PHYSICS FOR SCIENTISTS AND ENGINEERS

Prefixes for Powers of 10*

Multiple	Prefix	Abbreviation
10^{24}	yotta	Y
10^{21}	zetta	Z
10^{18}	exa	E
10^{15}	peta	P
10^{12}	tera	T
10^{9}	**giga**	**G**
10^{6}	**mega**	**M**
10^{3}	**kilo**	**k**
10^{2}	hecto	h
10^{1}	deka	da
10^{-1}	deci	d
10^{-2}	**centi**	**c**
10^{-3}	**milli**	**m**
10^{-6}	**micro**	**μ**
10^{-9}	**nano**	**n**
10^{-12}	**pico**	**p**
10^{-15}	femto	f
10^{-18}	atto	a
10^{-21}	zepto	z
10^{-24}	yocto	y

*Commonly used prefixes are in bold. All prefixes are pronounced with the accent on the first syllable.

The Greek Alphabet

Alpha	A	α	Nu	N	ν
Beta	B	β	Xi	Ξ	ξ
Gamma	Γ	γ	Omicron	O	o
Delta	Δ	δ	Pi	Π	π
Epsilon	E	ϵ, ε	Rho	P	ρ
Zeta	Z	ζ	Sigma	Σ	σ
Eta	H	η	Tau	T	τ
Theta	Θ	θ	Upsilon	Υ	υ
Iota	I	ι	Phi	Φ	ϕ
Kappa	K	κ	Chi	X	χ
Lambda	Λ	λ	Psi	Ψ	ψ
Mu	M	μ	Omega	Ω	ω

Terrestrial and Astronomical Data*

Acceleration of gravity at Earth's surface	g	$9.81\ \text{m/s}^2 = 32.2\ \text{ft/s}^2$
Radius of Earth R_E	R_E	6371 km = 3959 mi
Mass of Earth	M_E	5.97×10^{24} kg
Mass of the Sun		1.99×10^{30} kg
Mass of the moon		7.35×10^{22} kg
Escape speed at Earth's surface		11.2 km/s = 6.95 mi/s
Standard temperature and pressure (STP)		0°C = 273.15 K 1 atm = 101.3 kPa
Earth–moon distance†		3.84×10^{8} m = 2.39×10^{5} mi
Earth–Sun distance (mean)†		1.50×10^{11} m = 9.30×10^{7} mi
Speed of sound in dry air (at STP)		331 m/s
Speed of sound in dry air (20°C, 1 atm)		343 m/s
Density of dry air (STP)		$1.29\ \text{kg/m}^3$
Density of dry air (20°C, 1 atm)		$1.20\ \text{kg/m}^3$
Density of water (4°C, 1 atm)		$1000\ \text{kg/m}^3$
Heat of fusion of water (0°C, 1 atm)	L_f	333.5 kJ/kg
Heat of vaporization of water (100°C, 1 atm)	L_v	2.257 MJ/kg

* Additional data on the solar system can be found in Appendix B and at http://nssdc.gsfc.nasa.gov/planetary/planetfact.html.
† Center to center.

Mathematical Symbols

$=$	is equal to
$\equiv$	is defined by
$\neq$	is not equal to
$\approx$	is approximately equal to
$\sim$	is of the order of
$\propto$	is proportional to
$>$	is greater than
$\geq$	is greater than or equal to
$\gg$	is much greater than
$<$	is less than
$\leq$	is less than or equal to
$\ll$	is much less than
Δx	change in x
dx	differential change in x
$\lvert x \rvert$	absolute value of x
$\lvert \vec{v} \rvert$	magnitude of $\vec{v}$
$n!$	$n(n-1)(n-2)\ldots1$
Σ	sum
lim	limit
$\Delta t \to 0$	Δt approaches zero
$\frac{dx}{dt}$	derivative of x with respect to t
$\frac{\partial x}{\partial t}$	partial derivative of x with respect to t
$\int_{x_1}^{x_2} f(x)dx$	definite integral $= F(x)\Big\rvert_{x_1}^{x_2} = F(x_2) - F(x_1)$

Abbreviations for Units

A	ampere	H	henry	nm	nanometer (10^{-9} m)
Å	angstrom (10^{-10} m)	h	hour	pt	pint
atm	atmosphere	Hz	hertz	qt	quart
Btu	British thermal unit	in	inch	rev	revolution
Bq	becquerel	J	joule	R	roentgen
C	coulomb	K	kelvin	Sv	seivert
°C	degree Celsius	kg	kilogram	s	second
cal	calorie	km	kilometer	T	tesla
Ci	curie	keV	kilo-electron volt	u	unified mass unit
cm	centimeter	lb	pound	V	volt
dyn	dyne	L	liter	W	watt
eV	electron volt	m	meter	Wb	weber
°F	degree Fahrenheit	MeV	mega-electron volt	y	year
fm	femtometer, fermi (10^{-15} m)	Mm	megameter (10^{6} m)	yd	yard
ft	foot	mi	mile	μm	micrometer (10^{-6} m)
Gm	gigameter (10^{9} m)	min	minute	μs	microsecond
G	gauss	mm	millimeter	μC	microcoulomb
Gy	gray	ms	millisecond	Ω	ohm
g	gram	N	newton		

Some Conversion Factors

Length

1 m = 39.37 in = 3.281 ft = 1.094 yd

1 m = 10^{15} fm = 10^{10} Å = 10^{9} nm

1 km = 0.6214 mi

1 mi = 5280 ft = 1.609 km

1 lightyear = 1 $c \cdot$ y = 9.461×10^{15} m

1 in = 2.540 cm

Volume

1 L = 10^{3} cm^{3} = 10^{-3} m^{3} = 1.057 qt

Time

1 h = 3600 s = 3.6 ks

1 y = 365.24 d = 3.156×10^{7} s

Speed

1 km/h = 0.278 m/s = 0.6214 mi/h

1 ft/s = 0.3048 m/s = 0.6818 mi/h

Angle–angular speed

1 rev = 2π rad = 360°

1 rad = 57.30°

1 rev/min = 0.1047 rad/s

Force–pressure

1 N = 10^{5} dyn = 0.2248 lb

1 lb = 4.448 N

1 atm = 101.3 kPa = 1.013 bar = 76.00 cmHg = 14.70 lb/in^{2}

Mass

1 u = $[(10^{-3}\ \text{mol}^{-1})/N_A]$ kg = 1.661×10^{-27} kg

1 tonne = 10^{3} kg = 1 Mg

1 slug = 14.59 kg

1 kg weighs about 2.205 lb

Energy–power

1 J = 10^{7} erg = 0.7376 ft · lb = 9.869×10^{-3} L · atm

1 kW · h = 3.6 MJ

1 cal = 4.184 J = 4.129×10^{-2} L · atm

1 L · atm = 101.325 J = 24.22 cal

1 eV = 1.602×10^{-19} J

1 Btu = 778 ft · lb = 252 cal = 1054 J

1 horsepower = 550 ft · lb/s = 746 W

Thermal conductivity

1 W/(m · K) = 6.938 Btu · in/(h · ft^{2} · °F)

Magnetic field

1 T = 10^{4} G

Viscosity

1 Pa · s = 10 poise

SIXTH EDITION

PHYSICS FOR SCIENTISTS AND ENGINEERS

WITH MODERN PHYSICS

Paul A. Tipler
Gene Mosca

macmillan learning
Austin • Boston • New York • Plymouth

PT: For Claudia

GM: For Vivian

Senior Vice President:	Daryl Fox
Program Director, Physical Sciences:	Andrew Dunaway
Program Manager:	Lori Stover
Marketing Manager:	Nancy Bradshaw
Senior Developmental Editor:	Kharissia Pettus
Media Editor:	Jeanette Picerno
Editorial Assistants:	Janie Chan, Kathryn Treadway
Photo Editor:	Ted Szczepanski
Photo Researcher:	Dena Digilio Betz
Cover Designer:	Blake Logan
Text Designer:	Marsha Cohen/Parallelogram Graphics
Senior Project Editor:	Georgia Lee Hadler
Copy Editors:	Connie Parks, Trumbull Rogers
Illustrations:	Network Graphics
Illustration Coordinator:	Bill Page
Senior Workflow Project Supervisor:	Susan Wein
Production Supervisor:	Lawrence Guerra
Composition:	Lumina Datamatics, Inc.
Printing and Binding:	King Printing Co., Inc.

Library of Congress Control Number: 2007010418

ISBN-13: 978-1-319-36581-3
ISBN-10: 1-319-36581-7

Printed in the United States of America
3 4 5 6 28 27 26 25

Macmillan Learning
One New York Plaza
Suite 4600
New York, NY 10004-1562
www.macmillanlearning.com

w.h.freeman
Macmillan Learning

In 1946, William Freeman founded W. H. Freeman and Company and published Linus Pauling's *General Chemistry*, which revolutionized the chemistry curriculum and established the prototype for a Freeman text. W. H. Freeman quickly became a publishing house where leading researchers can make significant contributions to mathematics and science. In 1996, W. H. Freeman joined Macmillan and we have since proudly continued the legacy of providing revolutionary, quality educational tools for teaching and learning in STEM.

Contents in Brief

Thinkstock/Alamy

PART IV ELECTRICITY AND MAGNETISM

PART V LIGHT

PART VI MODERN PHYSICS: QUANTUM MECHANICS, RELATIVITY, AND THE STRUCTURE OF MATTER

APPENDICES

Contents

Extended

Courtesy of Rossignol Ski Company

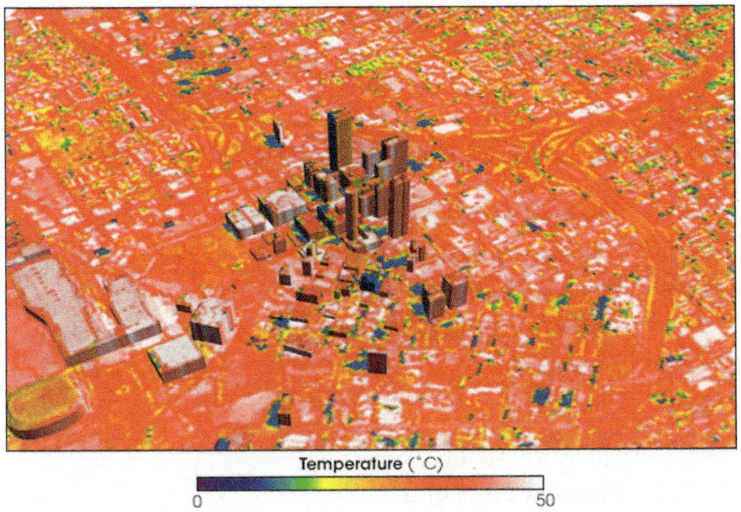

NASA/Goddard Space Flight Center Scientific Visualization Studio

Atlas Photo Bank/Photo Researchers, Inc.

NASA

The text you trust, with support like never before

For years, instructors have trusted Tipler's *Physics for Scientists and Engineers* to prepare physics, science, and engineering majors for future coursework. The text has long been respected for its rigorous emphasis on foundational skills, depth of coverage, and integration of calculus.

The new *Physics for Scientists and Engineers* Media Update makes it easier than ever for instructors to create lecture materials, engage students during class time, and assign homework exercises. These new resources can be easily edited, assigned, and integrated with your campus LMS to save time and effort for instructors and students alike. The text and its supplements are supported by Macmillan's personal customer service to fit your course needs and objectives.

New expanded suite of instructor resources:

- Image slides
- Clicker questions
- Lecture slides
- Instructor solutions manual
- Test bank

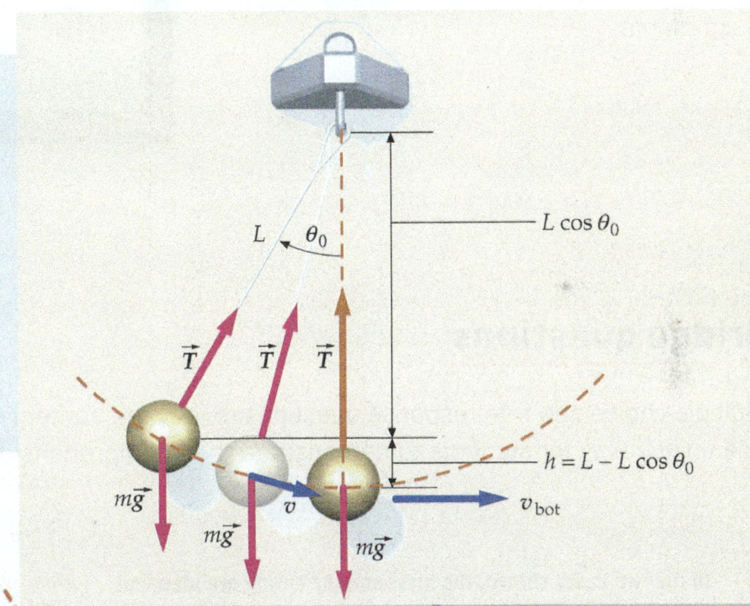

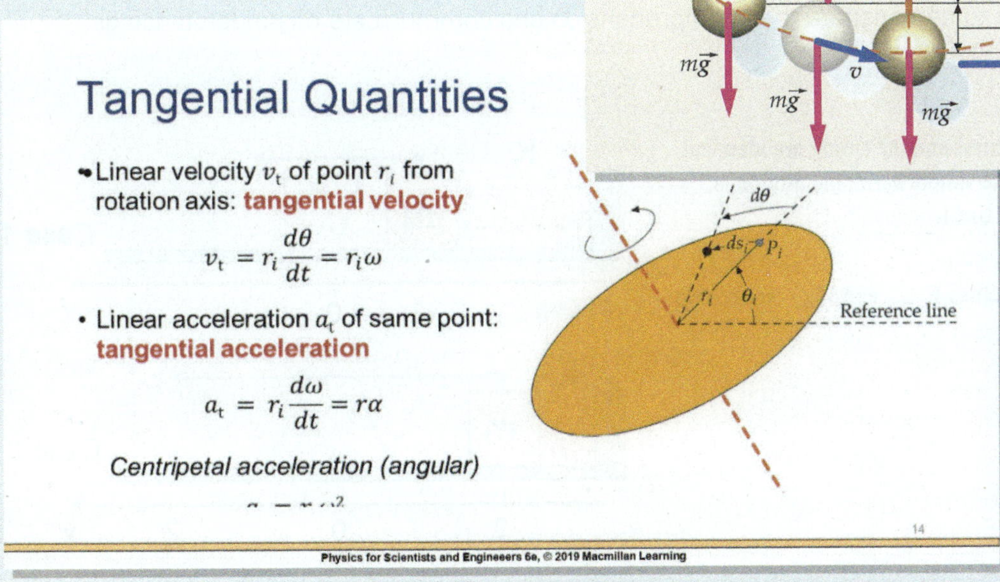

Prepare and Practice

Prelecture videos

Animated, narrated videos introduce core physics topics, laying the groundwork for conceptual understanding before lecture. Every chapter in the standard edition contains a series of 8-10 short videos; the activity can be assigned in its entirety or divided into smaller, more focused assignments.

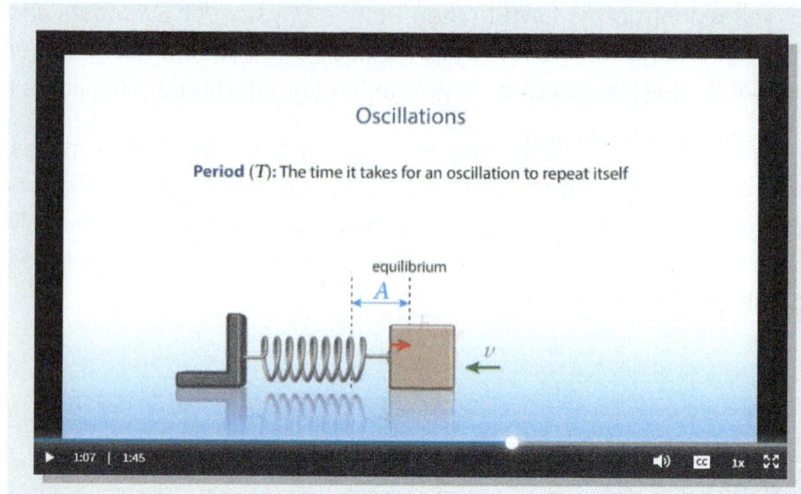

Bridge questions

Multiple choice and free response questions review the content covered in Prelecture videos and serve as a unique way for students to demonstrate their comprehension and communicate misunderstandings.

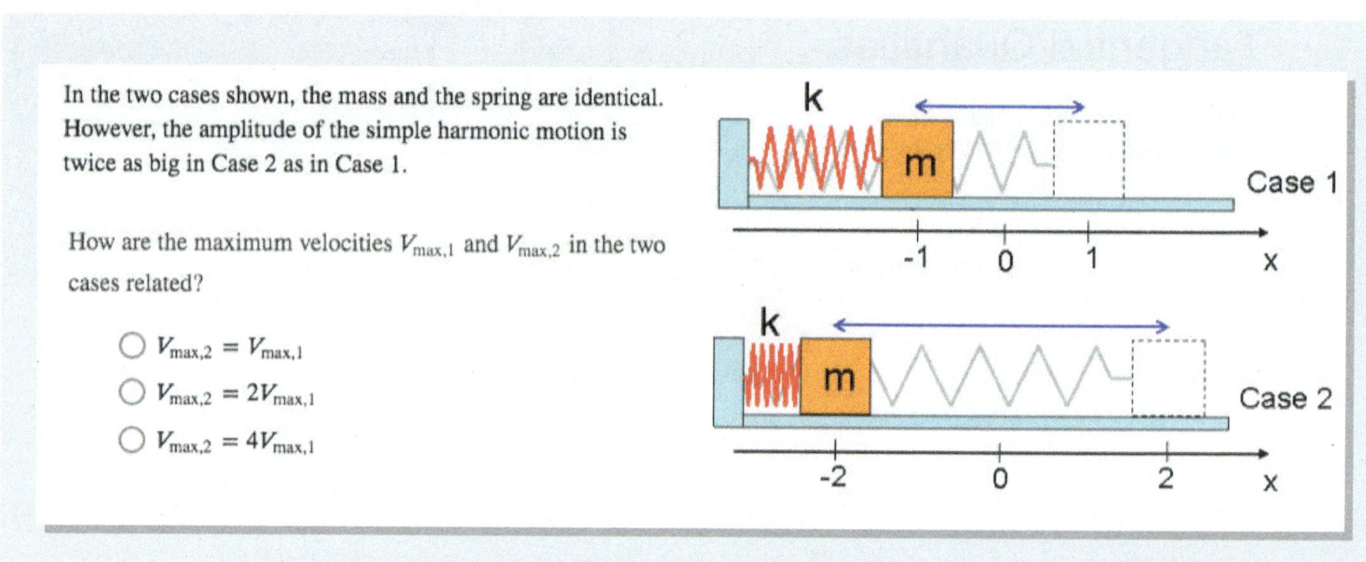

Every Problem Counts with SaplingPlus

Homework questions

Every homework question is built to test core competencies while coaching students through errors and common misconceptions.

Suppose Imani is on an amusement park ride that consists of a wide drum of radius R that rotates at ω rad/s and slowly increases its angle θ with the ground. Assume that Imani's back and heels are in contact with the wall of the drum.

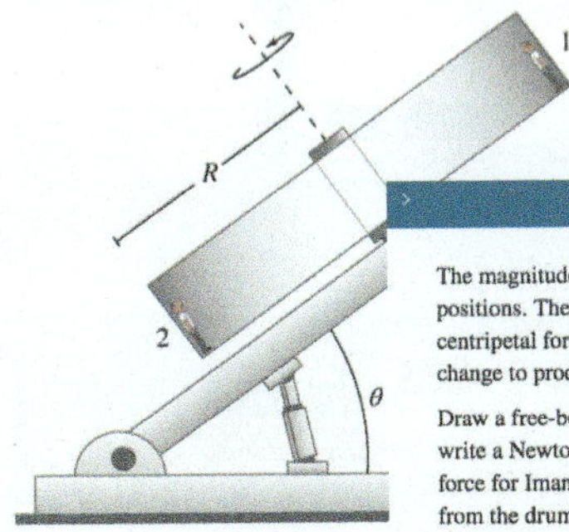

In the configuration shown, which forces contribute to the centripetal force on Imani?

- [x] the normal force applied by the drum wall
- [ ] the normal force applied by the drum floor
- [x] the gravitational force
- [ ] the centrifugal force
- [ ] the frictional force

In which position is the force of the drum's wall on Imani the greatest?

- ○ 1
- ● 2
- ○ neither

Feedback

The magnitude of the force of the drum's wall is not the same at both positions. The gravitational force component does not contribute to the centripetal force in the same way at both positions, so the normal force must change to produce the same centripetal force.

Draw a free-body diagram for Imani at both position 1 and position 2. Then, write a Newton's second law equation along the direction of the centripetal force for Imani at each position. Compare the magnitude of the normal force from the drum's wall in each position.

Wrong-answer-specific feedback targets and addresses students' misconceptions.

Hint

A centripetal force is a *net* force that generates uniform circular motion.

Use free-body diagrams and Newton's second law to compare the forces contributing to the centripetal force at each position.

Hints encourage critical thinking by providing suggestions without giving away the answer.

To determine in which position the normal force $\vec{F}_{N,wall}$ from the wall has the greatest magnitude, draw a free-body diagram and solve for $F_{N,wall}$ in each position using Newton's second law. Use a coordinate system in which the x-axis is parallel to the floor of the drum and the y-axis coincides with the drum's rotation axis.

In position 1, the normal force from the floor points in the positive y-direction, the normal force from the wall points in the negative x-direction, and the x-component $\vec{F}_{g,x}$ of the gravitational force points in the negative x-direction.

Write Newton's second law in the x-direction for Imani in position 1, then solve for $F_{N,wall}$.

$$-F_{N,wall} - F_{g,x} = -ma_c$$

$$F_{N,wall} = ma_c - F_{g,x}$$

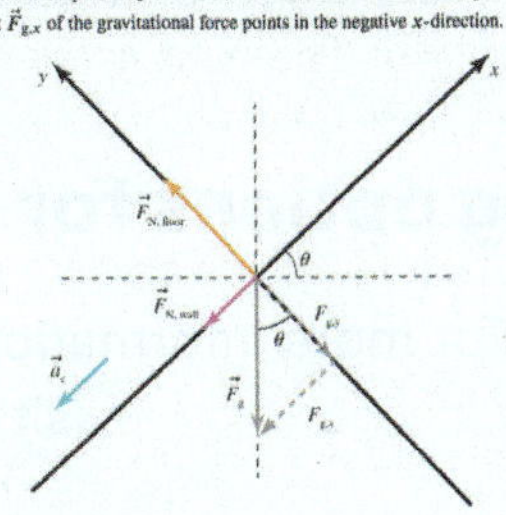

Detailed solutions provide an in-product study guide for every problem in SaplingPlus.

Custom Solutions and Ordering Options from Macmillan Learning Physics

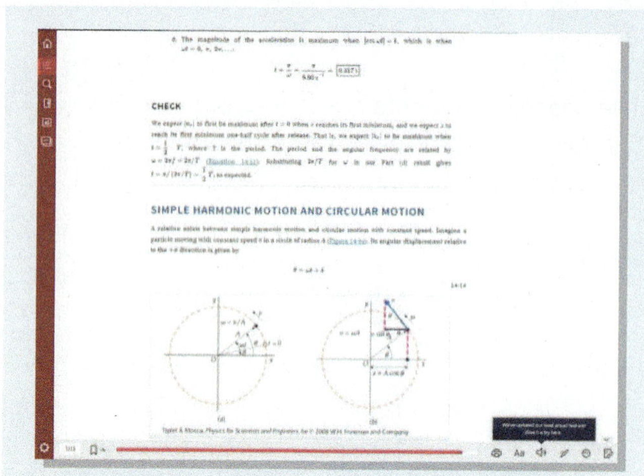

The new, enhanced e-book for Tipler's *Physics for Scientists and Engineers* is Macmillan's most affordable option for students. In partnership with VitalSource, the e-book contains highlighting and note-taking functionality, as well as offline reading and an accessible screen reader.

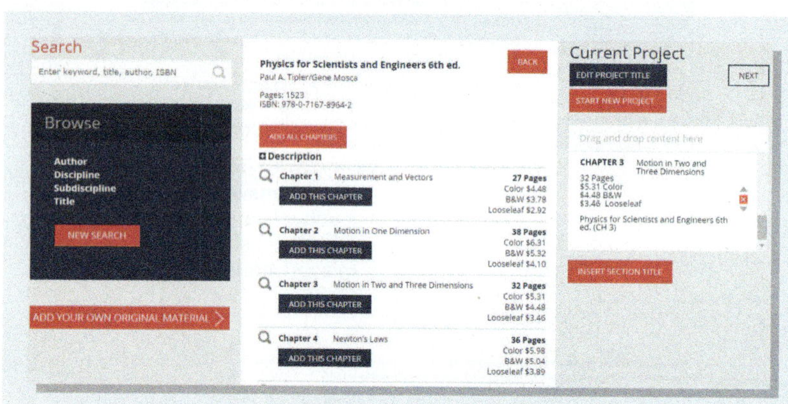

Custom texts and isolated volumes are easy to make in Freeman Select. Choose only the chapters you need and add your own content for your own custom text. Priced per page, it's an affordable solution for classes and students looking for a print solution. Try it for yourself at **macmillanlearning.com/freemanselect.**

Ordering options for *Physics for Scientists and Engineers 6e*:

For more information and ordering options, including rental options, visit **macmillanlearning.com/physics**.

E-book: 978-1-319-32171-0

E-book + SaplingPlus: 978-1-319-32169-7 (multi-term access)

Looseleaf: 978-1-319-36314-7

Looseleaf + SaplingPlus: 978-1-319-36875-3 (multi-term access)

Hardcover + SaplingPlus: 978-1-319-36828-9 (multi-term access)

Purposeful Classroom Engagement

iClicker

iClicker's market-leading student engagement solutions allow students to participate using laptops, mobile devices, or iClicker remotes—all in the same classroom. To learn more, visit **iclicker.com**.

Clicker questions for Tipler's *Physics for Scientists and Engineers* are available in SaplingPlus or on the Macmillan online catalog.

The iClicker app can be packaged with any book or digital version of *Physics for Scientists and Engineers* for an additional $5! Contact your local rep to learn more.

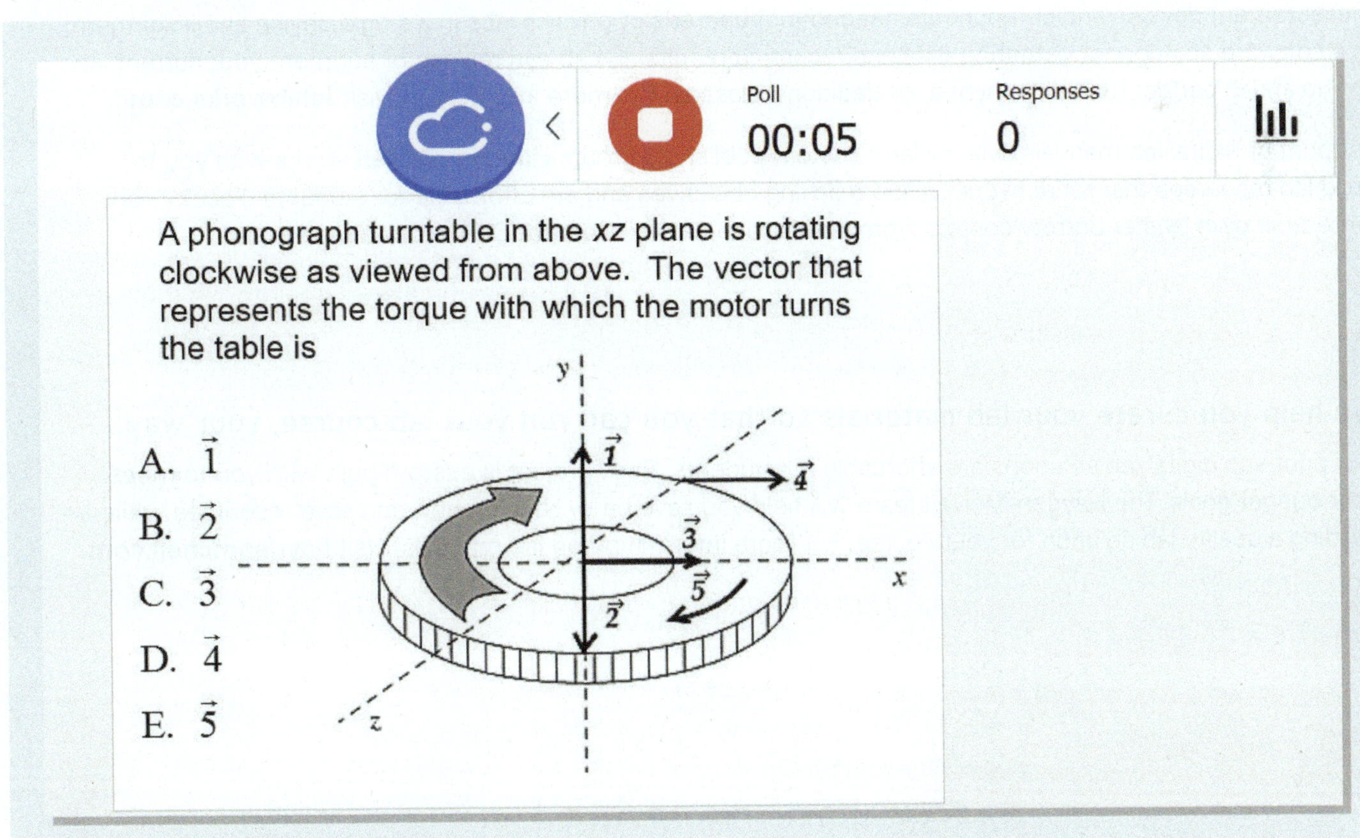

Lab Solutions From Macmillan Physics

Set Your Physics Lab Free

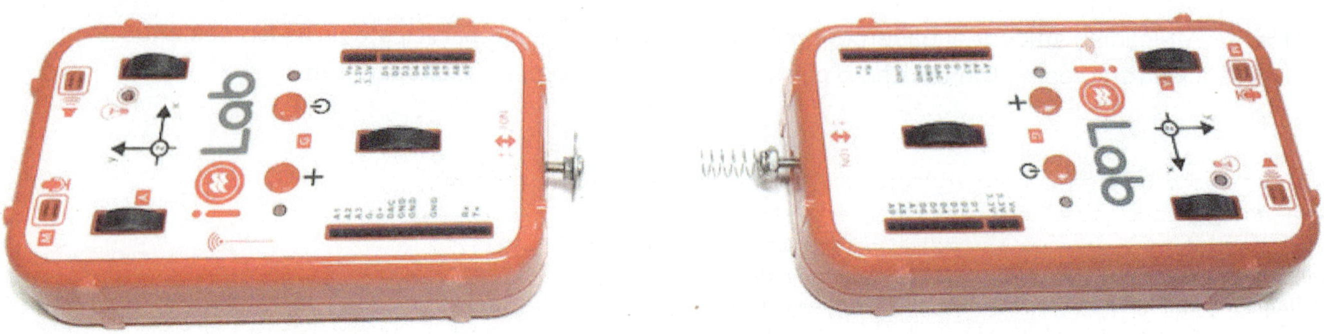

iOLab is a revolutionary new hardware and software solution for your physics lab. It combines all the measurement devices and components needed for hundreds of physics labs in a single device and links them to a software solution for gathering data and recording results. iOLab can replace or augment your current lab set up and is perfect for online, hybrid, or traditional classes. For more information, visit **iolabworks.com**.

For print or digital lab manuals, Macmillan's custom publishing group, **Hayden-McNeil** works with you to build lab resources that reflect your unique teaching objectives and are affordable for students. Faculty can write their own labs or borrow content from our Hayden-McNeil Content Collection to get started.

We help you curate your lab materials so that you can run your lab course, your way.

Our print and digital lab solutions are affordable for students. Each project is custom built with you to meet your budget goals. The Hayden-McNeil team will help you save time, save money, and save resources while building a quality lab solution for your course. For more information on lab manuals, visit **haydenmcneil.com**.

Acknowledgments

We are grateful to the many instructors, students, colleagues, and friends who have contributed to this edition and to earlier editions.

Anthony J. Buffa, professor emeritus at California Polytechnic State University in California, wrote many new end-of-chapter problems and edited the end-of-chapter problems sections. Laura Runkle wrote the Physics Spotlights. Richard Mickey revised the Math Review of the fifth edition, which is now the Math Tutorial of the sixth edition. David Mills, professor emeritus at the College of the Redwoods in California, extensively revised the Solutions Manual. We received invaluable help in creating text and checking the accuracy of text and problems from the following professors:

Thomas Foster
Southern Illinois University

Karamjeet Arya
San Jose State University

Mirley Bala
Texas A&M University—Corpus Christi

Michael Crivello
San Diego Mesa College

Carlos Delgado
Community College of Southern Nevada

David Faust
Mt. Hood Community College

Robin Jordan
Florida Atlantic University

Jerome Licini
Lehigh University

Dan Lucas
University of Wisconsin

Laura McCullough
University of Wisconsin, Stout

Jeannette Myers
Francis Marion University

Marian Peters
Appalachian State University

Todd K. Pedlar
Luther College

Paul Quinn
Kutztown University

Peter Sheldon
Randolph-Macon Woman's College

Michael G. Strauss
University of Oklahoma

Brad Trees
Ohio Wesleyan University

George Zober
Yough Senior High School

Patricia Zober
Ringgold High School

Many instructors and students have provided extensive and helpful reviews of one or more chapters of this edition. They have each made a fundamental contribution to the quality of this revision, and deserve our gratitude. We would like to thank the following reviewers:

Ahmad H. Abdelhadi
James Madison University

Edward Adelson
Ohio State University

Royal Albridge
Vanderbilt University

J. Robert Anderson
University of Maryland, College Park

Toby S. Anderson
Tennessee State University

Wickram Ariyasinghe
Baylor University

Yildirim Aktas
University of North Carolina, Charlotte

Eric Ayars
California State University

James Battat
Harvard University

Eugene W. Beier
University of Pennsylvania

Peter Beyersdorf
San Jose State University

Richard Bone
Florida International University

Juliet W. Brosing
Pacific University

Ronald Brown
California Polytechnic State University

Richard L. Cardenas
St. Mary's University

Troy Carter
University of California, Los Angeles

Alice D. Churukian
Concordia College

N. John DiNardo
Drexel University

Jianjun Dong
Auburn University

Fivos R Drymiotis
Clemson University

Mark A. Edwards
Hofstra University

James Evans
Broken Arrow Senior High

Nicola Fameli
University of British Columbia

N. G. Fazleev
University of Texas at Arlington

Thomas Furtak
Colorado School of Mines

Richard Gelderman
Western Kentucky University

Yuri Gershtein
Florida State University

Paolo Gondolo
University of Utah

Benjamin Grinstein
University of California, San Diego

Parameswar Hari
University of Tulsa

Joseph Harrison
University of Alabama—Birmingham

Patrick C. Hecking
Thiel College

Kristi R. G. Hendrickson
University of Puget Sound

Linnea Hess
Olympic College

Mark Hollabaugh
Normandale Community College

Daniel Holland
Illinois State University

Richard D. Holland II
Southern Illinois University

Eric Hudson
Massachusetts Institute of Technology

David C. Ingram
Ohio University

Colin Inglefield
Weber State University

Nathan Israeloff
Northeastern University

Donald J. Jacobs
California State University, Northridge

Erik L. Jensen
Chemeketa Community College

Colin P Jessop
University of Notre Dame

Ed Kearns
Boston University

Alice K. Kolakowska
Mississippi State University

Douglas Kurtze
Saint Joseph's University

Eric T. Lane
University of Tennessee at Chattanooga

Christie L. Larochelle
Franklin & Marshall College

Mary Lu Larsen
Towson University

Clifford L. Laurence
Colorado Technical University

Bruce W. Liby
Manhattan College

Ramon E. Lopez
Florida Institute of Technology

Ntungwa Maasha
Coastal Georgia Community Collegee and University Center

Jane H MacGibbon
University of North Florida

A. James Mallmann
Milwaukee School of Engineering

Rahul Mehta
University of Central Arkansas

R. A. McCorkle
University of Rhode Island

Linda McDonald
North Park University

Kenneth McLaughlin
Loras College

Eric R. Murray
Georgia Institute of Technology

Jeffrey S. Olafsen
University of Kansas

Richard P. Olenick
University of Dallas

Halina Opyrchal
New Jersey Institute of Technology

Russell L. Palma
Minnesota State University—Mankato

Todd K. Pedlar
Luther College

Daniel Phillips
Ohio University

Edward Pollack
University of Connecticut

Michael Politano
Marquette University

Robert L. Pompi
SUNY Binghamton

Damon A. Resnick
Montana State University

Richard Robinett
Pennsylvania State University

John Rollino
Rutgers University

Daniel V. Schroeder
Weber State University

Douglas Sherman
San Jose State University

Christopher Sirola
Marquette University

Larry K. Smith
Snow College

George Smoot
University of California at Berkeley

Zbigniew M. Stadnik
University of Ottawa

Kenny Stephens
Hardin-Simmons University

Daniel Stump
Michigan State University

Jorge Talamantes
California State University, Bakersfield

Charles G. Torre
Utah State University

Brad Trees
Ohio Wesleyan University

John K. Vassiliou
Villanova University

Theodore D. Violett
Western State College

Hai-Sheng Wu
Minnesota State University—Mankato

Anthony C. Zable
Portland Community College

Ulrich Zurcher
Cleveland State University

We also remain indebted to the reviewers of past editions. We would therefore like to thank the following reviewers, who provided immeasurable support as we developed the fourth and fifth editions:

Edward Adelson
The Ohio State University

Michael Arnett
Kirkwood Community College

Todd Averett
The College of William and Mary

Yildirim M. Aktas
University of North Carolina at Charlotte

Karamjeet Arya
San Jose State University

Alison Baski
Virginia Commonwealth University

William Bassichis
Texas A&M University

Joel C. Berlinghieri
The Citadel

Gary Stephen Blanpied
University of South Carolina

Frank Blatt
Michigan State University

Ronald Brown
California Polytechnic State University

Anthony J. Buffa
California Polytechnic State University

John E. Byrne
Gonzaga University

Wayne Carr
Stevens Institute of Technology

George Cassidy
University of Utah

Lay Nam Chang
Virginia Polytechnic Institute

I. V. Chivets
Trinity College, University of Dublin

Harry T. Chu
University of Akron

Alan Cresswell
Shippensburg University

Robert Coakley
University of Southern Maine

Robert Coleman
Emory University

Brent A. Corbin
UCLA

Andrew Cornelius
University of Nevada at Las Vegas

Mark W. Coffey
Colorado School of Mines

Peter P. Crooker
University of Hawaii

Jeff Culbert
London, Ontario

Paul Debevec
University of Illinois

Ricardo S. Decca
Indiana University-Purdue University

Robert W. Detenbeck
University of Vermont

N. John DiNardo
Drexel University

Bruce Doak
Arizona State University

Michael Dubson
University of Colorado at Boulder

John Elliott
University of Manchester, England

William Ellis
University of Technology — Sydney

Colonel Rolf Enger
U.S. Air Force Academy

John W. Farley
University of Nevada at Las Vegas

David Faust
Mount Hood Community College

Mirela S. Fetea
University of Richmond

David Flammer
Colorado School of Mines

Philip Fraundorf
University of Missouri, Saint Louis

Tom Furtak
Colorado School of Mines

James Garland
Retired

James Garner
University of North Florida

Ian Gatland
Georgia Institute of Technology

Ron Gautreau
New Jersey Institute of Technology

David Gavenda
University of Texas at Austin

Patrick C. Gibbons
Washington University

David Gordon Wilson
Massachusetts Institute of Technology

Christopher Gould
University of Southern California

Newton Greenberg
SUNY Binghamton

John B. Gruber
San Jose State University

Huidong Guo
Columbia University

Phuoc Ha
Creighton University

Richard Haracz
Drexel University

Clint Harper
Moorpark College

Michael Harris
University of Washington

Randy Harris
University of California at Davis

Tina Harriott
Mount Saint Vincent, Canada

Dieter Hartmann
Clemson University

Theresa Peggy Hartsell
Clark College

Kristi R.G. Hendrickson
University of Puget Sound

Michael Hildreth
University of Notre Dame

Robert Hollebeek
University of Pennsylvania

David Ingram
Ohio University

Shawn Jackson
The University of Tulsa

Madya Jalil
University of Malaya

Monwhea Jeng
University of California — Santa Barbara

James W. Johnson
Tallahassee Community College

Edwin R. Jones
University of South Carolina

Ilon Joseph
Columbia University

David Kaplan
University of California — Santa Barbara

William C. Kerr
Wake Forest University

John Kidder
Dartmouth College

Roger King
City College of San Francisco

James J. Kolata
University of Notre Dame

Boris Korsunsky
Northfield Mt. Hermon School

Thomas O. Krause
Towson University

Eric Lane
University of Tennessee, Chattanooga

Andrew Lang (graduate student)
University of Missouri

David Lange
University of California — Santa Barbara

Donald C. Larson
Drexel University

Paul L. Lee
California State University, Northridge

Peter M. Levy
New York University

Jerome Licini
Lehigh University

Isaac Leichter
Jerusalem College of Technology

William Lichten
Yale University

Robert Lieberman
Cornell University

Fred Lipschultz
University of Connecticut

Graeme Luke
Columbia University

Dan MacIsaac
Northern Arizona University

Edward McCliment
University of Iowa

Robert R. Marchini
The University of Memphis

Peter E. C. Markowitz
Florida International University

Daniel Marlow
Princeton University

Fernando Medina
Florida Atlantic University

Howard McAllister
University of Hawaii

John A. McClelland
University of Richmond

Laura McCullough
University of Wisconsin at Stout

M. Howard Miles
Washington State University

Matthew Moelter
University of Puget Sound

Eugene Mosca
U.S. Naval Academy

Carl Mungan
U.S. Naval Academy

Taha Mzoughi
Mississippi State University

Charles Niederriter
Gustavus Adolphus College

John W. Norbury
University of Wisconsin at Milwaukee

Aileen O'Donughue
St. Lawrence University

Jack Ord
University of Waterloo

Jeffry S. Olafsen
University of Kansas

Melvyn Jay Oremland
Pace University

Richard Packard
University of California

Antonio Pagnamenta
University of Illinois at Chicago

George W. Parker
North Carolina State University

John Parsons
Columbia University

Dinko Pocanic
University of Virginia

Edward Pollack
University of Connecticut

Robert Pompi
The State University of New York at Binghamton

Bernard G. Pope
Michigan State University

John M. Pratte
Clayton College and State University

Brooke Pridmore
Clayton State College

Yong-Zhong Qian
University of Minnesota

David Roberts
Brandeis University

Lyle D. Roelofs
Haverford College

R. J. Rollefson
Wesleyan University

Larry Rowan
University of North Carolina at Chapel Hill

Ajit S. Rupaal
Western Washington University

Todd G. Ruskell
Colorado School of Mines

Lewis H. Ryder
University of Kent, Canterbury

Andrew Scherbakov
Georgia Institute of Technology

Bruce A. Schumm
University of California, Santa Cruz

Cindy Schwarz
Vassar College

Mesgun Sebhatu
Winthrop University

Bernd Schuttler
University of Georgia

Murray Scureman
Amdahl Corporation

Marllin L. Simon
Auburn University

Scott Sinawi
Columbia University

Dave Smith
University of the Virgin Islands

Wesley H. Smith
University of Wisconsin

Kevork Spartalian
University of Vermont

Zbigniew M. Stadnik
University of Ottawa

G. R. Stewart
University of Florida

Michael G. Strauss
University of Oklahoma

Kaare Stegavik
University of Trondheim, Norway

Jay D. Strieb
Villanova University

Dan Styer
Oberlin College

Chun Fu Su
Mississippi State University

Jeffrey Sundquist
Palm Beach Community College — South

Cyrus Taylor
Case Western Reserve University

Martin Tiersten
City College of New York

Chin-Che Tin
Auburn University

Oscar Vilches
University of Washington

D. J. Wagner
Grove City College
Columbia University

George Watson
University of Delaware

Fred Watts
College of Charleston

David Winter

John A. Underwood
Austin Community College

John Weinstein
University of Mississippi

Stephen Weppner
Eckerd College

Suzanne E. Willis
Northern Illinois University

Frank L. H. Wolfe
University of Rochester

Frank Wolfs
University of Rochester

Roy C. Wood
New Mexico State University

Ron Zammit
California Polytechnic State University

Yuriy Zhestkov
Columbia University

Dean Zollman
Kansas State University

Fulin Zuo
University of Miami

Finally, we would like to thank our friends at W. H. Freeman and Company for their help and encouragement. Susan Brennan, Clancy Marshall, Kharissia Pettus, Georgia Lee Hadler, Susan Wein, Trumbull Rogers, Connie Parks, John Smith, Dena Digilio Betz, Ted Szczepanski, and Liz Geller were extremely generous with their creativity and hard work at every stage of the process.

We are also grateful for the contributions and help of our colleagues Larry Tankersley, John Ertel, Steve Montgomery, and Don Treacy.

About the Authors

Paul Tipler was born in the small farming town of Antigo, Wisconsin, in 1933. He graduated from high school in Oshkosh, Wisconsin, where his father was superintendent of the public schools. He received his B.S. from Purdue University in 1955 and his Ph.D. at the University of Illinois in 1962, where he studied the structure of nuclei. He taught for one year at Wesleyan University in Connecticut while writing his thesis, then moved to Oakland University in Michigan, where he was one of the original members of the physics department, playing a major role in developing the physics curriculum. During the next 20 years, he taught nearly all the physics courses and wrote the first and second editions of his widely used textbooks *Modern Physics* (1969, 1978) and *Physics* (1976, 1982). In 1982, he moved to Berkeley, California, where he now resides, and where he wrote *College Physics* (1987) and the third edition of *Physics* (1991). In addition to physics, his interests include music, hiking, and camping, and he is an accomplished jazz pianist and poker player.

Gene Mosca was born in New York City and grew up on Shelter Island, New York. He studied at Villanova University, the University of Michigan, and the University of Vermont, where he received his Ph.D. in physics. Gene recently retired from his teaching position at the U.S. Naval Academy, where as coordinator of the core physics course he instituted numerous enhancements to both the laboratory and classroom. Proclaimed by Paul Tipler "the best reviewer I ever had," Mosca became his coauthor beginning with the fifth edition of this book.

CHAPTER 1

THE NUMBER OF GRAINS OF SAND ON A BEACH MAY BE TOO GREAT TO COUNT, BUT WE CAN ESTIMATE THE NUMBER BY USING REASONABLE ASSUMPTIONS AND SIMPLE CALCULATIONS. *(Corbis.)*

Measurement and Vectors

? How many grains of sand are on your favorite beach? (See Example 1-7.)

We have always been curious about the world around us. Since the beginnings of recorded thought, we have sought to understand the bewildering diversity of events that we observe—the color of the sky, the change in sound of a passing car, the swaying of a tree in the wind, the rising and setting of the Sun, the flight of a bird or plane. This search for understanding has taken a variety of forms: one is religion, one is art, and one is science. Although the word *science* comes from the Latin verb meaning "to know," science has come to mean not merely knowledge but specifically knowledge of the natural world. Physics attempts to describe the fundamental nature of the universe and how it works. It is the science of matter and energy, space and time.

Like all science, physics is a body of knowledge organized in a specific and rational way. Physicists build, test, and connect models in an effort to describe, explain, and predict reality. This process involves hypotheses, repeatable experiments and observations, and new hypotheses. The end result is a set of fundamental principles and laws that describe the phenomena of the world around us.

These laws and principles apply both to the exotic—such as black holes, dark energy, and particles with names like leptoquarks and bosons—and to everyday life. As you will see, countless questions about our world can be answered with a basic knowledge of physics: Why is the sky blue? How do astronauts float in space? How do CD players work? Why does an oboe sound different from a flute? Why must a helicopter have two rotors? Why do metal objects feel colder than wood objects at the same temperature? How do moving clocks run slow?

In this book, you will learn how to apply the principles of physics to answer these, and many other questions. You will encounter the standard topics of physics, including mechanics, sound, light, heat, electricity, magnetism, atomic physics, and nuclear physics. You will also learn some useful techniques for solving physics problems. In the process, we hope you gain a greater awareness, appreciation, and understanding of the beauty of physics.

In this chapter, we'll begin by addressing some preliminary concepts that you will need throughout your study of physics. We'll briefly examine the nature of physics, establish some basic definitions, introduce systems of units and how to use them, and present an introduction to vector mathematics. We'll also look at the accuracy of measurements, significant figures, and estimations.

1-1 THE NATURE OF PHYSICS

The word physics comes from the Greek word meaning the knowledge of the natural world. It should come as no surprise, therefore, that the earliest recorded efforts to systematically assemble knowledge concerning motion came from ancient Greece. In Aristotle's (384–322 B.C.) system of natural philosophy, explanations of physical phenomena were deduced from assumptions about the world, rather than derived from experimentation. For example, it was a fundamental assumption that every substance had a "natural place" in the universe. Motion was thought to be the result of a substance trying to reach its natural place. Because of the agreement between the deductions of Aristotelian physics and the motions observed throughout the physical universe and the lack of experimentation that could overturn the ancient physical ideas, the Greek view was accepted for nearly two thousand years. It was the Italian scientist Galileo Galilei (1564–1642) whose brilliant experiments on motion established the absolute necessity of experimentation in physics. Within a hundred years, Isaac Newton had generalized the results of Galileo's experiments into his three spectacularly successful laws of motion, and the reign of the natural philosophy of Aristotle was over.

Experimentation during the next two hundred years brought a flood of discoveries—and raised a flood of new questions. Some of these discoveries involved electrical and thermal phenomena, and some involved the expansion and compression of gases. These discoveries and questions inspired the development of new models to explain them. By the end of the nineteenth century, Newton's laws for the motions of mechanical systems had been joined by equally impressive laws from James Maxwell, James Joule, Sadi Carnot, and others to describe electromagnetism and thermodynamics. The subjects that occupied physical scientists through the end of the nineteenth century—mechanics, light, heat, sound, electricity and magnetism—are usually referred to as *classical physics*. Because classical physics is what we need to understand the macroscopic world we live in, it dominates Parts I through V of this text.

The remarkable success of classical physics led many scientists to believe that the description of the physical universe was complete. However, the discovery of X rays by Wilhelm Röntgen in 1895 and of radioactivity by Antoine Becquerel and

Marie and Pierre Curie a few years later seemed to be outside the framework of classical physics. The theory of special relativity proposed by Albert Einstein in 1905 expanded the classical ideas of space and time promoted by Galileo and Newton. In the same year, Einstein suggested that light energy is quantized; that is, that light comes in discrete packets rather than being wavelike and continuous as was thought in classical physics. The generalization of this insight to the quantization of all types of energy is a central idea of quantum mechanics, one that has many amazing and important consequences. The application of special relativity, and particularly quantum theory, to extremely small systems such as atoms, molecules, and nuclei, has led to a detailed understanding of solids, liquids, and gases. This application is often referred to as *modern physics*. Modern physics is the subject of Part VI of this text.

While classical physics is the main subject of this book, from time to time in the earlier parts of the text we will note the relationship between classical and modern physics. For example, when we discuss velocity in Chapter 2, we will take a moment to consider velocities near the speed of light and briefly cross over to the relativistic universe first imagined by Einstein. After discussing the conservation of energy in Chapter 7, we will discuss the quantization of energy and Einstein's famous relation between mass and energy, $E = mc^2$. Four chapters later, in Chapter R, we will study the nature of space and time as revealed by Einstein in 1903.

1-2 UNITS

The laws of physics express relationships among physical quantities. **Physical quantities** are numbers that are obtained by measuring physical phenomena. For example, the length of this book is a physical quantity, as is the amount of time it takes for you to read this sentence and the temperature of the air in your classroom.

Measurement of any physical quantity involves comparing that quantity to some precisely defined standard, or **unit,** of that quantity. For example, to measure the distance between two points, we need a standard unit of distance, such as an inch, a meter, or a kilometer. The statement that a certain distance is 25 meters means that it is 25 times the length of the unit meter. It is important to include the unit, in this case meters, along with the number, 25, when expressing this distance because different units can be used to measure distance. To say that a distance is 25 is meaningless.

! When you use a number to describe a physical quantity, the number must always be accompanied by a unit.

Some of the most basic physical quantities—time, length, and mass—are defined by the processes of measuring them. The length of a pole, for example, is defined to be the number of some unit of length that is required to equal the length of the pole. A physical quantity is often defined using an **operational definition,** a statement that defines a physical quantity by the operation or procedure that should be carried out to measure the physical quantity. Other physical quantities are defined by describing how to calculate them from these fundamental quantities. The speed of an object, for example, is equal to a length divide by a time. Many of the quantities that you will be studying, such as velocity, force, momentum, work, energy, and power, can be expressed in terms of time, length, and mass. Thus, a small number of basic units are sufficient to express all physical quantities. These basic units are called base units, and the choice of base units determines a system of units.

Water clock used to measure time intervals in the thirteenth century. *(The Granger Collection.)*

THE INTERNATIONAL SYSTEM OF UNITS

In physics, it is important to use a consistent set of units. In 1960, an international committee established a set of standards for the scientific community called SI (for *Système International*). There are seven base quantities in the SI system. They are length, mass, time, electric current, thermodynamic temperature, amount of substance, and luminous intensity, and each base quantity has a base unit. The base SI unit of time is the second, the base unit of length is the meter, and the base unit of mass is the kilogram. Later, when you study thermodynamics and electricity, you will need to use the base SI units for temperature (the kelvin, K), for the amount of a substance (the mole, mol), and one for electrical current (the ampere, A). The seventh base SI unit, the candela (cd) for luminous intensity, we shall have no occasion to use in this book. Complete definitions of the SI units are given in Appendix A, along with commonly used units derived from these units.

Cesium fountain clock with developers Steve Jefferts and Dawn Meekhof. *(© 1999 Geoffrey Wheeler.)*

Time The unit of time, the **second** (s), was historically defined in terms of the rotation of Earth and was equal to $(1/60)(1/60)(1/24)$ of the mean solar day. However, scientists have observed that the rate of rotation of Earth is gradually slowing down. The second is now defined in terms of a characteristic frequency associated with the cesium atom. All atoms, after absorbing energy, emit light with frequencies and wavelengths characteristic of the particular element. There is a set of frequencies and wavelengths for each element, with a particular frequency and wavelength associated with each energy transition within the atom. As far as we know, these frequencies remain constant. The second is now defined so that the frequency of the light from a certain transition in cesium is exactly 9 192 631 770 cycles per second.

Length The **meter** (m) is the SI unit of length. Historically, this length was defined as one ten-millionth of the distance between the equator and the North Pole along the meridian through Paris (Figure 1-1). This distance proved to be difficult to measure accurately. So in 1889, the distance between two scratches on a bar made of platinum-iridium alloy held at a specified temperature was adopted as the new standard. In time, the precision of this standard also proved inadequate and other standards were created for the meter. Currently, the meter is determined using the speed of light through empty space, which is defined to be exactly 299 792 458 m/s. The meter, then, is the distance light travels through empty space in 1/(299 729 458) second. By using these definitions, the units of time and length are accessible to laboratories throughout the world.

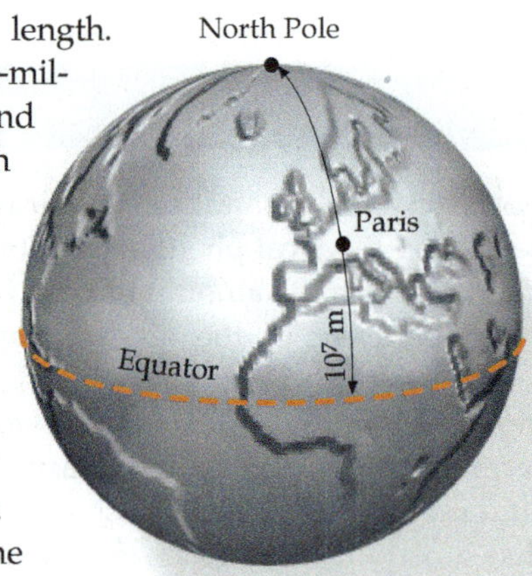

FIGURE 1-1 The meter was originally chosen so that the distance from the equator to the North Pole along the meridian through Paris would be 10 million meters (10 thousand kilometers).

Mass The SI unit of mass, the **kilogram** (kg) was once defined as the mass of one liter of water at 4°C. (A volume of one liter is equal to the volume of a cube 10 cm on an edge.) Like the standards for time and length, the kilogram standard has changed over time. The kilogram is now defined to be the mass of a specific platinum-iridium alloy cylinder. This cylinder, called the *standard body*, is kept at the International Bureau of Weights and Measures in Sèvres, France. A duplicate of the standard body is kept at the National Institute of Standards and Technology (NIST) in Gaithersburg, Maryland. We shall discuss the concept of mass in detail in Chapter 4, where we will see that the weight of an object at a given location is proportional to its mass. Thus, by comparing the weights of different objects of ordinary size with the weight of the standard body, the masses of the objects can be compared with each other.

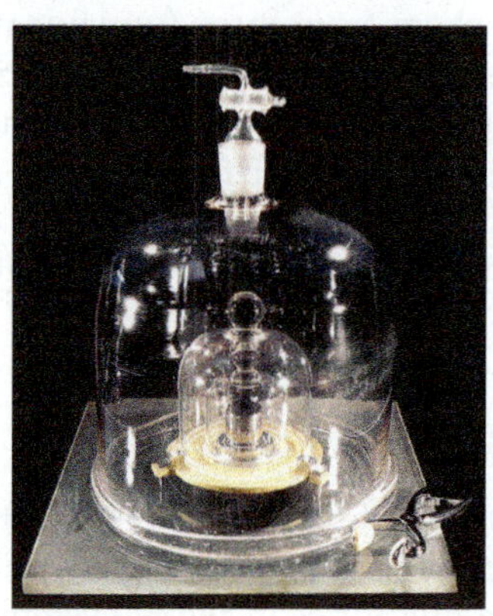

The standard body is the mass of a specific platinum-iridium alloy cylinder that is kept at the International Bureau of Weights and Measures in Sèvres, France. *(© BIPM; www.bipm.org.)*

Table 1-1 Prefixes for Powers of 10*

Multiple	Prefix	Abbreviation
10^{18}	exa	E
10^{15}	peta	P
10^{12}	tera	T
10^{9}	giga	G
10^{6}	mega	M
10^{3}	kilo	k
10^{2}	hecto	h
10^{1}	deka	da
10^{-1}	deci	d
10^{-2}	centi	c
10^{-3}	milli	m
10^{-6}	micro	μ
10^{-9}	nano	n
10^{-12}	pico	p
10^{-15}	femto	f
10^{-18}	atto	a

* The prefixes hecto (h), deka (da) and deci (d) are not multiples of 10^3 or 10^{-3} and are rarely used. The other prefix that is not a multiple of 10^3 or 10^{-3} is centi (c). The prefixes frequently used in this book are printed in red. Note that all prefix abbreviations for multiples 10^6 and higher are uppercase letters, all others are lowercase letters.

UNIT PREFIXES

Sometimes it is necessary to work with measurements that are much smaller or much larger than the standard SI units. In these situations, we can use other units that are related to the standard SI units by a multiple of ten. Prefixes are used to denote the different powers of ten. For example, the prefix "kilo" means 1000, or 10^3, while the prefix "micro" means 0.000 001, or 10^{-6}. Table 1-1 lists prefixes for common multiples of SI units. These prefixes can be applied to any SI unit; for example, 0.001 second is 1 millisecond (ms) and 1 000 000 watts is 1 megawatt (MW).

> **PRACTICE PROBLEM 1-1**
> Use prefixes to describe the following: (*a*) the delay caused by scrambling a cable television broadcast, which is about 0.000 000 3 second and (*b*) the circumference of Earth, which is about 40 000 000 meters.

OTHER SYSTEMS OF UNITS

In addition to SI, other systems of units are sometimes used. One such system is the *cgs system*. The fundamental units of the cgs system are the centimeter for length, the gram for mass, and the second for time. Other cgs units include the dyne (force) and the erg (work or energy).

The system of units with which you are probably most familiar is the U.S. customary system. In this system, the base unit of length is the foot and the base unit of time is the second. Also, a unit of force (the pound-force) rather than mass is considered a base unit. You will see in Chapter 4 that mass is a better choice for a fundamental unit than force, because mass is an intrinsic property of an object, independent of its location. The base U.S. customary units are now defined in terms of the base SI units.

1-3 CONVERSION OF UNITS

Because different systems of units are in use, it is important to know how to covert from one unit to another unit. When physical quantities are added, subtracted, multiplied, or divided in an algebraic equation, the unit can be treated like any other algebraic quantity. For example, suppose you want to find the distance traveled in 3 hours (h) by a car moving at a constant rate of 80 kilometers per hour (km/h). The distance is the product of the speed v and the time t:

$$x = vt = \frac{80\text{ km}}{\cancel{\text{h}}} \times 3\,\cancel{\text{h}} = 240\text{ km}$$

We cancel the unit of time, the hours, just as we would any algebraic quantity to obtain the distance in the proper unit of length, the kilometer. This method of treating units makes it easy to convert from one unit of distance to another. Now, suppose we want to convert the units in our answer from kilometers (km) to miles (mi). First, we need to find the relationship between kilometers and miles, which is 1 mi = 1.609 km (see either the front pages or Appendix A). Then, we divide each side of this equality by 1.609 km to obtain

$$\frac{1\text{ mi}}{1.609\text{ km}} = 1$$

Notice that the relationship is a ratio equal to 1. A ratio such as (1 mi)/(1.609 km) is called a **conversion factor,** which is a ratio equal to 1 and expresses a quantity expressed in some unit or units divided by its equal expressed some different unit or units. Because any quantity can be multiplied by 1 without changing its value, we can multiply the original quantity by the conversion factor to convert the units:

> ! If the units of the quantity and the conversion factor do not combine to give the desired final units, the conversion has not been properly carried out.

$$240\text{ km} = 240\,\cancel{\text{km}} \times \frac{1\text{ mi}}{1.609\,\cancel{\text{km}}} = 149\text{ mi}$$

By writing out the units explicitly and canceling them, you do not need to think about whether you multiply by 1.609 or divide by 1.609 to change kilometers to miles, because the units tell you whether you have chosen the correct or incorrect factor.

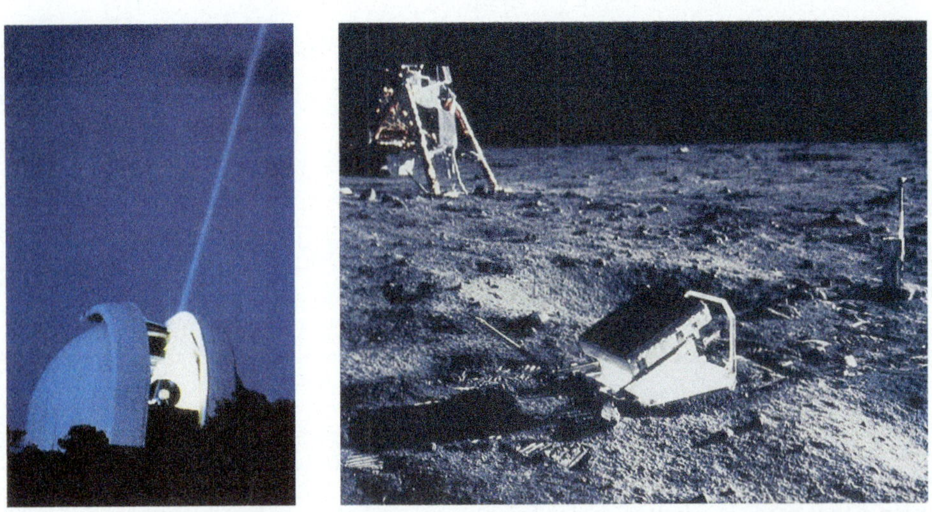

(a) (b)

(*a*) Laser beam from the Macdonald Observatory used to measure the distance to the moon. The distance can be measured within a few centimeters by measuring the time required for the beam to go to the moon and back after reflecting off a mirror (*b*) placed on the moon by the Apollo 14 astronauts. *(a, McDonald Observatory; b, Bruce Coleman).*

Example 1-1 Using Conversion Factors

Your employer sends you on a trip to a foreign country where the road signs give distances in kilometers and the automobile speedometers are calibrated in kilometers per hour. If you drive 90 km/h, how fast are you going in meters per second and in miles per hour?

PICTURE First we have to find the appropriate conversion factors for hours to seconds and kilometers to meters. We can use the facts that 1000 m = 1 km and 1 h = 60 min = 3600 s. The quantity 90 km/h is multiplied by the conversion factors, so the unwanted units cancel. (Each conversion factor has the value 1, so the value of the speed is not changed.) To convert to miles per hour, we use the conversion factor 1 mi/1.609 km.

SOLVE

1. Multiply 90 km/h by the conversion factors 1 h/3600 s and 1000 m/1 km to convert km to m and h to s:

$$\frac{90\ \cancel{\text{km}}}{\text{h}} \times \frac{1\ \cancel{\text{h}}}{3600\ \text{s}} \times \frac{1000\ \text{m}}{1\ \cancel{\text{km}}} = \boxed{25\ \text{m/s}}$$

2. Multiply 90 km/h by 1 mi/1.609 km:

$$\frac{90\ \cancel{\text{km}}}{\text{h}} \times \frac{1\ \text{mi}}{1.609\ \cancel{\text{km}}} = \boxed{56\ \text{mi/h}}$$

CHECK Notice that the final units in each step are correct. If you had not set up the conversion factors correctly, for example if you multiplied by 1 km/1000 m instead of 1000 m/1 km, the final units would not be correct.

TAKING IT FURTHER Step 1 can be shortened by writing 1 h/3600 s as 1 h/3.6 ks and canceling the prefixes in ks and km. That is,

$$\frac{90\ \text{km}}{\cancel{\text{h}}} \times \frac{1\ \cancel{\text{h}}}{3.6\ \text{ks}} = \boxed{25\ \text{m/s}}$$

Canceling these prefixes is equivalent to dividing the numerator and the denominator by 1000.

You may find it helpful to memorize the conversion results in Example 1-1. These results are

$$25\ \text{m/s} = 90\ \text{km/h} \approx (60\ \text{mi/h})$$

Knowing these values can provide you with a quick way to convert speeds to units you are more familiar with.

(Eunice Harris/Photo Researchers.)

1-4 DIMENSIONS OF PHYSICAL QUANTITIES

Recall that a physical quantity includes both a number and a unit. The unit tells the standard that is used for the measurement and the number gives the comparison of the quantity to the standard. To tell *what* you are measuring, however, you need to state the *dimension* of the physical quantity. Length, time, and mass are all dimensions. The distance d between two objects has dimensions of length. We express this relation as $[d] = \text{L}$, where $[d]$ represents the dimension of the distance d and L represents the dimension of length. All dimensions are represented by upper-case roman (nonitalic) letters. The letters T and M represent the dimensions of time and mass, respectively. The dimensions of a number of quantities can be written in terms of these fundamental dimensions. For example, the area A of a surface is found by multiplying one length by another. Because area is the product of two lengths, it is said to have the dimensions of length multiplied by length, or length squared, written $[A] = \text{L}^2$. In this equation, $[A]$ represents the dimension of the quantity A and L represents the dimension of length. Speed has the dimensions of length divided by time, or L/T. The dimensions of other quantities such as force or energy are written in terms of the fundamental quantities of length, time, and mass. Adding or subtracting two physical quantities makes sense only if the quantities have the same dimensions. For example, we cannot add an area to a speed to obtain a meaningful sum. For the equation

$$A = B + C$$

the quantities A, B, and C must all have the same dimensions. The addition of B and C also requires that these quantities be in the same units. For example, if B is an area of 500 in.^{2} and C is 4 ft^2, we must either convert B into square feet or C into square inches in order to find the sum of the two areas.

You can often find mistakes in a calculation by checking the dimensions or units of the quantities in your result. Suppose, for example, that you mistakenly use the formula $A = 2\pi r$ for the area of a circle. You can see immediately that this cannot be correct because $2\pi r$ has dimensions of length whereas area must have dimensions of length squared.

! Evaluating the dimensions of an expression will tell you only if the dimensions are correct, not whether the entire expression is correct. While expressing the area of a circle, for instance, dimensional analysis will not tell you the correct expression is πr^2 or $2\pi r^2$. (The correct expression is πr^2.)

Example 1-2 Dimensions of Pressure

The pressure P in a fluid in motion depends on its density ρ and its speed v. Find a simple combination of density and speed that gives the correct dimensions of pressure.

PICTURE Using Table 1-2, we can see that pressure has the dimensions M/(LT^2), density is M/L^3, and speed is L/T. In addition, both the dimensions of pressure and density have mass in the numerator, whereas the dimensions of speed do not contain mass. Therefore, the expression must involve multiplying or dividing dimensions of density and dimensions of speed to obtain the unit of mass in the dimensions of pressure. To find out the exact relationship, we can start by dividing the dimensions of pressure by those of density, and then evaluate the result with respect to the dimensions for speed.

SOLVE

1. Divide the dimensions of pressure by those of density to obtain an expression with no M in it:

$$\frac{[P]}{[\rho]} = \frac{\mathrm{M/LT^2}}{\mathrm{M/L^3}} = \frac{\mathrm{L^2}}{\mathrm{T^2}}$$

2. By inspection, we note that the result has dimensions of v^2. The dimensions of pressure are thus the same as the dimensions of density multiplied by speed squared:

$$[P] = [\rho][v^2] = \frac{\mathrm{M}}{\mathrm{L^3}} \times \left(\frac{\mathrm{L}}{\mathrm{T}}\right)^2 = \frac{\mathrm{M}}{\mathrm{L^3}} \times \frac{\mathrm{L^2}}{\mathrm{T^2}} = \boxed{\frac{\mathrm{M}}{\mathrm{LT^2}}}$$

CHECK Divide the dimensions of pressure by the dimensions of speed squared and the result is the dimensions of density $[P]/[v^2] = (\mathrm{M/LT^2})/(\mathrm{L^2/T^2}) = \mathrm{M/L^3} = [\rho]$.

1-5 SIGNIFICANT FIGURES AND ORDER OF MAGNITUDE

Many of the numbers in science are the result of measurement and are therefore known only to within a degree of experimental uncertainty. The magnitude of the uncertainty, which depends on both the skill of the experimenter and the apparatus used, often can only be estimated. A rough indication of the uncertainty in a measurement is inferred by the number of digits used. For example, if a tag on a table in a furniture store states that a table is 2.50 m long, it is saying that its length is close to, but not exactly, 2.50 m. The rightmost digit, the 0, is uncertain. If we use a tape measure with millimeter markings and measured the table length carefully, we might estimate that we could measure the length to ±0.6 mm of its true length. We would indicate this precision when giving the length by using four digits, such as 2.503 m. A reliably known digit (other than a zero used to locate the decimal point) is called a **significant figure.** The

Table 1-2 Dimensions of Physical Quantities

Quantity	Symbol	Dimension
Area	A	$\mathrm{L^2}$
Volume	V	$\mathrm{L^3}$
Speed	v	L/T
Acceleration	a	$\mathrm{L/T^2}$
Force	F	$\mathrm{ML/T^2}$
Pressure (F/A)	p	$\mathrm{M/LT^2}$
Density (M/V)	ρ	$\mathrm{M/L^3}$
Energy	E	$\mathrm{ML^2/T^2}$
Power (E/T)	P	$\mathrm{ML^2/T^3}$

number 2.50 has three significant figures; 2.503 m has four. The number 0.00130 has three significant figures. (The first three zeroes are not significant figures but are merely markers to locate the decimal point.) The number 2300. has four significant figures, but the number 2300 (the same as 2300. but without the decimal point) could have as few as two or as many as four significant figures. The number of significant figures in numbers with trailing zeros and no decimal point is ambiguous.

Suppose, for example, that you measure the area of a circular playing field by pacing off the radius and using the formula for the area of a circle, $A = \pi r^2$. If you estimate the radius to be 8 m and use a 10-digit calculator to compute the area, you obtain $\pi(8\text{ m})^2 = 201.0619298\text{ m}^2$. The digits after the decimal point give a false indication of the accuracy with which you know the area. To determine the appropriate numbers of significant figures for calculations involving multiplication and division, you can follow this general rule:

> When multiplying or dividing quantities, the number of significant figures in the final answer is no greater than that in the quantity with the fewest significant figures.

In the previous example, the radius is known to only one significant figure, so the area is also known only to one significant figure, or 200 m^2. This number indicates that the area is likely somewhere between 150 m^2 and 250 m^2.

The accuracy of the sum or difference of measurements is only as good as the accuracy of the *least* accurate of the measurements. A general rule is

> When adding or subtracting quantities, the number of decimal places in the answer should match that of the term with the smallest number of decimal places.

CONCEPT CHECK 1-1

How many significant figures does the number 0.010457 have?

! Exact values have an unlimited number of significant figures. For example, a value determined by counting, such as 2 tables, has no uncertainty and is an exact value. In addition, the conversion factor 1 m/100 cm is an exact value because 1 m *is* exactly equal to 100 cm.

! When you work with numbers that have uncertainties, you should be careful not to include more digits than the certainty of measurement warrants.

Example 1-3 Significant Figures

Subtract 1.040 from 1.21342.

PICTURE The first number, 1.040, has only three significant figures beyond the decimal point, whereas the second, 1.21342, has five. According to the rule stated for the addition and subtraction of numbers, the difference can have only three figures beyond the decimal point.

SOLVE

Sum the numbers, keeping only three digits beyond the decimal point: $1.21342 - 1.040 = 0.173\cancel{42} = \boxed{0.173}$

CHECK The answer cannot be more accurate than the least accurate number or 1.040. The answer has the same number of significant figures beyond the decimal point as 1.040.

TAKING IT FURTHER In this example, the given numbers have four and six significant figures, but the difference has only three significant figures. Most examples and exercises in this book will be done with data to two, three, or occasionally four significant figures.

PRACTICE PROBLEM 1-2 Apply the appropriate rule for significant figures to calculate the following: (*a*) 1.58×0.03, (*b*) $1.4 + 2.53$, (*c*) $2.456 - 2.453$

SCIENTIFIC NOTATION

When we work with very large or very small numbers we can show significant figures more easily by using scientific notation. In this notation, the number is written as a product of a number between 1 and 10 and a power of 10, such as 10^2 (= 100) or 10^3 (= 1000). For example, the number 12000000 is written 1.2×10^7; the distance from Earth to the Sun, which is about 150000000000 m, is written 1.5×10^{11} m. We have assumed that none of the trailing zeros in this number are significant. If two of the trailing zeros were significant we could express this unambiguously by writing the number as 1.500×10^{11} m. The number 11 in 10^{11} is called the exponent. For numbers smaller than 1, the exponent is negative. For example, $0.1 = 10^{-1}$, and $0.0001 = 10^{-4}$. The diameter of a virus, which is about 0.000 000 01 m, is written 1×10^{-8} m. Notice that by writing numbers in this form, you can easily identify the number of significant figures. For example, 1.5×10^{11} m contains two significant figures (1 and 5).

See **Math Tutorial** *for more information on* **Exponents**

PRACTICE PROBLEM 1-3

Apply the appropriate rule for significant figures to calculate $2.34 \times 10^2 + 4.93$.

Use the following Problem-Solving Strategy to do calculations with numbers in scientific notation.

PROBLEM-SOLVING STRATEGY

Scientific Notation

PICTURE If the numbers involved in a calculation are very large or very small, you may want to rewrite these numbers in scientific notation. This notation often makes it easier for you to determine the number of significant figures that a number has and makes it easier for you to perform calculations.

SOLVE Use these items to solve problems that involve scientific notation.

1. When numbers in scientific notation are multiplied, the exponents are added; when numbers in scientific notation are divided, the exponents are subtracted.

   Example: $10^2 \times 10^3 = 100 \times 1{,}000 = 100000 = 10^5$

   Example: $\dfrac{10^2}{10^3} = \dfrac{100}{1000} = \dfrac{1}{10} = 10^{-1}$

2. In scientific notation, 10^0 is defined to be 1. To see why, suppose we divide 1000 by 1000.

   Example: $\dfrac{1000}{1000} = \dfrac{10^3}{10^3} = 10^{3-3} = 10^0 = 1$

3. Be careful when adding or subtracting numbers written in scientific notation when their exponents do not match.

   Example: $(1.200 \times 10^2) + (8 \times 10^{-1}) = 120.0 + 0.8 = 120.8$

4. To find the sum without converting both numbers into ordinary decimal form, rewrite either of the numbers so that its power of 10 is the same as that of the other.

   Example: $(1200 \times 10^{-1}) + (8 \times 10^{-1}) = 1208 \times 10^{-1} = 120.8$

5. When raising a power to another power, the exponents are multiplied.

   Example: $(10^2)^4 = 10^2 \times 10^2 \times 10^2 \times 10^2 = 10^8$

CHECK Make sure that when you convert numbers smaller than one into scientific notation, the exponent is negative. You should also remember when exponents are added, subtracted, or multiplied, because performing the wrong operation can cause your answer to be inaccurate by powers of 10.

TAKING IT FURTHER During a calculation, avoid entering intermediate results via keyboard entry. Instead store these results in the calculator memory. If you must enter intermediate results via the keyboard, key in one or two additional (non significant) digits, called *guard digits*. This practice serves to minimize round-off errors.

! All exponents are dimensionless and have no units.

Example 1-4 How Much Water?

A liter (L) is the volume of a cube that is 10 cm by 10 cm by 10 cm. If you drink 1 L (exact) of water, how much volume in cubic centimeters and in cubic meters would it occupy in your stomach?

PICTURE The volume V of a cube of edge length ℓ is ℓ^3. The volume in cubic centimeters is found directly from $\ell = 10$ cm. To find the volume in cubic meters, convert cm^3 to m^3 using the conversion factor 1 cm $= 10^{-2}$ m.

SOLVE

1. Calculate the volume in cm^3:

$$V = \ell^3 = (10\text{ cm})^3 = 1000\text{ cm}^3 = \boxed{10^3\text{ cm}^3}$$

2. Convert to m^3:

$$10^3\text{ cm}^3 = 10^3\text{ cm}^3 \times \left(\frac{10^{-2}\text{ m}}{1\text{ cm}}\right)^3$$

Notice that the conversion factor (which equals 1) can be raised to the third power without changing its value, enabling us to cancel units.

$$= 10^3\text{ cm}^3 \times \frac{10^{-6}\text{ m}^3}{1\text{ cm}^3} = \boxed{10^{-3}\text{ m}^3}$$

CHECK Notice that the answers are in cubic centimeters and cubic meters. These answers are consistent with volume having dimensions of length cubed. Also note that the physical quantity 10^3 is greater than the physical quantity 10^{-3}, which is consistent with a meter being larger than a centimeter.

Example 1-5 Counting Atoms

In 12.0 g of carbon, there are $N_A = 6.02 \times 10^{23}$ carbon atoms (Avogadro's number). If you could count 1 atom per second, how long would it take to count the atoms in 1.00 g of carbon? Express your answer in years.

PICTURE We need to find the total number of atoms to be counted, N, and then use the fact that the number counted equals the counting rate R multiplied by the time t.

SOLVE

1. The time is related to the total number of atoms N, and the rate of counting $R = 1$ atom/s:

$$N = Rt$$

2. Find the number of carbon atoms in 1.00 g:

$$N = \frac{6.02 \times 10^{23}\text{ atoms}}{12.0\text{ g}} = 5.02 \times 10^{22}\text{ atoms}$$

3. Calculate the number of seconds it takes to count these at 1 per second:

$$t = \frac{N}{R} = \frac{5.02 \times 10^{22}\text{ atoms}}{1\text{ atom/s}} = 5.02 \times 10^{22}\text{ s}$$

4. Calculate the number n of seconds in a year:

$$n = \frac{365\text{ d}}{1.00\text{ y}} \times \frac{24\text{ h}}{1\text{ d}} \times \frac{3600\text{ s}}{1\text{ h}} = 3.15 \times 10^7\text{ s/y}$$

5. Use the conversion factor 3.15×10^7 s/y (a handy quantity to remember) to convert the answer in step 3 to years:

$$t = 5.02 \times 10^{22}\text{ s} \times \frac{1.00\text{ y}}{3.15 \times 10^7\text{ s}}$$

$$= \frac{5.02}{3.15} \times 10^{22-7}\text{ y} = \boxed{1.59 \times 10^{15}\text{ y}}$$

CHECK The answer can be checked by estimation. If you need approximately 10^{22} seconds to count the number of atoms in a gram of carbon and there are approximately 10^7 seconds in a year, then you would need $10^{22}/10^7 = 10^{15}$ y.

TAKING IT FURTHER The time required is about 100000 times the current known value for the age of the universe.

PRACTICE PROBLEM 1-4 How long would it take for 5 billion (5×10^9) people to count the atoms in 1 g of carbon?

Table 1-3 The Universe by Orders of Magnitude

Size or Distance	(m)	Mass	(kg)	Time Interval	(s)
Proton	10^{-15}	Electron	10^{-30}	Time for light to cross nucleus	10^{-23}
Atom	10^{-10}	Proton	10^{-27}	Period of visible light radiation	10^{-15}
Virus	10^{-7}	Amino acid	10^{-25}	Period of microwaves	10^{-10}
Giant amoeba	10^{-4}	Hemoglobin	10^{-22}	Half-life of muon	10^{-6}
Walnut	10^{-2}	Flu virus	10^{-19}	Period of highest audible sound	10^{-4}
Human being	10^{0}	Giant amoeba	10^{-8}	Period of human heartbeat	10^{0}
Highest mountain	10^{4}	Raindrop	10^{-6}	Half-life of free neutron	10^{3}
Earth	10^{7}	Ant	10^{-4}	Period of Earth's rotation	10^{3}
Sun	10^{9}	Human being	10^{2}	Period of Earth's revolution around the Sun	10^{7}
Distance from Earth to the Sun	10^{11}	Saturn V rocket	10^{6}	Lifetime of human being	10^{9}
Solar system	10^{13}	Pyramid	10^{10}	Half-life of plutonium-239	10^{12}
Distance to nearest star	10^{16}	Earth	10^{24}	Lifetime of mountain range	10^{15}
Milky Way galaxy	10^{21}	Sun	10^{30}	Age of Earth	10^{17}
Visible universe	10^{26}	Milky Way galaxy	10^{41}	Age of universe	10^{18}
		Universe	10^{52}		

ORDER OF MAGNITUDE

In doing rough calculations, we sometimes round off a number to the nearest power of 10. Such a number is called an **order of magnitude.** For example, the height of an ant might be 8×10^{-4} m or approximately 10^{-3} m. We would say that the order of magnitude of an ant's height is 10^{-3} m. Similarly, though the typical height h of most people is about 2 m, we might round that off further and say that $h \sim 10^{0}$ m, where the symbol $\sim$ means "is the order of magnitude of." By saying $h \sim 10^{0}$ m we do not mean that a typical height is really 1 m but that it is closer to 1 m than to 10 m or to 10^{-1} m. We might say that a human being is three orders of magnitude taller than an ant, meaning that the ratio of heights is about 1000 to 1. An order of magnitude does not provide any digits that are reliably known, so it has no significant figures. Table 1-3 gives some order-of-magnitude values for a variety of sizes, masses, and time intervals encountered in physics.

In many cases, the order of magnitude of a quantity can be estimated using plausible assumptions and simple calculations. The physicist Enrico Fermi was a master at using order-of-magnitude estimations to generate answers for questions that seemed impossible to calculate because of lack of information. Problems like these are often called **Fermi questions.** The following examples are Fermi questions.

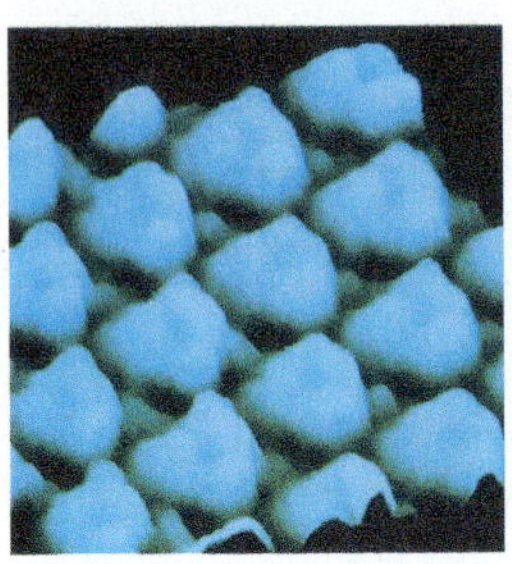

Benzene molecules of the order of 10^{-10} m in diameter as seen in a scanning electron microscope. *(IBM Research, Almaden Research Center.)*

The diameter of the Andromeda galaxy is of the order of 10^{21} m. *(Smithsonian Institution.)*

Distances familiar in our everyday world. The height of the woman is of the order of 10^{0} m and that of the mountain is of the order of 10^{4} m. *(Kent and Donnan Dannon/Photo Researchers.)*

Example 1-6 Burning Rubber

What thickness of rubber tread is worn off the tire of your automobile as it travels 1 km (0.6 mi)?

PICTURE Let's assume the tread thickness of a new tire is 1 cm. This estimation may be off by a factor of two or so, but 1 mm is certainly too small and 10 cm is too large. Because tires have to be replaced after about 60000 km (about 37000 mi), we will also assume that the tread is completely worn off after 60000 km. In other words, the rate of wear is 1 cm of tire per 60000 km of travel.

(Corbis.)

SOLVE

Use 1 cm wear per 60000 km travel to compute the thickness worn after 1 km of travel:

$$\frac{1\text{ cm wear}}{60000\text{ km travel}} = \frac{1.7 \times 10^{-5}\text{ cm wear}}{1\text{ km travel}} \approx \boxed{2 \times 10^{-7}\text{ m wear per km of travel}}$$

CHECK If you multiply 1.7×10^{-5} cm/km by 60000 km, you will get approximately 1 cm, which is the thickness of tread on a new tire.

TAKING IT FURTHER Atoms have diameters equal to about 2×10^{-10} m. Thus, the thickness worn off for each kilometer of travel is about 1000 atomic diameters thick.

Example 1-7 How Many Grains

Context-Rich

You have been placed on detention for falling asleep in class. Your teacher says you can get off detention early by estimating the number of grains of sand on a beach. You decide to give it your best shot.

PICTURE First, you make some assumptions about the size of the beach and the size of each grain of sand. Let's assume the beach is about 500 m long, 100 m wide, and 3 m deep. Searching the Internet, you learn that the diameter of a grain varies from 0.04 mm to 2 mm. You assume that each grain is a 1-mm-diameter sphere. Let's also assume that the grains are so tightly packed that the volume of the space between the grains is negligible compared to the volume of the sand itself.

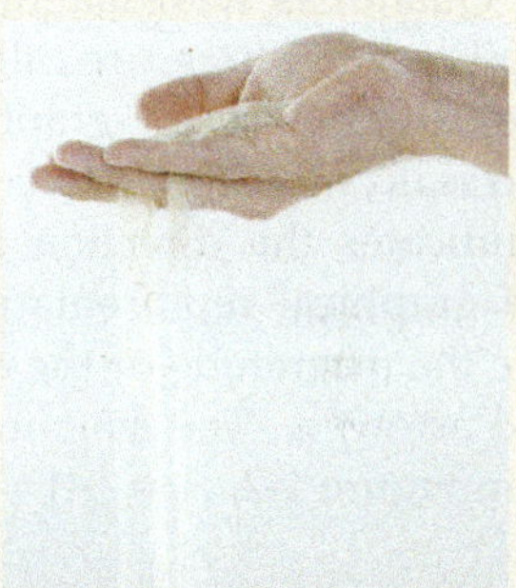

(Corbis.)

SOLVE

1. The volume V_B of the beach is equal to the number N of grains times the volume V_G of a single grain:

$$V_B = NV_G$$

2. Using the formula for the volume of a sphere, find the volume of a single grain of sand:

$$V_G = \frac{4}{3}\pi R^3$$

3. Solve for the number of grains. The numbers in our assumptions are specified to only one significant figure, so the answer will be expressed with one significant figure:

$$V_B = NV_G = N\frac{4}{3}\pi R^3$$

so

$$N = \frac{3V_B}{4\pi R^3} = \frac{3(500\text{ m})(100\text{ m})(3\text{ m})}{4\pi(0.5 \times 10^{-3}\text{ m})^3} = 2.9 \times 10^{14} \approx \boxed{3 \times 10^{14}}$$

CHECK To check the answer, divide the volume of the beach by the number of grains the beach holds. The result is $1.5 \times 10^5\text{ m}^3/3 \times 10^{14}$ grains $= 5 \times 10^{-10}\text{ m}^3$/grain. This result is the estimated volume of one grain of sand or $4/[3\pi(5 \times 10^{-4}\text{ m})^3]$.

TAKING IT FURTHER The volume of the space between grains can be found by first filling a one-liter container with dry sand, and then slowly pouring water into the container until the sand is saturated with water. If we assume that one-tenth of a liter of water is needed to fully saturate the sand in the container, the actual volume of the sand in the one-liter container is only nine-tenths of a liter. Our estimate of the number of grains on the beach is too high. Taking into account that the sand actually occupies only, say, 90 percent of the volume of its container, the number of grains on the beach would be only 90 percent of the value obtained in step 3 of our solution.

PRACTICE PROBLEM 1-5 How many grains of sand are on a 2-km stretch of beach that is 500 m wide? *Hint: Assume that the sand is* 3.00 m *deep and the diameter of one grain of sand is* 1.00 mm.

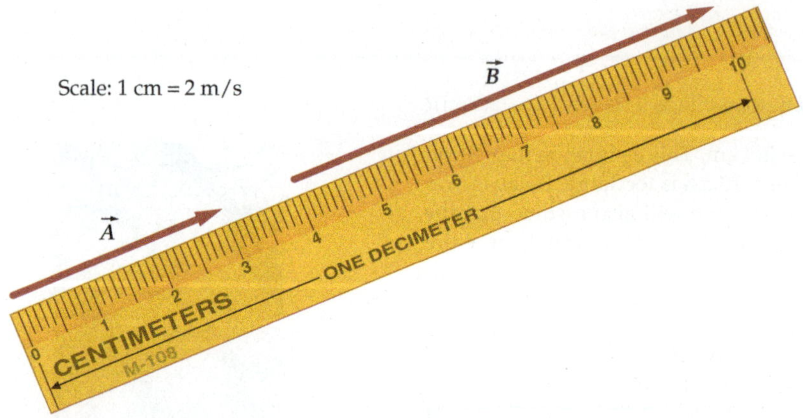

FIGURE 1-2 Velocity vectors $\vec{A}$ and $\vec{B}$ have magnitudes of 6 m/s and 12 m/s, respectively. The arrows representing them are drawn using the scale 1 cm = 2 m/s, so the arrows are drawn 3 and 6 cm long.

1-6 VECTORS

If an object moves in a straight line, we can describe its motion by describing how far or how fast it moves, and whether it moves to the left or right of the origin. But when we look at the motion of an object that is moving in two or three dimensions, we need more than just plus and minus signs to indicate direction. Quantities that have magnitude and direction, such as velocity, acceleration, and force, are called **vectors.** Quantities with magnitude but no associated direction, such as speed, mass, volume, and time, are called **scalars.**

We represent a vector graphically using an arrow. The length of the arrow, drawn to scale, indicates the magnitude of the vector quantity. The direction of the arrow indicates the direction of the vector quantity. Figure 1-2, for example, shows a graphical representation of two velocity vectors. One velocity vector has twice the magnitude of the other. We denote vectors by italic letters with an overhead arrow, $\vec{A}$. The magnitude of $\vec{A}$ is written $|\vec{A}|$, $\|\vec{A}\|$, or simply A. For the vectors in Figure 1-2, $A = |\vec{A}| = 6$ m/s and $B = |\vec{B}| = 12$ m/s.

! While working with vectors, you should always include an arrow over the letter to indicate a vector quantity. The letter without the arrow represents only the magnitude of the vector quantity. Note that the magnitude of a vector is never negative.

1-7 GENERAL PROPERTIES OF VECTORS

Like scalar quantities, vector quantities can be added, subtracted, and multiplied. However, manipulating vectors algebraically requires taking into account their direction. In this section, we will examine some of the general properties of vectors and how to work with them (multiplication of two vectors will be discussed in Chapters 6 and 9). Throughout most of the discussion, we will consider displacement vectors—vectors that represent change of position—because they are the most basic of vectors. However, keep in mind that the properties apply to all vectors, not just displacement vectors.

BASIC DEFINITIONS

If an object moves from location A to location B, we can represent its **displacement** with an arrow pointing from A to B, as shown in Figure 1-3*a*. The length of the arrow represents the distance, or magnitude, between the two locations. The direction of the arrow represents the direction from A to B. A displacement vector is a directed straight-line segment from the initial location to the final location that represents the change in position of an object. It does not necessarily represent the

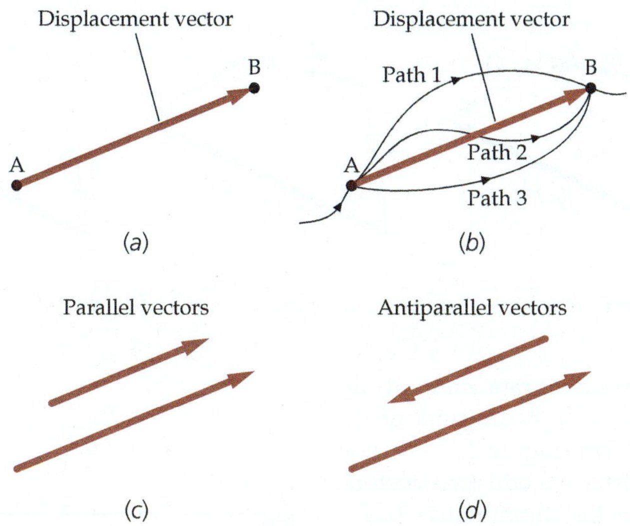

FIGURE 1-3 (*a*) shows a displacement vector from a point A to a point B; (*b*) shows the same displacement vector with three different paths between the two points; (*c*) shows the same displacement vector next to a second displacement vector that is parallel but a different length; (*d*) shows the same displacement vector next to a vector that is antiparallel (the head and tail are reversed) and a different length.

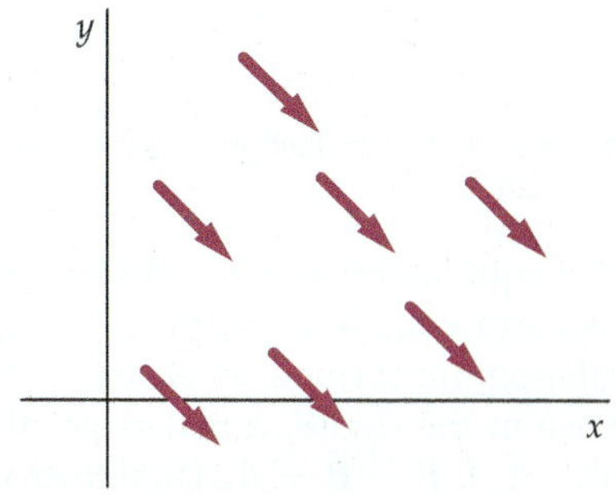

FIGURE 1-4 Vectors are equal if their magnitudes and directions are the same. All vectors in this figure are equal.

actual path the object takes. For example, in Figure 1-3*b*, the same displacement vector corresponds to all three paths between points A and B.

If two displacement vectors have the same direction, as shown in Figure 1-3*c*, they are **parallel.** If they have opposite directions, as shown in Figure 1-3*d*, they are **antiparallel.** If two vectors have *both* the same magnitude and the same direction, they are said to be equal. Graphically, this means that they have the same length and are parallel to each other. A vector can be drawn at different locations as long as it is drawn with the correct magnitude (length) and in the correct direction. Thus, all the vectors in Figure 1-4 are equal. In addition, vectors do not depend on the *coordinate system* used to represent them (except for position vectors, which are introduced in Chapter 3). Two or three mutually perpendicular coordinate axes form a coordinate system.

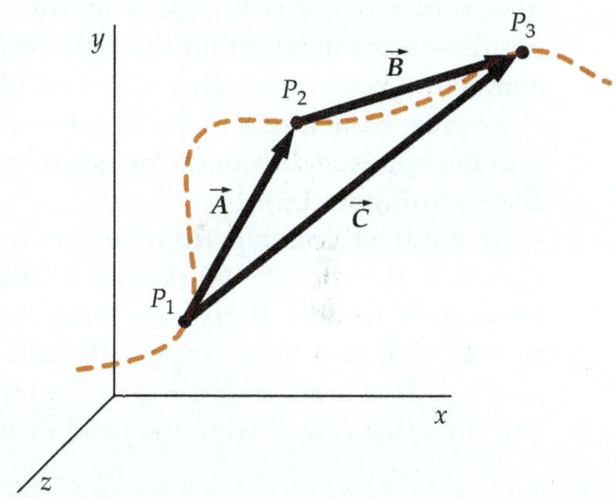

FIGURE 1-5

ADDITION AND SUBTRACTION OF VECTORS

Suppose you decide to take a hike along a trail through a forest. Figure 1-5 shows your path as you move from point P_1 to a second point P_2 and then to a third point P_3. The vector $\vec{A}$ represents your displacement from P_1 to P_2, while $\vec{B}$ represents your displacement from P_2 to P_3. Note that these displacement vectors depend only on the endpoints and not on the actual path taken. Your *net* displacement from P_1 to P_3, is a new vector, labeled $\vec{C}$ in the figure, and is called the sum of the two successive displacements $\vec{A}$ and $\vec{B}$:

$$\vec{C} = \vec{A} + \vec{B} \qquad 1\text{-}1$$

The sum of two vectors is called the sum, **vector sum,** or **resultant.**

The plus sign in Equation 1-1 refers to a process called *vector addition.* We find the sum using a geometric process that takes into account both the magnitudes and the directions of the quantities. To add two displacement vectors graphically, we draw the second vector $\vec{B}$ with its tail $\vec{B}$ at the head of the first vector $\vec{A}$ (Figure 1-6). The resultant vector is then drawn from the tail of the first to the head of the second. This method of adding vectors is called the **head-to-tail method.**

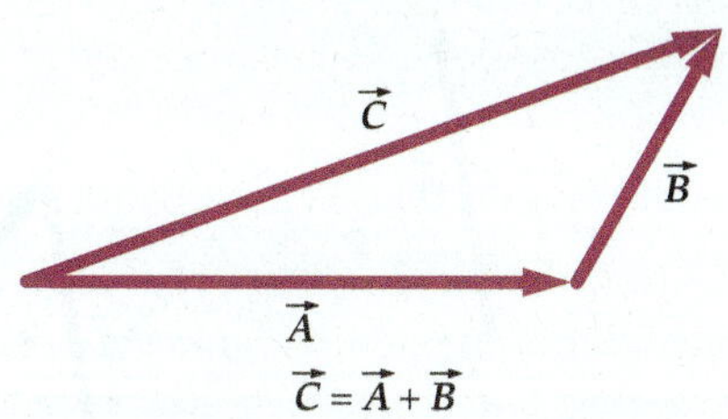

FIGURE 1-6 Head-to-tail method of vector addition.

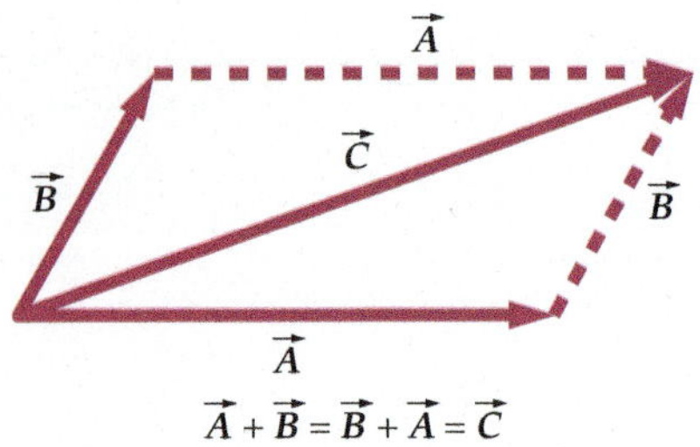

FIGURE 1-7 Parallelogram method of vector addition.

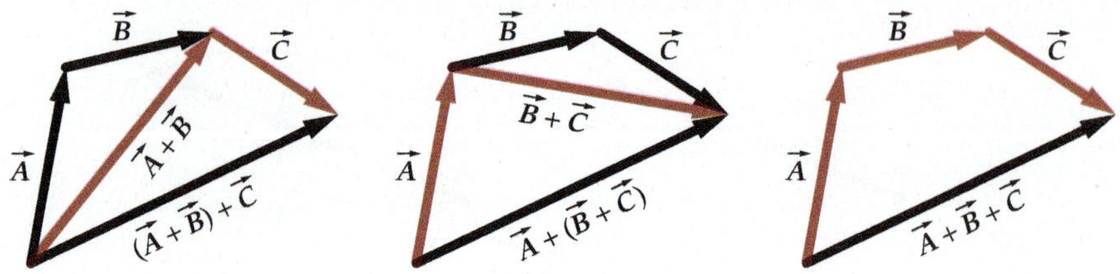

FIGURE 1-8 Vector addition is associative. That is, $(\vec{A}+\vec{B})+\vec{C}=\vec{A}+(\vec{B}+\vec{C})$.

An equivalent way of adding vectors, called the **parallelogram method,** involves drawing $\vec{B}$ so that it is tail-to-tail with $\vec{A}$ (Figure 1-7). A diagonal of the parallelogram formed by $\vec{A}$ and $\vec{B}$ then equals $\vec{C}$ as shown (Figure 1-7). As you can see in the figure, it makes no difference in which order we add two vectors; that is, $\vec{A}+\vec{B}=\vec{B}+\vec{A}$. Therefore, vector addition obeys the commutative law.

To add more than two vectors—for example, $\vec{A}$, $\vec{B}$, and $\vec{C}$—we first add two vectors (Figure 1-8), and then add the third vector to the vector sum of the first two. The order in which the vectors are grouped before adding does not matter; that is $(\vec{A}+\vec{B})+\vec{C}=\vec{A}+(\vec{B}+\vec{C})$. This reveals that like the addition of ordinary numbers, vector addition is associative.

> **!** C does not equal $A+B$ unless $\vec{A}$ and $\vec{B}$ are in the same direction. That is, $\vec{C}=\vec{A}+\vec{B}$ does not imply that $C=A+B$.

If vectors $\vec{A}$ and $\vec{B}$ are equal in magnitude and opposite in direction, then the vector $\vec{C}=\vec{A}+\vec{B}$ is a vector with a magnitude of zero. This can be shown by using the head-to-tail method of vector addition to graphically construct the sum $\vec{A}+\vec{B}$. Any vector with a magnitude of zero is called the zero vector $\vec{0}$. The direction of a vector with zero magnitude has no meaning, so in this book we will not use vector notation for the zero vector. That is, we will use 0 rather than $\vec{0}$ to denote the zero vector. If $\vec{A}+\vec{B}=0$, then $\vec{B}$ is said to be the negative of $\vec{A}$ and vice versa. Note that $\vec{B}$ is the negative of $\vec{A}$ if $\vec{B}$ has the same magnitude as $\vec{A}$ but is in the opposite direction. The negative of $\vec{A}$ is written $-\vec{A}$, so if $\vec{A}+\vec{B}=0$, then $\vec{B}=-\vec{A}$ (Figure 1-9).

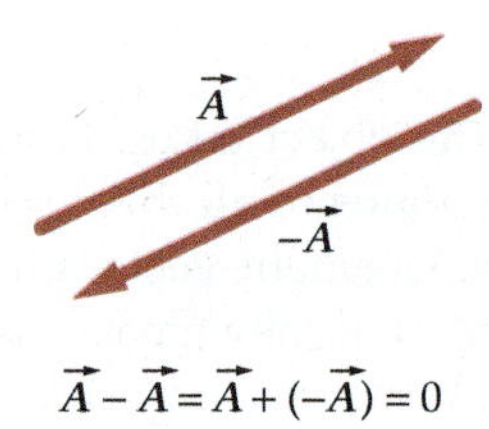

FIGURE 1-9

To subtract vector $\vec{B}$ from vector $\vec{A}$, add the negative of $\vec{B}$ to $\vec{A}$. The result is $\vec{C}=\vec{A}-\vec{B}=\vec{A}+(-\vec{B})$ (Figure 1-10*a*). An alternative method of subtracting $\vec{B}$ from $\vec{A}$ is to add $\vec{B}$ to both sides of the equation $\vec{C}=\vec{A}+(-\vec{B})$ to obtain $\vec{B}+\vec{C}=\vec{A}$, and then graphically add $\vec{B}$ and $\vec{C}$ to get $\vec{A}$ using the head-to-tail method. This is accomplished by first drawing $\vec{A}$ and$\vec{B}$ tail-to-tail (Figure 1-10*b*), and then drawing $\vec{C}$ from the head of $\vec{B}$ to the head of $\vec{A}$.

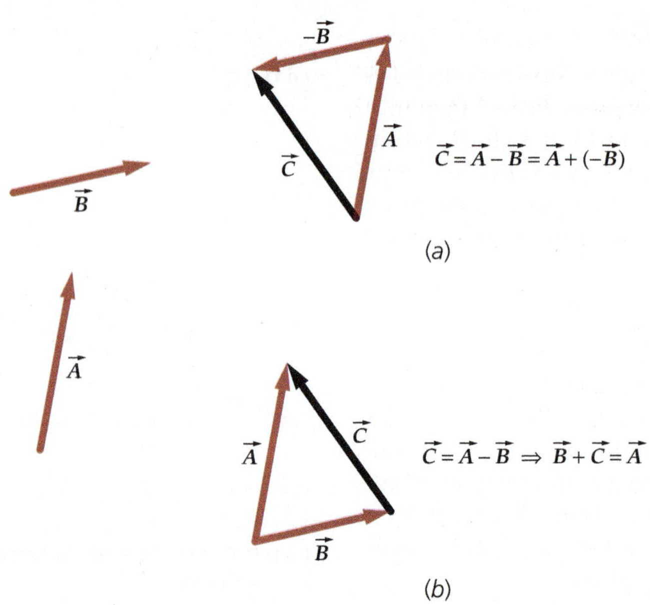

FIGURE 1-10 Alternative ways of subtracting vectors. Let $\vec{C}=\vec{A}-\vec{B}$. (*a*) To obtain $\vec{C}$, we add $-\vec{B}$ to $\vec{A}$. (*b*) To obtain $\vec{C}$, we first draw $\vec{A}$ and $\vec{B}$ with their tails together. Then, $\vec{C}$ is the vector we add to $\vec{B}$ to get $\vec{A}$.

Example 1-8 Your Displacement — *Conceptual*

You walk 3.00 km due east and then 4.00 km due north. Determine your resultant displacement by adding these two displacement vectors graphically.

PICTURE Your displacement is the vector from your initial position to your final position. You can add the two individual displacement vectors graphically to find the resultant displacement. To accurately draw the resultant, you must use a scale such as 1 cm on the drawing = 1 km on the ground.

SOLVE

1. Let $\vec{A}$ and $\vec{B}$ represent displacements of 3.00 km due east and 4.00 km due north, respectively, and let $\vec{C} = \vec{A} + \vec{B}$. Draw $\vec{A}$ and $\vec{B}$ with the tail of $\vec{B}$ at the head of $\vec{A}$, and with $\vec{C}$ drawn from the tail of $\vec{A}$ to the head of $\vec{B}$ (Figure 1-11). Use the scale 1 cm = 1 km. Include axes indicating the directions north and east.
2. Determine the magnitude and direction of $\vec{C}$ using your diagram, the scale 1 cm = 1 km, and a protractor.

   The arrow representing $\vec{C}$ has a length of 5.00 cm, so the magnitude of $\vec{C}$ is 5.00 km. The direction of $\vec{C}$ is approximately 53° north of east.

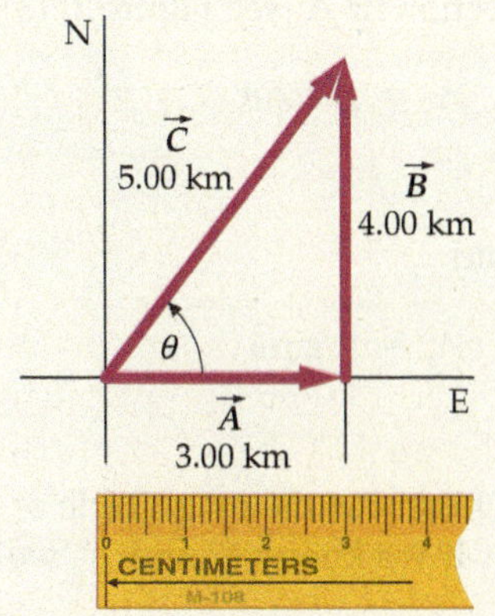

FIGURE 1-11

CHECK The distance traveled is 3.00 km + 4.00 km = 7.00 km and the magnitude of the net displacement is 5 km. This is consistent with the adage "the shortest distance between two points is a straight line." Also, if you go 3 km east and 4 km north, you should expect to be somewhat more than 45° north of east from your starting point.

TAKING IT FURTHER A vector is described by its magnitude and its direction. Your resultant displacement is therefore a vector of length 5.00 km in a direction approximately 53° north of east.

MULTIPLYING A VECTOR BY A SCALAR

The expression $3\vec{A}$, where $\vec{A}$ is an arbitrary vector, represents the sum $\vec{A} + \vec{A} + \vec{A}$. That is, $\vec{A} + \vec{A} + \vec{A} = 3\vec{A}$. (In like manner, $(-\vec{A}) + (-\vec{A}) + (-\vec{A}) = 3(-\vec{A}) = -3\vec{A}$.) More generally, the vector $\vec{A}$ multiplied by a scalar s is the vector $\vec{B} = s\vec{A}$, where $\vec{B}$ has magnitude $|s|A$. $\vec{B}$ is in the same direction as $\vec{A}$ if s is positive and is in the opposite direction if s is negative. The dimensions of $s\vec{A}$ are those of s multiplied by those of A. (In addition, to divide $\vec{A}$ by a scalar s, you multiply $\vec{A}$ by $1/s$.)

COMPONENTS OF VECTORS

We can add or subtract vectors algebraically by first breaking down the vectors into their components. The **component** of a vector in a given direction is the projection of the vector onto an axis in that direction. We can find the components of a vector by drawing perpendicular lines from the ends of the vector to the axis, as shown in Figure 1-12. The process of finding the x, y, and z components of a

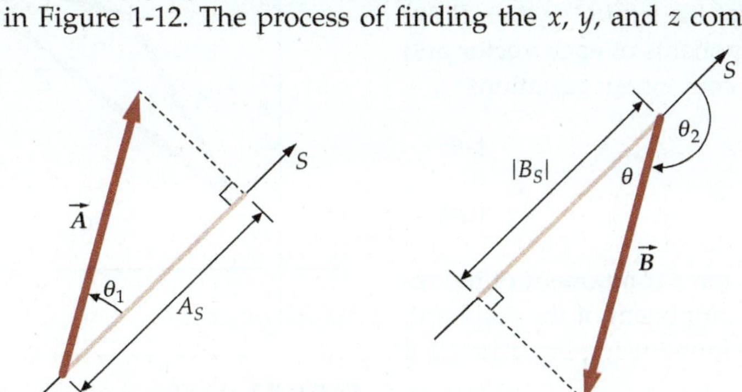

$A_S = A\cos\theta_1$ (*a*)

$B_S = B\cos\theta_2 = -B\cos\theta$ (*b*)

FIGURE 1-12 The component of a vector in a specified direction is equal to the magnitude of the vector times the cosine of the angle between the direction of the vector and the specified direction. The component of the vector $\vec{A}$ in the $+S$ direction is A_S, and A_S is positive. The component of the vector $\vec{B}$ in the $+S$ direction is B_S, and B_S is negative.

vector is called **resolving the vector** into its components. The components of a vector along the x, y, and z directions, illustrated in Figure 1-13 for a vector in the xy plane, are called the rectangular (or Cartesian) components. Note that the components of a vector *do* depend on the coordinate system used, although the vector itself does not.

We can use right-triangle geometry to find the rectangular components of a vector. If θ is the angle measured counterclockwise* from the $+x$ direction to the direction of $\vec{A}$ (see Figure 1-13), then

$$A_x = A\cos\theta \qquad \text{1-2}$$

x COMPONENT OF A VECTOR

and

$$A_y = A\sin\theta \qquad \text{1-3}$$

y COMPONENT OF A VECTOR

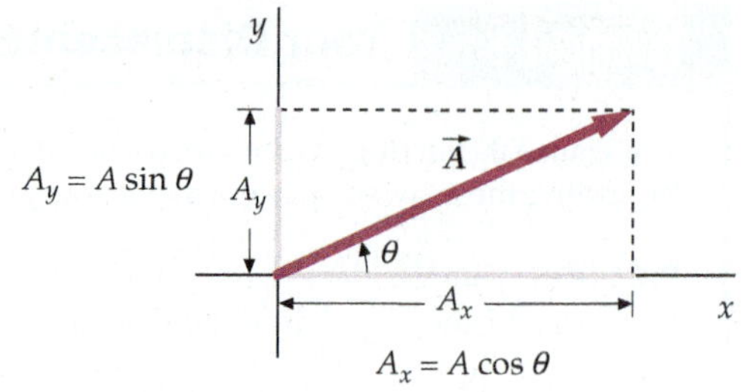

FIGURE 1-13 The rectangular components of a vector. θ is the angle between the direction of the vector and the $+x$ direction. The angle is positive if it is measured counterclockwise from the $+x$ direction, as shown.

where A is the magnitude of $\vec{A}$.

If we know A_x and A_y we can find the angle θ from

$$\tan\theta = \frac{A_y}{A_x} \qquad \theta = \tan^{-1}\frac{A_y}{A_x} \qquad \text{1-4}$$

and the magnitude A from the Pythagorean theorem:

$$A = \sqrt{A_x^2 + A_y^2} \qquad \text{1-5}a$$

In three dimensions,

$$A = \sqrt{A_x^2 + A_y^2 + A_z^2} \qquad \text{1-5}b$$

Components can be positive or negative. The x component of a vector is positive if the x coordinate of an ant as it walks from the tail to the head of the vector increases. Thus, if $\vec{A}$ points in the positive x direction, then A_x is positive, and if $\vec{A}$ points in the negative x direction, then A_x is negative.

It is important to note that in Equation 1-4, the inverse tangent function is multiple valued. This issue is clarified in Example 1-9.

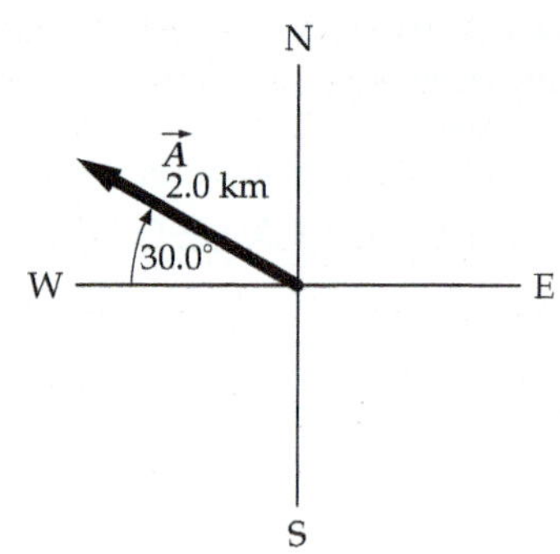

FIGURE 1-14

PRACTICE PROBLEM 1-6

A car travels 20.0 km in a direction 30.0° north of west. Let east be the $+x$ direction and north be the $+y$ direction, as in Figure 1-14. Find the x and y components of the displacement vector of the car.

Once we have resolved a vector into its components, we can manipulate the individual components. Consider two vectors $\vec{A}$ and $\vec{B}$ that lie in the xy plane. The rectangular components of each vector and those of the sum $\vec{C} = \vec{A} + \vec{B}$ are shown in Figure 1-15. We see that the rectangular components of each vector and those of the sum $\vec{C} = \vec{A} + \vec{B}$ are equivalent to the two component equations

$$C_x = A_x + B_x \qquad \text{1-6}a$$

and

$$C_y = A_y + B_y \qquad \text{1-6}b$$

In other words, the sum of the x components equals the x component of the resultant, and the sum of the y components equals the y component of the resultant. The angle and magnitude of the resultant vector can be found using Equations 1-4 and 1-5a, respectively.

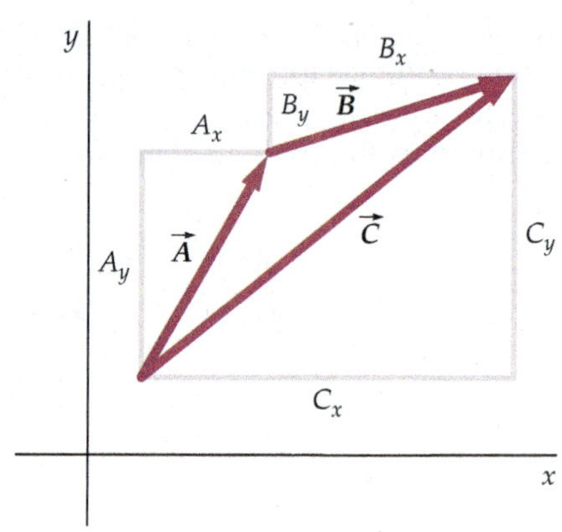

FIGURE 1-15

* This assumes the $+y$ direction is 90° counterclockwise from the $+x$ direction.

Example 1-9 A Treasure Map

Context-Rich

You are working at a tropical resort, and are setting up a treasure hunt activity for the guests. You've been given a map and told to follow its directions in order to bury a "treasure" at a specific location. You don't want to waste time walking around the island, because you want to finish early and go surfing. The directions are to walk 3.00 km headed 60.0° north of due east and then 4.00 km headed 40.0° north of due west. Where should you head and how far must you walk to get the job done quickly? Find your answer (*a*) graphically and (*b*) using components.

PICTURE In both cases you need to find your resultant displacement. In Part (*a*), use the head-to-tail method of vector addition and solve for the resultant vector graphically. You can do this by drawing each of the displacements to scale and then measuring the resultant displacement directly from your sketch. For Part (*b*), you will need to resolve the vectors into their individual components and then use the components to find the resultant displacement.

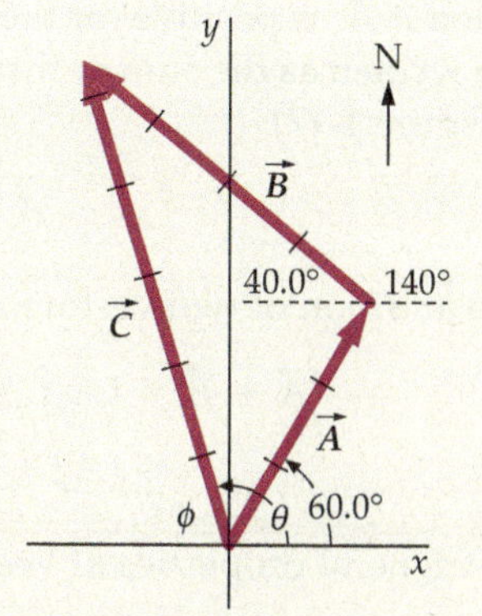

FIGURE 1-16

SOLVE

(*a*) 1. Draw a vector-addition diagram to scale (Figure 1-16). First draw coordinate axes, with the $+x$ direction toward the east and the $+y$ direction toward the north. Next, starting at the origin draw the first displacement vector $\vec{A}$ 3.00 cm long at 60.0° north of east. Beginning at the head of $\vec{A}$, draw the second vector $\vec{B}$ 4.00 cm long at 40.0° north of west. (You will need a protractor to measure the angles.) Then draw the resultant vector $\vec{C}$ from the tail of $\vec{A}$ to the head of $\vec{B}$:

2. Measure the length of $\vec{C}$. Using a protractor, measure the angle between the direction of $\vec{C}$ and the $+x$ direction:

$\vec{C}$ is about 5.40 cm long. Thus, the magnitude of the resultant displacement is $\boxed{5.40\text{ km}}$. The angle ϕ made between $\vec{C}$ and due west is about 73.2°. Therefore, you should walk 5.40 km headed $\boxed{73.2°\text{ north of west}}$.

(*b*) 1. To solve using components, let $\vec{A}$ denote the first displacement and choose the $+x$ direction toward the east and the $+y$ direction toward the north. Compute A_x and A_y from Equations 1-2 and 1-3:

$$A_x = (3.00\text{ km})\cos 60° = 1.50\text{ km}$$
$$A_y = (3.00\text{ km})\sin 60° = 2.60\text{ km}$$

2. Similarly, compute the components of the second displacement $\vec{B}$. The angle between the direction of $\vec{B}$ and the $+x$ direction is $180.0° - 40.0° = 140°$:

$$B_x = (4.00\text{ km})\cos 140° = -3.06\text{ km}$$
$$B_y = (4.00\text{ km})\sin 140° = +2.57\text{ km}$$

3. The components of the resultant displacement $\vec{C} = \vec{A} + \vec{B}$ are found by addition:

$$C_x = A_x + B_x = 1.50\text{ km} - 3.06\text{ km} = -1.56\text{ km}$$
$$C_y = A_y + B_y = 2.60\text{ km} + 2.57\text{ km} = 5.17\text{ km}$$

4. The Pythagorean theorem gives the magnitude of $\vec{C}$:

$$C^2 = C_x^2 + C_y^2 = (-1.56\text{ km})^2 + (5.17\text{ km})^2 = 29.2\text{ km}^2$$
$$C = \sqrt{29.2\text{ km}^2} = \boxed{5.40\text{ km}}$$

5. The ratio of C_y to C_x equals the tangent of the angle θ between $\vec{C}$ and the positive x direction. Be careful, the value you are seeking may be 180° larger than the value returned by your calculator for the inverse tangent:

$$\tan\theta = \frac{C_y}{C_x} \quad \text{so}$$
$$\theta = \tan^{-1}\frac{5.17\text{ km}}{-1.56\text{ km}} = \tan^{-1}(-3.31)$$
$$= \text{either } -73.2° \quad \text{or} \quad (-73.2° + 180°)$$
$$= \text{either } -73.2° \quad \text{or} \quad +107°$$

6. Because C_y is positive and C_x is negative we know to select the value for θ in the second quadrant:

$$\theta = \boxed{107°\text{ counterclockwise from east}}$$
$$\phi = \boxed{73.2°\text{ north of west}}$$

CHECK Step 4 of Part (*b*) gives the magnitude as 5.40 km and step 6 gives the direction as 73.2° north of west. This agrees with the results in Part (*a*) within the accuracy of our measurement.

TAKING IT FURTHER To specify a vector, you need to specify either the magnitude *and* direction, or both components. In this example, the magnitude and direction was specifically asked for.

UNIT VECTORS

A **unit vector** is a *dimensionless* vector with magnitude exactly equal to 1. The vector $\hat{A} = \vec{A}/A$ is an example of a unit vector that points in the direction of $\vec{A}$. The circumflex, or hat, denotes that it is a unit vector. Unit vectors that point in the positive x, y, and z directions are convenient for expressing vectors in terms of their rectangular components. These unit vectors are usually written $\hat{\imath}$, $\hat{\jmath}$, and $\hat{k}$, respectively. For example, the vector $A_x\hat{\imath}$ has magnitude $|A_x|$ and points in the $+x$ direction if A_x is positive (or the $-x$ direction if A_x is negative). A general vector $\vec{A}$ can be written as the sum of three vectors, each of which is parallel to a coordinate axis (Figure 1-17):

$$\vec{A} = A_x\hat{\imath} + A_y\hat{\jmath} + A_z\hat{k} \qquad 1\text{-}7$$

The addition of two vectors $\vec{A}$ and $\vec{B}$ can be written in terms of unit vectors as

$$\begin{aligned}\vec{A} + \vec{B} &= (A_x\hat{\imath} + A_y\hat{\jmath} + A_z\hat{k}) + (B_x\hat{\imath} + B_y\hat{\jmath} + B_z\hat{k}) \\ &= (A_x + B_x)\hat{\imath} + (A_y + B_y)\hat{\jmath} + (A_z + B_z)\hat{k} \qquad 1\text{-}8\end{aligned}$$

The general properties of vectors are summarized in Table 1-4.

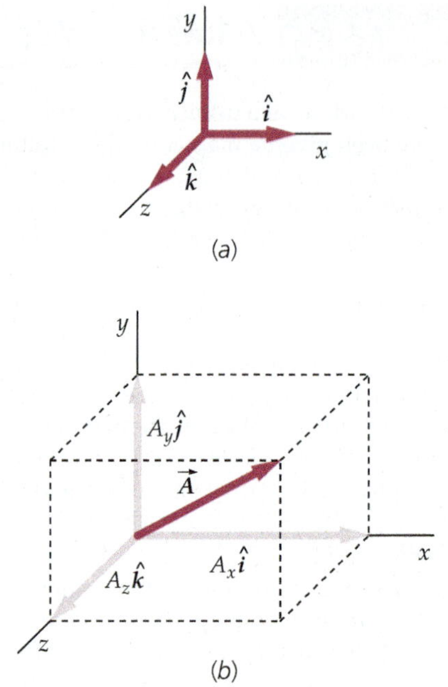

FIGURE 1-17 (*a*) The unit vectors $\hat{\imath}$, $\hat{\jmath}$, and $\hat{k}$ in a rectangular coordinate system. (*b*) The vector $\vec{A}$ in terms of the unit vectors: $\vec{A} = A_x\hat{\imath} + A_y\hat{\jmath} + A_z\hat{k}$.

PRACTICE PROBLEM 1-7

Given vectors $\vec{A} = (4.00\text{ m})\hat{\imath} + (3.00\text{ m})\hat{\jmath}$ and $\vec{B} = (2.00\text{ m})\hat{\imath} - (3.00\text{ m})\hat{\jmath}$, find (*a*) A, (*b*) B, (*c*) $\vec{A} + \vec{B}$, and (*d*) $\vec{A} - \vec{B}$.

Table 1-4 Properties of Vectors

Property	Explanation	Figure	Component Representation
Equality	$\vec{A} = \vec{B}$ if $\lvert\vec{A}\rvert = \lvert\vec{B}\rvert$ and their directions are the same		$A_x = B_x$ $A_y = B_y$ $A_z = B_z$
Addition	$\vec{C} = \vec{A} + \vec{B}$	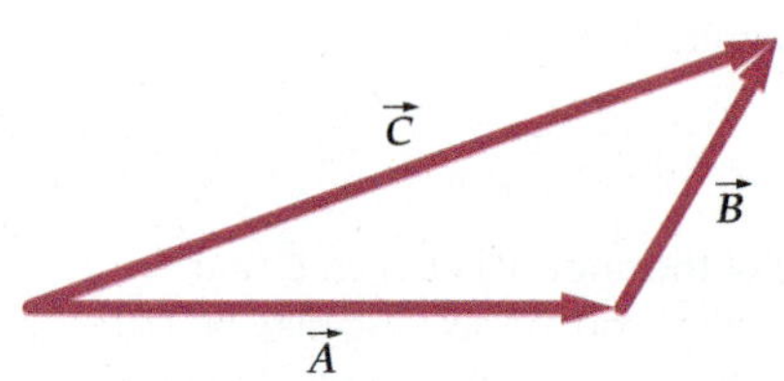	$C_x = A_x + B_x$ $C_y = A_y + B_y$ $C_z = A_z + B_z$
Negative of a vector	$\vec{A} = -\vec{B}$ if $\lvert\vec{B}\rvert = \lvert\vec{A}\rvert$ and their directions are opposite		$A_x = -B_x$ $A_y = -B_y$ $A_z = -B_z$
Subtraction	$\vec{C} = \vec{A} - \vec{B}$	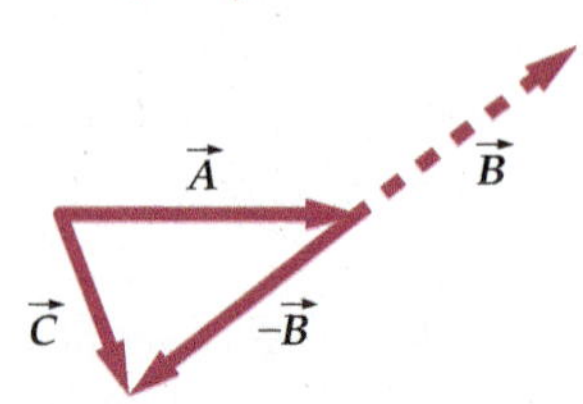	$C_x = A_x - B_x$ $C_y = A_y - B_y$ $C_z = A_z - B_z$
Multiplication by a scalar	$\vec{B} = s\vec{A}$ has magnitude $\lvert\vec{B}\rvert = \lvert s\rvert\lvert\vec{A}\rvert$ and has the same direction as $\vec{A}$ if s is positive or $-\vec{A}$ if s is negative	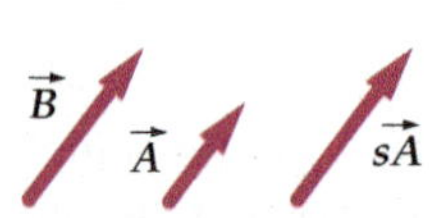	$B_x = sA_x$ $B_y = sA_y$ $B_z = sA_z$

Physics Spotlight

The 2005 Leap Second

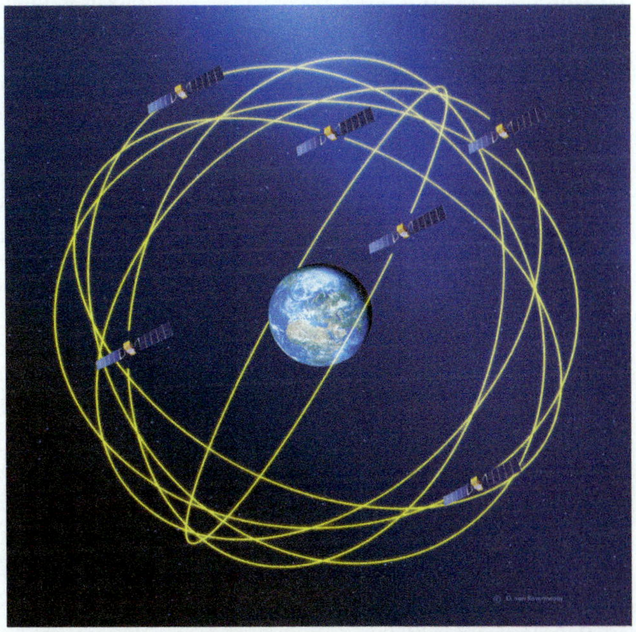

The global positioning system (GPS) requires that there be 24 satellites in primary service at least 70 percent of the time. Each primary satellite has an orbital period of 1/2 a sidereal day (1 sidereal day = ~23 h 56 min) and an orbital radius about 4 times the radius of Earth. There are 6 equally spaced orbital planes, each of which is inclined 55° with respect to the equatorial plane of Earth, and each of these planes contains 4 primary satellites. In addition, there are several other GPS satellites that serve as in-orbit spares in the event that one or more of the primary satellites fails. At the time of this writing (May 2006) there are 29 operational satellites in orbit. *(Detlev Van Ravenswaay/Photo Researchers.)*

The calendar year 2005 was longer—by exactly one second, known officially as a "leap second." This adjustment was necessary to synchronize two systems of keeping time, one based on Earth's rotation and the other based on a select group of atomic clocks.

Throughout history, timekeeping has been related to the position of the Sun in the sky, a factor determined by Earth's rotation on its axis and around the Sun. This astronomical time, now called Universal Time (UT1), assumed that the rate of Earth's rotation was uniform. But as more accurate methods of measurement were developed, it became evident that there were slight irregularities in the rotation rate of Earth. This meant that there would also be some variability in the scientific standard unit for time, the second, as long as its definition—(1/60)(1/60)(1/24) of a mean solar day—depended on astronomical time.

In 1955 the National Physical Laboratory in Britain developed the first cesium atomic clock, a device of far greater accuracy than any clock formerly in existence. Timekeeping could now be independent of astronomical observations, and a much more precise definition of the second could be given based on the frequency of radiation emitted in the transition between two energy levels of the cesium-133 atom. However, the more familiar UT1 continues to be of importance for systems such as navigation and astronomy. Thus it is important that atomic time and UT1 be synchronized.

According to the National Physical Laboratory, UK, "The solution adopted [for synchronization] was to construct an atomic time scale called Coordinated Universal Time ... (UTC) as the basis of international timekeeping. It combines all the regularity of atomic time with most of the convenience of UT1, and many countries have adopted it as the legal basis for time."* The International Bureau of Weights and Measures in Sèvres, France, takes data from select time laboratories around the world, including the U.S. Naval Observatory in Washington, DC, to provide the international standard UTC.

When slight differences accrue between UTC and UT1 because Earth's rotation varies slightly (usually slowing) over time, a leap second is added to close the gap. The concept is similar to the way that leap years are used to correct the calendar. A year is not exactly 365 days, but rather 365.242 days. To account for this, an extra day is added to the calendar every four years and designated February 29.

Since 1972 when the world shifted to atomic timekeeping, 23 leap seconds have been added to UTC. By international agreement, a leap second is added whenever the difference between UT1 and UTC approaches 0.9 seconds. The International Earth Rotation and Reference Systems (IERS) Service, through its center at the Paris Observatory, announces the need for a leap second months in advance.

In a year without any leap second, the last second of the year would be 23:59:59 UTC on December 31, while the first second of the new year would be 00:00:00 UTC on January 1 of the new year. But for 2005 a leap second was added at 23:59:59 UTC on December 31, so that atomic clocks read 23:59:60 UTC before changing to all zeros.

* http://www.npl.co.uk/time/leap_second.html

Summary

TOPIC	RELEVANT EQUATIONS AND REMARKS
1. Units	**Physical quantities** are numbers that are obtained by taking measurements of physical objects. Operational definitions specify operations or procedures that, if followed, define physical quantities. The magnitude of a physical quantity is expressed as a number times a unit.
2. Base Units	The base units in the SI system are the meter (m), the second (s), the kilogram (kg), the kelvin (K), the ampere (A), the mole (mol), and the candela (cd). The unit(s) of every physical quantity can be expressed in terms of these base units.
3. Units in Equations	Units in equations are treated just like any other algebraic quantity.
4. Conversion	**Conversion factors,** which are always equal to 1, provide a convenient method for converting from one kind of unit to another.
5. Dimensions	The terms of an equation must have the same dimensions.
6. Scientific Notation	For convenience, very small and very large numbers are generally written as a number between 1 and 10 times a power of 10.
7. Exponents	
Multiplication	When multiplying two numbers, the exponents are added.
Division	When dividing two numbers, the exponents are subtracted.
Raising to a power	When a number containing an exponent is itself raised to a power, the exponents are multiplied.
8. Significant Figures	
Multiplication and division	The number of significant figures in the result of multiplication or division is *no greater than* the least number of significant figures in any of the numbers.
Addition and subtraction	The result of addition or subtraction of two numbers has no significant figures beyond the last decimal place where both of the numbers being added or subracted have significant figures.
9. Order of Magnitude	A number rounded to the nearest power of 10 is called an order of magnitude. The order of magnitude of a quantity can often be estimated using plausible assumptions and simple calculations.
10. Vectors	
Definition	Vectors are quantities that have both magnitude and direction. Vectors add like displacements.
Components	The component of a vector in a direction in space is the projection of the vector on an axis in that direction. If $\vec{A}$ makes an angle θ with the positive x direction, its x and y components are $A_x = A\cos\theta$ (1-2) $A_y = A\sin\theta$ (1-3)
Magnitude	$A = \sqrt{A_x^2 + A_y^2}$ (1-5*a*)
Adding vectors graphically	Two vectors may be added graphically by drawing them with the tail of the second arrow at the head of the first arrow. The arrow representing the resultant vector is drawn from the tail of the first vector to the head of the second.
Adding vectors using components	If $\vec{C} = \vec{A} + \vec{B}$, then $C_x = A_x + B_x$ (1-6*a*) and $C_y = A_y + B_y$ (1-6*b*)
Unit vectors	A vector $\vec{A}$ can be written in terms of unit vectors $\hat{i}$, $\hat{j}$, and $\hat{k}$, which are dimensionless, have unit magnitude and lie along the x, y, and z axes, respectively $\vec{A} = A_x\hat{i} + A_y\hat{j} + A_z\hat{k}$ (1-7)

Answers to Concept Checks

1-1 5

Answers to Practice Problems

1-1 (*a*) 300 ns; (*b*) 40 Mm

1-2 (*a*) 0.05, (*b*) 3.9, (*c*) 0.003

1-3 2.39×10^2

1-4 3.2×10^5 y

1-5 $\approx 6 \times 10^{15}$

1-6 $A_x = -17.3$ km, $A_y = 10.0$ km

1-7 (*a*) $A = 5.00$ m, (*b*) $B = 3.61$ m, (*c*) $\vec{A} + \vec{B} = (6.00\text{ m})\hat{i}$, (*d*) $\vec{A} - \vec{B} = (2.00\text{ m})\hat{i} + (6.00\text{ m})\hat{j}$

Problems

In a few problems, you are given more data than you actually need; in a few other problems, you are required to supply data from your general knowledge, outside sources, or informed estimate.

Interpret as significant all digits in numerical values that have trailing zeroes and no decimal points.

- • Single-concept, single-step, relatively easy
- •• Intermediate-level, may require synthesis of concepts
- ••• Challenging, for advanced students
- SSM Solution is in the *Student Solutions Manual*
- Consecutive problems that are shaded are paired problems.

CONCEPTUAL PROBLEMS

1 • Which of the following *is not* one of the base quantities in the SI system? (*a*) mass, (*b*) length, (*c*) energy, (*d*) time, (*e*) All of the above are base quantities. SSM

2 • In doing a calculation, you end up with m/s in the numerator and m/s^2 in the denominator. What are your final units? (*a*) m^2/s^3, (*b*) 1/s, (*c*) s^3/m^2, (*d*) s, (*e*) m/s

3 • The prefix giga means (*a*) 10^3, (*b*) 10^6, (*c*) 10^9, (*d*) 10^{12}, (*e*) 10^{15}.

4 • The prefix mega means (*a*) 10^{-9}, (*b*) 10^{-6}, (*c*) 10^{-3}, (*d*) 10^6, (*e*) 10^9.

5 • Show that there are 30.48 cm per foot. How many centimeters are there in one mile? SSM

6 • The number 0.000 513 0 has ______ significant figures. (*a*) one, (*b*) three, (*c*) four, (*d*) seven, (*e*) eight

7 • The number 23.0040 has ______ significant figures. (*a*) two, (*b*) three, (*c*) four, (*d*) five, (*e*) six

8 • Force has dimensions of mass times acceleration. Acceleration has dimensions of speed divided by time. Pressure is defined as force divided by area. What are the dimensions of pressure? Express pressure in terms of the SI base units kilogram, meter, and second.

9 • True or false: Two quantities must have the same dimensions in order to be multiplied.

10 • A vector has a negative x component and a positive y component. Its angle measured counterclockwise from the positive x axis is (*a*) between zero and 90 degrees, (*b*) between 90 and 180 degrees, (*c*) more than 180 degrees.

11 • A vector $\vec{A}$ points in the $+x$ direction. Show graphically at least three choices for a vector $\vec{B}$ such that $\vec{B} + \vec{A}$ points in the $+y$ direction. SSM

12 • A vector $\vec{A}$ points in the $+y$ direction. Show graphically at least three choices for a vector $\vec{B}$ such that $\vec{B} - \vec{A}$ points in the $+x$ direction.

13 • Is it possible for three equal-magnitude vectors to add to zero? If so, sketch a graphical answer. If not, explain why not. SSM

ESTIMATION AND APPROXIMATION

14 • The angle subtended by the moon's diameter at a point on Earth is about 0.524° (Figure 1-18). Use this information and the fact that the moon is about 384 Mm away to find the diameter of the moon. *Hint: The angle can be determined from the diameter of the moon and the distance to the moon.*

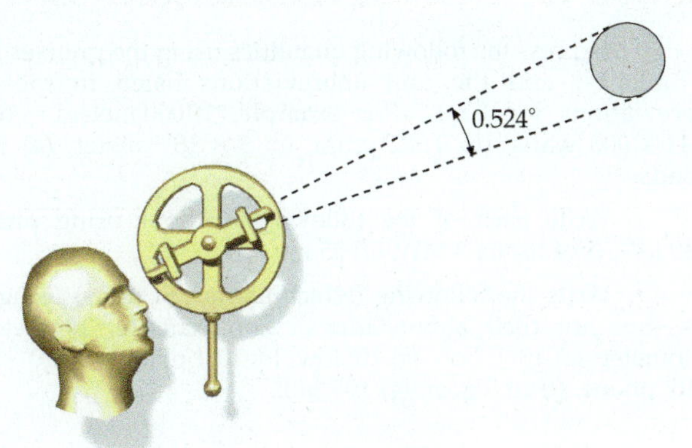

FIGURE 1-18 Problem 14

15 • **BIOLOGICAL APPLICATION** Some good estimates about the human body can be made if it is assumed that we are made mostly of water. The mass of a water molecule is 29.9×10^{-27} kg. If the mass of a person is 60 kg, estimate the number of water molecules in that person. SSM

16 •• **ENGINEERING APPLICATION** In 1989, IBM scientists moved atoms with a scanning tunneling microscope (STM). One of the first STM images seen by the general public was of the letters IBM spelled with xenon atoms on a nickel surface. The letters IBM were 15 xenon atoms across. If the space between the centers of adjacent xenon atoms is 5 nm (5×10^{-9} m), estimate how many times "IBM" could be written across this 8.5-inch page.

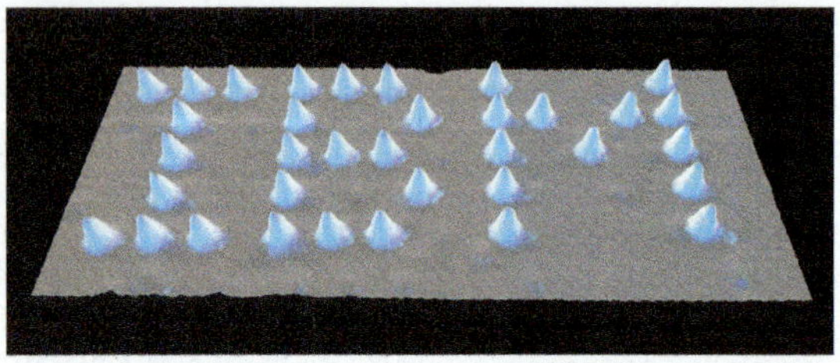

(By permission of IBM Reasearch, Almaden Research Center.)

17 •• There is an environmental debate over the use of cloth versus disposable diapers. (*a*) If we assume that between birth and 2.5 y of age, a child uses 3 diapers per day, estimate the total number of disposable diapers used in the United States per year. (*b*) Estimate the total landfill volume due to these diapers, assuming that 1000 kg of waste fills about 1 m³ of landfill volume. (*c*) How many square miles of landfill area at an average height of 10 m is needed for the disposal of diapers each year?

18 •• (*a*) Estimate the number of gallons of gasoline used per day by automobiles in the United States and the total amount of money spent on it. (*b*) If 19.4 gal of gasoline can be made from one barrel of crude oil, estimate the total number of barrels of oil imported into the United States per year to make gasoline. How many barrels per day is this?

19 •• **ENGINEERING APPLICATION** A megabyte (MB) is a unit of computer memory storage. A CD has a storage capacity of 700 MB and can store approximately 70 min of high-quality music. (*a*) If a typical song is 5 min long, how many megabytes are required for each song? (*b*) If a page of printed text takes approximately 5 kilobytes, estimate the number of novels that could be saved on a CD. SSM

UNITS

20 • Express the following quantities using the prefixes listed in Table 1-1 and the unit abbreviations listed in the table Abbreviations for Units. For example, 10 000 meters = 10 km. (*a*) 1 000 000 watts, (*b*) 0.002 gram, (*c*) 3×10^{-6} meter, (*d*) 30 000 seconds.

21 • Write each of the following without using prefixes: (*a*) 40 μW, (*b*) 4 ns, (*c*) 3 MW, (*d*) 25 km.

22 • Write the following (which are not SI units) using prefixes (but not their abbreviations). For example, 10^3 meters = 1 kilometer: (*a*) 10^{-12} boo, (*b*) 10^9 low, (*c*) 10^{-6} phone, (*d*) 10^{-18} boy, (*e*) 10^6 phone, (*f*) 10^{-9} goat, (*g*) 10^{12} bull.

23 •• In the following equations, the distance x is in meters, the time t is in seconds, and the velocity v is in meters per second. What are the SI units of the constants C_1 and C_2? (*a*) $x = C_1 + C_2 t$, (*b*) $x = \frac{1}{2}C_1 t^2$, (*c*) $v^2 = 2C_1 x$, (*d*) $x = C_1 \cos C_2 t$, (*e*) $v^2 = 2C_1 v - (C_2 x)^2$ SSM

24 •• If x is in feet, t is in milliseconds, and v is in feet per second, what are the units of the constants C_1 and C_2 in each part of Problem 23?

CONVERSION OF UNITS

25 • **MULTISTEP** From the original definition of the meter in terms of the distance along a meridian from the equator to the North Pole, find in meters (*a*) the circumference of Earth and (*b*) the radius of Earth. (*c*) Convert your answers for (*a*) and (*b*) from meters into miles.

26 • The speed of sound in air is 343 m/s. What is the speed of a supersonic plane that travels at twice the speed of sound? Give your answer in kilometers per hour and miles per hour.

27 • A basketball player is 6 ft $10\frac{1}{2}$ in tall. What is his height in centimeters?

28 • Complete the following: (*a*) 100 km/h = ______ mi/h, (*b*) 60 cm = ______ in., (*c*) 100 yd = ______ m.

29 • The main span of the Golden Gate Bridge is 4200 ft. Express this distance in kilometers.

30 • Find the conversion factor to convert from miles per hour into kilometers per hour.

31 • Complete the following: (*a*) 1.296×10^5 km/h² = ______ km/(h·s), (*b*) 1.296×10^5 km/h² = ______ m/s², (*c*) 60 mi/h = ______ ft/s, (*d*) 60 mi/h = ______ m/s.

32 • There are 640 acres in a square mile. How many square meters are there in one acre?

33 •• **CONTEXT-RICH** You are a delivery person for the Fresh Aqua Spring Water Company. Your truck carries 4 pallets. Each pallet carries 60 cases of water. Each case of water has 24 one-liter bottles. The dolly you use to carry the water into the stores has a weight limit of 250 lb. (*a*) If a milliliter of water has a mass of 1 g, and a kilogram has a weight of 2.2 lb, what is the weight, in pounds, of all the water in your truck? (*b*) How many full cases of water can you carry on the cart? SSM

34 •• A right circular cylinder has a diameter of 6.8 in. and a height of 2 ft. What is the volume of the cylinder in (*a*) cubic feet, (*b*) cubic meters, (*c*) liters?

35 •• In the following, x is in meters, t is in seconds, v is in meters per second, and the acceleration a is in meters per second squared. Find the SI units of each combination: (*a*) v^2/x, (*b*) $\sqrt{x/a}$, (*c*) $\frac{1}{2}at^2$. SSM

DIMENSIONS OF PHYSICAL QUANTITIES

36 • What are the dimensions of the constants in each part of Problem 23?

37 • The law of radioactive decay is $N(t) = N_0 e^{-\lambda t}$, where N_0 is the number of radioactive nuclei at $t = 0$, $N(t)$ is the number remaining at time t, and λ is a quantity known as the decay constant. What is the dimension of λ?

38 •• The SI unit of force, the kilogram-meter per second squared (kg·m/s²) is called the newton (N). Find the dimensions and the SI units of the constant G in Newton's law of gravitation $F = Gm_1 m_2/r^2$.

39 •• The magnitude of the force (F) that a spring exerts when it is stretched a distance x from its unstressed length is governed by Hooke's law, $F = kx$. (*a*) What are the dimensions of the *force constant*, k? (*b*) What are the dimensions and SI units of the quantity kx^2?

40 •• Show that the product of mass, acceleration, and speed has the dimensions of power.

41 •• The momentum of an object is the product of its velocity and mass. Show that momentum has the dimensions of force multiplied by time. SSM

42 •• What combination of force and one other physical quantity has the dimensions of power?

43 •• When an object falls through air, there is a drag force that depends on the product of the cross sectional area of the object and the square of its velocity, that is, $F_{air} = CAv^2$, where C is a constant. Determine the dimensions of C. SSM

44 •• Kepler's third law relates the period of a planet to its orbital radius r, the constant G in Newton's law of gravitation ($F = Gm_1m_2/r^2$), and the mass of the Sun M_s. What combination of these factors gives the correct dimensions for the period of a planet?

SCIENTIFIC NOTATION AND SIGNIFICANT FIGURES

45 • Express as a decimal number without using powers of 10 notation: (*a*) 3×10^4, (*b*) 6.2×10^{-3}, (*c*) 4×10^{-6}, (*d*) 2.17×10^5. SSM

46 • Write the following in scientific notation: (*a*) 1 345 100 m = ______ km, (*b*) 12 340 kW = ______ MW, (*c*) 54.32 ps = ______ s, (*d*) 3.0 m = ______ mm.

47 • Calculate the following, round off to the correct number of significant figures, and express your result in scientific notation: (*a*) $(1.14)(9.99 \times 10^4)$, (*b*) $(2.78 \times 10^{-8}) - (5.31 \times 10^{-9})$, (*c*) $12\pi/(4.56 \times 10^{-3})$, (*d*) $27.6 + (5.99 \times 10^2)$. SSM

48 • Calculate the following, round off to the correct number of significant figures, and express your result in scientific notation: (*a*) $(200.9)(569.3)$, (*b*) $(0.000\,000\,513)(62.3 \times 10^7)$, (*c*) $28\,401 + (5.78 \times 10^4)$, (*d*) $63.25/(4.17 \times 10^{-3})$.

49 • **BIOLOGICAL APPLICATION** A cell membrane has a thickness of 7.0 nm. How many cell membranes would it take to make a stack 1.0 in. high? SSM

50 •• **ENGINEERING APPLICATION** A circular hole of radius 8.470×10^{-1} cm must be cut into the front panel of a display unit. The *tolerance* is 1.0×10^{-3} cm, which means the actual hole cannot differ by more than this quantity from the desired radius. If the actual hole is larger than the desired radius by the allowed tolerance, what is the difference between the actual area and the desired area of the hole?

51 •• **ENGINEERING APPLICATION** A square peg must be made to fit through a square hole. If you have a square peg that has an edge length of 42.9 mm, and the square hole has an edge length of 43.2 mm, (*a*) what is the area of the space available when the peg is in the hole? (*b*) If the peg is made rectangular by removing 0.10 mm of material from one side, what is the area available now? SSM

VECTORS AND THEIR PROPERTIES

52 • **MULTISTEP** A vector that is 7.0 units long and a vector that is 5.5 units long are added. Their sum is a vector 10.0 units long. (*a*) Show graphically at least one way that the vectors can be added. (*b*) Using your sketch in Part (*a*), determine the angle between the original two vectors.

53 • Determine the x and y components of the following three vectors in the xy plane. (*a*) A 10-m displacement vector that makes an angle of 30° clockwise from the $+y$ direction. (*b*) A 25-m/s velocity vector that makes an angle of 40° counterclockwise from the $-x$ direction. (*c*) A 40-lb force vector that makes an angle of 120° counterclockwise from the $-y$ direction. SSM

54 • Rewrite the following vectors in terms of their magnitude and angle (counterclockwise from the $+x$ direction). (*a*) A displacement vector with an x component of +8.5 m and a y component of −5.5 m. (*b*) A velocity vector with an x component of −75 m/s and a y component of +35 m/s. (*c*) A force vector with a magnitude of 50 lb that is in the third quadrant with an x component whose magnitude is 40 lb.

55 • **CONCEPTUAL** You walk 100 m in a straight line on a horizontal plane. If this walk took you 50 m east, what are your possible north or south movements? What are the possible angles that your walk made with respect to due east?

56 • **ESTIMATION** The final destination of your journey is 300 m due east of your starting point. The first leg of this journey is the walk described in Problem 55, and the second leg in also a walk along a single straight-line path. Estimate graphically the length and heading for the second leg of your journey.

57 •• Given the following vectors: $\vec{A} = 3.4\hat{i} + 4.7\hat{j}$, $\vec{B} = (-7.7)\hat{i} + 3.2\hat{j}$, and $\vec{C} = 5.4\hat{i} + (-9.1)\hat{j}$. (*a*) Find the vector $\vec{D}$, in unit vector notation, such that $\vec{D} + 2\vec{A} - 3\vec{C} + 4\vec{B} = 0$. (*b*) Express your answer in Part (*a*) in terms of magnitude and angle with the $+x$ direction.

58 •• Given the following force vectors: $\vec{A}$ is 25 lb at an angle of 30° clockwise from the $+x$ axis, and $\vec{B}$ is 42 lb at an angle of 50° clockwise from the $+y$ axis. (*a*) Make a sketch and visually estimate the magnitude and angle of the vector $\vec{C}$ such that $2\vec{A} + \vec{C} - \vec{B}$ results in a vector with a magnitude of 35 lb pointing in the $+x$ direction. (*b*) Repeat the calculation in Part (*a*) using the method of components and compare your result to the estimate in (*a*).

59 •• Calculate the unit vector (in terms of $\hat{i}$ and $\hat{j}$) in the direction opposite to the direction of each of vectors $\vec{A}$, $\vec{B}$ and $\vec{C}$ in Problem 57. SSM

60 •• Unit vectors $\hat{i}$ and $\hat{j}$ are directed east and north, respectively. Calculate the unit vector (in terms of $\hat{i}$ and $\hat{j}$) in the following directions. (*a*) northeast, (*b*) 70° clockwise from the $-y$ axis, (*c*) southwest

GENERAL PROBLEMS

61 • The Apollo trips to the moon in the 1960s and 1970s typically took 3 days to travel the Earth–moon distance once they left Earth orbit. Estimate the spacecraft's average speed in kilometers per hour, miles per hour, and meters per second. SSM

62 • On many of the roads in Canada the speed limit is 100 km/h. What is this speed limit in miles per hour?

63 • If you could count \$1.00 per second, how many years would it take to count 1.00 billion dollars?

64 • (*a*) The speed of light in vacuum is 186 000 mi/s = 3.00×10^8 m/s. Use this fact to find the number of kilometers in a mile. (*b*) The weight of 1.00 ft³ of water is 62.4 lb, and 1.00 ft = 30.5 cm. Use this information and the fact that 1.00 cm³ of water has a mass of 1.00 g to find the weight in pounds of a 1.00-kg mass.

65 • The mass of one uranium atom is 4.0×10^{-26} kg. How many uranium atoms are there in 8.0 g of pure uranium?

66 •• During a thunderstorm, a total of 1.4 in. of rain falls. How much water falls on one acre of land? (1 mi² = 640 acres.) Express your answer in (*a*) cubic inches, (*b*) cubic feet, (*c*) cubic meters, and (*d*) kilograms. Note that the density of water is 1000 kg/m³.

67 •• An iron nucleus has a radius of 5.4×10^{-15} m and a mass of 9.3×10^{-26} kg. (*a*) What is its mass per unit volume in kg/m^3? (*b*) If Earth had the same mass per unit volume, what would be its radius? (The mass of Earth is 5.98×10^{24} kg.)

68 •• **ENGINEERING APPLICATION** The Canadian Norman Wells Oil Pipeline extends from Norman Wells, Northwest Territories, to Zama, Alberta. The 8.68×10^5-m-long pipeline has an inside diameter of 12 in. and can be supplied with oil at 35 L/s. (*a*) What is the volume of oil in the pipeline if it is full at some instant in time? (*b*) How long would it take to fill the pipeline with oil if it is initially empty?

69 •• The astronomical unit (AU) is defined as the mean center-to-center distance from Earth to the Sun, namely 1.496×10^{11} m. The parsec is the radius of a circle for which a central angle of 1 s intercepts an arc of length 1 AU. The light-year is the distance that light travels in 1 y. (*a*) How many parsecs are there in one astronomical unit? (*b*) How many meters are in a parsec? (*c*) How many meters in a light-year? (*d*) How many astronomical units in a light-year? (*e*) How many light-years in a parsec?

70 •• If the average density of the universe is at least 6×10^{-27} kg/m^3, then the universe will eventually stop expanding and begin contracting. (*a*) How many electrons are needed in each cubic meter to produce the critical density? (*b*) How many protons per cubic meter would produce the critical density? ($m_e = 9.11 \times 10^{-31}$ kg; $m_p = 1.67 \times 10^{-27}$ kg.)

71 ••• **CONTEXT-RICH, ENGINEERING APPLICATION, SPREADSHEET** You are an astronaut doing physics experiments on the moon. You are interested in the experimental relationship between distance fallen, y, and time elapsed, t, of falling objects dropped from rest. You have taken some data for a falling penny, which is represented in the table below.

(*a*) y (m)	10	20	30	40	50
(*b*) t (s)	3.5	5.2	6.0	7.3	7.9

You expect that a general relationship between distance y and time t is $y = Bt^C$, where B and C are constants to be determined experimentally. To accomplish this, create a log-log plot of the data: (*a*) graph $\log(y)$ vs. $\log(t)$, with $\log(y)$ the ordinate variable and $\log(t)$ the abscissa variable. (*b*) Show that if you take the log of each side of your expected relationship, you get $\log(y) = \log(B) + C\log(t)$. (*c*) By comparing this linear relationship to the graph of the data, estimate the values of B and C. (*d*) If you drop a penny, how long should it take to fall 1.0 m? (*e*) In the next chapter, we will show that the expected relationship between y and t is $y = \frac{1}{2}at^2$, where a is the acceleration of the object. What is the acceleration of objects dropped on the moon? SSM

72 ••• **SPREADSHEET** A particular company's stock prices vary with the market and with the company's type of business, and can be very unpredictable, but people often try to look for mathematical patterns where they may not belong. Corning is a materials-engineering company located in upstate New York. Below is a table of the price of Corning stock on August 3, for every 5 years from 1981 to 2001. Assume that the price follows a power law: price (in \$) $= Bt^C$ where t is expressed in years. (*a*) Evaluate the constants B and C (see methods suggested for the previous problem). (*b*) According to the power law, what should the price of Corning stock have been on August 3, 2000? (It was actually \$82.83!)

(*a*) Price (dollars)	2.10	4.19	9.14	10.82	16.85
(*b*) Years since 1980	1	6	11	16	21

73 ••• **ENGINEERING APPLICATION** The Super-Kamiokande neutrino detector in Japan is a large transparent cylinder filled with ultra-pure water. The height of the cylinder is 41.4 m and the diameter is 39.3 m. Calculate the mass of the water in the cylinder. Does this match the claim posted on the official Super-K Web site that the detector uses 50,000 tons of water? SSM

74 ••• **CONTEXT-RICH** You and a friend are out hiking across a large flat plain and decide to determine the height of a distant mountain peak, and also the horizontal distance from you to the peak (Figure 1-19). In order to do this, you stand in one spot and determine that the sightline to the top of the peak is inclined at 7.5° above the horizontal. You also make note of the heading to the peak at that point: 13° east of north. You stand at the original position, and your friend hikes due west for 1.5 km. He then sights the peak and determines that its sightline has a heading of 15° east of north. How far is the mountain from your position, and how high is its summit above your position?

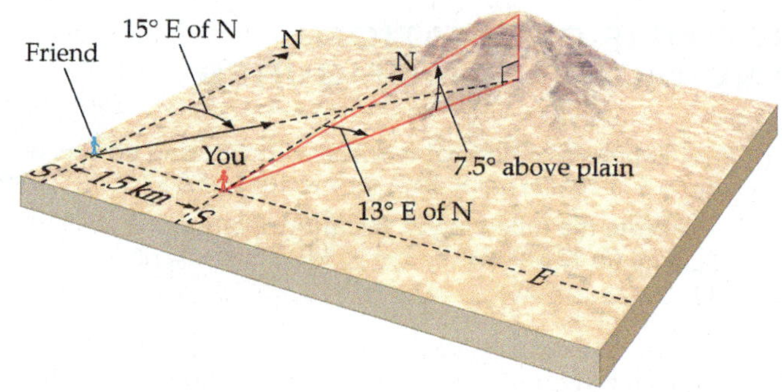

FIGURE 1-19 Problem 74

75 ••• **SPREADSHEET** The table below gives the periods T and orbit radii r for the motions of four satellites orbiting a dense, heavy asteroid. (*a*) These data can be fitted by the formula $T = Cr^n$. Find the values of the constants C and n. (*b*) A fifth satellite is discovered to have a period of 6.20 y. Find the radius for the orbit of this satellite, which fits the same formula.

(*a*) Period T, y	0.44	1.61	3.88	7.89
(*b*) Radius r, Gm	0.088	0.208	0.374	0.600

76 ••• **MULTISTEP** The period T of a simple pendulum depends on the length L of the pendulum and the acceleration of gravity g (dimensions L/T^2). (*a*) Find a simple combination of L and g that has the dimensions of time. (*b*) Check the dependence of the period T on the length L by measuring the period (time for a complete swing back and forth) of a pendulum for two different values of L. (*c*) The correct formula relating T to L and g involves a constant that is a multiple of π, and cannot be obtained by the dimensional analysis of Part (*a*). It can be found by experiment as in Part (*b*) if g is known. Using the value $g = 9.81$ m/s^2 and your experimental results from Part (*b*), find the formula relating T to L and g.

77 ••• A sled at rest is suddenly pulled in three horizontal directions at the same time but it goes nowhere. Paul pulls to the northeast with a force of 50 lb. Johnny pulls at an angle of 35° south of due west with a force of 65 lb. Connie pulls with a force to be determined. (*a*) Express the boys' two forces in terms of the usual unit vectors. (b) Determine the third force (from Connie), expressing it first in component form and then as a magnitude and angle (direction).

78 ••• You spot a plane that is 1.50 km north, 2.5 km east, and at an altitude 5.0 km above your position. (*a*) How far from you is the plane? (*b*) At what angle from due north (in the horizontal plane) are you looking? (*c*) Determine the plane's position vector (from your location) in terms of the unit vectors, letting $\hat{i}$ be toward the east direction, $\hat{j}$ be toward the north direction, and $\hat{k}$ be in vertically upward. (*d*) At what elevation angle (above the horizontal plane of Earth) is the airplane?

PART I MECHANICS

CHAPTER 2

MOTION IN ONE DIMENSION IS MOTION ALONG A STRAIGHT LINE LIKE THAT OF A CAR ON A STRAIGHT ROAD. THIS DRIVER ENCOUNTERS STOPLIGHTS AND DIFFERENT SPEED LIMITS ON HER COMMUTE ALONG A STRAIGHT HIGHWAY TO SCHOOL. *(Medio Images/Getty Images.)*

? How can she estimate her arrival time? (See Example 2-3.)

Motion in One Dimension

Imagine a car speeding down a highway. There are a number of ways in which you could describe the car's motion to someone else. For example, you could describe the change in the car's position as it travels from one point to another, how fast the car is moving and the direction in which it travels, and whether the car is speeding up or slowing down as it moves. These basic descriptors of motion—known as displacement, velocity, and acceleration—are an essential part of physics. In fact, the attempt to describe the motion of objects gave birth to physics more than 400 years ago.

The study of motion, and the related concepts of force and mass, is called **mechanics.** We begin our investigation into motion by examining **kinematics,** the branch of mechanics that deals with the characteristics of motion. You will need to understand kinematics to understand the rest of this book. Motion permeates all of physics, and an understanding of kinematics is needed to understand how force and mass effect motion. Starting in Chapter 4, we look at **dynamics,** which relates motion, force, and mass.

We study the simplest case of kinematics in this chapter — motion along a straight line. We will develop the models and tools you will need to describe motion in one dimension, and introduce the precise definitions of words commonly used to describe motion, such as displacement, speed, velocity, and acceleration. We will also look at the special case of straight-line motion when acceleration is constant. Finally, we consider the ways in which integration can be used to describe motion. In this chapter, moving objects are restricted to motion along a straight line. To describe such motion, it is not necessary to use the full vector notation developed in Chapter 1. A + or − sign are all that is needed to specify direction along a straight line.

2-1 DISPLACEMENT, VELOCITY, AND SPEED

In a horse race, the winner is the horse whose nose first crosses the finish line. One could argue that all that really matters during the race is the motion of that single point on the horse, and that the size, shape, and motion of the rest of the horse is unimportant. In physics, this type of simplification turns out to be useful for examining the motion of other objects as well. We can often describe the motion of an object by describing the motion of a single point of the object. For example, as a car moves in a straight line along a road, you could describe the motion of the car by examining the motion of a single point on the side of the car. An object that can be represented in this idealized manner is called a **particle.** In kinematics, any object can be considered a particle as long as we are not interested in its size, shape, or internal motion. For example, we can consider cars, trains, and rockets particles. Earth and other planets can also be thought of as particles as they move around the Sun. Even people and galaxies can be treated as particles.

(Bettmann/Corbis.)

POSITION AND DISPLACEMENT

To describe the motion of a particle, we need to be able to describe the **position** of the particle and how that position changes as the particle moves. For one-dimensional motion, we often choose the x axis as the line along which the motion takes place. For example, Figure 2-1 shows a student on a bicycle at position x_i at time t_i. At a later time, t_f, the student is at position x_f. The change in the student's position, $x_f - x_i$, is called a **displacement.** We use the Greek letter Δ (uppercase delta) to indicate the change in a quantity; thus, the change in x can be written as

$$\Delta x = x_f - x_i \qquad 2\text{-}1$$

DEFINITION—DISPLACEMENT

It is important to recognize the difference between displacement and distance traveled. The distance traveled by a particle is the length of the path a particle takes from its initial position to its final position. Distance is a scalar quantity and is always indicated by a positive number. Displacement is the *change in position* of the particle. It is positive if the change in position is in the direction of increasing x (the $+x$ direction), and negative if it is in the $-x$ direction. Displacement can be represented by vectors, as shown in Chapter 1. We will use the full vector notation developed in Chapter 1 when we study motion in two and three dimensions in Chapter 3.

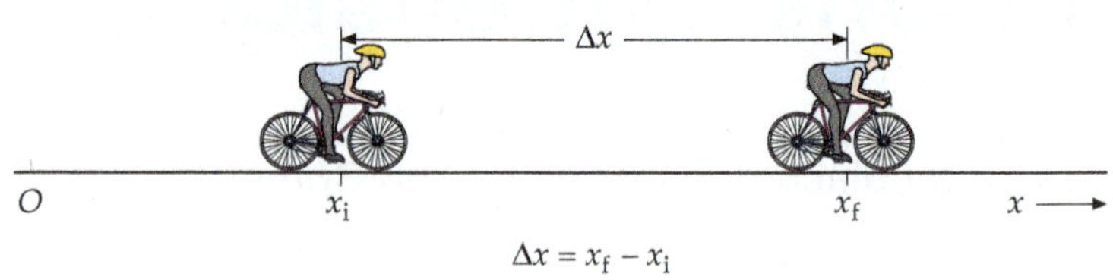

FIGURE 2-1 A student on a bicycle is moving in a straight line. A coordinate axis consists of a line along the path of the bicycle. A point on this line is chosen to be the origin O. Other points on it are assigned a number x, the value of x being proportional to its distance from O. The numbers assigned to points to the right of O are positive as shown, and those assigned to points to the left of O are negative. When the bicycle travels from point x_i to point x_f, its displacement is $\Delta x = x_f - x_i$.

! The notation Δx (read "delta x") stands for a single quantity that is the change in x. Δx is not a product of Δ and x any more than cos θ is a product of cos and θ. By convention, the change in a quantity is always its final value minus its initial value.

Example 2-1 Distance and Displacement of a Dog

You are playing a game of catch with a dog. The dog is initially standing near your feet. Then he jogs 20 feet in a straight line to retrieve a stick, and carries the stick 15 feet back toward you before lying on the ground to chew on the stick. (*a*) What is the total distance the dog travels? (*b*) What is the net displacement of the dog? (*c*) Show that the net displacement for the trip is the sum of the sequential displacements that make up the trip.

PICTURE The total distance, s, is determined by summing the individual distances the dog travels. The displacement is the dog's final position minus the dog's initial position. The dog leaves your side at time 0, gets the stick at time 1, and lies down to chew it at time 2.

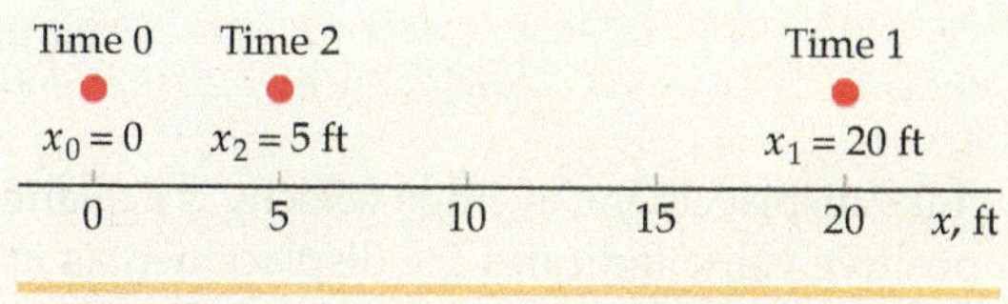

FIGURE 2-2 The red dots represent the dog's position at different times.

SOLVE

(*a*) 1. Make a diagram of the motion (Figure 2-2). Include a coordinate axis:

2. Calculate the total distance traveled:

$s_{02} = s_{01} + s_{12} = (20\text{ ft}) + (15\text{ ft}) = \boxed{35}\text{ ft}$

(The subscripts indicate the time intervals, where s_{01} is the distance traveled during the interval from time 0 to time 1, and so forth.)

(*b*) The net displacement is found from its definition, $\Delta x = x_f - x_i$, where $x_i = x_0 = 0$ is the dog's initial position. Five feet from the initial position or $x_f = x_2 = 5$ ft is the dog's final position:

$\Delta x_{02} = x_2 - x_0 = 5\text{ ft} - 0\text{ ft} = \boxed{5\text{ ft}}$

where Δx_{02} is the displacement during the interval from time 0 to time 2.

(*c*) The net displacement is also found by adding the displacement for the first leg to the displacement for the second leg.

$\Delta x_{01} = x_1 - x_0 = 20\text{ ft} - 0\text{ ft} = 20\text{ ft}$

$\Delta x_{12} = x_2 - x_1 = 5\text{ ft} - 20\text{ ft} = -15\text{ ft}$

adding, we obtain

$\Delta x_{01} + \Delta x_{12} = (x_1 - x_0) + (x_2 - x_1) = x_2 - x_0 = \Delta x_{02}$

so

$\Delta x_{02} = \Delta x_{01} + \Delta x_{12} = 20\text{ ft} - 15\text{ ft} = \boxed{5\text{ ft}}$

CHECK The magnitude of the displacement for any part of the trip is never greater than the total distance traveled for that part. The magnitude of the Part-(*b*) result (5 ft) is less than the Part-(*a*) result (35 ft), so the Part-(*b*) result is plausible.

TAKING IT FURTHER The total distance traveled for a trip is always equal to the sum of the distances traveled for the individual legs of the trip. The total or net displacement for a trip is always equal to the sum of the displacements for the individual legs of the trip.

AVERAGE VELOCITY AND SPEED

We often are interested in the speed something is moving. The **average speed** of a particle is the total distance traveled by the particle divided by the total time from start to finish:

$$\text{Average speed} = \frac{\text{total distance}}{\text{total time}} = \frac{s}{\Delta t} \qquad 2\text{-}2$$

DEFINITION—AVERAGE SPEED

Because the total distance and total time are both always positive, the average speed is always positive.

Although speed is a useful idea, it does not reveal anything about the direction of motion because neither the total distance nor the total time has an associated direction. A more useful quantity is one that describes both how fast and in what direction an object moves. The term used to describe this quantity is *velocity*. The **average velocity,** $v_{\text{av }x}$, of a particle is defined as the ratio of the displacement Δx to the time interval Δt:

$$v_{\text{av }x} = \frac{\Delta x}{\Delta t} = \frac{x_\text{f} - x_\text{i}}{t_\text{f} - t_\text{i}} \qquad (\text{so } \Delta x = v_{\text{av }x}\,\Delta t) \tag{2-3}$$

DEFINITION—AVERAGE VELOCITY

Like displacement, average velocity is a quantity that may be positive or negative. A positive value indicates the displacement is in the $+x$ direction. A negative value indicates the displacement is in the $-x$ direction. The dimensions of velocity are L/T and the SI unit of velocity is meters per second (m/s). Other common units include kilometers per hour (km/h), feet per second (ft/s), and miles per hour (mi/h).

Figure 2-3 is a graph of a particle's position as a function of time. Each point represents the position x of a particle at a particular time t. A straight line connects points P_1 and P_2 and forms the hypotenuse of the triangle having sides $\Delta x = x_2 - x_1$ and $\Delta t = t_2 - t_1$. Notice that the ratio $\Delta x/\Delta t$ is the line's **slope,** which gives us a geometric interpretation of average velocity:

The average velocity for the interval between $t = t_1$ and $t = t_2$ is the slope of the straight line connecting the points (t_1, x_1) and (t_2, x_2) on an x versus t graph.

GEOMETRIC INTERPRETATION OF AVERAGE VELOCITY

See* Math Tutorial *for more information on

Linear Equations

Notice that the average velocity depends on the time interval on which it is based. In Figure 2-3, for example, the smaller time interval indicated by t_2' and P_2' gives a larger average velocity, as shown by the greater steepness of the line connecting points P_1 and P_2'.

The definitions of average speed and average velocity are the most basic of the kinematic parameters. You will need to know these definitions and the definitions that appear later in this chapter to effectively solve kinematics problems.

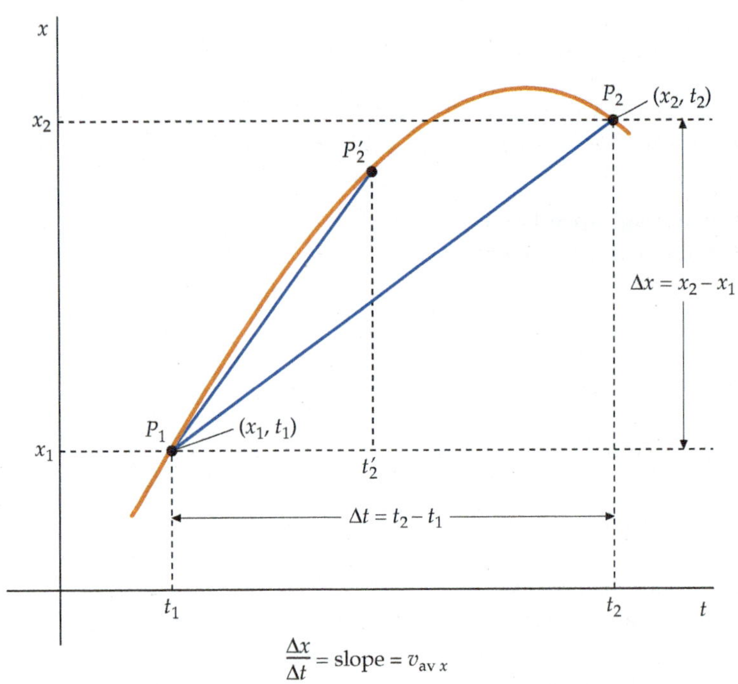

FIGURE 2-3 Graph of x versus t for a particle moving in one dimension. Each point on the curve represents the position x at a particular time t. We have drawn a straight line through points (x_1, t_1) and (x_2, t_2). The displacement $\Delta x = x_2 - x_1$ and the time interval $\Delta t = t_2 - t_1$ between these points are indicated. The straight line between P_1 and P_2 is the hypotenuse of the triangle having sides Δx and Δt, and the ratio $\Delta x/\Delta t$ is its slope. In geometric terms, the slope is a measure of the line's steepness.

Example 2-2 Average Speed and Velocity of the Dog

The dog that you were playing catch with in Example 2-1 jogged 20.0 ft away from you in 1.0 s to retrieve the stick and ambled back 15.0 ft in 1.5 s (Figure 2-4). Calculate (*a*) the dog's average speed, and (*b*) the dog's average velocity for the total trip.

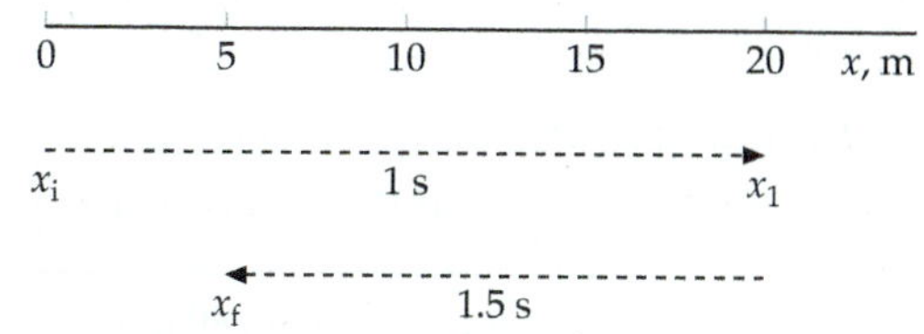

FIGURE 2-4

PICTURE We can solve this problem using the definitions of average speed and average velocity, noting that average *speed* is the total *distance* divided by the total time Δt, whereas the average *velocity* is the net *displacement* divided by Δt:

SOLVE

(*a*) 1. The dog's average speed equals the total distance divided by the total time:

$$\text{Average speed} = \frac{s}{\Delta t}$$

2. Calculate the total distance traveled and the total time:

$$s = s_1 + s_2 = 20.0\text{ ft} + 15.0\text{ ft} = 35.0\text{ ft}$$
$$\Delta t = (t_1 - t_i) + (t_f - t_2) = 1.0\text{ s} + 1.5\text{ s} = 2.5\text{ s}$$

3. Use s and Δt to find the dog's average speed:

$$\text{Average speed} = \frac{35.0\text{ ft}}{2.5\text{ s}} = \boxed{14\text{ ft/s}}$$

(*b*) 1. The dog's average velocity is the ratio of the net displacement Δx to the time interval Δt:

$$v_{\text{av }x} = \frac{\Delta x}{\Delta t}$$

2. The dog's net displacement is $x_f - x_i$, where $x_i = 0.0$ ft is the initial position of the dog and $x_f = 5.0$ ft is the dog's final position:

$$\Delta x = x_f - x_i = 5.0\text{ ft} - 0.0\text{ ft} = 5.0\text{ ft}$$

3. Use Δx and Δt to find the dog's average velocity:

$$v_{\text{av }x} = \frac{\Delta x}{\Delta t} = \frac{5.0\text{ ft}}{2.5\text{ s}} = \boxed{2.0\text{ ft/s}}$$

CHECK An Internet search reveals a greyhound can have an average speed of approximately 66 ft/s (45 mi/h), so our dog should easily be able to jog 14 ft/s (9.5 mi/h). A Part-(*a*) result greater than 66 ft/s would not be plausible.

TAKING IT FURTHER Note that the dog's speed is greater than the dog's average velocity because the total distance traveled is greater than the magnitude of the total displacement. Also, note that the total displacement is the sum of the individual displacements. That is, $\Delta x = \Delta x_1 + \Delta x_2 = (20.0\text{ ft}) + (-15.0\text{ ft}) = 5.0\text{ ft}$, which is the Part (*b*), step 2 result.

Example 2-3 Driving to School

It normally takes you 10 min to travel 5.0 mi to school along a straight road. You leave home 15 min before class begins. Delays caused by a broken traffic light slow down traffic to 20 mi/h for the first 2.0 mi of the trip. Will you be late for class?

PICTURE You need to find the total time that it will take you to travel to class. To do so, you must find the time $\Delta t_{2\text{ mi}}$ that you will be driving at 20 mi/h, and the time $\Delta t_{3\text{ mi}}$ for the remainder of the trip, during which you are driving at your usual velocity.

SOLVE

1. The total time equals the time to travel the first 2.0 mi plus the time to travel the remaining 3.0 mi:

$$\Delta t_{\text{tot}} = \Delta t_{2\text{ mi}} + \Delta t_{3\text{ mi}}$$

2. Using $\Delta x = v_{\text{av }x}\,\Delta t$, solve for the time taken to travel 2.0 mi at 20 mi/h:

$$\Delta t_{2\text{ mi}} = \frac{\Delta x_1}{v_{\text{av }x}} = \frac{2.0\text{ mi}}{20\text{ mi/h}} = 0.10\text{ h} = 6.0\text{ min}$$

3. Using $\Delta x = v_{\text{av }x}\,\Delta t$, relate the time taken to travel 3 mi at the usual velocity:

$$\Delta t_{3\text{ mi}} = \frac{\Delta x_2}{v_{\text{av }x}} = \frac{3.0\text{ mi}}{v_{\text{usual }x}}$$

4. Using $\Delta x = v_{\text{av }x}\,\Delta t$, solve for $v_{\text{usual }x}$, the velocity needed for you to travel the 5.0 mi in 10 min:

$$v_{\text{usual }x} = \frac{\Delta x_{\text{tot}}}{\Delta t_{\text{usual}}} = \frac{5.0\text{ mi}}{10\text{ min}} = 0.50\text{ mi/min}$$

5. Using the results from steps 3 and 4, solve for $\Delta t_{3\,\text{mi}}$:

$$\Delta t_{3\,\text{mi}} = \frac{\Delta x_2}{v_{\text{usual}\,x}} = \frac{3.0\text{ mi}}{0.50\text{ mi/min}} = 6.0\text{ min}$$

6. Solve for the total time:

$$\Delta t_{\text{tot}} = \Delta t_{2\,\text{mi}} + \Delta t_{3\,\text{mi}} = 12\text{ min}$$

7. The trip takes 12 min with the delay, compared to the usual 10 min. Because you wisely allowed yourself 15 min for the trip, $\boxed{\textit{you will not be late for class}}$.

CHECK Note that 20 mi/h = 20 mi/60 min = 1.0 mi/3.0 min. Traveling the entire 5.0 miles at one mile every three minutes, it would take 15 minutes for the trip to school. You allowed yourself 15 minutes for the trip, so you would get there on time even if you traveled at the slow speed of 20 mi/h for the entire 5.0 miles.

Example 2-4 A Train-Hopping Bird

Two trains 60 km apart approach each other on parallel tracks, each moving at 15 km/h. A bird flies back and forth between the trains at 20 km/h until the trains pass each other. How far does the bird fly?

PICTURE In this problem, you are asked to find the total distance flown by the bird. You are given the bird's speed, the trains' speeds, and the initial distance between the trains. At first glance, it might seem like you should find and sum the distances flown by the bird each time it moves from one train to the other. However, a much simpler approach is to use the facts that the time t the bird is flying is the time taken for the trains to meet. The total distance flown is the bird's speed multiplied by the time the bird is flying. Therefore, we can approach this problem by first writing an equation for the quantity to be found, the total distance s flown by the bird.

SOLVE

1. The total distance s_{bird} traveled by the bird equals its speed times the time of flight:

$$s_{\text{bird}} = (\text{average speed})_{\text{bird}} \times t = (\text{speed})_{\text{av bird}} \times t$$

2. The time t that the bird is in the air is the time taken for one of the trains to travel half the initial distance D separating the trains. (Because the trains are traveling at the same speed, each train travels half of the 60 km, which is 30 km, before they meet.):

$$\tfrac{1}{2}D = (\text{speed})_{\text{av train}} \times t$$

so

$$t = \frac{D}{2(\text{speed})_{\text{av train}}}$$

3. Substitute the step-2 result for the time into the step-1 result. The initial separation of the two trains is $D = 60$ km. The total distance traveled by the bird is therefore:

$$s_{\text{bird}} = (\text{speed})_{\text{av bird}}\, t = (\text{speed})_{\text{av bird}} \frac{D}{2(\text{speed})_{\text{av train}}}$$

$$= 20\text{ km/h}\,\frac{60\text{ km}}{2\,(15\text{ km/h})} = \boxed{40\text{ km}}$$

CHECK The speed of each train is three-fourths the speed of the bird, so the distance traveled by one of the trains will be three-fourths the distance the bird travels. Each train travels 30 km. Because 30 km is three-fourths of 40 km, our result of 40 km for the distance the bird travels is very plausible.

INSTANTANEOUS VELOCITY AND SPEED

Suppose your average velocity for a long trip was 60 km/h. Because this value is an average, it does not convey any information about how your velocity changed during the trip. For example, there may have been some parts of the journey where you were stopped at a traffic light and other parts where you went faster to make up time. To learn more about the details of your motion, we have to look at the velocity at any given instant during the trip. On first consideration, defining the velocity of a particle at a single instant might seem impossible. At a given instant, a particle is at a single point. If it is at a single point, how can it be moving? If it is not moving, how can it have a velocity? This age-old paradox is resolved when we realize that observing and defining motion requires that we look at the position of the object at more than one instant of time. For example, consider the graph of position versus time in Figure 2-5. As we consider successively shorter time intervals beginning at t_P, the average velocity for the interval approaches the slope of the tangent at t_P. We define the slope of this tangent as the **instantaneous velocity,** $v_x(t)$, at t_P. This tangent is the limit of the ratio $\Delta x/\Delta t$ as Δt, and therefore Δx, approaches zero. So we can say:

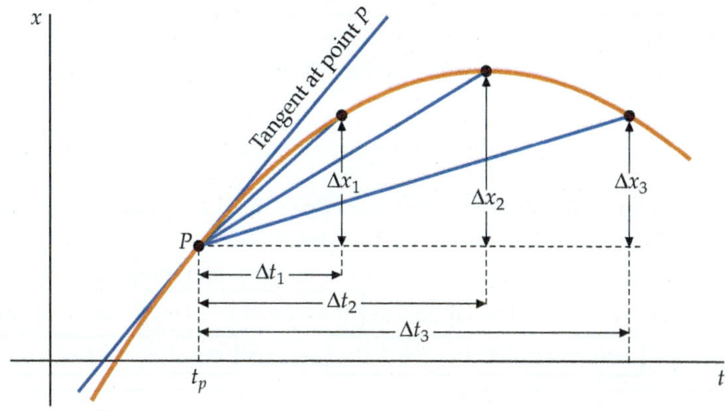

FIGURE 2-5 Graph of x versus t. Note the sequence of successively smaller time intervals, $\Delta t_3, \Delta t_2, \Delta t_1, \ldots$. The average velocity of each interval is the slope of the straight line for that interval. As the time intervals become smaller, these slopes approach the slope of the tangent to the curve at point t_P. The slope of this tangent line is defined as the instantaneous velocity at time t_P.

The instantaneous velocity v_x is the limit of the ratio $\Delta x/\Delta t$ as Δt approaches zero.

$$v_x(t) = \lim_{\Delta t \to 0} \frac{\Delta x}{\Delta t}$$

$$= \text{slope of the line tangent to the } x\text{-versus-}t \text{ curve} \qquad 2\text{-}4$$

DEFINITION—INSTANTANEOUS VELOCITY

In calculus, this limit is called the **derivative** of x with respect to t and is written dx/dt. Using this notation, Equation 2-4 becomes:

$$v_x(t) = \lim_{\Delta t \to 0} \frac{\Delta x}{\Delta t} = \frac{dx}{dt} \qquad 2\text{-}5$$

See **Math Tutorial** *for more information on*

Differential Calculus

A line's slope may be positive, negative, or zero; consequently, instantaneous velocity (in one-dimensional motion) may be positive (x increasing), negative (x decreasing), or zero (no motion). For an object moving with constant velocity, the object's instantaneous velocity is equal to its average velocity. The position versus time graph of this motion (Figure 2-6) will be a straight line whose slope equals the velocity.

The instantaneous velocity is a vector, and the magnitude of the instantaneous velocity is the **instantaneous speed.** Throughout the rest of the text, we shall use "velocity" in place of "instantaneous velocity" and "speed" in place of "instantaneous speed," except when emphasis or clarity is better served by the use of the adjective "instantaneous."

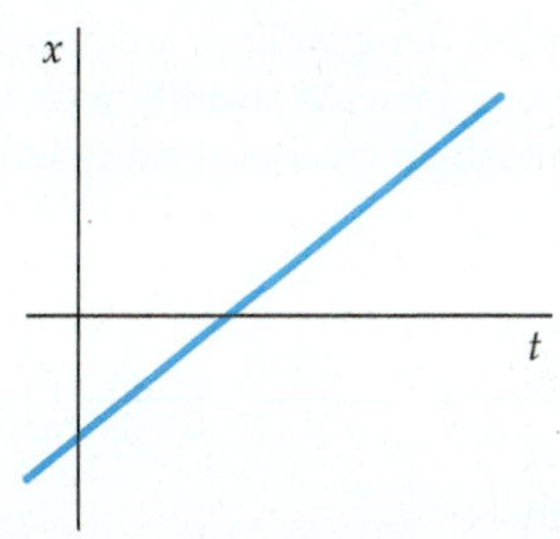

FIGURE 2-6 The position-versus-time graph for a particle moving at constant velocity.

Example 2-5 Position of a Particle as a Function of Time

Try It Yourself

The position of a particle as a function of time is given by the curve shown in Figure 2-7. Find the instantaneous velocity at time $t = 1.8$ s. When is the velocity greatest? When is it zero? Is it ever negative?

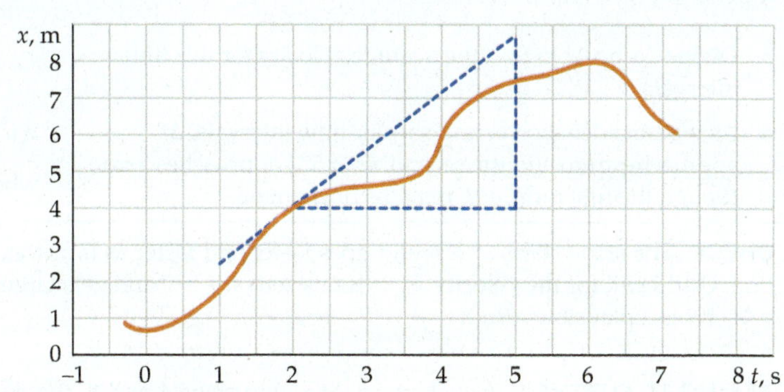

FIGURE 2-7

PICTURE In Figure 2-7, we have sketched the line tangent to the curve at $t = 1.8$ s. The tangent line's slope is the instantaneous velocity of the particle at the given time. You can measure the slope of the tangent line directly off this figure.

SOLVE

Cover the column to the right and try these on your own before looking at the answers.

Steps	Answers
1. Find the values x_1 and x_2 for the points on the tangent line at times $t_1 = 2.0$ s and $t_2 = 5.0$ s.	$x_1 \approx 4.0$ m, $x_2 \approx 8.5$ m
2. Compute the slope of the tangent line from these values. This slope equals the instantaneous velocity at $t = 2.0$ s.	$v_x = \text{slope} \approx \dfrac{8.5\text{ m} - 4.0\text{ m}}{5.0\text{ s} - 2.0\text{ s}} = \boxed{1.5\text{ m/s}}$

3. From the figure, the tangent line is steepest, and, therefore, the slope is greatest at about $t = 4.0$ s. The velocity is $\boxed{\text{greatest at } t \approx 4.0\text{ s}}$. The slope and the velocity both are $\boxed{\text{zero at } t = 0.0 \text{ and } t = 6.0\text{ s}}$ and are $\boxed{\text{negative for } t < 0.0 \text{ and } t > 6.0\text{ s}}$.

CHECK The position of the particle changes from about 1.8 m at 1.0 s to 4.0 m at 2.0 s, so the average velocity for the interval from 1.0 s to 2.0 s is 2.2 m/s. This is the same order of magnitude as the value for the instantaneous velocity at 1.8 s, so the step-2 result is plausible.

PRACTICE PROBLEM 2-1 Estimate the average velocity of this particle between $t = 2.0$ s and $t = 5.0$ s.

Example 2-6 A Stone Dropped from a Cliff

The position of a stone dropped from a cliff is described approximately by $x = 5t^2$, where x is in meters and t is in seconds. The $+x$ direction is downwards and the origin is at the top of the cliff. Find the velocity of the stone during its fall as a function of time t.

PICTURE We can compute the velocity at some time t by computing the derivative dx/dt directly from the definition in Equation 2-4. The corresponding curve giving x versus t is shown in Figure 2-8. Tangent lines are drawn at times t_1, t_2, and t_3. The slopes of these tangent lines increase steadily with increasing time, indicating that the instantaneous velocity increases steadily with time.

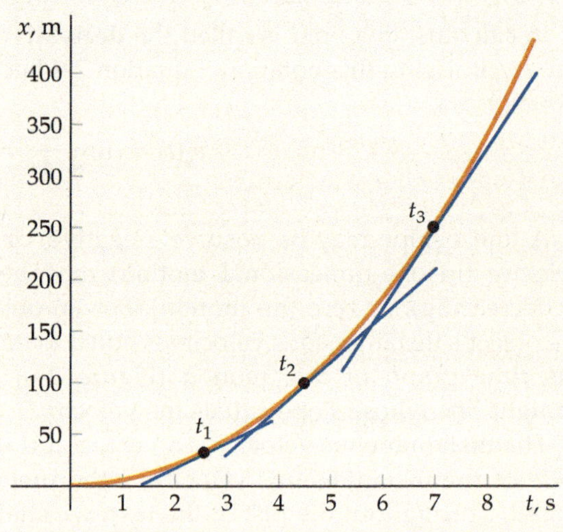

FIGURE 2-8

SOLVE

1. By definition the instantaneous velocity is
$$v_x(t) = \lim_{\Delta t \to 0} \frac{\Delta x}{\Delta t} = \lim_{\Delta t \to 0} \frac{x(t + \Delta t) - x(t)}{\Delta t}$$

2. We compute the displacement Δx from the position function $x(t)$:
$$x(t) = 5t^2$$

3. At a later time $t + \Delta t$, the position is $x(t + \Delta t)$, given by:
$$x(t + \Delta t) = 5(t + \Delta t)^2 = 5[t^2 + 2t\,\Delta t + (\Delta t)^2] = 5t^2 + 10t\,\Delta t + 5(\Delta t)^2$$

4. The displacement for this time interval is thus:
$$\Delta x = x(t + \Delta t) - x(t) = [5t^2 + 10t\,\Delta t + 5(\Delta t)^2] - 5t^2 = 10t\,\Delta t + 5(\Delta t)^2$$

5. Divide Δx by Δt to find the average velocity for this time interval:
$$v_{\text{av}\,x} = \frac{\Delta x}{\Delta t} = \frac{10t\,\Delta t + 5(\Delta t)^2}{\Delta t} = 10t + 5\,\Delta t$$

6. As we consider shorter and shorter time intervals, Δt approaches zero and the second term $5\Delta t$ approaches zero, though the first term, $10t$, remains unchanged:
$$v_x(t) = \lim_{\Delta t \to 0} \frac{\Delta x}{\Delta t} = \lim_{\Delta t \to 0}(10t + 5\,\Delta t) = \boxed{10t}$$
where v_x is in m/s and t is in s.

CHECK The stone starts at rest and goes faster and faster as it moves in the positive direction. Our result for the velocity, $v_x = 10t$, is zero at $t = 0$ and gets larger as t increases. Thus, $v_x = 10t$ is a plausible result.

TAKING IT FURTHER If we had set $\Delta t = 0$ in steps 4 and 5, the displacement would be $\Delta x = 0$, in which case the ratio $\Delta x/\Delta t$ would be undefined. Instead, we leave Δt as a variable until the final step, when the limit $\Delta t \to 0$ is well defined.

To find derivatives quickly, we use rules based on the limiting process above (see Appendix Table A-4). A particularly useful rule is

$$\text{If} \quad x = Ct^n, \quad \text{then} \quad \frac{dx}{dt} = Cnt^{n-1} \qquad \text{2-6}$$

where C and n are any constants. Using this rule in Example 2-6, we have $x = 5t^2$ and $v_x = dx/dt = 10t$, in agreement with our previous results.

2-2 ACCELERATION

When you step on your car's gas pedal or brake, you expect your velocity to change. An object whose velocity changes is said to be accelerating. **Acceleration** is the rate of change of velocity with respect to time. The **average acceleration,** $a_{\text{av}\,x}$, for a particular time interval Δt is defined as the change in velocity, Δv, divided by that time interval:

$$a_{\text{av}\,x} = \frac{\Delta v_x}{\Delta t} = \frac{v_{\text{f}\,x} - v_{\text{i}\,x}}{t_\text{f} - t_\text{i}} \qquad (\text{so } \Delta v_x = a_{\text{av}\,x}\,\Delta t) \qquad \text{2-7}$$

DEFINITION—AVERAGE ACCELERATION

Notice that acceleration has the dimensions of velocity (L/T) divided by time (T), which is the same as length divided by time squared (L/T^2). The SI unit is meters per second squared, m/s^2. Furthermore, like displacement and velocity, acceleration is a vector quantity. For one-dimensional motion, we can use + and − to indicate the direction of the acceleration. Equation 2-7 tells us that for $a_{\text{av}\,x}$ to be positive, Δv_x must be positive, and for $a_{\text{av}\,x}$ to be negative, Δv_x must be negative.

Instantaneous acceleration is the limit of the ratio $\Delta v_x/\Delta t$ as Δt approaches zero. On a plot of velocity versus time, the instantaneous acceleration at time t is the slope of the line tangent to the curve at that time:

$$a_x = \lim_{\Delta t \to 0} \frac{\Delta v_x}{\Delta t}$$

$$= \text{slope of the line tangent to the } v\text{-versus-}t \text{ curve} \qquad \text{2-8}$$

DEFINITION—INSTANTANEOUS ACCELERATION

Thus, acceleration is the derivative of velocity v_x with respect to time, dv_x/dt. Because velocity is the derivative of the position x with respect to t, acceleration is the second derivative of x with respect to t, d^2x/dt^2. We can see the reason for this notation when we write the acceleration as dv_x/dt and replace v_x with dx/dt:

$$a_x = \frac{dv_x}{dt} = \frac{d(dx/dt)}{dt} = \frac{d^2x}{dt^2} \qquad \text{2-9}$$

Notice that when the time interval becomes extremely small, the average acceleration and the instantaneous acceleration become equal. Therefore, we will use the word acceleration to mean "instantaneous acceleration."

It is important to note that the sign of an object's acceleration does not tell you whether the object is speeding up or slowing down. To determine this, you need to compare the signs of both the velocity and the acceleration of the object. If v_x and a_x are both positive, v_x is positive and becoming more positive so the speed is increasing. If v_x and a_x are both negative, v_x is negative and becoming more negative so the speed is again increasing. When v_x and a_x have opposite signs, the object is slowing down. If v_x is positive and a_x is negative, v_x is positive but is becoming less positive

! Deceleration does not mean the acceleration is negative. Deceleration does mean that v_x and a_x have opposite signs.

CONCEPT CHECK 2-1

You are following the car in front of you at high speed when the driver of the car in front of you brakes hard, bringing his car to a stop to avoid running over a huge pothole. Three tenths of a second after you see the brake lights on the lead car flash, you too brake hard. Assume that the two cars are initially traveling at the same speed, and that once both cars are braking hard, they lose speed at the same rate. Does the distance between the two cars remain constant while the two cars are both braking hard?

so the speed is decreasing. If v_x is negative and a_x is positive, v_x is negative but is becoming less negative so the speed is again decreasing. In summary, if v_x and a_x have the same sign, the speed is increasing; if v_x and a_x have opposite signs, the speed is decreasing. When an object is slowing down, we sometimes say it is decelerating.

If acceleration remains zero, there is no change in velocity over time—velocity is constant. In this case, the plot of x versus t is a straight line. If acceleration is nonzero and constant, as in Example 2-13 , then velocity varies linearly with time and x varies quadratically with time.

Example 2-7 A Fast Cat

A cheetah can accelerate from 0 to 96 km/h (60 mi/h) in 2.0 s, whereas a Corvette requires 4.5 s. Compute the average accelerations for the cheetah and Corvette and compare them with the free-fall acceleration, $g = 9.81\ \text{m/s}^2$.

(Gunther Ziesler/Peter Arnold.)

PICTURE Because we are given the initial and final velocities, as well as the change in time for both the cat and the car, we can simply use Equation 2-7 to find the acceleration for each object.

SOLVE

1. Convert 96 km/h into a velocity of m/s: $96\,\frac{\text{km}}{\text{h}}\left(\frac{1\ \text{h}}{3600\ \text{s}}\right)\left(\frac{1000\ \text{m}}{1\ \text{km}}\right) = 26.7\ \text{m/s}$

2. Find the average acceleration from the information given:

$$\text{cat}\quad a_{\text{av}\,x} = \frac{\Delta v_x}{\Delta t} = \frac{26.7\ \text{m/s} - 0}{2.0\ \text{s}} = 13.3\ \text{m/s}^2 = \boxed{13\ \text{m/s}^2}$$

$$\text{car}\quad a_{\text{av}\,x} = \frac{\Delta v_x}{\Delta t} = \frac{26.7\ \text{m/s} - 0}{4.5\ \text{s}} = 5.93\ \text{m/s}^2 = \boxed{5.9\ \text{m/s}^2}$$

3. To compare the result with the acceleration due to gravity, multiply each by the conversion factor $1g/9.81\ \text{m/s}^2$:

$$\text{cat}\quad 13.3\ \text{m/s}^2 \times \frac{1g}{9.81\ \text{m/s}^2} = 1.36g = \boxed{1.4g}$$

$$\text{car}\quad 5.93\ \text{m/s}^2 \times \frac{1g}{9.81\ \text{m/s}^2} = 0.604g = \boxed{0.60g}$$

CHECK Because the car takes slightly more than twice as long as the cheetah to accelerate to the same velocity, it makes sense that the car's acceleration is slightly less than half that of the cat's.

TAKING IT FURTHER To reduce round-off errors, calculations are carried out using values with at least three digits even though the answers are reported using only two significant digits. These extra digits used in the calculations are called *guard digits.*

PRACTICE PROBLEM 2-2 A car is traveling at 45 km/h at time $t = 0$. It accelerates at a constant rate of 10 km/(h·s). (*a*) How fast is it traveling at $t = 2.0$ s? (*b*) At what time is the car traveling at 70 km/h?

Example 2-8 Velocity and Acceleration as Functions of Time

The position of a particle is given by $x = Ct^3$, where C is a constant. Find the dimensions of C. In addition, find both the velocity and the acceleration as functions of time.

PICTURE We can find the velocity by applying $dx/dt = Cnt^{n-1}$ (Equation 2-6) to the position of the particle, where n in this case equals 3. Then, we repeat the process to find the time derivative of velocity.

SOLVE

1. The dimensions of x and t are L and T, respectively:

$$C = \frac{x}{t^3} \Rightarrow [C] = \frac{[x]}{[t]^3} = \boxed{\frac{\text{L}}{\text{T}^3}}$$

2. We find the velocity by applying $dx/dt = Cnt^{n-1}$ (Equation 2-6):

$$x = Ct^n = Ct^3$$

$$v_x = \frac{dx}{dt} = Cnt^{n-1} = C3t^{3-1} = \boxed{3Ct^2}$$

3. The time derivative of velocity gives the acceleration: $a = \frac{dv_x}{dt} = Cnt^{n-1} = 3C(2)(t^{2-1}) = \boxed{6Ct}$

CHECK We can check the dimensions of our answers. For velocity, $[v_x] = [C][t^2] = (L/T^3)(T^2) = L/T$. For acceleration, $[a_x] = [C][t] = (L/T^3)(T) = L/T^2$.

PRACTICE PROBLEM 2-3 If a car starts from rest at $x = 0$ with constant acceleration a_x, its velocity v_x depends on a_x and the distance traveled x. Which of the following equations has the correct dimensions and therefore could possibly be an equation relating x, a_x, and v_x?

(*a*) $v_x = 2a_x x$ (*b*) $v_x^2 = 2a_x/x$ (*c*) $v_x = 2a_x x^2$ (*d*) $v_x^2 = 2a_x x$

MOTION DIAGRAMS

In studying physics, you will often wish to estimate the direction of the acceleration vector from a description of the motion. Motion diagrams can help. In a motion diagram the moving object is drawn at a sequence of equally spaced time intervals. For example, suppose you are on a trampoline and, following a high bounce, you are falling back toward the trampoline. As you descend, you keep going faster and faster. A motion diagram of this motion is shown in Figure 2-9*a*. The dots represent your position at equally spaced time intervals, so the space between successive dots increases as your speed increases. The numbers placed next to the dots are there to indicate the progression of time and an arrow representing your velocity is drawn next to each dot. The direction of each arrow represents the direction of your velocity at that instant, and the length of the arrow represents how fast you are going. Your acceleration vector* is in the direction that your velocity vector is changing—which is downward. In general, if the velocity arrows get longer as time progresses, then the acceleration is in the same direction of the velocity. On the other hand, if the velocity arrows get shorter as time progresses (Figure 2-9*b*), the acceleration is in the direction opposite to that of the velocity. Figure 2-9*b* is a motion diagram of your motion as you rise toward the ceiling following a bounce on the trampoline.

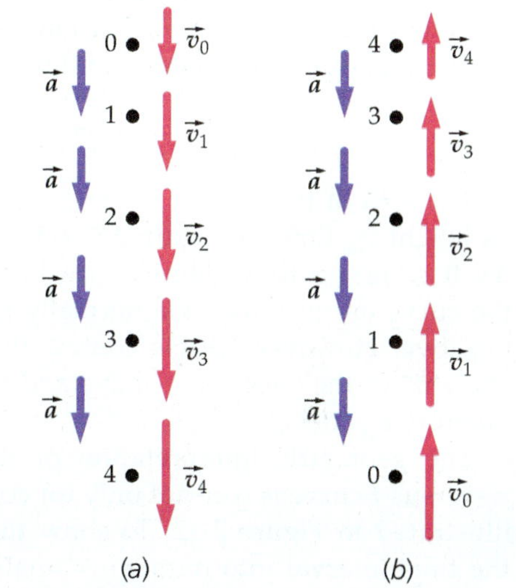

FIGURE 2-9 Motion diagrams. The time intervals between successive dots are identical. (*a*) The velocity vector is increasing, so the acceleration is in the direction of the velocity vector. (*b*) The velocity vector is decreasing, so the acceleration is in the direction opposite to that of the velocity vector.

2-3 MOTION WITH CONSTANT ACCELERATION

The motion of a particle that has nearly constant acceleration is found in nature. For example, near Earth's surface all unsupported objects fall vertically with nearly constant acceleration (provided air resistance is negligible). Other examples of near constant acceleration might include a plane accelerating along a runway for takeoff, and the motion of a car braking for a red light or speeding up at a green light. For a moving particle, the final velocity v_x equals the initial velocity plus the change in velocity, and the change in velocity equals the average acceleration multiplied by the time. That is,

$$v_x = v_{0x} + \Delta v = v_{0x} + a_{\mathrm{av}\,x}\,\Delta t \qquad \text{2-10}$$

If a particle has constant acceleration a_x, it follows that the instantaneous acceleration and the average acceleration are equal. That is,

$$a_x = a_{\mathrm{av}\,x} \quad (a_x \text{ is constant}) \qquad \text{2-11}$$

Because situations involving nearly constant acceleration are common, we can use the equations for acceleration and velocity to derive a special set of **kinematic equations** for problems involving one-dimensional motion at constant acceleration.

* The velocity vector and the acceleration vector were introduced in Chapter 1 and are further developed in Chapter 3.

DERIVING THE CONSTANT-ACCELERATION KINEMATIC EQUATIONS

Suppose a particle moving with constant acceleration a_x has a velocity v_{0x} at time $t_0 = 0$, and v_x at some later time t. Combining Equations 2-10 and 2-11, we have

$$v_x = v_{0x} + a_x t \qquad (a_x \text{ is constant}) \qquad 2\text{-}12$$

CONSTANT ACCELERATION: $v_x(t)$

A v_x-versus-t plot (Figure 2-10) of this equation is a straight line. The line's slope is the acceleration a_x.

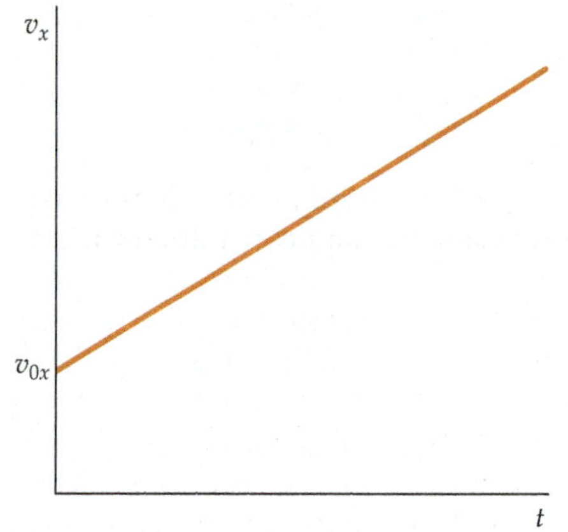

FIGURE 2-10 Graph of velocity versus time for constant acceleration.

To obtain an equation for the position x as a function of time, we first look at the special case of motion with a constant velocity $v_x = v_{0x}$ (Figure 2-11). The change in position Δx during an interval of time Δt is

$$\Delta x = v_{0x}\,\Delta t \qquad (a_x = 0)$$

The area of the shaded rectangle under the v_x-versus-t curve (Figure 2-11*a*) is its height v_{0x} times its width Δt, so the area under the curve is the displacement Δx. If v_{0x} is negative (Figure 2-11*b*), both the displacement Δx and the area under the curve are negative. We normally think of area as a quantity that cannot be negative. However, in this context that is not the case. If v_{0x} is negative, the "height" of the curve is negative and the "area under the curve" is the negative quantity $v_{0x}\,\Delta t$.

The geometric interpretation of the displacement as the area under the v_x-versus-t curve is true not only for constant velocity, but it is true in general, as illustrated in Figure 2-12. To show that this statement is true, we first divide the time interval into numerous small intervals, Δt_1, Δt_2, and so on. Then, we draw a set of rectangles as shown. The area of the rectangle corresponding to the ith time interval Δt_i (shaded in the figure) is $v_i\,\Delta t_i$, which is approximately equal

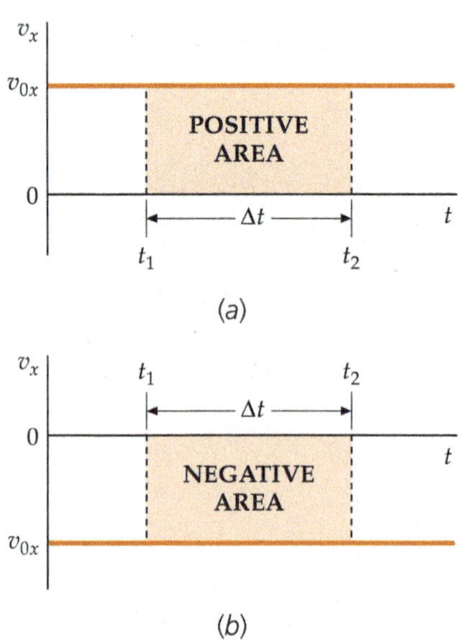

FIGURE 2-11 Motion with constant velocity.

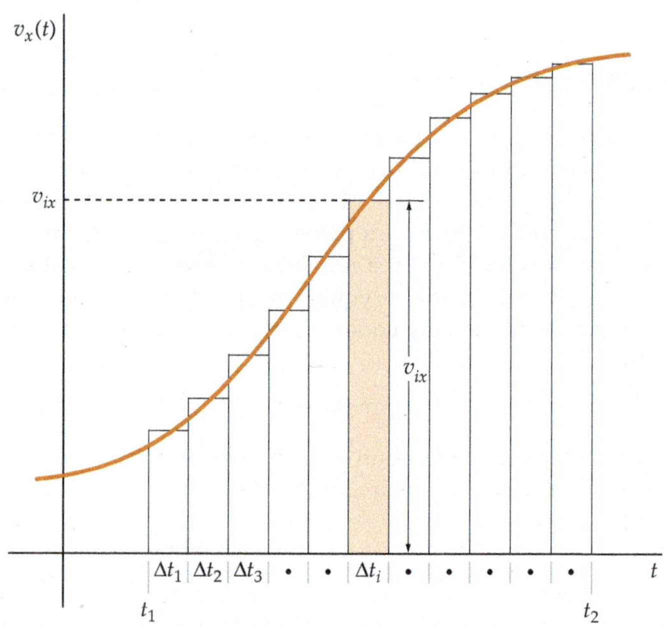

FIGURE 2-12 Graph of a general $v_x(t)$-versus-t curve. The total displacement from t_1 to t_2 is the area under the curve for this interval, which can be approximated by summing the areas of the rectangles.

to the displacement Δx_i during the interval Δt_i. The sum of the rectangular areas is therefore approximately the sum of the displacements during the time intervals and is approximately equal to the total displacement from time t_1 to t_2. We can make the approximation as accurate as we wish by putting enough rectangles under the curve, each rectangle having a sufficiently small value for Δt. For the limit of smaller and smaller time intervals (and more and more rectangles), the resulting sum approaches the area under the curve, which in turn equals the displacement. The displacement Δx is thus the area under the v_x-versus-t curve.

For motion with constant acceleration (Figure 2-13*a*), Δx is equal to the area of the shaded region. This region is divided into a rectangle and a triangle of areas $v_{1x}\,\Delta t$ and $\frac{1}{2}a_x(\Delta t)^2$, respectively, where $\Delta t = t_2 - t_1$. It follows that

$$\Delta x = v_{1x}\,\Delta t + \tfrac{1}{2}a_x(\Delta t)^2 \qquad \text{2-13}$$

If we set $t_1 = 0$ and $t_2 = t$, then Equation 2-13 becomes

$$x - x_0 = v_{0x}t + \tfrac{1}{2}a_x t^2 \qquad \text{2-14}$$

CONSTANT ACCELERATION: $x(t)$

where x_0 and v_{0x} are the position and velocity at time $t = 0$, and $x = x(t)$ is the position at time t. The first term on the right, $v_{0x}t$, is the displacement that would occur if a_x were zero, and the second term, $\frac{1}{2}a_x t^2$, is the additional displacement due to the constant acceleration.

We next use Equations 2-12 and 2-14 to obtain two additional kinematic equations for constant acceleration. Solving Equation 2-12 for t, and substituting for t, in Equation 2-14 gives

$$\Delta x = v_{0x}\frac{v_x - v_{0x}}{a_x} + \frac{1}{2}a_x\left(\frac{v_x - v_{0x}}{a_x}\right)^2$$

Multiplying both sides by $2a_x$ we obtain

$$2a_x\,\Delta x = 2v_{0x}(v_x - v_{0x}) + (v_x - v_{0x})^2$$

Simplifying and rearranging terms gives

$$v_x^2 = v_{0x}^2 + 2a_x\,\Delta x \qquad \text{2-15}$$

CONSTANT ACCELERATION: $v_x(x)$

The definition of average velocity (Equation 2-3) is:

$$\Delta x = v_{\text{av}\,x}\,\Delta t$$

where $v_{\text{av}\,x}\,\Delta t$ is the area under the horizontal line at height $v_{\text{av}\,x}$ in Figure 2-13*b* and Δx is the area under the v_x versus t curve in Figure 2-13*a*. We can see that if $v_{\text{av}\,x} = \frac{1}{2}(v_{1x} + v_{2x})$, the area under the line at height v_{av} in Figure 2-13*a* and the area under the v_x versus t curve in Figure 2-13*b* will be equal. Thus,

$$v_{\text{av}\,x} = \tfrac{1}{2}(v_{1x} + v_{2x}) \qquad \text{2-16}$$

CONSTANT ACCELERATION: $v_{\text{av}\,x}$ AND v_x

For motion with constant acceleration, the average velocity is the mean of the initial and final velocities.

For an example of an instance where Equation 2-16 is not applicable, consider the motion of a runner during a 10.0-km run that takes 40.0 min to complete. The

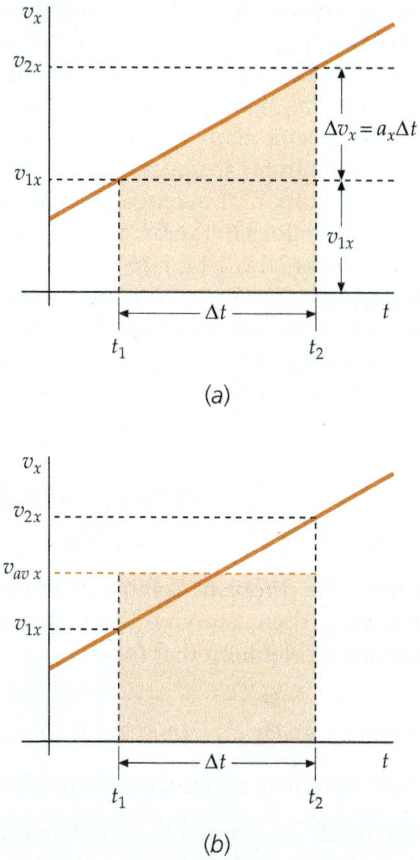

FIGURE 2-13 Motion with constant acceleration.

"It goes from zero to 60 in about 3 seconds." (*© Sydney Harris.*)

! Equation 2-16 is applicable only for time intervals during which the acceleration remains constant.

average velocity for the run is 0.250 km/min, which we compute using the definition of average velocity ($v_{\text{av}\,x} = \Delta x/\Delta t$). The runner starts from rest ($v_{1x} = 0$), and during the first one or two seconds his velocity increases rapidly, reaching a constant value v_{2x} that is sustained for the remainder of the run. The value of v_{2x} is just slightly greater than 0.250 km/min, so Equation 2-16 gives a value of about 0.125 km/min for the average velocity, a value almost 50% below the value given by the definition of average velocity. Equation 2-16 is not applicable because the acceleration does not remain constant for the entire run.

Equations 2-12, 2-14, 2-15, and 2-16 can be used to solve kinematics problems involving one-dimensional motion with constant acceleration. The choice of which equation or equations to use for a particular problem depends on what information you are given in the problem and what you are asked to find. Equation 2-15 is useful, for example, if we want to find the final velocity of a ball dropped from rest at some height x and we are not interested in the time the fall takes.

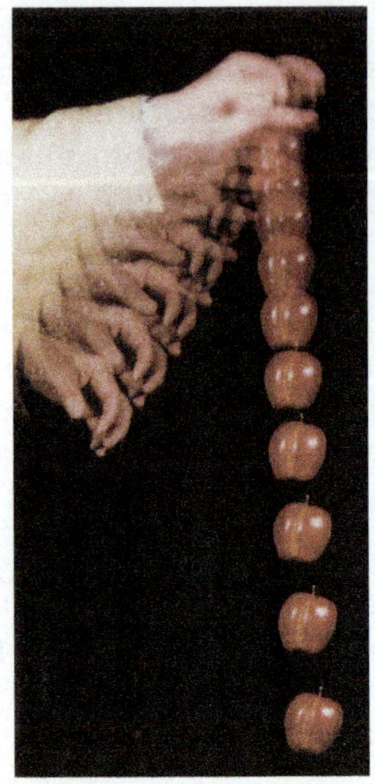

A falling apple captured by strobe photography at 60 flashes per second. The acceleration of the apple is indicated by the widening of the spaces between the images. *(Estate of Harold E. Edgerton/Palm Press.)*

USING THE CONSTANT-ACCELERATION KINEMATIC EQUATIONS

Review the Problem-Solving Strategy for solving problems using kinematic equations. Then, examine the examples involving one-dimensional motion with constant acceleration that follow.

PROBLEM-SOLVING STRATEGY

1-D Motion with Constant Acceleration

PICTURE Determine if a problem is asking you to find the time, distance, velocity, or acceleration for an object.

SOLVE Use the following steps to solve problems that involve one-dimensional motion and constant acceleration.

1. Draw a figure showing the particle in its initial and final positions. Include a coordinate axis and label the initial and final coordinates of the position. Indicate the + and − directions for the axis. Label the initial and final velocities, and label the acceleration.
2. Select one of the constant-acceleration kinematic equations (Equations 2-12, 2-14, 2-15, and 2-16). Substitute the given values into the selected equation and, if possible, solve for the desired value.
3. If necessary, select another of the constant-acceleration kinematic equations, substitute the given values into it, and solve for the desired value.

CHECK You should make sure that your answers are dimensionally consistent and the units of the answers are correct. In addition, check to make sure the magnitudes and signs of your answers agree with your expectations.

Problems with one object We will begin by looking at several examples that involve the motion of a single object.

Example 2-9 A Car's Stopping Distance

On a highway at night you see a stalled vehicle and brake your car to a stop. As you brake, the velocity of your car decreases at a constant rate of (5.0 m/s)/s. What is the car's stopping distance if your initial velocity is (*a*) 15 m/s (about 34 mi/h) or (*b*) 30 m/s?

PICTURE Use the Problem-Solving Strategy that precedes this example. The car is drawn as a dot to indicate a particle. We choose the direction of motion as $+x$ direction, and we choose the initial position $x_0 = 0$. The initial velocity is $v_{0x} = +15\text{ m/s}$ and the final velocity $v_x = 0$. Because the velocity is decreasing, the acceleration is negative. It is $a_x = -5.0\text{ m/s}^2$. We seek the distance traveled, which is the magnitude of the displacement Δx. We are neither given nor asked for the time, so $v_x^2 = v_{0x}^2 + 2a_x\,\Delta x$ (Equation 2-15) will provide a one-step solution.

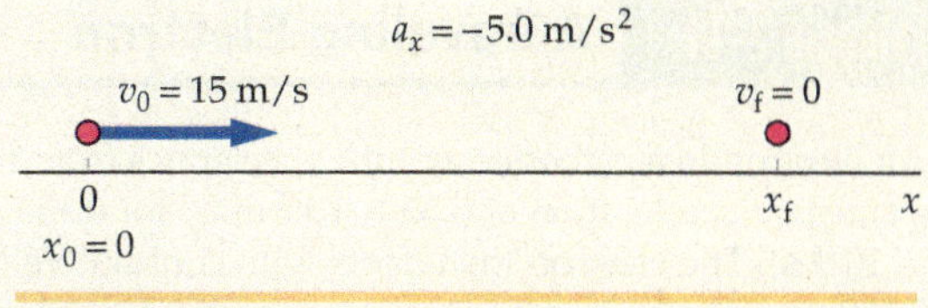

FIGURE 2-14

SOLVE

(*a*) 1. Draw the car (as a small dot) in its initial and final positions (Figure 2-14). Include a coordinate axis and label the drawing with the kinematic parameters.

2. Using Equation 2-15, calculate the displacement Δx:

$$v_x^2 = v_{0x}^2 + 2a_x\,\Delta x$$
$$0 = (15\text{ m/s})^2 + 2(-5.0\text{ m/s}^2)\Delta x$$
$$\Delta x = 22.5\text{ m} = \boxed{23\text{ m}}$$

(*b*) Substitute an initial speed of 30 m/s into the expression for Δx obtained in Part (*a*) (see Figure 2-14):

$$v_x^2 = v_{0x}^2 + 2a_x\,\Delta x$$
$$0 = (30\text{ m/s})^2 + 2(-5.0\text{ m/s}^2)\Delta x$$
$$\Delta x = \boxed{90\text{ m}}$$

CHECK The car's velocity decreases by 5.0 m/s each second. If its initial velocity is 15 m/s, it would take 3.0 s for it to come to rest. During the 3.0 s, it has an average velocity of half 15 m/s, so it would travel $\frac{1}{2}(15\text{ m/s})(3.0\text{ s}) = 23\text{ m}$. This confirms our Part-(*a*) result. Our Part-(*b*) result can be confirmed in the same manner.

Example 2-10 Stopping Distance

Try It Yourself

In the situation described in Example 2-9, (*a*) how much time does it take for the car to stop if its initial velocity is 30 m/s, and (*b*) how far does the car travel in the last second?

PICTURE Use the Problem-Solving Strategy that precedes Example 2-9. (*a*) In this part of the problem, you are asked to find the time it takes the car to stop. You are given the initial velocity $v_{0x} = 30\text{ m/s}$. From Example 2-9, you know the car has an acceleration $a_x = -5.0\text{ m/s}^2$. A relationship between time, velocity, and acceleration is given by Equation 2-12. (*b*) Because the car's velocity decreases by 5.0 m/s each second, the velocity 1.0 s before the car stops must be 5.0 m/s. Find the average velocity during the last second and use that to find the distance traveled.

SOLVE

Cover the column to the right and try these on your own before looking at the answers.

Steps	Answers
(*a*) 1. Draw the car (as a small dot) in its initial and final positions (Figure 2-15). Include a coordinate axis and label the drawing with the kinematic parameters.	
2. Use Equation 2-12 to find the total stopping time Δt.	$\Delta t = \boxed{6.0\text{ s}}$
(*b*) 1. Draw the car (as a small dot) in its initial and final positions (Figure 2-16). Include a coordinate axis.	
2. Find the average velocity during the last second from $v_{\text{av}\,x} = \frac{1}{2}(v_{\text{i}\,x} + v_{\text{f}\,x})$.	$v_{\text{av}\,x} = 2.5\text{ m/s}$
3. Compute the distance traveled from $\Delta x = v_{\text{av}\,x}\,\Delta t$.	$\Delta x = v_{\text{av}\,x}\,\Delta t = \boxed{2.5\text{ m}}$

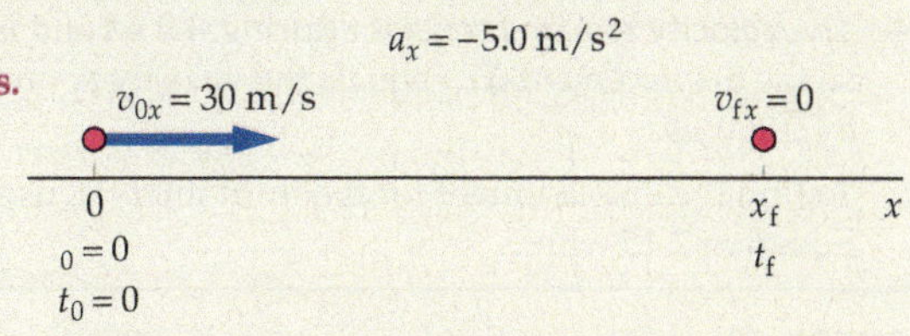

FIGURE 2-15

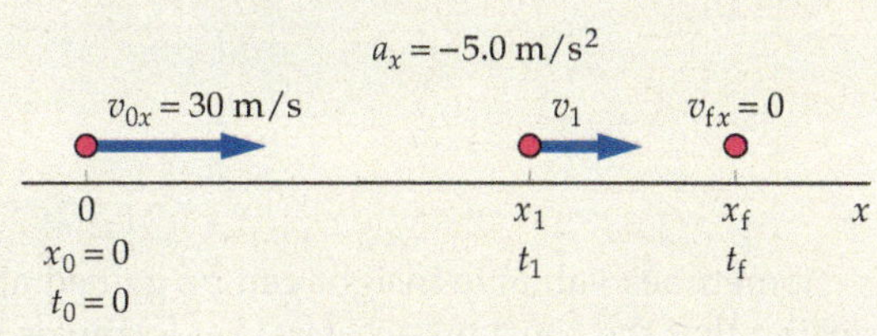

FIGURE 2-16

CHECK We would not expect the car to travel very far during the last second because it is moving relatively slowly. The Part-(*b*) result of 2.5 m is a plausible result.

Example 2-11 A Traveling Electron — *Try It Yourself*

An electron in a cathode-ray tube accelerates from rest with a constant acceleration of $5.33 \times 10^{12}\ \mathrm{m/s^2}$ for $0.150\ \mu s$ ($1\ \mu s = 10^{-6}$ s). The electron then drifts with constant velocity for $0.200\ \mu s$. Finally, it slows to a stop with an acceleration of $-2.67 \times 10^{13}\ \mathrm{m/s^2}$. How far does the electron travel?

The two-mile-long linear accelerator at Stanford University used to accelerate electrons and positrons in a straight line to nearly the speed of light. Cross section of the accelerator's electron beam as shown on a video monitor. *(Stanford Limear Accelerator, U.S. Department of Energy.)*

PICTURE The equations for constant acceleration do not apply to the full duration of the electron's motion because the acceleration changes twice during that time. However, we can divide the electron's motion into three intervals, each with a different constant acceleration, and use the final position and velocity for the first interval as the initial conditions for the second interval, and the final position and velocity for the second interval as the initial conditions for the third. Apply the Problem-Solving Strategy that precedes Example 2-9 to each of the three constant-acceleration intervals. We will choose the origin to be at the electron's starting position, and the $+x$ direction to be the direction of motion.

SOLVE

Cover the column to the right and try these on your own before looking at the answers.

Steps

1. Draw the electron in its initial and final positions for each constant-acceleration interval (Figure 2-17). Include a coordinate axis and label the drawing with the kinematic parameters.

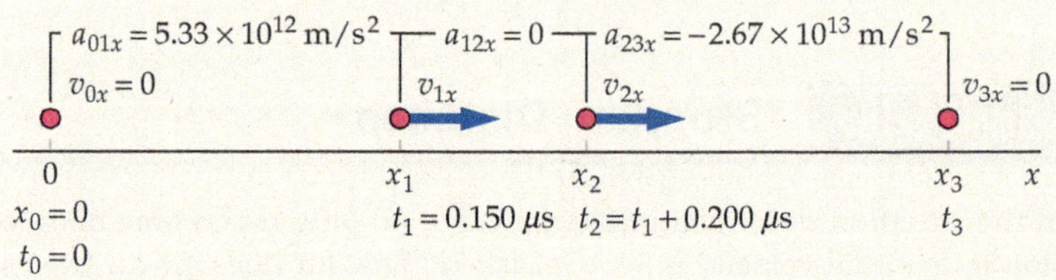

FIGURE 2-17

Answers

2. Set $v_{0x} = 0$ (because the electron starts from rest), use Equations 2-12 and 2-14 to find position x_1 and velocity v_{1x} at the end of the first $0.150\text{-}\mu s$ interval.
   $x_1 = 6.00$ cm, $v_{1x} = 8.00 \times 10^5\ \mathrm{m/s}$
3. The acceleration is zero during the second interval, so the velocity remains constant.
   $v_{2x} = v_{1x} = 8.00 \times 10^5\ \mathrm{m/s}$
4. The velocity remains constant during the second interval, so the displacement Δx_{12} equals the velocity v_{1x} multiplied by $0.200\ \mu s$.
   $\Delta x_{12} = 16.0$ cm, so $x_2 = 22.0$ cm
5. To find the displacement for the third interval, use Equation 2-15 with $v_{3x} = 0$.
   $\Delta x_{23} = 1.20$ cm, so $x_3 =$ **23.2 cm**

CHECK The average velocities are large, but the time intervals are small. Thus, the distances traveled are modest as we might expect.

Sometimes valuable insight can be gained about the motion of an object by asserting that the constant-acceleration formulas still apply even when the acceleration is not constant. The results are then estimates and not exact calculations. This is the case in the following example.

Example 2-12 The Crash Test

Context-Rich

In a crash test that you are performing, a car traveling 100 km/h (about 62 mi/h) hits an immovable concrete wall. What is the acceleration of the car during the crash?

(© 1994 General Motors Corporation, all rights reserved GM Archives.)

PICTURE In this example, different parts of the vehicle will have different accelerations as the car crumples to a halt. The front bumper stops virtually instantly while the rear bumper stops some time later. We will solve for the acceleration of a part of the car that is in the passenger compartment and out of the crumple zone. A bolt holding the driver's seat belt to the floor is such a part. We do not really expect the acceleration of this bolt to be constant. We need additional information to solve this problem–either the stopping distance or the stopping time. We can estimate the stopping distance using common sense. Upon impact, the center of the car will certainly move forward less than half the length of the car. We will choose 0.75 m as a reasonable estimate of the distance the center of the car will move during the crash. Because the problem neither asks for nor provides the time, we will use the equation $v_x^2 = v_{0x}^2 + 2a_x\,\Delta x$.

SOLVE

1. Draw the bolt (as a small circle) at the center of the car at its initial and final positions (Figure 2-18). Include a coordinate axis and label the drawing with the kinematic parameters.

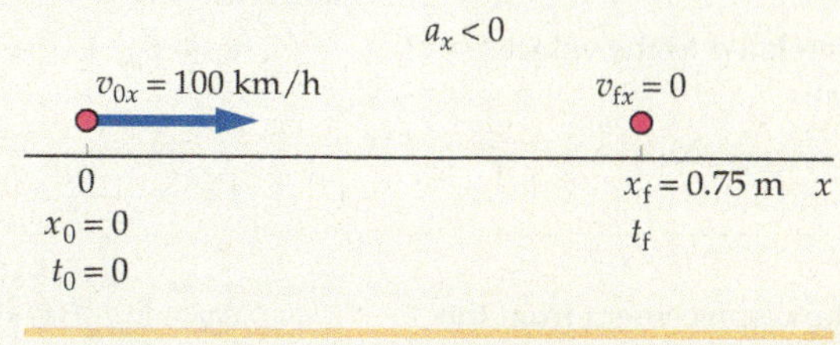

FIGURE 2-18

2. Convert the velocity from km/h to m/s.

$$(100\ \text{km/h}) \times \left(\frac{1\ \text{h}}{60\ \text{min}}\right) \times \left(\frac{1\ \text{min}}{60\ \text{s}}\right) \times \left(\frac{1000\ \text{m}}{1\ \text{km}}\right) = 27.8\ \text{m/s}$$

3. Using $v_x^2 = v_{0x}^2 + 2a_x\,\Delta x$, solve for the acceleration:

$$v_x^2 = v_{0x}^2 + 2a_x\,\Delta x$$

so

$$a_x = \frac{v_x^2 - v_{0x}^2}{2\Delta x} = \frac{0^2 - (27.8\ \text{m/s})^2}{2(0.75\ \text{m})}$$

4. Complete the calculation of the acceleration:

$$a_x = -\frac{(27.8\ \text{m/s})^2}{1.5\ \text{m}} = -514\ \text{m/s}^2 \approx \boxed{-500\ \text{m/s}^2}$$

CHECK The magnitude of the acceleration is about 50 times greater than the acceleration caused by the car breaking hard on a dry concrete road. The result is plausible because a large acceleration is expected for a high-speed head-on crash into an immovable object.

PRACTICE PROBLEM 2-4 Estimate the stopping time of the car.

Free-Fall Many practical problems deal with objects in free-fall, that is, objects that fall freely under the influence of gravity only. All objects in free-fall with the same initial velocity move identically. As shown in Figure 2-19, an apple and a feather, simultaneously released from rest in a large vacuum chamber, fall with identical motions. Thus, we know that the apple and the feather fall with the same acceleration. The magnitude of this acceleration, designated by g, has the approximate value $a = g \approx 9.81\ \text{m/s}^2 = 32.2\ \text{ft/s}^2$. If downward is designated as the $+y$ direction, then $a_y = +g$; if upward is designated as the $+y$ direction, then $a_y = -g$.

FIGURE 2-19 In a vacuum the apple and the feather, released simultaneously from rest, fall with identical motion. *(James Sugar/Black Star.)*

! Because g is the *magnitude* of the acceleration, g is *always* positive.

Example 2-13 The Flying Cap

Upon graduation, a joyful physics student throws her cap straight upward with an initial speed of 14.7 m/s. Given that its acceleration has a magnitude of 9.81 m/s^2 and is directed downward (we neglect air resistance), (*a*) how long does it take for the cap to reach its highest point? (*b*) What is the distance to the highest point? (*c*) Assuming the cap is caught at the same height from which it was released, what is the total time the cap is in flight?

PICTURE When the cap is at its highest point, its instantaneous velocity is zero. (When a problem specifies that an object is "at its highest point," translate this condition into the mathematical condition $v_y = 0$.)

SOLVE

(*a*) 1. Make a sketch of the cap in its initial position and again at its highest point (Figure 2-20). Include a coordinate axis and label the origin and the two specified positions of the cap.

2. The time is related to the velocity and acceleration:

$$v_y = v_{0y} + a_y t$$

3. Set $v_y = 0$ and solve for t:

$$t = \frac{0 - v_{0y}}{a_y} = \frac{-14.7\ \text{m/s}}{-9.81\ \text{m/s}^2} = \boxed{1.50\ \text{s}}$$

(*b*) We can find the displacement from the time t and the average velocity:

$$\Delta y = v_{\text{av}\,y} t = \tfrac{1}{2}(v_{0y} + v_y)\Delta t$$
$$= \tfrac{1}{2}(14.7\ \text{m/s} + 0)(1.50\ \text{s}) = \boxed{11.0\ \text{m}}$$

(*c*) 1. Set $y = y_0$ in Equation 2-14 and solve for t:

$$\Delta y = v_{0y}t + \tfrac{1}{2}a_y t^2$$
$$0 = (v_{0y} + \tfrac{1}{2}a_y t)t$$

2. There are two solutions for t when $y = y_0$. The first corresponds to the time at which the cap is released, the second to the time at which the cap is caught:

$$t = 0 \quad \text{(first solution)}$$
$$t = -\frac{2v_{0y}}{a_y} = -\frac{2(14.7\ \text{m/s})}{-9.81\ \text{m/s}^2} = \boxed{3.00\ \text{s}} \quad \text{(second solution)}$$

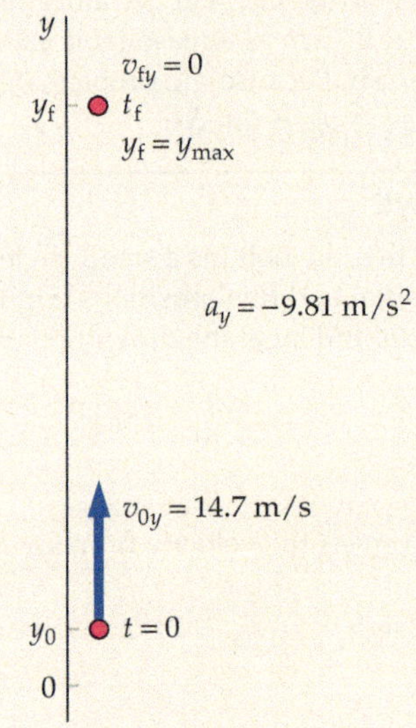

FIGURE 2-20

CHECK On the way up, the cap loses speed at the rate of 9.81 m/s each second. Because its initial speed is 14.7 m/s, we expect it rise for more than 1.00 s, but less than 2.00 s. Thus, a rise time of 1.50 s is quite plausible.

TAKING IT FURTHER On the plot of velocity versus time (Figure 2-21*b*) note that the slope is the same at all times, including the instant that $v_y = 0$. The slope is equal to the instantaneous acceleration, which is a constant -9.81 m/s^2. On the plot of height versus time (Figure 2-21*a*), note that the rise time equals the fall time. In reality, the cap will not have a constant acceleration because air resistance has a significant effect on a light object like a cap. If air resistance is not negligible, the fall time will exceed the rise time.

PRACTICE PROBLEM 2-5 Find $y_{max} - y_0$ using Equation 2-15. Find the velocity of the cap when it returns to its starting point.

PRACTICE PROBLEM 2-6 What is the velocity of the cap at the following points in time? (*a*) 0.100 s before it reaches its highest point; (*b*) 0.100 s after it reaches its highest point. (*c*) Compute $\Delta v_y / \Delta t$ for this 0.200-s-long time interval.

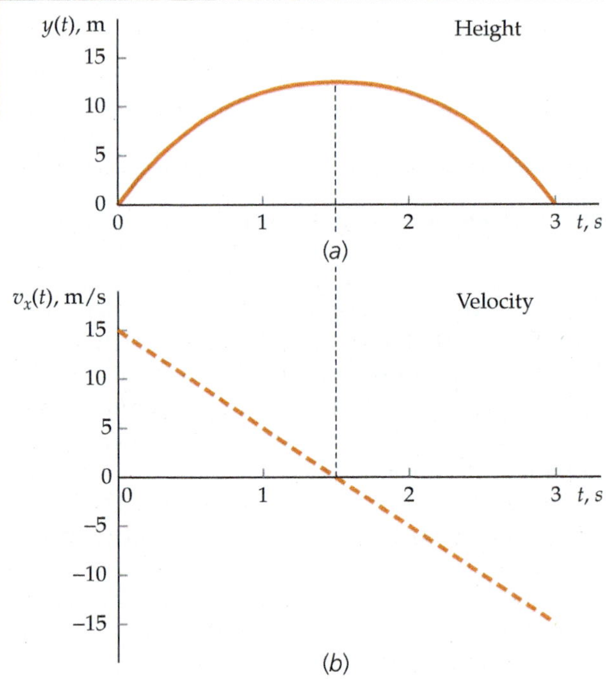

FIGURE 2-21 The height and velocity graphs are drawn one above the other so that both the height and the velocity can be observed at each instant of time.

Problems with two objects We now give some examples of problems involving two objects moving with constant acceleration.

Example 2-14 Catching a Speeding Car

A car is speeding at a constant 25 m/s (= 90 km/h ≈ 56 mi/h) in a school zone. A police car starts from rest just as the speeder passes by it and accelerates at a constant rate of 5.0 m/s². (*a*) When does the police car catch the speeding car? (*b*) How fast is the police car traveling when it catches up with the speeder?

PICTURE To determine when the two cars will be at the same position, we write the positions of the speeder x_S and of the police car x_P as functions of time and solve for the time t_c when $x_S = x_P$. Once we determine when the police car will catch up to the speeder, we can determine the velocity of the police car when it catches up to the speeder using the equation $v_x = a_x t$.

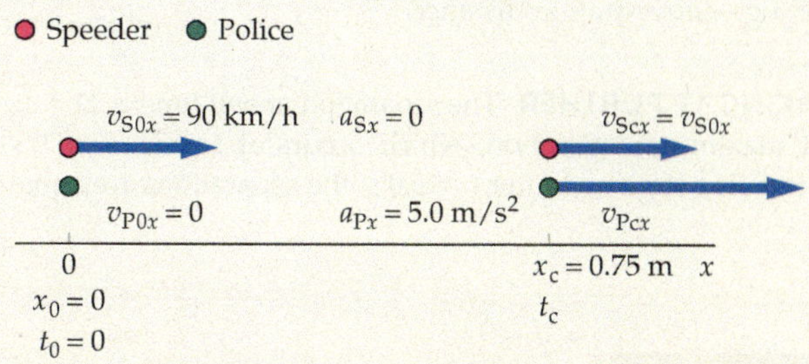

FIGURE 2-22 The speeder and the police car have the same position at $t = 0$ and again at $t = t_c$.

SOLVE

(*a*) 1. Draw the two cars at their initial positions (at $t = 0$) and again at their final positions (at $t = t_c$) (Figure 2-22). Include a coordinate axis and label the drawing with the kinematic parameters.

2. Write the position functions for the speeder and the police car:

$$x_S = v_{Sx}t \text{ and } x_P = \tfrac{1}{2}a_{Px}t^2$$

3. Set $x_S = x_P$ and solve for the time t_c, for $t_c > 0$:

$$v_{Sx}t_c = \tfrac{1}{2}a_{Px}t_c^2 \Rightarrow v_{Sx} = \tfrac{1}{2}a_{Px}t_c \quad t_c \neq 0$$

$$t_c = \frac{2v_{Sx}}{a_{Px}} = \frac{2(25\text{ m/s})}{5.0\text{ m/s}^2} = \boxed{10\text{ s}}$$

(*b*) The velocity of the police car is given by $v_x = v_{0x} + a_x t$, with $v_{0x} = 0$:

$$v_{Px} = a_{Px}t_c = (5.0\text{ m/s}^2)(10\text{ s}) = \boxed{50\text{ m/s}}$$

CHECK Notice that the final velocity of the police car in (*b*) is exactly twice that of the speeder. Because the two cars covered the same distance in the same time, they must have had the same average velocity. The speeder's average velocity, of course, is 25 m/s. For the police car to start from rest, maintain a constant acceleration, and have an average velocity of 25 m/s, it must reach a final velocity of 50 m/s.

PRACTICE PROBLEM 2-7 How far have the cars traveled when the police car catches the speeder?

Example 2-15 The Police Car — *Try It Yourself*

How fast is the police car in Example 2-14 traveling when it is 25 m behind the speeding car?

PICTURE The speed is given by $v_P = a_x t_1$, where t_1 is the time at which $x_S - x_P = 25$ m.

SOLVE

Cover the column to the right and try these on your own before looking at the answers.

Steps

1. Sketch an x-versus-t graph showing the positions of the two cars (Figure 2-23). On this graph identify the distance $D = x_S - x_P$ between the two cars at a given instant.

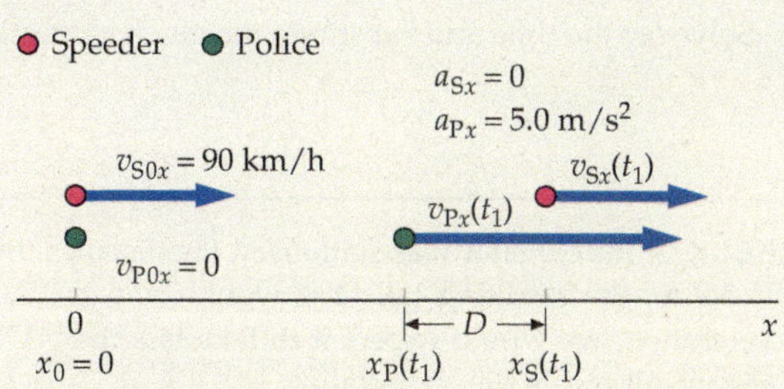

FIGURE 2-23

	Answers
2. Using the equations for x_P and x_S from Example 2-14, solve for t_1 when $x_S - x_P = 25$ m. We expect two solutions, one shortly after the start time and one shortly before the speeder is caught.	$t_1 = (5 \pm \sqrt{15})$ s
3. Use $v_{P1} = a_{Px}t_1$ to compute the speed of the police car when $x_S - x_P = 25$ m.	$v_{P1} = \boxed{5.64\text{ m/s}}$ and $\boxed{44.4\text{ m/s}}$

CHECK We see from Figure 2-24 that the distance between the cars starts at zero, increases to a maximum value, and then decreases. We would expect two speeds for a given separation distance.

TAKING IT FURTHER The separation at any time is $D = x_S - x_P = v_{Sx}t - \frac{1}{2}a_{Px}t^2$. At maximum separation, which occurs at $t = 5.0$ s, $dD/dt = 0$. At equal time intervals before and after $t = 5.0$ s, the separations are equal.

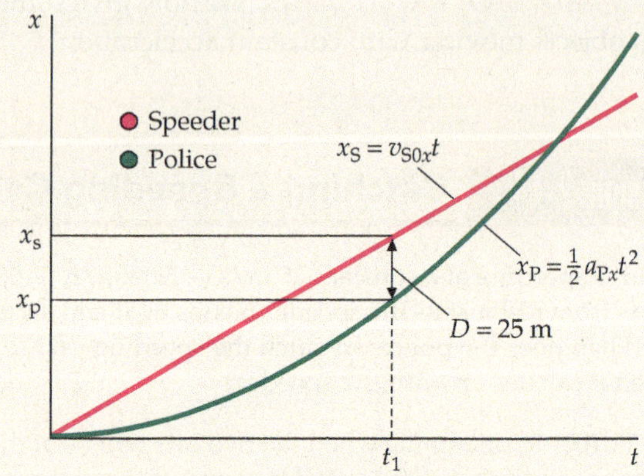

FIGURE 2-24

Example 2-16 A Moving Elevator

While standing in an elevator, you see a screw fall from the ceiling. The ceiling is 3.0 m above the floor. How long does it take the screw to hit the floor if the elevator is moving upward and gaining speed at a constant rate of 4.0 m/s² at the instant the screw leaves the ceiling?

PICTURE When the screw hits the floor, the positions of the screw and the floor are equal. Equate the these positions and solve for the time.

SOLVE

1. Draw a diagram showing the initial and final positions of the screw and the elevator floor (Figure 2-25). Include a coordinate axis and label the drawing with the kinematic parameters. The screw and the floor have the same initial velocity, but different accelerations. Choose the origin to be the initial position of the floor, and designate "upward" as the positive y direction. The screw hits the floor at time t_f:

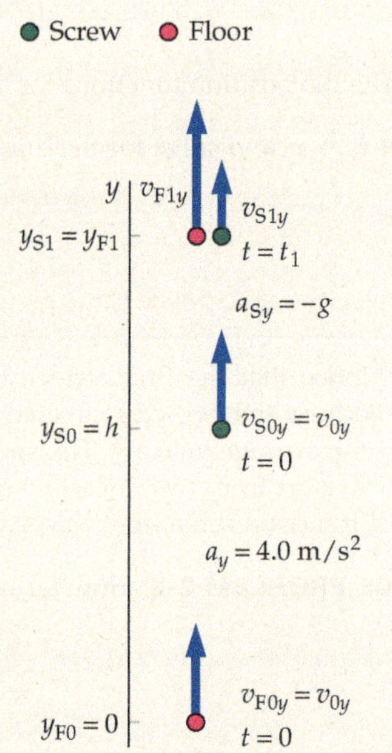

FIGURE 2-25 The y axis is fixed to the building.

2. Write equations specifying the *position* y_F of the elevator floor and the *position* y_S of the screw as functions of time. The screw and the elevator have the same initial velocity v_{0y}:

$$y_F - y_{F0} = v_{F0y}t + \tfrac{1}{2}a_{Fy}t^2$$
$$y_F - 0 = v_{0y}t + \tfrac{1}{2}a_{Fy}t^2$$
$$y_S - y_{S0} = v_{S0y}t + \tfrac{1}{2}a_{Sy}t^2$$
$$y_S - h = v_{0y}t + \tfrac{1}{2}(-g)t^2$$

3. Equate the expressions for y_S and y_F at $t = t_f$ and simplify:

$$y_S = y_F$$
$$h + v_{0y}t_f - \tfrac{1}{2}gt_f^2 = v_{0y}t_f + \tfrac{1}{2}a_{Fy}t_f^2$$
$$h - \tfrac{1}{2}gt_f^2 = \tfrac{1}{2}a_{Fy}t_f^2$$

4. Solve for the time and substitute the given values:

$$h = \tfrac{1}{2}(a_F + g)t_f^2 \quad \text{so}$$
$$t_f = \sqrt{\frac{2h}{a_F + g}} = \sqrt{\frac{2(3.0\text{ m})}{4.0\text{ m/s}^2 + 9.81\text{ m/s}^2}} = 0.659\text{ s} = \boxed{0.66\text{ s}}$$

CHECK If the elevator was stationary, the distance the screw falls is given by $h = \frac{1}{2}gt_f^2$. With $h = 3.0$ m, the resulting fall time would be $t_f = 0.78$ s. Because of the elevator's upward acceleration, we would expect it to take less than 0.78 s for the screw to hit the floor. Our 0.66-s result meets this expectation.

Example 2-17 The Moving Elevator

Try It Yourself

Consider the elevator and screw in Example 2-16. Assume the velocity of the elevator is 16 m/s upward when the screw separates from the ceiling. (*a*) How far does the elevator rise while the screw is in freefall? What is the displacement of the screw during freefall? (*b*) What are the velocity of the screw and the velocity of the elevator at impact?

PICTURE The time of flight of the screw is obtained in the solution of Example 2-16. Use this time to solve Parts (*a*) and (*b*).

SOLVE

Cover the column to the right and try these on your own before looking at the answers.

Steps	Answers
(*a*) 1. Using Equation 2-13, find the distance the floor rises between $t = 0$ and $t = t_f$, where t_f is calculated in step 4 of Example 2-16.	$\Delta y_F = v_{Fi}t_f + \frac{1}{2}a_F t_f^2 = \boxed{11.4\text{ m}}$
2. Between $t = 0$ and $t = t_f$, the displacement of the screw is less then that of the floor by 3.0 m.	$\Delta y_S = \boxed{+8.4\text{ m}}$
(*b*) Using $v_y = v_{0y} + a_y t$ (Equation 2-12), find the velocities of the screw and of the floor at impact.	$v_{Sy} = v_{Siy} - gt_f = \boxed{9.5\text{ m/s}}$ $v_{Fy} = v_{Fiy} + a_{Fy}t_f = \boxed{19\text{ m/s}}$

CHECK The Part-(*b*) results (velocity of the screw and the velocity of the floor at impact) are both positive indicating both velocities are directed upward. For impact to occur, the floor must be moving upward faster then the screw so it can catch up with the screw. This result is consistent with our Part-(*b*) results.

TAKING IT FURTHER The screw strikes the floor 8.4 m above its position when it leaves the ceiling. At impact, the velocity of the screw relative to the building is positive (upward). Relative to the building, the screw is still rising when it and the floor come in contact.

2-4 INTEGRATION

In this section, we use integral calculus to derive the equations of motion. A concise treatment of calculus can be found in the Math Tutorial.

To find the velocity from a given acceleration, we note that the velocity is the function $v_x(t)$ whose time derivative is the acceleration $a_x(t)$:

$$\frac{dv_x(t)}{dt} = a_x(t)$$

If the acceleration is constant, the velocity is that function of time which, when differentiated, equals this constant. One such function is

$$v_x = a_x t \qquad a_x \text{ is constant}$$

More generally, we can add any constant to $a_x t$ without changing the time derivative. Calling this constant c, we have

$$v_x = a_x t + c$$

When $t = 0$, $v_x = c$. Thus, c is the velocity v_{0x} at time $t = 0$.

Similarly, the position function $x(t)$ is that function whose derivative is the velocity:

$$\frac{dx}{dt} = v_x = v_{0x} + a_x t$$

We can treat each term separately. The function whose derivative is the constant v_{0x} is $v_{0x}t$ plus any constant. The function whose derivative is $a_x t$ is $\frac{1}{2}at^2$ plus any constant. Writing x_0 for the combined arbitrary constants, we have

$$x = x_0 + v_{0x}t + \tfrac{1}{2}a t_x^2$$

When $t = 0$, $x = x_0$. Thus, x_0 is the position at time $t = 0$.

Whenever we find a function from its derivative, we must include an arbitrary constant in the general form of the function. Because we go through the integration process twice to find $x(t)$ from the acceleration, two constants arise. These constants are usually determined from the velocity and position at some given time, which is usually chosen to be $t = 0$. They are therefore called the **initial conditions.** A common problem, called the **initial-value problem,** takes the form "given $a_x(t)$ and the initial values of x and v_x, find $x(t)$." This problem is particularly important in physics because the acceleration of a particle is determined by the forces acting on it. Thus, if we know the forces acting on a particle and the position and velocity of the particle at some particular time, we can find its position and velocity at all other times.

A function $F(t)$ whose derivative (with respect to t) equals the function $f(t)$ is called the **antiderivative** of $f(t)$. (Because $v_x = dx/dt$ and $a_x = dv_x/dt$, x is the antiderivative of v_x and v_x is the antiderivative of a_x.) Finding the antiderivative of a function is related to the problem of finding the area under a curve.

In deriving Equation 2-14 it was shown that the change in position Δx is equal to the area under the velocity-versus-time curve. To show this (see Figure 2-12), we first divided the time interval into numerous small intervals, Δt_1, Δt_2, and so on. Then, we drew a set of rectangles as shown. The area of the rectangle corresponding to the ith time interval Δt_i (shaded in the figure) is $v_{ix}\,\Delta t_i$, which is approximately equal to the displacement Δx_i during the interval Δt_i. The sum of the rectangular areas is therefore approximately the sum of the displacements during the time intervals and is approximately equal to the total displacement from time t_1 to t_2. Mathematically, we write this as

$$\Delta x \approx \sum_i v_{ix}\,\Delta t_i$$

For the limit of smaller and smaller time intervals (and more and more rectangles), the resulting sum approaches the area under the curve, which in turn equals the displacement. The limit of the sum as Δt approaches zero (and the number of rectangles approaches infinity) is called an **integral** and is written

$$\Delta x = x(t_2) - x(t_1) = \lim_{\Delta t \to 0}\left(\sum_i v_{ix}\,\Delta t_i\right) = \int_{t_1}^{t_2} v_x\,dt \qquad \text{2-17}$$

See **Math Tutorial** *for more information on*

Integrals

It is helpful to think of the integral sign $\int$ as an elongated S indicating a sum. The limits t_1 and t_2 indicate the initial and final values of the integration variable t.

The process of computing an integral is called **integration.** In Equation 2-17, v_x is the derivative of x, and x is the antiderivative of v_x. This is an example of the fundamental theorem of calculus, whose formulation in the seventeenth century greatly accelerated the mathematical development of physics. If

$$f(t) = \frac{dF(t)}{dt}, \quad \text{then} \quad F(t_2) - F(t_1) = \int_{t_1}^{t_2} f(t)\,dt \qquad \text{2-18}$$

FUNDAMENTAL THEOREM OF CALCULUS

The antiderivative of a function is also called the indefinite integral of the function and is written without limits on the integral sign, as in

$$x = \int v_x\,dt$$

Finding the function x from its derivative v_x (that is, finding the antiderivative) is also called integration. For example, if $v_x = v_{0x}$, a constant, then

$$x = \int v_{0x}\,dt = v_{0x}t + x_0$$

where x_0 is the arbitrary constant of integration. We can find a general rule for the integration of a power of t from Equation 2-6, which gives the general rule for the derivative of a power. The result is

$$\int t^n\,dt = \frac{t^{n+1}}{n+1} + C,\ n \neq -1 \qquad \text{2-19}$$

where C is an arbitrary constant. This equation can be checked by differentiating the right side using the rule of Equation 2-6. (For the special case $n = -1$, $\int t^{-1}dt = \ln t + C$, where $\ln t$ is the natural logarithm of t.)

Because $a_x = dv_x/dt$, the change in velocity for some time interval can similarly be interpreted as the area under the a_x-versus-t curve for that interval. This change is written

$$\Delta v_x = \lim_{\Delta t \to 0}\left(\sum_i a_{ix}\,\Delta t_i\right) = \int_{t_1}^{t_2} a_x\,dt \qquad \text{2-20}$$

We can now derive the constant-acceleration equations by computing the indefinite integrals of the acceleration and velocity. If a_x is constant, we have

$$v_x = \int a_x dt = a_x \int dt = v_{0x} + a_x t \qquad \text{2-21}$$

where we have expressed the product of a_x and the constant of integration as v_{0x}. Integrating again, and writing x_0 for the constant of integration, gives

$$x = \int (v_{0x} + a_x t)dt = x_0 + v_{0x}t + \tfrac{1}{2}a_x t^2 \qquad \text{2-22}$$

It is instructive to derive Equations 2-21 and 2-22 using definite integrals instead of indefinite ones. For constant acceleration, Equation 2-20, with $t_1 = 0$, gives

$$v_x(t_2) - v_x(0) = a_x \int_0^{t_2} dt = a_x(t_2 - 0)$$

where the time t_2 is arbitrary. Because it is arbitrary, we can set $t_2 = t$ to obtain

$$v_x = v_{0x} + a_x t$$

where $v_x = v_x(t)$ and $v_{0x} = v_x(0)$. To derive Equation 2-22, we substitute $v_{0x} + a_x t$ for v_x in Equation 2-17 with $t_1 = 0$. This gives

$$x(t_2) - x(0) = \int_0^{t_2} (v_{0x} + a_x t)dt$$

This integral is equal to the area under the v_x-versus-t curve (Figure 2-26). Evaluating the integral and solving for x gives

$$x(t_2) - x(0) = \int_0^{t_2} (v_{0x} + a_x t)dt = v_{0x}t + \tfrac{1}{2}a_x t^2 \Big|_0^{t_2} = v_{0x}t_2 + \tfrac{1}{2}a_x t_2^2$$

where t_2 is arbitrary. Setting $t_2 = t$, we obtain

$$x = x_0 + v_{0x}t + \tfrac{1}{2}a_x t^2$$

where $x = x(t)$ and $x_0 = x(0)$.

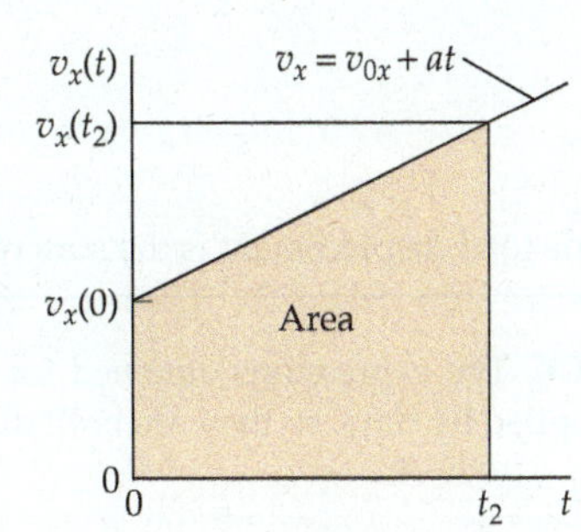

FIGURE 2-26 The area under the v_x-versus-t curve equals the displacement $\Delta x = x(t_2) - x(0)$.

The definition of average velocity is $\Delta x = v_{\text{av}\,x}\,\Delta t$ (Equation 2-3). In addition, $\Delta x = \int_{t_1}^{t_2} v_x\,dt$ (Equation 2-17). Equating the right sides of these equations and solving for $v_{\text{av}\,x}$ gives

$$v_{\text{av}\,x} = \frac{1}{\Delta t}\int_{t_1}^{t_2} v_x\,dt \qquad \text{2-23}$$

ALTERNATIVE DEFINITION OF AVERAGE VELOCITY

where $\Delta t = t_2 - t_1$. Equation 2-23 is mathematically equivalent to the definition of average velocity, so either equation can serve as a definition of average velocity.

Example 2-18 A Coasting Boat

A Shelter Island ferryboat moves with constant velocity $v_{0x} = 8.0\text{ m/s}$ for 60 s. It then shuts off its engines and coasts. Its coasting velocity is given by $v_x = v_{0x}t_1^2/t^2$, where $t_1 = 60$ s. What is the displacement of the boat for the interval $0 < t < \infty$?

(Gene Mosca.)

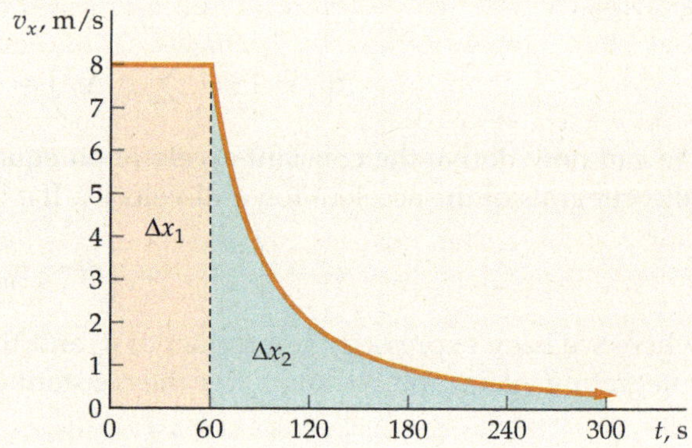

FIGURE 2-27

PICTURE The velocity function for the boat is shown in Figure 2-27. The total displacement is calculated as the sum of the displacement Δx_1 during the interval $0 < t < t_1 = 60$ s and the displacement Δx_2 during the interval $t_1 < t < \infty$.

SOLVE

1. The velocity of the boat is constant during the first 60 s; thus the displacement is simply the velocity times the elapsed time:

$$\Delta x_1 = v_{0x}\,\Delta t = v_{0x}t_1 = (8.0\text{ m/s})(60\text{ s}) = 480\text{ m}$$

2. The remaining displacement is given by the integral of the velocity from $t = t_1$ to $t = \infty$. We use Equation 2-17 to calculate the integral:

$$\Delta x_2 = \int_{t_1}^{\infty} v_x\,dt = \int_{t_1}^{\infty}\frac{v_{0x}t_1^2}{t^2}\,dt = v_{0x}t_1^2\int_{t_1}^{\infty} t^{-2}\,dt$$

$$= v_{0x}t_1^2\,\frac{t^{-1}}{-1}\bigg|_{t_1}^{\infty} = -v_{0x}t_1^2\left(\frac{1}{\infty} - \frac{1}{t_1}\right)$$

$$= -(0 - v_0t_1) = (8\text{ m/s})(60\text{ s}) = 480\text{ m}$$

3. The total displacement is the sum of the displacements found above:

$$\Delta x = \Delta x_1 + \Delta x_2 = 480\text{ m} + 480\text{ m} = \boxed{960\text{ m}}$$

CHECK The expressions obtained for the displacements in both steps 1 and 2 are velocity multiplied by time, so they are both dimensionally correct.

TAKING IT FURTHER Note that the area under the v_x-versus-t curve (Figure 2-27) is finite. Thus, even though the boat never stops moving, it travels only a finite distance. A better representation of the velocity of a coasting boat might be the exponentially decreasing function $v_x = v_{0x}e^{-b(t-t_1)}$, where b is a positive constant. In that case, the boat would also coast a finite distance in the interval $t_1 \le t \le \infty$.

Linear Accelerators

Linear accelerators are instruments that accelerate electrically charged particles to high speeds along a long, straight track to collide with a target. Large accelerators can impart very high kinetic energies (on the order of billions of electron volts) to charged particles that serve as probes for studying the fundamental particles of matter and the forces that hold them together. (The energy required to remove an electron from an atom is on the order of one electron volt.) In the two-mile-long linear accelerator at Stanford University, electromagnetic waves boost the speed of electrons or positrons as they move through an evacuated copper pipe. When the high-speed particles collide with a target, several different kinds of subatomic particles are produced along with X rays and gamma rays. These particles then pass into devices called particle detectors.

The beige cylinder in the background is the linear accelerator at the heart of the Naval Academy Tandem Accelerator Laboratory. A beam of high-speed protons travels from the accelerator to the target area in the foreground. *(Gene Mosca.)*

Through experiments with such accelerators, physicists have determined that protons and neutrons, once thought to be the ultimate particles of the nucleus, are themselves composed of more fundamental particles called quarks. Another group of particles known as leptons, which include electrons, neutrinos, and a few other particles, have also been identified. Most large accelerator research centers such as the Fermi National Accelerator Laboratory in Batavia, Illinois, use a series of linear and circular accelerators to achieve higher particle speeds. As the speed of a particle approaches the speed of light, the energy required to accelerate it to that speed approaches infinity.

Although the big accelerators may have a high profile, thousands of linear accelerators are used worldwide for a host of practical applications. One of the most common applications is the cathode ray tube (CRT) of a television set or computer monitor. In a CRT, electrons from the cathode (a heated filament) are accelerated in a vacuum toward a positively charged anode. Electromagnets control the direction of the electron beam onto the inside of a screen coated with a phosphor, a material that emits light when struck by electrons. The kinetic energy of electrons in a CRT ranges to a maximum of about 30,000 electron volts. The speed of an electron that has this kinetic energy is about one third of the speed of light.

In the field of medicine, linear accelerators about a thousand times more powerful than a CRT are used for radiation treatment of cancer. "The linear accelerator uses microwave technology (similar to that used for radar) to accelerate electrons in a part of the accelerator called the 'wave guide,' then allows these electrons to collide with a heavy metal target. As a result of the collisions, high-energy x-rays are scattered from the target. A portion of these x-rays is collected and then shaped to form a beam that matches the patient's tumor."*

Other applications of accelerators include the production of radioisotopes for tracers in medicine and biology, sterilization of surgical tools, and analysis of materials to determine their composition. For example, in a technique called particle-induced X-ray emission (PIXE), an ion beam, often consisting of protons, causes target atoms to emit X rays that identify the type of atoms present. This technique has been applied to the study of archeological materials and variety of other types of samples.

* The American College of Radiology and the Radiological Society of North America, http//www.radiologyinfo.org/content/therapy/linear_accelerator.htm.

SUMMARY

Displacement, velocity, and acceleration are important *defined* kinematic quantities.

TOPIC	RELEVANT EQUATIONS AND REMARKS	
1. Displacement	$\Delta x = x_2 - x_1$	2-1
Graphical interpretation	Displacement is the area under the v_x-versus-t curve.	
2. Velocity		
Average velocity	$v_{\text{av}\,x} = \dfrac{\Delta x}{\Delta t}$ or $v_{\text{av}\,x} = \dfrac{1}{\Delta t}\displaystyle\int_{t_1}^{t_2} v_x\,dt$	2-3, 2-23
Instantaneous velocity	$v_x(t) = \lim\limits_{\Delta t \to 0} \dfrac{\Delta x}{\Delta t} = \dfrac{dx}{dt}$	2-5
Graphical interpretation	The instantaneous velocity is the slope of the x-versus-t curve.	
3. Speed		
Average speed	average speed $= \dfrac{\text{total distance}}{\text{total time}} = \dfrac{s}{t}$	2-2
Instantaneous speed	Instantaneous speed is the magnitude of the instantaneous velocity speed $= \lvert v_x \rvert$	
4. Acceleration		
Average acceleration	$a_{\text{av}\,x} = \dfrac{\Delta v_x}{\Delta t}$	2-7
Instantaneous acceleration	$a_x = \dfrac{dv_x}{dt} = \dfrac{d^2x}{dt^2}$	2-9
Graphical interpretation	The instantaneous acceleration is the slope of the v_x-versus-t curve.	
Acceleration due to gravity	The acceleration of an object near the surface of Earth in free-fall under the influence of gravity alone is directed downward and has magnitude $g = 9.81\ \text{m/s}^2 = 32.2\ \text{ft/s}^2$	
5. Kinematic equations for constant acceleration		
Velocity	$v_x = v_{0x} + a_x t$	2-12
Average velocity	$v_{\text{av}\,x} = \frac{1}{2}(v_{0x} + v_x)$	2-16
Displacement in terms of $v_{\text{av}\,x}$	$\Delta x = x - x_0 = v_{\text{av}\,x} t = \frac{1}{2}(v_{0x} + v_x)t$	
Displacement as a function of time	$\Delta x = x - x_0 = v_{0x} t + \frac{1}{2} a_x t^2$	2-14
v_x^2 as a function of Δx	$v_x^2 = v_{0x}^2 + 2a_x \Delta x$	2-15
6. Displacement and velocity as integrals	Displacement is represented graphically as the area under the v_x-versus-t curve. This area is the integral of v_x over time from some initial time t_1 to some final time t_2 and is written $\Delta x = \lim\limits_{\Delta t \to 0} \sum_i v_{i\,x} \Delta t_i = \displaystyle\int_{t_1}^{t_2} v_x\,dt$	2-17
	Similarly, change in velocity is represented graphically as the area under the a_x-versus-t curve: $\Delta v_x = \lim\limits_{\Delta t \to 0} \sum_i a_{i\,x} \Delta t_i = \displaystyle\int_{t_1}^{t_2} a_x\,dt$	2-20

Answers to Concept Checks

2-1 No. The distance between cars will not remain constant. Instead, the distance will continuously decrease. When you first begin breaking, the speed of your car is greater than the speed of the car in front. That is because the car in front began breaking 0.3 s earlier. Because the cars lose speed at the same rate, the speed of your car will remain greater than the speed of the car in front throughout the braking period.

Answers to Practice Problems

2-1	1.2 m/s
2-2	(*a*) 65 km/h (*b*) 2.5 s
2-3	Only (*d*) has the same dimensions on both sides of the equation. Although we cannot obtain the exact equation from dimensional analysis, we can often obtain the functional dependence.
2-4	54 ms
2-5	(a) and (*b*) $y_{max} - y_0 = 11.0$ m (*c*) -14.7 m/s; notice that the final speed is the same as the initial speed
2-6	(*a*) $+0.981$ m/s (*b*) -0.981 m/s (*c*) $[(-0.981 \text{ m/s}) - (+0.981 \text{ m/s}]/(0.200 \text{ s}) = -9.81 \text{ m/s}^2$
2-7	250 m

PROBLEMS

In a few problems, you are given more data than you actually need; in a few other problems, you are required to supply data from your general knowledge, outside sources, or informed estimates.

Interpret as significant all digits in numerical values that have trailing zeroes and no decimal points.

For all problems, use $g = 9.81$ m/s² for the free-fall acceleration due to gravity and neglect friction and air resistance unless instructed to do otherwise.

- • Single-concept, single-step, relatively easy
- •• Intermediate-level, may require synthesis of concepts
- ••• Challenging
- SSM Solution is in the Student Solutions Manual
- Consecutive problems that are shaded are paired problems.

CONCEPTUAL PROBLEMS

1 • What is the average velocity over the "round trip" of an object that is launched straight up from the ground and falls straight back down to the ground?

2 • An object thrown straight up falls back and is caught at the same place it is launched from. Its time of flight is T; its maximum height is H. Neglect air resistance. The correct expression for its average speed for the entire flight is (*a*) H/T, (*b*) 0, (*c*) $H/(2T)$, (*d*) $2H/T$.

3 • Using the information in the previous question, what is its average speed just for the first half of the trip? What is its average velocity for the second half of the trip? (Answer in terms of H and T.)

4 • Give an everyday example of one-dimensional motion where (*a*) the velocity is westward and the acceleration is eastward, and (*b*) the velocity is northward and the acceleration is northward.

5 • Stand in the center of a large room. Call the direction to your right "positive," and the direction to your left "negative." Walk across the room along a straight line, using a constant acceleration to quickly reach a steady speed along a straight line in the negative direction. After reaching this steady speed, keep your velocity negative but make your acceleration positive. (*a*) Describe how your speed varied as you walked. (*b*) Sketch a graph of x versus t for your motion. Assume you started at $x = 0$. (*c*) Directly under the graph of Part (*b*), sketch a graph of v_x versus t. SSM

6 • True/false: The displacement *always* equals the product of the average velocity and the time interval. Explain your choice.

7 • Is the statement "for an object's velocity to remain constant, its acceleration *must* remain zero" true or false? Explain your choice.

8 •• **MULTISTEP** Draw careful graphs of the position and velocity and acceleration over the time period $0 \le t \le 30$ s for a cart that, in succession, has the following motion. The cart is moving at the constant speed of 5.0 m/s in the $+x$ direction. It passes by the origin at $t = 0.0$ s. It continues on at 5.0 m/s for 5.0 s, after which it gains speed at the constant rate of 0.50 m/s each second for 10.0 s. After gaining speed for 10.0 s, the cart loses speed at the constant rate of 0.50 m/s for the next 15.0 s.

9 • True/false: Average velocity *always* equals one-half the sum of the initial and final velocities. Explain your choice.

10 • Identical twin brothers standing on a horizontal bridge each throw a rock straight down into the water below. They throw rocks at exactly the same time, but one hits the water before the other. How can this be? Explain what they did differently. Ignore any effects due to air resistance.

11 •• Dr. Josiah S. Carberry stands at the top of the Sears Tower in Chicago. Wanting to emulate Galileo, and ignoring the safety of the pedestrians below, he drops a bowling ball from the top of the tower. One second later, he drops a second bowling ball. While the balls are in the air, does their separation (*a*) increase over time, (*b*) decrease, (*c*) stay the same? Ignore any effects due to air resistance. SSM

12 •• Which of the position-versus-time curves in Figure 2-28 best shows the motion of an object (*a*) with positive acceleration, (*b*) with constant positive velocity, (*c*) that is always at rest, and

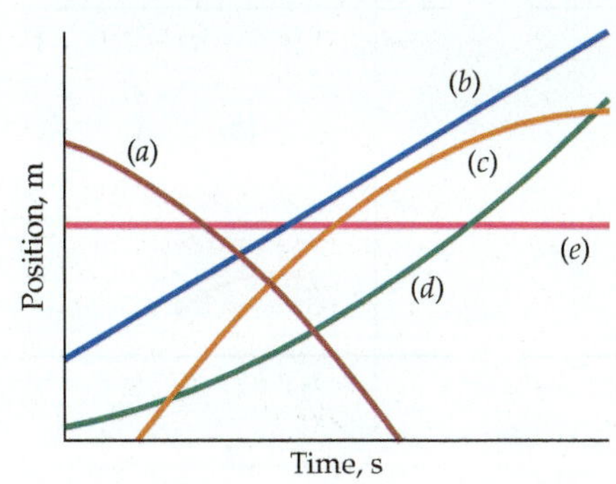

FIGURE 2-28 Problem 12

(*d*) with negative acceleration? (There may be more than one correct answer for each part of the problem.)

13 •• Which of the velocity-versus-time curves in Figure 2-29 best describes the motion of an object (*a*) with constant positive acceleration, (*b*) with positive acceleration that is decreasing with time, (*c*) with positive acceleration that is increasing with time, and (*d*) with no acceleration? (There may be more than one correct answer for each part of the problem.) SSM

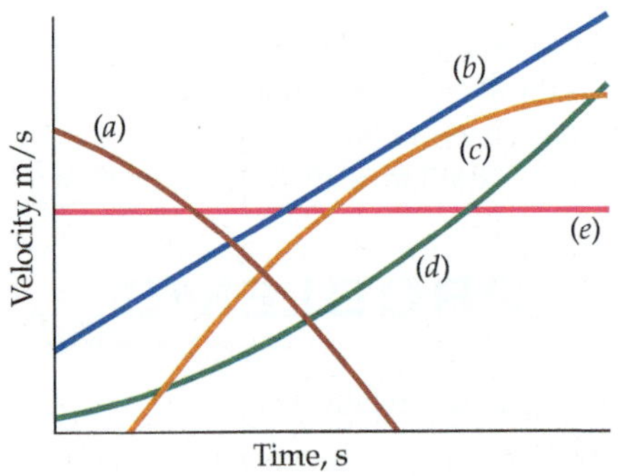

FIGURE 2-29 Problem 13

14 •• The diagram in Figure 2-30 tracks the location of an object moving in a straight line along the x axis. Assume that the object is at the origin at $t = 0$. Of the five times shown, which time (or times) represents when the object is (*a*) farthest from the origin, (*b*) at rest for an instant, (*c*) in the midst of being at rest for a while, and (*d*) moving away from the origin?

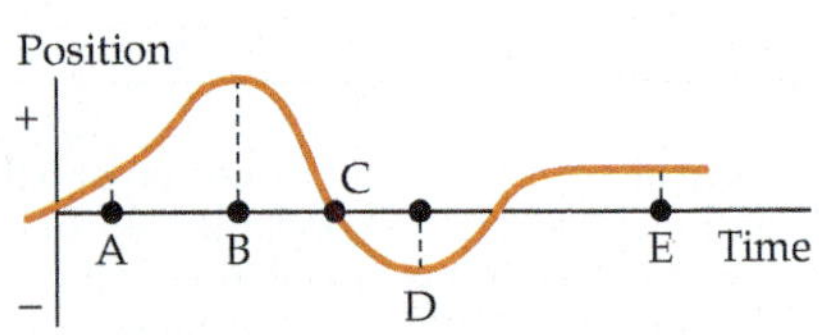

FIGURE 2-30 Problems 14 and 15

15 •• An object moves along a straight line. Its position-versus-time graph is shown in Figure 2-30. At which time or times is its (*a*) speed at a minimum, (*b*) acceleration positive, and (*c*) velocity negative? SSM

16 •• For each of the four graphs of x versus t in Figure 2-31 answer the following questions. (*a*) Is the velocity at time t_2 greater than, less than, or equal to the velocity at time t_1? (*b*) Is the speed at time t_2 greater than, less than, or equal to the speed at time t_1?

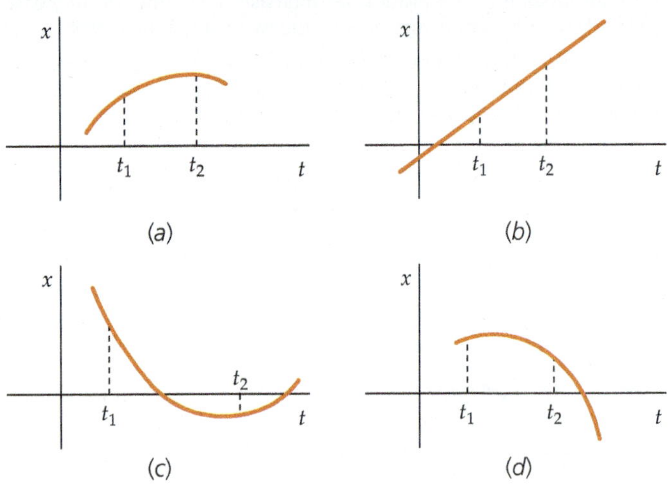

FIGURE 2-31 Problem 16

17 •• True/false:

(*a*) If the acceleration of an object is always zero, then it cannot be moving.
(*b*) If the acceleration of an object is always zero, then its x-versus-t curve must be a straight line.
(*c*) If the acceleration of an object is nonzero at an instant, it may be momentarily at rest at that instant.

Explain your reasoning for each answer. If an answer is *true*, give an example.

18 •• A hard-thrown tennis ball is moving horizontally when it bangs into a vertical concrete wall at perpendicular incidence. The ball rebounds straight back off the wall. Neglect any effects due to gravity for the small time interval described here. Assume that toward the wall is the $+x$ direction. What are the directions of its velocity and acceleration (*a*) just before hitting the wall, (*b*) at maximum impact, and (*c*) just after leaving the wall?

19 •• A ball is thrown straight up. Neglect any effects due to air resistance. (*a*) What is the velocity of the ball at the top of its flight? (*b*) What is its acceleration at that point? (*c*) What is different about the velocity and acceleration at the top of the flight if instead the ball impacts a horizontal ceiling very hard and then returns. SSM

20 •• An object that is launched straight up from the ground, reaches a maximum height H, and falls straight back down to the ground, hitting it T seconds after launch. Neglect any effects due to air resistance. (*a*) Express the average speed for the entire trip as a function of H and T. (*b*) Express the average speed for the same interval of time as a function of the initial launch speed v_0.

21 •• A small lead ball is thrown directly upward. True or false: (Neglect any effects due to air resistance.) (*a*) The magnitude of its acceleration decreases on the way up. (*b*) The direction of its acceleration on its way down is opposite to the direction of its acceleration on its way up. (*c*) The direction of its velocity on its way down is opposite to the direction of its velocity on its way up.

22 •• At $t = 0$, object A is dropped from the roof of a building. At the same instant, object B is dropped from a window 10 m below the roof. Air resistance is negligible. During the descent of B to the ground, the distance between the two objects (*a*) is proportional to t, (*b*) is proportional to t^2, (*c*) decreases, (*d*) remains 10 m throughout.

23 •• **CONTEXT-RICH** You are driving a Porsche that accelerates uniformly from 80.5 km/h (50 mi/h) at $t = 0.00$ s to 113 km/h (70 mi/h) at $t = 9.00$ s. (*a*) Which graph in Figure 2-32 best describes the velocity of your car? (*b*) Sketch a position-versus-time graph showing the location of your car during these nine seconds, assuming we let its position x be zero at $t = 0$.

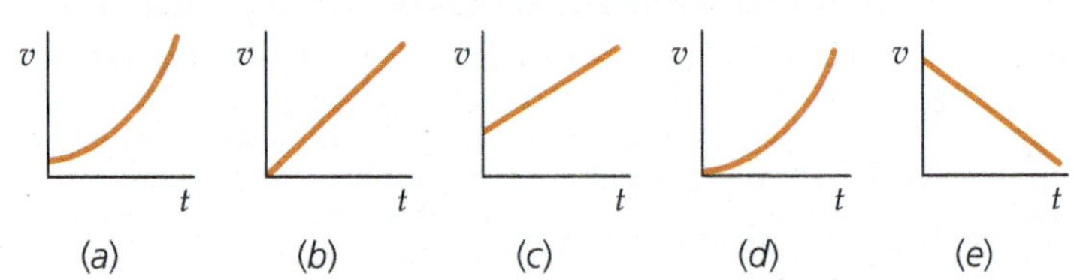

FIGURE 2-32 Problem 23

24 •• A small heavy object is dropped from rest and falls a distance D in a time T. After it has fallen for a time $2T$, what will be its (*a*) fall distance from its initial location, (*b*) its speed, and (*c*) its acceleration? (Neglect air resistance.)

25 •• In a race, at an instant when two horses are running right next to each other and in the same direction (the $+x$ direction), Horse A's instantaneous velocity and acceleration are $+10$ m/s and $+2.0$ m/s^2, respectively, and horse B's instantaneous velocity and acceleration are $+12$ m/s and -1.0 m/s^2, respectively. Which horse is passing the other at this instant? Explain.

26 •• True or false: (*a*) The equation $x - x_0 = v_{0x}t + \frac{1}{2}a_x t^2$ is always valid for particle motion in one dimension. (*b*) If the velocity at a given instant is zero, the acceleration at that instant must also be zero. (*c*) The equation $\Delta x = v_{av}\,\Delta t$ holds for all particle motion in one dimension.

27 •• If an object is moving in a straight line at constant acceleration, its instantaneous velocity halfway through any time interval is (*a*) greater than its average velocity, (*b*) less than its average velocity, (*c*) equal to its average velocity, (*d*) half its average velocity, (*e*) twice its average velocity.

28 •• A turtle, seeing his owner put some fresh lettuce on the opposite side of his terrarium, begins to accelerate (at a constant rate) from rest at time $t = 0$, heading directly toward the food. Let t_1 be the time at which the turtle has covered half the distance to his lunch. Derive an expression for the ratio of t_2 to t_1, where t_2 is the time at which the turtle reaches the lettuce.

29 •• The positions of two cars in parallel lanes of a straight stretch of highway are plotted as functions of time in the Figure 2-33. Take positive values of x as being to the right of the origin. Qualitatively answer the following: (*a*) Are the two cars ever side by side? If so, indicate that time (those times) on the axis. (*b*) Are they always traveling in the same direction, or are they moving in opposite directions for some of the time? If so, when? (*c*) Are they ever traveling at the same velocity? If so, when? (*d*) When are the two cars the farthest apart? (*e*) Sketch (no numbers) the velocity versus time curve for each car. SSM

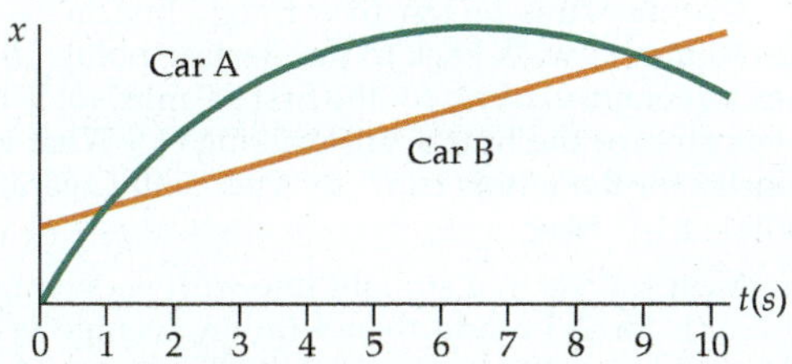

FIGURE 2-33 Problem 29

30 •• A car driving at constant velocity passes the origin at time $t = 0$. At that instant, a truck, at rest at the origin, begins to accelerate uniformly from rest. Figure 2-34 shows a qualitative plot of the velocities of truck and car as functions of time. Compare their displacements (from the origin), velocities, and accelerations at the instant that their curves intersect.

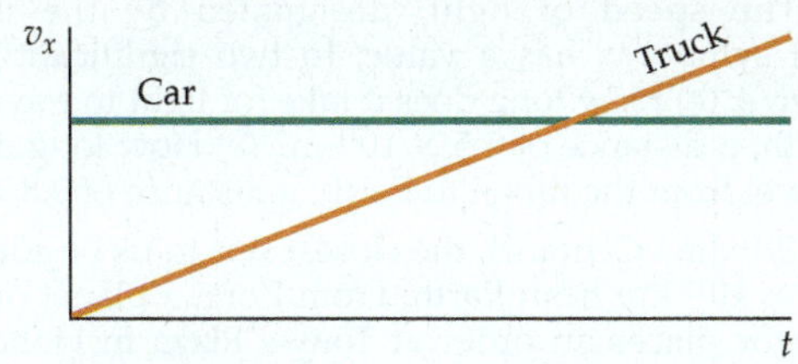

FIGURE 2-34 Problem 30

31 •• Reginald is out for a morning jog, and during the course of his run on a straight track, he has a velocity that depends upon time as shown in Figure 2-35. That is, he begins at rest, and ends at rest, peaking at a maximum velocity v_{max} at an arbitrary time t_{max}. A second runner, Josie, runs throughout the time interval $t = 0$ to $t = t_f$ at a constant speed v_J, so that each has the same displacement during the time interval. (*Note:* t_f is NOT twice t_{max}, but represents an arbitrary time.) What is the relation between v_J *and* v_{max}?

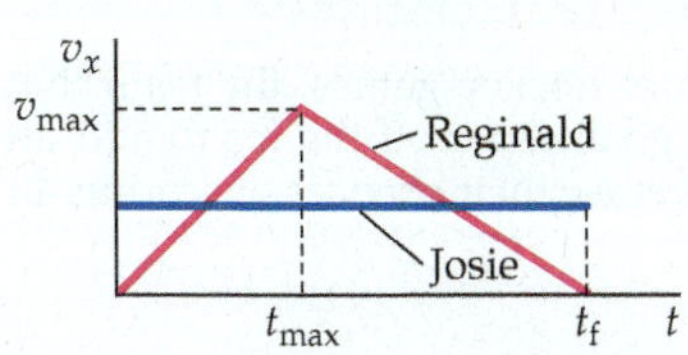

FIGURE 2-35 Problem 31

32 •• Which graph (or graphs), if any, of v_x versus t in Figure 2-36 best describes the motion of a particle with (*a*) positive velocity and increasing speed, (*b*) positive velocity and zero acceleration, (*c*) constant nonzero acceleration, and (*d*) a speed decrease?

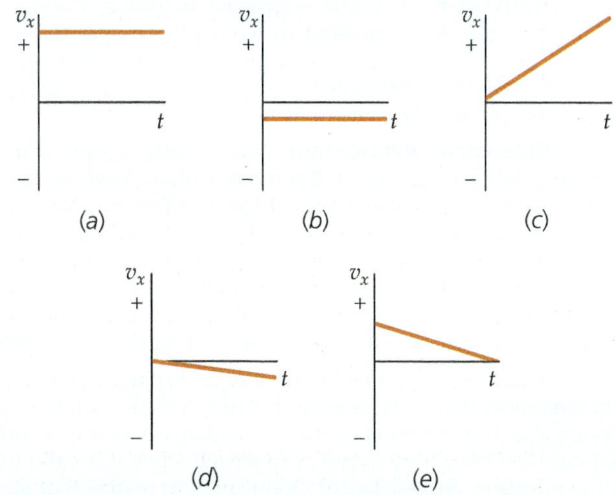

FIGURE 2-36 Problems 32 and 33

33 •• Which graph (or graphs), if any, of v_x versus t in Figure 2-36 best describes the motion of a particle with (*a*) negative velocity and increasing speed, (*b*) negative velocity and zero acceleration, (*c*) variable acceleration, and (*d*) increasing speed?

34 •• Sketch a v-versus-t curve for each of the following conditions: (*a*) Acceleration is zero and constant while velocity is not zero. (*b*) Acceleration is constant but not zero. (*c*) Velocity and acceleration are both positive. (*d*) Velocity and acceleration are both negative. (*e*) Velocity is positive and acceleration is negative. (*f*) Velocity is negative and acceleration is positive. (*g*) Velocity is momentarily zero but the acceleration is not zero.

35 •• Figure 2-37 shows nine graphs of position, velocity, and acceleration for objects in motion along a straight line. Indicate the graphs that meet the following conditions: (*a*) Velocity is constant,

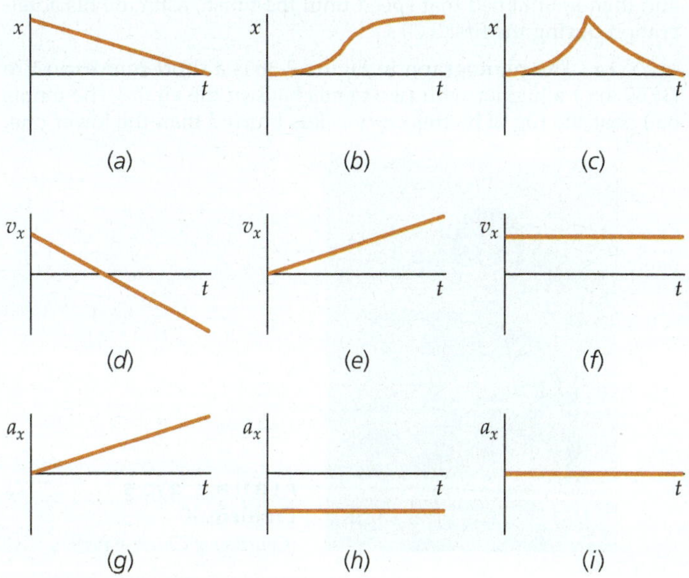

FIGURE 2-37 Problem 35

(*b*) velocity reverses its direction, (*c*) acceleration is constant, and (*d*) acceleration is not constant. (*e*) Which graphs of position, velocity, and acceleration are mutually consistent?

ESTIMATION AND APPROXIMATION

36 • **CONTEXT-RICH** While engrossed in thought about the scintillating lecture just delivered by your physics professor you mistakenly walk directly into the wall (rather than through the open lecture hall door). Estimate the magnitude of your average acceleration as you rapidly come to a halt.

37 • **BIOLOGICAL APPLICATION** Occasionally, people can survive falling large distances if the surface they land on is soft enough. During a traverse of the Eiger's infamous Nordvand, mountaineer Carlos Ragone's rock anchor gave way and he plummeted 500 feet to land in snow. Amazingly, he suffered only a few bruises and a wrenched shoulder. Assuming that his impact left a hole in the snow 4.0 ft deep, estimate his average acceleration as he slowed to a stop (that is, while he was impacting the snow). SSM

38 •• When we solve free-fall problems near Earth, it's important to remember that air resistance may play a significant role. If its effects are significant, we may get answers that are wrong by orders of magnitude if we ignore it. How can we tell when it is valid to ignore the effects of air resistance? One way is to realize that air resistance increases with increasing speed. Thus, as an object falls and its speed *increases*, its downward acceleration *decreases*. Under these circumstances, the object's speed will approach, as a limit, a value called its *terminal speed*. This terminal speed depends upon such things as the mass and cross-sectional area of the body. Upon reaching its terminal speed, its acceleration is zero. For a "typical" skydiver falling through the air, a typical terminal speed is about 50 m/s (roughly 120 mph). At half its terminal speed, the skydiver's acceleration will be about $\frac{3}{4}g$. Let us take half the terminal speed as a reasonable "upper bound" beyond which we should not use our constant acceleration free-fall relationships. Assuming the skydiver started from rest, (*a*) estimate how far, and for how long, the skydiver falls before we can no longer neglect air resistance. (*b*) Repeat the analysis for a Ping-Pong ball, which has a terminal speed of about 5.0 m/s. (*c*) What can you conclude by comparing your answers for Parts (*a*) and (*b*)?

39 •• **BIOLOGICAL APPLICATION** On June 14, 2005, Asafa Powell of Jamaica set a world's record for the 100-m dash with a time $t = 9.77$ s. Assuming he reached his maximum speed in 3.00 s, and then maintained that speed until the finish, estimate his acceleration during the first 3.00 s.

40 •• The photograph in Figure 2-38 is a short-time exposure (1/30 s) of a juggler with two tennis balls in the air. (*a*) The tennis ball near the top of its trajectory is less blurred than the lower one. Why is that? (*b*) Estimate the speed of the ball that he is just releasing from his right hand. (*c*) Determine how high the ball should have gone above the launch point and compare it to an estimate from the picture. *Hint: You have a built-in distance scale if you assume some reasonable value for the height of the juggler.*

FIGURE 2-38 Problem 40
(Courtesy of Chuck Adler.)

41 •• A rough rule of thumb for determining the distance between you and a lightning strike is to start counting the seconds that elapse ("one-Mississippi, two-Mississippi, . . .") until you hear the thunder (sound emitted by the lightning as it rapidly heats the air around it). Assuming the speed of sound is about 750 mi/h, (*a*) estimate how far away is a lightning strike if you counted about 5 s until you heard the thunder. (*b*) Estimate the uncertainty in the distance to the strike in Part (*a*). Be sure to explain your assumptions and reasoning. *Hint: The speed of sound depends on the air temperature, and your counting is far from exact!*

SPEED, DISPLACEMENT, AND VELOCITY

42 • **ENGINEERING APPLICATION** (*a*) An electron in a television tube travels the 16-cm distance from the grid to the screen at an average speed of 4.0×10^7 m/s. How long does the trip take? (*b*) An electron in a current-carrying wire travels at an average speed of 4.0×10^{-5} m/s. How long does it take to travel 16 cm?

43 • A runner runs 2.5 km, in a straight line, in 9.0 min and then takes 30 min to walk back to the starting point. (*a*) What is the runner's average velocity for the first 9.0 min? (*b*) What is the average velocity for the time spent walking? (*c*) What is the average velocity for the whole trip? (*d*) What is the average speed for the whole trip? SSM

44 • A car travels in a straight line with an average velocity of 80 km/h for 2.5 h and then with an average velocity of 40 km/h for 1.5 h. (*a*) What is the total displacement for the 4.0-h trip? (*b*) What is the average velocity for the total trip?

45 • One busy air route across the Atlantic Ocean is about 5500 km. The now-retired Concord, a supersonic jet capable of flying at twice the speed of sound, was used for traveling such routes. (*a*) Roughly how long did it take for a one-way flight? (Use 343 m/s for the speed of sound.) (*b*) Compare this time to the time taken by a subsonic jet flying at 0.90 times the speed of sound.

46 • The speed of light, designated by the universally recognized symbol c, has a value, to two significant figures, of 3.0×10^8 m/s. (*a*) How long does it take for light to travel from the Sun to Earth, a distance of 1.5×10^{11} m? (*b*) How long does it take light to travel from the moon to Earth, a distance of 3.8×10^8 m?

47 • Proxima Centauri, the closest star to us besides our own Sun, is 4.1×10^{13} km from Earth. From Zorg, a planet orbiting this star, a Gregor places an order at Tony's Pizza in Hoboken, New Jersey, communicating by light signals. Tony's fastest delivery craft travels at $1.00 \times 10^{-4}c$ (see Problem 46). (*a*) How long does it take Gregor's order to reach Tony's Pizza? (*b*) How long does Gregor wait between sending the signal and receiving the pizza? If Tony's has a "1000-years-or-it's-free" delivery policy, does Gregor have to pay for the pizza? SSM

48 • A car making a 100-km journey travels 40 km/h for the first 50 km. How fast must it go during the second 50 km to average 50 km/h?

49 •• **CONTEXT-RICH** Late in ice hockey games, the team that is losing sometimes "pulls" their goalkeeper off the ice to add an additional offensive player and increase their chances of scoring. In

such cases, the goalie on the opposing team might have an opportunity to score into the unguarded net 55.0 m away. Suppose you are the goaltender for your university team and are in just such a situation. You launch a shot (in hopes of getting your first career goal) on the frictionless ice. You eventually hear a disappointing "clang" as the puck strikes a goalpost (instead of going in!) exactly 2.50 s later. In this case, how fast did the puck travel? You should assume 343 m/s for the speed of sound.

50 •• Cosmonaut Andrei, your co-worker at the International Space Station, tosses a banana at you at a speed of 15 m/s. At exactly the same instant, you fling a scoop of ice cream at Andrei along exactly the same path. The collision between banana and ice cream produces a banana split 7.2 m from your location 1.2 s after the banana and ice cream were launched. (*a*) How fast did you toss the ice cream? (*b*) How far were you from Andrei when you tossed the ice cream? (Neglect any effects due to gravity.)

51 •• Figure 2-39 shows the position of a particle as a function of time. Find the average velocities for the time intervals *a*, *b*, *c*, and *d* indicated in the figure.

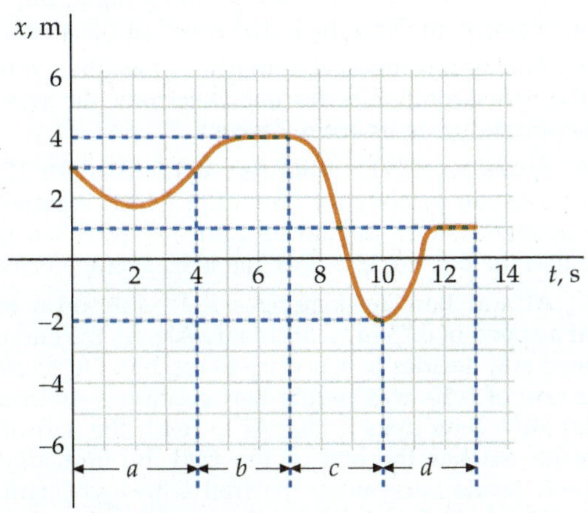

FIGURE 2-39 Problem 51

52 •• **ENGINEERING APPLICATION** It has been found that, on average, galaxies are moving away from Earth at a speed that is proportional to their distance from Earth. This discovery is known as Hubble's law, named for its discoverer, astrophysicist Sir Edwin Hubble. He found that the recessional speed v of a galaxy a distance r from Earth is given by $v = Hr$, where $H = 1.58 \times 10^{-18}\ \text{s}^{-1}$ is called the Hubble constant. What are the expected recessional speeds of galaxies (*a*) 5.00×10^{22} m from Earth, and (*b*) 2.00×10^{25} m from Earth? (*c*) If the galaxies at each of these distances had traveled at their expected recessional speeds, how long ago would they have been at our location?

53 •• The cheetah can run as fast as 113 km/h, the falcon can fly as fast as 161 km/h, and the sailfish can swim as fast as 105 km/h. The three of them run a relay with each covering a distance L at maximum speed. What is the average speed of this relay team for the entire relay? Compare this average speed with the numerical average of the three individual speeds. Explain carefully why the average speed of the relay team is *not* equal to the numerical average of the three individual speeds. SSM

54 •• Two cars are traveling along a straight road. Car A maintains a constant speed of 80 km/h and car B maintains a constant speed of 110 km/h. At $t = 0$, car B is 45 km behind car A. (*a*) How much farther will car A travel before car B overtakes it? (*b*) How much ahead of A will B be 30 s after it overtakes A?

55 •• **MULTISTEP** A car traveling at a constant speed of 20 m/s passes an intersection at time $t = 0$. A second car traveling at a constant speed of 30 m/s in the same direction passes the same intersection 5.0 s later. (*a*) Sketch the position functions $x_1(t)$ and $x_2(t)$ for the two cars for the interval $0 \leq t \leq 20\,\text{s}$. (*b*) Determine when the second car will overtake the first. (*c*) How far from the intersection will the two cars be when they pull even? (*d*) Where is the first car when the second car passes the intersection? SSM

56 •• **BIOLOGICAL APPLICATION** Bats use echolocation to determine their distance from objects they cannot easily see in the dark. The time between the emission of a high-frequency sound pulse (a click) and the detection of its echo is used to determine such distances. A bat, flying at a constant speed of 19.5 m/s in a straight line toward a vertical cave wall, makes a single clicking noise and hears the echo 0.15 s later. Assuming that she continued flying at her original speed, how close was she to the wall when she received the echo? Assume a speed of 343 m/s for the speed of sound.

57 ••• **ENGINEERING APPLICATION** A submarine can use *sonar* (sound traveling through water) to determine its distance from other objects. The time between the emission of a sound pulse (a "ping") and the detection of its echo can be used to determine such distances. Alternatively, by measuring the time between *successive* echo receptions of a *regularly timed set* of pings, the submarine's *speed* may be determined by comparing the time between echoes to the time between pings. Assume you are the sonar operator in a submarine traveling at a constant velocity underwater. Your boat is in the eastern Mediterranean Sea, where the speed of sound is known to be 1522 m/s. If you send out pings every 2.00 s, and your apparatus receives echoes reflected from an undersea cliff every 1.98 s, how fast is your submarine traveling?

ACCELERATION

58 • A sports car accelerates in third gear from 48.3 km/h (about 30 mi/h) to 80.5 km/h (about 50 mi/h) in 3.70 s. (*a*) What is the average acceleration of this car in m/s^2? (*b*) If the car maintained this acceleration, how fast would it be moving one second later?

59 • An object is moving along the x axis. At $t = 5.0$ s, the object is at $x = +3.0$ m and has a velocity of +5.0 m/s. At $t = 8.0$ s, it is at $x = +9.0$ m and its velocity is −1.0 m/s. Find its average acceleration during the time interval $5.0\ \text{s} < t < 8.0\ \text{s}$. SSM

60 •• A particle moves along the x axis with velocity $v_x = (8.0\ \text{m/s}^2)t - 7.0\ \text{m/s}$. (*a*) Find the average acceleration for two different one-second intervals, one beginning at $t = 3.0$ s and the other beginning at $t = 4.0$ s. (*b*) Sketch v_x versus t over the interval $0 < t < 10$ s. (*c*) How do the instantaneous accelerations at the middle of each of the two time intervals specified in Part (*a*) compare to the average accelerations found in Part (*a*)? Explain.

61 •• **MULTISTEP** The position of a certain particle depends on time according to the equation $x(t) = t^2 - 5.0t + 1.0$, where x is in meters if t is in seconds. (*a*) Find the displacement and average velocity for the interval $3.0\,\text{s} \leq t \leq 4.0\,\text{s}$. (*b*) Find the general formula for the displacement for the time interval from t to $t + \Delta t$. (*c*) Use the limiting process to obtain the instantaneous velocity for any time t. SSM

62 •• The position of an object as a function of time is given by $x = At^2 - Bt + C$, where $A = 8.0\ \text{m/s}^2$, $B = 6.0$ m/s, and $C = 4.0$ m. Find the instantaneous velocity and acceleration as functions of time.

63 ••• The one-dimensional motion of a particle is plotted in Figure 2-40. (*a*) What is the average acceleration in each of the intervals *AB*, *BC*, and *CE*? (*b*) How far is the particle from its starting point after 10 s? (*c*) Sketch the displacement of the particle as a function of time; label the instants *A*, *B*, *C*, *D*, and *E* on your graph. (*d*) At what time is the particle traveling most slowly?

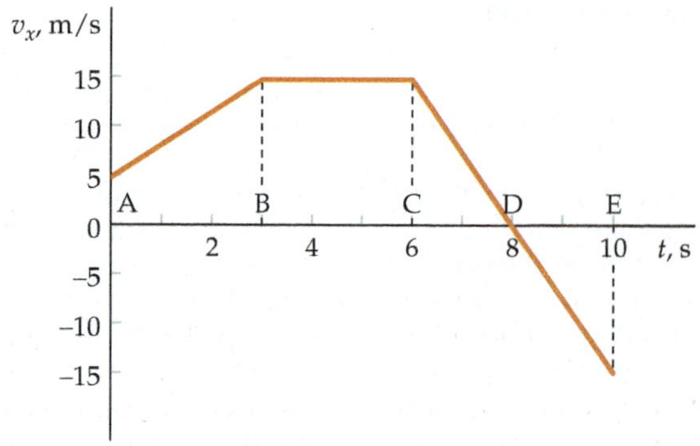

FIGURE 2-40 Problem 63

CONSTANT ACCELERATION AND FREE-FALL

64 • An object projected vertically upward with initial speed v_0 attains a maximum height h above its launch point. Another object projected up with initial speed $2v_0$ from the same height will attain a maximum height of (*a*) $4h$, (*b*) $3h$, (*c*) $2h$, (*d*) h. (Air resistance is negligible.)

65 • A car traveling along the x axis starts from rest at $x = 50$ m and accelerates at a constant rate of 8.0 m/s². (*a*) How fast is it going after 10 s? (*b*) How far has it gone after 10 s? (*c*) What is its average velocity for the interval $0 \le t \le 10$ s?

66 • An object traveling along the x axis with an initial velocity of +5.0 m/s has a constant acceleration of +2.0 m/s². When its speed is 15 m/s, how far has it traveled?

67 • An object traveling along the x axis at constant acceleration has a velocity of +10 m/s when it is at $x = 6.0$ m and of +15 m/s when it is at $x = 10.0$ m. What is its acceleration? SSM

68 • The speed of an object traveling along the x axis increases at the constant rate of +4.0 m/s each second. At $t = 0.0$ s, its velocity is +1.0 m/s and its position is +7.0 m. How fast is it moving when its position is +8.0 m, and how much time has elapsed from the start at $t = 0.0$ s?

69 •• A ball is launched directly upward from ground level with an initial speed of 20 m/s. (Air resistance is negligible.) (*a*) How long is the ball in the air? (*b*) What is the greatest height reached by the ball? (*c*) How many seconds after launch is the ball 15 m above the release point?

70 •• In the Blackhawk landslide in California, a mass of rock and mud fell 460 m down a mountain and then traveled 8.00 km across a level plain. It has been theorized that the rock and mud moved on a cushion of water vapor. Assume that the mass dropped with the free-fall acceleration and then slid horizontally, losing speed at a constant rate. (*a*) How long did the mud take to drop the 460 m? (*b*) How fast was it traveling when it reached the bottom? (*c*) How long did the mud take to slide the 8.00 km horizontally?

71 •• A load of bricks is lifted by a crane at a steady velocity of 5.0 m/s when one brick falls off 6.0 m above the ground. (*a*) Sketch the position of the brick $y(t)$ versus time, from the moment it leaves the pallet until it hits the ground. (*b*) What is the greatest height the brick reaches above the ground? (*c*) How long does it take to reach the ground? (*d*) What is its speed just before it hits the ground? SSM

72 •• A bolt comes loose from underneath an elevator that is moving upward at a constant speed of 6.0 m/s. The bolt reaches the bottom of the elevator shaft in 3.0 s. (*a*) How high above the bottom of the shaft was the elevator when the bolt came loose? (*b*) What is the speed of the bolt when it hits the bottom of the shaft?

73 •• An object is dropped from rest at a height of 120 m. Find the distance it falls during its final second in the air.

74 •• An object is released from rest at a height h. During the final second of its fall, it traverses a distance of 38 m. Determine h.

75 •• A stone is thrown vertically downward from the top of a 200-m cliff. During the last half second of its flight, the stone travels a distance of 45 m. Find the initial speed of the stone. SSM

76 •• An object is released from rest at a height h. It travels $0.4h$ during the first second of its descent. Determine the average velocity of the object during its entire descent.

77 •• A bus accelerates from rest at 1.5 m/s² for 12 s. It then travels at constant velocity for 25 s, after which it slows to a stop with an acceleration of magnitude 1.5 m/s². (*a*) What is the total distance that the bus travels? (*b*) What is its average velocity?

78 •• Al and Bert are jogging side-by-side on a trail in the woods at a speed of 0.75 m/s. Suddenly Al sees the end of the trail 35 m ahead and decides to speed up to reach it. He accelerates at a constant rate of 0.50 m/s² while Bert continues on at a constant speed. (*a*) How long does it take Al to reach the end of the trail? (*b*) Once he reaches the end of the trail, he immediately turns around and heads back along the trail with a constant speed of 0.85 m/s. How long does it take him to meet up with Bert? (*c*) How far are they from the end of the trail when they meet?

79 •• You have designed a rocket to be used to sample the local atmosphere for pollution. It is fired vertically with a constant upward acceleration of 20 m/s². After 25 s, the engine shuts off and the rocket continues rising (in freefall) for a while. (Air resistance is negligible.) The rocket eventually stops rising and then falls back to the ground. You want to get a sample of air that is 20 km above the ground. (*a*) Did you reach your height goal? If not, what would you change so that the rocket reaches 20 km? (*b*) Determine the total time the rocket is in the air. (*c*) Find the speed of the rocket just before it hits the ground.

80 •• A flowerpot falls from a windowsill of an apartment that is on the tenth floor of an apartment building. A person in an apartment below, coincidentally in possession of a high-speed high-precision timing system, notices that it takes 0.20 s for the pot to fall past his window, which is 4.0-m from top to bottom. How far above the top of the window is the windowsill from which the pot fell? (Neglect any effects due to air resistance.)

81 •• In a classroom demonstration, a glider moves along an inclined track with constant acceleration. It is projected from the low end of the track with an initial velocity. After 8.00 s have elapsed, it is 100 cm from the low end and is moving along the track at a velocity of −15 cm/s. Find the initial velocity and the acceleration. SSM

82 •• A rock dropped from a cliff covers one-third of its total distance to the ground in the last second of its fall. Air resistance is negligible. How high is the cliff?

83 •• A typical automobile under hard braking loses speed at a rate of about 7.0 m/s^2; the typical reaction time to engage the brakes is 0.50 s. A local school board sets the speed limit in a school zone such that all cars should be able to stop in 4.0 m. (*a*) What maximum speed does this imply for an automobile in this zone? (*b*) What fraction of the 4.0 m is due to the reaction time? SSM

84 •• Two trains face each other on adjacent tracks. They are initially at rest, and their front ends are 40 m apart. The train on the left accelerates rightward at 1.0 m/s^2. The train on the right accelerates leftward at 1.3 m/s^2. (*a*) How far does the train on the left travel before the front ends of the trains pass? (*b*) If the trains are each 150 m in length, how long after the start are they completely past one another, assuming their accelerations are constant?

85 •• Two stones are dropped from the edge of a 60-m cliff, the second stone 1.6 s after the first. How far below the top of the cliff is the second stone when the separation between the two stones is 36 m?

86 •• A motorcycle officer hidden at an intersection observes a car driven by an oblivious driver who ignores a stop sign and continues through the intersection at constant speed. The police officer takes off in pursuit 2.0 s after the car has passed the stop sign. She accelerates at 4.2 m/s^2 until her speed is 110 km/h, and then continues at this speed until she catches the car. At that instant, the car is 1.4 km from the intersection. (a) How long did it take for the officer to catch up to the car? (*b*) How fast was the car traveling?

87 •• At $t = 0$, a stone is dropped from the top of a cliff above a lake. Another stone is thrown downward 1.6 s later from the same point with an initial speed of 32 m/s. Both stones hit the water at the same instant. Find the height of the cliff.

88 •• A passenger train is traveling at 29 m/s when the engineer sees a freight train 360 m ahead of his train traveling in the same direction on the same track. The freight train is moving at a speed of 6.0 m/s. (*a*) If the reaction time of the engineer is 0.40 s, what is the minimum (constant) rate at which the passenger train must lose speed if a collision is to be avoided? (*b*) If the engineer's reaction time is 0.80 s and the train loses speed at the minimum rate described in Part (*a*), at what rate is the passenger train approaching the freight train when the two collide? (*c*) For both reaction times, how far will the passenger train have traveled in the time between the sighting of the freight train and the collision?

89 •• **BIOLOGICAL APPLICATION** The click beetle can project itself vertically with an acceleration of about 400g (an order of magnitude more than a human could survive!). The beetle jumps by "unfolding" its 0.60-cm-long legs. (*a*) How high can the click beetle jump? (*b*) How long is the beetle in the air? (Assume constant acceleration while in contact with the ground and neglect air resistance.)

90 •• An automobile accelerates from rest at 2.0 m/s^2 for 20 s. The speed is then held constant for 20 s, after which there is an acceleration of -3.0 m/s^2 until the automobile stops. What is the total distance traveled?

91 •• Consider measuring the free-fall motion of a particle (neglect air resistance). Before the advent of computer-driven data-logging software, these experiments typically employed a wax-coated tape placed vertically next to the path of a dropped electrically conductive object. A spark generator would cause an arc to jump between two vertical wires through the falling object and through the tape, thereby marking the tape at fixed time intervals Δt. Show that the change in height during successive time intervals for an object falling from rest follows *Galileo's Rule of Odd Numbers*: $\Delta y_{21} = 3\Delta y_{10}$, $\Delta y_{32} = 5\Delta y_{10}, \ldots$, where Δy_{10} is the change in y during the first interval of duration Δt, Δy_{21} is the change in y during the second interval of duration Δt, etc. SSM

92 •• Starting from rest, a particle travels along the x axis with a constant acceleration of $+3.0$ m/s^2. At a time 4.0 s following its start, it is at $x = -100$ m. At a time 6.0 s later it has a velocity of $+15$ m/s. Find its position at this later time.

93 •• If it were possible for a spacecraft to maintain a constant acceleration indefinitely, trips to the planets of the Solar System could be undertaken in days or weeks, while voyages to the nearer stars would only take a few years. (*a*) Using data from the tables at the back of the book, find the time it would take for a one-way trip from Earth to Mars (at Mars' closest approach to Earth). Assume that the spacecraft starts from rest, travels along a straight line, accelerates halfway at 1g, flips around, and decelerates at 1g for the rest of the trip. (*b*) Repeat the calculation for a 4.1×10^{13}-km trip to Proxima Centauri, our nearest stellar neighbor outside of the Sun. (See Problem 47.) SSM

94 •• The Stratosphere Tower in Las Vegas is 1137 ft high. It takes 1 min, 20 s to ascend from the ground floor to the top of the tower using the high-speed elevator. The elevator starts and ends at rest. Assume that it maintains a constant upward acceleration until it reaches its maximum speed, and then maintains a constant acceleration of equal magnitude until it comes to a stop. Find the magnitude of the acceleration of the elevator. Express this acceleration magnitude as a multiple of g (the acceleration due to gravity).

95 •• A train pulls away from a station with a constant acceleration of 0.40 m/s^2. A passenger arrives at a point next to the track 6.0 s after the end of the train has passed the very same point. What is the slowest constant speed at which she can run and still catch the train? On a single graph, plot the position versus time curves for both the train and the passenger.

96 ••• Ball A is dropped from the top of a building of height h at the same instant that ball B is thrown vertically upward from the ground. When the balls collide, they are moving in opposite directions, and the speed of A is twice the speed of B. At what height does the collision occur?

97 ••• Solve Problem 96 if the collision occurs when the balls are moving in the same direction and the speed of A is 4 times that of B.

98 ••• Starting at one station, a subway train accelerates from rest at a constant rate of 1.00 m/s^2 for half the distance to the next station, then slows down at the same rate for the second half of the journey. The total distance between stations is 900 m. (*a*) Sketch a graph of the velocity v_x as a function of time over the full journey. (*b*) Sketch a graph of the position as a function of time over the full journey. Place appropriate numerical values on both axes.

99 ••• A speeder traveling at a constant speed of 125 km/h races past a billboard. A patrol car pursues from rest with constant acceleration of (8.0 km/h)/s until it reaches its maximum speed of 190 km/h, which it maintains until it catches up with the speeder. (*a*) How long does it take the patrol car to catch the speeder if it starts moving just as the speeder passes? (*b*) How far does each car travel? (*c*) Sketch $x(t)$ for each car. SSM

100 ••• When the patrol car in Problem 99 (traveling at 190 km/h) is 100 m behind the speeder (traveling at 125 km/h), the speeder sees the police car and slams on his brakes, locking the wheels. (*a*) Assuming that each car can brake at 6.0 m/s^2 and that the driver of the police car brakes instantly as she sees the brake lights of the speeder (reaction time = 0.0 s), show that the cars collide. (*b*) At what time after the speeder applies his brakes do the two cars collide? (*c*) Discuss how reaction time would affect this problem.

101 ••• Leadfoot Lou enters the "Rest-to-Rest" auto competition, in which each contestant's car begins and ends at rest, covering a fixed distance L in as short a time as possible. The intention is to demonstrate driving skills, and to find which car is the best at the *total combination* of speeding up and slowing down. The course is designed so that maximum speeds of the cars are never reached. (*a*) If Lou's car maintains an acceleration (magnitude) of a during speedup, and maintains a deceleration (magnitude) of $2a$ during braking, at what fraction of L should Lou move his foot from the gas pedal to the brake? (*b*) What fraction of the total time for the trip has elapsed at that point? (*c*) What is the fastest speed Lou's car ever reaches? Neglect Lou's reaction time, and answer in terms of a and L.

102 ••• A physics professor, equipped with a rocket backpack, steps out of a helicopter at an altitude of 575 m with zero initial velocity. (Neglect air resistance.) For 8.0 s, she falls freely. At that time, she fires her rockets and slows her rate of descent at 15 m/s² until her rate of descent reaches 5.0 m/s. At this point, she adjusts her rocket engine controls to maintain that rate of descent until she reaches the ground. (*a*) On a single graph, sketch her acceleration and velocity as functions of time. (Take upward to be positive.) (*b*) What is her speed at the end of the first 8.0 s? (*c*) What is the duration of her slowing-down period? (*d*) How far does she travel while slowing down? (*e*) How much time is required for the entire trip from the helicopter to the ground? (*f*) What is her average velocity for the entire trip?

INTEGRATION OF THE EQUATIONS OF MOTION

103 • The velocity of a particle is given by $v_x(t) = (6.0\ \text{m/s}^2)t + (3.0\ \text{m/s})$. (*a*) Sketch v versus t and find the area under the curve for the interval $t = 0$ to $t = 5.0$ s. (*b*) Find the position function $x(t)$. Use it to calculate the displacement during the interval $t = 0$ to $t = 5.0$ s. SSM

104 • Figure 2-41 shows the velocity of a particle versus time. (*a*) What is the magnitude, in meters, represented by the area of the shaded box? (*b*) Estimate the displacement of the particle for the two 1-s intervals, one beginning at $t = 1.0$ s and the other at $t = 2.0$ s. (*c*) Estimate the average velocity for the interval $1.0\ \text{s} \le t \le 3.0\ \text{s}$. (*d*) The equation of the curve is $v_x = (0.50\ \text{m/s}^3)t^2$. Find the displacement of the particle for the interval $1.0\ \text{s} \le t \le 3.0\ \text{s}$ by integration and compare this answer with your answer for Part (*b*). Is the average velocity equal to the mean of the initial and final velocities for this case?

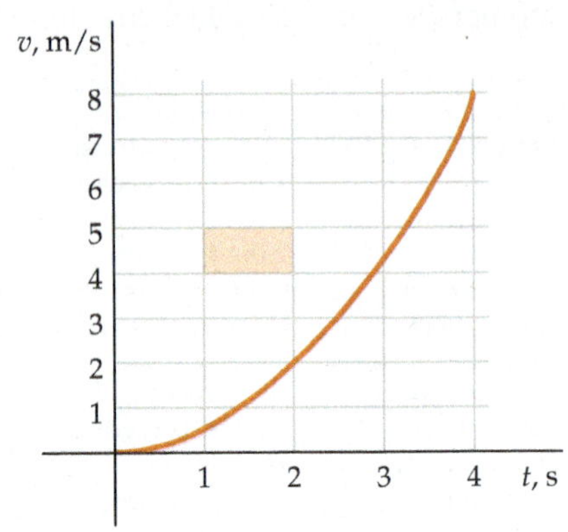

FIGURE 2-41 Problem 104

105 •• The velocity of a particle is given by $v_x = (7.0\ \text{m/s}^3)t^2 - 5.0\ \text{m/s}$. If the particle is at the origin at $t_0 = 0$, find the position function $x(t)$.

106 •• Consider the velocity graph in Figure 2-42. Assuming $x = 0$ at $t = 0$, write correct algebraic expressions for $x(t)$, $v_x(t)$, and $a_x(t)$ with appropriate numerical values inserted for all constants.

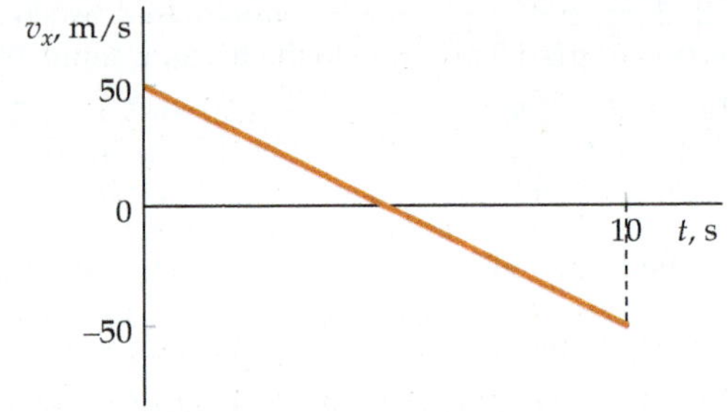

FIGURE 2-42 Problem 106

107 ••• Figure 2-43 shows the acceleration of a particle versus time. (*a*) What is the magnitude, in m/s, of the area of the shaded box? (*b*) The particle starts from rest at $t = 0$. Estimate the velocity at $t = 1.0$ s, 2.0 s, and 3.0 s by counting the boxes under the curve. (*c*) Sketch the curve v_x versus t from your results for Part (*b*); then estimate how far the particle travels in the interval $t = 0$ to $t = 3.0$ s.

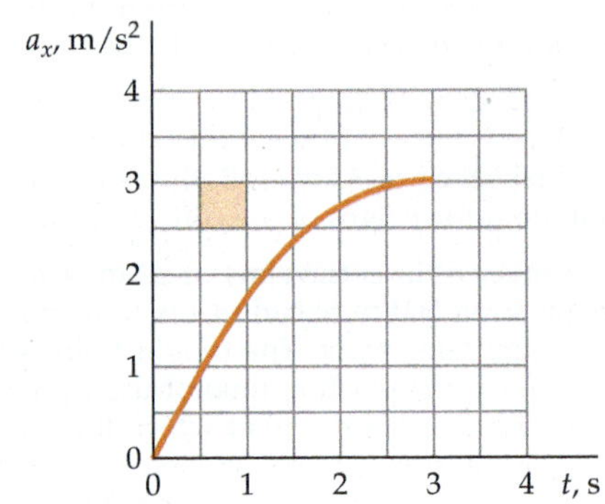

FIGURE 2-43 Problem 107

108 ••• Figure 2-44 is a graph of v_x versus t for a particle moving along a straight line. The position of the particle at time $t = 0$ is $x_0 = 5.0$ m. (*a*) Find x for various times t by counting boxes, and sketch x as a function of t. (*b*) Sketch a graph of the acceleration a_x as a function of the time t. (*c*) Determine the displacement of the particle between $t = 3.0$ s and 7.0 s.

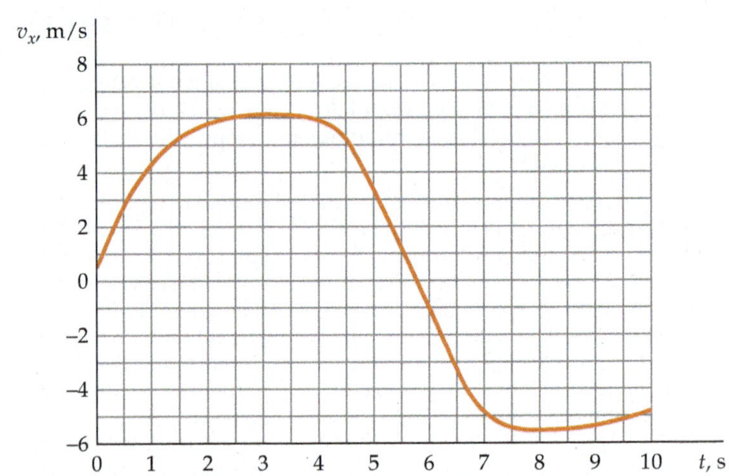

FIGURE 2-44 Problem 108

109 ••• **CONCEPTUAL** Figure 2-45 shows a plot of x versus t for an object moving along a straight line. For this motion, sketch graphs (using the same t axis) of (*a*) v_x as a function of t, and (*b*) a_x as a function of t. (*c*) Use your sketches to qualitatively compare the time(s) when the object is at its largest distance from the origin to the time(s) when its speed is greatest. Explain why the times are *not* the same. (*d*) Use your sketches to qualitatively compare the time(s) when the object is moving fastest to the time(s) when its acceleration is the largest. Explain why the times are *not* the same. SSM

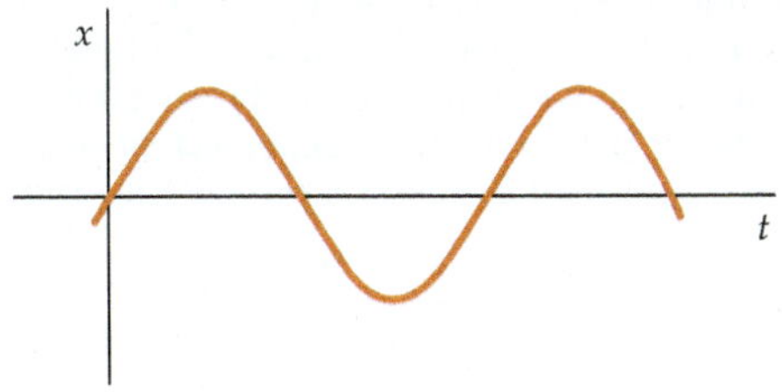

FIGURE 2-45 Problem 109

110 ••• **MULTISTEP** The acceleration of a certain rocket is given by $a_x = bt$, where b is a positive constant. (*a*) Find the position function $x(t)$ if $x = x_0$ and $v_x = v_{0x}$ at $t = 0$. (*b*) Find the position and velocity at $t = 5.0$ s if $x_0 = 0$, $v_{0x} = 0$ and $b = 3.0\ \text{m/s}^3$. (*c*) Compute the average velocity of the rocket between $t = 4.5$ s and 5.5 s at $t = 5.0$ s if $x_0 = 0$, $v_{0x} = 0$ and $b = 3.0\ \text{m/s}^3$. Compare this average velocity with the instantaneous velocity at $t = 5.0$ s.

111 ••• In the time interval from 0.0 s to 10.0 s, the acceleration of a particle traveling in a straight line is given by $a_x = (0.20\ \text{m/s}^3)t$. Let to the right be the $+x$ direction. The particle initially has a velocity to the right of 9.5 m/s and is located 5.0 m to the left of the origin. (*a*) Determine the velocity as a function of time during the interval; (*b*) determine the position as a function of time during the interval; (*c*) determine the average velocity between $t = 0.0$ s and 10.0 s, and compare it to the average of the instantaneous velocities at the start and ending times. Are these two averages equal? Explain. SSM

112 ••• Consider the motion of a particle that experiences a variable acceleration given by $a_x = a_{0x} + bt$, where a_{0x} and b are constants and $x = x_0$ and $v_x = v_{0x}$ at $t = 0$. (*a*) Find the instantaneous velocity as a function of time. (*b*) Find the position as a function of time. (*c*) Find the average velocity for the time interval with an initial time of zero and arbitrary final time t. (*d*) Compare the average of the initial and final velocities to your answer to Part (*c*). Are these two averages equal? Explain.

GENERAL PROBLEMS

113 ••• **CONTEXT-RICH** You are a student in a science class that is using the following apparatus to determine the value of g. Two photogates are used. (*Note:* You may be familiar with photogates in everyday living. You see them in the doorways of some stores. They are designed to ring a bell when someone interrupts the beam while walking through the door.) One photogate is located at the edge of a table that is 1.00 m above the floor, and the second photogate is located directly below the first, at a height 0.500 m above the floor. You are told to drop a marble through these gates, releasing it from rest a negligible distance above the upper gate. The upper gate starts a timer as the ball passes through its beam. The second photogate stops the timer when the ball passes through its beam. (*a*) Prove that the experimental magnitude of free-fall acceleration is given by $g_{exp} = (2\Delta y)/(\Delta t)^2$, where Δy is the vertical distance between the photogates and Δt is the fall time. (*b*) For your setup, what value of Δt would you expect to measure, assuming g_{exp} is the standard value ($9.81\ \text{m/s}^2$)? (*c*) During the experiment, a slight error is made. Instead of locating the first photogate even with the top of the table, your not-so-careful lab partner locates it 0.50 cm lower than the top of the table. However, she does manage to properly locate the second photogate at a height of 0.50 m above the floor. However, she releases the marble from the same height that it was released from when the photogate was 1.00 m above the floor. What value of g_{exp} will you and your partner determine? What percentage difference does this represent from the standard value of g?

114 ••• **MULTISTEP** The position of a body oscillating on a spring is given by $x = A \sin \omega t$, where A and ω (lower case Greek omega) are constants, $A = 5.0$ cm, and $\omega = 0.175\ \text{s}^{-1}$. (*a*) Plot x as a function of t for $0 \leq t \leq 36\,\text{s}$. (*b*) Measure the slope of your graph at $t = 0$ to find the velocity at this time. (*c*) Calculate the average velocity for a series of intervals, beginning at $t = 0$ and ending at $t = 6.0$, 3.0, 2.0, 1.0, 0.50, and 0.25 s. (*d*) Compute dx/dt to find the velocity at time $t = 0$. (*e*) Compare your results in Parts (*c*) and (*d*) and explain why your Part-(*c*) results approach your Part-(*d*) result.

115 ••• **CONCEPTUAL** Consider an object that is attached to a horizontally oscillating piston. The object moves with a velocity given by $v = B \sin(\omega t)$, where B and ω (lower case Greek omega) are constants and ω is in s^{-1}. (*a*) Explain why B is equal to the maximum speed v_{max}. (*b*) Determine the acceleration of the object as a function of time. Is the acceleration constant? (*c*) What is the maximum acceleration (magnitude) in terms of ω and v_{max}. (*d*) At $t = 0$, the object's position is known to be x_0. Determine the position as a function of time in terms of t, ω, x_0, and v_{max}. SSM

116 ••• Suppose the acceleration of a particle is a function of x, where $a_x(x) = (2.0\ \text{s}^{-2})x$. (*a*) If the velocity is zero when $x = 1.0$ m, what is the speed when $x = 3.0$ m? (*b*) How long does it take the particle to travel from $x = 1.0$ m to $x = 3.0$ m.

117 ••• A rock falls through water with a continuously decreasing acceleration. Assume that the rock's acceleration as a function of *velocity* has the form $a_y = g - bv_y$ where b is a positive constant. (The $+y$ direction is directly downward.) (*a*) What are the SI units of b? (*b*) Prove mathematically that if the rock is released from rest at time $t = 0$, the acceleration will depend exponentially on *time* according to $a_y(t) = ge^{-bt}$. (*c*) What is the terminal speed for the rock in terms of g and b? (See Problem 38 for an explanation of the phenomenon of *terminal speed*.) SSM

118 ••• A small rock sinking through water (see Problem 117) experiences an exponentially decreasing acceleration given by $a_y(t) = ge^{-bt}$, where b is a positive constant that depends on the shape and size of the rock and the physical properties of the water. Based upon this, find expressions for the velocity and position of the rock as functions of time. Assume that its initial position and velocity are both zero and that the $+y$ direction is directly downward.

119 ••• **SPREADSHEET** The acceleration of a skydiver jumping from an airplane is given by $a_y = g - bv_y^2$, where b is a positive constant that depends on the skydiver's cross-sectional area and the density of the surrounding atmosphere she is diving through. The $+y$ directions is directly downward. (*a*) If her initial speed is zero when stepping from a hovering helicopter, show that her speed as a function of time is given by $v_y(t) = v_t \tanh(t/T)$, where v_t is the terminal speed (see Problem 38) given by $v_t = \sqrt{g/b}$, and $T = v_t/g$ is a time-scale parameter. (*b*) What fraction of the terminal speed is the speed at $t = T$. (*c*) Use a **spreadsheet** program to graph $v_y(t)$ as a function of time, using a terminal speed of 56 m/s (use this value to calculate b and T). Does the resulting curve make sense?

120 ••• **APPROXIMATION** Imagine that you are standing at a wishing well, wishing that you knew how deep the surface of the water was. Cleverly, you make your wish. Then you take a penny from your pocket and drop it into the well. Exactly three seconds after you dropped the penny, you hear the sound it made when it struck the water. If the speed of sound is 343 m/s, how deep is the well? Neglect any effects due to air resistance.

121 ••• **CONTEXT-RICH** You are driving a car at the 25-mi/h speed limit when you observe the light at the intersection 65 m in front of you turn yellow. You know that at that particular intersection the light remains yellow for exactly 5.0 s before turning red. After you think for 1.0 s, you then accelerate the car at a constant rate. You somehow manage to pass your 4.5-m-long car completely through the 15.0-m-wide intersection just as the light turns red, thus narrowly avoiding a ticket for being in an intersection when the light is red. Immediately after passing through the intersection, you take your foot off the accelerator, relieved. However, down the road you are pulled over for speeding. You assume that you were ticketed for the speed of your car as it exited the intersection. Determine this speed and decide whether you should fight this ticket in court. Explain.

122 ••• For a spherical celestial object of radius R, the acceleration due to gravity g at a distance x from the *center* of the object is $g = g_0R^2/x^2$, where g_0 is the acceleration due to gravity at the object's surface and $x > R$. For the moon, take $g_0 = 1.63$ m/s^2 and $R = 3200$ km. If a rock is released from rest at a height of $4R$ above the lunar surface, with what speed does the rock impact the moon? *Hint: Its acceleration is a function of position and increases as the object falls. So do not use constant acceleration free-fall equations, but go back to basics.*

CHAPTER 3

SAILBOATS DO NOT TRAVEL IN STRAIGHT LINES TO THEIR DESTINATIONS, BUT INSTEAD MUST "TACK" BACK AND FORTH ACROSS THE WIND. THIS BOAT MUST SAIL EAST, THEN SOUTH, AND THEN EAST AGAIN, IN ITS JOURNEY TO A SOUTHEASTERN PORT. *(PhotoDisc/Getty Images.)*

? How can we calculate the boat's displacement and its average velocity? (See Example 3-1.)

Motion in Two and Three Dimensions

The motion of a sailboat tacking into the wind or the path of a home-run ball as it flies out of a ballpark cannot be fully described using the equations we presented in Chapter 2. Instead, to describe these motions, we must extend the idea of motion in one dimension discussed in Chapter 2 to two and three dimensions. To do this, we must revisit the concept of vectors and look at how they can be used to analyze and describe motion in more than one dimension.

In this chapter, we will discuss the displacement, velocity, and acceleration vectors in further detail. In addition, we will discuss two specific types of motion: projectile motion and circular motion. The material in this chapter presumes you are familiar with the material that introduces vectors in Sections 6 and 7 of Chapter 1. You are encouraged to review these sections before proceeding in this chapter.

3-1 DISPLACEMENT, VELOCITY, AND ACCELERATION

In Chapter 2, the concepts of displacement, velocity, and acceleration were used to describe the motion of an object moving in a straight line. Now we use the concept of vectors to extend these characteristics of motion in two and three dimensions.

POSITION AND DISPLACEMENT VECTORS

The **position vector** of a particle is a vector drawn from the origin of a coordinate system to the location of the particle. For a particle in the xy plane at the point with coordinates (x, y), the position vector $\vec{r}$ is

$$\vec{r} = x\hat{i} + y\hat{j} \qquad 3\text{-}1$$

DEFINITION—POSITION VECTOR

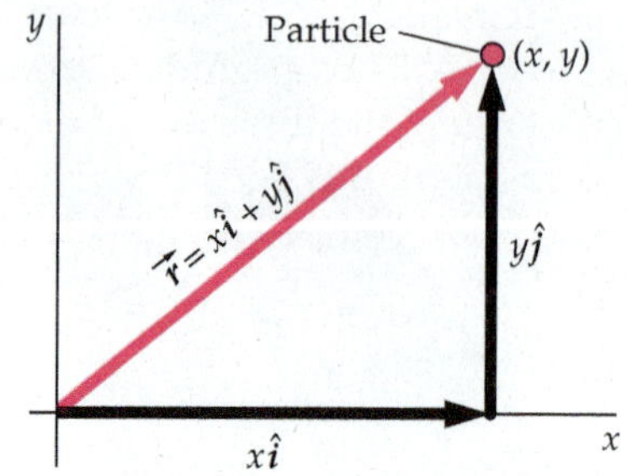

FIGURE 3-1 The x and y components of the position vector $\vec{r}$ for a particle are the x and y (Cartesian) coordinates of the particle.

Note that x and y components of the position vector $\vec{r}$ are the Cartesian coordinates (Figure 3-1) of the particle.

Figure 3-2 shows the actual path or trajectory of the particle. At time t_1, the particle is at P_1, with position vector $\vec{r}_1$; by time t_2, the particle has moved to P_2, with position vector $\vec{r}_2$. The particle's change in position is the displacement vector $\Delta\vec{r}$:

$$\Delta\vec{r} = \vec{r}_2 - \vec{r}_1 \qquad 3\text{-}2$$

DEFINITION—DISPLACEMENT VECTOR

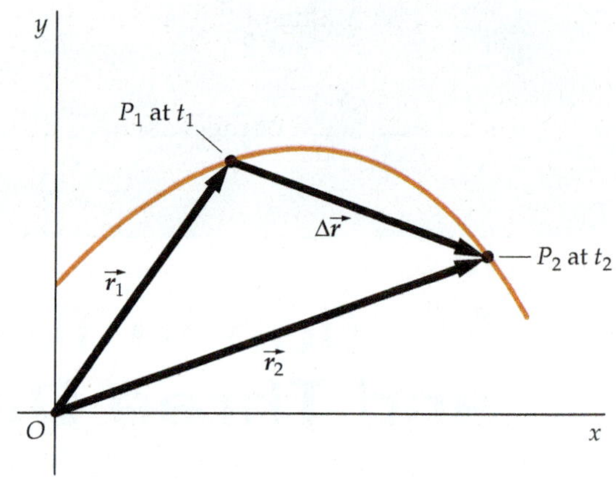

FIGURE 3-2 The displacement vector $\Delta\vec{r}$ is the difference in the position vectors, $\Delta\vec{r} = \vec{r}_2 - \vec{r}_1$. Equivalently, $\Delta\vec{r}$ is the vector that, when added to the initial position vector $\vec{r}_1$, yields the final position vector $\vec{r}_2$. That is, $\vec{r}_1 + \Delta\vec{r} = \vec{r}_2$.

Using unit vectors, we can rewrite this displacement as

$$\Delta\vec{r} = \vec{r}_2 - \vec{r}_1 = (x_2 - x_1)\hat{i} + (y_2 - y_1)\hat{j} = \Delta x\hat{i} + \Delta y\hat{j} \qquad 3\text{-}3$$

! Do not confuse the trajectory in graphs of x versus y with the curve in the x-versus-t plots of Chapter 2.

VELOCITY VECTORS

Recall that average velocity is defined as displacement divided by the elapsed time. The result of the displacement vector divided by the elapsed time interval $\Delta t = t_2 - t_1$ is the **average-velocity vector:**

$$\vec{v}_{av} = \frac{\Delta\vec{r}}{\Delta t} \qquad 3\text{-}4$$

DEFINITION—AVERAGE-VELOCITY VECTOR

The average velocity vector and the displacement vector are in the same direction.

The magnitude of the displacement vector is less than the distance traveled along the curve unless the particle moves along a straight line and never reverses its direction. However, if we consider smaller and smaller time intervals (Figure 3-3), the magnitude of the displacement approaches the distance along the curve, and the angle between $\Delta\vec{r}$ and the tangent to the curve at the beginning of the interval approaches zero. We define the **instantaneous-velocity vector** as the limit of the average-velocity vector as Δt approaches zero:

$$\vec{v} = \lim_{\Delta t \to 0} \frac{\Delta\vec{r}}{\Delta t} = \frac{d\vec{r}}{dt} \qquad 3\text{-}5$$

DEFINITION—INSTANTANEOUS-VELOCITY VECTOR

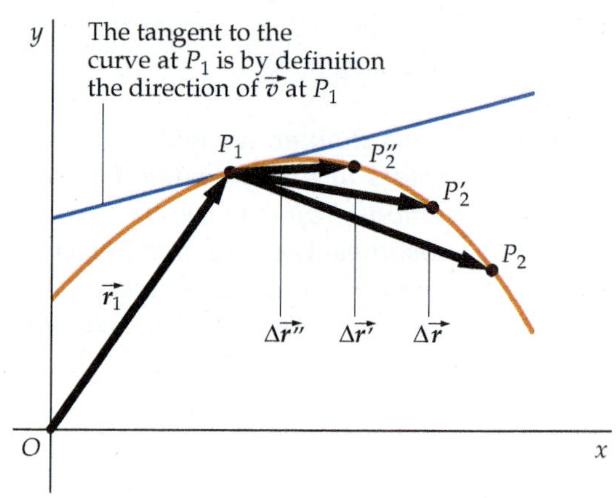

FIGURE 3-3 As the time interval decreases, the angle between direction of $\Delta\vec{r}$ and the tangent to the curve approaches zero.

The instantaneous-velocity vector is the derivative of the position vector with respect to time. Its magnitude is the speed and its direction is along the line tangent to the curve in the direction of motion of the particle.

To calculate the derivative in Equation 3-5, we write the position vectors in terms of their components (Equation 3-1):

$$\Delta\vec{r} = \vec{r}_2 - \vec{r}_1 = (x_2 - x_1)\hat{i} + (y_2 - y_1)\hat{j} = \Delta x\hat{i} + \Delta y\hat{j}$$

Then

$$\vec{v} = \lim_{\Delta t\to 0}\frac{\Delta\vec{r}}{\Delta t} = \lim_{\Delta t\to 0}\frac{\Delta x\hat{i} + \Delta y\hat{j}}{\Delta t} = \lim_{\Delta t\to 0}\left(\frac{\Delta x}{\Delta t}\right)\hat{i} + \lim_{\Delta t\to 0}\left(\frac{\Delta y}{\Delta t}\right)\hat{j}$$

or

$$\vec{v} = \frac{dx}{dt}\hat{i} + \frac{dy}{dt}\hat{j} = v_x\hat{i} + v_y\hat{j} \qquad \text{3-6}$$

where $v_x = dx/dt$ and $v_y = dy/dt$ are the x and y components of the velocity.

The magnitude of the velocity vector is given by:

$$v = \sqrt{v_x^2 + v_y^2} \qquad \text{3-7}$$

and the direction of the velocity is given by

$$\theta = \tan^{-1}\frac{v_y}{v_x} \qquad \text{3-8}$$

! Do not trust your calculator to always give the correct value for θ when using Equation 3-8. Most calculators will return the correct value for θ if v_x is positive. If v_x is negative, however, you will need to add 180° (π rad) to the value returned by the calculator.

See **Math Tutorial** *for more information on* **Trigonometry**

Example 3-1 The Velocity of a Sailboat

A sailboat has coordinates $(x_1, y_1) = (130\text{ m}, 205\text{ m})$ at $t_1 = 60.0$ s. Two minutes later, at time t_2, it has coordinates $(x_2, y_2) = (110\text{ m}, 218\text{ m})$. (*a*) Find the average velocity $\vec{v}_{av}$ for this two-minute interval. Express $\vec{v}_{av}$ in terms of its rectangular components. (*b*) Find the magnitude and direction of this average velocity. (*c*) For $t \geq 20.0$ s, the position of a second sailboat as a function of time is $x(t) = b_1 + b_2 t$ and $y(t) = c_1 + c_2/t$, where $b_1 = 100$ m, $b_2 = 0.500$ m/s, $c_1 = 200$ m, and $c_2 = 360$ m · s. Find the instantaneous velocity as a function of time t, for $t \geq 20.0$ s.

PICTURE The initial and final positions of the first sailboat are given. Because the motion of the boat is in two dimensions, we need to express the displacement, average velocity, and instantaneous velocity as vectors. Then we can use Equations 3-5 through 3-8 to obtain the requested values.

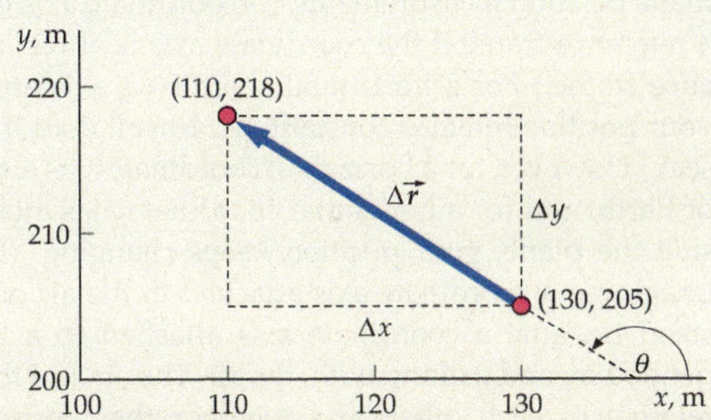

FIGURE 3-4

SOLVE

(*a*) 1. Draw a coordinate system (Figure 3-4) and draw the displacement of the sailboat. Draw the average-velocity vector (it and the displacement vector are in the same direction):

2. The x and y components of the average velocity $\vec{v}_{av}$ are calculated directly from their definitions:

$$\vec{v}_{av} = v_{x\,av}\hat{i} + v_{y\,av}\hat{j}$$

where

$$v_{x\,av} = \frac{\Delta x}{\Delta t} = \frac{110\text{ m} - 130\text{ m}}{120\text{ s}} = -0.167\text{ m/s}$$

$$v_{y\,av} = \frac{\Delta y}{\Delta t} = \frac{218\text{ m} - 205\text{ m}}{120\text{ s}} = 0.108\text{ m/s}$$

so

$$\vec{v}_{av} = \boxed{-(0.167\text{ m/s})\hat{i} + (0.108\text{ m/s})\hat{j}}$$

(*b*) 1. The magnitude of $\vec{v}_{av}$ is found from the Pythagorean theorem:

$$v_{av} = \sqrt{(v_{x\,av})^2 + (v_{y\,av})^2} = \boxed{0.199\text{ m/s}}$$

2. The ratio of $v_{y\,av}$ to $v_{x\,av}$ gives the tangent of the angle θ between $\vec{v}_{av}$ and the $+x$ direction (we add 180° to the value of −33.0° returned by the calculator because v_x is negative):

$$\tan\theta = \frac{v_{y\,av}}{v_{x\,av}}$$

so

$$\theta = \tan^{-1}\frac{v_{y\,av}}{v_{x\,av}} = \tan^{-1}\frac{0.108\text{ m/s}}{-0.167\text{ m/s}} = -33.0° + 180° = \boxed{147°}$$

(*c*) We find the instantaneous velocity $\vec{v}$ by calculating dx/dt and dy/dt: $\vec{v} = \frac{dx}{dt}\hat{i} + \frac{dy}{dt}\hat{j} = b_2\hat{i} - c_2 t^{-2}\hat{j} = \boxed{(0.500\text{ m/s})\hat{i} - \frac{360\text{ m}\cdot\text{s}}{t^2}\hat{j}}$

CHECK The magnitude of $\vec{v}_{av}$ is greater than the absolute value of either its x or its y component. With t in seconds, the units for the y component of $\vec{v}$ in Part (*c*) are $\text{m}\cdot\text{s/s}^2 = \text{m/s}$, which are appropriate units for velocity.

RELATIVE VELOCITY

If you are sitting in an airplane that is moving with a velocity of 500 mi/h toward the east, your velocity is the same as that of the airplane. This velocity might be your velocity relative to the surface of Earth, or it might be your velocity relative to the air outside the airplane. (These two relative velocities would be very different if the plane were flying in a jet stream.) In addition, your velocity relative to the airplane itself is zero.

The surface of Earth, the air outside the plane, and the plane itself are frames of reference. A **frame of reference** (or reference frame) is an extended object or collection of objects whose parts are at rest relative to each other. To specify the velocity of an object requires that you specify the frame of reference that the velocity is relative to.

Midair refueling. Each plane is nearly at rest relative to the other, though both are moving with very large velocities relative to Earth. *(Novastock/Dembinsky Photo Associates.)*

We use coordinate axes that are attached to reference frames to make position measurements. (A coordinate axis is said be attached to a reference frame if the coordinate axis is at rest relative to the reference frame.) For a horizontal coordinate axis attached to the plane, your position remains constant. (At least it does if you remain in your seat.) However, for a horizontal coordinate axis attached to the surface of Earth, and for a horizontal coordinate axis attached to the air outside the plane, your position keeps changing. (If you have trouble imagining a coordinate axis attached to the air outside the plane, instead imagine a coordinate axis attached to a balloon that is suspended in, and drifting with, the air. The air and the balloon are at rest relative to each other, and together they form a single reference frame.)

If a particle p moves with velocity $\vec{v}_{pA}$ relative to reference frame A, which is in turn moving with velocity $\vec{v}_{AB}$ relative to reference frame B, the velocity $\vec{v}_{pB}$ of the particle relative to reference frame B is related to $\vec{v}_{pA}$ and $\vec{v}_{AB}$ by

$$\vec{v}_{pB} = \vec{v}_{pA} + \vec{v}_{AB} \qquad 3\text{-}9$$

For example, if a person p is on a railroad car C that is moving with velocity $\vec{v}_{CG}$ relative to the ground G (Figure 3-5*a*), and the person is walking with velocity $\vec{v}_{pC}$ (Figure 3-5*b*) relative to the car, then the velocity of the person relative to the ground is the vector sum of these two velocities: $\vec{v}_{pG} = \vec{v}_{pC} + \vec{v}_{CG}$ (Figure 3-5*c*).

The velocity of object A relative to object B is equal in magnitude and opposite in direction to the velocity of object B relative to object A. For example, $\vec{v}_{pC}$ is equal to $-\vec{v}_{Cp}$, where $\vec{v}_{pC}$ is the velocity of the person relative to the car, and $\vec{v}_{Cp}$ is the velocity of the car relative to the person.

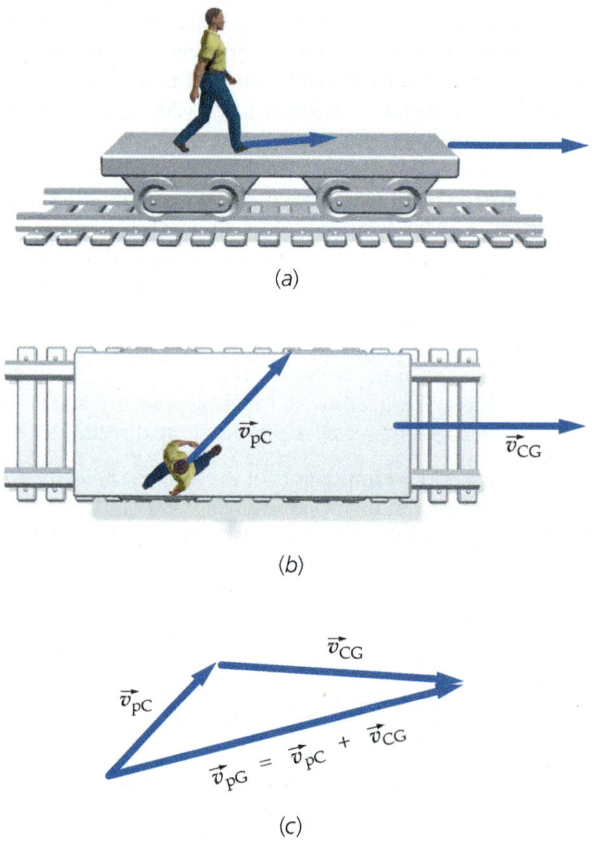

FIGURE 3-5 The velocity of the person relative to the ground is equal to the velocity of the person relative to the car plus the velocity of the car relative the ground.

PROBLEM-SOLVING STRATEGY

Relative Velocity

PICTURE The first step in solving a relative-velocity problem is to identify and label the relevant reference frames. Here, we will call them reference frame A and reference frame B.

SOLVE

1. Using $\vec{v}_{pB} = \vec{v}_{pA} + \vec{v}_{AB}$ (Equation 3-9), relate the velocity of the moving object (particle p) relative to frame A to the velocity of the particle relative to frame B.
2. Sketch a vector addition diagram for the equation $\vec{v}_{pB} = \vec{v}_{pA} + \vec{v}_{AB}$. Use the head-to-tail method of vector addition. Include coordinate axes on the sketch.
3. Solve for the desired quantity. Use trigonometry where appropriate.

CHECK Make sure that you solve for the velocity or position of the moving object relative to the proper reference frame.

! The order of the subscripts used when denoting relative velocity vectors is very important. When using relative velocity vectors, be very careful to write the subscripts in a consistent order.

Example 3-2 A Flying Plane

A pilot wishes to fly a plane due north relative to the ground. The airspeed of the plane is 200 km/h and the wind is blowing from west to east at 90 km/h. (*a*) In which direction should the plane head? (*b*) What is the ground speed of the plane?

PICTURE Because the wind is blowing toward the east, a plane headed due north will drift off course toward the east. To compensate for this crosswind, the plane must head west of due north. The velocity of the plane relative to the ground $\vec{v}_{pG}$ is equal to the velocity of the plane relative to the air $\vec{v}_{pA}$ plus the velocity of the air relative to the ground $\vec{v}_{AG}$.

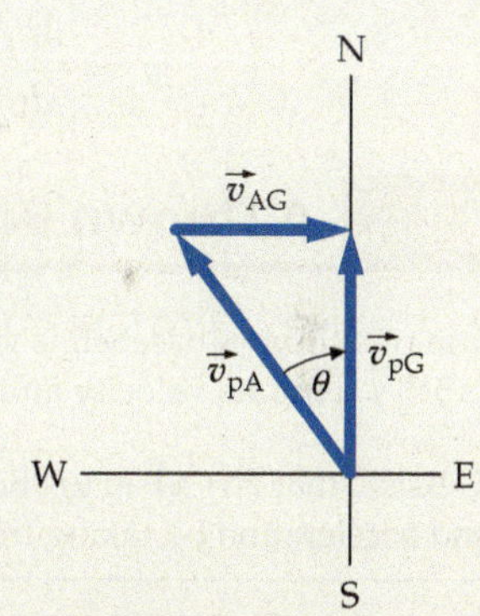

FIGURE 3-6

SOLVE

(a) 1. The velocity of the plane relative to the ground is given by Equation 3-9:

$$\vec{v}_{pG} = \vec{v}_{pA} + \vec{v}_{AG}$$

2. Make a velocity addition diagram (Figure 3-6) showing the addition of the vectors in step 1. Include direction axes:

3. The sine of the angle θ between the velocity of the plane relative to the air and due north equals the ratio of v_{AG} and v_{pA}:

$$\sin\theta = \frac{v_{AG}}{v_{pA}} = \frac{90\text{ km/h}}{200\text{ km/h}} = \frac{9}{20}$$

so

$$\theta = \sin^{-1}\frac{9}{20} = \boxed{27^\circ \text{ west of north}}$$

(*b*) Because $\vec{v}_{AG}$ and $\vec{v}_{pG}$ are mutually perpendicular, we can use the Pythagorean theorem to find the magnitude of $\vec{v}_{pG}$:

$$v_{pA}^2 = v_{pG}^2 + v_{AG}^2$$

so

$$v_{pG} = \sqrt{v_{pA}^2 - v_{AG}^2}$$

$$= \sqrt{(200\text{ km/h})^2 - (90\text{ km/h})^2} = \boxed{180\text{ km/h}}$$

CHECK Heading directly into the 90 km/h wind would result in a ground speed of 200 km/h − 90 km/h = 110 km/h. The Part-(*b*) result of 180 km/h is greater than 110 km/h and less than 200 km/h, as expected.

ACCELERATION VECTORS

The **average-acceleration vector** is the ratio of the change in the instantaneous-velocity vector, $\Delta\vec{v}$, to the elapsed time interval Δt:

$$\vec{a}_{av} = \frac{\Delta\vec{v}}{\Delta t} \qquad 3\text{-}10$$

DEFINITION—AVERAGE-ACCELERATION VECTOR

The **instantaneous-acceleration vector** is the limit of this ratio as Δt approaches zero; in other words, it is the derivative of the velocity vector with respect to time:

$$\vec{a} = \lim_{\Delta t \to 0} \frac{\Delta\vec{v}}{\Delta t} = \frac{d\vec{v}}{dt} \qquad 3\text{-}11$$

DEFINITION—INSTANTANEOUS-ACCELERATION VECTOR

To calculate the instantaneous acceleration, we express $\vec{v}$ in rectangular coordinates:

$$\vec{v} = v_x\hat{i} + v_y\hat{j} + v_z\hat{k} = \frac{dx}{dt}\hat{i} + \frac{dy}{dt}\hat{j} + \frac{dz}{dt}\hat{k}$$

Then

$$\vec{a} = \frac{dv_x}{dt}\hat{i} + \frac{dv_y}{dt}\hat{j} + \frac{dv_z}{dt}\hat{k} = \frac{d^2x}{dt^2}\hat{i} + \frac{d^2y}{dt^2}\hat{j} + \frac{d^2z}{dt^2}\hat{k}$$

$$= a_x\hat{i} + a_y\hat{j} + a_z\hat{k} \qquad 3\text{-}12$$

where the components of $\vec{a}$ are

$$a_x = \frac{dv_x}{dt}, \quad a_y = \frac{dv_y}{dt}, \quad a_z = \frac{dv_z}{dt}.$$

Example 3-3 A Thrown Baseball

The position of a thrown baseball is given by $\vec{r} = [1.5\text{ m} + (12\text{ m/s})t]\hat{i} + [(16\text{ m/s})t - (4.9\text{ m/s}^2)t^2]\hat{j}$. Find its velocity and acceleration as functions of time.

PICTURE Recall that $\vec{r} = x\hat{i} + y\hat{j}$ (Equation 3-1). We can find the x and y components of the velocity and acceleration by taking the time derivatives of x and y.

SOLVE

1. Find the x and y components of $\vec{r}$:

$$x = 1.5\text{ m} + (12\text{ m/s})t$$
$$y = (16\text{ m/s})t - (4.9\text{ m/s}^2)t^2$$

2. The x and y components of the velocity are found by differentiating x and y:

$$v_x = \frac{dx}{dt} = 12\text{ m/s}$$
$$v_y = \frac{dy}{dt} = (16\text{ m/s}) - 2(4.9\text{ m/s}^2)t$$

3. We differentiate v_x and v_y to obtain the components of the acceleration:

$$a_x = \frac{dv_x}{dt} = 0$$
$$a_y = \frac{dv_y}{dt} = -9.8\text{ m/s}^2$$

4. In vector notation, the velocity and acceleration are

$$\vec{v} = \boxed{(12\text{ m/s})\hat{i} + [16\text{ m/s} - (9.8\text{ m/s}^2)t]\hat{j}}$$
$$\vec{a} = \boxed{(-9.8\text{ m/s}^2)\hat{j}}$$

CHECK The units that accompany the quantities for velocity and acceleration are m/s and m/s^2, respectively. Our step 4 results for velocity and acceleration have the correct units of m/s and m/s^2.

For a vector to be constant, both its magnitude and direction must remain constant. If either magnitude or direction changes, the vector changes. Thus, if a car rounds a curve in the road at constant speed, it is accelerating because the velocity is changing due to the change in direction of the velocity vector.

Example 3-4 Rounding a Curve

A car is traveling east at 60 km/h. It rounds a curve, and 5.0 s later it is traveling north at 60 km/h. Find the average acceleration of the car.

PICTURE We can calculate the average acceleration from its definition, $\vec{a}_{av} = \Delta\vec{v}/\Delta t$. To do this, we first calculate $\Delta\vec{v}$, which is the vector that when added to $\vec{v}_i$, results in $\vec{v}_f$.

SOLVE

1. The average acceleration is the change in velocity divided by the elapsed time. To find $\vec{a}_{av}$, we first find the change in velocity:
$$\vec{a}_{av} = \frac{\Delta\vec{v}}{\Delta t}$$
2. To find $\Delta\vec{v}$, we first specify $\vec{v}_i$ and $\vec{v}_f$. Draw $\vec{v}_i$ and $\vec{v}_f$ (Figure 3-7*a*), and draw the vector addition diagram (Figure 3-7*b*) corresponding to $\vec{v}_f = \vec{v}_i + \Delta\vec{v}$:
3. The change in velocity is related to the initial and final velocities:
$$\vec{v}_f = \vec{v}_i + \Delta\vec{v}$$
4. Substitute these results to find the average acceleration:
$$\vec{a}_{av} = \frac{\vec{v}_f - \vec{v}_i}{\Delta t} = \frac{60\text{ km/h }\hat{j} - 60\text{ km/h }\hat{i}}{5.0\text{ s}}$$
5. Convert 60 km/h to meters per second:
$$60\text{ km/h} \times \frac{1\text{ h}}{3600\text{ s}} \times \frac{1000\text{ m}}{1\text{ km}} = 16.7\text{ m/s}$$
6. Express the acceleration in meters per second squared:
$$\vec{a}_{av} = \frac{\vec{v}_f - \vec{v}_i}{\Delta t} = \frac{16.7\text{ m/s }\hat{j} - 16.7\text{ m/s }\hat{i}}{5.0\text{ s}}$$
$$= \boxed{-3.4\text{ m/s}^2\hat{i} + 3.4\text{ m/s}^2\hat{j}}$$

CHECK The eastward component of the velocity decreases from 60 km/h to zero, so we expect a negative acceleration component in the *x* direction. The northward component of the velocity increases from zero to 60 km/h, so we expect a positive acceleration component in the *y* direction. Our step 6 result meets both of these expectations.

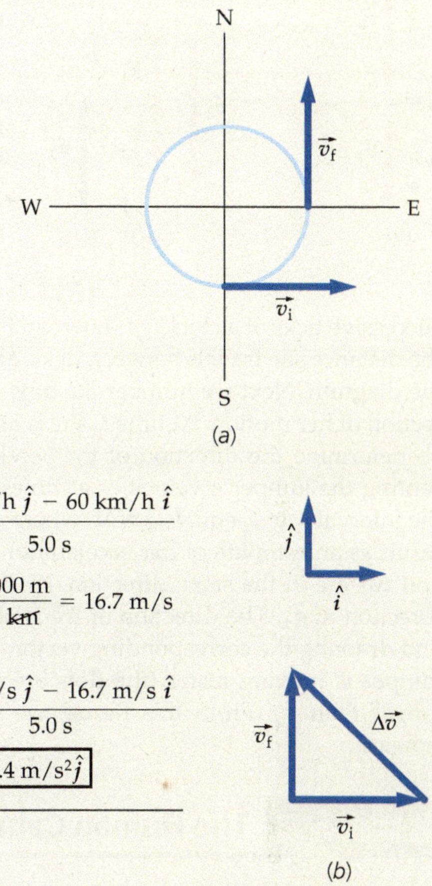

FIGURE 3-7

TAKING IT FURTHER Note that the car is accelerating even though its speed remains constant.

PRACTICE PROBLEM 3-1 Find the magnitude and direction of the average acceleration vector.

The motion of an object traveling in a circle at constant speed is an example of motion in which the direction of the velocity changes even though its magnitude, the speed, remains constant.

> ! Do not assume that the acceleration of an object is zero just because the object is traveling at constant speed. For the acceleration to be zero, neither the magnitude nor the direction of the velocity vector can be changing.

THE DIRECTION OF THE ACCELERATION VECTOR

In the next few chapters, you will need to determine the direction of the acceleration vector from a description of the motion. To see how this is done, consider a bungee jumper as she slows down prior to reversing direction at the lowest point of her jump. To find the direction of her acceleration as she loses speed during the last stages of her descent, we draw a series of dots representing her position at

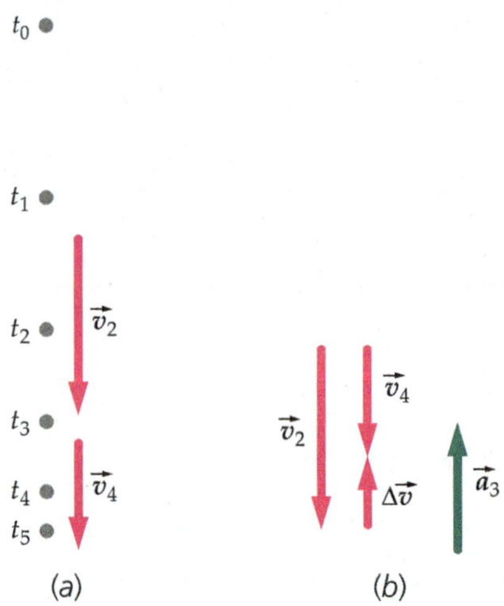

FIGURE 3-8 (*a*) A motion diagram of a bungee jumper losing speed as she descends. The dots are drawn at successive ticks of a clock. (*b*) We draw vectors $\vec{v}_2$ and $\vec{v}_4$ starting from the same point. Then, we draw $\Delta\vec{v}$ from the head of $\vec{v}_2$ to the head of $\vec{v}_4$ to obtain the graphical expression of the relation $\vec{v}_2 + \Delta\vec{v} = \vec{v}_4$. The acceleration $\vec{a}_3$ is in the same direction as $\Delta\vec{v}$.

successive ticks of a clock, as shown in Figure 3-8*a*. The faster she moves, the greater the distance she travels between ticks, and the greater the space between the dots in the diagram. Next we number the dots, starting with zero and increasing in the direction of her motion. At time t_0 she is at dot 0, at time t_1 she is at dot 1, and so forth. To determine the direction of the acceleration at time t_3, we draw vectors representing the jumper's velocities at times t_2 and t_4. The average acceleration during the interval t_2 to t_4 equals $\Delta\vec{v}/\Delta t$, where $\Delta\vec{v} = \vec{v}_4 - \vec{v}_2$ and $\Delta t = t_4 - t_2$. We use this result as an estimate of her acceleration at time t_3. That is, $\vec{a}_3 \approx \Delta\vec{v}/\Delta t$. Because $\vec{a}_3$ and $\Delta\vec{v}$ are in the same direction, by finding the direction of $\Delta\vec{v}$ we also find the direction of $\vec{a}_3$. The direction of $\Delta\vec{v}$ is obtained by using the relation $\vec{v}_2 + \Delta\vec{v} = \vec{v}_4$ and drawing the corresponding vector addition diagram (Figure 3-8*b*). Because the jumper is moving faster (the dots are farther apart) at t_2 than at t_4, we draw $\vec{v}_2$ longer than $\vec{v}_4$. From this figure, we can see that $\Delta\vec{v}$, and thus $\vec{a}_3$, is directed upward.

CONCEPT CHECK 3-1

Figure 3-9 is a motion diagram of the bungee jumper before, during, and after time t_6, when she momentarily come to rest at the lowest point in her descent. During the part of her ascent shown, she is moving upward with increasing speed. Use this diagram to determine the direction of the jumper's acceleration (*a*) at time t_6 and (*b*) at time t_9.

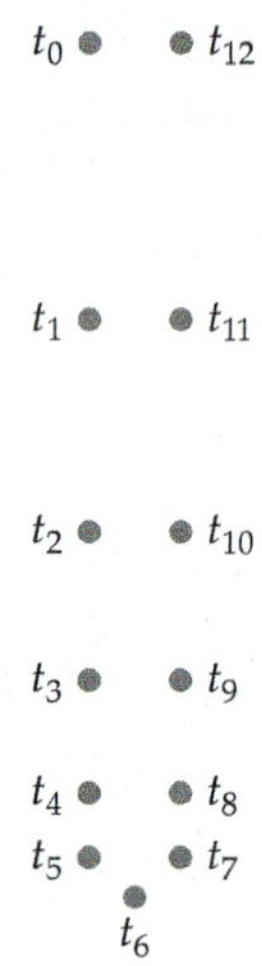

FIGURE 3-9 The dots for the bungee jumper's ascent are drawn to the right of those for her descent so that they do not overlap each other. Her motion, however, is straight down and then straight up.

Example 3-5 The Human Cannonball

Context-Rich

You are asked to substitute for an ill performer in a circus that is sponsored by your school. The job, should you choose to accept it, is to get shot out of a cannon. Never afraid to accept a challenge, you accept. The barrel of the cannon is inclined an angle of 60° above the horizontal. Your physics teacher offers you extra credit on the next exam if you successfully use a motion diagram to estimate the direction of your acceleration during the ascending portion of the flight.

PICTURE During the ascending portion of the flight, you travel in a curved path with decreasing speed. To estimate the direction of your acceleration you use $\vec{a}_{av} = \Delta\vec{v}/\Delta t$ and estimate the direction of $\Delta\vec{v}$. To estimate the direction of $\Delta\vec{v}$, we draw a motion diagram and then make a sketch of the relation $\vec{v}_i + \Delta\vec{v} = \vec{v}_f$.

SOLVE

1. Make a motion diagram (Figure 3-10*a*) of your motion during the ascending portion of the flight. Because your speed decreases as you ascend, the spacing between adjacent dots on your diagram decreases as you rise:
2. Pick a dot on the motion diagram and draw a velocity vector on the diagram for both the preceding and the following dot. These vectors should be drawn tangent to your trajectory.

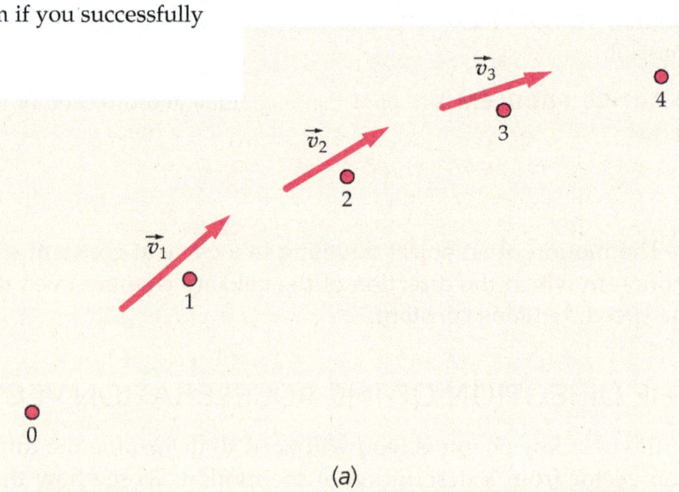

FIGURE 3-10

3. Draw the vector addition diagram (Figure 3-10*b*) of the relation $\vec{v}_i + \Delta\vec{v} = \vec{v}_f$. Begin by drawing the two velocity vectors from the same point. These vectors have the same magnitude and direction as the vectors drawn for step 2. Then, draw the $\Delta\vec{v}$ vector from the head of $\vec{v}_i$ to the head of $\vec{v}_f$.
4. Draw the estimated acceleration vector in the same direction as $\Delta\vec{v}$, but not the same length (because $\vec{a} \approx \Delta\vec{v}/\Delta t$).

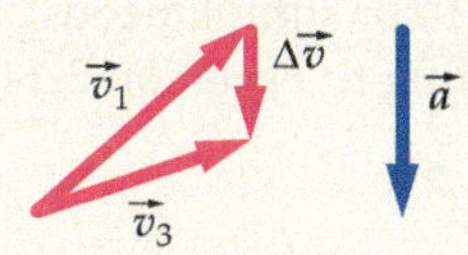

(*b*)

FIGURE 3.10 *(continued)*

CHECK During the ascent, the upward component of the velocity is decreasing, so we expect $\Delta\vec{v}$ to have a downward vertical component. Our step 3 result satisfies this expectation.

TAKING IT FURTHER The process of finding the direction of the acceleration using a motion diagram is not precise. Therefore, the result is an estimate of the direction of the acceleration, as opposed to a precise determination.

CONCEPT CHECK 3-2

Use a motion diagram to estimate the direction of the acceleration in Example 3-5 during the descending portion of your flight.

3-2 SPECIAL CASE 1: PROJECTILE MOTION

In a home run hit or a field goal kick, the ball follows a particular curved path through the air. This type of motion, known as **projectile motion,** occurs when an object (the projectile) is launched into the air and is allowed to move freely. The projectile might be a ball, a dart, water shooting out of a fountain, or even a human body during a long jump. If air resistance is negligible, then the projectile is said to be in free-fall. For objects in free-fall near the surface of Earth, the acceleration is the downward acceleration due to gravity.

Figure 3-11 shows a particle launched with initial speed v_0 at angle θ_0 above the horizontal. Let the launch point be at (x_0, y_0); y is positive upward and x is positive to the right. The initial velocity $\Delta\vec{v}_0$ then has components

$$v_{0x} = v_0 \cos\theta_0 \qquad \text{3-13}a$$

$$v_{0y} = v_0 \sin\theta_0 \qquad \text{3-13}b$$

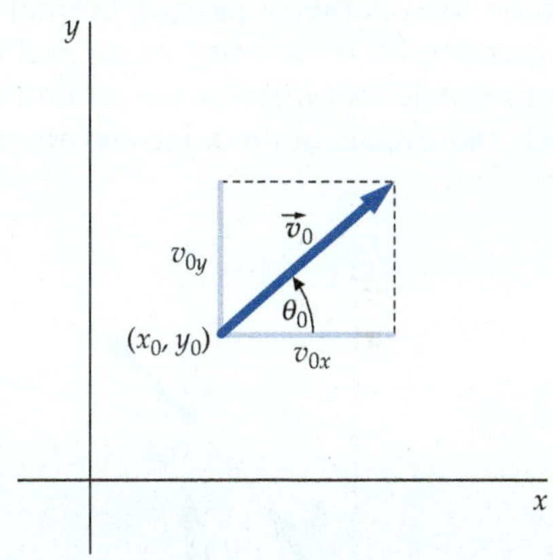

FIGURE 3-11 The components of $\vec{v}_0$ are $v_{0x} = v_0 \cos\theta_0$ and $v_{0y} = v_0 \sin\theta_0$, where θ_0 is the angle above the horizontal of $\vec{v}_0$.

In the absence of air resistance, the acceleration $\vec{a}$ is constant. The projectile has no horizontal acceleration, so the only acceleration is the free-fall acceleration $\vec{g}$, directed downward:

$$a_x = 0 \qquad \text{3-14}a$$

and

$$a_y = -g \qquad \text{3-14}b$$

Because the acceleration is constant, we can use the kinematic equations for constant acceleration presented in Chapter 2. The x component of the velocity $\vec{v}$ is constant because no horizontal acceleration exists:

$$v_x = v_{0x} \qquad \text{3-15}a$$

The y component of the velocity varies with time according to $v_y = v_{0y} + a_y t$ (Equation 2-12), with $a_y = -g$:

$$v_y = v_{0y} - gt \qquad \text{3-15}b$$

Notice that v_x does not depend on v_y and v_y does not depend on v_x: *The horizontal and vertical components of projectile motion are independent.* Dropping a ball from a desktop and projecting a second ball horizontally at the same time can demonstrate the independence of v_x and v_y, as shown in Figure 3-12. Notice that the two balls strike the floor simultaneously.

According to Equation 2-14, the displacements x and y are given by

$$x(t) = x_0 + v_{0x}t \qquad \text{3-16}a$$

$$y(t) = y_0 + v_{0y}t - \tfrac{1}{2}gt^2 \qquad \text{3-16}b$$

The notation $x(t)$ and $y(t)$ simply emphasizes that x and y are functions of time. If the y component of the initial velocity is known, the time t for which the particle is at height y can be found from Equation 3-16b. The horizontal position at that time can then be found using Equation 3-16a. (Equations 3-14 to 3-16 are expressed in vector form immediately preceding Example 3-10.)

The general equation for the path $y(x)$ of a projectile can be obtained from Equations 3-16 by eliminating the variable t. Choosing $x_0 = 0$ and $y_0 = 0$, we obtain $t = x/v_{0x}$ from Equation 3-16a. Substituting this into Equation 3-16b gives

$$y(x) = v_{0y}\left(\frac{x}{v_{0x}}\right) - \frac{1}{2}g\left(\frac{x}{v_{0x}}\right)^2 = \left(\frac{v_{0y}}{v_{0x}}\right)x - \left(\frac{g}{2v_{0x}^2}\right)x^2$$

Substituting for the velocity components using $v_{0x} = v_0 \cos\theta_0$ and $v_{0y} = v_0 \sin\theta_0$ yields

$$y(x) = (\tan\theta_0)x - \left(\frac{g}{2v_0^2\cos^2\theta_0}\right)x^2 \qquad \text{3-17}$$

PATH OF PROJECTILE

for the projectile's path. This equation is of the form $y = ax + bx^2$, which is the equation for a parabola passing through the origin. Figure 3-13 shows the path of a projectile with its velocity vector and components at several points. The path is for a projectile that impacts the ground at P. The horizontal distance $|\Delta x|$ between launch and impact at the same elevation is the **horizontal range** R.

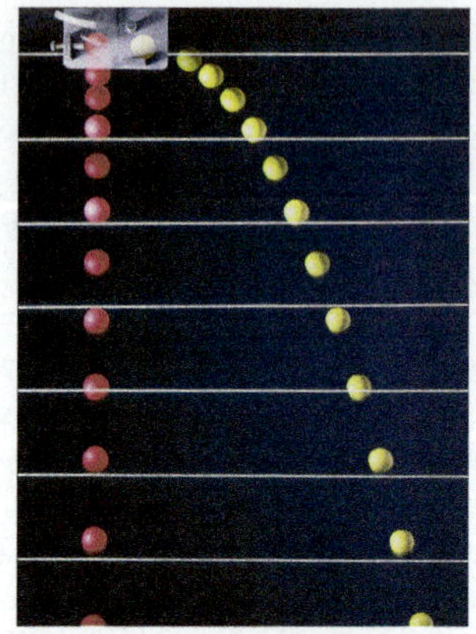

FIGURE 3-12 The red ball is released from rest at the instant the yellow ball rolls off the tabletop. The positions of the two balls are shown at successive equal time intervals. The vertical motion of the yellow ball is identical with the vertical motion of the red ball, thus demonstrating that the vertical motion of the yellow ball is independent of its horizontal motion. *(Richard Megna/Fundamental Photographs.)*

! Do not think that the velocity of a projectile is zero when the projectile is at the highest point in its trajectory. At the highest point in the trajectory v_y is zero, but the projectile may still be moving horizontally.

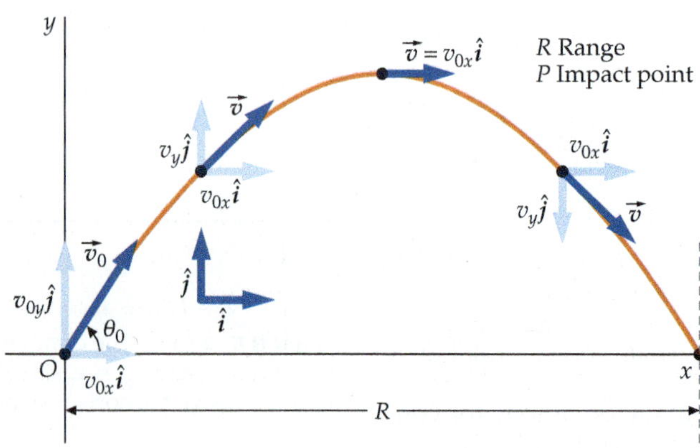

FIGURE 3-13 The path of a projectile, showing velocity components at different times.

Example 3-6 A Cap in the Air

A delighted physics graduate throws her cap into the air with an initial velocity of 24.5 m/s at 36.9° above the horizontal. The cap is later caught by another student. Find (a) the total time the cap is in the air, and (b) the total horizontal distance traveled. (Ignore effects of air resistance.)

PICTURE We choose the origin to be the initial position of the cap so that $x_0 = y_0 = 0$. We assume it is caught at the same height. The total time the cap is in the air is found by setting $y(t) = 0$ in $y(t) = y_0 + v_{0y}t - \frac{1}{2}gt^2$ (Equation 3-16b). We can then use this result in $x(t) = x_0 + v_{0x}t$ (Equation 3-16a) to find the total horizontal distance traveled.

SOLVE

(*a*) 1. Setting $y = 0$ in Equation 3-16*b*:

$$y = v_{0y}t - \tfrac{1}{2}gt^2$$
$$0 = t(v_{0y} - \tfrac{1}{2}gt)$$

2. There are two solutions for t:

$$t_1 = 0 \text{ (initial time)}$$
$$t_2 = \frac{2v_{0y}}{g}$$

3. Use trigonometry to relate v_{0y} to v_0 and θ_0 (see Figure 3-11):

$$v_{0y} = v_0 \sin\theta_0$$

4. Substitute for v_{0y} in the step 2 result to find the total time t_2:

$$t_2 = \frac{2v_{0y}}{g} = \frac{2v_0 \sin\theta_0}{g} = \frac{2(24.5\ \text{m/s})\sin 36.9°}{9.81\ \text{m/s}^2} = \boxed{3.00\ \text{s}}$$

(*b*) Use the value for the time in step 4 to calculate the total horizontal distance traveled:

$$x = v_{0x}t_2 = (v_0\cos\theta_0)t_2 = (24.5\ \text{m/s})\cos 36.9°(3.00\ \text{s}) = \boxed{58.8\ \text{m}}$$

CHECK If the cap traveled at a constant speed of 24.5 m/s for 3.00 s it would have traveled a distance of 73.5 m. Because it was launched at an angle, its horizontal speed was less than 24.5 m/s, so we expect its distance traveled to be less than 73.5 m. Our Part-(*b*) result of 58.8 m meets this expectation.

TAKING IT FURTHER The vertical component of the initial velocity of the cap is 14.7 m/s, the same as that of the cap in Example 2-13 (Chapter 2), where the cap was thrown straight up with $v_0 = 14.7$ m/s. The time the cap is in the air is also the same as in Example 2-13. Figure 3-14 shows the height y versus t for the cap. This curve is identical to Figure 2-20*a* (Example 2-13) because the caps each have the same vertical acceleration and vertical velocity. Figure 3-14 can be reinterpreted as a graph of y versus x if its time scale is converted to a distance scale, as shown in the figure. This is accomplished by multiplying the time values by 19.6 m/s. This works because the cap moves at (24.5 m/s) cos 36.9° = 19.6 m/s horizontally. The curve y versus x is a parabola (as is the curve y versus t).

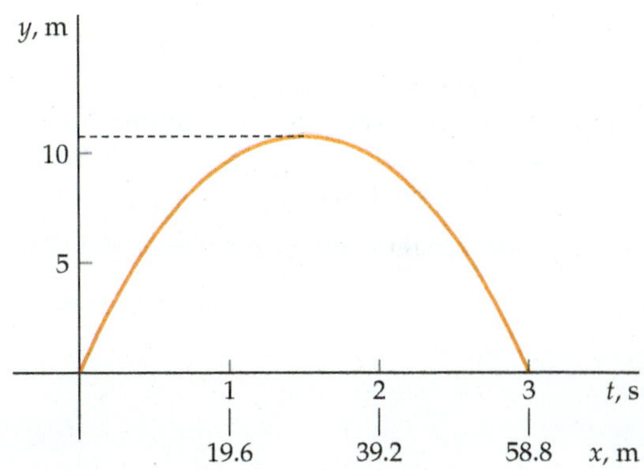

FIGURE 3-14 A plot of y versus t and of y versus x.

Example 3-7 A Supply Drop

A helicopter drops a supply package to flood victims on a raft on a swollen lake. When the package is released, the helicopter is 100 m directly above the raft and flying at a velocity of 25.0 m/s at an angle $\theta_0 = 36.9°$ above the horizontal. (*a*) How long is the package in the air? (*b*) How far from the raft does the package land? (*c*) If the helicopter continues at constant velocity, where is the helicopter when the package lands? (Ignore effects of air resistance.)

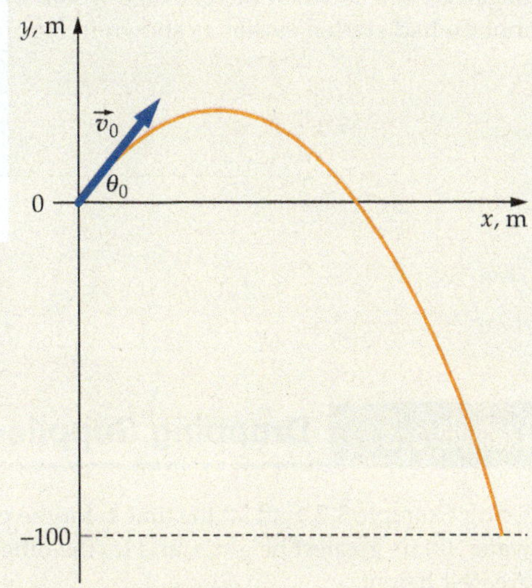

FIGURE 3-15 The parabola intersects the $y = -100$ m line twice, but only one of those times is greater than zero.

PICTURE The time in the air depends only on the vertical motion. Using $y(t) = y_0 + v_{0y}t - \tfrac{1}{2}gt^2$ (Equation 3-16*b*), you can solve for the time. Choose the origin to be at the location of the package when it is released. The initial velocity of the package is the velocity of the helicopter. The horizontal distance traveled by the package is given by $x(t) = v_{0x}t$ (Equation 3-16*a*), where t is the time the package is in the air.

SOLVE

(*a*) 1. Sketch the trajectory of the package during the time it is in the air. Include coordinate axes as shown in Figure 3-15:

2. To find the time of flight, write $y(t)$ for motion with constant acceleration, then set $y_0 = 0$ and $a_y = -g$ in the equation:

$$y = y_0 + v_{0y}t + \tfrac{1}{2}a_yt^2$$
$$y = 0 + v_{0y}t - \tfrac{1}{2}gt^2 = v_{0y}t - \tfrac{1}{2}gt^2$$

3. The solution to the quadratic equation $ax^2 + bx + c = 0$ is given by the quadratic formula:

$$x = \frac{-b \pm \sqrt{b^2 - 4ac}}{2a}$$

Using this, solve the quadratic equation from step 2 for t:

$y = v_{0y}t - \frac{1}{2}gt^2$

so

$0 = \frac{1}{2}gt^2 - v_{0y}t + y$

and

$$t = \frac{v_{0y} \pm \sqrt{v_{0y}^2 - 2gy}}{g}$$

4. Solve for the time when $y = -100$ m. First, solve for v_{0y}, then use the value for v_{0y} to find t.

$v_{0y} = v_0 \sin\theta_0 = (25.0\ \text{m/s}) \sin 36.9° = 15.0\ \text{m/s}$

so

$$t = \frac{15.0\ \text{m/s} \pm \sqrt{(15.0\ \text{m/s})^2 - 2(9.81\ \text{m/s}^2)(-100\ \text{m})}}{9.81\ \text{m/s}^2}$$

so

$t = -3.24$ s or $t = 6.30$ s

Because the package is released at $t = 0$, the time of impact cannot be negative. Hence:

$t = \boxed{6.30\ \text{s}}$

(*b*) 1. At impact the package has traveled a horizontal distance x, where x is the horizontal velocity times the time of flight. First solve for the horizontal velocity:

$v_{0x} = v_0 \cos\theta_0 = (25.0\ \text{m/s}) \cos 36.9° = 20.0\ \text{m/s}$

2. Next substitute for v_{0x} in $x = x_0 + v_{0x}t$ (Equation 3-16*a*) to find x.

$x = v_{0x}t = (20.0\ \text{m/s})(6.30\ \text{s}) = \boxed{126\ \text{m}}$

(*c*) The coordinates of the helicopter at the time of impact are

$x_h = v_{0x}t = (20.0\ \text{m/s})(6.30\ \text{s}) = 126\ \text{m}$

$y_h = y_{h0} + v_{h0}t = 0 + (15.0\ \text{m/s})(6.30\ \text{s}) = 94.4\ \text{m}$

At impact, the helicopter is

$\boxed{\text{194 m directly above the package}}$.

CHECK The helicopter is directly above the package when the package hits the water (and at all other times before then). This is because the horizontal velocities of the package and the helicopter were equal at release, and the horizontal velocities of both remain constant during flight.

TAKING IT FURTHER The positive time is appropriate because it corresponds to a time after the package is dropped (which occurs at $t = 0$). The negative time is when the package would have been at $y = -100$ m if its motion had started earlier as shown in Figure 3-16.

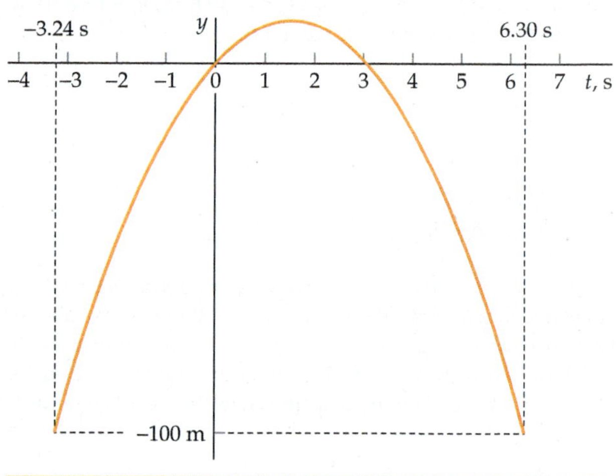

FIGURE 3-16

Example 3-8 Dropping Supplies

Try It Yourself

Using Example 3-7, find (*a*) the time t_1 for the package to reach its greatest height h above the water, (*b*) its greatest height h, and (*c*) the time t_2 for the package to fall to the water from its greatest height.

PICTURE The time t_1 is the time at which the vertical component of the velocity is zero. Using $v_y(t) = v_{0y} - gt$ (Equation 3-15*b*) solve for t_1.

SOLVE

Cover the column to the right and try these on your own before looking at the answers.

Steps	Answers
(a) 1. Write $v_y(t)$ for the package.	$v_y(t) = v_{0y} - gt$
2. Set $v_y(t_1) = 0$ and solve for t_1.	$t_1 = \boxed{1.53\text{ s}}$
(b) 1. Find $v_{y\,\text{av}}$ during the time the package is moving up.	$v_{y\,\text{av}} = 7.505\text{ m/s}$
2. Use $v_{y\,\text{av}}$ to find the distance traveled up. Then find h.	$\Delta y = 11.48\text{ m}$, so $h = \boxed{111\text{ m}}$
(c) Find the time for the package to fall a distance h.	$t_2 = \boxed{4.77\text{ s}}$

CHECK Note that $t_1 + t_2 = 6.30$ s, in agreement with Example 3-7. Also, note that t_1 is less than t_2. This is as expected because the package rises a distance of 12 m but falls a distance of 112 m.

PRACTICE PROBLEM 3-2 Solve Part (b) of Example 3-8 using $y(t)$ (Equation 3-16b) instead of finding $v_{y\,\text{av}}$.

HORIZONTAL RANGE OF A PROJECTILE

The horizontal range R of a projectile can be written in terms of its initial speed and initial angle above the horizontal. As in the preceding examples, we find the horizontal range by multiplying the x component of the velocity by the total time that the projectile is in the air. The total flight time T is obtained by setting $y = 0$ and $t = T$ in $y = v_0t - \frac{1}{2}gt^2$ (Equation 3-16b).

$$0 = v_{0y}T - \tfrac{1}{2}gT^2 \qquad T > 0$$

Dividing through by T gives

$$v_{0y} - \tfrac{1}{2}gT = 0$$

The flight time of the projectile is thus

$$T = \frac{2v_{0y}}{g} = \frac{2v_0}{g}\sin\theta_0$$

To find the horizontal range R, we substitute T for t in $x(t) = v_{0x}t$ (Equation 3-16a) to obtain

$$R = v_{0x}T = (v_0\cos\theta_0)\left(\frac{2v_0}{g}\sin\theta_0\right) = \frac{2v_0^2}{g}\sin\theta_0\cos\theta_0$$

This can be further simplified by using the trigonometric identity:

$$\sin 2\theta = 2\sin\theta\cos\theta$$

Thus,

$$R = \frac{v_0^2}{g}\sin 2\theta_0 \qquad \text{3-18}$$

HORIZONTAL RANGE OF A PROJECTILE

PRACTICE PROBLEM 3-3

Use Equation 3-18 to verify the answer Part (b) of Example 3-6.

Equation 3-18 is useful if you want to find the range for several projectiles that have equal initial speeds. For this case, this equation shows how the range depends on θ. Because the maximum value of $\sin 2\theta$ is 1, and because $\sin 2\theta = 1$ when $\theta = 45°$, the range is greatest when $\theta = 45°$. Figure 3-17 shows graphs of the vertical heights versus the horizontal distances for projectiles with an initial speed of 24.5 m/s and several different initial angles. The angles drawn are 45°, which has the maximum range, and pairs of angles at equal amounts above and below 45°. Notice that the paired angles have the same range. The green curve has an initial angle of 36.9°, as in Example 3-6.

In many practical applications, the initial and final elevations may not be equal, or other considerations are important. For example, in the shot put, the ball ends its flight when it hits the ground, but it is projected from an initial height of about 2 m above the ground. This condition causes the horizontal displacement to be at a maximum at an angle somewhat lower than 45°, as shown in Figure 3-18. Studies of the best shot-putters show that maximum horizontal displacement occurs at an initial angle of about 42°.

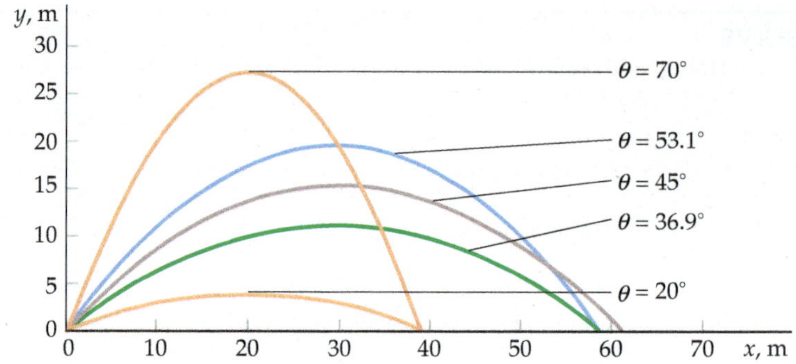

FIGURE 3-17 The initial speed is the same for each trajectory.

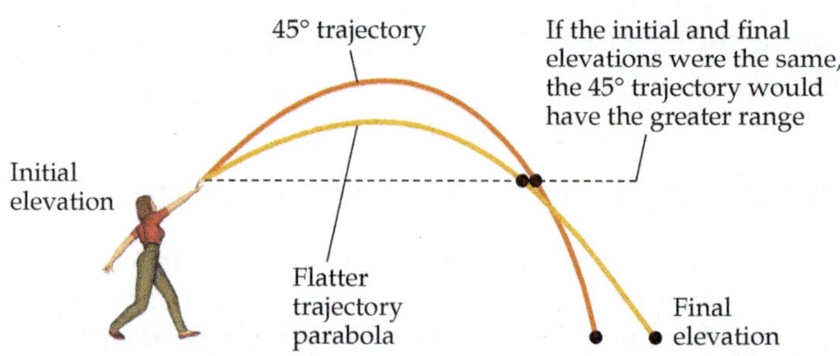

FIGURE 3-18 If a projectile lands at an elevation lower than the initial elevation, the maximum horizontal displacement is achieved when the projection angle is somewhat lower than 45°.

Example 3-9 To Catch a Thief

A police officer chases a master jewel thief across city rooftops. They are both running when they come to a gap between buildings that is 4.00 m wide and has a drop of 3.00 m (Figure 3-19). The thief, having studied a little physics, leaps at 5.00 m/s, at an angle of 45.0° above the horizontal, and clears the gap easily. The police officer did not study physics and thinks he should maximize his horizontal velocity, so he leaps horizontally at 5.00 m/s. (*a*) Does he clear the gap? (*b*) By how much does the thief clear the gap?

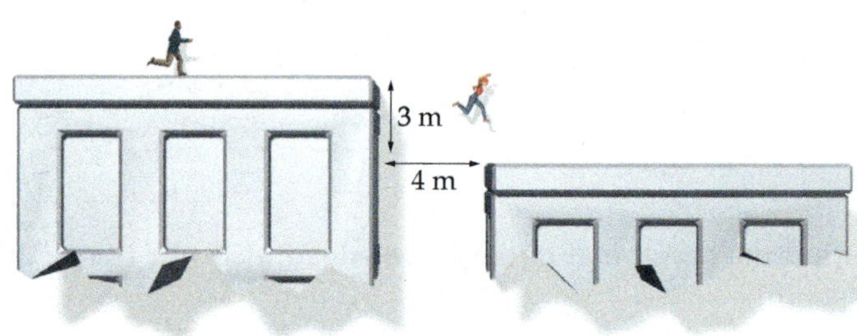

FIGURE 3-19

PICTURE Assuming they both clear the gap, the total time in the air depends only on the vertical aspects of the motion. Choose the origin at the launch point, with upward positive so that Equations 3-16 apply. Use Equation 3-16*b* for $y(t)$ and solve for the time when $y = -3.00$ m for $\theta_0 = 0$ and again for $\theta_0 = 45.0°$. The horizontal distances traveled are the values of x at these times.

SOLVE

(*a*) 1. Write $y(t)$ for the police officer and solve for t when $y = -3.00$ m.

$$y = y_0 + v_{y0}t - \tfrac{1}{2}gt^2 \quad -3.00\text{ m} = 0 + 0 - \tfrac{1}{2}(9.81\text{ m/s}^2)t^2$$

$$t = 0.782\text{ s}$$

2. Substitute this into the equation for $x(t)$ and find the horizontal distance traveled during this time.

$$x = x_0 + v_{x0}t$$

$$x = 0 + (5.00\text{ m/s})(0.782\text{ s})$$

$$x = \boxed{3.91\text{ m}}$$

Because 3.91 m is less than 4.00 m, it appears the police officer fails to make it across the gap between buildings.

(*b*) 1. Write $y(t)$ for the thief and solve for t when $y = -3.00$ m. $y(t)$ is a quadratic equation with two solutions, but only one of its solutions is acceptable.

$y = y_0 + v_{y0}t - \frac{1}{2}gt^2$ $-3.00\text{ m} = 0 + (5.00\text{ m/s})\sin 45.0° - \frac{1}{2}(9.81\text{ m/s}^2)t^2$

$t = -0.500$ s or $t = 1.22$ s

She must land at a time after she leaps, so

$t = 1.22$ s

2. Find the horizontal distance covered for the positive value of t.

$x = x_0 + v_{0x}t = 0 + (5.00\text{ m/s})\cos 45°(1.22\text{ s}) = 4.31\text{ m}$

3. Subtract 4.0 m from this distance.

$4.31\text{ m} - 4.00\text{ m} = \boxed{0.31\text{ m}}$

CHECK The policeman's horizontal speed remains 5.00 m/s during his flight. So, the policeman travels the 4.00 m to the next building in 4.00 m/(5.00 m/s) = 0.800 s. Because our Part-(*a*), step-1 result is less than 0.800 s, we know he falls below the second roof before reaching the second building—in agreement with our Part-(*a*), step-2 result.

TAKING IT FURTHER By modeling the policeman as a particle, we found that he was slightly below the second roof at impact. However, we cannot conclude that he did not complete the jump because he is not a particle. It is likely that he would raise his feet enough for them to clear the edge of the second roof.

PROJECTILE MOTION IN VECTOR FORM

For projectile motion, we have $a_x = 0$ and $a_y = -g$ (Equations 3-14*a* and 3-14*b*), where the $+y$ direction is directly upward. To express these equations in vector form, we multiply both sides of each equation by the appropriate unit vector and then add the two resulting equations. That is, $a_x\hat{i} = 0\hat{i}$ plus $a_y\hat{j} = -g\hat{j}$ gives

$$a_x\hat{i} + a_y\hat{j} = -g\hat{j} \quad \text{or} \quad \vec{a} = \vec{g} \qquad \text{3-14}c$$

where $\vec{g}$ is the free-fall acceleration vector. The magnitude of $\vec{g}$ is $g = 9.81\text{ m/s}^2$ (at sea level and at 45° latitude).

Combining equations $v_x = v_{0x}$ and $v_y = v_{0y} - gt$ in like manner gives

$$\vec{v} = \vec{v}_0 + \vec{g}t \quad (\text{or } \Delta\vec{v} = \vec{g}t) \qquad \text{3-15}c$$

where $\vec{v} = v_x\hat{i} + v_y\hat{j}$, $\vec{v}_0 = v_{0x}\hat{i} + v_{0y}\hat{j}$, and $\vec{g} = -g\hat{j}$. Repeating the process, this time for Equations $x(t) = x_0 + v_{0x}t$ and $y(t) = y_0 + v_{0y}t - \frac{1}{2}gt^2$, gives

$$\vec{r} = \vec{r}_0 + \vec{v}_0t + \tfrac{1}{2}\vec{g}t^2 \quad (\text{or } \Delta\vec{r} = \vec{v}_0t + \tfrac{1}{2}\vec{g}t^2) \qquad \text{3-16}c$$

where $\vec{r} = x\hat{i} + y\hat{j}$ and $\vec{r}_0 = x_0\hat{i} + y_0\hat{j}$. The vector forms of the kinematic equations (Equations 3-15*c* and 3-16*c*) are useful for solving a number of problems, including the following example.

Example 3-10 The Ranger and the Monkey

A park ranger with a tranquilizer dart gun intends to shoot a monkey hanging from a branch. The ranger points the barrel directly at the monkey, not realizing that the dart will follow a parabolic path that will pass below the present position of the creature. The monkey, seeing the gun discharge, immediately lets go of the branch and drops out of the tree, expecting to avoid the dart. (*a*) Show that the monkey will be hit regardless of the initial speed of the dart as long as this speed is great enough for the dart to travel the horizontal distance to the tree. Assume the reaction time of the monkey is negligible. (*b*) Let $\vec{v}_{d0}$ be the initial velocity of the dart relative to the monkey. Find the velocity of the dart *relative to the monkey* at an arbitrary time t during the dart's flight.

PICTURE In this example, both the monkey and the dart exhibit projectile motion. To show that the dart hits the monkey, we have to show that at some time t, the dart and the monkey have the same coordinates, regardless of the initial speed of the dart. To do this, we apply

Equation 3-16*c* to both the monkey and the dart. For Part (*b*), we can use Equation 3-15*c*, keeping in mind the relative reference frames.

SOLVE

(*a*) 1. Apply Equation 3-16*c* to the monkey at arbitrary time *t*:
$\Delta\vec{r}_{m} = \frac{1}{2}\vec{g}t^2$
(The initial velocity of the monkey is zero.)

2. Apply Equation 3-16*c* to the dart at arbitrary time *t*:
$\Delta\vec{r}_{d} = \vec{v}_{d0}t + \frac{1}{2}\vec{g}t^2$
where $\vec{v}_{d0}$ is the velocity of the dart as it leaves the gun.

3. Make a sketch of the monkey, the dart, and the gun, as shown in Figure 3-20. Show the dart and the monkey at their initial locations and at their locations a time *t* later. On the figure draw a vector representing each term in the step-1 and -2 results:

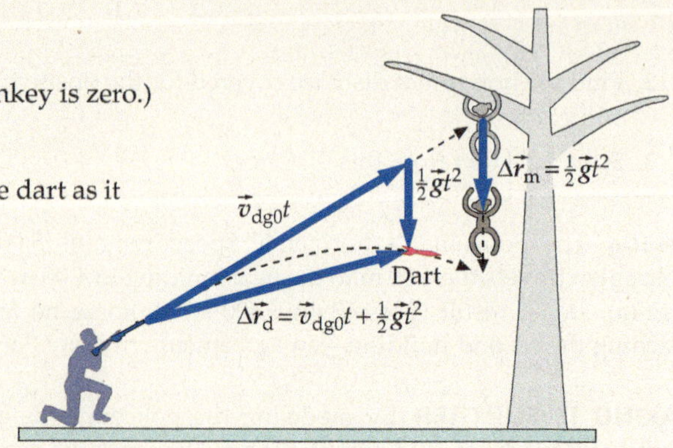

FIGURE 3-20

4. Note that at time *t* the dart and the monkey both are a distance $\frac{1}{2}gt^2$ below the line of sight of the gun:
The dart will strike the monkey when the dart reaches the monkey's line of fall.

(*b*) 1. The velocity of the dart relative to the monkey equals the velocity of the dart relative to the gun plus the velocity of the gun relative to the monkey:
$\vec{v}_{dm} = \vec{v}_{dg} + \vec{v}_{gm}$

2. The velocity of the gun relative to the monkey is the negative of the velocity of the monkey relative to the gun:
$\vec{v}_{dm} = \vec{v}_{dg} - \vec{v}_{mg}$

3. Using $\vec{v} = \vec{v}_0 + \vec{g}t$ (Equation 3-15*c*), express both the velocity of the dart relative to the gun and the velocity of the monkey relative to the gun:
$\vec{v}_{dg} = \vec{v}_{dg0} + \vec{g}t$
$\vec{v}_{mg} = \vec{g}t$

4. Substitute these expressions into the Part-(*b*), step-2 result:
$\vec{v}_{dm} = (\vec{v}_{dg0} + \vec{g}t) - (\vec{g}t) = \boxed{\vec{v}_{dg0}}$

CHECK The Part-(*a*), step-4 and the Part-(*b*), step-4 results are in agreement with each other. They agree that the dart will strike the monkey if the dart reaches the monkey's line of fall before the monkey lands on the ground.

TAKING IT FURTHER Relative to the falling monkey, the dart moves with constant speed v_{dg0} in a straight line. The dart strikes the monkey at time $t = L/v_{d0}$, where L is the distance from the muzzle of the gun to the initial position of the monkey.

In a familiar lecture demonstration, a target is suspended by an electromagnet. When the dart leaves the gun, the circuit to the magnet is broken and the target falls. The demonstration is then repeated using a different initial velocity of the dart. For a large value of v_{dg0}, the target is hit very near its original height, and for some lesser value of v_{dg0} it is hit just before it reaches the floor.

PRACTICE PROBLEM 3-4 A hockey puck at ice level is struck such that it misses the net and just clears the top of the Plexiglas wall of height $h = 2.80$ m. The flight time at the moment the puck clears the wall is $t_1 = 0.650$ s, and the horizontal position is $x_1 = 12.0$ m. (*a*) Find the initial speed and direction of the puck. (*b*) When does the puck reach its maximum height? (*c*) What is the maximum height of the puck?

3-3 SPECIAL CASE 2: CIRCULAR MOTION

Figure 3-21 shows a pendulum bob swinging back and forth in a vertical plane. The path of the bob is a segment of a circular path. Motion along a circular path, or a segment of a circular path, is called **circular motion.**

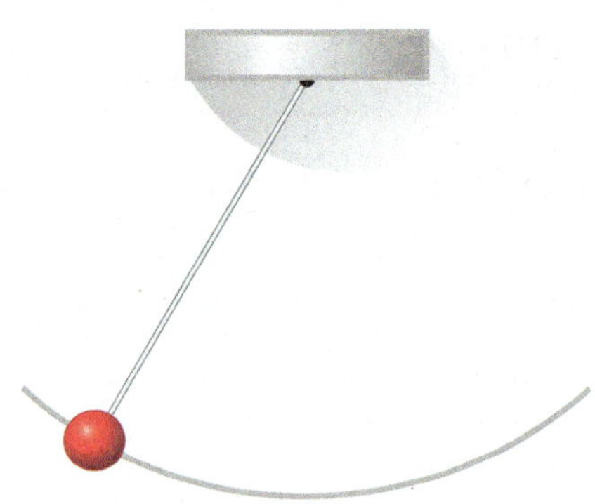

FIGURE 3-21 A pendulum bob swings along a circular arc centered at the point of support of the string.

Example 3-11 A Swinging Pendulum *Conceptual*

Consider the motion of the pendulum bob shown in Figure 3-21. Using a motion diagram (see Example 3-5), find the direction of the acceleration vector when the bob is swinging from left to right and (*a*) on the descending portion of the path, (*b*) passing through the lowest point on the path, and (*c*) on the ascending portion of the path.

PICTURE As the bob descends, it both gains speed and changes direction. The acceleration is related to the change in velocity by $\vec{a} \approx \Delta\vec{v}/\Delta t$. The direction of the acceleration at a point can be estimated by constructing a vector addition diagram for the relation $\vec{v}_i + \Delta\vec{v} = \vec{v}_f$ to find the direction of $\Delta\vec{v}$, and thus the direction of the acceleration vector.

SOLVE

(*a*) 1. Make a motion diagram for a full left-to-right swing of the bob (Figure 3-22*a*). The spacing between dots is greatest at the lowest point where the speed is greatest.

2. Pick the dot at t_2 on the descending portion of the motion and draw a velocity vector on the diagram for both the preceding and the following dot (the dots at t_1 and t_3). The velocity vectors should be drawn tangent to the path and with lengths proportional to the speed (Figure 3-22*a*).

3. Draw the vector addition diagram (Figure 3-22*b*) for the relation $\vec{v}_i + \Delta\vec{v} = \vec{v}_f$. On this diagram draw the acceleration vector. Because $\vec{a} \approx \Delta\vec{v}/\Delta t$, $\vec{a}$ is in the same direction as $\Delta\vec{v}$.

(*b*) Repeat steps 2 and 3 (Figure 3-22*c*) for the point at t_4, the lowest point on the path.

(*c*) Repeat steps 2 and 3 (Figure 3-22*d*) for the point at t_6, a point on the ascending portion of the path.

CHECK At the lowest point (at t_4) the horizontal component of $\vec{v}$ is a maximum, so we expect the horizontal component of $\vec{a}$ to be zero. Near the lowest point, the upward component of $\vec{v}$ is negative just prior to $t = t_4$ and is positive just after $t = t_4$, so the upward component of $\vec{v}$ is increasing at $t = t_4$. This means we expect the upward component of $\vec{a}$ to be positive at $t = t_4$. The acceleration vector in Figure 3-22*c* is in agreement with both of these expectations.

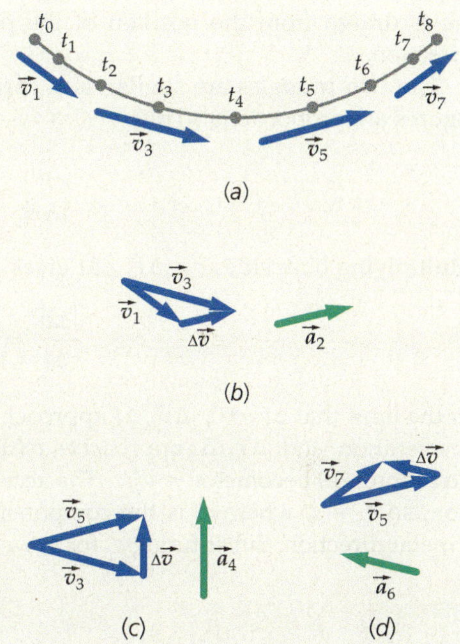

FIGURE 3-22

In Example 3-11, we saw that the acceleration vector is directed straight upward at the lowest point of the pendulum's swing (Figure 3-23)—toward point P at the center of the circle. Where the speed is increasing (on the descending portion), the acceleration vector has a component in the direction of the velocity vector as well as a component in the direction toward P. Where the speed is decreasing, the acceleration vector has a component in the direction opposite to the direction of the velocity vector, as well as a component in the direction toward P.

As a particle moves along a circular arc, the direction from the particle toward P (toward the center of the circle) is called the **centripetal direction** and the direction of the velocity vector is called the **tangential direction.** In Figure 3-23, the acceleration vector at the lowest point of the pendulum bob's path is in the centripetal direction. At virtually all other points along the path, the acceleration vector has both a tangential component and a centripetal component.

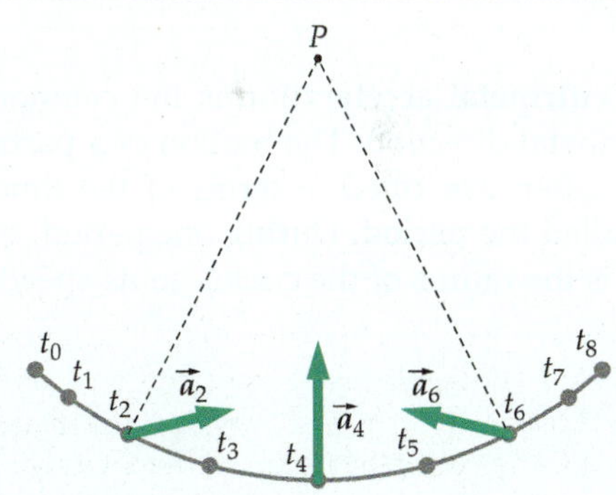

FIGURE 3-23

UNIFORM CIRCULAR MOTION

Motion in a circle at constant speed is called **uniform circular motion.** Even though the speed of a particle moving in uniform circular motion is constant, the moving particle is accelerating. To find an expression for the acceleration of a particle during uniform circular motion, we will extend the method used in Example 3-11 to relate the acceleration to the speed and the radius of the circle. The position and velocity vectors for a particle moving in a circle at constant speed are shown in Figure 3-24. The angle $\Delta\theta$ between $\vec{v}(t)$ and $\vec{v}(t + \Delta t)$ is equal to the angle between $\vec{r}(t)$ and $\vec{r}(t + \Delta t)$ because $\vec{r}$

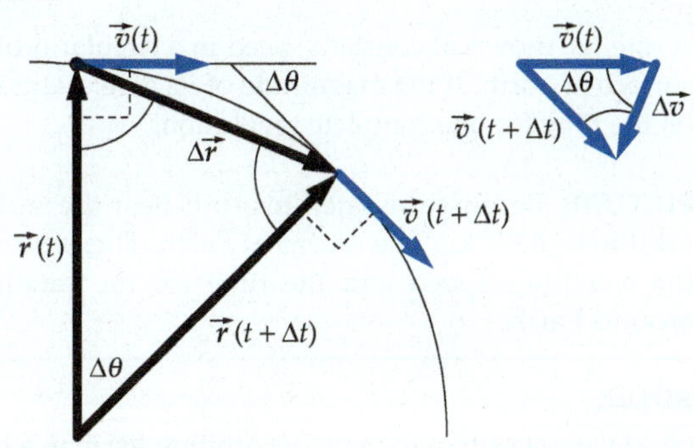

FIGURE 3-24

and $\vec{v}$ both rotate through the same angle $\Delta\theta$ during time Δt. An isosceles triangle is formed by the two velocity vectors and the $\Delta\vec{v}$ vector, and a second isosceles triangle is formed by the two position vectors and the $\Delta\vec{r}$ vector.

To find the direction of the acceleration vector we examine the triangle formed by the two velocity vectors and the $\Delta\vec{v}$ vector. The sum of the angles of any triangle is 180° and the base angles of any isosceles triangle are equal. In the limit that Δt approaches zero, $\Delta\theta$ also approaches zero, so in this limit the two base angles each approach 90°. This means that in the limit that $\Delta t \to 0$, $\Delta\vec{v}$ is perpendicular to $\vec{v}$. If $\Delta\vec{v}$ is drawn from the position of the particle, then it points in the centripetal direction.

The two triangles are similar, and corresponding lengths of similar geometric figures are proportional. Thus,

$$\frac{|\Delta\vec{v}|}{|\Delta\vec{r}|} = \frac{v}{r}$$

Multiplying both sides by $|\Delta\vec{r}|\ /\Delta t$ gives

$$\frac{|\Delta\vec{v}|}{\Delta t} = \frac{v}{r}\frac{|\Delta\vec{r}|}{\Delta t} \qquad \text{3-19}$$

In the limit that $\Delta t \to 0$, $|\Delta\vec{v}|/\Delta t$ approaches a, the magnitude of the instantaneous acceleration, and $|\Delta\vec{r}|/\Delta t$ approaches v (the speed). Thus, in the limit that $\Delta t \to 0$, Equation 3-19 becomes $a = v^2/r$. The acceleration vector is in the centripetal direction, so $a_c = a$, where a_c is the component of the acceleration vector in the centripetal direction. Substituting a_c for a, we have

$$a_c = \frac{v^2}{r} \qquad \text{3-20}$$

CENTRIPETAL ACCELERATION

Centripetal acceleration is the component of the acceleration vector in the centripetal direction. The motion of a particle moving in a circle with constant speed is often described in terms of the time T required for one complete revolution, called the **period.** During one period, the particle travels a distance of $2\pi r$ (where r is the radius of the circle), so its speed v is related to r and T by

$$v = \frac{2\pi r}{T} \qquad \text{3-21}$$

Example 3-12 A Satellite's Motion

A satellite moves at constant speed in a circular orbit about the center of Earth and near the surface of Earth. If the magnitude of its acceleration is $g = 9.81\ \text{m/s}^2$, find (*a*) its speed, and (*b*) the time for one complete revolution.

PICTURE Because the satellite orbits near the surface of Earth, we take the radius of the orbit to be 6370 km, the radius of Earth. Then, we can use Equations 3-20 and 3-21 to find the satellite's speed and the time for the satellite to make one complete revolution around Earth.

SOLVE

(*a*) Make a sketch of the satellite orbiting Earth in a low-Earth orbit (Figure 3-25). Include the velocity and acceleration vectors:

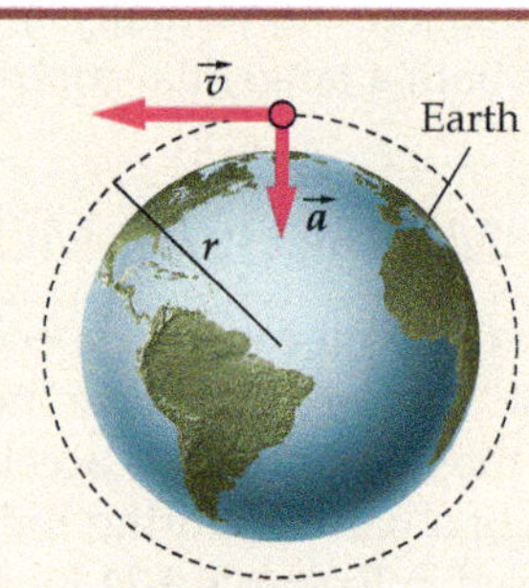

FIGURE 3-25 A satellite in a circular, low-Earth orbit.

Set the centripetal acceleration v^2/r equal to g and solve for the speed v:

$$a_c = \frac{v^2}{r} = g, \quad \text{so}$$

$$v = \sqrt{rg} = \sqrt{(6370\text{ km})(9.81\text{ m/s}^2)}$$

$$= \boxed{7.91\text{ km/s} = 17{,}700\text{ mi/h} = 4.91\text{ mi/s}}$$

(*b*) Use Equation 3-21 to get the period T:

$$T = \frac{2\pi r}{v} = \frac{2\pi(6370\text{ km})}{7.91\text{ km/s}} = \boxed{5060\text{ s} = 84.3\text{ min}}$$

CHECK It is well known that the orbital period for satellites orbiting just above the atmosphere of Earth is about 90 min, so the Part-(*b*) result of 84.3 min is close to what we would expect.

TAKING IT FURTHER For actual satellites in orbit a few hundred kilometers above Earth's surface, the orbital radius r is a few hundred kilometers greater than 6370 km. As a result, the acceleration is slightly less than 9.81 m/s² because of the decrease in the gravitational force with distance from the center of Earth. Orbits just above Earth's atmosphere are referred to as "low Earth orbits." Many satellites, including the Hubble telescope and the International Space Station, are in low Earth orbits. Information about these and other satellites can be found at www.nasa.gov.

PRACTICE PROBLEM 3-5 A car rounds a curve of radius 40 m at 48 km/h. What is its centripetal acceleration?

TANGENTIAL ACCELERATION

A particle moving in a circle with *varying speed* has a component of acceleration tangent to the circle, a_t, as well as the radially inward centripetal acceleration, v^2/r. For general motion along a curve, we can treat a portion of the curve as an arc of a circle (Figure 3-26). The particle then has centripetal acceleration $a_c = v^2/r$ toward the center of curvature, and if the speed v is changing, it has a tangential acceleration given by

$$a_t = \frac{dv}{dt} \qquad \text{3-22}$$

TANGENTIAL ACCELERATION

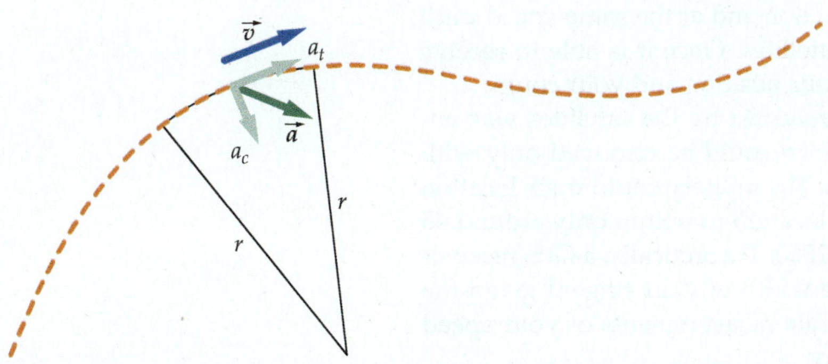

FIGURE 3-26 The tangential and centripetal acceleration components of a particle moving along a curved path.

PRACTICE PROBLEM 3-6

You are in a cart on a rollercoaster track that is entering a loop-the-loop. At the instant you are one fourth the way through the loop-the-loop your cart is going straight up at 20 m/s, and is losing speed at 5.0 m/s². The radius of curvature of the track is 25 m. What are your centripetal and tangential acceleration components at that instant?

Physics Spotlight

GPS: Vectors Calculated While You Move

If you fly to another city and rent a car to reach your destination, you might rent a Global Positioning System (GPS) navigation computer with the car. Many people use GPS navigators, but not everyone knows that these computers constantly calculate vectors for you.

Twenty-four GPS satellites orbit Earth at an altitude of 11,000 mi.* At most times and places, at least three satellites are visible (above the horizon). In many cases, four or more satellites are visible. Each satellite broadcasts a continual stream that includes its identification, information about its orbit, and a time marker that is precise to one billionth of a second.† The satellite's internal clocks and orbits are checked by a ground station that can send correction information.

A GPS receiver listens for the signals from the satellites. When it can get a lock on three or more satellite signals, it calculates how far away each satellite is by the difference between the satellite's time marker and the time on the receiver's clock when the marker is detected. From the known orbits of each satellite and the distance to each satellite, the receiver can triangulate its position. A calculation from three satellites will give the longitude and latitude of the receiver. A calculation from four satellites will also give altitude.

The navigation systems used in automobiles obtain information from GPS satellites and use the information to calculate the position and velocity of the car. At times, they calculate the car's displacement vector using a procedure called *dead reckoning*. *(Andrew Fox/Corbis.)*

But where do vectors come in? The receiver does not just triangulate its position once—that would give a point position. The receiver constantly listens for the satellites and calculates changes in the receiver position from changes in the triangulation results. It calculates any changes in distance and direction from the last known position. Within a very short time it has taken several readings, enough to calculate the velocity of your travel. The result? A speed in a particular direction—a velocity vector—is always part of the receiver's calculations.

But this vector is not there just to draw a pretty line on the screen for you. There are times when it is not possible to get a good reading on a receiver. Perhaps you have driven under a bridge or through a tunnel. If the GPS receiver is unable to lock onto a meaningful signal, it will start from your last known position. It will then use your last known velocity and direction to calculate a *dead reckoning*. It will assume that you are continuing in the same direction and at the same speed until it is able to get a reliable signal from enough satellites. Once it is able to receive good signals again, it will make corrections to your position and your course.

When GPS was pioneered, the time signal broadcast by the satellites was encoded with distortion, or *selective availability*, which could be removed only with decoding receivers enabled for defense purposes. The military could track location to within six meters, while civilians could track location to within only around 45 meters.** That coding was turned off in the year 2000. Theoretically, a GPS receiver would be able to tell your position down to the width of your finger,‡ given the right signals, and give equally precise and accurate measurements of your speed and direction—all from at least 11,000 miles away.

---

* The actual number of operational satellites varies. It is more than twenty-four, in case of satellite failure. "Block II Satellite Information." *ftp://tycho.usno.navy.mil/pub/gps/gpsb2.txt* United States Naval Observatory. March 2006.

† "GPS: The Role of Atomic Clocks—It Started with Basic Research." *http://www.beyonddiscovery.org/content/view.page.asp?I=464* Beyond Discovery. The National Academy of Sciences. March 2006.

** "Comparison of Positions With and Without Selective Availability: Full 24 Hour Data Sets." *http://www.ngs.noaa.gov/FGCS/info/sans_SA/compare/ERLA.htm* National Geodetic Survey. March, 2006.

‡ "Differential GPS: Advanced Concepts." *http://www.trimble.com/gps/advanced1.html* Trimble. March, 2006.

Summary

TOPIC	RELEVANT EQUATIONS AND REMARKS	
1. Kinematic Vectors		
Position vector	The position vector $\vec{r}$ points from the origin of the coordinate system to the particle.	
Instantaneous-velocity vector	The velocity vector $\vec{v}$ is the rate of change of the position vector. Its magnitude is the speed, and it points in the direction of motion. $$\vec{v} = \lim_{\Delta t \to 0} \frac{\Delta \vec{r}}{\Delta t} = \frac{d\vec{r}}{dt}$$	3-5
Instantaneous-acceleration vector	$$\vec{a} = \lim_{\Delta t \to 0} \frac{\Delta \vec{v}}{\Delta t} = \frac{d\vec{v}}{dt}$$	3-11
2. Relative Velocity	If a particle p moves with velocity $\vec{v}_{pA}$ relative to reference frame A, which is in turn moving with velocity $\vec{v}_{AB}$ relative to reference frame B, the velocity of p relative to B is $$\vec{v}_{pB} = \vec{v}_{pA} + \vec{v}_{AB}$$	3-9
3. Projectile Motion with No Air Resistance	The $+x$ direction is horizontal and the $+y$ direction is upward for the equations in this section.	
Independence of motion	In projectile motion, the horizontal and vertical motions are independent.	
Acceleration	$a_x = 0$ and $a_y = -g$	
Dependence on time	$v_x(t) = v_{0x} + a_x t$ and $v_y(t) = v_{0y} + a_y t$	2-12
	$\Delta x = v_{0x}t + \frac{1}{2}a_x t^2$ and $\Delta y = v_{0y}t + \frac{1}{2}a_y t^2$	2-14
	where $v_{x0} = v_0 \cos\theta_0$ and $v_{y0} = v_0 \sin\theta_0$. Alternatively, $\Delta\vec{v} = \vec{g}t$ and $\Delta\vec{r} = \vec{v}_0 t + \frac{1}{2}\vec{g}t^2$	3-15*c*, 3-16*c*
Horizontal displacement	The horizontal displacement is found by multiplying v_{0x} by the total time the projectile is in the air.	
4. Circular Motion		
Centripetal acceleration	$$a_c = \frac{v^2}{r}$$	3-20
Tangential acceleration	$$a_t = \frac{dv}{dt}$$ where v is the *speed*.	3-22
Period	$$v = \frac{2\pi r}{T}$$	3-21

Answers to Concept Checks

3-1 (*a*) upward, (*b*) upward

3-2 vertically downward

Answers to Practice Problems

3-1 $a_{av} = 4.7\ \text{m/s}^2$ at 45° west of north

3-2
$$\begin{aligned} y(t) &= y_0 + v_{0y}t - \tfrac{1}{2}gt^2 \\ &= 0 + (25.0\ \text{m/s})\sin 36.9°(1.43\ \text{s}) \\ &\quad + \tfrac{1}{2}(9.81\ \text{m/s}^2)(1.43\ \text{s})^2 \\ &= 11.48\ \text{m} \\ \therefore h &= 111\ \text{m} \end{aligned}$$

3-3 $$R = \frac{v_0^2}{g}\sin 2\theta_0 = \frac{(24.5\ \text{m/s})^2}{9.81\ \text{m/s}^2}\sin(2\times 36.9°) = 58.8°$$

3-4 (*a*) $\vec{v}_0 = 20.0$ m/s at $\theta_0 = 22.0°$, (*b*) $t = 0.764$ s, (*c*) $v_{y\,av}t = 2.86$ m

3-5 $4.44\ \text{m/s}^2$

3-6 $a_c = 16\ \text{m/s}^2$ and $a_t = -5.0\ \text{m/s}^2$

Problems

In a few problems, you are given more data than you actually need; in a few other problems, you are required to supply data from your general knowledge, outside sources, or informed estimate.

Interpret as significant all digits in numerical values that have trailing zeros and no decimal points.

- • Single-concept, single-step, relatively easy
- •• Intermediate-level, may require synthesis of concepts
- ••• Challenging
- SSM Solution is in the *Student Solutions Manual*
- Consecutive problems that are shaded are paired problems.

CONCEPTUAL PROBLEMS

1 • Can the magnitude of the displacement of a particle be less than the distance traveled by the particle along its path? Can its magnitude be more than the distance traveled? Explain. SSM

2 • Give an example in which the distance traveled is a significant amount, yet the corresponding displacement is zero. Can the reverse be true? If so, give an example.

3 • What is the average velocity of a batter who hits a home run (from when he hits the ball to when he touches home plate after rounding the bases)?

4 • A baseball is hit so its initial velocity upon leaving the bat makes an angle of 30° above the horizontal. It leaves that bat at a height of 1.0 m above the ground and lands untouched for a single. During its flight, from just after it leaves the bat to just before it hits the ground, describe how the angle between its velocity and acceleration vectors changes. Neglect any effects due to air resistance.

5 • If an object is moving toward the west at some instant, in what direction is its acceleration? (*a*) north, (*b*) east, (*c*) west, (*d*) south, (*e*) may be any direction

6 • Two astronauts are working on the lunar surface to install a new telescope. The acceleration due to gravity on the moon is only 1.64 m/s^2. One astronaut tosses a wrench to the other astronaut, but the speed of throw is excessive and the wrench goes over her colleague's head. When the wrench is at the highest point of its trajectory, (*a*) its velocity and acceleration are both zero, (*b*) its velocity is zero but its acceleration is nonzero, (*c*) its velocity is nonzero but its acceleration is zero, (*d*) its velocity and acceleration are both nonzero, (*e*) insufficient information is given to choose between any of the previous choices.

7 • The velocity of a particle is directed toward the east while the acceleration is directed toward the northwest, as shown in Figure 3-27. The particle is (*a*) speeding up and turning toward the north, (*b*) speeding up and turning toward the south, (*c*) slowing down and turning toward the north, (*d*) slowing down and turning toward the south, (*e*) maintaining constant speed and turning toward the south.

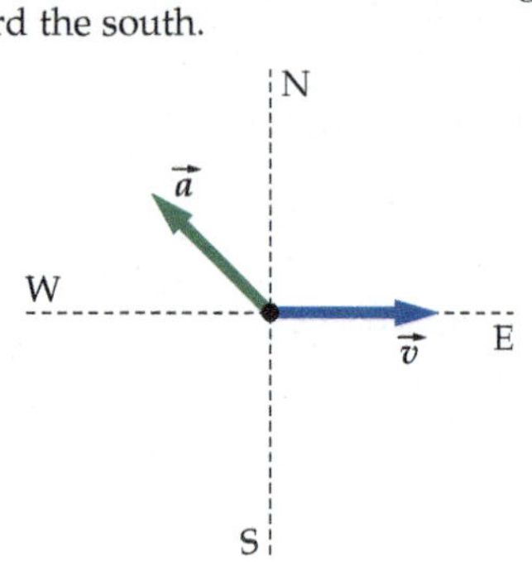

FIGURE 3-27 Problem 7

8 • Assume you know the position vectors of a particle at two points on its path, one earlier and one later. You also know the time it took the particle to move from one point to the other. Then, you can compute the particle's (*a*) average velocity, (*b*) average acceleration, (*c*) instantaneous velocity, and (*d*) instantaneous acceleration.

9 • Consider the path of a moving particle. (*a*) How is the velocity vector related geometrically to the path of the particle? (*b*) Sketch a curved path and draw the velocity vector for the particle for several positions along the path.

10 • The acceleration of a car is zero when it is (*a*) turning right at a constant speed, (*b*) driving up a long straight incline at constant speed, (*c*) traveling over the crest of a hill at constant speed, (*d*) bottoming out at the lowest point of a valley at constant speed, (*e*) speeding up as it descends a long straight decline.

11 • Give examples of motion in which the directions of the velocity and acceleration vectors are (*a*) opposite, (*b*) the same, and (*c*) mutually perpendicular. SSM

12 • How is it possible for a particle moving at constant speed to be accelerating? Can a particle with constant velocity be accelerating at the same time?

13 •• Imagine throwing a dart straight upward so that it sticks into the ceiling. After it leaves your hand, it steadily slows down as it rises before it sticks. (*a*) Draw the dart's velocity vector at times t_1 and t_2, where t_1 and t_2 occur after it leaves your hand but before it hits the ceiling and $t_2 - t_1$ is small. From your drawing, find the direction of the change in velocity $\Delta\vec{v} = \vec{v}_2 - \vec{v}_1$, and thus the direction of the acceleration vector. (*b*) After it has stuck in the ceiling for a few seconds, the dart falls down to the floor. As it falls it speeds up, of course, until it hits the floor. Repeat Part (*a*) to find the direction of its acceleration vector as it falls. (*c*) Now imagine tossing the dart horizontally. What is the direction of its acceleration vector after it leaves your hand, but before it strikes the floor? SSM

14 • As a bungee jumper approaches the lowest point in her descent, the rubber cord holding her stretches and she loses speed as she continues to move downward. Assuming that she is dropping straight down, make a motion diagram to find the direction of her acceleration vector as she slows down by drawing her velocity vectors at times t_1 and t_2, where t_1 and t_2 are two instants during the portion of her descent in which she is losing speed and $t_2 - t_1$ is small. From your drawing find the direction of the change in velocity $\Delta\vec{v} = \vec{v}_2 - \vec{v}_1$, and thus the direction of the acceleration vector.

15 • After reaching the lowest point in her jump at time t_{low}, a bungee jumper moves upward, gaining speed for a short time until gravity again dominates her motion. Draw her velocity vectors at times t_1 and t_2, where $t_2 - t_1$ is small and $t_1 < t_{low} < t_2$. From your drawing find the direction of the change in velocity $\Delta\vec{v} = \vec{v}_2 - \vec{v}_1$, and thus the direction of the acceleration vector at time t_{low}.

16 • A river is 0.76 km wide. The banks are straight and parallel (Figure 3-28). The current is 4.0 km/h and is parallel to the banks. A boat has a maximum speed of 4.0 km/h in still water. The pilot of the boat

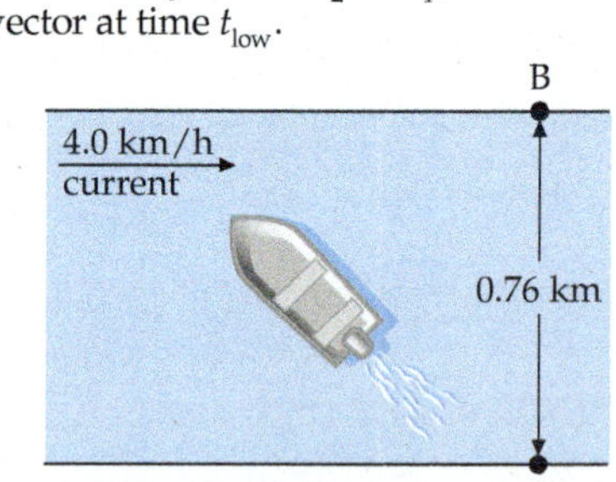

FIGURE 3-28 Problem 16

wishes to go on a straight line from A to B, where the line AB is perpendicular to the banks. The pilot should (*a*) head directly across the river, (*b*) head 53° upstream from the line AB, (*c*) head 37° upstream from the line AB, (*d*) give up—the trip from A to B is not possible with a boat of this limited speed, (*e*) do none of the above.

17 • During a heavy rain, the drops are falling at a constant velocity and at an angle of 10° west of the vertical. You are walking in the rain and notice that only the top surfaces of your clothes are getting wet. In what direction are you walking? Explain. SSM

18 • In Problem 17, what is your walking speed if the speed of the drops relative to the ground is 5.2 m/s?

19 • True or false (ignore any effects due to air resistance):

(*a*) When a projectile is fired horizontally, it takes the same amount of time to reach the ground as an identical projectile dropped from rest from the same height.
(*b*) When a projectile is fired from a certain height at an upward angle, it takes longer to reach the ground than does an identical projectile dropped from rest from the same height.
(*c*) When a projectile is fired horizontally from a certain height, it has a higher speed upon reaching the ground than does an identical projectile dropped from rest from the same height.

20 • A projectile is fired at 35° above the horizontal. Any effects due to air resistance are negligible. At the highest point in its trajectory, its speed is 20 m/s. The initial velocity had a horizontal component of (*a*) 0, (*b*) (20 m/s) cos 35°, (*c*) (20 m/s) sin 35°, (*d*) (20 m/s)/cos 35°, (*e*) 20 m/s.

21 • A projectile is fired at 35° above the horizontal. Any effects due to air resistance are negligible. The initial velocity of the projectile in Problem 20 has a vertical component that is (*a*) less than 20 m/s, (*b*) greater than 20 m/s, (*c*) equal to 20 m/s, (*d*) cannot be determined from the data given. SSM

22 • A projectile is fired at 35° above the horizontal. Any effects due to air resistance are negligible. The projectile lands at the same elevation of launch, so the vertical component of the impact velocity of the projectile is (*a*) the same as the vertical component of its initial velocity in both magnitude and sign, (*b*) the same as the vertical component of its initial velocity in magnitude but opposite in sign, (*c*) less than the vertical component of its initial velocity in magnitude but with the same sign, (*d*) less than the vertical component of its initial velocity in magnitude but with the opposite sign.

23 • Figure 3-29 represents the parabolic trajectory of a projectile going from A to E. Air resistance is negligible. What is the direction of the acceleration at point B? (*a*) up and to the right, (*b*) down and to the left, (*c*) straight up, (*d*) straight down, (*e*) the acceleration of the ball is zero

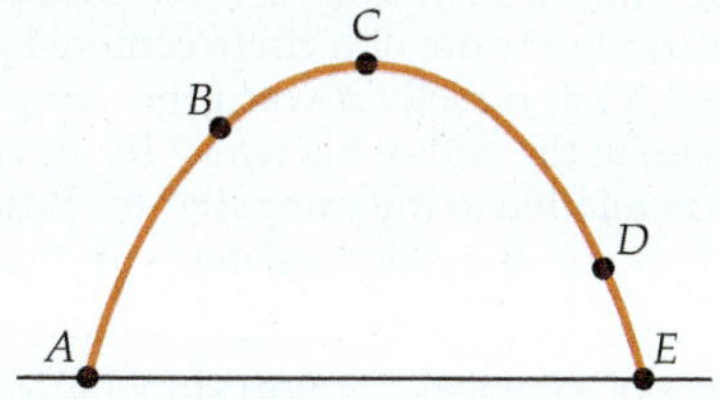

FIGURE 3-29 Problems 23 and 24

24 • Figure 3-29 represents the trajectory of a projectile going from *A* to *E*. Air resistance is negligible. (*a*) At which point(s) is the speed the greatest? (*b*) At which point(s) is the speed the least? (*c*) At which two points is the speed the same? Is the velocity also the same at these points?

25 • True or false:

(*a*) If an object's speed is constant, then its acceleration must be zero.
(*b*) If an object's acceleration is zero, then its speed must be constant.
(*c*) If an object's acceleration is zero, its velocity must be constant.
(*d*) If an object's speed is constant, then its velocity must be constant.
(*e*) If an object's velocity is constant, then its speed must be constant. SSM

26 • The initial and final velocities of a particle are as shown in Figure 3-30. Find the direction of the average acceleration.

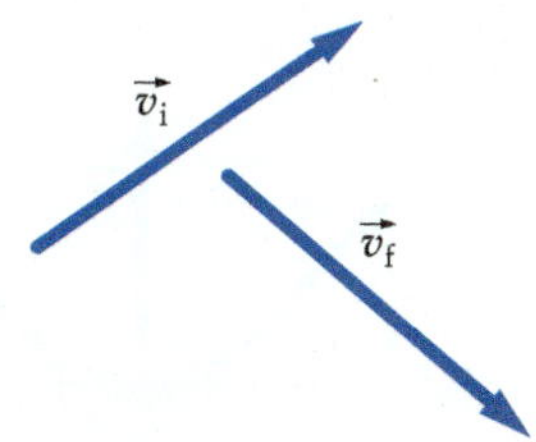

FIGURE 3-30 Problem 26

27 •• The automobile path shown in Figure 3-31 is made up of straight lines and arcs of circles. The automobile starts from rest at point A. After it reaches point B, it travels at constant speed until it reaches point E. It comes to rest at point F. (*a*) At the middle of each segment (AB, BC, CD, DE, and EF), what is the direction of the velocity vector? (*b*) At which of these points does the automobile have a nonzero acceleration? In those cases, what is the direction of the acceleration? (*c*) How do the magnitudes of the acceleration compare for segments BC and DE?

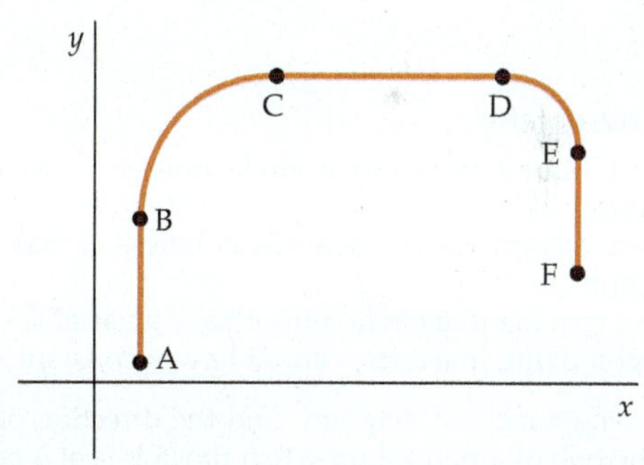

FIGURE 3-31 Problem 27

28 •• Two cannons are pointed directly toward each other, as shown in Figure 3-32. When fired, the cannonballs will follow the trajectories shown—P is the point where the trajectories cross each other. If we want the cannonballs to hit each other, should the gun crews fire cannon A first, cannon B first, or should they fire simultaneously? Ignore any effects due to air resistance.

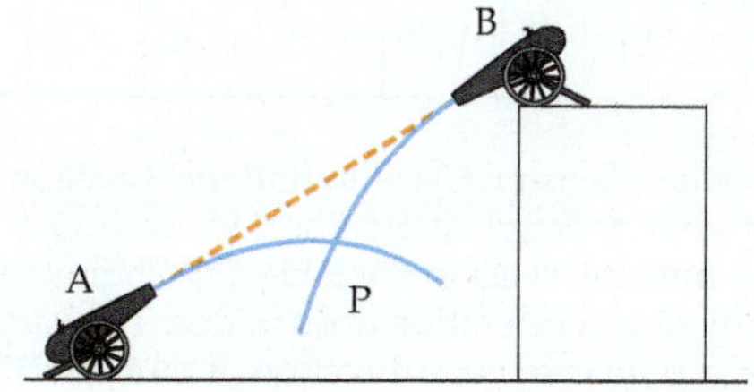

FIGURE 3-32 Problem 28

29 •• Galileo wrote the following in his *Dialogue concerning the two world systems:* "Shut yourself up . . . in the main cabin below decks on some large ship, and . . . hang up a bottle that empties drop by drop into a wide vessel beneath it. When you have observed [this] carefully . . . have the ship proceed with any speed you like, so long as the motion is uniform and not fluctuating this way and that . . . . The droplets will fall as before into the vessel beneath without dropping towards the stern, although while the drops are in the air the ship runs many spans." Explain this quotation.

30 •• A man swings a stone attached to a rope in a horizontal circle at constant speed. Figure 3-33 represents the path of the rock looking down from above. (*a*) Which of the vectors could represent the velocity of the stone? (*b*) Which could represent the acceleration?

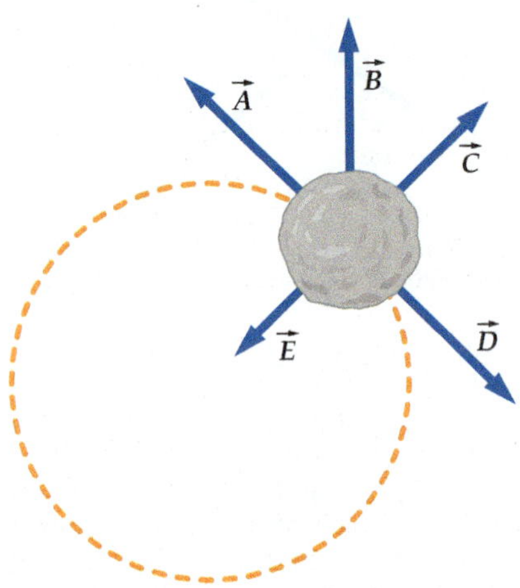

FIGURE 3-33 Problem 30

31 •• True or false:

(*a*) An object cannot move in a circle unless it has centripetal acceleration.
(*b*) An object cannot move in a circle unless it has tangential acceleration.
(*c*) An object moving in a circle cannot have a variable speed.
(*d*) An object moving in a circle cannot have a constant velocity.

32 •• Using a motion diagram, find the direction of the acceleration of the bob of a pendulum when the bob is at a point where it is just reversing its direction.

33 •• CONTEXT-RICH During your rookie bungee jump, your friend records your fall using a camcorder. By analyzing it frame by frame, he finds that the y component of your velocity is (recorded every 1/20 of a second) as follows:

t (s)	12.05	12.10	12.15	12.20	12.25	12.30	12.35	12.40	12.45
v_y (m/s)	−0.78	−0.69	−0.55	−0.35	−0.10	0.15	0.35	0.49	0.53

(*a*) Draw a motion diagram. Use it to find the direction and relative magnitude of your average acceleration for each of the eight successive 0.050-s time intervals in the table. (*b*) Comment on how the y component of your acceleration does or does not vary in sign and magnitude as you reverse your direction of motion. SSM

ESTIMATION AND APPROXIMATION

34 •• CONTEXT-RICH Estimate the speed in mph with which water comes out of a garden hose using your past observations of water coming out of garden hoses and your knowledge of projectile motion.

35 •• CONTEXT-RICH You won a contest to spend a day with a baseball team during spring training. You are allowed to try to hit some balls thrown by a pitcher. Estimate the acceleration during the hit of a fastball thrown by a major league pitcher when you hit the ball squarely—straight back at the pitcher. You will need to make reasonable choices for ball speeds, both just before and just after the ball is hit, and of the contact time of the ball with the bat.

36 •• Estimate how far you can throw a ball if you throw it (*a*) horizontally while standing on level ground, (*b*) at $\theta = 45°$ above horizontal while standing on level ground, (*c*) horizontally from the top of a building 12 m high, (*d*) at $\theta = 45°$ above horizontal from the top of a building 12 m high. Ignore any effects due to air resistance.

37 •• In 1978, Geoff Capes of Great Britain threw a heavy brick a horizontal distance of 44.5 m. Find the approximate speed of the brick at the highest point of its flight, neglecting any effects due to air resistance. Assume the brick landed at the same height it was launched.

POSITION, DISPLACEMENT, VELOCITY, AND ACCELERATION VECTORS

38 • A wall clock has a minute hand with a length of 0.50 m and an hour hand with a length of 0.25 m. Take the center of the clock as the origin, and use a Cartesian coordinate system with the positive x axis pointing to 3 o'clock and the positive y axis pointing to 12 o'clock. Using unit vectors $\hat{i}$ and $\hat{j}$, express the position vectors of the tip of the hour hand ($\vec{A}$) and the tip of the minute hand ($\vec{B}$) when the clock reads (*a*) 12:00, (*b*) 3:00, (*c*) 6:00, (*d*) 9:00.

39 • In Problem 38, find the displacements of the tip of each hand (that is, $\Delta\vec{A}$ and $\Delta\vec{B}$) when the time advances from 3:00 p.m. to 6:00 p.m. SSM

40 • In Problem 38, write the vector that describes the displacement of a fly if it quickly goes from the tip of the minute hand to the tip of the hour hand at 3:00 p.m.

41 • CONCEPTUAL, APPROXIMATION A bear, awakening from winter hibernation, staggers directly northeast for 12 m and then due east for 12 m. Show each displacement graphically and graphically determine the single displacement that will take the bear back to her cave to continue her hibernation.

42 • A scout walks 2.4 km due east from camp, then turns left and walks 2.4 km along the arc of a circle centered at the campsite, and finally walks 1.5 km directly toward the camp. (*a*) How far is the scout from camp at the end of his walk? (*b*) In what direction is the scout's position relative to the campsite? (*c*) What is the ratio of the final magnitude of the displacement to the total distance walked.

43 • The faces of a cubical storage cabinet in your garage have 3.0-m-long edges that are parallel to the xyz coordinate planes. The cube has one corner at the origin. A cockroach, on the hunt for crumbs of food, begins at that corner and walks along three edges until it is at the far corner. (*a*) Write the roach's displacement using the set of $\hat{i}$, $\hat{j}$, and $\hat{k}$ unit vectors, and (*b*) find the magnitude of its displacement. SSM

44 • **CONTEXT-RICH** You are the navigator of a ship at sea. You receive radio signals from two transmitters A and B, which are 100 km apart, one due south of the other. The direction finder shows you that transmitter A is at a heading of 30° south of east from the ship, while transmitter B is due east. Calculate the distance between your ship and transmitter B.

45 • A stationary radar operator determines that a ship is 10 km due south of him. An hour later the same ship is 20 km due southeast. If the ship moved at constant speed and always in the same direction, what was its velocity during this time?

46 • A particle's position coordinates (x, y) are (2.0 m, 3.0 m) at $t = 0$; (6.0 m, 7.0 m) at $t = 2.0$ s; and (13 m, 14 m) at $t = 5.0$ s. (*a*) Find the magnitude of the average velocity from $t = 0$ to $t = 2.0$ s. (*b*) Find the magnitude of the average velocity from $t = 0$ to $t = 5.0$ s.

47 • A particle moving at a velocity of 4.0 m/s in the $+x$ direction is given an acceleration of 3.0 m/s^2 in the $+y$ direction for 2.0 s. Find the final speed of the particle. SSM

48 • Initially, a swift-moving hawk is moving due west with a speed of 30 m/s; 5.0 s later it is moving due north with a speed of 20 m/s. (*a*) What are the magnitude and direction of $\Delta\vec{v}_{av}$ during this 5.0-s interval? (*b*) What are the magnitude and direction of $\vec{a}_{av}$ during this 5.0-s interval?

49 • At $t = 0$, a particle located at the origin has a velocity of 40 m/s at $\theta = 45°$. At $t = 3.0$ s, the particle is at $x = 100$ m and $y = 80$ m and has a velocity of 30 m/s at $\theta = 50°$. Calculate (*a*) the average velocity, and (*b*) the average acceleration of the particle during this 3.0-s interval.

50 •• At time zero, a particle is at $x = 4.0$ m and $y = 3.0$ m and has velocity $\vec{v} = (2.0\text{ m/s})\hat{i} + (-9.0\text{ m/s})\hat{j}$. The acceleration of the particle is constant and is given by $\vec{a} = (4.0\text{ m/s}^2)\hat{i} + (3.0\text{ m/s}^2)\hat{j}$. (*a*) Find the velocity at $t = 2.0$ s. (*b*) Express the position vector at $t = 4.0$ s in terms of $\hat{i}$ and $\hat{j}$. In addition, give the magnitude and direction of the position vector at this time.

51 •• A particle has a position vector given by $\vec{r} = (30t)\hat{i} + (40t - 5t^2)\hat{j}$, where r is in meters and t is in seconds. Find the instantaneous-velocity and instantaneous-acceleration vectors as functions of time t. SSM

52 •• A particle has a constant acceleration of $\vec{a} = (6.0\text{ m/s}^2)\hat{i} + (4.0\text{ m/s}^2)\hat{j}$. At time $t = 0$, the velocity is zero and the position vector is $\vec{r}_0 = (10\text{ m})\hat{i}$. (*a*) Find the velocity and position vectors as functions of time t. (*b*) Find the equation of the particle's path in the xy plane and sketch the path.

53 •• Starting from rest at a dock, a motor boat on a lake heads north while gaining speed at a constant 3.0 m/s^2 for 20 s. The boat then heads west and continues for 10 s at the speed that it had at 20 s. (*a*) What is the average velocity of the boat during the 30-s trip? (*b*) What is the average acceleration of the boat during the 30-s trip? (*c*) What is the displacement of the boat during the 30-s trip? SSM

54 •• Starting from rest at point A, you ride your motorcycle north to point B 75.0 m away, increasing speed at a steady rate of 2.00 m/s^2. You then gradually turn toward the east along a circular path of radius 50.0 m at constant speed from B to point C, until your direction of motion is due east at C. You then continue eastward, slowing at a steady rate of 1.00 m/s^2 until you come to rest at point D. (*a*) What is your average velocity and acceleration for the trip from A to D? (*b*) What is your displacement during your trip from A to C? (*c*) What distance did you travel for the entire trip from A to D?

RELATIVE VELOCITY

55 •• A plane flies at an airspeed of 250 km/h. A wind is blowing at 80 km/h toward the direction 60° east of north. (*a*) In what direction should the plane head in order to fly due north relative to the ground? (*b*) What is the speed of the plane relative to the ground?

56 •• A swimmer *heads* directly across a river, swimming at 1.6 m/s relative to the water. She arrives at a point 40 m downstream from the point directly across the river, which is 80 m wide. (*a*) What is the speed of the river current? (*b*) What is the swimmer's speed relative to the shore? (*c*) In what direction should the swimmer head in order to arrive at the point directly opposite her starting point?

57 •• A small plane departs from point A heading for an airport 520 km due north at point B. The airspeed of the plane is 240 km/h and there is a steady wind of 50 km/h blowing directly toward the southeast. Determine the proper heading for the plane and the time of flight. SSM

58 •• Two boat landings are 2.0 km apart on the same bank of a stream that flows at 1.4 km/h. A motorboat makes the round trip between the two landings in 50 min. What is the speed of the boat relative to the water?

59 •• **ENGINEERING APPLICATION, CONTEXT-RICH** During a radio-controlled model-airplane competition, each plane must fly from the center of a 1.0-km-radius circle to any point on the circle and back to the center. The winner is the plane that has the shortest round-trip time. The contestants are free to fly their planes along any route as long as the plane begins at the center, travels to the circle, and then returns to the center. On the day of the race, a steady wind blows out of the north at 5.0 m/s. Your plane can maintain an air speed of 15 m/s. Should you fly your plane upwind on the first leg and downwind on the trip back, or across the wind, flying east and then west? Optimize your chances by calculating the round-trip time for both routes using your knowledge of vectors and relative velocities. With this prerace calculation, you can determine the best route and have a major advantage over the competition!

60 •• **CONTEXT-RICH** You are piloting a small plane that can maintain an air speed of 150 kt (knots, or nautical miles per hour) and you want to fly due north (azimuth = 000°) relative to the ground. (*a*) If a wind of 30 kt is blowing from the east (azimuth = 090°), calculate the heading (azimuth) you must ask your copilot to maintain. (*b*) At that heading, what will be your ground speed?

61 •• Car A is traveling east at 20 m/s toward an intersection. As car A crosses the intersection, car B starts from rest 40 m north of the intersection and moves south steadily gaining speed at 2.0 m/s^2. Six seconds after A crosses the intersection find (*a*) the position of B relative to A, (*b*) the velocity of B relative to A, (*c*) the acceleration of B relative to A. *Hint: Let the unit vectors $\hat{i}$ and $\hat{j}$ be toward the east and north, respectively, and express your answers using $\hat{i}$ and $\hat{j}$.* SSM

62 •• While walking between gates at an airport, you notice a child running along a moving walkway. Estimating that the child runs at a constant speed of 2.5 m/s relative to the surface of the walkway, you decide to try to determine the speed of the walkway itself. You watch the child run on the entire 21-m walkway in one direction, immediately turn around, and run back to his starting point. The entire trip takes a total elapsed time of 22 s. Given this information, what is the speed of the moving walkway relative to the airport terminal?

63 •• Ben and Jack are shopping in a department store. Ben leaves Jack at the bottom of the escalator and walks east at a speed of 2.4 m/s. Jack gets on the escalator, which is inclined at an angle of 37° above the horizontal, and travels eastward and upward at a speed of 2.0 m/s. (*a*) What is the velocity of Ben relative to Jack? (*b*) At what speed should Jack walk up the escalator so that he is always directly above Ben (until he reaches the top)? SSM

64 ••• A juggler traveling in a train on level track throws a ball straight up, relative to the train, with a speed of 4.90 m/s. The train has a velocity of 20.0 m/s due east. As observed by the juggler, (*a*) what is the ball's total time of flight, and (*b*) what is the displacement of the ball during its rise? According to a friend standing on the ground next to the tracks, (*c*) what is the ball's initial speed, (*d*) what is the angle of the launch, and (*e*) what is the displacement of the ball during its rise?

CIRCULAR MOTION AND CENTRIPETAL ACCELERATION

65 • What is the magnitude of the acceleration of the tip of the minute hand of the clock in Problem 38? Express it as a fraction of the magnitude of free-fall acceleration g.

66 • **CONTEXT-RICH** You are designing a centrifuge to spin at a rate of 15,000 rev/min. (*a*) Calculate the maximum centripetal acceleration that a test-tube sample held in the centrifuge arm 15 cm from the rotation axis must withstand. (*b*) It takes 1 min, 15 s for the centrifuge to spin up to its maximum rate of revolution from rest. Calculate the *magnitude* of the tangential acceleration of the centrifuge while it is spinning up, assuming that the tangential acceleration is constant.

67 ••• Earth rotates on its axis once every 24 hours, so that objects on its surface execute uniform circular motion about the axis with a period of 24 hours. Consider only the effect of this rotation on the person on the surface. (Ignore Earth's orbital motion about the Sun.) (*a*) What is the speed and what is the magnitude of the acceleration of a person standing on the equator? (Express the magnitude of this acceleration as a percentage of g.) (*b*) What is the direction of the acceleration vector? (*c*) What is the speed and what is the magnitude of the acceleration of a person standing on the surface at 35°N latitude? (*d*) What is the angle between the direction of the acceleration of the person at 35°N and the direction of the acceleration of the person at the equator if both persons are at the same longitude? SSM

68 •• Determine the acceleration of the moon toward Earth, using values for its mean distance and orbital period from the Terrestrial and Astronomical Data table in this book. Assume a circular orbit. Express the acceleration as a fraction of g.

69 •• (*a*) What are the period and speed of a person on a carousel if the person has an acceleration with a magnitude of 0.80 m/s^2 when she is standing 4.0 m from the axis? (*b*) What are her acceleration magnitude and speed if she then moves to a distance of 2.0 m from the carousel center and the carousel keeps rotating with the same period?

70 ••• Pulsars are neutron stars that emit X rays and other radiation in such a way that we on Earth receive pulses of radiation from the pulsars at regular intervals equal to the period that they rotate. Some of these pulsars rotate with periods as short as 1 ms! The Crab Pulsar, located inside the Crab Nebula in the constellation Orion, has a period currently of length 33.085 ms. It is estimated to have an equatorial radius of 15 km, an average radius for a neutron star. (*a*) What is the value of the centripetal acceleration of an object on the surface at the equator of the pulsar? (*b*) Many pulsars are observed to have periods that lengthen slightly with time, a phenomenon called "spin down." The rate of slowing of the Crab Pulsar is 3.5×10^{-13} s per second, which implies that if this rate remains constant, the Crab Pulsar will stop spinning in 9.5×10^{10} s (about 3000 years from today). What is the tangential acceleration of an object on the equator of this neutron star?

71 ••• **BIOLOGICAL APPLICATION** Human blood contains plasma, platelets, and blood cells. To separate the plasma from other components, centrifugation is used. Effective centrifugation requires subjecting blood to an acceleration of 2000g or more. In this situation, assume that blood is contained in test tubes that are 15 cm long and are full of blood. These tubes ride in the centrifuge tilted at an angle of 45.0° above the horizontal (Figure 3-34). (*a*) What is the distance of a sample of blood from the rotation axis of a centrifuge rotating at 3500 rpm, if it has an acceleration of 2000g? (*b*) If the blood at the center of the tubes revolves around the rotation axis at the radius calculated in Part (*a*), calculate the accelerations experienced by the blood at each end of the test tube. Express all accelerations as multiples of g.

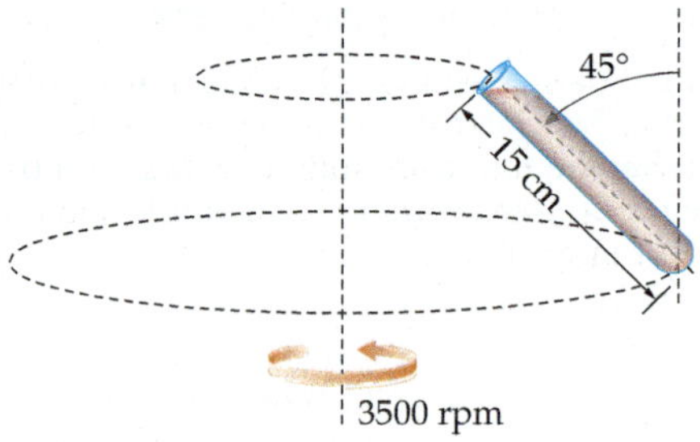

FIGURE 3-34 Problem 71

PROJECTILE MOTION AND PROJECTILE RANGE

72 • While trying out for the position of pitcher on your high school baseball team, you throw a fastball at 87 mi/h toward home plate, which is 18.4 m away. How far does the ball drop due to effects of gravity by the time it reaches home plate? (Ignore any effects due to air resistance.)

73 • A projectile is launched with speed v_0 at an angle of θ_0 above the horizontal. Find an expression for the maximum height it reaches above its starting point in terms of v_0, θ_0, and g. (Ignore any effects due to air resistance.)

74 •• A cannonball is fired with initial speed v_0 at an angle 30° above the horizontal from a height of 40 m above the ground. The projectile strikes the ground with a speed of $1.2v_0$. Find v_0. (Ignore any effects due to air resistance.)

75 •• In Figure 3-35, what is the minimum initial speed of the dart if it is to hit the monkey before the monkey hits the ground, which is 11.2 m below the initial position of the monkey, if x = 50 m and h = 10 m? (Ignore any effects due to air resistance.)

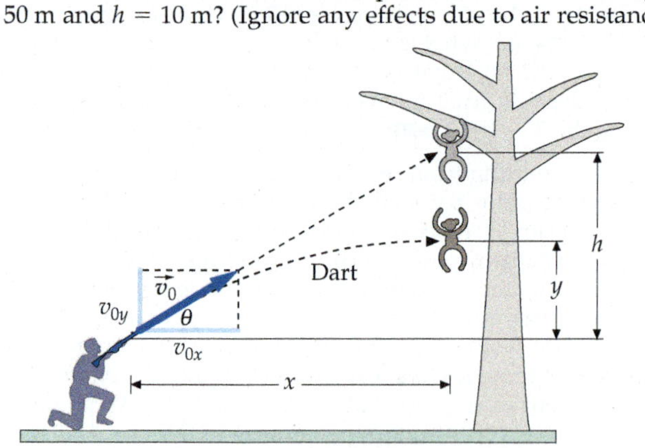

FIGURE 3-35 Problem 75

76 •• A projectile is launched from ground level with an initial speed of 53 m/s. Find the launch angle (the angle the initial velocity vector is above the horizontal) such that the maximum height of the projectile is equal to its horizontal range. (Ignore any effects due to air resistance.)

77 •• A ball launched from ground level lands 2.44 s later on a level field 40.0 m away from the launch point. Find the magnitude of the initial velocity vector and the angle it is above the horizontal. (Ignore any effects due to air resistance.) SSM

78 •• Consider a ball that is launched from ground level with initial speed v_0 at an angle θ_0 above the horizontal. If we consider its speed to be v at some height h above the ground, show that for a given value of h, v is independent of θ_0. (Ignore any effects due to air resistance.)

79 ••• At $\frac{1}{2}$ of its maximum height, the speed of a projectile is $\frac{3}{4}$ of its initial speed. What was its launch angle? (Ignore any effects due to air resistance.)

80 •• A cargo plane is flying horizontally at an altitude of 12 km with a speed of 900 km/h when a large crate falls out of the rear loading ramp. (Ignore any effects due to air resistance.) (*a*) How long does it take the crate to hit the ground? (*b*) How far horizontally is the crate from the point where it fell off when it hits the ground? (*c*) How far is the crate from the aircraft when the crate hits the ground, assuming that the plane continues to fly with the same velocity?

81 •• Wile E. Coyote (*Carnivorus hungribilus*) is chasing the Roadrunner (*Speedibus cantcatchmi*) yet again. While running down the road, they come to a deep gorge, 15.0 m straight across and 100 m deep. The Roadrunner launches himself across the gorge at a launch angle of 15° above the horizontal, and lands with 1.5 m to spare. (*a*) What was the Roadrunner's launch speed? (*b*) Wile E. Coyote launches himself across the gorge with the same initial speed, but at a different launch angle. To his horror, he is short of the other lip by 0.50 m. What was his launch angle? (Assume that it was less than 15°.) SSM

82 •• A cannon barrel is elevated 45° above the horizontal. It fires a ball with a speed of 300 m/s. (*a*) What height does the ball reach? (*b*) How long is the ball in the air? (*c*) What is the horizontal range of the cannon ball? (Ignore any effects due to air resistance.)

83 •• A stone thrown horizontally from the top of a 24-m tower hits the ground at a point 18 m from the base of the tower. (Ignore any effects due to air resistance.) (*a*) Find the speed with which the stone was thrown. (*b*) Find the speed of the stone just before it hits the ground. SSM

84 •• A projectile is fired into the air from the top of a 200-m cliff above a valley (Figure 3-36). Its initial velocity is 60 m/s at 60° above the horizontal. Where does the projectile land? (Ignore any effects due to air resistance.)

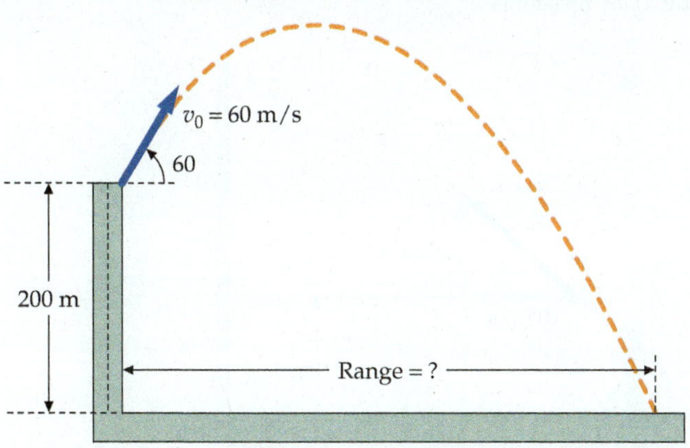

FIGURE 3-36 Problem 84

85 •• The range of a cannonball fired horizontally from a cliff is equal to the height of the cliff. What is the direction of the velocity vector of the projectile as it strikes the ground? (Ignore any effects due to air resistance.)

86 •• An archer fish launches a droplet of water from the surface of a small lake at an angle of 60° above the horizontal. He is aiming at a juicy spider sitting on a leaf 50 cm to the east and on a branch 25 cm above the water surface. The fish is trying to knock the spider into the water so that the fish can eat the spider. (*a*) What must the speed of the water droplet be for the fish to be successful? (*b*) When it hits the spider, is the droplet rising or falling?

87 •• **CONTEXT-RICH** You are trying out for the position of place-kicker on a professional football team. With the ball teed 50.0 m from the goalposts with a crossbar 3.05 m off the ground, you kick the ball at 25.0 m/s and 30° above the horizontal. (*a*) Is the field goal attempt good? (*b*) If so, by how much does it clear the bar? If not, by how much does it go under the bar? (*c*) How far behind the plane of the goalposts does the ball land? SSM

88 •• The speed of an arrow fired from a compound bow is about 45.0 m/s. (*a*) A Tartar archer sits astride his horse and launches an arrow into the air, elevating the bow at an angle of 10° above the horizontal. If the arrow is 2.25 m above the ground at launch, what is the arrow's horizontal range? Assume that the ground is level, and ignore any effects due to air resistance. (*b*) Now assume that his horse is at full gallop, moving in the same direction that the archer will fire the arrow. Also assume that the archer elevates the bow at the same elevation angle as in Part (*a*) and fires. If the horse's speed is 12.0 m/s, what is the arrow's horizontal range now?

89 The roof of a two-story house makes an angle of 30° with the horizontal. A ball rolling down the roof rolls off the edge at a speed of 5.0 m/s. The distance to the ground from that point is 7.0 m. (*a*) How long is the ball in the air? (*b*) How far from the base of the house does it land? (*c*) What is its speed and direction just before landing?

90 •• Compute $dR/d\theta_0$ from $R = (v_0^2/g)\sin(2\theta_0)$ and show that setting $dR/d\theta_0 = 0$ gives $\theta_0 = 45°$ for the maximum range.

91 •• In a science fiction short story written in the 1970s, Ben Bova described a conflict between two hypothetical colonies on the moon—one founded by the United States and the other by the USSR. In the story, colonists from each side started firing bullets at each other, only to find to their horror that their rifles had large enough muzzle velocities so that the bullets went into orbit. (*a*) If the magnitude of free-fall acceleration on the moon is 1.67 m/s^2, what is the maximum range of a rifle bullet with a muzzle velocity of 900 m/s? (Assume the curvature of the surface of the moon is negligible.) (*b*) What would the muzzle velocity have to be to send the bullet into a circular orbit just above the surface of the moon?

92 ••• On a level surface, a ball is launched from ground level at an angle of 55° above the horizontal, with an initial speed of 22 m/s. It lands on a hard surface, and bounces, reaching a peak height of 75% of the height it reached on its first arc. (Ignore any effects due to air resistance.) (*a*) What is the maximum height reached in its first parabolic arc? (*b*) How far horizontally from the launch point did it strike the ground the first time? (*c*) How far horizontally from the launch point did the ball strike the ground the second time? Assume the horizontal component of the velocity remains constant during the collision of the ball with the ground. *Hint: You cannot assume that the angle with which the ball leaves the ground after the first collision is the same as the initial launch angle.*

93 ••• In the text, we calculated the range for a projectile that lands at the same elevation from which it is fired as $R = (v_0^2/g)\sin 2\theta_0$. A golf ball hit from an elevated tee at 45.0 m/s and an angle of 35.0° lands on a green 20.0 m below the tee (Figure 3-37). (Ignore any effects due to air resistance.) (*a*) Calculate the range using the equation $R = (v_0^2/g)\sin 2\theta_0$ even though the ball is hit from an elevated tee. (*b*) Show that the range for the more general problem (Figure 3-37) is given by $R = \left(1 + \sqrt{1 - \dfrac{2gy}{v_0^2 \sin^2\theta_0}}\right)\dfrac{v_0^2}{2g}\sin 2\theta_0$, where y is the height of the green above the tee. That is, $y = -h$. (*c*) Compute the range using this formula. What is the percentage error in neglecting the elevation of the green?

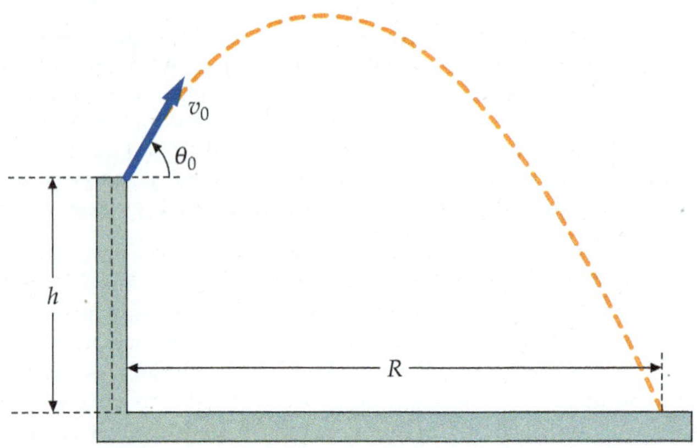

FIGURE 3-37 Problem 93

94 ••• **MULTISTEP** In the text, we calculated the range for a projectile that lands at the same elevation from which it is fired as $R = (v_0^2/g)\sin 2\theta_0$ if the effects of the air resistance are negligible. (*a*) Show that for the same conditions the change in the range for a small change in free-fall acceleration g, and the same initial speed and angle, is given by $\Delta R/R = -\Delta g/g$. (*b*) What would be the length of a homerun at a high altitude where g is 0.50% less than at sea level if the homerun at sea level traveled 400 ft?

95 ••• **MULTISTEP, APPROXIMATION** In the text, we calculated the range for a projectile that lands at the same elevation from which it is fired as $R = (v_0^2/g)\sin 2\theta_0$ if the effects of the air resistance are negligible. (*a*) Show that for the same conditions the change in the range for a small change in launch speed, and the same initial angle and free-fall acceleration, is given by $\Delta R/R = 2\Delta v_0/v_0$. (*b*) Suppose a projectile's range was 200 m. Use the formula in Part (*a*) to estimate its increase in range if the launch speed were increased by 20.0%. (*c*) Compare your answer in (*b*) to the increase in range by calculating the increase in range directly from $R = (v_0^2/g)\sin 2\theta_0$. If the results for Parts (*b*) and (*c*) are different, is the estimate too small or large, and why? SSM

96 ••• A projectile, fired with unknown initial velocity, lands 20.0 s later on the side of a hill, 3000 m away horizontally and 450 m vertically above its starting point. (Ignore any effects due to air resistance.) (*a*) What is the vertical component of its initial velocity? (*b*) What is the horizontal component of its initial velocity? (*c*) What was its maximum height above its launch point? (*d*) As it hit the hill, what speed did it have and what angle did its velocity make with the vertical?

97 ••• A projectile is launched over level ground at an initial elevation angle of θ. An observer standing at the launch site sees the projectile at the point of its highest elevation and measures the angle ϕ shown in Figure 3-38. Show that $\tan\phi = \frac{1}{2}\tan\theta$. (Ignore any effects due to air resistance.) SSM

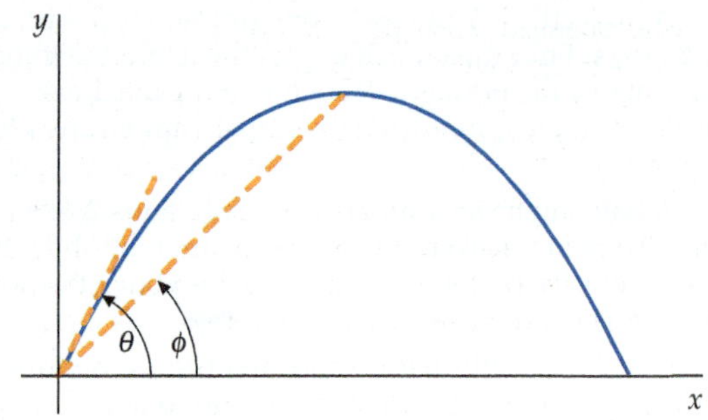

FIGURE 3-38 Problem 97

98 ••• A toy cannon is placed on a ramp that has a slope of angle ϕ. (*a*) If the cannonball is projected up the hill at an angle of θ_0 above the horizontal (Figure 3-39) and has a muzzle speed of v_0, show that the range R of the cannonball (as measured along the ramp) is given by

$$R = \frac{2v_0^2\cos^2\theta_0(\tan\theta_0 - \tan\phi)}{g\cos\phi}$$

Ignore any effects due to air resistance.

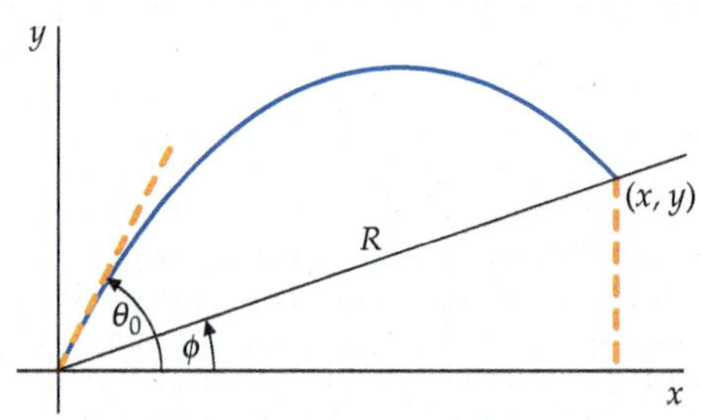

FIGURE 3-39 Problem 98

99 ••• A rock is thrown from the top of a 20-m-high building at an angle of 53° above the horizontal. (*a*) If the horizontal range of the throw is equal to the height of the building, with what speed was the rock thrown? (*b*) How long is it in the air? (*c*) What is the velocity of the rock just before it strikes the ground? (Ignore any effects due to air resistance.)

100 ••• A woman throws a ball at a vertical wall 4.0 m away (Figure 3-40). The ball is 2.0 m above ground when it leaves the woman's hand with an initial velocity of 14 m/s at 45°, as shown. When the ball hits the wall, the horizontal component of its velocity is reversed; the vertical component remains unchanged. (*a*) Where does the ball hit the ground? (*b*) How long was the ball in the air before it hit the wall? (*c*) Where did the ball hit the wall? (*d*) How long was the ball in the air after it left the wall? Ignore any effects due to air resistance.

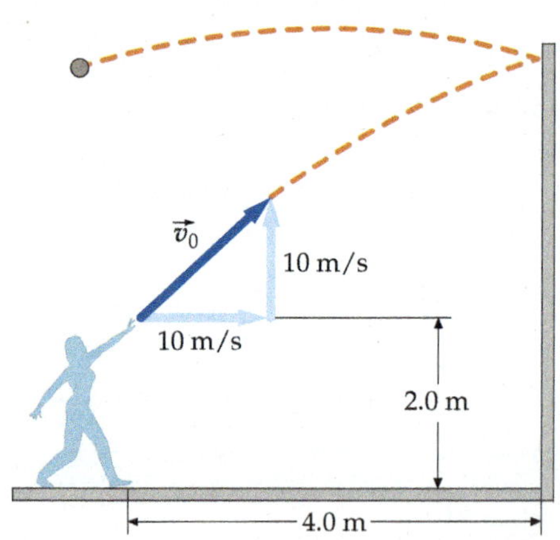

FIGURE 3-40 Problem 100

101 ••• **ENGINEERING APPLICATION** Catapults date from thousands of years ago, and were used historically to launch everything from stones to horses. During a battle in what is now Bavaria, inventive artillerymen from the united German clans launched giant *spaetzle* from their catapults toward a Roman fortification whose walls were 8.50 m high. The catapults launched the *spaetzle* projectiles from a height of 4.00 m above the ground and a distance of 38.0 m from the walls of the fortification at an angle of 60.0 degrees above the horizontal (Figure 3-41). If the projectiles were to hit the top of the wall, splattering the Roman soldiers atop the wall with pulverized pasta, (*a*) what launch speed was necessary? (*b*) How long were the *spaetzle* in the air? (*c*) At what speed did the projectiles hit the wall? Ignore any effects due to air resistance.

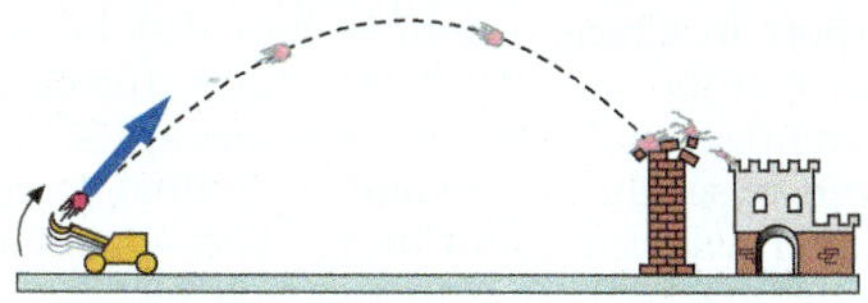

FIGURE 3-41 Problem 101

102 ••• The distance from the pitcher's mound to home plate is 18.4 m. The mound is 0.20 m above the level of the field. A pitcher throws a fastball with an initial speed of 37.5 m/s. At the moment the ball leaves the pitcher's hand, it is 2.30 m above the mound. (*a*) What should the angle between $\vec{v}_0$ and the horizontal be so that the ball crosses the plate 0.70 m above ground? (Ignore any effects due to air resistance.) (*b*) With what speed does the ball cross the plate? SSM

103 ••• You are watching your friend play hockey. During the course of the game, he strikes the puck in such a way that, when it is at its highest point, it just clears the surrounding 2.80-m-high Plexiglas wall that is 12.0 m away. Find (*a*) the vertical component of its initial velocity, (*b*) the time it takes to reach the wall, and (*c*) the horizontal component of its initial velocity, and its initial speed and angle. (Ignore any effects due to air resistance.)

104 ••• Carlos is on his trail bike, approaching a creek bed that is 7.0 m wide. A ramp with an incline of 10° has been built for daring people who try to jump the creek. Carlos is traveling at his bike's maximum speed, 40 km/h. (*a*) Should Carlos attempt the jump or brake hard? (*b*) What is the minimum speed a bike must have to make this jump? Assume equal elevations on either side of the creek. (Ignore any effects due to air resistance.)

105 ••• If a bullet that leaves the muzzle of a gun at 250 m/s is to hit a target 100 m away at the level of the muzzle (1.7 m above the level ground), the gun must be aimed at a point above the target. (*a*) How far above the target is that point? (*b*) How far behind the target will the bullet strike the ground? (Ignore any effects due to air resistance.) SSM

GENERAL PROBLEMS

106 •• During a do-it-yourself roof repair project, you are on the roof of your house and accidentally drop your hammer. The hammer then slides down the roof at constant speed of 4.0 m/s. The roof makes an angle of 30° with the horizontal, and its lowest point is 10 m from the ground. (*a*) How long after leaving the roof does the hammer hit the ground? (*b*) What is the horizontal distance traveled by the hammer between the instant it leaves the roof and the instant it hits the ground? (Ignore any effects due to air resistance.)

107 •• A squash ball typically rebounds from a surface with 25% of the speed with which it initially struck the surface. Suppose a squash ball is served in a shallow trajectory, from a height above the ground of 45 cm, at a launch angle of 6.0° degrees above the horizontal, and at a distance of 12 m from the front wall. (*a*) If it strikes the front wall exactly at the top of its parabolic trajectory, determine how high above the floor the ball strikes the wall. (*b*) How far horizontally from the wall does it strike the floor, after rebounding? (Ignore any effects due to air resistance.)

108 •• A football quarterback throws a pass at an angle of 36.5°. He releases the pass 3.50 m behind the line of scrimmage. His receiver left the line of scrimmage 2.50 s earlier, goes straight downfield at a constant speed of 7.50 m/s. In order that the pass land gently in the receiver's hands without the receiver breaking stride, with what speed must the quarterback throw the pass? Assume that the ball is released at the same height it is caught and that the receiver is straight downfield from the quarterback at the time of release. (Ignore any effects due to air resistance.)

109 •• Suppose a test pilot is able to safely withstand an acceleration of up to 5.0 times the acceleration due to gravity (that is, remain conscious and alert enough to fly). During the course of maneuvers, he is required to fly the plane in a horizontal circle at its top speed of 1900 mi/h. (*a*) What is the radius of the smallest circle he will be able to safely fly the plane in? (*b*) How long does it take him to go halfway around this minimum-radius circle?

110 •• A particle moves in the xy plane with constant acceleration. At $t = 0$ the particle is at $\vec{r}_1 = (4.0\text{ m})\hat{i} + (3.0\text{ m})\hat{j}$, with velocity $\vec{v}_1$. At $t = 2.0$ s, the particle has moved to $\vec{r}_2 = (10\text{ m})\hat{i} - (2.0\text{ m})\hat{j}$ and its velocity has changed to $\vec{v}_2 = (5.0\text{ m/s})\hat{i} - (6.0\text{ m/s})\hat{j}$. (*a*) Find $\vec{v}_1$. (*b*) What is the acceleration of the particle? (*c*) What is the velocity of the particle as a function of time? (*d*) What is the position vector of the particle as a function of time?

111 •• Plane A is flying due east at an air speed of 400 mph. Directly below, at a distance of 4000 ft, plane B is headed due north, flying at an air speed of 700 mph. Find the velocity vector of plane B relative to A. SSM

112 •• A diver steps off the cliffs at Acapulco, Mexico, 30.0 m above the surface of the water. At that moment, he activates his rocket-powered backpack horizontally, which gives him a constant horizontal acceleration of 5.00 m/s^2 but does not affect his vertical motion. (*a*) How long does he take to reach the surface of the water? (*b*) How far out from the base of the cliff does he enter the water, assuming the cliff is vertical? (*c*) Show that his flight path is a straight line. (Ignore any effects of air resistance.)

113 •• A small steel ball is projected horizontally off the top of a long flight of stairs. The initial speed of the ball is 3.0 m/s. Each step is 0.18 m high and 0.30 m wide. Which step does the ball strike first? SSM

114 •• Suppose you can throw a ball a maximum horizontal distance L when standing on level ground. How far can you throw it from the top of a building of height h if you throw it at (*a*) 0°, (*b*) 30°, (*c*) 45°? (Ignore any effects due to air resistance.)

115 ••• Darlene is a stunt motorcyclist in a traveling circus. For the climax of her show, she takes off from the ramp at angle θ_0, clears a ditch of width L, and lands on an elevated ramp (height h) on the other side (Figure 3-42). (*a*) For a given height h, find the minimum necessary takeoff speed v_{min} needed to make the jump successfully. (*b*) What is v_{min} for $L = 8.0$ m, $\theta = 30°$, and $h = 4.0$ m?

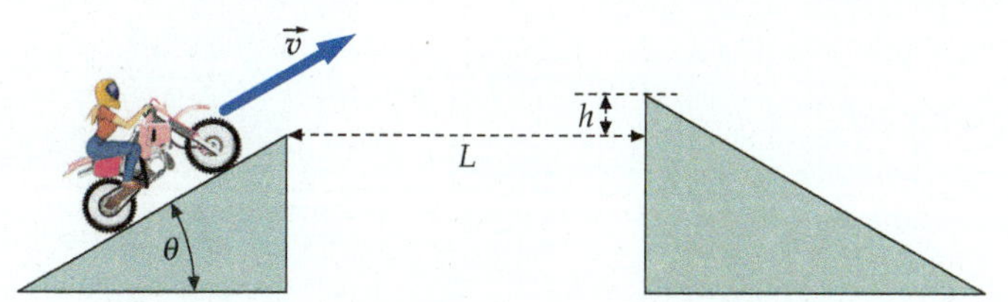

FIGURE 3-42 Problem 115

(*c*) Show that regardless of her takeoff speed, the maximum height of the platform is $h < L \tan \theta_0$. Interpret this result physically. (Neglect any effects due to air resistance and treat the rider and the bike as if they were a single particle.)

116 ••• A small boat is headed for a harbor 32 km directly northwest of its current position when it is suddenly engulfed in heavy fog. The captain maintains a compass bearing of northwest and a speed of 10 km/h relative to the water. The fog lifts 3.0 h later and the captain notes that he is now exactly 4.0 km south of the harbor. (*a*) What was the average velocity of the current during those 3.0 h? (*b*) In what direction should the boat have been heading to reach its destination along a straight course? (*c*) What would its travel time have been if it had followed a straight course?

117 •• Galileo showed that, if any effects due to air resistance are ignored, the ranges for projectiles (on a level field) whose angles of projection exceed or fall short of 45° by the same amount are equal. Prove Galileo's result. SSM

118 •• Two balls are thrown with equal speeds from the top of a cliff of height h. One ball is thrown at an angle of α above the horizontal. The other ball is thrown at an angle of β below the horizontal. Show that each ball strikes the ground with the same speed, and find that speed in terms of h and the initial speed v_0. (Ignore any effects due to air resistance.)

119 •• In his car, a driver tosses an egg vertically from chest height so that the peak of its path is just below the ceiling of the passenger compartment, which is 65 cm above his release point. He catches the egg at the same height that he releases it. If you are a roadside observer and measure the horizontal distance between catch and release points to be 19 m, (*a*) how fast is the car moving? (*b*) In your reference frame, at what angle above the horizontal was the egg thrown?

120 ••• A straight line is drawn on the surface of a 16-cm-radius turntable from the center to the perimeter. A bug crawls along this line from the center outward as the turntable spins counterclockwise at a constant 45 rpm. Its walking speed relative to the turntable is a steady 3.5 cm/s. Let its initial heading be in the positive x direction. As the bug reaches the edge of the turntable (still traveling at 3.5 cm/s radially, relative to the turntable), what are the x and y components of the velocity of the bug?

121 ••• On a windless day, a stunt pilot is flying his vintage World War I Sopwith Camel from Dubuque, Iowa, to Chicago, Illinois, for an air show. Unfortunately, he is unaware that his plane's ancient magnetic compass has a serious problem in that what it records as "north" is in fact 16.5° east of true north. At one moment during his flight, the airport in Chicago notifies him that he is, in reality, 150 km due west of the airport. He then turns due east, according to his plane's compass, and flies for 45 minutes at 150 km/h. At that point, he expects to see the airport and begin final descent. What is the plane's actual distance from Chicago and what should be the pilot's heading if he is to fly directly toward Chicago?

122 ••• **ENGINEERING APPLICATION, CONTEXT-RICH** A cargo plane in flight lost a package because somebody forgot to close the rear cargo doors. You are on the team of safety experts trying to analyze what happened. From the point of takeoff, while climbing to altitude, the airplane traveled in a straight line and at a constant speed of 275 mi/h at an angle of 37° above the horizontal. During the ascent, the package slid off the back ramp. You found the package in a field a distance of 7.5 km from the takeoff point. To complete the investigation you need to know exactly how long after takeoff the package slid off the back ramp of the plane. (Consider the sliding speed to be negligible.) Calculate the time at which the package fell off the back ramp. (Ignore any effects due to air resistance.)

CHAPTER 4

THIS AIRPLANE IS ACCELERATING AS IT HEADS DOWN THE RUNWAY BEFORE TAKEOFF. NEWTON'S LAWS RELATE AN OBJECT'S ACCELERATION TO ITS MASS AND THE FORCES ACTING ON IT. *(John Neubauer/FPG/Getty.)*

? If you were a passenger on this plane, how might you use Newton's laws to determine the plane's acceleration? (See Example 4-9.)

Newton's Laws

Now that we have studied how objects move in one, two, and three dimensions, we can ask the questions "Why do objects start to move?" and "What causes a moving object to change speed or change direction?"

These questions occupied the mind of Sir Isaac Newton, who was born in 1642, the year Galileo died. As a student at Cambridge, where he was later a mathematics professor, Newton studied the work of Galileo and Kepler. He wanted to figure out why planets move in ellipses at speeds dependent on their distance to the Sun, and even why the solar system stays together at all. During his lifetime, he developed both his law of gravitation, which we will examine in Chapter 12, and his three basic laws of motion that form the basis of classical mechanics.

Newton's laws relate the forces objects exert on each other, and relate any change in the motion of an object to the forces that act on it. Newton's laws of motion are the tools that enable us to analyze a wide range of mechanical phenomena. Although we may already have an intuitive idea of a force as a push or a pull, like that exerted by our muscles or by stretched rubber bands and springs, Newton's laws allow us to refine our understanding of forces.

In this chapter, we describe Newton's three laws of motion and begin using them to solve problems involving objects in motion and at rest.

4-1 NEWTON'S FIRST LAW: THE LAW OF INERTIA

Give a shove to a piece of ice on a counter top: It slides, and then slows to a stop. If the counter is wet, the ice will travel farther before stopping. A piece of dry ice (frozen carbon dioxide) riding on a cushion of carbon dioxide vapor slides quite far with little change in velocity. Before Galileo, it was thought that a steady force, such as a push or pull, was always needed to keep an object moving with constant velocity. But Galileo, and later Newton, recognized that the slowing of objects in everyday experience is due to the force of friction. If friction is reduced, the rate of slowing is reduced. A water slick or a cushion of gas is especially effective at reducing friction, allowing the object to slide a great distance with little change in velocity. Galileo reasoned that, if we could remove all external forces including friction from an object, then the velocity of the object would never change—a property of matter he described as **inertia.** This conclusion, restated by Newton as his first law, is also called the **law of inertia.**

Friction is greatly reduced by a cushion of air that supports the hovercraft. *(Jose Dupont/Explorer/Photo Researchers.)*

A modern wording of Newton's first law is

> **First law.** An object at rest stays at rest *unless* acted on by an external force. An object in motion continues to travel with constant speed in a straight line *unless* acted on by an external force.
>
> NEWTON'S FIRST LAW

INERTIAL REFERENCE FRAMES

Newton's first law makes no distinction between an object at rest and an object moving with constant (nonzero) velocity. Whether an object remains at rest or remains moving with constant velocity depends on the reference frame in which the object is observed. Suppose you are a passenger on an airplane that is flying along a straight path at constant altitude and you carefully place a tennis ball on your seat tray (which is horizontal). Relative to the plane, the ball will remain at rest as long as the plane continues to fly at constant velocity relative to the ground. Relative to the ground, the ball remains moving with the same velocity as the plane (Figure 4-1*a*).

Now, suppose that the pilot opens the throttle and the plane suddenly accelerates forward (relative to the ground). You will then observe that the ball on your tray suddenly starts to roll toward the rear of the plane, accelerating (relative to the plane) even though there is no horizontal force acting on it (Figure 4-1*b*). In this accelerating reference frame of the plane, Newton's first-law statement does not apply. Newton's first-law statement applies *only* in reference frames known as inertial reference frames. In fact, Newton's first law gives us the criterion for determining if a reference frame is an inertial frame:

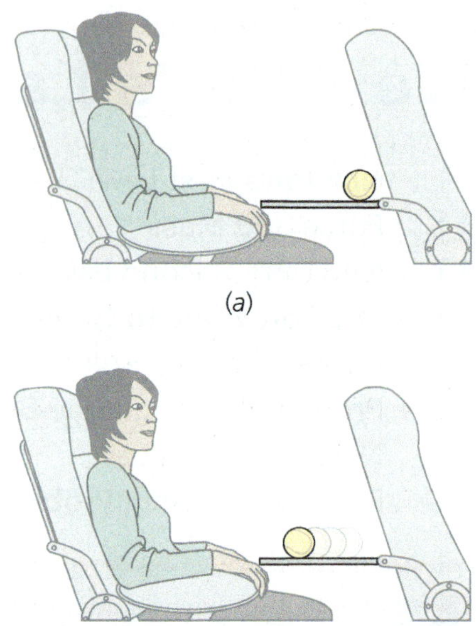

FIGURE 4-1 The plane is flying horizontally in a straight path at constant speed when you place a tennis ball on the tray. (*a*) The plane continues to fly at constant velocity (relative to the ground) and the ball remains at rest on the tray. (*b*) The pilot suddenly opens the throttle and plane rapidly gains speed (relative to the ground). The ball does not gain speed as quickly as the plane, so it accelerates (relative to the plane) toward the back of the plane.

> If no forces act on an object, any reference frame for which the acceleration of the object remains zero is an **inertial reference frame.**
>
> INERTIAL REFERENCE FRAME

Both the plane, when cruising at constant velocity, and the ground are, to a good approximation, inertial reference frames. Any reference frame moving with constant velocity relative to an inertial reference frame is also an inertial reference frame.

A reference frame attached to the ground is not quite an inertial reference frame because of the small acceleration of the ground due to the rotation of Earth and

> ! Newton's first-law statement applies *only* to inertial reference frames.

the small acceleration of Earth itself due to its revolution around the Sun. However, these accelerations are of the order of 0.01 m/s^2 or less, so to a good approximation, a reference frame attached to the surface of Earth is an inertial reference frame.

4-2 FORCE AND MASS

Using Newton's first law and the concept of inertial reference frames, we can define a **force** as an external influence or action on an object that causes the object to change velocity, that is, to accelerate relative to an inertial reference frame. (We assume that no other forces are acting on the object.) Force is a vector quantity. It has both magnitude (the size or strength of the force) and direction.

Forces are exerted on objects by other objects, and forces that are due to one object being physically touched by a second object are known as **contact forces.** Common examples of contact forces include hitting a ball with a bat, your hand pulling on a fishing line, your hands pushing on a shopping cart, and the force of friction between your sneakers and the floor. Note that in each case there is direct physical contact between the object applying the force and the object to which the force is applied. Other forces act on an object without direct physical contact with a second object. These forces, referred to as **action-at-a-distance forces,** include the gravitational force, the magnetic force, and the electric force.

THE FUNDAMENTAL INTERACTIONS OF NATURE

Forces are interactions between particles. Traditionally, physicists explain all interactions observed in nature in terms of four basic interactions that occur between elementary particles (see Figure 4-2):

1. The gravitational interaction—the long-range interaction between particles due to their mass. It is believed by some that the gravitational interaction involves the exchange of hypothetical particles called gravitons.
2. The electromagnetic interaction—the long-range interaction between electrically charged particles involving the exchange of photons.
3. The weak interaction—the extremely short-range interaction between subnuclear particles involving the exchange or production of W and Z bosons. The electromagnetic and weak interactions are now viewed as a single unified interaction called the electroweak interaction.
4. The strong interaction—the long-range interaction between hadrons, which themselves consist of quarks, that binds protons and neutrons together to form the atomic nuclei. It involves the exchange of mesons between hadrons, or gluons between quarks.

(a) (b) (c) (d)

FIGURE 4-2 (*a*) The magnitude of the gravitational force between Earth and an object near Earth's surface is the weight of the object. The gravitational interaction between the Sun and the other planets is responsible for keeping the planets in their orbits around the Sun. Similarly, the gravitational interaction between Earth and the moon keeps the moon in its nearly circular orbit around Earth. The gravitational forces exerted by the moon and the Sun on the oceans of Earth are responsible for the diurnal and semidiurnal tides. Mont-Saint-Michel, France, shown in the photo, is an island when the tide is in. (*b*) The electromagnetic interaction includes both the electric and the magnetic forces. A familiar example of the electric interaction is the attraction between small bits of paper and a comb that is electrified after being run through hair. The lightning bolts above the Kitt Peak National Observatory, shown in the photo, are the result of the electromagnetic interaction. (*c*) The strong nuclear interaction between elementary particles called hadrons, which include protons and neutrons, the constituents of atomic nuclei. This interaction results from the interaction of quarks, which are the building blocks of hadrons, and is responsible for holding nuclei together. The hydrogen bomb explosion shown here illustrates the strong nuclear interaction. (*d*) The weak nuclear interaction between leptons (which include electrons and muons) and between hadrons (which include protons and neutrons). This false-color cloud chamber photograph illustrates the weak interaction between a cosmic ray muon (green) and an electron (red) knocked out of an atom. *((a) Cotton Coulson/Woodfin Camp and Assoc.; (b) Gary Ladd; (c) Los Alamos National Lab; (d) Science Photo Library/Photo Researchers.)*

The everyday forces that we observe between macroscopic objects are due to either the gravitational or the electromagnetic interactions. Contact forces, for example, are actually electromagnetic in origin. They are exerted between the surface molecules of the objects in contact. Action-at-a-distance forces are due to the fundamental interactions of gravity and electromagnetism. These two forces act between particles that are separated in space. Although Newton could not explain how forces act through empty space, later scientists introduced the concept of a field, which acts as an intermediary agent. For example, we consider the attraction of Earth by the Sun in two steps. The Sun creates a condition in space that we call the gravitational field. This field then exerts a force on Earth. Similarly, Earth produces a gravitational field that exerts a force on the Sun. Your weight is the force exerted by the gravitational field of Earth on you. When we study electricity and magnetism (Chapters 22–31) we will study electric fields, which are produced by electrical charges, and magnetic fields, which are produced by electrical charges in motion. The strong and weak interactions are discussed in Chapter 42.

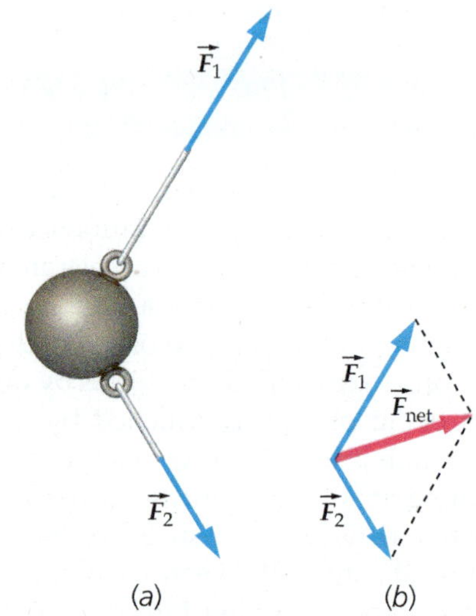

FIGURE 4-3 (*a*) The forces $\vec{F}_1$ and $\vec{F}_2$ pull on the sphere. (*b*) The effect of the two forces is as if a single force $\vec{F}_{net} = \vec{F}_1 + \vec{F}_2$ acts on the sphere instead of the two distinct forces $\vec{F}_1$ and $\vec{F}_2$.

CONCEPT CHECK 4-1

Is the net force an actual force?

COMBINING FORCES

If two or more individual forces simultaneously act on an object, the result is as if a single force, equal to the vector sum of the individual forces, acts in place of the individual forces. (That forces combine this way is called the **principle of superposition.**) The vector sum of the individual forces on an object is called the **net force $\vec{F}_{net}$** on the object. That is,

$$\vec{F}_{net} = \vec{F}_1 + \vec{F}_2 + \cdots$$

where $\vec{F}_1, \vec{F}_2, \ldots$ are the individual forces. Figure 4-3 shows an object being pulled in two directions by ropes. The effect is as if a single force equal to the net force acts on the object.

The SI unit of force is the newton (N). The newton is defined in the next section. One newton is equal to the weight of a modest-sized apple.

MASS

Objects intrinsically resist being accelerated. Imagine kicking both a soccer ball and a bowling ball. The bowling ball resists being accelerated much more than does the soccer ball, as would be evidenced by your sore toes. This intrinsic property is called the object's **mass.** It is a measure of the object's inertia. The greater an object's mass, the more the object resists being accelerated.

As noted in Chapter 1, the object chosen as the international standard for mass is a platinum-iridium alloy cylinder carefully preserved at the International Bureau of Weights and Measures at Sèvres, France. The mass of this standard object is 1 **kilogram** (kg), the SI unit of mass.

A convenient standard unit for mass in atomic and nuclear physics is the **unified atomic mass unit** (u), which is defined as one-twelfth of the mass of the carbon-12 (^{12}C) atom. The unified atomic mass unit is related to the kilogram by

$$1\ \text{u} = 1.660\,540 \times 10^{-27}\ \text{kg}$$

The concept of mass is defined as a constant of proportionality in Newton's second law. To measure the mass of an object, we compare its mass with a standard mass, such as the 1-kg standard kept at Sèvres. The comparison is accomplished using Newton's second law, and a procedure for doing this is described in Section 4-3 immediately following Example 4-1.

4-3 NEWTON'S SECOND LAW

Newton's first law tells us what happens when there is *no* force acting on an object. But what happens when there are forces exerted on the object? Consider again a block of ice sliding with constant velocity on a smooth, *frictionless* surface. If you push on the ice, you exert a force $\vec{F}$ that causes the ice's velocity to change. The harder you push, the greater the resulting acceleration $\vec{a}$. The acceleration, $\vec{a}$, of any object is directly proportional to the net force $\vec{F}_{net}$ exerted on it, and the reciprocal of the mass of the object is the proportionality constant. In addition, the acceleration vector and the net force vector are in the same direction. Newton summarized these observations in his second law of motion:

> **Second law.** The acceleration of an object is directly proportional to the net force acting on it, and the reciprocal of the mass of the object is the constant of proportionality. Thus,
>
> $$\vec{a} = \frac{\vec{F}_{net}}{m}, \quad \text{where} \quad \vec{F}_{net} = \sum \vec{F} \qquad 4\text{-}1$$
>
> NEWTON'S SECOND LAW

A net force on an object causes it to accelerate. It is a matter of cause and effect. The net force is the cause and the effect is the acceleration.*

A net force of 1 newton gives a 1-kg mass an acceleration of $1\ m/s^2$, so

$$1\ N = (1\ kg)(1\ m/s^2) = 1\ kg \cdot m/s^2 \qquad 4\text{-}2$$

Thus, a force of 2 N gives a 2-kg mass an acceleration of $1\ m/s^2$, and so on.

In the U.S. customary system, the unit of force is the *pound* (lb), where $1\ lb \approx 4.45\ N$,† and the unit of mass is the *slug*. The pound is defined to be the force required to produce an acceleration of $1\ ft/s^2$ on a mass of 1 slug:

$$1\ lb = 1\ slug \cdot ft/s^2$$

It follows that $1\ slug \approx 14.6\ kg$.

The Equation 4-1 is frequently expressed:

$$\vec{F}_{net} = m\vec{a}$$

and we will express it this way most of the time.

! Newton's second-law statement, like Newton's first–law statement, can be applied *only* in inertial reference frames.

CONCEPT CHECK 4-2

Is $m\vec{a}$ a force?

Example 4-1 A Sliding Ice-Cream Carton

A force exerted by a stretched rubber band (see Figure 4-4) produces an acceleration of $5.0\ m/s^2$ on an ice-cream carton of mass $m_1 = 1.0\ kg$. When a force exerted by an identical rubber band stretched by the same amount is applied to a carton of ice cream of mass m_2, it produces an acceleration of $11\ m/s^2$. (*a*) What is the mass of the second carton of ice cream? (*b*) What is the magnitude of the force exerted by the rubber band on the carton?

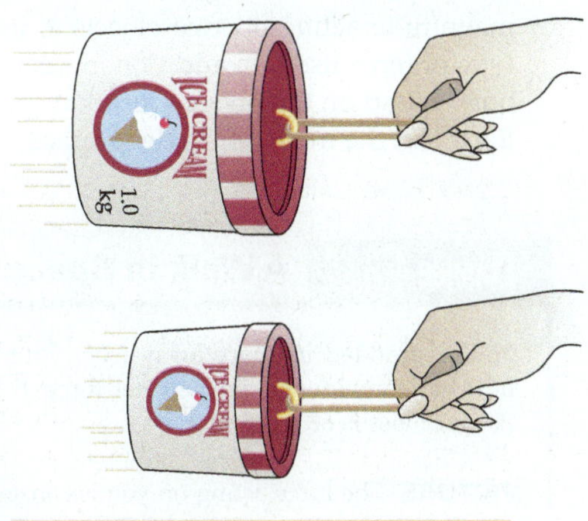

FIGURE 4-4

PICTURE We can apply Newton's second law, $\Sigma\vec{F} = m\vec{a}$, to each object and solve for the mass of the ice-cream carton and the magnitude of the force. The magnitudes of the forces exerted by the rubber bands are equal.

SOLVE

(*a*) 1. Apply $\Sigma\vec{F} = m\vec{a}$ to each object. There is only one force on each object, and we only need to consider magnitudes of the vector quantities:

$$F_1 = m_1 a_1 \quad \text{and} \quad F_2 = m_2 a_2$$

* Newton's second law relates the net force and the acceleration. Not everyone agrees that the net force is the cause and the acceleration is the effect.

† The pound we are talking about is the pound force (i.e., 1 pound force is exactly equal to 4.448 221 615 260 5 N). There is also a pound mass, which is exactly equal to 0.453 592 37 kg.

2. Because the applied forces are equal in magnitude, the ratio of the masses is equal to the reciprocal of the ratio of the accelerations: $F_1 = F_2 \Rightarrow m_1a_1 = m_2a_2$ and $\dfrac{m_2}{m_1} = \dfrac{a_1}{a_2}$

3. Solve for m_2 in terms of m_1, which is 1.0 kg: $m_2 = \dfrac{a_1}{a_2}m_1 = \dfrac{5.0\ \text{m/s}^2}{11\ \text{m/s}^2}(1.0\ \text{kg}) = \boxed{0.45\ \text{kg}}$

(*b*) The magnitude F_1 is found by using the mass and acceleration of either object: $F_1 = m_1a_1 = (1.0\ \text{kg})(5.0\ \text{m/s}^2) = \boxed{5.0\ \text{N}}$

CHECK A mass of 0.45 kg is a plausible mass for a carton of ice cream. One kilogram weighs about 2.2 pounds. So, the carton weighs about one pound and is the size of a one-pint carton.

PRACTICE PROBLEM 4-1 A net force of 3.0 N produces an acceleration of 2.0 m/s² on an object of unknown mass. What is the mass of the object?

To describe mass quantitatively, we can apply identical forces to two masses and compare their accelerations. If a force of magnitude F produces acceleration of magnitude a_1 when applied to an object of mass m_1, and an identical force produces acceleration of magnitude a_2 when applied to an object of mass m_2, then $m_1a_1 = m_2a_2$ (or $m_2/m_1 = a_1/a_2$). That is

If $F_1 = F_2$ then

$$\frac{m_2}{m_1} = \frac{a_1}{a_2} \qquad 4\text{-}3$$

COMPARING MASSES

This definition agrees with our intuitive idea of mass. If a force is applied to an object and a force of equal magnitude is applied to a second object, then the object with more mass will accelerate less. The ratio a_1/a_2 produced by forces of equal magnitude acting on two objects is independent of the magnitude, direction, or type of force used. In addition, mass is an intrinsic property of an object that does not depend on the object's location—it remains the same whether the object is on Earth, on the moon, or in deep space.

Example 4-2 A Walk in Space — *Context-Rich*

You're stranded in space away from your spaceship. Fortunately, you have a propulsion unit that provides a constant net force $\vec{F}$ for 3.0 s. After 3.0 s, you have moved 2.25 m. If your mass is 68 kg, find $\vec{F}$.

The propulsion unit (not shown) is pushing the astronaut to the right. *(NASA/Science Source/Photo Researchers.)*

PICTURE The force acting on you is constant, so your acceleration is also constant. We can use the kinematic equations of Chapters 2 and 3 to find $\vec{a}$, and then obtain the force from $\Sigma\vec{F} = m\vec{a}$. We choose the $+x$ direction to be in the direction of $\vec{F}$ (Figure 4-5), so $\vec{F} = F_x\hat{i}$ and $F_x = ma_x$.

SOLVE

1. To find the acceleration, we use Equation 2-14 with $v_0 = 0$: $\Delta x = v_0t + \frac{1}{2}a_xt^2 = 0 + \frac{1}{2}a_xt^2$

$$a_x = \frac{2\Delta x}{t^2} = \frac{2(2.25\ \text{m})}{(3.0\ \text{s})^2} = 0.50\ \text{m/s}^2$$

$$\vec{a} = a_x\hat{i} = 0.50\ \text{m/s}^2\hat{i}$$

2. Because $\vec{F}$ is the *net* force, $\Sigma \vec{F}_i = \vec{F}$. Therefore, we substitute $\vec{a} = 0.50\text{ m/s}^2\hat{i}$ and $m = 68\text{ kg}$ into this equation to find the force:

$$\vec{F} = ma_x\hat{i} = (68\text{ kg})(0.50\text{ m/s}^2)\hat{i}$$
$$= \boxed{34\text{ N}\hat{i}}$$

CHECK The acceleration is 0.50 m/s^2, which is about 5% of $g = 9.81\text{ m/s}^2$. This value seems plausible. If the magnitude of the acceleration were equal to g you would move a lot farther than 2.25 m in 3 s.

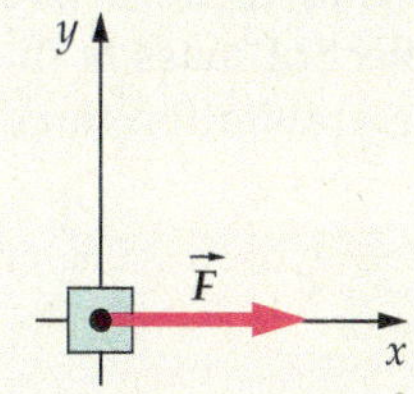

FIGURE 4-5

Example 4-3 A Particle Subjected to Two Forces

A particle of mass 0.400 kg is subjected simultaneously to two forces $\vec{F}_1 = -2.00\text{ N}\hat{i} - 4.00\text{ N}\hat{j}$ and $\vec{F}_2 = -2.60\text{ N}\hat{i} + 5.00\text{ N}\hat{j}$ (Figure 4-6). If the particle is at the origin and starts from rest at $t = 0$, find (*a*) its position $\vec{r}$ and (*b*) its velocity $\vec{v}$ at $t = 1.60\text{ s}$.

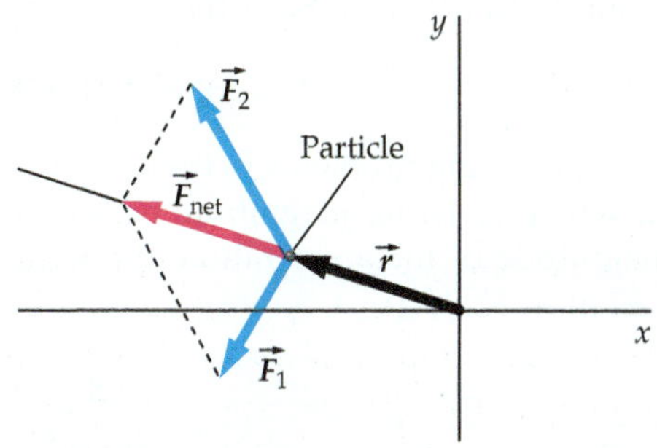

FIGURE 4-6 The acceleration is in the direction of the net force. The particle is released from rest at the origin. Following release, it moves in the direction of the net force, which is also the direction of the acceleration vector.

PICTURE Apply $\Sigma\vec{F} = m\vec{a}$ to find the acceleration. Once the acceleration is known, we can use the kinematic equations of Chapters 2 and 3 to determine the particle's position and velocity as functions of time.

SOLVE

(*a*) 1. Write the general equation for the position vector $\vec{r}$ as a function of time t for constant acceleration $\vec{a}$ in terms of $\vec{r}_0$, $\vec{v}_0$, and $\vec{a}$, and substitute $\vec{r}_0 = \vec{v}_0 = 0$.

$$\vec{r} = \vec{r}_0 + \vec{v}_0 t + \tfrac{1}{2}\vec{a}t^2 = 0 + 0 + \tfrac{1}{2}\vec{a}t^2$$
$$= \tfrac{1}{2}\vec{a}t^2$$

2. Use $\Sigma\vec{F} = m\vec{a}$ to write the acceleration $\vec{a}$ in terms of the resultant force $\Sigma\vec{F}$ and the mass m.

$$\vec{a} = \frac{\Sigma\vec{F}}{m}$$

3. Compute $\Sigma\vec{F}$ from the given forces.

$$\Sigma\vec{F} = \vec{F}_1 + \vec{F}_2$$
$$= (-2.00\text{ N}\,\hat{i} - 4.00\text{ N}\,\hat{j}) + (-2.60\text{ N}\,\hat{i} + 5.00\text{ N}\,\hat{j})$$
$$= -4.60\text{ N}\,\hat{i} + 1.00\text{ N}\,\hat{j}$$

4. Find the acceleration $\vec{a}$.

$$\vec{a} = \frac{\Sigma\vec{F}}{m} = -11.5\text{ m/s}^2\,\hat{i} + 2.50\text{ m/s}^2\,\hat{j}$$

5. Find the position $\vec{r}$ for a general time t.

$$\vec{r} = \tfrac{1}{2}\vec{a}t^2 = \tfrac{1}{2}a_x t^2\hat{i} + \tfrac{1}{2}a_y t^2\hat{j} = (-5.75\text{ m/s}^2\,\hat{i} + 1.25\text{ m/s}^2\,\hat{j})t^2$$

6. Find $\vec{r}$ at $t = 1.60\text{ s}$.

$$\vec{r} = \boxed{-14.7\text{ m}\,\hat{i} + 3.20\text{ m}\,\hat{j}}$$

(*b*) Write the velocity $\vec{v}$ by taking the time derivative of the step-5 result. Evaluate the velocity at $t = 1.6\text{ s}$.

$$\vec{v}(t) = \frac{d\vec{r}}{dt} = 2(-5.75\text{ m/s}^2\,\hat{i} + 1.25\text{ m/s}^2\,\hat{j})t$$
$$\vec{v}(1.6\text{ s}) = \boxed{-18.4\text{ m/s}\,\hat{i} + 4.00\text{ m/s}\,\hat{j}}$$

CHECK The position, velocity, acceleration, and net force vectors all have negative x components and positive y components. This is as expected for motion starting from rest at the origin and moving with constant acceleration.

4-4 THE FORCE DUE TO GRAVITY: WEIGHT

If you drop an object near Earth's surface, it accelerates toward Earth. If air resistance is negligible, all objects fall with the same acceleration, called the free-fall acceleration $\vec{g}$. The force causing this acceleration is the **gravitational force** ($\vec{F}_g$) exerted by Earth on the object. The *weight* of the object is the magnitude of the gravitational force on it. If the gravitational force is the *only* force acting on an object, the object is said to be in **freefall.** We can apply Newton's second law ($\Sigma\vec{F} = m\vec{a}$) to an object of mass m that is in freefall with acceleration $\vec{g}$ to obtain an expression for the gravitation force $\vec{F}_g$:

$$\vec{F}_g = m\vec{g} \qquad \text{4-4}$$

WEIGHT

Because $\vec{g}$ is the same for all objects, it follows that the gravitational force on an object is proportional to its mass. Near Earth, the vector $\vec{g}$ is the gravitational force per unit mass exerted by the planet Earth on any object and is called the **gravitational field** of Earth. Near the surface of Earth, the magnitude of $\vec{g}$ has the value

$$g = 9.81\,\text{N/kg} = 9.81\,\text{m/s}^2 \qquad \text{4-5}$$

When working problems in the U.S. customary system, we substitute F_g/g for mass m, where F_g is the magnitude of the gravitational force, in pounds, and g is the magnitude of the acceleration due to gravity in feet per second squared. Because $9.81\,\text{m} = 32.2\,\text{ft}$,

$$g = 32.2\,\text{ft/s}^2 \qquad \text{4-6}$$

Careful measurements show that near Earth $\vec{g}$ varies with location. $\vec{g}$ points toward the center of Earth and, at points above the surface of Earth, the magnitude of $\vec{g}$ varies inversely with the square of the distance to the center of Earth. Thus, an object weighs slightly less at very high altitudes than it does at sea level. The gravitational field also varies slightly with latitude because Earth is not exactly spherical but is slightly flattened at the poles. Thus weight, unlike mass, is not an intrinsic property of an object. Although the weight of an object varies from place to place because of variations in g, these variations are too small to be noticed in most practical applications on or near the surface of Earth.

! Weight is not an intrinsic property of an object.

An example should help clarify the difference between mass and weight. Consider a bowling ball near the moon. Its weight is the magnitude of the gravitational force exerted on it by the moon, but that force is a mere sixth of the magnitude of the gravitational force exerted on the bowling ball when it is similarly positioned on Earth. The ball weighs about one-sixth as much on the moon, and lifting the ball on the moon requires one-sixth the force. However, because the mass of the ball is the same on the moon as on Earth, throwing the ball horizontally at a specified speed requires the same force on the moon as on Earth.

Although the weight of an object may vary from one place to another, at any particular location the weight of the object is proportional to its mass. Thus, we can conveniently compare the masses of two objects at a given location by comparing their weights.

Our sensation of the gravitational force on us comes from other forces that balance it. When you sit on a chair, you feel a force exerted by the chair that balances the gravitational force on you and prevents you from accelerating toward the floor. When you stand on a spring scale, your feet feel the force exerted by the scale. The scale is calibrated to read the magnitude of the force it exerts (by the compression of its springs) to balance the gravitational force on you. The magnitude of this force is called your **apparent weight.** If there is no force to balance your weight, as in free-fall, your apparent weight is zero. This condition, called **weightlessness,** is experienced by astronauts in orbiting satellites. A satellite in a circular orbit near the surface of Earth is accelerating toward Earth. The only force acting on the satellite is that of gravity, so it is in free-fall. Astronauts in the satellite are also in free-fall. The only force on them is the gravitational force on them, which produces the acceleration $\vec{g}$. Because there is no force balancing the force of gravity, the astronauts have zero apparent weight.

Example 4-4 An Accelerating Student

The net force acting on a 130-lb student has a magnitude of 25.0 lb. What is the magnitude of her acceleration?

PICTURE Apply Newton's second law and solve for the acceleration. The mass can be found from the student's weight.

SOLVE

According to Newton's second law, the student's acceleration is the force divided by her mass, and her mass is equal to her weight divided by g:

$$a = \frac{F_{\text{net}}}{m} = \frac{F_{\text{net}}}{F_g/g} = \frac{25.0 \text{ lb}}{(130 \text{ lb})/(32.2 \text{ ft/s}^2)} = \boxed{6.19 \text{ ft/s}^2}$$

CHECK The force is slightly less than one-fifth of her weight, so we expect the acceleration to be slightly less than one-fifth of g. $(32.2 \text{ ft/s}^2)/5 = 6.44 \text{ ft/s}^2$, and 6.19 ft/s^2 is slightly less than 6.44 ft/s^2, so the result is plausible.

TAKING IT FURTHER Rearranging the equation in the solution gives

$$m = \frac{F_{\text{net}}}{a} = \frac{F_g}{g}$$

This arrangement reveals that you can solve for her acceleration without first solving for the mass. For any object, the ratio of F_{net} to a equals the ratio of F_g to g.

PRACTICE PROBLEM 4-2 What force is needed to give an acceleration of 3.0 ft/s^2 to a 5.0-lb block?

4-5 CONTACT FORCES: SOLIDS, SPRINGS, AND STRINGS

Many forces are exerted by one body in contact with another. In this section, we will examine some of the more common contact forces.

SOLIDS

If a surface is pushed against, it pushes back. Consider the ladder leaning against a wall shown in Figure 4-7. At the region of contact, the ladder pushes against the wall with a horizontal force, compressing the distance between the molecules in the surface of the wall. Like mattress springs, the compressed molecules in the wall push back on the ladder with a horizontal force. Such a force, *perpendicular* to the contacting surfaces, is called a **normal force** (the word *normal* means perpendicular). The wall bends slightly in response to a load, though this is rarely noticeable to the unaided eye.

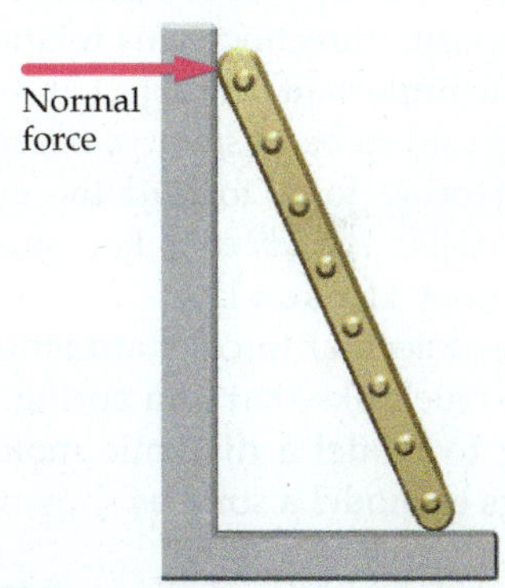

FIGURE 4-7 The wall supports the ladder by pushing on the ladder with a force normal to the wall.

Normal forces can vary over a wide range of magnitudes. A horizontal tabletop, for instance, will exert an upward normal force on any object resting on it. As long as the table doesn't break, this normal force will balance the downward gravitational force on the object. Furthermore, if you press down on the object, the magnitude of the upward normal force exerted by the table will increase, countering the extra force, thus preventing the object from accelerating downward.

In addition, surfaces in contact can exert forces on each other that are *parallel* to the contacting surfaces. Consider the large block on the floor shown in Figure 4-8. If the block is pushed sideways with a gentle enough force, it will not slide. The surface of the floor exerts a force back on the block, opposing its tendency to slide in the direction of the push. However, if the block is pushed sideways with a sufficiently large force, it will start to slide. To keep the block sliding, it is necessary to continue to push it. If the sideways push is not sustained, the contact force will slow the motion of the box until it stops. A component of a contact force that opposes sliding, or

FIGURE 4-8 The man is pushing on a block. The frictional force exerted by the floor on the block opposes its sliding motion or its tendency to slide.

the tendency to slide, is called a **frictional force;** it acts parallel to the contacting surfaces. (Frictional forces are treated more extensively in Chapter 5.)

A construction dumpster (Figure 4-9*a*) is situated on a road with a steep incline. Gravity pulls the dumpster downward, so to prevent the dumpster from moving, the road must exert an upward force $\vec{F}$ of equal magnitude on the dumpster (Figure 4-9*b*). The force $\vec{F}$ is a contact force by the road on the dumpster. A contact force such as this one is often thought of as two distinct forces, one, called the normal force $\vec{F}_n$, directed perpendicular to the road surface, and a second, called the frictional force $\vec{f}$, that is directed parallel to the road surface. The frictional force opposes any tendency of the dumpster to slide down the hill.

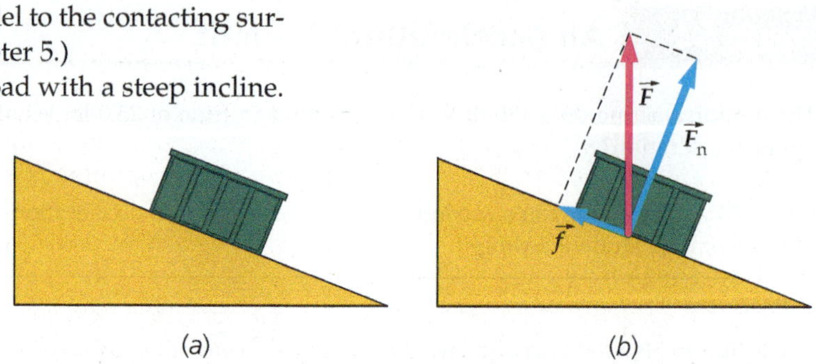

FIGURE 4-9 (*a*) A dumpster is parked on a steep incline. (*b*) The contact force by the road on the dumpster is represented either as the single force $\vec{F}$, or as a superposition of a normal force $\vec{F}_n$ and a frictional force $\vec{f}$.

SPRINGS

When a spring is stretched from its unstressed length by a distance x, the force it exerts is found experimentally to be

$$F_x = -kx \qquad 4\text{-}7$$

HOOKE'S LAW

where the positive constant k, called the **force constant** (or spring constant), is a measure of the stiffness of the spring (Figure 4-10). A negative value of x means the spring has been compressed a distance $|x|$ from its unstressed length. The negative sign in Equation 4-7 signifies that when the spring is stretched (or compressed) in one direction, the force it exerts is in the opposite direction. This relation, known as **Hooke's law,** turns out to be quite important. An object at rest under the influence of forces that balance is said to be in *static equilibrium.* If a small displacement results in a net restoring force toward the equilibrium position, the equilibrium is called *stable equilibrium.* For small displacements, nearly all restoring forces obey Hooke's law.

The molecular force of attraction between atoms in a molecule or solid varies much like that of a spring. We can therefore use two masses on a spring to model a diatomic molecule, or a set of masses connected by springs to model a solid as shown in Figure 4-11.

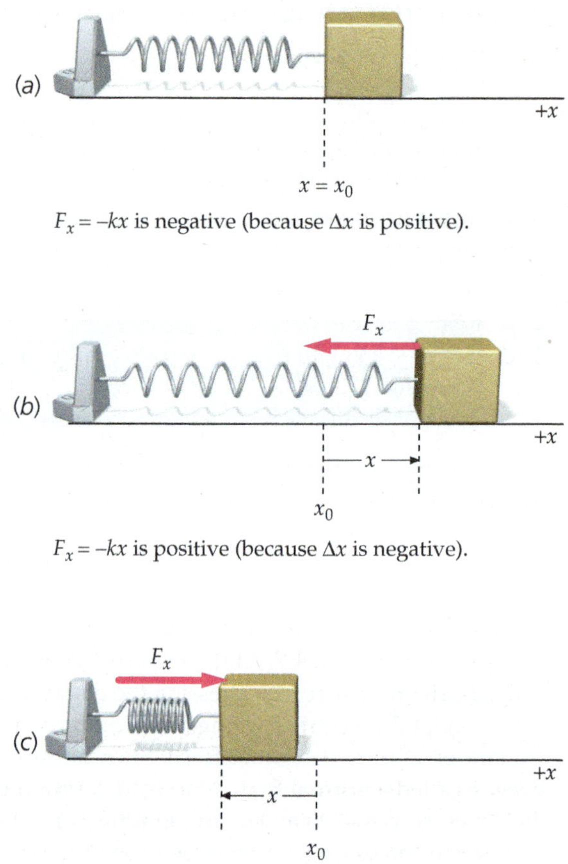

FIGURE 4-10 A horizontal spring. (*a*) When the spring is unstressed, it exerts no force on the block. (*b*) When the spring is stretched so that x is positive, it exerts a force of magnitude kx in the $-x$ direction. (*c*) When the spring is compressed so that x is negative, the spring exerts a force of magnitude $k|x|$ in the $+x$ direction.

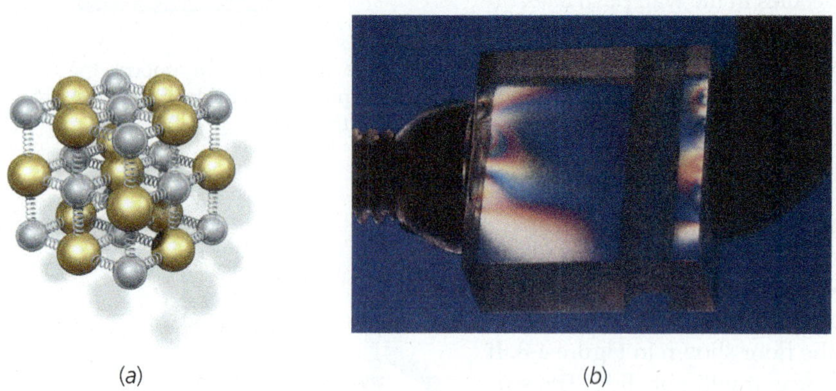

FIGURE 4-11 (*a*) Model of a solid consisting of atoms connected to each other by springs. The springs are very stiff (large force constant) so that when a weight is placed on the solid, its deformation is not visible. However, compression such as that produced by the clamp on the plastic block in (*b*) leads to stress patterns that are visible when viewed with polarized light. *((b) Fundamental Photographs.)*

Example 4-5 The Slam Dunk

A 110-kg basketball player hangs on the rim following a slam dunk (Figure 4-12). Prior to dropping to the floor, he hangs motionless with the front of the rim deflected down a distance of 15 cm. Assume the rim can be approximated by a spring and calculate the force constant k.

FIGURE 4-12 *(AFP-Getty Images.)*

PICTURE Because the acceleration of the player is zero, the net force exerted on him must also be zero. The upward force exerted by the rim balances his weight. Let $y = 0$ be the original position of the rim and choose down to be the $+y$ direction. Then the displacement of the rim y_e is positive, the weight $F_{gy} = mg$ is positive, and the force $F_y = -ky_e$ exerted by the rim is negative.

SOLVE

1. Apply $\Sigma F_y = m\vec{a}_y$ to the player. The acceleration of the player is zero:

$$\Sigma F_y = ma_y$$
$$F_{gy} + F_y = 0$$

2. Use Hooke's law (Equation 4-7) to find F_y:

$$F_y = -ky_e$$

3. Substitute expressions or values for the force components in step 1 and solve for k:

$$F_{gy} + F_y = 0$$
$$mg + (-ky_e) = 0$$
$$k = \frac{mg}{y_e} = \frac{(110\text{ kg})(9.8\text{ N/kg})}{0.15\text{ m}}$$
$$= \boxed{7.2 \times 10^3\text{ N/m}}$$

CHECK The weight of any object, in newtons, is almost ten times larger than the object's mass in kilograms. Thus, the weight is more than 1000 N. A deflection of only 0.10 m would mean k would be 10,000 N/m, so getting $k = 7200$N/m for a deflection of 0.15 m seems about right.

TAKING IT FURTHER Although a basketball rim doesn't look much like a spring, the rim is sometimes suspended by a hinge with a spring that is distorted when the front of the rim is pulled down. As a result, the upward force the rim exerts on the player's hands is proportional to the rim front's displacement and oppositely directed. Note that we used N/kg for the units of g so that kg cancels, giving N/m for the units of k. We can use either 9.81 N/kg or 9.81 m/s² for g, whichever is more convenient, because 1 N/kg = 1 m/s².

PRACTICE PROBLEM 4-3 A 4.0-kg bunch of bananas is suspended motionless from a spring balance whose force constant is 300 N/m. By how much is the spring stretched?

PRACTICE PROBLEM 4-4 A spring of force constant 400 N/m is attached to a 3.0-kg block that rests on a horizontal air track that renders friction negligible. What extension of the spring is needed to give the block an acceleration of 4.0 m/s² upon release?

PRACTICE PROBLEM 4-5 An object of mass m oscillates at the end of an ideal spring of force constant k. The time for one complete oscillation is the period T. Assuming that T depends on m and k, use dimensional analysis to find the form of the relationship $T = f(m, k)$, ignoring numerical constants. This is most easily found by looking at the units. Note that the units of k are N/m = (kg·m/s²)/m = kg/s², and the units of m are kg.

STRINGS

Strings (ropes) are used to pull things. We can think of a string as a spring with such a large force constant that the extension of the string is negligible. Strings are flexible, however, so unlike springs, they cannot push things. Instead, they flex or bend sharply. The magnitude of the force that one segment of a string exerts on an adjacent segment is called **tension,** T. It follows that if a string pulls on an object, the magnitude of the force on the object equals the tension. The concept of tension in a string or rope is further developed in Section 4-8.

4-6 PROBLEM SOLVING: FREE-BODY DIAGRAMS

Imagine a sled being pulled across icy ground by a sled dog. The dog pulls on a rope attached to the sled (Figure 4-13*a*) with a horizontal force causing the sled to gain speed. We can think of the sled and rope together as a single particle. What forces act on the sled-rope particle? Both the dog and the ice touch the sled-rope, so we know that the dog and the ice exert contact forces on it. We also know that Earth exerts a gravitational force on the sled-rope (the sled-rope's weight). Thus, a total of three forces act on the sled-rope (assuming that friction is negligible):

1. The gravitational force on the sled-rope $\vec{F}_g$.
2. The contact force $\vec{F}_n$ exerted by the ice on the runners. (Without friction, the contact force is directed normal to the ice.)
3. The contact force $\vec{F}$ exerted by the dog on the rope.

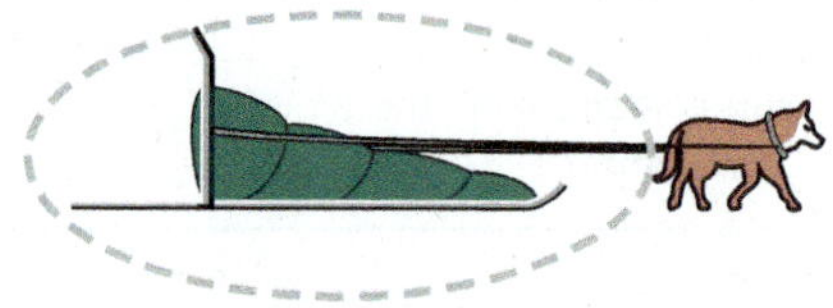
(a)

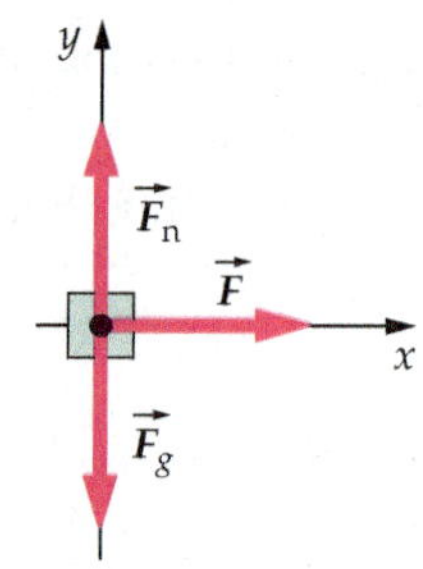

(b)

FIGURE 4-13 (*a*) A dog pulling a sled. The first step in problem solving is to isolate the system to be analyzed. In this case, the closed dashed curve represents the boundary between the sled-rope object and its surroundings. (*b*) The forces acting on the sled in Figure 4.13*a*.

A diagram that shows schematically all the forces acting on a system, such as Figure 4-13*b*, is called a **free-body diagram.** It is called a free-body diagram because the body (object) is drawn free from its surroundings.

Drawing the force vectors on a free-body diagram to scale requires that we first determine the direction of the acceleration vector using kinematic methods. We know the object is moving to the right with increasing speed. It follows from kinematics that its acceleration vector is in the direction that the velocity vector is changing—the forward direction. Note that $\vec{F}_n$ and $\vec{F}_g$ in the diagram have equal magnitudes. We know the magnitudes are equal because the vertical component of the acceleration is zero. As a qualitative check on the plausibility of our free-body diagram, we draw a vector-addition diagram (Figure 4-14) verifying that the vector sum of the forces is in the same direction as the acceleration vector.

We can now apply Newton's second law to determine the x and y components of the net force on the sled-rope particle. The x component of Newton's second law gives

$$\Sigma F_x = F_{nx} + F_{gx} + F_x = ma_x$$
$$0 + 0 + F = ma_x$$

or

$$a_x = \frac{F}{m}$$

The y component of Newton's second law gives

$$\Sigma F_y = F_{ny} + F_{gy} + F_y = ma_y$$
$$F_n - F_g + 0 = 0$$

or

$$F_n = F_g$$

Thus, the sled-rope particle has an acceleration in the $+x$ direction of F/m and the magnitude of the vertical force $\vec{F}_n$ exerted by the ice is $F_n = F_g = mg$.

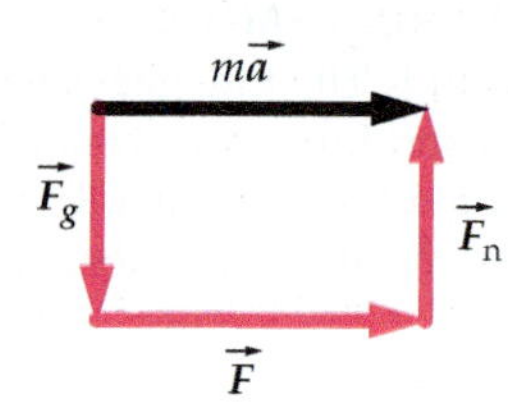

FIGURE 4-14 The vector sum of the forces in the free-body diagram is equal the mass times the acceleration vector.

PROBLEM-SOLVING STRATEGY

Applying Newton's Second Law

PICTURE Make sure you identify all of the forces acting on a particle. Then, determine the direction of the acceleration vector of the particle, if possible. Knowing the direction of the acceleration vector will help you choose the best coordinate axes for solving the problem.

SOLVE

1. Draw a neat diagram that includes the important features of the problem.
2. Isolate the object (particle) of interest, and identify each force that acts on it.
3. Draw a free-body diagram showing each of these forces.
4. Choose a suitable coordinate system. If the direction of the acceleration vector is known, choose a coordinate axis parallel to that direction. For objects sliding along a surface, choose one coordinate axis parallel to the surface and the other perpendicular to it.
5. Apply Newton's second law, $\Sigma\vec{F} = m\vec{a}$, usually in component form.
6. Solve the resulting equations for the unknowns.

CHECK Make sure your results have the correct units and seem plausible. Substituting extreme values into your symbolic solution is a good way to check your work for errors.

Example 4-6 A Dogsled Race

During your winter break, you enter a dogsled race in which students replace the dogs. Wearing cleats for traction, you begin the race by pulling on a rope attached to the sled with a force of 150 N at 25° above the horizontal. The mass of the sled–passenger–rope particle is 80 kg and there is negligible friction between the sled runners and the ice. Find (*a*) the acceleration of the sled and (*b*) the magnitude of the normal force exerted by the surface on the sled.

PICTURE Three forces act on the particle: its weight $\vec{F}_g$, which acts downward; the normal force $\vec{F}_n$, which acts upward; and the force with which you pull the rope $\vec{F}$, directed 25° above the horizontal. Because the forces are not all parallel to a single line, we study the system by applying Newton's second law to the x and y directions separately.

SOLVE

(*a*) 1. Sketch a free-body diagram (Figure 4-15*b*) of the sled-passenger-rope particle. Include a coordinate system with one of the coordinate axes in the direction of the acceleration. The particle moves to the right with increasing speed, so we know the acceleration is also to the right:

2. *Note:* Use the head-to-tail method of vector addition to verify that the sum of the forces on the free-body diagram can be in the direction of the acceleration (Figure 4-16):

3. Apply Newton's second law to the particle. Write out the equation in both vector and component form:

$$\vec{F}_n + \vec{F}_g + \vec{F} = m\vec{a} \quad \text{or}$$
$$F_{nx} + F_{gx} + F_x = ma_x$$
$$F_{ny} + F_{gy} + F_y = ma_y$$

4. Express the x components of $\vec{F}_n$, $\vec{F}_g$, and $\vec{F}$:

$$F_{nx} = 0, \quad F_{gx} = 0, \quad \text{and} \quad F_x = F\cos\theta$$

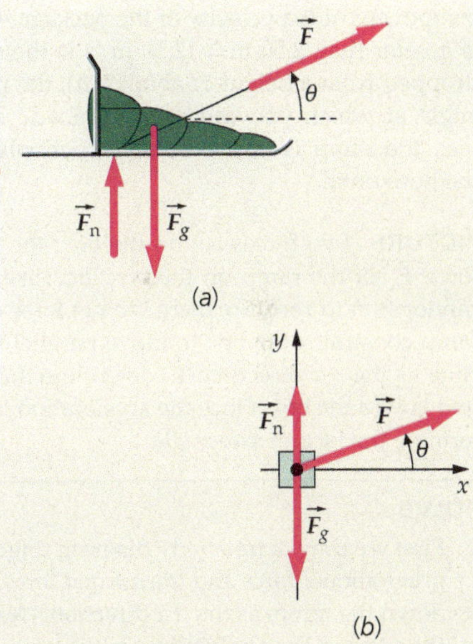

FIGURE 4-15

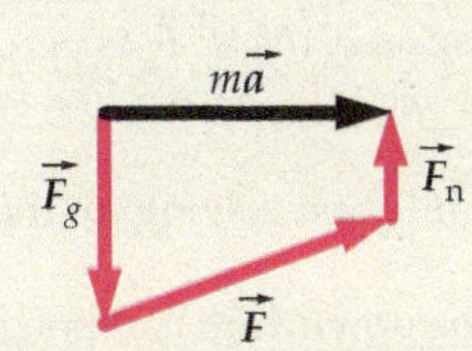

FIGURE 4-16 The vector sum of the forces in the free-body diagram is equal the mass times the acceleration vector.

5. Substitute the step-4 results into the x component equation in step 3. Then solve for the acceleration a_x:

$$\Sigma F_x = 0 + 0 + F\cos\theta = ma_x \quad \text{so}$$
$$a_x = \frac{F\cos\theta}{m} = \frac{(150\text{ N})\cos 25°}{80\text{ kg}} = \boxed{1.7\text{ m/s}^2}$$

(*b*) 1. Express the y component of $\vec{a}$:

$$a_y = 0$$

2. Express the y components of $\vec{F}_n$, $\vec{F}_g$, and $\vec{F}$:

$$F_{ny} = F_n,\quad F_{gy} = -mg, \quad \text{and} \quad F_y = F\sin\theta$$

3. Substitute the Part-(*b*), steps-1 and -2 results into the y component equation in Part-(*a*), step-3. Then solve for F_n:

$$\Sigma F_y = F_n - mg + F\sin\theta = 0$$
$$F_n = mg - F\sin\theta$$
$$= (80\text{ kg})(9.81\text{ N/kg}) - (150\text{ N})\sin 25° = \boxed{7.2\times 10^2\text{ N}}$$

CHECK Note that only the x component of $\vec{F}$, which is $F\cos\theta$, causes the object to accelerate. We expect the acceleration to be less if the rope is not horizontal. Also, we expect the normal force exerted by the ice to counter less than the full weight of the object because part of the weight is countered by the force exerted by the rope.

PRACTICE PROBLEM 4-6 If $\theta = 25°$, what is the maximum of the magnitude of the force $\vec{F}$ that can be applied to the rope without lifting the sled off the surface?

Example 4-7 Unloading a Truck *Context-Rich*

You are working for a big delivery company, and must unload a large, fragile package from your truck, using a delivery ramp (Figure 4-17). If the downward component of the velocity of the package when it reaches the bottom of the ramp is greater than 2.50 m/s (2.50 m/s is the speed an object would have if it were dropped from a height of about 1 ft), the package will break. What is the largest angle at which you can safely unload? The ramp is 1.00 m high, has rollers (i.e., the ramp is approximately frictionless), and is inclined at an angle θ to the horizontal.

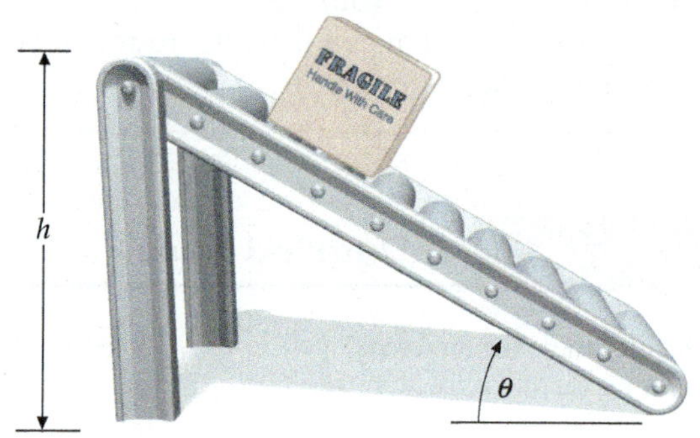

FIGURE 4-17

PICTURE Two forces act on the box, the gravitational force $\vec{F}_g$ and the normal force $\vec{F}_n$ of the ramp on the box. Because these forces are not antiparallel, they cannot sum to zero. So, there is a net force on the box causing it to accelerate. The ramp constrains the box to move parallel to its surface. We choose down the incline as the $+x$ direction. To determine the acceleration, we apply Newton's second law to the box. Once the acceleration is known, we can use kinematics to determine the largest safe angle.

SOLVE

1. First we draw a free-body diagram (Figure 4-18). Two forces act on the package, the gravitational force and the normal force. We choose the direction of the acceleration, down the ramp, as the $+x$ direction. (*Note:* The angle between $\vec{F}_g$ and the $-y$ direction is the same as the angle between the horizontal and the incline as we see from the free-body diagram.) We can also see that $F_{gx} = F_g\sin\theta$.

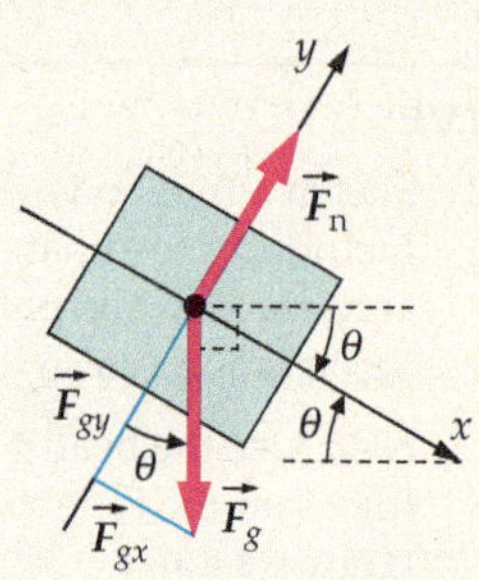

FIGURE 4-18

2. To find a_x we apply Newton's second law ($\Sigma F_x = ma_x$) to the package. (*Note:* $\vec{F}_n$ is perpendicular to the x axis and $F_g = mg$.)

$$F_{nx} + F_{gx} = ma_x \quad \text{where}$$
$$F_{nx} = 0 \quad \text{and} \quad F_{gx} = F_g\sin\theta = mg\sin\theta$$

3. Substituting and solving for the acceleration gives:

$$0 + mg\sin\theta = ma_x \quad \text{so} \quad a_x = g\sin\theta$$

4. Relate the downward component of the velocity of the box to its velocity component v_x in the x direction:

$$v_d = v_x\sin\theta$$

5. The velocity component v_x is related to the displacement Δx along the ramp by the kinematic equation:

$$v_x^2 = v_{0,x}^2 + 2a_x\Delta x$$

6. Substituting for a_x in the kinematic equation (step 5) and setting $v_{0,x}$ to zero gives:

$$v_x^2 = 2g\sin\theta\,\Delta x$$

7. From Figure 4-17, we can see that when Δx equals the length of the ramp, $\Delta x \sin\theta = h$, where h is the height of the ramp: $v_x^2 = 2gh$
8. Solve for v_d using the step-4 result and the expression for v_x from step 7: $v_d = \sqrt{2gh}\sin\theta$
9. Solve for the maximum angle: $2.50\text{ m/s} = \sqrt{2(9.81\text{ m/s}^2)(1.00\text{ m})}\sin\theta_{max}$

$\therefore\ \theta_{max} = \boxed{34.4°}$

CHECK At an angle of 34.4°, the downward component of the velocity will be slightly greater then half the speed that the box would have if it were dropped from a height of 1.00 m.

TAKING IT FURTHER The acceleration down the incline is constant and equal to $g\sin\theta$. In addition, the speed v at the bottom depends upon h but not upon the angle θ.

PRACTICE PROBLEM 4-7 Apply $\Sigma F_y = ma_y$ to the package and show that $F_n = mg\cos\theta$.

Example 4-8 Picture Hanging

Try It Yourself

A picture weighing 8.0 N is supported by two wires with tensions T_1 and T_2, as shown in Figure 4-19. Find each tension.

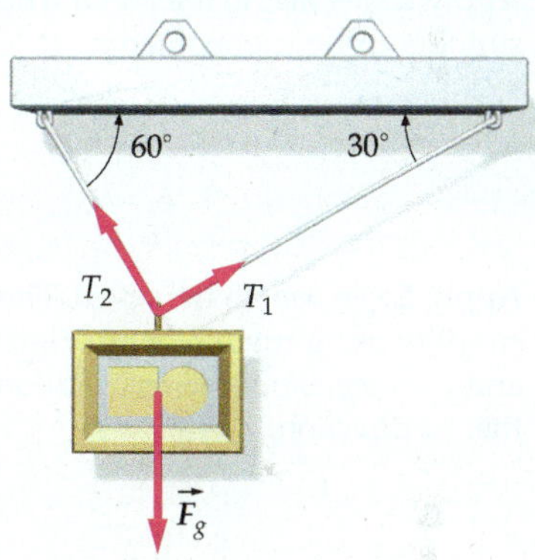

FIGURE 4-19

PICTURE Because the picture does not accelerate, the net force acting on it must be zero. The three forces acting on the picture (the gravitational force $\vec{F}_g$, and the tension forces $\vec{T}_1$ and $\vec{T}_2$) must therefore sum to zero.

SOLVE

Cover the column to the right and try these on your own before looking at the answers.

Steps:	Answers
1. Draw a free-body diagram for the picture (Figure 4-20). On your diagram show the x and y components of each tension force.	
2. Apply $\Sigma\vec{F} = m\vec{a}$ in vector form to the picture.	$\vec{T}_1 + \vec{T}_2 + \vec{F}_g = m\vec{a}$
3. Resolve each force into its x and y components. This gives you two equations for the two unknowns T_1 and T_2. The acceleration is zero.	$T_{1x} + T_{2x} + F_{gx} = 0$ $T_1\cos 30° - T_2\cos 60° + 0 = 0$ and $T_{1y} + T_{2y} + F_{gy} = 0$ $T_1\sin 30° + T_2\sin 60° - F_g = 0$
4. Solve the x component equation for T_2.	$T_2 = T_1\dfrac{\cos 30°}{\cos 60°}$
5. Substitute your result for T_2 (from step 4) into the y component equation and solve for T_1.	$T_1\sin 30° + \left(T_1\dfrac{\cos 30°}{\cos 60°}\right)\sin 60° - F_g = 0$ $T_1 = 0.50F_g = \boxed{4.0\text{ N}}$
6. Use your result for T_1 to find T_2.	$T_2 = T_1\dfrac{\cos 30°}{\cos 60°} = \boxed{6.9\text{ N}}$

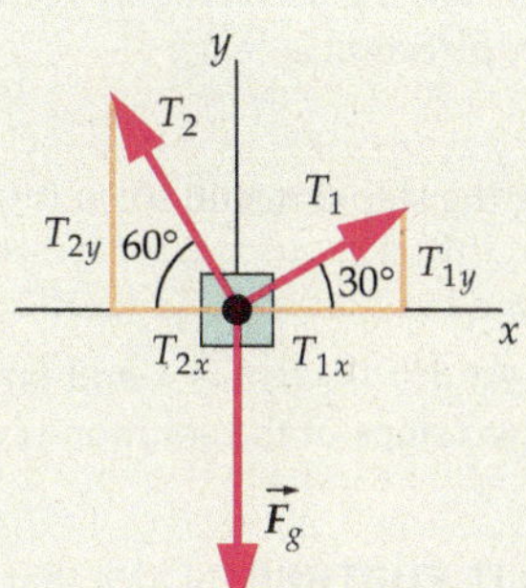

FIGURE 4-20

CHECK The more vertical of the two wires supports the greater share of the load, as you might expect. Also, we see that $T_1 + T_2 > 8$ N. The "extra" force is due to the wires pulling to the right and left.

Example 4-9 An Accelerating Jet Plane

As your jet plane speeds down the runway on takeoff, you decide to determine its acceleration, so you take out your yo-yo and note that when you suspend it, the string makes an angle of 22.0° with the vertical (Figure 4-21*a*). (*a*) What is the acceleration of the plane? (*b*) If the mass of the yo-yo is 40.0 g, what is the tension in the string?

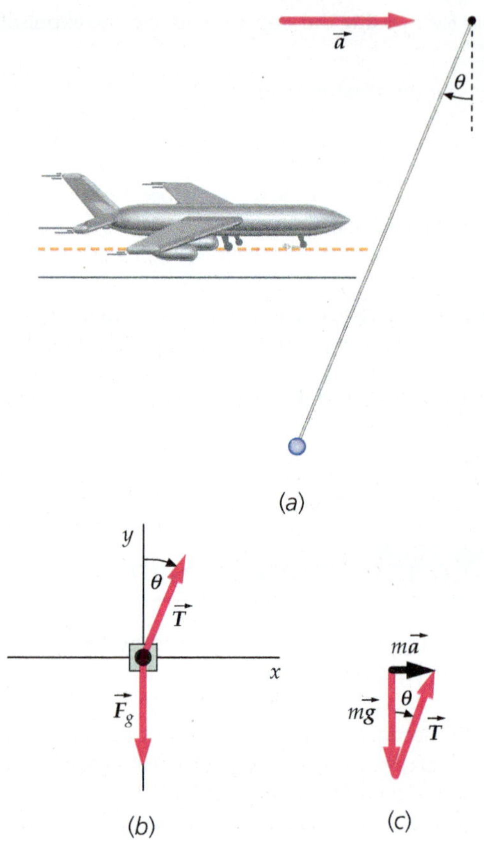

FIGURE 4-21

PICTURE Both the yo-yo and plane have the same acceleration. The net force on the yo-yo is in the direction of its acceleration—to the right. This force is supplied by the horizontal component of the tension force $\vec{T}$. The vertical component of $\vec{T}$ balances the gravitational force $\vec{F}_g$ on the yo-yo. We choose a coordinate system in which the $+x$ direction is in the direction of the acceleration vector $\vec{a}$ and the $+y$ direction is vertically upward. Writing Newton's second law for both the x and y directions gives two equations to determine the two unknowns, a and T.

SOLVE

(*a*) 1. Draw a free-body diagram for the yo-yo (Figure 4-21*b*). Choose the $+x$ direction to be the direction of the yo-yo's acceleration vector.

2. Apply $\Sigma F_x = ma_x$ to the yo-yo. Then simplify using trigonometry:

$$T_x + F_{gx} = ma_x$$

$$T\sin\theta + 0 = ma_x$$

or

$$T\sin\theta = ma_x$$

3. Apply $\Sigma F_y = ma_y$ to the yo-yo. Then, simplify using trigonometry (Figure 4-21) and $F_g = mg$. Since the acceleration is in the $+x$ direction, $a_y = 0$:

$$T_y + F_{gy} = ma_y$$

$$T\cos\theta - mg = 0$$

or

$$T\cos\theta = mg$$

4. Divide the step-2 result by the step 3 result and solve for the acceleration. Because the acceleration vector is in the $+x$ direction, $a = a_x$:

$$\frac{T\sin\theta}{T\cos\theta} = \frac{ma_x}{mg} \quad \text{so} \quad \tan\theta = \frac{a_x}{g} \quad \text{and}$$

$$a_x = g\tan\theta = (9.81\ \text{m/s}^2)\tan 22.0° = \boxed{3.96\ \text{m/s}^2}$$

(*b*) Using the step-3 result, solve for the tension:

$$T = \frac{mg}{\cos\theta} = \frac{(0.0400\ \text{kg})(9.81\ \text{m/s}^2)}{\cos 22.0°} = \boxed{0.423\ \text{N}}$$

CHECK At $\theta = 0$, $\cos\theta = 1$ and $\tan\theta = 0$. Substituting these values into the expressions in the last two steps of the solution gives and $a_x = 0$ and $T = mg$, as expected.

TAKING IT FURTHER Notice that for the Part-(*b*) result T is greater than the gravitational force on the yo-yo ($mg = 0.392\,\text{N}$) because the cord not only keeps the yo-yo from falling but also accelerates it in the horizontal direction. Here we use the units m/s² for g (instead of N/kg) because we are calculating acceleration.

PRACTICE PROBLEM 4-8 For what acceleration magnitude a would the tension in the string be equal to 3.00 mg? What is θ in this case?

Our next example is the application of Newton's second law to objects that are at rest relative to a reference frame that is itself accelerating.

Example 4-10 "Weighing" Yourself in an Elevator

Suppose that your mass is 80 kg, and you are standing on a scale fastened to the floor of an elevator. The scale measures force and is calibrated in newtons. What does the scale read when (*a*) the elevator is rising with upward acceleration of magnitude a; (*b*) the elevator is descending with downward acceleration of magnitude a'; (*c*) the elevator is rising at 20 m/s and its speed is decreasing at a rate of $8.0\,\text{m/s}^2$?

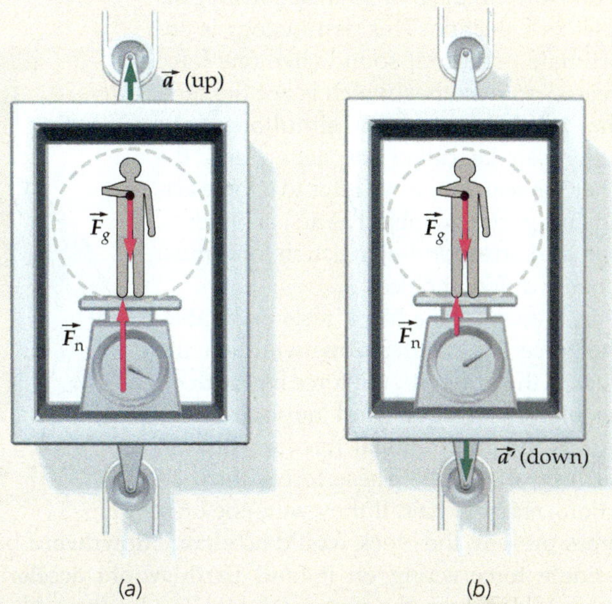

FIGURE 4-22

PICTURE The scale reading is the magnitude of the normal force exerted by the scale on you (Figure 4-22). Because you are at rest relative to the elevator, you and the elevator have the same acceleration. Two forces act on you: the downward force of gravity, $\vec{F}_g = m\vec{g}$, and the upward normal force from the scale, F_n. The sum of these forces gives you the observed acceleration. We choose upward to be the $+y$ direction.

SOLVE

(*a*) 1. Draw a free-body diagram of yourself (Figure 4-23):

2. Apply $\Sigma F_y = ma_y$: $\quad F_{ny} + F_{gy} = ma_y$
$$F_n - mg = ma_y$$

3. Solve for F_n. This is the reading on the scale (your apparent weight): $\quad F_n = mg + ma_y = m(g + a_y)$

4. $a_y = +a$: $\quad F_n = \boxed{m(g + a)}$

(*b*) $a_y = -a'$. Substitute for a_y in the Part-(*a*), step-3 result: $\quad F_n = m(g + a_y) = \boxed{m(g - a')}$

(*c*) The velocity is positive but decreasing, so the acceleration is negative. Thus, $a_y = -8.0\,\text{m/s}^2$. Substitute into the Part-(*a*), step-3 result: $\quad F_n = m(g + a_y) = (80\text{ kg})(9.81\text{ m/s}^2 - 8.0\text{ m/s}^2)$
$$= 144.8\text{ N} = \boxed{1.40 \times 10^2\text{ N}}$$

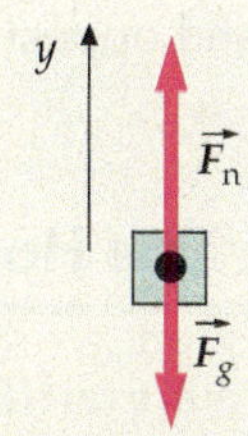

FIGURE 4-23

CHECK Independent of whether the elevator is ascending or descending, if its acceleration is upward you would expect to "feel heavier" and expect your apparent weight to be greater than mg. This is in keeping with the Part-(*a*) result. If its acceleration is downward you would expect to "feel lighter" and expect your apparent weight to be less than mg. The results for Parts (*b*) and (*c*) are in agreement with these expectations.

PRACTICE PROBLEM 4-9 A descending elevator comes to a stop with an acceleration of magnitude $4.00\,\text{m/s}^2$. If your mass is 70.0 kg and you are standing on a force scale in the elevator, what does the scale read as the elevator is stopping?

4-7 NEWTON'S THIRD LAW

Newton's third law describes an important property of forces: forces always occur in pairs. For example, if a force is exerted on some object A, there must be another object B exerting the force. Newton's third law states that these forces are equal in magnitude and opposite in direction. That is, if object A exerts a force on object B, then B exerts an equally strong but oppositely directed force on A.

> **Third law.** When two bodies interact, the force $\vec{F}_{BA}$ exerted by object B on object A is equal in magnitude and opposite in direction to the force $\vec{F}_{AB}$ exerted by object A on object B. Thus,
>
> $$\vec{F}_{BA} = -\vec{F}_{AB} \qquad 4\text{-}8$$
>
> NEWTON'S THIRD LAW

Each pair of forces is called a Newton's third-law **(N3L) pair.** It is common to refer to one force in the pair as an action and the other as a reaction. This terminology is unfortunate because it sounds like one force "reacts" to the other, which is not the case. The two forces occur simultaneously. Either can be called the action and the other the reaction. If we refer to a force acting on a particular object as an action force, then the corresponding reaction force must act on a different object.

In Figure 4-24, a block rests on a table. The force $\vec{F}_{\text{gEB}}$ acting downward on the block is the gravitational force by Earth on the block. An equal and opposite force $\vec{F}_{\text{gBE}}$ is the gravitational force exerted on Earth by the block. These forces form an action–reaction pair. If they were the only forces present, the block would accelerate downward because it would have only a single force acting on it (and Earth would accelerate upward for the same reason). However, the upward force $\vec{F}_{\text{nTB}}$ by the table on the block balances the gravitational force on the block. In addition, there is a downward force $\vec{F}_{\text{nBT}}$ by the block on the table. The forces $\vec{F}_{\text{nBT}}$ and $\vec{F}_{\text{nTB}}$ form a Newton's third law pair and, thus, are equal and opposite.

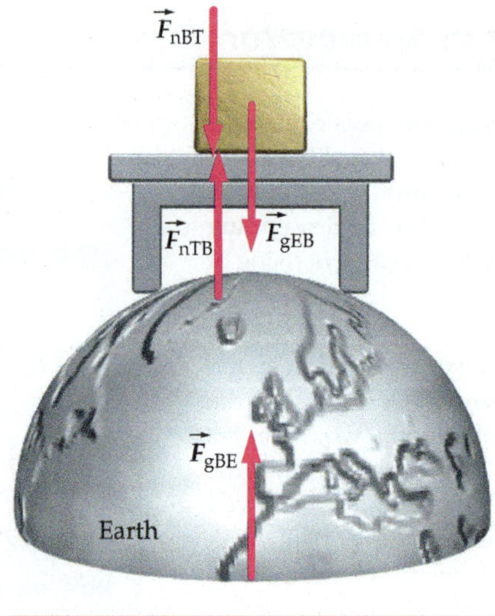

FIGURE 4-24

> **!** Newton's third-law force pairs are *always* equal and opposite.

> **!** No two external forces acting on *the same object* can ever constitute a Newton's third-law pair.

CONCEPT CHECK 4-3

Do the forces $\vec{F}_{\text{gBE}}$ and $\vec{F}_{\text{nBT}}$ in Figure 4-24 form a Newton's third-law pair?

Example 4-11 The Horse Before the Cart — *Conceptual*

A horse refuses to pull a cart (Figure 4-25*a*). The horse reasons, "according to Newton's third law, whatever force I exert on the cart, the cart will exert an equal and opposite force on me, so the net force will be zero and I will have no chance of accelerating the cart." What is wrong with this reasoning?

PICTURE Because we are interested in the motion of the cart, we draw a simple diagram for it (Figure 4-25*b*). The force exerted by the horse on the cart is labeled $\vec{F}_{\text{HC}}$. (This force is actually exerted on the harness. Because the harness is attached to the cart, we consider it a part of the cart.) Other forces acting on the cart are the gravitational force of Earth on the cart $\vec{F}_{g\text{EC}}$, the normal force of the pavement on the cart $\vec{F}_{\text{n PC}}$ and the frictional force exerted by the pavement on the cart, labeled $\vec{f}_{\text{PC}}$.

SOLVE

1. Draw a free-body diagram for the cart (see Figure 4-25*c*). Because the cart does not accelerate vertically, the vertical forces must sum to zero. The horizontal forces are $\vec{F}_{\text{HC}}$ to the right and $\vec{f}_{\text{PC}}$ to the left. The cart will accelerate to the right if $|\vec{F}_{\text{HC}}|$ is greater than $|\vec{f}_{\text{PC}}|$.

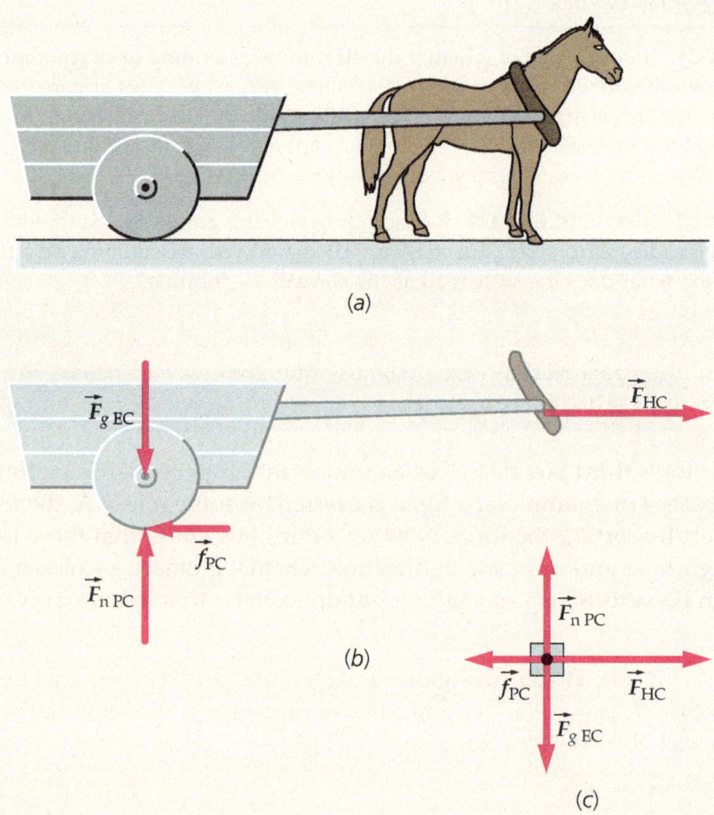

FIGURE 4-25

2. Note that the reaction force to $\vec{F}_{HC}$, which we call $\vec{F}_{CH}$, is exerted on the horse, not on the cart (Figure 4-25*d*). It has no effect on the motion of the cart, but it does affect the motion of the horse. If the horse is to accelerate to the right, there must be a force $\vec{F}_{PH}$ (to the right) exerted on the horse's hooves by the pavement that is greater in magnitude than $\vec{F}_{CH}$.

Because the reaction force to $\vec{F}_{HC}$ is exerted on the horse, it has no effect on the motion of the cart. This is the flaw in the horse's reasoning.

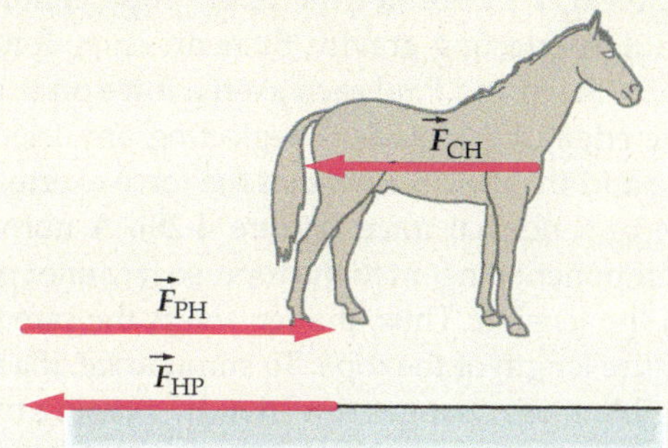

FIGURE 4-25
(continued) (*d*)

CHECK All forces on the cart have C for a rightmost subscript, and all forces on the horse have H for a rightmost subscript. Thus, we have not drawn both forces of a Newton's third-law force pair on either the horse or the cart.

TAKING IT FURTHER This example illustrates the importance of drawing a simple diagram when solving mechanics problems. Had the horse done so, he would have seen that he need only push back hard against the pavement so that the pavement will push him forward.

CONCEPT CHECK 4-4

As you stand facing a friend, place your palms against your friend's palms and push. Can your friend exert a force on you if you do not exert a force back? Try it.

4-8 PROBLEM SOLVING: PROBLEMS WITH TWO OR MORE OBJECTS

In some problems, the motions of two (or more) objects are influenced by the interactions between the objects. For example, such objects might touch each other, or they might be connected to each other by a string or spring.

The tension in a string or rope is the magnitude of the force that one segment of the rope exerts on a neighboring segment. The tension can vary throughout the length of the rope. For a rope dangling from a girder at the ceiling of a gymnasium, the tension is greatest at points near the ceiling because a short segment of rope next to the ceiling has to support the weight of all the rope below it. For the problems in this book, however, you can almost always assume that the masses of strings and ropes are negligible, so variations in tension due to the weight of a string or rope can be neglected. Conveniently, this also means that you may assume that variations in the tension due to the acceleration of a string or rope can also be neglected.

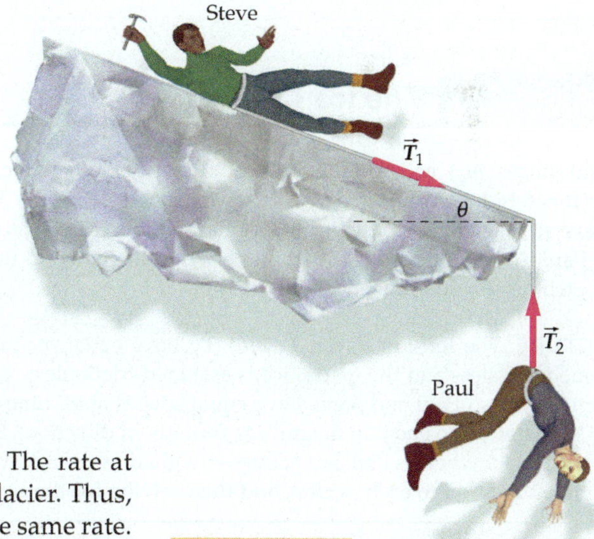

FIGURE 4-26

Consider, for example, the motion of Steve and Paul in Figure 4-26. The rate at which Paul descends equals the rate at which Steve slides along the glacier. Thus, their speeds remain equal. If Paul gains speed, Steve gains speed at the same rate. That is, their tangential accelerations remain equal. (The tangential acceleration of a particle is the component of the acceleration that is tangent to the path of the motion of the particle.)

The free-body diagram of a segment of the rope attached to Steve, where Δm_s is the segment's mass, is shown in Figure 4-27. Applying Newton's second law to the segment gives $T - T' = \Delta m_s a_x$. If the mass of the segment is negligible, then $T = T'$. To give a segment of negligible mass any finite acceleration, a net force of only a negligible magnitude is needed. (That is, only a negligible difference in tension is needed to give a rope segment of negligible mass any finite acceleration.)

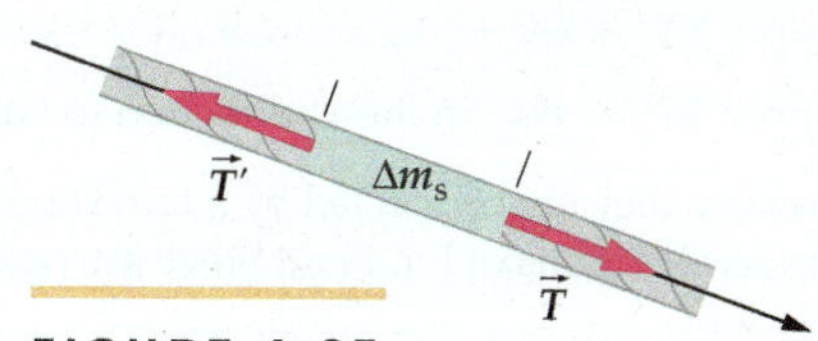

FIGURE 4-27

Next, we consider the entire rope connecting Steve and Paul. Neglecting gravity, there are three forces acting on the rope. Steve and Paul each exert a force on it, as does the ice at the edge of the glacier. Neglecting any friction between the ice and the rope means that the force exerted by the ice is always a normal force (Figure 4-28). A normal force has no component tangent to the rope, so it cannot produce a change in the tension. Thus, the tension is the same throughout the entire length of the rope. To summarize, if a taut rope of negligible mass changes direction by passing over a frictionless surface, the tension is the same throughout the rope. The following box summarizes the steps for solving such problems.

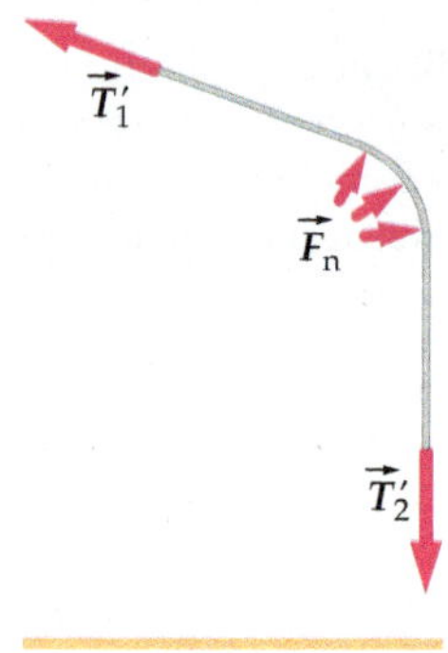

FIGURE 4-28

CONCEPT CHECK 4-5

Suppose that instead of passing over the edge of a glacier, the rope passed around a pulley with frictionless bearings as shown in Figure 4-29. Would the tension then be the same throughout the length of the rope?

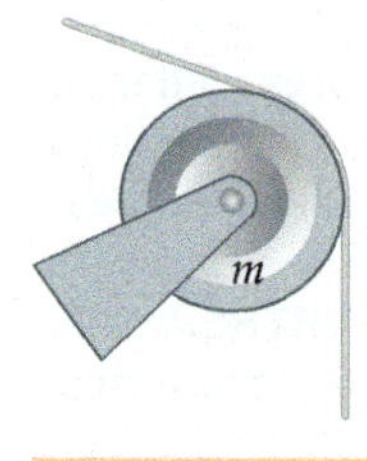

FIGURE 4-29

PROBLEM-SOLVING STRATEGY

Applying Newton's Laws to Problems with Two or More Objects

PICTURE Remember to draw a separate free-body diagram for each object. The unknowns can be obtained by solving simultaneous equations.

SOLVE

1. Draw a separate free-body diagram for each object. Use a separate coordinate system for each object. Remember, if two objects touch, the forces they exert on each other are equal and opposite (Newton's third law).
2. Apply Newton's second law to each object.
3. Solve the resultant equations, together with any equations describing interactions and constraints, for the unknown quantities.

CHECK Make sure your answer is consistent with the free-body diagrams that you have created.

Example 4-12 The Ice Climbers

Paul (mass m_P) accidentally falls off the edge of a glacier as shown in Figure 4-26. Fortunately, he is connected by a long rope to Steve (mass m_S), who has a climbing ax. Before Steve sets his ax to stop them, Steve slides without friction along the ice, attached by the rope to Paul. Assume there is no friction between the rope and the glacier. Find the acceleration of each person and the tension in the rope.

PICTURE The tension forces $\vec{T}_1$ and $\vec{T}_2$ have equal magnitudes because the rope is assumed massless and the glacial ice is assumed frictionless. The rope does not stretch or become slack, so Paul and Steve have equal speeds at all times. Their accelerations $\vec{a}_S$ and $\vec{a}_P$ must therefore be equal in magnitude (but not in direction). Steve accelerates down the face of the glacier whereas Paul accelerates straight downward. We can solve this problem by applying $\Sigma\vec{F} = m\vec{a}$ to each person, and then solving for the accelerations and the tension.

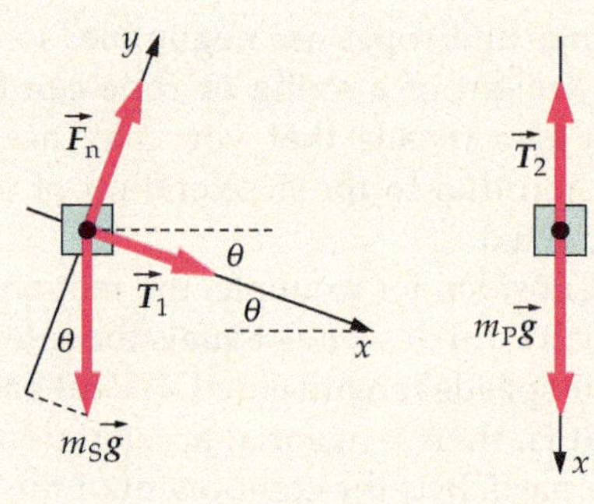

FIGURE 4-30

SOLVE

1. Draw separate free-body diagrams for Steve and Paul (Figure 4-30). Put axes x and y on Steve's diagram, choosing the direction of Steve's acceleration as the $+x$ direction. Choose the direction of Paul's acceleration as the $+x'$ direction.
2. Apply $\Sigma F_x = ma_x$ in the x direction to Steve:

   $F_{nx} + T_{1x} + m_S g_x = m_S a_{Sx}$

3. Apply $\Sigma F_{x'} = ma_{x'}$ in the x' direction to Paul:

   $T_{2x'} + m_P g_{x'} = m_P a_{Px'}$

4. Because they are connected by a taut rope that does not stretch, the accelerations of Paul and Steve are related. Express this relation:

   $a_{Px'} = a_{Sx} = a_t$

   a_t stands for the acceleration component in the *tangential* direction. (The direction of the motion.)

5. Because the rope is of negligible mass and slides over the ice with negligible friction, the forces $\vec{T}_1$ and $\vec{T}_2$ are simply related. Express this relation:

$$T_2 = T_1 = T$$

6. Substitute the steps-4 and -5 results into the step-2 and step-3 equations:

$$T + m_S g \sin\theta = m_S a_t$$
$$-T + m_P g = m_P a_t$$

7. Solve the step-6 equations for the acceleration by eliminating T and solving for a_t:

$$\boxed{a_t = \frac{m_S \sin\theta + m_P}{m_S + m_P} g}$$

8. Substitute the step-7 result into either step-6 equation and solve for T:

$$\boxed{T = \frac{m_S m_P}{m_S + m_P}(1 - \sin\theta) g}$$

CHECK If m_P is very much greater than m_S, we expect the acceleration to be approximately g and the tension to be approximately zero. Taking the limit as m_S approaches 0 does indeed give $a_t = g$ and $T = 0$ for this case. If m_P is much less than m_S, we expect the acceleration to be approximately $g \sin\theta$ (see step 3 of Example 4-7) and the tension to be zero. Taking the limit as m_P approaches 0 in steps 7 and 8, we indeed obtain at $a_t = g \sin\theta$ and $T = 0$. At an extreme value of the inclination ($\theta = 90°$) we again check our answers. Substituting $\theta = 90°$ in steps 7 and 8, we obtain $a_t = g$ and $T = 0$. This seems right since Steve and Paul would be in free-fall for $\theta = 90°$.

TAKING IT FURTHER In Step 1 we chose down the incline and straight down to be positive to keep the solution as simple as possible. With this choice, when Steve moves in the $+x$ direction (down the surface of the glacier), Paul moves in the $+x'$ direction (straight downward).

PRACTICE PROBLEM 4-10 (*a*) Find the acceleration if $\theta = 15°$ and if the masses are $m_S = 78\,\text{kg}$ and $m_P = 92\,\text{kg}$. (*b*) Find the acceleration if these two masses are interchanged.

Example 4-13 Building a Space Station

You are an astronaut constructing a space station, and you push on a box of mass m_1 with force $\vec{F}_{A1}$. The box is in direct contact with a second box of mass m_2 (Figure 4-31). (*a*) What is the acceleration of the boxes? (*b*) What is the magnitude of the force each box exerts on the other?

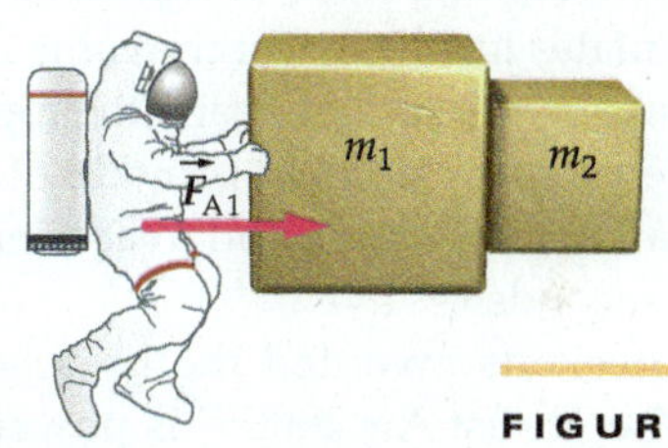

FIGURE 4-31

PICTURE Force $\vec{F}_{A1}$ is a contact force and only acts on box 1. Let $\vec{F}_{21}$ be the force exerted by box 2 on box 1, and $\vec{F}_{12}$ be the force exerted by box 1 on box 2. In accord with Newton's third law, these forces are equal and opposite ($\vec{F}_{21} = -\vec{F}_{12}$), so $F_{21} = F_{12}$. Apply Newton's second law to each box separately. The motions of the two boxes are identical, so the accelerations $\vec{a}_1$ and $\vec{a}_2$ are equal.

SOLVE

(*a*) 1. Draw free-body diagrams for the two boxes (Figure 4-32).

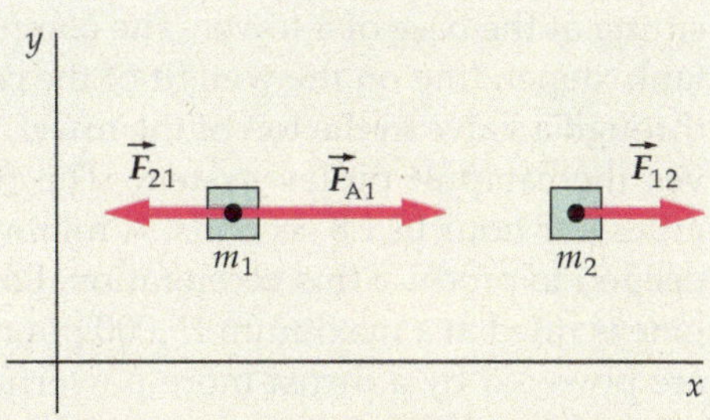

FIGURE 4-32

2. Apply $\Sigma\vec{F} = m\vec{a}$ to box 1.

$$F_{A1} - F_{21} = m_1 a_{1x}$$

3. Apply $\Sigma\vec{F} = m\vec{a}$ to box 2.

$$F_{12} = m_2 a_{2x}$$

4. Express both the relation between the two accelerations and the relation between the magnitudes of the forces the blocks exert on each other. The accelerations are equal because the speeds are equal at all times, so the rate of change of the speeds are equal. The forces are equal in magnitude because the forces constitute a N3L force pair:

$$a_{2x} = a_{1x} = a_x$$
$$F_{21} = F_{12} = F$$

5. Substitute these back into the step-2 and step-3 results and solve for a_x.

$$\boxed{a_x = \frac{F_{A1}}{m_1 + m_2}}$$

(*b*) Substitute your expression for a_x into either the step-2 or the step-3 result and solve for F.

$$\boxed{F = \frac{m_2}{m_1 + m_2} F_{A1}}$$

CHECK Note that the result in step 5 is the same as if the force $\vec{F}_{A1}$ had acted on a single mass equal to the sum of the masses of the two boxes. In fact, because the two boxes have the same acceleration, we can consider them to be a single particle with mass $m_1 + m_2$.

Physics Spotlight

Roller Coasters and the Need for Speed

Roller coasters have fascinated people since the spectacular Promenades Aeriennes (Aerial Walks) opened in Paris in 1817.* Until recently, though, ride designers were stuck with one major limitation—the ride needed to start at the top of a large hill.

In the 1970s, Anton Schwartzkopf, a German amusement park designer, was inspired by planes taking off from an aircraft carrier. In 1976, the Shuttle Loop roller coaster opened. A multiton weight was cranked to the top of a tower near the roller coaster. One end of a cable was attached to the weight, while the other was hooked to the coaster, in order to pull it. The weight dropped and quickly pulled the cable with the attached train of cars. In less than 3 seconds, the coaster train accelerated to 60 miles per hour.

At the same time, Schwartzkopf came up with a second catapult-style method of launching a coaster. A 5-ton flywheel was spun at high speed. A cable was connected to the coaster and the fly wheel. In less than 3 seconds, the coaster train, with up to 28 passengers, accelerated to speeds of 60 miles per hour. Both of these methods pioneered the use of catapult-style launches in roller coasters.†

Two new methods of launching roller coasters have allowed roller coasters to travel at even faster speeds. Intamin AG has created a *hydraulic*, or liquid-driven, system to pull the cable. The car alone for the Top Thrill Dragster weighs 12,000 pounds. Eighteen passengers are usually along for the ride, as well. The car is weighed as it passes over sensors, and a computer calculates just how fast the cable needs to go in order to catapult the car and passengers so they reach the top of the 420-foot first hill. Then, the liquid-filled motors quickly provide of up to 10,000 horsepower to wind the cable at up to 500 rpm, and to accelerate the coaster car to 120 miles per hour in 4 seconds.‡

Stan Checketts invented the first *pneumatic*, or compressed-air, roller coaster. The Thrust Air 2000™ is powered by a single, very large blast of air. The eight-passenger car is weighed as it passes over sensors. Then, four compressors swing into action. They pump air into a storage tank sitting at the base of a tower. The compressed air is measured into a shot tank, depending on the weight of the car. Finally, the air rapidly releases through a valve in the top of the tower, pushing against a piston that drives the catapult pulley system. The fully loaded car accelerates to 80 miles per hour in 1.8 seconds. A minimum of 40,000 pounds of thrust is needed to produce this acceleration. For comparison, a single F-15 jet engine is rated at a maximum 29,000 pounds of thrust.§ Roller coasters now are powered by a thrust more powerful than a jet engine. Something to think about the next time you pass by an amusement park.

The Hypersonic XLC at Paramount's King's Dominion Amusement Park, Virginia, the world's first compressed-airlaunched roller coaster, goes from 0 to 80 miles per hour in 1.8 seconds. *(Courtesy of King's Dominion Amusement Park.)*

* Cartmell, Robert, *The Incredible Scream Machine: A History of the Roller Coaster.* Bowling Green State University Popular Press, Bowling Green Ohio. 1987.

† "The Tidal Wave" http://www.greatamericaparks.com/tidalwave.html Marriott Great America Parks, 2006; Cartmell, op. cit.

‡ Hitchcox, Alan L. "Want Thrills? Go with Hydraulics." Hydraulics and Pneumatics, July 2005.

§ Goldman, Lea. "Newtonian Nightmare." *Forbes*, 7/23/2001. Vol. 168, Issue 2; "The F-100 Engine." http://www.pratt-whitney.com/prod_mil_f100.asp Pratt & Whitney, March 2006.

SUMMARY

1. Newton's laws of motion are fundamental laws of nature that serve as the basis for our understanding of mechanics.
2. Mass is an *intrinsic* property of an object.
3. Force is an important *derived* dynamic quantity.

TOPIC	RELEVANT EQUATIONS AND REMARKS
1. Newton's Laws	
First law	An object at rest stays at rest *unless* acted on by an external force. An object in motion continues to travel with constant velocity *unless* acted on by an external force. (Reference frames in which these statement hold are called inertial reference frames.)
Second law	The acceleration of an object is directly proportional to the net force acting on it. The reciprocal of the mass of the object is the proportionality constant. Thus $\vec{F}_{\text{net}} = m\vec{a}, \quad \text{where} \quad \vec{F}_{\text{net}} = \Sigma\vec{F}$ 4-1
Third law	When two bodies interact, the force $\vec{F}_{\text{BA}}$ exerted by object B on object A is equal in magnitude and opposite in direction to the force $\vec{F}_{\text{AB}}$ exerted by object A on object B: $\vec{F}_{\text{BA}} = -\vec{F}_{\text{AB}}$ 4-8
2. Inertial Reference Frames	Our statements of Newton's first and second laws are valid only in inertial reference frames. Any reference frame that is moving with constant velocity relative to an inertial reference frame is itself an inertial reference frame, and any reference frame that is accelerating relative to an inertial frame is not an inertial reference frame. Earth's surface is, to a good approximation, an inertial reference frame.
3. Force, Mass, and Weight	
Force	Force is defined in terms of the acceleration it produces on a given object. A force of 1 newton (N) is that force which produces an acceleration of 1 m/s^2 on a mass of 1 kilogram (kg).
Mass	Mass is an intrinsic property of an object. It is the measure of the inertial resistance of the object to acceleration. Mass does not depend on the location of the object. Applying identical forces to each of two objects and measuring their respective accelerations allows the masses of two objects to be compared. The ratio of the masses of the objects is equal to the inverse ratio of the accelerations produced: $\frac{m_2}{m_1} = \frac{a_1}{a_2}$
Gravitational Force	The gravitational force $\vec{F}_g$ on an object near the surface of Earth is the force of gravitational attraction exerted by Earth on the object. It is proportional to the gravitational field $\vec{g}$ (which is equal to the free-fall acceleration), and the mass m of the object is the proportionality constant: $\vec{F}_g = m\vec{g}$ 4-4 The weight of an object is the magnitude of the gravitational force on the object.
4. Fundamental Forces	All the forces observed in nature can be explained in terms of four basic interactions:
	1. The gravitational interaction
	2. The electromagnetic interaction
	3. The weak interaction*
	4. The strong nuclear interaction (also called the *hadronic force*)
5. Contact Forces	Contact forces of support and friction and those exerted by springs and strings are due to molecular forces that arise from the basic electromagnetic force.
Hooke's law	When an unstressed spring is compressed or extended by a small amount Δx, the restoring force it exerts is proportional to Δx: $F_x = -k\Delta x$ 4-7

* The electromagnetic and weak interactions are now viewed as the electroweak interaction.

Answers to Concept Checks

4-1 No, the net force is not an actual force. It is the vector sum of the actual forces.

4-2 No, it is the net force that causes the acceleration of the mass.

4-3 No, they do not.

4-4 No. Doing so would be contrary to Newton's third law.

4-5 No. Doing away with friction in the bearing is one thing, but the pulley still has mass. A difference in tension is needed in order to change the rate of rotation of the pulley wheel. Pulleys with non-negligible mass are studied in Chapter 8.

Answers to Practice Problems

4-1 1.5 kg

4-2 0.47 lb

4-3 13 cm

4-4 3.0 cm

4-5 $T = C\sqrt{m/k}$ where C is some dimensionless constant. The correct expression for the period, as we will see in Chapter 14, is $T = 2\pi\sqrt{m/k}$.

4-6 1.9 kN

4-7 Applying Newton's second law (for y components), we see from the free-body diagram (Figure 4-18) that $\Sigma F_y = ma_y \Rightarrow F_n - F_g \cos\theta = 0$, where we have used that a_y equals zero. Thus, $F_n = F_g \cos\theta$.

4-8 $a = 27.8\ \text{m/s}^2, \theta = 70.5°$

4-9 967 N

4-10 (*a*) $a_t = 0.66g$, (*b*) $a_t = 0.60g$

Problems

In a few problems, you are given more data than you actually need; in a few other problems, you are required to supply data from your general knowledge, outside sources, or informed estimate.

Interpret as significant all digits in numerical values that have trailing zeros and no decimal points.

For all problems, use $g = 9.81\ \text{m/s}^2$ for the free-fall acceleration due to gravity and neglect friction and air resistance unless instructed to do otherwise.

- • Single-concept, single-step, relatively easy
- •• Intermediate-level, may require synthesis of concepts
- ••• Challenging
- SSM Solution is in the *Student Solutions Manual*
- Consecutive problems that are shaded are paired problems.

CONCEPTUAL PROBLEMS

1 • While on a very smooth level transcontinental plane flight, your coffee cup sits motionless on your tray. Are there forces acting on the cup? If so, how do they differ from the forces that would be acting on the cup if it sat on your kitchen table at home?

2 • You are passing another car on a highway and determine that, relative to you, the car you pass has an acceleration $\vec{a}$ toward the west. However, the driver of the other car is maintaining a constant speed and direction relative to the road. Is the reference frame of your car an inertial one? If not, in which direction (east or west) is your car accelerating relative to the other car?

3 • CONTEXT-RICH You are riding in a limousine that has opaque windows that do not allow you to see outside. The car is on a flat horizontal plain, so the car can accelerate by speeding up, slowing down, or turning. Equipped with just a small heavy object on the end of a string, how can you use it to determine if the limousine is changing either speed or direction? Can you determine the limousine's velocity? SSM

4 •• If only a single nonzero force acts on an object, does the object accelerate relative to all inertial reference frames? Is it possible for such an object to have zero velocity in some inertial reference frame and not in another? If so, give a specific example.

5 •• A baseball is acted upon by a single known force. From this information alone, can you tell in which direction the baseball is moving relative to some reference frame? Explain.

6 •• A truck moves directly away from you at constant velocity (as observed by you while standing in the middle of the road). It follows that (*a*) no forces act on the truck, (*b*) a constant net force acts on the truck in the direction of its velocity, (*c*) the net force acting on the truck is zero, (*d*) the net force acting on the truck is its weight.

7 • ENGINEERING APPLICATION Several space probes have been launched that are now far out in space. *Pioneer 10*, for example, was launched in the 1970s and is still moving away from the Sun and its planets. Is the mass of *Pioneer 10* changing? Which of the known fundamental forces continue to act on it? Does it have a net force on it?

8 •• ENGINEERING APPLICATION Astronauts in apparent weightlessness during their stay on the International Space Station must carefully monitor their masses because significant loss of body mass is known to cause serious medical problems. Give an example of how you might design equipment to measure the mass of an astronaut on the orbiting space station.

9 •• CONTEXT-RICH You are riding in an elevator. Describe two situations in which your apparent weight is greater than your true weight. SSM

10 •• Suppose you are in a train moving at constant velocity relative to the ground. You toss a ball to your friend several seats in front of you. Use Newton's second law to explain why you cannot use your observations of the tossed ball to determine the train's velocity relative to the ground.

11 •• Explain why, of the fundamental interactions, gravitational interaction is the main concern in our everyday lives. One other on this list also plays an increasingly significant role in our rapidly advancing technology. Which one is that? Why are the others not obviously important?

12 •• Give an example of an object that has three forces acting on it, and (*a*) accelerates, (*b*) moves at constant (nonzero) velocity, and (*c*) remains at rest.

13 •• Suppose a block of mass m_1 rests on a block of mass m_2 and the combination rests on a table as shown in Figure 4-33. Tell the name of the force and its category (contact versus action-at-a-distance) for each of the following forces: (*a*) force exerted by m_1 on m_2, (*b*) force exerted by m_2 on m_1, (*c*) force exerted by m_2 on the table, (*d*) force exerted by the table on m_2, (*e*) force exerted by Earth on m_2. Which, if any, of these forces constitute a Newton's third-law pair of forces? SSM

FIGURE 4-33 Problem 13

14 •• CONTEXT-RICH You yank a fish you have just caught on your line upward from rest into your boat. Draw a free-body diagram of the fish after it has left the water and as it gains speed as it rises. In addition, tell the type (tension, spring, gravity, normal, friction, etc.) and category (contact versus action-at-a-distance) of each force on your diagram. Which, if any, pairs of the forces on your diagram constitute a Newton's third-law pair? Can you tell the relative magnitudes of the forces on your diagram from the information given? Explain.

15 •• If you gently set a fancy plate on the table, it will not break. However if you drop it from a height, it might very well break. Discuss the forces that act on the plate (as it contacts the table) in both these situations. Use kinematics and Newton's second law to describe what is different about the second situation that causes the plate to break?

16 •• For each of the following forces, give what produces it, what object it acts on, its direction, and the reaction force. (*a*) The force you exert on your briefcase as you hold it while standing at the bus stop. (*b*) The normal force on the soles of your feet as you stand barefooted on a horizontal wood floor. (*c*) The gravitational force on you as you stand on a horizontal floor. (*d*) The horizontal force exerted on a baseball by a bat as the ball is hit straight up the middle toward center field for a single.

17 •• For each case, identify the force (including its direction) that causes the acceleration. (*a*) A sprinter at the very start of the race. (*b*) A hockey puck skidding freely but slowly coming to rest on the ice. (*c*) A long fly ball at the top of its arc. (*d*) A bungee jumper at the very bottom of her descent.

18 • True or false:

(*a*) If two external forces that are both equal in magnitude and opposite in direction act on the same object, the two forces can never be a Newton's third-law pair.

(*b*) The two forces of a Newton's third-law pair are equal only if the objects involved are not accelerating.

19 •• An 80-kg man on ice skates is pushing his 40-kg son, also on skates, with a force of 100 N. Together, they move across the ice steadily gaining speed. (*a*) The force exerted by the boy on his father is (1) 200 N, (2) 100 N, (3) 50 N, or (4) 40 N. (*b*) How do the magnitudes of the two accelerations compare? (*c*) How do the directions of the two accelerations compare?

20 •• A girl holds a stone in her hand and can move it up or down or keep it still. True or false: (*a*) The force exerted by her hand on the rock is always the same magnitude as the force of gravity on the stone. (*b*) The force exerted by her hand on the rock is the reaction force to the force of gravity on the stone. (*c*) The force exerted by her hand on the stone is always the same magnitude as the force on her hand by the stone, but in the opposite direction. (*d*) If the girl moves her hand down at a constant speed, then her upward force on the stone is less than the force of gravity on the stone. (*e*) If the girl moves her hand downward but slows the stone to rest, then the force of the stone on the girl's hand is the same magnitude as the force of gravity on the stone.

21 •• A 2.5-kg object hangs at rest from a string attached to the ceiling. (*a*) Draw a free-body diagram of the object, indicate the reaction force to each force drawn and tell what object the reaction force acts on. (*b*) Draw a free-body diagram of the string, indicate the reaction force to each force drawn, and tell what object each reaction force acts on. Do not neglect the mass of the string. SSM

22 •• (*a*) Which of the free-body diagrams in Figure 4-34 represents a block sliding down a frictionless inclined surface? (*b*) For the correct diagram, label the forces and tell which are contact forces and which are action-at-a-distance forces. (*c*) For each force in the correct diagram, identify the reaction force, the object it acts on and its direction.

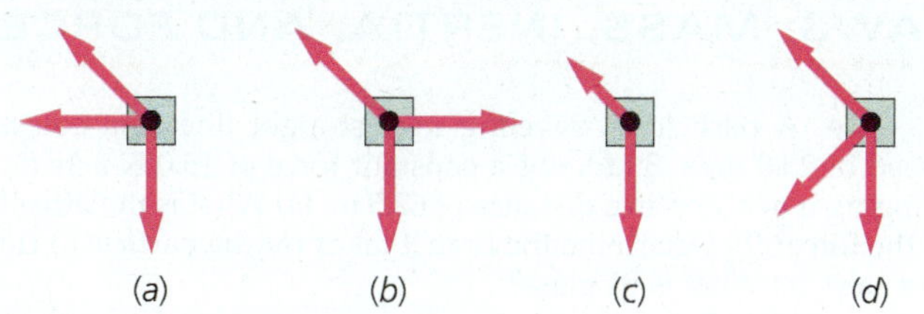

FIGURE 4-34 Problem 22

23 •• A wooden box on the floor is pressed against a compressed, horizontal spring that is attached to a wall. The horizontal floor beneath the box is frictionless. Draw the free-body diagram of the box in the following cases. (*a*) The box is held at rest against the compressed spring. (*b*) The force holding the box against the spring no longer exists, but the box is still in contact with the spring. (*c*) When the box no longer has contact with the spring.

24 •• Imagine yourself seated on a wheeled desk chair at your desk. Consider friction forces between the chair and the floor to be negligible. However, the friction forces between the desk and the floor are not negligible. When sitting at rest, you decide you need another cup of coffee. You push horizontally against the desk, and the chair rolls backward away from the desk. (*a*) Draw a free-body diagram of yourself during the push and clearly indicate which force was responsible for your acceleration. (*b*) What is the reaction force to the force that caused your acceleration? (*c*) Draw the free-body diagram of the desk and explain why it did not accelerate. Does this violate Newton's third law? Explain.

25 ••• The same (net) horizontal force F is applied for a fixed time interval Δt to each of two objects, having masses m_1 and m_2, that sit on a flat, frictionless surface. (Let $m_1 > m_2$.) (*a*) Assuming the two objects are initially at rest, what is the ratio of their accelerations during the time interval, in terms of F, m_1 and m_2? (*b*) What is the ratio of their speeds v_1 and v_2 at the end of the time interval? (*c*) How far apart are the two objects (and which is ahead) at the end of the time interval?

ESTIMATION AND APPROXIMATION

26 •• **CONCEPTUAL** Most cars have four springs attaching the body to the frame, one at each wheel position. Devise an experimental method of estimating the force constant of one of the springs using your known weight and the weights of several of your friends. Assume the 4 springs are identical. Use the method to estimate the force constant of your car's springs.

27 •• Estimate the force exerted on the goalie's glove by the puck when he catches a hard slap shot for a save. SSM

28 •• A baseball player slides into second base during a steal attempt. Assuming reasonable values for the length of the slide, the speed of the player at the beginning of the slide, and the speed of the player at the end of the slide, estimate the average force of friction acting on the player.

29 •• **ENGINEERING APPLICATION** A race car skidding out of control manages to slow down to 90 km/h before crashing head-on into a brick wall. Fortunately, the driver is wearing a safety harness. Using reasonable values for the mass of the driver and the stopping distance, estimate the average force exerted on the driver by the safety harness, including its direction. Neglect any effects of frictional forces on the driver by the seat.

NEWTON'S FIRST AND SECOND LAWS: MASS, INERTIA, AND FORCE

30 • A particle is traveling in a straight line at a constant speed of 25.0 m/s. Suddenly, a constant force of 15.0 N acts on it, bringing it to a stop in a distance of 62.5 m. (*a*) What is the direction of the force? (*b*) Determine the time it takes for the particle to come to a stop. (*c*) What is its mass?

31 • An object has an acceleration of 3.0 m/s^2 when a single force of magnitude F_0 acts on it. (*a*) What is the magnitude of its acceleration when the magnitude of this force is doubled? (*b*) A second object has an acceleration magnitude of 9.0 m/s^2 under the influence of a single force of magnitude F_0. What is the ratio of the mass of the second object to that of the first object? (*c*) If the two objects are glued together to form a composite object, what acceleration magnitude will a single force of magnitude F_0 acting on the composite object produce?

32 • A tugboat tows a ship with a constant force of magnitude F_1. The increase in the ship's speed during a 10-s interval is 4.0 km/h. When a second tugboat applies an additional constant force of magnitude F_2 in the same direction, the speed increases by 16 km/h during a 10-s interval. How do the magnitudes of F_1 and F_2 compare? (Neglect the effects of water resistance and air resistance.)

33 • A single constant force of magnitude 12 N acts on a particle of mass m. The particle starts from rest and travels in a straight line a distance of 18 m in 6.0 s. Find m.

34 • A net force of $(6.0\text{ N})\hat{i} - (3.0\text{ N})\hat{j}$ acts on a 1.5 kg object. Find the acceleration $\vec{a}$.

35 •• A bullet of mass 1.80×10^{-3} kg moving at 500 m/s impacts a tree stump and penetrates 6.00 cm into the wood before coming to rest. (*a*) Assuming that the acceleration of the bullet is constant, find the force (including direction) exerted by the wood on the bullet. (*b*) If the same force acted on the bullet and it had the same impact speed but half the mass, how far would it penetrate into the wood? SSM

36 •• A cart on a horizontal, linear track has a fan attached to it. The cart is positioned at one end of the track, and the fan is turned on. Starting from rest, the cart takes 4.55 s to travel a distance of 1.50 m. The mass of the cart plus fan is 355 g. Assume that the cart travels with constant acceleration. (*a*) What is the net force exerted on the cart–fan combination? (*b*) Mass is added to the cart until the total mass of the cart–fan combination is 722 g, and the experiment is repeated. How long does it take for the cart, starting from rest, to travel 1.50 m now? Ignore the effects due to friction.

37 •• A horizontal force of magnitude F_0 causes an acceleration of magnitude 3.0 m/s^2 when it acts on an object of mass m sliding on a frictionless surface. Find the magnitude of the acceleration of the same object in the circumstances shown in Figure 4-35*a* and 4-35*b*.

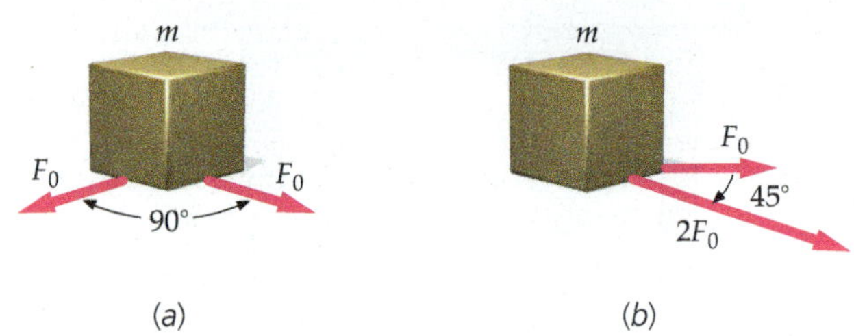

FIGURE 4-35 Problem 37

38 •• Al and Bert stand in the middle of a large frozen lake (frictionless surface). Al pushes on Bert with a force of 20 N for 1.5 s. Bert's mass is 100 kg. Assume that both are at rest before Al pushes Bert. (*a*) What is the speed that Bert reaches as he is pushed away from Al? (*b*) What speed does Al reach if his mass is 80 kg?

39 •• If you push a block whose mass is m_1 across a frictionless floor with a horizontal force of a magnitude F_0, the block has an acceleration of 12 m/s^2. If you push on a different block whose mass is m_2 with a horizontal force of magnitude F_0, its acceleration is 3.0 m/s^2. (*a*) What acceleration will a horizontal force of magnitude F_0 give to a single block with mass $m_2 - m_1$? (*b*) What acceleration will a horizontal force of magnitude F_0 give to a single block with mass $m_2 + m_1$?

40 •• **MULTISTEP** To drag a 75.0-kg log along the ground at constant velocity, your tractor has to pull it with a horizontal force of 250 N. (*a*) Draw the free-body diagram of the log. (*b*) Use Newton's laws to determine the force of friction on the log. (*c*) What is the normal force of the ground on the log? (*d*) What horizontal force must you exert if you want to give the log an acceleration of 2.00 m/s^2 assuming the force of friction does not change. Redraw the log's free-body diagram for this situation.

41 •• A 4.0-kg object is subjected to two constant forces, $\vec{F}_1 = (2.0\text{ N})\hat{i} + (-3.0\text{ N})\hat{j}$ and $\vec{F}_2 = (4.0\text{ N})\hat{i} - (11\text{ N})\hat{j}$. The object is at rest at the origin at time $t = 0$. (*a*) What is the object's acceleration? (*b*) What is its velocity at time $t = 3.0$ s? (*c*) Where is the object at time $t = 3.0$ s?

MASS AND WEIGHT

42 • On the moon, the acceleration due to the effect of gravity is only about 1/6 of that on Earth. An astronaut whose weight on Earth is 600 N travels to the lunar surface. His mass, as measured on the moon, will be (*a*) 600 kg, (*b*) 100 kg, (*c*) 61.2 kg, (*d*) 9.81 kg, (*e*) 360 kg.

43 • Find the weight of a 54-kg student in (*a*) newtons, and (*b*) pounds.

44 • Find the mass of a 165-lb engineer in kilograms.

45 •• **ENGINEERING APPLICATION** To train astronauts to work on the moon, where the free-fall acceleration is only about 1/6 of that on Earth, NASA submerges them in a tank of water. If an astronaut who is carrying a backpack, air conditioning unit, oxygen supply, and other equipment, has a total mass of 250 kg, determine the following quantities. (*a*) her weight including backpack, etc. on Earth, (*b*) her weight on the moon, (*c*) the required upward buoyancy force of the water during her training for the moon's environment on Earth. SSM

46 •• It is the year 2075 and space travel is common. A physics professor brings his favorite teaching demonstration with him to the moon. The apparatus consists of a very smooth (frictionless) horizontal table and an object to slide on it. On Earth, when the professor attaches a spring (force constant 50 N/m) to the object and pulls horizontally so the spring stretches 2.0 cm, the object accelerates at 1.5 m/s^2. (*a*) Draw the free-body diagram of the object and use it and Newton's laws to determine the object's mass. (*b*) What would the object's acceleration be under identical conditions on the moon?

FREE-BODY DIAGRAMS: STATIC EQUILIBRIUM

47 • **ENGINEERING APPLICATION, MULTISTEP** A 35.0-kg traffic light is supported by two wires as in Figure 4-36. (*a*) Draw the light's free-body diagram and use it to answer the following question qualitatively: Is the tension in wire 2 greater than or less than the tension in wire 1? (*b*) Verify your answer by applying Newton's laws and solving for the two tensions.

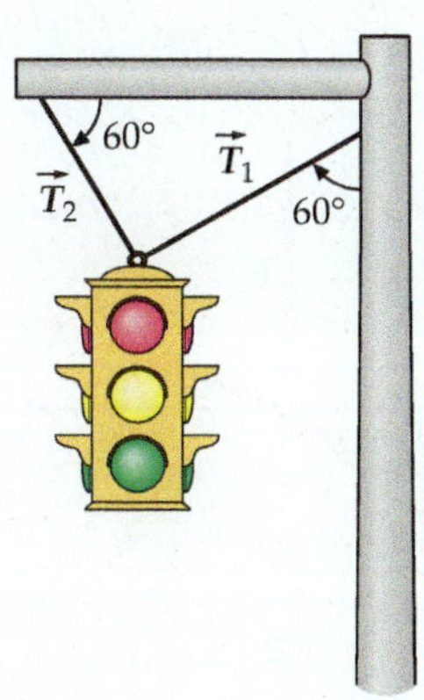

FIGURE 4-36 Problem 47

48 • A 42.6-kg lamp is hanging from wires as shown in Figure 4-37. The ring has negligible mass. The tension T_1 in the vertical wire is (*a*) 209 N, (*b*) 418 N, (*c*) 570 N, (*d*) 360 N, (*e*) 730 N.

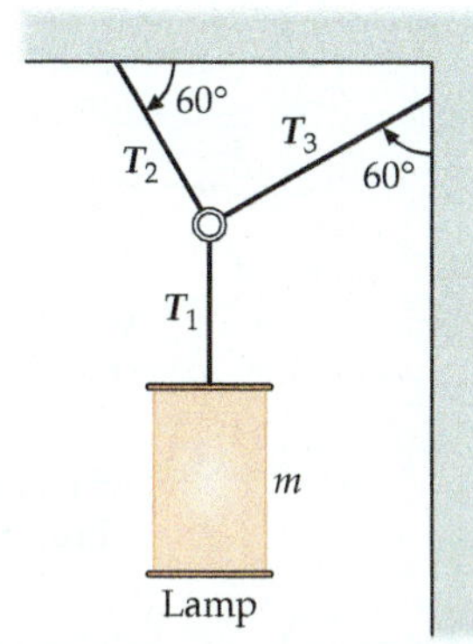

FIGURE 4-37 Problem 48

49 •• In Figure 4-38*a*, a 0.500-kg block is suspended at the midpoint of a 1.25-m-long string. The ends of the string are attached to the ceiling at points separated by 1.00 m. (*a*) What angle does the string make with the ceiling? (*b*) What is the tension in the string? (*c*) The 0.500-kg block is removed and two 0.250-kg blocks are attached to the string such that the lengths of the three string segments are equal (Figure 4-38*b*). What is the tension in each segment of the string? SSM

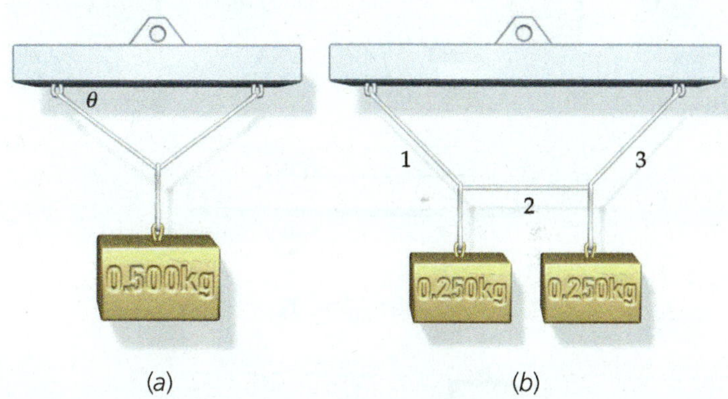

FIGURE 4-38 Problem 49

50 •• A ball weighing 100-N is shown suspended from a system of cords (Figure 4-39). What are the tensions in the horizontal and angled cords?

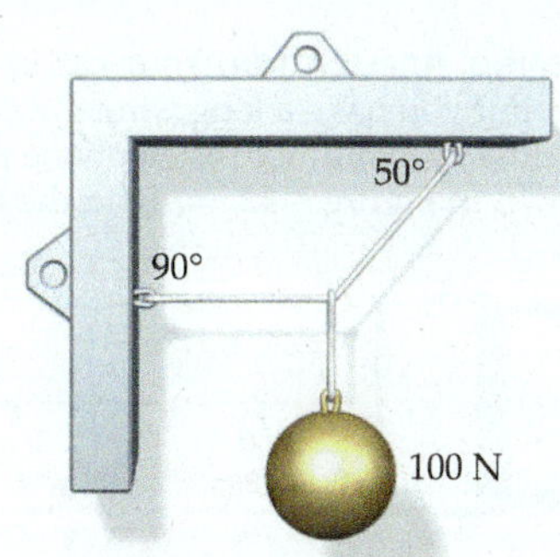

FIGURE 4-39 Problem 50

51 •• A 10-kg object on a frictionless table is subjected to two horizontal forces, $\vec{F}_1$ and $\vec{F}_2$, with magnitudes $F_1 = 20$ N and $F_2 = 30$ N, as shown in Figure 4-40. Find the third horizontal force $\vec{F}_3$ that must be applied so that the object is in static equilibrium. SSM

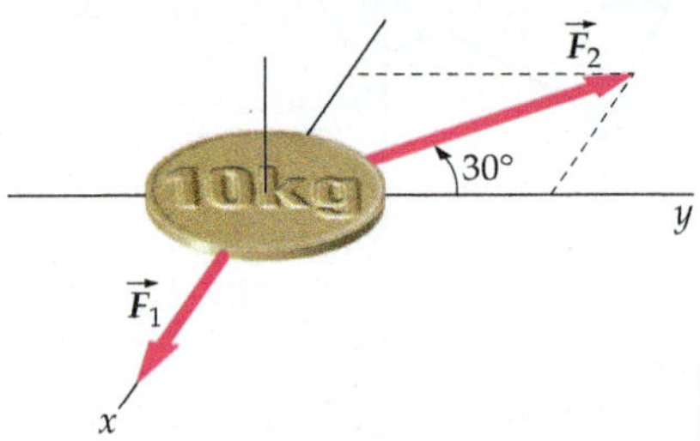

FIGURE 4-40
Problem 51

52 •• For the systems to be in equilibrium in Figure 4-41*a*, Figure 4-41*b*, and Figure 4-41*c*, find the unknown tensions and masses.

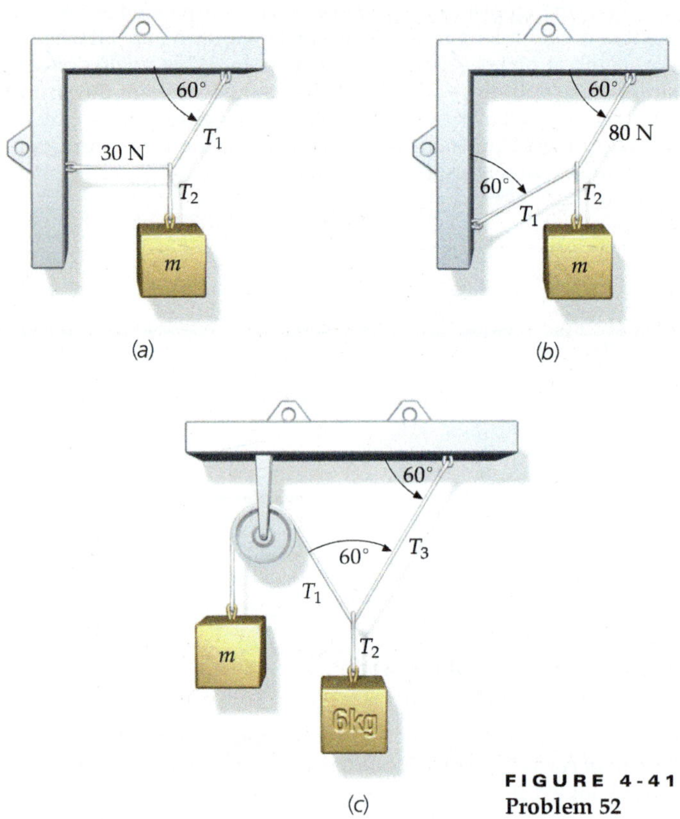

FIGURE 4-41
Problem 52

53 •• **Engineering Application** Your car is stuck in a mud hole. You are alone, but you have a long, strong rope. Having studied physics, you tie the rope tautly to a telephone pole and pull on it sideways, as shown in Figure 4-42. (*a*) Find the force exerted by the rope on the car when the angle θ is 3.00° and you are pulling with a force of 400 N, but the car does not move. (*b*) How strong must the rope be if it takes a force of 600 N to move the car when θ is 4.00°?

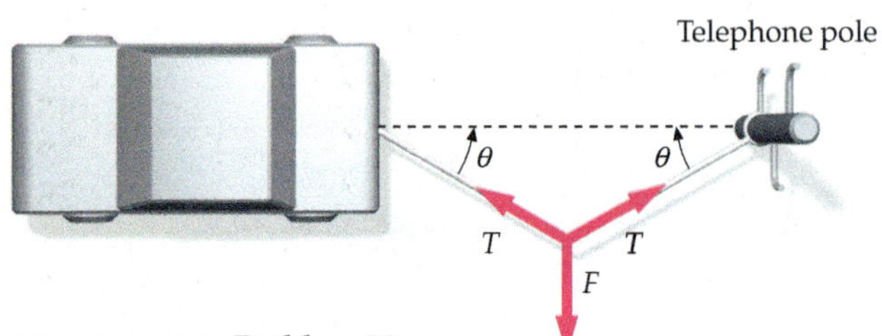

FIGURE 4-42 **Problem 53**

54 ••• **Engineering Application, Multistep** Balloon arches are often seen at festivals or celebrations; they are made by attaching helium-filled balloons to a rope that is fixed to the ground at each end. The lift from the balloons raises the structure into the arch shape. Figure 4-43*a* shows the geometry of such a structure: *N* balloons are attached at equally spaced intervals along a massless rope of length *L*, which is attached to two supports at its ends. Each balloon provides a lift force of magnitude *F*. The horizontal and vertical coordinates of the point on the rope where the *i*th balloon is attached are x_i and y_i, and T_i is the tension in the *i*th segment. (Note segment 0 is the segment between the point of attachment and the first balloon, and segment *N* is the segment between the last balloon and the other point of attachment). (*a*) Figure 4-43*b* shows a free-body diagram for the *i*th balloon. From this diagram, show that the horizontal component of the force T_i (call it T_H) is the same for all the string segments. (*b*) By considering the vertical component of the forces, use Newton's laws to derive the following relationship between the tension in the *i*th and $(i-1)$th segments: $T_{i-1}\sin\theta_{i-1} - T_i\sin\theta_i = F$. (*c*) Show that $\tan\theta_0 = -\tan\theta_{N+1} = NF/2T_H$. (*d*) From the diagram and the two expressions above, show that $\tan\theta_i = (N - 2i)F/2T_H$ and that $y_i = \frac{L}{N+1}\sum_{j=0}^{i-1}\cos\theta_j$, $y_i = \frac{L}{N+1}\sum_{j=0}^{i-1}\sin\theta_j$.

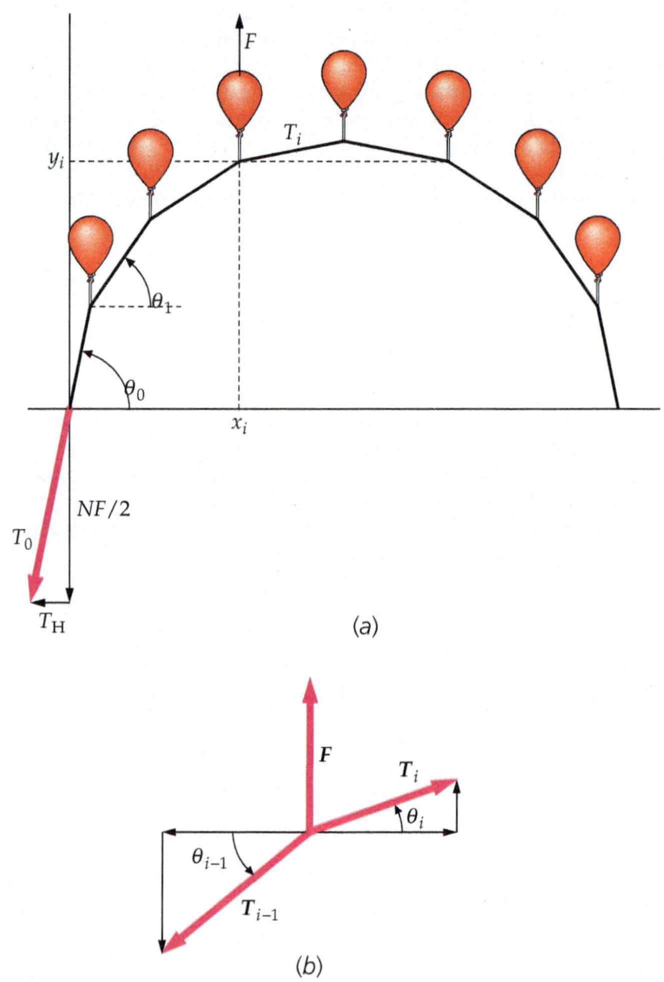

FIGURE 4-43 **Problem 54**

55 ••• **Engineering Application, Spreadsheet** (*a*) Consider a numerical solution to Problem 54. Write a **spreadsheet** program to make a graph of the shape of a balloon arch. Use the following parameters: $N = 10$ (balloons), each providing a lift force $F = 1.0$ N and each attached to a rope of length $L = 11$ m, with a horizontal component of tension $T_H = 10$ N. How far apart are the two points of attachment? How high is the arch at its highest point? (*b*) Note that we have not specified the spacing between the supports—it is determined by the other parameters. Vary T_H while keeping the other parameters the same until you create an arch that has a spacing of 8.0 m between the supports. What is T_H then? As you increase T_H, the arch should get flatter and more spread out. Does your spreadsheet model show this?

FREE-BODY DIAGRAMS: INCLINED PLANES AND THE NORMAL FORCE

56 • A large box whose mass is 20.0 kg rests on a frictionless floor. A mover pushes on the box with a force of 250 N at an angle 35.0° below the horizontal. Draw the box's free-body diagram and use it to determine the acceleration of the box.

57 • A 20.0-kg box rests on a frictionless ramp with a 15.0° slope. The mover pulls on a rope attached to the box to pull it up the incline (Figure 4-44). If the rope makes an angle of 40.0° with the horizontal, what is the smallest force F the mover will have to exert to move the box up the ramp?

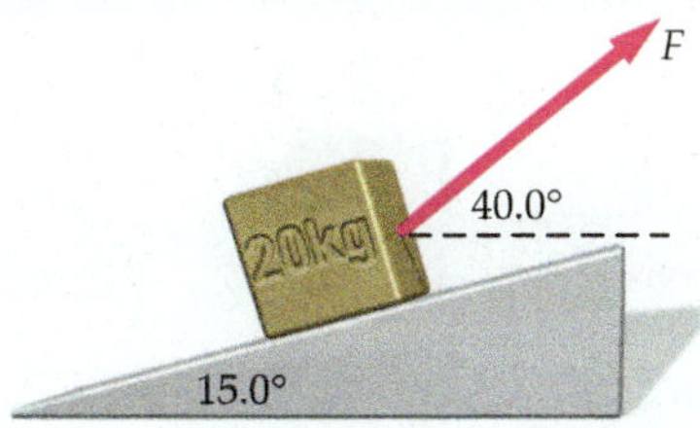

FIGURE 4-44 Problem 57

58 • In Figure 4-45, the objects are attached to spring scales calibrated in newtons. Give the reading(s) of the balance(s) in each case, assuming that both the scales and the strings are massless.

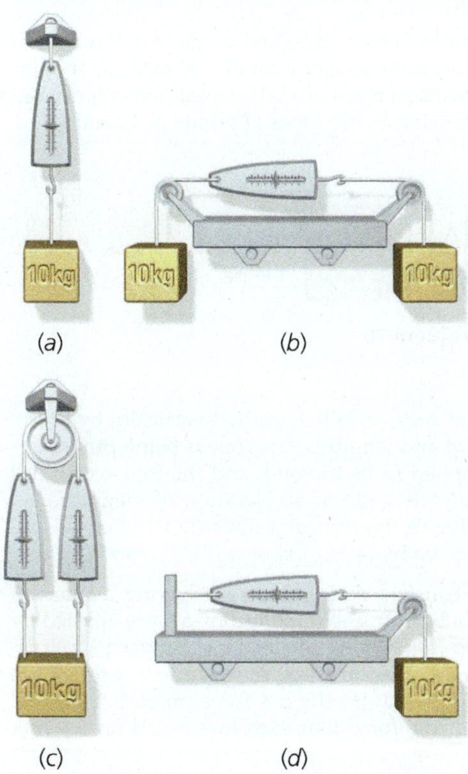

FIGURE 4-45 Problem 58

59 •• A box is held in position on a frictionless incline by a cable (Figure 4-46). (*a*) If $\theta = 60°$ and $m = 50$ kg, find the tension in the cable and the normal force exerted by the incline. (*b*) Find the tension as a function of θ and m, and check your result for plausibility in the special cases of $\theta = 0°$ and $\theta = 90°$.

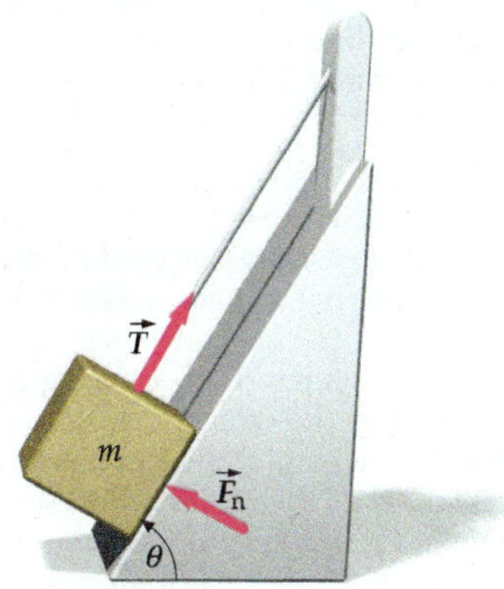

FIGURE 4-46 Problem 59

60 •• A horizontal force of 100 N pushes a 12-kg block up a frictionless incline that makes an angle of 25° with the horizontal. (*a*) What is the normal force that the incline exerts on the block? (*b*) What is the magnitude of acceleration of the block?

61 •• A 65-kg student weighs himself by standing on a force scale mounted on a skateboard that is rolling down an incline, as shown in Figure 4-47. Assume there is no friction so that the force exerted by the incline on the skateboard is normal to the incline. What is the reading on the scale if $\theta = 30°$? SSM

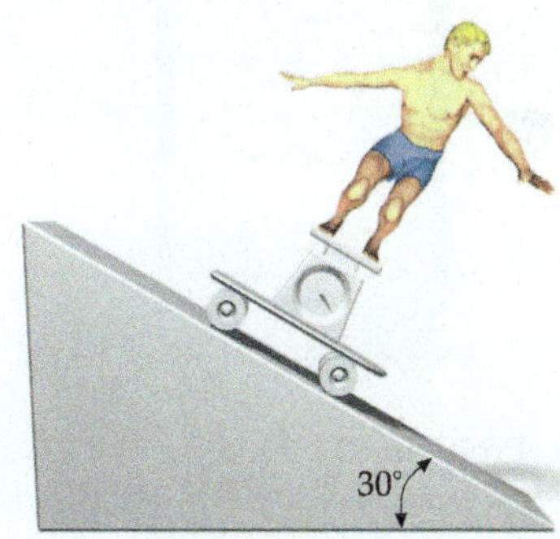

FIGURE 4-47 Problem 61

62 •• A block of mass m slides across a frictionless floor and then up a frictionless ramp (Figure 4-48). The angle of the ramp is θ and the speed of the block before it starts up the ramp is v_0. The block will slide up to some maximum height h above the floor before stopping. Show that h is independent of m and θ by deriving an expression for h in terms of v_0 and g.

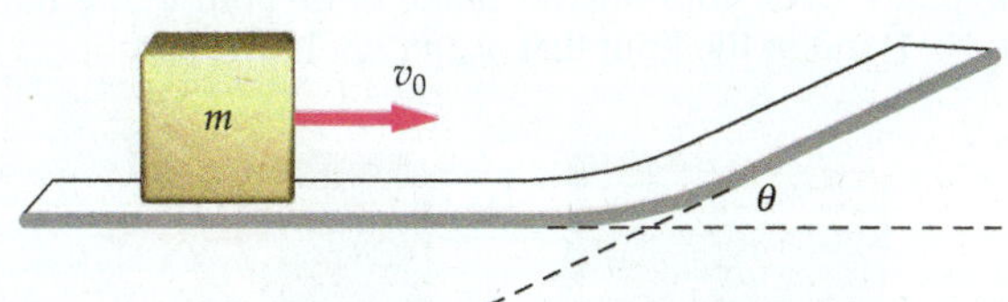

FIGURE 4-48 Problem 62

FREE-BODY DIAGRAMS: ELEVATORS

63 • **CONCEPTUAL** (*a*) Draw the free-body diagram (with accurate relative force magnitudes) for an object that is hung by a rope from the ceiling of an elevator that is ascending but slowing. (*b*) Repeat Part (*a*) but for the situation in which the elevator is descending and speeding up. (*c*) Can you tell the difference between the two diagrams? Explain why the diagrams do not tell anything about the object's velocity. SSM

64 • A 10.0-kg block is suspended from the ceiling of an elevator by a cord rated to withstand a tension of 150 N. Shortly after the elevator starts to ascend, the cord breaks. What was the minimum acceleration of the elevator when the cord broke?

65 •• A 2.0-kg block hangs from a spring scale calibrated in newtons that is attached to the ceiling of an elevator (Figure 4-49). What does the scale read when (*a*) the elevator is ascending with a constant speed of 30 m/s; (*b*) the elevator is descending with a constant speed of 30 m/s; (*c*) the elevator is ascending at 20 m/s and gaining speed at a rate of 3.0 m/s²? (*d*) Suppose that from $t = 0$ to $t = 5.0$ s, the elevator ascends at a constant speed of 10 m/s. Its speed is then steadily reduced to zero during the next 4.0 s, so that it is at rest at $t = 9.0$ s. Describe the reading of the scale during the interval $0 < t < 9.0$ s.

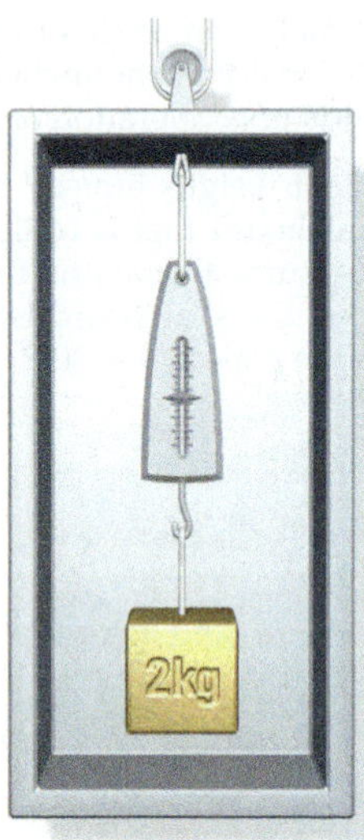

FIGURE 4-49 Problem 65

FREE-BODY DIAGRAMS: SEVERAL OBJECTS AND NEWTON'S THIRD LAW

66 •• **CONCEPTUAL** Two boxes of mass m_1 and m_2 connected by a massless string are being pulled along a horizontal frictionless surface by the tension force in a second string, as shown in Figure 4-50. (*a*) Draw the free-body diagram of both boxes separately and show that $T_1/T_2 = m_1/(m_1 + m_2)$. (*b*) Is this result plausible? Explain. Does your answer make sense both in the limit that $m_2/m_1 \gg 1$ and in the limit that $m_2/m_1 \ll 1$? Explain.

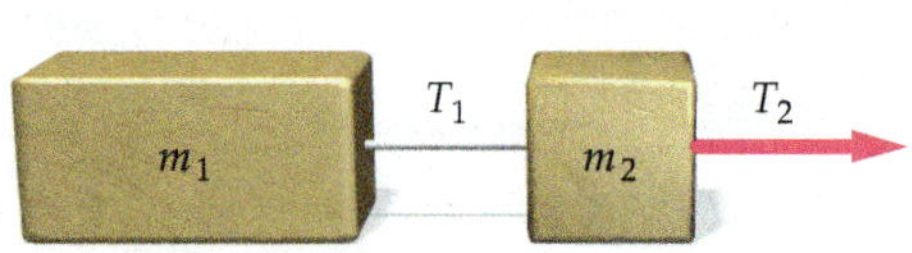

FIGURE 4-50 Problem 66

67 •• A box of mass $m_2 = 3.5$ kg rests on a frictionless horizontal shelf and is attached by strings to boxes of masses $m_1 = 1.5$ kg and $m_3 = 2.5$ kg as shown in Figure 4-51. Both pulleys are frictionless and massless. The system is released from rest. After it is released, find (*a*) the acceleration of each of the boxes, and (*b*) the tension in each string.

FIGURE 4-51 Problem 67

68 •• Two blocks are in contact on a frictionless horizontal surface. The blocks are accelerated by a single horizontal force $\vec{F}$ applied to one of them (Figure 4-52). Find the acceleration and the contact force of block 1 on block 2 (*a*) in terms of F, m_1 and m_2, and (*b*) for the specific values $F = 3.2$ N, $m_1 = 2.0$ kg and $m_2 = 6.0$ kg.

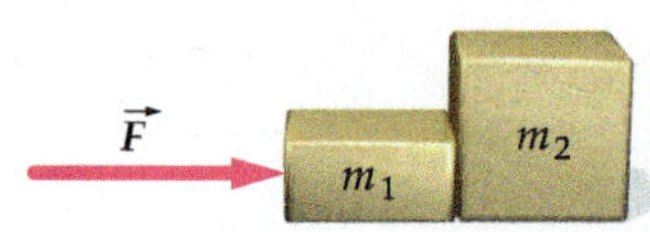

FIGURE 4-52 Problem 68

69 •• Repeat Problem 68, but with the two blocks interchanged. Are your answers for this problem the same as in Problem 68? Explain.

70 •• Two 100-kg boxes are dragged along a horizontal frictionless surface at a constant acceleration of 1.00 m/s², as shown in Figure 4-53. Each rope has a mass of 1.00 kg. Find the magnitude of the force $\vec{F}$ and the tension in the ropes at points A, B, and C.

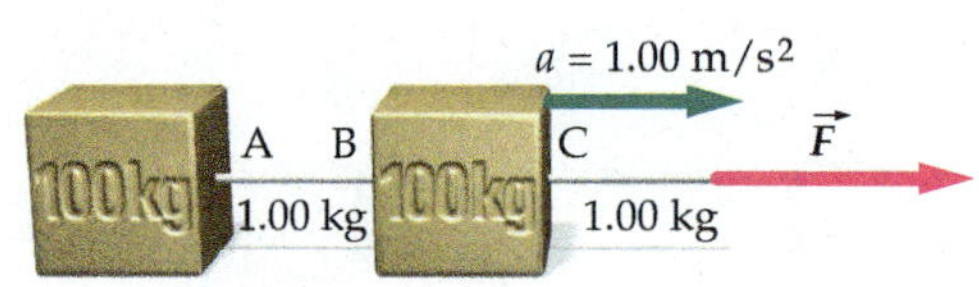

FIGURE 4-53 Problem 70

71 •• A block of mass m is being lifted vertically by a uniform rope of mass M and length L. The rope is being pulled upward by a force applied to its top end, and the rope and block are accelerating upward with an acceleration of magnitude a. Show that the tension in the rope at a distance x (where $x < L$) above the block is given by $(a + g)[m + (x/L)M]$. SSM

72 •• A chain consists of 5 links, each having a mass of 0.10 kg. The chain is being pulled upward by a force applied by your hand to its top link, giving the chain an upward acceleration of 2.5 m/s². Find (*a*) the force magnitude F exerted on the top link by your hand; (*b*) the net force on each link; and (*c*) the magnitude of the force that each link exerts on the link below it.

73 •• **MULTISTEP** A 40.0-kg object supported by a vertical rope. The rope, and thus the object, is then accelerated from rest upward so that it attains a speed of 3.50 m/s in 0.700 s. (*a*) Draw the object's free-body diagram with the relative lengths of the vectors showing the relative magnitudes of the forces. (*b*) Use the free-body diagram and Newton's laws to determine the tension in the rope. SSM

74 •• **ENGINEERING APPLICATION, MULTISTEP** A 15000-kg helicopter is lowering a 4000-kg truck to the ground by a cable of fixed length. The truck, helicopter, and cable are descending at 15.0 m/s and must be slowed to 5.00 m/s in the next 50.0 m of descent to prevent damaging the truck. Assume a constant rate of slowing. (*a*) Draw the free-body diagram of the truck. (*b*) Determine the tension in the cable. (*c*) Determine the lift force on the helicopter blades.

75 •• Two objects are connected by a massless string, as shown in Figure 4-54. The incline and the massless pulley are frictionless. Find the acceleration of the objects and the tension in the string (*a*) in terms of θ, m_1, and m_2, and for (*b*) $\theta = 30°$ and $m_1 = m_2 = 5.0\,\text{kg}$.

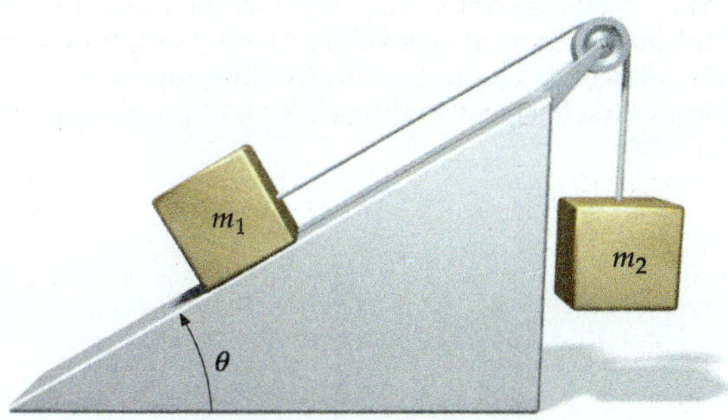

FIGURE 4-54 Problem 75

76 •• **ENGINEERING APPLICATION** During a stage production of *Peter Pan*, the 50-kg actress playing Peter has to fly in vertically (descend). To be in time with the music, she must, starting from rest, be lowered a distance of 3.2 m in 2.2 s at a constant acceleration. Backstage, a smooth surface sloped at 50° supports a counterweight of mass *m*, as shown in Figure 4-55. Show the calculations that the stage manager must perform to find (*a*) the mass of the counterweight that must be used and (*b*) the tension in the wire.

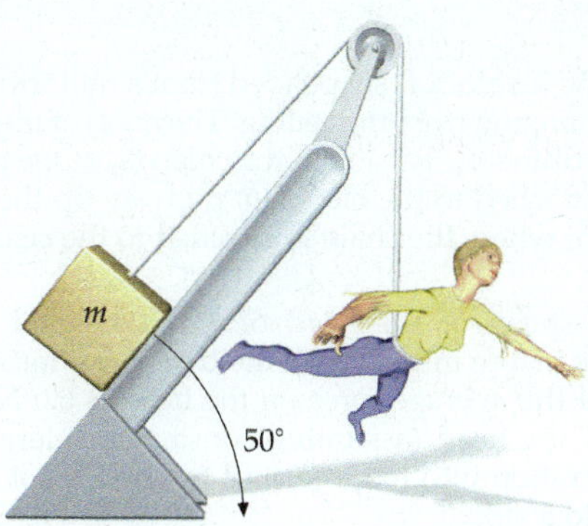

FIGURE 4-55 Problem 76

77 •• An 8.0-kg block and a 10-kg block, connected by a rope that passes over a frictionless peg, slide on frictionless incline, (Figure 4-56). (*a*) Find the acceleration of the blocks and the tension in the rope. (*b*) The two blocks are replaced by two others of masses m_1 and m_2 such that there is no acceleration. Find whatever information you can about the masses of these two new blocks.

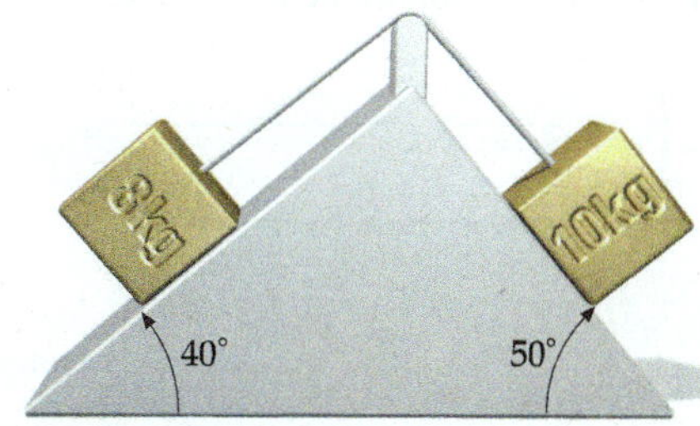

FIGURE 4-56 Problem 77

78 •• A heavy rope of length 5.0 m and mass 4.0 kg lies on a frictionless horizontal table. One end is attached to a 6.0-kg block. The other end of the rope is pulled by a constant horizontal 100-N force. (*a*) What is the acceleration of the system? (*b*) Give the tension in the rope as a function of position along the rope.

79 •• A 60-kg housepainter stands on a 15-kg aluminum platform. The platform is attached to a rope that passes through an overhead pulley, which allows the painter to raise herself and the platform (Figure 4-57). (*a*) With what force *F* must she pull down on the rope to accelerate herself and the platform upward at a rate of $0.80\,\text{m/s}^2$? (*b*) When her speed reaches 1.0 m/s, she pulls in such a way that she and the platform go up at a constant speed. What force is she exerting on the rope now? (Ignore the mass of the rope.) SSM

FIGURE 4-57 Problem 79

80 ••• Figure 4-58 shows a 20-kg block sliding on a 10-kg block. All surfaces are frictionless and the pulley is massless and frictionless. Find the acceleration of each block and the tension in the string that connects the blocks.

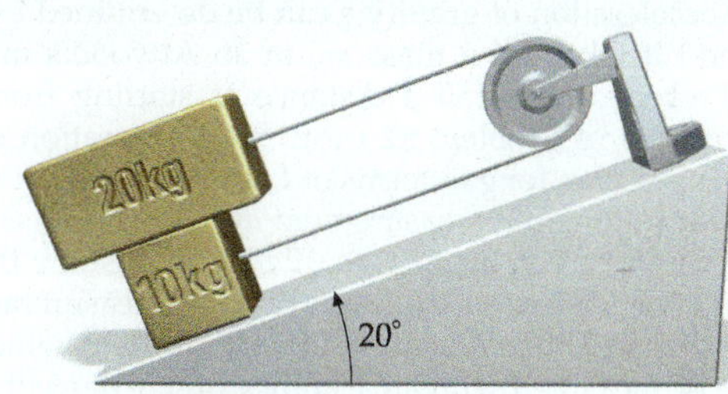

FIGURE 4-58 Problem 80

81 ••• A 20-kg block with a pulley attached slides along a frictionless ledge. It is connected by a massless string to a 5.0-kg block via the arrangement shown in Figure 4-59. Find (*a*) the acceleration of each block, and (*b*) the tension in the connecting string.

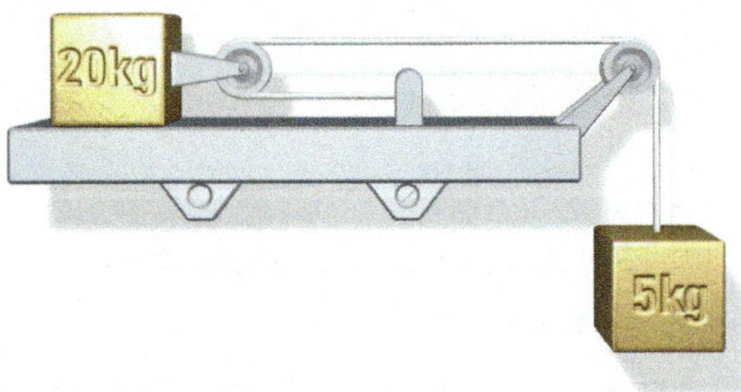

FIGURE 4-59 Problem 81

82 •• **MULTISTEP** The apparatus in Figure 4-60 is called an *Atwood's machine* and is used to measure the free-fall acceleration g by measuring the acceleration of the two blocks connected by a string over a pulley. Assume a massless, frictionless pulley and a massless string. (*a*) Draw the free-body diagram of each block. (*b*) Use the free-body diagrams and Newton's laws to show that the magnitude of the acceleration of either block and the tension in the string are $a = (m_1 - m_2)g/(m_1 + m_2)$ and $T = 2m_1m_2g/(m_1 + m_2)$. (*c*) Do these expressions give plausible results if $m_1 = m_2$, in the limit that $m_1 \gg m_2$ and in the limit that $m_1 \ll m_2$? Explain.

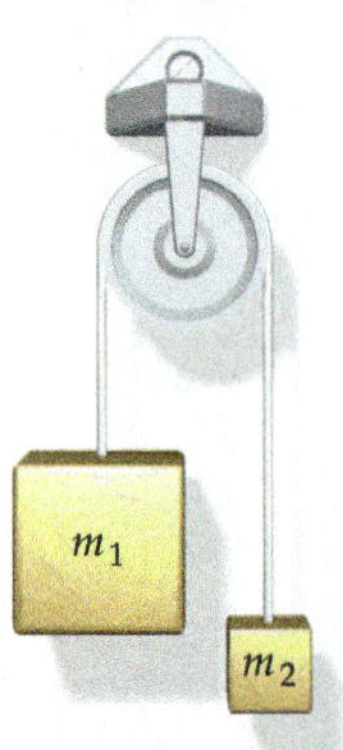

FIGURE 4-60 Problems 82 and 83

83 •• If one of the masses of the Atwood's machine in Figure 4-60 is 1.2 kg, what should be the other mass so that the displacement of either mass during the first second following release is 0.30 m? Assume a massless, frictionless pulley and a massless string.

84 ••• The acceleration of gravity g can be determined by measuring the time t it takes for a mass m_2 in an Atwood's machine described in Problem 82 to fall a distance L, starting from rest. (*a*) Using the results of Problem 82 (note the acceleration is constant), find an expression for g in terms of L, t, m_1, and m_2. (*b*) Show that a small error in the time measurement dt, will lead to an error in g by an amount dg given by $dg/g = -2dt/t$. (*c*) Assume that the only significant uncertainty in the experimental measurements is the time of fall. If $L = 3.00$ m and m_1 is 1.00 kg, find the value of m_2 such that g can be measured with an accuracy of ± 5 percent with a time measurement that is accurate to ± 0.1 s.

GENERAL PROBLEMS

85 •• A pebble of mass m rests on the block of mass m_2 of the ideal Atwood's machine in Figure 4-60. Find the force exerted by the pebble on the block of mass m_2.

86 •• A simple accelerometer can be made by suspending a small massive object from a string attached to a fixed point on an accelerating object. Suppose such an accelerometer is attached to point P on the ceiling of an automobile traveling in a straight line on a flat surface at constant acceleration. Due to the acceleration, the string will make an angle θ with the vertical. (*a*) Show that the magnitude of the acceleration a is related to the angle θ by $a = g\tan\theta$. (*b*) Suppose the automobile brakes steadily to rest from 50 km/h over a distance of 60 m. What angle will the string make with the vertical? Will the suspended object be positioned below and ahead or below and behind point P during the braking?

87 •• **ENGINEERING APPLICATION** The mast of a sailboat is supported at the bow and stern by stainless steel wires, the forestay and backstay, anchored 10 m apart (Figure 4-61). The 12.0-m-long mast weighs 800 N and stands vertically on the deck of the boat. The mast is positioned 3.60 m behind where the forestay is attached. The tension in the forestay is 500 N. Find the tension in the backstay and the force that the mast exerts on the deck. SSM

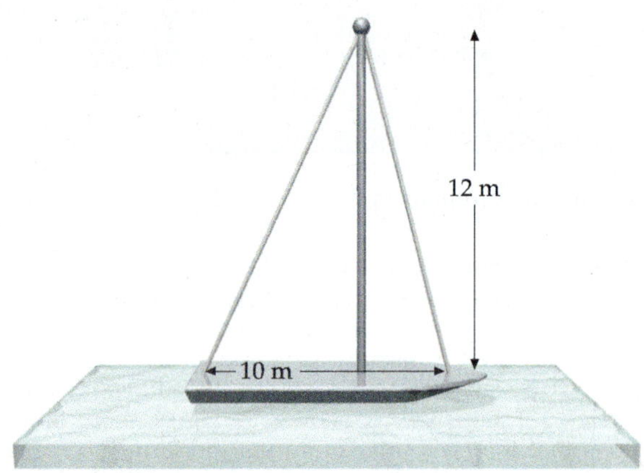

FIGURE 4-61 Problem 87

88 •• A 50-kg block is suspended from a uniform 1.5-m-long chain that is hanging from the ceiling. The mass of the chain itself is 20 kg. Determine the tension in the chain (*a*) at the point where the chain is attached to the block, (*b*) midway up the chain, and (*c*) at the point where the chain is attached to the ceiling.

89 •• The speed of the head of a red headed woodpecker reaches 5.5 m/s before impact with the tree. If the mass of the head is 0.060 kg and the average force on the head is 6.0 N, find (*a*) the acceleration of the head (assuming constant acceleration), (*b*) the depth of penetration into the tree, and (*c*) the time it takes for the head to come to a stop.

90 •• **MULTISTEP** A frictionless surface is inclined at an angle of 30.0° to the horizontal. A 270-g block on the ramp is attached to a 75.0-g block using a pulley, as shown in Figure 4-62. (*a*) Draw two free-body diagrams, one for the 270-g block and the other for the 75.0-g block. (*b*) Find the tension in the string and the acceleration of the 270-g block. (*c*) The 270-g block is released from rest.

FIGURE 4-62 Problem 90

How long does it take for it to slide a distance of 1.00 m along the surface? Will it slide up the incline, or down the incline?

91 •• A box of mass m_1 is pulled along a frictionless horizontal surface by a horizontal force $\vec{F}$ that is applied to the end of a rope of mass m_2 (see Figure 4-63). Neglect any sag of the rope. (*a*) Find the acceleration of the rope and block, assuming them to be one object. (*b*) What is the net force acting on the rope? (*c*) Find the tension in the rope at the point where it is attached to the block.

FIGURE 4-63 Problem 91

92 •• A 2.0-kg block rests on a frictionless wedge that has a 60° incline and an acceleration $\vec{a}$ to the right such that the mass remains stationary relative to the wedge (Figure 4-64). (*a*) Draw the free-body diagram of the block and use it to determine the magnitude of the acceleration. (*b*) What would happen if the wedge were given an acceleration larger than this value? Smaller than this value?

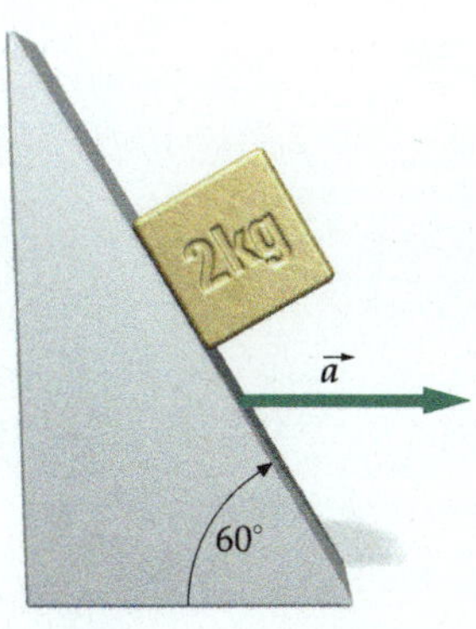

FIGURE 4-64 Problem 92

93 ••• The masses attached to each side of an ideal Atwood's machine consist of a stack of five washers, each of mass m, as shown in Figure 4-65. The tension in the string is T_0. When one of the washers is removed from the left side, the remaining washers accelerate and the tension decreases by 0.300 N. (*a*) Find m. (*b*) Find the new tension and the acceleration of each mass when a second washer is removed from the left side. SSM

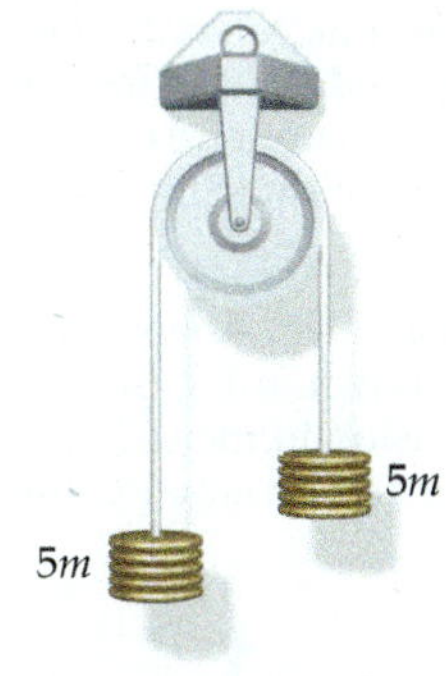

FIGURE 4-65 Problems 93 and 94

94 •• Consider the ideal Atwood's machine in Figure 4-65. When N washers are transferred from the left side to the right side, the right side descends 47.1 cm in 0.40 s. Find N.

95 •• Blocks of mass m and $2m$ are on a horizontal frictionless surface (Figure 4-66). The blocks are connected by a horizontal string. In addition, forces $\vec{F}_1$ and $\vec{F}_2$ are applied as shown. (*a*) If the forces shown are constant, find the tension in the connecting string. (*b*) If the magnitudes of the forces vary with time as $F_1 = Ct$ and $F_2 = 2Ct$, where C equals to 5.00 N/s and t is time, find the time t_0 at which the tension in the string equals to 10.0 N.

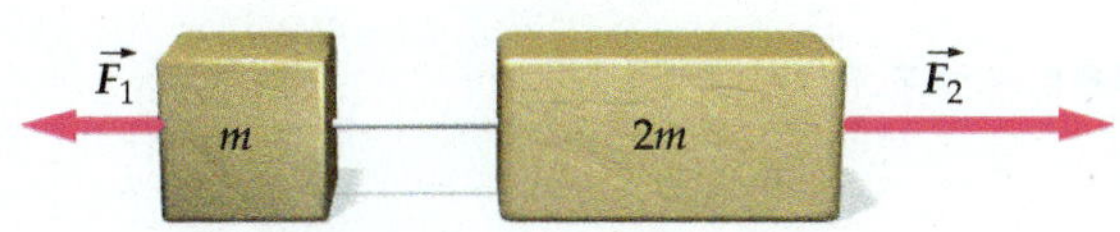

FIGURE 4-66 Problem 95

96 •• Elvis Presley has supposedly been sighted numerous times since his death on August 16, 1977. The following is a chart of what Elvis's weight would be if he were sighted on the surfaces of other objects in our solar system. Use the chart to determine: (*a*) Elvis's mass on Earth, (*b*) Elvis's mass on Pluto, and (*c*) the free-fall acceleration on Mars. (*d*) Compare the free-fall acceleration on Pluto to the free-fall acceleration on the moon.

Planet	Elvis's Weight (N)
Mercury	431
Venus	1031
Earth	1133
Mars	431
Jupiter	2880
Saturn	1222
Pluto	58
Moon	191

97 ••• **CONTEXT-RICH** As a prank, your friends have kidnapped you in your sleep, and transported you out onto the ice covering a local pond. When you wake up you are 30.0 m from the nearest shore. The ice is so slippery (i.e. frictionless) that you cannot seem to get yourself moving. You realize that you can use Newton's third law to your advantage, and choose to throw the heaviest thing you have, one boot, in order to get yourself moving. Take your weight to be 595 N. (*a*) What direction should you throw your boot so that you will most quickly reach the shore? (*b*) If you throw your 1.20-kg boot with an average force of 420 N, and the throw takes 0.600 s (the time interval over which you apply the force), what is the magnitude of the force that the boot exerts on you? (Assume constant acceleration.) (*c*) How long does it take you to reach shore, including the short time in which you were throwing the boot?

98 ••• The pulley of an ideal Atwood's machine is given an upward acceleration a, as shown in Figure 4-67. Find the acceleration of each mass and the tension in the string that connects them. The speeds of the two blocks are not equal in this situation.

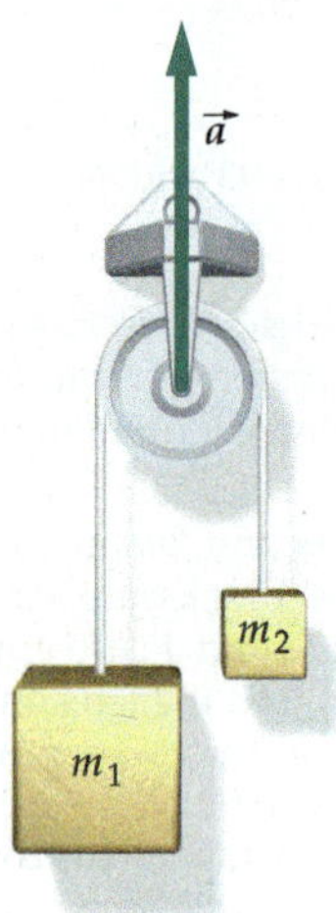

FIGURE 4-67 Problem 98

99 •• **ENGINEERING APPLICATION, CONTEXT-RICH, SPREADSHEET** You are working for an automotive magazine and putting a certain new automobile (mass 650 kg) through its paces. While accelerating from rest, its onboard computer records its velocity as a function of time as follows:

v_x (m/s):	0	10	20	30	40	50
t (s):	0	1.8	2.8	3.6	4.9	6.5

(*a*) Using a **spreadsheet,** find the average acceleration of the five time intervals and graph the velocity versus time and acceleration versus time for this car. (*b*) Where on the graph of velocity versus time is the net force on the car highest and lowest? Explain your reasoning. (*c*) What is the average net force on the car over the whole trip? (*d*) From the graph of velocity versus time, estimate the total distance covered by the car.

CHAPTER 5

DAYTONA INTERNATIONAL SPEEDWAY, THE "WORLD CENTER OF RACING," FEATURES A 2.5-MILE TRI-OVAL TRACK, WHICH HAS FOUR-STORY, 31-DEGREE-HIGH BANKED CURVES. IN THE DAYTONA 500 RACE THE STOCK CARS TRAVEL THROUGH THE CURVES AT SPEEDS CLOSE TO 200 MPH. SURPRISINGLY, THE FIERY CRASHES THAT THE DAYTONA 500 IS FAMOUS FOR, WITH THEIR ACCOMPANYING INJURIES AND FATALITIES, ARE USUALLY NOT CAUSED BY SKIDDING ON THE CURVES. *(PhotoDisc/Getty Images.)*

Additional Applications of Newton's Laws

In Chapter 4, we introduced Newton's laws and applied them to situations where action was restricted to straight-line motion and frictional forces were introduced. We now will consider some more general applications and how Newton's laws can be used to explain innumerable properties of the world in which we live.

In this chapter, we will extend the application of Newton's laws to motion along curved paths, and we will analyze the effects of resistive forces such as friction and air drag. We will also introduce the concept of the center of mass of a system of particles and show how modeling the system as a single particle located at the center of mass can result in being able to predict the bulk motion of such a system.

? What factors determine how fast a car can go through a curve without skidding? (See Example 5-12.)

5-1 FRICTION

If you shove a book that is resting on a desktop, the book will probably skid across the desktop. If the desktop is long enough, the book will eventually skid to a stop. This happens because a frictional force is exerted by the desktop on the book in a direction opposite to the book's velocity. This force, which acts on the surface of the book in contact with the desktop, is known as a *frictional force*. Frictional forces are a necessary part of our lives. Without friction our ground-based transportation system, from walking to automobiles, could not function. Friction allows you to start walking, and once you are already moving, friction allows you to change either your speed or direction. Friction allows you to start, steer, and stop a car. Friction holds a nut on a screw, a nail in wood, and a knot in a piece of rope. However, as important as friction is, it is often not desirable. Friction causes wear whenever moving pieces of machinery are in contact, and large amounts of time and money are spent trying to reduce such effects.

Friction is a complex, incompletely understood phenomenon that arises from the attraction between the molecules of one surface and the molecules on a second surface in close contact. The nature of this attraction is electromagnetic—the same as the molecular bonding that holds an object together. This short-ranged attractive force becomes negligible at distances of only a few atomic diameters.

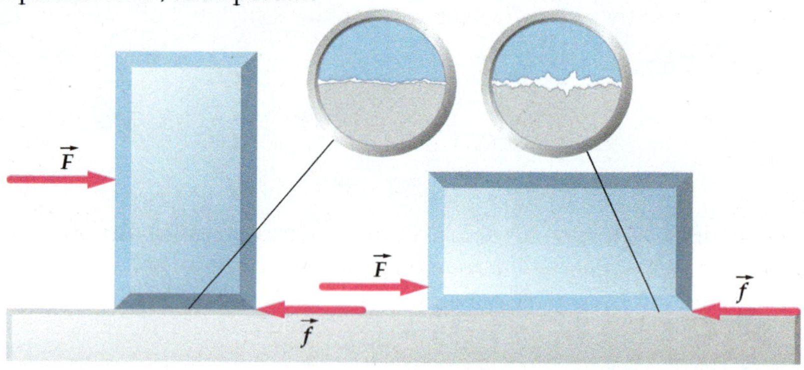

FIGURE 5-1 The microscopic area of contact between box and floor is only a small fraction of the macroscopic area of the box's bottom surface. The microscopic area of contact is proportional to the normal force exerted between the surfaces. If the box rests on its side, the macroscopic area is increased, but the force per unit area is decreased, so the microscopic area of contact is unchanged. Whether the box is upright or on its side, the same horizontal applied force F is required to keep it sliding at constant speed.

As shown in Figure 5-1, ordinary objects that look smooth and feel smooth are rough and bumpy at the microscopic (atomic) scale. This is the case even if the surfaces are highly polished. When surfaces come into contact, they touch only at prominences, called *asperities*, shown in Figure 5-1. The normal force exerted by a surface is exerted at the tips of these asperities where the normal force per unit area is very large, large enough to flatten the tips of the asperities. If the two surfaces are pressed together more strongly, the normal force increases and so does this flattening, resulting in a larger microscopic contact area. Under a wide range of conditions the microscopic area of contact is proportional to the normal force. The frictional force is proportional to the microscopic contact area; so, like the microscopic contact area, it is proportional to the normal force.

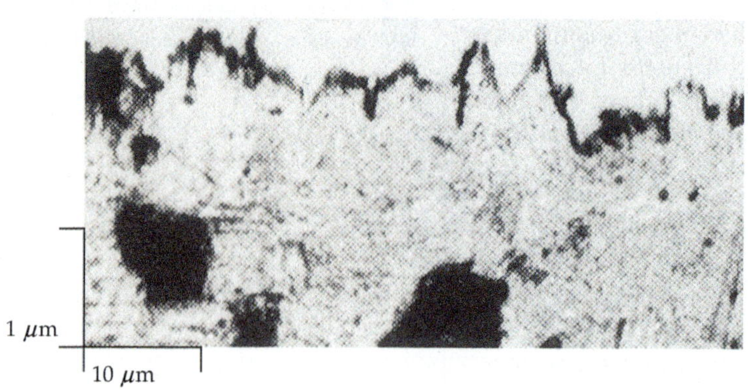

Magnified section of a polished steel surface showing surface irregularities. The irregularities are about $= 5 \times 10^{-7}$m high, a height that corresponds to several thousand atomic diameters. *(From F. P. Bowden and D. Tabor,* Lubrication of Solids, *Oxford University Press, 2000.)*

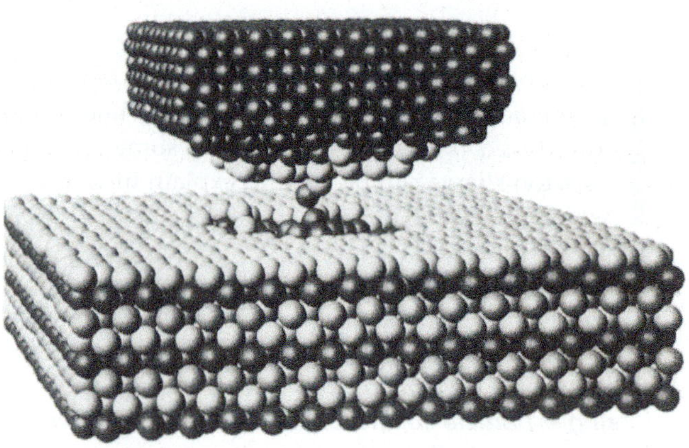

The computer graphic shows gold atoms (bottom) adhering to the fine point of a nickel probe (top) that has been in contact with the gold surface. *(Uzi Landman and David W. Leudtke/Georgia Institute of Technology.)*

STATIC FRICTION

Suppose you apply a small horizontal force $\vec{F}$ (Figure 5-2) to a large box resting on the floor. The box may not move noticeably because the force of static friction $\vec{f}_s$, exerted by the floor on the box, balances the force you apply. **Static friction** is the frictional force that acts when there is no sliding between the two surfaces in contact—it is the force that keeps the box from sliding. The force of static friction, which opposes the applied force on the box, can vary in magnitude from zero to some maximum value $f_{s\,max}$, depending on how hard you push. That is, as you push on the box, the opposing force of static friction increases to remain equal in magnitude to the applied force until the magnitude of the applied force exceeds $f_{s\,max}$. Data show that $f_{s\,max}$ is proportional to the strength of the forces pressing the two surfaces together. That is, $f_{s\,max}$ is proportional to the magnitude of the normal force exerted by one surface on the other:

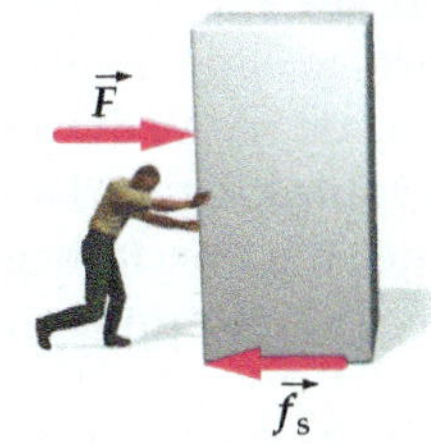

FIGURE 5-2

$$f_{s\,max} = \mu_s F_n \qquad 5\text{-}1$$

STATIC FRICTION RELATION

where the proportionality constant μ_s, is the **coefficient of static friction.** This coefficient depends on what materials the surfaces in contact are made of as well as the temperatures of the surfaces. If you exert a horizontal force with a magnitude that is less than or equal to $f_{s\,max}$ on the box, the static frictional force will just balance this horizontal force and the box will remain at rest. If you exert a horizontal force even slightly greater than $f_{s\,max}$ on the box, then the box will begin to slide. Thus, we can write Equation 5-1 as:

$$f_s \leq \mu_s F_n \qquad 5\text{-}2$$

The direction of the static frictional force is such that it opposes the tendency of the box to slide.

! Equation 5-2 is an inequality because the magnitude of the force of static friction ranges from zero up to $f_{s\,max}$.

! If the horizontal force you exert on the box is toward the left, then the static frictional force is toward the right. The static frictional force always opposes any tendency to slide.

KINETIC FRICTION

If you push the box in Figure 5-2 hard enough, it will slide across the floor. As it slides, the floor exerts a force of **kinetic friction** $\vec{f}_k$ (also called sliding friction) that opposes the motion. To keep the box sliding with constant velocity, you must exert a force on the box that is equal in magnitude and opposite in direction to the force of kinetic friction exerted by the floor.

Like the magnitude of a maximum static frictional force, the magnitude of a kinetic frictional force f_k is proportional to the microscopic contact area and the strength of the forces pressing the two surfaces together. That is, f_k is proportional to the normal force F_n one surface exerts on the other:

$$f_k = \mu_k F_n \qquad 5\text{-}3$$

KINETIC FRICTION RELATION

where the proportionality constant μ_k, is the **coefficient of kinetic friction.** The coefficient of kinetic friction depends on what materials the surfaces in contact are made of as well as the temperature of the contacting surfaces. Unlike static friction, the force of kinetic friction is independent of the magnitude of the applied horizontal force. Experiments show that μ_k is approximately constant for a wide range of speeds.

Figure 5-3 shows a plot of the frictional force exerted on the box by the floor as a function of the applied force. The force of friction balances the applied force until the box starts to slide, which occurs when the applied force exceeds $\mu_s F_n$ by an infinitesimal amount. As the box slides, the frictional force remains equal to $\mu_k F_n$. For any given contacting surfaces, μ_k is less than μ_s. This means you have to push harder to get the box to begin sliding than to keep it sliding at constant speed. Table 5-1 lists some approximate values of μ_s and μ_k for various pairs of surfaces.

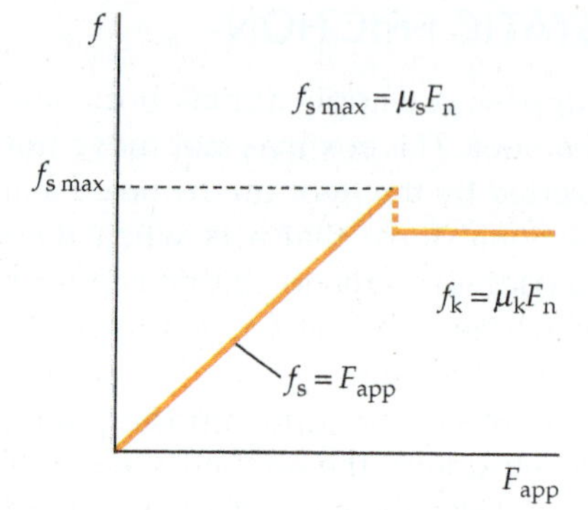

FIGURE 5-3

Table 5-1 Approximate Values of Frictional Coefficients

Materials	μ_s	μ_k
Steel on steel	0.7	0.6
Brass on steel	0.5	0.4
Copper on cast iron	1.1	0.3
Glass on glass	0.9	0.4
Teflon on Teflon	0.04	0.04
Teflon on steel	0.04	0.04
Rubber on concrete (dry)	1.0	0.80
Rubber on concrete (wet)	0.30	0.25
Waxed ski on snow (0°C)	0.10	0.05

ROLLING FRICTION

When a perfectly rigid wheel rolls *at constant speed* along a perfectly rigid horizontal road without slipping, no frictional force slows its motion. However, because real tires and roads continually deform (Figure 5-4) and because the tread and the road are continually peeled apart, in the real world the road exerts a force of **rolling friction** $\vec{f}_r$ that opposes the motion. To keep the wheel rolling with constant velocity, you must exert a force on the wheel that is equal in magnitude and opposite in direction to the force of rolling friction exerted on the wheel by the road.

The **coefficient of rolling friction** μ_r is the ratio of the magnitudes of the rolling frictional force f_r and the normal force F_n:

$$f_r = \mu_r F_n \qquad \text{5-4}$$

ROLLING FRICTION RELATION

where μ_r depends on the nature of the surfaces in contact and the composition of the wheel and road. Typical values of μ_r are 0.01 to 0.02 for rubber tires on concrete and 0.001 to 0.002 for steel wheels on steel rails. Coefficients of rolling friction are typically less than coefficients of kinetic friction by one to two orders of magnitude. Rolling friction is considered to be negligible in this book, except where it is specifically stated that it is significant.

FIGURE 5-4 As the car moves down the highway, the rubber flexes radially inward where the tread initiates contact with the pavement, and flexes radially outward where the tread loses contact with the road. The tire is not perfectly elastic, so the forces exerted on the tread by the pavement that flex the tread inward are greater than those exerted on the tread by the pavement as the tread flexes back as it leaves the pavement. This imbalance of forces results in a force opposing the rolling of the tire. This force is called a rolling frictional force. The more the tire flexes, the greater the rolling frictional force.

SOLVING PROBLEMS INVOLVING STATIC, KINETIC, AND ROLLING FRICTION

The following examples illustrate how to solve problems involving static and kinetic friction. The guidelines for approaching these types of problems are as follows:

PROBLEM-SOLVING STRATEGY

Solving Problems Involving Friction

PICTURE Determine which types of friction are involved in solving a problem. Objects experience static friction when no sliding exists between the surfaces of the objects that are in contact. The force of static friction opposes the tendency of the surfaces to slide on each other. The maximum static frictional force $f_{s\,max}$ is equal to the product of the normal force and the coefficient of static friction. If two surfaces are sliding against each other, they experience kinetic frictional forces (unless the problem states that one of the surfaces is frictionless). Rolling friction occurs because a rolling object and the surface that the object is rolling on continually deform and the object and the surface are continually peeling apart.

SOLVE

1. Construct a free-body diagram with the y axis normal to (and the x axis parallel to) the contacting surfaces. The direction of the frictional force is such that it opposes slipping, or the tendency to slip.
2. Apply $\Sigma F_y = ma_y$ and solve for the normal force F_n.
   If the friction is *kinetic* or *rolling*, relate the frictional and normal forces using $f_k = \mu_k F_n$ or $f_r = \mu_r F_n$, respectively.
   If the friction is *static*, relate the frictional and normal forces using $f_s \leq \mu_s F_n$ (or $f_{s\,max} = \mu_s F_n$).
3. Apply $\Sigma F_x = ma_x$ to the object and solve for the desired quantity.

CHECK In making sure that your answer makes sense, remember that coefficients of friction are dimensionless and that you must account for all forces (for example, tensions in ropes).

Example 5-1 A Game of Shuffleboard

A cruise-ship passenger uses a shuffleboard cue to push a shuffleboard disk of mass 0.40 kg horizontally along the deck so that the disk leaves the cue with a speed of 8.5 m/s. The disk then slides a distance of 8.0 m before coming to rest. Find the coefficient of kinetic friction between the disk and the deck.

PICTURE The force of kinetic friction is the only horizontal force acting on the disk after it separates from the cue. The acceleration is constant, because the frictional force is constant. We can find the acceleration using the constant-acceleration equations of Chapter 2 and relate the acceleration to μ_k using $\Sigma F_x = ma_x$.

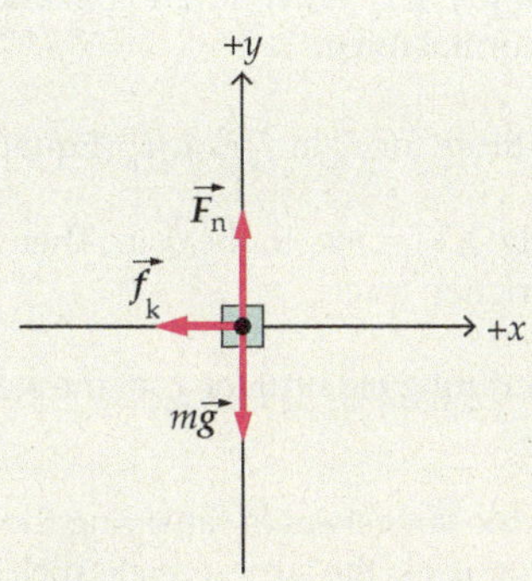

FIGURE 5-5

SOLVE

1. Draw a free-body diagram for the disk after it leaves the cue (Figure 5-5). Choose as the $+x$ direction the direction of the disk's velocity:

2. The coefficient of kinetic friction relates the magnitudes of the frictional and normal forces:

$$f_k = \mu_k F_n$$

3. Apply $\Sigma F_y = ma_y$ to the disk. Solve for the normal force. Then, using the relationship from step 2, solve for the frictional force:

$$\Sigma F_y = ma_y$$
$$F_n - mg = 0 \Rightarrow F_n = mg$$
$$\text{so} \quad f_k = \mu_k mg$$

4. Apply $\Sigma F_x = ma_x$ to the disk. Using the step-3 result, solve for the acceleration:

$$\Sigma F_x = ma_x$$
$$-f_k = ma_x$$
$$\text{so} \quad -\mu_k mg = ma_x \quad \text{so} \quad a_x = -\mu_k g$$

5. The acceleration is constant. Relate it to the total distance traveled and the initial velocity using $v_x^2 = v_{0x}^2 + 2a_x\Delta x$ (Equation 2-15). Using the step-4 result, solve for μ_k:

$$v_x^2 = v_{0x}^2 + 2a_x\Delta x \Rightarrow 0 = v_{0x}^2 - 2\mu_k g\,\Delta x$$

$$\text{so}\quad \mu_k = \frac{v_{0x}^2}{2g\,\Delta x} = \frac{(8.5\text{ m/s})^2}{2(9.81\text{ m/s}^2)(8.0\text{ m})} = \boxed{0.46}$$

CHECK The value obtained for μ_k is dimensionless and within the range of values for other materials listed in Table 5-1, so it is plausible.

TAKING IT FURTHER Note that the acceleration and the coefficient of friction are independent of the mass m. The greater the mass, the harder it is to stop the disk, but a greater mass is accompanied by a greater normal force, and thus a greater frictional force. The net result is that the mass has no effect on the acceleration (or the stopping distance).

Example 5-2 A Sliding Coin

A hardcover book (Figure 5-6) is resting on a tabletop with its front cover facing upward. You place a coin on this cover and very slowly open the book until the coin starts to slide. The angle θ_{max} (known as the *angle of repose*) is the angle the front cover makes with the horizontal just as the coin starts to slide. Find the coefficient of static friction μ_s between the book cover and the coin in terms of θ_{max}.

FIGURE 5-6 *(Ramón Rivera-Moret.)*

PICTURE The forces acting on the coin are the gravitational force $F_g = mg$, the normal force F_n, and the frictional force f. Because the coin is on the verge of sliding (but not yet sliding), the frictional force is a static frictional force directed up the incline. Because the coin remains stationary, its acceleration is zero. We use Newton's second law to relate this acceleration to the forces on the coin, and then solve for the fictional force.

SOLVE

1. Draw a free-body diagram for the coin when the book cover inclined at angle θ, where $\theta \le \theta_{max}$ (Figure 5-7). Draw the y axis normal to the book cover:

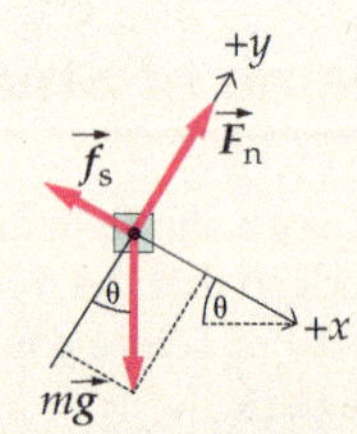

FIGURE 5-7

2. The coefficient of static friction relates the frictional and normal forces:

$$f_s \le \mu_s F_n$$

3. We apply $\Sigma F_y = ma_y$ to the coin and solve for the normal force:

$$\Sigma F_y = ma_y$$
$$F_n - mg\cos\theta = 0 \Rightarrow F_n = mg\cos\theta$$

4. Substitute for F_n in $f_s \le \mu_s F_n$ (Equation 5-1):

$$f_s \le \mu_s F_n \Rightarrow f_s \le \mu_s mg\cos\theta$$

5. Apply $\Sigma F_x = ma_x$ to the coin. Then solve for the friction force:

$$\Sigma F_x = ma_x$$
$$-f_s + mg\sin\theta = 0 \Rightarrow f_s = mg\sin\theta$$

6. Substituting $mg\sin\theta$ for f_s in the step-4 result gives:

$$mg\sin\theta \le \mu_s mg\cos\theta \Rightarrow \tan\theta \le \mu_s$$

7. θ_{max}, the largest angle satisfying the condition $\tan\theta \le \mu_s$, is the largest angle such that the coin does not slide:

$$\boxed{\mu_s = \tan\theta_{max}}$$

CHECK The coefficient of friction is dimensionless, and so is the tangent function. Also, for $0 < \theta_{max} < 45°$, $\tan\theta_{max}$ is between zero and one. One would expect the coin to slide before the angle reached 45° and one would expect the coefficient of static friction to be between zero and one. Thus, the step-7 result is plausible.

PRACTICE PROBLEM 5-1 The coefficient of static friction between a car's tires and the road on a particular day is 0.70. What is the steepest angle of inclination of the road for which the car can be parked with all four wheels locked and not slide down the hill?

Erosion due to a stream cutting across a beach. Even though the edge weaves in and out, the angle of the slope remains constant. The angle of the slope is the angle of repose for the granular material. *(David R. Bailey.)*

CONCEPT CHECK 5-1

The car in Practice Problem 5-1 is parked at the steepest angle of inclination with all four wheels locked. Would the car slide down the incline if only two of the wheels are locked?

Example 5-3 Pulling a Sled

Two children sitting on a sled at rest in the snow ask you to pull them. You oblige by pulling on the sled's rope, which makes an angle of 40° with the horizontal (Figure 5-8). The children have a combined mass of 45 kg and the sled has a mass of 5.0 kg. The coefficients of static and kinetic friction are $\mu_s = 0.20$ and $\mu_k = 0.15$ and the sled is initially at rest. Find both the magnitude of the frictional force exerted by the snow on the sled and the acceleration of the children and sled if the tension in the rope is (*a*) 100 N and (*b*) 140 N.

FIGURE 5-8 (*Jean-Claude LeJeune/Stock Boston.*)

PICTURE First, we need to find out whether the frictional force is static or kinetic. To do this, we see if the given tension forces satisfy the relation $f_s \leq \mu_s F_n$. Once we have done that, we can select the correct expression for the frictional force, and solve the corresponding expression for f.

SOLVE

(*a*) 1. Draw a free-body diagram for the sled (Figure 5-9):

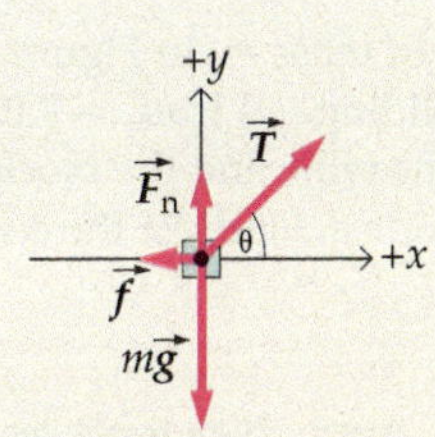

FIGURE 5-9

2. Write down the static friction relation. If this relation is satisfied, the sled does not begin sliding:

$$f_s \leq \mu_s F_n$$

3. Apply $\Sigma F_y = ma_y$ to the sled and solve for the normal force:

$$\Sigma F_y = ma_y$$

$$F_n + T\sin\theta - mg = 0 \Rightarrow F_n = mg - T\sin\theta$$

4. Apply $\Sigma F_x = ma_x$ (with $a_x = 0$) to the sled and solve for the static frictional force:

$$\Sigma f_x = ma_x$$

$$-f_s + T\cos\theta = 0 \Rightarrow f_s = T\cos\theta$$

5. Substitute the step-4 and step-5 results into the step-2 result:

$$T\cos\theta \le \mu_s(mg - T\sin\theta)$$

6. Check to see if the given tension of 100 N satisfies the nonslip condition (the step-3 inequality):

$$(100\text{ N})\cos 40° \le 0.20[(50\text{ kg})(9.81\text{ N/kg}) - (100\text{ N})\sin 40°]$$
$$77\text{ N} \le 85\text{ N}$$

The inequality *is* satisfied, thus the sled *is not* sliding.

7. Because the sled is not sliding, the frictional force is that of static friction. To find the frictional force, use the step-3 expression for f_s:

$$a_x = 0$$
$$f_s = T\cos\theta = (100\text{ N})\cos 40° = \boxed{77\text{ N}}$$

(*b*) 1. Check the step-4 result from Part (*a*) with $T = 140$ N. If the relation is satisfied, the sled does not slide:

$$(140\text{ N})\cos 40° \le 0.20[(50\text{ kg})(9.81\text{ N/kg}) - (140\text{ N})\sin 40°]$$
$$107\text{ N} \le 80\text{ N}$$

The inequality *is not* satisfied, thus the sled *is* sliding.

2. Because the sled is sliding, the friction is kinetic friction, where $f_k = \mu_k F_n$. In Part-(*a*), step 3 we applied $\Sigma F_y = ma_y$ to the sled and found $F_n = mg - T\sin\theta$. Using these results, solve for the kinetic frictional force:

$$f_k = \mu_k F_n$$
$$f_k = \mu_k(mg - T\sin\theta)$$
$$= 0.15[(50\text{ kg})(9.81\text{ N/kg}) - (140\text{ N})\sin 40°]$$
$$= \boxed{60\text{ N}}$$

3. Apply $\Sigma F_x = ma_x$ to the sled and solve for the frictional force. Then substitute the Part-(*b*), step-2 result for f_k and solve for the acceleration:

$$\Sigma F_x = ma_x$$
$$-f_k + T\cos\theta = ma_x \Rightarrow a_x = \frac{-f_k + T\cos\theta}{m}$$
$$\text{so}\quad a_x = \frac{(-60\text{ N}) + (140\text{ N})\cos 40°}{50\text{ kg}} = \boxed{0.94\text{ m/s}^2}$$

CHECK We expect a_x to be greater than or equal to zero, so we expect the magnitude of the frictional force to be less than or equal to the x component of the tension force. In Part (*a*) the magnitude of the frictional force and x component of the tension force both equal 77 N, and in Part (*b*) the magnitude of the frictional force equals 60 N and the x component of the tension force is 140 N cos 40° = 107 N.

TAKING IT FURTHER Note two important points about this example: (1) the normal force is less than the weight of the children and the sled. This is so because the vertical component of the tension helps the ground counter the gravitational force; and (2) in Part (*a*), the force of static friction is less than $\mu_s F_n$.

PRACTICE PROBLEM 5-2 What is the maximum force you can pull the rope at the specified angle without the sled beginning to slide?

Example 5-4 A Sliding Block

The block of mass m_2 in Figure 5-10 has been adjusted so that the block of mass m_1 is on the verge of sliding. (*a*) If $m_1 = 7.0$ kg and $m_2 = 5.0$ kg, what is the coefficient of static friction between the table and the block? (*b*) With a slight nudge, the blocks move with acceleration of magnitude a. Find a if the coefficient of kinetic friction between the table and the block is $\mu_k = 0.54$.

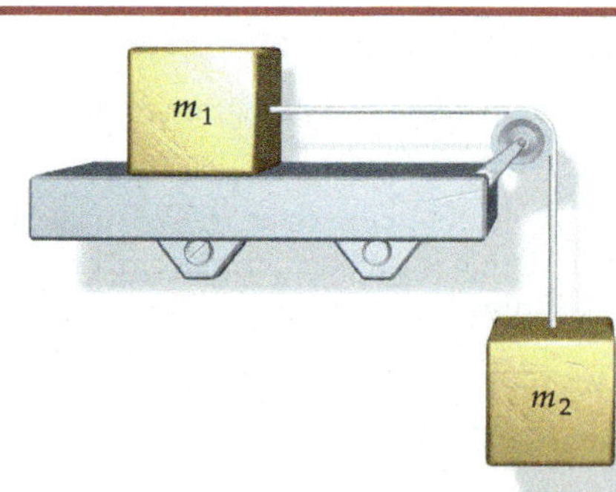

FIGURE 5-10

PICTURE Apply Newton's second law to each block. By neglecting the masses of both the rope and the pulley, and by neglecting friction in the pulley bearing, the tension has the same magnitude throughout the rope, so $T_1 = T_2 = T$. Because the rope remains taut but does not stretch, the accelerations have the same magnitude, so $a_1 = a_2 = a$.

To find the coefficient of static friction μ_s, as requested in Part (*a*), set the force of static friction on m_1 equal to its maximum value $f_{s\,max} = \mu_s F_n$ and set the acceleration equal to zero.

SOLVE

(*a*) 1. Draw a free-body diagram for each block (Figure 5-11). Choose the $+x$ and $+x'$ directions to be the same as the directions of the accelerations of blocks 1 and 2, respectively. That is, the $+x$ direction is to the right and the $+x'$ direction is vertically downward:

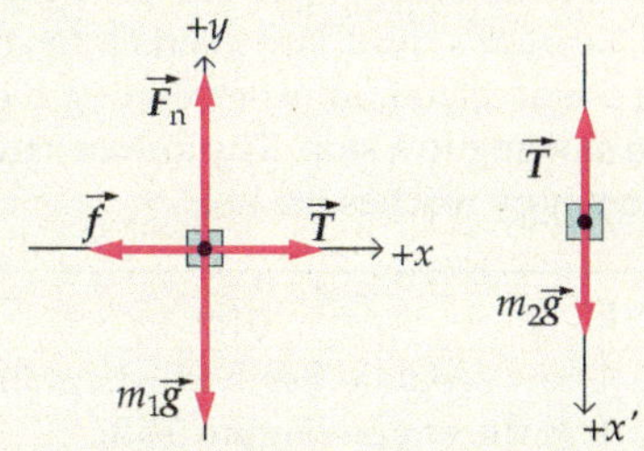

FIGURE 5-11

2. Apply $\Sigma F_y = ma_y$ to block 1 and solve for the normal force. Then solve for the static frictional force.

$\Sigma F_y = m_1 a_{1y}$

$F_n - m_1 g = 0 \Rightarrow F_n = m_1 g$

so

$f_{s\max} = \mu_s F_n$ so $f_{s\max} = \mu_s m_1 g$

3. Apply $\Sigma F_x = ma_x$ to block 1 and solve for the frictional force. Then substitute into the step-2 result.

$\Sigma F_x = m_1 a_{1x}$

$T - f_{s\max} = 0 \Rightarrow T = f_{s\max}$

so

$T = \mu_s m_1 g$

4. Apply $\Sigma F_x = ma_x$ to block 2 and solve for the tension. Then substitute into the step-3 result.

$\Sigma F_{x'} = m_2 a_{2x'} \Rightarrow m_2 g - T = 0$

so

$T = m_2 g$ and $m_2 g = \mu_s m_1 g$

5. Solve the step-4 result for μ_s.

$$\mu_s = \frac{m_2}{m_1} = \frac{5.0\text{ kg}}{7.0\text{ kg}} = \boxed{0.71}$$

(*b*) 1. During sliding, the frictional force is kinetic and the accelerations have the same magnitude a. Relate the kinetic frictional force f_k to the normal force. The normal force was found in step 2 of Part (*a*).

$f_k = \mu_k F_n$

so

$f_k = \mu_k m_1 g$

$|\vec{a}_1| = a_{1x} = a$ and $|\vec{a}_2| = a_{2x'} = a$

2. Apply $\Sigma F_x = ma_x$ to block 1. Then substitute for the frictional force using the result from step 1 of Part (*b*).

$\Sigma F_x = m_1 a_{1x} \Rightarrow T - f_k = m_1 a$

so

$T - \mu_k m_1 g = m_1 a$

3. Apply $\Sigma F_{x'} = ma_{x'}$ to block 2.

$\Sigma F_{x'} = m_2 a_{2x'} \Rightarrow m_2 g - T = m_2 a$

4. Add the equations in steps 2 and 3 of Part (*b*) and solve for a.

$$a = \frac{m_2 - \mu_k m_1}{m_1 + m_2} g = \boxed{1.0\text{ m/s}^2}$$

CHECK Note that if $m_1 = 0$ the expression for the acceleration reduces to $a = g$ as one would expect.

PRACTICE PROBLEM 5-3 What is the tension in the rope when the blocks are sliding?

Example 5-5 The Runaway Buggy *Context-Rich*

A runaway baby buggy is sliding without friction across a frozen pond toward a hole in the ice (Figure 5-12). You race after the buggy on skates. As you grab it, you and the buggy are moving toward the hole at speed v_0. The kinetic coefficient of friction between your skates and the ice as you turn out the blades to brake is μ_k. D is the distance between the buggy and the hole at the instant you reach the buggy, m_B is the mass of the buggy (including its precious cargo), and m_Y is your mass. (*a*) What is the lowest value of D such that you stop the buggy before it reaches the hole in the ice? (*b*) What force do you exert on the buggy as you slow it?

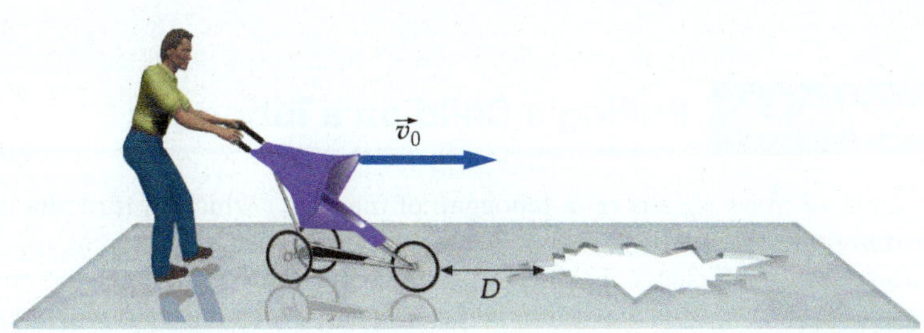

FIGURE 5-12

PICTURE Initially, you and the buggy are moving toward the hole with speed v_0, which we take to be in the $+x$ direction. If you exert a force $\vec{F}_{YB}$ on the buggy, the buggy, in accord with Newton's third law, exerts a force $\vec{F}_{BY}$ on you. Apply Newton's second law to determine the acceleration. After finding the acceleration, find the distance D the buggy travels while slowing to a stop. The lowest value of D is that for which your speed reaches zero just as the buggy reaches the hole.

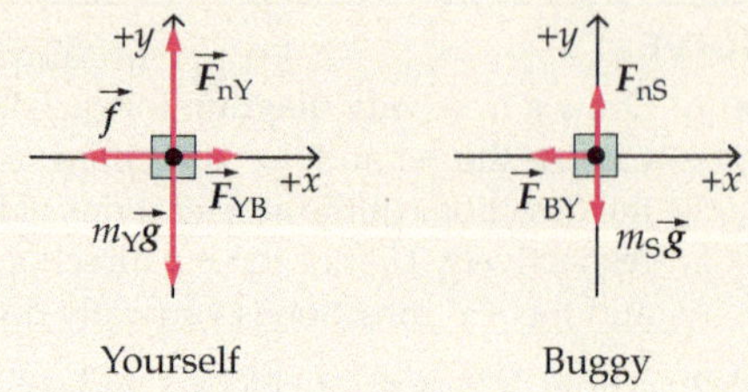

FIGURE 5-13

SOLVE

(*a*) 1. Draw separate free-body diagrams for yourself and the buggy (Figure 5-13).

2. To find the frictional force of the ice on you, you need to first find the normal force of the ice on you:

$$f_{IYk} = \mu_k F_{IYn}$$

3. Apply $\Sigma F_y = ma_y$ to yourself and solve first for the normal force and then for the frictional force:

$$\Sigma F_y = ma_y \Rightarrow F_{IYn} - m_Y g = 0 \quad (a_y = 0)$$

and $f_{IYk} = \mu_k F_{IYn}$ so $f_{IYk} = \mu_k m_Y g$

4. Apply $\Sigma F_x = ma_x$ to yourself. Then substitute in the step-3 result:

$$\Sigma F_x = ma_x \Rightarrow F_{BY} - f_{IYk} = m_Y a_x$$

so $F_{BY} - \mu_k m_Y g = m_Y a_x$

5. Apply $\Sigma F_x = ma_x$ to the buggy:

$$\Sigma F_x = m_B a_x \Rightarrow F_{YB} = m_B a_x$$

6. $\vec{F}_{BY}$ and $\vec{F}_{YB}$ form an N3L force pair, so they are equal in magnitude:

$$F_{BY} = F_{YB}$$

7. Add the step-4 and step-5 results and use $F_{BY} - F_{YB} = 0$ to simplify:

$$-F_{YB} + (F_{BY} - \mu_s m_Y g) = m_B a_x + m_Y a_x$$
$$0 - \mu_s m_Y g = m_B a_x + m_Y a_x$$

8. Solve the step-7 result for a_x:

$$a_x = -\frac{\mu_k m_Y}{m_Y + m_B} g \quad (a_x \text{ is negative, as expected.})$$

9. Substitute the step-8 result into a kinematic equation and solve for the magnitude of the displacement D:

$$v_x^2 = v_{0x}^2 + 2a_x \Delta x \Rightarrow 0 = v_0^2 + 2a_x D$$

so

$$D = \frac{-v_0^2}{2a_x} = \boxed{\left(1 + \frac{m_B}{m_Y}\right)\frac{v_0^2}{2\mu_k g}}$$

(*b*) F_{YB} can be found by combining the results for steps 5 and 8:

$$F_{YB} = m_B|a_x| = \boxed{\frac{\mu_k m_B g}{1 + (m_Y/m_B)}}$$

CHECK For large values of m_B/m_Y, D is large, as expected.

TAKING IT FURTHER The minimum value of D is proportional to v_0^2 and inversely proportional to μ_k. Figure 5-14 shows the stopping distance D versus initial velocity squared for values of m_B/m_Y equal to 0.1, 0.3, and 1.0, with $\mu_k = 0.5$. Note that the larger the mass ratio m_B/m_Y, the greater the distance D needed to stop for a given initial velocity. This is akin to braking to a stop in a car that is pulling a trailer that does not have its own brakes. The mass of the trailer increases the stopping distance for a given speed.

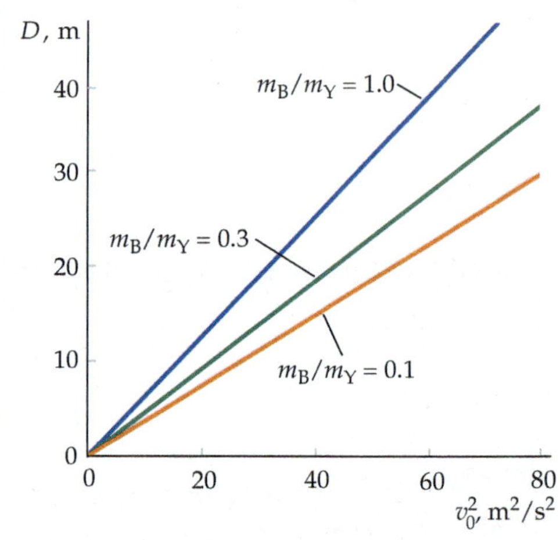

FIGURE 5-14

Example 5-6 Pulling a Child on a Toboggan

Try It Yourself

A child of mass m_C sits on a toboggan of mass m_T, which in turn sits on a frictionless frozen pond (Figure 5-15). The toboggan is pulled with a horizontal applied force $\vec{F}_{ap}$ as shown. The coefficients of static and sliding friction between the child and toboggan are μ_s and μ_k. (*a*) Find the maximum value of F_{ap} for which the child will not slide relative to the toboggan. (*b*) Find the acceleration of the toboggan and the acceleration of the child when F_{ap} is greater than this value.

FIGURE 5-15

PICTURE The only force accelerating the child forward is the frictional force exerted by the toboggan on the child. In Part (*a*) the challenge is to find F_{ap} when this frictional force is static and maximum. To do this, apply $\Sigma\vec{F} = m\vec{a}$ to the child and solve for the acceleration when the static frictional force is maximum. Then apply $\Sigma\vec{F} = m\vec{a}$ to the toboggan and solve for F_{ap}. In Part (*b*), we follow a parallel procedure. However, in Part (*b*) the minimum value of F_{ap} is given and we solve for the acceleration of the toboggan.

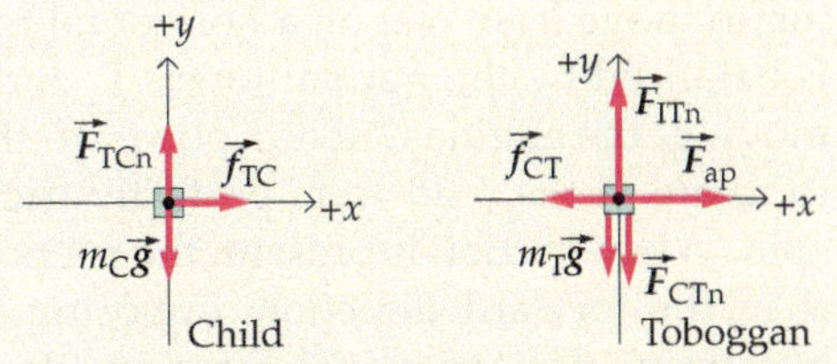

FIGURE 5-16 The force $\vec{F}_{ITn}$ is the normal force exerted by the ice on the toboggan.

SOLVE

Cover the column to the right and try these on your own before looking at the answers.

Steps	Answers
(*a*) 1. Draw one free-body diagram for the child and another for the toboggan (Figure 5-16).	
2. Apply $\Sigma F_x = ma_x$ to the toboggan:	$\Sigma F_{Tx} = m_T a_{Tx} \Rightarrow F_{ap} - f_{CTs\,max} = m_T a_{Tx}$
3. Apply $\Sigma F_y = ma_y$ to the child and solve for the normal force. Then apply $f_{x\,max} = \mu_s F_n$ and solve for the frictional force.	$\Sigma F_{Cy} = m_C a_y \Rightarrow F_{TCn} - m_C g = 0$ so $F_{TCn} = m_C g$ $f_{s\,max} = \mu F_n \Rightarrow f_{TCs\,max} = \mu_s F_{TCn}$ so $f_{TCs\,max} = \mu_s m_C g$
4. Apply $\Sigma F_x = ma_x$ to the child and solve for the acceleration.	$\Sigma F_{Cx} = m_C a_{Cx} \Rightarrow f_{TCs\,max} = m_C a_{Cx}$ so $\mu_s m_C g = m_C a_{Cx} \Rightarrow a_{Cx} = \mu_s g$
5. Equate the magnitudes of the forces in each N3L force pair appearing in the two free-body diagrams. In addition, express the relation between the accelerations due to the nonslipping constraint.	$F_{TCn} = F_{CTn}$ and $f_{TCs\,max} = f_{CTs\,max}$ and $a_{Cx} = a_{Tx} = a_x$
6. Substitute the step-4 and step-5 results into the step-2 result and solve for F_{ap}.	$F_{ap} - \mu_s m_C g = m_T \mu_s g$ so $F_{ap} = \boxed{(m_C + m_T)\mu_s g}$
(*b*) 1. Equate the magnitudes of each N3L force pair and express the relation between the accelerations if the child is slipping on the toboggan.	$F_{TCn} = F_{CTn} = F_n$ and $f_{TCk} = f_{CTk} = f_k$ and $a_{Cx} < a_{Tx}$
2. Solve for the magnitude of the kinetic frictional force using the result from step 3 of Part (*a*) for the normal force.	$f_k = \mu_k F_n$ so $f_k = \mu_k m_C g$
3. Apply $\Sigma F_x = ma_x$ to the child and solve for *her acceleration.*	$\Sigma F_{Cx} = m_C a_{Cx} \Rightarrow f_k = m_C a_{Cx}$ so $\mu_k m_C g = m_C a_{Cx} \Rightarrow a_{Cx} = \boxed{\mu_k g}$
4. Apply $\Sigma F_x = ma_x$ to the toboggan. Using the result from step 2 of Part (*b*), solve for *its acceleration.*	$\Sigma F_{Tx} = m_T a_{Tx} \Rightarrow F_{ap} - f_k = m_T a_{Tx}$ so $F_{ap} - \mu_k m_C g = m_T a_{Tx} \Rightarrow a_{Tx} = \boxed{\dfrac{F_{ap} - \mu_k m_C g}{m_T}}$

CHECK The Part-(*a*) result is what we expect if the child does not slip on the toboggan. If we model the child and the toboggan as a single particle and apply Newton's second law to it, we obtain $F_{ap} = (m_C + m_T)a_x$. If we substitute $\mu_s(m_C + m_T)g$ (our Part-(*a*), step-6 result) for F_{ap}, we get $\mu_s(m_C + m_T)g = (m_C + m_T)a_x$. Dividing both sides by the sum of the masses gives $a_x = \mu_s g$, our Part-(*a*), step-3 result. Thus, our Part (*a*) effort is consistent with modeling the sled and child as a single particle.

TAKING IT FURTHER In this example, the frictional force does not oppose the motion of the child, it causes it. Without friction, the girl would slip back relative to the toboggan. However, even though the child moves, or tends to move, backward (leftward) *relative to the toboggan,* she moves forward *relative to the ice*. Friction forces oppose relative motion, or tendency toward relative motion, between two surfaces in contact. In addition, relative to the child, the toboggan slips, or tends to slip, forward. The friction force on the toboggan is directed rearward, opposing this slipping forward, or tendency toward slipping forward.

! Note that frictional forces *do not always* oppose motion. However, frictional forces between contacting surfaces *do always* oppose relative motion, or the tendency toward relative motion, between the two contacting surfaces.

FRICTION, CARS, AND ANTILOCK BRAKES

Figure 5-17 shows the forces acting on a front-wheel-drive car that is just starting to move from rest on a horizontal road. The gravitational force F_g on the car is balanced by the normal forces F_n and F_n' exerted on the tires. To start the car moving, the engine delivers power to the axle that makes the front wheels start to rotate. If the road were perfectly frictionless, the front wheels would merely spin. When friction is present, the frictional force exerted by the road on the tires is in the forward direction, opposing the tendency of the tire surface to slip backward. This frictional force enables the acceleration needed for the car to start moving forward. If the power delivered by the engine is not so great that the surface of the tire slips on the surface of the road, the wheels will roll without slipping and that part of the tire tread touching the road remains at rest relative to the road. (The part in contact with the road continuously changes as the tire rolls.) The friction between the road and the tire tread is then that of static friction. The largest frictional force that the tire can exert on the road (and that the road can exert on the tire) is $\mu_s F_n$.

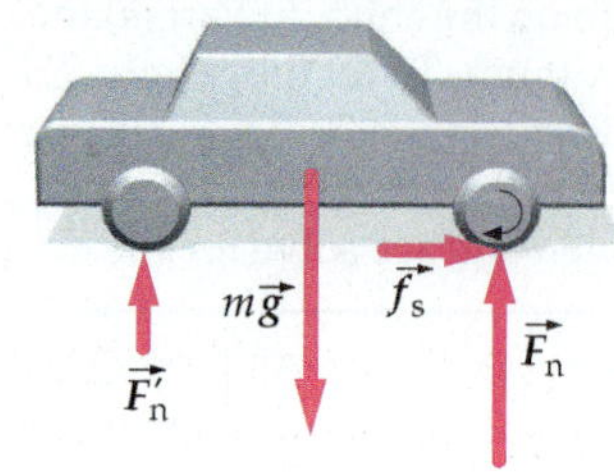

FIGURE 5-17 Forces acting on a car with front-wheel drive that is accelerating from rest. The normal force $\vec{F}_n$ on the front tires is usually larger than the normal force $\vec{F}_n'$ on the rear tires because typically the engine of the car is mounted near the front of the car. There is no drag force from the air shown because the car is just starting to move. There would be a rearward-directed rolling frictional force on all wheels, but that force has been ignored.

For a car moving in a straight line with speed v relative to the road, the center of each of its wheels also moves with speed v, as shown in Figure 5-18. If a wheel is rolling without slipping, its top is moving faster than v, whereas its bottom is moving slower than v. However, *relative to the car*, each point on the perimeter of the wheel moves in a circle with the same speed v. Moreover, the speed of the point on the tire momentarily in contact with the ground is zero *relative to the ground*. (Otherwise, the tire would be skidding along the road.)

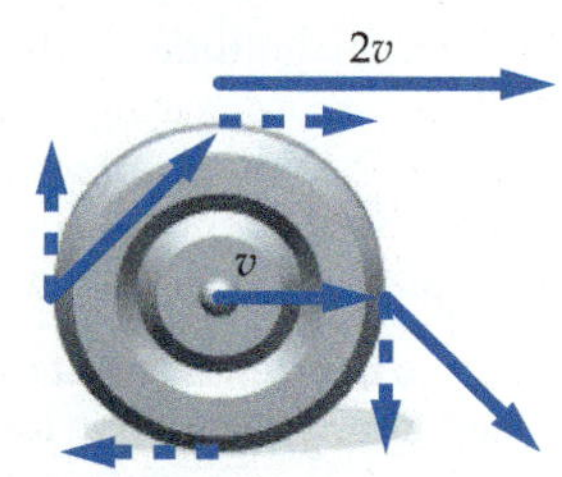

FIGURE 5-18 In this figure, dashed lines represent velocities relative to the body of the car; solid lines represent velocities relative to the ground.

If the engine supplies excessive power, the tire will slip and the wheels will spin. Then the force that accelerates the car is the force of kinetic friction, which is less than the maximum force of static friction. If we are stuck on ice or snow, our chances of getting free are better if we avoid spinning the wheels by using a light touch on the accelerator pedal. Similarly, when braking a car to a stop, the force exerted by the road on the tires may be either static friction or kinetic friction, depending on how hard the brakes are applied. If the brakes are applied so hard that the wheels lock, the tires will skid along the road and the stopping force will be that of kinetic friction. If the brakes are applied less forcefully, so that no slipping occurs between the tires and the road, the stopping force will be that of static friction.

When wheels do lock and tires skid, two undesirable things happen. The minimum stopping distance is increased and the ability to control the direction of the motion of the car is greatly diminished. Obviously, this loss of directional control can have dire consequences. Antilock braking systems (ABS) in cars are designed to prevent the wheels from locking even if the brakes are applied hard. These systems have wheel-speed sensors. If the control unit senses that a wheel is about to lock, the module signals the brake pressure modulator to drop, hold, and then restore the pressure to that wheel up to 15 times per second. This varying of pressure is much like "pumping" the brake, but with the ABS system, the wheel that is locking is the only one being pumped. This is called *threshold braking*. With threshold braking, maximum friction for stopping is maintained.

Example 5-7 The Effect of Antilock Brakes

A car is traveling at 30 m/s along a horizontal road. The coefficients of friction between the road and the tires are $\mu_s = 0.50$ and $\mu_k = 0.40$. How far does the car travel before stopping if (*a*) the car is braked with an antilock braking system so that threshold braking is sustained, and (*b*) the car is braked hard with no antilock braking system so that the wheels lock? (*Note:* Skidding heats up the tires and the coefficients of friction vary with changes in temperature. These temperature effects are neglected in this example.)

PICTURE The force that stops a car when it brakes without skidding is the force of static friction exerted by the road on the tires (Figure 5-19). We use Newton's second law to solve for the frictional force and the car's acceleration. Kinematics is then used to find the stopping distance.

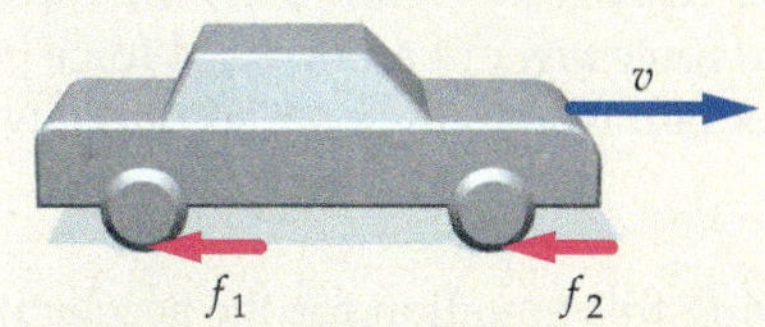

FIGURE 5-19

SOLVE

(*a*) 1. Draw a free-body diagram for the car (Figure 5-20). Treat all four wheels as if they were a single point of contact with the ground. Assume further that the brakes are applied to all four wheels. The $\vec{f}$ in the free-body diagram is the total of the frictional forces on individual wheels:

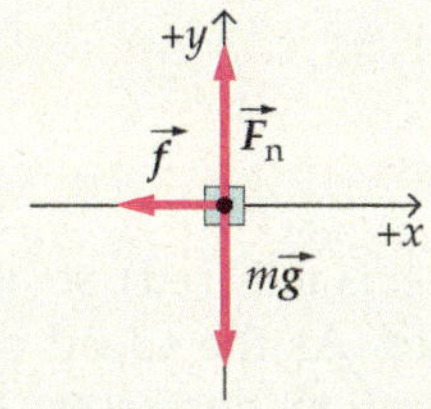

FIGURE 5-20

2. Assuming that the acceleration is constant, we use Equation 2-15 to relate the stopping distance Δx to the initial velocity v_{0x}:

$$v_x^2 = v_{0x}^2 + 2a_x\,\Delta x$$

When $v_x = 0$, $\Delta x = -\dfrac{v_{0x}^2}{2a_x}$

3. Apply $\Sigma F_y = ma_y$ to the car and solve for the normal force. Then apply $f_{x\max} = \mu_s F_n$ and solve for the frictional force:

$$\Sigma F_y = ma_y \;\Rightarrow\; F_n - mg = 0 \quad \text{so} \quad F_n = mg$$

$$f_{s\max} = \mu_s F_n \quad \text{so} \quad f_{s\max} = \mu_s mg$$

4. Apply $\Sigma F_x = ma_x$ to the car and solve for the acceleration:

$$\Sigma F_x = ma_x \;\Rightarrow\; -f_{s\max} = ma_x$$

Substituting $\mu_s mg$ for $f_{s\max}$ gives $-\mu_s mg = ma_x \;\Rightarrow\; a_x = -\mu_s g$

5. Substituting for a_x in the equation for Δx in step 2 gives the stopping distance:

$$\Delta x = -\frac{v_{0x}^2}{2a_x} = \frac{v_{0x}^2}{2\mu_s g} = \frac{(30\text{ m/s})^2}{2(0.50)(9.81\text{ m/s}^2)} = \boxed{0.92 \times 10^2\text{ m}}$$

(*b*) 1. When the wheels lock, the force exerted by the road on the car is that of kinetic friction. Using reasoning similar to that in Part (*a*), we obtain for the acceleration:

$$a_x = -\mu_k g$$

2. The stopping distance is then:

$$\Delta x = -\frac{v_{0x}^2}{2a_x} = \frac{v_{0x}^2}{2\mu_k g} = \frac{(30\text{ m/s})^2}{2(0.40)(9.81\text{ m/s}^2)} = \boxed{1.1 \times 10^2\text{ m}}$$

CHECK The calculated displacements are both positive as expected. In addition, the antilock brake system significantly shortens the stopping distance of the car as expected.

TAKING IT FURTHER Notice that the stopping distance is more than 20% greater when the wheels are locked. Also note that the stopping distance is independent of the car's mass—the stopping distance is the same for a subcompact car as for a large truck—provided the coefficients of friction are the same. The tires heat up dramatically when skidding occurs. This produces a change in μ_k that was not taken into account in this solution.

5-2 DRAG FORCES

When an object moves through a fluid such as air or water, the fluid exerts a **drag force** or retarding force that opposes the motion of the object. The drag force depends on the shape of the object, the properties of the fluid, and the speed of the object relative to the fluid. Unlike ordinary friction, the drag force increases as the speed of the object increases. At very low speeds, the drag force is approximately proportional to the speed of the object; at higher speeds, it is more nearly proportional to the square of the speed.

Consider an object dropped from rest and falling under the influence of the force of gravity, which we assume to be constant. The magnitude of the drag force is

$$F_d = bv^n \qquad \text{5-5}$$

DRAG FORCE RELATION

where b is a constant.

The drag force on the parachute slows this dragster. *(IHRA/Live Nation.)*

As shown in Figure 5-21, the forces acting on the object are a constant downward force mg and an upward force bv^n. If we take downward as the $+y$ direction, we obtain from Newton's second law

$$mg - bv^n = ma_y \tag{5-6}$$

Solving this equation for the acceleration gives:

$$a_y = g - \frac{b}{m}v^n \tag{5-7}$$

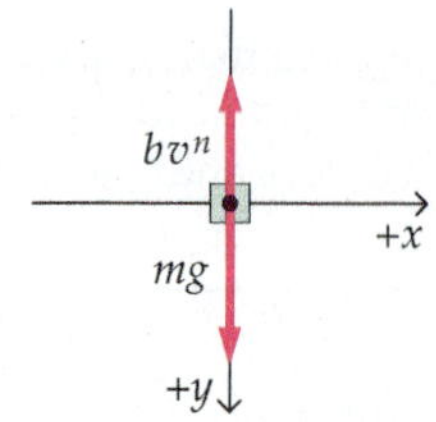

FIGURE 5-21 Free-body diagram showing forces on an object falling through the air.

The speed is zero at $t = 0$ (the instant the object is released), so at $t = 0$ the drag force is zero and the acceleration is g downward. As the speed of the object increases, the drag force increases so the acceleration decreases. Eventually, the speed is great enough for the magnitude of the drag force bv^n to approach the force of gravity mg. In this limit, the acceleration approaches zero and the speed approaches the **terminal speed** v_T. At terminal speed, the drag force balances the weight force and the acceleration is zero. Setting v equal to v_T and a_y equal to zero in Equation 5-6, we obtain

$$bv_T^n = mg$$

Solving for the terminal speed, we get

$$v_T = \left(\frac{mg}{b}\right)^{1/n} \tag{5-8}$$

The larger the constant b, the smaller the terminal speed. Cars are designed to minimize b to reduce the effect of wind resistance. A parachute, on the other hand, is designed to maximize b so that the terminal speed will be small. For example, the terminal speed of a skydiver before release of the parachute is about 200 km/h, which is about 60 m/s. When the parachute opens, the drag force rapidly increases, becoming greater than the force of gravity. As a result, the skydiver experiences an upward acceleration while falling; that is, the speed of the descending skydiver decreases. As the speed of the skydiver decreases, the drag force decreases and the speed approaches a new terminal speed of about 20 km/h.

(Stuart Williams/Dembinsky Photo Associates.)

(Joe McBride/Stone.)

Example 5-8 Terminal Speed

A 64.0-kg skydiver falls with a terminal speed of 180 km/h with her arms and legs outspread. (*a*) What is the magnitude of the upward drag force F_d on the skydiver? (*b*) If the drag force is equal to bv^2, what is the value of b?

PICTURE We use Newton's second law to solve for the drag force in Part (*a*), and then substitute in the appropriate values to find b in Part (*b*).

SOLVE

(*a*) 1. Draw a free-body diagram (Figure 5-22).

2. Apply $\Sigma F_y = ma_y$. Because the skydiver is moving with constant velocity, the acceleration is zero:

$$\Sigma F_y = ma_y \Rightarrow F_d - mg = 0$$

so $F_d = mg = (64.0\text{ kg})(9.81\text{ N/kg}) = \boxed{628\text{ N}}$

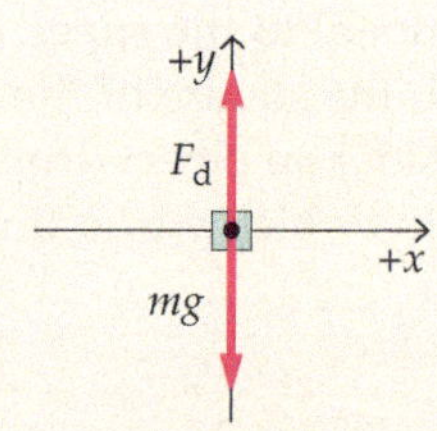

FIGURE 5-22

(*b*) 1. To find b we set $F_d = bv^2$:

$$F_d = mg = bv^2$$

$$\text{so} \quad b = \frac{mg}{v^2}$$

2. Convert the speed to m/s, then calculate b:

$$180\ \text{km/h} = \frac{180\ \text{km}}{1\ \text{h}} \times \frac{1\ \text{h}}{3.6\ \text{ks}} = 50.0\ \text{m/s}$$

$$b = \frac{(64.0\ \text{kg})(9.81\ \text{m/s}^2)}{(50.0\ \text{m/s})^2} = \boxed{0.251\ \text{kg/m}}$$

CHECK The units obtained for b are kg/m. To check that these units are correct, we multiply kg/m by $(\text{m/s})^2$ to get kg m/s². These units are units of force (because one newton is defined as a kg m/s²). We expected these units to be units of force because the drag force is given by bv^2.

TAKING IT FURTHER The conversion factor 1 h/3.6 ks appears in the last step of the solution. This conversion factor is exact because 1 h is exactly equal to 3600 s. Consequently, in converting her speed from km/h to m/s, three-figure accuracy is maintained. This is so even though we did not write the conversion factor as 1.00 h/3.60 ks.

5-3 MOTION ALONG A CURVED PATH

Objects often do not move in a straight line: a car rounding a curve is an example, as is a satellite orbiting Earth.

Consider a satellite moving in a circular orbit around Earth as shown in Figure 5-23. At an altitude of 200 km, the gravitational force on the satellite is just slightly less than at Earth's surface. Why does the satellite not fall toward Earth? Actually, the satellite does "fall." But because the surface of Earth is curved, the satellite does not get closer to the surface of Earth. If the satellite were not accelerating, it would move from point P_1 to P_2 in some time t. Instead, it arrives at point P_2' on its circular orbit. In a sense, the satellite "falls" the distance h shown in Figure 5-23. If t is small, P_2 and P_2' are nearly on a radial line. In that case, we can calculate h from the right triangle of sides vt, r, and $r + h$: Because $r + h$ is the hypotenuse of the right triangle, the Pythagorean theorem gives:

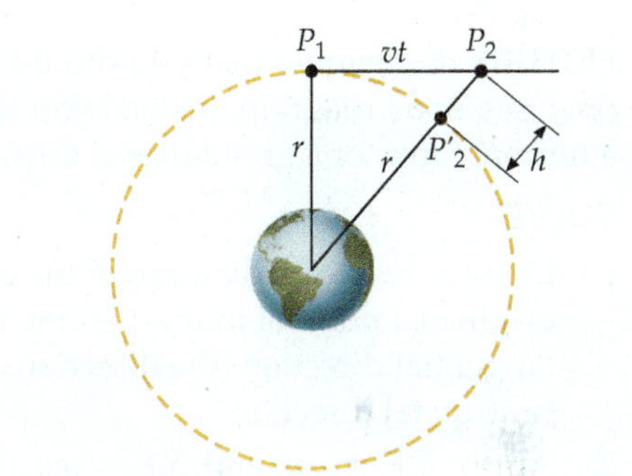

FIGURE 5-23 The satellite is moving with speed v in a circular orbit of radius r about Earth. If the satellite did not accelerate toward Earth, it would move from point P_1 to P_2 along a straight line. Because of its acceleration, it instead falls a distance h. For a sufficiently short time t, the acceleration is essentially constant, so $h = \frac{1}{2}at^2 = \frac{1}{2}(v^2/r)t^2$.

$$(r + h)^2 = (vt)^2 + r^2$$

$$r^2 + 2hr + h^2 = v^2t^2 + r^2$$

or

$$h(2r + h) = v^2t^2$$

For very short times, h will be much less than r, so we can neglect h compared with $2r$. Then

$$2rh \approx v^2t^2$$

or

$$h \approx \frac{1}{2}\left(\frac{v^2}{r}\right)t^2$$

Comparing this with the constant-acceleration expression $h = \frac{1}{2}at^2$, we see that the magnitude of the acceleration of the satellite is

$$a = \frac{v^2}{r}$$

which is the expression for centripetal acceleration established in Chapter 3. From Figure 3-24, we see this acceleration is directed toward the center of the circular orbit. By applying Newton's second law along the direction of the acceleration

See **Math Tutorial** *for more information on* **Trigonometry** vector, we find that the magnitude of the net force causing the acceleration is related to the magnitude of the acceleration by:

$$F_{net} = m\frac{v^2}{r}$$

Figure 5-24*a* shows a ball on the end of a string, the other end of which is attached to a fixed support. The ball is traveling in a horizontal circle of radius r at constant speed v. Consequently, the acceleration of the ball has magnitude v^2/r.

As we saw in Chapter 3, a particle moving with constant speed v in a circle of radius r (Figure 5-24*a*) has an acceleration of magnitude $a = v^2/r$ directed toward the center of the circle (the centripetal direction). The net force acting on an object is always in the same direction as the acceleration vector, so the net force (Figure 5-24*b*) on an object moving in a circle at constant speed is also in the centripetal direction. A net force in the centripetal direction is sometimes referred to as the **centripetal force.** It may be due to a string, spring, or other contact force such as a normal or frictional force; it may be an action-at-a-distance type of force such as a gravitational force; or it may be any combination of these. It is always directed inward—toward the center of curvature of the path.

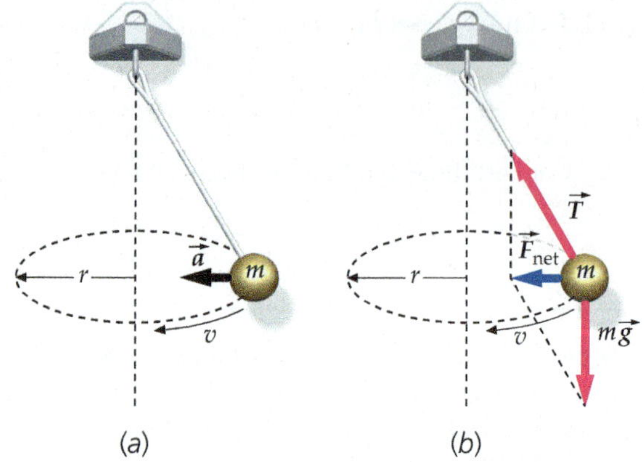

FIGURE 5-24 A ball suspended from a string moves in a horizontal circle at constant speed. (*a*) The acceleration vector is in the centripetal direction (toward the center of the circular path). Acceleration in the centripetal direction is called centripetal acceleration. (*b*) Two forces act on the ball, the tension force exerted by the string and the gravitational force. These two forces sum so that the net force is in the centripetal direction. A net force in the centripetal direction is sometimes referred to as the centripetal force.

! The centripetal force is not actually a force. It is merely a name for the component of net force toward the center of curvature of the path. Like the net force, the centripetal force does not belong on a free-body diagram. Only actual forces belong on free-body diagrams.

PROBLEM-SOLVING STRATEGY

Solving Motion Along A Curved Path Problems

PICTURE Remember that you should never label a force as a centripetal force on a free-body diagram. Instead you should label a force as a tension force, or a normal force, or a gravitational force, and so forth.

SOLVE

1. Draw a free-body diagram of the object. Include coordinate axes with the origin at a point of interest on the path. Draw one coordinate axis in the tangential direction (the direction of motion) and a second in the centripetal direction.
2. Apply $\sum F_c = ma_c$ and $\sum F_t = ma_t$ (Newton's second law in component form).
3. Substitute $a_c = v^2/r$ and $a_t = dv/dt$, where v is the speed.
4. If the object moves in a circle of radius r with constant speed v, use $v = 2\pi r/T$, where T is the time for one revolution.

CHECK Make sure that your answers are in accordance with the fact that the direction of the centripetal acceleration is always toward the center of curvature and perpendicular to the direction of the velocity vector.

Example 5-9 Swinging a Pail

You swing a pail containing mass m of water in a vertical circle of radius r (Figure 5-25). If the speed is v_{top} at the top of the circle, find (*a*) the force F_{PW} exerted by the pail on the water at the top of the circle, and (*b*) the minimum value of v_{top} for the water to remain in the pail. (*c*) What is the force exerted by the pail on the water at the bottom of the circle, where the speed of the pail is v_{bot}?

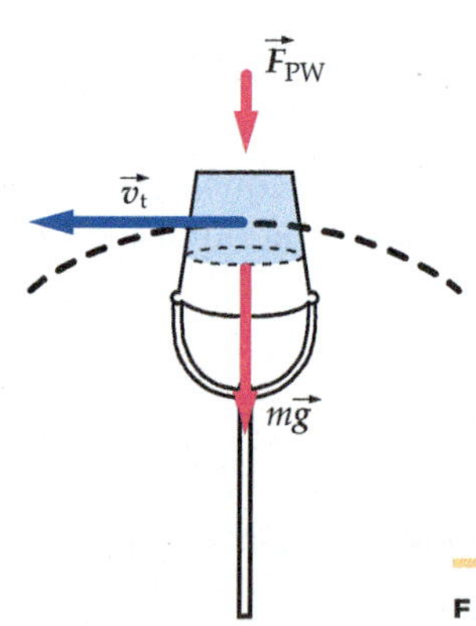

FIGURE 5-25

PICTURE When the bucket is at the top of its circle, both the force of gravity on the water and the contact force of the pail on the water are in the centripetal direction (downward). At the bottom of the circle, the contact force of the pail on the water must be greater than the force of gravity on the water to provide a net force in the centripetal direction (upward). We can apply Newton's second law to find the force exerted by the pail on the water at these points. Because the water moves in a circular path, it always has an acceleration component equal to v^2/r toward the center of the circle.

SOLVE

(*a*) 1. Draw free-body diagrams for the water at the top and bottom of the circle (Figure 5-26). Choose the $+r$ direction to be toward the center of the circle in each case.

2. Apply $\Sigma F_r = ma_r$ to the water as it passes through the top of the circle with speed v_{top}. Solve for the force F_{PW} exerted by the pail on the water:

$$\Sigma F_r = ma_r$$

$$F_{PW} + mg = m\frac{v_{top}^2}{r} \Rightarrow F_{PW} = \boxed{m\left(\frac{v_{top}^2}{r} - g\right)}$$

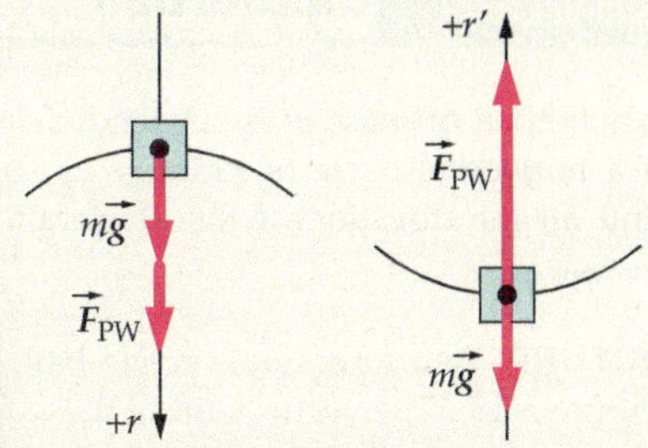

FIGURE 5-26

(*b*) The pail can push on the water, but not pull on it. The minimum force it can exert on the water is zero. Set $F_{PW} = 0$ and solve for the minimum speed:

$$0 = m\left(\frac{v_{top,min}^2}{r} - g\right) \Rightarrow v_{top,min} = \boxed{\sqrt{rg}}$$

(*c*) Apply $\Sigma F_r = ma_r$ to the water as it passes through the bottom of its path with speed v_{bot}. Solve for F_{PW}:

$$\Sigma F_r = ma_r$$

$$F_{PW} - mg = \frac{mv_{bot}^2}{r} \Rightarrow F_{PW} = \boxed{m\left(\frac{v_{bot}^2}{r} + g\right)}$$

CHECK In the Part-(*c*) result, when $v_{bot} = 0$, $F_{PW} = mg$. This is as expected.

TAKING IT FURTHER Note that there is no arrow labeled "centripetal force" in the free-body diagram. Centripetal force is not a kind of force exerted by some agent; it is just the name for the component of the resultant force in the centripetal direction.

PRACTICE PROBLEM 5-4 Estimate (*a*) the minimum speed at the top of the circle and (*b*) the maximum period of revolution that will keep you from getting wet if you swing a pail of water in a vertical circle at constant speed.

Example 5-10 Playing Ape

You step off the limb of a tree clinging to a 30-m-long vine that is attached to another limb at the same height and 30 m distant. Assuming air resistance is negligible, how fast are you gaining speed at the instant the vine makes an angle of 25° with the vertical during your descent?

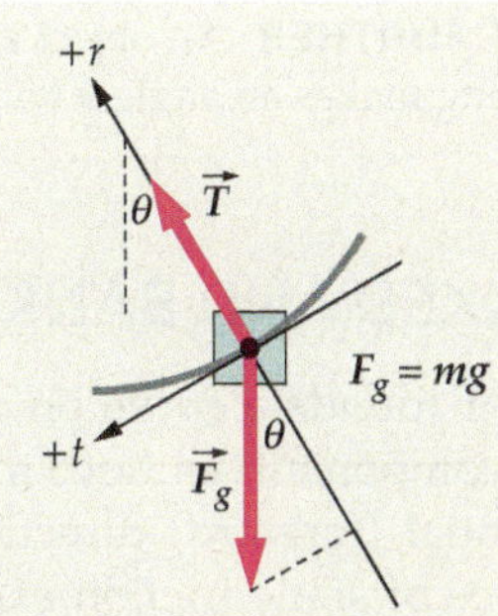

FIGURE 5-27 The $+t$ direction is in the direction of the velocity vector.

PICTURE Model this situation as a rope of negligible mass with one end attached to a limb and the other to a particle of mass m. Apply Newton's second law to the mass. The tangential acceleration is the rate of change of speed, so solve for the tangential acceleration.

SOLVE

1. Sketch a free-body diagram of the object (Figure 5-27). Let the $+r$ direction be toward the center of the path and the $+t$ direction be in the direction of the velocity:

2. Apply $\Sigma F_t = ma_t$ and use the free-body diagram to find expressions for the force components:

$$\Sigma F_t = ma_t$$

$$T_t + F_{gt} = ma_t$$

$$0 + mg\sin\theta = ma_t$$

3. Solve for a_t:

$$a_t = g\sin\theta = (9.81\ \text{m/s}^2)\sin 25° = \boxed{4.1\ \text{m/s}}$$

CHECK You expect your rate of change of speed to be positive because your speed would be increasing as long as you are descending. In addition, you expect that you will not be gaining speed at a rate equal to or greater than $g = 9.81\ \text{m/s}^2$. The step-3 result meets these expectations.

PRACTICE PROBLEM 5-5 What will your rate of change of speed be at the instant the vine is vertical and you are passing through the lowest point in your arc?

Example 5-11 Tetherball

Try It Yourself

A tetherball of mass m is suspended from a length of rope and travels at constant speed v in a horizontal circle of radius r as shown. The rope makes an angle θ with the vertical. Find (*a*) the direction of the acceleration, (*b*) the tension in the rope, and (*c*) the speed of the ball.

PICTURE Two forces act on the ball; the gravitational force and the tension in the rope (Figure 5-28). The vector sum of these forces is in the direction of the acceleration vector.

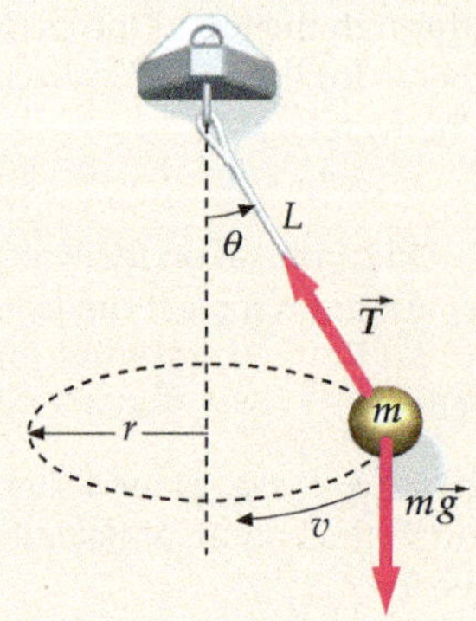

FIGURE 5-28

SOLVE

Cover the column to the right and try these on your own before looking at the answers.

Steps	Answers
(*a*) 1. The ball is moving in a horizontal circle at constant speed. The acceleration is in the centripetal direction.	The acceleration is horizontal and directed from the ball toward the center of the circle it is moving in.
(*b*) 1. Draw a free-body diagram for the ball (Figure 5-29). Choose as the $+x$ direction the direction of the ball's acceleration (toward the center of the circular path).	
2. Apply $\Sigma F_y = ma_y$ to the ball and solve for the tension T.	$\Sigma F_y = ma_y \Rightarrow T\cos\theta - mg = 0$ so $\boxed{T = \dfrac{mg}{\cos\theta}}$
(*c*) 1. Apply $\Sigma F_x = ma_x$ to the ball.	$\Sigma F_x = ma_x \Rightarrow T\sin\theta = m\dfrac{v^2}{r}$
2. Substitute $mg/\cos\theta$ for T and solve for v.	$\dfrac{mg}{\cos\theta}\sin\theta = m\dfrac{v^2}{r} \Rightarrow g\tan\theta = \dfrac{v^2}{r}$ so $v = \boxed{\sqrt{rg\tan\theta}}$

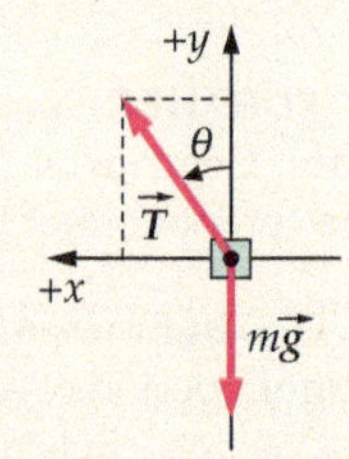

FIGURE 5-29

CHECK As $\theta \rightarrow 90°$, $\cos\theta \rightarrow 0$ and $\tan\theta \rightarrow \infty$. In the results for Parts (*b*) and (*c*), the expressions T and v both approach infinity as $\theta \rightarrow 90°$. This is what anyone who has played tetherball would expect. For θ to even approach 90°, the ball would have to move very fast.

TAKING IT FURTHER An object attached to a string and moving in a horizontal circle so that the string makes an angle θ with the vertical is called a *conical pendulum*.

UNBANKED AND BANKED CURVES

When a car rounds a curve on a horizontal road, the force components in both the centripetal and the tangential (forward) directions are provided by the force of static friction exerted by the road on the tires of the car. If the car travels at constant speed, then the forward component of the frictional force is balanced by the rearward-directed forces of air drag and rolling friction. The forward component of the static frictional force remains zero if air drag and rolling friction are both negligible and if the speed of the car remains constant.

If a curved road is not horizontal but banked, the normal force of the road will have a component in the centripetal direction. The *banking angle* is usually chosen so that no friction is needed for a car to complete the curve at the specified speed.

As a car rounds the curve, the tire is distorted by the frictional force exerted by the road. *(David Allio/Icon SMI/Corbis.)*

In 1993 a descent probe containing instruments went deep into the Jovian atmosphere—to the surface of Jupiter. The fully assembled probe was tested at accelerations up to 200gs in this large centrifuge at Sandia National Laboratories. *(Sandia National Laboratory.)*

Example 5-12 Rounding a Banked Curve

A curve of radius 30.0 m is banked at an angle θ. That is, the normal to the road surface makes an angle of 30.0° with the vertical. Find θ such that a car can round the curve at 40.0 km/h even if the road is coated with ice, making the road essentially frictionless.

PICTURE In this case, only two forces act on the car: the force of gravity and the normal force $\vec{F}_n$ (Figure 5-30). Because the road is banked, the normal force has a horizontal component that causes the car's centripetal acceleration. The vector sum of the two force vectors is in the direction of the acceleration. We can apply Newton's second law and then solve for θ.

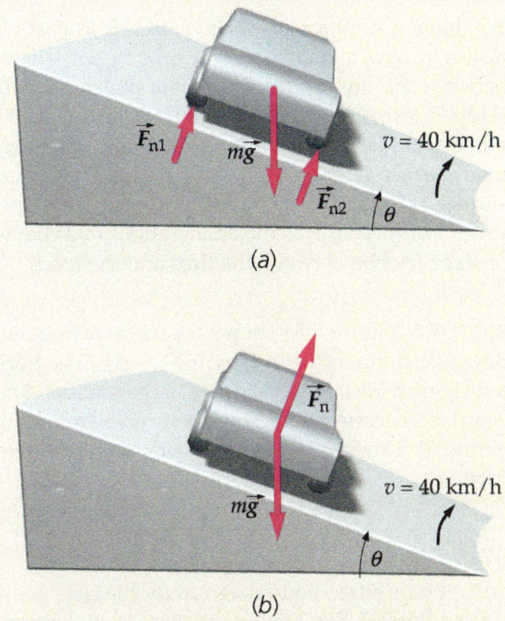

FIGURE 5-30

SOLVE

1. Draw a free-body diagram for the car (Figure 5-31)

2. Apply $\Sigma F_y = ma_y$ to the car:

$$\Sigma F_y = ma_y$$
$$F_n \cos\theta - mg = 0 \Rightarrow F_n \cos\theta = mg$$

3. Apply $\Sigma F_x = ma_x$ to the car.

$$\Sigma F_x = ma_x \Rightarrow F_n \sin\theta = m\frac{v^2}{r}$$

4. Divide the step-3 result by the step-2 result, then solve for θ:

$$\frac{\sin\theta}{\cos\theta} = \frac{mv^2}{rmg} \Rightarrow \tan\theta = \frac{v^2}{rg}$$
$$\theta = \tan^{-1}\frac{[(40.0\ \text{km/h})(1\ \text{h}/3.6\ \text{ks})]^2}{(30.0\ \text{m})(9.81\ \text{m/s}^2)}$$
$$= \boxed{22.8°}$$

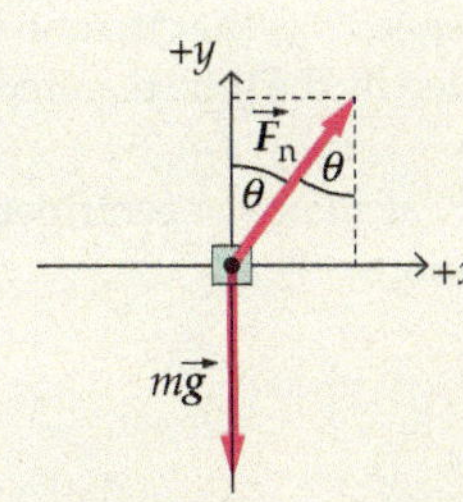

FIGURE 5-31

CHECK The banking angle is 22.8°. It is plausible because 30.0 m is a very small radius for a highway turn. For comparison, the turns at the Daytona International Speedway have a radius of 300 m and a banking angle of 31°.

TAKING IT FURTHER The banking angle θ depends on v and r, but not the mass m; θ increases with increasing v, and decreases with increasing r. When the banking angle, speed, and radius satisfy $\tan\theta = v^2/rg$, the car rounds the curve smoothly, with no tendency to slide either inward or outward. If the car speed is greater than $\sqrt{rg}\tan\theta$, the road must exert a static frictional force down the incline if the car is stay on the road. This force has a horizontal component, which provides the additional centripetal force needed to keep r from increasing. If the car speed is less than $\sqrt{rg}\tan\theta$, the road must exert a frictional force up the incline for the car is to stay on the road.

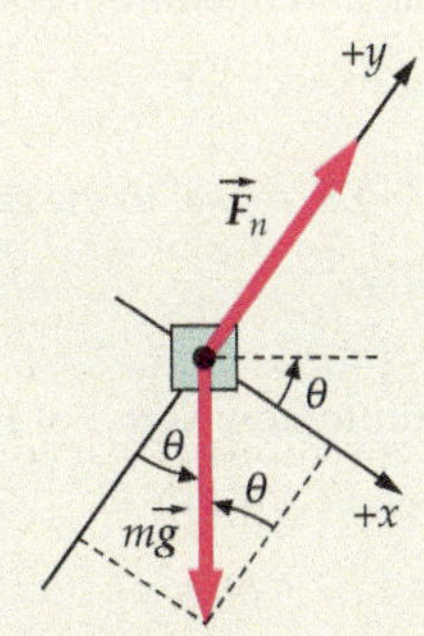

FIGURE 5-32

ALTERNATIVE SOLUTION In the preceding solution, we followed the guidelines to choose one of the coordinate axis directions to be the direction of the acceleration vector, the centripetal direction. However, the solution is no more difficult if we choose one of the axis directions to be down the incline. This choice is taken in the following solution.

1. Draw a free-body diagram for the car (Figure 5-32). The $+x$ direction is down the incline and the $+y$ direction is the normal direction.

2. Apply $\Sigma F_x = ma_x$ to the car:

$$\Sigma F_x = ma_x \Rightarrow mg\sin\theta = ma_x$$

3. Draw a sketch and use trigonometry to obtain an expression for a_x in terms of a and θ (Figure 5-33):

$$a_x = a\cos\theta$$

4. Substitute the step-3 result into the step-2 result. Then substitute v^2/r for a and solve for θ:

$$mg\sin\theta = ma\cos\theta$$
$$g\sin\theta = \frac{v^2}{r}\cos\theta$$
$$\tan\theta = \frac{v^2}{rg} \Rightarrow \theta = \boxed{\tan^{-1}\frac{v^2}{rg}}$$

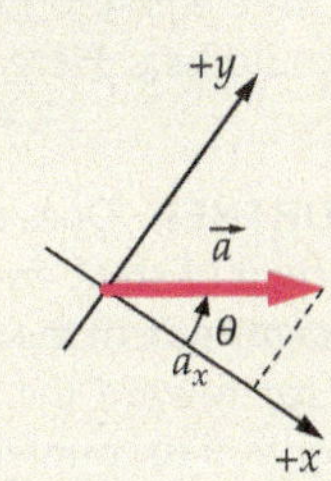

FIGURE 5-33

PRACTICE PROBLEM 5-6 Find the component of the acceleration normal to the road surface.

Example 5-13 A Road Test

Context-Rich

You have a summer job with NASCAR as part of an automobile tire testing team. You are testing a new model of racing tires to see whether or not the coefficient of static friction between the tires and dry concrete pavement is 0.90 as claimed by the manufacturer. In a skidpad test, a racecar is able to travel at constant speed in a circle of radius 45.7 m in 15.2 s without skidding. Assume air drag and rolling friction are negligible and assume that the road surface is horizontal. In a skidpad test a car travels in a circle on a flat, horizontal surface (a skidpad) at the maximum possible speed v without skidding. (*a*) What was its speed v? (*b*) What was the acceleration? (*c*) What was the minimum value for the coefficient of static friction between the tires and the road?

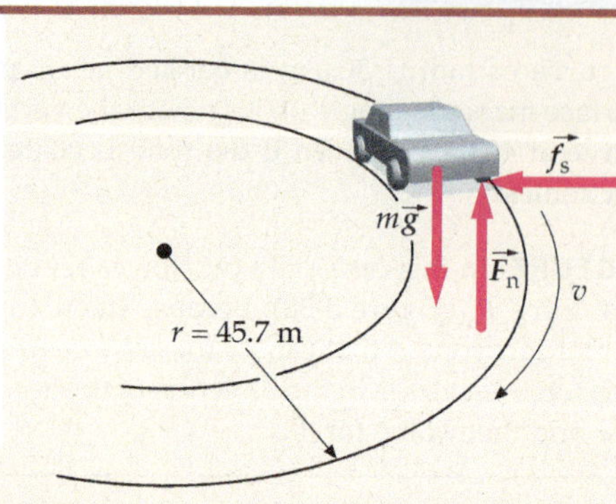

FIGURE 5-34

PICTURE Figure 5-34 shows the forces acting on the car. The normal force $\vec{F}_n$ balances the downward force due to gravity $\vec{F}_g = m\vec{g}$. The horizontal force is the force of static friction, which provides the centripetal acceleration. The faster the car travels, the greater the centripetal acceleration. The speed can be found from the circumference of the circle and the period T. This speed puts a lower limit on the maximum value of the coefficient of static friction.

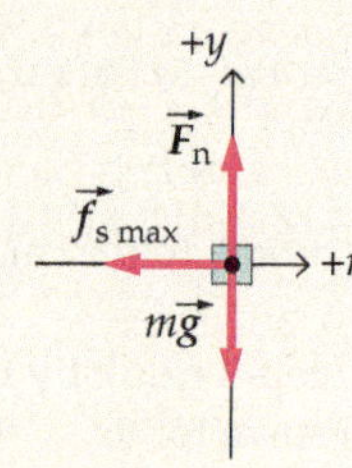

FIGURE 5-35

SOLVE

(*a*) 1. Draw a free-body diagram for the car (Figure 5-35). The $+r$ direction is away from the center of curvature.

2. The speed v is the circumference of the circle divided by the time required for one revolution:

$$v = \frac{2\pi r}{T} = \frac{2\pi(45.7\text{ m})}{15.2\text{ s}} = \boxed{18.9\text{ m/s}}$$

(*b*) Use v to calculate the centripetal and tangential accelerations:

$$a_c = \frac{v^2}{r} = \frac{(18.9\text{ m/s})^2}{(45.7\text{ m})} = \boxed{7.81\text{ m/s}^2}$$

$$a_t = \frac{dv}{dt} = \boxed{0}$$

The acceleration is 7.81 m/s² in the centripetal direction.

(*c*) 1. Apply $\Sigma F_y = ma_y$ to the car. Solve for the normal and maximum frictional force:

$$\Sigma F_y = ma_y$$

$$F_n - mg = 0 \quad \text{so} \quad F_n = mg$$

and $f_{s\max} = \mu_s F_n = \mu_s mg$

2. Apply $\Sigma F_r = ma_r$ to the car:

$$\Sigma F_r = ma_r$$

$$-f_{s\max} = m\left(-\frac{v^2}{r}\right) \Rightarrow f_{s\max} = m\frac{v^2}{r}$$

3. Substitute from step 1 of Part (*c*) and solve for μ_s:

$$\mu_s mg = m\frac{v^2}{r} \Rightarrow \mu_s g = \frac{v^2}{r}$$

$$\mu_s = \frac{v^2}{rg} = \frac{(18.9\text{ m/s})^2}{(45.7\text{ m})(9.81\text{ m/s}^2)} = \boxed{0.796}$$

CHECK If μ_s were equal to 1.00, the inward force would be equal to mg and the centripetal acceleration would be g. Here μ_s is about 0.80 and the centripetal acceleration is about $0.80g$.

TAKING IT FURTHER Does the result of the skidpad test support the manufacturer's claim that the coefficient of static friction is 0.90? It does support the manufacturer's claim. In calculating the magnitude of the frictional force, we accounted for the frictional force needed to accelerate the car toward the center of curvature, but we neglected to account for the frictional force required to counter the effects of air drag and rolling friction. A speed of 18.9 m/s equals 42.3 mi/h, a speed at which air drag is definitely significant.

* 5-4 NUMERICAL INTEGRATION: EULER'S METHOD

If a particle moves under the influence of a *constant* force, its acceleration is constant and we can find its velocity and position from the constant-acceleration kinematic formulas in Chapter 2. But consider a particle moving through space where the force on it, and therefore its acceleration, depends on its position and its velocity. The position, velocity, and acceleration of the particle at one instant determine the position and velocity the next instant, which then determines its acceleration at that instant. The actual position, velocity, and acceleration of an object all change continuously with time. We can approximate this by replacing the continuous time variations with small time steps of duration Δt. The simplest approximation is to assume constant acceleration during each step. This approximation is called **Euler's method.** If the time interval is sufficiently short, the change in acceleration during the interval will be small and can be neglected.

Let x_0, v_{0x}, and a_{0x} be the known position, velocity, and acceleration of a particle at some initial time t_0. If we neglect any change in velocity during the time interval, the new position is given by

$$x_1 = x_0 + v_{0x}\,\Delta t$$

Similarly, if we assume constant acceleration during Δt, the velocity at time $t_1 = t_0 + \Delta t$ is given by

$$v_1 = v_{0x} + a_{0x}\,\Delta t$$

We can use the values x_1 and v_{1x} to compute the new acceleration a_{1x} using Newton's second law, and then use x_1, v_{1x}, and a_{1x} to compute x_2 and v_{2x}.

$$x_2 = x_1 + v_{1x}\,\Delta t$$
$$v_2 = v_1 + a_{1x}\,\Delta t$$

The connection between the position and velocity at time t_n and time $t_{n+1} = t_n + \Delta t$ is given by

$$x_{n+1} = x_n + v_{nx}\,\Delta t \qquad \text{5-9}$$

and

$$v_{n+1} = v_{nx} + a_{nx}\,\Delta t \qquad \text{5-10}$$

To find the velocity and position at some time t, we therefore divide the time interval $t - t_0$ into a large number of smaller intervals Δt and apply Equations 5-9 and 5-10, beginning at the initial time t_0. This involves a large number of simple, repetitive calculations that are most easily done on a computer. The technique of breaking the time interval into small steps and computing the acceleration, velocity, and position at each step using the values from the previous step is called *numerical integration.*

To illustrate the use of numerical integration, let us consider a problem in which a skydiver is dropped from rest at some height under the influences of both gravity and a drag force that is proportional to the square of the speed. We will find the velocity v_x and the distance traveled x as functions of time.

The equation describing the motion of an object of mass m dropped from rest is Equation 5-6 with $n = 2$:

$$mg - bv^2 = ma_x$$

where down is the positive direction. The acceleration is thus

$$a_x = g - \frac{b}{m}v^2 \qquad \text{5-11}$$

	A	B	C	D
1	Δt =	0.5	s	
2	x0 =	0	m	
3	v0 =	0	m/s	
4	a0 =	9.81	m/s^2	
5	vt =	60	m/s	
6				
7	t	x	v	a
8	(s)	(m)	(m/s)	(m/s^2)
9	0.00	0.0	0.00	9.81
10	0.50	0.0	4.91	9.74
11	1.00	2.5	9.78	9.55
12	1.50	7.3	14.55	9.23
13	2.00	14.6	19.17	8.81
14	2.50	24.2	23.57	8.30
15	3.00	36.0	27.72	7.72
41	16.00	701.0	59.55	0.15
42	16.50	730.7	59.62	0.16
43	17.00	760.6	59.68	0.10
44	17.50	790.4	59.74	0.09
45	18.00	820.3	59.78	0.07
46	18.50	850.2	59.82	0.06
47	19.00	880.1	59.85	0.05
48	19.50	910.0	59.87	0.04
49	20.00	939.9	59.89	0.04
50				

(a)

	A	B	C	D
1	Δt =	0.5	s	
2	x0 =	0	m	
3	v0 =	0	m/s	
4	a0 =	9.81	m/s^2	
5	vt =	60	m/s	
6				
7	t	x	v	a
8	(s)	(m)	(m/s)	(m/s^2)
9	0	=B2	=B3	=$B$4*(1-C9^2/$B$5^2)
10	=A9+$B$1	=B9+C9*$B$1	=C9+D9*$B$1	=$B$4*(1-C10^2/$B$5^2)
11	=A10+$B$1	=B10+C10*$B$1	=C10+D10*$B$1	=$B$4*(1-C11^2/$B$5^2)
12	=A11+$B$1	=B11+C11*$B$1	=C11+D11*$B$1	=$B$4*(1-C12^2/$B$5^2)
13	=A12+$B$1	=B12+C12*$B$1	=C12+D12*$B$1	=$B$4*(1-C13^2/$B$5^2)
14	=A13+$B$1	=B13+C13*$B$1	=C13+D13*$B$1	=$B$4*(1-C14^2/$B$5^2)
15	=A14+$B$1	=B14+C14*$B$1	=C14+D14*$B$1	=$B$4*(1-C15^2/$B$5^2)
41	=A40+$B$1	=B40+C40*$B$1	=C40+D40*$B$1	=$B$4*(1-C41^2/$B$5^2)
42	=A41+$B$1	=B41+C41*$B$1	=C41+D41*$B$1	=$B$4*(1-C42^2/$B$5^2)
43	=A42+$B$1	=B42+C42*$B$1	=C42+D42*$B$1	=$B$4*(1-C43^2/$B$5^2)
44	=A43+$B$1	=B43+C43*$B$1	=C43+D43*$B$1	=$B$4*(1-C44^2/$B$5^2)
45	=A44+$B$1	=B44+C44*$B$1	=C44+D44*$B$1	=$B$4*(1-C45^2/$B$5^2)
46	=A45+$B$1	=B45+C45*$B$1	=C45+D45*$B$1	=$B$4*(1-C46^2/$B$5^2)
47	=A46+$B$1	=B46+C46*$B$1	=C46+D46*$B$1	=$B$4*(1-C47^2/$B$5^2)
48	=A47+$B$1	=B47+C47*$B$1	=C47+D47*$B$1	=$B$4*(1-C48^2/$B$5^2)
49	=A48+$B$1	=B48+C48*$B$1	=C48+D48*$B$1	=$B$4*(1+C49^2/$B$5^2)
50				

(b)

FIGURE 5-36 (*a*) Spreadsheet to compute the position and speed of a skydiver with air drag proportional to v^2. (*b*) The same Excel spreadsheet displaying the formulas rather than the values.

It is convenient to write the constant b/m in terms of the terminal speed v_T. Setting $a_x = 0$ in Equation 5-11, we obtain

$$0 = g - \frac{b}{m}v_T^2$$

$$\frac{b}{m} = \frac{g}{v_T^2}$$

Substituting g/v_T^2 for b/m in Equation 5-11 gives

$$a_x = g\left(1 - \frac{v^2}{v_T^2}\right) \qquad \text{5-12}$$

The acceleration at time t_n is calculated using the values x_n and v_{nx}.

To solve Equation 5-12 numerically, we need to use numerical values for g and v_t. A reasonable terminal speed for a skydiver is 60.0 m/s. If we choose $x_0 = 0$ for the initial position, the initial values are $x_0 = 0$, $v_0 = 0$, and $a_{0x} = g = 9.81\text{ m/s}^2$. To find the velocity v_x and position x after some time, say, $t = 20.0$ s, we divide the time interval $0 < t < 20.0$ s into many small intervals Δt and apply Equations 5-9, 5-10, and 5-12. We do this by using a computer spreadsheet (or by writing a computer program) as shown in Figure 5-36. This spreadsheet has $\Delta t = 0.5$ s, and the computed values at $t = 20$ s are $v = 59.89$ m/s and $x = 939.9$ m. Figure 5-37 shows graphs of v_x versus t and x versus t plotted from these data.

But how accurate are our computations? We can estimate the accuracy by running the program again using a smaller time interval. If we use $\Delta t = 0.25$ s, one-half of the value we originally used, we obtain $v = 59.86$ m/s and $x = 943.1$ m at $t = 20$ s. The difference in v is about 0.05 percent and that in x is about 0.3 percent. These are our estimates of the accuracy of the original computations.

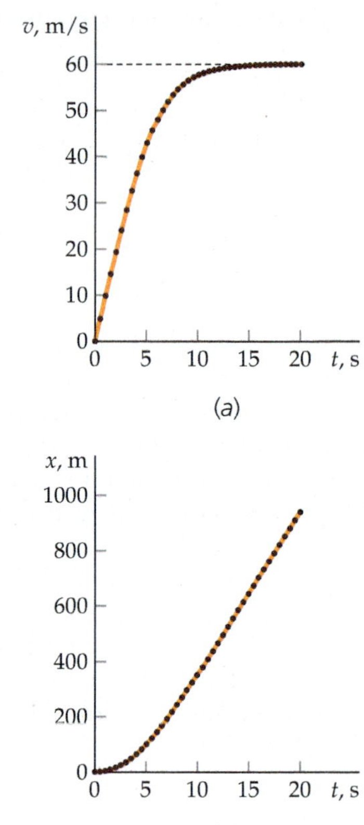

FIGURE 5-37 (*a*) Graph of v versus t for a skydiver, found by numerical integration using $\Delta t = 0.5$ s. The horizontal dashed line is the terminal speed $v_t = 60$ m/s. (*b*) Graph of x versus t using $\Delta t = 0.5$ s.

Because the difference between the value of a_{avx} for some time interval Δt and the value of a_x at the beginning of the interval becomes smaller as the time interval becomes shorter, we might expect that it would be better to use very short time intervals, say, $\Delta t = 0.000\,000\,001$ s. But there are two reasons for not using extremely short intervals. First, the shorter the time interval, the larger the number of calculations that are required and the longer the program takes to run. Second, the computer keeps only a fixed number of digits at each step of the calculation, so that at each step there is a round-off error. These round-off errors add up. The larger the number of calculations, the more significant the total round-off errors become. When we first decrease the time interval, the accuracy improves because a_i more nearly approximates a_{av} for the interval. However, as the time interval is decreased further, the round-off errors build up and the accuracy of the computation decreases. A good rule of thumb to follow is to use no more than about 10^5 time intervals for the typical numerical integration.

5-5 THE CENTER OF MASS

If you throw a ball into the air, the ball follows a smooth parabolic path. But if you toss a baton in the air (Figure 5-38), the motion of the baton is more complicated. Each end of the baton moves in a different direction, and both ends move in a different way than the middle. However, if you look at the motion of the baton more closely, you will see that there is one point on the baton that moves in a parabolic path, even though the rest of the baton does not. This point, called the **center of mass,** moves as if all the baton's mass were concentrated at that point and all external forces were applied there.

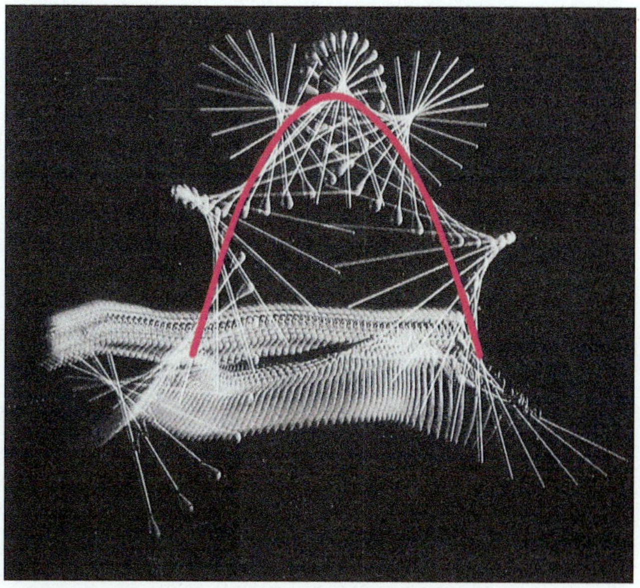

FIGURE 5-38 A multiflash photo of a baton thrown into the air. *(Estate of Harold E. Edgerton/Palm Press.)*

To determine the center of mass of an object, it is helpful to think of the object as a system of particles. Consider, for example, a simple system that consists of two point particles located on the x axis at positions x_1 and x_2 (Figure 5-39). If the particles have masses m_1 and m_2, then the center-of-mass is located on the x axis at position x_{cm}, defined by

$$Mx_{cm} = m_1x_1 + m_2x_2 \qquad 5\text{-}13$$

where $M = m_1 + m_2$ is the total mass of the system. If we choose the position of the origin and the $+x$ direction such that the position of m_1 is at the origin and that of m_2 is on the positive x axis, then $x_1 = 0$ and $x_2 = d$, where d is the distance between the particles. The center of mass is then given by

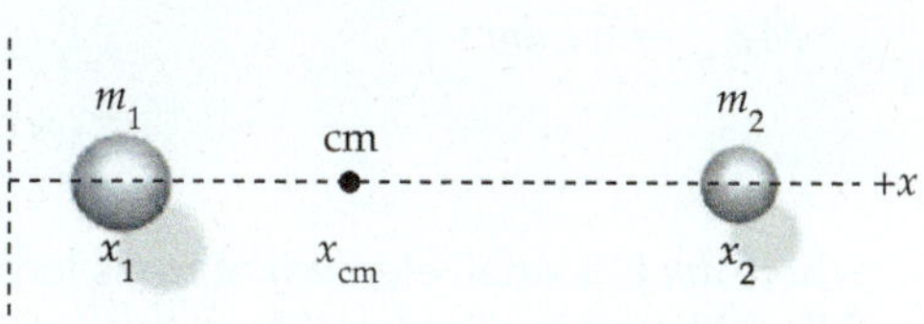

FIGURE 5-39

$$Mx_{cm} = m_1x_1 + m_2x_2 = m_1(0) + m_2d$$

$$x_{cm} = \frac{m_2}{M}d = \frac{m_2}{m_1 + m_2}d \qquad 5\text{-}14$$

In the case of just two particles, the center of mass lies at some point on the line between the particles; if the particles have equal masses, then the center of mass is midway between them (Figure 5-40*a*). If the two particles are of unequal mass, then the center of mass is closer to the more massive particle (Figure 5-40*b*).

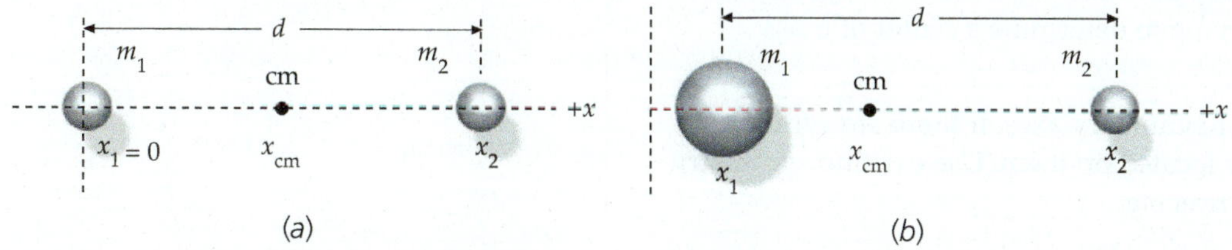

FIGURE 5-40

PRACTICE PROBLEM 5-7

A 4.0-kg mass is at the origin and a 2.0-kg mass is on the x axis at $x = 6.0$ cm. Find x_{cm}.

We can generalize from two particles in one dimension to a system of many particles in three dimensions. For N particles in three dimensions,

$$Mx_{cm} = m_1x_1 + m_2x_2 + m_3x_3 + \cdots + m_Nx_N$$

Using more concise notation, this is written

$$Mx_{cm} = \sum_i m_ix_i \qquad 5\text{-}15$$

where again $M = \sum_i m_i$ is the total mass of the system. Similarly, in the y and z directions,

$$My_{cm} = \sum_i m_iy_i \qquad 5\text{-}16$$

and

$$Mz_{cm} = \sum_i m_iz_i \qquad 5\text{-}17$$

In vector notation, $\vec{r}_i = x_i\hat{i} + y_i\hat{j} + z_i\hat{k}$ is the position vector of the ith particle. The **position of the center of mass,** $\vec{r}_{cm}$, is defined by

$$M\vec{r}_{cm} = m_1\vec{r}_1 + m_2\vec{r}_2 + \cdots = \sum_i m_i\vec{r}_i \qquad 5\text{-}18$$

DEFINITION: CENTER OF MASS

where $\vec{r}_{cm} = x_{cm}\hat{i} + y_{cm}\hat{j} + z_{cm}\hat{k}$.

Now let us consider extended objects, such as balls, baseball bats, and even cars. We can think of objects such as these as a system containing a very large number of particles, with a continuous distribution of mass. For highly symmetric objects, the center of mass is at the *center of symmetry*. For example, the center of mass of a uniform sphere or a uniform cylinder is located at its *geometric center*. For an object with a line or plane of symmetry, the center of mass lies somewhere along that line or plane. To find the position center of mass of an object, we replace the sum in Equation 5-18 with an integral:

$$M\vec{r}_{cm} = \int \vec{r}\,dm \qquad 5\text{-}19$$

CENTER OF MASS, CONTINUOUS OBJECT

where dm is a small element of mass located at position $\vec{r}$, as shown in Figure 5-41. (We will examine in detail how this integral is set up after Example 5-15.)

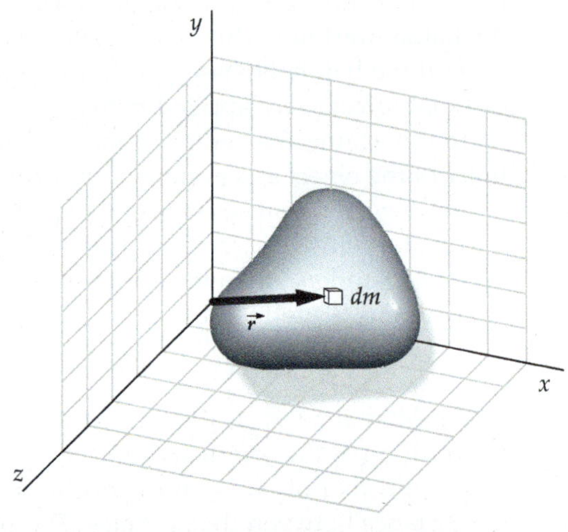

FIGURE 5-41 A mass element dm located at position $\vec{r}$ is used for finding the center of mass by integration.

PROBLEM-SOLVING STRATEGY

Solving Center-of-Mass Problems

PICTURE Determining centers of mass often simplifies determinations of the motions of an object or system of objects. Drawing a sketch of the object or system of objects is useful when trying to determine a center of mass.

SOLVE

1. Check the mass distribution for symmetry axes. If there are symmetry axes, the center of mass will be located on them. Use existing symmetry axes as coordinate axes where feasible.

2. Check to see if the mass distribution is composed of highly symmetric subsystems. If so, then calculate the centers of mass of the individual subsystems, and then calculate the center of mass of the system by treating each subsystem as a point particle at its center of mass.
3. If the system contains one or more point particles, place the origin at the location of a point particle. (If the ith particle is at the origin, then $\vec{r}_i = 0$.)

CHECK Make sure your center-of-mass determinations make sense. In many cases, the center of mass of an object is located near the more massive and larger part of the object. The center of mass of a multi-object system or an object such as a hoop may not be located within or on any object.

Example 5-14 The Center of Mass of a Water Molecule

A water molecule consists of an oxygen atom and two hydrogen atoms. An oxygen atom has a mass of 16.0 unified mass units (u), and each hydrogen atom has a mass of 1.00 u. The hydrogen atoms are each at an average distance of 96.0 pm (96.0×10^{-12} m) from the oxygen atom, and are separated from one another by an angle of 104.5°. Find the center of mass of a water molecule.

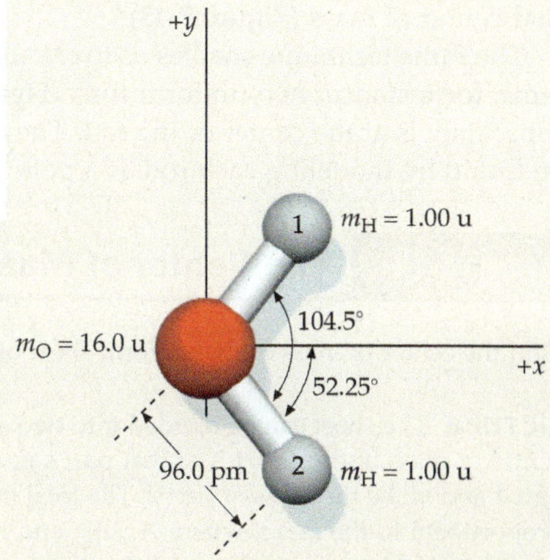

FIGURE 5-42

PICTURE We can simplify the calculation by selecting a coordinate system such that the origin is located at the oxygen atom, with the x axis bisecting the angle between the hydrogen atoms (Figure 5-42). Then, given the symmetries of the molecule, the center of mass will be on the x axis, and the line from the oxygen atom to each hydrogen atom will make an angle of 52.25°.

SOLVE

1. The location of the center of mass is given by its coordinates, x_{cm} and y_{cm} (Equations 5-15 and 5-16):

$$x_{cm} = \frac{\sum m_i x_i}{M}, \quad y_{cm} = \frac{\sum m_i y_i}{M}$$

2. Writing these out explicitly gives:

$$x_{cm} = \frac{m_{H1}x_{H1} + m_{H2}x_{H2} + m_O x_O}{m_{H1} + m_{H2} + m_O}$$

$$y_{cm} = \frac{m_{H1}y_{H1} + m_{H2}y_{H2} + m_O y_O}{m_{H1} + m_{H2} + m_O}$$

3. We have chosen the origin to be the location of the oxygen atom, so both the x and y coordinates of the oxygen atom are zero. The x and y coordinates of the hydrogen atoms are calculated from the 52.25° angle each hydrogen makes with the x axis:

$$x_O = y_O = 0$$

$$x_{H1} = 96.0 \text{ pm} \cos 52.25° = 58.8 \text{ pm}$$

$$x_{H2} = 96.0 \text{ pm} \cos(-52.25°) = 58.8 \text{ pm}$$

$$y_{H1} = 96.0 \text{ pm} \sin 52.25° = 75.9 \text{ pm}$$

$$y_{H2} = 96.0 \text{ pm} \sin(-52.25°) = -75.9 \text{ pm}$$

4. Substituting the coordinate and mass values into step 2 gives x_{cm}:

$$x_{cm} = \frac{(1.00 \text{ u})(58.8 \text{ pm}) + (1.00 \text{ u})(58.8 \text{ pm}) + (16.0 \text{ u})(0)}{1.00 \text{ u} + 1.00 \text{ u} + 16.0 \text{ u}} = 6.53 \text{ pm}$$

$$x_{cm} = \frac{(1.00 \text{ u})(75.9 \text{ pm}) + (1.00 \text{ u})(-75.9 \text{ pm}) + (16.0 \text{ u})(0)}{1.00 \text{ u} + 1.00 \text{ u} + 16.0 \text{ u}} = 0.00 \text{ pm}$$

5. The center of mass is on the x axis:

$$\vec{r}_{cm} = x_{cm}\hat{i} + y_{cm}\hat{j} = \boxed{6.53 \text{ pm } \hat{i} + 0.00\hat{j}}$$

CHECK That $y_{cm} = 0$ can be seen from the symmetry of the mass distribution. Also, the center of mass is very close to the relatively massive oxygen atom, as expected.

TAKING IT FURTHER The distance 96 pm is read "ninety-six picometers," where pico is the prefix for 10^{-12}.

Note that we could also have solved Example 5-14 by first finding the center of mass of just the two hydrogen atoms. For a system of three particles Equation 5-18 is

$$M\vec{r}_{cm} = m_1\vec{r}_1 + m_2\vec{r}_2 + m_3\vec{r}_3$$

The first two terms on the right side of this equation are related to the center of mass of the first two particles $\vec{r}\,'_{cm}$:

$$m_1\vec{r}_1 + m_2\vec{r}_2 = (m_1 + m_2)\vec{r}\,'_{cm}$$

The center of mass of the three-particle system can then be written

$$M\vec{r}_{cm} = (m_1 + m_2)\vec{r}\,'_{cm} + m_3\vec{r}_3$$

So we can first find the center of mass for two of the particles, the hydrogen atoms, for example, and then replace them with a single particle of total mass $m_1 + m_2$ at that center of mass (Figure 5-43).

The same technique enables us to calculate centers of mass for more complex systems, for instance, two uniform rods (Figure 5-44). The center of mass of each rod separately is at the center of the rod. The center of mass of the two-rod system can be found by modeling each rod as a point particle at its individual center of mass.

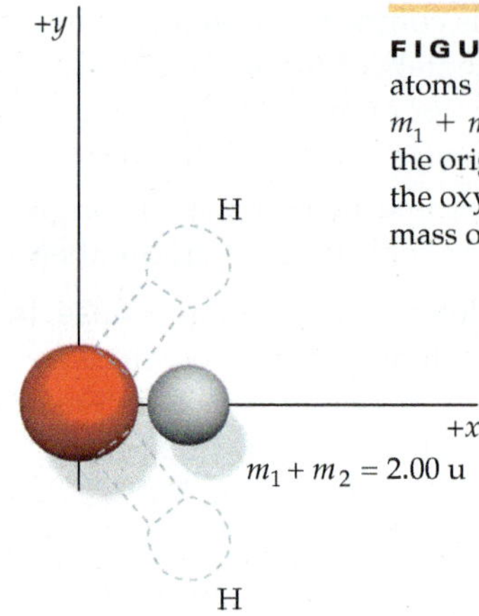

FIGURE 5-43 Example 5-14 with the two hydrogen atoms replaced by a single particle of mass $m_1 + m_2 = 2.00\,u$ on the x axis at the center of mass of the original atoms. The center of mass then falls between the oxygen atom at the origin and the calculated center of mass of the two hydrogen atoms.

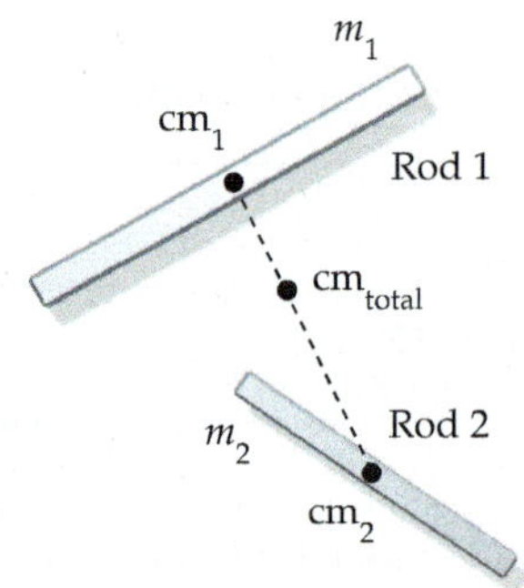

FIGURE 5-44

Example 5-15 The Center of Mass of a Plywood Sheet

Find the center of mass of the uniform sheet of plywood shown in Figure 5-45*a*.

PICTURE The sheet can be divided into two symmetrical parts (Figure 5-45*b*). The center of mass of each part is at that part's geometric center. Let m_1 be the mass of part 1 and m_2 be the mass of part 2. The total mass is $M = m_1 + m_2$. The masses are proportional to the areas, where A_1, A_2, and A are the areas of the pieces of mass m_1, m_2, and M, respectively.

SOLVE

Steps

1. Write the x and y coordinates of the center of mass in terms of m_1 and m_2.
$$x_{cm} = \frac{1}{M}(m_1x_{cm1} + m_2x_{cm2})$$
$$y_{cm} = \frac{1}{M}(m_1y_{cm1} + m_2y_{cm2})$$

2. Substitute area ratios for the mass ratios.
$$x_{cm} = \frac{A_1}{A}x_{cm1} + \frac{A_2}{A}x_{cm2}$$
$$y_{cm} = \frac{A_1}{A}y_{cm1} + \frac{A_2}{A}y_{cm2}$$

3. Calculate the areas and the ratios of the areas, using the values from Figure 5-45*b*.
$$A_1 = 0.32\ \text{m}^2; \qquad A_2 = 0.040\ \text{m}^2$$
$$\frac{A_1}{A} = \frac{8.0}{9.0} \qquad \frac{A_2}{A} = \frac{1.0}{9.0}$$

4. Write the x and y coordinates of the center-of-mass coordinates for each part by inspection of the figure.
$$x_{cm1} = 0.40\ \text{m}, \quad y_{cm1} = 0.20\ \text{m}$$
$$x_{cm2} = 0.70\ \text{m}, \quad y_{cm2} = 0.50\ \text{m}$$

5. Substitute these results into the step-2 result to calculate x_{cm} and y_{cm}.
$$x_{cm} = \boxed{0.43\ \text{m}}, \quad y_{cm} = \boxed{0.23\ \text{m}}$$

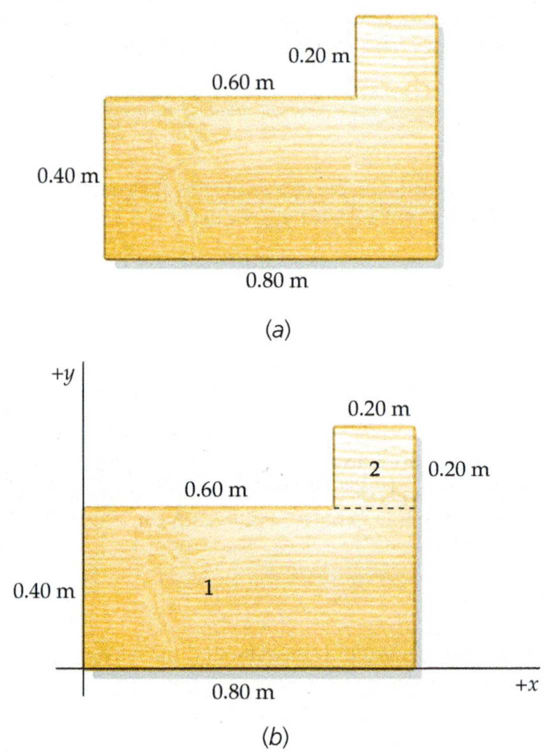

FIGURE 5-45

CHECK As expected, the center of mass of the system is very near the center of mass of part 1 (because $m_1 = 8m_2$.)

TAKING IT FURTHER Drawing an axis through the geometric centers of parts 1 and 2 and placing the origin at the geometric center of part 1, would have made locating the center of mass considerably easier.

* FINDING THE CENTER OF MASS BY INTEGRATION

In this section, we find the center of mass by integration (Equation 5-19):

$$\vec{r}_{\text{cm}} = \frac{1}{M}\int \vec{r}\, dm$$

We will start by finding the center of mass of a uniform thin rod to illustrate the technique for setting up the integration.

Uniform rod We first choose a coordinate system. A good choice for a coordinate system is one with an x axis through the length of the rod, with the origin at one end of the rod (Figure 5-46). Shown on the figure is a mass element dm of length dx a distance x from the origin. Equation 5-19 thus gives

$$\vec{r}_{\text{cm}} = \frac{1}{M}\int \vec{r}\, dm = \frac{1}{M}\int x\hat{i}\, dm$$

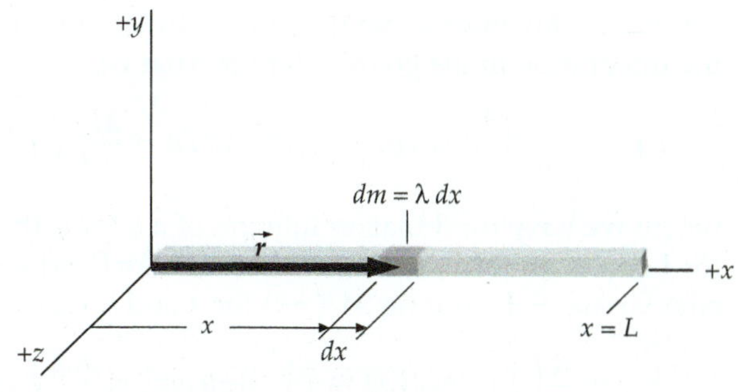

FIGURE 5-46

The mass is distributed on the x axis along the interval $0 \leq x \leq L$. To integrate dm along the mass distribution means the limits of the integral are 0 and L. (We integrate in the direction of increasing x.) The ratio dm/dx is the mass per unit length λ, so $dm = \lambda\, dx$:

$$\vec{r}_{\text{cm}} = \frac{1}{M}\hat{i}\int x\, dm = \frac{1}{M}\hat{i}\int_0^L x\lambda\, dx \qquad \text{5-20}$$

where

$$M = \int dm = \int_0^L \lambda\, dx \qquad \text{5-21}$$

Because the rod is uniform, λ is constant and can be factored from each of the integrals in Equations 5-20 and 5-21, giving

$$\vec{r}_{\text{cm}} = \frac{1}{M}\lambda\hat{i}\int_0^L x\, dx = \frac{1}{M}\lambda\hat{i}\frac{L^2}{2} \qquad \text{5-22}$$

and

$$M = \lambda\int_0^L dx = \lambda L \qquad \text{5-23}$$

Solving Equation 5-23 for λ gives $\lambda = M/L$. Thus, for a uniform rod the mass per unit length is equal to the total mass divided by the total length. Substituting λL for M in Equation 5-22, we complete the calculation and obtain the expected result

$$\vec{r}_{\text{cm}} = \frac{1}{\lambda L}\lambda\hat{i}\frac{L^2}{2} = \frac{1}{2}L\hat{i}$$

Semicircular hoop In calculating the center of mass of a uniform semicircular hoop of radius R, a good choice of coordinate axes is one with the origin at the center and with the y axis bisecting the semicircle (Figure 5-47). To find the center of mass, we use $M\vec{r}_{\text{cm}} = \int \vec{r}\, dm$ (Equation 5-19,) where $\vec{r} = x\hat{i} + y\hat{j}$. The semicircular mass distribution suggests using polar coordinates,* for which $x = r\cos\theta$ and $y = r\sin\theta$. The distance of the points on the semicircle from the origin is $r = R$. With these substitutions, we have

$$\vec{r}_{\text{cm}} = \frac{1}{M}\int (x\hat{i} + y\hat{j})\, dm = \frac{1}{M}\int R(\cos\theta\,\hat{i} + \sin\theta\,\hat{j})\, dm$$

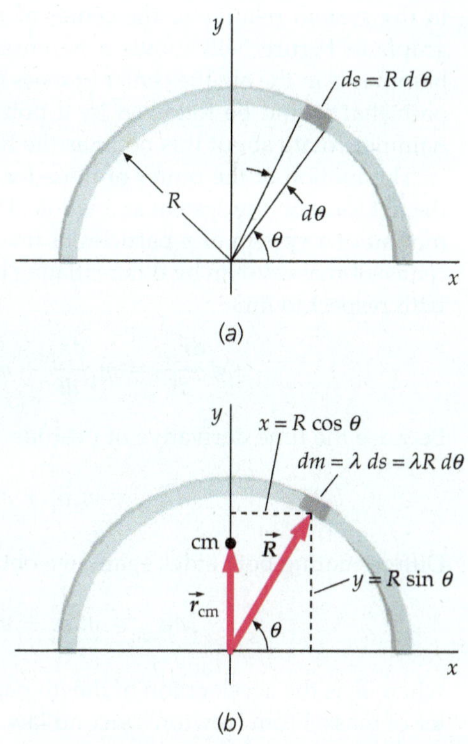

FIGURE 5-47 Geometry for calculating the center of mass of a semicircular hoop by integration.

* In polar coordinates the coordinates of a point are r and θ, were r is the magnitude of the position vector $\vec{r}$ and θ is the angle that the position vector makes with the $+x$ direction.

Next we express dm in terms of $d\theta$. First, the mass element dm has length $ds = R\,d\theta$, so

$$dm = \lambda\, ds = \lambda R\, d\theta$$

where $\lambda = dm/ds$ is the mass per unit length. Thus, we have

$$\vec{r}_{\text{cm}} = \frac{1}{M}\int R(\cos\theta\,\hat{i} + \sin\theta\,\hat{j})\lambda R\,d\theta$$

Evaluating this integral involves integrating dm along the semicircular mass distribution. This means that $0 \le \theta \le \pi$. Integrating in the direction of increasing θ, the integration limits go from 0 to π. That is,

$$\vec{r}_{\text{cm}} = \frac{1}{M}\int_0^{\pi} R(\cos\theta\hat{i} + \sin\theta\hat{j})\lambda R\,d\theta = \frac{\lambda R^2}{M}\left(\hat{i}\int_0^{\pi}\cos\theta\,d\theta + \hat{j}\int_0^{\pi}\sin\theta\,d\theta\right)$$

where we have used that an integral of a sum is the sum of the integrals. Because the hoop is uniform, we know that $M = \lambda\pi R$, where πR is the length of the semicircular arc. Substituting $M/(\pi R)$ for λ and integrating gives

$$\vec{r}_{\text{cm}} = \frac{R}{\pi}\left(\hat{i}\int_0^{\pi}\cos\theta\,d\theta + \hat{j}\int_0^{\pi}\sin\theta\,d\theta\right) = \frac{R}{\pi}\left(\hat{i}\,\sin\theta\Big|_0^{\pi} - \hat{j}\,\cos\theta\Big|_0^{\pi}\right) = \frac{2}{\pi}R\hat{j}$$

The center of mass is on the y axis a distance of $2R/\pi$ from the origin. Curiously, it is outside of the material of the semicircular hoop.

See **Math Tutorial** *for more information on* **Integrals**

MOTION OF THE CENTER OF MASS

The motion of any object or system of particles can be described in terms of the motion of the center of mass plus the motion of individual particles in the system relative to the center of mass. The multiple image photograph in Figure 5-48 shows a hammer thrown into the air. While the hammer is in the air, the center of mass follows a parabolic path, the same path that would be followed by a point particle. The other parts of the hammer rotate about this point as the hammer moves through the air.

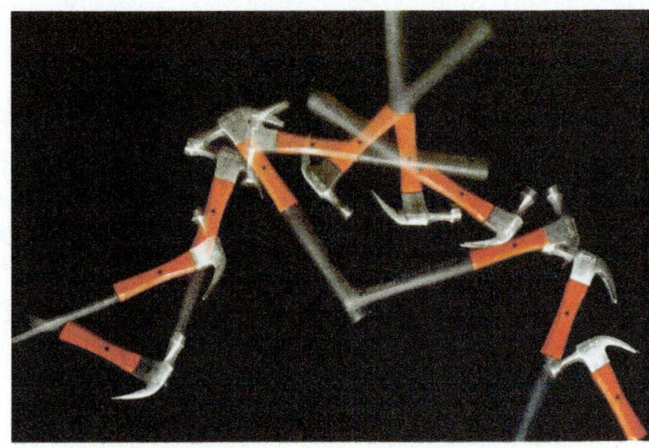

FIGURE 5-48 The center of mass (the black dot) of the hammer moves in a smooth parabolic path. *(Loren Winters/Visual Unlimited.)*

The motion of the center of mass for a system of particles is related to the net force on the system as a whole. We can show this by examining the motion of a system of n particles of total mass M. First, we find the velocity of the center-of-mass system by differentiating both sides of Equation 5-18 ($M\vec{r}_{\text{cm}} = \sum m_i\vec{r}_i$) with respect to time:

$$M\frac{d\vec{r}_{\text{cm}}}{dt} = m_1\frac{d\vec{r}_1}{dt} + m_2\frac{d\vec{r}_2}{dt} + \cdots = \sum_i m_i\frac{d\vec{r}_i}{dt}$$

Because the time derivative of position is velocity, this gives

$$M\vec{v}_{\text{cm}} = m_1\vec{v}_1 + m_2\vec{v}_2 + \cdots = \sum_i m_i\vec{v}_i \qquad \text{5-24}$$

Differentiating both sides again, we obtain the accelerations:

$$M\vec{a}_{\text{cm}} = m_1\vec{a}_1 + m_2\vec{a}_2 + \cdots = \sum_i m_i\vec{a}_i \qquad \text{5-25}$$

where $\vec{a}_i$ is the acceleration of the ith particle and $\vec{a}_{\text{cm}}$ is the acceleration of the center of mass. From Newton's second law, $m_i\vec{a}_i$ equals the sum of the forces acting on the ith particle, so

$$\sum_i m_i\vec{a}_i = \sum_i \vec{F}_i$$

where the sum on the right is the sum of all the forces acting on each and every particle in the system. Some of these forces are *internal* forces (exerted on a particle

Throwing the hammer. If the ball moves in a horizontal circle at constant speed, its acceleration vector points in the centripetal direction (toward the center of the circle). The net force on the ball is in the direction of the acceleration vector. The centripetal component of the net force is supplied by the person pulling the handle inward. *(Pete Saloutos/Corbis.)*

in the system by some other particle in the system) and others are *external* forces (exerted on a particle in the system by something external to the system). Thus,

$$M\vec{a}_{cm} = \sum_i \vec{F}_{i\,int} + \sum_i \vec{F}_{i\,ext} \qquad 5\text{-}26$$

According to Newton's third law, forces come in equal and opposite pairs. Therefore, for each internal force acting on a particle in the system there is an equal and opposite internal force acting on some other particle in the system. When we sum all the internal forces, each third-law force pair sums to zero, so $\sum \vec{F}_{i\,int} = 0$. Equation 5-22 then becomes

$$\vec{F}_{net\,ext} = \sum_i \vec{F}_{i\,ext} = M\vec{a}_{cm} \qquad 5\text{-}27$$

NEWTON'S SECOND LAW FOR A SYSTEM

That is, the net external force acting on the system equals the total mass M times the acceleration of the center of mass $\vec{a}_{cm}$. Thus,

The center of mass of a system moves like a particle of mass $M = \sum m_i$ under the influence of the net external force acting on the system.

This theorem is important because it describes the motion of the center of mass for *any* system of particles: The center of mass moves exactly like a single point particle of mass M acted on by only the external forces. The individual motion of a particle in the system is typically much more complex and is not described by Equation 5-27. The hammer thrown into the air in Figure 5-48 is an example. The only external force acting is gravity, so the center of mass of the hammer moves in a simple parabolic path, as would a point particle. However, Equation 5-27 does not describe the rotational motion of the head of the hammer about the center of mass.

If a system has a zero net external force acting on it, then $\vec{a}_{cm} = 0$. In this case the center of mass either remains at rest or moves with constant velocity. The internal forces and motion may be complex, but the motion of the center of mass is simple. Further, if the component of the net next force in a given direction, say the x direction, remains zero, then a_{cmx} remains zero and v_{cmx} remains constant. An example of this is a projectile in the absence of air drag. The net external force on the projectile is the gravitational force. This force acts straight downward, so its component in any horizontal direction remains zero. It follows that the horizontal component of the velocity of the center of mass remains constant.

CONCEPT CHECK 5-2

A cylinder rests on a sheet of paper on a table (Figure 5-49). You pull on the paper causing the paper to slide to the right. This results in the cylinder rolling leftward *relative to the paper*. How does the center of mass of the cylinder move *relative to the table*?

FIGURE 5-49

Example 5-16 An Exploding Projectile

A projectile is fired into the air over level ground on a trajectory that would result in it landing 55 m away. However, at its highest point it explodes into two fragments of equal mass. Immediately following the explosion one fragment has a momentary speed of zero and then falls straight down to the ground. Where does the other fragment land? Neglect air resistance.

PICTURE Let the projectile be the system. Then, the forces of the explosion are all internal forces. Because the only *external* force acting on the system is that due to gravity, the center of mass, which is midway between the two fragments, continues on its parabolic path as if there had been no explosion (Figure 5-50).

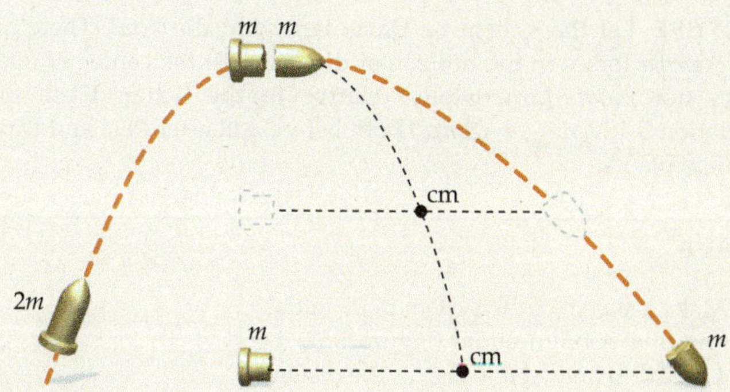

FIGURE 5-50

SOLVE

1. Let $x = 0$ be the initial position of the projectile. The landing positions x_1 and x_2 of the fragments are related to the final position of the center of mass by:

$$(2m)x_{cm} = mx_1 + mx_2$$
$$\text{or} \quad 2x_{cm} = x_1 + x_2$$

2. At impact, $x_{cm} = R$ and $x_1 = 0.5R$, where $R = 55$ m is the horizontal range for the unexploded projectile. Solve for x_2:

$$x_2 = 2x_{cm} - x_1 = 2R - 0.5R = 1.5R$$
$$= 1.5(55\text{ m}) = \boxed{83\text{ m}}$$

CHECK Fragment #1 was pushed backwards by the explosive forces, so fragment #2 was pushed forward by an equal but opposite force. As expected, fragment #2 impacts the ground at a distance farther from the launch point than the projectile would have impacted had it not exploded into two pieces.

TAKING IT FURTHER In Figure 5-51, height versus distance is plotted for exploding projectiles when fragment #1 has a horizontal velocity of half of the initial horizontal velocity. The center of mass follows a normal parabolic trajectory, as it did in the original example in which fragment #1 falls straight down. If both fragments have the same vertical component of velocity after the explosion, they land at the same time. If just after the explosion the vertical component of the velocity of one fragment is less than that of the other, the fragment with the smaller vertical-velocity component will hit the ground first. As soon as it does, the ground exerts a force on it and the net external force on the system is no longer just the gravitational force. From that instant on, our analysis is invalid.

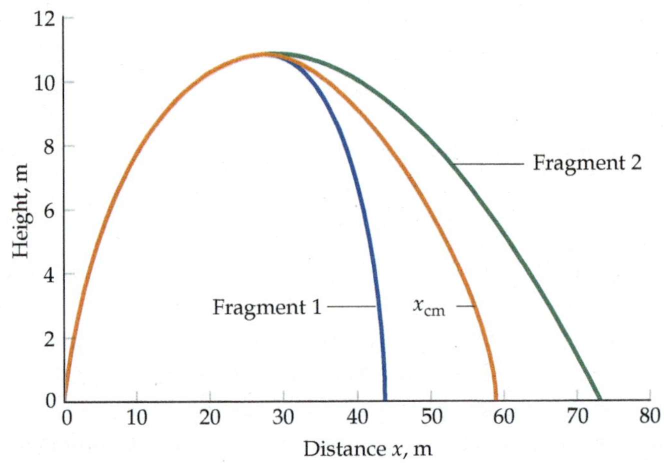

FIGURE 5-51

PRACTICE PROBLEM 5-8 If the fragment that falls straight down has twice the mass of the other fragment, how far from the launch position does the lighter fragment land?

Example 5-17 Changing Places in a Rowboat

Pete (mass 80 kg) and Dave (mass 120 kg) are in a rowboat (mass 60 kg) on a calm lake. Dave is near the bow of the boat, rowing, and Pete is at the stern, 2.0 m from the center. Dave gets tired and stops rowing. Pete offers to row, so after the boat comes to rest they change places. How far does the boat move as Pete and Dave change places? (Neglect any horizontal force exerted by the water.)

PICTURE Let the system be Dave, Pete, and the boat. There are no *external* forces in the horizontal direction, so the center of mass does not move horizontally relative to the water. Flesh out Equation 5-15 ($Mx_{cm} = \Sigma m_i x_i$) both before and after Pete and Dave change places.

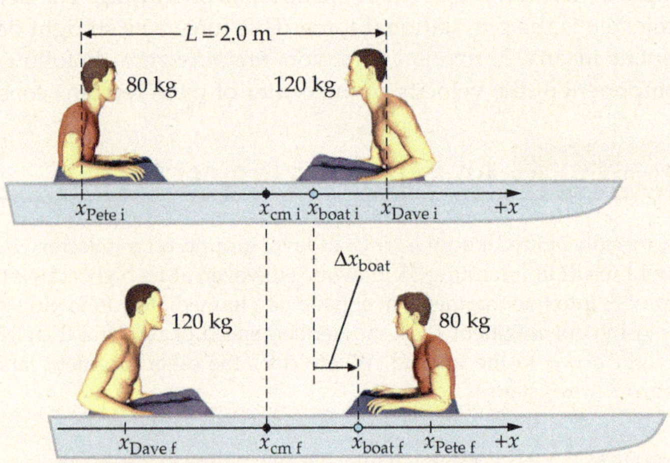

FIGURE 5-52 Pete and Dave changing places viewed from the reference frame of the water. The blue dot is the center of mass of the boat and the black dot is the center of mass of the Pete–Dave–boat system.

SOLVE

1. Make a sketch of the system in its initial and final configurations (Figure 5-52). Let $L = 2.0$ m and let $d = \Delta x_{boat}$, the distance the boat moves forward when Pete and Dave switch places:

2. Flesh out $Mx_{cm} = \Sigma m_i x_i$ both before and after Pete and Dave change places. The coordinate axis measures positions in the reference frame of the water:

$$Mx_{\text{cm i}} = m_{\text{Pete}}x_{\text{Pete i}} + m_{\text{Dave}}x_{\text{Dave i}} + m_{\text{boat}}x_{\text{boat i}}$$

and

$$Mx_{\text{cm f}} = m_{\text{Pete}}x_{\text{Pete f}} + m_{\text{Dave}}x_{\text{Dave f}} + m_{\text{boat}}x_{\text{boat f}}$$

3. Subtract the third step-2 equation from the second step-2 equation. Then substitute 0 for Δx_{cm}, $d + L$ for Δx_{Pete}, $d - L$ for Δx_{Dave} and d for Δx_{boat}:

$$M\Delta x_{\text{cm}} = m_{\text{Pete}}\,\Delta x_{\text{Pete}} + m_{\text{Dave}}\,\Delta x_{\text{Dave}} + m_{\text{boat}}\,\Delta x_{\text{boat}}$$
$$0 = m_{\text{Pete}}(d + L) + m_{\text{Dave}}(d + L) + m_{\text{boat}}d$$

4. Solve for d:

$$d = \frac{(m_{\text{Dave}} - m_{\text{Pete}})}{m_{\text{Dave}} + m_{\text{Pete}} + m_{\text{boat}}}L = \frac{(120\text{ kg} - 80\text{ kg})}{120\text{ kg} + 80\text{ kg} + 60\text{ kg}}(2.0\text{ m}) = \boxed{0.31\text{ m}}$$

CHECK Dave's mass is greater than Pete's mass, so when they changed places their center of mass moved toward the stern of the boat. The boat's center of mass had to move in the opposite direction for the center of mass of the Dave–Pete–boat system to remain stationary. The step-4 result is the displacement of the boat. It is positive as expected.

Example 5-18 A Sliding Block

A wedge of mass m_2 sits at rest on a scale, as shown in Figure 5-53. A small block of mass m_1 slides down the frictionless incline of the wedge. Find the scale reading while the block slides. The wedge does not slide on the scale.

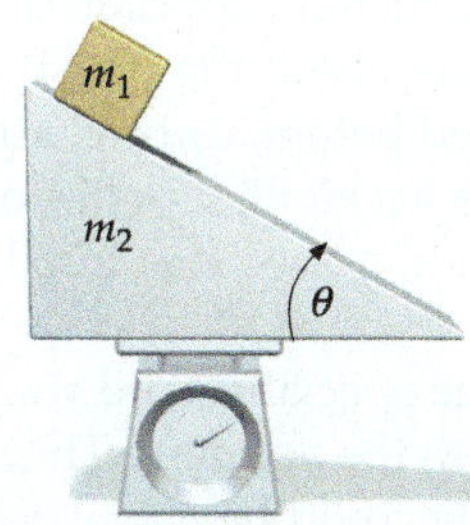

FIGURE 5-53

PICTURE We choose the wedge plus block to be the system. Because the block accelerates down the wedge, the center of mass of the system has acceleration components to the right and downward. The external forces on the system are the gravitational forces on the block and wedge, the static frictional force f_s of the scale on the wedge, and the normal force F_n exerted by the scale on the wedge. The scale reading is equal to the magnitude of F_n.

SOLVE

1. Draw a free-body diagram for the wedge–block system (Figure 5-54):

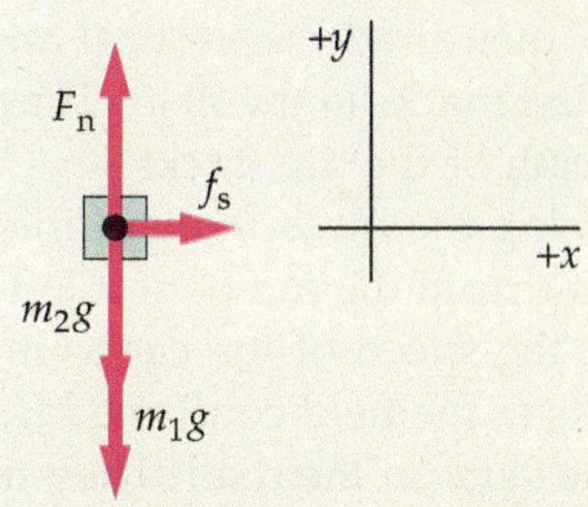

FIGURE 5-54

2. Write the vertical component of Newton's second law for the system and solve for F_n:

$$F_n - m_1g - m_2g = Ma_{cmy} = (m_1 + m_2)a_{cmy}$$
$$F_n = (m_1 + m_2)g + (m_1 + m_2)a_{cmy}$$

3. Using Equation 5-21, express a_{cmy} in terms of the acceleration of the block a_{1y}:

$$Ma_{cmy} = m_1a_{1y} + m_2a_{2y}$$
$$(m_1 + m_2)a_{cmy} = m_1a_{1y} + 0$$
$$a_{cmy} = \frac{m_1}{m_1 + m_2}a_{1y}$$

4. From Example 4-7, a block sliding down a stationary frictionless incline has acceleration $g \sin\theta$ down the incline. Use trigonometry to find the y component of this acceleration and use it to find a_{cmy}:

so $a_{1y} = -a_1 \sin\theta$, where $a_1 = g\sin\theta$
$$a_{1y} = -(g\sin\theta)\sin\theta = -g\sin^2\theta$$

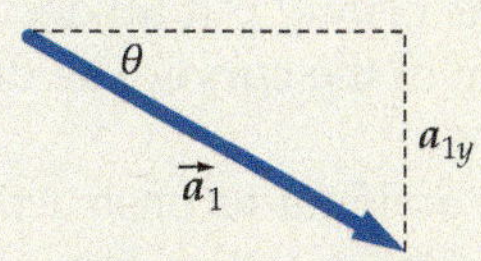

FIGURE 5-55

5. Substitute for a_{1y} in the step-3 result:

$$a_{cmy} = \frac{m_1}{m_1 + m_2}a_{1y} = -\frac{m_1}{m_1 + m_2}g\sin^2\theta$$

6. Substitute for a_{cmy} in the step-2 result and solve for F_n:

$$F_n = (m_1 + m_2)g + (m_1 + m_2)a_{cmy}$$
$$= (m_1 + m_2)g - m_1g\sin^2\theta = [m_1(1 - \sin^2\theta) + m_2]g$$
$$= \boxed{(m_1\cos^2\theta + m_2)g}$$

CHECK For $\theta = 0$, $\cos\theta = 1$, the step-5 result is, as expected, equal to the sum of the two weights. For $\theta = 90°$, $\cos\theta = 0$, and the step-5 result is, as expected, equal to the weight of the wedge alone.

PRACTICE PROBLEM 5-9 Find the force component F_x exerted on the wedge by the scale.

Physics Spotlight

Accident Reconstruction—Measurements and Forces

A car being towed after an accident. *(Mikael Karlsson.)*

Four teenagers drove to a Halloween haunted house that was located out in the countryside. As their car started into a curve, the driver saw a deer in the middle of the road. Frantic swerving and braking put the car into a slide. It skidded off the edge of the gently banked curve, flew over the narrow ditch, and landed in a newly harvested field below the road, where it skidded to a stop in the loose dirt.

Thanks to airbags and safety belts, no one was killed. All were taken to the hospital. The car was towed. But the accident investigation was not finished until a question was answered—was the car speeding?

Accident reconstruction specialists investigated the scene, and used information about the physics of an accident to determine what happened before, during, and immediately after that accident.* After this accident, a police officer, a specialist hired by the driver's insurance company, and a specialist hired by the county road department all looked at the scene.

The first thing the specialists did was measure and photograph everything that might be pertinent to the accident. They measured the road, so that the curve's radius and bank angle could be calculated and compared to the information at the county roads office. They measured the tire marks on the road, and in the field. They used a *drag sled* to determine the coefficient of kinetic friction for the field.† They measured the vertical and horizontal distance from the edge of the road to the first marks in the field. They measured the angle of the road to horizontal along the path of the tire tracks.

Using the measurement information, they calculated a simplified trajectory of the car from the moment it left the road until it landed in the field. This trajectory gave the speed of the car as it left the roadway. Their calculations using the skid marks in the field confirmed that speed. Finally, they calculated the starting speed of the skid on the road. They used the coefficient of kinetic friction of the road, as it was clear that the wheels had been locked, and not spinning.

They concluded that the car had been under the speed limit for the road, but, like most cars, had been going faster than the marked advisory speed for the curve.‡ The county put up deer warning signs, and installed a guard rail along the outer edge of the curve. The driver was ticketed for failure to maintain control of the vehicle.

Not all accident reconstruction is so simple or straightforward. Many accidents involve obstacles, other cars, or tires of the wrong size for the vehicle. Others may involve looking at the physics inside the car to determine whether or not seat belts were worn, or who was driving. But all accident reconstructions start with measurements to determine the forces at work during the accident.

---

* The International Association of Accident Reconstruction Specialists. http://www.iaars.org/ March 2006.

† Marks, Christopher C. O'N., *Pavement Skid Resistance Measurement and Analysis in the Forensic Context*. International Conference on Surface Friction. 2005, Christchurch. http://www.surfacefriction.org.nz.

‡ Chowdhury, Mashrur A., Warren, Davey L., "Analysis of Advisory Speed Setting Criteria." *Public Roads*, 00333735, Dec. 91. Vol. 55, Issue 3.

SUMMARY

Friction and drag forces are complex phenomena empirically approximated by simple equations. For a particle to move in a curved path at constant speed the net force is directed toward the center of curvature. The center of mass of a system moves as if the system were a single point particle with the net force on the system acting on it.

TOPIC	RELEVANT EQUATIONS AND REMARKS	
1. **Friction**	Two objects in contact exert frictional forces on each other. These forces are parallel to the contacting surfaces and directed so as to oppose sliding or tendency to slide.	
Static friction	$f_s \leq \mu_s F_n$	5-2
	where F_n is the normal force of contact and μ_s is the coefficient of static friction.	
Kinetic friction	$f_k = \mu_k F_n$	5-3
	where μ_k is the coefficient of kinetic friction. The coefficient of kinetic friction is slightly less than the coefficient of static friction.	
Rolling friction	$f_r = \mu_r F_n$	5-4
	where μ_r is the coefficient of rolling friction.	
2. **Drag Forces**	When an object moves through a fluid, it experiences a drag force that opposes its motion. The drag force increases with increasing speed. If the body is dropped from rest, its speed increases. As it does, the magnitude of the drag force comes closer and closer to the magnitude of the force of gravity, so the net force, and thus the acceleration, approaches zero. As the acceleration approaches zero, the speed approaches a constant value called its terminal speed. The terminal speed depends on both the shape of the body and on the medium through which it falls.	
3. **Motion Along a Curved Path**	A particle moving along an arbitrary curve can be considered to be moving along a circular arc for a short time interval. Its instantaneous acceleration vector has a component $a_c = v^2/r$ toward the center of curvature of the arc and a component $a_t = dv/dt$ that is tangential to the arc. If the particle is moving along a circular path of radius r at constant speed v, $a_t = 0$ and the speed, radius, and period T are related by $2\pi r = vT$.	
4. ***Numerical Integration: Euler's Method**	To estimate the position x and velocity v at some time t, we first divide the interval from zero to t into a large number of small intervals, each of length Δt. The initial acceleration a_0 is then calculated from the initial position x_0 and velocity v_0. The position x_1 and velocity v_1 a time Δt later are estimated using the relations	
	$x_{n+1} = x_n + v_n \Delta t$	5-9
	and	
	$v_{n+1} = v_n + a_n \Delta t$	5-10
	with $n = 0$. The acceleration a_{n+1} is calculated using the values for x_{n+1} and v_{n+1} and the process is repeated. This continues until estimations for the position and velocity at time t are calculated.	
5. **Center of Mass**		
Center of mass for a system of particles	The center of mass of a system of particles is defined to be the point whose coordinates are given by:	
	$Mx_{cm} = \sum_i m_i x_i$	5-15
	$My_{cm} = \sum_i m_i y_i$	5-16
	$Mz_{cm} = \sum_i m_i z_i$	5-17

TOPIC	RELEVANT EQUATIONS AND REMARKS	
Center of mass for continuous objects	If the mass is continuously distributed, the center of mass is given by: $M\vec{r}_{cm} = \int \vec{r}\,dm$	5-19
Position, velocity, and acceleration for the center of mass of a system of particles	$M\vec{r}_{cm} = m_1\vec{r}_1 + m_2\vec{r}_2 + \cdots$	5-18
	$M\vec{v}_{cm} = m_1\vec{v}_1 + m_2\vec{v}_2 + \cdots$	5-20
	$M\vec{a}_{cm} = m_1\vec{a}_1 + m_2\vec{a}_2 + \cdots$	5-21
Newton's second law for a system	$\vec{F}_{net\,ext} = \sum_i \vec{F}_{i\,ext} = M\vec{a}_{cm}$	5-23

Answers to Concept Checks

5-1 Yes, the car would slide down the incline.

5-2 It accelerates to the right, because the net external force acting on the cylinder is the frictional force to the right exerted on it by the paper. Try it. The cylinder may *appear* to move to the left, because relative to the paper it rolls leftward. However, relative to the table, which serves as an inertial reference frame, it moves to the right. If you mark the table with the original position of the cylinder, you will observe the center of mass move to the right *during the time the cylinder remains in contact with the moving paper*.

Answers to Practice Problems

5-1 35°

5-2 1.1×10^2 N

5-3 $T = m_2(g - a) = 44$ N

5-4 (*a*) Assuming $r \approx 1$ m, we find $v_{t,min} \approx 3$ m/s, (*b*) $T = 2\pi r/v \approx 2$ s

5-5 Zero. At that instant you are no longer gaining speed and not yet losing speed. Your rate of change of speed is momentarily zero.

5-6 1.60 m/s²

5-7 $x_{cm} = 2.0$ cm

5-8 $2R = 1.1 \times 10^2$ m

5-9 $F_x = m_1 g \sin\theta \cos\theta$

PROBLEMS

In a few problems, you are given more data than you actually need; in a few other problems, you are required to supply data from your general knowledge, outside sources, or informed estimate.

Interpret as significant all digits in numerical values that have trailing zeros and no decimal points.

- • Single-concept, single-step, relatively easy
- •• Intermediate-level, may require synthesis of concepts
- ••• Challenging
- SSM Solution is in the *Student Solutions Manual*
- Consecutive problems that are shaded are paired problems.

CONCEPTUAL PROBLEMS

1 • Various objects lie on the bed of a truck that is moving along a straight horizontal road. If the truck gradually speeds up, what force acts on the objects to cause them to speed up too? Explain why some of the objects might stay stationary on the floor while others might slip backward on the floor. SSM

2 • Blocks made of the same material but differing in size lie on the bed of a truck that is moving along a straight horizontal road. All of the blocks will slide if the truck's acceleration is sufficiently great. How does the minimum acceleration at which a small block slips compare with the minimum acceleration at which a much heavier block slips?

3 • A block of mass m rests on a plane that is inclined at an angle θ with the horizontal. It follows that the coefficient of static friction between the block and plane is (*a*) $\mu_s \geq g$, (*b*) $\mu_s = \tan\theta$, (*c*) $\mu_s \leq \tan\theta$, (*d*) $\mu_s \geq \tan\theta$.

4 • A block of mass m is at rest on a plane that is inclined at an angle of 30° with the horizontal, as shown in Figure 5-56. Which of the following statements about the magnitude of the static frictional force f_s is necessarily true? (*a*) $f_s > mg$, (*b*) $f_s > mg\cos 30°$, (*c*) $f_s = mg\cos 30°$, (*d*) $f_s = mg\sin 30°$, (*e*) None of these statements is true.

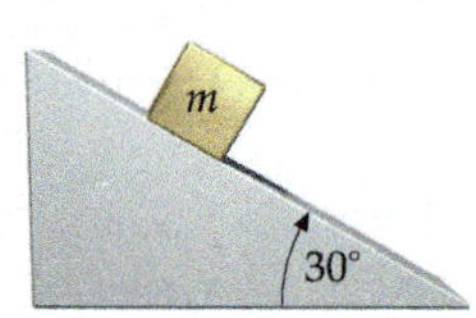

FIGURE 5-56 Problem 4

5 •• On an icy winter day, the coefficient of friction between the tires of a car and a roadway is reduced to one-quarter of its value on a dry day. As a result, the maximum speed $v_{max\,dry}$ at which the car can safely negotiate a curve of radius R is reduced. The new value for this speed is (*a*) $v_{max\,dry}$, (*b*) $0.71v_{max\,dry}$, (*c*) $0.50v_{max\,dry}$, (*d*) $0.25v_{max\,dry}$, (*e*) reduced by an unknown amount depending on the car's mass.

6 •• If it is started properly on the frictionless inside surface of a cone (Figure 5-57), a block is capable of maintaining uniform circular motion. Draw the free-body diagram of the block and identify clearly which force (or forces, or force components) is responsible for the centripetal acceleration of the block.

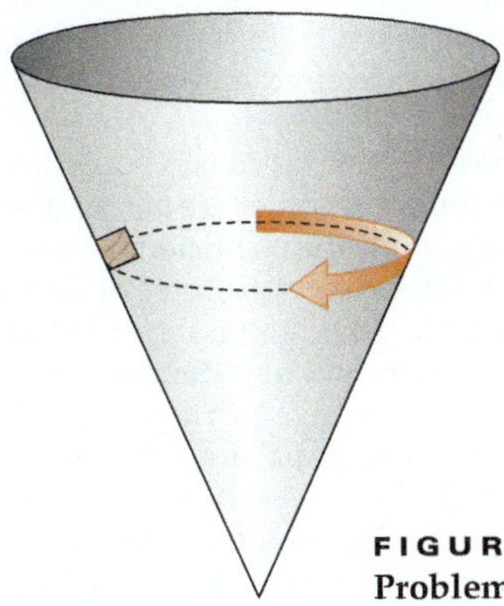

FIGURE 5-57 Problem 6

7 •• Here is an interesting experiment that you can perform at home: take a wooden block and rest it on the floor or some other flat surface. Attach a rubber band to the block and pull gently on the rubber band in the horizontal direction. Keep your hand moving at constant speed. At some point, the block will start moving, but it will not move smoothly. Instead, it will start moving, stop again, start moving again, stop again, and so on. Explain why the block moves this way. (The start-stop motion is sometimes called "stick-slip" motion.)

8 • Viewed from an inertial reference frame, an object is seen to be moving in a circle. Which, if any, of the following statements must be true. (*a*) A nonzero net force acts on the object. (*b*) The object cannot have a radially outward force acting on it. (*c*) At least one of the forces acting on the object must point directly toward the center of the circle.

9 •• A particle is traveling in a vertical circle at constant speed. One can conclude that the magnitude of its _____ is (are) constant. (*a*) velocity, (*b*) acceleration, (*c*) net force, (*d*) apparent weight

10 •• You place a lightweight piece of iron on a table and hold a small kitchen magnet above the iron at a distance of 1.00 cm. You find that the magnet cannot lift the iron, even though there is obviously a force between the iron and the magnet. Next, holding the magnet in one hand and the piece of iron in the other, with the magnet 1.00 cm above the iron, you simultaneously drop them from rest. As they fall, the magnet and the piece of iron bang into each other before hitting the floor. (*a*) Draw free-body diagrams illustrating all of the forces on the magnet and the iron for each demonstration. (*b*) Explain why the magnet and iron move closer together while they are falling, even though the magnet cannot lift the piece of iron when it is sitting on the table.

11 ••• The following question is an excellent "braintwister," invented by Boris Korsunsky:* Two identical blocks are attached by a massless string running over a pulley, as shown in Figure 5-58.

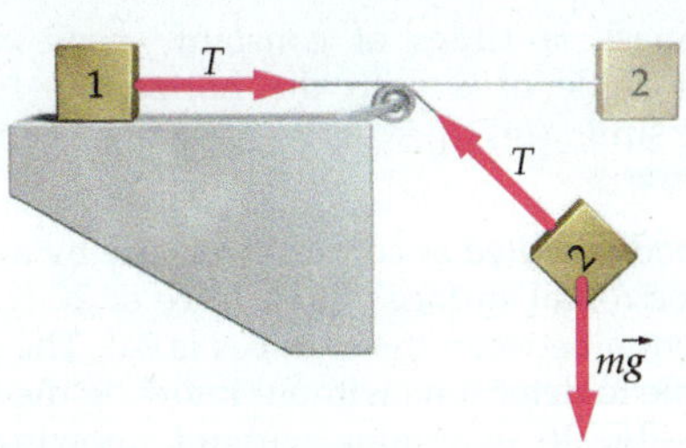

FIGURE 5-58 Problem 11

* Boris Korsunsky, "Braintwisters for Physics Students," *The Physics Teacher*, **33**, 550 (1995).

The rope initially runs over the pulley at the rope's midpoint, and the surface that block 1 rests on is frictionless. Blocks 1 and 2 are initially at rest when block 2 is released with the string taut and horizontal. Will block 1 hit the pulley before or after block 2 hits the wall? (Assume that the initial distance from block 1 to the pulley is the same as the initial distance from block 2 to the wall.) There is a very simple solution. SSM

12 •• In class, most professors do the following experiment while discussing the conditions under which air drag can be neglected while analyzing free-fall. First, a flat piece of paper and a small lead weight are dropped next to each other, and clearly the paper's acceleration is less than that of the lead weight. Then, the paper is crumpled into a small wad and the experiment repeated. Over the distance of a meter or two, it is clear the acceleration of the paper is now very close to that of the lead weight. To your dismay, the professor calls on you to explain why the paper's acceleration changed so dramatically. Repeat your explanation here!

13 •• CONTEXT-RICH Jim decides to attempt to set a record for terminal speed in skydiving. Using the knowledge he has gained from a physics course, he makes the following plans. He will be dropped from as high an altitude as possible (equipping himself with oxygen), on a warm day and go into a "knife" position, in which his body is pointed vertically down and his hands are pointed ahead. He will outfit himself with a special sleek helmet and rounded protective clothing. Explain how each of these factors helps Jim attain the record. SSM

14 •• CONTEXT-RICH You are sitting in the passenger seat in a car driving around a circular, horizontal, flat racetrack at a high speed. As you sit there, you "feel" a "force" pushing you toward the outside of the track. What is the true direction of the force acting on you, and where does it come from? (Assume that you do not slide across the seat.) Explain the *sensation* of an "outward force" on you in terms of the Newtonian perspective.

15 • The mass of the moon is only about 1% of that of Earth. Therefore, the force that keeps the moon in its orbit around Earth (*a*) is much smaller than the gravitational force exerted on the moon by Earth, (*b*) is much greater than the gravitational force exerted on the moon by Earth, (*c*) actually is the gravitational force exerted on the moon by Earth, (*d*) cannot be answered yet, because we have not yet studied Newton's law of gravity. SSM

16 •• A block is sliding on a frictionless surface along a loop-the-loop, as in Figure 5-59*a*. The block is moving fast enough so that it never loses contact with the track. Match the points along the track to the appropriate free-body diagrams in Figure 5-59*b*.

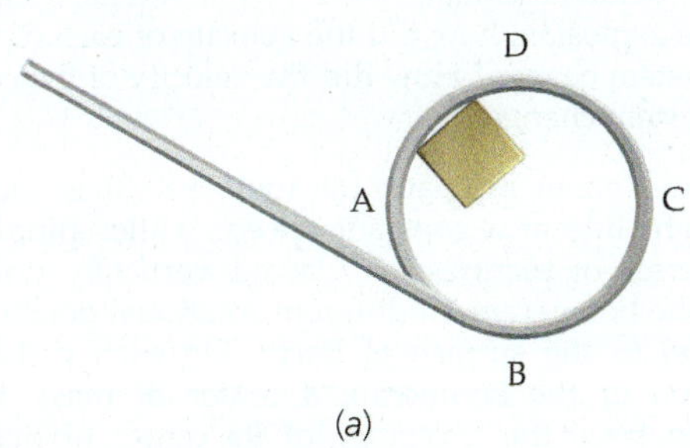

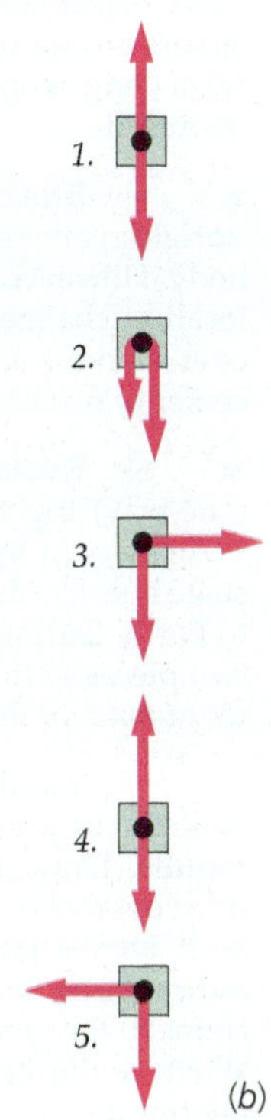

FIGURE 5-59 Problem 16

17 •• (*a*) A rock and a feather held at the same height above the ground are simultaneously dropped. During the first few milliseconds following release, the drag force on the rock is smaller than the drag force on the feather, but later on during the fall the *opposite* is true. Explain. (*b*) In light of this result, explain how the rock's acceleration can be so obviously larger than that of the feather. *Hint: Draw a free-body diagram of each object.* SSM

18 •• Two pucks of masses m_1 and m_2 are lying on a frictionless table and are connected by a massless spring of force constant k. A horizontal force F_1 directed away from m_2 is then exerted on m_1. What is the magnitude of the resulting acceleration of the center of mass of the two-puck system? (*a*) F_1/m_1, (*b*) $F_1/(m_1 + m_2)$, (*c*) $(F_1 + kx)/(m_1 + m_2)$, where x is the amount the spring is stretched, (*d*) $(m_1 + m_2)F_1/m_1m_2$

19 •• The two pucks in Problem 18 lie unconnected on a frictionless table. A horizontal force F_1 directed away from m_2 is then exerted on m_1. How does the magnitude of the resulting acceleration of the center of mass of the two-puck system compare to the magnitude acceleration of m_1? Explain your reasoning.

20 •• If only external forces can cause the center of mass of a system of particles to accelerate, how can a car on level ground ever accelerate? We normally think of the car's engine as supplying the force needed to accelerate the car, but is this true? Where does the external force that accelerates the car come from?

21 •• When you push on the brake pedal to slow down a car, a brake pad is pressed against the rotor so that the friction of the pad slows the wheel's rotation. However, the friction of the pad against the rotor *cannot be* the force that slows the car down, because it is an internal force (both the rotor and the wheel are parts of the car, so any forces between them are purely internal to the system). What is the external force that slows down the car? Give a detailed explanation of how this force operates.

22 •• Give an example of each of the following: (*a*) a three-dimensional object that has no matter at its center of mass, (*b*) a solid object whose center of mass is outside of it, (*c*) a solid sphere whose center of mass does not lie at its geometrical center, (*d*) a long wooden stick whose center of mass does not lie at its middle.

23 •• **BIOLOGICAL APPLICATION** When you are standing upright, your center of mass is located within the volume of your body. However, as you bend over (say to pick up a package), its location changes. Approximately where is it when you are bent over at right angles and what change in your body caused the center of mass location to change? Explain. SSM

24 •• **ENGINEERING APPLICATION** Early on their three-day (one-way) trip to the moon, the Apollo team (late 1960s to early 1970s) would explosively separate the lunar ship from the third-stage booster (that provided the final "boost") while still fairly close to Earth. During the explosion, how did the velocity of each of the two pieces of the system change? How did the velocity of the center of mass of the system change?

25 •• You throw a boomerang and for a while it "flies" horizontally in a straight line at a constant speed, while spinning rapidly. Draw a series of pictures, as viewed vertically down from overhead, of the boomerang in different rotational positions as it moves parallel to the surface of Earth. On each picture, indicate the location of the boomerang's center of mass and connect the dots to trace the trajectory of its center of mass. What is the center of mass's acceleration during this part of the flight?

ESTIMATION AND APPROXIMATION

26 •• **ENGINEERING APPLICATION** To determine the aerodynamic drag on a car, automotive engineers often use the "coast-down" method. The car is driven on a long, flat road at some convenient speed (60 mi/h is typical), shifted into neutral, and allowed to coast to a stop. The time that it takes for the speed to drop by successive 5-mi/h intervals is measured and used to compute the net force slowing the car down. (*a*) One day, a group measured that a Toyota Tercel with a mass of 1020 kg coasted down from 60.0 mi/h to 55.0 mi/h in 3.92 s. Estimate the average net force slowing the car down in this speed range. (*b*) If the coefficient of rolling friction for this car is known to be 0.020, what is the force of rolling friction that is acting to slow it down? Assuming that the only two forces acting on the car are rolling friction and aerodynamic drag, what is the average drag force acting on the car? (*c*) The drag force has the form $\frac{1}{2}C\rho Av^2$, where A is the cross-sectional area of the car facing into the air, v is the car's speed, ρ is the density of air, and C is a dimensionless constant of order-of-magnitude 1. If the cross-sectional area of the car is 1.91 m^2, determine C from the data given. (The density of air is 1.21 kg/m^3; use 57.5 mi/h for the speed of the car in this computation.)

27 •• Using dimensional analysis, determine the units and dimensions of the constant b in the retarding force bv^n if (*a*) $n = 1$ and (*b*) $n = 2$. (*c*) Newton showed that the air resistance of a falling object with a circular cross section should be approximately $\frac{1}{2}\rho\pi r^2v^2$, where $\rho = 1.20$ kg/m^3, the density of air. Show that this is consistent with your dimensional analysis for part (*b*). (*d*) Find the terminal speed for a 56.0-kg skydiver; approximate his cross-sectional area as a disk of radius 0.30 m. The density of air near the surface of Earth is 1.20 kg/m^3. (*e*) The density of the atmosphere decreases with height above the surface of Earth; at a height of 8.0 km, the density is only 0.514 kg/m^3. What is the terminal velocity at this height? SSM

28 •• Estimate the terminal velocity of an average sized raindrop and a golf-ball-sized hailstone. *Hint: See Problems 26 and 27.*

29 •• Estimate the minimum coefficient of static friction needed between a car's tires and the pavement in order to complete a left turn at a city street intersection at the posted straight-ahead speed limit of 25 mph and on narrow inner-city streets. Comment on the wisdom of attempting such a turn at that speed.

30 •• Estimate the widest stance you can take when standing on a dry, icy surface. That is, how wide can you safely place your feet and not slip into an undesired "split"? Let the coefficient of static friction of rubber on ice be roughly 0.25.

FRICTION

31 • A block of mass m slides at constant speed down a plane inclined at an angle of θ with the horizontal. It follows that (*a*) $\mu_k = mg\sin\theta$, (*b*) $\mu_k = \tan\theta$, (*c*) $\mu_k = 1 - \cos\theta$, (*d*) $\mu_k = \cos\theta - \sin\theta$. SSM

32 • A block of wood is pulled at constant velocity by a horizontal string across a horizontal surface with a force of 20 N. The coefficient of kinetic friction between the surfaces is 0.3. The force of friction is (*a*) impossible to determine without knowing the mass of the block, (*b*) impossible to determine without knowing the speed of the block, (*c*) 0.30 N, (*d*) 6.0 N, (*e*) 20 N.

33 • A block weighing 20-N rests on a horizontal surface. The coefficients of static and kinetic friction between the surface and the

block are $\mu_s = 0.80$ and $\mu_k = 0.60$. A horizontal string is then attached to the block and a constant tension T is maintained in the string. What is the magnitude of the force of friction acting on the block if (*a*) $T = 15$ N, (*b*) $T = 20$ N? SSM

34 • A block of mass m is pulled at a constant velocity across a horizontal surface by a string as shown in Figure 5-60. The magnitude of the frictional force is (*a*) $\mu_k mg$, (*b*) $T \cos\theta$, (*c*) $\mu_k(T - mg)$, (*d*) $\mu_k T \sin\theta$, or (*e*) $\mu_k(mg - T \sin\theta)$.

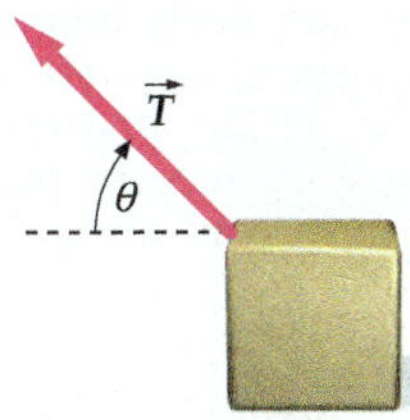

FIGURE 5-60 Problem 34

35 • A 100-kg crate rests on a thick-pile carpet. A weary worker then pushes on the crate with a horizontal force of 500 N. The coefficients of static and kinetic friction between the crate and the carpet are 0.600 and 0.400, respectively. Find the magnitude of the frictional force exerted by the carpet on the crate. SSM

36 • A box weighing 600 N is pushed along a horizontal floor at constant velocity with a force of 250 N parallel to the floor. What is the coefficient of kinetic friction between the box and the floor?

37 • The coefficient of static friction between the tires of a car and a horizontal road is 0.60. Neglecting air resistance and rolling friction, (*a*) what is the magnitude of the maximum acceleration of the car when it is braked? (*b*) What is the shortest distance in which the car can stop if it is initially traveling at 30 m/s? SSM

38 • The force that accelerates a car along a flat road is the frictional force exerted by the road on the car's tires. (*a*) Explain why the acceleration can be greater when the wheels do not slip. (*b*) If a car is to accelerate from 0 to 90 km/h in 12 s, what is the minimum coefficient of friction needed between the road and tires? Assume that the drive wheels support exactly half the weight of the car.

39 •• A 5.00-kg block is held at rest against a vertical wall by a horizontal force of 100 N. (*a*) What is the frictional force exerted by the wall on the block? (*b*) What is the minimum horizontal force needed to prevent the block from falling if the static coefficient of friction between the wall and the block is 0.400?

40 •• A tired and overloaded student is attempting to hold a large physics textbook wedged under his arm, as shown in Figure 5-61. The textbook has a mass of 3.2 kg, while the coefficient of static friction of the textbook against the student's underarm is 0.320 and the coefficient of static friction of the book against the student's shirt is 0.160. (*a*) What is the minimum horizontal force that the student must apply to the textbook to prevent it from falling? (*b*) If the student can only exert a force of 61 N, what is the acceleration of the textbook as it slides from under his arm? The coefficient of kinetic friction of arm against textbook is 0.200, while that of shirt against textbook is 0.090.

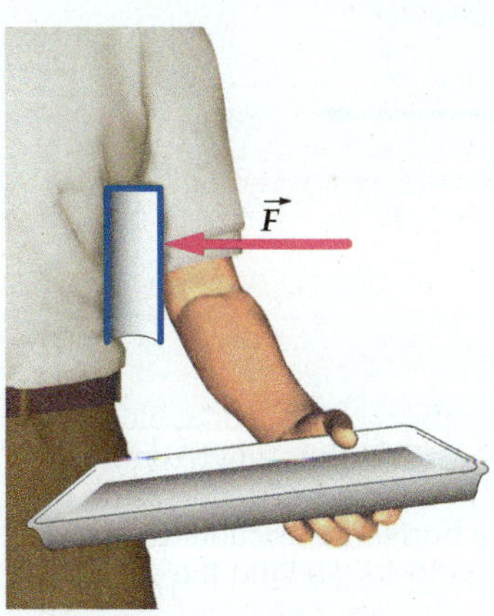

FIGURE 5-61 Problem 40

41 •• **ENGINEERING APPLICATION** You are racing in a rally on a snowy day when the temperature is near the freezing point. The coefficient of static friction between a car's tires and an icy road is 0.080. Your crew boss is concerned about some of the hills on the course and wants you to think about switching to studded tires. To address the issue, he wants to compare the actual hill angles on the course to see which of them your car can negotiate. (*a*) What is the angle of the steepest incline that a vehicle with four-wheel drive can climb at constant speed? (*b*) Given that the hills are icy, what is the steepest possible hill angle for the same four-wheel drive car to descend at constant speed?

42 •• A 50-kg box that is resting on a level floor must be moved. The coefficient of static friction between the box and the floor is 0.60. One way to move the box is to push down on the box at an angle θ below the horizontal. Another method is to pull up on the box at an angle θ above the horizontal. (*a*) Explain why one method requires less force than the other. (*b*) Calculate the minimum force needed to move the box by each method if $\theta = 30°$ and compare the answer with the results when $\theta = 0°$.

43 •• A block of mass $m_1 = 250$ g is at rest on a plane that makes an angle of $\theta = 30°$ with the horizontal. The coefficient of kinetic friction between the block and the plane is 0.100. The block is attached to a second block of mass $m_2 = 200$ g that hangs freely by a string that passes over a frictionless, massless pulley (Figure 5-62). When the second block has fallen 30.0 cm, what will be its speed? SSM

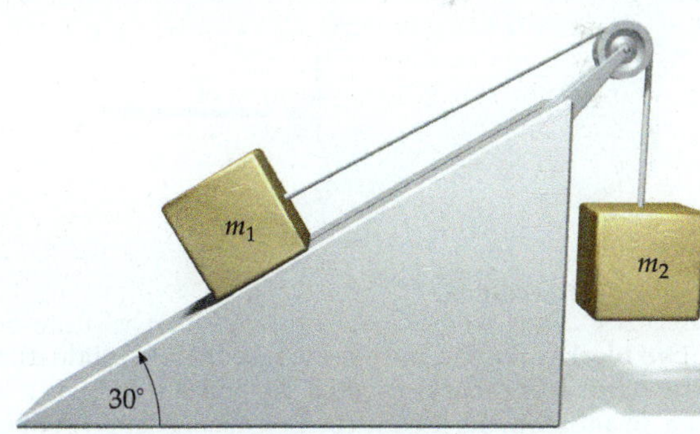

FIGURE 5-62 Problems 43, 44, 45

44 •• In Figure 5-62, $m_1 = 4.0$ kg and the coefficient of static friction between the block and the incline is 0.40. (*a*) Find the range of possible values for m_2 for which the system will be in static equilibrium. (*b*) Find the frictional force on the 4.0-kg block if $m_2 = 1.0$ kg.

45 •• In Figure 5-62, $m_1 = 4.0$ kg, $m_2 = 5.0$ kg, and the coefficient of kinetic friction between the inclined plane and the 4.0-kg block is $\mu_k = 0.24$. Find the magnitude of the acceleration of the masses and the tension in the cord.

46 •• A 12-kg turtle rests on the bed of a zookeeper's truck, which is traveling down a country road at 55 mi/h. The zookeeper spots a deer in the road, and slows to a stop in 12 s. Assuming constant acceleration, what is the minimum coefficient of static friction between the turtle and the truck bed surface that is needed to prevent the turtle from sliding?

47 •• A 150-g block is projected up a ramp with an initial speed of 7.0 m/s. The coefficient of kinetic friction between the ramp and the block is 0.23. (*a*) If the ramp is inclined 25° with the horizontal, how far along the surface of the ramp does the block slide before coming to a stop? (*b*) The block then slides back down the ramp. What is the minimum coefficient of static friction between the block and the ramp if the block is not to slide back down the ramp? SSM

48 •• An automobile is going up a 15° grade at a speed of 30 m/s. The coefficient of static friction between the tires and the road is 0.70. (*a*) What minimum distance does it take to stop the car? (*b*) What minimum distance would it take to stop if the car were going down the grade?

49 •• **ENGINEERING APPLICATION** A rear-wheel-drive car supports 40 percent of its weight on its two drive wheels and has a coefficient of static friction of 0.70 with a horizontal straight road. (*a*) Find the vehicle's maximum acceleration. (*b*) What is the shortest possible time in which this car can achieve a speed of 100 km/h? (Assume the engine can provide unlimited power.)

50 •• You and your best pal make a friendly bet that you can place a 2.0-kg box against the side of a cart, as in Figure 5-63, and that the box will not fall to the ground, even though you guarantee to use no hooks, ropes, fasteners, magnets, glue, or adhesives of any kind. When your friend accepts the bet, you begin pushing the cart in the direction shown in the figure. The coefficient of static friction between the box and the cart is 0.60. (*a*) Find the minimum acceleration for which you will win the bet. (*b*) What is the magnitude of the frictional force in this case? (*c*) Find the force of friction on the box if the acceleration is twice the minimum needed for the box not to fall. (*d*) Show that, for a box of any mass, the box will not fall if the magnitude of the forward acceleration is $a \geq g/\mu_s$, where μ_s is the coefficient of static friction.

FIGURE 5-63 Problem 50

51 •• Two blocks attached by a string (Figure 5-64) slide down a 10° incline. Block 1 has mass $m_1 = 0.80$ kg and block 2 has mass $m_2 = 0.25$ kg. In addition, the kinetic coefficients of friction between the blocks and the incline are 0.30 for block 1 and 0.20 for block 2. Find (*a*) the magnitude of the acceleration of the blocks, and (*b*) the tension in the string.

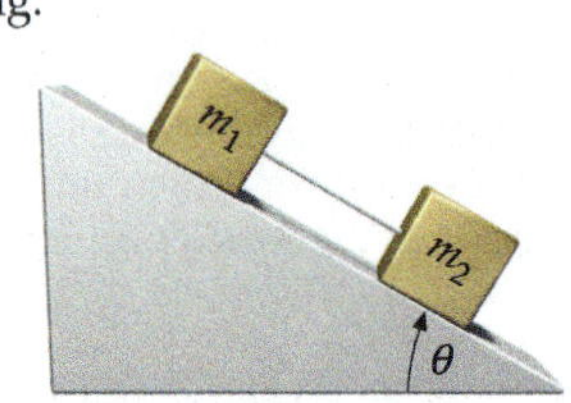

FIGURE 5-64 Problems 51 and 52

52 •• Two blocks of masses m_1 and m_2 are sliding down an incline as shown in Figure 5-64. They are connected by a massless *rod*. The coefficients of kinetic friction between the block and the surface are μ_1 for block 1 and μ_2 for block 2. (*a*) Determine the acceleration of the two blocks. (*b*) Determine the force that the rod exerts on each of the two blocks. Show that these forces are both 0 when $\mu_1 = \mu_2$ and give a simple, nonmathematical argument why this is true.

53 •• A block of mass m rests on a horizontal table (Figure 5-65). The block is pulled by a massless rope with a force $\vec{F}$ at an angle θ. The coefficient of static friction is 0.60. The minimum value of the force needed to move the block depends on the angle θ. (*a*) Discuss qualitatively how you would expect the magnitude of this force to depend on θ. (*b*) Compute the force for the angles $\theta = 0°, 10°, 20°, 30°, 40°, 50°$, and $60°$, and make a plot of F versus θ for $mg = 400$ N. From your plot, at what angle is it most efficient to apply the force to move the block? SSM

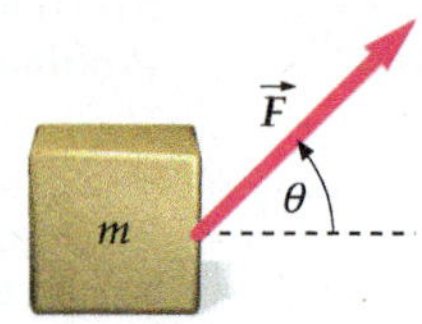

FIGURE 5-65 Problems 53 and 54

54 •• Consider the block in Figure 5-65. Show that, in general, the following results hold for a block of mass m resting on a horizontal surface whose coefficient of static friction is μ_s. (*a*) If you want to apply the *minimum* possible force to move the block, you should apply it with the force pulling upward at an angle $\theta = \tan^{-1}\mu_s$. (*b*) The minimum force necessary to start the block moving is $F_{min} = (\mu_s/\sqrt{1+\mu_s^2})mg$. (*c*) Once the block starts moving, if you want to apply the least possible force to keep it moving, should you keep the angle at which you are pulling the same, increase it, or decrease it?

55 •• Answer the questions in Problem 54, but for a force $\vec{F}$ that pushes down on the block at an angle θ *below* the horizontal.

56 •• A 100-kg mass is pulled along a frictionless surface by a horizontal force $\vec{F}$ such that its acceleration is $a_1 = 6.00\ \text{m/s}^2$ (Figure 5-66). A 20.0-kg mass slides along the top of the 100-kg mass and has an acceleration of $a_2 = 4.00\ \text{m/s}^2$. (It thus slides backward relative to the 100-kg mass.) (*a*) What is the frictional force exerted by the 100-kg mass on the 20.0-kg mass? (*b*) What is the net force acting on the 100-kg mass? What is the force F? (*c*) After the 20.0-kg mass falls off the 100-kg mass, what is the acceleration of the 100-kg mass? (Assume that the force F does not change.)

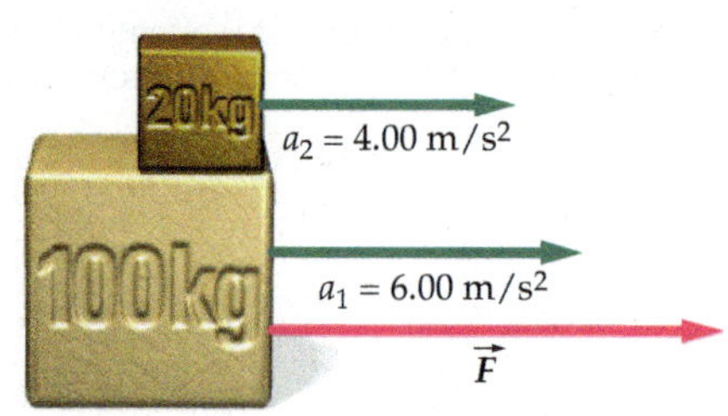

FIGURE 5-66 Problem 56

57 •• A 60-kg block slides along the top of a 100-kg block. The 60-kg block has an acceleration of 3.0 m/s² while a horizontal force of 320 N is applied to it, as shown in Figure 5-67. There is no friction between the 100-kg block and a horizontal frictionless surface, but there is friction between the two blocks. (*a*) Find the coefficient

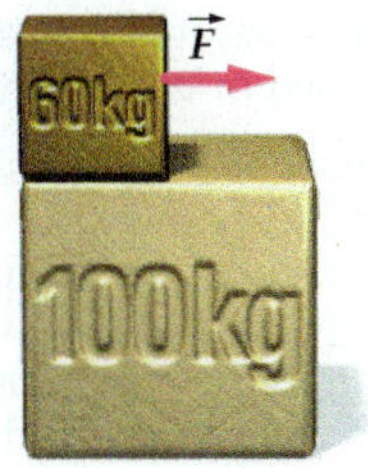

FIGURE 5-67 Problem 57

of kinetic friction between the blocks. (*b*) Find the acceleration of the 100-kg block during the time that the 60-kg block remains in contact.

58 •• The coefficient of static friction between a rubber tire and the road surface is 0.85. What is the maximum acceleration of a 1000-kg four-wheel-drive truck if the road makes an angle of 12° with the horizontal and the truck is (*a*) climbing, and (*b*) descending?

59 •• A 2.0-kg block sits on a 4.0-kg block that is on a frictionless table (Figure 5-68). The coefficients of friction between the blocks are $\mu_s = 0.30$ and $\mu_k = 0.20$. (*a*) What is the maximum horizontal force F that can be applied to the 4.0-kg block if the 2.0-kg block is not to slip? (*b*) If F has half this value, find the acceleration of each block and the force of friction acting on each block. (*c*) If F is twice the value found in (*a*), find the acceleration of each block.

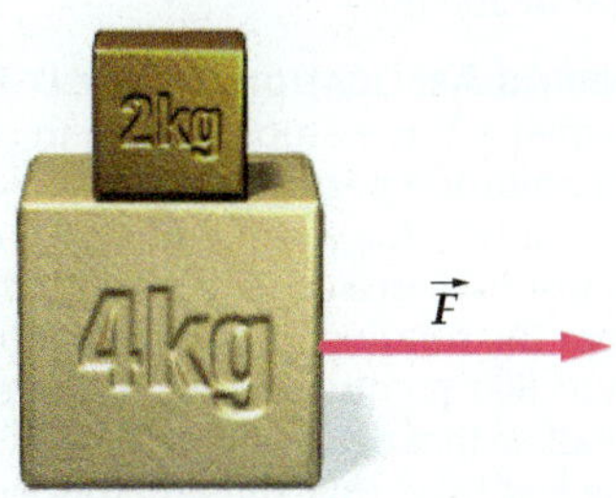

FIGURE 5-68 Problem 59

60 ••• A 10.0-kg block rests on a 5.0-kg bracket, as shown in Figure 5-69. The 5.0-kg bracket sits on a frictionless surface. The coefficients of friction between the 10.0-kg block and the bracket on which it rests are $\mu_s = 0.40$ and $\mu_k = 0.30$. (*a*) What is the maximum force F that can be applied if the 10.0-kg block is not to slide on the bracket? (*b*) What is the corresponding acceleration of the 5.0-kg bracket?

FIGURE 5-69 Problem 60

61 ••• You and your friends push a 75.0-kg greased pig up an aluminum slide at the county fair, starting from the low end of the slide. The coefficient of kinetic friction between the pig and the slide is 0.070. (*a*) All of you pushing together (parallel to the incline) manage to accelerate the pig from rest at the constant rate of 5.0 m/s^2 over a distance of 1.5 m, at which point you release the pig. The pig continues up the slide, reaching a maximum *vertical height* above its release point of 45 cm. What is the angle of inclination of the slide? (*b*) At the maximum height the pig turns around and begins to slip down the slide, how fast is it moving when it arrives at the low end of the slide?

62 ••• A 100-kg block on an inclined plane is attached to another block of mass m via a string, as in Figure 5-70. The coefficients of static and kinetic friction for the block and the incline are $\mu_s = 0.40$ and $\mu_k = 0.20$ and the plane is inclined 18° with horizontal. (*a*) Determine the range of values for m, the mass of the hanging block, for which the 100-kg block will not move unless disturbed, but if nudged, will slide down the incline. (*b*) Determine a range of values for m for which the 100-kg block will not move unless nudged, but if nudged will slide up the incline.

FIGURE 5-70 Problem 62

63 ••• A block of mass 0.50 kg rests on the inclined surface of a wedge of mass 2.0 kg, as in Figure 5-71. The wedge is acted on by a horizontal applied force $\vec{F}$ and slides on a frictionless surface. (*a*) If the coefficient of static friction between the wedge and the block is $\mu_s = 0.80$ and the wedge is inclined 35° with the horizontal, find the maximum and minimum values of the applied force for which the block does not slip. (*b*) Repeat part (*a*) with $\mu_s = 0.40$.

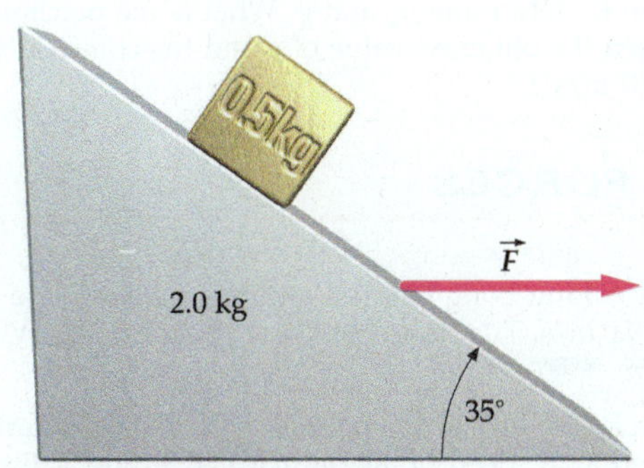

FIGURE 5-71 Problem 63

64 ••• **SPREADSHEET** In your physics lab, you and your lab partners push a block of wood with a mass of 10.0 kg (starting from rest), with a constant horizontal force of 70 N across a wooden floor. In the previous week's laboratory meeting, your group determined that the coefficient of kinetic friction was not exactly constant, but instead was found to vary with the object's speed according to $\mu_k = 0.11/(1 + 2.3 \times 10^{-4}v^2)^2$. Write a **spreadsheet** program using Euler's method to calculate and graph both the speed and the position of the block as a function of time from 0 to 10 s. Compare this result to the result you would get if you assumed the coefficient of kinetic friction had a constant value of 0.11.

65 ••• **MULTISTEP** In order to determine the coefficient of kinetic friction of a block of wood on a horizontal table surface, you are given the following assignment: Take the block of wood and give it an initial velocity across the surface of the table. Using a stopwatch, measure the time Δt it takes for the block to come to a stop and the total displacement Δx the block slides following the push. (*a*) Using Newton's laws and a free-body diagram of the block, show that the expression for the coefficient of kinetic friction is given by $\mu_k = 2\Delta x/[(\Delta t)^2 g]$. (*b*) If the block slides a distance of 1.37 m in 0.97 s, calculate μ_k. (*c*) What was the initial speed of the block?

66 •• **SPREADSHEET** (*a*) A block is sliding down an inclined plane. The coefficient of kinetic friction between the block and the plane is μ_k. Show that a graph of $a_x/\cos\theta$ versus $\tan\theta$ (where a_x is the acceleration down the incline and θ is the angle the plane is inclined with the horizontal) would be a straight line with slope g and intercept $-\mu_k g$. (*b*) The following data show the acceleration of a block sliding down an inclined plane as a function of the angle θ that the plane is inclined with the horizontal:*

θ (degrees)	Acceleration (m/s^2)
25.0	1.69
27.0	2.10
29.0	2.41
31.0	2.89
33.0	3.18
35.0	3.49
37.0	3.78
39.0	4.15
41.0	4.33
43.0	4.72
45.0	5.11

Using a **spreadsheet** program, graph these data and fit a straight line to them to determine μ_k and g. What is the percentage difference between the obtained value of g and the commonly specified value of 9.81 m/s^2?

DRAG FORCES

67 • A Ping-Pong ball has a mass of 2.3 g and a terminal speed of 9.0 m/s. The drag force is of the form bv^2. What is the value of b? SSM

68 • A small pollution particle settles toward Earth in still air. The terminal speed of the particle is 0.30 mm/s, the mass of the particle is 1.0×10^{-10} g and the drag force of the particle is of the form bv. What is the value of b?

69 •• A common classroom demonstration involves dropping basket-shaped coffee filters and measuring the time required for them to fall a given distance. A professor drops a single basket-shaped coffee filter from a height h above the floor, and records the time for the fall as Δt. How far will a stacked set of n identical filters fall during the same time interval Δt? Consider the filters to be so light that they instantaneously reach their terminal velocities. Assume a drag force that varies as the square of the speed and assume the filters are released oriented right-side up. SSM

* Data taken from Dennis W. Phillips, "Science Friction Adventure–Part II," *The Physics Teacher*, 553 (1990).

70 •• A skydiver of mass 60.0 kg can slow herself to a constant speed of 90 km/h by orienting her body horizontally, looking straight down with arms and legs extended. In this position, she presents the maximum cross-sectional area and thus maximizes the air-drag force on her. (*a*) What is the magnitude of the drag force on the skydiver? (*b*) If the drag force is given by bv^2, what is the value of b? (*c*) At some instant she quickly flips into a "knife" position, orienting her body vertically with her arms straight down. Suppose this reduces the value of b to 55 percent of the value in Parts (*a*) and (*b*). What is her acceleration at the instant she achieves the "knife" position?

71 •• **ENGINEERING APPLICATION, CONTEXT-RICH** Your team of test engineers is to release the parking brake so an 800-kg car will roll down a very long 6.0 percent grade in preparation for a crash test at the bottom of the incline. (On a 6.0 percent grade the change in altitude is 6.0 percent of the horizontal distance traveled.) The total resistive force (air drag plus rolling friction) for this car has been previously established to be $F_d = 100\text{ N} + (1.2\text{ N}\cdot\text{s}^2/\text{m}^2)v^2$, where v is the speed of the car. What is the terminal speed for the car rolling down this grade?

72 ••• **APPROXIMATION** Small, slowly moving spherical particles experience a drag force given by Stokes' law: $F_d = 6\pi\eta r v$, where r is the radius of the particle, v is its speed, and η is the coefficient of viscosity of the fluid medium. (*a*) Estimate the terminal speed of a spherical pollution particle of radius 1.00×10^{-5} m and density of 2000 kg/m^3. (*b*) Assuming that the air is still and that η is $1.80 \times 10^{-5}\text{ N}\cdot\text{s/m}^2$, estimate the time it takes for such a particle to fall from a height of 100 m.

73 ••• **ENGINEERING APPLICATION, CONTEXT-RICH** You have an environmental chemistry internship, and are in charge of a sample of air that contains pollution particles of the size and density given in Problem 72. You capture the sample in an 8.0-cm-long test tube. You then place the test tube in a centrifuge with the midpoint of the test tube 12 cm from the rotation axis of the centrifuge. You set the centrifuge to spin at 800 revolutions per minute. (*a*) Estimate the time you have to wait so that nearly all of the pollution particles settle to the end of the test tube. (*b*) Compare this to the time required for a pollution particle to fall 8.0 cm under the action of gravity and subject to the drag force given in Problem 72. SSM

MOTION ALONG A CURVED PATH

74 • A rigid rod with a 0.050-kg ball at one end rotates about the other end so the ball travels at constant speed in a vertical circle with a radius of 0.20 m. What is the maximum speed of the ball so that the force of the rod on the ball does not exceed 10 N?

75 • A 95-g stone is whirled in a horizontal circle on the end of an 85-cm-long string. The stone takes 1.2 s to make each complete revolution. Determine the angle that the string makes with the horizontal. SSM

76 •• A 0.20-kg stone is whirled in a horizontal circle on the end of an 0.80-m-long string. The string makes an angle of 20° with the horizontal. Determine the speed of the stone.

77 •• A 0.75-kg stone attached to a string is whirled in a horizontal circle of radius 35 cm as in the tetherball in Example 5-11. The string makes an angle of 30° with the vertical. (*a*) Find the speed of the stone. (*b*) Find the tension in the string.

78 •• **BIOLOGICAL APPLICATION** A pilot with a mass of 50 kg comes out of a vertical dive in a circular arc such that at the bottom of the arc her upward acceleration is $3.5g$. (*a*) How does the magnitude

of the force exerted by the airplane seat on the pilot at the bottom of the arc compare to her weight? (*b*) Use Newton's laws of motion to explain why the pilot might be subject to a blackout. This means that an above normal volume of blood "pools" in her lower limbs. How would an inertial reference frame observer describe the cause of the blood pooling?

79 •• An 80.0-kg airplane pilot pulls out of a dive by following, at a constant speed of 180 km/h, the arc of a circle whose radius is 300 m. (*a*) At the bottom of the circle, what are the direction and magnitude of his acceleration? (*b*) What is the net force acting on him at the bottom of the circle? (*c*) What is the force exerted on the pilot by the airplane seat?

80 •• An small object of mass m_1 moves in a circular path of radius r on a frictionless horizontal tabletop (Figure 5-72). It is attached to a string that passes through a small frictionless hole in the center of the table. A second object with a mass of m_2 is attached to the other end of the string. Derive an expression for r in terms of m_1, m_2, and the time T for one revolution.

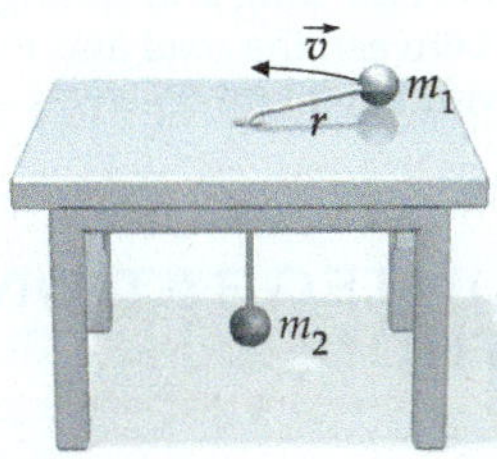

FIGURE 5-72 Problem 80

81 •• A block of mass m_1 is attached to a cord of length L_1, which is fixed at one end. The block moves in a horizontal circle on a frictionless tabletop. A second block of mass m_2 is attached to the first by a cord of length L_2 and also moves in a circle on the same frictionless tabletop, as shown in Figure 5-73. If the period of the motion is T, find the tension in each cord in terms of the given symbols. SSM

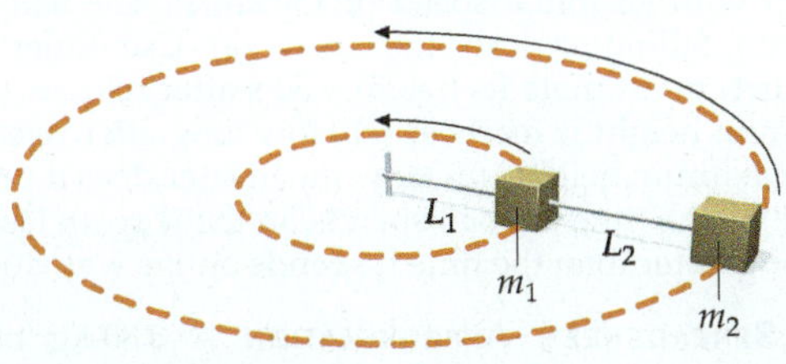

FIGURE 5-73 Problem 81

82 •• MULTISTEP A particle moves with constant speed in a circle of radius 4.0 cm. It takes 8.0 s to complete each revolution. (*a*) Draw the path of the particle to scale, and indicate the particle's position at 1.0-s intervals. (*b*) Sketch the displacement vectors for each interval. These vectors also indicate the directions for the average-velocity vectors for each interval. (*c*) Graphically find the magnitude of the change in the average velocity $|\Delta\vec{v}|$ for two consecutive 1-s intervals. Compare $|\Delta\vec{v}|/\Delta t$, measured in this way, with the magnitude of the instantaneous acceleration computed from $a_c = v^2/r$.

83 •• You are swinging your younger sister in a circle of radius 0.75 m, as shown in Figure 5-74. If her mass is 25 kg and you arrange it so she makes one revolution every 1.5 s, (*a*) what is the magnitude and direction of the force that must be exerted by you on her? (Model her as a point particle.) (*b*) What is the magnitude and direction of the force she exerts on you?

FIGURE 5-74 Problem 83 *(David de Lossy/The Image Bank.)*

84 •• The string of a conical pendulum is 50.0 cm long and the mass of the bob is 0.25 kg. (*a*) Find the angle between the string and the horizontal when the tension in the string is six times the weight of the bob. (*b*) Under those conditions, what is the period of the pendulum?

85 •• A 100-g coin sits on a horizontally rotating turntable. The turntable makes exactly 1.00 revolution each second. The coin is located 10 cm from the axis of rotation of the turntable. (*a*) What is the frictional force acting on the coin? (*b*) If the coin slides off the turntable when it is located more than 16.0 cm from the axis of rotation, what is the coefficient of static friction between the coin and the turntable?

86 •• A 0.25-kg tether ball is attached to a vertical pole by a 1.2-m cord. Assume the radius of the ball is negligible. If the ball moves in a horizontal circle with the cord making an angle of 20° with the vertical, (*a*) what is the tension in the cord? (*b*) What is the speed of the ball?

87 ••• A small bead with a mass of 100 g (Figure 5-75) slides without friction along a semicircular wire with a radius of 10 cm that rotates about a vertical axis at a rate of 2.0 revolutions per second. Find the value of θ for which the bead will remain stationary relative to the rotating wire.

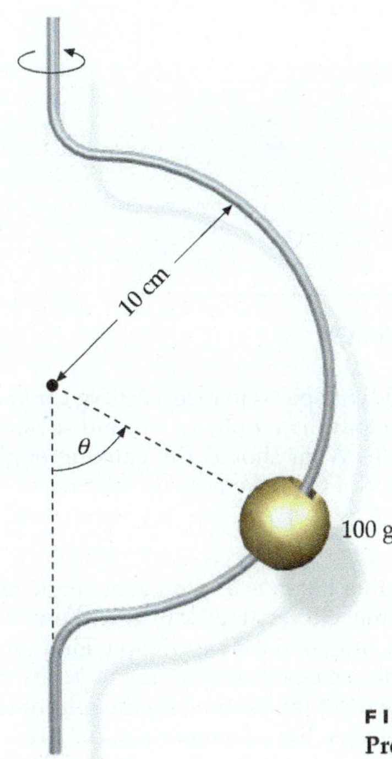

FIGURE 5-75
Problem 87

CENTRIPETAL FORCE

88 • A car speeds along the curved exit ramp of a freeway. The radius of the curve is 80.0 m. A 70.0-kg passenger holds the armrest of the car door with a 220-N force in order to keep from sliding across the front seat of the car. (Assume the exit ramp is not banked and ignore friction with the car seat.) What is the car's speed?

89 • The radius of curvature of the track at the top of a loop-the-loop on a roller-coaster ride is 12.0 m. At the top of the loop, the force that the seat exerts on a passenger of mass m is $0.40mg$. How fast is the roller-coaster car moving as it moves through the highest point of the loop. SSM

90 •• **ENGINEERING APPLICATION** On a runway of a decommissioned airport, a 2000-kg car travels at a constant speed of 100 km/h. At 100 km/h the air drag on the car is 500 N. Assume that rolling friction is negligible. (*a*) What is the force of static friction exerted on the car by the runway surface, and what is the minimum coefficient of static friction necessary for the car to sustain this speed? (*b*) The car continues to travel at 100 km/h, but now along a path with radius of curvature r. For what value of r will the angle between the static friction force vector and the velocity vector equal 45.0°, and for what value of r will it equal 88.0°? What is the minimum coefficient of static friction necessary for the car to hold this last radius of curvature without skidding?

91 •• Suppose you ride a bicycle in a 20-m-radius circle on a horizontal surface. The resultant force exerted by the surface on the bicycle (normal force plus frictional force) makes an angle of 15° with the vertical. (*a*) What is your speed? (*b*) If the frictional force on the bicycle is half its maximum possible value, what is the coefficient of static friction?

92 •• An airplane is flying in a horizontal circle at a speed of 480 km/h. The plane is banked for this turn, its wings tilted at an angle of 40° from the horizontal (Figure 5-76). Assume that a lift force acting perpendicular to the wings acts on the aircraft as it moves through the air. What is the radius of the circle in which the plane is flying?

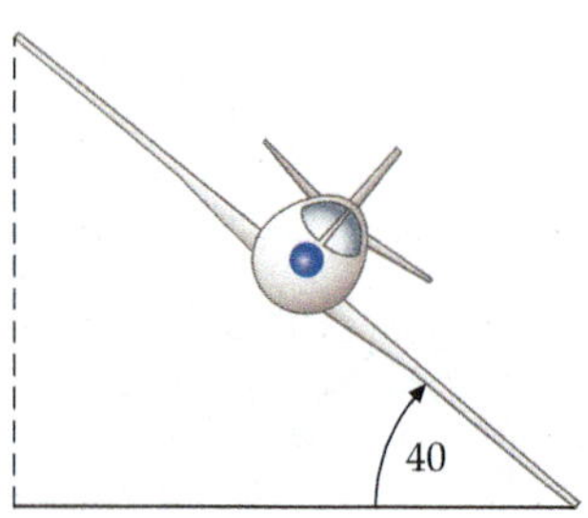

FIGURE 5-76 Problem 92

93 •• An automobile club plans to race a 750-kg car at the local racetrack. The car needs to be able to travel around several 160-m-radius curves at 90 km/h. What should the banking angle of the curves be so that the force of the pavement on the tires of the car is in the normal direction? *Hint: What does this requirement tell you about the frictional force?*

94 •• A curve of radius 150 m is banked at an angle of 10°. An 800-kg car negotiates the curve at 85 km/h without skidding. Neglect the effects of air drag and rolling friction. Find (*a*) the normal force exerted by the pavement on the tires, (*b*) the frictional force exerted by the pavement on the tires, (*c*) the minimum coefficient of static friction between the pavement and the tires.

95 •• On another occasion, the car in Problem 94 negotiates the curve at 38 km/h. Neglect the effects of air drag and rolling friction. Find (*a*) the normal force exerted on the tires by the pavement, and (*b*) the frictional force exerted on the tires by the pavement.

96 ••• **ENGINEERING APPLICATION** As a civil engineering intern during one of your summers in college, you are asked to design a curved section of roadway that meets the following conditions: When ice is on the road, and the coefficient of static friction between the road and rubber is 0.080, a car at rest must not slide into the ditch and a car traveling less than 60 km/h must not skid to the outside of the curve. Neglect the effects of air drag and rolling friction. What is the minimum radius of curvature of the curve and at what angle should the road be banked?

97 ••• **ENGINEERING APPLICATION** A curve of radius 30 m is banked so that a 950-kg car traveling at 40.0 km/h can round it even if the road is so icy that the coefficient of static friction is approximately zero. You are commissioned to tell the local police the range of speeds at which a car can travel around this curve without skidding. Neglect the effects of air drag and rolling friction. If the coefficient of static friction between the road and the tires is 0.300, what is the range of speeds you tell them?

* NUMERICAL INTEGRATION: EULER'S METHOD

* 98 •• **SPREADSHEET, APPROXIMATION** You are riding in a hovering hot air balloon when you throw a baseball straight down with an initial speed of 35.0 km/h. The baseball falls with a terminal speed of 150 km/h. Assuming air drag is proportional to the speed squared, use Euler's method (**spreadsheet**) to estimate the speed of the ball after 10.0 s. What is the uncertainty in this estimate? You drop a second baseball, this one is released from rest. How long does it take for it to reach 99 percent of its terminal speed? How far does it fall during this time?

* 99 •• **SPREADSHEET, APPROXIMATION** You throw a baseball straight up with an initial speed of 150 km/h. The ball's terminal speed when falling is also 150 km/h. (*a*) Use Euler's method (**spreadsheet**) to estimate its height 3.50 s after release. (*b*) What is the maximum height it reaches? (*c*) How long after release does it reach its maximum height? (*d*) How much later does it return to the ground? (*e*) Is the time the ball spends on the way up less than, the same as, or greater than the time it spends on the way down? SSM

* 100 ••• **SPREADSHEET, APPROXIMATION** A 0.80-kg block on a horizontal frictionless surface is held against a massless spring, compressing it 30 cm. The force constant of the spring is 50 N/m. The block is released and the spring pushes it 30 cm. Use Euler's method (**spreadsheet**) with $\Delta t = 0.0050$ s to estimate the time it takes for the spring to push the block the 30 cm. How fast is the block moving at this time? What is the uncertainty in this speed?

FINDING THE CENTER OF MASS

101 • Three point masses of 2.0 kg each are located on the x axis. One is at the origin, another at $x = 0.20$ m, and another at $x = 0.50$ m. Find the center of mass of the system.

102 • On a weekend archeological dig, you discover an old club-ax that consists of a symmetrical 8.0-kg stone attached to the end of a uniform 2.5-kg stick. You measure the dimensions of the club-ax as shown in Figure 5-77. How far is the center of mass of the club-ax from the handle end of the club-ax?

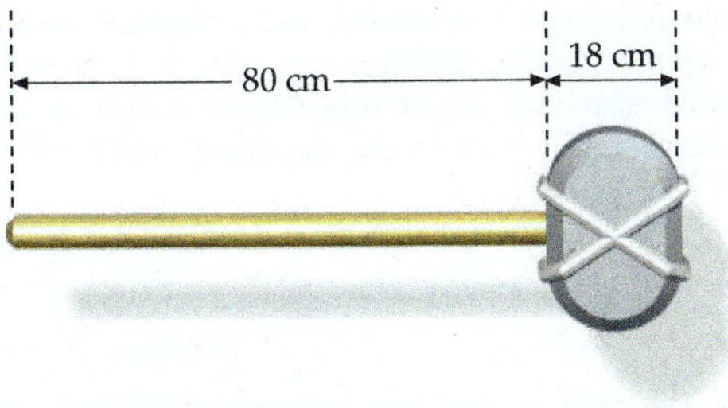

FIGURE 5-77 Problem 102

103 • Three balls A, B, and C, with masses of 3.0 kg, 1.0 kg, and 1.0 kg, respectively, are connected by massless rods, as shown in Figure 5-78. What are the coordinates of the center of mass of this system?

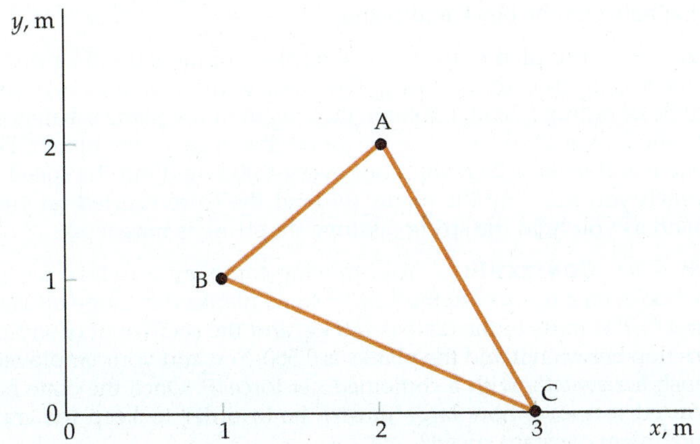

FIGURE 5-78 Problems 103 and 115

104 • By symmetry, locate the center of mass of a uniform sheet in the shape of an equilateral triangle with edges of length a. The triangle has one vertex on the y axis and the others at $(-a/2, 0)$ and $(+a/2, 0)$.

105 •• Find the center of mass of the uniform sheet of plywood in Figure 5-79. Consider this as a system of effectively two sheets, letting one have a "negative mass" to account for the cutout. Thus, one is a square sheet of 3.0-m edge length and mass m_1 and the second is a rectangular sheet measuring $1.0\text{ m} \times 2.0\text{ m}$ with a mass of $-m_2$. Let the coordinate origin be at the lower left corner of the sheet. SSM

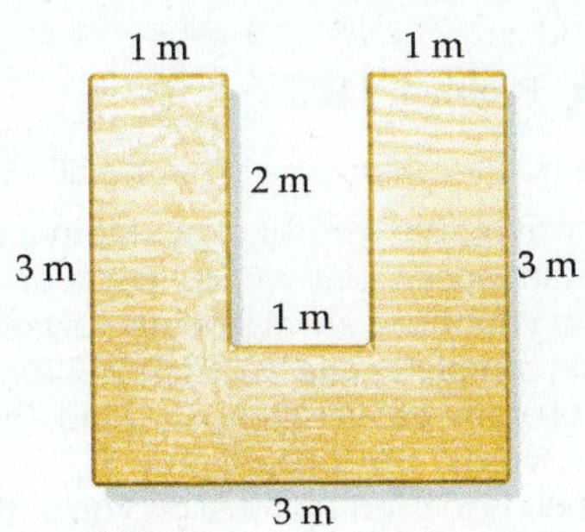

FIGURE 5-79 Problem 105

106 •• A can in the shape of a symmetrical cylinder with mass M and height H is filled with water. The initial mass of the water is M, the same mass as the can. A small hole is punched in the bottom of the can, and the water drains out. (*a*) If the height of the water in the can is x, what is the height of the center of mass of the can plus the water remaining in the can? (*b*) What is the minimum height of the center of mass as the water drains out?

107 •• Two identical thin uniform rods each of length L are glued together at the ends so that the angle at the joint is 90°. Determine the location of the center of mass (in terms of L) of this configuration relative to the origin taken to be at the joint. *Hint: You do not need the mass of the rods, but you should start by assuming a mass m and see that it cancels out.* SSM

108 ••• Repeat the analysis of Problem 107 with a general angle θ at the joint instead of 90°. Does your answer agree with the specific 90°-angle answer in Problem 107 if you set θ equal to 90°? Does your answer give plausible results for angles of zero and 180°?

* 109 •• **FINDING THE CENTER OF MASS BY INTEGRATION** Show that the center of mass of a uniform semicircular disk of radius R is at a point $4R/(3\pi)$ from the center of the circle.

* 110 •• Find the location of the center of mass of a nonuniform rod 0.40 m in length if its density varies linearly from 1.00 g/cm at one end to 5.00 g/cm at the other end. Specify the center-of-mass location relative to the less-massive end of the rod.

* 111 ••• You have a thin uniform wire bent into part of a circle that is described by a radius R and angle θ_m (see Figure 5-80). Show that the location of its center of mass is on the x axis and located a distance $x_{cm} = (R\sin\theta_m)/\theta_m$, where θ_m is expressed in radians. Check your answer by showing that this answer gives the physically expected limit for $\theta_m = 180°$. Verify that your answer gives you the result in the text (in the subsection Finding the Center of Mass by Integration) for the special case of $\theta_m = 90°$.

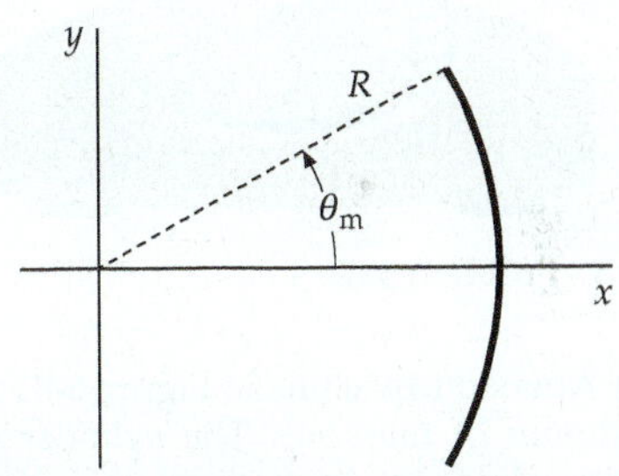

FIGURE 5-80 Problem 111

* 112 ••• A long, thin wire of length L has a linear mass density given by $A - Bx$, where A and B are positive constants and x is the distance from the more massive end. (*a*) A condition for this problem to be realistic is that $A > BL$. Explain why. (*b*) Determine x_{cm} in terms of L, A, and B. Does your answer makes sense if $B = 0$? Explain.

MOTION OF THE CENTER OF MASS

113 • Two 3.0-kg particles have velocities $\vec{v}_1 = (2.0\text{ m/s})\hat{i} + (3.0\text{ m/s})\hat{j}$ and $\vec{v}_2 = (4.0\text{ m/s})\hat{i} - (6.0\text{ m/s})\hat{j}$. Find the velocity of the center of mass of the system. SSM

114 • A 1500-kg car is moving westward with a speed of 20.0 m/s, and a 3000-kg truck is traveling east with a speed of 16.0 m/s. Find the velocity of the center of mass of the car–truck system.

115 • A force $\vec{F} = 12\text{ N }\hat{i}$ is applied to the 3.0-kg ball in Figure 5-78 in Problem 103. (No forces act on the other two balls.) What is the acceleration of the center of mass of the three-ball system?

116 •• A block of mass m is attached to a string and suspended inside an otherwise empty box of mass M. The box rests on a scale that measures the system's weight. (*a*) If the string breaks, does the reading on the scale change? Explain your reasoning. (*b*) Assume that the string breaks and the mass m falls with constant acceleration g. Find the magnitude and direction of the acceleration of the center of mass of the box–block system. (*c*) Using the result from (*b*), determine the reading on the scale while m is in free-fall.

117 •• The bottom end of a massless, vertical spring of force constant k rests on a scale and the top end is attached to a massless cup, as in Figure 5-81. Place a ball of mass m_b gently into the cup and ease it down into an equilibrium position where it sits at rest in the cup. (*a*) Draw the separate free-body diagrams for the ball and the spring. (*b*) Show that in this situation, the spring compression d is given by $d = m_b g/k$. (*c*) What is the scale reading under these conditions? SSM

FIGURE 5-81 Problem 117

118 ••• In the Atwood's machine in Figure 5-82 the string passes over a fixed cylinder of mass m_c. The cylinder does not rotate. Instead, the string slides on its frictionless surface. (*a*) Find the acceleration of the center of mass of the two-block–cylinder-string system. (*b*) Use Newton's second law for systems to find the force F exerted by the support. (*c*) Find the tension T in the string connecting the blocks and show that $F = m_c g + 2T$.

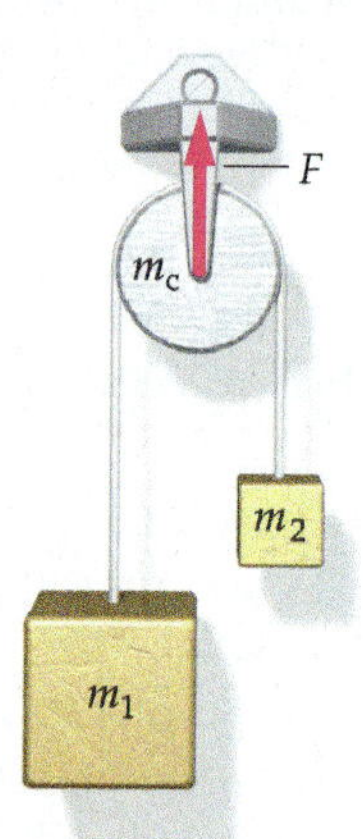

FIGURE 5-82 Problem 118

119 ••• Starting with the equilibrium situation in Problem 117, the whole system (scale, spring, cup, and ball) is now subjected to an upward acceleration of magnitude a (for example, in an elevator). Repeat the free-body diagrams and calculations in Problem 117?

GENERAL PROBLEMS

120 • In designing your new house in California, you are prepared for it to withstand a maximum horizontal acceleration of $0.50g$. What is the minimum coefficient of static friction between the floor and your prized Tuscan vase so that the vase does not slip on the floor under these conditions?

121 • A 4.5-kg block slides down an inclined plane that makes an angle of 28° with the horizontal. Starting from rest, the block slides a distance of 2.4 m in 5.2 s. Find the coefficient of kinetic friction between the block and plane.

122 •• You plan to fly a model airplane of mass 0.400 kg that is attached to a horizontal string. The plane will travel in a horizontal circle of radius 5.70 m. (Assume the weight of the plane is balanced by the upward "lift" force of the air on the wings of the plane.) The plane will make 1.20 revolutions every 4.00 s. (*a*) Find the speed at which you must fly the plane. (*b*) Find the force exerted on your hand as you hold the string (assume the string is massless).

123 •• **CONTEXT-RICH** Your moving company is to load a crate of books on a truck with the help of some planks that slope upward at 30°. The mass of the crate is 100 kg, and the coefficient of sliding friction between it and the planks is 0.500. You and your employees push *horizontally* with a combined net force $\vec{F}$. Once the crate has started to move, how large must F be in order to keep the crate moving at constant speed?

124 •• Three forces act on an object in static equilibrium (Figure 5-83). (*a*) If F_1, F_2, and F_3 represent the magnitudes of the forces acting on the object, show that $F_1/\sin\theta_{23} = F_2/\sin\theta_{31} = F_3/\sin\theta_{12}$. (*b*) Show that $F_1^2 = F_2^2 + F_3^2 + 2F_2F_3\cos\theta_{23}$.

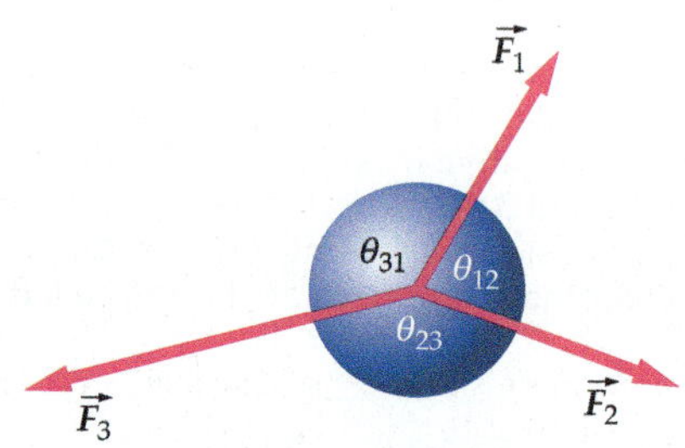

FIGURE 5-83 Problem 124

125 •• In a carnival ride, you sit on a seat in a compartment that rotates with constant speed in a vertical circle of radius 5.0 m. The ride is designed so your head always points toward the center of the circle. (*a*) If the ride completes one full circle in 2.0 s, find the direction and magnitude of your acceleration. (*b*) Find the slowest rate of rotation (in other words, the longest time T_m to complete one full circle) if the seat belt is to exert no force on you at the top of the ride.

126 •• A flat-topped toy cart moves on frictionless wheels, pulled by a rope under tension T. The mass of the cart is m_1. A load of mass m_2 rests on top of the cart with the coefficient of static friction μ_s between the cart and the load. The cart is pulled up a ramp that is inclined at angle θ above the horizontal. The rope is parallel to the ramp. What is the maximum tension T that can be applied without causing the load to slip?

127 ••• A sled weighing 200 N that is held in place by static friction, rests on a 15° incline. The coefficient of static friction between the sled and the incline is 0.50. *(a)* What is the magnitude of the normal force on the sled? *(b)* What is the magnitude of the static frictional force on the sled? *(c)* The sled is now pulled up the incline (Figure 5-84) at constant speed by a child walking up the incline ahead of the sled. The child weighs 500 N and pulls on the rope with a constant force of 100 N. The rope makes an angle of 30° with the incline and has negligible mass. What is the magnitude of the kinetic frictional force on the sled? *(d)* What is the coefficient of kinetic friction between the sled and the incline? *(e)* What is the magnitude of the force exerted on the child by the incline?

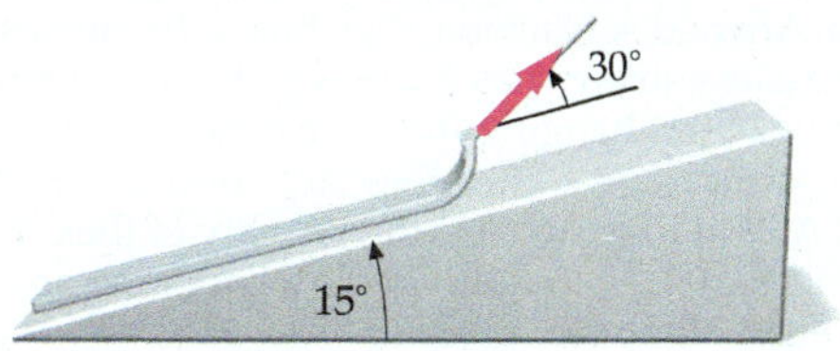

FIGURE 5-84 Problem 127

128 •• **ENGINEERING APPLICATION** In 1976, Gerard O'Neill proposed that large space stations be built for human habitation in orbit around Earth and the moon. Because prolonged free-fall has adverse medical effects, he proposed making the stations in the form of long cylinders and spinning them around the cylinder axis to provide the inhabitants with the sensation of gravity. One such O'Neill colony is to be built 5.0 miles long, with a diameter of 0.60 mi. A worker on the inside of the colony would experience a sense of "gravity," because he would be in an accelerated frame of reference due to the rotation. *(a)* Show that the "acceleration of gravity" experienced by the worker in the O'Neill colony is equal to his centripetal acceleration. *Hint: Consider someone "looking in" from outside the colony.* *(b)* If we assume that the space station is composed of several decks that are at varying distances (radii) from the axis of rotation, show that the "acceleration of gravity" becomes weaker the closer the worker gets to the axis. *(c)* How many revolutions per minute would this space station have to make to give an "acceleration of gravity" of 9.8 m/s^2 at the outermost edge of the station?

129 •• A child of mass m slides down a slide inclined at 30° in time t_1. The coefficient of kinetic friction between her and the slide is μ_k. She finds that if she sits on a small sled (also of mass m) with frictionless runners, she slides down the same slide in time $\frac{1}{2}t_1$. Find μ_k.

130 ••• The position of a particle of mass $m = 0.80$ kg as a function of time is given by $\vec{r} = x\hat{i} + y\hat{j} = (R\sin\omega t)\,\hat{i} + (R\cos\omega t)\,\hat{j}$, where $R = 4.0$ m and $\omega = 2\pi$ s^{-1}. *(a)* Show that the path of this particle is a circle of radius R, with its center at the origin of the xy plane. *(b)* Compute the velocity vector. Show that $v_x/v_y = -y/x$. *(c)* Compute the acceleration vector and show that it is directed toward the origin and has the magnitude v^2/R. *(d)* Find the magnitude and direction of the net force acting on the particle.

131 ••• **MULTISTEP** You are on an amusement park ride with your back against the wall of a spinning vertical cylinder. The floor falls away and you are held up by static friction. Assume your mass is 75 kg. *(a)* Draw a free-body diagram of yourself. *(b)* Use this diagram with Newton's laws to determine the force of friction on you. *(c)* If the radius of the cylinder is 4.0 m and the coefficient of static friction between you and the wall is 0.55. What is the minimum number of revolutions per minute necessary to prevent you from sliding down the wall? Does this answer hold only for you? Will other, more massive, patrons fall downward? Explain.

132 ••• An block of mass m_1 is on a horizontal table. The block is attached to a 2.5-kg block (m_2) by a light string that passes over a pulley at the edge of the table. The block of mass m_2 dangles 1.5 m above the ground (Figure 5-85). The string that connects them passes over a frictionless, massless pulley. This system is released from rest at $t = 0$ and the 2.5-kg block strikes the ground at $t = 0.82$ s. The system is now placed in its initial configuration and a 1.2-kg block is placed on top of the block of mass m_1. Released from rest, the 2.5-kg block now strikes the ground 1.3 s later. Determine the mass m_1 and the coefficient of kinetic friction between the block whose mass is m_1 and the table.

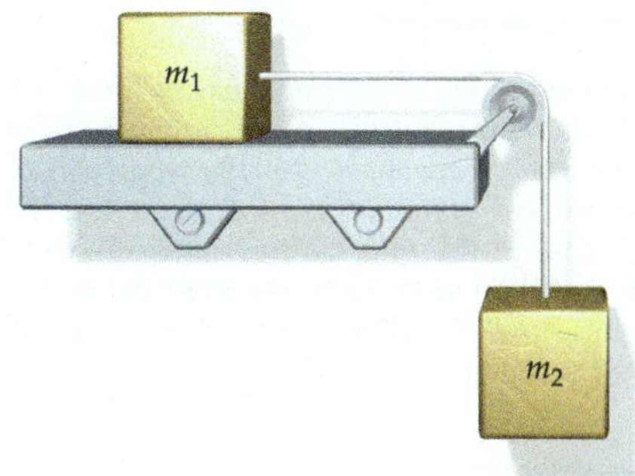

FIGURE 5-85 Problem 132

133 ••• Sally claims flying squirrels do not really fly; they jump and use the folds of skin that connect their forelegs and their back legs like a parachute to allow them to glide from tree to tree. Liz decides to test Sally's hypothesis by calculating the terminal speed of a falling outstretched flying squirrel. If the constant b in the drag force is proportional to the area of the object facing the air flow, use the results of Example 5-12 and some assumptions about the size of the squirrel to estimate its terminal (downward) speed. Is Sally's claim supported by Liz's calculation?

134 •• **BIOLOGICAL APPLICATION** After a parachutist jumps from an airplane (but before he pulls the rip cord to open his parachute), a downward speed of up to 180 km/h can be reached. When the parachute is finally opened, the drag force is increased by about a factor of 10, and this can create a large jolt on the jumper. Suppose this jumper falls at 180 km/h before opening his chute. *(a)* Determine the parachutist's acceleration when the chute is just opened, assuming his mass is 60 kg. *(b)* If rapid accelerations greater than 5.0g can harm the structure of the human body, is this a safe practice?

135 • Find the location of the center of mass of the Earth–moon system relative to the center of Earth. Is it inside or outside the surface of Earth?

136 •• A circular plate of radius R has a circular hole of radius $R/2$ cut out of it (Figure 5-86). Find the center of mass of the plate after the hole has been cut. *Hint: The plate can be modeled as two disks superimposed, with the hole modeled as a disk negative mass.*

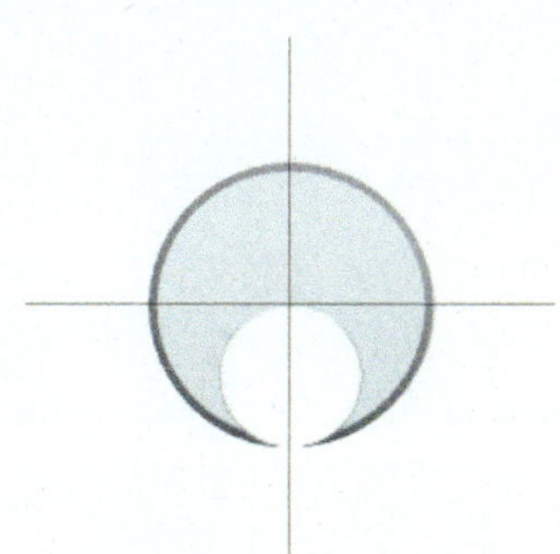

FIGURE 5-86 Problem 136

137 •• An unbalanced baton consists of a 50-cm-long uniform rod of mass 200 g. At one end there is a 10-cm-diameter uniform solid sphere of mass 500 g, and at the other end there is a 8.0-cm-diameter uniform solid sphere of mass 750 g. (The center-to-center distance between the spheres is 59 cm.) (*a*) Where, relative to the center of the light sphere, is the center of mass of this baton? (*b*) If this baton is tossed straight up (but spinning) so that its initial center of mass speed is 10.0 m/s, what is the velocity of the center of mass 1.5 s later? (*c*) What is the net external force on the baton while in the air? (*d*) What is the acceleration of the baton's center of mass 1.5 s following its release?

138 •• You are standing at the very rear of a 6.0-m-long, 120-kg raft that is at rest in a lake with its prow only 0.50 m from the end of the pier (Figure 5-87). Your mass is 60 kg. Neglect frictional forces between the raft and the water. (*a*) How far from the end of the pier is the center of mass of the you–raft system? (*b*) You walk to the front of the raft and then stop. How far from the end of the pier is the center of mass now? (*c*) When you are at the front of the raft, how far are you from the end of the pier?

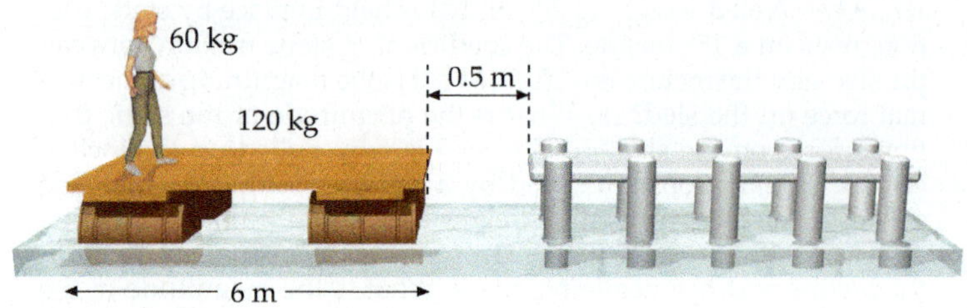

FIGURE 5-87 Problem 138

139 •• An Atwood's machine that has a frictionless massless pulley and massless strings has a 2.00-kg object hanging from one side and 4.00-kg object hanging from the other side. (*a*) What is the speed of each object 1.50 s after they are simultaneously released from rest? (*b*) At that time, what is the velocity of the center of mass of the two objects? (*c*) At that moment, what is the acceleration of the center of mass of the two objects?

CHAPTER 6

THE SNOW MELTS UNDER THE SKIES DUE TO THE KINETIC FRICTION BETWEEN THE SKIS AND THE SNOW. THE SKIER IS GLIDING DOWN THE MOUNTAIN ON A THIN LAYER OF LIQUID WATER. *(Courtesy of Rossignol Ski Company.)*

Work and Kinetic Energy

? How does the shape of the hill or the length of the path affect a skier's final speed at the finish line? (See Example 6-12.)

Thus far, we have analyzed motion by using concepts such as position, velocity, acceleration, and force. However, some types of motion are difficult to describe using Newton's laws directly. (A speed skier sliding down a curved slope is an example of this type of motion.) In this chapter and Chapter 7, we look at alternative methods for analyzing motion that involve two central concepts in science: energy and work. Unlike force, which is a vector physical quantity, energy and work are scalar physical quantities associated with particles and with systems of particles. As you will see, these new concepts provide powerful methods for solving a wide class of problems.

In this chapter, we explore the concept of work and how work is related to kinetic energy—the energy associated with the motion of objects. We also discuss the related concepts of power and center-of-mass work.

6-1 WORK DONE BY A CONSTANT FORCE

You may be used to thinking of work as anything that requires physical or mental exertion, such as studying for an exam, carrying a backpack, or riding a bike. But in physics, work is the transfer of energy by a force. If you stretch a spring by pulling on it with your hand (Figure 6-1), energy is transferred from you to the spring, and the energy transferred from you to the spring is equal to the work done by the force of your hand on the spring. The energy transferred to the spring can be evidenced if you let go of the spring and watch it rapidly contract and vibrate.

Work is a scalar quantity that can be positive, negative, or zero. The work done by object A on object B is positive if energy is transferred from A to B, and is negative if energy is transferred from B to A. If no energy is transferred, the work done is zero. In the case of you stretching a spring, the work done by you on the spring is positive because energy is transferred from you to the spring. However, suppose you move your hand so the spring slowly contracts to its unstressed state. During the contraction the spring loses energy—energy is transferred from the spring to you—and the work you do on the spring is negative.

It is commonly said that work is force times distance. Unfortunately, the statement "work is force times distance" is misleadingly simple. Work is done on an object by a force when the point of application of the force moves through a displacement. For a constant force, the work done equals the force component in the direction of the displacement times the magnitude of the displacement. For example, suppose you push a box along the ground with a constant horizontal force $\vec{F}$ in the direction of displacement $\Delta x\hat{i}$ (Figure 6-2*a*). Because the force acts on the box in the same direction as the displacement, the work W done by the force on the box is

$$W = F|\Delta x|$$

Now suppose you pull on a string attached to the box, such that force acts at an angle to the displacement, as shown in Figure 6-2*b*. In this case, the work done on the box by the force is given by the force component *in the direction* of the displacement times the magnitude of the displacement:

$$W = F_x\Delta x = F\cos\theta\,|\Delta x| \qquad \text{6-1}$$

WORK BY A CONSTANT FORCE

where F is the magnitude of the constant force, $|\Delta x|$ is the magnitude of the displacement of the point of application of the force, and θ is the angle between the directions of the force and displacement vectors. The displacement of the point of application of the force is identical to the displacement of any other point on the box because the box is rigid and moves without rotating. If you raise or lower a box by exerting a force $\vec{F}$ on it, you are doing work on the box. Let up be the positive y direction and let $\Delta y\hat{j}$ be the displacement of the box. The work done by you on the box is positive if Δy and F_y have the same signs and negative if they have opposite signs. But if you are simply holding the box in a fixed position then, according to the definition of work, you are not doing work *on the box* because the Δy is zero (Figure 6-3). In this case, the work you do on the box is zero, even though you are applying a force.

The SI unit of work is the **joule** (J), which equals the product of a newton and a meter:

$$1\,\text{J} = 1\,\text{N}\cdot\text{m} \qquad \text{6-2}$$

In the U.S. customary system, the unit of work is the foot-pound: $1\,\text{ft}\cdot\text{lb} = 1.356\,\text{J}$. Another convenient unit of work in atomic and nuclear physics is the **electron volt** (eV):

$$1\,\text{eV} = 1.602\times10^{-19}\,\text{J} \qquad \text{6-3}$$

FIGURE 6-1 Energy is transferred from the person to the spring as the spring lengthens. The energy transferred is equal to the work done by the person on the spring.

CONCEPT CHECK 6-1

For the contraction of the spring described immediately before this concept check, is the work done by the spring on the person positive or negative?

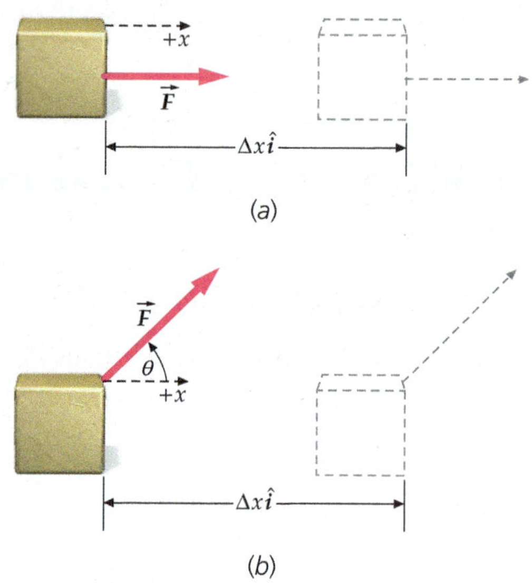

FIGURE 6-2

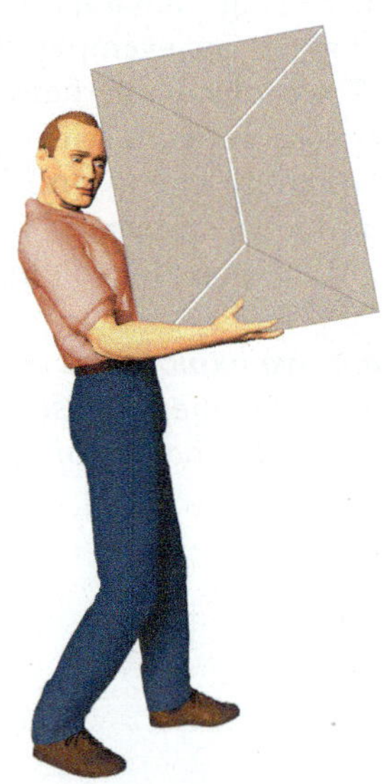

FIGURE 6-3

Commonly used multiples of eV are keV (10^3 eV) and MeV (10^6 eV). The work required to remove an electron from an atom is of the order of a few eV, whereas the work needed to remove a proton or a neutron from an atomic nucleus is of the order of several MeV.

PRACTICE PROBLEM 6-1

A force of 12 N is exerted on a box at an angle of $\theta = 20°$, as in Figure 6-2*b*. How much work is done by the force on the box as the box moves along the table a distance of 3.0 m?

If there are several forces that do work on *a system*, the total work is found by computing the work done by each force and adding each individual work together.

$$W_{\text{total}} = F_{1x}\,\Delta x_1 + F_{2x}\,\Delta x_2 + F_{3x}\,\Delta x_3 + \cdots \qquad 6\text{-}4$$

We model the system as a *particle* if the system moves so all the parts of the system undergo identical displacements. When several forces do work on such a *particle*, the displacements of the points of application of these forces are identical. Let the displacement of the point of application of any one of the forces be Δx. Then

$$W_{\text{total}} = F_{1x}\,\Delta x + F_{2x}\,\Delta x + \cdots = (F_{1x} + F_{2x} + \cdots)\Delta x = F_{\text{net}\,x}\,\Delta x \qquad 6\text{-}5$$

For a particle constrained to move along the x axis, the net force has only an x component. That is, $\vec{F}_{\text{net}} = F_{\text{net}\,x}\hat{i}$. Thus, for a particle, the x component of the net force times the displacement of any part of the object is equal to the total work done on the object.

Example 6-1 Loading with a Crane

A 3000-kg truck is to be loaded onto a ship by a crane that exerts an upward force of 31 kN on the truck. This force, which is strong enough to overcome the gravitational force and keep the truck moving upward, is applied over a distance of 2.0 m. Find (*a*) the work done on the truck by the crane, (*b*) the work done on the truck by gravity, and (*c*) the net work done on the truck.

PICTURE In Parts (*a*) and (*b*), the force acting on the truck is constant and the displacement is in a straight line, so we can use Equation 6-1, choosing the $+y$ direction as the direction of the displacement:

SOLVE

(*a*) 1. Sketch the truck at its initial and final positions, and choose the $+y$ direction to be the direction of the displacement (Figure 6-4):

2. Calculate the work done by the applied force:

$$W_{\text{app}} = F_{\text{app}\,y}\,\Delta y = (31\ \text{kN})(2.0\ \text{m}) = \boxed{62\ \text{kJ}}$$

(*b*) Calculate the work done by the force of gravity:
(*Note*: The vector $\vec{g}$ is directed downward and the $+y$ direction is upward. Consequently, $g_y = g\cos 180° = -g$.)

$$W_{\text{g}} = mg_y\,\Delta y = (3000\ \text{kg})(-9.81\ \text{N/kg})(2.0\ \text{m}) = \boxed{-59\ \text{kJ}}$$

(*c*) The net work done on the truck is the sum of the work done by each force:

$$W_{\text{net}} = W_{\text{app}\,y} + W_{\text{g}} = 62\ \text{kJ} + (-59\ \text{kJ}) = \boxed{3\ \text{kJ}}$$

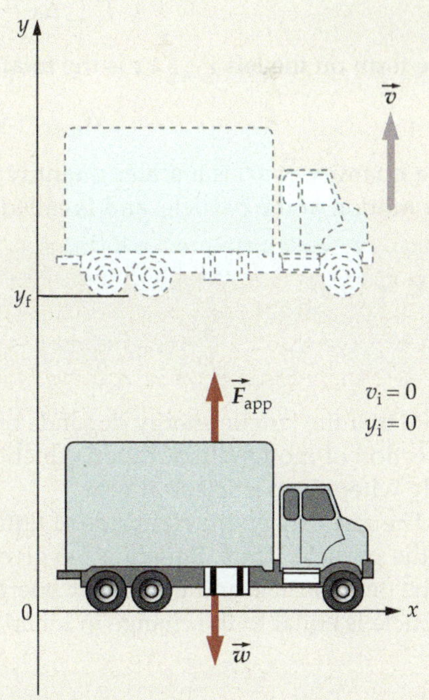

FIGURE 6-4

CHECK In Part (*a*), the force is applied in the same direction as the displacement, so we expect the work done on the truck to be positive. In Part (*b*), the force is applied in a direction opposite the displacement, so we expect the work done on the truck to be negative. Our results match these expectations.

TAKING IT FURTHER In Part (*c*), we also could have found the total work by first calculating the net force on the truck and then using Equation 6-5.

THE WORK–KINETIC-ENERGY THEOREM

Energy is one of the most important unifying concepts in science. All physical processes involve energy. The **energy** of a system is a measure of its ability to do work.

Different terms are used to describe the energy associated with different conditions or states. **Kinetic energy** is energy associated with motion. *Potential energy* is energy associated with the configuration of a system, such as the separation distance between two objects that attract each other. *Thermal energy* is associated with the random motion of the atoms, molecules, or ions within a system and is closely connected with the temperature of the system. In this chapter, we focus on kinetic energy. Potential energy and thermal energy are discussed in Chapter 7.

When forces do work on a particle, the result is a change in the energy associated with the motion of the particle—the kinetic energy. To evaluate the relationship between kinetic energy and work, let us look at what happens if a constant net force $\vec{F}_{\text{net}}$ acts on a particle of mass m that moves along the x axis. Applying Newton's second law, we see that

$$F_{\text{net}\,x} = ma_x$$

If the net force is constant, the acceleration is constant, and we can relate the displacement to the initial speed v_{i} and final speed v_{f} by using the constant-acceleration kinematic equation (Equation 2-16)

$$v_{\text{f}}^2 = v_{\text{i}}^2 + 2a_x\,\Delta x$$

Solving this for a_x gives

$$a_x = \frac{1}{2\Delta x}(v_{\text{f}}^2 - v_{\text{i}}^2)$$

Substituting for a_x in $F_{\text{net}\,x} = ma_x$ and then multiplying both sides by Δx, gives

$$F_{\text{net}\,x}\,\Delta x = \tfrac{1}{2}mv_{\text{f}}^2 - \tfrac{1}{2}mv_{\text{i}}^2$$

The term on the left $F_{\text{net}\,x}\,\Delta x$ is the total work done on the particle. Thus

$$W_{\text{total}} = \tfrac{1}{2}mv_{\text{f}}^2 - \tfrac{1}{2}mv_{\text{i}}^2 \qquad \text{6-6}$$

The quantity $\frac{1}{2}mv^2$ is a scalar quantity that represents the energy associated with the motion of the particle, and is called the **kinetic energy** K of the particle:

$$K = \tfrac{1}{2}mv^2 \qquad \text{6-7}$$

DEFINITION—KINETIC ENERGY

Note that the kinetic energy depends on only the particle's speed and mass, not its direction of motion. In addition, kinetic energy can never be negative, and is zero only when the particle is at rest.

The quantity on the right side of Equation 6-6 is the change in the kinetic energy of the particle. Thus, Equation 6-6 gives us a relationship between the total work done on a particle and the kinetic energy of the particle. The total work done on a particle is equal to the change in kinetic energy of the particle:

$$W_{\text{total}} = \Delta K \qquad \text{6-8}$$

WORK–KINETIC-ENERGY THEOREM

This result is known as the **work–kinetic-energy theorem.** This theorem tells us that when W_{total} is positive, the kinetic energy increases, which means the particle is moving faster at the end of the displacement than at the beginning. When W_{total} is negative, the kinetic energy decreases. When W_{total} is zero, the kinetic energy does not change, which means the particle's speed is unchanged.

Because total work on a particle is equal to its change in kinetic energy, we can see that the units of energy are the same as those of work. Three commonly used units of energy are the joule (J), the foot-pound (ft-lb), and the electron volt (eV).

The derivation of the work–kinetic-energy theorem presented here is valid only if the net force remains constant. However, as you will see later in this chapter, this theorem is valid even when the net force varies and the motion is not along a straight line.

! Note that kinetic energy depends on the *speed* of the particle, not the velocity. If the velocity changes direction, but not magnitude, the kinetic energy remains the same.

PROBLEM-SOLVING STRATEGY

Solving Problems Involving Work and Kinetic Energy

PICTURE The way you choose the $+y$ direction or $+x$ direction can help you to easily solve a problem that involves work and kinetic energy.

SOLVE

1. Draw the particle first at its initial position and second at its final position. For convenience, the object can be represented as a dot or a box. Label the initial and final positions of the object.
2. Put one or more coordinate axes on the drawing.
3. Draw arrows for the initial and final velocities, and label them appropriately.
4. On the initial-position drawing of the particle, place a vector for each force acting on it. Accompany each vector with a suitable label.
5. Calculate the total work done on the particle by the forces and equate this total to the change in the particle's kinetic energy.

CHECK Make sure you pay attention to negative signs during your calculations. For example, values for work done can be positive or negative, depending on the direction of the displacement relative to the direction of the force.

Example 6-2 Force on an Electron

In a television picture tube*, electrons are accelerated by an electron gun. The force that accelerates the electron is an electric force due to the electric field in the gun. An electron is accelerated from rest by an electron gun to a kinetic energy of 2.5 keV over a distance of 2.5 cm. Find the force on the electron, assuming it to be both constant and in the direction of the electron's motion.

PICTURE The electron can be modeled as a particle. Its initial and final kinetic energies are both given, and the electric force is the only force acting on it. Apply the work–kinetic-energy theorem and solve for the force.

* A television picture tube is a type of cathode-ray tube.

SOLVE

1. Make a drawing of the electron in its initial and final positions. Include the displacement, the initial and final speeds, and the force (Figure 6-5):

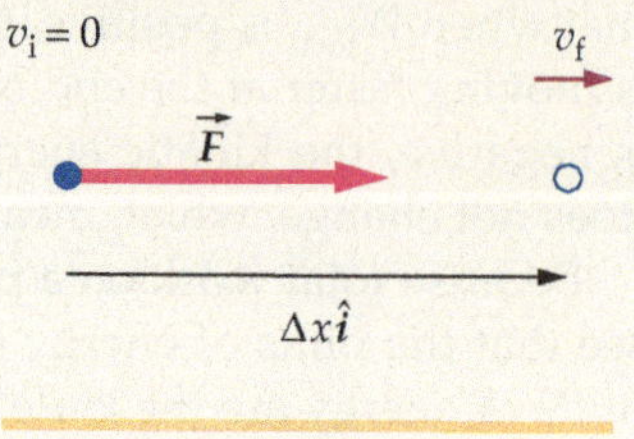

FIGURE 6-5

2. Set the work done equal to the change in kinetic energy:

$$W_{total} = \Delta K$$
$$F_x \Delta x = K_f - K_i$$

3. Solve for the force using the conversion factor $1.6 \times 10^{-19}\,J = 1.0\,eV$:

$$F_x = \frac{K_f - K_i}{\Delta x} = \frac{2500\,eV - 0}{0.025\,m} \times \frac{1.6 \times 10^{-19}\,J}{1.0\,eV}$$
$$= \boxed{1.6 \times 10^{-14}\,N}$$

CHECK The mass of an electron is only 9.1×10^{-31} kg. Thus, it is no surprise that such a small force would give it a large speed and thus a noticeable change in kinetic energy.

TAKING IT FURTHER (*a*) $1\,J = 1\,N \cdot m$, so $1\,J/m = 1\,N$. (*b*) 1 eV is the kinetic energy acquired by a particle of charge $-e$ (an electron, for example) when it is accelerated from the $-$ terminal to the $+$ terminal of a 1-V battery through a vacuum.

Example 6-3 A "Dogsled" Race

During your winter break you enter a "dogsled" race across a frozen lake. This is a race where each sled is pulled by a person, not by dogs. To get started you pull the sled (total mass 80 kg) with a force of 180 N at 40° above the horizontal. Find (*a*) the work you do, and (*b*) the final speed of the sled after it moves $\Delta x = 5.0$ m, assuming that it starts from rest and there is no friction.

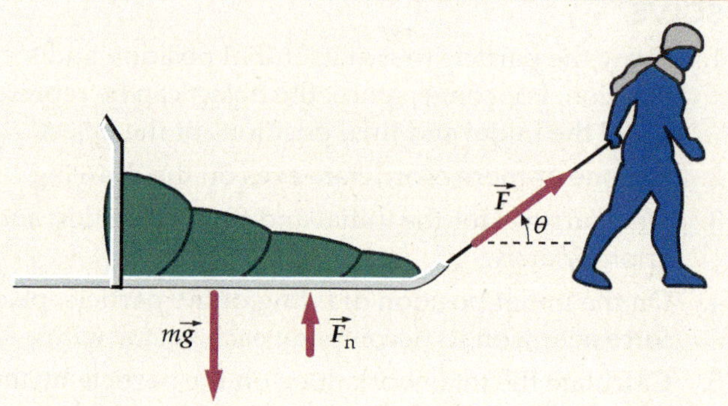

FIGURE 6-6

PICTURE The work done by you is $F_x \Delta x$, where we choose the direction of the displacement as the positive *x* direction. This is also the *total* work done on the sled because the other forces, mg and F_n, have no *x* components. The final speed of the sled can be found by applying the work–kinetic-energy theorem to the sled. Calculate the work done by each force on the sled (Figure 6-6) and equate the total work to the change in kinetic energy of the sled.

SOLVE

(*a*) 1. Sketch the sled both in its initial position and in its position after moving the 5.0 m. Draw the *x* axis in the direction of the motion (Figure 6-7).

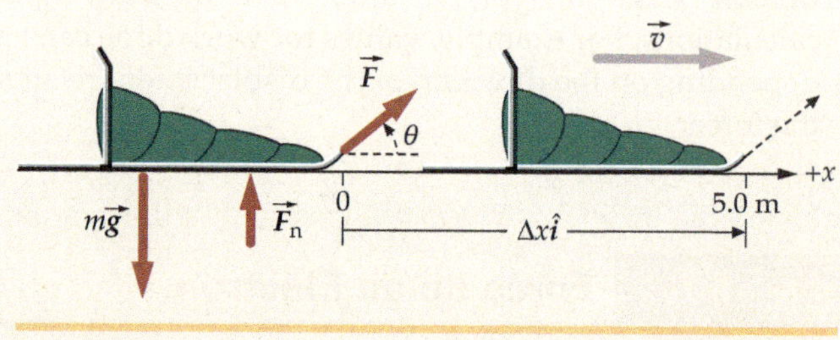

FIGURE 6-7

2. The work done by you on the sled is $F_x \Delta x$. This is the total work done on the sled. The other two forces each act perpendicular to the *x* direction (see Figure 6-7), so they do zero work:

$$W_{total} = W_{you} = F_x \Delta x = F \cos\theta \, \Delta x$$
$$= (180\,N)(\cos 40°)(5.0\,m) = 689\,J$$
$$= \boxed{6.9 \times 10^2\,J}$$

(*b*) Apply the work–kinetic-energy theorem to the sled and solve for the final speed:

$$W_{total} = \tfrac{1}{2}mv_f^2 - \tfrac{1}{2}mv_i^2$$
$$v_f^2 = v_i^2 + \frac{2W_{total}}{m}$$
$$= 0 + \frac{2(689\,J)}{80\,kg} = 17.2\,m^2/s^2$$
$$v_f = \sqrt{17.2\,m^2/s^2} = 4.151\,m/s = \boxed{4.2\,m/s}$$

CHECK In Part (*b*) we used that $1\ \mathrm{J/kg} = 1\ \mathrm{m^2/s^2}$. This is correct because

$$1\ \mathrm{J/kg} = 1\ \mathrm{N \cdot m/kg} = (1\ \mathrm{kg \cdot m/s^2}) \cdot \mathrm{m/kg} = 1\ \mathrm{m^2/s^2}$$

TAKING IT FURTHER The square root of 17.2 is 4.147, which rounds off to 4.1. However, the correct answer to Part (*b*) is 4.2 m/s. It is correct because it is calculated by taking the square root of 17.235 999 970 178 (the value stored in my calculator after executing the calculation of v_f).

PRACTICE PROBLEM 6-2 What is the magnitude of the force you exert if the 80-kg sled starts with a speed of 2.0 m/s and its final speed is 4.5 m/s after you pull it through a distance of 5.0 m while keeping the angle at 40°?

See **Math Tutorial** *for more information on* **Integrals**

6-2 WORK DONE BY A VARIABLE FORCE–STRAIGHT-LINE MOTION

Many forces vary with position. For example, a stretched spring exerts a force proportional to the distance it is stretched. In addition, the gravitational force Earth exerts on a spaceship varies inversely with the square of the center-to-center distance between the two bodies. How can we calculate the work done by forces like these?

Figure 6-8 shows the plot of a *constant* force F_x as a function of position x. Notice that the work done by the force on a particle whose displacement is Δx is represented by the area under the force-versus-position curve—indicated by the shading in Figure 6-8. We can approximate a variable force by a series of essentially constant forces (Figure 6-9). For each small displacement interval Δx_i, the force is approximately constant. Therefore the work done is approximately equal to the *area* of the rectangle of height F_{xi} and width Δx_i. The work W done by a variable force is then equal to the sum of the areas of an increasingly large number of these rectangles in the limit that the width of each individual rectangle approaches zero:

$$W = \lim_{\Delta x_i \to 0} \sum_i F_{xi}\,\Delta x_i = \text{area under the } F_x\text{-versus-}x \text{ curve} \qquad \text{6-9}$$

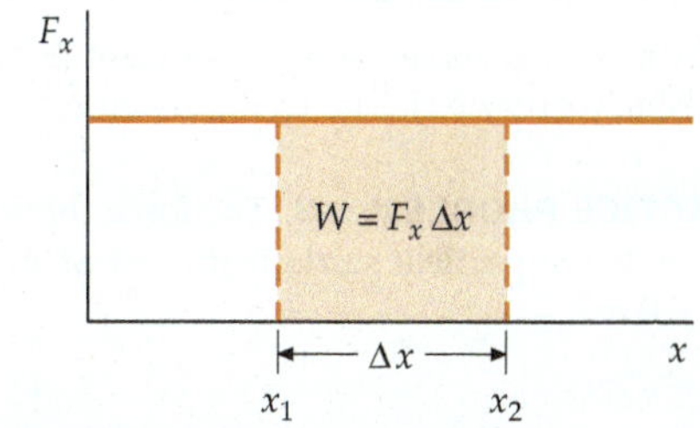

FIGURE 6-8 The work done by a constant force is represented graphically as the area under the F_x-versus-x curve.

This limit is the integral of $F_x dx$ over the interval from x_1 to x_2. So the work done by a variable force F_x acting on a particle as it moves from x_1 to x_2 is

$$W = \int_{x_1}^{x_2} F_x\,dx = \text{area under the } F_x\text{-versus-}x \text{ curve} \qquad \text{6-10}$$

WORK BY A VARIABLE FORCE—STRAIGHT-LINE MOTION

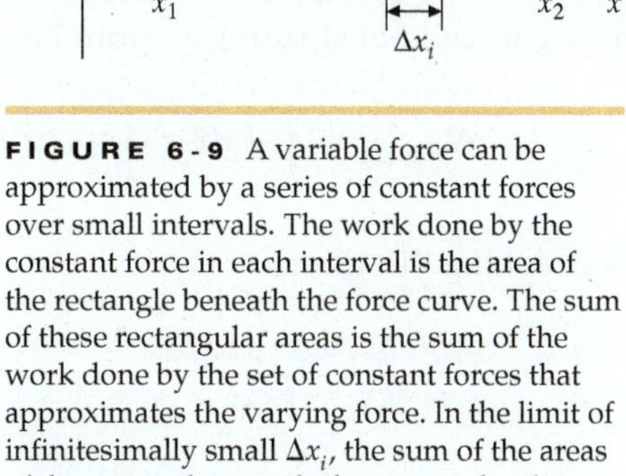

FIGURE 6-9 A variable force can be approximated by a series of constant forces over small intervals. The work done by the constant force in each interval is the area of the rectangle beneath the force curve. The sum of these rectangular areas is the sum of the work done by the set of constant forces that approximates the varying force. In the limit of infinitesimally small Δx_i, the sum of the areas of the rectangles equals the area under the complete force curve.

If the force plotted in Figure 6-9 is the net force on the particle, then each term $F_{xi}\,\Delta x_i$ in the sum in Equation 6-9 represents the total work done on the particle by a constant force as the particle undergoes the incremental displacement Δx_i. Thus, $F_{xi}\,\Delta x_i$ is equal to the change in kinetic energy ΔK_i of the particle during incremental displacement Δx_i (see Equation 6-8). In addition, the total change in the kinetic energy ΔK of the particle during the total displacement is equal to the sum of the incremental changes in kinetic energy. It follows that the total work W_{total} done on the particle for the total displacement equals the change in kinetic energy for the total displacement. Therefore, $W_{\text{total}} = \Delta K$ (Equation 6-8) holds for variable forces as well as for constant forces.

Example 6-4 Work Done by a Varying Force

A force $\vec{F} = F_x\hat{i}$ varies with x, as shown in Figure 6-10. Find the work done by the force on a particle as the particle moves from $x = 0.0$ m to $x = 6.0$ m.

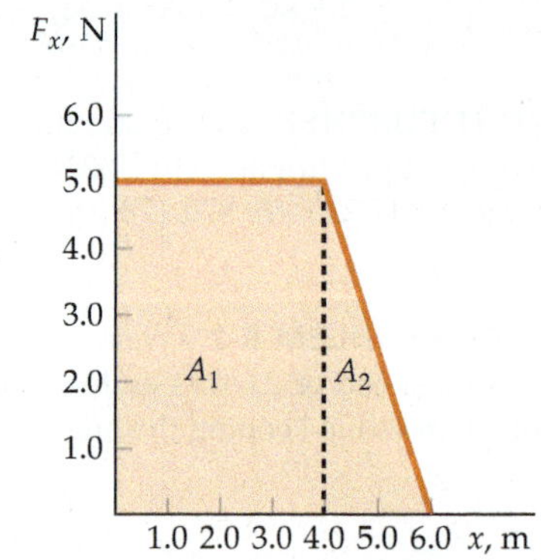

FIGURE 6-10

PICTURE The work done is the area under the curve from $x = 0.0$ m to $x = 6.0$ m. Because the curve consists of straight-line segments, the easiest approach is to break the area into two segments, one consisting of a rectangle (area A_1) and the other consisting of a right triangle (A_2), and then use the geometric formulas for area to find the work. (The alternative approach is to set up and execute an integration, as is done in Example 6-5.)

SOLVE

1. We find the work done by calculating the area under the F_x-versus-x curve: $W = A_{\text{total}}$

2. This area is the sum of the two areas shown. The area of a triangle is one half the altitude times the base:

$$W = A_{\text{total}} = A_1 + A_2$$
$$= (5.0\text{ N})(4.0\text{ m}) + \tfrac{1}{2}(5.0\text{ N})(2.0\text{ m})$$
$$= 20\text{ J} + 5.0\text{ J} = \boxed{25\text{ J}}$$

CHECK If the force were a constant 5.0 N over the entire 6.0 m, the work would be (5.0 N)(6.0 m) = 30 J. The step-2 result of 25 J is slightly less than the 30 J, as expected.

PRACTICE PROBLEM 6-3 The force shown is the only force that acts on a particle of mass 3.0 kg. If the particle starts from rest at $x = 0.0$ m, how fast is it moving when it reaches $x = 6.0$ m?

WORK DONE BY A SPRING THAT OBEYS HOOKE'S LAW

Figure 6-11 shows a block on a horizontal frictionless surface connected to a spring. If the spring is stretched or compressed, the spring exerts a force on the block. Recall from Equation 4-7 that the force exerted by the spring on the block is given by

$$F_x = -kx \qquad \text{(Hooke's law)} \qquad 6\text{-}11$$

where the k is a positive constant and x is the extension of the spring. If the spring is extended, then x is positive and the force component F_x is negative. If the spring is compressed, then x is negative and the force component F_x is positive.

Because the force varies with x, we can use Equation 6-10 to calculate the work done by the spring force on the block as the block undergoes a displacement from $x = x_i$ to $x = x_f$. (Besides the spring force, two other forces act on the block; the force of gravity, $m\vec{g}$, and the normal force of the table, $\vec{F}_n$. However, each of these forces does no work because neither has a component in the direction of the displacement. The only force that does work on the block is the spring force.) Substituting F_x from Equation 6-11 into Equation 6-10, we get

$$W_{\text{by spring}} = \int_{x_i}^{x_f} F_x dx = \int_{x_i}^{x_f} (-kx)dx = -k\int_{x_i}^{x_f} x\,dx = -k\left(\frac{x_f^2}{2} - \frac{x_i^2}{2}\right) \qquad 6\text{-}12$$

Rearranging this gives:

$$W_{\text{by spring}} = \tfrac{1}{2}kx_i^2 - \tfrac{1}{2}kx_f^2 \qquad 6\text{-}13$$

WORK BY A SPRING FORCE

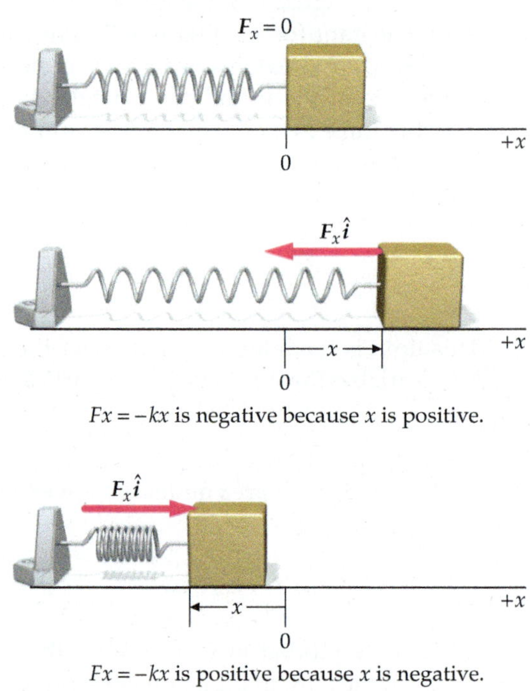

FIGURE 6-11 A horizontal spring. (*a*) When the spring is unstretched, it exerts no force on the block. (*b*) When the spring is stretched so that x is positive, it exerts a force of magnitude kx in the $-x$ direction. (*c*) When the spring is compressed so that x is negative, the spring exerts a force of magnitude $k|x|$ in the $+x$ direction.

The integral in Equation 6-12 can also be computed using geometry to calculate the area under the curve (Figure 6-12*a*). This gives

$$W_{\text{by spring}} = A_1 + A_2 = |A_1| - |A_2| = \tfrac{1}{2}kx_1^2 - \tfrac{1}{2}kx_2^2$$

which is identical to Equation 6-13.

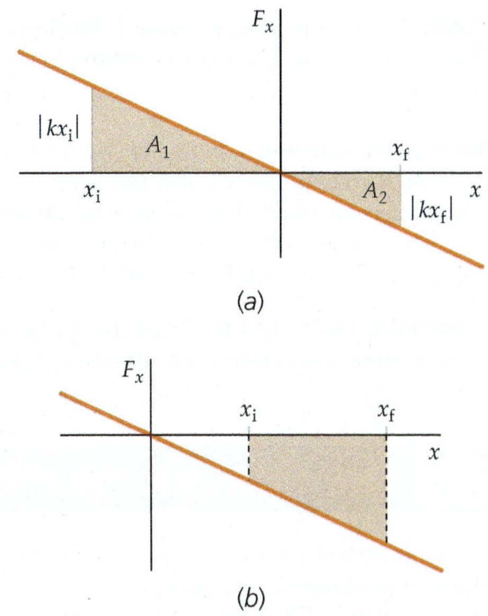

FIGURE 6-12

PRACTICE PROBLEM 6-4

Using geometry, calculate the area under the curve shown in Figure 6-12*b* and show you get an expression identical to that shown in Equation 6-13.

Suppose you pull on an initially relaxed spring (Figure 6-13), stretching it to a final extension x_f. How much work does the force exerted on the spring by your hand $\vec{F}_{SH}$ do? The force by your hand on the spring is equal to kx. (It is equal and opposite to the force by the spring on your hand.) As x increases from 0 to x_f, the force on the spring increases linearly from $F_{SHx} = 0$ to $F_{SHx} = kx_f$, and so has an average value* of $\frac{1}{2}kx_f$. The work done by this force is equal to the product of this average value and x_f. Thus, the work W done on the spring by your hand is given by

$$W = \tfrac{1}{2}kx_f^2$$

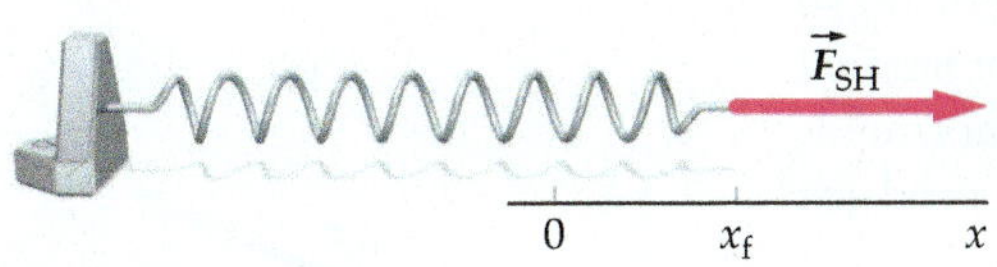

FIGURE 6-13

Example 6-5 Work Done on a Block by a Spring

A 4.0-kg block on a frictionless table is attached to a horizontal spring with $k = 400$ N/m. The spring is initially compressed 5.0 cm (Figure 6-14). Find (*a*) the work done on the block by the spring as the block moves from $x = x_1 = -5.0$ cm to its equilibrium position $x = x_2 = 0.0$ cm, and (*b*) the speed of the block at $x_2 = 0.0$ cm.

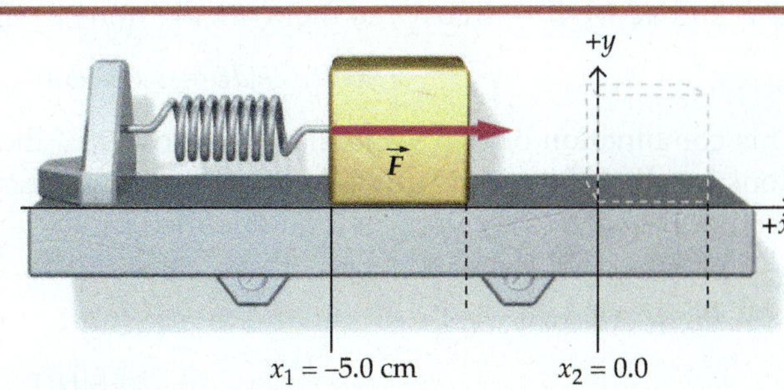

FIGURE 6-14

PICTURE Make a graph of F_x versus x. The work done on the block as it moves from x_1 to x_2 equals the area under the F_x-versus-x curve between these limits, shaded in Figure 6-15, which can be calculated by integrating the force over the distance. The work done equals the change in kinetic energy, which is simply its final kinetic energy because the initial kinetic energy is zero. The speed of the block at $x = 0.0$ cm is found from the kinetic energy of the block.

SOLVE

(*a*) The work W done on the block by the spring is the integral of $F_x dx$ from x_1 to x_2:

$$W = \int_{x_1}^{x_2} F_x\,dx = \int_{x_1}^{x_2} -kx\,dx = -k\int_{x_1}^{x_2} x\,dx$$

$$= -\tfrac{1}{2}kx^2\Big|_{x_1}^{x_2} = -\tfrac{1}{2}k(x_2^2 - x_1^2)$$

$$= -\tfrac{1}{2}(400\ \text{N/m})[(0.000\ \text{m})^2 - (0.050\ \text{m})^2]$$

$$= \boxed{0.50\ \text{J}}$$

(*b*) Apply the work–kinetic-energy theorem to the block and solve for v_2:

$$W_{\text{total}} = \tfrac{1}{2}mv_2^2 - \tfrac{1}{2}mv_1^2$$

so

$$v_2^2 = v_1^2 + \frac{2W_{\text{total}}}{m} = 0 + \frac{2(0.50\ \text{J})}{4.0\ \text{kg}} = 0.25\ \text{m}^2/\text{s}^2$$

$$v_2 = \boxed{0.50\ \text{m/s}}$$

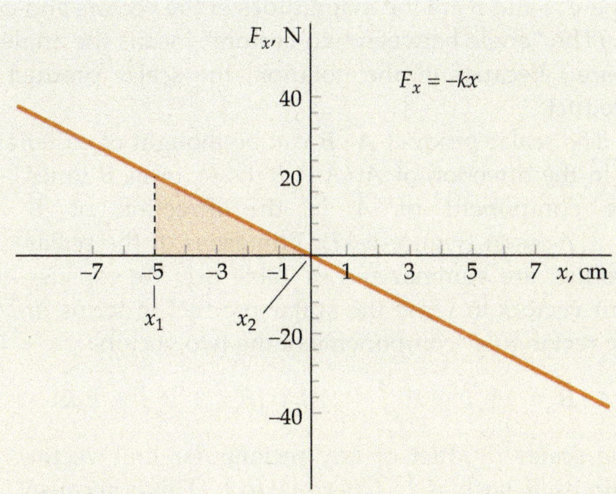

FIGURE 6-15

* Typically, an average value refers to an average over time. In this case, it refers to an average over position.

CHECK The work done is positive. The force and displacement are in the same direction, so this is as expected. The work is positive, so we expect the kinetic energy, and thus the speed, to increase. Our results fulfill this expectation.

TAKING IT FURTHER Note that we could *not* have solved this example by first applying Newton's second law to find the acceleration, and then using the constant-acceleration kinematic equations. This is so because the force exerted by the spring on the block, $F_x = -kx$, varies with position. Thus, the acceleration also varies with position. Therefore, the constant-acceleration kinematic equations are not in play.

PRACTICE PROBLEM 6-5 Find the speed of the 4.0-kg block when it reaches $x = 3.0$ cm if it starts from $x = 0.0$ cm with velocity $v_x = 0.50$ m/s.

6-3 THE SCALAR PRODUCT

Work is based on the component of force in the direction of an object's displacement. For straight-line motion, it is easy to calculate the component of the force in the direction of the displacement. However, in situations involving motion along a curved path, the force and the displacement can point in any direction. For these situations we can use a mathematical operation known as the *scalar* or *dot product* to determine the component of a given force in the direction of the displacement. The scalar product involves multiplying one vector by a second vector to produce a scalar.

Consider the particle moving along the arbitrary curve shown in Figure 6-16*a*. The component $F_{\|}$ in Figure 6-16*b* is related to the angle ϕ between the directions of $\vec{F}$ and $d\vec{\ell}$ by $F_{\|} = F\cos\phi$, so the work dW done by $\vec{F}$ for the displacement $d\vec{\ell}$ is

$$dW = F_{\|}d\ell = F\cos\phi\, d\ell$$

This combination of two vectors and the cosine of the angle between their directions is called the **scalar product** of the vectors. The scalar product of two general vectors $\vec{A}$ and $\vec{B}$ is written $\vec{A}\cdot\vec{B}$ and defined by

$$\vec{A}\cdot\vec{B} = AB\cos\phi \qquad 6\text{-}14$$

DEFINITION—SCALAR PRODUCT

where A and B are the magnitudes of the vectors and ϕ is the angle between $\vec{A}$ and $\vec{B}$. (The "angle between two vectors" means the angle between their directions in space.) Because of the notation, the scalar product is also known as the **dot product.**

The scalar product $\vec{A}\cdot\vec{B}$ can be thought of either as A times the component of $\vec{B}$ in the direction of $\vec{A}$ ($A \times B\cos\phi$), or as B times the component of $\vec{A}$ in the direction of $\vec{B}$ ($B \times A\cos\phi$) (Figure 6-17). Properties of the scalar product are summarized in Table 6-1. We can use unit vectors to write the scalar product in terms of the rectangular components of the two vectors:

$$\vec{A}\cdot\vec{B} = (A_x\hat{i} + A_y\hat{j} + A_z\hat{k})\cdot(B_x\hat{i} + B_y\hat{j} + B_z\hat{k})$$

The scalar product of any rectangular unit vector with itself, such as $\hat{i}\cdot\hat{i}$, is equal to 1. (This is because $\hat{i}\cdot\hat{i} = |\hat{i}||\hat{i}|\cos(0) = 1 \times 1 \times \cos(0) = 1$.) Thus, a term like $A_x\hat{i}\cdot B_x\hat{i} = A_xB_x\hat{i}\cdot\hat{i}$ is equal to A_xB_x. Also, because the unit vectors $\hat{i}$, $\hat{j}$, and $\hat{k}$ are mutually perpendicular, the scalar product of one of them

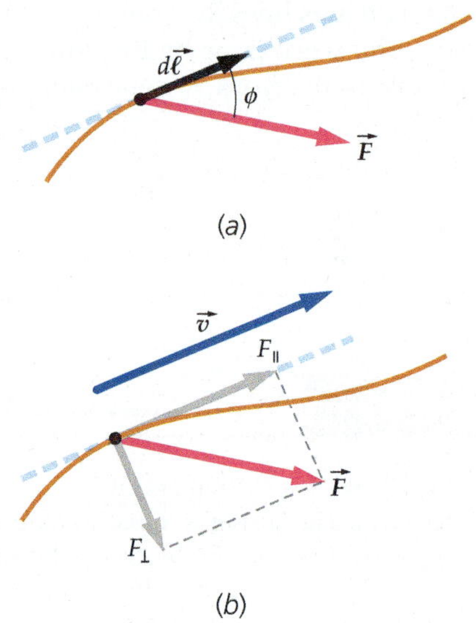

FIGURE 6-16 (*a*) A particle moving along an arbitrary curve in space. (*b*) The perpendicular component of the force $F_{\perp}$ changes the direction of the particle's motion, but not its speed. The tangential, or parallel, component $F_{\|}$ changes the particle's speed, but not its direction of motion. $F_{\|}$ is equal to the mass m times the tangential acceleration dv/dt. The parallel component of the force does work $F_{\|}d\ell$ and the perpendicular component does no work.

Table 6-1 Properties of Scalar Products

If	Then
$\vec{A}$ and $\vec{B}$ are perpendicular,	$\vec{A}\cdot\vec{B} = 0$ (because $\phi = 90°$, $\cos\phi = 0$)
$\vec{A}$ and $\vec{B}$ are parallel,	$\vec{A}\cdot\vec{B} = AB$ (because $\phi = 0°$, $\cos\phi = 1$)
$\vec{A}\cdot\vec{B} = 0$,	Either $\vec{A} = 0$ or $\vec{B} = 0$ or $\vec{A} \perp \vec{B}$
Furthermore,	
$\vec{A}\cdot\vec{A} = A^2$	Because $\vec{A}$ is parallel to itself
$\vec{A}\cdot\vec{B} = \vec{B}\cdot\vec{A}$	Commutative rule of multiplication
$(\vec{A} + \vec{B})\cdot\vec{C} = \vec{A}\cdot\vec{C} + \vec{B}\cdot\vec{C}$	Distributive rule of multiplication

with any other one of them, such as $\hat{i}\cdot\hat{j}$, is zero. (This is because $\hat{i}\cdot\hat{j} = |\hat{i}||\hat{j}|\cos(90°) = 1 \times 1 \times \cos(90°) = 0$.) Thus, any term like $A_x\hat{i}\cdot B_y\hat{j}$ (called a *cross term*) is equal to zero. The result is

$$\vec{A}\cdot\vec{B} = A_xB_x + A_yB_y + A_zB_z \qquad 6\text{-}15$$

The component of a vector in a specific direction can be written as the scalar product of the vector and the unit vector in that direction. For example, the component A_x is found from

$$\vec{A}\cdot\hat{i} = (A_x\hat{i} + A_y\hat{j} + A_z\hat{k})\cdot\hat{i} = A_x \qquad 6\text{-}16$$

This result provides an algebraic procedure for obtaining a component equation, given a vector equation. That is, multiplying both sides of the vector equation $\vec{A} + \vec{B} = \vec{C}$ by $\hat{i}$ gives $(\vec{A} + \vec{B})\cdot\hat{i} = \vec{C}\cdot\hat{i}$, which in turn gives $A_x + B_x = C_x$.

The rule for differentiating a dot product is

$$\frac{d}{dt}(\vec{A}\cdot\vec{B}) = \frac{d\vec{A}}{dt}\cdot\vec{B} + \vec{A}\cdot\frac{d\vec{B}}{dt} \qquad 6\text{-}17$$

This rule is analogous with the rule for differentiating the product of two scalars. The rule for differentiating a dot product can be obtained by differentiating both sides of Equation 6-15.

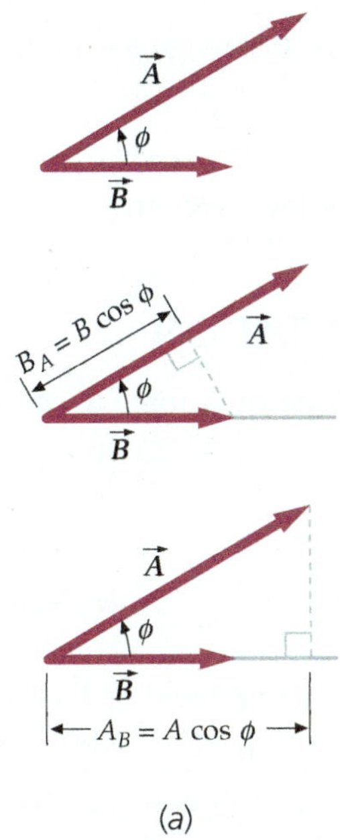

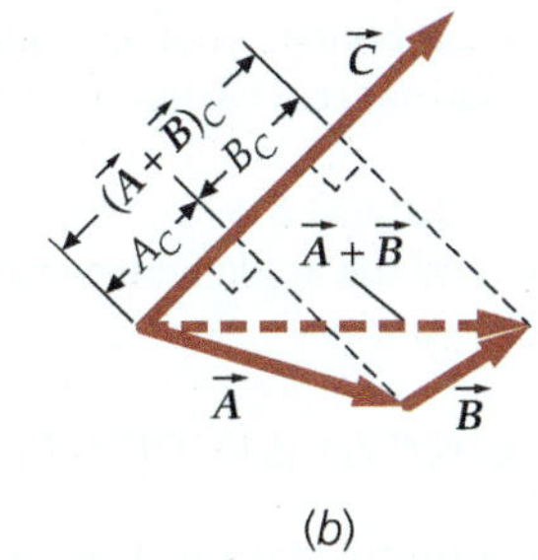

FIGURE 6-17 (*a*) The scalar product $\vec{A}\cdot\vec{B}$ is the product of A and the projection of $\vec{B}$ on $\vec{A}$, or the product of B and the projection of $\vec{A}$ on $\vec{B}$. That is, $\vec{A}\cdot\vec{B} = AB\cos\phi = AB_A = BA_B$. (*b*) $(\vec{A} + \vec{B})\cdot\vec{C}$ equals $(\vec{A} + \vec{B})_C C$ (the projection of $\vec{A} + \vec{B}$ in the direction of $\vec{C}$ times C). However, $(\vec{A} + \vec{B})_C = A_C + B_C$, so $(\vec{A} + \vec{B})\cdot\vec{C} = (A_C + B_C)C = A_C C + B_C C = \vec{A}\cdot\vec{C} + \vec{B}\cdot\vec{C}$. That is, for the scalar product, multiplication is distributive over addition.

Example 6-6 Using the Scalar Product

(*a*) Find the angle between the vectors $\vec{A} = (3.00\hat{i} + 2.00\hat{j})\,\text{m}$ and $\vec{B} = (4.00\hat{i} - 3.00\hat{j})\,\text{m}$ (Figure 6-18). (*b*) Find the component of $\vec{A}$ in the direction of $\vec{B}$.

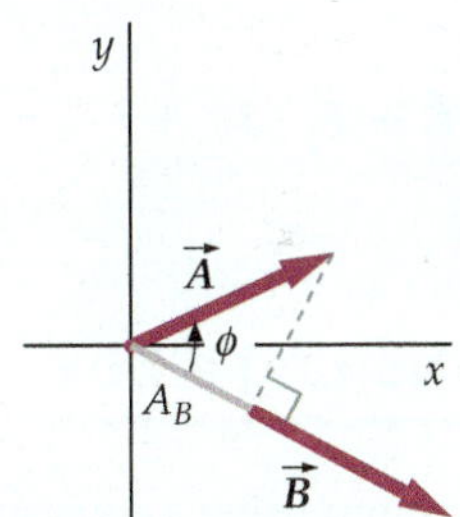

FIGURE 6-18

PICTURE For Part (a), we find the angle ϕ from the definition of the scalar product. Because we are given the components of the vectors, we first determine the scalar product and the values of A and B. Then, we use these values to determine the angle ϕ. For Part (*b*), the component of $\vec{A}$ in the direction of $\vec{B}$ is found from the scalar product $\vec{A}\cdot\hat{B}$, where $\hat{B} = \vec{B}/B$.

SOLVE

(*a*) 1. Write the scalar product of $\vec{A}$ and $\vec{B}$ in terms of A, B, and $\cos\phi$ and solve for $\cos\phi$:

$$\vec{A}\cdot\vec{B} = AB\cos\phi, \text{ so}$$
$$\cos\phi = \frac{\vec{A}\cdot\vec{B}}{AB}$$

2. Find $\vec{A}\cdot\vec{B}$ from the components of $\vec{A}$ and $\vec{B}$:

$$\vec{A}\cdot\vec{B} = A_xB_x + A_yB_y = (3.00\text{ m})(4.00\text{ m}) + (2.00\text{ m})(-3.00\text{ m})$$
$$= 12.0\text{ m}^2 - 6.00\text{ m}^2 = 6.0\text{ m}^2$$

3. The magnitudes of the vectors are obtained from the scalar product of the vector with itself:

$$\vec{A}\cdot\vec{A} = A^2 = A_x^2 + A_y^2 = (3.00\text{ m})^2 + (2.00\text{ m})^2 = 13.0\text{ m}^2$$
$$\text{so } A = \sqrt{13.0}\text{ m}$$
and
$$\vec{B}\cdot\vec{B} = B^2 = B_x^2 + B_y^2 = (4.00\text{ m})^2 + (-3.00\text{ m})^2 = 25.0\text{ m}^2$$
$$\text{so } B = 5.00\text{ m}$$

4. Substitute these values into the equation in step 1 for $\cos\phi$ to find ϕ:

$$\cos\phi = \frac{\vec{A}\cdot\vec{B}}{AB} = \frac{6.0\text{ m}^2}{(\sqrt{13}\text{ m})(5.00\text{ m})} = 0.333$$
$$\phi = \boxed{71°}$$

(*b*) The component of $\vec{A}$ in the direction of $\vec{B}$ is the scalar product of $\vec{A}$ with the unit vector $\hat{B} = \vec{B}/B$:

$$A_B = \vec{A}\cdot\hat{B} = \vec{A}\cdot\frac{\vec{B}}{B} = \frac{\vec{A}\cdot\vec{B}}{B} = \frac{6.0\text{ m}^2}{5.00\text{ m}} = \boxed{1.2\text{ m}}$$

CHECK The component of $\vec{A}$ in the direction of $\vec{B}$ is $A\cos\phi = (\sqrt{13}\ \text{m})\cos 71° = 1.2\ \text{m}$. This answer verifies our Part-(*b*) result.

PRACTICE PROBLEM 6-6 (*a*) Find $\vec{A}\cdot\vec{B}$ for $\vec{A} = (3.0\hat{i} + 4.0\hat{j})\,\text{m}$ and $\vec{B} = (2.0\hat{i} + 8.0\hat{j})\,\text{m}$. (*b*) Find *A*, *B*, and the angle between $\vec{A}$ and $\vec{B}$ for these vectors.

WORK IN SCALAR-PRODUCT NOTATION

In scalar-product notation, the work dW done by a force $\vec{F}$ on a particle over an infinitesimal displacement $d\vec{\ell}$ is

$$dW = F_{\|}\,d\ell = F\cos\phi\,d\ell = \vec{F}\cdot d\vec{\ell} \qquad 6\text{-}18$$

INCREMENTAL WORK

where $d\ell$ is the magnitude of $d\vec{\ell}$ and $F_{\|}$ is the component of $\vec{F}$ in the direction of $d\vec{\ell}$. The work done on the particle as it moves from point 1 to point 2 is

$$W = \int_1^2 \vec{F}\cdot d\vec{\ell} \qquad 6\text{-}19$$

THE DEFINITION OF WORK

(If the force remains constant, the work can be expressed $W = \vec{F}\cdot\vec{\ell}$, where $\vec{\ell}$ is the displacement. In Chapter 3, the displacement is denoted $\Delta\vec{r} = \Delta x\hat{i} + \Delta y\hat{j}$; $\vec{\ell}$ and $\Delta\vec{r}$ are different symbols for the same quantity.)

When several forces $\vec{F}_i$ act on a particle whose displacement is $d\vec{\ell}$, the total work done on it is

$$dW_{\text{total}} = \vec{F}_1\cdot d\vec{\ell} + \vec{F}_2\cdot d\vec{\ell} + \cdots = (\vec{F}_1 + \vec{F}_2 + \cdots)\cdot d\vec{\ell} = (\Sigma\vec{F}_i)\cdot d\vec{\ell} \qquad 6\text{-}20$$

Example 6-7 Pushing a Box

You push a box up a ramp using a constant horizontal 100-N force $\vec{F}$. For each distance of 5.00 m along the ramp, the box gains 3.00 m of height. Find the work done by $\vec{F}$ for each 5.00 m the box moves along the ramp (*a*) by directly computing the scalar product from the components of $\vec{F}$ and $\vec{\ell}$, where $\vec{\ell}$ is the displacement, (*b*) by multiplying the product of the magnitudes of $\vec{F}$ and $\vec{\ell}$ by $\cos\phi$, where ϕ is the angle between the direction of $\vec{F}$ and the direction of $\vec{\ell}$, (*c*) by finding $F_{\|}$ (the component of $\vec{F}$ in the direction of $\vec{\ell}$) and multiplying it by ℓ (the magnitude of $\vec{\ell}$), and (*d*) by finding $\ell_{\|}$ (the component of $\vec{\ell}$ in the direction of $\vec{F}$) and multiplying it by the magnitude of the force.

PICTURE Draw a sketch of the box in its initial and final positions. Place coordinate axes on the sketch with the *x* axis horizontal. Express the force and displacement vectors in component form and take the scalar product. Then find the component of the force in the direction of the displacement, and vice versa.

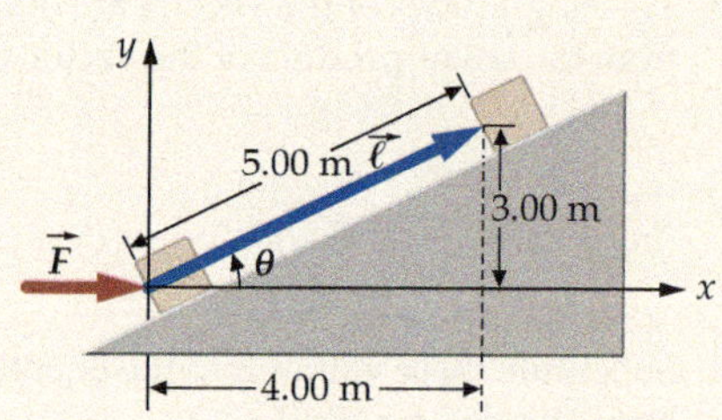

FIGURE 6-19

SOLVE

(*a*) 1. Draw a sketch of the situation (Figure 6-19).

2. Express $\vec{F}$ and $\vec{\ell}$ in component form and take the scalar product:

$$\vec{F} = (100\hat{i} + 0\hat{j})\,\text{N}$$
$$\vec{\ell} = (4.00\hat{i} + 3.00\hat{j})\,\text{m}$$
$$W = \vec{F}\cdot\vec{\ell} = F_x\Delta x + F_y\Delta y = (100\ \text{N})(4.00\ \text{m}) + 0(3.00\ \text{m})$$
$$= \boxed{4.00\times 10^2\ \text{J}}$$

(*b*) Calculate $F\ell\cos\phi$, where ϕ is the angle between the directions of the two vectors as shown. Equate this expression with the Part-(*a*) result and solve for $\cos\phi$. Then solve for the work:

$\vec{F}\cdot\vec{\ell} = F\ell\cos\phi$ and $\vec{F}\cdot\vec{\ell} = F_x\Delta x + F_y\Delta y$

so

$$\cos\phi = \frac{F_x\Delta x + F_y\Delta y}{F\ell} = \frac{(100\text{ N})(4.00\text{ m}) + 0}{(100\text{ N})(5.00\text{ m})} = 0.800$$

and

$$W = F\ell\cos\phi = (100\text{ N})(5.00\text{ m})0.800 = \boxed{4.00\times10^2\text{ J}}$$

(*c*) Find $F_{\|}$ and multiply it by ℓ:

$$F_{\|} = F\cos\phi = (100\text{ N})0.800 = 80.0\text{ N}$$

$$W = F_{\|}\ell = (80.0\text{ N})(5.00\text{ m}) = \boxed{4.00\times10^2\text{ J}}$$

(*d*) Multiply F and $\ell_{\|}$, where $\ell_{\|}$ is the component of $\vec{\ell}$ in the direction of $\vec{F}$:

$$\ell_{\|} = \ell\cos\phi = (5.00\text{ m})0.800 = 4.00\text{ m}$$

$$W = F\ell_{\|} = (100\text{ N})(4.00\text{ m}) = \boxed{4.00\times10^2\text{ J}}$$

CHECK The four distinct calculations give the same result for the work.

TAKING IT FURTHER For this problem, computing the work is easiest using the procedure in Part (*a*). For other problems, the procedure in Part (*b*), Part (*c*), or Part (*d*) may be the easiest. You need to be competent in using all four procedures. (The more problem-solving tools you have at your disposal, the better.)

Example 6-8 A Displaced Particle

Try It Yourself

A particle undergoes a displacement $\vec{\ell} = (2.00\hat{i} - 5.00\hat{j})$ m. During this displacement a constant force $\vec{F} = (3.00\hat{i} + 4.00\hat{j})$ N acts on the particle. Find (*a*) the work done by the force, and (*b*) the component of the force in the direction of the displacement.

PICTURE The force is constant, so the work W can be found by computing $W = \vec{F}\cdot\vec{\ell} = F_x\Delta x + F_y\Delta y$. Combining this with the relation $\vec{F}\cdot\vec{\ell} = F_{\|}\ell$, we can find the component of $\vec{F}$ in the direction of the displacement.

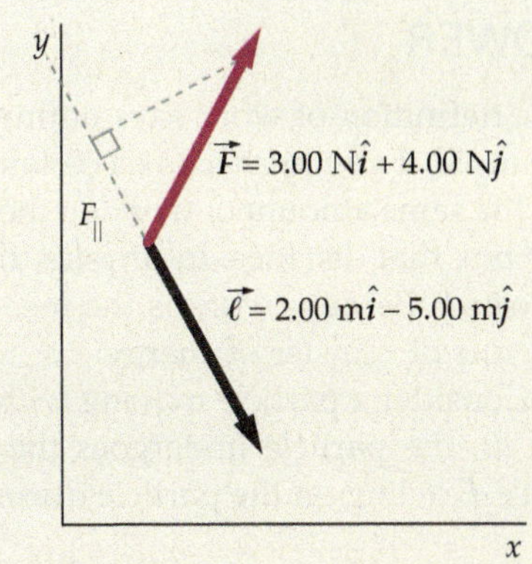

FIGURE 6-20

SOLVE

Cover the column to the right and try these on your own before looking at the answers.

Steps	Answers
(*a*) 1. Make a sketch showing $\vec{F}$, $\vec{\ell}$, and $F_{\|}$ (Figure 6-20).	
2. Compute the work done W.	$W = \vec{F}\cdot\vec{\ell} = \boxed{-14.0\text{ J}}$
(*b*) 1. Compute $\vec{\ell}\cdot\vec{\ell}$ and use your result to find ℓ.	$\vec{\ell}\cdot\vec{\ell} = 29.0\text{ m}^2$, so $\ell = \sqrt{29.0}$ m
2. Using $\vec{F}\cdot\vec{\ell} = F_{\|}\ell$, solve for $F_{\|}$.	$F_{\|} = \vec{F}\cdot\vec{\ell}/\ell = \boxed{-2.60\text{ N}}$

CHECK From Figure 6-20, we see that the angle between $\vec{F}$ and $\vec{\ell}$ is between 90° and 180°, so we expect both $F_{\|}$ and the work to be negative. Our results are in agreement with this expectation.

TAKING IT FURTHER Nowhere in the wording of either the problem statement or the solution of Example 6-8 does it say that the motion of the particle is along any particular path. Because the force is constant, the solution depends on the net displacement $\vec{\ell}$, but not on the path taken. The path could be a straight-line path or a curved path (Figure 6-21) and not a word in the solution would have to be changed.

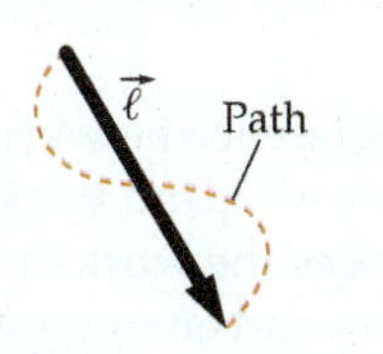

FIGURE 6-21

PRACTICE PROBLEM 6-7 Find the magnitude of $\vec{F}$ and the angle ϕ between $\vec{F}$ and $\vec{\ell}$.

Example 6-9 Differentiating a Scalar Product

Show that $\vec{a} \cdot \vec{v} = \frac{1}{2}d(v^2)/dt$, where $\vec{a}$ the acceleration, $\vec{v}$ the velocity, and v the speed.

PICTURE Note that $v^2 = \vec{v} \cdot \vec{v}$, so the rule for differentiating scalar products can be used here.

SOLVE

Apply the rule for differentiating scalar products (Equation 6-17) to the scalar product $\vec{v} \cdot \vec{v}$:

$$\frac{d}{dt}v^2 = \frac{d}{dt}(\vec{v} \cdot \vec{v}) = \frac{d\vec{v}}{dt} \cdot \vec{v} + \vec{v} \cdot \frac{d\vec{v}}{dt} = 2\frac{d\vec{v}}{dt} \cdot \vec{v} = 2\vec{a} \cdot \vec{v}$$

$$\text{so} \quad \vec{a} \cdot \vec{v} = \frac{1}{2}\frac{d}{dt}v^2$$

CHECK Speed v has dimensions of length over time, so dv^2/dt has dimensions of length squared over time cubed. Acceleration $\vec{a}$ has dimensions of length over time squared, so $\vec{a} \cdot \vec{v}$ has dimensions of length squared over time cubed. Thus, both sides of $\vec{a} \cdot \vec{v} = \frac{1}{2}d(v^2)/dt$ have the same dimensions (length squared over time cubed).

TAKING IT FURTHER This example involves only kinematic parameters, so the resulting relation is a strictly kinematic relation. The equation $\vec{a} \cdot \vec{v} = \frac{1}{2}d(v^2)/dt$ has unrestricted validity (unlike some kinematic equations we have studied that are valid only if the acceleration remains constant).

From Example 6-9 we have the kinematic relation

$$\vec{a} \cdot \vec{v} = \frac{1}{2}\frac{d}{dt}v^2 = \frac{d}{dt}\left(\frac{1}{2}v^2\right) \qquad \text{6-21}$$

In Section 6-4, this equation is used to derive the work–kinetic-energy theorem for particles moving along curved paths under the influence of forces that are not necessarily constant.

POWER

The definition of work says nothing about how long it takes to do the work. For example, if you push a box a certain distance up a hill with a constant velocity, you do the same amount of work on the box regardless of how long it takes you to push the box that distance. In physics, the rate at which a force does work is called the **power** P. Because work is the measure of energy transferred by a force, power is the rate of transfer of energy.

Consider a particle moving with instantaneous velocity $\vec{v}$. In a short time interval dt, the particle undergoes the displacement $d\vec{\ell} = \vec{v}\,dt$. The work done by a force $\vec{F}$ acting on the particle during this time interval is

$$dW = \vec{F} \cdot d\vec{\ell} = \vec{F} \cdot \vec{v}\,dt$$

The power is then

$$P = \frac{dW}{dt} = \vec{F} \cdot \vec{v} \qquad \text{6-22}$$

POWER BY A FORCE

Note the difference between power and work. Two motors that lift a given load a given distance expend the same amount of energy, but the power is greater for the force that does the work in the least time.

Like work and energy, power is a scalar quantity. The SI unit of power, one joule per second, is called a **watt** (W):

$$1\text{ W} = 1\text{ J/s}$$

In the U.S. customary system, the unit of energy is the foot-pound and the unit of power is the foot-pound per second. A commonly used multiple of this unit, called a horsepower (hp), is defined as

$$1\text{ hp} = 550\text{ ft}\cdot\text{lb/s} \approx 746\text{ W}$$

The product of a unit of power and a unit of time is a unit of energy. Electric companies charge for energy, not power, usually by the kilowatt-hour (kW · h). A kilowatt-hour of energy is the energy transferred in 1 hour at the constant rate of 1 kilowatt, or

$$1\text{ kW}\cdot\text{h} = (10^3\text{ W})(3600\text{ s}) = 3.6 \times 10^6\text{ W}\cdot\text{s} = 3.6\text{ MJ}$$

Example 6-10 The Power of a Motor

A small motor is used to operate a lift that raises a load of bricks weighing 500 N to a height of 10 m in 20 s (Figure 6-22) at constant speed. The lift weighs 300 N. What is the power output of the motor?

PICTURE Because the acceleration is zero, the magnitude of the upward force $\vec{F}$ exerted by the motor is equal to the weight of the lift plus the weight of the bricks. The rate the motor does work is the power.

SOLVE

The power is given by $\vec{F}\cdot\vec{v}$:

$$P = \vec{F}\cdot\vec{v} = Fv\cos\phi = Fv\cos(0) = Fv$$

$$= (800\text{ N})\frac{10\text{ m}}{20\text{ s}} = \boxed{4.0 \times 10^2\text{ W}}$$

CHECK The work done by the force is (800 N)(10 m) = 8000 J. This work took 20 s to do, so we expect the power to be the 8000 J/20 s = 4.0×10^2 W. Our result is in perfect agreement with this.

TAKING IT FURTHER (1) The lift could not actually operate at constant speed. The bricks and lift will have to initially be brought up to speed (because they are starting from rest.) The power output will exceed 400 W during this speedup interval. In addition, the power output will be less than 400 W as the lift slows to a stop at the top. The average power output of the motor during the lift is 400 W (and the power provided by the force of gravity is −400 W). (2) A power of 400 W is slightly more than $\frac{1}{2}$ hp.

PRACTICE PROBLEM 6-8 Find the average power output of the motor needed to raise the bricks and lift to a height of 10 m in 40 s. What is the work done by the force of the motor? What is the work done by the force of gravity?

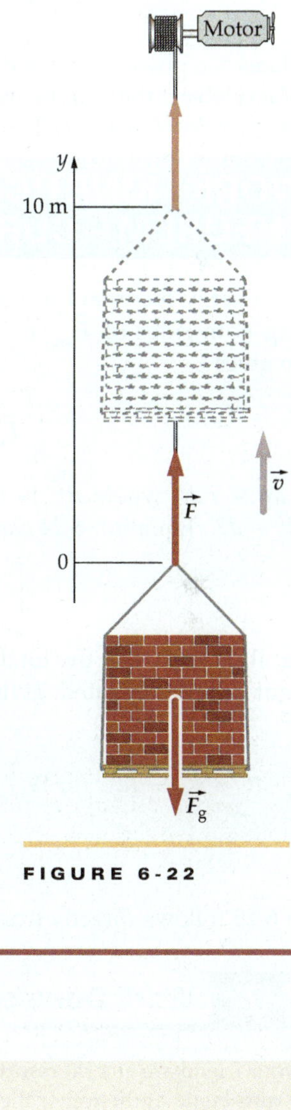

FIGURE 6-22

Example 6-11 Power and Kinetic Energy

Show that the power delivered by the net force acting on a particle equals the rate at which the kinetic energy of the particle is changing.

PICTURE The power delivered by the net force P_{net} equals $\vec{F}_{net}\cdot\vec{v}$. Show that $\vec{F}_{net}\cdot\vec{v} = dK/dt$, where $K = \frac{1}{2}mv^2$.

SOLVE

1. Substitute for $\vec{F}_{net}$ using Newton's second law: $\vec{F}_{net}\cdot\vec{v} = m\vec{a}\cdot\vec{v}$

2. The product $\vec{a}\cdot\vec{v}$ is related to the time derivative of v^2 by $2\vec{a}\cdot\vec{v} = d(v^2)/dt$ (Equation 6–21): $\frac{d}{dt}v^2 = \frac{d}{dt}(\vec{v}\cdot\vec{v}) = 2\vec{a}\cdot\vec{v}$

3. Substitute the step-2 result into the step-1 result: $\vec{F}_{net} \cdot \vec{v} = m\vec{a} \cdot \vec{v} = m\frac{1}{2}\frac{d}{dt}v^2$

4. The mass m is constant, so it and the fraction $\frac{1}{2}$ can be moved inside the argument of the derivative: $\vec{F}_{net} \cdot \vec{v} = \frac{d}{dt}\left(\frac{1}{2}mv^2\right)$

5. The argument of the derivative is the kinetic energy K: $\boxed{P_{net} = \vec{F}_{net} \cdot \vec{v} = \frac{dK}{dt}}$

CHECK The joule is the unit for energy, so dK/dt has units of joules per second, or watts. The watt is the unit for power, so $P_{net} = dK/dt$ is dimensionally consistent.

From Example 6-11 we have

$$P_{net} = \vec{F}_{net} \cdot \vec{v} = \frac{dK}{dt} \qquad 6\text{-}23$$

which relates the power delivered by the net force to the rate of change of kinetic energy of any object that can be modeled as a particle.

6-4 WORK–KINETIC-ENERGY THEOREM—CURVED PATHS

The work–energy theorem for motion along a curved path can be established by integrating both sides of $\vec{F}_{net} \cdot \vec{v} = dK/dt$ (Equation 6-23). Integrating both sides over time gives

$$\int_1^2 \vec{F}_{net} \cdot \vec{v}\,dt = \int_1^2 \frac{dK}{dt}dt \qquad 6\text{-}24$$

Because $d\vec{\ell} = \vec{v}\,dt$, where $d\vec{\ell}$ is the displacement during time dt, and because $(dK/dt)\,dt = dK$, Equation 6-24 can be expressed

$$\int_1^2 \vec{F}_{net} \cdot d\vec{\ell} = \int_1^2 dK$$

The integral on the left is the total work, W_{total}, done on the particle. The integral on the right can be evaluated, giving

$$\int_1^2 \vec{F}_{net} \cdot d\vec{\ell} = K_2 - K_1 \quad (\text{or } W_{total} = \Delta K) \qquad 6\text{-}25$$

WORK–KINETIC-ENERGY THEOREM

Equation 6-25 follows directly from Newton's second law of motion.

Example 6-12 Work Done on a Skier

Context-Rich

You and your friend are at a ski resort with two ski runs, a beginner's run and an expert's run. Both runs begin at the top of the ski lift and end at a finish line at the bottom of the same lift. Let h be the vertical descent for both runs. The beginner's run is longer and less steep than the expert's run. You and your friend, who is a much better skier than you, are testing some experimental frictionless skis. To make things interesting, you offer a wager that if she takes the expert's run and you take the beginner's run, her speed at the finish line will not be greater than your speed at the finish line. Forgetting that you are studying physics, she accepts the bet. The conditions are that you both start from rest at the top of the lift and both of you coast for the entire trip. Who wins the bet? (Assume air drag is negligible.)

(PhotoDisc/Getty.)

PICTURE Because you and your friend are coasting on the skis, you both can be modeled as particles. (The work–kinetic-energy theorem works only for particles.) Two forces act on each of you, a weight force and a normal force.

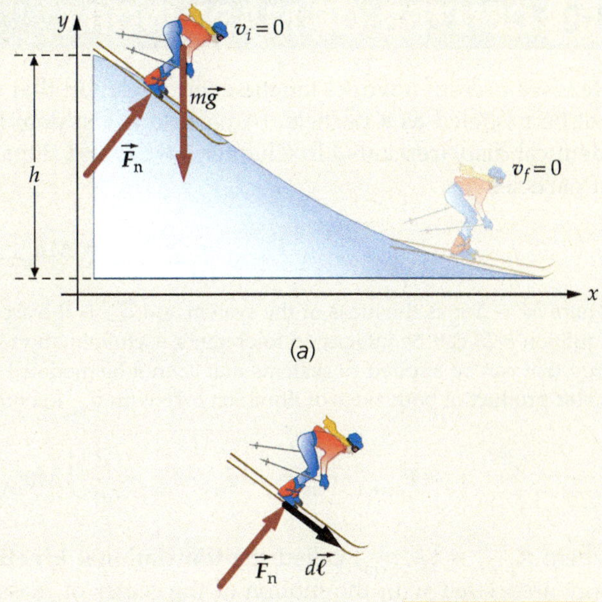

FIGURE 6-23

SOLVE

1. Make a sketch of yourself and draw the two force vectors on the sketch (Figure 6-23*a*). Also include coordinate axes. The work–kinetic-energy theorem, with $v_1 = 0$, relates the final speed v_f to the total work.
2. The final speed is related to the final kinetic energy, which in turn is related to the total work by the work–kinetic-energy theorem: $W_{total} = \frac{1}{2}mv_f^2 - \frac{1}{2}mv_i^2$
3. For each of you, the total work is the work done by the normal force plus the work done by the gravitational force: $W_{total} = W_n + W_g$
4. The force $m\vec{g}$ on you is constant, but the force $\vec{F}_n$ is not constant. First we calculate the work done by $\vec{F}_n$. Calculate the work dW_n done on you by $\vec{F}_n$ for an infinitesimal displacement $d\vec{\ell}$ (Figure 6-23*b*) at an arbitrary location along the run: $dW_n = \vec{F}_n \cdot d\vec{\ell} = F_n \cos\phi\, d\ell$
5. Find the angle ϕ between the directions of $\vec{F}_n$ and $d\vec{\ell}$. The displacement $d\vec{\ell}$ is tangent to the slope: $\phi = 90°$
6. Calculate the work done by $\vec{F}_n$ for the entire run: $W_n = \int F_n \cos 90°\, d\ell = \int (0)\, d\ell = 0$
7. The force of gravity $\vec{F}_g$ is constant, so the work done by gravity is $W_g = \vec{F}_g \cdot \vec{\ell}$, where $\vec{\ell}$ (Figure 6-24) is the net displacement from the top to the bottom of the lift: $W_g = m\vec{g} \cdot \vec{\ell} = -mg\hat{j} \cdot (\Delta x\hat{i} + \Delta y\hat{j}) = -mg\,\Delta y$
8. The skier is descending the hill, so Δy is negative. From Figure 6-23*a*, we see that $\Delta y = -h$: $\Delta y = -h$
9. Substituting gives: $W_g = mgh$
10. Apply the work–kinetic-energy theorem to find v_f: $W_n + W_g = \Delta K$
11. The final speed depends only on h, which is the same for both runs. Both of you will have the same final speeds. $0 + mgh = \frac{1}{2}mv_f^2 - 0$ so $v_f = \sqrt{2gh}$

YOU WIN! (The bet was that she would not be going faster than you.)

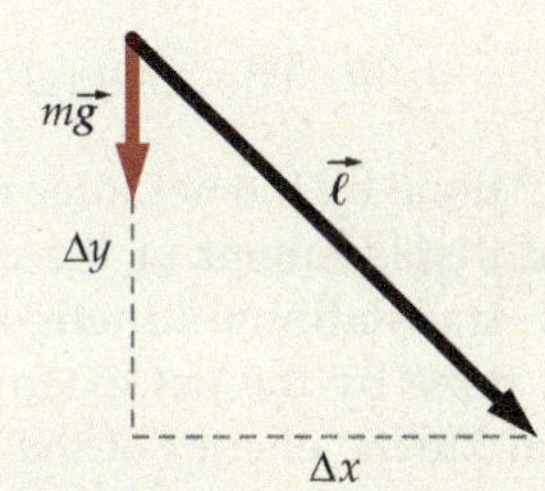

FIGURE 6-24

CHECK The force driving your motion is the gravitational force. This force is proportional to the mass, so the work done by it is proportional to the mass. Because the kinetic energy is also proportional to the mass, the mass cancels out of the work–kinetic-energy equation. Thus, we expect the final speed to be independent of mass. Our result is independent of mass as expected.

TAKING IT FURTHER Your friend on the steeper trail will cross the finish line in less time, but that was not the bet. What was shown here is that the work done by the gravitational force equals mgh. It does not depend upon the shape of the hill or upon the length of the path taken. It depends only upon the mass m and the vertical drop h between the starting point and the finishing point.

* 6-5 CENTER-OF-MASS WORK

Here we present a work–kinetic-energy relation that works for systems that cannot be modeled as a particle. (A particle is a system for which all parts undergo identical displacements.) In Chapter 5 we found (Equation 5-23) that for a system of particles

$$\vec{F}_{\text{net ext}} = \sum \vec{F}_{i\text{ext}} = M\vec{a}_{\text{cm}} \qquad 6\text{-}26$$

where $M = \Sigma m_i$ is the mass of the system and $\vec{a}_{\text{cm}}$ is the acceleration of the center of mass. Equation 6-26 can be integrated to obtain a useful equation involving work and kinetic energy that can be applied to systems that cannot be modeled as a particle. First, we take the scalar product of both sides of Equation 6-26 with $\vec{v}_{\text{cm}}$ to obtain

$$\vec{F}_{\text{net ext}} \cdot \vec{v}_{\text{cm}} = M\vec{a}_{\text{cm}} \cdot \vec{v}_{\text{cm}} = \frac{d}{dt}\left(\tfrac{1}{2}Mv_{\text{cm}}^2\right) = \frac{dK_{\text{trans}}}{dt} \qquad 6\text{-}27$$

where $K_{\text{trans}} = \frac{1}{2}Mv_{\text{cm}}^2$, called the **translational kinetic energy**, is the kinetic energy associated with the motion of the center of mass. Multiplying both sides of Equation 6-27 by dt and then integrating gives

$$\int_1^2 \vec{F}_{\text{net ext}} \cdot d\vec{\ell}_{\text{cm}} = \Delta K_{\text{trans}} \qquad 6\text{-}28$$

CENTER-OF-MASS WORK–TRANSLATIONAL-KINETIC-ENERGY RELATION

where $d\vec{\ell}_{\text{cm}} = \vec{v}_{\text{cm}}dt$. The integral $\int_1^2 \vec{F}_{\text{net ext}} \cdot d\vec{\ell}_{\text{cm}}$ is referred to as the **center-of-mass work*** done by the net force on a system of particles, and $d\vec{\ell}_{\text{cm}} = \vec{v}_{\text{cm}}dt$ is the incremental displacement of the center of mass. Equation 6-28 is the **center-of-mass work–translational-kinetic-energy relation.** Stated in words "The center-of-mass work done by the net external force on a system equals the change in the translational kinetic energy of the system. Although Equation 6-28 looks like the equation for the work–kinetic-energy theorem (Equation 6-25), there are some important differences. The center-of-mass work–translational-kinetic-energy relation deals only with the displacement and speed of the center of mass of the system, so when using this relation, we ignore the motion of any part of the system relative to the center-of-mass reference frame. (A center-of-mass reference frame is a nonrotating reference frame† that moves with the center of mass.) This allows us to calculate the bulk motion of the system without knowing all the internal details of the system.

For a system that moves as a particle (with all parts having the same velocity) the center-of-mass work–translational-kinetic-energy relation reduces to the work–kinetic-energy theorem (Equation 6-25).

It is also sometimes useful to refer to the center-of-mass work done by a single force. The center-of-mass work W_{cm} done by any particular force $\vec{F}$ is given by

$$W_{\text{cm}} = \int_1^2 \vec{F} \cdot d\vec{\ell}_{\text{cm}} \qquad 6\text{-}29$$

* Center-of-mass work is also called pseudowork.

† A nonrotating reference frame is a frame that is not rotating relative to an inertial reference frame.

Example 6-13 Two Pucks and a String

Two identical pucks on an air table are connected by a length of string (see Figure 6-25). The pucks, each of mass m, are initially at rest in the configuration shown. A constant force of magnitude F accelerates the system toward the right. After the point of application P of the force has moved a distance d, the pucks collide and stick together. What is the speed of the pucks immediately following the collision?

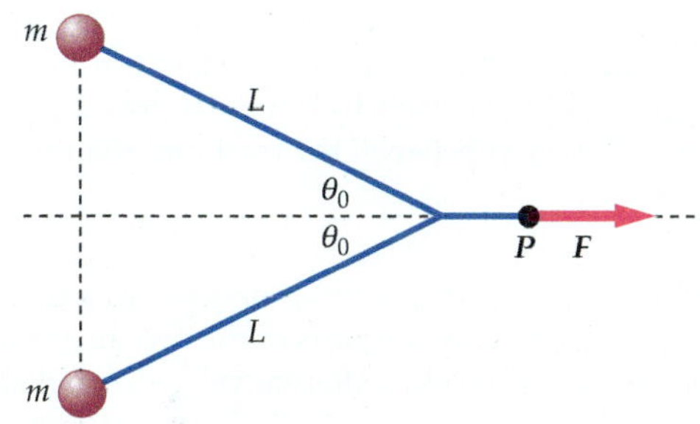

FIGURE 6-25

PICTURE Let the system be the two pucks and the string. Apply the center-of-mass work–kinetic-energy relation to the system. Following the collision the speed of each puck equals the speed of the center of mass. (The pucks can move without friction on the air table.)

SOLVE

1. Make a drawing showing the system initially, and after it has moved distance d (Figure 6-26):

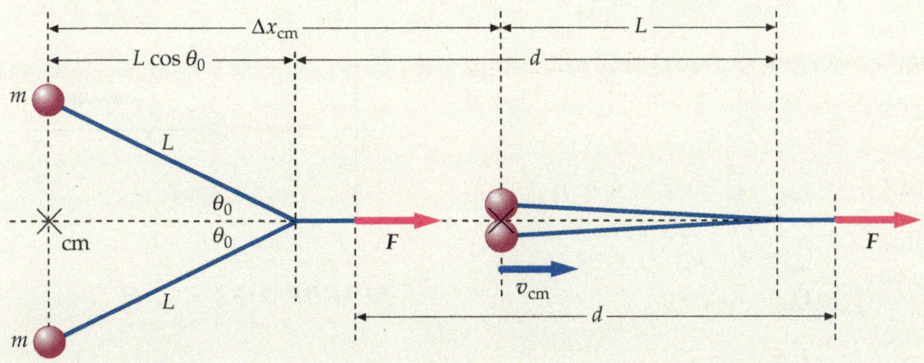

FIGURE 6-26 As the center of mass moves the distance Δx_{cm}, the point of application of the force $\vec{F}$ moves the distance d.

2. Apply the center-of-mass-work–translational-kinetic-energy relation to the system. The net force on the system is $\vec{F} = F\hat{i}$:

$$\int_i^f \vec{F}_{net\,ext} \cdot d\vec{\ell}_{cm} = \Delta K_{trans}$$

$$\int_i^f F\hat{i} \cdot dx_{cm}\hat{i} = K_{trans\,f} - K_{trans\,i}$$

$$F\int_i^f dx_{cm} = K_{trans\,f} - 0$$

$$F\Delta x_{cm} = \tfrac{1}{2}(2m)v_{cm}^2 = mv_{cm}^2$$

3. Find Δx_{cm} in terms of d and L. Figure 6-26 makes the calculation of Δx_{cm} fairly straightforward:

$$\Delta x_{cm} + L = L\cos\theta_0 + d$$

$$\text{so}\quad \Delta x_{cm} = d - L(1 - \cos\theta_0)$$

4. Substitute the step-3 result into the step-2 result and solve for v_{cm}:

$$F\Delta x_{cm} = mv_{cm}^2$$

$$F[d - L(1 - \cos\theta_0)] = mv_{cm}^2$$

$$\text{so}\quad v_{cm} = \boxed{\sqrt{\frac{F[d - L(1 - \cos\theta_0)]}{m}}}$$

CHECK If the initial angle θ_0 is zero, the system can be modeled as a particle and the work–kinetic-energy theorem can be used. This would give $Fd = \frac{1}{2}(2m)v^2 = mv^2$, or $v = \sqrt{Fd/m}$. Our step-4 result gives the very same expression for the speed if $\theta_0 = 0$.

TAKING IT FURTHER (1) In this example, the displacement of the center of mass Δx_{cm} is less than the displacement d of the point of application of the force $\vec{F}$. As a result, the center-of-mass work done by the force is less than the work Fd done by the force. (2) The pucks lose kinetic energy when they collide and stick together. This energy appears as some other form of energy, such as thermal energy. The conservation of energy is discussed further in Chapter 7.

Example 6-14 Stopping Distance

To avoid an accident, the driver of a 1000-kg car moving at 90 km/h on a straight horizontal road steps on the brakes with maximum force. The antilock braking system (ABS) is not working, so the wheels lock and the tires skid as the car comes to a stop. The kinetic coefficient of friction between the road and the tires is 0.80. How far does the car travel during the skid?

PICTURE The car cannot be modeled as a particle. The points of application of the kinetic frictional forces are the parts of the tires in contact with the road surface. The high points of the contacting surfaces alternatively stick and slip. Therefore, the car cannot be modeled as a particle during the skid. The center-of-mass work–translational-kinetic-energy relation applied to the car enables us to calculate the stopping distance.

SOLVE

1. Write the center-of-mass work–translational-kinetic-energy relation. We need to solve for the displacement of the center of mass of the car:

$$\int_1^2 \vec{F}_{\text{net ext}} \cdot d\vec{\ell}_{\text{cm}} = \Delta K_{\text{trans}}$$

2. Draw a free-body diagram of the car during the skid (Figure 6-27):

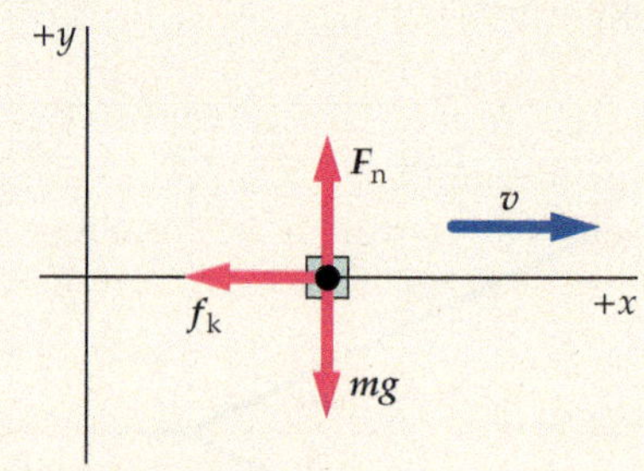

FIGURE 6-27

3. The vertical acceleration is zero, so the normal force and the gravitational force sum to zero. The net external force on the car is equal to the frictional force. Solve for the net force on the car:

$$\vec{F}_{\text{net}} = \vec{F}_{\text{n}} + m\vec{g} + \vec{f}_{\text{k}} = \vec{f}_{\text{k}}$$

so

$$F_{\text{net}} = f_{\text{k}} = \mu_{\text{k}} F_{\text{n}} = \mu_{\text{k}} mg$$

$$\vec{F}_{\text{net}} = -\mu_{\text{k}} mg\hat{i}$$

4. Apply the center-of-mass work–translational-kinetic-energy relation to the car:

$$\int_1^2 \vec{F}_{\text{net}} \cdot d\vec{\ell}_{\text{cm}} = \Delta K_{\text{trans}}$$

$$\int_1^2 -\mu_{\text{s}} mg\hat{i} \cdot dx_{\text{cm}}\hat{i} = K_{\text{trans}\,2} - K_{\text{trans}\,1}$$

$$-\mu_{\text{s}} mg \int_1^2 dx_{\text{cm}} = 0 - K_{\text{trans}\,1}$$

$$-\mu_{\text{s}} mg(x_{\text{cm}\,2} - x_{\text{cm}\,1}) = -\tfrac{1}{2}mv_{\text{cm}\,1}^2$$

5. Solve for the displacement, but first convert the initial speed from km/h to m/s:

$$v_{\text{cm}\,1} = 90\ \text{km/h} \cdot \frac{1\ \text{h}}{(3.6\ \text{ks})} = 25\ \text{m/s}$$

so

$$\Delta x_{\text{cm}} = x_{\text{cm}\,2} - x_{\text{cm}\,1} = \frac{v_{\text{cm}\,1}^2}{2\mu_{\text{k}} g}$$

$$\Delta x_{\text{cm}} = \frac{(25\ \text{m/s})^2}{2 \cdot (0.80)(9.81\ \text{m/s}^2)} = \boxed{40\ \text{m}}$$

CHECK We would expect the stopping distance to increase with initial speed, and to decrease with increasing coefficient of friction. The step-5 expression for Δx_{cm} meets these expectations.

TAKING IT FURTHER The translational kinetic energy of the car is dissipated as thermal energy of the tires and of the pavement. The dissipation of kinetic energy into thermal energy by kinetic friction is discussed further in Chapter 7.

Physics Spotlight

Coasters and Baggage and Work (Oh My!)

Baggage transfer methods at some major airports have a lot in common with roller coasters. High rates of change of acceleration for long periods of time are bad for both coaster passengers and baggage items. Both must move swiftly without unwanted jerking and halting.

Some roller coaster cars (and some baggage carriers) gain kinetic energy because of the work done on them by constant forces exerted on them by banks of *linear induction motors* (LIMs). A LIM is an electromagnetic method of providing force without moving parts.* The main reason for the use of linear induction motors is the flexibility of applying force at calculated locations during the travel of the coaster train or the baggage buggy. The roller coaster and the baggage buggies run on tracks that use sensors to determine the speed of the vehicles, and communicate that speed to the controllers for the motors. The LIMs can be turned off when the vehicle has reached the right speed. In both cases, some LIMs are also wired to act as brakes on the vehicles, exerting forces on them in opposition to their direction of travel.

The speeds of the buggies transporting luggage between terminals at Heathrow International Airport are controlled by LIMs. *(Vanderlande Industries.)*

Speed—The Ride is a roller coaster launched from the NASCAR Café in the Sahara Hotel and Casino in Las Vegas. The design firm, Ingenieurbuero Stengel GmbH, specified 88 motors in three locations along the track. The first bank of motors launches the coaster train. The 6-car, 24-passenger coaster train is smoothly accelerated to 45 mi/h in 2.0 s. It swoops around a corner and plunges 25 ft underground before rising and going through a clothoid loop-the-loop.† After it goes through the loop-the-loop, forces exerted on it by the second bank of LIMs quadruple its kinetic energy in 2.0 s.‡ The roller coaster glides along Las Vegas Boulevard, and races two hundred feet up a near-vertical incline. For safety's sake, a series of LIMs near the top of the incline can slow the train, if necessary. The coaster train then runs backward through the entire coaster route. As it returns to the station, the LIMs in the station act as brakes, and bring the train to a stop.

Other than the forces from the LIMs, the forces acting on the coaster train are gravity, friction, and the normal force. Each of the cars in the coaster train travels over the same path, although the starting and ending points for each car are not the same. The maximum acceleration of any passenger is $3.5g$. This is not excessive—the momentary acceleration caused by being hit on the head with a pillow can go above $20g$.§

Heathrow International Airport often transfers luggage between Terminals One and Four. The terminals are more than 1.0 km apart and are separated by a runway. Each piece of luggage is loaded onto a small buggy that rides on rails. (The speeds of the buggies are controlled by LIMs mounted on the tracks.) The buggy goes down a steep incline to reach the level of a tunnel, 20 m underground. It travels through the tunnel at 30 km/h, and is kept at that speed by regularly spaced LIMs. At the end of the tunnel, the buggy climbs up into the appropriate floor of the other terminal. When you transfer between flights at a large airport, remember that your luggage may well be going on its own special ride.

* "Whoa! Linear motors blast Vegas coaster straight up." *Machine Design*, May 4, 2000. Vol. 28; "Sectors" EI-WHS http://www.eiwhs.co.uk/sectors.asp April 2006; "Baggage Handling Case Study." Force Engineering http://www.force.co.uk/bagcase.htm, April 2006; "Leisure Rides." Force Engineering, http://www.force.co.uk/leishome.htm April 2006.

† "Roller coaster constructor Werner Stengel receives honorary doctorate at Göteborg University." Göteborg University Faculty of Science. http://www2.science.gu.se/english/werner_stengel.shtml April 2006.

‡ "Speed Facts." Sahara Hotel and Casino, http://www.saharavegas.com/thrills/facts.html April 2006.

§ Exponent Failure Analysis Associates. *Investigation of Amusement Park and Roller Coaster Injury Likelihood and Severity:* 48. http://www.emerson-associates.com/safety/articles/ExponentReport.pdf April 2006.

SUMMARY

1. Work, kinetic energy, and power are important derived dynamic quantities.
2. The work–kinetic-energy theorem is an important relation derived from Newton's laws applied to a particle. (In this context, a particle is a perfectly rigid object that moves without rotating.)
3. The scalar product of vectors is a mathematical definition that is useful throughout physics.

TOPIC	RELEVANT EQUATIONS AND REMARKS	
1. Work	$W = \int_1^2 \vec{F} \cdot d\vec{\ell} = \int_1^2 F_{\parallel} d\ell$ (definition)	
Constant force	$W = \vec{F} \cdot \vec{\ell} = F_{\parallel}\ell = F\ell_{\parallel} = F\ell\cos\theta$	
Constant force—straight-line motion	$W = F_x \Delta x = F\lvert\Delta x\rvert\cos\theta$	
Variable force—straight-line motion	$W = \int_{x_1}^{x_2} F_x dx =$ area under the F_x-versus-x curve	
2. Kinetic Energy	$K = \frac{1}{2}mv^2$ (definition)	
3. Work–Kinetic-Energy Theorem	$W_{\text{total}} = \Delta K = \frac{1}{2}mv_f^2 - \frac{1}{2}mv_i^2$	
4. Scalar or Dot Product	$\vec{A} \cdot \vec{B} = AB\cos\phi$ (definition)	
In terms of components	$\vec{A} \cdot \vec{B} = A_xB_x + A_yB_y + A_zB_z$	
Unit vector times vector	$\vec{A} \cdot \hat{i} = A_x$	
Derivative product rule	$\frac{d}{dt}(\vec{A} \cdot \vec{B}) = \frac{d\vec{A}}{dt} \cdot \vec{B} + \vec{A} \cdot \frac{d\vec{B}}{dt}$	
5. Power	$P = \frac{dW}{dt} = \vec{F} \cdot \vec{v}$	
6. Center-of-Mass Work–Translational-Kinetic-Energy Relation	$\int_1^2 \vec{F}_{\text{net ext}} \cdot d\vec{\ell}_{\text{cm}} = \Delta K_{\text{trans}}$ This relation is a useful problem-solving tool if for systems that cannot be modeled as a particle.	6-2
Center-of-Mass Work	$W_{\text{cm}} = \int_1^2 \vec{F} \cdot d\vec{\ell}_{\text{cm}}$	6-3
Translational Kinetic Energy	$K_{\text{trans}} = \frac{1}{2}Mv_{\text{cm}}^2$, where $M = \Sigma m_i$	

Answers to Concept Checks

6-1 The work being done by the force is negative.

Answers to Practice Problems

6-1 34 J

6-2 1.7×10^2 N

6-3 4.1 m/s

6-4 The region of interest is below the x axis, so the "area under the curve" is negative. The "area under the curve" is $-(\lvert A_1\rvert - \lvert A_2\rvert)$, where A_1 and A_2 are shown in Figure 6-28. The work done by the spring is equal to

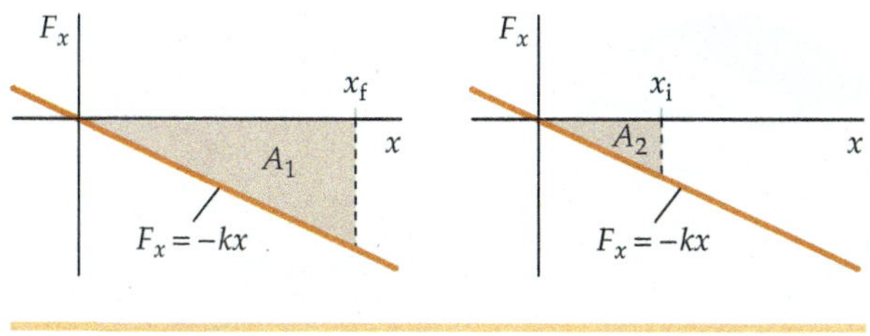

FIGURE 6-28

the "area under the curve," and the area of a triangle is one-half the altitude times the base. Thus,

$$W_{\text{by spring}} = -(|A_1| - |A_2|) = -(\tfrac{1}{2}\cdot kx_f \cdot x_f - \tfrac{1}{2}\cdot kx_i \cdot x_i) = \boxed{\tfrac{1}{2}kx_i^2 - \tfrac{1}{2}kx_f^2}$$

which is identical to Equation 6-13.

6-5 0.40 m/s

6-6 (*a*) 38 m^2, (*b*) $A = 5.0$ m, $B = 8.2$ m, $\phi = 23°$

6-7 $F = 5.00$ N, $\phi = 121°$

6-8 $P = 2.0 \times 10^2$ W, $W = 8.0 \times 10^3$ J, $W = -8.0 \times 10^3$ J

PROBLEMS

In a few problems, you are given more data than you actually need; in a few other problems, you are required to supply data from your general knowledge, outside sources, or informed estimates.

Interpret as significant all digits in numerical values that have trailing zeros and no decimal points.

For all problems, use $g = 9.81$ m/s² for the free-fall acceleration due to gravity and neglect friction and air resistance unless instructed to do otherwise.

- • Single-concept, single-step, relatively easy
- •• Intermediate-level, may require synthesis of concepts
- ••• Challenging
- SSM Solution is in the *Student Solutions Manual*
- Consecutive problems that are shaded are paired problems.

CONCEPTUAL PROBLEMS

1 • True or false: (*a*) If the net or total work done on a particle was not zero, then its speed must have changed. (*b*) If the net or total work done on a particle was not zero, then its velocity must have changed. (*c*) If the net or total work done on a particle was not zero, then its direction of motion could not have changed. (*d*) No work is done by the forces acting on a particle if it remains at rest. (*e*) A force that is always perpendicular to the velocity of a particle never does work on the particle.

2 • You push a heavy box in a straight line along the top of a rough horizontal table. The box starts at rest and ends at rest. Describe the work done on it (including sign) by each force acting on it and the net work done on it.

3 • You are riding on a Ferris wheel that is rotating at constant speed. True or false: During any fraction of a revolution: (*a*) None of the forces acting on you does work on you. (*b*) The total work done by all forces acting on you is zero. (*c*) There is zero net force on you. (*d*) You are accelerating.

4 • By what factor does the kinetic energy of a particle change if its speed is doubled but its mass is cut in half?

5 • Give an example of a particle that has constant kinetic energy but is accelerating. Can a non-accelerating particle have a changing kinetic energy? If so, give an example.

6 • An particle initially has kinetic energy *K*. Later it is found to be moving in the opposite direction with three times its initial speed. What is the kinetic energy now? (*a*) *K*, (*b*) 3*K*, (*c*) 23*K*, (*d*) 9*K*, (*e*) −9*K*

7 • How does the work required to stretch a spring 2.0 cm from its unstressed length compare with the work required to stretch it 1.0 cm from its unstressed length? SSM

8 • A spring is first stretched 2.0 cm from its unstressed length. It is then stretched an additional 2.0 cm. How does the work required for the second stretch compare to the work required for the first stretch (give a ratio of second to first)?

9 • The dimension of power is (*a*) $M\cdot L^2\cdot T^2$, (*b*) $M\cdot L^2/T$, (*c*) $M\cdot L^2/T^2$, (*d*) $M\cdot L^2/T^3$.

10 • Show that the SI units of the force constant of a spring can be written as kg/s^2.

11 • True or false: (*a*) The gravitational force cannot do work on an object, because it is not a contact force. (*b*) Static friction can never do work on an object. (*c*) As a negatively charged electron is removed from a positively charged nucleus, the force on the electron does work that as a positive value. (*d*) If a particle is moving along a circular path, the total work being done on it is necessarily zero.

12 •• A hockey puck has an initial velocity in the $+x$ direction on a horizontal sheet of ice. Qualitatively sketch the force-versus-position graph for the (constant) horizontal force that would need to act on the puck to bring it to rest. Assume that the puck is located at $x = 0$ when the force begins to act. Show that the sign of the area under the curve agrees with the sign of the change in the puck's kinetic energy and interpret this in terms of the work–kinetic-energy theorem.

13 •• True or false: (*a*) The scalar product cannot have units. (*b*) If the scalar product of two nonzero vectors is zero, then they are parallel. (*c*) If the scalar product of two nonzero vectors is equal to the product of their magnitudes, then the two vectors are parallel. (*d*) As an object slides up an incline, the sign of the scalar product of the force of gravity on it and its displacement is negative. SSM

14 •• (*a*) Must the scalar product of two perpendicular unit vectors always be zero? If not, give an example. (*b*) An object has a velocity $\vec{v}$ at some instant. Interpret $\sqrt{\vec{v}\cdot\vec{v}}$ physically. (*c*) A ball rolls off a horizontal table. What is the scalar product between its velocity and its acceleration the instant after it leaves the table? Explain. (*d*) In Part (*c*), what is the sign of the scalar product of its velocity and acceleration the instant before it impacts the floor?

15 •• You lift a package vertically upward a distance L in time Δt. You then lift a second package that has twice the mass of the first package vertically upward the same distance while providing the same power as required for the first package. How much time does lifting the second package take (answer in terms of Δt)?

16 •• There are lasers that output more than 1.0 GW of power. A typical large modern electric generation plant typically outputs 1.0 GW of electrical power. Does this mean the laser outputs a huge amount of energy? Explain. *Hint: These high-power lasers are pulsed on and off, so they are not outputting power for very long time intervals.*

17 •• You are driving a car that accelerates from rest on a level road without spinning its wheels. Use the center-of-mass work–translational-kinetic-energy relation and free-body diagrams to clearly explain which force (or forces) is (are) directly responsible for the gain in translational kinetic energy of both you and the car. *Hint: The relation refers to external forces only, so the car's engine is not the answer. Pick your "system" correctly for each case.* SSM

ESTIMATION AND APPROXIMATION

18 •• (*a*) Estimate the work done on you by gravity as you take an elevator from the ground floor to the top of the Empire State Building, a building 102 stories high. (*b*) Estimate the amount of work the normal force of the floor did on you. *Hint: The answer is not zero.* (*c*) Estimate the average power of the force of gravity.

19 •• ENGINEERING APPLICATION, CONTEXT-RICH The nearest stars, apart from the Sun, are light-years away from Earth. If we are to investigate these stars, our space ships will have to travel at an appreciable fraction of the speed of light. (*a*) You are in charge of estimating the energy required to accelerate a 10,000-kg capsule from rest to 10 percent of the speed of light in one year. What is the minimum amount of energy that is required? Note that at velocities approaching the speed of light, the kinetic energy formula $\frac{1}{2}mv^2$ is not correct. However, it gives a value that is within 1% of the correct value for speeds up to 10% of the speed of light. (*b*) Compare your estimate to the amount of energy that the United States uses in a year (about 5×10^{20} J). (*c*) Estimate the minimum average power required of the propulsion system.

20 •• The mass of the Space Shuttle orbiter is about 8×10^4 kg and the period of its orbit is 90 min. Estimate the kinetic energy of the orbiter and the work done on it by gravity between launch and orbit. (Although the force of gravity decreases with altitude, this effect is small in low-Earth orbit. Use this fact to make the necessary approximation; you do not need to do an integral.) The orbits are about 250 miles above the surface of Earth.

21 • CONTEXT-RICH Ten inches of snow have fallen during the night, and you must shovel out your 50-ft-long driveway (Figure 6-29). Estimate how much work you do on the snow by completing this task. Make a plausible guess of any value(s) needed (the width of the driveway, for example), and state the basis for each guess.

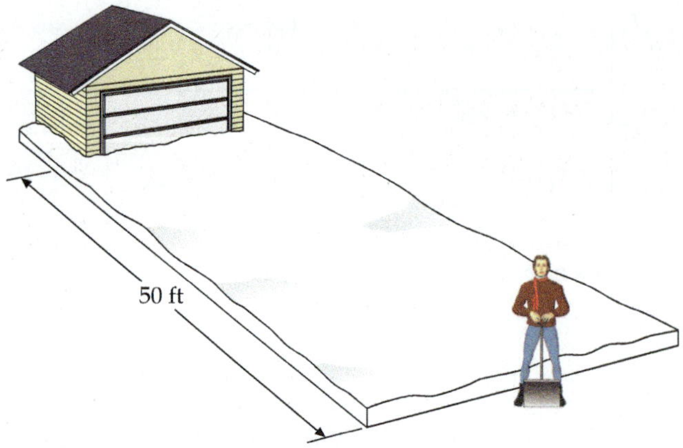

FIGURE 6-29 Problem 21

WORK, KINETIC ENERGY, AND APPLICATIONS

22 • A 15-g piece of space junk has a speed of 1.2 km/s. (*a*) What is its kinetic energy? (*b*) What is its kinetic energy if its speed is halved? (*c*) What is its kinetic energy if its speed is doubled?

23 • Find the kinetic energy of (*a*) a 0.145-kg baseball moving with a speed of 45.0 m/s, and (*b*) a 60.0-kg jogger running at a steady pace of 9.00 min/mi.

24 • A 6.0-kg box is raised a distance of 3.0 m from rest by a vertical applied force of 80 N. Find (*a*) the work done on the box by the applied force, (*b*) the work done on the box by gravity, and (*c*) the final kinetic energy of the box.

25 • A constant 80-N force acts on a 5.0 kg box. The box initially is moving at 20 m/s in the direction of the force, and 3.0 s later the box is moving at 68 m/s. Determine both the work done by this force and the average power delivered by the force during the 3.0-s interval.

26 •• You run a race with a friend. At first you each have the same kinetic energy, but she is running faster than you are. When you increase your speed by 25 percent, you are running at the same speed she is. If your mass is 85 kg, what is her mass?

27 •• A 3.0-kg particle moving along the x axis has a velocity of +2.0 m/s as it passes through the origin. It is subjected to a single force, F_x, that varies with position, as shown in Figure 6-30. (*a*) What is the kinetic energy of the particle as it passes through the origin? (*b*) How much work is done by the force as the particle moves from $x = 0.0$ m to $x = 4.0$ m? (*c*) What is the speed of the particle when it is at $x = 4.0$ m? SSM

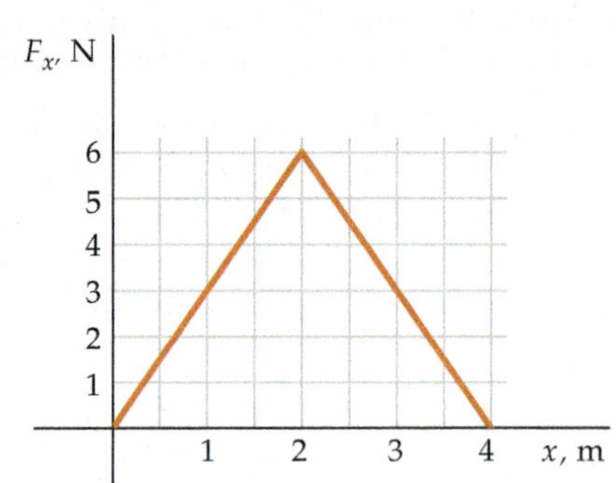

FIGURE 6-30 Problem 27

28 •• A 3.0-kg object moving along the x axis has a velocity of 2.4 m/s as it passes through the origin. It is acted on by a single force, F_x, that varies with x, as shown in Figure 6-31. (*a*) Find the work done by the force from $x = 0.0$ m to $x = 2.0$ m. (*b*) What is the kinetic energy of the object at $x = 2.0$ m? (*c*) What is the speed of the object at $x = 2.0$ m? (*d*) What is the work done on the object from $x = 0.0$ to $x = 4.0$ m? (*e*) What is the speed of the object at $x = 4.0$ m?

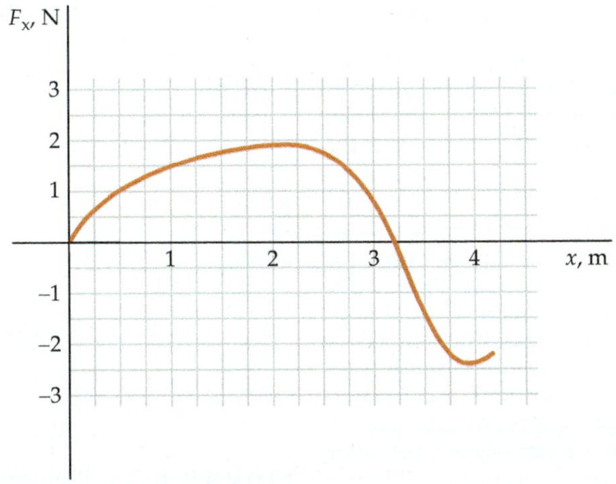

FIGURE 6-31 Problem 28

29 •• One end of a light spring (force constant k) is attached to the ceiling, the other end is attached to an object of mass m. The spring initially is vertical and unstressed. You then "ease the object down" to an equilibrium position a distance h below its initial position. Next, you repeat this experiment, but instead of easing the object down, you release it, with the result that it falls a distance H below the initial position before momentarily stopping. (*a*) Show that $h = mg/k$. (*b*) Use the work–kinetic-energy theorem to show that $H = 2h$. Try this experiment on your own.

30 •• A force F_x acts on a particle that has a mass of 1.5 kg. The force is related to the position x of the particle by the formula $F_x = Cx^3$, where $C = 0.50$ if x is in meters and F_x is in newtons. (*a*) What are the SI units of C? (*b*) Find the work done by this force as the particle moves from $x = 3.0$ m to $x = 1.5$ m. (*c*) At $x = 3.0$ m, the force points opposite the direction of the particle's velocity (speed is 12.0 m/s). What is its speed at $x = 1.5$ m? Can you tell its direction of motion at $x = 1.5$ m using only the work–kinetic-energy theorem? Explain.

31 •• You have a vacation cabin that has a nearby solar (black) water container used to provide a warm outdoor shower. For a few days last summer, your pump went out and you had to personally haul the water up the 4.0 m from the pond to the tank. Suppose your bucket has a mass of 5.0 kg and holds 15.0 kg of water when it is full. However, the bucket has a hole in it, and as you moved it vertically at a constant speed v, water leaked out at a constant rate. By the time you reached the top, only 5.0 kg of water remained. (*a*) Write an expression for the mass of the bucket plus water as a function of the height above the pond surface. (*b*) Find the work done by you on the bucket for each 5.0 kg of water delivered to the tank.

32 •• A 6.0-kg block slides 1.5 m down a frictionless incline that makes an angle of 60° with the horizontal. (*a*) Draw the free-body diagram of the block, and find the work done by each force when the block slides 1.5 m (measured along the incline). (*b*) What is the total work done on the block? (*c*) What is the speed of the block after it has slid 1.5 m, if it starts from rest? (*d*) What is its speed after 1.5 m, if it starts with an initial speed of 2.0 m/s?

33 •• **ENGINEERING APPLICATION** You are designing a jungle-vine–swinging sequence for the latest Tarzan movie. To determine his speed at the low point of the swing and to make sure it does not exceed mandatory safety limits, you decide to model the system of Tarzan + vine as a pendulum. Assume your model consists of a particle (Tarzan, mass 100 kg) hanging from a light string (the vine) of length ℓ attached to a support. The angle between the vertical and the string is written as ϕ. (*a*) Draw a free-body diagram for the object on the end of the string (Tarzan on the vine). (*b*) An infinitesimal distance along the arc (along which the object travels) is $\ell d\phi$. Write an expression for the total work dW_{total} done on the particle as it traverses that distance for an arbitrary angle ϕ. (*c*) If the $\ell = 7.0$ m, and if the particle starts from rest at an angle 50°, determine the particle's kinetic energy and speed at the low point of the swing using the work–kinetic-energy theorem. SSM

34 •• *Simple machines* are frequently used for reducing the amount of force that must be supplied to perform a task such as lifting a heavy weight. Such machines include the screw, block-and-tackle systems, and levers, but the simplest of the simple machines is the inclined plane. In Figure 6-32, you are raising a heavy box to the height of the truck bed by pushing it up an inclined plane (a ramp). (*a*) The *mechanical advantage MA* of the inclined plane is defined as the ratio of the magnitude of the force it would take to lift the block straight up (at constant speed) to the magnitude of the force it would take to push it up the ramp (at constant speed). If the plane is frictionless, show that $MA = 1/\sin\theta = L/H$, where H is the height of the truck bed and L is the length of the ramp. (*b*) Show that the work you do by moving the block into the truck is the same whether you lift it straight up or push it up the frictionless ramp.

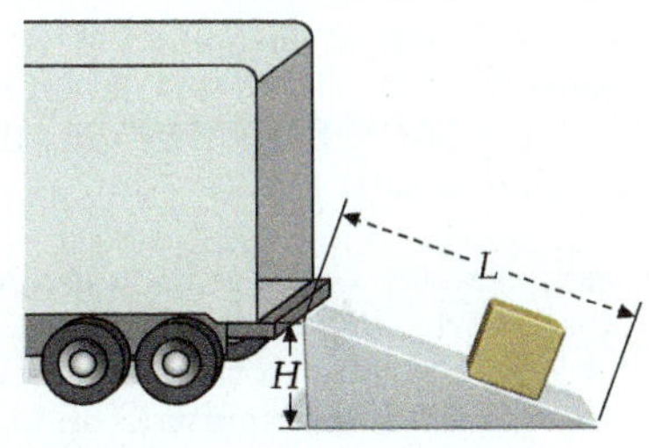

FIGURE 6-32 Problem 34

35 •• Particle a has mass m, is initially located on the positive x axis at $x = x_0$ and is subject to a repulsive force F_x from particle b. The location of particle b is fixed at the origin. The force F_x is inversely proportional to the square of the distance x between the particles. That is, $F_x = A/x^2$, where *A is a positive constant*. Particle a is released from rest and allowed to move under the influence of the force. Find an expression for the work done by the force on a as a function of x. Find both the kinetic energy and speed of a as x approaches infinity.

36 • You exert a force of magnitude F on the free end of the rope. (*a*) If the load moves up a distance h, through what distance does the point of application of the force move? (*b*) How much work is done by the rope on the load? (*c*) How much work do you do on the rope? (*d*) The mechanical advantage (defined in Problem 34) of this system is the ratio F/F_g, where F_g is the weight of the load. What is this mechanical advantage?

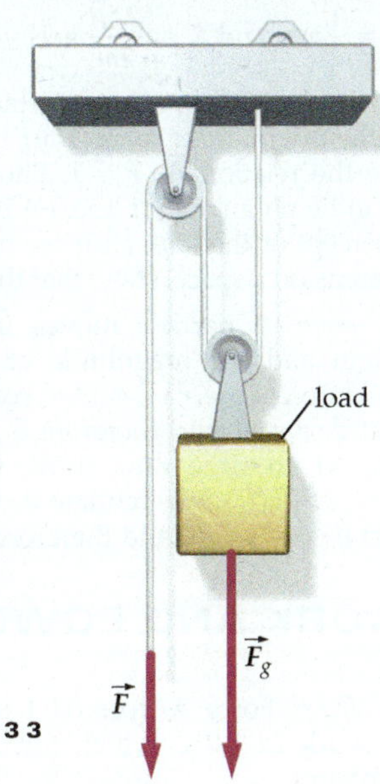

FIGURE 6-33 Problem 36

SCALAR (DOT) PRODUCTS

37 • What is the angle between the vectors $\vec{A}$ and $\vec{B}$ if $\vec{A} \cdot \vec{B} = -AB$?

38 • Two vectors $\vec{A}$ and $\vec{B}$ each have magnitudes of 6.0 m and the angle between their directions is 60°. Find $\vec{A} \cdot \vec{B}$.

39 • Find $\vec{A} \cdot \vec{B}$ for the following vectors: *(a)* $\vec{A} = 3\hat{i} - 6\hat{j}$, $\vec{B} = -4\hat{i} + 2\hat{j}$; *(b)* $\vec{A} = 5\hat{i} + 5\hat{j}$, $\vec{B} = 2\hat{i} - 4\hat{j}$; and *(c)* $\vec{A} = 6\hat{i} + 4\hat{j}$, $\vec{B} = 4\hat{i} - 6\hat{j}$.

40 • Find the angles between the vectors $\vec{A}$ and $\vec{B}$ given: *(a)* $\vec{A} = 3\hat{i} - 6\hat{j}$, $\vec{B} = -4\hat{i} + 2\hat{j}$; *(b)* $\vec{A} = 5\hat{i} + 5\hat{j}$, $\vec{B} = 2\hat{i} - 4\hat{j}$; and *(c)* $\vec{A} = 6\hat{i} + 4\hat{j}$, $\vec{B} = 4\hat{i} - 6\hat{j}$.

41 • A 2.0-kg particle is given a displacement of $\Delta\vec{r} = (3.0\text{ m})\hat{i} + (3.0\text{ m})\hat{j} + (-2.0\text{ m})\hat{k}$. During the displacement, a constant force $\vec{F} = (2.0\text{ N})\hat{i} - (1.0\text{ N})\hat{j} + (1.0\text{ N})\hat{k}$ acts on the particle. *(a)* Find the work done by $\vec{F}$ for this displacement. *(b)* Find the component of $\vec{F}$ in the direction of this displacement.

42 •• *(a)* Find the unit vector that is in the same direction as the vector $\vec{A} = 2.0\hat{i} - 1.0\hat{j} - 1.0\hat{k}$. *(b)* Find the component of the vector $\vec{A} = 2.0\hat{i} - 1.0\hat{j} - 1.0\hat{k}$ in the direction of the vector $\vec{B} = 3.0\hat{i} + 4.0\hat{j}$.

43 •• *(a)* Given two nonzero vectors $\vec{A}$ and $\vec{B}$, show that if $|\vec{A} + \vec{B}| = |\vec{A} - \vec{B}|$, then $\vec{A} \perp \vec{B}$. *(b)* Given a vector $\vec{A} = 4\hat{i} - 3\hat{j}$, find a vector in the xy plane that is perpendicular to $\vec{A}$ and has a magnitude of 10. Is this the only vector that satisfies the specified requirements? Explain. SSM

44 •• Unit vectors $\hat{A}$ and $\hat{B}$ are in the xy plane. They make angles of θ_1 and θ_2, respectively, with the $+x$ axis. *(a)* Use trigonometry to find the x and y components of the two vectors directly. (Your answer should be in terms of the angles.) *(b)* By considering the scalar product of $\hat{A}$ and $\hat{B}$, show that $\cos(\theta_1 - \theta_2) = \cos\theta_1 \cos\theta_2 + \sin\theta_1 \sin\theta_2$.

45 •• In Chapter 8, we shall introduce a new vector for a particle, called its *linear momentum*, symbolized by $\vec{p}$. Mathematically, it is related to the mass m and velocity $\vec{v}$ of the particle by $\vec{p} = m\vec{v}$. *(a)* Show that the particle's kinetic energy K can be expressed as $K = \dfrac{\vec{p} \cdot \vec{p}}{2m}$. *(b)* Compute the linear momentum of a particle of mass 2.5 kg that is moving at a speed of 15 m/s at an angle of 25° clockwise from the $+x$ axis in the xy plane. *(c)* Compute its kinetic energy using both $K = \frac{1}{2}mv^2$ and $K = \dfrac{\vec{p} \cdot \vec{p}}{2m}$ and verify that they give the same result.

46 ••• *(a)* Let $\vec{A}$ be a constant vector in the xy plane with its tail at the origin. Let $\vec{r} = x\hat{i} + y\hat{j}$ be a vector in the xy plane that satisfies the relation $\vec{A} \cdot \vec{r} = 1$. Show that the points with coordinates (x, y) lie on a straight line. *(b)* If $\vec{A} = 2\hat{i} - 3\hat{j}$, find the slope and y intercept of the line. *(c)* If we now let $\vec{A}$ and $\vec{r}$ be vectors in three-dimensional space, show that the relation $\vec{A} \cdot \vec{r} = 1$ specifies a plane.

47 ••• A particle moves in a circle that is centered at the origin and the magnitude of its position vector $\vec{r}$ is constant. *(a)* Differentiate $\vec{r} \cdot \vec{r} = r^2 =$ constant with respect to time to show that $\vec{v} \cdot \vec{r} = 0$, and therefore $\vec{v} \perp \vec{r}$. *(b)* Differentiate $\vec{v} \cdot \vec{r} = 0$ with respect to time and show that $\vec{a} \cdot \vec{r} + v^2 = 0$, and therefore $a_r = -v^2/r$. *(c)* Differentiate $\vec{v} \cdot \vec{v} = v^2$ with respect to time to show that $\vec{a} \cdot \hat{v} = dv/dt$, and therefore $a_t = dv/dt$. SSM

WORK AND POWER

48 • Force A does 5.0 J of work in 10 s. Force B does 3.0 J of work in 5.0 s. Which force delivers greater power, A or B? Explain.

49 • **MULTISTEP** A single force of 5.0 N in the $+x$ direction acts on an 8.0-kg object. *(a)* If the object starts from rest at $x = 0$ at time $t = 0$, write an expression for the power delivered by this force as a function of time. *(b)* What is the power delivered by this force at time $t = 3.0$ s?

50 • Find the power delivered by a force $\vec{F}$ acting on a particle that moves with a velocity $\vec{v}$, where *(a)* $\vec{F} = (4.0\text{ N})\hat{i} + (3.0\text{ N})\hat{k}$ and $\vec{v} = (6.0\text{ m/s})\hat{i}$; *(b)* $\vec{F} = (6.0\text{ N})\hat{i} - (5.0\text{ N})\hat{j}$ and $\vec{v} = -(5.0\text{ m/s})\hat{i} + (4.0\text{ m/s})\hat{j}$; and *(c)* $\vec{F} = (3.0\text{ N})\hat{i} + (6.0\text{ N})\hat{j}$ and $\vec{v} = (2.0\text{ m/s})\hat{i} + (3.0\text{ m/s})\hat{j}$.

51 •• **ENGINEERING APPLICATION** You are in charge of installing a small food-service elevator (called a *dumbwaiter* in the food industry) in a campus cafeteria. The elevator is connected by a pulley system to a motor, as shown in Figure 6-34. The motor raises and lowers the dumbwaiter. The mass of the dumbwaiter is 35 kg. In operation, it moves at a speed of 0.35 m/s upward, without accelerating (except for a brief initial period, which we can neglect, just after the motor is turned on). Electric motors typically have an efficiency of 78%. If you purchase a motor with an efficiency of 78%, what minimum power rating should the motor have? Assume that the pulleys are frictionless. SSM

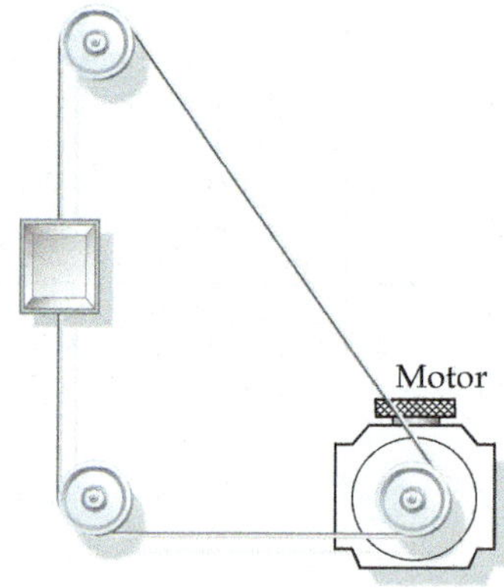

FIGURE 6-34 Problem 51

52 •• A cannon placed at the edge of a cliff of height H fires a cannonball directly upward with an initial speed v_0. The cannonball rises, falls back down (missing the cannon by a small margin), and lands at the foot of the cliff. Neglecting air resistance, calculate the velocity $\vec{v}$ as a function of time, and show explicitly that the integral of $\vec{F}_{net} \cdot \vec{v}$ over the time that the cannonball spends in flight is equal to the change in the kinetic energy of the cannonball over the same time.

53 •• A particle of mass m moves from rest at $t = 0$ under the influence of a single constant force $\vec{F}$. Show that the power delivered by the force at any time t is $P = F^2t/m$.

54 •• A 7.5-kg box is being lifted by means of a light rope that is threaded through a single, light, frictionless pulley that is attached to the ceiling. *(a)* If the box is being lifted at a *constant speed* of 2.0 m/s, what is the power delivered by the person pulling on the rope? *(b)* If the box is lifted, at *constant acceleration*, from rest on the floor to a height of 1.5 m above the floor in 0.42 s, what average power is delivered by the person pulling on the rope?

*CENTER OF MASS WORK AND CENTER OF MASS TRANSLATIONAL KINETIC ENERGY

55 ••• **ENGINEERING APPLICATION, CONTEXT-RICH, SPREADSHEET** You have been asked to test drive a car and study its actual performance relative to its specifications. This particular car's engine is rated at 164 hp. This value is the *peak* rating, which means that it is capable, at most, of providing energy at the rate of 164 hp to the drive wheels. You determine that the car's mass (including test equipment and driver on board) is 1220 kg. *(a)* When cruising at a constant 55.0 mi/h, your onboard engine-monitoring computer determines that the engine is producing 13.5 hp. From previous coasting experiments, it has been determined that the coefficient of rolling friction on the car is 0.0150. Assume that the drag force on the car varies as the square of the car's speed. That is, $F_d = Cv^2$. (*a*) What is the value of the constant, C? (*b*) Considering the peak power, what is the maximum speed (to the nearest 1 mi/h) that you would expect the car could attain? (This problem can be done by hand analytically, but it can be done more easily and quickly using a **graphing calculator** or **spreadsheet.**)

56 •• CONTEXT-RICH, CONCEPTUAL As you drive your car along a country road at night, a deer jumps out of the woods and stands in the middle of the road ahead of you. This occurs just as you are passing from a 55-mi/h zone to a 50-mi/h zone. At the 50-mi/h speed-limit sign, you slam on the car's brakes, causing them to lock up, and skid to a stop inches from the startled deer. As you breathe a sigh of relief, you hear the sound of a police siren. The policeman proceeds to write you a ticket for driving 56 mi/h in 50-mi/h zone. Because of your preparation in physics, you are able to use the 25-m-long skid marks that your car left behind as evidence that you were not speeding. What evidence do you present? In formulating your answer, you will need to know the coefficient of kinetic friction between automobile tires and dry concrete (see Table 5-1).

GENERAL PROBLEMS

57 • APPROXIMATION In February 2002, a total of 60.7 billion kW·h of electrical energy was generated by nuclear power plants in the United States. At that time, the population of the United States was about 287 million people. If the average American has a mass of 60 kg, and if 25% of the entire energy output of all nuclear power plants was diverted to supplying energy for a single giant elevator, estimate the height h to which the entire population of the country could be lifted by the elevator. In your calculations, assume also that g is constant over the entire height h.

58 • ENGINEERING APPLICATION One of the most powerful cranes in the world operates in Switzerland. It can slowly raise a 6000-t load to a height of 12.0 m. (Note that 1 t = one tonne is sometimes called a metric ton. It is a unit of mass, not force, and is equal to 1000 kg.) (*a*) How much work is done by the crane during this lift? (*b*) If it takes 1.00 min to lift this load to this height at constant velocity, and the crane is 20 percent efficient, find the total (gross) power rating of the crane.

59 • In Austria, there once was a 5.6-km-long ski lift. It took about 60 min for a gondola to travel up its length. If there were 12 gondolas going up, each with a cargo of mass 550 kg, and if there were 12 empty gondolas going down, and the angle of ascent was 30°, estimate the power P the engine needed to deliver in order to operate the ski lift.

60 •• ENGINEERING APPLICATION To complete your master's degree in physics, your advisor has you design a small, linear accelerator capable of emitting protons, each with a kinetic energy of 10.0 keV. (The mass of a single proton is 1.67×10^{-27} kg.) In addition, 1.00×10^9 protons per second must reach the target at the end of the 1.50-m-long accelerator. (*a*) What the average power must be delivered to the stream of protons? (*b*) What force (assumed constant) must be applied to each proton? (*c*) What speed does each proton attain just before it strikes the target, assuming the protons start from rest?

61 ••• The four strings pass over the bridge of a violin, as shown in Figure 6-35. The strings make and angle of 72.0° with the normal to the plane of the instrument on either side of the bridge. The resulting total normal force pressing the bridge into the violin is 1.00×10^3 N. The length of the strings from the bridge to the peg to which each is attached is 32.6 cm. (*a*) Determine the tension in the strings, assuming the tension is the same for each string. (*b*) One of the strings is plucked out a distance of 4.00 mm, as shown. Make a free-body diagram showing all of the forces acting on the segment of the string in contact with the finger (not shown), and determine the force pulling the segment back to its equilibrium position. Assume the tension in the string remains constant during the pluck. (*c*) Determine the work done on the string in plucking it out that distance. Remember that the net force pulling the string back to its equilibrium position is changing as the string is being pulled out, but assume that the magnitudes of the tension forces remain constant.

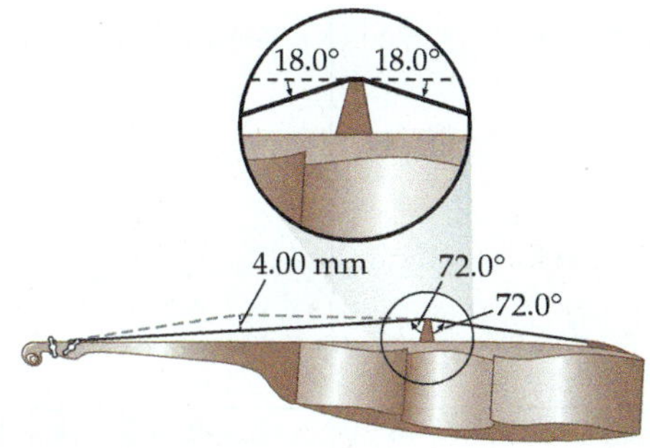

FIGURE 6-35 Problem 61

62 •• The magnitude of the single force acting on a particle of mass m is given by $F = bx^2$, where b is a constant. The particle starts from rest. After it travels a distance L, determine its (*a*) kinetic energy and (*b*) speed.

63 •• A single horizontal force in the $+x$ direction acts on a cart of mass m. The cart starts from rest at $x = 0$, and the speed of the cart increases with x as $v = Cx$, where C is a constant. (*a*) Find the force acting on the cart as a function of x. (*b*) Find the work done by the force in moving the cart from $x = 0$ to $x = x_1$. SSM

64 ••• A force $\vec{F} = (2.0\ \text{N/m}^2)x^2\hat{i}$ is applied to a particle initially at rest in the xy plane. Find the work done by this force on the particle and the final speed of the particle as it moves along a path that is (*a*) in a straight line from point (2.0 m, 2.0 m) to point (2.0 m, 7.0 m) and (*b*) in a straight line from point (2.0 m, 2.0 m) to point (5.0 m, 6.0 m). The given force is the only force doing work on the particle.

65 •• A particle of mass m moves along the x axis. Its position varies with time according to $x = 2t^3 - 4t^2$, where x is in meters and t is in seconds. Find (*a*) the velocity and acceleration of the particle as functions of t, (*b*) the power delivered to the particle as a function of t, and (*c*) the work done by the net force from $t = 0$ to $t = t_1$.

66 •• A 3.0-kg particle starts from rest at $x = 0.050$ m and moves along the x axis under the influence of a single force $F_x = 6.0 + 4.0x - 3.0x^2$, where F_x is in newtons and x is in meters. (*a*) Find the work done by the force as the particle moves from $x = 0.050$ m to $x = 3.0$ m. (*b*) Find the power delivered to the particle as it passes through the point $x = 3.0$ m.

67 •• The initial kinetic energy imparted to a 0.0200-kg bullet is 1200 J. (*a*) Assuming it accelerated down a 1.00-m-long rifle barrel, estimate the average power delivered to it during the firing. (*b*) Neglecting air resistance, find the range of this projectile when it is fired at an angle such that the range equals the maximum height attained.

68 •• The force F_x acting on a 0.500-kg particle is shown as a function of x in Figure 6-36. (*a*) From the graph, calculate the work done by the force when the particle moves from $x = 0.00$ to the following values of x: −4.00, −3.00, −2.00, −1.00, +1.00, +2.00,

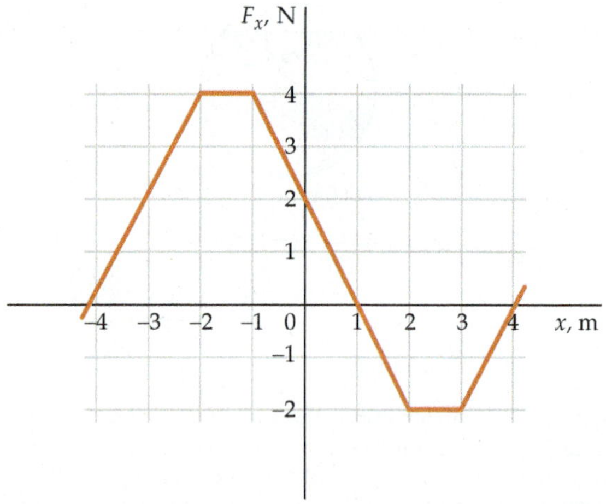

FIGURE 6-36 Problem 68

+3.00, and +4.00 m. (*b*) If it starts with a velocity of 2.00 m/s in the +*x* direction, how far will the particle go in that direction before stopping?

69 •• (*a*) Repeat Problem 68(*a*) for the force F_x shown in Figure 6-37. (*b*) If the object starts at the origin moving to the right with a kinetic energy of 25.0 J, how much kinetic energy does it have at $x = 4.00$ m.

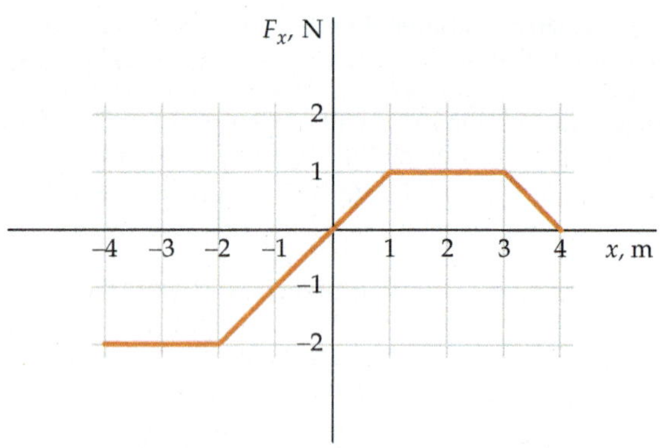

FIGURE 6-37 Problem 69

70 •• A box of mass M is at rest at the bottom of a frictionless inclined plane (Figure 6-38). The box is attached to a string that pulls with a constant tension T. (*a*) Find the work done by the tension T as the box moves through a distance x along the plane. (*b*) Find the speed of the box as a function of x. (*c*) Determine the power delivered by the tension in the string as a function of x.

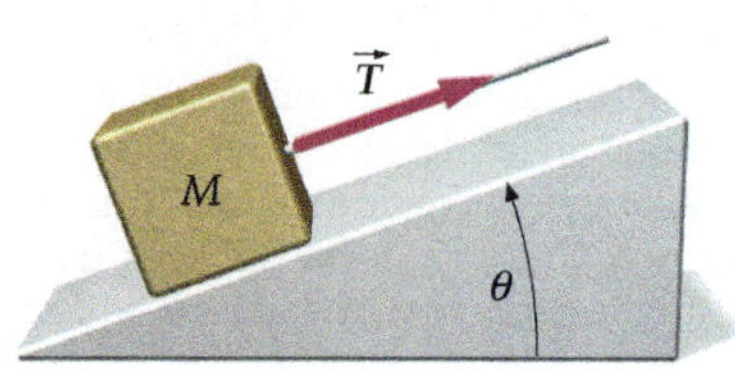

FIGURE 6-38 Problem 70

71 ••• A force acting on a particle in the xy plane at coordinates (x, y) is given by $\vec{F} = (F_0/r)(y\hat{i} - x\hat{j})$, where F_0 is a positive constant and r is the distance of the particle from the origin. (*a*) Show that the magnitude of this force is F_0 and that its direction is perpendicular to $\vec{r} = x\hat{i} + y\hat{j}$. (*b*) Find the work done by this force on a particle that moves once around a circle of radius 5.0 m that is centered at the origin. SSM

72 ••• A force acting on a 2.0-kg particle in the xy plane at coordinates (x, y) is given by $\vec{F} = -(b/r^3)(x\hat{i} + y\hat{j})$, where b is a positive constant and r is the distance from the origin. (*a*) Show that the magnitude of the force is inversely proportional to r^2, and that its direction is antiparallel (opposite) to the radius vector $\vec{r} = x\hat{i} + y\hat{j}$. (*b*) If $b = 3.0\ \text{N}\cdot\text{m}^2$, find the work done by this force as the particle moves from (2.0 m, 0.0 m), to (5.0 m, 0.0 m) along a straight-line path. (*c*) Find the work done by this force on a particle moving once around a circle of radius $r = 7.0$ m that is centered at the origin.

73 ••• A block of mass m on a horizontal frictionless tabletop is attached by a swivel to a spring that is attached to the ceiling (Figure 6-39). The vertical distance between the top of the block and the ceiling is y_0, and the horizontal position is x. When the block is at $x = 0$, the spring, which has force constant k, is completely unstressed. (*a*) What is F_x, the x component of the force on the block due to the spring, as a function of x? (*b*) Show that F_x is proportional to x^3 for sufficiently small values of $|x|$. (*c*) If the block is released from rest at $x = x_0$, where $|x_0| \ll y_0$, what is its speed when it reaches $x = 0$?

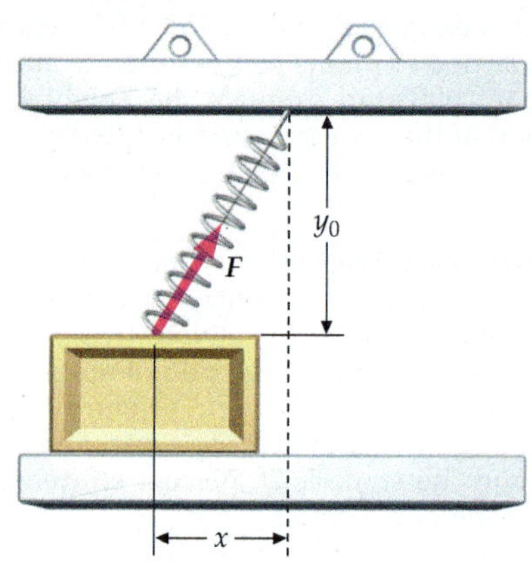

FIGURE 6-39 Problem 73

74 •• Two horses pull a large crate at constant speed across a barn floor by means of two light steel cables. A large box of mass 250 kg sits on the crate (Figure 6-40). As the horses pull, the cables are parallel to the horizontal floor. The coefficient of friction between the crate and the barn floor is 0.25. (*a*) What is the work done by each horse if the box is moved a distance of 25 m? (*b*) What is the tension in each cable if the angle between each cable and the direction the crate moves is 15°?

FIGURE 6-40 Problem 74

CHAPTER 7

AS A ROLLER COASTER RUSHES THROUGH ITS TWISTS AND TURNS, ENERGY IS TRANSFERRED IN DIFFERENT WAYS. ELECTRICAL POTENTIAL ENERGY PURCHASED FROM THE POWER COMPANY IS TRANSFORMED INTO GRAVITATIONAL POTENTIAL ENERGY AS THE CARS AND PASSENGERS ARE RAISED TO THE HIGHEST POINT OF THE RIDE. AS THE ROLLER-COASTER CARS PLUMMET DOWN THE STEEP INCLINE, THIS GRAVITATIONAL POTENTIAL ENERGY IS TRANSFORMED INTO KINETIC ENERGY AND THERMAL ENERGY—INCREASING THE TEMPERATURE OF BOTH THE CARS AND THEIR SURROUNDINGS BY A SMALL AMOUNT. *(Michael S. Lewis/Corbis.)*

Conservation of Energy

When work is done by one system on another, energy is transferred between the two systems. For example, when you pull a sled, energy from you goes partly into the kinetic energy of the sled and partly into the thermal energy that arises from the friction between the sled and the snow. At the same time, the internal chemical energy of your body decreases. The net result is the transfer of the internal chemical energy of your body to the external kinetic energy of the sled plus the thermal energy of the sled and the snow. This transfer of energy highlights one of the most important principles in science, *the law of conservation of energy,* which states that the total energy of a system and its surroundings does not change. Whenever the energy of a system changes, we can account for the change by the appearance or disappearance of energy somewhere else.

In this chapter, we continue the study of energy begun in Chapter 6 by describing and applying the law of conservation of energy and examining the energy associated with several different states, including potential energy and thermal energy. We also discuss that energy changes for a system are often not continuous, but occur in discrete "bundles" or "lumps" called quanta. *Although a quantum of energy in a macroscopic system typically is so small that it goes unnoticed, its presence has profound consequences for microscopic systems such as atoms and molecules.*

? How can we use the concept of energy transformation to determine how high the cars must be when they start their descent for them to complete the loop-the-loop? (See Example 7-8.)

7-1 POTENTIAL ENERGY

In Chapter 6, we showed that the total work done on a *particle* equals the change in its kinetic energy. However, sometimes a particle is part of a *system* consisting of two or more particles, and we need to examine the external work done on the system.* Often, the energy transferred to such a system by the work done by external forces on the system does not go into increasing the total kinetic energy *of the system.* Instead, the energy transferred is stored as **potential energy**—energy associated with the relative positions of different parts of the system. The configuration of a system is the way the different parts of the system are positioned relative to each other. Potential energy is an energy associated with system configuration, whereas kinetic energy is an energy associated with motion.

For example, consider a pile driver whose driver is suspended some distance h above a pile (a long, slender column). When the driver is released, it falls—gaining kinetic energy until it smashes into the pile, driving the pile into the ground. The driver is then raised back up to its initial height and released again. Each time the driver is raised from its lowest to its highest position, a gravitational force does work on it given by $-mgh$, where m is the mass of the driver. A second force, the force provided by the lifting agent, acts on the driver. As the driver is raised, the force exerted by the lifting agent does work on the driver that has a positive value. During the raising of the driver, these two work values sum to zero. We know they sum to zero because during the lift, the driver can be modeled as a particle, so the work–kinetic-energy theorem (Equation 6-8) tells us that the total work done on the driver is equal to the change in its kinetic energy—which is zero.

Consider lifting a barbell of mass m to a height h. The barbell starts at rest and ends at rest, so the net change in the kinetic energy of the barbell is zero. The barbell can be modeled as a particle during the lift, so the work–kinetic-energy theorem tells us that the total work done on the barbell is zero. There are two forces on the barbell, the force of gravity and the force of your hands. The gravitational force on the barbell is $m\vec{g}$, and the work done on the barbell by this force during the lift is $-mgh$. Because we know that the total work done on the barbell is zero, it follows that the work done on the barbell by the force of your hands is $+mgh$.

Consider the barbell and planet Earth to be a *system* of two particles (Figure 7-1). (You are not part of this system.) The external forces acting *on the Earth–barbell system* are the three forces exerted on it by you. These forces are the contact force by your hands *on the barbell*, the contact force by your feet *on the floor*, and the gravitational force by you *on Earth*. The gravitational force on Earth by you is equal and opposite to the gravitational force on you by Earth. (The gravitational forces you and the barbell attract each other with are negligible.) The barbell moves through a displacement of one or two meters, but displacements of the floor and planet Earth are negligibly small, so the force exerted on the barbell by your hands is the only one of the three external forces that does work on the Earth–barbell system. Thus, the total work done on this system by all three external forces is $+mgh$ (the work done on the barbell by your hands). The energy transferred to the system by this work is stored as *gravitational potential energy*, energy associated with the position of the barbell relative to Earth (energy associated with the height of the barbell above the floor).

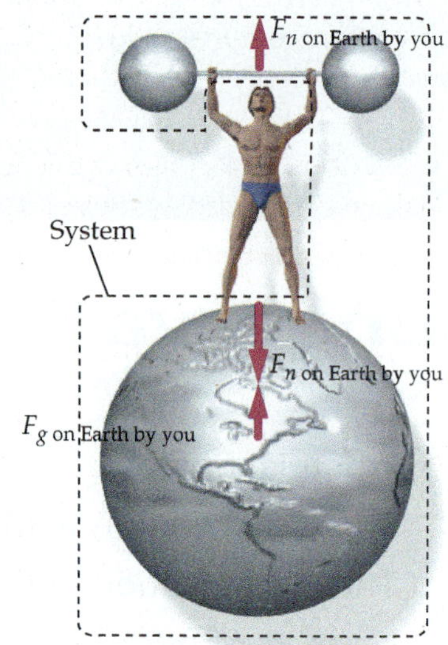

FIGURE 7-1

Another system that stores energy associated with its configuration is a spring. If you stretch or compress a spring, energy associated with the length of the spring is stored as *elastic potential energy*. Consider the spring shown in Figure 7-2 as the system. You compress the spring, pushing it with equal and opposite forces $\vec{F}_1$ and $\vec{F}_2$. These forces sum to zero, so the net force on the spring remains zero. Thus, there is no change in the kinetic energy of the spring. The energy transfer associated with the work you do on the spring is stored not as kinetic energy, but as

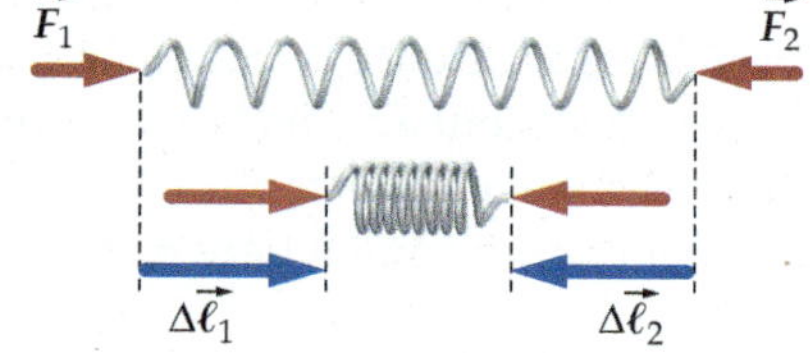

FIGURE 7-2 The spring is compressed by external forces $\vec{F}_1$ and $\vec{F}_2$. Both forces do work on the spring as they compress it. These work values are positive, so the elastic potential energy of the spring increases as it is compressed.

* Systems of particles are discussed more thoroughly in Chapter 8.

elastic potential energy. The configuration of this system has been changed, as evidenced by the change in the length of the spring. The total work done on the spring is positive because both $\vec{F}_1$ and $\vec{F}_2$ do positive work. (The work done by $\vec{F}_1$ is positive because both $\vec{F}_1$ and the displacement $\Delta\vec{\ell}_1$ are in the same direction. The same can be said for $\vec{F}_2$ and $\Delta\vec{\ell}_2$.)

CONSERVATIVE AND NONCONSERVATIVE FORCES

When you ride a ski lift to the top of a hill of height h, the work done on you by gravity is $-mgh$, where m is your mass. When you ski down the hill to the bottom, the work done by gravity is $+mgh$, independent of the shape of the hill (as you saw in Example 6-12). The total work done on you by gravity during the round-trip up and down the hill is zero and is independent of the path you take. In a situation such as this, where the total work done on an object by a force depends on only the initial and final positions of the object, and not the path taken, the force doing the work is called a **conservative force.**

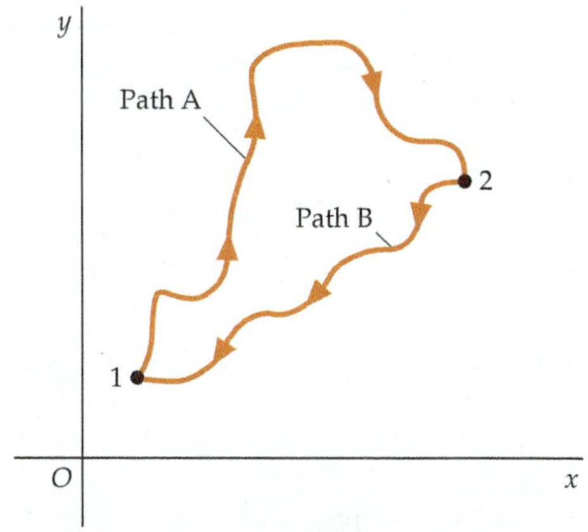

FIGURE 7-3 Two paths in space connecting the points 1 and 2. If the work done by a conservative force along path A from 1 to 2 is $+W$, the work done on the return trip along path B must be $-W$, because the roundtrip work is zero. When traversing path B from 1 to 2, the force is the same at each point, but the displacement is opposite to that when going on path B from 2 to 1. Thus, the work done along path B from 1 to 2 must also be W. It follows that the work done as a particle goes from point 1 to 2 is the same along any path connecting the two points.

> The work done by a conservative force on a particle is independent of the path taken as the particle moves from one point to another.
>
> DEFINITION—CONSERVATIVE FORCE

From Figure 7-3 we see that this definition implies that:

> A force is conservative if the work it does on a particle is zero when the particle moves around *any* closed path, returning to its initial position.
>
> ALTERNATIVE DEFINITION—CONSERVATIVE FORCE

In the ski-lift example, the force of gravity, exerted by Earth on you, is a conservative force, because the total work done by gravity on you during the round-trip is zero, independent of the path you take. Both the gravitational force on an object and the force exerted by a spring of negligible mass on an object are conservative forces. (If a spring's mass is negligible then its kinetic energy is also negligible.) Any spring in this book has negligible mass unless otherwise stated.

Not all forces are conservative. A force is said to be **nonconservative** if it does not meet the definition for conservative forces. Suppose, for example, that you push a block across a table along a straight line from point A to point B and back again, so that the block ends up at A, its original position. Friction opposes the block's motion, so the force you push on the block with is in the direction of motion and the value of the work done by the push is positive on both legs of the round-trip. The total work done by the push does *not* equal zero. Thus, the push is an example of a nonconservative force.

As another example, consider the force $\vec{F}$ a donkey exerts on a pole as the donkey pulls the pole around the circle at constant speed. As the donkey walks, $\vec{F}$ is continuously doing work whose value is positive. The point of application (point P) of $\vec{F}$ returns to the same position each time the donkey completes one pass around the circle, so the work done by $\vec{F}$ is not equal to zero each time P completes one trip around a closed path (the circle). We can thus conclude that $\vec{F}$ is a nonconservative force.

A machine for pumping water. The donkey exerts force $\vec{F}$ on the pole at P, the point of application of the force. *(O. Alamany and E. Vicens/Corbis.)*

If the work done around *any* particular closed path is not zero, we can conclude that the force is nonconservative. However, we can conclude the force is conservative only if the work is zero around *all* possible closed paths. Because there are infinitely many possible closed paths, it is impossible to calculate the work done for each one. Therefore, finding a single closed path along which the work done by a particular force *is not zero* is sufficient for showing that the force is nonconservative, but is of limited use when investigating whether a given force is conservative. In more advanced physics courses, more sophisticated mathematical methods for evaluating whether a force is conservative are studied.

Example 7-1 Integral Around a Closed Path

To calculate the work done by a force $\vec{F}$ around a closed curve (path) C, we evaluate $\oint_C \vec{F}\cdot d\vec{\ell}$, where the circle on the integral sign means that the integration is evaluated for one complete trip around C. For $\vec{F} = Ax\hat{i}$, calculate $\oint_C \vec{F}\cdot d\vec{\ell}$ for the path C shown in Figure 7-4.

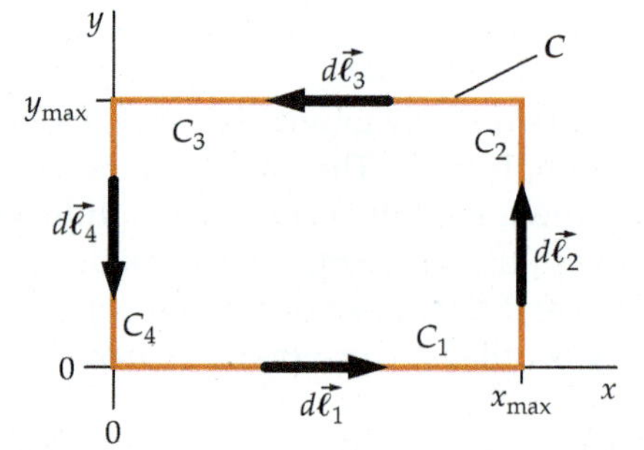

FIGURE 7-4

PICTURE The path C consists of four straight segments. Evaluate $d\vec{\ell} = dx\hat{i} + dy\hat{j}$ on each segment and calculate $\int \vec{F}\cdot d\vec{\ell}$ separately for each of the four segments.

SOLVE

1. The integral around C is equal to the sum of the integrals along the segments that make up C:

$$\oint_C \vec{F}\cdot d\vec{\ell} = \int_{C_1} \vec{F}\cdot d\vec{\ell}_1 + \int_{C_2} \vec{F}\cdot d\vec{\ell}_2 + \int_{C_3} \vec{F}\cdot d\vec{\ell}_3 + \int_{C_4} \vec{F}\cdot d\vec{\ell}_4$$

2. On C_1, $dy = 0$, so $d\vec{\ell}_1 = dx\hat{i}$:

$$\int_{C_1} \vec{F}\cdot d\vec{\ell}_1 = \int_0^{x_{max}} Ax\hat{i}\cdot dx\hat{i} = A\int_0^{x_{max}} x\,dx = \tfrac{1}{2}Ax_{max}^2$$

3. On C_2, $dx = 0$ and $x = x_{max}$, so $d\vec{\ell}_2 = dy\hat{j}$ and $\vec{F} = Ax_{max}\hat{i}$:

$$\int_{C_2} \vec{F}\cdot d\vec{\ell}_2 = \int_0^{y_{max}} Ax_{max}\hat{i}\cdot dy\hat{j} = Ax_{max}\int_0^{y_{max}} \hat{i}\cdot\hat{j}\,dy = 0$$

($\hat{i}\cdot\hat{j} = 0$ because $\hat{i}$ and $\hat{j}$ are perpendicular.)

4. On C_3, $dy = 0$, so $d\vec{\ell}_3 = dx\hat{i}$:

$$\int_{C_3} \vec{F}\cdot d\vec{\ell}_3 = \int_{x_{max}}^{0} Ax\hat{i}\cdot dx\hat{i} = -A\int_0^{x_{max}} x\,dx = -\tfrac{1}{2}Ax_{max}^2$$

5. On C_4, $dx = 0$ and $x = 0$, so $d\vec{\ell}_4 = dy\hat{j}$ and $\vec{F} = 0$:

$$\int_{C_4} \vec{F}\cdot d\vec{\ell}_4 = \int_{y_{max}}^{0} 0\hat{i}\cdot dy\hat{j} = 0$$

6. Add the step-2, -3, -4, and -5 results:

$$\oint_C \vec{F}\cdot d\vec{\ell} = \tfrac{1}{2}Ax_{max}^2 + 0 - \tfrac{1}{2}Ax_{max}^2 + 0 = \boxed{0}$$

CHECK The force is described by Hooke's law (the force for a spring). Thus, it is conservative, so the integral of this force around any closed path is zero.

TAKING IT FURTHER The negative sign in step 4 appeared because the integration limits were reversed.

PRACTICE PROBLEM 7-1 For $\vec{F} = Bxy\hat{i}$, calculate $\oint_C \vec{F}\cdot d\vec{\ell}$ for the path C shown in Figure 7-4.

POTENTIAL-ENERGY FUNCTIONS

The work done by a conservative force on a particle does not depend on the path, but it does depend on the endpoints of the path. We can use this property to define the **potential-energy function** U that is associated with a conservative force. Let us return to the ski-lift example once again. Now consider yourself and Earth to be a *two-particle system.* (The ski lift is not part of this system.) When a ski lift raises you to the top of the hill, it does work mgh on the you–Earth system. This work is stored as the gravitational potential energy of the you–Earth system. When you ski down the hill, this potential energy is converted to the kinetic energy of your motion. Note that when you ski down the hill, the work done by gravity *decreases* the potential energy of the system. We define the potential-energy function U such that the work done by a conservative force equals the decrease in the potential-energy function:

$$W = \int_1^2 \vec{F} \cdot d\vec{\ell} = -\Delta U$$

or

$$\Delta U = U_2 - U_1 = -\int_1^2 \vec{F} \cdot d\vec{\ell} \qquad 7\text{-}1a$$

DEFINITION—POTENTIAL-ENERGY FUNCTION

This equation gives the change in potential energy due to a change in the configuration of the system as an object moves from point 1 to point 2.

For an infinitesimal displacement, $d\vec{\ell}$, the change in potential energy is given by

$$dU = -\vec{F} \cdot d\vec{\ell} \qquad 7\text{-}1b$$

See Math Tutorial *for more information on* Integrals

Gravitational potential energy Using Equation 7-1*b* we can calculate the potential-energy function associated with the gravitational force near the surface of Earth. For the force $\vec{F} = -mg\hat{j}$, we have

$$dU = -\vec{F} \cdot d\vec{\ell} = -(-mg\hat{j}) \cdot (dx\hat{i} + dy\hat{j} + dz\hat{k}) = +mg\,dy$$

where we have exploited the fact that $\hat{j} \cdot \hat{i} = \hat{j} \cdot \hat{k} = 0$ and $\hat{j} \cdot \hat{j} = 1$. Integrating, we obtain

$$U = \int mg\,dy = mgy + U_0$$

$$U = U_0 + mgy \qquad 7\text{-}2$$

GRAVITATIONAL POTENTIAL ENERGY NEAR EARTH'S SURFACE

where U_0, the arbitrary constant of integration, is the value of the potential energy at $y = 0$. Because only a change in the potential energy is defined, the actual value of U is not important. For example, if the gravitational potential energy of the Earth–skier system is chosen to be zero when the skier is at the bottom of the hill, its value when the skier is at a height h above that level is mgh. Or we could choose the potential energy to be zero when the skier is at point P halfway down the ski slope, in which case its value at any other point would be mgy, where y is the height of the skier above point P. On the lower half of the slope, the potential energy would then be negative.

! We are free to choose U to be zero at any convenient reference point.

PRACTICE PROBLEM 7-2

A 55-kg window washer stands on a platform 8.0 m above the ground. What is the potential energy U of the window-washer–Earth system if (*a*) U is chosen to be zero on the ground, (*b*) U is chosen to be zero 4.0 m above the ground, and (*c*) U is chosen to be zero 10 m above the ground?

Example 7-2 A Falling Bottle

A 0.350-kg bottle falls from rest from a shelf that is 1.75 m above the floor. Find the potential energy of the bottle–Earth system when the bottle is on the shelf and just before impact with the floor. Find the kinetic energy of the bottle just before impact.

PICTURE The work done by Earth on the bottle as it falls equals the negative of the change in the potential energy of the bottle–Earth system. Knowing the work, we can use the work–kinetic-energy theorem to find the kinetic energy.

SOLVE

1. Make a sketch showing the bottle on the shelf and again when it is about to impact the floor (Figure 7-5). Choose the potential energy of the bottle–Earth system to be zero when the bottle is on the floor, and place a y axis on the sketch with the origin at floor level:

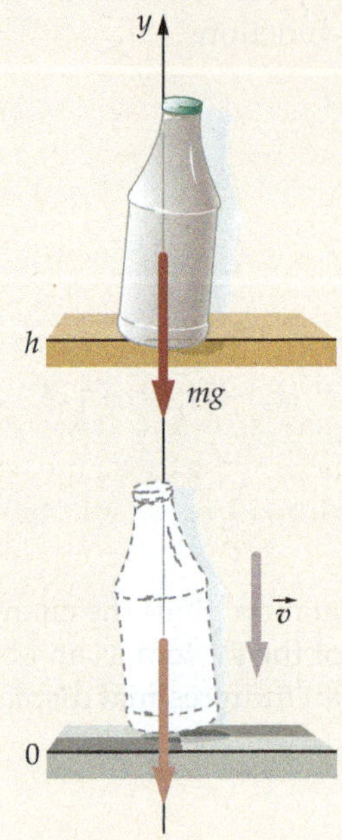

FIGURE 7-5

2. The only force doing work on the falling bottle is the force of gravity, so $W_{total} = W_g$. Apply the work–kinetic-energy theorem to the falling bottle:

$$W_{total} = W_g = \Delta K$$

3. The gravitational force exerted by Earth on the falling bottle is internal to the bottle–Earth system. It is also a conservative force, so the work done by it equals the negative of the change in the potential energy of the system:

$$W_g = -\Delta U = -(U_f - U_i) = -(mgy_f - mgy_i)$$
$$= mg(y_i - y_f) = mg(h - 0) = mgh$$

4. Substitute the step-3 result into the step-2 result and solve for the final kinetic energy. The original kinetic energy is zero:

$$mgh = \Delta K$$
$$mgh = K_f - K_i$$
$$K_f = K_i + mgh$$
$$= 0 + (0.350\text{ kg})(9.81\text{ N/kg})(1.75\text{ m})$$
$$= 6.01\text{ N}\cdot\text{m} = \boxed{6.01\text{ J}}$$

CHECK The units of the answer in step 4 are units of energy, because $1\text{ N}\cdot\text{m} = 1\text{ J}$.

TAKING IT FURTHER Potential energy is associated with the configuration of a *system of particles*, but we sometimes have systems, such as the bottle–Earth system in this example, in which only one particle moves (Earth's motion is negligible). For brevity, then, we sometimes refer to the potential energy of the bottle–Earth system as simply the potential energy of the bottle.

The gravitational potential energy of a system of particles in a uniform gravitational field is the same as if the entire mass of the system were concentrated at the system's center of mass. For such a system, let h_i be the height of the ith particle above some reference level. The gravitational potential energy of the system is then

$$U_g = \sum_i m_i g h_i = g \sum_i m_i h_i$$

where the sum is over all the particles in the system. By definition of the center of mass, the height of the center of mass of the system is given by

$$Mh_{cm} = \sum_i m_i h_i, \qquad \text{where } M = \sum_i m_i$$

Substituting Mh_{cm} for $\Sigma m_i h_i$ gives

$$U_g = Mgh_{cm} \qquad 7\text{-}3$$

GRAVITATIONAL POTENTIAL ENERGY OF A SYSTEM

Elastic potential energy Another example of a conservative force is that of a stretched (or compressed) spring of negligible mass. Suppose you pull a block attached to a spring from its equilibrium position at $x = 0$ to a new position at $x = x_1$ (Figure 7-6). The work done by the spring on the block is negative because the force exerted by the spring on the block and the displacement of the block are oppositely directed. If we then release the block, the force of the spring does positive work on the block as the block accelerates back toward its initial position. The total work done on the block by the spring as the block moves from $x = 0$ to $x = x_1$, and then back to $x = 0$, is zero. This result is independent of the size of x_1 (as long as the stretching is not so great as to exceed the elastic limit of the spring). The force exerted by the spring is therefore a conservative force. We can calculate the potential-energy function associated with this force from Equation 7-1*b*:

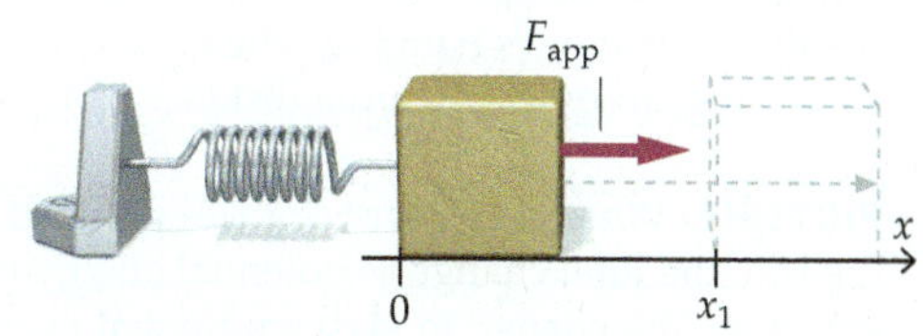

FIGURE 7-6 The applied force F_{app} pulls the block to the right, stretching the spring by x_1.

$$dU = -\vec{F} \cdot d\vec{\ell} = -F_x\,dx = -(-kx)dx = +kx\,dx$$

Then

$$U = \int kx\,dx = \frac{1}{2}kx^2 + U_0$$

where U_0 is the potential energy when $x = 0$, that is, when the spring is unstressed. Choosing U_0 to be zero gives

$$U = \tfrac{1}{2}kx^2 \qquad 7\text{-}4$$

POTENTIAL ENERGY OF A SPRING

The formula $U = \frac{1}{2}kx^2$ for the potential energy of a spring requires that the spring is relaxed if $x = 0$. Thus, the location of the point where $x = 0$ is not arbitrary when using the potential energy function $U = \frac{1}{2}kx^2$.

When we pull the block from $x = 0$ to $x = x_1$, the agent doing the pulling must exert an applied force on the block. If the block starts from rest at $x = 0$ and ends at rest at $x = x_1$, the change in its kinetic energy is zero. The work–energy theorem then tells us that the total work done on the block is zero. That is, $W_{app} + W_{spring} = 0$, or

$$W_{app} = -W_{spring} = \Delta U_{spring} = \tfrac{1}{2}kx_1^2 - 0 = \tfrac{1}{2}kx_1^2$$

The energy transferred from the agent doing the pulling to the block–spring system is equal to W_{app} and is stored as potential energy in the spring.

PRACTICE PROBLEM 7-3

A suspension spring on a Toyota Prius has a force constant of 11,000 N/m. How much energy is transferred to one of these springs when, starting from its relaxed length, it is compressed 30.0 cm?

FIGURE 7-7 *(Elio Castoria/APF/Getty Images.)*

Example 7-3 Potential Energy of a Basketball Player

A system consists of a 110-kg basketball player, the rim of a basketball hoop, and Earth. Assume that the potential energy of this system is zero when the player is standing on the floor and the rim is horizontal. Find the total potential energy of this system when the player is hanging on the front of the rim (much like that shown in Figure 7-7). Also, assume that the center of mass of the player is 0.80 m above the floor when he is standing and 1.30 m above the floor when he is hanging. The force constant of the rim is 7.2 kN/m and the front of the rim is displaced downward a distance of 15 cm. Assume the mass of the rim is negligible.

PICTURE When the player changes position from standing on the floor to hanging on the rim, the total change in potential energy is the change in gravitational potential energy plus the change in elastic potential energy stored in the stressed rim, whose potential energy can be measured just as if it were a spring: $U_s = \frac{1}{2}kx^2$. Choose 0.80 m above the floor as the reference point where $U_g = 0$.

SOLVE

1. Sketch the system, first in the initial configuration and again in the final configuration (Figure 7-8):

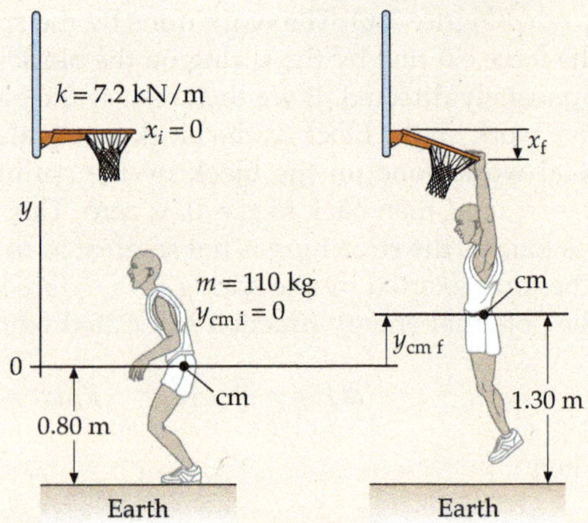

FIGURE 7-8 A basketball player jumps, grabs hold of the rim, and dangles from it.

2. The gravitational potential energy reference point where $U_g = 0$ is 0.80 m above the floor. Thus, $U_{gi} = 0$. The initial total potential energy equals zero:

$$U_{gi} = mgy_{\text{cm i}} = mg(0) = 0$$
$$U_{si} = \tfrac{1}{2}kx_i^2 = \tfrac{1}{2}k(0)^2 = 0$$
$$U_i = U_{gi} + U_{si} = 0$$

3. The total final potential energy is the sum of final gravitational potential energy and the final elastic potential energy of the rim:

$$U_f = U_{gf} + U_{sf} = mgy_{\text{cm f}} + \tfrac{1}{2}kx_f^2$$
$$= (110\text{ kg})(9.81\text{ N/kg})(0.50\text{ m}) + \tfrac{1}{2}(7.2\text{ kN/m})(0.15\text{ m})^2$$
$$= 540\text{ N}\cdot\text{m} + 81\text{ N}\cdot\text{m} = \boxed{6.2 \times 10^2\text{ J}}$$

CHECK The units check out if we use the definition of the joule. That definition is $1\text{ J} = 1\text{ N}\cdot\text{m}$.

TAKING IT FURTHER The front of the rim and player oscillate vertically immediately after the player grabs. However, they eventually come to rest with the front of the rim 15 cm below its initial position. The total potential energy is a minimum when the system is in equilibrium (Figure 7-9). Why this is so is explained near the end of Section 7-2.

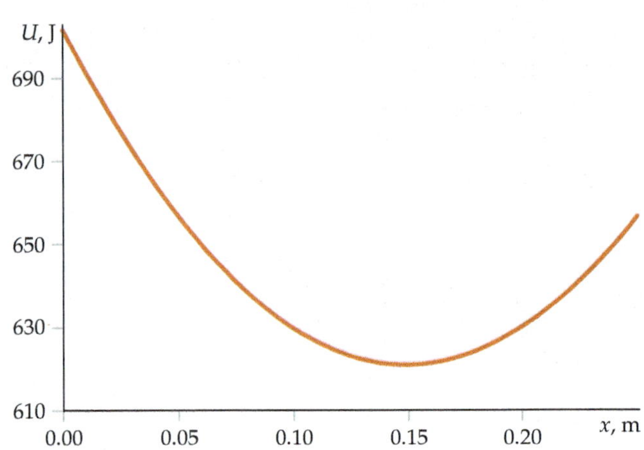

FIGURE 7-9 The graph shows the total potential energy $U = U_g + U_s$ as a function of the downward displacement of the rim.

PRACTICE PROBLEM 7-4 A 3.0-kg block is hung vertically from a spring with a force constant of 600 N/m. (*a*) By how much is the spring stretched? (*b*) How much potential energy is stored in the spring?

7-2 THE CONSERVATION OF MECHANICAL ENERGY

We are now ready to look at the relationship between kinetic energy and potential energy. Recall that the total work done on each particle in a system equals the change in the kinetic energy ΔK_i of that particle, so the total work done by all the forces W_{total} equals the change in the total kinetic energy of the system ΔK_{sys}:

$$W_{\text{total}} = \sum \Delta K_i = \Delta K_{\text{sys}} \qquad 7\text{-}5$$

Two sets of forces do work on a particle in a system: the external forces and the internal forces. Each internal force is either conservative or nonconservative. The total work done by all forces equals the work done by all external forces W_{ext}, plus the work done by all internal nonconservative forces W_{nc}, plus that done by all internal conservative forces W_{c}:

$$W_{\text{total}} = W_{\text{ext}} + W_{\text{nc}} + W_{\text{c}}$$

Rearranging gives

$$W_{\text{ext}} + W_{\text{nc}} = W_{\text{total}} - W_{\text{c}}$$

The negative of the total work done by all the conservative internal forces $-W_{\text{c}}$ equals the change in the potential energy of the system ΔU_{sys}

$$-W_{\text{c}} = \Delta U_{\text{sys}} \qquad 7\text{-}6$$

Substituting from Equations 7-5 and 7-6, we have

$$W_{\text{ext}} + W_{\text{nc}} = \Delta K_{\text{sys}} + \Delta U_{\text{sys}} \qquad 7\text{-}7$$

The right side of this equation can be simplified as

$$\Delta K_{\text{sys}} + \Delta U_{\text{sys}} = \Delta(K_{\text{sys}} + U_{\text{sys}}) \qquad 7\text{-}8$$

The sum of the kinetic energy K_{sys} and the potential energy U_{sys} is called the **total mechanical energy** E_{mech}:

$$E_{\text{mech}} = K_{\text{sys}} + U_{\text{sys}} \qquad 7\text{-}9$$

DEFINITION—TOTAL MECHANICAL ENERGY

Combining Equations 7-8 and 7-9, and then substituting into Equation 7-7 gives

$$W_{\text{ext}} = \Delta E_{\text{mech}} - W_{\text{nc}} \qquad 7\text{-}10$$

WORK–ENERGY THEOREM FOR SYSTEMS

The mechanical energy of a system of particles is conserved (E_{mech} = constant) if the total work done by all external forces and by all internal nonconservative forces is zero.

$$E_{\text{mech}} = K_{\text{sys}} + U_{\text{sys}} = \text{constant} \qquad 7\text{-}11$$

CONSERVATION OF MECHANICAL ENERGY

This is **conservation of mechanical energy,** and is the origin of the expression "conservative force."

If $E_{\text{mech i}} = K_i + U_i$ is the initial mechanical energy of the system and $E_{\text{mech f}} = K_f + U_f$ is the final mechanical energy of the system, conservation of mechanical energy implies that

$$E_{\text{mech f}} = E_{\text{mech i}} \quad (\text{or } K_f + U_f = K_i + U_i) \qquad 7\text{-}12$$

In other words, when the mechanical energy of a system is conserved, we can relate the final mechanical energy of the system to the initial mechanical energy of the system without considering the intermediate motion and the work done by the forces involved. Therefore, conservation of mechanical energy allows us to solve problems that might be difficult to solve using Newton's laws directly.

APPLICATIONS

Suppose that you are on skis and, starting at rest from a height h_i above the bottom of a hill, you coast down the hill. Assuming that friction and air drag are negligible, what is your speed as you pass by a marker on the hill a height h above the bottom of the hill?

The mechanical energy of the Earth–skier system is conserved because the only force doing work is the internal, conservative force of gravity. If we choose $U = 0$ at the bottom of the hill, the initial potential energy is mgh_i. This energy is also the total mechanical energy because the initial kinetic energy is zero. Thus,

$$E_{\text{mech i}} = K_i + U_i = 0 + mgh_i$$

As you pass the marker, the potential energy is mgh and the speed is v. Hence,

$$E_{\text{mech f}} = K_f + U_f = \tfrac{1}{2}mv^2 + mgh$$

Setting $E_{\text{mech f}} = E_{\text{mech i}}$, we find

$$\tfrac{1}{2}mv^2 + mgh = mgh_i$$

Solving for v gives

$$v = \sqrt{2g(h_i - h)}$$

Your speed is the same as if you had undergone free-fall, falling straight down through a distance $h_i - h$. However, by skiing down the hill, you travel a greater distance and take more time than you would if you were in free-fall and falling straight down.

PROBLEM-SOLVING STRATEGY

Solving Problems Involving Mechanical Energy

PICTURE Identify a system that includes the object (or objects) of interest and any other objects that interact with the object of interest by either a conservative or a kinetic-frictional force.

SOLVE

1. Make a sketch of the system and include labels. Include a coordinate axis (or axes) and show the system in its initial and final configurations. (Showing an intermediate configuration is often helpful also.) Objects may be represented as dots, just as is done in free-body diagrams.
2. Identify any external forces acting on the system that do work, and any internal nonconservative forces that do work. Also identify any internal conservative forces that do work.
3. Apply Equation 7-10 (the work–energy theorem for systems). For each internal conservative force doing work use a potential-energy function to represent the work done.

CHECK Make sure that you have accounted for the work done by all conservative and nonconservative forces in determining your answer.

Example 7-4 Kicking a Ball

Standing near the edge of the roof of a 12-m-high building, you kick a ball with an initial speed of $v_i = 16\text{ m/s}$ at an angle of 60° above the horizontal. Neglecting effects due to air resistance, find (*a*) how high above the height of the building the ball rises, and (*b*) its speed just before it hits the ground.

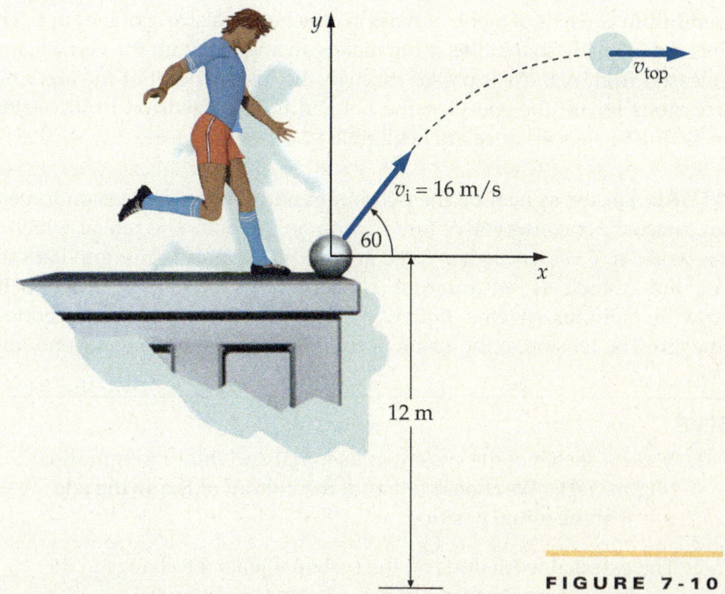

FIGURE 7-10

PICTURE We choose the ball and Earth as the system. We consider this system during the interval from just after the kick to just before impact with the ground. No external forces do work on the system, and no internal nonconservative forces do work, so the mechanical energy of the system is conserved. At the top of its flight, the ball is moving horizontally with a speed v_{top}, equal to the horizontal component of its initial velocity v_{ix}. We choose $y = 0$ at the roof of the building.

SOLVE

(*a*) 1. Make a sketch (Figure 7-10) of the trajectory. Include coordinate axes and show the initial position of the ball and its position at the top of its flight. Choose $y = 0$ at the roof of the building:

2. Apply the work–energy equation for systems. Choose the ball and Earth as the system. Following the kick and before impact with the ground no external forces do work and no nonconservative forces do work (we are neglecting the effects of air resistance):

$$W_{ext} = \Delta E_{mech} - W_{nc}$$
$$0 = \Delta E_{mech} - 0$$
$$\therefore E_{mech\,f} = E_{mech\,i}$$

3. The gravitational force does work on the system. We account for this work using the gravitational potential energy function mgy:

$$E_{mech\,top} = E_{mech\,i}$$
$$\tfrac{1}{2}mv_{top}^2 + mgy_{top} = \tfrac{1}{2}mv_i^2 + mgy_i$$
$$\tfrac{1}{2}mv_{top}^2 + mgh_{top} = \tfrac{1}{2}mv_i^2 + 0$$

4. Conservation of mechanical energy relates the height y_{top} above the roof of the building to the initial speed v_i and the speed at the top of its flight v_{top}:

$$E_{mech\,top} = E_{mech\,i}$$
$$\tfrac{1}{2}mv_{top}^2 + mgy_{top} = \tfrac{1}{2}mv_i^2 + mgy_i$$
$$\tfrac{1}{2}mv_{top}^2 + mgy_{top} = \tfrac{1}{2}mv_i^2 + 0$$

5. Solve for y_{top}:

$$y_{top} = \frac{v_i^2 - v_{top}^2}{2g}$$

6. The velocity at the top of its flight equals the x component of its initial velocity:

$$v_{top} = v_{ix} = v_i \cos\theta$$

7. Substitute the step-3 result into the step-2 result and solve for y_{top}:

$$y_{top} = \frac{v_i^2 - v_{top}^2}{2g} = \frac{v_i^2 - v_i^2\cos^2\theta}{2g} = \frac{v_i^2(1 - \cos^2\theta)}{2g}$$
$$= \frac{(16\text{ m/s})^2(1 - \cos^2 60°)}{2(9.81\text{ m s}^2)} = \boxed{9.8\text{ m}}$$

(*b*) 1. If v_f is the speed of the ball just before it hits the ground (where $y = y_f = -12$ m), its energy is expressed:

$$E_{mech\,f} = \tfrac{1}{2}mv_f^2 + mgy_f$$

2. Set the final mechanical energy equal to the initial mechanical energy:

$$\tfrac{1}{2}mv_f^2 + mgy_f = \tfrac{1}{2}mv_i^2 + 0$$

3. Solve for v_f, and set $y_f = -12$ m to find the final velocity:

$$v_f = \sqrt{v_i^2 - 2gy_f}$$
$$= \sqrt{(16\text{ m/s})^2 - 2(9.81\text{ m/s}^2)(-12\text{ m})}$$
$$= \boxed{22\text{ m/s}}$$

CHECK We would expect that the higher the building, the greater the speed at impact. The expression for v_f in step 3 of Part (*b*) reflects this expectation.

Example 7-5 A Pendulum

A pendulum consists of a bob of mass m attached to a string of length L. The bob is pulled aside so that the string makes an angle θ_0 with the vertical, and is released from rest. As it passes through the lowest point of the arc, find expressions for (*a*) the speed of the bob, and (*b*) the tension in the string. Effects due to air resistance are negligible.

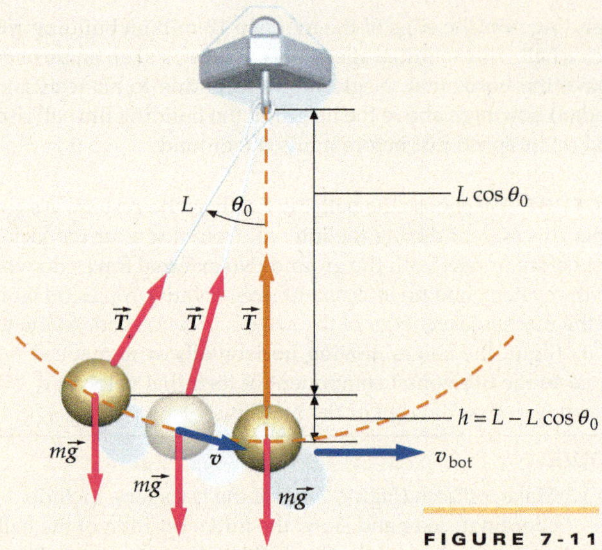

FIGURE 7-11

PICTURE Let the system be the pendulum and Earth. The tension force $\vec{T}$ is an internal nonconservative force acting on the bob. The rate at which $\vec{T}$ does work is $\vec{T}\cdot\vec{v}$. The other force acting on the bob is the gravitational force $m\vec{g}$, which is an internal conservative force. Use the work–energy theorem for systems (Equation 7-10) to find the speed at the bottom of the arc. The tension in the string is obtained using Newton's second law.

SOLVE

(*a*) 1. Make a sketch of the system in its initial and final configurations (Figure 7-11). We choose $y = 0$ at the bottom of the swing and $y = h$ at the initial position:

2. The external work done on the system equals the change in its mechanical energy minus the work done by internal nonconservative forces (Equation 7-10):

$$W_{ext} = \Delta E_{mech} - W_{nc}$$

3. There are no external forces acting on the system. The tension force is an internal nonconservative force:

$$W_{ext} = 0$$

$$W_{nc} = \int_1^2 \vec{T}\cdot d\vec{\ell}$$

4. The displacement increment $d\vec{\ell}$ equals the velocity times the time increment dt. Substitute into the step-3 result. The tension is perpendicular to the velocity, so $\vec{T}\cdot\vec{v} = 0$:

$$d\vec{\ell} = \vec{v}\,dt$$

$$\text{so}\quad W_{nc} = \int_1^2 \vec{T}\cdot d\vec{\ell} = \int_1^2 \vec{T}\cdot\vec{v}\,dt = 0$$

5. Substitute for W_{ext} and W_{nc} in the step-2 result. The bob initially is at rest:

$$W_{ext} = \Delta E_{mech} - W_{nc}$$
$$0 = \Delta E_{mech} - 0$$
$$\Delta E_{mech} = 0$$

6. Apply conservation of mechanical energy. The bob initially is at rest:

$$E_{mech\,f} = E_{mech\,i}$$
$$\tfrac{1}{2}mv_f^2 + mgy_f = \tfrac{1}{2}mv_i^2 + mgy_i$$
$$\tfrac{1}{2}mv_{bot}^2 + 0 = 0 + mgh$$

7. Conservation of mechanical energy thus relates the speed v_{bot} to the initial height $y_i = h$:

$$\tfrac{1}{2}mv_{bot}^2 = mgh$$

8. Solve for the speed v_{bot}:

$$v_{bot} = \sqrt{2gh}$$

9. To express speed in terms of the initial angle θ_0, we need to relate h to θ_0. This relation is illustrated in Figure 7-11:

$$L = L\cos\theta_0 + h$$
$$\text{so}\quad h = L - L\cos\theta_0 = L(1 - \cos\theta_0)$$

10. Substitute this value for h to express the speed at the bottom in terms of θ_0:

$$v_{bot} = \boxed{\sqrt{2gL(1 - \cos\theta_0)}}$$

(*b*) 1. When the bob is at the bottom of the circle, the forces on it are $m\vec{g}$ and $\vec{T}$. Apply $\Sigma F_y = ma_y$:

$$T - mg = ma_y$$

2. At the bottom, the bob has an acceleration v_{bot}^2/L in the centripetal direction (toward the center of the circle), which is upward:

$$a_y = \frac{v_{bot}^2}{L} = \frac{2gL(1 - \cos\theta_0)}{L} = 2g(1 - \cos\theta_0)$$

3. Substitute for a_y in the Part-(*b*), step-1 result and solve for T:

$$T = mg + ma_y = m(g + a_y) = m[g + 2g(1 - \cos\theta_0)]$$
$$= \boxed{(3 - 2\cos\theta_0)mg}$$

CHECK (1) The tension at the bottom is greater than the weight of the bob because the bob is accelerating upward. (2) Step 3 in Part (*b*) shows that for $\theta_0 = 0$, $T = mg$, the expected result for a stationary bob hanging from a string.

TAKING IT FURTHER (1) The rate at which a force does work is given by $\vec{F}\cdot\vec{v}$ (Equation 6-22). Step 4 of Part (*a*) reveals that the rate at which the tension force is doing work is zero. *Any* force that remains perpendicular to the velocity does zero work. (2) Step 8 of Part (*a*) shows that the speed at the bottom is the same as if the bob had dropped in free-fall from a height h. (3) The speed of the bob at the bottom of the arc can also be found using Newton's laws directly, but such a solution is more challenging because the tangential acceleration a_t varies with position, and therefore with time, so the constant-acceleration formulas do not apply. (4) If the string had not been included in the system then the W_{ext} would equal the work done by the tension force and W_{nc} would equal zero because there would be no internal nonconservative forces. The results would be identical.

Example 7-6 A Block Pushing a Spring *Try It Yourself*

A 2.0-kg block on a frictionless horizontal surface is pushed against a spring that has a force constant of 500 N/m, compressing the spring by 20 cm. The block is then released, and the force of the spring accelerates the block as the spring decompresses. The block then glides along the surface and then up a frictionless incline of angle 45°. How far up the incline does the block travel before momentarily coming to rest?

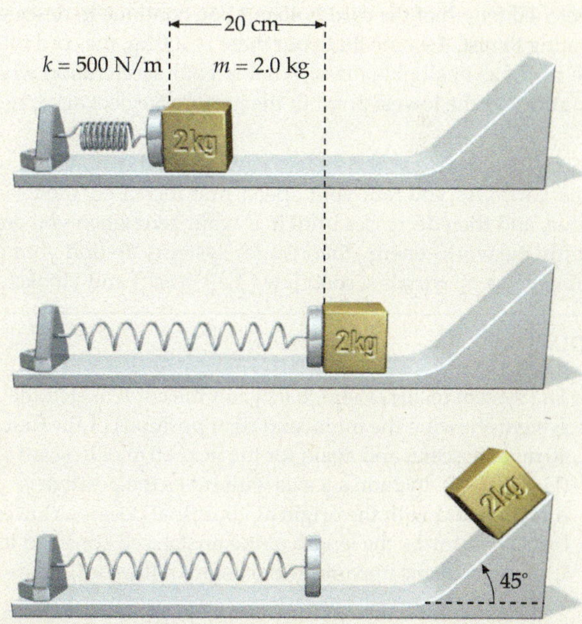

FIGURE 7-12

PICTURE Let the system include the block, the spring, Earth, the horizontal surface, the ramp, and the wall to which the spring is attached. After the block is released there are no external forces on this system. The only forces that do work are the force exerted by the spring on the block and the force of gravity, both of which are conservative. Thus, the total mechanical energy of the system is conserved. Find the maximum height h from the conservation of mechanical energy, and then the maximum distance up the incline s from $\sin 45° = h/s$.

SOLVE

Cover the column to the right and try these on your own before looking at the answers.

Steps	Answers
1. Choose the block, the spring, Earth, the horizontal surface, the ramp, and the wall to which the spring is attached. Sketch the system with both the initial and final configurations (Figure 7-12).	
2. Apply the work–energy theorem for systems. Following release no external forces act on the system and no internal nonconservative forces do work on the system:	$W_{ext} = \Delta E_{mech} - W_{nc}$ $0 = \Delta E_{mech} - 0$ $\therefore E_{mech\,f} = E_{mech\,i}$
3. Write the initial mechanical energy in terms of the compression distance x_i.	$E_{mech\,i} = U_{si} + U_{gi} + K_i = \frac{1}{2}kx_i^2 + 0 + 0$
4. Write the final mechanical energy in terms of the height h.	$E_{mech\,f} = U_{sf} + U_{gf} + K_f = 0 + mgh + 0$
5. Substitute into the step-3 result and solve for h.	$mgh = \frac{1}{2}kx_i^2$ $h = \dfrac{kx_i^2}{2mg} = 0.51\text{ m}$
6. Find the distance s from h and the angle of inclination (Figure 7-13).	$h = s \times \sin\theta$ $s = \boxed{0.72\text{ m}}$

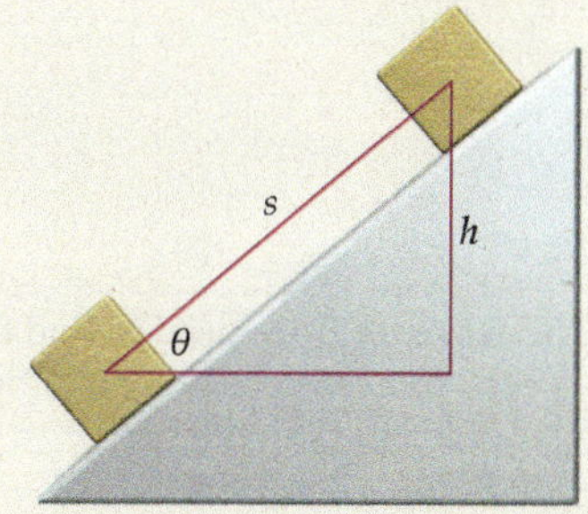

FIGURE 7-13

CHECK The expression for h in step 5 is plausible. It tells us by inspection that increasing x_i results in a larger maximum height, and using a larger mass results in a smaller maximum height.

TAKING IT FURTHER (1) In this problem, the initial mechanical energy of the system is the potential energy of the spring. This energy is transformed first into kinetic energy and then into gravitational potential energy. (2) The normal force $\vec{F}_n$ on the block always acts at right angles to the velocity, so $\vec{F}_n \cdot \vec{v} = 0$ at all times.

PRACTICE PROBLEM 7-5 Find the speed of the block just as it leaves the spring.

PRACTICE PROBLEM 7-6 How much work did the normal force on the block do?

Example 7-7 Bungee Jumping *Context-Rich*

You jump off a platform 134 m above the Nevis River. After you have free-fallen for the first 40 m, the bungee cord attached to your ankles starts to stretch. (The unstretched length of the cord is 40 m.) You continue to descend another 80 m before coming to rest. Assume that your mass is 100 kg, the cord follows Hooke's law, and the cord has negligible mass. What is your acceleration when you are momentarily at rest at the lowest point in the jump? (Neglect air drag.)

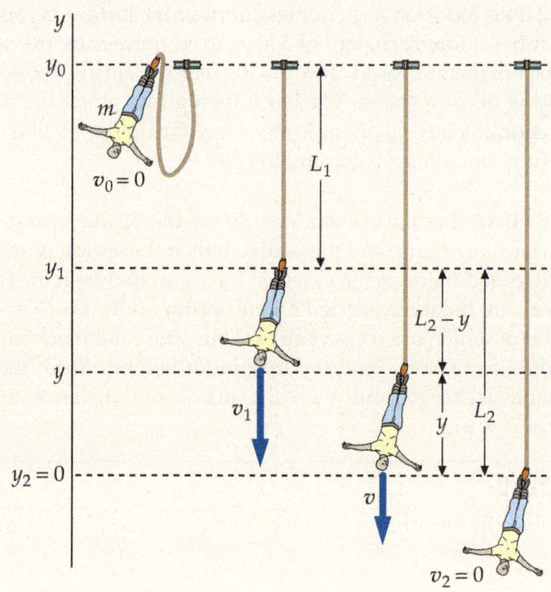

FIGURE 7-14

PICTURE Choose as the system everything mentioned in the problem statement, plus Earth. As you fall, your speed first increases, then reaches some maximum value, and then decreases until it is again zero when you are at your lowest point. Apply the work–energy theorem for systems. To find your acceleration at the bottom we use Newton's second law ($\Sigma F_x = ma_x$) and Hooke's law ($F_x = -kx$).

SOLVE

1. The system includes you, Earth and the cord. Sketch the system showing the initial and final positions of the first 40 m of descent, and again for the next 80 m of descent (Figure 7-14). Include a y axis with up as the positive y direction and with the origin at your final position (lowest). Let $L_1 = 40\,\text{m}$ be the length of the unstressed cord and let $L_2 = 80\,\text{m}$ be the maximum extension of the cord.

2. Apply the work–energy theorem for systems. There are no external forces and no internal nonconservative forces doing work:

$$W_{\text{ext}} = \Delta E_{\text{mech}} - W_{\text{nc}}$$
$$0 = \Delta E_{\text{mech}} - 0$$
$$\therefore E_{\text{mech f}} = E_{\text{mech i}}$$

3. Apply the step-2 result to the part of the descent that the cord is stretching. The extension of the cord is $L_2 - y$:

$$E_{\text{mech 3}} = E_{\text{mech 2}}$$
$$U_{g3} + U_{s3} + K_3 = U_{g2} + U_{s2} + K_2$$
$$mgy_3 + \tfrac{1}{2}k(L_2 - y_3)^2 + \tfrac{1}{2}mv_3^2 = mgy_2 + \tfrac{1}{2}ky_2^2 + \tfrac{1}{2}mv_2^2$$
$$0 + \tfrac{1}{2}kL_2^2 + 0 = mgL_2 + 0 + \tfrac{1}{2}mv_2^2$$
$$\tfrac{1}{2}kL_2^2 = mgL_2 + \tfrac{1}{2}mv_2^2$$

4. To solve for k, we need to find the kinetic energy at the end of the free-fall region. Apply the step-2 result again and solve for the kinetic energy:

$$E_{\text{mech 2}} = E_{\text{mech 1}}$$
$$U_{g2} + K_2 = U_{g1} + K_1$$
$$mgy_2 + \tfrac{1}{2}mv_2^2 = mgy_1 + \tfrac{1}{2}mv_1^2$$
$$mgL_2 + \tfrac{1}{2}mv_2^2 = mg(L_1 + L_2) + 0$$
$$\tfrac{1}{2}mv_2^2 = mgL_1$$

5. Substitute the step-4 result into the step-3 result and solve for k:

$$\tfrac{1}{2}kL_2^2 = mgL_2 + mgL_1$$

$$k = \frac{2mg(L_2 + L_1)}{L_2^2}$$

6. Apply Newton's second law when you are at the lowest point. First we construct a free-body diagram (Figure 7-15):

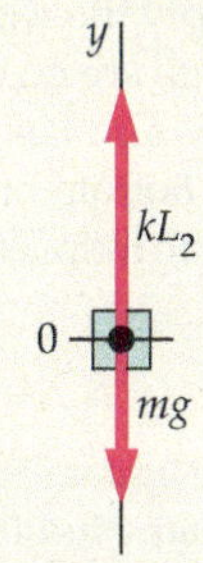

FIGURE 7-15

7. Apply Newton's second law and solve for the acceleration. Use the expression for k from step 5:

$$\Sigma F_y = ma_y$$

$$-mg + kL_2 = ma_y$$

$$a_y = -g + k\frac{L_2}{m} = -g + \frac{2mg(L_2 + L_1)}{L_2^2}\frac{L_2}{m}$$

$$= g\left(1 + 2\frac{L_1}{L_2}\right) = g\left(1 + 2\frac{40}{80}\right) = \boxed{2.0g}$$

CHECK We expect the acceleration at the bottom to be upward (the $+y$ direction) and our result is in agreement with that. Any time the velocity reverses directions, immediately following the reversal the velocity vector and the acceleration vector are in the same direction.

PRACTICE PROBLEM 7-7 As you fall, you gain speed until the upward pull of the cord equals the downward pull of gravity. What is your height above the lowest point when you achieve maximum speed?

Example 7-8 Back to the Future *Context-Rich*

Imagine that you have time-traveled back to the late 1800s and are watching your great-great-great-grandparents on their honeymoon taking a ride on the Flip Flap Railway, a Coney Island roller coaster with a circular loop-the-loop. The car they are in is about to enter the loop-the-loop when a 100-lb sack of sand falls from a construction-site platform and lands in the back seat of the car. No one is hurt, but the impact causes the car to lose 25 percent of its speed. The car started from rest at a point 2 times as high as the top of the circular loop. Neglect losses due to friction or air drag. Will their car make it over the top of the loop-the-loop without falling off?

PICTURE Let the system be the car, its contents, the track (including the loop-the-loop), and Earth. The car has to have enough speed at the top of the loop to maintain contact with the track. We can use the work–energy theorem for systems to determine the speed just before the sandbag hits the car, and we can use it again to determine the speed the car has at the top of the loop. Then we can use Newton's second law to determine the magnitude of the normal force, if any, exerted on the car by the track.

SOLVE

1. Choose the system to be the car, its contents, the track, and Earth. Draw a picture of the car and track, with the car at the starting point, at the bottom of the track, and again at the top of the loop (Figure 7-16):

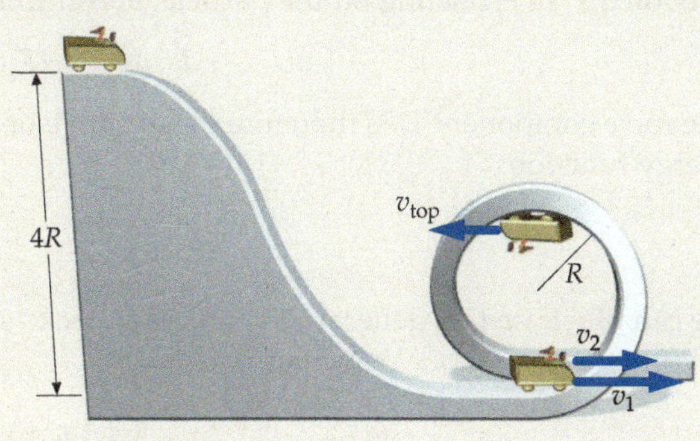

FIGURE 7-16

2. Apply Newton's second law to relate the speed at the top of the loop to the normal force:

$$F_n + mg = m\frac{v_{top}^2}{R}$$

3. Apply the work–energy theorem to the interval prior to impact. There are no external forces and no internal nonconservative forces do work. Find the speed just prior to impact. Measuring heights from the bottom of the loop, the initial height of $4R$, where R is the radius of the loop, is two times the height of the top of the loop:

$$W_{ext} = \Delta E_{mech} - W_{nc}$$
$$0 = \Delta E_{mech} - 0$$
$$\therefore E_{mech\,f} = E_{mech\,i}$$
$$U_0 + K_0 = U_1 + K_1$$
$$mg\,4R + 0 = 0 + \tfrac{1}{2}mv_1^2$$
$$\text{so} \quad v_1 = \sqrt{8Rg}$$

4. The impact with the sandbag results in a 25 percent decrease in speed. Find the speed after impact:

$$v_2 = 0.75v_1 = 0.75\sqrt{8Rg}$$

5. Apply the work–energy theorem to the interval following impact. Find the speed at the top of the loop-the-loop:

$$U_{top} + K_{top} = U_2 + K_2$$
$$mg2R + \tfrac{1}{2}mv_{top}^2 = 0 + \tfrac{1}{2}m(0.75^2 \cdot 8Rg)$$
$$\text{so} \quad v_{top}^2 = (0.75^2 \cdot 8 - 4)Rg = 0.5Rg$$

6. Substituting for v_{top}^2 in the step-2 result gives:

$$F_n + mg = m\frac{0.5Rg}{R}$$
$$F_n + mg = 0.5mg$$

7. Solve for F_n:

$$F_n = -0.5mg$$

8. F_n is the magnitude of the normal force. It cannot be negative:

Oops! The car has left the track.

CHECK A loss of 25 percent of your speed means losing almost 44 percent of your kinetic energy. The speed is the same as would be attained if the car started from rest at a height of $0.56 \times 4R = 1.12 \times 2R$ (12 percent higher than the height of the top of the loop). We should not be too surprised to find that the car left the track.

TAKING IT FURTHER Fortunately, there were safety devices to prevent the cars from falling, so your ancestors likely would have survived. The biggest concern for riders on the Flip Flap Railway was a broken neck. The Flip Flap Railway subjected riders to accelerations of up to 12g's during the loop-the-loop and was the last of the circular loop-the-loop roller coasters. Loop-the-loops on modern rides are higher than they are wide.

POTENTIAL ENERGY AND EQUILIBRIUM

We can gain insight into the motion of a system by looking at a graph of its potential energy versus the position of a particle in that system. For simplicity, we restrict our consideration to a particle constrained to move along a straight line—the x axis. To create such a graph, we first must find the relationship between the potential energy function and the force acting on the particle. Consider a conservative force $\vec{F} = F_x\hat{i}$ acting on the particle. Substituting this into Equation 7-1b gives

$$dU = -\vec{F}\cdot d\vec{\ell} = -F_x dx$$

The force component F_x is therefore the negative of the derivative* of the potential-energy function:

$$F_x = -\frac{dU}{dx} \qquad \text{7-13}$$

We can illustrate this general relation for a block–spring system by differentiating the function $U = \frac{1}{2}kx^2$. We obtain

$$F_x = -\frac{dU}{dx} = -\frac{d}{dx}\left(\frac{1}{2}kx^2\right) = -kx$$

Figure 7-17 shows a plot of $U = \frac{1}{2}kx^2$ versus x for a block and spring. The derivative of this function is represented graphically as the slope of the tangent line

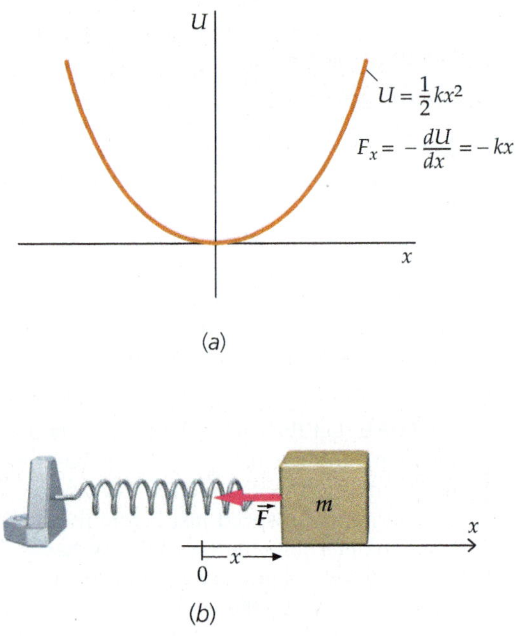

FIGURE 7-17 (a) Plot of the potential-energy function U versus x for an object on a spring. A minimum in the potential-energy curve is a point of stable equilibrium. Displacement in either direction results in a force directed toward the equilibrium position. (b) The applied force $\vec{F}_{app}$ pulls the block to the right, stretching the spring by x_1.

* The derivative in Equation 7-13 is replaced by the partial derivative with respect to x if the motion is not restricted to the x axis.

to the curve. The force is thus equal to the negative of the slope of the tangent line to the curve. At $x = 0$, the force $F_x = -dU/dx$ is zero and the block is in *equilibrium*, assuming no other forces are acting on it.

When x is positive in Figure 7-17*a*, the slope is positive and the force F_x is negative. When x is negative, the slope is negative and the force F_x is positive. In either case, the force is in the direction that will accelerate the block in the direction of decreasing potential energy. If the block is displaced slightly from $x = 0$, the force is directed back toward $x = 0$. The equilibrium at $x = 0$ is thus **stable equilibrium,** because a small displacement results in a restoring force that accelerates the particle back toward its equilibrium position.

! The potential energy function is a minimum at a point of stable equilibrium.

In stable equilibrium, a small displacement in any direction results in a restoring force that accelerates the particle back toward its equilibrium position.

CONDITION FOR STABLE EQUILIBRIUM

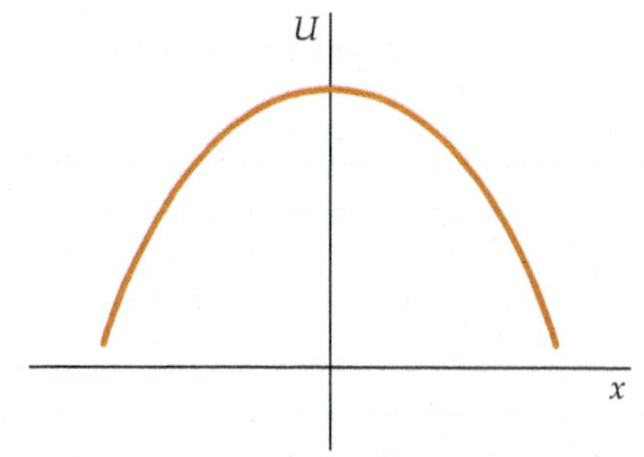

FIGURE 7-18 A particle with a potential energy shown by this potential-energy curve will be in unstable equilibrium at $x = 0$ because a displacement from $x = 0$ results in a force directed away from the equilibrium position.

Figure 7-18 shows a potential-energy curve with a maximum rather than a minimum at the equilibrium point $x = 0$. Such a curve could represent the potential energy of a space ship at the point between Earth and the moon where the gravitational pull on the ship by Earth is equal to the gravitational pull on the ship by the moon. (We are neglecting any gravitational pull from the Sun.) For this curve, when x is positive, the slope is negative and the force F_x is positive, and when x is negative, the slope is positive and the force F_x is negative. Again, the force is in the direction that will accelerate the particle toward lower potential energy, but this time the force is away from the equilibrium position. The maximum at $x = 0$ in Figure 7-18 is a point of **unstable equilibrium** because a small displacement results in a force that accelerates the particle away from its equilibrium position.

In unstable equilibrium, a small displacement results in a force that accelerates the particle away from its equilibrium position.

CONDITION FOR UNSTABLE EQUILIBRIUM

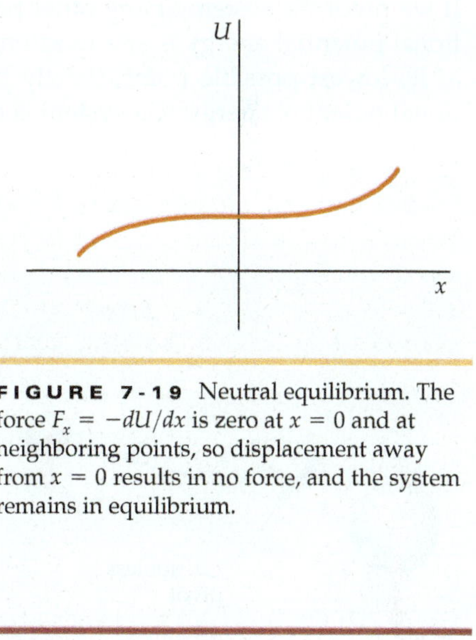

FIGURE 7-19 Neutral equilibrium. The force $F_x = -dU/dx$ is zero at $x = 0$ and at neighboring points, so displacement away from $x = 0$ results in no force, and the system remains in equilibrium.

Figure 7-19 shows a potential-energy curve that is flat in the region near $x = 0$. No force acts on a particle at $x = 0$, and hence the particle is at equilibrium; furthermore, there will be no resulting force if the particle is displaced slightly in either direction. This is an example of **neutral equilibrium.**

In neutral equilibrium, a small displacement in any direction results in zero force and the particle remains in equilibrium.

CONDITION FOR NEUTRAL EQUILIBRIUM

Example 7-9 Force and the Potential-Energy Function

In the region $-a < x < a$ the force on a particle is represented by the potential-energy function

$$U = -b\left(\frac{1}{a+x} + \frac{1}{a-x}\right)$$

where a and b are positive constants. (*a*) Find the force F_x in the region $-a < x < a$. (*b*) At what value of x is the force zero? (*c*) At the location where the force equals zero, is the equilibrium stable or unstable?

PICTURE The force is the negative of the derivative of the potential-energy function. The equilibrium is stable where the potential-energy function is a minimum and it is unstable where the potential-energy function is a maximum.

SOLVE

(*a*) Compute $F_x = -dU/dx$.

$$F_x = -\frac{d}{dx}\left[-b\left(\frac{1}{(a+x)} + \frac{1}{(a-x)}\right)\right] = \boxed{-b\left(\frac{1}{(a+x)^2} - \frac{1}{(a-x)^2}\right)}$$

(*b*) Set F_x equal to zero and solve for x.

$F_x = 0$ at $\boxed{x = 0}$

(*c*) Compute d^2U/dx^2. If it is positive at the equilibrium position, then U is a minimum and the equilibrium is stable. If it is negative, then U is a maximum and the equilibrium is unstable.

$$\frac{d^2U}{dx^2} = -2b\left(\frac{1}{(a+x)^3} + \frac{1}{(a-x)^3}\right)$$

At $x = 0$, $\dfrac{d^2U}{dx^2} = \dfrac{-4b}{a^3} < 0$

Thus, $\boxed{\text{unstable}}$ equilibrium.

CHECK If U is expressed in joules and x and a are expressed in meters, then b must be expressed in joule·meters and F_x must be expressed in newtons. Our Part-(*a*) result shows that F_x has the same units as those of Part (*b*) divided by m². That is, our expression for F_x has units of $\mathrm{J \cdot m/m^2 = J/m}$. Because $1\,\mathrm{J} = 1\,\mathrm{N \cdot m}$, our expression for F_x has units of newtons. Consequently, our Part-(*a*) result is dimensionally correct, and therefore is plausible.

TAKING IT FURTHER The potential-energy function in this example is for a particle under the influence of the gravitational forces exerted by two identical fixed-point masses, one at $x = -a$, the other at $x = +a$. The particle is located on the line joining the masses. Midway between the two masses the net force on the particle is zero. Otherwise, it is toward the closest mass.

We can use the result that a position of stable equilibrium is a potential-energy minimum to locate the center of mass experimentally. For example, two objects connected by a light rod will balance if the pivot is at the center of mass (Figure 7-20). If we pivot the system at any other point, the system will rotate until the gravitational potential energy is at a minimum, which occurs when the center of mass is at its lowest possible point directly below the pivot (Figure 7-21). (The gravitational potential energy of a system is given by $U_g = mgh_{cm}$ [Equation 7-3].)

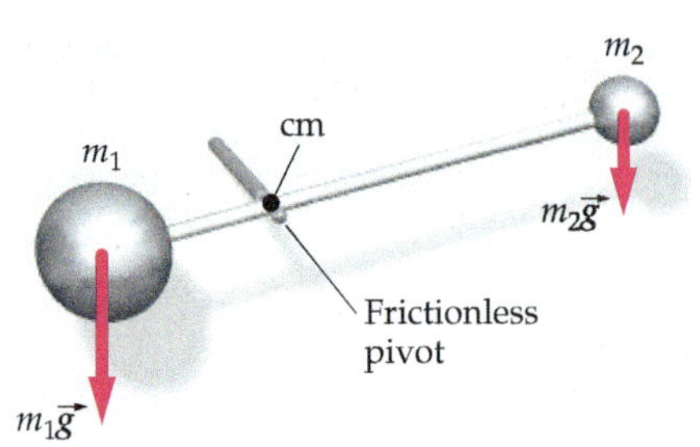

FIGURE 7-20

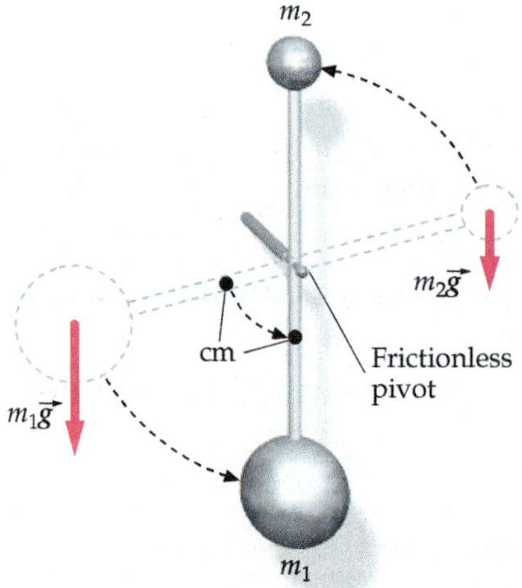

FIGURE 7-21

If we suspend any irregular object from a pivot, the object will hang so that its center of mass lies somewhere on the vertical line drawn directly downward from the pivot. Now suspend the object from another point and note where the vertical line through the pivot point now passes. The center of mass will lie at the intersection of the two lines (Figure 7-22).

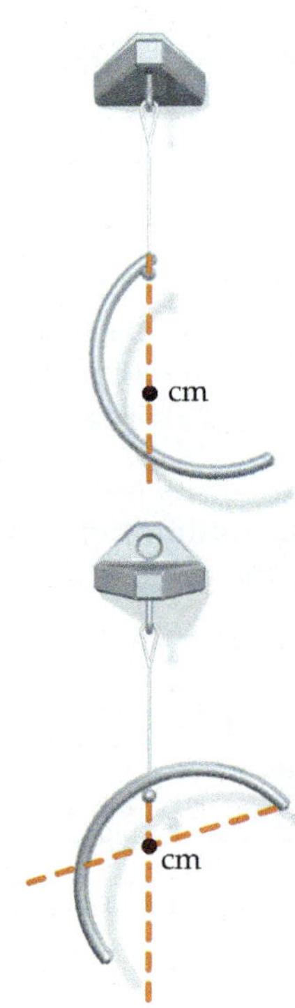

FIGURE 7-22 The center of mass of an irregular object can be found by suspending it first from one point and then from a second point.

For a system for which mechanical energy remains constant, graphs that plot both potential energy U and mechanical energy E are often useful. For example, Figure 7-23 is a plot of the potential-energy function

$$U = b\left(\frac{1}{a + x} + \frac{1}{a - x}\right)$$

which is the negative of the potential-energy function used in Example 7-9. Figure 7-23 contains plots of both this potential-energy function and the total mechanical energy E. The kinetic energy K for a specified value of x is represented by the distance that the total mechanical-energy line is above the potential-energy curve because $K = E - U$.

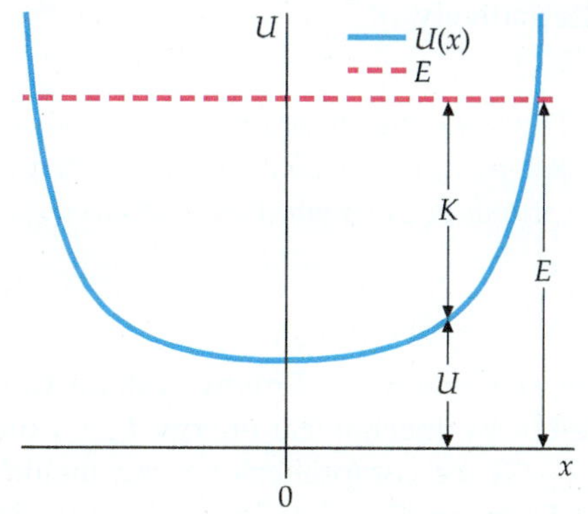

FIGURE 7-23 The potential energy U and the total mechanical energy E are plotted versus x. The sum of the kinetic energy K and the potential energy equals the total mechanical energy. That is, $K = E - U$.

7-3 THE CONSERVATION OF ENERGY

In the macroscopic world, dissipative nonconservative forces, such as kinetic friction, are always present to some extent. Such forces tend to decrease the mechanical energy of a system. However, any such decrease in mechanical energy is accompanied by a corresponding increase in thermal energy. (Consider that excessive automotive braking sometimes causes the temperature of the rotors or brake drums to increase to the point that the metal warps.) Another type of nonconservative force is that involved in the deformations of objects. When you bend a metal coat hanger back and forth, you do work on the coat hanger, but the work you do does not appear as mechanical energy. Instead, the coat hanger becomes warm. The work done in deforming the hanger is dissipated as thermal energy. Similarly, when a falling ball of modeling clay lands on the floor (thud), it becomes warmer as it deforms. The dissipated kinetic energy appears as thermal energy. For the clay–floor–Earth system, the total energy is the sum of the thermal energy and the mechanical energy. The total energy of the system is conserved even though neither the total mechanical energy nor the total thermal energy is individually conserved.

A third type of nonconservative force is associated with chemical reactions. When we include systems in which chemical reactions take place, the sum of mechanical energy plus thermal energy is not conserved. For example, suppose that you begin running from rest. You initially have no kinetic energy. When you begin to run, chemical energy stored in certain molecules in your muscles is transformed into kinetic energy and thermal energy. It is possible to identify and measure the chemical energy that is transformed into kinetic energy and thermal energy. In this case, the sum of mechanical energy, thermal energy, and chemical energy is conserved.

Even when thermal energy and chemical energy are included, the total energy of the system does not always remain constant, because energy can be converted to radiation energy, such as sound waves or electromagnetic waves. But *the increase or decrease in the total energy of a system can always be accounted for by the disappearance or appearance of energy outside the system.* This experimental result, known as the **law of conservation of energy,** is one of the most important laws in all of science. Let E_{sys} be the total energy of a given system, E_{in} be the energy that enters the system, and E_{out} be the energy that leaves the system. The law of conservation of energy then states

$$E_{in} - E_{out} = \Delta E_{sys} \qquad 7\text{-}14$$

LAW OF CONSERVATION OF ENERGY

Alternatively,

> The total energy of the universe is constant. Energy can be converted from one form to another, or transmitted from one region to another, but energy can never be created or destroyed.
>
> LAW OF CONSERVATION OF ENERGY

The total energy E of many systems from everyday life can be accounted for completely by mechanical energy E_{mech}, thermal energy E_{therm}, and chemical energy E_{chem}. To be comprehensive and include other possible forms of energy, such as electromagnetic or nuclear energy, we include E_{other}, and write

$$E_{\text{sys}} = E_{\text{mech}} + E_{\text{therm}} + E_{\text{chem}} + E_{\text{other}} \qquad \text{7-15}$$

THE WORK–ENERGY THEOREM

One way energy is transferred into or out of a system is for work to be done on the system by agents outside the system. In situations where this is the only method of energy transfer into or out of a system, the law of conservation of energy is expressed as:

$$W_{\text{ext}} = \Delta E_{\text{sys}} = \Delta E_{\text{mech}} + \Delta E_{\text{therm}} + \Delta E_{\text{chem}} + \Delta E_{\text{other}} \qquad \text{7-16}$$

WORK–ENERGY THEOREM

where W_{ext} is the work done on the system by external forces and ΔE_{sys} is the change in the system's total energy. This **work–energy theorem** for systems, which we call simply the work–energy theorem, is a powerful tool for studying a wide variety of systems. Note that if the system is just a single particle, its energy can only be kinetic. In that case, the work–energy theorem (Equation 7-16) reduces to the work–kinetic-energy theorem (Equation 6-8) studied in Chapter 6.

There are two methods for transferring energy into or out of a system. The second method is called heat. Heat is the transfer of energy due to a temperature difference. Exchanges of energy due to a temperature difference between a system and its surroundings are discussed in Chapter 18. In this chapter, the transfer of energy by heat is assumed to be negligible.

Example 7-10 Falling Clay — *Conceptual*

A ball of modeling clay with mass m is released from rest from a height h and falls to the perfectly rigid floor (thud). Discuss the application of the law of conservation of energy to (*a*) the system consisting of the clay ball alone, and (*b*) the system consisting of Earth, the floor, and the clay ball.

PICTURE Two forces act on the clay ball following its release: the force of gravity and the contact force of the floor. Because the floor does not move (it is rigid), the contact force it exerts on the clay ball does no work. There are no chemical or other energy changes, so we can neglect ΔE_{chem} and ΔE_{other}. (We neglect the sound energy radiated when the clay ball hits the floor.) Thus, the only energy transferred to or from the clay ball is the work done by the force of gravity.

SOLVE

(*a*) 1. Write the work–energy theorem for the clay ball:

$$W_{\text{ext}} = \Delta E_{\text{sys}} = \Delta E_{\text{mech}} + \Delta E_{\text{therm}} + \Delta E_{\text{chem}} + \Delta E_{\text{other}}$$
$$W_{\text{ext}} = \Delta E_{\text{sys}} = \Delta E_{\text{mech}} + E_{\text{therm}}$$

2. The two external forces on the system (the clay ball) are the force of gravity and the normal force exerted by the floor on the clay ball. However, the part of the ball in contact with the floor does not move, so the normal force on the ball by the floor does no work. Thus, the only work done on the clay ball is done by the force of gravity on the ball:

$$W_{ext} = mgh$$

3. Because the clay ball alone is our system, its mechanical energy is entirely kinetic, which is zero both initially and finally. Thus, the change in mechanical energy is zero:

$$\Delta E_{mech} = 0$$

4. Substitute mgh for W_{ext} and 0 for ΔE_{mech} in step 1:

$$W_{ext} = \Delta E_{mech} + E_{therm}$$

$$mgh = 0 + \Delta E_{therm}$$

so $\boxed{\Delta E_{therm} = mgh}$

(*Note:* If the floor were not perfectly rigid, the increase in thermal energy would be shared by the ball and the floor.)

(*b*) 1. No external forces act on the clay ball–Earth–floor system (the force of gravity and the force of the floor are now internal to the system), so there is no external work done:

$$W_{ext} = 0$$

2. Write the work–energy theorem with $W_{ext} = 0$:

$$W_{ext} = \Delta E_{sys} = \Delta E_{mech} + \Delta E_{therm}$$

$$0 = \Delta E_{mech} + \Delta E_{therm}$$

3. The initial mechanical energy of the clay ball–Earth system is the initial gravitational potential energy. The final mechanical energy is zero:

$$E_{mech\,i} = mgh$$

$$E_{mech\,f} = 0$$

4. The change in mechanical energy of the clay ball–Earth system is thus:

$$\Delta E_{mech} = 0 - mgh = -mgh$$

5. The work–energy theorem thus gives the same result as in Part (*a*):

$$\Delta E_{therm} = \boxed{-\Delta E_{mech} = mgh}$$

CHECK The Part-(*a*) and Part-(*b*) results are the same—that the thermal energy of the system increases by mgh. This is as one would expect.

TAKING IT FURTHER In Part (*a*), energy is transferred to the ball by the work done on it by the force of gravity. This energy appears as the kinetic energy of the ball before it impacts the floor and as thermal energy after impact. The ball warms slightly and the energy is eventually transferred to the surroundings. In Part (*b*), no energy is transferred to the ball–Earth–floor system. The original potential energy of the system is converted to kinetic energy of the ball just before it hits, and then into thermal energy.

PROBLEMS INVOLVING KINETIC FRICTION

When surfaces slide across each other, kinetic friction decreases the mechanical energy of the system and increases the thermal energy. Consider a block that begins with initial velocity v_i and slides along a board that is on a frictionless surface (Figure 7-24). The board is initially at rest. We choose the block and board to be our system, and $\Delta E_{chem} = \Delta E_{other} = 0$. No external work is done on this system. The work–energy theorem gives

$$0 = \Delta E_{mech} + \Delta E_{therm} \qquad 7\text{-}17$$

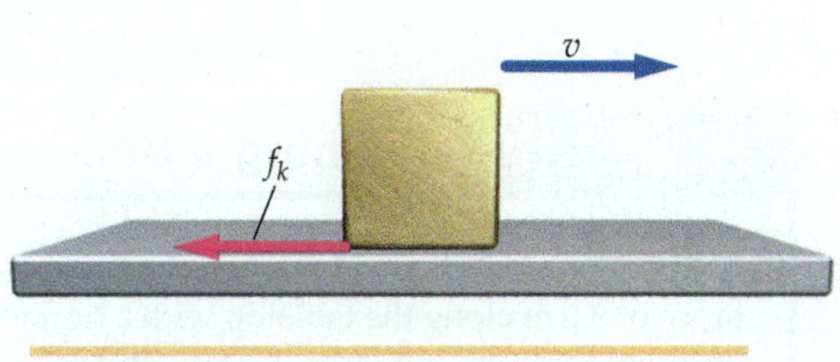

FIGURE 7-24

The change in mechanical energy is given by

$$\Delta E_{mech} = \Delta K_{block} + \Delta K_{board} = (\tfrac{1}{2}mv_f^2 - \tfrac{1}{2}mv_i^2) + (\tfrac{1}{2}MV_f^2 - 0) \qquad 7\text{-}18$$

where m is the mass of the block, M is the mass of the board, v is the speed of the block, and V is the speed of the board. We can relate this change in mechanical energy to the kinetic frictional force. If f_k is the magnitude of the frictional force on either the block or the board, Newton's second law applied to the block gives

$$-f_k = ma_x$$

where a_x is the acceleration of the block. Multiplying both sides by the displacement of the block Δx, we obtain

$$-f_k \Delta x = ma\Delta x \qquad 7\text{-}19$$

Solving the constant-acceleration formula $2a_x \Delta x = v_f^2 - v_i^2$ for $a_x \Delta x$ and substituting into Equation 7-19 gives

$$-f_k \Delta x = ma_x \Delta x = m\left(\tfrac{1}{2}v_f^2 - \tfrac{1}{2}v_i^2\right) = \tfrac{1}{2}mv_f^2 - \tfrac{1}{2}mv_i^2 \qquad 7\text{-}20$$

Equation 7-20 is just the center-of-mass work–translational-kinetic-energy relation (Equation 6-27) applied to the block. Applying this same relation to the board gives

$$f_k \Delta X = MA_X \Delta X = M\left(\tfrac{1}{2}V_f^2 - \tfrac{1}{2}V_i^2\right) = \tfrac{1}{2}MV_f^2 - 0 \qquad 7\text{-}21$$

where ΔX and A_x are the displacement and acceleration of the board. Adding Equations 7-20 and 7-21 gives

$$-f_k(\Delta x - \Delta X) = \left(\tfrac{1}{2}mv_f^2 - \tfrac{1}{2}mv_i^2\right) + \tfrac{1}{2}MV_f^2 \qquad 7\text{-}22$$

We note that $\Delta x - \Delta X$ is the distance s_{rel} that the block slides relative to the board, and that the right side of Equation 7-22 is the change in mechanical energy ΔE_{mech} of the block–board system. Substituting into Equation 7-22 gives

$$-f_k s_{rel} = \Delta E_{mech} \qquad 7\text{-}23$$

The decrease in mechanical energy of the block–board system is accompanied by a corresponding increase in the thermal energy of the system. This thermal energy appears both on the bottom surface of the block and on the upper surface of the board. Substituting ΔE_{therm} for $-\Delta E_{mech}$ (Equation 7-17), we obtain

$$f_k s_{rel} = \Delta E_{therm} \qquad 7\text{-}24$$

ENERGY DISSIPATED BY KINETIC FRICTION

where s_{rel} is the distance one contacting surface slides relative to the other contacting surface. Because the distance s_{rel} is the same in all frames of reference, Equation 7-24 is valid in all frames of reference, independent of whether they are inertial frames of reference or not.

Substituting this result into the work–energy theorem (with $\Delta E_{chem} = \Delta E_{other} = 0$), we obtain

$$W_{ext} = \Delta E_{mech} + \Delta E_{therm} = \Delta E_{mech} + f_k s_{rel} \qquad 7\text{-}25$$

WORK–ENERGY THEOREM WITH FRICTION

Example 7-11 Pushing a Box

A 4.0-kg box is initially at rest on a horizontal tabletop. You push the box a distance of 3.0 m along the tabletop with a horizontal force of 25 N. The coefficient of kinetic friction between the box and tabletop is 0.35. Find (*a*) the external work done on the block–table system, (*b*) the energy dissipated by friction, (*c*) the final kinetic energy of the box, and (*d*) the speed of the box.

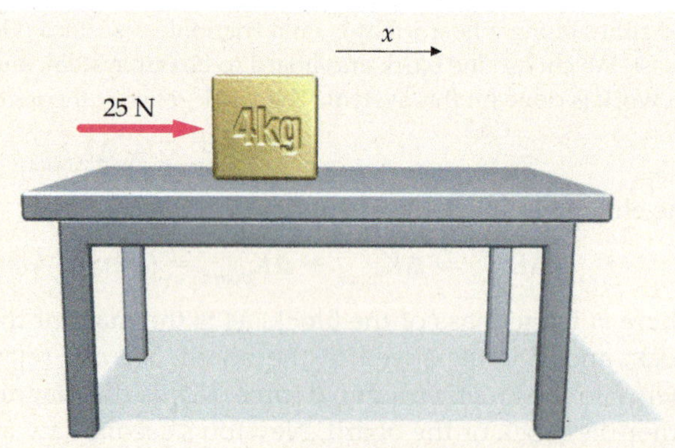

FIGURE 7-25

PICTURE The box plus table is the system (Figure 7-25). You are external to this system, so the force you push with is an external force. The final speed of the box is found from its final kinetic energy, which we find using the work–energy theorem with $\Delta E_{chem} = 0$ and $\Delta E_{therm} = f_k s_{rel}$. The energy of the system is increased by the external work. Some of the energy increase is kinetic energy and some is thermal energy.

SOLVE

(*a*) Four external forces are acting on the system. However, only one of them does work. The total external work done is the product of the push force and the distance traveled:

$$\Sigma W_{ext} = W_{\text{by you on block}} + W_{\text{by gravity on block}} + W_{\text{by gravity on table}} + W_{\text{by floor on table}}$$
$$= F_{push}\,\Delta x + 0 + 0 + 0 = (25\text{ N})(3.0\text{ m})$$
$$= \boxed{75\text{ J}}$$

(*b*) The energy dissipated by friction is $f_k\,\Delta x$ (the magnitude of the normal force equals mg):

$$\Delta E_{therm} = f_k\,\Delta x = \mu_k F_n\,\Delta x = \mu_k mg\,\Delta x$$
$$= (0.35)(4.0\text{ kg})(9.81\text{ N/kg})(3.0\text{ m})$$
$$= \boxed{41\text{ J}}$$

(*c*) 1. Apply the work–energy theorem to find the final kinetic energy:

$$W_{ext} = \Delta E_{mech} + \Delta E_{therm}$$

2. No internal conservative forces do work, so the change in potential energy ΔU is zero. Thus, the change in mechanical energy equals the change in kinetic energy:

$$\Delta E_{mech} = \Delta U + \Delta K = 0 + (K_f - 0) = K_f$$

3. Substitute this into the step-1 result, then use the values from Parts (*a*) and (*b*) to find K_f:

$$W_{ext} = K_f + \Delta E_{therm}$$
$$\text{so}\quad K_f = W_{ext} - \Delta E_{therm}$$
$$= 75\text{ J} - 41\text{ J} = \boxed{34\text{ J}}$$

(*d*) The final speed of the box is related to its kinetic energy. Solve for the final speed of the box:

$$K_f = \tfrac{1}{2}mv_f^2$$
$$\text{so}\quad v_f = \sqrt{\frac{2K_f}{m}} = \sqrt{\frac{2(34\text{ J})}{4.0\text{ kg}}} = \boxed{4.1\text{ m/s}}$$

CHECK Part of the energy transferred to the system by the pusher (you) ends up as kinetic energy and some of the energy ends up as thermal energy. As expected, the change in thermal energy (Part (*b*)) is positive and is less than the work done by the external force (Part (*a*)).

Example 7-12 A Moving Sled

Try It Yourself

A sled is coasting on a horizontal snow-covered surface with an initial speed of 4.0 m/s. If the coefficient of friction between the sled and the snow is 0.14, how far will the sled travel before coming to rest?

PICTURE We choose the sled and snow as our system and then apply the work–energy theorem.

SOLVE

Cover the column to the right and try these on your own before looking at the answers.

Steps / **Answers**

1. Sketch the system in its initial and final configurations (Figure 7-26).

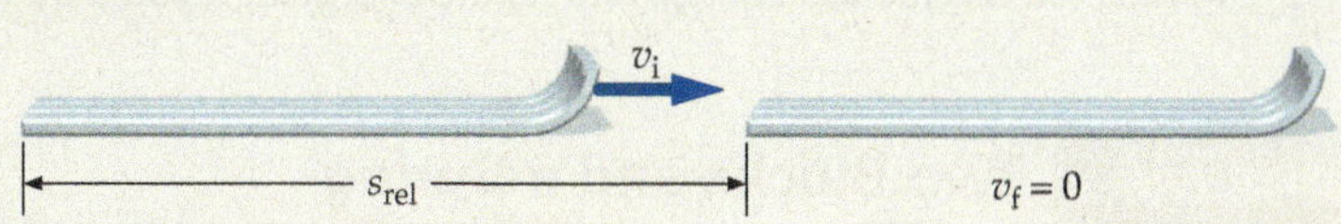

FIGURE 7-26

2. Apply the work–energy theorem. Relate the change in thermal energy to the frictional force.

$$W_{ext} = \Delta E_{mech} + \Delta E_{therm}$$
$$= (\Delta U + \Delta K) + f_k s_{rel}$$

3. Solve for f_k. The normal force is equal to mg.

$$f_k = \mu_k F_n = \mu_k mg$$

4. There are no external forces doing work on the system and there are no internal conservative forces doing work. Use these observations to eliminate two terms from the step-2 result.

$$W_{ext} = 0 \text{ and } \Delta U = 0$$
$$\text{so}\quad W_{ext} = \Delta U + \Delta K + f_k s_{rel}$$
$$0 = 0 + \Delta K + \mu_k mg\, s_{rel}$$

5. Express the change in kinetic energy in terms of the mass and the initial speed, and solve for s_{rel}.

$$s_{rel} = \frac{v^2}{2\mu_k g} = \boxed{5.8\text{ m}}$$

CHECK The expression for the displacement in step 5 is dimensionally correct. The coefficient of friction μ_k is dimensionless, and v^2/g has the dimension of length.

Example 7-13 A Playground Slide

A child of mass 40 kg goes down an 8.0-m-long slide inclined at 30° with the horizontal. The coefficient of kinetic friction between the child and the slide is 0.35. If the child starts from rest at the top of the slide, how fast is he traveling when he reaches the bottom?

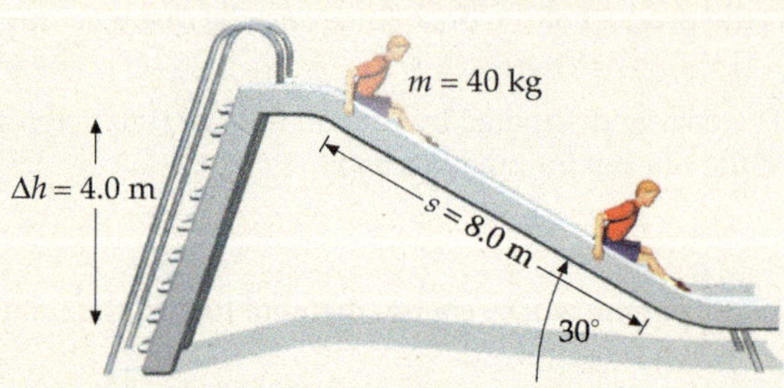

FIGURE 7-27

PICTURE As the child travels down the slide, some of his initial potential energy is converted into kinetic energy and, due to friction, some into thermal energy. We choose the child–slide–Earth as our system and apply the conservation of energy theorem.

SOLVE

1. Make a sketch of the child–slide–Earth system, showing both its initial and final configurations (Figure 7-27).
2. Write out the conservation-of-energy equation: $W_{ext} = \Delta E_{mech} + \Delta E_{therm} = (\Delta U + \Delta K) + f_k s_{rel}$
3. The initial kinetic energy is zero. The speed at the bottom is related to the final kinetic energy: $\Delta K = K_f - 0 = \frac{1}{2}mv_f^2$
4. There are no external forces acting on the system: $W_{ext} = 0$
5. The change in potential energy is related to the change in height Δh (which is negative): $\Delta U = mg\,\Delta h$
6. To find f_k we apply Newton's second law to the child. First we draw a free-body diagram (Figure 7-28):

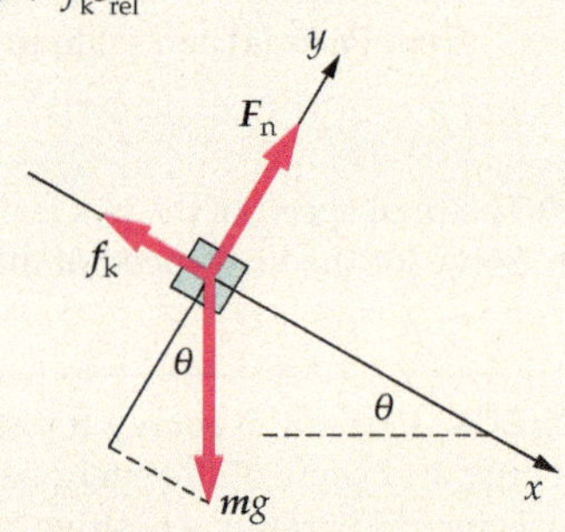

FIGURE 7-28

7. Next, we apply Newton's second law. The normal component of the acceleration is zero. To find F_n we take components in the normal direction. Then we solve for f_k using $f_k = \mu_k F_n$: $F_n - mg\cos\theta = 0$ so $f_k = \mu_k F_n = \mu_k mg\cos\theta$
8. We use trigonometry to relate $s = s_{rel}$ to Δh: $|\Delta h| = s\sin\theta$
9. Substituting into the step-2 result gives: $0 = mg\,\Delta h + \frac{1}{2}mv_f^2 + f_k s = -mgs\sin\theta + \frac{1}{2}mv_f^2 + \mu_k mg\cos\theta\, s$
10. Solving for v_f gives:

$$v_f^2 = 2gs(\sin\theta - \mu_k\cos\theta) = 2(9.81\text{ m/s}^2)(8.0\text{ m})(\sin 30° - 0.35\cos 30°)$$
$$= 30.9\text{ m}^2/\text{s}^2$$
$$\text{so}\quad v_f = \boxed{5.6\text{ m/s}}$$

CHECK Note that, as expected, the expression for v_f^2 in step 10 is independent of the mass of the child. This is expected because all forces acting on the child are proportional to the mass m.

PRACTICE PROBLEM 7-8 Use the bottom end of the slide as the reference level where the potential energy is zero. For the Earth–child–slide system, calculate (*a*) the initial mechanical energy, (*b*) the final mechanical energy, and (*c*) the energy dissipated by friction.

Example 7-14 Two Blocks and a Spring

A 4.0-kg block hangs by a light string that passes over a massless, frictionless pulley and is connected to a 6.0-kg block that rests on a shelf. The coefficient of kinetic friction is 0.20. The 6.0-kg block is pushed against a spring, compressing it 30 cm. The spring has a force constant of 180 N/m. Find the speed of the blocks after the 6.0-kg block is released and the 4.0-kg block has fallen a distance of 40 cm. (Assume the 6.0-kg block is initially 40 cm or more from the pulley.)

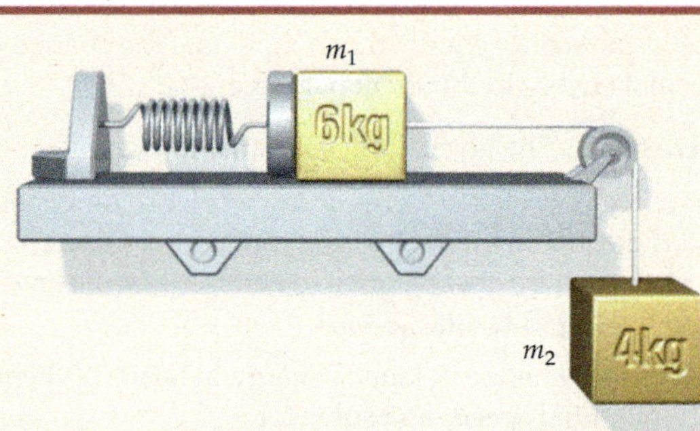

FIGURE 7-29 The system is everything shown plus Earth.

PICTURE The speed of the blocks is obtained from their final kinetic energy. Consider the system to be everything shown in Figure 7-29 plus Earth. This system has both gravitational and elastic potential energy. Apply the work–energy theorem to find the kinetic energy of the blocks. Then, use the kinetic energy of the blocks to solve for their speed.

SOLVE

1. The system is everything shown plus Earth. Write out the equation for the conservation of energy of the system.

$$W_{ext} = \Delta E_{mech} + \Delta E_{therm}$$
$$= (\Delta U_s + \Delta U_g + \Delta K) + f_k s_{rel}$$

2. Make a sketch of the system (Figure 7-30) in both the initial and final configurations:

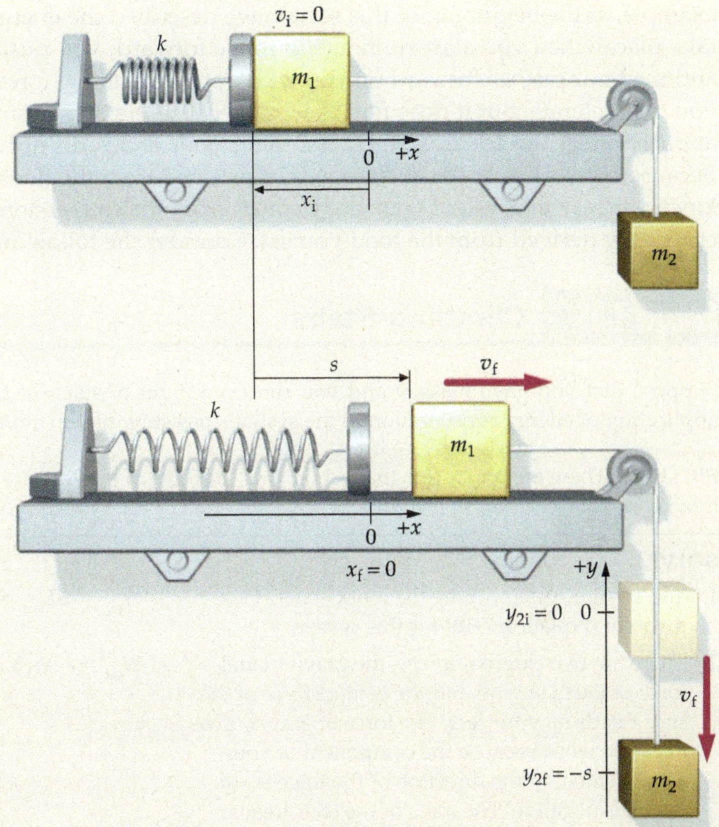

FIGURE 7-30 The system is everything shown plus Earth. The system is shown in both its initial and final configurations.

3. There are no external forces on the system.

$$W_{ext} = 0$$

4. The potential energy of the spring U_s depends on its force constant k and its extension x. (If the spring is compressed, x is negative.) The gravitational potential energy depends on the height of block 2:

$$U_s = \tfrac{1}{2}kx^2$$
$$U_s = mgy_2$$

5. Make a table of the mechanical-energy terms both initially, when the spring is compressed 30 cm, and finally, when each block has moved a distance $s = 40$ cm and the spring is unstressed. Let the gravitational potential energy of the initial configuration equal zero. Also, write down the difference (final minus initial) between each initial and final expression.

	Final	Initial	Difference
U_s	0	$\frac{1}{2}kx_i^2$	$-\frac{1}{2}kx_i^2$
U_g	$-m_2gs$	0	$-m_2gs$
K	$\frac{1}{2}(m_1 + m_2)v_f^2$	0	$\frac{1}{2}(m_1 + m_2)v_f^2$

6. Find an expression for f_k that includes μ_k.

$$f_k = \mu_k m_1 g$$

7. Substitute the results for steps 3–6 into the step-1 result.

$$0 = -\tfrac{1}{2}kx_i^2 - m_2gs + \tfrac{1}{2}(m_1 + m_2)v_f^2 + \mu_k m_1 gs$$

8. Solve the step-7 result for v_f^2, then substitute numerical values and solve for v_f:

$$v_f^2 = \frac{kx_i^2 + 2(m_2 - \mu_k m_1)gs}{m_1 + m_2}$$

so $v_f = \boxed{2.0\,\text{m/s}}$

CHECK If $m_2 = \mu_k = 0$, then the final speed does not depend on either g or μ_k (see step 8). This is as expected because m_2g is the gravitational force on m_2 pulling the system forward and $\mu_k m_1 g$ is the frictional force on m_1 opposing the forward motion. If these two forces sum to zero, the effects of gravity and friction do not affect the final speed.

TAKING IT FURTHER This solution assumes that the string remains taut at all times, which will be true if the acceleration of block 1 remains less than g, that is, if the net force on block 1 is less than $m_1 g = (6.0\text{ kg})(9.81\text{ N/kg}) = 59\text{ N}$. The force exerted by the spring on block 1 initially has the magnitude $kx_1 = (180\text{ N/m})(0.30\text{ m}) = 54\text{ N}$ and the frictional force initially has magnitude $f_k = \mu_k m_1 g = 0.20(59\text{ N}) = 12\text{ N}$. These forces combine to produce a net force of 42 N directed to the right. Because the spring's force decreases as block 1 moves following release, the acceleration of the 6.0-kg block will never exceed g and the string will remain taut.

PROBLEMS INVOLVING CHEMICAL ENERGY

Sometimes a system's internal chemical energy is converted into mechanical energy and thermal energy with no work being done on the system by external forces. For example, at the beginning of this section we described the energy conversions that take place when you start running. To move forward, you push back on the floor and the floor pushes forward on you with a static frictional force. This force causes you to accelerate, but it does not do work because the displacement of the point of application of the force is zero (assuming your shoes do not slip on the floor). Because no work is done, no energy is transferred from the floor to your body. The kinetic-energy increase of your body comes from the conversion of internal chemical energy derived from the food you eat. Consider the following example.

Example 7-15 Climbing Stairs — *Conceptual*

Suppose that you have mass m and you run up a flight of stairs of height h. Discuss the application of energy conservation to the system consisting of you alone.

(Corbis.)

PICTURE There are two forces that act on you: the force of gravity and the force of the stair treads on your feet. Apply the work–energy theorem to the system (you).

SOLVE

1. You are the system. Write the work–energy theorem (Equation 7-16) for this system:

$$W_{ext} = \Delta E_{sys} = \Delta E_{mech} + \Delta E_{therm} + \Delta E_{chem}$$

2. There are two external forces, the gravitational force of Earth on you and the contact force of the stair treads on your feet. The force of gravity does negative work because the component of your displacement in the direction of the force is $-h$, which is negative. The force of the stair treads does no work because the points of application, the soles of your feet, do not move while this force is applied:

$$W_{ext} = -mgh$$

3. You alone are the system. Because your configuration does not change (you remain upright), any change in your mechanical energy is entirely a change in your kinetic energy, which is the same initially and finally:

$$\Delta E_{mech} = 0$$

4. Substitute these results into the work–energy theorem:

$$W_{ext} = \Delta E_{mech} + \Delta E_{therm} + \Delta E_{chem} + \Delta E_{other}$$

so $-mgh = 0 + \Delta E_{therm} + \Delta E_{chem} + 0$

or $\Delta E_{chem} = -(mgh + \Delta E_{therm})$

CHECK We expect your chemical energy to decrease. According to the step-4 result, the change in chemical energy is negative as expected.

TAKING IT FURTHER If there were no change in thermal energy, then your chemical energy would decrease by mgh. Because the human body is relatively inefficient, the increase in thermal energy will be considerably greater than mgh. The decrease in stored chemical energy equals mgh plus some thermal energy. Any thermal energy is eventually transferred from your body to the surroundings.

CONCEPT CHECK 7-1

Discuss the energy conservation for the system consisting of both you and Earth.

Example 7-16 An Uphill Drive

You are driving a 1000-kg gasoline-powered car at a constant speed of 100 km/h (= 27.8 m/s = 62.2 mi/h) up a 10.0 percent grade (Figure 7-31). (A 10.0 percent grade means that the road rises 1.00 m for each 10.0 m of horizontal distance—that is, the angle of inclination θ is given by $\tan\theta = 0.100$.) (*a*) If the efficiency is 15.0 percent, what is the rate at which the chemical energy of the car–Earth–atmosphere system changes? (The efficiency is the fraction of the chemical energy consumed that appears as mechanical energy.) (*b*) What is the rate at which thermal energy is generated?

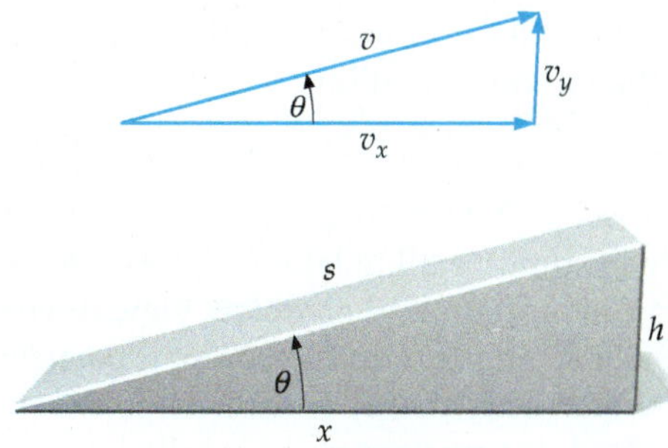

$\tan\theta = h/x \sim \sin\theta = h/s$

FIGURE 7-31

PICTURE Some of the chemical energy goes into increasing the potential energy of the car as it climbs the hill, and some goes into an increase in thermal energy, much of which is expelled by the car as exhaust. To solve this problem, we consider a system consisting of the car, the hill, the atmosphere, and Earth. We first need to find the rate of loss of the chemical energy. Then, we can apply the work–energy theorem to solve for the rate at which thermal energy is generated.

SOLVE

(*a*) 1. The rate of loss of chemical energy equals the absolute value of the change in chemical energy per unit time:

$$\text{Chemical energy loss rate} = \frac{|\Delta E_{\text{chem}}|}{\Delta t}$$

2. The increase in mechanical energy equals 15.0 percent of the decrease in chemical energy:

$$\Delta E_{\text{mech}} = 0.150|\Delta E_{\text{chem}}|$$

3. Solve for the loss rate of chemical energy:

$$\frac{|\Delta E_{\text{chem}}|}{\Delta t} = \frac{1}{0.150}\frac{\Delta E_{\text{mech}}}{\Delta t}$$

4. The car moves at constant speed, so $\Delta K = 0$ and $\Delta E_{\text{mech}} = \Delta U$. Relate the change in mechanical energy to the change in height Δh and substitute it into the step-3 result. (The chemical energy is decreasing):

$$\Delta E_{\text{mech}} = mg\,\Delta h$$

$$\text{so}\quad \frac{\Delta E_{\text{chem}}}{\Delta t} = -\frac{1}{0.150}\frac{mg\,\Delta h}{\Delta t}$$

5. Convert the changes to time derivatives. That is, take the limit of both sides as Δt approaches zero:

$$\frac{dE_{\text{chem}}}{dt} = -\frac{1}{0.150}\frac{mgdh}{dt}$$

6. The rate of change of h equals v_y, which is related to the speed v, as shown in Figure 7-31:

$$\frac{dh}{dt} = v_y = v\sin\theta$$

7. We can approximate $\sin\theta$ by $\tan\theta$ because the angle is small:

$$\sin\theta \approx \tan\theta = 0.100$$

8. Solve for the loss rate of chemical energy:

$$\frac{dE_{\text{chem}}}{dt} = -\frac{mg}{0.15}v\sin\theta \approx -\frac{(1000\text{ kg})(9.81\text{ N/kg})}{0.15}(27.8\text{ m/s})0.100$$

$$\approx -182\text{ kW}$$

$$\therefore\ -\frac{dE_{\text{chem}}}{dt} \approx \boxed{182\text{ kW}}$$

(*b*) 1. Write out the work–energy relation:

$$W_{\text{ext}} = \Delta E_{\text{mech}} + \Delta E_{\text{therm}} + \Delta E_{\text{chem}}$$

2. Set W_{ext} equal to zero, divide both sides by Δt, convert to derivatives, and solve for dE_{therm}/dt:

$$0 = \frac{dE_{\text{mech}}}{dt} + \frac{dE_{\text{therm}}}{dt} + \frac{dE_{\text{chem}}}{dt}$$

$$\text{so}\quad \frac{dE_{\text{therm}}}{dt} = -\frac{dE_{\text{mech}}}{dt} - \frac{dE_{\text{chem}}}{dt} = 0.150\frac{dE_{\text{chem}}}{dt} - \frac{dE_{\text{chem}}}{dt}$$

$$= -0.850\frac{dE_{\text{chem}}}{dt} = 0.850(182\text{ kW}) = \boxed{154\text{ kW}}$$

CHECK The relative size of the Part-(*a*) and Part-(*b*) results are as expected because it was given that the efficiency was only 15 percent.

TAKING IT FURTHER Gasoline-powered cars are typically only about 15 percent efficient. About 85 percent of the chemical energy of the gasoline goes to thermal energy, most of which is expelled out the exhaust pipe. Additional thermal energy is created by rolling friction and air resistance. The energy content of gasoline is about 31.8 MJ/L.

7-4 MASS AND ENERGY

In 1905, Albert Einstein published his special theory of relativity, a result of which is the famous equation

$$E = mc^2 \qquad \text{7-26}$$

where $c = 3.00 \times 10^8$ m/s is the speed of light in a vacuum. We will study this theory in some detail in later chapters. However, we use this equation here to present a more modern and complete view of energy conservation.

According to Equation 7-26, a particle or system of mass m has "rest" energy $E = mc^2$. This energy is intrinsic to the particle. Consider the positron—a particle emitted in a nuclear process called *beta decay*. Positrons and electrons have identical masses, but equal and opposite electrical charges. When a positron encounters an electron, electron–positron annihilation can occur. Annihilation is a process in which the electron and positron disappear and their energy appears as electromagnetic radiation. If the two particles are initially at rest, the energy of the electromagnetic radiation equals the rest energy of the electron plus the rest energy of the positron.

Energies in atomic and nuclear physics are usually expressed in units of electron volts (eV) or mega-electron-volts (1 MeV = 10^6 eV). A convenient unit for the masses of atomic particles is eV/c^2 or MeV/c^2. Table 7-1 lists rest energies (and therefore the masses) of some elementary particles and light nuclei. The rest energy of a positron plus the rest energy of an electron is 2(0.511 MeV), which is the energy of the electromagnetic radiation energy emitted upon annihilation of the electron and positron in a reference frame in which the electron and positron are initially at rest.

The rest energy of a *system* can consist of the potential energy of the system or other internal energies of the system, in addition to the intrinsic rest energies of the particles in the system. If the system at rest absorbs energy ΔE and remains at rest, its rest energy increases by ΔE and its mass increases by ΔM, where

$$\Delta M = \frac{\Delta E}{c^2} \qquad \text{7-27}$$

Table 7-1 Rest Energies* of Some Elementary Particles and Light Nuclei†

Particle	Symbol	Rest Energy (MeV)	
Electron	e^-	0.5110	
Positron	e^+	0.5110	
Proton	p	938.272	
Neutron	n	939.565	
Deuteron	d	1875.613	
Triton	t	2808.921	
Helion	h	2808.391	
Alpha particle	α	3727.379	

* Table values are from 2002 CODATA (except for the value for the triton).

† The proton, deuteron and triton are identical with the nuclei of ^{1}H, ^{2}H, and ^{3}H, respectively, and the helion and alpha particle are identical with the nuclei of ^{3}He and ^{4}He, respectively.

Consider two 1.00-kg blocks connected by a spring of force constant k. If we stretch the spring a distance x, the potential energy of the system increases by $\Delta U = \frac{1}{2}kx^2$. According to Equation 7-27, the mass of the system has also increased by $\Delta M = \Delta U/c^2$. Because c is such a large number, this increase in mass cannot be observed in macroscopic systems. For example, suppose $k = 800\text{ N/m}$ and $x = 10.0\text{ cm} = 0.100\text{ m}$. The potential energy of the spring system is then $\frac{1}{2}kx^2 = \frac{1}{2}(800\text{ N/m})(0.100\text{ m})^2 = 4.00\text{ J}$. The corresponding increase in mass of the system is $\Delta M = \Delta U/c^2 = 4.00\text{ J}/(3.00 \times 10^8\text{ m/s})^2 = 4.44 \times 10^{-17}\text{ kg}$. The fractional mass increase is given by

$$\frac{\Delta M}{M} = \frac{4.44 \times 10^{-17}\text{ kg}}{2.00\text{ kg}} = 2.22 \times 10^{-17}$$

which is much too small to be observed. However, in nuclear reactions, the energy changes are often a much, much larger fraction of the rest energy of the system. Consider the deuteron, which is the nucleus of deuterium, an isotope of hydrogen also called *heavy hydrogen*. The deuteron consists of a proton and neutron bound together. From Table 7-1 we see that the mass of the proton is $938.272\text{ MeV}/c^2$ and the mass of the neutron is $939.565\text{ MeV}/c^2$. The sum of these two masses is $1877.837\text{ MeV}/c^2$. But the mass of the deuteron is $1875.613\text{ MeV}/c^2$, which is less than the sum of the masses of the proton and neutron by $2.22\text{ MeV}/c^2$. Note that this mass difference is much greater than any uncertainties in the measurement of these masses, and the fractional mass difference of $\Delta M/M \approx 1.2 \times 10^{-3}$ is almost 14 orders of magnitude greater than the 2.2×10^{-17} discussed for the spring–blocks system.

Heavy water (deuterium oxide) molecules are produced in the primary cooling water of a nuclear reactor when neutrons collide with the hydrogen nuclei (protons) of the water molecules. If a slow moving neutron is captured by a proton, 2.22 MeV of energy are released in the form of electromagnetic radiation. Thus, the mass of a deuterium atom is $2.22\text{ MeV}/c^2$ less than the sum of the masses of an isolated ^{1}H atom and an isolated neutron. (The superscript 1 is the mass number of the isotope. So ^{1}H refers to protium, the isotope of hydrogen with no neutrons.)

This process can be reversed by breaking a deuteron into its constituent parts if at least 2.22 MeV of energy is transferred to the deuteron with electromagnetic radiation or by collisions with other energetic particles. Any transferred energy in excess of 2.22 MeV appears as kinetic energy of the resulting proton and neutron.

The energy needed to completely separate a nucleus into individual neutrons and protons is called the **binding energy.** The binding energy of a deuteron is 2.22 MeV. The deuteron is an example of a **bound system.** A system is bound if it does not have enough energy to spontaneously separate into separate parts. The rest energy of a bound system is less than the sum of the rest energies of its parts, so energy must be put into the system to break it apart. If the rest energy of a system is greater than the sum of the rest energies of its parts, the system is unbound. An example is uranium-236, which breaks apart or **fissions** into two smaller nuclei.* The sum of the masses of the resultant parts is less than the mass of the original nucleus. Thus, the mass of the system decreases, and energy is released.

In nuclear fusion, two very light nuclei such as a deuteron and a triton (the nucleus of the hydrogen isotope tritium) fuse together. The rest mass of the resultant nucleus is less than that of the original parts, and again energy is released. During a chemical reaction that liberates energy, such as burning coal, the mass decrease is of the order of $1\text{ eV}/c^2$ per atom. This is more than a million times smaller than the mass changes per nucleus in many nuclear reactions, and is not readily observable.

* Uranium-236, written ^{236}U, is made in a nuclear reactor when the stable isotope ^{235}U absorbs a neutron. This reaction is discussed in Chapter 34.

Example 7-17 Binding Energy

A hydrogen atom consisting of a proton and an electron has a binding energy of 13.6 eV. By what percentage is the mass of a proton plus the mass of an electron greater than that of the hydrogen atom?

PICTURE The mass of the proton m_p plus the mass of the electron m_e is equal to mass of the hydrogen atom plus the binding energy E_b divided by c^2. Thus, the fractional difference between $m_e + m_p$ and the mass of the hydrogen atom m_H is the ratio of E_b/c^2 to $m_e + m_p$.

SOLVE

1. The fractional difference (FD) in mass is the ratio of the binding energy E_b/c^2 to $m_e + m_p$:
   $$\text{FD} = \frac{(m_e + m_p) - m_H}{m_e + m_p} = \frac{E_b/c^2}{m_e + m_p} = \frac{13.6\text{ eV}/c^2}{m_e + m_p}$$
2. Obtain the rest masses of the proton and electron from Table 7-1:
   $$m_p = 938.28\text{ MeV}/c^2;$$
   $$m_e = 0.511\text{ MeV}/c^2$$
3. Add to find the sum of these masses:
   $$m_p + m_e = 938.79\text{ MeV}/c^2$$
4. The rest mass of the hydrogen atom is less than this by 13.6 eV/c^2. The fractional difference FD is:
   $$\text{FD} = \frac{13.6\text{ eV}/c^2}{938.79 \times 10^6\text{ eV}/c^2} = 1.45 \times 10^{-8} = \boxed{1.45 \times 10^{-6}\,\%}$$

CHECK The units work out. If we express all masses in units of eV/c^2, we get the fractional difference as a dimensionless number.

TAKING IT FURTHER This mass difference, $\Delta m = (m_e + m_p) - m_H$, is too small to be measured directly. However, binding energies can be accurately measured, so the mass difference Δm can be found from $E_b = (\Delta m)c^2$.

Example 7-18 Nuclear Fusion

Try It Yourself

In a typical nuclear fusion reaction, a triton (t) and a deuteron (d) fuse together to form an alpha particle (α) plus a neutron. The reaction is written d + t $\rightarrow \alpha$ + n. How much energy is released per deuteron produced for this fusion reaction?

PICTURE Because energy is released, the total rest energy of the initial particles must be greater than that of the final particles. This difference equals the energy released.

SOLVE

Cover the column to the right and try these on your own before looking at the answers.

Steps	Answers
1. Write down the rest energies of d and t from Table 7-1 and add to find the total initial rest energy.	$E_{\text{initial}} = 1875.613\text{ MeV} + 2808.921\text{ MeV} = 4684.534\text{ MeV}$
2. Do the same for α and n to find the final rest energy.	$E_{\text{final}} = 3727.379\text{ MeV} + 939.565\text{ MeV} = 4666.944\text{ MeV}$
3. Find the energy released from $E_{\text{released}} = E_{\text{initial}} - E_{\text{final}}$.	$E_{\text{released}} = 4684.534\text{ MeV} - 4666.944\text{ MeV} = \boxed{17.59\text{ MeV} \approx 17.6\text{ MeV}}$

CHECK The energy released is a small fraction of the initial energy. This fraction is 17.6 MeV/4685 MeV = 3.76×10^{-3}, which is the same order of magnitude as the fractional mass increase during the fusion of a proton and a neutron that was discussed at the beginning of this subsection on nuclear energy. Thus, 17.6 MeV is a plausible value for the energy release when a deuteron and helion fuse to form an alpha particle.

TAKING IT FURTHER This fusion reaction and other fusion reactions occur in the Sun. The energy that is released bathes Earth and is ultimately responsible for all life on the planet. The energy continuously emitted by the Sun is accompanied by a continuous decrease in the Sun's rest mass.

NONRELATIVISTIC (NEWTONIAN) MECHANICS AND RELATIVITY

As the speed of a particle approaches a significant fraction of the speed of light, Newton's second law breaks down, and we must modify Newtonian mechanics according to Einstein's theory of relativity. The criterion for the validity of Newtonian mechanics can also be stated in terms of the energy of a particle. In nonrelativistic (Newtonian) mechanics, the kinetic energy of a particle moving with speed v is

$$K = \tfrac{1}{2}mv^2 = \tfrac{1}{2}mc^2\frac{v^2}{c^2} = \tfrac{1}{2}E_0\frac{v^2}{c^2}$$

where $E_0 = mc^2$ is the rest energy of the particle. Solving for v/c gives

$$\frac{v}{c} = \sqrt{\frac{2K}{E_0}}$$

Nonrelativistic mechanics is valid if the speed of the particle is much less than the speed of light, or alternatively, the kinetic energy of a particle is much less than its rest energy.

PRACTICE PROBLEM 7-9

A low-Earth-orbit satellite has an orbital speed of $v \approx 5.0\ \text{mi/s} = 8.0\ \text{km/s}$. What fraction of the speed of light c is this speed? What speed, in mi/s, is equal to one percent of c?

7-5 QUANTIZATION OF ENERGY

When energy is put into a system that remains at rest, the *internal energy* of the system increases. (Internal energy is synonymous with rest energy. It is the total energy of the system less any kinetic energy associated with the motion of the system's center of mass.) While it might seem that we could change the internal energy of a bound system, like the solar system or a hydrogen atom, by any amount, this is found not to be true. This is particularly noticeable for microscopic systems, such as molecules, atoms, and atomic nuclei. The internal energy of a bound system can increase only by discrete increments.

If we have two blocks attached to a spring (Figure 7-32) and we stretch the spring by pulling the blocks further apart, we do work on the block–spring system, and its potential energy increases. If we then release the blocks, they oscillate back and forth. The energy of oscillation E—the kinetic energy of motion of the blocks plus the potential energy (due to the stretching of the spring)—equals the initial potential energy. In time, the energy of the system decreases because of various damping effects such as friction and air resistance. *As closely as we can measure*, the energy decreases continuously. All the energy is eventually dissipated and the energy of oscillation is zero.

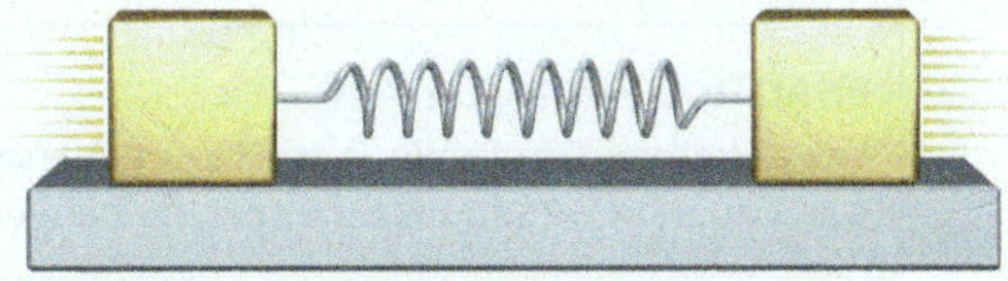

FIGURE 7-32

Now consider a diatomic molecule such as molecular oxygen, O_2. The force of attraction between the two oxygen atoms varies approximately linearly with the change in separation (for small changes), like that of two blocks connected by a spring. If a diatomic molecule is set oscillating with some energy E, the energy decreases with time as the molecule radiates or interacts with its surroundings, but careful measurements can show that the decrease is *not continuous*. The energy decreases in finite steps, and the lowest energy state, called the **ground state**, does not have zero energy. The vibrational energy of a diatomic molecule is said to be

quantized; that is, the molecule can absorb or release energies only in certain amounts, known as **quanta.**

When either blocks on a spring or diatomic molecules oscillate, the time for one oscillation is called the period T. The reciprocal of the period is the frequency of oscillation $f = 1/T$. We will see in Chapter 14 that the period and frequency of an oscillator do not depend on the energy of oscillation. As the energy decreases, the frequency remains the same. Figure 7-33 shows an **energy-level diagram** for an oscillator. The allowed energies are approximately equally spaced, and are given by*

$$E_n = (n + \tfrac{1}{2})hf \quad n = 0, 1, 2, 3, \ldots \qquad 7\text{-}28$$

where f is the frequency of oscillation and h is a fundamental constant of nature called Planck's constant:†

$$h = 6.626 \times 10^{-34}\ \text{J}\cdot\text{s} = 4.136 \times 10^{-15}\ \text{eV}\cdot\text{s} \qquad 7\text{-}29$$

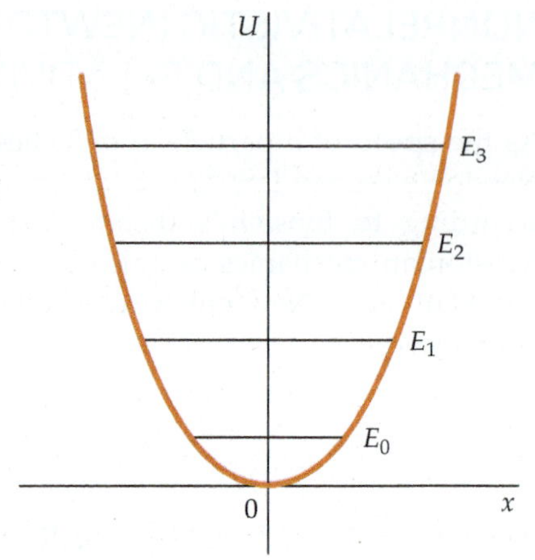

FIGURE 7-33

The integer n is called a **quantum number.** The lowest possible energy is the **ground state energy** $E_0 = \frac{1}{2}hf$.

Microscopic systems often gain or lose energy by absorbing or emitting electromagnetic radiation. By conservation of energy, if E_i and E_f are the initial and final energies of a system, the energy of the radiation emitted or absorbed is

$$E_{rad} = |E_f - E_i|$$

Because the system energies E_i and E_f are quantized, the radiated energy is also quantized.‡ The quantum of radiation is called a **photon.** The energy of a photon is given by

$$E_{photon} = hf \qquad 7\text{-}30$$

where f is the frequency of the electromagnetic radiation.§

As far as we know, all bound systems exhibit energy quantization. For macroscopic bound systems, the steps between energy levels are so small that they are unobservable. For example, typical oscillation frequencies for two blocks on a spring are 1 to 10 times per second. If $f = 10$ oscillations per second, the spacing between allowed levels is $hf = (6.626 \times 10^{-34}\ \text{J}\cdot\text{s})(10\ \text{s}^{-1}) \approx 7 \times 10^{-33}\ \text{J}$. Because the energy of a macroscopic system is of the order of 1 J, a quantum step of 10^{-33} J is too small to be noticed. To put it another way, if the energy of a system is 1 J, the value of n is of the order of 10^{32} and changes of one or two quantum units will not be observable.

! A typical energy for a diatomic molecule is 10^{-19} J. Thus, changes in the energy of oscillation are of the same order of magnitude as the energy of the molecule, and quantization is definitely not negligible.

PRACTICE PROBLEM 7-10

For a diatomic molecule, a typical frequency of vibration is 10^{14} vibrations per second. Use Equation 7-28 to find the spacing between the allowed energies.

* A diatomic molecule can also have rotational energy. The rotational energy is also quantized, but the energy levels are not equally spaced, and the lowest possible energy is zero. We will study rotational energy in Chapters 9 and 10.

† In 1900, the German physicist Max Planck introduced this constant during calculations to explain discrepancies between the theoretical curves and experimental data on the spectrum of blackbody radiation. The significance of Planck's constant was not appreciated by Planck or anyone else until Einstein postulated in 1905 that the energy of electromagnetic radiation is not continuous, but occurs in packets of size hf, where f is the frequency of the radiation.

‡ Historically, the quantization of electromagnetic radiation, as proposed by Max Planck and Albert Einstein, was the first "discovery" of energy quantization.

§ Electromagnetic radiation includes light, microwaves, radio waves, television waves, X rays, and gamma rays. These differ from one another in their frequencies.

Physics Spotlight

Blowing Warmed Air

Wind farms dot the Danish coast, the plains of the upper Midwest, and hills from California to Vermont. Harnessing the kinetic energy of the wind is nothing new. Windmills have been used to pump water, ventilate mines,* and grind grain for centuries.

Today, the most visible wind turbines run electrical generators. These turbines transform kinetic energy into electromagnetic energy. Modern turbines range widely in size, cost, and output. Some are very small, simple machines that cost under \$500/turbine, and put out less than 100 watts of power.† Others are complex behemoths that cost over \$2 million and put out as much as 2.5 MW/turbine.‡ All of these turbines take advantage of a widely available energy source—the wind.

The theory behind the windmill's conversion of kinetic energy to electromagnetic energy is straightforward. The moving air molecules push on the turbine blades, driving their rotational motion. The rotating blades then turn a series of gears. The gears, in turn, step up the rotation rate, and drive the rotation of a generator rotor. The generator sends the electromagnetic energy out along power lines.

But the conversion of the wind's kinetic energy to electromagnetic energy is not 100 percent efficient. The most important thing to remember is that it *cannot* be 100 percent efficient. If turbines converted 100 percent of the kinetic energy of the air into electrical energy, the air would leave the turbines with zero kinetic energy. That is, the turbines would stop the air. If the air were completely stopped by the turbine, it would flow around the turbine, rather than through the turbine.

So the theoretical efficiency of a wind turbine is a trade-off between capturing the kinetic energy of the moving air, and preventing most of the wind from flowing around the turbine. Propeller-style turbines are the most common, and their theoretical efficiency at transforming the kinetic energy of the air into electromagnetic energy varies from 30 percent to 59 percent.§ (The predicted efficiencies vary because of assumptions made about the way the air behaves as it flows through and around the propellers of the turbine.)

So even the most efficient turbine cannot convert 100 percent of the theoretically available energy. What happens? Upstream from the turbine, the air moves along straight streamlines. After the turbine, the air rotates and is turbulent. The rotational component of the air's movement beyond the turbine takes energy. Some dissipation of energy occurs because of the viscosity of air. When some of the air slows, there is friction between it and the faster moving air flowing by it. The turbine blades heat up, and the air itself heats up.° The gears within the turbine also convert kinetic energy into thermal energy through friction. All this thermal energy needs to be accounted for. The blades of the turbine vibrate individually—the energy associated with those vibrations cannot be used. Finally, the turbine uses some of the electricity it generates to run pumps for gear lubrication, and to run the yaw motor that moves the turbine blades into the most favorable position to catch the wind.

In the end, most wind turbines operate at between 10 and 20 percent efficiency.# They are still attractive power sources, because of the free fuel. One turbine owner explains, "The bottom line is we did it for our business to help control our future."**

A wind farm converting the kinetic energy of the air to electrical energy. *(Image Slate.)*

* Agricola, Georgius, *De Re Metallic.* (Herbert and Lou Henry Hoover, Transl.) Reprint Mineola, NY: Dover, 1950, 200–203.
† Conally, Abe, and Conally, Josie, "Wind Powered Generator," *Make*, Feb. 2006, Vol. 5, 90–101.
‡ "Why Four Generators May Be Better than One," *Modern Power Systems*, Dec. 2005, 30.
§ Gorban, A. N., Gorlov, A. M., and Silantyev, V. M., "Limits of the Turbine Efficiency for Free Fluid Flow." *Journal of Energy Resources Technology*, Dec. 2001, Vol. 123, 311–317.
° Roy, S. B., S. W. Pacala, and R. L. Walko. "Can Large Wind Farms Affect Local Meteorology?" *Journal of Geophysical Research (Atmospheres)*, Oct. 16, 2004, 109, D19101.
Gorban, A. N., Gorlov, A. M., and Silantyev, V. M., "Limits of the Turbine Efficiency for Free Fluid Flow." *Journal of Energy Resources Technology*, December 2001, Vol. 123, 311–317.
** Wilde, Matthew, "Colwell Farmers Take Advantage of Grant to Produce Wind Energy." *Waterloo-Cedar Falls Courier*, May 1, 2006, B1+.

Summary

1. The work–energy theorem and the conservation of energy are fundamental laws of nature that have applications in all areas of physics.
2. The conservation of mechanical energy is an important relation derived from Newton's laws for conservative forces. It is useful in solving many problems.
3. Einstein's equation $E = mc^2$ is a fundamental relation between mass and energy.
4. Quantization of energy is a fundamental property of bound systems.

TOPIC	RELEVANT EQUATIONS AND REMARKS	
1. Conservative Force	A force is conservative if the total work it does on a particle is zero when the particle moves along any path that returns it to its initial position. Alternatively, the work done by a conservative force on a particle is independent of the path taken by the particle as it moves from one point to another.	
2. Potential Energy	The potential energy of a system is the energy associated with the configuration of the system. The change in the potential energy of a system is defined as the negative of the work done by all internal conservative forces acting on the system.	
Definition	$\Delta U = U_2 - U_1 = -W = -\int_1^2 \vec{F} \cdot d\vec{\ell}$ $dU = -\vec{F} \cdot d\vec{\ell}$	7-1
Gravitational	$U = U_0 + mgy$	7-2
Elastic (spring)	$U = \frac{1}{2}kx^2$	
Conservative force	$F_x = -\frac{dU}{dx}$	7-13
Potential-energy curve	At a minimum on the curve of the potential-energy function versus the displacement, the force is zero and the system is in stable equilibrium. At a maximum, the force is zero and the system is in unstable equilibrium. A conservative force always tends to accelerate a particle toward a position of lower potential energy.	
3. Mechanical Energy	The sum of the kinetic and potential energies of a system is called the total mechanical energy $E_{mech} = K_{sys} + U_{sys}$	7-9
Work–Energy Theorem for Systems	The total work done on a system by external forces equals the change in mechanical energy of the system less the total work done by internal nonconservative forces: $W_{ext} = \Delta E_{mech} - W_{nc}$	7-10
Conservation of Mechanical Energy	If no external forces do work on the system, and if no internal nonconservative forces do work, then the mechanical energy of the system is constant: $K_f + U_f = K_i + U_i$	7-12
4. Total Energy of a System	The energy of a system consists of mechanical energy E_{mech}, thermal energy E_{therm}, chemical energy E_{chem}, and other types of energy E_{other}, such as sound radiation and electromagnetic radiation: $E_{sys} = E_{mech} + E_{therm} + E_{chem} + E_{other}$	7-15
5. Conservation of Energy		
Universe	The total energy of the universe is constant. Energy can be transformed from one form to another, or transmitted from one region to another, but energy can never be created or destroyed.	

TOPIC	RELEVANT EQUATIONS AND REMARKS
System	The energy of a system can be changed by work being done on the system and by energy transfer by heat. (These transfers include the emission or absorption of radiation.) The increase or decrease in the energy of the system can always be accounted for by the disappearance or appearance of some kind of energy somewhere else: $E_{in} - E_{out} = \Delta E_{sys}$ 7-14
Work–energy theorem	$W_{ext} = \Delta E_{sys} = \Delta E_{mech} + \Delta E_{therm} + \Delta E_{chem} + \Delta E_{other}$ 7-16
6. Energy Dissipated by Friction	For a system that has a surface that slides on a second surface, the energy dissipated by friction on both surfaces equals the increase in thermal energy of the system and is given by $f_k s_{rel} = \Delta E_{therm}$ 7-24 where s_{rel} is the distance one surface slides relative to the other.
7. Problem Solving	The conservation of mechanical energy and the work–energy theorem can be used as an alternative to Newton's laws to solve mechanics problems that require the determination of the speed of a particle as a function of its position.
8. Mass and Energy	A particle with mass m has an intrinsic rest energy E given by $E = mc^2$ 7-26 where $c = 3 \times 10^8$ m/s is the speed of light in a vacuum. A system with mass M also has a rest energy $E = Mc^2$.If a system gains or loses internal energy ΔE, it simultaneously gains or loses mass ΔM, where $\Delta M = \Delta E/c^2$.
Binding energy	The energy required to separate a bound system into its constituent parts is called its binding energy. The binding energy is ΔMc^2, where ΔM is the sum of the masses of the constituent parts, less the mass of the bound system.
9. Newtonian Mechanics and Special Relativity	If the speed of a particle approaches the speed of light c (when the kinetic energy of the particle is significant in comparison to its rest energy), Newtonian mechanics breaks down, and must be replaced by Einstein's special theory of relativity.
10. Energy Quantization	The internal energy of a bound system is found to have only a discrete set of possible values. For a system oscillating with frequency f, the allowed energy values are separated by an amount hf, where h is Planck's constant: $h = 6.626 \times 10^{-34}\,\text{J}\cdot\text{s}$ 7-29
Photons	Microscopic systems often exchange energy with their surroundings by emitting or absorbing electromagnetic radiation, which is also quantized. The quantum of energy of radiation is called the photon: $E_{photon} = hf$ 7-30 where f is the frequency of the electromagnetic radiation.

Answer to Concept Checks

7-1 On the you–Earth system no external work is done, so the total energy, which now includes gravitational potential energy, is conserved. The change in mechanical energy is mgh, so the work–energy theorem again gives $\Delta E_{chem} = -(mgh + \Delta E_{therm})$.

Answers to Practice Problems

7-1	$\oint_C \vec{F} \cdot d\vec{\ell} = -\frac{1}{2}Bx_{max}^2 y_{max}$
7-2	(*a*) 4.3 kJ, (b) 2.2 kJ, (*c*) −1.1 kJ
7-3	495 J
7-4	(*a*) 4.9 cm, (*b*) 0.72 J
7-5	3.16 m/s
7-6	None
7-7	53 m
7-8	(*a*) 1600 J, (*b*) 620 J, (*c*) 950 J
7-9	2.7×10^{-5}; 1.9×10^3 mi/s
7-10	$E_{n+1} - E_n = hf \approx (6.63 \times 10^{-34}\,\text{J}\cdot\text{s})(10^{14}\,\text{s}) \approx 6 \times 10^{-20}\,\text{J}$

Problems

In a few problems, you are given more data than you actually need; in a few other problems, you are required to supply data from your general knowledge, outside sources, or informed estimate.

Interpret as significant all digits in numerical values that have trailing zeros and no decimal points.

For all problems, use 9.81 m/s² for the free-fall acceleration and neglect friction and air resistance unless instructed to do otherwise.

• Single-concept, single-step, relatively easy

•• Intermediate-level, may require synthesis of concepts

••• Challenging

SSM Solution is in the *Student Solutions Manual*

Consecutive problems that are shaded are paired problems.

CONCEPTUAL PROBLEMS

1 • Two cylinders of unequal mass are connected by a massless cord that passes over a frictionless pulley (Figure 7-34). After the system is released from rest, which of the following statements are true? (U is the gravitational potential energy and K is the kinetic energy of the system.) (*a*) $\Delta U < 0$ and $\Delta K > 0$, (*b*) $\Delta U = 0$ and $\Delta K > 0$, (*c*) $\Delta U < 0$ and $\Delta K = 0$, (*d*) $\Delta U = 0$ and $\Delta K = 0$, (*e*) $\Delta U > 0$ and $\Delta K < 0$. SSM

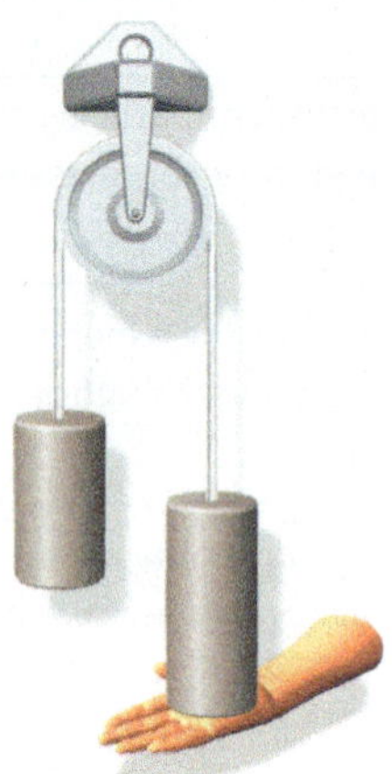

FIGURE 7-34
Problem 1

2 • Two stones are simultaneously thrown with the same initial speed from the roof of a building. One stone is thrown at an angle of 30° above the horizontal, the other is thrown horizontally. (Neglect effects due to air resistance.) Which statement below is true?

(*a*) The stones strike the ground at the same time and with equal speeds.
(*b*) The stones strike the ground at the same time with different speeds.
(*c*) The stones strike the ground at different times with equal speeds.
(*d*) The stones strike the ground at different times with different speeds.

3 • True or false:

(*a*) The total energy of a system cannot change.
(*b*) When you jump into the air, the floor does work on you, increasing your mechanical energy.
(*c*) Work done by frictional forces must always decrease the total mechanical energy of a system.
(*d*) Compressing a given spring 2.0 cm from its unstressed length takes more work than stretching it 2.0 cm from its unstressed length.

4 • As a novice ice hockey player (assume frictionless situation), you have not mastered the art of stopping except by coasting straight for the boards of the rink (assumed to be a rigid wall). Discuss the energy changes that occur as you use the boards to slow your motion to a stop.

5 • True or false (The particle in this question can move only along the x axis and is acted on by only one force, and $U(x)$ is the potential-energy function associated with this force.):

(*a*) The particle will be in equilibrium if it is at a location where $dU/dx = 0$.
(*b*) The particle will accelerate in the $-x$ direction if it is at a location where $dU/dx > 0$.
(*c*) The particle will both be in equilibrium and have constant speed if it is at a section of the x axis where $dU/dx = 0$ throughout the section.
(*d*) The particle will be in stable equilibrium if it is at a location where both $dU/dx = 0$ and $d^2U/dx^2 > 0$.
(*e*) The particle will be in neutral equilibrium if it is at a location where both $dU/dx = 0$ and $d^2U/dx^2 > 0$.

6 • Two knowledge seekers decide to ascend a mountain. Sal chooses a short, steep trail, while Joe, who weighs the same as Sal, chooses a long, gently sloped trail. At the top, they get into an argument about who gained more potential energy. Which of the following is true?

(*a*) Sal gains more gravitational potential energy than Joe.
(*b*) Sal gains less gravitational potential energy than Joe.
(*c*) Sal gains the same gravitational potential energy as Joe.
(*d*) To compare the gravitational potential energies, we must know the height of the mountain.
(*e*) To compare the gravitational potential energies, we must know the lengths of the two trails.

7 • True or false:

(*a*) Only conservative forces can do work.
(*b*) If only conservative forces act on a particle, the kinetic energy of the particle cannot change.
(*c*) The work done by a conservative force equals the change in the potential energy associated with that force.
(*d*) If, for a particle constrained to the x axis, the potential energy associated with a conservative force decreases as the particle moves to the right, then the force points to the left.
(*e*) If, for a particle constrained to the x axis, a conservative force points to the right, then the potential energy associated with the force increases as the particle moves to the left.

8 • Figure 7-35 shows the plot of a potential-energy function U versus x. (*a*) At each point indicated, state whether the x component of the force associated with this function is positive, negative, or zero. (*b*) At which point does the force have the greatest magnitude? (*c*) Identify any equilibrium points, and state whether the equilibrium is stable, unstable, or neutral.

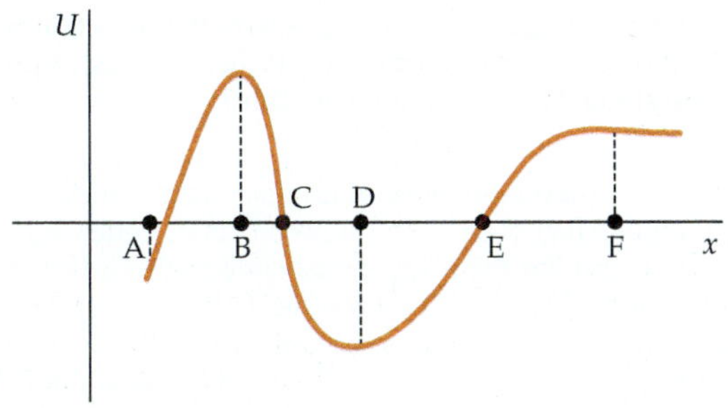

FIGURE 7-35 Problem 8

9 • Assume that, when the brakes are applied, a constant frictional force is exerted on the wheels of a car by the road. If that is so, then which of the following are necessarily true? (*a*) The distance the car travels before coming to rest is proportional to the speed of the car just as the brakes are first applied, (*b*) the car's kinetic energy diminishes at a constant rate, (*c*) the kinetic energy of the car is inversely proportional to the time that has elapsed since the application of the brakes, (*d*) none of the above

10 •• If a rock is attached to a massless, rigid rod and swung in a vertical circle (Figure 7-36) at a constant speed, the total mechanical energy of the rock–Earth system does not remain constant. The kinetic energy of the rock remains constant, but the gravitational potential energy is continually changing. Is the total work done on the rock equal to zero during all time intervals? Does the force by the rod on the rock ever have a nonzero tangential component?

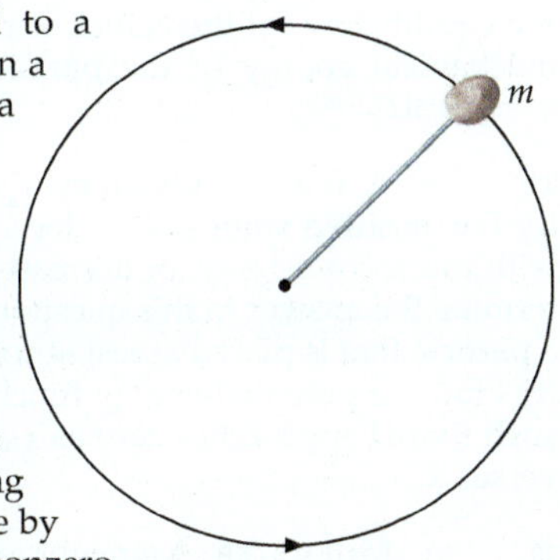

FIGURE 7-36 Problem 10

11 •• Use the rest energies given in Table 7-1 to answer the following questions. (*a*) Can the triton naturally decay into a helion? (*b*) Can the alpha particle naturally decay into helion plus a neutron? (*c*) Can the proton naturally decay into a neutron and a positron?

ESTIMATION AND APPROXIMATION

12 • Estimate (*a*) the change in your gravitational potential energy on taking an elevator from the ground floor to the top of the Empire State Building, (*b*) the average force exerted by the elevator on you during the trip, and (*c*) the average power delivered by that force. The building is 102 stories high. (You may need to estimate the time for the trip.)

13 • A tightrope walker whose mass is 50 kg walks across a tightrope held between two supports 10 m apart; the tension in the rope is 5000 N when she stands at the exact center of the rope. Estimate: (*a*) the sag in the tightrope when the acrobat stands in the exact center, and (*b*) the change in her gravitational potential energy from when she steps onto the tightrope to when she stands at its exact center.

14 •• **BIOLOGICAL APPLICATION** *The metabolic rate* is defined as the rate at which the body uses chemical energy to sustain its life functions. The average metabolic rate has been found to be proportional to the total skin surface area of the body. The surface area for a 5-ft, 10-in. male weighing 175 lb is about 2.0 m^2, and for a 5-ft, 4-in. female weighing 110 lb it is approximately 1.5 m^2. There is about a 1 percent change in surface area for every three pounds above or below the weights quoted here and a 1 percent change for every inch above or below the heights quoted. (*a*) Estimate your average metabolic rate over the course of a day using the following guide for metabolic rates (per square meter of skin area) for various physical activities: sleeping, 40 W/m^2; sitting, 60 W/m^2; walking, 160 W/m^2; moderate physical activity, 175 W/m^2; and moderate aerobic exercise, 300 W/m^2. How do your results compare to the power of a 100-W light bulb? (*b*) Express your average metabolic rate in terms of kcal/day (1 kcal = 4.19 kJ). (A kcal is the "food calorie" used by nutritionists.) (*c*) An estimate used by nutritionists is that each day the "average person" must eat roughly 12–15 kcal of food for each pound of body weight to maintain his or her weight. From the calculations in Part (*b*), are these estimates plausible?

15 •• **BIOLOGICAL APPLICATION** Assume that your maximum metabolic rate (the maximum rate at which your body uses its chemical energy) is 1500 W (about 2.7 hp). Assuming a 40 percent efficiency for the conversion of chemical energy into mechanical energy, estimate the following: (*a*) the shortest time you could run up four flights of stairs if each flight is 3.5 m high, (*b*) the shortest time you could climb the Empire State Building (102 stories high) using your Part-(*a*) result. Comment on the feasibility of you actually achieving your Part-(*b*) result. SSM

16 •• **ENGINEERING APPLICATION, CONTEXT-RICH** You are in charge of determining when the uranium fuel rods in a local nuclear power plant are to be replaced with fresh ones. To make this determination, you decide to estimate how much the mass of a core of a nuclear-fueled electric-generating plant is reduced per unit of electric energy produced. (*Note:* In such a generating plant the reactor core generates thermal energy, which is then transformed to electric energy by a steam turbine. It requires 3.0 J of thermal energy for each 1.0 J of electric energy produced.) What are your results for the production of (*a*) 1.0 J of thermal energy? (*b*) enough electric energy to keep a 100-W light bulb burning for 10.0 y? (*c*) electric energy at a constant rate of 1.0 GW for a year? (This is typical of modern plants.)

17 •• **ENGINEERING APPLICATION, MULTISTEP** The chemical energy released by burning a gallon of gasoline is approximately 1.3×10^5 kJ. Estimate the total energy used by all of the cars in the United States during the course of one year. What fraction does this represent of the total energy use by the United States in one year (currently about 5×10^{20} J)? SSM

18 •• **ENGINEERING APPLICATION** The maximum efficiency of a solar-energy panel in converting solar energy into useful electrical energy is currently about 12 percent. In a region such as the southwestern United States the solar intensity reaching Earth's surface is about 1.0 kW/m^2 on average during the day. Estimate the area that would have to be covered by solar panels in order to supply the energy requirements of the United States (approximately 5×10^{20} J/y) and compare it to the area of Arizona? Assume cloudless skies.

19 •• **ENGINEERING APPLICATION** Hydroelectric power plants convert gravitational potential energy into more useful forms by flowing water downhill through a turbine system to generate electric energy. The Hoover Dam on the Colorado River is 211 m high and generates 4×10^9 kW·h/y. At what rate (in L/s) must water be flowing through the turbines to generate this power? The density of water is 1.00 kg/L. Assume a total efficiency of 90.0 percent in converting the water's potential energy into electrical energy.

FORCE, POTENTIAL ENERGY, AND EQUILIBRIUM

20 • Water flows over Victoria Falls, which is 128 m high, at a rate of 1.4×10^6 kg/s. If half the potential energy of this water were converted into electric energy, how much electric power would be produced by these falls?

21 • A 2.0-kg box slides down a long, frictionless incline of angle 30°. It starts from rest at time $t = 0$ at the top of the incline at a height of 20 m above the ground. (*a*) What is the potential energy of the box relative to the ground at $t = 0$? (*b*) Use Newton's laws to find the distance the box travels during the interval $0.0\ \text{s} < t < 1.0\ \text{s}$ and its speed at $t = 1.0$ s. (*c*) Find the potential energy and the kinetic energy of the box at $t = 1.0$ s. (*d*) Find the kinetic energy and the speed of the box just as it reaches the ground at the bottom of the incline.

22 • A constant force $F_x = 6.0$ N is in the $+x$ direction. (*a*) Find the potential-energy function $U(x)$ associated with this force if $U(x_0) = 0$. (*b*) Find a function $U(x)$ such that $U(4.0\ \text{m}) = 0$. (*c*) Find a function $U(x)$ such that $U(6.0\ \text{m}) = 14$ J.

23 • A spring has a force constant of 1.0×10^4 N/m. How far must the spring be stretched for its potential energy to equal (*a*) 50 J, and (*b*) 100 J?

24 • (*a*) Find the force F_x associated with the potential-energy function $U = Ax^4$, where A is a constant. (*b*) At what value(s) of x does the force equal zero?

25 •• The force F_x is associated with the potential-energy function $U = C/x$, where C is a positive constant. (*a*) Find the force F_x as a function of x. (*b*) Is this force directed toward the origin or away from it in the region $x > 0$? Repeat the question for the region $x < 0$? (*c*) Does the potential energy U increase or decrease as x increases in the region $x > 0$? (*d*) Answer Parts (*b*) and (*c*) where C is a negative constant. SSM

26 •• The force F_y is associated with the potential-energy function $U(y)$. On the potential-energy curve for U versus y, shown in Figure 7-37, the segments AB and CD are straight lines. Plot F_y versus y. Include numerical values, with units, on both axes. These values can be obtained from the U versus y plot.

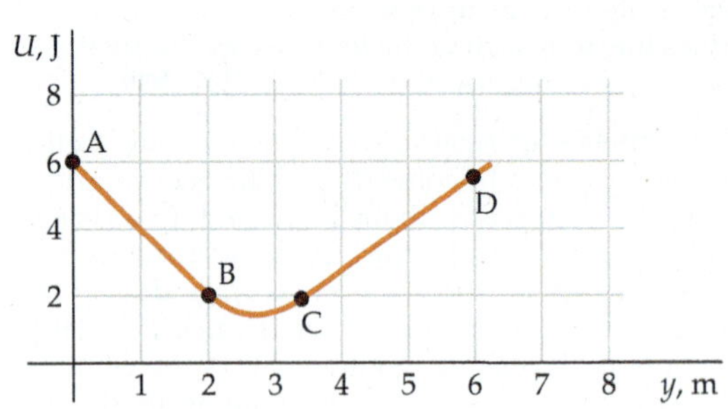

FIGURE 7-37 Problem 26

27 •• The force acting on an object is given by $F_x = a/x^2$. At $x = 5.0$ m, the force is known to point in the $-x$ direction and have a magnitude of 25 N. Determine the potential energy associated with this force as a function of x, assuming we assign a reference value of -10 J at $x = 2.0$ m for the potential energy.

28 •• The potential energy of an object constrained to the x axis is given by $U(x) = 3x^2 - 2x^3$, where U is in joules and x is in meters. (*a*) Determine the force F_x associated with this potential-energy function. (*b*) Assuming no other forces act on the object, at what positions is this object in equilibrium? (*c*) Which of these equilibrium positions are stable and which are unstable?

29 •• The potential energy of an object constrained to the x axis is given by $U(x) = 8x^2 - x^4$, where U is in joules and x is in meters. (*a*) Determine the force F_x associated with this potential-energy function. (*b*) Assuming no other forces act on the object, at what positions is this object in equilibrium? (*c*) Which of these equilibrium positions are stable and which are unstable? SSM

30 •• The net force acting on an object constrained to the x axis is given by $F_x(x) = x^3 - 4x$. (The force is in newtons and x in meters.) Locate the positions of unstable and stable equilibrium. Show that each position is stable or unstable by calculating the force one millimeter on either side of the locations.

31 •• The potential energy of a 4.0-kg object constrained to the x axis is given by $U = 3x^2 - x^3$ for $x \leq 3.0$ m and $U = 0$ for $x \geq 3.0$ m, where U is in joules and x is in meters, and the only force acting on this object is the force associated with this potential-energy function. (*a*) At what positions is this object in equilibrium? (*b*) Sketch a plot of U versus x. (*c*) Discuss the stability of the equilibrium for the values of x found in Part (*a*). (*d*) If the total mechanical energy of the particle is 12 J, what is its speed at $x = 2.0$ m?

32 •• A force is given by $F_x = Ax^{-3}$, where $A = 8.0\ \text{N}\cdot\text{m}^3$. (*a*) For positive values of x, does the potential energy associated with this force increase or decrease with increasing x? (You can determine the answer to this question by imagining what happens to a particle that is placed at rest at some point x and is then released.) (*b*) Find the potential-energy function U associated with this force such that U approaches zero as x approaches infinity. (*c*) Sketch U versus x.

33 •• **MULTISTEP** A straight rod of negligible mass is mounted on a frictionless pivot, as shown in Figure 7-38. Blocks having masses m_1 and m_2 are attached to the rod at distances ℓ_1 and ℓ_2. (*a*) Write an expression for the gravitational potential energy of the blocks–Earth system as a function of the angle θ made by the rod and the horizontal. (*b*) For what angle θ is this potential energy a minimum? Is the statement "systems tend to move toward a configuration of minimum potential energy" consistent with your result? (*c*) Show that if $m_1\ell_1 = m_2\ell_2$, the potential energy is the same for all values of θ. (When this holds, the system will balance at any angle θ. This result is known as *Archimedes' law of the lever*.) SSM

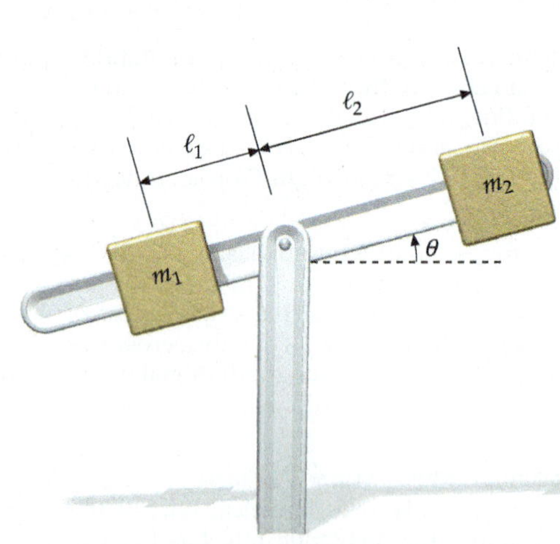

FIGURE 7-38 Problem 33

34 •• An Atwood's machine (Figure 7-39) consists of masses m_1 and m_2, and a pulley of negligible mass and friction. Starting from rest, the speed of the two masses is 4.0 m/s at the end of 3.0 s. At that time, the kinetic energy of the system is 80 J and each mass has moved a distance of 6.0 m. Determine the values of m_1 and m_2.

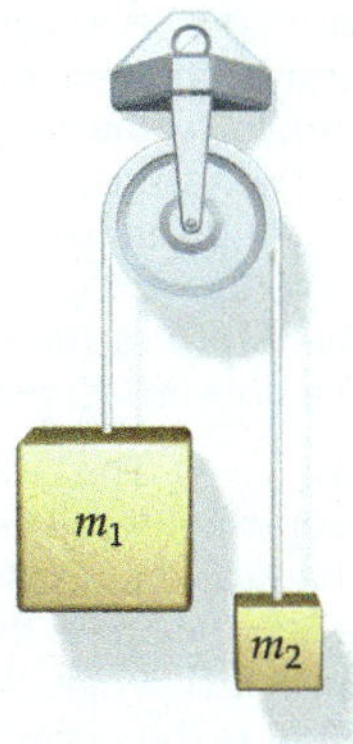

FIGURE 7-39 Problem 34

35 ••• **ENGINEERING APPLICATION, MULTISTEP** You have designed a novelty desk clock, as shown in Figure 7-40. You are worried that it is not ready for market because the clock itself might be in an unstable equilibrium configuration. You decide to apply your knowledge of potential energies and equilibrium conditions and analyze the situation. The clock (mass m) is supported by two light cables running over the two frictionless pulleys of negligible diameter, which are attached to counterweights that each have mass M. (*a*) Find the potential energy of the system as a function of the distance y. (*b*) Find the value of y for which the potential energy of the system is a minimum. (*c*) If the potential energy is a minimum, then the system is in equilibrium. Apply Newton's second law to the clock and show that it is in equilibrium (the forces on it sum to zero) for the value of y obtained for Part (*b*). (*d*) Finally, determine whether you are going to be able to market this gadget: is this a point of stable or unstable equilibrium?

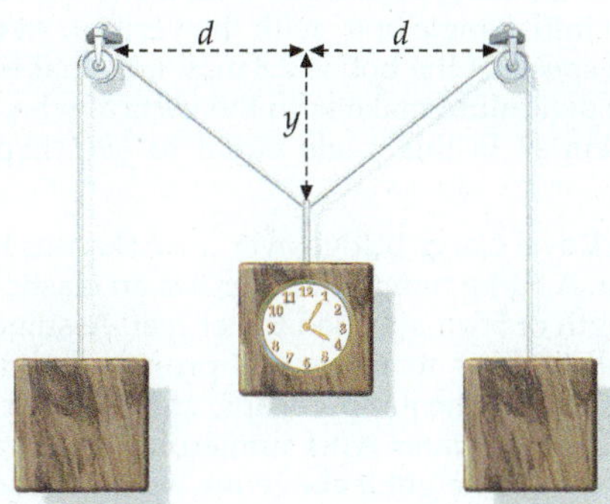

FIGURE 7-40 Problem 35

THE CONSERVATION OF MECHANICAL ENERGY

36 • A block of mass m on a horizontal frictionless tabletop is pushed against a horizontal spring, compressing it a distance x, and the block is then released. The spring propels the block along the tabletop, giving a speed v. The same spring is then used to propel a second block of mass $4m$, giving it a speed $3v$. What distance was the spring compressed in the second case? Express your answer in terms of x.

37 • A simple pendulum of length L with a bob of mass m is pulled aside until the bob is at a height $L/4$ above its equilibrium position. The bob is then released. Find the speed of the bob as it passes through the equilibrium position. Neglect any effects due to air resistance.

38 • A 3.0-kg block slides along a frictionless horizontal surface with a speed of 7.0 m/s (Figure 7-41). After sliding a distance of 2.0 m, the block makes a smooth transition to a frictionless ramp inclined at an angle of 40° to the horizontal. What distance along the ramp does the block slide before coming momentarily to rest?

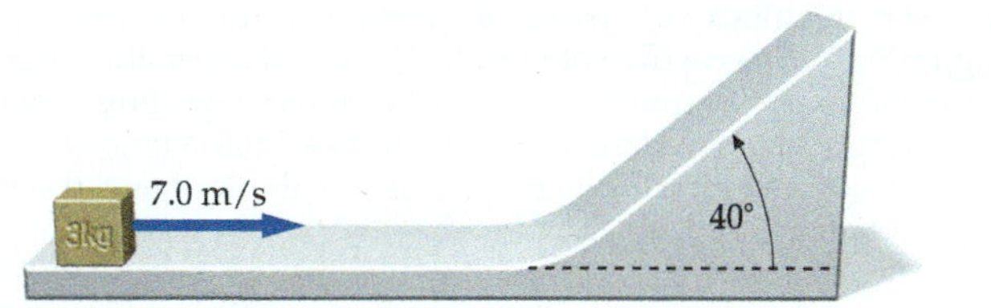

FIGURE 7-41 Problems 38 and 64

39 • The 3.00-kg object in Figure 7-42 is released from rest at a height of 5.00 m on a curved frictionless ramp. At the foot of the ramp is a spring of force constant 400 N/m. The object slides down the ramp and into the spring, compressing it a distance x before coming momentarily to rest. (*a*) Find x. (*b*) Describe the motion of the object (if any) after the block momentarily comes to rest?

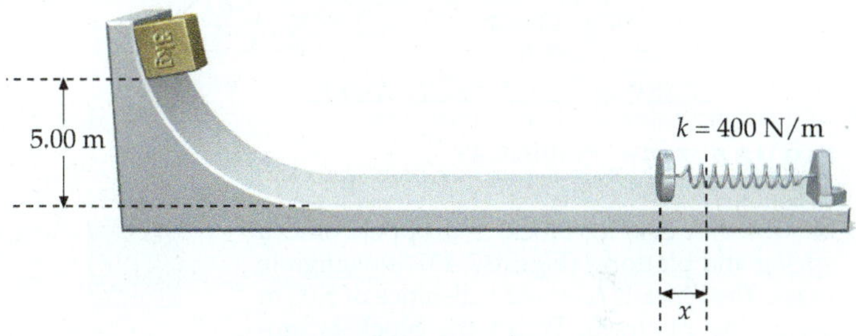

FIGURE 7-42 Problem 39

40 • **ENGINEERING APPLICATION, CONTEXT-RICH** You are designing a game for small children and want to see if the ball's maximum speed is sufficient to require the use of goggles. In your game, a 15.0-g ball is to be shot from a spring gun whose spring has a force constant of 600 N/m. The spring will be compressed 5.00 cm when in use. How fast will the ball be moving as it leaves the gun and how high will the ball go if the gun is aimed vertically upward? What would be your recommendation on the use of goggles?

41 • A 16-kg child on a 6.0-m-long playground swing moves with a speed of 3.4 m/s when the swing seat passes through its lowest point. What is the angle that the swing makes with the vertical when the swing is at its highest point? Assume that the effects due to air resistance are negligible, and assume that the child is not pumping the swing. SSM

42 •• The system shown in Figure 7-43 is initially at rest when the lower string is cut. Find the speed of the objects when they are momentarily at the same height. The frictionless pulley has negligible mass.

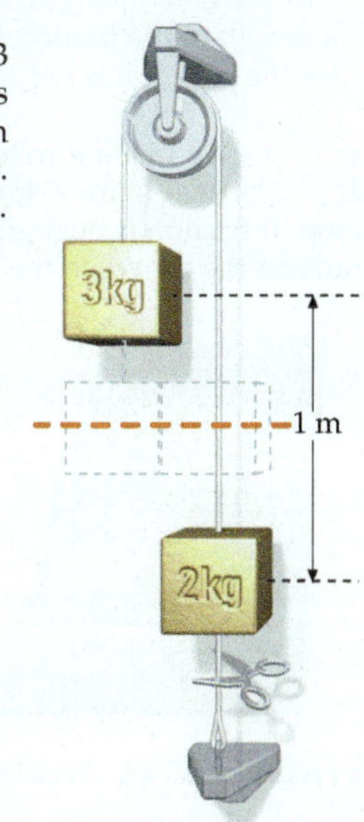

FIGURE 7-43 Problem 42

43 •• A block of mass m rests on an inclined plane (Figure 7-44). The coefficient of static friction between the block and the plane is μ_s. A gradually increasing force is pulling down on the spring (force constant k). Find the potential energy U of the spring (in terms of the given symbols) at the moment the block begins to move.

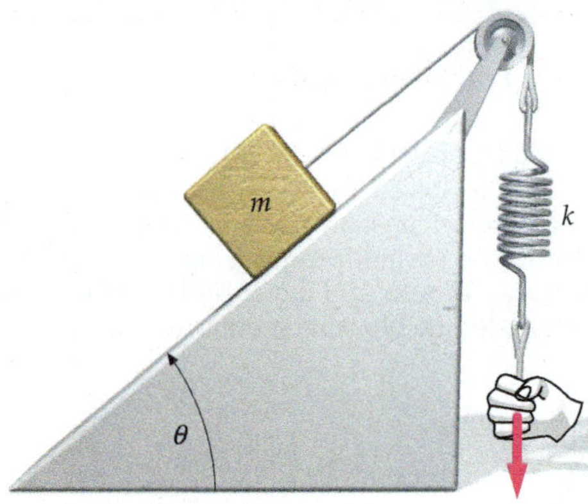

FIGURE 7-44 Problem 43

44 •• A 2.40-kg block is dropped onto a spring and platform (Figure 7-45) of negligible mass. The block is released a distance of 5.00 m above the platform. When the block is momentarily at rest, the spring is compressed by 25.0 cm. Find the speed of the block when the compression of the spring is only 15.0 cm.

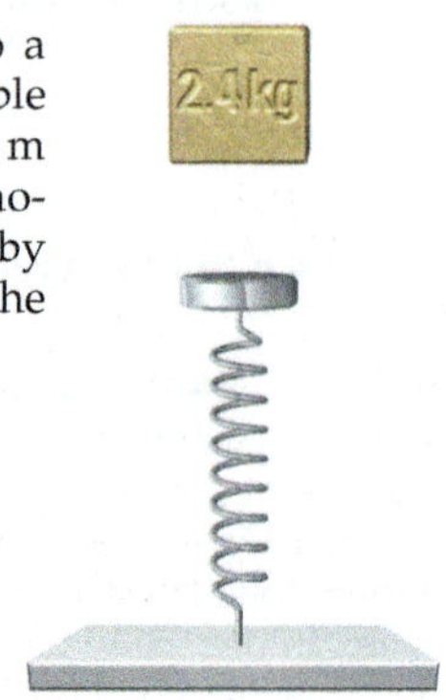

FIGURE 7-45 Problem 44

45 •• A ball at the end of a string moves in a vertical circle with constant mechanical energy E. What is the difference between the tension at the bottom of the circle and the tension at the top? SSM

46 •• A girl of mass m is taking a picnic lunch to her grandmother. She ties a rope of length R to a tree branch over a creek and starts to swing from rest at a point that is a distance $R/2$ lower than the branch. What is the minimum breaking tension for the rope if it is not to break and drop the girl into the creek?

47 •• A 1500-kg roller coaster car starts from rest at a height $H = 23.0$ m (Figure 7-46) above the bottom of a 15.0-m-diameter loop. If friction is negligible, determine the downward force of the rails on the car when the upside-down car is at the top of the loop.

FIGURE 7-46 Problem 47

48 •• A single roller-coaster car is moving with speed v_0 on the first section of track when it descends a 5.0-m-deep valley, then climbs to the top of a hill that is 4.5 m above the first section of track. Assume any effects of friction or of air resistance are negligible. (*a*) What is the minimum speed v_0 required if the car is to travel beyond the top of the hill? (*b*) Can we affect this speed by changing the depth of the valley to make the coaster pick up more speed at the bottom? Explain.

49 •• The Gravitron single-car roller coaster consists of a single loop-the-loop. The car is initially pushed, giving it just the right mechanical energy so the riders on the coaster will feel "weightless" when they pass through the top of the circular arc. How heavy will they feel when they pass through the bottom of the arc (that is, what is the normal force pressing up on them when they are at the bottom of the loop)? Express the answer as a multiple of mg (their actual weight). Assume any effects of friction or of air resistance are negligible.

50 •• A stone is thrown upward at an angle of 53° above the horizontal. Its maximum height above the release point is 24 m. What was the stone's initial speed? Assume any effects of air resistance are negligible.

51 •• A 0.17-kg baseball is launched from the roof of a building 12 m above the ground. Its initial velocity is 30 m/s at 40° above the horizontal. Assume any effects of air resistance are negligible. (*a*) What is the maximum height above the ground that the ball reaches? (*b*) What is the speed of the ball as it strikes the ground?

52 •• An 80-cm-long pendulum with a 0.60-kg bob is released from rest at an initial angle of θ_0 with the vertical. At the bottom of the swing, the speed of the bob is 2.8 m/s. (*a*) What is θ_0? (*b*) What angle does the pendulum make with the vertical when the speed of the bob is 1.4 m/s? Is this angle equal to $\frac{1}{2}\theta_0$? Explain why or why not.

53 •• The Royal Gorge bridge over the Arkansas River is 310 m above the river. A 60-kg bungee jumper has an elastic cord with an unstressed length of 50 m attached to her feet. Assume that, like an ideal spring, the cord is massless and provides a linear restoring force when stretched. The jumper leaps, and at at her lowest point she barely touches the water. After numerous ascents and descents, she comes to rest at a height h above the water. Model the jumper as a point particle and assume that any effects of air resistance are negligible. (*a*) Find h. (*b*) Find the maximum speed of the jumper.

54 •• A pendulum consists of a 2.0-kg bob attached to a light 3.0-m-long string. While hanging at rest with the string vertical, the bob is struck a sharp horizontal blow, giving it a horizonal velocity of 4.5 m/s. At the instant the string makes an angle of 30° with the vertical, what is (*a*) the speed, (*b*) the gravitational potential energy (relative to its value is at the lowest point), and (*c*) the tension in the string? (*d*) What is the angle of the string with the vertical when the bob reaches its greatest height?

55 •• A pendulum consists of a string of length L and a bob of mass m. The bob is rotated until the string is horizontal. The bob is then projected downward with the minimum initial speed needed to enable the bob to make a full revolution in the vertical plane. (*a*) What is the maximum kinetic energy of the bob? (*b*) What is the tension in the string when the kinetic energy is maximum? SSM

56 •• A child whose weight is 360 N swings out over a pool of water using a rope attached to the branch of a tree at the edge of the pool. The branch is 12 m above ground level and the surface of the water is 1.8 m below ground level. The child holds onto the rope at a point 10.6 m from the branch and moves back until the angle between the rope and the vertical is 23°. When the rope is in the vertical position, the child lets go and drops into the pool. Find the speed of the child just as he impacts the surface of the water.

(Model the child as a point particle attached to the rope 10.6 m from the branch.)

57 •• Walking by a pond, you find a rope attached to a stout tree limb that is 5.2 m above ground level. You decide to use the rope to swing out over the pond. The rope is a bit frayed, but supports your weight. You estimate that the rope might break if the tension is 80 N greater than your weight. You grab the rope at a point 4.6 m from the limb and move back to swing out over the pond. (Model yourself as a point particle attached to the rope 4.6 m from the limb.) (*a*) What is the maximum safe initial angle between the rope and the vertical at which it will not break during the swing? (*b*) If you begin at this maximum angle, and the surface of the pond is 1.2 m below the level of the ground, with what speed will you enter the water if you let go of the rope when the rope is vertical?

58 ••• A pendulum bob of mass m is attached to a light string of length L and is also attached to a spring of force constant k. With the pendulum in the position shown in Figure 7-47, the spring is at its unstressed length. If the bob is now pulled aside so that the string makes a *small* angle θ with the vertical and released, what is the speed of the bob as it passes through the equilibrium position? *Hint: Recall the small-angle approximations: if θ is expressed in radians, and if $|\theta| \ll 1$, then $\sin\theta \approx \tan\theta \approx \theta$ and $\cos\theta \approx 1 - \frac{1}{2}\theta^2$.*

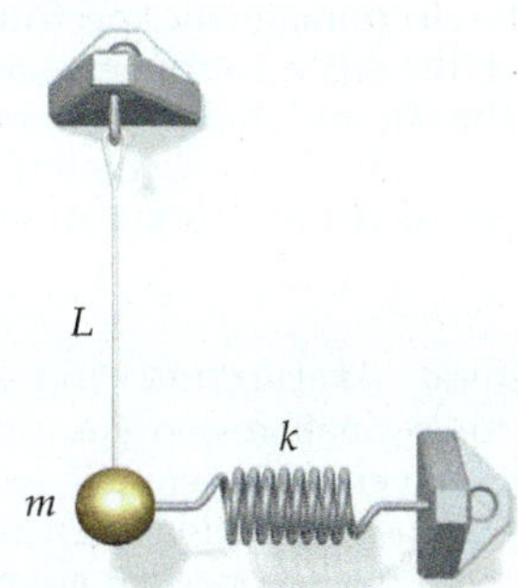

FIGURE 7-47 Problem 58

59 ••• A pendulum is suspended from the ceiling and attached to a spring fixed to the floor directly below the pendulum support (Figure 7-48). The mass of the pendulum bob is m, the length of the pendulum is L, and the force constant is k. The unstressed length of the spring is $L/2$ and the distance between the floor and ceiling is $1.5L$. The pendulum is pulled aside so that it makes an angle θ with the vertical and is then released from rest. Obtain an expression for the speed of the pendulum bob as the bob passes through a point directly below the pendulum support. SSM

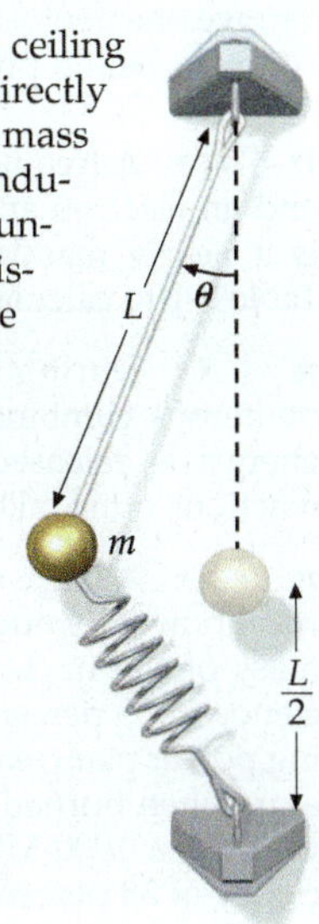

FIGURE 7-48 Problem 59

TOTAL ENERGY AND NONCONSERVATIVE FORCES

60 • In a volcanic eruption, 4.00 km^3 of mountain with an average density of 1600 kg/m^3 was raised an average height of 500 m. (*a*) What is the minimum amount of energy, in joules, that was released during this eruption? (*b*) The energy released by thermonuclear bombs is measured in megatons of TNT, where 1 megaton of TNT $= 4.2 \times 10^{15}$ J. Convert your answer for Part (*a*) to megatons of TNT.

61 • CONTEXT-RICH To work off a large pepperoni pizza you ate on Friday night, on Saturday morning you climb a 120-m-high hill. (*a*) Assuming a reasonable value for your mass, determine your increase in gravitational potential energy. (*b*) Where does this energy come from? (*c*) The human body is typically 20 percent efficient. How much energy was converted into thermal energy? (*d*) How much chemical energy is expended by you during the climb? Given that oxidation (burning) of a single slice of pepperoni pizza releases about 1.0 MJ (250 food calories) of energy, do you think one climb up the hill is enough?

62 • A 2000-kg car moving at an initial speed of 25 m/s along a horizontal road skids to a stop in 60 m. (*a*) Find the energy dissipated by friction. (*b*) Find the coefficient of kinetic friction between the tires and the road. (*Note*: When stopping without skidding and using conventional brakes, 100 percent of the kinetic energy is dissipated by friction within the brakes. With regenerative braking, such as that used in hybrid vehicles, only 70 percent of the kinetic energy is dissipated.)

63 • An 8.0-kg sled is initially at rest on a horizontal road. The coefficient of kinetic friction between the sled and the road is 0.40. The sled is pulled a distance of 3.0 m by a force of 40 N applied to the sled at an angle of 30° above the horizontal. (*a*) Find the work done by the applied force. (*b*) Find the energy dissipated by friction. (*c*) Find the change in the kinetic energy of the sled. (*d*) Find the speed of the sled after it has traveled 3.0 m.

64 •• Using Figure 7-41, suppose that the surfaces described are not frictionless and that the coefficient of kinetic friction between the block and the surfaces is 0.30. The block has an initial speed of 7.0 m/s and slides 2.0 m before reaching the ramp. Find (*a*) the speed of the block when it reaches the ramp, and (*b*) the distance that the block slides along the inclined surface before coming momentarily to rest. (Neglect any energy dissipated along the transition curve.)

65 •• The 2.0-kg block in Figure 7-49 slides down a frictionless curved ramp, starting from rest at a height of 3.0 m. The block then slides 9.0 m on a rough horizontal surface before coming to rest. (*a*) What is the speed of the block at the bottom of the ramp? (*b*) What is the energy dissipated by friction? (*c*) What is the coefficient of kinetic friction between the block and the horizontal surface? SSM

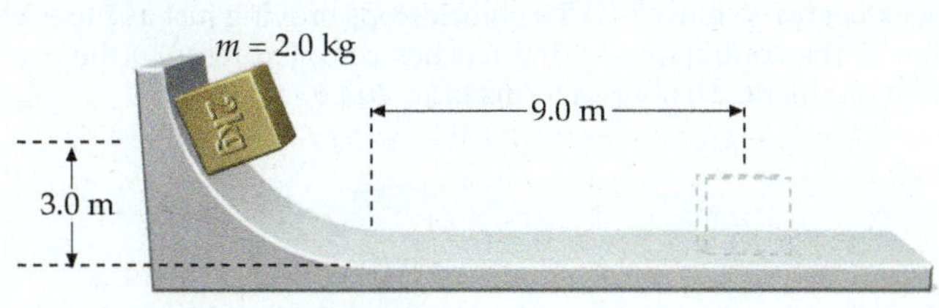

FIGURE 7-49 Problem 65

66 •• A 20-kg girl slides down a playground slide with a vertical drop of 3.2 m. When she reaches the bottom of the slide, her speed is 1.3 m/s. (*a*) How much energy was dissipated by friction? (*b*) If the slide is inclined at 20° with the horizontal, what is the coefficient of kinetic friction between the girl and the slide?

67 •• In Figure 7-50, the coefficient of kinetic friction between the 4.0-kg block and the shelf is 0.35. (*a*) Find the energy dissipated by friction when the 2.0-kg block falls a distance *y*. (*b*) Find the change in the mechanical energy E_{mech} of the two-block–Earth system during the time it takes the 2.0-kg block to fall a distance *y*. (*c*) Use your result for Part (*b*) to find the speed of either block after the 2.0-kg block falls 2.0 m.

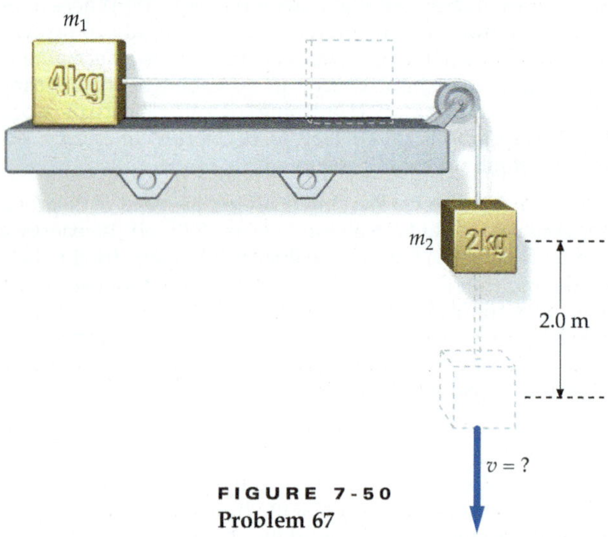

FIGURE 7-50 Problem 67

68 •• A small object of mass *m* moves in a horizontal circle of radius *r* on a rough table. It is attached to a horizontal string fixed at the center of the circle. The speed of the object is initially v_0. After completing one full trip around the circle, the speed of the object is $0.5v_0$. (*a*) Find the energy dissipated by friction during that one revolution in terms of *m*, v_0, and *r*. (*b*) What is the coefficient of kinetic friction? (*c*) How many more revolutions will the object make before coming to rest?

69 •• The initial speed of a 2.4-kg box traveling up a plane inclined 37° to the horizontal is 3.8 m/s. The coefficient of kinetic friction between the box and the plane is 0.30. (*a*) How far along the incline does the box travel before coming to a stop? (*b*) What is its speed when it has traveled half the distance found in Part (*a*)? SSM

70 ••• A block of mass *m* rests on a plane inclined at an angle *θ* with the horizontal (Figure 7-51). A spring with force constant *k* is attached to the block. The coefficient of static friction between the block and plane is μ_s. The spring is pulled upward along the plane very slowly. (*a*) What is the extension of the spring the instant the block begins to move? (*b*) The block stops moving just as the extension of the contracting spring reaches zero. Express μ_k (the coefficient of kinetic friction) in terms of μ_s and *θ*.

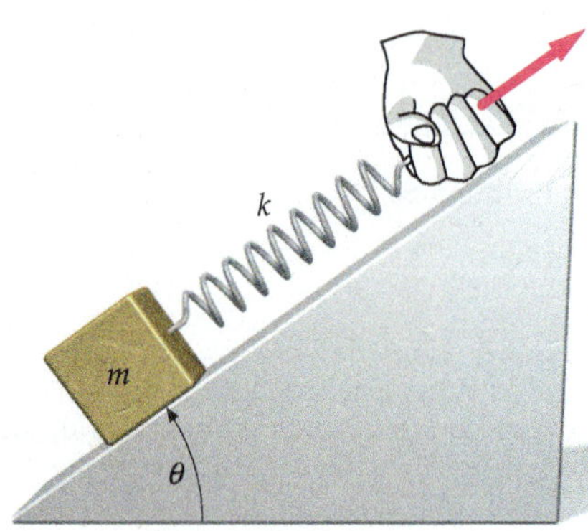

FIGURE 7-51 Problem 70

MASS AND ENERGY

71 • (*a*) Calculate the rest energy of 1.0 g of dirt. (*b*) If you could convert this energy completely into electrical energy and sell it for \$0.10/kW·h, how much money would you take in? (*c*) If you could power a 100-W light bulb with this energy, for how long could you keep the bulb lit?

72 • One kiloton of TNT, when detonated, yields an explosive energy of roughly 4×10^{12} J. How much less is the total mass of the bomb remnants after the explosion than before? If you could find and reassemble the pieces, would this loss of mass be noticeable?

73 • **CONCEPTUAL** Calculate your rest energy in both mega electron-volts and joules. If that energy could be converted completely to the kinetic energy of your car, estimate its speed. Use the nonrelativistic expression for kinetic energy and comment on whether or not your answer justifies using the nonrelativistic expression for kinetic energy.

74 • If a black hole and a "normal" star orbit each other, gases from the normal star falling into the black hole can have their temperature increased by millions of degrees due to frictional heating. When the gases are heated that much, they begin to radiate light in the X-ray region of the electromagnetic spectrum (high-energy light photons). Cygnus X-1, the second strongest known X-ray source in the sky, is thought to be one such binary system; it radiates at an estimated power of 4×10^{31} W. If we assume that 1.0 percent of the in-falling mass escapes as X ray energy, at what rate is the black hole gaining mass?

75 • **ENGINEERING APPLICATION** You are designing the fuel requirements for a small fusion electric-generating plant. Assume 33 percent conversion to electric energy. For the deuterium–tritium (D–T) fusion reaction in Example 7-18, calculate the number of reactions per second that are necessary to generate 1.00 kW of electric power. SSM

76 • Use Table 7-1 to calculate the energy needed to remove one neutron from a stationary alpha particle, leaving a stationary helion plus a neutron with a kinetic energy of 1.5 MeV.

77 • A free neutron can decay into a proton plus an electron and an electron antineutrino [an electron antineutrino (symbol $\bar{\nu}_e$) is a nearly massless elementary particle]: $n \rightarrow p + e^- + \bar{\nu}_e$. Use Table 7-1 to calculate the energy released during this reaction.

78 •• During one type of nuclear fusion reaction, two deuterons combine to produce an alpha particle. (*a*) How much energy is released during this reaction? (*b*) How many such reactions must take place per second to produce 1 kW of power?

79 •• A large nuclear power plant produces 1000 MW of electrical power by nuclear fission. (*a*) By how many kilograms does the mass of the nuclear fuel decrease in one year? (Assume an efficiency of 33 percent for a nuclear power plant.) (*b*) In a coal-burning power plant, each kilogram of coal releases 31 MJ of thermal energy when burned. How many kilograms of coal are needed each year for a 1000-MW coal-burning power plant? (Assume an efficiency of 38 percent for a coal-burning power plant.)

QUANTIZATION OF ENERGY

80 •• A mass on the end of a spring with a force constant of 1000 N/kg oscillates at a frequency of 2.5 oscillations per second. (*a*) Determine the quantum number, *n*, of the state it is in if it has a total energy of 10 J. (*b*) What is its ground state energy?

81 •• Repeat Problem 80, but consider instead an atom in a solid vibrating at a frequency of 1.00×10^{14} oscillations per second and having a total energy of 2.7 eV.

GENERAL PROBLEMS

82 • A block of mass m, starting from rest, is pulled up a frictionless inclined plane that makes an angle θ with the horizontal by a string parallel to the plane. The tension in the string is T. After traveling a distance L, the speed of the block is v_f. Derive an expression for work done by the tension force.

83 • A block of mass m slides with constant speed v down a plane inclined at angle θ with the horizontal. Derive an expression for the energy dissipated by friction during the time interval Δt.

84 • In particle physics, the potential energy associated with a pair of quarks bound together by the strong nuclear force is in one particular theoretical model written as the following function: $U(r) = -(\alpha/r) + kr$, where k and α are positive constants, and r is the distance of separation between the two quarks.* (*a*) Sketch the general shape of the potential-energy function. (*b*) What is a general form for the force each quark exerts on the other? (*c*) At the two extremes of very small and very large values of r, what does the force simplify to?

85 • **ENGINEERING APPLICATION, CONTEXT-RICH** You are in charge of "solar-energizing" your grandfather's farm. At the farm's location, an average of $1.0\,\text{kW/m}^2$ reaches the surface during the daylight hours on a clear day. If this could be converted at 25 percent efficiency to electric energy, how large a collection area would you need to run a 4.0-hp irrigation water pump during the daylight hours? SSM

86 •• **ENGINEERING APPLICATION** The radiant energy from the Sun that reaches Earth's orbit is $1.35\,\text{kW/m}^2$. (*a*) Even when the Sun is directly overhead and under dry desert conditions, 25 percent of this energy is absorbed and/or reflected by the atmosphere before it reaches Earth's surface. If the average frequency of the electromagnetic radiation from the Sun is 5.5×10^{14} Hz, how many photons per second would be incident upon a 1.0-m^2 solar panel? (*b*) Suppose the efficiency of the panels for converting the radiant energy to electrical energy and delivering it is a highly efficient 10.0 percent. How large a solar panel is needed to supply the needs of a 5.0-hp solar-powered car (assuming the car runs directly off the solar panel and not batteries) during a race in Cairo at noon on March 21? (*c*) Assuming a more-realistic efficiency of 3.3 percent and panels capable of rotating to be always perpendicular to the sunlight, how large an array of solar panels is needed to supply the power needs of the International Space Station (ISS)? The ISS requires about 110 kW of continuous electric power.

87 •• In 1964, after the 1250-kg jet-powered car *Spirit of America* lost its parachute and went out of control during a run at Bonneville Salt Flats, Utah, it left skid marks about 8.00 km long. (This earned a place in the Guinness Book of World Records for longest skid marks.) (*a*) If the car was moving initially at a speed of about 800 km/h, and was still going at about 300 km/h when it crashed into a brine pond, estimate the coefficient of kinetic friction μ_k. (*b*) What was the kinetic energy of the car 60 s after the skid began?

88 •• **ENGINEERING APPLICATION, CONTEXT-RICH** A T-bar tow is planned in a new ski area. At any one time, it will be required, to pull a maximum of 80 skiers up a 600-m slope inclined at 15° above the horizontal at a speed of 2.50 m/s. The coefficient of kinetic friction between the skiers skis and the snow is typically 0.060. As the manager of the facility, what motor power should you request of the construction contractor if the mass of the average skier is 75.0 kg. Assume you want to be ready for any emergency and will order a motor whose power rating is 50 percent larger than the bare minimum.

89 •• **MULTISTEP** A box of mass m on the floor is connected to a horizontal spring of force constant k (Figure 7-52). The coefficient of kinetic friction between the box and the floor is μ_k. The other end of the spring is connected to a wall. The spring is initially unstressed. If the box is pulled away from the wall a distance d_0 and released, the box slides toward the wall. Assume the box does not slide so far that the coils of the spring touch. (*a*) Obtain an expression for the distance d_1 the box slides before it first comes to a stop. (*b*) Assuming $d_1 > d_0$, obtain an expression for the speed of the box when it has slid a distance d_0 following the release. (*c*) Obtain the special value of μ_k such that $d_1 = d_0$.

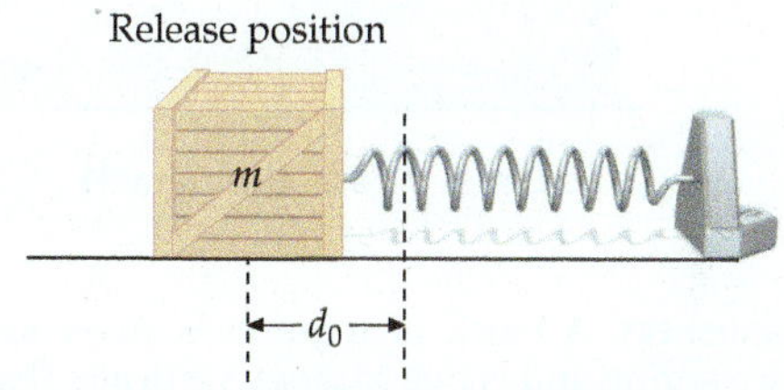

FIGURE 7-52 Problem 89

90 •• **ENGINEERING APPLICATION, CONTEXT-RICH** You operate a small grain elevator near Champaign, Illinois. One of your silos uses a bucket elevator that carries a full load of 800 kg through a vertical distance of 40 m. (A bucket elevator works with a continuous belt, like a conveyor belt.) (*a*) What is the power provided by the electric motor powering the bucket elevator when the bucket elevator ascends with a full load at a speed of 2.3 m/s? (*b*) Assuming the motor is 85 percent efficient, how much does it cost you to run this elevator, per day, assuming it runs 60 percent of the time between 7:00 a.m. and 7:00 p.m. with an average load of 85 percent of a full load? Assume the cost of electric energy in your location is 15 cents per kilowatt hour.

91 •• **ENGINEERING APPLICATION** To reduce the power requirement of elevator motors, elevators are counterbalanced with weights connected to the elevator by a cable that runs over a pulley at the top of the elevator shaft. Neglect any effects of friction in the pulley. If a 1200-kg elevator that carries a maximum load of 800 kg is counterbalanced with a mass of 1500 kg, (*a*) what is the power provided by the motor when the elevator ascends fully loaded at a speed of 2.3 m/s? (*b*) How much power is provided by the motor when the elevator ascends at 2.3 m/s without a load?

92 •• In old science fiction movies, writers attempted to come up with novel ways of launching spacecraft toward the moon. In one hypothetical case, a screenwriter envisioned launching a moon probe from a deep, smooth tunnel, inclined at 65.0° above the horizontal. At the bottom of the tunnel a very stiff spring designed to launch the craft was anchored. The top of the spring, when the spring is unstressed, is 30.0 m from the upper end of the tunnel. The screenwriter knew from his research that to reach the moon, the 318-kg probe should have a speed of at least 11.2 km/s when it exits the tunnel. If the spring is compressed by 95.0 m just before launch, what is the minimum value for its force constant to achieve a successful launch? Neglect friction with the tunnel walls and floor.

* This is known as the "Cornell potential," developed in *Physical Review Letters*. Make reference here also to the 2004 Nobel Prize to Gross, Wilczek, and Politzer.

93 •• In a volcanic eruption, a 2-kg piece of porous volcanic rock is thrown straight upward with an initial speed of 40 m/s. It travels upward a distance of 50 m before it begins to fall back to Earth. (*a*) What is the initial kinetic energy of the rock? (*b*) What is the increase in thermal energy due to air resistance during ascent? (*c*) If the increase in thermal energy due to air resistance on the way down is 70 percent of that on the way up, what is the speed of the rock when it returns to its initial position? SSM

94 •• A block of mass m starts from rest at a height h and slides down a frictionless plane inclined at angle θ with the horizontal, as shown in Figure 7-53. The block strikes a spring of force constant k. Find the distance the spring is compressed when the block momentarily stops.

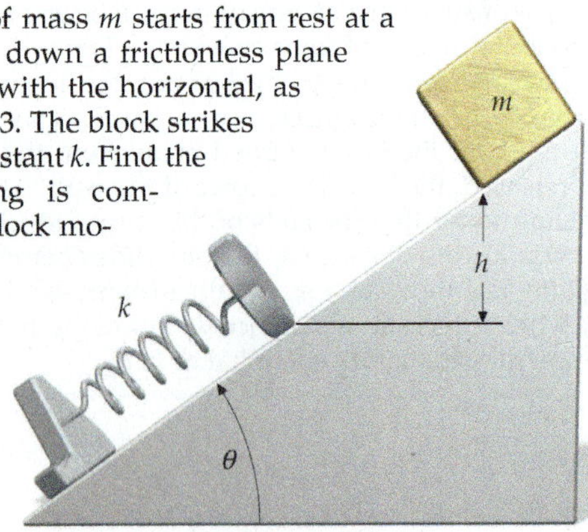

FIGURE 7-53 Problem 94

95 •• **SPREADSHEET** A block of mass m is suspended from a wall bracket by a spring and is free to move vertically (Figure 7-54). The $+y$ direction is downward and the origin is at the position of the block when the spring is unstressed. (*a*) Show that the potential energy as a function of position may be expressed as $U = \frac{1}{2}ky^2 - mgy$. (*b*) Using a **spreadsheet** program or graphing calculator, make a graph of U as a function of y, with $k = 2$ N/m and $mg = 1$ N. (*c*) Explain how this graph shows that there is a position of stable equilibrium for a positive value of y. Using the Part (*a*) expression for U, determine (symbolically) the value of y when the block is at its equilibrium position. (*d*) From the expression for U, find the net force acting on m at any position y. (*e*) The block is released from rest with the spring unstressed; if there is no friction, what is the maximum value of y that will be reached by the mass? Indicate y_{max} on your graph/spreadsheet. SSM

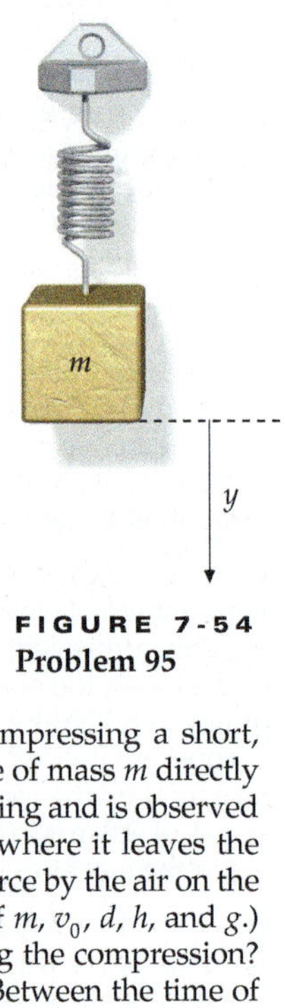

FIGURE 7-54 Problem 95

96 •• A spring-loaded gun is cocked by compressing a short, strong spring by a distance d. It fires a signal flare of mass m directly upward. The flare has speed v_0 as it leaves the spring and is observed to rise to a maximum height h above the point where it leaves the spring. After it leaves the spring, effects of drag force by the air on the flare are significant. (Express answers in terms of m, v_0, d, h, and g.) (*a*) How much work is done on the spring during the compression? (*b*) What is the value of the force constant k? (*c*) Between the time of firing and the time at which maximum elevation is reached, how much mechanical energy is dissipated into thermal energy?

97 •• **ENGINEERING APPLICATION, CONTEXT-RICH** Your firm is designing a new roller-coaster ride. The permit process requires the calculation of forces and accelerations at various important locations on the ride. Each roller-coaster car will have a total mass (including passengers) of 500 kg and will travel freely along the winding frictionless track shown in Figure 7-55. Points A, E, and G are on horizontal straight sections, all at the same height of 10 m above the ground. Point C is at a height of 10 m above the ground on an inclined section of slope angle 30°. Point B is at the crest of a hill, while point D is at ground level at the bottom of a valley; the radius of curvature at both of these points is 20 m. Point F is at the middle of a banked horizontal curve with a radius of curvature of 30 m, and at the same height as points A, E, and G. At point A the speed of the car is 12 m/s. (*a*) If the car is just barely to make it over the hill at point B, what must be the height of point B above the ground? (*b*) If the car is to just barely make it over the hill at point B, what should be the magnitude of the force exerted by the track on the car at that point? (*c*) What will be the acceleration of the car at point C? (*d*) What will be the magnitude and direction of the force exerted by the track on the car at point D? (*e*) What will be the magnitude and direction of the force exerted by the track on the car at point F? (*f*) At point G a constant braking force is to be applied to the car, bringing it to a halt in a distance of 25 m. What is the magnitude of this required braking force?

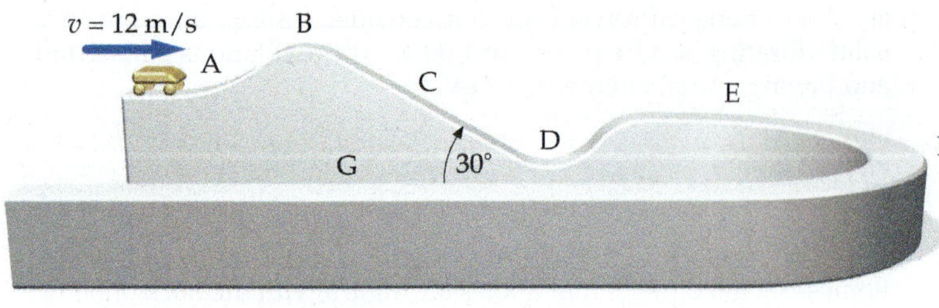

FIGURE 7-55 Problem 97

98 •• **ENGINEERING APPLICATION** The cable of a 2000-kg elevator has broken, and the elevator is moving downward at a steady speed of 1.5 m/s. A safety braking system that works on friction prevents the downward speed from increasing. (*a*) At what rate is the braking system converting mechanical energy to thermal energy? (*b*) While the elevator is moving downward at 1.5 m/s, the braking system fails and the elevator is in free-fall for a distance of 5.0 m before hitting the top of a large safety spring with force constant of 1.5×10^4 N/m. After the elevator hits the top of the spring, find the distance d that the spring is compressed before the elevator is brought to rest.

99 ••• **ENGINEERING APPLICATION, CONTEXT-RICH** To measure the combined force of friction (rolling friction plus air drag) on a moving car, an automotive engineering team you are on turns off the engine and allows the car to coast down hills of known steepness. The team collects the following data: (1) On a 2.87° hill, the car can coast at a steady 20 m/s. (2) On a 5.74° hill, the steady coasting speed is 30 m/s. The total mass of the car is 1000 kg. (*a*) What is the magnitude of the combined force of friction at 20 m/s (F_{20}) and at 30 m/s (F_{30})? (*b*) How much power must the engine deliver to drive the car on a level road at steady speeds of 20 m/s (P_{20}) and 30 m/s (P_{30})? (*c*) The maximum power the engine can deliver is 40 kW. What is the angle of the steepest incline up which the car can maintain a steady 20 m/s? (*d*) Assume that the engine delivers the same total useful work from each liter of gas, no matter what the speed. At 20 m/s on a level road, the car gets 12.7 km/L. How many kilometers per liter does it get if it goes 30 m/s instead? SSM

100 •• **ENGINEERING APPLICATION** (*a*) Calculate the kinetic energy of a 1200-kg car moving at 50 km/h. (*b*) If friction (rolling friction and air drag) results in a retarding force of 300 N at a speed of 50 km/h, what is the minimum energy needed to move the car a distance of 300 m at a constant speed of 50 km/h?

101 ••• A pendulum consists of a small bob of mass m attached to a string of length L. The bob is held to the side with the string horizontal (see Figure 7-56). Then the bob is released from rest. At the lowest point of the swing, the string catches on a thin peg a distance R above the lowest point. Show that R must be smaller than $2L/5$ if the string is to remain taut as the bob swings around the peg in a full circle.

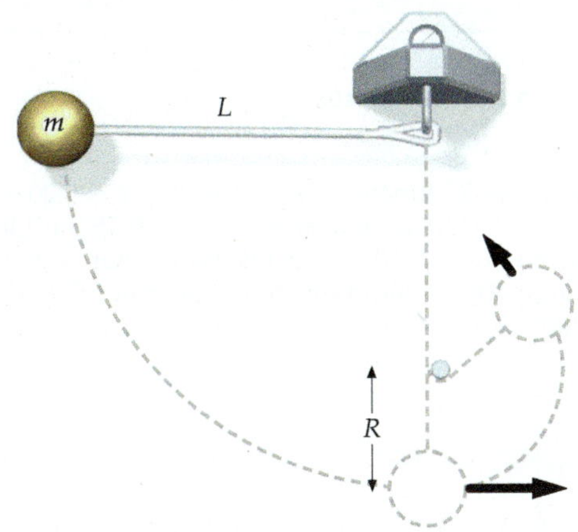

FIGURE 7-56 Problem 101

102 •• A 285-kg stunt boat is driven on the surface of a lake at a constant speed of 13.5 m/s toward a ramp, which is angled at 25.0° above the horizontal. The coefficient of friction between the boat bottom and the ramp's surface is 0.150, and the raised end of the ramp is 2.00 m above the water surface. (*a*) Assuming the engines are cut off when the boat hits the ramp, what is the speed of the boat as it leaves the ramp? (*b*) What is the speed of the boat when it strikes the water again? Neglect any effects due to air resistance.

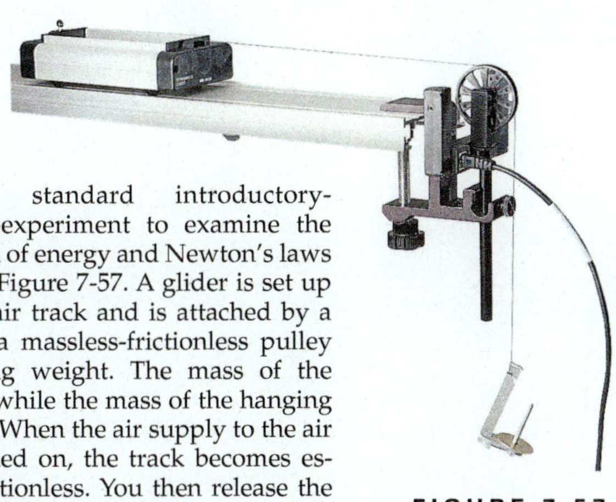

FIGURE 7-57 Problem 103

103 •• A standard introductory-physics lab-experiment to examine the conservation of energy and Newton's laws is shown in Figure 7-57. A glider is set up on a linear air track and is attached by a string over a massless-frictionless pulley to a hanging weight. The mass of the glider is M, while the mass of the hanging weight is m. When the air supply to the air track is turned on, the track becomes essentially frictionless. You then release the hanging weight and measure the speed of the glider after the weight has fallen a given distance (y). (*a*) To show that the measured speed is the speed predicted by theory, apply conservation of mechanical energy and calculate the speed as a function of y. (*b*) To verify this calculation, apply Newton's second and third laws directly by sketching a free-body diagram for each of the two masses and applying Newton's laws to find their accelerations. Then use kinematics to calculate the speed of the glider as a function of y.

104 •• **BIOLOGICAL APPLICATION** In one model of a person jogging, the energy expended is assumed to go into accelerating and decelerating the feet and the lower portions of the legs. If the jogging speed is v, then the maximum speed of the foot and lower leg is about $2v$. (From the moment a foot leaves the ground, to the moment it next contacts the ground, the foot travels nearly twice as far as the torso, so it must be going, on average, nearly twice as fast as the torso.) If the mass of the foot and lower portion of a leg is m, the energy needed to accelerate the foot and lower portion of a leg from rest to speed $2v$ is $\frac{1}{2}m(2v)^2 = 2mv^2$, and the same energy is needed to decelerate this mass back to rest for the next stride. Assume that the mass of the foot and lower portion of a man's leg is 5.0 kg and that he jogs at a speed of 3.0 m/s with 1.0 m between one footfall and the next. The energy he must provide to each leg in each 2.0 m of travel is $2mv^2$, so the energy he must provide to both legs during each second of jogging is $6mv^2$. Calculate the rate of the man's energy expenditure using this model, assuming that his muscles have an efficiency of 20 percent.

105 •• A high school teacher once suggested measuring the magnitude of free-fall acceleration by the following method: Hang a mass on a very fine thread (length L) to make a pendulum with the mass a height H above the floor when the mass is at its lowest point P. Pull the pendulum back so that the thread makes an angle θ_0 with the vertical. Just above point P, place a razor blade that is positioned to cut through the thread as the mass swings through point P. Once the thread is cut, the mass is projected horizontally, and hits the floor a horizontal distance D from point P. The idea was that the measurement of D as a function of θ_0 should somehow determine g. Apart from some obvious experimental difficulties, the experiment had one fatal flaw: D does not depend on g! Show that this is true, and that D depends only on the angle θ_0. SSM

106 ••• The bob of a pendulum of length L is pulled aside so that the string makes an angle θ_0 with the vertical, and the bob is then released. In Example 7-5, the conservation of energy was used to obtain the speed of the bob at the bottom of its swing. In this problem, you are to obtain the same result using Newton's second law. (*a*) Show that the tangential component of Newton's second law gives $dv/dt = -g\sin\theta$, where v is the speed and θ is the angle between the string and the vertical. (*b*) Show that v can be written $v = L\,d\theta/dt$. (*c*) Use this result and the chain rule for derivatives to obtain $\dfrac{dv}{dt} = \dfrac{dv}{d\theta}\dfrac{v}{L}$. (*d*) Combine the results of Parts (*a*) and (*c*) to obtain $v\,dv = -gL\sin\theta\,d\theta$. (*e*) Integrate the left side of the equation in Part (*d*) from $v = 0$ to the final speed v and the right side from $\theta = \theta_0$ to $\theta = 0$, and show that the result is equivalent to $v = \sqrt{2gh}$, where h is the original height of the bob above the bottom of its swing.

107 ••• **SPREADSHEET** A rock climber is rappelling down the face of a cliff when his hold slips and he slides down over the rock face, supported only by the bungee cord he attached to the top of the cliff. The cliff face is in the form of a smooth quarter-cylinder with height (and radius) $H = 300$ m (Figure 7-58). Treat the bungee

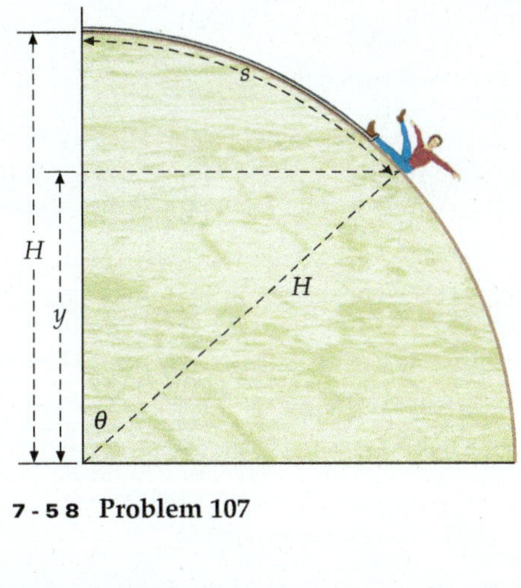

FIGURE 7-58 Problem 107

cord as a spring with force constant $k = 5.00\text{ N/m}$ and unstressed length $L = 60.0\text{ m}$. The climber's mass is 85.0 kg. (*a*) Using a **spreadsheet** program, make a graph of the rock climber's potential energy as a function of s, his distance from the top of the cliff *measured along the curved surface.* Use values of s between 60.0 m and 200 m. (*b*) His fall began when he was a distance $s_i = 60.0\text{ m}$ from the top of the cliff, and ended when he was a distance $s_f = 110\text{ m}$ from the top. Determine how much energy is dissipated by friction between the time he initially slipped and the time when he came to a stop.

108 ••• **SPREADSHEET** A block of wood (mass m) is connected to two massless springs, as shown in Figure 7-59. Each spring has unstressed length L and force constant k. (*a*) If the block is displaced a distance x, as shown, what is the change in the potential energy stored in the springs? (*b*) What is the magnitude of the force pulling the block back toward the equilibrium position? (*c*) Using a **spreadsheet** program or graphing calculator, make a graph of the potential energy U as a function of x for $0 \le x \le 0.20\text{ m}$. Assume $k = 1.0\text{ N/m}$, $L = 0.10\text{ m}$, and $m = 1.0\text{ kg}$. (*d*) If the block is displaced a distance $x = 0.10\text{ m}$ and released, what is its speed as it passes through the equilibrium point? Assume that the block is resting on a frictionless surface.

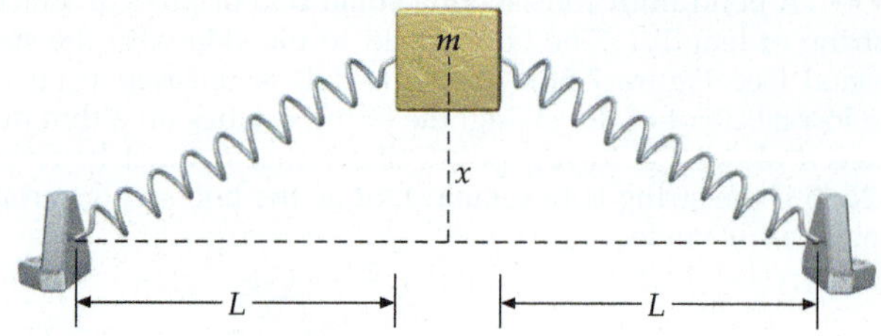

FIGURE 7-59 Problem 108

CHAPTER 8

DURING THE BRIEF PERIOD OF TIME THAT THE HEAD OF A GOLF CLUB IS IN CONTACT WITH A GOLF BALL, IT EXERTS A VERY LARGE FORCE ON THE BALL, SENDING THE BALL FLYING THROUGH THE AIR. THIS FORCE CAN BE ROUGHLY 10 000 TIMES THE WEIGHT OF THE BALL, GIVING THE BALL AN AVERAGE ACCELERATION OF APPROXIMATELY 10 000g FOR ABOUT ONE-HALF OF A MILLISECOND. *(Joe McNally/Getty Images.)*

Conservation of Linear Momentum

? If the golfer drives the ball 200 yards, how large a force does the clubface exert on the ball? (See Example 8-8.)

When the head of a golf club strikes a golf ball, the magnitude of the force exerted on the ball increases to a maximum value and then returns to zero as the ball leaves the club head. To describe how such a time-varying force affects the motion of an object it acts on, we need to introduce two new concepts: the impulse of a force and the momentum* of an object. Like energy, momentum is a conserved quantity. One of the most important principles in physics is *the law of conservation of momentum*, which says that the total momentum of a system and its surroundings does not change. Whenever the momentum of a system changes, we can account for the change by the appearance or disappearance of momentum somewhere else. By having these new ideas in our problem-solving toolbox, we can analyze collisions such as those that occur between golf clubs and golf balls, cars, and between subatomic particles in a nuclear reactor.

* In this chapter, the term *momentum* refers to linear momentum. (Angular momentum is developed in Chapter 10.)

In this chapter, we introduce the ideas of impulse and linear momentum, and show how integrating Newton's second law produces an important theorem known as the impulse–momentum theorem. We will also determine if the momentum of a system remains constant, and how to exploit constant momentum to solve problems involving collisions between objects. In addition, we examine a new reference frame, known as the center-of-mass reference frame, and explore situations in which a system has a continuously changing mass.

8-1 CONSERVATION OF LINEAR MOMENTUM

When Newton devised his second law, he considered the product of mass and velocity as a measure of an object's "quantity of motion." Today, we call the product of a particle's mass and velocity **linear momentum, $\vec{p}$:**

$$\vec{p} = m\vec{v} \qquad 8\text{-}1$$

DEFINITION—MOMENTUM OF A PARTICLE

The quantity $\vec{p}$ is called the *linear momentum* of a particle to distinguish linear momentum from *angular momentum*, which is presented in Chapter 10. However, when there is no need to distinguish between the two types, the adjective *linear* is often dropped and we just refer to the *momentum*. The plural of momentum is *momenta*.

Linear momentum is a vector quantity. It is the product of a vector (velocity) and a scalar (mass). Its magnitude is mv and it has the same direction as $\vec{v}$. The units of momentum are units of mass times speed, so the SI units of momentum are kg · m/s.

Momentum may be thought of as a measurement of the effort needed to bring a particle to rest. For example, a massive truck has more momentum than a small passenger car traveling at the same speed. It takes a greater force to stop the truck in a given time than it does to stop the car in the same time.

Using Newton's second law, we can relate the momentum of a particle to the resultant force acting on the particle. Differentiating Equation 8-1, we obtain

$$\frac{d\vec{p}}{dt} = \frac{d(m\vec{v})}{dt} = m\frac{d\vec{v}}{dt} = m\vec{a}$$

Then, substituting the force $\vec{F}_{\text{net}}$ for $m\vec{a}$,

$$\vec{F}_{\text{net}} = \frac{d\vec{p}}{dt} \qquad 8\text{-}2$$

Thus, the net force acting on a particle equals the time rate of change of the particle's momentum. (In his famous treatise *Principia* (1687), Isaac Newton presents the second law of motion in this form, rather than as $\vec{F}_{\text{net}} = m\vec{a}$.)

The total momentum $\vec{P}_{\text{sys}}$ of a system of particles is the vector sum of the momenta of the individual particles:

$$\vec{P}_{\text{sys}} = \sum_i m_i\vec{v}_i = \sum_i \vec{p}_i$$

According to Equation 5-20, $\Sigma m_i\vec{v}_i$ equals the total mass M times the velocity of the center of mass:

$$\vec{P}_{\text{sys}} = \sum_i m_i\vec{v}_i = M\vec{v}_{\text{cm}} \qquad 8\text{-}3$$

TOTAL MOMENTUM OF A SYSTEM

Differentiating this equation, we obtain

$$\frac{d\vec{P}_{sys}}{dt} = M\frac{d\vec{v}_{cm}}{dt} = M\vec{a}_{cm}$$

But according to Newton's second law for a system of particles, $M\vec{a}_{cm}$ equals the net external force acting on the system. Thus,

$$\sum_i \vec{F}_{ext} = \vec{F}_{net\,ext} = \frac{d\vec{P}_{sys}}{dt} \qquad 8\text{-}4$$

When the sum of the external forces acting on a system of particles remains zero, the rate of change of the total momentum remains zero and the total momentum of the system remains constant. That is,

$$\text{If } \Sigma\vec{F}_{ext} = 0, \text{ then } \vec{P}_{sys} = \sum_i m_i\vec{v}_i = M\vec{v}_{cm} = \text{constant} \qquad 8\text{-}5$$

CONSERVATION OF MOMENTUM

This result is known as the **law of conservation of momentum:**

> If the sum of the external forces on a system remains zero, the total momentum of the system remains constant.

This law is one of the most important in physics. It is more widely applicable than the law of conservation of *mechanical* energy because internal forces exerted by one particle in a system on another are often not conservative. The nonconservative internal forces can change the total mechanical energy of the system, though they effect no change of the system's total momentum. If the total momentum of a system remains constant, then the velocity of the center of mass of the system remains constant. The law of conservation of momentum is a vector relation, so it is valid component by component. For example, if the sum of the x components of the external forces on a system remains zero, then the x component of the total momentum of the system remains constant. That is,

$$\text{If } \Sigma F_{ext\,x} = 0, \text{ then } P_{sys\,x} = \text{constant} \qquad 8\text{-}6$$

CONSERVATION OF A COMPONENT OF MOMENTUM

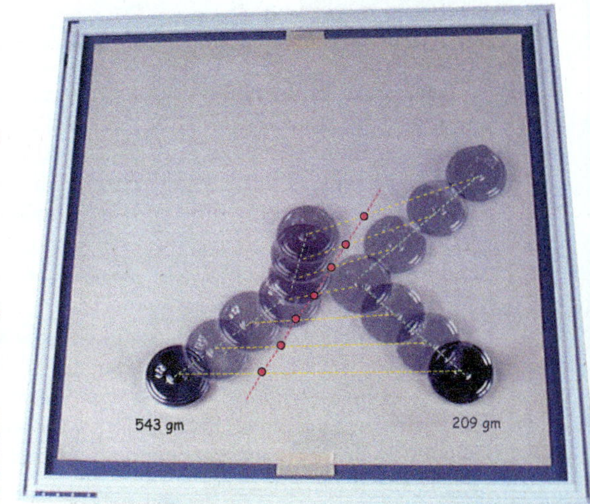

The two pucks are moving on an air cushion on a horizontal flat surface. (The hoses supplying the air are not shown.) The velocity of each puck changes in both magnitude and direction during the collision, but the velocity of the center of mass remains constant—unaffected by the internal forces of the collision. *(Courtesy of Daedalon Corporation.)*

PROBLEM-SOLVING STRATEGY

Finding Velocities Using Momentum Conservation (Equation 8-5)

PICTURE Determine that the net external force $\Sigma\vec{F}_{ext}$ (or $\Sigma F_{ext\,x}$) on the system is negligible for some interval of time. If the net force is determined not to be negligible, do not proceed.

SOLVE

1. Draw a sketch showing the system both before and after the time interval. Include coordinate axes and label the initial and final velocity vectors.
2. Equate the initial momentum to the final momentum. That is, write out the equation $m_1\vec{v}_{1i} + m_2\vec{v}_{2i} = m_1\vec{v}_{1f} + m_2\vec{v}_{2f}$ (or $m_1v_{1ix} + m_2v_{2ix} = m_1v_{1fx} + m_2v_{2fx}$).
3. Substitute the given information into the step-2 equation(s) and solve for the quantity of interest.

CHECK Make sure you include any minus signs that accompany velocity components because they influence your final answer.

Example 8-1 A Space Repair

During repair of the Hubble Space Telescope, an astronaut replaces a damaged solar panel during a spacewalk. Pushing the detached panel away into space, she is propelled in the opposite direction. The astronaut's mass is 60 kg and the panel's mass is 80 kg. Both the astronaut and the panel initially are at rest relative to the telescope. The astronaut then gives the panel a shove. After the shove it is moving at 0.30 m/s relative to the telescope. What is her subsequent velocity relative to the telescope? (During this operation the astronaut is tethered to the ship; for our calculations assume that the tether remains slack.)

(NASA.)

PICTURE Let us choose the system to be the astronaut and the panel, and let us choose the direction of motion of the panel to be the $-x$ direction. For this system, there are no external forces, so the momentum of the system is conserved. Because we know both the mass of the astronaut and the panel, the velocity of the astronaut can be found from the velocity of the panel using conservation of momentum. Because the total momentum of the system is initially zero, it remains zero.

SOLVE

1. Sketch a figure showing the system before and after the shove. Include velocity vectors (Figure 8-1):

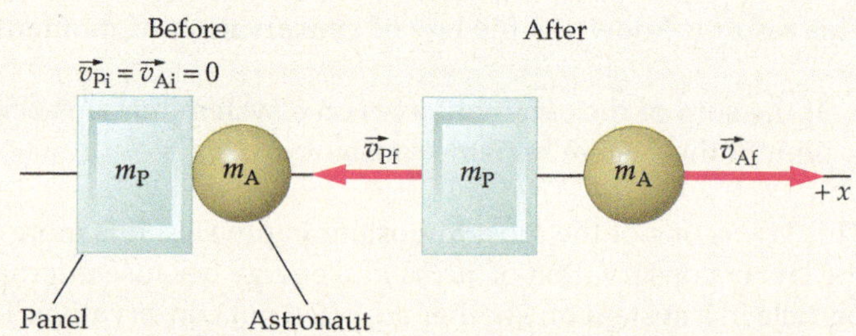

FIGURE 8-1

2. Apply Newton's second law to the system. There are no external forces on the system, so the momentum of the system remains constant:

$$\sum_i \vec{F}_{\text{ext}} = \frac{d\vec{P}_{\text{sys}}}{dt}$$

$$0 = \frac{d\vec{P}_{\text{sys}}}{dt},$$

$$\text{so} \quad \vec{P}_{\text{sys}} = \text{constant}$$

3. Equate the initial momentum of the system to the final momentum. Because the initial momentum is zero, the momentum of the system remains zero:

$$\vec{P}_{\text{sys i}} = \vec{P}_{\text{sys f}}$$

$$m_{\text{P}}\vec{v}_{\text{Pi}} + m_{\text{A}}\vec{v}_{\text{Ai}} = m_{\text{P}}\vec{v}_{\text{Pf}} + m_{\text{A}}\vec{v}_{\text{Af}}$$

$$0 + 0 = m_{\text{P}}\vec{v}_{\text{Pf}} + m_{\text{A}}\vec{v}_{\text{Af}}$$

4. Solve for the astronaut's velocity:

$$\vec{v}_{\text{Af}} = -\frac{m_{\text{P}}}{m_{\text{A}}}\vec{v}_{\text{Pf}} = -\frac{80\text{ kg}}{60\text{ kg}}(-0.30\text{ m/s})\hat{i} = \boxed{(0.40\text{ m/s})\hat{i}}$$

CHECK We expect the astronaut's velocity to be in the $+x$ direction because the velocity of the panel was in the $-x$ direction. Also, because her mass is less than the mass of the panel, we expect that her speed would be greater than the speed of the panel. Our result meets both these expectations.

TAKING IT FURTHER Although momentum is conserved, the mechanical energy of this system increased because chemical energy in the astronaut's muscles was transformed into kinetic energy.

PRACTICE PROBLEM 8-1 Find the final kinetic energy of the astronaut–panel system.

Example 8-2 A Runaway Railroad Car

A runaway 14,000-kg railroad car is rolling horizontally at 4.00 m/s toward a switchyard. As it passes by a grain elevator, 2000 kg of grain suddenly drops into the car. How long does it take the car to cover the 500-m distance from the elevator to the switchyard? Assume that the grain falls straight down and that slowing due to rolling friction or air drag is negligible.

PICTURE We can find the travel time we seek from the distance traveled and the speed of the car. Consider the car and the grain as our system (Figure 8-2). Let the direction of the car be the $+x$ direction. There are no external forces with nonzero x components, so the x component of the momentum is conserved. The final speed of the grain-filled car is found from its final momentum, which equals the car's initial momentum (the grain initially has zero momentum in the x direction). Let m_c and m_g be the masses of the car and grain, respectively.

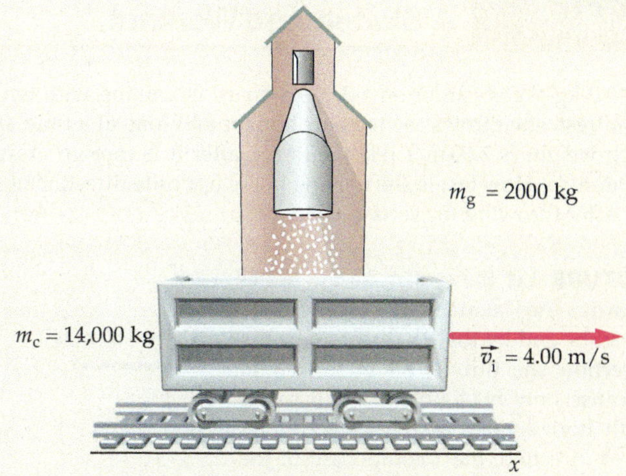

FIGURE 8-2

SOLVE

1. The time for the car to travel from the elevator to the yard is related to the distance to the yard d and the car's speed v_{fx} following the grain dump. We are looking for this time:

$$d = v_{fx}\,\Delta t$$

2. Sketch a free-body diagram (FBD) of the system consisting of the car, the grain already in the car, and the grain that is falling into the car (Figure 8-3). Include coordinate axes:

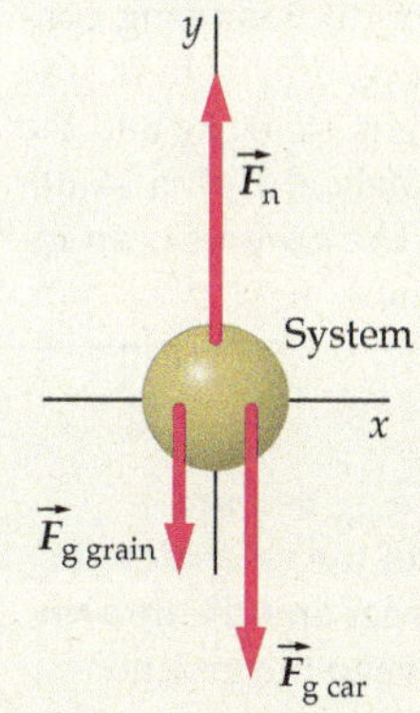

FIGURE 8-3 Three forces act on the system: the gravitational forces on the grain and the car, and the normal force of the track on the car.

3. The sum of the external forces acting on the grain–car system equals the rate of change of the momentum of the system (Equation 8-4):

$$\sum \vec{F}_{i\,ext} = \vec{F}_{g\,grain} + \vec{F}_{g\,car} + \vec{F}_n = \frac{d\vec{P}_{sys}}{dt}$$

4. Each of the external forces is vertical, so the x component of each is zero. Take the x component of each term in the step-3 result. The x component of the net external force is zero, so $P_{sys\,x}$ is constant:

$$F_{g\,grain\,x} + F_{g\,car\,x} + F_{nx} = \frac{dP_{sys\,x}}{dt}$$

$$0 + 0 + 0 = \frac{dP_{sys\,x}}{dt}$$

$$\therefore\ P_{sys\,fx} = P_{sys\,ix}$$

5. Make a sketch of the system before the collision and again after the collision (Figure 8-4):

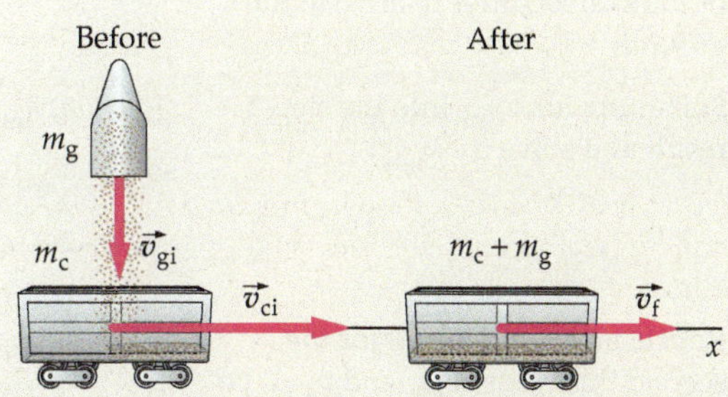

FIGURE 8-4

6. Apply conservation of momentum to relate the final velocity v_{fx} to the initial velocity v_{ix}. The x component of the system's momentum is conserved:

$$P_{sys\,fx} = P_{sys\,i}$$

$$(m_c + m_g)v_{fx} = m_c v_{ix} + m_g(0)$$

7. Solve for v_{fx}:

$$v_{fx} = \frac{m_c}{m_c + m_s} v_{ix}$$

8. Substitute the result for v_{fx} into step 1 and solve for the time:

$$\Delta t = \frac{d}{v_{fx}} = \frac{(m_c + m_s)d}{m_c v_{ix}}$$

$$= \frac{(14\,000\text{ kg} + 2000\text{ kg})(500\text{ m})}{(14\,000\text{ kg})(4.00\text{ m/s})}$$

$$= \boxed{1.43 \times 10^2\text{ s}}$$

CHECK The mass of the empty car is seven times greater than the mass of the grain, so we do not expect the grain to reduce the speed of the car by much. If the car continued at its initial speed of 4.00 m/s, the time to travel the 500 m would be (500 m)/(4.00 m/s) = 125 s. The step-8 result of 140 s is only slightly longer than 125 s, as expected.

PRACTICE PROBLEM 8-2 Suppose that there is a small vertical chute in the bottom of the car so that the grain leaks out at 10 kg/s. Now how long does it take the car to cover the 500 m?

Example 8-3 A Skateboard Workout

A 40.0-kg skateboarder on a 3.00-kg board is training with two 5.00-kg weights. Beginning from rest, she throws the weights horizontally, one at a time, from her board. The speed of each weight is 7.00 m/s relative to her after it is thrown. Assume the board rolls without friction. (*a*) How fast is she moving in the opposite direction after throwing the first weight? (*b*) After throwing the second weight?

PICTURE Let the system be the skateboard, the skateboarder, and the weights, and let the $-x$ direction be the direction she throws the first weight. Because only negligible external forces with horizontal components act along the x direction, the x component of the momentum of the system is conserved. We need to find the velocity of the skateboarder after throwing each weight (Figure 8-5). We can do this using conservation of momentum, where the mass m of each weight is 5.00 kg and the mass M of the skateboard plus skateboarder is 43.0 kg. The ground is an inertial reference frame.

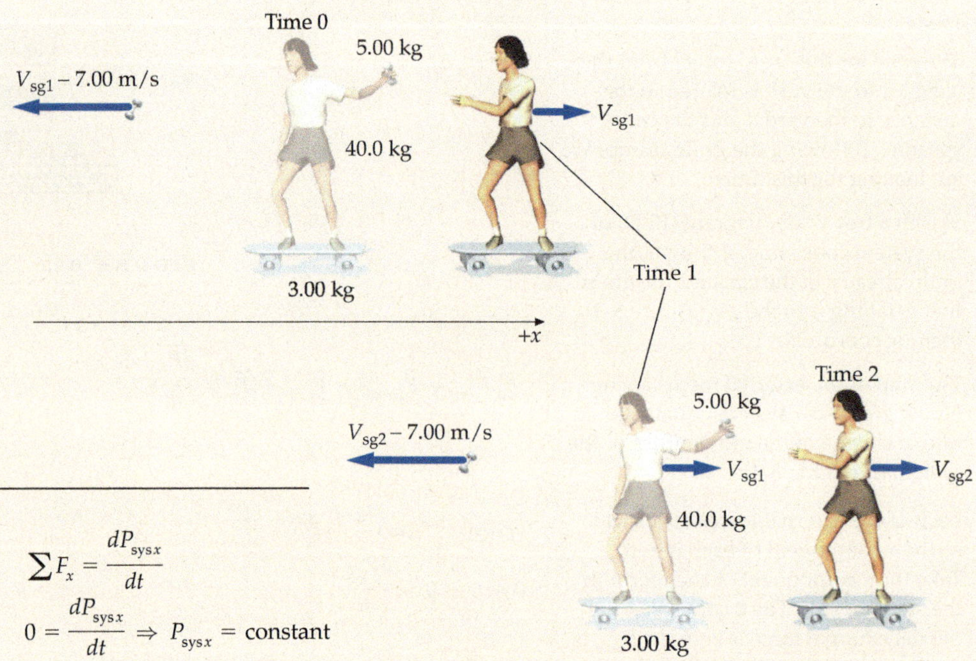

FIGURE 8-5 The numbers in the subscripts stand for times. Time 0 occurs just before the first throw, time 1 occurs between the two throws, and time 2 occurs following the second throw.

SOLVE

(*a*) 1. Let V_{sg1x} and v_{ws1x} be the x components of the velocities of the skateboarder and the thrown weight relative to the ground, respectively. Apply conservation of momentum for the first throw:

$$\sum F_x = \frac{dP_{sysx}}{dt}$$

$$0 = \frac{dP_{sysx}}{dt} \Rightarrow P_{sysx} = \text{constant}$$

$$\text{so} \quad P_{sys\,1x} = P_{sys\,0x}$$

$$(M + m)V_{sg1x} + mv_{wg1x} = 0$$

2. The velocity of the thrown weight relative to the ground equals the velocity of the weight relative to the skateboarder plus the velocity of the skateboarder relative to the ground:

$$v_{wg1x} = v_{ws\,1x} + V_{sg1x}$$

3. Substitute for v_{wg1x} into the step-1 result and solve for V_{sg1x}:

$$(M + m)V_{sg1x} + m(v_{ws1x} + V_{sg1x}) = 0$$

$$\text{so} \quad V_{sg1x} = -\frac{m}{M + 2m}v_{ws1x}$$

$$= -\frac{5.00\text{ kg}}{43.0\text{ kg} + 10.0\text{ kg}}(-7.00\text{ m/s}) = \boxed{0.660\text{ m/s}}$$

(*b*) 1. Repeat step 1 of Part (*a*) for the second throw. Let V_{sg2x} and $v_{w'g2x}$ be the x components of the respective velocities of the skateboarder and the second thrown weight relative to the ground:

$$P_{sys2x} = P_{sys1x}$$

$$MV_{sg2x} + mv_{w'g2x} = (M + m)V_{sg1x}$$

2. Repeat step 2 of Part (*a*) for the second throw.

$$v_{w'g2x} = v_{w's2x} + V_{sg2x}$$

3. Substitute for $v_{w'g2x}$ in the Part-(*b*), step-1 result and solve for V_{sg2x}:

$$MV_{sg2x} + m(v_{w's2x} + V_{sg2x}) = (M + m)V_{sg1x}$$

$$\text{so} \quad V_{sg2x} = \frac{(M + m)V_{sg1x} - mv_{w's2x}}{M + m} = V_{sg1x} - \frac{m}{M + m}v_{w's2x}$$

$$= 0.660\text{ m/s} - \frac{5.00\text{ kg}}{48.0\text{ kg}}(-7.00\text{ m/s}) = \boxed{1.39\text{ m/s}}$$

CHECK After the second throw, the mass of the skateboard and its cargo is 43.0 kg, which is 5.00 kg less than it was after the first throw. Because its mass was less for the second throw, we expect the increase in the speed of the skateboarder to be greater during the second throw. Our results show that her speed increases by 0.660 m/s during the first throw and by 1.39 m/s − 0.660 m/s = 0.73 m/s during the second throw, a small increase in the change in speed, as we expected.

TAKING IT FURTHER This example illustrates the principle of the rocket; a rocket moves forward by throwing its fuel out backward in the form of exhaust gases. As a rocket's mass lessens its acceleration increases, just as the skateboarder gains more speed with the second throw than she did with the first throw.

PRACTICE PROBLEM 8-3 How fast is the skateboarder moving if, starting from rest, she throws both weights simultaneously? The weights move at speed 7.00 m/s relative to her *after they are thrown*. Does she gain more speed by throwing them simultaneously or sequentially?

Example 8-4 Radioactive Decay

A thorium-227 nucleus (mass 227 u) at rest decays into a radium-223 nucleus (mass 223 u) by emitting an alpha particle (mass 4.00 u) (Figure 8-6). The kinetic energy of the α particle is measured to be 6.00 MeV. What is the kinetic energy of the recoiling radium nucleus?

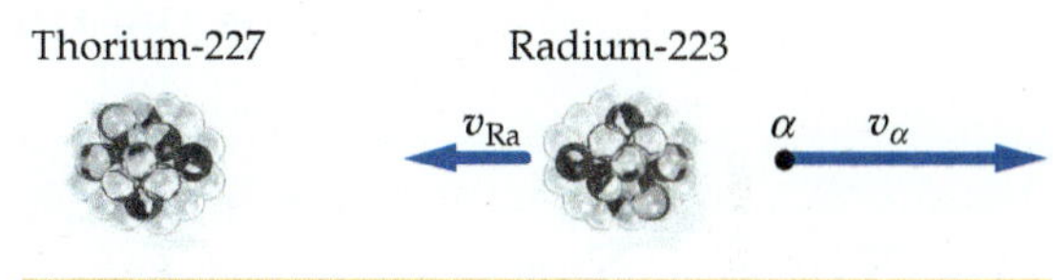

FIGURE 8-6

PICTURE A decay of a particle into two particles is like a collision run backwards in time. There are no external forces, so the momentum of the system is conserved. Recall that the kinetic energy of an object is $K = \frac{1}{2}mv^2$. Because the thorium nucleus before decay is at rest, its total momentum is zero. Therefore, we can relate the velocity of the radium nucleus to that of the alpha particle using conservation of momentum.

SOLVE

1. Write the kinetic energy of the radium nucleus K_{ra} in terms of its mass m_{ra} and speed v_{ra}.

$$K_{ra} = \tfrac{1}{2}m_{ra}v_{ra}^2$$

2. Write the kinetic energy of the alpha particle K_α in terms of its mass m_α and speed v_α.

$$K_\alpha = \tfrac{1}{2}m_\alpha v_\alpha^2$$

3. Use conservation of momentum to relate v_{ra} to v_α. The thorium nucleus was at rest, so the momentum of the system is zero:

$$m_\alpha v_\alpha = m_{ra}v_{ra}$$

4. Solve the step-1 and step-2 results the speeds v_{ra} and v_α, and substitute these expressions into the step-3 result.

$$K_{ra} = \tfrac{1}{2}m_{ra}v_{ra}^2 \qquad K_\alpha = \tfrac{1}{2}m_\alpha v_\alpha^2$$

$$v_{ra} = \left(\frac{2K_{ra}}{m_{ra}}\right)^{1/2} \qquad v_\alpha = \left(\frac{2K_\alpha}{m_\alpha}\right)^{1/2}$$

$$\text{so} \quad m_\alpha\left(\frac{2K_\alpha}{m_\alpha}\right)^{1/2} = m_{ra}\left(\frac{2K_{ra}}{m_{ra}}\right)^{1/2}$$

5. Solve the step-4 result for K_{ra}.

$$K_{ra} = \frac{m_\alpha}{m_{ra}}K_\alpha = \frac{4.00\text{ u}}{223\text{ u}}(6.00\text{ MeV}) = \boxed{0.107\text{ MeV}}$$

CHECK Let us check the step-5 result $K_{ra} = (m_\alpha/m_{ra})K_\alpha$ for several values of the ratio m_α/m_{ra}. If the two masses are equal, our result gives $K_{ra} = K_\alpha$, as expected. If $m_\alpha \ll m_{ra}$ then our result gives $K_{ra} \ll K_\alpha$, which means the kinetic energy of the alpha particle is much greater than that of the radium nuclei. This also means that the speed of the alpha particle is much greater than that of the radium nuclei, as expected.

TAKING IT FURTHER In this process, some of the rest energy of the thorium nucleus is converted into kinetic energy of the alpha particle and radium nucleus. The mass of the thorium nucleus is greater than the sum of the masses of the alpha particle and radium nucleus by $(K_\alpha + K_{ra})/c^2 = 6.11\text{ MeV}/c^2$.

8-2 KINETIC ENERGY OF A SYSTEM

If the net external force on a system of particles remains zero, then the total momentum of a system must remain constant; however, the total mechanical energy of the system can change. As we saw in the examples of the previous section, internal forces that cannot change the total momentum may be nonconservative and thus change the total mechanical energy of the system. There is an important theorem concerning the kinetic energy of a system of particles that allows us to treat the energy of complex systems more easily and gives us insight into energy changes within a system:

> The kinetic energy of a system of particles can be written as the sum of two terms: (1) the kinetic energy associated with the motion of the center of mass, $\frac{1}{2}Mv_{\text{cm}}^2$, where M is the total mass of the system; and (2) the kinetic energy associated with the motion of the particles of the system relative to the center of mass, $\Sigma\frac{1}{2}m_iu_i^2$, where $\vec{u}_i$ is the velocity of the ith particle relative to the center of mass.

THEOREM FOR THE KINETIC ENERGY OF A SYSTEM

So

$$K = \sum_i \tfrac{1}{2}m_iv_{\text{cm}}^2 + \sum_i \tfrac{1}{2}m_iu_i^2 = \tfrac{1}{2}Mv_{\text{cm}}^2 + K_{\text{rel}} \qquad \text{8-7}$$

KINETIC ENERGY OF A SYSTEM OF PARTICLES

***See* Math Tutorial *for more information on* Factoring**

where M is the total mass and K_{rel} is the kinetic energy of the particles *relative to the center of mass.*

To prove this theorem, recall that the kinetic energy K of a system of particles is the sum of the kinetic energies of the individual particles:

$$K = \sum_i K_i = \sum_i \tfrac{1}{2}m_iv_i^2 = \sum_i \tfrac{1}{2}m_i(\vec{v}_i \cdot \vec{v}_i)$$

where we have used the result that $v_i^2 = \vec{v}_i \cdot \vec{v}_i$. The velocity of the ith particle can be written as the sum of the velocity of the center of mass $\vec{v}_{\text{cm}}$ and the velocity of the ith particle relative to the center of mass $\vec{u}_i$:

$$\vec{v}_i = \vec{v}_{\text{cm}} + \vec{u}_i \qquad \text{8-8}$$

Substituting, we obtain

$$K = \sum_i \tfrac{1}{2}m_i(\vec{v}_i \cdot \vec{v}_i) = \sum_i \tfrac{1}{2}m_i(\vec{v}_{\text{cm}} + \vec{u}_i)\cdot(\vec{v}_{\text{cm}} + \vec{u}_i)$$

$$= \sum_i \tfrac{1}{2}m_i(v_{\text{cm}}^2 + u_i^2 + 2\vec{v}_{\text{cm}}\cdot\vec{u}_i)$$

We can write this as the sum of three terms:

$$K = \sum_i \tfrac{1}{2}m_iv_{\text{cm}}^2 + \sum_i \tfrac{1}{2}m_iu_i^2 + \vec{v}_{\text{cm}} \cdot \sum_i m_i\vec{u}_i$$

where in the term on the right we have factored $\vec{v}_{\text{cm}}$ from the sum ($\vec{v}_{\text{cm}}$ is a system parameter and does not change from particle to particle). The quantity $\Sigma m_i\vec{u}_i$ is equal to $M\vec{u}_{\text{cm}}$, where $\vec{u}_{\text{cm}}$ is the velocity of the center of mass relative to the center of mass. It follows that $\vec{u}_{\text{cm}}$ and thus $\Sigma m_i\vec{u}_i$ are equal to zero. (The velocity of anything relative to itself is always equal to zero.) It follows that $\Sigma m_i\vec{u}_i$ is equal to zero, so

$$K = \sum_i \tfrac{1}{2}m_iv_{\text{cm}}^2 + \sum_i \tfrac{1}{2}m_iu_i^2 = \tfrac{1}{2}Mv_{\text{cm}}^2 + K_{\text{rel}}$$

which completes the proof of Equation 8-7. If the net external force is zero, $\vec{v}_{cm}$ remains constant and the kinetic energy associated with bulk motion of the system $\frac{1}{2}Mv_{cm}^2$ does not change. Only the relative kinetic energy K_{rel} can change in an isolated system.

PRACTICE PROBLEM 8-4

Air-track glider A is moving at 1.0 m/s in the $+x$ direction along a frictionless horizontal air track. An identical glider, glider B, is parked on the track ahead of glider A. The mass of each glider is 1.0 kg, and the system consists of the two gliders. (*a*) What is the velocity of the center of mass, and what is the velocity of each glider, relative to the center of mass? (*b*) What is the kinetic energy of each glider relative to the center of mass? (*c*) What is the total kinetic energy relative to the center of mass? (*d*) The gliders collide and stick together. What then is the total kinetic energy relative to the center of mass?

8-3 COLLISIONS

A car crashes head-on into another car. A bat smashes into a baseball. A dart lands with a resounding "thunk" in the bull's-eye of a dartboard. Each of these is an example of a collision in which two objects approach and interact strongly for a very short time.

During the brief time of collision, any external forces on the two objects are usually much much weaker than the forces of interaction between the two objects. Thus, the colliding objects can usually be treated as an isolated system for the duration of the collision. During the collision the only significant forces are the internal interaction forces, which are equal and opposite. As a result, momentum is conserved. That is, the total momentum of the system the instant before the collision is equal to the total momentum the instant following the collision. In addition, the collision time is usually so short that the displacements of the colliding objects *during the collision* can be neglected.

When the total kinetic energy of the two-object system is the same after the collision as before, the collision is called an **elastic collision.** Otherwise, it is called an **inelastic collision.** An extreme case is the **perfectly inelastic collision,** during which all of the kinetic energy relative to the center of mass is converted to thermal or internal energy of the system, and the two objects share a common velocity (often because they stick together) at the end of the collision. We examine these different types of collisions in more detail later in this section.

IMPULSE AND AVERAGE FORCE

When two objects collide, they usually exert very large forces on each other for a very brief time. The force exerted by a baseball bat on a ball, for example, may be several thousand times the weight of the ball, but this enormous force is exerted for only a millisecond or so. Such forces are sometimes called *impulsive forces*. Figure 8-7 shows the time variation of the magnitude of a typical force exerted by one object on another during a collision. The force is large during much of the collision time interval $\Delta t = t_f - t_i$. For other times the force is negligibly small. The **impulse** $\vec{I}$ of a force $\vec{F}$ during time interval $\Delta t = t_f - t_i$ is a vector defined as

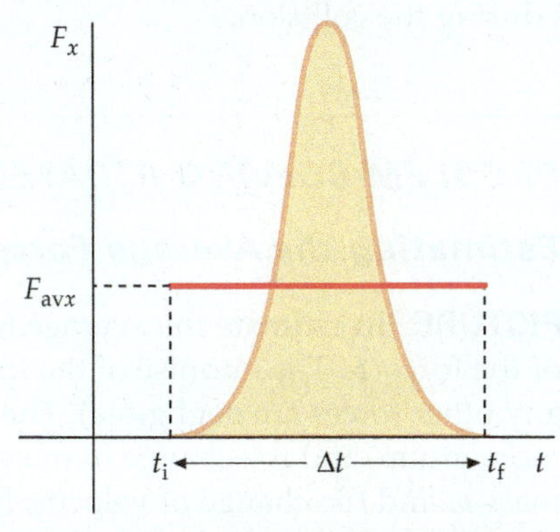

FIGURE 8-7 Typical time variation of force during a collision. The area under the F_x-versus-t curve is the x component of the impulse, I_x. F_{avx} is the average force for time interval Δt. The rectangular area $F_{avx}\,\Delta t$ is the same as the area under the F_x-versus-t curve.

$$\vec{I} = \int_{t_i}^{t_f} \vec{F}\,dt \qquad 8\text{-}9$$

DEFINITION—IMPULSE

The impulse is a measure of both the strength and the duration of the collision force. The x component of the impulse of the force is the area under its F_x-versus-t curve, and the S.I. units of impulse are newton seconds (N · s).

The net force $\vec{F}_{net}$ acting on a particle is related to the rate of change of momentum of the particle by Newton's second law: $\vec{F}_{net} = d\vec{p}/dt$. Taking the time integral of both sides of this equation gives

$$\int_{t_i}^{t_f} \vec{F}_{net}\,dt = \int_{t_i}^{t_f} \frac{d\vec{p}}{dt}\,dt = \vec{p}_f - \vec{p}_i$$

Recognizing the left side of this equation as the impulse of the net force, we have

$$\vec{I}_{net} = \Delta\vec{p} \qquad \text{8-10}$$

IMPULSE–MOMENTUM THEOREM FOR A PARTICLE

where $\vec{I}_{net} = \int_{t_i}^{t_f} \vec{F}_{net}\,dt$ and $\Delta\vec{p} = \vec{p}_f - \vec{p}_i$. Equation 8-10 is called the impulse–momentum theorem *for a particle*. Also, the net impulse on *a system* due to the external forces acting on the system equals the change in the total momentum of the system:

$$\vec{I}_{net\,ext} = \int_{t_i}^{t_f} \vec{F}_{net\,ext}\,dt = \Delta\vec{P}_{sys} \qquad \text{8-11}$$

IMPULSE–MOMENTUM THEOREM FOR A SYSTEM

By definition, the average of a force $\vec{F}$ for the interval $\Delta t = t_f - t_i$ is given by

$$\vec{F}_{av} = \frac{1}{\Delta t}\int_{t_i}^{t_f} \vec{F}\,dt = \frac{1}{\Delta t}\vec{I} \qquad \text{8-12}$$

AVERAGE FORCE

Rearranging gives

$$\vec{I} = \vec{F}_{av}\,\Delta t \qquad \text{8-13}$$

IMPULSE AND AVERAGE FORCE

The average force is the constant force that gives the same impulse as the actual force in the time interval Δt, as shown by the rectangle in Figure 8-7. The average net force can be calculated from the change in momentum if a collision time is known. This time can often be estimated using the displacement of one of the bodies during the collision.

PROBLEM-SOLVING STRATEGY

Estimating the Average Force

PICTURE To estimate the average force $\vec{F}_{av}$ we first estimate the impulse of the force $\vec{I}$. The impulse of the force equals the net impulse (assuming any other forces are negligible). The net impulse is equal to the change in momentum, and the change in momentum equals the product of the mass m and the change of velocity $\vec{v}_f - \vec{v}_i$. An estimate of the change in velocity can be gotten from estimates of both the collision time Δt and the displacement $\Delta\vec{r}$.

SOLVE

1. Calculate (or estimate) the impulse $\vec{I}$ and the time Δt. This estimate assumes that during the collision, the collision force on the object is very

large compared to all other forces on the object. This procedure works *only* if the displacement during the collision can be determined.

2. Draw a sketch showing the before and after position of the object. Add coordinate axes and label the pre- and postcollision velocities $\vec{v}_i$ and $\vec{v}_f$. In addition, label the displacement $\Delta\vec{r}$ during the collision.
3. Calculate the change in momentum of the object during a collision. The impulse on the object equals its change in momentum ($\vec{I} = \Delta\vec{p} = m\,\Delta\vec{v}$).
4. Use kinematics to estimate the collision time. This means using both $\vec{v}_{av} \approx \frac{1}{2}(\vec{v}_i + \vec{v}_f)$ and $\Delta\vec{r} = \vec{v}_{av}\,\Delta t$ to obtain $\Delta\vec{r} \approx \frac{1}{2}(\vec{v}_i + \vec{v}_f)\,\Delta t$, and then solving for Δt.
5. Use $\vec{F}_{av} = \vec{I}/\Delta t = m\,\Delta\vec{v}/\Delta t$ to calculate the average force (Equation 8-13).

CHECK Average force is a vector. Your result for average force should have the same direction as the change in velocity vector.

(Robert R. Edwards/BOB-E.)

Example 8-5 A Karate Collision

With an expert karate blow, you shatter a concrete block. Consider your hand to have a mass 0.70 kg, to be moving 5.0 m/s as it strikes the block, and to stop 6.0 mm beyond the point of contact. (*a*) What impulse does the block exert on your hand? (*b*) What is the approximate collision time and the average force the block exerts on your hand?

PICTURE The net impulse equals the change in momentum $\Delta\vec{p}$. We find $\Delta\vec{p}$ from the mass and velocity of your hand. The time of collision for Part (*b*) comes from the given displacement during the collision and by *estimating* the average velocity during the collision using a constant-acceleration kinematic formula.

SOLVE

(*a*) 1. Make a before and after sketch of your hand and the block. In the before picture, the edge of you hand is just reaching the block. Include a vertical coordinate axis on the sketch (Figure 8-8).

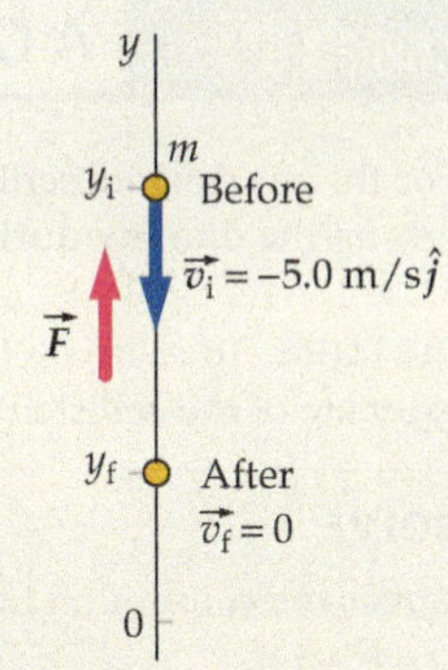

FIGURE 8-8

2. Set the impulse equal to the change in momentum: $\vec{I} = \Delta\vec{p} = m\,\Delta\vec{v}$

3. The initial velocity $\vec{v}_i$ is that of the hand just before it hits the block. The final velocity is zero: $\vec{v}_i = -5.0\,\mathrm{m/s}\,\hat{j}$, $\vec{v}_f = 0$

4. Substitute the values from step 3 into the equation from step 2 to find the impulse exerted by the block on your hand:

$$\vec{I} = m\,\Delta\vec{v} = (0.70\,\mathrm{kg})[0 - (-5.0\,\mathrm{m/s}\,\hat{j})] = 3.5\,\mathrm{kg\cdot m/s}\,\hat{j} = \boxed{3.5\,\mathrm{N\cdot s}\,\hat{j}}$$

(*b*) 1. The displacement equals the average velocity multiplied by the time. We estimate the average velocity by assuming constant acceleration. For constant a_y, $v_{av\,y} \approx \frac{1}{2}(v_{iy} + v_{fy})$:

$$\Delta y = v_{av}\,\Delta t \approx \tfrac{1}{2}(v_{iy} + v_{fy})\,\Delta t$$

2. Because we have chosen up as the $+y$ direction, both Δy and $v_{av\,y}$ are negative. Calculate Δt:

$$\Delta t \approx \frac{\Delta y}{\frac{1}{2}(v_{iy} + v_{fy})} = \frac{-0.006\,\mathrm{m}}{-2.5\,\mathrm{m/s}} = 0.0024\,\mathrm{s} = 2.4\,\mathrm{ms}$$

3. From Equation 8-12, the average force is the impulse divided by the collision time:

$$\vec{F}_{av} = \frac{\vec{I}}{\Delta t} = \frac{3.5\,\mathrm{N\cdot s}\,\hat{j}}{2.4\,\mathrm{ms}} = \boxed{1.5\,\mathrm{kN}\,\hat{j}}$$

CHECK The average force on your hand is in the $+y$ direction (upward). This is the same direction as the change in velocity vector, as expected. (The average force by the edge of your hand on the block is equal and opposite to the average force by the block on the hand.)

TAKING IT FURTHER Note that the average force relatively is large. Assuming the mass of a hand is about one kilogram, the average force is about 150 times the weight of a hand. The average collision force is much larger than the gravitational force on the hand during the collision.

Example 8-6 A Crumpled Car

A car equipped with an 80-kg crash-test dummy (Figure 8-9) drives into a massive concrete wall at 25 m/s (about 56 mi/h.) Estimate the displacement of the dummy during the crash.

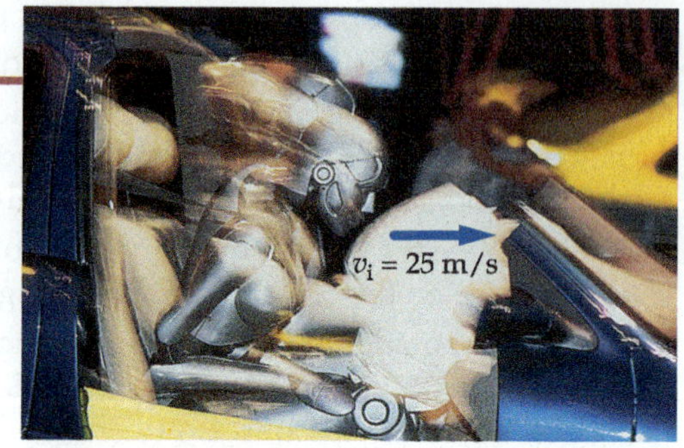

FIGURE 8-9 *(Romilly Lockyer/The Image Bank.)*

PICTURE The passenger compartment of a modern car is designed to remain rigid, whereas the front and rear ends of the car are designed to collapse upon impact. Assume the dummy is halfway between the front and rear bumpers and that the front end of the car completely collapses.

SOLVE

1. The front end of the car is the engine compartment, the radiator, the grill and the bumper. Estimate the fraction of the entire length of the car that the front end occupies.
   The front end is approximately 25% of the length of the car.
2. Estimate the displacement of the passenger compartment if the front end completely crumples.
   As the front end completely crumples, the displacement of the rest of the car, including the dummy, might be equal to 25 percent of the length of the car.
3. Estimate the length of a typical car.
   The length of a car is about 4.0 m (about 13 ft).
4. The length of the displacement is equal to the length of the front end.
   The length of the displacement is 25% of the length of the car, or about $\boxed{1.0\ \text{m}}$.

CHECK The dummy was 2.0 m from the wall at impact. Our result is half of that distance, which is plausible.

Example 8-7 A Crash Test

Try It Yourself

For the car crash described in Example 8-6, estimate the average force that the seat belt exerts on the dummy during the crash.

PICTURE To estimate the average force, calculate the impulse $\vec{I}$, and then divide it by an estimate of the collision time, Δt.

SOLVE

Cover the column on the right and try these on your own before looking at the answers.

Steps	Answers
1. Relate the average force to the impulse, and thus to the change in momentum.	$\vec{I} = \vec{F}_{av}\,\Delta t = \Delta\vec{p}$
2. Sketch and label a figure showing a representation of the dummy both before and after the crash (Figure 8-10).	
3. Find the change in the dummy's momentum. Let the forward direction of the car be the $+x$ direction.	$\Delta\vec{p} = m\vec{v}_f - m\vec{v}_i = -2000\ \text{N}\cdot\text{s}\ \hat{i}$
4. Relate the time to the displacement, assuming constant acceleration.	$\Delta x = v_{av}\,\Delta t$
5. Get the displacement of the dummy during the crash from step 4 of Example 8-6:	$\Delta x = 1.0\ \text{m}$
6. Estimate the average velocity and use it and the steps-4 and -5 results to find the time.	$\vec{v}_{av} \approx \frac{1}{2}(\vec{v}_f + \vec{v}_i) = 12.5\ \text{m/s}\ \hat{i}$, so $\Delta t = 0.080\ \text{s} = 80\ \text{ms}$
7. Substitute the step-3 and step-6 results into the step-1 result and solve for the force.	$\vec{F}_{av} = \boxed{-25\ \text{kN}\hat{i}}$

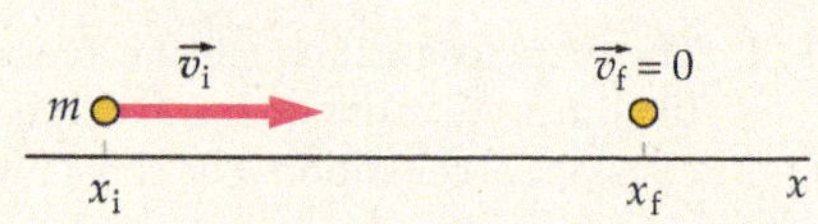

FIGURE 8-10

CHECK The average force is in the $-x$ direction, which is opposite to the forward direction of the car. This result is what is expected because the force must oppose the forward motion of the dummy.

TAKING IT FURTHER The magnitude of the average acceleration is $a_{av} = \Delta v/\Delta t \approx 300\ \text{m/s}^2$, or roughly $30g$. Such an acceleration means a net force about 30 times the weight of the dummy, clearly enough to cause serious injuries. An air bag increases the stopping time somewhat, which helps to reduce the force. In addition, the air bag allows the force to be distributed over a much larger area. In Figure 8-11, plot (*a*) shows the average force on the dummy as a function of the stopping distance. With no seat belt or air bag, you either fly though the windshield, or are stopped in a small fraction of a meter by the dashboard or steering wheel. Plot (*b*) shows the force as a function of the initial velocity for three stopping distances: 2.0 m, 1.5 m, and 1.0 m.

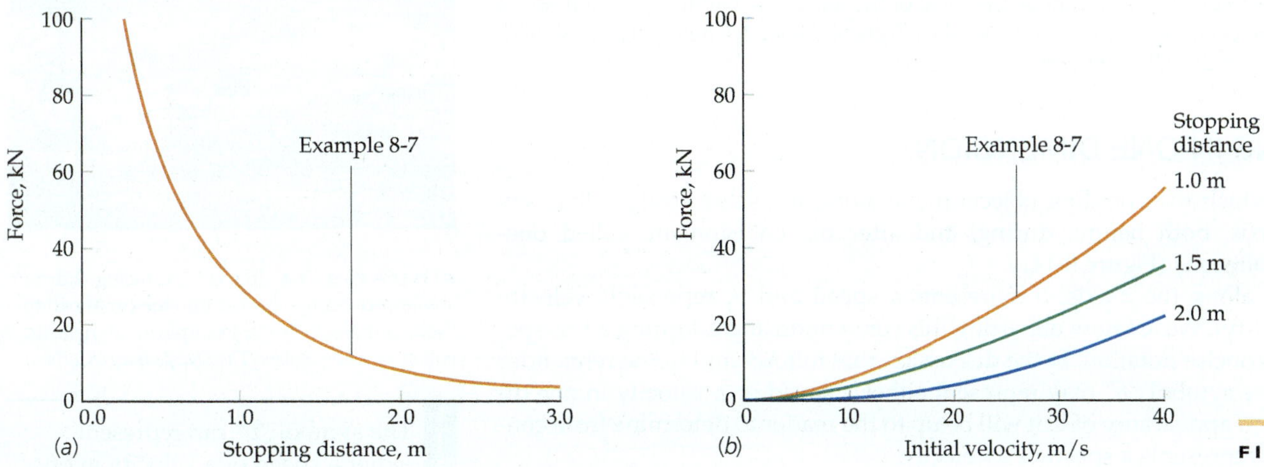

FIGURE 8-11

Example 8-8 Hitting a Golf Ball

You strike a golf ball with a driving iron. What are reasonable estimates for the magnitudes of the (*a*) impulse $\vec{I}$, (*b*) collision time Δt, and (*c*) average force $\vec{F}_{av}$? A typical golf ball has a mass $m = 45$ g and a radius $r = 2.0$ cm. For a typical drive, the range R is roughly 190 m (about 210 yd). Assume the ball leaves the ground at an angle $\theta_0 = 13°$ above the horizontal (Figure 8-12).

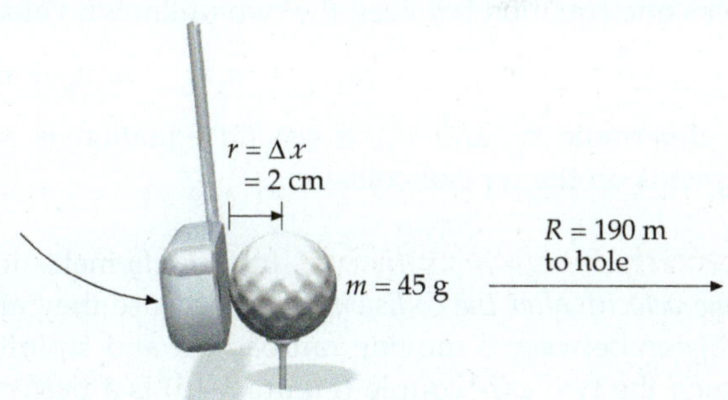

FIGURE 8-12

PICTURE Let v_0 denote the speed of the ball as it leaves the clubface. The impulse on the ball equals its change in momentum (mv_0) during the collision. We estimate v_0 from the range. We estimate the collision time from the displacement Δx and the average velocity $\frac{1}{2}(v_{ix} + v_{fx})$ during the collision, assuming constant acceleration. Taking $\Delta x = 2.0$ cm (half the diameter of the ball), the average force is then obtained from the impulse $\vec{I}$ and collision time Δt.

SOLVE

(*a*) 1. Set the impulse equal to the change in momentum of the ball:

$$I_x = F_{av\,x}\,\Delta t = \Delta p_x$$

2. Make a sketch showing the ball in both the pre- and postcollision positions (Figure 8-13):

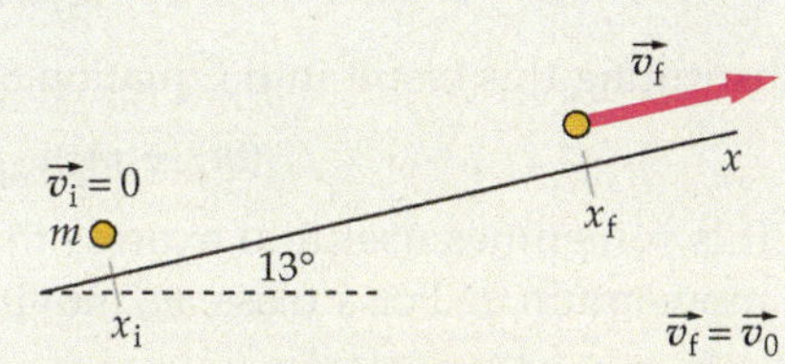

FIGURE 8-13

3. The speed v_f immediately after the collision is related to the range R, which is given by $R = (v_0^2/g)\sin 2\theta_0$ (Equation 2-23) with v_0 equal to the post-collision speed v_f:

$$R = \frac{v_f^2}{g}\sin 2\theta_0$$

4. Take $\theta_0 = 13°$ and calculate the initial speed for the projectile motion:

$$v_0 = \sqrt{\frac{Rg}{\sin 2\theta_0}} = \sqrt{\frac{(190\ \text{m})(9.81\ \text{m/s}^2)}{\sin 26°}} = 65.2\ \text{m/s}$$

5. Use this value of v_0 to calculate the impulse:

$$I_x = \Delta p_x = m(v_{0x} - 0) = (0.045\text{ kg})(65.2\text{ m/s}) = 2.93\text{ kg}\cdot\text{m/s} = \boxed{2.90\text{ N}\cdot\text{s}}$$

(*b*) Calculate the collision time Δt using $\Delta x = 2.0$ cm and $v_{\text{av}x} = \frac{1}{2}(v_{\text{i}x} + v_{\text{f}x})$:

$$\Delta t = \frac{\Delta x}{v_{\text{av}x}} = \frac{\Delta x}{\frac{1}{2}(0 + v_0)} = \frac{0.020\text{ m}}{\frac{1}{2}(65.2\text{ m/s})} = 6.13 \times 10^{-4}\text{ s} = \boxed{6.1 \times 10^{-4}\text{ s}}$$

(*c*) Use the calculated values of I_x and Δt to find the magnitude of the average force:

$$F_{\text{av}} = F_{\text{av}x} = \frac{I_x}{\Delta t} = \frac{2.93\text{ N}\cdot\text{s}}{6.13 \times 10^{-4}\text{ s}} = 4.78\text{ kN} = \boxed{4.8\text{ kN}}$$

CHECK The weight of the ball is mg, which is $(0.045\text{ kg})(9.81\text{ N/kg}) \approx 0.50$ N. We found the force of the golf club on the ball to be many times the weight of the ball, as expected.

TAKING IT FURTHER In this example, the force of the air on the ball has been left out of our analysis. However, for an actual golf shot the effects of the air are definitely *not* negligible, as any player with a slice can verify.

FIGURE 8-14 In stock car racing, a driver sometimes bumps his car into the car ahead to "send a message"—an example of an inelastic collision. *(Sam Sharpe/The Sharpe Image/Corbis.)*

COLLISIONS IN ONE DIMENSION

Collisions in which the colliding objects move along the same straight line, say along the x axis, both before, during, and after the collision are called one-dimensional collisions (Figure 8-14).

For motion along the x axis, v represents a speed and v_x represents velocity (a signed quantity). We are now departing this convention and adopting a less specific but more concise notation. In the discussion that follows and in the remainder of this book, the symbol "v" may represent either a speed or a velocity in one dimension. In each appearance of v, it will be up to the reader to determine from context whether v represents a speed or a velocity.

> ! The symbol "v" can represent either a speed or a velocity in one dimension. (In one dimension, velocity is a signed quantity.)

Consider an object of mass m_1 with initial velocity $v_{1\text{i}}$ approaching a second object of mass m_2 that is moving in the same direction with initial velocity $v_{2\text{i}}$. If $v_{2\text{i}} < v_{1\text{i}}$, the objects collide. Let $v_{1\text{f}}$ and $v_{2\text{f}}$ be their velocities after the collision. The two objects can be considered an isolated system. Conservation of momentum gives one equation between the two unknown velocities $v_{1\text{f}}$ and $v_{2\text{f}}$:

$$m_1v_{1\text{f}} + m_2v_{2\text{f}} = m_1v_{1\text{i}} + m_2v_{2\text{i}} \qquad \text{8-14}$$

To determine $v_{1\text{f}}$ and $v_{2\text{f}}$, a second equation is needed. That second equation depends on the type of collision.

Perfectly inelastic collisions In *perfectly* inelastic collisions, the objects *have the same velocity after the collision*, often because they are stuck together. A low-speed collision between a moving railroad car and an initially stationary railroad car in which the two cars couple (Figure 8-15) is a perfectly inelastic collision. For perfectly inelastic collisions, the final velocities are equal to each other and to the velocity of the center of mass:

$$v_{1\text{f}} = v_{2\text{f}} = v_{\text{cm}}$$

Substituting this result into Equation 8-14 gives

$$(m_1 + m_2)v_{\text{cm}} = m_1v_{1\text{i}} + m_2v_{2\text{i}} \qquad \text{8-15}$$

It is sometimes useful to express the kinetic energy, K, of a particle in terms of its momentum, p. For a mass, m, moving along the x axis with velocity, v, we have

$$K = \tfrac{1}{2}mv^2 = \frac{(mv)^2}{2m}$$

Because $p = mv$,

$$K = \frac{p^2}{2m} \qquad \text{8-16}$$

FIGURE 8-15 The engine bumps into the car causing the two to couple together—an example of a perfectly inelastic collision. *(Courtesy of Dick Tinder.)*

We can apply this to a perfectly inelastic collision where one object is initially at rest. The momentum of the system is that of the incoming object:

$$P_{sys} = p_{1i} = m_1 v_{1i}$$

The initial kinetic energy is

$$K_i = \frac{p^2_{sys}}{2m_1} \qquad \text{8-17}$$

After colliding, the objects move together as a single mass $m_1 + m_2$ with velocity v_{cm}. Momentum is conserved, so the final momentum equals P_{sys}. The final kinetic energy is then

$$K_f = \frac{p^2_{sys}}{2(m_1 + m_2)} \quad \text{(perfectly inelastic collisions)} \qquad \text{8-18}$$

Comparing Equations 8-17 and 8-18, we see that the final kinetic energy is less than the initial kinetic energy.

(Romilly Lockyer/ The Image Bank.)

Example 8-9 A Catch in Space

An astronaut of mass 60 kg is on a space walk to repair a communications satellite when he realizes he needs to consult the repair manual. You happen to have it with you, so you throw it to him with speed 4.0 m/s relative to the spacecraft. He is at rest relative to the spacecraft before catching the 3.0-kg book (Figure 8-16). Find (*a*) his velocity just after he catches the book, (*b*) the initial and final kinetic energies of the book–astronaut system, and (*c*) the impulse exerted by the book on the astronaut.

FIGURE 8-16

PICTURE This collision is a perfectly inelastic collision. So, following the catch, the book and astronaut move with the same final velocity. (*a*) We find the final velocity using conservation of momentum, as expressed in Equation 8-15. (*b*) The kinetic energies of the book and astronaut can be calculated directly from their masses and their initial and final velocities. (*c*) The impulse exerted by the book on the astronaut equals the change in momentum of the astronaut.

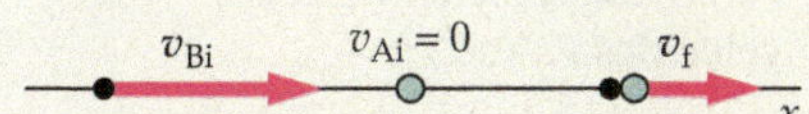

FIGURE 8-17

SOLVE

(*a*) 1. Make a drawing (Figure 8-17) showing the objects just before and just after the catch. Let the direction you throw the book be the $+x$ direction:

2. Use conservation of momentum to relate the final velocity of the system v_f to the initial velocities:

$$m_B v_{Bi} + m_A v_{Ai} = (m_A + m_B)v_f$$

3. Solve for v_f:

$$v_f = \frac{m_B v_B + m_A v_A}{m_B + m_A} = \frac{(3.0\,\text{kg})(4.0\,\text{m/s}) + (60\,\text{kg})(0\,\text{m/s})}{3.0\,\text{kg} + 60\,\text{kg}}$$
$$= 0.190\,\text{m/s} = \boxed{0.19\,\text{m/s}}$$

(*b*) 1. Because the astronaut is initially at rest, the initial kinetic energy of the book–astronaut system is the initial kinetic energy of the book:

$$K_{sys\,i} = K_{Bi} = \tfrac{1}{2}m_B v^2_{Bi} = \tfrac{1}{2}(3.0\,\text{kg})(4.0\,\text{m/s})^2 = \boxed{24\,\text{J}}$$

2. The final kinetic energy is the kinetic energy of the book and astronaut moving together at v_f:

$$K_{sys\,f} = \tfrac{1}{2}(m_B + m_A)v^2_f = 2(63\,\text{kg})(0.190\,\text{m/s})^2 = 1.14\,\text{J} = \boxed{1.1\,\text{J}}$$

(*c*) Set the impulse exerted on the astronaut equal to the change in momentum of the astronaut:

$$I_{by\,B\,on\,A} = \Delta p_A = m_A\,\Delta v_A = (60\,\text{kg})(0.190\,\text{m/s} - 0)$$
$$= 11.4\,\text{kg}\cdot\text{m/s} = \boxed{11\,\text{N}\cdot\text{s}}$$

CHECK The final velocity, the Part-(*a*), step-3 result, is equal to the velocity of the center of mass ($v_f = v_{cm}$). Before the collision, the book–astronaut system had both kinetic energy associated with the motion of the center of mass and kinetic energy relative to the center of mass. After the collision, the kinetic energy relative to the center of mass is equal to zero. As expected, the total kinetic energy of the system decreased.

TAKING IT FURTHER Most of the initial kinetic energy in this collision is lost by conversion to thermal energy. In addition, the impulse exerted by the book on the astronaut is equal and opposite to that exerted by the astronaut on the book, so the total change in momentum of the book–astronaut system is zero.

Example 8-10 A Ballistic Pendulum

In a feat of public marksmanship, you fire a bullet into a hanging wood block (Figure 8-18), which is a device known as a *ballistic pendulum*. The block, with bullet embedded, swings upward. Noting the height reached at the top of the swing, you immediately inform the crowd of the bullet's speed. How fast was the bullet traveling?

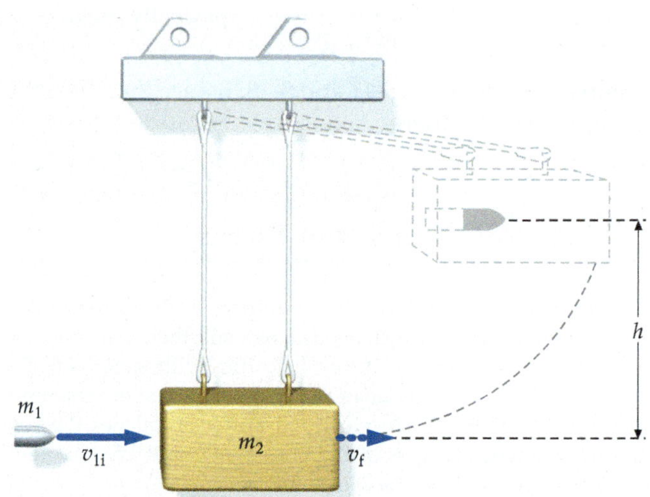

FIGURE 8-18

PICTURE Although the block moves upward after the collision, we can still consider this a one-dimensional collision, because the direction of the bullet and the block just after the collision is in the bullet's original direction of motion. The precollision velocity of the bullet v_{1i} is related to the postcollision velocity of the bullet–block system v_f by conservation of momentum. The speed v_f is related to the height, h, by conservation of mechanical energy. Let m_1 be the mass of the bullet and m_2 be the mass of the block.

SOLVE

1. Using conservation of mechanical energy *after* the collision, we relate the postcollision speed v_f to the maximum height h:

$$\tfrac{1}{2}(m_1 + m_2)v_f^2 = (m_1 + m_2)gh$$
$$v_f = \sqrt{2gh}$$

2. Using conservation of momentum *during* the collision we relate velocities v_{1i} and v_f:

$$m_1 v_{1i} = (m_1 + m_2)v_f$$
$$v_{1i} = \frac{m_1 + m_2}{m_1}v_f$$

3. Substituting for v_f in the step-2 result, we can solve for v_{1i}:

$$v_{1i} = \frac{m_1 + m_2}{m_1}v_f = \boxed{\frac{m_1 + m_2}{m_1}\sqrt{2gh}}$$

CHECK The mass of the block is much greater than the mass of the bullet. Thus, we expect the speed of the bullet to be much greater than the speed of the block after impact. Our step-2 result meets this expectation.

TAKING IT FURTHER We assumed that the time of the collision is so short that the displacement of the block during the collision is negligible. This assumption means the block had the postcollision speed v_f while still at the lowest point in the arc.

CONCEPT CHECK 8-1

Can this example be solved by equating the initial kinetic energy of the bullet with the potential energy of the block–bullet composite at maximum height? That is, is mechanical energy conserved both during the perfectly inelastic collision and during the rise of the pendulum?

Example 8-11 Collision with an Empty Box

Try It Yourself

You repeat your feat of Example 8-10, this time with an empty box as the target. The bullet strikes the box and passes through it completely. A laser ranging device indicates that the bullet emerged with half its initial velocity. Hearing this, you correctly report how high the target must have swung. How high did it swing?

PICTURE The height h is related to the box's speed v_2 just after colliding by the conservation of mechanical energy (Figure 8-19). This speed can be determined using conservation of momentum.

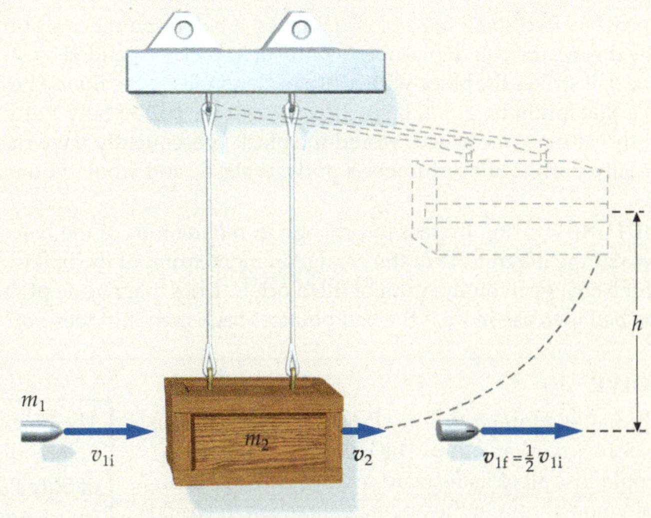

FIGURE 8-19

SOLVE

Cover the column to the right and try these on your own before looking at the answers.

Steps	Answers
1. Use conservation of mechanical energy to relate the final height h to the speed v_2 of the box just after the collision.	$m_2gh = \frac{1}{2}m_2v_2^2$
2. Use conservation of momentum to write an equation relating v_{1i} to the postcollision speed of the box, v_2.	$m_1v_{1i} = m_2v_2 + m_1(\frac{1}{2}v_{1i})$
3. Eliminate v_2 from the two equations and solve for h.	$h = \boxed{\frac{m_1^2v_{1i}^2}{8m_2^2g}}$

CHECK The quotient v^2/g has units of m^2/s^2 divided by m/s^2, which reduce to just m (meters). Thus, the step-3 result has appropriate units for height.

TAKING IT FURTHER The collision of the bullet and the box is an inelastic collision, but it is not a perfectly inelastic collision, because the two objects do not have the same velocity after the collision. Inelastic collisions also occur in microscopic systems. For example, when an electron collides with an atom, the atom is sometimes excited to a higher internal energy state. As a result, the total kinetic energy of the atom and the electron is less after the collision than before the collision.

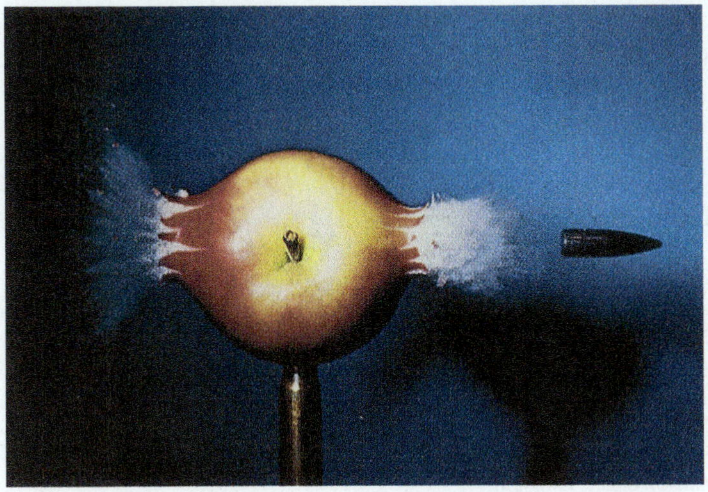

A bullet traveling 850 m/s collides inelastically with an apple, which disintegrates completely moments later. Exposure time is less than a millionth of a second. *(Estate of Harold E. Edgerton/ Palm Press Inc.)*

Example 8-12 Collisions with Putty — Conceptual

Mary has two small balls of equal mass, a ball of plumber's putty and a one of Silly Putty. She throws the ball of plumber's putty at a block suspended by strings shown in Figure 8-20. The ball strikes the block with a "thonk" and falls to the floor. The block subsequently swings to a maximum height h. If she had thrown the ball of Silly Putty (instead of the plumber's putty) at the same speed, would the block subsequently have risen to a height greater than h? Silly Putty, unlike plumber's putty, is elastic and would bounce back from the block.

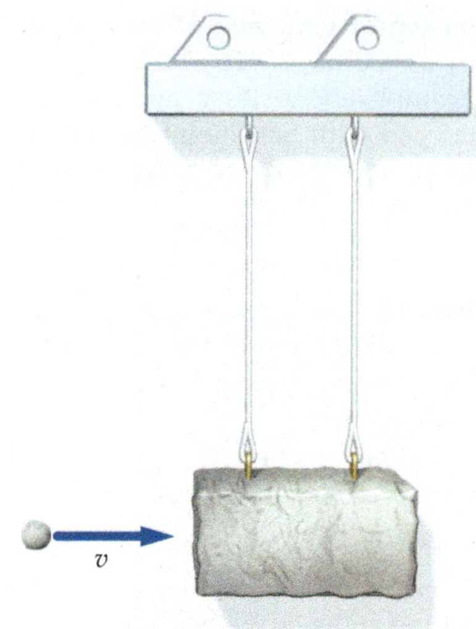

FIGURE 8-20

PICTURE During impact the change in momentum of the ball–block system is zero. The greater the magnitude of the change in momentum of the ball is, the greater the magnitude of the change in momentum of the block is. Does magnitude of the change in momentum of the ball increase more if the ball bounces back than if it does not?

SOLVE

The ball of plumber's putty loses a large fraction of its forward momentum. The ball of Silly Putty would lose all of its forward momentum and then gain momentum in the opposite direction. It would undergo a larger change in momentum than did the ball of plumber's putty.

The block would swing to a greater height after being struck with the ball of Silly Putty than it did after being struck with the ball of plumbers putty.

CHECK The block exerts a backward impulse on the ball of plumber's putty to slow the ball to a stop. The same backward impulse on the ball of Silly Putty would also bring it to a stop, and an additional backward impulse on the ball would give it momentum in the backward direction. Thus, the block exerts the larger backward impulse on the Silly-Putty ball. In accord with Newton's third law, the impulse of a ball on the block is equal and opposite to the impulse of the block on the ball. Thus, the Silly-Putty ball exerts the larger forward impulse on the block, giving the block a larger forward change in momentum.

Elastic collisions In elastic collisions, the kinetic energy of the system is the same before and after the collision. Elastic collisions are an ideal that is sometimes approached but never realized in the macroscopic world. If a ball dropped onto a concrete platform bounces back to its original height, then the collision between the ball and the concrete would be elastic. That situation has never been observed. At the microscopic level, elastic collisions are common. For example, the collisions between air molecules at the temperatures found on the surface of Earth are almost always elastic.

Figure 8-21 shows two objects before and after they have a one-dimensional head-on collision. Momentum is conserved during this collision so

$$m_1v_{1f} + m_2v_{2f} = m_1v_{1i} + m_2v_{2i} \qquad 8\text{-}19$$

The collision is elastic. Only for elastic collisions is the kinetic energy the same after the collision as before the collision. Therefore,

$$\tfrac{1}{2}m_1v_{1f}^2 + \tfrac{1}{2}m_2v_{2f}^2 = \tfrac{1}{2}m_1v_{1i}^2 + \tfrac{1}{2}m_2v_{2i}^2 \qquad 8\text{-}20$$

These two equations are sufficient to determine the final velocities of the two objects if given the initial velocities and the masses. However, the quadratic nature of Equation 8-20 often complicates the solution of simultaneous Equations 8-19 and 8-20. Such problems can be treated more easily if we express the velocity of the two

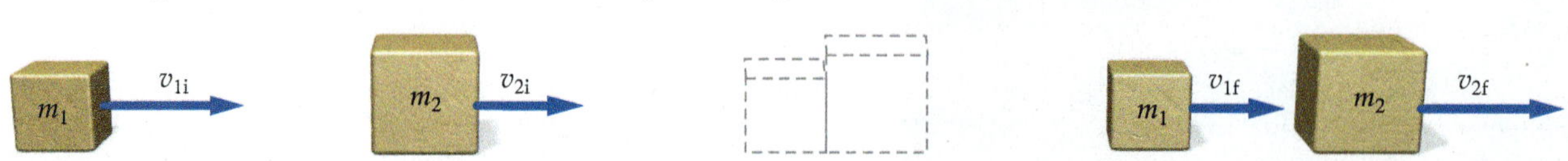

FIGURE 8-21 Closing (approaching) and separating (receding) in a head-on elastic collision.

objects relative to each other after the collision in terms of their relative velocity before the collision. Rearranging Equation 8-20 gives

$$m_2(v_{2f}^2 - v_{2i}^2) = m_1(v_{1i}^2 - v_{1f}^2)$$

Because $(v_{2f}^2 - v_{2i}^2) = (v_{2f} - v_{2i})(v_{2f} + v_{2i})$ and $(v_{1i}^2 - v_{1f}^2) = (v_{1i} - v_{1f})(v_{1i} + v_{1f})$, we have

$$m_2(v_{2f} - v_{2i})(v_{2f} + v_{2i}) = m_1(v_{1i} - v_{1f})(v_{1i} + v_{1f}) \qquad \text{8-21}$$

From conservation of momentum, we have that

$$m_1v_{1f} + m_2v_{2f} = m_1v_{1i} + m_2v_{2i}$$

Rearranging the equation for conservation of momentum (Equation 8-19) gives

$$m_2(v_{2f} - v_{2i}) = m_1(v_{1i} - v_{1f}) \qquad \text{8-22}$$

Dividing Equation 8-21 by Equation 8-22, we get

$$v_{2f} + v_{2i} = v_{1i} + v_{1f}$$

Rearranging yet again, we obtain

$$v_{1i} - v_{2i} = v_{2f} - v_{1f} \qquad \text{8-23}$$

RELATIVE VELOCITIES IN AN ELASTIC COLLISION

where $v_{1i} - v_{2i}$ is the *speed of approach* (closing speed) of the two particles before the collision and $v_{2f} - v_{1f}$ is the *speed of separation* following the collision (Figure 8-22). Equation 8-23 states:

In elastic collisions, the speed of separation equals the speed of approach.

Solving head-on elastic-collision problems is *always* easier using Equations 8-19 and 8-23 rather than Equations 8-19 and 8-20.

! Equation 8-23 is valid only if the initial and final kinetic energies are equal, so it applies *only* to elastic collisions.

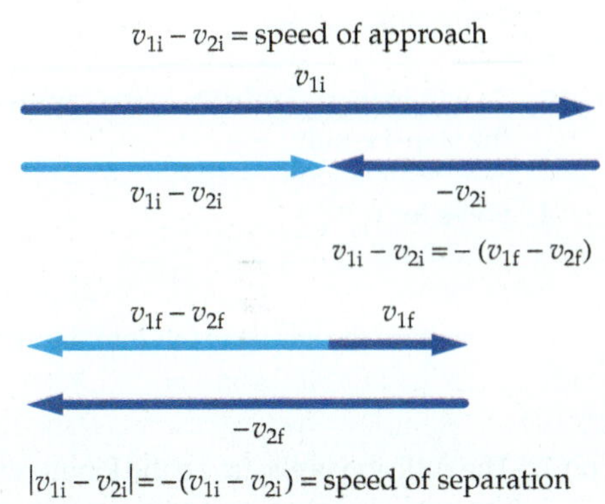

FIGURE 8-22

Example 8-13 Elastic Collision of Two Blocks

A 4.0-kg block moving to the right at 6.0 m/s undergoes an elastic head-on collision with a 2.0-kg block moving to the right at 3.0 m/s (Figure 8-23). Find their final velocities.

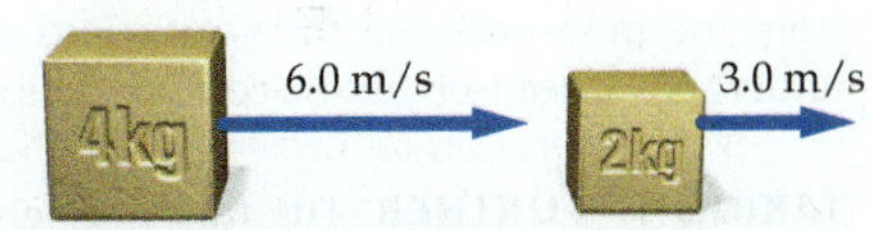

FIGURE 8-23

PICTURE Conservation of momentum and the equality of the initial and final kinetic energies (expressed as a reversal of relative velocities) give two equations for the two unknown final velocities. Let subscript 1 denote the 4.0-kg block, subscript 2 the 2.0-kg block.

SOLVE

1. Apply conservation of momentum and simplify to obtain an equation relating the two final velocities:
   $m_1v_{1i} + m_2v_{2i} = m_1v_{1f} + m_2v_{2f}$
   $(4.0\text{ kg})(6.0\text{ m/s}) + (2.0\text{ kg})(3.0\text{ m/s}) = (4.0\text{ kg})v_{1f} + (2.0\text{ kg})v_{2f}$
   so $2v_{1f} + v_{2f} = 15\text{ m/s}$
2. Because this is a head-on elastic collision, we can use Equation 8-23 to obtain a second equation:
   $v_{2f} - v_{1f} = v_{1i} - v_{2i}$
   $= 6.0\text{ m/s} - 3.0\text{ m/s} = 3.0\text{ m/s}$
3. Subtract the step-2 result from the step-1 result and solve for v_{1f}:
   $2v_{1f} + v_{1f} = 12\text{ m/s}$ so $v_{1f} = \boxed{4.0\text{ m/s}}$
4. Substitute into the step-2 result and solve for v_{2f}:
   $v_{2f} - 4.0\text{ m/s} = 3.0\text{ m/s}$ so $v_{2f} = \boxed{7.0\text{ m/s}}$

CHECK As a check, we calculate the initial and final kinetic energies.

$$K_i = \tfrac{1}{2}(4.0\text{ kg})(6.0\text{ m/s})^2 + \tfrac{1}{2}(2.0\text{ kg})(3.0\text{ m/s})^2 = 72\text{ J} + 9.0\text{ J} = 81\text{ J}$$
$$K_f = \tfrac{1}{2}(4.0\text{ kg})(4.0\text{ m/s})^2 + \tfrac{1}{2}(2.0\text{ kg})(7.0\text{ m/s})^2 = 32\text{ J} + 49\text{ J} = 81\text{ J}$$

The pre- and postcollision kinetic energies are equal, as expected.

Example 8-14 Elastic Collision of a Neutron and a Nucleus

A neutron of mass m_n and speed v_{ni} undergoes a head-on elastic collision with a carbon nucleus of mass m_C initially at rest (Figure 8-24). (*a*) What are the final velocities of both particles? (*b*) What fraction f of its initial kinetic energy does the neutron lose?

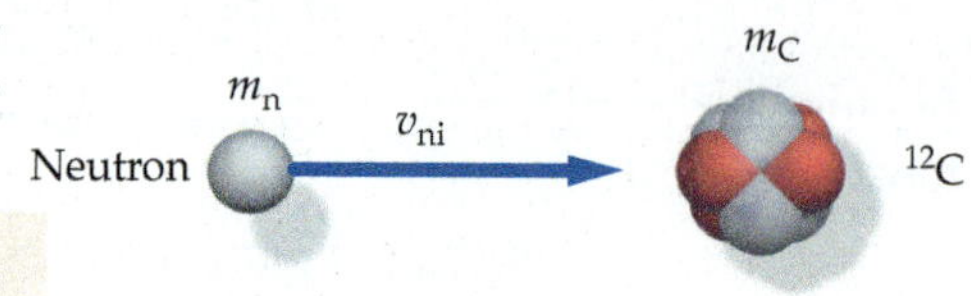

FIGURE 8-24

PICTURE Conservation of momentum and conservation of kinetic energy allow us to find the final velocities. Because the kinetic energy of the carbon nucleus is initially zero, its final kinetic energy equals the kinetic energy lost by the neutron.

SOLVE

(*a*) 1. Use conservation of momentum to obtain one relation for the final velocities:

$$m_n v_{ni} = m_n v_{nf} + m_C v_{Cf}$$

2. Use Equation 8-23 to equate the speeds of recession and approach:

$$v_{Cf} - v_{nf} = v_{ni} - v_{Ci} = v_{ni} - 0$$

so $$v_{Cf} = v_{ni} + v_{nf}$$

3. To eliminate v_{Cf}, substitute the expression for v_{Cf} from step 2 into the step-1 result:

$$m_n v_{ni} = m_n v_{nf} + m_C(v_{ni} + v_{nf})$$

4. Solve for v_{nf}:

$$\boxed{v_{nf} = -\frac{m_C - m_n}{m_n + m_C} v_{ni}}$$

5. Substitute the step-4 result into the step-2 result and solve for v_{Cf}:

$$v_{Cf} = v_{ni} - \frac{m_C - m_n}{m_n + m_C} v_{ni} = \boxed{\frac{2m_n}{m_n + m_C} v_{ni}}$$

(*b*) 1. The collision is elastic, so the kinetic energy lost by the neutron is the final kinetic energy of the carbon nucleus:

$$f = \frac{-\Delta K_n}{K_{ni}} = \frac{K_{Cf}}{K_{ni}} = \frac{\frac{1}{2}m_C v_{Cf}^2}{\frac{1}{2}m_n v_{ni}^2} = \frac{m_C}{m_n}\left(\frac{v_{Cf}}{v_{ni}}\right)^2$$

2. Solve the Part-(*a*), step-5 result for the ratio of the velocities; substitute into the Part-(*b*), step-1 result, and solve for the fractional energy loss of the neutron:

$$f = \frac{m_C}{m_n}\left(\frac{2m_n}{m_n + m_C}\right)^2 = \boxed{\frac{4m_n m_C}{(m_n + m_C)^2}}$$

CHECK Note that our calculated value for v_{nf} is negative. The neutron m_n bounces back from the more massive carbon nucleus m_C. This result is expected when a light particle undergoes a head-on elastic collision with a more massive particle that is initially at rest.

TAKING IT FURTHER The fractional energy loss for head-on collisions depends on the ratio of the masses (see Figure 8-25).

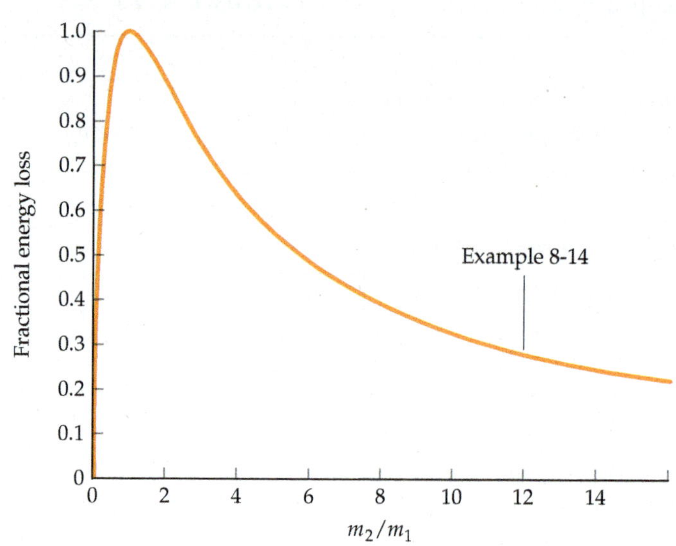

FIGURE 8-25 Fractional energy loss as a function of the ratio of the two masses. The maximum energy loss occurs when $m_1 = m_2$.

PRACTICE PROBLEM 8-5 Consider an elastic head-on collision between a moving object (object 1) and a second moving object of equal mass (object 2). Use Equations 8-19 and 8-23 to show that the two objects exchange velocities. That is, show that the final velocity of object 2 equals the initial velocity of object 1, and vice versa.

The final velocity of the incoming particle v_{1f} and that of the originally stationary particle v_{2f} are related to the initial velocity of the incoming particle by

$$v_{1f} = \frac{m_1 - m_2}{m_1 + m_2} v_{1i} \qquad \text{8-24}a$$

and

$$v_{2f} = \frac{2m_1}{m_1 + m_2} v_{1i} \qquad \text{8-24}b$$

These equations were derived in Example 8-14. Here we show they give plausible results for limiting values of the masses. When a very massive object (say a bowling ball) collides with a light stationary object (say a Ping-Pong ball), the massive object is essentially unaffected. Before the collision, the relative speed of approach is v_{1i}. If the massive object continues with a velocity that is essentially v_{1i} after the collision, the velocity of the smaller object must be $2v_{1i}$, so that the speed of recession is equal to the speed of approach. This result also follows from Equations 8-24*a* and 8-24*b* if we take m_2 to be much smaller than m_1, in which case $v_{1f} \approx v_{1i}$ and $v_{2f} \approx 2v_{1i}$, as expected.

The coefficient of restitution Most collisions lie somewhere between the extreme cases of elastic, in which the relative velocities are reversed, and perfectly inelastic, in which there is no relative velocity after the collision. The **coefficient of restitution** *e* is a measure of the elasticity of a collision. It is defined as the ratio of the speed of recession to the speed of approach.

$$e = \frac{v_{rec}}{v_{app}} = \frac{v_{1f} - v_{2f}}{v_{2i} - v_{1i}} \qquad \text{8-25}$$

DEFINITION—COEFFICIENT OF RESTITUTION

For an elastic collision, $e = 1$. For a perfectly inelastic collision, $e = 0$.

FIGURE 8-26
(Estate of Harold E. Edgerton/Palm Press Inc.)

PRACTICE PROBLEM 8-6
From the picture (Figure 8-26) of the golf club hitting the golf ball, estimate the coefficient of restitution of the golf ball–golf club interaction.

COLLISIONS IN TWO AND THREE DIMENSIONS

For one-dimensional collisions, the directions of the initial and final velocity vectors can be specified by a + or a −. For two- or three-dimensional collisions, this is not the case. During such a collision, momentum is conserved in each of the *x*, *y*, and *z* directions.

Inelastic collisions For collisions in two or three dimensions, the total initial momentum is the sum of the initial momentum vectors of each object involved in the collision. Because following a perfectly inelastic collision the two objects have the same final velocity and because momentum is conserved, we have $m_{1i}\vec{v}_{1i} + m_2\vec{v}_{2i} = (m_1 + m_2)\vec{v}_f$. Due to this relation we know that the velocity vectors, and thus the collision, takes place in a single plane. In addition, from the definition of the center of mass we know that $\vec{v}_f = \vec{v}_{cm}$.

Example 8-15 A Car–Truck Collision

You are at the wheel of a 1200-kg car traveling east through an intersection when a 3000-kg truck traveling north through the intersection crashes into your car, as shown in Figure 8-27. Your car and the truck remain stuck together after impact. The driver of the truck claims you were at fault because you were speeding. You look for evidence to disprove this claim. First, there are no skid marks, indicating that neither you nor the truck driver saw the accident coming and braked hard; second, the posted speed limit for the road on which you were driving is 80 km/h; third, the speedometer of the truck was smashed on impact, leaving the needle stuck at 50 km/h; and fourth, the wreck initially skidded from the impact zone at an angle of 59° north of east. Does this evidence support or undermine the claim that you were speeding?

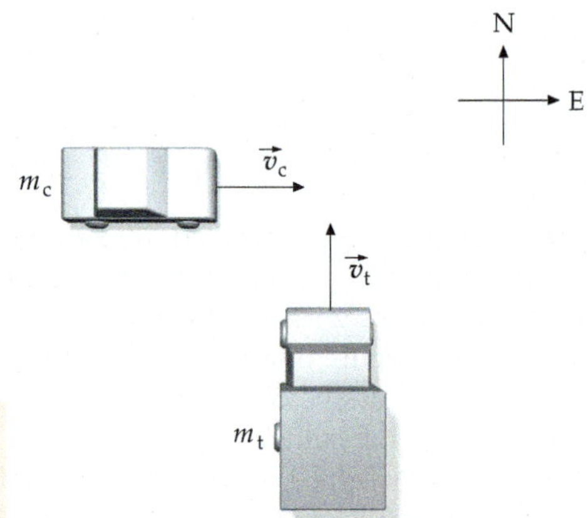

FIGURE 8-27

PICTURE We are given the masses of the vehicles and the velocity of the truck at impact. We know this is a perfectly inelastic collision because the car and truck stick together. Use conservation of momentum to determine your car's initial velocity.

SOLVE

1. Make a drawing (Figure 8-28) showing the two objects just before and just after the collision. Choose a coordinate system so that initially the car is traveling in the $+x$ direction and the truck is traveling in the $+y$ direction:

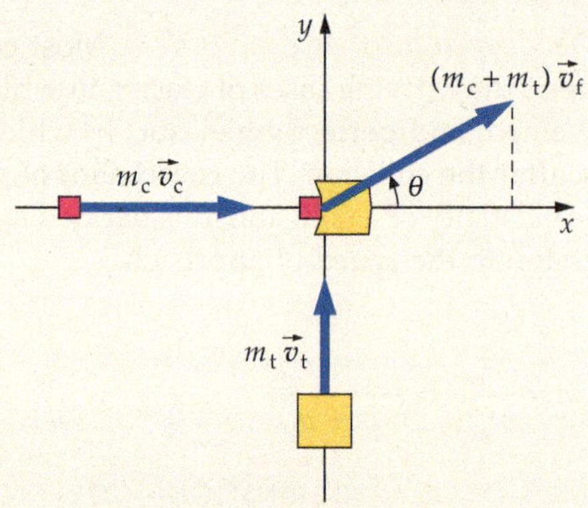

FIGURE 8-28

2. Write out the conservation of momentum equation in terms of masses and velocities:
$$m_c\vec{v}_c + m_t\vec{v}_t = (m_c + m_t)\vec{v}_f$$

3. Equate the x component of the initial momentum to the x component of the final momentum:
$$m_c v_c + 0 = (m_c + m_t)v_f\cos\theta$$

4. Equate the y component of the initial momentum to the y component of the final momentum:
$$0 + m_t v_t = (m_c + m_t)v_f\sin\theta$$

5. Eliminate v_f by dividing the y component equation by the x component equation:
$$\frac{m_t v_t}{m_c v_c} = \frac{\sin\theta}{\cos\theta} = \tan\theta$$
so
$$v_c = \frac{m_t v_t}{m_c \tan\theta} = \frac{(3000\text{ kg})(50\text{ km/h})}{(1200\text{ kg})\tan 59°} = \boxed{75\text{ km/h}}$$

6. Does this undermine the truck driver's claim that you were speeding?

> Because 75 km/h is less than the 80 km/h speed limit, the truck driver's claim is undermined by the careful application of physics.

CHECK The mass of the truck is 2.5 times the mass of the car. If the car were going 80 km/h, the truck's speed would be 5/8 that of the car, and the ratio of the magnitude of the momentum of the truck to that of the car would be $2.5 \times 5/8 = 1.56$. Because $\tan^{-1} 1.56 = 57°$ and 57° is slightly less than 59°, the step-6 result seems about right.

Next, we consider a three-dimensional inelastic collision in which the colliding objects do not share the same final velocity.

Example 8-16 A Glancing Collision

An object with mass m_1 and with an initial speed of 20 m/s undergoes an off-center collision with a second object of mass m_2. The second object is initially at rest. After the collision the first object is moving at 15 m/s at an angle of 25° with the direction of the initial velocity of the first object. In what direction is the second object moving?

PICTURE Momentum is conserved during this collision. It is a two-dimensional collision, so we equate the sum of the initial momentum vectors to the sum of the final momentum vectors and solve for the desired direction. (The problem does not indicate whether or not the collision is elastic, so we do not assume that it is.)

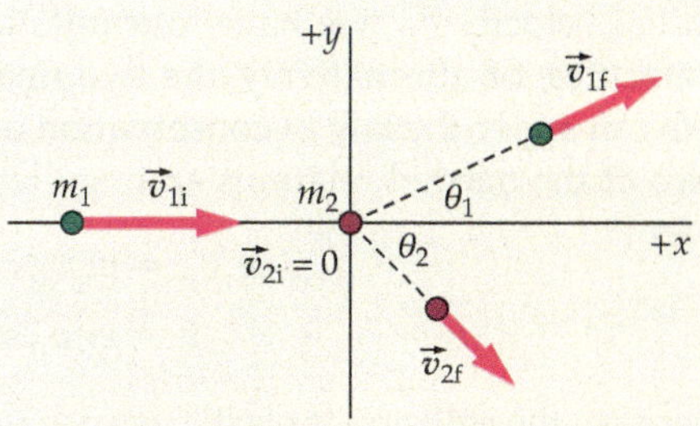

FIGURE 8-29

SOLVE

1. Sketch the two particles both before and after the collision (Figure 8-29). Choose the $+x$ direction to be the direction of the initial velocity of object 1. Include velocity vectors and labels in the sketch:

2. Write down the conservation of momentum relation, both in vector and in component form:

$$m_1\vec{v}_{1i} + 0 = m_1\vec{v}_{1f} + m_2\vec{v}_{2f}$$

or $\quad m_1v_{1ix} = m_1v_{1fx} + m_2v_{2fx}$

$\quad m_1v_{1iy} = m_1v_{1fy} + m_2v_{2fy}$

3. Express the component equations in terms of magnitudes and angles:

$$m_1v_{1i} = m_1v_{1f}\cos\theta_1 + m_2v_{2f}\cos\theta_2$$
$$0 = m_1v_{1f}\sin\theta_1 + m_2v_{2f}\sin\theta_2$$

4. In order to solve for θ_2 we exploit the relation $\tan\theta = \sin\theta/\cos\theta$. First we solve the step-3 results for the ratio $\sin\theta_2/\cos\theta_2$:

$$m_2v_{2f}\sin\theta_2 = -m_1v_{1f}\sin\theta_1$$
$$m_2v_{2f}\cos\theta_2 = m_1v_{1i} - m_1v_{1f}\cos\theta_1$$

so $$\frac{m_2v_{2f}\sin\theta_2}{m_2v_{2f}\cos\theta_2} = \frac{-m_1v_{1f}\sin\theta_1}{m_1v_{1i} - m_1v_{1f}\cos\theta_1}$$

and $$\tan\theta_2 = \frac{-\sin\theta_1}{\dfrac{v_{1i}}{v_{1f}} - \cos\theta_1}$$

5. Substitute in numbers and solve for θ_2:

$$\tan\theta_2 = \frac{-\sin 25°}{\dfrac{20}{15} - \cos 25°} = -0.990$$

$$\therefore \theta_2 = \boxed{-45°}$$

CHECK In step 1 we chose a coordinate system such that $\theta_1 = +25°$. We expected that θ_2 would be between zero and $-90°$. Our result that $\theta_2 = -45°$ meets this expectation.

TAKING IT FURTHER The problem did not specify either m_2 or v_{2f}, so you may have been surprised that solving for θ_2 was possible. It was possible because both initial and one final momentum vectors were fully specified in the problem statement, so the conservation of momentum relation (step 2) completely specifies the other final momentum vector. Once all four momentum vectors were fully specified, solving for the direction of the final momentum of particle 2 became possible.

Elastic collisions Elastic collisions in two and three dimensions are more complicated than those we have already covered. Figure 8-30 shows an off-center collision between an object of mass m_1 moving with velocity $\vec{v}_{1i}$ parallel to the x axis toward an object of mass m_2 that is initially at rest at the origin. This type of collision is often referred to as a *glancing* collision (as opposed to a head-on collision). The distance b between the centers measured perpendicular to the direction of $\vec{v}_{1i}$ is called the *impact parameter*. After the collision, object 1 moves off with velocity $\vec{v}_{1f}$, making an angle θ_1 with the direction of its initial velocity, and object 2 moves with velocity $\vec{v}_{2f}$, making an angle θ_2 with $\vec{v}_{1i}$. The final velocities depend on the impact parameter and on the type of force exerted by one object on the other.

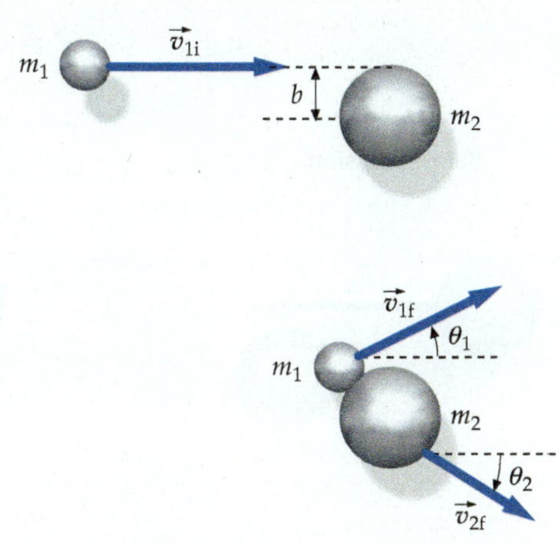

FIGURE 8-30 Off-center collision. The final velocities depend on the impact parameter b and on the type of force exerted by one object on the other.

Linear momentum is conserved, so we know that

$$\vec{P}_{sys} = m_1\vec{v}_{1i} = m_1\vec{v}_{1f} + m_2\vec{v}_{2f} \qquad \text{8-26}$$

We can see from this equation that the vector $\vec{v}_{2f}$ must lie in the plane formed by $\vec{v}_{1i}$ and $\vec{v}_{1f}$, which we will take to be the xy plane. Assuming that we know the initial velocity $\vec{v}_{1i}$, we have four unknowns: the x and y components of both final velocities; or alternatively, the two final speeds and the two angles of deflection. We can apply the law of conservation of momentum in component form to give us two of the needed relations among these quantities:

$$m_1v_{1i} = m_1v_{1f}\cos\theta_1 + m_2v_{2f}\cos\theta_2 \qquad \text{8-27}$$

$$0 = m_1v_{1f}\sin\theta_1 - m_2v_{2f}\sin\theta_2 \qquad \text{8-28}$$

Because the collision is elastic, we can use the conservation of kinetic energy to find a third relation:

$$\tfrac{1}{2}m_1v_{1i}^2 = \tfrac{1}{2}m_1v_{1f}^2 + m_2v_{2f}^2 \qquad \text{8-29}$$

We need an additional equation to be able to solve for the unknowns. The fourth relation depends on the impact parameter b and on the type of interacting force exerted by the colliding objects on each other. In practice, the fourth relation is often found experimentally, by measuring the angle of deflection or the angle of recoil. Such a measurement can then give us information about the type of interacting force between the bodies.

Now let us consider the interesting special case of the glancing elastic collision of two objects *of equal mass* when one is initially at rest (Figure 8-31*a*). If $\vec{v}_{1i}$ and $\vec{v}_{1f}$ are the initial and final velocities of object 1, and if $\vec{v}_{2f}$ is the final velocity of object 2, conservation of momentum gives

$$m\vec{v}_{1i} = m\vec{v}_{1f} + m\vec{v}_{2f}$$

or

$$\vec{v}_{1i} = \vec{v}_{1f} + \vec{v}_{2f}$$

These vectors form the triangle shown in Figure 8-31*b*. Because the collision is elastic,

$$\tfrac{1}{2}mv_{1i}^2 = \tfrac{1}{2}mv_{1f}^2 + \tfrac{1}{2}mv_{2f}^2$$

or

$$v_{1i}^2 = v_{1f}^2 + v_{2f}^2 \qquad \text{8-30}$$

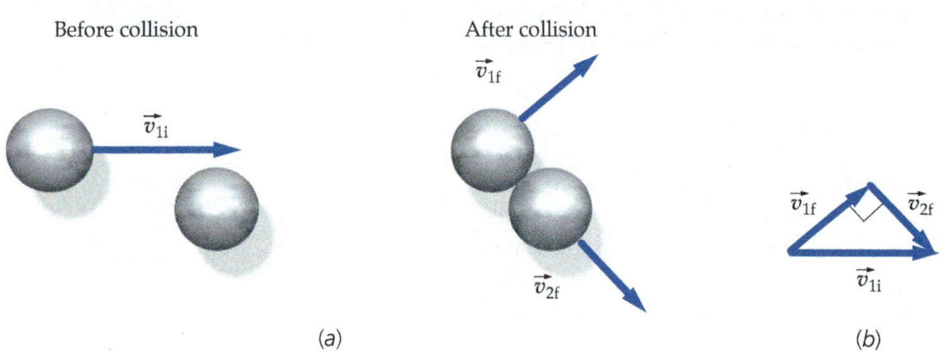

FIGURE 8-31 (*a*) Off-center elastic collision of two spheres of equal mass when one sphere is initially at rest. After the collision, the spheres move off at right angles to each other. (*b*) The velocity vectors for this collision form a right triangle.

(Joe Strunk/Visuals Unlimited.)

! The relation $v_{1i} - v_{2i} = v_{2f} - v_{1f}$ (Equation 8-23) is very useful for solving 1-dimensional elastic-collision problems. Do not think you can use this equation, or a vector form of the equation, to solve 2- or 3-dimensional elastic-collision problems. You cannot.

Multiflash photograph of an off-center elastic collision of two balls of equal mass. The dotted ball, entering from the left, strikes the striped ball, which is initially at rest. The final velocities of the two balls are perpendicular to each other. *(Berenice Abbot/Photo Researchers.)*

Equation 8-30 is the Pythagorean theorem for a right triangle formed by the vectors $\vec{v}_{1f}$, $\vec{v}_{2f}$, and $\vec{v}_{1i}$, with the hypotenuse of the triangle being $\vec{v}_{1i}$. So for this special case, the final velocity vectors $\vec{v}_{1f}$ and $\vec{v}_{2f}$ are perpendicular to each other, as shown in Figure 8-31*b*.

PRACTICE PROBLEM 8-7

In a friendly game of pool, the cue ball moving at speed v_0 glances off the initially stationary 2 ball. The collision is elastic, the 2 ball is at rest prior to the collision, and the cue ball is deflected 30° from its pre-collision path. What is the speed of the 2 ball following the collision? (The 2 ball and the cue ball have the same mass.)

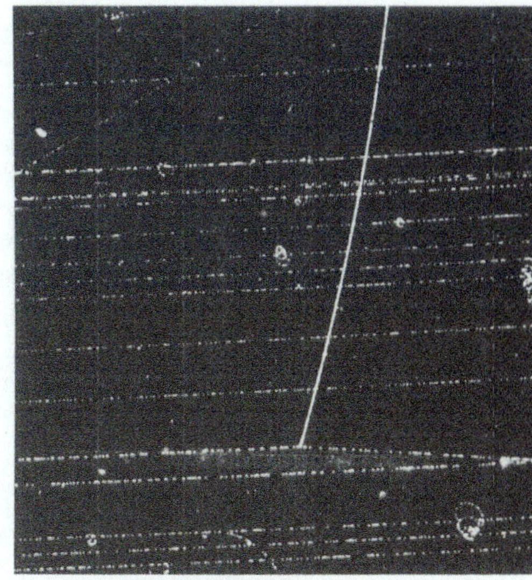

Proton–proton collision in a liquid-hydrogen bubble chamber. A proton entering from the left interacts with a stationary proton. The two then move off at right angles. The slight curvature of the tracks is due to a magnetic field. *(Brookhaven National Laboratory.)*

* 8-4 COLLISIONS IN THE CENTER-OF-MASS REFERENCE FRAME

If the net external force on a system remains zero, the velocity of the center of mass remains constant in any inertial reference frame. It is often convenient to do calculations in an alternate reference frame that moves with the center of mass. Relative to the original reference frame, called the laboratory reference frame, this reference frame moves with a constant velocity $\vec{v}_{cm}$ relative to the laboratory frame. A reference frame that moves at the same velocity as the center of mass is called the **center-of-mass reference frame.** If a particle has velocity $\vec{v}$ relative to the laboratory reference frame, then its velocity relative to the center-of-mass reference frame is $\vec{u} = \vec{v} - \vec{v}_{cm}$. Because the total momentum of a system equals the total mass times the velocity of the center of mass, the total momentum is also zero in the center-of-mass frame. Thus, the center-of-mass reference frame is also a **zero-momentum reference frame.**

The mathematics of collisions is greatly simplified when considered from the center-of-mass reference frame. The velocities of the particles in the center-of-mass frame are $\vec{u}_1$ and $\vec{u}_2$. The momenta, $m_1\vec{u}_1$ and $m_2\vec{u}_2$, of the two incoming objects are equal and opposite:

$$m_1\vec{u}_1 + m_2\vec{u}_2 = 0$$

After a perfectly inelastic collision, the objects remain at rest. However, for an elastic head-on collision the direction of each velocity vector is reversed without changing its magnitude. That is,

$$\vec{u}_{1i} = -\vec{u}_{1f} \quad \text{and} \quad \vec{u}_{2i} = -\vec{u}_{2f} \quad \text{(one-dimensional collision)}$$

Consider a simple two-particle system in a reference frame in which one particle of mass m_1 is moving with a velocity $\vec{v}_1$ and a second particle of mass m_2 is moving with a velocity $\vec{v}_2$ (Figure 8-32). In this frame, the velocity of the center of mass is

$$\vec{v}_{cm} = \frac{m_1\vec{v}_1 + m_2\vec{v}_2}{m_1 + m_2}$$

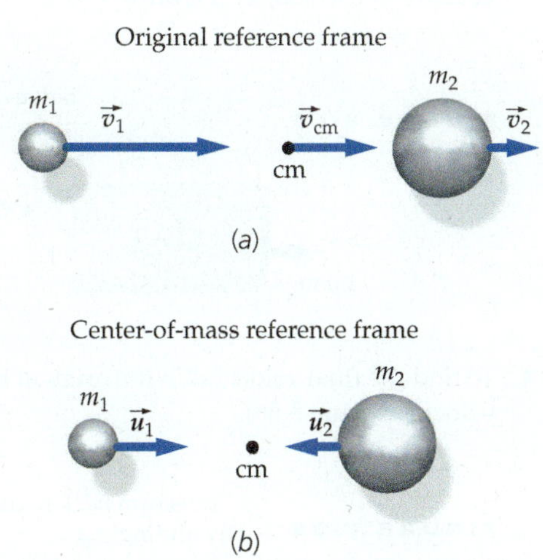

FIGURE 8-32 (*a*) Two particles viewed from a frame in which the center of mass has a velocity $\vec{v}_{cm}$. (*b*) The same two particles viewed from a reference frame for which the center of mass is at rest.

We can transform the velocities of each particle to its velocity in the center-of-mass reference frame by subtracting $\vec{v}_{cm}$ from it. Thus, the velocities of the particles in the center-of-mass frame are $\vec{u}_1$ and $\vec{u}_2$, given by

$$\vec{u}_1 = \vec{v}_1 - \vec{v}_{cm} \tag{8-31a}$$

and

$$\vec{u}_2 = \vec{v}_2 - \vec{v}_{cm} \tag{8-31b}$$

Example 8-17 The Elastic Collision of Two Blocks

Find the final velocities for the elastic head-on collision in Example 8-13 (in which a 4.0-kg block moving right at 6.0 m/s collides elastically with a 2.0-kg block moving right at 3.0 m/s) by transforming their velocities to the center-of-mass reference frame.

PICTURE We transform to the center-of-mass reference frame by first finding v_{cm} and subtracting it from each velocity. We then solve the collision by reversing the velocities and transforming back to the original frame.

SOLVE

1. Calculate the velocity of the center of mass v_{cm} (Figure 8-33):

$$v_{cm} = \frac{m_1 v_{1i} + m_2 v_{2i}}{m_1 + m_2} = \frac{(4.0\text{ kg})(6.0\text{ m/s}) + (2.0\text{ kg})(3.0\text{ m/s})}{4.0\text{ kg} + 2.0\text{ kg}} = 5.0\text{ m/s}$$

FIGURE 8-33

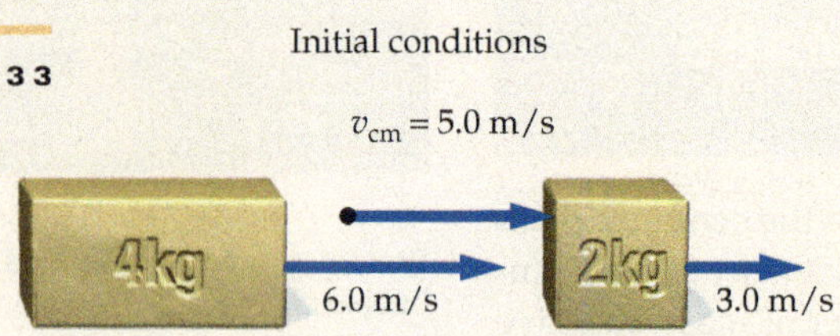

2. Transform the initial velocities to the center-of-mass reference frame by subtracting v_{cm} from the initial velocities (Figure 8-34):

$$u_{1i} = v_{1i} - v_{cm} = 6.0\text{ m/s} - 5.0\text{ m/s} = 1.0\text{ m/s}$$

$$u_{2i} = v_{2i} - v_{cm} = 3.0\text{ m/s} - 5.0\text{ m/s} = -2.0\text{ m/s}$$

FIGURE 8-34

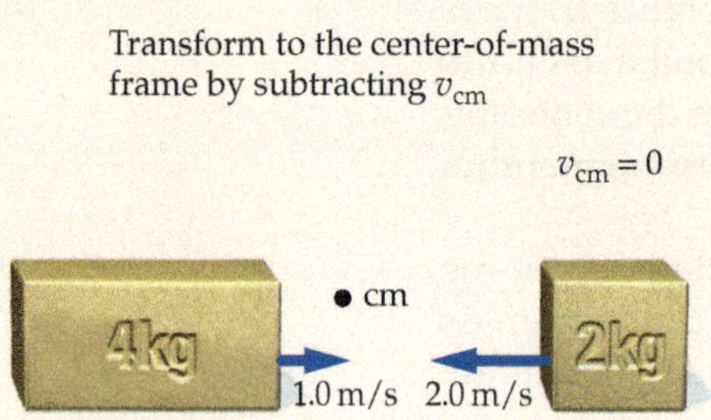

3. Solve the collision in the center-of-mass reference frame by reversing the velocity of each object (Figure 8-35):

$$u_{1f} = -u_{1i} = -1.0\text{ m/s}$$

$$u_{2f} = -u_{2i} = +2.0\text{ m/s}$$

FIGURE 8-35

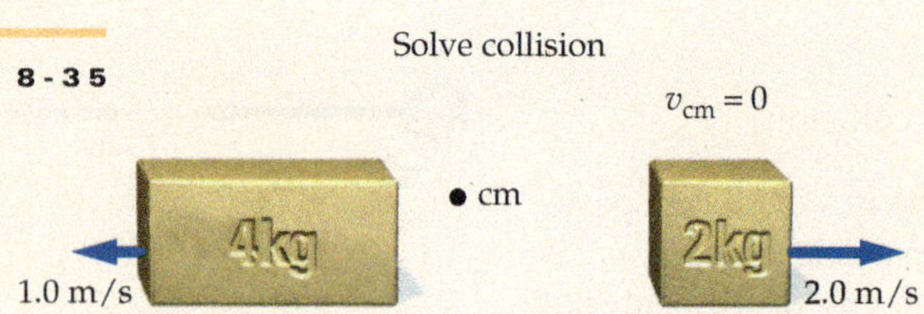

4. To find the final velocities in the original frame, add v_{cm} to each final velocity (Figure 8-36).

$$v_{1f} = u_{1f} + v_{cm} = -1.0\text{ m/s} + 5.0\text{ m/s} = \boxed{4.0\text{ m/s}}$$

$$v_{2f} = u_{2f} + v_{cm} = 2.0\text{ m/s} + 5.0\text{ m/s} = \boxed{7.0\text{ m/s}}$$

FIGURE 8-36

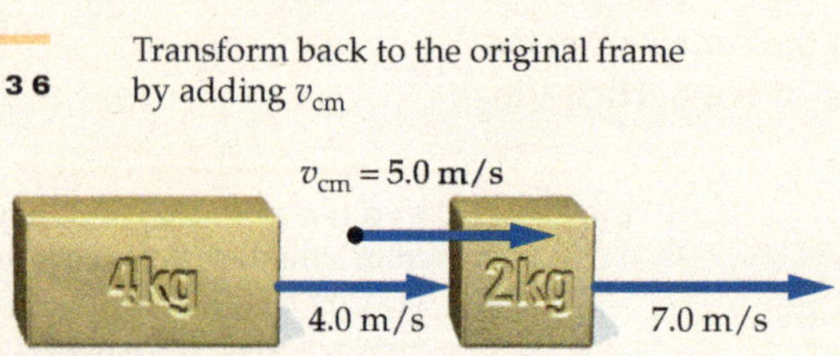

CHECK This result is the same result found in Example 8-13.

PRACTICE PROBLEM 8-8 Verify that the total momentum of the system both before the collision and after the collision is zero in the center-of-mass reference frame.

8-5 CONTINUOUSLY VARYING MASS AND ROCKET PROPULSION

A creative and important step in solving physics problems is specifying the system. In this section, we explore situations in which the system has a continuously changing mass. One example of such a system is a rocket. For a rocket, we specify the system to be the rocket plus any unspent fuel in it. As the spent fuel (the exhaust) spews out the back, the mass of the system decreases. Another example is the sand falling onto base of an hourglass (Figure 8-37). We specify the system as the sand currently resting on the base. The mass of the system continues to increase as the sand on the base continues to accumulate.

FIGURE 8-37 *(Brand-X Pictures/PunchStock.)*

On the Jovian moon Io there is a large volcano. When the volcano erupts the speed of the effluence exceeds the escape speed of Io. Consequently, a stream of the effluence is projected into space. The material in the stream collides with and sticks to the surface of an asteroid passing through the stream. We now consider the effect of the impact of this material on the motion of the asteroid. Doing so involves developing an equation, that is, a version of Newton's second law for systems with continuously varying mass.

Suppose a continuous stream of matter moving at velocity $\vec{u}$ is impacting an object of mass M that is moving with velocity $\vec{v}$ (Figure 8-38). These impacting particles stick to the object, increasing its mass by ΔM during time Δt. In addition, during time Δt the velocity $\vec{v}$ changes by $\Delta\vec{v}$, as shown. Applying the impulse–momentum theorem to this system gives

$$\vec{F}_{\text{net ext}}\,\Delta t = \Delta\vec{P} = \vec{P}_{\text{f}} - \vec{P}_{\text{i}} = [(M + \Delta M)(\vec{v} + \Delta\vec{v})] - [M\vec{v} + \Delta M\vec{u}]$$

where the first term in square brackets is the momentum at time $t + \Delta t$ and the second term in square brackets is the momentum at time t. Rearranging terms gives

$$\vec{F}_{\text{net ext}}\,\Delta t = M\,\Delta\vec{v} + \Delta M\,(\vec{v} - \vec{u}) + \Delta M\,\Delta\vec{v} \qquad \text{8-32}$$

Dividing through Equation 8-32 by Δt gives

$$\vec{F}_{\text{net ext}} = M\frac{\Delta\vec{v}}{\Delta t} + \frac{\Delta M}{\Delta t}(\vec{v} - \vec{u}) + \frac{\Delta M}{\Delta t}\Delta\vec{v}$$

Taking the limit as $\Delta t \to 0$ (which also means as $\Delta M \to 0$ and as $\Delta\vec{v} \to 0$) gives

$$\vec{F}_{\text{net ext}} = M\frac{d\vec{v}}{dt} - \frac{dM}{dt}(\vec{v} - \vec{u}) + \frac{dM}{dt}(0)$$

Rearranging once again, we obtain

$$\vec{F}_{\text{net ext}} + \frac{dM}{dt}\vec{v}_{\text{rel}} = M\frac{d\vec{v}}{dt} \qquad \text{8-33}$$

NEWTON'S SECOND LAW—CONTINUOUSLY VARIABLE MASS

where $\vec{v}_{\text{rel}} = \vec{u} - \vec{v}$ is the velocity of the impacting material relative to the object. Note that except for the term $(dM/dt)\vec{v}_{\text{rel}}$, Equation 8-33 is the identical to the equation for Newton's second law for a system with constant mass.

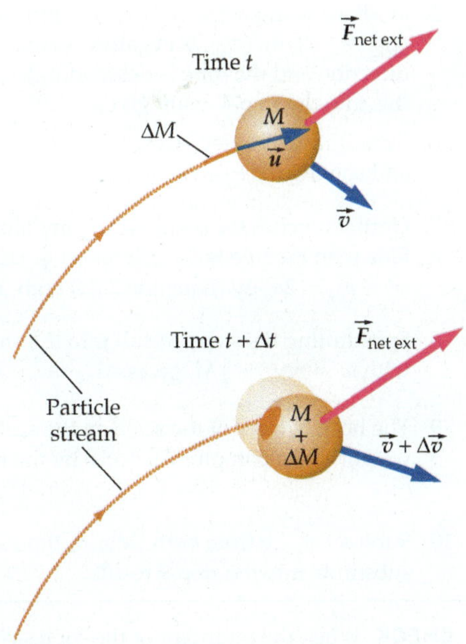

FIGURE 8-38 Particles in a continuous stream and moving at velocity $\vec{u}$ undergo perfectly inelastic collisions with an object of mass M moving at velocity $\vec{v}$. In addition, there is a net external force $\vec{F}_{\text{net ext}}$ acting on the object. The system is shown at time t and again at time $t + \Delta t$.

Example 8-18 A Falling Rope

A uniform rope of mass M and length L is held by one end, with the other end just touching the surface of a scale pan. The rope is released and begins to fall. Find the force of the rope on the scale pan just as the midpoint of the rope reaches the scale pan.

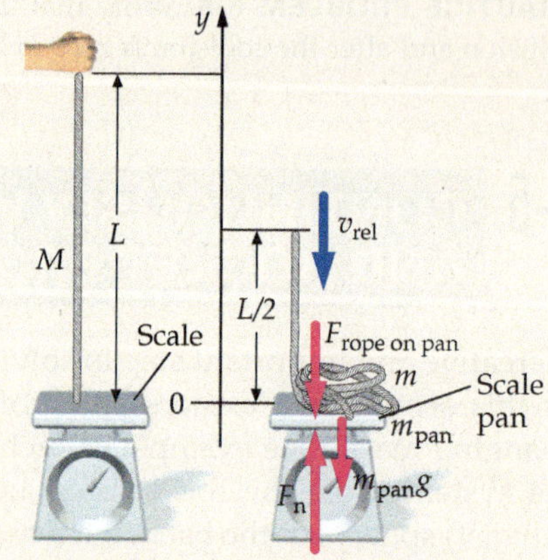

FIGURE 8-39 A very flexible rope of length L and mass M is released from rest and falls on the pan of a scale.

PICTURE Apply Equation 8-33 to the system consisting of the scale pan and that portion of the rope on the scale at time t. There are two external forces on that system, the force of gravity and the normal force exerted by the scale on the scale pan. The impact velocities of the different points along the falling rope depend upon their initial heights above the scale pan. The normal force exerted by the scale must both change the momentum of the rope impacting the scale, support the weight of both the pan and that portion of the rope already on the pan.

SOLVE

1. Draw a sketch of the situation (Figure 8-39). Include the initial configuration and the configuration at an arbitrary time later. Include a coordinate axis:

2. Express Equation 8-33 in component form. Let m denote the mass of the system (the pan plus that portion of the rope on the scale). The velocity of the system remains zero, so the dv_y/dt is zero:

$$F_{\text{net ext }y} + \frac{dm}{dt}v_{\text{rel }y} = m\frac{dv_y}{dt}$$
$$F_n - mg + \frac{dm}{dt}v_{\text{rel }y} = 0$$

3. Let dm denote the mass of the rope segment of length $d\ell$ that falls on the scale during time dt. Because the rope is uniform, the relation between dm and $d\ell$ is:

$$\frac{dm}{d\ell} = \frac{M}{L}$$

4. Solve for dm/dt by multiplying both sides of the step-3 result by $d\ell/dt$:

$$\frac{dm}{dt} = \frac{M}{L}\frac{d\ell}{dt}$$

5. $d\ell/dt$ is the impact speed of the segment, so $v_{\text{rel }y} = -d\ell/dt$. ($v_{\text{rel }y}$ is negative because up is the $+y$ direction and the rope is descending). Substituting this into the step-4 result gives:

$$\frac{dm}{dt} = -\frac{M}{L}v_{\text{rel }y}$$

6. Substituting the step-5 result into the step-2 result and solving for F_n gives:

$$F_n = mg + \frac{M}{L}v^2_{\text{rel }y}$$

7. Until it touches the scale, each point along the rope falls with the free-fall acceleration $\vec{g}$. Using $v_y^2 = v_{0y}^2 + 2a_y\,\Delta y$ (Equation 2-23) with $\Delta y = -L/2$ gives:

$$v^2_{\text{rel }y} = v^2_{\text{rel }y0} + 2a_y\,\Delta y = 0 + 2(-g)(-L/2) = gL$$

8. Substituting the step-7 result into the step-6 result, with $m = m_{\text{pan}} + \frac{1}{2}M$, gives:

$$F_n = \left(m_{\text{pan}} + \frac{M}{2}\right)g + \frac{M}{L}gL = m_{\text{pan}}g + \frac{3}{2}Mg$$

9. The normal force of the scale on the scale pan equals the weight of the pan plus the force by the rope on the pan:

$$F_n = m_{\text{pan}}g + F_{\text{by rope on pan}}$$
so $F_{\text{by rope on pan}} = F_n - m_{\text{pan}}g$

10. Subtract $m_{\text{pan}}g$ from both sides of the step-8 result and substitute into the step-9 result:

$$F_{\text{by rope on pan}} = \boxed{\frac{3}{2}Mg}$$

CHECK When the midpoint of the rope strikes the scale pan, the force of the rope currently on the scale pan is greater than $\frac{1}{2}Mg$ (the weight of the rope on the pan at that instant), as expected. We expect the force of the rope on the pan to be greater than $\frac{1}{2}Mg$ because the scale pan must both support the weight of the rope on it, and stop the momentum of the impacting rope.

PRACTICE PROBLEM 8-9 Find the force exerted by the scale pan on the rope (a) just before the upper end of the rope reaches the pan, and (b) just after the upper end of the rope comes to rest on the pan.

Rocket propulsion is a striking example of the conservation of momentum in action. We now derive the rocket equation (a special case of Equation 8-33). The mass of the rocket changes continuously as it burns fuel and expels exhaust gas. Consider a rocket moving straight up with velocity $\vec{v}$ relative to Earth, as shown in Figure 8-40. Assuming that the fuel is burned at a constant rate R, the rocket's mass at time t is

$$M = M_0 - Rt \qquad \text{8-34}$$

where M_0 is the initial mass of the rocket. The exhaust gases leave the rocket engine with velocity $\vec{u}_{ex}$ relative to the rocket, and the rate at which the fuel is burned is the rate at which the mass M decreases. We choose the rocket, including unspent fuel in it, as the system. Neglecting air drag, the only external force on the system is that of gravity. With $\vec{F}_{net\,ext} = M\vec{g}$ and $dM/dt = -R$, Equation 8-33 becomes the **rocket equation:**

$$M\vec{g} - R\vec{u}_{ex} = M\frac{d\vec{v}}{dt} \qquad \text{8-35}$$

ROCKET EQUATION

FIGURE 8-40 *(NASA/Superstock.)*

The quantity $-R\vec{u}_{ex}$ is the force exerted on the rocket by the exhaust gases. This force is called the **thrust** $\vec{F}_{th}$:

$$\vec{F}_{th} = -R\vec{u}_{ex} = -\left|\frac{dM}{dt}\right|\vec{u}_{ex} \qquad \text{8-36}$$

DEFINITION—ROCKET THRUST

The rocket is moving straight up, so we choose upward as the positive y direction and express Equation 8-35 in component form

$$-Mg - Ru_{ex\,y} = M\frac{dv_y}{dt}$$

The direction of $\vec{u}_{ex}$ is downward, so $u_{ex\,y} = -u_{ex}$. Substituting gives

$$-Mg + Ru_{ex} = M\frac{dv_y}{dt} \qquad \text{8-37}$$

Solving for dv_y/dt (the acceleration) gives

$$\frac{dv_y}{dt} = \frac{Ru_{ex}}{M} - g$$

where Ru_{ex}/M is the contribution of the thrust to the acceleration and $-g$ is the contribution of the gravitational force to the acceleration. Substituting for M from Equation 8-34 gives

$$\frac{dv_y}{dt} = \frac{Ru_{ex}}{M_0 - Rt} - g \qquad \text{8-38}$$

Equation 8-38 is solved by integrating both sides with respect to time. For a rocket starting at rest at $t = 0$, this gives

$$v_y = \int_0^{t_f}\left(\frac{Ru_{ex}}{M_0 - Rt} - g\right)dt = u_{ex}\int_0^{t_f}\frac{dt}{b - t} - \int_0^{t_f} g\,dt = -u_{ex}\ln\frac{b - t_f}{b} - gt_f$$

where $b = M_0/R$. Rearranging, after substituting t for t_f and M_0/R for b, gives

$$v_y = u_{ex}\ln\left(\frac{M_0}{M_0 - Rt}\right) - gt \qquad \text{8-39}$$

Example 8-19 Liftoff

The Saturn V rocket used in the Apollo moon-landing program had an initial mass M_0 of 2.85×10^6 kg, 73.0 percent of which was fuel, a burn rate R of 13.84×10^3 kg/s, and a thrust F_{th} of 34.0×10^6 N. Find (*a*) the exhaust speed relative to the rocket, (*b*) the burn time t_b, (*c*) the acceleration at liftoff, (*d*) the acceleration at just before burnout t_b, and (*e*) the final speed of the rocket.

PICTURE (*a*) The exhaust speed relative to the rocket can be found from the thrust and burn rate. (*b*) The mass of the rocket without any fuel is 73.0% of the initial mass. To find the burn time, you need to find the total mass of fuel burned, which is the initial mass minus the mass at burnout. (*c*) and (*d*) The acceleration is found from Equation 8-38. (*e*) The final speed is given by Equation 8-39.

SOLVE

(*a*) 1. Calculate u_{ex} from the given thrust and burn rate.

$$F_{th} = \left|\frac{dM}{dt}\right| u_{ex}$$

so $$u_{ex} = \frac{F_{th}}{|dM/dt|} = \frac{34.0 \times 10^6\ \text{N}}{13.84 \times 10^3\ \text{kg/s}} = \boxed{2.46\ \text{km/s}}$$

(*b*) 1. Calculate the mass M_b of the rocket at burnout (when it runs out of fuel).

$$M_b = 0.270M_0 = 7.70 \times 10^5\ \text{kg}$$

2. The mass of the fuel equals the burn rate multiplied by the burn time t_b.

$$M_{fuel} = Rt_b$$

so $$t_b = \frac{M_{fuel}}{R} = \frac{M_0 - M_b}{R} = \boxed{150\ \text{s}}$$

(*c*) Calculate dv_y/dt at $t = 0$ using Equation 8-38.

Initially,

$$\frac{dv_y}{dt} = \frac{u_{ex}}{M_0}\left|\frac{dM}{dt}\right| - g = \frac{2.46\ \text{km/s}}{2.85 \times 10^6\ \text{kg}}(13.84 \times 10^3\ \text{kg/s}) - 9.81\ \text{m/s}^2$$

$$= \boxed{2.14\ \text{m/s}^2}$$

(*d*) Calculate dv_y/dt at $t = t_b$ using Equation 8-38.

At burnout,

$$\frac{dv_y}{dt} = \frac{u_{ex}}{M_b}\left|\frac{dM}{dt}\right| - g = \frac{2.46\ \text{km/s}}{7.70 \times 10^5\ \text{kg}}(13.84 \times 10^3\ \text{kg/s}) - 9.81\ \text{m/s}^2$$

$$= \boxed{34.3\ \text{m/s}^2}$$

(*e*) Calculate the speed at $t = t_b$ using Equation 8-39.

$$v_y = u_{ex}\ln\left(\frac{M_0}{M_0 - Rt}\right) - gt = \boxed{1.75\ \text{km/s}}$$

CHECK At burnout, the mass being accelerated is 73 percent less than the mass being accelerated at liftoff. Therefore, we would expect the acceleration at burnout to be much larger than the initial acceleration. This is shown by our results in Parts (*c*) and (*d*).

TAKING IT FURTHER (1) The initial acceleration is small—only $0.21g$. At burnout (also called flameout), the rocket's acceleration has increased to $3.5g$. Immediately following burnout, the acceleration is $-g$. The speed of the rocket at burnout is 1.75 km/s = 6300 km/h ≈ 3900 mi/h. (2) The calculations for Parts (*d*) and (*e*) assume the rocket moved vertically upward and that g did not vary with altitude. In practice the actual rocket initially moved vertically upward, but it then gradually turned eastward.

Physics Spotlight

Pulse Detonation Engines: Faster (and Louder)

Experimental pulse detonation engine in Rutan Vari-Eze. *(Tim Anderson.)*

Liquid-fueled rocket engines need expensive, delicate pumps to compress the fuel to very high pressures in the combustion chamber. Most jet engines are gas-turbine engines, which have many moving parts with tight tolerances and high maintenance needs. Rocket and aeronautical engineers want an engine with higher fuel efficiency, few moving parts, and an ability to operate at a wide range of speeds.

The *pulse detonation engine* (PDE) may fulfill these requirements. The PDE is powered by *detonation* rather than *deflagration*.

Both detonation and deflagration are types of combustion. Deflagration propagates slower than the speed of sound by heating up the air around it. Fireworks, properly tuned automobile engines, and charcoal barbecues with too much starter fluid on them are deflagrations. Detonation propagates faster than the speed of sound—sometimes much faster, by a shock wave that compresses and ignites the air. Confined high explosives used in mining and demolition detonate; improperly tuned car engines can also have internal detonations.

In a PDE, a detonation tube is closed at one end, and open at the other for exhaust. Air and fuel are admitted into the closed end, and are ignited by a spark. This starts a deflagration. As the deflagration moves down the complex interior surface* of the detonation tube, it is compressed quickly and begins to detonate. Once the detonation begins, it propagates much faster than the speed of sound. Detonation wavefronts as fast as Mach 5 have been measured in various laboratories.† The exhaust leaves the open end of the tube very rapidly. Because the exhaust has such a high velocity, it has higher momentum than the same exhaust would have from a deflagration. This gives a greater thrust to the rocket for the same amount of fuel. Detonation has given double the impulse of deflagration, using the same fuel and apparatus.‡

The only moving parts in the PDE are the valves to admit the fuel/air mixture. The ignition can be supplied by an automobile spark plug, and the rest of the engine is just the detonation tube. It seems very simple at first glance. But combustion is a complex process, and the combustion in a PDE is over very quickly. To power a jet or rocket, the PDE needs many detonations per second, just as an automobile needs many combustion incidents per second to move. PDEs have been tested at 80 detonations per second for several minutes and hours, but ideally, PDEs would reach speeds of a few hundred detonations per second.§

But detonation is a violent process. It is extremely loud, and causes the engine to vibrate even more than existing jet and rocket engines do.° Excess vibration can be harmful to rockets and jets. The noise produced by existing PDEs is not practical for a vehicle with a human pilot or passenger. Finally, heavy tubes have been used to contain the detonation. The tubes need to be made of a material strong enough to withstand the detonations, but lightweight enough to fly.

By early 2006, no planes have flown with a PDE, but the idea of engines for jets and rockets that cost less, generate a wide range of thrusts, and have higher fuel efficiency is well worth pursuing.

---

* Paxson, D. E., Rosenthal, B. N., Sgondea, A., and Wilson, J., "Parametric Investigation of Thrust Augmentation by Ejectors on a Pulsed Detonation Tube." Paper presented at the 41st Joint Propulsion Conference, 2005, Tuscon, AZ.

† Borisov, A.A., Frolov, S.M., Netzer, D. W., and Roy, G. D., "Pulse Detonation Propulsion: Challenges, Current Status, and Future Perspective." *Progress in Energy and Combustion Science* 30 (2004) 545-672

‡ "Detonation Initiation and Impulse Measurement." Explosion Dynamics Laboratory: Pulse Detonation Engines, http://www.galcit.caltech.edu/EDL/projects/pde/pde.html May, 2006

§ Kandebo, Stanley W., "Taking the Pulse." *Aviation Week and Space Technology,* 160:10 Mar. 8, 2004, 32-33.

° Borisov et al. op. cit.

Summary

The conservation of momentum for an isolated system is a fundamental law of nature that has applications in all areas of physics.

TOPIC	RELEVANT EQUATIONS AND REMARKS	
1. Momentum		
Definition for a particle	$\vec{p} = m\vec{v}$	8-1
Kinetic energy of a particle	$K = \dfrac{p^2}{2m}$	8-16
Momentum of a system	$\vec{P}_{sys} = \sum_i m_i\vec{v}_i = M\vec{v}_{cm}$	8-3
Newton's second law for a system	$\vec{F}_{net\,ext} = \dfrac{d\vec{P}_{sys}}{dt}$	8-4
Law of conservation of momentum	If the net external force acting on a system remains zero, the total momentum of the system is conserved.	
2. Energy of a System		
Kinetic energy	The kinetic energy associated with the motion of the particles of a system relative to its center of mass is $K_{rel} = \Sigma\frac{1}{2}m_iu_i^2$, where u_i is the speed of the ith particle relative to the center of mass. $K = \frac{1}{2}Mv_{cm}^2 + K_{rel}$	8-7
3. Collisions		
Impulse	The impulse of a force is defined as the integral of the force over the time interval during which the force acts. $\vec{I} = \int_{t_i}^{t_f}\vec{F}\,dt$	8-9
Impulse–momentum theorem	$\vec{I}_{net} = \int_{t_i}^{t_f}\vec{F}_{net}dt = \Delta\vec{p}$	8-10
Average force	$\vec{F}_{av} = \dfrac{1}{\Delta t}\int_{t_i}^{t_f}\vec{F}\,dt = \dfrac{\vec{I}}{\Delta t}$ (so $\vec{I} = \vec{F}_{av}\Delta t$)	8-13
Elastic collisions	An elastic collision between two objects is one in which the sum of their kinetic energies is the same before and after the collision.	
Relative speeds of approach and separation	For an elastic collision, the speed of separation equals the speed of approach. For a *head-on* elastic collision, $v_{2f} - v_{1f} = v_{1i} - v_{2i}$	8-23
Perfectly inelastic collisions	Following a perfectly inelastic collision, the two objects stick together and move with the velocity of the center of mass.	
*Coefficient of restitution	The coefficient of restitution e is a measure of the elasticity. It is the ratio of the separation speed to the closing speed: $e = \dfrac{v_{2f} - v_{1f}}{v_{1i} - v_{2i}}$ For an elastic collision, $e = 1$; for a perfectly inelastic collision $e = 0$.	8-25
***4. Continuously Variable Mass**		
Newton's second law	$\vec{F}_{net\,ext} + \dfrac{dM}{dt}\vec{v}_{rel} = M\dfrac{d\vec{v}}{dt}$ where $R = \lvert dM/dt\rvert$ is the burn rate.	8-33
Rocket equation	$M\vec{g} - R\vec{u}_{ex} = M\dfrac{d\vec{v}}{dt}$	8-35
Thrust	$\vec{F}_{th} = -R\vec{u}_{ex} = -\left\lvert\dfrac{dM}{dt}\right\rvert\vec{u}_{ex}$	8-36

Answer to Concept Check

8-1 No

Answers to Practice Problems

8-1 8.4 J

8-2 140 s. The grain leaking out does not impart any momentum to the rest of the system. If the ground were frictionless and flat, all of the grain initially in the car would arrive at the switchyard along with the car.

8-3 1.32 m/s. She gains more speed by throwing them sequentially rather than simultaneously.

8-4 (*a*) $v_{cm} = 0.50$ m/s, $v_A = +0.50$ m/s, and $v_B = -0.50$ m/s, (*b*) $K_{A\,rel} = K_{B\,rel} = 0.125$ J, (*c*) $K_{rel} = 0.25$ J, (*d*) $K_{rel} = 0$

8-5 Momentum conservation implies $v_{1i} + v_{2i} = v_{1f} + v_{2f}$, elastic collision implies $v_{1i} - v_{2i} = v_{2f} - v_{1f}$. Together, these imply $v_{2f} = v_{1i}$ and $v_{1f} = v_{2i}$.

8-6 0.73

8-7 $\frac{1}{2}v_0$

8-8 Before: $P_{sys\,i} = (4.0\text{ kg})(1.0\text{ m/s}) + (2.0\text{ kg})(-2.0\text{ m/s}) = 0.0\text{ kg}\cdot\text{m/s}$

After: $P_{sys\,f} = (4.0\text{ kg})(-1.0\text{ m/s}) + (2.0\text{ kg})(2.0\text{ m/s}) = 0.0\text{ kg}\cdot\text{m/s}$

8-9 (*a*) $3Mg$, (*b*) Mg

Problems

In a few problems, you are given more data than you actually need; in a few other problems, you are required to supply data from your general knowledge, outside sources, or informed estimates.

Interpret as significant all digits in numerical values that have trailing zeros and no decimal points.

For all problems, use 9.81 m/s² for the free-fall acceleration and neglect friction and air resistance unless instructed to do otherwise.

• Single-concept, single-step, relatively easy

•• Intermediate-level, may require synthesis of concepts

••• Challenging

SSM Solution is in the *Student Solutions Manual*

Consecutive problems that are shaded are paired problems.

CONCEPTUAL PROBLEMS

1 • Show that if two particles have equal kinetic energies, the magnitudes of their momenta are equal only if they have the same mass. SSM

2 • Particle A has twice the momentum and four times the kinetic energy of particle B. What is the ratio of the mass of particle A to that of particle B? Explain your reasoning.

3 • Using SI units, show that the units of momentum squared divided by those of mass is equivalent to the joule.

4 • True or false:

(*a*) The total linear momentum of a system may be conserved even when the mechanical energy of the system is not.

(*b*) For the total linear momentum of a system to be conserved, there must be no external forces acting on the system.

(*c*) The velocity of the center of mass of a system changes only when there is a net external force on the system.

5 • If a bullet is fired due west, explain how conservation of linear momentum enables you to predict that the recoil of the rifle will be exactly due east. Is kinetic energy conserved here?

6 • A child jumps from a small boat to a dock. Why does she have to jump with more effort than she would need if she were jumping through an identical displacement, but from a boulder to a tree stump?

7 •• Much of the early research in rocket motion was done by Robert Goddard, physics professor at Clark College in Worcester, Massachusetts. A quotation from a 1920 editorial in the *New York Times* illustrates the public's opinion of his work: "That Professor Goddard with his 'chair' at Clark College and the countenance of the Smithsonian Institution does not know the relation between action and reaction, and the need to have something better than a vacuum against which to react—to say that would be absurd. Of course, he only seems to lack the knowledge ladled out daily in high schools."* The belief that a rocket needs something to push against was a prevalent misconception before rockets in space were commonplace. Explain why that belief is wrong. SSM

* On page 43 of the July 17, 1969, edition of the *New York Times* "A Correction" to their editorial of 1920 was printed. This commentary, which was published three days before man's first walk on the moon, stated that "it is now definitely established that a rocket can function in a vacuum as well as in an atmosphere. *The Times* regrets the error.

8 • Two identical bowling balls are moving with the same center-of-mass velocity, but one just slides down the alley without rotating, whereas the other rolls down the alley. Which ball has more kinetic energy?

9 • A philosopher tells you, "Changing motion of objects is impossible. Forces always come in equal but opposite pairs. Therefore, all forces cancel out. Because forces cancel, the momenta of objects can never be changed." Answer his argument.

10 • A moving object collides with a stationary object. Is it possible for both objects to be at rest immediately after the collision? (Assume any external forces acting on this two-object system are negligibly small.) Is it possible for one object to be at rest immediately after the collision? Explain.

11 • Several researchers in physics education claim that part of the cause of physical misconceptions among students comes from special effects they observe in cartoons and movies. Using the conservation of linear momentum, how would you explain to a class of high school physics students what is conceptually wrong with a superhero hovering at rest in midair while tossing massive objects such as cars at villains? Does this action violate the conservation of energy as well? Explain.

12 •• A struggling physics student asks, "If only external forces can cause the center of mass of a system of particles to accelerate, how can a car move? Doesn't the car's engine supply the force needed to accelerate the car?" Explain what external agent produces the force that accelerates the car, and explain how the engine makes that agent do so.

13 •• When we push on the brake pedal to slow down a car, a brake pad is pressed against the rotor so that the friction of the pad slows the rotation of the rotor and thus the rotation of the wheel. However, the friction of the pad against the rotor cannot be the force that slows the car down because it is an internal force—both the rotor and the wheel are parts of the car, so any forces between them are internal, not external forces. What external agent exerts the force that slows down the car? Give a detailed explanation of how this force operates.

14 •• Explain why a circus performer falling into a safety net can survive unharmed, while a circus performer falling from the same height onto the hard concrete floor suffers serious injury or death. Base your explanation on the impulse–momentum theorem.

15 •• In Problem 14, estimate the ratio of the collision time with the safety net to the collision time with the concrete for the performer falling from a height of 25 m. *Hint: Use the procedure outlined in step 4 of the Problem-Solving Strategy located in Section 8-3.* SSM

16 •• (*a*) Why does a drinking glass survive a fall onto a carpet but not a fall onto a concrete floor? (*b*) On many automobile race tracks, dangerous curves are surrounded by massive bails of hay. Explain how this setup reduces the chances of car damage and driver injury.

17 • True or false:

(*a*) Following any perfectly inelastic collision, the kinetic energy of the system is zero in all inertial reference frames.

(*b*) For a head-on elastic collision, the relative speed of recession equals the relative speed of approach.

(*c*) In a perfectly inelastic head-on collision with one object initially at rest, only some of the system's kinetic energy is dissipated.

(*d*) After a perfectly inelastic head-on collision along the east–west horizontal axis, the two objects are observed to be moving west. The initial total system momentum was therefore to the west.

18 •• Under what conditions can all the initial kinetic energy of an isolated system consisting of two colliding objects be lost in a collision? Explain how this result can be, and yet the momentum of the system can be conserved.

19 •• Consider a perfectly inelastic collision of two objects of equal mass. (*a*) Is the loss of kinetic energy greater if the two objects are moving in opposite directions, each moving at speed $v/2$, or if one of the two objects is initially at rest and the other has an initial speed of v? (*b*) In which of these situations is the percentage loss in kinetic energy the greatest?

20 •• A double-barreled pea shooter is shown in Figure 8-41. Air is blown into the left end of the pea shooter, and identical peas A and B are positioned inside each straw as shown. If the pea shooter is held horizontally while the peas are shot off, which pea, A or B, will travel farther after leaving the straw? Explain. (Base your explanation on the impulse–momentum theorem.)

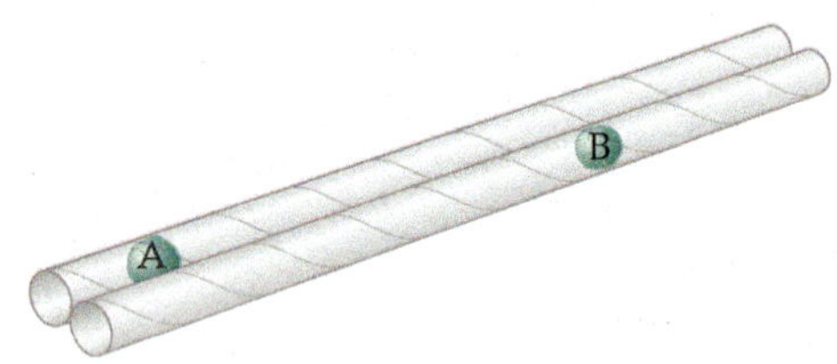

FIGURE 8-41 Problem 20

21 •• A particle of mass m_1 traveling with a speed v makes a head-on elastic collision with a stationary particle of mass m_2. In which scenario will the largest amount of energy be imparted to the particle of mass m_2? (*a*) $m_2 < m_1$, (*b*) $m_2 = m_1$, (*c*) $m_2 > m_1$, (*d*) none of the above

22 •• **ENGINEERING APPLICATION, CONTEXT-RICH** Suppose you are in charge of an accident reconstruction team that has reconstructed an accident in which a car was "rear-ended," causing the two cars to lock bumpers and skid to a halt. During the trial, you are on the stand as an expert witness for the prosecution and the lawyer for the defense claims that you wrongly neglected friction and the force of gravity during the fraction of a second while the cars collided. Defend your report. Why were you correct in ignoring these forces? You did not ignore these two forces in your skid analysis both before and after the collision. Can you explain to the jury why you did not ignore these two forces during the pre- and postcollision skids?

23 •• Nozzles for a garden hose are often made with a right-angle shape, as shown in Figure 8-42. If you open the nozzle to spray water out, you will find that the nozzle presses against your hand with a pretty strong force—much stronger than if you used a nozzle not bent into a right angle. Why is this situation true?

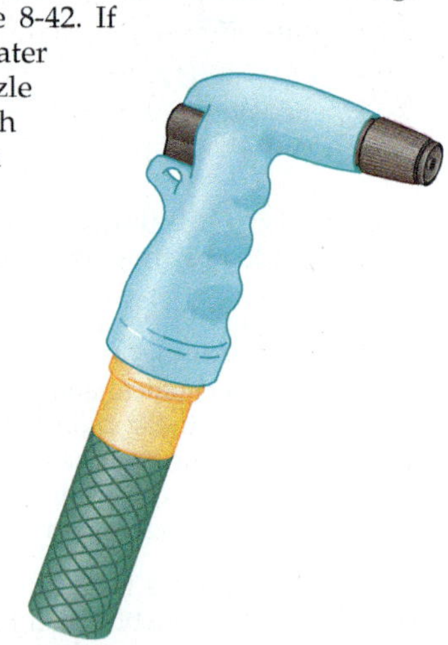

FIGURE 8-42 Problem 23

CONCEPTUAL PROBLEMS FROM OPTIONAL SECTIONS

24 • Describe a perfectly inelastic head-on collision between two stunt cars as viewed in the center-of-mass reference frame.

25 •• One air-hockey puck is initially at rest. An identical air-hockey puck collides with it, striking it with a glancing blow. Assume the collision is elastic and neglect any rotational motion of the pucks. Describe the collision in the center-of-mass frame of the pucks.

26 •• A baton with one end more massive than the other is tossed at an angle into the air. (*a*) Describe the trajectory of the center of mass of the baton in the reference frame of the ground. (*b*) Describe the motion of the two ends of the baton in the center-of-mass frame of the baton.

27 •• Describe the forces acting on a descending Lunar lander as it fires its retrorockets to slow down for a safe landing. (Assume the lander's mass loss during the rocket firing is not negligible.)

28 •• A railroad car rolling along by itself is passing by a grain elevator, which is dumping grain into it at a constant rate. (*a*) Does momentum conservation imply that the railroad car should be slowing down as it passes the grain elevator? Assume that the track is frictionless and perfectly level and that the grain is falling vertically. (*b*) If the car is slowing down, this situation implies that there is some external force acting on the car to slow it down. Where does this force come from? (*c*) After passing the elevator, the railroad car springs a leak, and grain starts leaking out of a vertical hole in its floor at a constant rate. Should the car speed up as it loses mass?

29 •• To show that even really intelligent people can make mistakes, consider the following problem, which was asked of a freshman class at Caltech on an exam (paraphrased): *A sailboat is sitting in the water on a windless day. In order to make the boat move, a misguided sailor sets up a fan in the back of the boat to blow into the sails to make the boat move forward. Explain why the boat will not move.* The idea was that the net force of the wind pushing the sail forward would be counteracted by the force pushing the fan back (Newton's third law). However, as one of the students pointed out to his professor, the sailboat *could* in fact move forward. Why is that? SSM

ESTIMATION AND APPROXIMATION

30 •• **ENGINEERING APPLICATION** A 2000-kg car traveling at 90 km/h crashes into an immovable concrete wall. (*a*) Estimate the time of collision, assuming that the center of the car travels halfway to the wall with constant acceleration. (Use any plausible length for the car.) (*b*) Estimate the average force exerted by the wall on the car.

31 •• In hand-pumped railcar races, a speed of 32.0 km/h has been achieved by teams of four people. A car that has a mass equal to 350 kg is moving at that speed toward a river when Carlos, the chief pumper, notices that the bridge ahead is out. All four people (each with a mass of 75.0 kg) simultaneously jump backward off the car with a velocity that has a horizontal component of 4.00 m/s relative to the car. The car proceeds off the bank and falls into the water a horizontal distance of 25.0 m from the bank. (*a*) Estimate the time of the fall of the railcar. (*b*) What is the horizontal component of the velocity of the pumpers when they hit the ground?

32 •• A wooden block and a gun are firmly fixed to opposite ends of a long glider mounted on a frictionless air track (Figure 8-43). The block and gun are a distance L apart. The system

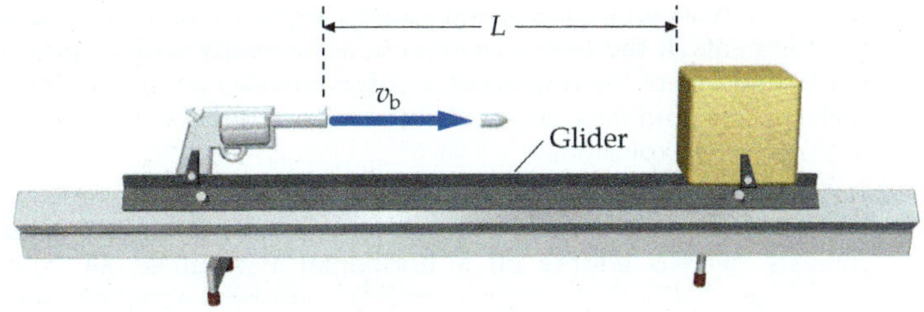

FIGURE 8-43 Problem 32

is initially at rest. The gun is fired and the bullet leaves the gun with a velocity v_b and impacts the block, becoming imbedded in it. The mass of the bullet is m_b and the mass of the gun–glider–block system is m_p. (*a*) What is the velocity of the glider immediately after the bullet leaves the gun? (*b*) What is the velocity of the glider immediately after the bullet comes to rest in the block? (*c*) How far does the glider move while the bullet is in transit between the gun and the block?

CONSERVATION OF LINEAR MOMENTUM

33 • Tyrone, a 85-kg teenager, runs off the end of a horizontal pier and lands on a free-floating 150-kg raft that was initially at rest. After he lands on the raft, the raft, with him on it, moves away from the pier at 2.0 m/s. What was Tyrone's speed as he ran off the end of the pier? SSM

34 •• A 55-kg woman contestant on a reality television show is at rest at the south end of a horizontal 150-kg raft that is floating in crocodile-infested waters. She and the raft are initially at rest. She needs to jump from the raft to a platform that is several meters off the north end of the raft. She takes a running start. When she reaches the north end of the raft she is running at 5.0 m/s relative to the raft. At that instant, what is her velocity relative to the water?

35 • A 5.0-kg object and a 10-kg object, both resting on a frictionless table, are connected by a massless compressed spring. The spring is released and the objects fly off in opposite directions. The 5.0-kg object has a velocity of 8.0 m/s to the left. What is the velocity of the 10-kg object?

36 • Figure 8-44 shows the behavior of a projectile just after it has broken up into three pieces. What was the speed of the projectile the instant before it broke up: (*a*) v_3, (*b*) $v_3/3$, (*c*) $v_3/4$, (*d*) $4v_3$, (*e*) $(v_1 + v_2 + v_3)/4$?

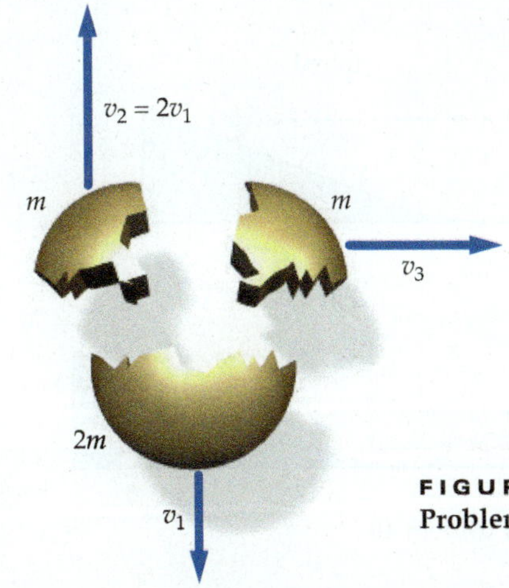

FIGURE 8-44 Problem 36

37 • A shell of mass m and speed v explodes into two identical fragments. If the shell was moving horizontally with respect to Earth, and one of the fragments is subsequently moving vertically with speed v, find the velocity $\vec{v}'$ of the other fragment immediately following the explosion.

38 •• For this week's physics lab, the experimental setup consists of two gliders on a horizontal frictionless air track (see Figure 8-45). Each glider supports a strong magnet centered on top of it, and the magnets are oriented so they attract each other. The mass of glider 1 and its magnet is 0.100 kg, and the mass of glider 2 and its magnet is 0.200 kg. You and your lab partners are instructed to take the origin to be at the left end of the track and to center glider 1 at $x_1 = 0.100$ m and glider 2 at $x_2 = 1.600$ m. Glider 1 is 10.0 cm long, and glider 2 is 20.0 cm long, and each glider has its center of mass at its geometric center. When the two gliders are released from rest, they will move toward each other and stick. (*a*) Predict the position of the center of each glider when they first touch. (*b*) Predict the velocity that the two gliders will continue to move with after they stick. Explain the reasoning behind this prediction for your lab partners.

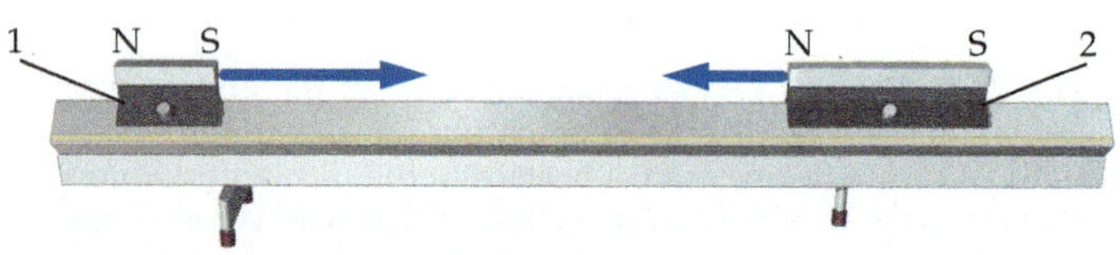

FIGURE 8-45 Problem 38

39 •• Bored, a boy shoots his pellet gun at a piece of cheese that sits on a massive block of ice. On one particular shot, his 1.2-g pellet gets stuck in the cheese, causing it to slide 25 cm before coming to a stop. If the muzzle velocity of the gun is known to be 65 m/s, and the cheese has a mass of 120 g, what is the coefficient of friction between the cheese and the ice?

40 ••• **MULTISTEP** A wedge of mass M is placed on a frictionless, horizontal surface, and a block of mass m is placed on the wedge, which also has a frictionless surface (Figure 8-46). The block's center of mass moves downward a distance h as the block slides from its initial position to the horizontal floor. (*a*) What are the speeds of the block and of the wedge as they separate from each other and go their own ways? (*b*) Check your calculation plausibility by considering the limiting case when $M \gg m$.

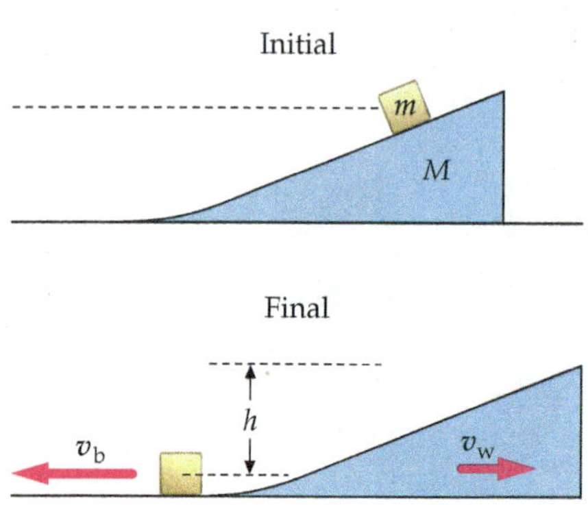

FIGURE 8-46 Problem 40

KINETIC ENERGY OF A SYSTEM OF PARTICLES

41 •• **MULTISTEP** A 3.0-kg block is traveling to the right (the $+x$ direction) at 5.0 m/s, and a second 3.0-kg block is traveling to the left at 2.0 m/s. (*a*) Find the total kinetic energy of the two blocks. (*b*) Find the velocity of the center of mass of the two-block system. (*c*) Find the velocity of each block relative to the center of mass. (*d*) Find the kinetic energy of the blocks relative to the center of mass. (*e*) Show that your answer for Part (*a*) is greater than your answer for Part (*d*) by an amount equal to the kinetic energy associated with the motion of the center of mass. SSM

42 •• Repeat Problem 41 with the second 3.0-kg block replaced by a 5.0-kg block moving to the right at 3.0 m/s.

IMPULSE AND AVERAGE FORCE

43 • You kick a soccer ball whose mass is 0.43 kg. The ball leaves your foot with an initial speed of 25 m/s. (*a*) What is the magnitude of the impulse associated with the force of your foot on the ball? (*b*) If your foot is in contact with the ball for 8.0 ms, what is the magnitude of the average force exerted by your foot on the ball? SSM

44 • A 0.30-kg brick is dropped from a height of 8.0 m. It hits the ground and comes to rest. (*a*) What is the impulse exerted by the ground on the brick during the collision? (*b*) If it takes 0.0013 s from the time the brick first touches the ground until it comes to rest, what is the average force exerted by the ground on the brick at impact?

45 • A meteorite that has a mass equal to 30.8 tonnes (1 tonne = 1000 kg) is exhibited in the American Museum of Natural History in New York City. Suppose that the kinetic energy of the meteorite as it hit the ground was 617 MJ. Find the magnitude of the impulse experienced by the meteorite up to the time its kinetic energy was halved (which took about $t = 3.0$ s). Also find the average force exerted on the meteorite during this time interval.

46 •• A 0.15-kg baseball traveling horizontally is hit by a bat and its direction is exactly reversed. Its velocity changes from $+20$ m/s to -20 m/s. (*a*) What is the magnitude of the impulse delivered by the bat to the ball? (*b*) If the baseball is in contact with the bat for 1.3 ms, what is the average force exerted by the bat on the ball?

47 •• A 60-g handball moving with a speed of 5.0 m/s strikes the wall at a 40° angle with the normal, and then bounces off with the same speed at the same angle with the normal. It is in contact with the wall for 2.0 ms. What is the average force exerted by the ball on the wall?

48 •• **ESTIMATION** You throw a 150-g ball straight up to a height of 40.0 m. (*a*) Use a reasonable value for the displacement of the ball while it is in your hand to estimate the time the ball is in your hand while you are throwing it. (*b*) Calculate the average force exerted by your hand while you are throwing it. (Is it okay to neglect the gravitational force on the ball while it is being thrown?)

49 •• A 0.060-g handball is thrown straight toward a wall with a speed of 10 m/s. It rebounds straight backward at a speed of 8.0 m/s. (*a*) What impulse is exerted on the wall? (*b*) If the ball is in contact with the wall for 3.0 ms, what average force is exerted on the wall by the ball? (*c*) The rebounding ball is caught by a player who brings it to rest. In the process, her hand moves back 0.50 m. What is the impulse received by the player? (*d*) What average force was exerted on the player by the ball?

50 •• A spherical 0.34-kg orange, 2.0 cm in radius, is dropped from the top of a 35-m-tall building. After the orange strikes the pavement, its shape is that of a 0.50-cm-thick pancake. Neglect air resistance and assume that the collision is completely inelastic. (*a*) How much time did the orange take to completely "squish" to a stop? (*b*) What average force did the pavement exert on the orange during the collision?

51 •• The pole vault landing pad at an Olympic competition contains what is essentially a bag of air that compresses from its "resting" height of 1.2 m down to 0.20 m as the vaulter is slowed to a stop. (*a*) What is the time interval during which a vaulter who has just cleared a height of 6.40 m slows to a stop? (*b*) What is the time interval if instead the vaulter is brought to rest by a 20-cm layer of sawdust that compresses to 5.0 cm when he lands? (*c*) Qualitatively discuss the difference in average force the vaulter experiences from the two different landing pads. That is, which landing pad would exert the least force on the vaulter and why?

52 ••• Large limestone caverns have been formed by dripping water. (*a*) If water droplets of 0.030 mL fall from a height of 5.0 m at a rate of 10 droplets per minute, what is the average force exerted on the limestone floor by the droplets of water during a 1.0-min period? (Assume the water does not accumulate on the floor.) (*b*) Compare this force to the weight of one water droplet.

COLLISIONS IN ONE DIMENSION

53 • A 2000-kg car traveling to the right at 30 m/s is chasing a second car of the same mass that is traveling in the same direction at 10 m/s. (*a*) If the two cars collide and stick together, what is their speed just after the collision? (*b*) What fraction of the initial kinetic energy of the cars is lost during this collision? Where does it go? SSM

54 • An 85-kg running back moving at 7.0 m/s makes a perfectly inelastic head-on collision with a 105-kg linebacker who is initially at rest. What is the speed of the players just after their collision?

55 • A 5.0-kg object with a speed of 4.0 m/s collides head-on with a 10-kg object moving toward it with a speed of 3.0 m/s. The 10-kg object stops dead after the collision. (*a*) What is the postcollision speed of the 5.0-kg object? (*b*) Is the collision elastic?

56 • A small superball of mass m moves with speed v to the right toward a much more massive bat that is moving to the left with speed v. Find the speed of the ball after it makes an elastic head-on collision with the bat.

57 •• A proton that has a mass m and is moving at 300 m/s undergoes a head-on elastic collision with a stationary carbon nucleus of mass $12m$. Find the velocities of the proton and the carbon nucleus after the collision.

58 •• A 3.0-kg block moving at 4.0 m/s has a head-on elastic collision with a stationary block of mass 2.0 kg. Use conservation of momentum and the fact that the relative speed of recession equals the relative speed of approach to find the velocity of each block after the collision. Check your answer by calculating the initial and final kinetic energies of each block.

59 •• A block of mass $m_1 = 2.0$ kg slides along a frictionless table with a speed of 10 m/s. Directly in front of it, and moving in the same direction with a speed of 3.0 m/s, is a block of mass $m_2 = 5.0$ kg. A massless spring that has a force constant $k = 1120$ N/m is attached to the second block, as in Figure 8-47. (*a*) What is the velocity of the center of mass of the system? (*b*) During the collision, the spring is compressed by a maximum amount Δx.

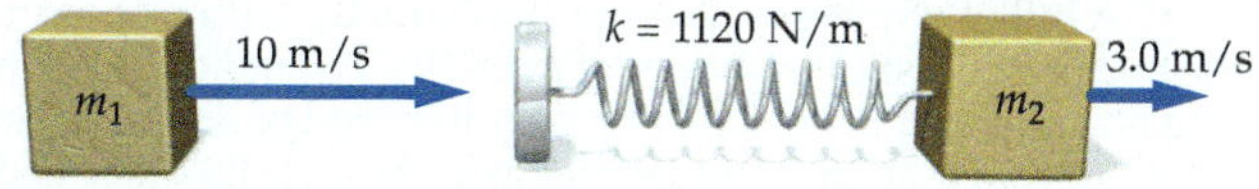

FIGURE 8-47 Problem 59

What is the value of Δx? (*c*) The blocks will eventually separate again. What are the velocities of the two blocks measured in the reference frame of the table, after they separate?

60 •• A bullet of mass m is fired vertically from below into a thin horizontal sheet of plywood of mass M that is initially at rest, supported by a thin sheet of paper (Figure 8-48). The bullet punches through the plywood, which rises to a height, H, above the paper before falling back down. The bullet continues rising to a height, h, above the paper. (*a*) Express the upward velocity of the bullet and the plywood immediately after the bullet exits the plywood in terms of h and H. (*b*) What is the speed of the bullet? (*c*) What is the mechanical energy of the system before and after the inelastic collision? (*d*) How much mechanical energy is dissipated during the collision?

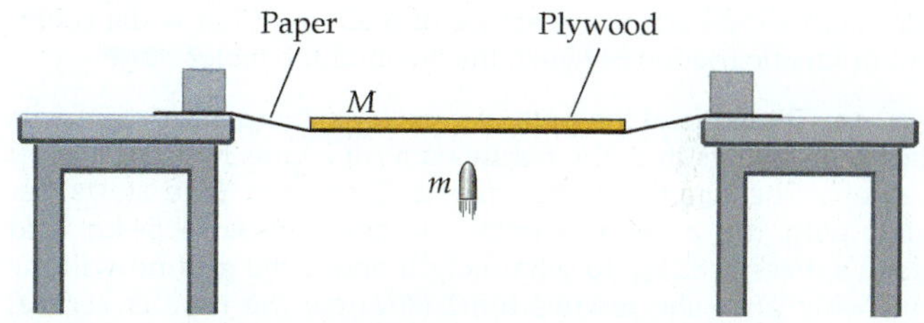

FIGURE 8-48 Problem 60

61 •• A proton of mass m is moving with initial speed v_0 directly toward the center of an α particle of mass $4m$, which is initially at rest. Both particles carry positive charge, so they repel each other. (The repulsive forces are sufficient to prevent the two particles from coming into direct contact.) Find the speed v_α of the α particle (*a*) when the distance between the two particles is a minimum, and (*b*) later when the two particles are far apart.

62 •• An electron collides elastically with a hydrogen atom that is initially at rest. Assume all the motion occurs along a straight line. What fraction of the electron's initial kinetic energy is transferred to the atom? (Take the mass of the hydrogen atom to be 1840 times the mass of an electron.)

63 •• A 16-g bullet is fired into the bob of a 1.5-kg ballistic pendulum (Figure 8-18). When the bob is at its maximum height, the strings make an angle of 60° with the vertical. The pendulum strings are 2.3 m long. Find the speed of the bullet prior to impact. SSM

64 •• Show that in a one-dimensional elastic collision, if the mass and velocity of object 1 are m_1 and v_{1i}, and if the mass and velocity of object 2 are m_2 and v_{2i}, then their final velocities v_{1f} and v_{2f} are given by

$$v_{1f} = \frac{m_1 - m_2}{m_1 + m_2}v_{1i} + \frac{2m_2}{m_1 + m_2}v_{2i}$$

and

$$v_{2f} = \frac{2m_1}{m_1 + m_2}v_{1i} + \frac{m_2 - m_1}{m_1 + m_2}v_{2i}$$

65 •• Investigate the plausibility of the results of Problem 64 by calculating the final velocities in the following limits: (*a*) When the two masses are equal, show that the particles "swap" velocities: $v_{1f} = v_{2i}$ and $v_{2f} = v_{1i}$. (*b*) If $m_2 \gg m_1$, and $v_{2i} = 0$, show that $v_{1f} \approx -v_{1i}$ and $v_{2f} \approx 0$. (*c*) If $m_1 \gg m_2$, and $v_{2i} = 0$, show that $v_{1f} \approx v_{1i}$ and $v_{2f} \approx 2v_{1i}$.

66 •• A bullet of mass m_1 is fired horizontally with a speed v_0 into the bob of a ballistic pendulum of mass m_2. The pendulum consists of a bob attached to one end of a very light rod of length L. The rod is free to rotate about a horizontal axis through its other end. The bullet is stopped in the bob. Find the minimum v_0 such that the bob will swing through a complete circle.

67 •• A bullet of mass m_1 is fired horizontally with a speed v into the bob of a ballistic pendulum of mass m_2 (Figure 18-19). Find the maximum height h attained by the bob if the bullet passes through the bob and emerges with a speed $v/3$.

68 •• A heavy wooden block rests on a flat table and a high-speed bullet is fired horizontally into the block, the bullet stopping in it. How far will the block slide before coming to a stop? The mass of the bullet is 10.5 g, the mass of the block is 10.5 kg, the bullet's impact speed is 750 m/s, and the coefficient of kinetic friction between the block and the table is 0.220. (Assume that the bullet does not cause the block to spin.)

69 •• A 0.425-kg ball with a speed of 1.30 m/s rolls across a level surface toward an open 0.327-kg box that is resting on its side. The ball enters the box, and the box (with the ball inside it) then slides across the surface a distance of 0.520 m. What is the coefficient of kinetic friction between the box and the table? SSM

70 •• Tarzan is in the path of a pack of stampeding elephants when Jane swings in to the rescue on a rope vine, hauling him off to safety. The length of the vine is 25 m, and Jane starts her swing with the rope horizontal. If Jane's mass is 54 kg, and Tarzan's mass is 82 kg, to what height above the ground will the pair swing after she rescues him? (Assume the rope is vertical when she grabs him.)

71 •• Scientists estimate that the meteorite responsible for the creation of Barringer Meteorite Crater in Arizona weighed roughly 2.72×10^5 tonnes (1 tonne = 1000 kg) and was traveling at a speed of 17.9 km/s. Take Earth's orbital speed to be about 30.0 km/s. (*a*) What should the direction of impact be if Earth's orbital speed is to be changed by the maximum possible amount? (*b*) Assuming the condition of collision in Part (*a*), estimate the maximum percentage change in Earth's orbital speed as a result of this collision. (*c*) What mass of an asteroid, having a speed equal to Earth's orbital speed, would be necessary to change Earth's orbital speed by 1.00%? SSM

72 ••• William Tell shoots an apple from his son's head. The speed of the 125-g arrow just before it strikes the apple is 25.0 m/s, and at the time of impact it is traveling horizontally. If the arrow sticks in the apple and the arrow/apple combination strikes the ground 8.50 m behind the son's feet, how massive was the apple? Assume the son is 1.85 m tall.

EXPLOSIONS AND RADIOACTIVE DECAY

73 •• The beryllium isotope ^{8}Be is unstable, decays into two α particles ($m_\alpha = 6.64 \times 10^{-27}$ kg), and releases 1.5×10^{-14} J of energy. Determine the velocities of the two α particles that arise from the decay of a ^{8}Be nucleus at rest, assuming that all the energy appears as kinetic energy of the particles. SSM

74 •• The light isotope, ^{5}Li, of lithium is unstable and breaks up spontaneously into a proton and an α particle. In this process, 3.15×10^{-13} J of energy are released, appearing as the kinetic energy of the two decay products. Determine the velocities of the proton and the α particle that arise from the decay of a ^{5}Li nucleus at rest. (*Note*: The masses of the proton and alpha particle are $m_p = 1.67 \times 10^{-27}$ kg and $m_\alpha = 4m_p = 6.64 \times 10^{-27}$ kg.)

75 ••• A 3.00-kg projectile is fired with an initial speed of 120 m/s at an angle of 30.0° with the horizontal. At the top of its trajectory, the projectile explodes into two fragments of masses 1.00 kg and 2.00 kg. At 3.60 s after the explosion, the 2.00-kg fragment lands on the ground directly below the point of explosion. (*a*) Determine the velocity of the 1.00-kg fragment immediately after the explosion. (*b*) Find the distance between the point of firing and the point at which the 1.00-kg fragment strikes the ground. (*c*) Determine the energy released in the explosion.

76 ••• The boron isotope ^{9}B is unstable and disintegrates into a proton and two α particles. The total energy released as kinetic energy of the decay products is 4.4×10^{-14} J. In one such event, with the ^{9}B nucleus at rest prior to decay, the velocity of the proton is measured as 6.0×10^6 m/s. If the two α particles have equal energies, find the magnitude and the direction of their velocities with respect to the direction of the proton.

COEFFICIENT OF RESTITUTION

77 • ENGINEERING APPLICATION, CONTEXT-RICH You are in charge of measuring the coefficient of restitution for a new alloy of steel. You convince your engineering team to accomplish this task by simply dropping a small ball onto a plate, both ball and plate made from the experimental alloy. If the ball is dropped from a height of 3.0 m and rebounds to a height of 2.5 m, what is the coefficient of restitution? SSM

78 • According to official racquetball rules, to be acceptable for tournament play, a ball must bounce to a height of between 173 and 183 cm when dropped from a height of 254 cm at room temperature. What is the acceptable range of values for the coefficient of restitution for the racquetball–floor system?

79 • A ball bounces to 80 percent of its original height. (*a*) What fraction of its mechanical energy is lost each time it bounces? (*b*) What is the coefficient of restitution of the ball–floor system?

80 •• A 2.0-kg object moving to the right at 6.0 m/s collides head-on with a 4.0-kg object that is initially at rest. After the collision, the 2.0-kg object is moving to the left at 1.0 m/s. (*a*) Find the velocity of the 4.0-kg object after the collision. (*b*) Find the energy lost in the collision. (*c*) What is the coefficient of restitution for these objects?

81 •• A 2.0-kg block moving to the right with a speed of 5.0 m/s collides with a 3.0-kg block that is moving in the same direction at 2.0 m/s, as in Figure 8-49. After the collision, the 3.0-kg block moves to the right at 4.2 m/s. Find (*a*) the velocity of the 2.0-kg block after the collision, and (*b*) the coefficient of restitution between the two blocks.

FIGURE 8-49 Problem 81

82 ••• CONTEXT-RICH To keep homerun records and distances consistent from year to year, organized baseball randomly checks the coefficient of restitution between new baseballs and wooden surfaces similar to that of an average bat. Suppose you are in charge of making sure that no "juiced" baseballs are produced. (*a*) In a random test, you find one that when dropped from 2.0 m rebounds 0.25 m. What is the coefficient of restitution for this ball? (*b*) What is the maximum distance home run shot you would expect from this

ball, neglecting any effects due to air resistance and making reasonable assumptions for bat speeds and incoming pitch speeds? Is this a "juiced" ball, a "normal" ball, or a "dead" ball?

83 •• **CONCEPTUAL** To make puck handling easy, hockey pucks are kept frozen until they are used in the game. (*a*) Explain why room-temperature pucks would be more difficult to handle on the end of a stick than a frozen puck. *Hint: Hockey pucks are made of rubber.* (b) A room-temperature puck rebounds 15 cm when dropped onto a wooden surface from 100 cm. If a frozen puck has only half the coefficient of restitution of a room-temperature one, predict how high the frozen puck would rebound under the same conditions. SSM

COLLISIONS IN MORE THAN ONE DIMENSION

84 •• In Section 8-3, it was proved by using geometry that when a particle elastically collides with another particle of equal mass that is initially at rest, the two postcollision velocities are perpendicular. Here we examine another way of proving this result that illustrates the power of vector notation. (*a*) Given that $\vec{A} = \vec{B} + \vec{C}$, square both sides of this equation (obtain the scalar product of each side with itself) to show that $A^2 = B^2 + C^2 + 2\vec{B}\cdot\vec{C}$. (*b*) Let the momentum of the initially moving particle be $\vec{P}$ and the momenta of the particles after the collision be $\vec{p}_1$ and $\vec{p}_2$. Write the vector equation for the conservation of linear momentum and square both sides (obtain the dot product of each side with itself). Compare it to the equation gotten from the elastic collision condition (kinetic energy is conserved) and finally show that these two equations imply that $\vec{p}_1\cdot\vec{p}_2 = 0$.

85 •• In a pool game, the cue ball, which has an initial speed of 5.0 m/s, makes an elastic collision with the eight ball, which is initially at rest. After the collision, the eight ball moves at an angle of 30° to the right of the original direction of the cue ball. Assume that the balls have equal mass. (*a*) Find the direction of motion of the cue ball immediately after the collision. (*b*) Find the speed of each ball immediately after the collision.

86 •• Object A, which has a mass m and a velocity $v_0\hat{i}$, collides with object B, which has a mass $2m$ and a velocity $\frac{1}{2}v_0\hat{j}$. Following the collision, object B has a velocity of $\frac{1}{4}v_0\hat{i}$. (*a*) Determine the velocity of object A after the collision. (*b*) Is the collision elastic? If not, express the change in the kinetic energy in terms of m and v_0.

87 •• A puck of mass 5.0 kg moving at 2.0 m/s approaches an identical puck that is stationary on frictionless ice. After the collision, the first puck leaves with a speed v_1 at 30° to the original line of motion; the second puck leaves with speed v_2 at 60°, as in Figure 8-50. (*a*) Calculate the speeds v_1 and v_2. (*b*) Was the collision elastic? SSM

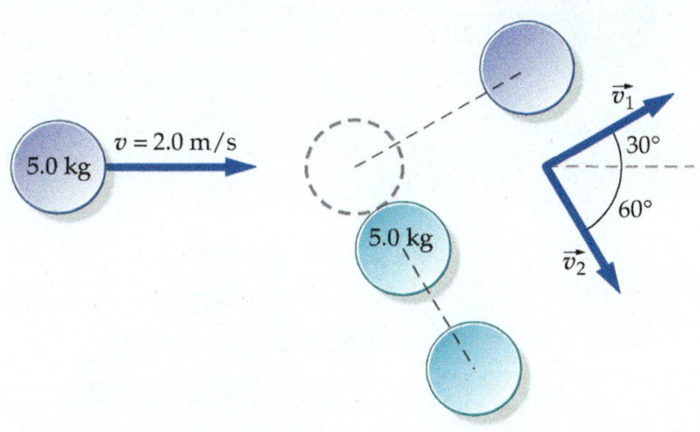

FIGURE 8-50 Problem 87

88 •• Figure 8-51 shows the result of a collision between two objects of unequal mass. (*a*) Find the speed v_2 of the larger mass after the collision; also find the angle θ_2. (*b*) Show that the collision is elastic.

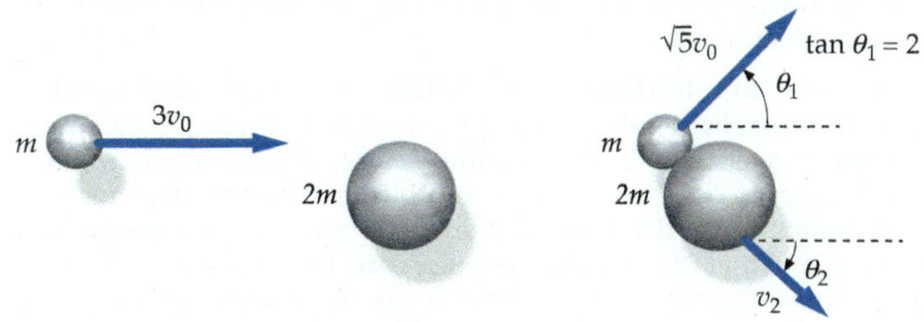

FIGURE 8-51 Problem 88

89 •• A 2.0-kg ball moving at 10 m/s makes an off-center collision with a 3.0-kg ball that is initially at rest. After the collision, the 2.0-kg ball is deflected at an angle of 30° from its original direction of motion and the 3.0-kg ball is moving at 4.0 m/s. Find the speed of the 2.0-kg ball and the direction of the 3.0-kg ball after the collision. *Hint:* $\sin^2\theta + \cos^2\theta = 1$.

90 •• A particle has initial speed v_0. It collides with a second particle with the same mass that is initially at rest. The first particle is deflected through an angle ϕ. Its speed after the collision is v. The second particle recoils, and its velocity makes an angle θ with the initial direction of the first particle. (*a*) Show that $\tan\theta = (v\sin\phi)/[(v_0 - v\cos\phi)]$. (*b*) Show that if the collision is elastic, then $v = v_0\cos\phi$.

* CENTER-OF-MASS REFERENCE FRAME

91 •• In the center-of-mass reference frame, a particle with mass m_1 and momentum p_1 makes an elastic head-on collision with a second particle of mass m_2 and momentum $p_2 = -p_1$. After the collision the first particle's momentum is p_1'. Write the total kinetic initial energy in terms of m_1, m_2, and p_1 and the total final energy in terms of m_1, m_2, and p_1', and show that $p_1' = \pm p_1$. If $p_1' = -p_1$, the particle is merely turned around by the collision and leaves with the speed it had initially. What is the situation for the $p_1' = +p_1$ solution?

92 •• **MULTISTEP** A 3.0-kg block is traveling in the $-x$ direction at 5.0 m/s, and a 1.0-kg block is traveling in the $+x$ direction at 3.0 m/s. (*a*) Find the velocity v_{cm} of the center of mass. (*b*) Subtract v_{cm} from the velocity of each block to find the velocity of each block in the center-of-mass reference frame. (*c*) After they make a head-on elastic collision, the velocity of each block is reversed (in the center-of-mass frame). Find the velocity of each block in the center-of-mass frame after the collision. (*d*) Transform back into the original frame by adding v_{cm} to the velocity of each block. (*e*) Check your result by finding the initial and final kinetic energies of the blocks in the original frame and comparing them.

93 •• Repeat Problem 92 with the second block having a mass of 5.0 kg and moving to the right at 3.0 m/s. SSM

* SYSTEMS WITH CONTINUOUSLY VARYING MASS: ROCKET PROPULSION

94 • **ENGINEERING APPLICATION** A rocket burns fuel at a rate of 200 kg/s and exhausts the gas at a speed of 6.00 km/s relative to the rocket. Find the magnitude of the thrust of the rocket.

95 •• **ENGINEERING APPLICATION** A rocket has an initial mass of 30,000 kg, 80 percent of which is the fuel. It burns fuel at a rate of 200 kg/s and exhausts its gas at a relative speed of 1.80 km/s. Find (*a*) the thrust of the rocket, (*b*) the time until burnout, and (*c*) the rocket's speed at burnout, assuming it moves straight upward near the surface of Earth. Assume g is constant and neglect any effects of air resistence.

96 •• **ENGINEERING APPLICATION** The *specific impulse* of a rocket propellant is defined as $I_{sp} = F_{th}/(Rg)$, where F_{th} is the thrust of the propellant, g is the magnitude of free-fall acceleration, and R is the rate at which the propellant is burned. The rate depends predominantly on the type and exact mixture of the propellant. (*a*) Show that the specific impulse has the dimension of time. (*b*) Show that $u_{ex} = gI_{sp}$, where u_{ex} is the relative speed of the exhaust. (*c*) What is the specific impulse (in seconds) of the propellant used in the Saturn V rocket of Example 8-19.

97 ••• **SPREADSHEET, ENGINEERING APPLICATION** The initial *thrust-to-weight ratio* τ_0 of a rocket is $\tau_0 = F_{th}/(m_0 g)$, where F_{th} is the rocket's thrust and m_0 the initial mass of the rocket, including the propellant. (*a*) For a rocket launched straight up from Earth's surface, show that $\tau_0 = 1 + (a_0/g)$, where a_0 is the initial acceleration of the rocket. For manned rocket flight, τ_0 cannot be made much larger than 4 for the comfort and safety of the astronauts. (As the rocket lifts off, the astronauts will feel that their weight is equal to τ_0 times their normal weight.) (*b*) Show that the final velocity of a rocket launched from Earth's surface can, in terms of τ_0 and I_{sp} (see Problem 96), be written as

$$v_f = gI_{sp}\left[\ln\left(\frac{m_0}{m_f}\right) - \frac{1}{\tau_0}\left(1 - \frac{m_f}{m_0}\right)\right]$$

where m_f is the mass of the rocket (not including the spent propellant). (*c*) Using a **spreadsheet** program or **graphing calculator,** graph v_f as a function of the mass ratio m_0/m_f for $I_{sp} = 250$ s and $\tau_0 = 2$ for values of the mass ratio from 2 to 10. (Note that the mass ratio cannot be less than 1.) (*d*) To lift a rocket into orbit, a final velocity after burnout of $v_f = 7.0$ km/s is needed. Calculate the mass ratio required of a single-stage rocket to do this, using the values of specific impulse and thrust ratio given in Part (*b*). For engineering reasons, it is difficult to make a rocket with a mass ratio much greater than 10. Can you see why multistage rockets are usually used to put payloads into orbit around Earth? SSM

98 •• **ENGINEERING APPLICATION** The height that a model rocket launched from Earth's surface can reach can be estimated by assuming that the burn time is short compared to the total flight time; the rocket is therefore in free-fall for most of the flight. (This estimate neglects the burn time in calculations of both time and displacement.) For a model rocket with specific impulse $I_{sp} = 100$ s, mass ratio $m_0/m_f = 1.20$, and initial thrust-to-weight ratio $\tau_0 = 5.00$ (these parameters are defined in Problems 96 and 97), estimate (*a*) the height the rocket can reach, and (*b*) the total flight time. (*c*) Justify the assumption used in the estimates by comparing the flight time from Part (*b*) to the time it takes to consume the fuel.

GENERAL PROBLEMS

99 • A 250-g model-train car traveling at 0.50 m/s links up with a 400-g car that is initially at rest. What is the speed of the cars immediately after they link up? Find the pre- and postcollision kinetic energies of the two-car system. SSM

100 • **MULTISTEP** A 250-g model-train car traveling at 0.50 m/s heads toward a 400-g car that is initially at rest. (*a*) Find the total kinetic energy of the two-car system. (*b*) Find the velocity of each car in the center-of-mass reference frame, and use these velocities to calculate the kinetic energy of the two-car system in the center-of-mass reference frame. (*c*) Find the kinetic energy associated with the motion of the center of mass of the system. (*d*) Compare your answer for Part (*a*) with the sum of your answers for Parts (*b*) and (*c*).

101 •• A 1500-kg car traveling north at 70 km/h collides at an intersection with a 2000-kg car traveling west at 55 km/h. The two cars stick together. (*a*) What is the total momentum of the system before the collision? (*b*) What are the magnitude and direction of the velocity of the wreckage just after the collision.

102 •• A 60-kg woman stands on the back of a 6.0-m-long, 120-kg raft that is floating at rest in still water. The raft is 0.50 m from a fixed pier, as shown in Figure 8-52. (*a*) The woman walks to the front of the raft and stops. How far is the raft from the pier now? (*b*) While the woman walks, she maintains a constant speed of 3.0 m/s relative to the raft. Find the total kinetic energy of the system (woman plus raft), and compare your answer with the kinetic energy if the woman walked at 3.0 m/s on a raft tied to the pier. (*c*) Where do these kinetic energies come from, and where do they go when the woman stops at the front of the raft? (*d*) On land, the woman puts a lead shot 6.0 m. Then, standing at the back of the raft, she aims forward, and puts the shot so that just after it leaves her hand, it has the same velocity relative to her as it did when she threw it from the ground. Approximately, where does her shot land?

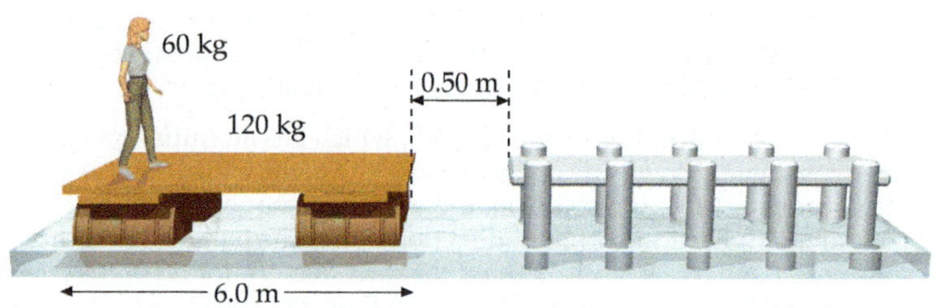

FIGURE 8-52 Problem 102

103 •• A 1.0-kg steel ball and a 2.0-m cord of negligible mass make up a simple pendulum that can pivot without friction about the point O, as in Figure 8-53. This pendulum is released from rest in a horizontal position, and when the ball is at its lowest point it strikes a 1.0-kg block sitting at rest on a shelf. Assume that the collision is perfectly elastic and that the coefficient of kinetic friction between the block and shelf is 0.10. (*a*) What is the velocity of the block just after impact? (*b*) How far does the block slide before coming to rest (assuming that the shelf is long enough)?

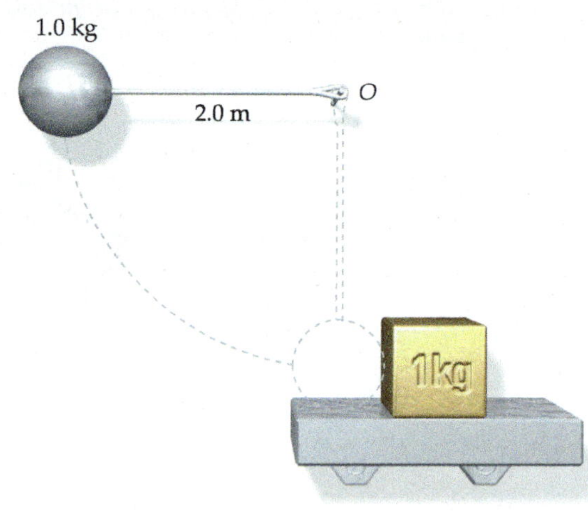

FIGURE 8-53 Problem 103

104 •• Figure 8-54 shows a World War I cannon mounted on a railcar and set so that it will project a shell at an angle of 30° above the horizontal. With the car initially at rest on a horizontal frictionless track, the cannon fires a 200-kg projectile at 125 m/s. (All values are for the frame of reference of the track.). (*a*) Will the vector momentum of the car–cannon–shell system be the same just before and just after the shell is fired? Explain your answer. (*b*) If the mass of the railcar plus cannon is 5000 kg, what will be the recoil velocity of the car along the track after the firing? (*c*) The shell is observed to rise to a maximum height of 180 m as it moves through its trajectory. At this point, its speed is 80.0 m/s. On the basis of this information, calculate the amount of thermal energy produced by air friction on the shell from the cannon's mouth to this maximum height.

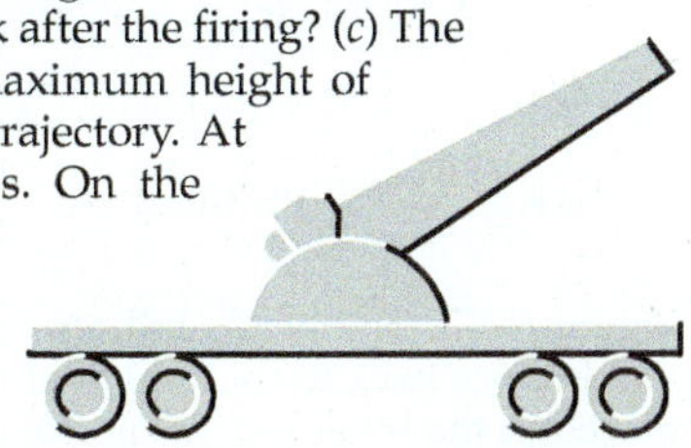

FIGURE 8-54 Problem 104

105 ••• **MULTISTEP** One popular, if dangerous, classroom demonstration involves holding a baseball an inch or so directly above a basketball that you are holding a few feet above a hard floor, and dropping the two balls simultaneously. The two balls will collide just after the basketball bounces from the floor; the baseball will then rocket off into the ceiling tiles, while the basketball will stop in midair. (*a*) Assuming that the collision of the basketball with the floor is elastic, what is the relation between the velocities of the balls just before they collide? (*b*) Assuming the collision between the two balls is elastic, use the result of Part (*a*) and the conservation of momentum and the conservation of energy to show that, if the basketball is three times as heavy as the baseball, the final velocity of the basketball will be zero. (This is approximately the true mass ratio, which is why the demonstration is so dramatic.) (*c*) If the speed of the baseball is v just before the collision, what is its speed just after the collision? SSM

106 ••• (*a*) In Problem 105, if we held a third ball above the baseball and basketball and wanted both the baseball and basketball to stop in midair, what should the ratio of the mass of the top ball to the mass of the baseball be? (*b*) If the speed of the top ball is v just before the collision, what is its speed just after the collision?

107 ••• In the "slingshot effect," the transfer of energy in an elastic collision is used to boost the energy of a space probe so that it can escape from the solar system. All speeds are relative to an inertial frame in which the center of the Sun remains at rest. Figure 8-55 shows a space probe moving at 10.4 km/s toward Saturn, which is moving at 9.6 km/s toward the probe. Because of the gravitational attraction between Saturn and the probe, the probe swings around Saturn and heads back in the opposite direction with speed v_f. (*a*) Assuming this collision to be a one-dimensional elastic collision with the mass of Saturn much much greater than that of the probe, find v_f. (*b*) By what factor is the kinetic energy of the probe increased? Where does this energy come from? SSM

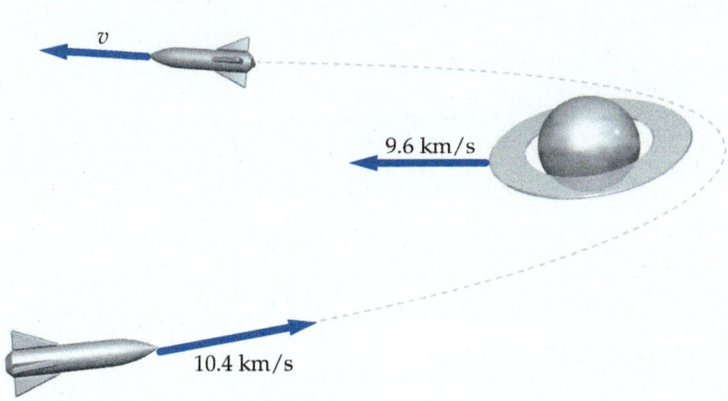

FIGURE 8-55 Problem 107

108 •• A 13-kg block is at rest on a level floor. A 400-g glob of putty is thrown at the block so that the putty travels horizontally, hits the block, and sticks to it. The block and putty slide 15 cm along the floor. If the coefficient of kinetic friction is 0.40, what is the initial speed of the putty?

109 ••• **CONTEXT-RICH** Your accident-reconstruction team has been hired by the local police to analyze the following accident. A careless driver rear-ended a car that was halted at a stop sign. Just before impact, the driver slammed on his brakes, locking the wheels. The driver of the struck car had his foot solidly on the brake pedal, locking his brakes. The mass of the struck car was 900 kg, and that of the initially moving vehicle was 1200 kg. On collision, the bumpers of the two cars meshed. Police determine from the skid marks that after the collision the two cars moved 0.76 m together. Tests revealed that the coefficient of kinetic friction between the tires and pavement was 0.92. The driver of the moving car claims that he was traveling at less than 15 km/h as he approached the intersection. Is he telling the truth? SSM

110 •• A pendulum consists of a compact 0.40-kg bob attached to a string of length 1.6 m. A block of mass m rests on a horizontal frictionless surface. The pendulum is released from rest at an angle of 53° with the vertical. The bob collides elastically with the block at the lowest point in its arc. Following the collision, the maximum angle of the pendulum with the vertical is 5.73°. Determine the mass m.

111 ••• A 1.00-kg block (mass m) and a second block (mass M) are both initially at rest on a frictionless inclined plane (Figure 8-56). Mass M rests against a spring that has a force constant of 11.0 kN/m. The distance along the plane between the two blocks is 4.00 m. The 1.00-kg block is released, making an elastic collision with the larger block. The 1.00-kg block then rebounds a distance of 2.56 m back up the inclined plane. The block of mass M momentarily comes to rest 4.00 cm from its initial position. Find M. SSM

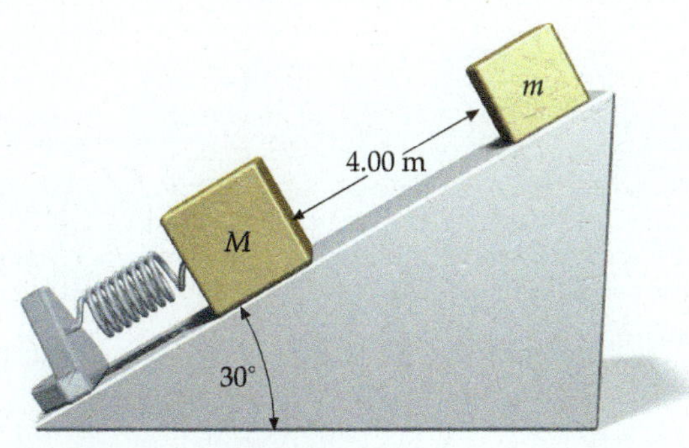

FIGURE 8-56 Problem 111

112 ••• A neutron of mass m makes an elastic head-on collision with a stationary nucleus of mass M. (*a*) Show that the kinetic energy of the nucleus after the collision is given by $K_{nucleus} = [4mM/(m + M)^2]K_n$, where K_n is the initial kinetic energy of the neutron. (*b*) Show that the *fractional* change in the kinetic energy of the neutron is given by

$$\frac{\Delta K_n}{K_n} = -\frac{4(m/M)}{(1 + [m/M])^2}$$

(*c*) Show that this expression gives plausible results both if $m \ll M$ and if $m = M$. What is the best stationary nucleus for the neutron to collide head-on with if the objective is to produce a maximum loss in the kinetic energy of the neutron?

113 ••• **ENGINEERING APPLICATION** The mass of a carbon nucleus is approximately 12 times the mass of a neutron. (*a*) Use the results of Problem 112 to show that after *N* head-on collisions of a neutron with carbon nuclei at rest, the kinetic energy of the neutron is approximately $0.716^N K_0$, where K_0 is its initial kinetic energy. (*b*) Neutrons emitted in the fission of a uranium nucleus have kinetic energies of about 2.0 MeV. For such a neutron to cause the fission of another uranium nucleus in a reactor, its kinetic energy must be reduced to about 0.020 eV. How many head-on collisions are needed to reduce the kinetic energy of a neutron from 2.0 MeV to 0.020 eV, assuming elastic head-on collisions with stationary carbon nuclei?

114 ••• **ENGINEERING APPLICATION** On average, a neutron actually loses only 63 percent of its energy in an elastic collision with a hydrogen atom (not 100 percent) and 11 percent of its energy in an elastic collision with a carbon atom (not 28 percent). (These numbers are an average over all types of collisions, not just head-on ones. Thus, the results are lower than the ones determined from analyses like that in Problem 112, because most collisions are not head-on.) Calculate the actual number of collisions, on average, needed to reduce the energy of a neutron from 2.0 MeV to 0.020 eV if the neutron collides with (*a*) stationary hydrogen atoms and (*b*) stationary carbon atoms.

115 ••• Two astronauts at rest face each other in space. One, who has mass m_1, throws a ball of mass m_b to the other, whose mass is m_2. The second astronuat catches the ball and throws it back to the first astronaut. Following each throw, the ball has a speed of v relative to the thrower. After each has made one throw and one catch, (*a*) how fast are the astronauts moving? (*b*) How much has the two-astronaut system's kinetic energy changed and where did this energy come from? SSM

116 ••• A stream of elastic glass beads, each with a mass of 0.50 g, comes out of a horizontal tube at a rate of 100 per second (see Figure 8-57). The beads fall a distance of 0.50 m to a balance pan and bounce back to their original height. How much mass must be placed in the other pan of the balance to keep the pointer at zero?

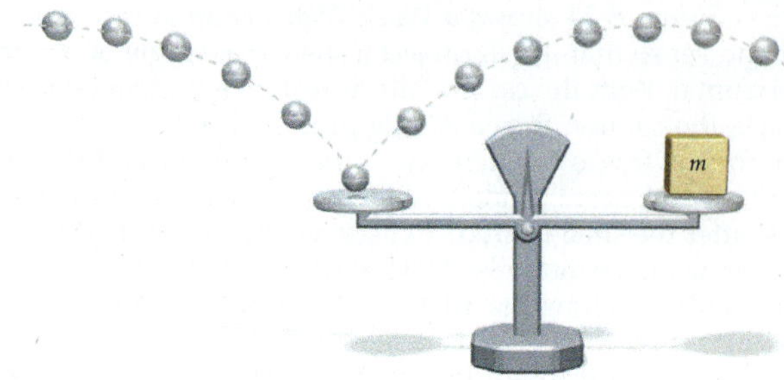

FIGURE 8-57 Problem 116

117 ••• A dumbbell consisting of two balls of mass m connected by a massless 1.00-m-long rod rests on a frictionless floor against a frictionless wall with one ball directly above the other. The center-to-center distance between the balls is equal to 1.00 m. The dumbbell then begins to slide down the wall, as in Figure 8-58. Find the speed of the bottom ball at the moment when it equals the speed of the top ball.

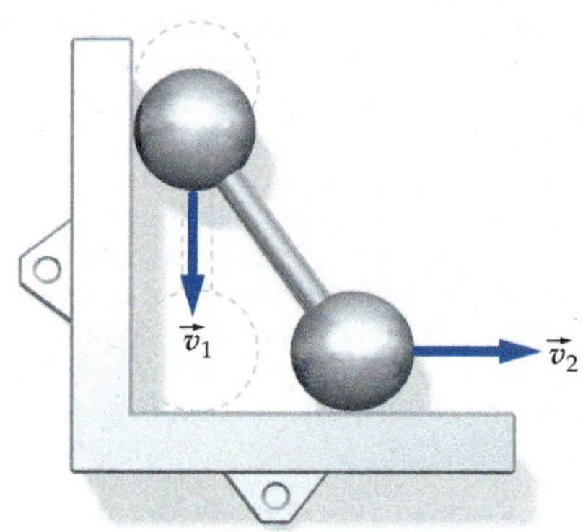

FIGURE 8-58 Problem 117

CHAPTER 9

THE LONDON EYE IS A 135-METER-HIGH OBSERVATION WHEEL THAT CARRIES A MAXIMUM OF 800 PASSENGERS.
(Ian Britton/FreeFoto.com.)

Rotation

? What is the torque required to stop the wheel so that the passengers travel a distance of 10 m as the wheel slows to a stop? (See Example 9-15.)

In Chapters 4 and 5, we explored Newton's laws. In Chapters 6 and 7, we examined the conservation of energy, and we studied the conservation of momentum in Chapter 8. We discovered tools (laws, theorems, and problem-solving techniques) that are useful in analyzing new situations and solving new problems in those chapters. We now continue to use those tools as we explore rotational motion.

Rotational motion is all around us. Earth rotates about its axis. Wheels, gears, propellers, motors, the drive shaft in a car, a CD in its player, a pirouetting ice skater, all rotate.

In this chapter, we consider rotation about an axis that is fixed in space, as in a merry-go-round, or about an axis that is moving without changing its direction in space, as in a rolling wheel on a car that is traveling in a straight line. The study of rotational motion is continued in Chapter 10 where more general examples of rotational motion are considered.

9-1 ROTATIONAL KINEMATICS: ANGULAR VELOCITY AND ANGULAR ACCELERATION

Every point of a rigid object rotating about a fixed axis moves in a circle whose center is on the axis and whose radius is the radial distance from the axis of rotation to that point. A radius drawn from the rotation axis to any point on the body sweeps out the same angle in the same time. Imagine a disk spinning about a fixed axis perpendicular to the disk and through its center (Figure 9-1). Let r_i be the distance from the center of the disk to the ith particle (Figure 9-2), and let θ_i be the angle measured counterclockwise from a fixed reference line in space to a radial line from the axis to the particle. As the disk rotates through an angle $d\theta$, the particle moves through a circular arc of directed length ds_i, such that

$$ds_i = r_i\, d\theta \qquad 9\text{-}1$$

where $d\theta$ is measured in radians. If counterclockwise is designated as the positive direction, then $d\theta$, θ_i, and ds_i, shown in Figure 9-2, are all positive. (If clockwise is designated the positive direction, they are all negative.) The angle θ_i, the directed length ds_i, and the distance r_i vary from particle to particle, but the ratio ds_i/r_i, called the **angular displacement** $d\theta$, is the same for all particles of the disk. For one complete revolution, the arc length s_i is $2\pi r_i$ and the angular displacement $\Delta\theta$ is

$$\Delta\theta = \frac{s_i}{r_i} = \frac{2\pi r_i}{r_i} = 2\pi \text{ rad} = 360° = 1 \text{ rev}$$

The time rate of change of the angle is the same for all particles of the disk, and is called the angular velocity ω of the disk. The instantaneous **angular velocity** ω is an angular displacement of short duration divided by the time. That is,

$$\omega = \frac{d\theta}{dt} \qquad 9\text{-}2$$

DEFINITION—ANGULAR VELOCITY

so ω is positive if $d\theta$ is positive and negative if $d\theta$ is negative. All points on the disk undergo the same angular displacement during the same time, so they all have the same angular velocity. The SI units of ω are rad/s. Because radians are dimensionless, the dimension of angular velocity is that of reciprocal time, T^{-1}. The magnitude of the angular velocity is called the **angular speed.** We often use revolutions per minute (rev/min or RPM) to specify the angular speed. To convert between revolutions, radians, and degrees, we use

$$1 \text{ rev} = 2\pi \text{ rad} = 360°$$

PRACTICE PROBLEM 9-1

A compact disk is rotating at 3000 rev/min. What is its angular speed in radians per second?

Angular acceleration is the rate of change of angular velocity. If the rotation rate of a rotating object increases, the angular speed $|\omega|$ increases. (If $|\omega|$ is increasing, and if the angular velocity ω is clockwise, then the change in the angular velocity $\Delta\omega$ is also clockwise.) The average angular acceleration vector

$$\alpha_{av} = \frac{\Delta\omega}{\Delta t} \qquad 9\text{-}3$$

DEFINITION—AVERAGE ANGULAR ACCELERATION

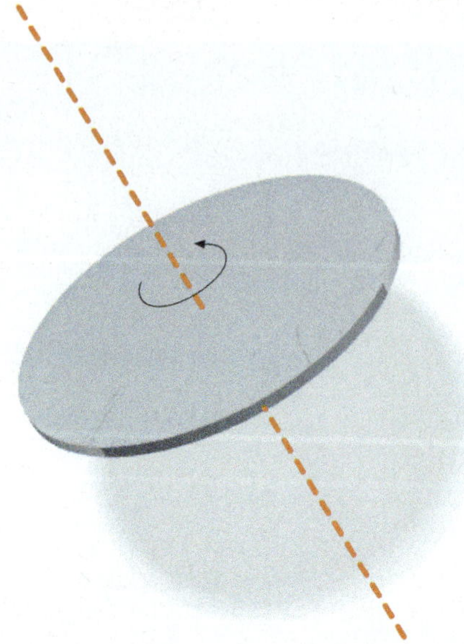

FIGURE 9-1

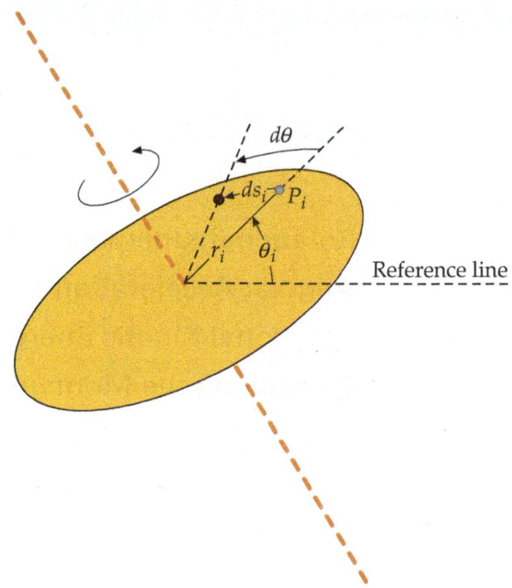

FIGURE 9-2

(Fred Habegger/Grant Heilman Photography, Inc.)

is always in the same direction as $\Delta\omega$. If the rotation rate decreases, then both $\Delta\omega$ and α_{av} are in the opposite direction to ω.

The instantaneous rate of change of angular velocity is called the **angular acceleration** α. That is,

$$\alpha = \frac{d\omega}{dt} = \frac{d^2\theta}{dt^2} \qquad \text{9-4}$$

DEFINITION—ANGULAR ACCELERATION

The SI units of α are rad/s^2. α is positive if ω is increasing, and α is negative if ω is decreasing.

The angular displacement θ, the angular velocity ω, and angular acceleration α are analogous to the linear displacement x, linear velocity v_x, and linear acceleration a_x in one-dimensional motion. If the angular acceleration α is constant, we can integrate both sides of $d\omega = \alpha\, dt$ (Equation 9-4) to obtain:

$$\omega = \omega_0 + \alpha t \qquad \text{9-5}$$

CONSTANT ANGULAR ACCELERATION

where the constant of integration ω_0 is the initial angular velocity. (Equation 9-5 is the rotational analog of the equation $v_x = v_{0x} + a_x t$.) Substituting $d\theta/dt$ in Equation 9-5, we obtain $d\theta = (\omega_0 + \alpha t)dt$. Integrating both sides of this equation gives

$$\theta = \theta_0 + \omega_0 t + \tfrac{1}{2}\alpha t^2 \qquad \text{9-6}$$

CONSTANT ANGULAR ACCELERATION

(which is the rotational analog of $x = x_0 + v_{0x}t + \frac{1}{2}a_x t^2$). By eliminating t from Equations 9-5 and 9-6, we get

$$\omega^2 = \omega_0^2 + 2\alpha(\theta - \theta_0) \qquad \text{9-7}$$

CONSTANT ANGULAR ACCELERATION

These constant-angular-acceleration kinematic equations have exactly the same form as the equations for constant linear acceleration developed in Chapter 2.

Star tracks in a time exposure of the night sky. *(David Malin/Anglo-Australian Telescope Board.)*

Example 9-1 A CD Player

A compact disk rotates from rest to 500 rev/min in 5.5 s. (*a*) What is its angular acceleration, assuming that it is constant? (*b*) How many revolutions does the disk make in 5.5 s? (*c*) How far does a point on the rim 6.0 cm from the center of the disk travel during the 5.5 s it takes to get to 500 rev/min?

PICTURE Part (*a*) is analogous to the one-dimensional linear problem of finding the acceleration, given the time and the final velocity. Part (*b*) is analogous to the one-dimensional linear problem of finding the displacement, given the time and the final velocity. Part (*c*), unlike Parts (*a*) and (*b*), involves both a linear quantity (distance traveled) and an angular quantity (angular displacement). Thus, Part (*c*) is not analogous to a one-dimensional angular problem.

SOLVE

(*a*) 1. The angular acceleration is related to the initial and final angular velocities:

$$\omega = \omega_0 + \alpha t = 0 + \alpha t$$

2. Solve for α:

$$\alpha = \frac{\omega}{t} = \frac{500\ \text{rev/min}}{5.5\ \text{s}} \times \frac{2\pi\ \text{rad}}{1\ \text{rev}} \times \frac{1\ \text{min}}{60\ \text{s}}$$

$$= 9.52\ \text{rad/s}^2 = \boxed{9.5\ \text{rad/s}^2}$$

(*b*) 1. The angular displacement is related to the time by Equation 9-6:

$$\theta - \theta_0 = \omega_0 t + \tfrac{1}{2}\alpha t^2 = 0 + \tfrac{1}{2}(9.52\ \text{rad/s}^2)(5.5\ \text{s})^2$$

$$= 144\ \text{rad}$$

2. Convert radians to revolutions:

$$144\ \text{rad} \times \frac{1\ \text{rev}}{2\pi\ \text{rad}} = 22.9\ \text{rev} = \boxed{23\ \text{rev}}$$

(*c*) The distance traveled Δs is r times the angular displacement in radians:

$$\Delta s = r\Delta\theta = (6.0\,\text{cm})(144\,\text{rad}) = 8.65\,\text{m} = \boxed{8.7\,\text{m}}$$

CHECK The average angular velocity is 250 rev/min. In 5.5 s, the compact disk rotates (250 rev/60 s)(5.5 s) = 23 rev.

TAKING IT FURTHER A compact disk is scanned by a laser that begins at the inner radius of 2.4 cm and moves to the outer radius of 6.0 cm. As the laser moves outward, the angular velocity of the disk decreases from 500 rev/min to 200 rev/min so that the linear (tangential) velocity of the disk at the point where the laser beam strikes remains constant.

PRACTICE PROBLEM 9-2 (*a*) Convert 500 rev/min to rad/s. (*b*) Check the result of Part (*b*) in the example using $\omega^2 = \omega_0^2 + 2\alpha(0 - \theta_0)$.

The linear velocity v_t of a particle on the disk is tangent to the circular path of the particle and has magnitude ds_i/dt. We can relate this tangential velocity to the angular velocity of the disk using Equations 9-1 and 9-2:

$$v_t = \frac{ds_i}{dt} = \frac{r_i d\theta}{dt} = r_i \frac{d\theta}{dt}$$

so

$$v_t = r_i\omega \qquad 9\text{-}8$$

Similarly, the tangential acceleration of a particle on the disk is dv_t/dt:

$$a_t = \frac{dv_t}{dt} = r_i\frac{d\omega}{dt}$$

so

$$a_t = r\alpha \qquad 9\text{-}9$$

Each particle of the disk also has a centripetal acceleration, which points inward along the radial line and has the magnitude

$$a_c = \frac{v_t^2}{r_i} = \frac{(r_i\omega)^2}{r_i}$$

so

$$a_c = r_i\omega^2 \qquad 9\text{-}10$$

! Equations that contain both linear and angular parameters, such as Equations 9-1, 9-8, 9-9, and 9-10, are valid only if the angle values are expressed in radians.

PRACTICE PROBLEM 9-3

A point on the rim of a compact disk is 6.00 cm from the axis of rotation. Find the tangential speed v_t, tangential acceleration a_t, and centripetal acceleration a_c of the point when the disk is rotating at a constant angular speed of 300 rev/min.

PRACTICE PROBLEM 9-4

Find the linear speed of a point on the CD in Example 9-1 at (*a*) $r = 2.40$ cm, when the disk rotates at 500 rev/min, and (*b*) $r = 6.00$ cm, when the disk rotates at 200 rev/min.

9-2 ROTATIONAL KINETIC ENERGY

The kinetic energy of a rigid object rotating about a fixed axis is the sum of the kinetic energies of the individual particles that collectively constitute the object. The kinetic energy of the ith particle, with mass m_i, is

$$K = \tfrac{1}{2}m_i v_i^2$$

Summing over all the particles and using $v_i = r_i\omega$ gives

$$K = \sum_i \left(\tfrac{1}{2} m_i v_i^2\right) = \tfrac{1}{2} \sum (m_i r_i^2 \omega^2) = \tfrac{1}{2}\left(\sum_i m_i r_i^2\right)\omega^2$$

The sum in the expression farthest to the right is the object's **moment of inertia** I for the axis of rotation.

$$I = \sum_i m_i r_i^2 \qquad \text{9-11}$$

MOMENT OF INERTIA DEFINED

The kinetic energy is thus

$$K = \tfrac{1}{2} I\omega^2 \qquad \text{9-12}$$

KINETIC ENERGY OF ROTATING OBJECT

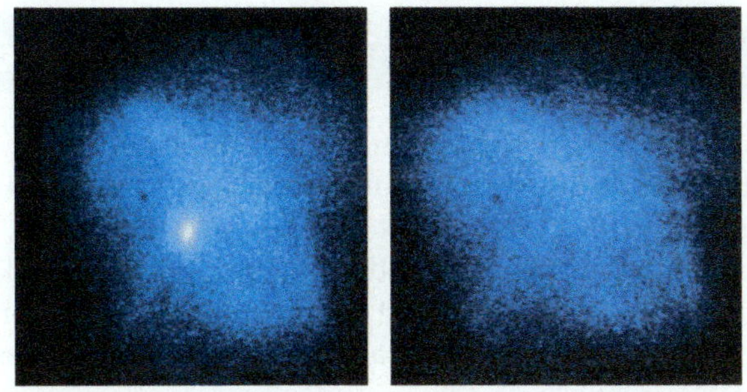

The Crab Pulsar is one of the fastest-rotating neutron stars known, but it is slowing down. It appears to blink on (left) and off (right) like the rotating lamp in a lighthouse, at the fast rate of about 30 times per second, but the period is increasing by about 10^{-5} s/y. This rate of loss in rotational energy is equivalent to the power output of 100,000 suns. The lost kinetic energy appears as light emitted by electrons accelerated in the magnetic field of the pulsar. *(David Malin/ Anglo-Australian Telescope Board.)*

Example 9-2 A Rotating System of Particles

An object consists of four point particles, each of mass m, connected by rigid massless rods to form a rectangle of edge lengths $2a$ and $2b$, as shown in Figure 9-3. The system rotates with angular speed ω about an axis in the plane of the figure through the center, as shown. (*a*) Find the kinetic energy of this object using Equations 9-11 and 9-12. (*b*) Check your result by individually calculating the kinetic energy of each particle and then taking their sum.

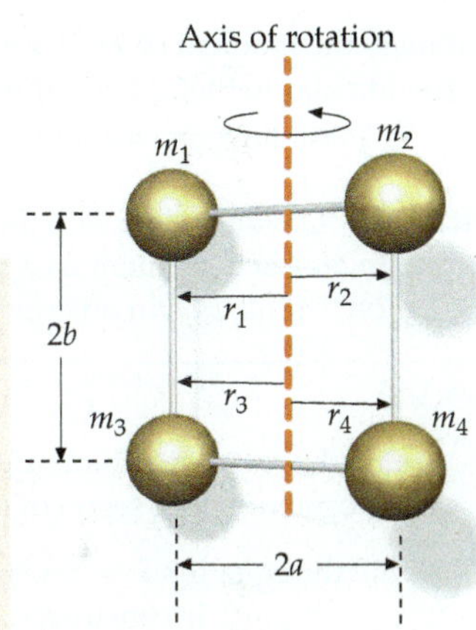

FIGURE 9-3

PICTURE Because we are given that the objects are point particles, we use Equation 9-11 to calculate I and then use Equation 9-12 to calculate K.

SOLVE

(*a*) 1. Apply the definition of moment of inertia $I = \Sigma m_i r_i^2$ (Equation 9-11), where r_i is the radial distance from the rotation axis to the particle of mass m_i:

$$I = \sum_i m_i r_i^2 = m_1 r_1^2 + m_2 r_2^2 + m_3 r_3^2 + m_4 r_4^2$$

2. The masses m_i and the distances r_i are given:

$$m_1 = m_2 = m_3 = m_4 = m$$
$$r_1 = r_2 = r_3 = r_4 = a$$

3. Substitution gives the moment of inertia:

$$I = ma^2 + ma^2 + ma^2 + ma^2 = 4ma^2$$

4. Using Equation 9-12, solve for the kinetic energy:

$$K = \tfrac{1}{2} I\omega^2 = \tfrac{1}{2} 4ma^2\omega^2 = \boxed{2ma^2\omega^2}$$

(*b*) 1. To find the kinetic energy of the *i*th particle, we must first find its speed:

$$K_i = \tfrac{1}{2} m_i v_i^2$$

2. The particles are all moving in circles of radius a. Find the speed of each particle:

$$v_i = r_i\omega = a\omega \quad (i = 1, \ldots, 4)$$

3. Substitute into the Part-(*b*), step-1 result:

$$K_i = \tfrac{1}{2} m_i v_i^2 = \tfrac{1}{2} ma^2\omega^2$$

4. Each particle has the same kinetic energy. Sum the kinetic energies to get the total:

$$K = \sum_{i=1}^{4} K_i = \tfrac{1}{2} m_1 v_1^2 + \tfrac{1}{2} m_2 v_2^2 + \tfrac{1}{2} m_3 v_3^2 + \tfrac{1}{2} m_4 v_4^2$$
$$= 4\left(\tfrac{1}{2} ma^2\omega^2\right) = 2ma^2\omega^2$$

5. Compare with the Part-(*a*) result:

The two calculations give the same result.

CHECK The fact that the two calculations give the same result is a plausibility check.

TAKING IT FURTHER Notice that I is independent of the length b. The moment of inertia depends only upon how far from the axis the masses are, and not where along the axis they are.

PRACTICE PROBLEM 9-5 Find the moment of inertia of this system for rotation about an axis parallel to the first axis but passing through two of the particles, as shown in Figure 9-4.

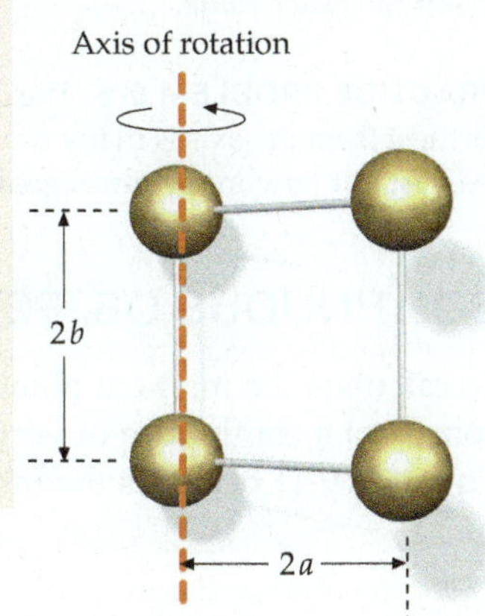

FIGURE 9-4

9-3 CALCULATING THE MOMENT OF INERTIA

The moment of inertia about an axis is a measure of the inertial resistance of the object to changes in its rotational motion about the axis. It is the rotational analog of mass. The farther an element of mass is from the axis, the greater its contribution to the moment of inertia about that axis. Thus, unlike the mass of an object, which is a property of the object itself, the moment of inertia depends on the location of the axis as well as the mass distribution of the object.

! The moment of inertia of an object about an axis depends both on the mass and on the distribution of the mass relative to the axis.

SYSTEMS OF DISCRETE PARTICLES

For systems consisting of discrete particles, we can compute the moment of inertia about a given axis directly from Equation 9-11. We can also use Equation 9-10 to obtain approximate values for the moment of inertia, as is done in the following example.

Example 9-3 Estimating the Moment of Inertia

Estimate the moment of inertia of a thin uniform rod of length L and mass M about an axis perpendicular to the rod and through one end. Execute this estimation by modeling the rod as three point masses, each point mass representing one-third of the rod.

PICTURE Divide the rod into three identical segments, each of mass $\frac{1}{3}M$ and length $\frac{1}{3}L$, and approximate each segment as a point mass located at its center of mass. Apply $I = \Sigma m_i r_i^2$ (Equation 9-11) to obtain an approximate value for I.

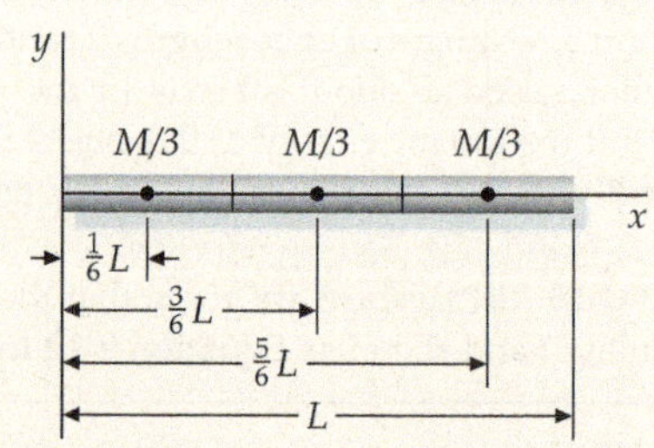

FIGURE 9-5

SOLVE

1. Sketch the rod divided into three segments and superpose the point-particle constructs at the center of each segment (Figure 9-5):
2. Apply the equation $I = \Sigma m_i r_i^2$ to the approximate system (the point-particle constructs): $I = \Sigma m_i r_i^2 \approx m_1 r_1^2 + m_2 r_2^2 + m_3 r_3^2$
3. The mass of each particle is $\frac{1}{3}M$, and the distances of the particles from the axis are $\frac{1}{6}L$, $\frac{3}{6}L$, and $\frac{5}{6}L$: $I \approx (\frac{1}{3}M)(\frac{1}{6}L)^2 + (\frac{1}{3}M)(\frac{3}{6}L)^2 + (\frac{1}{3}M)(\frac{5}{6}L)^2$

$$= \frac{1}{3}M\left(\frac{1 + 3^2 + 5^2}{6^2}\right)L^2 = \boxed{\frac{35}{108}ML^2}$$

CHECK The exact value for the moment of inertia of the rod about this axis is $\frac{1}{3}ML^2$. (The exact value is calculated in Example 9-4.) One-third is equal to 36/108, so our result is within 1% of the exact value.

PRACTICE PROBLEM 9-6 The contribution to the moment of inertia of the third of the rod farthest from the axis is many times greater than is the contribution of the third closest to the axis. About how many times greater is it?

CONTINUOUS OBJECTS

To calculate the moment of inertia for continuous objects, we imagine the object to consist of a continuum of very small mass elements. Thus, the finite sum $\Sigma m_i r_i^2$ in Equation 9-11 becomes the integral

$$I = \int r^2\,dm \qquad 9\text{-}13$$

where r is the radial distance from the axis to mass element dm. To evaluate this integral, we first express dm as a density times an element of length, area, or volume, as is done in the following examples.

Example 9-4 Moment of Inertia of a Thin Uniform Rod

Find the moment of inertia of a thin uniform rod of length L and mass M about an axis perpendicular to the rod and through one end.

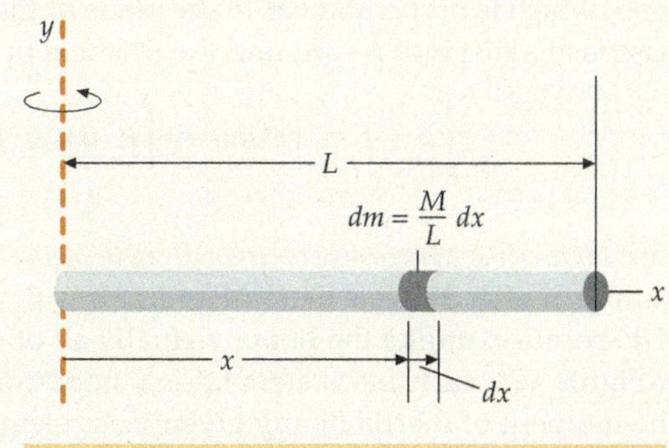

FIGURE 9-6

PICTURE Use $I = \int r^2 dm$ (Equation 9-13) to calculate the moment of inertia about the specified axis. The rod is uniform, which means that for any segment of the rod, the mass per unit length λ is equal to M/L.

SOLVE

1. Draw a sketch (Figure 9-6) showing the rod along the $+x$ axis with its end at the origin. To calculate I about the y axis, we choose a mass element dm at a distance x from the axis:
2. The moment of inertia about the y axis is given by the integral: $I = \int x^2\, dm$
3. To compute the integral, we first relate dm to dx. Express dm in terms of the linear mass density λ and dx: $dm = \lambda\, dx = \frac{M}{L}dx$
4. Substitute and perform the integration. We choose integration limits so that we integrate through the mass distribution in the direction of increasing x:

$$I = \int x^2\, dm = \int_0^L x^2 \frac{M}{L}dx = \frac{M}{L}\int_0^L x^2\, dx$$

$$= \frac{M}{L}\frac{1}{3}x^3\Big|_0^L = \frac{M}{L}\frac{L^3}{3} = \boxed{\frac{1}{3}ML^2}$$

CHECK This result is in good agreement with the approximate result obtained in Example 9-3.

TAKING IT FURTHER The moment of inertia about the z axis is also $\frac{1}{3}ML^2$, and that about the x axis is zero (assuming that all of the mass is a negligible distance from the x axis).

We can calculate I for continuous objects of various shapes, again using Equation 9-13 (Table 9-1). Some of these calculations are done here.

Table 9-1 Moments of Inertia of Uniform Bodies of Various Shapes

Thin cylindrical shell about axis 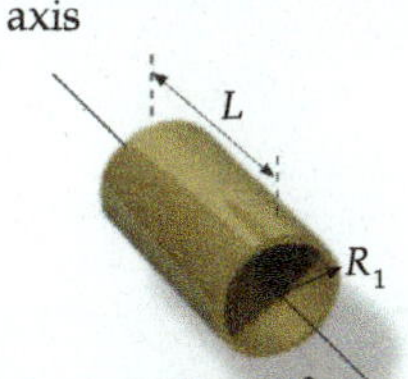 $I = MR^2$	Thin cylindrical shell about diameter through center  $I = \frac{1}{2}MR^2 + \frac{1}{12}ML^2$	Thin rod about perpendicular line through center 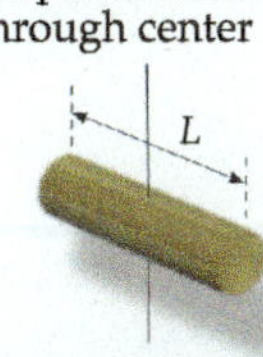 $I = \frac{1}{12}ML^2$	Thin spherical shell about diameter 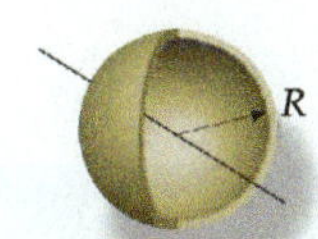 $I = \frac{2}{3}MR^2$
Solid cylinder about axis 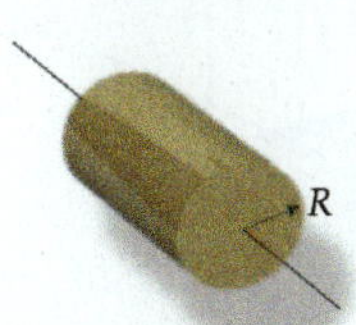 $I = \frac{1}{2}MR^2$	Solid cylinder about diameter through center 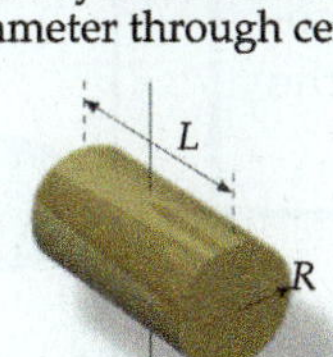 $I = \frac{1}{4}MR^2 + \frac{1}{12}ML^2$	Thin rod about perpendicular line through one end 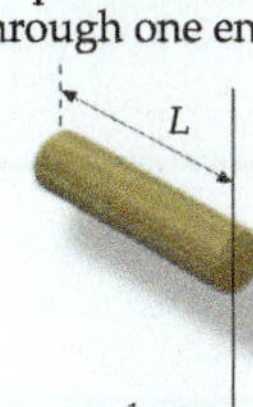 $I = \frac{1}{3}ML^2$	Solid sphere about diameter 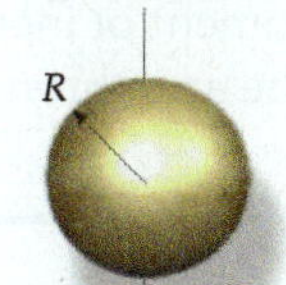 $I = \frac{2}{5}MR^2$
Hollow cylinder about axis 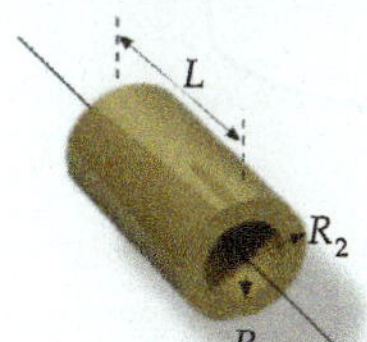 $I = \frac{1}{2}M(R_1^2 + R_2^2)$	Hollow cylinder about diameter through center 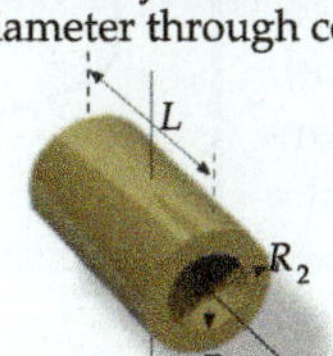 $I = \frac{1}{4}M(R_1^2 + R_2^2) + \frac{1}{12}ML^2$		Solid rectangular parallelepiped about axis through center perpendicular to face 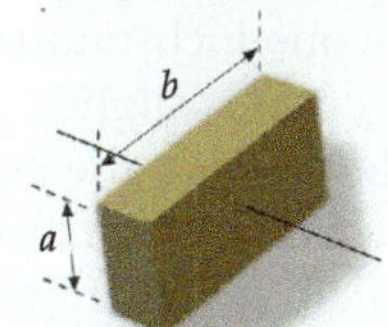 $I = \frac{1}{12}M(a^2 + b^2)$

*A disk is a cylinder whose length L is negligible. By setting $L = 0$, the above formulas for cylinders hold for disks.

* **Hoop about a perpendicular axis through its center** Assume that a hoop has mass M and radius R (Figure 9-7). The axis of rotation is the symmetry axis of the hoop, which is perpendicular to the plane of the hoop and through its center. All the mass is at a distance $r = R$, and the moment of inertia is

$$I = \int r^2\,dm = \int R^2\,dm = R^2 \int dm = MR^2$$

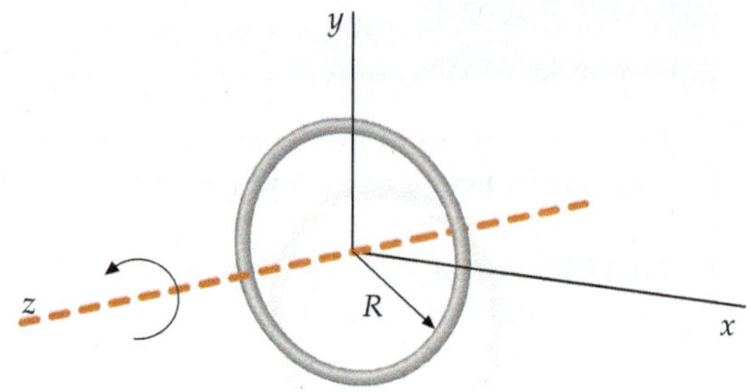

FIGURE 9-7

* **Uniform disk about a perpendicular axis through its center** For the case of a uniform disk that has mass M and radius R, we expect that I will be smaller than MR^2 because, unlike the hoop, virtually all of the mass is closer to the axis than R. In Figure 9-8, each mass element is a hoop (or ring) of radius r and thickness dr. The moment of inertia of any given mass element is $r^2 dm$. Because the disk is uniform, the mass per unit area σ is the same for any piece of it, so $\sigma = M/A$, where $A = \pi R^2$ is the area of the disk. Because the area of each hoop-shaped mass element is $dA = 2\pi r\,dr$, the mass of each element is

$$dm = \sigma dA = \frac{M}{A} 2\pi r\,dr$$

We thus have

$$I = \int r^2\,dm = \int_0^R r^2 \sigma\, 2\pi r\,dr = 2\pi\sigma \int_0^R r^3\,dr = 2\pi \frac{M}{A} \frac{r^4}{4}\Big|_0^R$$

$$= \frac{2\pi M}{A}\frac{R^4}{4} = \frac{\pi M}{2\pi R^2} R^4 = \frac{1}{2} MR^2$$

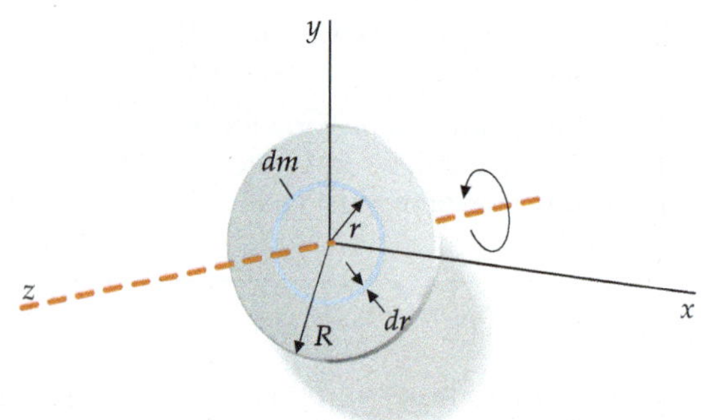

FIGURE 9-8

* **Uniform solid cylinder about its axis** We consider a cylinder to be a set of disks, each with mass dm and moment of inertia $dI = \frac{1}{2}(dm)R^2$ (Figure 9-9). The moment of inertia of the complete cylinder is then

$$I = \int \frac{1}{2} dm\,R^2 = \frac{1}{2} R^2 \int dm = \frac{1}{2} MR^2$$

where M is the total mass of the cylinder.

CONCEPT CHECK 9-1

Consider two identical uniform one-inch-diameter disks A and B. You drill a quarter-inch-diameter hole through the center of disk B. Which disk, A or B, then has the greater moment of inertia? (For each disk, consider only the moment of inertia about the axis through the center of, and perpendicular to, the disk.)

CONCEPT CHECK 9-2

Consider two uniform one-inch-radius disks A and B. The disks are identical, except that the density of B is slightly greater than that of A. You drill a quarter-inch-diameter hole through the center of disk B and find that the disks now have identical masses. Which disk, A or B, then has the greater moment of inertia? (For each disk, consider only the moment of inertia about the axis through the center of, and perpendicular to, the disk.)

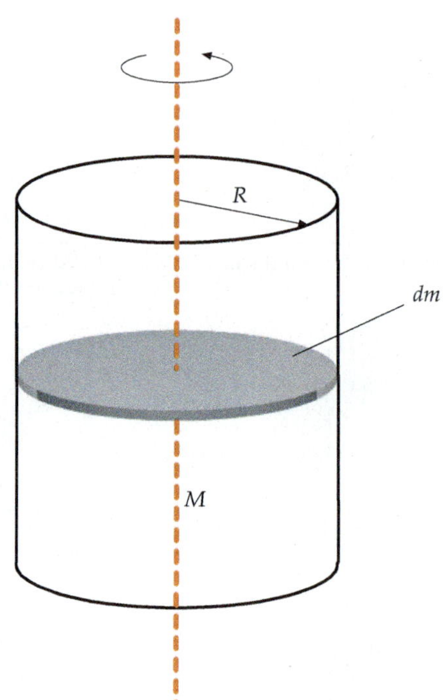

FIGURE 9-9

THE PARALLEL-AXIS THEOREM

We can often simplify the calculation of moments of inertia for various objects by using the **parallel-axis theorem,** which relates the moment of inertia about an axis through the center of mass to the moment of inertia about a second, parallel axis (Figure 9-10). Let I be the moment of inertia, and let I_{cm} be the moment of inertia about a parallel axis through the center of mass. In addition, let M be the total mass of the object and let h be the distance between the two axes. The parallel-axis theorem states that

$$I = I_{cm} + Mh^2 \tag{9-14}$$

PARALLEL-AXIS THEOREM

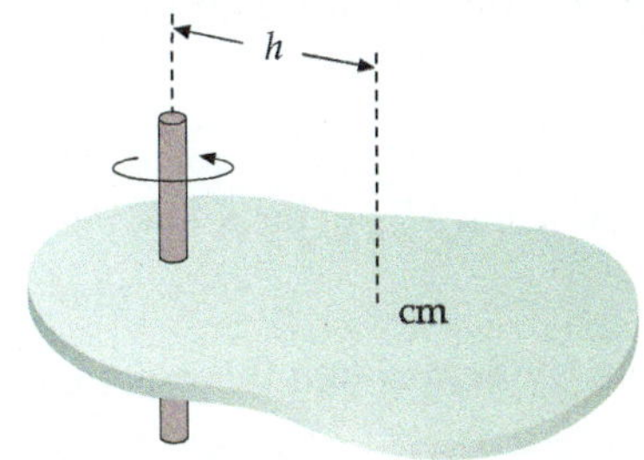

FIGURE 9-10 An object rotating about an axis parallel to an axis through the center of mass and a distance h from it.

Example 9-2 and the Practice Problem following it illustrate a special case of this theorem with $h = a, M = 4m$, and $I_{cm} = 4ma^2$.

Example 9-5 Applying the Parallel-Axis Theorem — *Try It Yourself*

A thin uniform rod of mass M and length L on the x axis (Figure 9-11) has one end at the origin. Using the parallel-axis theorem, find the moment of inertia about the y' axis, which is parallel to the y axis, and through the center of the rod.

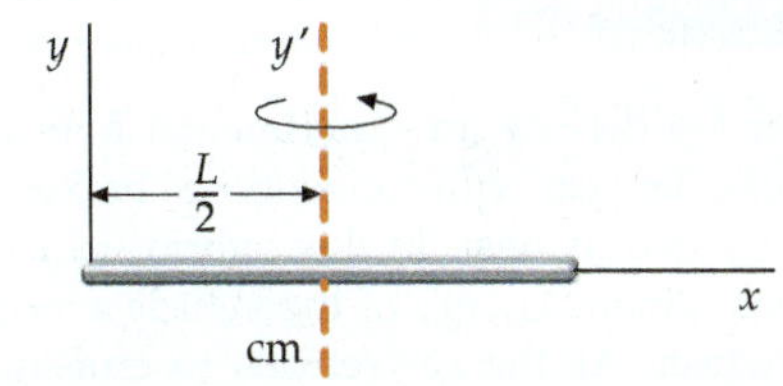

FIGURE 9-11

PICTURE Here you know that $I = \frac{1}{3}ML^2$ about one end (see Example 9-4) and want to find I_{cm}. Use the parallel-axis theorem with $h = \frac{1}{2}L$.

SOLVE

Cover the column to the right and try these on your own before looking at the answers.

Steps	Answers
1. Apply the parallel-axis theorem to write I about the end in terms of I_{cm}.	$I = I_{cm} + Mh^2$ $I_y = I_{y'} + M(\frac{1}{2}L)^2$
2. Substitute, using $\frac{1}{3}ML^2$ for I_y, I_{cm} for $I_{y'}$, and solve for I_{cm}.	$I_{cm} = I_y - Mh^2 = \frac{1}{3}ML^2 - M(\frac{1}{2}L)^2 = \boxed{\frac{1}{12}ML^2}$

CHECK Calculate the moment of inertia by direct integration. This calculation is the same as the calculation in Example 9-4 except that the integration limits are from $-\frac{1}{2}L$ to $+\frac{1}{2}L$. The result is

$$I = \int x^2\,dm = \frac{M}{L}\int_{-L/2}^{+L/2} x^2\,dx = \frac{M}{L}\frac{1}{3}x^3\Big|_{-L/2}^{+L/2} = \frac{M}{3L}\left(\frac{L^3}{8} + \frac{L^3}{8}\right) = \frac{1}{12}ML^2$$

which is the same as the step-2 result.

TAKING IT FURTHER The step-2 result is only 25% of the result gotten in Example 9-4, where the uniform rod is rotated about an axis through one end.

* PROOF OF THE PARALLEL-AXIS THEOREM

To prove the parallel-axis theorem, we start with an object (Figure 9-12) that is rotating about a fixed axis, one that does not pass through the center of mass. The kinetic energy K of such an object is given by $\frac{1}{2}I\omega^2$ (Equation 9-12), where I is the moment of inertia about the fixed axis. We saw in Chapter 8 (Equation 8-7) that the kinetic energy of a system can be written as the sum of its translational kinetic energy ($\frac{1}{2}Mv_{cm}^2$) and the kinetic energy relative to the center of mass. For an object that is rotating, the kinetic energy relative to its center of mass is $\frac{1}{2}I_{cm}\omega^2$, where I_{cm}

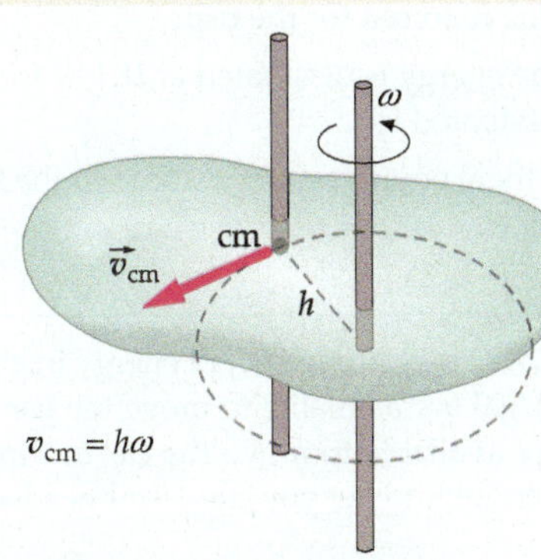

FIGURE 9-12

is the moment of inertia about the axis through the center of mass.* Thus, the total kinetic energy of the object is

$$K = \frac{1}{2}Mv_{cm}^2 + \frac{1}{2}I_{cm}\omega^2$$

The center of mass moves along a circular path of radius h, so $v_{cm} = h\omega$ Substituting $\frac{1}{2}I_{cm}\omega^2$ for K and $h\omega$ for v_{cm} gives

$$\frac{1}{2}I\omega^2 = \frac{1}{2}Mh^2\omega^2 + \frac{1}{2}I_{cm}\omega^2$$

Multiplying through this equation by $2/\omega^2$ gives

$$I = Mh^2 + I_{cm}$$

which completes the proof of the parallel-axis theorem.

PRACTICE PROBLEM 9-7

Using the parallel-axis theorem, show that when comparing the moments of inertia of an object about two parallel axes, the moment of inertia is less about the axis that is nearer to the center of mass.

Example 9-6 A Flywheel-Powered Car *Context-Rich*

You are driving an experimental hybrid vehicle that is designed for use in stop-and-go traffic. In a car with conventional brakes, each time you brake to a stop the kinetic energy is dissipated as heat. In this hybrid vehicle, the braking mechanism transforms the translational kinetic energy of the vehicle's motion into the rotational kinetic energy of a massive flywheel. As the car returns to cruising speed, this energy is transferred back into the translational kinetic energy of the car. The 100-kg flywheel is a hollow cylinder with an inner radius R_1 of 25.0 cm, an outer radius R_2 of 40.0 cm, and a maximum angular speed of 30,000 rev/min. On a dark and dreary night, the car runs out of gas 15.0 mi from home with the flywheel spinning at maximum speed. Is there sufficient energy stored in the flywheel for you and your nervous grandmother to make it home? (When driving at the minimum highway speed of 40.0 mi/h, air drag and rolling friction dissipate energy at 10.0 kW.)

PICTURE The kinetic energy is calculated directly from $K = \frac{1}{2}I\omega^2$.

SOLVE

1. The kinetic energy of rotation is	$K = \frac{1}{2}I\omega^2$
2. Calculate the moment of inertia of the hollow cylinder using an expression from Table 9-1:	$I = \frac{1}{2}m(R_1^2 + R_2^2) = 11.1\ \text{kg}\cdot\text{m}^2$
3. Convert ω to rad/s:	$\omega = 30{,}000\ \text{rev/min} = 3142\ \text{rad/s}$
4. Substitute these values to find the kinetic energy:	$K = \frac{1}{2}I\omega^2 = 54.9\ \text{MJ}$
5. Energy is dissipated at 10 kW at a speed of 40 mi/h. To find the energy dissipated during the 15-mi trip, we first need to find the time required for the trip:	$\Delta x = v\,\Delta t$, so $\Delta t = 1350\ \text{s}$
6. The energy is dissipated at 10 kW for 1350 s. The total energy dissipated is	13.5 MJ
7. Is there enough energy stored in the flywheel?	54.9 MJ are available and 13.5 MJ are dissipated. Yes, there is more than enough energy stored in the flywheel.

CHECK There are 130 MJ of energy in a gallon of gasoline. If the engine is 10% efficient, only 13 MJ/gal are available to move the car. The initial energy in the flywheel is more than the energy available to move the car in three gallons of gasoline. This energy should be more than enough to get you the 15 mi needed to get home.

* *Relative to its center of mass* means "relative to an inertial reference frame in which the center of mass is at least momentarily at rest."

Example 9-7 The Pivoted Rod

A uniform thin rod of length L and mass M, pivoted at one end as shown in Figure 9-13, is held horizontal and then released from rest. Assuming that effects due to friction and air resistance are negligible, find (*a*) the angular speed of the rod as it sweeps through the vertical position, and (*b*) the force exerted on the rod by the pivot at this instant. (*c*) What initial angular speed would be needed for the rod to just reach a vertical position at the top of its swing?

PICTURE We choose everything shown in Figure 9-13 plus Earth as the system. (*a*) As the rod swings down, the potential energy decreases and the kinetic energy increases. Because the pivot is frictionless, mechanical energy remains constant. The angular speed of the rod is then found from its rotational kinetic energy. (*b*) To find the force of the pivot we apply Newton's second law for a system to the rod. (*c*) As in Part (*a*), mechanical energy remains constant.

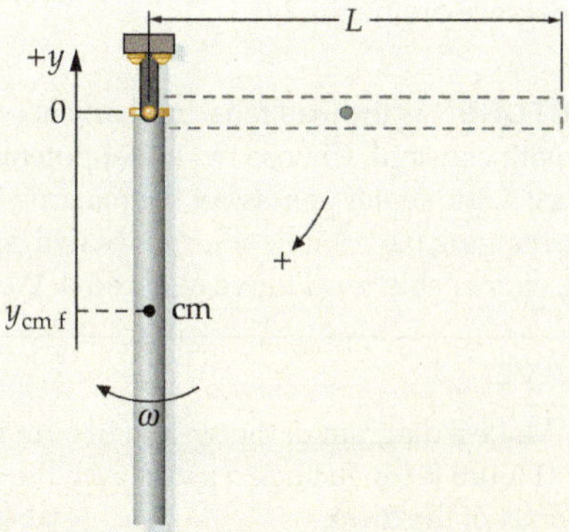

FIGURE 9-13

SOLVE

(*a*) 1. The diagram of the rod (Figure 9-13) shows both the initial and final configurations of the rod-Earth system. The origin of the y axis is at the same height as the rotation axis.

2. Apply conservation of mechanical energy to relate the initial and final mechanical energies:

$$K_f + U_f = K_i + U_i$$

$$\tfrac{1}{2}I\omega_f^2 + Mgy_{\text{cm f}} = \tfrac{1}{2}I\omega_i^2 + Mgy_{\text{cm i}}$$

$$\frac{1}{2}I\omega_f^2 + Mg\left(-\frac{L}{2}\right) = 0 + 0$$

3. Solve for ω_f.

$$\omega_f = \sqrt{\frac{MgL}{I}}$$

4. Obtain I from Table 9-1 and substitute into the step-3 result:

$$I = \tfrac{1}{3}ML^2 \quad \text{so} \quad \omega_f = \sqrt{\frac{MgL}{\tfrac{1}{3}ML^2}} = \boxed{\sqrt{\frac{3g}{L}}}$$

(*b*) 1. Draw a free-body diagram of the rod as it sweeps through the vertical position at the bottom of its swing (Figure 9-14).

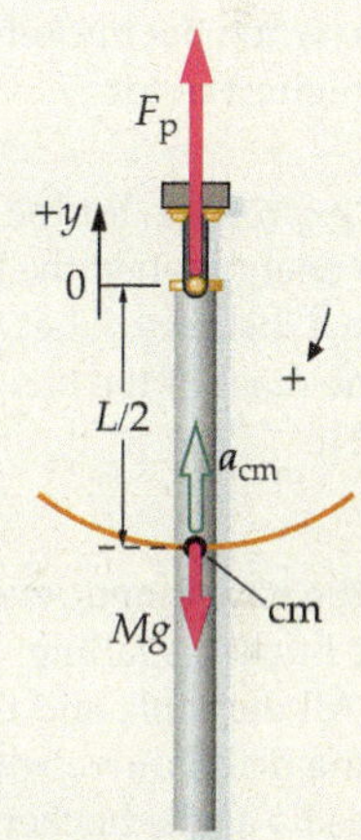

FIGURE 9-14

2. Apply Newton's second law for a system to the rod. At the bottom of the swing the acceleration of the center of mass is in the centripetal (upward) direction:

$$\Sigma F_{\text{ext}y} = Ma_{\text{cm}}$$

$$F_p - Mg = Ma_{\text{cm}}$$

3. Relate the acceleration of the center of mass to the angular speed using $a_c = r\omega^2$. Substitute the Part-(*a*), step-4 result for ω and solve for a_{cm}:

$$a_{\text{cm}} = r\omega^2$$

$$a_{\text{cm}} = \frac{L}{2}\frac{3g}{L} = \frac{3}{2}g$$

4. Substitute into the Part-(*b*), step-2 result and calculate F_p:

$$F_p = Mg + Ma_{\text{cm}} = Mg + M\tfrac{3}{2}g = \boxed{\tfrac{5}{2}Mg}$$

(*c*) 1. The initial angular velocity ω_i is related to the initial kinetic energy:

$$K_i = \tfrac{1}{2}I\omega_i^2$$

2. Make a diagram of the rod showing both the initial and final configuration (Figure 9-15). Include the same coordinate axis as in Part (*a*):

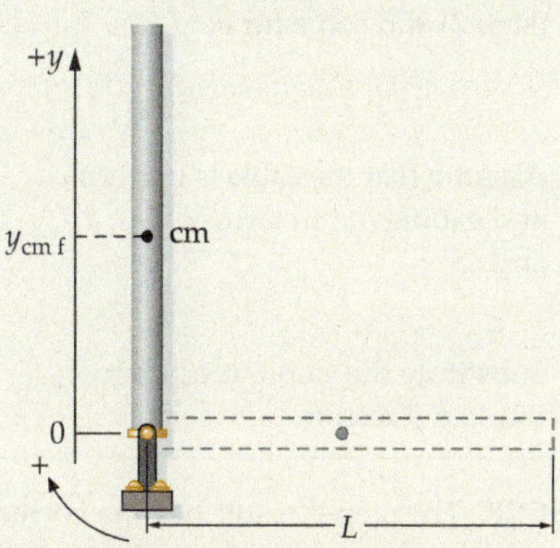

FIGURE 9-15

3. Apply conservation of mechanical energy to relate the initial kinetic energy to the final position:

$$K_f + U_f = K_i + U_i$$

$$\tfrac{1}{2}I\omega_f^2 + Mgy_{cm\,f} = \tfrac{1}{2}I\omega_i^2 + Mgy_{cm\,i}$$

$$0 + Mg\frac{L}{2} = \frac{1}{2}I\omega_i^2 + 0$$

4. Solve for the initial angular velocity:

$$\omega_i = \sqrt{\frac{MgL}{I}} = \sqrt{\frac{MgL}{\tfrac{1}{3}ML^2}} = \boxed{\sqrt{\frac{3g}{L}}}$$

CHECK It is no coincidence that the answers to Part (*a*) and Part (*c*) are identical. The decrease in both height and potential energy in Part (*a*) is equal to the increase in height and potential energy in Part (*c*). Thus, the increase in kinetic energy in Part (*a*) is equal to the decrease in kinetic energy in Part (*c*).

Example 9-8 A Winch and a Bucket

(D. S. Kerr/ Visuals Unlimited.)

A winch is at the top of a deep well. The drum of the winch has mass m_w and radius R. Virtually all its mass is concentrated a distance R from the axis. A cable wound around the drum suspends a bucket of water of mass m_b. The entire cable has mass m_c and length L. Just when you have the bucket at the highest point, your hand slips and the bucket falls back down the well, unwinding the winch cable as it goes. How fast is the bucket moving after it has fallen a distance d, where d is less than L? Assume that effects due to friction and air resistance are negligible.

PICTURE As the load falls, mechanical energy of the drum-cable-bucket of water-Earth system remain constant. Choose the initial potential energy to be zero. When the load has fallen a distance d, the center of mass of the hanging part of the cable has dropped a distance $d/2$. Because the hanging part of the cable moves with speed v and the cable does not stretch or become slack, the entire cable must move at speed v. We find v from the conservation of mechanical energy.

SOLVE

1. Make a diagram of the system in both its initial and its final configuration (Figure 9-16). Include a y axis with the origin at the height of the rotation axis of the drum.

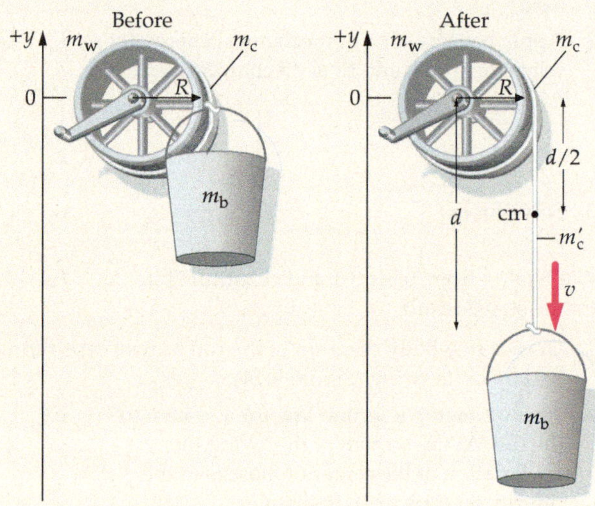

FIGURE 9-16

2. Apply conservation of mechanical energy. Choose the potential energy to be zero when the bucket of water is at the highest point:

$$U_f + K_f = U_i + K_i = 0 + 0 = 0$$

3. Write an expression for the total potential energy when the bucket has fallen a distance d. Let m'_c denote the mass of the hanging part of the cable:

$$U_f = U_{bf} + U_{cf} + U_{wf} = m_b g(-d) + m'_c g\left(-\frac{d}{2}\right) + 0 = -(m_b + \tfrac{1}{2}m'_c)gd$$

4. Express the total kinetic energy when the bucket is falling with speed v. All the cable and the entire mass of the drum move with the same speed v as the bucket:

$$K_f = K_{fc} + K_{fb} + K_{fw} = \tfrac{1}{2}m_c v^2 + \tfrac{1}{2}m_b v^2 + \tfrac{1}{2}m_w v^2 = \tfrac{1}{2}(m_c + m_b + m_w)v^2$$

5. Substitute into the conservation of mechanical energy equation (step 2) and solve for v:

$$-(m_b + \tfrac{1}{2}m'_c)gd + \tfrac{1}{2}(m_c + m_b + m_w)v^2 = 0$$

$$\text{so} \quad v = \sqrt{\frac{(2m_b + m'_c)gd}{(m_c + m_b + m_w)}}$$

6. Assume that the cable is uniform and express m'_c in terms of m_c, d, and L:

$$\frac{m'_c}{d} = \frac{m_c}{L} \Rightarrow m'_c = \frac{d}{L}m_c$$

7. Substitute the step-6 result into the step-5 result:

$$\boxed{v = \sqrt{\frac{(2m_b L + m_c d)gd}{(m_c + m_b + m_w)L}}}$$

CHECK The step-7 result has the correct dimensions for speed because acceleration times length has dimensions of length squared divided by time squared.

TAKING IT FURTHER Because the entire mass of the drum is moving at the same speed v, we can express its kinetic energy as $\frac{1}{2}m_w v^2$. However, we can also express it as $\frac{1}{2}I_w\omega^2$, where $I_w = m_w R^2$ and $\omega = v/R$. With these substitutions, $K_w = \frac{1}{2}I_w\omega^2 = \frac{1}{2}m_w R^2(v^2/R^2) = \frac{1}{2}m_w v^2$.

9-4 NEWTON'S SECOND LAW FOR ROTATION

To set a top spinning, you twist it. In Figure 9-17, a disk is set spinning by the forces $\vec{F}_1$ and $\vec{F}_2$ exerted at the edges of the disk in the tangential direction. The directions of these forces and their points of application are important. If the same forces are applied at the same points but in the radial direction (Figure 9-18*a*), the disk will not start to spin. In addition, if the same forces are applied in the tangential direction, but at points closer to the center of the disk (Figure 9-18*b*), the disk will not gain angular speed as quickly.

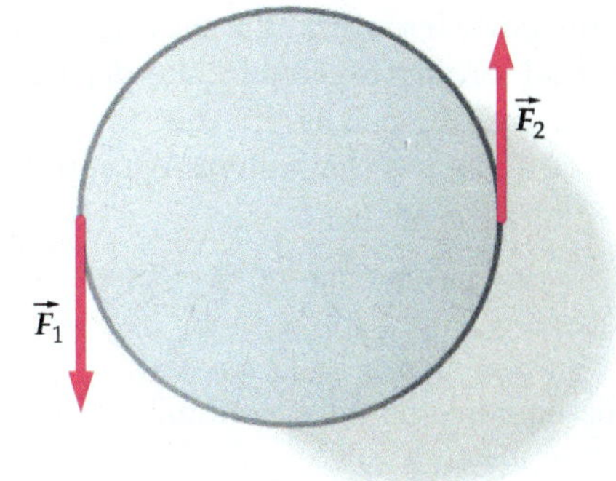

FIGURE 9-17

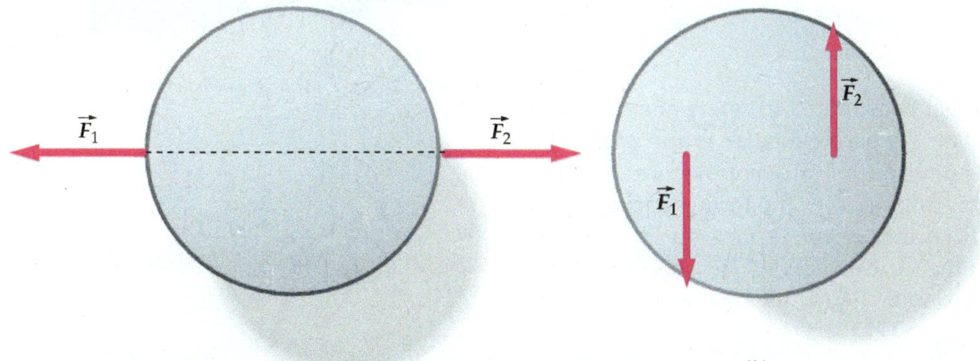

(*a*) (*b*)

FIGURE 9-18

Figure 9-19 shows a particle of mass m attached to one end of a massless rigid rod of length r. The rod is free to rotate about a fixed axis perpendicular to the rod and passing through the end of the rod at A. Consequently, the particle is constrained to move in a circle of radius r. A single force $\vec{F}$ is applied to the particle as shown. Applying Newton's second law to the particle and taking components in the tangential direction gives

$$F_t = ma_t$$

where $F_t = F\sin\phi$ is the tangential component of $\vec{F}$ and a_t is the tangential component of the acceleration. We wish to obtain an equation involving angular quantities. Substituting $r\alpha$ for a_t (Equation 9-9) and multiplying both sides by r gives

$$rF_t = mr^2\alpha \qquad 9\text{-}15$$

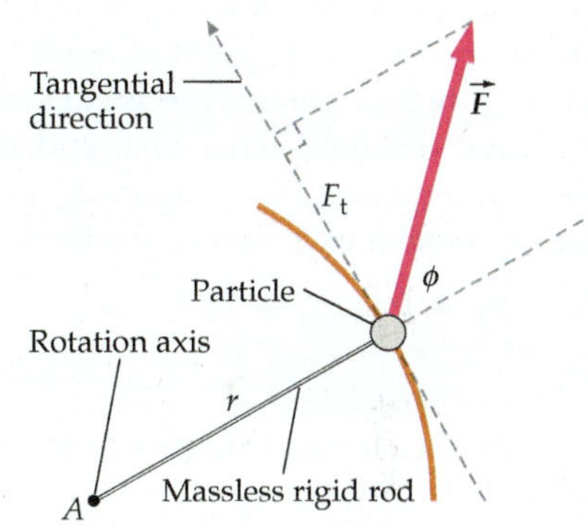

FIGURE 9-19

The product rF_t is the **torque** τ about the rotation axis associated with the force. That is,

$$\tau = F_t r \qquad 9\text{-}16$$

TORQUE ABOUT AN AXIS

(Torque about a point is defined as a vector quantity in Chapter 10. What we refer to as "torque about an axis" is the component of the torque vector parallel with the axis.)

Substituting τ for rF_t in Equation 9-15 gives

$$\tau = mr^2\alpha \qquad 9\text{-}17$$

A rigid object that rotates about a fixed axis is just a collection of individual particles, each of which is constrained to move in a circular path with the same angular velocity ω and acceleration α. Applying Equation 9-17 to the ith of these particles gives

$$\tau_{i\,\text{net}} = m_i r_i^2 \alpha$$

where $\tau_{i\,\text{net}}$ is the torque due to the net force on the ith particle. Summing both sides over all particles gives

$$\sum \tau_{i\,\text{net}} = \sum m_i r_i^2 \alpha = \left(\sum m_i r_i^2\right)\alpha = I\alpha \qquad 9\text{-}18$$

In Chapter 8, we saw that the net force acting on a system of particles is equal to the net *external* force acting on the system because the internal forces (those exerted

See **Math Tutorial** *for more information on* **Trigonometry**

by the particles within the system on one another) cancel in pairs. The treatment of internal torques exerted by the particles within a system on one another leads to a similar result, that is, the net torque acting on a system equals the net *external* torque acting on the system. We can thus write Equation 9-18 as

$$\tau_{\text{net ext}} = \sum \tau_{\text{ext}} = I\alpha \tag{9-19}$$

NEWTON'S SECOND LAW FOR ROTATION

This equation is the rotational analog of Newton's second law for linear motion ($\Sigma\vec{F} = m\vec{a}$).

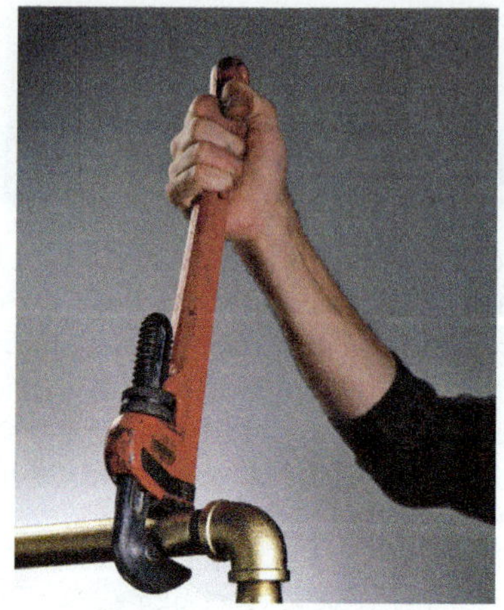

(*Richard Menga/ Fundamental Photographs.*)

CALCULATING TORQUES

Figure 9-20 shows a force $\vec{F}$ acting on an object constrained to rotate about a fixed axis A, not shown, which passes through O and is perpendicular to the page. The positive tangential direction is shown at the point of application of the force, and r is the radial distance of this point of application from axis A. The torque τ due to this force about axis A is $\tau = F_t r$ (Equation 9-16). In principle, the expression $F_t r$ is all that is needed to calculate torques. However, in practice, calculations are often simpler if alternative expressions for torque are used. From the figure, we can see that

$$F_t = F \sin\phi$$

where ϕ is the angle between the radial direction and the direction of the force. Thus, we can express the torque as $\tau = F_t r = (F\sin\phi)r$. The *line of action* of a force is the line through the point of application of the force that is parallel to the force. From Figure 9-21 we can see that $r\sin\phi = \ell$, where the **moment arm** ℓ is the perpendicular distance between A and the line of action. (The moment arm is also called the *lever arm.*) Consequently, the torque is also given by $\tau = F\ell$. Putting all three equivalent expressions for the torque in one place, we have

$$\tau = F_t r = F\sin\phi\, r = F\ell \tag{9-20}$$

EQUIVALENT EXPRESSIONS FOR TORQUE

The torque of a force about an axis is also called the *moment of the force* about the axis.

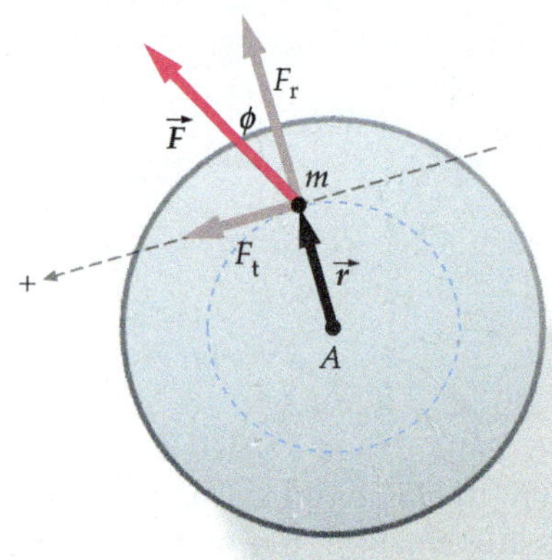

FIGURE 9-20 The force $\vec{F}$ produces a torque $F_t R$ about the center.

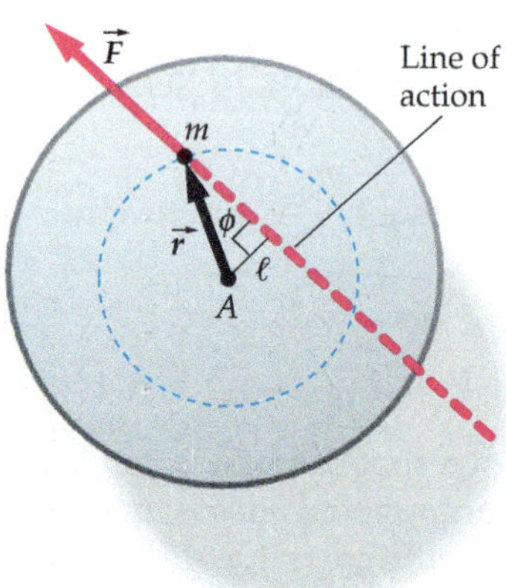

FIGURE 9-21 The force $\vec{F}$ produces a torque $F\ell$ about the center.

TORQUE DUE TO GRAVITY

We can model an extended object as an assembly of microscopic point particles, and there is a microscopic gravitational force on each particle. Each of these microscopic gravitational forces exerts a microscopic torque about a given axis, and the net gravitational torque on the object is the sum of these microscopic torques. The net gravitational torque can be calculated by considering the total gravitational force (the sum of the microscopic gravitational forces) to act at a single point-the **center of gravity.** Consider an object (Figure 9-22) constrained to rotate about a horizontal axis A coming out of the page. We choose the z axis of our coordinate system to coincide with axis A, and choose the x axis direction to be horizontal and the y axis vertical as shown. The torque on a particle of mass m_i due to gravity is $m_i g x_i$, where x_i is the moment arm of the force $m_i\vec{g}$. The net gravitational torque on

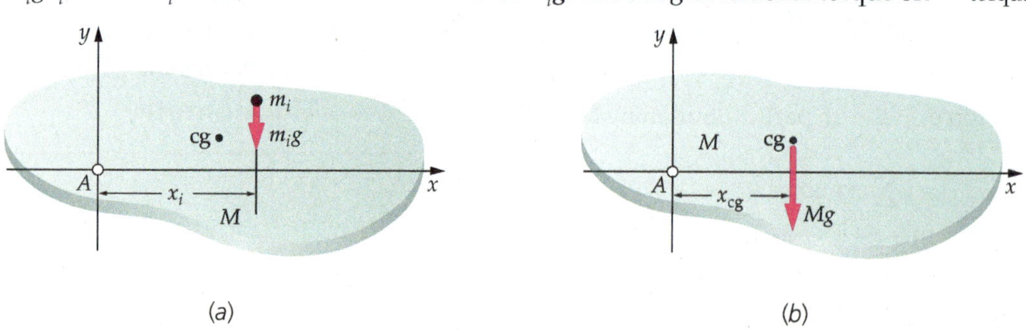

FIGURE 9-22 In a uniform gravitational field, the center of gravity coincides with the center of mass.

the object is the sum of the gravitational torques for all the particles that make up the object. That is, $\tau_{\text{grav net}} = \Sigma m_i g x_i$. If $\vec{g}$ is uniform (has the same magnitude and direction) throughout the region of space occupied by object, then g can be factored out of the sum. Factoring g out of the sum gives $\tau_{\text{grav net}} = (\Sigma m_i x_i)g$. You should recognize the sum in the parentheses as Mx_{cg} (see Equation 5-13). Substituting this for the sum gives

$$\tau_{\text{grav net}} = Mgx_{\text{cg}} \qquad \text{9-21}$$

TORQUE DUE TO GRAVITY

For any object in a uniform gravitational field, the center of gravity and the center of mass coincide.

The torque due to a uniform gravitational field is calculated as if the entire gravitational force is applied at the center of gravity.

9-5 APPLICATIONS OF NEWTON'S SECOND LAW FOR ROTATION

In this section, we give several applications of Newton's second law for rotation as expressed in Equation 9-19.

PROBLEM-SOLVING STRATEGY

Applying Newton's Second Law for Rotation

PICTURE Angular accelerations for rigid objects can be found by using free-body diagrams and Newton's second law for rotation, which is $\tau_{\text{net ext}} = \Sigma\tau_{\text{ext}} = I\alpha$. If $\tau_{\text{net ext}}$ is constant, then the constant angular acceleration equations apply. Time intervals and angular positions, velocities, and angular accelerations can then be determined using these equations.

SOLVE

1. Draw a free-body diagram with the object shown as a likeness of the object (not just as a dot).
2. Draw each force vector along the line of action of that force.
3. On the diagram indicate the positive direction (clockwise or counterclockwise) for rotations.

CHECK Make sure that the signs of your results are consistent with your choice for the positive directions of rotation.

Example 9-9 A Stationary Bike

To get some exercise without going anywhere, you set your bike on a stand so that the rear wheel is free to turn. As you pedal, the chain applies a force of 18 N to the rear sprocket wheel at a distance of $r_s = 7.0$ cm from the rotation axis of the wheel. Consider the wheel to be a hoop ($I = MR^2$) of radius $R = 35$ cm and mass $M = 2.4$ kg. What is the angular velocity of the wheel after 5.0 s?

PICTURE The angular velocity is found from the angular acceleration, which is found from Newton's second law for rotation. Because the forces are constant, the torques are also constant. Thus, the constant angular acceleration equations apply. Note that $\vec{F}$ acts in the direction of the chain, so the line of action of $\vec{F}$ is tangent to the sprocket wheel, and the moment arm is the radius r_s of the sprocket wheel (Figure 9-23).

FIGURE 9-23

SOLVE

1. The angular velocity is related to the angular acceleration and the time: $\omega = \omega_0 + \alpha t = 0 + \alpha t$
2. Apply Newton's second law for rotational motion to relate α to the net torque and the moment of inertia: $\tau_{net} = I\alpha$
3. The only torque acting on the system is that due to the applied force F with moment arm r_s: $\tau_{net} = Fr_s$
4. Substitute this value for the torque and $I = MR^2$ for the moment of inertia: $\alpha = \dfrac{\tau_{net}}{I} = \dfrac{Fr_s}{MR^2}$
5. Substitute into the step-1 result and solve for the angular velocity after 5.0 s: $\omega = \alpha t = \dfrac{Fr_s}{MR^2}t = \dfrac{(18\text{ N})(0.070\text{ m})}{(2.4\text{ kg})(0.35\text{ m})^2}5.0\text{ s} = 21.4\text{ rad/s} = \boxed{21\text{ rad/s}}$

CHECK The tangential speed of the rim is given by $R\omega = (0.36\text{ m})(21\text{ rad/s}) = 7.6\text{ m/s}$, which is a plausible speed. (A world-class runner can sprint at speeds in excess of 10 m/s.)

Example 9-10 A Uniform Rod, Pivoted at One End

A uniform thin rod of length L and mass M is pivoted at one end. It is held horizontal and released. Effects due to friction and air resistance are negligible. Find (*a*) the angular acceleration of the rod immediately following its release, and (*b*) the magnitude of the force F_A exerted on the rod by the pivot at this instant.

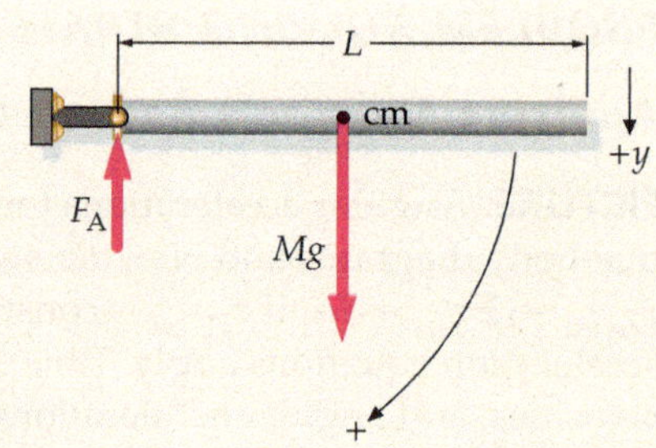

FIGURE 9-24

PICTURE The angular acceleration is found from Newton's second law for rotation (Equation 9-19). The force F_A is found from Newton's second law for a system (Equation 5-23). The tangential acceleration of the center of mass is related to the angular acceleration (Equation 9-6), and the centripetal acceleration of the center of mass is related to the angular speed (Equation 9-7).

SOLVE

(*a*) 1. Sketch a free-body diagram of the rod (Figure 9-24).

2. Write Newton's second law for rotation: $\Sigma\tau_{ext} = I\alpha$

3. Compute the torque due to gravity about the given axis. The rod is uniform so its center of gravity is at its center, a distance $L/2$ from the axis: $\tau_{grav} = Mg\dfrac{L}{2}$

4. Find the moment of inertia about the end of the rod from Table 9-1: $I = \frac{1}{3}ML^2$

5. Substitute these values into the step-2 equation to compute α: $\alpha = \dfrac{\tau_{grav}}{I} = \dfrac{Mg(L/2)}{(1/3)ML^2} = \boxed{\dfrac{3g}{2L}}$

(*b*) 1. Write Newton's second law for a system for the rod: $\Sigma F_{ext\,y} = Ma_{cm\,y}$
$Mg - F_A = Ma_{cm\,y}$

2. Use the relation $a_c = r\omega^2$ to find $a_{cm\,c}$. Immediately following release, $\omega = 0$: $a_{cm\,c} = r_{cm}\omega^2 = \dfrac{L}{2}\omega^2 = 0$

3. We now have two equations and three unknowns, α, $a_{cm\,y}$, and F_A. Use the relation $a_t = r\alpha$ to obtain a third equation, one relating $a_{cm\,y}$ to α. Then substitute the Part-(*a*), step- 5 result for α.
$a_t = r\alpha$
$a_{cm\,y} = a_{cm\,t} = r_{cm}\alpha = \dfrac{L}{2}\dfrac{3g}{2L} = \dfrac{3}{4}g$

4. Substitute the Part-(*b*), step-3 result into the Part-(*b*), step-1 result and solve for F_A: $Mg - F_A = M\frac{3}{4}g$
so $F_A = \boxed{\frac{1}{4}Mg}$

CHECK The axis exerts an upward force on the rod. Consequently, we expect that following release, the acceleration of the center of mass will be somewhat less than the free-fall acceleration g. Our Part-(*b*), step-3 result confirms this expectation.

TAKING IT FURTHER Just after the rod is released, the acceleration of the center of mass is directed straight down. Because the net external force and the acceleration of the center of mass are in the same direction, it follows that $\vec{F}_A$ does not have a horizontal component at this instant.

PRACTICE PROBLEM 9-8 A small pebble of mass $m \ll M$ is placed on top of the rod at its center. Just after the rod is released find (*a*) the acceleration of the pebble, and (*b*) the force it exerts on the rod.

NONSLIP CONDITIONS

In physics courses, there are many situations in which a taut string is in contact with a rotating pulley wheel. For the string not to slip on the pulley wheel, the parts of the string and the wheel that are in direct contact with each other must share the same tangential velocity. As a result,

$$v_t = R\omega \qquad \text{9-22}$$

NONSLIP CONDITION FOR v_t AND ω

where v_t is the tangential velocity of the string and $R\omega$ is the tangential velocity of the perimeter of the pulley wheel. The wheel has radius R and is rotating with angular velocity ω. Differentiating both sides of the nonslip condition (Equation 9-22) with respect to time gives

$$a_t = R\alpha \qquad \text{9-23}$$

a_t AND α UNDER NONSLIP CONDITIONS

where a_t is the tangential acceleration of the string and α is the angular acceleration of the wheel.

(*Fundamental Photographs.*)

Example 9-11 Tension in a String

An object of mass m is suspended from a light string that is wound around the rim of a pulley wheel that has moment of inertia I and radius R. The wheel bearing is frictionless and the string does not slip on the rim. The wheel is released from rest. It starts to rotate as the object descends and the string unwinds. Find the tension in the string and the acceleration of the object.

PICTURE The object descends with a downward acceleration a, while the wheel rotates with an angular acceleration α (Figure 9-25). We apply Newton's second law for rotation to the wheel to determine α, and Newton's second law to the object to obtain a. Relate a_t and α using the nonslip condition.

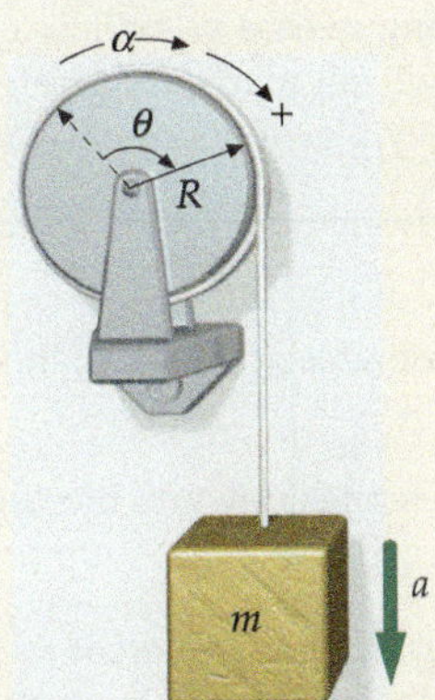

FIGURE 9-25

SOLVE

1. Draw a free-body diagram of the pulley wheel, drawing each force vector with its tail at the point of application of the force. Put labels on the diagram and indicate the positive rotational direction as shown in Figure 9-26:

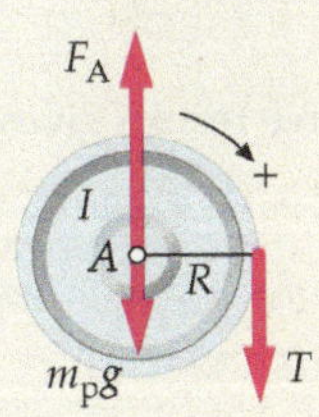

FIGURE 9-26

2. The only force that exerts a torque on the wheel is the tension T, which has moment arm R. Apply Newton's second law for rotational motion to relate T to the angular acceleration α:

$$\Sigma\tau_{\text{ext}} = I\alpha$$
$$TR = I\alpha$$

3. Draw a free-body diagram for the suspended object, and apply Newton's second law to relate T to the tangential acceleration a_t (Figure 9-27):

$$\Sigma F_{\text{ext}\,y} = ma_y$$
$$mg - T = ma_t$$

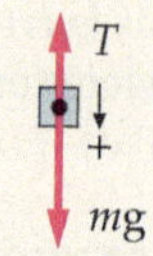

FIGURE 9-27

4. We have two equations for three unknowns, T, a_t, and α. A third equation is the relation between a_t and α, called the nonslip condition. (The tangential accelerations of the object, the string, and the perimeter of the wheel are all equal.)

$$a_t = R\alpha$$

5. We now have three equations enabling us to determine T, a_t, and α. To solve for T, substitute into the step-4 equation. Substitute for α using the step-2 result and for a_t using the step-3 result. Then solve for T.

$$\frac{mg - T}{m} = R\frac{TR}{I}$$

$$\text{so} \quad T = \boxed{\frac{mg}{1 + (mR^2/I)}}$$

6. Substitute this result for T into the step-3 result and solve for a_t. The object and the wheel perimeter gain speed at the same rate. Thus, a_t is the acceleration of the wheel perimeter:

$$mg - \frac{mg}{1 + (mR^2/I)} = ma_t$$

$$\text{so} \quad a_t = \boxed{\frac{1}{1 + (I/mR^2)}g}$$

CHECK Let us evaluate our results for a couple of limiting cases. If $I = 0$, the object should fall freely and the string should be slack; our results give $T = 0, a_t = g$, as expected. For very large values of I we expect the wheel to remain at rest. If $I \to \infty$, our equations give $T \to mg$ and $a_t \to 0$, as expected.

Example 9-12 Two Blocks and a Pulley I

Conceptual

The system shown in Figure 9-28 is released from rest. The mass of the pulley wheel is not negligible, but the friction in the bearing is negligible. The string does not slip on the pulley wheel. Given that $m_1 > m_2$, what can be determined about the tensions T_1 and T_2?

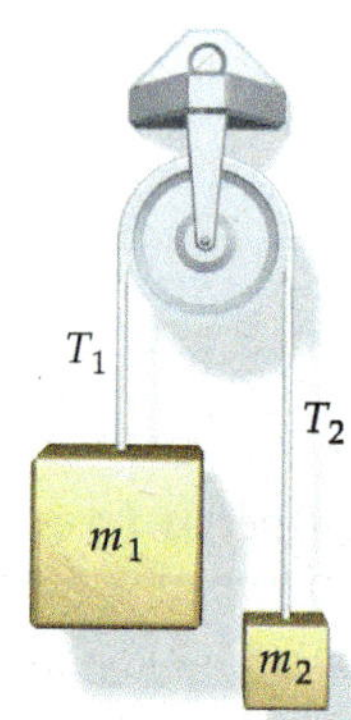

FIGURE 9-28

PICTURE Following release m_1 will accelerate downward, m_2 will accelerate upward, and the pulley wheel will angularly accelerate counterclockwise. Apply Newton's second law to each mass and apply Newton's second law for rotations to the pulley wheel.

SOLVE

1. Because m_1 accelerates downward, the net force on it must be downward:

$$m_1 g > T_1$$

2. Because m_2 accelerates upward, the net force on it must be upward:

$$T_2 > m_2 g$$

3. Because the angular acceleration of the pulley wheel is counterclockwise, the net torque on it must be counterclockwise. Because the two moment arms are equal, larger torque means larger tension:

$$\tau_1 > \tau_2 \quad \text{so} \quad T_1 > T_2$$

4. Combining the three results gives:

$$\boxed{m_1 g > T_1 > T_2 > m_2 g}$$

CHECK If T_1 were not greater than T_2, the angular acceleration of the pulley wheel would not be counterclockwise.

Example 9-13 Two Blocks and a Pulley II

Try It Yourself

Two blocks are connected by a string that passes over a pulley of radius R and moment of inertia I. The block of mass m_1 slides on a frictionless, horizontal surface; the block of mass m_2 is suspended from the string (Figure 9-29). Find the acceleration a of the blocks and the tensions T_1 and T_2. The string does not slip on the pulley.

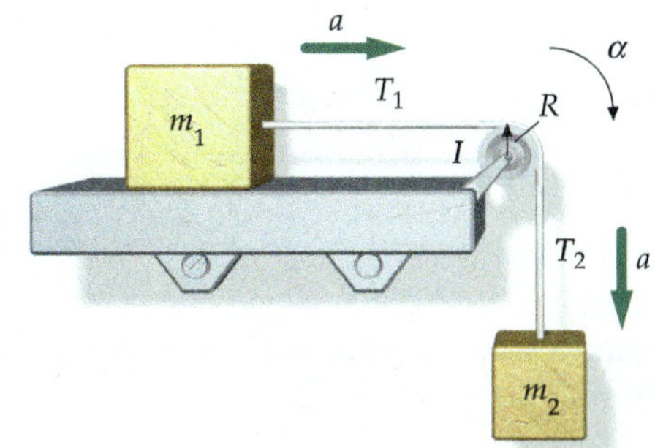

FIGURE 9-29

PICTURE The tensions T_1 and T_2 are not equal because the pulley has mass and because there is static friction between the string and the pulley (Figure 9-29). (If the two tensions were equal, the torque on the pulley by the string would be zero.) Note that T_2 exerts a clockwise torque and T_1 exerts a counterclockwise torque on the pulley. Use Newton's second law for each block, and Newton's second law for rotational motion for the pulley. Relate α and a using the nonslip condition.

SOLVE

Cover the column to the right and try these on your own before looking at the answers.

Steps	Answers
1. Draw a free-body diagram for each block and for the pulley, as shown in Figure 9-30. Note that the center of mass of the pulley wheel does not accelerate, so the support must exert a force on the axle F_s that balances the resultant of the gravitational force on the wheel and the forces exerted on it by the string.	
2. Apply Newton's second law to each block.	$T_1 = m_1a;\quad m_2g - T_2 = m_2a$
3. Apply Newton's second law for rotation to the pulley wheel.	$T_2R - T_1R = I\alpha$
4. We have three equations and four unknowns. To get a fourth equation, use the nonslip condition. The acceleration a of the blocks is equal to the tangential acceleration a_t of the string and pulley-wheel perimeter.	$a_t = R\alpha$ $a = a_t = R\alpha$
5. We have four equations and four unknowns, so the rest is algebra. Do the algebra and obtain expressions for a, T_1, and T_2. *Hint: To find a, obtain expressions for T_1 and T_2 from the step-2 results. Substitute these results into the step-3 result to obtain an equation with unknowns a and α. Use the step-4 result to eliminate α and solve for a.*	$a = \boxed{\dfrac{m_2}{m_1 + m_2 + (I/R)^2}g}$ $T_1 = m_1a = \boxed{\dfrac{m_2}{m_1 + m_2 + (I/R)^2}m_1g}$ $T_2 = m_2(g - a) = \boxed{\dfrac{m_1 + (I/R)^2}{m_1 + m_2 + (I/R)^2}m_2g}$

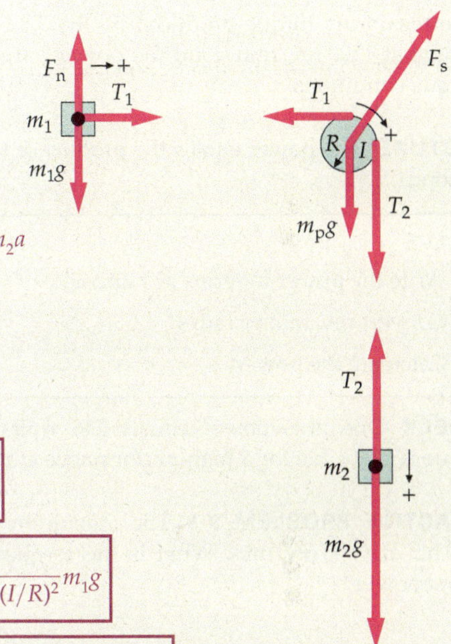

FIGURE 9-30

CHECK If $I = 0, T_1 = T_2$, and $a = m_2g/(m_1 + m_2)$, as expected. If $I \to \infty$, then $a \to 0, T_1 \to 0$, and $T_2 \to m_2g$, again as expected.

! Do not assume that the tension in a string passing over a pulley is the same on either side of the pulley. If it were, the string would not exert a torque on the pulley wheel and the wheel would not change its rotation rate. Use two distinct labels, such as T_1 and T_2, for the tensions in the string on opposite sides of the wheel.

POWER

When you spin up an object, you do work on it, increasing its kinetic energy. Consider the force $\vec{F}$ acting on a rotating object. As the object rotates through an angle $d\theta$, the point of application of the force moves a distance $ds = r\,d\theta$, and the force does work

$$dW = F_t\,ds = F_t r\,d\theta = \tau\,d\theta$$

where τ is the torque exerted by the force $\vec{F}$, and F_t is the tangential component of $\vec{F}$. The work dW done by a torque τ on an object when the object rotates through a small angle $d\theta$ is thus

$$dW = \tau\,d\theta \qquad 9\text{-}24$$

WORK

The rate at which the torque does work—the power input of the torque—is

$$P = \frac{dW}{dt} = \tau\frac{d\theta}{dt}$$

or

$$P = \tau\omega \qquad 9\text{-}25$$

POWER

Equations 9-24 and 9-25 are the rotational analogs of the linear equations $dW = F_{\|}\,d\ell$ and $P = F_{\|}v$.

Example 9-14 Torque Exerted by an Automobile Engine

The maximum torque produced by the 5.4-L V8 engine of a 2005 Ford GT is 678 N·m of torque at 4500 rev/min. Find the power output of the engine operating at these maximum torque conditions.

PICTURE The power equals the product of the torque and angular velocity (in radians per second).

SOLVE

1. Write the power in terms of τ and ω: $P = \tau\omega$
2. Convert rev/min to rad/s: $\omega = 4500\text{ rev/min} = 471\text{ rad/s}$
3. Calculate the power: $P = (678\text{ N}\cdot\text{m})\cdot(471\text{ rad/s}) = \boxed{320\text{ kW}}$

CHECK One horsepower equals 746 watts, so $320\text{ kW} \times 1\text{ hp}/0.746\text{ kW} = 429\text{ hp}$. This value is plausible for a high-performance automobile engine.

PRACTICE PROBLEM 9-9 The maximum power produced by the Ford GT engine is 500 hp at 6000 rev/min. What is the torque when the engine is operating at maximum horsepower?

There are many parallels between one-dimensional linear motion and rotational motion about a fixed axis. The similarities of the formulas can be seen in Table 9-2. The relations are the same, but the symbols are different.

Example 9-15 Stopping the Wheel

The specifications for the London Eye include that it be able to brake to a stop so that the passenger compartments move no more than 10 m during braking. The operating speed of the 135-m-diameter 1600-tonne wheel is 2.0 rev/h. (One tonne equals 1000 kg.) A picture of the wheel can be found at the beginning of this chapter. (*a*) Estimate the torque that is required to stop the wheel so the rim travels 10 m during the braking. (*b*) Assuming that the braking force is applied at the rim, what is the magnitude of the breaking force?

PICTURE The work done on the wheel is equal to its change in kinetic energy. Use $dW = \tau\,d\theta$ (Equation 9-24) to calculate the work in terms of the torque. Almost all of the mass is near the perimeter of the wheel (see the picture on the first page of this chapter). This suggests a way to estimate the moment of inertia. The braking force can be found from the torque.

SOLVE

(*a*) 1. Set the work done equal to the change in kinetic energy: $W = \Delta K$

2. Using $dW = \tau\, d\theta$ (Equation 9-24), relate the work to the torque and the angular displacement:

$$W = \tau\,\Delta\theta$$

3. Using $ds = r\,d\theta$ (Equation 9-2), relate the angular displacement to the stopping distance s:

$$s = r\,\Delta\theta \Rightarrow \Delta\theta = \frac{s}{r} = \frac{10\text{ m}}{67.5\text{ m}} = 0.148\text{ rad}$$

4. The mass is concentrated near the rim of the wheel, so $I \approx mr^2$:

$$I = mr^2 = (1.6 \times 10^6\text{ kg})(67.5\text{ m})^2$$
$$= 7.29 \times 10^9\text{ kg}\cdot\text{m}^2$$

5. Substitute into the step-1 result and solve for the torque. The initial angular velocity is 2.0 rev/h = 3.49×10^{-3} rad/s:

$$\tau\,\Delta\theta = 0 - \tfrac{1}{2}I\omega_0^2$$
$$\text{so}\quad \tau = -\frac{I\omega_0^2}{2\,\Delta\theta} = -\frac{(7.29 \times 10^9\text{ kg}\cdot\text{m}^2)(3.49 \times 10^{-3}\text{ rad/s})^2}{2\,(0.148\text{ rad})}$$
$$= \boxed{-3.0 \times 10^5\text{ N}\cdot\text{m}}$$

(*b*) 1. The line of action of the braking force is tangent to the rim, so the moment arm is equal to the radius of the wheel.

$$|\tau| = FR$$
$$F = \frac{|\tau|}{R} = \frac{3.0 \times 10^5\text{ N}\cdot\text{m}}{67.5\text{ m}} = \boxed{4.4 \times 10^3\text{ N}}$$

CHECK From the expression for the torque in step 5 of Part (*a*) we can see that τ is negative if $\Delta\theta$ is positive, and vice versa. This result is expected because the torque opposes the motion during braking.

TAKING IT FURTHER The braking force of 1.3×10^5 N is approximately equal to one-half ton.

Table 9-2 Analogs in Fixed-Axis Rotational and One-Dimensional Linear Motion

Rotational Motion		Linear Motion	
Angular displacement	$\Delta\theta$	Displacement	Δx
Angular velocity	$\omega = \frac{d\theta}{dt}$	Velocity	$v_x = \frac{dx}{dt}$
Angular acceleration	$\alpha = \frac{d\omega}{dt} = \frac{d^2\theta}{dt^2}$	Acceleration	$a_x = \frac{dv_x}{dt} = \frac{d^2x}{dt^2}$
Constant-angular-acceleration equations	$\omega = \omega_0 + \alpha t$	Constant-acceleration equations	$v_x = v_{0x} + a_x t$
	$\Delta\theta = \omega_{av}\,\Delta t$		$\Delta x = v_{av\,x}\,\Delta t$
	$\omega_{av} = \frac{1}{2}(\omega_0 + \omega)$		$v_{av\,x} = \frac{1}{2}(v_{0x} + v_x)$
	$\theta = \theta_0 + \omega_0 t + \frac{1}{2}\alpha t^2$		$x = x_0 + v_{0x}t + \frac{1}{2}a_x t^2$
	$\omega^2 = \omega_0^2 + 2\alpha\,\Delta\theta$		$v_x^2 = v_{0x}^2 + 2a_x\,\Delta x$
Torque	τ	Force	F_x
Moment of inertia	I	Mass	m
Work	$dW = \tau\,d\theta$	Work	$dW = F_x\,dx$
Kinetic energy	$K = \frac{1}{2}I\omega^2$	Kinetic energy	$K = \frac{1}{2}mv^2$
Power	$P = \tau\omega$	Power	$P = F_x v_x$
Angular momentum*	$L = I\omega$	Momentum	$p_x = mv_x$
Newton's second law	$\tau_{net} = I\alpha = \frac{dL}{dt}$	Newton's second law	$F_{net\,x} = ma_x = \frac{dp_x}{dt}$

*Angular momentum is introduced in Chapter 10.

9-6 ROLLING OBJECTS

ROLLING WITHOUT SLIPPING

When a spool rolls without slipping down an incline (Figure 9-31), the points of the spool in contact with the incline are instantaneously at rest and the spool rotates about a rotation axis through the contact points. This can be observed because rapid motion causes blurring, which means the part of the spool that is moving slowest is blurred the least. In Figure 9-32 a wheel of radius R is rolling without slipping along a flat surface. Point P on the wheel moves as shown with speed

$$v = r\omega \tag{9-26}$$

NONSLIP CONDITION FOR SPEED

where r is the radial distance from the rotation axis to point P. The center of mass of the wheel moves with speed

$$v_{cm} = R\omega \tag{9-27}$$

NONSLIP CONDITION FOR v_{cm}

For a point on the very top of the wheel, $r = 2R$, so the top of the wheel is moving at twice the speed of the center of the wheel.

Differentiating both sides of Equation 9-27 gives

$$a_{cm} = R\alpha \tag{9-28}$$

NONSLIP CONDITION FOR ACCELERATION

A falling yo-yo that is unwinding from a string—the top end of which is held fixed—follows the same nonslip conditions as the wheel.

A wheel of radius R is rolling without slipping along a straight path. As the wheel rotates through angle ϕ (Figure 9-33), the point of contact between the wheel and the surface moves a distance s that is related to ϕ by

$$s = R\phi \tag{9-29}$$

NONSLIP CONDITION FOR DISTANCE

If the wheel is rolling on a flat surface, the wheel's center of mass remains directly over the point of contact, so it also moves through distance $R\phi$.

In Chapter 8 we saw (Equation 8-7) that the kinetic energy of a system can be written as the sum of its translational kinetic energy ($\frac{1}{2}Mv_{cm}^2$) and the kinetic energy relative to the center of mass K_{rel}. For an object that is rotating, the kinetic energy relative to an inertial frame moving with center of mass $\frac{1}{2}I_{cm}\omega^2$. Thus, the total kinetic energy of the object is

$$K = \tfrac{1}{2}Mv_{cm}^2 + \tfrac{1}{2}I_{cm}\omega^2 \tag{9-30}$$

TOTAL KINETIC ENERGY OF A ROTATING OBJECT

FIGURE 9-31 A spool that has dots on it is rolling without slipping down an inclined meter stick. The shaft of the spool is in contact with the stick. The exposure time for this photo was long enough that the dots appear as streaks, with the length of the streaks increasing with distance from the rotation axis. *(Loren Winters/Visuals Unlimited.)*

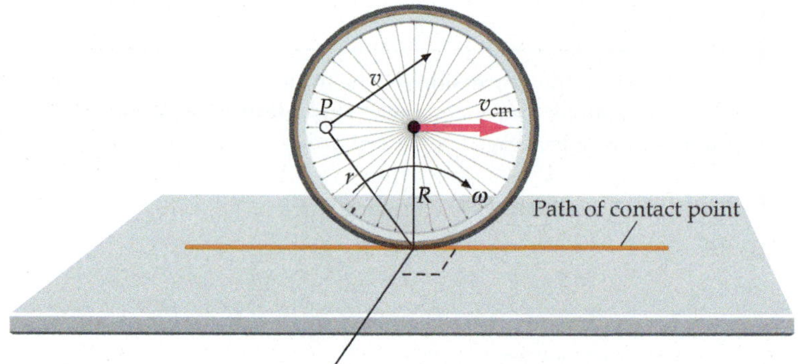

FIGURE 9-32 As the disk rolls to the right, point P moves upward and to the right. Point P reaches its greatest height when is passes over the center of the disk.

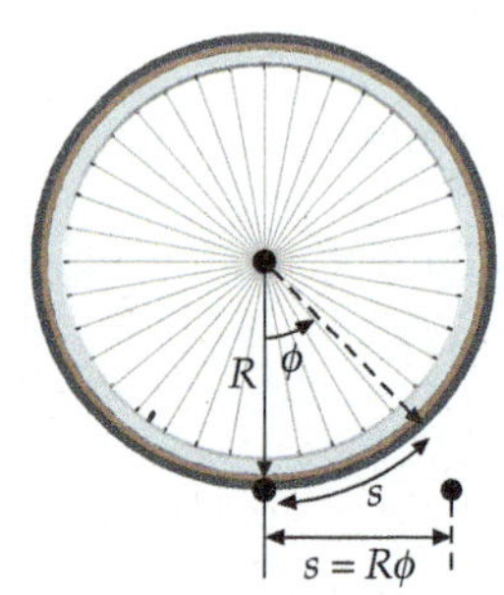

FIGURE 9-33

Remember, a rolling object has both translational and rotational kinetic energy.

Example 9-16 A Bowling Ball

A bowling ball that has an 11-cm radius and a 7.2-kg mass is rolling without slipping at 2.0 m/s on a horizontal ball return. It continues to roll without slipping up a hill to a height h before momentarily coming to rest and then rolling back down the hill. Model the bowling ball as a uniform sphere and find h.

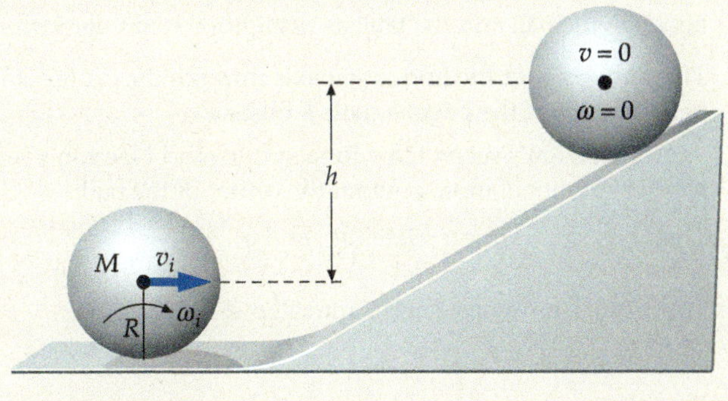

FIGURE 9-34

PICTURE No slipping occurs, so no energy is dissipated by kinetic friction. No external forces act on the ball–hill–Earth system, so no external forces do work on the system. The initial kinetic energy, which is the translational kinetic energy, $\frac{1}{2}Mv_{\text{cm}}^2$, plus the kinetic energy of rotation about the center of mass, $\frac{1}{2}I_{\text{cm}}\omega^2$, is converted to potential energy Mgh. Because the sphere rolls without slipping, the initial linear and angular speeds are related by $v_{\text{cm}} = R\omega$

SOLVE

1. Make a labeled sketch showing the ball in both its initial and final positions (Figure 9-34).
2. No external forces act on the system, so the work done by external forces is zero, and no slipping occurs, so no energy is dissipated by kinetic friction. Thus, mechanical energy is constant:

$$W_{\text{ext}} = \Delta E_{\text{mech}} + \Delta E_{\text{therm}}$$
$$0 = \Delta E_{\text{mech}} + 0$$

3. Apply conservation of mechanical energy with $U_i = 0$ and $K_f = 0$. Write the total initial kinetic energy K_i in terms of the speed v_{cm} and the angular speed ω_1:

$$U_f + K_f = U_i + K_i$$
$$Mgh + 0 = 0 + \tfrac{1}{2}Mv_{\text{cm i}}^2 + \tfrac{1}{2}I_{\text{cm}}\omega_i^2$$

4. Substitute from $\omega_i = v_{\text{cm i}}/R$ and $I_{\text{cm}} = \frac{2}{5}MR^2$ and solve for h:

$$Mgh = \frac{1}{2}Mv_{\text{cm i}}^2 + \frac{1}{2}\left(\frac{2}{5}MR^2\right)\frac{v_{\text{cm i}}^2}{R^2} = \frac{7}{10}Mv_{\text{cm i}}^2$$

$$\text{so} \quad h = \frac{7v_{\text{cm i}}^2}{10g} = 0.2854\text{ m} = \boxed{29\text{ cm}}$$

CHECK The height h is independent of the mass. This result is expected because both the kinetic energy and the potential energy are proportional to the mass.

TAKING IT FURTHER The height h is also independent of the radius R of the ball. This result is because $I_{\text{cm}} = \frac{2}{5}MR^2$ and $\omega_i = v_{\text{cm i}}/R$, so in the product $I_{\text{cm}}\omega_i^2$, the Rs cancel.

PRACTICE PROBLEM 9-10 Find the initial kinetic energy of the ball.

Example 9-17 Playing Pool

A cue stick strikes a cue ball horizontally at a point a distance d above the center of the ball (Figure 9-35). Find the value of d for which the cue ball will roll without slipping from the beginning. Express your answer in terms of the radius R of the ball.

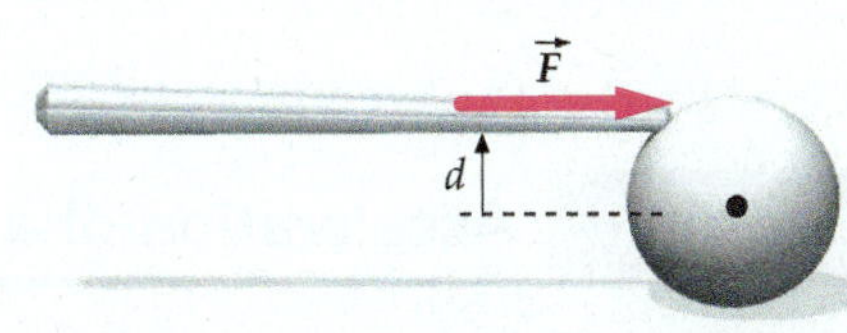

FIGURE 9-35

PICTURE The lines of action of the gravitational and normal forces both pass through the center of mass of the ball, and thus exert no torque about the center of mass. The frictional force exerted by the table is much smaller than the collision force of the cue stick, so its effects during the collision can be ignored. If the stick strikes the ball at the level of the ball's center, the ball initially translates with no rotation. If the stick strikes below the level of the center, the ball initially has backspin. However, if the stick strikes a certain distance d above the level of the center, the ball acquires just the right forward spin and forward translational motion to satisfy the nonslip condition. The value of d determines the torque-to-force ratio applied to the ball and hence the ball's angular acceleration to linear acceleration ratio. The linear acceleration a_{cm} is F/m, independent of d. For the ball to roll without slipping from the start, we find α and a_{cm}, then set $a_{\text{cm}} = R\alpha$ (the nonslip condition) to find d.

SOLVE

1. Sketch a free-body diagram of the ball (Figure 9-36). We are assuming that friction between the ball and the table is negligible, so do not include this frictional force:

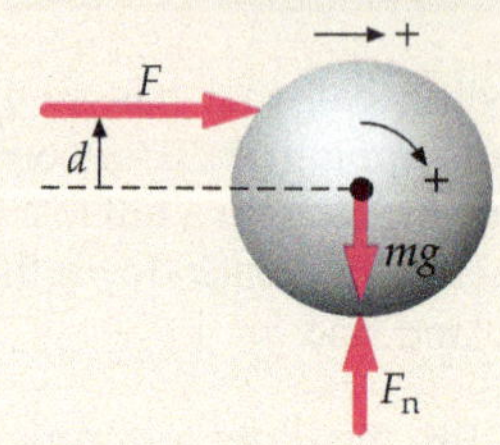

FIGURE 9-36

2. The torque about the horizontal axis through the center of the ball (and out of the page) equals F times d: $\tau = Fd$
3. Apply Newton's second law for a system and Newton's second law for rotational motion about the center of the ball: $F = ma_{cm}$ and $\tau = I_{cm}\alpha$
4. The nonslip condition relates a_{cm} and α: $a_{cm} = R\alpha$
5. Substitute from steps 2 and 3 into step 4: $\frac{F}{m} = R\frac{Fd}{I_{cm}}$
6. Find the moment of inertia from Table 9-1 and solve for d: $d = \frac{I_{cm}}{mR} = \frac{\frac{2}{5}mR^2}{mR} = \boxed{\frac{2}{5}R}$

CHECK The step-6 result is plausible because the value obtained for d is greater than zero and less than R, as expected.

TAKING IT FURTHER Striking the ball at a point either higher or lower than $2R/5$ from the center will result in the ball rolling *and* slipping (skidding). Skidding is often desirable in the game of pool. Rolling and slipping is discussed in the next subsection.

(Scott Goldsmith/ Stone/Getty.)

When an object rolls down an incline, its center of mass accelerates. The analysis of such a problem is simplified by an important theorem concerning the center of mass:

Newton's second law for rotation ($\tau = I\alpha$) holds in any inertial reference frame. It also holds in reference frames moving with the center of mass—even when the center of mass is accelerating—as long as the moment of inertia and all torques are computed about an axis through the center of mass. That is,

$$\tau_{\text{net cm}} = I_{cm}\alpha \qquad 9\text{-}31$$

Equation 9-31 is the same as Equation 9-19, except that in Equation 9-31 the torques and the moment of inertia are computed from a reference frame moving with the center of mass. When the center of mass is accelerating (a ball rolling down an incline, for example), the center-of-mass reference frame is a noninertial one, where we would not expect our equations for Newton's second law for rotation to be valid. Nevertheless, they are valid for this special case.

Example 9-18 Acceleration of a Ball That Is Rolling Without Slipping

A uniform solid ball of mass m and radius R rolls without slipping down a plane inclined at an angle ϕ above the horizontal. Find the frictional force and the acceleration of the center of mass.

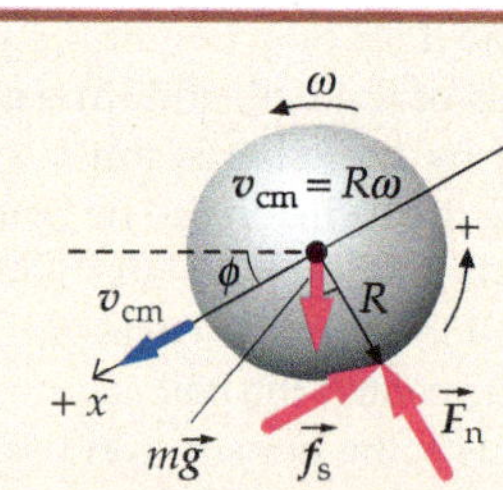

FIGURE 9-37

PICTURE From Newton's second law, the acceleration of the center of mass equals the net force divided by the mass. The forces acting are the gravitational force $m\vec{g}$ downward, the normal force $\vec{F}_n$, and the static frictional force $\vec{f}_s$ acting up the incline (Figure 9-37). As the object accelerates down the incline, its angular velocity must increase to maintain the nonslip condition. This angular acceleration requires a net external torque about the axis through the center of mass. We apply Newton's second law for rotation to find α. The nonslip condition relates α and a_{cm}.

SOLVE

1. Apply Newton's second law for a system in component form for the x axis:

$$\Sigma F_x = ma_{cm\,x}$$
$$mg\sin\phi - f_s = ma_{cm}$$

2. Apply Newton's second law for rotational motion about a horizontal axis passing through the center of mass and perpendicular to $\vec{v}_{cm}$. The moment arms for the normal and gravitational forces each equal zero, so they do not exert torques on the ball:

$$\Sigma \tau_i = I_{cm}\alpha$$
$$f_s R + 0 + 0 = I_{cm}\alpha$$

3. Relate a_{cm} and α using the nonslip condition:

$$a_{cm} = R\alpha$$

4. We now have three equations and three unknowns. Solve the step-1 result for f_s and the step-3 result for α, substitute for these quantities in the step-2 result, and solve for a_{cm}:

$$(mg\sin\phi - ma_{cm})R = I_{cm}\frac{a_{cm}}{R}$$
$$\text{so}\quad a_{cm} = \frac{g\sin\phi}{1 + \dfrac{I_{cm}}{mR^2}}$$

5. Substitute the step-4 result into the step-1 result and solve for f_s:

$$f_s = mg\sin\phi - ma_{cm} = mg\sin\phi - \frac{mg\sin\phi}{1 + \dfrac{I_{cm}}{mR^2}} = \frac{mg\sin\phi}{1 + \dfrac{mR^2}{I_{cm}}}$$

6. For a solid sphere, $I_{cm} = \frac{2}{5}mR^2$ (see Table 9-1). Substitute for I_{cm} in the step-4 and step-5 results:

$$a_{cm} = \frac{g\sin\phi}{1 + \frac{2}{5}} = \boxed{\frac{5}{7}g\sin\phi}$$
$$f_s = \frac{mg\sin\phi}{1 + \frac{5}{2}} = \boxed{\frac{2}{7}mg\sin\phi}$$

CHECK If the incline was frictionless the ball would not rotate and the acceleration would be $g\sin\phi$. With friction, we expect the acceleration to be less than $g\sin\phi$, which is evidenced by our first step-6 result.

TAKING IT FURTHER Because the ball rolls without slipping, the friction is static friction. Note that the result seems independent of the coefficient of static friction. However, we have assumed that the coefficient of static friction was large enough to prevent slipping.

The results of steps 5 and 6 in Example 9-18 apply to any round object with the center of mass at the geometric center that is rolling without slipping. For such objects, $I_{cm} = \beta mR^2$, where $\beta = \frac{2}{5}$ for a solid sphere, $\frac{1}{2}$ for a rolling solid cylinder, 1 for a thin hollow cylinder, and so forth. (These values of β are obtained from the expressions for I found in Table 9-1.) For such objects the step-5 and step-6 results can be expressed as

$$f_s = \frac{mg\sin\phi}{1 + \beta^{-1}} \qquad 9\text{-}32$$

$$a_{cm} = \frac{g\sin\phi}{1 + \beta} \qquad 9\text{-}33$$

The linear acceleration of any object rolling without slipping down an incline is less than $g\sin\phi$ because of the frictional force directed up the incline. Note that these accelerations are independent of the mass and the radius of the objects. That is, all uniform solid spheres will roll without slipping down an incline with the same acceleration. However, if we release a sphere, a cylinder, and a hoop at the top of an incline, and if they all roll without slipping, the sphere will reach the bottom first because it has the greatest acceleration. The cylinder will be second and the hoop last (Figure 9-38). A block that slides without friction down the incline will arrive at the bottom ahead of all three rolling objects. Perhaps surprisingly, a full can of soda pop that rolls without slipping will reach the bottom almost as fast

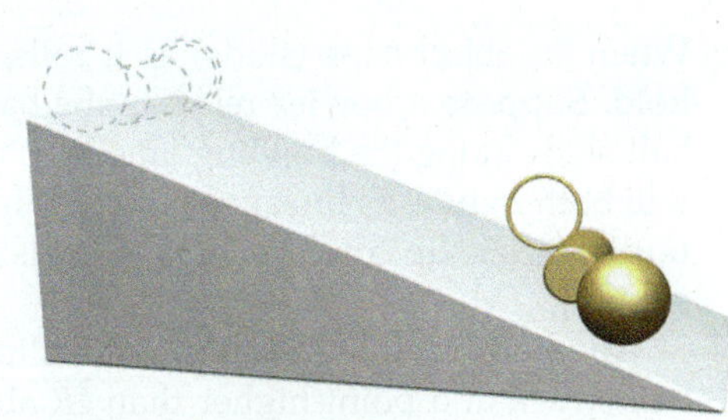

FIGURE 9-38

as the friction-free block. That is because the liquid in the can does not rotate with the can, so the effective moment of inertia of the can full of pop is just the moment of inertia of the metal can.

Static frictional forces do no work on the rolling objects, and if there is no slipping there is no dissipation of energy. Therefore, we use the conservation of mechanical energy to find the final speed of an object released from rest that is rolling without slipping down an incline. At the top of the incline, the total energy is the potential energy mgh. At the bottom, the total energy is kinetic energy. Conservation of mechanical energy therefore gives

$$\tfrac{1}{2}mv_{\text{cm}}^2 + \tfrac{1}{2}I_{\text{cm}}\omega^2 = mgh$$

We can use the nonslip condition to eliminate either v_{cm} or ω. Substituting $I_{\text{cm}} = \beta mR^2$ and $\omega = v_{\text{cm}}/R$, we obtain $\tfrac{1}{2}mv_{\text{cm}}^2 + \tfrac{1}{2}\beta mR^2(v_{\text{cm}}^2/R^2) = mgh$. Solving for v_{cm}^2 gives

$$v_{\text{cm}}^2 = \frac{2gh}{1+\beta} \qquad \text{9-34}$$

For a cylinder, with $\beta = \tfrac{1}{2}$, we obtain $v_{\text{cm}} = \sqrt{\tfrac{4}{3}gh}$. Note that this speed is independent of both the mass and the radius of the cylinder, and is less than $\sqrt{2gh}$ (the final speed if there were no friction so that the object just slides down the incline).

For an object rolling without slipping down an incline, the frictional force f_{s} is less than or equal to its maximum value; that is, $f_{\text{s}} \leq \mu_{\text{s}}F_{\text{n}}$, where $F_{\text{n}} = mg\cos\phi$. Substituting the expression from Equation 9-32 for the frictional force, we have

$$\frac{mg\sin\phi}{1+\beta^{-1}} \leq \mu_{\text{s}}mg\cos\phi$$

or

$$\tan\phi \leq (1+\beta^{-1})\mu_{\text{s}} \qquad \text{9-35}$$

For a uniform cylinder, $\beta = \tfrac{1}{2}$, and Equation 9-35 becomes $\tan\phi = 3\mu_{\text{s}}$. If the tangent of the angle of incline is greater than $(1+\beta^{-1})\mu_{\text{s}}$, the object will slip as it rolls down the incline.

PRACTICE PROBLEM 9-11 A uniform cylinder rolls down a plane inclined at $\phi = 50°$. What is the minimum value of the coefficient of static friction for which the cylinder will roll without slipping?

PRACTICE PROBLEM 9-12 For a uniform hoop of mass m that is rolling without slipping down an incline, (*a*) what is the force of friction, and (*b*) what is the maximum value of $\tan\phi$ for which the hoop will roll without slipping?

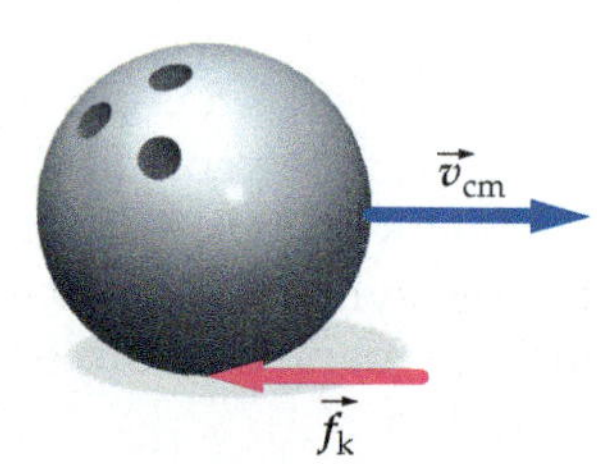

FIGURE 9-39 A bowling ball moving with no initial angular speed. The frictional force $\vec{f}_{\text{k}}$ exerted by the floor reduces the linear speed v_{cm} and increases the angular speed ω until $v_{\text{cm}} = R\omega$.

* ROLLING WITH SLIPPING

When an object slips (skids) as it rolls, the nonslip condition $v_{\text{cm}} = R\omega$ does not hold. Suppose a bowler releases the ball with no initial rotation ($\omega_0 = 0$). As the ball skids along the bowling lane, $v_{\text{cm}} > R\omega$. However, the kinetic frictional force will both reduce its linear speed v_{cm} (Figure 9-39) and increase its angular speed ω until the nonslip condition $v_{\text{cm}} = R\omega$ is reached, after which the ball rolls without slipping.

Another example of rolling with slipping is a ball with topspin, such as a cue ball struck at a point higher than $\tfrac{2}{5}R$ above the center (see Example 9-17) so that $v_{\text{cm}} < R\omega$. Then the kinetic frictional force both increases v_{cm} and decreases ω until the nonslip condition $v_{\text{cm}} = R\omega$ is reached (Figure 9-40).

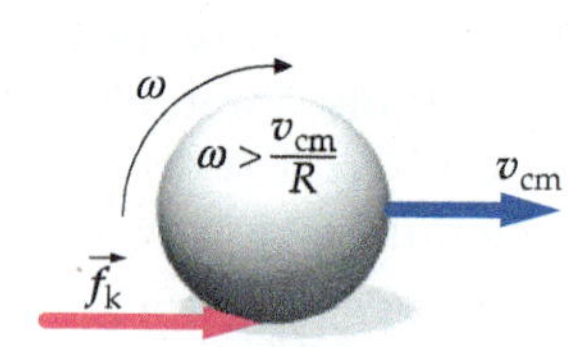

FIGURE 9-40 Ball with excess topspin. The frictional force accelerates the ball in the direction of motion.

Example 9-19 A Skidding Bowling Ball

A bowling ball of mass M and radius R is released at floor level so that at release it is moving horizontally with speed $v_0 = 5.0$ m/s and is not rotating. The coefficient of kinetic friction between the ball and the floor is $\mu_k = 0.080$. Find (*a*) the time the ball skids along the floor (after which it begins rolling without slipping), and (*b*) the distance the ball skids.

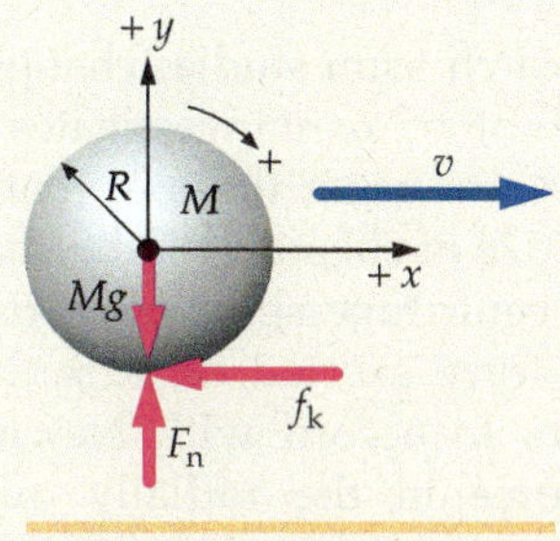

FIGURE 9-41

PICTURE During the skidding, $v_{cm} > R\omega$. We calculate v_{cm} and ω as functions of time, set v_{cm} equal to $R\omega$, and solve for the time. The linear and angular accelerations are found from $\Sigma F = ma$ and $\Sigma\tau_{cm} = I_{cm}\alpha$. Let the direction of motion be positive. There is slipping so the friction is kinetic (not static). This means that energy is dissipated by friction, so conservation of mechanical energy cannot be used to solve this problem.

SOLVE

(*a*) 1. Sketch a free-body diagram of the ball (Figure 9-41).

2. The net force on the ball is the force of kinetic friction f_k, which acts in the negative x direction. Apply Newton's second law:

$$\Sigma F_x = Ma_{cmx}$$
$$-f_k = Ma_{cmx}$$

3. The acceleration is in the negative x direction and $a_{cmy} = 0$. Find f_k by first finding F_n:

$$\Sigma F_y = Ma_{cmy} = 0 \Rightarrow F_n = Mg$$
$$\text{so}\quad f_k = \mu_k F_n = \mu_k Mg$$

4. Find the acceleration using the step-2 and step-3 results:

$$-\mu_k Mg = Ma_{cmx} \Rightarrow a_{cmx} = -\mu_k g$$

5. Relate the linear velocity to the constant acceleration and the time using a kinematic equation:

$$v_{cmx} = v_0 + a_{cmx}t = v_0 - \mu_k g t$$

6. Find α by applying Newton's second law for rotational motion to the ball. Compute the torques about the axis through the center of mass. Note that the free-body diagram has clockwise as positive:

$$\Sigma\tau = I_{cm}\alpha$$
$$\mu_k MgR + 0 + 0 = \tfrac{2}{5}MR^2\alpha$$
$$\text{so}\quad \alpha = \frac{5}{2}\frac{\mu_k g}{R}$$

7. Relate the angular velocity to the constant angular acceleration and the time using a kinematic equation:

$$\omega = \omega_0 + \alpha t = 0 + \alpha t = \frac{5}{2}\frac{\mu_k g}{R}t$$

8. Solve for the time t at which $v_{cm} = R\omega$:

$$v_{cm} = R\omega$$
$$(v_0 - \mu_k g t) = R\left(\frac{5}{2}\frac{\mu_k g}{R}t\right)$$
$$\text{so}\quad t = \frac{2v_0}{7\mu_k g} = \frac{2(5.0\ \text{m/s})}{7(0.080)(9.81\ \text{m/s}^2)} = \boxed{1.8\ \text{s}}$$

(*b*) The distance traveled while skidding is

$$\Delta x = v_0 t + \tfrac{1}{2}a_{cm}t^2 = v_0\left(\frac{2v_0}{7\mu_k g}\right) + \frac{1}{2}(-\mu_k g)\left(\frac{2v_0}{7\mu_k g}\right)^2 = \frac{12}{49}\frac{v_0^2}{\mu_k g}$$
$$= \frac{12}{49}\frac{(5.0\ \text{m/s})^2}{(0.080)(9.81\ \text{m/s}^2)} = \boxed{7.8\ \text{m}}$$

CHECK A bowling alley is about 60 feet, or 18 m, long. That the ball skids for 7.8 m, almost half way down the alley, is plausible.

TAKING IT FURTHER At a bowling alley, or on television, you can observe bowling balls skidding a considerable distance down the alley. The lanes at a well-maintained bowling alley are lightly oiled and very slick, so the balls skid for considerable distances.

PRACTICE PROBLEM 9-13 Find the speed of the bowling ball when it begins to roll without slipping. Does this speed depend on the value of μ_k?

PRACTICE PROBLEM 9-14 Find the total kinetic energy of the ball as it initiates rolling without slipping.

Physics Spotlight

Spindizzy—Ultracentrifuges

One research team studies changes in blood lipids as people change their diets.[*] Another team investigates the stability of viral proteins.[†] All the researchers are using the tool that won Theodor (The) Svedberg the 1926 Nobel Prize in chemistry—the analytical ultracentrifuge.[‡]

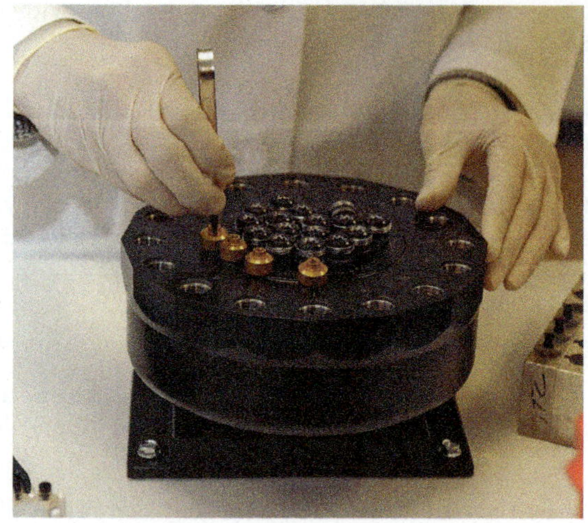

Sample tubes being loaded into an ultracentrifuge. *(Atherotech.)*

As a centrifuge spins, each particle suspended in the sample experiences a force exerted on it by its surroundings in the radially inward direction. In accord with Newton's third law, each particle exerts an equal force in the radially outward direction on its surroundings. Consequently, the particles *sediment*, or settle, to the outer region of the sample tube as the centrifuge spins. Larger, more-dense particles move most quickly to the end of the sample tube farthest from the rotation axis. This movement depends on several variables—the mass, density, and friction coefficients of the particles in the solution. The final result, *sedimentation equilibrium*, has layers, or strata, of particles arranged according to these variables. An ultracentrifuge is powerful enough to analyze complex molecules, and an ultracentrifuge that has windows into the sample chambers to measure changes in the absorption of ultraviolet light is called an analytical ultracentrifuge. These changes in absorption show the *sedimentation velocity*, or the speed at which different particles stratify. From the sedimentation velocities and sedimentation equilibrium, the purity, mass, shape, and composition of complex molecules can be calculated.

To analyze complex molecules, analytical ultracentrifuges must rotate extremely fast. Svedberg's first ultracentrifuge, built with an oil turbine rotor in 1924, spun at 12 000 rev/min, and generated an acceleration of $7000g$. In 1935, an ultracentrifuge was created that spun in a vacuum. This prevented delicate samples from being heated by the friction between air and the centrifuge.[§] This centrifuge was able to generate an acceleration of $150\,000g$ at the outer edge of the sample tubes. Today, analytical ultracentrifuges spin at 60 000 rev/min, with accelerations of $250\,000g$. Other types of ultracentrifuges can generate accelerations as large as $810\,000g$.[°]

The enormous accelerations and associated forces in these centrifuges cause high mechanical stresses on the rotors. At full speed, a single gram of material in an analytical ultracentrifuge sample tube has an apparent weight equal to 550 lb. In fact, the speed of the rotors is limited mostly by the tensile strength of the materials from which the rotors are made.[#] Over time, the stresses cause material fatigue in the ultracentrifuge. Caustic solutions can also make stress corrosion more likely in aluminum rotors. Rotor failures are catastrophic events[**] and have cracked cement block walls, shattered windows, and sent parts through the laboratory at high speeds.

Fortunately, ultracentrifuge rotor failures are very rare. The rotors in analytic ultracentrifuges are cast from strong materials—titanium, aluminum, or even carbon fiber compounds. The windows of analytic ultracentrifuges are made of optical-grade quartz or sapphire. Samples are carefully balanced in the rotors. Rotors are matched to particular ultracentrifuges and particular purposes. Finally, time logs are kept for each ultracentrifuge rotor. After a set number of hours, the rotors are replaced.

An early user of centrifuges was happy that "by means of the centrifuge, forces can be obtained many times greater than gravity, which otherwise exist only on the very largest planets."[††] Today, ultracentrifuges give researchers the effects of forces nearly 1000 times greater than the force of gravity at the surface of our Sun.

* Dragsted, L. O., Finne´ Nielsen, I.-L., Grønbæk, M., Hansen, A. S., Marckmann, P., and Nielsen, S. E., "Effect of Red Wine and Red Grape Extract on Blood Lipids, Haemostatic Factors, and Other Risk Factors for Cardiovascular Disease," *European Journal of Clinical Nutrition*, 2005 Vol. 59, 449–455.

† Chang, G.-G., Chang, H.-C., Chou, C.-Y., Hsu, W.-C., Lin, C.-H., and Lin, T.-Z., "Quaternary Structure of the Severe Acute Respiratory Syndrome (SARS) Coronavirus Main Protease." *Biochemistry*, Nov. 30, 2004, Vol. 43 No. 47, 14958–14970.

‡ Svedberg, Theodor, "The Ultracentrifuge," *1926 Nobel Lectures*. https://www.nobelprize.org/prizes/chemistry/1926/svedberg/lecture/May 2006.

§ Beams, H. W., "The Air Turbine Ultracentrifuge, Together with Some Results upon Ultracentrifuging the Eggs of *Fucus serratus*," *Journal of the Marine Biological Association*, Mar., 1937, Vol XXI, No. 2, 571–588.

° *Introduction to Analytical Ultracentrifugation*. https://www.beckman.com/resources/videos/products/introduction-to-auc

"Rotor Safety Guide." Beckman Coulter. https://www.beckman.com/resources/May, 2006.

** "Urgent Corrective Action Notice." http://www.ehs.cornell.edu/lrs/Centrifuge/letter.pdf May 2006; "Laboratory Safety Incidents—Explosions." American Industrial Hygiene Association. http://www2.umdnj.edu/eohssweb/aiha/accidents/explosion.htm May 2006.

†† Beams, op. cit., p. 571.

SUMMARY

1. Angular displacement, angular velocity, and angular acceleration are fundamental defined quantities in rotational kinematics.
2. Torque and moment of inertia are important derived dynamic concepts. Torque is a measure of the effect of a force in changing an object's rate of rotation. Moment of inertia is the measure of an object's inertial resistance to angular accelerations. The moment of inertia depends on the distribution of the mass relative to the rotation axis.
3. The parallel-axis theorem, which follows from the definition of the moment of inertia, often simplifies the calculation of I.
4. Newton's second law for rotation, $\Sigma\tau_{\text{ext}} = I\alpha$, is derived from Newton's second law and the definitions of τ, I, and α. It is an important relation for problems involving the rotation of a rigid object about an axis of fixed direction.

TOPIC	RELEVANT EQUATIONS AND REMARKS	
1. Angular Velocity and Angular Acceleration		
Angular velocity	$\omega = \dfrac{d\theta}{dt}$ (Definition)	9-2
Angular acceleration	$\alpha = \dfrac{d\omega}{dt}$ (Definition)	9-4
Tangential velocity	$v_{\text{t}} = r\omega$	9-8
Tangential acceleration	$a_{\text{t}} = r\alpha$	9-9
Centripetal acceleration	$a_{\text{c}} = \dfrac{v^2}{r} = r\omega^2$	9-10
2. Equations for Rotation with Constant Angular Acceleration	$\omega = \omega_0 + \alpha t$	9-5
	$\theta = \theta_0 + \omega_0 t + \frac{1}{2}\alpha t^2$	9-6
	$\omega^2 = \omega_0^2 + 2\alpha(\theta - \theta_0)$	9-7
3. Moment of Inertia		
System of particles	$I = \Sigma m_i r_i^2$ (Definition)	9-11
Continuous object	$I = \int r^2 dm$	9-13
Parallel-axis theorem	The moment of inertia about an axis a distance h from a parallel axis through the center of mass is $I = I_{\text{cm}} + Mh^2$ where I_{cm} is the moment of inertia about the axis through the center of mass and M is the total mass of the object.	9-14
4. Energy		
Kinetic energy for rotation about a fixed axis	$K = \frac{1}{2}I\omega^2$	9-12
Kinetic energy for a rotating object	$K = \frac{1}{2}Mv_{\text{cm}}^2 + \frac{1}{2}I_{\text{cm}}\omega^2$	9-30
Power	$P = \tau\omega$	9-25

TOPIC	RELEVANT EQUATIONS AND REMARKS	
5. Torque About an Axis	The torque due to a force equals the product of the tangential component of the force and the radial distance from the axis to the point of application of the force:	
	$\tau = F_t r = Fr\sin\phi = F\ell$	9-20
6. Newton's Second Law for Rotation	$\tau_{\text{net ext}} = \sum_i \tau_{i\text{ext}} = I\alpha$	9-19
	Newton's second law for rotation holds, even if the reference frame is noninertial, if the moment of inertia and the torques are calculated about an axis through the center of mass.	
7. Nonslip Conditions	If a string that is wrapped around a pulley wheel does not slip, the linear and angular quantities are related by	
	$v_t = R\omega$	9-22
	$a_t = R\alpha$	9-23
8. Rolling Objects		
Rolling without slipping	$v_{\text{cm}} = R\omega$	9-27
*Rolling with slipping	When an object slips while rolling, $v_{\text{cm}} \neq R\omega$. Kinetic friction then tends to change both v_{cm} and ω (increasing one while decreasing the other) until $v_{\text{cm}} = R\omega$, at which point rolling without slipping occurs.	

Answer to Concept Checks

9-1 Disk A has the greater moment of inertia. Mentally divide disk A into two parts, the part closer to the axis than one-eighth inch (Part 1), and the part farther from the axis than one-eighth inch (Part 2). Part 2 alone has the same mass and moment of inertia as does disk B, so the additional moment of inertial of Part 1 gives disk A the greater moment of inertia.

9-2 Disk B has the greater moment of inertia. The two disks have the same mass, but the mass of disk B is distributed farther from its axis than the mass of disk A is from its axis.

Answers to Practice Problems

9-1 314 rad/s

9-2 (*a*) 500 rev/min = 52.4 rad/s

9-3 $v_t = 1.88$ m/s, $a_t = 0$, $a_c = 59.2$ m/s^2

9-4 (*a*) 1.26 m/s, (*b*) 1.26 m/s

9-5 $I = 8ma^2$

9-6 Approximately 25 times greater

9-8 $a = 3g/4$ downward, (*b*) $F = mg/4$ downward

9-9 594 N · m

9-10 20 J

9-11 0.40

9-12 (*a*) $f = \frac{1}{2}mg\sin\phi$, (*b*) $\tan\phi \leq 2\mu_s$

9-13 $v_{\text{cm}} = \frac{5}{7}v_0$. No.

9-14 $K = \frac{5}{14}mv_0^2$

PROBLEMS

In a few problems, you are given more data than you actually need; in a few other problems, you are required to supply data from your general knowledge, outside sources, or informed estimate.

For all problems, use 9.81 m/s² for the free-fall acceleration and neglect friction and air resistance unless instructed to do otherwise.

Interpret as significant all digits in numerical values that have trailing zeros and no decimal points.

- • Single-concept, single-step, relatively easy
- •• Intermediate-level, may require synthesis of concepts
- ••• Challenging
- SSM Solution is in the *Student Solutions Manual*
- Consecutive problems that are shaded are paired problems.

CONCEPTUAL PROBLEMS

1 • Two points are on a disk that is turning about a fixed axis perpendicular to the disk and through its center at increasing angular velocity. One point is on the rim and the other point is halfway between the rim and the center. (*a*) Which point moves the greater distance in a given time? (*b*) Which point turns through the greater angle? (*c*) Which point has the greater speed? (*d*) Which point has the greater angular speed? (*e*) Which point has the greater tangential acceleration? (*f*) Which point has the greater angular acceleration? (*g*) Which point has the greater centripetal acceleration?

2 • True or false: (*a*) Angular speed and linear speed have the same dimensions. (*b*) All parts of a wheel rotating about a fixed axis must have the same angular speed. (*c*) All parts of a wheel rotating about a fixed axis must have the same angular acceleration. (*d*) All parts of a wheel rotating about a fixed axis must have the same centripetal acceleration.

3 • Starting from rest and rotating at constant angular acceleration, a disk takes 10 revolutions to reach an angular speed ω. How many additional revolutions at the same angular acceleration are required for it to reach an angular speed of 2ω? (*a*) 10 rev, (*b*) 20 rev, (*c*) 30 rev, (*d*) 40 rev, (*e*) 50 rev

4 • You are looking down from above at a merry-go-round and observe that it is rotating counterclockwise and that its rotation rate is slowing. If we designate counterclockwise as positive, what is the sign of the angular acceleration?

5 • Chad and Tara go for a ride on a merry-go-round. Chad sits on a pony that is 2.0 m from the rotation axis, and Tara sits on a pony 4.0 m from the axis. The merry-go-round is traveling counterclockwise and is speeding up. Does Chad or Tara have (*a*) the larger linear speed? (*b*) the larger centripetal acceleration? (*c*) the larger tangential acceleration?

6 • Disk B was identical to disk A before a hole was drilled though the center of disk B. Which disk has the largest moment of inertia about its symmetry axis center? Explain your answer.

7 • **CONTEXT-RICH** The pitcher in a baseball game has a blazing fastball. You have not been able to swing the bat in time to hit the ball. You are now just trying to make the bat connect with the ball, hit the ball foul, and avoid a strikeout. So you decide to take your coach's advice and grip the bat high rather than at the very end. This change should increase bat speed; thus, you will be able to swing the bat quicker and increase your chances of hitting the ball. Explain how this theory works in terms of the moment of inertia, angular acceleration, and torque of the bat. SSM

8 • (*a*) Is the direction of an object's angular velocity necessarily the same as the direction net torque on it? Explain. (*b*) If net torque and angular velocity are in opposite directions, what does that tell you about the angular speed? (*c*) Can the angular velocity be zero even if the net torque is not zero? If your answer is yes, give an example.

9 • A disk is free to rotate about a fixed axis. A tangential force applied a distance *d* from the axis causes an angular acceleration α. What angular acceleration is produced if the same force is applied a distance 2*d* from the axis? (*a*) α, (*b*) 2α, (*c*) $\alpha/2$, (*d*) 4α, (*e*) $\alpha/4$

10 • The moment of inertia of an object about an axis that does not pass through its center of mass is ______ the moment of inertia about a parallel axis through its center of mass: (*a*) always less than, (*b*) sometimes less than, (*c*) sometimes equal to, (*d*) always greater than.

11 • The motor of a merry-go-round exerts a constant torque on it. As it speeds up from rest, the power output of the motor (*a*) is constant, (*b*) increases linearly with the angular speed of the merry-go-round, (*c*) is zero, (*d*) none of the above. SSM

12 • A constant net torque acts on a merry-go-round from startup until it reaches its operating speed. During this time, the merry-go-round's kinetic energy (*a*) is constant, (*b*) increases linearly with angular speed, (*c*) increases quadratically as the square of the angular speed, (*d*) none of the above.

13 • **ENGINEERING APPLICATION** Most doors knobs are designed so the knob is located on the side opposite the hinges (rather than in the center of the door, for example). Explain why this practice makes doors easier to open.

14 • A wheel of radius *R* and angular speed ω is rolling without slipping toward the north on a flat, stationary surface. The velocity of the point on the rim that is (momentarily) in contact with the surface is (*a*) equal in magnitude to $R\omega$ and directed toward the north, (*b*) equal to in magnitude $R\omega$ and directed toward the south, (*c*) zero, (*d*) equal to the speed of the center of mass and directed toward the north, (*e*) equal to the speed of the center of mass and directed toward the south.

15 • A uniform solid cylinder and a uniform solid sphere have equal masses. Both roll on a horizontal surface without slipping. If their total kinetic energies are the same, then (*a*) the translational speed of the cylinder is greater than the translational speed of the sphere, (*b*) the translational speed of the cylinder is less than the translational speed of the sphere, (*c*) the translational speeds of the two objects are the same, (*d*), (*a*), (*b*), or (*c*) could be correct, depending on the radii of the objects.

16 • Two identical-looking 1.0-m-long pipes are each plugged with 10 kg of lead. In the first pipe, the lead is concentrated at the middle of the pipe, while in the second the lead is divided into two 5-kg masses placed at opposite ends of the pipe. The ends of the pipes are then sealed using four identical caps. Without opening either pipe, how could you determine which pipe has the lead at the ends?

17 •• Starting simultaneously from rest, a coin and a hoop roll without slipping down an incline. Which of the following statements is true? (*a*) The hoop reaches the bottom first. (*b*) The coin reaches the bottom first. (*c*) The coin and hoop arrive at the bottom simultaneously. (*d*) The race to the bottom depends on their relative masses. (*e*) The race to the bottom depends on their relative diameters.

18 •• For a hoop of mass *M* and radius *R* that is rolling without slipping, which is larger, its translational kinetic energy or its kinetic energy relative to the center of mass? (*a*) Its translational kinetic energy is larger. (*b*) Its kinetic energy relative to the center of mass is larger. (*c*) Both energies have the same magnitude. (*d*) The answer depends on the radius of the hoop. (*e*) The answer depends on the mass of the hoop.

19 •• For a disk of mass *M* and radius *R* that is rolling without slipping, which is larger, its translational kinetic energy or its kinetic energy relative to the center of mass? (*a*) Its translational kinetic energy is larger. (*b*) Its kinetic energy relative to the center of mass is larger. (*c*) Both energies have the same magnitude. (*d*) The answer depends on the radius of the disk. (*e*) The answer depends on the mass of the disk.

20 •• A perfectly rigid ball rolls without slipping along a perfectly rigid horizontal plane. Show that the frictional force acting on the ball must be zero. *Hint: Consider a possible direction for the action of the frictional force and what effects such a force would have on the velocity of the center of mass and on the angular velocity.*

21 •• A spool is free to rotate about a fixed axis, and a string wrapped around the axle of the spool causes the spool to rotate in a counterclockwise direction (Figure 9-42*a*). However, if the spool is set on a horizontal tabletop, the spool instead (given sufficient frictional force between the table and the spool) rotates in a clockwise

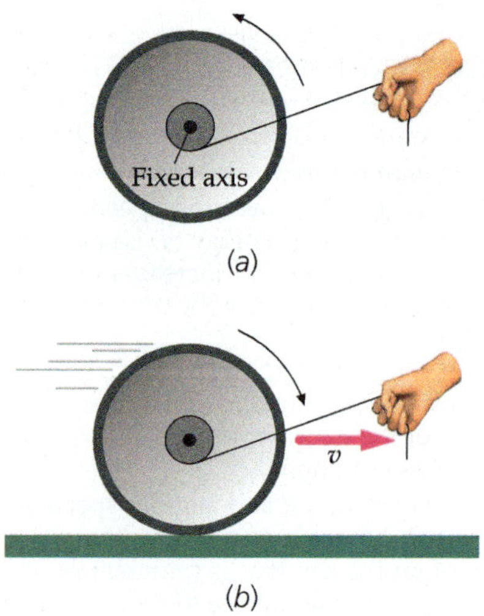

FIGURE 9-42 Problem 21

direction and rolls to the right (Figure 9-42*b*). By considering torque about the appropriate axes, show that these conclusions are both consistent with Newton's second law for rotations. SSM

22 •• You want to locate the center of gravity of an arbitrarily shaped flat object. You are told to suspend the object from a point, and to suspend a plumb line from the same point. Then draw a vertical line on the object to represent the plumb line. Next, you repeat the process using a different suspension point. The center of gravity will be at the intersection of the drawn lines. Explain the principle(s) behind this process.

ESTIMATION AND APPROXIMATION

23 • A baseball is thrown at 88 mi/h and with a spin rate of 1500 rev/min. If the distance between the pitcher's point of release and the catcher's glove is about 61 feet, estimate how many revolutions the ball makes between release and catch. Neglect any effects of gravity or air resistance on the ball's flight.

24 •• Consider the Crab Pulsar, discussed on page 293. Justify the statement that the loss in rotational energy is equivalent to the power output of 100 000 stars. The total power radiated by the Sun is about 4×10^{26} W. Assume that the pulsar has a mass that is 2×10^{30} kg, has a radius that is 20 km, is rotating at about 30 rev/s, and has a rotational period that is increasing at 10^{-5} s/y.

25 •• A 14-kg bicycle has 1.2-m-diameter wheels, each with a mass of 3.0 kg. The mass of the rider is 38 kg. Estimate the fraction of the total kinetic energy of the rider–bicycle system that is associated with rotation of the wheels.

26 •• Why does toast falling off a table always land jelly-side down? The question may sound silly, but it has been a subject of serious scientific enquiry. The analysis is too complicated to reproduce here, but R. D. Edge and Darryl Steinert showed that a piece of toast, pushed gently over the edge of a table until it tilts off, typically falls off the table when it makes an angle of about 30° with the horizontal (Figure 9-43) and at that instant has an angular speed of $\omega = 0.956\sqrt{g/\ell}$, where ℓ is the length of one edge of the piece of toast (assumed to be square).* Assuming that a piece of toast is jelly-side up, what side will it land on if it falls from a 0.500-m-high table? If it falls from a 1.00-m-high table? Assume that $\ell = 10.0$ cm. Ignore any forces due to air resistance.

* For readers interested in this problem and a host of others, we highly recommend Robert Erlich's wonderful book, *Why Toast Lands Jelly-Side Down: Zen and the Art of Physics Demonstrations.*

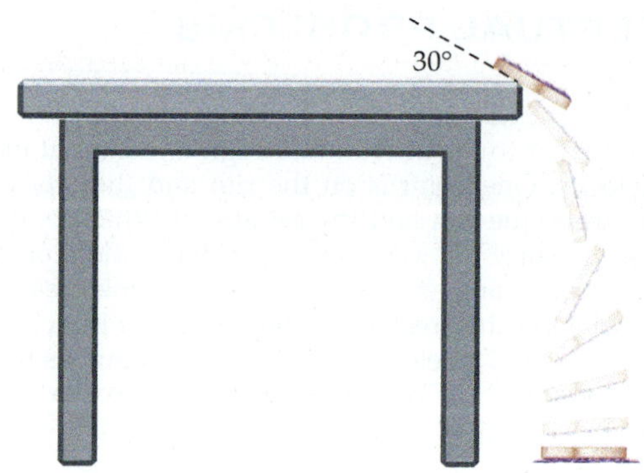

FIGURE 9-43 Problem 26

27 •• Consider your moment of inertia about a vertical axis through the center of your body, both when you are standing straight up with your arms flat against your sides, and when you are standing straight up holding your arms straight out to your sides. Estimate the ratio of the moment of inertia with your arms straight out to the moment of inertia with your arms flat against your sides.

ANGULAR VELOCITY, ANGULAR SPEED, AND ANGULAR ACCELERATION

28 • A particle moves with a constant speed of 25 m/s in a 90-m-radius circle. (*a*) What is its angular speed in radians per second about the center of the circle? (*b*) How many revolutions does it make in 30 s?

29 • A wheel released from rest is rotating with constant angular acceleration of 2.6 rad/s². At 6.0 s after the release: (*a*) What is its angular speed? (*b*) Through what angle has the wheel turned? (*c*) How many revolutions has it completed? (*d*) What is the linear speed and what is the magnitude of the linear acceleration of a point 0.30 m from the axis of rotation? SSM

30 • **MULTISTEP** When a turntable rotating at 33 rev/min is shut off, it comes to rest in 26 s. Assuming constant angular acceleration, find (*a*) the angular acceleration. During the 26 s, find (*b*) the average angular speed and (*c*) the angular displacement in revolutions.

31 • A 12-cm-radius disk that begins to rotate about its axis at $t = 0$, rotates with a constant angular acceleration of 8.0 rad/s². At $t = 5.0$ s, (*a*) what is the angular speed of the disk, and (*b*) what are the tangential and centripetal components of the acceleration of a point on the edge of the disk?

32 • A 12-m-radius Ferris wheel rotates once each 27 s. (*a*) What is its angular speed (in radians per second)? (*b*) What is the linear speed of a passenger? (*c*) What is the acceleration of a passenger?

33 • A cyclist accelerates uniformly from rest. After 8.0 s, the wheels have rotated 3.0 rev. (*a*) What is the angular acceleration of the wheels? (*b*) What is the angular speed of the wheels at the end of the 8.0 s?

34 • What is the angular speed of Earth in radians per second as it rotates about its axis?

35 • A wheel rotates through 5.0 rad in 2.8 s as it is brought to rest with constant angular acceleration. Determine the wheel's initial angular speed before braking began.

36 • A bicycle has 0.750-m-diameter wheels. The bicyclist accelerates from rest with constant acceleration to 24.0 km/h in 14.0 s. What is the angular acceleration of the wheels?

37 •• **ENGINEERING APPLICATION** The tape in a standard VHS videotape cassette has a total length of 246 m, which is enough for the tape to play for 2.0 h (Figure 9-44). As the tape starts, the full reel has a 45-mm outer radius and a 12-mm inner radius. At some point during the play, both reels have the same angular speed. Calculate this angular speed in radians per second and in revolutions per minute. *Hint: Between the two reels the tape moves at constant speed.* SSM

FIGURE 9-44 Problem 37 (© *Treë.*)

38 •• **CONTEXT-RICH** To start a lawn mower, you must pull on a rope wound around the perimeter of a flywheel. After you pull the rope for 0.95 s, the flywheel is rotating at 4.5 revolutions per second, at which point the rope disengages. This attempt at starting the mower does not work, however, and the flywheel slows, coming to rest 0.24 s after the disengagement. Assume constant acceleration during both spin-up and spin-down. (*a*) Determine the average angular acceleration during the 4.5-s spin-up and again during the 0.24-s spin-down. (*b*) What is the maximum angular speed reached by the flywheel? (*c*) Determine the ratio of the number of revolutions made during spin-up to the number made during spin-down.

39 ••• Mars orbits the Sun at a mean orbital radius of 228 Gm (1 Gm = 10^9 m) and has an orbital period of 687 d. Earth orbits the Sun at a mean orbital radius of 149.6 Gm. (*a*) The Earth–Sun line sweeps out an angle of 360° during one Earth-year. Approximately what angle is swept out by the Mars-Sun line during one Earth-year? (*b*) How frequently are Mars and the Sun in opposition (on diametrically opposite sides of Earth)?

CALCULATING THE MOMENT OF INERTIA

40 • A tennis ball has a mass of 57 g and a diameter of 7.0 cm. Find the moment of inertia about its diameter. Model the ball as a thin spherical shell.

41 • Four particles, one at each of the four corners of a square with 2.0-m-long edges, are connected by massless rods (Figure 9-45). The masses of the particles are $m_1 = m_3 = 3.0$ kg and $m_2 = m_4 = 4.0$ kg. Find the moment of inertia of the system about the z axis. SSM

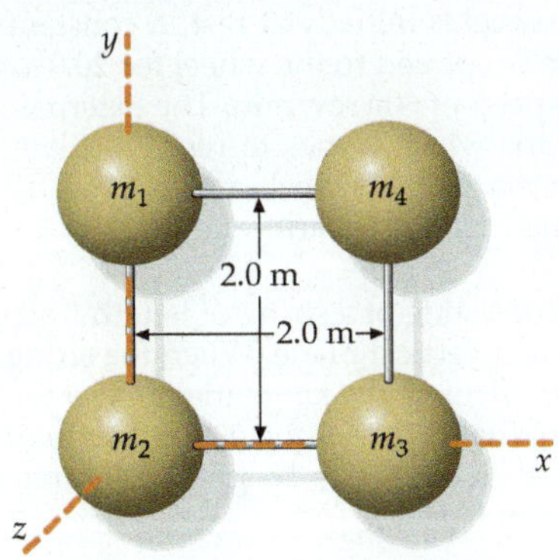

FIGURE 9-45 Problem 41

42 •• Use the parallel-axis theorem and the result for Problem 41 to find the moment of inertia of the four-particle system in Figure 9-45 about an axis that passes through the center of mass and is parallel with the z axis. Check your result by direct computation.

43 • For the four-particle system of Figure 9-45, (*a*) find the moment of inertia I_x about the x axis, which passes through m_2 and m_3, and (*b*) find the moment of inertia I_y about the y axis, which passes through m_1 and m_2.

44 • Determine the moment of inertia of a uniform solid sphere of mass M and radius R about an axis that is tangent to the surface of the sphere (Figure 9-46).

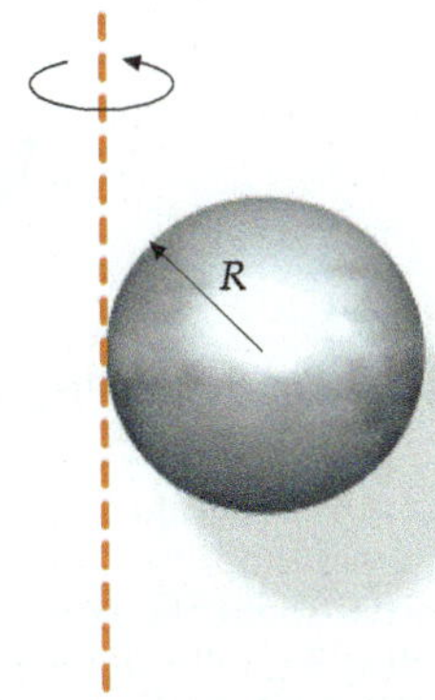

FIGURE 9-46 Problem 44

45 •• A 1.00-m-diameter wagon wheel consists of a thin rim having a mass of 8.00 kg and 6 spokes, each with a mass of 1.20 kg. Determine the moment of inertia of the wagon wheel about its axis.

46 •• **MULTISTEP** Two point masses m_1 and m_2 are separated by a massless rod of length L. (*a*) Write an expression for the moment of inertia I about an axis perpendicular to the rod and passing through it a distance x from mass m_1. (*b*) Calculate dI/dx and show that I is at a minimum when the axis passes through the center of mass of the system.

47 •• A uniform rectangular plate has mass m and edges of lengths a and b. (*a*) Show by integration that the moment of inertia of the plate about an axis that is perpendicular to the plate and passes through one corner is $\frac{1}{3}m(a^2 + b^2)$. (*b*) What is the moment of inertia about an axis that is perpendicular to the plate and passes through its center of mass?

48 •• **CONTEXT-RICH** In attempting to ensure a spot on the pep squad, you and your friend Corey research baton-twirling. Each of you is using "The Beast" as a model baton: two uniform spheres, each of mass 500 g and radius 5.00 cm, mounted at the ends of a 30.0-cm uniform rod of mass 60.0 g (Figure 9-47). You want to determine the moment of inertia I of "The Beast" about an axis perpendicular to the rod and passing through its center. Corey uses the approximation that the two spheres can be treated as point particles that are 20.0 cm from the axis of rotation, and that the mass of the rod is negligible. You, however, decide to do an exact calculation. (*a*) Compare the two results. (Give the percentage difference between them). (*b*) Suppose the spheres were replaced by two thin spherical shells, each of the same mass as the original solid spheres. Give a conceptual argument explaining how this replacement would, or would not, change the value of I.

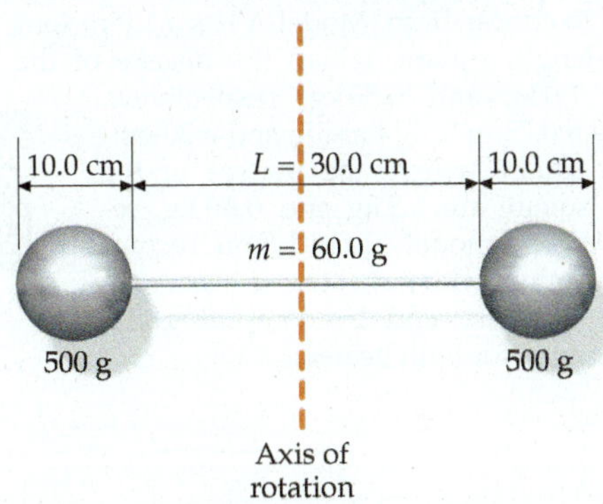

FIGURE 9-47 Problem 48

49 •• The methane molecule (CH_4) has four hydrogen atoms located at the vertices of a regular tetrahedron of edge length 0.18 nm, with the carbon atom at the center of the tetrahedron (Figure 9-48). Find the moment of inertia of this molecule for rotation about an axis that passes through the centers of the carbon atom and one of the hydrogen atoms.

FIGURE 9-48 Problem 49

50 •• A hollow cylinder has mass m, an outside radius R_2, and an inside radius R_1. Use integration to show that the moment of inertia about its axis is given by $I = \frac{1}{2}m(R_2^2 + R_1^2)$. *Hint: Review Section 9-3, where the moment of inertia is calculated for a solid cylinder by direct integration.*

51 •• **BIOLOGICAL APPLICATION** While slapping the water's surface with his tail to communicate danger, a beaver must rotate it about one of its narrow ends. Let us model the tail as a rectangle of uniform thickness and density (Figure 9-49). Estimate its moment of inertia about the line passing through its narrow end (dashed line). Assume that the tail measures 15 by 30 cm with a thickness of 1.0 cm and that the flesh has the density of water.

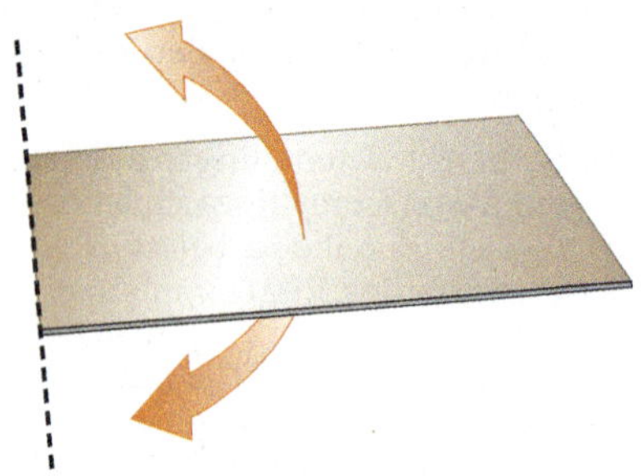

FIGURE 9-49 Problem 51

52 •• **CONTEXT-RICH** To prevent damage to her shoulders, your elderly grandmother wants to purchase the rug beater (Figure 9-50) with the lowest moment of inertia about its grip end. Knowing you are taking physics, she asks your advice. There are two models to choose from. Model A has a 1.0-m-long handle on a 40-cm-edge-length square, where the masses of the handle and square are 1.0 kg and 0.50 kg, respectively. Model B has a 0.75-m-long handle and a 30-cm-edge-length square, where the masses of the handle and square are 1.5 kg and 0.60 kg, respectively. Which model should you recommend? Determine which beater is easier to swing from the very end by computing the moment of inertia for both beaters.

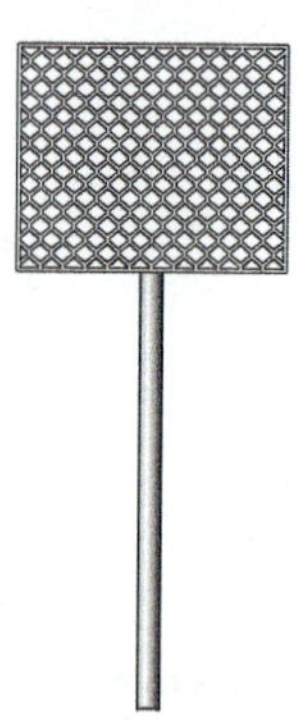

FIGURE 9-50 Problem 52

53 ••• Use integration to show that the moment of inertia of a thin spherical shell of radius R and mass m about an axis through its center is $2mR^2/3$. SSM

54 ••• According to one model, the density of Earth varies with the distance r from the center of Earth as $\rho = C[1.22 - (r/R)]$, where R is the radius of Earth and C is a constant. (*a*) Find C in terms of the total mass M and the radius R. (*b*) According to this model, what is the moment of inertia of Earth about an axis through its center. (See Problem 53.)

55 ••• Use integration to determine the moment of inertia about its axis of a uniform right circular cone of height H, base radius R, and mass M.

56 ••• Use integration to determine the moment of inertia of a thin uniform disk of mass M and radius R about an axis in the plane of the disk and passing through its center. Check your answer by referring to Table 9-1.

57 ••• **ENGINEERING APPLICATION, CONTEXT-RICH** An advertising firm has contacted your engineering firm to create a new advertisement for a local ice-cream stand. The owner of this stand wants to add rotating solid cones (painted to look like ice-cream cones, of course) to catch the eye of travelers. Each cone will rotate about an axis parallel to its base and passing through its apex. The actual size of the cones is to be decided upon, and the owner wonders if it would be more energy-efficient to rotate smaller cones than larger ones. He asks your firm to write a report showing the determination of the moment of inertia of a homogeneous right circular cone of height H, base radius R, and mass M. What is the result of your report?

TORQUE, MOMENT OF INERTIA, AND NEWTON'S SECOND LAW FOR ROTATION

58 • **ENGINEERING APPLICATION, CONTEXT-RICH** A firm wants to determine the amount of frictional torque in their current line of grindstones, so they can redesign them to be more energy efficient. To do this, they ask you to test the best-selling model, which is basically a disk-shaped grindstone of mass 1.70 kg and radius 8.00 cm that operates at 730 rev/min. When the power is shut off, you time the grindstone and find it takes 31.2 s for it to stop rotating. (*a*) What is the angular acceleration of the grindstone? (Assume constant angular acceleration.) (*b*) What is the frictional torque exerted on the grindstone?

59 • A 2.5-kg 11-cm-radius cylinder, initially at rest, is free to rotate about the axis of the cylinder. A rope of negligible mass is wrapped around it and pulled with a force of 17 N. Assuming that the rope does not slip, find (*a*) the torque exerted on the cylinder by the rope, (*b*) the angular acceleration of the cylinder, and (*c*) the angular speed of the cylinder after 0.50 s. SSM

60 •• A grinding wheel is initially at rest. A constant external torque of 50.0 N·m is applied to the wheel for 20.0 s, giving the wheel an angular speed of 600 rev/min. The external torque is then removed, and the wheel comes to rest 120 s later. Find (*a*) the moment of inertia of the wheel, and (*b*) the frictional torque, which is assumed to be constant.

61 •• A pendulum consisting of a string of length L attached to a bob of mass m swings in a vertical plane. When the string is at an angle θ to the vertical, (*a*) calculate the tangential acceleration of the bob using $\Sigma F_t = ma_t$. (*b*) What is the torque exerted about the pivot point? (*c*) Show that $\Sigma\tau = I\alpha$ with $a_t = L\alpha$ gives the same tangential acceleration as found in Part (*a*).

62 ••• A uniform rod of mass M and length L is pivoted at one end and hangs as in Figure 9-51 so that it is free to rotate without friction about its pivot. It is struck a sharp horizontal blow a distance x below the pivot, as shown. (*a*) Show that, just after the rod is struck, the speed of the center of mass of the rod is given by $v_0 = 3xF_0\,\Delta t/(2ML)$, where F_0 and Δt are the average force and duration, respectively, of the blow. (*b*) Find the horizontal component of the force exerted by the pivot on the rod, and show that this force component is zero if $x = \frac{2}{3}L$. This point (the point of impact when the horizontal component of the pivot force is zero) is called the *center of percussion* of the rod-pivot system.

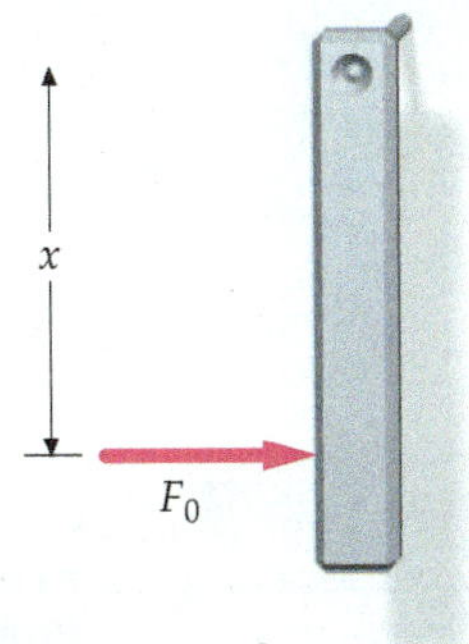

FIGURE 9-51 Problem 62

63 ••• **MULTISTEP** A uniform horizontal disk of mass M and radius R is spinning about the vertical axis through its center with an angular speed ω. When the spinning disk is dropped onto a horizontal tabletop, kinetic-frictional forces on the disk oppose its spinning motion. Let μ_k be the coefficient of kinetic friction between the disk and the tabletop. (*a*) Find the torque $d\tau$ exerted by the force of friction on a circular element of radius r and width dr. (*b*) Find the total torque exerted by friction on the disk. (*c*) Find the time required for the disk to stop rotating.

ENERGY METHODS, INCLUDING KINETIC ENERGY DUE TO ROTATION

64 • The particles in Figure 9-52 are connected by a very light rod. They rotate about the y axis at 2.0 rad/s. (*a*) Find the speed of each particle, and use it to calculate the kinetic energy of this system directly from $\Sigma\frac{1}{2}m_iv_i^2$. (*b*) Find the moment of inertia about the y axis, calculate the kinetic energy from $K = \frac{1}{2}I\omega^2$, and compare your result with your Part-(*a*) result.

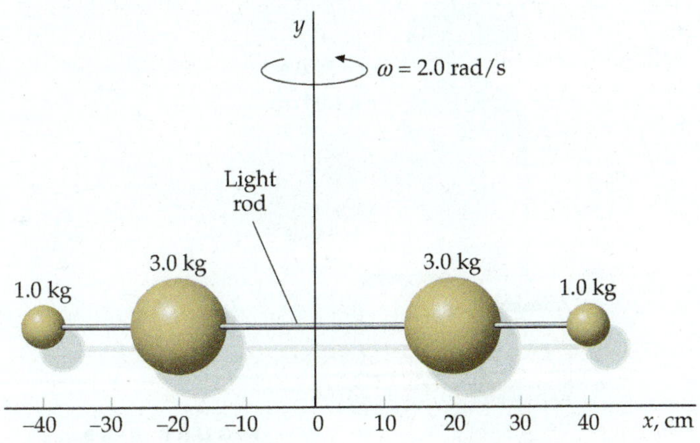

FIGURE 9-52 Problem 64

65 • A 1.4-kg 15-cm-diameter solid sphere is rotating about its diameter at 70 rev/min. (*a*) What is its kinetic energy? (*b*) If an additional 5.0 mJ of energy are added to the kinetic energy, what is the new angular speed of the sphere? SSM

66 •• Calculate the kinetic energy of Earth due to its spinning about its axis, and compare your answer with the kinetic energy of the orbital motion of Earth's center of mass about the Sun. Assume Earth to be a homogeneous sphere of mass 6.0×10^{24} kg and radius 6.4×10^{6} m. The radius of Earth's orbit is 1.5×10^{11} m.

67 •• A 2000-kg block is lifted at a constant speed of 8.0 cm/s by a steel cable that passes over a massless pulley to a motor-driven winch (Figure 9-53). The radius of the winch drum is 30 cm. (*a*) What is the tension in the cable? (*b*) What torque does the cable exert on the winch drum? (*c*) What is the angular speed of the winch drum? (*d*) What power must be developed by the motor to drive the winch drum? SSM

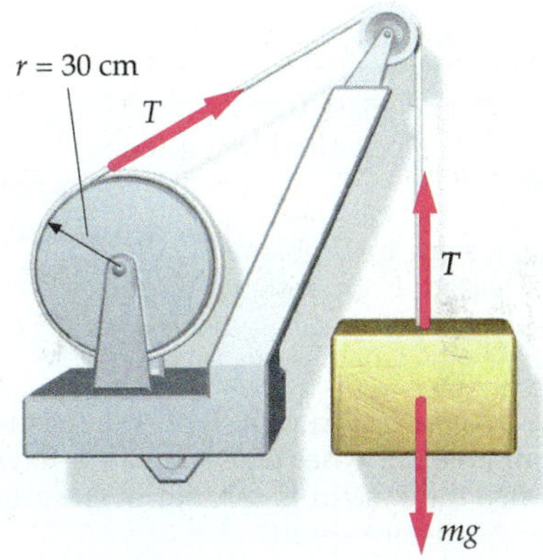

FIGURE 9-53 Problem 67

68 •• A uniform disk that has a mass M and a radius R can rotate freely about a fixed horizontal axis that passes through its center and is perpendicular to the plane of the disk. A small particle that has a mass m is attached to the rim of the disk at the top, directly above the pivot. The system is gently nudged, and the disk begins to rotate. As the particle passes through its lowest point, (*a*) what is the angular speed of the disk, and (*b*) what force is exerted by the disk on the particle?

69 •• A uniform 1.5-m-diameter ring is pivoted at a point on its perimeter so that it is free to rotate about a horizontal axis that is perpendicular to the plane of the ring. The ring is released with the center of the ring at the same height as the axis (Figure 9-54). (*a*) If the ring was released from rest, what was its maximum angular speed? (*b*) What minimum angular speed must it be given at release if it is to rotate a full 360°?

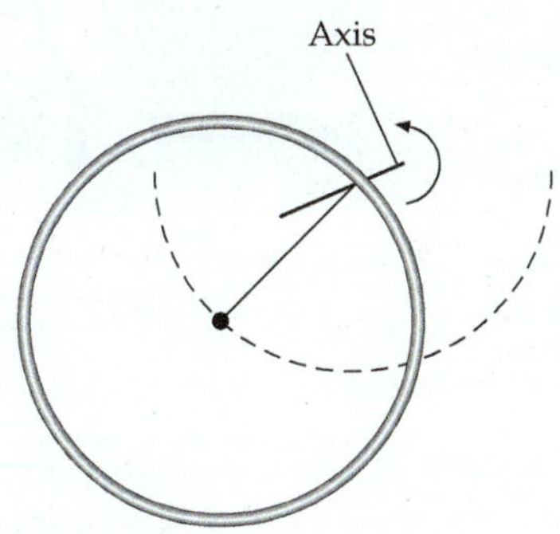

FIGURE 9-54 Problem 69

70 •• **ENGINEERING APPLICATION, CONTEXT-RICH** You set out to design a car that uses the energy stored in a flywheel consisting of a uniform 100-kg cylinder of radius R that has a maximum angular speed of 400 rev/s. The flywheel must deliver an average of 2.00 MJ of energy for each kilometer of distance. Find the smallest value of R for which the car can travel 300 km without the flywheel needing to be reenergized.

PULLEYS, YO-YOS, AND HANGING THINGS

71 •• The system shown in Figure 9-55 consists of a 4.0-kg block resting on a frictionless horizontal ledge. This block is attached to a string that passes over a pulley, and the other end of the string is attached to a hanging 2.0-kg block. The pulley is a uniform disk of radius 8.0 cm and mass 0.60 kg. Find the acceleration of each block and the tension in the string. SSM

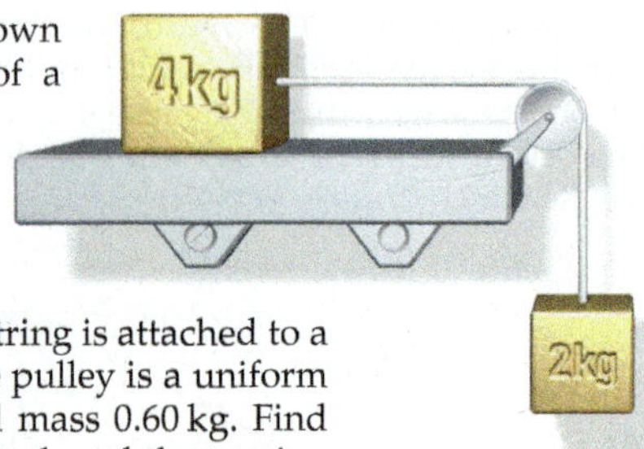

FIGURE 9-55
Problems 71–74

72 •• For the system in Problem 71, the 2.0-kg block is released from rest. (*a*) Find the speed of the block after it falls a distance of 2.5 m. (*b*) What is the angular speed of the pulley at this instant?

73 •• For the system in Problem 71, if the (frictionless) ledge were adjustable in angle, at what angle would it have to be tilted upward so that once the system is set into motion the blocks will continue to move at constant speed?

74 •• In the system shown in Figure 9-55, there is a 4.0-kg block resting on a horizontal ledge. The coefficient of kinetic friction between the ledge and the block is 0.25. The block is attached to a string that passes over a pulley, and the other end of the string is attached to a hanging 2.0-kg block. The pulley is a uniform disk of radius 8.0 cm and mass 0.60 kg. Find the acceleration of each block and the tensions in the segments of string between each block and the pulley.

75 •• A 1200-kg car is being raised over water by a winch. At the moment the car is 5.0 m above the water (Figure 9-56), the gearbox breaks, allowing the winch drum to spin freely as the car falls. During the car's fall, there is no slipping between the (massless) rope, the pulley wheel, and the winch drum. The moment of inertia of the winch drum is $320\ \text{kg}\cdot\text{m}^2$, and the moment of inertia of the pulley wheel is $4.00\ \text{kg}\cdot\text{m}^2$. The radius of the winch drum is 0.800 m, and the radius of the pulley is 0.300 m. Assume that the car starts to fall from rest. Find the speed of the car as it hits the water.

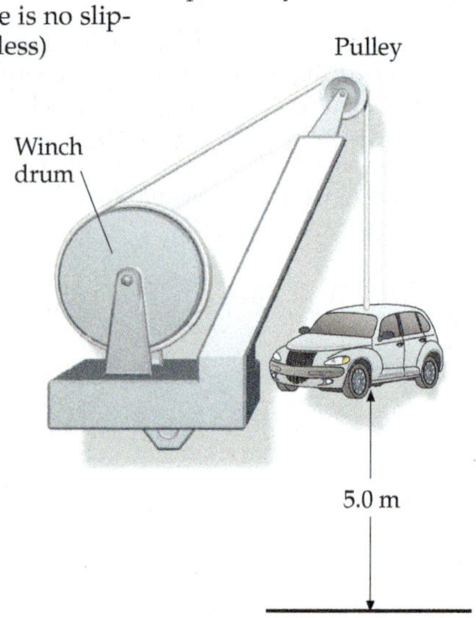

FIGURE 9-56
Problem 75

76 •• The system in Figure 9-57 is released from rest when the 30-kg block is 2.0 m above the ledge. The pulley is a uniform 5.0-kg disk with a radius of 10 cm. Just before the 30-kg block hits the ledge, find (*a*) its speed, (*b*) the angular speed of the pulley, and (*c*) the tensions in the strings. (*d*) Find the time of descent for the 30-kg block. Assume that the string does not slip on the pulley.

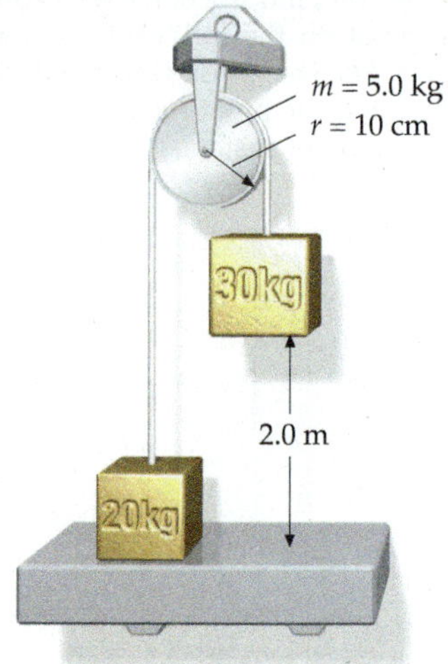

FIGURE 9-57
Problem 76

77 •• A uniform solid sphere of mass M and radius R is free to rotate about a horizontal axis through its center. A string is wrapped around the sphere and is attached to an object of mass m (Figure 9-58). Assume that the string does not slip on the sphere. Find (*a*) the acceleration of the object and (*b*) the tension in the string.

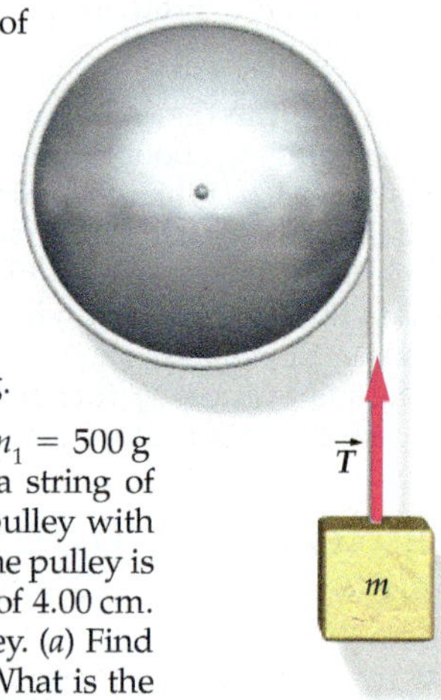

FIGURE 9-58
Problem 77

78 •• Two objects, of masses $m_1 = 500$ g and $m_2 = 510$ g, are connected by a string of negligible mass that passes over a pulley with frictionless bearings (Figure 9-59). The pulley is a uniform 50.0-g disk with a radius of 4.00 cm. The string does not slip on the pulley. (*a*) Find the accelerations of the objects. (*b*) What is the tension in the string between the 500-g block and the pulley? What is the tension in the string between the 510-g block and the pulley? By how much do these tensions differ? (*c*) What would your answers be if you neglected the mass of the pulley?

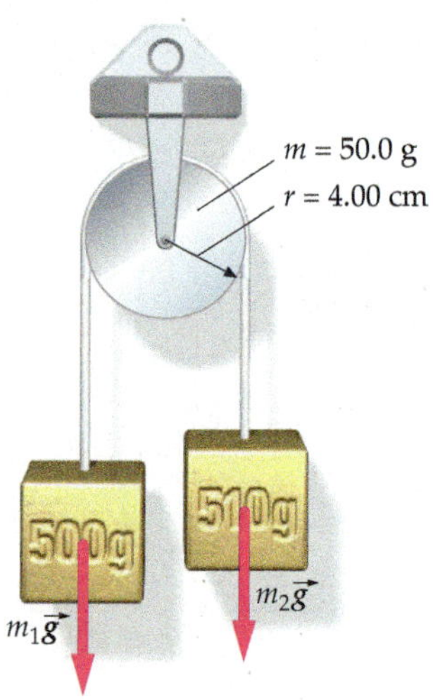

FIGURE 9-59
Problem 78

79 •• Two objects are attached to ropes that are attached to two wheels on a common axle, as shown in Figure 9-60. The two wheels are attached together so that they form a single rigid object. The moment of inertia of the rigid object is $40\ \text{kg}\cdot\text{m}^2$. The radii of the wheels are $R_1 = 1.2\ \text{m}$ and $R_2 = 0.40\ \text{m}$. (*a*) If $m_1 = 24\ \text{kg}$, find m_2 such that there is no angular acceleration of the wheels. (*b*) If 12 kg is placed on top of m_1, find the angular acceleration of the wheels and the tensions in the ropes. SSM

80 •• The upper end of the string wrapped around the cylinder in Figure 9-61 is held by a hand that is accelerated upward so that the center of mass of the cylinder does not move as the cylinder spins up. Find (*a*) the tension in the string, (*b*) the angular acceleration of the cylinder, and (*c*) the acceleration of the hand.

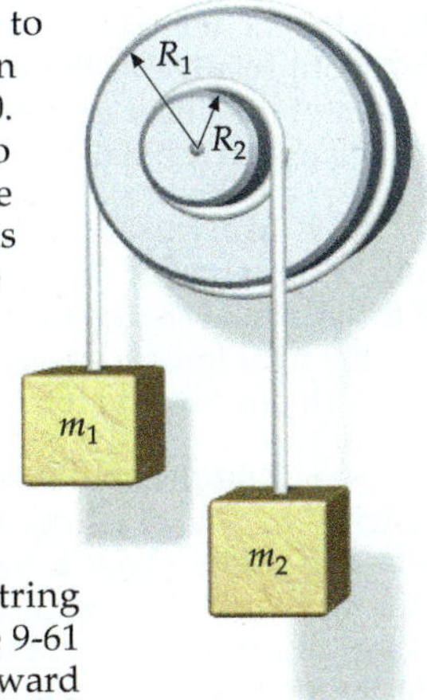

FIGURE 9-60 Problem 79

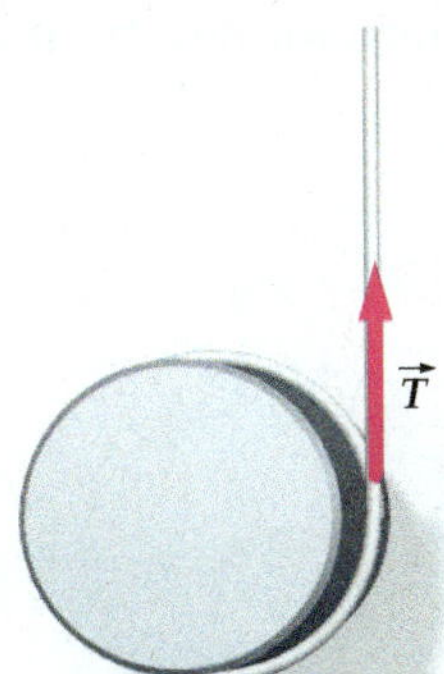

FIGURE 9-61 Problems 80 and 86

81 •• A uniform cylinder of mass m_1 and radius R is pivoted on frictionless bearings. A massless string wrapped around the cylinder is connected to a block of mass m_2 that is on a frictionless incline of angle θ as shown in Figure 9-62. The system is released from rest when the block is at a vertical distance h above the bottom of the incline. (*a*) What is the acceleration of the block? (*b*) What is the tension in the string? (*c*) What is the speed of the block as it reaches the bottom of the incline? (*d*) Evaluate your answers for the special case where $\theta = 90°$ and $m_1 = 0$. Are your answers what you would expect for this special case? Explain. SSM

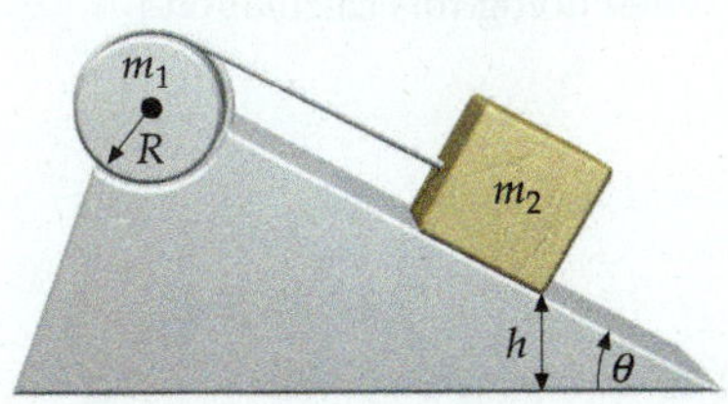

FIGURE 9-62 Problem 81

82 •• A device for measuring the moment of inertia of an object is shown in Figure 9-63. The circular platform is attached to a concentric drum of radius R, and the platform and the drum are free to rotate about a frictionless vertical axis. The string that is wound around the drum passes over a frictionless and massless pulley to a block of mass M. The block is released from rest, and the time t_1 it takes for it to drop a distance D is measured. The system is then rewound, the object whose moment of inertia I we wish to measure is placed on the platform, and the system is again released from rest. The time t_2 required for the block to drop the same distance D then provides the data needed to calculate I. Using $R = 10\ \text{cm}, M = 2.5\ \text{kg}, D = 1.8\ \text{m}, t_1 = 4.2\ \text{s}$, and $t_2 = 6.8\ \text{s}$, (*a*) find the moment of inertia of the platform-drum combination. (*b*) Find the moment of inertia of the platform-drum-object combination. (*c*) Use your results for Parts (*a*) and (*b*) to find the moment of inertia of the object.

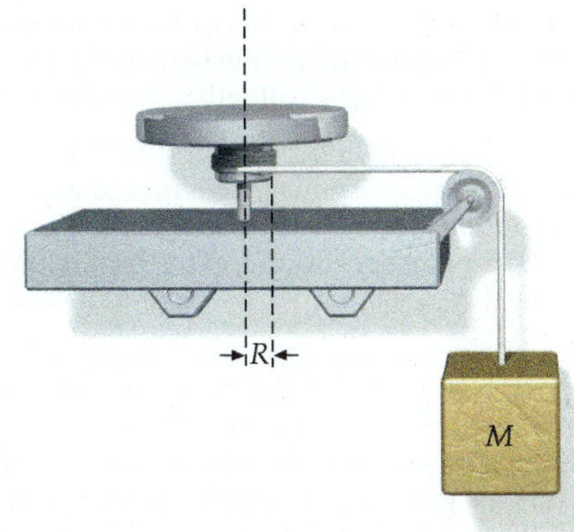

FIGURE 9-63 Problem 82

OBJECTS ROTATING AND ROLLING WITHOUT SLIPPING

83 • A homogeneous 60-kg cylinder of radius 18 cm is rolling without slipping along a horizontal floor at a speed of 15 m/s. What is the minimum amount of work that was required to give it this motion?

84 • An object is rolling without slipping. What percentage of its total kinetic energy is its translational kinetic energy if the object is (*a*) a uniform sphere, (*b*) a uniform cylinder, or (*c*) a hoop?

85 •• In 1993 a giant 400-kg yo-yo with a radius of 1.5 m was dropped from a crane at height of 57 m. One end of the string was tied to the top of the crane, so the yo-yo unwound as it descended. Assuming that the axle of the yo-yo had a radius of 0.10 m, estimate its linear speed at the end of the fall. SSM

86 •• A uniform cylinder of mass M and radius R has a string wrapped around it. The string is held fixed, and the cylinder falls vertically as shown in Figure 9-61. (*a*) Show that the acceleration of the cylinder is downward with a magnitude $a = 2g/3$. (*b*) Find the tension in the string.

87 •• A 0.10-kg yo-yo consisting of two solid disks, each of radius 10 cm, is joined by a massless rod of radius 1.0 cm. A string is wrapped around the rod. One end of the string is held fixed and is under tension as the yo-yo is released. The yo-yo rotates as it descends vertically. Find (*a*) the acceleration of the yo-yo, and (*b*) the tension T.

88 •• A uniform solid sphere rolls down an incline without slipping. If the linear acceleration of the center of mass of the sphere is $0.20g$, then what is the angle the incline makes with the horizontal?

89 •• A thin spherical shell rolls down an incline without slipping. If the linear acceleration of the center of mass of the shell is $0.20g$, what is the angle the incline makes with the horizontal?

90 •• A basketball rolls without slipping down an incline of angle θ. The coefficient of static friction is μ_s. Model the ball as a thin

spherical shell. Find (*a*) the acceleration of the center of mass of the ball, (*b*) the frictional force acting on the ball, and (*c*) the maximum angle of the incline for which the ball will roll without slipping.

91 •• A uniform solid cylinder of wood rolls without slipping down an incline of angle θ. The coefficient of static friction is μ_s. Find (*a*) the acceleration of the center of mass of the cylinder, (*b*) the frictional force acting on the cylinder, and (*c*) the maximum angle of the incline for which the cylinder will roll without slipping.

92 •• Released from rest at the same height, a thin spherical shell and solid sphere of the same mass m and radius R roll without slipping down an incline through the same vertical drop H (Figure 9-64). Each is moving horizontally as it leaves the ramp. The spherical shell hits the ground a horizontal distance L from the end of the ramp and the solid sphere hits the ground a distance L' from the end of the ramp. Find the ratio L'/L.

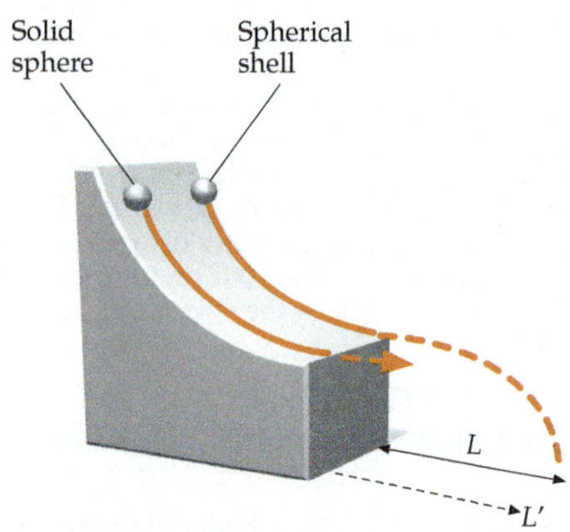

FIGURE 9-64 Problem 92

93 •• A uniform, thin cylindrical shell and a solid cylinder roll horizontally without slipping. The speed of the cylindrical shell is v. The solid cylinder and the hollow cylinder encounter an incline that they climb without slipping. If the maximum height they reach is the same, find the initial speed v' of the solid cylinder. SSM

94 •• A thin cylindrical shell and a solid sphere start from rest and roll without slipping down a 3.0-m-long inclined plane. The cylinder arrives at the bottom of the incline 2.4 s after the sphere. Determine the angle the incline makes with the horizontal.

95 •• A wheel has a thin 3.0-kg rim and four spokes, each of mass 1.2 kg. Find the kinetic energy of the wheel when it is rolling at 6.0 m/s on a horizontal surface.

96 ••• A uniform solid cylinder of mass M and radius R is at rest on a slab of mass m, which in turn rests on a horizontal, frictionless table (Figure 9-65). If a horizontal force $\vec{F}$ is applied to the slab, it accelerates and the cylinder rolls without slipping. Find the acceleration of the slab in terms of M, R, and F.

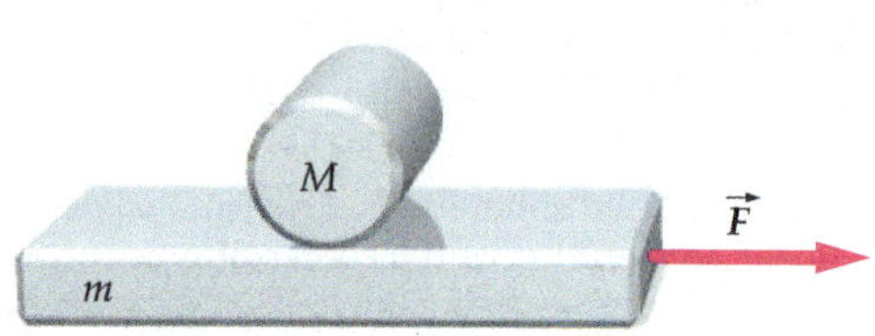

FIGURE 9-65 Problems 96–98

97 ••• (*a*) Find the angular acceleration of the cylinder in Problem 96. Is the cylinder rotating clockwise or counterclockwise? (*b*) What is the cylinder's linear acceleration (magnitude and direction) relative to the table? (*c*) What is the magnitude and direction of the linear acceleration of the center of mass of the cylinder relative to the slab?

98 ••• **MULTISTEP** If the force in Problem 96 acts over a distance d, in terms of the symbols given, find (*a*) the kinetic energy of the slab, and (*b*) the total kinetic energy of the cylinder. (*c*) Show that the total kinetic energy of the slab-cylinder system is equal to the work done by the force.

99 ••• **ENGINEERING APPLICATION** Two large gears that are being designed as part of a large machine are shown in Figure 9-66; each is free to rotate about a fixed axis through its center. The radius and moment of inertia of the smaller gear are 0.50 m and 1.0 kg · m², respectively, and the radius and moment of inertia of the larger gear are 1.0 m and 16 kg · m², respectively. The lever attached to the smaller gear is 1.0 m long and has a negligible mass. (*a*) If a worker will typically apply a force of 2.0 N to the end of the lever, as shown, what will be the angular accelerations of the two gears? (*b*) Another part of the machine (not shown) will apply a force tangentially to the outer edge of the larger gear to temporarily keep the gear system from rotating. What should the magnitude and direction of this force (clockwise or counterclockwise) be? SSM

FIGURE 9-66 Problem 99

100 ••• **ENGINEERING APPLICATION, CONTEXT-RICH** As the chief design engineer for a major toy company, you are in charge of designing a "loop-the-loop" toy for youngsters. The idea, as shown in Figure 9-67, is that a ball of mass m and radius r will roll down an inclined track and around the loop without slipping. The ball starts from rest at a height h above the tabletop that supports the whole track. The loop radius is R. Determine the minimum height h, in terms of R and r, for which the ball will remain in contact with the track during the whole of its loop-the-loop journey. (Do not neglect the size of the ball's radius when doing this calculation.)

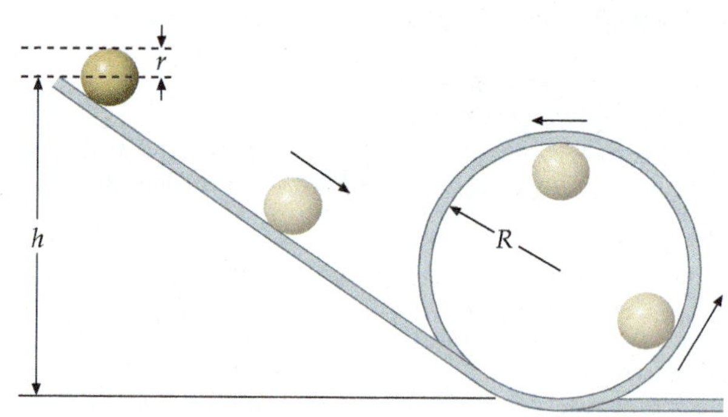

FIGURE 9-67 Problem 100

ROLLING WITH SLIPPING

101 •• A bowling ball of mass M and radius R is released so that at the instant it touches the floor it is moving horizontally with a speed v_0 and is not rotating. It slides for a time t_1 a distance s_1 before it begins to roll without slipping. (*a*) If μ_k is the coefficient of kinetic friction between the ball and the floor, find s_1, t_1, and the final speed v_1 of the ball. (*b*) Find the ratio of the final kinetic energy to the initial kinetic energy of the ball. (*c*) Evaluate s_1, t_1, and v_1 for $v_0 = 8.0$ m/s and $\mu_k = 0.060$.

102 •• **CONTEXT-RICH** During a game of pool, the cue ball (a uniform sphere of radius r) is at rest on the horizontal pool table (Figure 9-68). You strike the ball horizontally with your cue stick, which delivers a large horizontal force of magnitude F_0 for a short time. The stick strikes the ball at a point a vertical height h above the tabletop. Assume that the striking location is above the ball's center. Show that the ball's angular speed ω is related to the initial linear speed of its center of mass v_{cm} by $\omega = (5/2)v_{cm}(h - r)/r^2$. Estimate the ball's rotation rate just after the hit using reasonable estimates for h, r, and v_{cm}.

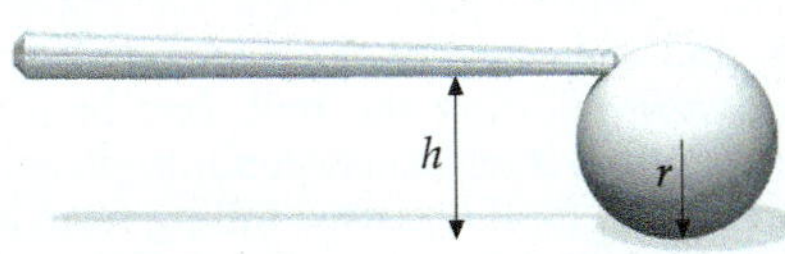

FIGURE 9-68 Problem 102

103 •• A uniform solid sphere is set rotating about a horizontal axis at an angular speed ω_0 and then is placed on the floor with its center of mass at rest. If the coefficient of kinetic friction between the sphere and the floor is μ_k, find the speed of the center of mass of the sphere when the sphere begins to roll without slipping.

104 •• A uniform solid ball that has a mass of 20 g and a radius of 5.0 cm rests on a horizontal surface. A sharp force is applied to the ball in a horizontal direction 9.0 cm above the horizontal surface. During impact the force increases linearly from 0.0 N to 40.0 kN in 1.0×10^{-4} s, and then it decreases linearly to 0.0 N in 1.0×10^{-4} s. (*a*) What is the speed of the ball just after impact? (*b*) What is the angular speed of the ball after impact? (*c*) What is the speed of the ball when it begins to roll without slipping? (*d*) How far does the ball travel along the surface before it begins to roll without slipping? Assume that $\mu_k = 0.50$.

105 •• A 0.16-kg billiard ball whose radius is 3.0 cm is given a sharp blow by a cue stick. The applied force is horizontal and the line of action of the force passes through the center of the ball. The speed of the ball just after the blow is 4.0 m/s, and the coefficient of kinetic friction between the ball and the billiard table is 0.60. (*a*) How long does the ball slide before it begins to roll without slipping? (*b*) How far does it slide? (*c*) What is its speed once it begins rolling without slipping? SSM

106 •• A billiard ball that is initially at rest is given a sharp blow by a cue stick. The force is horizontal and is applied at a distance $2R/3$ below the centerline, as shown in Figure 9-69. The speed of the ball just after the blow is v_0 and the coefficient of kinetic friction between the ball and the billiard table is μ_k. (*a*) What is the angular speed of the ball just after the blow? (*b*) What is the speed of the ball once it begins to roll without slipping? (*c*) What is the kinetic energy of the ball just after the hit?

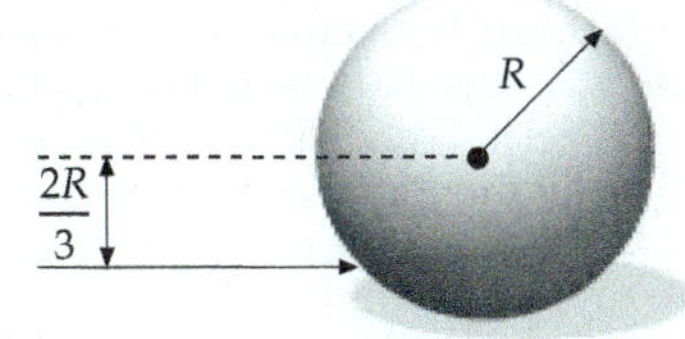

FIGURE 9-69 Problem 106

107 •• A bowling ball of radius R has an initial speed v_0 down the lane and a forward spin $\omega_0 = 3v_0/R$ just after its release. The coefficient of kinetic friction is μ_k. (*a*) What is the speed of the ball just as it begins rolling without slipping? (*b*) For how long a time does the ball slide before it begins rolling without slipping? (*c*) What distance does the ball slide down the lane before it begins rolling without slipping?

GENERAL PROBLEMS

108 •• The radius of a small playground merry-go-round is 2.2 m. To start it rotating, you wrap a rope around its perimeter and pull with a force of 260 N for 12 s. During this time, the merry-go-round makes one complete rotation. Neglect any effects of friction. (*a*) Find the angular acceleration of the merry-go-round. (*b*) What torque is exerted by the rope on the merry-go-round? (*c*) What is the moment of inertia of the merry-go-round?

109 •• A uniform 2.00-m-long stick is raised at an angle of 30° to the horizontal above a sheet of ice. The bottom end of the stick rests on the ice. The stick is released from rest. The bottom end of the stick remains in contact with the ice at all times. How far will the bottom end of the stick have traveled during the time the rest of the stick is falling to the ice? Assume that the ice is frictionless.

110 •• A uniform 5.0-kg disk has a radius of 0.12 m and is pivoted so that it rotates freely about its axis (Figure 9-70). A string wrapped around the disk is pulled with a force equal to 20 N. (*a*) What is the torque being exerted by this force about the rotation axis? (*b*) What is the angular acceleration of the disk? (*c*) If the disk starts from rest, what is its angular speed after 5.0 s? (*d*) What is its kinetic energy after the 5.0 s? (*e*) What is the angular displacement of the disk during the 5.0 s? (*f*) Show that the work done by the torque, $\tau\,\Delta\theta$, equals the kinetic energy.

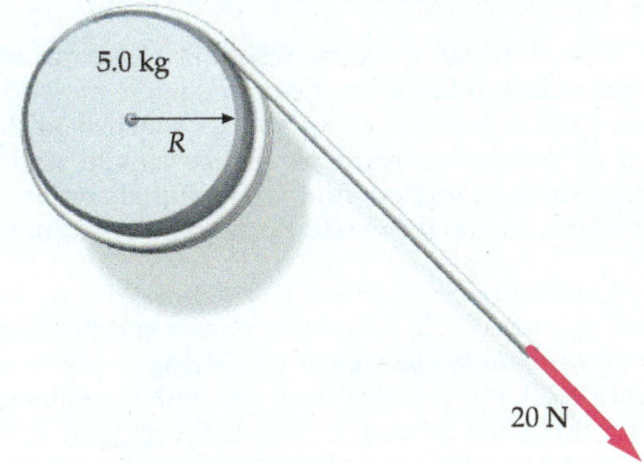

FIGURE 9-70 Problem 110

111 •• A uniform 0.25-kg thin rod that has an 80-cm length is free to rotate about a fixed horizontal axis perpendicular to and through one end of the rod. It is held horizontal and released. Immediately after it is released, what is (*a*) the acceleration of the center of the rod, and (*b*) the initial acceleration of the free end of the rod? (*c*) What is the speed of the center of mass of the rod when the rod is (momentarily) vertical?

112 •• A marble of mass M and radius R rolls without slipping down the track on the left from a height h_1, as shown in Figure 9-71. The marble then goes up the *frictionless* track on the right to a height h_2. Find h_2.

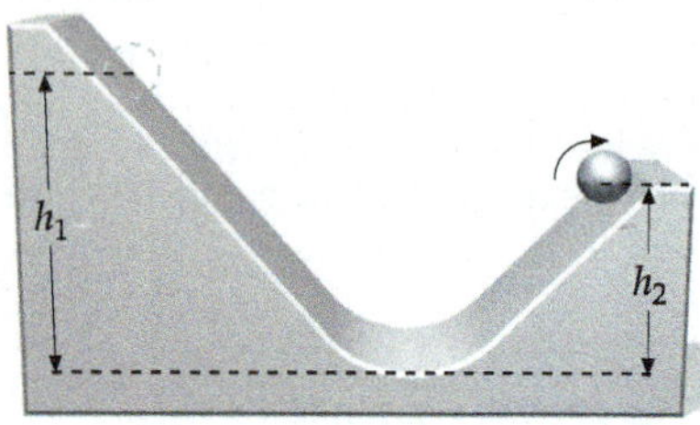

FIGURE 9-71 Problem 112

113 •• A uniform 120-kg disk with a radius equal to 1.4 m initially rotates with an angular speed of 1100 rev/min. A constant tangential force is applied at a radial distance of 0.60 m from the axis. (*a*) How much work must this force do to stop the wheel? (*b*) If the wheel is brought to rest in 2.5 min, what torque does the force produce? What is the magnitude of the force? (*c*) How many revolutions does the wheel make in these 2.5 min? SSM

114 •• A day-care center has a merry-go-round that consists of a uniform 240-kg circular wooden platform 4.00 m in diameter. Four children run alongside the merry-go-round and push tangentially along the platform's circumference until, starting from rest, the merry-go-round is spinning at 2.14 rev/min. During the spin-up: (*a*) If each child exerts a sustained force equal to 26 N, how far does each child run? (*b*) What is the angular acceleration of the merry-go-round? (*c*) How much work does each child do? (*d*) What is the increase in the kinetic energy of the merry-go-round?

115 •• A uniform 1.5-kg hoop with a 65-cm radius has a string wrapped around its circumference and lies flat on a horizontal frictionless table. The free end of the string is pulled with a constant horizontal force equal to 5.0 N and the string does not slip on the hoop. (*a*) How far does the center of the hoop travel in 3.0 s? (*b*) What is the angular speed of the hoop after 3.0 s?

116 •• A hand-driven grinding wheel is a uniform 60-kg disk with a 45-cm radius. It has a handle of negligible mass 65 cm from the rotation axis. A compact 25-kg load is attached to the handle when it is at the same height as the horizontal rotation axis. Ignoring the effects of friction, find (*a*) the initial angular acceleration of the wheel, and (*b*) the maximum angular speed of the wheel.

117 •• A uniform disk of radius R and mass M is pivoted about a horizontal axis parallel to its symmetry axis and passing through a point on its perimeter, so that it can swing freely in a vertical plane (Figure 9-72). It is released from rest with its center of mass at the same height as the pivot. (*a*) What is the angular speed of the disk when its center of mass is directly below the pivot? (*b*) What force is exerted by the pivot on the disk at this moment?

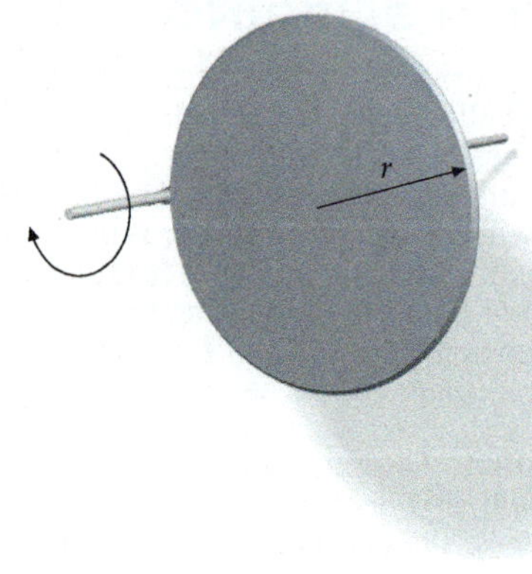

FIGURE 9-72 Problem 117

118 •• **ENGINEERING APPLICATION** The roof of the student dining hall at your college will be supported by high cross-braced wooden beams attached in the shape of an upside-down L (Figure 9-73). Each vertical beam is 10.0 ft high and 2.0 ft wide, and the horizontal cross-member is 8.0 ft long. The mass of the vertical beam is 350 kg, and the mass of the horizontal beam is 280 kg. As the workers were building the hall, one of these structures started to fall over before it was anchored into place. (Luckily they stopped it before it fell.) (*a*) If it started falling from an upright position, what was the initial angular acceleration of the structure? Assume that the bottom did not slide across the floor and that it did not fall out of plane; that is, during the fall, the structure remained in the vertical plane defined by the initial position of the structure. (*b*) What would be the magnitude of the initial linear acceleration of the upper right corner of the horizontal beam? (*c*) What would the horizontal component of the initial linear acceleration be at this same location? (*d*) Estimate the structure's rotational speed at impact.

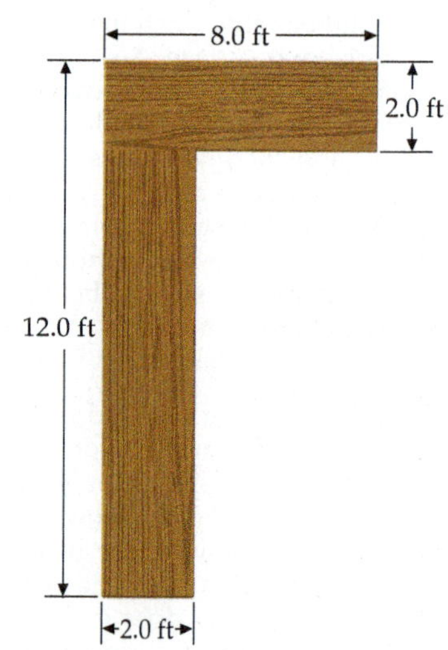

FIGURE 9-73 Problem 118

119 •• **CONTEXT-RICH** You are participating in league bowling with your friends. Time after time, you notice that your bowling ball rolls back to you without slipping on the flat section of track. When the ball encounters the slope that brings it up to the ball return, it is moving at 3.70 m/s. The length of the sloped part of the track is 2.50 m. The radius of the bowling ball is 11.5 cm. (*a*) What is

the angular speed of the ball before it encounters the slope? (*b*) If the speed with which the ball emerges at the top of the incline is 0.40 m/s, what is the angle (assumed constant) that the sloped section of the track makes with the horizontal? (*c*) What is the magnitude of the angular acceleration of the ball while it is on the slope? SSM

120 •• Figure 9-74 shows a hollow cylinder that has a length equal to 1.80 m, a mass equal to 0.80 kg, and radius equal to 0.20 m. The cylinder is free to rotate about a vertical axis that passes through its center and is perpendicular to the cylinder. Two objects are inside the cylinder. Each object has a mass equal to 0.20 kg and is attached to a spring that has a force constant k and an unstressed length equal to 0.40 m. The inside walls of the cylinder are frictionless. (*a*) Determine the value of the force constant if the objects are located 0.80 m from the center of the cylinder when the cylinder rotates at 24 rad/s. (*b*) How much work is required to bring the system from rest to an angular speed of 24 rad/s?

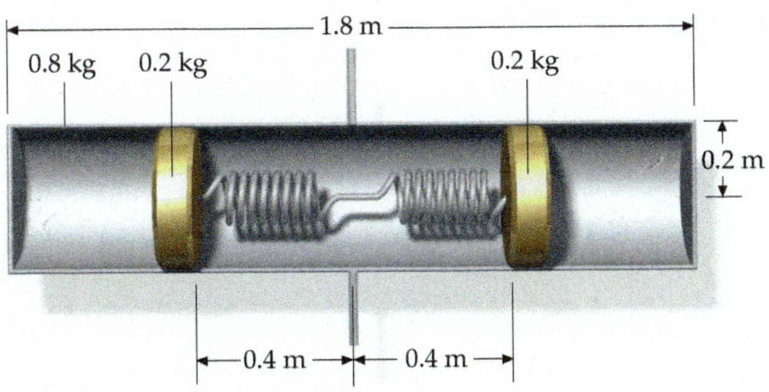

FIGURE 9-74 Problems 120 and 123

121 •• A popular classroom demonstration involves taking a meterstick and holding it horizontally at the 0.0-cm end with a number of pennies spaced evenly along its surface. If the hand is suddenly relaxed so that the meterstick pivots freely about the 0.0-cm mark under the influence of gravity, an interesting thing is seen during the first part of the stick's rotation: the pennies nearest the 0.0-cm mark remain on the meterstick, while those nearest the 100-cm mark are left behind by the falling meterstick. (This demonstration is often called the "faster than gravity" demonstration.) Suppose this demonstration is repeated without any pennies on the meterstick. (*a*) What would the initial acceleration of the 100.0-cm mark then be? (The initial acceleration is the acceleration just after the release.) (*b*) What point on the meterstick would then have an initial acceleration greater than g? SSM

122 •• A solid metal rod 1.5 m long is free to pivot without friction about a fixed horizontal axis perpendicular to the rod and passing through one of its ends. The rod is held in a horizontal position. Small coins, each of mass m, are placed on the rod 25 cm, 50 cm, 75 cm, 1 m, 1.25 m, and 1.5 m from the pivot. If the free end is now released, calculate the initial force exerted on each coin by the rod. Assume that the masses of the coins can be ignored in comparison to the mass of the rod.

123 •• Suppose that for the system described in Problem 120, the force constants are each 60 N/m. The system starts from rest and slowly accelerates until the masses are 0.80 m from the center of the cylinder. How much work was done in the process?

124 ••• A string is wrapped around a uniform solid cylinder of radius R and mass M that rests on a horizontal frictionless surface. (The string does not touch the surface because there is a groove cut in the surface to provide space for the string to clear.) The string is pulled horizontally from the top with force F. (*a*) Show that the magnitude of the angular acceleration of the cylinder is twice the magnitude of the angular acceleration needed for rolling without slipping, so that the bottom point on the cylinder slides backward against the table. (*b*) Find the magnitude and direction of the frictional force between the table and cylinder that would be needed for the cylinder to roll without slipping. What would be the magnitude of acceleration of the cylinder in this case?

125 ••• **SPREADSHEET** Let us calculate the position y of the falling load attached to the winch in Example 9-8 as a function of time by numerical integration. Let the $+y$ direction be straight downward. Then $v(y) = dy/dt$, or

$$t = \int_0^y \frac{1}{v(y')}dy' \approx \sum_{i=0}^{N} \frac{1}{v(y_i')}\Delta y'$$

where t is the time taken for the bucket to fall a distance y, $\Delta y'$ is a small increment of y', and $y' = N\Delta y'$. Hence, we can calculate t as a function of d by numerical summation. Make a graph of y versus t between 0 s and 2.00 s. Assume that $m_w = 10.0$ kg, $R = 0.50$ m, $m_b = 5.0$ kg, $L = 10.0$ m, and $m_c = 3.50$ kg. Use $\Delta y' = 0.10$ m. Compare this position to the position of the falling load if it were in free-fall. SSM

126 ••• Figure 9-75 shows a solid cylinder that has mass M and radius R to which a second solid cylinder that has mass m and radius r is attached. A string is wound about the smaller cylinder. The larger cylinder rests on a horizontal surface. The coefficient of static friction between the larger cylinder and the surface is μ_s. If a light tension is applied to the string in the vertical direction, the cylinder will roll to the left; if the tension is applied with the string horizontally to the right, the cylinder rolls to the right. Find the angle between the string and the horizontal that will allow the cylinder to remain stationary when a light tension is applied to the string.

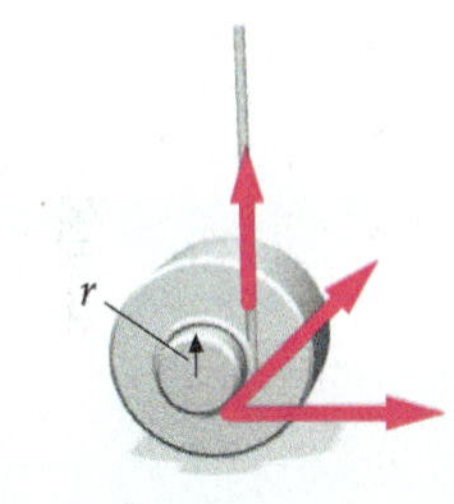

FIGURE 9-75 Problem 126

127 ••• In problems dealing with a pulley with a nonzero moment of inertia, the magnitude of the tensions in the ropes hanging on either side of the pulley are not equal. The difference in the tension is due to the static frictional force between the rope and the pulley; however, the static frictional force cannot be made arbitrarily large. Consider a massless rope wrapped partly around a cylinder through an angle $\Delta\theta$ (measured in radians). It can be shown that if the tension on one side of the pulley is T, while the tension on the other side is T' ($T' > T$), the maximum value of T' that can be maintained without the rope slipping is $T'_{max} = Te^{\mu_s \Delta\theta}$, where μ_s is the coefficient of static friction. Consider the Atwood's machine in Figure 9-76: the pulley has a radius $R = 0.15$ m, the moment of inertia is $I = 0.35\ \text{kg}\cdot\text{m}^2$, and the coefficient of static friction between the wheel and the string is $\mu_s = 0.30$. (*a*) If the tension on one side of the pulley is 10 N, what is the maximum tension on the other side that will prevent the rope from slipping on the pulley? (*b*) What is the acceleration of the blocks in this case? (c) If the mass of one of the hanging blocks is 1.0 kg, what is the maximum mass of the other block if, after the blocks are released, the pulley is to rotate without slipping? SSM

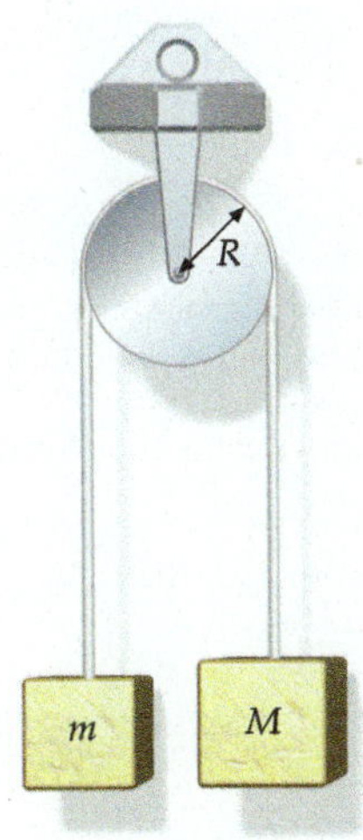

FIGURE 9-76 Problem 127

128 ••• A massive, uniform cylinder has a mass m and a radius R (Figure 9-77). It is accelerated by a tension force $\vec{T}$ that is applied through a rope wound around a light drum of radius r that is attached to the cylinder. The coefficient of static friction is sufficient for the cylinder to roll without slipping. (*a*) Find the frictional force. (*b*) Find the acceleration a of the center of the cylinder. (*c*) Show that it is possible to choose r so that a is greater than T/m. (*d*) What is the direction of the frictional force in the circumstances of Part (*c*)?

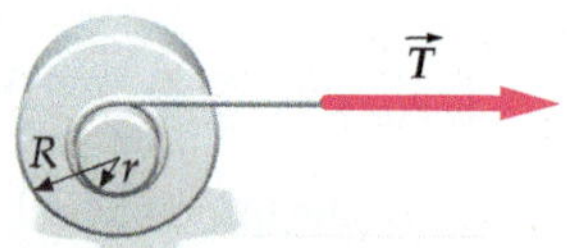

FIGURE 9-77 Problem 128

129 ••• A uniform rod that has a length L and a mass M is free to rotate about a horizontal axis through one end, as shown in Figure 9-78. The rod is released from rest at $\theta = \theta_0$. Show that the parallel and perpendicular components of the force exerted by the axis on the rod are given by $F_{\parallel} = \frac{1}{2}Mg(5\cos\theta - 3\cos\theta_0)$ and $F_{\perp} = \frac{1}{4}Mg\sin\theta$, where $F_{\parallel}$ is the component parallel with the rod and $F_{\perp}$ is the component perpendicular to the rod.

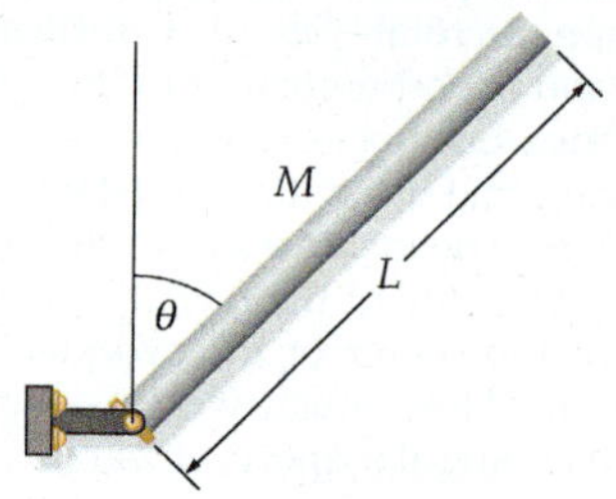

FIGURE 9-78 Problem 129

CHAPTER 10

THE HUBBLE SPACE TELESCOPE WAS PLACED IN ORBIT AND PUT INTO OPERATION IN 1990. ALMOST IMMEDIATELY A MAJOR FLAW IN ITS PRIMARY MIRROR WAS DISCOVERED. HOWEVER, IN 1993 A SERVICE SHIP VISITED THE TELESCOPE AND CORRECTED THE PROBLEM. SINCE THEN, THE HUBBLE HAS PROVIDED SPECTACULAR IMAGES OF THE UNIVERSE. THE IMPACT OF THESE IMAGES HAS BEEN TO ENRICH AND EXTEND OUR KNOWLEDGE ABOUT THE UNIVERSE. THE HUBBLE TELESCOPE IS AN EXTRAORDINARY SCIENTIFIC INSTRUMENT. *(NASA.)*

Angular Momentum

As with conservation of energy and conservation of linear momentum, conservation of angular momentum is one of the basic principles of physics. Experimental evidence shows that angular momentum is never created nor destroyed.

In this chapter, we extend our study of rotational motion to situations in which the direction of the axis of rotation may change. Angular velocity, angular acceleration, and torque are presented in Chapter 9. Here we begin by introducing the vector nature of these quantities and of angular momentum, which is the rotational analog of linear momentum. We then show that the net torque acting on a system equals the time rate of change of its angular momentum. Angular momentum is conserved in systems that have zero net external torque. Like conservation of linear momentum, conservation of angular momentum is a fundamental law of nature, relating to even atoms, molecules, subatomic particles, and photons.

? In order to aim the Hubble telescope in a new direction the telescope must rotate. How is this accomplished? (See Example 10-7.)

10-1 THE VECTOR NATURE OF ROTATION

In Chapter 9, we indicated the direction of rotation about an axis by assigning plus and minus signs to indicate the direction of the angular velocity, just as in Chapter 2 we used plus and minus signs to indicate the direction of the velocity in one-dimensional motion. However, plus and minus signs are *not* adequate to specify the direction of the angular velocity if the direction of the rotation axis is not fixed in space. This inadequacy is overcome by treating the angular velocity as a vector quantity $\vec{\omega}$ directed along the rotation axis. Consider the rotating disk in Figure 10-1. If the rotation is directed as shown, $\vec{\omega}$ is directed as shown; if the rotation direction is reversed, so is the direction of $\vec{\omega}$. The convention relating the direction of $\vec{\omega}$ with the direction of rotation is specified by a convention called the **right-hand rule.** You can obtain the direction of $\vec{\omega}$ by curling the fingers of your right hand in the direction of rotation (Figure 10-2); your thumb then points along the rotation axis in the direction of $\vec{\omega}$.

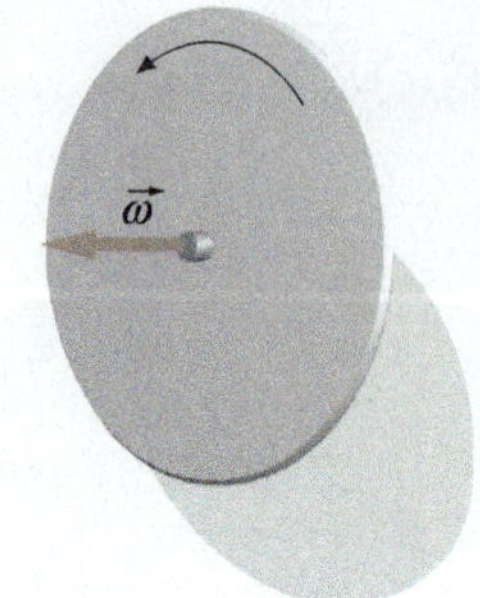

FIGURE 10-1

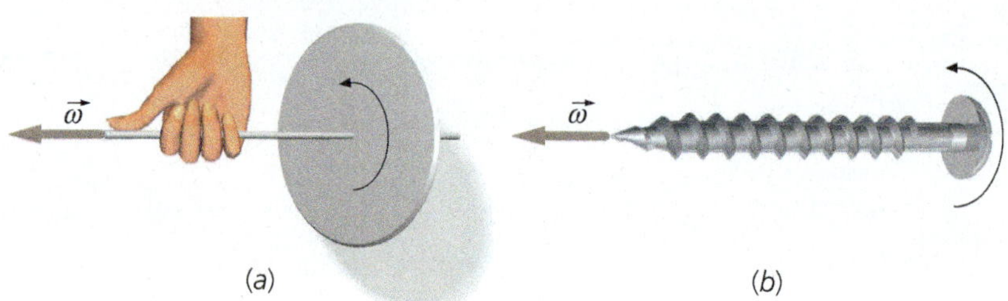

FIGURE 10-2 (*a*) When the fingers of the right hand curl in the direction of rotation, the thumb points in the direction of $\vec{\omega}$, (*b*) Looked at another way, the direction of $\vec{\omega}$ is that of the advance of a rotating right-hand screw.

In Chapter 9, we indicated the direction of torque about an axis by assigning plus and minus signs to indicate the direction of the torque. In this chapter, we define the torque $\vec{\tau}$ about a point as a vector quantity, and, as with $\vec{\omega}$, the direction of $\vec{\tau}$ is specified by a right hand rule. Figure 10-3 shows a force $\vec{F}$ acting on a particle at some position $\vec{r}$ relative to the origin O. The torque $\vec{\tau}$ about O exerted by this force is defined as a vector that is perpendicular to both $\vec{F}$ and $\vec{r}$ and has magnitude $Fr\sin\phi$, where ϕ is the angle between the directions of $\vec{F}$ and $\vec{r}$. If $\vec{F}$ and $\vec{r}$ are both perpendicular to the z axis, as in Figure 10-3, the torque vector $\vec{\tau}$ is parallel to the z axis. If $\vec{F}$ is applied to the rim of a disk of radius r, as shown in Figure 10-4, the torque vector has the magnitude Fr, and is along the axis of rotation in the direction shown.

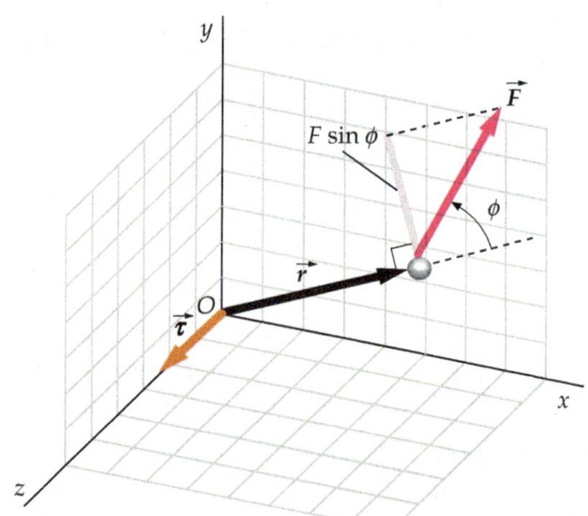

FIGURE 10-3 If $\vec{F}$ and $\vec{r}$ are both perpendicular to the z axis, then $\vec{\tau}$ is parallel with the z axis.

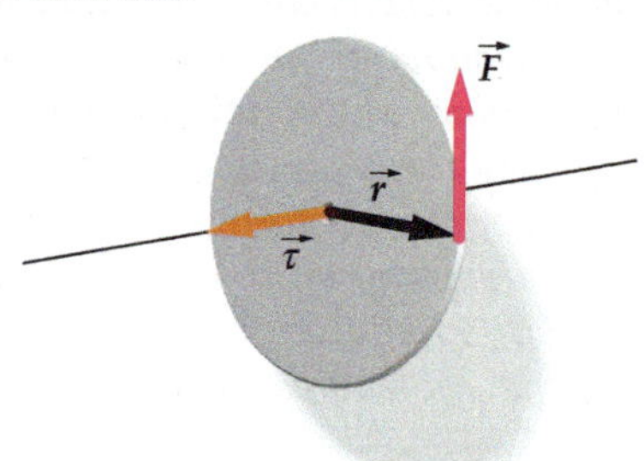

FIGURE 10-4

THE VECTOR PRODUCT

Torque is expressed mathematically as the **vector product** of $\vec{r}$ and $\vec{F}$:

$$\vec{\tau} = \vec{r} \times \vec{F} \qquad 10\text{-}1$$

(Because of the $\times$ used to indicate this type of multiplication, the vector product is also called the *cross product*.) The vector product of two vectors $\vec{A}$ and $\vec{B}$ is defined to be a vector $\vec{C} = \vec{A} \times \vec{B}$ whose magnitude equals the area of the parallelogram formed by $\vec{A}$ and $\vec{B}$ (Figure 10-5). The vector $\vec{C}$ is perpendicular to both $\vec{A}$ and $\vec{B}$ in the direction of the thumb of your right hand if you curl your fingers from the direction of $\vec{A}$ toward the direction of $\vec{B}$ (Figure 10-6). If ϕ is the angle between $\vec{A}$ and $\vec{B}$,* and $\hat{n}$ is a unit vector that is perpendicular to both $\vec{A}$ and $\vec{B}$ in the direction of $\vec{C}$, the vector product of $\vec{A}$ and $\vec{B}$ is

$$\vec{A} \times \vec{B} = AB\sin\phi\,\hat{n} \qquad 10\text{-}2$$

DEFINITION—VECTOR PRODUCT

It follows from the definition of the vector product that

$$\vec{A} \times \vec{A} = 0 \qquad 10\text{-}3$$

and

$$\vec{A} \times \vec{B} = -\vec{B} \times \vec{A} \qquad 10\text{-}4$$

Note that the order in which two vectors are multiplied in a vector product makes a difference. Below is a Problem-Solving Strategy that should help you work with the vector product.

* The angle between two vectors is the angle between their directions in space.

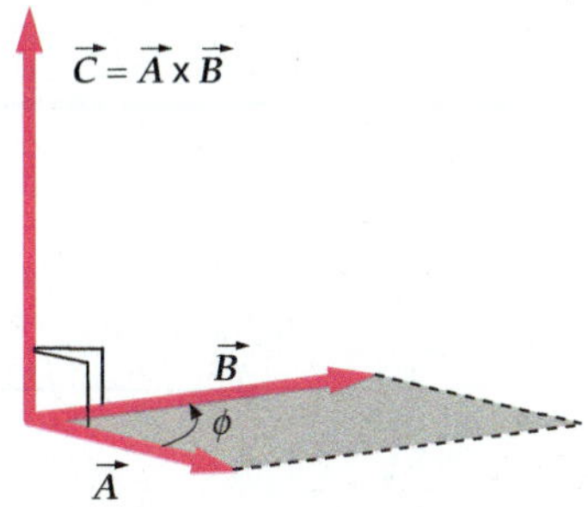

FIGURE 10-5 The vector product $\vec{A} \times \vec{B}$ is a vector $\vec{C}$ that is perpendicular to both $\vec{A}$ and $\vec{B}$, and has a magnitude $AB \sin\phi$, which equals the area of the parallelogram shown.

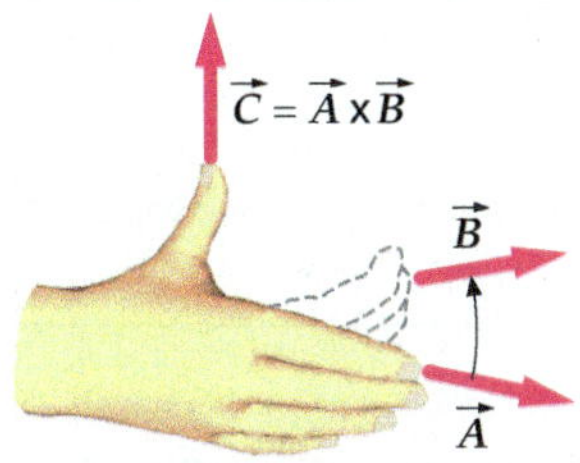

FIGURE 10-6 The direction of $\vec{A} \times \vec{B}$ is given by the right-hand rule when the fingers are rotated from the direction of $\vec{A}$ toward $\vec{B}$ through the angle ϕ.

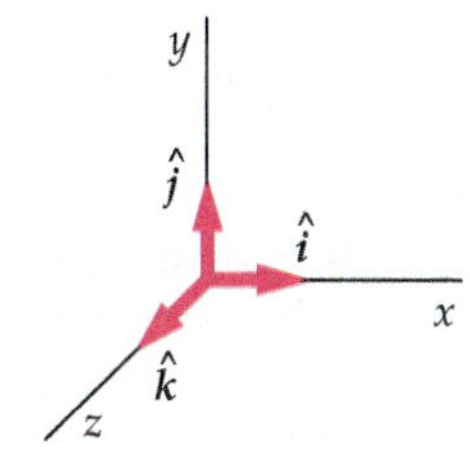

FIGURE 10-7

PROBLEM-SOLVING STRATEGY

Finding the Vector Product of Two Vectors

PICTURE At times it is easier to find a vector product of two vectors by using the equation $\vec{A} \times \vec{B} = AB \sin\phi\, \hat{n}$. At other times it is easier to find the vector product using the Cartesian components of the two vectors.

SOLVE

1. The vector product obeys a distributive law under addition:

$$\vec{A} \times (\vec{B} + \vec{C}) = \vec{A} \times \vec{B} + \vec{A} \times \vec{C} \qquad \text{10-5}$$

2. If $\vec{A}$ and $\vec{B}$ are functions of some variable such as t, the derivative of $\vec{A} \times \vec{B}$ follows the usual product rule for derivatives:

$$\frac{d}{dt}(\vec{A} \times \vec{B}) = \left(\vec{A} \times \frac{d\vec{B}}{dt}\right) + \left(\frac{d\vec{A}}{dt} \times \vec{B}\right) \qquad \text{10-6}$$

3. The unit vectors $\hat{i}$, $\hat{j}$, and $\hat{k}$ (Figure 10-7), which are mutually perpendicular, have vector products given by

$$\hat{i} \times \hat{j} = \hat{k},\ \hat{j} \times \hat{k} = \hat{i}, \text{ and } \hat{k} \times \hat{i} = \hat{j} \qquad \text{10-7}a$$

(Reversing the order of multiplication gives $\hat{j} \times \hat{i} = -\hat{k}$, $\hat{k} \times \hat{j} = -\hat{i}$ and $\hat{i} \times \hat{k} = -\hat{j}$, in accord with Equation 10-4. A tool for remembering this is shown in Figure 10-8. Furthermore,

$$\hat{i} \times \hat{i} = \hat{j} \times \hat{j} = \hat{k} \times \hat{k} = 0 \qquad \text{10-7}b$$

CHECK Make sure that your vector products make sense. For example, the vector product of two vectors is a vector and is perpendicular to each of the two vectors. In addition, check your work to make certain you did not inadvertently reverse the order of the two vectors being multiplied, and thus create a sign error.

TAKING IT FURTHER Any coordinate system for which Equations 10-7*a* and 10-7*b* hold is called a *right-handed coordinate system* (Figure 10-9). Only right-handed coordinate systems are used in this book.

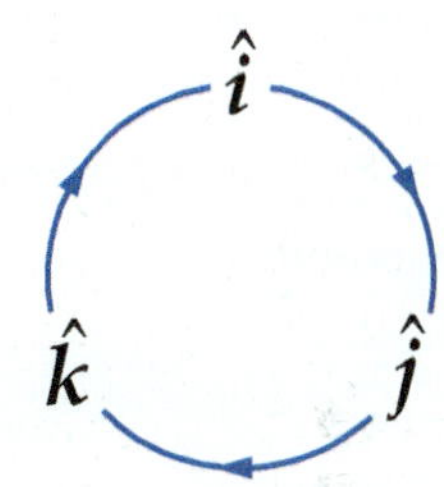

FIGURE 10-8 Taking the vector product by going around this figure in the direction of the arrows (clockwise) and the sign is positive ($\hat{i} \times \hat{j} = \hat{k}$). Going around against the arrows and the sign is negative ($\hat{i} \times \hat{k} = -\hat{j}$).

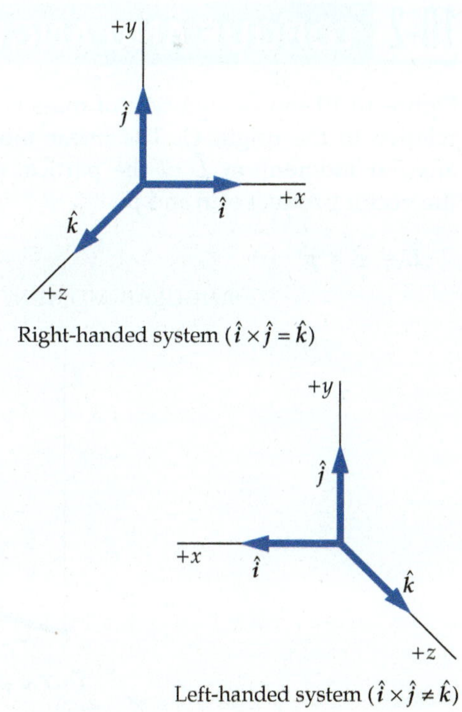

FIGURE 10-9 A right-handed and a left-handed coordinate system. In this book, only right-handed coordinate systems are used.

Example 10-1 Vector Products and Dot Products

If $\vec{A} = 3\hat{j}$, $\vec{A} \times \vec{B} = 9\hat{i}$, and $\vec{A} \cdot \vec{B} = 12$, find $\vec{B}$.

PICTURE Express $\vec{B}$ in terms of its Cartesian components and solve for each Cartesian component of $\vec{B}$ using the given information.

SOLVE

1. Express $\vec{B}$ in terms of its Cartesian components. The task is to solve for each of these components:
$$\vec{B} = B_x\hat{i} + B_y\hat{j} + B_z\hat{k}$$
2. We are given $\vec{A} \cdot \vec{B} = 12$. Evaluate $\vec{A} \cdot \vec{B}$ and simplify using Equation 6-15:
$$\vec{A} \cdot \vec{B} = 3\hat{j} \cdot (B_x\hat{i} + B_y\hat{j} + B_z\hat{k}) = 3B_x\hat{j} \cdot \hat{i} + 3B_y\hat{j} \cdot \hat{j} + 3B_z\hat{j} \cdot \hat{k}$$
$$= 0 + 3B_y + 0 = 3B_y$$
3. Set the step-2 result equal to 12 and solve for B_y:
$$3B_y = 12, \quad \text{so} \quad B_y = 4$$
4. We are given $\vec{A} \times \vec{B} = 9\hat{i}$. Evaluate $\vec{A} \times \vec{B}$ and simplify using Equations 10-7*a* and 10-7*b*:
$$\vec{A} \times \vec{B} = 3\hat{j} \times (B_x\hat{i} + B_y\hat{j} + B_z\hat{k}) = 3B_x\hat{j} \times \hat{i} + 3B_y\hat{j} \times \hat{j} + 3B_z\hat{j} \times \hat{k}$$
$$= 3B_x(-\hat{k}) + 3B_y(0) + 3B_z(\hat{i}) = 3B_z\hat{i} - 3B_x\hat{k}$$
5. Set the step-4 result equal to $9\hat{i}$ and solve for the remaining components of $\vec{B}$:
$$\vec{A} \times \vec{B} = 9\hat{i}$$
$$3B_z\hat{i} - 3B_x\hat{k} = 9\hat{i} \quad \text{so}$$
$$B_z = 3 \text{ and } B_x = 0$$
$$\therefore \vec{B} = 0\hat{i} + 4\hat{j} + 3\hat{k} = \boxed{4\hat{j} + 3\hat{k}}$$

CHECK The vector product of any two vectors is perpendicular to both vectors (except when the vector product is equal to zero). Because $\vec{A} \times \vec{B} = 9\hat{i}$, we expect $\vec{B}$ to be perpendicular to $\hat{i}$, which means we expect the x component of $\vec{B}$ to be zero. Our calculated value of $\vec{B}$ meets this expectation.

10-2 TORQUE AND ANGULAR MOMENTUM

Figure 10-10 shows a particle of mass m moving with a velocity $\vec{v}$ at a position $\vec{r}$ relative to the origin O. The linear momentum of the particle is $\vec{p} = m\vec{v}$. The **angular momentum** $\vec{L}$ of the particle relative to the origin O is defined to be the vector product of $\vec{r}$ and $\vec{p}$:

$$\vec{L} = \vec{r} \times \vec{p} \qquad 10\text{-}8$$

ANGULAR MOMENTUM OF A POINT PARTICLE DEFINED

See **Math Tutorial** *for more information on* **Trigonometry**

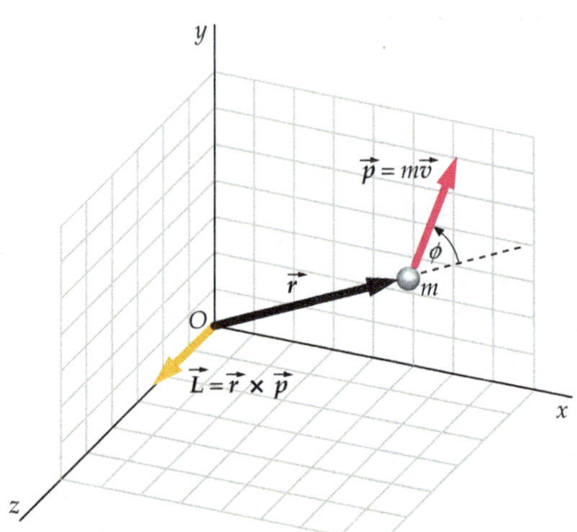

FIGURE 10-10

If $\vec{r}$ and $\vec{p}$ are both perpendicular to the z axis, as in Figure 10-10, $\vec{L}$ is parallel to the z axis and is given by $\vec{L} = \vec{r} \times \vec{p} = mvr \sin\phi\, \hat{k}$. Like torque, angular momentum is defined *relative to a point in space;* in this case the angular momentum is defined about the origin.

Figure 10-11 shows a particle of mass m attached to a circular disk of negligible mass moving in a circle in the xy plane with its center at the origin. The disk is spinning about the z axis with angular speed ω. The speed v of the particle and its angular speed are related by $v = r\omega$. The angular momentum of the particle relative to the center of the disk is

$$\vec{L} = \vec{r} \times \vec{p} = \vec{r} \times m\vec{v} = rmv \sin 90^\circ\, \hat{k} = rmv\hat{k} = mr^2\omega\hat{k} = mr^2\vec{\omega}$$

(*Note:* In this example the angular momentum vector is in the same direction as the angular velocity vector.)

Because mr^2 is the moment of inertia for a single particle about the z axis, we have

$$\vec{L} = mr^2\vec{\omega} = I\vec{\omega}$$

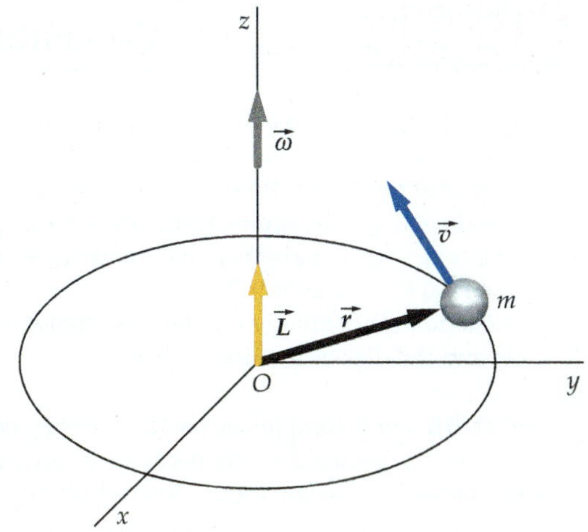

FIGURE 10-11

The angular momentum of this particle about a general point on the z axis is not parallel to the angular velocity vector. Figure 10-12 shows the angular momentum vector $\vec{L}'$ for the same particle attached to the same disk, but with $\vec{L}'$ computed about a point on the z axis that is not at the center of the circle. In this case, the angular momentum is not parallel to the angular velocity vector $\vec{\omega}$, which is parallel to the z axis.

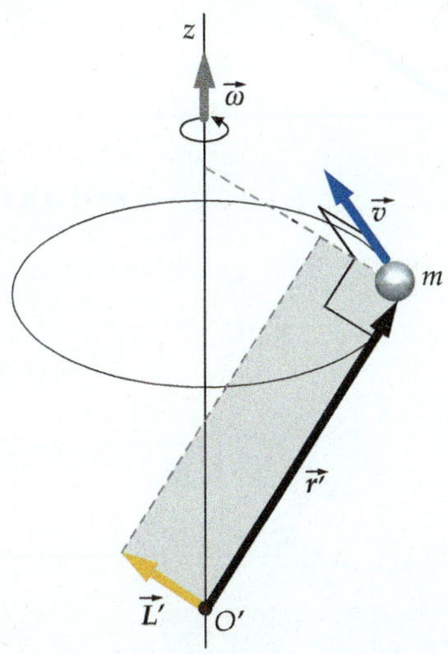

FIGURE 10-12

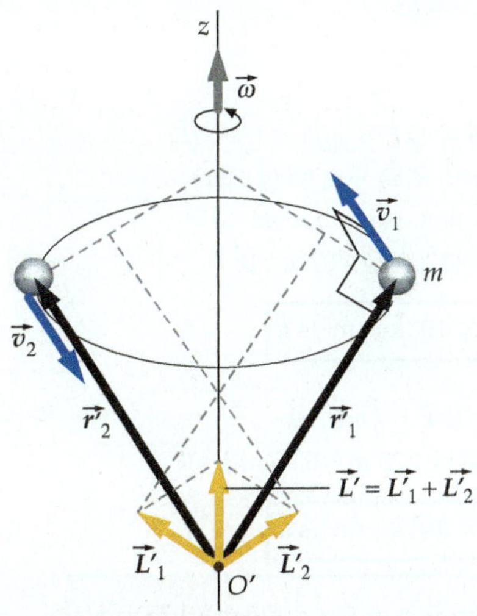

FIGURE 10-13

In Figure 10-13, we attach a second particle of equal mass to the spinning disk at a point diametrically opposite this first particle. The angular-momentum vectors $\vec{L}'_1$ and $\vec{L}'_2$ are shown relative to the same point O'. The total angular momentum $\vec{L}' = \vec{L}'_1 + \vec{L}'_2$ of the two-particle system is again parallel to the angular velocity vector $\vec{\omega}$. In this case, the axis of rotation, the z axis, passes through the center of mass of the two-particle system, and the mass distribution is symmetric about this axis. Such an axis is called a **symmetry axis.** For any system of particles that rotates about a symmetry axis, the total angular momentum (which is the sum of the angular momenta of the individual particles) is parallel to the angular velocity and is given by

$$\vec{L} = I\vec{\omega} \qquad \text{10-9}$$

ANGULAR MOMENTUM OF A SYSTEM ROTATING ABOUT A SYMMETRY AXIS

where I is a scalar.*

* In more advanced treatments Equation 10-9 is valid about any axis, but I is a tensor of rank 3.

Example 10-2 Angular Momentum About the Origin

Find the angular momentum about the origin for the following situations. (*a*) A car of mass 1200 kg moves in a circle of radius 20 m with a speed of 15 m/s. The circle is in the xy plane, centered at the origin. When viewed from a point on the positive z axis, the car moves counterclockwise. (*b*) The same car moves in the xy plane with velocity $\vec{v} = -(15\text{ m/s})\hat{i}$ along the line $y = y_0 = 20$ m parallel to the x axis. (*c*) A uniform disk in the xy plane of radius 20 m and mass 1200 kg rotates at 0.75 rad/s about its axis, which is also the z axis. When viewed from a point on the positive z axis, the disk rotates counterclockwise. Model the car as a point particle and the disk as a uniform disk.

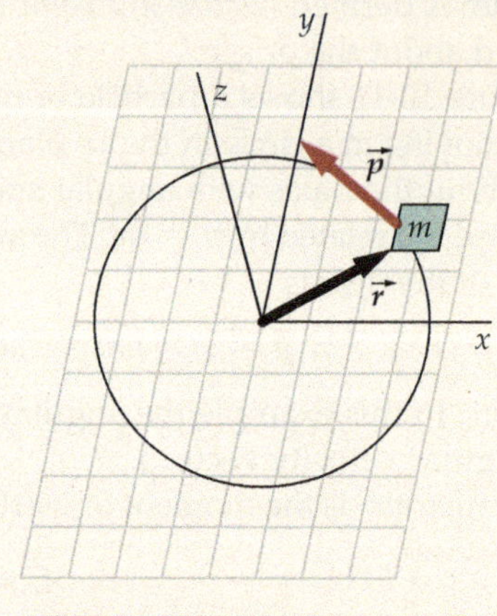

FIGURE 10-14

PICTURE For (*a*) and (*b*) we use $\vec{L} = \vec{r} \times \vec{p}$, because we are modeling the car as a point particle. For (*c*), we use $\vec{L} = I\vec{\omega}$ because we are modeling the disk as a rigid extended body—a disk. Draw a figure and apply the right-hand rule to find the direction of $\vec{L}$.

SOLVE

(*a*) $\vec{r}$ and $\vec{p}$ are perpendicular and $\vec{r} \times \vec{p}$ is in the $+z$ direction (Figure 10-14):

$$\vec{L} = \vec{r} \times \vec{p} = rmv \sin 90°\hat{k}$$
$$= (20\text{ m})(1200\text{ kg})(15\text{ m/s})\hat{k}$$
$$= \boxed{3.6 \times 10^5\text{ kg}\cdot\text{m}^2/\text{s}\ \hat{k}}$$

(*b*) 1. For the same car moving in the direction of decreasing x along the line $y = 20$ m, we express $\vec{r}$ and $\vec{p}$ in terms of unit vectors:

$$\vec{r} = x\hat{i} + y\hat{j} = x\hat{i} + y_0\hat{j}$$
$$\vec{p} = m\vec{v} = -mv\hat{i}$$

2. Now compute $\vec{r} \times \vec{p}$ (Figure 10-15):

$$\vec{L} = \vec{r} \times \vec{p} = (x\hat{i} + y_0\hat{j}) \times (-mv\hat{i})$$
$$= -xmv(\hat{i} \times \hat{i}) - y_0 mv(\hat{j} \times \hat{i})$$
$$= 0 - y_0 mv(-\hat{k}) = y_0 mv\hat{k}$$
$$= (20\text{ m})(1200\text{ kg})(15\text{ m/s})\hat{k}$$
$$= \boxed{3.6 \times 10^5\text{ kg}\cdot\text{m}^2/\text{s}\ \hat{k}}$$

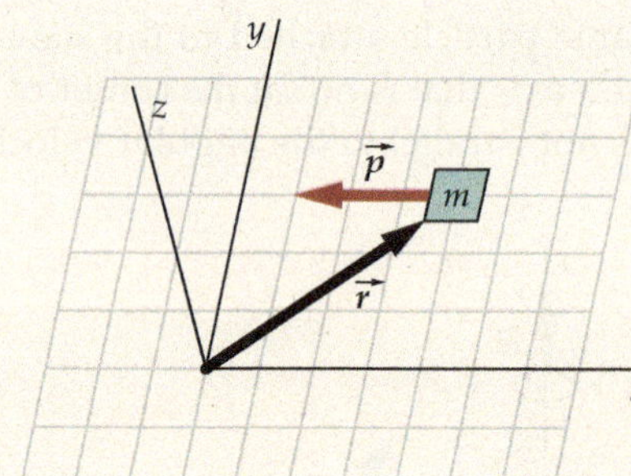

FIGURE 10-15

(*c*) Use $\vec{L} = I\vec{\omega}$ (Figure 10-16):

$$\vec{L} = I\vec{\omega} = I\omega\hat{k} = \tfrac{1}{2}mR^2\omega\hat{k}$$
$$= \tfrac{1}{2}(1200\text{ kg})(20\text{ m})^2(0.75\text{ rad/s})\hat{k}$$
$$= \boxed{1.8 \times 10^5\text{ kg}\cdot\text{m}^2/\text{s}\ \hat{k}}$$

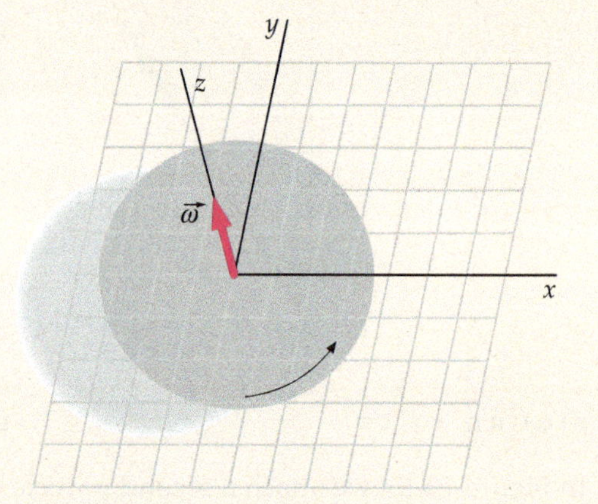

FIGURE 10-16

CHECK In Part (*c*), the velocity of a point on the rim is $v = R\omega = (20\text{ m})(0.75\text{ rad/s}) = 15$ m/s, the same as the velocity of the car in Parts (*a*) and (*b*). The angular momentum of the rotating disk is less than that of the car because virtually all of the mass of the disk is less than 20 m from the axis of rotation.

TAKING IT FURTHER The angular momentum of the car moving in a circle in Part (*a*) is the same as that of the car moving along a straight line in Part (*b*).

There are several additional results concerning torque and angular momentum for a system of particles. The first of these is

$$\vec{\tau}_{\text{net ext}} = \frac{d\vec{L}_{\text{sys}}}{dt} \qquad \text{10-10}$$

The net external torque about a fixed point acting on a system equals the rate of change of the angular momentum of the system about the same point.

NEWTON'S SECOND LAW FOR ANGULAR MOTION

In Equation 10-10, the net external torque about the point is the vector sum of the external torques about that point acting on the system. Integrating both sides of this equation with respect to time gives

$$\Delta\vec{L}_{sys} = \int_{t_i}^{t_f} \vec{\tau}_{net\,ext}\,dt \qquad 10\text{-}11$$

ANGULAR IMPULSE–ANGULAR-MOMENTUM EQUATION

Equation 10-11 is the rotational analog of $\Delta\vec{P}_{sys} = \int_{t_i}^{t_f} \vec{F}_{net\,ext}\,dt$ (Equation 8-11).

It is often useful to split the total angular momentum of a system about an arbitrary point O into orbital angular momentum and spin angular momentum:

$$\vec{L}_{sys} = \vec{L}_{orbit} + \vec{L}_{spin} \qquad 10\text{-}12$$

SPIN AND ORBITAL ANGULAR MOMENTUM

Earth has spin angular momentum due to its spinning motion about its rotational axis, and it has orbital angular momentum about the center of the Sun due to its orbital motion around the Sun (Figure 10-17). The total angular momentum of Earth relative to the center of the Sun is the vector sum of the spin and orbital angular momenta. $\vec{L}_{spin}$ is the angular momentum of a system about its center of mass, and $\vec{L}_{orbit}$ is the angular momentum that a point particle of mass M located at the center of mass and moving at the velocity of the center of mass would have. That is,

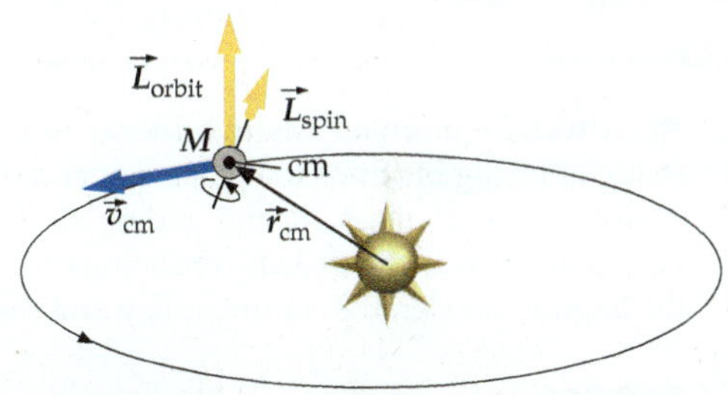

FIGURE 10-17 The total angular momentum of Earth about the center of the Sun is the sum of the orbital and the spin-angular-momentum vectors.

$$\vec{L}_{orbit} = \vec{r}_{cm} \times M\vec{v}_{cm} \qquad 10\text{-}13$$

DEFINITION: ORBITAL ANGULAR MOMENTUM

In Chapter 9, torques are computed about axes instead of about points. The relation between the torque about an axis and the torque about a point is straightforward. If point O is the origin and if force $\vec{F}$ exerts torque $\vec{\tau}$ about O, then τ_z (the z component of $\vec{\tau}$) is the torque of $\vec{F}$ about the z axis.

! Do not confuse torque about a point with torque about an axis. The torque of a force about the z axis is the z component of the torque of the force about any point on the z axis.

Taking components of vector products requires some care. If $\vec{\tau} = \vec{r} \times \vec{F}$, then

$$\vec{\tau}_z = \vec{r}_{rad} \times \vec{F}_{xy} \qquad 10\text{-}14$$

TORQUE ABOUT z AXIS

where $\vec{\tau}_z$, $\vec{r}_{rad}$ and $\vec{F}_{xy}$ (see Figure 10-18) are the vector components of $\vec{\tau}$, $\vec{r}$, and $\vec{F}$. The vector component in a given direction is the scalar component in that direction times the unit vector in that direction. For example, $\vec{\tau}_z = \tau_z\hat{k}$. Here, $\vec{r}_{rad}$ is the vector component of $\vec{r}$ in the positive radial direction (directly away from the z axis), and $\vec{F}_{xy}$ is the component of $\vec{F}$ perpendicular to the z axis, and thus parallel to the xy plane ($\vec{F}_{xy} = \vec{F} - F_z\hat{k}$). The relation between angular momentum about an axis and angular momentum about a point is also straightforward. If the angular momentum of a point particle about the origin is $\vec{L} = \vec{r} \times \vec{p}$, then the angular momentum of the particle about the z axis is

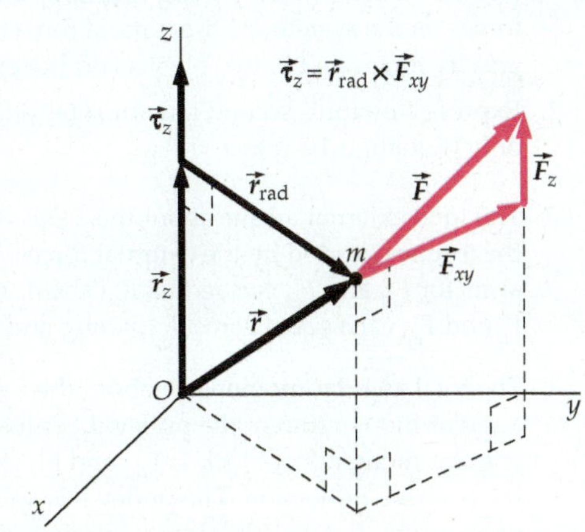

FIGURE 10-18

$$\vec{L}_z = \vec{r}_{rad} \times \vec{p}_{xy} \qquad 10\text{-}15$$

ANGULAR MOMENTUM ABOUT z AXIS

where $\vec{p}_{xy}$ is the component of the linear momentum $\vec{p}$ perpendicular to the z axis ($\vec{p}_{xy} = \vec{p} - p_z\hat{k}$). Taking the z vector components of both sides of Equation 10-10 gives

$$\vec{\tau}_{\text{net ext } z} = \frac{d\vec{L}_{\text{sys } z}}{dt} \qquad \text{10-16}$$

For a symmetric rigid object rotating about the z axis, $\vec{L}_{\text{sys } z} = I_z\vec{\omega}$, where I_z is the moment of inertia about the z axis. Substituting this into Equation 10-16 gives

$$\vec{\tau}_{\text{net ext } z} = \frac{d\vec{L}_{\text{sys } z}}{dt} = \frac{d}{dt}(I_z\vec{\omega}) = I_z\vec{\alpha} \qquad \text{10-17}$$

where the angular acceleration vector $\vec{\alpha}$ is defined by $\vec{\alpha} = d\vec{\omega}/dt$. (Equation 10-17 is the vector form of Equation 9-18.)

For a system of particles, the total angular momentum about the z axis equals the sum of the angular momenta about the z axis. In addition, the total torque about the z axis is the sum of the external torques about the z axis acting on the system.

(Dick Luria/Science Source/ Photo Researchers.)

Example 10-3 The Atwood's Machine Revisited

An Atwood's machine has two blocks with masses m_1 and m_2 ($m_1 > m_2$) connected by a string of negligible mass that passes over a pulley with frictionless bearings. The pulley is a uniform disk of mass M and radius R. The string does not slip on the pulley. Apply Equation 10-16 to the system consisting of the two blocks, the string, and the pulley, to find the angular acceleration of the pulley and the linear acceleration of the blocks.

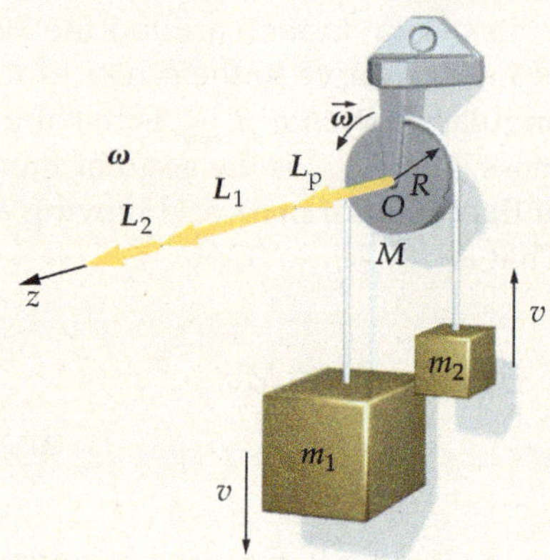

FIGURE 10-19

PICTURE Let the pulley and blocks be centered in the xy plane with the z axis out of the page and through the center of the pulley at point O, as shown in Figure 10-19. We compute the torques and angular momenta about the z axis and apply Newton's second law for angular motion (Equation 10-10). Because m_1 is greater than m_2, the disk will rotate counterclockwise, which means $\vec{\omega}$ is directed in the $+z$ direction. All the forces are in the xy plane, so all torques are parallel to the z axis. Also, all the velocities are in the xy plane, so all the angular-momentum vectors are also parallel with the z axis. Because the torque, angular velocity, and angular-momentum vectors are all parallel with the z axis, we can treat this as a one-dimensional problem with positive assigned to counterclockwise motion and negative to clockwise motion. The acceleration a of the blocks is related to the angular acceleration α of the pulley by the nonslip condition $a = R\alpha$.

SOLVE

1. Let the system be everything that moves. Draw a free-body diagram of the system (Figure 10-20). The only thing touching the system is the pulley bearings. The external forces on the system are the normal force of the pulley bearings on the pulley and the gravity forces on the two blocks and the pulley:
2. Express Newton's second law for rotation, z components only (Equation 10-16):
   $$\sum \tau_{\text{ext} z} = \frac{dL_z}{dt}$$
3. The total external torque about the z axis is the sum of the torques exerted by the external forces. The moment arms for F_{g1} and F_{g2} each equal R. (The moment arms of F_n and F_{gp} each equal zero.) $F_{g1} = m_1 g$ and $F_{g2} = m_2 g$:
   $$\sum \tau_{i\,\text{ext} z} = \tau_n + \tau_{gp} + \tau_{g1} + \tau_{g2}$$
   $$= 0 + 0 + m_1 gR - m_2 gR$$
4. The total angular momentum about the z axis equals the angular momentum of the pulley, $\vec{L}_p$, plus the angular momenta of block 1, $\vec{L}_1$, and block 2, $\vec{L}_2$, each in the positive z direction. The pulley has spin angular momentum, but no orbital angular momentum because its center of mass is at rest. Each block has orbital angular momentum, but no spin angular momentum.
   $$L_z = L_1 + L_2 + L_p$$
   $$= m_1 vR + m_2 vR + I\omega$$

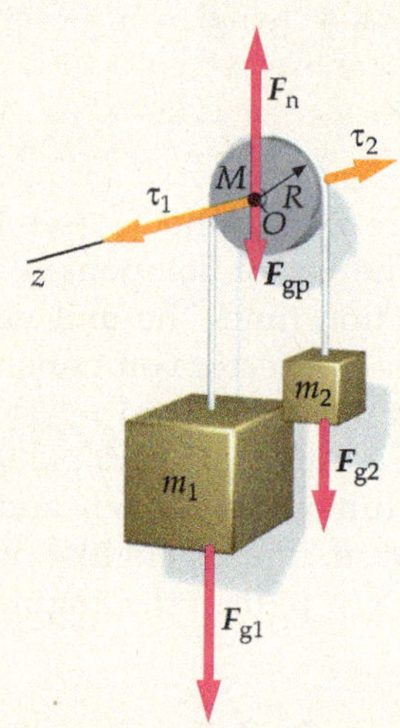

FIGURE 10-20

5. Substitute these results into Newton's second law for rotation in step 2:

$$\sum \tau_{\text{ext }z} = \frac{dL_z}{dt}$$

$$m_1gR - m_2gR = \frac{d}{dt}(m_1vR + m_2vR + I\omega)$$

$$m_1gR - m_2gR = (m_1 + m_2)Ra + I\alpha$$

6. Relate I to M and R, and use the nonslip condition to relate α to a and solve for both a and α:

$$m_1gR - m_2gR = (m_1 + m_2)Ra + \tfrac{1}{2}MR^2\frac{a}{R}$$

so $$a = \boxed{\frac{m_1 - m_2}{m_1 + m_2 + \frac{1}{2}M}g}$$

and $$\alpha = \frac{a}{R} = \boxed{\frac{m_1 - m_2}{m_1 + m_2 + \frac{1}{2}M}\frac{g}{R}}$$

CHECK The answers are dimensionally correct. Both the numerators and the denominators contain factors that have the dimensions of mass, so these factors do not contribute to the dimensions of the ratios. For the first answer, a and g have dimension L/T^2, and for the second answer, α and g/R have dimension T^{-2}. Both of these dimensions are what one would expect.

TAKING IT FURTHER (1) This problem could be solved by writing the tensions T_1 on the left and T_2 on the right and using $\tau = I\alpha$ (Equation 10-17) for the pulley and $\Sigma F_y = ma_y$ for each block. However, using angular momentum (Equation 10-16) is easier, and once you have solved for the acceleration, it is straightforward to solve for the two tensions. (2) Because $\vec{L}_2 = \vec{r}_2 \times m_2\vec{v}_2$ (Figure 10-21), the direction of $\vec{L}_2$ is gotten by applying the right-hand rule (Figure 10-6). And because $\vec{\tau}_2 = \vec{r}_2 \times \vec{F}_{g2}$ (Figure 10-21), the direction of $\vec{\tau}_2$ also is gotten by applying the right-hand rule.

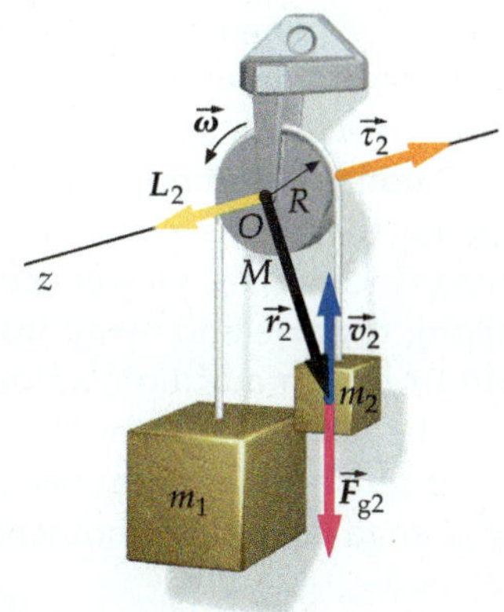

FIGURE 10-21

There are many problems in which the forces, position vectors, and velocities all remain perpendicular to a fixed axis, so the torques, angular velocities, and angular-momentum vectors all remain parallel with an axis of rotation that remains fixed in space. In such cases, we can assign positive and negative values to counterclockwise or clockwise rotations, as we did in Example 10-3, and treat the case like a one-dimensional problem. However, there are other situations, such as the motion of a gyroscope, where torque, angular velocity, and angular momentum must be treated as multidimensional vectors.

THE GYROSCOPE

A *gyroscope* is a common example of an object exhibiting motion in which its axis of rotation changes direction. Figure 10-22 shows a gyroscope consisting of a bicycle wheel that is free to turn on its axle. The axle is pivoted at a point a distance D from the center of the wheel, and the axle is free to rotate about the pivot in any direction. We can give a qualitative understanding of the complex motion of such a system by using Newton's second law for rotation,

$$\vec{\tau}_{\text{net}} = \frac{d\vec{L}}{dt} \quad (\text{or } \Delta\vec{L} \approx \vec{\tau}_{\text{net}}\Delta t) \tag{10-18}$$

along with the relations

$$\vec{\tau}_{\text{net}} = \vec{r}_{\text{cm}} \times M\vec{g}$$

and

$$\vec{L} = I_s\vec{\omega}_s$$

where M is the mass of the wheel-axle system, $\vec{r}_{\text{cm}}$ is the position of the center of mass relative to O, and I_s and ω_s are the moment of inertia and angular velocity of the wheel about its spin axis. (The torque about O on the system that is due to the normal force exerted by the support stand is zero, so the net torque about O is equal to the torque about O that is due to the gravitational force.)

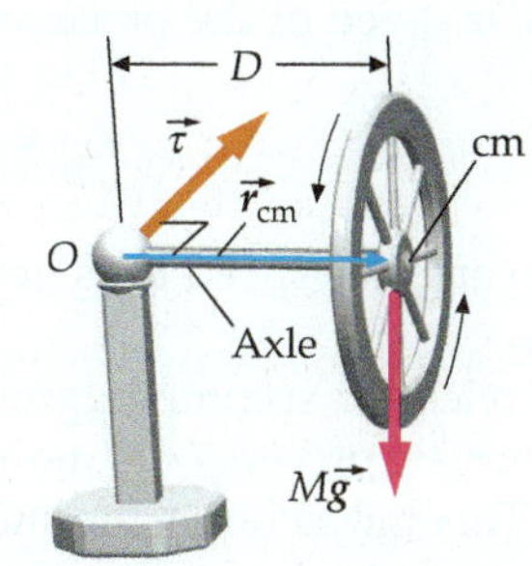

FIGURE 10-22

In accord with Equation 10-18, the *change* in the angular momentum of the system is in the same direction as the net torque acting on the system. We wish to describe the motion of the wheel-axle system after it is released from rest in the horizontal position shown in Figure 10-22, and we will do this first without the wheel spinning, and then with the wheel spinning rapidly. If the wheel *is not* spinning about its axle, Equation 10-18 predicts that upon release the wheel-axle system will simply tip downward, rotating about a horizontal axis that passes through O and is perpendicular to the axle. This prediction is based on the following reasoning. The torque vector is horizontal, perpendicular to the axle, and directed as shown in Figure 10-22. Both the wheel and the axle shaft are initially at rest, so the initial angular momentum $\vec{L}_i$ is zero. Consequently, the change in angular momentum $\Delta\vec{L} = \vec{L}_f - \vec{L}_i$ equals the final angular momentum $\vec{L}_f$, which, according to Equation 10-18, is in the same direction as the torque. The final angular velocity vector $\vec{\omega}_f$ is in the same direction as the final angular-momentum vector $\vec{L}_f$. If you place your right thumb in the direction of $\vec{\omega}_f$, your fingers will curl in the direction of the motion of the wheel-axle system.

If the wheel *is* spinning rapidly about its axle, Equation 10-18 predicts that upon release the wheel-axle system will slowly rotate about a vertical axis through O. This prediction is based on the following reasoning. The torque vector is horizontal, perpendicular to the axle, and directed into the page, just as before. The wheel is spinning clockwise as viewed from O, so the initial angular momentum $\vec{L}_i$ is directed along the axle and away from O. (The direction of $\vec{L}_i$ is obtained from the right-hand rule.) In addition, according to Equation 10-18, $\Delta\vec{L}$ is in the same direction as the net torque, which is initially directed into the page (Figure 10-23*a*). The final angular-momentum vector $\vec{L}_f$ is equal to the initial angular momentum plus the change in angular momentum. That is,

$$\vec{L}_f = \vec{L}_i + \Delta\vec{L}$$

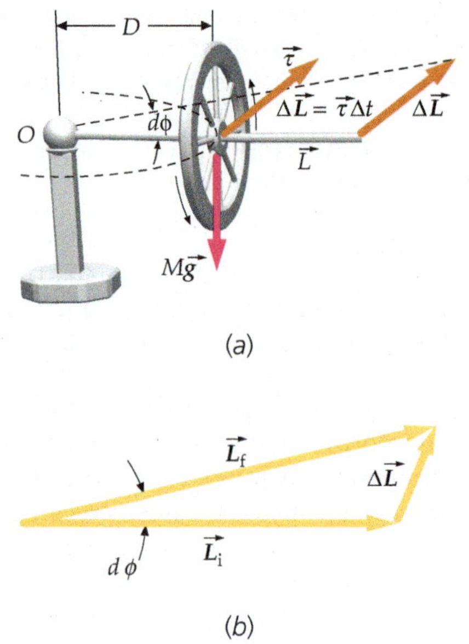

FIGURE 10-23

The direction of $\vec{L}_f$ is shown in the vector-addition diagram (Figure 10-23*b*). Because the wheel is spinning rapidly, and because the wheel contains much of the system's mass, the angular momentum of the wheel-axle system is dominated by the spin angular momentum of the wheel, which means that $\vec{L}$ is directed along the axle and away from O. Thus, Equation 10-18 predicts the center of mass of the system will rotate about a vertical axis through O in the direction that keeps the tip of the angular-momentum vector moving horizontally—and in the direction of the torque vector. This motion, which is always surprising when first encountered, is called **precession.** We can calculate the angular speed ω_p of precession. In a small time interval dt, the change in the angular momentum has a magnitude dL:

$$dL = \tau\, dt = MgD\, dt$$

where MgD is the magnitude of the torque about the pivot point. From Figure 10-23*b*, the angle $d\phi$ through which the axle moves is

$$d\phi = \frac{dL}{L} = \frac{\tau\, dt}{L} = \frac{MgD\, dt}{L}$$

The angular speed of the precession is thus

$$\omega_p = \frac{d\phi}{dt} = \frac{MgD}{L} = \frac{MgD}{I_s\omega_s} \qquad \text{10-19}$$

If the spin angular speed ω_s is very fast, then the precessional angular speed ω_p is very slow.

If you release a spinning gyroscope with its spin axis at rest, upon release this axis will initiate precessional motion with an up and down bouncing motion called *nutation.* This initial bouncing motion can be avoided by releasing the gyroscope with the spin axis already rotating with an initial angular velocity exactly equal to ω_p (see Equation 10-19).

10-3 CONSERVATION OF ANGULAR MOMENTUM

When the net external torque acting on a system about some point remains zero, we have

$$\vec{\tau}_{\text{net ext}} = \frac{d\vec{L}_{\text{sys}}}{dt} = 0$$

or

$$\vec{L}_{\text{sys}} = \text{constant} \qquad 10\text{-}20$$

Equation 10-20 is a statement of the **law of conservation of angular momentum.**

> If the net external torque acting on a system about some point is zero, the total angular momentum of the system about that point remains constant.
>
> CONSERVATION OF ANGULAR MOMENTUM

This law is the rotational analog of the law of conservation of linear momentum. If a system is isolated from its surroundings, so that there are no external forces or torques acting on it, three quantities are conserved: energy, linear momentum, and angular momentum. The law of conservation of angular momentum is a fundamental law of nature. There are many examples of the conservation of angular momentum in everyday life. Figure 10-24 and Figure 10-25 illustrate angular-momentum conservation in diving and ice skating. Even on the scale of atomic and nuclear physics, where Newtonian mechanics does not hold, the angular momentum of an isolated system is found to be constant over time.

Although conservation of angular momentum is a law, independent of Newton's laws of motion, the fact that the internal torques of a system cancel is suggested by Newton's third law. Consider the two particles shown in Figure 10-26. Let $\vec{F}_{12}$ be the force exerted by particle 1 on particle 2 and $\vec{F}_{21}$ be the force exerted by particle 2 on particle 1. By Newton's third law, $\vec{F}_{21} = -\vec{F}_{12}$. The sum of the torques exerted by these forces about the origin O is

$$\vec{\tau}_1 + \vec{\tau}_2 = \vec{r}_1 \times \vec{F}_{21} + \vec{r}_2 \times \vec{F}_{12} = \vec{r}_1 \times \vec{F}_{21} + \vec{r}_2 \times (-\vec{F}_{21}) = (\vec{r}_1 - \vec{r}_2) \times \vec{F}_{21}$$

The vector $\vec{r}_1 - \vec{r}_2$ is along the line joining the two particles. If $\vec{F}_{21}$ acts parallel to the line joining m_1 and m_2, $\vec{F}_{21}$ and $\vec{r}_1 - \vec{r}_2$ are either parallel or antiparallel and

$$(\vec{r}_1 - \vec{r}_2) \times \vec{F}_{21} = 0$$

If this is true for all the internal forces, the internal torques cancel in pairs.*

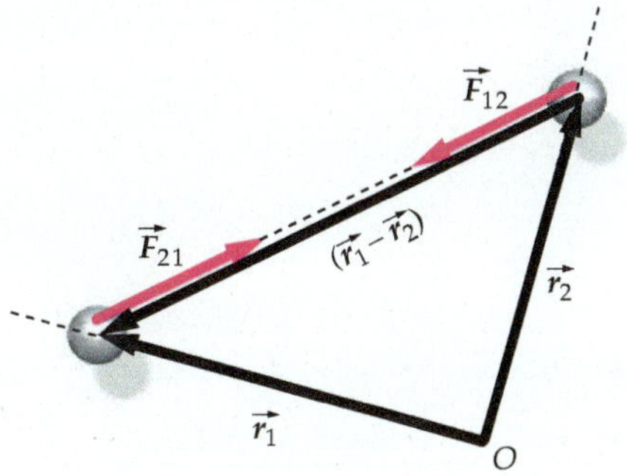

FIGURE 10-26

* Not all forces do come in pairs of equal and opposite forces. For example, the magnetic forces that moving charged particles exert on each other do not.

FIGURE 10-24 Multiflash photograph of a diver. The diver's center of mass moves along a parabolic path after he leaves the board. The angular momentum is provided by the initial external torque due to the force of the board, which does not pass through the diver's center of mass if he leans forward as he jumps. If the diver wanted to undergo one or more somersaults in the air, he would draw in his arms and legs, decreasing his moment of inertia to increase his angular velocity. *(© The Harold E. Edgerton 1992 Trust.)*

FIGURE 10-25 A spinning skater. Because the torque exerted by the ice is small, the angular momentum of the skater is approximately constant. When she reduces her moment of inertia by drawing in her arms, her angular velocity increases. *(Mike Powell/ Getty.)*

Example 10-4 A Rotating Disk

Disk 1 is rotating freely and has an angular velocity ω_i about a vertical axis that coincides with its symmetry axis, as shown in Figure 10-27. Its moment of inertia about this axis is I_1. It drops onto disk 2, of moment of inertia I_2, that is initially at rest. Disk 2 is centered on the same axis as disk 1 and is free to rotate about that axis. Because of kinetic friction, the two disks eventually attain a common angular velocity ω_f. Find ω_f.

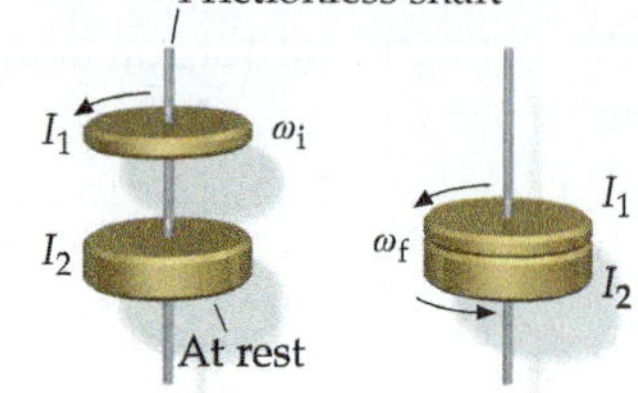

FIGURE 10-27

PICTURE We find the final angular velocity from the final angular momentum, which is equal to the initial angular momentum because there are no external torques acting on the two-disk system. The angular speed of the upper disk is reduced, while that of the lower disk is increased by the forces of kinetic friction. Because the direction of the rotation axis is fixed, the direction of the rotational motion can be specified by a + or − sign. Kinetic friction dissipates mechanical energy, so we expect that the total mechanical energy is decreased.

SOLVE

1. The final angular velocity is related to the initial angular velocity by conservation of angular momentum:
$$L_f = L_i$$
$$(I_1 + I_2)\omega_f = I_1\omega_i$$

2. Solve for the final angular velocity:
$$\boxed{\omega_f = \frac{I_1}{I_1 + I_2}\omega_i = \frac{1}{1 + (I_2/I_1)}\omega_i}$$

CHECK If $I_2 \ll I_1$, the collision should have little effect on the motion of disk 1. Our answer agrees, giving that as $(I_2/I_1) \to 0$, $\omega_f \to \omega_i$. (Read "$\to$" as "approaches".) If $I_2 \gg I_1$, then disk 1 should slow almost to a stop without causing disk 2 to rotate appreciably. Our answer again agrees, giving that as $(I_2/I_1) \to \infty$, $\omega_f \to 0$.

TAKING IT FURTHER Plates rotating at differing speeds engage in the drive trains of truck and automobiles. The photo shows such plates in a truck transmission.

The rotating plates in the transmission of a truck make inelastic collisions when engaged. *(Dick Luria/FPG International.)*

In the collision of the two disks in Example 10-4, mechanical energy is dissipated. We can see this by writing the energy in terms of the angular momentum. An object rotating with an angular velocity ω has kinetic energy

$$K = \tfrac{1}{2}I\omega^2 = \frac{(I\omega)^2}{2I}$$

Substituting L for $I\omega$ gives

$$K = \frac{L^2}{2I} \qquad \text{10-21}$$

(This result is analogous with that for linear motion $K = p^2/2m$, Equation 8-25.) The initial kinetic energy in Example 10-4 is

$$K_i = \frac{L_i^2}{2I_1}$$

and the final kinetic energy is

$$K_f = \frac{L_f^2}{2(I_1 + I_2)}$$

Because $L_f = L_i$, the ratio of the final to the initial kinetic energy is

$$\frac{K_f}{K_i} = \frac{I_1}{I_1 + I_2}$$

which is less than one. This interaction of the disks is analogous to a one-dimensional perfectly inelastic collision of two objects.

Example 10-5 Mud in Your Eye

Context-Rich

You and three of your friends have been bullied for many years by Gene, who has avoided taking physics classes. So you and your three friends who are taking advanced-placement physics decide to teach him a lesson using conservation of angular momentum. Here is your plan. A nearby park has a small merry-go-round (Figure 10-28) with a 3.0-m-diameter turntable that has a 130-kg · m² moment of inertia. You initially get all five of you to stand on the merry-go-round next to the rim while the merry-go-round is rotating at a modest 20 rev/min. When the signal is given, you and your friends will quickly walk to the center of the merry-go-round, leaving Gene near the rim. The merry-go-round will speed up, throwing Gene off and into the mud. (You plan to do this after a heavy rain.) Gene is very quick and very strong, so throwing him off will require that the centripetal acceleration of the rim be at least 4.0gs. Will this plan work? (Assume that each person has a mass of 60 kg.)

FIGURE 10-28

PICTURE By moving to the center of the merry-go-round, you and your friends are decreasing the moment of inertia of the students–merry-go-round system. No external torques about the axis act on the system (assume effects due to friction and air resistance to be negligible), so the angular momentum about the axis remains constant. The angular momentum is the moment of inertia times the angular velocity, so a decrease in the moment of inertia means an increase in the angular velocity. The angular velocity can be used to find the centripetal acceleration at the rim. Because the direction of the rotation axis is fixed, the direction of the rotational motion can be specified by a + or − sign.

SOLVE

1. The centripetal acceleration depends on the angular speed ω and the radius R:

$$a_c = \omega^2 R$$

2. Angular momentum is conserved. For rotations about a fixed axis, $L = I\omega$:

$$L_f = L_i$$
$$I_f\omega_f = I_i\omega_i$$

3. The moment of inertia of the system is the sum of the moments of inertia of each person plus that of the merry-go-round. Each person has mass $m = 60$ kg:

$$I_i = 5mR^2 + I_{mgr} = 5(60\text{ kg})(1.5\text{ m})^2 + 130\text{ kg}\cdot\text{m}^2 = 805\text{ kg}\cdot\text{m}^2$$

4. To find the final moment of inertia, assume that you and your friends are 30 cm ($\approx$1 ft) from the center:

$$I_f = mR^2 + 4mr^2 + I_{mgr} = (60\text{ kg})(1.5\text{ m})^2 + 4(60\text{ kg})(0.3\text{ m})^2 + 130\text{ kg}\cdot\text{m}^2 = 287\text{ kg}\cdot\text{m}^2$$

5. Using conservation of angular momentum, solve for the final angular velocity:

$$\omega_f = \frac{I_i}{I_f}\omega_i = \frac{805\text{ kg}\cdot\text{m}^2}{287\text{ kg}\cdot\text{m}^2}20\text{ rev/min} = 56.2\text{ rev/min} = 5.88\text{ rad/s}$$

6. Solve for the centripetal acceleration of the rim:

$$a_c = \omega^2 R = (5.88\text{ rad/s})^2(1.5\text{ m}) = 51.9\text{ m/s}^2$$

7. Convert to g's:

$$a_c = 51.9\text{ m/s}^2 \times \frac{1g}{9.81\text{ m/s}^2} = 5.29g = \boxed{5.3g}$$

8. Does Gene end up in the mud?

Success! The acceleration is much greater than 4.0g, so Gene flies off and lands in the mud.

CHECK At the rim of the merry-go-round the four friends are five times farther from the axis than they are after walking to the center. Thus, their contribution to the final moment of inertia of the system is 1/25 of their contribution to the initial moment of inertia. To conserve angular momentum, this large reduction in the moment of inertia will be accompanied by a compensating increase in the angular speed. Our step-5 results show that the angular speed increased from 20 rev/min to 56 rev/min.

TAKING IT FURTHER The linear speed of the rotating merry-go-round is greatest at the rim and decreases to zero at the center. At the rim, everyone is moving in a circle. As the four friends walk toward the center, they are stepping onto a part of the merry-go-round that is moving more slowly than they are, so the frictional force of their feet on the merry-go-round has a component in the tangential direction that speeds up the merry-go-round. Also, the merry-go-round exerts an equal and opposite frictional force on the feet of the four friends, slowing their motion in the tangential direction. The static frictional forces exerted by their feet, result in a net torque on the merry-go-round, increasing its angular momentum about the rotation axis. The equal and opposite static frictional forces on the feet of the friends exert torque in the opposite direction on the friends, so the torques associated with these forces decreases their angular momentum about the axis. The two net torques are equal and opposite, as are the associated angular-momentum changes. Thus, the angular momentum of the students–merry-go-round system remains constant.

The moment of inertia of the students–merry-go-round system decreases as the friends walk toward the center. Thus, the system's moment of inertia decreases while its angular momentum remains constant. As a result, we can see from Equation 10-21 that the kinetic energy of the students–merry-go-round system increases. The energy for this kinetic energy increase comes from the internal energy of the friends. Walking radially inward, like walking up a steep incline, requires the expenditure of internal energy.

Example 10-6 Another Ride on the Merry-go-Round

Try It Yourself

A 25-kg child in a playground runs with an initial speed of 2.5 m/s along a path *tangent* to the rim of a merry-go-round, whose radius is 2.0 m. The merry-go-round, which is initially at rest, has a moment of inertia of 500 kg · m² The child then jumps on (Figure 10-29). Find the final angular velocity of the child and the merry-go-round together.

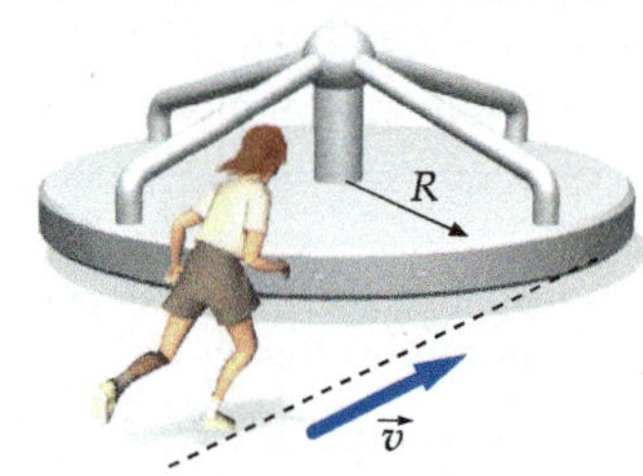

FIGURE 10-29

PICTURE Once the child's feet leave the ground, no external torques about the rotation axis act on the child–merry-go-round system, hence, the total angular momentum of the system about the rotation axis is conserved. The initial angular speed of the merry-go-round is zero. Because the direction of the rotation axis is fixed, the direction of the rotational motion can be specified by a + or − sign.

SOLVE

Cover the column at the right and solve it yourself before looking at the answers.

Steps	Answers
1. Write an expression for the initial angular momentum of the child–merry-go-round system. The initial angular momentum of the merry-go-round is zero. The child has mass m and speed v_i in the tangential direction just before making contact with the merry-go-round. Model the child as a point particle.	$L_i = \lvert\vec{r}_{child} \times m\vec{v}_i\rvert = Rmv_i$
2. Write an expression for the total final angular momentum of the child–merry-go-round system in terms of the final angular velocity ω_f.	$L_f = I_{sys}\omega_f = (mR^2 + I_m)\omega_f$
3. Set your expressions in steps 1 and 2 equal and solve for ω_f.	$\omega_f = \dfrac{mvR}{mR^2 + I_m} = \boxed{0.21\ \text{rad/s}}$

CHECK The final speed of the child is $\omega_f R$ = (0.21 rad/s)(2.0 m) = 0.42 m/s. As expected, this speed is much less then the child's initial speed of 2.5 m/s.

PRACTICE PROBLEM 10-1 Calculate the initial and final kinetic energies of the child–merry-go-round system.

An astronaut examines the reaction wheel of the Hubble Space Telescope. *(NASA/Goddard Space Flight Center.)*

The Hubble Space Telescope is aimed by regulating the spin rates of 45-kg reaction wheels arranged off-axis from each other and spinning at up to 3000 rpm. Software-controlled changes in the spin rates exchange angular momentum between the flywheels and the rest of the satellite. The changes in the angular momentum of the rest of the satellite cause it to slew to different orientations. This aiming mechanism can achieve and hold a target to within 0.005 arcsec, which is equivalent to holding a flashlight beam in Los Angeles on a dime in San Francisco.

Example 10-7 Spinning the Wheel — *Conceptual*

You are sitting on a stool on a frictionless turntable holding a bicycle wheel (Figure 10-30). Initially, neither the wheel nor the turntable is spinning. Following instructions from your teacher, you hold the spin axis of the wheel vertical with one hand, and with your other hand you set the wheel spinning counterclockwise (as viewed from above). Surprise! When you start the wheel spinning one way, the turntable, the stool, you, and the axis of the wheel start rotating in the opposite direction. After a few seconds, you use your free hand to brake the spinning motion of the wheel. You are surprised again when you, the stool, and the wheel axis cease rotating as the wheel ceases spinning. Explain.

FIGURE 10-30 As she starts to spin the wheel clockwise, which way does she start to rotate?

PICTURE Because the turntable is frictionless, there are no torques on the student–turntable–stool–wheel system about the axis of the turntable. Thus, the angular momentum of the system about the turntable axis remains constant.

SOLVE

Initially, the entire system is at rest, so its total angular momentum is zero. As it spins up, the wheel acquires an upward-directed spin angular momentum. The total angular momentum of the system must remain zero.

Thus, the orbital angular momentum acquired by the wheel about the turntable axis, plus the angular momentum acquired by you, the stool, and the turntable about the turntable axis, is equal in magnitude to the spin angular momentum of the wheel, but is directed downward. A downward-directed angular momentum means you start rotating clockwise (as viewed from above). As you brake the wheel, its upward-directed spin angular momentum decreases to zero. To keep the total angular momentum of the system equal to zero, the entire system must slow to a stop as the wheel ceases spinning.

CHECK This situation is analogous to that of a person walking on flatbed cart that has frictionless wheel bearings and is on a smooth horizontal track. As the person walks forward, the cart moves backward, but when the person stops walking forward, the cart stops moving backward, as the principle of conservation of linear momentum predicts.

TAKING IT FURTHER The Hubble Space Telescope is aimed at different targets using four wheels mounted on the telescope. The wheels are spun up and spun down by computer-controlled electric motors. As a result the telescope is able to aim at the targets specified in the instructions to the computers.

Example 10-8 Pulling Through a Hole

A particle of mass m moves with speed v_0 in a circle of radius r_0 on a frictionless tabletop. The particle is attached to a string that passes through a hole in the table, as shown in Figure 10-31. The string is slowly pulled downward until the particle is a distance r_f from the hole, after which the particle moves in a circle of radius r_f. (*a*) Find the final velocity in terms of r_0, v_0 and r_f. (*b*) Find the tension when the particle is moving in a circle of radius r in terms of m, r, and the angular momentum L. (*c*) Calculate the work done on the particle by the tension force $\vec{T}$ by integrating $\vec{T}\cdot d\vec{\ell}$. Express your answer in terms of r and L_0.

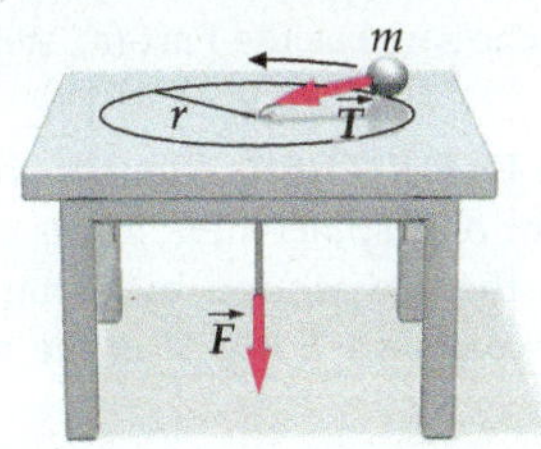

FIGURE 10-31

PICTURE The speed of the particle is related to its angular momentum. The net torque is equal to the rate of change of the angular momentum. Because the net force acting on the particle is the tension force $\vec{T}$ exerted by the string, which is always directed toward the hole, the torque about the vertical axis through the hole is zero. Thus, the angular momentum about this axis remains constant.

SOLVE

(*a*) Conservation of angular momentum relates the final speed to the initial speed and the initial and final radii:

$$L_f = L_0$$

$$mv_f r_f = mv_0 r_0$$

$$\text{so} \quad \boxed{v_f = \frac{r_0}{r_f}v_0}$$

(*b*) 1. Apply Newton's second law to relate T to v and r. Because the particle is being pulled in slowly, the acceleration is virtually the same as if the particle were moving in a circle:

$$T \approx m\frac{v^2}{r}$$

2. Obtain a relation between L, r, and v using the definition of angular momentum. Because the particle is being pulled in slowly, $|\beta| \ll 1$ (Figure 10-32*a*):

$$\vec{L} = \vec{r} \times \vec{p}$$

$$L = rmv\cos\beta \approx rmv \quad (|\beta| \ll 1, \text{so } \cos\beta \approx 1)$$

3. Eliminate v by solving the Part-(*b*), step-2 result for v and then substituting into the Part-(*b*), step-1 result:

$$T = m\frac{v^2}{r} = \frac{m}{r}\left(\frac{L}{mr}\right)^2 = \boxed{\frac{L^2}{mr^3}}$$

(*c*) 1. Make a drawing of the particle as it moves closer to the hole (Figure 10-32*b*). When the particle undergoes displacement $d\vec{\ell}$, its distance r from the axis changes by dr. Because r is decreasing, dr is negative. Thus:

$$dr = -|dr|$$

2. Write $dW = \vec{T}\cdot d\vec{\ell}$ in terms of T and dr:

$$dW = \vec{T}\cdot d\vec{\ell} = T\,d\ell\cos\phi$$

Because $\quad |dr| = d\ell\cos\phi,$

$$dW = T|dr| = -T\,dr$$

3. Integrate from r_0 to r_f after substituting for T from the Part-(*b*), step-3 result:

$$W = -\int_{r_0}^{r_f} T\,dr = -\int_{r_0}^{r_f}\frac{L^2}{mr^3}\,dr$$

$$= -\frac{L^2}{m}\int_{r_0}^{r_f} r^{-3}\,dr = -\frac{L^2}{m}\frac{r^{-2}}{-2}\bigg|_{r_0}^{r_f}$$

$$= \boxed{\frac{L^2}{2m}\left(\frac{1}{r_f^2} - \frac{1}{r_0^2}\right)}$$

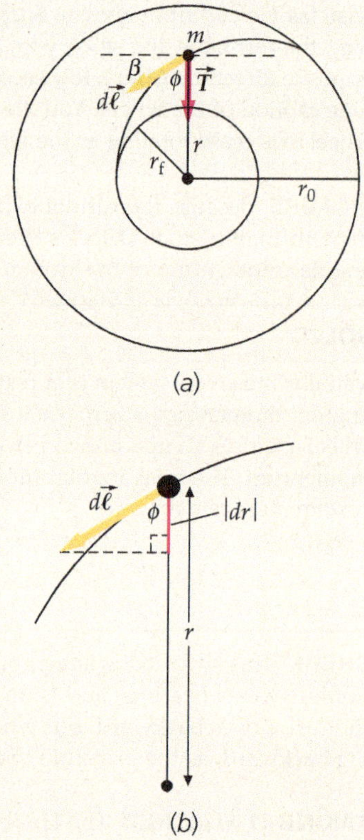

(*a*) (*b*)

FIGURE 10-32

CHECK Note that work must be done to pull the string downward. Because r_f is less than r_0, the work is positive. This work is equal to the increase in kinetic energy. We can calculate the change in kinetic energy of the particle directly. Using $K = L^2/2I$, with $L_0 = L_f = L$, and $I = mr^2$, the change in kinetic energy is $K_f - K_i = (L^2/2mr_f^2) - (L^2/2mr_0^2) = (L^2/2m)(r_f^{-2} - r_0^{-2})$, which is the same as the Part-(*c*), step-3 result found by direct integration.

TAKING IT FURTHER The increment of work dW can also be obtained by expressing the increment of displacement $d\vec{\ell}$ as $d\vec{r}$, the change in the position vector $\vec{r}$. The dot product $\vec{T}\cdot d\vec{r}$ is then expanded by using components which give $dW = \vec{T}\cdot d\vec{r} = T_r\,dr = -T\,dr$. In this expansion $T_r = -T$ is the radial component of $\vec{T}$ and dr is the radial component of $d\vec{r}$.

PRACTICE PROBLEM 10-2 At what final radius r_N would the tension be N times the tension at initial radius r_0?

In Figure 10-33 a puck on a frictionless plane is given an initial speed v_0. The puck is attached to a string that wraps around a vertical post. This situation looks similar to Example 10-8, but it is not the same. There is no agent that does work on the puck, nor is there any mechanism for energy dissipation. Thus, mechanical energy must be conserved. Because $K = L^2/(2I)$, where L is the magnitude of the angular momentum about the axis of the post, is constant and I decreases as r_0 decreases, L must also decrease. Note that the tension force does not act toward the axis of the post. The tension force on the puck produces a torque vector $\vec{\tau}$ about the axis of the post in the downward direction, which reduces the puck's angular-momentum vector $\vec{L}$ about the axis, which is in the upward direction.

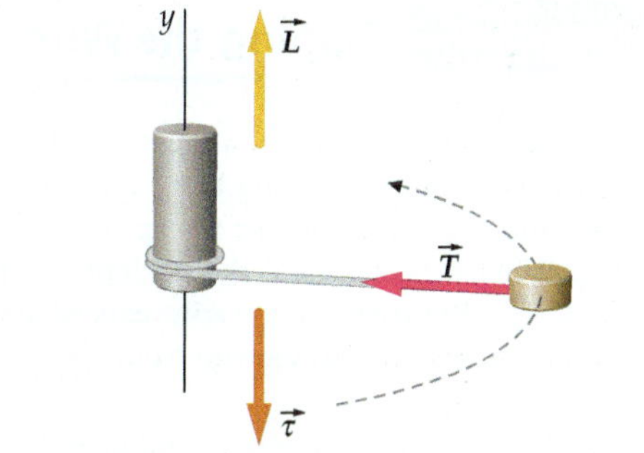

FIGURE 10-33

Example 10-9 The Ballistic Pendulum Revisited

A thin rod of mass M and length d hangs vertically from a pivot attached to one end. A piece of clay of mass m and moving horizontally at speed v hits the rod a distance x from the pivot and sticks to it (Figure 10-34). Find the ratio of the clay–rod system's kinetic energy just after the collision to its kinetic energy just before the collision.

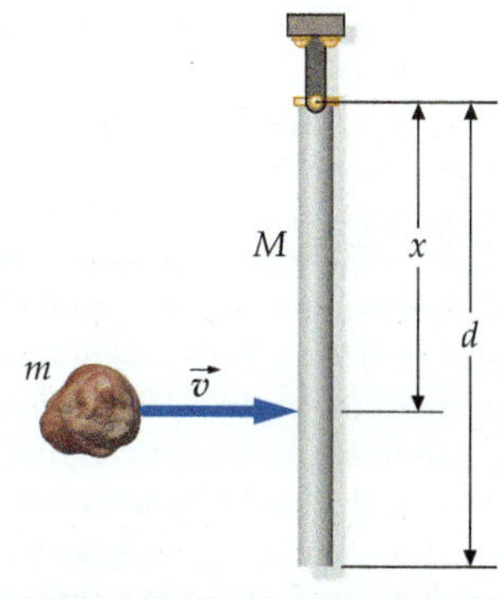

FIGURE 10-34

PICTURE The collision is inelastic, so we do not expect mechanical energy to be constant. During the collision, the pivot exerts a large force on the rod, so the linear momentum of the rod–clay system is not conserved. However, there are no external torques about the horizontal axis perpendicular to the page and through the pivot point, so angular momentum of the system about this axis is conserved. The kinetic energy after the inelastic collision can be written in terms of the angular momentum L_{sys} and the moment of inertia I_f of the combined clay–rod system. Conservation of angular momentum allows you to relate L_{sys} to the mass m and velocity v of the clay. Model the clay as a point particle.

SOLVE

1. Before the collision the kinetic energy of the system is that of the moving clay ball.

$$K_i = \tfrac{1}{2}mv^2$$

2. After the collision it is that of the swinging clay–rod object. Write the kinetic energy after the collision in terms of the angular momentum L_{sys} and the moment of inertia I_f of the clay–rod system.

$$K_f = \frac{L_{sys}^2}{2I_f}$$

3. During the collision, angular momentum is conserved. Write the angular momentum L_{sys} in terms of m, v, and x. Before impact the angular momentum of the rod is zero.

$$L_{sys} = |\vec{r} \times m\vec{v}| = mvx$$

where $\vec{r}$ is the vector from the axis to the clay and $\vec{v}$ is the velocity of the clay before impact.

4. Write I_f in terms of m, x, M, and d.

$$I_f = mx^2 + \tfrac{1}{3}Md^2$$

5. Substitute these expressions for L_{sys} and I_f into your equation for K_f.

$$K_f = \frac{L_{sys}^2}{2I_f} = \frac{(mvx)^2}{2(mx^2 + \frac{1}{3}Md^2)}$$

$$= \frac{3}{2}\,\frac{m^2x^2v^2}{(3mx^2 + Md^2)}$$

6. Divide the kinetic energy after the collision by the initial kinetic energy.

$$\frac{K_f}{K_i} = \frac{\dfrac{3}{2}\,\dfrac{m^2x^2v^2}{(3mx^2 + Md^2)}}{\frac{1}{2}mv^2} = \boxed{\frac{1}{1 + \dfrac{Md^2}{3mx^2}}}$$

CHECK Md^2 and mx^2 obviously have the same dimensions, so the step-6 result has no dimensions, as is expected for a ratio of two energies. In addition, the ratio K_f/K_i is between zero and one, as expected for an inelastic collision. In the limit that $M/m \to \infty$, $K_f/K_i \to 0$, and in the limit that $M/m \to 0$, $K_f/K_i \to 1$. Both of these limiting values of K_f/K_i meet expectations.

TAKING IT FURTHER This example is the rotational analog of the ballistic pendulum discussed in Example 8-10. In that example, we used conservation of linear momentum to determine the kinetic energy of the pendulum after the collision.

Example 10-10 Tipping the Wheel

Conceptual

A student sitting on a stool that rests on a turntable with frictionless bearings (Figure 10-35*a*) is holding a rapidly spinning bicycle wheel. The rotation axis of the wheel is initially horizontal, and the magnitude of the spin-angular-momentum vector of the spinning wheel is $L_{\text{wheel i}}$. What will happen if the student suddenly tips the axle of the wheel (Figure 10-35*b*) so that after the rotation the spin axis of the wheel is vertical and the wheel is spinning counterclockwise (when viewed from above)?

PICTURE The turntable–stool–student–wheel system is free to rotate about a vertical axis through the center of the turntable. Because the turntable is frictionless, there can be no external torques about this axis. Thus, the angular momentum of the system about this axis remains constant.

SOLVE

Rotating the axle changes the direction, but not the magnitude, of the spin angular momentum of the wheel. The final spin angular momentum of the wheel is directed upward. The initial angular momentum of the turntable–stool–student–wheel system about the vertical axis of the turntable is zero. Thus, the final angular momentum of the system about the same vertical axis is also zero. Following the rotation of the axle, the spin angular momentum of the wheel is counterclockwise (when viewed from above) and equal in magnitude to $L_{\text{wheel i}}$. Conservation of angular momentum dictates that the remaining angular momentum of the system about the vertical axis of the turntable must be clockwise, and equal in magnitude to $L_{\text{wheel i}}$.

The turntable, stool, and student will be rotating clockwise with an angular momentum about the vertical axis of the turntable of magnitude $L_{\text{wheel i}}$.

CHECK The student exerts an upward torque on the spinning wheel when she tips it upward. (Due to the vector product definition of torque, an upward torque requires horizontal forces.) The wheel exerts an equal and opposite torque (also horizontal forces) on the student, causing her to rotate clockwise.

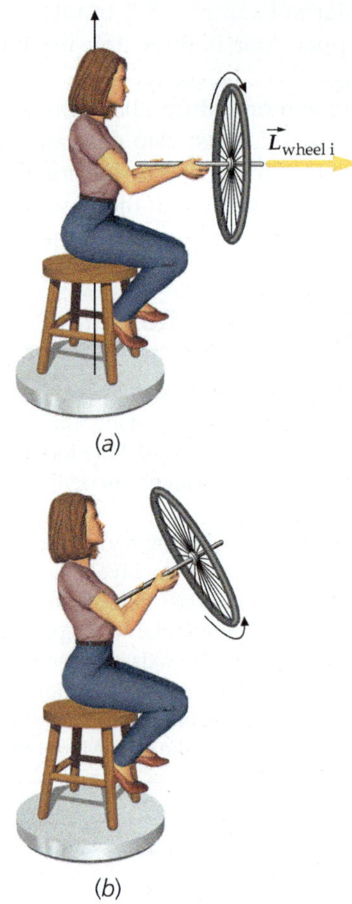

FIGURE 10-35

PROOFS OF EQUATIONS 10-10, 10-12, 10-13, 10-14, AND 10-15

Proof of Equation 10-10 We now show that Newton's second law implies that the rate of change of the angular momentum of a point particle equals the net torque acting on the particle. If more than one force acts on a particle, then the net torque relative to the origin O is the sum of the torques due to each force:

$$\vec{\tau}_{\text{net}} = \vec{r} \times \vec{F}_1 + \vec{r} \times \vec{F}_2 + \cdots = \vec{r} \times \sum_i \vec{F}_i = \vec{r} \times \vec{F}_{\text{net}}$$

According to Newton's second law, the net force on a particle equals the rate of change of the particle's linear momentum $d\vec{p}/dt$. Thus

$$\vec{\tau}_{\text{net}} = \vec{r} \times \vec{F}_{\text{net}} = \vec{r} \times \frac{d\vec{p}}{dt} \qquad \text{10-22}$$

We now compare this expression with the expression for the time rate of change of the particle's angular momentum. The definition of the angular momentum of a particle (Equation 10-8) is

$$\vec{L} = \vec{r} \times \vec{p}$$

We can compute $d\vec{L}/dt$ using the product rule for derivatives:

$$\frac{d\vec{L}}{dt} = \frac{d}{dt}(\vec{r} \times \vec{p}) = \left(\frac{d\vec{r}}{dt} \times \vec{p}\right) + \left(\vec{r} \times \frac{d\vec{p}}{dt}\right)$$

CONCEPT CHECK 10-1

The bicycle wheel is spinning counterclockwise (if viewed from above) with its spin axis vertical when it is handed to the student who is on the non-rotating turntable. In what direction will the turntable rotate as the student rotates the axis of the wheel toward the horizontal?

The second term from the right is zero because $\vec{p} = m\vec{v}$ and $\vec{v} = d\vec{r}/dt$, so

$$\frac{d\vec{r}}{dt} \times \vec{p} = \vec{v} \times m\vec{v} = 0$$

because the vector product of two vectors in the same direction is zero. Thus

$$\frac{d\vec{L}}{dt} = \vec{r} \times \frac{d\vec{p}}{dt}$$

Substituting $\vec{\tau}_{\text{net}}$ for $\vec{r} \times (d\vec{p}/dt)$ (from Equation 10-22) gives

$$\vec{\tau}_{\text{net}} = \frac{d\vec{L}}{dt} \qquad \text{10-23}$$

The net torque acting on a system of particles is the sum of the net torques on the individual particles. The generalization of Equation 10-23 to a system of particles is then

$$\vec{\tau}_{\text{net sys}} = \sum_i \vec{\tau}_{\text{net}\,i} = \sum_i \frac{d\vec{L}_i}{dt} = \frac{d}{dt}\sum_i \vec{L}_i = \frac{d\vec{L}_{\text{sys}}}{dt}$$

In this equation, the sum of the torques may include internal as well as external torques. The sum of the internal torques equals zero, so

$$\vec{\tau}_{\text{net ext}} = \frac{d\vec{L}_{\text{sys}}}{dt} \qquad \text{10-10}$$

NEWTON'S SECOND LAW FOR ANGULAR MOTION

Proofs of Equations 10-12 and 10-13 We now show that the angular momentum of a system of particles can be written as the sum of the orbital angular momentum and the spin angular momentum.

Figure 10-36 shows a system of particles. The angular momentum $\vec{L}_i$ of the ith particle about arbitrary point O is given by

$$\vec{L}_i = \vec{r}_i \times \vec{p}_i = \vec{r}_i \times m_i\vec{v}_i \qquad \text{10-24}$$

and the angular momentum of the system about O is

$$\vec{L} = \Sigma\vec{L}_i = \Sigma(\vec{r}_i \times m_i\vec{v}_i)$$

The angular momentum about the center of mass is given by

$$\vec{L}_{\text{cm}} = \Sigma(\vec{r}_i' \times m_i\vec{u}_i)$$

where $\vec{r}_i'$ and $\vec{u}_i$ are the position and velocity, respectively, of the ith particle relative to the center of mass. It can be seen from the figure that

$$\vec{r}_i = \vec{r}_{\text{cm}} + \vec{r}_i'$$

Differentiating both sides gives

$$\vec{v}_i = \vec{v}_{\text{cm}} + \vec{u}_i$$

Substituting these into Equation 10-24, we have

$$\vec{L}_i = \vec{r}_i \times m_i\vec{v}_i = (\vec{r}_{\text{cm}} \times \vec{r}_i') \times m_i(\vec{v}_{\text{cm}} + \vec{u}_i)$$

Expanding the right side, we obtain

$$\vec{L}_i = (\vec{r}_{\text{cm}} \times m_i\vec{v}_{\text{cm}}) + (\vec{r}_{\text{cm}} \times m_i\vec{u}_i) + (m_i\vec{r}_i' \times \vec{v}_{\text{cm}}) + (\vec{r}_i' \times m_i\vec{u}_i)$$

Summing both sides and factoring common terms out of the sums gives

$$\vec{L}_{\text{sys}} = \Sigma\vec{L}_i = \vec{r}_{\text{cm}} \times (\Sigma m_i)\vec{v}_{\text{cm}} + \vec{r}_{\text{cm}} \times (\Sigma m_i\vec{u}_i) + (\Sigma m_i\vec{r}_i') \times \vec{v}_{\text{cm}} + \Sigma(\vec{r}_i' \times m_i\vec{u}_i)$$

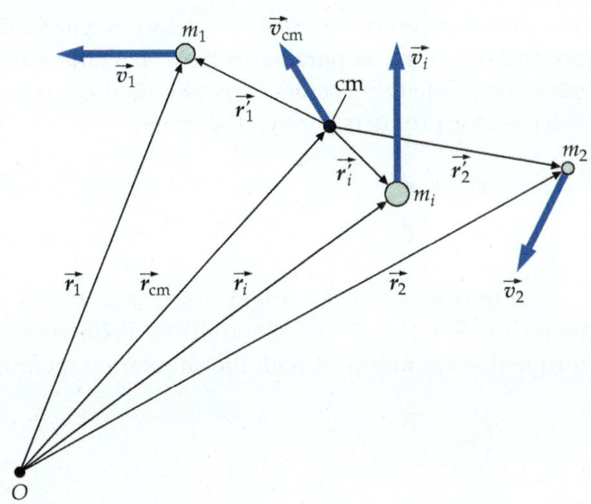

FIGURE 10-36

Because $\Sigma m_i\vec{r}_i'$ and $\Sigma m_i\vec{u}_i$ are both zero, and because $\Sigma m_i = M$ and $\Sigma(\vec{r}_i' \times m_i\vec{u}_i) = \vec{L}_{cm}$, we have $\vec{L}_{sys} = \vec{r}_{cm} \times M\vec{v}_{cm} + \vec{L}_{cm}$ or

$$\vec{L}_{sys} = \vec{L}_{orbit} + \vec{L}_{spin} \qquad 10\text{-}12$$

where $\vec{L}_{orbit} = \vec{v}_{cm} \times M\vec{v}_{cm}$ and

$$\vec{L}_{spin} = \vec{L}_{cm} = \Sigma(\vec{r}_i' \times m_i\vec{u}_i) \qquad 10\text{-}25$$

DEFINITION: SPIN ANGULAR MOMENTUM

Proofs of Equations 10-14 and 10-15 We now take the z components of the vectors for the torque and the angular momentum about a point to obtain the formulas for the torque and angular momentum about a fixed axis. The angular momentum of a particle about the origin is $\vec{L} = \vec{r} \times \vec{p}$, so finding the z component of the angular momentum means finding the z component of the product $\vec{r} \times \vec{p}$. To do this, we express $\vec{r}$ and $\vec{p}$ as

$$\vec{r} = \vec{r}_{rad} + \vec{r}_z \quad \text{and} \quad \vec{p} = \vec{p}_{xy} + \vec{p}_z$$

where $\vec{r}_{rad}$, $\vec{r}_z$, $\vec{p}_{xy}$, and $\vec{p}_z$ are vector components (Figure 10-37) of $\vec{r}$ and $\vec{p}$.

Substituting for $\vec{r}$ and $\vec{p}$ gives

$$\vec{L} = \vec{r} \times \vec{p} = (\vec{r}_{rad} + \vec{r}_z) \times (\vec{p}_{xy} \times \vec{p}_z)$$

and expanding the right side, we have

$$\vec{L} = (\vec{r}_{rad} \times \vec{p}_{xy}) + (\vec{r}_{rad} \times \vec{p}_z) + (\vec{r}_z \times \vec{p}_{xy}) + (\vec{r}_z \times \vec{p}_z)$$

The vector product of any two vectors is perpendicular to both vectors, so the product $\vec{r}_{rad} \times \vec{p}_{xy}$ is parallel to the z axis. In each of the other three products at least one of the two vectors is parallel to the z axis, so the z component of each of these vector products is zero. Therefore,

$$\vec{L}_z = \vec{r}_{rad} \times \vec{p}_{xy} \qquad 10\text{-}14$$

ANGULAR MOMENTUM ABOUT z AXIS

The torque about the origin associated with a force acting on the particle is given by $\vec{\tau} = \vec{r} \times \vec{F}$ (Equation 10-1). Following the same procedure with the torque that we followed with the angular momentum gives

$$\vec{\tau}_z = \vec{r}_{rad} \times \vec{F}_{xy} \qquad 10\text{-}15$$

TORQUE ABOUT z AXIS

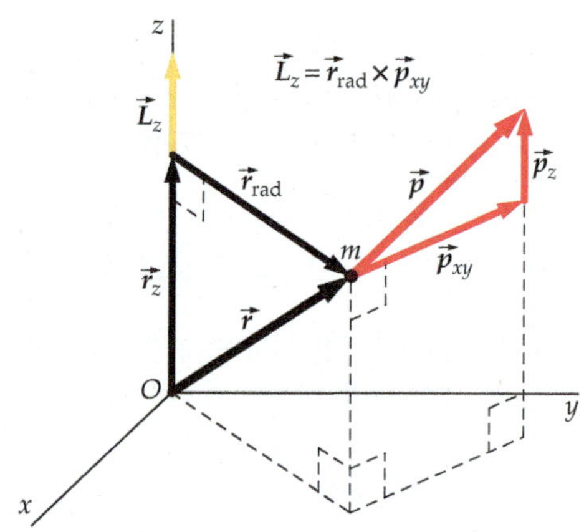

FIGURE 10-37 The vector components $\vec{r}_{rad}$, $\vec{r}_z$, $\vec{p}_{xy}$, and $\vec{p}_z$ of $\vec{r}$ and $\vec{p}$ that are used for calculating the angular momentum about the z axis $\vec{L}_z$.

* 10-4 QUANTIZATION OF ANGULAR MOMENTUM

Angular momentum plays an important role in the description of atoms, molecules, nuclei, and elementary particles. If a particle is bound to one or more other particles, the particle is said to be a *bound particle.* The planets, the asteroids, the comets, and the Sun make up a *bound system,* called the solar system, and Earth is bound to the solar system. Like energy, the angular momentum of bound systems is **quantized,** that is, changes in angular momentum occur only in discrete amounts.

The angular momentum of a particle due to its orbital motion is its orbital angular momentum. The magnitude of the orbital angular momentum L of a bound particle can have only the values

$$L = \sqrt{\ell(\ell + 1)}\,\hbar \qquad \ell = 0, 1, 2, \ldots \tag{10-26}$$

where $\hbar$ (read "h-bar") is the **fundamental unit of angular momentum,** which is related to Planck's constant h:

$$\hbar = \frac{h}{2\pi} = 1.05 \times 10^{-34}\,\text{J}\cdot\text{s} \tag{10-27}$$

The component of orbital angular momentum along any direction in space is also quantized and can have only the values $\pm m\hbar$, where m is a nonnegative integer that is less than or equal to ℓ. For example, if $\ell = 2$, m can equal 2, 1, or 0.

Because the quantum of angular momentum $\hbar$ is so small, the quantization of angular momentum is not noticed in the macroscopic world. Consider a particle of mass 1.00 g $= 1.00 \times 10^{-3}$ kg moving in a circle of radius 1.00 cm with a period of 1.00 s. Its orbital angular momentum is

$$L = mvr = mr^2\omega = mr^2\frac{2\pi}{T} = (1.00 \times 10^{-3}\,\text{kg})(1.00 \times 10^{-2}\,\text{m})^2\frac{2\pi}{1.00\,\text{s}}$$
$$= 6.28 \times 10^{-7}\,\text{J}\cdot\text{s}$$

If we divide by $\hbar$, we obtain

$$\frac{L}{\hbar} = \frac{6.28 \times 10^{-7}\,\text{J}\cdot\text{s}}{1.05 \times 10^{-34}\,\text{J}\cdot\text{s}} = 6.00 \times 10^{27}$$

Thus, this typical macroscopic angular momentum contains 6.00×10^{27} units of the fundamental unit of angular momentum. Even if we could measure L to one part in a billion, we would never observe the quantization of this macroscopic angular-momentum value.

The quantization of orbital angular momentum leads to the quantization of rotational kinetic energy. Consider a molecule rotating about its center of mass with angular momentum L (Figure 10-38). Let I be its moment of inertia. Its kinetic energy is

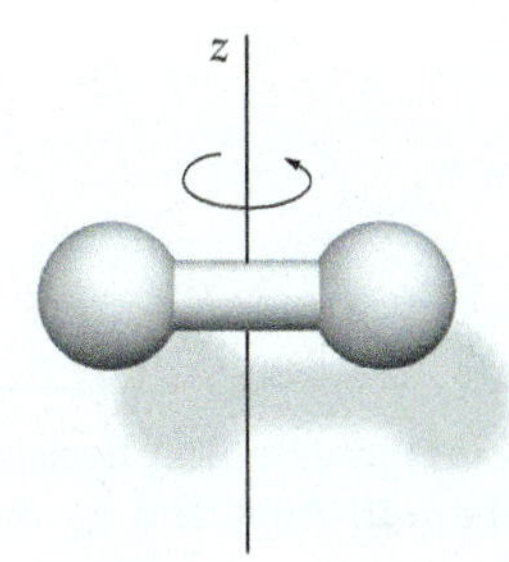

FIGURE 10-38 Model of a rigid diatomic molecule rotating about the z axis.

$$K = \frac{L^2}{2I} \tag{10-28}$$

But L^2 is quantized to the values $L^2 = \ell(\ell + 1)\hbar^2$, with $\ell = 0, 1, 2, \ldots$. Thus, the kinetic energy is quantized to the values K_ℓ given by

$$K_\ell = \frac{L^2}{2I} = \frac{\ell(\ell + 1)\hbar^2}{2I} = \ell(\ell + 1)E_{0\text{r}} \tag{10-29a}$$

where

$$E_{0\text{r}} = \frac{\hbar^2}{2I} \tag{10-29b}$$

Figure 10-39 shows an energy-level diagram for a rotating molecule with constant moment of inertia I. Note that, unlike the energy levels for a vibrating system (Section 7-4), the rotational energy levels are not equally spaced, and the lowest level is zero.

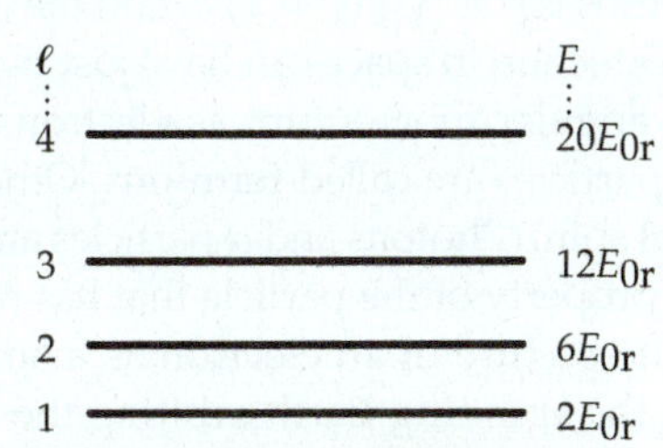

FIGURE 10-39 Energy-level diagram for a rotating molecule.

Example 10-11 Rotational Energy Levels

The characteristic rotational energy E_{0r} (Equation 10-29*b*) for the rotation of the N_2 molecule is 2.48×10^{-4} eV. Using this information, find the separation distance of the two nitrogen atoms.

PICTURE The characteristic rotational energy depends on the moment of inertia, and the moment of inertia depends on the separation distance.

SOLVE

1. The characteristic rotational energy is related to the moment of inertia (see Equation 10-29*b*):

$$E_{0r} = \frac{\hbar^2}{2I}$$

2. Model each nitrogen atom as a point mass at the center of its nucleus. The N_2 molecule then is modeled as two point masses rotating about the center of mass of the molecule. Calculate the moment of inertia about the axis through the center of mass and perpendicular to the line joining the molecules:

$$I = m_1 r_1^2 + m_2 r_2^2$$

3. The distance of each molecule from the center of mass is half of the separation distance d:

$$r_1 = r_2 = \frac{d}{2} \quad \text{and} \quad m_1 = m_2 = m$$

4. Calculate the moment of inertial in terms of d and m:

$$I = m\frac{d^2}{4} + m\frac{d^2}{4} = \frac{1}{2}md^2$$

5. Substitute for I in the step-1 result. The mass of a nitrogen atom is 14.00 u. (Atomic masses can be found in Appendix C):

$$E_{0r} = \frac{\hbar^2}{2I} = \frac{\hbar^2}{md^2}$$

where $m = (14.00\text{ u})(1.661 \times 10^{-27}\text{ kg/u}) = 2.325 \times 10^{-26}\text{ kg}$

and $E_{0r} = (2.48 \times 10^{-4}\text{ eV})(1.602 \times 10^{-19}\text{ J/eV}) = 3.973 \times 10^{-23}\text{ J}$

6. Solve for d:

$$d = \frac{\hbar}{\sqrt{mE_{0r}}} = \frac{1.055 \times 10^{-34}\text{ J}\cdot\text{s}}{\sqrt{(2.325 \times 10^{-26}\text{ kg})(3.973 \times 10^{-23}\text{ J})}}$$

$$= 1.097 \times 10^{-10}\text{ m} = \boxed{0.110\text{ nm}}$$

CHECK In 1911, the British physicist Ernest Rutherford (1871–1935) found the diameter of a nucleus to be $\sim 10^{-6}$ nm, and the diameter of an atom to be typically ~ 0.1 nm, which is approximately the same as our step-6 result. Is it plausible that the separation distance between atoms in N_2 is about equal to the diameter of an isolated atom? Yes. In a nitrogen molecule, the valence electrons are shared by the two atoms. This process, called covalent bonding, is discussed further in Chapter 37.

Stable matter contains just three kinds of particles: electrons, protons, and neutrons. In addition to orbital angular momentum, each of these particles also has an intrinsic angular momentum called **spin.** The spin angular momentum of a particle, like its mass and electric charge, is a fundamental property of the particle that cannot be changed. The magnitude of the spin angular-momentum vector for electrons, protons, and neutrons is $s = \sqrt{\frac{1}{2}(\frac{1}{2} + 1)}\,\hbar$, and the component of the spin angular momentum along any direction in space can have just two values: $+\frac{1}{2}\hbar$ and $-\frac{1}{2}\hbar$. Particles with the same spin angular momentum as electrons are called "spin-one-half" particles. Spin-one-half particles are called **fermions.** Other particles, called **bosons,** have zero spin or integral spin. (Photons and α particles are examples of bosons.) Curiously, spin is a quantum property of the particle that has nothing to do with the motion of the particle.

The picture of an electron as a spinning ball that orbits the nucleus in an atom (like the spinning Earth orbiting the Sun) is often a useful visualization. However, the angular momentum of a spinning ball can be increased or decreased, whereas the spin of the electron is a fixed property like its charge and mass. Furthermore, as far as we know, electrons are point particles that have no size. In addition, electrons do not orbit the nucleus as the planets orbit the Sun. The quantum mechanical model of an atom allows us to calculate the probability that an electron will be found in some specified volume of space.

Physics Spotlight

As the World Turns: Atmospheric Angular Momentum

Cloud patterns suggest two adjacent low pressure cells. *(SeaWiFS Project, NASA/Goddard Space Flight Center, and ORBIMAGE.)*

Air has measurable density, which varies both with the available moisture and the altitude within the atmosphere. Air also has measurable speed. Most surface winds are local, but masses of air higher in the atmosphere have measurable global circulation.

Over the years, increased computing power[*] has allowed scientists to calculate the total atmospheric angular momentum (AAM) for Earth. These calculations are available at the Special Bureau for the Atmosphere of the Global Geophysical Fluids Center.[†] They are based on measurements from meteorological services of several countries. Most measurements made between 10 and 50 kilometers high, in the upper *troposphere* and the *stratosphere,* are made with the help of weather balloons. AAM is calculated using the magnitude and direction of winds at various heights—wind vectors—and AAM Units, where 1 AAM Unit $= 10^{25}\ \mathrm{kg \cdot m^2/s}$.[‡]

In the past few decades, the length of day (LOD) for Earth also has been measured to great precision,[#] calculated by astronomical measurements, and reckoned in solar time corrected for polar wobbling (UT1). The measurements use a combination of satellite laser ranging, very-long-baseline interferometry, recent data from GPS, and the Doppler Orbitography and Radio Positioning Integrated by Satellite (DORIS) system. Variations in LOD of tenths of a millisecond are routinely reported. That value is less than one part in a hundred million.

When variations in the LOD and the AAM are compared, they have striking similarity.[°] Both AAM and LOD have weekly, monthly, seasonal, yearly, and multiyear variations.[§] What is more, they correlate with each other to 95.4 percent and 98.02 percent,[¶] depending on the model for AAM. These correlations are not accidental. The spin angular momentum of the entire Earth–atmosphere system is conserved. Both Earth's spin angular momentum and AAM are in the same direction—from west to east. This means that when AAM speeds up, the angular momentum of Earth itself (excluding the atmosphere) slows down, and LOD increases.

The strongest confirmation of this result is El Niño weather patterns.[**] During El Niños, the southern Pacific Ocean warms and subtropical westerly winds speed up, while tropical easterly winds slow down. These wind patterns increase AAM. In 1984,[††] measurements at Goddard Space Flight Center showed that the day had lengthened by over a millisecond during El Niño. In 1997, the day grew by four-tenths of a millisecond[‡‡] during the El Niño event. When the AAM decreases, Earth speeds up, and days shorten. AAM is the strongest cause of variation in LOD for Earth. Other causes include solar flares, volcanic eruptions, and even core–mantle friction.[##]

Because precise measurements of Earth's AAM and LOD can be made, predictions about what changes in Earth's atmosphere, such as an increase in carbon dioxide,[°°] will mean for the angular momentum of the atmosphere can be made as well as studies of AAM of other planets.[§§]

---

[*] Marcus, S. L., et al., "Detection and Modeling of Nontidal Oceanic Effects on Earth's Rotation Rate," *Science,* Sept. 11, 1998, Vol. 281, 1656–1659.

[†] "GGFC Special Bureau for the Atmosphere," *International Earth Rotation and Reference Systems Service* https://www.iers.org/IERS/EN/Organization/ProductCentres/GGFC/SBAtmosphere/atmosphere.html as of June 2019.

[‡] Huang, H.-P., Weickmann, K. M., and Rosen, R. D., "Unusual Behaviour in Atmospheric Angular Momentum during the 1965 and 1972 El Niños," *Journal of Climate,* Aug. 2003, Vol. 16, 2526–2539.

[#] Chao, B. F. et al., "Space Geodesy Monitors Mass Transports in Global Geophysical Fluids," *Eos, Transactions, American Geophysical Union,* May 30, 2000, Vol. 81, 247+; "Universal Time (UT1) and Length of Day (LOD.) http://www.iers.org/MainDisp.csl?pid=95-97

[°] Marcus et al., op. cit. "Studies of Atmospheric Angular Momentum," *NOAA-CIRES Climate Diagnostics Center,* https://www.esrl.noaa.gov/psd/psd1/review/Chap04/sec3.html as of June, 2019.

[§] Barnes, R. T. H., et al., "Atmospheric Angular Momentum Fluctuations, Length-of-Day Changes, and Polar Motion," *Proceedings of the Royal Society of London A,* May 9, 1983, Vol. 387, 31–73.

[¶] Koot, L., De Viron, O., and Dehant, V., "Atmospheric Angular Momentum Time Series: Characterization of Their Internal Noise and Creation of a Combined Series," *Proceedings of the Journées 2004 Systèmes de Référence Spatio-Temporels,* N. Capitaine, Ed., Observatoire de Paris, 2005, 138–139.

[**] Huang, H.-P., et al., op. cit.

[††] Simon, C., "The Pull of El Niño: Sluggish Rotation and Longer Days," *Science News,* Jan. 14, 1984, Vol. 125, 20

[‡‡] Monastersky, R., "El Niño Shifts Earth's Momentum," *Science News,* Jan 17, 1998, Vol. 153, 45.

[##] Marcus, S. L., et al., op. cit.

[°°] Rosen, R. D., and Gutowski, W. J., "Response of Zonal Winds and Atmospheric Angular Momentum to a Doubling Of CO2," *Journal of Climate,* Dec. 1992, Vol. 5, 1391–1404.

[§§] Zhu, Xun, "Dynamics in Planetary Atmospheric Physics: Comparative Studies of Equatorial Superrotation for Venus, Titan, and Earth," *Johns Hopkins APL Technical Digest,* 2005, Vol. 26, 164–174.

Summary

1. Angular momentum is an important derived dynamic quantity in macroscopic physics. In microscopic physics, spin angular momentum is an intrinsic, fundamental property of elementary particles.
2. Conservation of angular momentum is a fundamental law of nature.
3. Quantization of angular momentum is a fundamental law of nature.

TOPIC	RELEVANT EQUATIONS AND REMARKS	
1. Vector Nature of Rotation	Right-hand rules are used to obtain the direction of the angular velocity and the torque.	
Angular velocity $\vec{\omega}$	The direction of the angular velocity $\vec{\omega}$ is along the axis of rotation in the direction given by the right-hand rule.	
Torque $\vec{\tau}$	$\vec{\tau} = \vec{r} \times \vec{F}$	10-1
2. Vector Product	$\vec{A} \times \vec{B} = AB\sin\phi\,\hat{n}$	10-2
	where ϕ is the angle between the vectors and $\hat{n}$ is a unit vector perpendicular to the plane of $\vec{A}$ and $\vec{B}$ in the direction given by the right-hand rule as $\vec{A}$ is rotated into $\vec{B}$.	
Properties	$\vec{A} \times \vec{A} = 0$	10-3
	$\vec{A} \times \vec{B} = -\vec{B} \times \vec{A}$	10-4
	$\frac{d}{dt}(\vec{A} \times \vec{B}) = \left(\vec{A} \times \frac{d\vec{B}}{dt}\right) + \left(\frac{d\vec{A}}{dt} \times \vec{B}\right)$	10-6
	$\hat{i} \times \hat{j} = \hat{k},\quad \hat{j} \times \hat{k} = \hat{i},\quad \text{and} \quad \hat{k} \times \hat{i} = \hat{j}$	10-7*a*
	$\hat{i} \times \hat{i} = \hat{j} \times \hat{j} = \hat{k} \times \hat{k} = 0$	10-7*b*
3. Angular Momentum		
For a point particle	$\vec{L} = \vec{r} \times \vec{p}$	10-8
For a system rotating about a symmetry axis	$\vec{L} = I\vec{\omega}$	10-9
For any system	The angular momentum about any point O is the angular momentum about the center of mass (spin angular momentum) plus the angular momentum associated with center-of-mass motion about O (orbital angular momentum).	
	$\vec{L} = \vec{L}_{\text{orbit}} + \vec{L}_{\text{spin}} = \vec{r}_{\text{cm}} \times M\vec{v}_{\text{cm}} + \sum_i (\vec{r}_i' \times m_i\vec{u}_i)$	10-12
Newton's second law for angular motion	$\vec{\tau}_{\text{net ext}} = \frac{d\vec{L}}{dt}$	10-10
Conservation of angular momentum	If the net external torque remains zero, the angular momentum of the system is conserved. (If the component of the net external torque in a given direction remains zero, the component of the angular momentum of the system in that direction remains conserved.)	
Kinetic energy of an object rotating about a fixed axis	$K = \frac{1}{2}I\omega^2 = \frac{L^2}{2I}$	10-21
Quantization of angular momentum	The magnitude of the orbital angular momentum of a bound particle can have only the values $L = \sqrt{\ell(\ell + 1)}\,\hbar \qquad \ell = 0, 1, 2, \ldots$	
*Quantization of any component of orbital angular momentum	The component of orbital angular momentum of a bound particle along any direction in space is also quantized and can have only the values $\pm m\hbar$, where m is a nonnegative integer that is less than or equal to ℓ.	
Spin	Electrons, protons, and neutrons have an intrinsic angular momentum called spin.	

Answers to Concept Check

10-1 Counterclockwise, if viewed from above.

Answers to Practice Problems

10-1 $K_i = 78.2\,\text{J}$, $K_f = 13.0\,\text{J}$

10-2 $r_N = r_0/\sqrt[3]{N}$

Problems

In a few problems, you are given more data than you actually need; in a few other problems, you are required to supply data from your general knowledge, outside sources, or informed estimates.

For all problems, use 9.81 m/s² for the free-fall acceleration and neglect friction and air resistance unless instructed to do otherwise.

Interpret as significant all digits in numerical values that have trailing zeros and no decimal points.

- • Single-concept, single-step, relatively easy
- •• Intermediate-level, may require synthesis of concepts
- ••• Challenging
- SSM Solution is in the *Student Solutions Manual*
- Consecutive problems that are shaded are paired problems.

CONCEPTUAL PROBLEMS

1 • True or false:
(*a*) If two vectors are exactly opposite in direction, their vector product must be zero.
(*b*) The magnitude of the vector product of two vectors is at a minimum when the vectors are perpendicular.
(*c*) Knowing the magnitude of the vector product of two nonzero vectors and the vectors' individual magnitudes uniquely determines the angle between them.

2 • Consider two nonzero vectors $\vec{A}$ and $\vec{B}$. Their vector product has the greatest magnitude if $\vec{A}$ and $\vec{B}$ are (*a*) parallel, (*b*) perpendicular, (*c*) antiparallel, (*d*) at an angle of 45° to each other.

3 • What is the angle between a force vector $\vec{F}$ and a torque vector $\vec{\tau}$ generated by $\vec{F}$?

4 • A point particle of mass m is moving with a constant speed v along a straight line that passes through point *P*. What can you say about the angular momentum of the particle relative to point *P*? (*a*) Its magnitude is mv. (*b*) Its magnitude is zero. (*c*) Its magnitude changes sign as the particle passes through point *P*. (*d*) Its magnitude increases as the particle approaches point *P*.

5 • A particle travels in a circular path, and point *P* is at the center of the circle. (*a*) If the particle's linear momentum $\vec{p}$ is doubled without changing the radius of the circle, how is the magnitude of its angular momentum about *P* affected? (*b*) If the radius of the circle is doubled but the speed of the particle is unchanged, how is the magnitude of its angular momentum about *P* affected? SSM

6 • A particle moves along a straight line at constant speed. How does the particle's angular momentum about any fixed point vary with time?

7 •• True or false: If the net torque on a rotating object is zero, the angular velocity of the object cannot change. If your answer is false, give an example of such a situation.

8 •• You are standing on the edge of a turntable with frictionless bearings that is initially rotating when you catch a ball that is moving in the same direction but faster than you are moving and on a line tangent to the edge of the turntable. Assume you do not move relative to the turntable. (*a*) Does the angular speed of the turntable increase, decrease, or remain the same during the catch? (*b*) Does the magnitude of your angular momentum (about the rotation axis of the table) increase, decrease, or remain the same during the catch? (*c*) How does the ball's angular momentum (about the rotation axis of the table) change after the catch? (*d*) How does the total angular momentum of the system, you–table–ball (above the rotation axis of the turntable), change after the catch?

9 •• If the angular momentum of a system about a fixed point *P* is constant, which one of the following statements must be true?
(*a*) No torque about *P* acts on any part of the system.
(*b*) A constant torque about *P* acts on each part of the system.
(*c*) Zero net torque about *P* acts on each part of the system.
(*d*) A constant external torque about *P* acts on the system.
(*e*) Zero net external torque about *P* acts on the system.

10 •• A block sliding on a frictionless table is attached to a string that passes through a narrow hole through the tabletop. Initially, the block is sliding with speed v_0 in a circle of radius r_0. A student under the table pulls slowly on the string. What happens as the block spirals inward? Give supporting arguments for your choice. (The term "angular momentum" refers to the angular momentum about a vertical axis through the hole.) (*a*) Its energy and angular momentum are conserved. (*b*) Its angular momentum is conserved and its energy increases. (*c*) Its angular momentum is conserved and its energy decreases. (*d*) Its energy is conserved and its angular momentum increases. (*e*) Its energy is conserved and its angular momentum decreases.

11 •• One way to tell if an egg is hardboiled or uncooked without breaking the egg is to lay the egg flat on a hard surface and try to spin it. A hardboiled egg will spin easily, an uncooked egg will not. However, once spinning, the uncooked egg may do something unusual: If you stop it with your finger, it may start spinning again. Explain the difference in the behavior of the two types of eggs. SSM

12 •• Explain why a helicopter with just one main rotor has a second smaller rotor mounted on a horizontal axis at the rear, as in Figure 10-40. Describe the resultant motion of the helicopter if this rear rotor fails during flight.

FIGURE 10-40 Problem 12
(Chris Sorenson/The Stock Market.)

13 •• The spin angular-momentum vector for a spinning wheel is parallel with its axle and is pointed east. To cause this vector to rotate toward the south, in which direction must a force be exerted on the east end of the axle? (*a*) up, (*b*) down, (*c*) north, (*d*) south, (*e*) east

14 •• **CONTEXT-RICH** You are walking toward the north, and in your left hand you are carrying a suitcase that contains a massive spinning wheel mounted on an axle attached to the front and back of the case. The angular velocity of the gyroscope points north. You now begin to turn to walk toward the east. As a result, the front end of the suitcase will (*a*) resist your attempt to turn it and will try to maintain its original orientation, (*b*) resist your attempt to turn and will pull to the west, (*c*) rise upward, (*d*) dip downward, (*e*) show no effect whatsoever.

15 •• **ENGINEERING APPLICATION** The angular momentum of the propeller of a small single-engine airplane points forward. The propeller rotates clockwise if viewed from behind. (*a*) Just after liftoff, as the nose of the plane tilts upward, the airplane tends to veer to one side. To which side does it tend to veer and why? (*b*) If the plane is flying horizontally and suddenly turns to the right, does the nose of the plane tend to veer upward or downward? Why? SSM

16 •• **CONTEXT-RICH, ENGINEERING APPLICATION** You have designed a car that is powered by the energy stored in a single flywheel with a spin angular momentum $\vec{L}$. In the morning, you plug the car into an electrical outlet and a motor spins the flywheel up to speed, adding a huge amount of rotational kinetic energy to it—energy that will be changed into translational kinetic energy of the car during the day. Having taken a physics course involving angular momentum and torques, you realize that problems would arise during various maneuvers of the car. Discuss some of these problems. For example, suppose the flywheel is mounted so that $\vec{L}$ points vertically upward when the car is on a horizontal road. What would happen as the car travels over a hilltop? Through a valley? Suppose the flywheel is mounted so that $\vec{L}$ points forward or to one side when the car is on a horizontal road. Then what would happen if the car attempts to turn to the left or right? In each case that you examine, consider the direction of the torque exerted on the car by the road.

17 •• You are sitting on a spinning piano stool with your arms folded. (*a*) When you extend your arms out to your sides, what happens to your kinetic energy? What is the cause of this change? (*b*) Explain what happens to your moment of inertia, angular speed, and angular momentum as you extend your arms. SSM

18 •• A uniform rod of mass M and length L rests on a horizontal frictionless table. A blob of putty of mass $m = M/4$ moves along a line perpendicular to the rod, strikes the rod near its end, and sticks to the rod. Describe qualitatively the subsequent motion of the rod and putty.

ESTIMATION AND APPROXIMATION

19 •• An ice-skater starts her pirouette with arms outstretched, rotating at 1.5 rev/s. Estimate her rotational speed (in revolutions per second) when she brings her arms flat against her body. SSM

20 •• Estimate the ratio of angular velocities for the rotation of a diver between the full tuck position and the full layout position.

21 •• The days on Mars and Earth are of nearly identical length. Earth's mass is 9.35 times Mars's mass, Earth's radius is 1.88 times Mars's radius, and Mars is on average 1.52 times farther away from the Sun than Earth is. The Martian year is 1.88 times longer than Earth's year. Assume that they are both uniform spheres and that their orbits about the Sun are circles. Estimate the ratio (Earth to Mars) of (*a*) their spin angular momenta, (*b*) their spin kinetic energies, (*c*) their orbital angular momenta, and (*d*) their orbital kinetic energies.

22 •• The polar ice caps contain about 2.3×10^{19} kg of ice. This mass contributes negligibly to the moment of inertia of Earth because it is located at the poles, close to the axis of rotation. Estimate the change in the length of the day that would be expected if the polar ice caps were to melt and the water were distributed uniformly over the surface of Earth.

23 •• A 2.0-g particle moves at a constant speed of 3.0 mm/s around a circle of radius 4.0 mm. (*a*) Find the magnitude of the angular momentum of the particle. (*b*) If $L = \sqrt{\ell(\ell + 1)}\,\hbar$, where ℓ is an integer, find the value of $\ell(\ell + 1)$ and the approximate value of ℓ. (*c*) By how much does ℓ change if the particle's speed increases by one-millionth of a percent, and nothing else changes? Use your result to explain why the quantization of angular momentum is not noticed in macroscopic physics. SSM

24 ••• Astrophysicists in the 1960s tried to explain the existence and structure of *pulsars*—extremely regular astronomical sources of radio pulses whose periods ranged from seconds to milliseconds. At one point, these radio sources were given the acronym LGM (Little Green Men), a reference to the idea that they might be signals from extraterrestrial civilizations. The explanation given today is no less interesting. Consider the following. Our Sun, which is a fairly typical star, has a mass of 1.99×10^{30} kg and a radius of 6.96×10^{8} m. Although it does not rotate uniformly, because it is not a solid body, its average rate of rotation is about 1 rev/25 d. Stars larger than the Sun can expire in spectacular explosions called *supernovae,* leaving behind a collapsed remnant of the star called a *neutron star.* These neutron-stars have masses comparable to the original masses of the stars but radii of only a few kilometers! The high rotation rate are due to the conservation of angular momentum during the collapses. These stars emit beams of radio waves. Because of the rapid angular speed of the stars, the beam sweeps past Earth at regular, very short, intervals. To produce the observed radio-wave pulses, the star has to rotate at rates that range from about 1 rev/s to 1000 rev/s. (*a*) Using data from the textbook, estimate the rotation rate of the Sun if it were to collapse into a neutron star of radius 10 km. The Sun is not a uniform sphere of gas, and its moment of inertia is given by $I = 0.059MR^2$. Assume that the neutron star is spherical and has a uniform mass distribution. (*b*) Is the rotational kinetic energy of the Sun greater or smaller after the collapse? By what factor does it change, and where does the energy go to or come from?

25 •• The moment of inertia of Earth about its spin axis is approximately 8.03×10^{37} kg·m^2. (*a*) Because Earth is nearly spherical, assume that the moment of inertia can be written as $I = CMR^2$, where C is a dimensionless constant, $M = 5.98 \times 10^{24}$ kg is the mass of Earth, and $R = 6370$ km is its radius. Determine C. (*b*) If Earth's mass were distributed uniformly, C would equal 2/5. From the value of C calculated in Part (*a*), is Earth's density greater near its center or near its surface? Explain your reasoning.

26 ••• Estimate Timothy Goebel's initial takeoff speed, rotational velocity, and angular momentum when he performs a quadruple lutz (Figure 10-41). Make any assumptions you think reasonable, but justify them. Goebel's mass is about 60 kg and the height of the jump is about 0.60 m, Note that his angular speed will

change quite a bit during the jump, as he begins with arms outstretched and then pulls them in. Your answer should be accurate to within a factor of 2, if you are careful.

FIGURE 10-41
Problem 26
(Chris Trotman/DUOMO/Corbis.)

THE VECTOR PRODUCT AND THE VECTOR NATURE OF TORQUE AND ROTATION

27 • A force of magnitude F is applied horizontally in the $-x$ direction to the rim of a disk of radius R, as shown in Figure 10-42. Write $\vec{F}$ and $\vec{r}$ in terms of the unit vectors $\hat{i}$, $\hat{j}$, and $\hat{k}$, and compute the torque produced by this force about the origin at the center of the disk. SSM

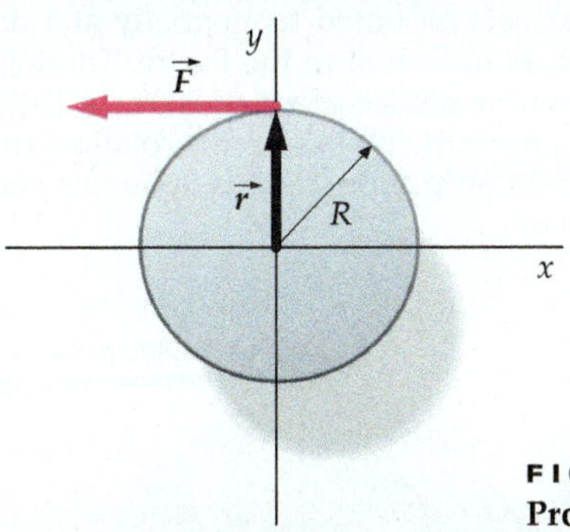

FIGURE 10-42
Problem 27

28 • Compute the torque about the origin of the gravitational force $\vec{F} = -mg\hat{j}$ acting on a particle of mass m located at $\vec{r} = x\hat{i} + y\hat{j}$, and show that this torque is independent of the y coordinate.

29 • Find $\vec{A} \times \vec{B}$ for the following choices: (*a*) $\vec{A} = 4\hat{i}$ and $\vec{B} = 6\hat{i} + 6\hat{j}$, (*b*) $\vec{A} = 4\hat{i}$ and $\vec{B} = 6\hat{i} + 6\hat{k}$, and (*c*) $\vec{A} = 2\hat{i} + 3\hat{j}$ and $\vec{B} = 3\hat{i} + 2\hat{j}$.

30 •• For each case in Problem 29, compute $|\vec{A} \times \vec{B}|$. Compare it to $|\vec{A}||\vec{B}|$ to estimate which of the pairs of vectors are closest to being perpendicular. Verify your answers by calculating the angle using the dot product.

31 •• A particle moves in a circle that is centered at the origin. The particle has position $\vec{r}$ and angular velocity $\vec{\omega}$. (*a*) Show that its velocity is given by $\vec{v} = \vec{\omega} \times \vec{r}$. (*b*) Show that its centripetal acceleration is given by $\vec{a}_c = \vec{\omega} \times \vec{v} = \vec{\omega} \times (\vec{\omega} \times \vec{r})$.

32 •• You are given three vectors and their components in the form: $\vec{A} = a_x\hat{i} + a_y\hat{j} + a_z\hat{k}$, $\vec{B} = b_x\hat{i} + b_y\hat{j} + b_z\hat{k}$, and $\vec{C} = c_x\hat{i} + c_y\hat{j} + c_z\hat{k}$. Show that the following equalities hold: $\vec{A}\cdot(\vec{B} \times \vec{C}) = \vec{C}\cdot(\vec{A} \times \vec{B}) = \vec{B}\cdot(\vec{C} \times \vec{A})$.

33 •• If $\vec{A} = 3\hat{j}$, $\vec{A} \times \vec{B} = 9\hat{i}$, and $\vec{A}\cdot\vec{B} = 12$, find $\vec{B}$.

34 •• If $\vec{A} = 4\hat{i}$, $B_z = 0$, $|\vec{B}| = 5$, and $\vec{A} \times \vec{B} = 12\hat{k}$, determine $\vec{B}$.

35 ••• Given three noncoplanar vectors, $\vec{A}$, $\vec{B}$, and $\vec{C}$, show that $\vec{A}\cdot(\vec{B} \times \vec{C})$ is the volume of the parallelepiped formed by the three vectors.

36 ••• Using the vector product, prove the *law of sines* for the triangle shown in Figure 10-43. That is, if A, B, and C are the lengths of each side of the triangle, show that $A/\sin a = B/\sin b = C/\sin c$.

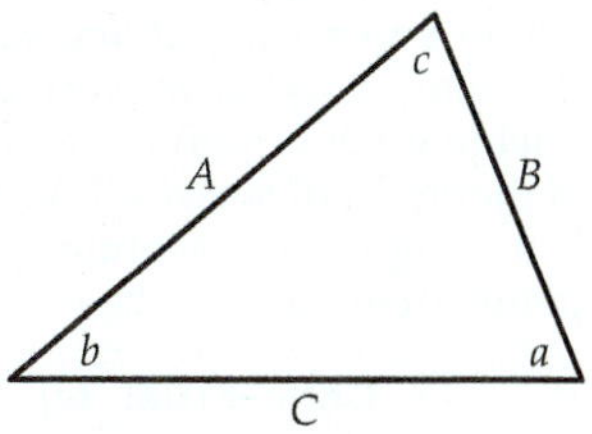

FIGURE 10-43
Problem 36

TORQUE AND ANGULAR MOMENTUM

37 • A 2.0-kg particle moves directly eastward at a constant speed of 4.5 m/s along an east–west line. (*a*) What is its angular momentum (including direction) about a point that lies 6.0 m north of the line? (*b*) What is its angular momentum (including direction) about a point that lies 6.0 m south of the line? (*c*) What is its angular momentum (including direction) about a point that lies 6.0 m directly east of the particle? SSM

38 • You observe a 2.0-kg particle moving at a constant speed of 3.5m/s in a clockwise direction around a circle of radius 4.0 m. (*a*) What is its angular momentum (including direction) about the center of the circle? (*b*) What is its moment of inertia about an axis through the center of the circle and perpendicular to the plane of the motion? (*c*) What is the angular velocity of the particle?

39 •• (*a*) A particle moving at constant velocity has zero angular momentum about a particular point. Use the definition of angular momentum to show that under this condition the particle is moving either directly toward or directly away from the point. (*b*) You are a right-handed batter and let a waist-high fastball go past you without swinging. What is the direction of the ball's angular momentum relative to your navel? (Assume that the ball travels in a straight, horizontal line as it passes you.)

40 •• A particle that has a mass m is traveling with a constant velocity $\vec{v}$ along a straight line that is a distance b from the origin O (Figure 10-44). Let dA be the area swept out by the position vector from O to the particle during a time interval dt. Show that dA/dt is constant and is equal to $L/2m$, where L is the magnitude of the angular momentum of the particle about the origin.

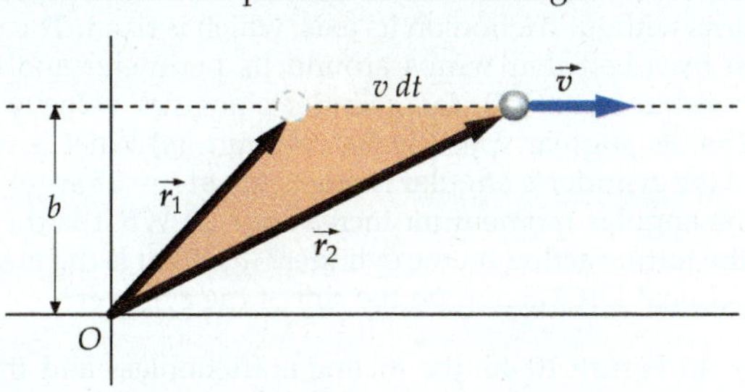

FIGURE 10-44 **Problem 40**

41 •• A 15-g coin that has a diameter equal to 1.5 cm is spinning at 10 rev/s about a fixed vertical axis. The coin is spinning on edge with its center directly above the point of contact with the tabletop. As you look down on the tabletop, the coin spins clockwise. (*a*) What is the angular momentum (including direction) of the coin about its center of mass? (To find the moment of inertia about the axis, see Table 9-1.) Model the coin as a cylinder of length L and take the limit as L approaches zero. (*b*) What is the coin's angular momentum (including direction) about a point on the tabletop 10 cm from the axis? (*c*) Now the coin's center of mass travels at 5.0 cm/s in a straight line east across the tabletop, while spinning the same way as in Part (*a*). What is the angular momentum (including direction) of the coin about a point on the line of motion of the center of mass? (*d*) When it is both spinning and sliding, what is the angular momentum of the coin (including direction) about a point 10 cm north of the line of motion of the center of mass?

42 •• **CONCEPTUAL** (*a*) Two stars of masses m_1 and m_2 are located at $\vec{r}_1$ and $\vec{r}_2$ relative to some origin O, as shown in Figure 10-45. They exert equal and opposite attractive gravitational forces on each other. For this two-star system, calculate the net torque exerted by these internal forces about the origin O and show that it is zero only if both forces lie along the line joining the particles. (*b*) The fact that the Newton's third-law pair of forces are not only equal and oppositely directed but also lie along the line connecting the two objects is sometimes called the strong form of Newton's third law. Why is it important to add that last phrase? *Hint: Consider what would happen to these two objects if the forces were offset from each other.*

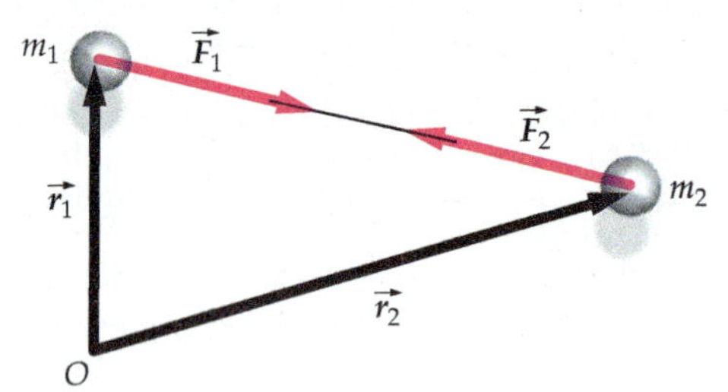

FIGURE 10-45 Problem 42

43 •• A 1.8-kg particle moves in a circle of radius 3.4 m. As you look down on the plane of its orbit, the particle is initially moving clockwise. If we call the clockwise direction positive, the particle's angular momentum relative to the center of the circle varies with time according to $L(t) = 10\,\mathrm{N\cdot m\cdot s} - (4.0\,\mathrm{N\cdot m})t$. (*a*) Find the magnitude and direction of the torque acting on the particle. (*b*) Find the angular velocity of the particle as a function of time.

44 •• **CONTEXT-RICH, ENGINEERING APPLICATION** You are designing a lathe motor, part of which consists of a uniform cylinder whose mass is 90 kg and whose radius is 0.40 m that is mounted so that it turns without friction on its axis, which is fixed. The cylinder is driven by a belt that wraps around its perimeter and exerts a constant torque. At $t = 0$, the cylinder's angular velocity is zero. At $t = 25$ s, its angular speed is 500 rev/min. (*a*) What is the magnitude of the cylinder's angular momentum at $t = 25$ s? (*b*) At what rate is the angular momentum increasing? (*c*) What is the magnitude of the torque acting on the cylinder? (*d*) What is the magnitude of the frictional force acting on the rim of the cylinder?

45 •• In Figure 10-46, the incline is frictionless and the string passes through the center of mass of each block. The pulley has a moment of inertia I and radius R. (*a*) Find the net torque acting on the system (the two masses, string, and pulley) about the center of the pulley. (*b*) Write an expression for the total angular momentum of the system about the center of the pulley. Assume the masses are moving with a speed v. (*c*) Find the acceleration of the masses by using your results for Parts (*a*) and (*b*) and by setting the net torque equal to the rate of change of the system's angular momentum. SSM

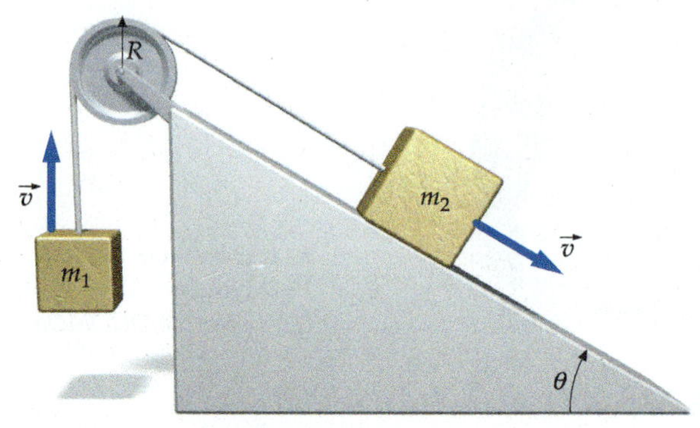

FIGURE 10-46 Problem 45

46 •• **CONTEXT-RICH, ENGINEERING APPLICATION** Figure 10-47 shows the rear view of a space capsule that was left rotating rapidly about its axis at 30 rev/min after a collision with another capsule. You are the flight controller and have just moments to tell the crew how to stop this rotation before they become ill from the rotation and the situation becomes dangerous. You know that they have access to two small jets mounted tangentially at a distance $R = 3.0$ m from the axis, as indicated in the figure. These jets can each eject 10 g/s of gas with a nozzle speed of 800 m/s. Determine the length of time these jets must run to stop the rotation. In flight, the moment of inertia of the ship about its axis (assumed constant) is known to be $4000\,\mathrm{kg\cdot m^2}$.

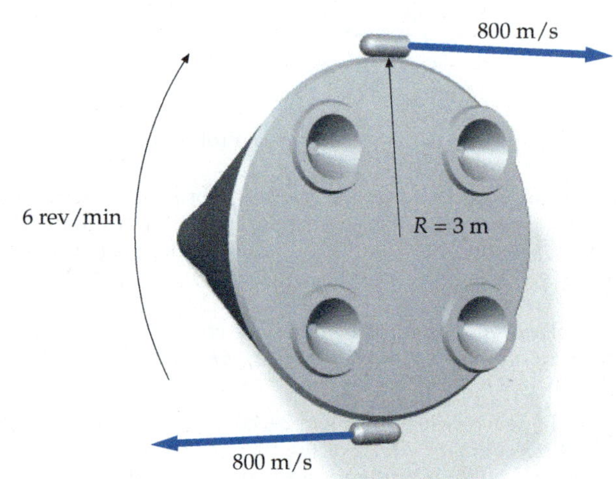

FIGURE 10-47 Problem 46

47 •• A projectile (mass M) is launched at an angle θ with an initial speed v_0. Considering the torque and angular momentum about the launch point, explicitly show that $dL/dt = \tau$. Ignore the effects of air resistance. (The equations for projectile motion are found in Chapter 3.)

CONSERVATION OF ANGULAR MOMENTUM

48 • A planet moves in an elliptical orbit about the Sun, with the Sun at one focus of the ellipse, as in Figure 10-48. (*a*) What is the torque about the center of the Sun due to the gravitational force of attraction of the Sun on the planet? (*b*) At position *A*, the planet has an orbital radius r_1 and is moving with a speed v_1 perpendicular to the line from the Sun to the planet. At position *B*, the planet has an orbital radius r_2 and is moving with speed v_2, again perpendicular to the line from the Sun to the planet. What is the ratio of v_1 to v_2 in terms of r_1 and r_2?

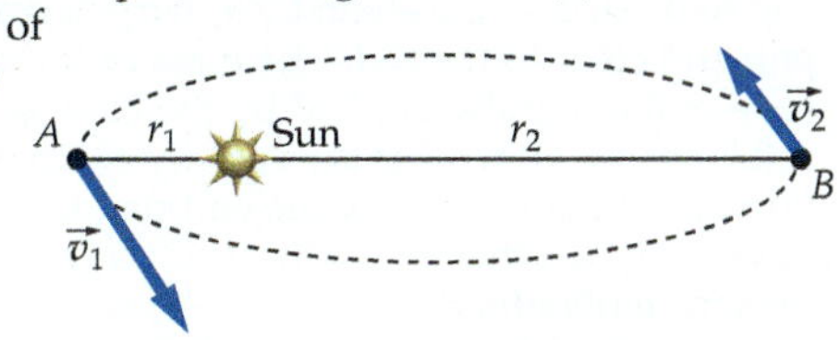

FIGURE 10-48 Problem 48

49 •• You stand on a frictionless platform that is rotating at an angular speed of 1.5 rev/s. Your arms are outstretched, and you hold a heavy weight in each hand. The moment of inertia of you, the extended weights, and the platform is 6.0 kg · m². When you pull the weights in toward your body, the moment of inertia decreases to 1.8 kg · m². (*a*) What is the resulting angular speed of the platform? (*b*) What is the change in kinetic energy of the system? (*c*) Where did this increase in energy come from? SSM

50 •• A small blob of putty of mass m falls from the ceiling and lands on the outer rim of a turntable of radius R and moment of inertia I_0 that is rotating freely with angular speed ω_0 about its vertical fixed-symmetry axis. (*a*) What is the postcollision angular speed of the turntable–putty system? (*b*) After several turns, the blob flies off the edge of the turntable. What is the angular speed of the turntable after the blob's departure?

51 •• A lazy Susan consists of a heavy plastic cylinder mounted on a frictionless bearing resting on a vertical shaft through its center. The cylinder has a radius $R = 15$ cm and mass $M = 0.25$ kg. A cockroach (mass $m = 0.015$ kg) is on the lazy Susan, at a distance of 8.0 cm from the center. Both the cockroach and the lazy Susan are initially at rest. The cockroach then walks along a circular path concentric with the axis of the lazy Susan at a constant distance of 8.0 cm from the axis of the shaft. If the speed of the cockroach with respect to the lazy Susan is 0.010 m/s, what is the speed of the cockroach with respect to the room? SSM

52 •• Two disks of identical mass but different radii (r and $2r$) are spinning on frictionless bearings at the same angular speed ω_0, but in opposite directions (Figure 10-49). The two disks are brought slowly together. The resulting frictional force between the surfaces eventually brings them to a common angular velocity. (*a*) What is the magnitude of that final angular velocity in terms of ω_0? (*b*) What is the change in rotational kinetic energy of the system? Explain.

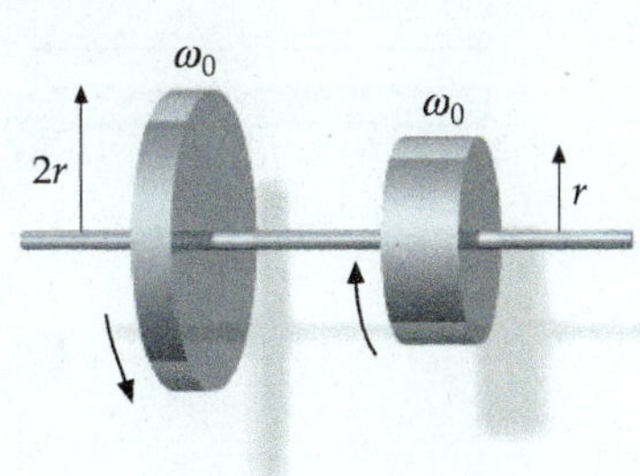

FIGURE 10-49 Problem 52

53 •• A block of mass m sliding on a frictionless table is attached to a string that passes through a narrow hole through the center of the table. The block is sliding with speed v_0 in a circle of radius r_0. Find (*a*) the angular momentum of the block, (*b*) the kinetic energy of the block, and (*c*) the tension in the string. (*d*) A student under the table now slowly pulls the string downward. How much work is required to reduce the radius of the circle from r_0 to $r_0/2$?

54 ••• A 0.20-kg point mass moving on a frictionless horizontal surface is attached to a rubber band whose other end is fixed at point *P*. The rubber band exerts a force whose magnitude is $F = bx$, where x is the length of the rubber band and b is an unknown constant. The rubber band force points inward toward *P*. The mass moves along the dotted line in Figure 10-50. When it passes point *A*, its velocity is 4.0 m/s, directed as shown. The distance *AP* is 0.60 m and *BP* is 1.0 m. (*a*) Find the speed of the mass at points *B* and *C*. (*b*) Find *b*.

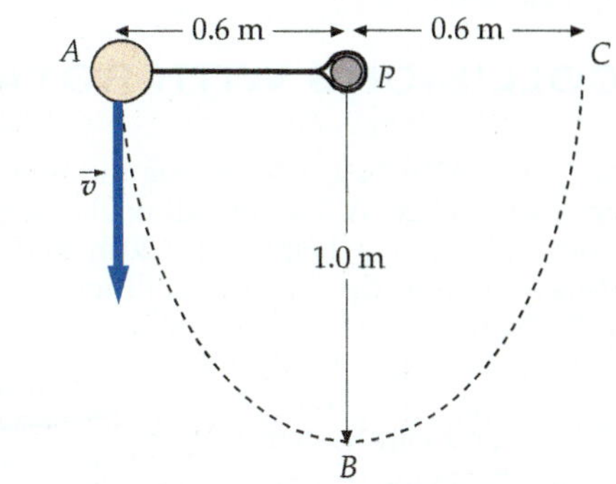

FIGURE 10-50 Problem 54

*QUANTIZATION OF ANGULAR MOMENTUM

55 •• The z component of the spin of an electron is $-\frac{1}{2}\hbar$, but the magnitude of the spin vector is $\sqrt{0.75}\,\hbar$. What is the angle between the electron's spin angular-momentum vector and the positive z axis? SSM

56 •• Show that the energy difference between one rotational state of a molecule and the next higher state is proportional to $\ell + 1$.

57 •• **CONTEXT-RICH, BIOLOGICAL APPLICATION** You work in a biochemical research lab, where you are investigating the rotational energy levels of the HBr molecule. After consulting the periodic chart, you know that the mass of the bromine atom is 80 times that of the hydrogen atom. Consequently, in calculating the rotational motion of the molecule, you assume, to a good approximation, that the Br nucleus remains stationary as the H atom (mass 1.67×10^{-27} kg) revolves around it. You also know that the separation between the hydrogen atom and bromine nucleus is 0.144 nm. Calculate (*a*) the moment of inertia of the HBr molecule about the bromine nucleus, and (*b*) the rotational energies for the bromine nucleus's *ground state* (lowest energy) $\ell = 0$, and the next two states of higher energy (called the first and second *excited states*) described by $\ell = 1$, and $\ell = 2$. SSM

58 ••• The equilibrium separation between the nuclei of the nitrogen molecule (N_2) is 0.110 nm and the mass of each nitrogen nucleus is 14.0 u, where u = 1.66×10^{-27} kg. For rotational energies, the total energy is due to rotational kinetic energy. (*a*) Approximate the nitrogen molecule as a rigid dumbbell of two equal point masses and calculate the moment of inertia about its center of mass. (*b*) Find the energy E_ℓ of the lowest three energy levels using $E_\ell = K_\ell = \ell(\ell + 1)\hbar^2/(2I)$. (*c*) Molecules emit a particle (or *quantum*) of light called a *photon* when they make a *transition* from a higher energy state to a lower one. Determine the energy of a photon emitted when a nitrogen molecule drops from the $\ell = 2$ to the $\ell = 1$ state. Visible-light photons each have between 2 and 3 eV of energy. Are these photons in the visible region?

59 ••• **CONCEPTUAL** Consider a *transition* from a lower energy state to a higher one—that is, the absorption of a quantum of energy, resulting in an increase in the rotational energy of an N_2 molecule (see Problem 58). Suppose such a molecule, initially in its ground rotational state, was exposed to photons, each with an energy equal to three times the energy of its first excited state. (*a*) Would the molecule be able to absorb this photon energy? Explain why or why not, and if it can, determine the energy level to which it goes. (*b*) To make a transition from its ground state to its second excited state requires how many times the energy of the first excited state?

COLLISIONS WITH ROTATIONS

60 •• A 16.0-kg, 2.40-m-long rod is supported at its midpoint on a knife edge. A 3.20-kg ball of clay drops from rest from a height of 1.20 m and makes a perfectly inelastic collision with the rod 0.90 m from the point of support (Figure 10-51). Find the angular momentum of the rod and clay system about the point of support immediately after the inelastic collision.

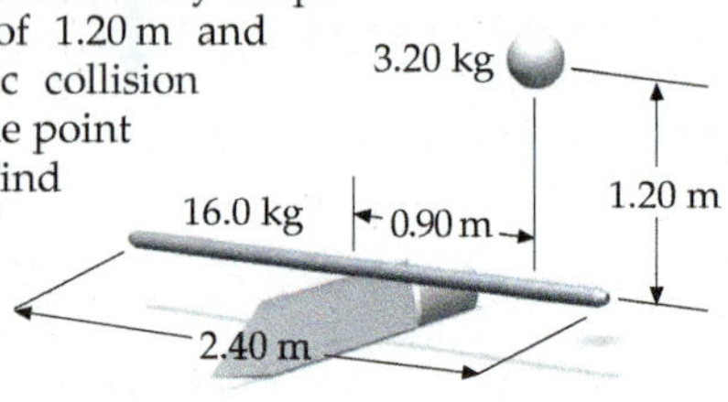

FIGURE 10-51 Problem 60

61 •• Figure 10-52 shows a thin, uniform bar of length L and mass M and a small blob of putty of mass m. The system is supported by a frictionless horizontal surface. The putty moves to the right with velocity $\vec{v}$, strikes the bar at a distance d from the center of the bar, and sticks to the bar at the point of contact. Obtain expressions for the velocity of the system's center of mass and for the angular speed following the collision. SSM

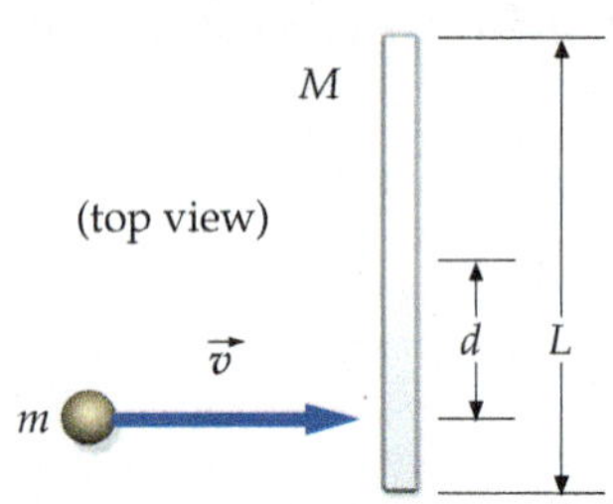

FIGURE 10-52 Problem 61

62 •• Figure 10-52 shows a thin, uniform bar whose length is L and mass is M and a compact hard sphere whose mass is m. The system is supported by a frictionless horizontal surface. The sphere moves to the right with velocity $\vec{v}$, and strikes the bar at a distance $\frac{1}{4}L$ from the center of the bar. The collision is elastic, and following the collision the sphere is at rest. Find the value of the ratio m/M.

63 •• Figure 10-53 shows a uniform rod whose length is L and mass is M pivoted at the top. The rod, which is initially at rest, is struck by a particle whose mass is m at a point $x = 0.8L$ below the pivot. Assume that the particle sticks to the rod. What must be the speed v of the particle so that following the collision the maximum angle between the rod and the vertical is 90°?

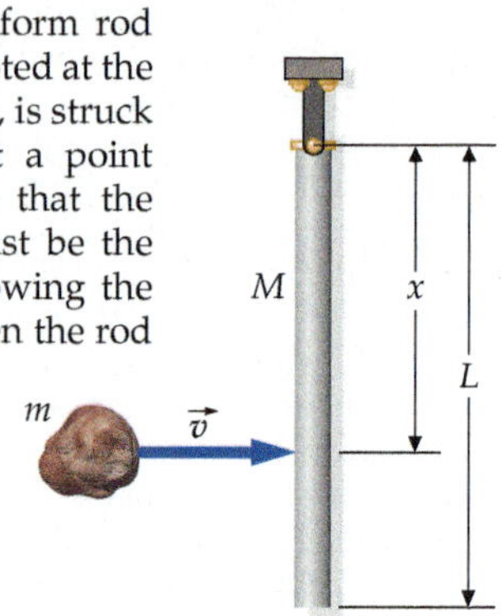

FIGURE 10-53 Problem 63

64 •• If, for the system of Problem 63, $L = 1.2$ m, $M = 0.80$ kg, $m = 0.30$ kg, and the maximum angle between the rod and the vertical following the collision is 60°, find the speed of the particle before impact.

65 •• A uniform rod is resting on a frictionless table when it is suddenly struck at one end by a sharp horizontal blow in a direction perpendicular to the rod. The mass of the rod is M and the magnitude of the impulse applied by the blow is J. Immediately after the rod is struck, (*a*) what is the velocity of the center of mass of the rod, (*b*) what is the velocity of the end that is struck, (*c*) and what is the velocity of the other end of the rod? (*d*) Is there a point on the rod that remains motionless?

66 •• A projectile of mass m_p is traveling at a constant velocity $\vec{v}_0$ toward a stationary disk of mass M and radius R that is free to rotate its axis (Figure 10-54). Before impact, the projectile is traveling along a line displaced a distance b below the axis. The projectile strikes the disk and sticks to point B. Model the projectile as a point mass. (*a*) Before impact, what is the total angular momentum L_0 of the disk–projectile system about the axis? Answer the following questions in terms of the symbols given at the start of this problem. (*b*) What is the angular speed ω of the disk–projectile system just after the impact? (*c*) What is the kinetic energy of the disk–projectile system after impact? (*d*) How much mechanical energy is lost in this collision?

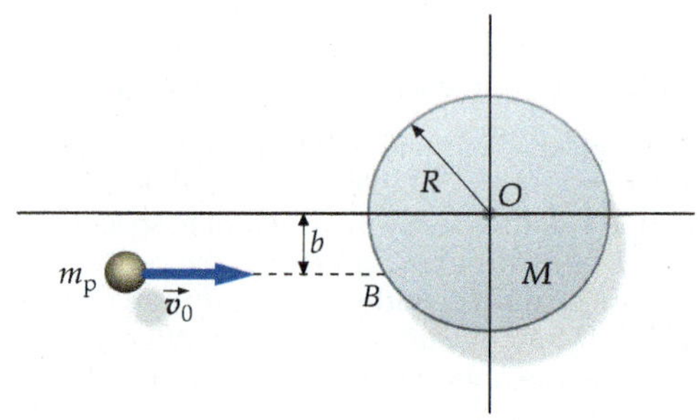

FIGURE 10-54 Problem 66

67 •• A uniform rod of length L_1 and mass M equal to 0.75 kg is attached to a hinge of negligible mass at one end and is free to rotate in the vertical plane (Figure 10-55). The rod is released from rest in the position shown. A particle of mass $m = 0.50$ kg is suspended from a thin string of length L_2 from the hinge. The particle sticks to the rod on contact. What should the ratio L_2/L_1 be so that $\theta_{max} = 60°$ after the collision? SSM

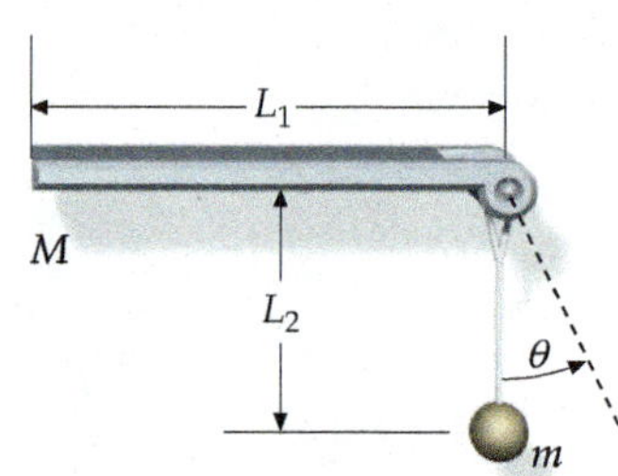

FIGURE 10-55 Problem 67

68 •• A uniform rod that has a length L_1 equal to 1.2 m and a mass M equal to 2.0 kg is attached to a hinge of negligible mass at one end and is free to rotate in the vertical plane (Figure 10-55). The rod is released from rest in the position shown. A particle whose mass is m is suspended from a thin string whose length L_2 is equal to 0.80 m from the hinge. The particle sticks to the rod on contact, and after the collision the rod continues to rotate until $\theta_{max} = 37°$. (*a*) Find m. (*b*) How much energy is dissipated during the collision? SSM

PRECESSION

69 •• A bicycle wheel that has a radius equal to 28 cm is mounted at the middle of a 50-cm-long axle. The tire and rim weigh 30 N. The wheel is spun at 12 rev/s, and the axle is then placed in a horizontal position with one end resting on a pivot. (*a*) What is the angular momentum due to the spinning of the wheel? (Treat the wheel as a hoop.) (*b*) What is the angular velocity of precession? (*c*) How long does it take for the axle to swing through 360° around the pivot? (*d*) What is the angular momentum associated with the motion of the center of mass, that is, due to the precession? In what direction is this angular momentum? SSM

70 •• A uniform disk, whose mass is 2.50 kg and radius is 6.40 cm, is mounted at the center of a 10.0-cm-long axle and spun at 700 rev/min. The axle is then placed in a horizontal position with one end resting on a pivot. The other end is given an initial horizontal velocity such that the precession is smooth with no nutation. (*a*) What is the angular velocity of precession? (*b*) What is the speed of the center of mass during the precession? (*c*) What is the acceleration (magnitude and direction) of the center of mass? (*d*) What are the vertical and horizontal components of the force exerted by the pivot on the axle?

GENERAL PROBLEMS

71 • A particle whose mass is 3.0 kg moves in the xy plane with velocity $\vec{v} = (3.0\text{ m/s})\hat{i}$ along the line $y = 5.3$ m. (*a*) Find the angular momentum $\vec{L}$ about the origin when the particle is at (12 m, 5.3 m). (*b*) A force $\vec{F} = (-3.9\text{ N})\hat{i}$ is applied to the particle. Find the torque about the origin due to this force as the particle passes through the point (12 m, 5.3 m). SSM

72 • The position vector of a particle whose mass is 3.0 kg is given by $\vec{r} = (4.0 + 3.0t^2)\hat{j}$, where $\vec{r}$ is in meters and t is in seconds. Determine the angular momentum and net torque about the origin acting on the particle.

73 •• Two ice-skaters, whose masses are 55 kg and 85 kg, hold hands and rotate about a vertical axis that passes between them, making one revolution in 2.5 s. Their centers of mass are separated by 1.7 m, and their center of mass is stationary. Model each skater as a point particle and find (*a*) the angular momentum of the system about their center of mass and (*b*) the total kinetic energy of the system.

74 •• A 2.0-kg ball attached to a string whose length is 1.5 m moves counterclockwise (as viewed from above) in a horizontal circle (Figure 10-56). The string makes an angle $\theta = 30°$ with the vertical. (*a*) Determine both the horizontal and vertical components of the angular momentum $\vec{L}$ of the ball about the point of support. (*b*) Find the magnitude of $d\vec{L}/dt$ and verify that it equals the magnitude of the torque exerted by gravity about the point of support.

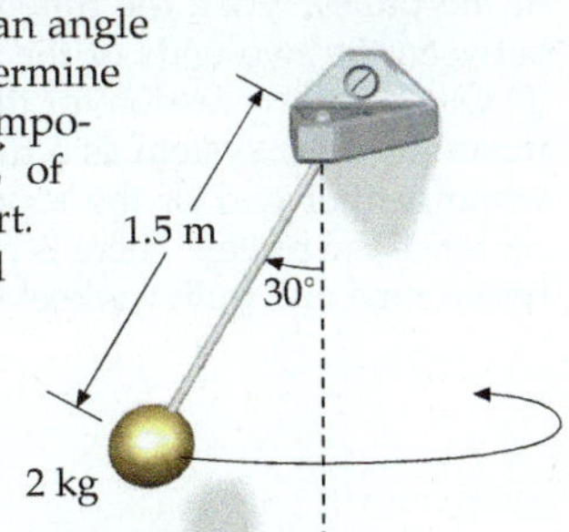

FIGURE 10-56 Problem 74

75 •• A compact object whose mass is m resting on a horizontal, frictionless surface is attached to a string that wraps around a vertical cylindrical post attached to the surface. Thus, when the object is set into motion, it follows a path that spirals inward. (*a*) Is the angular momentum of the object about the axis of the post conserved? Explain your answer. (*b*) Is the energy of the object conserved? Explain your answer. (*c*) If the speed of the object is v_0 when the unwrapped length of the string is r, what is its speed when the unwrapped length has shortened to $r/2$?

76 •• CONTEXT-RICH, ENGINEERING APPLICATION Figure 10-57 shows a hollow cylindrical tube that has a mass M, a length L, and a moment of inertia $ML^2/10$. Inside the cylinder are two disks, each of mass m and radius r, separated by a distance ℓ, and tied to a central post by a thin string. The system can rotate about a vertical axis through the center of the cylinder. You are designing this cylinder–disk apparatus to shut down the rotations by triggering an electronic "shutoff" signal (sent to the rotating motor) when the strings break as the disks hit the ends of the cylinder. During development, you notice that with the system rotating at some critical angular speed ω, the string suddenly breaks. When the disks reach the ends of the cylinder, they stick. Obtain expressions for the final angular speed and the initial and final kinetic energies of the system. Assume that the inside walls of the cylinder are frictionless.

77 •• CONTEXT-RICH, ENGINEERING APPLICATION Repeat Problem 76, but this time friction between the disks and the walls of the cylinder is not negligible. However, the coefficient of friction is not great enough to prevent the disks from reaching the ends of the cylinder. Can the final kinetic energy of the system be determined without knowing the coefficient of kinetic friction? SSM

78 •• CONTEXT-RICH, ENGINEERING APPLICATION Suppose that in Figure 10-57 $\ell = 0.60$ m, $L = 2.0$ m, $M = 0.80$ kg, and $m = 0.40$ kg. The string breaks when the system's angular speed approaches a critical angular speed ω_i at which time the tension in the string approaches 108 N. The masses then move radially outward until they undergo perfectly inelastic collisions with the ends of the cylinder. Determine the critical angular speed and the angular speed of the system after the inelastic collisions. Find the total kinetic energy of the system at the critical angular speed and again after the inelastic collisions. Assume that the inside walls of the cylinder are frictionless.

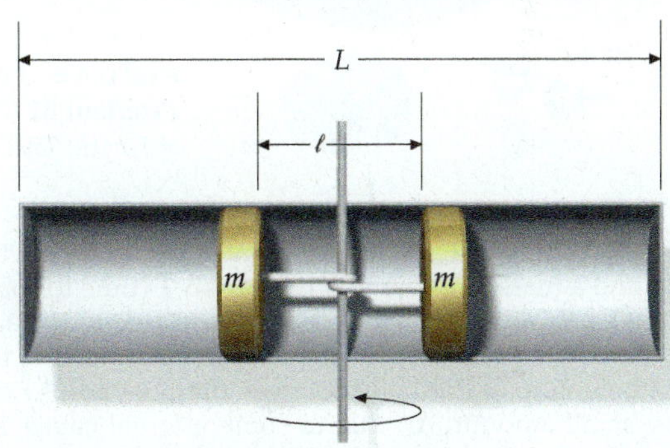

FIGURE 10-57 Problems 76, 77 and 78

79 •• Kepler's second law states: *The line from the center of the Sun to the center of a planet sweeps out equal areas in equal times.* Show that this law follows directly from the law of conservation of angular momentum and the fact that the force of gravitational attraction between a planet and the Sun acts along the line joining the centers of the two celestial objects. SSM

80 •• Consider a cylindrical turntable whose mass is M and radius is R, turning with an initial angular speed ω_1. (*a*) A parakeet of mass m, after hovering in flight above the outer edge of the turntable, gently lands on it and stays in one place on it, as

shown in Figure 10-58. What is the angular speed of the turntable after the parakeet lands? (*b*) Becoming dizzy, the parakeet jumps off (not flies off) with a velocity $\vec{v}$ relative to the turntable. The direction of $\vec{v}$ is tangent to the edge of the turntable and in the direction of its rotation. What will be the angular speed of the turntable afterwards? Express your answer in terms of the two masses m and M, the radius R, the parakeet speed v, and the initial angular speed ω_1.

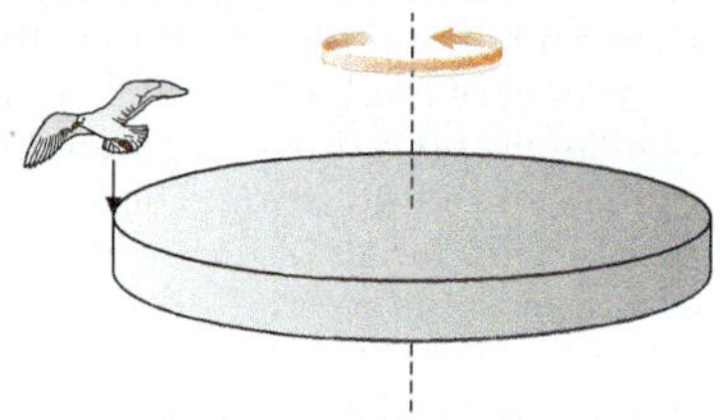

FIGURE 10-58 Problem 80

81 •• You are given a heavy but thin metal disk (like a coin, but larger; Figure 10-59) that has a mass of 0.500 kg and a radius of 0.125 m. (Objects like this are called *Euler disks*.) Placing the disk on a turntable, you spin the disk, on edge, about a vertical axis through a diameter of the disk and the center of the turntable. As you do this, you hold the turntable still with your other hand, letting it go immediately after you spin the disk. The turntable is a uniform solid cylinder with a radius equal to 0.250 m and a mass equal to 0.735 kg and rotates on a frictionless bearing. The disk has an initial angular speed of 30 rev/min. (*a*) The disk spins down and falls over, finally coming to rest on the turntable with its symmetry axis coinciding with the turntable. What is the final angular speed of the turntable? (*b*) What will be the final angular speed if the disk's symmetry axis ends up 0.100 m from the axis of the turntable?

FIGURE 10-59 Problem 81 *(Courtesy of Tangent Toy Co.)*

82 •• (*a*) Assuming Earth to be a homogeneous sphere that has a radius r and a mass m, show that the period T (time for one daily rotation) of Earth's rotation about its axis is related to its radius by $T = br^2$, where $b = (4/5)\pi\, m/L$. Here L is the magnitude of the spin angular momentum of Earth. (*b*) Suppose that the radius r changes by a very small amount, Δr, due to some internal cause, such as thermal expansion. Show that the fractional change in the period ΔT is given approximately by $\Delta T/T = 2\Delta r/r$. (*c*) By how many kilometers would r need to increase for the period to change by 0.25 d/y, (so that leap years would no longer be necessary)?

83 •• The term *precession of the equinoxes* refers to the fact that Earth's spin axis does not stay fixed but sweeps out a cone once every 26,000 y. (This explains why our pole star, Polaris, will not remain the pole star forever.) The reason for this instability is that Earth is a giant gyroscope. The spin axis of Earth precesses because of the torques exerted on it by the gravitational forces of the Sun and moon. The angle between the direction of Earth's spin axis and the normal to the ecliptic plane (the plane of Earth's orbit) is 22.5 degrees. Calculate an approximate value for this torque, given that the period of rotation of Earth is 1.00 d and its moment of inertia is $8.03 \times 10^{37}\ \text{kg}\cdot\text{m}^2$. SSM

84 •• As indicated in the text, according to the Standard Model of Particle Physics, electrons are pointlike particles having no spatial extent. (This assumption has been confirmed experimentally, and the radius of the electron has been shown to be less than 10^{-18} m.) The intrinsic spin of an electron could *in principle* be due to its rotation. Let us check to see if this conclusion is feasible. (*a*) Assuming that the electron is a uniform sphere whose radius is 1.00×10^{-18} m, what angular speed would be necessary to produce the observed intrinsic angular momentum of $\hbar/2$? (*b*) Using this value of angular speed, show that the speed of a point on the "equator" of a "spinning" electron would be moving faster than the speed of light. What is your conclusion about the spin angular momentum being analogous to a spinning sphere with spatial extent?

85 •• An interesting phenomenon occurring in certain *pulsars* (see Problem 24) is an event known as a "spin glitch," that is, a quick change in the spin rate of the pulsar due to a shift in mass location and a resulting rotational change in inertia. Imagine a pulsar whose radius is 10.0 km and whose period of rotation is 25.032 ms. The rotation period is observed to suddenly decrease from 25.032 ms to 25.028 ms. If that decrease was caused by a contraction of the star, by what amount would the pulsar radius have had to change?

86 ••• Figure 10-60 shows a pulley in the form of a uniform disk with a rope hanging over it. The circumference of the pulley is 1.2 m and its mass is 2.2 kg. The rope is 8.0 m long and its mass is 4.8 kg. At the instant shown in the figure, the system is at rest and the difference in height of the two ends of the rope is 0.60 m. (*a*) What is the angular speed of the pulley when the difference in height between the two ends of the rope is 7.2 m? (*b*) Obtain an expression for the angular momentum of the system as a function of time while neither end of the rope is above the center of the pulley. There is no slippage between rope and pulley wheel.

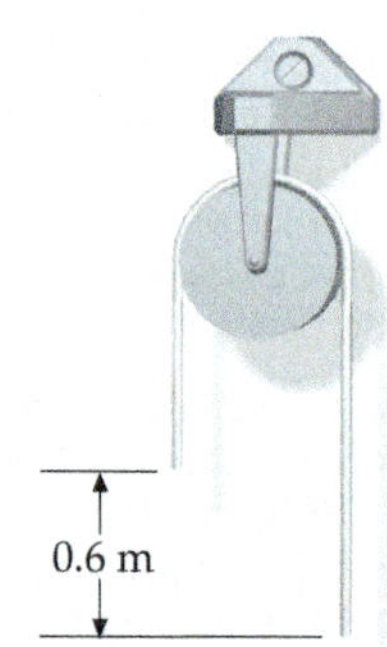

FIGURE 10-60 Problem 86

CHAPTER R

ABOARD THE SPACE SHUTTLE ORBITER, ASTRONAUTS REST IN BUNKS. THEY SLEEP STRAPPED IN SO THEY DO NOT FLOAT AROUND THE CABIN AND THEY DO NOT NEED SOFT MATTRESSES. WHEN THE SHUTTLE IS IN A LOW EARTH ORBIT, IT ORBITS EARTH ONCE EVERY 90 MIN AT A SPEED OF ABOUT 7 KM/S (4.9 MI/S). THIS SPEED IS A SMALL FRACTION OF THE SPEED OF LIGHT, WHICH IS 3.0×10^5 KM/S. *(NASA.)*

Special Relativity

? If an astronaut on a spaceship that is traveling at $0.6c$ relative to Earth takes a one-hour-long nap, does that nap take one hour according to observers on Earth? (See Example R-1.)

The theory of relativity consists of two rather different theories, the special theory and the general theory. The special theory, developed by Albert Einstein and others in 1905, concerns the comparison of measurements made in different inertial reference frames moving with constant velocity relative to one another. Its consequences, which can be derived with a minimum of mathematics, are applicable in a wide variety of situations encountered in physics and engineering. An application of the special theory can be seen in the development of the global positioning system (GPS), which is able to give your position coordinates (latitude, longitude, and altitude) to within a few meters. The system contains 24 satellites, each carrying an atomic clock and each broadcasting a time signal that can be picked up by any GPS receiver in the line of sight of the satellite. The relative arrival times of the signals from several satellites together with knowledge of the positions of the satellites enables the receiver to calculate its position coordinates. According to the special theory, the faster the speed of a clock, the slower it runs (Section R-3). The satellites are moving at about 3.9 km/s, and the effect of the clocks running more slowly at this speed is not negligible. However, the design of the system accounts for the slowing down of the orbiting clocks. (In addition, according to the general theory, the greater the gravitation potential energy of a clock, the faster it runs (Section 39-8). The design of the system also accounts for the speeding up of the clocks due to their high altitude.) The GPS is able to function only because it takes into account the effects of the special and general theories of relativity on the observed clock rates.

In this chapter, we concentrate on the special theory of relativity (often referred to as special relativity). You will see how this theory challenges our everyday experience of time and distance, as we describe the slowing down of moving clocks, the shortening of moving sticks, the relativity of simultaneity for events that occur in different locations, and the relativity of the momentum and energy relation.

Albert Michelson.
(AIP Emilio Segre Visual Archives.)

R-1 THE PRINCIPLE OF RELATIVITY AND THE CONSTANCY OF THE SPEED OF LIGHT

The principle of relativity can be stated as follows:

> It is impossible to devise an experiment that determines whether you are at rest or moving uniformly.
>
> POSTULATE I: THE PRINCIPLE OF RELATIVITY

Moving uniformly means moving at constant velocity relative to an inertial reference frame. For example, suppose that you are in your seat on board a high-speed airplane moving uniformly relative to the surface of Earth. If you drop your fork, it will fall to the floor in the same way that it would if the plane were parked on a runway. When the airplane is in flight, you can consider yourself and the airplane to be at rest and the surface of Earth below you to be moving. There is nothing to distinguish whether you and the plane are moving and the surface of Earth is at rest, or vice versa.

Any reference frame in which a particle with no forces acting on it moves with constant velocity is, by definition, an inertial reference frame.* The surface of Earth is, to a good approximation, an inertial reference frame. The airplane is also an inertial reference frame as long as it moves with constant velocity relative to the surface of Earth. As long as you remain seated or standing still on the airplane, you can consider yourself and the airplane to be at rest and the surface of Earth to be moving, or you can consider the surface of Earth to be at rest and yourself and the airplane to be moving.

Edward Morley.
(AIP Emilio Segre Visual Archives.)

In the nineteenth century, the existence of a preferred frame of reference that could be considered to be at rest was widely accepted. This reference frame was thought to be the reference frame of the *ether*, which is the medium filling all of space through which light was thought to propagate. (It was then accepted that light waves needed a medium to propagate through, just as it is now accepted that sound waves need air or some other medium through which to propagate.) The ether was considered to be the preferred "at rest" reference frame.

A carefully devised series of measurements to measure the orbital speed of Earth relative to the ether were carried out in 1887 by Albert Michelson and Edward Morley. These measurements were considered challenging because the

* Further discussion on inertial reference frames can be found in Chapter 4.

orbital speed of Earth is less than 1/10,000 the speed of light in vacuum. Much to the surprise of nearly everyone, the observations always found the speed of Earth relative to the ether to be zero. It was Albert Einstein who came up with a theory that was consistent with these observations. His explanation was that light is capable of traveling through empty space and that the ether was an unnecessary construct that did not exist. Einstein also postulated:

> The speed of light is independent of the speed of the light source.
>
> POSTULATE II

Here the *speed of light* refers to the speed at which light travels through the vacuum of empty space.

A consequence of Postulate II and the principle of relativity (Postulate I) is that all inertial observers measure the same value for the speed of light. (An inertial observer is one that remains at rest relative to an inertial reference frame.) To establish that all inertial observers measure the same value for the speed of light, we consider inertial observers A and B, where observer A is moving relative to observer B. The principle of relativity states that it is impossible to devise an experiment that determines whether an inertial observer is at rest or moving uniformly. If observer A measures a different value for the speed of light than observer B, then observers A and B could not both consider themselves to be at rest—a result in direct contradiction with the principle of relativity. Thus, a consequence of both the principle of relativity and Postulate II (that the speed of light is independent of the speed of the source) leads to the **constancy of the speed of light:**

Albert Einstein.
(Underwood & Underwood/Corbis.)

> The speed of light c is the same in any inertial reference frame.
>
> THE CONSTANCY OF THE SPEED OF LIGHT

That is, anything (not just light) that travels at speed c relative to one inertial reference frame travels at the same speed c relative to any inertial reference frame.

Suppose you are in your backyard here on Earth and Bob is on a spaceship moving away from you at half the speed of light $(\frac{1}{2}c)$ (Figure R-1). You point a flashlight in Bob's direction and turn it on. The light leaves the flashlight, traveling at speed c (relative to the flashlight), and passes by your neighbor Keisha, who is standing on the roof of her house next door. Keisha measures the speed of the light going by and finds it to be traveling at speed c. A few minutes later, the light travels past Bob and his spaceship. Like Keisha, Bob measures the speed of the light going by him and also finds it to be traveling at speed c. This surprises Bob because he expected the light to be traveling past him at speed $\frac{1}{2}c$ rather than at speed c—after all, Bob is moving at speed $\frac{1}{2}c$ relative to the source of the light (the flashlight in your backyard). Like many people, Bob finds the constancy of the speed of light to be counterintuitive. As a result, he is faced with a dilemma. Should he trust his measuring instruments or trust his intuition? It turns out that it is Bob's intuition that needs adjusting, not his instruments. Bob must change his concepts of both space and time.

Suppose that instead of pointing a flashlight, you point a high-speed particle beam in his direction, where by "high speed" we mean a speed very close to the speed of light, c. (A particle such as an electron or proton cannot travel at the speed of light, but it can travel at speeds that approach the speed of light.) If Keisha measures the particles going by her to be traveling at $0.9999c$ (relative to her), then how fast will Bob measure the particles going by him? Bob's intuition tells him that, because he is moving away from the source of the particles, they will be traveling

FIGURE R-1

past him at the slower speed of $0.4999c$, but that is not the case. When Bob measures the speed of the particles (relative to him) he finds it to be extremely close to $0.9999c$. (The actual value is $0.9997c$.)*

We tend to think of distances between cities as fixed. However, this too is not the case. According to a certain road map, the distance between Baltimore and Philadelphia is 160 km. However, if you travel from Baltimore to Philadelphia at a significant fraction of the speed of light, the distance between the two cities will be much shorter than it is if you travel at 100 km/h (62 mi/h). For someone driving at 100 km/h, the distance between Baltimore and Philadelphia is very close to 160 km. However, for someone traveling at a speed of $0.866c$ (relative to Earth's surface), the distance is only 80 km, and for someone traveling at $0.9999c$, the distance is only 2.2 km.

The fastest speed that a human being has ever traveled relative to Earth is only about 10 km/s $= 3.3 \times 10^{-5}c$. (During the Apollo missions to the moon, the capsule reached this speed on its return to Earth.) This speed is quite slow compared to the speed of light. For someone traveling at that speed from Baltimore to Philadelphia, the distance between those cities would be shorter than the distance for someone traveling at 100 km/h by less than the diameter of a human hair. The logic explaining how this is determined is presented in the next three sections.

R-2 MOVING STICKS

We wish to show that if a stick moves perpendicular to its length, its length does not change. We do this by showing that any increase or decrease in length would contradict the principle of relativity. Showing that a stick does not change its length may seem mundane. However, we show it because an immediate consequence is that moving clocks run slow.

Suppose that we have two identical metersticks, stick A and stick B. We verify that the sticks are identical in length by placing them side by side and comparing their lengths visually. We then give stick B to Bob just before he takes off on another trip in his spaceship. On this trip, Bob makes sure to hold the stick at right angles to the velocity of the spaceship relative to Earth. Stick A remains back on Earth with us. During the trip, is stick B shorter than stick A?

To answer this question we conduct a thought experiment. We attach felt-tipped marking pens to stick A, one at the 20-cm mark, the other at the 80-cm mark. Then Bob and his spaceship execute a flyby, during which Bob holds stick B out a porthole, keeping it at right angles to the ship's velocity. During the flyby, we hold up our stick (stick A), keeping it parallel with stick B. As the sticks pass by each other, two marks are drawn on stick B by the pens (Figure R-2). Bob then returns to Earth with stick B, the two sticks are again placed side by side (Figure R-3), and the distance between the two marks on stick B and the distance between the two marking pens on stick A are compared. Let us assume that a stick moving perpendicular to its length is shorter than is an identical stick that is stationary. Then, the distance between the two pens on A will be less than the distance between the two marks on B (Figure R-3)—clear evidence that during the flyby the moving stick (stick B) was shorter than the stationary stick. However, according to the principle of relativity it is equally valid to think of stick B as stationary and stick A as moving during the flyby. From this perspective, the same evidence (Figure R-3) demonstrates that the moving stick—now stick A—is longer than the stationary stick. Thus, our assumption—that a stick moving perpendicular to its length is shorter than an identical stationary stick—leads to a contradiction and must be rejected. The assumption that a stick moving perpendicular to its length is longer than is an iden-

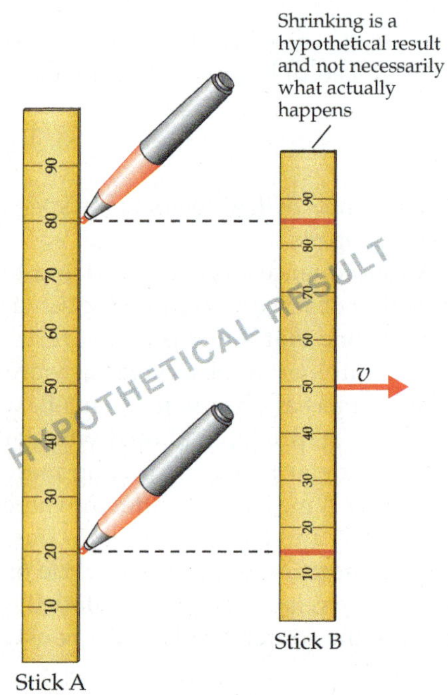

FIGURE R-2 During the flyby, marks like this would be made on stick B by marking pens attached to stick A, if the moving stick was shortened.

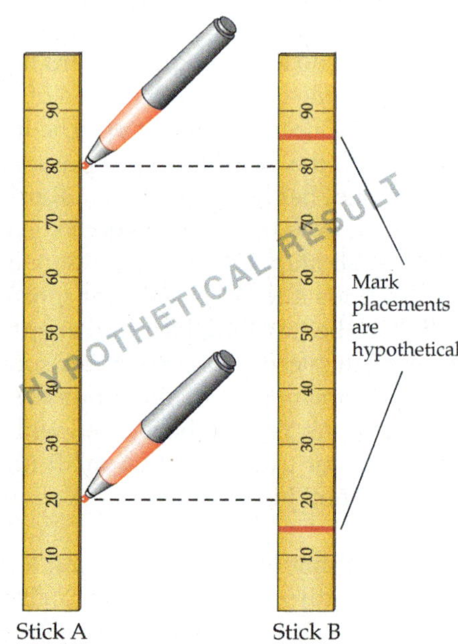

FIGURE R-3 If the distance between marks is greater than the distance between marker pens, this would demonstrate that stick B was shorter than stick A when the marks were made.

* Relative velocity in special relativity is covered in Chapter 39.

tical stationary stick also leads to a contradiction, as can be shown using an analogous argument. Thus, we conclude:

> A stick moving perpendicular to its length has the same length as an identical stick that is stationary.

This rule is established without any consideration of the material from which the two sticks are made. Thus, the rule does not reflect a property of sticks. Instead, it reflects a property of space.

The frame of reference in which the stick is at rest is called the **proper frame** or **rest frame** of the stick, and the length of a stick in its proper reference frame is called its **proper length** or **rest length.**

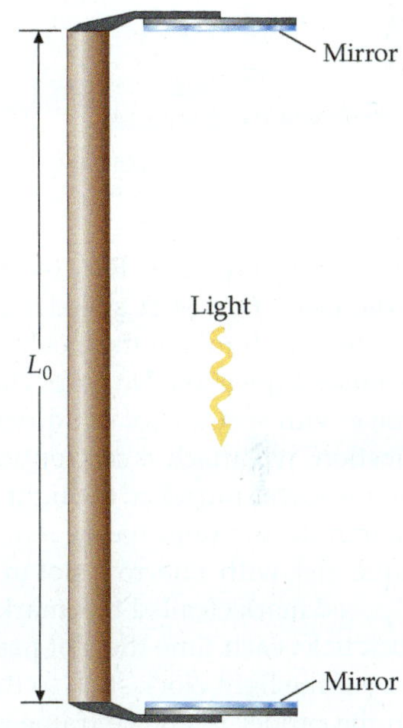

FIGURE R-4 The light clock ticks each time the light pulse reflects off the lower mirror.

R-3 MOVING CLOCKS

Clocks are used to measure time. In this section, we show that clocks moving at high speeds run slowly, so if a high-speed spaceship travels by us, we would observe that all the clocks on the ship run slower than our clocks. However, the people on the ship are free to consider themselves to be at rest and us to be moving, and they would observe that our clocks run slow compared to their clocks. Let us examine how these observations are consistent with the constancy of the speed of light and the principle of relativity.

We construct a clock, called a *light clock*, using a stick of proper length L_0 and two mirrors (Figure R-4). The two mirrors face each other, and a pulse of light bounces back and forth between them. Each time the light pulse strikes one of the mirrors, say the lower mirror, the clock is said to tick. Between successive ticks the light pulse travels a distance $2L_0$ in the proper reference frame of the clock. Thus, the time between ticks T_0 is related to L_0 by

$$2L_0 = cT_0 \qquad \text{R-1}$$

Next, we consider the time between ticks T of the same light clock, but this time we observe it from a reference frame in which the clock is moving perpendicular to the stick with speed v (Figure R-5). In this reference frame, the clock moves a distance vT between ticks and the light pulse moves a distance cT between ticks. The distance the pulse moves in traveling from the bottom mirror to the top mirror is $\sqrt{L_0^2 + (\frac{1}{2}vT)^2}$. The light pulse travels the same distance in traveling from the top mirror to the bottom mirror. Thus,

$$2\sqrt{L_0^2 + (\tfrac{1}{2}vT)^2} = cT \qquad \text{R-2}$$

Because the speed of light is the same in all inertial reference frames, we have used the same symbol c for the speed of light in Equations R-1 and R-2. Solving Equation R-1 for L_0 and substituting into Equation R-2 gives

$$\sqrt{(\tfrac{1}{2}cT_0)^2 + (\tfrac{1}{2}vT)^2} = \tfrac{1}{2}cT \qquad \text{R-3}$$

Solving for T gives

$$T = \frac{T_0}{\sqrt{1 - (v^2/c^2)}} \qquad \text{R-3}$$

TIME DILATION

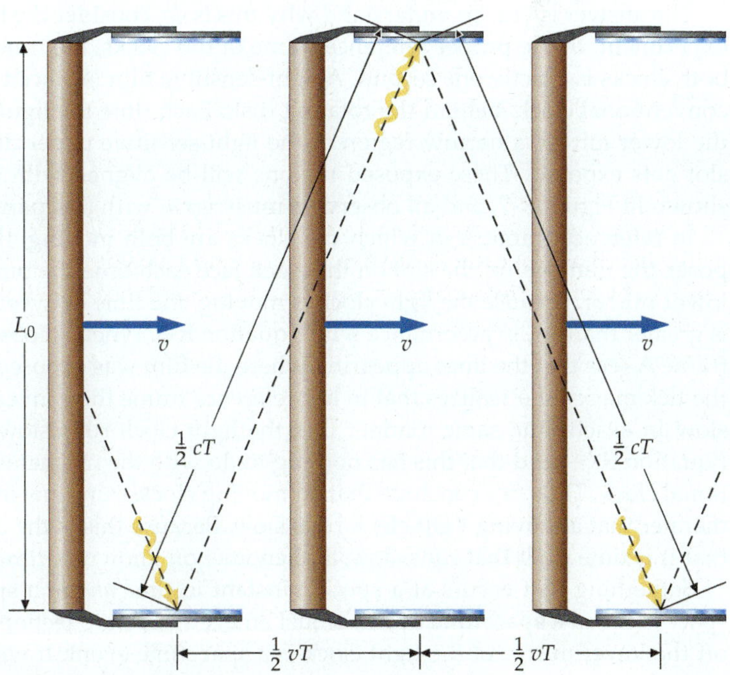

FIGURE R-5 The light clock moves with speed v.

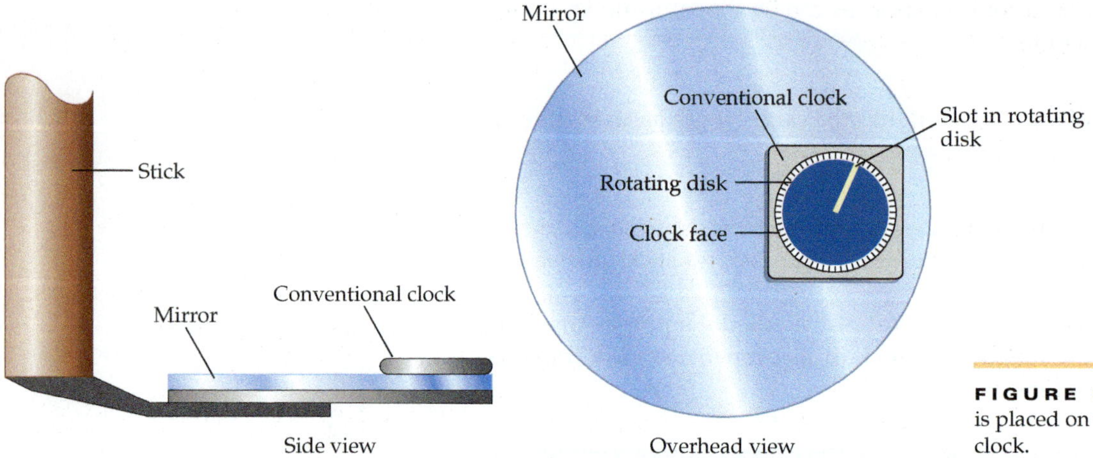

FIGURE R-6 A small conventional clock is placed on the lower mirror of the light clock.

According to Equation R-3, the time between ticks in the reference frame in which the clock moves at speed v is greater than the time between ticks in the proper reference frame of the clock.

This raises a question: Do other clocks run slow according to Equation R-3 when they move with speed v, or is Equation R-3 valid only for light clocks? To answer this question, we attach a conventional clock (with a conventional clock mechanism) to the lower mirror of the light clock (Figure R-6). The conventional clock has had the minute and hour hands removed. In place of a second hand, the clock has an opaque disk with a narrow slot to indicate the time. The clock's face contains 60 equal-spaced marks (called tick marks) around its perimeter—one for each second. The clock ticks each time the slot passes over one of the tick marks. We adjust the length L_0 of the light-clock stick so that the time between ticks of both clocks is the same in the proper reference frame of the clocks. Next, we synchronize the clocks so each tick of the light clock occurs simultaneously with a tick of the conventional clock. We then ask, "If the ticks of the two clocks occur simultaneously in the proper reference frame of the clocks, do they also occur simultaneously in a reference frame in which the clocks are moving at speed v?"

The answer is yes. To understand why this is so, consider the following thought experiment. In the proper reference frame of the clocks, the time between ticks of both clocks is exactly one second. A light-sensitive film is placed on the face of the conventional clock, behind the rotating disk. Each time the light pulse reflects off the lower mirror, a narrow region of the light-sensitive paper directly behind the slot gets exposed. These exposed regions will be aligned with the tick marks as shown in Figure R-7, and all observers must agree with this permanent record.

In reference frame A in which the clocks are both moving, the light pulse exposes the film behind the slot on the clock face each time the pulse reflects off the lower mirror. Because the light clock is moving, the time between these reflections is greater than 1 s, in accordance with Equation R-3. When an observer of reference frame A sees that the lines appearing where the film was exposed are aligned with the tick marks, she realizes that in her reference frame the conventional clock runs slow in exactly the same manner that the light clock runs slow—in accord with Equation R-3—and that this has nothing to do with the mechanism of the conventional clock. Thus, we conclude that all moving clocks run slow in exactly the same manner that a moving light clock runs slow. Because this is the case, we conclude that it is time itself that runs slow, a phenomenon known as **time dilation.**

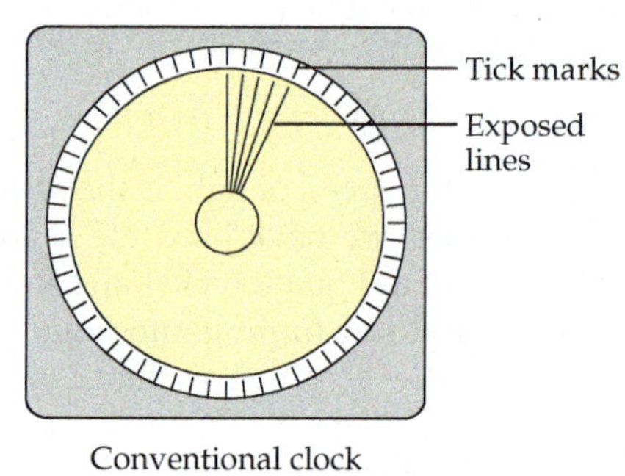

FIGURE R-7 The light striking the conventional clock that passes through the slot exposes the light-sensitive film behind the slot.

Something that occurs at a specific instant in time *and* at a specific location in space is called a **spacetime event,** or just an **event.** Each reflection of the light pulse off the lower mirror of the light clock is a spacetime event. If we call one of these reflections event 1, and the next reflection event 2, then the time between events 1 and 2 in a frame of reference in which the two events occur at the same location is called the **proper time interval** T_0 between the two events. Let T be the time between the same two events in a reference frame in which they occur at different locations. Equation R-3 relates the time T between two events to the proper time T_0 between the same two events.

Each time the light pulse reflects off the lower mirror, the slot (second hand) of the conventional clock is directly over a tick mark. In the proper frame of the two clocks, these two events—the arrival of the light pulse and the passing of the slot over a tick mark—occur at the same *time* and at the same place. Any two events that occur both at the same time and at the same place in one reference frame will occur both at the same time and at the same place in all reference frames. This is because such events can have lasting consequences—like producing lines on the light-sensitive film aligned with the tick marks on the clock face. We cannot have the marks aligned with the tick marks in one reference frame and not aligned with the tick marks in another reference frame. After all, there is only one clock face and one set of marks. This conclusion can be generalized into a principle called the principle of **invariance of coincidences:**

> If two events occur at the same time and at the same place in one reference frame, then they occur at the same time and at the same place in all reference frames.
>
> INVARIANCE OF COINCIDENCES

We can better visualize this principle by considering two automobiles passing through an intersection at the same time. The two events are (1) automobile A passes through the intersection, and (2) automobile B passes through the intersection. If these two events occur at the same time in one reference frame, then they must occur at the same time in all reference frames. Either a fender becomes dented or it does not. That is, if the automobiles collide, then there is no question that the two cars were in the intersection at the same time. The lasting evidence dictates that observers in all reference frames must agree on this fact. Any pair of events that occur at the same time *and* at the same location are referred to as a **spacetime coincidence.**

Example R-1 The Napping Astronauts *Context-Rich*

You work at space control and communicate regularly with astronauts in a spaceship traveling at $v = 0.600c$ relative to Earth. The astronauts sign off from space control, stating that they are going to nap for 1.00 h and then will call back. How long does their nap last according to you and other observers back on Earth?

PICTURE Clock S on the ship reads t_0 when the nap begins (a spacetime coincidence) and reads $t_0 + 1.00$ h when the nap ends (also a spacetime coincidence). Observers on the ship agree that, because clock S is stationary it does not run slow, so the nap lasted 1.00 h. In the reference frame of the ship, the two events (the beginning of the nap and the end of the nap) occur at the same location, so the time interval between the events is the proper time interval between them. You and other observers on Earth agree that clock S reads t_0 when the nap begins and it reads $t_0 + 1.00$ h when the nap ends. However, you and other observers on Earth also agree that because clock S is moving at speed v, it is running slow, so the nap lasted more than 1.00 h. In the reference frame of Earth, the ship is moving so the nap begins and ends at different locations. Therefore, in the reference frame of Earth, the time interval between the events is not the proper time interval between the events.

SOLVE

1. Event 1 is the beginning of the nap and event 2 is the end of the nap. Clock S on the ship advances 1.00 h between these events. Determine the proper time interval T_0 between these events:

$$T_0 = 1.00 \text{ h}$$

2. Find the time interval T between events 1 and 2 for you and other observers on Earth:

$$T = \frac{T_0}{\sqrt{1 - (v^2/c^2)}} = \frac{1.00 \text{ h}}{\sqrt{1 - \frac{(0.600c)^2}{c^2}}}$$

$$= \frac{1.00 \text{ h}}{\sqrt{1 - 0.360}} = \frac{1.00 \text{ h}}{\sqrt{0.640}} = \frac{1.00 \text{ h}}{0.800} = \boxed{1.25 \text{ h}}$$

CHECK The length of the nap is longer in the reference frame in which the napping person is moving, which is in accord with Equation R-3.

TAKING IT FURTHER Clock S is an unnecessary construct, as the astronauts themselves serve as clocks. What is necessary to realize is that the proper time between the beginning and the end of the nap is 1.00 h, so the time T *between* the same two events in a reference frame where the clocks (astronauts) are moving with speed v is given by Equation R-3.

PRACTICE PROBLEM R-1 A pion* has a mean proper lifetime of 26 ns ($1 \text{ ns} = 1 \times 10^{-9}$ s) (measured when the pion is at rest). What is the mean pion lifetime if measured when the pion is moving at $0.995c$?

PRACTICE PROBLEM R-2 A beam of pions (see Practice Problem R-1) moving at $0.995c$ passes point P. How far from P do the pions travel before only half of the pions in the beam remain?

R-4 MOVING STICKS AGAIN

In Section R-2 the length of a stick moving perpendicular to its length and the length of an identical stationary stick are compared and found to be equal. However, the technique used for this comparison works only if the velocity of the moving stick is perpendicular to the length of the stick. Here we apply a different technique to compare the length of a stick at rest to its length when the stick is moving parallel to its length.

A light clock is shown in its proper frame in Figure R-8. This clock ticks each time the light pulse reflects off the mirror on the left. In its proper reference frame, the length of the clock is L_0 and the time between ticks is $T_0 = 2L_0/c$ (Equation R-1). To find the length of the clock in a reference frame in which it is moving to the right at speed v we consider three sequential events:

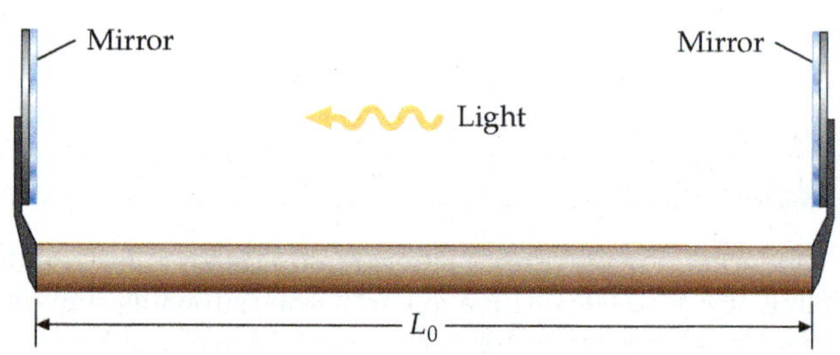

FIGURE R-8

Event 0 Light pulse reflects off the mirror at the left end.
Event 1 Light pulse reflects off the mirror at the right end.
Event 2 Light pulse reflects off at the mirror at the left end.

In Figure R-9, the clock is shown at the time of each of these events in a reference frame in which the clock moves to the right with speed v. (The clock is drawn lower down the page at later times to avoid visual overlap.) The times of occurrence for events 0, 1, and 2 in this reference frame are t_0', t_1', and t_2', respectively. In the time between events 0 and 1, the clock moves a distance $v(t_1' - t_0')$ and the light pulse travels a distance $c(t_1' - t_0')$. Thus,

$$c(t_1' - t_0') = L + v(t_1' - t_0') \qquad \text{R-4}$$

* A pion (short for pi meson) is a subatomic particle.

In the time between events 1 and 2 the clock moves a distance $v(t_2' - t_1')$ and the light pulse travels a distance $c(t_2' - t_1')$, so

$$c(t_2' - t_1') = L - v(t_2' - t_1') \qquad \text{R-5}$$

Eliminating t_1' by solving Equation R-4 for t_1', substituting the result into Equation R-5, and then solving for $t_2' - t_0'$ gives

$$t_2' - t_0' = \frac{2L/c}{1 - (v^2/c^2)} \qquad \text{R-6}$$

The time interval $t_2' - t_0'$ is related to the proper time interval $t_2 - t_0$ between events 0 and 2 (Equation R-3) by

$$t_2' - t_0' = \frac{t_2 - t_0}{\sqrt{1 - (v^2/c^2)}} \qquad \text{R-7}$$

where $t_2 - t_0 = 2L_0/c$ (Equation R-1). Substituting $2L_0/c$ for $t_2 - t_0$ gives

$$t_2' - t_0' = \frac{2L_0/c}{\sqrt{1 - (v^2/c^2)}} \qquad \text{R-8}$$

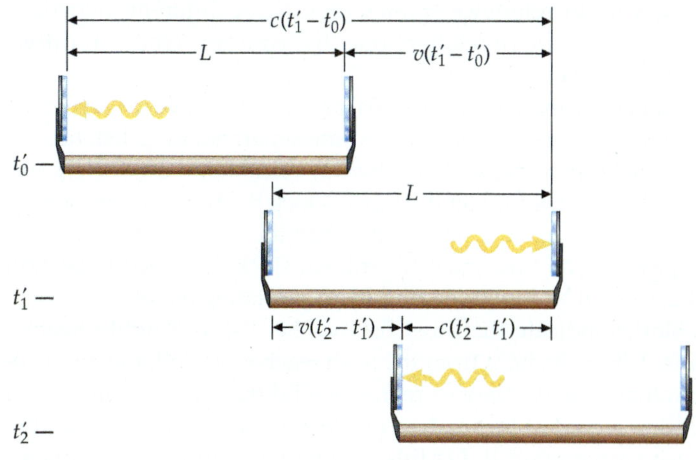

FIGURE R-9 A light clock moving to the right at speed v is shown at times t_0, t_1, and t_2.

Equating the right sides of Equations R-6 and R-8, and then solving for L gives

$$L = L_0\sqrt{1 - (v^2/c^2)} \qquad \text{R-9}$$

LENGTH CONTRACTION

Establishing this result did not involve any properties of the stick. Thus, Equation R-9 reflects the nature of space and time, and not the nature of the sticks.

Example R-2 The Length of a Railroad Car

Keisha is on a train that is moving at $0.80c$ relative to the station. She measures the length of the railroad car she is in, and finds its length to be 40 m. Bob is standing on the station platform as the train streaks by. Bob measures the time it takes for the car to pass him and multiplies this time by $0.80c$ to determine the length of the car. What is the length of the car according to Bob's calculation?

PICTURE The car is at rest in Keisha's reference frame, so 40 m is the proper length of the car.

SOLVE

The car is at rest in Keisha's frame, so 40 m is its proper length. In Bob's frame the train is moving at $0.8c$. Use Equation R-9 to find the length of the car in Bob's frame:

$$L = L_0\sqrt{1 - (v^2/c^2)} = 40\text{ m}\sqrt{1 - 0.80^2} = \boxed{24\text{ m}}$$

CHECK As expected, the car is shorter in the reference frame in which it is moving.

R-5 DISTANT CLOCKS AND SIMULTANEITY

We have established three useful relations: (1) that the length of a stick moving perpendicular to its length is the same as its rest length; (2) that the time T *between* two ticks of a moving clock is greater than the proper time between the two ticks of the same clock according to $T = T_0/\sqrt{1 - (v^2/c^2)}$; and (3) that the length L of a stick moving parallel to its length is less than its rest length L_0 according to $L = L_0\sqrt{1 - (v^2/c^2)}$. But in order to analyze events from the perspective of observers in reference frames moving at different velocities, we need one more relation, one that concerns clocks at different locations.

Clocks A and B (Figure R-10*a*) are at rest relative to each other, and in their rest frame the clocks are separated by a distance L_0. To synchronize these clocks there is a flashlamp on clock A and a light-sensitive film on the face of clock B. The alarm on clock A is set to energize the flashlamp when the second hand on clock A passes zero. Like the conventional clock described in Section R-3, clock B has only a second hand, a rotating opaque disk with a slot to indicate the time. Behind the disk is a light-sensitive film. When the light from the flash reaches clock B, the film is illuminated on the narrow region behind the slot. This provides a lasting record of the reading on clock B when the light from the flashlamp reaches it. Let this reading be t_1. In the rest frame of the clocks, the time for the light to travel at speed c from clock A to clock B is L_0/c, so when the light arrives at clock B, clock A reads L_0/c and clock B reads t_1. To synchronize the two clocks, we turn clock B back by $\Delta t = t_1 - L_0/c$.

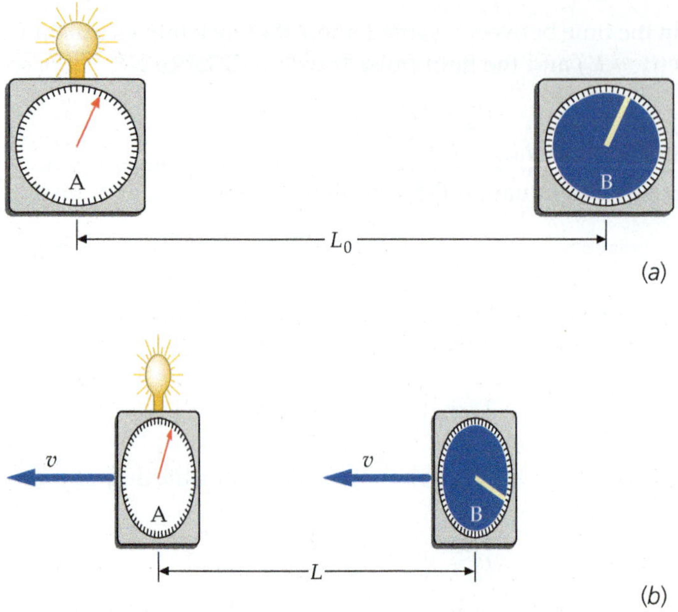

FIGURE R-10 (*a*) The clocks are synchronized in the reference frame in which they are at rest. (*b*) Are the clocks also synchronized in the reference frame in which they are moving with speed v parallel to the line joining them?

With the two clocks synchronized in their rest frame (frame 1), we then determine whether they are also synchronized in a reference frame (frame 2) in which they are moving at speed v parallel to the line joining them (shown in Figure R-10*b*). We reset the alarm to energize the flashlamp when clock A next reads zero. These two events—clock A reads zero and the lamp flashes—are a spacetime coincidence, so we know they occur simultaneously in all reference frames. Also, the light reaching clock B and clock B reading L_0/c, are a spacetime coincidence, so we know they occur simultaneously in all reference frames.

In frame 2, the distance L between the clocks is given by

$$L = L_0\sqrt{1 - (v^2/c^2)}$$

and clock B is moving toward the flashlamp. In this frame, the light traveling from clock A to clock B travels a distance $L - vt$, where t is the time required for the light to travel this distance. Thus, the time t, the distance L, and the speed v are related by

$$ct = L - vt$$

Solving for the time gives $t = L/(c + v)$.

Moving clocks run slow, so during time t the readings on both clocks advance not by t but $t\sqrt{1 - (v^2/c^2)}$, where $t = L/(c + v)$. This advance is equal to

$$\frac{L}{c + v}\sqrt{1 - (v^2/c^2)} = \frac{L_0\sqrt{1 - (v^2/c^2)}}{c + v}\sqrt{1 - (v^2/c^2)}$$

$$= \frac{L_0}{(c + v)}\frac{(c + v)(c - v)}{c^2} = \frac{L_0}{c} - \frac{vL_0}{c^2}$$

Thus, when the light arrives at clock B, clock B reads L_0/c and clock A reads $L_0/c - vL_0/c^2$. Therefore, in frame 2, clock B is ahead of clock A by vL_0/c^2:

> If two clocks that are moving with the same velocity are synchronized in their rest frame, then in a frame where they move with speed v parallel to the line joining them, the clock in the rear is ahead of the clock in front by vL_0/c^2.
>
> THE RELATIVITY OF SIMULTANEITY

In this case, L_0 is the distance between the clocks in their rest frame. It is also true that if two clocks are synchronized in their rest frame, they are also synchronized in any frame in which they are moving perpendicular to the line joining them. This condition follows from the symmetry of the situation. (For one thing, there is no way to state a rule specifying which of the two clocks is ahead.)

APPLYING THE RULES

Example R-3 A Train Through a Tunnel

A high-speed train is about to enter a tunnel through a mountain. The tunnel has a proper length of 1.2 km. The length of the train in the reference frame of the mountain is also 1.2 km, and the proper length of the train is 2.0 km. Clock A is attached to the mountain at the entrance to the tunnel, and clock B is attached to the mountain at the exit to the tunnel. In the reference frame of the mountain, at the instant the front of the train enters the tunnel both clocks read zero. (*a*) In the reference frame of the mountain, what is the speed of the train, and what is the reading of both clocks at the instant the front of the train exits the tunnel (Figure R-11*a*)? (*b*) In the reference frame of the train, what is the length of the tunnel, what is the reading of both clocks at the instant the front of the train enters the tunnel (Figure R-11*b*), and what is the reading of both clocks at the instant the front of the train exits the tunnel? (*c*) For a passenger on the train, how long does it take for the front of the train to pass through the tunnel?

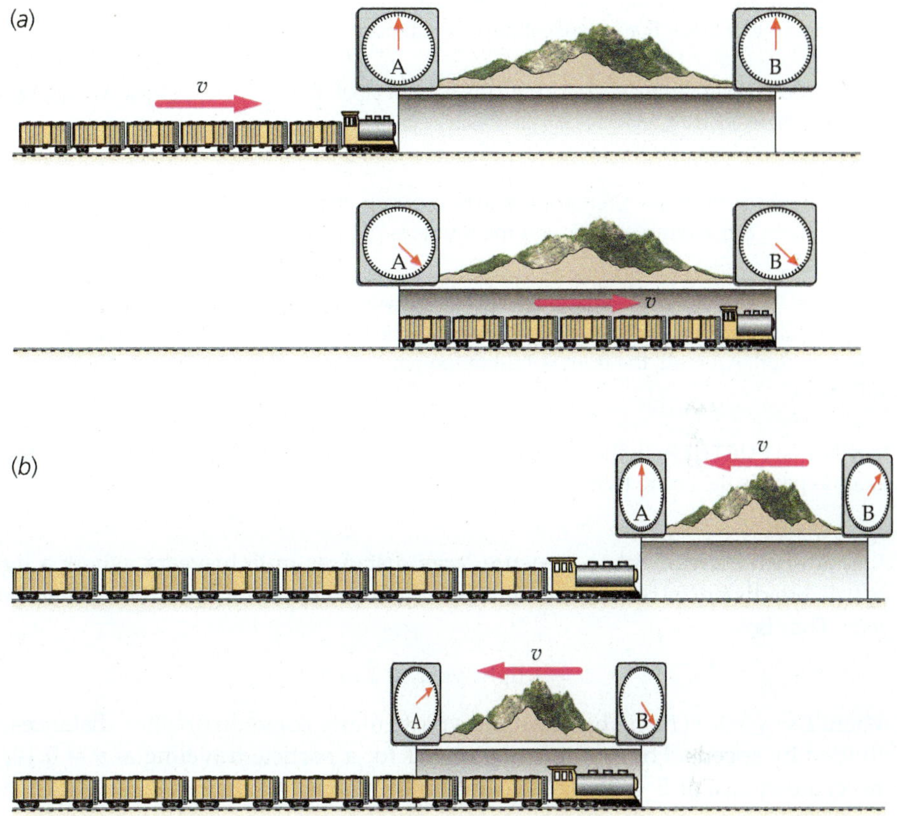

FIGURE R-11 (*a*) In the reference frame of the mountain, the train approaches the tunnel at high speed. The clocks are synchronized in this reference. (*b*) In the reference frame of the train, the tunnel (and the mountain) approaches at high speed. Clock B is ahead of clock A in this reference frame.

PICTURE The speed of the train and the length of the train are related by the length-contraction formula. Some of the clock readings in the two reference frames can be equated because they are event pairs that form spacetime coincidences. Other clock readings can be related by the relativity-of-simultaneity relation.

SOLVE

(*a*) 1. Using the length-contraction formula, solve for the speed of the train:

$$L = L_0\sqrt{1 - (v^2/c^2)}$$

$$1.2\text{ km} = 2.0\text{ km}\sqrt{1 - (v^2/c^2)}$$

$$\text{so}\quad v = 0.80c = 0.80(3.00 \times 10^8\text{ m/s}) = \boxed{2.4 \times 10^8\text{ m/s}}$$

2. The length of the tunnel equals its proper length, and because the clocks are not moving, they do not run slow. The reading on both clocks is the time t that it takes for the front of the train to travel the length of the tunnel:

$$L_{\text{tunnel }0} = vt$$

$$\text{so}\quad t = \frac{L_{\text{tunnel }0}}{v} = \frac{1.2 \times 10^3\text{ m}}{2.4 \times 10^8\text{ m/s}} = 5.0 \times 10^{-6}\text{ s} = \boxed{5.0\ \mu\text{s}}$$

3. The clocks are synchronized, so when the front of the train exits the tunnel, both clocks read 5 μs:

$$\text{Clock A reading} = \text{Clock B reading} = \boxed{5.0\ \mu\text{s}}$$

(*b*) 1. In this frame, the mountain is moving at 0.80*c*. Using the length-contraction formula, solve for the length of the tunnel:

$$L_{\text{tunnel}} = L_{\text{tunnel }0}\sqrt{1 - (v^2/c^2)} = 1.2\text{ km}\sqrt{1 - \frac{(0.80c)^2}{c^2}}$$

$$= 1.2\text{ km}\sqrt{1 - 0.80^2} = \boxed{0.72\text{ km} = 720\text{ m}}$$

2. The front of the train entering the tunnel and a zero reading on clock A are a spacetime coincidence:

$$\boxed{\text{Clock A reads zero}}$$

3. The two clocks are moving toward the train with clock B in the rear, so clock B is ahead of clock A by vL_0/c^2. When the train enters the tunnel, clock A reads zero, so clock B reads vL_0/c^2:

$$\text{Clock B reading} = \frac{vL_{\text{tunnel }0}}{c^2} = \frac{0.80L_{\text{tunnel }0}}{c}$$
$$= \frac{0.80(1.2 \times 10^3\ \text{m})}{3.0 \times 10^8\ \text{m/s}} = \boxed{3.2\ \mu\text{s}}$$

4. The front of the train exiting the tunnel and a 5.0-μs reading of clock B are a spacetime coincidence:

$$\text{Clock B reads} = \boxed{5.0\ \mu\text{s}}$$

5. Clock B is in the rear, so clock A lags behind clock B by vL_0/c^2.

$$\text{Clock A reading} = \text{clock B reading} - \frac{vL_{\text{tunnel }0}}{c^2}$$
$$= 5.0\ \mu\text{s} - 3.2\ \mu\text{s} = \boxed{1.8\ \mu\text{s}}$$

(*c*) For an observer in the reference frame of the train, the mountain is traveling at $0.80c$ and the tunnel is 720-m long:

$$L_{\text{tunnel}} = vt$$
$$\text{so}\quad t = \frac{L_{\text{tunnel}}}{v} = \frac{L_{\text{tunnel}}}{0.80c} = \frac{720\ \text{m}}{2.4 \times 10^8\ \text{m/s}} = \boxed{3.0\ \mu\text{s}}$$

CHECK Our Part-(*c*) result of 3.0 μs is less than the Part-(*a*), step-2 result of 5 μs. This result is as expected. After all, the train is 1.2-km long in Part (*a*) and the tunnel is only 720-m long in Part (*c*).

TAKING IT FURTHER In the reference frame of the train, the train is longer than the tunnel so at no time is the entire train within the tunnel.

CONCEPT CHECK R-1

Event 1 is the front of the train entering the tunnel, and event 2 is the front of the train exiting the tunnel. (*a*) In which reference frame do these two events occur at the same location? (*b*) What is the proper time interval between events 1 and 2?

It is often convenient to measure large distances in light-years, where a light-year is the distance traveled when traveling at the speed of light for a time of one year. That is,

$$1\ \text{light-year} = 1\ c \cdot \text{y}$$

where $1\ c \cdot \text{y} = c \cdot (1\ \text{y})$. This notation is particularly convenient when distances are divided by speeds. For example, the time T for a particle traveling at $v = 0.10c$ to travel a distance of $L = 25$ light-years is

$$T = \frac{L}{v} = \frac{25\ c \cdot \text{y}}{0.10c} = 250\ \text{y}$$

where the c's cancel.

PRACTICE PROBLEM R-3

In the reference frame of Earth, it takes 8 minutes for light to travel from the Sun to Earth, so the distance between the Sun and Earth is $8.3c \cdot \text{min}$. How many minutes does it take a particle from the Sun to reach Earth if the particle travels at $0.10c$?

R-6 RELATIVISTIC MOMENTUM, MASS, AND ENERGY

MOMENTUM AND MASS

In special relativity, both momentum and energy are conserved, just as they are in classic physics. The laws of conservation of momentum and energy are essential to analyzing the high-speed collisions that take place in high-energy physics laboratories. However, the classic equations for conserving momentum and energy are not adequate for the analysis of high-speed collisions. Here, we present the relativistically correct form of these conservation equations. The momentum of a particle moving with velocity v is given by

$$p = \frac{mv}{\sqrt{1 - (v^2/c^2)}} \qquad \text{R-10}$$

RELATIVISTIC MOMENTUM

where m is the mass of the particle.* The relativity of momentum is discussed further in Chapter 39.

ENERGY

In relativistic mechanics, as in classic mechanics, the net force on a particle is equal to the time rate of change of the momentum of the particle. Considering one-dimensional motion only, we have

$$F_{\text{net}} = \frac{dp}{dt} \qquad \text{R-11}$$

We wish to find an expression for the kinetic energy. To do this, we multiply both sides of Equation R-11 by the displacement $d\ell$. This gives

$$F_{\text{net}}\, d\ell = \frac{dp}{dt}\, d\ell \qquad \text{R-12}$$

where we identify the term on the left as the work and the term on the right as the change in kinetic energy dK. Substituting $v\,dt$ for $d\ell$ in the term on the right we obtain

$$dK = \frac{dp}{dt}\, v\, dt = v\, dp$$

Integrating both sides gives

$$K = \int_0^{p_f} v\, dp \qquad \text{R-13}$$

To evaluate this integral, we first change the integration variable from p to v. Using Equation R-10 and the quotient rule, we have

$$dp = d\left(\frac{v}{\sqrt{1 - (v^2/c^2)}}\right)$$

$$= \frac{[1 - (v^2/c^2)]^{1/2} dv - v\frac{1}{2}[1 - (v^2/c^2)]^{-1/2}\left(-\frac{2vdv}{c^2}\right)}{1 - (v^2/c^2)} = \frac{dv}{[1 - (v^2/c^2)]^{3/2}}$$

Substituting for dp in Equation R-13 gives

$$K = \int_0^{p_f} vdp = m\int_0^{v_f} \frac{vdv}{[1 - (v^2/c^2)]^{3/2}} = mc^2\left(\frac{1}{\sqrt{1 - (v_f^2/c_2)}} - 1\right)$$

so

$$K = \frac{mc^2}{\sqrt{1 - (v^2/c^2)}} - mc^2 \qquad \text{R-14}$$

(In this expression, because the only speed is v_f, the subscript f is not needed.)

---

* Equation R-10 is sometimes written $p = m_r v$, where m_r is called the *relativistic mass*: $m_r = m/\sqrt{1 - (v^2/c^2)}$. In the rest frame of the particle, $v = 0$ and $m_r = m$. (The mass m is sometimes called the *rest mass* to differentiate it from the relativistic mass.)

Defining $mc^2/\sqrt{1-(v^2/c^2)}$ as the **total relativistic energy** E, Equation R-14 can be written

$$E = K + mc^2 = \frac{mc^2}{\sqrt{1-(v^2/c^2)}} \qquad \text{R-15}$$

where mc^2, called the **rest energy** E_0, is energy the particle has when it is at rest.

By multiplying both sides of Equation R-10 by c and then dividing the resulting equation by Equation R-15, we obtain

$$\frac{v}{c} = \frac{pc}{E} \qquad \text{R-16}$$

which can be useful when trying to solve for the speed v. Eliminating v from Equations R-10 and R-16, and solving for E^2, gives

$$E^2 = p^2c^2 + m^2c^4 \qquad \text{R-17}$$

The relation between mass and energy is briefly discussed in Section 4 of Chapter 7.

See **Math Tutorial** *for more information on the* **Binomial Expansion**

Example R-4 Momentum and Energy

A proton has kinetic energy of 1100 MeV and a mass of 938 MeV/c^2. What is its momentum? What is its speed?

PICTURE Equations R-15 and R-17 relate the momentum to the total energy, the kinetic energy and the mass. Equation R-16 relates the speed to the momentum and the total energy.

SOLVE

1. The momentum is related to the total energy by Equation R-17, and the total energy is related to the kinetic energy by Equation R-15:

   $E^2 = p^2c^2 + m^2c^4$

   $E = K + mc^2$

2. Substitute for E in the first step-1 equation and solve for p^2c^2:

   $(K + mc^2)^2 = p^2c^2 + m^2c^4$

   so $p^2c^2 = (K + mc^2)^2 - m^2c^4$

3. Calculate the value of p^2c^2:

   $p^2c^2 = (1100\text{ MeV} + 938\text{ MeV})^2 - (938\text{ MeV})^2 = 3.27 \times 10^6\ (\text{MeV})^2$

4. Solve for p:

   $p = \sqrt{3.27 \times 10^6}\ \text{MeV}/c = 1.81 \times 10^3\ \text{MeV}/c = \boxed{1.8 \times 10^3\ \text{MeV}/c}$

5. Using Equation R-16 solve for the speed:

   $\frac{v}{c} = \frac{pc}{E} = \frac{pc}{K + mc^2} = \frac{1.81 \times 10^3\text{ MeV}}{1100\text{ MeV} + 938\text{ MeV}} = 0.888$

   so $\boxed{v = 0.888c = 0.89c}$

CHECK As expected, the speed is greater than zero and less than c. In addition, the kinetic energy is greater than the rest energy (938 MeV), so we expect the speed to be a significant fraction of c.

Example R-5 Colliding Particles — *Conceptual*

Two identical particles, each with mass m, travel in opposite directions, each with a total energy equal to twice its rest energy. They undergo a perfectly inelastic head-on collision and stick together to form a single particle of mass M. Find M.

PICTURE Use conservation of momentum to determine the speed of the mass-M particle. Use conservation of energy to find the mass of the mass-M particle.

SOLVE

1. The identical particles have the same mass, m, and the same total energy, E, so they have the same speed. They are traveling in opposite directions, so the momentum of one is equal and opposite to the momentum of the other. The total momentum of the two-particle system is zero.

   Conservation of momentum tells us that the momentum, and thus the speed, of the mass-M particle is zero.

2. The particle of mass M is at rest, so its total energy is equal to its rest energy Mc^2. For each mass-m particle, the total energy is twice the rest energy mc^2. *Conservation of energy* tells us that the total energy in the same before and after the collision.

   $E_f = E_i$

   $Mc^2 = 2mc^2 + 2mc^2$

   $\therefore \boxed{M = 4m}$

CHECK The kinetic energies of the two mass-m particles is transformed into the rest energy of the mass-M particle. Because the kinetic energy of each mass-m particle is equal to its rest energy, the mass of the mass-M particle is equal to $4m$.

! Do not think during an inelastic collision mass is conserved. It is not. Mass is proportional to rest energy. If kinetic energy is transformed into rest energy, then the mass increases.

Summary

1. The principle of relativity is a *fundamental law of physics.*
2. That the speed of light in a vacuum is independent of the speed of the source is a *fundamental law of physics.*

TOPIC	RELEVANT EQUATIONS AND REMARKS	
1. Postulates of Special Relativity		
Postulate I: Principle of relativity	It is impossible to devise an experiment that determines whether you are at rest or moving uniformly, where moving uniformly means moving at constant velocity relative to an inertial reference frame.	
Postulate II	The speed of light is independent of the speed of the source.	
Constancy of the speed of light	It follows that the speed of light is the same in any inertial reference frame.	
2. Moving Sticks	The length of a stick moving perpendicular to its length is equal to its proper length. The length of a stick moving with speed v parallel to its length is shorter than its proper length according to $$L = L_0\sqrt{1 - (v^2/c^2)}$$	R-9
3. Moving Clocks		
Time dilation	The time between ticks of a clock moving with speed v is longer than the proper time between ticks of the same clock by $$T = \frac{T_0}{\sqrt{1 - (v^2/c^2)}}$$	R-3
Relativity of simultaneity	If two clocks that move with the same velocity are synchronized in their rest frame, in a frame where they move with speed v parallel to the line joining them, the clock in the rear is ahead of the clock in front by vL_0/c^2, where L_0 is the distance between them in their rest frame. If two clocks that move with the same velocity are synchronized in their rest frame they are also synchronized in any frame where they are moving perpendicular to the line joining them.	

TOPIC	RELEVANT EQUATIONS AND REMARKS	
4. Spacetime Coincidence	If two events occur both at the same time *and* at the same place in one reference frame, they occur both at the same time and at the same place in any reference frame.	
5. Momentum, Mass, and Energy		
Momentum	The momentum of a particle is given by $$p = \frac{m_0 v}{\sqrt{1 - (v^2/c^2)}}$$	R-10
Kinetic energy	$$K = \left(\frac{1}{\sqrt{1 - (v^2/c^2)}} - 1\right)mc^2$$	R-14
Mass and energy	The total relativistic energy E of a particle equals its rest energy plus its kinetic energy. $$E = K + mc^2 = \frac{mc^2}{\sqrt{1 - (v^2/c^2)}}$$ where mc^2 is the rest energy E_0.	R-15
Momentum and Energy	$\frac{v}{c} = \frac{pc}{E}$ and $E^2 = p^2c^2 + m^2c^4$	R-16, R-17

Answers to Concept Checks

R-1 (*a*) The reference frame of the train, because both events occur at the front end of the train, (*b*) 3.0 μs

Answers to Practice Problems

R-1 260 ns

R-2 78 m

R-3 $(8.3\,c\cdot\text{min})/0.10c = (8.3\text{ min})/0.10 = 83\text{ min}$

Problems

In a few problems, you are given more data than you actually need; in a few other problems, you are required to supply data from your general knowledge, outside sources, or informed estimate.

Interpret as significant all digits in numerical values that have trailing zeros and no decimal points.

- • Single-concept, single-step, relatively easy
- •• Intermediate-level, may require synthesis of concepts
- ••• Challenging
- SSM Solution is in the *Student Solutions Manual*
- Consecutive problems that are shaded are paired problems.

CONCEPTUAL PROBLEMS

1 • CONTEXT-RICH You are standing on a corner when you see a friend drive past you in a car. Each of you is wearing a wristwatch. Both of you note the times when the car passes two different intersections and determine from your watch readings the time that elapses between the two events. Have either of you determined the proper time interval? Explain your answer.

2 • CONTEXT-RICH In Problem 1, suppose your friend in the car measures the width of the car door to be 90 cm. You also measure the width as he goes by you. (*a*) Does either of you measure the proper width of the door? Explain your answer. (*b*) How will your value for the door width compare to his? (1) Your value will be smaller. (2) Your value will be larger. (3) Your value will be the same. (4) You cannot compare the widths, as the answer depends on the car's speed.

3 • If event A occurs at a different location than event B in some reference frame, might it be possible for there to be a second reference frame in which they occur at the same location? If so, give an example. If not, explain why not. SSM

4 • If event A occurs prior to event B in some reference frame, might it be possible for there to be a second reference frame in which event B occurs prior to event A? If so, give an example. If not, explain why not.

5 • Two events are simultaneous in a reference frame in which they also occur at the same location. Are they simultaneous in all other reference frames? SSM

6 • Two inertial observers are in relative motion. Under what circumstances can they agree on the simultaneity of two different events?

7 • The approximate total energy of a particle of mass m moving at speed $v \ll c$ is (*a*) $mc^2 + \frac{1}{2}mv^2$, (*b*) mv^2, (*c*) cmv, (*d*) $\frac{1}{2}mc^2$.

8 • True or false:

(*a*) The speed of light is the same in all reference frames.
(*b*) The proper time interval is the shortest time interval between two events.
(*c*) Absolute motion can be determined by means of length contraction.
(*d*) The light-year is a unit of distance.
(*e*) For two events to form a spacetime coincidence, they must occur at the same place.
(*f*) If two events are not simultaneous in one reference frame, they cannot be simultaneous in any other reference frame.

9 •• (*a*) Show that pc has dimensions of energy. (*b*) There is a geometrical interpretation of Equation R-17 based on the Pythagorean theorem. Draw a picture of a triangle illustrating this interpretation.

10 •• A wad of putty of mass m_1 strikes and sticks to a second wad of putty of mass m_2, which is initially at rest. Do you expect that after the collision the combined putty mass will be (*a*) greater than, (*b*) less than, (*c*) the same as $m_1 + m_2$? Explain your answer.

11 •• **BIOLOGICAL APPLICATION** Many nuclei of atoms are unstable; for example, ^{14}C, an isotope of carbon, has a half-life of 5700 years. (By definition, the *half-life* is the time it takes for any given number of unstable particles to decay to half that number of particles.) This fact is used extensively for archeological and biological dating of old artifacts. Such unstable nuclei decay into several decay products, each with significant kinetic energy. Which of the following is true? (*a*) The mass of the unstable nucleus is larger than the sum of the masses of the decay products. (*b*) The mass of the unstable nucleus is smaller than the sum of the masses of the decay products. (*c*) The mass of the unstable nucleus is the same as the sum of the masses of the decay products. Explain your choice. SSM

12 •• **BIOLOGICAL APPLICATION** Positron emission tomography (PET) scans are common in modern medicine. During this procedure, positrons (a positron has the same mass but the opposite charge of an electron) are emitted by radioactive nuclei that have been introduced into the body. Assume that an emitted positron, traveling slowly (with negligible kinetic energy), collides with an electron traveling at the same slow speed in the opposite direction. They undergo annihilation and two quanta of light (photons) are formed. You are in charge of designing detectors to receive these photons and measure their energies. (*a*) Explain why you would expect these two photons to come off in exactly opposite directions. (*b*) In terms of the electron mass m_e, how much energy would each photon have? (1) less than mc^2, (2) greater mc^2, (3) exactly mc^2. Explain your choice.

ESTIMATION AND APPROXIMATION

13 •• In 1975, an airplane carrying an atomic clock flew back and forth at low altitude for 15 hours at an average speed of 140 m/s as part of a time-dilation experiment. The time on the clock was compared to the time on an atomic clock kept on the ground. What is the time difference between the atomic clock on the airplane and the atomic clock on the ground? (Ignore any effects that accelerations of the airplane have on the atomic clock that is on the airplane. Also assume that the airplane travels at costant speed.) SSM

14 •• (*a*) By making any necessary assumptions and finding certain stellar distances, estimate the speed at which a spaceship would have to travel for its passengers to make a trip to the nearest star (not the Sun!) and back to Earth in 1.0 Earth years, as measured by an observer on the spaceship. Assume that the passengers make the outgoing and return trips at constant speed, and ignore any effects due to the spaceship stopping and starting. (*b*) How much time would elapse on Earth during their round-trip? Include 2.0 Earth years for a low-speed exploration of the planets in the vicinity of the star.

15 •• (*a*) Compare the kinetic energy of a moving car to its rest energy. (*b*) Compare the total energy of a moving car to its rest energy. (*c*) Estimate the error made in computing the kinetic energy of a moving car using nonrelativistic expressions compared to the relativistically correct expressions. *Hint: Use of the binomial expansion may help.*

LENGTH CONTRACTION AND TIME DILATION

16 • The proper average (or mean) lifetime of a pion (a subatomic particle) is 2.6×10^{-8} s. (A neutral pion has a much shorter lifetime. See Chapter 41.) A beam of pions has a speed of $0.85c$ relative to a laboratory. (*a*) What would be their mean lifetime as measured in that laboratory? (*b*) On average, how far would they travel in that laboratory before they decay? (*c*) What would your answer to Part (*b*) be if you had neglected time dilation?

17 • In the reference frame of a pion in Problem 16, how far does the laboratory travel in 2.6×10^{-8} s? SSM

18 • The proper average (or mean) lifetime of a muon (a subnuclear particle) is 2.20 μs. Muons in a beam are traveling at $0.999c$ relative to a laboratory. (*a*) What is their lifetime as measured in that laboratory? (*b*) On average, how far do they travel in that laboratory before they decay?

19 • In the reference frame of the muon in Problem 18, how far does the laboratory travel in 2.20 μs?

20 • **CONTEXT-RICH** You have been posted to a remote region of space to monitor traffic. Toward the end of a quiet shift, a spacecraft goes by and you measure its length using a laser device. This device reports a length of 85.0 m. You flip open your handy reference catalog and identify the craft as a CCCNX-22, which has a proper length of 100 m. When you phone in your report, what speed should you give for this spacecraft?

21 • A spaceship travels from Earth to a star 95 light-years away at a speed of 2.2×10^8 m/s. How long does the spaceship take to get to the star (*a*) as measured on Earth and (*b*) as measured by a passenger on the a spaceship? SSM

22 • The average lifetime of a beam of subatomic particles called pions (see Problem 16 for details on these particles) traveling at high speed is measured to be 7.5×10^{-8} s. Their average lifetime at rest is known to be 2.6×10^{-8} s. How fast is this pion beam traveling?

23 • A meterstick moves with speed $0.80c$ relative to you in the direction parallel to the stick. (*a*) Find the length of the stick as measured by you. (*b*) How long does it take for the stick to pass you?

24 • Recall that the half-life is the time it takes for any given amount of unstable particles to decay to half that amount of particles. The proper half-life of a species of charged subatomic particles called pions is 1.80×10^{-8} s. (See Problem 16 for details on pions.) Suppose a group of these pions are produced in an accelerator and emerge with a speed of $0.998c$. How far do these particles travel in the accelerator's laboratory before half of them have decayed?

25 •• Your friend, who is the same age as you, travels to the star Alpha Centauri, which is 4.0 light-years away, and returns immediately. He claims that the entire trip took just 6.0 y. What was his speed? Ignore any accelerations of your friend's spaceship and assume that the spaceship traveled at the same speed during the entire trip. SSM

26 •• Two spaceships pass each other traveling in opposite directions. A passenger in ship A, knows that her ship is 100 m long. She notes that ship B is moving with a speed of $0.92c$ relative to ship A and that the length of B is 36 m. What are the lengths of the two spaceships as measured by a passenger in ship B?

27 •• Supersonic jets achieve maximum speeds of about $3.00 \times 10^{-6}c$. (*a*) By what percentage would a jet traveling at this speed contract in length? (*b*) During a time of exactly one year or 3.15×10^7 s on your clock, how much time would elapse on the pilot's clock? How many minutes are lost by the pilot's clock in one year of your time? Assume that you are on the ground and the pilot is flying at the specified speed for the entire year.

28 •• The proper mean lifetime of a muon (see Problems 18 and 19 for details regarding muons) is 2.20 μs. Consider a muon, created in Earth's upper atmosphere, speeding toward the surface 8.00 km below, at a speed of $0.980c$. (*a*) What is the likelihood that the muon will survive its trip to Earth's surface before decaying? The probability of a muon decaying is given by $P = 1 - e^{-\Delta t/\tau}$, where Δt is the time interval as measured in the reference frame in question. (*b*) Calculate the probability from the point of view of an observer moving with the muon. Show that the answer is the same from the point of view of an observer on Earth.

29 •• A spaceship commander traveling to the Magellanic Clouds travels at a uniform speed of $0.800c$. When leaving the Kuiper belt, whose outer edge is 50.0 AU from Earth (*Note:* 1 AU = 150,000,000 km and represents the average distance between Earth and the Sun; AU = astronomical unit), he sends a message to ground control in Houston, Texas, saying that he is fine. Fifteen minutes later (according to him), he realizes he has made a typo, so he sends a correction. How much time passes at Houston between the receipt of his initial message and the receipt of second message?

THE RELATIVITY OF SIMULTANEITY

Problems 30 through 34 refer to the following situation: Mary is a worker on a large space platform. She places clock A at point A and clock B at point B, which is 100 light-minutes from point A (Figure R-12). She also places a flashbulb at a point midway between points A and B. Jamal, a worker on a different platform, is standing next to clock C. Each clock immediately starts at zero when the flash reaches it. Mary's platform moves at speed of $0.600c$ to the left relative to Jamal's. As Mary's platform passes by, clock B, then the flashbulb, and then clock A pass directly over clock C—just missing it as they go by. As the flashbulb passes next to clock C, it flashes and clock C immediately starts at zero.

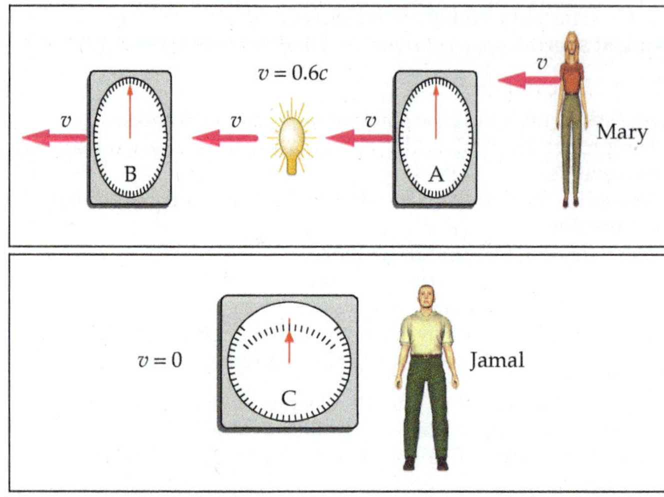

FIGURE R-12 Problems 30–34

30 •• According to Jamal: (*a*) What is the distance between the flashbulb and clock A? (*b*) How far does the flash travel to reach clock A? (*c*) How far does clock A travel while the flash is traveling from the flashbulb to it?

31 •• According to Jamal, how long does it take the flash to travel to clock A, and what does clock C read as the flash reaches clock A.

32 •• Show that clock C reads 100 min as the light flash reaches clock B, which is traveling away from clock C with speed $0.600c$.

33 •• According to Jamal, the reading on clock C advances from 25 min to 100 min between the reception of the flashes by clocks A and B in Problems 31 and 32. According to Jamal, how much will the reading on clock A advance during this 75-min interval?

34 •• According to Jamal, the advance of clock A calculated in Problem 33 is the amount that clock A leads clock B. Compare this result with vL_0/c^2, where $v = 0.600c$.

35 •• In an inertial reference frame *S*, event B occurs 2.00 μs after and 1.50 km distant from event A. How fast must an observer be moving along the line joining the two events so that the two events occur simultaneously? For an observer traveling fast enough, is it possible for event B to precede event A? SSM

36 •• A large flat space platform has an *x* axis painted on it. A firecracker explodes on the *x* axis at $x_1 = 480$ m, and a second firecracker explodes on the *x* axis 5.00 μs later at $x_2 = 1200$ m. In the reference frame of a train traveling alongside the *x* axis at speed *v* relative to the platform, these two explosions occur at the same place on that axis. What is the separation in time between the two explosions in the reference frame of the train?

37 •• Herb and Randy are twin jazz musicians who perform as a trombone-saxophone duo. At the age of twenty, however, Randy got an irresistible offer to perform on a star 15 light-years away. To celebrate his good fortune, he bought a new vehicle for the trip—a deluxe space-coupé that travels at $0.99c$. Each of the twins promises to practice diligently, so they can reunite afterward. However, Randy's gig goes so well that he stays for a full 10 years before returning to Herb. After their reunion, (*a*) how many years of practice will Randy have had, and (*b*) how many years of practice will Herb have had?

38 •• Al and Bert are twins. Al travels at $0.600c$ to Alpha Centauri (which is $4.00\ c \cdot y$ from Earth, as measured in the reference frame of Earth) and returns immediately. Each twin sends the other a light signal every 0.0100 y, as measured in his own reference frame. (*a*) At what rate does Bert receive signals as Al is moving away from him? (*b*) How many signals does Bert receive at this rate? (*c*) How many total signals are received by Bert before Al returns to Earth? (*d*) At what rate does Al receive signals as Bert is moving away from him? (*e*) How many signals does Al receive at this rate? (*f*) How many total signals are received by Al before Al returns to Earth? (*g*) Which twin is younger at the end of the trip and by how many years?

RELATIVISTIC ENERGY AND MOMENTUM

39 • Find the ratio of the total energy to the rest energy of a particle of mass m moving with speed (*a*) $0.100c$, (*b*) $0.500c$, (*c*) $0.800c$, and (*d*) $0.990c$.

40 • A proton of rest energy 938 MeV has a total energy of 1400 MeV. (*a*) What is its speed? (*b*) What is its momentum?

41 • How much energy would be required to accelerate a particle of mass m from rest to (*a*) $0.500c$, (*b*) $0.900c$, and (*c*) $0.990c$? Express your answers as multiples of the rest energy, mc^2. SSM

42 • If the kinetic energy of a particle equals its rest energy, what percentage error is made by using $p = mv$ for its momentum? Is the nonrelativistic expression always smaller or larger than the relativistically correct expression for momentum?

43 • What is the total energy of a proton whose momentum is $3mc$?

44 •• **SPREADSHEET, ESTIMATION** Using a spreadsheet program or graphing calculator, make a graph of the kinetic energy of a particle with rest energy of 100 MeV for speeds between 0 and c. On the same graph, plot $\frac{1}{2}mv^2$ by way of comparison. Using the graph, estimate what speed of the nonrelativistic expression is no longer a good approximation of the kinetic energy. As a suggestion, plot the energy in units of MeV and the speed in the dimensionless form v/c.

45 •• (*a*) Show that the speed v of a particle of mass m and total energy E is given by $v/c = [1 - ((mc^2)^2/E^2)]^{1/2}$, and that when E is much greater than mc^2, this can be approximated by $(v/c) \approx 1 - ((mc^2)/2E^2)$. Find the speed of an electron with kinetic energy of (*b*) 0.510 MeV, and (*c*) 10.0 MeV. SSM

46 •• Use the binomial expansion and Equation R-17 to show that when $pc \ll mc^2$, the total energy is given approximately by $E \approx mc^2 + (p^2/2m)$.

47 •• Derive the equation $E^2 = p^2c^2 + m^2c^4$ (Equation R-17) by eliminating v from Equations R-10 and R-16.

48 •• The rest energy of a proton is about 938 MeV. If its kinetic energy is also 938 MeV, find (*a*) its momentum, and (*b*) its speed.

49 •• What percentage error is made in using $\frac{1}{2}m_0v^2$ for the kinetic energy of a particle if its speed is (*a*) $0.10c$, and (*b*) $0.90c$?

GENERAL PROBLEMS

50 • A spaceship departs from Earth for the star Alpha Centauri, which is $4.0\ c \cdot y$ away in the reference frame of Earth. The spaceship travels at $0.75c$. How long does it take to get there (*a*) as measured on Earth, and (*b*) as measured by a passenger on the spaceship?

51 • The total energy of a particle is three times its rest energy. (*a*) Find v/c for the particle. (*b*) Show that its momentum is given by $p = \sqrt{8}mc$.

52 • A spaceship travels past Earth moving at $0.70c$ relative to Earth. Five minutes after the spaceship is closest to Earth, a message is sent from the control center at Houston, Texas, to the craft. (Ignore any effects of the rotational motion of Earth.) (*a*) How long does it take for the signal to arrive? (*b*) The spaceship and the control center agree on the time when the ship is closest to Earth. Five minutes after the message is received aboard the ship, a return message is sent by the ship back to Houston. What is the time interval in Houston between the time their message is sent, and the time the return message is received?

53 •• Particles called muons traveling at $0.99995c$ are detected at the surface of Earth. One of your fellow students claims that the detected muons might have originated from the Sun. Prove him wrong. (The proper mean lifetime of the muon is $2.20\ \mu s$.) SSM

54 •• (*a*) How tall is Mount Everest in a reference frame traveling with a cosmic ray muon that is traveling straight down, relative to Earth, at $0.99c$? Take the height of Mount Everest according to an Earth-based observer to be 8846 m. (*b*) How long does it take the muon to travel the height of the mountain from the reference frame traveling with the muon? (*c*) How long does it take the muon to travel the height of the mountain from the Earth-based reference frame?

55 ••• A gold nucleus has a radius of 3.00×10^{-14} m, and a mass of 197 AMU. (One AMU has a rest energy of 932 MeV.) During experiments at Brookhaven National Laboratory, these nuclei are routinely accelerated to a kinetic energy of 3.35×10^4 GeV. (*a*) How much less than the speed of light are they traveling? (*b*) At these energies, how long does it take them to travel 100 m in the laboratory reference frame?

56 ••• **APPROXIMATION** Consider the flight of a beam of neutrons produced in a nuclear reactor. These neutrons have kinetic energies of up to 1.00 MeV. The rest energy of a neutron is 939 MeV. (*a*) What is the speed of 1.00 MeV neutrons? Express your answer

in terms of v/c. (*b*) If the average lifetime of such a neutron is 15.0 min (in the laboratory reference frame), what is the maximum length of a beam of such neutrons (in a vacuum, in the absence of any interactions between the neutrons and other material)? Estimate this maximum range by calculating the length corresponding to five mean lifetimes. After five mean lifetimes, only e^{-5} or 0.007 (0.7%) of the neutrons remain. (*c*) Compare this range to the range of so-called "thermally moderated" neutrons, whose kinetic energies are around 0.025 eV. Express your answer as a percentage. That is, what percent of the 1.00 MeV-neutron range is the thermally moderated-neutron range? (Note that our assumption of a vacuum continues; however, in reality neutrons of this energy interact readily with matter, such as air or water, and "real" ranges are much shorter.)

57 ••• **CONTEXT-RICH** You and Ernie are trying to fit a 15-ft-long ladder into a 10-ft-long shed with doors at each end. You suggest to Ernie that you open the front door to the shed and that he run toward it with the ladder at a speed such that the length contraction of the ladder shortens it enough so that it fits in the shed. As soon as the back end of the ladder passes through the door, you will slam it shut. (*a*) What is the minimum speed at which Ernie must run to fit the ladder into the shed? Express it as a fraction of the speed of light. (*b*) As Ernie runs toward the shed at a speed of $0.866c$, he realizes that in the reference frame of the ladder, it is the *shed* that is shorter, not the ladder. How long is the shed in the rest frame of the ladder? (*c*) In the reference frame of the ladder, is there any instant that both ends of the ladder are simultaneously inside the shed? Examine this from the point of view of relativistic simultaneity.

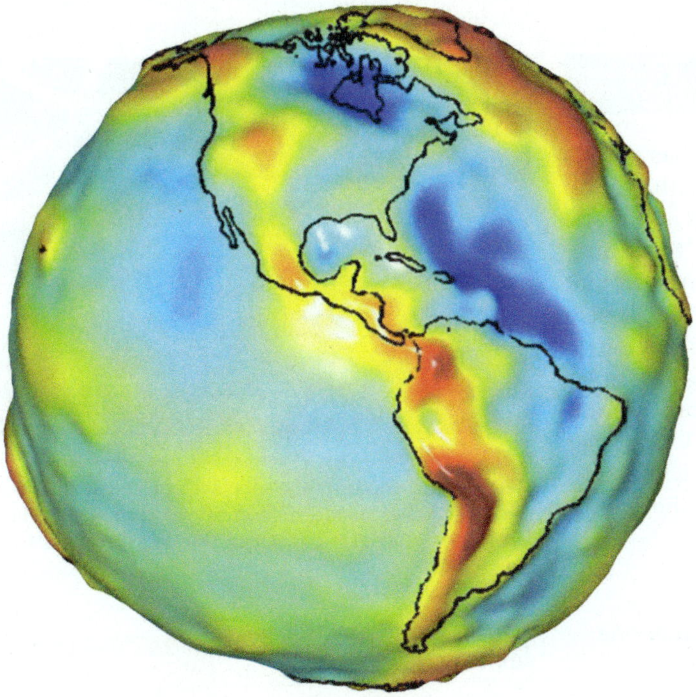

CHAPTER 11

THIS EXAGGERATED GRAVITY MAP OF THE WESTERN HEMISPHERE WAS PRODUCED AS PART OF THE GRACE MISSION, A MISSION JOINTLY UNDERTAKEN BY DLR OF GERMANY AND NASA OF THE UNITED STATES. THE STRENGTH OF THE GRAVITATIONAL FIELD VARIES SLIGHTLY FROM LOCATION TO LOCATION. (THE VARIATIONS ON THIS MAP ARE EXAGGERATED.) THE DATA FOR THIS MAP WERE ACQUIRED BY ACCURATELY MONITORING THE DISTANCE BETWEEN A PAIR OF ORBITING SATELLITES. *(NASA/University of Texas Center for Space Research.)*

Gravity

An understanding of gravity and its role in how celestial bodies move and interact, galaxies expand and contract, and black holes develop is well understood. The gravitational force exerted by Earth on us and on the objects around us is a fundamental part of our experience. It is gravity that binds us to Earth and keeps Earth and the other planets within the solar system. However, the variations in gravity are often too small to notice on the surface of Earth. But these minuscule variations should not be completely disregarded. Geophysicists have found ways to use these small variations in gravity to determine the location of oil and mineral deposits.

During the time of Newton, many believed that nature followed different rules in other parts of the universe than here on Earth. Newton's law of universal gravity, along with his three laws of motion, revealed that nature follows the same rules everywhere, and this revelation has had a profound effect on our view of the universe.

In this chapter, we use the tools of conservation of angular momentum, conservation of energy, Newton's laws of motion, and Newton's law of gravity to predict the motion of the planets and other celestial bodies, including satellites that we have put in space.

? Using two satellites, how might you use your understanding of gravity to detect a region of increased gravitation-field strength? (See Example 11-9.)

11-1 KEPLER'S LAWS

The nighttime sky with its myriad stars and shining planets has always fascinated people. Toward the end of the sixteenth century, the astronomer Tycho Brahe studied the motions of the planets and made observations that were considerably more accurate than those previously available. Using Brahe's data, Johannes Kepler discovered that the paths of the planets about the Sun are ellipses (Figure 11-1). He also showed that each planet moves faster when its orbit brings it closer to the Sun and slower when its orbit takes it farther away. Finally, Kepler developed a precise mathematical relation between the orbital period of a planet and its average distance from the Sun (Table 11-1). He stated these results in three empirical laws of planetary motion. Ultimately, these laws provided the basis for Newton's discovery of the law of gravity. Kepler's three laws follow.

A mechanical model of the solar system, called an *orrery*, in the collection of Historical Scientific Instruments at Harvard University. *(Collection of Historical Scientific Instruments, Harvard University.)*

Law 1. All planets move in elliptical orbits with the Sun at one focus.

An ellipse is the locus of points for which the sum of the distances from two fixed points, called foci F, is constant, as shown in Figure 11-2. Figure 11-3 shows a planet following an elliptical path with the Sun at one focus. Earth's orbit is nearly circular, with the distance to the Sun at perihelion (closest point) being 1.48×10^{11} m and at aphelion (farthest point) being 1.52×10^{11} m. The semimajor axis equals the mean of these two distances, which is 1.50×10^{11} m (93 million miles) for Earth's orbit. This mean distance defines the astronomical unit (AU):

$$1\text{ AU} = 1.50 \times 10^{11}\text{ m} = 93.0 \times 10^{6}\text{ mi} \qquad 11\text{-}1$$

The AU is used frequently in problems dealing with the solar system.

Table 11-1 Mean Orbital Radii and Orbital Periods for the Planets

Planet	Mean Radius r ($\times 10^{10}$ m)	Period T (y)
Mercury	5.79	0.241
Venus	10.8	0.615
Earth	15.0	1.00
Mars	22.8	1.88
Jupiter	77.8	11.9
Saturn	143	29.5
Uranus	287	84
Neptune	450	165
Pluto	590	248

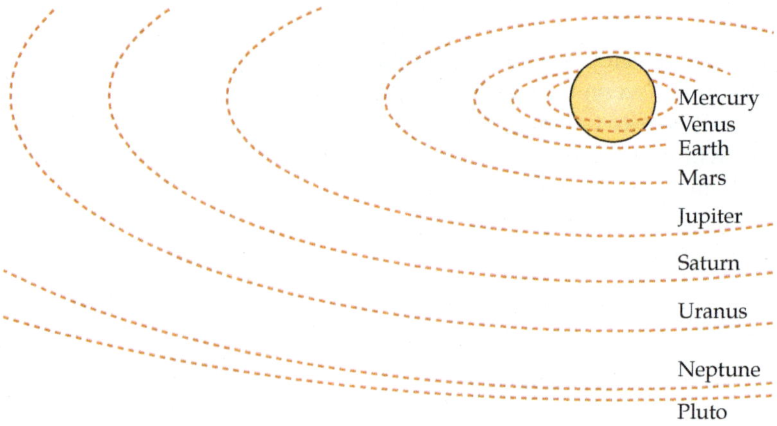

FIGURE 11-1 Orbits of the planets around the Sun. (The sizes are not to scale.) In 2006, the International Astronomical Union passed a new definition of planet that excludes Pluto and puts it in a new category of "dwarf planet."

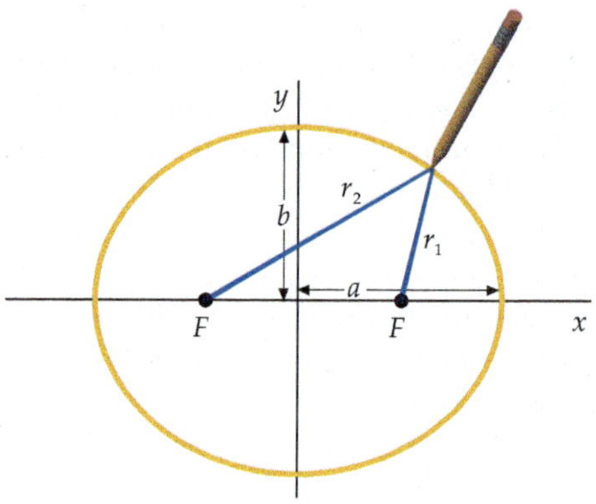

FIGURE 11-2 An ellipse is the locus of points for which $r_1 + r_2 =$ constant. The distance a is called the semimajor axis, and b is the semiminor axis. You can draw an ellipse with a piece of string by fixing each end at a focus F and using it to guide the pencil. Circles are special cases in which the two foci coincide.

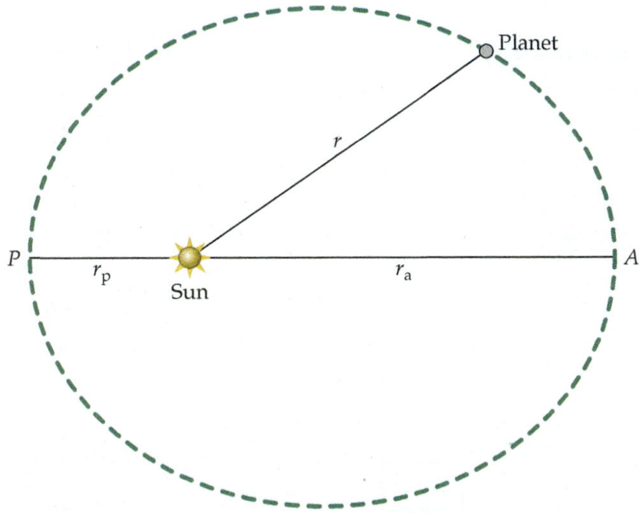

FIGURE 11-3 The elliptical path of a planet with the Sun at one focus. Point P, where the planet is closest to the Sun, is called the *perihelion*, and point A, where it is farthest from the Sun, is called the *aphelion*. The average distance between the planet and the Sun, defined as $(r_p + r_a)/2$, is equal to the semimajor axis. (The known planets travel along more circular paths than the orbit shown here.)

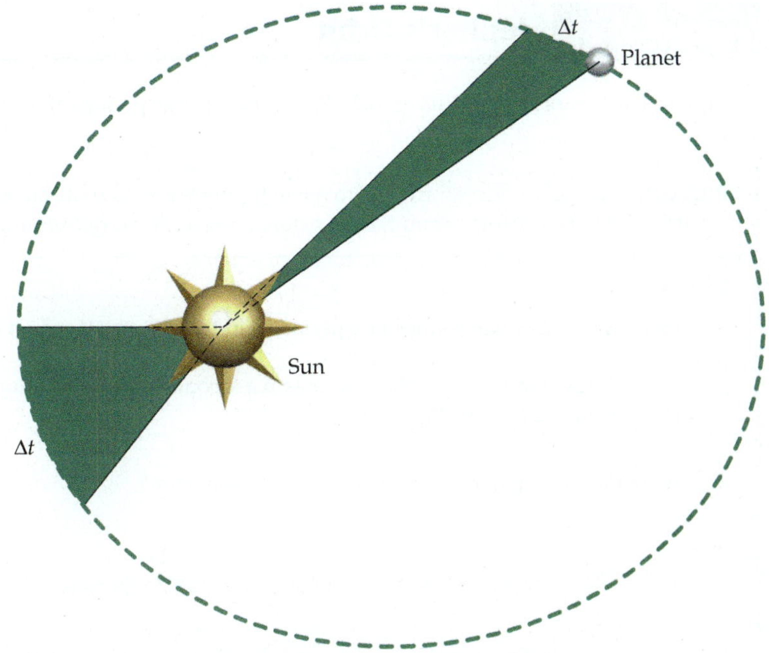

FIGURE 11-4 When a planet is close to the Sun, it moves faster than when it is farther away. The areas swept out by the line joining the centers of the Sun and the planet during a given time interval are equal.

> Law 2. A line joining any planet to the Sun sweeps out equal areas in equal times.

Figure 11-4 illustrates Kepler's second law, the law of equal areas. A planet moves so that the area swept out by the line joining the centers of the Sun and the planet during a given time interval is the same throughout the orbit. The law of equal areas is a consequence of the conservation of angular momentum, as we will see in the next section.

> Law 3. The square of the period of any planet is proportional to the cube of the semimajor axis of its orbit.

Kepler's third law relates the period of any planet to its mean distance from the Sun, which equals the semimajor axis of its elliptical path. In algebraic form, if r is the mean orbital radius* and T is the period of revolution, Kepler's third law states that

$$T^2 = Cr^3 \qquad 11\text{-}2$$

where the constant C has the same value for all the planets. This law is a consequence of the fact that the force exerted by the Sun on a planet varies inversely with the square of the distance from the Sun to the planet. We will demonstrate this in Section 11-2 for the special case of a circular orbit.

* By *mean orbital radius* we refer to the average of the perihelion and aphelion distances.

Example 11-1 Jupiter's Orbit

Jupiter's mean orbital radius is 5.20 AU. What is the period of Jupiter's orbit around the Sun?

PICTURE We use Kepler's third law to relate Jupiter's period to its mean orbital radius. The constant C can be obtained from Earth's known mean orbital radius and period.

SOLVE

1. Kepler's third law relates Jupiter's period T_J and mean orbital radius r_J:
$$T_J^2 = Cr_J^3$$

2. Apply Kepler's third law to Earth to obtain a second equation relating the same constant C to T_E and r_E:
$$T_E^2 = Cr_E^3$$

3. Divide the two equations, eliminating C, and solve for T_J:
$$\frac{T_J^2}{T_E^2} = \frac{r_J^3}{r_E^3}$$
so
$$T_J = T_E\left(\frac{r}{r_E}\right)^{3/2} = (1\text{ y})\left(\frac{5.20\text{ AU}}{1\text{ AU}}\right)^{3/2}$$
$$= \boxed{11.9\text{ y}}$$

CHECK The step-3 result agrees with the orbital period of Jupiter listed in Table 11-1.

TAKING IT FURTHER The periods of the planets Earth, Jupiter, Saturn, Uranus, and Neptune are plotted in Figure 11-5 as functions of their mean orbital radii. In Figure 11-5*a*, periods are plotted versus mean orbital radii. In Figure 11-5*b*, the squares of the periods are plotted versus the cubes of the mean orbital radii. Here, the points fall on a straight line.

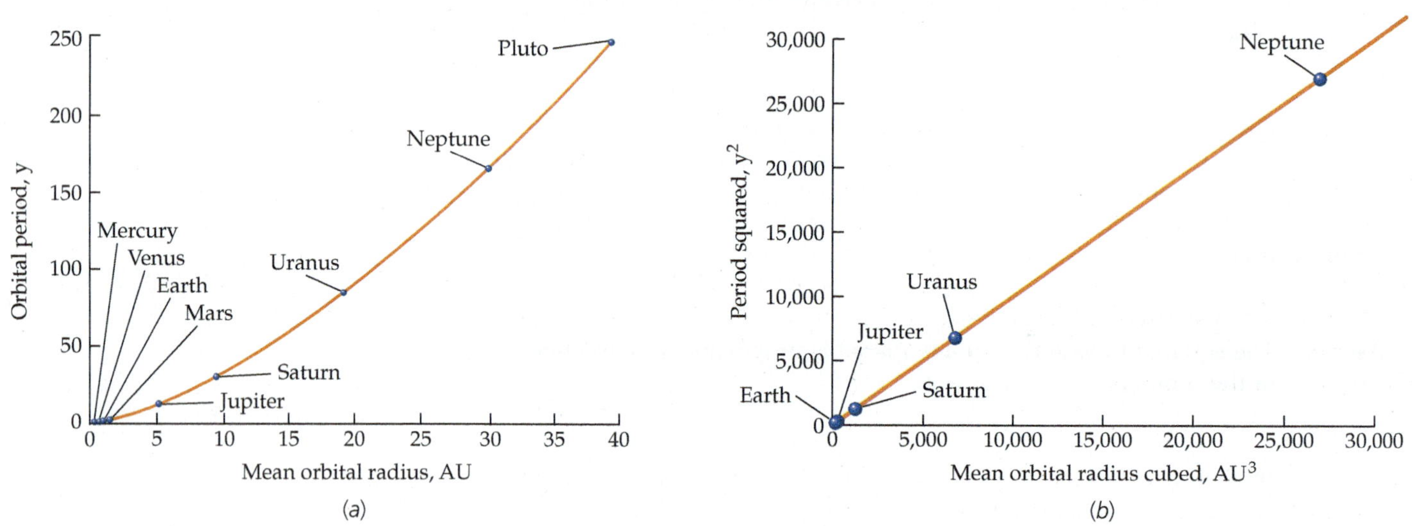

FIGURE 11-5

PRACTICE PROBLEM 11-1 The period of Neptune is 164.8 y. What is its mean orbital radius?

PRACTICE PROBLEM 11-2 If the logarithms of the periods of the planets Earth, Jupiter, Saturn, Uranus, and Neptune are plotted versus the logarithms of their mean orbital radii, the points fall on a curve. What is the shape of this curve?

11-2 NEWTON'S LAW OF GRAVITY

Although Kepler's laws were an important first step in understanding the motion of planets, they were nothing more than empirical rules obtained from the astronomical observations of Brahe. It remained for Newton to take the next giant step by attributing the acceleration of a planet in its orbit to a specific force exerted on it by the Sun. Using his second law, Newton proved that an attractive force that varies inversely with the square of the distance between the Sun and a planet would result in an elliptical orbit, as observed by Kepler. He then made the bold assumption that this attractive force acts between any two objects in the universe. Before Newton, it was not widely believed that the laws of physics observed on Earth were applicable to the heavenly bodies. Newton changed our understanding of the nature of the nonterrestrial realm by showing that the laws of physics apply equally well to both terrestrial and nonterrestrial objects. **Newton's law of gravity** postulates that there is a force of attraction between each pair of point particles that is proportional to the product of the masses of the particles and inversely proportional to the square of the distance separating them. Let m_1 and m_2 be the masses of point particles 1 and 2 (at positions $\vec{r}_1$ and $\vec{r}_2$, respectively) and $\vec{r}_{12}$ be the position of particle 2 relative to particle 1 (Figure 11-6*a*).

The gravitational force $\vec{F}_{12}$ exerted by particle 1 on particle 2 is then

$$\vec{F}_{12} = -\frac{Gm_1m_2}{r_{12}^2}\hat{r}_{12} \qquad 11\text{-}3$$

NEWTON'S LAW OF GRAVITY

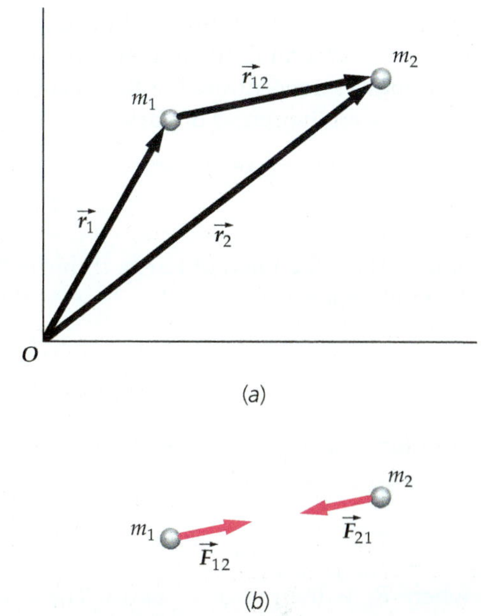

FIGURE 11-6 (*a*) Particles at $\vec{r}_1$ and $\vec{r}_2$. (*b*) The particles exert equal and opposite forces on each other.

where $\hat{r}_{12} = \vec{r}_{12}/r_{12}$ is a unit vector in the direction from 1 toward 2 and G is the **universal gravitational constant,** which has the value

$$G = 6.67 \times 10^{-11}\ \text{N}\cdot\text{m}^2/\text{kg}^2 \qquad 11\text{-}4$$

The force $\vec{F}_{21}$ exerted by 2 on 1 is equal and opposite to $\vec{F}_{12}$, in accord with Newton's third law (Figure 11-6*b*). The magnitude of the gravitational force exerted by a point particle of mass m_1 on a point particle of mass m_2 a distance r away is thus given by

$$F_g = \frac{Gm_1m_2}{r^2} \qquad 11\text{-}5$$

Newton published his theory of gravitation in 1686, but it was not until a century later that an accurate experimental determination of G was made by Henry Cavendish, as is discussed in Section 11-4.

We can use the known value of G to compute the gravitational attraction between two ordinary objects.

PRACTICE PROBLEM 11-3

Show that the gravitational force that attracts a 65-kg man to a 50-kg woman when they are 0.50 m apart is 8.7×10^{-7} N. (Model them as point particles for the purpose of this calculation.)

This calculation demonstrates that the gravitational force exerted by an object of ordinary size on another such object is so small as to be unnoticeable. For comparison, a mosquito weighs about 1×10^{-7} N so the force of attraction is equal to the weight of 9 mosquitos. The weight of a 50-kg woman is about 490 N—*half a billion* times greater than the force of attraction calculated in Practice Problem 11-3! Gravitational attraction is easily noticed only when at least one of the objects is astronomically massive. The gravitational attraction between the woman and Earth for example, is readily apparent.

To check the validity of the inverse-square nature of the gravitational force, Newton compared the acceleration of the moon in its orbit with the free-fall acceleration of objects near the surface of Earth (such as the legendary apple). He reasoned that the gravitational attraction due to Earth causes both accelerations. He first assumed that Earth and the moon could be treated as point particles with their total masses concentrated at their centers. The force on a particle of mass m a distance r from the center of Earth is

$$F_g = \frac{GM_E m}{r^2} \tag{11-6}$$

where M_E is the mass of Earth. If this is the only force acting on the particle, then its acceleration is

$$a = \frac{F_g}{m} = \frac{GM_E}{r^2} \tag{11-7}$$

For objects near the surface of Earth, $r \approx R_E$ and the free-fall acceleration is g:

$$g = \frac{GM_E}{R_E^2} \tag{11-8}$$

where R_E is the radius of Earth. The distance to the moon is about 60 times the radius of Earth ($r = 60R_E$). Substituting this into Equation 11-7 gives $a = g/60^2 = g/3600$, so the acceleration of the moon in its near-circular orbit is the free-fall acceleration g at the surface of Earth divided by 60^2. That is, the acceleration of the moon a_m should be $(9.81/3600)\ \text{m/s}^2$. The moon's acceleration can be calculated from its known distance from the center of Earth, $r = 3.84 \times 10^8$ m, and its known period $T = 27.3\ \text{d} = 2.36 \times 10^6$ s:

$$a_m = \frac{v^2}{r} = \frac{(2\pi r/T)^2}{r} = \frac{4\pi^2 r}{T^2} = \frac{4\pi^2(3.84 \times 10^8\ \text{m})}{(2.36 \times 10^6\ \text{s})^2} = 2.72 \times 10^{-3}\ \text{m/s}^2$$

Then

$$\frac{g}{a_m} = \frac{9.81\ \text{m/s}^2}{2.72 \times 10^{-3}\ \text{m/s}^2} = 3607 \approx 3600$$

In Newton's words, "I thereby compared the force requisite to keep the Moon in her orb with the force of gravity at the surface of the Earth, and found them answer pretty nearly."

The assumption that Earth and the moon can be treated as point particles in the calculation of the force on the moon is reasonable because the Earth-to-moon distance is large compared with the radius of either Earth or the moon, but such an assumption is certainly questionable when applied to an object near Earth's surface. After considerable effort, Newton was able to mathematically demonstrate that the force exerted by any object with a spherically symmetric mass distribution on a point mass either on or outside its surface is the same as if all the mass of the object were concentrated at its center. (This calculation is the subject of Section 11-5.) The proof involves integral calculus, which Newton developed to solve this problem.

Because $g = 9.81\ \text{m/s}^2$ is readily measured and the radius of Earth is known, Equation 11-8 can be used to determine the value of the product GM_E. Newton estimated the value of G from an estimation of the average density of Earth. When Cavendish determined G to within one percent some 100 years later by measuring the force between small spheres of known mass and separation, he called his experiment "weighing Earth." Knowing the value of G meant that the mass of the Sun and the mass of any planet with a satellite could be determined. The method for doing this is described in Section 11-4.

Earth as seen from *Apollo 11* orbiting the Moon on July 16, 1969. *(NASA.)*

Example 11-2 Falling to Earth

What is the free-fall acceleration of an object at the altitude of the space shuttle's orbit, about 400 km above Earth's surface?

(NASA.)

PICTURE The only force acting on an object in freefall is the force of gravity.

SOLVE

1. The free-fall acceleration is given by $a = F_g/m$: $\quad a = \dfrac{F_g}{m} = \dfrac{GmM_E/r^2}{m} = \dfrac{GM_E}{r^2}$

2. The distance r is related to the radius of Earth R_E and the altitude h: $\quad r = R_E + h = 6370\text{ km} + 400\text{ km} = 6770\text{ km}$

3. The acceleration is then:

$$a = \frac{GM_E}{r^2} = \frac{(6.67 \times 10^{-11}\text{ N}\cdot\text{m}^2/\text{kg}^2)(5.98 \times 10^{24}\text{ kg})}{(6.77 \times 10^6\text{ m})^2} = \boxed{8.70\text{ m/s}^2}$$

CHECK The altitude of 400 km is 6% of the radius of Earth (6370 km), and the free-fall acceleration of 8.70 m/s^2 is 11% less than 9.81 m/s^2. A free-fall acceleration that is only 11% less than 9.81 m/s^2 is plausible because the altitude is only 6% of Earth's radius.

TAKING IT FURTHER The acceleration of both the shuttle and the shuttle astronauts as they accelerate in their near circular orbit is 8.70 m/s^2.

CONCEPT CHECK 11-1

How is it that the astronauts in the orbiting shuttle are said to be weightless, even though the force of gravity on them is only 11 percent less than it is on Earth's surface?

The calculation in Example 11-2 can be simplified by using Equation 11-8 to eliminate GM_E from Equation 11-7. Then the acceleration at a distance r is

$$a = \frac{F_g}{m} = \frac{GM_E}{r^2} = g\frac{R_E^2}{r^2} \qquad 11\text{-}9$$

PRACTICE PROBLEM 11-4

At what distance h above the surface of Earth is the free-fall acceleration half its value at sea level?

MEASUREMENT OF G

The universal gravitational constant G was first measured in 1798 by Henry Cavendish, who used the apparatus shown in Figure 11-7. Cavendish's measurement of G has been repeated by other experimenters with various improvements and refinements. All measurements of G are difficult because of the extreme

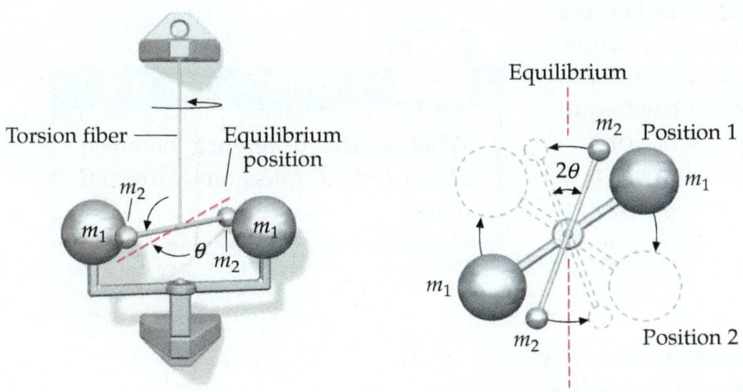

FIGURE 11-7 (*a*) Two small spheres, each of mass m_2, are at the ends of a light rod that is suspended by a fine fiber. Careful measurements determine the torque required to turn the fiber through a given angle. Two large spheres, each of mass m_1, are then placed near the small spheres. Because of the gravitational attraction of the large spheres of mass m_1 for the small spheres, the fiber is twisted through a very small angle θ from its equilibrium position. (*b*) The apparatus as seen from above. After the apparatus comes to rest, the positions of the large spheres are reversed, as shown by the dashed lines, so that they are at the same distance from the equilibrium position of the balance, but on the other side. If the apparatus is again allowed to come to rest, the fiber will turn through angle 2θ in response to the reversal of the torque. Once the torsion constant has been determined, the forces between the masses m_1 and m_2 can be determined from the measurement of this angle. Because the masses and their separations are known, G can be calculated. Cavendish obtained a value for G within about 1 percent of the currently accepted value given by Equation 11-4.

weakness of the gravitational attraction. Consequently, the value of G is known today only to about 1 part in 10,000. Although G was one of the first physical constants ever measured, it remains one of the least accurately known.

GRAVITATIONAL AND INERTIAL MASS

The property of an object that is responsible for the gravitational force the object exerts on another object, or for the gravitational force another object exerts on it, is its *gravitational* mass. On the other hand, *inertial* mass is the property of an object that measures the object's resistance to acceleration. We have used the same symbol m for these two properties because, experimentally, they are proportional. For convenience, units are judiciously defined to make the proportionality constant one. The fact that the gravitational force exerted on an object is proportional to its inertial mass is a characteristic unique to the force of gravity. One consequence is that all objects near the surface of Earth fall with the same acceleration if air resistance is neglected. The well-known story of Galileo dropping objects from the Leaning Tower of Pisa to demonstrate that the free-fall acceleration is the same for objects with different inertial masses is just one example of the excitement this discovery aroused in the sixteenth century.

We could easily imagine that the gravitational and inertial masses of an object were not the same. Suppose we write m_G for the gravitational mass and m for the inertial mass. The force exerted by Earth on an object near its surface would then be

$$F_g = \frac{GM_E m_G}{R_E^2} \qquad \text{11-10}$$

where M_E is the gravitational mass of Earth. The free-fall acceleration of the object near Earth's surface would then be

$$a = \frac{F_g}{m} = \left(\frac{GM_E}{R_E^2}\right)\frac{m_G}{m} \qquad \text{11-11}$$

If gravity were just another property of matter, like porosity or hardness, it might be reasonable to expect that the ratio m_G/m would depend on such things as the chemical composition of the object or its temperature. The free-fall acceleration would then be different for different objects. The experimental evidence, however, is that a is the same for all objects. Thus, we need not maintain the distinction between m_G and m, and can set $m_G = m$. We must keep in mind, however, that the equivalence of gravitational and inertial mass is an empirical law that is limited by the accuracy of experiment. Experiments testing this equivalence were carried out by Simon Stevin in the 1580s. Galileo publicized this law widely, and his contemporaries made considerable improvements in the experimental accuracy with which the law was established.

The most precise early comparisons of gravitational and inertial mass were made by Newton. Through experiments using simple pendulums rather than falling bodies, Newton was able to establish the equivalence between gravitational and inertial mass to an accuracy of about 1 part in 1000. Experiments comparing gravitational and inertial mass have improved steadily over the years. Their equivalence is now established to about 1 part in 5×10^{13}. Thus, the equivalence of gravitational and inertial mass is one of the best established of all physical laws. It is the basis for the principle of equivalence, which is the foundation of Einstein's general theory of relativity.

Gravitational torsional balance used in student labs for the measurement of G. A tiny angular deflection of the balance results in a large angular deflection of the laser beam that reflects from a mirror on the balance. *(Courtesy of Central Scientific Company.)*

! Free-fall acceleration is the same for all objects.

CONCEPT CHECK 11-2

What is the difference between gravitational mass and inertial mass?

DERIVATION OF KEPLER'S LAWS

Newton used his second law of motion to show that a particle moving under the influence of an attractive force that varies inversely with the square of the distance from a fixed point moves along a path the shape of a conic section (an ellipse, a parabola, or a hyperbola) with a focus located at the fixed point. He inferred from this result and Kepler's laws that the planets (and comets) are attracted to the center of the Sun by a force that varies inversely with the square of their distances from the center of the Sun. The parabolic and hyperbolic paths apply to objects that make one pass by the Sun and never return. Such orbits are not closed. The only closed orbits are those of objects that follow elliptical paths. Thus, Kepler's first law is a direct consequence of Newton's law of gravity. Kepler's second law, the law of equal areas, follows from the fact that the force exerted by the Sun on a planet is directed toward a force center—the center of the Sun. Such a force is called a **central force.** Figure 11-8*a* shows a planet moving in an elliptical orbit around the Sun. In time dt, the planet moves a distance $v\,dt$ and the radius vector $\vec{r}$ sweeps out the area shaded in the figure. This is half the area of the parallelogram formed by the vectors $\vec{r}$ and $\vec{v}\,dt$, which is $|\vec{r} \times \vec{v}\,dt|$. Thus, the area dA swept out by the radius vector $\vec{r}$ in time dt is given by

$$dA = \frac{1}{2}|\vec{r} \times \vec{v}\,dt| = \frac{|\vec{r} \times m\vec{v}|}{2m}\,dt$$

or

$$\frac{dA}{dt} = \frac{L}{2m} \qquad \text{11-12}$$

where $L = |\vec{r} \times m\vec{v}|$ is the magnitude of the orbital angular momentum of the planet about the Sun. The area dA swept out in a given time interval dt is therefore proportional to the magnitude of the orbital angular momentum L. Because the force on a planet is along the line from the planet to the Sun, it exerts no torque about the Sun. Thus, the orbital angular momentum of the planet is constant; that is, L is constant. Therefore, the rate at which the area is swept out is the same for all parts of the orbit, which is Kepler's second law. Also, the fact that L is constant means that $rv \sin \phi$ is constant. At aphelion and perihelion $\phi = 90°$ (Figure 11-8*b*), so $r_a v_a = r_p v_p$.

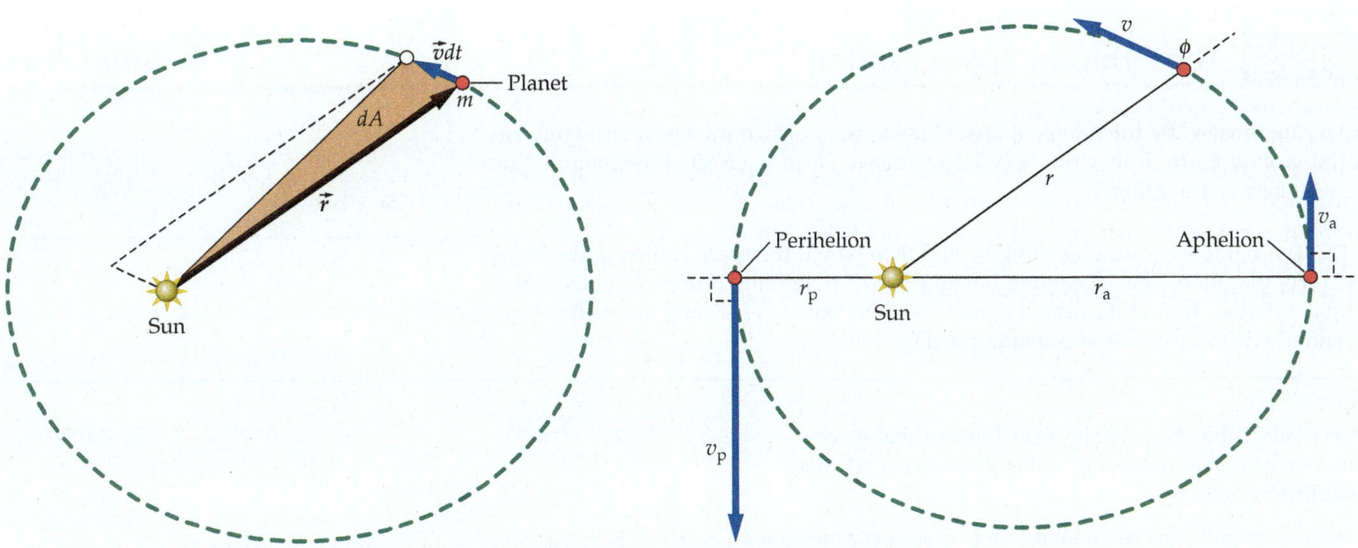

FIGURE 11-8 (*a*) The area dA swept out in time dt equals $\frac{1}{2}|\vec{r} \times \vec{v}\,dt| = \frac{1}{2m}L\,dt$, where $\vec{L} = \vec{r} \times m\vec{v}$. Because $\vec{L}$ remains constant, dA/dt remains constant. (*b*) The magnitude of the angular momentum, given by $L = mvr \sin \phi$, remains constant, so $rv \sin \theta$ remains constant. In addition, $\phi = 90°$ at both perihelion and aphelion, so $r_a v_a = r_p v_p$.

We now show that Newton's law of gravity implies Kepler's third law for the special case of a circular orbit. Consider a planet moving with speed v in a circular orbit of radius r about the Sun. The gravitational force on the planet by the Sun provides the centripetal acceleration v^2/r. Applying Newton's second law ($F = ma$) to the planet gives

$$\frac{GM_S M_P}{r^2} = M_P \frac{v^2}{r} \tag{11-13}$$

where M_S is the mass of the Sun and M_P is that of the planet. Solving for v^2 gives

$$v^2 = \frac{GM_S}{r} \tag{11-14}$$

Because the planet moves a distance $2\pi r$ in time T, its speed is related to the period by

$$v = \frac{2\pi r}{T} \tag{11-15}$$

Substituting $2\pi r/T$ for v in Equation 11-14, we obtain

$$\frac{4\pi^2 r^2}{T^2} = \frac{GM_S}{r}$$

or

$$T^2 = \frac{4\pi^2}{GM_S} r^3 \tag{11-16}$$

KEPLER'S THIRD LAW

Equation 11-16 is a version of Kepler's third law. Equation 11-16 is the same as Equation 11-2 with $C = 4\pi^2/GM_S$. Equation 11-16 also applies to the orbits of the satellites of any planet if we replace the mass of the Sun M_S with the mass of the planet.

(NASA.)

Example 11-3 The Orbiting Space Station

Context-Rich

You are trying to view the International Space Station (ISS), which travels in a roughly circular orbit around Earth. If its altitude is 385 km above Earth's surface, how long do you have to wait between sightings?

PICTURE The sightings occur only at night, and then only if the space station is above the horizon at your location. Thus, the minimum time between sightings is approximately equal to the orbital period. To find the orbital period we apply Newton's second law to the space station and use distance equals speed multiplied by time.

SOLVE

1. For a circular orbit, the orbital period T and orbital speed v can be related to the orbital radius r by using distance equals speed multiplied by time.
   $$2\pi r = vT$$
2. To obtain a second equation relating v and r we apply Newton's second law to the space station of mass m:
   $$F_g = ma$$
   $$\frac{GM_E m}{r^2} = m\frac{v^2}{r}$$
3. Substituting $2\pi r/T$ for v (from our step-1 result) gives:
   $$\frac{GM_E}{r^2} = \frac{4\pi^2 r}{T^2}$$

4. Solving for T^2, we obtain:

$$T^2 = \frac{4\pi^2}{GM_E} r^3$$

5. At an altitude $h = 385$ km, $r = R_E + h = 6760$ km. Substitute $r = R_E + h$ and solve for the period:

$$T^2 = \frac{4\pi^2}{GM_E}(R_E + h)^3$$

$$= \frac{4\pi^2}{(6.67 \times 10^{-11}\,\text{N}\cdot\text{m}^2/\text{kg}^2)(5.98 \times 10^{24}\,\text{kg})}(6.76 \times 10^6\,\text{m})^3$$

$$= 30.56 \times 10^6\,\text{s}^2$$

so $T = 5528\,\text{s} = \boxed{92.1\,\text{min}}$

CHECK An Internet search for "NASA orbital period ISS" tells us that the orbital period is 91.5 min, so our step-2 result is in the right ballpark. In addition, our step-3 result is Kepler's third law (Equation 11-16) for a satellite orbiting Earth.

TAKING IT FURTHER Kansas City is due west of New York City by 23 degrees of longitude. The plane of the near-circular orbit of the ISS, which is inclined ≈52° to the equatorial plane, does not rotate with Earth. If the ISS is directly over your home at time t, 92.1 min later it will be directly over a location 23.0° due west of your home. If the ISS passes over your home in New York City at midnight Eastern Time, you could tell your friend in Kansas City that it will pass over Kansas City at 12:32 a.m. Central Time (1:32 a.m. Eastern Time).

PRACTICE PROBLEM 11-5 How many degrees does Earth rotate in one hour? *Hint: How many degrees does Earth rotate in 24 h?*

PRACTICE PROBLEM 11-6 Find the radius of the circular orbit of a satellite that orbits Earth with a period of 1.00 d.

Because G is known, we can determine the mass of an astronomical object by measuring the orbital period T and the mean orbital radius r of a satellite orbiting it and by substituting these values into Equation 11-16. In establishing Equation 11-16, the mass of the satellite was assumed negligible compared to the mass of the central object. This means that the central object remains stationary as the satellite revolves around it. In fact, the central object and satellite both revolve around a common point, their center of mass. If the mass of the satellite is not assumed negligible, the result is

$$T^2 = \frac{4\pi^2}{G(M_1 + M_2)} r^3 \qquad \text{11-17}$$

where r is the center-to-center separation of the objects. (The derivation of Equation 11-17 for circular orbits is left to Problem 11-93. For the more general elliptical orbits, the math is more challenging, but the result is the same, only r is replaced by the mean of the maximum and minimum center-to-center distances between the objects.) If the mass of the satellite is not negligible, as is the case with most binary star systems, then only the sum of the masses is determined, as revealed by Equation 11-17. The moon, along with planets Mercury and Venus, have no natural satellites, so their masses were not well known until the 1960s when artificial satellites were first placed in orbit around them.

PRACTICE PROBLEM 11-7

The Martian moon Phobos has a period of 460 min and a mean orbital radius of 9400 km. What is the mass of Mars?

CONCEPT CHECK 11-3

Kepler's first law is that the planets all follow elliptical paths with the center of the Sun and a focus of each ellipse. Newton inferred from Kepler's first law that the planets are all attracted to the center of the Sun by a force that varies inversely with the square of the distance. What led him to this inference?

11-3 GRAVITATIONAL POTENTIAL ENERGY

Near the surface of Earth, the gravitational force exerted by Earth on an object is essentially uniform because the distance to the center of Earth, $r = R_E + h$, is always approximately R_E for $h \ll R_E$. The potential energy of an object near Earth's surface is $mgh = mg(r - R_E)$, where we have chosen $U = 0$ at Earth's surface, $r = R_E$. When we are far from the surface of Earth, we must take into account the fact that the gravitational force exerted by Earth is not uniform, but decreases as $1/r^2$. The general definition of potential energy (Equation 7-1) is

$$dU = -\vec{F}\cdot d\vec{\ell}$$

where $\vec{F}$ is a conservative force on a particle and $d\vec{\ell}$ is a general displacement of the particle. For the gravitational force $\vec{F}_g$ given by Equation 11-6 (Figure 11-9), we have

$$dU = -\vec{F}_g\cdot d\vec{\ell} = -(-F_g\hat{r})\cdot d\vec{\ell} = F_g\hat{r}\cdot d\vec{\ell} = \frac{GM_E m}{r^2}\,dr \qquad \text{11-18}$$

where we have used $\vec{F}_g = -F_g\hat{r}$ and $\hat{r}\cdot d\vec{\ell} = d\ell\cos\phi = dr$. Integrating both sides of Equation 11-18 we obtain

$$U = GM_E m\int r^{-2}\,dr = -\frac{GM_E m}{r} + U_0 \qquad \text{11-19}$$

where U_0 is a constant of integration. The expression for U is algebraically simplest if we choose $U_0 = 0$. Then

$$U(r) = -\frac{GMm}{r} \qquad \text{11-20}$$

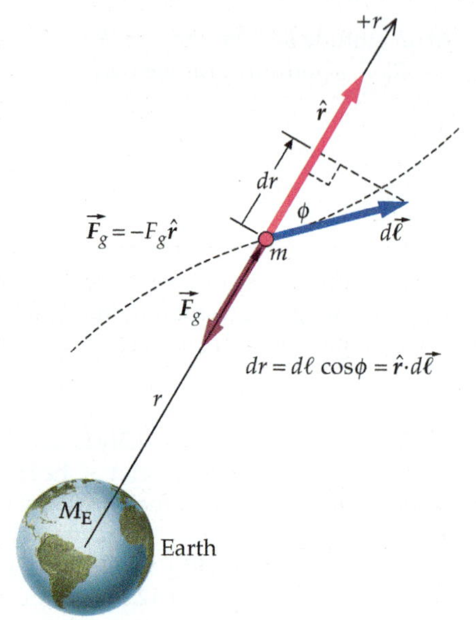

FIGURE 11-9 The distance r of the particle from Earth increases by dr when the particle undergoes displacement $d\vec{\ell}$. In the figure the length of $d\vec{\ell}$ has been exaggerated.

Thus, a choice of zero for U_0 means that U approaches zero as r approaches infinite. At first this may seem like a strange choice, because for finite values of r all values of U are negative. This just means, however, that the potential energy is maximum when Earth and the particle are at infinite separation. Negative potential energy is nothing new. When we use the potential-energy function $U = mgh$, where h is the height above some reference point on a tabletop, the potential energy is negative any time the particle of mass m is anywhere below the level of the tabletop. This reflects the fact that when the particle is below the level of the tabletop the potential energy is less than it is when the particle is at the level of the tabletop.

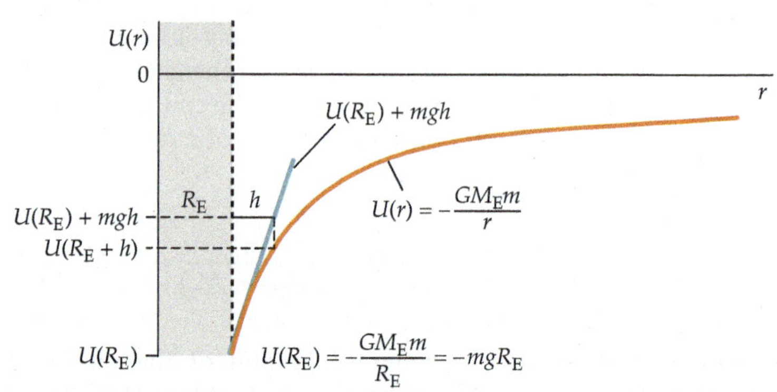

FIGURE 11-10

Figure 11-10 is a plot of $U(r)$ versus r for $U(r) = -GM_E m/r$ for $R_E \le r < \infty$. This plot begins at the negative value $U = -GM_E m/R_E$ at Earth's surface and increases as r increases, approaching zero as r approaches infinity. The slope of this curve at $r = R_E$ is $GM_E m/R_E^2 = mg$. (Recall that $g = GM_E m/R_E^2$.) The equation of the tangent line, drawn in blue, is $f(h) = U(R_E) + mgh$, where $h = r - R_E$ is the distance above Earth's surface. From the figure, you can see that for small h, $U(R_E) + mgh \approx U(r)$.

ESCAPE SPEED

Since the mid-1950s, the idea of escaping from Earth's gravity has changed from fantasy to reality. Space probes have been sent out to the far reaches of the solar system. Many of these probes orbit the Sun, while a few leave the solar system and drift on into outer space. We will see that a minimum initial speed, called the **escape speed**, is required for a projectile to escape from Earth.

If we project an object upward from the surface of Earth with some initial kinetic energy, the kinetic energy decreases and the potential energy increases as the object rises. But the maximum increase in potential energy is $GM_E m/R_E$. Therefore, this amount is the most that the kinetic energy can decrease. If the initial kinetic energy is greater than $GM_E m/R_E$, then the total energy E will be greater than zero (E_2 in Figure 11-11), and the object will still have some kinetic energy when r is very large (even if r approaches infinity). Thus, if the initial kinetic energy is greater than $GM_E m/R_E$, the object is said to escape from Earth. Because the potential energy at Earth's surface is $-GM_E m/R_E$, the total energy $E = K + U$ must be greater than or equal to zero in order for the object to escape Earth. The speed near Earth's surface corresponding to zero total energy is called the escape speed v_e. It is found from

$$K_f + U_f = K_i + U_i$$

$$0 + 0 = \frac{1}{2}mv_e^2 - \frac{GM_E m}{R_E}$$

so

$$v_e = \sqrt{\frac{2GM_E}{R_E}} \qquad \text{11-21}$$

ESCAPE SPEED

(NASA.)

Using $G = 6.67 \times 10^{-11}\,\text{N}\cdot\text{m}^2/\text{kg}^2$, $M_E = 5.98 \times 10^{24}$ kg, and $R_E = 6.37 \times 10^6$ m, we obtain

$$v_e = \sqrt{2(9.81\text{ m/s}^2)(6.37 \times 10^6\text{ m})} = 11.2\text{ km/s}$$

This speed is about 7 mi/s or 25,000 mi/h. An object that has this speed will escape Earth, but it will not escape the solar system because we have neglected the gravitational attraction of the Sun and other planets (see Problem 50).

The escape speed for a planet or moon relative to the thermal speeds of gas molecules determines the kind of atmosphere a planet or moon can have. The average kinetic energy of gas molecules, $(\frac{1}{2}mv^2)_{av}$, is proportional to the absolute temperature T (Chapter 18). Near the surface of Earth, the speeds of nearly all of the oxygen and nitrogen molecules are much much lower than the escape speed, so these gases are retained in our atmosphere. For the lighter molecules hydrogen and helium, however, a significant fraction of them have speeds greater than the escape speed. Hydrogen and helium gases are therefore not found in our atmosphere. The escape speed at the surface of the moon is 2.3 km/s, which can be calculated from Equation 11-21 using the mass and radius of the moon instead of M_E and R_E. This speed is considerably smaller than the escape speed for Earth and, in fact, is too small for any atmosphere to exist.

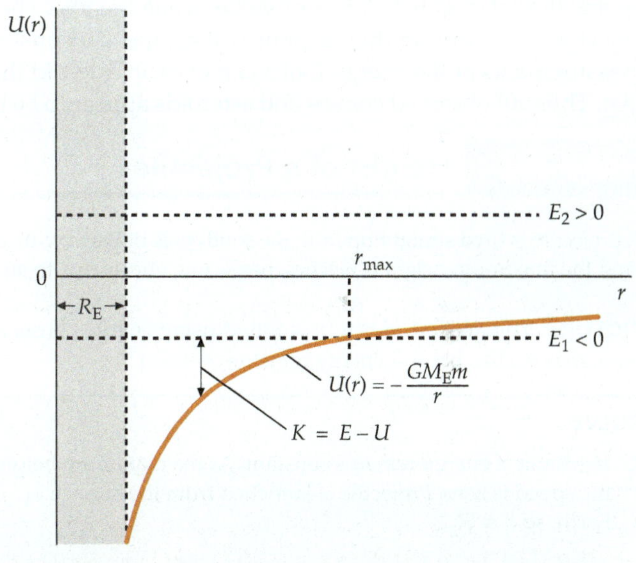

FIGURE 11-11 The kinetic energy of an object at a distance r from the center of Earth is $E - U(r)$. When the total energy is less than zero (E_1 in the figure), the kinetic energy K is zero at $r = r_{max}$ and the object is bound to Earth. When the total energy is greater than zero (E_2 in the figure), the object can escape Earth.

PRACTICE PROBLEM 11-8

Find the escape speed at the surface of Mercury, which has a mass $M = 3.31 \times 10^{23}$ kg and a radius $R = 2440$ km.

CLASSIFICATION OF ORBITS BY ENERGY

In Figure 11-11, two possible values for the total energy E are indicated on a graph of $U(r)$ versus r: E_1, which is negative, and E_2, which is positive. A negative total energy simply means that the kinetic energy at Earth's surface is less than $GM_E m/R_E$, so that $K + U$ is never greater than zero. From this figure we see that, if the total energy is negative, the total-energy line intersects the potential-energy curve at some maximum separation r_{max} and the system is bound. On the other hand, if the total energy is zero or positive, there is no such intersection and the system is unbound. The criteria for a bound or unbound system are simply stated:

If $E < 0$, the system is bound.
If $E \geq 0$, the system is unbound.

When E is negative, its absolute value $|E|$ is called the *binding energy*. The binding energy is the energy that must be added to the system to bring the total energy up to zero.

The potential energy of an object such as a planet or comet of mass m at a distance r from the Sun is

$$U(r) = -\frac{GM_S m}{r} \qquad \text{11-22}$$

where M_S is the mass of the Sun. The kinetic energy of the object is $\frac{1}{2}mv^2$. If the total energy, kinetic plus potential, is less than zero, then the orbit will be an ellipse (possibly a circle), and the object will be bound to the Sun. If, instead, the total energy is positive, then the orbit will be a hyperbola, and the object will make one trip around the Sun and leave the solar system, never to return. If the total energy is exactly zero, the orbit will be a parabola, and again the object will make one trip and then escape the solar system. To summarize, when the total energy is zero or positive the object is not bound to the Sun, but will escape. Curiously, there have not been any measurements of the energy E of a comet or an asteroid that are definitely nonnegative. Thus, all observed comets and asteroids appear to be bound to the solar system.

Example 11-4 Height of a Projectile

A projectile is fired straight up from the south pole of Earth with an initial speed $v_i = 8.0$ km/s. Find the maximum height it reaches, neglecting effects due to air resistance.

PICTURE The maximum height is found using energy conservation. We are neglecting air resistance, so mechanical energy remains constant.

SOLVE

1. Mechanical energy remains constant. At the maximum height the speed is zero. Projectile is launched from the surface of Earth, so $r_i = R_E$.

$$K_f + U_f = K_i + U_i$$

$$\frac{1}{2}mv_f^2 - \frac{GM_E m}{r_f} = \frac{1}{2}mv_i^2 - \frac{GM_E m}{r_i}$$

$$0 - \frac{GM_E m}{r_f} = \frac{1}{2}mv_i^2 - \frac{GM_E m}{R_E}$$

2. Multiply through by $-1/(GM_E m)$ and solve for r_f:

$$\frac{1}{r_f} = -\frac{v_i^2}{2GM_E} + \frac{1}{R_E}$$

$$= \frac{-(8000\ \text{m/s})^2}{2(6.67 \times 10^{-11}\ \text{N}\cdot\text{m}^2/\text{kg}^2)(5.98 \times 10^{24}\ \text{kg})} + \frac{1}{6.37 \times 10^6\ \text{m}}$$

$$= 7.68 \times 10^{-8}\ \text{m}^{-1}$$

so $r_f = 1/(7.68 \times 10^{-8}\ \text{m}^{-1}) = 1.30 \times 10^7\ \text{m}$

3. Solve for h_f, where $h_f = r_f - R_E$:

$$h = r_f - R_E = 1.30 \times 10^7\ \text{m} - 6.37 \times 10^6\ \text{m} = \boxed{6.7 \times 10^6\ \text{m}}$$

CHECK If g remained equal to 9.81 m/s², then the maximum height h could be calculated from $mgh = \frac{1}{2}mv_i^2$. Solving this for h gives $h = v_i^2/(2g) = (8000\ \text{m/s})^2/(19.6\ \text{m/s}^2) = 3.3 \times 10^6$ m. Our step-3 result is greater than this value—as expected.

TAKING IT FURTHER Our step-3 result is 4.5% greater than the radius of Earth.

Example 11-5 Speed of a Projectile

Try It Yourself

A projectile is fired straight up from the south pole of Earth with an initial speed $v_i = 15\,\text{km/s}$. Find the speed of the projectile when it is very far from Earth, neglecting effects due to air resistance.

PICTURE The maximum height is found using energy conservation. We are neglecting effects due to air resistance, so mechanical energy remains constant. The initial speed of 15 m/s is greater than the escape speed of 11.2 km/s, so the total energy of the projectile is positive and the projectile will retain some kinetic energy when it is very far from Earth.

SOLVE

Cover the column to the right and try these on your own before looking at the answers.

Steps	Answers
1. Mechanical energy remains constant. Note that $r_f \rightarrow \infty$, so $U_f \rightarrow 0$.	$\frac{1}{2}mv_f^2 + 0 = \frac{1}{2}mv_i^2 - \frac{GM_E m}{R_E}$
2. Solve for v_f^2.	$v_f^2 = v_i^2 - \frac{2GM_E}{R_E}$
3. Calculate v_f.	$v_f = \boxed{1.0 \times 10^4\,\text{m/s}}$

CHECK The initial speed is just slightly less than $\sqrt{2}$ times the escape speed, so the initial kinetic energy is almost twice that needed to escape with zero final speed. This means the final kinetic energy will be slightly less than that of the projectile if it were moving at the escape speed of 11 km/s. Our step-3 result of 10 km/s is, as expected, slightly less than 11 km/s.

TAKING IT FURTHER In Figure 11-12, the speed of the projectile in kilometers per second is plotted versus h/R_E, where h is the height above Earth's surface. At very large values of h/R_E, the speed of the projectile approaches the horizontal line $v = 10\,\text{km/s}$.

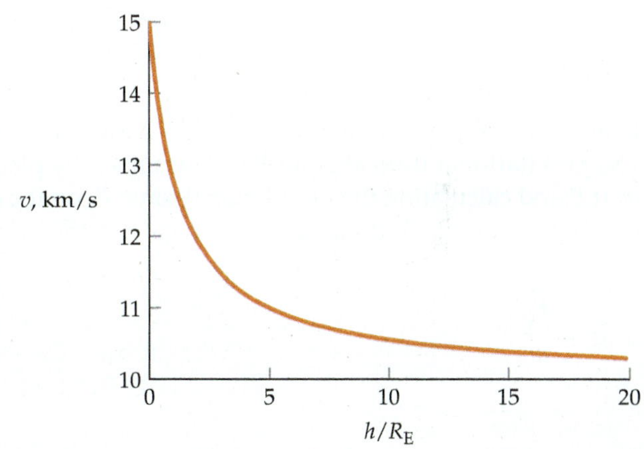

FIGURE 11-12

Example 11-6 Total Energy of a Satellite

Show that the total energy of a satellite in a circular orbit about Earth is equal to half of the potential energy.

PICTURE The total energy of a satellite is the sum of its potential and kinetic energies, $E = U + K$. Newton's second law allows us to relate the speed v of the satellite to its orbital radius r. The kinetic energy depends on the speed, so we can find the kinetic energy in terms of r.

SOLVE

1. Write the total energy equal to the sum of the potential energy and the kinetic energy.

$$E = K + U = \frac{1}{2}mv^2 - \frac{GM_E m}{r}$$

2. Apply Newton's second law to the satellite and solve for the square of the speed.

$$F = ma$$

$$\frac{GM_E m}{r^2} = m\frac{v^2}{r}$$

$$\text{so} \quad v^2 = \frac{GM_E}{r}$$

3. Substitute into the step-1 result and simplify.

$$E = \frac{1}{2}m\frac{GM_E}{r} - \frac{GM_E m}{r} = -\frac{GM_E m}{2r}$$

4. Compare the step-3 result with U in step 1.

$$E = -\frac{GM_E m}{2r} = \frac{1}{2}\left(-\frac{GM_E m}{r}\right) = \boxed{\frac{1}{2}U}$$

CHECK $E = K + U$, so $K = E - U$. Because K is positive, this means that E is greater than U. Because U is negative, $U//2$ is greater than U. Thus, our step-4 result correctly meets the expectation that E is greater than U.

PRACTICE PROBLEM 11-9 A satellite of mass 450 kg orbits Earth in a circular orbit at 6830 km above Earth's surface. The potential energy is zero at infinite separation from Earth. Find (*a*) the potential energy, (*b*) the kinetic energy, and (*c*) the total energy of the satellite.

(NASA.)

11-4 THE GRAVITATIONAL FIELD

The gravitational force exerted by a point particle of mass m_1 on a second point particle of mass m_2 a distance r_{12} away is given by

$$\vec{F}_{12} = -\frac{Gm_1m_2}{r_{12}^2}\hat{r}_{12}$$

where $\hat{r}_{12} = \vec{r}_{12}/r_{12}$ is a unit vector directed away from particle 1 toward particle 2. The **gravitational field** at point P is determined by placing a point particle of mass m at P and calculating the gravitational force $\vec{F}_g$ on it due to all other particles. The gravitational force $\vec{F}_g$ divided by the mass m is the gravitational field $\vec{g}$ at P:

$$\vec{g} = \frac{\vec{F}_g}{m} \qquad 11\text{-}23$$

DEFINITION—GRAVITATIONAL FIELD

The point P is called a **field point.** The gravitational field at a field point due to the masses of a collection of point particles is the vector sum of the fields due to the individual masses:

$$\vec{g} = \sum_i \vec{g}_i \qquad 11\text{-}24a$$

The locations of these point particles are called **source points.** To find the gravitational field at a field point due to a continuous object, we find the field $d\vec{g}$ due to a small element of volume with mass dm and integrate over the entire mass distribution of the object (the entire set of source points).

$$\vec{g} = \int d\vec{g} \qquad 11\text{-}24b$$

The gravitational field of Earth at a distance $r \geq R_E$ points toward Earth and has the magnitude $g(r)$ given by

$$g(r) = \frac{F_g}{m} = \frac{GM_E}{r^2} \qquad 11\text{-}25$$

GRAVITATIONAL FIELD OF EARTH

The following Problem-Solving Strategy and two examples involve calculations of the gravitational field produced by rather artificial distributions of mass. We present these here because the skills needed to accomplish these calculations are also needed in many other areas of physics. More specifically, these skills will be used extensively in Chapters 21 and 22 where the task at hand is to calculate the electric field produced by distributions of electric charge.

PROBLEM-SOLVING STRATEGY

Calculating a Gravitational Field

PICTURE Making a sketch of the mass or masses described by a problem is crucial in determining where the field point and source points are. These locations are needed to find both the magnitude and the direction of the gravitational field.

SOLVE

1. Draw a diagram that describes the situation given in the problem statement. Do not forget to identify the field point and the source points. The placement of these points must be accurate because their locations will help you solve the problem.
2. Determine r or the distance between the field point and the source points. You may have to use geometry or trigonometry to determine r.
3. Use the equation $g(r) = (GM/r^2)$ to determine the magnitude of the gravitational field. The direction can be obtained by using your diagram.

CHECK Do not forget that gravitational fields are vector fields, so your answers for these fields must include both their magnitudes and their directions.

Example 11-7 Gravitational Field of Two Point Particles

Two point particles, each of mass M, are fixed in position on the y axis at $y = +a$ and $y = -a$. Find the gravitational field at all points on the x axis as a function of x.

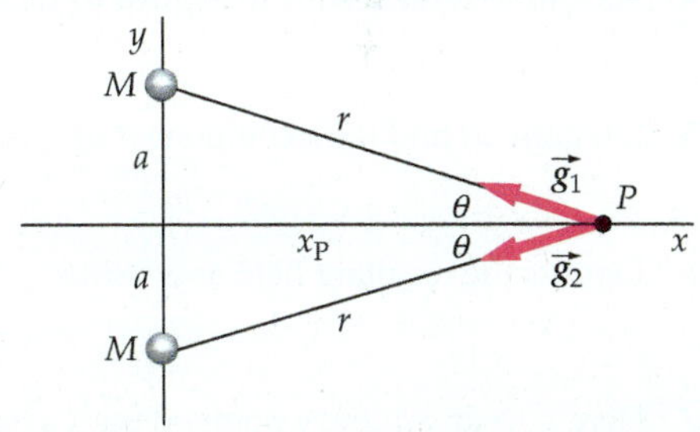

FIGURE 11-13 The particles are each located at a source point, and point P is a field point.

PICTURE Make a sketch of the two particles and the coordinate axis (Figure 11-13). Two particles of mass M each produce a gravitational field at point P located at $x = x_P$. The distance r between P and either particle is $\sqrt{x_P^2 + a^2}$. The resultant field $\vec{g}$ is the vector sum of the fields $\vec{g}_1$ and $\vec{g}_2$ due to each particle.

SOLVE

1. Calculate the magnitude of $\vec{g}_1$ and $\vec{g}_2$:

$$g_1 = g_2 = \frac{GM}{r^2}$$

2. The y component of the resultant field, the sum of g_{1y} and g_{2y}, is zero. The x component is the sum of g_{1x} and g_{2x}:

$$g_y = g_{1y} + g_{2y} = g_1 \sin\theta - g_2 \sin\theta = 0$$

$$g_x = g_{1x} + g_{2x} = g_1 \cos\theta + g_2 \cos\theta = 2g_1 \cos\theta = \frac{2GM}{r^2}\cos\theta$$

3. Express $\cos\theta$ in terms of x_P and r from the figure:

$$\cos\theta = \frac{x_P}{r}$$

4. Combining the last two results yields $\vec{g}$. To express $\vec{g}$ as a function of x_P, substitute $(x_P^2 + a^2)^{1/2}$ for r:

$$\vec{g} = g_x\hat{i} = -\frac{2GM}{r^2}\frac{x_P}{r}\hat{i} = -\frac{2GMx_P}{r^3}\hat{i} = -\frac{2GMx_P}{(x_P^2 + a^2)^{3/2}}\hat{i}$$

5. x_P is an arbitrary point on the x axis. For simplicity, we replace it with x:

$$\boxed{\vec{g} = -\frac{2GMx}{(x^2 + a^2)^{3/2}}\hat{i}}$$

CHECK: For $x < 0$, $\vec{g}$ is in the positive x direction and for $x > 0$, $\vec{g}$ is in the negative direction, as expected. If $x = 0$, we find that $\vec{g} = 0$; the fields $\vec{g}_1$ and $\vec{g}_2$ are equal and opposite at $x = 0$, and hence they cancel.

TAKING IT FURTHER For $x \gg a$, $\vec{g} \approx -(2GM/x^2)\hat{i}$. The field is the same as if a single particle of mass $2M$ were at the origin.

Example 11-8 Gravitational Field of a Uniform Rod

A thin uniform rod of mass M and length L is centered at the origin and lies along the x axis. Find the gravitational field due to the rod at all points on the x axis in the region $x > L/2$.

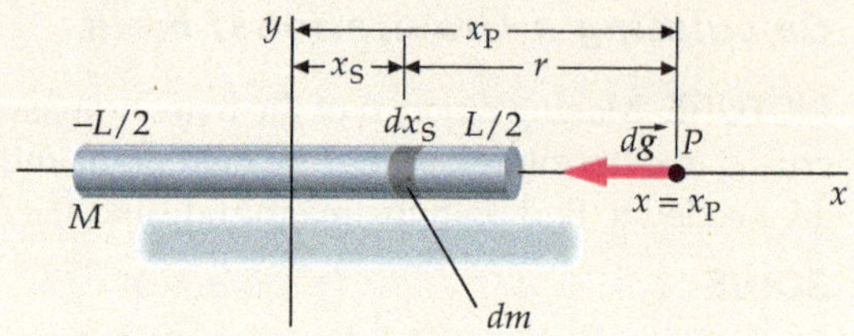

FIGURE 11-14 All points on x axis in the range $-L/2 < x < L/2$ are source points, and point P is a field point.

PICTURE Make a sketch of the rod (Figure 11-14). Identify a mass element dm of length dx_S at $x = x_S$, where $-L/2 < x_S < L/2$, and choose a field point P on the x axis at $x = x_P$, where $x_P > L/2$. Each mass element of the rod produces a gravitational field at P that points in the negative x direction. We can calculate the total field at P by integrating the field due to the mass element over the length of the rod.

SOLVE

1. Find the x component of the field at P due to the mass element dm:

$$dg_x = -\frac{G\,dm}{r^2}$$

2. Because the rod is uniform, the mass per unit length λ is constant and equal to the total mass divided by the total length. The mass dm of an element of length dx_S is equal to the mass per unit length times the length dx_S:

$$dm = \lambda\,dx$$

where $\lambda = \dfrac{M}{L}$

3. Write the distance r between dm and point P in terms of x_S and x_P:

$$r = x_P - x_S$$

4. Substitute these results to express dg in terms of x:

$$dg_x = -\frac{G\,dm}{r^2} = -\frac{G\lambda\,dx_S}{(x_P - x_S)^2}$$

5. Integrate to find the x component of the resultant field:

$$g_x = \int dg_x = -G\lambda\int_{-L/2}^{L/2}\frac{dx_S}{(x_P - x_S)^2} = -\frac{GM}{x_P^2 - (L/2)^2}$$

6. Express the resultant field as a vector:

$$\vec{g} = g_x\hat{i} = -\frac{GM}{x_P^2 - (L/2)^2}\hat{i}$$

7. Here x_P is an arbitrary point on the x axis in the region $x > L/2$. For simplicity, we replace it with x:

$$\boxed{\vec{g} = -\frac{GM}{x^2 - (L/2)^2}\hat{i} \qquad x > L/2}$$

CHECK For $x \gg L/2$, the field approaches that of a point particle of mass M, $\vec{g} = -(GM/x^2)\hat{i}$.

Example 11-9 A Gravity Map of Earth

Conceptual

Twin satellites, launched in March 2002, are making detailed measurements of Earth's gravitational field. They are in identical orbits, with one satellite directly in front of the other by about 220 km. The distance between the satellites is continuously monitored with micrometer accuracy using onboard microwave telemetry equipment. How does the distance between the two satellites change as the satellites approach a region of increased mass?

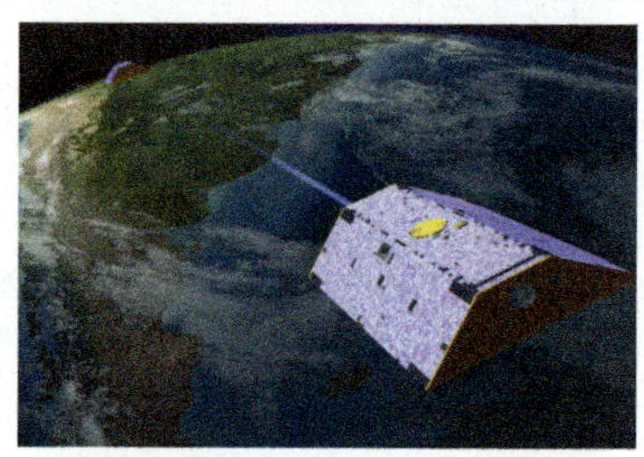

Twin satellites monitoring the distance between them and measuring variations in Earth's gravitational field. *(NASA and DLR under the NASA Earth System Science Pathfinder Program.)*

PICTURE Earth's gravitational-field strength varies because the mass of Earth is not uniformly distributed. For example, rock is denser than water, so the gravitational field is stronger over a region of dense rock than it is over water.

SOLVE

As the twin satellites approach a region where there is excess mass, the increased gravitational-field strength due to the excess mass pulls them forward (toward the excess mass). The pull on the leading satellite is greater than the pull on the trailing satellite because the leading satellite is closer to the excess mass. Consequently, the leading satellite is gaining speed more rapidly than is the trailing satellite. This results in an increase in the separation distance between the satellites. Thus, the separation distance increases as the satellites approach a region of increased mass.

The distance between them gets larger.

TAKING IT FURTHER A map of the gravitational field is also a map of the mass distribution both at and below the surface of Earth. The buildup of water in the western Pacific during an El Niño can be detected by mapping the gravitational field of Earth with the twin satellites. Gravitational maps provide information that is often useful in the search for underground resources, such as water and oil.

CONCEPT CHECK 11-4

When the twin satellites *straddle* a region of increased mass, with the leading satellite leaving the region and the trailing satellite entering the region, is the distance between the twin satellites changing? If so, is it increasing or decreasing?

$\vec{g}$ OF A SPHERICAL SHELL AND OF A SOLID SPHERE

One of Newton's motivations for developing calculus was to prove that the gravitational field outside a solid sphere is the same as if all the mass of the sphere were concentrated at its center. (This statement is correct only if the mass density of the sphere is uniform or if it varies only with distance from the center of the sphere.) A proof of this statement is in Section 11-5. Here, we merely discuss the consequences of this proof. We first consider a uniform thin spherical shell of mass M and radius R (Figure 11-15). We will show that the gravitational field due to the shell a distance r from the center of the shell is given by

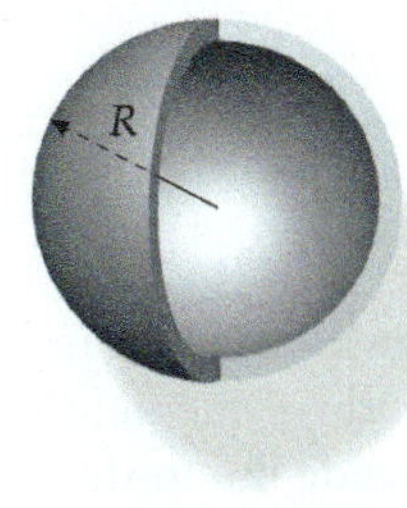

FIGURE 11-15 A uniform spherical shell of mass M and radius R.

$$\vec{g} = -\frac{GM}{r^2}\hat{r} \qquad r > R \qquad \text{11-26}a$$

$$\vec{g} = 0 \qquad r < R \qquad \text{11-26}b$$

GRAVITATIONAL FIELD OF A UNIFORM THIN SPHERICAL SHELL

From Figure 11-16, which shows a point mass m_0 inside a uniform spherical shell, we can understand the result that $\vec{g} = 0$ inside the shell. In this figure, the shell segments with masses m_1 and m_2 are proportional to the areas A_1 and A_2, respectively, and the areas A_1 and A_2 are proportional to the squares of radii r_1 and r_2, respectively. It follows that

$$\frac{m_1}{m_2} = \frac{A_1}{A_2} = \frac{r_1^2}{r_2^2} \quad \text{so} \quad \frac{m_1}{r_1^2} = \frac{m_2}{r_2^2}$$

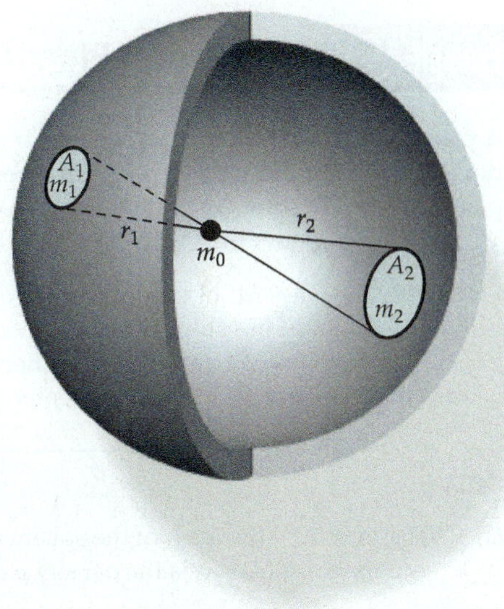

FIGURE 11-16 A point mass m_0 inside a uniform spherical shell feels no net force.

Because the gravitational force falls off inversely as the square of the distance, the force on m_0 due to the smaller mass m_1 on the left is exactly balanced by that due to the more distant, larger mass m_2 on the right.

The gravitational field outside a uniform solid sphere is a simple extension of Equation 11-26*a*. We merely consider the solid sphere to consist of a continuum of concentric uniform spherical shells. Because the field due to each shell is the same as if its mass were concentrated at the center of the shell, the field due to the entire sphere is the same as if the entire mass of the sphere were concentrated at its center:

$$g_r = -\frac{GM}{r^2} \qquad r > R \qquad \text{11-27}$$

This result holds whether or not the sphere has uniform density, as long as the density depends only on r.

$\vec{g}$ INSIDE A SOLID SPHERE

We now use Equations 11-26*a* and 11-26*b* to find the gravitational field inside a solid sphere of uniform density at a point a distance r from the center, where r is less than the radius R of the sphere. This would apply, for example, to finding the gravitational force on an object at the bottom of a deep mine shaft. As we have seen, the field inside a spherical shell is zero. Thus, in Figure 11-17 the mass of that part of the sphere outside r exerts no force at or inside r.

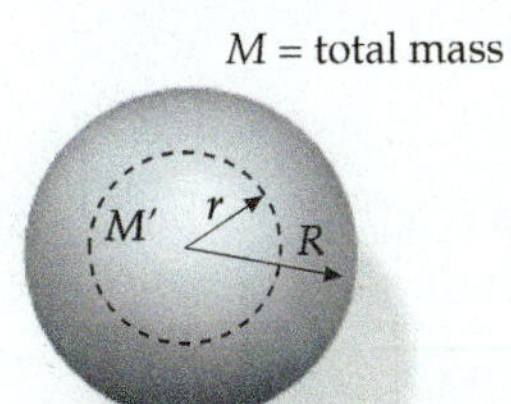

FIGURE 11-17 A uniform solid sphere of radius R and mass M. Only the mass M', which is inside the sphere of radius r, contributes to the gravitational field at the distance r.

Therefore, only the mass M' within the radius r contributes to the gravitational field at r. This mass produces a field equal to that of a point mass M' at the center of the sphere. For a uniform sphere, the fraction of the total mass of the sphere within radius r is equal to the ratio of the volume of a sphere of radius r to that of a sphere of radius R. Thus, if M is the total mass of the sphere, M' is given by

$$M' = \frac{\frac{4}{3}\pi r^3}{\frac{4}{3}\pi R^3}M = \frac{r^3}{R^3}M \qquad \text{11-28}$$

The gravitational field at the distance r is thus

$$g_r = -\frac{GM'}{r^2} = -\frac{GM}{r^2}\frac{r^3}{R^3} \qquad r \leq R$$

or

$$g_r = -\frac{GM}{R^3}r \qquad r \leq R \qquad \text{11-29}$$

The magnitude of the field is zero at the center and increases linearly with distance r inside the uniform sphere. Figure 11-18 shows a plot of the field g_r as a function of r for a solid sphere of uniform mass density.

See **Math Tutorial** *for more information on* **Geometry**

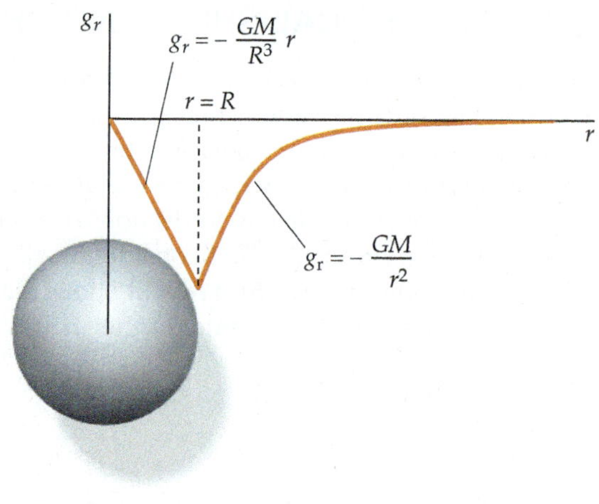

FIGURE 11-18 A plot of g_r versus r for a uniform solid sphere of mass M. The magnitude of the field increases linearly with r inside the sphere and decreases as $1/r^2$ outside the sphere.

Example 11-10 A Hollow Planet

A planet that has a hollow core consists of a uniform spherical shell with mass M, outer radius R, and inner radius $R/2$. (*a*) What amount of mass is closer than $\frac{3}{4}R$ to the center of the planet? (*b*) What is the gravitational field a distance $\frac{3}{4}R$ from the center?

PICTURE The mass M' of that part of the spherical shell that is closer to the center than $\frac{3}{4}R$ is the density times the volume of the spherical shell with outer radius $\frac{3}{4}R$ and inner radius $\frac{1}{2}R$. First, find the density and volume, then find the mass. The gravitational field at $r = \frac{3}{4}R$ is due only to the mass closer to the center than $\frac{3}{4}R$.

SOLVE

(*a*) 1. The mass M' (the mass of the spherical shell with outer radius $\frac{3}{4}R$ and inner radius $\frac{1}{2}R$) is the density ρ times the volume V':

$$M' = \rho V'$$

2. The density is the total mass M divided by the total volume V:

$$\rho = \frac{M}{V} = \frac{M}{\frac{4}{3}\pi R^3 - \frac{4}{3}\pi(\frac{1}{2}R)^3} = \frac{M}{\frac{7}{6}\pi R^3} = \frac{6M}{7\pi R^3}$$

3. Find the volume V' of the thick shell with outer radius $\frac{3}{4}R$ and inner radius $\frac{1}{2}R$:

$$V' = \frac{4}{3}\pi\left(\frac{3R}{4}\right)^3 - \frac{4}{3}\pi\left(\frac{R}{2}\right)^3 = \frac{19}{48}\pi R^3$$

4. Find the mass M':

$$M' = \rho V' = \frac{6M}{7\pi R^3}\frac{19}{48}\pi R^3 = \boxed{\frac{19}{56}M}$$

(*b*) The gravitational field at $r = \frac{3}{4}R$ is due only to the mass M':

$$\vec{g} = -\frac{GM'}{r^2}\hat{r} = -\frac{G\frac{19}{56}M}{(\frac{3}{4}R)^2}\hat{r} = \boxed{-\frac{38}{63}\frac{GM}{R^2}\hat{r}}$$

CHECK The volume V' (step 3) is less than half the volume V (step 2), so we expect M' to be less than half M. Our step-4 result meets this expectation.

Example 11-11 Radially Dependent Density

A solid sphere of radius R and mass M is spherically symmetric but not uniform. Its density ρ, defined as its mass per unit volume, is proportional to the distance r from the center for $r \le R$. That is, $\rho = Cr$ for $r \le R$, where C is a constant. (*a*) Find C. (*b*) Find $\vec{g}$ for all $r \ge R$. (*c*) Find $\vec{g}$ at $r = \frac{1}{2}R$.

PICTURE (*a*) You can find C by integrating the density over the volume of the sphere and setting the result equal to M. For a volume element, take a spherical shell of radius r and thickness dr (Figure 11-19). Its volume is $4\pi r^2\,dr$ and its mass is $dM = \rho\,dV = Cr(4\pi r^2\,dr)$. (*b*) The field outside the sphere ($r \ge R$) is the same as if the total mass M were at the center of the sphere. (*c*) The field at $r = \frac{1}{2}R$ is the same as if mass M' were at the center of the sphere, where M' is the amount of mass within the sphere of radius $\frac{1}{2}R$. The mass between $r = \frac{1}{2}R$ and $r = R$ produces zero field at $r = \frac{1}{2}R$.

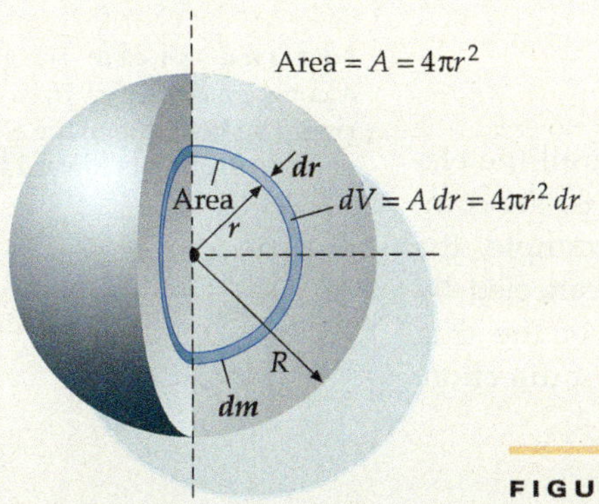

FIGURE 11-19

SOLVE

(*a*) 1. Integrate $dM = \rho\,dV$ to relate C to the mass M, where $dV = 4\pi r^2\,dr$. ($4\pi r^2$ is the area of a sphere of radius r so $4\pi r^2\,dr$ is the volume of a spherical shell of radius r and thickness dr):

$$M = \int dM = \int \rho\,dV$$
$$= \int_0^R Cr(4\pi r^2\,dr) = C\pi R^4$$

2. Solve for C in terms of the given quantities M and R.

$$\boxed{C = \frac{M}{\pi R^4}}$$

(*b*) Write an expression for the field outside the sphere in terms of the mass M, the distance r from the center, and the unit vector $\hat{\boldsymbol{r}}$. The unit vector $\hat{\boldsymbol{r}}$ is in the direction of increasing r:

$$\vec{g} = \boxed{-\frac{GM}{r^2}\hat{r}} \quad (r > R)$$

(*c*) 1. Compute the mass M' that is within the radius $\frac{1}{2}R$ by integrating $dm = \rho\,dV$ from $r = 0$ to $\frac{1}{2}R$ and use the value of C found in Part (*a*), step 2.

$$M' = \int \rho\,dV = \int_0^{R/2} Cr(4\pi r^2\,dr) = C\pi R^4/16$$
$$M' = \frac{M}{16}$$

2. Write an expression for the field at $r = \frac{1}{2}R$ in terms of M and R.

$$\vec{g} = -\frac{GM'}{r^2}\hat{r} = \boxed{-\frac{GM}{4R^2}\hat{r} \quad \text{at} \quad r = \tfrac{1}{2}R}$$

CHECK For a uniform sphere, Equation 11-29 gives the field at $r = \frac{1}{2}R$ as $g_r = -GM/(2R^2)$, which is twice as large as our Part-(*c*) result. We expected a larger value for a uniform sphere, because a uniform sphere has a larger fraction of its total mass in the region $0 < r < \frac{1}{2}R$ than does the sphere in Example 11-11.

TAKING IT FURTHER Note that the units for C are kg/m^4, so the units for ρ are kg/m^3, which is mass per volume.

* 11-5 FINDING THE GRAVITATIONAL FIELD OF A SPHERICAL SHELL BY INTEGRATION

We now derive the equation for the gravitational field of a uniform thin spherical shell. First, we find the gravitational field on the axis of a thin ring of uniform mass. We then apply our result to a thin spherical shell, which we treat as a continuum of thin coaxial rings.

Figure 11-20 shows a thin ring of total mass m and radius a and a field point P on the axis of the ring a distance x from its center. We choose an element with mass dm on the ring that is small enough to be considered a point particle. The distance from the element to P is s, and the line joining the element and P makes an *angle* α with the axis of the ring.

The field at P, which is due to the element dm, is toward the element and has magnitude dg given by

$$dg = \frac{G\,dm}{s^2}$$

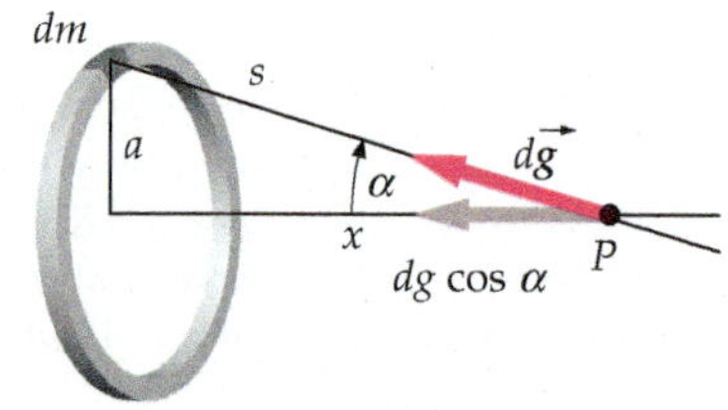

FIGURE 11-20 The gravitational field at a point P a distance x from a thin uniform ring. The field due to the element with a mass equal to dm points toward the element.

From the symmetry of the figure, we can see that when we sum over all the elements of the ring, the net field will be along the axis of the ring; that is, the components of $\vec{g}$ perpendicular to the x axis will sum to zero. For example, the perpendicular component of the field shown in the figure will be canceled by the perpendicular component of the field due to another element of the ring directly opposite the one shown. The net field will therefore be in the $-x$ direction. The x component of the field due to the element dm is

$$dg_x = -dg\cos\alpha = -\frac{G\,dm}{s^2}\cos\alpha$$

We obtain the total field by integrating both sides of this equation:

$$g_x = -\int \frac{G\cos\alpha}{s^2}\,dm$$

See Math Tutorial *for more information on* Integrals

Because s and α are the same for all points on the ring, they are constants as far as this integration is concerned. Thus,

$$g_x = -\frac{G\cos\alpha}{s^2}\int dm = -\frac{Gm}{s^2}\cos\alpha \qquad \text{11-30}$$

where $m = \int dm$ is the total mass of the ring.

We now use this result to calculate the gravitational field of a thin uniform spherical shell of mass M and radius R at a point a distance r from the center of the shell. We first consider the case in which the field point P is outside the shell, as in Figure 11-21. By symmetry, the field must be directed toward the center of the spherical shell. We choose for our element of mass the strip shown, which can be considered to be a thin ring of mass dM. The field due to this strip is given by Equation 11-30 with m replaced by dM:

$$dg_r = -\frac{G\,dM}{s^2}\cos\alpha \qquad \text{11-31}$$

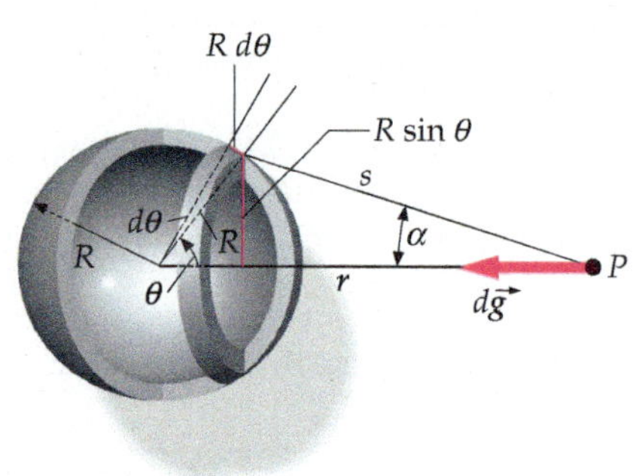

FIGURE 11-21 A uniform thin spherical shell of radius R and total mass M. The strip shown can be considered to be a ring of width $R\,d\theta$ and circumference $2\pi R\sin\theta$.

The mass dM is proportional to the area of the strip dA, which equals the circumference times the width. The radius of the strip is $R\sin\theta$, so the circumference is $2\pi R\sin\theta$. The width is $R\,d\theta$. If M is the total mass of the shell and $A = 4\pi R^2$ is its total area, the mass of the strip of area dA is

$$dM = \frac{M}{A}dA = \frac{M}{4\pi R^2}2\pi R^2\sin\theta\,d\theta = \frac{1}{2}M\sin\theta\,d\theta \qquad \text{11-32}$$

Substituting this result into Equation 11-31 gives

$$dg_r = -\frac{G\,dM}{s^2}\cos\alpha = -\frac{GM\sin\theta\,d\theta}{2s^2}\cos\alpha \qquad \text{11-33}$$

The right-hand term of Equation 11-33 contains three variables (s, θ, and α). Before integrating this term we must express it as a function of a single variable. It turns out to be easiest if we express it in terms of s. By the law of cosines, we have

$$s^2 = r^2 + R^2 - 2rR\cos\theta$$

Differentiating gives

$$2s\,ds = +2rR\sin\theta\,d\theta$$

so

$$\sin\theta\,d\theta = \frac{s\,ds}{rR}$$

An expression for cos α can be obtained by again applying the law of cosines to the same triangle. We have

$$R^2 = s^2 + r^2 - 2sr\cos\alpha$$

so

$$\cos\alpha = \frac{s^2 + r^2 - R^2}{2sr}$$

Substituting these results into Equation 11-33 gives

$$dg_r = -\frac{GM\sin\theta\,d\theta}{2s^2}\cos\alpha = -\frac{GM}{2s^2}\left(\frac{s\,ds}{rR}\right)\frac{s^2 + r^2 - R^2}{2sr}$$
$$= -\frac{GM\,ds}{4s^2r^2R}(s^2 + r^2 - R^2) = -\frac{GM}{4r^2R}\left(1 + \frac{r^2 - R^2}{s^2}\right)ds \qquad \text{11-34}$$

To find the field at P, we integrate over the entire shell. The integration limits for this step depend on whether the field point P lies outside the shell or inside it. For P outside the shell, s varies from $r - R$ (at $\theta = 0$) to $s = r + R$ (at $\theta = 180°$), so the field due to the entire shell is found by integrating from $s = r - R$ to $s = r + R$.

$$g_r = -\frac{GM}{4r^2R}\int_{r-R}^{r+R}\left(1 + \frac{(r-R)(r+R)}{s^2}\right)ds = -\frac{GM}{4r^2R}\left[s - \frac{(r-R)(r+R)}{s}\right]_{r-R}^{r+R}$$

Substituting the upper and lower limits for s yields $4R$ for the quantity in the square brackets. Thus,

$$g_r = -\frac{GM}{r^2} \qquad \text{for} \quad r > R$$

which is the same result as for Equation 11-26*a*.

If the field point P is inside the shell (Figure 11-22), the calculation is identical except that s now varies from $R - r$ to $R + r$. Thus,

$$g_r = -\frac{GM}{4r^2R}\left[s - \frac{(r-R)(r+R)}{s}\right]_{R-r}^{R+r}$$

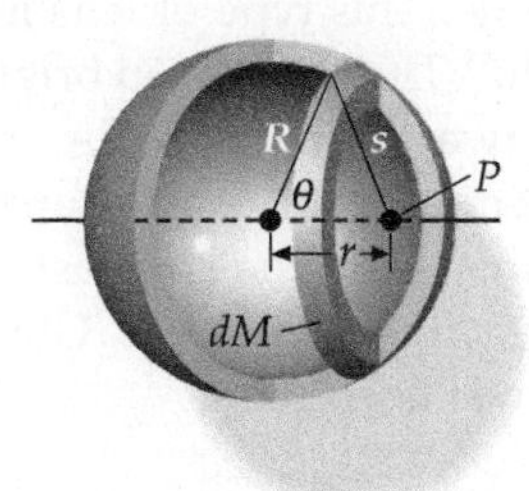

FIGURE 11-22

Substituting these upper and lower limits for s yields $g_r = 0$. Therefore,

$$g_r = 0 \qquad \text{for} \quad r < R$$

which is the same as Equation 11-26*b*.

Applying these results to find the gravitational field due to a uniform spherical shell of finite thickness is the topic of Problem 11-99.

Physics Spotlight

Gravitational Lenses: A Window on the Universe

In 1919, Arthur Eddington took photographs during a 1919 solar eclipse which showed stars where they should not be. The starlight had been "bent" by the mass of the Sun. This verified a key prediction of Einstein's 1915 theory of general relativity, namely, space is warped by massive objects. The degree of curvature of space depends on the object's mass.

The bent light was mostly a curiosity for many years after 1919. Years later, many astronomers started studying quasars, star-sized objects that gave off more light than most galaxies. In 1979, twin images of a distant quasar were seen. These images had been formed when the quasar's light was bent by a cluster of galaxies[*] between the quasar and Earth.

The central reddish-white blobs are giant elliptical galaxies, and the thin blue light ringing these blobs comes from galaxies twice as far away and directly behind the giant elliptical galaxies. The light from the more distant galaxies is distorted into circular shapes by the gravitation field of the giant elliptical galaxies. *(NASA, ESA, A. Bolton (Harvard-Smithsonian CfA), & the SLACS Team.)*

Clusters of galaxies are massive objects. Space in and around them warps, and thus curves the light from distant objects that travels through and near them on its journey toward Earth. The region of warped space near a massive object is called a *gravitational lens*. Gravitational lenses can brighten the light of distant objects, just as light shining through a drop of water can be brightened. Gravitational lenses are now used to study very distant quasars and galaxies. Because the lenses magnify faint light, they help determine the age and expansion of the universe.[†]

Calculations using gravitational lenses are based on the images of the distant object. To get a precise description for an object, the distance, mass, and shape of the intervening gravitational lens must be determined. A lens formed by a uniform circular mass directly between the distant object and Earth would create a uniform circular image—an *Einstein ring*—with easily calculated values.[‡,#] But gravitational lenses create multiple[°] or strangely distorted images much more often than they create perfect rings. The strongest lenses are formed by clusters of galaxies.[§] These galaxies are difficult to model, and their detectable energy cannot account for all of the distorting mass. Calculations show that the galaxies must have a large halo of unseen mass. Gravitational lenses confirm that most of the universe's mass is *dark matter*, matter that does not emit detectable energy.[¶,**]

A weak lens, or *microlens*, does not create multiple images of a distant object, but brightens the image of a known object for a short time. This type of lens is formed by a massive compact halo object (macho), which passes between the known object and Earth.[††] The changes in brightness reveal much about the shape and mass of the macho. One microlens was determined to be a red dwarf, orbited by the smallest planet known outside of our solar system.[‡‡]

Gravitational lenses have enabled the discovery of objects from galaxies[##] produced less than a billion years after the start of the universe, to the smallest known objects outside of our solar system. These lenses have defined questions about matter and energy in the universe. Gravitational lenses have come full circle—the curve of light is now used to measure things that have not yet been seen.

* Walsh, D., Carswell, R. F., and Weymann, R.J., "0957 + 561 A, B - Twin Quasistellar Objects or Gravitational Lens," *Nature*, May 31, 1979, Vol. 279, 381–384.
† Irion, R., "Through a Lens, Deeply," *Science*, Jan. 24, 2003, Vol 299, 500+.
‡ Greene, Katie, "Ring Around the Galaxy," *Science News*, Nov. 25, 2005, 342.
\# Liu, C., "The Quest for the Golden Lens," *Natural History*, Sept. 2003, 64–66.
° Rusin, D., Kochanek, C. S., Norbury, M., Falco, E. E., Impey, C. D., Lehár, J., McLeod, B. A., Rix, H.-W., Keeton, C. R., Muñoz, J. A., and Peng, C. Y., "B1359+154: A Six-Image Lens Produced by a $z \simeq 1$ Compact Group of Galaxies," *Astrophysical Journal*, Aug. 20, 2001, Vol. 557, 594–604.
§ Abell, G. O., Corwin, H. G., and Olowin, R. P., 1989, *A Catalog of Rich Clusters of Galaxies*, The Astrophysical Journal Supplement Series, 1989, Vol. 70, 1–138.
¶ Koopmans, L.V.E., and Blandford, R. D., "Gravitational Lenses," *Physics Today*, June 2004, 45–51.
** Seife, Charles, "The Intelligent Noncosmologist's Guide to Spacetime," *Science*, May 2002, Vol. 296, 1418–1421.
†† Mancini, L., Jetzer, Ph., and Scarpetta, G., "Compact Dark Objects and Gravitational Microlensing towards the Large Magellanic Cloud," in *Highlights in Condensed Matter Physics*, A. Avella et al., eds. New York: American Institute of Physics, 2003, 339–347.
‡‡ Cowan, Ron, "Tiny Planet Orbits Faraway Star," *Science News*, Feb. 25, 2006, 126.
\#\# Cowan, Ron, "A Galaxy that Goes the Distance?" *Science News*, Apr. 24, 2004, 270.

Summary

1. Kepler's laws are *empirical* observations. They can also be derived from Newton's laws of motion and Newton's law of gravity.
2. Newton's law of gravity is a *fundamental law* of physics, and G is a fundamental *universal* physical constant.
3. The gravitational potential energy of a two-particle system, relative to $U = 0$ at infinite separation, is given by $U = 2Gm_1m_2/r$. If the system is bound, its total energy is negative.
4. The gravitational field is a *fundamental physical concept* that describes the condition in space set up by a mass distribution.

TOPIC	RELEVANT EQUATIONS AND REMARKS	
1. Kepler's Three Laws	Law 1. All of the planets move in elliptical orbits with the Sun at one focus.	
	Law 2. A line joining any planet to the Sun sweeps out equal areas in equal times.	
	Law 3. The square of the period of any planet is proportional to the cube of the planet's mean distance from the Sun:	
	$T^2 = Cr^3$	11-2
	where C has almost the same value for all planets; from Newton's law of gravity, C can be shown to be $4\pi^2/[G(M_S + M_p)]$. If $M_S \gg M_p$, this can be expressed as	
	$T^2 = \frac{4\pi^2}{GM_S}r^3$	11-16
	Kepler's laws can be derived from Newton's law of gravity. The first and third of Kepler's laws follow from the fact that the force exerted by the Sun on the planets varies inversely as the square of the separation distance. The second law follows from the fact that the force exerted by the Sun on a planet is along the line joining them, so the orbital angular momentum of the planet is conserved. Kepler's laws also hold for any object orbiting another in an inverse-square gravitational field, such as a satellite orbiting a planet.	
2. Newton's Law of Gravity	Every point particle exerts an attractive force on every other point particle that is proportional to the masses of the two particles and inversely proportional to the square of the distance separating them:	
	$\vec{F}_{12} = -\frac{Gm_1m_2}{r_{12}^2}\hat{r}_{12}$	11-3
Universal gravitational constant	$G = 6.67 \times 10^{-11}\ \mathrm{N \cdot m^2/kg^2}$	11-4
3. Gravitational Potential Energy	The gravitational potential energy U for a system consisting of a particle of mass m outside a spherically symmetric object of mass M and at a distance r from its center is	
	$U(r) = -\frac{GMm}{r}$	11-20
	This potential-energy function approaches zero as r approaches infinity.	
4. Mechanical Energy	The mechanical energy E for a system consisting of a particle of mass m outside a spherically symmetric object of mass M and at a distance r from its center is	
	$E = \frac{1}{2}mv^2 - \frac{GMm}{r}$	
Escape speed	For a given value of r, the speed of the particle for which $E = 0$ is called the escape speed v_e. That is, if $v = v_e$, then $E = 0$.	
5. Classification of Orbits	If $E < 0$, the system is bound and the orbit is an ellipse (or circle, which is a type of ellipse). If $E \geq 0$, the system is unbound and the orbit is a hyperbola (or a parabola for $E = 0$).	

TOPIC	RELEVANT EQUATIONS AND REMARKS	
6. Gravitational Field		
Definition	$\vec{g} = \dfrac{\vec{F}_g}{m}$	11-23
Due to Earth	$\vec{g}(r) = \dfrac{\vec{F}_g}{m} = -\dfrac{GM_E}{r^2}\hat{r} \quad (r \geq R_E)$	11-29
Due to a thin, spherical shell	Outside the shell, the gravitational field is the same as if all the mass of the shell were concentrated at the center. The field inside the shell is zero.	
	$\vec{g} = -\dfrac{GM}{r^2}\hat{r} \quad \text{for} \quad r > R$	11-26*a*
	$\vec{g} = 0 \quad \text{for} \quad r < R$	11-26*b*

Answers to Concept Checks

11-1 An orbiting astronaut is said to be weightless because both he and the orbiting shuttle are in free-fall with the same acceleration, so if an astronaut would stand on a scale that is attached to the shuttle, the scale would read zero. An orbiting astronaut is not actually weightless because we have defined weight to be the magnitude of the gravitational force.

11-2 The property of an object responsible for the gravitational force it exerts on another object, or for the gravitational force another object exerts on it, is its *gravitational* mass. On the other hand, the property of an object that measures its inertial resistance to acceleration is its *inertial* mass.

11-3 Using his second law, Newton proved that an attractive force that varies inversely with the square of the distance between the Sun and a planet would result in an elliptical orbit with the center of the Sun at a focus of the ellipse.

11-4 The distance between the satellites is decreasing.

Answers to Practice Problems

11-1 30.1 AU

11-2 A straight line that has a slope of 1.5

11-3 8.67×10^{-7} N

11-4 2640 km

11-5 Earth rotates 360° in 24 h, so in 1 h Earth rotates 15°. Because of Earth's orbital motion, the direction of the Sun relative to Earth changes by (1/365.25) rev each 24 h. As a result, Earth rotates through 1 rev in about 4 min less than 24 h. The time for 1 rev is called the sidereal day.

11-6 $r = 6.63R_E = 4.22 \times 10^7$ m = 26,200 mi. If such a satellite is in orbit over the equator and moves in the same direction as the rotation of Earth, it appears stationary relative to Earth. Many communication satellites are "parked" in such orbits, called *geosynchronous orbits*.

11-7 6.45×10^{23} kg $= 0.108\, M_E$

11-8 $v_e = 4.25$ km/s

11-9 (Note that $r = R_E + h = 13{,}200$ km.)
(*a*) $U = -13.6 \times 10^9$ J, (*b*) $K = 6.80 \times 10^9$ J,
(*c*) $E = -6.80 \times 10^9$ J

PROBLEMS

In a few problems, you are given more data than you actually need; in a few other problems, you are required to supply data from your general knowledge, outside sources, or informed estimates.

For all problems, use 9.81 m/s² for the free-fall acceleration and neglect friction and air resistance unless instructed to do otherwise.

Interpret as significant all digits in numerical values that have trailing zeros and no decimal points.

- • Single-concept, single-step, relatively easy
- •• Intermediate-level, may require synthesis of concepts
- ••• Challenging
- SSM Solution is in the *Student Solutions Manual*
- Consecutive problems that are shaded are paired problems.

CONCEPTUAL PROBLEMS

1 • True or false:

(*a*) For Kepler's law of equal areas to be valid, the force of gravity must vary inversely with the square of the distance between a given planet and the Sun.
(*b*) The planet closest to the Sun has the shortest orbital period.
(*c*) Venus's orbital speed is larger than the orbital speed of Earth.
(*d*) The orbital period of a planet allows accurate determination of that planet's mass. SSM

2 • If the mass of a small Earth-orbiting satellite is doubled, the radius of its orbit can remain constant if the speed of the satellite (*a*) increases by a factor of 8, (*b*) increases by a factor of 2, (*c*) does not change, (*d*) is reduced by a factor of 8, (*e*) is reduced by a factor of 2.

3 • During what season in the northern hemisphere does Earth attain its maximum orbital speed about the Sun? In what season does it attain its minimum orbital speed? *Hint: Earth is at perihelion in early January.* SSM

4 • Haley's comet is in a highly elliptical orbit about the Sun with a period of about 76 y. Its last closest approach to the Sun occurred in 1987. In what years of the twentieth century was it traveling at its fastest or slowest orbital speed about the Sun?

5 • Venus has no natural satellites. However, artificial satellites have been placed in orbit around it. To use one of their orbits to determine the mass of Venus, what orbital parameters would you have to measure? How would you then use these parameters to do the mass calculation?

6 • A majority of the asteroids are in approximately circular orbits in a "belt" between Mars and Jupiter. Do they all have the same orbital period about the Sun? Explain.

7 • At the moon's surface, the acceleration due to the gravity of the moon is a. At a distance from the moon's center equal to four times the radius of the moon, the acceleration due to the gravity of the moon is (*a*) $16a$, (*b*) $a/4$, (*c*) $a/3$, (*d*) $a/16$, (*e*) none of the above. SSM

8 • At a depth equal to half the radius of Earth, the acceleration due to gravity is about (*a*) g (*b*) $2g$ (*c*) $g/2$, (*d*) $g/4$, (*e*) $g/8$, (*f*) You cannot determine the answer based on the data given.

9 •• Two stars orbit their common center of mass as a *binary* star system. If each of their masses were doubled, what would have to happen to the distance between them in order to maintain the same gravitational force? The distance would have to (*a*) remain the same, (*b*) double, (*c*) quadruple, (*d*) be reduced by a factor of 2, (*e*) You cannot determine the answer based on the data given.

10 •• **CONTEXT-RICH** If you had been working for NASA in the 1960s and planning the trip to the moon, you would have determined that a unique location exists somewhere between Earth and the moon, where a spaceship is, for an instant, truly weightless. (Consider only the moon, Earth, and the Apollo spaceship, and neglect other gravitational forces.) Explain this phenomenon and explain whether this location is closer to the moon, midway on the trip, or closer to Earth.

11 •• Suppose the escape speed from a planet is only slightly larger than the escape speed from Earth, yet the planet is considerably larger than Earth. How would the planet's (average) density compare to Earth's (average) density? (*a*) It must be denser, (*b*) It must be less dense. (*c*) It must be the same density. (*d*) You cannot determine the answer based on the data given. SSM

12 •• Suppose that, using a telescope in your backyard, you discovered a distant object approaching the Sun, and were able to determine both its distance from the Sun and its speed. How would you be able to predict whether the object will remain "bound" to the solar system, or if it is an interstellar interloper that would come in, turn around, and escape, never to return?

13 •• **CONTEXT-RICH, ENGINEERING APPLICATION** Near the end of their useful lives, several large Earth-orbiting satellites have been maneuvered so they burn up as they enter Earth's atmosphere. These maneuvers have to be done carefully so large fragments do not impact populated land areas. You are in charge of such a project. Assuming a satellite of interest has on-board propulsion, in what direction would you fire the rockets for a short burn time to start this downward spiral? What would happen to the kinetic energy, gravitational potential energy, and total mechanical energy following the burn as the satellite came closer and closer to Earth? SSM

14 •• **ENGINEERING APPLICATION** During a trip back from the moon, the Apollo spacecraft fires its rockets to leave its lunar orbit. Then, it coasts back to Earth where it enters the atmosphere at high speed, survives a blazing reentry, and parachutes safely into the ocean. In what direction do you fire the rockets to initiate this return trip? Explain the changes in kinetic energy, gravitational potential, and total mechanical energy that occur to the spacecraft from the beginning to the end of this journey.

15 •• Explain why the gravitational field inside a solid sphere of uniform mass is directly proportional to r rather than inversely proportional to r.

16 •• In the movie *2001: A Space Odyssey*, a spaceship containing two astronauts is on a long-term mission to Jupiter. A model of their ship could be a uniform pencil-like rod (containing the propulsion systems) with a uniform sphere (the crew habitat and flight deck) attached to one end (Figure 11-23). The design is such that the radius of the sphere is much smaller than the length of the rod. At a location a few meters away from the ship, at point P on the perpendicular bisector of the rodlike section, what would be the direction of the gravitational field due to the ship alone (that is, assuming all other gravitational fields are negligible)? Explain your answer. At a large distance from the ship, what would be the dependence of the ship's gravitational field on the distance from the ship?

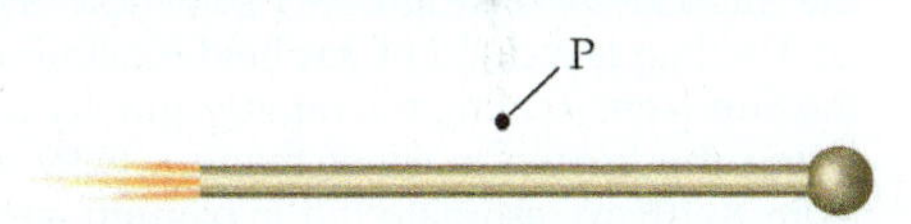

FIGURE 11-23 Problem 16

ESTIMATION AND APPROXIMATION

17 • Estimate the mass of our galaxy (the Milky Way) if the Sun orbits the center of the galaxy with a period of 250 million years at a mean distance of 30,000 $c \cdot y$. Express the mass in terms of multiples of the solar mass M_S. (Neglect the mass farther from the center than the Sun, and assume that the mass closer to the center than the Sun exerts the same force on the Sun, as would a point particle of the same mass located at the center of the galaxy.) SSM

18 •• Besides studying samples of the lunar surface, the Apollo astronauts had several ways of determining that the moon is *not* made of green cheese. Among these ways are measurements of the gravitational acceleration at the lunar surface. Estimate the gravitational acceleration at the lunar surface if the moon were, in fact, a solid block of green cheese and compare your answer to the known value of the gravitational acceleration at the lunar surface.

19 •• **CONTEXT-RICH, ENGINEERING APPLICATION** You are in charge of the first manned exploration of an asteroid. You are concerned that, due to the weak gravitation field and resulting low escape speed, tethers might be required to bind the explorers to the surface of the asteroid. Therefore, if you do not wish to use tethers, you have to be careful about which asteroids to choose to explore. Estimate the largest radius the asteroid can have that would still allow you to escape its surface by jumping. Assume spherical geometry and reasonable rock density.

20 ••• One of the great discoveries in astronomy in the past decade is the detection of planets outside the solar system. Since 1996, more than 100 planets have been detected orbiting stars other than the Sun. While the planets themselves cannot be seen directly, telescopes can detect the small periodic motion of the star as the star and planet orbit around their common center of mass. (This is measured using the *Doppler effect*, which is discussed in Chapter 15.) Both the period of this motion and the variation in the speed of the star over the course of time can be determined observationally. The mass of the star is found from its observed luminance and from the theory of stellar structure. *Iota Draconis* is the eighth brightest star in the constellation Draco. Observations show that a planet, with an orbital period of 1.50 y, is orbiting this star. The mass of Iota Draconis is $1.05M_{\rm Sun}$. (*a*) Estimate the size (in AU) of the semimajor axis of this planet's orbit. (*b*) The radial speed of the star is observed to vary by 592 m/s. Use conservation of momentum to find the mass of the planet. Assume the orbit is circular, we are observing the orbit edge-on, and no other planets orbit Iota Draconis. Express the mass as a multiple of the mass of Jupiter.

21 ••• One of the biggest unresolved problems in the theory of the formation of the solar system is that, while the mass of the Sun is 99.9 percent of the total mass of the solar system, it carries only about 2 percent of the total angular momentum. The most widely accepted theory of solar system formation has as its central hypothesis the collapse of a cloud of dust and gas under the force of gravity, with most of the mass forming the Sun. However, because the net angular momentum of this cloud is conserved, a simple theory would indicate that the Sun should be rotating much more rapidly than it currently is. In this problem, you are to show why it is important that most of the angular momentum was somehow transferred to the planets. (*a*) The Sun is a cloud of gas held together by the force of gravity. If the Sun were rotating too rapidly, gravity could not hold it together. Using the known mass of the Sun (1.99×10^{30} kg) and its radius (6.96×10^{8} m), estimate the maximum angular speed that the Sun can have if it is to stay intact. What is the period of rotation corresponding to this rotation rate? (*b*) Calculate the orbital angular momentum of Jupiter and of Saturn from their masses (318 and 95.1 Earth masses, respectively), mean distances from the Sun (778 and 1430 million km, respectively), and orbital periods (11.9 and 29.5 y, respectively). Compare them to the experimentally measured value of the Sun's angular momentum of $1.91 \times 10^{41}\ \text{kg}\cdot\text{m}^2/\text{s}$. (*c*) If we were to somehow transfer all of Jupiter's and Saturn's angular momentum to the Sun, what would be the Sun's new rotational period? The Sun is not a uniform sphere of gas, and its moment of inertia is given by the formula $I = 0.059MR^2$. Compare this to the maximum rotational period of Part (*a*).

KEPLER'S LAWS

22 • The new comet Alex-Casey has a very elliptical orbit with a period of 127.4 y. If the closest approach of Alex-Casey to the Sun is 0.1 AU, what is its greatest distance from the Sun?

23 • The radius of Earth's orbit is 1.496×10^{11} m and that of Uranus is 2.87×10^{12} m. What is the orbital period of Uranus?

24 • The asteroid Hektor, discovered in 1907, is in a nearly circular orbit of radius 5.16 AU about the Sun. Determine the period of this asteroid.

25 •• One of the so-called "Kirkwood gaps" in the asteroid belt occurs at an orbital radius at which the period of the orbit is half that of Jupiter's. The reason there is a gap for orbits of this radius is because of the periodic pulling (by Jupiter) that an asteroid experiences at the same location with every *other* orbit around the Sun. Repeated tugs from Jupiter of this kind would eventually change the orbit of such an asteroid. Therefore, all asteroids that would otherwise have orbited at this radius have presumably been cleared away from the area due to this resonance phenomenon. How far from the Sun is this particular 2:1 resonance "Kirkwood" gap? SSM

26 •• The tiny Saturnian moon, Atlas, is locked into what is known as an orbital resonance with another moon, Mimas, whose orbit lies outside that of Atlas. The ratio between periods of these orbits is 3:2, that is, for every 3 orbits of Atlas, Mimas completes 2 orbits. Thus, Atlas, Mimas and Saturn are aligned at intervals equal to two orbital periods of Atlas. If Mimas orbits Saturn at a radius of 186,000 km, what is the radius of Atlas's orbit?

27 •• The asteroid Icarus, discovered in 1949, was so named because its highly eccentric elliptical orbit brings it close to the Sun at perihelion. The eccentricity e of an ellipse is defined by the relation $r_{\rm p} = a(1 - e)$, where $r_{\rm p}$ is the perihelion distance and a is the semimajor axis. Icarus has an eccentricity of 0.83 and a period of 1.1 y. (*a*) Determine the semimajor axis of the orbit of Icarus. (*b*) Determine the perihelion and aphelion distances of the orbit of Icarus.

28 •• **CONTEXT-RICH, ENGINEERING APPLICATION, BIOLOGICAL APPLICATION** A manned mission to Mars and its attendant problems due to the extremely long time the astronauts would spend weightless and without supplies space have been extensively discussed. To examine this issue in a simple way, consider one possible trajectory for the spacecraft: the "Hohmann transfer orbit." This orbit consists of an elliptical orbit tangent to the orbit of Earth at its perihelion and tangent to the orbit of Mars at its aphelion. Given that Mars has a mean distance from the Sun of 1.52 times the mean Sun–Earth distance, calculate the time spent by the astronauts during the outbound part of the trip to Mars. Many adverse biological effects (such as muscle atrophy and decreased bone density) have been observed in astronauts returning from near-Earth orbit after only a few months in space. As the flight doctor, are there any health concerns that you should be aware of?

29 •• **ESTIMATION** Kepler determined distances in the solar system from his data. For example, he found the relative distance from the Sun to Venus (as compared to the distance from the Sun to Earth) as follows. Because Venus's orbit is closer to the Sun than is Earth's orbit, Venus is a morning or evening star—its position in the sky is never very far from the Sun (Figure 11-24). If we suppose

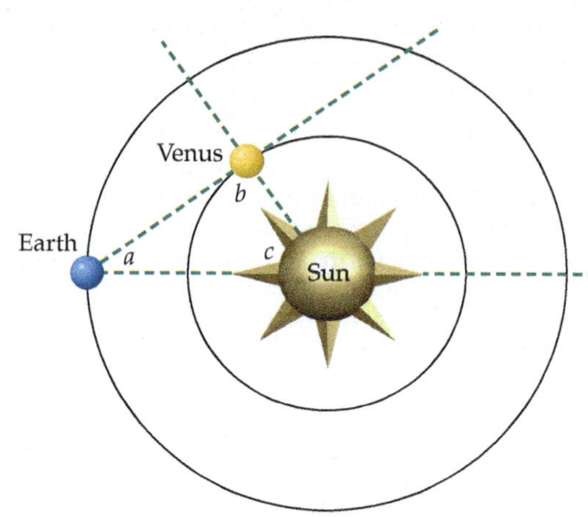

FIGURE 11-24 Problem 29

the orbit of Venus is a perfect circle, then consider the relative orientation of Venus, Earth, and the Sun at maximum extension, that is, when Venus is farthest from the Sun in the sky. (*a*) Under this condition, show that angle *b* in Figure 11-24 is 90°. (*b*) If the maximum elongation angle *a* between Venus and the Sun is 47°, what is the distance between Venus and the Sun in AU? (*c*) Use this result to estimate the length of a Venusian "year." SSM

30 •• At apogee, the center of the moon is 406,395 km from the center of Earth and at perigee, the moon is 357,643 km from the center of Earth. What is the orbital speed of the moon at perigee and at apogee? The mass of Earth is 5.98×10^{24} kg.

NEWTON'S LAW OF GRAVITY

31 • Jupiter's satellite Europa orbits Jupiter with a period of 3.55 d at an average orbital radius of 6.71×10^8 m. (*a*) Assuming that the orbit is circular, determine the mass of Jupiter from the data given. (*b*) Another satellite of Jupiter, Callisto, orbits at an average radius of 18.8×10^8 m with an orbital period of 16.7 d. Show that these data are consistent with an inverse-square force law for gravity (*Note:* Do NOT use the value of *G* anywhere in Part (*b*)). SSM

32 • **Biological Application** Some people think that shuttle astronauts are "weightless" because they are "beyond the pull of Earth's gravity." In fact, this is completely untrue. (*a*) What is the magnitude of the gravitational field in the vicinity of a shuttle orbit? A shuttle orbit is about 400 km above the ground. (*b*) Given the answer in Part (*a*), explain why shuttle astronauts suffer from adverse biological affects such as muscle atrophy even though they are not actually "weightless"?

33 • The mass of Saturn is 5.69×10^{26} kg. (*a*) Find the period of its moon Mimas, whose mean orbital radius is 1.86×10^8 m. (*b*) Find the mean orbital radius of its moon Titan, whose period is 1.38×10^6 s. SSM

34 • Calculate the mass of Earth from the period of the moon, $T = 27.3$ d; its mean orbital radius, $r_m = 3.84 \times 10^8$ m; and the known value of *G*.

35 • Suppose you leave the solar system and arrive at a planet that has the same mass-to-volume ratio as Earth but has 10 times Earth's radius. What would you weigh on this planet compared with what you weigh on Earth?

36 • Suppose that Earth retained its present mass but was somehow compressed to half its present radius. What would be the value of *g* at the surface of this new, compact planet?

37 • A planet orbits a massive star. When the planet is at perihelion, it has a speed of 5.0×10^4 m/s and is 1.0×10^{15} m from the star. The orbital radius increases to 2.2×10^{15} m at aphelion. What is the planet's speed at aphelion?

38 • What is the magnitude of the gravitational field at the surface of a neutron star whose mass is 1.60 times the mass of the Sun and whose radius is 10.5 km?

39 •• The speed of an asteroid is 20 km/s at perihelion and 14 km/s at aphelion. (*a*) Determine the ratio of the aphelion to perihelion distances. (*b*) Is this asteroid farther from the Sun or closer to the Sun than Earth, on average? Explain.

40 •• A satellite that has a mass of 300 kg moves in a circular orbit 5.00×10^7 m above Earth's surface. (*a*) What is the gravitational force on the satellite? (*b*) What is the speed of the satellite? (*c*) What is the period of the satellite?

41 •• A superconducting gravity meter can measure changes in gravity on the order $\Delta g/g = 1.00 \times 10^{-11}$. (*a*) You are hiding behind a tree holding the meter, and your 80-kg friend approaches the tree from the other side. How close to you can your friend come before the meter detects a change in *g* due to his presence? (*b*) You are in a hot air balloon and are using the gravity meter to determine the rate of ascent (assume the balloon has constant acceleration). What is the smallest change in altitude that results in a detectable change in the gravitational field of Earth? SSM

42 •• Suppose that the attractive interaction between a star of mass *M* and a planet of mass $m \ll M$ is of the form $F = KMm/r$, where *K* is the gravitational constant. What would be the relation between the radius of the planet's circular orbit and its period?

43 •• Earth's radius is 6370 km and the moon's radius is 1738 km. The acceleration of gravity at the surface of the moon is 1.62 m/s². What is the ratio of the average density of the moon to that of Earth? SSM

GRAVITATIONAL AND INERTIAL MASS

44 • The weight of a standard object defined as having a mass of exactly 1.00.... kg is measured to be 9.81 N. In the same laboratory, a second object weighs 56.6 N. (*a*) What is the mass of the second object? (*b*) Is the mass you determined in Part (*a*) gravitational or inertial mass?

45 • **Estimation** The *Principle of Equivalence* states that the free-fall acceleration of any object in a gravitational field is independent of the mass of the object. This can be deduced from the law of universal gravitation, but how well does it hold up experimentally? The Roll–Krotkov–Dicke experiment performed in the 1960s indicates that the free-fall acceleration is independent of mass to at least 1 part in 10^{12}. Suppose two objects are simultaneously released from rest in a uniform gravitational field. Also, suppose one of the objects falls with a constant acceleration of exactly 9.81 m/s², while the other falls with a constant acceleration that is greater than 9.81 m/s² by one part in 10^{12}. How far will the first object have fallen when the second object has fallen 1.00 mm farther than the first object has? Note that this estimate provides only an upper bound on the difference in the accelerations; most physicists believe that there is no difference in the accelerations.

GRAVITATIONAL POTENTIAL ENERGY

46 • (*a*) If we take the potential energy of a 100-kg object and Earth are zero when the two are separated by an infinite distance, what is the potential energy when the object is at the surface of Earth? (*b*) Find the potential energy of the same object at a height above Earth's surface equal to Earth's radius. (*c*) Find the escape speed for a body projected from this height. SSM

47 • Knowing that the acceleration of gravity on the moon is 0.166 times that on Earth and that the moon's radius is $0.273\,R_E$, find the escape speed for a projectile leaving the surface of the moon. SSM

48 •• What initial speed would a particle need to be given at the surface of Earth if it is to have a final speed that is equal to its escape speed when it is very far from Earth? Neglect any effects due to air resistance.

49 •• **Context-Rich, Engineering Application** While preparing its budget for the next fiscal year, NASA wants to report to the nation a rough estimate of the cost (per kilogram) of launching a modern satellite into near-Earth orbit. You are chosen for this task, because you know both physics and accounting. (*a*) Determine the energy, in kW · h, necessary to place a 1.0-kg object in low-Earth orbit. In low-Earth orbit, the height of the object above the surface of Earth is much smaller than Earth's radius. Take the orbital height to be 300 km. (*b*) If this energy can be obtained at a typical electrical energy rate of \$0.15/kW · h, what is the minimum cost of launching a 400-kg satellite into low-Earth orbit? Neglect any effects due to air resistance.

50 •• The science fiction writer Robert Heinlein once said, "If you can get into orbit, then you're halfway to anywhere." Justify this statement by comparing the minimum energy needed to place a satellite into low Earth orbit (h = 400 km) to that needed to set it completely free from the bonds of Earth's gravity. Neglect any effects due to air resistance.

51 •• An object is dropped from rest from a height of 4.0×10^6 m above the surface of Earth. If there is no air resistance, what is its speed when it strikes Earth? SSM

52 •• An object is projected straight upward from the surface of Earth with an initial speed of 4.0 km/s. What is the maximum height it reaches?

53 •• A particle is projected from the surface of Earth with a speed twice the escape speed. When it is very far from Earth, what is its speed?

54 ••• When we calculate escape speeds, we usually do so with the assumption that the object from which we are calculating escape speed is isolated. This is, of course, generally not true in the solar system. Show that the escape speed at a point near a system that consists of two stationary massive spherical objects is equal to the square root of the sum of the squares of the escape speeds from each of the two objects considered individually.

55 ••• Calculate the minimum necessary speed, relative to Earth, for a projectile launched from the surface of Earth to escape the solar system. The answer will depend on the direction of launch. Explain the choice of direction you would make for the direction of the launch in order to minimize the necessary launch speed relative to Earth. Neglect Earth's rotational motion and effects due to air resistance.

56 ••• An object is projected vertically from the surface of Earth at less than the escape speed. Show that the maximum height reached by the object is $H = R_E H'/(R_E - H')$, where H' is the height that it would reach if the gravitational field were constant. Neglect any effects due to air resistance.

GRAVITATIONAL ORBITS

57 •• A 100-kg spacecraft is in a circular orbit about Earth at a height $h = 2R_E$. (*a*) What is the orbital period of the spacecraft? (*b*) What is the spacecraft's kinetic energy? (*c*) Express the angular momentum L of the spacecraft about the center of Earth in terms of the kinetic energy K and find the numerical value of L.

58 •• **ESTIMATION** The orbital period of the moon is 27.3 d, the average center-to-center distance between the moon and Earth is 3.82×10^8 m, the length of an Earth year 365.25 d, and the average center-to-center distance between Earth and the Sun is 1.50×10^{11} m. Use this data to estimate the ratio of the mass of the Sun to the mass of Earth. Compare this estimation to the measured ratio of 3.33×10^5. List some neglected factors that might account for any discrepancy.

59 •• Many satellites orbit Earth at maximum altitudes above Earth's surface of 1000 km or less. *Geosynchronous* satellites, however, orbit at an altitude of 35 790 km above Earth's surface. How much more energy is required to launch a 500-kg satellite into a geosynchronous orbit than into an orbit 1000 km above the surface of Earth? SSM

60 ••• **CONTEXT-RICH, ENGINEERING APPLICATION** The idea of a spaceport orbiting Earth is an attractive proposition for launching probes and/or manned missions to the outer planets of the solar system. Suppose such a "platform" has been constructed, and orbits Earth at a distance of 450 km above Earth's surface. Your research team is launching a lunar probe into an orbit that has its perigee at the spaceport's orbital radius, and its apogee at the moon's orbital radius. (*a*) To launch the probe successfully, first determine the orbital speed for the platform. (*b*) Next, determine the necessary speed relative to the platform that is necessary to launch the probe so that it attains the desired orbit. Assume that any effects due to the gravitational pull of the moon on the probe are negligible. In addition, assume that the launch takes place in a negligible amount of time. (*c*) You have the probe designed to radio back when it has reached apogee. How long after launch should you expect to receive this signal from the probe (neglect the second or so delay for the transit time of the signal back to the platform)?

THE GRAVITATIONAL FIELD ($\vec{g}$)

61 • A 3.0-kg space probe experiences a gravitational force of $12\ \text{N}\hat{i}$ as it passes through point P. What is the gravitational field at point P?

62 • The gravitational field at some point is given by $\vec{g} = 2.5 \times 10^{-6}\ \text{N/kg}\ \hat{j}$. What is the gravitational force on a 0.0040 kg object located at that point?

63 •• A point particle of mass m is on the x axis at $x = L$ and an identical point particle is on the y axis at $y = L$. (*a*) What is the direction of the gravitational field at the origin? (*b*) What is the magnitude of this field? SSM

64 •• Five objects, each of mass M, are equally spaced on the arc of a semicircle of radius R, as in Figure 11-25. An object of mass m is located at the center of curvature of the arc. (*a*) If M is 3.0 kg, m is 2.0 kg, and R is 10 cm, what is the gravitational force on the particle of mass m due to the five objects? (*b*) If the object whose mass is m is removed, what is the gravitational field at the center of curvature of the arc?

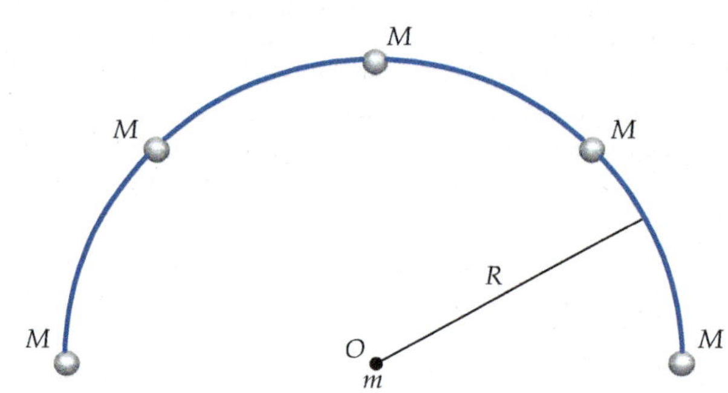

FIGURE 11-25 Problem 64

65 •• A point particle of mass $m_1 = 2.0$ kg is at the origin and a second point particle of mass $m_2 = 4.0$ kg is on the x axis at $x = 6.0$ m. Find the gravitational field $\vec{g}$ at (*a*) $x = 2.0$ m, and (*b*) $x = 12$ m. (*c*) Find the point on the x axis for which $g = 0$.

66 •• Show that on the x axis, the maximum value of g for the field of Example 11-7 occurs at points $x = \pm a/\sqrt{2}$.

67 ••• A nonuniform thin rod of length L lies on the x axis. One end of the rod is at the origin, and the other end is at $x = L$. The rod's mass per unit length λ varies as $\lambda = Cx$, where C is a constant. (Thus, an element of the rod has mass $dm = \lambda\, dx$.) (*a*) Determine the total mass of the rod. (*b*) Determine the gravitational field due to the rod on the x axis at $x = x_0$, where $x_0 > L$. SSM

68 •• A uniform thin rod of mass M and length L lies on the positive x axis with one end at the origin. Consider an element of the rod of length dx, and mass dm, at point x, where $0 < x < L$. (*a*) Show that this element produces a gravitational field at a point x_0 on the x axis in the region $x_0 > L$ is given by $dg_x = -\frac{GM}{L(x_0 - x)^2}dx$. (*b*) Integrate this result over the length of the rod to find the total gravitational field at the point x_0 due to the rod. (*c*) Find the gravitational force on a point particle of mass m_0 at x_0. (*d*) Show that for $x_0 \gg L$, the field of the rod approximates the field of a point particle of mass M at $x = 0$.

THE GRAVITATIONAL FIELD ($\vec{g}$) DUE TO SPHERICAL OBJECTS

69 • A uniform thin spherical shell has a radius of 2.0 m and a mass of 300 kg. What is the gravitational field at the following distances from the center of the shell: (*a*) 0.50 m, (*b*) 1.9 m, (*c*) 2.5 m?

70 • A uniform thin spherical shell has a radius of 2.00 m and a mass of 300 kg and its center is located at the origin of a coordinate system. Another uniform thin spherical shell with a radius of 1.00 m and a mass of 150 kg is inside the larger shell, with its center at 0.600 m on the x axis. What is the gravitational force of attraction between the two shells?

71 •• Two widely separated solid spheres, S_1 and S_2, each have radius R and mass M. Sphere S_1 is uniform, whereas the density of S_2 is given by $\rho(r) = C/r$, where r is the distance from its center. If the gravitational field strength at the surface of S_1 is g_1, what is the gravitational field strength at the surface of S_2? SSM

72 •• Two widely separated uniform solid spheres, S_1 and S_2, have equal masses, but different radii, R_1 and R_2. If the gravitational field strength on the surface of S_1 is g_1, what is the gravitational field strength on the surface of S_2?

73 •• Two concentric uniform thin spherical shells have masses M_1 and M_2 and radii a and $2a$, as in Figure 11-26. What is the magnitude of the gravitational force on a point particle of mass m (not shown) located (*a*) a distance $3a$ from the center of the shells, (*b*) a distance $1.9a$ from the center of the shells, (*c*) a distance $0.9a$ from the center of the shells?

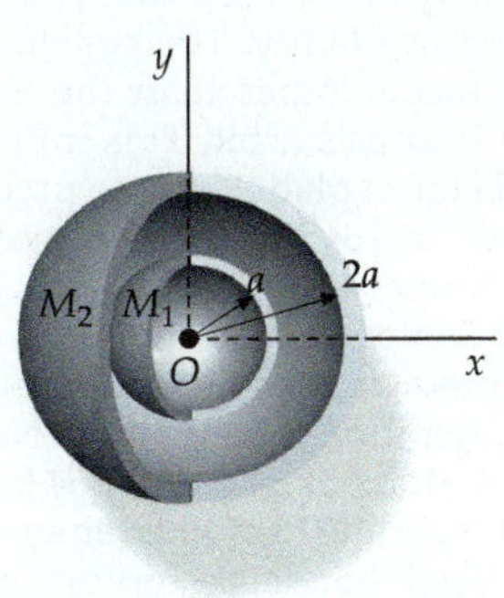

FIGURE 11-26 Problem 73

74 •• The inner spherical shell in Problem 73 is shifted so that its center is now on the x axis at $x = 0.8a$. What is the magnitude of the gravitational force on a particle of point mass m located on the x axis at (*a*) $x = 3a$? (*b*) $x = 1.9a$? (*c*) $x = 0.9a$?

75 •• Suppose you are standing on a spring scale in an elevator that is descending at costant speed in a mine shaft located on the equator. Model Earth as a homogeneous sphere.

(*a*) Show that the force on you due to Earth's gravity alone is proportional to your distance from the center of the planet.

(*b*) Assume that the mine shaft is located on the equator and is vertical. Do not neglect Earth's rotational motion. Show that the reading on the spring scale is proportional to your distance from the center of the planet. SSM

76 •• CONTEXT-RICH Suppose Earth were a nonrotating uniform sphere. As a reward for earning the highest lab grade, your physics professor chooses your laboratory team to participate in a gravitational experiment at a deep mine on the equator. This mine has an elevator shaft going 15.0 km into Earth. Before making the measurement, you are asked to predict the decrease in the weight of a team member, who weighs 800 N at the surface of Earth, when she is at the bottom of the shaft. The density of Earth's crust actually increases with depth. Is your answer higher or lower than the actual experimental result?

77 •• A solid sphere of radius R has its center at the origin. It has a uniform mass density ρ_0, except that the sphere has a spherical cavity in it of radius $r = \frac{1}{2}R$ centered at $x = \frac{1}{2}R$, as in Figure 11-27. Find the gravitational field at points on the x axis for $|x| > R$. *Hint: The cavity may be thought of as a sphere of mass* $m = (4/3)\pi r^3 \rho_0$ *plus a sphere of "negative" mass* $-m$. SSM

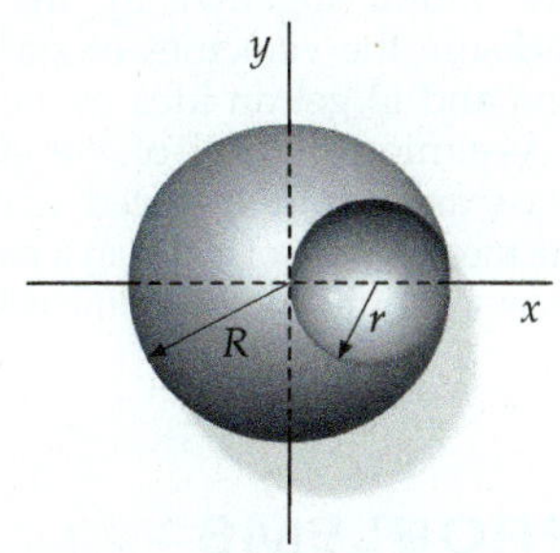

FIGURE 11-27 Problem 77

78 ••• For the sphere with the cavity in Problem 77, show that the gravitational field is uniform throughout the cavity, and find its magnitude and direction there.

79 ••• A straight, smooth tunnel is dug through a uniform spherical planet of mass density ρ_0. The tunnel passes through the center of the planet and is perpendicular to the planet's axis of rotation, which is fixed in space. The planet rotates with a constant angular speed ω, so objects in the tunnel have no apparent weight. Find the required angular speed of the planet ω.

80 ••• The density of a sphere is given by $\rho(r) = C/r$. The sphere has a radius of 5.0 m and a mass of 1.0×10^{11} kg. (*a*) Determine the constant C. (*b*) Obtain expressions for the gravitational field for the regions (1) $r > 5.0$ m, and (2) $r < 5.0$ m.

81 ••• A small-diameter hole is drilled into the sphere of Problem 80 toward the center of the sphere to a depth of 2.0 m below the sphere's surface. A small mass is dropped from the surface into the hole. Determine the speed of the small mass when it strikes the bottom of the hole. SSM

82 ••• **CONTEXT-RICH, ENGINEERING APPLICATION** As a geologist for a mining company, you are working on a method for determining possible locations of underground ore deposits. Assume that where the company owns land the crust of Earth is 40.0 km thick and has a density of about 3000 kg/m^3. Suppose a spherical deposit of heavy metals with a density of 8000 kg/m^3 and radius of 1000 m is centered 2000 m below the surface. You propose to detect it by determining its effect on the local surface value of g. Find $\Delta g/g$ at the surface directly above this deposit, where Δg is the increase in the gravitational field due to the deposit.

83 ••• Two identical spherical cavities are made in a lead sphere of radius R. The cavities each have a radius $R/2$. They touch the outside surface of the sphere and its center as in Figure 11-28. The mass of a solid uniform lead sphere of radius R is M. Find the force of attraction on a point particle of mass m located a distance d from the center of the lead sphere as shown. SSM

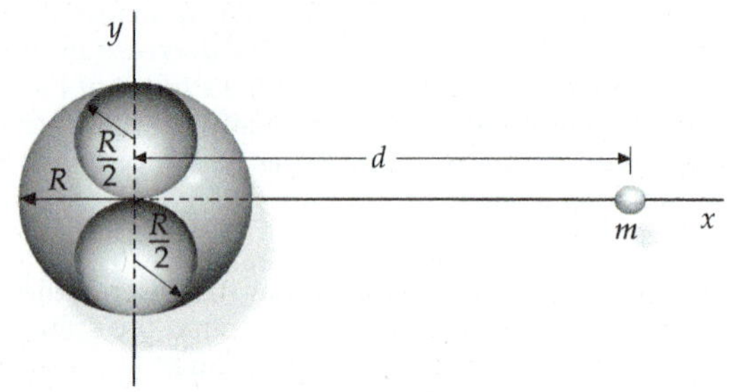

FIGURE 11-28 Problem 83

84 ••• A globular cluster is a roughly spherical collection of up to millions of stars bound together by the force of gravity. Astronomers can measure the velocities of stars in the cluster to study its composition and to get an idea of the mass distribution within the cluster. Assuming that all of the stars have approximately the same mass and are distributed uniformly within the cluster, show that the mean speed of a star in a circular orbit around the center of the cluster should increase linearly with its distance from the center.

GENERAL PROBLEMS

85 • The mean distance of Pluto from the Sun is 39.5 AU. Calculate the period of Pluto's orbital motion.

86 • Calculate the mass of Earth using the known values of G, g, and R_E.

87 •• The force exerted by Earth on a particle of mass m a distance r $(r > R_E)$ from the center of Earth has the magnitude mgR_E^2/r^2, where $g = GM_E/R_E^2$. (*a*) Calculate the work you must do to move the particle from distance r_1 to distance r_2. (*b*) Show that when $r_1 = R_E$ and $r_2 = R_E + h$, the result can be written as $W = mgR_E^2[(1/R_E) - 1/(R_E + h)]$. (*c*) Show that when $h \ll R_E$, the work is given approximately by $W = mgh$.

88 •• The average density of the moon is $\rho = 3340$ kg/m^3. Find the minimum possible period T of a spacecraft orbiting the moon.

89 •• A *neutron star* is a highly condensed remnant of a massive star in the last phase of its evolution. It is composed of neutrons (hence the name), because the star's gravitational force causes electrons and protons to "coalesce" into the neutrons. Suppose at the end of its current phase, the Sun collapsed into a neutron star (it cannot actually do this because it does not have enough mass) of radius 12.0 km, without losing any mass in the process. (*a*) Calculate the ratio of the gravitational acceleration at the surface of the Sun following the collapse compared to the value at the surface of the Sun today. (*b*) Calculate the ratio of the escape speed from the surface of the neutron-Sun to the Sun's value today. SSM

90 •• **CONTEXT-RICH, ENGINEERING APPLICATION** Suppose the Sun could collapse into a neutron star of radius 12.0 km, as in Problem 89. Your research team is in charge of sending a probe from Earth to study the transformed Sun, and the probe needs to end up in a circular orbit 4500 km from the neutron-Sun's center. (*a*) Calculate the orbital speed of the probe. (*b*) Later on, plans call for construction of a permanent spaceport in that same orbit. To transport equipment and supplies, scientists on Earth need you to determine the escape speed for rockets launched from the spaceport (relative to the spaceport) in the direction of the spaceport's orbital velocity at takeoff time. What is that speed, and how does it compare to the escape speed at the surface of Earth?

91 •• A satellite is circling the moon (radius 1700 km) close to the surface at a speed v. A projectile is launched vertically up from the moon's surface at the same initial speed v How high will the projectile rise?

92 •• *Black holes* are objects whose gravitational field is so strong that not even light can escape. One way of thinking about this is to consider a spherical object whose density is so large that the escape speed at its surface is greater than the speed of light, c. If a star's radius is smaller than a value called the *Schwarzschild radius* R_S, then the star will be a black hole, that is, light originating from its surface cannot escape. (*a*) For a nonrotating black hole, the Schwarzschild radius depends only upon the mass of the black hole. Show that it is related to that mass M by $R_S = (2GM)/c^2$. (*b*) Calculate the value of the Schwarzschild radius for a black hole whose mass is ten solar masses.

93 •• In a binary star system, two stars follow circular orbits about their common center of mass. If the stars have masses m_1 and m_2 and are separated by a distance r, show that the period of rotation is related to r by $T^2 = 4\pi^2 r^3/[G(m_1 + m_2)]$.

94 •• Two particles of masses m_1 and m_2 are released from rest at a large separation distance. Find their speeds v_1 and v_2 when their separation distance is r. The initial separation distance is given as large, but large is a relative term. Relative to what distance is it large?

95 •• Uranus, the seventh planet in the solar system, was first observed in 1781 by William Herschel. Its orbit was then analyzed in terms of Kepler's laws. By the 1840s, observations of Uranus clearly indicated that its true orbit was different from the Keplerian calculation by an amount that could not be accounted for by observational uncertainty. The conclusion was that there must be another influence other than the Sun and the known planets lying inside Uranus's orbit. This influence was hypothesized to be due to an eighth planet, whose predicted orbit was described independently in 1845 by two astronomers: John Adams (no relation to the former president of the United States) and Urbain LeVerrier. In September of 1846, John Galle, searching in the sky at the place predicted by Adams and LeVerrier, made the first observation of Neptune. Uranus and Neptune are in orbit about the Sun with periods of 84.0 and 164.8 years, respectively. To see the effect that Neptune had on Uranus, determine the ratio of the gravitational force between Neptune and Uranus to that between Uranus and the Sun, when Neptune and Uranus are at their closest approach to one another (i.e., when aligned with the

Sun). The masses of the Sun, Uranus, and Neptune are 333,000; 14.5; and 17.1 times that of Earth, respectively. SSM

96 •• It is believed that there is a "supermassive" black hole at the center of our galaxy. One datum that leads to this conclusion is the important recent observation of stellar motion in the vicinity of the galactic center. If one such star moves in an elliptical orbit with a period of 15.2 years and has a semimajor axis of 5.5 light-days (the distance light travels in 5.5 days), what is the mass around which the star moves in its Keplerian orbit?

97 •• Four identical planets are arranged in a square, as shown in Figure 11-29. If the mass of each planet is M and the edge length of the square is a, what must their speed be if they are to orbit their common center under the influence of their mutual attraction? SSM

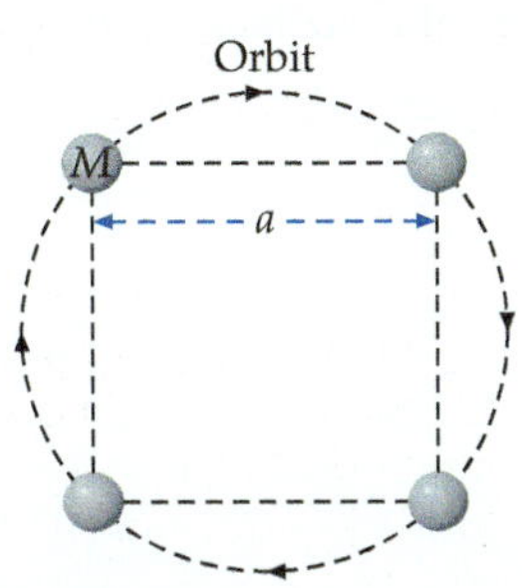

FIGURE 11-29 Problem 97

98 •• A hole is drilled from the surface of Earth to its center, as in Figure 11-30. Ignore Earth's rotation and any effects due to air resistance, and model Earth as a uniform sphere. (*a*) How much work is required to lift a particle of mass m from the center of Earth to Earth's surface? (*b*) If the particle is dropped from rest at the surface of Earth, what is its speed when it reaches the center of Earth? (*c*) What is the escape speed for a particle projected from the center of Earth? Express your answers in terms of m, g, and R_E.

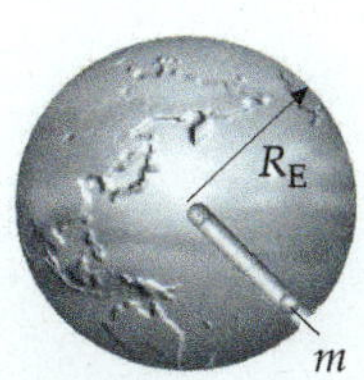

FIGURE 11-30 Problem 98

99 •• A thick spherical shell of mass M and uniform density has an inner radius R_1 and an outer radius R_2. Find the gravitational field g_r as a function of r for $0 < r < \infty$. Sketch a graph of g_r versus r.

100 •• (*a*) A thin uniform ring of mass M and radius R lies in the $x = 0$ plane and is cenered at the origin. Sketch a plot of the gravitational field g_x versus x for all points on the x axis. (*b*) At what point, or points, on the axis is the magnitude of g_x a maximum?

101 ••• Find the magnitude of the gravitational field that is at a distance r from an infinitely long uniform thin rod whose mass per unit length is λ.

102 ••• One question in early planetary science was whether each of the rings of Saturn were solid or were, instead, composed of individual chunks, each in its own orbit. The issue could be resolved by an observation in which astronomers would measure the speed of the inner and outer portions of the ring. If the inner portion of the ring moved more slowly than the outer portion, then the ring was solid; if the opposite was true, then it was actually composed of separate chunks. Let us see how this results from a theoretical viewpoint. Let the radial width of a given ring (there are many) be Δr, the average distance of that ring from the center of Saturn be represented by R, and the average speed of that ring be v_{avg}. (*a*) If the ring is solid, show that the *difference* in speed between its outermost and innermost portions, Δv, is given by the expression $\Delta v = v_{out} - v_{in} \approx v_{avg}(\Delta r/R)$. Here, v_{out} is the speed of the outermost portion of the ring, and v_{in} is the speed of the innermost portion. (*b*) If, however, the ring is composed of many small chunks, show that $\Delta v \approx -\frac{1}{2}(v_{avg}(\Delta r/R))$. (Assume that $\Delta r \ll R$.)

103 ••• Find the gravitational potential energy of the thin rod in Example 11-8 and a point particle of mass m_0 that is on the x axis at $x = x_0$ where $x_0 \geq \frac{1}{2}L$. (*a*) Show that the potential energy shared by an element of the rod of mass dm (shown in Figure 11-14) and the point particle of mass m_0, is given by

$$dU = -\frac{Gm_0\,dm}{x_0 - x_s} = \frac{GMm_0}{L(x_0 - x_s)}dx_s$$

where $U = 0$ at $x_0 = \infty$. (*b*) Integrate your result for Part (*a*) over the length of the rod to find the total potential energy for the system. Generalize your function $U(x_0)$ to any place on the x axis in the region $x > L/2$ by replacing x_0 with a general coordinate x and write it as $U(x)$. (*c*) Compute the force on m_0 at a general point x using $F_x = -dU/dx$ and compare your result with m_0g, where g is the field at x_0 calculated in Example 11-8. SSM

104 ••• A uniform sphere of mass M is located near a thin, uniform rod of mass m and length L, as in Figure 11-31. Find the gravitational force of attraction exerted by the sphere on the rod.

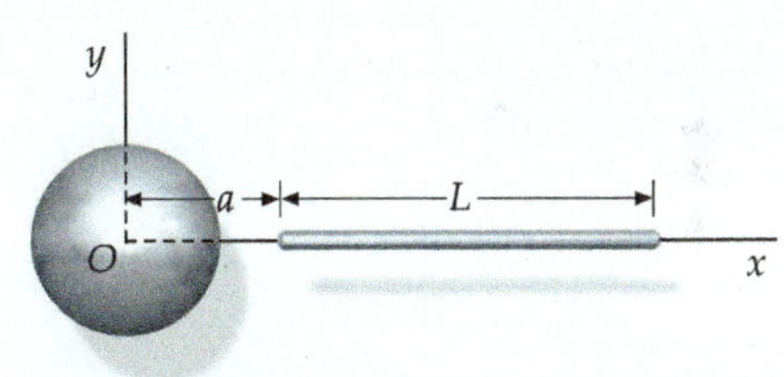

FIGURE 11-31 Problem 104

105 ••• A thin uniform 20-kg rod with a length equal to 5.0 m is bent into a semicircle. What is the gravitational force exerted by the rod on a 0.10-kg point mass located at the center of curvature of the circular arc?

106 ••• Both the Sun and the moon exert gravitational forces on the oceans of Earth, causing tides. (*a*) Show that the ratio of the force exerted on a point particle on the surface of Earth by the Sun to that exerted by the moon is $M_S r_m^2/M_m r_S^2$. Here M_S and M_m represent the masses of the Sun and moon, and r_S and r_m are the distances of the particle from Earth to the Sun and Earth to the moon, respectively. Evaluate this ratio numerically. (*b*) Even though the Sun exerts a much greater force on the oceans than does the moon, the moon has a greater effect on the tides because it is the *difference* in the force from one side of Earth to the other that is important. Differentiate the expression $F = Gm_1m_2/r^2$ to calculate the change in F due to a small change in r. Show that $dF/F = -2\,dr/r$. (*c*) The oceanic *tidal bulge* (that is, the elongation of the liquid water of the oceans causing two opposite high and two opposite low spots) is caused by the difference in gravitational force on the oceans from one side of Earth to the other. Show that for a small difference in

distance compared to the average distance, the ratio of the differential gravitational force exerted by the Sun to the differential gravitational force exerted by the moon on Earth's oceans is given by $\Delta F_S/\Delta F_m \approx (M_S r_m^3)/(M_m r_S^3)$. Calculate this ratio. What is your conclusion? Which object, the moon or the Sun, is the main cause of the tidal stretching of the oceans on Earth?

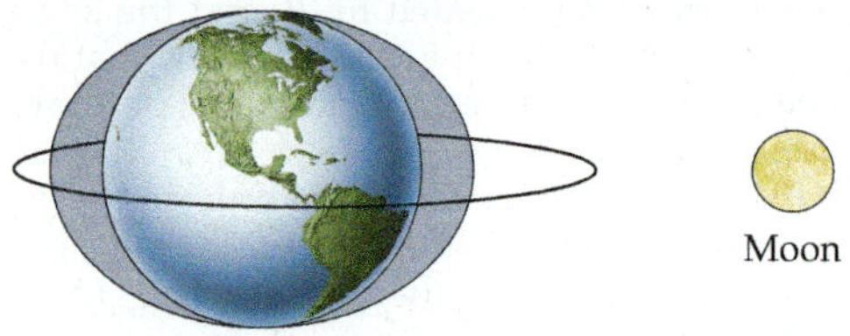

FIGURE 11-32 Problem 106 The lunar tidal bulges (exaggerated here) are due to the difference in the gravitational pull of the moon on opposite sides of Earth.

107 ••• **CONTEXT-RICH, ENGINEERING APPLICATION** United Federation Starship *Excelsior* drops two small robotic probes toward the surface of a neutron star for exploration. The mass of the star is the same as that of the Sun, but the star's diameter is only 10 km. The robotic probes are linked together by a 1.0-m-long steel cord (which includes communication lines between the two probes), and are dropped vertically (that is, one always above the other). The ship hovers at rest above the star's surface. As the Chief of Materials Engineering on the ship, you are concerned that the communication between the two probes, a crucial aspect of the mission, will not survive. (*a*) Outline your briefing session to the mission commander and explain the existence of a "stretching force" that will try to pull the robots apart as they fall toward the planet. (See Problem 106 for hints.) (*b*) Assume that the cord in use has a breaking tension of 25 kN and that the robots each have a mass of 1.0 kg. How close will the robots be to the surface of the star before the cord breaks?

CHAPTER 12

LARGE FORCES AND TORQUES ARE OFTEN EXERTED ON CONSTRUCTION CRANES SUCH AS THIS ONE. CRANES HAVE TO BE BOTH RIGID AND WELL ANCHORED IF THEY ARE TO WITHSTAND SUCH FORCES AND TORQUES AND NOT COLLAPSE. *(Eric M. Anderson/Tower Cranes of America, Inc.)*

Static Equilibrium and Elasticity

? Tower cranes are a part of the landscape of cities around the world. The model shown has a maximum reach of 81 m. Counterweights are used to counterbalance the load and prevent the crane from tipping over. (See Example 12-5.)

In this chapter, we study the forces and torques needed to keep extended objects static (stationary). For example, the forces exerted by the cables of a suspension bridge must be known so the cables can be designed to be strong enough to support the bridge. Similarly, cranes must be designed so that they do not topple over when lifting a weight.

In this chapter, we study the equilibrium of rigid bodies and then briefly consider the deformations and elastic forces that arise when real solids are under stress.

12-1 CONDITIONS FOR EQUILIBRIUM

A necessary condition for a particle at rest to remain at rest is that the net force acting on it remains zero. Similarly, a necessary condition for the center of mass of a rigid object to remain at rest is that the net force acting on the object remains zero. A rigid object can rotate, even when its center of mass is at rest, but then the object is not in *static equilibrium.* Therefore, a second necessary condition for a rigid object to remain in static equilibrium is that the net torque acting on it about *any* axis must remain zero. This condition gives us the option to choose any point or any axis for calculating torques, an option that greatly simplifies the solution of most static problems.

The two necessary conditions for a rigid body to be in static equilibrium are as follows:

1. The net external force acting on the body must remain zero:

$$\Sigma\vec{F} = 0 \qquad \text{12-1}$$

2. The net external torque about any point must remain zero:

$$\Sigma\vec{\tau} = 0 \qquad \text{12-2}$$

CONDITIONS FOR EQUILIBRIUM

12-2 THE CENTER OF GRAVITY

In Section 4 of Chapter 9, the center of gravity is introduced in terms of torques about an axis. Here we introduce the center of gravity in terms of torques about a point. Figure 12-1*a* shows a rigid object in static equilibrium and a point O. We consider the object to be composed of many small mass elements. The force of gravity on the ith small mass element is $\vec{F}_{gi}$, and the total force of gravity on the object is $\vec{F}_g = \Sigma\vec{F}_{gi}$. If $\vec{r}_i$ is the position vector of the ith particle relative to O, then $\vec{\tau}_i = \vec{r}_i \times \vec{F}_{gi}$, where $\vec{\tau}_i$ is the torque due to $\vec{F}_{gi}$ about O. The net gravitational torque about O is then $\vec{\tau}_{\text{net}} = \Sigma(\vec{r}_i \times \vec{F}_{gi})$. Conveniently, the net torque due to gravity about a point can be calculated as if the entire force of gravity $\vec{F}_g$ were applied at a single point, the **center of gravity** (see Figure 12-1*b*). That is,

$$\vec{\tau}_{\text{net}} = \vec{r}_{\text{cg}} \times \vec{F}_g \qquad \text{12-3}$$

CENTER OF GRAVITY

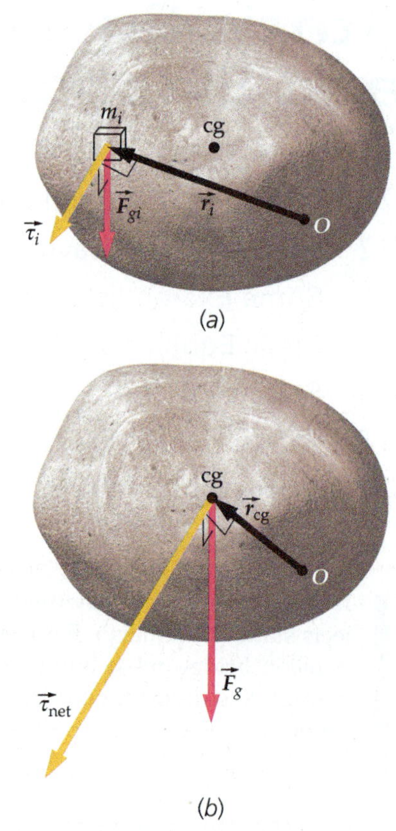

FIGURE 12-1 The direction of the torque is gotten by applying the right-hand rule for cross products. (*a*) $\vec{\tau}_i$ is the torque about O due to the gravitational force $\vec{F}_{gi}$ on the ith element of mass. (*b*) The net gravitational torque $\vec{\tau}$ about O can be calculated by considering the total gravitational force $\vec{F}_g$ to be applied at a point called the center of gravity.

where $\vec{r}_{\text{cg}}$ is the position vector of the center of gravity relative to O.

If the gravitational field $\vec{g}$ is uniform over the object (as is nearly always the case for objects of less than astronomical size), we can write $\vec{F}_{gi} = m_i\vec{g}$. Summing both sides of this gives $\vec{F}_g = M\vec{g}$, where $M = \Sigma m_i$ is the mass of the object. The net torque is the sum of the individual torques. That is,

$$\vec{\tau}_{\text{net}} = \sum_i(\vec{r}_i \times \vec{F}_{gi}) = \sum_i(\vec{r}_i \times m_i\vec{g}) = \sum_i(m_i\vec{r}_i \times \vec{g})$$

Factoring $\vec{g}$ from the term on the right gives

$$\vec{\tau}_{\text{net}} = \left(\sum_i m_i\vec{r}_i\right) \times \vec{g}$$

and substituting for $\Sigma m_i \vec{r}_i$ using the definition of center of mass ($M\vec{r}_{cm} = \Sigma m_i \vec{r}_i$), we obtain

$$\vec{\tau}_{net} = M\vec{r}_{cm} \times \vec{g} = \vec{r}_{cm} \times M\vec{g} = \vec{r}_{cm} \times \vec{F}_g \qquad 12\text{-}4$$

Equations 12-3 and 12-4 are valid for any choice of the point O only if $\vec{r}_{cg} = \vec{r}_{cm}$. That is, the center of gravity and the center of mass coincide if the object is in a uniform gravitational field.

If O is directly above the center of gravity, then $\vec{r}_{cg}$ and $\vec{F}_g$ are both in the same direction (downward), so $\vec{\tau}_{net} = \vec{r}_{cg} \times \vec{F}_g = 0$. For example, when a mobile is suspended with its center of gravity directly below its suspension point, the net torque on the mobile about the suspension point is zero, so it is in static equilibrium.

The center of gravity of the Calder mobile is directly below the suspension point. *(Copyright © 2002 Estate of Alexander Calder/ Artists Rights Society (ARS), New York.)*

12-3 SOME EXAMPLES OF STATIC EQUILIBRIUM

For most examples and problems in this chapter, all the forces are perpendicular to the z axis. It is therefore best in such problems to calculate torques about an axis parallel to the z axis (rather than about a point). For figures in this chapter, the z axis is typically perpendicular to the page, and out of the page is frequently chosen as $+z$ direction. Calculating torques about the z axis and choosing out of the page as the $+z$ direction is equivalent to choosing counterclockwise as positive and clockwise as negative. (If into the page is chosen as the $+z$ direction, then clockwise is positive and counterclockwise is negative.)

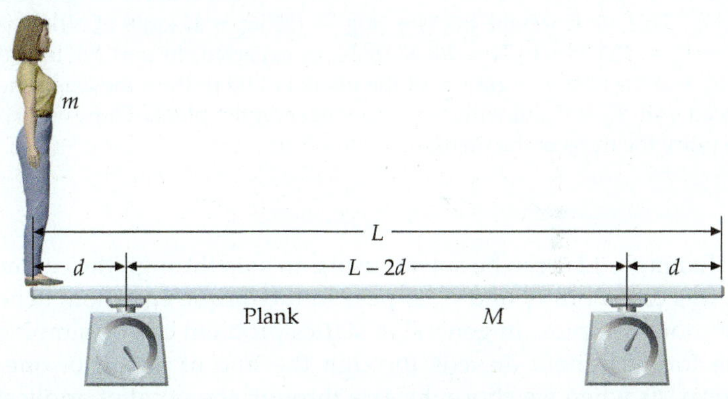

FIGURE 12-2

Example 12-1 Walking the Plank

A uniform plank of length $L = 3.00$ m and mass $M = 35$ kg is supported by scales a distance $d = 0.50$ m from the ends of the board, as shown in Figure 12-2. (*a*) Find the reading on the scales when Mary, whose mass $m = 45$ kg, stands on the left end of the plank. (*b*) Sergio climbs onto the plank and walks toward Mary, who jumps to the floor when the plank starts to tip. Sergio keeps walking all the way to the left end of the plank, and when he gets there the scale supporting the right end of the plank reads zero. Find Sergio's mass.

PICTURE The readings on the scales are the magnitudes of the forces Mary and Sergio exert on the boards. To find these magnitudes, we apply the two conditions for equilibrium to the system consisting of Mary plus the plank.

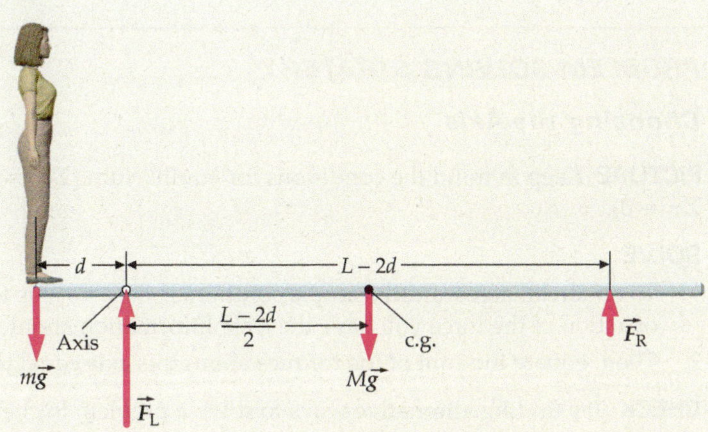

FIGURE 12-3

SOLVE

(*a*) 1. Draw a free-body diagram of the system consisting of Mary and the plank (Figure 12-3). Forces $\vec{F}_L$ and $\vec{F}_R$ are the forces exerted by the left and right scales.

2. Set the net force equal to zero, taking upward as positive:

$$\Sigma F_y = 0$$
$$F_L + F_R - Mg - mg = 0$$

3. Calculate the net torque about the axis directed out of the page (making counterclockwise positive) and through the point of application of $\vec{F}_L$:

$$\Sigma\tau = F_L(0) + F_R(L - 2d) - Mg\frac{L - 2d}{2} + mgd$$

4. Set the net torque equal to zero and solve for F_R:

$$0 = F_R(L - 2d) - Mg\frac{L - 2d}{2} + mgd$$

$$\text{so}\quad F_R = \left(\frac{1}{2}M - \frac{d}{L - 2d}m\right)g$$

5. Substitute this result for F_R into the step-2 result and solve for F_L:

$$F_L = Mg + mg - \left(\frac{1}{2}M - \frac{d}{L - 2d}m\right)g = \left(\frac{1}{2}M + \frac{L - d}{L - 2d}m\right)g$$

6. Substitute numerical values to obtain numerical values for the forces:

$$F_R = \left(\frac{1}{2}35\text{ kg} - \frac{0.50\text{ m}}{1.5\text{ m}}45\text{ kg}\right)(9.81\text{N/kg})$$

$$= 61.3\text{ N} = \boxed{61\text{ N}}$$

$$F_L = \left(\frac{1}{2}35\text{ kg} + \frac{2.5\text{ m}}{2.0\text{ m}}45\text{kg}\right)(9.81\text{ N/kg})$$

$$= 723\text{ N} = \boxed{7.2 \times 10^2\text{ N}}$$

(*b*) Using the Part-(*a*), step-4 result, set $F_R = 0$ and solve for m:

$$0 = \left(\frac{1}{2}M - \frac{d}{L - 2d}m\right)g$$

$$\text{so}\quad m = \frac{L - 2d}{2d}M = \frac{2.0\text{ m}}{1.0\text{ m}}35\text{ kg} = \boxed{70\text{ kg}}$$

CHECK In part (*a*), the sum $F_L + F_R$ should equal Mary's weight plus the weight of the plank. This total weight is $(M + m)g = (35\text{ kg} + 45\text{ kg})(9.81\text{ N/kg}) = 7.8 \times 10^2\text{ N}$. Also, $F_L + F_R = 723\text{ N} + 61\text{ N} = 7.8 \times 10^2\text{ N}$, as expected. In part (*b*), Sergio is 0.50 m from the axis, and the center of gravity of the plank is 1.00 m from the axis, when the system is balanced with $F_R = 0$ and with Mary no longer on the plank. Thus, we expect Sergio's mass to be twice the mass of the plank.

Example 12-1 can be solved using an axis through the center of the plank, but in this case both F_L and F_R appear in the torque equation, hence the algebra is a bit more complex. In general, a statics problem can be simplified by computing the torques about an axis through the line of action of one of the unknown forces, as when we chose the axis through the point of application of force F_L in Example 12-1.

PROBLEM-SOLVING STRATEGY

Choosing the Axis

PICTURE Keep in mind the conditions for equilibrium ($\Sigma\vec{F} = 0$ and $\Sigma\tau = 0$).

SOLVE

1. To obtain an algebraically simple solution, choose an axis through the line of action of the force you have the least information about.
2. Then, equate the sum of the torques about this axis equal to zero.

CHECK Try finding alternative ways to solve a problem to check the plausibility of your solution.

Example 12-2 Force on an Elbow

You hold a 6.0-kg weight in your hand with your forearm making a 90° angle with your upper arm, as shown in Figure 12-4. Your biceps muscle exerts an upward force $\vec{F}_m$ that acts 3.4 cm from the pivot point O at the elbow joint. Model the forearm and hand as a 30.0-cm-long uniform rod with a mass of 1.0 kg. (*a*) Find the magnitude of $\vec{F}_m$ if the distance from the weight to the pivot point (elbow joint) is 30 cm, and (*b*) find the magnitude and direction of the force exerted on the elbow joint by the upper arm.

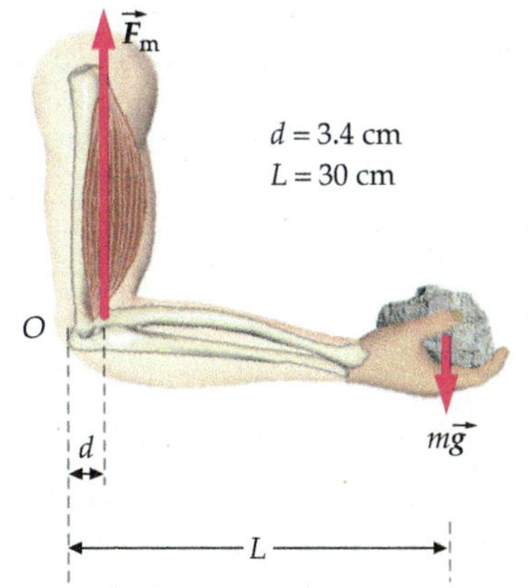

FIGURE 12-4

PICTURE To find the two forces, apply the two conditions for static equilibrium ($\Sigma\vec{F} = 0$ and $\Sigma\tau = 0$) to the forearm.

SOLVE

(*a*) 1. Draw a free-body diagram of the forearm (Figure 12-5). Model the forearm as a horizontal rod.

2. The force we know least about is the force of the upper arm on the elbow joint $\vec{F}_{ua}$ (we know neither its magnitude nor its direction). Apply $\Sigma\tau = 0$ about an axis directed out of the page and through the point of application of $\vec{F}_{ua}$:

$$F_{ua}(0) - m_h g\frac{L}{2} + F_m d - mgL = 0$$

so

$$F_m = \left(\frac{1}{2}m_h + m\right)g\frac{L}{d}$$

$$= \left(\frac{1}{2}(1.0\text{ kg}) + 6.0\text{ kg}\right)(9.81\text{ N/kg})\frac{30\text{ cm}}{3.4\text{ cm}}$$

$$= 563\text{ N} = \boxed{5.6 \times 10^2\text{ N}}$$

(*b*) Apply $\Sigma F_x = 0$ and $\Sigma F_y = 0$ to obtain $\vec{F}_{ua}$:

$$F_{uax} + 0 + 0 + 0 = 0$$

and $F_{uay} + F_m - m_h g - mg = 0$

so $F_{uax} = 0$

and $F_{uay} = (m + m_h)g - F_m$

$$= (7.0\text{ kg})(9.81\text{ N/kg}) - 563\text{ N}$$

$$= -494\text{ N}$$

Therefore,

$$\vec{F}_{ua} = \boxed{4.9 \times 10^2\text{ N, down}}$$

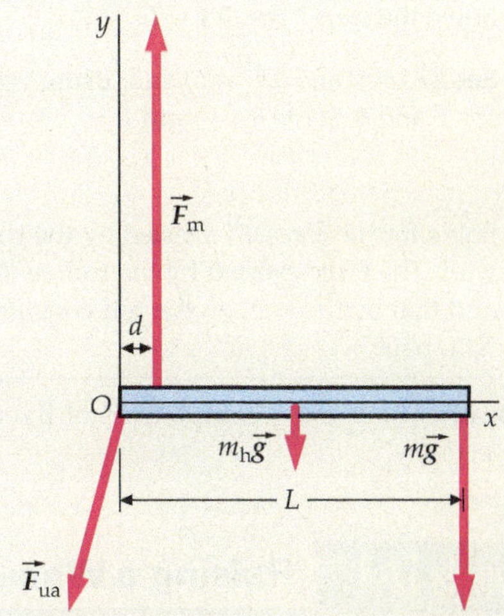

FIGURE 12-5

CHECK F_{ua} can be found in one step by choosing the axis to be through the point where the biceps attaches to the forearm. Setting net torque equal to zero gives $F_{ua}(3.4\text{ cm}) + F_m(0) - (6.0\text{ kg})(9.81\text{ N/kg})(30.0\text{ cm} - 3.4\text{ cm}) - (1.0\text{ kg})(9.81\text{ N/kg})(15.0\text{ cm} - 3.4\text{ cm}) = 0$. This equation yields $F_{ua} = 4.9 \times 10^2$ N, the same as our Part-(*b*) result.

TAKING IT FURTHER (1) The force that must be exerted by the muscle is 9.6 times the weight of the object! In addition, as the muscle pulls upward, the upper arm must push downward to keep the forearm in equilibrium. The force exerted by the upper arm is 8.4 times greater than the object's weight. (2) This example and plausibility check show that we can choose the axis to be wherever it is convenient for our calculation.

Example 12-3 Hanging a Sign *Try It Yourself*

The manager of the campus bookstore has ordered a new 20-kg sign to hang in front of the store, from the end of a rod that will be attached to the wall by a wire (Figure 12-6). The manager needs to know how strong a wire is needed. She knows you are taking a physics course, so she asks you to calculate the tension in the wire. She is also concerned about how much force the rod puts on the wall, so she asks you to calculate that as well. The rod has a length of 2.0 m and a mass of 4.0 kg, and the wire is attached to a point on the wall 1.0 m above the rod.

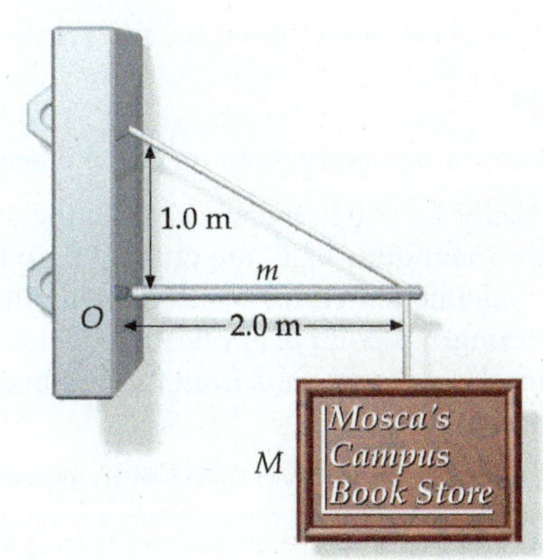

FIGURE 12-6

PICTURE The conditions for the rod to be in equilibrium are $\Sigma F_x = 0$, $\Sigma F_y = 0$, and $\Sigma\tau = 0$, where the torques are to be calculated about an axis through the line of action of the force we have the least information about. The force exerted by the rod on the wall is equal but opposite to the force exerted by the wall on the rod.

SOLVE

Cover the column to the right and try these on your own before looking at the answers.

Steps	Answers
1. Draw a free-body diagram for the rod (Figure 12-7).	
2. Set $\Sigma\tau = 0$ about an axis perpendicular to the page and through point O, which is on the line of action of the force of the wall on the rod:	$TL\sin\theta - MgL - mg\frac{L}{2} = 0$ so $T = \frac{(M + \frac{1}{2}m)g}{\sin\theta}$
3. Use trigonometry to solve for θ:	$\theta = \tan^{-1}\frac{1}{2} = 26.6°$
4. Solve the step-2 result for T:	$T = 483\text{ N} = \boxed{4.8 \times 10^2\text{ N}}$
5. Set $\Sigma F_x = 0$ and $\Sigma F_y = 0$ and, using your values for T and θ, solve for F_x and F_y:	$F_x + T_x = 0$ $F_y + T_y - Mg - mg = 0$ so $F_x = 432\text{ N}, F_y = 19.2\text{ N}$
6. Solve for the force $\vec{F}'$ exerted by the rod on the wall. The force exerted by the rod on the wall and that by the wall on the rod constitute a N3L pair:	$\vec{F}' = -\vec{F} = \boxed{-4.3 \times 10^2\text{ N}\hat{i} - 19\text{ N}\hat{j}}$

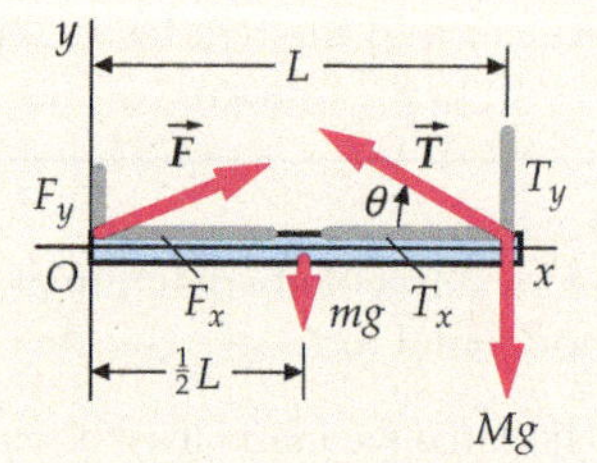

FIGURE 12-7

CHECK The x and y components of the force on the wall by the rod are both negative, as expected.

Example 12-4 Raising a Wheel

Try It Yourself

A wheel of mass M and radius R (Figure 12-8) rests on a horizontal surface against a step of height $h(h < R)$. The wheel is to be raised over the step by a horizontal force $\vec{F}$ applied to the axle of the wheel, as shown. Find the minimum force F_{min} necessary to raise the wheel over the step.

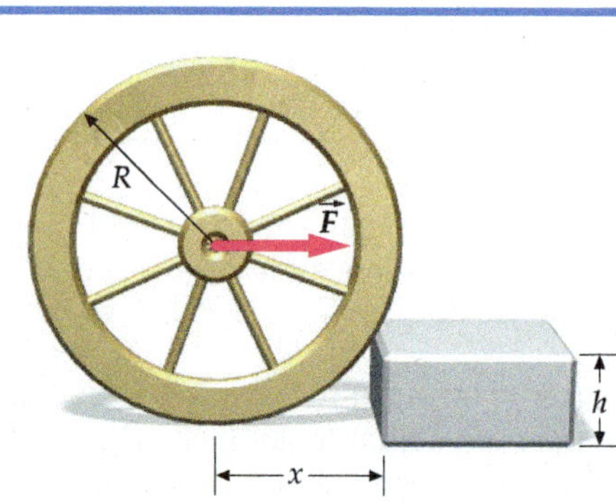

FIGURE 12-8

PICTURE If the magnitude of F is less than F_{min}, the surface at the bottom of the wheel exerts an upward normal force on the wheel. If F is increased, this normal force decreases. Apply the conditions for static equilibrium to find the value of F that will hold the wheel in place when the normal force is zero.

SOLVE

Cover the column to the right and try these on your own before looking at the answers.

Steps	Answers
1. Draw a free-body diagram of the wheel (Figure 12-9).	
2. Apply $\Sigma\tau = 0$ to the wheel. Both the direction and the magnitude of $\vec{F}'$ are unknown, so follow the guidelines and calculate torques about an axis through its point of application. Obtain expressions for the moment arms from the free-body diagram and solve for F_{min}:	$\Sigma\tau = F_{min}(R - h) - Mgx = 0$ so $F_{min} = \frac{Mgx}{R - h}$
3. Use the Pythagorean theorem to express x in terms of h and R:	$x = \sqrt{h(2R - h)}$
4. Substitute $\sqrt{h(2R - h)}$ for x to obtain an expression for F_{min}:	$F_{min} = \boxed{\frac{\sqrt{h(2R - h)}}{R - h}Mg}$

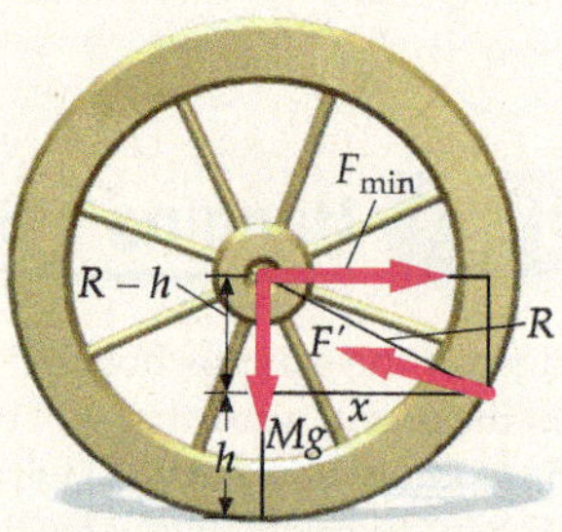

FIGURE 12-9

CHECK We evaluate F_{min} for the limiting cases of $h = 0$ and $h = R$. There is no curb for $h = 0$, so we expect F_{min} to equal zero, and for $h = R$ we expect that no force is big enough to get the wheel to roll up the step. Our step-4 result gives $F_{min} = 0$ if $h = 0$, as expected, and gives $F_{min} \to \infty$ as $h \to R$, again as expected.

TAKING IT FURTHER Applying $\Sigma\tau = 0$ about the axis through the center of the wheel shows that $\vec{F}'$ is directed toward the wheel's center. (Otherwise, there would be a nonzero net torque about the axis.)

Example 12-5 Balancing a Crane

Figure 12-10 shows a standard K-10000 tower crane. The horizontal members extending to either side of the tower are called *jibs.* The tower is 12 m across. The forward jib is 80 m long with a mass of $m_{FJ} = 80$ t (1 t = 1 tonne = 1000 kg). The counterweight (cw) jib is 44 m long with mass of $m_{cwJ} = 31$ t, the fixed counterweight has a mass of $m_{Fcw} = 100$ t, the outer mobile cw has a mass of $m_{OMcw} = 40$ t, the inner mobile cw has a mass of $m_{IMcw} = 83$ t, and the tower has a mass of $m_T = 100$ t. A load with a mass of $m_L = 100$ t is suspended from the center of the forward jib. Is the crane balanced or unbalanced? If unbalanced, would you move the load toward or away from the tower in order to balance the load?

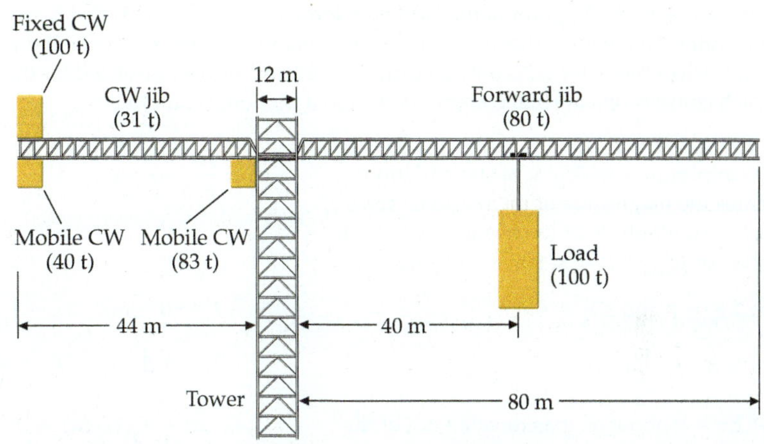

FIGURE 12-10 CW stands for counterweight.

PICTURE The crane is balanced if the center of gravity, and thus the center of mass, is inside the tower. Model each jib as a uniform rod and each counterweight as a point mass. Calculate the x component of the center of mass of the crane and load, where the $+x$ direction it to the right in Figure 12-10. If the center of gravity is within the tower, the crane is balanced.

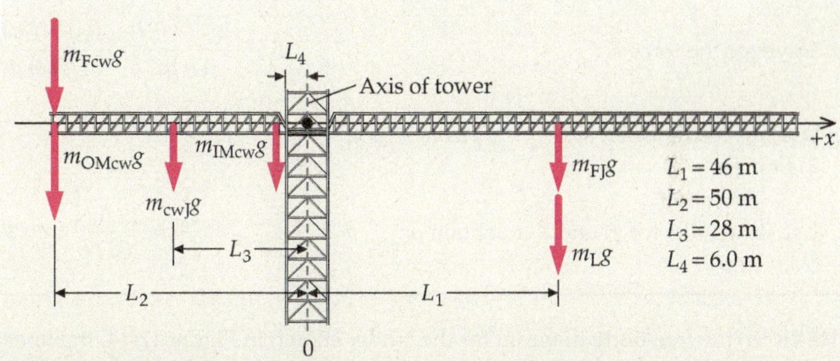

FIGURE 12-11

SOLVE

1. Draw a free-body diagram of the crane and load (Figure 12-11). Draw the x axis with the origin at the center of the tower.

2. Calculate the center of mass of the system:

$$Mx_{cm} = m(m_{FJ} + m_L)L_1 + m_T(0) - (m_{Fcw} + m_{OMcw})L_2 - m_{cwJ}L_3 - m_{IMcw}L_4$$

$$\text{so} \quad x_{cm} = \frac{(180\text{ t})(46\text{ m}) + 0 - (140\text{ t})(50\text{ m}) - (31\text{ t})(28\text{ m}) - (83\text{ t})(6.0\text{ m})}{180\text{ t} + 100\text{ t} + 140\text{ t} + 31\text{ t} + 83\text{ t}}$$

$$= -0.16\text{ m}$$

3. If the center of mass is outside the tower, the crane is unbalanced:

The center of gravity is 16 cm to the left of the axis of the tower. The center of mass is within the tower so **the crane is balanced.**

CHECK The left end of the cw jib is at $x = -50$ m, and the right end of the forward jib is at $x = 86$ m. Thus, the step-2 result is plausible because it is within the range $-50\text{ m} \le x \le +86\text{ m}$. This is a somewhat crude plausibility check, but if the step-2 result was not within this range, the result would definitely not be plausible.

TAKING IT FURTHER The tower is fastened to a rotating platform that is firmly anchored to a massive concrete base.

Example 12-6 A Leaning Ladder

A uniform 5.0-m ladder weighing 60 N leans against a frictionless vertical wall, as shown in Figure 12-12. The foot of the ladder is 3.0 m from the wall. What is the minimum coefficient of static friction necessary between the ladder and the floor if the ladder is not to slip?

PICTURE There are three conditions for the ladder to be in equilibrium: $\Sigma F_x = 0$, $\Sigma F_y = 0$, and $\Sigma\tau = 0$. Apply these along with $f_s \le \mu_s F_n$ to solve for the minimum value of μ_s needed to prevent slipping.

SOLVE

1. Draw a free-body diagram of the ladder as shown in Figure 12-13. The forces acting on the ladder are the force due to gravity $\vec{F}_g$, the force $\vec{F}_1$ exerted by the wall (because the wall is frictionless, it exerts only a normal force), and the force exerted by the floor, which consists of a normal component F_n and a frictional component f_s.

2. The minimum coefficient of static friction relates the magnitude of the frictional force f_s and the magnitude of the normal force F_n. To solve for $\mu_{s\min}$, we first solve for f_s and F_n:
$$\mu_s \ge \frac{f_s}{F_n} \quad \text{so} \quad \mu_{s\min} = \frac{f_s}{F_n}$$

3. Set $\Sigma F_x = 0$ and $\Sigma F_y = 0$:
$$f_s - F_1 = 0 \quad \text{and} \quad F_n - F_g = 0$$

4. Solve for f_s and F_n:
$$f_s = F_1 \quad \text{and} \quad F_n = F_g = 60\text{ N}$$

5. Set $\Sigma\tau = 0$ about an axis directed out of the page and through the foot of the ladder, the point of application of the force we know the least about:
$$F_1(4.0\text{ m}) - F_g(1.5\text{ m}) = 0$$

6. Solve for the force F_1:
$$F_1 = \frac{F_g(1.5\text{ m})}{4.0\text{ m}} = \frac{(60\text{ N})(1.5\text{ m})}{4.0\text{ m}} = 22.5\text{ N}$$

7. Use this result for F_1, and $f_s = F_1$ from step 4, to find f_s:
$$f_s = F_1 = 22.5\text{ N}$$

8. Use the results for f_s and F_n to obtain $\mu_{s\min}$ from step 2:
$$\mu_{s\min} = \frac{f_s}{F_n} = \frac{22.5\text{ N}}{60\text{ N}} = 0.375 = \boxed{0.38}$$

CHECK In the free-body diagram for the ladder shown in Figure 12-14, the lines of action of $\vec{F}_g$ and $\vec{F}_1$ intersect at point P. This means the torques about P due to $\vec{F}_g$ and $\vec{F}_1$ must both be equal to zero. Because the sum of all the torques about point P must equal zero, we know the torque about P due to $\vec{F}_2$ must also equal zero. This means the line of action of $\vec{F}_2$ must pass through point P as well. Consequently, $\tan\theta' = 4.0\text{ m}/1.5\text{ m} = F_n/f_s$, which means $f_s/F_n = 1.5/4.0 = 0.375$. This value of f_s/F_n is the same as that obtained in step 8.

! If an object is in static equilibrium under the influence of three forces, where the lines of action of any two of the forces intersect at a point, the lines of action of all three forces will intersect at that same point.

PRACTICE PROBLEM 12-1
Show that if an object is in static equilibrium under the influence of three forces, where the lines of action of two of the forces intersect at a point, the lines of action of all three forces will intersect at that same point.

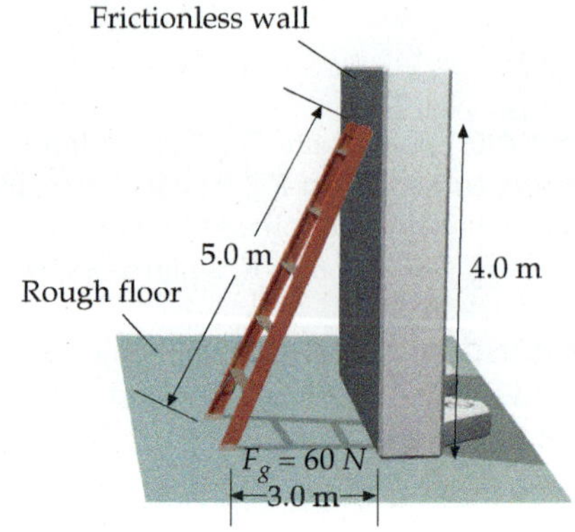

FIGURE 12-12

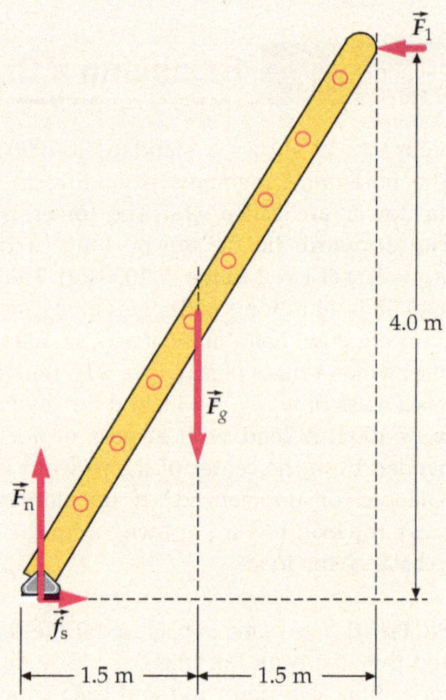

FIGURE 12-13

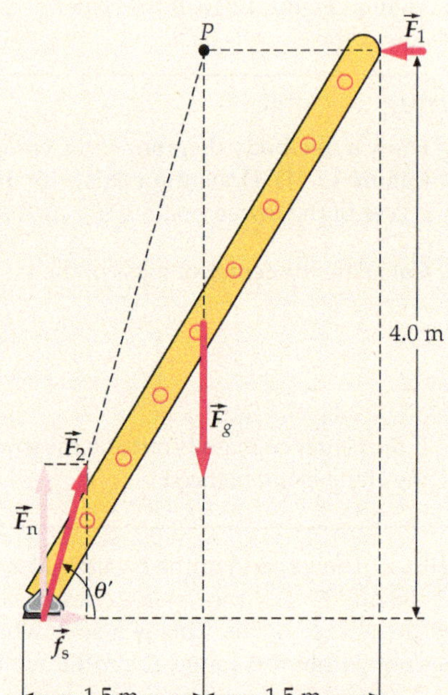

FIGURE 12-14

COUPLES

The forces $\vec{F}_n$ and $\vec{F}_g$ in Figure 12-13 of Example 12-6 are equal in magnitude, opposite in direction, and are not collinear. Such a pair of forces, called a **couple,** tends to produce an angular acceleration, but its net force is zero. The forces $\vec{f}_s$ and $\vec{F}_1$ in Figure 12-13 also constitute a couple. Figure 12-15 shows a couple consisting of forces $\vec{F}_1$ and $\vec{F}_2$ a distance D apart. The torque produced by this couple about an arbitrary point O is

$$\vec{\tau} = \vec{r}_1 \times \vec{F}_1 + \vec{r}_2 \times \vec{F}_2 = \vec{r}_1 \times \vec{F}_1 + \vec{r}_2 \times (-\vec{F}_1) = (\vec{r}_1 - \vec{r}_2) \times \vec{F}_1 \qquad 12\text{-}5$$

This result does not depend on the choice of the point O.

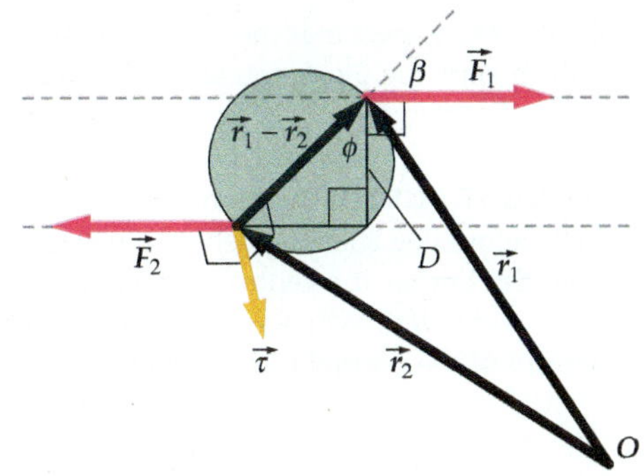

FIGURE 12-15 The torque $\vec{\tau}$ produced by the two forces is directed into the page—perpendicular to the plane containing the two forces. D is the perpendicular distance between the lines of action of the two forces.

> The torque produced by a couple is the same about all points in space.

The magnitude of the torque exerted by a couple is

$$\tau = FD \qquad 12\text{-}6$$

where F is the magnitude of either force and D is the perpendicular distance between the lines of action of the two forces.

PRACTICE PROBLEM 12-2

Show that the magnitude of $(\vec{r}_1 - \vec{r}_2) \times \vec{F}_1$ (see Equation 12-5) is FD (see Equation 12-6), where D (shown in Figure 12-15) is the distance between the lines of action of the two forces and F is the magnitude of either force.

Example 12-7 Tipping the Block

In visiting a marble quarry you notice half of a \$100 bill (Figure 12-16) sticking out from under a block of marble of mass m, height H, and with a square cross section of edge-length L. You try to retrieve the \$100 bill, but it is stuck. To retrieve it, you push the block with a horizontal force a distance h above the floor. How hard do you have to push to cause the block to tip up (slightly), thus freeing the \$100 bill? (Assume friction is sufficient to prevent the block from slipping.)

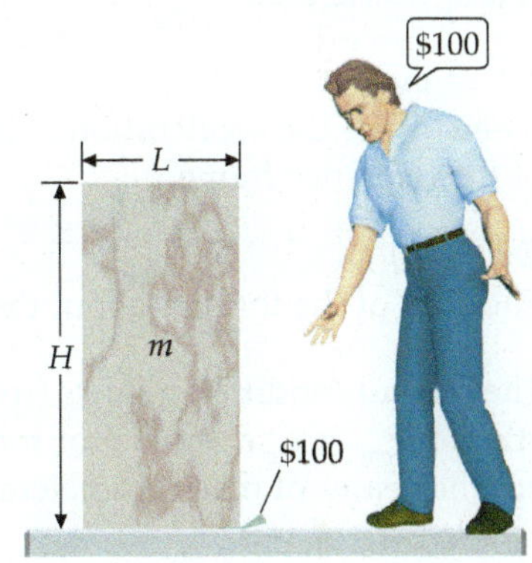

FIGURE 12-16

PICTURE Assume you are pushing so hard that if you pushed just slightly harder, the block would start to tip up. Draw a free-body diagram of the block, and apply the conditions for equilibrium. If there are any couples, use Equation 12-6 to calculate the magnitude of the torque.

SOLVE

1. Assume the block is on the verge of tipping up, and draw a free-body diagram of the block (Figure 12-17). Draw the normal force at the left edge of the block (see the remark at the end of this example).
2. To relate the forces, apply $\Sigma F_x = ma_x$ and $\Sigma F_y = ma_y$ to the block, where $a_x = a_y = 0$:
   $\Sigma F_x = 0 \Rightarrow F_{app} = f_s$
   and $\Sigma F_y = 0 \Rightarrow mg = F_n$
3. Identify any couples:
   $\vec{F}_{app}$ and $\vec{f}_s$ form couple 1
   and $m\vec{g}$ and $\vec{F}_n$ form couple 2
4. Choose counterclockwise as positive and, using Equation 12-6, calculate the torque due to each couple:
   $\tau_1 = +F_{app}h$ and $\tau_2 = -mg\frac{1}{2}L$
5. Using $\Sigma\tau = 0$, solve for F_{app}:
   $F_{app}h - mg\frac{1}{2}L = 0$
   so $\boxed{F_{app} = \dfrac{L}{2h}mg}$

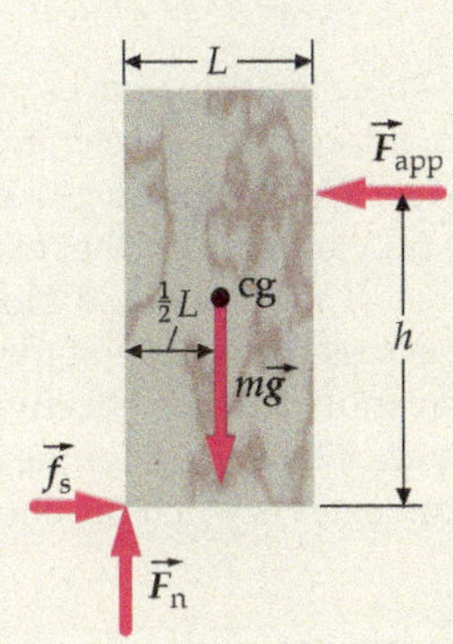

FIGURE 12-17

CHECK We expect that the higher up on the block you push, the more gently you need to push to start the block rotating. The step-6 result meets this expectation. That is, as h increases F_{app} decreases.

TAKING IT FURTHER The normal force is uniformly distributed across the bottom of the block before you start to push on the block. When you do push on the block, the harder you push, the farther the centroid (the effective center) of the distribution of the normal force moves to the left. When you push so hard that the block is on the verge of tipping up, the centroid of the normal force is at the left edge of the bottom of the block.

CONCEPT CHECK 12-1

There is a minimal-effort solution to this example. There is a specific axis choice such that the first step-6 equation is immediately obtained by setting the sum of the torques about the axis equal to zero. What is the axis choice used in the minimal-effort solution?

12-4 STATIC EQUILIBRIUM IN AN ACCELERATED FRAME

By an accelerated frame, we mean a reference frame that is accelerating, but is not rotating, relative to an inertial reference frame. The net force on an object that remains at rest relative to an accelerated reference frame is not equal to zero. An object at rest relative to the accelerated frame has the same acceleration as the frame. The two conditions for an object to be in static equilibrium in an accelerated reference frame follow:

1. $\Sigma\vec{F} = m\vec{a}_{cm}$
   where $\vec{a}_{cm}$ is the acceleration of the center of mass, which is also the acceleration of the reference frame.
2. $\Sigma\vec{\tau}_{cm} = 0$
   The sum of the torques about the center of mass must be zero.

The second condition follows from the fact that Newton's second law for rotation, $\Sigma\vec{\tau}_{cm} = I_{cm}\vec{\alpha}$, holds for torques about the center of mass whether or not the center of mass is accelerating.*

(a)

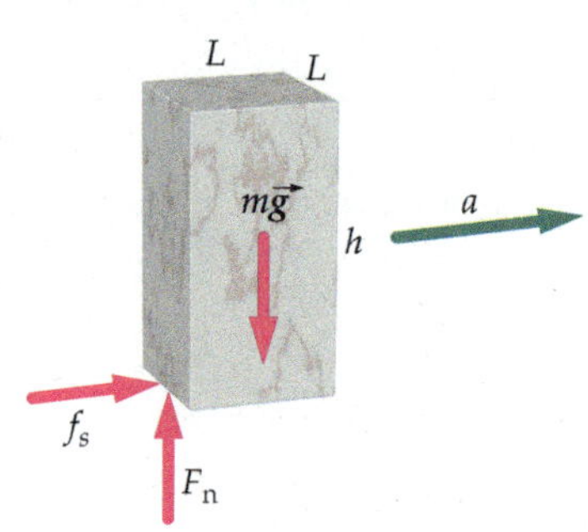

(b)

FIGURE 12-18

Example 12-8 Moving the Block

A truck (Figure 12-18*a*) carries a uniform block of marble of mass m, height h, and square cross section of edge-length L. What is the greatest acceleration the truck can have without the block tipping over? Assume that the block tips before it slides.

PICTURE There are three forces on the block, a gravitational force, a static frictional force, and a normal force. The acceleration of the block is due to the frictional force, as shown in Figure 12-18*b*. This force exerts a counterclockwise torque about the center of mass of the block. The only other force that exerts a torque about the center of mass of the block is the normal force. If the truck and block are not accelerating, the normal force is distributed uniformly across the bottom of the block. If the magnitude of the acceleration is small, this distribution shifts and the effective point of application of the normal force† moves to the left to provide a balancing torque about the center of mass. The greatest balancing torque this force can exert is when the effective normal force is at the edge of the base of the block, as shown.

* See the discussion surrounding Equation 9-30.

† By "effective point of application" of the normal force, we mean the point where the entire normal force can be considered to be applied for the purpose of calculating the torque exerted by the force.

SOLVE

1. Draw a free-body diagram of the block (Figure 12-19).

2. Apply $\Sigma F_y = ma_{\text{cm}y}$ to the block, and then solve for the normal force: $F_n - mg = 0 \quad \text{so} \quad F_n = mg$

3. Apply $\Sigma F_x = ma_{\text{cm}x}$ to the block: $f_s = ma$

4. Apply $\Sigma \tau_{\text{cm}} = 0$: $f_s\frac{h}{2} - F_n d = 0, \text{ where } d \le \frac{1}{2}L$

5. If $d = \frac{1}{2}L$, then the acceleration is maximum. Substitute $\frac{1}{2}L$ for d, ma_{max} for f_s, and mg for F_n, and solve for a_{max}: $ma_{\text{max}}\frac{h}{2} - mg\frac{L}{2} = 0 \quad \text{so} \quad a_{\text{max}} = \boxed{\frac{L}{h}g}$

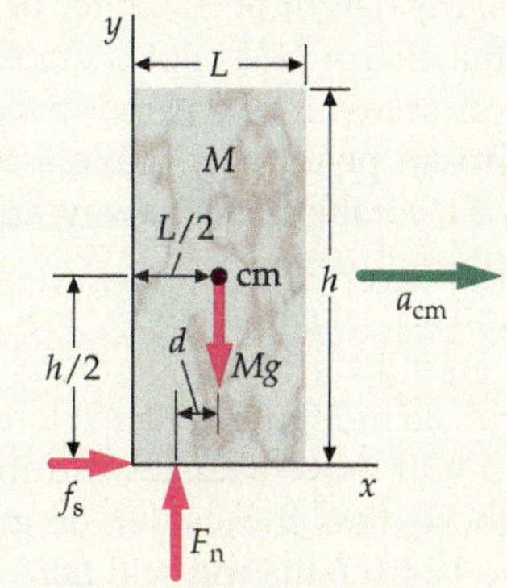

FIGURE 12-19

CHECK One would expect a_{max} to be larger for a short wide block (small h and large L) than for a tall narrow block (large h and small L). Our step-5 result meets this expectation.

12-5 STABILITY OF ROTATIONAL EQUILIBRIUM

There are three categories of rotational equilibrium for an object: stable, unstable, or neutral. **Stable rotational equilibrium** occurs when the torques that arise from a small angular displacement of the object from equilibrium tend to rotate the object back toward its equilibrium orientation. Stable equilibrium is illustrated in Figure 12-20*a*. When the cone is tipped slightly as shown, the resulting gravitational torque about the pivot point tends to restore the cone to its original orientation. Note that this slight tipping lifts the center of gravity, increasing the gravitational potential energy.

Unstable rotational equilibrium, illustrated in Figure 12-20*b*, occurs when the torques that arise from a small angular displacement of the object tends to rotate the object even farther away from its equilibrium orientation. A slight tipping of the cone causes it to fall over because the torque due to the gravitational force tends to rotate it away from its original orientation. Here the rotation lowers the center of gravity and decreases the gravitational potential energy.

The cone resting on a horizontal surface in Figure 12-20*c* illustrates **neutral rotational equilibrium.** If the cone is rolled slightly, there is no torque that tends

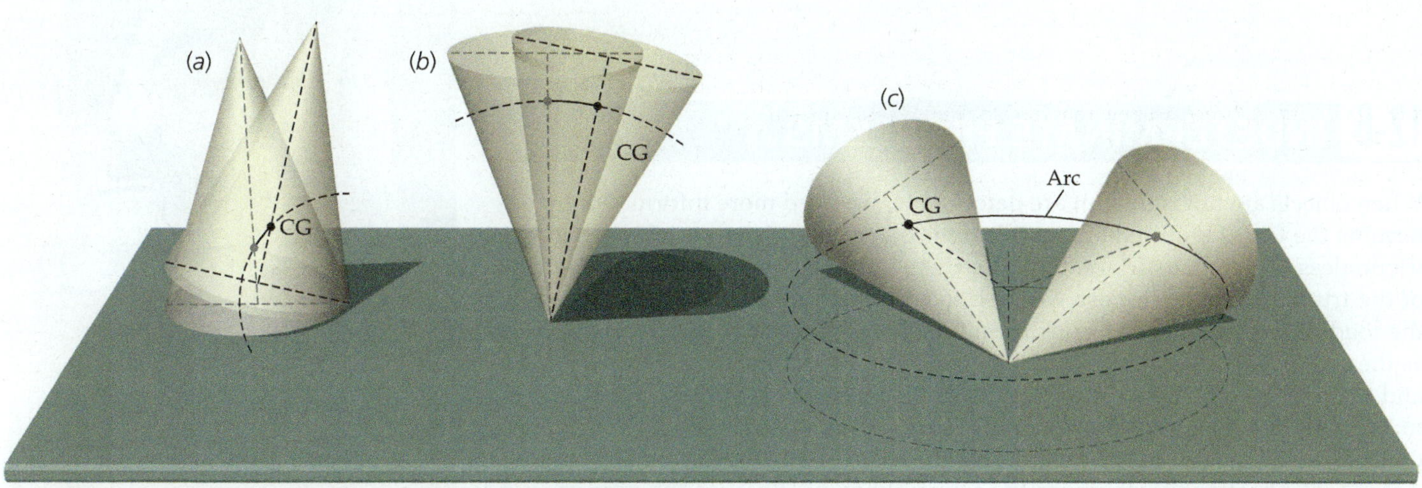

FIGURE 12-20 If slight rotation raises the center of gravity, as in (*a*), the equilibrium is stable. If a slight rotation lowers the center of gravity, as in (*b*), the equilibrium is unstable. If a slight rotation neither raises nor lowers the center of gravity, as in (*c*), the equilibrium is neutral.

to rotate it either back toward, or away from, its original orientation. As the cone rotates, the height of its center of gravity remains unchanged, so the gravitational potential energy does not change.

In summary, if a system is rotated slightly from an equilibrium orientation, the equilibrium position is stable if the system returns to its original orientation, unstable if it rotates farther away, and neutral if there are no torques tending to rotate it in either direction.

Because "rotated slightly" is a relative term, stability is also relative. One example of equilibrium may be more or less stable than another. A rod is balanced on one end, as in Figure 12-21*a*. Here, if the disturbance is very small (Figure 12-21*b*), the rod will move back toward its original position, but if the disturbance is great enough so that the center of gravity no longer lies over the base of support (Figure 12-21*c*), the rod will fall.

We can improve the stability of a system by either lowering the center of gravity or widening the base of support. Figure 12-22 shows a nonuniform rod that is loaded so that its center of gravity is near one end. If it stands on its heavy end so that the center of gravity is low (Figure 12-22*a*), it is much more stable than if it stands on the other end so that the center of gravity is high (Figure 12-22*b*).

In Figure 12-23, the system is stable for any angular displacement because the resulting torque always rotates the system back toward its equilibrium position.

Standing or walking upright is difficult for a human because the center of gravity is high, and must be kept over a relatively small base of support, the feet. Human infants take about a year to learn to walk. A four-footed creature has a much easier time, in part because its base of support is wider and its center of gravity is lower. Newborn kittens can walk almost immediately.

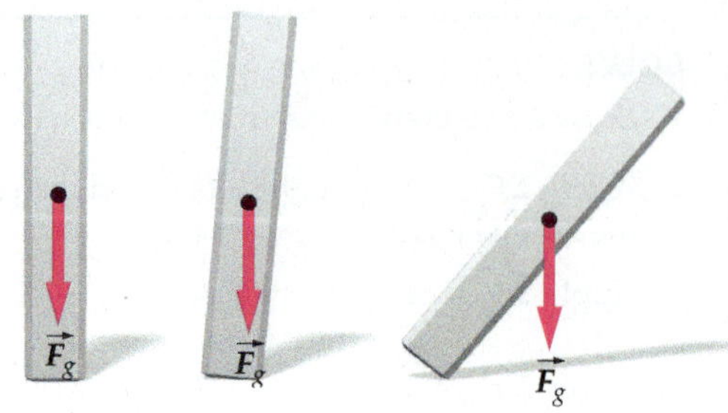

FIGURE 12-21 Stability of equilibrium is relative. If the rod in (*a*) is rotated slightly, as in (*b*), it returns to its original equilibrium position as long as the center of gravity lies over the base of support. (*c*) If the rotation is too great, the center of gravity is no longer over the base of support, and the rod falls over.

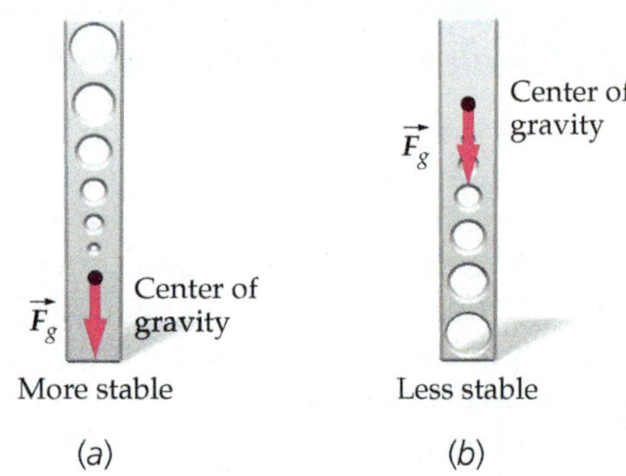

FIGURE 12-22 When a nonuniform rod rests on its heavy end (so its center of gravity is low), as in (*a*), the equilibrium is more stable than when its center of gravity is high, as in (*b*).

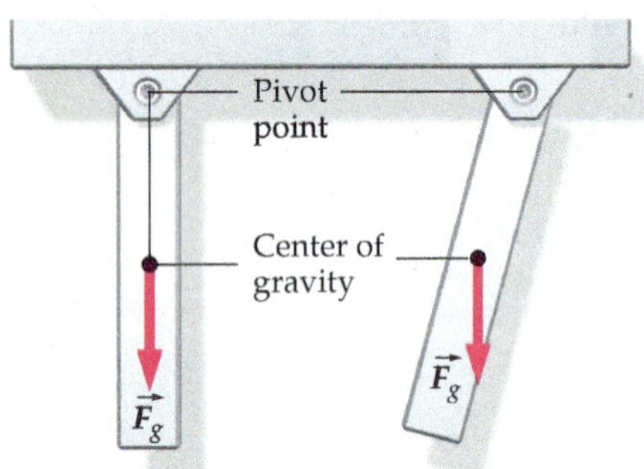

FIGURE 12-23

The toddler is relatively unstable (in comparison to a kitten). *(Photodisk.)*

12-6 INDETERMINATE PROBLEMS

When objects are not rigid, but are deformable, we need more information to determine the forces required for equilibrium. Consider a pickup truck resting on a frictionless horizontal surface. Suppose there is a very heavy object on one side of the trunk bed, and suppose we wish to calculate the normal force exerted by the road on each of the four truck tires. Consider the truck plus the heavy object as the system and assume we know the location of each tire, the system's weight, and the location of the center of gravity. Is this knowledge sufficient to allow us to calculate the magnitudes of the four normal forces? The answer to this question is no. The magnitude of each normal force is unknown, so we need four independent equations to solve for the four unknowns. Because the system is in equilibrium, the conditions for equilibrium can only supply us with three independent equations. Let the road surface be in the xy plane. The first condition for equilibrium is that the sum of the external forces is equal to zero. This provides only one equation ($\Sigma F_z = 0$) because all the forces are vertical. The second condition for equilibrium is that the sum of the external torques about any point is equal to zero. This provides two additional equations, $\Sigma\tau_x = 0$ and $\Sigma\tau_y = 0$. The reason there are no vertical torque components is because torque vectors are cross products ($\vec{\tau} = \vec{r} \times \vec{F}$), and the direction of a cross product is perpendicular to each vector in the product. Because the forces on the truck are all vertical, all the torque vectors are horizontal.

There are two external forces acting on the truck: the force of gravity and the normal forces by the road on the tires. Let the road surface be in the xy plane. If we choose the contact point of one of the tires with the road as our origin, the torque exerted by all the forces about that point has both x and y components. All the forces are vertical, so all the torque vectors must be horizontal. There are no z components because there are no horizontal forces. We thus obtain two equations by setting the net torque equal to zero, and a third equation by setting the net vertical force equal to zero. We need another equation to find the force exerted by the road on each of the four tires. Because we do not have another equation at our disposal, the forces cannot be determined. If we let air out of one of the tires and pump up another tire to a greater pressure, the car remains in equilibrium, but the force exerted on each tire changes. Clearly, the forces on the tires in this problem are not determined by the information given. The tires are not rigid bodies. To some extent, every object is deformable.

12-7 STRESS AND STRAIN

If a solid object is subjected to forces that tend to stretch, shear, or compress the object, its shape changes. If the object returns to its original shape when the forces are removed, it is said to be **elastic.** Most objects are elastic for forces up to a certain maximum, called the **elastic limit.** If the forces exceed the elastic limit, the object does not return to its original shape but is permanently deformed.

Figure 12-24 shows a solid bar subjected to a stretching or **tensile force** F acting equally to the right and to the left. The bar is in equilibrium, but the forces acting on it tend to increase its length. The fractional change in the length $\Delta L/L$ of a segment of the bar is called the **strain**:

$$\text{Strain} = \frac{\Delta L}{L} \qquad 12\text{-}7$$

The ratio of the force F to the cross-sectional area A is called the **tensile stress**:

$$\text{Stress} = \frac{F}{A} \qquad 12\text{-}8$$

Figure 12-25 shows a graph of stress versus strain for a typical solid bar. The graph is linear until point A. Up to this point, known as the proportional limit, the strain is proportional to the stress. The result that strain varies linearly with stress is known as Hooke's law. Point B in Figure 12-25 is the elastic limit of the material. If the bar is stretched beyond this point, it is permanently deformed. If an even greater stress is applied, the material eventually breaks, shown happening at point C. The ratio of stress to strain in the linear region of the graph is a constant called **Young's modulus** Y:

$$Y = \frac{\text{Stress}}{\text{Strain}} = \frac{F/A}{\Delta L/L} \qquad 12\text{-}9$$

YOUNG'S MODULUS DEFINED

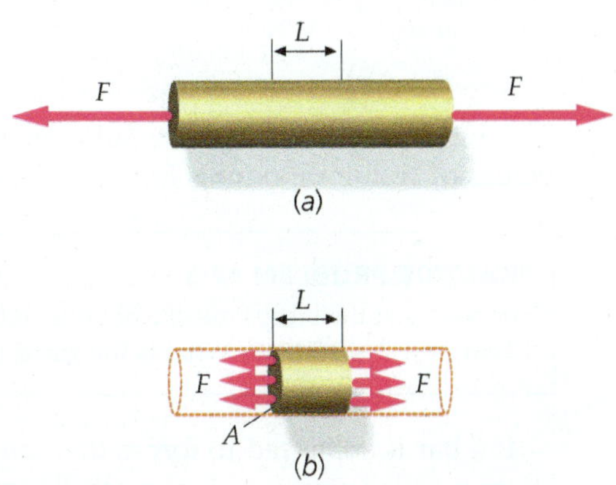

FIGURE 12-24 (*a*) A solid bar subjected to stretching forces of magnitude F acting on each other. (*b*) A small section of the bar of length L. The elements of the bar to the left and right of this section exert forces on the section. These forces are distributed equally over the cross-sectional area. The force per unit area is the stress.

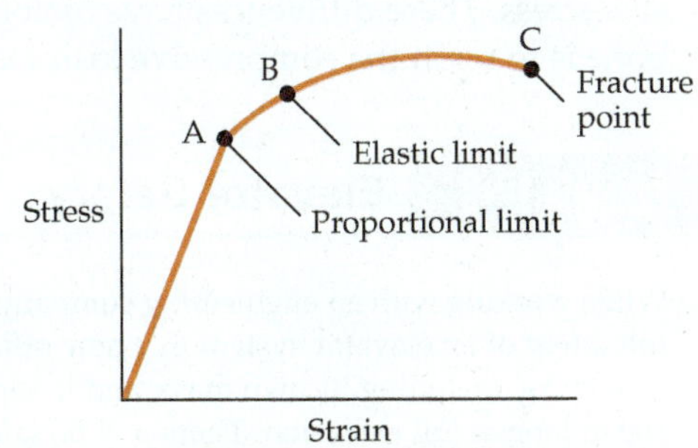

FIGURE 12-25 A graph of stress versus strain. Up to point A, the strain is proportional to the stress. Beyond the elastic limit at point B, the bar will not return to its original length when the stress is removed. At point C, the bar fractures.

Table 12-1 Young's Modulus Y and Strengths of Various Materials†

Material	Y, GN/m^2‡	Tensile strength, MN/m^2	Compressive strength, MN/m^2
Aluminum	70	90	
Bone			
Tensile	16	200	
Compressive	9		270
Brass	90	370	
Concrete	23	2	17
Copper	110	230	
Iron (wrought)	190	390	
Lead	16	12	
Steel	200	520	520

† These values are representative. Actual values for particular samples may differ.

‡ $1\ GN = 10^3\ MN = 1 \times 10^9\ N$.

See **Math Tutorial** *for more information on* **Direct and Inverse Proportions**

The dimensions of Young's modulus are those of force divided by area. Approximate values of Young's modulus for various materials are listed in Table 12-1.

PRACTICE PROBLEM 12-3

Suppose that the biceps muscle of your right arm has a maximum cross-sectional area of $12\ cm^2 = 1.2 \times 10^{-3}\ m^2$. What is the stress in the muscle if it exerts a force of 300 N?

If a bar is subjected to forces that tend to compress it rather than stretch it, the stress is called **compressive stress.** For many materials, Young's modulus for compressive stress is the same as that for tensile stress. Note that for compressive strain, ΔL, in Equation 12-7 is then taken to be the *decrease* in the length of the bar. If the tensile or compressive stress is too great, the bar breaks. The stress at which breakage occurs is called the **tensile strength,** or in the case of compression, the **compressive strength.** Approximate values of the tensile and compressive strengths of various materials are listed in Table 12-1. Note from the table that the compressive strength of bone is greater than the tensile strength. Also note that, for bone, Young's modulus is significantly larger for tensile stress than for compressive stress. These differences have biological significance, because the major job of bone is to resist the compressive load exerted by contracting muscles.

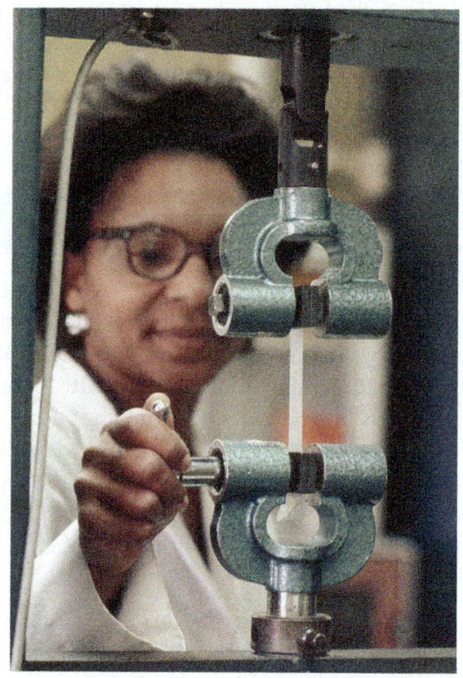

To measure the tensile strength, the rod is stretched until it breaks. *(Vince Streano/CORBIS.)*

Example 12-9 Elevator Safety

Context-Rich

While working with an engineering company during the summer, you are assigned to check the safety of an elevator system in a new office building. The elevator has a maximum load of 1000 kg, including its own mass, and is supported by a steel cable 3.0 cm in diameter and 300 m long at full extension. There will be safety concerns if the steel stretches more than 3.0 cm. Your job is to determine whether or not the elevator is safe as planned, given a maximum acceleration of the system of $1.5\ m/s^2$.

PICTURE L is the length of the unstressed cable, F is the magnitude of the force acting on it, and A is its cross-sectional area. The stretch in the cable ΔL is related to Young's modulus by $Y = (F/A)/(\Delta L/L)$. From Table 12-1, we find the numerical value of Young's modulus for steel, $Y = 2.0 \times 10^{11}\ N/m^2$.

SOLVE

1. The amount the cable is stretched, ΔL, is related to Young's modulus:

$$Y = \frac{F/A}{\Delta L/L} \quad \text{so} \quad \Delta L = \frac{FL}{AY}$$

2. To find the force acting on the cable, we apply Newton's second law to the elevator. There are two forces on the elevator, the force F of the cable and the gravitational force:

$$F - mg = ma_y$$

$$\text{so} \quad F_{max} = m(g + a_{y\,max}) = (1000\ \text{kg})(9.81\ \text{N/kg} + 1.5\ \text{N/kg}) = 1.13 \times 10^4\ \text{N}$$

3. Substitute into the step-1 result and obtain the maximum amount of stretch:

$$\Delta L = \frac{F_{max}L}{AY} = \frac{F_{max}L}{\pi r^2 Y} = \frac{(1.13 \times 10^4\ \text{N})(300\ \text{m})}{\pi(0.015\ \text{m})^2(2.0 \times 10^{11}\ \text{N/m}^2)} = 2.40\ \text{cm}$$

4. Report your results to your boss:

According to my calculations, the most the cable will stretch is 2.4 cm, only 20 percent less than the 3.0-cm limit. However, in reading the footnote to the table, I note that the values given for Young's modulus are representative values, and that actual values vary from sample to sample. I recommend that you consult an engineer and get a professional evaluation.

CHECK Is the step-3 expression for ΔL dimensionally correct? Young's modulus has the dimensions of force per unit area, so AY has the dimensions of force. Thus, the dimension of F_{max} in the numerator cancels the dimension of AY in the denominator. The expression has the dimensions of length and is dimensionally correct.

PRACTICE PROBLEM 12-4 A 1.5-m-long wire has a cross-sectional area of 2.4 mm². It is hung vertically and stretches 0.32 mm when a 10-kg block is attached to it. Find (*a*) the stress, (*b*) the strain, and (*c*) Young's modulus for the wire.

In Figure 12-26, a force F_s is applied tangentially to the top of a block of Jello. Such a force is called a **shear force.** The ratio of the shear force F_s to the area A is called the **shear stress**:

$$\text{Shear stress} = \frac{F_s}{A} \qquad \text{12-10}$$

A shear stress tends to deform an object, as shown in Figure 12-26. The ratio $\Delta X/L$ is called the **shear strain**:

$$\text{Shear strain} = \frac{\Delta X}{L} = \tan\theta \qquad \text{12-11}$$

where θ is the shear angle shown in the figure. The ratio of the shear stress to the shear strain is called the **shear modulus** M_s:

$$M_s = \frac{\text{Shear stress}}{\text{Shear strain}} = \frac{F_s/A}{\Delta X/L} = \frac{F_s/A}{\tan\theta} \qquad \text{12-12}$$

DEFINITION—SHEAR MODULUS

The shear modulus is also known as the **torsion modulus.** The torsion modulus is approximately constant for small stresses, which implies that the shear strain varies linearly with the shear stress. This observation is known as Hooke's law for torsional stress. In a torsion balance, such as that used in Cavendish's apparatus for measuring the universal gravitational constant G, the torque (which is related to the stress) is proportional to the angle of twist (which equals the strain for small angles). Approximate values of the shear modulus for various materials are listed in Table 12-2.

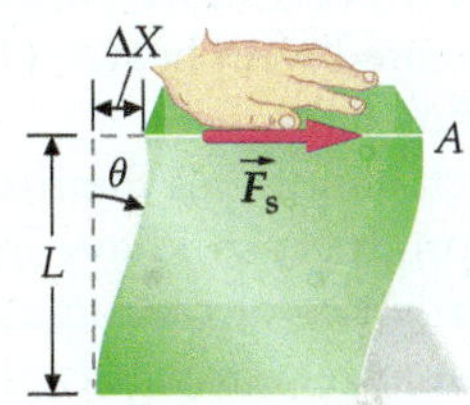

FIGURE 12-26 The application of the horizontal force $\vec{F}_s$ to the Jello causes a shear stress defined as the force per unit area. The ratio $\Delta X/L = \tan\theta$ is the shear strain and A is the horizontal cross-sectional area of the Jello.

Table 12-2 Approximate Values of the Shear Modulus M_s of Various Materials

Material	M_s, GN/m²
Aluminum	30
Brass	36
Copper	42
Iron	70
Lead	5.6
Steel	84
Tungsten	150

Physics Spotlight

Carbon Nanotubes: Small and Mighty

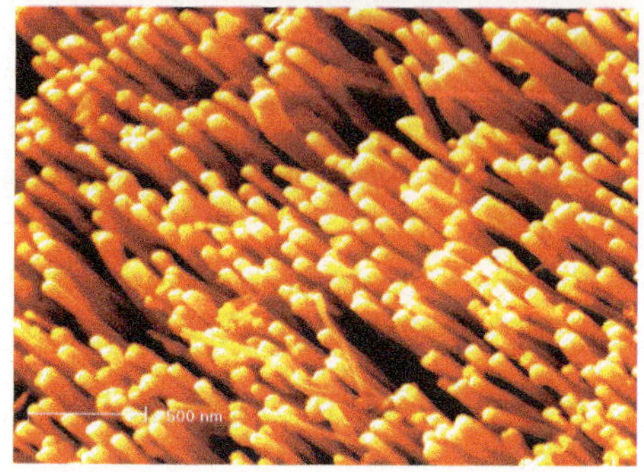

Carbon nanotubes are produced in large numbers. *(Courtesy of Prof. Zhong Lin Wang, Georgia Tech.)*

The most common form of pure carbon is graphite—slippery, strong sheets one atom thick. The lattice of carbon atoms in graphite is arranged in a hexagonal pattern, much like the hexagonal pattern of chicken wire. Carbon-atom lattices can also form tubes that are a few nanometers in diameter, and a few micrometers long. Because of their small size, they have been dubbed *nanotubes*. Nanotube walls can contain a single layer of atoms, or they can consist of many nested tubes, in a multiwalled tube.

Nanotubes can have different properties, depending on the orientation of the lattice, and the diameter of the tube. More than 300 different types of carbon nanotubes have been identified. Each production method[*] makes between 10 and 50 different types of nanotubes at once.[†] Isolating a large pure group of nanotubes is a difficult process.[‡,#] Most nanotubes are sold in batches containing between 65 and 95 percent nanotubes. The cost is per gram, and purer types are more expensive. (The remaining impurities are different forms of carbon.) Nanotubes differ dramatically[°] from carbon fibers currently used in composites. Carbon fibers are a specialized type of manufactured graphite, but they are not hollow tubes.

Because nanotubes are so small, new methods of measuring their tensile strength and Young's modulus were created.[§,¶] Young's modulus for single-walled nanotubes was measured[**] at an average of 1.25 TN/m^2, with a range[††] of 0.32 to 1.47 TN/m^2. These values are more than five times stronger than steel by volume, and many more times stronger by weight. Multiwalled carbon nanotubes have a greater variation in Young's modulus,[‡‡] from 270 GN/m^2 to 950 GN/m^2, and their tensile strength ranges from 11 GN/M^2 to 63 GN/m^2. Carbon nanotubes have higher tensile strength and a much higher Young's modulus than Kevlar™ fibers,[##] at an equivalent weight. Nanotubes are the stiffest materials known, and have the highest tensile strength known.

Carbon nanotubes can not only take stress with a low rate of strain, but can also exert strong stresses. Carbon nanotubes have recently been shown to exert pressures of 40.53 GN/m^2 on metal crystals trapped within them, when the tubes are radiated and treated with heat.[°°] (This is about one-tenth the pressure at Earth's core!) As the tubes shrink, they squeeze the metal into very fine hairs.

Although nanotubes themselves are extremely strong, threads,[§§] fibers,[¶¶] and ribbons[***] spun from them are not as strong. But these products still have high tensile strengths and high Young's moduli. Much of the strength of nanotubes comes from the regular lattice of carbon atoms. If defects exist in that lattice, the nanotube is weakened.[†††] (This also explains the wide variation in tested strengths of carbon nanotubes.) Because billions of nanotubes are required for even medium-scale applications, statistically these larger applications cannot have the same strength per unit volume (or per unit mass) as individual nanotubes.[‡‡‡] Still, even a small amount of relatively pure nanotubes can add strength and stiffness to existing materials. By weight, 5 percent nanotubes in a composite can more than double the tensile strength and stiffness of the composite.[###] Wherever strong, lightweight materials are needed, carbon nanotubes have a bright future.

---

* Guice, C., "Dynamics of Nanotube Synthesis," *Penn State McNair Journal 2003,* 115–119.

† Kumar, Satish, in Goho, Alexandra, "Nice Threads: The Golden Secret Behind Spinning Carbon-Nanotube Fibers," *Science News*, June 5, 2004, p. 363+.

‡ Benavides, J. M., "Method for Manufacturing of High Quality Carbon Nanotubes,." U.S. Patent 7,008,605, Mar. 7, 2006.

\# Harutyunyan et al., "Method of Purifying Nanotubes and Nanofibers Using Electromagnetic Radiation," U. S. Patent 7,014,737, Mar 21, 2006.

° Ajayan, P. M., Charlier, J.-C., and Rinzle, A. G., "Carbon Nanotubes: From Macromolecules to Nanotechnology," *Proceedings of the National Academy of Sciences*, Dec. 7, 1999, 14199–14200.

§ Pasquali et al., "Method for Determining the Length of Single-Walled Carbon Nanotubes," U.S. Patent 6,962,092, Nov. 8, 2005.

¶ Yu, M.-F., Lourie, O., Dyer, M., Motini, K, Kelly, T., and Ruoff, R., "Strength and Breaking Mechanism of Multiwalled Carbon Nanotubes Under Tensile Load," *Science*, Jan. 28, 2000, 637–640.

** Krishan, A., Dujardin, E., Ebbesen, T. W., Treacy, M. M. J., and Yianilos, P. N., "Young's Modulus of Single-Walled Nanotubes," *Physical Review Letters B*, Nov. 15, 1998, 14013–1409.

†† Yu, M.-F., Files, B., Arepalli, S., and Ruoff, R., "Tensile Loading of Ropes of Single Wall Carbon Nanotubes and Their Mechanical Properties," *Physical Review Letters*, June 12, 2000, 5552–5555.

‡‡ Yu et al., op. cit.

\## Tang, Benjamin, "Fiber Reinforced Polymer Composites Applications in USA DOT - Federal Highway Administration," *First Korea/USA Road Workshop Proceedings*, Jan. 28–29, 1997 http://www.fhwa.dot.gov/bridge/frp/frp197.htm

°° Sun, L., Bahhart, F., Krasheninnikov, A. V., Rodriguez-Manzo, J. A., Terrones, M., and Ajayan, P. M., "Carbon Nanotubes as High-Pressure Cylinders and Nanoextruders," *Science*, May 26, 2006, 1199–1202.

§§ Ericson et al., "Macroscopic, Neat, Single-Walled Carbon Nanotube Fibers," *Science*, Sept. 3, 2004, 1447–1450.

¶¶ Li, Y.-L., Kinloch, I. A., and Windle, A. H., "Direct Spinning of Carbon Nanotube Fibers from Chemical Vapor Deposition Synthesis," *Science*, Apr. 9, 2004, 276–278.

*** Zhang et al., "Strong, Transparent, Multifunctional Carbon Nanotube Sheets," *Science*, Aug. 19, 2005, 1215–1219.

††† Mielke, S., Diego, T., Zhang, S., Li, J.-L., Xiao, S., Car, R., Ruoff, R., Schatz, G., and Belytschko, T., "The Role of Vacancy Defects and Holes in the Fracture of Carbon Nanotubes," *Chemical Physical Letters*, Apr. 16, 2004, 413–420.

‡‡‡ Pugno, N., "On the Strength of the Carbon Nanotube-Based Space Elevator Cable: From Nano- to Mega-Mechanics," *Journal of Physics: Condensed Matter, Special Issue: Nanoscience and Nanotechnology*, (Prepublication), July 2006.

\### Andrews, D., Jacques, D., Rao, A. M., Rantell, T., Derbyshire, F., Chen, Y., Chen, J., and Haddon, R. C., "Nanotube Composite Carbon Fibers," *Applied Physics Letters*, Aug. 30, 1999, 1329–1331.

Summary

TOPIC	RELEVANT EQUATIONS AND REMARKS	
1. Equilibrium of a Rigid Object		
Conditions	1. The net external force acting on the object must be zero: $\Sigma\vec{F} = 0$	12-1
	2. The net external torque about any point must be zero: $\Sigma\vec{\tau} = 0$	12-2
	(The sum of the torques about any axis also equals zero.)	
Stability	The equilibrium of an object can be classified as stable, unstable, or neutral. An object resting on some surface will be in equilibrium if its center of gravity lies over its base of support. Stability can be improved by lowering the center of gravity or by increasing the width of the base.	
2. Center of Gravity	The force of gravity exerted on the various parts of an object can be replaced by a single force, the total gravitational force, acting at the center of gravity: $\vec{\tau}_{net} = \sum_i(\vec{r}_i \times \vec{F}_{gi}) = \vec{r}_{cg} \times \vec{F}_g$	12-3
	For an object in a uniform gravitational field, the center of gravity coincides with the center of mass.	
3. Couples	A pair of equal and opposite forces constitutes a couple. The torque produced by a couple is the same about any point in space. $\vec{\tau} = (\vec{r}_1 - \vec{r}_2) \times \vec{F}_1$ so $\tau = FD$	12-5, 12-6
	where D is the distance between the lines of action of the forces.	
4. Accelerated Reference Frame	The conditions for static equilibrium in an accelerated reference frame are 1. $\Sigma\vec{F} = m\vec{a}_{cm}$, where $\vec{a}_{cm}$ is the acceleration of the center of mass, which is also the acceleration of the reference frame. 2. $\Sigma\vec{\tau}_{cm} = 0$ The sum of the external torques about the center of mass must be zero.	
5. Stress and Strain		
Young's modulus	$Y = \frac{\text{Stress}}{\text{Strain}} = \frac{F/A}{\Delta L/L}$	12-9
Shear modulus	$M_s = \frac{\text{Shear stress}}{\text{Shear strain}} = \frac{F_s/A}{\Delta X/L} = \frac{F_s/A}{\tan\theta}$	12-12

Answer to Concept Check

12-1 A horizontal axis along the lowest edge of the block on the side opposite the side that is being pushed (Figure 12-17). (The torques about this axis due to the normal and frictional forces are both equal to zero.)

Answers to Practice Problems

12-1 Forces $\vec{F}_1$, $\vec{F}_2$, and $\vec{F}_3$ act on the object. The object is in equilibrium, so the torques due to these forces about any point must sum to zero. Let P be the point of intersection of the lines of action of forces $\vec{F}_1$ and $\vec{F}_2$. Then the torques about P due to $\vec{F}_1$ and $\vec{F}_2$ each must equal zero, so the torque about P due to $\vec{F}_3$ must also equal zero. It follows that the line of action of $\vec{F}_3$ must pass through point P.

12-2 The angle between $\vec{r}_1 - \vec{r}_2$ and $\vec{F}_1$ is β (Figure 12-15), so $|\vec{\tau}| = |(\vec{r}_1 - \vec{r}_2) \times \vec{F}_1| = |\vec{r}_1 - \vec{r}_2|F\sin\beta$. Because $D = |\vec{r}_1 - \vec{r}_2|\sin\beta$, $|\vec{\tau}| = FD$.

12-3 Stress $= F/A = 2.5 \times 10^5\ \text{N/m}^2$. The maximum stress that can be exerted is approximately the same for all human muscles. Greater forces can be exerted by muscles with greater cross-sectional areas.

12-4 (*a*) $4.1 \times 10^7\ \text{N/m}^2$, (*b*) 2.1×10^{-4}, (*c*) $190\ \text{GN/m}^2$

Problems

In a few problems, you are given more data than you actually need; in a few other problems, you are required to supply data from your general knowledge, outside sources, or informed estimate.

Interpret as significant all digits in numerical values that have trailing zeros and no decimal points.

For all problems, use $9.81\ \text{m/s}^2$ for the free-fall acceleration and neglect friction and air resistance unless instructed to do otherwise.

• Single-concept, single-step, relatively easy

•• Intermediate-level, may require synthesis of concepts

••• Challenging

SSM Solution is in the *Student Solutions Manual*

Consecutive problems that are shaded are paired problems.

CONCEPTUAL PROBLEMS

1 • True or false: SSM

(*a*) $\sum_i \vec{F}_i = 0$ is sufficient for static equilibrium to exist.

(*b*) $\sum_i \vec{F}_i = 0$ is necessary for static equilibrium to exist.

(*c*) In static equilibrium, the net torque about any point is zero.

(*d*) An object in equilibrium cannot be moving.

2 • True or false:

(*a*) The center of gravity is always at the geometric center of a body.

(*b*) The center of gravity must be located inside an object.

(*c*) The center of gravity of a baton is located between the two ends.

(*d*) The torque produced by the force of gravity about the center of gravity is always zero.

3 • The horizontal bar in Figure 12-27 will remain horizontal if (*a*) $L_1 = L_2$ and $R_1 = R_2$, (*b*) $M_1R_1 = M_2R_2$, (*c*) $M_2R_1 = M_1R_2$ (*d*) $L_1M_1 = L_2M_2$, (*e*) $R_1L_1 = R_2L_2$.

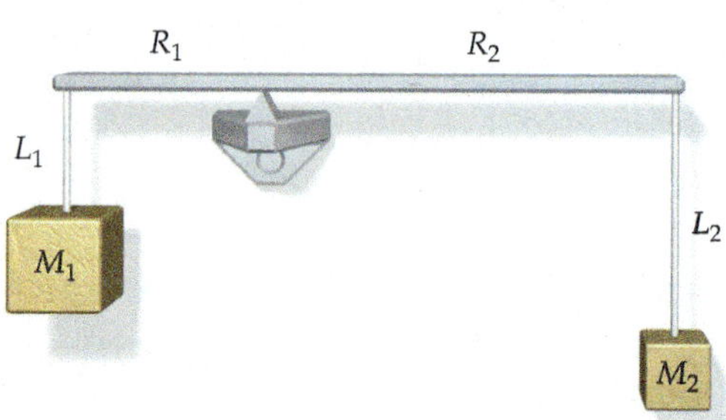

FIGURE 12-27 Problem 3

4 • Sit in a chair with your back straight. Now try to stand up without leaning forward. Explain why you cannot do it.

5 • **ENGINEERING APPLICATION** You have a job digging holes for posts to support signs for a Louisiana restaurant (called Mosca's). Explain why the higher above the ground a sign is mounted, the farther the posts should extend into the ground.

6 • A father (mass M) and his son (mass m) begin walking out toward opposite ends of a balanced see-saw. As they walk, the see-saw stays exactly horizontal. What can be said about the relationship between the father's speed V and the son's speed v?

7 • Travel mugs that people might set on the dashboards of their cars are often made with broad bases and relatively narrow mouths. Why would travel mugs be designed with this shape, rather than have the roughly cylindrical shape that mugs normally have?

8 •• **ENGINEERING APPLICATION** The sailors in the photo are using a technique called "hiking out." What purpose does positioning themselves in this way serve? If the wind were stronger, what would they need to do to keep their craft stable?

Sailors who are hiking out. *(Peter Andrews/Reuters/Corbis.)*

9 •• An aluminum wire and a steel wire of the same length L and diameter D are joined end-to-end to form a wire of length $2L$. One end of wire is then fastened to the ceiling and an object of mass M is attached to the other end. Neglecting the mass of the wires, which of the following statements is true? (*a*) The aluminum portion will stretch by the same amount as the steel portion. (*b*) The tensions in the aluminum portion and the steel portion are equal. (*c*) The tension in the aluminum portion is greater than that in the steel portion. (*d*) none of the above SSM

ESTIMATION AND APPROXIMATION

10 •• A large crate weighing 4500 N rests on four 12-cm-high blocks on a horizontal surface (Figure 12-28). The crate is 2.0 m long, 1.2 m high, and 1.2 m deep. You are asked to lift one end of the crate using a long steel pry bar. The fulcrum on the pry bar is 10 cm from the end that lifts the crate. Estimate the length of the bar you will need to lift the end of the crate.

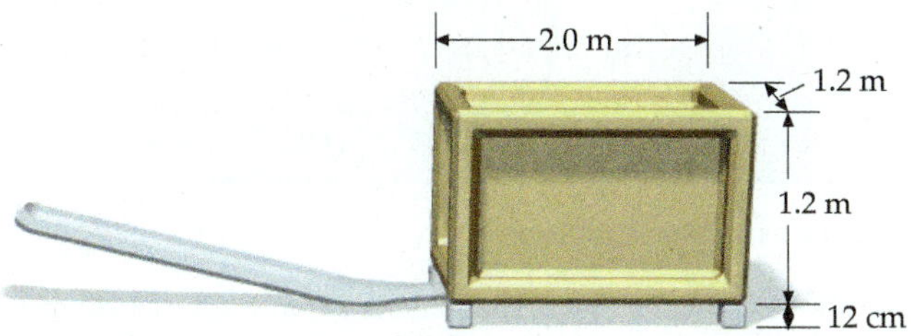

FIGURE 12-28 Problem 10

11 •• **ENGINEERING APPLICATION** Consider an atomic model for Young's modulus. Assume that a large number of atoms are arranged in a cubic array, with each atom at a corner of a cube and each atom a distance a from its six nearest neighbors. Imagine that each atom is attached to its six nearest neighbors by little springs each with spring constant k. (*a*) Show that this material, if stretched, will have a Young's modulus $Y = k/a$. (*b*) Using Table 12-1 and assuming that $a \approx 1.0$ nm, estimate a typical value for the "atomic spring constant" k in a metal. SSM

12 •• By considering the torques about the centers of the ball joints in your shoulders, estimate the force your deltoid muscles (those muscles on top of the shoulder) must exert on your upper arm to keep your arm held out and extended at shoulder level. Then, estimate the force they must exert when you hold a 10-lb weight out to the side at arm's length.

CONDITIONS FOR EQUILIBRIUM

13 • Your crutch is pressed against the sidewalk with a force $\vec{F}_c$ along its own direction, as in Figure 12-29. This force is balanced by the normal force $\vec{F}_n$ and a frictional force $\vec{f}_s$. (*a*) Show that when the force of friction is at its maximum value, the coefficient of friction is related to the angle θ by $\mu_s = \tan\theta$. (*b*) Explain how this result applies to the forces on your foot when you are not using a crutch. (*c*) Explain why it is advantageous to take short steps when walking on slippery surfaces?

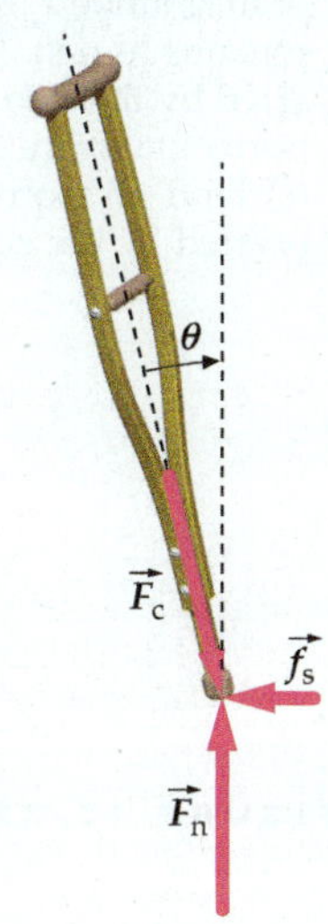

FIGURE 12-29 Problem 13

14 •• A thin uniform rod of mass M is suspended horizontally by two vertical wires. One wire is at the left end of the rod, and the other wire is 2/3 of the length of the rod from the left end. (*a*) Determine the tension in each wire. (*b*) An object is now hung by a string attached to the right end of the rod. When this happens, it is noticed that the rod remains horizontal but the tension in the wire on the left vanishes. Determine the mass of the object.

THE CENTER OF GRAVITY

15 • An automobile has 58 percent of its weight on the front wheels. The front and back wheels on each side are separated by 2.0 m. Where is the center of gravity located?

STATIC EQUILIBRIUM

16 • Figure 12-30 shows a lever of negligible mass with a vertical force F_{app} being applied to lift a load F. The *mechanical advantage* of the lever is defined as $M = F/F_{app\,min}$, where $F_{app\,min}$ is the smallest force necessary to lift the load F. Show that for this simple lever system, $M = x/X$, where x is the moment arm (distance to the pivot) for the applied force, and X is the moment arm for the load.

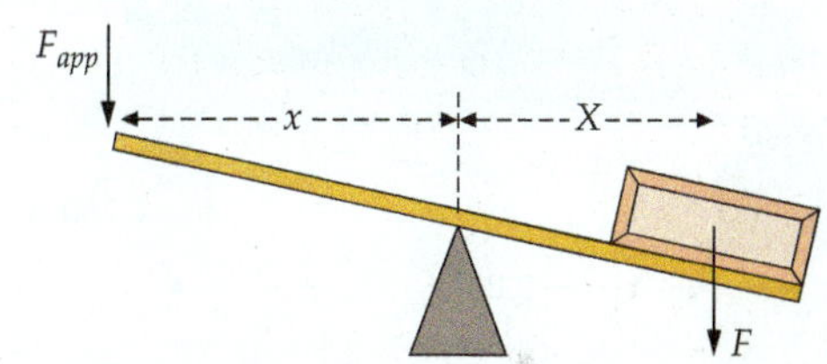

FIGURE 12-30 Problem 16

17 • **ENGINEERING APPLICATION** Figure 12-31 shows a 25-foot sailboat. The mast is a uniform 120-kg pole that is supported on the deck and held fore and aft by wires as shown. The tension in the *forestay* (wire leading to the bow) is 1000 N. Determine the tension in the *backstay* (wire leading aft) and the normal force that the deck exerts on the mast. (Assume that the frictional force the deck exerts on the mast to be negligible.) SSM

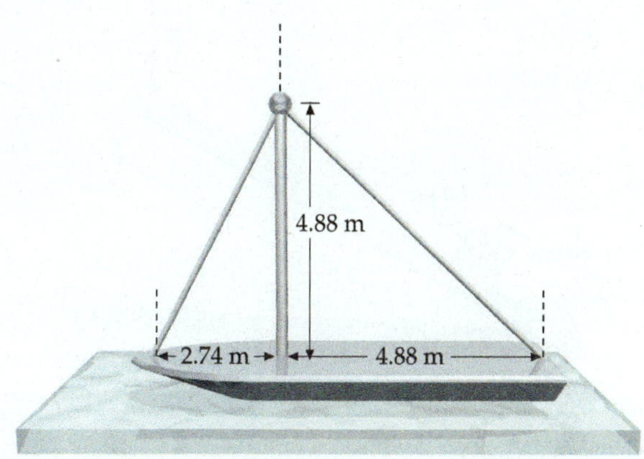

FIGURE 12-31 Problem 17

18 •• A uniform 10.0-m beam of mass 300 kg extends over a ledge as in Figure 12-32. The beam is not attached, but simply rests on the surface. A 60.0-kg student intends to position the beam so that he can walk to the end of it. What is the maximum distance the beam can extend past end of the ledge and still allow him to perform this feat?

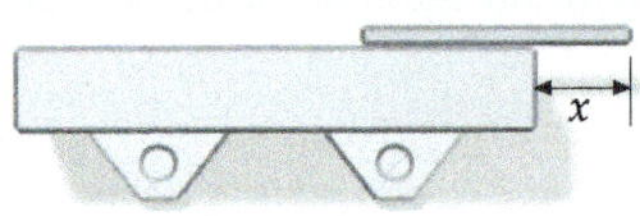

FIGURE 12-32 Problem 18

19 •• **BIOLOGICAL APPLICATION** A *gravity board* is a convenient and quick way to determine the location of the center of gravity of a person. It consists of a horizontal board supported by a fulcrum at one end and a scale at the other end. To demonstrate this in class, your physics professor calls on you to lie horizontally on the board with the top of your head directly above the fulcrum point as shown in Figure 12-33. The fulcrum is 2.00 m from the scale. In preparation for this experiment, you had accurately weighed yourself and determined your mass to be 70.0 kg. When you are at rest on the gravity board, the scale advances 250 N beyond its reading when the board is there by itself. Use this data to determine the location of your center of gravity relative to your feet. SSM

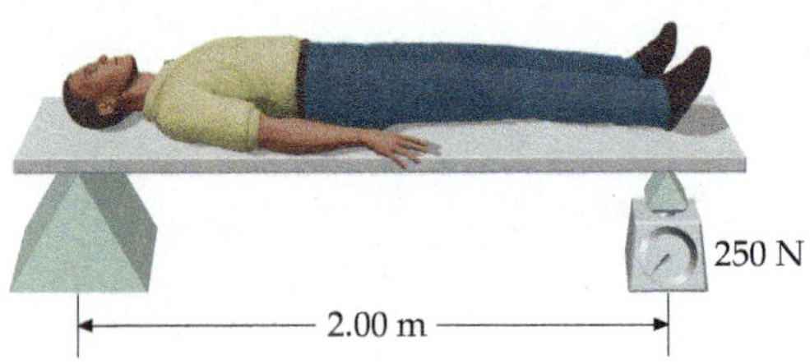

FIGURE 12-33 Problem 19

20 •• A stationary 3.0-m board of mass 5.0 kg is hinged at one end. A force $\vec{F}$ is applied vertically at the other end, and the board makes a 30° angle with the horizontal. A 60-kg block rests on the board 80 cm from the hinge as shown in Figure 12-34. (*a*) Find the magnitude of the force $\vec{F}$. (*b*) Find the force exerted by the hinge. (*c*) Find the magnitude of the force $\vec{F}$ as well as the force exerted by the hinge, if $\vec{F}$ is exerted, instead, at right angles to the board.

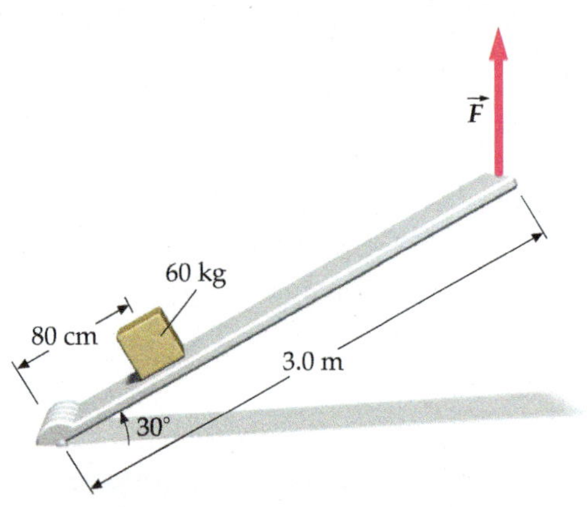

FIGURE 12-34 Problem 20

21 •• A cylinder of mass M is supported by a frictionless trough formed by a plane inclined at 30° to the horizontal on the left and one inclined at 60° on the right, as shown in Figure 12-35. Find the force exerted by each plane on the cylinder.

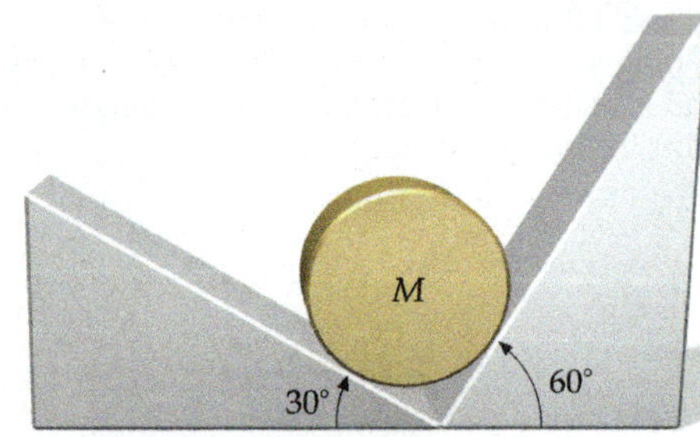

FIGURE 12-35 Problem 21

22 •• A uniform 18-kg door that is 2.0 m high by 0.80 m wide is hung from two hinges that are 20 cm from the top and 20 cm from the bottom. If each hinge supports half the weight of the door, find the magnitude and direction of the horizontal components of the forces exerted by the two hinges on the door.

23 •• Find the force exerted on the strut by the hinge at A for the arrangement in Figure 12-36 if (*a*) the strut is weightless, and (*b*) the strut weighs 20 N.

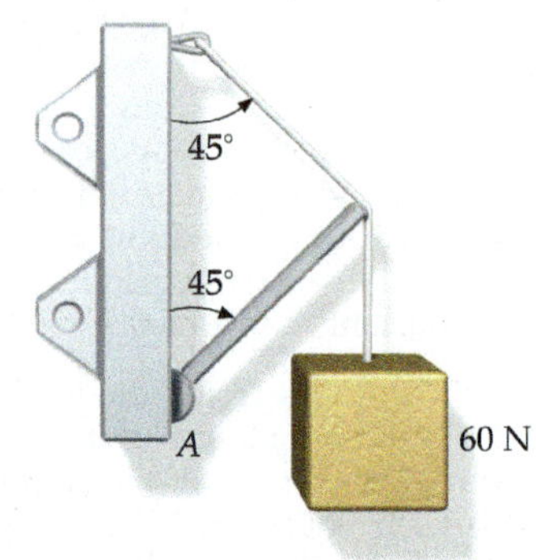

FIGURE 12-36 Problem 23

24 •• Julie has been hired to help paint the trim of a building, but she is not convinced of the safety of the apparatus. A 5.0-m plank is suspended horizontally from the top of the building by ropes attached at each end. Julie knows from previous experience that the ropes being used will break if the tension exceeds 1.0 kN. Her 80-kg boss dismisses Julie's worries and begins painting while standing 1.0 m from the end of the plank. If Julie's mass is 60 kg and the plank has a mass of 20 kg, over what range of positions can Julie stand to join her boss without causing the ropes to break?

25 •• A cylinder of mass M and radius R rolls against a step of height h, as shown in Figure 12-37. When a horizontal force of magnitude F is applied to the top of the cylinder, the cylinder remains at rest. (*a*) Find an expression for the normal force exerted by the floor on the cylinder. (*b*) Find an expression for the horizontal force exerted by the edge of the step on the cylinder. (*c*) Find an expression for the vertical component of the force exerted by the edge of the step on the cylinder. SSM

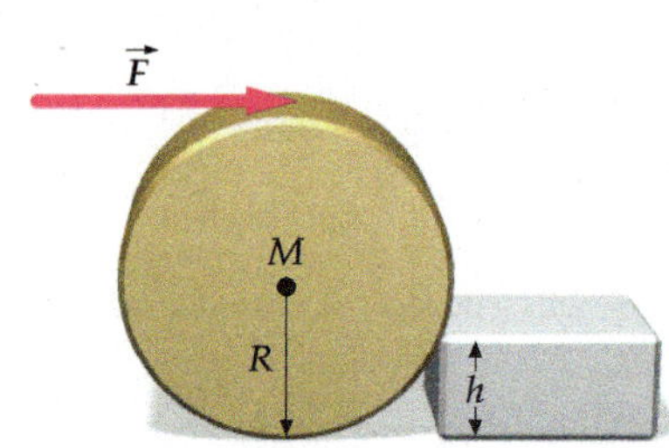

FIGURE 12-37 Problem 25

26 •• For the cylinder in Problem 25, find an expression for the minimum magnitude of the horizontal force $\vec{F}$ that will roll the cylinder over the step if the cylinder does not slide on the edge.

27 •• **CONTEXT-RICH** Figure 12-38 shows a hand holding an epee, a weapon used in the sport of fencing, which you are taking as a physical education elective. The center of mass of your epee is 24 cm from the pommel (the end of the epee at the grip). You have weighed it so you know that the epee's mass is 0.700 kg and its full length is 110 cm. (*a*) At the beginning of a match you hold it straight out in static equilibrium. Find the total force exerted by your hand on the epee. (*b*) Find the torque exerted by your hand on the epee. (*c*) Your hand, being an extended object, actually exerts its force along the length of the epee grip. Model the total force exerted by your hand as two oppositely directed forces whose lines of action are separated by the width of your hand (taken to be 10.0 cm). Find the magnitudes and directions of these two forces.

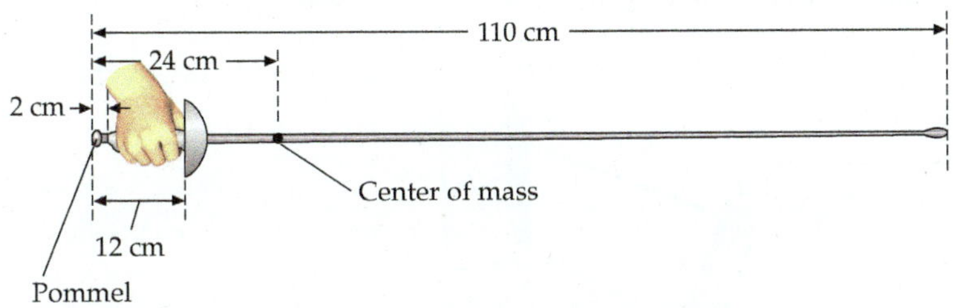

FIGURE 12-38 Problem 27

28 •• A large gate weighing 200 N is supported by hinges at the top and bottom, and is further supported by a wire, as shown in Figure 12-39. (*a*) What must the tension in the wire be for the force on the upper hinge to have no horizontal component? (*b*) What is the horizontal force on the lower hinge? (*c*) What are the vertical forces on the hinges?

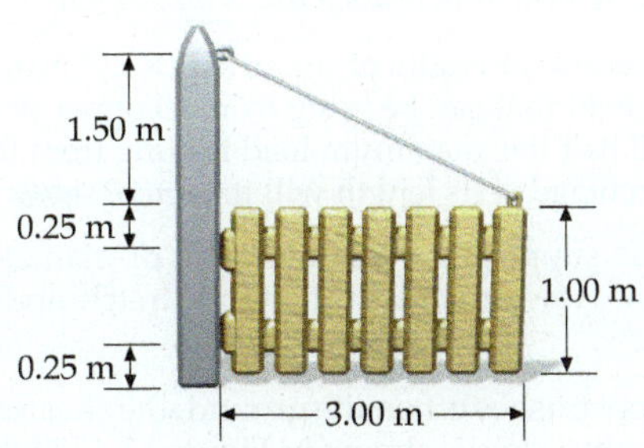

FIGURE 12-39 Problem 28

29 •• **CONTEXT-RICH** On a camping trip, you moor your boat at the end of a dock in a river that is rapidly flowing to the right. The boat is anchored to the dock by a chain 5.0 m long, as shown in Figure 12-40. A 100-N weight is suspended from the center in the chain. This will allow the tension in the chain to change as the force of the current which pulls the boat away from the dock and to the right varies. The drag force by the water on the boat depends on the speed of the water. You decide to apply the principles of statics you learned in physics class. (Ignore the weight of the chain.) The drag force on the boat is 50 N. (*a*) What is the tension in the chain? (*b*) How far is the boat from the dock? (*c*) The maximum tension the chain can sustain is 500 N. What is the minimum drag force on the boat that would snap the chain?

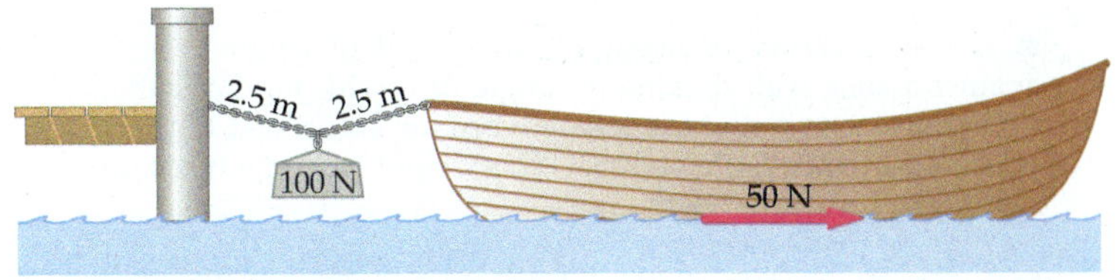

FIGURE 12-40 Problem 29

30 •• Romeo takes a uniform 10-m ladder and leans it against the smooth (frictionless) wall of the Capulet residence. The ladder's mass is 22 kg and the bottom rests on the ground 2.8 m from the wall. When Romeo, whose mass is 70 kg, gets 90 percent of the way to the top, the ladder begins to slip. What is the coefficient of static friction between the ground and the ladder?

31 •• Two 80-N forces are applied to opposite corners of a rectangular plate, as shown in Figure 12-41. (*a*) Find the torque produced by this couple using the Equation 12-6. (*b*) Show that the result is the same as if you determine the torque about the lower left-hand corner. SSM

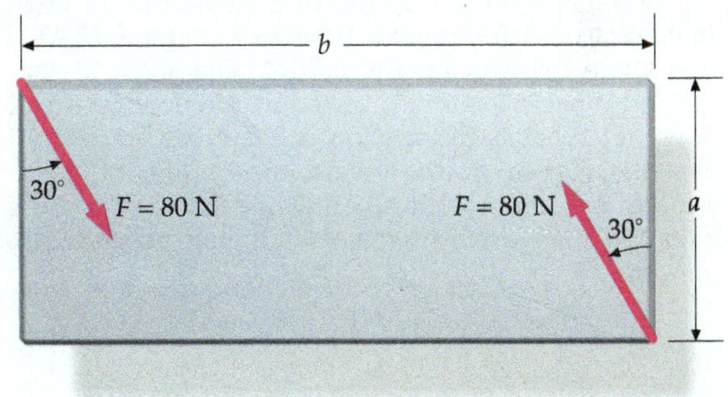

FIGURE 12-41 Problem 31

32 •• A uniform cube of side a and mass M rests on a horizontal surface. A horizontal force $\vec{F}$ is applied to the top of the cube, as in Figure 12-42. This force is not sufficient to move or tip the cube. (*a*) Show that the force of static friction exerted by the surface and the applied force constitute a couple, and find the torque exerted by the couple. (*b*) The torque exerted by the couple is balanced by the torque exerted by the couple consisting of the normal force on the cube and the gravitational force on the cube. Use this fact to find the effective point of application of the normal force when $F = Mg/3$. (*c*) Find the greatest magnitude of $\vec{F}$ for which the cube will not tip. (Assuming the cube does not slip.)

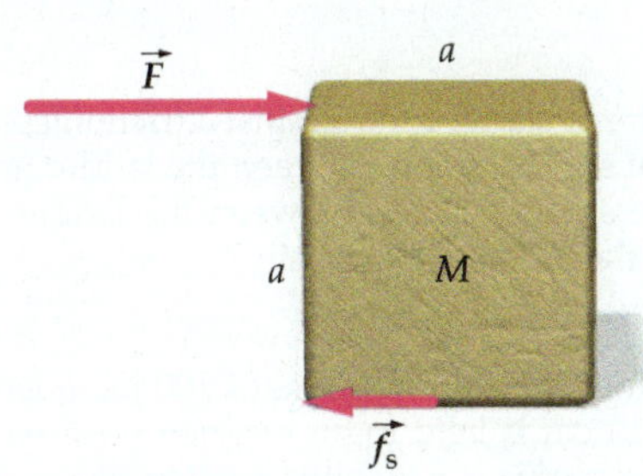

FIGURE 12-42 Problem 32

33 •• A ladder of negligible mass and of length L leans against a slick wall making an angle of θ with the horizontal floor. The coefficient of friction between the ladder and the floor is μ_s. A man climbs the ladder. What height h can he reach before the ladder slips? SSM

34 •• A uniform ladder of length L and mass m leans against a frictionless vertical wall, making an angle of 60° with the horizontal. The coefficient of static friction between the ladder and the ground is 0.45. If your mass is four times that of the ladder, how high can you climb before the ladder begins to slip?

35 •• A ladder of mass m and length L leans against a frictionless vertical wall, so that it makes an angle θ with the horizontal. The center of mass of the ladder is a height h above the floor. A force F directly away from the wall pulls on the ladder at its midpoint. Find the minimum coefficient of static friction μ_s for which the top end of the ladder will separate from the wall before the lower end begins to slip.

36 •• A 900-N man sits on top of a stepladder of negligible mass that rests on a frictionless floor as in Figure 12-43. There is a cross brace halfway up the ladder. The angle at the apex is $\theta = 30°$. (*a*) What is the force exerted by the floor on each leg of the ladder? (*b*) What is the tension in the cross brace? (*c*) If the cross brace is moved down toward the bottom of the ladder (maintaining the same angle θ), will its tension be the same, greater, or less than when it was at its higher position? Explain your answer.

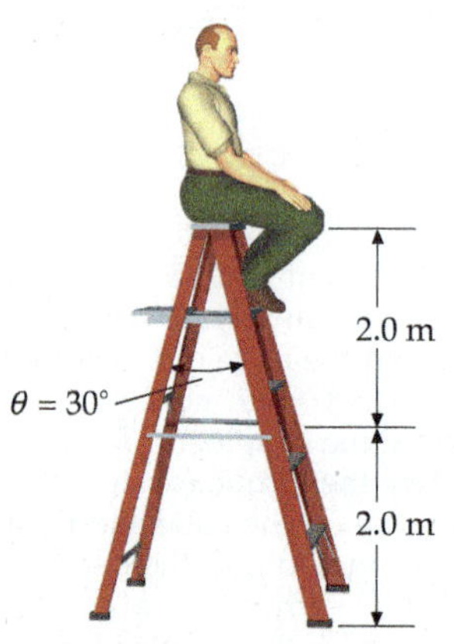

FIGURE 12-43 Problem 36

37 •• A uniform ladder rests against a frictionless vertical wall. The coefficient of static friction between the ladder and the floor is 0.30. What is the smallest angle between the ladder and the horizontal such that the ladder will not slip?

38 ••• A uniform log with a mass of 100 kg, a length of 4.0 m, and a radius of 12 cm is held in an inclined position, as shown in Figure 12-44. The coefficient of static friction between the log and the horizontal surface is 0.60. The log is on the verge of slipping to the right. Find the tension in the support wire and the angle the wire makes with the vertical wall.

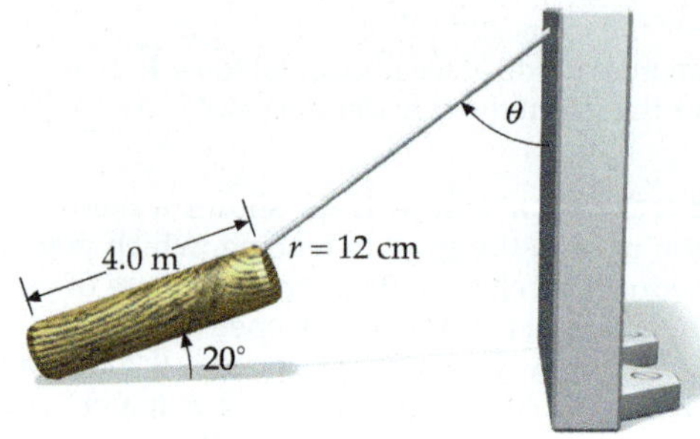

FIGURE 12-44 Problem 38

39 ••• A tall, uniform, rectangular block sits on an inclined plane, as shown in Figure 12-45. A cord attached to the top of the block prevents it from falling down the incline. What is the maximum angle θ for which the block will not slide on the incline? Assume the block has a height-to-width ratio, b/a, of 4.0 and the coefficient of static friction between it and the incline is $\mu_s = 0.80$. SSM

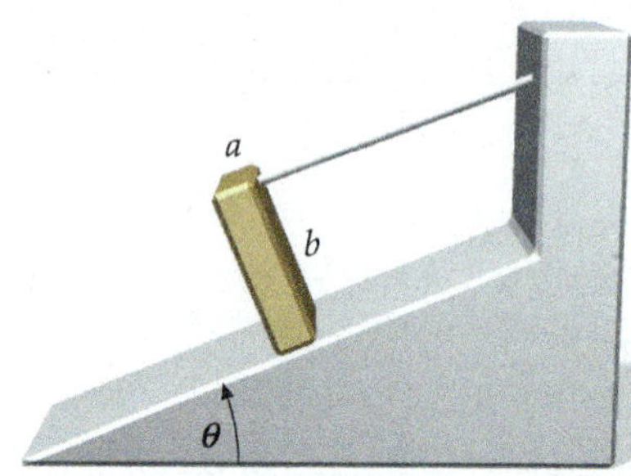

FIGURE 12-45 Problem 39

STRESS AND STRAIN

40 • A 50-kg ball is suspended from a steel wire of length 5.0 m and radius 2.0 mm. By how much does the wire stretch?

41 • Copper has a tensile strength of about $3.0 \times 10^8\ \mathrm{N/m^2}$. (*a*) What is the maximum load that can be hung from a copper wire of diameter 0.42 mm? (*b*) If half this maximum load is hung from the copper wire, by what percentage of its length will it stretch? SSM

42 • A 4.0-kg mass is supported by a steel wire of diameter 0.60 mm and length 1.2 m. How much will the wire stretch under this load?

43 • As a runner's foot pushes off on the ground, the shearing force acting on an 8.0-mm-thick sole is shown in Figure 12-46. If the force of 25 N is distributed over an area of 15 cm², find the angle of shear θ, given that the shear modulus of the sole is $1.9 \times 10^5\ \mathrm{N/m^2}$. SSM

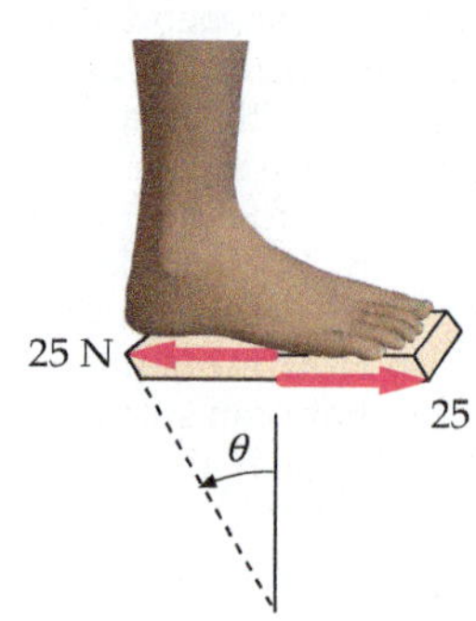

FIGURE 12-46 Problem 43

44 •• A steel wire of length 1.50 m and diameter 1.00 mm is joined to an aluminum wire of identical dimensions to make a composite wire of length 3.00 m. Find the resulting change in length of this composite wire if an object with a mass of 5.00 kg is hung vertically from one of its ends. (Neglect any effects the masses of two wires have on the changes in their lengths.)

45 •• Equal but opposite forces of magnitude F are applied to both ends of a thin wire of length L and cross-sectional area A. Show that if the wire is modeled as a spring, the force constant k is given by $k = AY/L$ and the potential energy stored in the wire is $U = \frac{1}{2}F\Delta L$, where Y is Young's modulus and ΔL is the amount the wire has stretched. SSM

46 •• The steel E string of a violin is under a tension of 53.0 N. The diameter of the string is 0.200 mm and the length under tension is 35.0 cm. Find (*a*) the unstretched length of this string, and (*b*) the work needed to stretch the string.

47 •• **ENGINEERING APPLICATION, SPREADSHEET** During a materials science experiment on the Young's modulus of rubber, the teaching assistant supplies you and your team with a rubber strip that is rectangular in cross section. She tells you to first measure the cross section dimensions, and you find their values are 3.0 mm × 1.5 mm. The lab write-up calls for the rubber strip to be suspended vertically and various (known) masses attached to its lower end. Your team obtains the following data for the length of the strip as a function of the load (mass) suspended from the end of the strip:

Load, kg	0.0	0.10	0.20	0.30	0.40	0.50
Length, cm	5.0	5.6	6.2	6.9	7.8	8.8

(*a*) Use a **spreadsheet** or **graphing calculator** to find Young's modulus for the rubber strip over this range of loads. *Hint: It is probably best to plot F/A versus $\Delta L/L$. Why?*

(*b*) Find the energy stored in the strip when the load is 0.15 kg. (See Problem 45.)

(*c*) Find the energy stored in the strip when the load is 0.30 kg. Is it twice as much as your answer to Part (*b*)? Explain.

48 •• A large mirror is hung from a nail, as shown in Figure 12-47. The supporting steel wire has a diameter of 0.20 mm and an unstretched length of 1.7 m. The distance between the points of support at the top of the mirror's frame is 1.5 m. The mass of the mirror is 2.4 kg. How much will the distance between the nail and the mirror increase due to the stretching of the wire as the mirror is hung?

FIGURE 12-47 Problem 48

49 •• Two masses, M_1 and M_2, are supported by wires that have equal lengths when unstretched. The wire supporting M_1 is an aluminum wire 0.70 mm in diameter, and the one supporting M_2 is a steel wire 0.50 mm in diameter. What is the ratio M_1/M_2 if the two wires stretch by the same amount?

50 •• A 0.50-kg ball is attached to one end of an aluminum wire that has a diameter of 1.6 mm and an unstretched length of 0.70 m. The other end of the wire is fixed to the top of a post. The ball rotates about the post in a horizontal plane at a rotational speed such that the angle between the wire and the horizontal is 5.0°. Find the tension in the wire and the increase in its length due to the tension in the wire.

51 •• An elevator cable is to be made of a new type of composite developed by Acme Laboratories. In the lab, a sample of the cable that is 2.00 m long and has a cross-sectional area of 0.200 mm^2 fails under a load of 1000 N. The actual cable used to support the elevator will be 20.0 m long and have a cross-sectional area of 1.20 mm^2. It will need to support a load of 20,000 N safely. Will it? SSM

52 •• If a material's density remains constant when it is stretched in one direction, then (because its total volume remains constant), its length must decrease in one or both of the other directions. Take a rectangular block of length x, width y, and depth z, and pull on it so that its new length $x' = x + \Delta x$. If $\Delta x \ll x$ and $\Delta y/y = \Delta z/z$, show that $\Delta y/y = -\frac{1}{2}\Delta x/x$.

53 •• You are given a wire with a circular cross section of radius r and a length L. If the wire is made from a material whose density remains constant when it is stretched in one direction, then show that $\Delta r/r = -\frac{1}{2}\Delta L/L$, assuming that $\Delta L \ll L$. (See Problem 52.) SSM

54 ••• For most materials listed in Table 12-1, the tensile strength is two to three orders of magnitude lower than Young's modulus. Consequently, most of these materials will break before their strain exceeds 1 percent. Of man-made materials, nylon has about the greatest extensibility—it can take strains of about 0.2 before breaking. But spider silk beats anything man-made. Certain forms of spider silk can take strains on the order of 10 before breaking! (*a*) If such a thread has a circular cross-section of radius r_0 and unstretched length L_0, find its new radius r when stretched to a length $L = 10L_0$. (Assume that the density of the thread remains constant as it stretches.) (*b*) If the Young's modulus of the spider thread is Y, calculate the tension needed to break the thread in terms of Y and r_0.

GENERAL PROBLEMS

55 • **BIOLOGICAL APPLICATION** A standard bowling ball weighs 16 pounds. You wish to hold a bowling ball in front of you, with your elbow bent at a right angle. Assume that your biceps attaches to your forearm at 2.5 cm out from the elbow joint, and that your biceps muscle pulls vertically upward, that is, it acts at right angles to the forearm. Also assume that the ball is held 38 cm out from the elbow joint. Let the mass of your forearm be 5.0 kg and assume its center of gravity is located 19 cm out from the elbow joint. How much force must your biceps muscle apply to the forearm in order to hold out the bowling ball at desired angle? SSM

56 •• **BIOLOGICAL APPLICATION, CONTEXT-RICH** A biology laboratory at your university is studying the location of a person's center of gravity as a function of her or his body weight. They pay well and you decide to volunteer. The location of your center of gravity when standing erect is to be determined by having you lie on a uniform board (mass of 5.00 kg, length 2.00 m)

supported by two scales, as shown in Figure 12-48. If your height is 188 cm and the left scale reads 445 N while the right scale reads 400 N, where is your center of gravity relative to your feet? Assume the scales are both exactly the same distance from the two ends of the board, are separated by 178 cm, and are set to each read zero before you get on the platform.

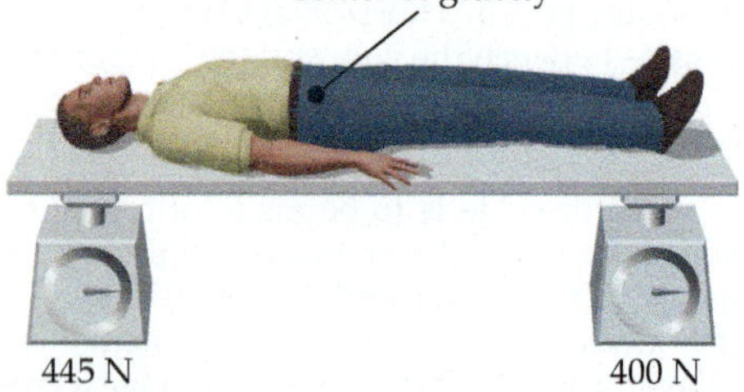

FIGURE 12-48 Problem 56

57 •• Figure 12-49 shows a mobile consisting of four objects hanging on three rods of negligible mass. Find the values of the unknown masses of the objects if the mobile is to balance. *Hint: Find the mass m_1 first.*

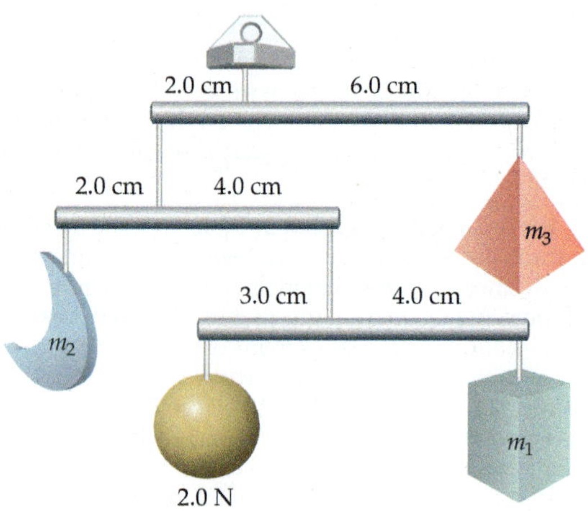

FIGURE 12-49 Problem 57

58 •• **ENGINEERING APPLICATION, CONTEXT-RICH** Steel construction beams, with an industry designation of "W12 × 22," have a weight of 22 pounds per foot. A new business in town has hired you to place its sign on a 4.0 m long steel beam of this type. The design calls for the beam to extend outward horizontally from the front brick wall (Figure 12-50). It is to be held in place by a 5.0-m-long steel cable. The cable is attached to one end of the beam and to the wall above the point at which the beam is in contact with the wall. During an initial stage of construction, the beam is *not* to be bolted to the wall, but to be held in place solely by friction. (*a*) What is the minimum coefficient of friction between the beam and the wall for the beam to remain in static equilibrium? (*b*) What is the tension in the cable in this case?

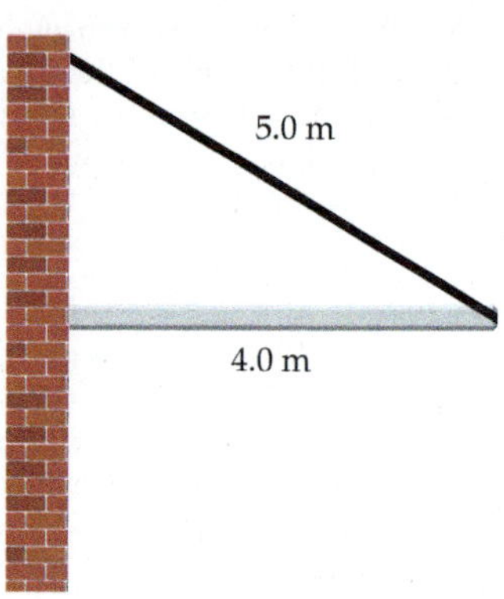

FIGURE 12-50 Problem 58

59 ••• Consider a rigid 2.5-m-long beam (Figure 12-51) that is supported by a fixed 1.25-m-high post through its center and pivots on a frictionless bearing at its center atop the vertical 1.25-m-high post. One end of the beam is connected to the floor by a spring that has a force constant $k = 1250$ N/m. When the beam is horizontal, the spring is vertical and unstressed. If an object is hung from the opposite end of the beam, the beam settles into an equilibrium position where it makes an angle of 17.5° with the horizontal. What is the mass of the object? SSM

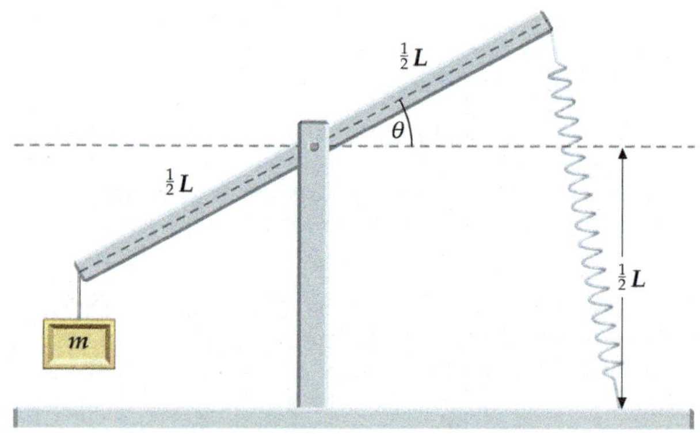

FIGURE 12-51 Problem 59

60 •• A rope and pulley system, called a *block and tackle*, is used to raise an object of mass M (Figure 12-52) at constant speed. When the end of the rope moves downward through a distance L, the height of the lower pulley is increased by h. (*a*) What is the ratio L/h? (*b*) Assume that the mass of the block and tackle is negligible and that the pulley bearings are frictionless. Show that $FL = Mgh$ by applying the work–energy principle to the block–tackle object.

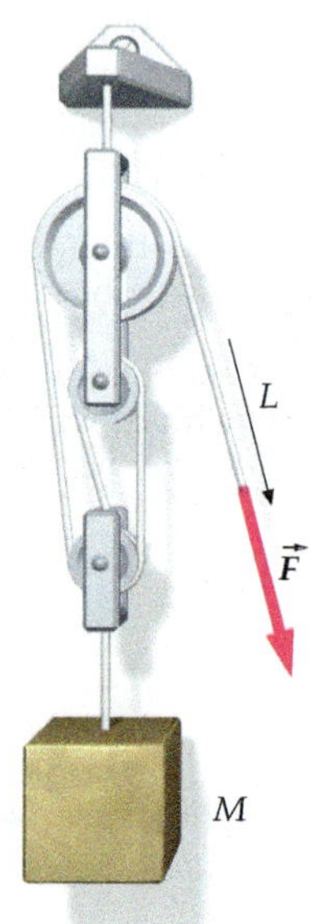

FIGURE 12-52 Problem 60

61 •• A plate of mass M in the shape of an equilateral triangle is suspended from one of its corners, and a mass m is suspended from another of its corners. If the base of the triangle makes an angle of 6.0° with the horizontal, what is the ratio m/M?

62 •• A standard six-sided pencil is placed on a notebook (Figure 12-53). Find the minimum coefficient of static friction μ_s such that, if the upper cover is raised, the pencil rolls down the incline rather than sliding.

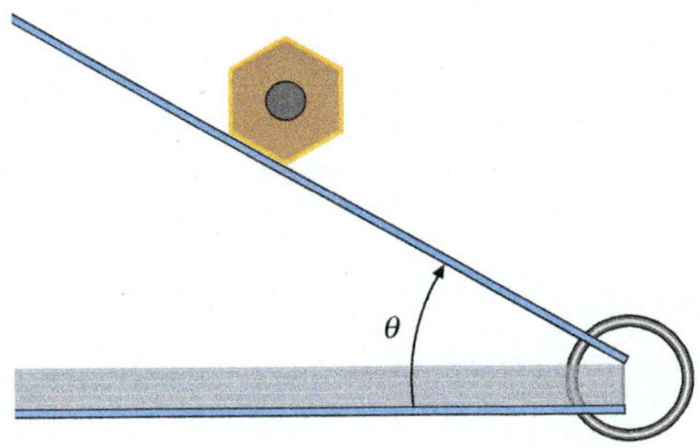

FIGURE 12-53 Problem 62

63 •• An 8.0-kg box that has a uniform density and is twice as tall as it is wide rests on the floor of a truck. What is the maximum coefficient of static friction between the box and floor so that the box will slide toward the rear of the truck rather than tip when the truck accelerates forward on a level road?

64 •• A balance scale has unequal arms. The scale is balanced with a 1.50-kg block on the left pan and a 1.95-kg block on the right pan (Figure 12-54). If the 1.95-kg block is removed from the right pan and the 1.50-kg block is then moved to the right pan, what mass on the left pan will balance the scale?

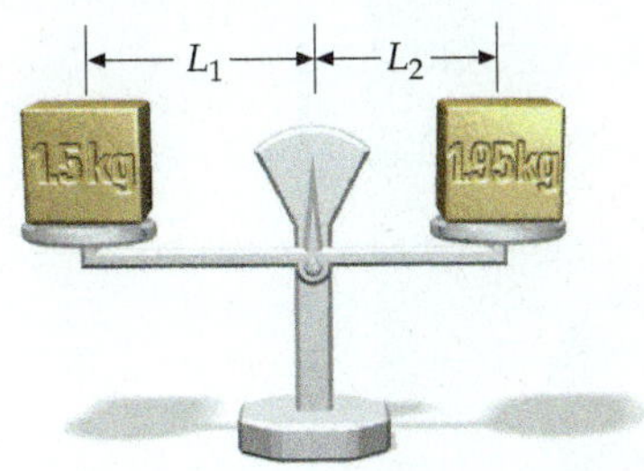

FIGURE 12-54 Problem 64

65 •• A cube leans against a frictionless wall, making an angle of θ with the floor, as shown in Figure 12-55. Find the minimum coefficient of static friction μ_s between the cube and the floor that is needed to prevent the cube from slipping. SSM

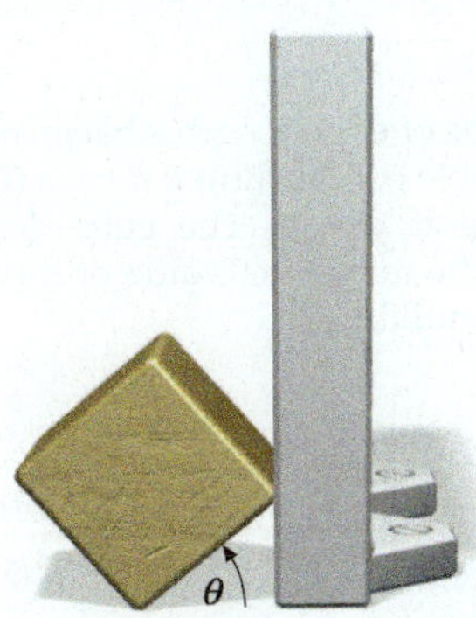

FIGURE 12-55 Problem 65

66 •• Figure 12-56 shows a 5.00-kg, 1.00-m-long rod hinged to a vertical wall and supported by a thin wire. The wire and rod each make angles of 45° with the vertical. When a 10.0-kg block is suspended from the midpoint of the rod, the tension T in the supporting wire is 52.0 N. If the wire will break when the tension exceeds 75 N, what is the maximum distance from the hinge at which the block can be suspended?

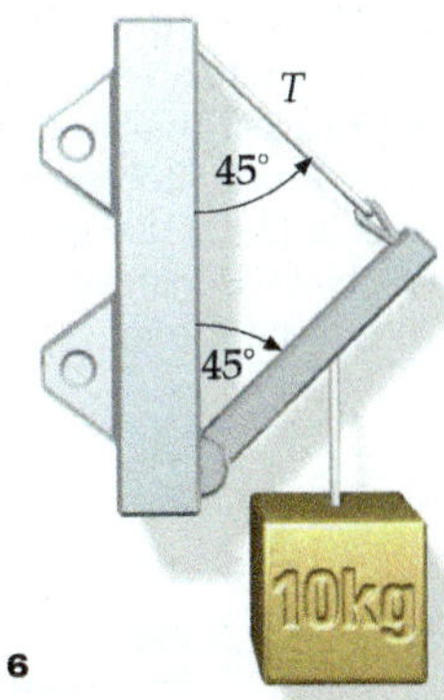

FIGURE 12-56 Problem 66

67 •• Figure 12-57 shows a 20.0-kg ladder leaning against a frictionless wall and resting on a frictionless horizontal surface. To keep the ladder from slipping, the bottom of the ladder is tied to the wall by a thin wire. When no one is on the ladder, the tension in the wire is 29.4 N. (The wire will break if the tension exceeds 200 N.) (*a*) If an 80.0-kg person climbs halfway up the ladder, what force will be exerted by the ladder against the wall? (*b*) How far from the bottom end of the ladder can an 80.0-kg person climb? SSM

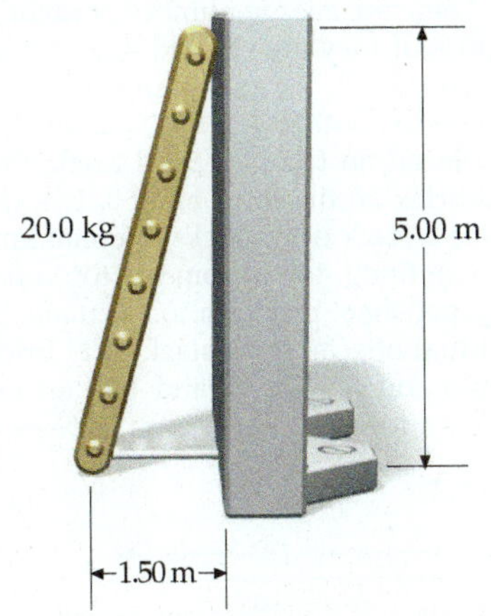

FIGURE 12-57 Problem 67

68 •• A 360-kg object is supported on a wire attached to a 15-m-long steel bar that is pivoted at a vertical wall and supported by a cable as shown in Figure 12-58. The mass of the bar is 85 kg. With the cable attached to the bar 5.0 m from the lower end, as shown, what are the tension in the cable and the force exerted by the wall on the steel bar?

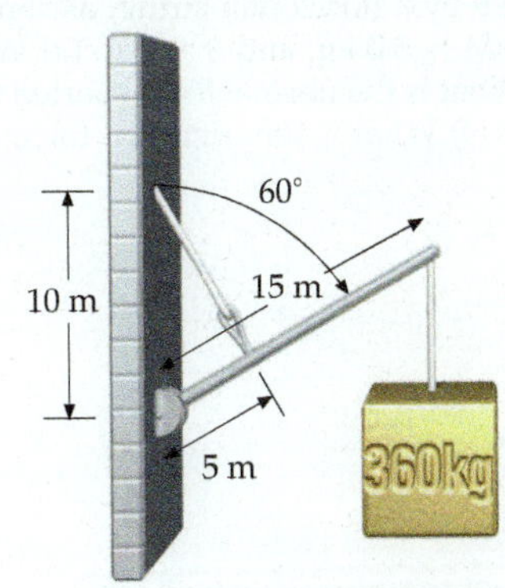

FIGURE 12-58 Problem 68

69 •• Repeat Problem 63 if the truck accelerates up a hill that makes an angle of 9.0° with the horizontal.

70 •• A thin uniform rod 60 cm long is balanced 20 cm from one end when an object whose mass is ($2m + 2.0$ grams) is at the end nearest the pivot and an object of mass m is at the opposite end (Figure 12-59*a*). Balance is again achieved if the object whose mass is ($2m + 2.0$ grams) is replaced by the object of mass m and no object is placed at the other end of the rod (Figure 12-59*b*). Determine the mass of the rod.

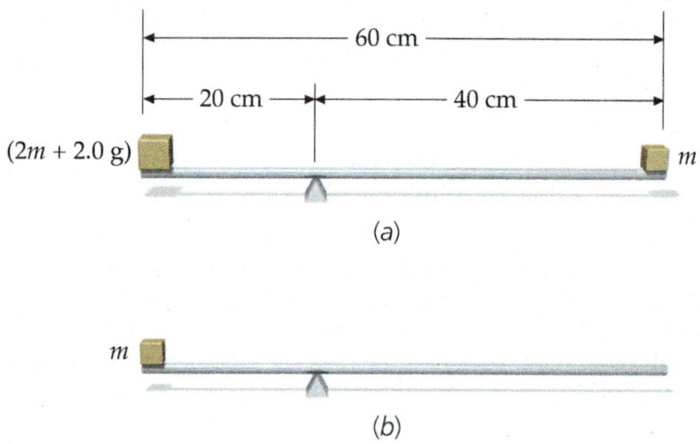

FIGURE 12-59 Problem 70

71 ••• **SPREADSHEET** There are a large number of identical uniform bricks, each of length L. If they are stacked one on top of another lengthwise (see Figure 12-60), the maximum offset that will allow the top brick to rest on the bottom brick is $L/2$. (*a*) Show that if this two-brick stack is placed on top of a third brick, the maximum offset of the second brick on the third brick is $L/4$. (*b*) Show that, in general, if you have a stack of N bricks, the maximum offset of the $(n - 1)$th brick (counting down from the top) on the nth brick is L/n. (*c*) Write a **spreadsheet** program to calculate total offset (the sum of the individual offsets) for a stack of N bricks, and calculate this for $L = 20$ cm and $N = 5, 10$, and 100. (*d*) Does the sum of the individual offsets approach a finite limit as $N \to \infty$? If so, what is that limit? SSM

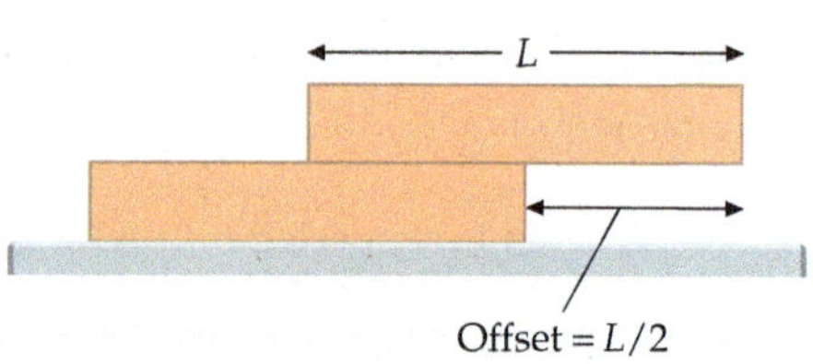

FIGURE 12-60 Problem 71

72 •• A uniform sphere of radius R and mass M is held at rest on an inclined plane of angle θ by a horizontal string, as shown in Figure 12-61. Let $R = 20$ cm, $M = 3.0$ kg, and $\theta = 30°$. (*a*) What is the tension in the string? (*b*) What is the normal force exerted on the sphere by the inclined plane? (*c*) What is the frictional force acting on the sphere?

FIGURE 12-61 Problem 72

73 ••• The legs of a tripod make equal angles of 90° with each other at the apex, where they join together. A 100-kg block hangs from the apex. What are the compressional forces in the three legs?

74 ••• Figure 12-62 shows a 20-cm-long uniform beam resting on a cylinder that has a radius of 4.0-cm. The mass of the beam is 5.0 kg, and that of the cylinder is 8.0 kg. The coefficient of static friction between beam and cylinder is zero, whereas the coefficients of static friction between the cylinder and the floor, and between the beam and the floor, are *not* zero. Are there any values for these coefficients of static friction such that the system is in static equilibrium? If so, what are these values? If not, explain why none exist.

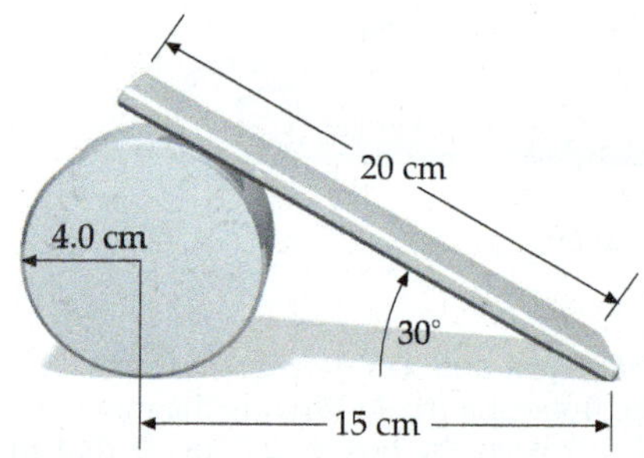

FIGURE 12-62 Problem 74

75 ••• Two solid smooth (frictionless) spheres of radius r are placed inside a cylinder of radius R, as in Figure 12-63. The mass of each sphere is m. Find the force exerted by the bottom of the cylinder on the bottom sphere, the force exerted by the wall of the cylinder on each sphere, and the force exerted by one sphere on the other. Express all forces in terms of m, R, and r. SSM

FIGURE 12-63 Problem 75

76 ••• A solid cube of edge-length a balanced atop a cylinder of diameter d is in unstable equilibrium if $d \ll a$ (Figure 12-64), and is in stable equilibrium if $d \gg a$. The cube does not slip on the cylinder. Determine the minimum value of the ratio d/a for which the cube is in stable equilibrium.

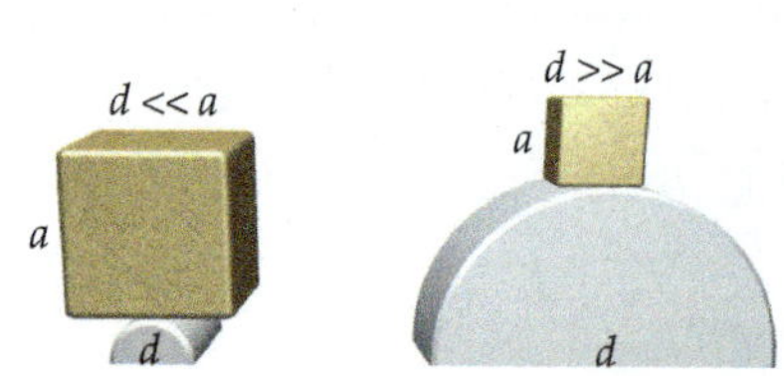

FIGURE 12-64 Problem 76

CHAPTER 13

CANAL BOATS IN SCOTLAND ARE ABLE TO TRAVEL BETWEEN THE FORTH & CLYDE CANAL AND THE UNION CANAL THANKS TO THE FALKIRK WHEEL. EACH OF THE TWO GONDOLAS OF THE WHEEL LIFTS 300 METRIC TONS [ONE METRIC TON (1 TONNE) IS EQUAL TO 1000 kg] OF WATER AND MAY ACCOMMODATE UP TO FOUR 20-m-LONG BOATS AT ANY ONE TIME. A VERY SMALL TORQUE AND A VERY SMALL AMOUNT OF ENERGY ARE REQUIRED TO ROTATE THIS MASSIVE WHEEL. *(Powered by Light/Alan Spencer/Alamy.)*

Fluids

Consider the air that fills our lungs, the blood that flows through our bodies, and even the rain that falls on us as we hurry to and from class. Air, blood, and rainwater are all fluids. It may seem odd to think of air as a fluid, but fluids include both liquids and gases. Liquids flow until they occupy the lowest possible regions of the space in which they are contained, be it a plastic bottle, a lock in a canal, or behind a dam. Unlike liquids, gases expand to fill their containers. To better understand fluid behavior is to better understand much about our own bodies and our interactions with the world around us.

Civil engineers employ their knowledge of fluids to design dams, which are thicker at the bottom than at the top. Automotive and aerospace engineers use wind tunnels to observe the flow of air around cars and aircraft to help them evaluate the aerodynamic aspects of particular vehicles. Blood-pressure gauges are used by medical professionals to measure the pressure of our blood.

We begin this chapter by studying fluids at rest, taking into account the density of and the pressure in fluids, as well as buoyancy and Archimedes' principle. Then, we study steady-state flow while emphasizing laminar flow.

? Why does it not require a large torque and a large amount of energy to rotate such a massive wheel? (See Example 13-8.)

13-1 DENSITY

In a gas, the average distance between two molecules is large compared with the size of a molecule. The molecules have little influence on one another except during their brief collisions. In a liquid or solid, the molecules are close together and exert forces on one another that are comparable to the forces that bind atoms into molecules. Molecules in a liquid form temporary short-range bonds that are continually broken and reformed due to the proximity of the molecules as they bump into each other. These bonds hold the liquid together; if the bonds were not present, the liquid would immediately evaporate and the molecules would escape as a vapor. The strength of the bonds in a liquid depends on the type of molecule that makes up the liquid. For example, the bonds between helium molecules are very weak and, for this reason, helium does not liquefy at atmospheric pressure unless the temperature is 4.2 K or lower. The ratio of the mass of an object to its volume is called its *average density*:

$$\text{Average density} = \frac{\text{Mass}}{\text{Volume}}$$

DEFINITION—AVERAGE DENSITY

If the mass of substance within a small element of volume dV is dm, then the density of the substance at the location of the volume element is

$$\rho = \frac{dm}{dV} \qquad 13\text{-}1$$

DEFINITION—DENSITY

where ρ (the lowercase Greek letter rho) is used to denote density. Because the gram was originally defined as the mass of one cubic centimeter of liquid water, the density of liquid water in cgs (centimeter–gram–second) units is 1 g/cm^3. Converting to SI units, we obtain for the density of water

$$\rho_{\text{w}} = \frac{1\text{ g}}{\text{cm}^3} \times \frac{\text{kg}}{10^3\text{ g}} \times \left(\frac{100\text{ cm}}{1\text{ m}}\right)^3 = 1000\text{ kg/m}^3 \qquad 13\text{-}2$$

Precise measurements of density must take temperature into account, because the densities of most solids and liquids, including water, vary with temperature. Equation 13-2 gives the maximum value for the density of water, which occurs at 4°C. Table 13-1 lists the densities of some common substances.

A convenient unit of volume for fluids is the **liter** (L):

$$1\text{ L} = 10^3\text{ cm}^3 = 10^{-3}\text{ m}^3$$

In terms of this unit, the density of water at 4°C is 1.00 kg/L = 1.00 g/mL. When the average density of a solid object is greater than that of water, it sinks in water, and when a solid object's average density is less than the density of water, it floats. The ratio of the density of a substance to that of a reference substance, usually water, is its **specific gravity.** For example, the specific gravity of aluminum is 2.7, meaning that a volume of aluminum has 2.7 times the mass of an equal volume of water.

Table 13-1 Densities of Selected Substances

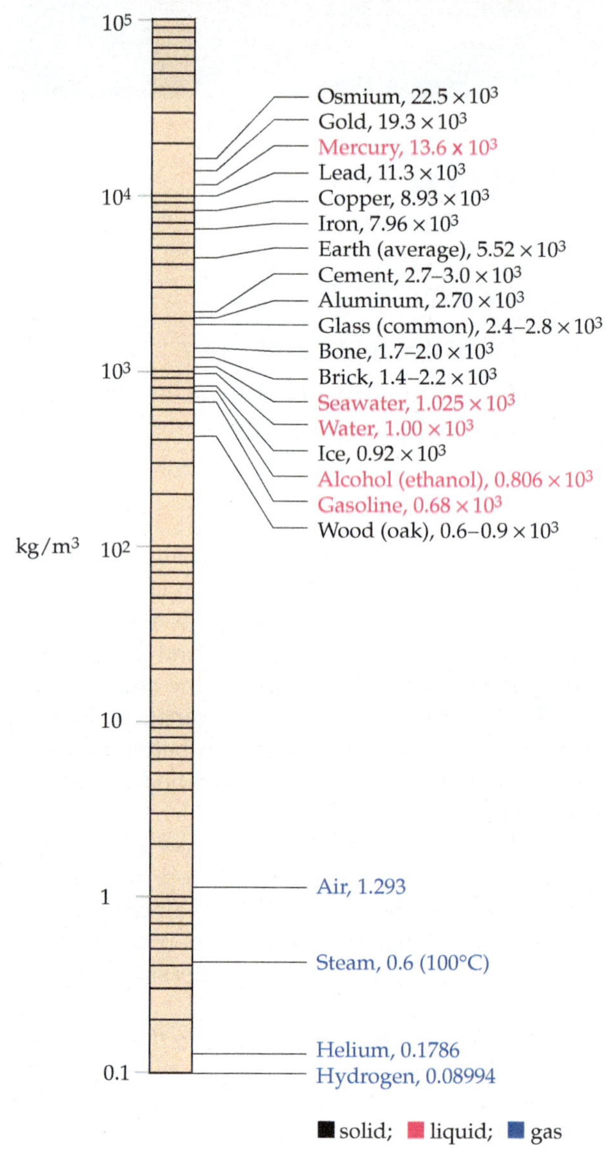

The density values exceed five orders of magnitude.

The specific gravities of objects that sink when submerged in water range from 1 to about 22.5 (for the densest element, osmium).

Most solids and liquids expand only slightly when heated, and contract only slightly when subjected to an increase in external pressure. Because these changes in volume are relatively small, we often treat the densities of solids and liquids as approximately independent of temperature and pressure. The density of a gas, on the other hand, depends strongly on the pressure and temperature, so these variables must be specified when reporting the densities of gases. By convention, the standard conditions for the measurement of physical properties are atmospheric pressure at sea level and a temperature of 0°C. The densities for the substances listed in Table 13-1 are for these conditions. Note that the densities of liquids and solids are considerably greater than those of gases. For example, the density of liquid water is about 800 times that of air under standard conditions.

Example 13-1 Calculating Density

A flask is filled to the brim with 200-mL of water at 4.0°C. When the flask is heated to 80.0°C, 6.0 g of water spill out. What is the density of water at 80°C? (Assume that the expansion of the flask is negligible.)

PICTURE The density of water at 80°C is $\rho' = m'/V$, where $V = 0.200\text{ L} = 200\text{ cm}^3$ is the volume of the flask, and m' is the mass remaining in the flask after 6.0 g spill out. We find m' by first finding the mass of water originally in the flask.

SOLVE

1. Let ρ' be the density of water at 80°C and let m' be the mass of the water remaining in the flask of volume $V = 200$ mL. Relate ρ' to m' using the definition of density:
$$\rho' = \frac{m'}{V}$$
2. Calculate the original mass m of water in the flask at 4.0°C using $\rho = 1.00$ kg/L and the definition of density:
$$m = \rho V = (1.00\text{ kg/L})(0.200\text{ L}) = 0.200\text{ kg}$$
3. Calculate the mass of water remaining after 6 g spill out:
$$m' = m - 6\text{ g} = 0.200\text{ kg} - 0.006\text{ kg} = 0.194\text{ kg}$$
4. Use this value of m' to find the density of water at 80°C:
$$\rho' = \frac{m'}{V} = \frac{0.194\text{ kg}}{0.200\text{ L}} = \boxed{0.970\text{ kg/L}}$$

CHECK The density of water at 4.0°C is 1.00 kg/L. The density of water is greatest at 4.0°C so we expect the density of water at 80°C to be less than 1.00 kg/L. Our step-3 result confirms this expectation.

PRACTICE PROBLEM 13-1 A solid metal cube 8.00 cm on an edge has a mass of 4.08 kg. (*a*) What is the average density of the cube? (*b*) If the cube is made from a single element listed in Table 13-1, what is the element?

PRACTICE PROBLEM 13-2 A gold brick is 5.0 cm × 10 cm × 20 cm. What is its mass?

13-2 PRESSURE IN A FLUID

When a fluid such as water is in contact with a solid surface, the fluid exerts a force normal (perpendicular) to the surface at each point on the surface. The force per unit area is called the **pressure** P of the fluid:

$$P = \frac{F}{A} \qquad 13\text{-}3$$

DEFINITION—PRESSURE

The SI unit of pressure is the newton per square meter (N/m^2), which is called the **pascal** (Pa):

$$1 \text{ Pa} = 1 \text{ N/m}^2 \qquad 13\text{-}4$$

> Force is a vector quantity, but pressure is a scalar quantity. (Pressure is the magnitude of the force per unit area.)

In the U.S. customary system, pressure is usually given in pounds per square inch (lb/in^2). Another common unit of pressure is the atmosphere (atm), which approximately equals the air pressure at sea level. One atmosphere is defined as exactly 101.325 kilopascals (kPa), which is about 14.70 lb/in^2:

$$1 \text{ atm} = 101.325 \text{ kPa} \approx 14.70 \text{ lb/in}^2 \qquad 13\text{-}5$$

Other units of pressure in common use are discussed later in this chapter.

If the pressure on an object increases, the ratio of the increase in pressure, ΔP, to the fractional decrease in volume, $(-\Delta V/V)$, is called the **bulk modulus:**

$$B = -\frac{\Delta P}{\Delta V/V} \qquad 13\text{-}6$$

DEFINITION—BULK MODULUS

Like other elastic moduli (Young's modulus and shear modulus were introduced in Section 12-7), bulk modulus is a ratio of stress to strain, with ΔP the stress and $-\Delta V/V$ the strain. (All stable materials decrease in volume when subjected to an increase in external pressure. Thus, the negative sign in Equation 13-6 means that B is always positive.)

The more difficult it is to compress a material, the smaller is the fractional volume decrease $-\Delta V/V$ for a given pressure increase ΔP, and hence the greater the bulk modulus. The **compressibility** is the reciprocal of the bulk modulus. (The easier it is to compress a material, the larger the compressibility.) Liquids, gases, and solids all have a bulk modulus. Because liquids and solids are relatively incompressible, they have large values of B, and these values are relatively independent of temperature and pressure. Gases, on the other hand, are easily compressed, and their values for B depend strongly on pressure and temperature. Table 13-2 charts values for the bulk modulus of various materials.

As any scuba diver knows, the pressure in a lake or ocean increases with depth. Similarly, the pressure of the atmosphere decreases with altitude. For a liquid, whose density is approximately constant throughout, the pressure increases linearly with depth. We can see this by considering a column of liquid of cross-sectional area A, as shown in Figure 13-1. To support the weight of the liquid in the column of height Δh, the pressure at the bottom of the column must be greater than the pressure at the top. The weight of the liquid in the column is

$$F_g = mg = (\rho V)g = \rho A\,\Delta h g$$

where ρ and V are the density and volume of the liquid. If P_0 is the pressure at the top and P is the pressure at

Table 13-2 Approximate Values for the Bulk Modulus B of Various Materials

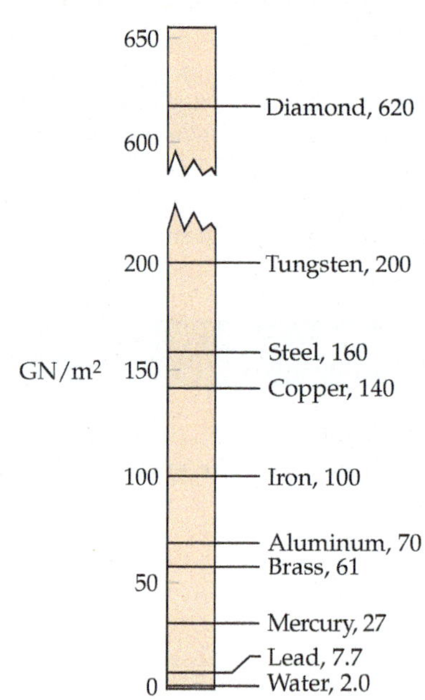

FIGURE 13-1

the bottom, the net upward force exerted by this pressure difference is $PA - P_0A$. Setting this net upward force equal to the weight of the column, we obtain

$$PA - P_0A = (\rho A\,\Delta h g)$$

or

$$P = P_0 + \rho g\,\Delta h \quad (\rho \text{ constant}) \qquad 13\text{-}7$$

PRACTICE PROBLEM 13-3

How far below the surface of a lake is a scuba diver if the pressure is equal to 2.00 atm? (The pressure at the surface is 1.00 atm.)

(Corbis.)

Example 13-2 Force on a Dam

A rectangular dam 30 m wide supports a body of water to a depth of 25 m. Find the total horizontal force on the dam due to both water and air pressure.

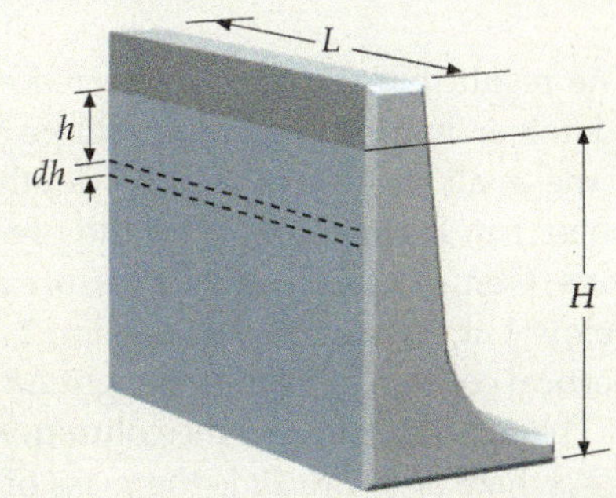

FIGURE 13-2

PICTURE Because the pressure varies with depth, we cannot merely multiply the pressure times the area of the dam to find the force exerted by the water. Instead we can consider the force exerted on a strip of surface of length $L = 30$ m, height dh, and area $dA = L\,dh$ at a depth h (Figure 13-2), and then integrate from $h = 0$ to $h = H = 25$ m. The water pressure at depth h is $P_{at} + \rho gh$, where P_{at} is the atmospheric pressure. Neglect any variation in air pressure over the 25-m height of the dam.

SOLVE

1. Express the force dF of the water on the element of length L and height dh in terms of the pressure $P_{at} + \rho hg$ on the dam by the water:

$$dF = P\,dA = (P_{at} + \rho gh)L\,dh$$

2. Integrate from $h = 0$ to $h = H$ to find the horizontal component of the force of the water on the dam:

$$F = \int_{h=0}^{h=H} dF = \int_0^H (P_{at} + \rho gh)L\,dh = P_{at}LH + \frac{1}{2}\rho gLH^2$$

3. The downstream surface of the dam is not vertical. Sketch an edge-on view (Figure 13-3) of a horizontal strip across the downstream side of the surface, a strip of length L and width ds. Let dh be the height of the strip:

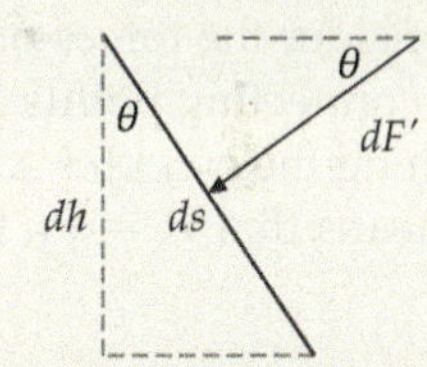

FIGURE 13-3

4. Relate the force dF' exerted on this strip by the air to the pressure of the air and the area of the strip:

$$dF' = P_{at}\,dA = P_{at}L\,ds$$

5. Express the horizontal component of dF'_x in terms of dh:

$$dF'_x = dF\cos\theta = P_{at}L\,ds\cos\theta = P_{at}L\,dh$$

6. Integrate from $h = 0$ to $h = H$ to find the horizontal component of the force of the air on the downstream side of the dam:

$$F' = \int_{h=0}^{h=H} dF' = \int_0^H P_{at}L\,dh = P_{at}LH$$

7. The net horizontal force on the dam is $F - F'_x$:

$$F - F'_x = (P_{at}LH + \tfrac{1}{2}\rho gLH^2) - P_{at}LH = \tfrac{1}{2}\rho gLH^2$$
$$= \tfrac{1}{2}(1000\text{ kg/m}^3)(9.81\text{ N/kg}))(30\text{ m})(25\text{ m})^2$$
$$= 9.20 \times 10^7\text{ N} = \boxed{9.2 \times 10^7\text{ N}}$$

CHECK The net horizontal force on the dam is independent of air pressure, as expected. This is as expected because the pressure of the air on the surface of the water increases the pressure throughout the water by one atmosphere, and the air presses on the downstream side of the dam a pressure of one atmosphere.

TAKING IT FURTHER Dams typically are thicker at the bottom than at the top because the pressure on the dam increases with the depth of the water.

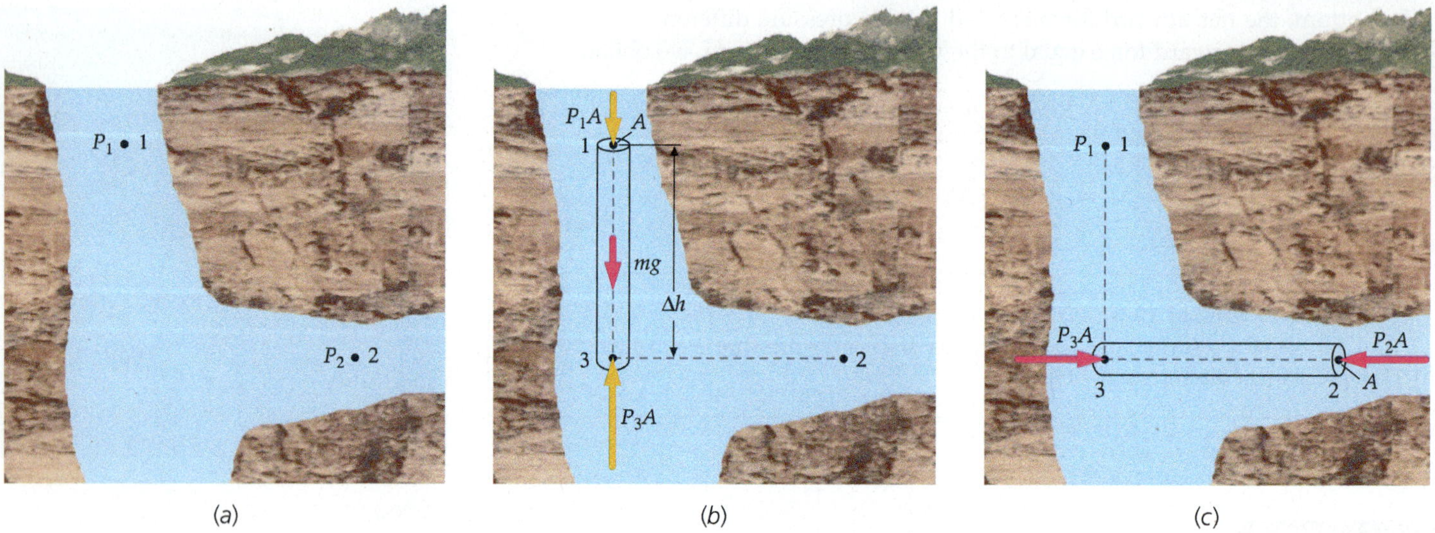

FIGURE 13-4

The result that the pressure increases linearly with depth holds for a liquid in any container, independent of the shape of the container. Furthermore, the pressure is the same at all points at the same depth. We can see this by comparing the pressure at point 1 in Figure 13-4*a* with the pressure at point 2, which is inside an underwater cave. First, we compare the pressure at points 1 and 3, where point 3 is a point directly below 1 at the same depth as point 2 (Figure 13-4*b*). Consider the vertical forces on the vertical column of water of height Δh and cross-sectional area A between points 1 and 3. The upward force on the column, P_3A, balances the two downward forces P_1A and mg, where $m = \rho A\,\Delta h$ is the mass of the water in the column ($A\,\Delta h$ is the volume of the column). That is, $P_3A = P_1A + (\rho A\,\Delta hg)$. Dividing both sides by A gives

$$P_3 = P_1 + \rho g\,\Delta h$$

Next consider the forces on the horizontal cylinder of water, also of cross-sectional area A, connecting points 2 and 3 (Figure 13-4*c*). There are two forces with components along the cylinder's axis, P_3A and P_2A. The fact that these forces balance each other means that $P_3 = P_2$. It follows that

$$P_2 = P_1 + \rho g\,\Delta h$$

If we increase the pressure in a container of water by pressing down on the top surface with a piston, the increase in pressure is the same throughout the liquid. This holds for both liquids and gasses and is known as **Pascal's principle,** named after Blaise Pascal (1623–1662):

> A pressure change applied to a confined fluid is transmitted undiminished to every point in the fluid and to the walls of the container.
>
> PASCAL'S PRINCIPLE

A common application of Pascal's principle is the hydraulic lift shown in Figure 13-5.

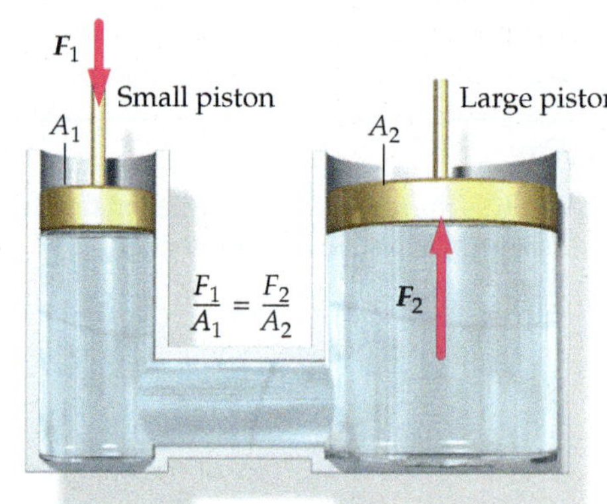

FIGURE 13-5 Hydraulic lift. A small force F_1 on the small piston produces an increase in pressure F_1/A_1 that is transmitted by the liquid to the large piston. Because the pressure changes are the same throughout the fluid, the forces exerted on the pistons are related by $F_2/A_2 = F_1/A_1$. Because the area of the large piston is much greater than that of the small piston, the force on the large piston $F_2 = (A_2/A_1)F_1$ is much greater than F_1.

Example 13-3 A Hydraulic Lift

The large piston in a hydraulic lift has a radius of 20 cm. What force must be applied to the small piston of radius 2.0 cm to raise a car of mass 1500 kg?

PICTURE The pressure P times the area A_2 of the large piston must equal the weight mg of the car. The force F_1 that must be exerted on the small piston is this pressure times the area A_1 (Figure 13-5).

SOLVE

1. The force F_1 is the pressure P times the area A_1: $F_1 = PA_1$

2. The pressure P times the area A_2 equals the weight of the car: $PA_2 = mg$ so $P = \frac{mg}{A_2}$

3. Substitute this result for P into the step-1 result and calculate F_1:

$$F_1 = PA_1 = \frac{mg}{A_2}A_1 = mg\frac{A_1}{A_2} = mg\frac{\pi r_1^2}{\pi r_2^2}$$

$$= (1500\text{ kg})(9.81\text{ N/kg})\left(\frac{2.0\text{ cm}}{20\text{ cm}}\right)^2$$

$$= 147\text{ N} = \boxed{150\text{ N}}$$

CHECK The radii differ by a factor of 10, so the areas differ by a factor of $10^2 = 100$. Thus, the forces also differ by a factor of 100.

Figure 13-6 shows water in a container that has sections of different shapes. At first glance, it might seem that the pressure at the bottom of section 3, the section containing the most water, would be greatest and that water would therefore be forced to a greater height in section 2, which is the section with the least water. But that is not observed, a result known as the **hydrostatic paradox.** The pressure depends only on the depth of the water, not on the shape of the container, so at the same depth the pressure is the same in all parts of the container, a finding that can be shown experimentally. Although the water in section 4 of the container weighs more than that in section 2, the portion of the water in section 4 that is not above the opening at the bottom is supported by the horizontal shelf of the section. In fact, the water above the opening at the bottom of section 5 weighs less than the water above an opening of the same size at the bottom of section 1. However, the horizontal shelf of section 5 exerts a downward force on the water—exactly compensating for the shortfall of weight.

! Pressure depends only on the depth of the water, not on the shape of a container. So the pressure is the same for all parts of the container that are at the same depth.

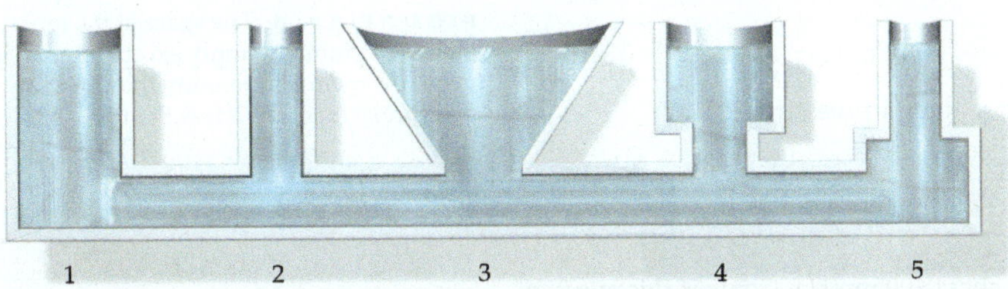

FIGURE 13-6 The hydrostatic paradox. The water level is the same regardless of the shape of the vessel. The weight of those portions of the water not above an opening is supported by the sides of the containers.

We can use the fact that the pressure increases linearly with the depth of a liquid to measure unknown pressures. Figure 13-7 shows a simple pressure gauge, the open-tube manometer. The top of the tube is open to the atmosphere at pressure P_{at}. The other end of the tube is at pressure P, which is to be measured. The difference $P - P_{at}$, called the **gauge pressure** P_{gauge}, is equal to ρgh, where ρ is the density of the liquid in the tube. The pressure you measure in your automobile tire is gauge pressure. When the tire is entirely flat, the gauge pressure is zero, and the absolute pressure of the air remaining in the tire is atmospheric pressure. The absolute pressure P is obtained from the gauge pressure by adding atmospheric pressure to it:

$$P = P_{gauge} + P_{at} \qquad 13\text{-}8$$

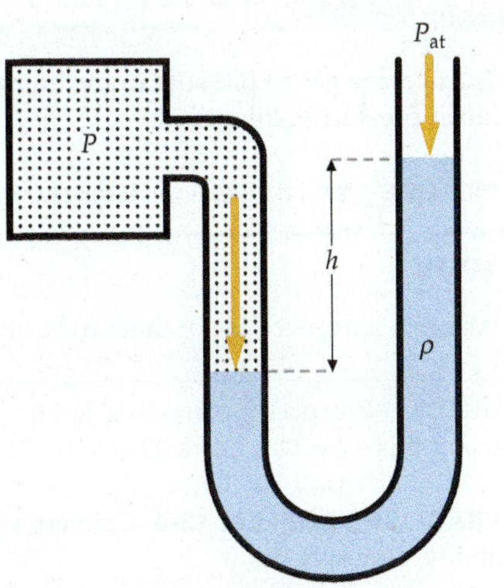

FIGURE 13-7 Open-tube manometer for measuring an unknown pressure P. The difference $P - P_{at}$ equals ρgh.

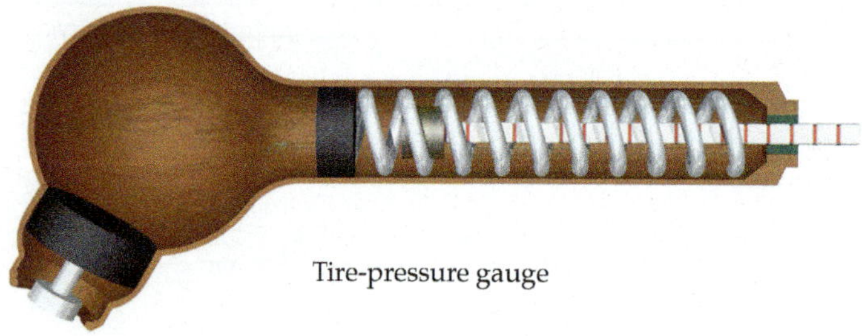

Tire-pressure gauge. The piston pushes the rod to the right until the force of the spring plus the force due to atmospheric pressure balances the force due to the air pressure in the tire.

Checking the tire pressure. *(Vanessa Vick/ Photo Researchers, Inc.)*

Figure 13-8 shows a mercury barometer, which is used to measure atmospheric pressure. The top end of the tube has been closed off and evacuated so that the pressure there is zero. The other end is submerged in a pool of mercury that is open to the atmosphere at pressure P_{at}. The pressure P_{at} is $\rho g h$, where ρ is the density of mercury.

> **PRACTICE PROBLEM 13-4**
> At 0°C, the density of mercury is 13.595×10^3 kg/m³. What is the height of the mercury column in a barometer if the pressure P is exactly 1 atm = 101.325 kPa?

In practice, pressure is often measured in millimeters of mercury, a unit called the **torr,** after the Italian physicist Evangelista Torricelli, or in inches of mercury (written inHg). The various units of pressure are related as follows:

$$\begin{aligned} 1.00 \text{ atm} &= 760 \text{ mmHg} = 760 \text{ torr} \\ &= 29.9 \text{ inHg} = 101 \text{ kPa} = 14.7 \text{ lb/in}^2 \end{aligned} \qquad 13\text{-}9$$

Other units commonly used on weather maps are the **bar** and the **millibar,** which are defined as follows:

$$1 \text{ bar} = 10^3 \text{ millibars} = 100 \text{ kPa} \qquad 13\text{-}10$$

A pressure of 1 atm is about 1.01 percent greater than a pressure of 1 bar.

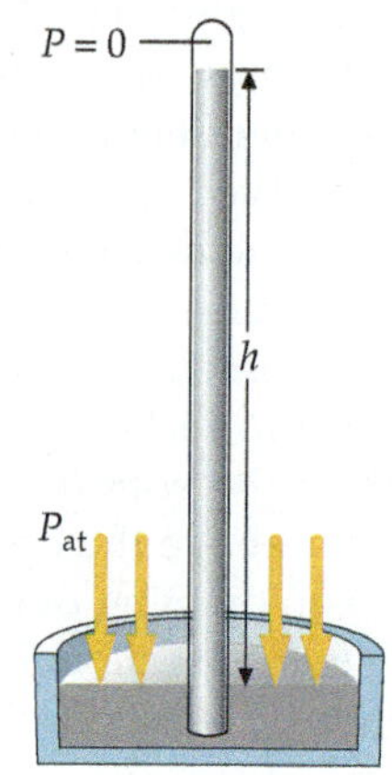

FIGURE 13-8 The space at the top of the mercury column is empty except for mercury vapor. At room temperature the vapor pressure of mercury is less than 10^{-5} atm.

Example 13-4 Blood Pressure in the Aorta

The average gauge pressure in the human aorta is about 100 mmHg. Convert this average blood pressure to kilopascals.

PICTURE Use a conversion factor obtained from Equation 13-9:

SOLVE

We use a conversion factor that can be obtained from Equation 13-9:

$$P = 100 \text{ mmHg} \times \frac{101 \text{ kPa}}{760 \text{ mmHg}} = \boxed{13.3 \text{ kPa}}$$

CHECK We expect the pressure to be a small fraction of 1.00 atm. A pressure of 13.3 kPa meets this expectation as 1.00 atm = 101 kPa.

PRACTICE PROBLEM 13-5 Convert a pressure of 45.0 kPa to (*a*) millimeters of mercury, and (*b*) atmospheres.

The relation between pressure and altitude (or depth) is more complicated for a gas than for a liquid. The density of a liquid is essentially constant, whereas the density of a gas is approximately proportional to the pressure. As you go up from the surface of Earth, pressure in a column of air decreases, just as the pressure would decrease as you go up from the bottom in a column of water. But the decrease in air pressure is not linear with distance.

Example 13-5 The Law of Atmospheres

The assumption that the density of air is proportional to the pressure predicts that the pressure decreases exponentially with altitude. Use this assumption to verify this prediction and calculate the altitude at which the pressure is one-half of its value at sea level.

PICTURE Apply Newton's second law to an element of air at altitude y and vertical thickness dy to find an expression for the change in pressure over the change in altitude dy. Integrate this expression, taking into account that the density is proportional to the pressure.

SOLVE

1. Draw a thin horizontal disk-shaped element of air at altitude y, vertical thickness dy, cross-sectional area A, and mass dm. Draw and label all the forces on this element (Figure 13-9):

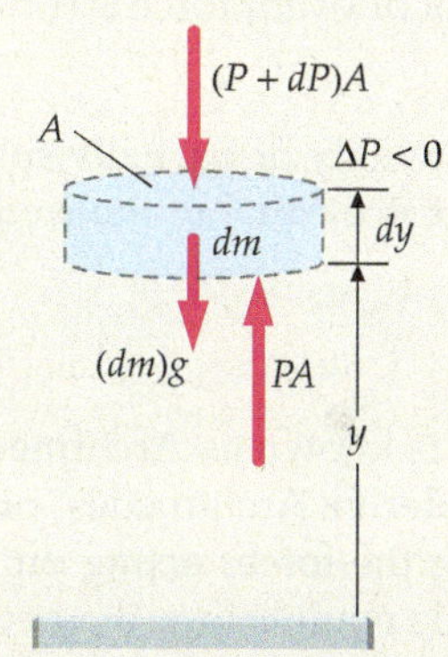

FIGURE 13-9 The pressure below the thin disk-shaped element of air is greater than the pressure above it. This pressure difference produces an upward force on the disk that balances the downward force of gravity on it.

2. Apply Newton's second law to the disk. The acceleration is zero so the sum of the forces equal zero:

$$PA - (P + dP)A - (dm)g = 0$$

3. Simplify the equation and substitute $\rho A\,dy$ for dm:

$$-A\,dP - \rho g A\,dy = 0$$

so $$dP = -\rho g\,dy$$

4. We are assuming that the density is proportional to pressure, and we know the density ρ_0 and pressure P_0 at sea level ($y = 0$):

$$\frac{\rho}{P} = \frac{\rho_0}{P_0}$$

5. Substitute for ρ in the step-3 result and divide both sides by P to separate variables:

$$dP = -P\frac{\rho_0}{P_0}g\,dy$$

so $$\frac{dP}{P} = -\frac{\rho_0}{P_0}g\,dy$$

6. Integrate from $y = 0$ to $y = y_f$. Let $P = P_f$ be the pressure at altitude y_f:

$$\int_{P_0}^{P_f}\frac{dP}{P} = -\frac{\rho_0}{P_0}g\int_0^{y_f}dy$$

so $$\ln\frac{P_f}{P_0} = -\frac{\rho_0}{P_0}gy_f$$

7. Solve for P_f. Then substitute P for P_f and y for y_f:

$$P_f = P_0e^{-(\rho_0/P_0)gy_f} \quad \text{or} \quad \boxed{P = P_0e^{-(\rho_0/P_0)gy}}$$

8. Solve for the height h at which $P = \frac{1}{2}P_0$. Look up the density of air at 1-atm pressure in Table 13-1:

$$\tfrac{1}{2}P_0 = P_0e^{-(\rho_0/P_0)gh} \Rightarrow h = \frac{P_0}{\rho_0 g}\ln 2$$

so $$h = \frac{(1.01 \times 10^5\ \text{Pa})\ln 2}{(1.29\ \text{kg/m}^3)(9.81\ \text{N/kg})} = \boxed{5.5\ \text{km}}$$

CHECK An altitude of 5.5 km is about 18,000 ft. We know that many people suffer from the effects of oxygen deprivation at this altitude. That people would suffer at half the normal air pressure is not surprising.

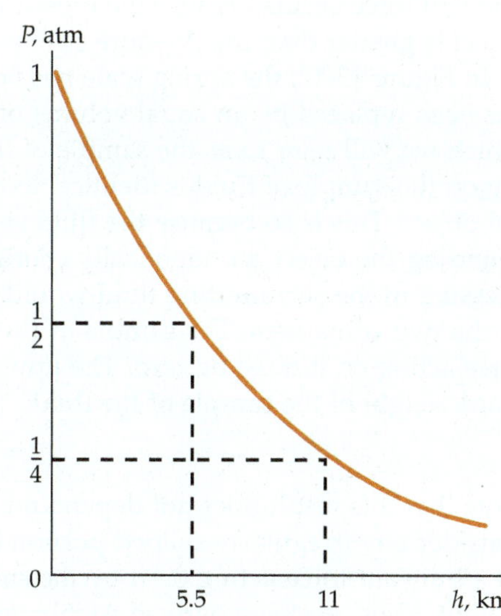

FIGURE 13-10 Variation in pressure with height above Earth's surface. For each 5.5-km increase in height, the pressure decreases by half.

TAKING IT FURTHER The step-7 result reveals that air pressure decreases exponentially with altitude. This conclusion means that air pressure decreases by a constant fraction for a given increase in height, as shown in Figure 13-10. At a height of about 5.5 km, the air pressure is half its value at sea level. If we go up another 5.5 km to an altitude of 11 km (a typical altitude for airliners), the pressure is again halved so that it is one-fourth its value at sea level, and so on. At the high altitudes at which commercial jets fly, the cabins must be pressurized. The density of air is approximately proportional to the pressure, so the density of air decreases with altitude. Less oxygen is available on a mountain top than at normal elevations. As a result, exercising in the Rockies is difficult, and climbing in the Himalayas is dangerous.

13-3 BUOYANCY AND ARCHIMEDES' PRINCIPLE

If a dense object submerged in water is weighed by suspending it from a spring scale (Figure 13-11*a*), the apparent weight of the object when submerged (the reading on the scale) is less than the weight of the object. This difference exists because the water exerts an upward force that partially balances the force of gravity. This upward force is even more evident when we submerge a piece of cork. When completely submerged, the cork experiences an upward force from the water pressure that is greater than the force of gravity, so when released it accelerates up toward the surface. The force exerted by a fluid on a body wholly or partially submerged in it is called the **buoyant force.** It is equal to the weight of the fluid displaced by the body. (The definition of buoyant force is further refined later in this section.)

> A body wholly or partially submerged in a fluid is buoyed up by a force equal to the weight of the displaced fluid.
>
> ARCHIMEDES' PRINCIPLE

This result is known as **Archimedes' principle.**

We can derive Archimedes' principle from Newton's laws by considering the forces acting on a portion of a fluid and noting that in static equilibrium the net force must be zero. Figure 13-11*b* shows the vertical forces acting on an object being weighed while submerged. These forces are the force of gravity $\vec{F}_g$ acting down, the force of the spring scale $\vec{F}_s$ acting up, a force $\vec{F}_1$ acting down because of the fluid pressure on the top surface of the object, and a force $\vec{F}_2$ acting up because of the fluid pressing on the bottom surface of the object. Because the spring scale reads a force less than the weight of the object, the magnitude of force $\vec{F}_2$ must be greater than the magnitude of force $\vec{F}_1$. The vector sum of these two forces is equal to the buoyant force $\vec{B} = \vec{F}_1 + \vec{F}_2$ (Figure 13-11*c*). The buoyant force occurs because the pressure of the fluid on the bottom surface of the object is greater than the pressure on the top surface of the object.

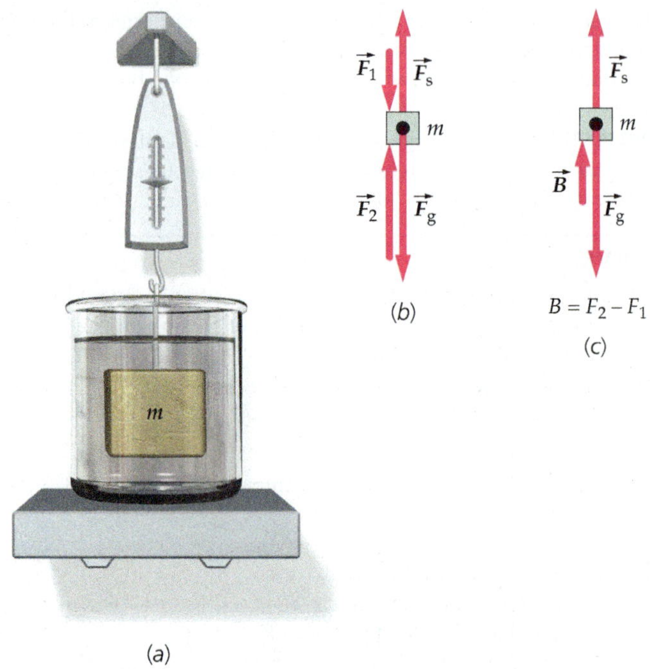

FIGURE 13-11 (*a*) Weighing an object submerged in a fluid. (*b*) Free-body diagram showing the weight $\vec{F}_g$, the force $\vec{F}_s$ of the spring, and the forces $\vec{F}_1$ and $\vec{F}_2$ that are exerted on the object by the surrounding fluid. (*c*) The buoyant force $\vec{B}$, where $\vec{B} = \vec{F}_1 + \vec{F}_2$, is the total force exerted by the fluid on the object.

In Figure 13-12, the spring scale has been eliminated and the submerged object has been replaced by an equal volume of the fluid (outlined by the dashed lines), which we will refer to as the sample of fluid. The buoyant force $\vec{B} = \vec{F}_1 + \vec{F}_2$ acting on the sample of fluid is identical to the buoyant force that acted on our original object. This is so because the fluid surrounding the sample and the fluid surrounding the object are identically configured; there is no reason to suppose the pressure in the surrounding fluid would not be the same at corresponding points in the two containers. The sample of fluid is in equilibrium, so we know the net force acting on it must be zero. The upward buoyant force thus equals the downward weight of the sample of the fluid:

$$B = F_{gf} \qquad 13\text{-}11$$

Note that this result does not depend on the shape of the submerged object. If we consider any irregularly shaped portion of a static fluid as our sample, there will be a buoyant force acting on it by the surrounding fluid that exactly supports its weight. Thus, we have derived Archimedes' principle.

Archimedes (287–212 B.C.E.) had been given the task of determining whether a crown (actually a wreath) made for King Hieron II was of pure gold or had been adulterated with some cheaper metal such as silver, and to do this without destroying the crown. For Archimedes, the problem was to determine if the density

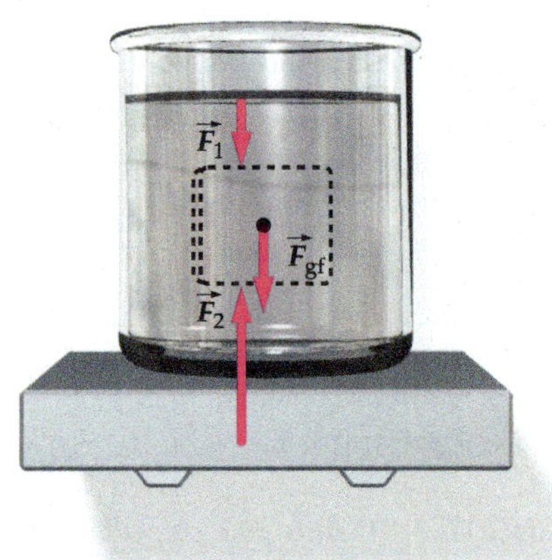

FIGURE 13-12 Figure 13-11 with the submerged body replaced by an equal volume of fluid. The forces $\vec{F}_1$ and $\vec{F}_2$, due to the pressure of the fluid, are the same as in Figure 13-11. The magnitude of the buoyant force is thus equal to the weight F_{gf} of the displaced fluid.

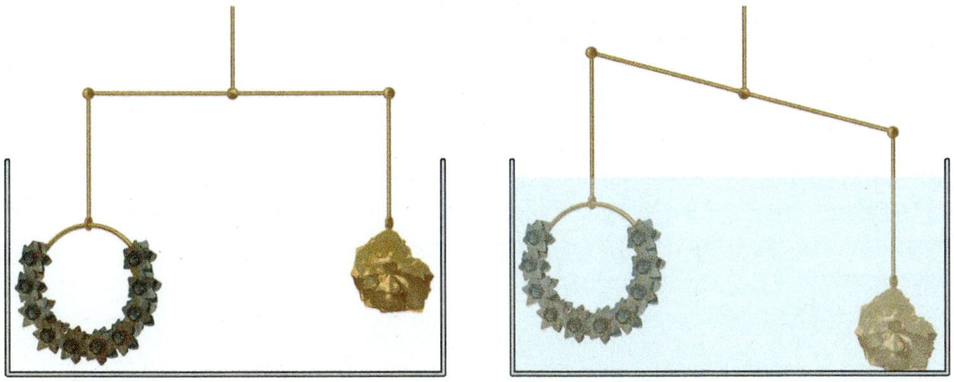

(*a*) Crown and gold nugget have equal weight.

(*b*) Crown displaces more water than does the gold nugget.

FIGURE 13-13 (*a*) The crown and the gold nugget have equal weight. (*b*) The balance tips because the wreath displaces more water than the gold nugget.

of the irregularly shaped crown was the same as the density of gold. As the story goes, he came upon the solution while sinking himself into a bathtub and immediately rushed home, running naked through the streets of Syracuse shouting "Eureka!" ("I have found it!"). This flash of insight preceded Newton's laws, which we used to derive Archimedes' principle, by some 1900 years. What Archimedes had found was a simple and accurate way to compare the density of the crown with the density of gold by using a balance. He placed the balance over a large basin, suspended the crown from one arm of the balance and an equal mass of pure gold from the other arm. He then added water to the basin (Figure 13-13*a*) submerging the crown and the pure gold. The balance tilted, with the crown rising (Figure 13-13*b*) —indicating that the buoyant force on the crown was greater than that on the pure gold because the volume of water displaced by the crown was greater than that displaced by the pure gold. The crown was less dense than the pure gold.

The apparent weight $F_{g\,app}$ of an object submerged in a fluid is the difference between its weight F_g and the magnitude of the buoyant force B:

$$F_{g\,app} = F_g - B \qquad \text{13-12}$$

The hot-air balloon and the boat both need buoyancy to float. *(Richard Hamilton Smith/ CORBIS.)*

PROBLEM-SOLVING STRATEGY

Solving Problems Using Archimedes' Principle

PICTURE Carefully read the problem statement to determine the situation. Sketching a picture of the situation is often helpful.

SOLVE

1. Apply Archimedes' principle to relate the buoyant force to the weight of the displaced fluid.
2. Apply Newton's second law to the object and solve for the desired quantity.

CHECK Verify that your answer is plausible.

Example 13-6 Is It Really Gold? *Context-Rich*

Your friend is concerned about a gold ring she bought on a recent trip. The ring was expensive, and she would like to know whether it is really made of gold or of something else. You decide to help her, using your knowledge of physics. You weigh the ring and find that it has a weight of 0.158 N. Using a string, you suspend the ring from the scale and, with the ring submerged in water, weigh it again to find a new reading of 0.150 N. Is the ring pure gold?

PICTURE If the ring is pure gold, its density (relative to that of water) is 19.3 (see Table 13-1). Using Archimedes' principle as a guide, determine the density of the ring relative to the density of water.

SOLVE

1. The weight F_g of the ring equals its density ρ_R times its volume V times g. The buoyant force B on the ring (when submerged) equals the density of water ρ_w times Vg:

$$F_g = \rho_R V g$$
$$B = \rho_w V g$$

2. Divide the first equation by the second to relate the ratio of the weight to the buoyant force to the ratio of the density to the density of water:

$$\frac{F_g}{B} = \frac{\rho_R V g}{\rho_w V g} = \frac{\rho_R}{\rho_w}$$

3. In accord with Newton's second law, B equals the weight minus the apparent weight when submerged:

$$F_{g\,app} = F_g - B \Rightarrow B = F_g - F_{g\,app}$$

4. Substitute for B in step 2:

$$\frac{F_g}{F_g - F_{g\,app}} = \frac{\rho_R}{\rho_w}$$

5. Solve for the ratio ρ_R/ρ_w:

$$\frac{\rho_R}{\rho_w} = \frac{F_g}{F_g - F_{g\,app}} = \frac{0.158\text{ N}}{0.158\text{ N} - 0.150\text{ N}} = \frac{0.158\text{ N}}{0.008\text{ N}}$$

6. The denominator has one significant figure, so the ratio of the densities is determined to one significant figure:

$$\frac{\rho_R}{\rho_w} = \frac{0.158\text{ N}}{0.008\text{ N}} = 19.3 = 20$$

The 2 in the 20 is a significant digit, but the 0 is not.

7. Compare the ratio of the densities with the ratio of the density of gold to the density of water, which is 19.3:

According to the measurement, the ratio of the densities is 2×10^1.

The ring may be pure gold, but the measurement is not accurate enough to be certain.

CHECK The uncertainty in the result is large, which is expected. When two numbers that are almost equal are subtracted there are fewer significant figures in the result than in the original numbers.

TAKING IT FURTHER A scale capable of much greater accuracy is needed to make a more certain determination.

PRACTICE PROBLEM 13-6 A block of an unknown material weighs 3.00 N and has an apparent weight of 1.89 N when submerged in water. What is the material?

PRACTICE PROBLEM 13-7 A piece of lead (specific gravity = 11.3) weighs 80.0 N in air. What does it weigh when submerged in water?

REVISITING THE BUOYANT FORCE

The density of the block shown in Figure 13-14 is greater than the density of the surrounding fluid, and both the block and the scale pan are completely submerged in the fluid. The gravitation force on the block is its weight, $\vec{F}_g$, and the scale is adjusted, so it reads zero when the block is not being supported by the pan (Figure 13-14*b*). If the block is on the pan (Figure 13-14*a*), the scale reading is equal to the magnitude of the apparent weight $F_{g\,app}$ of the block. When the block is on the pan, the fluid is in direct contact with the entire surface of the block—except for those regions of the bottom surface of the block that are in direct contact with the pan. We assume the pan surface is not perfectly flat, but instead has some high and some low regions, and that the pan is in direct contact with the bottom surface of the block only

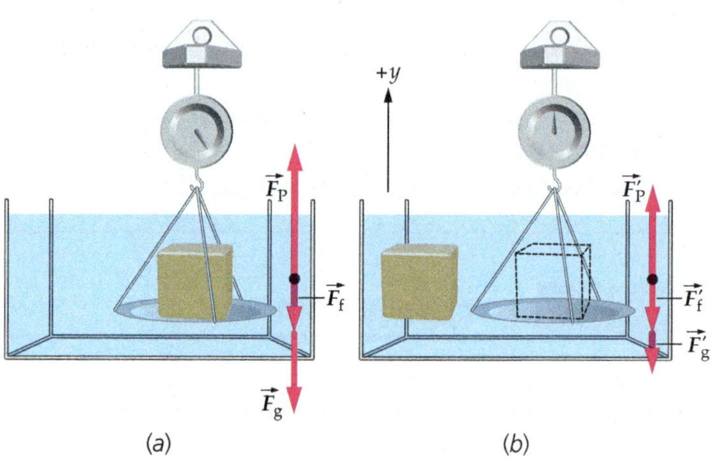

FIGURE 13-14

at the high regions of the pan. (At the low regions of the pan, there is fluid between the pan and the block.) We now analyze this situation to show that the scale reading is equal to the weight of the block less the weight of an equal volume of fluid.

While resting on the pan, the net force exerted by the fluid on the block $\vec{F}_f$ is a combination of the downward force of the fluid on the top surface of the block and the upward force of the fluid on those regions of the bottom surface of the block that are in direct contact with the fluid. (We have drawn $\vec{F}_f$ downward. However, if the fluid were in direct contact with a large enough area of the bottom surface of the block $\vec{F}_f$ would be upward.) The two other vertical forces acting on the block are the gravitational force $\vec{F}_g$ and the upward force $\vec{F}_P$ exerted on the block by the pan at the regions of direct contact between the block and the pan.

In Figure 13-14*b*, the block has been moved off the pan, and in its place is a sample of fluid of identical size and shape (outlined by the dashed lines). The same regions of the surface of the pan are in direct contact with the bottom surface of this sample of fluid as were in direct contact with the block before it was moved. The forces acting on the fluid sample are the forces acting on it by the surrounding fluid $\vec{F}'_f$, the upward force on it by the pan $\vec{F}'_P$, and the gravitational force $\vec{F}'_g$. The forces $\vec{F}_f$ and $\vec{F}'_f$ are equal because, at every point where the sample and the surrounding fluid are in direct contact, the pressure of the surrounding fluid is the same as it was at the same point before the block was moved off the pan.

When the submerged block rests on the pan, the block is in equilibrium, so

$$F_{fy} + F_{Py} + F_{gy} = 0 \qquad \text{or} \qquad F_{fy} + F_P - F_g = 0$$

and when the block is moved off the pan, the fluid sample in its place is in equilibrium, so

$$F'_{fy} + F'_P - F'_g = 0$$

Subtracting these equations, exploiting that $F'_{fy} = F_{fy}$ and rearranging gives

$$F_P - F'_P = F_g - F'_g$$

where $F_P - F'_P$ is the decrease in the scale reading when the block is moved off the pan. Thus, $F_P - F'_P$ is the apparent weight of the submerged block. That is,

$$F_{g\,app} = F_g - F'_g$$

It is common parlance to refer to $F_g - F_{g\,app}$ as the buoyant force B. Rearranging gives

$$B = F_g - F_{g\,app} = F'_g$$

This is the same expression for the buoyant force as is in Equation 13-12, which was established with the fluid in direct contact with 100 percent of the surface of the submerged object.

Example 13-7 Measuring the Fat

You decide to enroll in a fitness program. To determine your initial fitness, at the first meeting your percentage of body fat is measured. Your percentage of body fat can be estimated by measuring your body density (the average density of your body). Fat is less dense than muscle or bone. Assume the average density of fat is 0.90×10^3 kg/m^3 and the average density of lean tissue (everything except fat) is 1.1×10^3 kg/m^3. Measuring your body density involves measuring your apparent weight while you are submerged in water with the air completely exhaled from your lungs. (In practice, the amount of air remaining in the lungs is estimated and corrected for.) Suppose that your apparent weight when submerged in water is 5 percent of your weight. What percentage of your body mass is fat?

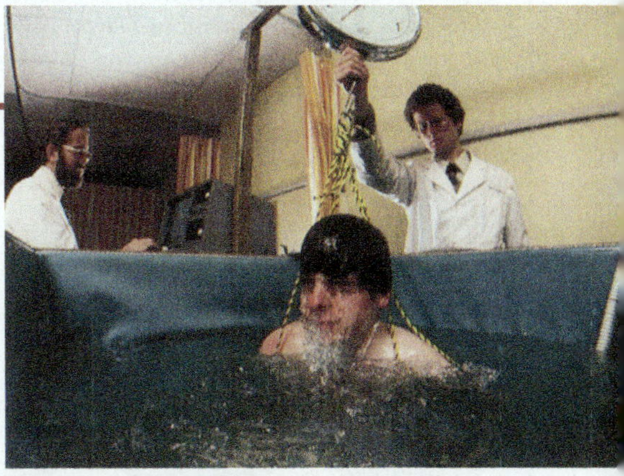

To determine the percentage of fat in this man's body, his density is measured by weighing him while he is submerged in the water. *(David Burnett/ Woodfin Camp and Assoc.)*

PICTURE For the person, the total volume equals the volume of the fat plus the volume of the lean, and the total mass equals the mass of the fat plus the mass of the lean. The volume and average density are related to the mass by $m = \rho V$. The fraction of fat equals the mass of the fat divided by the total mass and the fraction of lean equals the mass of the lean divided by the total mass. Also, the fraction of fat plus the fraction of lean equals 1.

SOLVE

1. Using Equations 13-2 and 13-3, find the ratio of your body's density to the density of water:
$$\frac{\rho}{\rho_{water}} = \frac{F_g}{F_g - F_{g\,app}} = \frac{F_g}{F_g - 0.05F_g} = 1.05$$
2. Your total body volume equals the volume of fat plus the volume of lean tissue:
$$V_{tot} = V_{fat} + V_{lean}$$
3. Because mass equals density times volume, volume equals mass divided by density. Substitute the corresponding mass-to-density ratio for each volume in the step-2 result:
$$\frac{m_{tot}}{\rho} = \frac{m_{fat}}{\rho_{fat}} + \frac{m_{lean}}{\rho_{lean}}$$
4. The mass of fat is $f_{fat}m_{tot}$, where f_{fat} is the fraction of fat, and the mass of lean is $f_{lean}m_{tot}$, where f_{lean} is the fraction of lean. Substitute for m_{fat} and m_{lean} in the step-3 result:
$$\frac{m_{tot}}{\rho} = \frac{f_{fat}m_{tot}}{\rho_{fat}} + \frac{f_{lean}m_{tot}}{\rho_{lean}}$$
5. The fraction of fat plus the fraction of lean tissue equals 1:
$$f_{fat} + f_{lean} = 1$$
6. Divide both sides of the step-4 result by m_{tot} and substitute $1 - f_{fat}$ for f_{lean}:
$$\frac{1}{\rho} = \frac{f_{fat}}{\rho_{fat}} + \frac{(1 - f_{fat})}{\rho_{lean}}$$
7. Solve the step-6 result for f_{fat}:
$$f_{fat} = \frac{1 - (\rho_{lean}/\rho)}{1 - (\rho_{lean}/\rho_{fat})}$$
8. Using the step-1 result, substitute for ρ in the step-7 result and solve for f_{fat}:
$$f_{fat} = \frac{1 - (\rho_{lean}/1.05\rho_{water})}{1 - (\rho_{lean}/\rho_{fat})} = \frac{1 - (1.1/1.05)}{1 - (1.1/0.90)} = 0.21$$
9. Convert to a percentage:
$$100\% \times f_{fat} = \boxed{21\%}$$

CHECK You are an adult male and are not overweight. The charts inform you that for adult males, a body-fat percentage between 18% and 25% is acceptable. Thus, the step-9 result is a plausible result.

PRACTICE PROBLEM 13-8 If Ed's apparent weight when submerged is zero, what is his body-fat percentage?

Example 13-8 Which Weighs the Most? *Conceptual*

Consider five identical beakers (Figure 13-15). The water level in each beaker is up to the point of overflowing. A toy boat is floating on the surface of Beaker *A*. A second toy boat, one that tipped over and sank to the bottom, is in Beaker *B*. An ice cube is floating on the surface of Beaker C. The block of wood submerged in Beaker *D* is tethered to the bottom by a thread and dab of superglue. (There is nothing in Beaker *E* except water.) The two boats, the ice cube, and the block of wood have equal masses. The density of the submerged boat is twice that of water, and the density of the block of wood is half that of water. Each beaker rests on a scale. Rank the readings on the scales from most to least.

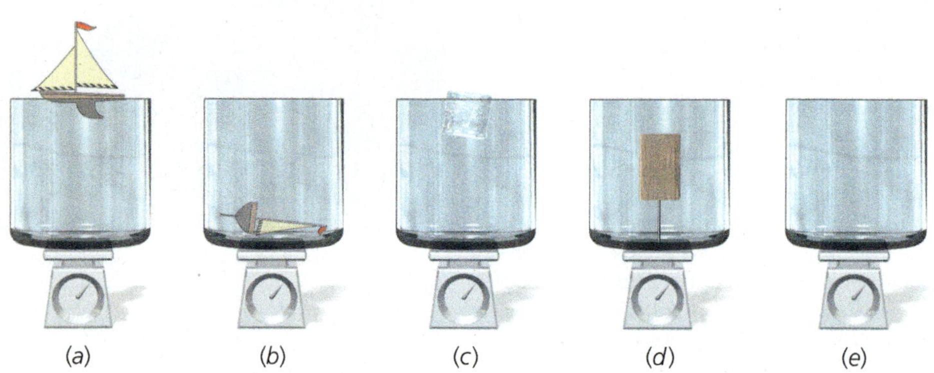

FIGURE 13-15

PICTURE The reading of each scale is equal to the total weight of the system resting on it. In each case, the system consists of the beaker, the water in the beaker, and any object submerged in, or floating on, the water. Beaker *E* contains the most water. The buoyant force on an object is equal to the weight of the fluid it displaces, a submerged object displaces its own volume of water, and a floating object displaces its own weight of water. The amount of water in each of the beakers equals the amount of water in Beaker *E* less the amount of water displaced by a floating or submerged object.

SOLVE

1. Let $F_{g\,A}$ be the reading on the scale under Beaker *A*, let $F_{g\,B}$ be the reading on the scale under Beaker *B*, and so on. In addition, let $F_{g\,obj}$ be the weight of each of the objects (they have identical weights). A floating object displaces its own weight of water. Calculate the weight of the water displaced by each floating object:

   The weight of the water displaced by the floating boat is $F_{g\,obj}$, and the weight displaced by the floating ice cube is also $F_{g\,obj}$.

2. A submerged object displaces its own volume of water. Calculate the weight of the water displaced by each submerged object:

   The weight of the water displaced by the submerged wooden block is $2F_{g\,obj}$ and the weight displaced by the submerged boat is $\frac{1}{2}F_{g\,obj}$.

3. In each case, the system consists of a beaker, the water in the beaker, and any object submerged in, or floating on, the water. In addition, the weight of water in each beaker equals the weight of water in Beaker *E* less the weight of water displaced by an object. Calculate the total weight of the system on each scale:

Beaker	Weight
A	$F_{g\,sys} = (F_{g\,E} - F_{g\,obj}) + F_{g\,obj} = F_{g\,E}$
B	$F_{g\,sys} = (F_{g\,E} - \frac{1}{2}F_{g\,obj}) + F_{g\,obj} = F_{g\,E} + \frac{1}{2}F_{g\,obj}$
C	$F_{g\,sys} = (F_{g\,E} - F_{g\,obj}) + F_{g\,obj} = F_{g\,E}$
D	$F_{g\,sys} = (F_{g\,E} - 2F_{g\,obj}) + F_{g\,obj} = F_{g\,E} - F_{g\,obj}$
E	$F_{g\,sys} = (F_{g\,E} - 0) + 0 = F_{g\,E}$

4. The reading on the scale equals the weight of the system $F_{g\,sys}$.

   The reading on the scale with the sunken boat has the highest reading, the reading on the scale with the submerged block has the lowest reading, and the readings on the other three scales are equal.

CHECK The water pressure at the bottom of each beaker is the same because the water has the same depth in all of the beakers. This means that the water pushes down on the bottoms of Beakers *A*, *C*, *D*, and *E* with identical forces. Thus, the readings on the scales under Beakers *A*, *C*, and *E* are identical. For Beaker *D*, in addition to the water pushing down the thread pulls up, so the reading on the scale under Beaker *D* is less than the readings on the scales under Beakers *A*, *C*, and *E*. For Beaker *B*, the submerged boat pushes down on the bottom of the beaker with a force per unit area that exceeds the water pressure there, so the reading on the scale under Beaker *B* is largest of them all.

TAKING IT FURTHER The Falkirk Wheel is perfectly balanced for the same reason that the readings on the scales under Beakers *A*, *C*, and *E* are equal. As long as the depths of the water in the two gondolas remain equal, and the boats in the gondolas (if any) remain floating, the wheel will remain in perfect balance. Because it remains in balance, only a small effort is needed to rotate the wheel.

The tallest iceberg seen in the North Atlantic extended about 168 m above sea level, about the same height as the Washington Monument. This iceberg was sighted in Melville Bay, Greenland, in 1957. *(Courtesy of the U.S. Coast Guard International Ice Patrol.)*

Example 13-9 An Iceberg

Find the fraction of the volume of an iceberg that is below sea level.

PICTURE Let V be the volume of the iceberg and V_{sub} be the volume that is submerged. The weight of the iceberg is $\rho_{IB}Vg$ and the buoyant force due to the seawater is $\rho_{SW}V_{sub}g$. The densities of ice and seawater are found in Table 13-1.

SOLVE

1. Because the iceberg is in equilibrium, the buoyant force equals its weight:

$$F_g = B$$
$$\rho_{IB} V g = \rho_{SW} V_{sub} g$$

2. Solve for V_{sub}/V:

$$f = \frac{V_{sub}}{V} = \frac{\rho_{IB}}{\rho_{SW}} = \frac{0.92 \times 10^3\ \text{kg/m}^3}{1.025 \times 10^3\ \text{kg/m}^3} = 0.898 = \boxed{0.90}$$

CHECK We have all seen an ice cube floating in fresh water. The great majority of the ice cube is submerged. We expect pretty much the same for an iceberg floating in seawater, and our step-2 result agrees with our expectation. Because the density of seawater is 2 to 3% more than the density of fresh water, ice floats a bit higher in seawater than in fresh water.

If we replace ρ_{SW} in the preceding calculation with ρ_f, the density of the fluid, we can determine the submerged fraction of an object floating in any fluid. From Example 13-9, the fraction of a floating object of uniform density ρ that is submerged equals the ratio of its density to the density of the fluid.

$$\frac{V_{sub}}{V} = \frac{\rho}{\rho_f} \qquad \text{13-13}$$

Smoke from a burning cigarette. At first the smoke rises in a regular stream, but the simple streamlined flow quickly becomes turbulent and the smoke begins to swirl irregularly. *(Estate of Harold E. Edgerton.)*

13-4 FLUIDS IN MOTION

The behavior of a fluid in motion can be complex. Consider, for example, the rise of smoke from a burning cigarette. At first the smoke rises in a regular stream of warm gas, but the simple streamlined flow quickly becomes turbulent and the smoke begins to swirl irregularly. Turbulent flow is very difficult to describe, even qualitatively. If no turbulence exists, the fluid flows along streamlines. Sophisticated computer programs that simulate streamlines of the air flowing around objects are of great value to automotive design engineers.

Figure 13-16 shows a tube full of fluid. The tube contains a tapered section with decreasing cross-sectional area. The fluid is flowing without turbulence from left to right, and the shaded portion on the left depicts the fluid that passes through cross-sectional surface 1 during time Δt. If the density and speed of the

Streamlined body designs can greatly reduce the drag forces on moving objects such as automobiles or airplanes. *(Takeski Takahara/Photo Researchers, Inc.)*

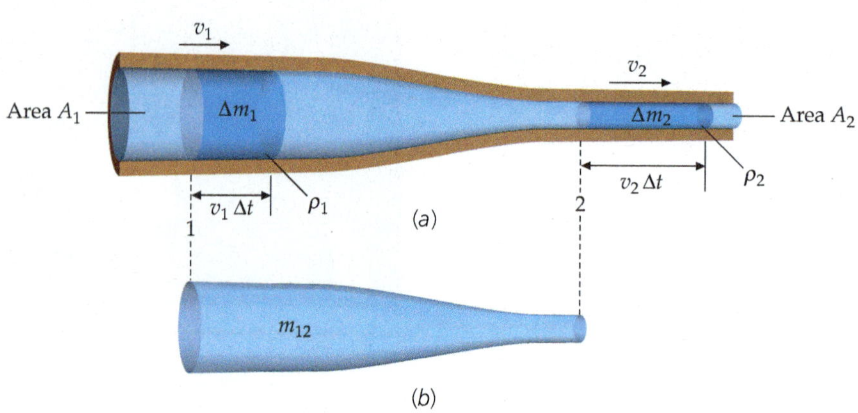

FIGURE 13-16

fluid at this surface are ρ_1 and v_1 and the area of this surface is A_1, then the mass Δm_1 flowing through surface 1 is given by

$$\Delta m_1 = \rho_1 \, \Delta V_1 = \rho_1 A_1 v_1 \, \Delta t$$

where $\Delta V_1 = A_1 v_1 \, \Delta t$ is the volume of the fluid flowing through surface 1 during time Δt. The quantity $I_M = \rho A v$ is called the **mass flow rate.** The dimensions of I_M are mass divided by time. The mass Δm_2 of fluid that passes through surface 2 during the same time Δt and depicted by the shaded portion on the right is given by

$$\Delta m_2 = \rho_2 A_2 v_2 \, \Delta t$$

where ρ_2, v_2, and A_2 are, respectively, the density and speed of the fluid at surface 2, and the cross-sectional area of surface 2.

If the mass flow rate through surface 1 is greater than the mass flow rate through surface 2, then fluid is entering the region between surfaces 1 and 2 faster than it is leaving the region, so the mass of fluid in the region is increasing. The rate at which mass enters the region minus the rate at which mass leaves the region is equal to the rate of change of the accumulated mass in the region. That is,

$$I_{M1} - I_{M2} = dm_{12}/dt \qquad 13\text{-}14$$

CONTINUITY EQUATION

where I_{M1} and I_{M2} are the mass flow rates through surfaces 1 and 2, respectively, and m_{12} is the accumulated mass between surfaces 1 and 2. Equation 13-14 is called the **continuity equation.** Flow in which the motion of the fluid is constant (does not change as time increases) at all points is called **steady-state flow.** If the flow is steady state, then dm_{12}/dt in Equation 13-14 equals zero. In steady-state flow mass flow rate is the same through all cross-sectional surfaces. In addition, the mass flow rate is constant.

The quantity $I_V = Av$ is called the **volume flow rate.** The dimensions of I_V are volume divided by time. In the flow of an incompressible fluid, the instantaneous volume flow rate is the same through any cross-sectional surface perpendicular to the flow. In addition, if the flow is steady-state flow, the volume flow rate is constant:

$$I_V = Av \qquad 13\text{-}15$$

VOLUME FLOW RATE

In an incompressible fluid, the density is equal to a single fixed value throughout the fluid. Liquids are almost always considered incompressible because their densities, to an excellent approximation, do not vary.

PRACTICE PROBLEM 13-9

Blood flows in an aorta of radius 1.0 cm at 30 cm/s. What is the volume flow rate?

PRACTICE PROBLEM 13-10

Blood flows from a large artery of radius 0.30 cm, where its speed is 10 cm/s, into a region where the radius has been reduced to 0.20 cm because of thickening of the arterial walls (arteriosclerosis). What is the speed of the blood in the narrower region?

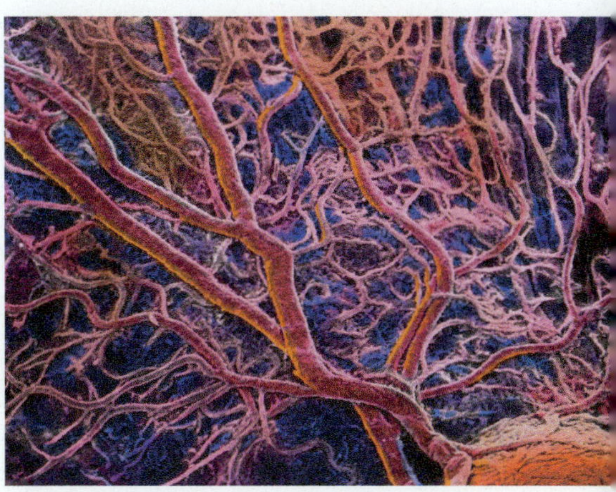

The larger arteries branch out into smaller arteries, which in turn branch out into smaller arteries, and so forth. *(P. Motta/ Photo Researchers Inc.)*

THE BERNOULLI EQUATION

The Bernoulli equation relates the pressure, elevation, and speed of an incompressible **inviscid** fluid in steady **streamlined flow.** Inviscid means without **viscosity,** which is the property of a fluid that causes it to resist flowing. During streamlined flow, the particles of the fluid move along **streamlines,** which are straight or smoothly curved paths that do not intersect. The Bernoulli equation can be derived by applying Newton's second law to a small parcel of the fluid moving along a streamline. As a parcel enters a region of reduced pressure the parcel gains speed because the pressure behind the parcel pushing it forward exceeds the pressure in front of the parcel opposing its motion.

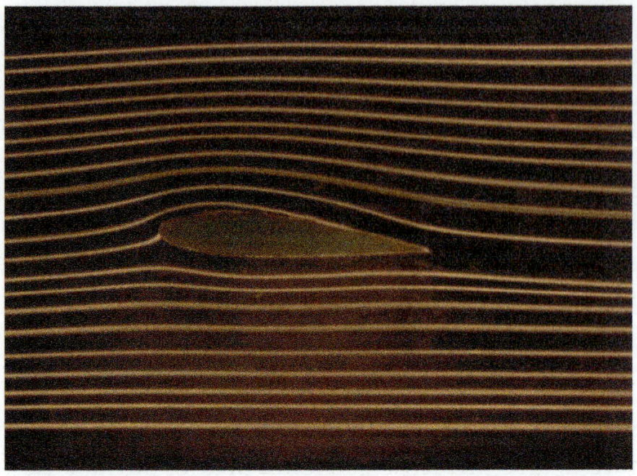

The streamlines are made visible by using smoke trails. In streamlined flow the particles of the fluid follow smoothly curved lines. *(Holger Babinsky. 2003 Phys. Educ. 38 497-503.)*

Applying Newton's second law to a small parcel of air (Figure 13-17) of mass m moving along a horizontal streamline gives

$$F = m\frac{dv}{dt}$$

The fluid has density ρ, and the parcel has area A and width $\Delta\ell$, so the volume and mass of the parcel are $A\,\Delta\ell$ and $m = \rho A\,\Delta\ell$. The force F is due to the pressure P behind the parcel and the slightly different pressure $P + \Delta P$ in front of it. This force is given by

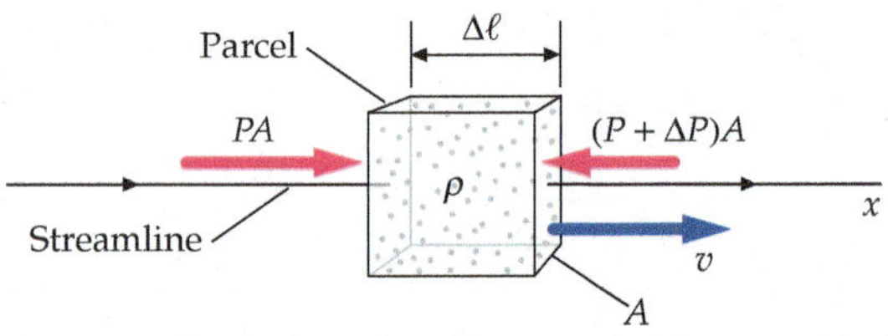

FIGURE 13-17 The small parcel moves along a streamline into a region of reduced pressure.

$$F = (P)(A) - (P + \Delta P)A = -A\,\Delta P$$

The parcel is small, so the pressure difference ΔP can be accurately expressed using the differential approximation

$$\frac{\Delta P}{\Delta\ell} = \frac{dP}{dx} \quad \text{so} \quad \Delta P = \frac{dP}{dx}\Delta\ell$$

Substituting for F and m in Newton's second law, gives

$$-A\frac{dP}{dx}\Delta\ell = \rho A\,\Delta\ell\,\frac{dv}{dt}$$

Simplifying this equation, we obtain

$$-dP = \rho\frac{dv}{dt}dx$$

Because $dx/dt = v$, this becomes

$$dP = -\rho v\,dv$$

Integrating both sides gives

$$\int_{P_1}^{P_2} dP = -\rho\int_{v_1}^{v_2} v\,dv$$

where the density ρ was factored from the integral on the right. Factoring ρ from the integral restricts the validity of the results to situations where the density remains constant. Evaluating the integrals gives

$$P_2 - P_1 = \tfrac{1}{2}\rho v_1^2 - \tfrac{1}{2}\rho v_2^2$$

Rearranging gives the Bernoulli equation for flow along a horizontal streamline,

$$P_2 + \tfrac{1}{2}\rho v_2^2 = P_1 + \tfrac{1}{2}\rho v_1^2 \qquad \text{13-16}$$

The Bernoulli equation for flow along a streamline that is not horizontal is derived in Problem 13-63. The result is

$$P_2 + \rho g h_2 + \tfrac{1}{2}\rho v_2^2 = P_1 + \rho g h_1 + \tfrac{1}{2}\rho v_1^2 \qquad \text{13-17}a$$

THE BERNOULLI EQUATION

where h_1 and h_2 are the initial and final heights, respectively.

The Bernoulli equation can be restated

$$P + \rho g h + \tfrac{1}{2}\rho v^2 = \text{constant} \qquad \text{13-17}b$$

THE BERNOULLI EQUATION

! The Bernoulli equation relates pressure and speed between two points that are on a single streamline of an inviscid fluid.

A special application of the Bernoulli equation is for a fluid at rest. Then $v_1 = v_2 = 0$, and we obtain

$$P_1 - P_2 = \rho g h_2 - \rho g h_1 = \rho g\, \Delta h$$

This is the same as Equation 13-7.

Example 13-10 Torricelli's Law

A large tank of water, open at the top, has a small hole through its side a distance h below the surface of the water. Find the speed of the water as it flows out the hole.

PICTURE The streamlines begin at the top of the water and continue through the small hole. We apply the Bernoulli equation to points a and b in Figure 13-18. Because the diameter of the hole is much smaller than the diameter of the tank, we can neglect the speed of the water at the top (point a).

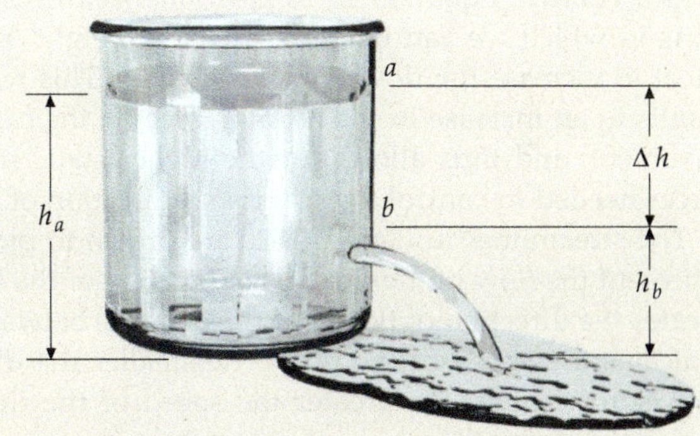

FIGURE 13-18

SOLVE

1. The Bernoulli equation with $v_a = 0$ gives: $P_a + \rho g h_a + 0 = P_b + \rho g h_b + \tfrac{1}{2}\rho v_b^2$
2. The pressure at point a and at point b is the same, P_{at}, because both points are open to the atmosphere: $P_a = P_{at}$ and $P_b = P_{at}$, so $P_{at} + \rho g h_a + 0 = P_{at} + \rho g h_b + \tfrac{1}{2}\rho v_b^2$
3. Solve the step-2 result for the speed v_b of the water flowing from the hole: $v_b^2 = 2g(h_a - h_b) = 2g\,\Delta h$, so $v_b = \boxed{\sqrt{2g\,\Delta h}}$

CHECK We can solve this directly using conservation of mechanical energy with the water and Earth as the system. The mass m of water that flows out of the hole in the short time Δt equals the mass of water that has "disappeared" from the top of the tank. Thus, the decrease in potential energy is $mg\,\Delta h$, while the increase in kinetic energy is $\tfrac{1}{2}mv^2$. Equating these and solving for v gives the step-3 result.

PRACTICE PROBLEM 13-11 If the water flowing out of the hole is directed vertically upward, how high does it rise?

CONCEPT CHECK 13-1

In an Olympic swimming pool the water is continuously cycled through a filter by a pump. The water reenters the pool through underwater nozzles. The stream of water exiting the nozzles extends almost the entire length of the pool before dissipating. As the water in the stream loses speed, does its pressure increase as the Bernoulli equation (Equation 13-16) seems to indicate?

In Example 13-10, the water emerges from the hole with a speed equal to the speed it would have if it dropped in free-fall a distance h. This finding is known as *Torricelli's law.*

In Figure 13-19, water is shown flowing through a horizontal pipe that has a constricted section. Because both sections of the pipe are at the same elevation, $h_2 = h_1$, in Equation 13-17*a*. Then the Bernoulli equation becomes

$$P + \frac{1}{2}\rho v^2 = \text{constant} \qquad 13\text{-}18$$

When the fluid moves into the constriction, the area A gets smaller, so the speed v must get larger because Av remains constant. But because $P + \frac{1}{2}\rho v^2$ is constant, when the speed becomes larger, the pressure must become less. Thus, the pressure in the constriction is reduced.

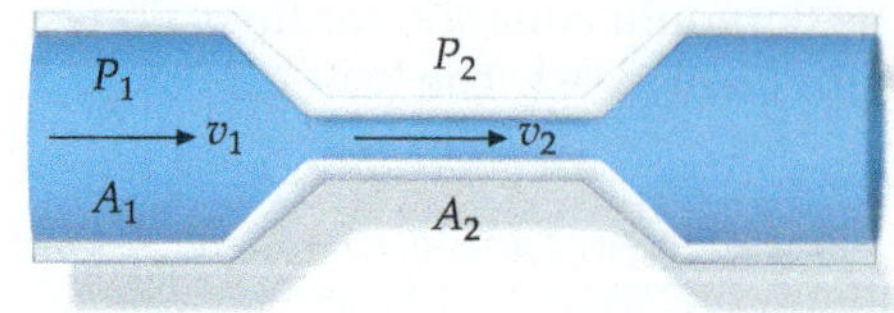

FIGURE 13-19 Constriction in a pipe carrying a moving fluid. The pressure is lower in the narrow section of the pipe where the fluid is moving faster.

> As air, or another fluid, passes through a constriction, its speed increases and its pressure drops.
>
> VENTURI EFFECT

This result is often referred to as the **venturi effect** and the constriction is referred to as a **venturi.** Equation 13-18 is an important result that applies to many situations in which we can ignore changes in height. Racing cars exploit the venturi effect to increase the downforce on the car. This reduction results in an increase in the normal force on the car by the pavement, and thus allows for the higher static frictional forces needed to control the speed and direction of the car.

The streamlines in Figure 13-20 are drawn to pictorially represent the flow of the fluid. The direction of the lines indicates the direction of flow and the distance between lines indicates the speed of the flow. The smaller the distances between the lines, the greater the speed of the fluid. For horizontal flow, where the speed increases, the pressure decreases, so a decrease in the distance between streamlines is accompanied by a decrease in pressure.

The wing on this Formula 1 racing car deflects the air upward, increasing downforce on the car for better control at high speeds. In addition, an underbody venturi reduces the pressure under the car body. The downforce associated with this pressure reduction is called the ground effect. *(Schlegelmilch/Corbis.)*

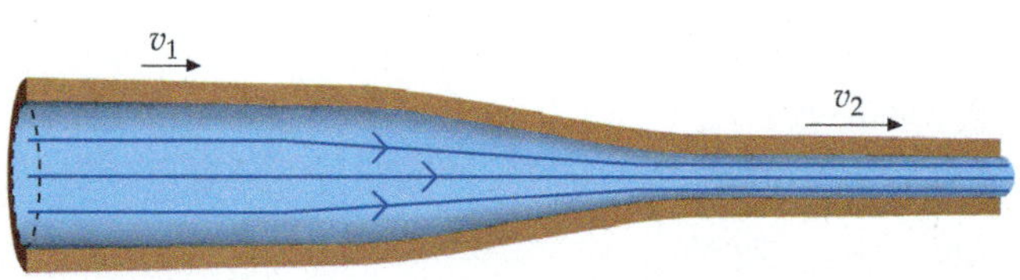

FIGURE 13-20

FIGURE 13-21 When the bulb of an atomizer is squeezed, the air is forced through the constriction in the horizontal tube, which reduces the pressure there below atmospheric pressure. Because of the resulting pressure difference, the liquid in the jar, which is open to the atmosphere is pumped up through the vertical tube, enters the air stream, and emerges from the nozzle. A similar effect occurs in the carburetor of a gasoline lawnmower engine.

CONCEPT CHECK 13-2

In the atomizer shown in Figure 13-21, the horizontal tube is constricted at the point where the vertical tube joins it. Is the constriction functional, or is it there because the vertical tube is narrower than the horizontal tube? Explain.

Example 13-11 A Venturi Meter

A *venturi meter*, used to measure the flow rate of an incompressible inviscid fluid, is shown in Figure 13-22. The fluid of density ρ_F passes through a pipe of cross-sectional area A_1 that has a constriction of cross-sectional area A_2. Because the fluid gains speed as it enters the constricted section, the pressure in the constricted section is less than in the other portions of the pipe. The two parts of the pipe are connected with a U-tube manometer partially filled with a liquid of density ρ_L. The pressure difference is measured by the difference in the levels of the liquid in the U-tube, Δh. Express the velocity v_1 in terms of the measured height Δh and the known quantities ρ_F, ρ_L, and $r = A_1/A_2$.

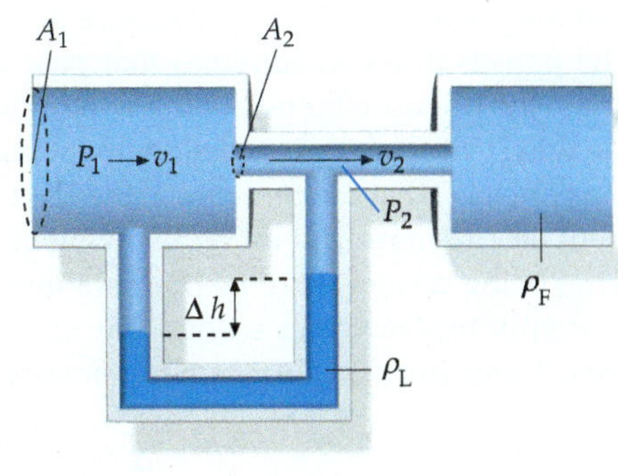

FIGURE 13-22 A venturi meter.

PICTURE The pressures P_1 and P_2 in the two regions are related to the speeds v_1 and v_2 by the Bernoulli equation. The pressure difference is related to the height Δh. You can express v_2 in terms of v_1 and the areas A_1 and A_2 by the continuity equation.

SOLVE

1. Write the Bernoulli equation for constant elevation for the two regions.

$$P_1 + \tfrac{1}{2}\rho_F v_1^2 = P_2 + \tfrac{1}{2}\rho_F v_2^2$$

2. Write the continuity equation for the two regions, and solve for v_2 in terms of v_1 and r, where $r = A_1/A_2$.

$$v_2 A_2 = v_1 A_1$$

$$\text{so} \quad v_2 = \frac{A_1}{A_2} v_1 = r v_1$$

3. Substitute your result for v_2 into the equation in step 1 and obtain an equation for $P_1 - P_2$.

$$P_1 - P_2 = \tfrac{1}{2}\rho_F(v_2^2 - v_1^2) = \tfrac{1}{2}\rho_F(r^2 - 1)v_1^2$$

4. Write $P_1 - P_2$ in terms of the difference in height Δh of the liquid in the arms of the U-tube. This pressure difference equals the pressure drop in the column of height Δh of the liquid, less the pressure drop and that in the column of the same height of the fluid.

$$P_1 - P_2 = \rho_L g\,\Delta h - \rho_F g\,\Delta h = (\rho_L - \rho_F) g\,\Delta h$$

5. Equate the two expressions for $P_1 - P_2$ and solve for v_1 in terms of Δh.

$$\tfrac{1}{2}\rho_F(r^2 - 1)v_1^2 = (\rho_L - \rho_F) g\,\Delta h$$

$$\text{so} \quad v_1 = \boxed{\sqrt{\frac{2(\rho_L - \rho_F) g\,\Delta h}{\rho_F(r^2 - 1)}}}$$

CHECK Let us check the dimensions of the expression for v_1 in step 5. A ratio of two densities is dimensionless, as is the ratio of two areas, r. Therefore, the dimension of the expression for v_1 is the same as the dimension of $\sqrt{2g\,\Delta h}$. The dimension of g is length divided by time squared, so the dimension of gh is length squared divided by time squared. Thus, the dimension of the square root of gh is length divided by time—the dimension of speed. The step-5 result is dimensionally correct.

PRACTICE PROBLEM 13-12 Find v_1 if $\Delta h = 3$ cm, $r = 4$, the fluid is air ($\rho_F = 1.29$ kg/m^3), and the liquid in the U-tube portion of the venturi meter is water ($\rho_w = 10^3$ kg/m^3).

Air is a compressible fluid, so the calculation in Practice Problem 13-12 is not as accurate as the calculation in the Example 13-10. Strictly speaking, the Bernoulli equation and the continuity equation hold only for incompressible fluids.

An airplane wing is an airfoil (Figure 13-23) that in normal circumstances causes streamlines to curve—following the curve of the airfoil surfaces. (The streamlines fail to follow the surfaces during an undersirable event occurance a stall.) During our analysis of how the curved streamlines produce lift (the upward force on the wing), we will neglect any variations in pressure due to the effects of gravity on the air. In addition, the analysis will be done in a frame of reference moving with the wing.

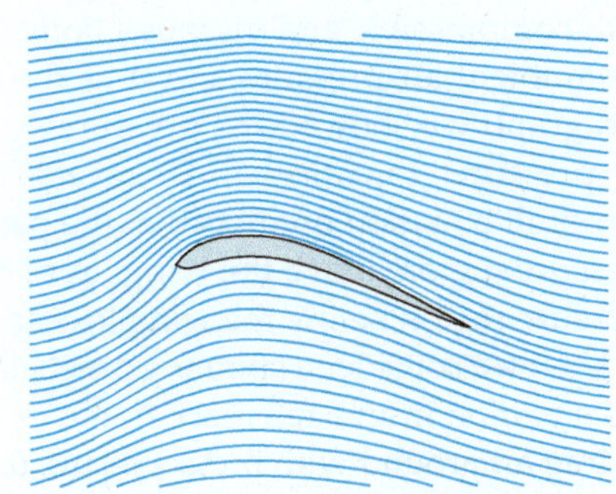

FIGURE 13-23 The purpose of an airfoil is to cause the streamlines to curve. Under normal conditions the streamlines will follow the curve of the airfoil. The airfoil shown is very thin, like the wing of a raptor. It is very efficient in creating lift.

A parcel of air in moving along a curved streamline is accelerating in the centripetal direction—toward the center of curvature of the streamline. For the streamlines above the wing, the direction of this acceleration is more or less downward. Thus, the net force on the parcel is downward. This means the air pressure just above the parcel is greater than the air pressure just below the parcel. As a result, the air pressure is greater at points far above the wing than at points next to the upper surface of the wing. The pressure far above the wing is the ambient air pressure, so we can conclude that the pressure at the upper surface of the wing is less than the ambient air pressure. The parcels in the streamlines that pass below the wing also accelerate downward. Thus, the pressure far below the wing is less than the pressure at the lower surface of the wing. The pressure far below the wing is equal to the ambient air pressure, so we can conclude that the pressure at the lower surface of the wing is greater than the ambient air pressure. The lift on the wing is due to the pressure immediately below the wing being greater than the pressure immediately above the wing.

Applying Newton's second law ($F = ma$) to a parcel of air (Figure 13-24) with area A and thickness Δr, moving at speed v, we get

$$(P + \Delta P)(A) - PA = (\rho A\,\Delta r)\frac{v^2}{r}$$

where r is the distance of the parcel from the center of curvature of the streamline, ρ is the density of air, P is the pressure of the parcel surface closest to the center of curvature, and $P + \Delta P$ is the pressure at the opposite surface of the parcel. Simplifying and rearranging this equation gives

$$\frac{\Delta P}{\Delta r} = \rho\frac{v^2}{r}$$

As Δr approaches zero this becomes

$$\frac{\partial P}{\partial r} = \rho\frac{v^2}{r} \qquad 13\text{-}19$$

The derivative in this equation is a partial derivative because it represents the rate of change of pressure perpendicular to the streamlines. The pressure also varies along the direction tangent to the streamlines.

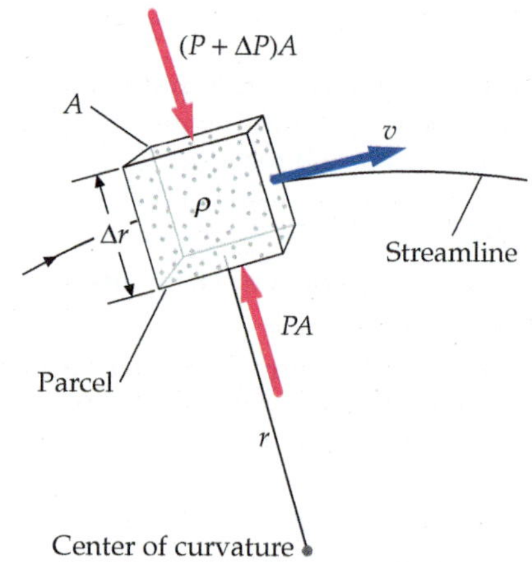

FIGURE 13-24 The centripetal force on a parcel moving along a curved streamline is due to a pressure difference across the parcel.

We now show how the pressure varies with position along a streamline. Consider a small parcel of air moving along a streamline that passes over the wing. When the parcel is far in front of the wing, it is at ambient pressure. As the parcel moves into the region above the wing, it moves into a region of lower pressure. The parcel gains speed as it enters this region because the pressure behind the parcel pushing it forward is greater than the pressure in front of it pushing it backward. Suppose point 1 is far in front of the wing and point 2 is directly above the wing and on the same streamline. The pressures at these two points are then related by the Bernoulli equation:

$$P_1 + \tfrac{1}{2}\rho v_1^2 = P_2 + \tfrac{1}{2}\rho v_2^2 \qquad 13\text{-}20$$

(The Bernoulli equation is only approximately valid in this context, because air is both compressible and viscous.) Point 1 is far in front of the wing, so P_1 is equal to the ambient air pressure and v_1 is the speed of the air at point 1. In the previous paragraph we showed the pressure P_2 above the wing to be less than the ambient air pressure. Equation 13-20 reveals that if $P_2 < P_1$, then $v_2 < v_1$. That is, the Bernoulli equation predicts that the parcels of air gain speed as they enter the low-pressure region above the wing. In addition, the parcels of air that enter the high-pressure region below the wing lose speed, as predicted by the Bernoulli equation.

Now we consider a spinning baseball moving through still air. Figure 13-25 shows this situation from a reference frame moving with the ball. As the ball spins, it tends to drag air around with it. As a result, the streamlines are curved as shown. Is the air pressure greater immediately above the ball or immediately below it? (As before, we are neglecting any changes in air pressure due to the effects of gravity.) The air pressure is greater immediately below the ball, as we shall explain. The surrounding air exerts pressure forces on the packets of air moving along the streamlines. If the streamlines are curved, there are resultant pressure forces on the packets in the centripetal direction. Thus, as with the wing, the pressure immediately above the ball is

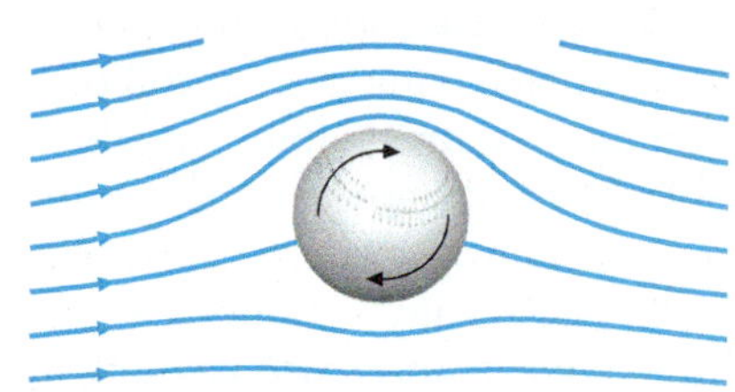

FIGURE 13-25 The ball is moving from right to left, so in the reference frame of the ball, the air moves from left to right, as shown.

significantly less than the pressure far above the ball—which is equal to the ambient air pressure. The streamlines just below the ball are almost straight, so the pressure just below the ball is almost the same as the pressure far below the ball, which is equal to the ambient air pressure. Thus, pressure immediately below the ball is greater than the pressure immediately above the ball. As a result the air exerts an upward force on the ball. The ball will not fall as fast as it would if gravity alone were acting on it.

Although the Bernoulli equation is very useful for qualitative descriptions of many features of fluid flow, such descriptions are often grossly inaccurate when compared with the quantitative results of experiments. Prominent reasons for the discrepancies are that gases like air are hardly incompressible, and liquids like water are hardly inviscid, which invalidates the assumptions made in deriving the Bernoulli equation. In addition, it is often difficult to maintain steady, streamlined flow without turbulence, and the introduction of turbulence can greatly affect the results.

* VISCOUS FLOW

According to Bernoulli's equation, when a fluid flows steadily through a long, narrow, horizontal pipe of constant cross section, the pressure along the pipe will be constant. In practice, however, we observe a pressure drop as we move along the direction of the flow. Put another way, a pressure difference is required to push a fluid through a horizontal pipe. This pressure difference is needed because the fluid flows in very thin layers, and the thin layer of fluid in contact with the pipe is held stationary by forces exerted on it by the pipe. As we move away from the surface of the pipe, the speed of each successive layer is slightly greater than the speed of the layer next to it on one side, and each layer is held back by forces exerted on it by the slightly slower-moving layer next to it on one side (just as each lamina is pulled forward by forces exerted on it by the slightly faster-moving lamina next to it on the other side). These forces between adjacent layers are called **viscous forces.** As a result of viscous forces, the velocity of the fluid is not constant across the diameter of the pipe. Instead, it is greatest near the center of the pipe and approaches zero where the fluid is in contact with the walls of the pipe (Figure 13-26). Let P_1 be the pressure at point 1 and P_2 be that at point 2, a distance L downstream from point 1. The pressure drop $\Delta P = P_1 - P_2$ is proportional to the volume flow rate:

$$\Delta P = P_1 - P_2 = I_V R \qquad 13\text{-}21$$

DEFINITION: RESISTANCE

where $I_V = vA$ is the volume flow rate and the proportionality constant R is the *resistance* to flow, which depends on the length of the pipe L, the radius r, and the viscosity of the fluid.

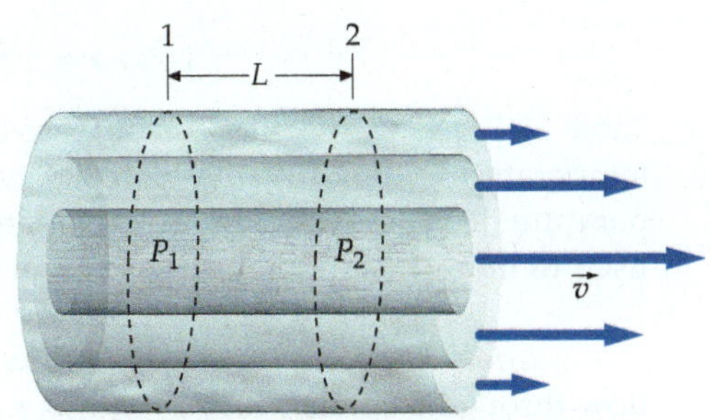

FIGURE 13-26 When a viscous fluid flows through a pipe, the speed is greatest at the center of the pipe. At the walls of the pipe, the speed of the fluid approaches zero.

Example 13-12 Resistance to Blood Flow

Blood flows from the aorta through the major arteries, the small arteries, the capillaries, and the veins until it reaches the right atrium. During the course of that flow, the (gauge) pressure drops from about 100 torr to zero. If the volume flow rate is 800 mL/s, find the total resistance of the circulatory system.

PICTURE The resistance is related to the pressure drop and volume flow rate by Equation 13-21. We can use Equation 13-9 to convert from torr to kPa.

SOLVE

Write the resistance in terms of the pressure drop and volume flow rate, and convert all terms to SI units:

$$R = \frac{\Delta P}{I_V} = \frac{100\text{ torr}}{0.800\text{ L/s}} \times \frac{101\text{ kPa}}{760\text{ torr}} \times \frac{1\text{ L}}{10^3\text{ cm}^3} \times \frac{1\text{ cm}^3}{10^{-6}\text{ m}^3}$$

$$= 16.61\text{ kPa}\cdot\text{s/m}^3 = \boxed{16.6\text{ kPa}\cdot\text{s/m}^3}$$

CHECK We could have used 1 Pa = 1 N/m^2 to write the result as 16.6 kN · s/m^5. The dimensions of resistance are the dimensions of pressure divided by the dimensions of volume flow rate. The dimensions of pressure are the dimensions of force divided by the dimension of length squared, and the dimensions of volume flow rate are the dimension of length cubed divided by the dimension of time. Thus, the dimensions of resistance are the dimensions of pressure times the dimension of time divided by the dimension of length raised to the fifth power. The units of the result have the correct dimensions, so the result is plausible.

To define the coefficient of viscosity of a fluid, we consider a fluid that is confined between two rectangular parallel plates, each of area A, separated by a distance z, as shown in Figure 13-27. The upper plate is pulled at a constant speed v by a force $\vec{F}$, while the bottom plate is held at rest. A force is needed to pull the upper plate because the fluid next to the plate exerts a viscous drag force opposing its motion. The fluid flows in thin layers, or **lamina,** and the motion is called **laminar flow.** The speed of the lamina in contact with the upper plate is v, the speed of the lamina in contact with the lower plate approaches zero, and the speeds of the laminas increase linearly with distance from the lower plate. The force $\vec{F}$ on the upper plate is found to be directly proportional to v and A and inversely proportional to the plate separation z. The proportionality constant is the **coefficient of viscosity** η:

$$F = \eta \frac{vA}{z} \qquad 13\text{-}22$$

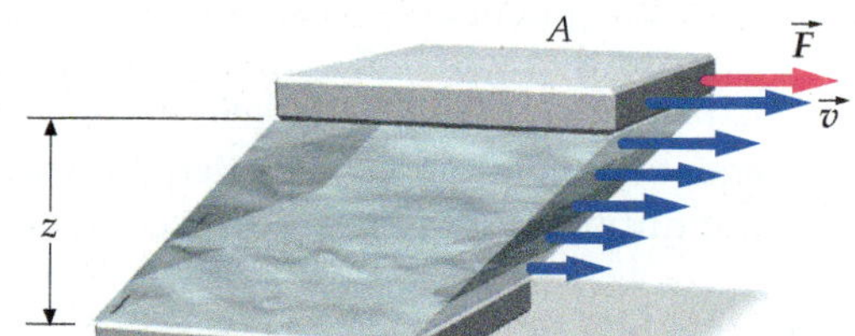

FIGURE 13-27 Two plates of equal area with a viscous fluid between them. When the upper plate is moved relative to the lower one, each layer of fluid exerts a drag force on the adjacent layers. The force needed to pull the upper plate is directly proportional to v and the area A, and inversely proportional to z, the separation between the plates.

The SI unit of viscosity is the N · s/m^2 = Pa · s. An older cgs unit still in common use is the **poise,** named after the French physicist Jean Poiseuille. These units are related by

$$1 \text{ Pa} \cdot \text{s} = 10 \text{ poise} \qquad 13\text{-}23$$

Table 13-3 lists the coefficients of viscosity for several fluids at various temperatures. Typically, the viscosity of a liquid increases as the temperature decreases. Thus, in cold climates, a less viscous grade of oil is used to lubricate automobile engines in the winter than in summer.

Table 13-3 Coefficients of Viscosity for Various Fluids

Fluid	t, °C	η, mPa · s
Water	0	1.8
	20	1.00
	60	0.65
Blood (whole)	37	4.0
Engine oil (SAE 10W)	30	200
Glycerin	0	10,000
	20	1,410
	60	81
Air	20	0.018

Poiseuille's law The resistance to flow R in Equation 13-21 for steady flow through a circular tube of radius r can be shown to be

$$R = \frac{8\eta L}{\pi r^4} \qquad 13\text{-}24$$

Equations 13-21 and 13-24 can be combined to give the pressure drop over a length L of a circular tube of radius r:

$$\Delta P = \frac{8\eta L}{\pi r^4} I_V \qquad 13\text{-}25$$

POISEUILLE'S LAW

Equation 13-25 is known as **Poiseuille's law.** Note the inverse r^4 dependence of the pressure drop. If the radius of the tube is halved, the pressure drop for a given volume flow rate is increased by a factor of 16; or a pressure 16 times as great is needed to pump the fluid through the tube at the original volume flow rate. Thus, for example, if the diameter of a person's blood vessels or arteries is reduced for some reason, either the volume flow rate of the blood is greatly reduced, or the blood pressure must escalate to maintain the volume flow rate. For water flowing through a long garden hose open at one end and connected to a constant pressure source at the other end, the pressure drop is fixed. It equals the difference in pressure between that at the water source and atmospheric pressure at the open end. The volume flow rate is then proportional to the fourth power of the radius. Thus, the volume flow rate increases by a factor of more than 5 when you switch to a three-quarter-inch-diameter hose from a half-inch-diameter hose. That is so because $(0.75/0.50)^4 = 5.1$.

Poiseuille's law applies only to the laminar flow of a fluid of constant viscosity. In some fluids, viscosity changes with velocity, violating Poiseuille's law. Blood, for example, is a complex fluid consisting of solid particles of various shapes suspended in a liquid. Red blood cells are disk-shaped objects that are randomly oriented at low speeds, but at high speeds tend to become oriented to facilitate the flow. Thus, the viscosity of blood decreases as the flow speed increases, so Poiseuille's law cannot be strictly applied. Nevertheless, Poiseuille's law is a good approximation that is very useful for obtaining a qualitative understanding of blood flow.

In Chapter 25 the flow of electrical current I through metal wires is studied. One of the basic relations in that chapter is Ohm's law, $\Delta V = IR$, where ΔV is the potential difference and R is the electrical resistance of the wire. As we shall see, Ohm's law is analogous to Poiseuille's law, $\Delta P = I_V R$.

TURBULENCE: REYNOLDS NUMBER

When the flow speed of a fluid becomes sufficiently great, laminar flow breaks down and turbulence sets in. The critical speed above which the flow through a tube is turbulent depends on the density and viscosity of the fluid and on the radius of the tube. The flow of a fluid can be characterized by a dimensionless number called the **Reynolds number,** N_R, which is defined by

$$N_R = \frac{2r\rho v}{\eta} \qquad \text{13-26}$$

where v is the average speed of the fluid. Experiments have shown that the flow will be laminar if the Reynolds number is less than about 2000 and turbulent if it is greater than 3000. Between these values, the flow is unstable and may change from one type to the other.

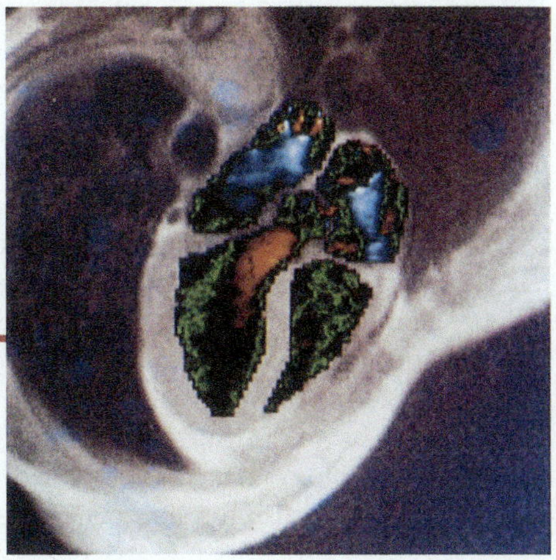

False-color view of turbulence of blood flowing into and out of the heart as seen by magnetic resonance imaging (MRI). Systolic ejection from the left ventricle into the aorta is seen in red, and diastolic filling of the ventricles in blue. *(Pickler International.)*

Example 13-13 Blood Flow in the Aorta

Calculate the Reynolds number for blood flowing at 30 cm/s through an aorta of radius 1.0 cm. Assume that blood has a viscosity of 4.0 mPa · s and a density of 1060 kg/m³.

PICTURE Because N_R is dimensionless, we can use any system of units as long as we are consistent.

SOLVE

Write Equation 13-26 for the Reynolds number, expressing each quantity in SI units:

$$N_R = \frac{2r\rho}{\eta} = \frac{2(0.010\text{ m})(1060\text{ kg/m}^3)(0.30\text{ m/s})}{4.0 \times 10^{-3}\text{ Pa}\cdot\text{s}}$$

$$= 1590 = \boxed{1.6 \times 10^3}$$

CHECK Because the Reynolds number is less than 2000, this flow will be laminar rather than turbulent. We expect the flow of blood to be nonturbulent, so our result is plausible.

Physics Spotlight

Automotive Aerodynamics: Ride with the Wind

The shape and finish of a car's body can reduce drag and increase fuel economy. As a result, many recent passenger cars have a half-teardrop shape in profile. However, the curve of the air flow over the top of this type of car creates a region of low pressure at the top of the car. This low pressure provides lift, which reduces the normal force upon the road. This makes it harder for the driver to safely maneuver curves. Lift is proportional to the square of the speed. At speeds achieved during car races, the lift can be significant. The lift results in a loss of traction, which on curves can determine the outcome of a race.

Automotive engineers call an increase in normal force "downforce," which is negative lift. In order to increase speeds on curves, different racing teams use different methods to increase the downforce on their cars. Formula 1 and Indy cars use large airfoils shaped like upside down airplane wings to create low pressure underneath the cars that will give increases in downforce. Airfoils were first introduced to Indy cars in 1972. The one-lap record speed increased by 20 mph that year.[*] Indy cars also contour the bottoms of their cars, to reduce the pressure underneath their cars. At racing speeds, the downforce-to-weight ratio can exceed one at racing speeds.[†,‡] Some race cars have even used fans to pull air quickly under the car's body,[#] although most racing rules do not allow them today.

The most visible modifications to NASCAR cars are aerodynamic. They have stiff skirts on the sides, and a very low spoiler in front. A spoiler extends the width of the car in back, and on some tracks, a roof spoiler is required. The spoilers increase the drag of the car, so the car does not go faster than safety concerns demand. In rare cases, aero flaps open on the roof of the car when cars drift rapidly sideways or backwards. In fact, an entire class of accidents has been prevented because these flaps were introduced in 1994.[°]

In 1994, Formula 1 banned the use of bottom contouring in cars, and required that the car bottoms be flattened. The intent was to reduce the speed of the races, as two drivers had died in crashes.[§] Teams had two weeks to implement these rules, and figure out how to maintain as much "downforce" as possible.[¶] All these teams use computation fluid dynamics (CFD) modeling programs, as well as scale wind tunnels for testing ideas before implementing them.

Race teams are not the only groups to use wind tunnels and CFD programs to test their designs. A group at the Georgia Tech Research Institute modeled the turbulence behind semitrucks, and tested their models in a small-scale wind tunnel. They found that by adding a system of slits and air compressors, they could decrease a truck's drag by up to 35 percent at highway speeds.[**] In road tests,[††] the system was shown to improve overall fuel economy by 8 to 9 percent.[‡‡] Some of the same technology that has been used to make cars go very quickly may soon be used to save over a billion gallons of gasoline a year.

At highway speeds aerodynamic drag is reduced because of the jets of air along the back edge of the truck body of this prototype. *(Courtesy of Georgia Institute of Technology.)*

* Katz, J., *Race Car Aerodynamics: Designing for Speed*, 2nd ed. Cambridge, MA: Bentley, 2006, 4.
† Simanaitis, D., "Technology Update: Automotive Aerodynamics," *Road and Track*, June 2002, 84+.
‡ Robertson, C., quoted in "Fast Cars," *Nova*, PBS. Aug. 19, 1997. http://www.pbs.org/wgbh/nova/transcripts/2208fast.html as of June 2006.
\# Fuller, M. J., "A Brief History of Sports Car Racing," *Mulsanne's Corner*, http://www.mulsannescorner.com/history.htm 1996, as of June 2006.
° Katz, J., op. cit., 191.
§ Butler, R., "Not So Fast!" *Professional Engineering*, Nov. 9, 2005, 37–38.
¶ Zeimelis, K., and Wenz, C., "Science in the Fast Lane," *Nature*, Oct. 14, 2004, 736–738.
** Weiss, P., "Aircraft Trick May Give Big Rigs a Gentle Lift," *Science News*, Oct. 28, 2000, 279.
†† Toon, John, "Low-Drag Trucks: Aerodynamic Improvements and Flow Control System Boost Fuel Efficiency in Heavy Trucks," *Georgia Institute of Technology Research News*, Jan. 5, 2004. http://gtresearchnews.gatech.edu/newsrelease/truckfuel.htm
‡‡ Weiss, P., "Thrifty Trucks Go with the Flow," *Science News*, Jan. 29, 2005, 78.

Summary

1. Density, specific gravity, and pressure are defined quantities that are important in fluid statics and dynamics.
2. Pascal's principle, Archimedes' principle and the Bernoulli equation are derived from Newton's laws.

*3. The venturi effect is a special case of Bernoulli's equation.

4. A transverse pressure gradient always accompanies curved streamlines.

*5. Poiseuille's law accounts for pressure drops due to viscosity; Reynolds number is used to predict whether flow is laminar or turbulent.

TOPIC	RELEVANT EQUATIONS AND REMARKS		
1. Density	The density of a substance is the ratio of its mass to its volume:		
	$\rho = \frac{dm}{dV}$		13-1
	The densities of most solids and liquids are approximately independent of temperature and pressure, whereas those of gases depend strongly on these quantities.		
2. Specific Gravity	The specific gravity of a substance is the ratio of its density to that of another substance, usually water.		
3. Pressure	$P = \frac{F}{A}$		13-3
Units	$1\ \text{Pa} = 1\ \text{N/m}^2$		13-4
	$1\ \text{atm} = 760\ \text{mmHg} = 760\ \text{torr} = 29.9\ \text{inHg} = 101.325\ \text{kPa} = 14.7\ \text{lb/in.}^2$		13-9
	$1\ \text{bar} = 10^3\ \text{millibars} = 100\ \text{kPa}$		13-10
Gauge pressure	Gauge pressure is the difference between the absolute pressure and atmospheric pressure:		
	$P = P_{\text{gauge}} + P_{\text{at}}$		
In a static liquid	$P = P_0 + \rho g\,\Delta h$ (ρ constant)		13-7
In a gas	In a gas such as air, pressure decreases exponentially with altitude.		
Bulk modulus	$B = -\frac{\Delta P}{\Delta V/V}$		13-6
4. Pascal's Principle	Pressure changes applied to a confined fluid are transmitted undiminished to every point in the fluid and to the walls of the container.		
5. Archimedes' Principle	A body wholly or partially submerged in a fluid is buoyed up by a force equal to the weight of the displaced fluid.		
***6. Fluid Flow**			
Mass flow rate and continuity equation	$I_{\text{M}} = \rho Av$	Mass flow rate	
	$I_{\text{M1}} - I_{\text{M2}} = dm_{12}/dt$	Continuity equation	13-14
Volume flow rate and continuity equation for an incompressible fluid	$I_{\text{V}} = Av$	Volume flow rate	
	$A_1v_1 = A_2v_3$	Incompressible fluid	
Bernoulli equation	Along a streamline of a nonviscous, incompressible fluid undergoing steady flow:		
	$P + gh + \frac{1}{2}\rho v^2 = \text{constant}$		13-17*b*
Venturi effect	As air, or another fluid, passes through a constriction its speed increases and its pressure drops		

TOPIC	RELEVANT EQUATIONS AND REMARKS	
Resistance to fluid flow	$\Delta P_2 = I_V R$	13-21
Coefficient of viscosity	$\eta = \dfrac{F/A}{v/z}$	13-22
Poiseuille's law for viscous flow	$\Delta P = RI_V = \dfrac{8\eta L}{\pi r^4} I_V$	13-25
Laminar flow, turbulent flow, and the Reynolds number	The flow will be laminar if the Reynolds number N_R is less than about 2000 and turbulent if it is greater than 3000, where N_R is given by N_R $N_R = \dfrac{2r\rho v}{\eta}$	13-26

Answers to Concept Checks

13-1 No. The water packets are not slowing because they are entering a region of higher pressure. Instead they are slowing because of viscous drag forces on them. The Bernoulli equation is valid only if viscous forces are negligible.

13-2 The constriction is functional. The narrow region is a venturi. When the bulb is vigorously squeezed, the pressure in the constricted region drops below atmospheric pressure due to the venturi effect. This reduces the pressure in the vertical tube so the air pressure on the top of the liquid in the reservoir is able to push the liquid up the vertical tube and into the horizontal air stream.

Answers to Practice Problems

13-1 (*a*) 7.97 kg/L, (*b*) iron

13-2 19 kg

13-3 With $P_0 = 1.00$ atm $= 101$ kPa, $P = 2.00$ atm, $\rho = 1000\ \text{kg/m}^3$, and $g = 9.81$ N/kg, we have $\Delta h = \Delta P/\rho g = 10.3$ m. The pressure at a depth of 10.3 m is twice that at the surface.

13-4 $h = P/\rho g = 0.760\ \text{m} = 760\ \text{mm}$

13-5 (*a*) 338 mmHg, (*b*) 0.444 atm

13-6 The specific gravity of the material is 2.7, which is the specific gravity of aluminum. The material is aluminum.

13-7 72.9 N

13-8 45 percent

13-9 $I_V = vA = 9.4 \times 10^{-5}\ \text{m}^3/\text{s}$. It is customary to give the pumping rate of the heart in liters per minute. Using $1\ \text{m}^3 = 1000$ L and 1 min = 60 s, we have $I_V = 5.7$ L/min.

13-10 If v_1 and v_2 are the initial and final speeds and A_1 and A_2 are the initial and final areas, Equation 13-15 gives

$$v_2 = \frac{A_1}{A_2}v_1 = \frac{\pi(0.30\ \text{cm})^2}{\pi(0.20\ \text{cm})^2}(10\ \text{cm/s}) = 23\ \text{cm/s}$$

13-11 The water shoots upward a distance h; that is, to the same level as the surface of the water in the tank.

13-12 5.51 m/s

Problems

In a few problems, you are given more data than you actually need; in a few other problems, you are required to supply data from your general knowledge, outside sources, or informed estimate.

Interpret as significant all digits in numerical values that have trailing zeros and no decimal points.

- • Single-concept, single-step, relatively easy
- •• Intermediate-level, may require synthesis of concepts
- ••• Challenging
- SSM Solution is in the *Student Solutions Manual*
- Consecutive problems that are shaded are paired problems.

CONCEPTUAL PROBLEMS

1 • If the gauge pressure is doubled, the absolute pressure will be (*a*) halved, (*b*) doubled, (*c*) unchanged, (*d*) increased by a factor greater than 2, (*e*) increased by a factor less than 2.

2 • Two spherical objects differ in size and mass. Object A has a mass that is eight times the mass of object B. The radius of object A is twice the radius of object B. How do their densities compare? (*a*) $\rho_A > \rho_B$, (*b*) $\rho_A < \rho_B$, (*c*) $\rho_A = \rho_B$, (*d*) Not enough information is given to compare their densities.

3 • Two objects differ in density and mass. Object A has a mass that is eight times the mass of object B. The density of object A is four times the density of object B. How do their volumes compare? (*a*) $V_A = \frac{1}{2}V_B$, (*b*) $V_A = V_B$, (*c*) $V_A = 2V_B$, (*d*) Not enough information is given to compare their volumes. **SSM**

4 • A sphere is constructed by gluing together two hemispheres. The density of each hemisphere is uniform, but the density of one is greater than the density of the other. True or false: The average density of the sphere is the numerical average of the two different densities. Clearly explain your reasoning.

5 • **BIOLOGICAL APPLICATION, CONTEXT-RICH** In several jungle adventure movies, the hero and heroine escape the bad guys by hiding underwater for extended periods of time. To do this, they breathe through long vertical hollow reeds. Imagine that in one movie, the water is so clear that to be safely hidden, the two are at a depth of 15 m. As a science consultant to the movie producers, you tell them that this depth is not realistic and the knowledgeable viewer will laugh during this scene. Explain why this is so.

6 •• Two objects are balanced as in Figure 13-28. The objects have identical volumes but different masses. Assume all the objects in the figure are denser than water and thus none will float. Will the equilibrium be disturbed if the entire system is completely immersed in water? Explain your reasoning.

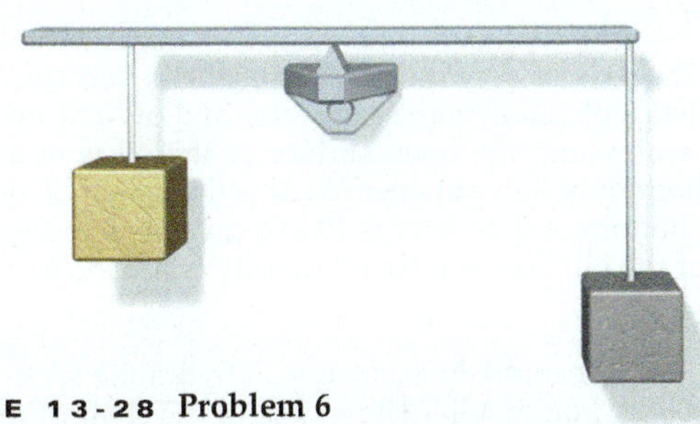

FIGURE 13-28 Problem 6

7 •• A solid 200-g block of lead and a solid 200-g block of copper are completely submerged in a aquarium filled with water. Each block is suspended just above the bottom of the aquarium by a thread. Which of the following is true? **SSM**

(*a*) The buoyant force on the lead block is greater than the buoyant force on the copper block.
(*b*) The buoyant force on the copper block is greater than the buoyant force on the lead block.
(*c*) The buoyant force is the same on both blocks.
(*d*) More information is needed to choose the correct answer.

8 •• A 20-cm^3 block of lead and a 20-cm^3 block of copper are completely submerged in an aquarium filled with water. Each is suspended just above the bottom of the aquarium by a thread. Which of the following is true?

(*a*) The buoyant force on the lead block is greater than the buoyant force on the copper block.
(*b*) The buoyant force on the copper block is greater than the buoyant force on the lead block.
(*c*) The buoyant force is the same on both blocks.
(*d*) More information is needed to choose the correct answer.

9 •• Two bricks are completely submerged in water. Brick 1 is made of lead and has rectangular dimensions of 2″ × 4″ × 8″. Brick 2 is made of wood and has rectangular dimensions of 1″ × 8″ × 8″. True or false: The buoyant force on brick 2 is larger than the buoyant force on brick 1.

10 •• Figure 13-29 shows an object called a "Cartesian diver." The diver consists of a small tube, open at the bottom, with an air bubble at the top, inside a closed plastic soda bottle that is partly filled with water. The diver normally floats, but sinks when the bottle is squeezed hard. (*a*) Explain why this happens. (*b*) Explain the physics behind how a submarine can "silently" sink vertically simply by allowing water to flow into empty tanks near its keel. (*c*) Explain why a person floating in water will oscillate up and down on the water's surface as he or she breathes in and out.

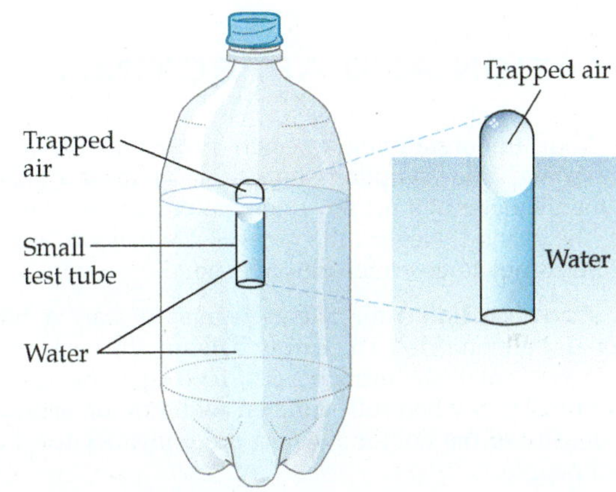

FIGURE 13-29 Problem 10

11 •• A certain object has a density just slightly less than that of water so that it floats almost completely submerged. However, the object is more compressible than water. What happens if the floating object is given a slight downward push? Explain.

12 •• In Example 13-11, the fluid is accelerated to a greater speed as it enters the narrow part of the pipe. Identify the forces that act on the fluid at the entrance to the narrow region to produce this acceleration.

13 •• An upright glass of water is accelerating to the right along a flat, horizontal surface. What is the origin of the force that produces the acceleration on a small element of water in the middle of the glass? Explain by using a diagram. *Hint: The water surface will not remain level as long as the glass of water is accelerating. Draw a free-body diagram of the small element of water.* **SSM**

14 •• You are sitting in a boat floating on a very small pond. You take the anchor out of the boat and drop it into the water. Does the water level in the pond rise, fall, or remain the same? Explain your answer.

15 •• A horizontal pipe narrows from a diameter of 10 cm at location A to 5.0 cm at location B. For a nonviscous incompressible fluid flowing without turbulence from location A to location B, how do the flow speeds v (in m/s) compare at the two locations? (*a*) $v_A = v_B$, (*b*) $v_A = \frac{1}{2}v_B$, (*c*) $v_A = \frac{1}{4}v_B$, (*d*) $v_A = 2v_B$, (*e*) $v_A = 4v_B$

16 •• A horizontal pipe narrows from a diameter of 10 cm at location A to 5.0 cm at location B. For a nonviscous incompressible fluid flowing without turbulence from location A to location B, how do the pressures P compare at the two locations? (*a*) $P_A = P_B$, (*b*) $P_A = \frac{1}{2}P_B$, (*c*) $P_A = \frac{1}{4}P_B$, (*d*) $P_A = 2P_B$, (*e*) $P_A = 4P_B$, (f) There is not enough information to compare the pressures quantitatively.

17 •• **BIOLOGICAL APPLICATION** Figure 13-30 is a diagram of a prairie dog tunnel. The geometry of the two entrances are such that entrance 1 is surrounded by a mound and entrance 2 is surrounded by flat ground. Explain how the tunnel remains ventilated, and indicate in which direction air will flow through the tunnel. SSM

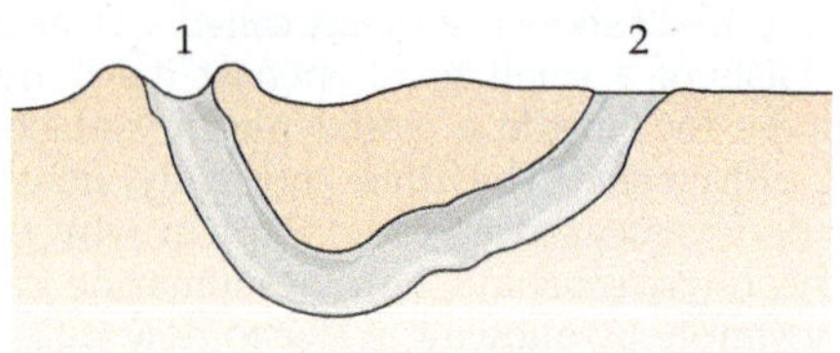

FIGURE 13-30 Problem 17

ESTIMATION AND APPROXIMATION

18 •• Your undergraduate research project involves atmospheric sampling. The sampling device has a mass of 25.0 kg. Estimate the diameter of a helium-filled balloon required to lift the device off the ground. Neglect the mass of the balloon "skin" and the small buoyancy force on the device itself.

19 ••• **CONTEXT-RICH** Your friend wants to start a business giving hot-air balloon rides. The empty balloon, the basket and the occupants have a total maximum mass of 1000 kg. If the balloon has a diameter of 22.0 m when fully inflated with hot air, estimate the required density of the hot air. Neglect the buoyancy force on the basket and people.

DENSITY

20 • Find the mass of a solid lead sphere with a radius equal to 2.00 cm.

21 • Consider a room measuring 4.0 m × 5.0 m × 4.0 m. Under normal atmospheric conditions at Earth's surface, what would be the mass of the air in the room? SSM

22 • An average neutron star has approximately the same mass as the Sun, but is compressed into a sphere of radius roughly 10 km. What would be the approximate mass of a teaspoonful of matter that dense?

23 •• A 50.0-g ball consists of a plastic spherical shell and a water-filled core. The shell has an outside diameter equal to 50.0 mm and an inside diameter equal to 20.0 mm. What is the density of the plastic?

24 •• A 60.0-mL flask is filled with mercury at 0°C (Figure 13-31). When the temperature increases to 80°C, 1.47 g of mercury spills out of the flask. Assuming that the volume of the flask stays constant, find the change in the density of mercury at 80°C if its density at 0°C is 13645 kg/m^3.

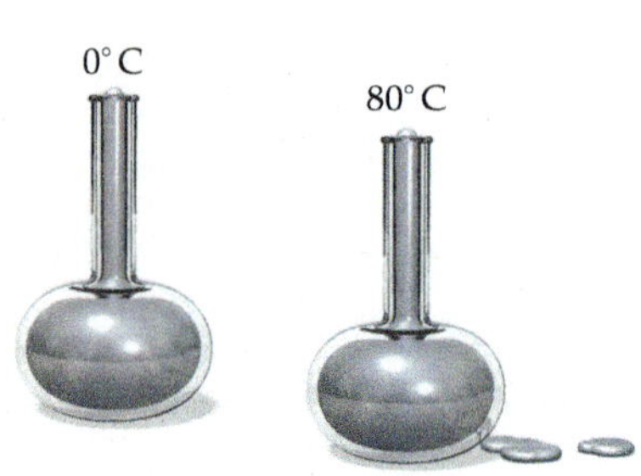

FIGURE 13-31 Problem 24

25 •• One sphere is made of gold and has a radius r_{Au} and another sphere is made of copper and has a radius r_{Cu}. If the spheres have equal mass, what is the ratio of the radii, r_{Au}/r_{Cu}?

26 ••• Since 1983, the U.S. Mint has coined pennies that are made out of zinc with a copper cladding. The mass of this type of penny is 2.50 g. Model the penny as a uniform cylinder of height 1.23 mm and radius 9.50 mm. Assume the copper cladding is uniformly thick on all surfaces. If the density of zinc is 7140 kg/m^3 and that of copper is 8930 kg/m^3, what is the thickness of the copper cladding?

PRESSURE

27 • Barometer readings are commonly given in inches of mercury (inHg). Find the pressure in inches of mercury equal to 101 kPa.

28 • The pressure on the surface of a lake is P_{at} = 101 kPa. (*a*) At what depth is the pressure $2P_{at}$? (*b*) If the pressure at the top of a deep pool of mercury is P_{at}, at what depth is the pressure $2P_{at}$?

29 • **BIOLOGICAL APPLICATION** When at cruising altitude, a typical airplane cabin will have an air pressure equivalent to an altitude of about 2400 m. During the flight, ears often equilibrate, so that the air pressure inside the inner ear equalizes with the air pressure outside the plane. The Eustachian tubes allow for this equalization, but can become clogged. If an Eustachian tube is clogged, pressure equalization may not occur on descent and the air pressure inside an inner ear may remain equal to the pressure at 2400 m. In that case, by the time the plane lands and the cabin is repressurized to sea-level air pressure, what is the net force on one ear drum due to this pressure difference, assuming the ear drum has an area of 0.50 cm^2?

30 • The axis of a cylindrical container is vertical. The container is filled with equal masses of water and oil. The oil floats on top of the water, and the open surface of the oil is at a height *h* above the bottom of the container. What is the height, *h*, if the pressure at the bottom of the water is 10 kPa greater than the pressure at the top of the oil? Assume the oil density is 875 kg/m^3.

31 • **ENGINEERING APPLICATION** A hydraulic lift is used to raise a 1500-kg automobile. The radius of the shaft of the lift is 8.00 cm and the radius of the compressor's piston is 1.00 cm. How much force must be applied to the piston to raise the automobile? *Hint: The shaft of the lift is the other piston.* SSM

32 • **ENGINEERING APPLICATION** A 1500-kg car rests on four tires, each of which is inflated to a gauge pressure of 200 kPa. If the four tires support the car's weight equally, what is the area of contact of each tire with the road?

33 •• What pressure increase is required to compress the volume of 1.00 kg of water from 1.00 L to 0.99 L? Could this compression occur in the ocean, where the maximum depth is about 11 km? Explain.

34 •• When a woman in high-heeled shoes takes a step, she momentarily places her entire weight on one heel of her shoe. If her mass is 56.0 kg and if the area of the heel is 1.00 cm^2, what is the pressure exerted on the floor by the heel? Compare your answer to the pressure exerted by one foot of an elephant on a flat floor. Assume the elephant's mass is 5000 kg, that he has all four feet equally distributed on the floor, and that each foot has an area of 400 cm^2.

35 •• In the seventeenth century, Blaise Pascal performed the experiment shown in Figure 13-32. A wine barrel filled with water was coupled to a long tube. Water was added to the tube until the barrel burst. The radius of the barrel's lid was 20 cm and the height of the water in the tube was 12 m. (*a*) Calculate the force exerted on the lid due to the pressure increase. (*b*) If the tube had an inner radius of 3.0 mm, what mass of water in the tube caused the pressure that burst the barrel?

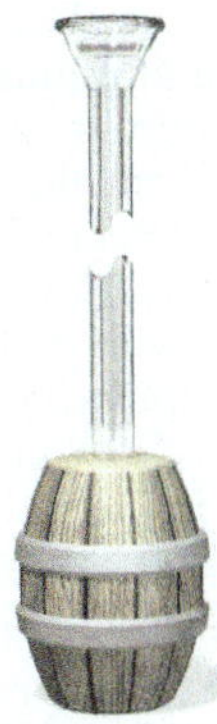

FIGURE 13-32
Problem 35

36 •• **BIOLOGICAL APPLICATION** Blood plasma flows from a bag through a tube into a patient's vein, where the blood pressure is 12 mmHg. The specific gravity of blood plasma at 37°C is 1.03. What is the minimum elevation of the bag so that the plasma flows into the vein?

37 •• **BIOLOGICAL APPLICATION** Many people have imagined that if they were to float the top of a flexible snorkel tube out of the water, they would be able to breathe through it while walking underwater (Figure 13-33). However, they generally do not take into account just how much water pressure opposes the expansion of the chest and the inflation of the lungs. Suppose you can just breathe while lying on the floor with a 400-N (90-lb) weight on your chest. How far below the surface of the water could your chest be for you still to be able to breathe, assuming your chest has a frontal area of 0.090 m^2?

FIGURE 13-33
Problem 37

38 •• **ENGINEERING APPLICATION** In Example 13-3, a 150-N force is applied to a small piston to lift a car that weighs 15 000 N. Demonstrate that this does not violate the law of conservation of energy by showing that, when the car is lifted some distance h, the work done by the 150-N force acting on the small piston equals the work done on the car by the large piston.

39 •• A 5.00-kg lead sinker is accidentally dropped overboard by fishermen in a boat directly above the deepest portion of the Marianas trench, near the Philippines. By what percentage does the volume of the sinker change, by the time it settles on the trench bottom, which is 10.9 km below the surface?

40 ••• The volume of a cone of height h and base radius r is $V = \pi r^2 h/3$. A jar in the shape of a cone of height 25 cm has a base with a radius equal to 15 cm. The jar is filled with water. Then its lid (the base of the cone) is screwed on and the jar is turned over so its lid is horizontal. (*a*) Find the volume and weight of the water in the jar. (*b*) Assuming the pressure inside the jar at the top of the cone is equal to 1 atm, find the excess force exerted by the water on the base of the jar, where by excess force we mean the force minus the force exerted by air pressure on the outside of the base of the jar. Explain how this force can be greater than the weight of the water in the jar.

BUOYANCY

41 • A 500-g piece of copper, with specific gravity of 8.96, is suspended from a spring scale and is submerged in water (Figure 13-34). What force does the spring scale read?

FIGURE 13-34
Problem 41

42 • When a certain rock is suspended from a spring scale, the scale-display reads 60 N. However, when the suspended stone is submerged in water, the display reads 40 N. What is the density of the rock?

43 • A block of an unknown material weighs 5.00 N in air and 4.55 N when submerged in water. (*a*) What is the density of the material? (*b*) From what material is the block likely to have been made? SSM

44 • A solid piece of metal weighs 90.0 N in air and 56.6 N when submerged in water. What is the density of this metal?

45 •• A homogeneous solid object floats on water, with 80.0 percent of its volume below the surface. When placed in a second liquid, the same object floats on that liquid with 72.0 percent of its volume below the surface. Determine the density of the object and the specific gravity of the liquid.

46 •• A 5.00-kg iron block is suspended from a spring scale and is submerged in a fluid of unknown density. The spring scale reads 6.16 N. What is the density of the fluid?

47 •• A large piece of cork weighs 0.285 N in air. When held submerged underwater by a spring scale, as shown in Figure 13-35, the spring scale reads 0.855 N. Find the density of the cork.

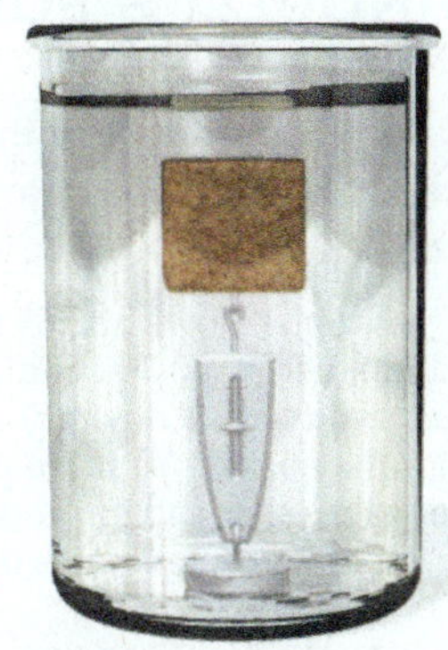

FIGURE 13-35
Problem 47

48 •• A helium balloon lifts a basket and cargo with a total weight of 2000 N under standard conditions, at which the density of air is 1.29 kg/m^3 and the density of helium is 0.178 kg/m^3. What is the minimum volume of the balloon?

49 •• An object has "neutral buoyancy" when its density equals that of the liquid in which it is submerged, which means that it neither floats nor sinks. If the average density of an 85-kg diver is 0.96 kg/L, what mass of lead should the dive master suggest be added to give the diver neutral buoyancy? SSM

50 •• A 1.00-kg beaker containing 2.00 kg of water rests on a scale. A 2.00-kg block of aluminum (density 2.70×10^3 kg/m^3) suspended from a spring scale is submerged in the water, as in Figure 13-36. Find the readings of both scales.

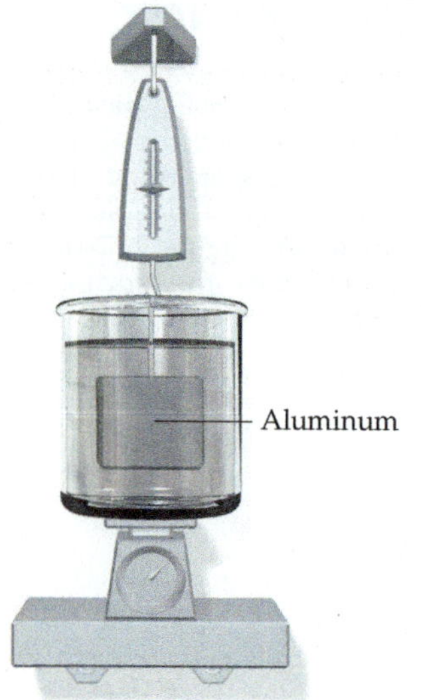

FIGURE 13-36 Problem 50

51 •• **ENGINEERING APPLICATION** When cracks form at the base of a dam, the water seeping into the cracks exerts a buoyant force that tends to lift the dam. As a result, the dam can topple. Estimate the buoyant force exerted on a 2.0-m-thick by 5.0-m-long dam wall by water seeping into cracks at its base. The water level in the lake is 5.0 m above the cracks.

52 •• **ENGINEERING APPLICATION, CONTEXT-RICH** Your team is in charge of launching a large helium weather balloon that is spherical in shape, and whose radius is 2.5 m and total mass is 15 kg (balloon plus helium plus equipment). (*a*) What is the initial upward acceleration of the balloon when it is released from sea level? (*b*) If the drag force on the balloon is given by $F_D = \frac{1}{2}\pi r^2 \rho v^2$, where r is the balloon radius, ρ is the density of air, and v the balloon's ascension speed, calculate the terminal velocity of the ascending balloon.

53 ••• **ENGINEERING APPLICATION** A ship sails from seawater (specific gravity 1.025) into freshwater, and therefore sinks slightly. When its 600,000-kg load is removed, it returns to its original level. Assuming that the sides of the ship are vertical at the water line, find the mass of the ship before it was unloaded. SSM

CONTINUITY AND BERNOULLI'S EQUATION

Note: **For the problems in this section, assume laminar nonviscous steady-state flow in all cases unless otherwise indicated.**

54 • Water flows at 0.65 m/s through a 3.0-cm-diameter hose that terminates in a 0.30-cm-diameter nozzle. Assume laminar nonviscous steady-state flow. (*a*) At what speed does the water pass through the nozzle? (*b*) If the pump at one end of the hose and the nozzle at the other end are at the same height, and if the pressure at the nozzle is 1.0 atm, what is the pressure at the pump outlet?

55 • Water is flowing at 3.00 m/s in a horizontal pipe under a pressure of 200 kPa. The pipe narrows to half its original diameter. (*a*) What is the speed of flow in the narrow section? (*b*) What is the pressure in the narrow section? (*c*) How do the volume flow rates in the two sections compare? SSM

56 •• The pressure in a section of horizontal pipe with a diameter of 2.00 cm is 142 kPa. Water flows through the pipe at 2.80 L/s. If the pressure at a certain point is to be reduced to 101 kPa by constricting a section of the pipe, what should the diameter of the constricted section be?

57 •• **BIOLOGICAL APPLICATION** Blood flows at 30 cm/s in an aorta of radius 9.0 mm. (*a*) Calculate the volume flow rate in liters per minute. (*b*) Although the cross-sectional area of a capillary is much smaller than that of the aorta, there are many capillaries, so their total cross-sectional area is much larger. If all the blood from the aorta flows into the capillaries and the speed of flow through the capillaries is 1.0 mm/s, calculate the total cross-sectional area of the capillaries. Assume laminar nonviscous steady-state flow. SSM

58 •• Water flows through a 1.0-m-long conical section of pipe that joins a cylindrical pipe of radius 0.45 m, on the left, to a cylindrical pipe of radius 0.25 m, on the right. If the water flows into the 0.45-m pipe with a speed of 1.50 m/s, and if we assume laminar nonviscous steady-state flow, (*a*) what is the speed of flow in the 0.25-m pipe? (*b*) What is the speed of flow at a position x in the conical section, if x is the distance measured from the left-hand end of the conical section of pipe?

59 •• **ENGINEERING APPLICATION** The $8 billion, 800-mile-long Alaskan Pipeline has a maximum volume flow rate of 240,000 m^3 of oil per day. Most of the pipeline has a radius of 60.0 cm. Find the pressure P' at a point where the pipe has a 30.0-cm radius. Take the pressure in the 60.0-cm radius sections to be P = 180 kPa and the density of oil to be 800 kg/m^3. Assume laminar nonviscous steady-state flow.

60 •• Water flows through a Venturi meter like that in Example 13-11 with a pipe diameter of 9.50 cm and a constriction diameter of 5.60 cm. The U-tube manometer is partially filled with mercury. Find the volume flow rate of the water if the difference in the mercury level in the U-tube is 2.40 cm.

61 •• **ENGINEERING APPLICATION, CONTEXT-RICH** Horizontal flexible tubing for carrying cooling water extends through a large electromagnet used in your physics experiment at Fermi National Accelerator Laboratory. A minimum volume flow rate of 0.0500 L/s through the tubing is necessary to keep your magnet cool. Within the magnet volume, the tubing has a circular cross section of radius 0.500 cm. In regions outside the magnet, the tubing widens to a radius of 1.25 cm. You have attached pressure sensors to measure differences in pressure between the 0.500- and 1.25-cm sections. The lab technicians tell you that if the flow rate in the system drops below 0.050 L/s, the magnet is in danger of overheating and that you should install an alarm to sound a warning when the flow rate drops below that level. What is the critical pressure difference at which you should program the sensors to send the alarm signal (and is this a minimum, or maximum, pressure difference)? Assume laminar nonviscous steady-state flow. SSM

62 •• Figure 13-37 shows a Pitot-static tube, a device used for measuring the speed of a gas. The inner pipe faces the incoming fluid, while the ring of holes in the outer tube is parallel to the gas flow. Show that the speed of the gas is given by $v_2 = 2gh(\rho_L - \rho_g)/\rho_g$, where ρ_L is the density of the liquid used in the manometer and ρ_g is the density of the gas.

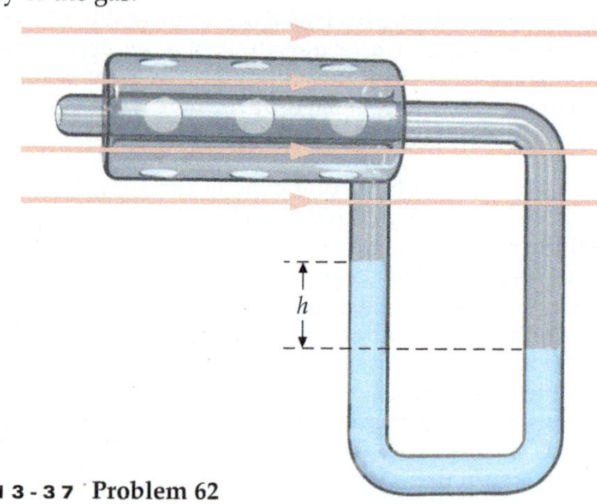

FIGURE 13-37 Problem 62

63 ••• Derive the Bernoulli equation in more generality than is done in the text, that is, allow for the fluid to change elevation during its movement. Using the work–energy theorem, show that when changes in elevation are allowed, Equation 13-16 becomes $P_1 + \rho g h_1 + \frac{1}{2}\rho v_1^2 = P_2 + \rho g h_2 + \frac{1}{2}\rho v_2^2$ (Equation 13-17). SSM

64 ••• A large keg of height H and cross-sectional area A_1 is filled with root beer. The top is open to the atmosphere. There is a spigot opening of area A_2, which is much smaller than A_1, at the bottom of the keg. (*a*) Show that when the height of the root beer is h, the speed of the root beer leaving the spigot is approximately $\sqrt{2gh}$. (*b*) Show that if $A_2 \ll A_1$, the rate of change of the height h of the root beer is given by $dh/dt = -(A_2/A_1)(2gh)^{1/2}$. (*c*) Find h as a function of time if $h = H$ at $t = 0$. (*d*) Find the total time needed to drain the keg if $H = 2.00$ m, $A_1 = 0.800\ \text{m}^2$, and $A_2 = 1.00 \times 10^{-4} A_1$. Assume laminar nonviscous flow.

65 •• A siphon is a device for transferring a liquid from one container to another. The tube shown in Figure 13-38 must be filled to start the siphon, but once this has been done, fluid will flow through the tube until the liquid surfaces in the containers are at the same level. (*a*) Using Bernoulli's equation, show that the speed of water in the tube is $v = \sqrt{2gd}$. (*b*) What is the pressure at the highest part of the tube?

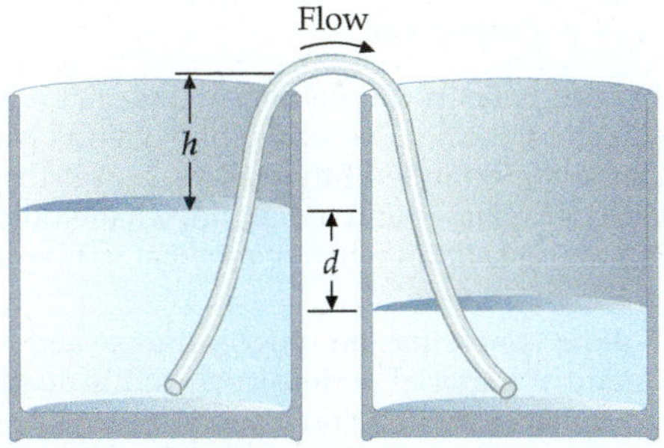

FIGURE 13-38 Problem 65

66 •• A fountain designed to spray a column of water 12 m into the air has a 1.0-cm-diameter nozzle at ground level. The water pump is 3.0 m below the ground. The pipe to the nozzle has a diameter of 2.0 cm. Find the pump pressure necessary if the fountain is to operate as designed. (Assume laminar nonviscous steady-state flow.)

67 •• Water at 20°C exits a circular tap moving straight down with a flow rate of 10.5 cm³/s. (*a*) If the diameter of the tap is 1.20 cm, what is the speed of the water? (*b*) As the fluid falls from the tap, the stream of water narrows. What is the new diameter of the stream at a point 7.50 cm below the tap? Assume that the stream still has a circular cross section and neglect any effects of drag forces acting on the water. (*c*) If turbulent flows are characterized by Reynolds numbers above 2300 or so, how far does the water have to fall before it becomes turbulent? Does this match your everyday observations?

68 •• **ENGINEERING APPLICATION, CONTEXT-RICH** To better fight fires in your seaside community, the local fire brigade has asked you to construct a pump system to draw seawater from the ocean to the top of the steep cliff adjacent to the water where most of the homes are. If the cliff is 12.0 m high, and the pump is capable of producing a gauge pressure of 150 kPa, how much water (in L/s) can be pumped using a hose with a radius of 4.00 cm?

69 ••• **MULTISTEP** In Figure 13-39, H is the depth of the liquid and h is the distance from the surface of the liquid to the pipe inserted in the tank's side. (*a*) Find the distance x at which the water strikes the ground after exiting the pipe as a function of h and H. (*b*) Show that, for a given value of H, there are two values of h (whose average value is $\frac{1}{2}H$), both of which give the same distance x. (*c*) Show that for a given value of H, x is a maximum when $h = \frac{1}{2}H$. Find the maximum value for x as a function of H.

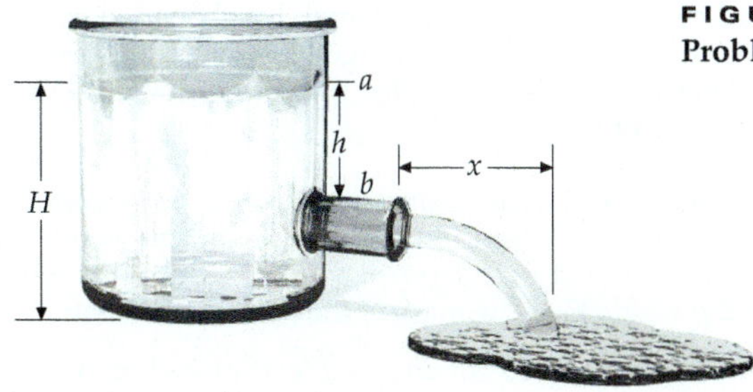

FIGURE 13-39 Problem 69

*VISCOUS FLOW

70 • Water flows through a horizontal 25.0-cm-long tube with an inside diameter of 1.20 mm at 0.300 mL/s. Find the pressure difference required to drive this flow if the viscosity of water is 1.00 mPa · s. Assume laminar flow.

71 • Find the diameter of a tube that would give double the flow rate for the pressure difference in Problem 70.

72 • **BIOLOGICAL APPLICATION** Blood takes about 1.00 s to pass through a 1.00-mm-long capillary in the human circulatory system. If the diameter of the capillary is 7.00 μm and the pressure drop is 2.60 kPa, find the viscosity of blood. Assume laminar flow.

73 • An abrupt transition occurs at Reynolds numbers of about 3×10^5, where the drag on a sphere moving through a fluid abruptly decreases. Estimate the speed at which this transition occurs for a baseball, and comment on whether it should play a role in the physics of the game. SSM

74 •• A horizontal pipe of radius 1.5 cm and length 25 m is connected to the output that can sustain an output gauge pressure of 10 kPa. What is the speed of 20°C water flowing through the pipe? If the temperature of the water is 60°C, what is the speed of the water in the pipe?

75 •• A very large tank is filled to a depth of 250 cm with oil that has a density of 860 kg/m³ and a viscosity of 180 mPa · s. If the container walls are 5.00 cm thick, and a cylindrical hole of radius 0.750 cm is bored through the base of the container, what is the initial volume flow rate (in L/s) of the oil through the hole?

76 ••• The drag force on a moving sphere at a very low Reynolds number is given by $F_D = 6\pi\eta a v$, where η is the viscosity of the surrounding fluid and a is the radius of the sphere. (This relation is called Stokes' law.) Using this information, find the terminal speed of ascent for a spherical 1.0-mm-diameter carbon dioxide bubble rising in a carbonated beverage ($\rho = 1.1$ kg/L and $\eta = 1.8$ mPa · s). How long should it take for this bubble to rise 20 cm (the height of the drinking glass)? Is this length of time consistent with your observations?

GENERAL PROBLEMS

77 • Several teenagers swim toward a rectangular, wooden raft that is 3.00 m wide and 2.00 m long. If the raft is 9.00 cm thick, how many 75.0-kg teenage boys can stand on top of the raft without the raft becoming submerged? Assume the wood density is 650 kg/m³. SSM

78 • A thread attaches a 2.7-g Ping-Pong ball to the bottom of a beaker. When the beaker is filled with water so that the ball is totally submerged, the tension in the thread is 7.0 mN. Determine the diameter of the ball.

79 • Seawater has a bulk modulus of 2.30×10^9 N/m^2. Find the difference in density of seawater at a depth where the pressure is 800 atm as compared to the density at the surface which is 1025 kg/m^3. Neglect any effects due to changes in either temperature or salinity.

80 • A solid cube with 0.60-m edge length is suspended from a spring balance. When the cube is submerged in water, the spring balance reads 80 percent of the reading for when the cube is in air. Determine the density of the cube.

81 •• A 1.5-kg block of wood floats on water with 68 percent of its volume submerged. A lead block is placed on the wood, fully submerging the wood to a depth where the lead remains entirely out of the water. Find the mass of the lead block. SSM

82 •• A Styrofoam cube, 25 cm on an edge, is placed on one pan of a balance. The balance is in equilibrium when a 20-g mass of brass is placed on the other pan. Find the mass of the Styrofoam cube. Neglect the buoyant force of the brass mass, but do not neglect the buoyant force of the air on the Styrofoam cube.

83 •• A spherical shell of copper with an outer diameter of 12.0 cm floats on water with half its volume above the water's surface. Determine the inner diameter of the shell. The cavity inside the spherical shell is empty.

84 •• A 200-mL beaker that is half-filled with water is on the left pan of a balance, and a sufficient amount of sand is placed on the right pan to bring the balance to equilibrium. A cube 4.0 cm on an edge that is attached to a string is then lowered into the water until the cube is completely submerged, but not touching the bottom of the beaker. A piece of brass of mass m is then added to the right pan to restore equilibrium. What is m?

85 •• **ENGINEERING APPLICATION, CONTEXT-RICH** Crude oil has a viscosity of about 0.800 Pa·s at normal temperature. You are the chief design engineer in charge of constructing a 50.0-km horizontal pipeline that connects an oil field to a tanker terminal. The pipeline is to deliver oil at the terminal at a rate of 500 L/s, and the flow through the pipeline is to be laminar. Assuming that the density of crude oil is 700 kg/m^3, estimate the diameter of the pipeline that should be used. SSM

86 •• Water flows through the pipe in Figure 13-40 and exits to the atmosphere at the right end of section C. The diameter of the pipe is 2.00 cm at A, 1.00 cm at B, and 0.800 cm at C. The gauge pressure in the pipe at the center of section A is 1.22 atm and the flow rate is 0.800 L/s. The vertical pipes are open to the air. Find the level (above the flow midline as shown) of the liquid-air interfaces in the two vertical pipes. Assume laminar nonviscous flow.

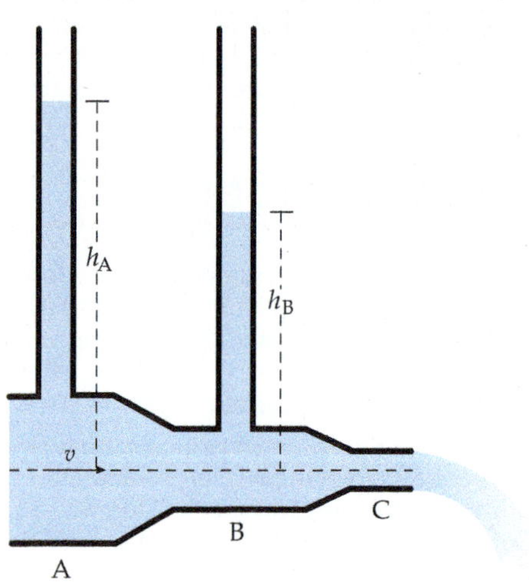

FIGURE 13-40
Problem 98

87 •• **ENGINEERING APPLICATION, CONTEXT-RICH** You are employed as a tanker truck driver for the summer. Heating oil is delivered to customers for winter usage by your large tanker truck. The delivery hose has a 1.00-cm radius. The specific gravity of the oil is 0.875, and its coefficient of viscosity is 200 mPa·s. What is the minimum time it will take you to fill a customer's 55-gal oil drum, if laminar flow through the hose must be maintained? SSM

88 •• A U-tube is filled with water until the liquid level reaches 28 cm above the bottom of the tube (Figure 13-41a). Oil, which has a specific gravity of 0.78, is now poured into one arm of the U-tube until the level of the water in the other arm of the tube reaches 34 cm above the bottom of the tube (Figure 13-41b). Find the levels of the oil–water and oil–air interfaces in the other arm of the tube.

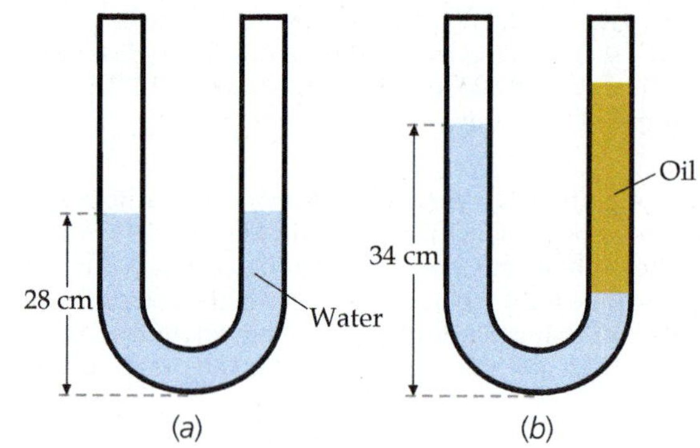

FIGURE 13-41 Problem 88

89 •• A helium balloon can just lift a load that weighs 750 N and has a negligible volume. The skin of the balloon has a mass of 1.5 kg. (*a*) What is the volume of the balloon? (*b*) If the volume of the balloon were twice that found in Part (*a*), what would be the initial acceleration of the balloon when released at sea level carrying a load weighing 900 N? SSM

90 •• A hollow sphere has an inner radius R and an outer radius $2R$. It is made of material of density ρ_0 and is floating in a liquid of density $2\rho_0$. The interior is now completely filled with material of density ρ' such that the sphere just floats completely submerged. Find ρ'.

91 •• According to *the law of atmospheres*, the fractional decrease in atmospheric pressure is proportional to the change in altitude. This law can be expressed as the differential equation $dP/P = -C dh$, where C is a positive constant. (*a*) Show that $P(h) = P_0 e^{-Ch}$, where P_0 is the pressure at $h = 0$, is a solution of the differential equation. (*b*) Given that the pressure 5.5 km above sea level is half that at sea level, find the constant C.

92 •• **ENGINEERING APPLICATION** A submarine has a total mass of 2.40×10^6 kg, including crew and equipment. The vessel consists of two parts, the pressure hull, which has a volume of 2.00×10^3 m^3, and the ballast tanks, which have a volume of 4.00×10^2 m^3. When the boat cruises on the surface, the ballast tanks are filled with air at atmospheric pressure; to cruise below the surface, seawater must be admitted into the tanks. (*a*) What fraction of the submarine's volume is above the water surface when the tanks are filled with air? (*b*) How much water must be admitted into the tanks to give the submarine neutral buoyancy? Neglect the mass of any air in the tanks and use 1.025 as the specific gravity of seawater.

93 ••• **BIOLOGICAL APPLICATION** Most species of fish have expandable sacs, commonly known as "swim bladders," that enable fish to rise in the water by filling the bladders with oxygen collected by their gills and to sink by emptying the bladders into the surrounding water. A freshwater fish has an average density equal to 1.05 kg/L when its swim bladder is empty. How large must the volume of oxygen in the fish's swim bladder be if the fish is to have neutral buoyancy? The fish has a mass of 0.825 kg. Assume the density of oxygen in the bladder is equal to air density at standard temperature and pressure.

PART II OSCILLATIONS AND WAVES

CHAPTER 14

MONSTER TRUCKS CAN POWER OVER JUST ABOUT ANYTHING, BUT WHAT KEEPS THESE GIANT TRUCKS FROM THROWING THEIR DRIVERS RIGHT OUT OF THEIR SEATS? MONSTER TRUCKS HAVE MONSTER-SIZE SHOCK ABSORBERS. THESE GIANT SHOCK ABSORBERS HELP DAMPEN THE OSCILLATION OF THE VEHICLE, PROVIDING A SMOOTHER RIDE AS THE OPERATOR DRIVES OVER TOUGH TERRAIN OR EVEN OTHER TRUCKS. *(Jeff Greenberg/Photoedit.)*

Oscillations

? How does a mechanic installing monster truck shock absorbers determine which size shock absorber to use? (See Example 14-13.)

We discuss oscillatory motion in this chapter. The kinematics of motion with constant acceleration is presented in Chapters 2 and 3. In this chapter, the kinematics and dynamics of motion with acceleration that is proportional to displacement from equilibrium is presented. The word "oscillate" means to swing back and forth. Oscillation occurs when a system is disturbed from a position of stable equilibrium. Many familiar examples exist: surfers bob up and down waiting for the right wave, clock pendulums swing back and forth, and the strings and reeds of musical instruments vibrate.

Other, less familiar, examples are the oscillations of air molecules in a sound wave and the oscillations of electric currents in radios, television sets, and metal detectors. In addition, many other devices rely on oscillatory motion to function.

In this chapter, we deal mostly with the most fundamental type of oscillatory motion—simple harmonic motion. We also consider both damped and driven oscillations.

14-1 SIMPLE HARMONIC MOTION

A common, very important, and very basic kind of oscillatory motion is **simple harmonic motion** such as the motion of a solid object attached to a spring (Figure 14-1). In equilibrium, the spring exerts no force on the object. When the object is displaced an amount x from its equilibrium position, the spring exerts a force $-kx$, as given by Hooke's law:*

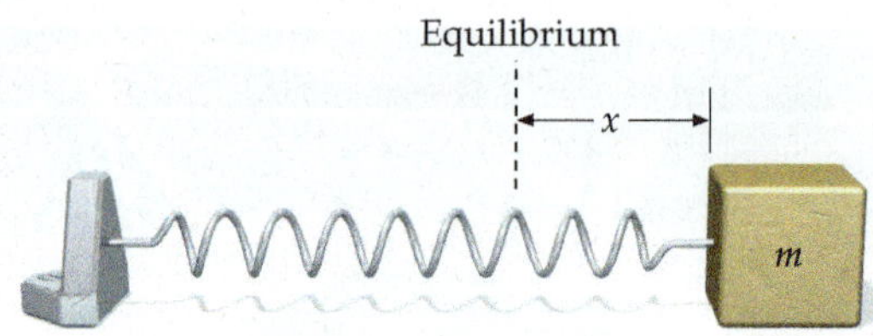

FIGURE 14-1 An object and spring on a frictionless surface. The displacement x, measured from the equilibrium position, is positive if the spring is stretched and negative if the spring is compressed.

$$F_x = -kx \qquad 14\text{-}1$$

LINEAR RESTORING FORCE

where k is the force constant of the spring, a measure of the spring's stiffness. The minus sign indicates that the force is a restoring force; that is, it is opposite to the direction of the displacement from the equilibrium position. Combining Equation 14-1 with Newton's second law ($F_x = ma_x$), we have

$$-kx = ma_x$$

or

$$a_x = -\frac{k}{m}x \quad \left(\text{or} \quad \frac{d^2x}{dt^2} = -\frac{k}{m}x\right) \qquad 14\text{-}2$$

The acceleration is proportional to the displacement and the minus sign indicates that the acceleration and the displacement are oppositely directed. This relation is the defining characteristic of simple harmonic motion and can be used to identify systems that will exhibit it:

> In simple harmonic motion, the acceleration, and thus the net force, are both proportional to, and oppositely directed from, the displacement from the equilibrium position.
>
> CONDITIONS FOR SIMPLE HARMONIC MOTION

The time it takes for a displaced object to execute a complete cycle of oscillatory motion—from one extreme to the other extreme and back—is called the **period** T. The reciprocal of the period is the **frequency** f, which is the number of cycles per unit of time:

$$f = \frac{1}{T} \qquad 14\text{-}3$$

The unit of frequency is the cycle per second (cy/s), which is called a **hertz** (Hz). For example, if the time for one complete cycle of oscillation is 0.25 s, the frequency is 4.0 Hz.

Figure 14-2 shows how we can experimentally obtain x versus t for a mass on a spring. The general equation for such a curve is

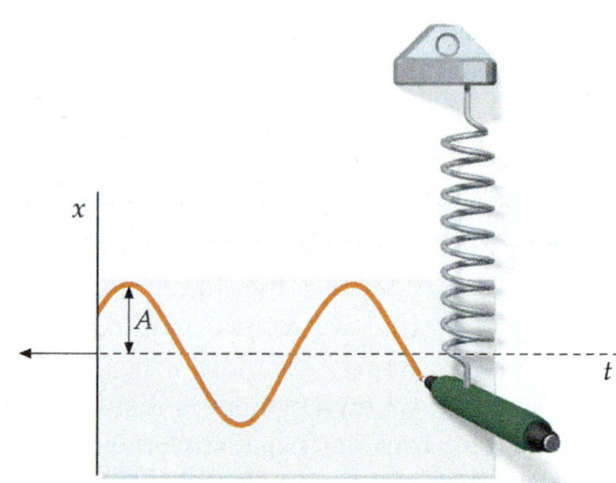

FIGURE 14-2 A marking pen is attached to a mass on a spring, and the paper is pulled to the left. As the paper moves with constant speed, the pen traces out the displacement x as a function of time t. (Here, we have chosen x to be positive when the spring is compressed.)

$$x = A\cos(\omega t + \delta) \qquad 14\text{-}4$$

POSITION IN SIMPLE HARMONIC MOTION

where A, ω, and δ are constants. The maximum displacement x_{max} from equilibrium is called the **amplitude** A. The argument of the cosine function, $\omega t + \delta$, is

* Hooke's law is introduced in Chapter 4, Section 5.

called the **phase** of the motion, and the constant δ is called the **phase constant,** which equals the phase at $t = 0$. [Note that $\cos(\omega t + \delta) = \sin(\omega t + \delta + [\pi/2])$; thus, whether the equation is expressed as a cosine function or a sine function simply depends on the phase of the oscillation at $t = 0$.] If we have just one oscillating system, we can always choose $t = 0$ so that $\delta = 0$. If we have two systems oscillating with the same frequency but with different phases, we can choose $\delta = 0$ for one of them. The equations for the two systems are then

$$x_1 = A_1 \cos(\omega t)$$

and

$$x_2 = A_2 \cos(\omega t + \delta)$$

If the phase difference δ is 0 or an integer times 2π, then the systems are said to be *in phase.* If the phase difference δ is π or an odd integer times π, then the systems are said to be 180° *out of phase.*

We can show that Equation 14-4 is a solution of Equation 14-2 by differentiating x twice with respect to time. The first derivative of x gives the velocity v_x:

$$v_x = \frac{dx}{dt} = -\omega A \sin(\omega t + \delta) \qquad 14\text{-}5$$

VELOCITY IN SIMPLE HARMONIC MOTION

Differentiating velocity with respect to time gives the acceleration:

$$a_x = \frac{dv_x}{dt} = \frac{d^2x}{dt^2} = -\omega^2 A \cos(\omega t + \delta) \qquad 14\text{-}6$$

Substituting x for $A \cos(\omega t + \delta)$ (see Equation 14-4) gives

$$a_x = -\omega^2 x \qquad 14\text{-}7$$

ACCELERATION IN SIMPLE HARMONIC MOTION

Comparing $a_x = -\omega^2 x$ (Equation 14-7) with $a_x = -(k/m)x$ (Equation 14-2), we see that $x = A\cos(\omega t + \delta)$ is a solution of $d^2x/dt^2 = -(k/m)x$ (Equation 14-2) if

$$\omega = \sqrt{\frac{k}{m}} \qquad 14\text{-}8$$

The amplitude A and the phase constant δ can be determined from the initial position x_0 and the initial velocity v_{0x} of the system. Setting $t = 0$ in $x = A\cos(\omega t + \delta)$ gives

$$x_0 = A \cos \delta \qquad 14\text{-}9$$

Similarly, setting $t = 0$ in $v_x = dx/dt = -A\omega \sin(\omega t + \delta)$ gives

$$v_{0x} = -A\omega \sin \delta \qquad 14\text{-}10$$

By using these equations, we can determine A and δ in terms of x_0, v_{0x}, and ω.

The period T is the shortest time interval satisfying the relation

$$x(t) = x(t + T)$$

for all t. Substituting into this relation using $x(t) = A\cos(\omega t + \delta)$ (Equation 14-4) gives

$$\begin{aligned} A\cos(\omega t + \delta) &= A\cos[\omega(t + T) + \delta] \\ &= A\cos(\omega t + \delta + \omega T) \end{aligned}$$

The swaying of the Citicorp Building in New York City during high winds is reduced by this tuned-mass damper mounted on an upper floor. It consists of a 400-ton sliding block connected to the building by a spring. The force constant is chosen so that the natural frequency of the spring-block system is the same as the natural sway frequency of the building. Set into motion by winds, the building and damper oscillate 180° out of phase with each other, thereby significantly reducing the swaying. *(Citibank.)*

See
Math Tutorial *for more information on*
Trigonometry

The cosine (and sine) function repeats in value when the phase increases by 2π, so

$$\omega T = 2\pi \quad \text{or} \quad \omega = 2\pi\left(\frac{1}{T}\right)$$

The constant ω is called the **angular frequency.** It has units of radians per second and dimensions of inverse time, the same as angular speed, which is also designated by ω. Substituting $2\pi/T$ for ω in Equation 14-4 gives

$$x = A\cos\left(2\pi\frac{t}{T} + \delta\right)$$

We can see by inspection that each time t increases by T, the ratio t/T increases by 1, the phase increases by 2π, and one cycle of the motion is completed.

The frequency is related to the angular frequency by

$$\omega = 2\pi\frac{1}{T} = 2\pi f \qquad \text{14-11}$$

Because $\omega = \sqrt{k/m}$, the frequency and period of an object on a spring are related to the force constant k and the mass m by

$$f = \frac{1}{T} = \frac{1}{2\pi}\sqrt{\frac{k}{m}} \qquad \text{14-12}$$

The frequency increases with increasing k (spring stiffness) and decreases with increasing mass. Equation 14-12 provides a way to measure the inertial mass of an astronaut in a "weightless" environment.

PRACTICE PROBLEM 14-1

A 0.80-kg object is attached to a spring that has a force constant $k = 400$ N/m. (*a*) Find the frequency and period of motion of the object when it is displaced from equilibrium and then released. (*b*) Repeat Part (*a*) except with a 1.6-kg object attached to the spring in place of the 0.80-kg object. *Hint: Review Example 14-4 first.*

PROBLEM-SOLVING STRATEGY

Solving Simple Harmonic Motion Problems

PICTURE Choose the origin of the x axis at the equilibrium position. For a spring, choose the $+x$ direction so that x is positive if the spring is extended.

SOLVE Do not use the kinematic equations for constant acceleration. Instead, use the equations developed for simple-harmonic motion.

CHECK Make sure your calculator is in the appropriate mode (degrees or radians) when evaluating trigonometric functions and their arguments.

! At just 1 rad/s off resonance, the amplitude drops by a factor of 20. This is not surprising, because the width $\Delta\omega$ of the resonance is only 0.0957 rad/s.

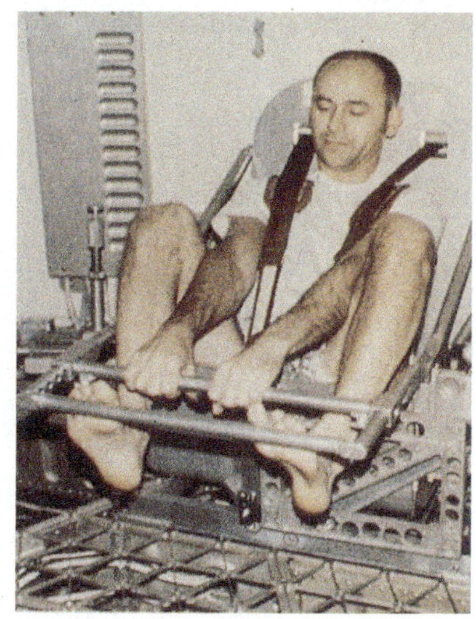

Astronaut Alan L. Bean measures his body mass during the second Skylab mission by sitting in a seat attached to a spring and oscillating back and forth. The total mass of the astronaut plus the seat is related to his frequency of vibration by Equation 14-12. *(NASA.)*

Example 14-1 Riding the Waves

You are sitting on a surfboard that is riding up and down on some swells. The board's vertical displacement y is given by

$$y = (1.2\text{ m})\cos\left(\frac{1}{2.0\text{ s}}t + \frac{\pi}{6}\right)$$

(*a*) Find the amplitude, angular frequency, phase constant, frequency, and period of the motion. (*b*) Where is the surfboard at $t = 1.0$ s? (*c*) Find the velocity and acceleration as functions of time t. (*d*) Find the initial values of the position, velocity, and acceleration of the surfboard.

Surfers waiting. *(David Pu'u/CORBIS.)*

PICTURE We find the quantities asked for in (*a*) by comparing the equation of motion

$$y = (1.2\text{ m})\cos\left(\frac{1}{2.0\text{ s}}t + \frac{\pi}{6}\right)$$

with the standard equation for simple harmonic motion, Equation 14-4. The velocity and acceleration are found by differentiating $y(t)$.

SOLVE

(*a*) 1. Compare this equation with $y = A\cos(\omega t + \delta)$ (Equation 14-4) to get A, ω, and δ:

$$y = (1.2\text{ m})\cos\left(\frac{1}{2.0\text{ s}}t + \frac{\pi}{6}\right)$$

$$A = \boxed{1.2\text{ m}} \quad \omega = \boxed{0.50\text{ rad/s}} \quad \delta = \boxed{\pi/6\text{ rad}}$$

2. The frequency and period are found from ω:

$$f = \frac{\omega}{2\pi} = \frac{0.50\text{ rad/s}}{2\pi} = 0.0796\text{ Hz} = \boxed{0.080\text{ Hz}}$$

$$T = \frac{1}{f} = \frac{1}{0.0796\text{ Hz}} = 12.6\text{ s} = \boxed{13\text{ s}}$$

(*b*) Set $t = 1.0$ s to find the surfboard's position above mean sea level:

$$y = (1.2\text{ m})\cos\left[(0.50\text{ rad/s})(1.0\text{ s}) + \frac{\pi}{6}\right] = \boxed{0.62\text{ m}}$$

(*c*) The velocity and acceleration are obtained from the position by differentiation with respect to time:

$$v_y = \frac{dy}{dt} = \frac{d}{dt}[A\cos(\omega t + \delta)] = -\omega A\sin(\omega t + \delta)$$

$$= -(0.50\text{ rad/s})(1.2\text{ m})\sin\left[(0.50\text{ rad/s})t + \frac{\pi}{6}\right]$$

$$= \boxed{-(0.60\text{ m/s})\sin\left[(0.50\text{ rad/s})t + \frac{\pi}{6}\right]}$$

$$a_y = \frac{dv_y}{dt} = \frac{d}{dt}[-\omega A\sin(\omega t + \delta)] = -\omega^2 A\cos(\omega t + \delta)$$

$$= -(0.50\text{ rad/s})^2(1.2\text{ m})\cos\left[(0.50\text{ rad/s})t + \frac{\pi}{6}\right]$$

$$= \boxed{-(0.30\text{ m/s}^2)\cos\left[(0.50\text{ rad/s})t + \frac{\pi}{6}\right]}$$

(*d*) Set $t = 0$ to find y_0, v_{0y}, and a_{0y}:

$$y_0 = (1.2\text{ m})\cos\frac{\pi}{6} = 1.04 = \boxed{1.0\text{ m}}$$

$$v_{0y} = -(0.60\text{ m/s})\sin\frac{\pi}{6} = \boxed{-0.30\text{ m/s}}$$

$$a_{0y} = -(0.30\text{ m/s}^2)\cos\frac{\pi}{6} = \boxed{-0.26\text{ m/s}^2}$$

CHECK We can check the plausibility of the Part-(*d*) results using $a_y = -\omega^2 y$ (Equation 14-7) at $t = 0$, with $y = 1.04$ m and $\omega = 0.50$ rad/s. Substituting into Equation 14-7 gives $a_{0y} = -\omega^2 y_0 = -(0.50\text{ rad/s})^2(1.04\text{ m}) = -0.26\text{ m/s}^2$, the same as the third Part-(*d*) result.

Figure 14-3 shows two identical masses attached to identical springs and resting on a horizontal frictionless surface. The spring attached to object 2 is stretched 10 cm and the spring attached to object 1 is stretched 5 cm. If they are released at the same time, which object reaches the equilibrium position first?

According to Equation 14-12, the period depends only on k and m and not on the amplitude. Because k and m are the same for both systems, the periods are the same. Thus, the objects reach the equilibrium position at the same time. The second object has twice as far to go to reach equilibrium, but it will also have twice the speed at any given instant. Figure 14-4 shows a sketch of the position functions for the two objects. This sketch illustrates an important general property of simple harmonic motion:

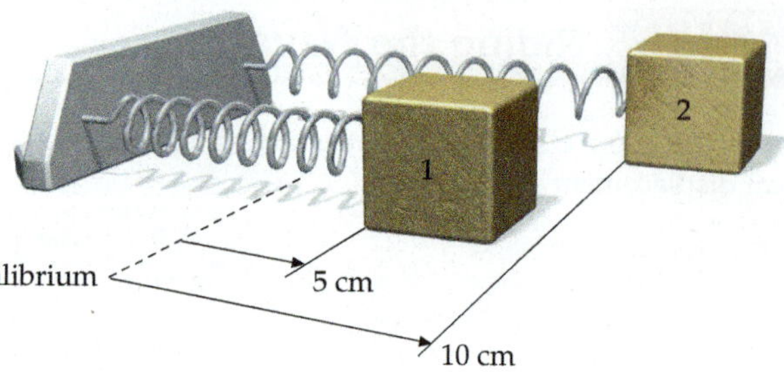

FIGURE 14-3 Two identical mass-spring systems.

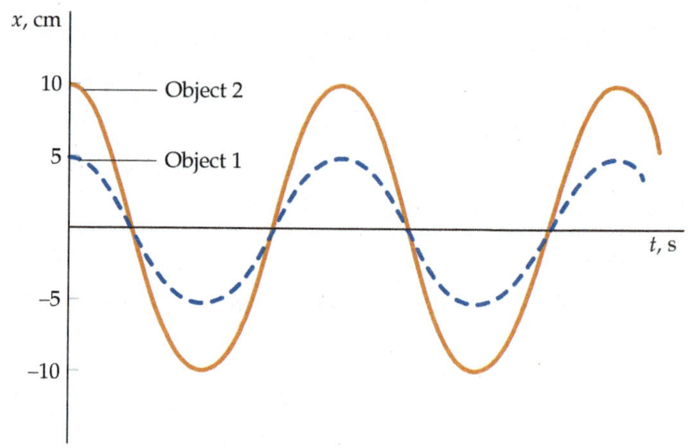

FIGURE 14-4 Plots of x versus t for the systems in Figure 14-3. Both reach their equilibrium positions at the same time.

> The frequency (and thus the period) of simple harmonic motion is independent of the amplitude.

The fact that the frequency in simple harmonic motion is independent of the amplitude has important consequences in many fields. In music, for example, it means that when a note is struck on the piano, the pitch (which corresponds to the frequency) does not depend on how loudly the note is played (which corresponds to the amplitude).† If changes in amplitude had a large effect on the frequency, then musical instruments would be unplayable.

Example 14-2 An Oscillating Object

An object oscillates with angular frequency $\omega = 8.0$ rad/s. At $t = 0$, the object is at $x = 4.0$ cm with an initial velocity $v_x = -25$ cm/s. (*a*) Find the amplitude and phase constant for the motion. (*b*) Write x as a function of time.

PICTURE The initial position and velocity give us two equations from which to determine the amplitude A and the phase constant δ.

SOLVE

(*a*) 1. The initial position and velocity are related to the amplitude and phase constant. The position is given by Equation 14-4. The velocity is found by taking the derivative with respect to time:

$$x = A\cos(\omega t + \delta) \quad \text{and}$$

$$v_x = \frac{dx}{dt} = -\omega A \sin(\omega t + \delta)$$

2. At $t = 0$ the position and velocity are:

$$x_0 = A\cos\delta \quad \text{and} \quad v_{0x} = -\omega A \sin\delta$$

3. Divide these equations to eliminate A:

$$\frac{v_{0x}}{x_0} = \frac{-\omega A\sin\delta}{A\cos\delta} = -\omega\tan\delta$$

4. Substituting numerical values yields δ:

$$\tan\delta = -\frac{v_{0x}}{\omega x_0} \quad \text{so}$$

$$\delta = \tan^{-1}\left(-\frac{v_{0x}}{\omega x_0}\right) = \tan^{-1}\left[-\frac{-25\text{ cm/s}}{(8.0\text{ rad/s})(4.0\text{ cm})}\right]$$

$$= 0.663\text{ rad} = \boxed{0.66\text{ rad}}$$

† For many musical instruments, there is a slight dependence of frequency on amplitude. The vibration of an oboe reed, for example, is not exactly simple harmonic; thus its pitch depends slightly on how hard it is blown. This effect can be corrected for by a skilled musician.

5. The amplitude can be found using either the x_0 or v_{0x} equation. Here we use x_0: $A = \dfrac{x_0}{\cos\delta} = \dfrac{4.0\text{ cm}}{\cos 0.663} = \boxed{5.1\text{ cm}}$

(*b*) Comparing with Equation 14-4 yields x: $x = \boxed{(5.1\text{ cm})\cos[(8.0\text{ s}^{-1})t + 0.66]}$

CHECK To see if the Part-(*b*) result ($x = (5.1\text{ cm})\cos[(8.0\text{ s}^{-1})t + 0.66]$) is plausible, we set t equal to zero and see if $x = 4.0$ cm. That is, $x = (5.1\text{ cm})\cos[(0) + 0.66] = 4.0$ cm. Thus, the Part-(*b*) result is plausible.

If the phase constant δ is 0, Equations 14-4, 14-5, and 14-6 then become

$$x = A\cos\omega t \qquad 14\text{-}13a$$

$$v_x = -\omega A\sin\omega t \qquad 14\text{-}13b$$

and

$$a_x = -\omega^2 A\cos\omega t \qquad 14\text{-}13c$$

These functions are plotted in Figure 14-5.

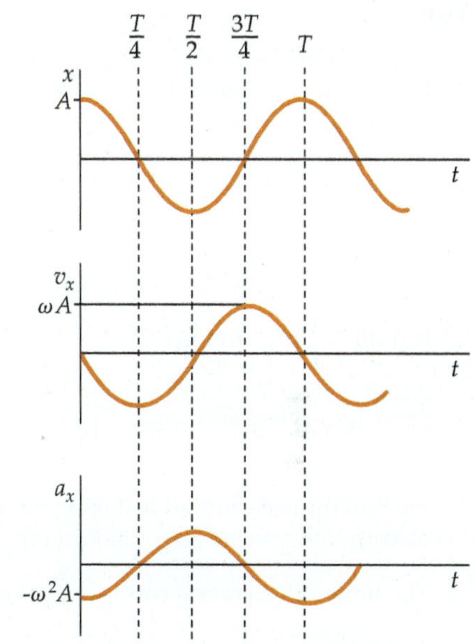

FIGURE 14-5 Plots of x, v_x and a_x as functions of time t for $\delta = 0$. At $t = 0$, the displacement is maximum, the velocity is zero, and the acceleration is negative and equal to $-\omega^2 A$. The velocity becomes negative as the object moves back toward its equilibrium position. After one quarter-period ($t = T/4$), the object is at equilibrium, $x = 0$, $a_x = 0$, and the velocity has its minimum value of $-\omega A$. At $t = T/2$, the displacement is $-A$, the velocity is again zero, and the acceleration is $+\omega^2 A$. At $t = 3T/4$, $x = 0$, $a_x = 0$, and $v_x = +\omega A$.

Example 14-3 A Block on a Spring — *Try It Yourself*

A 2.00-kg block is attached to a spring as in Figure 14-1. The force constant of the spring is $k = 196$ N/m. The block is held a distance 5.00 cm from the equilibrium position and is released at $t = 0$. (*a*) Find the angular frequency ω, the frequency f and the period T. (*b*) Write x as a function of time.

PICTURE For Part (*a*) use Equations 14-8 and 14-12 For Part (*b*) use Equation 14-4.

SOLVE

Cover the column to the right and try these on your own before looking at the answers.

Steps	Answers
(*a*) 1. Calculate ω from $\omega = \sqrt{k/m}$.	$\omega = \boxed{9.90\text{ rad/s}}$
2. Use your result to find f and T.	$f = \boxed{1.58\text{ Hz}}$ $T = \boxed{0.635\text{ s}}$
3. Find A and δ from the initial conditions.	$A = 5.00$ cm $\delta = 0.00$
(*b*) Write $x(t)$ using your results for A, ω, and δ.	$x = \boxed{(5.00\text{ cm})\cos[(9.90\text{ s}^{-1})t]}$

CHECK The block was released from rest, so we expect the velocity at $t = 0$ to be zero. To verify that our Part-(*b*) result is correct, we take the derivative of the expression $x = (5.00\text{ cm})\cos[(9.90\text{ s}^{-1})t]$ and evaluate it at $t = 0$. That is, $v_x(t) = dx/dt = -(4.95\text{ cm/s})\sin[(9.90\text{ s}^{-1})t]$. Evaluating this at $t = 0$ gives $v_x(0) = -(4.95\text{ cm/s})\sin(0) = 0$, as expected.

Example 14-4 Speed and Acceleration of an Object on a Spring

Consider an object on a spring whose position is given by $x = (5.00\text{ cm})\cos(9.90\text{ s}^{-1}t)$. (*a*) What is the maximum speed of the object? (*b*) When does this maximum speed first occur after $t = 0$? (*c*) What is the maximum of the acceleration of the object? (*d*) When does the maximum of the magnitude of the acceleration first occur after $t = 0$?

PICTURE Because the object is released from rest, $\delta = 0$, and the position, velocity, and acceleration are given by Equations 14-13*a*, *b*, and *c*.

SOLVE

(*a*) 1. Equation 14-13*a*, with $\delta = 0$, gives the position. We get the velocity by taking the derivative with respect to time:

$$x = A\cos\omega t$$

$$\text{so}\quad v_x = \frac{dx}{dt} = -\omega A\sin\omega t$$

2. Maximum speed v occurs when $|\sin\omega t| = 1$:

$$v = \omega A|\sin\omega t|$$

$$\text{so}\quad v_{max} = \omega A = (9.90\text{ rad/s})(5.00\text{ cm}) = \boxed{49.5\text{ cm/s}}$$

(*b*) 1. $|\sin\omega t| = 1$ first occurs when $\omega t = \pi/2$:

$$|\sin\omega t| = 1 \Rightarrow \omega t = \frac{\pi}{2}, \frac{3\pi}{2}, \frac{5\pi}{2}, \ldots$$

2. Solve for t when $\omega t = \pi/2$:

$$t = \frac{\pi}{2\omega} = \frac{\pi}{2(9.90\text{ s}^{-1})} = \boxed{0.159\text{ s}}$$

(*c*) 1. We find the acceleration by taking the derivative of the velocity, obtained in step 1 of Part (*a*):

$$a_x = \frac{dv_x}{dt} = -\omega^2 A\cos\omega t$$

2. Maximum acceleration corresponds to $\cos\omega t = -1$.

$$a_{max} = \omega^2 A = (9.90\text{ rad/s})^2(5.00\text{ cm}) = \boxed{490\text{ cm/s}^2 \approx \tfrac{1}{2}g}$$

(*d*) The magnitude of the acceleration is maximum when $|\cos\omega t| = 1$, which is when $\omega t = 0, \pi, 2\pi, \ldots$:

$$t = \frac{\pi}{\omega} = \frac{\pi}{9.90\text{ s}^{-1}} = \boxed{0.317\text{ s}}$$

CHECK We expect $|a_x|$ to first be maximum after $t = 0$ when x reaches its first minimum, and we expect x to reach its first minimum one-half cycle after release. That is, we expect $|a_x|$ to be maximum when $t = \frac{1}{2}T$, where T is the period. The period and the angular frequency are related by $\omega = 2\pi f = 2\pi/T$ (Equation 14-11). Substituting $2\pi/T$ for ω in our Part-(*d*) result gives $t = \pi/(2\pi/T) = \frac{1}{2}T$, as expected.

SIMPLE HARMONIC MOTION AND CIRCULAR MOTION

A relation exists between simple harmonic motion and circular motion with constant speed. Imagine a particle moving with constant speed v in a circle of radius A (Figure 14-6*a*). Its angular displacement relative to the $+x$ direction is given by

$$\theta = \omega t + \delta \qquad 14\text{-}14$$

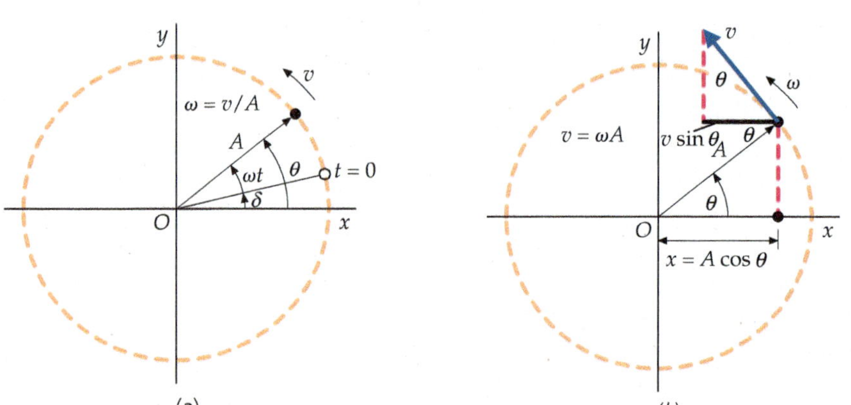

FIGURE 14-6 A particle moves in a circular path with constant speed. (*a*) Its x component of position describes simple harmonic motion, and (*b*) its x component of velocity describes the velocity of the simple harmonic motion.

where δ is the angular displacement at time $t = 0$ and $\omega = v/A$ is the angular speed of the particle. The x component of the particle's position (Figure 14-6*b*) is

$$x = A\cos\theta = A\cos(\omega t + \delta)$$

which is the same as Equation 14-4 for simple harmonic motion.

When a particle moves with constant speed in a circle, its projection onto a diameter of the circle moves with simple harmonic motion (see Figure 14-6).

The speed of a particle moving in a circle is $r\omega$, where r is the radius. For the particle in Figure 14-6*b*, $r = A$, so its speed is $A\omega$. The projection of the velocity vector onto the x axis gives $v_x = -v\sin\theta$. Substituting for v and θ gives

$$v_x = -v\sin\theta = -\omega A\sin(\omega t + \delta)$$

which is the same as Equation 14-5 for simple harmonic motion. The relation between circular motion and simple harmonic motion is nicely demonstrated by the image of the bubble trail produced by a rotating boat propeller.

Bubbles foaming off the edge of a rotating propeller that is moving through water produce a sinusoidal pattern. *(Institute for Marine Dynamics.)*

14-2 ENERGY IN SIMPLE HARMONIC MOTION

When an object on a spring undergoes simple harmonic motion, the system's potential energy and kinetic energy vary with time. Their sum, the total mechanical energy $E = K + U$, is constant. Consider an object a distance x from equilibrium, acted on by a restoring force $-kx$. The system's potential energy is

$$U = \tfrac{1}{2}kx^2$$

This is Equation 7-4. For simple harmonic motion, $x = A\cos(\omega t + \delta)$. Substituting gives

$$U = \tfrac{1}{2}kA^2\cos^2(\omega t + \delta) \qquad 14\text{-}15$$

POTENTIAL ENERGY IN SIMPLE HARMONIC MOTION

The kinetic energy of the system is

$$K = \tfrac{1}{2}mv^2$$

where m is the object's mass and v is its speed. For simple harmonic motion, $v_x = -\omega A\sin(\omega t + \delta)$. Substituting gives

$$K = \tfrac{1}{2}m\omega^2A^2\sin^2(\omega t + \delta)$$

Then using $\omega^2 = k/m$,

$$K = \tfrac{1}{2}kA^2\sin^2(\omega t + \delta) \qquad 14\text{-}16$$

KINETIC ENERGY IN SIMPLE HARMONIC MOTION

The total mechanical energy E is the sum of the potential and kinetic energies:

$$E = U + K = \tfrac{1}{2}kA^2\cos^2(\omega t + \delta) + \tfrac{1}{2}k\,A\sin^2(\omega t + \delta)$$
$$= \tfrac{1}{2}kA^2[\cos^2(\omega t + \delta) + \sin^2(\omega t + \delta)]$$

Because $\sin^2(\omega t + \delta) + \cos^2(\omega t + \delta) = 1$,

$$E = U + K = \tfrac{1}{2}kA^2 \qquad 14\text{-}17$$

TOTAL MECHANICAL ENERGY IN SIMPLE HARMONIC MOTION

This equation reveals an important general property of simple harmonic motion:

> The total mechanical energy in simple harmonic motion is proportional to the square of the amplitude.

For an object at its maximum displacement, the total energy is all potential energy. As the object moves toward its equilibrium position, the kinetic energy of the system increases and its potential energy decreases. As the object moves through its equilibrium position, the kinetic energy of the object is maximum, the potential energy of the system is zero, and the total energy is kinetic.

As the object moves past the equilibrium point, its kinetic energy begins to decrease, and the potential energy of the system increases until the object again stops momentarily at its maximum displacement (now in the other direction). At all times, the sum of the potential and kinetic energies is constant. Figure 14-7*b* and *c* show plots of U and K versus time. These curves have the same shape except that one is zero when the other is maximum. Their average values over one or more cycles are equal, and because $U + K = E$, their average values are given by

$$U_{av} = K_{av} = \tfrac{1}{2}E \qquad 14\text{-}18$$

In Figure 14-8, the potential energy U is graphed as a function of x. The total energy E is constant and is therefore plotted as a horizontal line. This line intersects the potential-energy curve at $x = A$ and $x = -A$. At these two points, called the **turning points,** oscillating objects reverse direction and head back toward the equilibrium position. Because $U \leq E$, the motion is restricted to $-A \leq x \leq +A$.

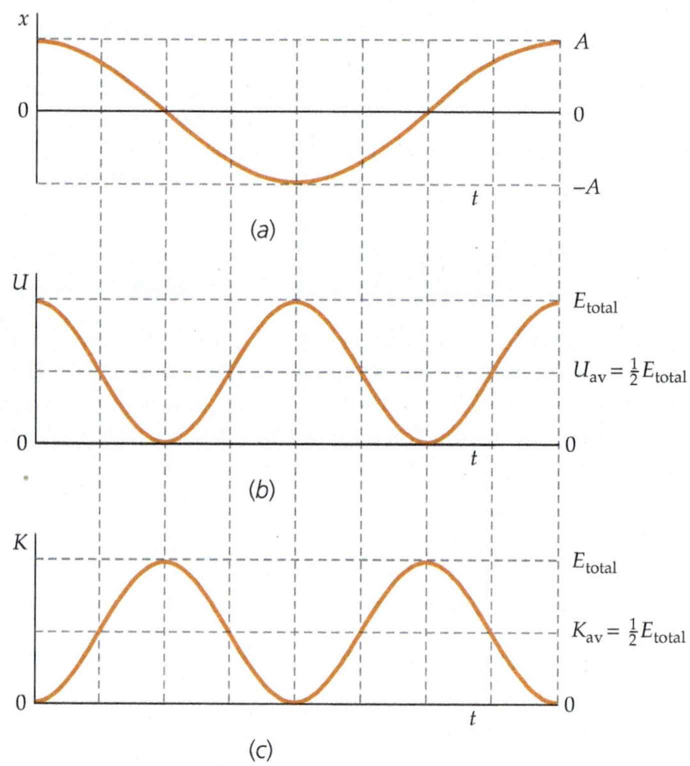

FIGURE 14-7 Plots of x, U, and K versus t.

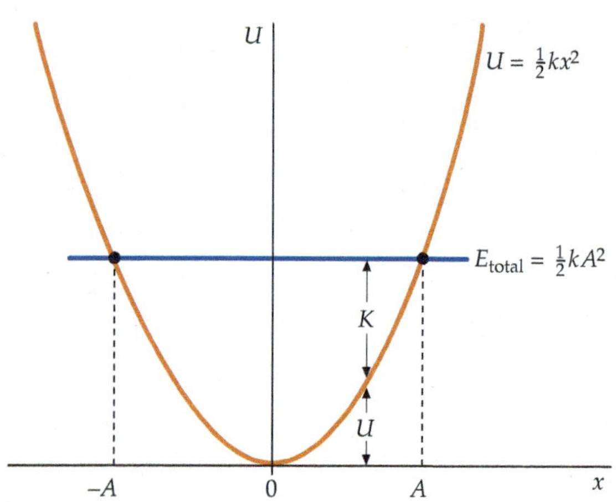

FIGURE 14-8 The potential-energy function $U = \tfrac{1}{2}kx^2$ for an object of mechanical mass m on a (massless) spring of force constant k. The horizontal blue line represents the total mechanical energy E_{total} for an amplitude of A. The kinetic energy K is represented by the vertical distance $K = E_{total} - U$. $E_{total} \geq U$, so the motion is restricted to $-A \leq x \leq +A$.

Example 14-5 Energy and Speed of an Oscillating Object

A 3.0-kg object attached to a spring oscillates with an amplitude of 4.0 cm and a period of 2.0 s. (*a*) What is the total energy? (*b*) What is the maximum speed of the object? (*c*) At what position x_1 is the speed equal to half its maximum value?

PICTURE (*a*) The total energy can be found from the amplitude and the force constant, and the force constant can be found from the mass and period. (*b*) The maximum speed occurs when the kinetic energy equals the total energy. (*c*) We can relate the position to the speed by using conservation of energy.

SOLVE

(*a*) 1. Write the total energy E in terms of the force constant k and amplitude A:

$$E = \tfrac{1}{2}kA^2$$

2. The force constant is related to the period and mass:

$$k = m\omega^2 = m\left(\frac{2\pi}{T}\right)^2$$

3. Substitute the given values to find E:

$$E = \frac{1}{2}kA^2 = \frac{1}{2}m\left(\frac{2\pi}{T}\right)^2 A^2 = \frac{1}{2}(3.0\text{ kg})\left(\frac{2\pi}{2.0\text{ s}}\right)^2(0.040\text{ m})^2$$

$$= 2.37 \times 10^{-2}\text{ J} = \boxed{2.4 \times 10^{-2}\text{ J}}$$

(*b*) To find v_{max}, set the kinetic energy equal to the total energy and solve for v:

$$\tfrac{1}{2}mv_{max}^2 = E$$

$$\text{so} \quad v_{max} = \sqrt{\frac{2E}{m}} = \sqrt{\frac{2(2.37 \times 10^{-2}\text{ J})}{3.0\text{ kg}}} = 0.126\text{ m/s} = \boxed{0.13\text{ m/s}}$$

(*c*) 1. Conservation of energy relates the position x to the speed v:

$$E = \tfrac{1}{2}mv^2 + \tfrac{1}{2}kx^2$$

2. Substitute $v = \frac{1}{2}v_{max}$ and solve for x_1. It is convenient to find x in terms of E and then write $E = \frac{1}{2}kA^2$ to obtain an expression for x in terms of A:

$$E = \tfrac{1}{2}m(\tfrac{1}{2}v_{max})^2 + \tfrac{1}{2}kx_1^2 = \tfrac{1}{4}(\tfrac{1}{2}mv_{max}^2) + \tfrac{1}{2}kx_1^2 = \tfrac{1}{4}E + \tfrac{1}{2}kx_1^2$$

$$\text{so} \quad \tfrac{1}{2}kx_1^2 = E - \tfrac{1}{4}E = \tfrac{3}{4}E$$

$$\text{and} \quad x_1 = \pm\sqrt{\frac{3E}{2k}} = \pm\sqrt{\frac{3}{2k}\left(\frac{1}{2}kA^2\right)} = \pm\frac{\sqrt{3}}{2}A$$

$$= \pm\frac{\sqrt{3}}{2}(4.0\text{ cm}) = \boxed{\pm 3.5\text{ cm}}$$

CHECK As expected, the result for Part (*c*), step 2 has two values, one with the spring extended, the other with the spring compressed. In addition, we expected these values to be equal, except for the sign. Further, the positive result is less than 4.0 cm (the amplitude is 4.0 cm), as expected.

PRACTICE PROBLEM 14-2 Calculate ω for this example and find v_{max} from $v_{max} = \omega A$.

PRACTICE PROBLEM 14-3 An object of mass 2.00 kg is attached to a spring that has a force constant 40.0 N/m. The object is moving at 25.0 cm/s when it is at its equilibrium position. (*a*) What is the total energy of the object? (*b*) What is the amplitude of the motion?

* GENERAL MOTION NEAR EQUILIBRIUM

Simple harmonic motion typically occurs when a particle is displaced slightly from a position of stable equilibrium. Figure 14-9 is a graph of the potential energy U versus x for a force that has a position of stable equilibrium and a position of unstable equilibrium. As discussed in Chapter 7, the potential-energy maximum at x_2 on Figure 14-9 corresponds to unstable equilibrium, whereas the minimum at x_1 corresponds to stable equilibrium. Many smooth curves with a minimum as in Figure 14-9 can be closely approximated near the minimum by a parabola. The dashed curve in this figure is a parabolic curve that approximately fits U near the

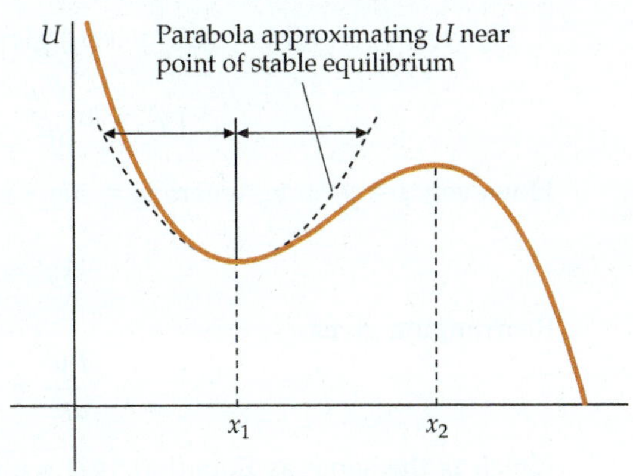

FIGURE 14-9 Plot of U versus x for a force that has a position of stable equilibrium (x_1) and a position of unstable equilibrium (x_2).

stable equilibrium point. The general equation for a parabola that has a minimum at point x_1 can be written

$$U = A + B(x - x_1)^2 \qquad 14\text{-}19$$

where A and B are constants. The constant A is the value of U at the equilibrium position $x = x_1$. The force is related to the potential energy curve by $F_x = -dU/dx$. Then

$$F_x = -\frac{dU}{dx} = -2B(x - x_1)$$

If we set $2B = k$, this equation reduces to

$$F_x = -\frac{dU}{dx} = -k(x - x_1) \qquad 14\text{-}20$$

According to Equation 14-20, the force is proportional to the displacement from equilibrium and oppositely directed, so the motion will be simple harmonic. Figure 14-9 shows a graph of this system's potential energy function $U(x)$, which has a position of stable equilibrium at $x = x_1$. Figure 14-10 shows a potential-energy function that has a position of stable equilibrium at $x = 0$. The system for this function is a small particle of mass m oscillating back and forth at the bottom of a frictionless spherical bowl.

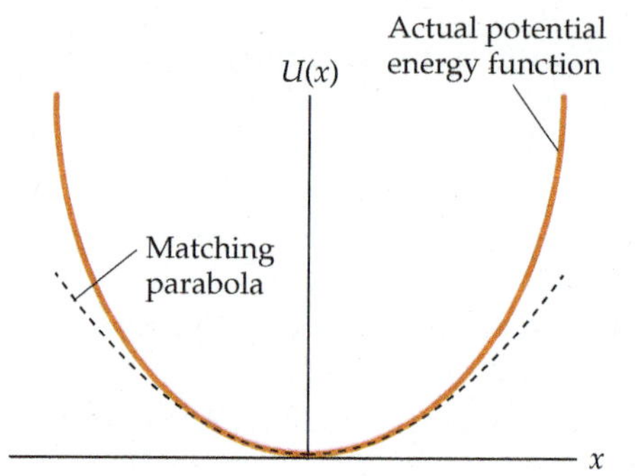

FIGURE 14-10 Plot of U versus x for a small particle oscillating back and forth at the bottom of a spherical bowl.

14-3 SOME OSCILLATING SYSTEMS

OBJECT ON A VERTICAL SPRING

When an object hangs from a vertical spring, there is a downward force mg in addition to the force of the spring (Figure 14-11). If we choose downward as the positive y direction, then the spring's force on the object is $-ky$, where y is the extension of the spring. The net force on the object is then

$$\Sigma F_y = -ky + mg \qquad 14\text{-}21$$

We can simplify this equation by changing to a new variable $y' = y - y_0$, where $y_0 = mg/k$ is the amount the spring is stretched when the object is in equilibrium. Substituting $y' + y_0$ for y gives

$$\Sigma F_y = -k(y' + y_0) + mg$$

But $ky_0 = mg$, so

$$\Sigma F_y = -ky' \qquad 14\text{-}22$$

Newton's second law ($\Sigma F_y = ma_y$) gives

$$-ky' = m\frac{d^2y}{dt^2}$$

However, $y = y' + y_0$, where $y_0 = mg/k$ is a constant. Thus $d^2y/dt^2 = d^2y'/dt^2$, so

$$-ky' = m\frac{d^2y'}{dt^2}$$

Rearranging gives

$$\frac{d^2y'}{dt^2} = -\frac{k}{m}y'$$

which is the same as Equation 14-2 with y' replacing x. It has the now familiar solution

$$y' = A\cos(\omega t + \delta)$$

where $\omega = \sqrt{k/m}$.

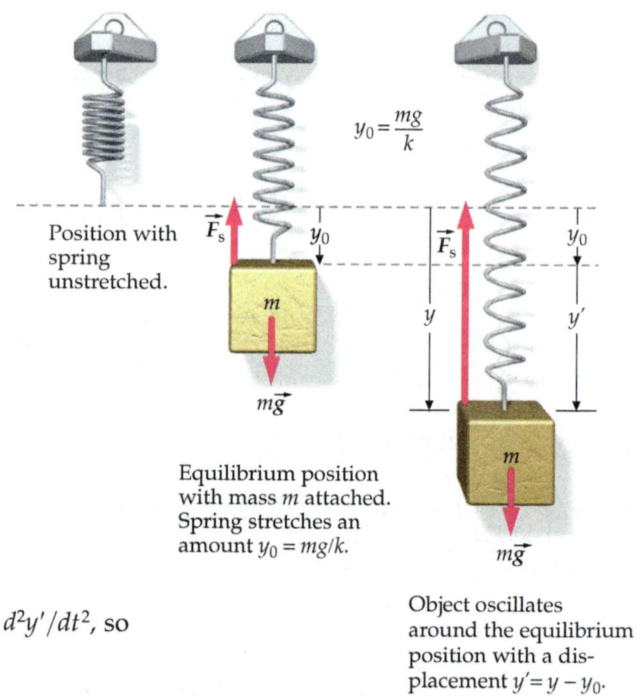

FIGURE 14-11 The Newton's second law equation for the motion of a mass on a vertical spring is greatly simplified if the displacement (y') is measured from the equilibrium position of the spring with the mass attached.

Thus, the effect of the gravitational force mg is merely to shift the equilibrium position from $y = 0$ to $y' = 0$. When the object is displaced from this equilibrium position by the amount y', the net force is $-ky'$. The object oscillates about this equilibrium position with an angular frequency $\omega = \sqrt{k/m}$, the same angular frequency as that for an object on a horizontal spring.

A force is conservative if the work done by it is independent of the path. Both the force of the spring and the force of gravity are conservative, and the sum of these forces (Equations 14-21 and 14-22) also is conservative. The potential energy function U associated with the sum of these forces is the negative of the work done plus an arbitrary integration constant. That is,

$$U = -\int -ky'dy' = \tfrac{1}{2}ky'^2 + U_0$$

where the integration constant U_0 is the value of U at the equilibrium position ($y' = 0$). Thus,

$$U = \tfrac{1}{2}ky'^2 + U_0 \qquad \text{14-23}$$

Example 14-6 Paper Springs

Context-Rich

You are showing your nieces how to make paper party decorations using paper springs. One niece makes a paper spring. The spring is stretched 8 cm and has a single sheet of colored paper suspended from it. You want the decorations to bounce at approximately 1.0 cy/s. How many sheets of colored paper should be used for the decoration on that spring if it is to bounce at 1.0 cy/s?

A paper spring (under construction). *(Rhoda Peacher.)*

PICTURE The frequency depends on the ratio of the force constant to the suspended mass (Equation 14-12), and you do not know either the force constant or the mass. However, Hooke's law (Equation 14-1) can be used to find the required ratio from the information given.

SOLVE

1. Write the frequency in terms of the force constant k and the mass M (Equation 14-12), where M is the mass of N sheets. We need to find N:

$$f = \frac{\omega}{2\pi} = \frac{1}{2\pi}\sqrt{\frac{k}{M}}$$

2. The spring stretches a distance of $y_0 = 8.0$ cm when a single sheet of mass m is suspended:

$$ky_0 = mg \quad \text{so} \quad \frac{k}{m} = \frac{g}{y_0}$$

3. The mass of N sheets equals N times the mass of a single sheet:

$$M = Nm$$

4. Using the step-2 and step-3 results, solve for k/M:

$$\frac{k}{M} = \frac{k}{Nm} = \frac{1}{N}\frac{g}{y_0}$$

5. Substitute the step-4 result into the step-1 result and solve for N:

$$f = \frac{1}{2\pi}\sqrt{\frac{k}{M}} = \frac{1}{2\pi}\sqrt{\frac{1}{N}\frac{g}{y_0}}$$

$$\text{so} \quad N = \frac{g}{(2\pi f)^2 y_0} = \frac{9.81\ \text{m/s}^2}{4\pi^2(1.0\ \text{Hz})^2(0.080\ \text{m})} = 3.1$$

Three sheets are needed.

CHECK Three or more sheets of construction paper seems plausible. Fifty or 100 sheets would likely wreck a paper spring.

TAKING IT FURTHER Note that we did not need to use the value of m or k in this example because the frequency depends on the ratio k/m, which equals g/y_0. In addition, we have neglected the mass of the spring itself. Its mass is probably not negligible compared to the mass of a few sheets of construction paper, so our step-5 result is an approximate result.

PRACTICE PROBLEM 14-4 How much is the paper spring stretched when a decoration made from three sheets of paper is suspended from it and the paper is in equilibrium?

Example 14-7 A Bead on a Block

A block securely attached to a spring oscillates vertically with a frequency of 4.00 Hz and an amplitude of 7.00 cm. A tiny bead is placed on top of the oscillating block just as it reaches its lowest point. Assume that the bead's mass is so small that its effect on the motion of the block is negligible. At what displacement from the equilibrium position does the bead lose contact with the block?

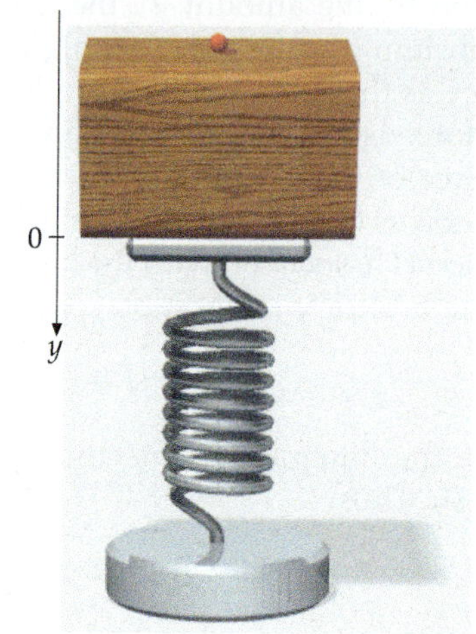

FIGURE 14-12

PICTURE The forces on the bead are its weight mg downward and the upward normal force exerted by the block. The magnitude of this normal force changes as the acceleration changes. As the block moves upward *from equilibrium*, its acceleration and the acceleration of the bead are *downward* and increasing in magnitude. When the acceleration reaches g downward, the normal force will be zero. If the block's downward acceleration becomes even slightly larger, the bead will leave the block.

SOLVE

1. Draw a sketch of the system (Figure 14-12). Include a y coordinate axis with its origin at the equilibrium position and with down as the positive direction:
2. We are looking for the value of y when the acceleration is g downward. Use Equation 14-7:
$$a_y = -\omega^2 y$$
$$g = -\omega^2 y$$
3. Substitute $2\pi f$ for ω and solve for y:
$$g = -(2\pi f)^2 y$$
$$\text{so} \quad y = -\frac{g}{(2\pi f)^2} = -\frac{9.81\ \text{m/s}^2}{[2\pi(4.00\ \text{Hz})]^2} = -0.0155\ \text{m} = \boxed{-1.55\ \text{cm}}$$

CHECK The bead leaves the block when y is negative, which is when the bead is above the equilibrium position because down was chosen as the positive y direction. This is as expected.

THE SIMPLE PENDULUM

A simple pendulum consists of a string of length L and a bob of mass m. When the bob is released from an initial angle ϕ_0 with the vertical, it swings back and forth with some period T. The units of length, mass, and g, are m, kg, and m/s^2, respectively. If we divide L by g, the meters cancel and we are left with seconds squared, suggesting the form $\sqrt{L/g}$. If the formula for the period contains the mass, then the unit kg must be canceled by some other quantity. But there is no combination of L and g that can cancel mass units. So the period cannot depend on the mass of the bob. Because the initial angle ϕ_0 is dimensionless, we cannot tell whether or not it is a factor in the period. We will see below that for small ϕ_0, the period is given by $T = 2\pi\sqrt{L/g}$.

CONCEPT CHECK 14-1

We might expect the period of a simple pendulum to depend on the mass m of a pendulum bob, the length L of the pendulum, the acceleration due to gravity g, and the initial angle ϕ_0. Find a simple combination of some or all of these quantities that gives the correct dimensions for the period.

A Foucault pendulum at the University of Louisville. In 1851, Leon Foucault suspended a 67-m-long pendulum from the ceiling of the Pantheon in Paris. Because of the rotation of Earth about its axis, the Pantheon rotates about the pendulum. (If the Pantheon were at the North Pole, it would rotate once every 24 hours.) The observation of the building rotating about the plane of the pendulum captured the imagination of the world. *(Courtesy of John Kielkopf/University of Louisville.)*

The forces on the bob are its weight $m\vec{g}$ and the string tension $\vec{T}$ (Figure 14-13). At an angle ϕ with the vertical, the weight has components $mg\cos\phi$ along the string and $mg\sin\phi$ tangential to the circular arc in the direction of decreasing ϕ. Using tangential components, Newton's second law ($\Sigma F_t = ma_t$) gives

$$-mg\sin\phi = m\frac{d^2s}{dt^2} \qquad 14\text{-}24$$

where the arc length s is related to the angle ϕ by $s = L\phi$. Repeatedly differentiating both sides of $s = L\phi$ gives

$$\frac{d^2s}{dt^2} = L\frac{d^2\phi}{dt^2}$$

Substituting $Ld^2\phi/dt^2$ into Equation 14-24 for d^2s/dt^2 and rearranging gives

$$\frac{d^2\phi}{dt^2} = -\frac{g}{L}\sin\phi \qquad 14\text{-}25$$

Note that the mass m does not appear in Equation 14-25—the motion of a pendulum does not depend on its mass. For small ϕ, $\sin\phi \approx \phi$, and

$$\frac{d^2\phi}{dt^2} \approx -\frac{g}{L}\phi \qquad \phi \ll 1 \qquad 14\text{-}26$$

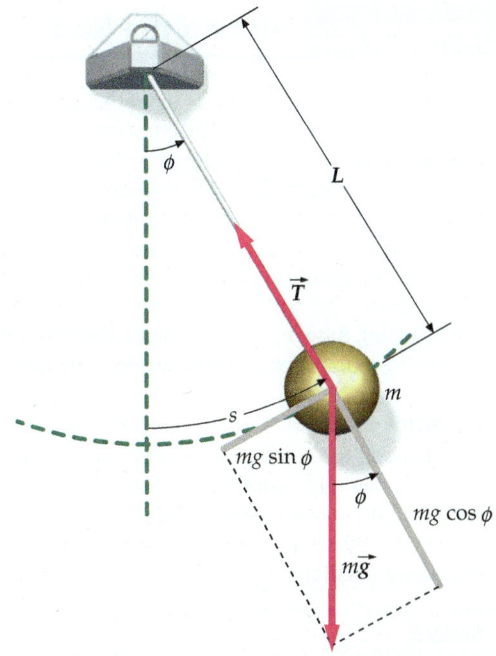

FIGURE 14-13 Forces on a pendulum bob.

Equation 14-26 is of the same form as Equation 14-2 for an object on a spring. Thus, the motion of a pendulum approximates simple harmonic motion for small angular displacements.

Equation 14-26 can be written

$$\frac{d^2\phi}{dt^2} = -\omega^2\phi, \qquad \text{where } \omega^2 = \frac{g}{L} \qquad 14\text{-}27$$

> ! ω is the angular frequency—not the angular speed—of the motion of the pendulum.

The period of the motion is thus

$$T = \frac{2\pi}{\omega} = 2\pi\sqrt{\frac{L}{g}} \qquad \text{(for small oscillations)} \qquad 14\text{-}28$$

PERIOD OF A SIMPLE PENDULUM

The solution of Equation 14-27 is

$$\phi = \phi_0\cos(\omega t + \delta)$$

where ϕ_0 is the maximum angular displacement.

According to Equation 14-28, the greater the length of a pendulum, the greater the period, which is consistent with experimental observation. The period and therefore the frequency are independent of the amplitude of oscillation (as long as the amplitude is small). This statement is a general feature of simple harmonic motion.

PRACTICE PROBLEM 14-5

Find the period of a simple pendulum of length 1.00 m undergoing small oscillations.

The acceleration due to gravity can be measured using a simple pendulum undergoing small oscillations. We need only measure the length L and period T of the pendulum, and using Equation 14-28, solve for g. (To measure T, we usually measure the time for n oscillations and then divide by n, which minimizes measurement error.)

Example 14-8 Timing the Run *Conceptual*

In a physics lab on one-dimensional kinematics, Liz and Bob are tasked with measuring the time it takes for a glider released from rest on an inclined 2.00-m-long air track to travel various distances. (An air track is a virtually frictionless track.) They tilt the track by putting a 2.0-cm-thick notebook under the legs at one end of the track. They release the glider from the middle of the track and find the time for it to accelerate half the length of the track to be 4.8 s. They then release the glider from the high end of the track and find that the time it takes for the glider to accelerate the entire length of the track is 4.8 s—the same time it took to accelerate half the length of the track. Thinking that the times for the two distances cannot be equal, they repeat both measurements, only to obtain the same results. Confused, they ask the instructor for an explanation. Can you think of a plausible explanation?

PICTURE If the track were perfectly straight, the acceleration would be the same everywhere along the track and the time for the glider to accelerate the entire length of the track, starting from rest, would be greater than the time for it to accelerate only half the length of the track. If the track sagged slightly, however, then the acceleration would be greatest at the high end of the track where the slope is steepest. What would the assumption that the track is sagging predict?

SOLVE

1. Suppose the track sags slightly, in such a way that the track forms a circular arc whose center of curvature is directly above the low end of the track:

   If the track sags as supposed, then the glider would move like the bob of a simple pendulum of length $L = R$, where R is the radius of curvature of the track.

2. The period T of a pendulum is independent of amplitude for small amplitudes:

   The times measured by Liz and Bob would equal $\frac{1}{4}$ the period T of the pendulum, given by Equation 14-28. Because the period of a pendulum is independent of amplitude (for small amplitudes), the times measured by Liz and Bob would be expected to be equal.

CHECK Is the amplitude of the pendulum sufficiently small when the glider is released from the high end of the track? It is if the R is much greater than 2.00 m. Equation 14-28 tells us that the length of the pendulum is given by $L = gT^2/(4\pi^2)$. Substituting $4 \times (4.8\text{ s})$ for T gives $R = L = 92$ m, justifying the supposition that the amplitudes were small.

This clock keeps time by using a torsional oscillator. *(Courtesy of Bill Master/Alibaba. http://yuning.en.alibaba.com.)*

Pendulum in an accelerated reference frame Figure 14-14*a* shows a simple pendulum suspended from the ceiling of a boxcar that has acceleration $\vec{a}_0$, relative to the ground, to the right, and $\vec{a}$ is the acceleration of the bob relative to the ground. Applying Newton's second law to the bob gives

$$\Sigma\vec{F} = \vec{T} + m\vec{g} = m\vec{a} \tag{14-29}$$

If the bob remains at rest relative to the boxcar, then $\vec{a} = \vec{a}_0$ and

$$\Sigma F_x = T\sin\theta_0 = ma_0$$
$$\Sigma F_y = T\cos\theta_0 - mg = 0$$

where θ_0 is the equilibrium angle. Thus, θ_0 is given by $\tan\theta_0 = a_0/g$. If the bob is

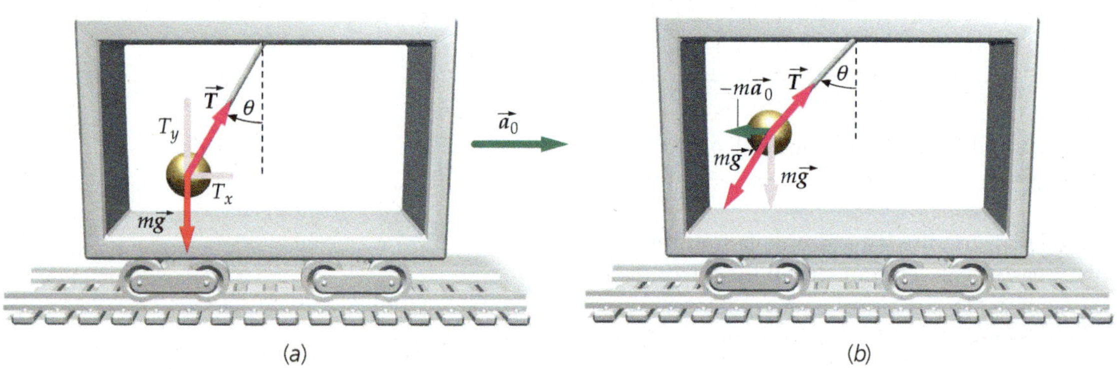

FIGURE 14-14 (*a*) Simple pendulum in apparent equilibrium in an accelerating boxcar. Forces are those as seen from a separate stationary frame. (*b*) Forces on the bob as seen in the accelerated frame. Adding the pseudoforce $-m\vec{a}_0$ is equivalent to replacing $\vec{g}$ by $\vec{g}'$.

moving relative to the boxcar, then $\vec{a}' = \vec{a} - \vec{a}_0$, where $\vec{a}'$ is the acceleration of the bob relative to the boxcar. Substituting for $\vec{a}$ in Equation 14-29 gives

$$\Sigma\vec{F} = \vec{T} + m\vec{g} = m(\vec{a}' + \vec{a}_0)$$

Subtracting $m\vec{a}_0$ from both sides of this equation and rearranging terms gives

$$\vec{T} + m\vec{g}' = m\vec{a}'$$

where $\vec{g}' = \vec{g} - \vec{a}_0$. Thus, by replacing $\vec{g}$ by $\vec{g}'$ and $\vec{a}$ by $\vec{a}'$ in Equation 14-29 we can solve for the motion of the bob relative to the boxcar. The vectors $\vec{T}$ and $m\vec{g}'$ are shown in Figure 14-14*b*. If the string breaks so that $\vec{T} = 0$, then our equation gives $\vec{a}' = \vec{g}'$, which means that $\vec{g}'$ is the free-fall acceleration in the reference frame of the boxcar. If the bob is displaced slightly from equilibrium, it will oscillate with a period T given by Equation 14-28 with g replaced by g'.

PRACTICE PROBLEM 14-6

A simple pendulum of length 1.00 m is in a boxcar that is accelerating horizontally with acceleration $a_0 = 3.00\ \text{m/s}^2$. Find g' and the period T.

Large-Amplitude oscillations When the amplitude of a pendulum's oscillation becomes large, its motion continues to be periodic, but it is no longer a simple harmonic. For an angular amplitude ϕ_0, the period can be shown to be given by

$$T = T_0\left[1 + \frac{1}{2^2}\sin^2\frac{1}{2}\phi_0 + \frac{1}{2^2}\left(\frac{3}{4}\right)^2\sin^4\frac{1}{2}\phi_0 + \cdots\right] \qquad 14\text{-}30$$

PERIOD FOR LARGE-AMPLITUDE OSCILLATIONS

where $T_0 = 2\pi\sqrt{L/g}$ is the period for very small amplitudes. Figure 14-15 shows T/T_0 as a function of amplitude ϕ_0.

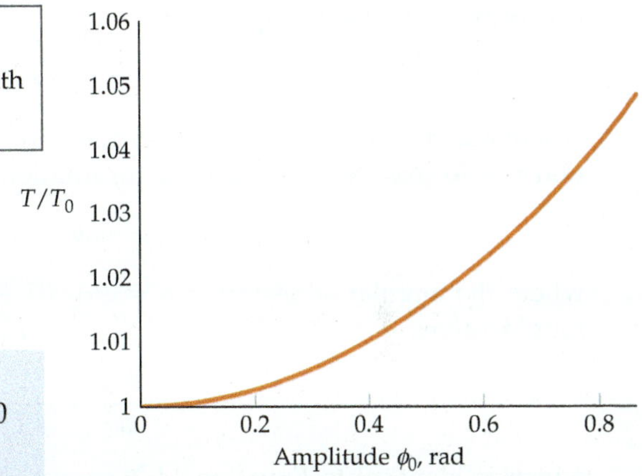

FIGURE 14-15 Note that the values on the vertical axis range from 1 to 1.06. Over a range of ϕ from 0 to 0.8 rad (46°), the period varies by about 5 percent.

Example 14-9 A Pendulum Clock

Try It Yourself

A simple pendulum clock is calibrated to keep accurate time at an amplitude of $\phi_0 = 10.0°$. When the amplitude has decreased to the point where it is very small, does the clock gain or lose time? About how much time will the clock gain or lose in one day if the amplitude remains very small.

PICTURE To calculate the period when the angular amplitude is 10°, retain only the first correction term to Equation 14-30. That is, use

$$T \approx T_0\left[1 + \frac{1}{2^2}\sin^2\frac{1}{2}\phi_0\right]$$

This equation provides sufficient accuracy because 10° is a fairly small amplitude. The amplitude of the pendulum slowly decreases due to the effects of air drag.

SOLVE

Cover the column to the right and try these on your own before looking at the answers.

Steps	Answers
1. Use Equation 14-30 to determine if T_0 is greater or less than T.	T decreases as ϕ_0 decreases, so **the clock gains time.**
2. Use Equation 14-30 to find the percentage change $[(T - T_0)/T] \times 100\%$ for $\phi = 10°$. Use only the first correction term.	0.190%
3. Find the number of minutes in a day.	There are 1440 minutes in a day.
4. Combine steps 2 and 3 to find the change in the number of minutes in a day.	The gain is **2.73 min/d**

CHECK The first correction term in Equation 14-30 is $\frac{1}{4}\sin^2(10.0°/2) = 1.90 \times 10^{-3}$, so $T = 1.00190T_0$ and $(T - T_0)/T = (1.00190T_0 - T_0)/1.00190T_0 = 0.00190$. This value agrees with our step-2 result.

TAKING IT FURTHER To avoid this gain, pendulum-clock mechanisms are designed to keep the amplitude fairly constant.

* THE TORSIONAL OSCILLATOR

A system that undergoes rotational oscillations in a variation of simple-harmonic motion is called a **torsional oscillator.** Figure 14-16 shows a torsional oscillator consisting of a solid disk suspended from a steel wire. If the angular displacement of the disk from the equilibrium position is ϕ then the wire exerts a linear restoring torque τ on the disk given by

$$\tau = -\kappa\phi \qquad \text{14-31}$$

where κ is the **torsional constant** of the wire. Substituting $-\kappa\phi$ for τ in the equation $\tau = I\alpha$ (Newton's second law for rotational motion) gives

$$-\kappa\phi = I\alpha$$

where the angular acceleration $\alpha = d^2\phi/dt^2$. Substituting $d^2\phi/dt^2$ for α and rearranging gives

$$\frac{d^2\phi}{dt^2} = -\frac{\kappa}{I}\phi \qquad \text{14-32}$$

which is identical to Equation 14-2, except with I in place of m, κ in place of k, and ϕ in place of x. Thus, the solution to Equation 14-32 can be written by directly substituting into Equation 14-4. Doing so gives

$$\phi = \phi_0 \cos(\omega t + \delta) \qquad \text{14-33}$$

where $\omega = \sqrt{\kappa/I}$ is the angular frequency—and not the angular speed—of the motion. The period is therefore

$$T = \frac{2\pi}{\omega} = 2\pi\sqrt{\frac{I}{\kappa}} \qquad \text{14-34}$$

PERIOD OF A TORSIONAL OSCILLATOR

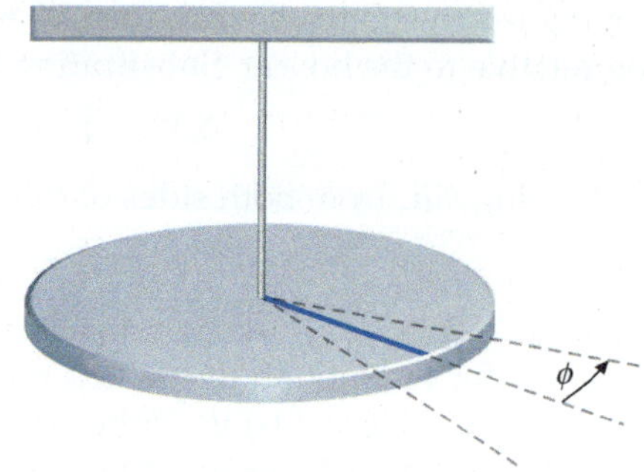

FIGURE 14-16 This torsional pendulum consists of a solid disk suspended by a steel wire.

All mechanical clocks keep time because the period of the oscillating part of the mechanism remains constant. The period of any pendulum changes with changes in amplitude. However, the driving mechanism of a pendulum clock maintains the amplitude at a constant value. *(Richard Menga/ Fundamental Photographers.)*

* THE PHYSICAL PENDULUM

A rigid object free to rotate about a horizontal axis that is not through its center of mass will oscillate when displaced from equilibrium. Such a system is called a **physical pendulum.** Consider a plane figure with a rotation axis a distance D from the figure's center of mass and displaced from equilibrium by the angle ϕ (Figure 14-17). The torque about the axis has a magnitude $MgD\sin\phi$. For sufficiently small values of ϕ, we can simplify our expression for the torque using the small-angle approximation ($\sin\phi = \phi$). Thus, for small angles the torque is a linear restoring torque given by

$$\tau = -MgD\phi \qquad \text{14-35}$$

Comparing this with $\tau = -\kappa\phi$ (Equation 14-31), we can see that for small angular displacements the physical pendulum is a torsional oscillator with a torsional constant given by

$$\kappa = MgD$$

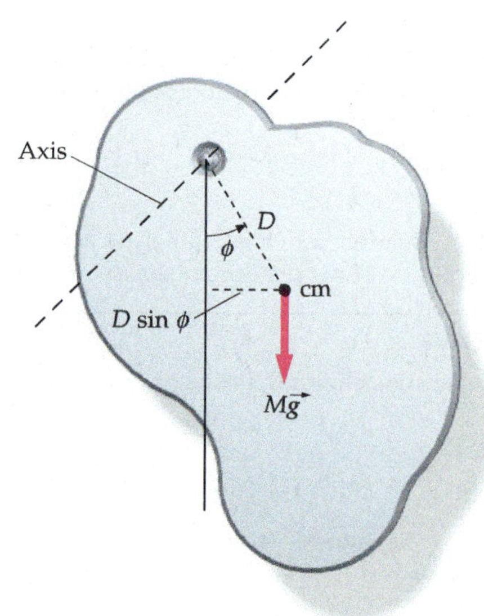

FIGURE 14-17 A physical pendulum.

Thus, the motion of the pendulum is described by Equation 14-33 with $\kappa = MgD$. The period is therefore

$$T = \frac{2\pi}{\omega} = 2\pi\sqrt{\frac{I}{MgD}} \qquad 14\text{-}36$$

PERIOD OF A PHYSICAL PENDULUM

! The period of a physical pendulum depends on the distribution of the mass, but not on the total mass M. The moment of inertia I is proportional to M, so the ratio I/M is independent of M.

For large amplitudes, the period is given by Equation 14-30, with T_0 given by Equation 14-36. For a simple pendulum of length L, the moment of inertia is $I = ML^2$ and $D = L$. Then, Equation 14-36 gives $T = 2\pi\sqrt{ML^2/(MgL)} = 2\pi\sqrt{L/g}$, the same as Equation 14-28.

Example 14-10 A Comfortable Pace

Context-Rich

You claim that the pace of a comfortable walk can be predicted if we model each leg as a physical pendulum. Your teacher is skeptical about this claim and asks you to back it up. Is your claim correct?

PICTURE A simple model of each leg is that of a uniform rod pivoted at one end. Each leg swings back and forth once every two steps, so the time required to walk 10 steps is $5T$, where T is the period of the "pendulum." How long will it take you to complete 10 steps at a leisurely pace if your prediction is correct? Model your leg as a 0.90-m-long uniform rod pivoted about an axis through one end.

SOLVE

1. Draw and label a uniform thin rod pivoted about one end (Figure 14-18):

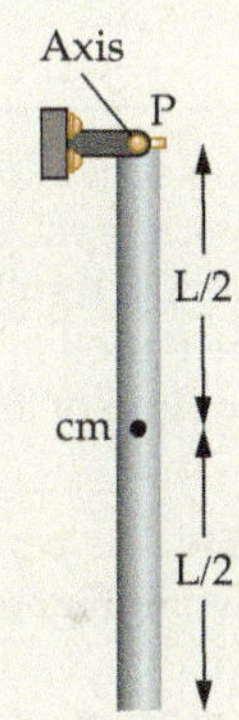

FIGURE 14-18 The distance between the rotation axis and the center of mass is $L/2$.

2. The period of a physical pendulum is given by $T = 2\pi\sqrt{I/MgD}$ (Equation 14-36):
$$T = 2\pi\sqrt{\frac{I}{MgD}}$$

3. I about the end is found in Table 9-1 and D is half the length of the rod:
$$I = \tfrac{1}{3}ML^2 \quad \text{and} \quad D = \tfrac{1}{2}L$$

4. Substitute the expressions for I and D to find T:
$$T = 2\pi\sqrt{\frac{\frac{1}{3}ML^2}{Mg(\frac{1}{2}L)}} = 2\pi\sqrt{\frac{2L}{3g}}$$

5. The length $L = 0.90$ m and the time for 10 steps is $5T$:
$$5T = 5\cdot 2\pi\sqrt{\frac{2L}{3g}} = 10\pi\sqrt{\frac{2(0.90\text{ m})}{3(9.81\text{ m/s}^2)}} = 7.8\text{ s}$$

6. My hypothesis has merit. My hip joint is about 90 cm above the floor and it took me almost 6.7 s to complete 10 leisurely steps. The upper half of my leg is more massive than the lower half, so modeling my leg as a uniform rod is not completely appropriate. In addition, what is a leisurely pace is subject to interpretation.

CHECK Long-legged animals, like elephants and giraffes, seem to walk at a slow, lumbering pace, and short-legged animals, like mice and sandpipers, walk at a fast pace. This conclusion is predicted by this model, because the period of a long pendulum is greater than that of a short pendulum.

Example 14-11 A Swinging Rod

A uniform rod of mass M and length L is free to swing about a horizontal axis perpendicular to the rod and a distance x from the rod's center. Find the period of oscillation for small angular displacements of the rod.

PICTURE The period is given by Equation 14-36. The center of mass is at the center of the rod, so the distance from the center of mass to the rotation axis is x (Figure 14-19). The moment of inertia of a uniform rod can be found from the parallel-axis theorem $I = I_{cm} + MD^2$ (Equation 9-13), where I_{cm} can be found in Table 9-1.

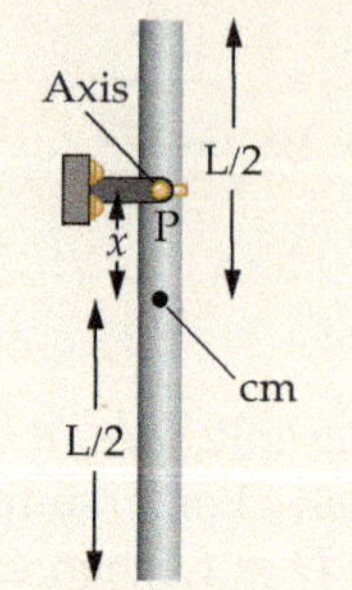

FIGURE 14-19 The distance between the rotation axis and the center of mass is x.

SOLVE

1. The period is given by Equation 14-36: $T = 2\pi\sqrt{\dfrac{I}{MgD}}$
2. $D = x$, and the moment of inertia is given by the parallel-axis theorem. The moment of inertia about a parallel axis through the center of mass is found in Table 9-1: $D = x$; $I = I_{cm} + MD^2 = \frac{1}{12}ML^2 + Mx^2$
3. Substitute these values to find T: $T = 2\pi\sqrt{\dfrac{I}{MgD}} = 2\pi\sqrt{\dfrac{(\frac{1}{12}ML^2 + Mx^2)}{Mgx}}$

$$= \boxed{2\pi\sqrt{\dfrac{(\frac{1}{12}L^2 + x^2)}{gx}}}$$

CHECK As $x \rightarrow 0$, $T \rightarrow \infty$ as expected. (If the rotation axis of the rod passes through its center of mass, we do not expect gravity to exert a restoring torque.) Also, if $x = L/2$ we get $T = 2\pi\sqrt{2L/3g}$, the same result as found in step 4 of Example 14-10. In addition, if $x \gg L$ the expression for the period approaches $T = 2\pi\sqrt{x/g}$, which is the expression for the period of a simple pendulum of length x (Equation 14-28).

TAKING IT FURTHER The period T versus distance x from the center of mass for a rod of length 1.00 m is shown in Figure 14-20.

PRACTICE PROBLEM 14-7 Show that the step-3 expression for the period gives the same period for $x = L/6$ as for $x = L/2$.

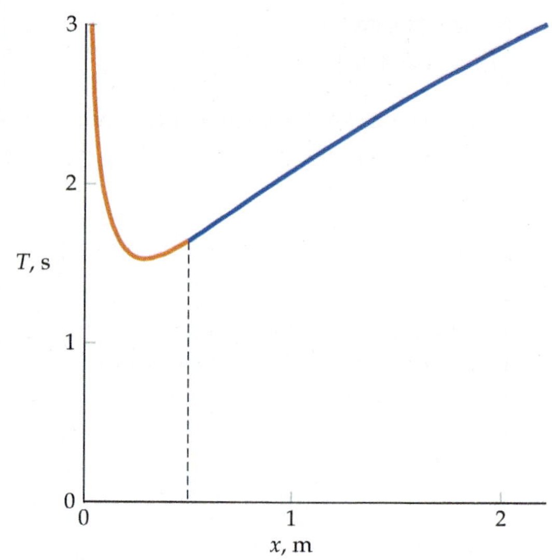

FIGURE 14-20 Plot of the period versus the distance from the pivot to the center of mass. For $x > 0.5$ m the pivot point is beyond the end of the rod.

Example 14-12 The Swinging Rod Revisited

Try It Yourself

Find the value of x in Example 14-11 for which the period is a minimum.

PICTURE At the value of x for which T is a minimum, $dT/dx = 0$.

SOLVE

Cover the column to the right and try these on your own before looking at the answers.

Steps	Answers
1. The period, given by the Example 14-11 result, is $T = 2\pi\sqrt{Z/g}$, where $Z = (\frac{1}{12}L^2 + x^2)/x$. Find the period both as x approaches zero and as x approaches infinity.	$T = 2\pi\sqrt{\dfrac{(\frac{1}{12}L^2 + x^2)}{gx}} = 2\pi\sqrt{\dfrac{Z}{g}}$ where $Z = (\frac{1}{12}L^2 + x^2)/x$. As $x \rightarrow 0$, $Z \rightarrow \infty$, and $T \rightarrow \infty$. As $x \rightarrow \infty$, $Z \rightarrow \infty$, and $T \rightarrow \infty$.

2. The period goes to infinity as x approaches zero and as x approaches infinity. Somewhere in the range $0 < x < \infty$ the period is a minimum. To find the minimum, evaluate dT/dx, set it equal to zero, and solve for x.

$$\frac{dT}{dx} = \frac{dT}{dZ}\frac{dZ}{dx} = \frac{\pi}{\sqrt{g}} Z^{-1/2}\frac{dZ}{dx}$$

$Z > 0$ throughout the range $0 < x < \infty$, so

$$\frac{dT}{dx} = 0 \Rightarrow \frac{dZ}{dx} = 0.$$

$$\frac{dZ}{dx} = 0 \Rightarrow x = \boxed{\frac{L}{\sqrt{12}} = 0.289L}$$

CHECK We expect an answer between 0 and $0.5L$. The step-2 result of $x = 0.289L$ meets that expectation.

14-4 DAMPED OSCILLATIONS

Left to itself, a spring or a pendulum eventually stops oscillating because the mechanical energy is dissipated by frictional forces. Such motion is said to be **damped.** If the damping is large enough, as, for example, a pendulum submerged in molasses, the oscillator fails to complete even one cycle of oscillation. Instead, it just moves toward the equilibrium position with a speed that approaches zero as the object approaches the equilibrium position. This type of motion is referred to as **overdamped.** If the damping is small enough that the system oscillates with an amplitude that decreases slowly with time—like a child on a playground swing when a parent stops providing a push each cycle—the motion is said to be **underdamped.** Motion with the minimum damping for nonoscillatory motion is said to be **critically damped.** (With any less damping, the motion would be underdamped.)

Underdamped motion The damping force exerted on an oscillator such as the one shown in Figure 14-21*a* can be represented by the empirical expression

$$\vec{F}_d = -b\vec{v}$$

where b is a constant. Such a system is said to be *linearly damped.* The discussion here is for linearly damped motion. Because the damping force is opposite to the direction of motion, it does negative work and causes the mechanical energy of

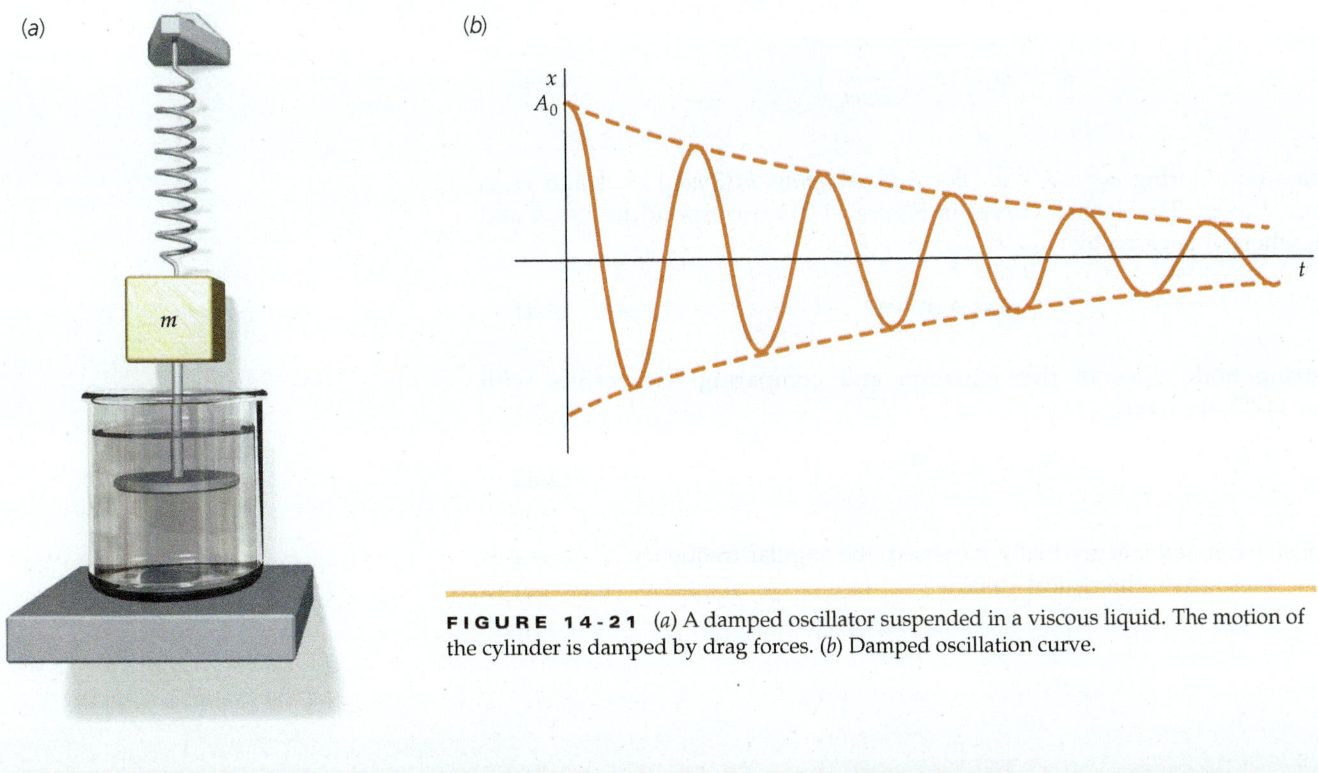

FIGURE 14-21 (*a*) A damped oscillator suspended in a viscous liquid. The motion of the cylinder is damped by drag forces. (*b*) Damped oscillation curve.

the system to decrease. This energy is proportional to the square of the amplitude (Equation 14-17), and the square of the amplitude decreases exponentially with increasing time. That is,

$$A^2 = A_0^2 e^{-t/\tau} \qquad \text{14-37}$$

DEFINITION—TIME CONSTANT

where A is the amplitude, A_0 is the amplitude at $t = 0$, and τ is the decay time or time constant. The **time constant** is the time for the energy to change by a factor of e^{-1}.

The motion of a damped system can be obtained from Newton's second law. For an object of mass m on a spring that has a force constant k, the net force is $-kx - b(dx/dt)$. Setting the net force equal to the mass times the acceleration d^2x/dt^2, we obtain

$$-kx - b\frac{dx}{dt} = m\frac{d^2x}{dt^2}$$

which we rearrange to appear as

$$m\frac{d^2x}{dt^2} + b\frac{dx}{dt} + kx = 0 \qquad \text{14-38}$$

DIFFERENTIAL EQUATION FOR A DAMPED OSCILLATOR

The exact solution of this equation can be found using standard methods for solving differential equations. The solution for the underdamped case is

$$x = A_0 e^{-(b/2m)t}\cos(\omega' t + \delta) \qquad \text{14-39}$$

where A_0 is the initial amplitude. The frequency ω' is related to the natural frequency ω_0 (the frequency with no damping) by

$$\omega' = \omega_0\sqrt{1 - \left(\frac{b}{2m\omega_0}\right)^2} \qquad \text{14-40}$$

For a mass on a spring $\omega_0 = \sqrt{k/m}$. For *weak damping*, $b/(2m\omega_0) \ll 1$ and ω' is nearly equal to ω_0. The dashed curves in Figure 14-21*b* correspond to $x = A$ and $x = -A$, where A is given by

$$A = A_0 e^{-(b/2m)t} \qquad \text{14-41}$$

By squaring both sides of this equation and comparing the results with Equation 14-37, we have

$$\tau = \frac{m}{b} \qquad \text{14-42}$$

If the *damping constant* b is gradually increased, the angular frequency ω' decreases until it becomes zero at the critical value

$$b_c = 2m\omega_0 \qquad \text{14-43}$$

When b is greater than or equal to b_c, the system does not oscillate. If $b > b_c$, the system is overdamped. The smaller b is, the more rapidly the object returns to equilibrium. If $b = b_c$, the system is said to be critically damped and the object returns to equilibrium (without oscillation) very rapidly. Figure 14-22 shows plots of the displacement versus time for a critically damped and an overdamped oscillator. We often use critical damping when we want a system to avoid oscillations and yet return to equilibrium quickly.

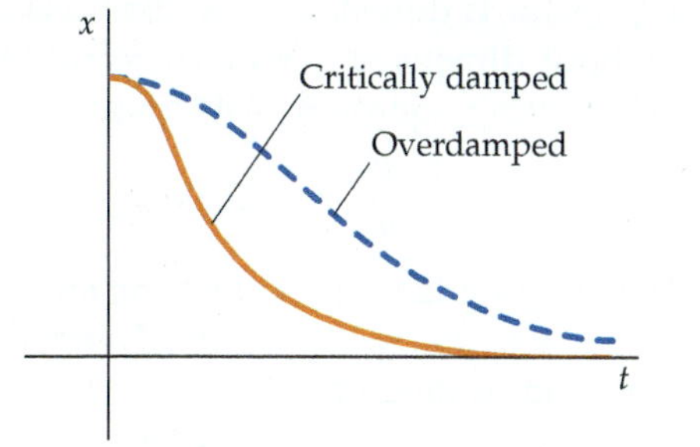

FIGURE 14-22 Plots of displacement versus time for a critically damped and an overdamped oscillator, each released from rest.

Example 14-13 Sprung Mass of a Passenger Car

The sprung mass of an automobile is the mass that is supported by the springs. (It does not include the mass of the wheels, axles, brakes, and so on.) A passenger car has a sprung mass of 1100 kg and an unsprung mass of 250 kg. If the four shock absorbers are removed, the car bounces up and down on its springs with a frequency of 1.0 Hz. What is the damping constant provided by the four shocks if the car, with shocks, is to return to equilibrium as quickly as possible without passing it after hitting a speed bump?

PICTURE Because the car returns to equilibrium as quickly as possible without passing it, we know the car is a critically damped oscillator. Use $b_c = 2m\omega_0$ (Equation 14-43) to solve for the damping constant for critical damping.

SOLVE

1. The damping constant for critical damping is related to the natural frequency by $b_c = 2m\omega_0$ (Equation 14-43): $b_c = 2m\omega_0$
2. With the tires in contact with the pavement, only the inertia of the sprung mass enters the picture: $m = 1100\ \text{kg}$
3. The natural frequency ω_0 is given in the problem statement: $\omega_0 = 1.0\ \text{Hz}$
4. Calculate the damping constant: $b = b_c = 2(1100\ \text{kg})/(1.0\ \text{Hz}) = \boxed{2.2 \times 10^3\ \text{kg/s}}$

CHECK The damping force is given by $\vec{F} = -b\vec{v}$, so bv has SI units of newtons. Our step-4 value for b has units of kg/s, so bv has units of (kg/s)(m/s) = kg·m/s², which are the SI units for mass times acceleration. Thus, kg/s are appropriate units for b.

TAKING IT FURTHER The optimal shock absorber for any vehicle is a shock absorber that has a damping constant such that the oscillations are critically damped. Thus, the optimal choice for the critical damping constant b_c is determined by the sprung mass of the vehicle and the force constant k of the suspension springs.)

Weights are placed in automobile wheels when the wheels are "balanced." The purpose of balancing the wheels is to prevent vibrations that will drive oscillations of the wheel assembly. *(David Wrobel/ Visuals Unlimited.)*

Because the energy of an oscillator is proportional to the square of its amplitude, the energy of an underdamped oscillator (averaged over a cycle) also decreases exponentially with time:

$$E = \tfrac{1}{2}m\omega^2A^2 = \tfrac{1}{2}m\omega^2(A_0e^{-(b/2m)t})^2 = \tfrac{1}{2}m\omega^2A_0^2e^{-(b/m)t} = E_0e^{-t/\tau} \qquad 14\text{-}44$$

where $E_0 = \frac{1}{2}m\omega^2A_0^2$ and $\tau = m/b$.

A damped oscillator is often described by its Q factor (for quality factor),

$$Q = \omega_0\tau \qquad 14\text{-}45$$

DEFINITION—Q FACTOR

The Q factor is dimensionless. (Because ω_0 has dimensions of reciprocal time, $\omega_0\tau$ is without dimension.) We can relate Q to the fractional energy loss per cycle. Differentiating Equation 14-44 gives

$$\frac{dE}{dt} = -(1/\tau)E_0e^{-t/\tau} = -(1/\tau)E \quad \text{or} \quad \frac{dE}{E} = -\frac{dt}{\tau}$$

If the damping is weak so that the energy loss per cycle is a small fraction of the energy E, we can replace dE by ΔE and dt by the period T. Then $|\Delta E|/E$ in one cycle (one period) is given by

$$\left(\frac{|\Delta E|}{E}\right)_{\text{cycle}} = \frac{T}{\tau} = \frac{2\pi}{\omega_0\tau} = \frac{2\pi}{Q} \qquad \text{14-46}$$

so

$$Q = \frac{2\pi}{(|\Delta E|/E)_{\text{cycle}}} \qquad \frac{|\Delta E|}{E} \ll 1 \qquad \text{14-47}$$

PHYSICAL INTERPRETATION OF Q FOR WEAK DAMPING

Q is thus inversely proportional to the fractional energy loss per cycle.

Example 14-14 Making Music

When middle C on a piano (frequency 262 Hz) is struck, it loses half its energy after 4.00 s. (*a*) What is the decay time τ? (*b*) What is the Q factor for this piano wire? (*c*) What is the fractional energy loss per cycle?

PICTURE (*a*) We use $E = E_0e^{-t/\tau}$ and set E equal to $\frac{1}{2}E_0$. (*b*) The Q value can then be found from the decay time and the frequency.

SOLVE

(*a*) 1. Set the energy at time $t = 4.00$ s equal to half the original energy:

$$E = E_0e^{-t/\tau} \quad \text{so} \quad \tfrac{1}{2}E_0 = E_0e^{-(4.00\text{ s}/\tau)}$$
$$\tfrac{1}{2} = e^{-(4.00\text{ s}/\tau)}$$

2. Solve for the time τ by taking the natural log of both sides:

$$\ln\frac{1}{2} = -\frac{4.00\text{ s}}{\tau}$$

$$\text{so} \quad \tau = \frac{4.00\text{ s}}{\ln 2} = 5.771 = \boxed{5.77\text{ s}}$$

(*b*) Calculate Q from τ and ω_0:

$$Q = \omega_0\tau = 2\pi f\tau$$
$$= 2\pi(262\text{ Hz})(5.771\text{ s}) = 9.500 \times 10^3 = \boxed{9.50 \times 10^3}$$

(*c*) The fractional energy loss in a cycle is given by Equation 14-46 and the frequency $f = 1/T$:

$$\left(\frac{|\Delta E|}{E}\right)_{\text{cycle}} = \frac{T}{\tau} = \frac{2\pi}{\omega_0\tau} = \frac{1}{f\tau} = \frac{1}{(262\text{ Hz})(5.771\text{ s})}$$
$$= 6.614 \times 10^{-4} = \boxed{6.61 \times 10^{-4}}$$

CHECK Q can also be calculated from $Q = 2\pi/(\Delta E/E)_{\text{cycle}} = 2\pi/(6.61 \times 10^{-4}) = 9.50 \times 10^3$. Note that the fractional energy loss after 4.00 s is not just the number of cycles (4.00 × 262) times the fractional energy loss per cycle, because the energy decreases exponentially, not linearly.

TAKING IT FURTHER Figure 14-23 shows the relative amplitude A/A_0 versus time and the relative energy E/E_0 versus time for the oscillation of a piano string after middle C is struck. After 4.00 s, the amplitude has decreased to about 0.7 times its initial value, and the energy, which is proportional to the amplitude squared, drops to about half its initial value.

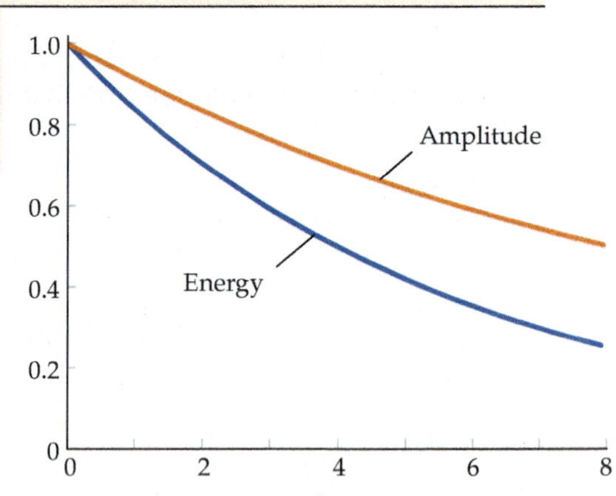

FIGURE 14-23 Plots of A/A_0 and E/E_0 for a struck piano string.

Note that the value of Q in Example 14-4 is relatively large. You can estimate τ and Q for various oscillating systems. Tap a crystal wine glass and see how long it rings. The longer it rings, the greater the value of τ and Q and the lower the damping. Glass beakers from the laboratory may also have a high Q. Try tapping a plastic cup. How does the damping compare to that of the glass beaker?

In terms of Q, the exact frequency of an underdamped oscillator is

$$\omega' = \omega_0\sqrt{1 - \left(\frac{b}{2m\omega_0}\right)^2} = \omega_0\sqrt{1 - \frac{1}{4Q^2}} \qquad \text{14-48}$$

Because b is quite small (and Q is quite large) for a weakly damped oscillator (Example 14-14), we see that ω' is nearly equal to ω_0.

We can understand much of the behavior of a weakly damped oscillator by considering its energy. The power dissipated by the damping force equals the instantaneous rate of change of the total mechanical energy

$$P = \frac{dE}{dt} = \vec{F}_d \cdot \vec{v} = -b\vec{v} \cdot \vec{v} = -bv^2 \qquad \text{14-49}$$

For a weakly damped oscillator with linear damping, the total mechanical energy decreases slowly with time. The average kinetic energy per cycle equals half the total energy

$$\left(\frac{1}{2}mv^2\right)_{\text{av}} = \frac{1}{2}E \quad \text{or} \quad (v^2)_{\text{av}} = \frac{E}{m}$$

If we substitute $(v^2)_{\text{av}} = E/m$ for v^2 in Equation 14-49, we have

$$\frac{dE}{dt} = -bv^2 \approx -b(v^2)_{\text{av}} = -\frac{b}{m}E \qquad \text{14-50}$$

Rearranging Equation 14-50 gives

$$\frac{dE}{E} = -\frac{b}{m}dt$$

which upon integration gives

$$E = E_0 e^{-(b/m)t} = E_0 e^{-t/\tau}$$

which is Equation 14-44.

By pumping the swing, the young woman is transferring her internal energy into the mechanical energy of the oscillator. *(Eye Wire/ Getty.)*

14-5 DRIVEN OSCILLATIONS AND RESONANCE

To keep a damped system going indefinitely, mechanical energy must be put into the system. When this is done, the oscillator is said to be *driven* or *forced*. When Mom (or Dad) kept your swing going by pushing on it once each cycle, she was driving an oscillator. Likewise, when you keep a swing going by "pumping," you are driving an oscillator. If the driving mechanism puts energy into the system at a greater rate than it is dissipated, the system's mechanical energy increases with time, and the amplitude increases. If the driving mechanism puts energy in at the same rate it is being dissipated, the amplitude remains constant over time. The motion of the oscillator is then said to be steady-state motion.

Figure 14-24 shows a system consisting of an object on a spring that is being driven by moving the point of support up and down with simple harmonic motion of frequency ω. At first the motion is complicated, but eventually steady-state motion is reached in which the system oscillates with the same

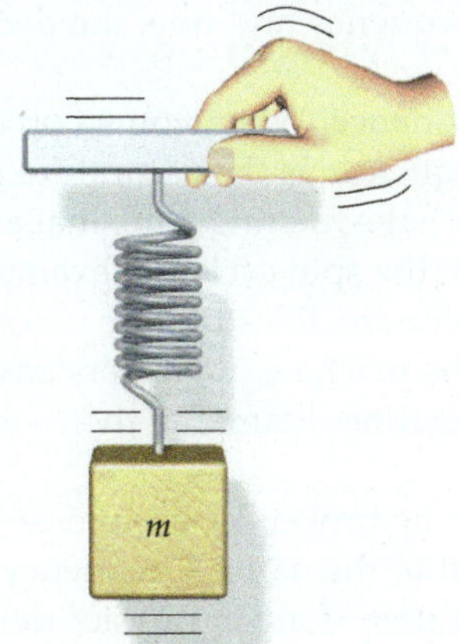

FIGURE 14-24 An object on a vertical spring can be driven by moving the support up and down.

frequency as that of the driver and with a constant amplitude and, therefore, at constant energy. In the steady state, the energy put into the system per cycle by the driving force equals the energy dissipated per cycle due to the damping.

The amplitude, and therefore the energy, of a system in the steady state depends not only on the amplitude of the driving force, but also on its frequency. The **natural frequency** of an oscillator, ω_0, is its frequency when no driving or damping forces are present. (In the case of a spring, for example, $\omega_0 = \sqrt{k/m}$.) If the driving frequency is sufficiently close to the natural frequency of the system, the system will oscillate with a relatively large amplitude. For example, if the support in Figure 14-24 oscillates at a frequency close to the natural frequency of the mass-spring system, the mass will oscillate with a much greater amplitude than it would if the support oscillates at significantly higher or lower frequencies. This phenomenon is called **resonance.** When the driving frequency equals the natural frequency of the oscillator, the energy per cycle transferred to the oscillator is maximum. The natural frequency of the system is thus called the **resonance frequency.** (Mathematically, the angular frequency ω is more convenient to use than the frequency $f(f = \omega/2\pi)$. Because ω and f are proportional, most statements concerning angular frequency also hold for frequency. In verbal descriptions, we usually omit the word angular when the omission will not cause confusion.) Figure 14-25 shows plots of the average power delivered to an oscillator as a function of the driving frequency for two different values of damping. These curves are called **resonance curves.** When the damping is weak (large Q), the width of the peak of the resonance curve is correspondingly narrow, and we speak of the resonance as being sharp. For strong damping, the resonance curve is broad. The width of each resonance curve $\Delta\omega$, indicated in the figure, is the width at half the maximum height. For weak damping, the ratio of the width of the resonance to the resonant frequency can be shown to equal the reciprocal of the Q factor (see Problem 106):

$$\frac{\Delta\omega}{\omega_0} = \frac{1}{Q} \qquad 14\text{-}51$$

RESONANCE WIDTH FOR WEAK DAMPING

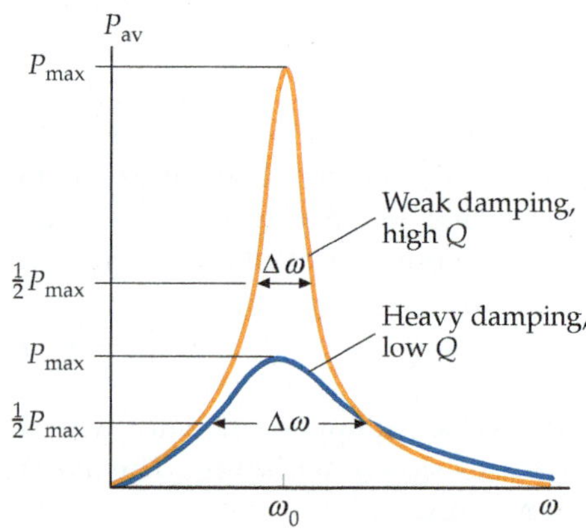

FIGURE 14-25 Resonance for an oscillator. The width $\Delta\omega$ of the resonance peak for a high-Q oscillator (the orange curve) is small compared to the natural frequency of ω_0. The resonance peak of the low-Q oscillator (the blue curve) with the same natural frequency has a width that is considerably larger that that for the high-Q oscillator.

Thus, the Q factor is a direct measure of the sharpness of resonance.

You can do a simple experiment to demonstrate resonance. Hold a meterstick at one end between two fingers so that it acts like a pendulum. (If a meterstick is not available, use whatever is convenient. A golf club works fine.) Release the stick from some initial angular displacement and observe the natural frequency of its motion. Then, move your hand back and forth horizontally, driving it at its natural frequency. Even if the amplitude of the motion of your hand is small, the stick will oscillate with a substantial amplitude. Now move your hand back and forth at a frequency two or three times the natural frequency and note the decrease in amplitude of the oscillating stick.

There are many familiar examples of resonance. When you sit on a swing, you learn intuitively to pump with the same frequency as the natural frequency of the swing. Many machines vibrate because they have rotating parts that are not in perfect balance. (Observe a washing machine in the spin cycle, for example.) If such a machine is attached to a structure that can vibrate, the structure becomes a driven oscillatory system that is set in motion by the machine. Engineers pay great attention to balancing the rotary parts of such machines, damping their vibrations, and isolating them from building supports.

A crystal goblet with weak damping can be broken by an intense sound wave at a frequency equal to or very nearly equal to the natural frequency of vibration of the goblet. The breaking of the goblet is often done in physics demonstrations using an audio oscillator, a loudspeaker and an amplifier.

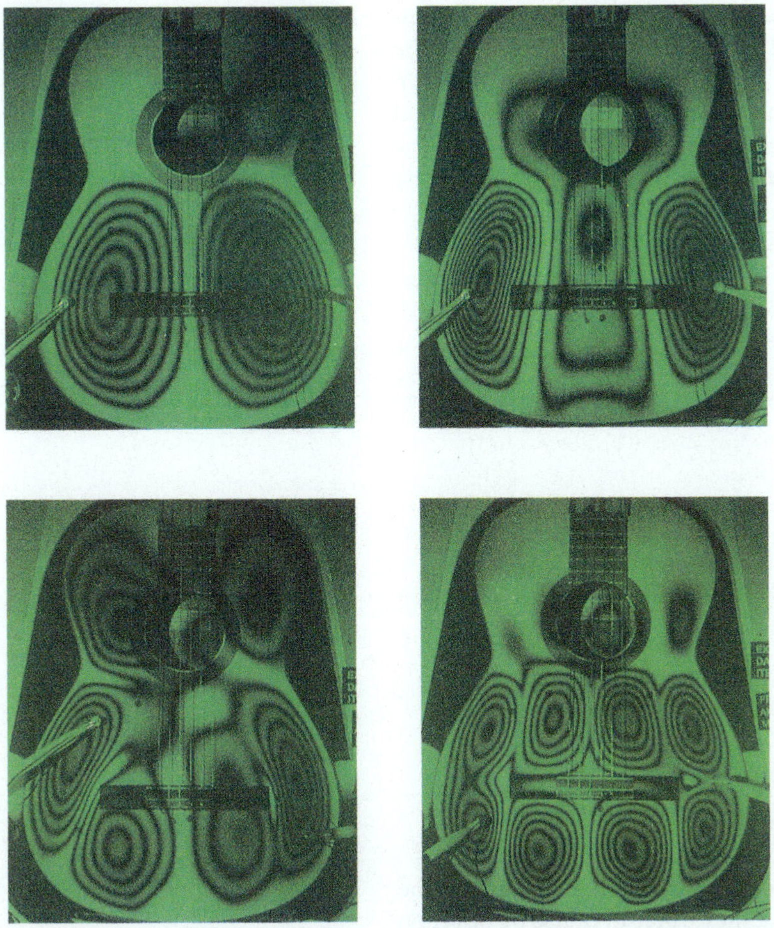

Extended objects have more than one resonance frequency. When plucked, a guitar string transmits its energy to the body of the guitar. The body's oscillations, coupled to those of the air mass it encloses, produce the resonance patterns shown. *(Royal Swedish Academy of Music.)*

* MATHEMATICAL TREATMENT OF RESONANCE

We can treat a driven oscillator mathematically by assuming that, in addition to the restoring force and a damping force, the oscillator is subject to an external driving force that varies harmonically with time:

$$F_{ext} = F_0 \cos \omega t \qquad 14\text{-}52$$

where F_0 and ω are the amplitude and angular frequency of the driving force. This frequency is generally not related to the natural angular frequency of the system ω_0.

Newton's second law applied to an object that has a mass m attached to a spring that has a force constant k and subject to a damping force $-bv_x$ and an external force $f_0 \cos \omega t$ gives

$$\Sigma F_x = ma_x$$

$$-kx - bv_x + F_0 \cos \omega t = m\frac{d^2x}{dt^2}$$

where we have used $a_x = d^2x/dt^2$. Substituting $m\omega_0^2$ for k (Equation 14-8) and rearranging gives

$$m\frac{d^2x}{dt^2} + b\frac{dx}{dt} + m\omega_0^2 x = F_0 \cos \omega t \qquad 14\text{-}53$$

DIFFERENTIAL EQUATION FOR A DRIVEN OSCILLATOR

We now discuss the general solution of Equation 14-53 qualitatively. It consists of two parts, the **transient solution** and the **steady-state solution.** The transient part of the solution is identical to that for a damped oscillator given in Equation 14-39. The constants in this part of the solution depend on the initial conditions. Over time, this part of the solution becomes negligible because of the exponential decrease of the amplitude. We are then left with the steady-state solution, which can be written as

$$x = A\cos(\omega t - \delta) \tag{14-54}$$

POSITION FOR A DRIVEN OSCILLATOR

where the angular frequency ω is the same as that of the driving force. The amplitude A is given by

$$A = \frac{F_0}{\sqrt{m^2(\omega_0^2 - \omega^2)^2 + b^2\omega^2}} \tag{14-55}$$

AMPLITUDE FOR A DRIVEN OSCILLATOR

and the phase constant δ is given by

$$\tan\delta = \frac{b\omega}{m(\omega_0^2 - \omega^2)} \tag{14-56}$$

PHASE CONSTANT FOR A DRIVEN OSCILLATOR

Comparing Equations 14-52 and 14-54, we can see that the displacement and the driving force oscillate with the same frequency, but they differ in phase by δ. When the driving frequency ω approaches zero, δ approaches zero, as can be seen from Equation 14-56. At resonance, ω equals ω_0 and δ equals 90°, and when ω is much greater than ω_0, δ approaches 180°. At the beginning of this chapter, the displacement of a particle undergoing simple harmonic motion is written $x = A\cos(\omega t + \delta)$ (Equation 14-4). This equation is identical to Equation 14-54 except for the sign preceding the phase constant δ. The phase of a driven oscillator always lags behind the phase of the driving force. The negative sign in Equation 14-54 ensures that δ is always positive (rather than always negative).

In your simple experiment to drive a meterstick by moving your hand back and forth (see discussion immediately following Equation 14-51), you should note that at resonance the oscillation of your hand is neither in phase nor 180° out of phase with the oscillation of the stick. If you move your hand back and forth at a frequency several times the natural frequency of the pendulum, the stick's steady-state motion will be almost 180° out of phase with your hand.

The velocity of the object in the steady state is obtained by differentiating x with respect to t:

$$v_x = \frac{dx}{dt} = -\omega A\sin(\omega t - \delta)$$

At resonance, $\delta = \pi/2$, and the velocity is in phase with the driving force:

$$v_x = -\omega A\sin\left(\omega t - \frac{\pi}{2}\right) = +\omega A\cos\omega t$$

Thus, at resonance, the object is always moving in the direction of the driving force, as would be expected for maximum power input. The velocity amplitude ωA is maximum at $\omega = \omega_0$.

! At resonance, the object is always moving in the direction of the driving force, as would be expected for maximum power input.

Example 14-15 An Object on a Spring

Try It Yourself

An object of mass 1.5 kg on a spring that has a force constant equal to 600 N/m loses 3.0% of its energy in each cycle. The same system is driven by a sinusoidal force with a maximum value of $F_0 = 0.50$ N. (*a*) What is Q for this system? (*b*) What is the resonance (angular) frequency? (*c*) If the driving frequency is slowly varied through resonance, what is the width $\Delta\omega$ of the resonance? (*d*) What is the amplitude at resonance? (*e*) What is the amplitude if the driving frequency $\omega = 19$ rad/s?

PICTURE The energy loss per cycle is only 3.0%, so the damping is weak. We can find Q from $Q = 2\pi/(\Delta E/E)_{\text{cycle}}$ (Equation 14-47) and then use this result and $\Delta\omega/\omega_0 = 1/Q$ (Equation 14-51) to find the width $\Delta\omega$ of the resonance. The resonance frequency is the natural frequency. The amplitude both at resonance and off resonance can be found from Equation 14-55, with the damping constant calculated from Q using $Q = \omega_0\tau$ (Equation 14-45) and $\tau = m/b$ (Equation 14-42).

SOLVE

Cover the column to the right and try these on your own before looking at the answers.

Steps	Answers
(*a*) The damping is weak. Relate Q to the fractional energy loss using $Q = 2\pi/(\Delta E/E)_{\text{cycle}}$ (Equation 14-47):	$Q \approx \dfrac{2\pi}{(\lvert\Delta E\rvert/E)_{\text{cycle}}} = \dfrac{2\pi}{0.030} = \boxed{210}$
(*b*) The resonance frequency is the natural frequency of the system:	$\omega_0 = \sqrt{\dfrac{k}{m}} = \boxed{20\text{ rad/s}}$
(*c*) Relate the width of the resonance $\Delta\omega$ to Q using $\Delta\omega/\omega_0 = 1/Q$ (Equation 14-51):	$\Delta\omega = \dfrac{\omega_0}{Q} = \boxed{0.096\text{ rad/s}}$
(*d*) 1. Write an expression for the amplitude A for any driving frequency ω (Equation 14-55):	$A(\omega) = \dfrac{F_0}{\sqrt{m^2(\omega_0^2 - \omega^2)^2 + b^2\omega^2}}$
2. Set ω equal to ω_0 to calculate A at resonance:	$A(\omega_0) = \dfrac{F_0}{b\omega_0}$
3. Use $Q = \omega_0\tau$ (Equation 14-45) and $\tau = m/b$ (Equation 14-42) to relate the damping constant b to Q:	$b = \dfrac{m\omega_0}{Q} = 0.144$ kg/s
4. Use the results of the previous two steps to calculate the amplitude at resonance:	$A(\omega_0) = \dfrac{F_0}{b\omega_0} = \boxed{17\text{ cm}}$
(*e*) Calculate the amplitude for $\omega = 19$ rad/s. (We omit the units to simplify the equation. Because all quantities are in SI units, A will be in meters.)	$A(19) = \dfrac{0.5}{\sqrt{1.5^2(20^2 - 19^2)^2 + 0.144^2(19)^2}} = \boxed{0.85\text{ cm}}$

CHECK At a frequency just 1 rad/s below the 20 rad/s resonance frequency, the amplitude drops by a factor of 20. This is not surprising, because the width $\Delta\omega$ of the resonance is only 0.096 rad/s.

TAKING IT FURTHER Off resonance the term $b^2\omega^2$ is negligible compared with the other term in the denominator of the expression for A. When $\omega - \omega_0$ is more than several times the width $\Delta\omega$, as it was in this example, we can neglect the $b^2\omega^2$ term and calculate A from $A \approx F_0/[m(\omega_0^2 - \omega^2)]$. Figure 14-26 shows the amplitude versus driving frequency ω. Note that the horizontal scale is over a small range of ω.

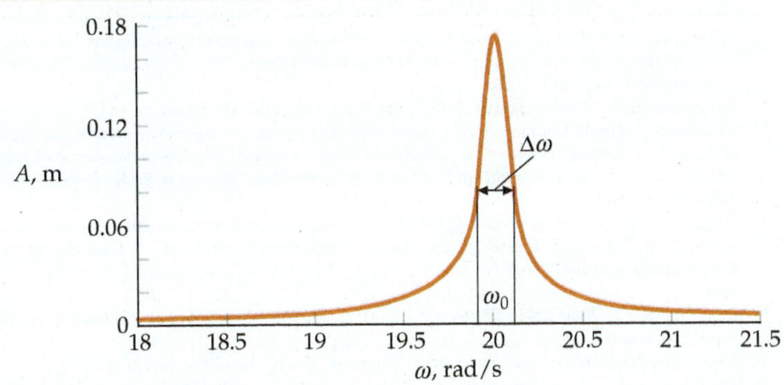

FIGURE 14-26

Physics Spotlight

Moving to the Beat: Millennium Bridge

Massive damped oscillators were attached under the walkway shortly after this suspension bridge opened. The oscillators were put there to prevent the excessive swaying that was driven by lateral forces exerted by the footsteps of the walkers. *(Alamy.)*

The three-span London Millennium footbridge opened in June of 2000. Between 80,000 and 100,000 people crossed the suspension bridge during the course of the day. As the bridge became crowded with up to 2000 people on it at any one time,[*] it began to sway from side to side. Soon the lateral swaying was so strong that many people had to hold onto the handrails.[†] The "Wobbly Bridge"[‡] was closed three days after its opening and did not reopen until February 2002. The footbridge was designed to withstand extremely strong winds, as well as hits from heavy barges on both piers. However, the lateral motion was a shock to the designers and engineers. After several months of study, researchers concluded that walking has a lateral component of force, as well as vertical and forward/backward components.

The typical cadence of a walking person is such that his or her left foot strikes the walkway at approximately one-second intervals. The same is true, of course, for the right foot. When someone steps forward onto his or her left foot, nearly 25 N of force is directed to the left; the force is directed to the right for the right foot.[#] At each step, a left or right lateral force is exerted on the walkway, so the lateral forces shake the walkway with a frequency of 1 Hz. Unfortunately, the two lowest natural frequencies of sideways motion for the 144-meter-long center span were 0.5 Hz and 1.0 Hz,[°] and the 100-meter-long southern span had a natural vibration mode at 0.8 Hz. The footsteps of the crowd drove the motion of the bridge. When the crowd was small, the combined force of the footsteps was not enough to cause motion. But after more than 200 people were on the bridge,[§] the natural damping of the bridge was not high enough to resist the combined force of the crowd's footsteps pushing the bridge sideways.

The swaying increased because of human reaction to the sideways motion. Calculations show that the maximum lateral acceleration was between $0.2g$ and $0.3g$,[¶] enough to cause people to lose their balance. An instinctive method of regaining balance on a moving surface is to walk so that the timing of footsteps matches the motion of the surface. This resonant walking increased the amplitude of the motion.

Measurements were made of test crowds on the bridge, which led to the solution of a series of dampers. Eight tuned-mass dampers and 37 viscous dampers were installed to reduce the lateral swaying. The tuned-mass dampers are 2.5-ton steel blocks suspended on pendulums. They reduce lateral sway by vibrating 180° out of phase with the bridge.[**] The viscous dampers are similar to the shock absorbers used to dampen vertical oscillations in automobiles; they work by moving a piston back and forth through a viscous fluid. The main lateral damping is performed by 37 viscous dampers.[††] Additional mass dampers were installed to dampen any vertical oscillations. During the final tests before reopening, the peak measured accelerations on the bridge dropped by 97%, from $0.25g$ to $0.006g$.[‡‡] The bridge has had no swaying problems since reopening.

Any[##] bridge with a lateral vibration mode below 1.3 Hz is susceptible to oscillation caused by the footsteps of a crowd.[°°] Several different types of bridges had exhibited lateral swaying under pedestrian loads, including a cable-stayed bridge in Japan[§§] and footbridges in Paris and Ottawa. Even highway bridges have shown the same behavior.[¶¶] Because of the London Millennium footbridge, engineers have been motivated to look at vibration in a new manner.

---

[*] Dallard, P., et al., "The London Millennium Footbridge," *The Structural Engineer*, Nov. 20, 2001, Vol. 79, No. 22, 17–33.

[†] Smith, Michael, "Bouncing Bridge May Be Closed 'for Weeks,'" *The Telegraph*, Jun. 13, 2000. https://www.telegraph.co.uk/news/uknews/1342600/Bouncing-bridge-may-be-closed-for-weeks.html as of July 2006.

[‡] Binney, Magnus, "Throwing a Wobbly," *The Times*, Oct. 31, 2000, Features, 16.

[#] "Oscillation," *The Millennium Bridge—Challenge*. Arup Engineering. http://www.arup.com/MillenniumBridge/challenge/oscillation.html as of July 2006.

[°] Fitzpatrick, T., *Linking London: The Millennium Bridge*. London: The Royal Academy of Engineering, June 2001.

[§] Roberts, T. M., "Lateral Pedestrian Excitation of Footbridge," *Journal of Bridge Engineering*, Jan./Feb. 2005, Vol. 10, No. 1, 107–112s.

[¶] Dallard et al., op. cit.

[**] "Elegant, Filigran, and Not Moving," GERB Vibration Control Systems. http://gerb.com/images/both/projektbeispiele/pdf/millenium_bridge_en.pdf as of July 2006.

[††] Taylor, D. P., "Damper Retrofit of the London Millenium Footbridge—A Case Study in Biodynamic Design," Taylor Devices. http://www.taylordevices.com/papers/damper/damper.pdf as of July 2006.

[‡‡] Ibid.

[##] *Structural Safety 2000-2001: Thirteenth Report of SCOSS—The Standing Committee on Structural Safety*. London: Standing Committee on Structural Safety. May 2001, 24–26. http://www.scoss.org.uk/publications/rtf/13Report.pdf as of July 2006.

[°°] "Designing Footbridges with Eurocodes," *Eurocode News*, Mar. 2004, No. 2, 6.

[§§] Nakamura, S.-I., "Model for Lateral Excitation of Footbridges by Synchronous Walking," *Journal of Structural Engineering*, Jan. 2004, 32–37.

[¶¶] Fitzpatrick, op. cit.

Summary

1. Simple harmonic motion occurs whenever the restoring force is proportional to the displacement from equilibrium. It has wide application in the study of oscillations, waves, electrical circuits, and molecular dynamics.
2. Resonance is an important phenomenon in many areas of physics. It occurs when the frequency of the driving force is close to the natural frequency of the oscillating system.

TOPIC	RELEVANT EQUATIONS AND REMARKS	
1. Simple Harmonic Motion	In simple harmonic motion, the acceleration (and thus the net force) is both proportional to, and oppositely directed from, the displacement from the equilibrium position. $F_x = -kx = ma_x$	14-1
Position function	$x = A\cos(\omega t + \delta)$	14-4
Angular frequency	$\omega = 2\pi f = \frac{2\pi}{T}$	14-11
Mechanical energy	$E = K + U = \frac{1}{2}kA^2$	14-17
Circular motion	If a particle moves in a circle with constant speed, the projection of the particle onto a diameter of the circle moves in simple harmonic motion.	
General motion near equilibrium	If an object is given a small displacement from a position of stable equilibrium, it typically oscillates about this position with simple harmonic motion.	
2. Natural Frequencies for Various Systems		
Mass on a spring	$\omega = \sqrt{\frac{k}{m}}$	14-8
Simple pendulum	$\omega = \sqrt{\frac{g}{L}}$	14-27
Torsional oscillator	$\omega = \sqrt{\frac{\kappa}{I}}$ where I is the moment of inertia and κ is the torsional constant. For small oscillations of a physical pendulum, $\kappa = MgD$, where D is the distance of the center of mass from the rotation axis.	14-33
3. Damped Oscillations	In the oscillations of real systems, the motion is damped because of dissipative forces. If the damping is greater than some critical value, the system does not oscillate when disturbed, but merely returns to its equilibrium position. The motion of a weakly damped system is nearly simple harmonic with an amplitude that decreases exponentially with time.	
Frequency	$\omega' = \omega_0\sqrt{1 - \frac{1}{4Q^2}}$	14-48
Energy	$E = E_0 e^{-t/\tau}$	14-44
Amplitude	$A = A_0 e^{-(1/2)t/\tau}$	14-41
Decay time	$\tau = \frac{m}{b}$	14-42
Q factor defined	$Q = \omega_0 \tau$	14-45
Q factor for weak damping	$Q \approx \frac{2\pi}{(\lvert\Delta E\rvert/E)_{\text{cycle}}} \quad \left(\frac{\lvert\Delta E\rvert}{E}\right)_{\text{cycle}} \ll 1$	14-47
4. Driven Oscillations	When an underdamped ($b < b_c$) system is driven by an external sinusoidal force $F_{\text{ext}} = F_0\cos\omega t$, the system oscillates with a frequency ω equal to the driving frequency and an amplitude A that depends on the driving frequency.	

TOPIC	RELEVANT EQUATIONS AND REMARKS	
Resonance frequency	$\omega = \omega_0$	
Resonance width for weak damping	$\frac{\Delta\omega}{\omega_0} = \frac{1}{Q}$	14-51
*Position function	$x = A\cos(\omega t - \delta)$	14-54
*Amplitude	$A = \frac{F_0}{\sqrt{m^2(\omega_0^2 - \omega^2)^2 + b^2\omega^2}}$	14-55
*Phase constant	$\tan\delta = \frac{b\omega}{m(\omega_0^2 - \omega^2)}$	14-56

Answer to Concept Check

14-1 $\sqrt{L/g}$

Answers to Practice Problems

14-1 (*a*) $f = 3.6$ Hz, $T = 0.28$ s, (*b*) $f = 2.5$ Hz, $T = 0.40$ s

14-2 $\omega = 3.1$ rad/s, $v_{max} = 0.13$ m/s

14-3 (*a*) $E = \frac{1}{2}mv_{max}^2 = 0.0625$ J (*b*) $A = \sqrt{2E_{total}/k} = 5.59$ cm

14-4 24 cm

14-5 2.01 s

14-6 $g' = 10.3\ \text{m/s}^2$, $T = 1.96$ s

14-7 $T = \sqrt{\frac{2L}{3g}}$ for $x = L/6$ and for $x = L/2$

Problems

In a few problems, you are given more data than you actually need; in a few other problems, you are required to supply data from your general knowledge, outside sources, or informed estimate.

Interpret as significant all digits in numerical values that have trailing zeros and no decimal points.

- • Single-concept, single-step, relatively easy
- •• Intermediate-level, may require synthesis of concepts
- ••• Challenging
- SSM Solution is in the *Student Solutions Manual*
- Consecutive problems that are shaded are paired problems.

CONCEPTUAL PROBLEMS

1 • True or false:
(*a*) For a simple harmonic oscillator, the period is proportional to the square of the amplitude.
(*b*) For a simple harmonic oscillator, the frequency does not depend on the amplitude.
(*c*) If the net force on a particle undergoing one-dimensional motion is proportional to, and oppositely directed from, the displacement from equilibrium, the motion is simple harmonic.

2 • If the amplitude of a simple harmonic oscillator is tripled, by what factor is the energy changed?

3 •• An object attached to a spring exhibits simple harmonic motion with an amplitude of 4.0 cm. When the object is 2.0 cm from the equilibrium position, what percentage of its total mechanical energy is in the form of potential energy? (*a*) one-quarter, (*b*) one-third, (*c*) one-half, (*d*) two-thirds, (*e*) three-quarters SSM

4 •• An object attached to a spring exhibits simple harmonic motion with an amplitude of 10.0 cm. How far from equilibrium will the object be when the system's potential energy is equal to its kinetic energy? (*a*) 5.00 cm, (*b*) 7.07 cm, (*c*) 9.00 cm, (*d*) The distance cannot be determined from the data given.

5 •• Two identical systems each consist of a spring with one end attached to a block and the other end attached to a wall. The springs are horizontal, and the blocks are supported from below by a frictionless horizontal table. The blocks are oscillating in simple harmonic motions such that the amplitude of the motion of block A is four times as large as the amplitude of the motion of block B. How do their maximum speeds compare? (*a*) $v_{A\,max} = v_{B\,max}$, (*b*) $v_{A\,max} = 2v_{B\,max}$, (*c*) $v_{A\,max} = 4v_{B\,max}$, (*d*) This comparison cannot be done by using the data given.

6 •• Two systems each consist of a spring with one end attached to a block and the other end attached to a wall. The springs are horizontal, and the blocks are supported from below by a frictionless horizontal table. The identical blocks are oscillating in simple harmonic motions with equal amplitudes. However, the force constant of spring A is four times as large as the force constant of spring B. How do their maximum speeds compare? (*a*) $v_{A\,max} = v_{B\,max}$, (*b*) $v_{A\,max} = 2v_{B\,max}$, (*c*) $v_{A\,max} = 4v_{B\,max}$, (*d*) This comparison cannot be done by using the data given.

7 •• Two systems each consist of a spring with one end attached to a block and the other end attached to a wall. The identical springs are horizontal, and the blocks are supported from below by a frictionless horizontal table. The blocks are oscillating in simple harmonic motions with equal amplitudes. However, the mass of block A is four times as large as the mass of block B. How do their maximum speeds compare? (*a*) $v_{A\,max} = v_{B\,max}$, (*b*) $v_{A\,max} = 2v_{B\,max}$, (*c*) $v_{A\,max} = \frac{1}{2}v_{B\,max}$, (*d*) This comparison cannot be done by using the data given. SSM

8 •• Two systems each consist of a spring with one end attached to a block and the other end attached to a wall. The identical springs are horizontal, and the blocks are supported from below by a frictionless horizontal table. The blocks are oscillating in simple harmonic motions with equal amplitudes. However, the mass of block A is four times as large as the mass of block B. How do the magnitudes of their maximum accelerations compare? (*a*) $a_{A\max} = a_{B\max}$, (*b*) $a_{A\max} = 2a_{B\max}$, (*c*) $a_{A\max} = \frac{1}{2}a_{B\max}$, (*d*) $a_{A\max} = \frac{1}{4}a_{B\max}$, (*e*) This comparison cannot be done by using the data given.

9 •• In general physics courses, the mass of the spring in simple harmonic motion is usually neglected because its mass is usually much smaller than the mass of the object attached to it. However, this is not always the case. If you neglect the mass of the spring when it is not negligible, how will your calculation of the system's period, frequency, and total energy compare to the actual values of these parameters? Explain. SSM

10 •• Two mass–spring systems oscillate with periods T_A and T_B. If $T_A = 2T_B$ and the systems' springs have identical force constants, it follows that the systems' masses are related by (*a*) $m_A = 4m_B$, (*b*) $m_A = m_B/\sqrt{2}$, (*c*) $m_A = m_B/2$, (*d*) $m_A = m_B/4$.

11 •• Two mass–spring systems oscillate at frequencies f_A and f_B. If $f_A = 2f_B$ and the systems' springs have identical force constants, it follows that the systems' masses are related by (*a*) $m_A = 4m_B$, (*b*) $m_A = m_B/\sqrt{2}$, (*c*) $m_A = m_B/2$, (*d*) $m_A = m_B/4$.

12 •• Two mass–spring systems *A* and *B* oscillate so that their total mechanical energies are equal. If $m_A = 2m_B$, which expression best relates their amplitudes? (*a*) $A_A = A_B/4$, (*b*) $A_A = A_B/\sqrt{2}$, (*c*) $A_A = A_B$, (*d*) Not enough information is given to determine the ratio of the amplitudes.

13 •• Two mass–spring systems *A* and *B* oscillate so that their total mechanical energies are equal. If the force constant of spring *A* is two times as large as the force constant of spring *B*, then which expression best relates their amplitudes? (*a*) $A_A = A_B/4$, (*b*) $A_A = A_B/\sqrt{2}$, (*c*) $A_A = A_B$, (*d*) Not enough information is given to determine the ratio of the amplitudes. SSM

14 •• The length of the string or wire supporting a pendulum bob increases slightly when the temperature of the string or wire increases. How does this affect a clock operated by a simple pendulum?

15 •• A lamp hanging from the ceiling of the club car in a train oscillates with period T_0 when the train is at rest. The period will be (match left and right columns)

1. greater than T_0 when	A. The train moves horizontally at constant velocity.
2. less than T_0 when	B. The train rounds a curve at constant speed.
3. equal to T_0 when	C. The train climbs a hill at constant speed.
	D. The train goes over the crest of a hill at constant speed.

16 •• Two simple pendulums are related as follows. Pendulum A has a length L_A and a bob of mass m_A; pendulum B has a length L_B and a bob of mass m_B. If the period of A is twice that of B, then (*a*) $L_A = 2L_B$ and $m_A = 2m_B$, (*b*) $L_A = 4L_B$ and $m_A = m_B$, (*c*) $L_A = 4L_B$, whatever the ratio m_A/m_B, (*d*) $L_A = \sqrt{2}L_B$, whatever the ratio m_A/m_B.

17 •• Two simple pendulums are related as follows. Pendulum A has a length L_A and a bob of mass m_A; pendulum B has a length L_B and a bob of mass m_B. If the frequency of A is one-third the frequency of B, then (*a*) $L_A = 3L_B$ and $m_A = 3m_B$, (*b*) $L_A = 9L_B$ and $m_A = m_B$, (*c*) $L_A = 9L_B$, regardless of the ratio m_A/m_B, (*d*) $L_A = \sqrt{3}L_B$ regardless of the ratio m_A/m_B. SSM

18 •• Two simple pendulums are related as follows. Pendulum A has a length L_A and a bob of mass m_A; pendulum B has a length L_B and a bob of mass m_B. They have the same period. If the only difference between their motions is that the amplitude of A's motion is twice the amplitude of B's motion, then (*a*) $L_A = L_B$ and $m_A = m_B$, (*b*) $L_A = 2L_B$ and $m_A = m_B$, (*c*) $L_A = L_B$ whatever the ratio m_A/m_B, (*d*) $L_A = \frac{1}{2}L_B$, whatever the ratio m_A/m_B.

19 •• True or false:

(*a*) The mechanical energy of a damped, undriven oscillator decreases exponentially with time.
(*b*) Resonance for a damped, driven oscillator occurs when the driving frequency exactly equals the natural frequency.
(*c*) If the *Q* factor of a damped oscillator is high, then its resonance curve will be narrow.
(*d*) The decay time τ for a spring–mass oscillator with linear damping is independent of its mass.
(*e*) The *Q* factor for a driven spring–mass oscillator with linear damping is independent of its mass.

20 •• Two damped spring–mass oscillating systems have identical spring and damping constants. However, system A's mass m_A is four times system B's mass m_B. How do their decay times compare? (*a*) $\tau_A = 4\tau_B$, (*b*) $\tau_A = 2\tau_B$, (*c*) $\tau_A = \tau_B$, (*d*) Their decay times cannot be compared, given the information provided.

21 •• Two damped spring–mass oscillating systems have identical spring constants and decay times. However, system A's mass m_A is twice system B's mass m_B. How do their damping constants, *b*, compare? (*a*) $b_A = 4b_B$, (*b*) $b_A = 2b_B$, (*c*) $b_A = b_B$, (*d*) $b_A = \frac{1}{2}b_B$, (*e*) Their decay times cannot be compared, given the information provided.

22 •• Two damped, driven spring–mass oscillating systems have identical driving forces as well as identical spring and damping constants. However, the mass of system A is four times the mass of system B. Assume both systems are very weakly damped. How do their resonant frequencies compare? (*a*) $\omega_A = \omega_B$ (*b*) $\omega_A = 2\omega_B$, (*c*) $\omega_A = \frac{1}{2}\omega_B$, (*d*) $\omega_A = \frac{1}{4}\omega_B$, (*e*) Their resonant frequencies cannot be compared, given the information provided.

23 •• Two damped, driven spring–mass oscillating systems have identical masses, driving forces, and damping constants. However, system A's force constant k_A is four times system B's force constant k_B. Assume they are both very weakly damped. How do their resonant frequencies compare? (*a*) $\omega_A = \omega_B$, (*b*) $\omega_A = 2\omega_B$, (*c*) $\omega_A = \frac{1}{2}\omega_B$, (*d*) $\omega_A = \frac{1}{4}\omega_B$, (*e*) Their resonant frequencies cannot be compared, given the information provided. SSM

24 •• Two damped, driven simple-pendulum systems have identical masses, driving forces, and damping constants. However, system A's length is four times system B's length. Assume they are both very weakly damped. How do their resonant frequencies compare? (*a*) $\omega_A = \omega_B$, (*b*) $\omega_A = 2\omega_B$, (*c*) $\omega_A = \frac{1}{2}\omega_B$, (*d*) $\omega_A = \frac{1}{4}\omega_B$, (*e*) Their resonant frequencies cannot be compared, given the information provided.

ESTIMATION AND APPROXIMATION

25 • Estimate the width of a typical grandfather clock's cabinet relative to the width of the pendulum bob, presuming the desired motion of the pendulum is simple harmonic. SSM

26 • A small punching bag for boxing workouts is approximately the size and weight of a person's head and is suspended from a very short rope or chain. Estimate the natural frequency of oscillations of such a punching bag.

27 •• For a child on a swing, the amplitude drops by a factor of $1/e$ in about eight periods if no additional mechanical energy is given to the system. Estimate the *Q* factor for this system.

28 •• (*a*) Estimate the natural period of oscillation for swinging your arms as you walk, when your hands are empty. (*b*) Now estimate the natural period of oscillation when you are carrying a heavy briefcase. (*c*) Observe other people while they walk. Do your estimates seem reasonable?

SIMPLE HARMONIC MOTION

***Note:* Unless otherwise specified, assume that all objects in this section are in simple harmonic motion.**

29 • The position of a particle is given by $x = (7.0\text{ cm})\cos 6\pi t$, where t is in seconds. What are (*a*) the frequency, (*b*) the period, and (*c*) the amplitude of the particle's motion? (*d*) What is the first time after $t = 0$ that the particle is at its equilibrium position? In what direction is it moving at that time?

30 • What is the phase constant δ in $x = A\cos(\omega t + \delta)$ (Equation 14-4) if the position of the oscillating particle at time $t = 0$ is (*a*) 0, (*b*) $-A$, (*c*) A, and (*d*) $A/2$?

31 • A particle of mass m begins at rest from $x = +25$ cm and oscillates about its equilibrium position at $x = 0$ with a period of 1.5 s. Write expressions for (*a*) the position x as a function of t, (*b*) the velocity v_x as a function of t, and (*c*) the acceleration a_x as a function of t. SSM

32 •• Find (*a*) the maximum speed, and (*b*) the maximum acceleration of the particle in Problem 29. (*c*) What is the first time that the particle is at $x = 0$ and moving to the right?

33 •• Work Problem 31 for when the particle is initially at $x = 25$ cm and moving with velocity $v_0 = +50$ cm/s.

34 •• The period of a particle that is oscillating in simple harmonic motion is 8.0 s and its amplitude is 12 cm. At $t = 0$, it is at its equilibrium position. Find the distance the particle travels during the intervals (*a*) $t = 0$ to $t = 2.0$ s, (*b*) $t = 2.0$ s to $t = 4.0$ s, (*c*) $t = 0$ to $t = 1.0$ s, and (*d*) $t = 1.0$ s to $t = 2.0$ s.

35 •• The period of a particle oscillating in simple harmonic motion is 8.0 s. At $t = 0$, the particle is at rest at $x = A = 10$ cm. (*a*) Sketch x as a function of t. (*b*) Find the distance traveled in the first, second, third, and fourth second after $t = 0$.

36 •• **ENGINEERING APPLICATION, CONTEXT-RICH** Military specifications often call for electronic devices to be able to withstand accelerations of up to $10g$ ($10g = 98.1\text{ m/s}^2$). To make sure that your company's products meet this specification, your manager has told you to use a "shaking table," which can vibrate a device at controlled and adjustable frequencies and amplitudes. If a device is placed on the table and made to oscillate at an amplitude of 1.5 cm, what should you adjust the frequency to in order to test for compliance with the $10g$ military specification?

37 •• The position of a particle is given by $x = 2.5\cos \pi t$, where x is in meters and t is in seconds. (*a*) Find the maximum speed and maximum acceleration of the particle. (*b*) Find the speed and acceleration of the particle when $x = 1.5$ m. SSM

38 ••• (*a*) Show that $A_0\cos(\omega t + \delta)$ can be written as $A_s\sin(\omega t) + A_c\cos(\omega t)$, and determine A_s and A_c in terms of A_0 and δ. (*b*) Relate A_c and A_s to the initial position and velocity of a particle undergoing simple harmonic motion.

SIMPLE HARMONIC MOTION AS RELATED TO CIRCULAR MOTION

39 • A particle moves at a constant speed of 80 cm/s in a circle of radius 40 cm centered at the origin. (*a*) Find the frequency and period of the x component of its position. (*b*) Write an expression for the x component of the particle's position as a function of time t, assuming that the particle is located on the $+y$ axis at time $t = 0$. SSM

40 • A particle moves in a 15-cm-radius circle centered at the origin and completes 1.0 rev every 3.0 s. (*a*) Find the speed of the particle. (*b*) Find its angular speed ω. (*c*) Write an equation for the x component of the particle's position as a function of time t, assuming that the particle is on the $-x$ axis at time $t = 0$.

ENERGY IN SIMPLE HARMONIC MOTION

41 • A 2.4-kg object on a frictionless hoizontal surface is attached to one end of a horizontal spring of force constant $k = 4.5$ kN/m. The other end of the spring is held stationary. The spring is stretched 10 cm from equilibrium and released. Find the system's total mechanical energy.

42 • Find the total energy of a system consisting of a 3.0-kg object on a frictionless horizontal surface oscillating with an amplitude of 10 cm and a frequency of 2.4 Hz at the end of a horizontal spring.

43 • A 1.50-kg object on a frictionless horizontal surface oscillates at the end of a spring (force constant $k = 500$ N/m). The object's maximum speed is 70.0 cm/s. (*a*) What is the system's total mechanical energy? (*b*) What is the amplitude of the motion? SSM

44 • A 3.0-kg object on a frictionless horizontal surface oscillating at the end of a spring that has a force constant equal to 2.0 kN/m has a total mechanical energy of 0.90 J. (*a*) What is the amplitude of the motion? (*b*) What is the maximum speed?

45 • An object on a frictionless horizontal surface oscillates at the end of a spring with an amplitude of 4.5 cm. Its total mechanical energy is 1.4 J. What is the force constant of the spring?

46 •• A 3.0-kg object on a frictionless horizontal surface oscillates at the end of a spring with an amplitude of 8.0 cm. Its maximum acceleration is 3.5 m/s². Find the total mechanical energy.

SIMPLE HARMONIC MOTION AND SPRINGS

47 • A 2.4-kg object on a frictionless horizontal surface is attached to the end of a horizontal spring that has a force constant $k = 4.5$ kN/m. The spring is stretched 10 cm from equilibrium and released. What are (*a*) the frequency of the motion, (*b*) the period, (*c*) the amplitude, (*d*) the maximum speed, and (*e*) the maximum acceleration? (*f*) When does the object first reach its equilibrium position? What is its acceleration at this time?

48 • A 5.00-kg object on a frictionless horizontal surface is attached to one end of a horizontal spring that has a force constant $k = 700$ N/m. The spring is stretched 8.00 cm from equilibrium and released. What are (*a*) the frequency of the motion, (*b*) the period, (*c*) the amplitude, (*d*) the maximum speed, and (*e*) the maximum acceleration? (*f*) When does the object first reach its equilibrium position? What is its acceleration at this time?

49 • A 3.0-kg object on a frictionless horizonal surface is attached to one end of a horizontal spring and oscillates with an amplitude $A = 10$ cm and a frequency $f = 2.4$ Hz. (*a*) What is the force constant of the spring? (*b*) What is the period of the motion? (*c*) What is the maximum speed of the object? (*d*) What is the maximum acceleration of the object? SSM

50 • An 85.0-kg person steps into a car of mass 2400 kg, causing it to sink 2.35 cm on its springs. If started into vertical oscillation, and assuming no damping, at what frequency will the car and passenger vibrate on these springs?

51 • A 4.50-kg object oscillates on a horizontal spring with an amplitude of 3.80 cm. The object's maximum acceleration is 26.0 m/s². Find (*a*) the force constant of the spring, (*b*) the frequency of the object, and (*c*) the period of the motion of the object.

52 •• An object of mass m is suspended from a vertical spring of force constant 1800 N/m. When the object is pulled down 2.50 cm from equilibrium and released from rest, the object oscillates at 5.50 Hz. (*a*) Find m. (*b*) Find the amount the spring is stretched from its unstressed length when the object is in equilibrium. (*c*) Write expressions for the displacement x, the velocity v_x and the acceleration a_x as functions of time t.

53 •• An object is hung on the end of a vertical spring and is released from rest with the spring unstressed. If the object falls 3.42 cm before first coming to rest, find the period of the resulting oscillatory motion.

54 •• A suitcase of mass 20 kg is hung from two bungee cords, as shown in Figure 14-27. Each cord is stretched 5.0 cm when the suitcase is in equilibrium. If the suitcase is pulled down a little and released, what will be its oscillation frequency?

FIGURE 14-27 Problem 54

55 •• A 0.120-kg block is suspended from a spring. When a small pebble of mass 30 g is placed on the block, the spring stretches an additional 5.0 cm. With the pebble on the block, the block oscillates with an amplitude of 12 cm. (*a*) What is the frequency of the motion? (*b*) How long does the block take to travel from its lowest point to its highest point? (*c*) What is the net force on the pebble when it is at the point of maximum upward displacement?

56 •• Referring to Problem 55, find the maximum amplitude of oscillation at which the pebble will remain in contact with the block.

57 •• An object of mass 2.0 kg is attached to the top of a vertical spring that is anchored to the floor. The unstressed length of the spring is 8.0 cm and the length of the spring when the object is in equilibrium is 5.0 cm. When the object is resting at its equilibrium position, it is given a sharp downward blow with a hammer so that its initial speed is 0.30 m/s. (*a*) To what maximum height above the floor does the object eventually rise? (*b*) How long does it take for the object to reach its maximum height for the first time? (*c*) Does the spring ever become unstressed? What minimum initial velocity must be given to the object for the spring to be unstressed at some time?

58 ••• **ENGINEERING APPLICATION** A winch cable has a cross-sectional area of 1.5 cm^2 and a length of 2.5 m. Young's modulus for the cable is 150 GN/m^2. A 950-kg engine block is hung from the end of the cable. (*a*) By what length does the cable stretch? (*b*) If we treat the cable as a simple spring, what is the oscillation frequency of the engine block at the end of the cable?

SIMPLE PENDULUM SYSTEMS

59 • Find the length of a simple pendulum if its frequency for small amplitudes is 0.75 Hz. SSM

60 • Find the length of a simple pendulum if its period for small amplitudes is 5.0 s.

61 • What would the period of the pendulum in Problem 60 be if the pendulum were on the moon, where the acceleration due to gravity is one-sixth that on Earth?

62 • If the period of a 70.0-cm-long simple pendulum is 1.68 s, what is the value of g at the location of the pendulum?

63 • A simple pendulum that is set up in the stairwell of a 10-story building consists of a heavy weight suspended on a 34.0-m-long wire. What is the period of oscillation?

64 •• Show that the total energy of a simple pendulum undergoing oscillations of small amplitude ϕ_0 (in radians) is $E \approx \frac{1}{2}mgL\phi_0^2$. *Hint: Use the approximation* $\cos\phi \approx 1 - \frac{1}{2}\phi^2$ *for small* ϕ.

65 ••• A simple pendulum of length L is attached to a massive cart that slides without friction down a plane inclined at angle θ with the horizontal, as shown in Figure 14-28. Find the period of oscillation for small oscillations of this pendulum. SSM

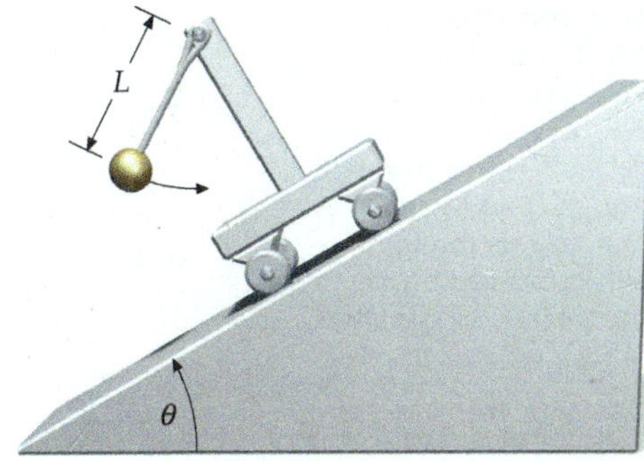

FIGURE 14-28 Problem 65

66 ••• The bob at the end of a simple pendulum of length L is released from rest from an angle ϕ_0. (*a*) Model the pendulum's motion as simple harmonic motion, and find its speed as it passes through $\phi = 0$ by using the small angle approximation. (*b*) Using the conservation of energy, find this speed exactly for any angle (not just small angles). (*c*) Show that your result from Part (*b*) agrees with the approximate answer in Part (*a*) when ϕ_0 is small. (*d*) Find the difference between the approximate and exact results for $\phi_0 = 0.20$ rad and $L = 1.0$ m. (*e*) Find the difference between the approximate and exact results for $\phi_0 = 1.20$ rad and $L = 1.0$ m.

* PHYSICAL PENDULUMS

67 • A thin 5.0-kg uniform disk with a 20-cm radius is free to rotate about a fixed horizontal axis perpendicular to the disk and passing through its rim. The disk is displaced slightly from equilibrium and released. Find the period of the subsequent simple harmonic motion. SSM

68 • A circular hoop that has a 50-cm radius is hung on a narrow horizontal rod and allowed to swing in the plane of the hoop. What is the period of its oscillation, assuming that the amplitude is small?

69 • A 3.0-kg plane figure is suspended at a point 10 cm from its center of mass. When it is oscillating with small amplitude, the period of oscillation is 2.6 s. Find the moment of inertia I about an axis perpendicular to the plane of the figure through the pivot point.

70 •• **ENGINEERING APPLICATION, CONTEXT-RICH, CONCEPTUAL** You have designed a cat door that consists of a square piece of plywood that is 1.0 in. thick and 6.0 in. on a side, and is hinged at its top. To make sure the cat has enough time to get through it safely, the door should have a natural period of at least 1.0 s. Will your design work? If not, explain qualitatively what you would need to do to make it meet your requirements.

71 •• You are given a meterstick and asked to drill a small diameter hole through it so that, when the stick is pivoted about a horizontal axis through the hole, the period of the pendulum will be a minimum. Where should you drill the hole?

72 •• Figure 14-29 shows a uniform disk that has a radius $R = 0.80$ m, a mass of 6.00 kg, and a small hole a distance d from the disk's center that can serve as a pivot point. (*a*) What should be the distance d so that the period of this physical pendulum is 2.50 s? (*b*) What should be the distance d so that this physical pendulum will have the shortest possible period? What is this shortest possible period?

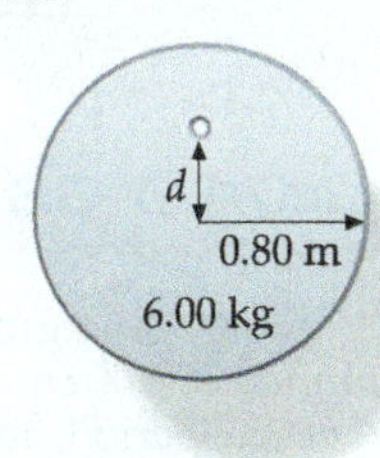

FIGURE 14-29 Problem 72

73 ••• Points P_1 and P_2 on a plane object (Figure 14-30), are distances h_1 and h_2 respectively, from the center of mass. The object oscillates with the same period T when it is free to rotate about an axis through P_1 and when it is free to rotate about an axis through P_2. Both of these axes are perpendicular to the plane of the object. Show that $h_1 + h_2 = gT^2/(4\pi^2)$, where $h_1 \neq h_2$. SSM

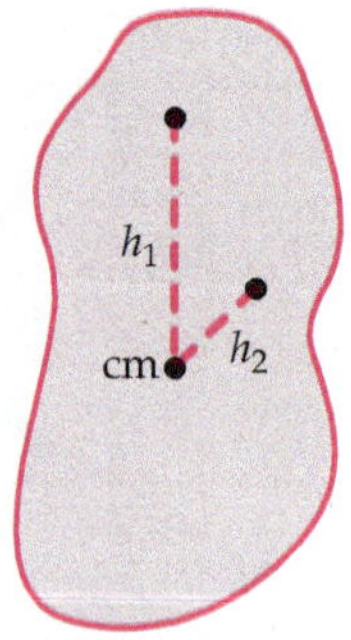

FIGURE 14-30
Problem 73

74 ••• A physical pendulum consists of a spherical bob of radius r and mass m suspended from a rigid rod of negligible mass, as in Figure 14-31. The distance from the center of the sphere to the point of support is L. When r is much less than L, such a pendulum is often treated as a simple pendulum of length L. (*a*) Show that the period for small oscillations is given by $T = T_0\sqrt{1 + (2r^2/5L^2)}$, where $T_0 = 2\pi\sqrt{L/g}$ is the period of a simple pendulum of length L. (*b*) Show that when r is smaller than L, the period can be approximated by $T \approx T_0(1 + r^2/5L^2)$. (*c*) If $L = 1.00$ m and $r = 2.00$ cm, find the error in the calculated value when the approximation $T = T_0$ is used for this period. How large must the radius of the bob be for the error to be 1.00 percent?

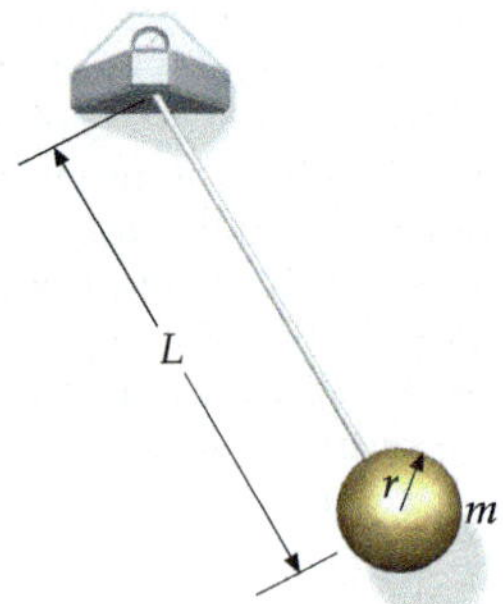

FIGURE 14-31
Problem 74

75 ••• Figure 14-32 shows the pendulum of a clock in your grandmother's house. The uniform rod of length $L = 2.00$ m has a mass $m = 0.800$ kg. Attached to the rod is a uniform disk of mass $M = 1.20$ kg and radius 0.150 m. The clock is constructed to keep perfect time if the period of the pendulum is exactly 3.50 s. (*a*) What should the distance d be so that the period of this pendulum is 2.50 s? (*b*) Suppose that the pendulum clock loses 5.00 min/d. To make sure your grandmother will not be late for her quilting parties, you decide to adjust the clock back to its proper period. How far and in what direction should you move the disk to ensure that the clock will keep perfect time?

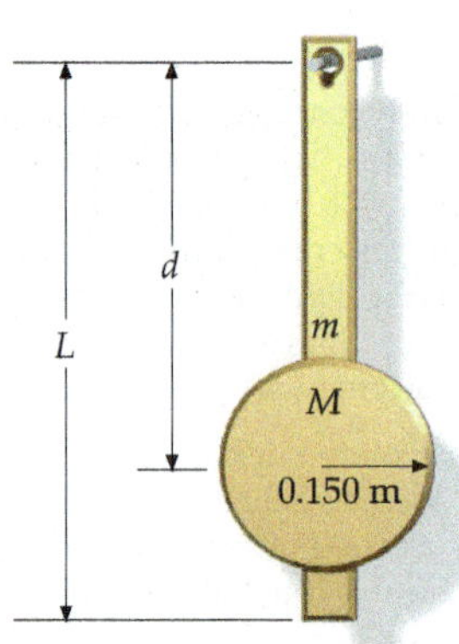

FIGURE 14-32
Problem 75

DAMPED OSCILLATIONS

76 • A 2.00-kg object oscillates on a spring with an initial amplitude of 3.00 cm. The force constant of the spring is 400 N/m. Find (*a*) the period, and (*b*) the total initial energy. (*c*) If the energy decreases by 1.00 percent per period, find the linear damping constant b and the Q factor.

77 •• Show that the ratio of the amplitudes for two successive oscillations is constant for a linearly damped oscillator. SSM

78 •• An oscillator has a period of 3.00 s. Its amplitude decreases by 5.00 percent during each cycle. (*a*) By how much does its mechanical energy decrease during each cycle? (*b*) What is the time constant τ? (*c*) What is the Q factor?

79 •• A linearly damped oscillator has a Q factor of 20. (*a*) By what fraction does the energy decrease during each cycle? (*b*) Use Equation 14-40 to find the percentage difference between ω' and ω_0. *Hint: Use the approximation* $(1 + x)^{1/2} \approx 1 + \frac{1}{2}x$ *for small* x.

80 •• A linearly damped mass-spring system oscillates at 200 Hz. The time constant of the system is 2.0 s. At $t = 0$, the amplitude of oscillation is 6.0 cm and the energy of the oscillating system is 60 J. (*a*) What are the amplitudes of oscillation at $t = 2.0$ s and $t = 4.0$ s? (*b*) How much energy is dissipated in the first 2-s interval and in the second 2-s interval?

81 •• **ENGINEERING APPLICATION** Seismologists and geophysicists have determined that the vibrating Earth has a resonance period of 54 min and a Q factor of about 400. After a large earthquake, Earth will "ring" (continue to vibrate) for up to 2 months. (*a*) Find the percentage of the energy of vibration lost to damping forces during each cycle. (*b*) Show that after n periods the vibrational energy is given by $E_n = (0.984)^n E_0$, where E_0, is the original energy. (*c*) If the original energy of vibration of an earthquake is E_0, what is the energy after 2.0 d? SSM

82 ••• A pendulum that is used in your physics laboratory experiment has a length of 75 cm and a compact bob with a mass equal to 15 g. To start the bob oscillating, you place a fan next to it that blows a horizontal stream of air on the bob. While the fan is on, the bob is in equilibrium when the pendulum is displaced by an angle of 5.0° from the vertical. The speed of the air from the fan is 7.0 m/s. You turn the fan off, and allow the pendulum to oscillate. (*a*) Assuming that the drag force due to the air is of the form $-bv$, predict the decay time constant τ for this pendulum. (*b*) How long will it take for the pendulum's amplitude to reach 1.0°?

83 ••• **ENGINEERING APPLICATION, CONTEXT-RICH** You are in charge of monitoring the viscosity of oils at a manufacturing plant and you determine the viscosity of an oil by using the following method: The viscosity of a fluid can be measured by determining the decay time of oscillations for an oscillator that has known properties and operates while immersed in the fluid. As long as the speed of the oscillator through the fluid is relatively small, so that turbulence is not a factor, the drag force of the fluid on a sphere is proportional to the sphere's speed v relative to the fluid: $F_d = 6\pi a\eta v$, where η is the viscosity of the fluid and a is the sphere's radius. Thus, the constant b is given by $6\pi a\eta$. Suppose your apparatus consists of a stiff spring that has a force constant equal to 350 N/cm and a gold sphere (radius 6.00 cm) hanging on the spring. (*a*) What viscosity of an oil do you measure if the decay time for this system is 2.80 s? (*b*) What is the Q factor for your system? SSM

DRIVEN OSCILLATIONS AND RESONANCE

84 • A linearly damped oscillator loses 2.00 percent of its energy during each cycle. (*a*) What is its Q factor? (*b*) If its resonance frequency is 300 Hz, what is the width of the resonance curve $\Delta\omega$ when the oscillator is driven?

85 • Find the resonance frequency for each of the three systems shown in Figure 14-33.

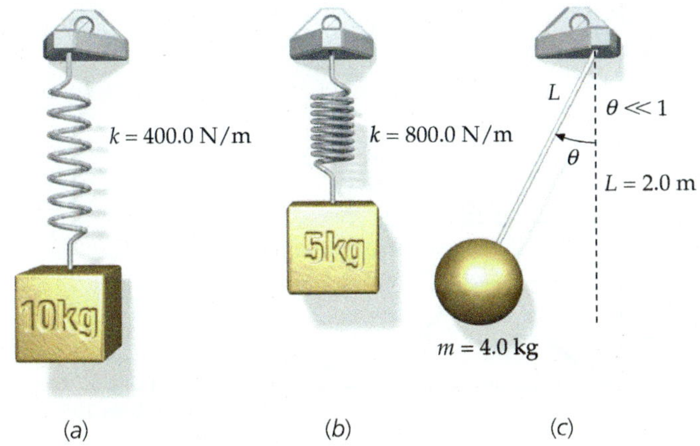

FIGURE 14-33 Problem 85

86 •• A damped oscillator loses 3.50 percent of its energy during each cycle. (*a*) How many cycles elapse before half of its original energy is dissipated? (*b*) What is its Q factor? (*c*) If the natural frequency is 100 Hz, what is the width of the resonance curve when the oscillator is driven by a sinusoidal force?

87 •• A 2.00-kg object oscillates on a spring that has a force constant equal to 400 N/m. The linear damping constant has a value of $b = 2.00$ kg/s. The system is driven by a sinusoidal force of maximum value 10.0 N and angular frequency $\omega = 10.0$ rad/s. (*a*) What is the amplitude of the oscillations? (*b*) If the driving frequency is varied, at what frequency will resonance occur? (*c*) What is the amplitude of oscillation at resonance? (*d*) What is the width of the resonance curve $\Delta\omega$? SSM

88 •• **ENGINEERING APPLICATION, CONTEXT-RICH** Suppose you have the same apparatus that is described in Problem 83 and the same gold sphere hanging from a weaker spring that has a force constant of only 35.0 N/cm. You have studied the viscosity of ethylene glycol with this device, and found that ethylene glycol has a viscosity value of 19.9 mPa · s. Now you decide to drive this system with an external oscillating force. (*a*) If the magnitude of the driving force for the device is 0.110 N and the device is driven at resonance, how large would be the amplitude of the resulting oscillation? (*b*) If the system were not driven, but were allowed to oscillate, what percentage of its energy would it lose per cycle?

GENERAL PROBLEMS

89 • **MULTISTEP** A particle's displacement from equilibrium is given by $x(t) = 0.40 \cos(3.0t + \pi/4)$, where x is in meters and t is in seconds. (*a*) Find the frequency f and period T of its motion. (*b*) Find an expression for the velocity of the particle as a function of time. (*c*) What is its maximum speed?

90 • **ENGINEERING APPLICATION** An astronaut arrives at a new planet, and gets out his simple device to determine the gravitational acceleration there. Prior to his arrival, he noted that the radius of the planet was 7550 km. If his 0.500-m-long simple pendulum has a period of 1.0 s, what is the mass of the planet?

91 •• A pendulum clock keeps perfect time on Earth's surface. In which case will the error be greater: if the clock is placed in a mine of depth h, or if the clock is elevated to a height h? Prove your answer and assume that $h \ll R_E$.

92 •• Figure 14-34 shows a pendulum of length L with a bob of mass M. The bob is attached to a spring that has a force constant k, as shown. When the bob is directly below the pendulum support, the spring is unstressed. (*a*) Derive an expression for the period of this oscillating system for small-amplitude vibrations. (*b*) Suppose that $M = 1.00$ kg and L is such that in the absence of the spring the period is 2.00 s. What is the force constant k if the period of the oscillating system is 1.00 s?

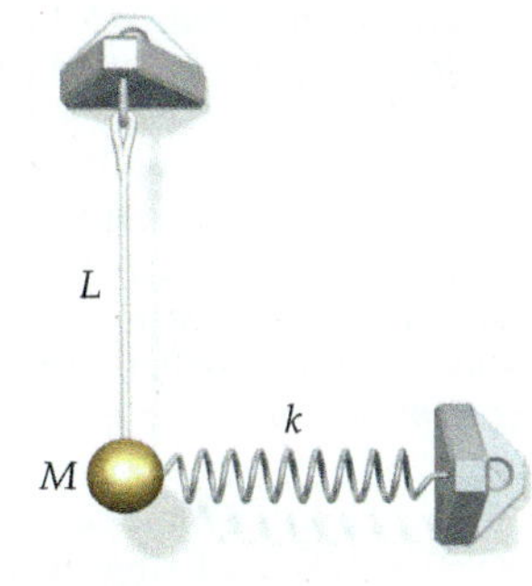

FIGURE 14-34 Problem 92

93 •• A block that has a mass equal to m_1 is supported from below by a frictionless horizontal surface. The block, which is attached to the end of a horizontal spring that has a force constant k, oscillates with an amplitude A. When the spring is at its greatest extension and the block is instantaneously at rest, a second block of mass m_2 is placed on top of it. (*a*) What is the smallest value for the coefficient of static friction μ_s such that the second object does not slip on the first? (*b*) Explain how the total mechanical energy E, the amplitude A, the angular frequency ω, and the period T of the system are affected by the placing of m_2 on m_1, assuming that the coefficient of friction is great enough to prevent slippage. SSM

94 •• A 100-kg box hangs from the ceiling of a room—suspended from a spring with a force constant of 500 N/m. The unstressed length of the spring is 0.500 m. (*a*) Find the equilibrium position of the box. (*b*) An identical spring is stretched and attached to the ceiling and the box, and is parallel with the first spring. Find the frequency of the oscillations when the box is released. (*c*) What is the new equilibrium position of the box once it comes to rest?

95 •• **ENGINEERING APPLICATION** The acceleration due to gravity g varies with geographical location because of Earth's rotation and because Earth is not exactly spherical. This was first discovered in the seventeenth century, when it was noted that a pendulum clock carefully adjusted to keep correct time in Paris lost about 90 s/d near the equator. (*a*) Show by using the differential approximation that a small change in the acceleration of gravity Δg produces a small change in the period ΔT of a pendulum given by $\Delta T/T \approx -\frac{1}{2}\Delta g/g$. (*b*) How large a change in g is needed to account for a 90-s/d change in the period?

96 •• A small block that has a mass equal to m_1 rests on a piston that is vibrating vertically with simple harmonic motion described by the formula $y = A \sin \omega t$. (*a*) Show that the block will leave the piston if $\omega^2 A > g$. (*b*) If $\omega^2 A = 3g$ and $A = 15$ cm, at what time will the block leave the piston?

97 •• Show that for the situations in Figure 14-35*a* and 14-35*b*, the object oscillates with a frequency $f = (1/2\pi)\sqrt{k_{eff}/m}$, where k_{eff} is given by (*a*) $k_{eff} = k_1 + k_2$, and (*b*) $1/k_{eff} = 1/k_1 + 1/k_2$. *Hint: Find the magnitude of the net force F on the object for a small displacement x and write* $F = -k_{eff}x$. *Note that in Part(b) the springs stretch by different amounts, the sum of which is x.* SSM

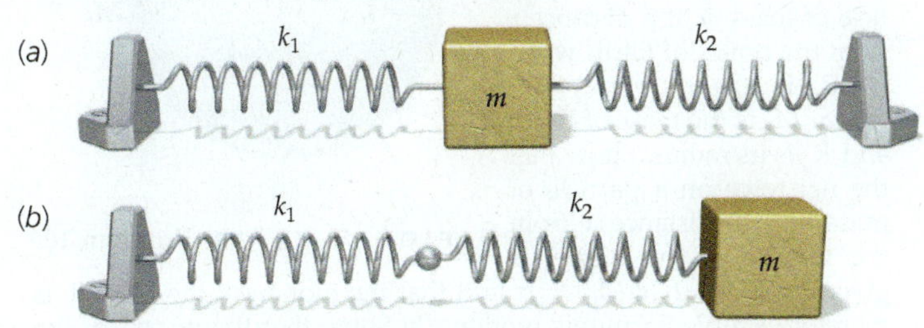

FIGURE 14-35 Problem 97

98 •• **CONTEXT-RICH** During an earthquake, a horizontal floor oscillates horizontally in approximately simple harmonic motion. Assume it oscillates at a single frequency with a period of 0.80 s. (*a*) After the earthquake, you are in charge of examining the video of the floor motion and discover that a box on the floor started to slip

when the amplitude reached 10 cm. From your data, determine the coefficient of static friction between the box and the floor (*b*) If the coefficient of friction between the box and floor were 0.40, what would be the maximum amplitude of vibration before the box would slip?

99 •• If we attach two blocks that have masses m_1 and m_2 to either end of a spring that has a force constant k and set them into oscillation by releasing them from rest with the spring stretched, show that the oscillation frequency is given by $\omega = (k/\mu)^{1/2}$, where $\mu = m_1m_2/(m_1 + m_2)$ is the reduced mass of the system.

100 •• In one of your chemistry labs, you determine that one of the vibrational modes of the HCl molecule has a frequency of 8.969×10^{13} Hz. Using the result of Problem 99, find the "effective spring constant" between the H atom and the Cl atom in the HCl molecule.

101 •• If a hydrogen atom in HCl were replaced by a deuterium atom (forming DCl) in Problem 100, what would be the new vibration frequency of the molecule? Deuterium consists of 1 proton and 1 neutron.

102 ••• **SPREADSHEET** A block of mass m resting on a horizontal table is attached to a spring that has a force constant k, as shown in Figure 14-36. The coefficient of kinetic friction between the block and the table is μ_k. The spring is unstressed if the block is at the origin ($x = 0$), and the $+x$ direction is to the right. The spring is stretched a distance A, where $kA > \mu_k mg$, and the block is released. (*a*) Apply Newton's second law to the block to obtain an equation for its acceleration d^2x/dt^2 for the first half-cycle, during which the block is moving to the left. Show that the resulting equation can be written as $d^2x'/dt^2 = -\omega^2x'$, where $\omega = \sqrt{k/m}$ and $x' = x - x_0$, with $x_0 = \mu_k mg/k = \mu_k g/\omega^2$. (*b*) Repeat Part (*a*) for the second half-cycle as the block moves to the right, and show that $d^2x''/dt^2 = -\omega^2x''$, where $x'' = x + x_0$ and x_0 has the same value. (*c*) Use a **spreadsheet program** to graph the first five half-cycles for $A = 10x_0$. Describe the motion, if any, after the fifth half-cycle.

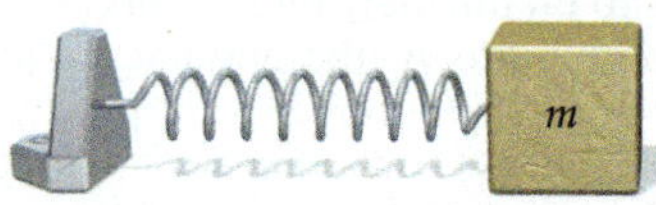

FIGURE 14-36 Problem 102

103 ••• Figure 14-37 shows a uniform solid half-cylinder of mass M and radius R resting on a horizontal surface. If one side of this cylinder is pushed down slightly and then released, the half-cylinder will oscillate about its equilibrium position. Determine the period of this oscillation. SSM

FIGURE 14-37 Problem 103

104 ••• A straight tunnel is dug through Earth, as shown in Figure 14-38. Assume that the walls of the tunnel are frictionless. (*a*) The gravitational force exerted by Earth on a particle of mass m at a distance r from the center of Earth when $r < R_E$ is $F_r = -(GmM_E/R_E^3)r$, where M_E is the mass of Earth and R_E is its radius. Show that the net force on a particle of mass m at a distance x from the middle of the tunnel is given by $F_x = -(GmM_E/R_E^3)x$ and that the motion of the particle is therefore simple harmonic motion. (*b*) Show that the period of the motion is independent of the length of the tunnel and is given by $T = 2\pi\sqrt{R_E/g}$ (c) Find its numerical value in minutes.

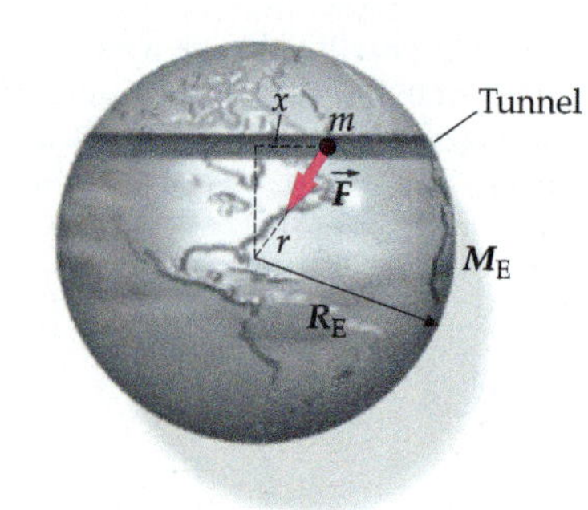

FIGURE 14-38 Problem 104

105 ••• **MULTISTEP** In this problem, derive the expression for the average power delivered by a driving force to a driven oscillator (Figure 14-39).

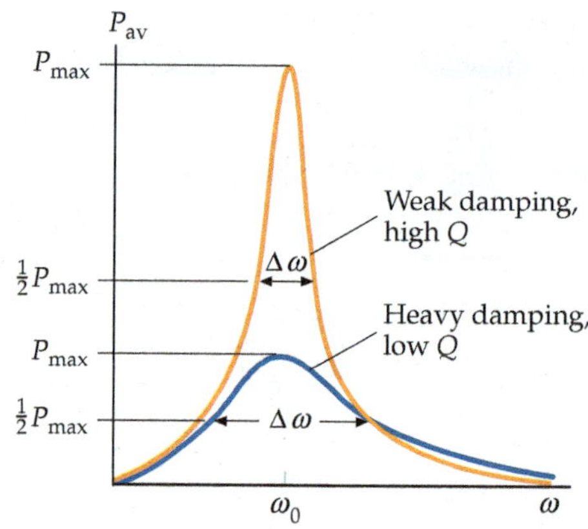

FIGURE 14-39 Problem 105

(*a*) Show that the instantaneous power input of the driving force is given by $P = Fv = -A\omega F_0 \cos \omega t \sin(\omega t - \delta)$.
(*b*) Use the identity $\sin(\theta_1 - \theta_2) = \sin\theta_1 \cos\theta_2 - \cos\theta_1 \sin\theta_2$ to show that the equation in Part (*a*) can be written as $P = A\omega F_0 \sin\delta \cos^2 \omega t \sin - A\omega F_0 \cos\delta \cos\omega t \sin\omega t$.
(*c*) Show that the average value of the second term in your result for Part (*b*) over one or more periods is zero, and that therefore $P_{av} = \frac{1}{2}A\omega F_0 \sin\delta$.
(*d*) From Equation 14-56 for $\tan\delta$, construct a right triangle in which the side opposite the angle δ is $b\omega$ and the side adjacent is $m(\omega_0^2 - \omega^2)$, and use this triangle to show that

$$\sin\delta = \frac{b\omega}{\sqrt{m^2(\omega_0^2 - \omega^2)^2 + b^2\omega^2}} = \frac{b\omega A}{F_0}$$ SSM

(*e*) Use your result for Part (*d*) to eliminate ωA from your result for Part (*c*), so that the average power input can be written as

$$P_{av} = \frac{1}{2}\frac{F_0^2}{b}\sin^2\delta = \frac{1}{2}\left[\frac{b\omega^2F_0^2}{m^2(\omega_0^2 - \omega^2)^2 + b^2\omega^2}\right]$$

106 ••• **MULTISTEP** In this problem, you are to use the result of Problem 105 to derive Equation 14-51. At resonance, the denominator of the fraction in brackets in Problem 105(*e*) is $b^2\omega_0^2$ and P_{av} has its maximum value. For a sharp resonance, the variation in ω in the numerator in this equation can be neglected. Then, the power input will be half its maximum value at the values of ω, for which the denominator is $2b^2\omega_0^2$.
(*a*) Show that ω then satisfies $m^2(\omega - \omega_0)^2(\omega + \omega_0)^2 \approx b^2\omega_0^2$.
(*b*) Using the approximation $\omega + \omega_2 \approx 2\omega_0$, show that $\omega - \omega_0 \approx \pm b/2m$.
(*c*) Express b in terms of Q.
(*d*) Combine the results of Part (*b*) and Part (*c*) to show that there are two values of ω for which the power input is half that at resonance and that they are given by

$$\omega_1 = \omega_0 - \frac{\omega_0}{2Q} \quad \text{and} \quad \omega_2 = \omega_0 - \frac{\omega_0}{2Q}$$

Therefore, $\omega_2 - \omega_1 = \Delta\omega = \omega_0/Q$, which is equivalent to Equation 14-51.

107 ••• **SPREADSHEET** The Morse potential, which is often used to model interatomic forces, can be written in the form $U(r) = D(1 - e^{-\beta(r-r_0)})^2$, where r is the distance between the two atomic nuclei. (*a*) Using a **spreadsheet program** or **graphing calculator,** make a graph of the Morse potential using $D = 5.00$ eV, $\beta = 0.20\ \text{nm}^{-1}$, and $r_0 = 0.750$ nm. (*b*) Determine the equilibrium separation and "force constant" for small displacements from equilibrium for the Morse potential. (*c*) Determine an expression for the oscillation frequency for a *homonuclear* diatomic molecule (that is, two of the same atoms), where the atoms each have mass m.

CHAPTER 15

THE CREW OF AN NOAA (NATIONAL OCEANIC AND ATMOSPHERIC ADMINISTRATION) VESSEL DEPLOYS A DART (DEEP-OCEAN ASSESSMENT AND REPORTING OF TSUNAMIS) BUOY IN THE NORTH PACIFIC. THE DECEMBER 2004 INDIAN OCEAN EARTHQUAKE (ALSO KNOWN AS THE SUMATRA-ANDAMAN EARTHQUAKE) AND THE RESULTING TSUNAMI CAUSED THE LOSS OF HUNDREDS OF THOUSANDS OF LIVES. TSUNAMI DETECTION DEVICES SUCH AS THE DART CAN HELP PREVENT THIS TYPE OF CATASTROPHIC LOSS BY PREDICTING WHEN THE GIANT WAVES WILL HIT LAND. *(Courtesy of NOAA and the Harbor Branch Oceanographic Institution.)*

Traveling Waves

15-1 Simple Wave Motion
15-2 Periodic Waves
15-3 Waves in Three Dimensions
15-4 Waves Encountering Barriers
15-5 The Doppler Effect

? Why do tsunami waves travel so much faster than ocean surface waves? (See Example 15-2.)

In Chapter 14, we looked at oscillatory motion and things that move with repeating patterns. In this chapter, we are still concerned with oscillation, but we explore the physics of waves. Waves travel through various media, such as water, air, and land, and travel through space where there is no medium in which to travel. Think of ocean waves, music, earthquakes, and sunlight. Waves *do* transport energy and momentum, but do *not* transport matter.

The study of wave motion has resulted in many fascinating inventions. Police radar guns and garage door openers both employ electromagnetic waves to achieve very different goals—determining the speed of motorists and opening doors from several meters away. Sonographic equipment, which uses ultrasonic waves, allows medical professionals to obtain remarkable images such as a fetus in its mother's uterus. An understanding of how waves act when confronted with obstacles helps performance-hall architects to create the best acoustic setting for concerts.

In this chapter, we discuss simple wave motion. We examine periodic waves, particularly harmonic waves. We also discuss how waves move in three dimensions and explore what happens when waves encounter obstacles. Finally, we look at the Doppler effect and discuss its relevance to the world around us.

15-1 SIMPLE WAVE MOTION

TRANSVERSE AND LONGITUDINAL WAVES

A mechanical wave is caused by a disturbance in a medium. For example, when a taut string is plucked, the disturbance produced travels along the string as a wave. The disturbance in this case is the change in shape of the string from its equilibrium shape. Its propagation arises from the interaction of each string segment with the adjacent segments. The segments of the string move in the direction transverse to (perpendicular to) the string as the pulses propagate back and forth along the string. Waves such as these, in which the motion of the medium (the string) is perpendicular to the direction of propagation of the disturbance, are called **transverse** waves (Figure 15-1). Waves in which the motion of the medium is along (parallel to) the direction of propagation of the disturbance are called **longitudinal** waves (Figure 15-2). Sound waves are examples of longitudinal waves. When sound travels through a medium (a gas, a liquid, or a solid) the molecules of the medium oscillate (move back and forth) along the line of propagation, alternately compressing and rarefying (expanding) the medium.

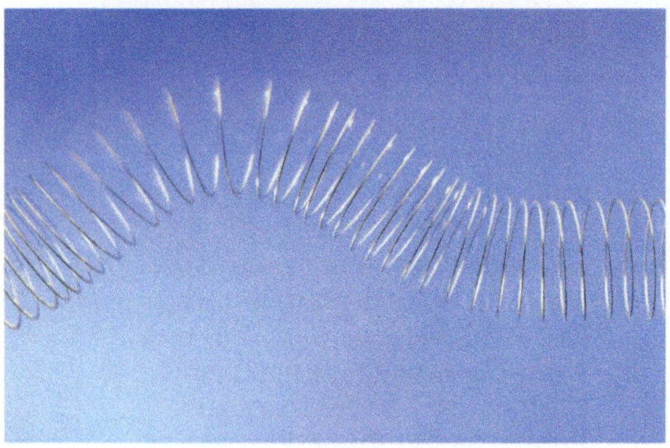

(a)

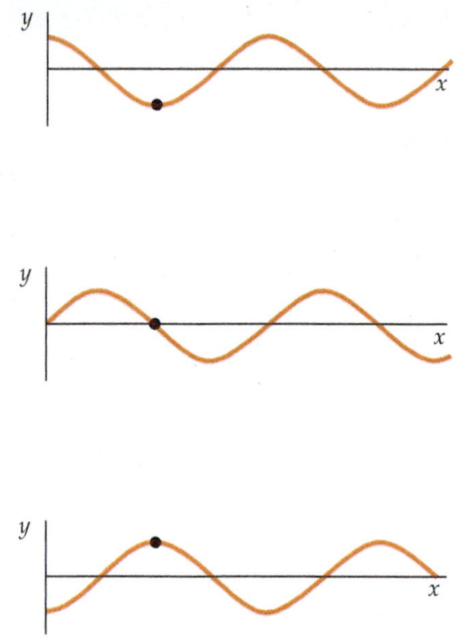

(b)

FIGURE 15-1 (*a*) Transverse wave pulse on a spring. The motion of the propagating medium is perpendicular to the direction of motion disturbance. (*b*) Three successive drawings of a transverse wave on a string traveling to the right. An element of the string (the black dot) moves up and down as the wave crests and troughs travel to the right. *(Richard Menga/Fundamental Photographs.)*

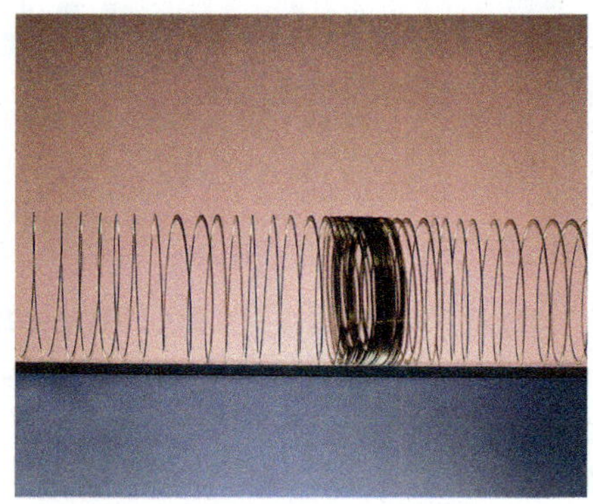

FIGURE 15-2 Longitudinal wave pulse on a spring. The disturbance is parallel with the direction of the motion of the wave. *(Richard Menga/Fundamental Photographs.)*

WAVE PULSES

Figure 15-3*a* shows a pulse on a string at time $t = 0$. The shape of the string at this instant can be represented by some function $y = f(x)$. At some later time (Figure 15-3*b*), the pulse is farther down the string. In a new coordinate system with origin O' that moves to the right with the same speed as the pulse, the pulse is stationary. The string is described in this frame by $f(x')$ for all times. The x coordinates of the two reference frames are related by

$$x' = x - vt$$

so $f(x') = f(x - vt)$. Thus, the shape of the string in the original frame is

$$y = f(x - vt) \qquad \text{wave moving in the } +x \text{ direction} \qquad 15\text{-}1$$

The same line of reasoning for a pulse moving to the left leads to

$$y = f(x + vt) \qquad \text{wave moving in the } -x \text{ direction} \qquad 15\text{-}2$$

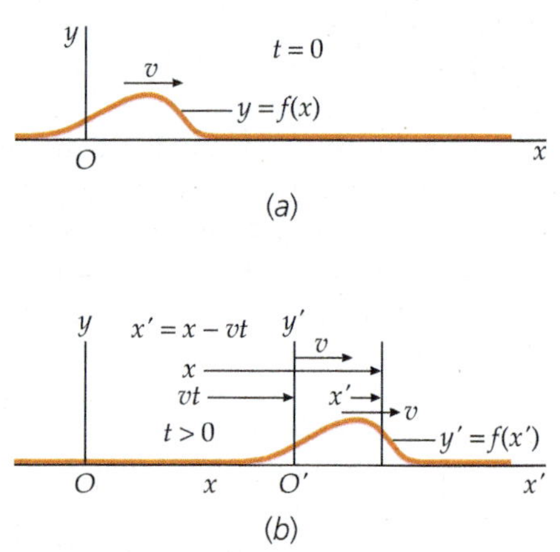

FIGURE 15-3

In both expressions, v is the speed of propagation of the wave. (Because v is a speed and not a velocity, it is always a positive quantity.) The function $y = f(x - vt)$ is called a **wave function.** For waves on a string, the wave function represents the transverse displacement of the string. For sound waves in air, the wave function can be the longitudinal displacement of the air molecules, or the pressure of the air. These wave functions are solutions of a differential equation called the *wave equation,* which can be derived using Newton's laws.

SPEED OF WAVES

A general property of waves is that their speed relative to the medium depends on the properties of the medium, but is independent of the motion of the source of the waves. For example, the speed of a sound from a car horn depends only on the properties of air and not on the motion of the car.

For wave pulses on a rope, we can demonstrate that the greater the tension, the faster the propagation of the waves. Furthermore, waves propagate faster in a light rope than in a heavy rope under the same tension. If F_T is the tension (we use F_T rather than T for tension because we use T for the period) and μ is the linear mass density (mass per unit length), then the wave speed is

$$v = \sqrt{\frac{F_T}{\mu}} \qquad 15\text{-}3$$

SPEED OF WAVES ON A STRING

Example 15-1 Inchy Runs for His Life

Inchy, an inchworm, is inching along a cotton clothesline (Figure 15-4). The 25-m-long clothesline has a mass of 1.0 kg and is kept taut by a hanging object of mass 10 kg, as shown. Vivian is hanging up her swimsuit 5.0 m from one end when she sees Inchy 2.5 cm from the opposite end. She plucks the line sending a terrifying 3.0-cm-high pulse toward Inchy. If Inchy crawls at 1.0 in/s, will he get to the end of the clothesline before the pulse reaches him?

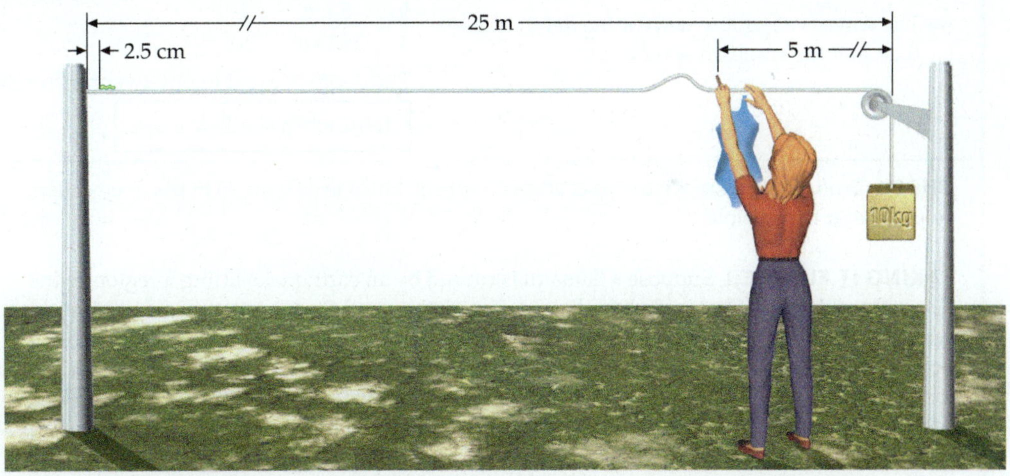

FIGURE 15-4

PICTURE We need to know how fast the wave travels. To find the wave speed we use the formula $v = \sqrt{F_T/\mu}$. Let $m_s = 1.0$ kg be the mass of the string and let $m = 10$ kg be the mass of the hanging object.

SOLVE

1. The speed of the pulse is related to the tension F_T and mass density μ: $\quad v = \sqrt{\frac{F_T}{\mu}}$

2. Express the mass density and tension in terms of the given parameters: $\quad \mu = \frac{m_s}{L}$ and $F_T = mg$

3. Substitute these values to calculate the speed: $\quad v = \sqrt{\frac{F_T}{\mu}} = \sqrt{\frac{mgL}{m_s}} = \sqrt{\frac{(10\text{ kg})(9.81\text{ m/s}^2)(25\text{ m})}{1.0\text{ kg}}} = 49.5\text{ m/s}$

4. Use this speed to find the time for the pulse to travel the 20 m to the far end: $\quad \Delta t = \frac{\Delta x}{v} = \frac{20\text{ m}}{49.5\text{ m/s}} = 0.40\text{ s}$

5. Find the time it takes Inchy to travel the 2.5 cm to the end traveling at 1.0 in/s:

$$\Delta t' = \frac{\Delta x'}{v'} = \frac{2.5\text{ cm}}{1\text{ in/s}} \times \frac{1\text{ in}}{2.54\text{ cm}} = 0.98\text{ s}$$

$\Delta t' > \Delta t$, so Inchy does not beat the pulse.

CHECK The pulse travels at 49 m/s and Inchy travels at 1.0 in/s = 0.025 m/s. The pulse travels almost 2000 times faster than the inchworm. No wonder Inchy does not beat the pulse.

PRACTICE PROBLEM 15-1 Show that the units of $\sqrt{F_T/\mu}$ are m/s when F_T is in newtons and μ is in kg/m.

! While the wave pulse in Example 15-1 moves to the left at 49 m/s, the particles that make up the string do not. Instead, they move first up and then down as the pulse passes by them.

Example 15-2 The Speed of a Shallow Gravity Wave

Surface ocean waves are possible because of gravity and are called *gravity waves.* Gravity waves are called shallow waves if the water depth is less than half a wavelength. The wave speed for gravity waves depends on the depth and is given by $v = \sqrt{gh}$, where h is the depth. A gravity wave in the open ocean, where the depth is 5.0 km, has a wavelength of 100 km. (*a*) What is the wave speed of the wave? (*b*) Is the wave a shallow wave?

PICTURE Use $v = \sqrt{gh}$ to calculate the wave speed. Check to see if the depth is greater than half the specified wavelength.

SOLVE

(*a*) Using $v = \sqrt{gh}$, calculate the wave speed:

$$v = \sqrt{gh} = \sqrt{(9.81\text{ m/s}^2)(5000\text{ m})} = \boxed{221\text{ m/s} = 797\text{ km/h}}$$

(*b*) The wave is a shallow wave if the depth is less than half the specified wavelength:

$$\frac{h}{\lambda} = \frac{5\text{ km}}{100\text{ km}} = \frac{1}{20}$$

The depth is equal to one-twentieth of the wavelength, so the wave is $\boxed{\text{definitely a shallow wave.}}$

CHECK Tsunamis are known to travel at speeds of 800 km/h (~500 mi/h) in the open ocean, so our result is plausible.

TAKING IT FURTHER Suppose a tsunami is caused by an earthquake lifting a region of the ocean floor that is 50 km wide by a height of a meter or so. Such a tsunami would have a wavelength of ~100 km, and the height of the wave would be only a meter or so in the open ocean. Tsunamis travel so fast in the open ocean because they have wavelengths that are longer than the depth of the ocean. Typical ocean waves have wavelengths of a 100 m or less, which is much less then the depth of the open ocean. These waves are deep-water waves, and deep-water waves travel much more slowly than do shallow-water waves. In really shallow water, like the water very near the shore, other factors must be considered when calculating the speed of the waves.

For sound waves in a fluid such as air or water, the speed v is given by

$$v = \sqrt{\frac{B}{\rho}} \qquad \text{15-4}$$

where ρ is the equilibrium density of the medium and B is the bulk modulus* (Equation 13-6). Comparing Equations 15-3 and 15-4, we can see that, in general, the speed of waves depends on an elastic property of the medium (the tension for string waves and the bulk modulus for sound waves) and on an inertial property of the medium (the linear mass density or the volume mass density).

* The bulk modulus is the negative of the ratio of the change in pressure to the fractional change in volume (Chapter 13):

$$B = -\frac{\Delta P}{\Delta V/V}$$

For sound waves in a gas such as air, the bulk modulus* is proportional to the pressure, which in turn is proportional to the density ρ and to the absolute temperature T of the gas. The ratio B/ρ is thus independent of density and is merely proportional to the absolute temperature T. In Chapter 17, we show that, in this case, Equation 15-4 is equivalent to

$$v = \sqrt{\frac{\gamma RT}{M}} \qquad 15\text{-}5$$

SPEED OF SOUND IN A GAS

In this equation, T is the absolute temperature measured in kelvins (K), which is related to the Celsius temperature t_C by

$$T = t_C + 273.15 \qquad 15\text{-}6$$

The dimensionless constant γ depends on the kind of gas. For diatomic molecules, such as O_2 and N_2, γ has the value 7/5. Because O_2 and N_2 comprise 98 percent of the atmosphere, 7/5 is also the γ value for air. (For gases composed of monatomic molecules such as He, γ has the value 5/3.)† The constant R is the universal gas constant

$$R = 8.3145\ \text{J/(mol}\cdot\text{K)} \qquad 15\text{-}7$$

and M is the molar mass of the gas (that is, the mass of one mole of the gas), which for air is

$$M = 29.0 \times 10^{-3}\ \text{kg/mol}$$

Example 15-3 Speed of Sound in Air

Try It Yourself

The spring track season at a school in the Northeast starts in early April when air temperatures are around 13.0°C. By the end of the season, the weather has warmed and temperatures are then around 33.0°C. Calculate the speed of sound as it leaves the starter's pistol in air at (*a*) 13.0°C and (*b*) 33.0°C. Of course, runners should always leave the blocks at the sight of smoke from the pistol, instead of waiting for the sound of the shot to reach them.

PICTURE The speeds at the specified temperatures can be obtained using Equation 15-5, using 7/5 for the value of γ (for a diatomic gas), and using 29.0×10^{-3} kg/mol for M.

SOLVE

Cover the column to the right and try these on your own before looking at the answers.

Steps	Answers
(*a*) 1. Use Equation 15-5 ($v = \sqrt{\gamma RT/M}$) and given values to solve for the speed at 13.0°C. (Be sure to convert the temperature to kelvins.)	$v_a = \sqrt{\dfrac{\gamma RT_a}{M}} = \boxed{339\ \text{m/s}}$
(*b*) 1. From Equation 15-5, we can see that v is proportional to $\sqrt{T}$. Use this proportionality to express the ratio of the speed at 33.0°C to the speed at 13.0°C:	$\dfrac{v_b}{v_a} = \sqrt{\dfrac{T_b}{T_a}}$
2. Calculate v at 33.0°C:	$v_b = \boxed{351\ \text{m/s}}$

CHECK The Part-(*b*) result is larger than the Part-(*a*) result. This is what is expected because the speed of sound increases with increasing temperature.

TAKING IT FURTHER We see from this example that the speed of sound in air is about 343 m/s at 20°C. (This temperature is commonly referred to as room temperature.)

PRACTICE PROBLEM 15-2 For helium, $M = 4.00 \times 10^{-3}$ kg/mol and $\gamma = 5/3$. What is the speed of sound waves in helium gas at 20.0°C?

* The **isothermal bulk modulus,** which describes changes that occur at constant temperature, differs from the **adiabatic bulk modulus,** which describes changes that have no heat transfer. For sound waves at audible frequencies, the changes in the pressure occur too rapidly for appreciable heat transfer, so the appropriate bulk modulus is the adiabatic bulk modulus.

† These values of γ for monatomic and diatomic gases are established in Section 9 of Chapter 18.

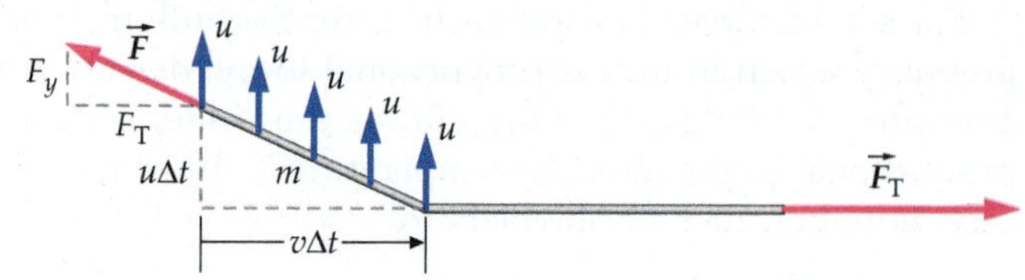

FIGURE 15-5 As the end of the string moves upward at constant speed u, the point where the string changes from horizontal to inclined moves to the right at the wave speed v.

Derivation of v for waves on a string Equation 15-3 ($v = \sqrt{F_T/\mu}$) can be obtained by applying the impulse-momentum theorem to the motion of a string. Suppose you are holding one end of a long taut string with tension F_T and uniform mass per unit length μ. (The other end of the string is attached to a distant wall.) Suddenly, you begin to move your hand upward at a constant speed u. After a short time, the string appears as shown in Figure 15-5, with the rightmost point of the inclined segment of the string moving to the right at the wave speed v and the entire inclined segment moving upward at speed u. By applying the impulse-momentum theorem ($\vec{F}_{av}\,\Delta t = \Delta\vec{p}$) to the string, we obtain

$$F_y\,\Delta t = mu - 0 \qquad \text{15-8}$$

where F_y is the upward component of the force of your hand on the string, m is the mass of the inclined segment, and Δt is the time that your hand has been moving upward. The two triangles in the figure are similar, so

$$\frac{F_y}{F_T} = \frac{u\,\Delta t}{v\,\Delta t} \qquad \text{or} \qquad F_y = \frac{u}{v}F_T$$

Substituting for F_y in Equation 15-8 gives

$$\frac{u}{v}F_T\,\Delta t = (\mu v\,\Delta t)u$$

where $\mu v\,\Delta t$ has been substituted for m. Solving for v gives

$$v = \sqrt{\frac{F_T}{\mu}}$$

which is the expression for the wave speed that is given in Equation 15-3.

In the following discussion we show that this result is true not only for a wave pulse shaped like that shown in Figure 15-5, but for pulses with a wide variety of shapes.

* THE WAVE EQUATION

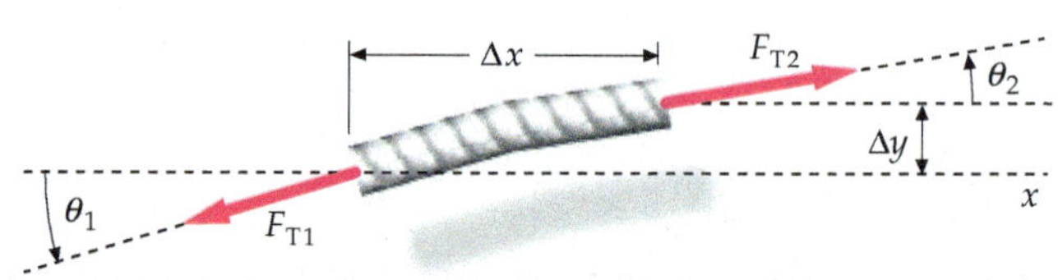

FIGURE 15-6 Segment of a stretched string used for the derivation of the wave equation. The net vertical force on the segment is $F_{T2}\sin\theta_2 - F_{T1}\sin\theta_1$, where F is the tension in the string. The wave equation is derived by applying Newton's second law to the segment.

We can apply Newton's second law to a segment of the string to derive a differential equation known as the wave equation, which relates the spatial derivatives of $y(x,t)$ to its time derivatives. Figure 15-6 shows one segment of a string. We consider only small angles θ_1 and θ_2. Then the length of the segment is approximately Δx and its mass is $m = \mu\,\Delta x$, where μ is the string's mass per unit length. First, we show that, for small vertical displacements, the net horizontal force on a segment is zero and the tension is uniform and constant. The net force in the horizontal direction is zero. That is,

$$\Sigma F_x = F_{T2}\cos\theta_2 - F_{T1}\cos\theta_1 = 0$$

where θ_2 and θ_1 are the angles shown and F_T is the tension in the string. Because the angles are assumed to be small, we may approximate $\cos\theta$ by 1 for each angle. Then, the net horizontal force on the segment can be written

$$\Sigma F_x = F_{T2} - F_{T1} = 0$$

Thus,

$$F_{T2} = F_{T1} = F_T$$

The segment moves vertically, and the net force in this direction is

$$\Sigma F_y = F_T\sin\theta_2 - F_T\sin\theta_1$$

Because the angles are assumed to be small, we may approximate $\sin\theta$ by $\tan\theta$ for each angle. Then the net vertical force on the string segment can be written

$$\sum F_y = F_T(\sin\theta_2 - \sin\theta_1) \approx F_T(\tan\theta_2 - \tan\theta_1)$$

The tangent of the angle made by the string with the horizontal is the slope of the line tangent to the string. The slope S is the first derivative of $y(x,t)$ with respect to x for constant t. A derivative of a function of two variables with respect to one of the variables with the other held constant is called a **partial derivative.** The partial derivative of y with respect to x is written $\partial y/\partial x$. Thus, we have

$$S = \tan\theta = \frac{\partial y}{\partial x}$$

Then

$$\sum F_y = F_T(S_2 - S_1) = F_T\,\Delta S$$

where S_1 and S_2 are the slopes of either end of the string segment and ΔS is the change in the slope. Setting this net force equal to the mass $\mu\,\Delta x$ times the acceleration $\partial^2 y/\partial t^2$ gives

$$F_T\,\Delta S = \mu\Delta x\frac{\partial^2 y}{\partial t^2} \quad \text{or} \quad F_T\frac{\Delta S}{\Delta x} = \mu\frac{\partial^2 y}{\partial t^2} \qquad \text{15-9}$$

In the limit as $\Delta x \rightarrow 0$, we have

$$\lim_{\Delta x \to 0}\frac{\Delta S}{\Delta x} = \frac{\partial S}{\partial x} = \frac{\partial}{\partial x}\frac{\partial y}{\partial x} = \frac{\partial^2 y}{\partial x^2}$$

Thus, in the limit as $\Delta x \rightarrow 0$, Equation 15-9 becomes

$$\frac{\partial^2 y}{\partial x^2} = \frac{\mu}{F_T}\frac{\partial^2 y}{\partial t^2} \qquad \text{15-10}a$$

Equation 15-10*a* is the **wave equation** for a taut string.

We now show that the wave equation is satisfied by any function $x - vt$. Let $\alpha = x - vt$ and consider any wave function

$$y = y(x - vt) = y(\alpha)$$

We use y' for the derivative of y with respect to α. Then, by the chain rule for derivatives,

$$\frac{\partial y}{\partial x} = \frac{dy}{d\alpha}\frac{\partial \alpha}{\partial x} = y'\frac{\partial \alpha}{\partial x} \quad \text{and} \quad \frac{\partial y}{\partial t} = \frac{dy}{d\alpha}\frac{\partial \alpha}{\partial t} = y'\frac{\partial \alpha}{\partial t}$$

Because

$$\frac{\partial \alpha}{\partial x} = \frac{\partial(x - vt)}{\partial x} = 1 \quad \text{and} \quad \frac{\partial \alpha}{\partial t} = \frac{\partial(x - vt)}{\partial t} = -v$$

we have

$$\frac{\partial y}{\partial x} = y' \quad \text{and} \quad \frac{\partial y}{\partial t} = -vy'$$

Taking the second derivatives, we obtain

$$\frac{\partial^2 y}{\partial x^2} = y'' \quad \text{and} \quad \frac{\partial^2 y}{\partial t^2} = -v\frac{\partial y'}{\partial t} = -v\frac{dy'}{d\alpha}\frac{\partial \alpha}{\partial t} = +v^2 y''$$

Thus,

$$\frac{\partial^2 y}{\partial x^2} = \frac{1}{v^2}\frac{\partial^2 y}{\partial t^2} \qquad \text{15-10}b$$

WAVE EQUATION

The same result (Equation 15-10b) can be obtained for any function of $x + vt$ as well. Comparing Equations 15-10a and 15-10b, we see that the speed of propagation of the wave is $v = \sqrt{F_T/\mu}$, which is Equation 15-3.

Example 15-4 Harmonic Wave Function

In the following section, harmonic waves are defined by the wave function $y(x,t) = A\sin(kx - \omega t)$, where $v = \omega/k$. Show that this wave function satisfies Equation 15-10b by explicitly calculating the second derivatives.

PICTURE We can show this by explicitly calculating $\partial^2 y/\partial x^2$ and $\partial^2 y/\partial t^2$, where $y = A\sin(kx - \omega t)$, and substituting into Equation 15-10b.

SOLVE

1. Calculate the second partial derivative of y with respect to x:

$$\frac{\partial y}{\partial x} = \frac{\partial}{\partial x}[A\sin(kx - \omega t)] = A\cos(kx - \omega t)\frac{\partial(kx - \omega t)}{\partial x} = kA\cos(kx - \omega t)$$

$$\frac{\partial^2 y}{\partial x^2} = \frac{\partial}{\partial x}\frac{\partial y}{\partial x} = \frac{\partial}{\partial x}kA\cos(kx - \omega t) = -kA\sin(kx - \omega t)\frac{\partial(kx - \omega t)}{\partial x}$$

$$= -k^2A\sin(kx - \omega t)$$

2. Similarly, calculate the second partial derivative of y with respect to t:

$$\frac{\partial y}{\partial t} = \frac{\partial}{\partial t}[A\sin(kx - \omega t)] = A\cos(kx - \omega t)\frac{\partial(kx - \omega t)}{\partial t} = -\omega A\cos(kx - \omega t)$$

$$\frac{\partial^2 y}{\partial t^2} = \omega A\sin(kx - \omega t)\frac{\partial y(kx - \omega t)}{\partial t} = -\omega^2 A\sin(kx - \omega t)$$

3. Substituting these results in Equation 15-10b gives:

$$-k^2A\sin(kx - \omega t) = \frac{1}{v^2}[-\omega^2 A\sin(kx - \omega t)]$$

$$\text{or}\quad A\sin(kx - \omega t) = \frac{\omega^2/k^2}{v^2}A\sin(kx - \omega t)$$

4. The two sides of the step-3 result are equal to each other, provided $(\omega^2/k^2)/v^2 = 1$:

$A\sin(kx - \omega t)$ is a solution to the wave equation (Equation 15-10b), provided $(\omega^2/k^2)/v^2 = 1$. That is, provided $v = \omega/k$.

CHECK Any function of the form $y(x - vt)$ satisfies the wave equation (Equation 15-10b). The function $y = A\sin(kx - \omega t)$ is of the $y(x - vt)$, provided $v = \omega/k$. To show that this function is in the proper form, we substitute kv for ω to obtain

$$y = A\sin(kx - \omega t) = A\sin(kx - kvt) = A\sin(k[x - vt])$$

which is of the form $y(x - vt)$.

PRACTICE PROBLEM 15-3 Show that any function $y(kx + \omega t)$ satisfies Equation 15-10b, provided $v = \omega/k$.

Derivation of v for sound waves The speed of sound is given by $v = \sqrt{B/\rho}$ (Equation 15-4), where B and ρ are the bulk modulus and density of the medium, respectively. This equation can be obtained by applying the impulse-momentum theorem to the motion of the air in a long cylinder (Figure 15-7) with a piston at one end and with the other end open to the atmosphere. Suddenly, you begin to move the piston to the right at constant speed u. After a short time, Δt, the piston has moved a distance $u\,\Delta t$ and all the air within distance $v\,\Delta t$ from the initial position of the piston is moving to the right with speed u. By applying the impulse-momentum theorem ($\vec{F}_{av}\,\Delta t = \Delta\vec{p}$) to the air in the cylinder we obtain

$$F\,\Delta t = mu - 0 \qquad 15\text{-}11$$

where m is the mass of the air moving with speed u and F is the net force on the air in the cylinder. The air was initially at rest. The net force F is related to the pressure increase ΔP of the air near the moving piston by

$$F = A\,\Delta P$$

where A is the cross-sectional area of the cylinder.

The bulk modulus of the air is given by

$$B = -\frac{\Delta P}{\Delta V/V} \quad \text{so} \quad \Delta P = -B\frac{\Delta V}{V} = -B\frac{-Au\,\Delta t}{Av\,\Delta t} = B\frac{u}{v}$$

where $Au\Delta t$ is the volume swept out by the piston and $Av\Delta t$ is the initial volume of the air that is now moving with speed u. Substituting for F in Equation 15-11 gives

$$A\Delta P\Delta t = mu \quad \text{or} \quad AB\frac{u}{v}\Delta t = (\rho Av\Delta t)u$$

where $\rho Av\,\Delta t$ has been substituted for m. Solving for v gives

$$v = \sqrt{\frac{B}{\rho}}$$

which is the same as the expression for v in Equation 15-4.

A wave equation for sound waves can be derived using Newton's laws. In one dimension, this equation is

$$\frac{\partial^2 s}{\partial x^2} = \frac{1}{v_s^2}\frac{\partial^2 s}{\partial t^2}$$

where s is the displacement of the medium in the x direction and v_s is the speed of sound in the medium.

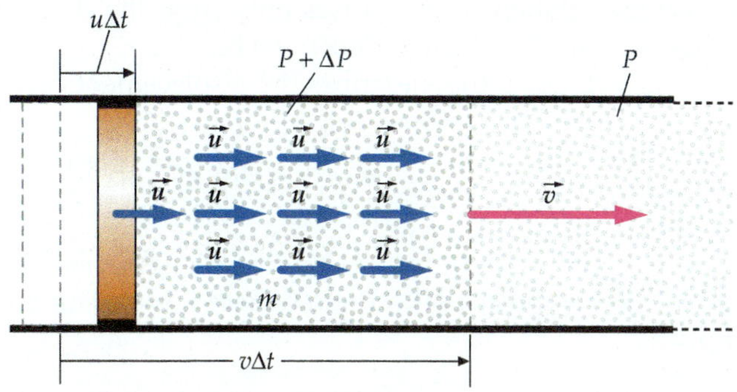

FIGURE 15-7 The air near the piston is moving to the right at the same constant speed u as the piston. The right edge of this pressure pulse moves to the right with the wave speed v. The pressure in the pulse is higher than the pressure in the rest of the cylinder by ΔP.

15-2 PERIODIC WAVES

If one end of a long taut string is shaken back and forth in periodic motion, then a **periodic wave** is generated. If a periodic wave is traveling along a taut string or any other medium, each point along the medium oscillates with the same period.

HARMONIC WAVES

Harmonic waves are the most basic type of periodic waves. All waves, whether they are periodic or not, can be modeled as a superposition of harmonic waves. Consequently, an understanding of harmonic wave motion can be generalized to form an understanding of any type of wave motion. If a **harmonic wave** is traveling through a medium, each point of the medium oscillates in simple harmonic motion.

If one end of a string is attached to a vibrating tuning fork that is moving up and down with simple harmonic motion, a sinusoidal wave train propagates along the string. This wave train is a harmonic wave. As shown in Figure 15-8, the shape of the string is that of a sinusoidal function. The minimum distance after which the wave repeats (the distance between crests, for example) in this figure is called the **wavelength** λ.

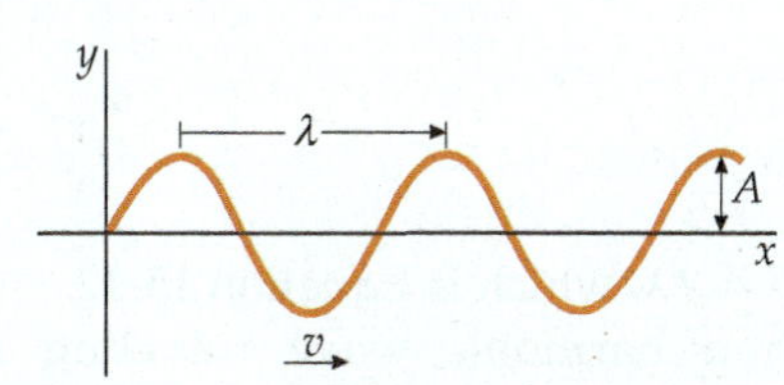

FIGURE 15-8 Harmonic wave at some instant in time. A is the amplitude and λ is the wavelength. For a wave on a string, this figure can be obtained by taking a high-speed photographic snapshot of the string.

As the wave propagates along the string, each point on the string moves up and down—perpendicular to the direction of propagation—in simple harmonic motion with the frequency f of the tuning fork. During one period T of this motion the wave moves a distance of one wavelength, so its speed is given by

$$v = \frac{\lambda}{T} = f\lambda \qquad \text{15-12}$$

where we have used the relation $T = 1/f$.

Because the relation $v = f\lambda$ arises only from the definitions of wavelength and frequency, it applies to all periodic waves.

The sine function that describes the displacements in Figure 15-8 is

$$y(x) = A\sin\left(2\pi\frac{x}{\lambda} + \delta\right)$$

where A is the amplitude, λ is the wavelength, and δ is a phase constant that depends on the choice of the origin (where $x = 0$). This equation is expressed more simply as

$$y(x) = A\sin(kx + \delta) \qquad 15\text{-}13$$

where k, called the **wave number,** is given by

$$k = \frac{2\pi}{\lambda} \qquad 15\text{-}14$$

Note that k has units of m^{-1}. (Because the angle must be in radians, we sometimes write the units of k as rad/m.) When dealing with a single harmonic wave we usually choose the location of the origin so that $\delta = 0$.

For a wave traveling in the direction of increasing x with speed v, replace x in Equation 15-13 with $x - vt$ (see "Wave Pulses" in Section 15-1). With δ equal to zero, this gives

$$y(x,t) = A\sin k(x - vt) = A\sin(kx - kvt)$$

or

$$y(x,t) = A\sin(kx - \omega t) \qquad 15\text{-}15$$

HARMONIC WAVE FUNCTION

where

$$\omega = kv \qquad 15\text{-}16$$

is the angular frequency, and the argument of the sine function, $(kx - \omega t)$, is called the **phase.** The angular frequency is related to the frequency f and period T by

$$\omega = 2\pi f = \frac{2\pi}{T} \qquad 15\text{-}17$$

Substituting $\omega = 2\pi f$ into Equation 15-16 and using $k = 2\pi/\lambda$, we obtain

$$2\pi f = kv = \frac{2\pi}{\lambda}v$$

or $v = f\lambda$, which is Equation 15-12.

If a harmonic wave traveling along a string is described by $y(x,t) = A\sin(kx - \omega t)$, the velocity of a point on the string at a fixed value of x is

$$v_y = \frac{\partial y}{\partial t} = \frac{\partial}{\partial t}[A\sin(kx - \omega t)] = -\omega A\cos(kx - \omega t) \qquad 15\text{-}18$$

TRANSVERSE VELOCITY

The acceleration of this point is given by $\partial^2 y/\partial t^2$.

Example 15-5 A Harmonic Wave on a String

The wave function $y(x,t) = (0.030 \text{ m}) \times \sin[(2.2 \text{ m}^{-1})x - (3.5 \text{ s}^{-1})t]$ is for a harmonic wave on a string. (*a*) In what direction does this wave travel and what is its speed? (*b*) Find the wavelength, frequency, and period of this wave. (*c*) What is the maximum displacement of any point on the string? (*d*) What is the maximum speed of any point on the string?

PICTURE (*a*) To find the direction of travel, express $y(x,t)$ as either a function of $(x - vt)$ or as a function of $(x + vt)$ and use Equations 15-1 and 15-2. To find the wave speed, use $\omega = kv$ (Equation 15-16). (*b*) The wavelength, frequency, and period can be found from the wave number k and the angular frequency ω. (*c*) The maximum displacement of a point on the string is the amplitude A. (*d*) The velocity of a point on the string is $\partial y/\partial t$.

SOLVE

(*a*) 1. The given wave function is of the form $y(x,t) = A\sin(kx - \omega t)$. Using $\omega = kv$ (Equation 15-16), write the wave function as a function of $x - vt$. Then, use Equations 15-1 and 15-2 to find the direction of travel:

$y(x,t) = A\sin(kx - \omega t)$ and $\omega = kv$

so $y(x,t) = A\sin(kx - kvt) = A\sin[k(x - vt)]$

The wave travels in the $\boxed{+x \text{ direction.}}$

2. Because the form is $y = A\sin(kx - \omega t)$, we know A as well as both ω and k. Use these to calculate the speed:

$$v = \frac{\lambda}{T} = \frac{\lambda}{2\pi}\frac{2\pi}{T} = \frac{\omega}{k} = \frac{3.5 \text{ s}^{-1}}{2.2 \text{ m}^{-1}} = 1.59 \text{ m/s}$$
$$= \boxed{1.6 \text{ m/s}}$$

(*b*) The wavelength λ is related to the wave number k, and the period T and frequency f are related to ω:

$$\lambda = \frac{2\pi}{k} = \frac{2\pi}{2.2 \text{ m}^{-1}} = 2.86 \text{ m} = \boxed{2.9 \text{ m}}$$

$$T = \frac{2\pi}{\omega} = \frac{2\pi}{3.5 \text{ s}^{-1}} = 1.80 \text{ s} = \boxed{1.8 \text{ s}}$$

$$f = \frac{1}{T} = \frac{1}{1.80 \text{ s}} = 0.557 \text{ Hz} = \boxed{0.56 \text{ Hz}}$$

(*c*) The maximum displacement of a string segment is the amplitude A:

$$A = \boxed{0.030 \text{ m}}$$

(*d*) 1. Compute $\partial y/\partial t$ to find the velocity of a point on the string:

$$v_y = \frac{\partial y}{\partial t} = (0.030 \text{ m})\frac{\partial[\sin(2.2 \text{ m}^{-1}x - 3.5 \text{ s}^{-1}t)]}{\partial t}$$
$$= (0.030 \text{ m})(-3.5 \text{ s}^{-1})\cos(2.2 \text{ m}^{-1}x - 3.5 \text{ s}^{-1}t)$$
$$= -(0.105 \text{ m/s})\cos(2.2 \text{ m}^{-1}x - 3.5 \text{ s}^{-1}t)$$

2. The maximum transverse speed occurs when the cosine function has the value of ± 1:

$$v_{y,\text{max}} = 0.105 \text{ m/s} = \boxed{0.11 \text{ m/s}}$$

CHECK We have included the units explicitly to show how they work out. They serve as a plausibility check. Often we will omit the units for brevity.

Energy transfer via waves on a string Consider again a string attached to a tuning fork. As the fork vibrates, it transfers energy to the segment of the string attached to it. For example, as the fork moves upward from its equilibrium position it stretches the adjacent string segment slightly—increasing its elastic potential energy. In addition, the fork slows as it moves upward from it equilibrium, so it slows the string segment closest to it. This decreases the kinetic energy of the segment. As a wave moves along the string, energy is transferred from one segment to the next in a similar manner.

Power is the rate of energy transfer. We can calculate the power by considering work done by the force that one segment of the string exerts on a neighboring segment. The rate of work done by this force is the power. Figure 15-9 shows a harmonic wave moving to the right along a string segment. That is, we assume a wave function of the form

$$y(x,t) = A\sin(kx - \omega t) \qquad \text{15-19}$$

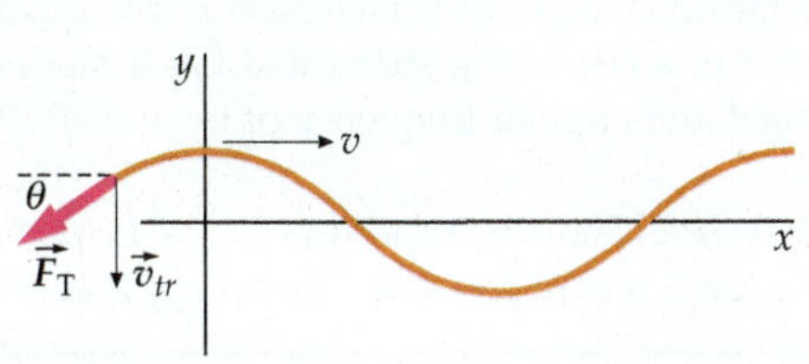

FIGURE 15-9 The tension force $\vec{F}_T$ has a component in the direction of the transverse velocity $\vec{v}_{tr}$, so at this instant the force is doing work on the end of the string that has a positive value.

The tension force $\vec{F}_T$ on the left end of the segment is directed tangent to the string, as shown. To calculate the power transferred by this force, we use the formula $P = \vec{F}_T \cdot \vec{v}_{tr}$ (Equation 6-16), where F_T is the tension and $\vec{v}_{tr}$, the transverse velocity, is the velocity of the end of the segment. To obtain an expression for the power, we first express the vectors in component form. That is, $\vec{F}_T = F_{Tx}\hat{i} + F_{Ty}\hat{j}$ and $\vec{v}_{tr} = v_y\hat{j}$. Taking the scalar product gives $P = F_{Ty}v_y$. We obtain v_y by differentiating Equation 15-18. From the figure, we see that $F_{Ty} = -F_T \sin\theta \approx -F_T \tan\theta$, where we have used the small angle approximation sin $\approx \tan\theta$. Because $\tan\theta$ is the slope of line tangent to the string, we have $\tan\theta = \partial y/\partial x$. Thus

$$P = F_{Ty}v_y \approx -F_T v_y \tan\theta = -F_T \frac{\partial y}{\partial t}\frac{\partial y}{\partial x} \qquad 15\text{-}20$$

Applying Equation 15-20 to a harmonic wave (by taking derivatives of Equation 15-19) gives

$$P = -F_T[-\omega A\cos(kx - \omega t)][kA\cos(kx - \omega t)] = F_T\omega kA^2\cos^2(kx - \omega t)$$

Using $v = \sqrt{F_T/\mu}$ (Equation 15-3) and $v = \omega/k$ (Equation 15-16), we substitute for F_T and the leading k to obtain

$$P = \mu v\omega^2A^2\cos^2(kx - \omega t) \qquad 15\text{-}21$$

where v is the wave speed. The average power at any location x is then

$$P_{av} = \tfrac{1}{2}\mu v\omega^2A^2 \qquad 15\text{-}22$$

because the average value of $\cos^2(kx - \omega t)$ is $\frac{1}{2}$. This average is taken over an entire period T of the motion with x held constant.

The energy travels along a taut string at an average speed equal to the wave speed v, so the average energy $(\Delta E)_{av}$ flowing past point P_1 during time Δt (Figure 15-10*a* and Figure 15-10*b*) is

$$(\Delta E)_{av} = P_{av}\,\Delta t = \tfrac{1}{2}\mu v\omega^2A^2\Delta t$$

This energy is distributed over a length $\Delta x = v\,\Delta t$, so the average energy in length Δx is

$$(\Delta E)_{av} = \tfrac{1}{2}\mu\omega^2A^2\,\Delta x \qquad 15\text{-}23$$

Note that like the average power, the average energy per unit length is proportional to the square of the amplitude of the wave.

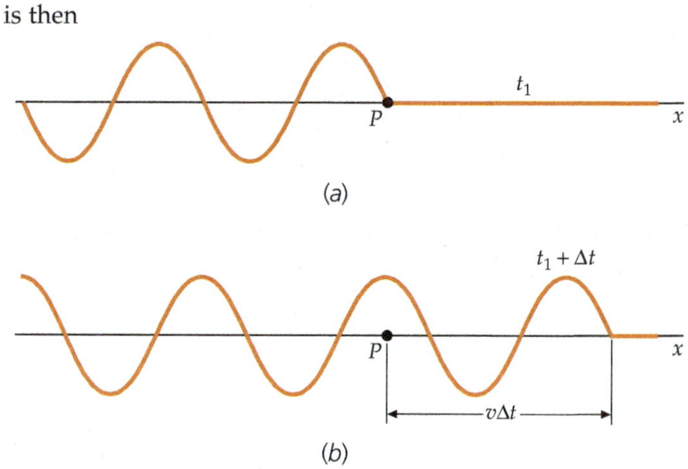

FIGURE 15-10 The wave has reached point P at time t_1. During time Δt, the wave advanced past point P a distance $v\,\Delta t$.

Example 15-6 Average Total Energy of a Wave on a String

A harmonic wave of wavelength 25 cm and amplitude 1.2 cm moves along a 15-m-long segment of a 60-m-long string that has a mass of 320 g and a tension of 12 N. (*a*) What is the speed and angular frequency of the wave? (*b*) What is the average total energy of the wave?

PICTURE The wave speed is $v = \sqrt{F_T/\mu}$, where F_T is given and $\mu = m/L$. We find ω from $\omega = 2\pi f$, where $f = v/\lambda$. The energy is found using $(\Delta E)_{av} = \frac{1}{2}\mu\omega^2A^2\,\Delta x$ (Equation 15-23).

SOLVE

(*a*) 1. The speed is related to the tension and mass density: $v = \sqrt{\dfrac{F_T}{\mu}}$ and $\mu = \dfrac{m}{L}$

2. Calculate the wave speed: $v = \sqrt{\dfrac{F_T L}{m}} = \sqrt{\dfrac{(12\text{ N})(60\text{ m})}{(0.32\text{ kg})}} = 47.4\text{ m/s} = \boxed{47\text{ m/s}}$

3. The angular frequency is found from the frequency, which is found from the speed and wavelength:

$$\omega = 2\pi f \quad \text{and} \quad v = f\lambda,$$

$$\text{so} \quad \omega = 2\pi\frac{v}{\lambda} = 2\pi\frac{47.4\ \text{m/s}}{0.25\ \text{m}} = 1190\ \text{rad/s}$$

$$= \boxed{1200\ \text{rad/s}}$$

(*b*) The average total energy of a harmonic wave on the string is given by $(\Delta E)_{av} = \frac{1}{2}\mu\omega^2 A^2 \Delta x$ (Equation 15-23):

$$(\Delta E)_{av} = \frac{1}{2}\mu\omega^2 A^2\,\Delta x = \frac{1}{2}\frac{m}{L}\omega^2 A^2\,\Delta x$$

$$= \frac{1}{2}\frac{0.32\ \text{kg}}{60\ \text{m}}(1190\ \text{s}^{-1})^2(0.012\ \text{m})^2(15\ \text{m})$$

$$= 8.19\ \text{J} = \boxed{8.2\ \text{J}}$$

CHECK The units for the average energy in the Part-(*b*) result are given by

$$1\frac{\text{kg}\cdot\text{s}^{-2}\,\text{m}^3}{\text{m}} = 1\frac{\text{kg}\cdot\text{m}^2}{\text{s}^2} = 1\ \text{N}\cdot\text{m} = 1\ \text{J}$$

where we have used that $1\ \text{N} = 1\ \text{kg}\cdot\text{m/s}^2$. The units work out, so the Part-(*b*) result is plausible.

PRACTICE PROBLEM 15-4 Calculate the average rate at which energy is transmitted along the string.

HARMONIC SOUND WAVES

Harmonic sound waves can be generated by a tuning fork or loudspeaker that is vibrating with simple harmonic motion. The vibrating source causes the air molecules next to it to oscillate with simple harmonic motion about their equilibrium positions. These molecules collide with neighboring molecules, causing them to oscillate, which in turn collide with their neighboring molecules, causing them to oscillate, and so forth, thereby propagating the sound wave. Equation 15-15 describes a harmonic sound wave if the wave function $y(x,t)$ is replaced by $s(x,t)$, which represents the displacements of the molecules from their equilibrium positions. Thus,

$$s(x,t) = s_0 \sin(kx - \omega t) \qquad 15\text{-}24$$

These displacements are along the direction of propagation of the wave, and lead to variations in the density and pressure of the air. Figure 15-11 shows the displacement of air molecules and the density variations caused by a sound wave at some fixed time. The pressure is maximum where the density is maximum. We see from this figure

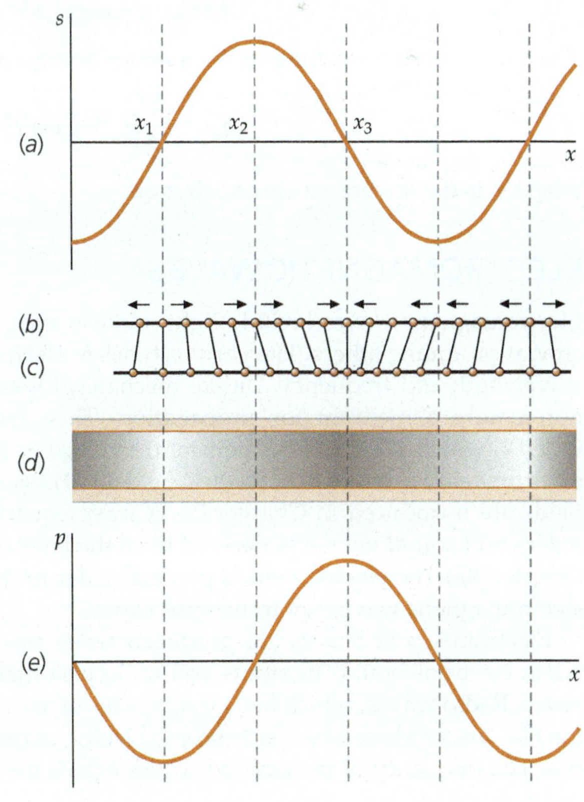

FIGURE 15-11 (*a*) Displacement from equilibrium of air molecules in a harmonic sound wave versus position at some instant. Points x_1 and x_3 are points of zero displacement. (*b*) Some representative molecules equally spaced at their equilibrium positions 1/4 cycle earlier. The arrows indicate the directions of their velocities at that instant. (*c*) Molecules near points x_1, x_2, and x_3 after the sound wave arrives. Just to the left of x_1, the displacement is negative, indicating that the gas molecules are displaced to the left, away from point x_1, at this time. Just to the right of x_1, the displacement is positive, indicating that the molecules are displaced to the right, which is again away from point x_1. So at point x_1, the density is a minimum because the gas molecules on both sides are displaced away from that point. At point x_3, the density is a maximum because the molecules on both sides of that point are displaced toward point x_3. At point x_2, the density does not change because the gas molecules on both sides of that point have equal displacements in the same direction. (*d*) Density of the air at this instant. The density is maximum at x_3 and minimum at x_1, which are both points of zero displacement. It is equal to the equilibrium value at point x_2, which is a maximum in displacement. (*e*) Pressure change, which is proportional to the density change, versus position. The pressure change and displacement (position change) are 90° out of phase.

that the density wave, and thus the pressure wave, is 90° out of phase with the displacement wave. (In the arguments of sine or cosine functions, we will always express phase angles in radians. However, in verbal descriptions, we usually say that "two waves are 90° out of phase" rather than "two waves are out of phase by $\pi/2$ rad.") Where the displacement s is zero, the density, and thus the pressure, is either maximum or minimum, and where the displacement is a maximum or a minimum, the density, and thus the pressure, is equal to its equilibrium value. A displacement wave given by Equation 15-24 thus implies a pressure wave given by

$$p = p_0 \sin\left(kx - \omega t - \frac{\pi}{2}\right) = -p_0 \cos(kx - \omega t) \qquad \text{15-25}$$

where p stands for the pressure minus the local equilibrium pressure, and p_0, the maximum value of p, is called the pressure amplitude. It can be shown that the pressure amplitude p_0 is related to the displacement amplitude s_0 by

$$p_0 = \rho \omega v s_0 \qquad \text{15-26}$$

where v is the speed of propagation and ρ is the equilibrium density of the gas. Thus, as a harmonic sound wave travels through air, the displacement of air molecules, the pressure, and the density all vary sinusoidally with the frequency of the vibrating source.

PRACTICE PROBLEM 15-5

Sound of frequencies from about 20 Hz to about 20,000 Hz are audible to humans (although many people have rather limited hearing above 15,000 Hz). If the speed of sound in air is 343 m/s, what are the wavelengths that correspond to the highest and lowest audible frequencies?

Energy of sound waves The average energy of a harmonic sound wave in a volume element ΔV is given by Equation 15-23 with A replaced by s_0 and $\mu\,\Delta x$, replaced by $\rho\,\Delta V$, where ρ is the equilibrium density of the medium.

$$(\Delta E)_{\text{av}} = \tfrac{1}{2}\rho\omega^2 s_0^2\,\Delta V \qquad \text{15-27}$$

The energy per unit volume is the average energy density η_{av}:

$$\eta_{\text{av}} = \frac{\Delta E_{\text{av}}}{\Delta V} = \frac{1}{2}\rho\omega^2 s_0^2 \qquad \text{15-28}$$

where η is the lowercase Greek letter eta.

ELECTROMAGNETIC WAVES

Electromagnetic waves include light, radio waves, X rays, gamma rays, and microwaves, among others. The various types of electromagnetic waves differ only in wavelength and frequency. Unlike mechanical waves, electromagnetic waves do not require a medium for propagation. They travel through a vacuum with speed c, which is a universal constant, $c \approx 3.00 \times 10^8$ m/s. The wave function for electromagnetic waves is an electric field $\vec{E}(x,t)$ associated with the wave. (Electric fields are introduced in Chapter 21. A wave equation, similar to those for string waves and sound waves, is derived from the laws of electricity and magnetism in Chapter 30.) The electric field is perpendicular to the direction of propagation, so electromagnetic waves are transverse waves.

Electromagnetic waves are produced when free electric charges accelerate or when electrons bound to atoms and molecules make transitions to lower energy states. Radio waves, which have frequencies of about 1 MHz for AM and 100 MHz for FM, are produced by macroscopic electric currents oscillating in radio antennas. The frequency of the emitted waves equals the frequency of oscillation of the charges. Light waves, which have frequencies of the order of 10^{14} Hz, are generally produced by atomic or molecular transitions involving bound electrons. The spectrum of electromagnetic waves is discussed in Chapter 31.

(David Sacks/ The Image Bank/ Getty.)

15-3 WAVES IN THREE DIMENSIONS

Figure 15-12 shows two-dimensional circular waves on the surface of water in a ripple tank. These waves are generated by drops of water striking the surface. The wave crests form concentric circles called **wavefronts.** For a point source of sound, the waves move out in three dimensions, and the wavefronts are concentric spherical surfaces.

FIGURE 15-12 Circular wavefronts diverging from a point source in a ripple tank. *(PhotoDisc/Getty Images.)*

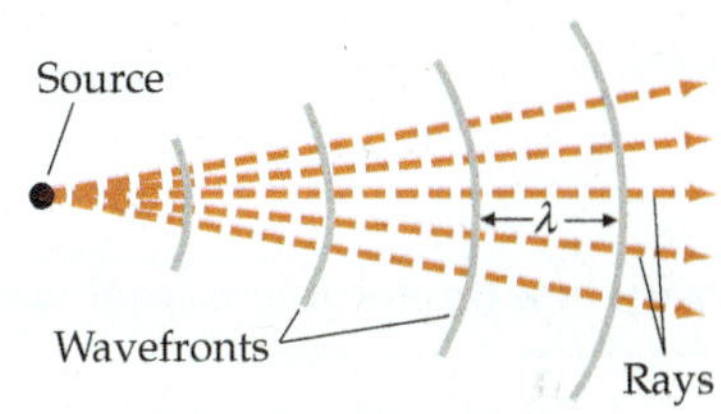

FIGURE 15-13 The motion of wavefronts can be represented by rays drawn perpendicular to the wavefronts. For a point source, the rays are radial lines diverging from the source.

The motion of any set of wavefronts can be indicated by rays, which are directed lines perpendicular to the wavefronts (Figure 15-13). For circular or spherical waves, the rays are radial lines.

In a homogeneous medium, such as air at constant density, the wavefronts travels in straight lines in the direction of the rays, much like a beam of particles. At a great distance from a point source, a sufficiently small section of the wavefront can be approximated by a flat surface (a plane), and the rays are approximately parallel lines; such a wave is called a **plane wave** (Figure 15-14). The two-dimensional analog of a plane wave is a *line wave,* which is a small part of a circular wavefront at a great distance from the source. Line waves can also be produced in a ripple tank by a line source, as in Figure 15-15.

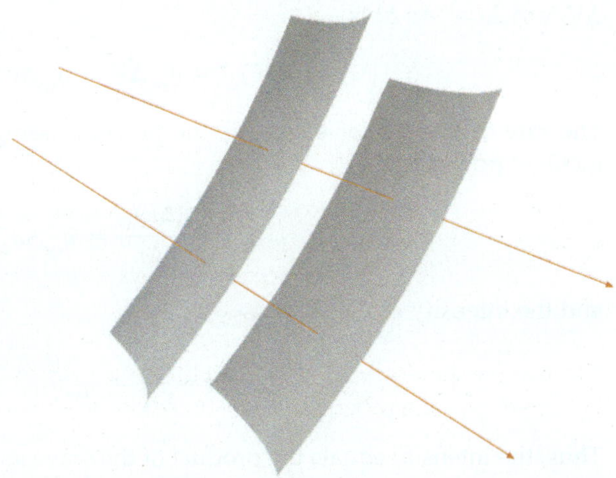

FIGURE 15-14 Plane waves. At great distances from a point source, the wavefronts are approximately parallel planes, and the rays are approximately parallel lines perpendicular to the wavefronts.

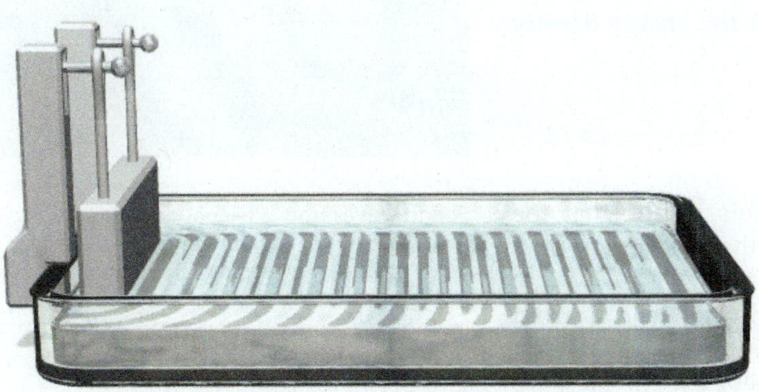

FIGURE 15-15 A two-dimensional analog of a plane wave can be generated in a ripple tank by a flat board that oscillates up and down in the water to produce the wavefronts, which are straight lines.

WAVE INTENSITY

If a point source emits waves uniformly in all directions, then the energy at a distance r from the source is distributed uniformly on a spherical surface of radius r and area $A = 4\pi r^2$. If P_{av} is the average power emitted by the source, then the average power per unit area at a distance r from the source is $P_{av}/(4\pi r^2)$. The average power per unit area that is incident perpendicular to the direction of propagation is called the **intensity:**

$$I = \frac{P_{av}}{A} \tag{15-29}$$

INTENSITY DEFINED

The SI units of intensity are watts per square meter (W/m²). At a distance r from a point source, the intensity is

$$I = \frac{P_{av}}{4\pi r^2} \tag{15-30}$$

INTENSITY DUE TO A POINT SOURCE

The intensity of a three-dimensional wave varies inversely with the square of the distance from a point source.

There is a simple relation between the intensity of a wave and the energy density in the medium through which it propagates. Figure 15-16 shows a spherical wave that has just reached the radius r_1. The volume inside the radius r_1 contains energy because the particles in that region are oscillating. The region outside r_1 contains no energy because the wave has not yet reached it. After a short time Δt, the wave moves out a short distance $\Delta r = v\,\Delta t$ past r_1. The average energy in the spherical shell of surface area A, thickness $v\,\Delta t$, and volume $\Delta V = A\,\Delta r = Av\,\Delta t$ is

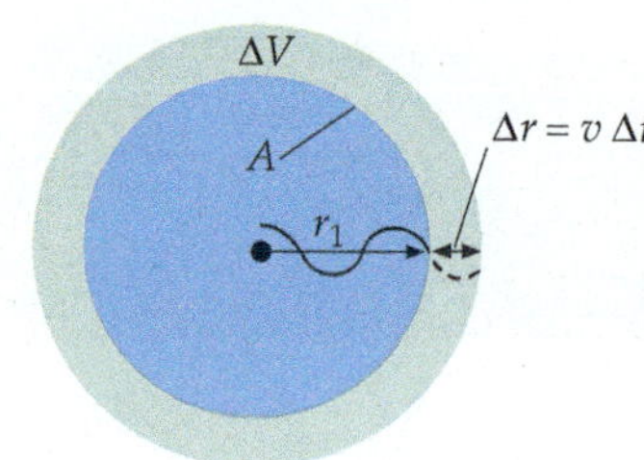

Volume of shell = $\Delta V = A\,\Delta r = Av\,\Delta t$

FIGURE 15-16

$$(\Delta E)_{av} = \eta_{av}\,\Delta V = \eta_{av} Av\,\Delta t$$

The rate of transfer of energy is the power passing into the shell. The average incident power is

$$P_{av} = \frac{(\Delta E)_{av}}{\Delta t} = \eta_{av} Av$$

and the intensity of the wave is

$$I = \frac{P_{av}}{A} = \eta_{av} v \tag{15-31}$$

Thus, the intensity equals the product of the wave speed v and the average energy density η_{av}. Substituting $\eta_{av} = \frac{1}{2}\rho\omega^2 s_0^2$ from Equation 15-28 for the energy density in a harmonic sound wave, we obtain

$$I = \eta_{av} v = \frac{1}{2}\rho\omega^2 s_0^2 v = \frac{1}{2}\frac{p_0^2}{\rho v} \tag{15-32}$$

Sound waves from a telephone handset spreading out in the air. The waves have been made visible by sweeping out the space in front of the handset with a light source whose brightness is controlled by a microphone. *(From Winston E. Kock, Lasers and Holography, 1978, Dover Publications, New York.)*

where we have used $s_0 = p_0/(\rho\omega v)$ from Equation 15-26. This result—that the intensity of a sound wave is proportional to the square of the amplitude—is a general property of harmonic waves.

The human ear can accommodate a large range of sound-wave intensities, from about 10^{-12} W/m² (which is usually taken to be the threshold of hearing) to about 1 W/m² (an intensity great enough to stimulate pain in most people). The pressure amplitudes that correspond to these extreme intensities are about 3×10^{-5} Pa for the hearing threshold and 30 Pa for the pain threshold. (Recall that a pascal is a newton per square meter.) These very small pressure variations add to or subtract from the normal atmospheric pressure of about 101.3 kPa.

Example 15-7 A Loudspeaker

A loudspeaker diaphragm 30 cm in diameter is vibrating at 1.0 kHz with an amplitude of 0.020 mm. Assuming that the air molecules in the vicinity have the same amplitude of vibration, find (*a*) the pressure amplitude immediately in front of the diaphragm, (*b*) the sound intensity immediately in front of the diaphragm, and (*c*) the acoustic power being radiated. (*d*) If the sound is radiated uniformly into the forward hemisphere, find the intensity at 5.0 m from the loudspeaker.

PICTURE (*a*) and (*b*) The pressure amplitude is calculated directly from $p_0 = \rho\omega v s_0$ (Equation 15-26), and the intensity from $I = \frac{1}{2}\rho\omega^2 s_0^2 v$ (Equation 15-32). (*c*) The power radiated is the intensity times the area of the diaphragm. (*d*) The area of a hemisphere of radius r is $2\pi r^2$. We can use Equation 15-29 with $A = 2\pi r^2$.

SOLVE

(*a*) Equation 15-26 relates the pressure amplitude to the displacement amplitude, frequency, wave velocity, and air density:

$$p_0 = \rho\omega v s_0 = (1.29\ \text{kg/m}^3)2\pi(10^3\ \text{Hz})(343\ \text{m/s})(2.0 \times 10^{-5}\ \text{m})$$
$$= 55.6\ \text{N/m}^2 = \boxed{56\ \text{Pa}}$$

(*b*) Equation 15-32 relates the intensity to these same known quantities:

$$I = \tfrac{1}{2}\rho\omega^2 s_0^2 v = \tfrac{1}{2}(1.29\ \text{kg/m}^3)[2\pi(1.0\ \text{kHz})]^2(2.0 \times 10^{-5}\ \text{m})^2(343\ \text{m/s})$$
$$= 3.494\ \text{W/m}^2 = \boxed{3.5\ \text{W/m}^2}$$

(c) The power is the intensity times the area of the diaphragm:

$$P_{\text{av}} = IA = (3.494\ \text{W/m}^2)\pi(0.15\ \text{m})^2 = 0.247\ \text{W} = \boxed{0.25\ \text{W}}$$

(*d*) Calculate the intensity at $r = 5.0$ m, assuming uniform radiation into the forward hemisphere:

$$I = \frac{P_{\text{av}}}{A} = \frac{0.247\ \text{W}}{2\pi(5.0\ \text{m})^2} = 1.57 \times 10^{-3}\ \text{W/m}^2 = \boxed{1.6\ \text{mW/m}^2}$$

CHECK The Part-(*d*) result is smaller than the Part-(*b*) result, as expected. (We expect the intensity to be greatest immediately in front of the diaphragm.)

TAKING IT FURTHER The assumption of uniform radiation in the forward hemisphere is not a very good one because the wavelength in this case [$\lambda = v/f = (343\ \text{m/s})/(1000\ \text{s}^{-1}) = 34.3$ cm] is not large compared with the speaker diameter. There is also some radiation in the backward direction, as can be observed if you stand behind a loudspeaker.

Loudspeakers at a rock concert may put out more than 100 times as much power as the speaker in this example.

* ***Intensity level and loudness*** Our perception of loudness is not proportional to the intensity. However, our perception of loudness varies logarithmically with intensity to a good approximation. We therefore use a logarithmic scale to describe the **intensity level** β of a sound wave, which is measured in **decibels** (dB) and defined by

$$\beta = (10\ \text{dB}) \log \frac{I}{I_0} \qquad 15\text{-}33$$

DEFINITION—INTENSITY LEVEL IN dB

See **Math Tutorial** *for more information on* **Exponents and Logarithms**

where *log* refers to a base-10 logarithm. The decibel is a dimensionless number, like the radian. Typically, we write Equation 15-33 without explicitly writing the units. That is, we write it as $\beta = 10 \log(I/I_0)$. Here I is the intensity of the sound and I_0 is a reference level, which usually is taken to be the threshold of hearing:

$$I_0 = 10^{-12}\ \text{W/m}^2 \qquad 15\text{-}34$$

THRESHOLD OF HEARING

On this scale, the threshold of hearing ($I = 10^{-12}$ W/m^2) corresponds to an intensity level of $\beta = 10\log(10^{-12}/10^{-12}) = 0$ dB and the pain threshold ($I = 1$ W/m^2) corresponds to $\beta = 10\log(1/10^{-12}) = 10\log 10^{12} = 120$ dB. Thus, the range of sound intensities from 10^{-12} W/m^2 to 1 W/m^2 corresponds to intensity levels from 0 dB to 120 dB. Table 15-1 lists the intensity levels of some common sounds.

Table 15-1 Intensity and Intensity Level of Some Common Sounds

Source	I/I_0	dB	Description
	10^0	0	Hearing threshold
Normal breathing	10^1	10	Barely audible
Rustling leaves	10^2	20	
Soft whisper (at 5 m)	10^3	30	Very quiet
Library	10^4	40	
Quiet office	10^5	50	Quiet
Normal conversation (at 1 m)	10^6	60	
Busy traffic	10^7	70	
Noisy office with machines; average factory	10^8	80	
Heavy truck (at 15 m); Niagara Falls	10^9	90	Constant exposure endangers hearing
Old subway train	10^{10}	100	
Construction noise (at 3 m)	10^{11}	110	
Rock concert with amplifiers (at 2 m); jet takeoff (at 60 m)	10^{12}	120	Pain threshold
Pneumatic riveter; machine gun	10^{13}	130	
Jet takeoff (nearby)	10^{15}	150	
Large rocket engine (nearby)	10^{18}	180	

Example 15-8 Soundproofing

A sound absorber attenuates the sound *intensity level* by 30 dB. By what factor is the *intensity* changed?

PICTURE Inspect Table 15-1 to see the change in intensity for every 10-dB change in intensity level. Can you discern the pattern?

SOLVE

1. From Table 15-1, we can see that for every 10-dB decrease in the intensity level, the intensity changes by a factor of 1/10.

Thus, if the sound level decreases by 30 dB, then the intensity changes by a factor of $10^{-1} \times 10^{-1} \times 10^{-1} = \boxed{10^{-3}.}$

CHECK We can compare this result with the result gotten by directly using Equation 15-33. That is, $\beta_2 - \beta_1 = 10\log(I_2/I_0) - 10\log(I_1/I_0) = 10\log(I_2/I_1)$. Solving for I_2 gives $I_2 = 10^{(\beta_2-\beta_1)/10}I_1$. Substituting -30 for $\beta_2 - \beta_1$ gives $I_2 = 10^{-3}I_1$, thus verifying our previous result.

CONCEPT CHECK 15-1

When his radio breaks, Chuck purchases a new one that produces twice as much acoustic power as the old one. His expectation is that his new radio will sound twice as loud as his old radio. Will he be disappointed? Explain.

The sensation of loudness depends on the frequency as well as the intensity of a sound. Figure 15-17 is a plot of intensity level versus frequency for sounds of equal loudness to the human ear. (In this figure, the frequency is plotted on a logarithmic scale to display the wide range of frequencies from 20 Hz to 10 kHz.) We observe from this plot that the human ear is most sensitive at about 4 kHz for all intensity levels.

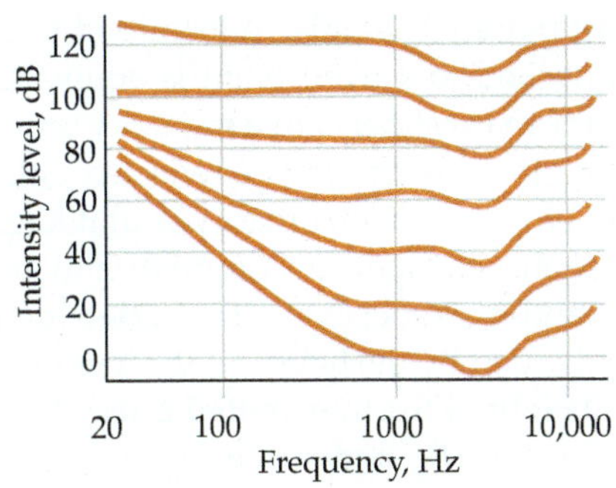

FIGURE 15-17 Intensity level versus frequency for sounds perceived to be of equal loudness. The lowest curve is below the threshold for hearing of all but about one percent of the population. The second lowest curve is approximately the hearing threshold for about 50 percent of the population.

Example 15-9 Barking Dogs

A barking dog delivers about 1.0 mW of acoustic power. (*a*) If this power is uniformly distributed in all directions, what is the sound intensity level at a distance of 5.0 m? (*b*) What would be the intensity level of two dogs, each 5.0 m away, barking at the same time if each delivered 1.0 mW of power?

PICTURE The intensity level is found from the intensity, which is found from $I = P_{av}/(4\pi r^2)$. For two dogs, the intensities are added.

SOLVE

(*a*) 1. The intensity level β is related to the intensity I. Thus, we must first calculate the intensity I:

$$\beta = 10 \log \frac{I}{I_0}$$

2. Using $I = P_{av}/(4\pi r^2)$, calculate the intensity at $r = 5.0$ m:

$$I_1 = \frac{P_{1\,av}}{4\pi r^2} = \frac{1.0 \times 10^{-3}\ \text{W}}{4\pi(5.0\ \text{m})^2} = 3.18 \times 10^{-6}\ \text{W/m}^2$$

3. Use your result to find the intensity level at 5 m:

$$\beta_1 = 10 \log \frac{I_1}{I_0} = 10 \log \frac{3.18 \times 10^{-6}}{1 \times 10^{-12}} = \boxed{65.0\ \text{dB}}$$

(*b*) If I_1 is the intensity for one dog barking, the intensity for two dogs barking is $I_2 = 2I_1$:

$$\beta_2 = 10 \log \frac{I_2}{I_0} = 10 \log \frac{2I_1}{I_0} = 10\left(\log 2 + \log \frac{I_1}{I_0}\right)$$

$$= 10 \log 2 + \beta_1 = 3.01 + 65.0 = \boxed{68.0\ \text{dB}}$$

CHECK If the Part-(*b*) result is correct then whenever the intensity is doubled, the intensity level increases by ~3 dB. To see if this checks out we divide 65 dB by 3 dB to get 21.7, so doubling the threshold intensity 21.7 times should give an intensity of $I_1 \approx 3 \times 10^{-6}\ \text{W/m}^2$. That is, $2^{21.7} I_0$ should equal about $3 \times 10^{-6}\ \text{W/m}^2$. Multiplying $1 \times 10^{-12}\ \text{W/m}^2$ by $2^{21.7}$ gives $3.4 \times 10^{-6}\ \text{W/m}^2$, so our Part-(*b*) result is plausible.

15-4 WAVES ENCOUNTERING BARRIERS

REFLECTION, TRANSMISSION, AND REFRACTION

When a wave is incident on a boundary that separates two regions of differing wave speed, part of the wave is reflected and part is transmitted. Figure 15-18*a* shows a pulse on a light string that is attached to a heavier string (one with a slower wave speed). In this case, the pulse reflected at the boundary is inverted. If the second string is lighter than the first (Figure 15-18*b*), then the reflected pulse is upright. (By upright we mean

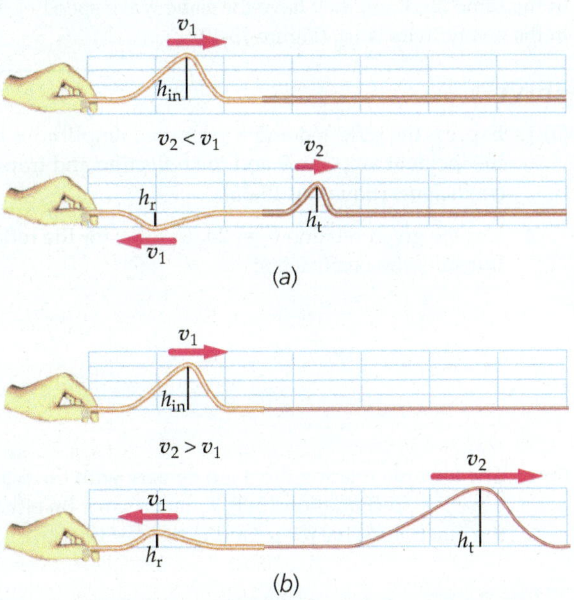

FIGURE 15-18 The leading edges of the pulses are steeper than the trailing edges because the end of the string was raised more quickly than it was lowered. (*a*) A wave pulse traveling on a string attached to a more massive string in which the wave speed is half as large. The reflected pulse is inverted, whereas the transmitted pulse is not. (*b*) A wave pulse traveling on a string attached to a less massive string in which the wave speed is twice as large. In this case, the reflected pulse is not inverted.

it has the same orientation as the incident pulse.) The pulse transmitted to the second string is always upright. A string attached to a fixed point is equivalent to a string being attached to a second string with an extremely large mass per unit length, so for an incident pulse on a string attached to a fixed point the reflected pulse is inverted. Conversely, if the string is tied to a string of smaller mass per unit length, the reflected pulse is upright. The heights of the incident, transmitted, and reflected pulses, shown in Figure 15-18, are h_{in}, h_t, and h_r, respectively. The **reflection coefficient** r is the height of the reflected pulse divided by the height of the incident pulse, and the **transmission coefficient** τ is the height of the transmitted pulse divided by the height of the incident pulse. That is, $r = h_r/h_{in}$ and $\tau = h_t/h_{in}$, where heights h_r, h_{in}, and h_t are shown in Figure 15-18. The expressions for r and τ, are

$$r = \frac{v_2 - v_1}{v_2 + v_1} \quad \text{and} \quad \tau = \frac{2v_2}{v_2 + v_1} \qquad 15\text{-}35$$

REFLECTION AND TRANSMISSION COEFFICIENTS

These expressions for the reflection and transmission coefficients r and τ are known as the *Fresnel relations*. They can be derived by requiring that the tension, the height of the string and the slope of the string all remain continuous at the point where the mass per unit length is discontinuous. (Newton's third law requires that the slope be continuous.) Note that τ is never negative and that r is negative if $v_2 < v_1$. This implies the transmitted pulse is never inverted and the reflected pulse is inverted if $v_2 < v_1$. In addition, the Fresnel relations are valid for both light waves and sound waves.

Example 15-10 Two Soldered Wires

Two wires of different linear mass densities are soldered together end-to-end and then stretched under a tension F_T (the tension is the same in both wires). The wave speed in the first wire is twice that in the second. A harmonic wave traveling in the first wire is incident on the junction of the wires. (*a*) If the amplitude of the incident wave is A, what are the amplitudes of the reflected and transmitted waves? (*b*) What is the ratio μ_2/μ_1 of the mass densities of the wires? (*c*) What fraction of the incident average power is reflected at the junction and what fraction is transmitted?

PICTURE To calculate the amplitudes of the reflected and transmitted waves, use $A_r = rA$ and $A_t = \tau A$, where A_r and A_t are the amplitudes of the reflected and transmitted waves, respectively, and r and τ are the reflection and transmission coefficients given in Equation 15-35. Each power is expressed using $P_{av} = \frac{1}{2}\mu v \omega^2 A^2$ (Equation 15-22). The incident, reflected and transmitted waves all share the same frequency. Because the reflected wave and incident wave are in the same medium, they have the same wave speed v_1. We are given that the wave speed v_2 in the second wire is $\frac{1}{2}v_1$ (Figure 15-19).

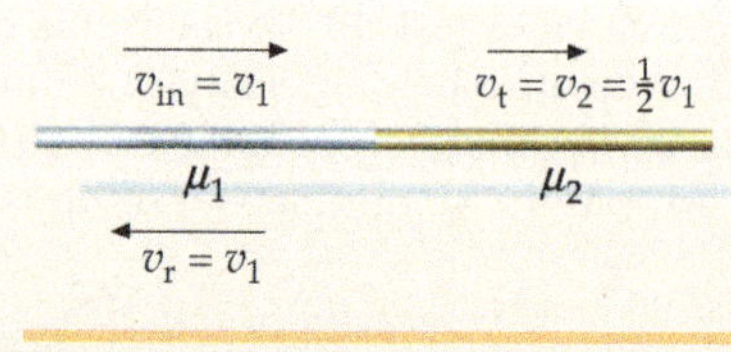

FIGURE 15-19

SOLVE

(*a*) 1. Express the reflected and transmitted amplitudes in terms of the incident amplitude and the reflection and transmission coefficients (Equation 15-35):

$$A_r = rA \quad \text{and} \quad A_t = \tau A$$

2. Use the given relation $v_1 = 2v_2$ to solve for the reflection and transmission coefficients:

$$r = \frac{v_2 - v_1}{v_2 + v_1} = \frac{v_2 - 2v_2}{v_2 + 2v_2} = -\frac{1}{3}$$

$$\tau = \frac{2v_2}{v_2 + v_1} = \frac{2v_2}{v_2 + 2v_2} = \frac{2}{3}$$

$$\text{so} \quad A_r = \boxed{-\frac{1}{3}A} \quad \text{and} \quad A_t = \boxed{\frac{2}{3}A}$$

(*b*) 1. The formula relating the mass density with the wave speed is $v = \sqrt{F_T/\mu}$ (Equation 15-3). F_T is the same on either side of the junction. Solve for μ_2 and μ_1:

$$v_1^2 = \frac{F_T}{\mu_1} \quad \text{and} \quad v_2^2 = \frac{F_T}{\mu_2}$$

$$\text{so} \quad \mu_1 = \frac{F_T}{v_1^2} \quad \text{and} \quad \mu_2 = \frac{F_T}{v_2^2}$$

2. Divide μ_2 by μ_1 and use the given information that $v_1 = 2v_2$:

$$\frac{\mu_2}{\mu_1} = \frac{v_1^2}{v_2^2} = \frac{(2v_2)^2}{v_2^2} = \boxed{4}$$

(*c*) 1. Write expressions for the incident, reflected, and transmitted power using $P_{av} = \frac{1}{2}\mu v\omega^2 A^2$ (Equation 15-22):

$$P_{\text{in av}} = \tfrac{1}{2}\mu_1\omega^2 A^2 v_1$$

$$P_{\text{r av}} = \tfrac{1}{2}\mu_1\omega^2 A_r^2 v_1$$

$$P_{\text{t av}} = \tfrac{1}{2}\mu_2\omega^2 A_t^2 v_2$$

2. Substitute the Part-(*a*) results into the expressions for the reflected and transmitted power:

$$P_{\text{r av}} = \tfrac{1}{2}\mu_1\omega^2\left(-\frac{1}{3}A\right)^2 v_1 = \frac{1}{18}\mu_1\omega^2 A^2 v_1$$

$$P_{\text{t av}} = \tfrac{1}{2}\mu_2\omega^2\left(\frac{2}{3}A\right)^2 v_2 = \frac{2}{9}\mu_2\omega^2 A^2 v_2$$

3. Obtain expressions for P_r/P_{in} and for P_t/P_{in}:

$$\frac{P_{\text{r av}}}{P_{\text{in av}}} = \frac{\frac{1}{18}\mu_1\omega^2 A^2 v_1}{\frac{1}{2}\mu_1\omega^2 A^2 v_1} = \boxed{\frac{1}{9}}$$

$$\frac{P_{\text{t av}}}{P_{\text{in av}}} = \frac{\frac{2}{9}\mu_2\omega^2 A^2 v_2}{\frac{1}{2}\mu_1\omega^2 A^2 v_1} = \frac{4}{9}\frac{\mu_2}{\mu_1}\frac{v_2}{v_1}$$

4. Simplify using the Part-(*b*) result and the given relation $v_1 = 2v_2$:

$$\frac{P_{\text{t av}}}{P_{\text{in av}}} = \frac{4}{9}4\frac{v_2}{2v_2} = \boxed{\frac{8}{9}}$$

CHECK The fractional power reflected plus the fractional power transmitted equals one, as one would expect.

TAKING IT FURTHER The reflected wave is inverted relative to the incident wave, so it is 180° out of phase with it. A negative amplitude corresponds to a phase shift of 180°.

PRACTICE PROBLEM 15-6 Repeat Example 15-10 except with $v_2 = 2v_1$.

Energy conservation gives another relation between the reflection and transmission coefficients. This relation, established in Problem 15-70 is given by

$$1 = r^2 + \frac{v_1}{v_2}\tau^2 \qquad 15\text{-}36$$

where r^2 is the fraction of the incident power that is reflected and $(v_1/v_2)\tau^2$ is the fraction that is transmitted.

PRACTICE PROBLEM 15-7

Show that the values of r and τ for the wires in Example 15-10 satisfy Equation 15-36.

In three dimensions, a boundary between two regions of differing wave speed is a surface. Figure 15-20 shows a ray incident on such a boundary surface. This example could be an ultrasonic pressure wave in air striking a solid or liquid surface. The reflected ray makes an angle with the normal to the surface equal to that of the incident ray, as shown.

The transmitted ray is bent toward or away from the normal—depending on whether the wave speed in the second medium is less or greater than that in the incident medium. The bending of the transmitted ray is called **refraction.** When the wave speed in the second medium is greater than that in the incident medium (as occurs when a light wave in glass or water is refracted into the air), the ray

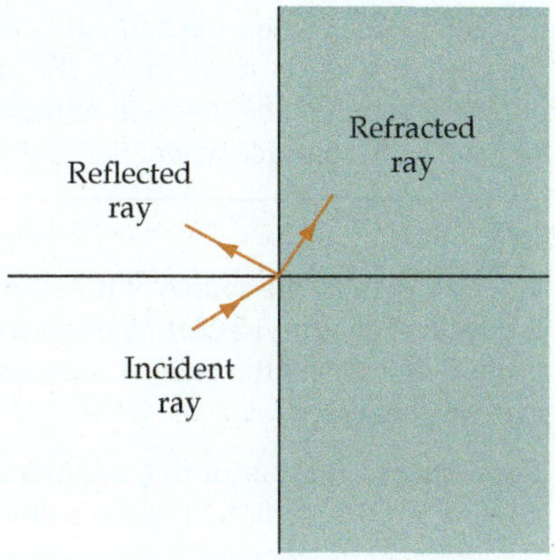

FIGURE 15-20 A wave striking a boundary surface between two media in which the wave speed differs. Part of the wave is reflected and part is transmitted. The change in direction of the transmitted (refracted) ray is called refraction.

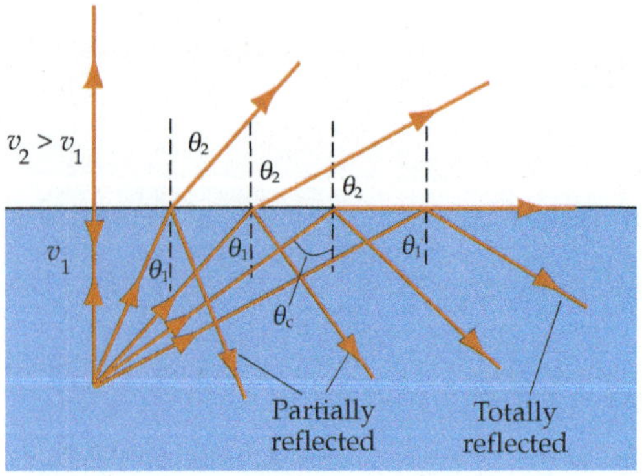

FIGURE 15-21 Light from a source in the water is bent away from the normal when it enters the air. For angles of incidence above a critical angle θ_C, there is no transmitted ray, a condition known as total internal reflection.

describing the direction of propagation is bent away from the normal, as shown in Figure 15-21. As the angle of incidence is increased, the angle of refraction increases, until a critical angle of incidence is reached for which the angle of refraction is 90°. For incident angles greater than the critical angle, there is no refracted ray, a phenomenon known as **total internal reflection.**

The amount of energy reflected from a surface depends on the surface. Rigid flat walls, floors, and ceilings make good reflectors for sound waves, whereas porous and less rigid materials, such as cloth in draperies and furniture coverings, absorb much of the incident sound. The reflection of sound waves plays an important role in the design of a lecture hall, a library, or a music auditorium. If a lecture hall has many flat reflecting surfaces, speech is difficult to understand because of the many echoes that simultaneously arrive at the listener's ear. Absorbent material is often placed on the walls and ceiling to reduce such reflections. In a concert hall, a reflecting shell is placed behind the orchestra, and reflecting panels are hung from the ceiling to reflect and direct the sound back toward the listeners.

(Courtesy of Davies Symphony Hall.)

Example 15-11 Balloon Hearing Aid — *Conceptual*

A popular physics demonstration uses a weather balloon filled with carbon dioxide. If you place the balloon between yourself and a sound source, the sound gets louder. Why is that?

PICTURE The molar mass of carbon dioxide is greater than the effective molar mass of air. Thus, sound travels faster in air than it does in carbon dioxide at atmospheric pressure. To "see" why the sound gets louder when the balloon is between you and the source of the sound, draw a diagram of the rays of sound as they pass through the balloon. The rays will refract (bend) when they are transmitted through a surface where the speed of sound changes.

SOLVE

1. Trace a ray from the source of the sound through the top half of the balloon (Figure 15-22*a*). The ray will refract toward the normal as it enters the balloon, and away from the normal as it exits the balloon:
2. Repeat step 1 for four or five additional rays, including some that go through the lower half of the balloon (Figure 15-22*b*).
3. Use the diagram to explain why the sound is louder when the balloon is between you and the source of sound: The sound is loudest in the region where the rays intersect.

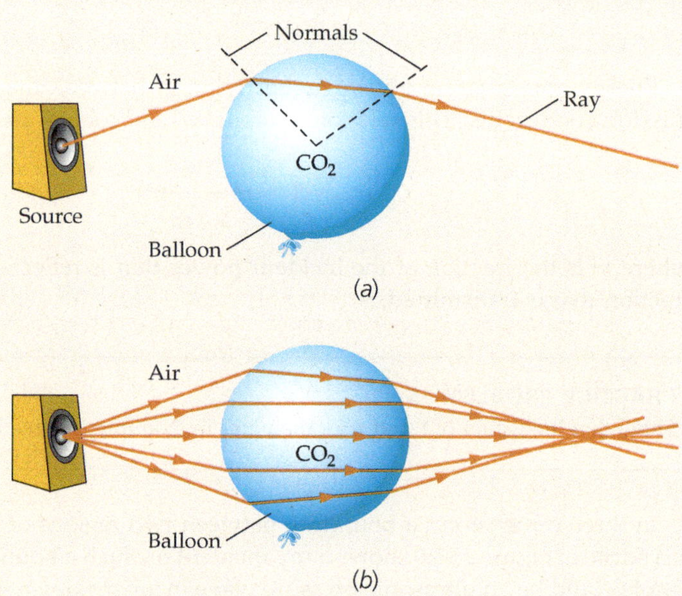

FIGURE 15-22

CHECK The balloon is to sound as a magnifying glass is to light. In glass, light travels more slowly than it does in air, just as in CO_2 sound travels more slowly than it does in air.

DIFFRACTION

If a wavefront is partially blocked by an obstacle, the unblocked part of the wavefront bends behind the obstacle. This bending of the wavefronts is called **diffraction.** Almost all of the diffraction occurs for that part of the wavefront that passes within a few wavelengths of the edge of the obstacle. For the parts of the wavefront that pass farther than a few wavelengths from the edge, diffraction is negligible and the wave propagates in straight lines in the direction of the incident rays. When wavefronts encounter a barrier with an aperture (hole) only a few wavelengths across, the part of the wavefronts passing through the aperture all pass within a few wavelengths of an edge. Thus, flat wavefronts bend and spread out and become spherical or circular (Figure 15-23). In contrast, for a beam of *particles* falling upon a barrier with an aperture, the part of the beam passing through the aperture does so with no change in the direction of the particles (Figure 15-24). Diffraction is one of the key characteristics that distinguish waves from particles. We will discuss how diffraction arises when we study the interference and diffraction of light in Chapter 35.

FIGURE 15-23 Plane waves in a ripple tank meeting a barrier with an opening that is approximately one wavelength wide. Beyond the barrier are circular waves that are concentric about the opening, much as if there were a point source at the opening. *(Fundamental Photographers.)*

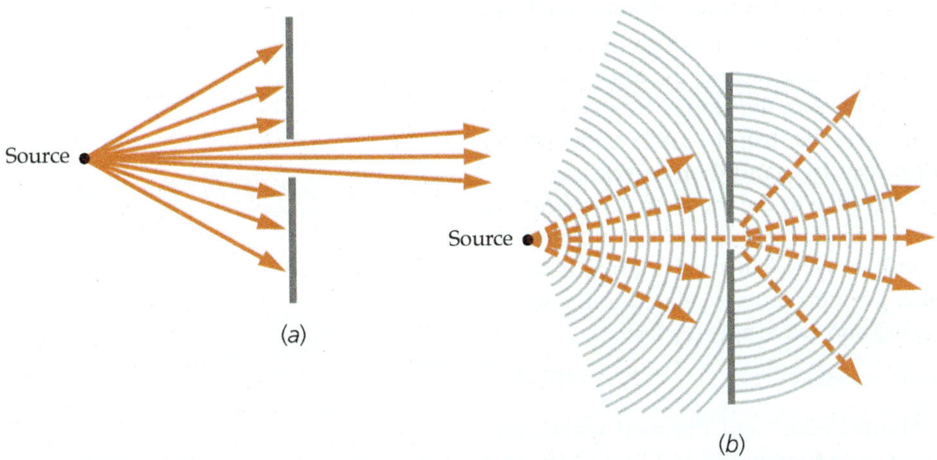

FIGURE 15-24 Comparison of particles and waves passing through a narrow opening in a barrier. (*a*) Transmitted particles are confined to a narrow-angle beam. (*b*) Transmitted waves spread out (radiate widely) from the aperture, which acts like a point source of circular waves.

Although waves passing through an aperture always bend, or diffract, to some extent, the amount of diffraction depends on whether the wavelength is small or large relative to the width of the aperture. If the wavelength is greater than or equal to the width aperture, as in Figure 15-25, the diffraction effects are large, and the waves spread out as they pass through the aperture—as if the waves were originating from a point source. On the other hand, if the wavelength is small relative to the aperture, the effect of diffraction is small, as shown in Figure 15-25. Near the edges of the aperture the wavefronts are distorted and the waves appear to bend slightly. For the most part, however, the wavefronts are not affected and the waves propagate in straight lines, much like a beam of particles. The approximation that waves propagate in straight lines in the direction of the rays with no diffraction is known as the **ray approximation.** Wavefronts are distorted *near* the edges of any obstacle blocking part of the wavefronts. By *near* we mean within a few wavelengths of the edges.

Because the wavelengths of audible sound (which range from a few centimeters to several meters) are generally large compared with apertures and obstacles (doors or windows, and people, for example), diffraction of sound waves is a phenomenon that is regularly observed. On the other hand, the wavelengths of visible light 4×10^{-7} to 7×10^{-7} m) are so small compared with the size of ordinary objects and apertures that the diffraction of light is not easily noticed; light appears to travel in straight lines. Nevertheless, the diffraction of light is an important phenomenon that we study in detail in Chapter 35.

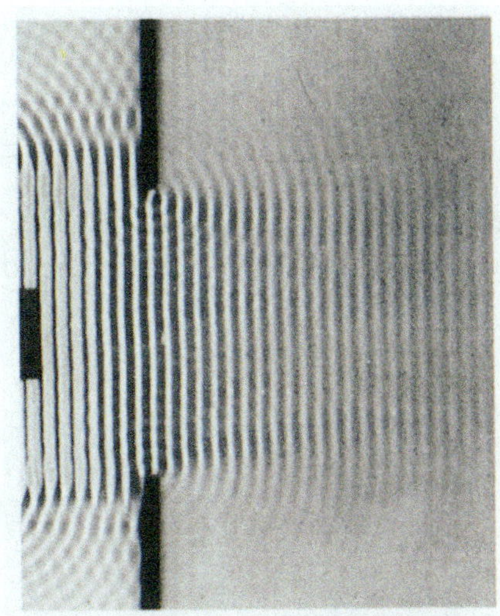

FIGURE 15-25 Plane waves in a ripple tank meeting a barrier with an opening width that is large compared to λ. The wave continues in the forward direction, with only a small amount of spreading into the regions to either side of the opening. *(Fundamental Photographers.)*

Diffraction places a limitation on how accurately small objects can be located by reflecting waves off them and on how well details of the objects can be resolved. Waves are not reflected appreciably from objects smaller than the wavelength, so detail cannot be observed on a scale smaller than the wavelength used. If waves of wavelength λ are used to locate an object, then its position can be known only to within an uncertainty of one wavelength.

Sound waves with frequencies above 20,000 Hz are called **ultrasonic waves.** Because of their very small wavelengths, narrow beams of ultrasonic waves can be sent out and reflected from small objects. Bats can emit and detect frequencies up to about 120 kHz, corresponding to a wavelength of 2.8 mm, which they use to locate small prey such as moths. Echolocation systems, called sonar (from *so*und and *na*vigation *r*anging), are used to detect the outlines of submerged objects with sound waves. The frequency used by commercially available fish finders ranges from about 25 to 200 kHz, and porpoises produce echolocation clicks in the same frequency range. In medicine, ultrasonic waves are used for diagnostic purposes. Ultrasonic waves are passed through the human body and information about the frequency and intensity of the transmitted and reflected waves is processed to construct a three-dimensional picture of the body's interior, called a sonogram.

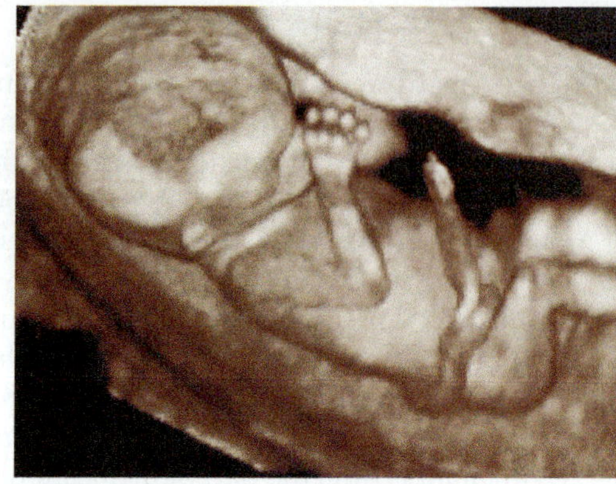

(GE Medical Systems/Photo Researchers, Inc.)

15-5 THE DOPPLER EFFECT

If a wave source and a receiver are moving relative to each other, the received frequency is not the same as the frequency of the source. If they are moving closer together, the received frequency is greater than the source frequency; and if they are moving farther apart, the received frequency is less than the source frequency. This is called the **Doppler effect.** A familiar example is the drop in pitch of the sound of the horn of an approaching car as the car passes by—and then recedes.

In the following discussion, all motions are relative to the medium. Consider the source moving with speed u_s, shown in Figure 15-26*a* and *b*, and a stationary receiver. The source has frequency f_s (and period $T_s = 1/f_s$). The received frequency f_r, the number of wave crests passing the receiver per unit time, is related to the wavelength λ (the distance between successive crests) and wave speed v by

$$f_r\lambda = v \quad \text{(stationary receiver)} \qquad 15\text{-}37$$

A wave crest leaves the source at time t_1 (Figure15-26*c*) and the next wave crest leaves the source at time t_2. The time between these two events is $T_s = t_2 - t_1$, and during this time the source and the crest leaving the source at time t_1 travel distances u_sT_s and vT_s, respectively. Consequently, at time t_2, the distance between the source and the crest leaving at time t_1 equals the wavelength λ. Behind the source $\lambda = \lambda_b = (v + u_s)T_s$, and in front of the source $\lambda = \lambda_f = (v - u_s)T_s$, provided $u_s < v$. (If $u_s \geq v$, no wavefronts reach the region ahead of the source.) We can express both λ_b and λ_f as

$$\lambda = (v \pm u_s)T_s = \frac{v \pm u_s}{f_s} \qquad 15\text{-}38$$

where the minus sign is used if $\lambda = \lambda_f$ and the plus sign is used if $\lambda = \lambda_b$. We have substituted $1/f_s$ for T_s. Substituting for λ in Equation 15-37 and rearranging gives

$$f_r = \frac{v}{\lambda} = \frac{v}{v \pm u_s} f_s \quad \text{(stationary receiver)} \qquad 15\text{-}39$$

(a) (b)

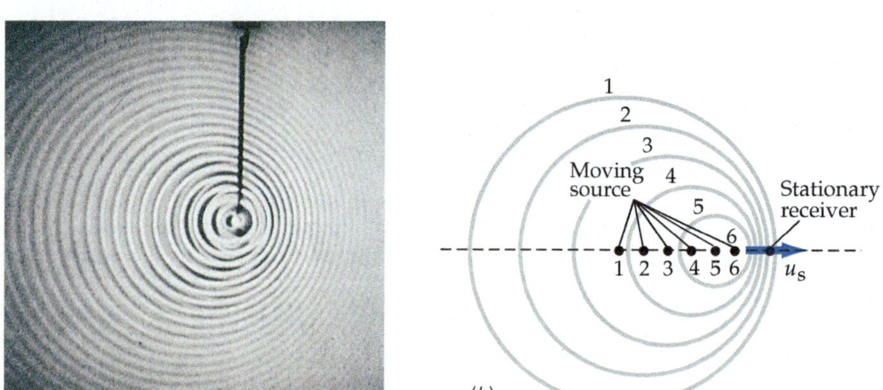

(c)

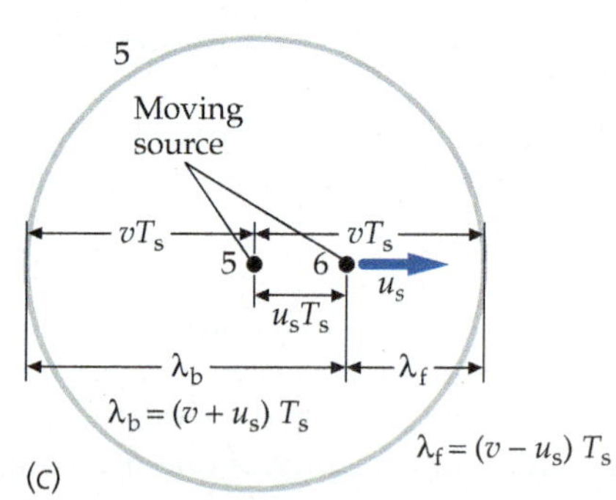

FIGURE 15-26 (*a*) Waves in a ripple tank produced by a point source moving to the right. The wavefronts are closer together in front of the source and farther apart behind the source. (*b*) Successive wavefronts emitted by a point source moving with speed u_s to the right. The numbers of the wavefronts correspond to the positions of the source when the wave was emitted. (*c*) The source vibrates one cycle in time T_s. During time T_s the source moves a distance u_sT_s and the fifth wavefront travels a distance vT_s. In front of the source the wavelength $\lambda_f = (v - u_s)T_s$, while behind the source $\lambda_b = (v + u_s)T_s$. *(Educational Development Center.)*

When the receiver moves relative to the medium, the received frequency is different simply because the receiver moves past more or fewer wave crests in a given time. Let T_r denote the time between arrivals of successive crests for a receiver moving with speed u_r. Then, during the time between the arrivals of two successive crests, each crest will have traveled a distance vT_r, and during the same time the receiver will have traveled a distance u_rT_r. If the receiver moves in the direction opposite to that of the wave (Figure 15-27), then during time T_r, the distance a crest moves plus the distance the receiver moves equals the wavelength. That is, $vT_r + u_rT_r = \lambda$, or $T_r = \lambda/(v + u_r)$. [If the receiver moves in the same direction as the wave, then $vT_r - \lambda = u_rT_r$, so $T_r = \lambda/(v - u_r)$]. Because $f_r = \lambda/T_r$, we have

$$f_r = \frac{1}{T_r} = \frac{v \pm u_r}{\lambda} \qquad \text{15-40}$$

where, if the receiver moves in the same direction as the wave, the received frequency is lower, so we choose the negative sign. If the receiver moves in the direction opposite to that of the wave, the frequency is higher, so we choose the positive sign. Substituting for λ from Equation 15-38, we obtain

$$f_r = \frac{v \pm u_r}{v \pm u_s} f_s \qquad \text{15-41}a$$

The correct choices for the plus or minus signs are most easily determined by remembering that the frequency tends to increase both when the source moves toward the receiver and when the receiver moves toward the source. For example, if the receiver is moving toward the source, the plus sign is selected in the numerator, which tends to increase the received frequency; if the source is moving away from the receiver, the plus sign is selected in the denominator so that the equation predicts a decrease in the received frequency. Equation 15-41*a* appears more symmetric, and thus is easier to remember, if expressed in the form

$$\frac{f_r}{v \pm u_r} = \frac{f_s}{v \pm u_s} \qquad \text{15-41}b$$

It can be shown (see Problem 83) that if both u_s and u_r are much smaller than the wave speed v, then the shift in frequency $\Delta f = f_r - f_s$ is given approximately by

$$\frac{\Delta f}{f_s} \approx \pm\frac{u}{v} \qquad (u \ll v) \qquad \text{15-42}$$

where $u = u_s \pm u_r$ is the speed of the source relative to the receiver.

In a reference frame in which the medium is moving (for example, the reference frame of the ground if air is the medium and if there is a wind blowing), the wave speed v is replaced by $v' = v \pm u_w$, where u_w is the speed of the wind relative to the ground.

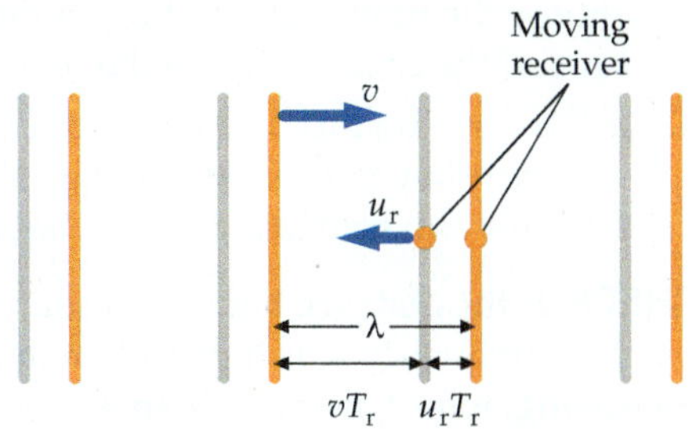

FIGURE 15-27 The time between arrivals of wave crests at the receiver is T_r. The wave crests are represented by orange lines when a wave crest reaches the receiver, and they are represented by grey lines when the next crest reaches the receiver. During time T_r the receiver travels the distance u_rT_r, while the wave crest travels the distance vT_r.

! Equations 15-37 through 15-42 are valid only in the reference frame of the propagating medium.

PROBLEM-SOLVING STRATEGY

Solving Problems Involving Doppler Shift

PICTURE Solving problems involving the Doppler shift means using the equation

$$f_r = \frac{v \pm u_r}{v \pm u_s} f_s$$

(Equation 15-41*a*).

SOLVE

1. Find the speed of the source u_s and of the receiver u_r in the reference frame of the propagating medium.
2. Find the directions of the motions of the source and receiver in the same reference frame.
3. Substitute values into Equation 15-41*a*. Both the source moving toward the receiver and the receiver moving toward the source tend to increase the received frequency. Thus, if the source is moving toward the receiver,

choose the minus sign in the denominator, and if the receiver is moving toward the source, choose the plus sign in the numerator.

4. If the wave bounces off a reflector before reaching the receiver, treat the reflector first as a receiver and apply Equation 15-41*a*, then treat the reflector as a source and apply Equation 15-41*a* once again.

CHECK If the distance between a source and receiver is decreasing, then the received frequency f_r is higher than the source frequency f_s. If this distance is increasing, then f_r is lower than f_s.

Example 15-12 Sounding the Horn

The frequency of a car horn is 400 Hz. If the horn is honked as the car moves with a speed $u_s = 34$ m/s (about 122 km/h) through still air toward a stationary receiver, find (*a*) the wavelength of the sound passing the receiver, and (*b*) the frequency received. Take the speed of sound in air to be 343 m/s. (*c*) Find the wavelength of the sound passing the receiver and find the frequency received if the car is stationary as the horn is honked and a receiver moves with a speed $u_r = 34$ m/s toward the car.

PICTURE (*a*) The waves in front of the source are compressed, so we use the minus sign in $\lambda = (v \pm u_s)/f_s$ (Equation 15-38). (*b*) We calculate the received frequency using $f_r = [(v \pm u_r)/(v \pm u_s)]f_s$ (Equation 15-41*a*). (*c*) For a moving receiver, we use the same equations as in Parts (*a*) and (*b*).

SOLVE

(*a*) Using Equation 15-38, calculate the wavelength in front of the car. In front of the source the wavelength is shorter, so choose the sign accordingly:

$$\lambda = \frac{v - u_s}{f_s} = \frac{343\text{ m/s} - 34\text{ m/s}}{400\text{ Hz}} = 0.773\text{ m} = \boxed{0.77\text{ m}}$$

(*b*) Using Equation 15-41*a* with $u_r = 0$, solve for the received frequency:

$$f_r = \frac{v \pm u_r}{v \pm u_s} f_s = \frac{v}{v - u_s} f_s = \left(\frac{343}{343 - 34}\right)(400\text{ Hz}) = 444\text{ Hz} = \boxed{440\text{ Hz}}$$

(*c*) 1. Using Equation 15-38 with $u_s = 0$, calculate the wavelength in front of the source:

$$\lambda = \frac{v \pm u_s}{f_s} = \frac{343\text{ m/s}}{400\text{ Hz}} = 0.858\text{ m} = \boxed{0.86\text{ m}}$$

2. The received frequency is given by Equation 15-41*a* with $u_s = 0$. The source is approaching the receiver, so the frequency is shifted upward. Choose the sign accordingly:

$$f_r = \frac{v \pm u_r}{v \pm u_s} f_s = \frac{v + u_r}{v} f_s = \left(1 + \frac{u_r}{v}\right) f_s = \left(1 + \frac{34}{343}\right)(400\text{ Hz}) = \boxed{440\text{ Hz}}$$

CHECK The receiver is moving at about 10% of the speed of sound and the frequency received is about 10% higher than the frequency of the source, which is plausible. (Caution, though, this works only if the source is at rest.)

TAKING IT FURTHER The frequency f_r can also be obtained using Equation 15-40.

PRACTICE PROBLEM 15-8 As a train moving at 90 km/h is approaching a stationary listener on a windless day, it sounds its horn, which has a frequency of 630 Hz. (*a*) What is the wavelength of the sound waves in front of the train? (*b*) What frequency is heard by the listener? (Use 343 m/s for the speed of sound.)

Example 15-13 The Speed of the Wave

Context-Rich

You work for an insurance company. An asteroid crashing into the ocean generates a tsunami. When the waves strikes land a 10-m-high wave does a lot of damage. Your boss wants to know how fast the big waves were moving. Knowing that you took a physics course, he asks you to find out. All you have to go on is the audio from a tape recorder that was found in a tree after the waves receded. The audio on the tape has a siren in the background, and in between blasts from the local warning siren is a faint echo of the siren. You measure the frequencies of the sound produced by the siren and its echo and find the siren had a frequency of 4000 Hz, but the echo had a frequency of 4080 Hz. How fast was the big wave approaching?

PICTURE You check with the weather service and find there was no wind at the time the tsunami hit. In addition, the reported temperature is 20°C, so the speed of sound was 343 m/s. First, apply the Doppler-effect equation (Equation 15-41*a*) to calculate the frequency of the sound received by the tsunami in terms of the speed u of the big wave. Apply the equation again, this time considering the big wave as the source of the sound and the tape recorder as the receiver. Assume that the tape recorder was not moving.

SOLVE

1. Apply the Doppler-effect equation with $u_s = 0$ to relate the frequency f_r received by the big wave to the speed u of the big wave:

$$f_r = \frac{v \pm u_r}{v \pm u_s} f_s = \frac{v + u}{v} f_s \qquad f_r = \frac{v + u}{v} f_s$$

2. Apply the Doppler-effect equation, this time with $u_r = 0$, to relate the frequency f_r' received by the tape recorder to the speed of the big wave. Use the step-1 result f_r as the frequency of the big wave as a source of sound:

$$f_r' = \frac{v \pm u_r}{v \pm u_s} f_s' = \frac{v}{v - u} f_r \qquad f_r' = \frac{v}{v - u} f_r$$

3. We now have two equations and two unknowns. Substitute the step-1 result into the step-2 result and simplify:

$$f_r' = \frac{v}{v - u} f_r = \frac{v}{v - u}\,\frac{v + u}{v} f_s \qquad f_r' = \frac{v + u}{v - u} f_s$$

4. Solve for the speed u:

$$u = \frac{f_r' - f_s}{f_r' - f_s} v = \frac{4400\text{ Hz} - 4000\text{ Hz}}{4400\text{ Hz} + 4000\text{ Hz}} 343\text{ m/s} = \boxed{16.3\text{ m/s}}$$

CHECK Sixteen meters per second is about twice as fast as a person can sprint under ideal conditions. Having seen video of tsunamis striking the shore, this seems like a plausible speed.

Another familiar example of the Doppler effect is the radar used by police to measure the speed of a car. Electromagnetic waves emitted by the radar transmitter strike the moving car. The car acts as both a moving receiver and a moving source as the waves reflect off it back to the radar receiver. Equation 15-41*a* is not valid for electromagnetic waves. Electromagnetic waves require the use of the relativistic Doppler-effect formula. (The relativistic Doppler effect is discussed following Example 15-14.) It turns out that if $u \ll c$, where c is the speed of light, Equation 15-42 is valid for electromagnetic waves.

Example 15-14 Police Radar

Try It Yourself

The radar unit in a police car sends out electromagnetic waves that travel at the speed of light c. The electric current in the antenna of the radar unit oscillates at frequency f_s. The waves reflect from a speeding car moving away from the police car at speed u relative to the police car. There is a frequency difference of Δf between f_s and f_r', the frequency received at the police car. Find u in terms of f_s and Δf.

PICTURE The radar wave strikes the speeding car at frequency f_r. This frequency is less than f_s because the car is moving away from the source. The frequency shift is given by $\Delta f/f = \pm u/v$ (Equation 15-42) with $v = c$. The car then acts as a moving source emitting waves of frequency f_r. The police unit detects waves of frequency $f_x' < f_r$ because the source (the speeding car) is moving away from the police car. The frequency difference is $f_x' - f_s$.

SOLVE

Cover the column to the right and try these on your own before looking at the answers.

Steps	Answers
1. The radar unit must be able to determine the speed based only on what it transmits and what it detects.	The radar unit must determine u in terms of f_s and f_r'. We solve $\Delta f/f = \pm u/v$ (Equation 15-42) for u in terms of f_s and $\Delta f = f_r' - f_s$.
2. The frequency difference Δf is the frequency difference $\Delta f_1 = f_r - f_s$ plus the frequency difference $\Delta f_2 = f - f_r$.	$\Delta f = \Delta f_1 + \Delta f_2$
3. Using Equation 15-42 with $v = c$, substitute for the frequency differences in step 2.	$\Delta f = -\frac{u}{c} f_s - \frac{u}{c} f_r = -\frac{u}{c}(f_s + f_r)$
4. Again using Equation 15-42, solve for f_r in terms of f_s.	$\frac{\Delta f_1}{f_s} = -\frac{u}{c}$ so $f_r = \left(1 - \frac{u}{c}\right) f_s$

5. Substitute your step-4 result into your step-3 result and simplify. $\Delta f = -\frac{u}{c}\left(2 - \frac{u}{c}\right)f_s$

6. Compared to 2, u/c is negligible. Use this to simplify the step-5 result and solve for u in terms of Δf and f_s. $\Delta f \approx -2\frac{u}{c}f_s$ so $u = -\frac{\Delta f}{2f_s}c = \boxed{\frac{|\Delta f|}{2f_s}c}$

CHECK The step-6 result is a dimensionless ratio times the speed of light, so it has the correct dimensions for a speed. Thus, the dimensions of the step-6 result are plausible.

TAKING IT FURTHER The difference in frequency between two waves of nearly equal frequency is easy to detect because the two waves interfere to produce a wave whose amplitude oscillates with frequency $|\Delta f|$, which is called the beat frequency. Interference and beats are discussed in Chapter 16.

PRACTICE PROBLEM 15-9 Calculate Δf if $f_s = 1.50 \times 10^9$ Hz, $c = 3.00 \times 10^8$ m/s, and $u = 50.0$ m/s.

The Doppler shift and relativity We see from Example 15-12 (and Equations 15-39, 15-40, and 15-41) that the magnitude of the Doppler shift in frequency depends on whether it is the source or the receiver that is moving relative to the medium. For sound, these two situations are physically different. For example, if you move relative to still air, you feel air rushing past you. In your reference frame, there is a wind. For sound waves in air, therefore, we can tell whether the source or receiver is moving by noting if there is a wind in the reference frame of the source or the receiver. However, light and other electromagnetic waves propagate through empty space in which there is no propagating medium. There is no "wind" to tell us whether the source or receiver is moving. According to Einstein's theory of relativity, absolute motion cannot be detected, and all observers measure the same speed c for light, independent of their motion relative to the source. Thus Equation 15-41 cannot be correct for the Doppler shift for light. Two modifications must be made in calculating the relativistic Doppler effect for light. First, the speed of waves passing a receiver is c, which is independent of the motion of the receiver. Second, the time interval between the emission of successive wave crests, which is $T_s = 1/f_s$ in the reference frame of the source, is different in the reference frame of the receiver when the two reference frames are in relative motion, because of relativistic time dilation and length contraction (Equations R-9 and R-3). (We discuss the relativistic Doppler effect in Chapter 39.) The result is that the frequency received depends only on the relative speed of approach (or recession) u, and is related to the frequency emitted by

$$f_r = \sqrt{\frac{c \pm u}{c \pm u}}f_s \qquad \text{15-43}$$

Choose the signs that give an up-shift in frequency when the source and receiver are approaching, and vice versa. Again, when $u \ll c$, $\Delta f/f_s \approx \pm u/c$, as given by Equation 15-42.

SHOCK WAVES

During our derivations of the Doppler-shift expressions, we assumed that the speed u of the source was less than the wave speed v. If a source moves with speed greater than the wave speed, then there will be no waves in front of the source. Instead, the waves pile up behind the source to form a shock wave. In the case of sound waves, this shock wave is heard as a sonic boom when it arrives at the receiver.

Figure 15-28 shows a source originally at point P_1 moving to the right with speed u. After some time t, the wave emitted from point P_1 has traveled a distance vt. The source has traveled a distance ut and will be at point P_2. The line from

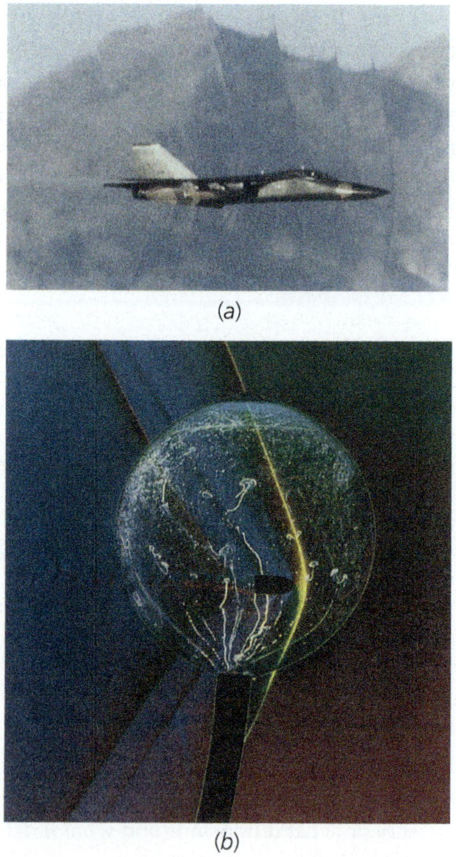

(*a*) Shock waves from a supersonic airplane. *(Sandia National Laboratory.)* (*b*) Shock waves produced by a bullet traversing a helium balloon. *(Estate of Harold E. Edgerton/ Palm Press Inc.)*

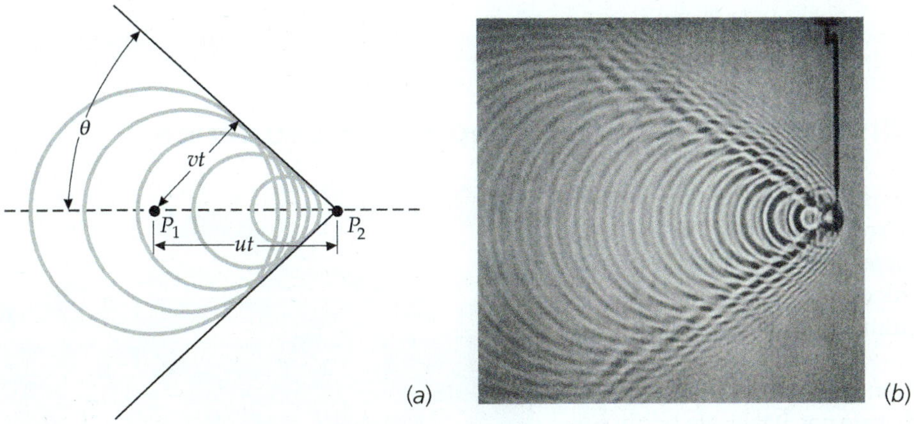

FIGURE 15-28 (*a*) Source moving with a speed u that is greater than the wave speed v. The envelope of the wavefronts forms a cone with the source at the apex. (*b*) Waves in a ripple tank produced by a source moving with a speed $u > v$. *(Educational Development Center.)*

this new position of the source to the wavefront emitted when the source was at P_1 makes an angle θ, called the **Mach angle,** with the path of the source, given by

$$\sin\theta = \frac{vt}{ut} = \frac{v}{u} \qquad 15\text{-}44$$

Thus, the shock wave is confined to a cone that narrows as u increases. The ratio of the source speed u to the wave speed v is called the **Mach number:**

$$\text{Mach number} = \frac{u}{v} \qquad 15\text{-}45$$

Equation 15-44 also applies to the electromagnetic radiation called *Cerenkov radiation,* which is given off when a charged particle moves in a medium with speed u that is greater than the speed of light v in that medium. (According to the special theory of relativity, it is impossible for a particle to move faster than c, the speed of light in vacuum. In a medium such as glass however, electrons and other particles can move faster than the speed of light in that medium.) The blue glow surrounding the fuel elements of a nuclear reactor is an example of Cerenkov radiation.

Example 15-15 A Sonic Boom — *Try It Yourself*

A supersonic plane flying due east at an altitude of 15 km passes directly over point P. The sonic boom is heard at point P when the plane is 22 km east of point P. What is the speed of the supersonic plane?

PICTURE The speed of the plane is related to the sine of the Mach angle (Equation 15-2). Draw a picture so the sine of the Mach angle can be calculated.

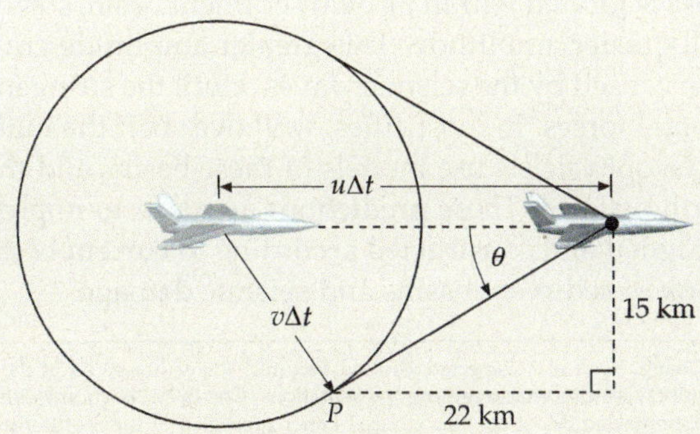

FIGURE 15-29 In the time that the plane moves distance $u\,\Delta t$, the sound moves distance $v\,\Delta t$.

SOLVE

Cover the column to the right and try these on your own before looking at the answers.

Steps	Answers
1. Sketch the position of the plane (Figure 15-29) both at the instant the sonic boom is heard at point P and at the instant that sound was produced. Label the distance the sound travels $v\,\Delta t$, and the distance the plane travels $u\,\Delta t$.	
2. From your sketch and Equation 15-44, calculate u:	$\tan\theta = \dfrac{15\text{ km}}{22\text{ km}}$ so $\theta = 34.3°$ $\sin\theta = \dfrac{v\,\Delta t}{u\,\Delta t} = \dfrac{v}{u}$ so $u = \dfrac{v}{\sin\theta} = 609\text{ m/s} = \boxed{610\text{ m/s}}$

CHECK The speed of sound is 343 m/s, so 610 m/s is plausible for a supersonic speed.

Physics Spotlight

All Shook Up: Sediment Basins and Earthquake Resonance

The damage to buildings constructed on water-logged loose gravel, sand, dirt and clay is greater than the damage to those constructed on hard rock. *(Roger Ressmeyer/CORBIS.)*

On April 18, 1906, the city of San Francisco was devastated by a powerful earthquake. All the buildings in the lowest area collapsed. These buildings were built on water-logged *unconsolidated sediments*—loose gravel, sand, dirt, and clay. Some buildings even sank one or more stories into the ground as the shaking liquefied the loose sediment. Buildings up on rocky Nob Hill and Russian Hill fared better.

Cities located on unconsolidated sediments and near major faults are more vulnerable to earthquake damage than those that are not. If they are partly surrounded by rocky hills or mountains, the danger increases. Vulnerable cities include Seattle,[*,†] Istanbul,[‡] Rome,[#] Los Angeles,[°] San Francisco,[§] and Taipei.[¶]

Unconsolidated sediments are at much greater risk for seismic shaking than rock. When an earthquake occurs, some of the earthquake's energy is sent forth in seismic waves. These waves cause the ground to vibrate over a wide range of frequencies. In solid rock, the waves vibrate and have relatively small amplitudes. The looser the rock or sediment, the more the propagation speed decreases and the amplitude increases.[**] In loose gravel, the waves vibrate more slowly and have greater amplitude. In water-logged sediments, the waves vibrate still more slowly and have much greater amplitude. If you tap a bowl of gelatin sharply on the side, you can hear the sound of tapping the bowl. If it is a metal or glass bowl, the sound will have a frequency of hundreds of hertz. But the gelatin attenuates and scatters the higher frequencies, and resonates at lower frequencies. The same principle underlies the vulnerability of many cities to earthquakes.[††]

Unfortunately, the resonance frequencies of most buildings are closer to the resonance frequencies of the seismic waves in loose sediments.[‡‡] Thus, not only do the sediments vibrate with greater amplitude, but they vibrate most strongly at frequencies that are closer to the resonance frequencies of buildings. This problem was clearly recognized in the State report on the San Francisco earthquake of 1906.[##] Buildings located in areas of unconsolidated sediment were far more damaged than those that were located on higher, firmer ground.

The situation is worst for those cities that are built on sediments that are partially surrounded by areas of hard rock. There the waves resonate within the basin of sediments with large amplitudes. This was true in 1906, when the town of Santa Rosa sustained great damage, even though it was farther from the epicenter of the quake than other, less damaged towns. Santa Rosa is located within its own sediment basin, surrounded by rock.[°°] The basin resonance causes the sediments to vibrate with still greater amplitude. This greater amplitude creates greater damage. Usually, the damage is from the horizontal acceleration caused by the seismic waves. Until the stringent earthquake codes of the 1970s, buildings were not built to withstand horizontal forces. In most cities, well over half the buildings date from before the stringent codes were adopted.

Geophysicists use models of these basins and their sediments to predict areas that are likely to sustain high damage from earthquakes.[§§] These predictions are used to improve codes or to require that bridges,[¶¶] breakwaters,[***] and buildings[†††] are designed and constructed according to current best practices for hazard reduction. The next time you shake a bowl of gelatin, think of sediment basins and seismic damage.

---

* Chang, S., et al., "Expected Ground Failure," Paper presented at the Seattle Fault Earthquake Scenario Conference, 2005. Seattle: Earthquake Engineering Research Institute. http://seattlescenario.eeri.org/presentations/Ch%202%20Ground%20Failure%20-%20Chang.pdf

† Pierepiekarz, M. et al., "Buildings," Paper presented at the Seattle Fault Earthquake Scenario Conference, 2005. Seattle: Earthquake Engineering Research Institute. http://seattlescenario.eeri.org/presentations/Ch%205%20Buildings%20-%20Pierepiekarz.pdf

‡ Erdik, M. *Earthquake Vulnerability of Buildings and a Mitigation Strategy: Case of Istanbul.* World Bank. http://info.worldbank.org/etools/docs/library/114715/istanbul03/docs/istanbul03/06erdik3-n%5B1%5D.pdf as of June 2006.

\# Perkins, S. "Rome at Risk: Seismic Shaking Could Be Long and Destructive," *Science News*, Feb. 25, 2006, 115.

° Perkins, S., "Portrait of Destruction," *Science News*, May 21, 2005, 325.

§ Zoback, M. L., "The 1906 Earthquake—Lessons Learned, Lessons Forgotten, and Future Directions," Paper presented at the American Geophysical Union Meeting, San Francisco, Dec. 5–9, 2005. http://www.ucmp.berkeley.edu/museum/events/shortcourse2006/zoback/

¶ Altenburger, E., "Earthquake Hazards in Taiwan—The September 1999 Chichi Earthquake," *FOCUS on Geography*, Winter 2004, 1–8.

** O'Connell, D. R.H., "Replications of Apparent Nonlinear Seismic Response with Linear Wave Propagation Models," *Science*, Mar. 26, 1999, Vol. 283, No. 5410, p. 2045–2050.

†† Page, R. A., Blume, J. A., and Joyner, W. B., "Earthquake Shaking and Damage to Buildings." *Science*, Aug. 22, 1975, Vol. 189., No. 4203, p. 601–608.

‡‡ Seed, H. B., et al., "Soil Conditions and Building Damage in 1967 Caracas Earthquake," *Journal of Soil Mechanics Division of American Society of Civil Engineers*, 1972, Vol. 98, No. 8, 787–806.

\#\# Lawson, A., et al., *Report of the State Earthquake Investigation Commission*. 1908. Washington, DC: Carnegie Institution.

°° Sloan, D., "Portrait of a Tectonic Landscape," *Bay Nature*, Spring 2006, Vol. 6, No. 2, 24–27.

§§ United States Geological Survey, "1906 Ground Motion Simulations," Earthquake Hazards Program. http://earthquake.usgs.gov/regional/nca/1906/simulations/, as of June 2006.

¶¶ Treyger, S., Jones, M., and Orsolini, G., "Suspending the Big One," *Roads and Bridges*, May 2004, 22–25.

*** Banijamali, B., "Rubble Mounds Feel the Rumble," *Dredging and Port Construction*, June 2005, 35–41.

††† Gonchar, J., "One Project, but Many Seismic Solutions," *Architectural Record*, May 2006, 167–174.

Summary

1. In wave motion, energy and momentum are transported from one point in space to another without the transport of matter.
2. The relation $v = f\lambda$ holds for all harmonic waves.

TOPIC	RELEVANT EQUATIONS AND REMARKS	
1. Transverse and Longitudinal Waves	In transverse waves, such as waves on a string, the disturbance is perpendicular to the direction of propagation. In longitudinal waves, such as sound waves, the disturbance is along the direction of propagation.	
2. Speed of Waves	The wave speed v is independent of the motion of the wave source. The speed of a wave relative to the medium depends on the density and elastic properties of the medium.	
Waves on a string	$v = \sqrt{F_T/\mu}$	15-3
Sound waves	$v = \sqrt{B/\rho}$	15-4
Sound waves in a gas	$v = \sqrt{\gamma RT/M}$	15-5
	where T is the absolute temperature,	
	$T = t_C + 273.15$	15-6
	R is the universal gas constant,	
	$R = 8.314\ \text{J}/(\text{mol}\cdot\text{K})$	15-7
	M is the molar mass of the gas, which for air is 29.0×10^{-3} kg/mol, and γ is a constant that depends on the kind of gas. For a diatomic gas such as air, $\gamma = 7/5$. For a monatomic gas such as helium, $\gamma = 5/3$.	
Electromagnetic waves	The speed of electromagnetic waves in vacuum is a universal constant $c = 3.00 \times 10^8$ m/s	
***3. Wave Equation**	$\frac{\partial^2 y}{\partial x^2} = \frac{1}{v^2}\frac{\partial^2 y}{\partial t^2}$	15-10*b*
4. Harmonic Waves		
Wave function	$y(x,t) = A \sin(kx \pm \omega t)$	15-15
	where A is the amplitude, k is the wave number, and ω is the angular frequency. Use the minus sign for a wave traveling in the $+x$ direction, and the plus sign for a wave traveling in the $-x$ direction.	
Wave number	$k = \frac{2\pi}{\lambda}$	15-14
Angular frequency	$\omega = 2\pi f = \frac{2\pi}{T}$	15-17
Speed	$v = f\lambda = \omega/k$	15-12, 15-16
Energy	The energy in a harmonic wave is proportional to the square of the amplitude.	
Power for harmonic waves on a string	$P_{av} = \frac{1}{2}\mu v \omega^2 A^2$	15-22
5. Harmonic Sound Waves	Sound waves can be considered to be either displacement waves or pressure waves. The human ear is sensitive to sound waves of frequencies from about 20 Hz to 20 kHz. In a harmonic sound wave, the pressure and displacement are 90° out of phase.	
Amplitudes	The pressure and displacement amplitudes are related by	
	$p_0 = \rho\omega v s_0$	15-26
	where ρ is the density of the medium.	
Energy density	$\eta_{av} = \frac{(\Delta E)_{av}}{\Delta V} = \frac{1}{2}\rho\omega^2 s_0^2$	15-28

TOPIC	RELEVANT EQUATIONS AND REMARKS	
6. Intensity	The intensity of a wave is the average power per unit area. $$I = \frac{P_{av}}{A}$$	15-29
Average intensity I of a sound wave	$$I = \eta_{av} v = \frac{1}{2}\rho\omega^2 s_0^2 v = \frac{1}{2}\frac{p_0^2}{\rho v}$$	15-32
*Intensity level β in dB	Sound intensity levels are measured on a logarithmic scale. $$\beta = (10\ \text{dB}) \log \frac{I}{I_0}$$ where $I_0 = 10^{-12}\ \text{W/m}^2$ is taken as the threshold of hearing.	15-33
7. Reflection and Refraction	When a wave is incident on a boundary surface that separates two regions of differing wave speed, part of the wave is reflected and part is transmitted.	
	The reflection and transmission coefficients are $$r = \frac{v_2 - v_1}{v_2 + v_1} \quad \text{and} \quad \tau = \frac{2v_2}{v_2 + v_1}$$	15-35
8. Diffraction	If a wavefront is partially blocked by an obstacle, the unblocked part of the wavefront diffracts (bends) into the region behind the obstacle.	
Ray approximation	If a wavefront is partially blocked by an obstacle, almost all of the diffraction occurs for that part of the wavefront that passes within a few wavelengths of the edge. For those parts of the wavefront that pass farther from the edge than a few wavelengths, diffraction is negligible and the wave propagates in straight lines in the direction of the incident rays.	
9. Doppler Effect	When a sound source and receiver are in relative motion, the received frequency f_r is higher than the frequency of the source f_s if their separation is decreasing, and lower if their separation is increasing.	
Moving source	$$\lambda = \frac{v \pm u_s}{f_s}$$	15-38[4]
Moving receiver	$$f_r = \frac{v \pm u_r}{\lambda}$$	15-40[3]
Either source or receiver moving	$$f_r = \frac{v \pm u_r}{v \pm u_s} f_s \quad \text{or} \quad \frac{f_r}{v \pm u_r} = \frac{f_s}{v \pm u_s}$$ Choose the signs that give an up-shift in frequency for an approaching source or receiver, and vice versa.	15-41[3]
Small speeds of source or receiver	$$\frac{\Delta f}{f_s} \approx \pm \frac{u}{v} \quad (u \ll v), \quad \text{where} \quad u_s = u_s \pm u_r,$$	15-42[3]
Relativistic Doppler shift	$$f_r = \sqrt{\frac{c \pm u}{c \pm u}} f_s$$ Choose the signs that give an up-shift in frequency for an approaching source or receiver, and vice versa.	15-43
10. Shock Waves	When the source speed is greater than the wave speed, the waves behind the source are confined to a cone of angle θ given by	
Mach angle	$$\sin\theta = \frac{v}{u}$$	15-44
Mach number	$$\text{Mach number} = \frac{u}{v}$$	15-45

Answer to Concept Check

15-1 Chuck will be disappointed. Twice the acoustic power will produce twice the *intensity* a given distance from the radio, not twice the *intensity level*.

Answers to Practice Problems

15-1 $\sqrt{\dfrac{\text{N}}{\text{kg/m}}} = \sqrt{\dfrac{\text{kg}\cdot\text{m/s}^2}{\text{kg/m}}} = \sqrt{\dfrac{\text{kg}\cdot\text{m}^2/\text{s}^2}{\text{kg}}} = \sqrt{\text{m}^2/\text{s}^2} = \text{m/s}$

15-2 1.01 km/s

15-3 $\dfrac{\partial^2 y}{\partial x^2} = k^2\dfrac{d^2y}{d^2\beta}$ and $\dfrac{\partial^2 y}{\partial t^2} = \omega^2\dfrac{d^2y}{d^2\beta}$, where $\beta = kx + \omega t$. Thus $k^2 = \dfrac{\omega^2}{v^2} \Rightarrow \omega = kv$

15-4 26 W

15-5 $\lambda = 17$ m at 20 Hz, 17 mm at 20,000 Hz

15-6 (*a*) $A_r = +\frac{1}{3}A$ and $A_r = \frac{4}{3}A$, (*b*) $\dfrac{\mu_2}{\mu_1} = \dfrac{1}{4}$, (*c*) $P_r/P_{in} = 1/9$ and $P_r/P_{in} = 8/9$

15-7 $1 = \left(-\dfrac{1}{3}\right)^2 + 2\left(\dfrac{2}{3}\right)^2 = \dfrac{1}{9} + 2\dfrac{4}{9} = 1$

15-8 (*a*) $\lambda = 0.5$ m, (*b*) $f_r = 680$ Hz

15-9 $\Delta f = 500$ Hz

Problems

Interpret as significant all digits in numerical values that have trailing zeros and no decimal points.

Use 343 m/s as the speed of sound for air, unless otherwise indicated.

In intensity level problems involving the hearing threshold, the reference intensity is exactly 1×10^{-12} W/m² by convention. It is assumed to be accurate to an infinite number of significant figures. Therefore, the number of significant figures in the answers is determined only by those of the data.

- • Single-concept, single-step, relatively easy
- •• Intermediate-level, may require synthesis of concepts
- ••• Challenging
- SSM Solution is in the *Student Solutions Manual*
- Consecutive problems that are shaded are paired problems.

CONCEPTUAL PROBLEMS

1 • A rope hangs vertically from the ceiling. A pulse is sent up the rope. Does the pulse travel faster, slower, or at a constant speed as it moves toward the ceiling? Explain your answer. SSM

2 • A pulse on a horizontal taut string travels to the right. If the rope's mass per unit length decreases to the right, what happens to the speed of the pulse as it travels to the right? (*a*) It slows down. (*b*) It speeds up. (*c*) Its speed is constant. (*d*) You cannot tell from the information given.

3 • As a sinusoidal wave travels past a point on a taut string the arrival time between successive crests is measured to be 0.20 s. Which of the following is true? (*a*) The wavelength of the wave is 5.0 m. (*b*) The frequency of the wave is 5.0 Hz. (*c*) The velocity of propagation of the wave is 5.0 m/s. (*d*) The wavelength of the wave is 0.20 m. (*e*) There is not enough information to justify any of these statements.

4 • Two harmonic waves on identical strings differ only in amplitude. Wave A has an amplitude that is twice the amplitude of wave B. How do the energies of these waves compare? (*a*) $E_A = E_B$, (*b*) $E_A = 2E_B$, (*c*) $E_A = 4E_B$, (*d*) There is not enough information to compare their energies.

5 • True or false: The rate that energy is transported by a harmonic wave is proportional to the square of the amplitude of the wave.

6 • Musical instruments produce sounds of widely varying frequencies. Which sound waves have the longer wavelengths? (*a*) The lower frequencies. (*b*) The higher frequencies. (*c*) All frequencies have the same wavelength. (*d*) There is not enough information to compare the wavelengths of the different frequency sounds.

7 • In Problem 6, which sound waves have the larger speeds? (*a*) The lower frequency sounds. (*b*) The higher frequency sounds. (*c*) All frequencies have the same wave speed. (*d*) There is not enough information to compare their speeds.

8 • Sound travels at 343 m/s in air and 1500 m/s in water. A sound of 256 Hz is made under water, but you hear the sound while walking along the side of the pool. In the air, the frequency is (*a*) the same, but the wavelength of the sound is shorter, (*b*) higher, but the wavelength of the sound stays the same, (*c*) lower, but the wavelength of the sound is longer, (*d*) lower, and the wavelength of the sound is shorter, (*e*) the same, and the wavelength of the sound stays the same.

9 • While out on patrol, the battleship *Rodger Young* hits a mine, begins to burn, and ultimately explodes. Sailor Abel jumps into the water and begins swimming away from the doomed ship, while Sailor Baker gets into a life raft. Comparing their experiences later, Abel tells Baker, "I was swimming underwater, and heard a big explosion from the ship. When I surfaced, I heard a second explosion. What do you think it could be?" Baker says, "I think it was your imagination—I only heard one explosion." Explain why Baker only heard one explosion, while Abel heard two.

10 • True or false: A 60-dB sound has twice the intensity of a 30-dB sound.

11 • At a given location, two harmonic sound waves have the same displacement amplitude, but the frequency of sound A is twice the frequency of sound B. How do their average energy densities compare? (*a*) The average energy density of A is twice the average energy density of B. (*b*) The average energy density of A is four times the average energy density of B. (*c*) The average energy density of A is 16 times the average energy density of B. (*d*) You cannot compare the average energy densities from the data given. SSM

12 • At a given location, two harmonic sound waves have the same frequency, but the amplitude of sound A is twice the amplitude of sound B. How do their average energy densities compare? *(a)* The average energy density of A is twice the average energy density of B. *(b)* The average energy density of A is four times the average energy density of B. *(c)* The average energy density of A is 16 times the average energy density of B. *(d)* You cannot compare the average energy densities from the data given.

13 • What is the ratio of the intensity of normal conversation to the sound intensity of a soft whisper (at a distance of 5.0 m)? *(a)* 10^3, *(b)* 2, *(c)* 10^{-3}, *(d)* 1/2. *Hint: See Table 15-1.*

14 • What is the ratio of the intensity level of normal conversation to the sound intensity level of a soft whisper (at a distance of 5.0 m)? *(a)* 10^3, *(b)* 2, *(c)* 10^{-3}, *(d)* 1/2. *Hint: See Table 15-1.*

15 • To increase the sound intensity level by 20 dB requires the sound intensity to increase by what factor? *(a)* 10, *(b)* 100, *(c)* 1000, *(d)* 2

16 • You are using a hand-held sound level meter to measure the intensity level of the roars produced by a lion prowling in the high grass. To decrease the measured sound intensity level by 20 dB requires the lion move away from you until its distance from you has increased by what factor? *(a)* 10, *(b)* 100, *(c)* 1000, *(d)* You cannot tell the required distance from the data given.

17 • One end of a very light (but strong) thread is attached to an end of a thicker and denser cord. The other end of the thread is fastened to a sturdy post and you pull the other end of the cord so the thread and cord are taut. A pulse is sent down the thicker, denser cord. True or false:

(a) The pulse that is reflected back from the thread-cord attachment point is inverted compared to the initial incoming pulse.
(b) The pulse that continues past the thread-cord attachment point is not inverted compared to the initial incoming pulse.
(c) The pulse that continues past the thread-cord attachment point has an amplitude that is smaller than the pulse that is reflected.

18 • Light traveling in air strikes a glass surface at a 45° incident angle. True or false:

(a) The angle between the reflected light ray and the incident ray is 90°.
(b) The angle between the reflected light ray and the refracted light ray is less than 90°.

19 • Sound waves in air encounter an open 1.0-m-wide door into a classroom. Due to the effects of diffraction, the sound of which frequency is least likely to be heard by all the students in the room–assuming the room is full? *(a)* 600 Hz, *(b)* 300 Hz, *(c)* 100 Hz, *(d)* All of the sounds are equally likely to be heard in the room. *(e)* Diffraction depends on wavelength not frequency, so you cannot tell from the data given. SSM

20 • Microwave radiation in modern microwave ovens has a wavelength on the order of centimeters. Would you expect significant diffraction if this radiation was aimed at a 1.0-m-wide door? Explain.

21 •• Stars often occur in pairs revolving around their common center of mass. If one of the stars is a black hole, it is invisible. Explain how the existence of such a black hole might be inferred by measuring the Doppler frequency shift of the light observed from the other, visible star. SSM

22 •• Figure 15-30 shows a wave pulse at time $t = 0$ moving to the right. *(a)* At this particular time, which segments of the string are moving up? *(b)* Which segments are moving down? *(c)* Is there any segment of the string at the pulse that is instantaneously at rest? Answer these questions by sketching the pulse at a slightly later time and a slightly earlier time to see how the segments of the string are moving.

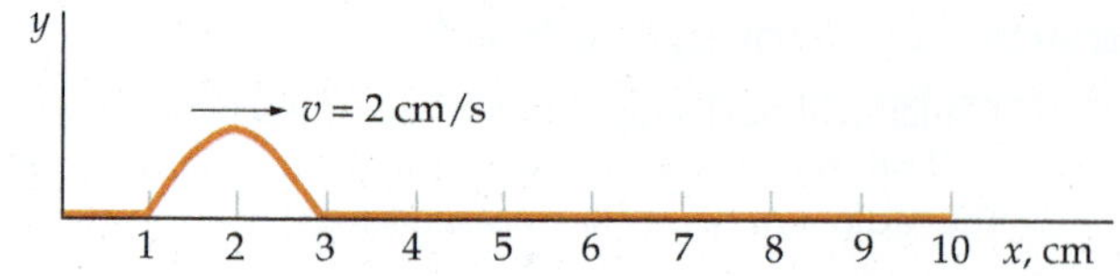

FIGURE 15-30 Problems 22, 23

23 •• Make a sketch of the velocity of each string segment versus position for the pulse shown in Figure 15-30.

24 •• An object of mass m hangs on a very light rope that is connected to the ceiling. You pluck the rope just above the object, and a wave pulse travels up to the ceiling and back. Compare the round-trip time for such a wave pulse to the round-trip time of a wave pulse on the same rope if an object of mass $9m$ is hung on the rope instead. (Assume that the rope does not stretch, that is, that the mass-to-ceiling distance is the same in each case.)

25 •• The speed of sound in water is greater than the speed of sound in air. The explosion of a depth charge beneath the surface of a body of water is recorded by a helicopter hovering above the water's surface, as shown in Figure 15-31. Along which path—A, B, or C—will the sound wave take the least time to reach the helicopter? Explain why you chose the path you did.

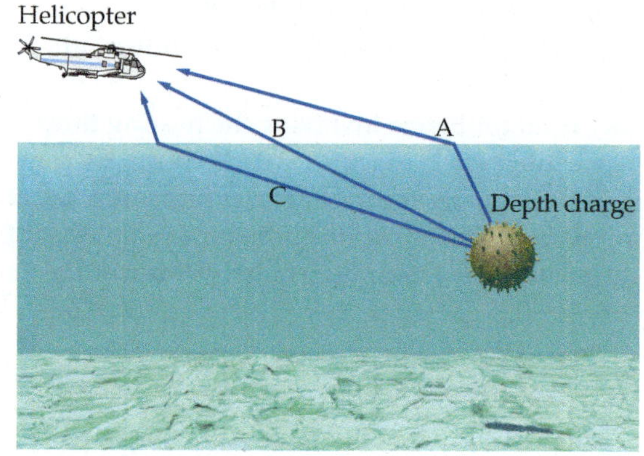

FIGURE 15-31 Problem 25

26 •• Does a speed of Mach 2 at an altitude 60,000 feet mean the same as a speed of Mach 2 near ground level? Explain clearly.

ESTIMATION AND APPROXIMATION

27 •• Many years ago, Olympic 100-m dashes were started by the sound from a starter's pistol, with the starter positioned several meters down the track, just on the inside of the track. (Today, the pistol that is used is often only a trigger, which is used to electronically activate speakers behind each sprinter's starting blocks. This method avoids the problem of one runner hearing the sound before the other runners.) Estimate the time advantage the runner at the inside lane (relative to the runner at the outside lane of eight runners) would have if all runners started when they heard the sound from the starter's pistol.

28 •• Estimate the speed of the bullet as it passes through the helium balloon in Figure 15-32. *Hint: A protractor would be beneficial.*

FIGURE 15-32 Problem 28 *(Estate of Harold E. Edgerton/Palm Press Inc.)*

29 •• The new student townhouses at a local college are in the form of a semicircle half-enclosing the track field. To estimate the speed of sound in air, an ambitious physics student stood at the center of the semicircle and clapped his hands rhythmically at a frequency at which he could not hear the echo of the clap, because the echo reached him at the same time as his next clap. This frequency was about 2.5 claps/s. Once he established this frequency, he paced off the distance to the townhouses, which was 30 double strides. Assuming that the length of each stride is equal to half his height (5 ft, 11 in.), estimate the speed of sound in air using these data. How far off is your estimation from the commonly accepted value of 343 m/s?

SPEED OF WAVES

30 • (*a*) The bulk modulus of water is $2.00 \times 10^9\ \text{N/m}^2$. Use this value to find the speed of sound in water. (*b*) The speed of sound in mercury is 1410 m/s. What is the bulk modulus of mercury ($\rho = 13.6 \times 10^3\ \text{kg/m}^3$)?

31 • Calculate the speed of sound waves in hydrogen gas ($M = 2.00$ g/mol and $\gamma = 1.40$) at $T = 300$ K.

32 • A 7.00-m-long string has a mass of 100 g and is under a tension of 900 N. What is the speed of a transverse wave pulse on this string?

33 •• (*a*) Compute the derivative of the speed of a wave on a string with respect to the tension dv/dF_T and show that the differentials dv and dF_T obey $dv/v = \frac{1}{2}\,dF_T/F_T$. (*b*) A wave moves with a speed of 300 m/s on a string that is under a tension of 500 N. Using the differential approximation, estimate how much the tension must be changed to increase the speed to 312 m/s. (*c*) Calculate ΔF_T exactly and compare it to the differential approximation result in Part (*b*). Assume that the string does not stretch with the increase in tension. SSM

34 •• (*a*) Compute the derivative of the speed of sound in air with respect to the absolute temperature, and show that the differentials dv and dT obey $dv/v = \frac{1}{2}\,dT/T$. (*b*) Use this result to estimate the percentage change in the speed of sound when the temperature changes from 0°C to 27°C. (*c*) If the speed of sound is 331 m/s at 0°C, estimate its value at 27°C using the differential approximation. (*d*) How does this approximation compare with the result of an exact calculation?

35 ••• Derive a convenient formula for the speed of sound in air at temperature t in degrees Celsius. Begin by writing the temperature as $T = T_0 + \Delta T$, where $T_0 = 273$ K and corresponds to 0°C and $\Delta T = t$, which is the Celsius temperature. The speed of sound is a function of T, $v(T)$. To a first-order approximation, you can write $v(T) \approx v(T_0) + (dv/dT)_{T_0}\,\Delta T$, where $(dv/dT)_{T_0}$ is the derivative evaluated at $T = T_0$. Compute this derivative, and show that the result leads to

$$v = (331\ \text{m/s})(1 + (t/2T_0)) = (331 + 0.606t)\ \text{m/s}$$

THE WAVE EQUATION

36 • Show explicitly that the following functions satisfy the wave equations $\partial^2 y/\partial x^2 = (1/v^2)\,\partial^2 y/\partial t^2$: (*a*) $y(x,t) = k(x + vt)^3$, (*b*) $y(x,t) = Ae^{ik(x-vt)}$, where A and k are constants and $i = \sqrt{-1}$, and (*c*) $y(x,t) = \ln[k(x - vt)]$.

37 • Show that the function $y = A \sin kx \cos \omega t$ satisfies the wave equation.

HARMONIC WAVES ON A STRING

38 • One end of a 6.0-m-long string is moved up and down with simple harmonic motion at a frequency of 60 Hz. If the wave crests travel the length of the string in 0.50 s, find the wavelength of the waves on the string.

39 • A harmonic wave on a string that has a mass per unit length of 0.050 kg/m and a tension of 80 N has an amplitude of 5.0 cm. Each point on the string moves with simple harmonic motion at a frequency of 10 Hz. What is the power carried by the wave propagating along the string? SSM

40 • A 2.00-m-long rope has a mass of 0.100 kg. The tension is 60.0 N. An oscillator at one end sends a harmonic wave with an amplitude of 1.00 cm down the rope. The other end of the rope is terminated so all of the energy of the wave is absorbed and none is reflected. What is the frequency of the oscillator if the power transmitted is 100 W?

41 •• The wave function for a harmonic wave on a string is $y(x,t) = (1.00\ \text{mm}) \sin(62.8\ \text{m}^{-1}x + 314\ \text{s}^{-1}t)$. (*a*) In what direction does this wave travel, and what is the wave's speed? (*b*) Find the wavelength, frequency, and period of this wave. (*c*) What is the maximum speed of any point on the string?

42 •• A harmonic wave on a string with a frequency of 80 Hz and an amplitude of 0.025 m travels in the $+x$ direction with a speed of 12 m/s. (*a*) Write a suitable wave function for this wave. (*b*) Find the maximum speed of a point on the string. (*c*) Find the maximum acceleration of a point on the string.

43 •• A 200-Hz harmonic wave with an amplitude equal to 1.2 cm moves along a 40-m-long string that has a mass of 0.120 kg and a tension of 50 N. (*a*) What is the average total energy of the waves on a 20-m-long segment of string? (*b*) What is the power transmitted past a given point on the string?

44 •• On a real string, some of the energy of a wave dissipates as the wave travels down the string. Such a situation can be described by a wave function whose amplitude $A(x)$ depends on x: $y = A(x) \sin(kx - \omega t)$, where $A(x) = A_0 e^{-bx}$. What is the power transported by the wave as a function of x, where $x > 0$?

45 •• Power is to be transmitted along a taut string by means of transverse harmonic waves. The wave speed is 10 m/s and the linear mass density of the string is 0.010 kg/m. The power source oscillates with an amplitude of 0.50 mm. (*a*) What average power is transmitted along the string if the frequency is 400 Hz? (*b*) The power transmitted can be increased by increasing the tension in the string, the frequency of the source, or the amplitude of the waves. By how much would each of these quantities have to increase to cause an increase in power by a factor of 100 if it is the only quantity changed? SSM

46 ••• Two very long strings are tied together at the point $x = 0$. In the region $x < 0$, the wave speed is v_1, while in the region $x > 0$, the speed is v_2. A sinusoidal wave is incident on the knot from the left ($x < 0$); part of the wave is reflected and part is transmitted. For $x < 0$, the displacement of the wave is described by $y(x,t) = A \sin(k_1 x - \omega t) + B \sin(k_1 x + \omega t)$, while for $x > 0$, $y(x,t) = C \sin(k_2 x - \omega t)$, where $\omega/k_1 = v_1$ and $\omega/k_2 = v_2$. (*a*) If we assume that both the wave function y and its first spatial derivative $\partial y/\partial x$ must be continuous at $x = 0$, show that $C/A = 2v_2/(v_1 + v_2)$, and that $B/A = (v_1 - v_2)/(v_1 + v_2)$. (*b*) Show that $B^2 + (v_1/v_2)C^2 = A^2$.

HARMONIC SOUND WAVES

47 • A sound wave in air produces a pressure variation given by $p(x,t) = 0.75 \cos\left[\frac{\pi}{2}(x - 343t)\right]$, where p is in pascals, x is in meters, and t is in seconds. Find (*a*) the pressure amplitude, (*b*) the wavelength, (*c*) the frequency, and (*d*) the wave speed.

48 • (*a*) Middle C on the musical scale has a frequency of 262 Hz. What is the wavelength of this note in air? (*b*) The frequency of the C an octave above middle C is twice that of middle C. What is the wavelength of this note in air?

49 • The density of sound is 1.29 kg/m^3. (*a*) What is the displacement amplitude for a sound wave with a frequency of 100 Hz and a pressure amplitude of 1.00×10^{-4} atm? (*b*) The displacement amplitude of a sound wave of frequency 300 Hz is 1.00×10^{-7} m. What is the pressure amplitude of this wave? SSM

50 • The density of sound is 1.29 kg/m^3. (*a*) What is the displacement amplitude of a sound wave that has a frequency of 500 Hz at the pain-threshold pressure amplitude of 29.0 Pa? (*b*) What is the displacement amplitude of a sound wave that has the same pressure amplitude as the wave in Part (*a*), but has a frequency of 1.00 kHz?

51 • A typical loud sound wave that has a frequency of 1.00 kHz has a pressure amplitude of about 1.00×10^{-4} atm. (*a*) At $t = 0$, the pressure is a maximum at some point x_1. What is the displacement at that point at $t = 0$? (*b*) Assuming the density of air is 1.29 kg/m^3, what is the maximum value of the displacement at any time and place?

52 • An octave represents a change in frequency by a factor of 2. Over how many octaves can a typical person hear?

53 •• **BIOLOGICAL APPLICATION** In the oceans, whales communicate by sound transmission through the water. A whale emits a sound of 50.0 Hz to tell a wayward calf to catch up to the pod. The speed of sound in water is about 1500 m/s. (*a*) How long does it take the sound to reach the calf if he is 1.20 km away? (*b*) What is the wavelength of this sound in the water? (*c*) If the whales are close to the surface, some of the sound energy might refract out into the air. What would be the frequency and wavelength of the sound in the air?

WAVES IN THREE DIMENSIONS: INTENSITY

54 • A spherical sinusoidal source radiates sound uniformly in all directions. At a distance of 10.0 m, the sound intensity level is 1.00×10^{-4} W/m^2. (*a*) At what distance from the source is the intensity 1.00×10^{-6} W/m^2? (*b*) What power is radiated by this source?

55 • **ENGINEERING APPLICATION** A loudspeaker at a rock concert generates a sound that has an intensity level equal to 1.00×10^{-2} W/m^2 at 20.0 m and has a frequency of 1.00 kHz Assume that the speaker spreads its energy uniformly in three dimensions. (*a*) What is the total acoustic power output of the speaker? (*b*) At what distance will the sound intensity be at the pain threshold of 1.00 W/m^2? (*c*) What is the sound intensity at 30.0 m? SSM

56 •• When a pin of mass 0.100 g is dropped from a height of 1.00 m, 0.050 percent of its energy is converted into a sound pulse that has a duration of 0.100 s. (*a*) Estimate how far away the dropped pin can be heard if the minimum audible intensity is 1.00×10^{-11} W/m^2. (*b*) Your result in Part (*a*) is much too large in practice, because of background noise. If you assume the intensity must be at least 1.00×10^{-8} W/m^2 for the sound to be heard, estimate how far away the dropped pin can be heard. (In both parts, assume that the intensity is $P/4\pi r^2$.)

*INTENSITY LEVEL

57 • What is the intensity level in decibels of a sound wave that has an intensity equal to (*a*) 1.00×10^{-10} W/m^2 and (*b*) 1.00×10^{-2} W/m^2? SSM

58 • What is the intensity of a sound wave if, at a particular location, the intensity level is (*a*) $\beta = 10$ dB and (*b*) $\beta = 3.0$ dB?

59 • At a certain distance, the sound intensity level of a dog's bark is 50 dB. At that same distance, the sound intensity of a rock concert is 10,000 times that of the dog's bark. What is the sound intensity level of the rock concert?

60 • What fraction of the acoustic power of a noise would have to be eliminated to lower its sound intensity level from 90 to 70 dB?

61 •• A spherical source radiates sound uniformly in all directions. At a distance of 10 m, the sound intensity level is 80 dB. (*a*) At what distance from the source is the intensity level 60 dB? (*b*) What power is radiated by this source?

62 •• Harry and Sally are sitting on opposite sides of a circus tent when an elephant trumpets a loud blast. If Harry experiences a sound intensity level of 65 dB and Sally experiences only 55 dB, what is the ratio of the distance between Sally and the elephant to the distance between Harry and the elephant?

63 •• Three noise sources produce intensity levels of 70 dB, 73 dB, and 80 dB, when acting separately. When the sources act together, the resultant intensity is the sum of the individual intensities. (*a*) Find the sound intensity level in decibels when the three sources act at the same time. (*b*) Discuss the effectiveness of eliminating the two less intense sources in reducing the intensity level of the noise.

64 •• Show that if two people are different distances away from a sound source, the difference $\Delta\beta$ between the intensity levels reaching the people, in decibels, will always be the same, no matter the power radiated by the source.

65 ••• Everyone at a party is talking equally loudly. One person is talking to you and the sound intensity level at your location is 72 dB. Assuming that all 38 people at the party are at the same distance from you as the person who you are talking to, find the sound intensity level at your location.

66 ••• When a violinist pulls the bow across a string, the force with which the bow is pulled is fairly small, about 0.60 N. Suppose the bow travels across the A string, which vibrates at 440 Hz, at 0.50 m/s. A listener 35 m from the performer hears a sound of 60-dB intensity. Assuming that the sound radiates uniformly in all directions, with what efficiency is the mechanical energy of bowing converted to sound energy?

67 ••• The noise intensity level at some location in an empty classroom is 40 dB. When 100 students are writing during an exam, the noise level at that location increases to 60 dB. Assuming that the noise produced by each student contributes an equal amount of acoustic power, find the noise intensity level in the room after 50 students have left. SSM

STRING WAVES EXPERIENCING SPEED CHANGES

68 • A 3.00-m-long piece of string, with a mass of 25.0 g, is tied to 4.00 m of heavy twine with a mass of 75.0 g, and the combination is put under a tension of 100 N. If a transverse pulse is sent down the less dense string, determine the reflection and transmission coefficients at the junction point.

69 • Consider a taut string, with a mass per unit length μ_1, carrying transverse wave pulses that are incident upon a point where the string connects to a second string, with a mass per unit length μ_2. (*a*) Show that if $\mu_2 = \mu_1$, then the reflection coefficient r equals zero and the transmission coefficient τ equals $+1$. (*b*) Show that if $\mu_2 \gg \mu_1$, then $r \approx -1$ and $\tau \approx 0$. (*c*) Show that if $\mu_2 \ll \mu_1$, then $r \approx +1$ and $\tau \approx +2$. SSM

70 •• Verify the validity of $1 = r^2 + (v_1/v_2)\tau^2$ (Equation 15-36) by substituting the expressions for r and τ into it.

71 ••• Consider a taut string that has a mass per unit length μ_1 carrying transverse wave pulses of the form $y = f(x - v_1 t)$ that are incident upon a point P where the string connects to a second string with mass per unit length μ_2. Derive $1 = r^2 + (v_1/v_2)\tau^2$ by equating the power incident on point P to the power reflected at P plus the power transmitted at P.

THE DOPPLER EFFECT

In Problems 72 through 75, assume that the source emits sound at a frequency of 200 Hz. Assume also that the sound travels through still air at 343 m/s.

72 • A sound source is moving at 80 m/s toward a stationary listener that is standing in still air. (*a*) Find the wavelength of the sound in the region between the source and the listener. (*b*) Find the frequency heard by the listener.

73 • Consider the situation described in Problem 72 from the reference frame of the source. In this frame, the listener and the air are moving toward the source at 80 m/s and the source is at rest. (*a*) At what speed, relative to the source, is the sound traveling in the region between the source and the listener? (*b*) Find the wavelength of the sound in the region between the source and the listener. (*c*) Find the frequency heard by the listener.

74 • A sound source is moving away from the stationary listener at 80 m/s. (*a*) Find the wavelength of the sound waves in the region between the source and the listener. (*b*) Find the frequency heard by the listener.

75 • The listener is moving at 80 m/s away from the stationary source that is at rest relative to the air. Find the frequency heard by the listener.

76 •• CONTEXT-RICH You have made the trek to observe a Space Shuttle landing. Near the end of its descent, the ship is traveling at Mach 2.50 at an altitude of 5000 m. (*a*) What is the angle that the shock wave makes with the line of flight of the shuttle? (*b*) How far are you from the shuttle by the time you hear its shock wave, assuming the shuttle maintains both a constant heading and a constant 5000-m altitude after flying directly over your head?

77 •• ENGINEERING APPLICATION The SuperKamiokande neutrino detector of Japan is a water tank the size of a 14-story building. When neutrinos collide with electrons in the water, most of their energy is transferred to the electrons. As a consequence, the electrons then fly off at speeds that approach *c*. The neutrino is counted by detecting the shock wave, called Cerenkov radiation, that is produced when the high-speed electrons travel through the water at speeds greater than the speed of light in water. If the maximum angle of the Cerenkov shock-wave cone is 48.75°, what is the speed of light in water?

78 •• ENGINEERING APPLICATION, CONTEXT-RICH You are in charge of calibrating the radar guns for a local police department. One such device emits microwaves at a frequency of 2.00 GHz. During the trials, these waves are reflected from a car moving directly away from the stationary emitter. You detect a frequency difference (between the received microwaves and the ones sent out) of 293 Hz. Find the speed of the car.

79 •• ENGINEERING APPLICATION, CONTEXT-RICH The Doppler effect is routinely used to measure the speed of winds in storm systems. As the manager of a weather monitoring station in the Midwest, you are using a Doppler radar system that has a frequency of 625 MHz to bounce a radar pulse off of the raindrops in a swirling thunderstorm system 50 km away. You measure the reflected radar pulse to be up-shifted in frequency by 325 Hz. Assuming the wind is headed directly toward you, how fast are the winds in the storm system moving? *Hint: The radar system can only measure the component of the wind velocity along its "line of sight."* SSM

80 •• ENGINEERING APPLICATION A stationary destroyer is equipped with sonar that sends out 40 MHz pulses of sound. The destroyer receives reflected pulses back from a submarine directly below with a time delay of 80 ms at a frequency of 39.958 MHz. If the speed of sound in seawater is 1.54 km/s, (*a*) what is the depth of the submarine? (*b*) What is its vertical speed?

81 •• A police radar unit transmits microwaves of frequency 3.00×10^{10} Hz, and their speed in air is 3.00×10^{8} m/s. Suppose a car is receding from the stationary police car at a speed of 140 km/h. (*a*) What is the frequency difference between the transmitted signal and the signal received from the receding car? (*b*) Suppose the police car is, instead, moving at a speed of 60 km/h in the same direction as the other vehicle. What is the difference in frequency between the emitted and the reflected signals?

82 •• BIOLOGICAL APPLICATION, CONTEXT-RICH In modern medicine, the Doppler effect is routinely used to measure the rate and direction of blood flow in arteries and veins. High-frequency "ultrasound" (sound at frequencies above the human hearing range) is typically employed. Suppose you are in charge of measuring the blood flow in a vein (located in the lower leg of an older patient) that returns blood upward to the heart. Her varicose veins indicate that perhaps the one-way valves in the vein are not working properly and that the blood is "pooling" in the veins and perhaps even that the blood flow is backward toward her feet. Employing sound that has a frequency of 50.0 kHz, you point the sound source from above her thigh region down toward her feet and measure the sound reflected from that vein area to be lower than 50.0 kHz. (*a*) Was your diagnosis of the valve condition correct? If so, explain. (*b*) Estimate the instrument's frequency difference capability to enable you to measure speeds down to 1.00 mm/s. Take the speed of sound in flesh to be the same as that in water, 1500 m/s.

83 •• A sound source of frequency f_s moves with speed u_s relative to still air toward a receiver who is moving away from the source with speed u_r relative to still air. (*a*) Write an expression for the received frequency f'_r. (*b*) Use the result that $(1 - x)^{-1} \approx 1 + x$ to show that if both u_s and u_r are small compared to v, then the received frequency is approximately

$$f'_r = \left(1 + \frac{u_{rel}}{v}\right) f_s$$

where $u_{rel} = u_s - u_r$ is the velocity of the source relative to the receiver. SSM

84 •• To study the Doppler shift on your own, you take an electronic tone generating device that is set to a frequency of middle C (262 Hz) to a campus wishing well known as "The Abyss." When you hold the device at arm's length (1.0 m), you measure its intensity level to be 80.0 dB. You then drop the tuner down the hole, listening to its sound as it falls. After the tuner has fallen for 5.50 s, what frequency do you hear?

85 •• You are in a hot-air balloon carried along by a 36-km/h wind and have a sound source with you that emits a sound of 800 Hz as it approaches a tall building. (*a*) What is the frequency of the sound heard by an observer at the window of this building? (*b*) What is the frequency of the reflected sound heard by you?

86 •• A car is approaching a reflecting wall. A stationary observer behind the car hears a sound of frequency 745 Hz from the car horn and a sound of frequency 863 Hz from the wall. (*a*) How fast is the car traveling? (*b*) What is the frequency of the car horn? (*c*) What frequency does the car driver hear reflected from the wall?

87 •• The driver of a car traveling at 100 km/h toward a vertical wall briefly sounds the horn. Exactly 1.00 s later she hears the echo and notes that its frequency is 840 Hz. How far from the wall was the car when the driver sounded the horn and what is the frequency of the horn?

88 •• You are on a transatlantic flight traveling due west at 800 km/h. An experimental plane flying at Mach 1.6 and 3.0 km to the north of your plane is also on an east-to-west course. What is the distance between the two planes when you hear the sonic boom from the experimental plane?

89 ••• The Hubble space telescope has been used to determine the existence of planets orbiting around distant stars. A planet orbiting a star will cause the star to "wobble" with the same period as the planet's orbit. Because of this wobble, light from the star will be Doppler-shifted up and down periodically. Estimate the maximum and minimum wavelengths of light of nominal wavelength 500 nm emitted by the Sun that is Doppler-shifted by the motion of the Sun due to the planet Jupiter.

GENERAL PROBLEMS

90 • At time $t = 0$, the shape of a wave pulse on a string is given by the function $y(x,0) = 0.120\ \text{m}^3/((2.00\ \text{m})^2 + x^2)$, where x is in meters. (*a*) Sketch $y(x, 0)$ versus x. (*b*) Give the wave function $y(x,t)$ at a general time t if the pulse is moving in the $+x$ direction with a speed of 10.0 m/s and if the pulse is moving in the $-x$ direction with a speed of 10.0 m/s.

91 • A whistle that has a frequency of 500 Hz moves in a circle of radius 1.00 m at 3.00 rev/s. What are the maximum and minimum frequencies heard by a stationary listener in the plane of the circle and 5.00 m away from its center? SSM

92 • Ocean waves move toward the beach with a speed of 8.90 m/s and a crest-to-crest separation of 15.0 m. You are in a small boat anchored off shore. (*a*) At what frequency do the wave crests reach your boat? (*b*) You now lift anchor and head out to sea at a speed of 15.0 m/s. At what frequency do the wave crests reach your boat now?

93 •• A 12.0-m-long wire that has an 85.0-g mass is under a tension of 180 N. A pulse is generated at the left end of the wire, and 25.0 ms later a second pulse is generated at the right end of the wire. Where do the pulses first meet?

94 •• You are parked on the shoulder of a highway. Find the speed of a car in which the tone of the car's horn drops by 10 percent as it passes you. (In other words, the total drop in frequency between the "approach" value and the "recession" value is 10%.)

95 •• A loudspeaker driver 20.0 cm in diameter is vibrating at 800 Hz with an amplitude of 0.0250 mm. Assuming that the air molecules in the vicinity have the same amplitude of vibration, find (*a*) the pressure amplitude immediately in front of the driver, (*b*) the sound intensity, and (*c*) the acoustic power being radiated by the front surface of the driver. SSM

96 •• A plane, harmonic, sound wave in air has an amplitude of 1.00 μm and an intensity of 10.0 m W/m^2. What is the frequency of the wave?

97 •• Water flows at 7.0 m/s in a pipe of radius 5.0 cm. A plate with area equal to the cross-sectional area of the pipe is suddenly inserted to stop the flow. Find the force exerted on the plate. Take the speed of sound in water to be 1.4 km/s. *Hint: When the plate is inserted, a pressure wave propagates through the water at the speed of sound, v_s. The mass of water brought to a stop in time Δt is the water in a length of pipe equal to $v_s\,\Delta t$.*

98 •• A high-speed flash photography setup meant to capture a picture of a bullet exploding a soap bubble is shown in Figure 15-33. The shock wave from the bullet is to be detected by a microphone that will trigger the flash. The microphone is placed on a track that is parallel to and 0.350 m below the path of the bullet. The track is used to adjust the position of the microphone. If the bullet is traveling at 1.25 times the speed of sound, how far back from the soap bubble must the microphone be set to trigger the flash? (Assume that the flash itself is instantaneous once the microphone is triggered.)

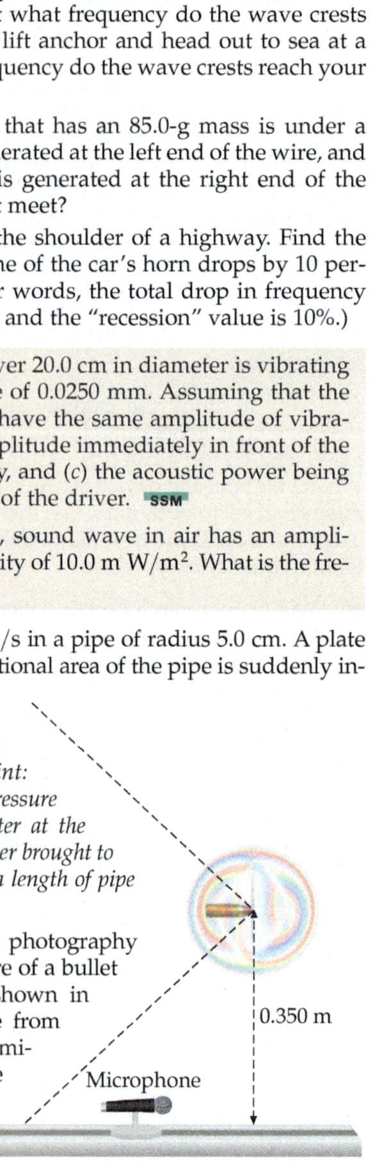

FIGURE 15-33 Problem 98

99 •• A column of precision marchers keeps in step by listening to the band positioned at the head of the column. The beat of the music is for 100 paces/min. A television camera shows that only the marchers at the front and the rear of the column are actually in step. The marchers in the middle section are striding forward with the left foot when those at the front and rear are striding forward with the right foot. The marchers are so well trained, however, that they are all certain that they are in proper step with the music. How long is the column?

100 •• **BIOLOGICAL APPLICATION** A bat flying toward a stationary obstacle at 12.0 m/s emits brief, high-frequency sound pulses at a repetition frequency of 80.0 Hz. What is the interval between the arrival times of the reflected pulses heard by the bat?

101 •• Laser ranging to the moon is done routinely to accurately determine the Earth-moon distance. However, to determine the distance accurately, corrections must be made for the average speed of light in Earth's atmosphere, which is 99.997 percent of the speed of light in vacuum. Assuming that Earth's atmosphere is 8.00 km high, estimate the length of the correction.

102 •• A tuning fork attached to a taut string generates transverse waves. The vibration of the fork is perpendicular to the string. Its frequency is 400 Hz and the amplitude of its oscillation is 0.50 mm. The string has a linear mass density of 0.010 kg/m and is under a tension of 1.0 kN. Assume that there are no waves reflected at the far end of the string. (*a*) What are the period and frequency of waves on the string? (*b*) What is the speed of the waves? (*c*) What are the wavelength and wave number? (*d*) What is a suitable wave function for the waves on the string? (*e*) What is the maximum speed and acceleration of a point on the string? (*f*) At what minimum average rate must energy be supplied to the fork to keep it oscillating at a steady amplitude?

103 ••• A long rope with a mass per unit length of 0.100 kg/m is under a constant tension of 10.0 N. A motor drives one end of the rope with transverse simple harmonic motion at 5.00 cycles per second and an amplitude of 40.0 mm. (*a*) What is the wave speed? What is the wavelength? (*c*) What is the maximum transverse linear momentum of a 1.00-mm segment of the rope? (*d*) What is the maximum net force on a 1.00-mm segment of the rope?

104 ••• In this problem, you will derive an expression for the potential energy of a segment of a string carrying a traveling wave (Figure 15-34). The potential energy of a segment equals the work done by the tension in stretching the string, which is $\Delta U = F_T(\Delta\ell - \Delta x)$, where F_T is the tension, $\Delta\ell$ is the length of the stretched segment, and Δx is its original length. (*a*) Use the binomial expansion to show that $\Delta\ell - \Delta x \approx \frac{1}{2}(\Delta y/\Delta x)^2\,\Delta x$, and therefore $\Delta U \approx \frac{1}{2}F_T(\Delta y/\Delta x)^2\,\Delta x$. (*b*) Compute $\partial y/\partial x$ from the wave function $y(x, t) = A\sin(kx - \omega t)$ (Equation 15-15) and show that $\Delta U \approx \frac{1}{2}F_T k^2 A^2\cos^2(kx - \omega t)\,\Delta x$.

$$\Delta\ell = \sqrt{(\Delta x)^2 + (\Delta y)^2} = \Delta x[1 + (\Delta y/\Delta x)^2]^{1/2}$$

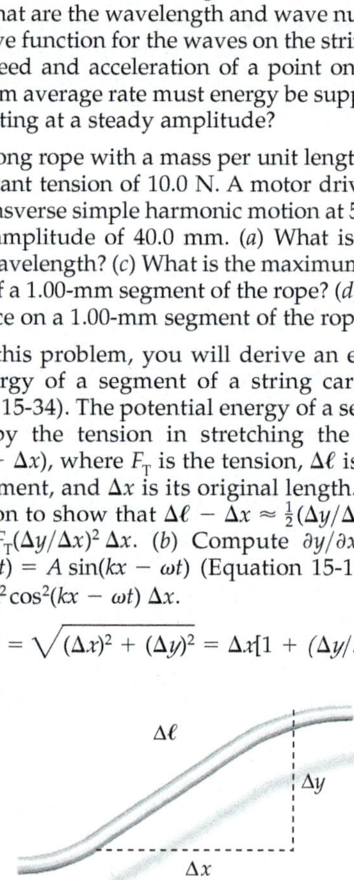

FIGURE 15-34 Problem 104

CHAPTER

16

COMPOSED OF OVER 6134 PIPES OF WIDELY VARYING SIZES, THIS ORGAN IS CAPABLE OF NOTES RANGING FROM A C THAT IS BELOW THE LOWEST C ON A PIANO AND HAS A FREQUENCY OF ONLY 16 Hz TO A NOTE THAT IS A FULL OCTAVE AND A THIRD HIGHER THAN A PIANO'S HIGHEST NOTE AND HAS A FREQUENCY OF 10,548 Hz. *(Ted Soqui/Corbis.)*

Superposition and Standing Waves

? What is the length of the organ pipe that produces the 16-Hz note? (See Example 16-9.)

To get a clear understanding of simple wave motion, in Chapter 15 we examined the movement of a sequence of disturbances through a medium. However, if you have been to the ocean, perhaps you have observed what happens when these disturbances collide and cut across each other. When two or more waves overlap in space, their individual disturbances superimpose, adding algebraically, to create a resultant wave. For the case of harmonic waves, overlapping waves of the same frequency produce sustained wave patterns in space.

The Walt Disney Concert Hall in Los Angeles, California, which houses the pipe organ shown here, is an engineering and acoustic marvel. Structural and civil engineers worked to establish structural integrity of the Frank Gehry–designed organ and to ensure that the organ is strong enough to withstand earthquakes. Acoustical engineers created models for acoustic testing. One such model, scaled to one-tenth actual size, even included felt-covered lead figures to represent audience members. (Sound waves at 10 times the normal frequency—and one-tenth of the normal wavelength—were used to test the design.)

Our study of waves does not end with this chapter, though. We will continue our examination of waves in Chapter 34 where the wave nature of electrons and other material objects are integral to our understanding of quantum physics.

In this chapter, we begin with the superposition of wave pulses on a string and then consider the superposition and interference of harmonic waves. We examine the phenomenon of beats and study standing waves, which occur when harmonic waves are confined in space. Finally, we consider the analysis of complex musical tones.

16-1 SUPERPOSITION OF WAVES

Figure 16-1*a* shows two small-amplitude wave pulses of different durations moving in opposite directions on a string. The shape of the string when they overlap can be found by adding the displacements that would be produced by each pulse separately. The **principle of superposition** is a property of wave motion, which states:

> When two or more waves overlap, the resultant wave is the algebraic sum of the individual waves.
>
> PRINCIPLE OF SUPERPOSITION

That is, when there are two pulses on the string, the total wave function is the algebraic sum of the individual wave functions. While the principle of superposition holds for many waves, it does not hold for all waves. For example, the principle of superposition does not hold if the sum of two displacements exceeds the proportional limit* of the medium. Throughout the discussions that follow, we assume that the principle of superposition holds.

In the special case of two pulses that are identical except that one is inverted relative to the other, as in Figure 16-1*b*, there is an instant when the pulses exactly overlap and add to zero. At this instant the string is horizontal. A short time later the individual pulses emerge, each continuing in its original direction. That is, they exit the overlap region looking exactly as they did prior to entering the overlap region.

> ! After two wave pulses traveling in opposite directions "collide," they each continue moving with the same speed, size, and shape that they had before the "collision."

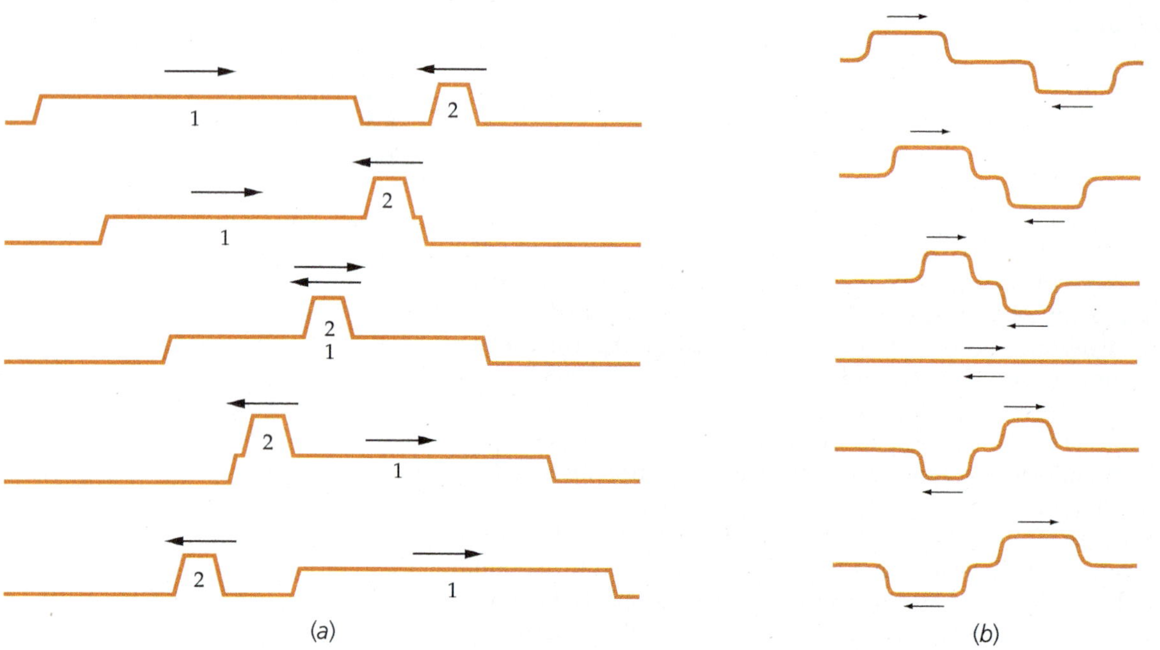

FIGURE 16-1 Wave pulses moving in opposite directions on a string. The shape of the string when the pulses overlap is found by adding the displacements due to each separate pulse. (*a*) Superposition of two pulses having displacements in the same direction (upward). The figure shows the shape of the string at equal time intervals of duration Δt. Each pulse travels the length of pulse 2 during time Δt. (*b*) Superposition of two pulses having equal displacements in opposite directions. Here the algebraic addition of the displacement amounts to the subtraction of the magnitudes.

* The proportional limit of an elastic material is the maximum strain for which stress is proportional to strain. Stress and strain are discussed in Section 8 of Chapter 12.

Example 16-1 Colliding Pulses *Conceptual*

An upright pulse on a taut string moves to the right, while an inverted pulse of the same size and shape moves to the left. When these pulses overlap there is an instant when the string is flat and no pulses can be seen. This is all in accord with the principle of superposition. The question is, why do the pulses reappear and continue on following the collision?

PICTURE The displacement of each point on the string is zero at the instant the string is flat, but is the velocity of each point zero at that instant? For an upright pulse, the string in the leading edge of the pulse is moving upward and the string in the trailing edge is moving downward. For an inverted pulse the opposite is true: the string in the leading edge is moving downward and the string in the trailing edge is moving upward.

SOLVE

1. Plot both the position and the velocity of the string versus the position along the string before the pulses overlap (Figure 16-2). For an upright pulse, the string in the leading edge is moving upward and the string in the trailing edge is moving downward. For an inverted pulse, the opposite is true; the string in the leading edge is moving downward and the string in the trailing edge is moving upward.

2. This time plot both the position and the velocity of the string versus the position along the string at the instant the pulses completely overlap (Figure 16-3).

3. Is the velocity zero at all points on the string at the instant the string is flat? In step 1, the velocity profiles of the string are identical for the two pulses, so when the two pulses overlap, the displacements add to zero, but the velocities do not add to zero. The pulses reform after they overlap because the string is moving and has inertia. Thus, it does not stay flat.

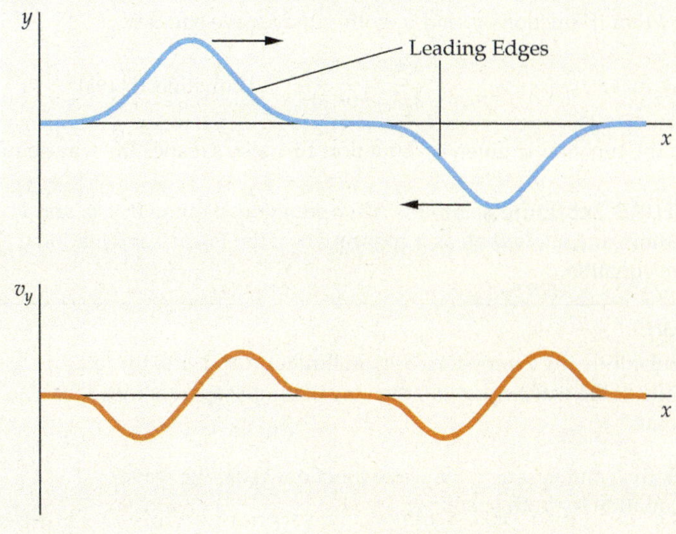

FIGURE 16-2

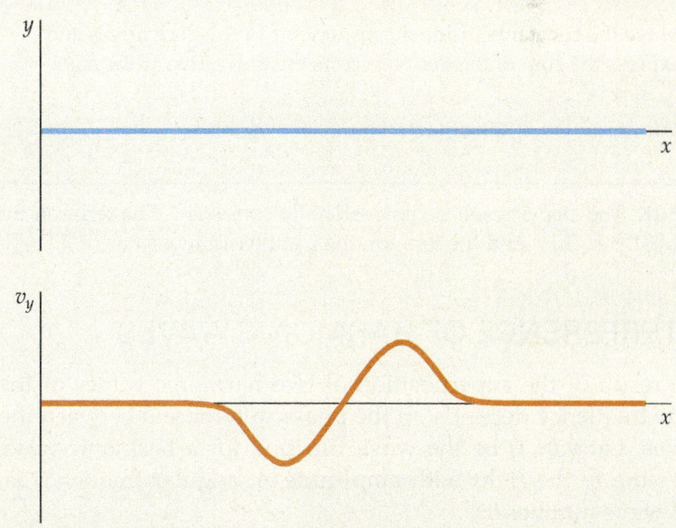

FIGURE 16-3

* SUPERPOSITION AND THE WAVE EQUATION

The principle of superposition follows from the fact that the wave equation (Equation 15-10*b*) is linear for small transverse displacements. That is, the function $y(x, t)$ and its derivatives occur only to the first power. The defining property of a linear equation is that if y_1 and y_2 are two solutions of the equation, then the linear combination

$$y_3 = C_1 y_1 + C_2 y_2 \tag{16-1}$$

where C_1 and C_2 are any constants, is also a solution. The linearity of the wave equation can be shown by the direct substitution of y_3 into the wave equation. The result is the mathematical statement of the principle of superposition. If any two waves satisfy a wave equation, then their algebraic sum also satisfies the same wave equation.

Example 16-2 Superposition and the Wave Equation

Show that if functions y_1 and y_2 both satisfy wave equation

$$\frac{\partial^2 y}{\partial x^2} = \frac{1}{v^2}\frac{\partial^2 y}{\partial t^2} \qquad \text{(Equation 15-10}b\text{)}$$

then the function y_3 given by Equation 16-1 also satisfies the wave equation.

PICTURE Substitute y_3 into the wave equation, assume that y_1 and y_2 each satisfy the wave equation, and show that, as a consequence, the linear combination $C_1y_1 + C_2y_2$ satisfies the wave equation.

SOLVE

1. Substitute the expression for y_3 in Equation 16-1 into the left side of the wave equation, then break it into separate terms for y_1 and y_2:

$$\frac{\partial^2 y_3}{\partial x^2} = \frac{\partial^2}{\partial x^2}(C_1y_1 + C_2y_2) = C_1\frac{\partial^2 y_1}{\partial x^2} + C_2\frac{\partial^2 y_2}{\partial x^2}$$

2. Both y_1 and y_2 satisfy the wave function. Write the wave equation for both y_1 and y_2:

$$\frac{\partial^2 y_1}{\partial x^2} = \frac{1}{v^2}\frac{\partial^2 y_1}{\partial t^2} \quad \text{and} \quad \frac{\partial^2 y_2}{\partial x^2} = \frac{1}{v^2}\frac{\partial^2 y_2}{\partial t^2}$$

3. Substitute the step-2 results into the step-1 result and factor out any common terms:

$$\frac{\partial^2 y_3}{\partial x^2} = C_1\frac{1}{v^2}\frac{\partial^2 y_1}{\partial t^2} + C_2\frac{1}{v^2}\frac{\partial^2 y_2}{\partial t^2} = \frac{1}{v^2}\left(C_1\frac{\partial^2 y_1}{\partial t^2} + C_2\frac{\partial^2 y_2}{\partial t^2}\right)$$

4. Move the constants inside the arguments of the derivatives and express the sum of the derivatives as the derivative of the sum:

$$\frac{\partial^2 y_3}{\partial x^2} = \frac{1}{v^2}\left(\frac{\partial^2 C_1y_1}{\partial t^2} + \frac{\partial^2 C_2y_2}{\partial t^2}\right) = \frac{1}{v^2}\frac{\partial^2}{\partial t^2}(C_1y_1 + C_2y_2)$$

5. The argument of the time derivative in step 4 is y_3:

$$\therefore \boxed{\frac{\partial^2 y_3}{\partial x^2} = \frac{1}{v^2}\frac{\partial^2 y_3}{\partial t^2}}$$

CHECK The step-5 result is dimensionally consistent. The term on the left has dimensions of $[\text{L}]/[\text{L}]^2 = [\text{L}]^{-1}$ and the term on the right has dimensions of $\{[\text{T}]^2/[\text{L}]^2\}\{[\text{L}]/[\text{T}]^2\} = [\text{L}]^{-1}$.

INTERFERENCE OF HARMONIC WAVES

The result of the superposition of two harmonic waves of the same frequency depends on the phase difference δ between the waves. Let $y_1(x, t)$ be the wave function for a harmonic wave traveling to the right with amplitude A, angular frequency ω, and wave number k:

$$y_1 = A\sin(kx - \omega t) \qquad 16\text{-}2$$

For this wave function, we have chosen the phase constant to be zero.* If we have another harmonic wave also traveling to the right with the same amplitude, frequency, and wave number, then the general equation for its wave function can be written

$$y_2 = A\sin(kx - \omega t + \delta) \qquad 16\text{-}3$$

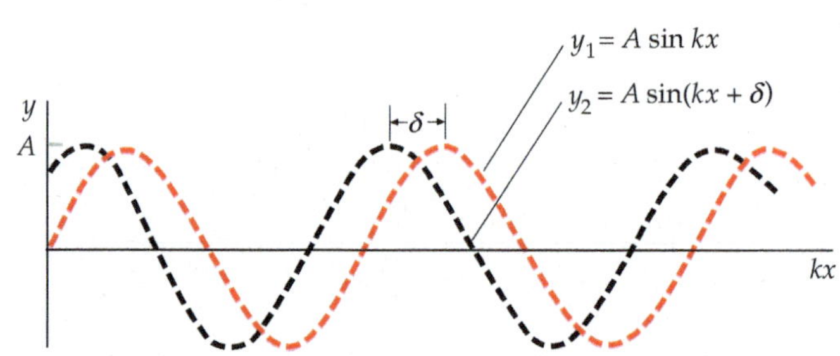

FIGURE 16-4 Displacement versus position at (a given instant) for two harmonic waves having the same amplitude, frequency, and wavelength, but differing in phase by δ.

where δ is the phase constant. The two waves described by Equations 16-2 and 16-3 differ in phase by δ. Figure 16-4 shows a plot of the two wave functions versus position at time $t = 0$. The resultant wave is the sum

$$y_1 + y_2 = A\sin(kx - \omega t) + A\sin(kx - \omega t + \delta) \qquad 16\text{-}4$$

We can simplify Equation 16-4 by using the trigonometric identity

$$\sin\theta_1 + \sin\theta_2 = 2\cos\tfrac{1}{2}(\theta_1 - \theta_2)\sin\tfrac{1}{2}(\theta_1 + \theta_2) \qquad 16\text{-}5$$

See Math Tutorial *for more information on* Trigonometry

* This choice is convenient but not mandatory. If, for example, we chose $t = 0$ when the displacement was maximum at $x = 0$, we would write $y_1 = A\cos(kx - \omega t) = A\sin(kx - \omega t + \frac{1}{2}\pi)$.

For this case, $\theta_1 = kx - \omega t$ and $\theta_2 = kx - \omega t + \delta$, so that

$$\tfrac{1}{2}(\theta_1 - \theta_2) = -\tfrac{1}{2}\delta$$

and

$$\tfrac{1}{2}(\theta_1 + \theta_2) = kx - \omega t + \tfrac{1}{2}\delta$$

Thus, Equation 16-4 becomes

$$y_1 + y_2 = [2A\cos\tfrac{1}{2}\delta]\sin(kx - \omega t + \tfrac{1}{2}\delta) \qquad 16\text{-}6$$

SUPERPOSITION OF TWO WAVES OF THE SAME AMPLITUDE AND FREQUENCY

where we have used $\cos(-\frac{1}{2}\delta) = \cos\frac{1}{2}\delta$. We see that the result of the superposition of two harmonic waves having the same wave number k and frequency ω is a harmonic wave having wave number k and frequency ω. The resultant wave has amplitude $2A\cos\frac{1}{2}\delta$ and a phase equal to half the difference between the phases of the original waves. The phenomenon of two or more waves of the same, or almost the same, frequency superposing to produce an observable pattern in the intensity is called **interference.** In this example, the intensity, which is proportional to the square of the amplitude, is uniform. If the two waves are in phase, then $\delta = 0$, $\cos 0 = 1$, and the amplitude of the resultant wave is $2A$. The interference of two waves in phase is called **constructive interference** (Figure 16-5). If the two waves are 180° out of phase, then $\delta = \pi$, $\cos(\frac{1}{2}\delta) = 0$, and the amplitude of the resultant wave is zero. The interference of two waves 180° out of phase is called **destructive interference** (Figure 16-6).

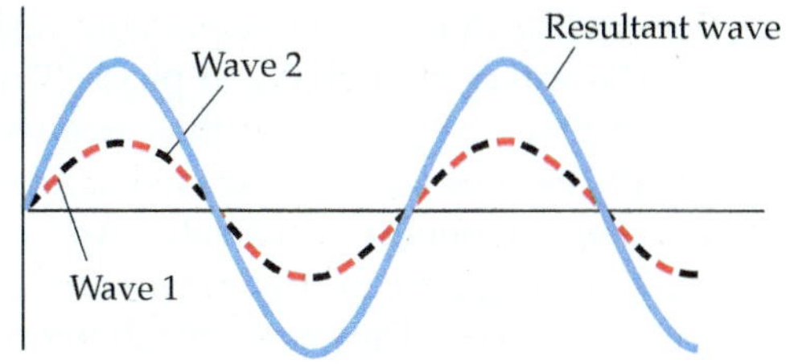

FIGURE 16-5 Constructive interference. If two harmonic waves of the same frequency are in phase, the amplitude of the resultant wave is the sum of the amplitudes of the individual waves. Waves 1 and 2 are identical, so they appear as a single harmonic wave. Wave 1 is shown as a red dashed curve and Wave 2 is shown as a black dashed curve.

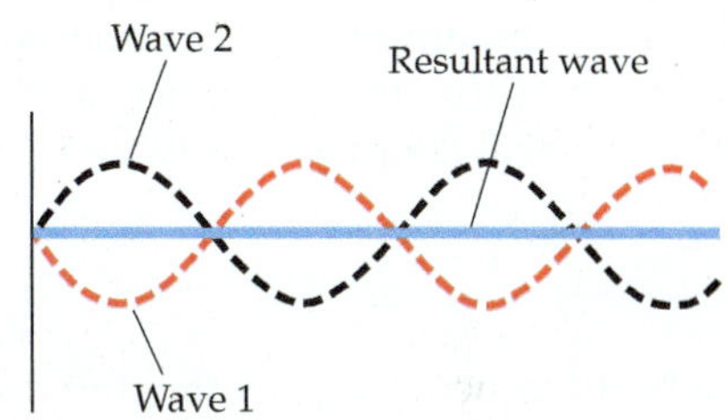

FIGURE 16-6 Destructive interference. If two harmonic waves of the same frequency differ in phase by 180°, the amplitude of the resultant wave is the difference between the amplitudes of the individual waves. If the original waves have equal amplitudes, they cancel completely.

PRACTICE PROBLEM 16-1

Two waves with the same frequency, wavelength, and amplitude are traveling in the same direction. (*a*) If they differ in phase by 90.0° and each has an amplitude of 4.00 cm, what is the amplitude of the resultant wave? (*b*) For what phase difference δ will the resultant amplitude be equal to 4.0 cm?

Beats The interference of two sound waves with slightly different frequencies produces the interesting phenomenon known as **beats.** Consider two sound waves that have angular frequencies of ω_1 and ω_2 and the same pressure amplitude p_0. What do we hear? At a fixed point, the spatial dependence of the wave merely contributes a phase constant, so we can neglect it. The pressure at the ear due to either wave acting alone will be a simple harmonic function of the type

$$p_1 = p_0 \sin\omega_1 t$$

and

$$p_2 = p_0 \sin\omega_2 t$$

where we have chosen sine functions, rather than cosine functions for convenience, and have assumed that the waves are in phase at time $t = 0$. Using the trigonometry identity

$$\sin\theta_1 + \sin\theta_2 = 2\cos\tfrac{1}{2}(\theta_1 - \theta_2)\sin\tfrac{1}{2}(\theta_1 + \theta_2)$$

for the sum of two sine functions, we obtain for the resultant wave

$$p = p_0\sin\omega_1 t + p_0\sin\omega_2 t = 2p_0\cos\tfrac{1}{2}(\omega_1 - \omega_2)t\,\sin\tfrac{1}{2}(\omega_1 + \omega_2)t$$

If we write $\omega_{av} = (\omega_1 + \omega_2)/2$ for the average angular frequency and $\Delta\omega = \omega_1 - \omega_2$ for the difference in angular frequencies, the resultant wave function is

$$p = 2p_0\cos(\tfrac{1}{2}\Delta\omega t)\sin\omega_{av}t = 2p_0\cos(2\pi\tfrac{1}{2}\Delta f t)\sin 2\pi f_{av}t \qquad 16\text{-}7$$

where $\Delta f = \Delta\omega/(2\pi)$ and $f_{av} = \omega_{av}/(2\pi)$.

Figure 16-7 shows a plot of pressure variations as a function of time. The waves are initially in phase. Thus, they add constructively at time $t = 0$. Because their frequencies differ, the waves gradually become out of phase, and at time t_1 they are 180° out of phase and interfere destructively.* An equal time interval later (time t_2 in the figure), the two waves are again in phase and interfere constructively. The greater the difference in the frequencies of the two waves, the more rapidly they oscillate in and out of phase.

When two tuning forks vibrate with equal amplitudes and with almost equal frequencies f_1 and f_2, the tone that we hear has a frequency of $f_{av} = (f_1 + f_2)/2$ and an amplitude of $2p_0 \cos(2\pi \frac{1}{2} \Delta f\, t)$. (For some values of t the amplitude is negative. Because $-\cos\theta = \cos(\theta + \pi)$, a change in the sign of the amplitude is equivalent to a 180° phase change.) The amplitude oscillates with the frequency $\frac{1}{2}\Delta f$. Because the sound intensity is proportional to the square of the amplitude, the sound is loud whenever the amplitude function is either a maximum or a minimum. Thus, the frequency of this variation in intensity, called the **beat frequency,** is twice $\frac{1}{2}\Delta f$:

$$f_{beat} = \Delta f \qquad 16\text{-}8$$

BEAT FREQUENCY

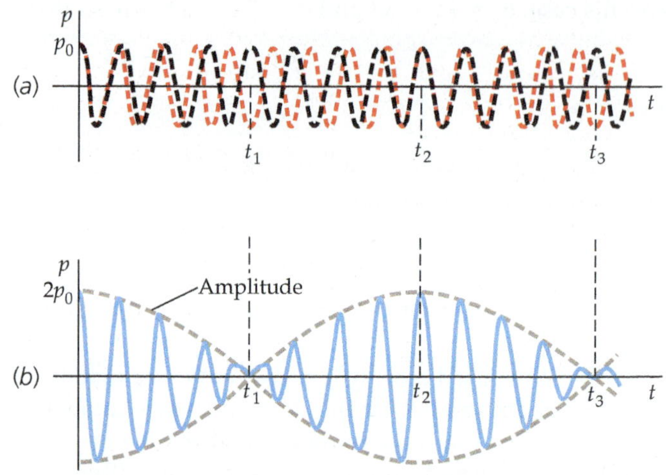

FIGURE 16-7 Beats. (*a*) Two harmonic waves of different but nearly equal frequencies that are in phase at $t = 0$ are 180° out of phase at some later time t_1. At a still later time, t_2, they are back in phase. (*b*) The resultant of the two waves shown in (*a*). The frequency of the resultant wave is about the same as the frequencies of the original waves, but the amplitude is modulated as indicated. The intensity is maximum at times 0 and t_2, and zero at times t_1 and t_3.

The beat frequency equals the difference in the individual frequencies of the two waves. If we simultaneously strike two tuning forks having the frequencies 241 Hz and 243 Hz, we will hear a pulsating tone at the average frequency of 242 Hz that has a maximum intensity at half-second intervals; that is, the beat frequency is 2 Hz. The ear can detect beats with beat frequencies of up to about 15 to 20 per second. Above this frequency, the fluctuations in loudness are too rapid to be distinguished.

The phenomenon of beats is often used to compare an unknown frequency with a known frequency, as when a tuning fork is used to tune a piano string. Pianos are tuned by simultaneously ringing the tuning fork and striking a key, while at the same time adjusting the tension of the piano string until the beats are far apart, indicating that the difference in frequency of the two sound generators is very small.

Example 16-3 Tuning a Guitar

When a 440-Hz (concert A) tuning fork is struck simultaneously with the playing of the A string of a slightly out-of-tune guitar, 3.00 beats per second are heard. The guitar string is tightened a little to increase its frequency. As the guitar string is slowly tightened, you hear the beat frequency slowly increase. What was the initial frequency of the guitar string (the frequency before it was tightened)?

PICTURE Because 3.00 beats per second were heard initially, the initial frequency of the guitar string was either 437 Hz or 443 Hz. The greater the difference between the frequency of the string and the frequency of the tuning fork, the greater the beat frequency. The frequency of the string increases with an increase in the tension.

SOLVE

1. Because the beat frequency increases as the tension increases, the initial frequency must have been 443 Hz: $f = f_A + f_{beat} = 440\text{ Hz} + 3.00\text{ Hz} = \boxed{443\text{ Hz}}$

CHECK The answer has the correct number of significant figures.

The tuning fork has been struck and a string has been plucked. Listening to the beats the man is tightening (or loosening) the string in order to bring the beat frequency to zero. As the beat frequency approaches zero the string frequency approaches the tuning-fork frequency. *(Ray Malace Photography.)*

* Complete cancellation occurs only when the pressure amplitudes of the two waves are equal.

Phase difference due to path difference A common cause of a phase difference between two waves is different path lengths between the sources of the waves and the point where the interference occurs. Suppose that two sources oscillate in phase (positive crests leave the sources at the same time) and emit harmonic waves of the same frequency and wavelength. Now consider a point in space for which the path lengths to the two sources differ. If the path difference is one wavelength, as is the case in Figure 16-8*a*, or any other integral number of wavelengths, the interference is constructive. If the path difference is one-half of a wavelength or an odd number of half wavelengths, as in Figure 16-8*b*, the maximum of one wave at the same time as the minimum of the other and the interference is destructive.

The wave functions for waves from two sources oscillating in phase can be written

$$p_1 = p_0 \sin(kx_1 - \omega t)$$

and

$$p_2 = p_0 \sin(kx_2 - \omega t)$$

The phase difference for these two wave functions is

$$\delta = (kx_2 - \omega t) - (kx_1 - \omega t) = k(x_2 - x_1) = k\,\Delta x$$

Using $k = 2\pi/\lambda$, we have

$$\delta = k\,\Delta x = 2\pi\frac{\Delta x}{\lambda} \qquad 16\text{-}9$$

PHASE DIFFERENCE DUE TO PATH DIFFERENCE

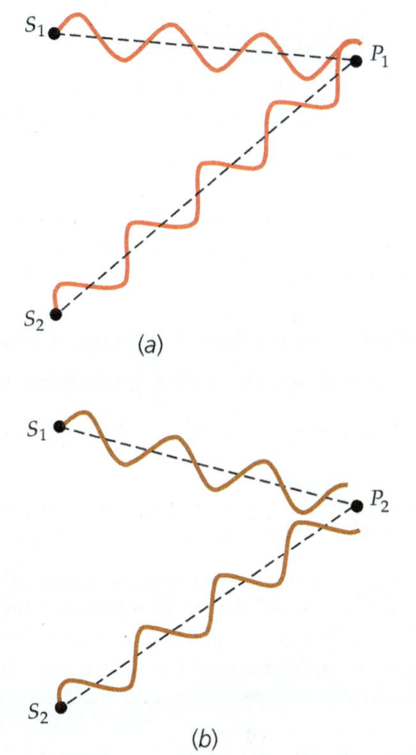

FIGURE 16-8 Waves from two sources S_1 and S_2 are in phase when they meet at a point P_1. (*a*) When the path difference is one wavelength λ, the waves are in phase at P_1 and therefore interfere constructively. (*b*) When the path difference is $\frac{1}{2}\lambda$, the waves at P_2 are out of phase by 180° and therefore interfere destructively. If the waves are of equal amplitude at P_2, they will cancel completely at this point.

Example 16-4 A Resultant Sound Wave

Two identical loudspeakers are driven in phase by a common audio oscillator. At a point 5.00 m from one speaker cone and 5.17 m from the other, the amplitude of the sound from each is p_0. Find the amplitude of the resultant wave at that point if the frequency of the sound waves is (*a*) 1000 Hz, (*b*) 2000 Hz, and (*c*) 500 Hz. (Use 340 m/s for the speed of sound.)

PICTURE The amplitude of the resultant wave due to superposition of two waves differing in phase by δ is given by $A = 2p_0 \cos\frac{1}{2}\delta$ (Equation 16-6), where p_0 is the amplitude of either wave and $\delta = 2\pi\,\Delta x/\lambda$ (Equation 16-9) is the phase difference. We are given the path difference, $\Delta x = 5.17\text{ m} - 5.00\text{ m} = 0.17\text{ m}$, so all that is needed is the wavelength λ.

SOLVE

(*a*) 1. The wavelength equals the speed divided by the frequency. Calculate λ for $f = 1000$ Hz:

$$\lambda = \frac{v}{f} = \frac{340\text{ m/s}}{1000\text{ Hz}} = 0.340\text{ m}$$

2. For $\lambda = 0.340$ m, the given path difference ($\Delta x = 0.17$ m) is $\frac{1}{2}\lambda$, so we expect destructive interference. Use this value of λ and $A = 2p_0 \cos\frac{1}{2}\delta$ (Equation 16-6), to calculate the phase difference δ, and then use δ to calculate the amplitude A:

$$\delta = 2\pi\frac{\Delta x}{\lambda} = 2\pi\frac{0.17\text{ m}}{0.340\text{ m}} = \pi$$

$$\text{so}\quad A = 2p_0\cos\frac{1}{2}\delta = 2p_0\cos\frac{\pi}{2} = \boxed{0.0\text{ m}}$$

(*b*) 1. Calculate λ for $f = 2000$ Hz:

$$\lambda = \frac{v}{f} = \frac{340\text{ m/s}}{2000\text{ Hz}} = 0.170\text{ m}$$

2. For $\lambda = 0.170$ m, the path difference equals λ, so we expect constructive interference. Calculate the phase difference and amplitude:

$$\delta = 2\pi\frac{\Delta x}{\lambda} = 2\pi\frac{0.170\text{ m}}{0.17\text{ m}} = 2\pi$$

$$\text{so}\quad A = 2p_0\cos\tfrac{1}{2}\delta = 2p_0\cos\pi = \boxed{-2p_0}$$

(*c*) 1. Calculate λ for $f = 500$ Hz: $\lambda = \dfrac{v}{f} = \dfrac{340\text{ m/s}}{500\text{ Hz}} = 0.680\text{ m}$

2. Calculate the phase difference and amplitude: $\delta = 2\pi\dfrac{\Delta x}{\lambda} = 2\pi\dfrac{0.17\text{ m}}{0.680\text{ m}} = \dfrac{\pi}{2}$

so $A = 2p_0 \cos\dfrac{1}{2}\delta = 2p_0\cos\dfrac{\pi}{4} = \boxed{\sqrt{2}\,p_0}$

CHECK Each of the three answers is between $-2p_0$ and $+2p_0$, so the answers are within the expected range.

TAKING IT FURTHER In Part (*b*), *A* is found to be negative. Equation 16-6 can be written $y_1 + y_2 = A'\sin\left(kx - \omega t + \dfrac{\delta}{2}\right)$, which can also be written $y_1 + y_2 = -A'\sin\left(kx - \omega t + \dfrac{\delta}{2} + \pi\right)$. A phase shift of $\pi = 180°$ is equivalent to multiplying by -1.

Example 16-5 Sound Intensity of Two Loudspeakers

The two identical loudspeakers in Example 16-4 are now turned to face each other at a distance of 180 cm. In addition, they are now driven at 686 Hz. Locate the points between the speakers along a line joining them for which the sound intensity is (*a*) maximum, and (*b*) minimum. (Neglect the variation in intensity with distance from either speaker, and use 343 m/s for the speed of sound.)

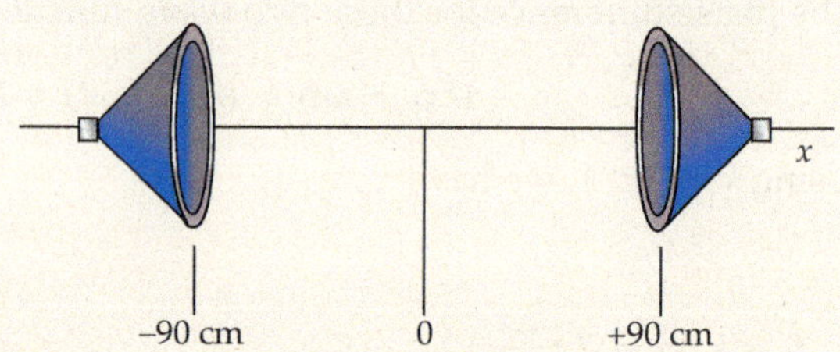

FIGURE 16-9 The two loudspeakers are on the *x* axis with $x = 0$ midway between them.

PICTURE We choose the origin to be at the midpoint between the speakers (Figure 16-9). Because the origin is equidistant from the speakers, it is a point of maximum intensity. When we move a distance *x* from the origin toward one of the speakers, the path difference between us and the two speakers is $2x$. The intensity will be maximum at points where $2x = 0, \lambda, 2\lambda, 3\lambda, \ldots$, and minimum where $2x = \frac{1}{2}\lambda, \frac{3}{2}\lambda, \frac{5}{2}\lambda, \ldots$.

SOLVE

(*a*) 1. The intensity will be maximum when $2x$ equals an integral number of wavelengths: $2x = 0, \pm\lambda, \pm 2\lambda, \pm 3\lambda, \ldots$

2. Calculate the wavelength: $\lambda = \dfrac{v}{f} = \dfrac{343\text{ m/s}}{686\text{ Hz}} = 0.500\text{ m} = 50.0\text{ cm}$

3. Solve for *x* using the calculated wavelength: $x = 0, \pm\frac{1}{2}\lambda, \pm\lambda, \pm\frac{3}{2}\lambda, \ldots = \boxed{0, \pm 25.0\text{ cm}, \pm 50.0\text{ cm}, \pm 75.0\text{ cm}}$

(*b*) 1. The intensity will be minimum when $2x$ equals an odd number of half wavelengths: $2x = \pm\frac{1}{2}\lambda, \pm\frac{3}{2}\lambda, \pm\frac{5}{2}\lambda, \ldots$

2. Solve for *x* using the calculated wavelength: $x = \pm\frac{1}{4}\lambda, \pm\frac{3}{4}\lambda, \pm\frac{5}{4}\lambda, \ldots = \boxed{\pm 12.5\text{ cm}, \pm 37.5\text{ cm}, \pm 62.5\text{ cm}, \pm 87.5\text{ cm}}$

CHECK The answers for Parts (*a*) and (*b*) complement each other, with the intensity minima located halfway between the intensity maxima, as expected.

TAKING IT FURTHER The maxima and minima will be relative maxima and relative minima, because at each maxima (and minima) the amplitude from the nearer speaker will be slightly greater than that from the farther speaker. Only seven terms were used for the maxima and only eight terms for the minima, because any additional terms would not be in the region between the two speakers.

Figure 16-10*a* shows the wave pattern in a ripple tank produced by two point sources that are oscillating in phase. Each source produces waves with circular wavefronts. The circular wavefronts shown all have the same phase (they are all crests) and are separated by one wavelength. We can construct a similar pattern with a compass by drawing circular arcs representing the wave crests from each source at some particular instant of time (Figure 16-10*b*). Where the crests from each source overlap, the waves interfere constructively. At these points, the path lengths from the two sources are either equal or they differ by an integral number of wavelengths. The dashed lines indicate the points that are both equidistant from the sources or whose path differences from the sources are one wavelength, two wavelengths, or three wavelengths. At each point along any of these lines the interference is constructive, so these are lines of interference maxima. Between the lines of interference maxima are lines of interference minima. On a line of interference minima, the path length from any point on the line to each of the two sources differs by an odd number of half wavelengths. Throughout the region where the two waves are superposed, the amplitude of the resultant wave is given by $A = 2p_0 \cos\frac{1}{2}\delta$, where p_0 is the amplitude of each wave separately and δ is related to the path difference Δr by $\delta = 2\pi\,\Delta r/\lambda$ (Equation 16-9).

Figure 16-11 shows the intensity I of the resultant wave from two sources as a function of path difference Δx. At points where the interference is constructive, the amplitude of the resultant wave is twice that of either wave alone, and because the intensity is proportional to the square of the amplitude, the intensity is $4I_0$, where I_0 is the intensity due to either source alone. At points of destructive interference, the intensity is zero. The average intensity, shown by the dashed line at $2I_0$ in the figure, is twice the intensity due to either source alone, a result required by the conservation of energy. The interference of the waves from the two sources thus redistributes the energy in space. The interference of two sound sources can be demonstrated by driving two separated speakers with the same amplifier (so that they are always in phase) fed by an audio-signal generator. Moving about the room, one can detect by ear the positions of constructive and destructive interference.* This demonstration is best done in a room called an *anechoic chamber,* where reflections (echoes) off the walls of the room are minimized.

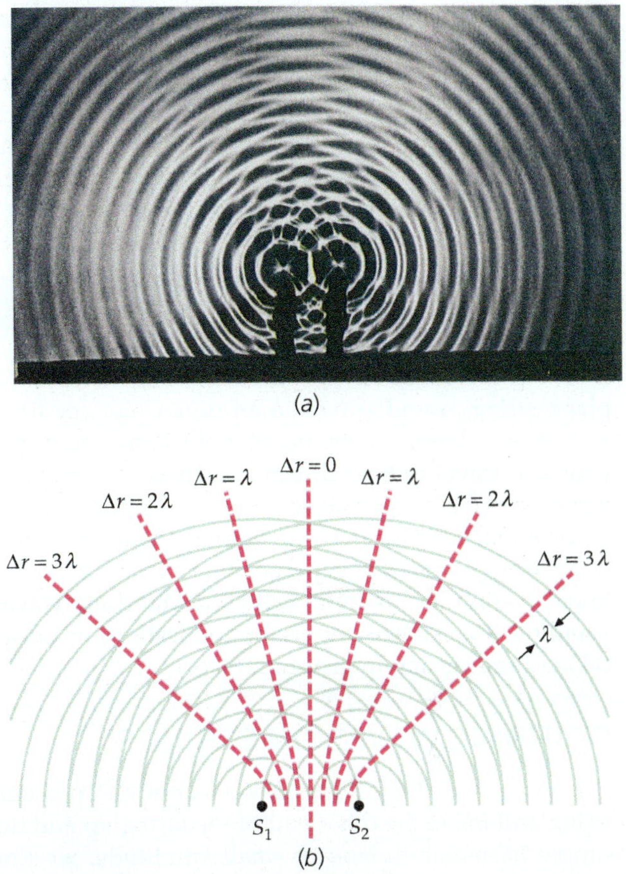

FIGURE 16-10 (*a*) Water waves in a ripple tank produced by two sources oscillating in phase. (*b*) Drawing of wave crests for the sources in (*a*). The dashed lines indicate points for which the path difference is an integral number of wavelengths. *(Part (a) Berenice Abbott, 8J 1328/Photo Researchers.)*

Coherence Two sources need not be in phase to produce an interference pattern. Consider two sources that are 180° out of phase. (Two speakers that are in phase can be made to be out of phase by 180° merely by switching the leads to one of the speakers.) The intensity pattern is the same as that in Figure 16-11, except that the locations of the maxima and minima are interchanged. At points for which the distance differs by an integral number of wavelengths, the interference is destructive because the waves are 180° out of phase. At points where the path difference is an odd number of half wavelengths, the waves are now in phase because the 180° phase difference of the sources is offset by the 180° phase difference due to the path difference.

Similar interference patterns will be produced by any two sources whose phase difference remains constant. Two sources that remain in phase or maintain a constant phase difference are said to be **coherent.** Coherent sources of water waves in a ripple tank are easy to produce by driving both sources with the same motor. Coherent sound sources are obtained by driving two speakers with the same signal source and amplifier.

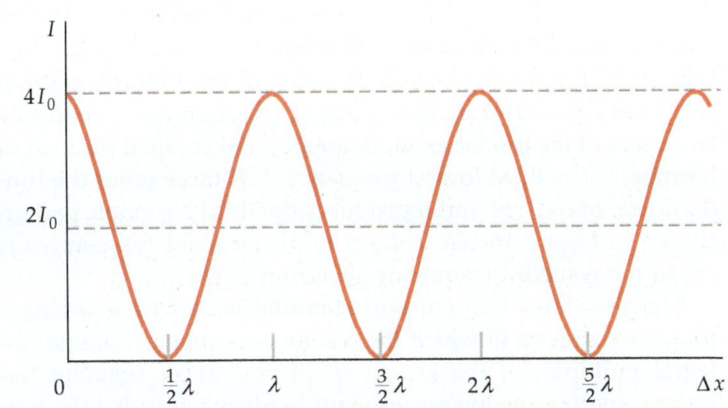

FIGURE 16-11 Intensity versus path difference for two sources that are in phase. I_0 is the intensity due to each source individually.

Wave sources whose difference in phase is not constant, but varies randomly, are said to be **incoherent sources.** There are many examples of incoherent sources, such as two speakers driven by different amplifiers or two violins played by

* In this demonstration, the sound intensity will be not quite zero at the points of destructive interference because of sound reflections from the walls or objects in the room.

different violinists. For incoherent sources, the interference at a particular point varies rapidly back and forth from constructive to destructive, and no interference pattern is sustained long enough to be observed. The resultant intensity of waves from two or more incoherent sources is simply the sum of the intensities due to the individual sources.

16-2 STANDING WAVES

If waves are confined in space, like the waves on a piano string, sound waves in an organ pipe, or light waves in a laser, reflections at both ends cause the waves to travel in both directions. These superposing waves interfere in accordance with the principle of superposition. For a given string or pipe, there are certain frequencies for which superposition results in a stationary vibration pattern called a **standing wave.** Standing waves have important applications in musical instruments and in quantum theory.

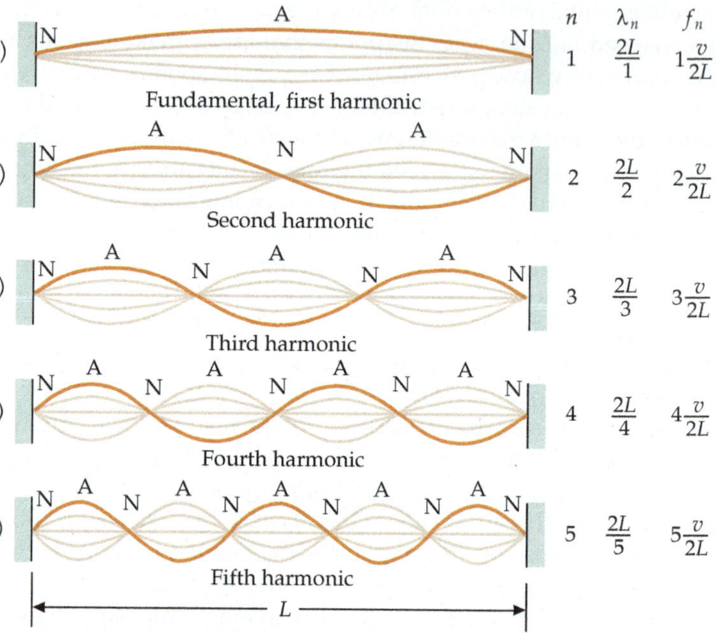

FIGURE 16-12 Standing waves on a string that is fixed at both ends. Antinodes are labeled A and nodes are labeled N. The nth harmonic has n antinodes, where $n = 1, 2, 3, \ldots$.

STANDING WAVES ON STRINGS

String fixed at both ends If we fix one end of a taut flexible string and move the other end of the string up and down with simple harmonic motion of small amplitude, we find that at certain frequencies, standing-wave patterns such as those shown in Figure 16-12 are produced. The frequencies that produce these patterns are called the **resonance frequencies** of the string system. Each such frequency, with its accompanying wave function, is called a **mode of vibration.** The lowest resonance frequency is called the **fundamental** frequency f_1. It produces the standing-wave pattern shown in Figure 16-12*a*, which is called the **fundamental mode** of vibration or the **first harmonic.** The second lowest frequency f_2 produces the pattern shown in Figure 16-12*b*. This mode of vibration has a frequency twice that of the fundamental frequency and is called the second harmonic. The third lowest frequency f_3 is three times the fundamental frequency, and it produces the third harmonic pattern shown in Figure 16-12*c*. The set of all resonant frequencies is called the **resonant frequency spectrum** of the string.

Many systems that support standing waves have resonant frequency spectra in which the resonant frequencies are not integral multiples of the lowest frequency. In all resonant frequency spectra, the lowest resonant frequency is called the fundamental frequency (or just the fundamental), the next lowest resonant frequency is called the first **overtone,** the next lowest the second overtone, and so forth. This terminology has its roots in music. Only if each resonant frequency is an integral multiple of the fundamental frequency are they referred to as harmonics.

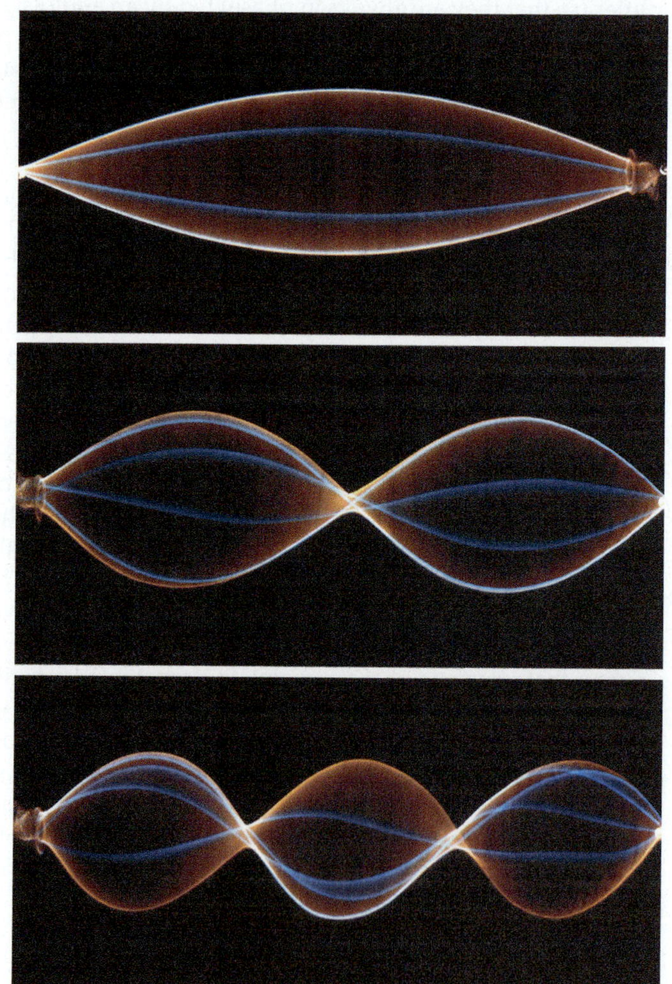

Standing waves on a string made to oscillate by a vibrator attached to the left end of the string. These standing waves occur only at specific frequencies. *(Richard Megna/Fundamental Photographs, New York.)*

We note from Figure 16-12 that for each harmonic there are certain points on the string (the midpoint in Figure 16-12*b*, for example) that do not move. Such points are called **nodes.** Midway between each adjacent pair of nodes is a point of maximum amplitude of vibration called an **antinode.** A fixed end of the string is, of course, a node. (If one end is attached to a tuning fork or other vibrator rather than being fixed, it will still be approximately a node because the amplitude of the vibration at that end is so much smaller than the amplitude at the antinodes.) We note that the first harmonic has one antinode, the second harmonic has two antinodes, and so on.

! Not all resonant frequencies are called harmonics. Only frequencies that are a part of a resonant frequency spectrum that is composed of integral multiples of the fundamental (lowest) frequency are referred to as harmonics.

We can relate the resonance frequencies to the wave speed in the string and the length of the string. The distance between a node and the nearest antinode is one-fourth of the wavelength. Therefore, the length of the string L equals one-half the wavelength in the fundamental mode of vibration (Figure 16-13) and, as Figure 16-12 reveals, L equals two half-wavelengths for the second harmonic, three half-wavelengths for the third harmonic, and so forth. In general, if λ_n is the wavelength of the nth harmonic, we have

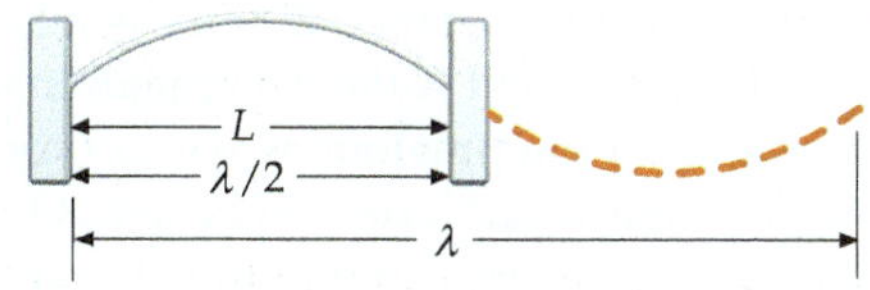

FIGURE 16-13 For the first harmonic of a taut string fixed at both ends, $\lambda = 2L$.

$$L = n\frac{\lambda_n}{2} \qquad n = 1, 2, 3, \ldots \qquad 16\text{-}10$$

STANDING-WAVE CONDITION, BOTH ENDS FIXED

This result is known as the **standing-wave condition.** We can find the frequency of the nth harmonic from the fact that the wave speed v equals the frequency f_n times the wavelength. Thus,

$$f_n = \frac{v}{\lambda_n} = \frac{v}{2L/n} \qquad n = 1, 2, 3, \ldots$$

or

$$f_n = n\frac{v}{2L} = nf_1 \qquad n = 1, 2, 3, \ldots \qquad 16\text{-}11$$

RESONANCE FREQUENCIES, BOTH ENDS FIXED

where $f_1 = v/(2L)$ is the fundamental frequency.

We can understand standing waves in terms of resonance. Consider a string of length L that is attached at one end to a vibrator (Figure 16-14) and is fixed at the other end. The first wave crest sent out by the vibrator travels down the string a distance L to the fixed end, where it is reflected and inverted. It then travels back a distance L and is again reflected and inverted at the vibrator. The total time for the round-trip is $2L/v$. If this time equals the period of the vibrator, the twice-reflected wave crest exactly overlaps the second wave crest produced by the vibrator, and the two crests interfere constructively, producing a crest with twice the original amplitude. The combined wave crest travels down the string and back and is added to by the third crest produced by the vibrator, increasing the amplitude threefold, and so on. Thus, the vibrator is in resonance with the string. The wavelength is equal to $2L$ and the frequency is equal to $v/(2L)$.

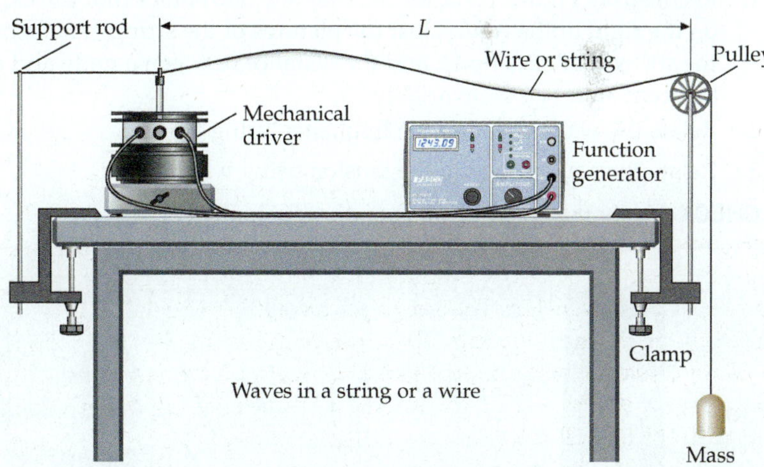

FIGURE 16-14 The mechanical driver sends waves down the string. The waves reflect off the pulley.

Resonance also occurs at other vibrator frequencies. The vibrator is in resonance with the string if the time it takes for the first wave crest to travel the distance $2L$ is any integer n times the period T of the vibrator. That is, if $2L/v = nT_n$, where $2L/v$ is the round-trip time for a wave crest. Thus,

$$f_n = \frac{1}{T_n} = n\frac{v}{2L} \qquad n = 1, 2, 3, \ldots$$

is the condition for resonance. This result is the same result we found by fitting an integral number of half-wavelengths into the distance L. Various damping effects, such as the loss of energy during reflection and air drag on the string, put a limit on the maximum amplitude that can be reached.

The resonance frequencies given by Equation 16-11 are also called the **natural frequencies** of the string. When the frequency of the vibrator is not one of the natural frequencies of the vibrating string, standing waves are not produced. After the first wave travels the distance $2L$ and is reflected from the fork, it differs in phase from the wave being generated at the vibrator (Figure 16-15). When this resultant wave has traveled the distance $2L$ and is again reflected at the vibrator, it will differ in phase from the next wave generated. In some cases, the new resultant wave will superpose with the previous wave to produce a wave of greater amplitude, in other cases the new amplitude will be less. On the average, the amplitude will neither increase nor decrease, but will remain on the order of the amplitude of the first wave generated, which is the amplitude of the vibrator. This amplitude is very small compared with the amplitudes attained at resonance frequencies.

The resonance of standing waves is analogous to the resonance of a simple harmonic oscillator with a harmonic driving force. However, a vibrating string does not have just one natural frequency, but a sequence of natural frequencies that are integral multiples of the fundamental frequency. This sequence is called a **harmonic series.**

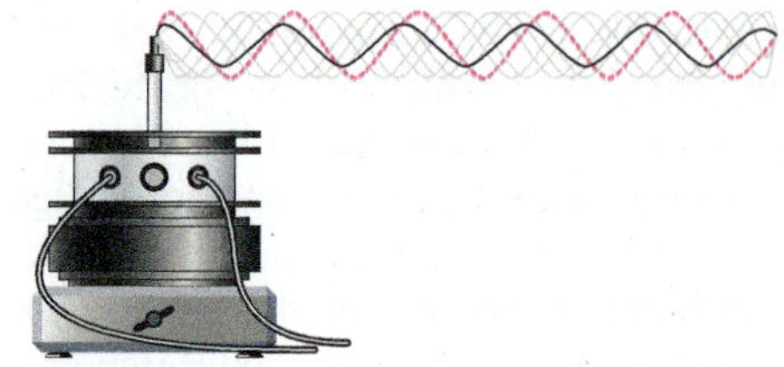

FIGURE 16-15 Waves on a string produced by a mechanical wave driver whose frequency is not in resonance with the natural frequencies of the string. The wave leaving the wave driver for the first time (dashed red line) is not in phase with the waves that have been reflected two or more times (gray lines), and these waves are not in phase with each other, so there is no buildup in amplitude. The resultant wave (black line) has about the same amplitude as the individual waves, which is about the amplitude of the driver.

PROBLEM-SOLVING STRATEGY

Using the Standing-Wave Condition to Solve Problems

PICTURE You should not bother to memorize Equation 16-11. Just sketch Figure 16-12 to remind yourself of the standing-wave condition, $\lambda_n = 2L/n$, and then use $v = f_n\lambda_n$.

SOLVE

1. Reconstruct Figure 16-12 for the first few harmonics (not the expression on the right of the figure, just the pictures of the string). At each end of the string there is a node, and the distance between a node and an adjacent antinode is invariably $\frac{1}{4}\lambda$.
2. Relate the wave speed to the frequency using $v = f\lambda$.
3. Relate the wave speed to the tension using $v = \sqrt{F_T/\mu}$.

CHECK Verify that your results are dimensionally correct.

Standing waves generated by 45 mi/h winds generated standing waves in the Tacoma Narrows suspension bridge, leading to its collapse on November 7, 1940, just four months after it had been opened for traffic. *(University of Washington.)*

Example 16-6 Give Me an A

A string is stretched between two fixed supports 0.700 m apart and the tension is adjusted until the fundamental frequency of the string is concert A, 440 Hz. What is the speed of transverse waves on the string?

PICTURE The wave speed equals the frequency times the wavelength. For a string fixed at both ends, in the fundamental mode there is a single antinode in the middle of the string. Thus, the length of the string equals one-half wavelength.

SOLVE

1. The wave speed is related to the frequency and wavelength We are given the fundamental frequency f_1: $v = f_1\lambda_1$
2. Use Figure 16-12 to relate the wavelength of the fundamental to the length of the string: $\lambda_1 = 2L$
3. Use this wavelength and the given frequency to find the speed: $v = f_1\lambda_1 = f_1 2L = 2f_1L = 2(440\text{ Hz})(0.700\text{ m}) = \boxed{616\text{ m/s}}$

CHECK To check the plausibility of this answer, we check the units. The unit for frequency is the hertz, where 1 Hz = 1 cy/s, or just 1 s^{-1} (because a cycle is dimensionless). Thus, 1 Hz times 1 m equals 1 m/s, which are the correct units for speed.

PRACTICE PROBLEM 16-2 The speed of transverse waves on a stretched string is 200 m/s. If the string is 5.0 m long, find the frequencies of the fundamental and the second and third harmonics.

Example 16-7 Testing Piano Wire

Context-Rich

You have a summer job at a music shop, helping the owner build instruments. He asks you to test a new wire for possible use in pianos. He tells you that the 3.00-m-long wire has a linear mass density of 0.00250 kg/m, and he has found two adjacent resonant frequencies at 252 Hz and at 336 Hz. He wants you to determine the fundamental frequency of the wire and determine whether or not the wire is a good choice for piano strings. You know that safety issues start to arise if the tension in the wire gets above 700 N.

A technician uses a micrometer to measure the diameter of a piano wire. *(Courtesy of Buck Rogers/Craftsmen Piano Rebuilders North Attleboro, MA.)*

PICTURE The tension F_T is found from $v = \sqrt{F_T/\mu}$, where the speed v can be found from $v = f\lambda$ using any harmonic. The wavelength for the fundamental is twice the length of the wire. To find the fundamental frequency let 252 Hz be the frequency of the nth harmonic. Then $f_n = nf_1$ and $f_{n+1} = (n+1)f_1$, where $f_{n+1} = 336$ Hz. We can solve these two equations for f_1.

SOLVE

1. The tension is related to the wave speed: $v = \sqrt{F_T/\mu}$ so $F_T = \mu v^2$
2. The wave speed is related to the wavelength and frequency: $v = f\lambda$
3. Use Figure 16-12 to relate the wavelength of the fundamental to the length of the wire: $\lambda_1 = 2L$
4. Use the step-2 and step-3 results to relate the speed v to the fundamental frequency f_1: $v = f_1\lambda_1 = f_1 \times 2L = 2f_1L$
5. Substitute into the step-1 result to find the tension: $F_T = \mu v^2 = 4\mu f_1^2L^2$
6. The consecutive harmonics f_n and f_{n+1} are related to the fundamental frequency f_1: $nf_1 = 252$ Hz; $(n+1)f_1 = 336$ Hz
7. Dividing these equations eliminates f_1 and allows us to determine n: $\dfrac{n}{n+1} = \dfrac{252\text{ Hz}}{336\text{ Hz}} = 0.750 \Rightarrow n = 3$
8. Solve for f_1: $f_n = nf_1$ so $f_1 = \dfrac{f_n}{n} = \dfrac{f_3}{3} = \dfrac{252\text{ Hz}}{3} = 84.0\text{ Hz}$
9. Using the step-5 result, solve for F_T: $F_T = 4\mu f_1^2L^2 = 4(0.00250\text{ kg/m})(84.0\text{ Hz})^2(3.00\text{ m})^2 = 635\text{ N}$
10. Is the tension safe? **The tension is less than the 700-N safety limit. The wire is safe to use.**

CHECK That the tension is the same order of magnitude as the safety limit makes the answer plausible.

String fixed at one end, free at the other Figure 16-16 shows a string that has one end fixed and one end attached to a ring that is free to slide up and down on a friction-free pole. The vertical motion of the ring is driven by the vertical component of the tension force (we are neglecting any effects of gravity). Ideally, we let the mass of the ring approach zero. Then the vertical motion of the end of the string that is attached to the ring is unconstrained, so it is said to be a *free end.* Any finite vertical force by the string on the massless ring would give the ring an infinite acceleration. However, the acceleration of the ring will remain finite as long as the tangent to the string at the point where it attaches to the ring remains parallel to the string's equilibrium position. For a string oscillating in a standing wave, the antinodes are the only points where the tangent to the string remains parallel to the string's equilibrium position. It follows that there is an antinode at the end of the string attached to the ring.

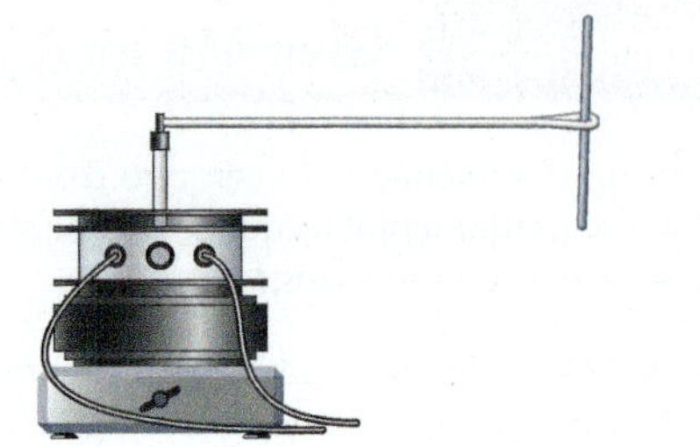

FIGURE 16-16 An approximation of a string fixed at one end and free at the other end can be produced by connecting the "free" end of the string to a ring that is free to move on a post. The end attached to the mechanical wave driver is approximately fixed because the amplitude of the driver is very small.

In the fundamental mode of vibration for a string fixed at one end and free at the other, there is a node at the fixed end and an antinode at the free end, so $L = \frac{1}{4}\lambda$ (Figure 16-17). (Recall that distance from a node to an adjacent antinode is equal to one-quarter wavelength.)

In each mode of vibration shown in Figure 16-18 there is an odd number of quarter-wavelengths in the length L. That is, $L = n\frac{1}{4}\lambda_n$, where $n = 1, 3, 5, \ldots$. The standing-wave condition can thus be written

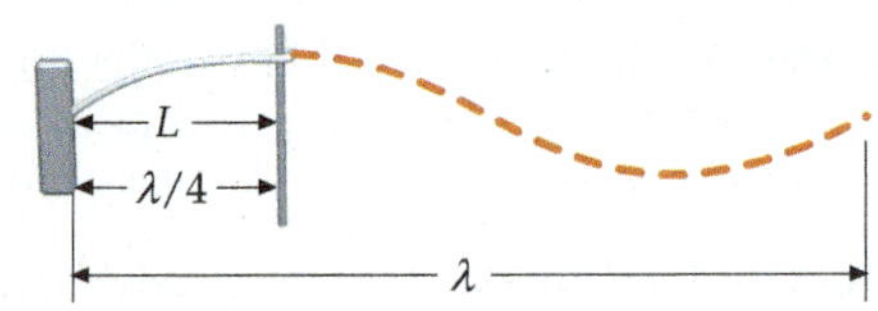

FIGURE 16-17 For the first harmonic of a taut string fixed at one end and free at the other end, $\lambda = 4L$.

$$L = n\frac{\lambda_n}{4} \qquad n = 1, 3, 5, \ldots \qquad \text{16-12}$$

STANDING-WAVE CONDITION, ONE END FREE

so $\lambda_n = 4L/n$. The resonance frequencies are therefore given by

$$f_n = \frac{v}{\lambda_n} = n\frac{v}{4L} = nf_1 \qquad n = 1, 3, 5, \ldots \qquad \text{16-13}$$

RESONANCE FREQUENCIES, ONE END FREE

where

$$f_1 = \frac{v}{4L} \qquad \text{16-14}$$

is the fundamental frequency. The natural frequencies of this system occur in the ratios 1:3:5:7:..., which means that all the even harmonics are missing.

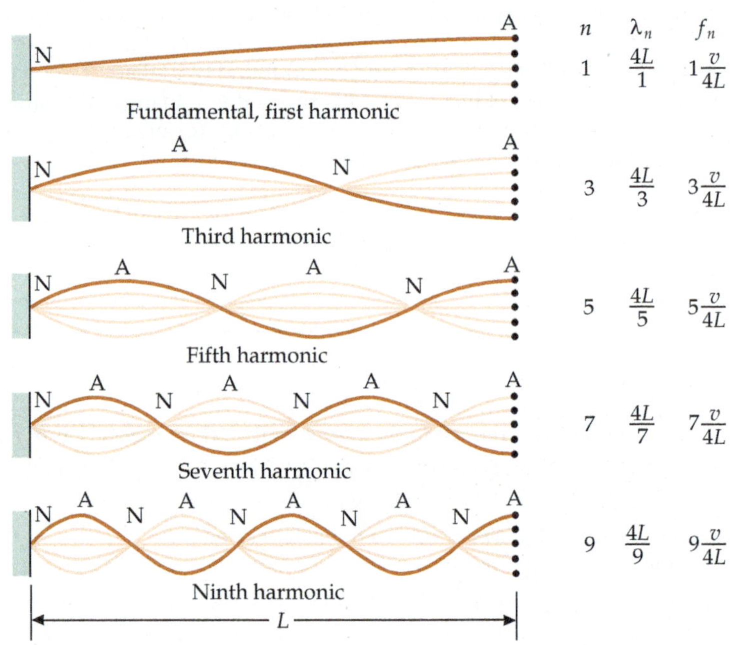

FIGURE 16-18 Standing waves on a string fixed at only one end. An antinode exists at the free end.

Wave functions for standing waves If a string vibrates in its nth mode, each point on the string moves with simple harmonic motion. Its displacement $y_n(x, t)$ is given by

$$y_n(x, t) = A_n(x)\cos(\omega_n t + \delta_n)$$

where ω_n is the angular frequency, δ_n is the phase constant, which depends on the initial conditions, and $A_n(x)$ is the amplitude, which depends on the position x of the point. The function $A_n(x)$ is the shape of the string when $\cos(\omega_n t + \delta_n) = 1$ (the instant that the vibration has its maximum displacement). The amplitude of a string vibrating in its nth mode is described by

$$A_n(x) = A_n \sin k_n x \qquad \text{16-15}$$

where $k_n = 2\pi/\lambda_n$ is the wave number. The wave function for a standing wave in the nth harmonic can thus be written

$$y_n(x, t) = A_n \sin(k_n x) \cos(\omega_n t + \delta_n) \qquad \text{16-16}$$

It is useful to remember the two conditions necessary for standing-wave motion, which are as follows:

1. Each point on the string either remains at rest or oscillates in simple harmonic motion. (Those points remaining at rest are the nodes.)
2. Any two oscillating points on the string oscillate either in phase or 180° out of phase.

NECESSARY CONDITIONS FOR A STANDING-WAVE MOTION ON A LENGTH OF STRING

Example 16-8 Standing Waves

Try It Yourself

(*a*) The wave functions for two waves that have equal amplitude, frequency, and wavelength, but that travel in opposite directions, are given by $y_1 = y_0 \sin(kx - \omega t)$ and $y_2 = y_0 \sin(kx + \omega t)$. Show that the superposition of these two waves is a standing wave. (*b*) A standing wave on a string that is fixed at both ends is given by $y(x, t) = (0.024\text{ m}) \sin(52.3\text{ m}^{-1}\, x) \cos(480\text{ s}^{-1}\, t)$. Find the speed of waves on the string and find the distance between adjacent nodes for the standing waves.

PICTURE To show that the superposition of the two given waves is a standing wave is to show that the algebraic sum of y_1 and y_2 can be written in the form of $y_n(x, t) = A_n \sin(k_n x) \cos(\omega_n t + \delta_n)$ (Equation 16-16). To find the wave speed and the wavelength, we compare the given wave function with Equation 16-16 and identify the wave number and angular frequency. Knowing these, we can determine the wavelength and wave speed.

SOLVE

Cover the column to the right and try these on your own before looking at the answers.

Steps	Answers
(*a*) 1. Write Equation 16-16. If the sum of y_1 and y_2 can be written in this form, then the superposition of the two traveling waves is a standing wave:	$y(x, t) = A \sin kx \cos \omega t$
2. Add the two wave functions and use the trigonometric identity $\sin\theta_1 + \sin\theta_2 = 2\sin\frac{1}{2}(\theta_1 + \theta_2)\cos\frac{1}{2}(\theta_1 - \theta_2)$.	$y = y_0 \sin(kx - \omega t) + y_0 \sin(kx + \omega t)$ $= 2y_0 \sin kx \cos \omega t$ This is of the form given by Equation 16-16 (with $A = 2y_0$), so the superposition is a standing wave.
(*b*) 1. Identify the wave number and the angular frequency:	$k = 52.3\text{ m}^{-1}$, $\omega = 480\text{ s}^{-1}$
2. Calculate the speed from $v = \omega/k$:	$v = 9.18\text{ m/s}$
3. Find the wavelength $\lambda = 2\pi/k$, and use it to find the distance between adjacent nodes:	$\frac{\lambda}{2} = 6.01\text{ cm}$

CHECK Anyone would expect that the superposition of a wave traveling to the right and an otherwise identical wave traveling to the left would not be a traveling wave. (If it were a traveling wave, which way would it be traveling?) Thus, we are not surprised that the superposition of the two traveling waves is a standing wave.

STANDING SOUND WAVES

An organ pipe is a familiar example of the use of standing waves in air columns. In the flue-type organ pipe, a stream of air is directed against the sharp edge of an opening (point A in Figure 16-19). The complicated swirling motion of the air near the edge sets up vibrations in the air column. The resonance frequencies of the pipe depend on the length of the pipe and on whether the top is stopped (closed) or open.

In an open organ pipe, the pressure does not vary appreciably near each open end. (It remains at atmospheric pressure.) Because the pressure just beyond the ends does not vary appreciably, there is a pressure node near each end. If the sound wave in the tube is a one-dimensional wave, which is largely correct if the tube diameter is much smaller than the wavelength, then the pressure node is extremely close to the open end of the tube. In practice, however, the pressure node lies slightly beyond the open end of the tube. The effective length of the pipe is $L_{\text{eff}} = L + \Delta L$, where ΔL is the end correction, which is somewhat smaller than the tube diameter. The standing-wave condition for this system is the same as that for a string fixed at both ends, where L is replaced by L_{eff} (the effective length of the tube), and all the same equations apply.

In a stopped organ pipe (open at one end, closed at the other), there is a pressure node near the opening (point A in Figure 16-19) and a pressure antinode at the closed end. The standing-wave condition for this system is the same as that for a string with one end fixed and one end free. The effective length of the tube is equal to an odd integer times $\lambda/4$. That is, the wavelength of the fundamental mode is four times the effective length of the tube, and only the odd harmonics are present.

As we saw in Chapter 15, a sound wave can be thought of as either a pressure wave or a displacement wave. The pressure and displacement variations in a sound wave are 90° out of phase. Thus, in a standing sound wave, the pressure nodes are displacement antinodes and vice versa. Near the open end of an organ pipe there is a pressure node and a displacement antinode, whereas at a stopped end there is a pressure antinode and a displacement node.

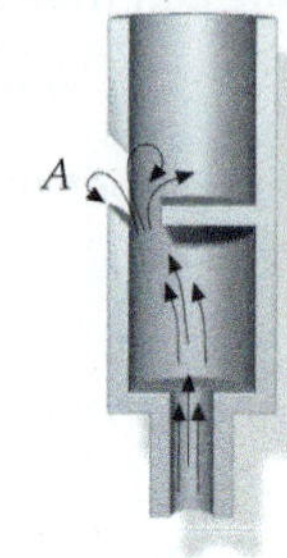

FIGURE 16-19 Cutaway view of a section of a flue-type organ pipe. Air is blown against the edge, causing a swirling motion of the air near point A that excites standing waves in the pipe. There is a pressure node near point A, which is open to the atmosphere.

Example 16-9 Standing Sound Waves in an Air Column: I — *Try It Yourself*

An unstopped (open at both ends) organ pipe has an effective length equal to 1.00 m. (*a*) If the speed of sound is 343 m/s, what are the allowed frequencies and wavelengths for standing sound waves in this pipe? (*b*) The speed of sound in helium is 975 m/s. What are the allowed frequencies for standing sound waves in this pipe if it is filled with and surrounded by helium?

PICTURE There is a displacement antinode (and a pressure node) at each end. Therefore, the effective length of the pipe is equal to an integral number of half-wavelengths.

SOLVE

Cover the column to the right and try these on your own before looking at the answers.

Steps	Answers
(*a*) 1. Using Figure 16-12, determine the wavelength of the fundamental mode:	$\lambda_1 = 2L_{\text{eff}} = 2.00\text{ m}$
2. Use $v = f\lambda$ to calculate the fundamental frequency f_1:	$f_1 = \dfrac{v}{\lambda_1} = 172\text{ Hz}$
3. Write expressions for the frequencies f_n and wavelengths λ_n of the other harmonics in terms of n:	$f_n = nf_1 = \boxed{n(172\text{ Hz}) \quad n = 1, 2, 3, \ldots}$ $\lambda_n = \dfrac{2L}{n} = \boxed{(2.00\text{ m})/n \quad n = 1, 2, 3, \ldots}$
(*b*) 1. Repeat Part (*a*) to calculate the resonant frequency spectrum of the helium-filled organ pipe:	$f_n = nf_1 = n\dfrac{v}{\lambda_1} = n\dfrac{v}{2L} = n\dfrac{975\text{ m/s}}{2.00\text{ m}}$ $= \boxed{n(488\text{ Hz}) \quad n = 1, 2, 3, \ldots}$

CHECK The product of the two Part-(*a*), step-3 results does not depend on n. (The n's cancel when you take the product.) This is as expected because the product is equal to the wave speed, which does not depend on frequency or wavelength.

PRACTICE PROBLEM 16-3 The longest organ pipe is the one with a fundamental frequency that is equal to 16 Hz, the lowest frequency audible to humans. What is the length of an unstopped organ pipe that has a fundamental frequency of 16.0 Hz?

CONCEPT CHECK 16-1

Why does your voice change pitch when you speak after inhaling the contents of a helium-filled balloon?

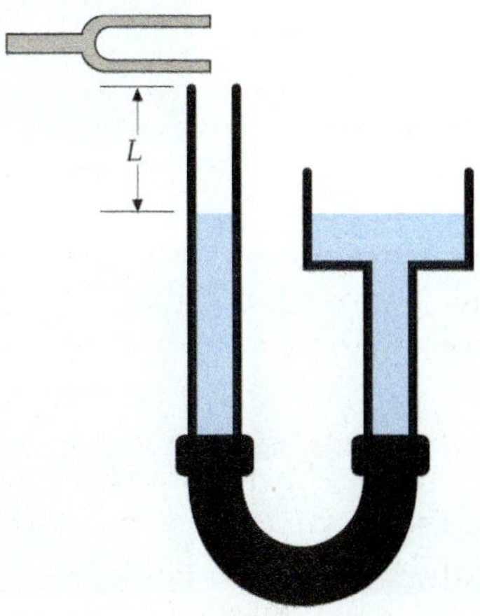

FIGURE 16-20 The length of the air column in the cylinder on the left is varied by moving the reservoir on the right up or down. The two cylinders are connected by a flexible hose.

Example 16-10 Standing Sound Waves in an Air Column: II

When a tuning fork of frequency 500 Hz is held above a tube that is partly filled with water, as in Figure 16-20, resonances are found when the water level is at distances L = 16.0, 50.5, 85.0, and 119.5 cm from the top of the tube. (*a*) What is the speed of sound in air? (*b*) How far from the open end of the tube is the displacement antinode?

PICTURE Standing sound waves of frequency 500 Hz are excited in the air column whose length L can be adjusted (by adjusting the water level). The air column is stopped at one end, open at the other. Thus, at resonance, the number of quarter-wavelengths in the effective length L_{eff} of the tube is equal to an odd integer (Figure 16-21). A displacement node exists at the surface of the water and a displacement antinode exists a short distance ΔL above the open end of the tube. Because the frequency is fixed, so is the wavelength. The speed is then found from $v = f\lambda$, where f is 500 Hz.

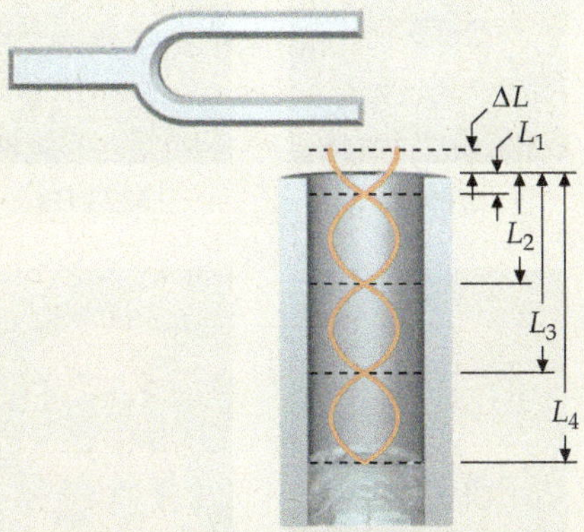

FIGURE 16-21 A displacement node exists at the surface of the water and a displacement antinode exists a distance ΔL above the top of the cylinder.

SOLVE

(*a*) 1. The speed of sound in air is related to the frequency and wavelength: $v = f\lambda$

2. Resonance occurs each time the water level is at the location of a displacement node (see Figure 16-21). That is, when the length L changes by half a wavelength: $L_{n+1} = L_n + \dfrac{\lambda}{2} \quad n = 1, 2, 3, 4$

3. The distance between successive levels is found from the data given in the problem: $L_{n+1} - L_n = L_4 - L_3 = 119.5\text{ cm} - 85.0\text{ cm} = 34.5\text{ cm}$
so $\lambda = 2(34.5\text{ cm}) = 69.0\text{ cm} = 0.690\text{ m}$

4. Substitute the values of f and λ to determine v: $v = f\lambda = (500\text{ Hz})(0.690\text{ m}) = \boxed{345\text{ m/s}}$

(*b*) There will be a displacement antinode one-quarter wavelength above the displacement node at the surface of the water. Thus, the distance from the highest water level supporting resonance and the displacement antinode above the opening of the tube is one-quarter wavelength: $\tfrac{1}{4}\lambda = L_1 + \Delta L$
so $\Delta L = \tfrac{1}{4}\lambda - L_1 = \tfrac{1}{4}(69.0\text{ cm}) - (16.0\text{ cm})$
$= \boxed{1.25\text{ cm}}$

CHECK As expected, the wave speed (step 4) is approximately equal to the speed of sound in air at room temperature.

Most musical wind instruments are much more complicated than simple cylindrical tubes. The conical tube, which is the basis for the oboe, bassoon, English horn, and saxophone, has a complete harmonic series with its fundamental wavelength equal to twice the length of the cone. Brass instruments are combinations of cones and cylinders. The analysis of these instruments is extremely complex. The fact that they have nearly harmonic series is a triumph of educated trial and error rather than mathematical calculation.

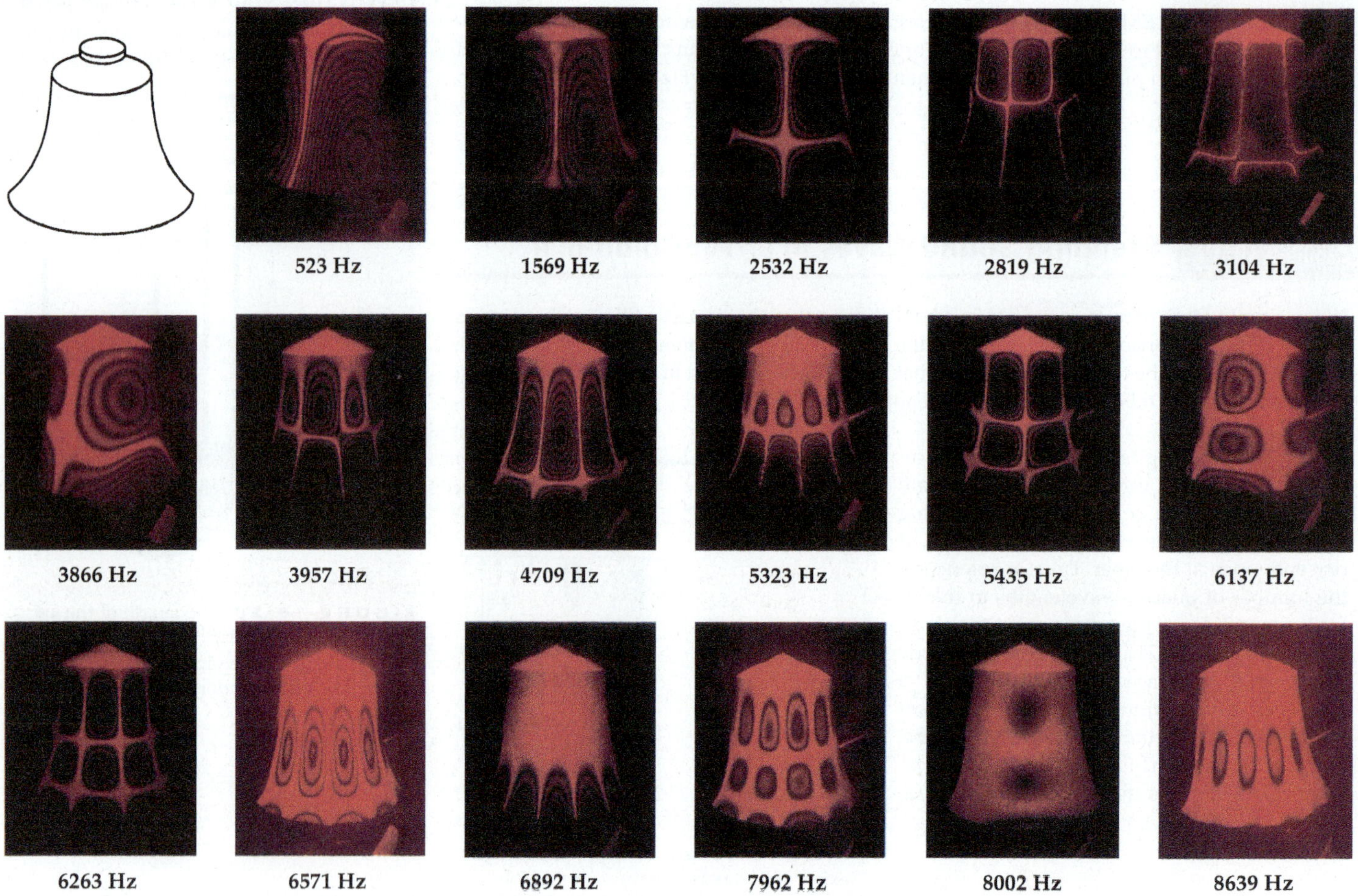

Holographic interferograms showing standing waves in a handbell. The "bull's eyes" locate the antinodes. *(Professor Thomas D. Rossing, Northern Illinois University, DeKalb.)*

* 16-3 ADDITIONAL TOPICS

THE SUPERPOSITION OF STANDING WAVES

As we saw in the preceding section, there is a set of natural resonance frequencies that produce standing waves for sound waves in air columns or vibrating strings that are fixed at one or both ends. For example, for a string fixed at both ends, the frequency of the fundamental mode of vibration is $f_1 = v/(2L)$, where L is the length of the string and v is the wave speed and the wave function is Equation 16-16:

$$y_1(x, t) = A_1 \sin k_1 x \cos(\omega_1 t + \delta_1)$$

In general, a vibrating system does not vibrate in a single harmonic mode. Instead, the motion consists of a superposition of several of the allowed harmonics. The wave function is a linear combination of the harmonic wave functions:

$$y(x, t) = \sum_n A_n \sin(k_n x) \cos(\omega_n t + \delta_n) \qquad 16\text{-}17$$

where $k_n = 2\pi/\lambda_n$, $\omega_n = 2\pi f_n$, and A_n and δ_n are constants. The constants A_n and δ_n depend on the initial positions and velocities of the points on the string. If a harp string, for example, is plucked at the center and released, as in Figure 16-22, the

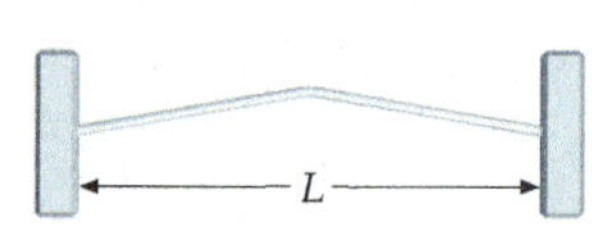

FIGURE 16-22 A string plucked at the center. When it is released, its vibration is a linear superposition of standing waves.

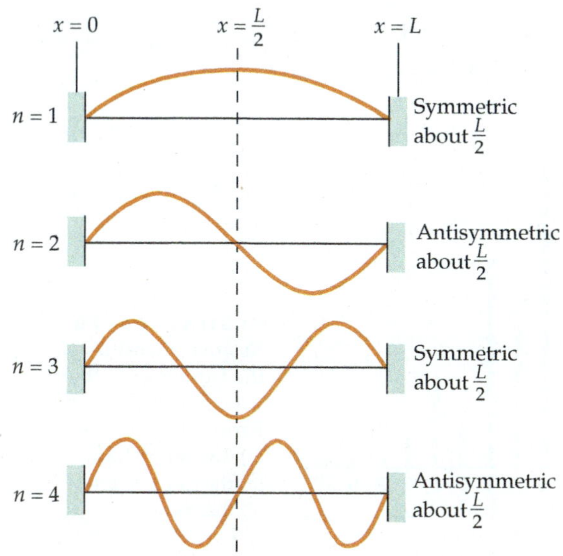

FIGURE 16-23 The first four harmonics for a string fixed at both ends. The odd harmonics are symmetrical about the center of the string, whereas the even harmonics are not. When a string is plucked at the center, it vibrates only in its odd harmonics.

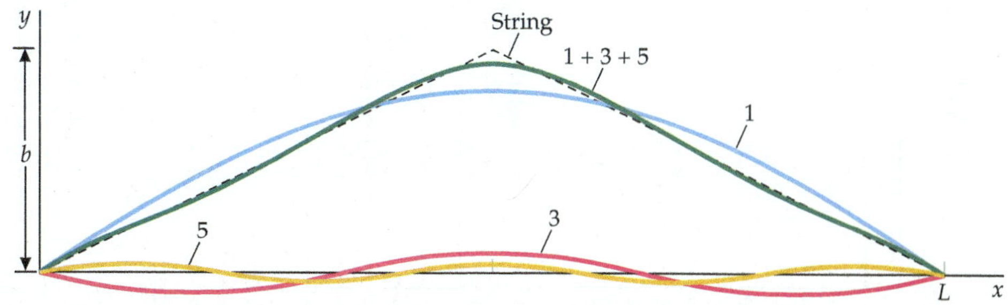

FIGURE 16-24 Approximating the shape of a string plucked at the center, as in Figure 16-22, using harmonics. The green line is an approximation of the original shape of the string based on the first three odd harmonics. The height of the string is exaggerated in this drawing to show the relative amplitudes of the harmonics. Most of the energy is associated with the fundamental, but there is some energy in the third, fifth, and other odd harmonics.

initial shape of the string is symmetric about the point $x = \frac{1}{2}L$ and the initial velocity is zero throughout the length of the string. The motion of the string after it has been released will remain symmetric about $x = \frac{1}{2}L$. Only the odd harmonics, which are also symmetric about $x = \frac{1}{2}L$, will be excited. The even harmonics, which are antisymmetric about $x = \frac{1}{2}L$, are not excited; that is, the constant A_n is zero for all even values of n. The shapes of the first four harmonics are shown in Figure 16-23. Most of the energy of the plucked string is associated with the fundamental, but small amounts of energy are associated with the third, fifth, and other odd harmonic modes. Figure 16-24 shows an approximation to the initial shape of the string using the superposition of only the first three odd harmonics.

(Corbis.)

HARMONIC ANALYSIS AND SYNTHESIS

When a clarinet and an oboe play the same note, say, concert A, they sound quite different. Both notes have the same **pitch,** a physiological sensation of the highness or lowness of the note that is strongly correlated with frequency. However, the notes differ in what is called **tone quality.** The principal reason for the difference in tone quality is that, although both the clarinet and oboe are producing vibrations at the same fundamental frequency, each instrument is also producing harmonics whose relative intensities depend on the instrument and how it is played. If the sound produced by each instrument were entirely at the fundamental frequency of the instrument, they would sound identical.

Figure 16-25 shows plots of the pressure variations versus time for the sound from a tuning fork, a clarinet, and an oboe, each playing the same note. These patterns are called **waveforms.** The waveform for the sound from the tuning fork is nearly a pure sine wave, but those from the clarinet and the oboe are clearly more complex.

Waveforms can be analyzed in terms of the harmonics that constitute them by means of **harmonic analysis.** (Harmonic analysis is also called **Fourier analysis** after the French mathematician J.B.J. Fourier, who developed the

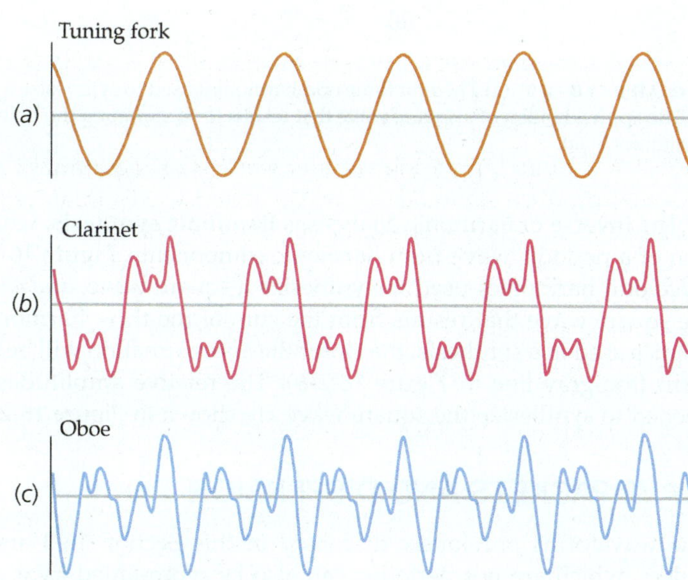

FIGURE 16-25 Waveforms of (*a*) a tuning fork, (*b*) a clarinet, and (*c*) an oboe, each at a fundamental frequency of 440 Hz and at approximately the same intensity.

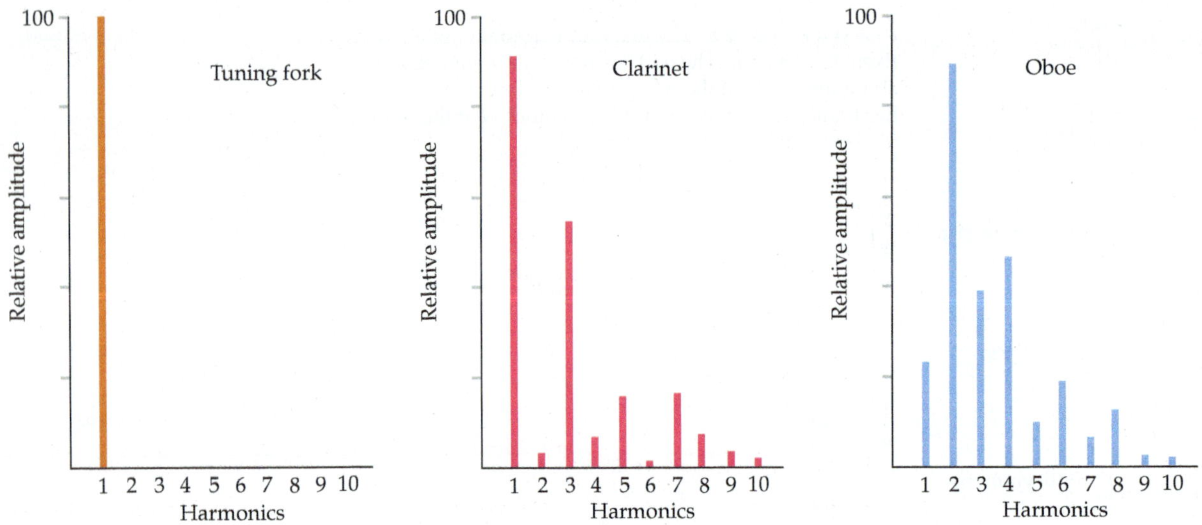

FIGURE 16-26 Relative intensities of the harmonics in the waveforms shown in Figure 16-25 for (*a*) the tuning fork, (*b*) the clarinet, and (*c*) the oboe.

techniques for analyzing periodic functions.) Figure 16-26 shows a plot of the relative intensities of the harmonics of the waveforms in Figure 16-25. The waveform of the sound from the tuning fork contains only the fundamental frequency. The waveform for the sound from the clarinet contains the fundamental, large amounts of the third, fifth, and seventh harmonics, and lesser amounts of the second, fourth, and sixth harmonics. For the sound from the oboe, there is more intensity in the second, third and fourth harmonics than in the fundamental.

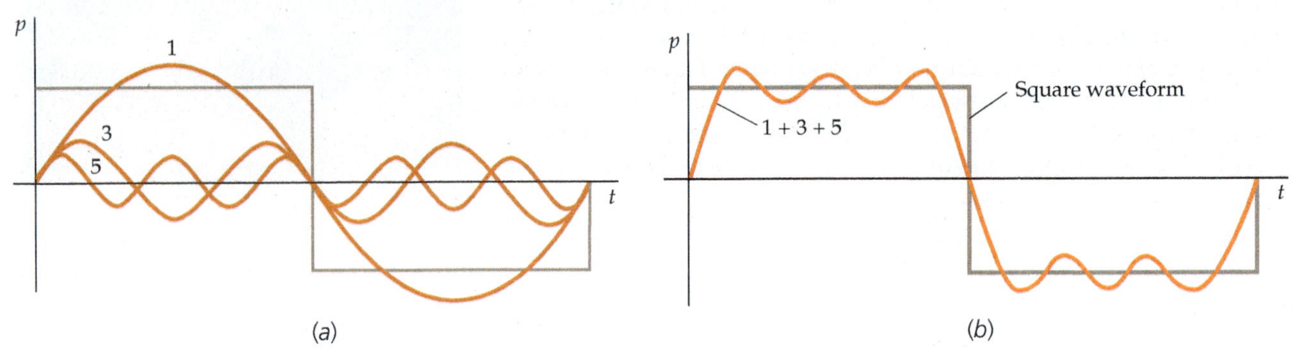

FIGURE 16-27 (a) The first three odd harmonics used to synthesize a square wave. (*b*) The approximation of a square wave that results from summing the first three odd harmonics in (*a*).

The inverse of harmonic analysis is **harmonic synthesis,** which is the construction of a periodic wave from harmonic components. Figure 16-27*a* shows the first three odd harmonics used to synthesize a square wave, and Figure 16-27*b* shows the square wave that results from the sum of the three harmonics. The more harmonics used in a synthesis, the closer the approximation will be to the actual waveform (the gray line in Figure 16-27*b*). The relative amplitudes of the harmonics needed to synthesize the square wave are shown in Figure 16-28.

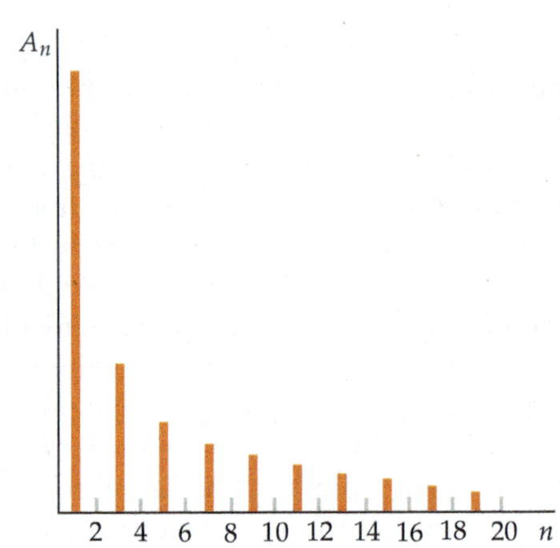

FIGURE 16-28 Relative amplitudes A_n of the first 10 harmonics needed to synthesize a square wave. The more harmonics that are used, the closer the approximation is to the square wave.

WAVE PACKETS AND DISPERSION

The waveforms previously discussed in this Section 16-3 are periodic in time. Pulses, which are not periodic, can also be represented by a group of harmonic waves of different frequencies. However, the synthesis of an isolated pulse requires a continuous distribution of frequencies rather than a discrete set of harmonics, as in Figure 16-28. Such a group is called a **wave packet.** The characteristic feature of a wave pulse is that it has a beginning and an end, whereas a harmonic wave repeats over and over. If the duration Δt of the pulse is very short, the range of frequencies $\Delta\omega$ needed to describe the pulse is very large. The general relation between Δt and $\Delta\omega$ is

$$\Delta\omega\,\Delta t \sim 1 \qquad 16\text{-}18$$

where the tilde (~) means "of the order of magnitude of."

The exact value of this product depends on just how the quantities $\Delta\omega$ and Δt are defined. For any reasonable definitions, $\Delta\omega$ and $1/\Delta t$ have the same order of magnitude. A wave pulse produced by a source of short duration Δt, like the crack of a bat on a ball, has a narrow width in space $\Delta x = v\,\Delta t$, where v is the wave speed. Each harmonic wave of frequency ω has a wave number $k = \omega/v$. A range of frequencies $\Delta\omega$ implies a range of wave numbers $\Delta k = \Delta\omega/v$. Substituting $v\,\Delta k$ for $\Delta\omega$ in Equation 16-18 gives $v\,\Delta k\,\Delta t \sim 1$, or

$$\Delta k\,\Delta x \sim 1 \qquad 16\text{-}19$$

Example 16-11 Estimating $\Delta\omega$ and Δk

In Example 15-1 a wave pulse on a long clothesline is moving at 100 m/s. (*a*) If the width of the pulse is 1.00 m, what is the duration of the pulse? That is, how long does it take for the pulse to travel past a point on the clothesline? (*b*) The pulse can be considered as a superposition of harmonic waves. What is the range of frequencies of these harmonic waves? (*c*) What is the range of wave numbers?

PICTURE To find the duration of the pulse, we use distance equals speed times the time. To find the range of frequencies and the range of wave numbers, we use $\Delta\omega\,\Delta t \sim 1$ and $\Delta k\,\Delta x \sim 1$ (Equations 16-18 and 16-19).

SOLVE

(*a*) The duration of the pulse is the time it takes to pass a point on the clothesline:

$$L = v\,\Delta t \quad \text{so} \quad \Delta t = \frac{L}{v} = \frac{1.00\text{ m}}{100\text{ m/s}} = \boxed{0.0100\text{ s}}$$

(*b*) To find the range of frequencies, we use $\Delta\omega\,\Delta t \sim 1$ (Equation 16-18):

$$\Delta\omega\,\Delta t \sim 1 \quad \text{so} \quad \Delta\omega \sim \frac{1}{\Delta t} = \frac{1}{0.0100\text{ s}} = \boxed{100\text{ s}^{-1}}$$

(*c*) To find the range of wave numbers, we use $\Delta k\,\Delta x \sim 1$ (Equation 16-19):

$$\Delta k\,\Delta x \sim 1 \quad \text{so} \quad \Delta k \sim \frac{1}{\Delta x} = \frac{1}{1.00\text{ m}} = \boxed{1.00\text{ m}^{-1}}$$

CHECK We know that $k = \omega/v$, so a range of frequencies $\Delta\omega$ implies a range of wave numbers $\Delta k = \Delta\omega/v$. Dividing our Part-(*b*) result by the wave speed v, we obtain $(100\text{ s}^{-1})/(100\text{ m/s}) = 1\text{ m}^{-1}$. This value is our Part-(*c*) result.

If a wave packet is to maintain its shape as it travels, all of the component harmonic waves that make up the packet must travel with the same speed. This occurs if the speed of the component waves in a given medium is independent of frequency or wavelength. Such a medium is called a **nondispersive medium.** Air is, to an excellent approximation, a nondispersive medium for sound waves, but solids and liquids are not. (Probably the most familiar example of dispersion is the formation of a rainbow, which is due to the fact that the velocity of light waves in water depends slightly on the frequency of the light, so the different colors, corresponding to different frequencies, have slightly different angles of refraction.)

When the wave speed in a dispersive medium depends only slightly on the frequency (or wavelength), a wave packet changes shape very slowly as it travels, and it covers a considerable distance as a recognizable entity. But the speed of the packet, called the **group velocity,** is not the same as the (average) speed of the individual component harmonic waves, called the **phase velocity.** (By the speed of an individual harmonic wave we mean the speed of its wavefronts. Because wavefronts are lines or surfaces of constant phase, their speed is called the phase velocity of the wave.)

Physics Spotlight

Echoes of Silence: Acoustical Architecture

Architectural acoustics deals with the ways that sound energy reflects, reverberates, and absorbs within a venue. Computer modeling of spaces has allowed acoustic engineers to design flexible spaces,[*, †] while taking into account the different needs for listening to lectures, theater, and several types of music. In general, the goal is to make the sound uniform, audible, and intelligible at each seat.

The baffles hanging from the ceiling and attached to the walls above the doorways are there to absorb sound. Their surfaces are made of acoustically dead material, like felt. *(Courtesy of Perdue Acoustics.)*

There should not be any whole-room standing waves in the listening room.[‡] Whole-room standing waves make certain frequencies harder to hear for people in the seats near nodes, and key frequencies too loud for people seated near the antinodes. Rooms that are designed so that whole-room standing waves are reduced have long walls that are not parallel to each other, and ceilings and floors that are also nonparallel.

If listeners are seated an average of 50 feet from the main sound source, well under one percent of the sound energy can go directly into their ears,[#] and nearly all the sound energy that reaches listeners will be reflected sound. The reflections must be clean and energetic enough to give the listener a reasonable total volume. Timing the reflections is also important. If a reflection up to 15 decibels below the source level reaches a listener's ear more than 60 milliseconds after the source sound, it will be perceived as an echo.[°, §] If reflections louder than the source occur in the first 30 milliseconds, they may also be perceived as echoes. Echoes detract from the intelligibility of speech, and make music sound fuzzy. Late reflections arriving 50 ms or more after the source should be avoided.

Reflectors should be closer than 50 feet to each listener. This is a problem for open-air venues surrounded by tall buildings.[¶] Many older venues have nonstructural plaster work. These structures provide early reflections to listeners. Newer venues often use multiple speakers along the walls and ceiling. Chandeliers and panels suspended from high ceilings also reflect sound. Vaulted and detailed ceilings disperse the sound into many small, unenergetic reflections.

Acoustic absorbers are used to lower the ambient noise energy within a room. The materials for both the reflective structures and the absorbent structures are carefully tailored to the venue, because most materials have different *absorption coefficients* at different frequencies.[**] The absorption coefficient is a measure of the fraction of the sound energy that is absorbed, rather than reflected or transmitted. Window glass, for instance, has absorption coefficients of 0.35 at 125 Hz and 0.04 at 4 kHz. Indoor/outdoor carpet has absorption coefficients of 0.01 at 125 Hz and 0.65 at 4 kHz. Different materials must be used for both absorption and reflection to give a full-spectrum response at each seat.

Too much absorption gives rooms a dead feeling, and gives people claustrophobia.[††] Reverberation, or chaotic sound energy, gives rooms a warm feeling. Reverberation time, the measure of how quickly chaotic noise dissipates, is used as a measure of how lively a room sounds. Reverberation times of venues vary according to the purpose of the venue.

---

* Orfali, and Ahnert, op. cit.
† "Gallagher Bluedorn Performing Arts Center," Acoustic Dimensions, http://www.acousticdimensions.com/profiles/gb_uni.htm
‡ Everest, F. Alton, *Master Handbook of Acoustics*, 4th ed., New York: McGraw-Hill, 2001, 320
\# Noxon, A., "Auditorium Acoustics 101," *Church & Worship Technology*, April 2002, 22+.
° Everest, op. cit., 356.
§ Noxon, A., "Auditorium Acoustics 102," *Church & Worship Technology*, May 2002, 24+.
¶ Orfali, W., and Ahnert, W., "Measurments (sic) and Verification in Two Mosques in Saudi Arabia and Jordan," paper presented at the 151st Meeting of the Acoustical Society of America, Providence, RI, June 1–5, 2006, http://scitation.aip.org/confst/ASA/data/5/1aAA9.pdf
** Everest, op. cit., 585–587.
†† Freiheit, R., "Historic Recording Gives Choir 'Alien' Feeling: In Anechoic Space, No One Can Hear You Sing," paper presented at the ASA/Noise Conference 2005 Minneapolis, http://www.acoustics.org/press/150th/Freiheit.html

Summary

1. The principle of superposition, which holds for all electromagnetic waves in empty space, for waves on a flexible taut string in the small-angle approximation, and for sound waves of small amplitude, follows from the linearity of the corresponding wave equations.
2. Interference is an important wave phenomenon that applies to all coherent superposing waves. It follows from the principle of superposition. Diffraction and interference distinguish wave motion from particle motion.
3. The standing-wave conditions can be recalled by sketching a string or tube and drawing waves that have displacement nodes at a fixed or stopped end, and displacement antinodes at a free or open end.

TOPIC	RELEVANT EQUATIONS AND REMARKS	
1. Superposition and Interference	The superposition of two harmonic waves of equal amplitude, wave number, and frequency but phase difference δ, results in a harmonic wave of the same wave number and frequency, but differing in phase and amplitude from each of the two waves $y = y_1 + y_2 = y_0 \sin(kx - \omega t) + y_0 \sin(k - \omega t + \delta) = [2y_0 \cos\frac{1}{2}\delta] \sin(kx - \omega t + \frac{1}{2}\delta)$	16-6
Constructive interference	If waves are in phase or differ in phase by an integer times 2π, then the amplitudes of the waves add and the interference is constructive.	
Destructive interference	If waves differ in phase by π or by an odd integer times π, then the amplitudes subtract and the interference is destructive.	
Beats	Beats are the result of the interference of two waves of slightly different frequencies. The beat frequency equals the difference in the frequencies of the two waves: $f_{\text{beat}} = \Delta f$	16-8
Phase difference δ due to path difference Δx	$\delta = k\,\Delta x = 2\pi \frac{\Delta x}{\lambda}$	16-9
2. Standing Waves	Standing waves occur for certain frequencies and wavelengths when waves are confined in space. If they occur, then each point of the system oscillates in simple harmonic motion and any two points not at nodes move either in phase or 180° out of phase.	
Wavelength	The distance between a node and an adjacent antinode is a quarter-wavelength.	
String fixed at both ends	For a string fixed at both ends, there is a node at each end so that an integral number of half-wavelengths must fit into the length of the string. The standing-wave condition in this case is $L = n\frac{\lambda_n}{2} \quad n = 1, 2, 3, \ldots$	16-10
Standing-wave function for a string fixed at both ends	The allowed waves form a harmonic series, with the frequencies given by $f_n = \frac{v}{\lambda_n} = n\frac{v}{\lambda_1} = n\frac{v}{2L} = nf_1 \quad n = 1, 2, 3, \ldots$ where $f_1 = v/2L$ is the lowest frequency, called the fundamental.	16-18
Organ pipe open at both ends	Standing sound waves in the air in a pipe that is open at both ends have a pressure node (and a displacement antinode) near each end so that the standing-wave condition is the same as for a string fixed at both ends.	
String fixed at one end and free at the other	For a string with one end fixed and one end free, there is a node at the fixed end and an antinode at the free end, so that an integral number of quarter-wavelengths must fit into the length of the string. The standing-wave condition in this case is $L = n\frac{\lambda_n}{4} \quad n = 1, 3, 5, \ldots$	16-12
	Only the odd harmonics are present. Their frequencies are given by $f_n = \frac{v}{\lambda_n} = n\frac{v}{\lambda_1} = n\frac{v}{4L} = nf_1 \quad n = 1, 3, 5, \ldots$ where $f_1 = v/4L$.	16-13

TOPIC	RELEVANT EQUATIONS AND REMARKS	
Organ pipe open at one end and stopped at the other	Standing sound waves in a pipe that is open at one end and stopped at the other end have a displacement antinode at the open end and a displacement node at the stopped end. The standing wave condition is the same as for a string fixed at one end.	
Wave Functions for Standing Waves	$y_n(x, t) = A_n \sin(k_n x) \cos(\omega_n t + \delta_n)$	16-16
	where $k_n = 2\pi/\lambda_n$ and $\omega_n = 2\pi f_n$. The necessary conditions for standing waves on a string are 1. Each point on the string either remains at rest or oscillates with simple harmonic motion. (Those points remaining at rest are nodes.) 2. The motions of any two points on the string that are not nodes oscillate either in phase or 180° out of phase.	
***3. Superposition of Standing Waves**	A vibrating system typically does not vibrate in a single harmonic mode, but in a superposition of the allowed harmonic modes.	
***4. Harmonic Analysis and Synthesis**	Sounds of different tone quality contain different mixtures of harmonics. The analysis of a particular tone in terms of its harmonic content is called harmonic analysis. Harmonic synthesis is the construction of a tone by the addition of harmonics.	
***5. Wave Packets**	A wave pulse can be represented by a continuous distribution of harmonic waves. The range of frequencies $\Delta\omega$ is related to the width in time Δt, and the range of wave numbers Δk is related to the width in space Δx.	
Frequency and time ranges	$\Delta\omega\, \Delta t \sim 1$	16-18
Wave number and space ranges	$\Delta k\, \Delta x \sim 1$	16-19
***6. Dispersion**	In a nondispersive medium, the phase velocity is independent of frequency, and a pulse (wave packet) travels without change in shape. In a dispersive medium, the phase velocity varies with frequency, and the pulse changes shape as it moves. The pulse moves with a velocity called the group velocity of the packet.	

Answer to Concept Check

16-1 Your voice changes pitch because the fundamental frequency of your throat and mouth cavity is increased, just like the resonant frequency of the organ pipe in Example 16-9 increased when it was filled with helium.

Answers to Practice Problems

16-1 (*a*) 5.66 cm, (*b*) 120° or 240°

16-2 $f_1 = 20$ Hz, $f_2 = 40$ Hz, $f_3 = 60$ Hz

16-3 About 10.7 m ≈ 35 ft

Problems

In a few problems, you are given more data than you actually need; in a few other problems, you are required to supply data from your general knowledge, outside sources, or informed estimate.

Interpret as significant all digits in numerical values that have trailing zeros and no decimal points.

Use 343 m/s as the speed of sound for air, unless otherwise indicated.

- • Single-concept, single-step, relatively easy
- •• Intermediate-level, may require synthesis of concepts
- ••• Challenging
- SSM Solution is in the *Student Solutions Manual*
- Consecutive problems that are shaded are paired problems.

CONCEPTUAL PROBLEMS

1 • Two rectangular wave pulses are traveling in opposite directions along a string. At $t = 0$, the two pulses are as shown in Figure 16-29. Sketch the wave functions for $t = 1.0$, 2.0, and 3.0 s. SSM

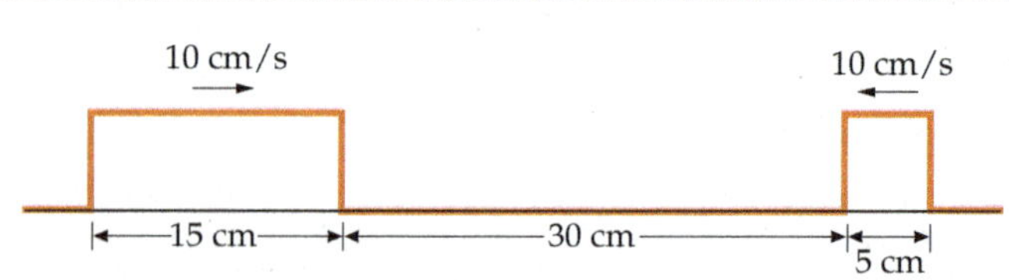

FIGURE 16-29 Problems 1, 2

2 • Repeat Problem 1 for the case in which the pulse on the right is inverted.

3 • Beats are produced by the superposition of two harmonic waves if (*a*) their amplitudes and frequencies are equal, (*b*) their amplitudes are the same but their frequencies differ slightly, (*c*) their frequencies are equal but their amplitudes differ slightly.

4 • Two tuning forks are struck and the sounds from each reach your ears at the same time. One sound has a frequency of 256 Hz, and the second sound has a frequency of 258 Hz. The underlying "hum" frequency that you hear is (*a*) 2 Hz, (*b*) 256 Hz, (*c*) 258 Hz, (*d*) 257 Hz.

5 • In Problem 4, the beat frequency is (*a*) 2 Hz, (*b*) 256 Hz, (*c*) 258 Hz, (*d*) 257 Hz.

6 • **CONTEXT-RICH** As a graduate student, you are teaching your first physics lecture while the professor is away. To demonstrate interference of sound waves, you have set up two speakers that are driven coherently and in phase by the same frequency generator on the front desk. Each speaker generates sound with a 2.4-m wavelength. One student in the front row says she hears a very low volume (loudness) of the sound from the speakers compared to the volume of the sound she hears when only one speaker is generating sound. What could be the difference in the distance between her and each of the two speakers? (*a*) 1.2 m, (*b*) 2.4 m, (*c*) 4.8 m, (*d*) You cannot determine the difference in distances from the data given.

7 • In Problem 6, determine the longest wavelength for which a student would hear "extra loud" sound due to constructive interference, assuming this student is located so that one speaker is 3.0 m farther from her than the other speaker.

8 • Consider standing waves in an organ pipe. True or false:

(*a*) In a pipe open at both ends, the frequency of the third harmonic is three times that of the first harmonic.
(*b*) In a pipe open at both ends, the frequency of the fifth harmonic is five times that of the fundamental.
(*c*) In a pipe that is open at one end and stopped at the other, the even harmonics are not excited.

Explain your choices.

9 • Standing waves result from the superposition of two waves that have (*a*) the same amplitude, frequency, and direction of propagation, (*b*) the same amplitude and frequency and opposite directions of propagation, (*c*) the same amplitude, slightly different frequencies, and the same direction of propagation, (*d*) the same amplitude, slightly different frequencies, and opposite directions of propagation.

10 • If you blow air over the top of a fairly large drinking straw you can hear a fundamental frequency due to a standing wave being set up in the straw. What happens to the fundamental frequency, (*a*) if while blowing, you cover the bottom of the straw with your fingertip? (*b*) if while blowing you cut the straw in half with a pair of scissors? (*c*) Explain your answers to Parts (*a*) and (*b*).

11 • An organ pipe that is open at both ends has a fundamental frequency of 400 Hz. If one end of this pipe is now stopped, the fundamental frequency is (*a*) 200 Hz, (*b*) 400 Hz, (*c*) 546 Hz, (*d*) 800 Hz. SSM

12 •• A string fixed at both ends resonates at a fundamental frequency of 180 Hz. Which of the following will reduce the fundamental frequency to 90 Hz? (*a*) Double the tension and double the length. (*b*) Halve the tension and keep the length and the mass per unit length fixed. (*c*) Keep the tension and the mass per unit length fixed and double the length. (*d*) Keep the tension and the mass per unit length fixed and halve the length.

13 •• **ENGINEERING APPLICATION** Explain how you might use the resonance frequencies of an organ pipe to estimate the temperature of the air in the pipe. SSM

14 •• In the fundamental standing-wave pattern of an organ pipe stopped at one end, what happens to the wavelength, frequency, and speed of the sound needed to create the pattern if the air in the pipe becomes significantly colder? Explain your reasoning.

15 •• (*a*) When a guitar string is vibrating in its fundamental mode, is the wavelength of the sound it produces in air typically the same as the wavelength of the standing wave on the string? Explain. (*b*) When an organ pipe is in any one of its standing-wave modes, is the wavelength of the traveling sound wave it produces in air typically the same as the wavelength of the standing sound wave in the pipe? Explain. SSM

16 •• Figure 16-30 is a photograph of two pieces of very finely woven silk placed one on top of the other. Where the pieces overlap, a series of light and dark lines are seen. This moiré pattern can also be seen when a scanner is used to copy photos from a book or newspaper. What causes the moiré pattern, and how is it similar to the phenomenon of interference?

FIGURE 16-30 **Problem 16** (*Courtesy of Chuck Adler.*)

17 •• When a musical instrument consisting of drinking glasses, each partially filled to a different height with water, is struck with a small mallet, each glass produces a different frequency of sound wave. Explain how this instrument works.

18 •• **ENGINEERING APPLICATION** During an organ recital, the air compressor that drives the organ pipes suddenly fails. An enterprising physics student in the audience tries to help by replacing the compressor with a pressurized tank of nitrogen gas. What effect, if any, will the nitrogen gas have on the frequency output of the organ pipes? What effect, if any, would helium gas have on the frequency output of the organ pipes?

19 •• The constant γ for helium (and all monatomic gases) is 1.67. If a man inhales helium and then speaks, his voice has a high pitch and becomes cartoon-like. Why?

ESTIMATION AND APPROXIMATION

20 • It is said that a powerful opera singer can hit a high note with sufficient intensity to shatter an empty wine glass by causing the air in it to resonate at the frequency of her voice. Estimate the frequency necessary to obtain a standing wave in an 8.0-cm-high glass. (The 8.0 cm does not include the height of the stem.) Approximately how many octaves above middle C (262 Hz) is this? *Hint: To go up one octave means to double the frequency.*

21 • Estimate how accurately you can tune a piano string to a tuning fork of known frequency using only your ears, the tuning fork, and a wrench. Explain your answer.

22 •• The shortest pipes used in organs are 7.5 cm long. (*a*) Estimate the fundamental frequency of a pipe this long that is open at both ends. (*b*) For such a pipe, estimate the harmonic number n of the highest-frequency harmonic that is within the audible range. (The audible range of human hearing is about 20 to 20,000 Hz.)

23 •• **BIOLOGICAL APPLICATION** Estimate the resonant frequencies that are in the audible range of human hearing of the human ear canal. Treat the canal as an air column open at one end, stopped at the other end, and with a length of 1.00 in. How many resonant frequencies lie in this range? Human hearing has been found experimentally to be the most sensitive at frequencies of about 3, 9, and 15 kHz. How do these frequencies compare to your calculations?

SUPERPOSITION AND INTERFERENCE

24 • Two harmonic waves traveling on a string in the same direction both have a frequency of 100 Hz, a wavelength of 2.0 cm, and an amplitude of 0.020 m. In addition, they overlap each other. What is the amplitude of the resultant wave if the original waves differ in phase by (*a*) $\pi/6$, and (*b*) $\pi/3$?

25 • Two harmonic waves having the same frequency, wave speed, and amplitude are traveling in the same direction and in the same propagating medium. In addition, they overlap each other. If they differ in phase by $\pi/2$ and each has an amplitude of 0.050 m, what is the amplitude of the resultant wave? SSM

26 • Two audio speakers facing in the same direction oscillate in phase at the same frequency. They are separated by a distance equal to one-third of a wavelength. Point P is in front of both speakers, on the line that passes through their centers. The amplitude of the sound at P due to either speaker acting alone is A. What is the amplitude (in terms of A) of the resultant wave at point P?

27 • Two compact sources of sound oscillate in phase with a frequency of 100 Hz. At a point 5.00 m from one source and 5.85 m from the other, the amplitude of the sound from each source separately is A. (*a*) What is the phase difference of the two waves at that point? (*b*) What is the amplitude (in terms of A) of the resultant wave at that point?

28 • With a drawing program or a compass, draw circular arcs of radius 1 cm, 2 cm, 3 cm, 4 cm, 5 cm, 6 cm, and 7 cm centered at each of two points (P_1 and P_2) a distance $d = 3.0$ cm apart. Draw smooth curves through the intersections corresponding to points N centimeters farther from P_1 than from P_2 for $N = +2, +1, 0, -1$, and -2, and label each curve with the corresponding value of N. There are two additional such curves you can draw, one for $N = +3$ and one for $N = -3$. If identical sources of coherent in-phase 1.0-cm-wavelength waves were placed at points P_1 and P_2, the waves would interfere constructively along each of the smooth curves.

29 • Two speakers separated by some distance emit sound waves of the same frequency. At some point P, the intensity due to each speaker separately is I_0. The distance from P to one of the speakers is $\frac{1}{2}\lambda$ longer than that from P to the other speaker. What is the intensity at P if (*a*) the speakers are coherent and in phase, (*b*) the speakers are incoherent, and (*c*) the speakers are coherent and 180° out of phase? SSM

30 • Two speakers separated by some distance emit sound waves of the same frequency. At some point P' the intensity due to each speaker separately is I_0. The distance from P' to one of the speakers is one wavelength longer than that from P' to the other speaker. What is the intensity at P' if (*a*) the speakers are coherent and in phase, (*b*) the speakers are incoherent, and (*c*) the speakers are coherent and out of phase?

31 •• A transverse harmonic wave with a frequency equal to 40.0 Hz propagates along a taut string. Two points 5.00 cm apart are out of phase by $\pi/6$. (*a*) What is the wavelength of the wave? (*b*) At a given point on the string, how much does the phase change in 5.00 ms? (*c*) What is the wave speed?

32 •• **BIOLOGICAL APPLICATION** It is thought that the brain determines the direction of the source of a sound by sensing the phase difference between the sound waves striking the eardrums. A distant source emits sound of frequency 680 Hz. When you are directly facing a sound source there is no phase difference. Estimate the phase difference between the sounds received by your ears when you are facing 90° away from the direction of the source.

33 •• Sound source A is located at $x = 0, y = 0$, and sound source B is located at $x = 0, y = 2.4$ m. The two sources radiate coherently and in phase. An observer at $x = 15$ m, $y = 0$ notes that as he takes a few steps from $y = 0$ in either the $+y$ or $-y$ direction, the sound intensity diminishes. What is the lowest frequency, and the next to lowest frequency of the sources that can account for that observation? SSM

34 •• Suppose that the observer in Problem 33 finds himself at a point of minimum intensity at $x = 15$ m, $y = 0$. What is then the lowest frequency and next to lowest frequency of the sources that can account for this observation?

35 ••• **SPREADSHEET** Two harmonic water waves of equal amplitudes but different frequencies, wave numbers, and speeds are traveling in the same direction. In addition, they are superposed on each other. The total displacement of the wave can be written as $y(x, t) = A[\cos(k_1x - \omega_1t) + \cos(k_2x - \omega_2t)]$, where $\omega_1/k_1 = v_1$ (the speed of the first wave) and $\omega_2/k_2 = v_2$ (the speed of the second wave). (*a*) Show that $y(x, t)$ can be written in the form $y(x, t) = Y(x, t)\cos(k_{av}x - \omega_{av}t)$, where $\omega_{av} = (\omega_1 + \omega_2)/2$, $k_{av} = (k_1 + k_2)/2$, $Y(x, t) = 2A\cos[(\Delta k/2)x - (\Delta\omega/2)t]$, $\Delta\omega = \omega_1 - \omega_2$, and $\Delta k = k_1 - k_2$. The factor $Y(x, t)$ is called the *envelope* of the wave. (*b*) Let $A = 1.00$ cm, $\omega_1 = 1.00$ rad/s, $k_1 = 1.00\ \text{m}^{-1}$, $\omega_2 = 0.900$ rad/s, and $k_2 = 0.800\ \text{m}^{-1}$. Using a **spreadsheet program** or **graphing calculator,** make a plot of $y(x, t)$ versus x at $t = 0.00$ s for $0 < x < 5.00$ m. (*c*) Using a **spreadsheet program** or **graphing calculator,** make three plots of $Y(x, t)$ versus x for $-5.00\ \text{m} < x < 5.00$ m on the same graph. Make one plot for $t = 0.00$ s, the second for $t = 5.00$ s, and the third for $t = 10.00$ s. Estimate the speed at which the envelope moves from the three plots, and compare this estimate with the speed obtained using $v_{envelope} = \Delta\omega/\Delta k$. SSM

36 ••• Two coherent point sources are in phase and are separated by a distance d. An interference pattern is detected along a line parallel to the line through the sources and a large distance D from the sources, as shown in Figure 16-31. (*a*) Show that the path difference Δs from the two sources to some point on the line at an angle θ is given, approximately, by $\Delta s \approx d\sin\theta$. *Hint: Assume that $D \gg d$, so the lines from the sources to P are approximately parallel* (Figure 16-31*b*). (*b*) Show that the two waves interfere constructively at P if $\Delta s = m\lambda$, where $m = 0, 1, 2, \ldots$. (That is, show there is an interference maximum at P if $\Delta s = m\lambda$, where $m = 0, 1, 2, \ldots$.) (*c*) Show that the distance y_m from the central maximum (at $y = 0$) to the mth interference maximum at P is given by $y_m = D\tan\theta_m$, where $d\sin\theta_m = m\lambda$.

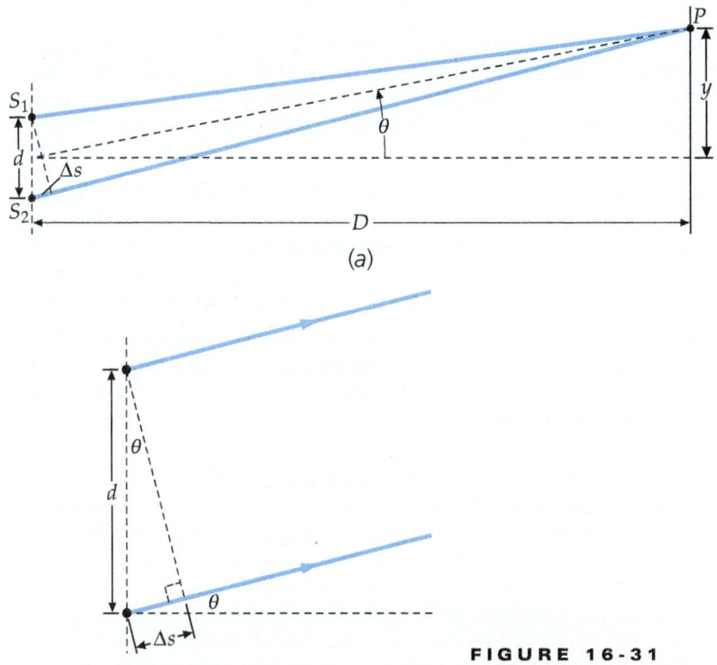

FIGURE 16-31
Problem 36

37 •• Two sound sources radiating in phase at a frequency of 480 Hz interfere such that maxima are heard at angles of 0° and 23° from a line perpendicular to that joining the two sources. The listener is at a large distance from the line through both sources, and no additional maxima are heard at angles in the range $0° < \theta < 23°$. Find the separation d between the two sources, and any other angles at which intensity maxima will be heard. (Use the result of Problem 36.)

38 •• Two loudspeakers are driven in phase by an audio amplifier at a frequency of 600 Hz. The speakers are on the y axis, one at $y = +1.00$ m and the other at $y = -1.00$ m. A listener, starting at $(x, y) = (D, 0)$, where $D \gg 2.00$ m, walks in the $+y$ direction along the line $x = D$. (See Problem 36.) (*a*) At what angle θ will she first hear a minimum in the sound intensity? (θ is the angle between the positive x axis and the line from the origin to the listener.) (*b*) At what angle will she first hear a maximum in the sound intensity (after $\theta = 0$)? (*c*) How many maxima can she possibly hear if she keeps walking in the same direction?

39 ••• Two sound sources driven in phase by the same amplifier are 2.00 m apart on the y axis, one at $y = +1.00$ m and the other at $y = -1.00$ m. At points large distances from the y axis, constructive interference is heard at angles with the x axis of $\theta_0 = 0.000$ rad, $\theta_1 = 0.140$ rad, and $\theta_2 = 0.283$ rad, and at no angles in between (see Figure 16-31). (*a*) What is the wavelength of the sound waves from the sources? (*b*) What is the frequency of the sources? (*c*) At what other angles is constructive interference heard? (*d*) What is the smallest angle for which the sound waves cancel? SSM

40 ••• The two sound sources from Problem 39 are now driven 90° out-of-phase, but at the same frequency as in Problem 39. At what angles are constructive and destructive interference heard?

41 •• **ENGINEERING APPLICATION** An astronomical radio telescope consists of two antennas separated by a distance of 200 m. Both antennas are tuned to the frequency of 20 MHz. The signals from each antenna are fed into a common amplifier, but one signal first passes through a phase selector that delays its phase by a chosen amount so that the telescope can "look" in different directions (Figure 16-32). When the phase delay is zero, plane radio waves that are incident vertically on the antennas produce signals that add constructively at the amplifier. What should the phase delay be so that signals coming from an angle $\theta = 10°$ with the vertical (in the plane formed by the vertical and the line joining the antennas) will add constructively at the amplifier? *Hint: Radio waves travel at* 3.00×10^8 *m/s.*

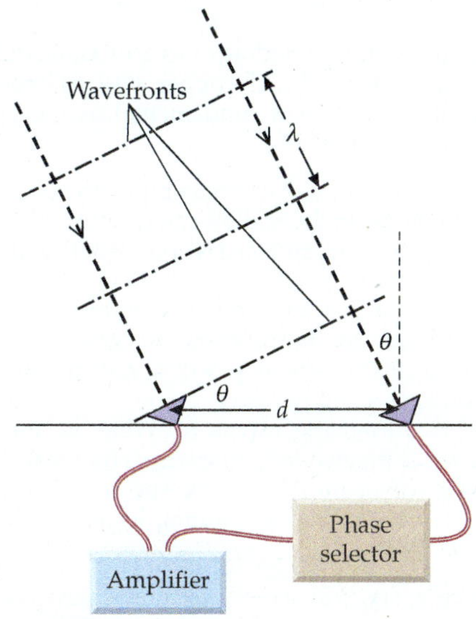

FIGURE 16-32
Problem 41

BEATS

42 • When two tuning forks are struck simultaneously, 4.0 beats per second are heard. The frequency of one fork is 500 Hz. (*a*) What are the possible values for the frequency of the other fork? (*b*) A piece of wax is placed on the 500-Hz fork to lower its frequency slightly. Explain how the measurement of the new beat frequency can be used to determine which of your answers to Part (*a*) is the correct frequency of the second fork.

43 ••• **ENGINEERING APPLICATION** A stationary police radar gun emits microwaves at 5.00 GHz. When the gun is aimed at a car, it superposes the transmitted and reflected waves. Because the frequencies of these two waves differ, beats are generated, with the speed of the car proportional to the beat frequency. The speed of the car, 83 mi/h, appears on the display of the radar gun. Assuming the car is moving along the line-of-sight of the police officer, and using the Doppler-shift equations, (*a*) show that, for a fixed radar-gun frequency, the beat frequency is proportional to the speed of the car. *Hint: Car speeds are tiny compared to the speed of light.* (*b*) What is the beat frequency in this case? (*c*) What is the calibration factor for this radar gun? That is, what is the beat frequency generated per mi/h of speed? SSM

STANDING WAVES

44 • A string fixed at both ends is 3.00 m long. It resonates in its second harmonic at a frequency of 60.0 Hz. What is the speed of transverse waves on the string?

45 • A string 3.00 m long and fixed at both ends is vibrating in its third harmonic. The maximum displacement of any point on the string is 4.00 mm. The speed of transverse waves on this string is 50.0 m/s. (*a*) What are the wavelength and frequency of this standing wave? (*b*) Write the wave function for this standing wave.

46 • Calculate the fundamental frequency for an organ pipe, with an effective length equal to 10 m, that is (*a*) open at both ends, and (*b*) stopped at one end.

47 • A 5.00-g, 1.40-m-long flexible wire has a tension of 968 N and is fixed at both ends. (*a*) Find the speed of transverse waves on the wire. (*b*) Find the wavelength and frequency of the fundamental. (*c*) Find the frequencies of the second and third harmonics. SSM

48 • A taut, 4.00-m-long rope has one end fixed and the other end free. (The free end is attached to a long, light string.) The speed of waves on the rope is 20.0 m/s. (*a*) Find the frequency of the fundamental. (*b*) Find the second harmonic. (*c*) Find the third harmonic.

49 • A steel piano wire without windings has a fundamental frequency of 200 Hz. When it is wound with copper wire, its linear mass density is doubled. What is its new fundamental frequency, assuming that the tension is unchanged?

50 • What is the greatest length that an organ pipe can have in order that its fundamental note be in the audible range (20 to 20,000 Hz) if (*a*) the pipe is stopped at one end, and (*b*) it is open at both ends?

51 •• The wave function $y(x, t)$ for a certain standing wave on a string that is fixed at both ends is given by $y(x, t) = 4.20 \sin(0.200x) \cos(300t)$, where y and x are in centimeters and t is in seconds. A standing wave can be considered as the superposition of two traveling waves. (*a*) What are the wavelength and frequency of the two traveling waves that make up the specified standing wave? (*b*) What is the speed of these waves on this string? (*c*) If the string is vibrating in its fourth harmonic, how long is it? SSM

52 •• The wave function $y(x, t)$ for a certain standing wave on a string that is fixed at both ends is given by $y(x, t) = (0.0500 \text{ m}) \sin(2.50 \text{ m}^{-1} x) \cos(500 \text{ s}^{-1} t)$. A standing wave can be considered as the superposition of two traveling waves. (*a*) What are the speed and amplitude of the two traveling waves that result in the specified standing wave? (*b*) What is the distance between successive nodes on the string? (*c*) What is the shortest possible length of the string?

53 •• A 1.20-m-long pipe is stopped at one end. Near the open end, there is a loudspeaker that is driven by an audio oscillator whose frequency can be varied from 10.0 to 5000 Hz. (Neglect any end corrections.) (*a*) What is the lowest frequency of the oscillator that will produce resonance within the tube? (*b*) What is the highest frequency of the oscillator that will produce resonance within the tube? (*c*) How many different frequencies of the oscillator will produce resonance within the tube?

54 •• A 460-Hz tuning fork causes resonance in the tube depicted in Figure 16-33 when the length L of the air column above the water is 18.3 and 55.8 cm. (*a*) Find the speed of sound in air. (*b*) What is the end correction to adjust for the fact that the antinode does not occur exactly at the open end of the tube?

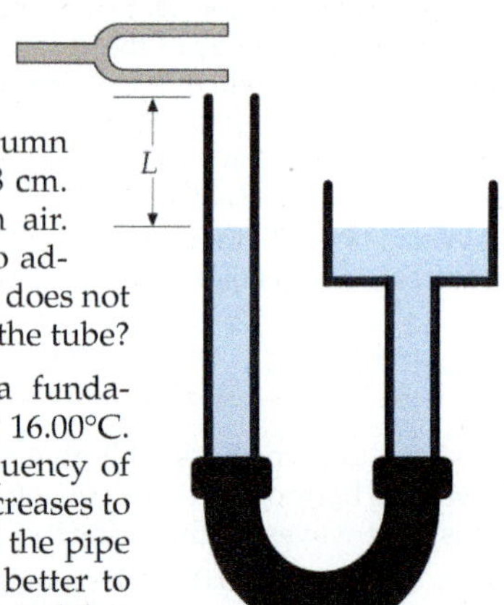

FIGURE 16-33
Problem 54

55 •• An organ pipe has a fundamental frequency of 440.0 Hz at 16.00°C. What will the fundamental frequency of the pipe be if the temperature increases to 32.00°C (assuming the length of the pipe remains constant)? Would it be better to construct organ pipes from a material that expands substantially as the temperature increases, or should the pipes be made of material that maintains the same length at all normal temperatures? SSM

56 •• According to theory, the end correction for a pipe is approximately $\Delta L = 0.3186D$, where D is the pipe diameter. Find the actual length of a pipe open at both ends that will produce a middle C (256 Hz) as its fundamental mode for pipes of diameter $D = 1.00$ cm, 10.0 cm, and 30.0 cm.

57 •• Assume a 40.0-cm-long violin string has a mass of 1.20 g and is vibrating in its fundamental mode* at a frequency of 500 Hz. (*a*) What is the wavelength of the standing wave on the string? (*b*) What is the tension in the string? (*c*) Where should you place your finger to increase the fundamental frequency to 650 Hz?

* A bowed string does not vibrate in a single mode. Thus, the conditions described in this problem statement are not completely accurate.

58 •• The G string on a violin is 30.0 cm long. When played without fingering, it vibrates in its fundamental mode* at a frequency of 196 Hz. The next higher notes on its C-major scale are A (220 Hz), B (247 Hz), C (262 Hz), and D (294 Hz). How far from the end of the string must a finger be placed to play each of these notes?

59 •• A string that has a linear mass density of 4.00×10^{-3} kg/m is under a tension of 360 N and is fixed at both ends. One of its resonance frequencies is 375 Hz. The next higher resonance frequency is 450 Hz. (*a*) What is the fundamental frequency of this string? (*b*) Which harmonics have the given frequencies? (*c*) What is the length of the string?

60 •• A string fixed at both ends has successive resonances with wavelengths of 0.54 m for the nth harmonic and 0.48 m for the $(n + 1)$th harmonic. (*a*) Which harmonics are these? (*b*) What is the length of the string?

61 •• The strings of a violin are tuned to the tones G, D, A, and E, which are separated by a fifth from one another. That is, $f(\text{D}) = 1.5f(\text{G})$, $f(\text{A}) = 1.5f(\text{D}) = 440$ Hz, and $f(\text{E}) = 1.5f(\text{A})$. The distance between the bridge at the scroll and the bridge over the body, the two fixed points on each string, is 30.0 cm. The tension on the E string is 90.0 N. (*a*) What is the linear mass density of the E string? (*b*) To prevent distortion of the instrument over time, it is important that the tension on all strings be the same. Find the linear mass densities of the other strings. SSM

62 •• On a cello, like most other stringed instruments, the positioning of the fingers by the player determines the fundamental frequencies of the strings. Suppose that one of the strings on a cello is tuned to play a middle C (262 Hz) when played at its full length. By what fraction must that string be shortened in order to play a note that is the interval of a third higher (namely, an E (330 Hz)? How about a fifth higher or a G (392 Hz)?

63 •• To tune your violin, you first tune the A string to the correct pitch of 440 Hz, and then you bow both it and an adjoining string simultaneously, all the while listening for beats. While bowing the A and E strings, you hear a beat frequency of 3.00 Hz and note that the beat frequency increases as the tension on the E string is increased. (The E string is to be tuned to 660 Hz.) (*a*) Why are beats produced by these two strings when bowed simultaneously? (*b*) What is the frequency of the E string vibration when the beat frequency is 3.00 Hz?

64 •• A 2.00-m-long string fixed at one end and free at the other end (the free end is fastened to the end of a long, light thread) is vibrating in its third harmonic with a maximum amplitude of 3.00 cm and a frequency 100 Hz. (*a*) Write the wave function for this vibration. (*b*) Write a function for the kinetic energy of a segment of the string of length dx, at a point a distance x from the fixed end, as a function of time t. At what times is this kinetic energy maximum? What is the shape of the string at these times? (*c*) Find the maximum kinetic energy of the string by integrating your expression for Part (*b*) over the total length of the string.

65 •• CONTEXT-RICH A commonly used physics experiment that examines resonances of transverse waves on a string is shown in Figure 16-34. A weight is attached to the end of a string draped over a pulley; the other end of the string is attached to a mechanical oscillator that moves up and down at a frequency f that remains fixed throughout the demonstration. The length L between the oscillator and the pulley is fixed, and the tension is equal to the gravitational force on the weight. For certain values of the tension, the string resonates. Assume the string does not stretch or shrink as the tension is varied. You are in charge of setting up this apparatus for a lecture demonstration. (*a*) Explain why only certain discrete values of the tension result in standing waves on the string. (*b*) Do you need to increase or decrease the tension to produce a standing wave with an additional antinode? Explain. (*c*) Prove your reasoning in Part (*b*) by showing that the values for the tension F_{Tn} for the nth

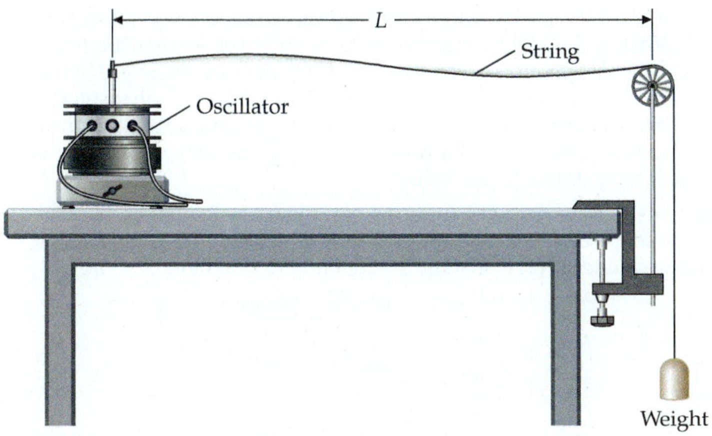

FIGURE 16-34 Problem 65

standing-wave mode are given by $F_{Tn} = 4L^2f^2\mu/n^2$, and thus the F_{Tn} is inversely proportional to n^2. (*d*) For your particular setup to fit onto the lecture table, you chose $L = 1.00$ m, $f = 80.0$ Hz, and $\mu = 0.750$ g/m. Calculate how much tension is needed to produce each of the first three modes (standing waves) of the string. SSM

*HARMONIC ANALYSIS

66 • A guitar string is given a light pluck at its midpoint. A microphone on your computer detects the sound and a program on the computer determines that most of the subsequent sound consists of a 100-Hz tone accompanied by a bit of sound with a 300-Hz tone. What are the two dominant standing-wave modes on the string?

*WAVE PACKETS

67 •• A tuning fork with natural frequency f_0 begins vibrating at time $t = 0$ and is stopped after a time interval, Δt. The waveform of the sound at some later time is shown (Figure 16-35) as a function of x. Let N be an estimate of the number of cycles in this waveform. (*a*) If Δx is the length in space of this wave packet, what is the range in wave numbers Δk of the packet? (*b*) Estimate the average value of the wavelength λ in terms of N and Δx. (*c*) Estimate the average wave number k in terms of N and Δx. (*d*) If Δt is the time it takes the wave packet to pass a point in space, what is the range in angular frequencies $\Delta\omega$ of the packet? (*e*) Express f_0 in terms of N and Δt. (*f*) The number N is uncertain by about ±1 cycle. Use Figure 16-35 to explain why. (*g*) Show that the uncertainty in the wave number due to the uncertainty in N is $2\pi/\Delta x$. SSM

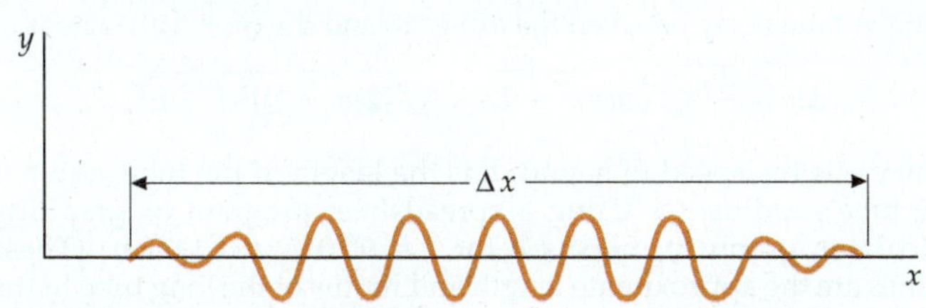

FIGURE 16-35 Problem 67

GENERAL PROBLEMS

68 •• A 35-m-long string has a linear mass density of 0.0085 kg/m and is under a tension of 18 N. Find the frequencies of the lowest four harmonics (*a*) if the string is fixed at both ends, and (*b*) if the string is fixed at one end and free at the other. (That is, if the free end is attached to a long string of negligible mass.)

69 •• **CONTEXT-RICH, ENGINEERING APPLICATION** Working for a small gold mining company, you stumble across an abandoned mine shaft that, because of decaying wood shoring, looks too dangerous to explore in person. To measure its depth, you employ an audio oscillator of variable frequency. You determine that successive resonances are produced at frequencies of 63.58 and 89.25 Hz. Estimate the depth of the shaft.

70 •• A 5.00-m-long string that is fixed at one end and attached to a long string of negligible mass at the other end is vibrating in its fifth harmonic, which has a frequency of 400 Hz. The amplitude of the motion at each antinode is 3.00 cm. (*a*) What is the wavelength of this wave? (*b*) What is the wave number? (*c*) What is the angular frequency? (*d*) Write the wave function for this standing wave.

71 •• The wave function for a standing wave on a string is described by $y(x, t) = (0.020)\sin(4\pi x)\cos(60\pi t)$, where y and x are in meters and t is in seconds. Determine the maximum displacement and maximum speed of a point on the string at (*a*) $x = 0.10$ m, (*b*) $x = 0.25$ m, (*c*) $x = 0.30$ m, and (*d*) $x = 0.50$ m. SSM

72 •• A 2.5-m-long string that has a mass of 0.10 kg is fixed at both ends and is under a tension of 30 N. When the nth harmonic is excited, there is a node 0.50 m from one end. (*a*) What is n? (*b*) What are the frequencies of the first three harmonics of this string?

73 •• An organ pipe is such that under normal conditions its fundamental frequency is 220 Hz. It is placed in an atmosphere of sulfur hexafluoride (SF_6) at the same temperature and pressure. The molar mass of air is 29.0×10^{-3} kg/mol and the molar mass of SF_6 is 146×10^{-3} kg/mol. What is the fundamental frequency of the organ pipe when it is in an atmosphere of SF_6?

74 •• During a lecture demonstration of standing waves, one end of a string is attached to a device that vibrates at 60 Hz and produces transverse waves of that frequency on the string. The other end of the string passes over a pulley, and the tension is varied by attaching weights to that end. The string has approximate nodes next to both the vibrating device and the pulley. (*a*) If the string has a linear mass density of 8.0 g/m and is 2.5 m long from the vibrating device to the pulley, what must be the tension for the string to vibrate in its fundamental mode? (*b*) Find the tension necessary for the string to vibrate in its second, third, and fourth harmonics.

75 •• Three successive resonance frequencies in an organ pipe are 1310, 1834, and 2358 Hz. (*a*) Is the pipe closed at one end or open at both ends? (*b*) What is the fundamental frequency? (*c*) What is the effective length of the pipe? SSM

76 •• During an experiment studying the speed of sound in air using an audio oscillator and a tube that is open at one end and stopped at the other end, a particular resonant frequency is found to have nodes roughly 6.94 cm apart. The oscillator's frequency is increased, and the next resonant frequency found has nodes 5.40 cm apart. (*a*) What are the two resonant frequencies? (*b*) What is the fundamental frequency? (*c*) Which harmonics are these two modes? The speed of sound is 343 m/s.

77 •• A standing wave on a rope is represented by the wave function $y(x, t) = (0.020)\sin(\frac{1}{2}\pi x)\cos(40\pi t)$, where x and y are in meters and t is in seconds. (*a*) Write wave functions for two traveling waves that, when superimposed, produce this standing-wave pattern. (*b*) What is the distance between the nodes of the standing wave? (*c*) What is the maximum speed of the rope at $x = 1.0$ m? (*d*) What is the maximum acceleration of the rope at $x = 1.0$ m?

78 •• **SPREADSHEET** Two traveling-wave pulses on a string are represented by the wave functions

$$y_1(x, t) = \frac{0.020}{2.0 + (x - 2.0t)^2} \quad \text{and} \quad y_2(x, t) = \frac{-0.020}{2.0 + (x + 2.0t)^2}$$

where x is in meters and t is in seconds. (*a*) Using a **spreadsheet program** or **graphing calculator,** make a separate graph of each wave function as a function of x at $t = 0$ and again at $t = 1.0$ s, and describe the behavior of each as time increases. For each graph make your plot for $-5.0\text{ m} < x < +5.0\text{ m}$. (*b*) Graph the resultant wave function at $t = -1.0$ s, at $t = 0.0$ s, and at $t = 1.0$ s.

79 ••• Three waves that have the same frequency, wavelength, and amplitude are traveling along the x axis. The three waves are described by the following wave functions: $y_1(x, t) = (5.00\text{ cm})\sin(kx - \omega t - \frac{1}{3}\pi)$, $y_2(x, t) = (5.00\text{ cm})\sin(kx - \omega t)$, and $y_3(x, t) = (5.00\text{ cm})\sin(kx - \omega t + \frac{1}{3}\pi)$, where x is in meters and t is in seconds. The resultant wave function is given by $y_3(x, t) = A\sin(kx - \omega t + \delta)$. What are the values of A and δ?

80 ••• A harmonic pressure wave produced by a distant source is traveling through your vicinity, and the wavefronts that travel through your vicinity are vertical planes. Let the $+x$ direction be to the east and the $+y$ direction be toward the north. The wave function for the wave is $p(x, y, t) = A\cos(k_x x + k_y y - \omega t)$. Show that the direction in which the wave is traveling makes an angle $\theta = \tan^{-1}(k_y/k_x)$ with the $+x$ direction and that the wave speed is $v = \omega\sqrt{k_x^2 + k_y^2}$.

81 •• The speed of sound in air is proportional to the square root of the absolute temperature T (Equation 15-5). (*a*) Show that if the air temperature changes by a small amount, the fractional change in the fundamental frequency of an organ pipe is approximately equal to half the fractional change in the absolute temperature. That is, show that $\Delta f/f \approx \frac{1}{2}\Delta T/T$, where f is the frequency at absolute temperature T and Δf is the change in frequency when the temperature changes by ΔT. (Ignore any change in the length of the pipe due to thermal expansion.) (*b*) Suppose that an organ pipe that is stopped at one end has a fundamental frequency of 200.0 Hz when the temperature is 20.00°C. Use the approximate result from Part (*a*) to determine the pipe's fundamental frequency when the temperature is 30.00°C. (*c*) Compare your Part-(*b*) result to what you would get using exact calculations. (Ignore any change in the length of the pipe due to thermal expansion.) SSM

82 •• The pipe in Figure 16-36 is kept filled with natural gas [methane (CH_4)]. The pipe is punctured by a line of small holes 1.00 cm apart down its entire 2.20-m length. A speaker forms the closure on one end of the pipe, and a solid piece of metal closes the other end. What frequency is being played in this picture? The speed of sound in low-pressure methane at room temperature is about 460 m/s.

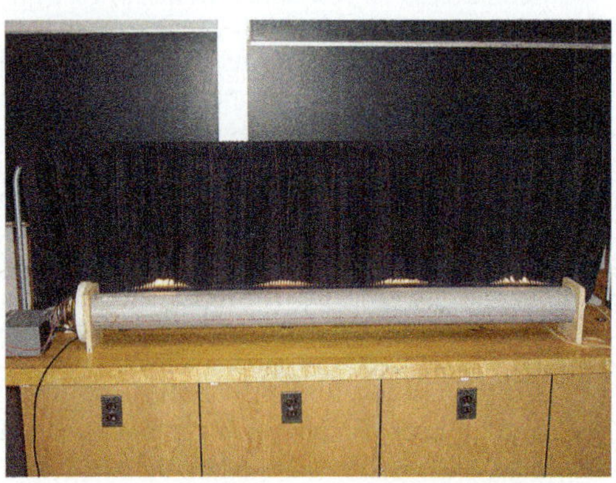

FIGURE 16-36 Problem 82 *(University of Michigan Demonstration Laboratory.)*

83 •• **CONTEXT-RICH** Assume that your clarinet is entirely filled with helium and that before you start to play you fill your lungs with helium. You pick up the clarinet and play it as though you were trying to play a B-flat, which has a frequency of 277 Hz. The frequency of 277 Hz is the natural resonance frequency of this clarinet with all finger holes closed and when filled with air. What frequency do you actually hear?

84 ••• A 2.00-m-long wire that is fixed at both ends is vibrating in its fundamental mode. The tension in the wire is 40.0 N and the mass of the wire is 0.100 kg. At the midpoint of the wire, the amplitude is 2.00 cm. (*a*) Find the maximum kinetic energy of the wire. (*b*) What is the kinetic energy of the wire at the instant the transverse displacement is given by $y = 0.0200\sin(\frac{\pi}{2}x)$, where y is in meters if x is in meters, for $0.00\text{ m} \le x \le 2.00\text{ m}$? (*c*) For what value of x is the average value of the kinetic energy per unit length the greatest? (*d*) For what value of x does the elastic potential energy per unit length have its maximum value?

85 ••• **SPREADSHEET** In principle, a wave with almost any arbitrary shape can be expressed as a sum of harmonic waves of different frequencies. (*a*) Consider the function defined by

$$f(x) = \frac{4}{\pi}\left(\frac{\cos x}{1} - \frac{\cos 3x}{3} + \frac{\cos 5x}{5} - \cdots\right)$$

$$= \frac{4}{\pi}\sum_{n=0}^{\infty}(-1)^n\frac{\cos[(2n+1)x]}{2n+1}$$

Write a **spreadsheet program** to calculate this series using a finite number of terms, and make three graphs of the function in the range $x = 0$ to $x = 4\pi$. To create the first graph, for each value of x that you plot, approximate the sum from $n = 0$ to $n = \infty$ with the first term of the sum. To create the second and third graphs, use only the first five terms and the first ten terms, respectively. This function is sometimes called the *square wave*. (*b*) What is the relation between this function and Leibnitz's series for π,

$$\frac{\pi}{4} = 1 - \frac{1}{3} + \frac{1}{5} - \frac{1}{7} + \cdots$$ SSM

86 ••• **SPREADSHEET** Write a **spreadsheet program** to calculate and graph the function

$$y(x) = \frac{4}{\pi}\left(\sin x - \frac{\sin 3x}{9} + \frac{\sin 5x}{25} - \cdots\right)$$

$$= \frac{4}{\pi}\sum_{n}\frac{(-1)^n\sin(2n+1)x}{(2n+1)^2}$$

for $0 \le x \le 4\pi$. Use only the first 25 terms in the sum for each value of x that you plot.

87 ••• **SPREADSHEET** If you clap your hands at the end of a long, cylindrical tube, the echo you hear back will not sound like the handclap; instead, you will hear what sounds like a whistle, initially at a very high frequency, but descending rapidly down to almost nothing. This "culvert whistler" is easily explained if you think of the sound from the clap as a single compression radiating outward from the hands. The echoes of the handclap arriving at your ear have traveled along different paths through the tube, as shown in Figure 16-37. The first echo to arrive travels straight down and straight back along the tube, while the second echo reflects once off of the center of the tube going out, and again going back, the third echo reflects twice at points 1/4 and 3/4 of the distance, and so on. The tone of the sound you hear reflects the frequency at which these echoes reach your ears. (*a*) Show that the time delay between the nth echo and the $(n+1)$th echo is

$$\Delta t_n = \frac{2}{v}\left(\sqrt{(2n)^2r^2 + L^2} - \sqrt{[2(n-1)]^2r^2 + L^2}\right)$$

where v is the speed of sound, L is the length of the tube, and r is the tube's radius. (*b*) Using a **spreadsheet program** or **graphing calculator,** graph Δt_n versus n for $L = 90.0$ m, $r = 1.00$ m. (These values are the approximate length and radius of the long tube in the San Francisco Exploratorium.) Go to at least $n = 100$. (*c*) From your graph, explain why the frequency decreases over time. What are the highest and lowest frequencies you will hear in the whistler?

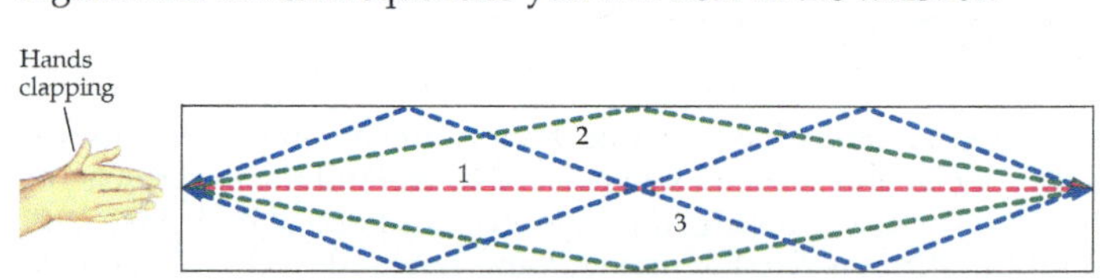

FIGURE 16-37 Problem 87

CHAPTER 17

WHEN BEN FRANKLIN WENT TO PARIS, HE SAW THE FIRST KNOWN MANNED HOT-AIR BALLOON FLIGHT. PEOPLE HAVE BEEN FLYING HOT-AIR BALLOONS EVER SINCE. *(Corbis.)*

Temperature and Kinetic Theory of Gases

? Why does the balloon rise when the air inside it is heated? (See Example 17-7.)

Even very small children have a basic understanding of hot and cold, but what is temperature? What is it a measurement of? In Chapter 17, we begin our study of temperature.

A pilot, a hot air balloonist, and a scuba diver must all have a good working understanding of air and water temperatures as they plan their flights and dives. Pilots and balloonists need to be aware of how changes in air temperature affect air density as well as wind patterns. Scuba divers know that changes in body temperature affect how much air they will use over the course of a dive. They also understand the importance of equalizing the pressure on their bodies and the gases within their bodies. For the diver, the pilot, and the balloonist, the importance of how gases behave in relation to temperature is vital. Thus, we begin our study of thermodynamics with a discussion of temperature and an examination of the ideal-gas law.

In this chapter, we show that a consistent temperature scale can be defined in terms of the properties of gases that have low densities, and that temperature is a measure of the average internal molecular kinetic energy of an object.

17-1 THERMAL EQUILIBRIUM AND TEMPERATURE

Our sense of touch can usually tell us if an object is hot or cold. We know that to make a cold object warmer, we can place it in contact with a hot object, and to make a hot object cooler, we can place it in contact with a cold object.

When an object is heated or cooled, some of its physical properties change. If a solid or liquid is heated, its volume usually increases. If a gas is heated and its pressure is kept constant, its volume increases. However, if a gas is heated and its volume is kept constant, its pressure increases. If an electrical conductor is heated, its electrical resistance changes. (This property is discussed in Chapter 25.) A physical property that changes with temperature is called a **thermometric property.** A change in a thermometric property indicates a change in the temperature of the object.

Suppose that we place a warm copper bar in close contact with a cold iron bar so that the copper bar cools and the iron bar warms. We say that the two bars are in **thermal contact.** The copper bar contracts slightly as it cools, and the iron bar expands slightly as it warms. This process eventually stops and the lengths of the bars remain constant. The two bars are then in **thermal equilibrium** with each other.

Suppose instead that we place the warm copper bar in a cool running stream of water. The bar cools until it stops contracting, at the point at which the bar and the water are in thermal equilibrium. Next, we place a cold iron bar in the stream, near but not touching the copper bar. The iron bar will warm until the iron bar and the water are also in thermal equilibrium. If we remove the bars and place them in thermal contact with each other, we find that their lengths do not change. They are in thermal equilibrium with each other. Although it is common sense, there is no logical way to deduce this fact, which is called the **zeroth law of thermodynamics** (Figure 17-1):

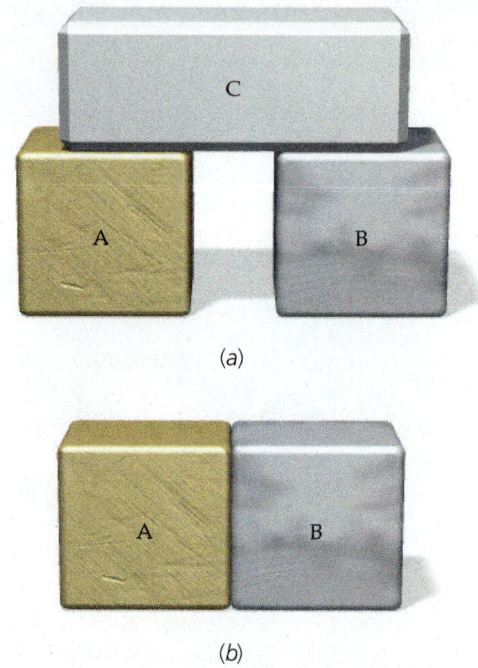

FIGURE 17-1 The zeroth law of thermodynamics. (*a*) Systems A and B are in thermal contact with system C, but not with each other. When A and B are each in thermal equilibrium with C, they are in thermal equilibrium with each other, which can be checked by placing them in contact with each other as in Part (*b*).

> If two objects are in thermal equilibrium with a third object, then all three of the objects are in thermal equilibrium with each other.
>
> ZEROTH LAW OF THERMODYNAMICS

Two objects are defined to have the same *temperature* if they are in thermal equilibrium with each other. The zeroth law, as we will see, enables us to define a temperature scale.

THE CENTIGRADE AND FAHRENHEIT TEMPERATURE SCALES

Any thermometric property can be used to establish a temperature scale. The common mercury thermometer consists of a glass bulb and tube containing a fixed amount of mercury.* When this thermometer is put in contact with a warmer object, the mercury expands, increasing the length of the mercury column (the glass expands too, but by a negligible amount). We can create a scale along the glass tube by using the following procedure. First, the thermometer is placed in ice and water in equilibrium† at a pressure of 1 atm. When the thermometer is in thermal equilibrium with the ice water, the top of the mercury column is marked on the glass tube. This mark represents the **ice-point** temperature (also called the **normal freezing point** of water). Next, the thermometer is placed in boiling water at a pressure

* Mercury is highly toxic. Today, alcohol is commonly used in thermometers.

† Water and ice in equilibrium provide a constant-temperature bath. When ice is placed in warm water, the water cools as some of the ice melts. Thermal equilibrium is eventually reached and no more ice melts. If the water/ice system is heated slightly, some more of the ice melts, but the temperature of the system does not change as long as some ice remains.

of 1 atm. When the thermometer is in thermal equilibrium with the boiling water, the top of the mercury column is marked. This mark represents the **steam-point temperature** (also called the **normal boiling point** of water).

The **centigrade temperature scale** defines the ice-point temperature as zero degrees centigrade (0°C) and the steam-point temperature as 100°C. The space between the 0-degree and 100-degree marks is divided into 100 equal intervals (degrees). Degree markings are also extended below and above these points. If L_t is the length of the mercury column, the centigrade temperature t_C is given by

$$t_C = \frac{L_t - L_0}{L_{100} - L_0} \times 100° \qquad 17\text{-}1$$

where L_0 is the length of the mercury column when the thermometer is in an ice bath and L_{100} is its length when the thermometer is in a steam bath. The normal temperature of the human body measured on the centigrade scale is about 37°C.

A shortcoming of the centigrade scale is that it depends on the thermometric property of some material, such as mercury. An improvement is the Celsius scale, discussed in Section 17-2, which is in close agreement with the centigrade scale. (So close is the agreement between these two scales that many refer to the Celsius scale as the centigrade scale.)

Historically, the **Fahrenheit temperature scale** (which is widely used in the United States) defines the ice-point temperature as 32°F and the steam-point temperature as 212°F.* To convert temperatures between Fahrenheit and centigrade scales, we note there are 100 centigrade degrees and 180 Fahrenheit degrees between the ice and steam points. A temperature change of one centigrade degree therefore equals a change of 1.8 = 9/5 Fahrenheit degrees. To convert a temperature from one scale to the other, we must also take into account the fact that the zero temperatures of the two scales are not the same. The general relation between a Fahrenheit temperature t_F and centigrade temperature t_C is

$$t_C = \tfrac{5}{9}(t_F - 32°) \qquad (\text{or } t_F = \tfrac{9}{5}t_C + 32°) \qquad 17\text{-}2$$

FAHRENHEIT–CENTIGRADE CONVERSION

Today, we define the Fahrenheit scale using Equation 17-2, with t_C the *Celsius* temperature.

Example 17-1 Converting Fahrenheit and Celsius Temperatures

Vivian measures her ill six-month-old son's temperature with a Celsius thermometer and finds it to be 40.0°C. She then telephone's the doctor for advice. When she gives the doctor the baby's temperature, the doctor asks, "What is that in Fahrenheit?" She does the conversion using Equation 17-2 and says, "102°F." Did she convert from Celsius to Fahrenheit correctly?

PICTURE Solve for t_F by using $t_C = \frac{5}{9}(t_F - 32°)$ (Equation 17-2), where $t_C = 40.0°$.

SOLVE

1. Solve $t_C = \frac{5}{9}(t_F - 32°)$ (Equation 17-2) for t_F in terms of t_C: $t_F = \frac{9}{5}t_C + 32°$
2. Substitute $t_C = 40.0°\text{C}$: $t_F = \frac{9}{5}(40.0°) + 32° = \boxed{104°\text{F}}$

Vivian's estimate is off by 2°F.

CHECK The temperature of 40°C is 0.4 of the way between 0°C and 100°C, and the temperature of 72°F is 0.4 of the way between 0°F and 180°F. Thus, we expect the Fahrenheit temperature to be 72°F + 32°F = 104°F, which verifies our step-2 result.

PRACTICE PROBLEM 17-1 (*a*) Find the Celsius temperature equivalent to 68°F. (*b*) Find the Fahrenheit temperature equivalent to −40°C.

* When the German physicist Daniel Fahrenheit devised his temperature scale, he wanted all measurable temperatures to be positive. He originally chose 0°F for the coldest temperature he could obtain with a mixture of ice and salt water and 96°F (a convenient number with many factors for subdivision) for the temperature of the human body. He then modified his scale slightly to make the ice-point and steam-point temperatures whole numbers. This modification resulted in the average temperature of the human body being between 98° and 99°F.

Other thermometric properties can be used to set up thermometers and construct temperature scales. Figure 17-2 shows a bimetallic strip consisting of two different metals bonded together. When the strip is heated or cooled, it bends to accommodate the difference in the thermal expansion of the two metals. Figure 17-3 shows a thermometer consisting of a bimetallic coil with a pointer attached to indicate the temperature. When the thermometer is heated, the coil bends and the pointer moves. Like mercury thermometers, it is calibrated by dividing the interval between the ice point and the steam point into 100 centigrade degrees (or 180 Fahrenheit degrees).

FIGURE 17-2 A bimetallic strip. When heated or cooled, the two metals expand or contract by different amounts, causing the strip to bend.

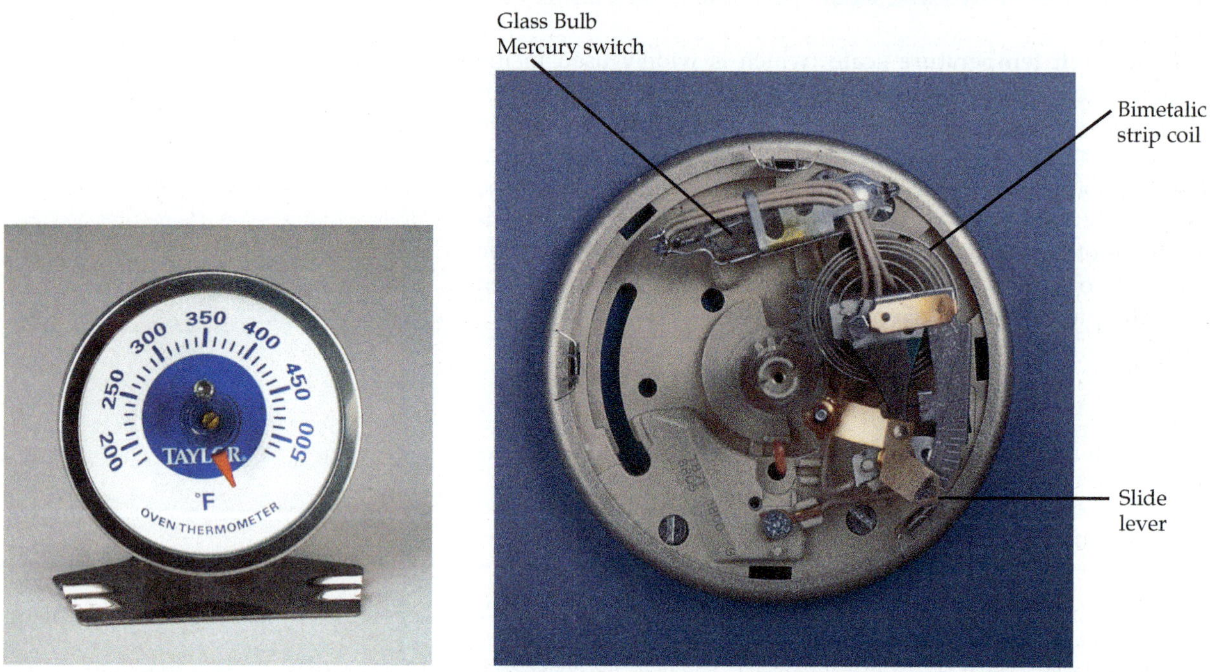

(*a*) (*b*)

FIGURE 17-3 (*a*) A thermometer using a bimetallic strip in the form of a coil. (The red pointer is attached to one end of the coil.) When the temperature of the coil increases, the needle rotates clockwise because the outer metal expands more than the inner metal. (*b*) A home thermostat controls the central air conditioner. When the air gets warmer, the coil expands, the glass bulb mounted on it tilts, and mercury in the tube slides to close an electrical switch, turning on the air conditioning. A slide lever (at the lower right), used to rotate the coil mount, is used to set the desired temperature. The circuit will be broken when the cooler air causes the bimetallic coil to contract. *((a) Courtesy of Taylor Precision Products. (b) Richard Menga/Fundamental Photographs.)*

17-2 GAS THERMOMETERS AND THE ABSOLUTE TEMPERATURE SCALE

When different types of centigrade thermometers are calibrated in ice water and steam, they agree (by definition) at 0°C and 100°C, but they give slightly different readings at points in between. Discrepancies increase markedly above the steam point and below the ice point. However, in one group of thermometers, gas thermometers, measured temperatures agree closely with each other, even far from the calibration points. In a **constant-volume gas thermometer,** the gas volume is kept constant, and change in gas pressure is used to indicate a change in temperature (Figure 17-4). An ice-point pressure P_0 and steam-point pressure P_{100} are determined by placing the thermometer in ice–water and water–steam baths, and the interval between them is divided into 100 equal degrees (for the centigrade scale). If the pressure is P_t in a bath whose temperature is to be determined, that temperature in degrees centigrade is defined to be

$$t_C = \frac{P_t - P_0}{P_{100} - P_0} \times 100°C \qquad 17\text{-}3$$

CONSTANT-VOLUME CENTIGRADE GAS THERMOMETER

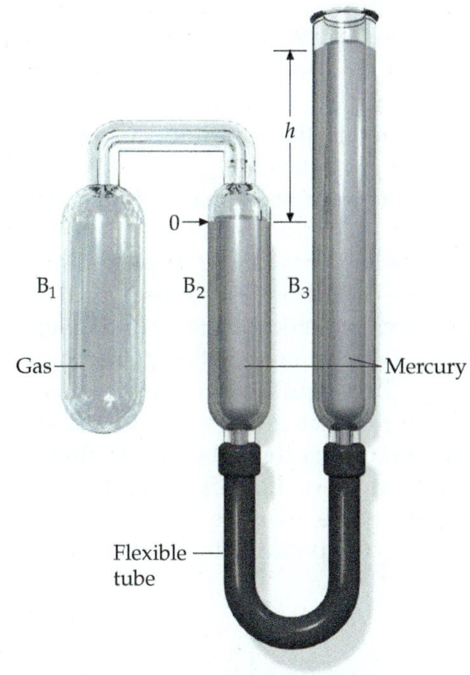

FIGURE 17-4 A constant-volume gas thermometer. The volume is kept constant by raising or lowering tube B_3 so that the mercury in tube B_2 remains at the zero mark. The temperature is chosen to be proportional to the pressure of the gas in tube B_1, which is indicated by the height h of the mercury column in tube B_3.

Suppose we measure a specific temperature, say the boiling point of sulfur at 1-atm pressure, using four constant-volume gas thermometers, each containing one of four gases: air, hydrogen, nitrogen, or oxygen. The thermometers are calibrated, meaning values for P_{100} and P_0 are determined for each. Each thermometer is then immersed in boiling sulfur, and when it is in thermal equilibrium with the sulfur, the pressure in the thermometer is measured. Next, the temperature is calculated using Equation 17-3. Will this process give the same result for each of the four thermometers? Surprisingly perhaps, the answer is yes. All four thermometers measure the same temperature as long as the density of the gas in each is sufficiently low.

One measure of the density of the gas in the thermometer is its pressure at the steam point, P_{100}. If we vary the amount of gas in a constant-volume gas thermometer, by either adding or removing gas, we change both P_{100} and P_0. As a result, each time the amount of gas is varied, the thermometer must be recalibrated. Figure 17-5 shows the results of measurements of the boiling point of sulfur using four constant-volume gas thermometers, each filled with air, hydrogen, nitrogen, or oxygen. For each thermometer the measured temperature is plotted as a function of the steam-point pressure P_{100} of the thermometer. As the amount of a gas is reduced, its density and the steam-point pressure both decrease. We see that when low densities of gas are used (small P_{100}), the thermometers are in close agreement. In the limit as gas density approaches zero, all gas thermometers give the same value for the temperature of boiling sulfur. This low-density temperature measurement is independent of the properties of any particular gas. Of course, there is nothing special about the boiling point of sulfur. Constant-volume gas thermometers that have low densities of gas are in agreement at any temperature. Thus, constant-volume gas thermometers that contain low densities of gas can be used to define temperature.

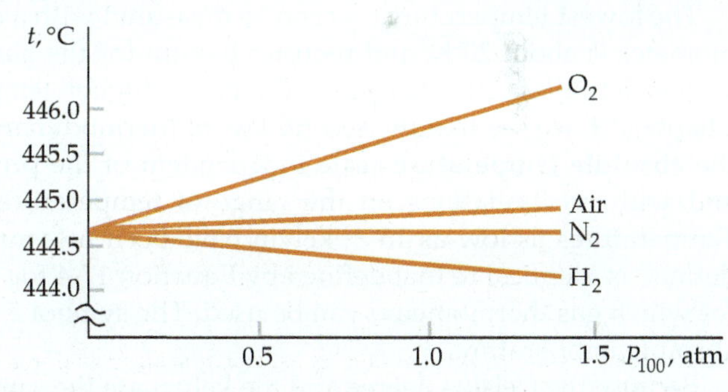

FIGURE 17-5 Temperature of the boiling point of sulfur measured with constant-volume gas thermometers filled with various gases. Increasing or decreasing the amount of gas in the thermometer varies the pressure P_{100} at the steam point of water. As the amount of gas is reduced, the temperature of the boiling point of sulfur measured by all the thermometers approaches the value 444.60°C. Note that the temperature axis shows a range of temperatures from 444°C to 446°C.

Now consider a series of temperature measurements using a constant-volume gas thermometer that has a very small but fixed amount of gas. According to Equation 17-3, the pressure in the thermometer P_t varies linearly with the measured temperature t_C. Figure 17-6 shows a plot of gas pressure versus measured

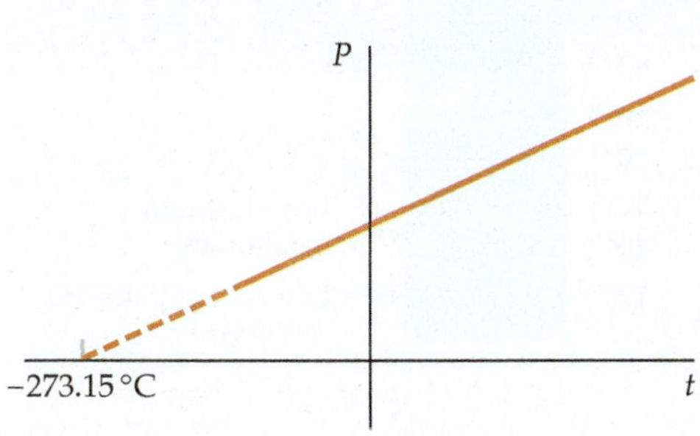

FIGURE 17-6 Plot of pressure versus temperature for a gas, as measured by a constant-volume gas thermometer. When extrapolated to zero pressure, the plot intersects the temperature axis at the value −273.15°C.

temperature in a constant-volume gas thermometer. When we extrapolate this straight line to zero gas pressure, the temperature approaches −273.15°C. This limit is the same no matter what kind of gas is used.

A reference state that is much more precisely reproducible than either the ice or steam points is the **triple point of water**—the unique temperature and pressure at which water, water vapor, and ice coexist in equilibrium (see Figure 17-7). This equilibrium state occurs at 4.58 mmHg and 0.01°C. The **ideal-gas temperature scale** is defined so that the temperature of the triple-point state is 273.16 kelvins (K). The temperature T of any other state is defined to be proportional to the pressure in a constant-volume gas thermometer:

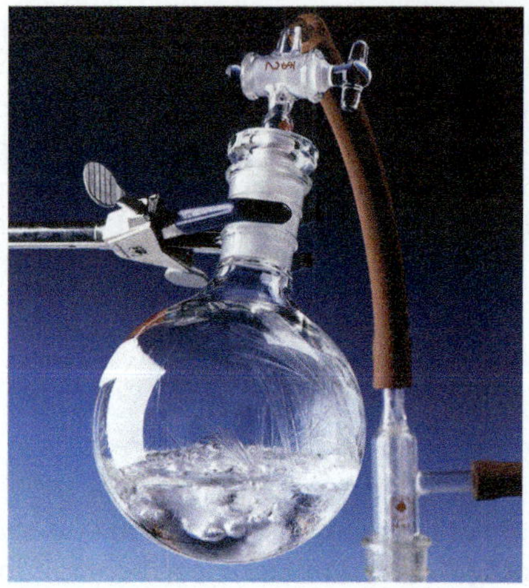

FIGURE 17-7 Water at its triple point. The spherical flask contains liquid water, ice, and water vapor in thermal equilibrium. *(Richard Menga/Fundamental Photographs.)*

$$T = \frac{P}{P_3} T_3 \qquad 17\text{-}4$$

CONSTANT-VOLUME IDEAL-GAS–TEMPERATURE THERMOMETER

where P is the observed pressure of the gas in the thermometer, P_3 is the pressure when the thermometer is immersed in a water–ice–vapor bath at its triple point, and T_3 = 273.16 K (the triple-point temperature). The value of P_3 depends on the amount of gas in the thermometer.

! The ideal-gas temperature scale, defined by Equation 17-4, has the advantage that any measured temperature does not depend on the properties of the particular gas that is used, but depends only on the general properties of gases.

! Note that the SI temperature unit, the kelvin, is not a degree and is not accompanied by a degree symbol.

The Celsius degree is a degree unit that is the same size as the kelvin, but the zero point of the **Celsius scale** differs from the zero point of the ideal-gas temperature scale. By definition, zero on the Celsius scale corresponds to an ideal-gas temperature of exactly 273.15 K.

The lowest temperature that can be measured with a constant-volume gas thermometer is about 20 K, and requires helium for the gas. Below this temperature helium liquefies; all other gases liquefy at higher temperatures (Table 17-1). In Chapter 19, we see that the second law of thermodynamics can be used to define the **absolute temperature scale** independent of the properties of any substance, and with no limitations on the range of temperatures that can be measured. Temperatures as low as 10^{-10} kelvin have been measured. The absolute scale so defined is identical to that defined by Equation 17-4 for the range of temperatures for which gas thermometers can be used. The symbol T is used when referring to absolute temperature.

Because the Celsius degree and the kelvin are the same size, temperature *differences* are the same on both the Celsius scale and the absolute temperature scale (also called the **Kelvin scale**). That is, a temperature *change* of 1 K is identical to a temperature *change* of 1°C. The two scales differ only in the choice of zero temperature. To convert from degrees Celsius to kelvins, we merely add 273.15:*

$$T = t_C + 273.15\text{ K} \qquad 17\text{-}5$$

CELSIUS–ABSOLUTE CONVERSION

Although the Celsius and Fahrenheit scales are convenient for everyday use, the absolute scale is much more convenient for scientific purposes, partly because many formulas are more simply expressed using it, and partly because the absolute temperature can be given a fundamental interpretation.

* For most purposes, we can round off the temperature of absolute zero to −273°C.

Table 17-1 The Temperatures of Various Places and Phenomena

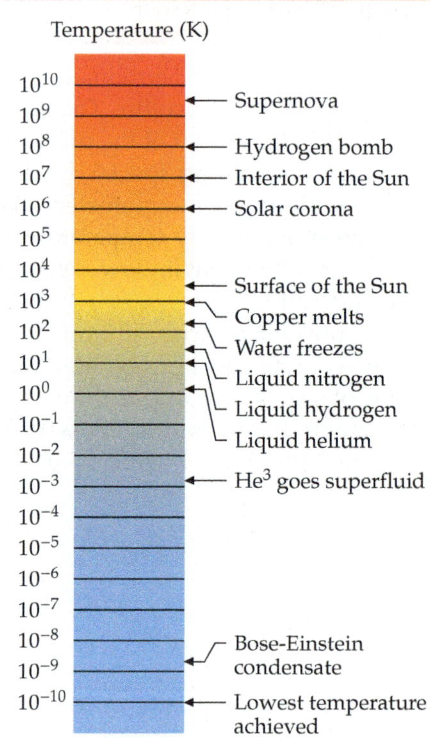

Example 17-2 Converting from Kelvin to Fahrenheit

The "high-temperature" superconductor $YBa_2Cu_3O_7$ becomes superconducting when the temperature is decreased to 92 K. Find the superconducting threshold temperature in degrees Fahrenheit.

PICTURE First, convert to degrees Celsius, then to kelvins.

SOLVE

1. Convert from kelvins to degrees Celsius: $T = t_C + 273.15$
   so $92 = t_C + 273.15 \Rightarrow t_C = -181.15°C$
2. To find the Fahrenheit temperature we use $t_C = \frac{5}{9}(t_F - 32°)$ (Equation 17-2): $t_C = \frac{5}{9}(t_F - 32°)$
   so $-181.15° = \frac{5}{9}(t_F - 32°) \Rightarrow t_F = \boxed{-294°F}$

CHECK A temperature of 92 K is closer to 0 K than it is to 273 K, so we should expect the Fahrenheit temperature to be considerably less than 32°F. Our result meets this expectation.

(U.S. Dept. of Energy.)

17-3 THE IDEAL-GAS LAW

The properties of gas samples that have low densities led to the definition of the ideal-gas temperature scale. If we compress such a gas while keeping its temperature constant, the pressure increases. Similarly, if a gas expands at constant temperature, its pressure decreases. To a good approximation, the product of the pressure and volume of a gas sample that has a low density is constant at a constant temperature. This result was discovered experimentally by Robert Boyle (1627–1691), and is known as **Boyle's law:**

$$PV = \text{constant} \qquad \text{(constant temperature)}$$

See **Math Tutorial** *for more information on* **Direct and Inverse Proportions**

A more general law exists that reproduces Boyle's law as a special case. According to Equation 17-4, the absolute temperature of a gas sample that has a low density is proportional to its pressure at constant volume. In addition, the absolute temperature of a gas sample that has a low density is proportional to its volume at constant pressure. This result was discovered experimentally by Jacques Charles (1746–1823) and Joseph Gay-Lussac (1778–1850). We can combine these two results by stating

$$PV = CT \qquad \text{17-6}$$

where C is a constant that has positive value. We can see that this constant is proportional to the number of molecules of the gas sample by considering the following. Suppose that we have two containers that have identical volumes, each holding the same amount of the same kind of gas at the same temperature and pressure. If we consider the two containers as one system, we have twice the amount of gas at twice the volume, but at the same temperature and pressure. We have thus doubled the quantity $PV/T = C$ by doubling the amount of gas. We can therefore write C as a constant k times the number N of molecules in the gas:

$$C = kN$$

Equation 17-6 then becomes

$$PV = NkT \qquad \text{17-7}$$

The constant k is called **Boltzmann's constant.** It is found experimentally to have the same value for any kind of gas:

$$k = 1.381 \times 10^{-23}\ \text{J/K} = 8.617 \times 10^{-5}\ \text{eV/K} \qquad \text{17-8}$$

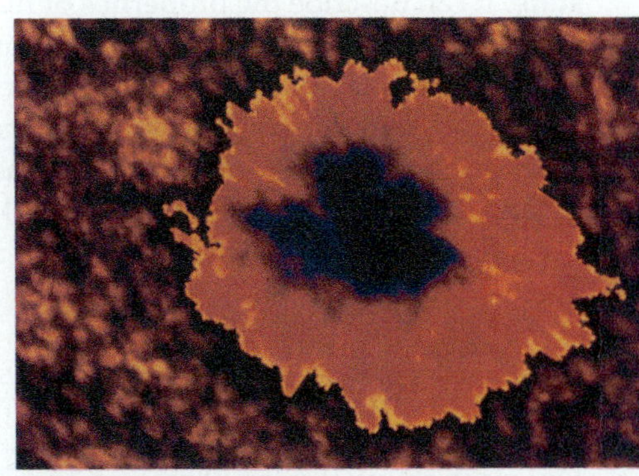

Sunspots appear on the surface of the Sun when streams of gases slowly erupt from deep inside the star. The solar "flower" is 10,000 miles in diameter. The temperature variation, indicated by computer-enhanced color changes, is not fully understood. The central portion of the sunspot is cooler than the outer regions, as indicated by the dark area. The temperature at the Sun's core is of the order of 10^7 K, whereas at the surface the temperature is only about 6000 K. *(NASA.)*

An amount of gas is often expressed in moles. A **mole** (mol) of any substance is the amount of that substance that contains **Avogadro's number,** N_A, of particles (such as atoms or molecules). Avogadro's number is defined as the number of carbon atoms in exactly 12 g (1 mol) of ^{12}C:

$$N_A = 6.022 \times 10^{23} \text{ mol}^{-1} \quad 17\text{-}9$$

AVOGADRO'S NUMBER

If we have n moles of a substance, then the number of molecules is

$$N = nN_A \quad 17\text{-}10$$

Equation 17-7 is then

$$PV = nN_A kT = nRT \quad 17\text{-}11$$

where $R = N_A k$ is called the **universal gas constant.** Its value, which is the same for all gases, is

$$R = N_A k = 8.314 \text{ J/(mol}\cdot\text{K)} = 0.08206 \text{ L}\cdot\text{atm/(mol}\cdot\text{K)} \quad 17\text{-}12$$

Figure 17-8 shows plots of $PV/(nT)$ versus the pressure P for several gases. For all gases, $PV/(nT)$ is nearly constant over a large range of pressures. Even oxygen, which varies the most in this graph, changes by only about 1 percent between 0 and 5 atm. An **ideal gas** is defined as a gas for which $PV/(nT)$ is constant for all pressures. The pressure, volume, and temperature of an ideal gas are related by

$$PV = nRT \quad 17\text{-}13$$

IDEAL-GAS LAW

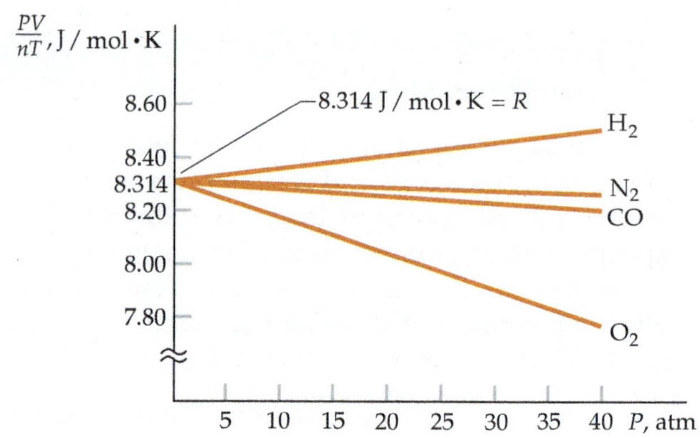

FIGURE 17-8 Plot of PV/nT versus P for real gases. In these plots, varying the amount of gas varies the pressure. The ratio PV/nT approaches the same value, 8.314 J/(mol • K), for all gases as we reduce their densities, and thereby their pressures, of the gases. This value is the universal gas constant R.

Equation 17-13, which relates the variables P, V, and T, is known as the ideal-gas law, and is an example of an **equation of state.** It can describe the properties of real gases that have low densities (and therefore low pressures). Corrections must be made to this equation if higher densities of gases are used. In Chapter 20, we discuss another equation of state, the van der Waals equation, which includes such corrections. For any density of gas, there is an equation of state relating P, V, and T for a given amount of gas. Thus, the state of a given amount of gas is completely specified by knowledge of any two of the three **state variables** P, V, and T.

PARTIAL PRESSURES

Dry air is about 21 percent oxygen and 79 percent nitrogen. Scuba divers often use oxygen-enriched air (called nitrox) because it extends the length of time for a dive. For very deep dives, a mixture of oxygen and helium (called heliox) is used because this mixture reduces the chance that a diver will suffer from nitrogen narcosis.

If we have a confined mixture of two or more gases, and if the mixture is sufficiently dilute (so each gas can be modeled as an ideal gas), then we can think of each gas as occupying the entire volume of the container. This is because the volume of the individual molecules of the gas is negligible compared to the volume of the empty space surrounding them. The total pressure exerted by the mixture is the sum of the individual pressures, called **partial pressures,** exerted by each of the individual gases in the mixture. Furthermore, the partial pressure of each gas in the mixture is the pressure it would exert if it alone occupied the container. This result—the total pressure is the sum of the partial pressures—is called the **law of partial pressures.**

PROBLEM-SOLVING STRATEGY

Dilute Gases

PICTURE A dilute gas is one for which the ideal-gas model gives sufficiently accurate results. The variables are pressure, volume, temperature, mass, and/or the amount of substance (number of moles).

SOLVE

1. Apply the ideal-gas law, $PV = nRT$, to each dilute gas. Be sure to use the absolute temperature and the absolute pressure.
2. For a mixture of dilute gases, the ideal-gas law applies to each gas in the mixture, the volume of each gas in the mixture is the volume of the container, and the pressure of each gas is the partial pressure of that gas. The pressure of the mixture is the sum of the partial pressures of the constituent gases.
3. Additional useful relations are $R = N_A k$, $N = nN_A$, and $m = nM$, where k is the Boltzmann constant, N is the number of molecules, m is the mass of the gas, and M is the molar mass.
4. Solve for the desired quantity.

CHECK The pressure, volume, and temperature can never be negative.

Example 17-3 Mixing the Gases

A 20-L tank of oxygen is at a pressure of $0.30P_{at}$, and a 30-L tank of nitrogen is at a pressure of $0.60P_{at}$. The temperature of each gas is 300 K. The oxygen is then transferred into the 30-L tank containing the nitrogen, where the two mix. What is the pressure of the mixture if its temperature is 300 K?

PICTURE The final volume of both gases is 30 L. The initial temperatures of both gases are equal. Thus, we can use Boyle's law ($P_iV_i = P_fV_f$) to find the partial pressure of each gas in the mixture. Then, we use the law of partial pressures to find the pressure of the mixture.

SOLVE

1. The pressure of the mixture is the sum of the partial pressures of the two gases:
$$P = P_{O_2} + P_{N_2}$$
2. The initial and final temperatures of the gases are the same. So, by using Boyle's law, we find the partial pressures of the gases:
$$P_iV_i = P_fV_f \Rightarrow P_f = \frac{V_i}{V_f}P_i$$
3. The final volume of the oxygen is 30 L (as is the final volume of the nitrogen):
$$P_{O_2} = \frac{V_i}{V_f}P_i = \frac{20\text{ L}}{30\text{ L}}0.30P_{at} = 0.20P_{at}$$
$$P_{N_2} = \frac{V_i}{V_f}P_i = \frac{30\text{ L}}{30\text{ L}}0.60P_{at} = 0.60P_{at}$$
4. The pressure is the sum of the partial pressures:
$$P = P_{O_2} + P_{N_2} = 0.20P_{at} + 0.60P_{at} = \boxed{0.80P_{at}}$$

CHECK We expect an increase in pressure in the 30-L tank when the oxygen is transferred into it. This expectation is met with our final result ($0.80P_{at}$ represents an increase in pressure of $0.20P_{at}$).

Example 17-4 Volume of an Ideal Gas

What volume is occupied by 1.00 mol of an ideal gas at a temperature of 0.00°C and a pressure of 1.00 atm?

PICTURE Use the ideal-gas law to determine the volume occupied by the ideal gas.

SOLVE

We can find the volume using the ideal-gas law, with $T = 273$ K:

$$V = \frac{nRT}{P} = \frac{(1.00\ \text{mol})[0.0821\ \text{L}\cdot\text{atm/(mol}\cdot\text{K)}](273.15\ \text{K})}{1.00\ \text{atm}} = \boxed{22.4\ \text{L}}$$

CHECK Note that by writing R in L·atm/(mol·K), we can write P in atmospheres to get V in liters.

PRACTICE PROBLEM 17-2 Find (*a*) the number of moles n, and (*b*) the number of molecules N in 1.00 cm^3 of a gas at 0.00°C and 1.00 atm.

CONCEPT CHECK 17-1

Two identically sized dorm rooms in a suite, Toni's and Keisha's, are connected by an open door. Toni's room, which is air-conditioned, is 5.0°C cooler than Keisha's room. Whose room has more air in it?

The temperature of 0°C = 273.15 K and the pressure of 1 atm are often referred to as **standard temperature and pressure (STP),** or just **standard conditions.** We see from Example 17-4 that under standard conditions, 1 mol of an ideal gas occupies a volume of 22.4 L.

Figure 17-9 shows plots of P versus V at several constant temperatures T. These curves are called **isotherms.** The isotherms for an ideal gas are hyperbolas. For a fixed amount of gas, we can see from the ideal-gas law (Equation 17-13) that the quantity PV/T is constant. Using the subscripts 1 for the initial values and 2 for the final values, we have

$$\frac{P_2V_2}{T_2} = \frac{P_1V_1}{T_1} \qquad \text{17-14}$$

IDEAL-GAS LAW FOR FIXED AMOUNT OF GAS

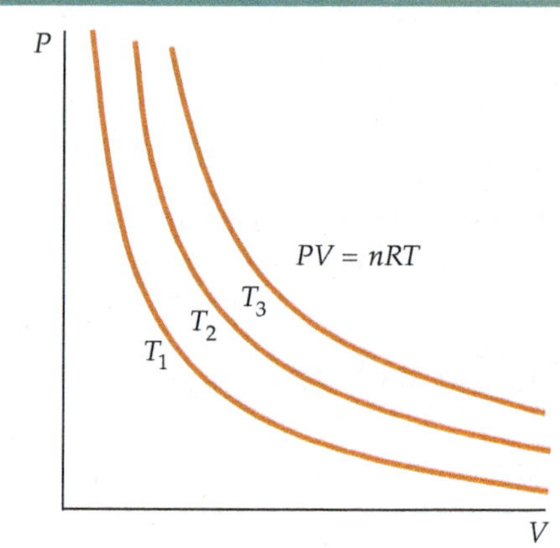

FIGURE 17-9 Isotherms on the PV diagram for a gas. For an ideal gas, these curves are hyperbolas given by $PV = nRT$. (The generic equation for a hyperbola that asymptotically approaches the coordinate axes is xy = constant.)

Example 17-5 Heating and Compressing a Gas

A gas has a volume of 2.00 L, a temperature of 30.0°C, and a pressure of 1.00 atm. When the gas is heated to 60.0°C and compressed to a volume of 1.50 L, what is its new pressure?

PICTURE Because the amount of gas is fixed, the pressure can be found using Equation 17-14. Let subscripts 1 and 2 refer to the initial and final states, respectively.

SOLVE

1. Express the pressure P_2 in terms of P_1 and the initial and final volumes and temperatures:

$$\frac{P_1V_1}{T_1} = \frac{P_2V_2}{T_2}$$

$$P_2 = \frac{T_2V_1}{T_1V_2}P_1$$

2. Calculate the initial and final absolute temperatures:

$$T_1 = 273.15 + 30.0 = 303.15\ \text{K}$$

$$T_2 = 273.15 + 60.0 = 333.15\ \text{K}$$

3. Substitute numerical values in step 1 to find P_2:

$$P_2 = \frac{(333.15\ \text{K})(2.00\ \text{L})}{(303.15\ \text{K})(1.50\ \text{L})}(1.00\ \text{atm}) = \boxed{1.47\ \text{atm}}$$

CHECK Heating a gas and compressing a gas each tend to increase the pressure. Thus, we expect the pressure to exceed the starting pressure of 1.00 atm. Our result of 1.47 atm meets this expectation.

PRACTICE PROBLEM 17-3 How many moles of gas are in the system described in this example?

The mass per mole of a substance is called its **molar mass,** M. (The terms *molecular weight or molecular mass* are also sometimes used.) The molar mass of ^{12}C is, by definition, 12 g/mol or 0.012 kg/mol. Molar masses of the elements are given in Appendix C. The molar mass of an element represents the average of the molar masses of the isotopes of that element—with the average weighted by the relative abundance of those isotopes on Earth. The molar mass of a compound such as CO_2 is the sum of the molar masses of the elements in the molecule. The molar mass of carbon is 12.011 g/mol and the molar mass of oxygen is 15.999 g/mol. Thus, the molar mass of CO_2 is 12.011 g/mol + 2 × 15.999 g/mol = 44.009 g/mol.

CONCEPT CHECK 17-2

If the temperature is decreased at constant pressure, what happens to the volume?

Example 17-6 The Mass of a Hydrogen Atom

The molar mass of hydrogen is 1.008 g/mol. What is the average mass of a hydrogen atom in a glass of H_2O (water)?

PICTURE Let m be the mass of a hydrogen atom. Because there are N_A hydrogen atoms in a mole of hydrogen, the molar mass M is given by $M = mN_A$. We can use this to solve for m.

SOLVE

The average mass of a hydrogen atom is the molar mass divided by Avogadro's number:

$$m = \frac{M}{N_A} = \frac{1.008\text{ g/mol}}{6.022 \times 10^{23}\text{ atoms/mol}} = \boxed{1.674 \times 10^{-24}\text{ g/atom}}$$

CHECK The calculated mass of the hydrogen atom is, as expected, many many orders of magnitude less than the molar mass of about 1 g/atom.

TAKING IT FURTHER The three isotopes of hydrogen are protium ^1H, deuterium ^2H, and tritium ^3H. The relative abundance of ^1H in naturally occurring hydrogen is 99.985%.

Example 17-7 Floating a Hot-Air Balloon

A small hot-air balloon has a volume of 15.0 m^3 and is open at the bottom. The air inside the balloon is at an average temperature of 75°C, while the air next to the balloon has a temperature of 24°C, and a pressure, on average, of 1.00 atm. The balloon is tethered to prevent it from rising, and the tension in the tether is 10.0 N. Use 0.0290 kg/mol for the molar mass of air. (Neglect the gravitational force on the fabric of the balloon.) What is the pressure, on average, inside the balloon?

PICTURE Three forces act on the balloon and its contents, the buoyant force of the surrounding air, the tension force of the tether, and the gravitational force of Earth. The net force is the sum of these three forces. The buoyant force is equal to the weight of the displaced air (Archimedes' principle). The pressure, temperature, and volume of the gas are related by $PV = nRT$. The mass m of the air is equal to the number of moles n times the molar mass M of air.

SOLVE

(*a*) 1. The net force on the system (the balloon and air inside it) is zero. Sketch a free-body diagram (Figure 17-10) of the system:

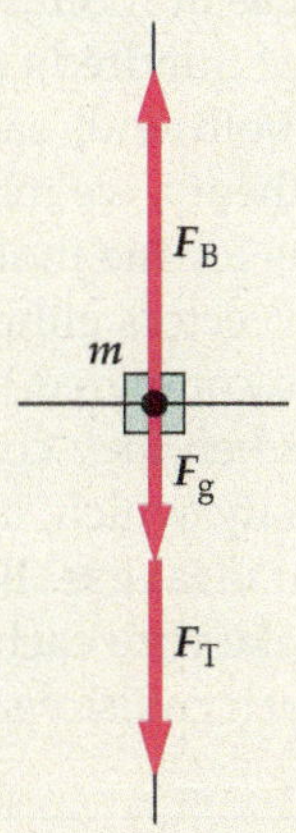

FIGURE 17-10

2. Apply Newton's second law to the system.

$$\Sigma\vec{F} = m\vec{a}$$

$$\vec{F}_T + \vec{F}_B + \vec{F}_g = 0 \Rightarrow F_B = F_T + F_g$$

3. The buoyant force is equal to the weight of a volume V of the air next to the balloon, where V is the volume of the air inside the balloon. Let ρ_1 be the average density of the air next to the balloon, and let ρ_2 be the average density of the air within the balloon:

$$F_B = \rho_1 Vg$$

$$F_g = \rho_2 Vg$$

4. Substitute from step 2 into the step-1 result:

$$F_T = F_B - F_g$$

$$F_T = \rho_1 Vg - \rho_2 Vg = (\rho_1 - \rho_2)Vg$$

5. The mass of a sample of air in the balloon is the number of moles n times the molar mass M of air:

$$\rho_1 = \frac{m_1}{V} = \frac{n_1 M}{V}$$

$$\rho_2 = \frac{m_2}{V} = \frac{n_2 M}{V}$$

6. Substitute from step 4 into the step-3 result:

$$F_T = \left(\frac{n_1 M}{V} - \frac{n_2 M}{V}\right)Vg \Rightarrow \frac{F_T}{Mg} = n_1 - n_2$$

7. Using the ideal-gas law ($PV = nRT$), substitute for n_1 and n_2:

$$\frac{F_T}{Mg} = \frac{P_1 V}{RT_1} - \frac{P_2 V}{RT_2}$$

8. Solve for P_2:

$$P_2 = \left(\frac{P_1}{T_1} - \frac{F_T R}{MgV}\right)T_2$$

9. Solve for the value of P_2. The temperature in kelvins equals 273 plus the temperature in degrees Celsius, 1 atm = 101.3 kPa and $R = 8.314\ \text{J/(mol}\cdot\text{K)}$:

$$P_2 = \left(\frac{1.013 \times 10^5}{297} - \frac{10.0 \times 8.314}{0.0290 \times 9.81 \times 15.0}\right)348$$

$$= \boxed{1.12 \times 10^5\ \text{Pa}} = \boxed{1.10\ \text{atm}}$$

CHECK To produce a net upward force on the fabric of the balloon, the air inside the balloon must be at a higher pressure than the air outside the balloon. Thus, our result of 1.10 atm for the average pressure inside the balloon is plausible.

TAKING IT FURTHER The pressure at the opening at the bottom of the balloon is the same as the pressure in the surrounding air at that altitude. In a static fluid, the pressure decreases with increasing altitude, and the greater the density of the fluid, the greater the rate of decrease in pressure with altitude. The air inside the balloon is less dense than the outside air. Thus, inside the balloon the pressure decrease from the opening to the top of the balloon is less in the air inside the balloon than it is in the air adjacent to the balloon.

17-4 THE KINETIC THEORY OF GASES

The description of the behavior of a gas in terms of the macroscopic state variables P, V, and T can be related to simple averages of microscopic quantities, such as the mass and speed of the molecules in the gas. The resulting theory, called **the kinetic theory of gases,** provides a detailed model of dilute gases.

From the point of view of kinetic theory, a confined gas consists of a large number of rapidly moving particles. In a monatomic gas, like helium and neon, these particles are single atoms, but in polyatomic gases, like oxygen and carbon dioxide, the particles are molecules. In kinetic theory, it is common practice to refer to the constituent particles of a gas as molecules. (This is the practice, even though referring to a single atom as a molecule is something of a misnomer.) We shall follow this practice in the discussions that follow.

In a gas at room temperature, a very large number of molecules are moving at speeds of hundreds of meters per second. These molecules are making elastic collisions, both with each other and with the walls of a container. In the context of kinetic theory, we may neglect any effects due to gravity, so there are no preferred positions for the molecules in the container,* and no preferred directions for their velocity vectors either. The molecules are separated, on average, by distances that are large compared with their diameters. They also exert no forces on each other except when they collide. (This assumption is equivalent to assuming a very low gas density, which, as we saw in the last section, is the same as assuming that the gas is an ideal gas. Because momentum is conserved, the collisions that the molecules make with each other have no effect on the total momentum in any direction. Thus, such collisions can be neglected.)

* Because of gravity, the density of molecules at the bottom of the container is slightly greater than at the top. As discussed in Chapter 13, the density of air decreases by half at a height of about 5.5 km, so the variation over a normal sized container is negligible.

CALCULATING THE PRESSURE EXERTED BY A GAS

The pressure that a gas exerts on its container is due to collisions between gas molecules and the container walls. This pressure is a force per unit area and, by Newton's second law, this force is the rate of change of momentum of the gas molecules colliding with the wall.

Consider a rectangular container of volume V containing N gas molecules, each of mass m moving with a speed v. Let us calculate the force exerted by these molecules on the right-hand wall, which is perpendicular to the x axis and has area A. The molecules hitting this wall in a time interval Δt are those that are within distance $|v_x|\,\Delta t$ of the wall (Figure 17-11) and are moving to the right. Thus, the number of molecules hitting the wall during time Δt is the number per unit volume N/V multiplied by the volume $A|v_x|\,\Delta t$ multiplied by $\frac{1}{2}$ because, on average, only half the molecules are moving to the right. That is, during time Δt

$$\text{Number of molecules that hit the wall} = \frac{1}{2}\frac{N}{V}|v_x|\,\Delta t\,A$$

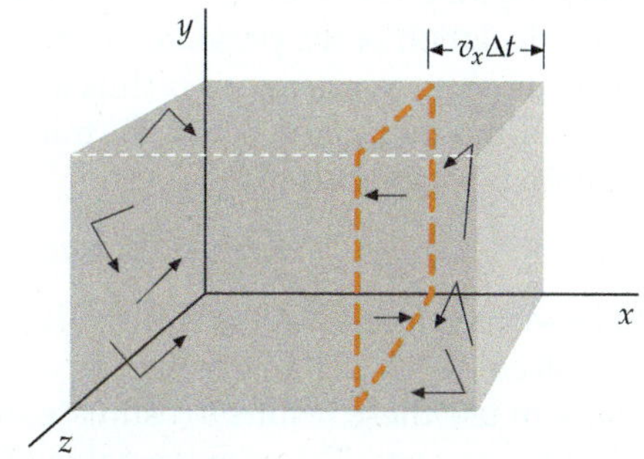

FIGURE 17-11 Gas molecules in a rectangular container. The molecules that both are moving to the right and are closer than $v_x\Delta t$ to the right wall will hit the right wall during time Δt.

The x component of momentum of a molecule is $+mv_x$ before it hits the wall, and $-mv_x$ after an elastic collision with the wall. The change in momentum has the magnitude $2mv_x$. The magnitude of the total change in momentum $|\Delta\vec{p}|$ of all molecules during a time interval Δt is $2m|v_x|$ multiplied by the number of molecules that hit the wall during this interval:

$$|\Delta\vec{p}| = (2m|v_x|) \times \left(\frac{1}{2}\frac{N}{V}|v_x|\,\Delta t\,A\right) = \frac{N}{V}mv_x^2 A\,\Delta t \qquad 17\text{-}15$$

The magnitude of the force exerted by the wall on the molecules, and the magnitude of the force exerted by the molecules on the wall, is the ratio $|\Delta\vec{p}|/\Delta t$. The pressure is the magnitude of this force divided by the area A:

$$P = \frac{F}{A} = \frac{1}{A}\frac{|\Delta\vec{p}|}{\Delta t} = \frac{N}{V}mv_x^2$$

or

$$PV = Nmv_x^2 \qquad 17\text{-}16$$

To allow for the fact that all the molecules in a container do not have the same speed, we merely replace v_x^2 with its average value $(v_x^2)_{av}$. Then, writing Equation 17-16 in terms of the kinetic energy $\frac{1}{2}mv_x^2$ associated with motion along the x axis, we have

$$PV = 2N\left(\tfrac{1}{2}mv_x^2\right)_{av} \qquad 17\text{-}17$$

THE MOLECULAR INTERPRETATION OF TEMPERATURE

Comparing Equation 17-17 with $PV = NkT$ (Equation 17-7), which was obtained experimentally for any gas that has a low density, we can see that

$$NkT = 2N\left(\tfrac{1}{2}mv_x^2\right)_{av} \quad \text{or} \quad \left(\tfrac{1}{2}mv_x^2\right)_{av} = \tfrac{1}{2}kT \qquad 17\text{-}18$$

Thus, the average kinetic energy associated with motion along the x axis is $\frac{1}{2}kT$. But there is nothing special about the x direction. Consequently,

$$(v_x^2)_{av} = (v_y^2)_{av} = (v_z^2)_{av} \quad \text{and} \quad (v^2)_{av} = (v_x^2)_{av} + (v_y^2)_{av} + (v_z^2)_{av} = 3(v_x^2)_{av}$$

Writing $(v_x^2)_{av} = \frac{1}{3}(v^2)_{av}$ and $K_{\text{trans av}}$ for the average translational kinetic energy of the molecules, Equation 17-18 becomes

$$K_{\text{trans av}} = \left(\tfrac{1}{2}mv^2\right)_{av} = \tfrac{3}{2}kT \qquad 17\text{-}19$$

AVERAGE TRANSLATIONAL KINETIC ENERGY OF A MOLECULE

In addition to translational kinetic energy, the molecules may also have rotational or vibrational kinetic energy. However, only the translational kinetic energy is relevant to the calculation of the pressure exerted by a gas on the walls of its container.

The absolute temperature is thus a measure of the average translational kinetic energy of the molecules. The total translational kinetic energy of n moles of a gas containing N molecules is

$$K_{trans} = N\left(\tfrac{1}{2}mv^2\right)_{av} = \tfrac{3}{2}NkT = \tfrac{3}{2}nRT \qquad 17\text{-}20$$

where we have used $Nk = nN_A k = nR$. Thus, the translational kinetic energy is $\frac{3}{2}kT$ per molecule and $\frac{3}{2}RT$ per mole.

We can use these results to estimate the order of magnitude of the speeds of the molecules in a gas. The average value of v^2 is, by Equation 17-19,

$$(v^2)_{av} = \frac{3kT}{m} = \frac{3N_A kT}{N_A m} = \frac{3RT}{M}$$

where $M = N_A m$ is the molar mass. The square root of $(v^2)_{av}$ is referred to as the **root-mean-square** (rms) speed:

$$v_{rms} = \sqrt{(v^2)_{av}} = \sqrt{\frac{3kT}{m}} = \sqrt{\frac{3RT}{M}} \qquad 17\text{-}21$$

ROOT-MEAN-SQUARE SPEED OF A MOLECULE

Note that Equation 17-21 is similar to Equation 15-5 for the speed of sound in a gas:

$$v_{sound} = \sqrt{\frac{\gamma RT}{M}} \qquad 17\text{-}22$$

where $\gamma = 1.4$ for air. This is not surprising because a sound wave in air is a pressure disturbance propagated by collisions between air molecules.

Example 17-8 The rms Speed of Gas Molecules

Oxygen gas (O_2) has a molar mass of about 32.0 g/mol, and hydrogen gas (H_2) has a molar mass of about 2.00 g/mol. Calculate (*a*) the rms speed of an oxygen molecule when the temperature is 300 K, and (*b*) the rms speed of a hydrogen molecule at the same temperature.

PICTURE We find v_{rms} using Equation 17-21. For the units to work out right, we use $R = 8.314\ \text{J/(mol}\cdot\text{K)}$, and we express the molecular masses of O_2 and H_2 in kg/mol.

SOLVE

(*a*) 1. Substitute the given values into Equation 17-21:

$$v_{rms} = \sqrt{\frac{3RT}{M}} = \sqrt{\frac{3(8.314\ \text{J/mol}\cdot\text{K})(300\ \text{K})}{0.0320\ \text{kg/mol}}}$$
$$= 483.56\ \text{m/s} = \boxed{484\ \text{m/s}}$$

(*b*) 1. Repeat the calculation with $M = 0.00200$ kg/mol:

$$v_{rms} = \sqrt{\frac{3RT}{M}} = \sqrt{\frac{3(8.314\ \text{J/mol}\cdot\text{K})(300\ \text{K})}{0.00200\ \text{kg/mol}}}$$
$$= 1934\ \text{m/s} = \boxed{1.93\times10^3\ \text{m/s}}$$

CHECK Because v_{rms} is inversely proportional to $\sqrt{M}$ (Equation 17-21), and the molar mass of hydrogen is one-sixteenth that of oxygen, the rms speed of hydrogen is four times that of oxygen. Our calculations are consistent with the mass ratio because 1930/484 = 4.00.

TAKING IT FURTHER The rms speed of oxygen molecules is 484 m/s = 1080 mi/h, about 1.4 times the speed of sound in air, which at 300 K is about 343 m/s.

PRACTICE PROBLEM 17-4 Find the rms speed of nitrogen molecules (M = 28 g/mol) at 300 K.

THE EQUIPARTITION THEOREM

We have seen that the average kinetic energy associated with translational motion in any direction is $\frac{1}{2}kT$ per molecule (Equation 17-19) or, equivalently, $\frac{1}{2}RT$ per mole, where k is the Boltzmann constant and R is the universal gas constant. If the energy of a molecule associated with its motion in one direction is momentarily increased, say, by a collision between the molecule and a moving piston during a compression of the gas, collisions between that molecule and other molecules will quickly redistribute the added energy. When the gas is again in equilibrium, energy will be equally partitioned among the translational kinetic energies associated with motion in the x, y, and z directions. This sharing of the energy equally among the three terms in the translational kinetic energy is a special case of the **equipartition theorem,** a result that follows from classic statistical mechanics. Each component of position and momentum (including angular position and angular momentum) that appears as a squared term in the expression for the energy of the system is called a **degree of freedom.** Typical degrees of freedom are associated with the kinetic energy of translation, rotation, and vibration, and with the potential energy of vibration. The equipartition theorem states that:

> When a substance is in equilibrium, there is an average energy of $\frac{1}{2}kT$ per molecule or $\frac{1}{2}RT$ per mole associated with each degree of freedom.
>
> EQUIPARTITION THEOREM

In Chapter 18, we use the equipartition theorem to relate the measured heat capacities of gases to their molecular structure.

Example 17-9 Mixing the Gases — *Conceptual*

A thermally insulated tank is divided into two 20-L sections by a partition. One 20-L section contains a mole of nitrogen at 300 K and the other 20-L section contains a mole of helium at 320 K. The partition is removed and the gases are allowed to mix. For the mixture, is the partial pressure of the nitrogen gas less than, equal to, or greater than the partial pressure of the helium gas? Is the final temperature of the mixture less than, equal to, or greater than 310 K?

PICTURE The tank is insulated, so the energy of its contents remains fixed. Any energy gained by the molecules of nitrogen is lost by the molecules of helium. After mixing, the temperature of each gas is equal to the temperature of the mixture, and the temperature of each gas is proportional to its translational kinetic energy. Helium is monatomic and nitrogen is diatomic. Thus, we should expect the energy gained by the nitrogen molecules to end up as rotational kinetic energy as well as translational kinetic energy.

SOLVE

1. After mixing, the volume, temperature, and number of moles are the same for both gases. The ideal-gas law relates the volume, temperature, partial pressure, and number of moles of each gas.

   The ideal-gas law implies that, for each gas, the partial pressure is completely specified by the volume, temperature, and number of moles. The volume, temperature, and number of moles are the same for both gases, so **the partial pressures are the same** as well.

2. The tank is insulated, so the total energy of the two gases remains constant during the mixing.

   The tank is thermally insulated, so any energy gained by the nitrogen molecules is lost by the helium atoms. That is, the average increase in energy of a nitrogen molecule is equal to the average decrease in energy of a helium atom.

3. The final temperature of both gases is the same as the temperature of the mixture.

   After mixing, the temperature is the same for each gas, so the average translational kinetic energy is the same for the molecules of each gas.

4. Nitrogen is a diatomic gas and helium is a monatomic gas, so the nitrogen has more degrees of freedom than does helium. Some of the energy gained by the nitrogen will end up as an increase in rotational kinetic energy.

The decrease in translational kinetic energy of the helium atoms is equal to the increase in translational kinetic energy PLUS the increase of rotational kinetic energy of the nitrogen molecules.

5. The change in temperature in each gas is proportional to the change in translational kinetic energy of each gas.

The decrease in temperature of the helium gas is greater than the increase in temperature of the nitrogen gas. The final temperature is less than 310 K.

CHECK If the two gases had both been monatomic gases, the final temperature would have been equal to 310 K. This is so even if the atomic masses of the two substances were very different.

MEAN FREE PATH

The average speed of molecules in a gas at normal pressures is several hundred meters per second, yet if somebody across the room from you opens a perfume bottle, you do not detect the odor for several minutes. The reason for the time delay is that the perfume molecules do not travel directly toward you, but instead travel a zigzag path due to collisions with the air molecules. The average distance λ traveled by a molecule between collisions is called its **mean free path.** (The reason you smell the perfume at all is due to air currents (convection). The time for a perfume molecule to diffuse across a room is of the order of weeks.)

The mean free path of a gas molecule is related to its size, to the size of the surrounding gas molecules, and to the density of the gas. Consider one gas molecule of radius r_1 moving with speed v through a region of stationary molecules (Figure 17-12). The moving molecule will collide with another molecule of radius r_2 if the centers of the two molecules come within a distance $d = r_1 + r_2$ from each other. (If all the molecules are the same type, then d is the molecular diameter.) As the molecule moves, it will collide with any molecule whose center is in a circle of radius d (Figure 17-13). In some time t, the molecule moves a distance vt and collides with every molecule in the cylindrical volume $\pi d^2 vt$. The number of molecules in this volume is $n_V \pi d^2 vt$, where the number density $n_V = N/V$ is the number of molecules per unit volume. (After each collision, the direction of the molecule changes, so the path actually zigs and zags.) The total path length divided by the number of collisions is the mean free path:

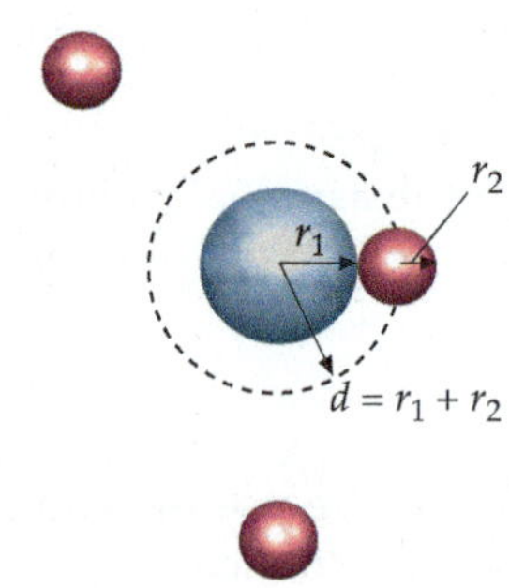

FIGURE 17-12 Model of a molecule (center sphere) moving in a gas. The molecule of radius r_1 will collide with any molecule of radius r_2 if their centers are a distance $d = r_1 + r_2$ apart, which is any molecule whose center is on a sphere of radius $d = r_1 + r_2$ centered about the molecule.

$$\lambda = \frac{vt}{n_V \pi d^2 vt} = \frac{1}{n_V \pi d^2}$$

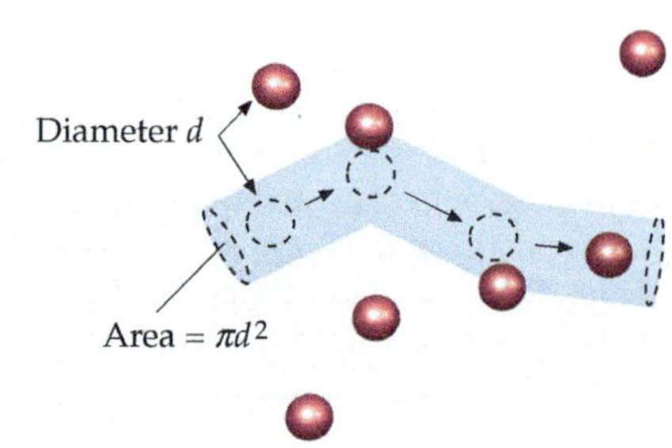

FIGURE 17-13 Model of a molecule moving with speed v in a gas of similar molecules. The motion is shown during time t. The molecule of diameter d will collide with any similar molecule whose center is in a cylinder of volume $\pi d^2 vt$. In this picture, all collisions are assumed to be elastic and all but one of the molecules are assumed to be at rest.

This calculation of the mean free path assumes that all but one of the gas molecules are stationary, which is not a realistic situation. When the motion of all the molecules is taken into account, the correct expression for the mean free path is given by

$$\lambda = \frac{1}{\sqrt{2}\, n_V \pi d^2} \tag{17-23}$$

MEAN FREE PATH OF A MOLECULE

The average time between collisions is called the **collision time** τ. The reciprocal of the collision time, $1/\tau$, is equal to the average number of collisions per second, or the **collision frequency.** If v_{av} is the average speed, then the average distance traveled between collisions is

$$\lambda = v_{av}\tau \tag{17-24}$$

Example 17-10 Mean Free Path of a CO Molecule in Air *Context-Rich*

The local poison control center wants to know more about carbon monoxide and how it spreads through a room. You are asked (*a*) to calculate the mean free path of a carbon monoxide molecule, and (*b*) to estimate the mean time between collisions. The molar mass of carbon monoxide is 28.0 g/mol. Assume that the CO molecule is traveling in air at 300 K and 1.00 atm, and that the diameters of both CO molecules and air molecules are 3.75×10^{-10} m.

PICTURE (*a*) Because d is given, we can find λ from $\lambda = 1/(\sqrt{2}\, n_V \pi d^2)$ using the ideal-gas law ($PV = NkT$) to find $n_V = N/V$. (*b*) We can estimate the collision time by using v_{rms} for the average speed.

SOLVE

(*a*) 1. Write λ in terms of the number density n_V and the molecular diameter d:

$$\lambda = \frac{1}{\sqrt{2}\, n_V \pi d^2}$$

2. Use the ideal-gas law ($PV = NkT$) to calculate $n_V = N/V$:

$$n_V = \frac{N}{V} = \frac{P}{kT} = \frac{101.3 \times 10^3\ \text{Pa}}{(1.381 \times 10^{-23}\ \text{J/K})(300\ \text{K})} = 2.446 \times 10^{25}\ \text{molecules/m}^3$$

3. Substitute this value of n_V and the given value of d to calculate λ:

$$\lambda = \frac{1}{\sqrt{2}\, n_V \pi d^2} = \frac{1}{\sqrt{2}(2.451 \times 10^{25}/\text{m}^3)\pi(3.75 \times 10^{-10}\ \text{m}^2)^2}$$

$$= 6.5428 \times 10^{-8}\ \text{m} = \boxed{6.54 \times 10^{-8}\ \text{m}}$$

(*b*) 1. Write τ in terms of the mean free path λ:

$$\tau = \frac{\lambda}{v_{av}}$$

2. Estimate v_{av} by calculating v_{rms}:

$$v_{rms} = \sqrt{\frac{3RT}{M}} = \sqrt{\frac{3(8.3145\ \text{J/[mol}\cdot\text{K]})(300\ \text{K})}{0.0280\ \text{kg/mol}}} = 517.0\ \text{m/s}$$

3. Use $v_{av} \approx v_{rms}$ to estimate τ:

$$\tau = \frac{\lambda}{v_{av}} \approx \frac{\lambda}{v_{rms}} = \frac{6.530 \times 10^{-8}\ \text{m}}{517.0\ \text{m/s}} = \boxed{1.27 \times 10^{-10}\ \text{s}}$$

CHECK The mean free path [the Part-(*a*), step-3 result] is 174 times the molecular diameter $d = 3.75 \times 10^{-10}$ m. If the calculated value of the mean free path were less than the molecular diameter, then we would look for a mistake in our calculation.

TAKING IT FURTHER The collision frequency is about $1/\tau \approx 8 \times 10^9$ collisions per second. (That is a lot of collisions for one second.)

* THE DISTRIBUTION OF MOLECULAR SPEEDS

We would not expect all of the molecules in a gas to have the same speed. The calculation of the temperature of a gas allows us to calculate the mean square speed (the square of the rms speed), and therefore the average translational kinetic energy of molecules in a gas, but it does not yield any details about the *distribution* of molecular speeds. Before we consider this problem, we discuss the idea of distribution functions in general with some elementary examples from common experience.

Distribution functions Suppose that a teacher gave a 25-point quiz to a large number N of students. To describe the results, the teacher might give the average score, but this would not be a complete description. If all the students received a score of 12.5, for example, that would be quite different from half the students receiving 25 and the other half zero, but the average score would be the same in both cases. A complete description of the results would be to give the number of students n_i who received a score s_i for all the scores received. Alternatively, one could give the fraction of the students $f_i = n_i/N$ who received the score s_i. Both n_i and f_i, which are functions of the variable s, are called **distribution functions.** The fractional distribution is somewhat more convenient to use. The probability that one of the N students selected at random received the score s_i equals the total number of students who received that score n_i divided by N, that is, the probability equals f_i. Note that

$$\sum_i f_i = \sum_i \frac{n_i}{N} = \frac{1}{N}\sum_i n_i$$

and because $\Sigma n_i = N$,

$$\sum_i f_i = 1 \qquad \text{17-25}$$

DEFINITION: NORMALIZATION CONDITION

Equation 17-25 is called the **normalization condition** for fractional distributions.

To find the average score, we add all the scores and divide by N. Because each score s_i was obtained by $n_i = Nf_i$ students, this is equivalent to

$$s_{av} = \frac{1}{N}\sum_i n_i s_i = \sum_i s_i f_i \qquad \text{17-26}$$

Similarly, the average of any function $g(s)$ is defined by

$$g(s)_{av} = \frac{1}{N}\sum_i g(s_i)n_i = \sum_i g(s_i)f_i \qquad \text{17-27}$$

AVERAGE OF $g(s)$

In particular, the average of the square of the scores is

$$(s^2)_{av} = \frac{1}{N}\sum_i s_i^2 n_i = \sum_i s_i^2 f_i \qquad \text{17-28}$$

where $(s^2)_{av}$ is called the *mean square score* and the square root of $(s^2)_{av}$ is called the **root-mean-square score** s_{rms}:

$$s_{rms} = \sqrt{(s^2)_{av}} \qquad \text{17-29}$$

DEFINITION: ROOT MEAN SQUARE OF s

A possible distribution function is shown in Figure 17-14. For this distribution, the most probable score (obtained by the most students) is 16, the average score is 14.2, and the rms score is 14.9.

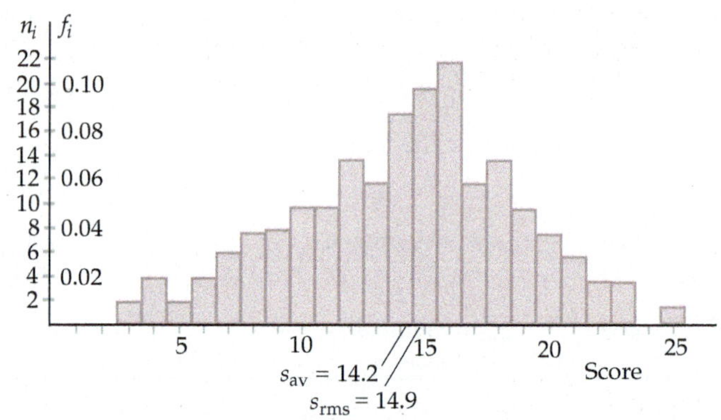

FIGURE 17-14 Grade distribution for a 25-point quiz given to 200 students. n_i is the number of students receiving grade s_i and $f_i = n_i/N$ is the fraction of students receiving grade s_i. The most probable score is 16.

Example 17-11 Making the Grade

Fifteen students took a 25-point quiz. Their scores are 25, 22, 22, 20, 20, 20, 18, 18, 18, 18, 18, 15, 15, 15, and 10. Find the average score and the rms score.

PICTURE The distribution function for this problem is $n_{25} = 1, n_{22} = 2, n_{20} = 3, n_{18} = 5, n_{15} = 3,$ and $n_{10} = 1$. To find the average score, we use $s_{av} = N^{-1}\Sigma n_i s_i$ (Equation 17-26). To find the rms score, we use $(s^2)_{av} = N^{-1}\Sigma s_i^2 n_i$ (Equation 17-28) and then take the square root.

SOLVE

1. By definition, s_{av} is

$$s_{av} = \frac{1}{N}\sum_i n_i s_i = \frac{1}{15}[1(25) + 2(22) + 3(20) + 5(18) + 3(15) + 1(10)] = \frac{1}{15}(274) = 18.27 = \boxed{18.3}$$

2. To calculate s_{rms}, first find the average of s^2:

$$(s^2)_{av} = \frac{1}{N}\sum_1 n_i s_i^2 = \frac{1}{15}[1(25)^2 + 2(22)^2 + 3(20)^2 + 5(18)^2 + 3(15)^2 + 1(10)^2] = \frac{1}{15}(5188) = 345.9$$

3. Take the square root of $(s^2)_{av}$:

$$s_{rms} = \sqrt{(s^2)_{av}} = \boxed{18.6}$$

CHECK The average and rms scores differ by only 1 or 2 percent. In addition, the rms value is greater than the average value. The fact that the rms value is always greater than (or equal to) the mean is explained in the discussion following Equation 17-34*b*.

Now consider the case of a continuous distribution, for example, the distribution of heights in a population. For any finite number N, the number of people who are *exactly* 2 m tall is zero. If we assume that height can be determined to any desired accuracy, there are an infinite number of possible heights, so the probability is zero that anybody has any one particular (exact) height. Therefore, we divide the heights into intervals Δh (for example, Δh might be 1 cm or 0.5 cm) and ask what fraction of people has heights that fall in any particular interval. For very large N, this number is proportional to the size of the interval, provided the interval is sufficiently small. We define the distribution function $f(h)$ as the fraction of the number of people with heights in the interval between h and $h + \Delta h$. Then for N people, $Nf(h)\,\Delta h$ is the number of people whose height is between h and $h + \Delta h$. Figure 17-15 shows a possible height distribution.

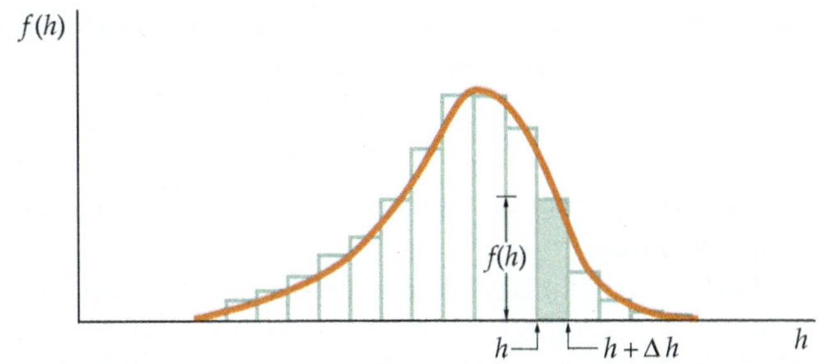

FIGURE 17-15 A possible height distribution function. The fraction of the number of heights between h and $h + \Delta h$ equals the shaded area $f(h)\,\Delta h$. The histogram can be approximated by a continuous curve as shown.

The fraction of people who have heights in a given interval Δh is the area $f(h)\,\Delta h$. If N is very large, we can choose Δh to be very small, and the histogram will approximate a continuous curve. We can therefore consider the distribution function $f(h)$ to be a continuous function, write the interval as dh, and replace the sums in Equations 17-25 through 17-28 by integrals:

$$\int f(h)\,dh = 1 \qquad 17\text{-}30$$

NORMALIZATION CONDITION

$$h_{av} = \int h f(h)\,dh \qquad 17\text{-}31$$

$$[g(h)]_{av} = \int g(h) f(h)\,dh \qquad 17\text{-}32$$

AVERAGE VALUE OF $g(h)$

where $g(h)$ is an arbitrary function of h. Thus,

$$(h^2)_{av} = \int h^2 f(h)\,dh \qquad 17\text{-}33$$

The probability of a person selected at random having a height between h and $h + dh$ is $f(h)\,dh$. A useful quantity characterizing a distribution is the **standard deviation** σ, defined by

$$\sigma^2 = [(x - x_{av})^2]_{av} \qquad 17\text{-}34a$$

STANDARD DEVIATION σ

Expanding $(x - x_{av})^2$, we obtain

$$\sigma^2 = [x^2 - 2xx_{av} + x_{av}^2]_{av} = (x^2)_{av} - 2x_{av}x_{av} + x_{av}^2 = (x^2)_{av} - x_{av}^2$$

or

$$\sigma^2 = (x^2)_{av} - x_{av}^2 \qquad 17\text{-}34b$$

The standard deviation is a measure of the spread of the values about the average value. For most distributions there will be few values that differ from x_{av} by more than a few multiples of σ. For the familiar bell-shaped distribution (called a normal distribution), 68.3 percent of the values are expected to fall within 1σ of x_{av} (i.e., between $x_{av} - \sigma$ and $x_{av} + \sigma$).

In Example 17-11, we found that the rms value was greater than the average value. This is a general feature for any distribution (unless all the values are identical, in which case $\sigma = 0$ and $x_{rms} = x_{av}$). From the definition of rms

People walking down a city street. Consider the various heights of the people you see. *(Getty Images/PhotoAlto.)*

(Equation 17-29), we have $x_{rms}^2 = (x^2)_{av}$. By substituting x_{rms}^2 for $(x^2)_{av}$ in Equation 17-34*b*, we obtain

$$\sigma^2 = x_{rms}^2 - x_{av}^2$$

Because σ^2 and x_{rms} are always positive, x_{rms} must always be greater than $|x_{av}|$.

For the familiar bell-shaped distribution (called a normal distribution), 68.3 percent of the values fall within $x_{av} \pm \sigma$, 95.5 percent fall within $x_{av} \pm 2\sigma$, and 99.7 percent fall within $x_{av} \pm 3\sigma$. (This is known as the 68–95–99.7 rule.)

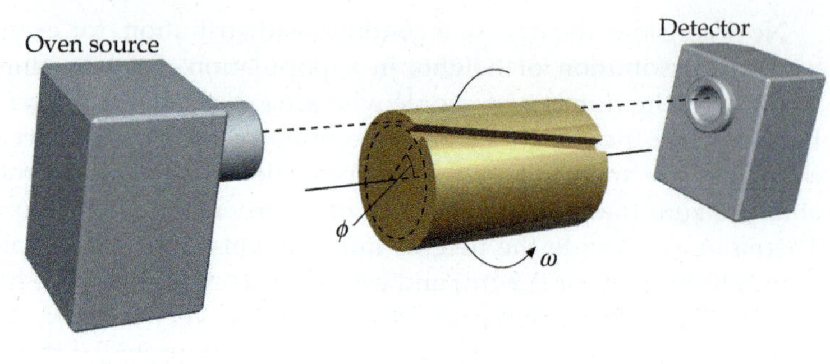

FIGURE 17-16 Schematic diagram of the apparatus for determining the speed distribution of the molecules of a gas. A substance is vaporized in an oven and the vapor molecules are allowed to escape through a hole in the oven wall into a vacuum chamber. The molecules are collimated into a narrow beam by a series of slits (not shown). The beam is aimed at a detector that counts the number of molecules that are incident on it in a given period of time. A rotating cylinder stops most of the beam. Small slits in the cylinder (only one of which is depicted here) allow the passage of molecules that have a narrow range of speeds that is determined by the angular speed of the rotating cylinder. Counting the number of molecules that reach the detector for each of a large number of angular speeds, gives a measure of the number of molecules in each range of speeds.

The Maxwell–Boltzmann distribution The distribution of the molecular speeds of a gas can be measured directly using the apparatus illustrated in Figure 17-16. In Figure 17-17, these speeds are shown for two different temperatures. The quantity $f(v)$ in Figure 17-17 is called the **Maxwell–Boltzmann speed distribution function.** In a gas of N molecules, the number with speeds in the range between v and $v + dv$ is dN, given by

$$dN = Nf(v)\,dv \qquad 17\text{-}35$$

The fraction $dN/N = f(v)\,dv$ in a particular range dv is illustrated by the shaded region in the figure. The Maxwell–Boltzmann speed distribution function can be derived using statistical mechanics. The result is

$$f(v) = \frac{4}{\sqrt{\pi}}\left(\frac{m}{2kT}\right)^{3/2} v^2 e^{-mv^2/(2kT)} \qquad 17\text{-}36$$

MAXWELL–BOLTZMANN SPEED DISTRIBUTION FUNCTION

The most probable speed v_{max}, that speed for which $f(v)$ is maximum, is given by

$$v_{max} = \sqrt{\frac{2\,kT}{m}} = \sqrt{\frac{2\,RT}{M}} \qquad 17\text{-}37$$

MOST PROBABLE SPEED

The rms speed is given by $v_{rms} = \sqrt{3RT/M}$ (Equation 17–21). Comparing Equation 17-37 with Equation 17-21, we see that the most probable speed is slightly less than the rms speed.

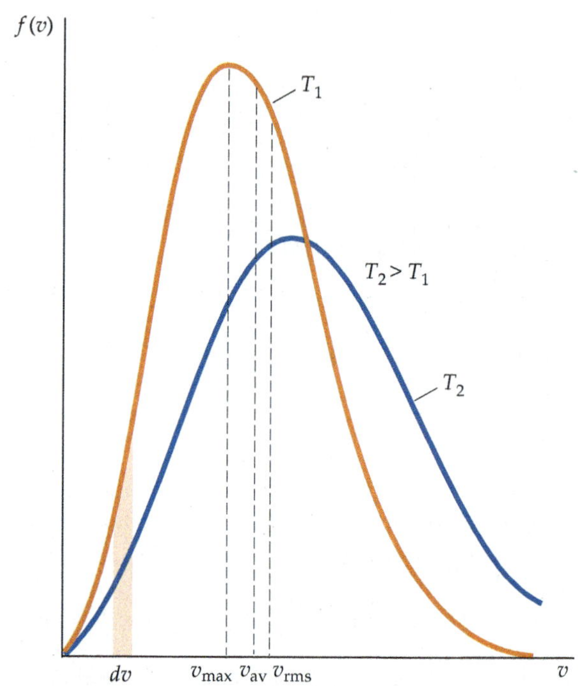

FIGURE 17-17 Distributions of molecular speeds in a gas at two temperatures, T_1 and $T_2 > T_1$. The shaded area $f(v)\,dv$ equals the fraction of the number of molecules having a particular speed in a narrow range of speeds dv. The mean speed v_{av} and the rms speed v_{rms} are both slightly greater than the most probable speed v_{max}.

Example 17-12 Using the Maxwell–Boltzmann Distribution

Calculate the mean square speed (the average value of v^2) for the molecules in a gas using the Maxwell–Boltzmann distribution function.

PICTURE The average value of v^2 is calculated from $(h^2)_{av} = \int h^2 f(h)\,dh$ (Equation 17-33), with v replacing h and $f(v)$ given by Equation 17-36.

SOLVE

1. By definition, $(v^2)_{av}$ is

$$(v^2)_{av} = \int_0^\infty v^2 f(v)\, dv$$

2. Use Equation 17-36 for $f(v)$:

$$(v^2)_{av} = \int_0^\infty v^2 \frac{4}{\sqrt{\pi}}\left(\frac{m}{2kT}\right)^{3/2} v^2 e^{-mv^2/(2kT)}\, dv = \frac{4}{\sqrt{\pi}}\left(\frac{m}{2kT}\right)^{3/2} \int_0^\infty v^4 e^{-mv^2/(2kT)}\, dv$$

3. The integral in step 2 can be found in standard integral tables:

$$\int_0^\infty v^4 e^{-mv^2/(2kT)}\, dv = \frac{3}{8}\sqrt{\pi}\left(\frac{2kT}{m}\right)^{5/2}$$

4. Use this result to calculate $(v^2)_{av}$:

$$(v^2)_{av} = \frac{4}{\sqrt{\pi}}\left(\frac{m}{2kT}\right)^{3/2} \frac{3}{8}\sqrt{\pi}\left(\frac{2kT}{m}\right)^{5/2} = \boxed{\frac{3kT}{m}}$$

CHECK Our result agrees with $v_{rms} = \sqrt{3kT/m}$ from Equation 17-21.

In Example 17-8 we found that the rms speed of hydrogen molecules is about 1.93 km/s. This is about 17 percent of the escape speed at the surface of Earth, which we found to be 11.2 km/s in Section 11-3. So why is there no free hydrogen in Earth's atmosphere? As we can see from Figure 17-17, a considerable fraction of the molecules of a gas in equilibrium have speeds greater than the rms speed. When the rms speed of the molecules of a particular gas is as great as 15 to 20 percent of the escape speed for a planet, enough of the molecules have speeds greater than the escape speed so that most of the gas does not remain in the atmosphere of that planet very long before escaping. Thus, there is virtually no hydrogen gas in Earth's atmosphere. The rms speed of oxygen molecules, on the other hand, is about one-fourth that of hydrogen molecules, which makes it only about 4 percent of the escape speed at the surface of Earth. Therefore, only a negligible fraction of the oxygen molecules have speeds greater than the escape speed, and oxygen remains in Earth's atmosphere.

CONCEPT CHECK 17-3

As any low-temperature physicist knows, liquid nitrogen is much cheaper than liquid helium. One reason for this is that while nitrogen is the most common constituent of the atmosphere, only minute amounts of helium are found in the atmosphere. (Helium is found in natural gas deposits.) Why are only minute amounts of helium found in the atmosphere?

The energy distribution The Maxwell–Boltzmann speed distribution as given by Equation 17-36 can also be written as a translational-kinetic-energy distribution. We write the number of molecules with translation kinetic energy E in the range between E and $E + dE$ as

$$dN = NF(E)\, dE$$

where $F(E)$ is the energy distribution function. This will be the same number as given by Equation 17-36, with the energy E related to the speed v by $E = \frac{1}{2}mv^2$. Then

$$dE = mv\, dv$$

and

$$Nf(v)\, dv = NF(E)\, dE$$

We can write

$$f(v)\, dv = Cv^2 e^{-mv^2/(2kT)}\, dv = Cve^{-E/(kT)} v\, dv = C\left(\frac{2E}{m}\right)^{1/2} e^{-E/(kT)} \frac{dE}{m}$$

where $C = (4/\sqrt{\pi})[m/(2kT)]^{3/2}$ (from Equation 17-36). The translational-kinetic-energy distribution function $F(E)$ is thus given by

$$F(E) = \frac{4}{\sqrt{\pi}}\left(\frac{m}{2kT}\right)^{3/2}\left(\frac{2}{m}\right)^{1/2}\frac{1}{m}E^{1/2}e^{-E/(kT)}$$

Simplifying, we obtain the **Maxwell–Boltzmann energy distribution function:**

$$F(E) = \frac{2}{\sqrt{\pi}}\left(\frac{1}{kT}\right)^{3/2} E^{1/2} e^{-E/(kT)} \qquad \text{17-38}$$

MAXWELL–BOLTZMANN ENERGY DISTRIBUTION FUNCTION

See Math Tutorial for more information on Differential Calculus

In the language of statistical mechanics, the energy distribution is considered to be the product of two factors: one factor is called the **density of states** and is proportional to $\sqrt{E}$, the other factor is the probability of a state being occupied, which is $e^{-E/(kT)}$, and is called the **Boltzmann factor.**

Physics Spotlight

Molecular Thermometers

Molecular thermometers show changes in temperature by changes in the molecules themselves. Molecular thermometers can be simple and inexpensive and are targets of recent intensive research.

Mood rings feature liquid crystals[*] that show changes in the wearer's finger temperature by changing color. Many liquid crystals have *thermochromic* properties—they change color with temperature. These liquid crystals are made up of twisted molecules. As the temperature changes, the twist of each molecule tightens or loosens, which changes how the liquid crystal absorbs and reflects light. Some liquid crystals are sensitive to changes as small as 0.1°C.[†] Any particular liquid crystal will show color changes over a small temperature range, usually less than 7°C. However, different liquid crystals can be encased in different compartments within strips to allow sensing of a range of temperatures from −30°C to 120°C. These strip thermometers are used as aquarium thermometers, as well as fever thermometers.[‡] Liquid crystal thermometers provide real-time monitoring,[#] and are an inexpensive option when cost is a concern.[°]

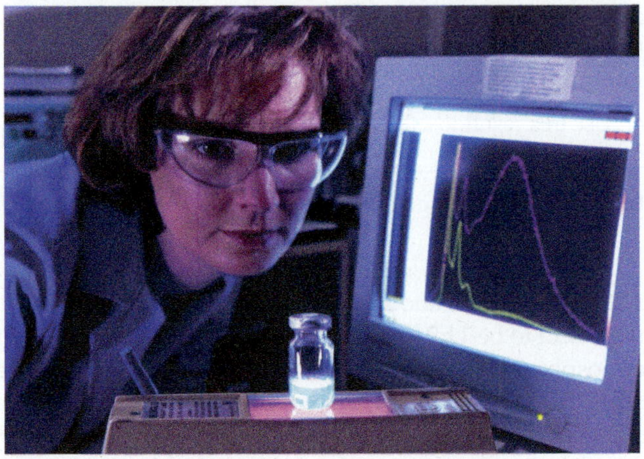

A scientist views a sample that is fluorescing in two wavelengths. It is fluorescing because it is being illuminated from below by ultraviolet light. The ratio of the two wavelengths is a sensitive thermometric property of the material. *(LeRoy N. Sanchez/Los Alamos National Laboratory.)*

Fluorescent thermometers can be useful in monitoring temperature changes of very small computer chips during manufacturing processes as well as monitoring temperature changes in automobiles during manufacturing and aerodynamic testing.[§] When these thermometers are bathed in ultraviolet light, most respond by fluorescing (emitting light) in two wavelengths. The ratio of these two wavelengths is related to the temperature. A recently developed fluorescent thermometer is accurate to 0.05°C, and indicates temperatures ranging from 25°C to 140°C[¶] by changes in the ratio of wavelengths of emitted light. As the ratio changes, the visible color changes with the temperature.

But there are times that molecular thermometers must measure temperatures that are higher than 150°C. In that case, objects that need to be measured can be coated with a powder that phosphoresces, or briefly shines, when excited with light. The length of time the phosphors shine depends upon the temperature of the coated object. Phosphorescence time duration has been used by the steel industry to determine whether steel is at the right temperature for the formation of desired alloys. Phosphor thermometry allows measurement of 700°C steel within 3°C.[**] Conventional thermometry had errors of as much as 40°C. The accuracy could save the steel industry up to $70 million per year.

Real-time thermometers cannot tell what has happened in the past. The highest, or endpoint, cooking temperature that meat reaches is an important measurement for food safety. The endpoint cooking temperature (EPT) determines whether food-related illness is likely to occur. Unfortunately, it is not possible to measure the EPT once the meat has cooled. But the ratio of three large molecules in beef allows determination of the EPT within 2°C,[††] even if the beef has been frozen and thawed since cooking. It may soon be possible to tell whether precooked meats delivered to nursing homes and schools have reached a safe temperature before cooling.

Because of the wide number of applications, from inexpensive temperature monitoring to detection of past temperatures and real-time industrial monitoring, molecular thermometry has a glowing future.

* James, B. G., "Heat Sensitive Novelty Device," *U.S. Patent 3,802,945*, Apr. 9, 1974.
† White, M. A., and LeBlanc, M., "Thermochromism in Commercial Products," *Journal of Chemical Education*, Sept. 1999, Vol. 76, 1201–1205.
‡ Krause, B. F., "Accuracy and Response Time Comparisons of Four Skin Temperature-Monitoring Devices," *Nurse Anesthesia*, June 1993, Vol. 4, 55–61.
# Dart, R. C., et al., "Liquid Crystal Thermometry for Continuous Temperature Measurement in Emergency Department Patients," *Annals of Emergency Medicine*, Dec. 1985, Vol. 14, 1188–1190.
° Manandhar, N., et al., "Liquid Crystal Thermometry for the Detection of Neonatal Hypothermia in Nepal," *Journal of Tropical Pediatrics*, Feb. 1998, Vol. 55, 15+.
§ Chandrasekharan, N., and Kelly, L., "Fluorescent Molecular Thermometers Based on Monomer/Exciplex Interconversion," *The Spectrum*, Sept. 2002, Vol. 15, No. 3, 1–7.
¶ Hanson, T., "Laboratory Scientists Develop Novel Fluorescent Thermometer," *Los Alamos National Laboratory News*, Sept. 4, 2004. http://www.lanl.gov/news/index.php?fuseaction=nb.story&story_id=5007&nb_date=2004-04-15 as of July 2006.
** "Thermometry for the Steel Industry," *Thermographic Phosphor Sensing Applications*, Oak Ridge National Laboratory. http://www.ornl.gov/sci/phosphors/galv.htm as of July 2006.
†† Miller, D. R., and Keeton, J. T., "Verification of Safe Cooking Endpoints in Beef by Multiple Antigen Elisa," *2004 Beef Cattle Research In Texas Publication*. http://animalscience.tamu.edu/ANSC/beef/bcrt/2004/miller_3.pdf as of July 2006.

Summary

TOPIC	RELEVANT EQUATIONS AND REMARKS	
1. Centigrade and Fahrenheit Scale	On the centigrade scale, the ice point is defined to be 0°C and the steam point is 100°C. On the Fahrenheit scale, the ice point is 32°F and the steam point is 212°F. Temperatures on the Fahrenheit and centigrade scales are related by $t_C = \frac{5}{9}(t_F - 32°)$	17-2
2. Gas Thermometers	Gas thermometers have the property that they all agree with each other in the measurement of any temperature as long as the density of the gas is very low. The ideal-gas temperature T (in kelvins) is defined by $T = \frac{P}{P_3}T_3$ where P is the observed pressure of the gas in the thermometer, P_3 is the pressure when the thermometer is immersed in a water–ice–vapor bath at its triple point, and $T_3 = 273.16$ K (the triple-point temperature).	17-4
3. Celsius Scale	The Celsius temperature t_C is related to the ideal-gas temperature T in kelvins by $t_C = T + 273.15\text{ K}$	17-5
4. Ideal Gas	At low densities, all gases obey the ideal-gas law.	
Equation of state	$PV = nRT$	17-13
Universal gas constant	$R = N_A k = 8.314\text{ J/(mol}\cdot\text{K)}$ $= 0.08206\text{ L}\cdot\text{atm/(mol}\cdot\text{K)}$	17-12
Boltzmann's constant	$k = 1.381 \times 10^{-23}\text{ J/K} = 8.617 \times 10^{-5}\text{ eV/K}$	17-8
Avogadro's number	$N_A = 6.022 \times 10^{23}\text{ mol}^{-1}$	17-9
Equation for a fixed amount of gas	A form of the ideal-gas law that is useful for solving problems involving a fixed amount of gas is $\frac{P_2V_2}{T_2} = \frac{P_1V_1}{T_1}$	17-14
5. Kinetic Theory of Gases		
Molecular interpretation of temperature	The absolute temperature T is a measure of the average molecular translational kinetic energy.	
Equipartition theorem	When a system is in equilibrium, there is an average energy of $\frac{1}{2}kT$ per molecule ($\frac{1}{2}RT$ per mole) associated with each degree of freedom.	
Average kinetic energy	For an ideal gas, the average translational kinetic energy of the molecules is $K_{\text{trans av}} = \left(\frac{1}{2}mv^2\right)_{\text{av}} = \frac{3}{2}kT$	17-19
Total translational kinetic energy	The total translational kinetic energy of n moles of a gas containing N molecules is given by $K_{\text{trans}} = N\left(\frac{1}{2}mv^2\right)_{\text{av}} = \frac{3}{2}NkT = \frac{3}{2}nRT$	17-20
rms speed of molecules	The rms speed of a molecule of a gas is related to the absolute temperature by $v_{\text{rms}} = \sqrt{(v^2)_{\text{av}}} = \sqrt{\frac{3kT}{m}} = \sqrt{\frac{3RT}{M}}$ where m is the mass of the molecule and M is the molar mass.	17-21
Mean free path	The mean free path λ of a molecule is related to its diameter d and the number of molecules per unit volume n_V by $\lambda = \frac{1}{\sqrt{2}\, n_V \pi d^2}$	17-23

TOPIC	RELEVANT EQUATIONS AND REMARKS	
*6. **Maxwell–Boltzmann Speed Distribution**	$f(v) = \frac{4}{\sqrt{\pi}}\left(\frac{m}{2kT}\right)^{3/2} v^2 e^{-mv^2/(2kT)}$	17-36
Maxwell–Boltzmann Energy Distribution	$F(E) = \frac{2}{\sqrt{\pi}}\left(\frac{1}{kT}\right)^{3/2} E^{1/2} e^{-E/(kT)}$	17-38

Answers to Concept Checks

17-1 Toni's room has more air in it.

17-2 It decreases.

17-3 The rms speed of helium is about 12 percent of the escape speed from Earth's surface. Thus there are enough helium molecules with speeds above escape speed for the helium to slowly escape Earth.

Answers to Practice Problems

17-1 (*a*) 20°C, (*b*) −40°F

17-2 (*a*) $n = 4.47 \times 10^{-5}$ mol, (*b*) $N = 2.69 \times 10^{19}$ molecules

17-3 $n = 0.0804$ mol

17-4 5.2×10^2 m/s

Problems

In a few problems, you are given more data than you actually need; in a few other problems, you are required to supply data from your general knowledge, outside sources, or informed estimate.

Interpret as significant all digits in numerical values that have trailing zeros and no decimal points.

- • Single-concept, single-step, relatively easy
- •• Intermediate-level, may require synthesis of concepts
- ••• Challenging
- SSM Solution is in the *Student Solutions Manual*
- Consecutive problems that are shaded are paired problems.

CONCEPTUAL PROBLEMS

1 • True or false:

(*a*) The zeroth law of thermodynamics states that two objects in thermal equilibrium with each other must be in thermal equilibrium with a third object.

(*b*) The Fahrenheit and Celsius temperature scales differ only in the choice of the **ice-point** temperature.

(*c*) The Celsius degree and the kelvin are the same size.

2 • How can you determine if two objects are in thermal equilibrium with each other when putting them into physical contact with each other would have undesirable effects? (For example, if you put a piece of sodium in contact with water there would be a violent chemical reaction.)

3 • "Yesterday I woke up and it was 20°F in my bedroom," said Mert to his old friend Mort. "That's nothing," replied Mort. "My room was −5.0°C." Who had the colder room, Mert or Mort? SSM

4 • Two identical vessels contain different ideal gases at the same pressure and temperature. It follows that (*a*) the number of gas molecules is the same in both vessels, (*b*) the total mass of gas is the same in both vessels, (*c*) the average speed of the gas molecules is the same in both vessels, (*d*) None of the above.

5 • Figure 17-18 shows a plot of volume V versus absolute temperature T for a process that takes a fixed amount of an ideal gas from point A to point B. What happens to the pressure of the gas during this process? SSM

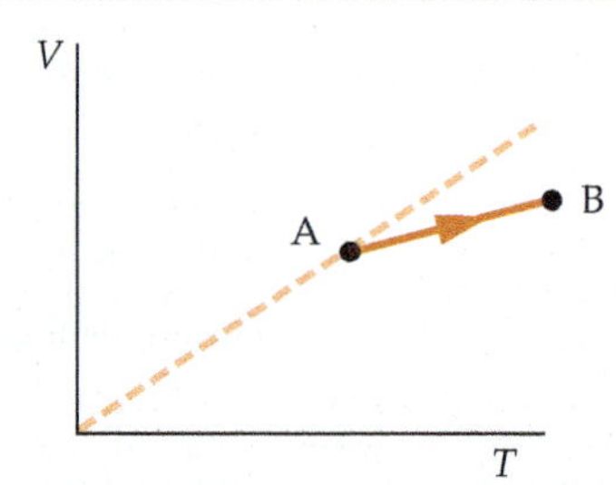

FIGURE 17-18 Problem 5

6 • Figure 17-19 shows a plot of pressure P versus absolute temperature T for a process that takes a sample of an ideal gas from point A to point B. What happens to the volume of the gas during this process?

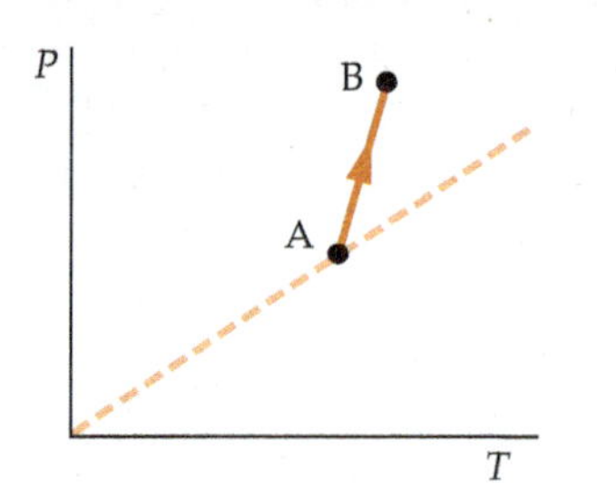

FIGURE 17-19 Problem 6

7 • If a vessel contains equal amounts, by mass, of helium and argon, which of the following are true?

(*a*) The partial pressure exerted by each of the two gases on the walls of the container is the same.
(*b*) The average speed of a helium atom is the same as that of an argon atom.
(*c*) The number of helium atoms and argon atoms in the vessel are equal.
(*d*) None of the above.

8 • By what factor must the absolute temperature of a gas be increased to double the rms speed of its molecules?

9 • Two different gases are at the same temperature. What can be said about the average translational kinetic energies of the molecules? What can be said about the rms speeds of the gas molecules?

10 • A vessel holds a mixture of helium (He) and methane (CH_4). The ratio of the rms speed of the He atoms to that of the CH_4 molecules is (*a*) 1, (*b*) 2, (*c*) 4, (*d*) 16.

11 • True or false: If the pressure of a fixed amount of gas increases, the temperature of the gas must increase.

12 • Why might the Celsius and Fahrenheit scales be more convenient than the absolute scale for ordinary, nonscientific purposes?

13 • An astronomer claims that the temperature at the center of the Sun is about 10^7 degrees. Do you think that this temperature is in kelvins, degrees Celsius, or does it not matter?

14 • Imagine that you have a fixed amount of ideal gas in a container that expands to maintain constant pressure. If you double the absolute temperature of the gas, the average speed of the molecules (*a*) remains constant, (*b*) doubles, (*c*) quadruples, (*d*) increases by a factor of $\sqrt{2}$.

15 • Suppose that you compress an ideal gas to half its original volume, while also halving its absolute temperature. During this process, the pressure of the gas (*a*) halves, (*b*) remains constant, (*c*) doubles, (*d*) quadruples. SSM

16 • The average translational kinetic energy of the molecules of a gas depends on (*a*) the number of moles and the temperature, (*b*) the pressure and the temperature, (*c*) the pressure only, (*d*) the temperature only.

17 •• Which speed is greater, the speed of sound in a gas or the rms speed of the molecules of the gas? Justify your answer, using the appropriate formulas, and explain why your answer is intuitively plausible. SSM

18 •• Imagine that you increase the temperature of a gas while holding its volume fixed. Explain in terms of molecular motion why the pressure of the gas on the walls of its container increases.

19 •• Imagine that you compress a gas while holding it at a fixed temperature (perhaps by immersing the container in cool water). Explain in terms of molecular motion why the pressure of the gas on the walls of its container increases.

20 •• Oxygen has a molar mass of 32 g/mol, and nitrogen has a molar mass of 28 g/mol. The oxygen and nitrogen molecules in a room have:

(*a*) equal average translational kinetic energies, but the oxygen molecules have a larger average speed than the nitrogen molecules have.
(*b*) equal average translational kinetic energies, but the oxygen molecules have a smaller average speed than the nitrogen molecules have.
(*c*) equal average translational kinetic energies and equal average speeds.
(*d*) equal average speeds, but the oxygen molecules have a larger average translational kinetic energy than the nitrogen molecules have.
(*e*) equal average speeds, but the oxygen molecules have a smaller average translational kinetic energy than the nitrogen molecules have.
(*f*) None of the above.

21 •• Liquid nitrogen is relatively cheap, while liquid helium is relatively expensive. One reason for the difference in price is that while nitrogen is the most common constituent of the atmosphere, only small traces of helium can be found in the atmosphere. Use ideas from this chapter to explain why it is that only small traces of helium can be found in the atmosphere. SSM

ESTIMATION AND APPROXIMATION

22 • Estimate the total number of air molecules in your classroom.

23 •• Estimate the density of dry air at sea level on a warm summer day.

24 •• A stoppered test tube that has a volume of 10.0 mL has 1.00 mL of water at its bottom. The water has a temperature of 100°C and is initially at a pressure of 1.00 atm. The test tube is held over a flame until the water has completely boiled away. Estimate the final pressure inside the test tube.

25 •• In Chapter 11, we found that the escape speed at the surface of a planet of radius R is $v_e = \sqrt{2gR}$, where g is the acceleration due to gravity at the surface of the planet. If the rms speed of a gas is greater than about 15 to 20 percent of the escape speed of a planet, virtually all of the molecules of that gas will escape the atmosphere of the planet.

(*a*) At what temperature is v_{rms} for O_2 equal to 15 percent of the escape speed for Earth?
(*b*) At what temperature is v_{rms} for H_2 equal to 15 percent of the escape speed for Earth?
(*c*) Temperatures in the upper atmosphere reach 1000 K. How does this help account for the low abundance of hydrogen in Earth's atmosphere?
(*d*) Compute the temperatures for which the rms speeds of O_2 and H_2 are equal to 15 percent of the escape speed at the surface of the moon, where g is about one-sixth of its value on Earth and R = 1738 km. How does this account for the absence of an atmosphere on the moon? SSM

26 •• The escape speed for gas molecules in the atmosphere of Mars is 5.0 km/s and the surface temperature of Mars is typically 0°C. Calculate the rms speeds for (*a*) H_2, (*b*) O_2, and (*c*) CO_2 at this temperature. (*d*) Are H_2, O_2, and CO_2 likely to be found in the atmosphere of Mars?

27 •• The escape speed for gas molecules in the atmosphere of Jupiter is 60 km/s and the surface temperature of Jupiter is typically −150°C. Calculate the rms speeds for (*a*) H_2, (*b*) O_2, and (*c*) CO_2 at this temperature. (*d*) Are H_2, O_2, and CO_2 likely to be found in the atmosphere of Jupiter? SSM

Jupiter as seen from about twelve million miles. Because the escape speed at the surface of Jupiter is about 600 km/s, Jupiter easily retains hydrogen in its atmosphere. (*Jet Propulsion Laboratory/NASA.*)

28 •• Estimate the average pressure on the front wall of a racquetball court, due to the collisions of the ball with the wall during a game. Use any reasonable numbers for the mass of the ball, its typical speed, and the dimensions of the court. Is the average pressure from the ball significant compared to that from the air?

29 •• To a first approximation, the Sun consists of a gas of equal numbers of protons and electrons. (The masses of these particles can be found in Appendix B.) The temperature at the center of the Sun is about 1×10^7 K, and the density of the Sun is about $1 \times 10^5\ \mathrm{kg/m^3}$. Because the temperature is so high, the protons and electrons are separate particles (rather than being joined together to form hydrogen atoms). (*a*) Estimate the pressure at the center of the Sun. (*b*) Estimate the rms speeds of the protons and the electrons at the center of the Sun.

30 •• **CONTEXT-RICH, ENGINEERING APPLICATION** You are designing a vacuum chamber for fabricating reflective coatings. Inside this chamber, a small sample of metal will be vaporized so that its atoms travel in straight lines (the effects of gravity are negligible during the brief time of flight) to a surface where they land to form a very thin film. The sample of metal is 30 cm from the surface to which the metal atoms will adhere. How low must the pressure in the chamber be so that the metal atoms only rarely collide with air molecules before they land on the surface?

31 ••• **BIOLOGICAL APPLICATION** In normal breathing conditions, approximately 5 percent of each exhaled breath is carbon dioxide. Given this information, and neglecting any difference in water-vapor content, estimate the typical difference in mass between an inhaled breath and an exhaled breath. SSM

TEMPERATURE SCALES

32 • A certain ski wax is rated for use between −12° and −7.0°C. What is this temperature range on the Fahrenheit scale?

33 • The melting point of gold is 1945.4°F. Express this temperature in degrees Celsius. SSM

34 • A weather report indicates that the temperature is expected to drop by 15.0°C over the next four hours. By how many degrees on the Fahrenheit scale will the temperature drop?

35 • The length of the column of mercury in a thermometer is 4.00 cm when the thermometer is immersed in ice water at 1 atm of pressure, and 24.0 cm when the thermometer is immersed in boiling water at 1 atm of pressure. Assume that the length of the mercury column varies linearly with temperature. (*a*) Sketch a graph of the length of the mercury column versus temperature (in degrees Celsius). (*b*) What is the length of the column at room temperature (22.0°C)? (*c*) If the mercury column is 25.4 cm long when the thermometer is immersed in a chemical solution, what is the temperature of the solution?

36 • The temperature of the interior of the Sun is about 1.0×10^7 K. What is this temperature in (*a*) Celsius degrees, (*b*) Fahrenheit degrees?

37 • The boiling point of nitrogen, N_2, is 77.35 K. Express this temperature in degrees Fahrenheit.

38 • The pressure of a constant-volume gas thermometer is 0.400 atm at the ice point and 0.546 atm at the steam point. (*a*) Sketch a graph of pressure versus Celsius temperature for this thermometer. (*b*) When the pressure is 0.100 atm, what is the temperature? (*c*) What is the pressure at 444.6°C (the boiling point of sulfur)?

39 • A constant-volume gas thermometer reads 50.0 torr at the triple point of water. (*a*) Sketch a graph of pressure versus absolute temperature for this thermometer. (*b*) What will be the pressure when the thermometer measures a temperature of 300 K? (*c*) What ideal-gas temperature corresponds to a pressure of 678 torr? SSM

40 • A constant-volume gas thermometer has a pressure of 30.0 torr when it reads a temperature of 373 K. (*a*) Sketch a graph of pressure versus absolute temperature for this thermometer. (*b*) What is its triple-point pressure P_3? (*c*) What temperature corresponds to a pressure of 0.175 torr?

41 • At what temperature do the Fahrenheit and Celsius temperature scales give the same reading?

42 • Sodium melts at 371 K. What is the melting point of sodium on the Celsius and Fahrenheit temperature scales?

43 • The boiling point of oxygen at 1.00 atm is 90.2 K. What is the boiling point of oxygen at 1.00 atm on the Celsius and Fahrenheit scales?

44 •• On the Réaumur temperature scale, the melting point of ice is 0°R and the boiling point of water is 80°R. Derive expressions for converting temperatures on the Réaumur scale to the Celsius and Fahrenheit scales.

45 ••• **ENGINEERING APPLICATION** A thermistor is a solid-state device widely used in a variety of engineering applications. Its primary characteristic is that its electrical resistance varies greatly with temperature. Its temperature dependence is given approximately by $R = R_0 e^{B/T}$, where R is in ohms (Ω), T is in kelvins, and R_0 and B are constants that can be determined by measuring R at calibration points such as the ice point and the steam point. (*a*) If $R = 7360\,\Omega$ at the ice point and $153\,\Omega$ at the steam point, find R_0 and B. (*b*) What is the resistance of the thermistor at $t = 98.6$°F? (*c*) What is the rate of change of the resistance with temperature (dR/dT) at the ice point and the steam point? (*d*) At which temperature is the thermistor most sensitive? SSM

THE IDEAL-GAS LAW

46 • An ideal gas in a cylinder fitted with a piston (Figure 17-20) is held at fixed pressure. If the temperature of the gas increases from 50° to 100°C, by what factor does the volume change?

FIGURE 17-20 Problems 46, 71

47 • A 10.0-L vessel contains gas at a temperature of 0.00°C and a pressure of 4.00 atm. How many moles of gas are in the vessel? How many molecules? SSM

48 •• A pressure as low as 1.00×10^{-8} torr can be achieved using an oil diffusion pump. How many molecules are there in $1.00\ \mathrm{cm^3}$ of a gas at this pressure if its temperature is 300 K?

49 •• You copy the following paragraph from a Martian physics textbook: "1 *snorf* of an ideal gas occupies a volume of 1.35 *zaks*. At a temperature of 22 *glips*, the gas has a pressure of 12.5 *klads*. At a temperature of −10 *glips*, the same gas now has a pressure of 8.7 *klads*." Determine the temperature of absolute zero in *glips*.

50 •• A motorist inflates the tires of her car to a gauge pressure of 180 kPa on a day when the temperature is –8.0°C. When she arrives at her destination, the tire pressure has increased to 245 kPa. What is the temperature of the tires if we assume that (*a*) the tires do not expand, or (*b*) that the tires expand so the volume of the enclosed air increases by 7 percent?

51 •• A room is 6.0 m by 5.0 m by 3.0 m. (*a*) If the air pressure in the room is 1.0 atm and the temperature is 300 K, find the number of moles of air in the room. (*b*) If the temperature increases by 5.0 K and the pressure remains constant, how many moles of air leave the room?

52 •• Imagine that 10.0 g of liquid helium, initially at 4.20 K, evaporate into an empty balloon that is kept at 1.00-atm pressure. What is the volume of the balloon at (*a*) 25.0 K, and (*b*) 293 K?

53 •• A closed container with a volume of 6.00 L holds 10.0 g of liquid helium at 25.0 K and enough air to fill the rest of its volume at a pressure of 1.00 atm. The helium then evaporates and the container warms to room temperature (293 K). What is the final pressure inside the container?

54 •• An automobile tire is filled to a gauge pressure of 200 kPa when its temperature is 20°C. (Gauge pressure is the difference between the actual pressure and atmospheric pressure.) After the car has been driven at high speeds, the tire temperature increases to 50°C. (*a*) Assuming that the volume of the tire does not change and that air behaves as an ideal gas, find the gauge pressure of the air in the tire. (*b*) Calculate the gauge pressure if the tire expands so the volume of the enclosed air increases by 10 percent.

55 •• After nitrogen (N_2) and oxygen (O_2), the most abundant molecule in Earth's atmosphere is water, H_2O. However, the fraction of H_2O molecules in a given volume of air varies dramatically, from practically zero percent under the driest conditions to a high of 4 percent where it is very humid. (*a*) At a given temperature and pressure, would air be denser when its water vapor content is large or small? (*b*) What is the difference in mass, at room temperature and atmospheric pressure, between a cubic meter of air with no water vapor molecules and a cubic meter of air in which 4 percent of the molecules are water vapor molecules? SSM

56 •• A scuba diver is 40 m below the surface of a lake, where the temperature is 5.0°C. He releases an air bubble that has a volume of 15 cm³. The bubble rises to the surface, where the temperature is 25°C. Assume that the air in the bubble is always in thermal equilibrium with the surrounding water, and assume that there is no exchange of molecules between the bubble and the surrounding water. What is the volume of the bubble right before it breaks the surface? *Hint: Remember that the pressure also changes.*

57 •• **Engineering Application** A hot-air balloon is open at the bottom. The balloon, which has a volume of 446 m³, is filled with air that has an average temperature of 100°C. The air outside the balloon has a temperature of 20.0°C and a pressure of 1.00 atm. How large a payload (including the envelope of the balloon itself) can the balloon lift? Use 29.0 g/mol for the molar mass of air. (Neglect the volume of both the payload and the envelope of the balloon.) SSM

58 ••• A helium balloon is used to lift a load of 110 N. The weight of the envelope of the balloon is 50.0 N and the volume of the helium when the balloon is fully inflated is 32.0 m³. The temperature of the air is 0°C and the atmospheric pressure is 1.00 atm. The balloon is inflated with a sufficient amount of helium gas that the net upward force on the balloon and its load is 30.0 N. Neglect any effects due to the changes of temperature as the altitude changes. (*a*) How many moles of helium gas are contained in the balloon? (*b*) At what altitude will the balloon be fully inflated? (*c*) Does the balloon ever reach the altitude at which it is fully inflated? (*d*) If the answer to Part (*c*) is "Yes," what is the maximum altitude attained by the balloon?

KINETIC THEORY OF GASES

59 • (*a*) One mole of argon gas is confined to a 1.0-liter container at a pressure of 10 atm. What is the rms speed of the argon atoms? (*b*) Compare your answer to the rms speed for helium atoms under the same conditions. SSM

60 • Find the total translational kinetic energy of the molecules of 1.0 L of oxygen gas at a temperature of 0.0°C and a pressure of 1.0 atm.

61 • Estimate the rms speed and the average kinetic energy of a hydrogen atom in a gas at a temperature of 1.0×10^7 K. (At this temperature, which is approximately the temperature in the interior of a star, hydrogen atoms are ionized and become protons.)

62 • Liquid helium has a temperature of only 4.20 K and is in equilibrium with its vapor at atmospheric pressure. Calculate the rms speed of a helium atom in the vapor at this temperature, and comment on the result.

63 • Show that the mean free path for a molecule in an ideal gas at temperature T and pressure P is given by $\lambda = kT/(\sqrt{2}\,P\pi d^2)$.

64 •• **Engineering Application** State-of-the-art vacuum equipment can attain pressures as low as 7.0×10^{-11} Pa. Suppose that a chamber contains helium at this pressure and at room temperature (300 K). Estimate the mean free path and the collision time for helium in the chamber. Assume the diameter of a helium atom is 1.0×10^{-10} m.

65 •• Oxygen (O_2) is confined to a cube-shaped container 15 cm on an edge at a temperature of 300 K. Compare the average kinetic energy of a molecule of the gas to the change in its gravitational potential energy if it falls 15 cm (the height of the container). SSM

*THE DISTRIBUTION OF MOLECULAR SPEEDS

66 •• Use calculus to show that $f(v)$, given by Equation 17-36, has its maximum value at a speed $v = \sqrt{2kT/m}$.

67 •• The fractional distribution function $f(v)$ is defined in Equation 17-36. Because $f(v)\,dv$ gives the fraction of molecules that have speeds in the range between v and $v + dv$, the integral of $f(v)\,dv$ over all the possible ranges of speeds must equal 1. Given that the integral $\int_0^\infty v^2 e^{-av^2}\,dv = \sqrt{(\pi/4)}a^{-3/2}$, show that $\int_0^\infty f(v)\,dv = 1$, where $f(v)$ is given by Equation 17-36. SSM

68 •• Given that the integral $\int_0^\infty v^3 e^{-av^2}\,dv = (1/2a^2)$, calculate the average speed v_{av} of molecules in a gas using the Maxwell–Boltzmann distribution function.

69 •• **Multistep** The translational kinetic energies of the molecules of a gas are distributed according to the Maxwell–Boltzmann energy distribution, Equation 17-38. (*a*) Determine the most probable value of the translational kinetic energy (in terms of the temperature *T*) and compare this value to the average value. (*b*) Sketch a graph of the translational kinetic energy distribution [$f(E)$ versus E] and label the most probable energy and the average energy. (Do not worry about calibrating the vertical scale of the graph.) (*c*) Your teacher says, "Just looking at the graph $f(E)$ versus E allows you to see that the average translational kinetic energy is considerably greater than the most probable translational kinetic energy." What feature(s) of the graph support her claim?

GENERAL PROBLEMS

70 • Find the temperature at which the rms speed of a molecule of hydrogen gas equals 343 m/s.

71 •• (*a*) If 1.0 mol of a gas in a cylindrical container occupies a volume of 10 L at a pressure of 1.0 atm, what is the temperature of the gas in kelvins? (*b*) The cylinder is fitted with a piston so that the volume of the gas (Figure 17-20) can vary. When the gas is heated at constant pressure, it expands to a volume of 20 L. What is the temperature of the gas in kelvins? (*c*) Next, the volume is fixed at 20 L, and the gas's temperature is increased to 350 K. What is the pressure of the gas now?

72 •• **MULTISTEP** (*a*) The volume per molecule of a gas is the reciprocal of the number density (the number of molecules per unit volume). Find the average volume per molecule for dry air at room temperature and atmospheric pressure. (*b*) Take the cube root of your answer to Part (*a*) to obtain a rough estimate of the average distance d between air molecules. (*c*) Find or estimate the average diameter D of an air molecule, and compare it to your answer to Part (*b*). (*d*) Sketch the molecules in a cube-shaped volume of air, with the edge length of the cube equal to $3d$. Make your figure to scale and place the molecules in what you think is a typical configuration. (*e*) Use your picture to explain why the mean free path of an air molecule is much greater than the average distance between molecules.

73 •• **CONCEPTUAL** The Maxwell–Boltzmann distribution applies not just to gases, but also to the molecular motions within liquids. The fact that not all molecules have the same speed helps us understand the process of evaporation. (*a*) Explain in terms of molecular motion why a drop of water becomes cooler as molecules evaporate from the drop's surface. (Evaporative cooling is an important mechanism for regulating our body temperatures, and is also used to cool buildings in hot, dry locations.) (*b*) Use the Maxwell–Boltzmann distribution to explain why even a slight increase in temperature can greatly increase the rate at which a drop of water evaporates. SSM

74 •• A cubic metal box that has 20-cm-long edges contains air at a pressure of 1.0 atm and a temperature of 300 K. The box is sealed so that the enclosed volume remains constant, and it is heated to a temperature of 400 K. Find the force due to the internal air pressure on each wall of the box.

75 •• **ENGINEERING APPLICATION** In attempting to create liquid hydrogen for fuel, one of the proposals is to convert plain old water (H_2O) into H_2 and O_2 gases by *electrolysis*. How many moles of each of these gases result from the electrolysis of 2.0 L of water? SSM

76 •• A 40-cm-long hollow cylinder of negligible mass rests on its side on a horizontal frictionless table. The cylinder is divided into two equal sections by a vertical nonporous membrane. One section contains nitrogen and the other contains oxygen. The pressure of the nitrogen is twice that of the oxygen. How far will the cylinder move if the membrane breaks?

77 •• A cylinder of fixed volume contains a mixture of helium gas (He) and hydrogen gas (H_2) at a temperature T_1 and pressure P_1. If the temperature is doubled to $T_2 = 2T_1$, the pressure would also double, except for the fact that at this temperature the H_2 is essentially 100 percent dissociated into H_1. In reality, at pressure $P_2 = 2P_1$ the temperature is $T_2 = 3T_1$. If the mass of the hydrogen in the cylinder is m, what is the mass of the helium in the cylinder?

78 •• The mean free path for O_2 molecules at a temperature of 300 K and at 1.00-atm pressure is 7.10×10^{-8} m. Use these data to estimate the size of an O_2 molecule.

79 •• **ENGINEERING APPLICATION** Current experiments in atomic trapping and cooling can create low-density gases of rubidium and other atoms with temperatures in the nanokelvin (10^{-9} K) range. These atoms are trapped and cooled using magnetic fields and lasers in ultrahigh vacuum chambers. One method that is used to measure the temperature of a trapped gas is to turn the trap off and measure the time it takes for molecules of the gas to fall a given distance. Consider a gas of rubidium atoms at a temperature of 120 nK. Calculate how long it would take an atom traveling at the rms speed of the gas to fall a distance of 10.0 cm if (*a*) it were initially moving directly downward, and (*b*) if it were initially moving directly upward. Assume that the atom does not collide with any other atoms along its trajectory. SSM

80 ••• A cylinder is filled with 0.10 mol of an ideal gas at standard temperature and pressure, and a 1.4-kg piston seals the gas in the cylinder (Figure 17-21) with a frictionless seal. The trapped column of gas is 2.4 m high. The piston and cylinder are surrounded by air, also at standard temperature and pressure. The piston is released from rest and starts to fall. The motion of the piston ceases after the oscillations stop with the piston and the trapped air in thermal equilibrium with the surrounding air. (*a*) Find the height of the gas column. (*b*) Suppose that the piston is pushed down below its equilibrium position by a small amount and then released. Assuming that the temperature of the gas remains constant, find the frequency of vibration of the piston.

FIGURE 17-21 Problem 80

81 ••• **SPREADSHEET, MULTISTEP** To solve this problem, you will use a **spreadsheet** to study the distribution of molecular speeds in a gas. Figure 17-22 should help you get started. (*a*) Enter the values for constants R, M, and T, as shown. Then in column A, enter values of speeds ranging from 0 to 1200 m/s, in increments of 1 m/s. (This spreadsheet will be long.) In cell B7, enter the formula for the Maxwell–Boltzmann fractional speed distribution. This formula contains parameters v, R, M and T. Substitute A7 for v, B$1 for R, B$2 for M and B$3 for T. Then use the FILL DOWN command to enter the formula in the cells below B7. Create a graph of $f(v)$ versus v using the data in columns A and B. (*b*) Explore how the graph changes as you increase and decrease the temperature, and describe the results. (*c*) Add a third column in which each cell contains the cumulative sum of all $f(v)$ values, multiplied by the interval size dv (which equals 1), in the rows above and including the row in question. What is the physical interpretation of the numbers in this column? (*d*) For nitrogen gas at 300 K, what percentage of the molecules has speeds less than 200 m/s? (*e*) For nitrogen gas at 300 K, what percentage of the molecules has speeds greater than 700 m/s?

	A	B	C
1	R =	8.31	J/mol-K
2	M =	0.028	kg/mol
3	T =	300	K
4			
5	**v**	**f(v)**	**sum f(v)dv**
6	(m/s)	(s/m)	(unitless)
7	0	0	0
8	1	3.0032E-08	3.00325E-08
9	2	1.2013E-07	1.5016E-07
10	3	2.7028E-07	4.20441E-07
11	4	4.8048E-07	9.0092E-07
12	5	7.5071E-07	1.65163E-06

FIGURE 17-22 Problem 81 (Only the first few rows of the spreadsheet are shown.)

CHAPTER

18

THE MEN'S LITTLE 500 BICYCLE RACE HAS BEEN HELD AT INDIANA UNIVERSITY SINCE 1951. (THE CLIMATIC SCENE OF THE 1979 MOVIE *BREAKING AWAY* FEATURED THIS RACE WHEN A TEAM OF LOCALS, THE CUTTERS, COMPETED.) ONLY IU STUDENTS ARE PERMITTED TO COMPETE IN THIS 200-LAP 50-MILE EVENT. THIS PICTURE IS THE 2006 RACE. (THE RACE WAS WON BY ALPHA TAU OMEGA). *(AJ Mast/Icon SMI/Corbis.)*

Heat and the First Law of Thermodynamics

? Pumping up a bicycle tire requires how much work on the air in order to compress it? (See Example 18-13.)

The relation between heating a system, doing work on the system, and the change in the internal energy of the system is the basis of the first law of thermodynamics. In Part I of this book, we discussed motion; now, we consider the role that heat plays in the generation of motion—whether it is the movement of people hurrying to catch a bus, the cyclic motion of pistons in a car engine, or even the drips of water sliding down a glass of cold lemonade on a hot day.

For years, the power generated by heating has been harnessed. From early steam engines to internal combustion automobile engines to jet engines, engineers have been finding ways to improve the performance of their machines to get the most energy out of them. Even athletes today train and eat to optimize their performance on the field, essentially treating their bodies like other mechanical engines.

In this chapter, we define heat capacity, and examine how heating a sample can cause either an increase in its temperature or a change in its phase (from solid to liquid, for example). We then examine the relationship between changes in the internal energy of a system, the energy transferred to the system via heat and work, and express the law of conservation of energy for systems as the first law of thermodynamics. Finally, we shall see how the heat capacity of a system is related to its molecular structure.

18-1 HEAT CAPACITY AND SPECIFIC HEAT

Heat is the transfer of energy due to a difference in temperature. During the seventeenth century, Galileo, Newton, and other scientists generally supported the theory of the ancient Greek atomists who considered thermal energy to be a manifestation of molecular motion. During the next century, methods were developed for making quantitative measurements of the amount of energy transferred because of differences in temperature, and it was found that if objects are in thermal contact, the amount of energy that is released by one object equals the amount that is absorbed by the other object. This discovery led to a theory of in which heat was modeled as a conserved material substance. In this theory, an invisible fluid called "caloric" flowed out of one object and into another, and this caloric could be neither created nor destroyed.

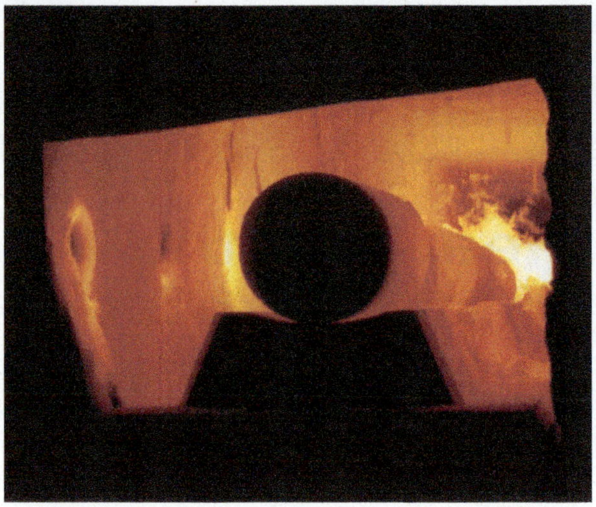

Steel ingots in a twin-tube tunnel furnace. The three 53-cm-diameter carbon–steel ingots seen here have been heated for about 7 hours to approximately 1340°C. Each 3200-kg ingot sits on a furnace car that transports it through the 81-m-long furnace, which is divided into 12 separate heating zones so that the temperature of the ingot is increased gradually to prevent cracking. The ingots, glowing a yellow-whitish color, exit the furnace to be milled into large, heavy-walled pipes. *(Phoenix Pipe & Tube/ Lana Berkovich.)*

The caloric theory reigned until the nineteenth century, when it was observed that kinetic friction between objects could produce an unlimited transfer of energy between objects, deposing of the idea that caloric was a substance present in a fixed amount. The modern theory of heat did not emerge until the 1840s, when James Joule (1818–1889) demonstrated that when a viscous liquid is stirred with a paddle, the increase or decrease of a given amount of thermal energy was always accompanied by the decrease or increase of an equivalent quantity of mechanical energy. Thermal energy, therefore, is not itself conserved. Instead, thermal energy is a form of internal energy, and it is energy that is conserved.

When a warmer object is in thermal contact with a cooler object, the energy being transferred from the warmer object to the cooler object due to the difference in the temperatures of the two objects is called **heat.** The energy is no longer identified as heat once it has been transferred to the cooler object. Instead, it is identified as part of the internal energy of the cooler object. The **internal energy** of an object is its total energy in the center-of-mass reference frame of the object. In this book Q is the symbol for heat and E_{int} is the symbol for internal energy.

When energy is transferred to a substance by heating it, the temperature of the substance usually increases.* The amount of heat Q needed to increase the temperature of a sample of the substance is proportional to both the temperature change and the mass of the sample:

$$Q = \Delta E_{\text{int}} = C\,\Delta T = mc\,\Delta T \qquad 18\text{-}1$$

DEFINITION: HEAT CAPACITY

* An exception occurs during a change in phase, as when water freezes or evaporates. Changes of phase are discussed in Section 18-2.

where C is the **heat capacity,** which is defined as the change in internal energy required to increase the temperature of a sample by one degree. The **specific heat** capacity c is the heat capacity per unit mass:

$$c = \frac{C}{m} \qquad 18\text{-}2$$

DEFINITION: SPECIFIC HEAT

> **!** The term *heat capacity* does not mean that a body contains a certain amount of heat.

The term *specific heat* is short for specific heat capacity. The historical unit of heat, the **calorie,** was originally defined to be the amount of heat needed to increase the temperature of one gram of water one Celsius degree.* Because we now recognize that heat is a measure of energy transfer, we can define the calorie in terms of the SI unit of energy, the joule:

$$1 \text{ cal} = 4.184 \text{ J} \qquad 18\text{-}3$$

The U.S. customary unit of heat is the **Btu** (for British thermal unit), which was originally defined to be the amount of energy needed to increase the temperature of 1 pound of water by 1°F. The Btu is related to the calorie and to the joule by

$$1 \text{ Btu} = 252 \text{ cal} = 1.054 \text{ kJ} \qquad 18\text{-}4$$

The original definition of the calorie implies that the specific heat of water (in the liquid state) is†

$$c_{\text{water}} = 1 \text{ cal/(g}\cdot\text{K)} = 1 \text{ kcal/(kg}\cdot\text{K)} = 4.184 \text{ kJ/(kg}\cdot\text{K)} \qquad 18\text{-}5a$$

Similarly, from the definition of the Btu, the specific heat of water in U.S. customary units is

$$c_{\text{water}} = 1 \text{ Btu/(lb}\cdot{}^\circ\text{F)} \qquad 18\text{-}5b$$

The heat capacity per mole is called the **molar specific heat** c',

$$c' = \frac{C}{n}$$

where n is the number of moles. Because $C = mc$, the molar specific heat c' and specific heat c are related by

$$c' = \frac{C}{n} = \frac{mc}{n} = Mc \qquad 18\text{-}6$$

MOLAR SPECIFIC HEAT

where $M = m/n$ is the molar mass. Table 18-1 lists the specific heats and molar specific heats of some solids and liquids. Note that the molar heats of all the metals are about the same. We discuss the significance of the similar specific-heat values of metals in Section 18-7.

Table 18-1 Specific Heats and Molar Specific Heats of Some Solids and Liquids

Substance	c, kJ/kg·K	c, kcal/kg·K or Btu/lb·F°	c', J/mol·K
Aluminium	0.900	0.215	24.3
Bismuth	0.123	0.0294	25.7
Copper	0.386	0.0923	24.5
Glass	0.840	0.20	—
Gold	0.126	0.0301	25.6
Ice (−10°C)	2.05	0.49	36.9
Lead	0.128	0.0305	26.4
Silver	0.233	0.0558	24.9
Tungsten	0.134	0.0321	24.8
Zinc	0.387	0.0925	25.2
Alcohol (ethyl)	2.4	0.58	111
Mercury	0.140	0.033	28.3
Water	4.18	1.00	75.2
Steam (at 1 atm)	2.02	0.48	36.4

Liquids are in red typeface and gases are in blue typeface.

* The kilocalorie is then the amount of heat needed to increase the temperature of 1 kg of water by 1°C. The "calorie" used in measuring the energy equivalent of foods is actually the kilocalorie.

† Careful measurement shows that the specific heat of water varies by about 1 percent over the temperature range from 0° to 100°C. We will usually neglect this small variation.

1 line short

Example 18-1 Increasing the Temperature

A jewelry designer is creating gold charms. To make the charms, he must melt gold to fill molds. How much heat is needed to increase the temperature of 3.00 kg of gold from 22°C (room temperature) to 1063°C, the melting point of gold?

PICTURE The amount of heat Q needed to increase the temperature of the substance (gold) is proportional to the temperature change and to the mass of the substance.

SOLVE

1. The required heat is given by Equation 18-1, with $c = 0.126\ \text{kJ}/(\text{kg}\cdot\text{K})$ from Table 18-1:

$$Q = mc\,\Delta T = (3.00\ \text{kg})(0.126\ \text{kJ}/(\text{kg}\cdot\text{K}))(1041\ \text{K}) = \boxed{393\ \text{kJ}}$$

CHECK The problem asks for an amount of energy and the answer came out in joules, which are energy units.

TAKING IT FURTHER Note that we use $\Delta T = 1063°\text{C} - 22°\text{C} = 1041°\text{C} = 1041\ \text{K}$.

PRACTICE PROBLEM 18-1 A 2.0-kg aluminum block is originally at 10°C. If 36 kJ of energy are added to the block, what is its final temperature?

We see from Table 18-1 that the specific heat of liquid water is considerably larger than that of other common substances. Thus, water is an excellent material for storing thermal energy, as in a solar heating system. It is also an excellent coolant, as in a car engine. (The coolant in automotive engines is a mixture of water and ethylene glycol.)

Large bodies of water, such as lakes or oceans, tend to moderate fluctuations of the air temperature nearby because the bodies of water can absorb or release large quantities of heat while undergoing only very small changes in temperature. *(From Frank Press and Raymond Siever,* Understanding Earth, *3rd ed., W. H. Freeman and Company, 2001.)*

CALORIMETRY

To measure the specific heat of an object, we can first heat it to some known temperature, say the boiling point of water. Then, we transfer the object to a water bath of known mass and initial temperature. Finally, we measure the final equilibrium temperature of the system (the object, the water in the bath, and the water-bath container). If the system is thermally isolated from its surroundings (by insulating the container, for example), then the heat released by the object will equal the heat absorbed by the water and the container. This procedure is called **calorimetry,** and the insulated water container is called a **calorimeter.**

Let m be the mass of the object, let c be its specific heat, and let T_{io} be the initial temperature of the object. If T_f is the final temperature of both the object, the water and container, the heat released by the object is

$$Q_{out} = mc(T_{io} - T_f)$$

Similarly, if T_{iw} is the initial temperature of the water and container, then the heat absorbed by the water and container is

$$Q_{in} = m_w c_w (T_f - T_{iw}) - m_c c_c (T_f - T_{iw})$$

where m_w and $c_w = 4.18\ \text{kJ}/(\text{kg}\cdot\text{K})$ are the mass and specific heat of the water, and m_c and c_c are the mass and specific heat of the container. (Note that we have expressed the temperature differences so that they are both positive quantities. As a consequence, our expressions for Q_{in} and Q_{out} are both positive.) Setting these amounts of heat equal yields an equation that can be solved for the specific heat c of the object:

$$Q_{out} = Q_{in} \Rightarrow mc(T_{io} - T_f) = m_w c_w (T_f - T_{iw}) + m_c c_c (T_f - T_{iw}) \qquad 18\text{-}7$$

Because only temperature differences occur in Equation 18-7, and because the kelvin and the Celsius degree are the same size, it does not matter whether we use kelvins or Celsius degrees.

Example 18-2 Measuring Specific Heat

To measure the specific heat of lead, you heat 600 g of lead shot to 100.0°C and place it in an aluminum calorimeter of mass 200 g that contains 500 g of water initially at 17.3°C. If the final temperature of the mixture is 20.0°C, what is the specific heat of lead? (The specific heat of the aluminum container is 0.900 kJ/(kg · K).)

PICTURE We set the heat released by the lead equal to the heat absorbed by the water and container and solve for the specific heat of lead c_{Pb}.

SOLVE

1. Write the heat released by the lead in terms of its specific heat: $Q_{Pb} = m_{Pb} c_{Pb} |\Delta T_{Pb}|$
2. Write the heat absorbed by the water: $Q_w = m_w c_w \Delta T_w$
3. Write the heat absorbed by the container: $Q_c = m_c c_c \Delta T_c$
4. Set the heat released equal to the heats absorbed by the water and the container:
   $Q_{out} = Q_{in} \Rightarrow Q_{Pb} = Q_w + Q_c$
   $m_{Pb} c_{Pb} |\Delta T_{Pb}| = m_w c_w \Delta T_w + m_c c_c \Delta T_c$
   where $\Delta T_c = \Delta T_w = 2.7\text{ K}$ and $|\Delta T_{Pb}| = 80.0\text{ K}$
5. Solve for c_{Pb}:
   $$c_{Pb} = \frac{(m_w c_w + m_c c_c)\,\Delta T_w}{m_{Pb}|\Delta T_{Pb}|}$$
   $$= \frac{[(0.50\text{ kg})(4.18\text{ kJ/(kg}\cdot\text{K)}) + (0.20\text{ kg})(0.90\text{ kJ/(kg}\cdot\text{K)})](2.7\text{ K})}{(0.600\text{ kg})(80.0\text{ K})}$$
   $$= 0.128\text{ kJ/(kg}\cdot\text{K)} = \boxed{0.13\text{ kJ/(kg}\cdot\text{K)}}$$

CHECK As expected, the specific heat of lead is considerably less than that of water. (The specific heat of liquid water is 4.18 kJ/(kg · K).)

TAKING IT FURTHER The step-5 result is expressed to two figures because the temperature change of the water is known to only two figures.

PRACTICE PROBLEM 18-2 A solar home contains 1.00×10^5 kg of concrete (specific heat = 1.00 kJ/kg · K). How much heat is released by the concrete when it cools from 25°C to 20°C?

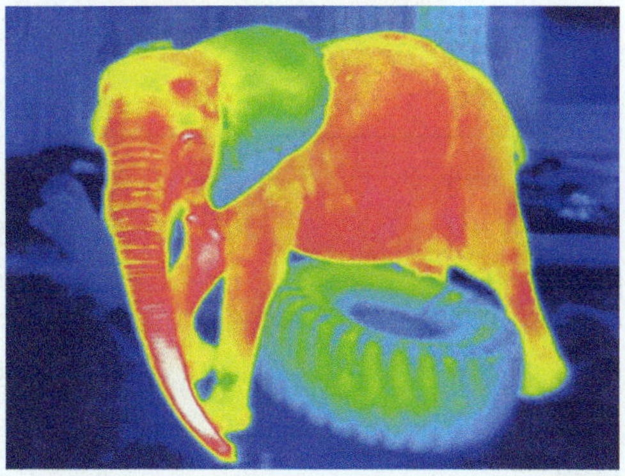

(*Edward Kinsman/ Photo Researchers, Inc.*)

18-2 CHANGE OF PHASE AND LATENT HEAT

If heat is absorbed by ice at 0°C, the temperature of the ice does not change. Instead, the ice melts. Melting is an example of a **phase change** or **change of state.** Common types of phase changes include fusion (liquid to solid), melting (solid to liquid), vaporization (liquid to vapor or gas), condensation (gas or vapor to liquid), and sublimation (solid directly to gas or vapor, such as solid carbon dioxide, or dry ice, changing to vapor). Other types of phase changes exist as well, such as the change of a solid from one crystalline form to another. For example, carbon graphite under intense pressure becomes a diamond.

Molecular theory can help us to understand why temperature remains constant during a phase change. The molecules in a liquid are close together and exert attractive forces on each other, whereas molecules in a gas are far apart. Because of this intermolecular attraction, it takes energy to remove molecules from a liquid to form a gas. Consider a pot of water sitting over a flame on the stove. At first, as the water is heated, the motions of its molecules increase and the temperature increases. When the temperature reaches the boiling point, the molecules can no longer increase their kinetic energy and remain in the liquid. As the liquid water vaporizes, the added energy goes into breaking the intermolecular attractions. That is, it goes into increasing the potential energy of the molecules rather than their kinetic energy. Because temperature is a measure of the average translational *kinetic* energy of molecules, the temperature does not change.

For a pure substance, a change in phase at a given pressure occurs only at a particular temperature. For example, pure water at a pressure of 1 atm changes from solid to liquid at 0°C (the normal melting point of water) and from liquid to gas at 100°C (the normal boiling point of water).

The energy required to melt a sample of a substance of mass m with no change in its temperature is proportional to the mass of the sample:

$$Q_f = mL_f \qquad 18\text{-}8$$

where L_f is called the **latent heat of fusion** of the substance. At a pressure of 1 atm, the latent heat of fusion for water is 333.5 kJ/kg = 79.7 kcal/kg. If the phase change is from liquid to gas, the heat required is

$$Q_v = mL_v \qquad 18\text{-}9$$

where L_v is the **latent heat of vaporization.** For water at a pressure of 1 atm, the latent heat of vaporization is 2.26 MJ/kg = 540 kcal/kg. Table 18-2 gives the melting and boiling points and the latent heats of fusion and vaporization, all at 1 atm, for various substances.

Although melting indicates that the ice has experienced a change in phase, the temperature of the ice does not change. *(From Donald Wink, Sharon Gislason, and Sheila McNicholas,* The Practice of Chemistry, *W. H. Freeman and Company, 2002.)*

Table 18-2 Melting Point (MP), Latent Heat of Fusion (L_f), Boiling Point (BP), and Latent Heat of Vaporization (L_v), all at 1 atm, for Various Substances

Substance	MP, K	L_f, kJ/kg	BP, K	L_v, kJ/kg
Alcohol, ethyl	159	109	351	879
Bromine	266	67.4	332	369
Carbon dioxide	—	—	194.6*	573*
Copper	1356	205	2839	4726
Gold	1336	62.8	3081	1701
Helium	—	—	4.2	21
Lead	600	24.7	2023	858
Mercury	234	11.3	630	296
Nitrogen	63	25.7	77.35	199
Oxygen	54.4	13.8	90.2	213
Silver	1234	105	2436	2323
Sulfur	388	38.5	717.75	287
Water (liquid)	273.15	333.5	373.15	2257
Zinc	692	102	1184	1768

* These values are for sublimation. Carbon dioxide does not have a liquid state at 1 atm.

Example 18-3 Changing Ice into Steam

How much heat is needed to change 1.5 kg of ice at −20°C and 1.0 atm into steam?

PICTURE The heat required to change the ice into steam consists of four parts: Q_1, the heat needed to increase the temperature of the ice from −20°C to 0°C; Q_2, the heat needed to melt the ice; Q_3, the heat needed to increase the temperature of the water from 0°C to 100°C; and Q_4, the heat needed to vaporize the water. In calculating Q_1 and Q_3, we assume that the specific heats are constant, with the values 2.05 kJ/kg·K for ice and 4.18 kJ/kg·K for water.

SOLVE

1. Use $Q_1 = mc\,\Delta T$ to find the heat needed to increase the temperature of the ice to 0°C:

$$Q_1 = mc\,\Delta T = (1.5\text{ kg})(2.05\text{ kJ/kg}\cdot\text{K})(20\text{ K})$$
$$= 61.5\text{ kJ} = 0.0615\text{ MJ}$$

2. Use L_f from Table 18-2 to find the heat Q_2 needed to melt the ice:

$$Q_2 = mL_f = (1.5\text{ kg})(333.5\text{ kJ/kg}) = 500\text{ kJ} = 0.500\text{ MJ}$$

3. Find the heat Q_3 needed to increase the temperature of the water from 0°C to 100°C:

$$Q_3 = mc\,\Delta T = (1.5\text{ kg})(4.18\text{ kJ/kg}\cdot\text{K})(100\text{ K}) = 627\text{ kJ} = 0.627\text{ MJ}$$
$$Q_3 = 627\text{ kJ} = 0.627\text{ MJ}$$

4. Use L_v from Table 18-2 to find the heat Q_4 needed to vaporize the water:

$$Q_4 = mL_V = (1.5\text{ kg})(2.26\text{ MJ/kg}) = 3.39\text{ MJ}$$

5. Sum your results to find the total heat Q:

$$Q = Q_1 + Q_2 + Q_3 + Q_4 = \boxed{4.6\text{ MJ}}$$

CHECK You may have observed that much less time is needed to bring a kettle full of water to a boil than the time needed to boil the kettle dry. This observation is consistent with the fact that our step-3 result is less than 20 percent of our step-4 result.

TAKING IT FURTHER Notice that most of the heat was needed to vaporize the water, and that the amount needed to melt the ice was a significant fraction of the heat needed to increase the temperature of the liquid water by 100°C. A graph of temperature versus time for the case in which the heat is absorbed at a constant rate of 1.0 kJ/s is shown in Figure 18-1. Note that it takes considerably longer to vaporize the water than it does to melt the ice or to increase the temperature of the water. When all of the water has vaporized, the temperature again increases as heat is absorbed.

PRACTICE PROBLEM 18-3 An 830-g piece of lead is heated to its melting point of 600 K. How much additional energy must be absorbed by the lead at 600 K to completely melt all 830 g?

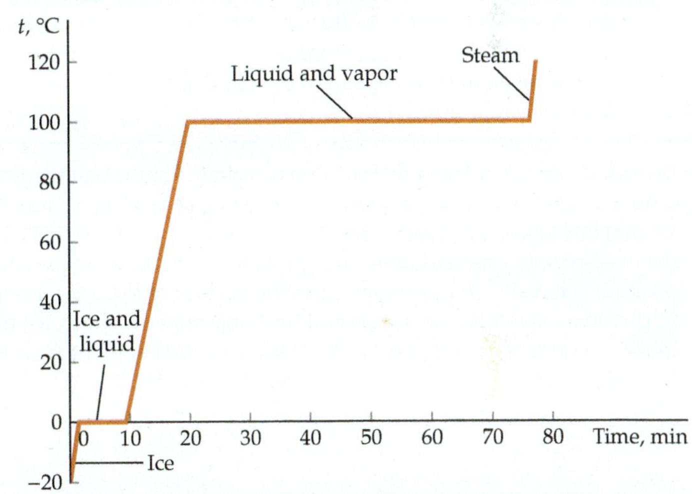

FIGURE 18-1 A 1.5-kg sample of water is heated from −20°C to 120°C at a constant rate of 60 kJ/min.

Example 18-4 A Cool Drink

Context-Rich

A 2.0-liter pitcher of lemonade has been sitting on the picnic table in the sunlight all day at 33°C. You pour 0.24 kg into a Styrofoam cup and add two ice cubes (each 0.025 kg at 0.0°C). (*a*) Assuming no heat is released to the surroundings, what is the final temperature of the lemonade? (*b*) What is the final temperature if you add six, instead of the two, ice cubes?

PICTURE We set the heat released by the lemonade equal to the heat absorbed by the ice cubes. Let T_f be the final temperature of the lemonade and water. We assume that lemonade has the same specific heat as water.

SOLVE

(*a*) 1. Write the heat absorbed by the lemonade in terms of the final temperature T_f:

$$Q_{out} = m_L c|\Delta T| = m_L c(T_{Li} - T_f)$$

2. Write the heat absorbed by the ice cubes and resulting water in terms of the final temperature:

$$Q_{in} = m_{ice}L_f + m_{ice}c\,\Delta T_w = m_{ice}L_f + m_{ice}c(T_f - T_{wi})$$

3. Set the heat released equal to the heat absorbed and solve for T_f:

$$Q_{out} = Q_{in}$$

$$m_L c(T_{Li} - T_f) = m_{ice}L_f + m_{ice}c(T_f - T_{wi})$$

$$\text{so}\quad T_f = \frac{(m_{ice}T_{wi} + m_L T_{Li})c - m_{ice}L_f}{(m_L + m_{ice})c}$$

$$= \frac{(0.050 \times 273.15 + 0.24 \times 306.15)4.18 - 0.050 \times 333.5}{0.29 \times 4.18}$$

$$= 286.7\text{ K} = \boxed{14°\text{C}}$$

(*b*) 1. For six ice cubes, $m_{ice} = 0.15$ kg. Find the final temperature as in step 3 of Part (*a*):

$$T_f = \frac{(m_{ice}T_{wi} + m_L T_{Li})c - m_{ice}L_f}{(m_L + m_{ice})c}$$

$$= \frac{(0.150 \times 273.15 + 0.24 \times 306.15)4.18 - 0.150 \times 333.5}{0.39 \times 4.18}$$

$$= 262.8\text{ K} = -10.4\text{ °C}$$

2. A final temperature below 0°C cannot be correct! No amount of ice at 0°C can decrease the temperature of warm lemonade to below 0°C. Our calculation is wrong because our assumption that all of the ice melts was wrong. Instead, the heat released by the lemonade as it cools from 32°C to 0°C is not enough to melt all of the ice. The final temperature is thus:

$$= \boxed{0°\text{C}}$$

CHECK Let us calculate how much ice is melted. For the lemonade to cool from 33°C to 0°C, it must release heat in the amount $Q = (0.24\text{ kg})(4.18\text{ kJ/kg}\cdot\text{K})(33\text{ K}) = 33.1\text{ kJ}$. The mass of ice that this amount of heat will melt is $m_{ice} = Q/L_f = 33.1\text{ kJ}/(333.5\text{ kJ/kg}) = 0.10\text{ kg}$. This is the mass of only four ice cubes. Adding more than four ice cubes does not decrease the temperature below 0°C. It merely increases the amount of ice in the ice–lemonade mixture at 0°C. In problems like this one, we should first find out how much ice must be melted to reduce the temperature of the liquid to 0°C. If less than that amount is added, we can proceed as in Part (*a*). If more ice is added, the final temperature is 0°C.

18-3 JOULE'S EXPERIMENT AND THE FIRST LAW OF THERMODYNAMICS

We can increase the temperature of a system by adding energy but we can also increase its temperature by doing work on it. Figure 18-2 is a diagram of the apparatus Joule used in a famous experiment in which he determined the amount of work needed to increase the temperature of one pound of water by one Fahrenheit degree. Here, the system is a thermally insulated container of water. Joule's apparatus converts the potential energy of falling weights into work done on the water by an attached paddle, as shown in the figure. Joule found that he could increase the temperature of 1.00 lb of water by 1.00°F by dropping 772 lb of attached weights a distance of one foot. Converting to modern units and using current values, Joule found that it takes about 4.184 J (the energy units adopted by the scientific community in 1948) to increase the temperature of 1 g of water by 1°C. The result that 4.184 J of mechanical energy is exactly equivalent to 1 cal of heat is known as the **mechanical equivalence of heat.**

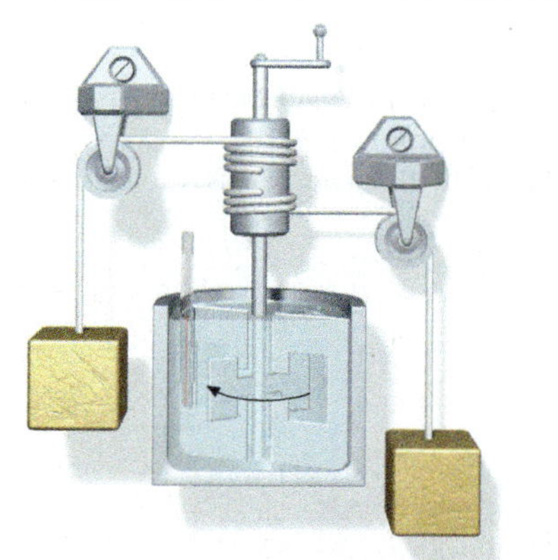

FIGURE 18-2 Schematic diagram for Joule's experiment. Insulating walls surround water. As the weights fall at constant speed, they turn a paddle wheel, which does work on the water. If friction is negligible, the work done by the paddle wheel on the water equals the loss of mechanical energy of the weights, which is determined by calculating the loss in the potential energy of the weights.

There are other ways of doing work on this system. For example, we could let gravity do the work by dropping the insulated container of water from some

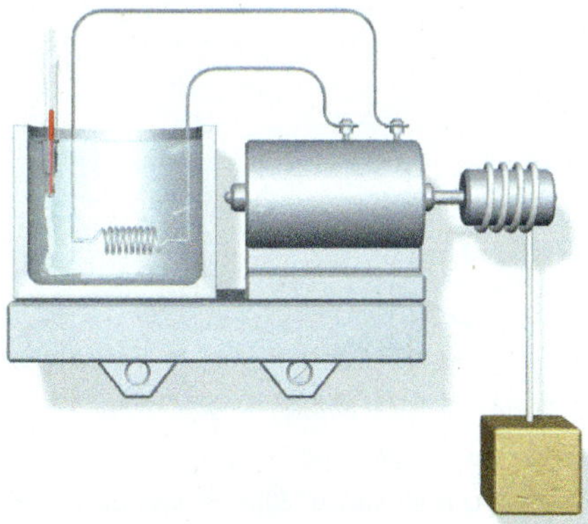

FIGURE 18-3 Another method of doing work on a thermally insulated container of water. Electrical work is done on the system by the generator, which is driven by the falling weight.

CONCEPT CHECK 18-1

Joule's experiment establishing the mechanical equivalence of heat involved the conversion of mechanical energy into internal energy. Give some examples of the internal energy of a system being converted into mechanical energy.

height h, letting the system make an inelastic collision with the ground, or we could do mechanical work to generate electricity and then use the electricity to heat the water (Figure 18-3). During all such experiments, the same amount of work is required to produce a given temperature change. By the conservation of energy, the work done equals the increase in the internal energy of the system.

Example 18-5 Warming Water by Dropping It

(*a*) At Niagara Falls, the water drops 50 m. If the decrease in the gravitational potential energy of the water is equal to the increase in the internal energy of the water, compute the increase in its temperature. (*b*) At Yosemite Falls, the water drops 740 m. If the decrease in the gravitational potential energy of the water is equal to the increase in the internal energy of the water, compute the increase in its temperature. (These temperature increases are not easily observed because as the water falls its temperature is affected by several other effects. For example it cools by evaporation and it is warmed as the air does work on it via the viscosity interaction.)

PICTURE The kinetic energy of the water just before it hits the bottom equals its original potential energy mgh. During the collision, this energy is converted into internal energy, which in turn causes an increase in temperature given by $mc\,\Delta T$.

SOLVE

(*a*) 1. Set the decrease in the potential energy equal to the increase in the internal energy:

$$mgh = mc\,\Delta T$$

2. Solve for the temperature change:

$$\Delta T = \frac{gh}{c} = \frac{(9.81\text{ N/kg})(50\text{ m})}{4.184\text{ kJ/kg}\cdot\text{K}} = 0.117\text{ K} = \boxed{0.12\text{ K}}$$

(*b*) Repeat the calculation with $h = 740$ m:

$$\Delta T = \frac{gh}{c} = \frac{(9.81\text{ N/kg})(740\text{ m})}{4.184\text{ kJ/kg}\cdot\text{K}} = 1.74\text{ K} = \boxed{1.7\text{ K}}$$

CHECK Yosemite Falls is 14.8 times higher than Niagara Falls, so the potential energy change of the Yosemite Falls water is 14.8 times larger than the potential energy change of Niagara Falls water. Thus, the change in temperature should be 14.8 times greater for Yosemite Falls water than for Niagara Falls water. Multiplying 0.117 K by 14.8 gives 1.73 K, which is very close to our Part-(*b*) result.

TAKING IT FURTHER These calculations illustrate one of the difficulties with Joule's experiment—a large amount of mechanical energy must be dissipated to produce a measurable change in the temperature of the water.

Suppose that we perform Joule's experiment, but replace the insulating walls of the container with conducting walls. We find that the work needed to produce a given change in the temperature of the system depends on how much heat is absorbed or released by the system by conduction through the walls. However, if we sum the work done on the system and the net heat absorbed by the system, the result is always the same for a given temperature change. That is, the sum of the heat transfer *into* the system and the work done *on* the system equals the change in the internal energy of the system. This result is the **first law of thermodynamics,** which is simply a statement of the conservation of energy.

Let W_{on} stand for the work done by the surroundings *on* the system. For example, suppose our system is a gas confined to a cylinder by a piston. If the piston compresses the gas, the surroundings do work on the gas and W_{on} is positive. (However, if the gas expands against the piston, the gas does work on the surroundings and W_{on} is negative.) Also, let Q_{in} stand for the heat transfer into the system. If heat is transferred into the system, then Q_{in} is positive; if heat is transferred out of the system, then Q_{in} is negative (Figure 18-4). Using these conventions, and denoting the internal energy by E_{int},* the first law of thermodynamics is written

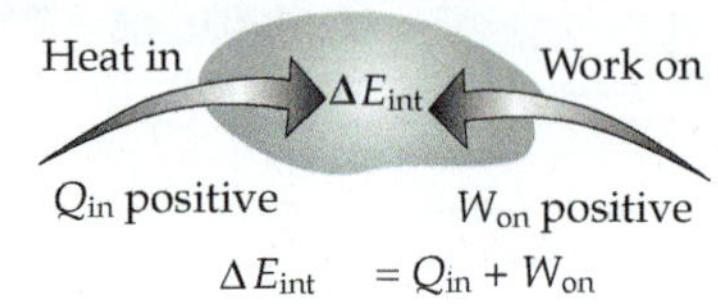

FIGURE 18-4 Sign convention for the first law of thermodynamics.

$$\Delta E_{int} = Q_{in} + W_{on} \tag{18-10}$$

The change in the internal energy of the system equals the heat transfer into the system plus the work done on the system.

FIRST LAW OF THERMODYNAMICS

Equation 18-10 is the same as the work–energy theorem $W_{ext} = \Delta E_{sys}$ of Chapter 7 (Equation 7-9), except we have added the heat term Q_{in} and called the energy of the system E_{int}.

Example 18-6 Stirring the Water

You do 25 kJ of work on a system consisting of 3.0 kg of water by stirring it with a paddle wheel. During this time, 15 kcal of heat is released by the system due to poor thermal insulation. What is the change in the internal energy of the system?

PICTURE We express all energies in joules and apply the first law of thermodynamics.

SOLVE

1. ΔE_{int} is found by using the first law of thermodynamics: $\Delta E_{int} = Q_{in} + W_{on}$
2. Heat is *released* by the system, thus the Q_{in} term is negative: $Q_{in} = -15\text{ kcal} = -(15\text{ kcal})\left(\dfrac{4.18\text{ kJ}}{1\text{ kcal}}\right) = -62.7\text{ kJ}$
3. The work is done *on* the system, thus W_{on} term is positive: $W_{on} = +25\text{ kJ}$
4. Substitute these quantities and solve for ΔE_{int}: $\Delta E_{int} = Q_{in} + W_{on} = (-62.7\text{ kJ}) + (+25\text{ kJ})$ $= -37.7\text{ kJ} = \boxed{-38\text{ kJ}}$

CHECK The heat loss exceeds the work gain, so the change in internal energy is negative.

It is important to understand that the internal energy E_{int} is a function of the state of the system, just as P, V, and T are functions of the state of the system. Consider a gas in some initial state (P_i, V_i). The temperature T_i can be determined by the equation of state. For example, if the gas is ideal, then $T_i = P_iV_i/(nR)$. The internal energy $E_{int\,i}$ also depends only on the state of the gas, which is determined

* Another commonly used symbol for internal energy is U.

by any two state variables such as P and V, P and T, or V and T. If we slowly heat the gas or cool the gas, do work on the gas, or let the gas do work, the gas will move through a sequence of states; that is, it will have different values of the state functions P, V, T, and E_{int}.

On the other hand, the heat Q and the work W are not functions of the state of the system. That is, there are no functions Q or W associated with any particular state of the gas. We could take the gas through a sequence of states beginning and ending at state (P_i, V_i) during which the gas did work that has a positive value and absorbed an equal amount of heat. Or we could take it through a different sequence of states such that work was done on the gas and heat was released from the gas. Heat is not something that is contained in a system. Rather, heat is a measure of the energy that is transferred from one system to another because of a difference in temperature. Work is a measure of the energy that is transferred from one system to another because the point of contact of a force exerted by one system on the other undergoes a displacement with a component that is parallel to the force.

For very small amounts of heat absorbed, work done, or changes in internal energy, it is customary to write Equation 18-10 as

$$dE_{int} = dQ_{in} + dW_{on} \qquad 18\text{-}11$$

In this equation, dE_{int} is the differential of the internal-energy function. However, neither dQ_{in} nor dW_{on} is a differential of any function. Instead, dQ_{in} merely represents a small amount of energy transferred to or from the system by heating or cooling, and dW_{on} represents a small amount energy transferred to or from the system by work being done on or by the system.

> ! If the gas is then returned to its original state (P_i, V_i), the temperature T and the internal energy E_{int} must equal their original values.

> ! It is correct to say that the internal energy of a system has increased, but it is not correct to say either that the work of a system has increased or that the heat of a system has increased.

18-4 THE INTERNAL ENERGY OF AN IDEAL GAS

The translational kinetic energy K of the molecules in an *ideal* gas is related to the absolute temperature T by Equation 17-20:

$$K = \tfrac{3}{2}nRT$$

where n is the number of moles of gas and R is the universal gas constant. If the internal energy of a gas is just this translational kinetic energy, then $E_{int} = K$, and

$$E_{int} = \tfrac{3}{2}nRT \qquad 18\text{-}12$$

Then, the internal energy will depend only on the temperature of the gas, and not on its volume or pressure. If the molecules have other types of energy in addition to translational kinetic energy, such as rotational energy, the internal energy will be greater than that given by Equation 18-12. But according to the equipartition theorem (Chapter 17, Section 4), the average energy associated with any degree of freedom will be $\frac{1}{2}RT$ per mole ($\frac{1}{2}kT$ per molecule), so again, the internal energy will depend only on the temperature and not on the volume or pressure.

We can imagine that the internal energy of a *real* gas might include other kinds of energy, which depend on the pressure and volume of the gas. Suppose, for example, that nearby gas molecules exert attractive forces on each other. Work is then required to increase the separation of the molecules. Then, if the average distance between the molecules is increased, the potential energy associated with the molecular attraction will increase. The internal energy of the gas will then depend on the volume of the gas as well as on its temperature.

Joule, using an apparatus like the one shown in Figure 18-5, performed a simple but interesting experiment to determine whether or not the internal energy of a gas depends on its volume. The compartment on the left in Figure 18-5 initially contains a gas and the compartment on the right has been evacuated. A stopcock that is

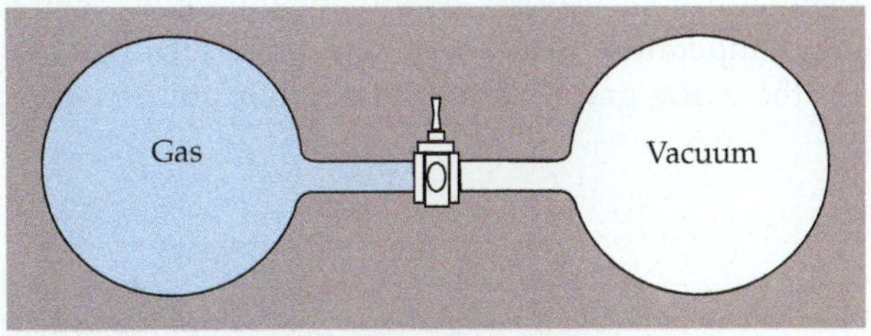

FIGURE 18-5 Free expansion of a gas. When the stopcock on the gas is opened, the gas expands rapidly into the evacuated chamber. Because no work is done on the gas and the whole system is thermally insulated, the initial and final internal energies of the gas are equal.

initially closed connects the two compartments. The whole system is thermally insulated from its surroundings by rigid walls so that no energy can be transferred into or out of the system by heating or cooling *and* no energy can be transferred by work being done on or by the gas. When the stopcock is opened, the gas rushes into the evacuated chamber. This process is called a **free expansion.** The gas eventually reaches thermal equilibrium with itself. Because no work has been done on the gas and no heat has been transferred to it, the final internal energy of the gas must equal its initial internal energy. If the gas molecules exert attractive forces on one another, the potential energy associated with these forces will increase as the volume increases. Because energy is conserved, the kinetic energy of translation will therefore decrease, which will result in a decrease in the temperature of the gas.

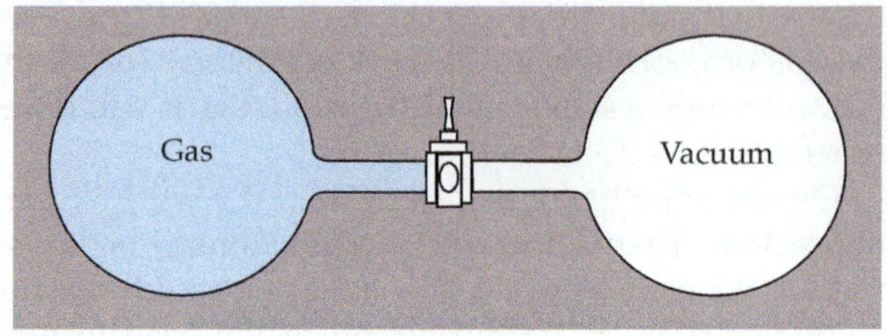

FIGURE 18-5 *(repeated)* Free expansion of a gas. When the stopcock on the gas is opened, the gas expands rapidly into the evacuated chamber. Because no work is done on the gas and the whole system is thermally insulated, the initial and final internal energies of the as are equal.

When Joule performed this experiment, he found the final temperature to be equal to the initial temperature. Subsequent experiments verified this result for gases at low densities. This result implies that for a gas at low density—that is, for an ideal gas—the temperature depends only on the internal energy, or as we usually think of it, the internal energy depends only on the temperature. However, if the experiment is done with a large amount of gas initially in the left compartment so that the initial density is high, then the temperature after expansion is slightly cooler than the temperature before the expansion. This result indicates that a small attraction exists between the gas molecules of a gas.

18-5 WORK AND THE DIAGRAM FOR A GAS

In many types of engines, a gas does work expanding against a moving piston. For example, in a steam engine, water is heated in a boiler to produce steam. The steam then does work as it expands and drives a piston. In an automobile engine, a mixture of gasoline vapor and air is ignited, causing it to combust. The resulting high temperatures and pressures cause the gas to expand rapidly, driving a piston and doing work. In this section, we see how we can mathematically describe the work done by an expanding gas.

QUASI-STATIC PROCESSES

Figure 18-6 shows an ideal gas confined in a container that has a tightly fitting piston that we assume to be frictionless. If the piston moves, the volume of the gas changes. The temperature or pressure or both must also change because these three variables are related by the equation of state $PV = nRT$. If we suddenly push in the piston to compress the gas, the pressure will initially be greater near the piston than far from it. The gas eventually will settle down to a new equilibrium pressure and temperature. We cannot determine such macroscopic variables as T, P, or E_{int} for the entire gas system until equilibrium is restored in the gas. However, if we

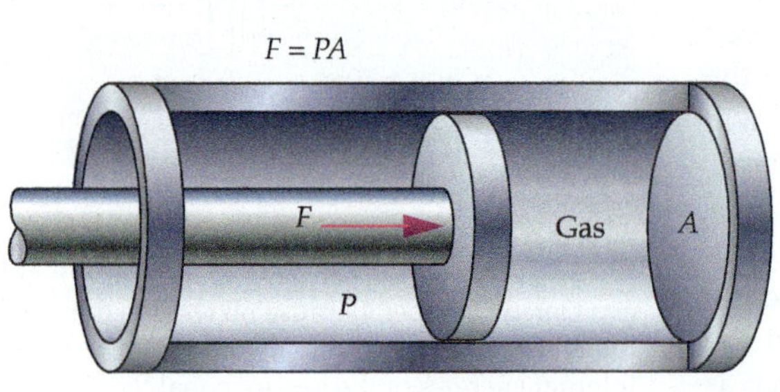

FIGURE 18-6 Gas confined in a thermally insulated cylinder with a movable piston. If the piston moves a distance dx, the volume of the gas changes by $dV = A\,dx$. The work done by the gas is $PA\,dx = P\,dV$, where P is the pressure.

move the piston slowly in small steps and allow equilibrium to be reestablished after each step, we can compress or expand the gas in such a way that the gas is never far from an equilibrium state. During this kind of process, called a **quasi-static process,** the gas moves through a series of equilibrium states. In practice, it is possible to approximate quasi-static processes fairly well.

Let us begin with a gas at a high pressure, and let it expand quasi-statically. The magnitude of the force F exerted by the gas on the piston is PA, where A is the area of the piston and P is the gas pressure. As the piston moves a small distance dx, the work done *by* the gas *on* the piston is

$$dW_{\text{by gas}} = F_x\,dx = PA\,dx = P\,dV \qquad \text{18-13}$$

where $dV = A\,dx$ is the increase in the volume of the gas. During the expansion the piston exerts a force of magnitude PA on the gas, but opposite in direction to the force of the gas on the piston. Thus, work done *by* the piston *on* the gas is just the negative of the work done *by* the gas

$$dW_{\text{on gas}} = -dW_{\text{by gas}} = -P\,dV \qquad \text{18-14}$$

Note that for an expansion, dV is positive, the gas does work on the piston, so $dW_{\text{on gas}}$ is negative, and for a compression, dV is negative, work is done on the gas, so $dW_{\text{on gas}}$ is positive.

The work done on the gas during an expansion or a compression from a volume of V_i to a volume of V_f is

$$W_{\text{on gas}} = -\int_{V_i}^{V_f} P\,dV \qquad \text{18-15}$$

WORK DONE ON A GAS

To calculate this work, we need to know how the pressure varies during the expansion or compression. The various possibilities can be illustrated most easily using a PV diagram.

PV DIAGRAMS

We can represent the states of a gas on a diagram of P versus V. Because by specifying both P and V we specify the state of the gas, each point on the PV diagram indicates a particular state of the gas. Figure 18-7 shows a PV diagram with a directed horizontal line representing a series of states that all have the same value of P. This line represents a *compression* at constant pressure. Such a process is called an **isobaric compression.** For a volume change of ΔV (ΔV is negative for a compression), we have

$$W_{\text{on}} = -\int_{V_i}^{V_f} P\,dV = -P\int_{V_i}^{V_f} dV = -P\,\Delta V = |P\,\Delta V|$$

which is equal to the shaded area under the curve (directed line) in the figure. For a compression, the work done on the gas is equal to the area under the P-versus-V curve. (For an expansion the work done on the gas is equal to the negative of the area under the P-versus-V curve.) Because pressures are often given in atmospheres and volumes are often given in liters, it is convenient to have a conversion factor between liter-atmospheres and joules:

$$1\text{ L}\cdot\text{atm} = (10^{-3}\text{ m}^3)(101.325 \times 10^3\text{ N/m}^2) \approx 101.3\text{ J} \qquad \text{18-16}$$

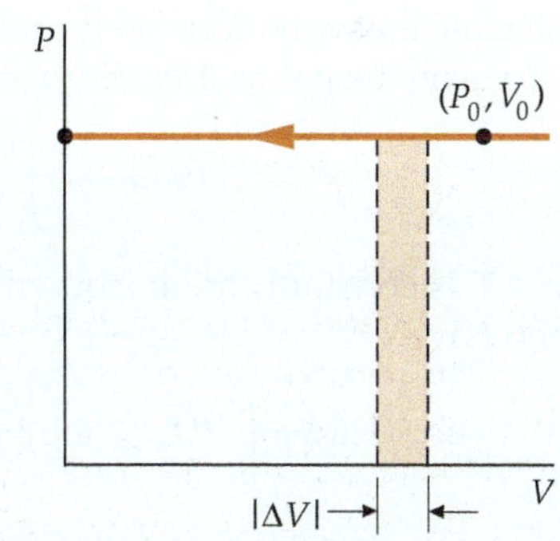

FIGURE 18-7 Each point on a PV diagram, such as (P_0, V_0), represents a particular state of the gas. The horizontal line represents states with a constant pressure P_0. The shaded area, $P_0|\Delta V|$, represents the work done on the gas as it is compressed an amount $|\Delta V|$.

PRACTICE PROBLEM 18-4

If 5.00 L of an ideal gas at a pressure of 2.00 atm is cooled so that it contracts at constant pressure until its volume is 3.00 L, what is the work done on the gas?

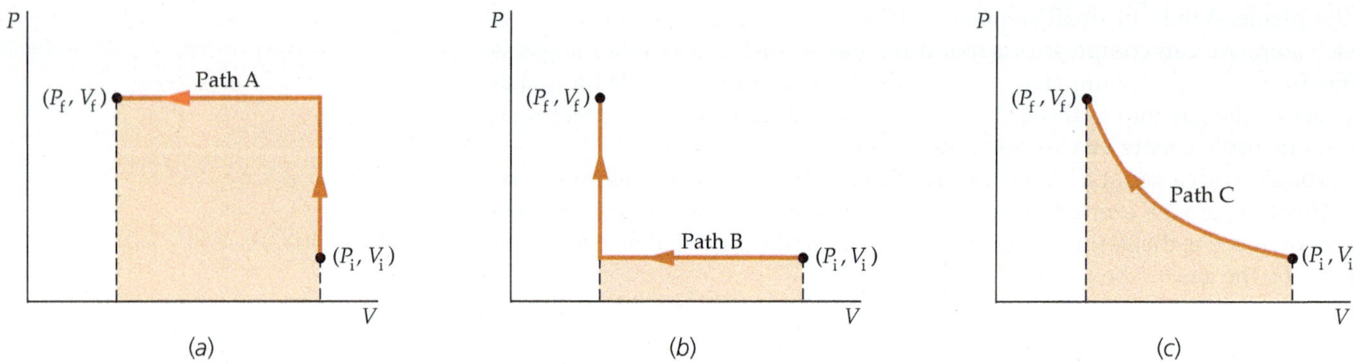

FIGURE 18-8 Three paths on *PV* diagrams connecting an initial state (P_i, V_i) and a final state (P_f, V_f). The corresponding shaded area indicates the work done on the gas along each path.

Figure 18-8 shows three different possible paths on a *PV* diagram for a gas that is initially in state (P_i, V_i) and is finally in state (P_f, V_f). We assume that the gas is ideal and have chosen the original and final states to have the same temperature so that $P_iV_i = P_fV_f = nRT$. Because the internal energy depends only on the temperature, the initial and final internal energies are the same also.

In Figure 18-8*a*, the gas is heated **isometrically** (at constant volume)* until its pressure is P_f, after which it is cooled isobarically (at constant pressure) until its volume is V_f. The work done on the gas along the constant-volume (vertical) part of path A is zero; along the constant pressure (horizontal) part of the path A, it is

$$P_f|V_f - V_i| = -P_f(V_f - V_i).$$

In Figure 18-8*b*, the gas is first cooled at constant pressure until its volume is V_f, after which it is heated at constant volume until its pressure is P_f. The work done on the gas along this path is $P_i|V_f - V_i| = -P_i(V_f - V_i)$, which is much less than that done along the path shown in Figure 18-8*a*, as can be seen by comparing the shaded regions in Figure 18-8*a* and Figure 18-8*b*.

In Figure 18-8*c*, path C represents an **isothermal** compression, meaning that the temperature remains constant. (Keeping the temperature constant during the compression requires that energy be transferred out of the gas via heat during the compression.) We can calculate the work done on the gas along path C by using $P = nRT/V$. Hence, the work done on the gas as it is compressed from V_i to V_f along path C is

$$W_{on} = -\int_{V_i}^{V_f} P\,dV = -\int_{V_i}^{V_f} \frac{nRT}{V}\,dV$$

See **Math Tutorial** *for more information on* **Logarithms**

Because *T* is constant for an isothermal process, we can factor it from the integral. We then have

$$W_{isothermal} = -nRT\int_{V_i}^{V_f} \frac{dV}{V} = -nRT\ln\frac{V_f}{V_i} = nRT\ln\frac{V_i}{V_f} \qquad 18\text{-}17$$

WORK DONE ON GAS DURING ISOTHERMAL COMPRESSION

We see that the amount of work done on the gas is different for each process illustrated. The change in internal energy of the gas depends on the initial and final states of the gas, but does not depend on the path taken. The change in internal energy equals the work done on the gas plus the heat transfer into the gas. Thus, we can see that because the work done is different for each process illustrated, the amount of heat transfer also must be different for each process. This discussion illustrates the fact that both the work done and the amount of heat transfer depend only on how a system transitions from one state to another, but the change in the internal energy of the system does not.

* Constant-volume processes are also called *isochoric processes or isovolumetric processes.*

PROBLEM-SOLVING STRATEGY

Calculating Work Done by an Ideal Gas During a Constrained Quasi-Static Process

PICTURE The increment of work done by a gas is equal to the pressure times the increment of volume. That is, $dW_{\text{by}} = P\,dV$. It follows that $W_{\text{by}} = \int_{V_i}^{V_f} P\,dV$. The constraint dictates how to evaluate this integral.

SOLVE

1. If the volume V is constant, then dV equals zero and $W_{\text{by}} = 0$.
2. If the pressure P is constant, then $W_{\text{by}} = P\int_{V_i}^{V_f} dV = P(V_f - V_i)$.
3. If the temperature T is constant, then $P = nRT/V$ and

$$W_{\text{by}} = \int_{V_i}^{V_f} \frac{nRT}{V}dV = nRT\int_{V_i}^{V_f}\frac{dV}{V} = nRT\ln\frac{V_f}{V_i}$$

4. If no energy is transferred to or from the gas via heat, then see Section 18-9.

CHECK If the volume increases then the W_{by} must be positive, and vice versa.

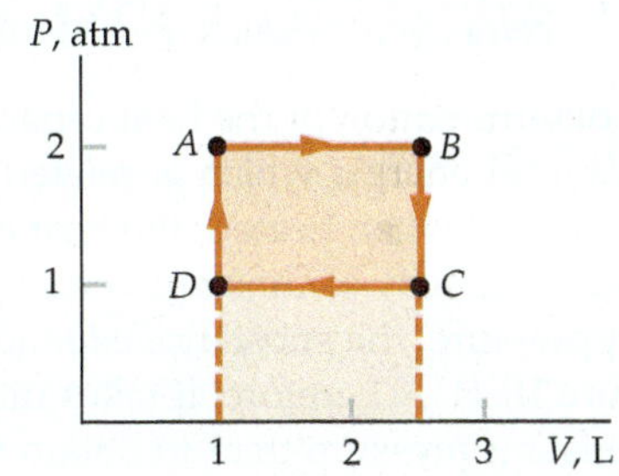

FIGURE 18-9

Example 18-7 Work Done on an Ideal Gas

An ideal gas undergoes a cyclic process from point A to point B to point C to point D and back to point A, as shown in Figure 18-9. The gas begins at a volume of 1.00 L and a pressure of 2.00 atm and expands at constant pressure until the volume is 2.50 L, after which it is cooled at constant volume until its pressure is 1.00 atm. It is then compressed at constant pressure until its volume is again 1.00 L, after which it is heated at constant volume until it is back in its original state. Find the total work done on the gas and the total amount of heat transfer into it during the cycle.

PICTURE We calculate the work done during each step. Because $\Delta E_{\text{int}} = 0$ for any complete cycle, the first law of thermodynamics implies that the total amount of heat transfer into the gas plus the total work done on the gas equals zero.

SOLVE

1. From point A to point B the process is a constant pressure expansion, so the work done on the gas has a negative value. The work done on the gas equals the negative of the shaded area under the AB curve, shown in Figure 18-10*a*:

$$W_{AB} = -P\,\Delta V = -P(V_B - V_A)$$
$$= -(2.00\text{ atm})(2.50\text{ L} - 1.00\text{ L})$$
$$= -3.00\text{ L}\cdot\text{atm}$$

2. Convert the units to joules:

$$W_{AB} = -3.00\text{ L}\cdot\text{atm}\times\frac{101.3\text{ J}}{1\text{ L}\cdot\text{atm}} = -304\text{ J}$$

3. From point B to point C (Figure 18-10) the gas cools at constant volume, so the work done is zero:

$$W_{BC} = 0$$

4. As the gas undergoes a constant pressure compression from point C to point D, the work done on it has a positive value. This work equals the area under the CD curve, shown in Figure 18-10*b*:

$$W_{CD} = -P\,\Delta V = -P(V_D - V_C)$$
$$= -(1.00\text{ atm})(1.00\text{ L} - 2.50\text{ L})$$
$$= 1.50\text{ L}\cdot\text{atm} = 152\text{ J}$$

5. As the gas is heated back to its original state A, the volume is again constant (Figure 18-10), so no work is done:

$$W_{DA} = 0$$

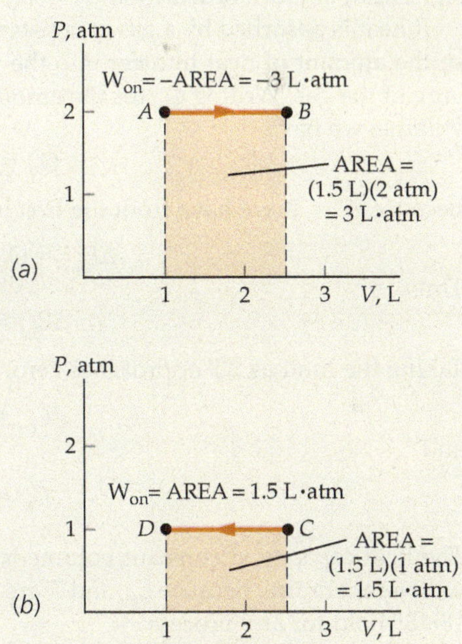

FIGURE 18-10 (*a*) The work done on the gas during the expansion from A to B is equal to the negative of the area under the curve. (*b*) The work done on the gas during the compression from C to D is equal to the area under the curve.

6. The total work done by the gas is the sum of the work done along each step:

$$W_{\text{total}} = W_{AB} + W_{BC} + W_{CD} + W_{DA}$$
$$= (-304\ \text{J}) + 0 + 152\ \text{J} + 0 = \boxed{-152\ \text{J}}$$

7. Because the gas is back in its original state, the total change in internal energy is zero:

$$\Delta E_{\text{int}} = 0$$

8. The amount of heat transfer into the gas is found from the first law of thermodynamics:

$$\Delta E_{\text{int}} = Q_{\text{in}} + W_{\text{on}}$$
$$\text{so}\quad Q_{\text{in}} = \Delta E_{\text{int}} - W_{\text{on}} = 0 - (-152\ \text{J}) = \boxed{152\ \text{J}}$$

CHECK We expect the net energy transferred to the gas to be zero for a cyclic process, which is the case in this example because the work done on the gas is −152 J and the amount of heat transfer into the gas is +152 J.

TAKING IT FURTHER The work done by the gas equals the negative of the work done on the gas, so the total work done by the gas during the cycle is +152 J. During the cycle, the gas absorbs 152 J of heat from its surroundings and does 152 J of work on its surroundings. This process leaves the gas in its initial state. The total work done by the gas equals the area enclosed by the cycle in Figure 18-9. Such cyclic processes have important applications for heat engines, as we will see in Chapter 19.

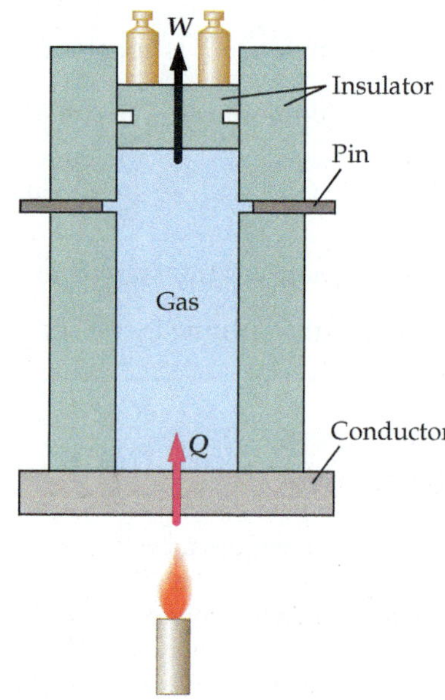

FIGURE 18-11 Heat is absorbed and the pressure remains constant. The gas expands, thus doing work on the piston.

18-6 HEAT CAPACITIES OF GASES

The determination of the heat capacity of a substance provides information about its internal energy, which is related to its molecular structure. For all substances that expand when heated, the heat capacity at constant pressure C_P is greater than the heat capacity at constant volume C_V. If heat is absorbed by a substance at constant pressure, the substance expands and does positive work on its surroundings (Figure 18-11). Therefore, it takes more heat to obtain a given temperature change at constant pressure than to obtain the same temperature change when heated at constant volume. The expansion is usually negligible for solids and liquids, so for them $C_P \approx C_V$. But a gas heated at constant pressure readily expands and does a significant amount of work, so $C_P - C_V$ is not negligible.

If heat is absorbed by a gas at constant volume, no work is done (Figure 18-12), so the amount of heat transfer into the gas equals the increase in the internal energy of the gas. Writing Q_V for the amount of heat transfer into the gas at constant volume, we have

$$Q_V = C_V\,\Delta T$$

Because $W = 0$, we have from the first law of thermodynamics,

$$\Delta E_{\text{int}} = Q_V + W = Q_V$$

Thus,

$$\Delta E_{\text{int}} = C_V\,\Delta T$$

Taking the limit as ΔT approaches zero, we obtain

$$dE_{\text{int}} = C_V\,dT \qquad \text{18-18}a$$

and

$$C_V = \frac{dE_{\text{int}}}{dT} \qquad \text{18-18}b$$

The heat capacity at constant volume is the rate of change of the internal energy with temperature. Because E_{int} and T are both state functions, Equations 18-18*a* and 18-18*b* hold for any process.

Now let us calculate the difference $C_P - C_V$ for an ideal gas. From the definition of C_P, the amount of heat transfer into the gas at constant pressure is

$$Q_P = C_P\,\Delta T$$

From the first law of thermodynamics,

$$\Delta E_{\text{int}} = Q_P + W_{\text{on}} = Q_P - P\,\Delta V$$

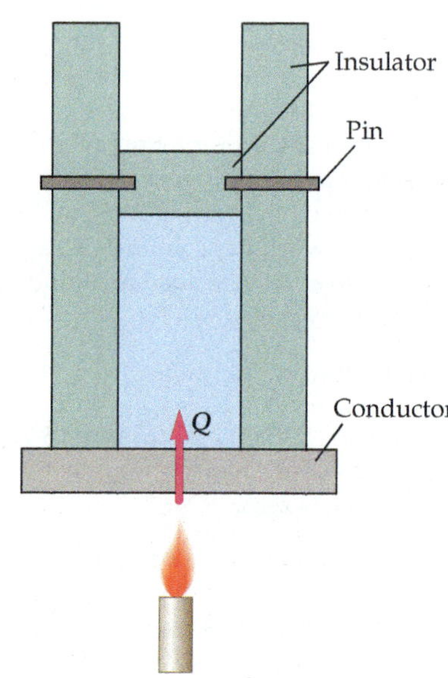

FIGURE 18-12 The piston is held in place by pins. Heat is absorbed at constant volume, so no work is done and all the heat is transferred into the internal energy of the gas.

Then

$$\Delta E_{int} = C_P \Delta T - P \Delta V \qquad \text{or} \qquad C_P \Delta T = \Delta E_{int} + P \Delta V$$

For infinitesimal changes, this becomes

$$C_P \, dT = dE_{int} + P \, dV$$

Using Equation 18-18*a* for dE_{int}, we obtain

$$C_P \, dT = C_V \, dT + P \, dV \qquad \text{18-19}$$

The pressure, volume, and temperature for an ideal gas are related by

$$PV = nRT$$

Taking the differentials of both sides of the ideal-gas law, we obtain

$$P \, dV + V \, dP = nR \, dT$$

For a constant-pressure process $dP = 0$, so

$$P \, dV = nR \, dT$$

Substituting $nR \, dT$ for $P \, dV$ in Equation 18-19 gives

$$C_P \, dT = C_V \, dT + nR \, dT = (C_V + nR) \, dT$$

Therefore,

$$C_P = C_V + nR \qquad \text{18-20}$$

which shows that, for an ideal gas, the heat capacity at constant pressure is greater than the heat capacity at constant volume by the amount nR.

Table 18-3 lists measured molar heat capacities c'_P and c'_V for several gases. Note from this table that the ideal-gas prediction, $c'_P - c'_V = R$, holds quite well for all gases. The table also shows that c'_V is approximately $1.5R$ for all monatomic gases, $2.5R$ for all diatomic gases, and greater than $2.5R$ for gases consisting of more complex molecules. We can understand these results by considering the molecular model of a gas (Chapter 17). The total translational kinetic energy of n moles of a gas is $K_{trans} = \frac{3}{2}nRT$ (Equation 17-20). Thus, if the internal energy of a gas consists of translational kinetic energy only, we have

$$E_{int} = \frac{3}{2}nRT \qquad \text{18-21}$$

Table 18-3 Molar Heat Capacities in of Various Gases at

Gas	c'_P	c'_V	c'_V/R	$c'_P - c'_V$	$(c'_P - c'_V)/R$
Monatomic					
He	20.79	12.52	1.51	8.27	0.99
Ne	20.79	12.68	1.52	8.11	0.98
Ar	20.79	12.45	1.50	8.34	1.00
Kr	20.79	12.45	1.50	8.34	1.00
Xe	20.79	12.52	1.51	8.27	0.99
Diatomic					
N_2	29.12	20.80	2.50	8.32	1.00
H_2	28.82	20.44	2.46	8.38	1.01
O_2	29.37	20.98	2.52	8.39	1.01
CO	29.04	20.74	2.49	8.30	1.00
Polyatomic					
CO_2	36.62	28.17	3.39	8.45	1.02
N_2O	36.90	28.39	3.41	8.51	1.02
H_2S	36.12	27.36	3.29	8.76	1.05

The heat capacities are then

$$C_V = \frac{dE_{int}}{dT} = \tfrac{3}{2}nR \qquad 18\text{-}22$$

C_V FOR AN IDEAL MONATOMIC GAS

and

$$C_P = C_V + nR = \tfrac{5}{2}nR \qquad 18\text{-}23$$

C_P FOR AN IDEAL MONATOMIC GAS

The results in Table 18-3 agree well with these predictions for monatomic gases, but for other gases, the heat capacities are greater than those predicted by Equations 18-22 and 18-23. The internal energy for a gas consisting of diatomic or more complicated molecules is evidently greater than $\frac{3}{2}nRT$. The reason is that such molecules can have other types of energy, such as rotational or vibrational energy, in addition to translational kinetic energy.

Example 18-8 Heating, Cooling, and Compressing an Ideal Gas

A system consisting of 0.32 mol of a monatomic ideal gas, with $c'_V = \frac{3}{2}RT$, occupies a volume of 2.2 L at a pressure of 2.4 atm, as represented by point A in Figure 18-13. The system is carried through a cycle consisting of three processes:

1. The gas is heated at constant pressure until its volume is 4.4 L at point B.
2. The gas is cooled at constant volume until the pressure decreases to 1.2 atm (point C).
3. The gas undergoes an isothermal compression back to point A.

(*a*) What is the temperature at points A, B, and C? (*b*) Find W, Q, and ΔE_{int} for each process and for the entire cycle.

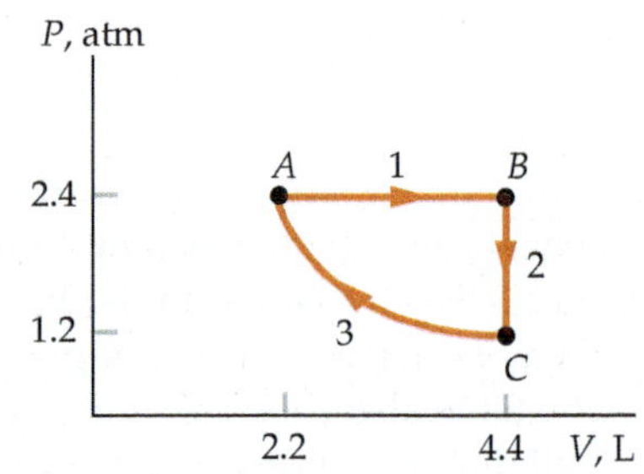

FIGURE 18-13 The total work done *on* the gas during one cycle is the negative of the area enclosed by the curves. The total work done *by* the gas during one cycle is the area enclosed by the curves.

PICTURE You can find the temperatures at all points from the ideal-gas law. You can find the work for each process by finding the area under the curve, and the heat transferred by using the given heat capacity and the initial and final temperatures for each process. In process 3, T is constant, so $\Delta E_{int} = 0$ and the heat absorbed by the gas plus the work done on the gas equals zero.

SOLVE

(*a*) Find the temperatures at points A, B, and C using the ideal-gas law:

$$T_C = T_A = \frac{P_A V_A}{nR} = \frac{(2.4\ \text{atm})(2.2\ \text{L})}{(0.32\ \text{mol})(0.08206\ \text{L}\cdot\text{atm}/(\text{mol}\cdot\text{K}))}$$

$$= 201\ \text{K} = \boxed{2.0 \times 10^2\ \text{K}}$$

$$T_B = \frac{P_B V_B}{nR} = \frac{P_A 2V_A}{nR} = 2\frac{P_A V_A}{nR} = 2T_A = 402\ \text{K} = \boxed{4.0 \times 10^2\ \text{K}}$$

(*b*) 1. For process 1, use $W_1 = -P_c\,\Delta V$ to calculate the work, and $C_P = \frac{5}{2}nR$ to calculate the heat Q_1. Then use W_1 and Q_1 to calculate $\Delta E_{int\,1}$:

$$W_1 = -P_A\,\Delta V = -P_A(V_B - V_A) = -(2.4\ \text{atm})(2.2\ \text{L})$$

$$= -5.28\ \text{L}\cdot\text{atm}\left(\frac{101.3\ \text{J}}{1\ \text{L}\cdot\text{atm}}\right) = -534.9\ \text{J} = \boxed{-0.53\ \text{kJ}}$$

$$Q_1 = C_P\,\Delta T = \tfrac{5}{2}nR\,\Delta T = \tfrac{5}{2}(0.32\ \text{mol})(8.314\ \text{J}/(\text{mol}\cdot\text{K}))(201\ \text{K})$$

$$= 1337\ \text{J} = \boxed{1.3\ \text{kJ}}$$

$$\Delta E_{int\,1} = Q_1 + W_1 = 1337\ \text{J} - 534.9\ \text{J} = 802\ \text{J} = \boxed{0.80\ \text{kJ}}$$

2. For process 2, use $C_V = \frac{3}{2}nR$ and $T_C - T_B$ from step 1 to find Q_2:

$$W_2 = \boxed{0}$$

Then, because $W_2 = 0$, $\Delta E_{int\,2} = Q_2$:

$$Q_2 = C_V\,\Delta T = \tfrac{3}{2}nR\,\Delta T = \tfrac{3}{2}(0.32\ \text{mol})[8.314\ \text{J}/(\text{mol}\cdot\text{K})](-201\ \text{K})$$

$$= -802\ \text{J} = \boxed{-0.80\ \text{kJ}}$$

$$\Delta E_{int\,2} = W_2 + Q_2 = 0 + (-802\ \text{J}) = \boxed{-0.80\ \text{kJ}}$$

3. Calculate W_3 from $W = -nRT\ln(V_A/V_C)$ (Equation 18-17) in the isothermal compression. Then, because $\Delta E_{\text{int }3} = 0$, $Q_3 = -W_3$:

$$W_3 = nRT_A \ln\frac{V_A}{V_C} = (0.32\text{ mol})[8.314\text{ J/(mol}\cdot\text{K)}](-201\text{ K})\ln 2.0$$
$$= 371\text{ J} = \boxed{0.37\text{ kJ}}$$
$$\Delta E_{\text{int }3} = \boxed{0}$$
$$Q_3 = \Delta E_{\text{int }3} - W_3 = -371\text{ J} = \boxed{-0.37\text{ kJ}}$$

4. Find the total work W, the total heat Q, and the total change ΔE_{int} by summing the quantities found in steps 2, 3, and 4:

$$W_{\text{total}} = W_1 + W_2 + W_3 = (-535\text{ J}) + 0 + 371\text{ J} = \boxed{-0.16\text{ kJ}}$$
$$Q_{\text{total}} = Q_1 + Q_2 + Q_3 = 1337\text{ J} + (-802\text{ J}) + (-371\text{ J})$$
$$= \boxed{0.16\text{ kJ}}$$
$$\Delta E_{\text{int total}} = \Delta E_{\text{int }1} + \Delta E_{\text{int }2} + \Delta E_{\text{int }3}$$
$$= 802\text{ J} + (-802\text{ J}) + 0 = \boxed{0.00\text{ kJ}}$$

CHECK The total change in internal energy is zero, as it must be for a cyclic process. The total work done on the gas plus the total heat absorbed by the gas equals zero.

TAKING IT FURTHER The total work done on the gas equals the area under the *CA* curve minus the area under the *AB* curve, which equals the negative of the area enclosed by the three curves in Figure 18-13.

HEAT CAPACITIES AND THE EQUIPARTITION THEOREM

According to the equipartition theorem stated in Section 4 of Chapter 17, the internal energy of n moles of a gas should equal $\frac{1}{2}nRT$ for each degree of freedom of the gas molecules. The heat capacity at constant volume of a gas should then be $\frac{1}{2}nR$ times the number of degrees of freedom of the molecules. From Table 18-2, nitrogen, oxygen, hydrogen, and carbon monoxide all have molar heat capacities at constant volume of about $\frac{5}{2}R$. Thus, the molecules in each of these gases have five degrees of freedom. About 1880, Rudolf Clausius speculated that these gases must consist of diatomic molecules that can rotate about two axes, giving them two additional degrees of freedom (Figure 18-14). The two degrees of freedom besides the three for translation are now known to be associated with their rotation about each of the two axes, x' and y', perpendicular to the line joining the atoms. The kinetic energy of a diatomic molecule is therefore

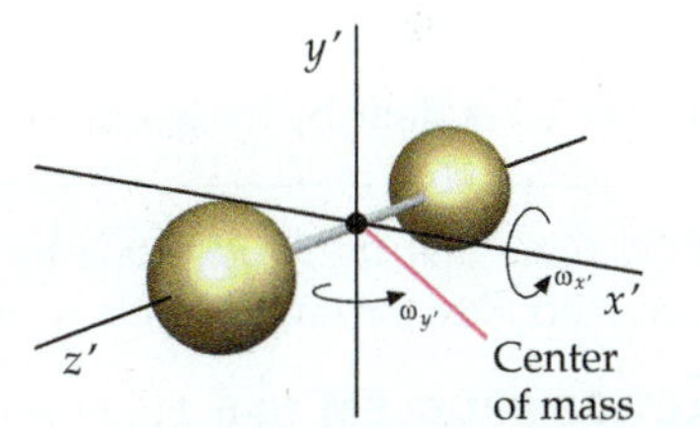

FIGURE 18-14 Rigid-dumbbell model of a diatomic molecule.

$$K = \tfrac{1}{2}mv_x^2 + \tfrac{1}{2}mv_y^2 + \tfrac{1}{2}mv_z^2 + \tfrac{1}{2}I_{x'}\omega_{x'}^2 + \tfrac{1}{2}I_{y'}\omega_{y'}^2$$

The total internal energy of n moles of such a gas is then

$$E_{\text{int}} = 5 \times \tfrac{1}{2}nRT = \tfrac{5}{2}nRT \qquad 18\text{-}24$$

and the heat capacity at constant volume is

$$C_V = \tfrac{5}{2}nR \qquad 18\text{-}25$$

Apparently, diatomic gases do not rotate about the line joining the two atoms—if they did, there would be six degrees of freedom and C_V would be $\frac{6}{2}nR = 3nR$, which is contrary to experimental results. Furthermore, monatomic gases do not rotate at all. We will see in Section 18-8 that these puzzling facts are easily explained when we take the quantization of angular momentum into account.

Example 18-9 Heating a Diatomic Ideal Gas

A sample consisting of 2.00 mol of oxygen gas at an initial pressure of 1.00 atm is heated from a temperature of 20.0°C to a temperature of 100.0°C. Assume that the sample can be modeled as an ideal gas. (*a*) What amount of heat transfer into the sample is required if the volume is kept constant during the heating? (*b*) What amount of heat transfer into the sample is required if the pressure is kept constant? (*c*) How much work does the gas do in Part (*b*)?

PICTURE The amount of heat transfer needed for constant-volume heating is $Q_V = C_V\,\Delta T$, where $C_V = \frac{5}{2}nR$ (because oxygen is a diatomic gas). For constant-pressure heating, $Q_P = C_P\,\Delta T$, where $C_P = C_V + nR$. Finally, the amount of work done by the gas equals the negative of the work done on the gas, which can be found from $\Delta E_{int} = Q_{in} + W_{on}$. (Alternatively, $W_{by} = P\,\Delta V$.)

SOLVE

(*a*) 1. Write the amount of heat transfer needed for constant volume in terms of C_V and ΔT:

$$Q_V = C_V\,\Delta T$$

2. Calculate the amount of heat transfer needed for $\Delta T = 80°\text{C} = 80\text{ K}$:

$$Q_V = C_V\,\Delta T = \tfrac{5}{2}nR\,\Delta T = \tfrac{5}{2}(2.00\text{ mol})[8.314\text{ J/(mol}\cdot\text{K)}](80.0\text{ K}) = \boxed{3.33\text{ kJ}}$$

(*b*) 1. Write the amount of heat transfer needed for constant pressure in terms of C_P and ΔT:

$$Q_P = C_P\,\Delta T$$

2. Calculate the heat capacity at constant pressure:

$$C_P = C_V + nR = \tfrac{5}{2}nR + nR = \tfrac{7}{2}nR$$

3. Calculate the amount of heat transfer needed at constant pressure for $\Delta T = 80\text{ K}$:

$$Q_P = C_P\,\Delta T = \tfrac{7}{2}(2.00\text{ mol})[8.314\text{ J/(mol}\cdot\text{K)}](80.0\text{ K}) = \boxed{4.66\text{ kJ}}$$

(*c*) 1. The work W_{on} can be found from the first law of thermodynamics:

$$\Delta E_{int} = Q_{in} + W_{on} \quad \text{so} \quad W_{on} = \Delta E_{int} - Q_{in}$$

2. The internal energy change equals the heat transferred at constant volume, which was calculated in Part (*a*):

$$\Delta E_{int} = Q_V = C_V\,\Delta T = \tfrac{5}{2}nR\,\Delta T$$

$$\text{and} \quad Q_P = C_P\,\Delta T = \tfrac{7}{2}nR\,\Delta T$$

$$\text{so} \quad W_{on} = \Delta E_{int} - Q_P = \tfrac{5}{2}nR\,\Delta T - \tfrac{7}{2}nR\,\Delta T = -nR\,\Delta T$$

$$= -(2.00\text{ mol})[8.314\text{ J/(mol}\cdot\text{K)}](80.0\text{ K}) = -1.33\text{ kJ}$$

3. The work done by the gas at constant pressure is then:

$$W_{by} = -W_{on} = \boxed{1.33\text{ kJ}}$$

CHECK Note that the work done by the gas in Part (*c*) has a positive value. This is as expected because the gas expands when heated at constant pressure.

PRACTICE PROBLEM 18-5 Find the initial and final volumes of this gas from the ideal-gas law, and use them to calculate the work done by the gas if the heat is added at constant pressure using $W_{by} = P\,\Delta V$.

Example 18-10 Vibrational Modes of Carbon Dioxide *Conceptual*

The carbon dioxide molecule consists of a carbon atom directly between two oxygen atoms. This molecule has three distinct modes of vibration. Sketch these modes in a reference frame where the center of mass of the molecule is at rest.

PICTURE If the molecule were not vibrating, the centers of the atoms would lie in a straight line. When vibrating the atoms can move both parallel and perpendicular to the line through their centers. There are two stretching modes in which the atoms move parallel to the line through their centers, and one bending mode where they move perpendicular to the line through their centers.

SOLVE

1. In the symmetric stretch mode (Figure 18-15a) the carbon atom remains stationary and the oxygen atoms oscillate 180° out of phase with each other. Can you see why this mode is sometimes referred to as the breathing mode?
2. In the asymmetric stretch mode (Figure 18-15*b*) the two oxygen atoms vibrate in phase with each other, but 180° out of phase with the motion of the carbon atom:
3. In the bending mode (Figure 18-15*c*) the two oxygen atoms vibrate in phase with each other, but 180° out of phase with the motion of the carbon atom:

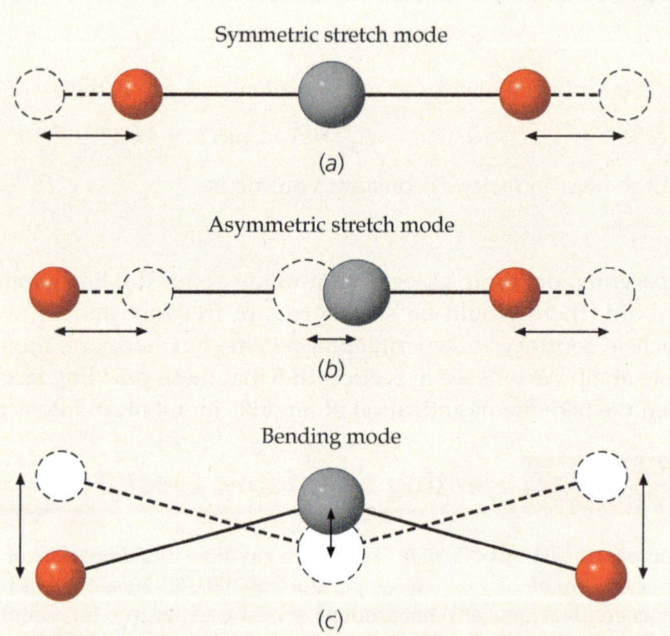

FIGURE 18-15

18-7 HEAT CAPACITIES OF SOLIDS

In Section 18-1, we noted that all of the metals listed in Table 18-1 have approximately equal molar specific heats. Most solids have molar heat capacities approximately equal to $3R$:

$$c' = 3R = 24.9\ \text{J/mol}\cdot\text{K} \qquad 18\text{-}26$$

This result is known as the **Dulong–Petit law.** We can understand this law by applying the equipartition theorem to the simple model for a solid shown in Figure 18-16. According to this model, a solid consists of a regular array of atoms in which each of the atoms has a fixed equilibrium position and is connected by springs to its neighbors. Each atom can vibrate in the x, y, and z directions. The total energy of an atom in a solid is

$$E = \tfrac{1}{2}mv_x^2 + \tfrac{1}{2}mv_y^2 + \tfrac{1}{2}mv_z^2 + \tfrac{1}{2}k_{\text{eff}}x^2 + \tfrac{1}{2}k_{\text{eff}}y^2 + \tfrac{1}{2}k_{\text{eff}}z^2$$

where k_{eff} is the effective force constant of the hypothetical springs. Each atom thus has six degrees of freedom. The equipartition theorem states that a substance in equilibrium has an average energy of $\frac{1}{2}RT$ per mole for each degree of freedom. Thus, the internal energy of a mole of a solid is

$$E_{\text{int m}} = 6 \times \tfrac{1}{2}RT = 3RT \qquad 18\text{-}27$$

which means that c' is equal to $3R$.

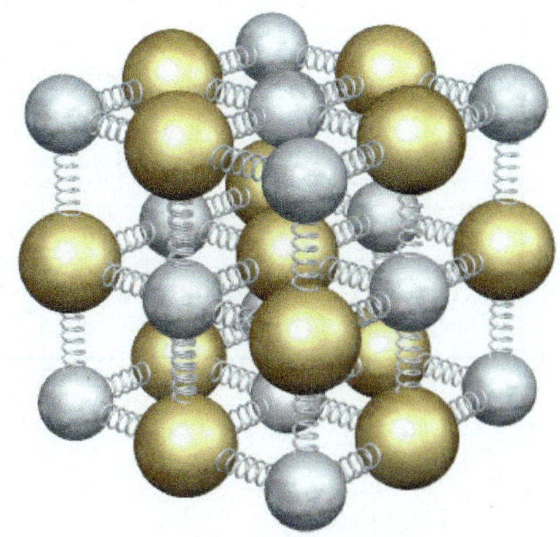

FIGURE 18-16 Model of a solid in which the atoms are connected to each other by springs. The internal energy of the molecule consists of the kinetic and potential energies of vibration.

Example 18-11 Using the Dulong–Petit Law

The molar mass of copper is 63.5 g/mol. Use the Dulong–Petit law to calculate the specific heat of copper.

PICTURE The Dulong–Petit law gives the molar specific heat of a solid, c'. The specific heat is then $c = c'/M$ (Equation 18-6), where M is the molar mass.

SOLVE

1. The Dulong–Petit law gives c' in terms of R: $\quad c' = 3R$

2. Using $M = 63.5$ g/mol for copper, the specific heat is: $\quad c = \dfrac{c'}{M} = \dfrac{3R}{M} = \dfrac{3(8.314\ \text{J/(mol}\cdot\text{K))}}{63.5\ \text{g/mol}}$

$$= 0.392\ \text{J/(g}\cdot\text{K)} = \boxed{0.392\ \text{kJ/(kg}\cdot\text{K)}}$$

CHECK This result differs from the measured value of 0.386 kJ/kg·K, given in Table 18-1, by less than 2 percent.

PRACTICE PROBLEM 18-6 The specific heat of a certain metal is measured to be 1.02 kJ/kg·K. (*a*) Calculate the molar mass of this metal, assuming that the metal obeys the Dulong–Petit law. (*b*) What is the metal?

18-8 FAILURE OF THE EQUIPARTITION THEOREM

Although the equipartition theorem had spectacular successes in explaining the heat capacities of gases and solids, it had equally spectacular failures. For example, if a diatomic gas molecule like the one in Figure 18-14 rotates about the line joining the atoms, there should be an additional degree of freedom. Similarly, if a diatomic molecule is not rigid, the two atoms should vibrate along the line joining them. We would then have two more degrees of freedom corresponding to kinetic and potential energies of vibration. But according to the measured values of the molar

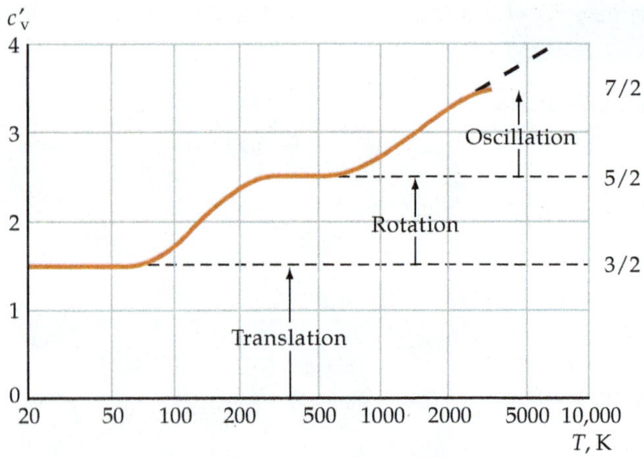

FIGURE 18-17 Temperature dependence of the molar heat capacity of H_2. (The curve is qualitative in those regions where c'_V is changing.) Ninety-five percent of H_2 molecules are dissociated into atomic hydrogen at 5000 K.

heat capacities in Table 18-3, diatomic gases apparently do not rotate about the line joining them, nor do they vibrate. The equipartition theorem does not give explanations for this consequence, or for the fact that monatomic atoms do not rotate about any of the three possible perpendicular axes in space. Furthermore, heat capacities are found to depend on temperature, contrary to the predictions of the equipartition theorem. The most spectacular case of the temperature dependence of heat capacity is that of H_2, as shown in Figure 18-17. At temperatures below 70 K, c'_V for H_2 is $\frac{3}{2}R$, which is the same as that for a gas of molecules that translate, but do not rotate or vibrate. At temperatures between 250 K and 700 K, $c'_V = \frac{5}{2}R$, which is that for molecules that translate and rotate but do not vibrate. And at temperatures above 700 K, the H_2 molecules begin to vibrate. However, the molecules dissociate before c'_V reaches $\frac{7}{2}R$. Finally, the equipartition theorem predicts a constant value of $3R$ for the heat capacity of solids. While this result holds for many, although not all, solids at high temperatures, it does not hold at very low temperatures.

The equipartition theorem fails because the energy is **quantized.** That is, a molecule can have only certain values of internal energy, as illustrated schematically by the energy-level diagram in Figure 18-18. The molecule can gain or lose energy only if the gain or loss takes it to another allowed level. For example, the energy that can be transferred between colliding gas molecules is of the order of kT, the typical thermal energy of a molecule. The validity of the equipartition theorem depends on the relative size of kT and the spacing of the allowed energy levels.

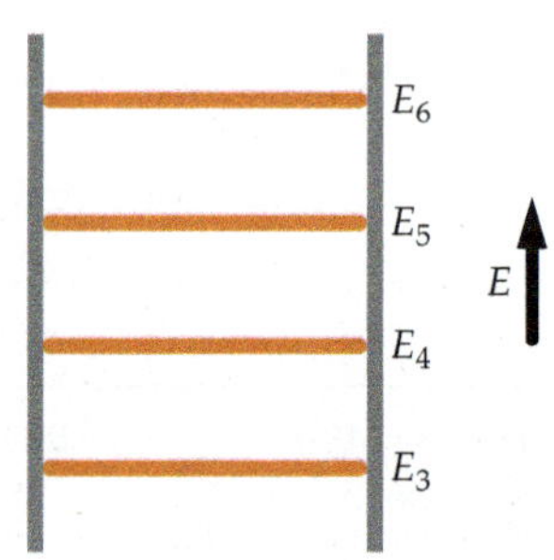

FIGURE 18-18 Energy-level diagram. A bound system can have only certain discrete energies.

> If the spacing of the allowed energy levels is large compared with kT, energy cannot be transferred by collisions and the classic equipartition theorem is not valid. If the spacing of the levels is much smaller than kT, energy quantization will not be noticed and the equipartition theorem will hold.
>
> CONDITIONS FOR THE VALIDITY OF THE EQUIPARTITION THEOREM

Consider the rotation of a molecule. The energy of rotation is

$$E = \frac{1}{2}I\omega^2 = \frac{(I\omega)^2}{2I} = \frac{L^2}{2I} \qquad \text{18-28}$$

where I is the moment of inertia of the molecule, ω is its angular velocity, and $L = I\omega$ is its angular momentum. In Section 10-5, we mentioned that angular momentum is quantized, and its magnitude is restricted to

$$L = \sqrt{\ell(\ell + 1)}\hbar \qquad \ell = 0, 1, 2, \ldots \qquad \text{18-29}$$

where $\hbar = h/(2\pi)$, and h is Planck's constant. The energy of a rotating molecule is therefore quantized to the values

$$E = \frac{L^2}{2I} = \frac{\ell(\ell + 1)\hbar^2}{2I} = \ell(\ell + 1)\, E_{0r} \qquad 18\text{-}30$$

where

$$E_{0r} = \frac{\hbar^2}{2I} \qquad 18\text{-}31$$

is characteristic of the energy gap between levels. If this energy is much less than kT, we expect classical physics and the equipartition theorem to hold. Let us define a critical temperature T_c by

$$kT_c = E_{0r} = \frac{\hbar^2}{2I} \qquad 18\text{-}32$$

If T is much greater than this critical temperature, then kT will be much greater than the spacing of the energy levels, which is of the order of kT_c, and we expect classical physics and the equipartition theorem to be valid. If T is less than or of the order of T_c, then kT will not be much greater than the energy-level spacing, and we expect classical physics and the equipartition theorem to break down. Let us estimate T_c for some cases of interest.

1. *Rotation of* H_2 *about an axis perpendicular to the line joining the H atoms and through the center of mass* (Figure 18-19): The moment of inertia of H_2 about the axis is

$$I_H = 2M_H\left(\frac{r_s}{2}\right)^2 = \frac{1}{2}M_H r_s^2$$

where M_H is the mass of an H atom and r_s is the separation distance. For hydrogen, $M_H = 1.67 \times 10^{-27}$ kg, and $r_s \approx 8 \times 10^{-11}$ m. The critical temperature is then

$$T_c = \frac{\hbar^2}{2kI} = \frac{\hbar^2}{kM_H r_s^2}$$
$$= \frac{(1.05 \times 10^{-34}\ \text{J}\cdot\text{s})^2}{(1.38 \times 10^{-23}\ \text{J/K})(1.67 \times 10^{-27}\ \text{kg})(8 \times 10^{-11}\ \text{m})^2} \approx 75\ \text{K}$$

As we see from Figure 18-17, this is approximately the temperature below which the rotational energy does not contribute to the heat capacity.

2. O_2: Because the mass of O_2 is about 16 times that of H_2, and the separation is about the same, the critical temperature for O_2 should be about (75/16) ≈ 4.6 K. For all temperatures for which O_2 exists as a gas, $T \gg T_c$, so kT is much greater than the energy level spacing. Consequently, we expect the equipartition theorem of classical physics to apply.

3. *Rotation of a monatomic gas*: Consider the He atom that has a nucleus that is composed of two protons and two neutrons and has two electrons. The mass of an electron is about 8000 times smaller than the mass of the He nucleus, but the radius of the nucleus is about 100,000 times smaller than the distance from the nucleus to an electron. Therefore, the moment of inertia of the He atom is almost entirely due to its two electrons. The distance from the He nucleus to one of its electrons is about half the separation distance of the H atoms in H_2, and the electron mass is about 2000 times smaller than that of the H nucleus. Thus, using $m_e = M_H/2000$ and $r = r_s/2$, we find the moment of inertia of the two electrons in He to be roughly

$$I_{He} = 2m_e r^2 \approx 2\frac{M_H}{2000}\left(\frac{r_s}{2}\right)^2 = \frac{I_H}{2000}$$

The critical temperature for He is thus about 2000 times that of H_2 or about 150,000 K. This is much higher than the dissociation temperature (the temperature at which electrons are stripped from their nuclei) for helium. So, the gap between

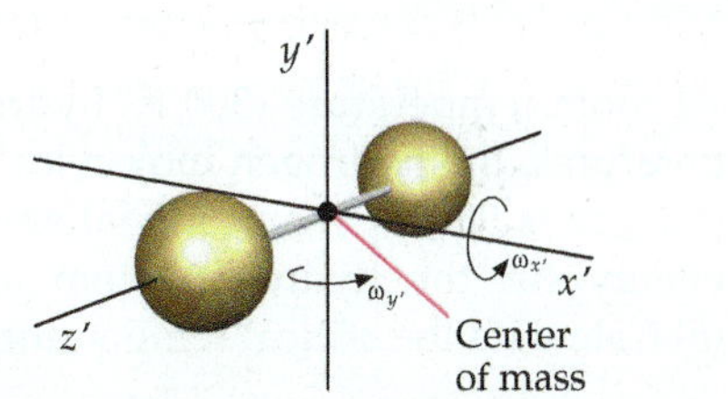

FIGURE 18-19 Rigid-dumbbell model of a diatomic molecule.

allowed energy levels is always much greater than kT, and the He molecules cannot be induced to rotate by collisions occurring in the gas. Other monatomic gases have slightly greater moments of inertia because they have more electrons, but their critical temperatures are still tens of thousands of kelvins. Therefore, their molecules also cannot be induced to rotate by collisions occurring in the gas.

4. *Rotation of a diatomic gas about the axis joining the atoms*: We see from our discussion of monatomic gases that the moment of inertia for a diatomic gas molecule about this axis will also be due mainly to the electrons and will be of the same order of magnitude as for a monatomic gas. Again, the critical temperature, T_c, calculated in order for this rotation to occur due to collisions between molecules in the gas, exceeds the gas's dissociation temperature, making rotation under those circumstances impossible.

It is interesting to note that the successes of the equipartition theorem in explaining the measured heat capacities of gases and solids led to the first real understanding of molecular structure in the nineteenth century, whereas its failures played an important role in the development of quantum mechanics in the twentieth century.

Example 18-12 Rotational Energy of the Hydrogen Atom

At room temperature (300 K) hydrogen gas is diatomic. However, at higher temperatures, the hydrogen molecules dissociate. At a temperature of 8000 K, hydrogen gas is 99.99 percent monatomic. (*a*) Estimate the lowest (nonzero) rotational energy for the hydrogen atom and compare it to kT at room temperature. (*b*) Calculate the critical temperature T_c for a gas of atomic hydrogen.

PICTURE From Equation 18-30, the lowest rotational energy is for $\ell = 1$. We use Equation 18-30 to determine the energy in terms of the moment of inertia. We can neglect the moment of inertia of the nucleus because its radius is 100,000 times smaller than the radius of the atom. Therefore, the moment of inertia for the atom is essentially the moment of inertia of the electron about the nucleus. Then $I = m_e r^2$, where $r = 5.29 \times 10^{-11}$ m is the distance from the nucleus to the electron.

SOLVE

(*a*) 1. The lowest energy greater than zero occurs for $\ell = 1$:

$$E_\ell = \frac{\ell(\ell + 1)\hbar^2}{2I} \qquad \ell = 0, 1, 2, \ldots$$

$$\text{so}\quad E_1 = \frac{1(1+1)\hbar^2}{2m_e r^2} = \frac{\hbar^2}{m_e r^2}$$

2. The numerical values are:

$\hbar = 1.05 \times 10^{-34}\ \text{J}\cdot\text{s}$
$m_e = 9.11 \times 10^{-31}$ kg
$r = 5.29 \times 10^{-11}$ m

3. Substitute the numerical values:

$$E_1 = \frac{\hbar^2}{m_e r^2} = \boxed{4.32 \times 10^{-18}\ \text{J}}$$

4. The value of kT at $T = 300$ K is

$$kT = (1.38 \times 10^{-23}\ \text{J/K})(300\ \text{K}) = 4.14 \times 10^{-21}\ \text{J}$$

5. Compare E_1 and kT:

$$\frac{E_1}{kT} = \frac{4.32 \times 10^{-18}\ \text{J}}{4.14 \times 10^{-21}\ \text{J}} \approx 10^3$$

$\boxed{E_1 \text{ is about three orders of magnitude larger than } kT.}$

(*b*) Set $kT_c = E_1$ and solve for T_c:

$$kT_c = E_1$$

$$T_c = \frac{E_1}{k} = \frac{4.32 \times 10^{-18}\ \text{J}}{1.38 \times 10^{-23}\ \text{J/K}} = \boxed{3.13 \times 10^5\ \text{K}}$$

CHECK The critical temperature of a hydrogen atom (~3×10^5 K) is so high that the atom would be ionized well before the critical temperature could be reached. This "explains" why no rotational degrees of freedom contribute to the heat capacity of hydrogen atoms.

18-9 THE QUASI-STATIC ADIABATIC COMPRESSION OF A GAS

A process in which no heat transfers into or out of a system is called an **adiabatic process.** Such a process occurs either when the system is extremely well insulated or when the process happens very quickly. Consider the quasi-static adiabatic compression of a gas in which the gas in a thermally insulated container is slowly compressed by a piston, which is thereby doing work on the gas. Because no heat is transferred to or from the gas, the work done on the gas equals the increase in the internal energy of the gas, and the temperature of the gas increases. The curve representing this process on a PV diagram is shown in Figure 18-20.

We can find the equation for the adiabatic curve for an ideal gas by using the equation of state ($PV = nRT$) and the first law of thermodynamics ($dE_{int} = dQ_m + dW_{on}$). The first law of thermodynamics gives

$$C_V dT = 0 + (-P\,dV) \qquad \text{18-33}$$

where we have used $dE_{int} = C_V dT$ (Equation 18-18a), $dQ_m = 0$ (the process is adiabatic), and $dW_{on} = -P\,dV$ (Equation 18-15). Then, substituting for P using $P = nRT/V$, we obtain

$$C_V dT = -nRT\frac{dV}{V}$$

Separating variables by dividing both sides by TC_V, we obtain

$$\frac{dT}{T} + \frac{nR}{C_V}\frac{dV}{V} = 0$$

Integration gives

$$\ln T + \frac{nR}{C_V}\ln V = \text{constant}$$

Simplifying,

$$\ln T + \frac{nR}{C_V}\ln V = \ln T + \ln V^{nR/C_V} = \ln(TV^{nR/C_V}) = \text{constant}$$

Thus,

$$TV^{nR/C_V} = \text{constant} \qquad \text{18-34}$$

where the constants in the two preceding equations are not the same constant. Equation 18-34 can be rewritten by noting that $C_P - C_V = nR$, so

$$\frac{nR}{C_V} = \frac{C_P - C_V}{C_V} = \frac{C_P}{C_V} - 1 = \gamma - 1 \qquad \text{18-35}$$

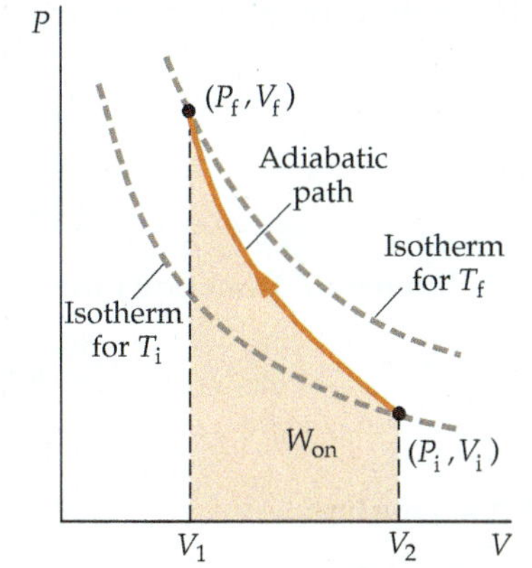

FIGURE 18-20 Quasi-static adiabatic compression of an ideal gas. The dashed lines are the isotherms for the initial and final temperatures. The curve connecting the initial and final states of the adiabatic compression is steeper than the isotherms because the temperature increases during the compression.

Clouds form if rising moist air cools due to adiabatic expansion of the air. Cooling causes water vapor to condense into liquid droplets. *(Will and Deni McIntyre/Photo Researchers.)*

where γ is the ratio of the heat capacities:

$$\gamma = \frac{C_P}{C_V} \qquad \text{18-36}$$

Therefore,

$$TV^{\gamma-1} = \text{constant} \qquad \text{18-37}$$

We can eliminate T from Equation 18-37 using $PV = nRT$. We then have

$$\frac{PV}{nR}V^{\gamma-1} = \text{constant}$$

or

$$PV^{\gamma} = \text{constant} \qquad \text{18-38}$$

QUASI-STATIC ADIABATIC PROCESS

Equation 18-38 relates P and V for adiabatic expansions and compressions. Solving $PV = nRT$ (the ideal gas equation) for V, then substituting the resulting expression for V into Equation 18-38, and then simplifying, gives

$$\frac{T^{\gamma}}{P^{\gamma-1}} = \text{constant} \qquad \text{18-39}$$

PRACTICE PROBLEM 18-7

Show that for quasi-static adiabatic process $T^{\gamma}/P^{\gamma-1} = \text{constant}$.

The work done on the gas in an adiabatic compression can be calculated from the first law of thermodynamics:

$$dE_{\text{int}} = dQ_{\text{in}} + dW_{\text{on}} \qquad \text{or} \qquad dW_{\text{on}} = dE_{\text{int}} - dQ_{\text{in}}$$

Because $dE_{\text{int}} = C_V dT$ and $dQ_{\text{in}} = 0$, we have

$$dW_{\text{on}} = C_V\, dT$$

Then

$$W_{\text{adiabatic}} = \int dW_{\text{on}} = \int C_V\, dT = C_V\, \Delta T \qquad \text{18-40}$$

ADIABATIC WORK ON IDEAL GAS

where we have assumed that C_V is constant.* We note that the work done on the gas depends only on the change in the temperature of the gas. During an adiabatic compression, work is done on the gas, and its internal energy and temperature increase. During a quasi-static adiabatic *expansion*, work is done *by* the gas, and the internal energy and temperature decrease.

We can use the ideal-gas law to write Equation 18-40 in terms of the initial and final values of the pressure and volume. If T_i is the initial temperature and T_f is the final temperature, we have for the work done

$$W_{\text{adiabatic}} = C_V\, \Delta T = C_V(T_f - T_i)$$

Using $PV = nRT$, we obtain

$$W_{\text{adiabatic}} = C_V\left(\frac{P_fV_f}{nR} - \frac{P_iV_i}{nR}\right) = \frac{C_V}{nR}(P_fV_f - P_iV_i)$$

Using Equation 18-35 to simplify this expression, we have

$$W_{\text{adiabatic}} = \frac{P_fV_f - P_iV_i}{\gamma - 1} \qquad \text{18-41}$$

* For an ideal gas, E_{int} is proportional to the absolute temperature, and therefore $C_v = dE_{\text{int}}/dT$ is a constant.

Example 18-13 Quasi-Static Adiabatic Compression of Air

A hand pump is used to inflate a bicycle tire to a gauge pressure of 482 kPa (about 70.0 lb/in.^{2}). (*a*) How much work must be done if each stroke of the pump is a quasi-static adiabatic process? Atmospheric pressure is 1.00 atm, the outdoor air temperature is 20°C, and the volume of the air in the tire remains constant at 1.00 L. (*b*) What is the pressure in the inflated tire after the pump is removed and the temperature of the air in the tire returns to 20°C?

PICTURE The work done is found from $\Delta E_{\text{int}} = Q_{\text{in}} + W_{\text{on}}$, with $Q_{\text{in}} = 0$. For an ideal gas, $\Delta E_{\text{int}} = C_V \Delta T$ (Equation 18-40). Because the process is both quasi-static and adiabatic, we know that $T^\gamma/P^{\gamma-1}$ = constant (Equation 18-39). (This relation yields the final temperature.) Find γ using $\gamma = C_P/C_V$, $C_P = C_V + nR$, and $C_V = \frac{5}{2}nR$ (Equations 18-36, 18-20, and 18-25). Let subscript 1 refer to initial values, and subscript 2 to final values. Then $P_1 = 1.00$ atm, $V_2 = 1.00$ L, and $T_1 = 20°\text{C} = 293$ K.

SOLVE

(*a*) 1. To find the work done, we apply the first law of thermodynamics. Because the compression is adiabatic, $Q_{\text{in}} = 0$:

$$\Delta E_{\text{int}} = Q_{\text{in}} + W_{\text{on}} = 0 + W_{\text{on}}$$

2. For an ideal gas, the change in internal energy is $C_V \Delta T$:

$$W = \Delta E_{\text{int}} = C_V \Delta T$$

3. For a diatomic gas, $C_V = \frac{5}{2}nR$:

$$W = C_V \Delta T = \tfrac{5}{2}nR\,\Delta T$$

4. The final temperature can be determined using $T^\gamma/P^{\gamma-1}$ = constant (Equation 18-39):

$$\frac{T_1^\gamma}{P_1^{\gamma-1}} = \frac{T_2^\gamma}{P_2^{\gamma-1}} \Rightarrow T_2 = \left(\frac{P_2}{P_1}\right)^{(\gamma-1)/\gamma} T_1$$

5. Find γ for a diatomic gas using Equations 18-36, 18-20 and 18-25:

$$\gamma = \frac{C_P}{C_V} = \frac{C_V + nR}{C_V} = 1 + \frac{nR}{C_V} = 1 + \frac{nR}{\frac{5}{2}nR} = \frac{7}{5} = 1.4$$

6. Solve for T_2. The given pressure is a gauge pressure, so add 1.00 atm = 101.3 kPa to the given pressure of 482 kPa:

$$T_2 = \left(\frac{P_2}{P_1}\right)^{(\gamma-1)/\gamma} T_1 = \left(\frac{583\text{ kPa}}{101.3\text{ kPa}}\right)^{0.4/1.4} 293\text{ K} = 483\text{ K}$$

$$P_2 = P_1\left(\frac{V_1}{V_2}\right)^\gamma = (1.00\text{ atm})\left(\frac{4.00\text{ L}}{2.00\text{ L}}\right)^{1.4} = \boxed{2.64\text{ atm}}$$

7. Calculate the work using the step-3 result. Use $PV = nRT$ (the ideal-gas law) to express nR in terms of P_2, V_2, and T_2:

$$W = \frac{5}{2}nR\,\Delta T = \frac{5}{2}\frac{P_2V_2}{T_2}(T_2 - T_1)$$

$$= \frac{5}{2}\frac{(583\text{ kPa})(1.00 \times 10^{-3}\text{ m}^3)}{483\text{ K}}(483\text{ K} - 293\text{ K}) = \boxed{634\text{ J}}$$

(*b*) The air in the tire cools at constant volume. Thus, $P_3/T_3 = P_2/T_3$, where P_3 and T_3 are the final pressure and temperature:

$$\frac{P_3}{T_3} = \frac{P_2}{T_2}, \quad \text{where} \quad T_3 = T_0 = 293\text{ K}$$

$$P_3 = \frac{T_3}{T_2}P_2 = \frac{293\text{ K}}{634\text{ K}}2.64\text{ atm} = \boxed{1.22\text{ atm}}$$

CHECK For the adiabatic compression the final temperature is higher than the initial temperature, as expected, and the work done on the gas is positive, as expected.

TAKING IT FURTHER (1) The work can also be calculated using $W_{\text{adiabatic}} = (P_fV_f - P_iV_i)/(\gamma - 1)$ (Equation 18-41), but using $W_{\text{adiabatic}} = C_V \Delta T$ is preferable because it is more directly connected to a principle (the first law of thermodynamics) and thus is easier to recall. (2) A real bicycle pump and tire are not insulated, so the process of pumping up the tire would not be even approximately adiabatic.

SPEED OF SOUND WAVES

We can use Equation 18-38 to calculate the adiabatic bulk modulus of an ideal gas, which is related to the speed of sound waves in air. We first compute the differential of both sides of PV^γ = constant (Equation 18-38):

$$P\,d(V^\gamma) + V^\gamma\,dP = 0$$

or

$$\gamma P V^{\gamma-1}\,dV + V^\gamma\,dP = 0$$

Then

$$dP = -\frac{\gamma P\,dV}{V}$$

Referring to Equation 13-6, the adiabatic bulk modulus* is then:

$$B_{\text{adiabatic}} = -\frac{dP}{dV/V} = \gamma P \qquad \text{18-42}$$

The speed of sound (Equation 15-4) is given by

$$v = \sqrt{\frac{B_{\text{adiabatic}}}{\rho}}$$

where the mass density ρ is related to the number of moles n and the molecular mass M by $\rho = m/V = nM/V$. Using the ideal-gas law, $PV = nRT$, we can eliminate V from the density

$$\rho = \frac{nM}{V} = \frac{nM}{nRT/P} = \frac{MP}{RT}$$

Using this result and γP for $B_{\text{adiabatic}}$, we obtain

$$v = \sqrt{\frac{B_{\text{adiabatic}}}{\rho}} = \sqrt{\frac{\gamma P}{MP/(RT)}} = \sqrt{\frac{\gamma RT}{M}}$$

which is Equation 15-5, the speed of sound in a gas.

* The bulk modulus, discussed in Chapter 13, is the negative ratio of the pressure change to the fractional change in volume, $B = -\Delta P/(\Delta V/V)$. The isothermal bulk modulus, which describes changes that occur at constant temperature, differs from the adiabatic bulk modulus, which describes changes with no heat transfer. For sound waves at audible frequencies, the pressure changes occur too rapidly for appreciable heat transfer, so the appropriate bulk modulus is the adiabatic bulk modulus.

Physics Spotlight

Respirometry: Breathing the Heat

Calorimetry, the study and measurement of heat transfer, helps to determine the full energy budget of systems. Wilber O. Atwater, the first director of experimental stations for the U.S. Department of Agriculture,* ambitiously decided to measure the energy budgets of people. This effort involved measuring and analyzing the food and water given to participants, measuring, analyzing, and burning the waste of the participants, and analyzing the temperature, chemistry, and moisture of a tiny room in which the participants lived.† This room was thermally insulated, and the interior was a copper box, which was lined with copper water tubes for careful measurement of heat released, and electric coils for temperature maintenance. Any change in temperature of the air in the room was due to energy coming from the people in the room. This energy was measured by the changes in temperature registered by the sensitive hanging thermometer within the box, and by changes in the temperature of the water pumped through the tubes lining the walls.‡

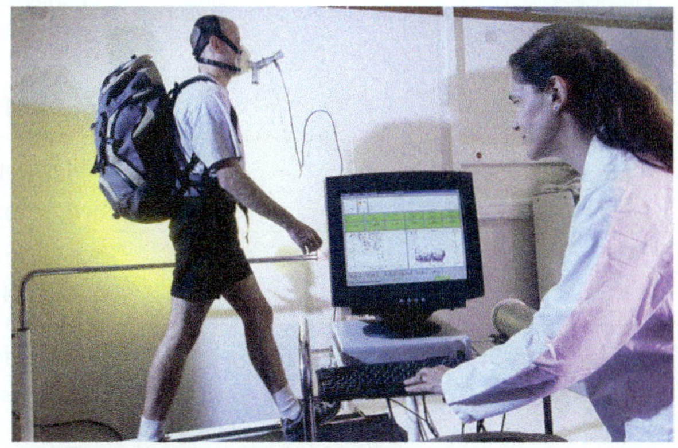

About 5 kcal of energy is expended for every 1 L of oxygen consumed. The oxygen consumption of this man is being monitored while he walks on a treadmill. *(Philippe Psaila/ Photo Researchers, Inc.)*

But although this copper room excelled at measuring the energy budgets of people at rest and in action, it was expensive and difficult to use. This led to indirect calorimetry by measuring respiration —*respirometry*. Further research showed that better than 95 percent# of human energy expenditure could reliably be calculated just by measuring the amounts of inhaled oxygen and exhaled carbon dioxide.° One figure that is frequently used today is 5 kcal/L of oxygen consumption.§ Depending on the measuring equipment, the volume of oxygen may be calculated from partial pressure of oxygen in the inhaled air, or it may be based on medically administered oxygen inhaled by the subject.

Respirometry is extremely useful, because it is the fastest measure of energy use by organisms. With appropriate modifications, respirometry is used for cattle,¶ poultry,** exotic animals,†† and even sewage sludge.‡‡ Recently, respirometry has been used to determine whether compost is mature enough to be added to soil. If the gas exchange rate of the compost is high, then bacterial activity is still high, and the compost is not fully mature.##

In medical care, respirometry is used to tailor nutritional therapy, especially for badly injured or very ill patients.°°, §§ In sports and fitness centers, handheld respirometers give rapid, accurate measurements of energy requirements for athletes and dieters¶¶ and are used to help patients, dieters, and athletes reach and maintain a healthy weight.

Finally, respirometry is used as a tool to help evaluate public policy and set nutrition standards. A study compared the calculations from two different nutrition standards to actual respirometry measurements of sedentary and active adults. One standard called for more energy than the participants actually used.*** As indirect calorimetry becomes less expensive, it is being used to help study the energy needs of people around the world.

* Swan, P., "100 Years Ago," *Nutrition Notes of the American Society for Nutritional Sciences*, June 2004, Vol. 40, No. 2, 4–5. http://www.asns.org/nnjun04a.pdf
† Atwater, W. O., *A Respiration Calorimeter with Appliances for the Direct Determination of Oxygen*. Washington, D. C.: Carnegie Institution, 1905.
‡ Morrison, P., and Morrison, P., "Laws of Calorie Counting," *Scientific American*, Aug. 2000, 93+.
# Ferrannini, E., "The Theoretical Bases of Indirect Calorimetry: A Review," *Metabolism*, Mar. 1988, Vol. 37, No. 3, 287–301.
° Mansell, P. I., and MacDonald, I. A., "Reappraisal of the Weir Equation for Calculation of Metabolic Rate," *AJP—Regulatory, Integrative and Comparative Physiology*, June 1990, Vol. 258, No. 6, R1347–R1354.
§ Food and Nutrition Board, *Dietary Reference Intakes for Energy, Carbohydrate, Fiber, Fat, Fatty Acids, Cholesterol, Protein, and Amino Acids*. Washington, D. C.: National Academies Press, 2005, 884.
¶ Mcleod, K., et al. "Effects of Brown Midrib Corn Silage on the Energy Balance of Dairy Cattle," *Journal of Dairy Science*, April 2000, Vol. 84, 885–895.
** "Animal Calorimetry," *Biomeasurements and Experimental Techniques for Avian Species*. http://web.uconn.edu/poultry/NE-127/NewFiles/Home2.html as of July 2006.
†† Schalkwyk, S. J., et al., "Gas Exchange of the Ostrich Embryo During Peak Metabolism in Relation to Incubator Design," *South African Journal of Animal Science*, 2002, Vol. 32, 122–129.
‡‡ Rai, C. L., et al., "Influence of Ultrasonic Disintegration on Sludge Growth Reduction and Its Estimation by Respirometry," *Environmental Science and Technology*, Nov. 2004, Vol. 38, No. 21, 5779–5785.
Seekings, B., "Field Test for Compost Maturity," *Biocycle*, July 1996, Vol. 37, No. 8, 72–75.
°° American Association for Respiratory Care, "Metabolic Measurement Using Indirect Calorimetry During Mechanical Ventilation—2004 Revision & Update," *Respiratory Care*, Sept. 2004, Vol. 49, No. 9, 1073–1079. http://www.guideline.gov/summary/summary.aspx?ss=15&doc_id=6515 as of July 2006.
§§ Steward, D., and Pridham, K., "Stability of Respiratory Quotient and Growth Outcomes of Very Low Birth Weight Infants," *Biological Research for Nursing*, Jan. 2001, Vol. 2, No. 3, 198–205.
¶¶ St-Onge, M., et al., "A New Hand-Held Indirect Calorimeter to Measure Postprandial Energy Expenditure," *Obesity Research*, April 2004, Vol. 12, No. 4, 704–709.
*** Alfonzo-González, G., et al., "Estimation of Daily Energy Needs with the FAO/WHO/UNU 1985 Procedures in Adults: Comparison to Whole-Body Indirect Calorimetry Measurements," *European Journal of Clinical Nurtrition*, Aug. 2004, Vol. 58, No. 8, 1125–1131.

Summary

1. The first law of thermodynamics, which is a statement of the conservation of energy, is a fundamental law of physics.
2. The equipartition theorem is a fundamental law of classical physics. It breaks down if the typical thermal energy kT is small compared to the spacing of quantized energy levels.

TOPIC	RELEVANT EQUATIONS AND REMARKS	
1. Heat	If energy is transferred from one system to another due to a temperature difference, the energy transferred is called heat.	
Calorie	The calorie, originally defined as the heat necessary to increase the temperature of 1 g of water by 1°C, is now defined to be exactly 4.184 joules.	
2. Heat Capacity	Heat capacity is the amount of heat needed to increase the temperature of a substance by one degree. $$C = \frac{Q}{\Delta T}$$	18-1
At constant volume	$$C_V = \frac{Q_V}{\Delta T}$$	
At constant pressure	$$C_P = \frac{Q_P}{\Delta T}$$	
Specific heat (heat capacity per unit mass)	$$c = \frac{C}{m}$$	18-2
Molar specific heat (heat capacity per mole)	$$c' = \frac{C}{n}$$	18-6
Heat capacity–internal energy relation	$$C_V = \frac{dE_{int}}{dT}$$	18-18*a*
Ideal gas	$$C_P - C_V = nR$$	18-20
Monatomic ideal gas	$$C_V = \tfrac{3}{2}nR$$	18-22
Diatomic ideal gas	$$C_V = \tfrac{5}{2}nR$$	18-25
3. Fusion and Vaporization	Both melting and vaporization occur at a constant temperature.	
Latent heat of fusion	The heat needed to melt a substance is the product of the mass of the substance and its latent heat of fusion L_f: $$Q_f = mL_f$$	18-8
L_f of water	$$L_f = 333.5\ \text{kJ/kg}$$	
Latent heat of vaporization	The heat needed to vaporize a liquid is the product of the mass of the liquid and its latent heat of vaporization, L_v: $$Q_v = mL_v$$	18-9
L_v of water	$$L_v = 2257\ \text{kJ/kg}$$	
4. First Law of Thermodynamics	The change in the internal energy of a system equals the energy transferred into the system via heat plus the energy transferred into the system via work: $$\Delta E_{int} = Q_{in} + W_{on}$$	18-10
5. Internal Energy E_{int}	The internal energy of a system is a property of the state of the system, as are the pressure, volume, and temperature. Heat and work are not properties of state.	
Ideal gas	E_{int} depends only on the temperature T.	

TOPIC	RELEVANT EQUATIONS AND REMARKS	
Monatomic ideal gas	$E_{int} = \frac{3}{2}nRT$	18-12
Internal energy related to heat capacity	$dE_{int} = C_V\,dT$	18-18*b*
6. Quasi-Static Process	A quasi-static process is one that occurs slowly so that the system moves through a series of equilibrium states.	
Isometric (isochoric)	$V = \text{constant}$	
Isobaric	$P = \text{constant}$	
Isothermal	$T = \text{constant}$	
Adiabatic	$Q = 0$	
Adiabatic, ideal gas	$TV^{\gamma-1} = \text{constant}$	18-37
	$PV^{\gamma} = \text{constant}$	18-38
	$T^{\gamma}/P^{\gamma-1} = \text{constant}$	18-39
	where $\gamma = C_P/C_V$	18-36
7. Work Done on a Gas	$W_{on} = -\int_{V_i}^{V_f} P\,dV = C_V\,\Delta T - Q_{in}$	18-10, 18-15, and 18-18
Isometric	$W_{on} = -\int_{V_i}^{V_f} P\,dV = 0 \qquad V_f = V_i$	
Isobaric	$W_{on} = -\int_{V_i}^{V_f} P\,dV = -P\int_{V_i}^{V_f} dV = -P\,\Delta V$	
Isothermal	$W_{isothermal} = -\int_{V_i}^{V_f} P\,dV = -nRT\int_{V_i}^{V_f}\frac{dV}{V} = nRT\ln\frac{V_i}{V_f}$	18-17
Adiabatic	$W_{adiabatic} = C_V\,\Delta T$	18-40
8. Equipartition Theorem	The equipartition theorem states that if a system is in equilibrium, there is an average energy of $\frac{1}{2}kT$ per molecule or $\frac{1}{2}RT$ per mole associated with each degree of freedom.	
Failure of the equipartition theorem	The equipartition theorem fails if the thermal energy ($\sim kT$) that can be transferred in collisions is smaller than the energy gap ΔE between quantized energy levels. For example, monatomic gas molecules cannot rotate, because the first nonzero energy permitted is much greater than kT.	
9. Dulong–Petit Law	The molar specific heat of most solids is $3R$. This is predicted by the equipartition theorem, assuming a solid atom has six degrees of freedom.	

Answer to Concept Check

18-1 A compressed spring in a dart gun is released, and its internal energy goes into the dart kinetic energy acquired by the dart. The compressed air in a tank is released, and the air is used to raise a car on a lift in a service station.

Answers to Practice Problems

18-1 30°C

18-2 500 kJ

18-3 20.5 kJ

18-4 405 J

18-5 $V_i = 48.0$ L, $V_f = 61.1$ L, $W = 13.1$ L·atm $= 1.33$ kJ

18-6 (*a*) $M = 24.4$ g/mol. (*b*) The metal must be magnesium, which has a molar mass of 24.3 g/mol

18-7 For a quasi-static process, $PV^{\gamma} = \text{constant}$. Solving the ideal-gas law for V gives $V = nRT/P$. Substituting nRT/P into the equation PV^{γ} gives $P(nRT/P)^{\gamma} = \text{constant}$. Rearranging gives $T^{\gamma}/P^{\gamma-1} = \text{constant}/(nR)$.

Problems

In a few problems, you are given more data than you actually need; in a few other problems, you are required to supply data from your general knowledge, outside sources, or informed estimate.

Interpret as significant all digits in numerical values that have trailing zeros and no decimal points.

Use 343 m/s as the speed of sound unless otherwise indicated.

- • Single-concept, single-step, relatively easy
- •• Intermediate-level, may require synthesis of concepts
- ••• Challenging
- SSM Solution is in the *Student Solutions Manual*
- Consecutive problems that are shaded are paired problems.

CONCEPTUAL PROBLEMS

1 • Object A has a mass that is twice the mass of object B, and object A has a specific heat that is twice the specific heat of object B. If equal amounts of heat are transferred to these objects, how do the subsequent changes in their temperatures compare? (*a*) $\Delta T_A = 4\Delta T_B$, (*b*) $\Delta T_A = 2\Delta T_B$, (*c*) $\Delta T_A = \Delta T_B$, (*d*) $\Delta T_A = \frac{1}{2}\Delta T_B$, (*e*) $\Delta T_A = \frac{1}{4}\Delta T_B$

2 • Object A has a mass that is twice the mass of object B. The temperature change of object A is equal to the temperature change of object B when the objects absorb equal amounts of heat. It follows that their specific heats are related by (*a*) $c_A = 2c_B$, (*b*) $2c_A = c_B$, (*c*) $c_A = c_B$, (*d*) none of the above.

3 • The specific heat of aluminum is more than twice the specific heat of copper. A block of copper and a block of aluminum have the same mass and temperature (20°C). The blocks are simultaneously dropped into a single calorimeter containing water at 40°C. Which statement is true when thermal equilibrium is reached? (*a*) The aluminum block is at a higher temperature than the copper block. (*b*) The aluminum block has absorbed less energy than the copper block. (*c*) The aluminum block has absorbed more energy than the copper block. (*d*) Both (*a*) and (*c*) are correct statements. SSM

4 • A block of copper is in a pot of boiling water and has a temperature of 100°C. The block is removed from the boiling water and immediately placed in an insulated container filled with a quantity of water that has a temperature of 20°C and the same mass as the block of copper. (The heat capacity of the insulated container is negligible.) The final temperature will be closest to (*a*) 40°C, (*b*) 60°C, (*c*) 80°C.

5 • You pour both a certain amount of water at 100°C and an equal amount of water at 20°C into an insulated container. The final temperature of the mixture will be (*a*) 60°C, (*b*) less than 60°C, (*c*) greater than 60°C.

6 • You pour some water at 100°C and some ice cubes at 0°C into an insulated container. The final temperature of the mixture will be (*a*) 50°C, (*b*) less than 50°C, but larger than 0°C, (*c*) 0°C, (*d*) You cannot tell the final temperature from the data given.

7 • You pour some water at 100°C and some ice cubes at 0°C into an insulated container. When thermal equilibrium is reached, you notice some ice remains and floats in liquid water. The final temperature of the mixture is (*a*) above 0°C, (*b*) less than 0°C, (*c*) 0°C, (*d*) You cannot tell the final temperature from the data given.

8 • Joule's experiment establishing the mechanical equivalence of heat involved the conversion of mechanical energy into internal energy. Give some everyday examples in which some of the internal energy of a system is converted into mechanical energy.

9 • Can a gas absorb heat while its internal energy does not change? If so, give an example. If not, explain why not.

10 • The equation $\Delta E_{int} = Q + W$ is the formal statement of the first law of thermodynamics. In this equation, the quantities Q and W, respectively, represent (*a*) the heat absorbed by the system and the work done by the system, (*b*) the heat absorbed by the system and the work done on the system, (*c*) the heat released by the system and the work done by the system, (*d*) the heat released by the system and the work done on the system.

11 • A real gas cools during a free expansion, while an ideal gas does not cool during a free expansion. Explain the reason for this difference. SSM

12 • An ideal gas that has a pressure of 1.0 atm and a temperature of 300 K is confined to half of an insulated container by a thin partition. The other half of the container is a vacuum. The partition is punctured and equilibrium is quickly reestablished. Which of the following is correct? (*a*) The gas pressure is 0.50 atm and the temperature of the gas is 150 K. (*b*) The gas pressure is 1.0 atm and the temperature of the gas is 150 K. (*c*) The gas pressure is 0.50 atm and the temperature of the gas is 300 K. (*d*) none of the above

13 • A gas consists of ions that repel each other. The gas undergoes a free expansion in which no heat is absorbed or released and no work is done. Does the temperature of the gas increase, decrease, or remain the same? Explain your answer.

14 • Two gas-filled rubber balloons that have equal volumes are located at the bottom of a dark, cold lake. The temperature of the water decreases with increasing depth. One balloon rises rapidly and expands adiabatically as it rises. The other balloon rises more slowly and expands isothermally. The pressure in each balloon remains equal to the pressure in the water just next to the balloon. Which balloon has the larger volume when it reaches the surface of the lake? Explain your answer.

15 • A gas changes its state quasi-statically from A to C along the paths shown in Figure 18-21. The work done by the gas is (*a*) greatest for path A → B → C, (*b*) least for path A → C, (*c*) greatest for path A → D → C, (*d*) the same for all three paths.

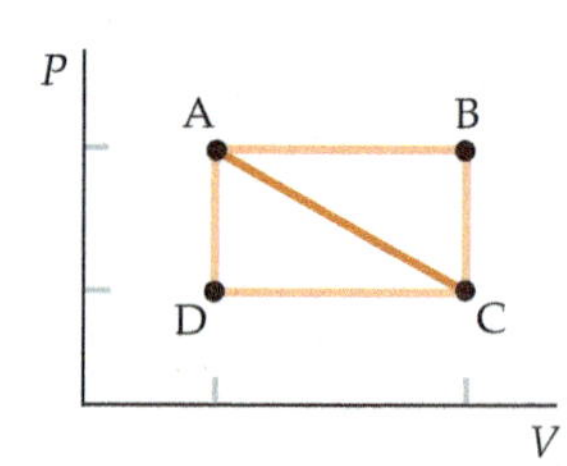

FIGURE 18-21 Problem 15

16 • When an ideal gas undergoes an adiabatic process, (*a*) no work is done by the system, (*b*) no heat is transferred to the system, (*c*) the internal energy of the system remains constant, (*d*) the amount of heat transfer into the system equals the amount of work done by the system.

17 • True or false:

(*a*) When a system can go from state 1 to state 2 by several different processes, the amount of heat absorbed by the system will be the same for all processes.
(*b*) When a system can go from state 1 to state 2 by several different processes, the amount of work done on the system will be the same for all processes.
(*c*) When a system goes from state 1 to state 2 by several different processes, the change in the internal energy of the system will be the same for all processes.
(*d*) The internal energy of a given amount of an ideal gas depends only on its absolute temperature.
(*e*) A quasi-static process is one in which the system is never far from being in equilibrium.
(*f*) For any substance that expands when heated, its C_P is greater than its C_V.

18 • The volume of a sample of gas remains constant while its pressure increases. (*a*) The internal energy of the system is unchanged. (*b*) The system does work. (*c*) The system absorbs no heat. (*d*) The change in internal energy must equal the heat absorbed by the system. (*e*) none of the above

19 •• When an ideal gas undergoes an isothermal process, (*a*) no work is done by the system, (*b*) no heat is absorbed by the system, (*c*) the heat absorbed by the system equals the change in the system's internal energy, (*d*) the heat absorbed by the system equals the work done by the system.

20 •• Consider the following series of sequential quasi-static processes that a system undergoes: (*1*) an adiabatic expansion, (*2*) an isothermal expansion, (*3*) an adiabatic compression, and (*4*) an isothermal compression that brings the system back to its original state. Sketch the series of processes on a *PV* diagram, and then sketch the series of processes on a *VT* diagram (in which volume is plotted as a function of temperature).

21 • An ideal gas in a cylinder is at pressure P and volume V. During a quasi-static adiabatic process, the gas is compressed until its volume has decreased to $V/2$. Then, in a quasi-static isothermal process, the gas is allowed to expand until its volume again has a value of V. What kind of process will return the system to its original state? Sketch the cycle on a graph. SSM

22 •• Metal A is denser than metal B. Which would you expect to have a higher heat capacity per unit mass—metal A or metal B? Why?

23 •• An ideal gas undergoes a process during which $P\sqrt{V}$ = constant and the volume of the gas decreases. Does its temperature increase, decrease, or remain the same during this process? Explain.

ESTIMATION AND APPROXIMATION

24 • **Engineering Application, Context-Rich** During the early stages of designing a modern electric generating plant, you are in charge of the team of environmental engineers. The new plant is to be located on the ocean and will use ocean water for cooling. The plant will produce electrical power at the rate of 1.00 GW. Because the plant will have an efficiency of one-third (typical of most modern plants), heat will be released to the cooling water at the rate of 2.00 GW. If environmental codes require that only water with a temperature increase of 15°F or less can be returned to the ocean, estimate the flow rate (in kg/s) of cooling water through the plant.

25 •• A typical microwave oven has a power consumption of about 1200 W. Estimate how long it should take to boil a cup of water in the microwave, assuming that 50 percent of the electrical power consumption goes into heating the water. How does this estimate correspond to everyday experience? SSM

26 •• A demonstration of the heating of a gas under adiabatic compression involves putting a small strip of paper into a large glass test tube, which is then sealed with a piston. If the piston compresses the trapped air very rapidly, the paper will catch fire. Assuming that the burning point of paper is 451°F, estimate the factor by which the volume of the air trapped by the piston must be reduced for this demonstration to work.

27 •• A small change in the volume of a liquid occurs when heating the liquid at constant pressure. Use the following data to estimate the fractional contribution this change makes to the heat capacity of water between 4.00°C and 100°C. The density of water at 4.00°C and 1.00 atm pressure is 1.000 g/cm^3. The density of liquid water at 100°C and 1.00 atm pressure is 0.9584 g/cm^3.

HEAT CAPACITY, SPECIFIC HEAT, LATENT HEAT

28 • **Engineering Application, Context-Rich** You designed a solar home that contains 1.00×10^5 kg of concrete (specific heat = 1.00 kJ/kg·K). How much heat is released by the concrete at night when it cools from 25.0°C to 20.0°C?

29 • How much heat must be absorbed by 60.0 g of ice at −10.0°C to transform it into 60.0 g of liquid water at 40.0°C? SSM

30 •• How much heat must be released by 0.100 kg of steam at 150°C to transform it into 0.100 kg of ice at 0.00°C?

31 •• A 50.0-g piece of aluminum at 20°C is cooled to −196°C by placing it in a large container of liquid nitrogen at that temperature. How much nitrogen is vaporized? (Assume that the specific heat of aluminum is constant over this temperature range.)

32 •• **Engineering Application, Context-Rich** You are supervising the creation of some lead castings for use in the construction industry. Each casting involves one of your workers pouring 0.500 kg of molten lead that has a temperature of 327°C into a cavity in a large block of ice at 0°C. How much liquid water should you plan on draining per hour if there are 100 workers who are able to each average one casting every 10.0 min?

CALORIMETRY

33 • **Engineering Application, Context-Rich** While spending the summer on your uncle's horse farm, you spend a week apprenticing with his farrier (a person who makes and fits horseshoes). You observe the way he cools a shoe after pounding the hot, pliable shoe into the correct size and shape. Suppose a 750-g iron horseshoe is taken from the farrier's fire, shaped, and at a temperature of 650°C, dropped into a 25.0-L bucket of water at 10.0°C. What is the final temperature of the water after the horseshoe and water arrive at equilibrium? Neglect any heating of the bucket and assume the specific heat of iron is 460 J/(kg·K). SSM

34 • The specific heat of a certain metal can be determined by measuring the temperature change that occurs when a piece of the metal is heated and then placed in an insulated container that is made of the same material and contains water. Suppose the piece of metal has a mass of 100 g and is initially at 100°C. The container has a mass of 200 g and contains 500 g of water at an initial temperature of 20.0°C. The final temperature is 21.4°C. What is the specific heat of the metal?

35 •• **BIOLOGICAL APPLICATION** During his many appearances at the Tour de France, champion bicyclist Lance Armstrong typically expended an average power of 400 W, 5.0 hours a day for 20 days. What quantity of water, initially at 24°C, could be brought to a boil if you could harness all of that energy?

36 •• A 25.0-g glass tumbler contains 200 mL of water at 24.0°C. If two 15.0-g ice cubes, each at a temperature of −3.00°C, are dropped into the tumbler, what is the final temperature of the drink? Neglect any heat transfer between the tumbler and the room.

37 •• A 200-g piece of ice at 0°C is placed in 500 g of water at 20°C. This system is in a container of negligible heat capacity and is insulated from its surroundings. (*a*) What is the final equilibrium temperature of the system? (*b*) How much of the ice melts? SSM

38 •• A 3.5-kg block of copper at a temperature of 80°C is dropped into a bucket containing a mixture of ice and water whose total mass is 1.2 kg. When thermal equilibrium is reached, the temperature of the water is 8.0°C. How much ice was in the bucket before the copper block was placed in it? (Assume that the heat capacity of the bucket is negligible.)

39 •• A well-insulated bucket of negligible heat capacity contains 150 g of ice at 0°C. (*a*) If 20 g of steam at 100°C is injected into the bucket, what is the final equilibrium temperature of the system? (*b*) Is any ice left after the system reaches equilibrium?

40 •• A calorimeter of negligible heat capacity contains 1.00 kg of water at 303 K and 50.0 g of ice at 273 K. (*a*) Find the final temperature T. (*b*) Find the final temperature T if the mass of ice is 500 g.

41 •• A 200-g aluminum calorimeter contains 600 g of water at 20.0°C. A 100-g piece of ice cooled to −20.0°C is placed in the calorimeter. (*a*) Find the final temperature of the system, assuming no heat is transferred to or from the system. (*b*) A 200-g piece of ice at −20.0°C is added. How much ice remains in the system after the system reaches equilibrium? (*c*) Would the answer for Part (*b*) change if both pieces of ice were added at the same time?

42 •• The specific heat of a 100-g block of a substance is to be determined. The block is placed in a 25-g copper calorimeter holding 60 g of water at 20°C. Then, 120 mL of water at 80°C are added to the calorimeter. When thermal equilibrium is reached, the temperature of the system is 54°C. Determine the specific heat of the block.

43 •• A 100-g piece of copper is heated in a furnace to a temperature t_C. The copper is then inserted into a 150-g copper calorimeter containing 200 g of water. The initial temperature of the water and calorimeter is 16.0°C, and the temperature after equilibrium is established is 38.0°C. When the calorimeter and its contents are weighed, 1.20 g of water are found to have evaporated. What was the temperature t_C? SSM

44 ••• A 200-g aluminum calorimeter contains 500 g of water at 20.0°C. Aluminum shot with a mass equal to 300 g is heated to 100.0°C and is then placed in the calorimeter. Find the final temperature of the system, assuming that there is no heat transfer to the surroundings.

FIRST LAW OF THERMODYNAMICS

45 • A diatomic gas does 300 J of work and also absorbs 2.50 kJ of heat. What is the change in internal energy of the gas?

46 • If a gas absorbs 1.67 MJ of heat while doing 800 kJ of work, what is the change in the internal energy of the gas?

47 • If a gas absorbs 84 J of heat while doing 30 J of work, what is the change in the internal energy of the gas?

48 •• A lead bullet initially at 30°C just melts upon striking a target. Assuming that all of the initial kinetic energy of the bullet goes into the internal energy of the bullet, calculate the impact speed of the bullet.

49 •• During a cold day, you can warm your hands by rubbing them together. Assume the coefficient of kinetic friction between your hands is 0.500, the normal force between your hands is 35.0 N, and that you rub them together at an average relative speed of 35.0 cm/s. (*a*) What is the rate at which mechanical energy is dissipated? (*b*) Assume further that the mass of each of your hands is 350 g, the specific heat of your hands is 4.00 kJ/kg·K, and that all the dissipated mechanical energy goes into increasing the temperature of your hands. How long must you rub your hands together to produce a 5.00°C increase in their temperature?

WORK AND THE *PV* DIAGRAM FOR A GAS

In Problems 50 through 53, the initial state of 1.00 mol of a dilute gas is P_1 = 3.00 atm, V_1 = 1.00 L, and $E_{int\,1}$ = 456 J, and its final state is P_2 = 2.00 atm, V_2 = 3.00 L, and $E_{int\,2}$ = 912 J.

50 • The gas is allowed to expand at constant pressure until it reaches its final volume. It is then cooled at constant volume until it reaches its final pressure. (*a*) Illustrate this process on a *PV* diagram and calculate the work done by the gas. (*b*) Find the heat absorbed by the gas during this process.

51 • The gas is first cooled at constant volume until it reaches its final pressure. It is then allowed to expand at constant pressure until it reaches its final volume. (*a*) Illustrate this process on a *PV* diagram and calculate the work done by the gas. (*b*) Find the heat absorbed by the gas during this process. SSM

52 •• The gas is allowed to expand isothermally until it reaches its final volume and its pressure is 1.00 atm. It is then heated at constant volume until it reaches its final pressure. (*a*) Illustrate this process on a *PV* diagram and calculate the work done by the gas. (*b*) Find the heat absorbed by the gas during this process.

53 •• The gas is heated and is allowed to expand such that it follows a single straight-line path on a *PV* diagram from its initial state to its final state. (*a*) Illustrate this process on a *PV* diagram and calculate the work done by the gas. (*b*) Find the heat absorbed by the gas during this process.

54 •• In this problem, 1.00 mol of a dilute gas initially has a pressure equal to 1.00 atm, a volume equal to 25.0 L, and an internal energy equal to 456 J. As the gas is slowly heated, the plot of its state on a *PV* diagram moves in a straight line to the final state. The gas now has a pressure equal to 3.00 atm, a volume equal to 75.0 L, and an internal energy equal to 912 J. Find the work done and the heat absorbed by the gas.

55 •• In this problem, 1.00 mol of the ideal gas is heated while its volume changes, so that $T = AP^2$, where A is a constant. The temperature changes from T_0 to $4T_0$. Find the work done by the gas.

56 •• **ENGINEERING APPLICATION, CONTEXT-RICH** A sealed, almost-empty spray paint can still contains a residual amount of the propellant: 0.020 mol of nitrogen gas. The can's warning label clearly states: "Do Not Dispose by Incineration." (*a*) Explain this warning and draw the *PV* diagram for the gas if, in fact, the can is subject to a high temperature. (*b*) You are in charge of testing the can. The manufacturer claims it can withstand an internal gas pressure of 6.00 atm before it breaks. The can is initially at room-temperature and standard pressure in your testing laboratory. You begin to heat it uniformly using a heating element that has a power output of 200 W. The can and element are in an insulating oven, and you can assume 1.0% of the heat released by the heating element is absorbed by the gas in the can. How long should you expect the heating element to remain on before the can bursts?

57 •• An ideal gas initially at 20°C and 200 kPa has a volume of 4.00 L. It undergoes a quasi-static, isothermal expansion until its pressure is reduced to 100 kPa. Find (*a*) the work done by the gas, and (*b*) the heat absorbed by the gas during the expansion. SSM

HEAT CAPACITIES OF GASES AND THE EQUIPARTITION THEOREM

58 • The heat capacity at constant volume of a certain amount of a monatomic gas is 49.8 J/K. (*a*) Find the number of moles of the gas. (*b*) What is the internal energy of the gas at $T = 300$ K? (*c*) What is the heat capacity at constant pressure of the gas?

59 •• The heat capacity at constant pressure of a certain amount of a diatomic gas is 14.4 J/K. (*a*) Find the number of moles of the gas. (*b*) What is the internal energy of the gas at $T = 300$ K? (*c*) What is the molar heat capacity of this gas at constant volume? (*d*) What is the heat capacity of this gas at constant volume? SSM

60 •• (*a*) Calculate the heat capacity per unit mass of air at constant volume and the heat capacity per unit mass of air at constant pressure. Assume that air has a temperature of 300 K and a pressure of 1.00×10^5 N/m^2. Also, assume that air is composed of 74.0 percent N_2 molecules (molecular weight 28.0 g/mol) and 26.0 percent O_2 molecules (molar mass of 32.0 g/mol) and that both components are ideal gases. (*b*) Compare your answer for the specific heat at constant pressure to the value listed in the *Handbook of Chemistry and Physics* of 1.032 kJ/kg · K.

61 •• In this problem, 1.00 mol of an ideal diatomic gas is heated at constant volume from 300 K to 600 K. (*a*) Find the increase in the internal energy of the gas, the work done by the gas, and the heat absorbed by the gas. (*b*) Find the same quantities if this gas is heated from 300 K to 600 K at constant pressure. Use the first law of thermodynamics and your results for Part (*a*) to calculate the work done by the gas. (*c*) Again calculate the work done in Part (*b*). This time calculate it by integrating the equation $dW = P\,dV$.

62 •• A diatomic gas is confined to a closed container of constant volume V_0 and at a pressure P_0. The gas is heated until its pressure triples. What amount of heat had to be absorbed by the gas in order to triple the pressure?

63 •• In this problem, 1.00 mol of air is confined in a cylinder with a piston. The confined air is maintained at a constant pressure of 1.00 atm. The air is initially at 0°C and has volume V_0. Find the volume after 13,200 J of heat are absorbed by the trapped air.

64 •• The heat capacity at constant pressure of a sample of a gas is greater than the heat capacity at constant volume by 29.1 J/K. (*a*) How many moles of the gas are present? (*b*) If the gas is monatomic, what are C_V and C_P? (*c*) What are the values of C_V and C_P at normal room temperatures?

65 •• Carbon dioxide (CO_2) at a pressure of 1.00 atm and a temperature of −78.5°C sublimates directly from a solid to a gaseous state without going through a liquid phase. What is the change in the heat capacity at constant pressure per mole of CO_2 when it undergoes sublimation? (Assume that the gas molecules can rotate but do not vibrate.) Is the change in the heat capacity positive or negative during sublimation? The CO_2 molecule is pictured in Figure 18-22. SSM

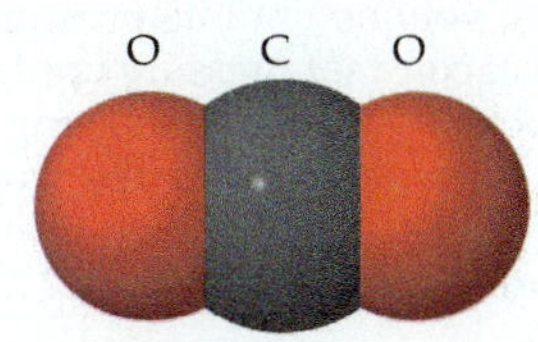

FIGURE 18-22 Problem 65

66 •• In this problem, 1.00 mol of a monatomic ideal gas is initially at 273 K and 1.00 atm. (*a*) What is the initial internal energy of the gas? (*b*) Find the work done by the gas when 500 J of heat are absorbed by the gas at constant pressure. What is the final internal energy of the gas? (*c*) Find the work done by the gas when 500 J of heat are absorbed by the gas at constant volume. What is the final internal energy of the gas?

67 •• List all of the degrees of freedom possible for a water molecule and estimate the heat capacity of water at a temperature very far above its boiling point. (Ignore the fact the molecule might dissociate at high temperatures.) Think carefully about all of the different ways in which a water molecule can vibrate.

HEAT CAPACITIES OF SOLIDS AND THE DULONG–PETIT LAW

68 • The Dulong–Petit law was originally used to determine the molar mass of a substance from its measured heat capacity. The specific heat of a certain solid substance is measured to be 0.447 kJ/kg · K. (*a*) Find the molar mass of the substance. (*b*) What element has this specific-heat value?

QUASI-STATIC ADIABATIC EXPANSION OF A GAS

69 •• A 0.500-mol sample of an ideal monatomic gas at 400 kPa and 300 K expands quasi-statically until the pressure decreases to 160 kPa. Find the final temperature and volume of the gas, the work done by the gas, and the heat absorbed by the gas if the expansion is (*a*) isothermal, and (*b*) adiabatic. SSM

70 •• A 0.500-mol sample of an ideal diatomic gas at 400 kPa and 300 K expands quasi-statically until the pressure decreases to 160 kPa. Find the final temperature and volume of the gas, the work done by the gas, and the heat absorbed by the gas if the expansion is (*a*) isothermal, and (*b*) adiabatic.

71 •• A 0.500-mol sample of helium gas expands adiabatically and quasi-statically from an initial pressure of 5.00 atm and temperature of 500 K to a final pressure of 1.00 atm. Find (*a*) the final temperature of the gas, (*b*) the final volume of the gas, (*c*) the work done by the gas, and (*d*) the change in the internal energy of the gas.

CYCLIC PROCESSES

72 •• A 1.00-mol sample of N_2 gas at 20.0°C and 5.00 atm is allowed to expand adiabatically and quasi-statically until its pressure equals 1.00 atm. It is then heated at constant pressure until its temperature is again 20.0°C. After it reaches a temperature of 20.0°C, it is heated at constant volume until its pressure is again 5.00 atm. It is then compressed at constant pressure until it is back to its original state. (*a*) Construct a *PV* diagram showing each process in the cycle. (*b*) From your graph, determine the work done by the gas during the complete cycle. (*c*) How much heat is absorbed (or released) by the gas during the complete cycle?

73 •• A 1.00-mol sample of an ideal diatomic gas is allowed to expand. This expansion is represented by the straight line from 1 to 2 in the *PV* diagram (Figure 18-23). The gas is then compressed isothermally. This compression is represented by the curved line from 2 to 1 in the *PV* diagram. Calculate the work per cycle done by the gas. SSM

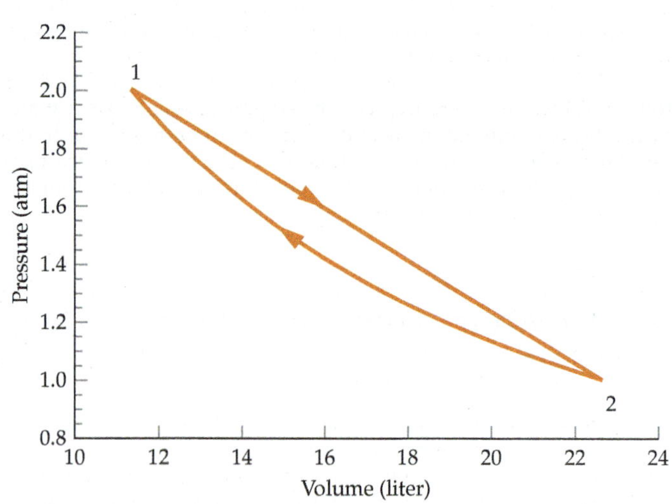

FIGURE 18-23 Problem 73

74 •• A 2.00-mol sample of an ideal monatomic gas has an initial pressure of 2.00 atm and an initial volume of 2.00 L. The gas is taken through the following quasi-static cycle: It is expanded isothermally until it has a volume of 4.00 L. Next, it is heated at constant volume until it has a pressure of 2.00 atm. It is then cooled at constant pressure until it is back to its initial state. (*a*) Show this cycle on a *PV* diagram. (*b*) Find the temperature at the end of each part of the cycle. (*c*) Calculate the heat absorbed and the work done by the gas during each part of the cycle.

75 ••• At point D in Figure 18-24, the pressure and temperature of 2.00 mol of an ideal monatomic gas are 2.00 atm and 360 K, respectively. The volume of the gas at point B on the *PV* diagram is three times that at point D and its pressure is twice that at point C. Paths AB and CD represent isothermal processes. The gas is carried through a complete cycle along the path DABCD. Determine the total amount of work done by the gas and the heat absorbed by the gas along each portion of the cycle. SSM

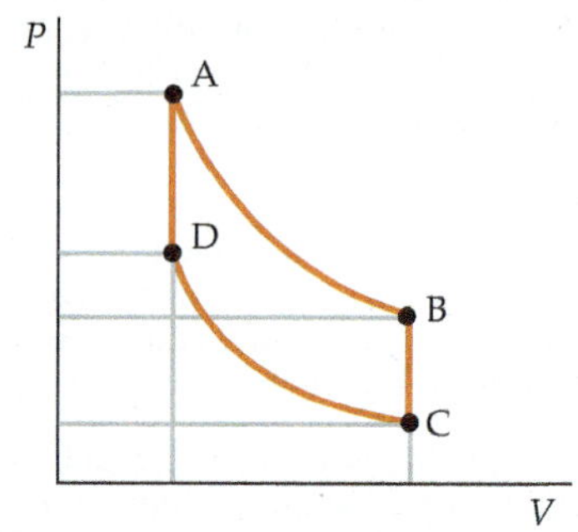

FIGURE 18-24 Problems 75, 76

76 ••• At point D in Figure 18-24, the pressure and temperature of 2.00 mol of an ideal diatomic gas are 2.00 atm and 360 K, respectively. The volume of the gas at point B on the *PV* diagram is three times that at point D and its pressure is twice that at point C. Paths AB and CD represent isothermal processes. The gas is carried through a complete cycle along the path DABCD. Determine the total amount of work done by the gas and the heat absorbed by the gas along each portion of the cycle.

77 ••• A sample consisting of n moles of an ideal gas is initially at pressure P_1, volume V_1, and temperature T_h. It expands isothermally until its pressure and volume are P_2 and V_2. It then expands adiabatically until its temperature is T_c and its pressure and volume are P_3 and V_3. It is then compressed isothermally until it is at a pressure P_4 and a volume V_4, which is related to its initial volume V_1 by $T_c V_4^{\gamma-1} = T_h V_1^{\gamma-1}$. The gas is then compressed adiabatically until it is back in its original state. (*a*) Assuming that each process is quasi-static, plot this cycle on a *PV* diagram. (This cycle is known as the Carnot cycle for an ideal gas.) (*b*) Show that the heat Q_h absorbed during the isothermal expansion at T_h is $Q_h = nRT_h \ln(V_2/V_1)$. (*c*) Show that the heat Q_c released by the gas during the isothermal compression at T_c is $Q_c = nRT_c \ln(V_3/V_4)$. (*d*) Using the result that $TV^{\gamma-1}$ is constant for a quasi-static adiabatic expansion, show that $V_2/V_1 = V_3/V_4$. (*e*) The efficiency of a Carnot cycle is defined as the net work done by the gas, divided by the heat absorbed Q_h by the gas. Using the first law of thermodynamics, show that the efficiency is $1 - Q_c/Q_h$. (*f*) Using your results from the previous parts of this problem, show that $Q_c/Q_h = T_c/T_h$.

GENERAL PROBLEMS

78 • During the process of quasi-statically compressing an ideal diatomic gas to one-fifth of its initial volume, 180 kJ of work are done on the gas. (*a*) If this compression is accomplished isothermally at room temperature (293 K), how much heat is released by the gas? (*b*) How many moles of gas are in this sample?

79 • The *PV* diagram in Figure 18-25 represents 3.00 mol of an ideal monatomic gas. The gas is initially at point A. The paths AD and BC represent isothermal changes. If the system is brought to point C along the path AEC, find (*a*) the initial and final temperatures of the gas, (*b*) the work done by the gas, and (*c*) the heat absorbed by the gas. SSM

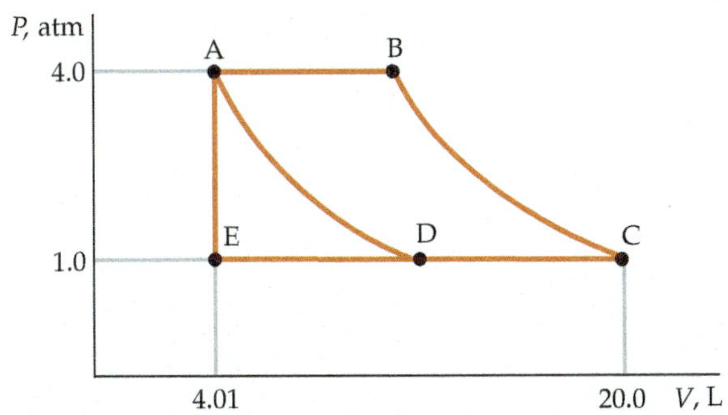

FIGURE 18-25 Problems 79, 80, 81, 82

80 •• The *PV* diagram in Figure 18-25 represents 3.00 mol of an ideal monatomic gas. The gas is initially at point A. The paths AD and BC represent isothermal changes. If the system is brought to point C along the path ABC, find (*a*) the initial and final temperatures of the gas, (*b*) the work done by the gas, and (*c*) the heat absorbed by the gas.

81 •• The *PV* diagram in Figure 18-25 represents 3.00 mol of an ideal monatomic gas. The gas is initially at point A. The paths AD and BC represent isothermal changes. If the system is brought to point C along the path ADC, find (*a*) the initial and final temperatures of the gas, (*b*) the work done by the gas, and (*c*) the heat absorbed by the gas.

82 •• Suppose that the paths AD and BC in Figure 18-25 represent adiabatic processes. What are the work done by the gas and the heat absorbed by the gas in following the path ABC?

83 •• **BIOLOGICAL APPLICATION, CONTEXT-RICH** As part of a laboratory experiment, you test the calorie content of various foods. Assume that when you eat these foods, 100% of the energy released by the foods is absorbed by your body. Suppose you burn a 2.50-g potato chip, and the resulting flame warms a small aluminum can of water. After burning the potato chip, you measure its mass to be 2.20 g. The mass of the can is 25.0 g, and the volume of water contained in the can is 15.0 mL. If the temperature increase in the water is 12.5°C, how many kilocalories (1 kcal = 1 dietary calorie) per 150-g serving of these potato chips would you estimate there are? Assume the can of water captures 50.0 percent of the heat released during the burning of the potato chip. (*Note:* Although the joule is the SI unit of choice in most thermodynamic situations, the food industry in the United States currently expresses the energy released during metabolism in terms of the "dietary calorie," which is our kilocalorie). SSM

84 •• **ENGINEERING APPLICATION** Diesel engines operate without spark plugs, unlike gasoline engines. The cycle that diesel engines undergo involves adiabatically compressing the air in a cylinder, and then fuel is injected. When the fuel is injected, if the air temperature inside the cylinder is above the fuel's flashpoint, the fuel–air mixture will ignite. Most diesel engines have compression ratios in the range from 14:1 to 25:1. For this range of compression ratios (which are the ratio of maximum to minimum volume), what is the range of maximum temperatures of the air in the cylinder, assuming the air is taken into the cylinder at 35°C? Most modern gasoline engines typically have compression ratios on the order of 8:1. Explain why you expect the diesel engine to require a better (more efficient) cooling system than its gasoline counterpart.

85 •• At very low temperatures, the specific heat of a metal is given by $c = aT + bT^3$. For copper, $a = 0.0108\ \text{J/kg}\cdot\text{K}^2$ and $b = 7.62 \times 10^{-4}\ \text{J/kg}\cdot\text{K}^4$. (*a*) What is the specific heat of copper at 4.00 K? (*b*) How much heat is required to heat copper from 1.00 to 3.00 K?

86 •• How much work must be done on 30.0 g of carbon monoxide (CO) at standard temperature and pressure to compress it to one-fifth of its initial volume if the process is (*a*) isothermal, (*b*) adiabatic?

87 •• How much work must be done on 30.0 g of carbon dioxide (CO_2) at standard temperature and pressure to compress it to one-fifth of its initial volume if the process is (*a*) isothermal, (*b*) adiabatic?

88 •• How much work must be done on 30.0 g of argon (Ar) at standard temperature and pressure to compress it to one-fifth of its initial volume if the process is (*a*) isothermal, (*b*) adiabatic?

89 •• A thermally insulated system consists of 1.00 mol of a diatomic gas at 100 K and 2.00 mol of a solid at 200 K that are separated by a rigid insulating wall. Find the equilibrium temperature of the system after the insulating wall is removed, assuming that the gas obeys the ideal-gas law and that the solid obeys the Dulong–Petit law. SSM

90 •• When an ideal gas undergoes a temperature change at constant volume, its internal energy change is given by the formula $\Delta E_{\text{int}} = C_V \Delta T$. However, this formula correctly gives the change in internal energy whether the volume remains constant or not. (*a*) Explain why this formula gives correct results for an ideal gas even when the volume changes. (*b*) Using this formula, along with the first law of thermodynamics, show that for an ideal gas $C_P = C_V + nR$.

91 •• An insulated cylinder is fitted with an insulated movable piston to maintain constant pressure. The cylinder initially contains 100 g of ice at −10°C. Heat is transferred to the ice at a constant rate by a 100-W heater. Make a graph showing the temperature of the ice/water/steam as a function of time starting at t_i when the temperature is −10°C and ending at t_f when the temperature is 110°C.

92 •• (*a*) In this problem, 2.00 mol of a diatomic ideal gas expands adiabatically and quasi-statically. The initial temperature of the gas is 300 K. The work done by the gas during the expansion is 3.50 kJ. What is the final temperature of the gas? (*b*) Compare your result to the result you would get if the gas were monatomic.

93 •• A vertical insulated cylinder is divided into two parts by a movable piston of mass m. The piston is initially held at rest. The top part of the cylinder is evacuated and the bottom part is filled with 1.00 mol of diatomic ideal gas at temperature 300 K. After the piston is released and the system comes to equilibrium, the volume occupied by gas is halved. Find the final temperature of the gas.

94 ••• According to the Einstein model of a crystalline solid, the internal energy per mole is given by $U = (3N_AkT_E)/(e^{T_E/T} - 1)$, where T_E is a characteristic temperature called the *Einstein temperature*, and T is the temperature of the solid in kelvins. Use this expression to show that a crystalline solid's molar heat capacity at constant volume is given by

$$c'_V = 3R\left(\frac{T_E}{T}\right)^2 \frac{e^{T_E/T}}{(e^{T_E/T} - 1)^2}$$

95 ••• (*a*) Use the results of Problem 94 to show that in the limit that $T \gg T_E$, the Einstein model gives the same expression for specific heat that the Dulong-Petit law does. (*b*) For diamond, T_E is approximately 1060 K. Integrate numerically $dE_{int} = c'_v\, dT$ to find the increase in the internal energy if 1.00 mol of diamond is heated from 300 K to 600 K. SSM

96 ••• Use the results of the Einstein model in Problem 94 to determine the molar internal energy of diamond ($T_E = 1060$ K) at 300 K and 600 K, and thereby the increase in internal energy as diamond is heated from 300 K to 600 K. Compare your result to that of Problem 95.

97 ••• During an isothermal expansion, an ideal gas at an initial pressure P_0 expands until its volume is twice its initial volume V_0. (*a*) Find its pressure after the expansion. (*b*) The gas is then compressed adiabatically and quasi-statically until its volume is V_0 and its pressure is $1.32P_0$. Is the gas monatomic, diatomic, or polyatomic? (*c*) How does the translational kinetic energy of the gas change in each stage of this process?

98 ••• If a hole is punctured in a tire, the gas inside will gradually leak out. Assume the following: The area of the hole is A; the tire volume is V; and the time, τ, it takes for most of the air to leak out of the tire can be expressed in terms of the ratio A/V, the temperature T, the Boltzmann constant k, and the initial mass m of the gas inside the tire. (*a*) Based on these assumptions, use dimensional analysis to find an estimate for τ. (*b*) Use the result of Part (*a*) to estimate the time it takes for a car tire with a nail hole punched in it to go flat.

CHAPTER 19

DIVERS CARRY TANKS OF AIR TO ALLOW THEM TO STAY UNDERWATER FOR LONG PERIODS OF TIME. *(Paul Springett/ Alamy.)*

The Second Law of Thermodynamics

? The probability that at a given instant the molecules of air in one of the tanks are all in the half of the tank opposite the hose connection is very very small. Just how small is it? (See Example 19-13.)

We are often asked to conserve energy. But according to the first law of thermodynamics, energy is always conserved. What then does it mean to conserve energy if the total amount of energy in the universe does not change regardless of what we do? The first law of thermodynamics does not tell the whole story. Energy is always conserved, but some forms of energy are more useful than others. The possibility or impossibility of putting energy to *use* is the subject of the second law of thermodynamics. Scientists and engineers are constantly trying to improve the efficiency of heat engines (devices that transform heat into work). In the power industry engineers strive to achieve higher efficiencies in transforming the thermal energy liberated by the burning of fossil fuels and the fission of uranium and plutonium into useful work.

In this chapter, we examine the second law of thermodynamics as it relates directly to heat engines as well as refrigerators. We also discuss an ideal heat engine—the Carnot engine. Irreversibility and entropy are also covered as they relate to the availability of energy, disorder, and probability.

Solar energy is directed toward the solar oven at the center by this circular array of reflectors at Barstow, California. *(Sandia National Laboratory.)*

19-1 HEAT ENGINES AND THE SECOND LAW OF THERMODYNAMICS

> No system can absorb heat from a single reservoir and convert it entirely into work without additional net changes in the system or its surroundings.
>
> SECOND LAW OF THERMODYNAMICS: KELVIN STATEMENT

A common example of the conversion of work into heat is movement with friction. For example, suppose you spend two minutes pushing a block this way and that way along a tabletop in a closed path, leaving the block in its initial position. Also, suppose that the block–table system is initially in thermal equilibrium with its surroundings. The work you do on the system is converted into internal energy of the system, and as a result the block–table system becomes warmer. Consequently, the system is no longer in thermal equilibrium with its surroundings. However, the system will transfer energy as heat to its surroundings until it returns to thermal equilibrium with those surroundings. Because the final and initial states of the system are the same, the first law of thermodynamics dictates that the energy transferred to the environment as heat equals the work done by you on the system. The reverse process never occurs—a block and table that are warm will never spontaneously cool by converting their internal energy into work that causes the block to push your hand around the table! Yet such an amazing occurrence would not violate the first law of thermodynamics or any other physical laws we have encountered so far. It does, however, violate the second law of thermodynamics. Thus, there is a lack of symmetry in the roles played by heat and work that is not evident from the first law. This lack of symmetry is related to the fact that some processes are *irreversible*.

Many other irreversible processes exist, seemingly quite different from one another, but all related to the second law. For example, heat transfer is an irreversible process. If we place a hot body in contact with a cold body, heat will transfer from the hot body to the cold body until they are at the same temperature. However, the reverse does not occur. Two bodies in contact at the same temperature remain at the same temperature; heat is not transferred from one to the other leaving one colder and the other warmer. This experimental fact gives us an equivalent statement of the second law of thermodynamics.

> A process whose only net result is to absorb heat from a cold reservoir and release the same amount of heat to a hot reservoir is impossible.
>
> SECOND LAW OF THERMODYNAMICS: CLAUSIUS STATEMENT

We will show in this chapter that the Kelvin and Clausius statements of the second law are equivalent.

The study of the efficiency of heat engines gave rise to the first clear statements of the second law. A **heat engine** is a cyclic device whose purpose is to convert as much heat into work as possible. Heat engines contain a **working substance** (water in a steam engine) that absorbs a quantity of heat Q_h from a high temperature reservoir, does work W on its surroundings, and releases heat Q_c as it returns to its initial state, where Q_h, W, and Q_c represent magnitudes and are never negative.

> ! Q_h, W, and Q_c represent magnitudes and are never negative.

The earliest heat engines were steam engines, invented in the eighteenth century for pumping water from coal mines. Today steam engines are used to generate electricity. In a typical steam engine, liquid water under several hundred atmospheres of pressure absorbs heat from a high temperature reservoir until it vaporizes at about 500°C (Figure 19-1). This steam expands against a piston (or turbine blades), doing work, and then exits at a much lower temperature. The steam is cooled further in the condenser where it condenses by releasing heat to a low temperature reservoir. The water is then pumped back into the boiler and heated again.

Figure 19-2 is a schematic diagram of the heat engine used in many automobiles—the internal-combustion engine. With the exhaust valve closed, a mixture of gasoline vapor and air enters the combustion chamber as the piston moves down during the intake stroke. The mixture is then compressed, after which it is ignited by a spark from the spark plug. The hot gases then expand, driving the piston down and doing work on it during the *power stroke.* The gases are then exhausted through the exhaust valve, and the cycle repeats.

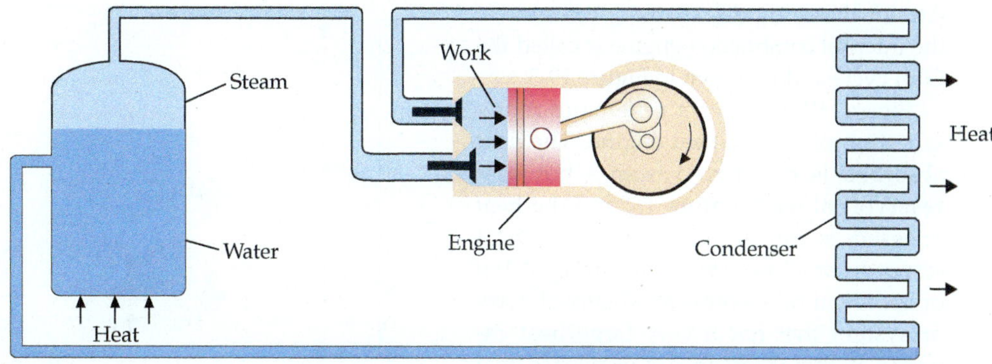

FIGURE 19-1 Schematic drawing of a steam engine. High-pressure steam does work on the piston.

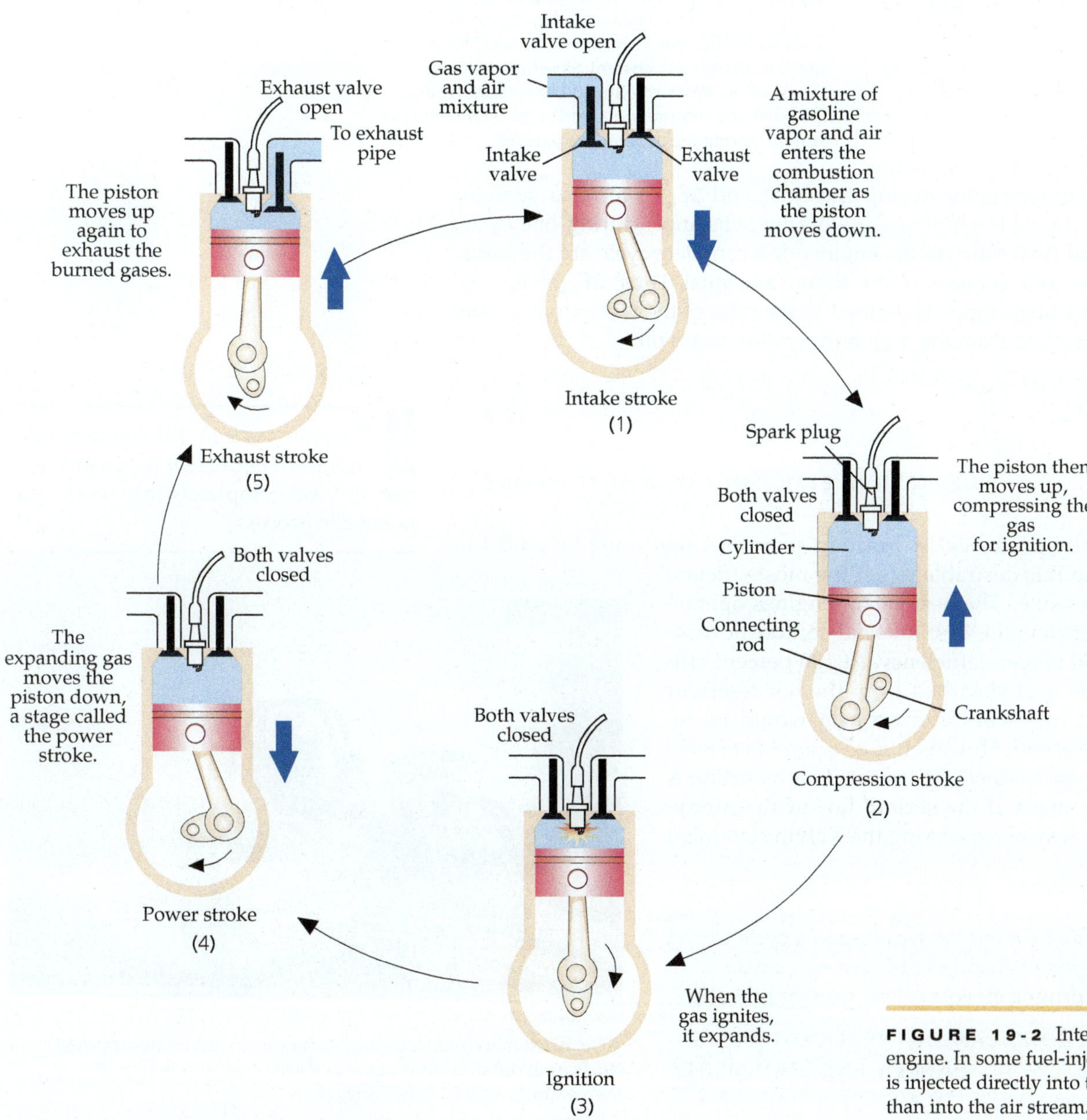

FIGURE 19-2 Internal-combustion engine. In some fuel-injected engines, the fuel is injected directly into the cylinder rather than into the air stream.

An idealized model of the processes in the internal combustion engine is called the **Otto cycle** and is shown in Figure 19-3.

Figure 19-4 shows a schematic representation of a basic heat engine. The heat absorbed is transferred from a hot **heat reservoir** at temperature T_h, and the heat released is transferred into a cold heat reservoir at a lower temperature T_c. A hot or cold heat reservoir is an idealized body or system that has a very large heat capacity so that it can absorb or release heat with no noticeable change in its temperature. In practice, burning fossil fuel often acts as the high-temperature reservoir, and the surrounding atmosphere or a lake often acts as the low-temperature reservoir. Applying the first law of thermodynamics ($\Delta E_{int} = Q_{in} + W_{on}$) to the heat engine gives

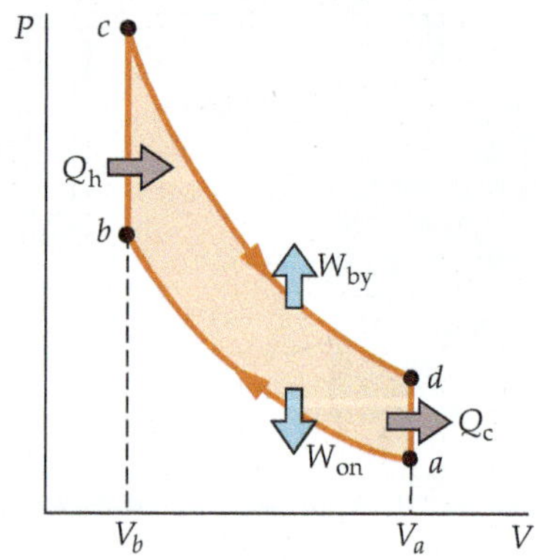

FIGURE 19-3 Otto cycle, representing the internal-combustion engine. The fuel–air mixture is adiabatically compressed from *a* to *b*. It is then heated (by combustion) at constant volume to *c*. The power stroke is represented by the adiabatic expansion from *c* to *d*. The cooling at constant volume from *d* to *a* represents the release of heat. The combustion products are exchanged for a fresh fuel–air mixture at constant pressure at step *a* (not shown). Work is done on the system during the adiabatic compression, and work is done by the system during the adiabatic expansion.

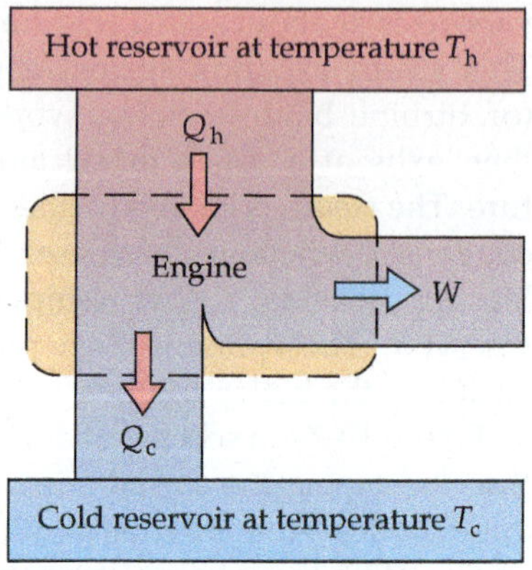

FIGURE 19-4 Schematic representation of a heat engine. The engine absorbs heat Q_h from a hot reservoir at a temperature T_h, does work W, and releases heat Q_c to a cold reservoir at a temperature T_c.

$$W = Q_h - Q_c \qquad 19\text{-}1$$

where W is the work done *by* the engine during one complete cycle, $Q_h - Q_c$ is the total heat transferred to the engine during one cycle, and ΔE_{int} is the change in internal energy of the engine (including the working substance) during one cycle. Because the initial and final states of the engine for a complete cycle are the same, the initial and final internal energies of the engine are equal. Thus, $\Delta E_{int} = 0$.

The efficiency ε of a heat engine is defined as the ratio of the work done by the engine to the heat absorbed from the high temperature reservoir:

$$\varepsilon = \frac{W}{Q_h} = \frac{Q_h - Q_c}{Q_h} = 1 - \frac{Q_c}{Q_h} \qquad 19\text{-}2$$

DEFINITION: EFFICIENCY OF A HEAT ENGINE

! The word *cycle* in this statement is important because *it* is possible to convert heat completely into work in a noncyclic process.

The heat Q_h is usually produced by burning some fuel that must be paid for, such as coal or oil, so it is desirable to get the most efficient use of the fuel as possible. The best steam engines operate near 40 percent efficiency; the best internal-combustion engines operate near 50 percent efficiency. At 100 percent efficiency ($\varepsilon = 1$), all the heat absorbed from the hot reservoir would be converted into work and no heat would be released to the cold reservoir. However, *it is impossible to make a heat engine with an efficiency of 100 percent.* This assertion is the **heat-engine statement of the second law of thermodynamics.** It is another way of expressing the Kelvin statement given earlier:

> It is impossible for a heat engine working in a cycle to produce *only the effect* of absorbing heat from a single reservoir and performing an equivalent amount of work.
>
> SECOND LAW OF THERMODYNAMICS: HEAT-ENGINE STATEMENT

An exhaust manifold feeds the header pipes seen on this top-fuel dragster in order for the engine to release heat and reduce its temperature. *(© 2002 Robert Briggs.)*

An ideal gas undergoing an isothermal expansion does just this. But after the expansion, the gas is not in its original state. To bring the gas back to its original state, work must be done on the gas, and some heat will be released.

The second law tells us that to do work with energy absorbed from a heat reservoir, we must have a colder reservoir available to absorb the energy that is not used by the engine to do work. If this were not true, we could design a ship that has a heat engine that is powered by simply extracting energy as heat from the ocean. Unfortunately, the lack of a colder reservoir, which would absorb heat from the engine, makes this enormous reservoir of energy unavailable for such use. (It is theoretically possible to run a heat engine between the warmer surface water of the ocean and the colder water at greater depths, but no practical scheme for using this temperature difference has yet emerged.) In order to convert heat at a single temperature into the energy that does work (with no other changes in the source or object), a separate cold reservoir must be used.

PROBLEM-SOLVING STRATEGY

Calculating the Work Done by a Heat Engine Operating in a Cycle

PICTURE A heat engine absorbs heat from a high-temperature heat reservoir, does work and releases heat to a low-temperature heat reservoir. Conservation of energy informs us that the heat absorbed by the engine per cycle equals the heat released by the engine per cycle plus the work done by the engine per cycle. The efficiency of a heat engine is defined as the ratio of the work done by the engine per cycle to the heat absorbed by the engine per cycle. The working substance for the engine is an ideal gas for virtually all calculations in this text.

SOLVE

1. For an integral number of cycles, the change in internal energy $\Delta E_{\text{int}} = 0$, so $Q_h = W + Q_c$.
2. The efficiency is given by $\varepsilon = W/Q_h$.
3. The work during a step in a cycle is given by $W_{\text{step}} = \int_{V_i}^{V_f} P\,dV$, where $P = nRT/V$.
4. The heat absorbed by the gas during a step is given by $C\,\Delta T$, where C is the heat capacity.

CHECK The work done, W, must be equal to $Q_h - Q_c$ if the engine completes an integral number of cycles.

Example 19-1 Efficiency of a Heat Engine

During each cycle, a heat engine absorbs 200 J of heat from a hot reservoir, does work, and releases 160 J to a cold reservoir. What is the efficiency of the engine?

PICTURE We use the definition of the efficiency of a heat engine $\varepsilon = W/Q_h$ (Equation 19-2).

SOLVE

1. The efficiency is the work done divided by the heat absorbed: $\varepsilon = \dfrac{W}{Q_h}$
2. The heat absorbed and the heat released are given: $Q_h = 200\text{ J}$ and $Q_c = 160\text{ J}$
3. The work is found from the first law: $W = Q_h - Q_c = 200\text{ J} - 160\text{ J} = 40\text{ J}$
4. Substitute the values of Q_h and W to calculate the efficiency: $\varepsilon = \dfrac{W}{Q_h} = \dfrac{40\text{ J}}{200\text{ J}} = 0.20 = \boxed{20\%}$

CHECK The efficiency is dimensionless. In this example, both W and Q_h are both expressed in joules, so the ratio is dimensionless, as expected.

PRACTICE PROBLEM 19-1 A heat engine has an efficiency of 35 percent. (*a*) How much work does it perform in a cycle if it absorbs 150 J of heat per cycle from the high-temperature reservoir? (*b*) How much heat is transferred to the low-temperature reservoir per cycle?

Example 19-2 The Otto Cycle

Try It Yourself

(*a*) Find the efficiency of the Otto cycle shown in Figure 19-3. (*b*) Express your answer in terms of the ratio of the volumes $r = V_a/V_b$.

PICTURE (*a*) To find ε, you need to find Q_h and Q_c. Heat transfer occurs only during the two constant-volume processes, b to c and d to a. You can thus find Q_h and Q_c, and therefore ε in terms of the temperatures T_a, T_b, T_c, and T_d. (*b*) The temperatures can be related to the volumes using $TV^{\gamma-1}$ = constant for adiabatic processes.

SOLVE

Cover the column to the right and try these on your own before looking at the answers.

Steps	Answers
(*a*) 1. Write the efficiency in terms of Q_h and Q_c:	$\varepsilon = 1 - \dfrac{Q_{\text{cold}}}{Q_{\text{hot}}} = 1 - \dfrac{Q_c}{Q_h}$
2. The release of heat occurs at constant volume from d to a. Write Q_c in terms of C_v and the temperatures T_a and T_d:	$Q_c = \lvert Q_{d\to a}\rvert = C_v\lvert T_a - T_d\rvert = C_v(T_d - T_a)$
3. The absorption of heat occurs at constant volume from b to c. Write Q_h in terms of C_v and the temperatures T_c and T_b:	$Q_h = Q_{b\to c} = C_v(T_c - T_b)$
4. Substitute these expressions Q_c and Q_h to find the efficiency in terms of the temperatures T_a, T_b, T_c, and T_d:	$\varepsilon = \boxed{1 - \dfrac{T_d - T_a}{T_c - T_b}}$
(*b*) 1. Relate T_c to T_d using $TV^{\gamma-1}$ = constant, and $V_a/V_c = r$:	$T_cV_c^{\gamma-1} = T_dV_d^{\gamma-1}$ $T_c = T_d\dfrac{V_d^{\gamma-1}}{V_c^{\gamma-1}} = T_dr^{\gamma-1}$
2. Relate T_b to T_a as in step 1:	$T_b = T_ar^{\gamma-1}$
3. Use these relations to eliminate T_c and T_b from ε in Part (*a*) so that ε is expressed in terms of r:	$\varepsilon = 1 - \dfrac{T_d - T_a}{T_dr^{\gamma-1} - T_ar^{\gamma-1}} = \boxed{1 - \dfrac{1}{r^{\gamma-1}}}$

CHECK The Part-(*b*) result is a dimensionless number, as expected. In addition, the expression for ε is between 0 and 1, and approaches 0 as r approaches 1, as expected.

TAKING IT FURTHER The ratio r (volume before compression/volume after compression) is called the compression ratio.

19-2 REFRIGERATORS AND THE SECOND LAW OF THERMODYNAMICS

A **refrigerator** is essentially a heat engine run backward (Figure 19-5*a*). The refrigerator's engine absorbs heat from the interior of the refrigerator (cold reservoir) and releases heat to the surroundings (hot reservoir) (Figure 19-5*b*). Experience shows that such a transfer always requires that work be done on the refrigerator—a result known as the **refrigerator statement of the second law of thermodynamics,** which is another way of expressing the Clausius statement:

FIGURE 19-5 (*a*) Schematic representation of a refrigerator. Work W is done on the refrigerator and it absorbs heat Q_c from a cold reservoir and releases heat Q_h. (*b*) An actual refrigerator.

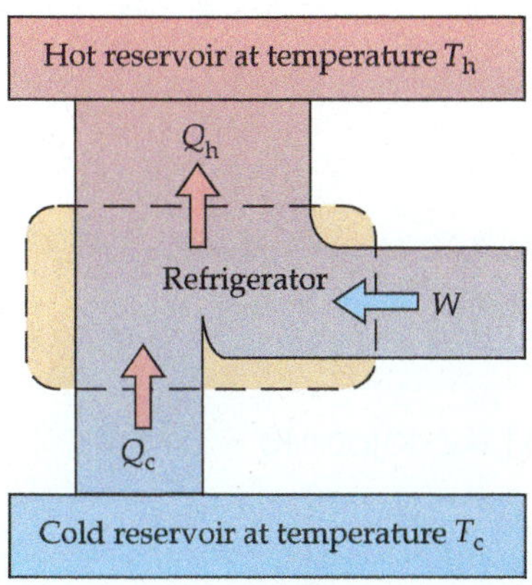

(*a*)

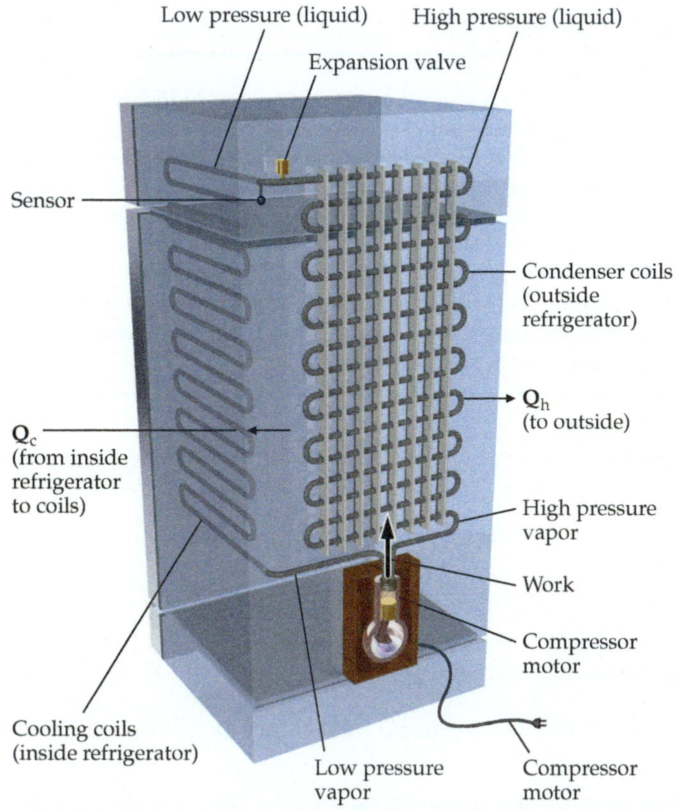

(*b*)

It is impossible for a refrigerator working in a cycle to produce *only the effect* of absorbing heat from a cold object and releasing the same amount of heat to a hot object.

SECOND LAW OF THERMODYNAMICS: REFRIGERATOR STATEMENT

Were the preceding statement not true, we could cool our homes in the summer with refrigerators that released heat to the outdoors without using electricity or any other energy source.

A measure of a refrigerator's performance is the ratio Q_c/W of the heat absorbed from the low-temperature reservoir to the work done on the refrigerator. (This work equals the electrical energy that comes from the wall outlet.) The ratio Q_c/W is called the **coefficient of performance** (COP):

$$\text{COP} = \frac{Q_c}{W} \qquad 19\text{-}3$$

DEFINITION: COEFFICIENT OF PERFORMANCE (REFRIGERATOR)

The greater the COP, the better the refrigerator. Typical refrigerators have coefficients of performance of about 5 or 6. [In the United States, the coefficient of performance of air conditioners is called the seasonal energy efficiency ratio (SEER).]* In terms of this ratio, the refrigerator statement of the second law states that the COP of a refrigerator cannot be infinite.

(Anderson Ross/PhotoDisk/Getty.)

* The SEER of an air conditioner is equal to the seasonal average of Q_c/W, with Q_c in BTUs and W in watt-hours.

Example 19-3 Making Ice Cubes

Context-Rich

You have half an hour before guests start arriving for your party when you suddenly realize that you forgot to buy ice for drinks. You quickly put 1.00 L of water at 10.0°C into your ice cube trays and pop them into the freezer. Will you have ice in time for your guests? The label on your refrigerator states that the appliance has a coefficient of performance of 5.5 and a power rating of 550 W. You estimate that only 10 percent of the electrical power contributes to the cooling and freezing of the water.

PICTURE Work equals power times time. We are given the power, so we need to find the work to determine the time. The work W is related to Q_c by COP = W/Q_c (Equation 19-3). To find Q_c we calculate how much heat must be released by the water.

SOLVE

1. The time needed is related to the power available and the work required:
$$P = W/\Delta t \Rightarrow \Delta t = W/P$$

2. The work is related to the coefficient of performance and the heat absorbed:
$$\text{COP} = \frac{Q_c}{W}$$

3. The heat Q_c absorbed from the inside of the refrigerator equals the heat Q_{cool} to be absorbed from the water to cool the water plus the heat Q_{freeze} to be absorbed from the water to freeze the water:
$$Q_c = Q_{cool} + Q_{freeze}$$

4. The release of heat needed to cool 1.00 L of water (mass 1 kg) by 10°C is:
$$Q_{cool} = mc\,\Delta T = (1.00\text{ kg})[4.18\text{ kJ/(kg}\cdot\text{K)}](10.0\text{ K}) = 41.8\text{ kJ}$$

5. The release of heat needed to freeze 1 L of water into ice cubes is:
$$Q_{freeze} = mL_f = (1.00\text{ kg})(333.5\text{ kJ/kg}) = 333.5\text{ kJ}$$

6. Add these heats to obtain Q_c:
$$Q_c = 41.8\text{ kJ} + 333.5\text{ kJ} = 375\text{ kJ}$$

7. Substitute Q_c into step 2 to find the work W:
$$W = \frac{Q_c}{\text{COP}} = \frac{375\text{ kJ}}{5.5} = 68.2\text{ kJ}$$

8. Use this value of W and 55 W for the available power to find the time Δt:
$$\Delta t = \frac{W}{P} = \frac{68.2\text{ kJ}}{55\text{ J/s}} = 1.24\text{ ks} \times \frac{1\text{ min}}{60\text{ s}} = 20.7\text{ min}$$

Your guests will have ice.

CHECK Twenty minutes is a short, but feasible, time to freeze one liter of water.

PRACTICE PROBLEM 19-2 A refrigerator has a coefficient of performance of 4.0. How much heat per cycle is absorbed by the hot reservoir if 200 kJ of heat per cycle are released by the cold reservoir?

EQUIVALENCE OF THE HEAT-ENGINE AND REFRIGERATOR STATEMENTS

The heat-engine and refrigerator statements (that is, the Kelvin and Clausius statements, respectively) of the second law of thermodynamics seem quite different, but they are actually equivalent. The heat-engine statement is, "It is impossible for a heat engine working in a cycle to produce *only the effect* of absorbing heat from a single reservoir and performing an equivalent amount of work," whereas the refrigerator statement is, "It is impossible for a refrigerator working in a cycle to produce *only the effect* of absorbing heat from a cold object and releasing the same amount of heat to a hot object." We can prove the equivalence of these statements by showing that if either statement is assumed to be false, then the other must also be false. We will use a numerical example to show that if the heat-engine statement is false, then the refrigerator statement is false.

Figure 19-6*a* shows an ordinary refrigerator that uses 50 J of work to absorb 100 J of heat from a cold reservoir and release 150 J of heat to a hot reservoir. Suppose the heat-engine statement of the second law were not true. Then, a "perfect" heat engine could absorb 50 J of heat from the hot reservoir and do 50 J of work with 100 percent efficiency. We could use this perfect heat engine to absorb 50 J of heat from the hot reservoir and do 50 J of work (Figure 19-6*b*) on the ordinary refrigerator. Then, the combination of the perfect heat engine and the ordinary refrigerator would be a perfect refrigerator, transferring 100 J of energy as heat from the cold reservoir to the hot reservoir without requiring any work, as illustrated in Figure 19-6*c*. This violates the refrigerator statement of the second law. Thus, if the heat-engine statement is false, the refrigerator statement is also false. Similarly, if a perfect refrigerator existed, it could be used in conjunction with an ordinary heat engine to construct a perfect heat engine. Thus, if the refrigerator statement is false, the heat-engine statement is also false. It then follows that if one statement is true, the other is also true. Therefore, the heat-engine statement and the refrigerator statement are equivalent.

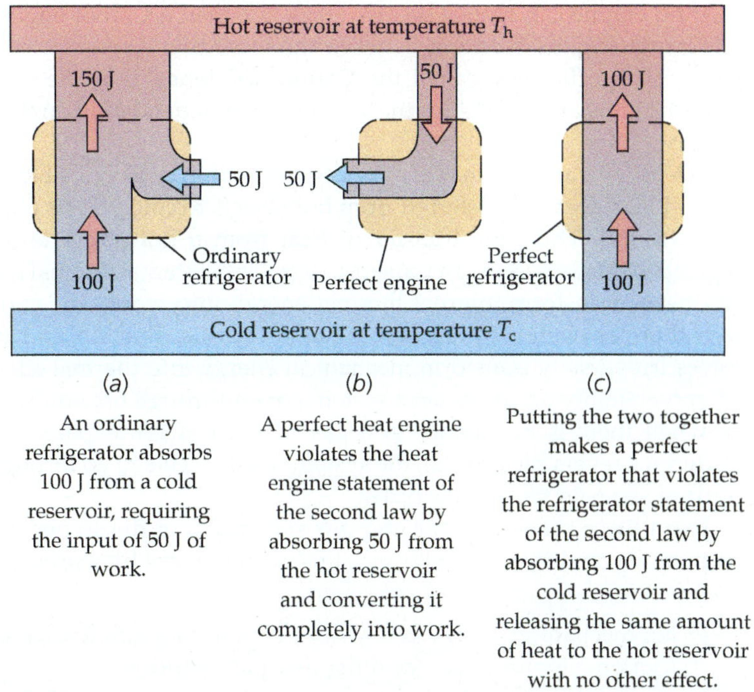

FIGURE 19-6 Demonstration of the equivalence of the heat-engine and refrigerator statements of the second law of thermodynamics.

19-3 THE CARNOT ENGINE

According to the second law of thermodynamics, it is impossible for a heat engine working between two heat reservoirs to be 100 percent efficient. What, then, is the maximum possible efficiency for such an engine? A young French engineer, Sadi Carnot answered this question in 1824, before either the first or the second law of thermodynamics had been established. Carnot found that a *reversible engine* is the most efficient engine that can operate between any two given reservoirs. This result is known as the Carnot theorem:

> No engine working between two given heat reservoirs can be more efficient than a reversible engine working between these two reservoirs.
>
> CARNOT THEOREM

A reversible engine working in a cycle between two heat reservoirs is called a **Carnot engine,** and its cycle is called a **Carnot cycle.** Figure 19-7 illustrates the Carnot theorem with a numerical example worked out in the figure caption.

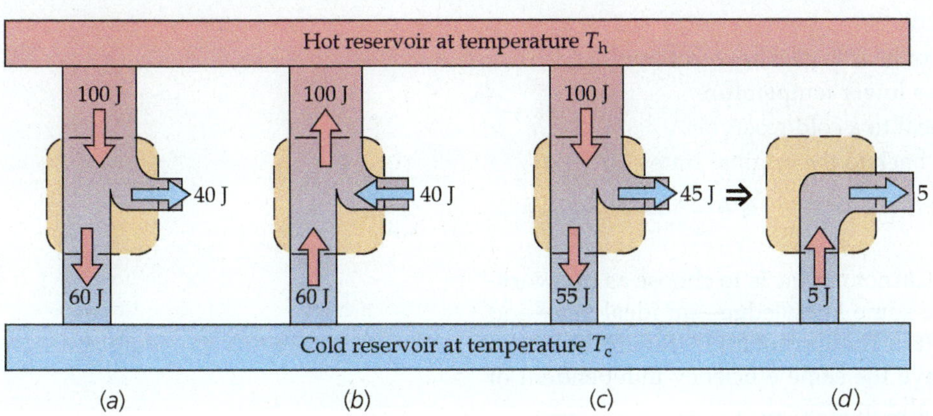

FIGURE 19-7 Illustration of the Carnot theorem. (*a*) A reversible heat engine with 40 percent efficiency absorbs 100 J from a hot reservoir, does 40 J work, and releases 60 J to the cold reservoir. (*b*) If the same engine runs backwards as a refrigerator, 40 J of work are done to absorb 60 J from the cold reservoir and release 100 J to the hot reservoir. (*c*) An assumed heat engine working between the same two reservoirs with an efficiency of 45 percent, which is greater than that of the reversible engine in Part (*a*). (*d*) The net effect of running the engine in Part (*c*) in conjunction with the refrigerator in Part (*b*) is the same as that of a perfect heat engine that absorbs 5 J from the cold reservoir and converts it completely into work with no other effect, violating the second law of thermodynamics. Thus, the reversible engine in Part (*a*) is the most efficient engine that can operate between these two reservoirs.

If no engine can have a greater efficiency than a Carnot engine, it follows that all Carnot engines working between the same two reservoirs have the same efficiency. This efficiency, called the **Carnot efficiency,** must be independent of the working substance of the engine, and thus can depend only on the temperatures of the reservoirs.

Let us look at what makes a process reversible or irreversible. According to the second law, heat is transferred from hot objects to cold objects and never the other way around. Thus, the transfer of heat from a hot object to a cold one is *not* reversible. Also, friction can transform work into internal thermal energy, but friction can never transform internal thermal energy into work. The conversion of work into thermal energy by friction is *not* reversible. Friction and other dissipative forces irreversibly transform mechanical energy into thermal energy. A third type of irreversibility occurs when a system passes through nonequilibrium states, such as when there is turbulence in a gas or when a gas explodes. For a system to undergo a reversible process, the system must be able to go through the same equilibrium states in the reverse order.

From these considerations and our statements of the second law of thermodynamics, we can list some conditions that are necessary for a process to be reversible:

1. No mechanical energy is transformed into internal thermal energy by friction, viscous forces, or other dissipative forces.
2. Energy transfer as heat only occurs between objects with an infinitesimal difference in temperature.
3. The process must be quasi-static so that the system is always at or infinitesimally near an equilibrium state.

CONDITIONS FOR REVERSIBILITY

Any process that violates any of the preceding conditions is irreversible. Most processes we observe in nature are irreversible. To have a reversible process, great care must be taken to eliminate frictional and other dissipative forces and to make the process quasi-static. Because this can never be completely accomplished, a reversible process is an idealization similar to the idealization of motion without friction in mechanics problems. Reversibility can, nevertheless, be closely approximated in practice.

We can now understand the features of a Carnot cycle, which is a reversible cycle between two heat reservoirs. Because all heat transfer must be done isothermally in order for the process to be reversible, the heat absorbed from the hot reservoir must be absorbed isothermally. The next step is a quasi-static, adiabatic expansion to the lower temperature of the cold reservoir. Next, heat is released isothermally to the cold reservoir. Finally, there is a quasi-static, adiabatic compression to the higher temperature of the hot reservoir. The Carnot cycle thus consists of four reversible steps:

1. A quasi-static, isothermal absorption of heat from a hot reservoir
2. A quasi-static, adiabatic expansion to a lower temperature
3. A quasi-static, isothermal release of heat to a cold reservoir
4. A quasi-static, adiabatic compression back to the original state

STEPS IN A CARNOT CYCLE

One way to calculate the efficiency of a Carnot engine is to choose as the working substance a material of which we have some knowledge—an ideal gas—and then explicitly calculate the work done on it over a Carnot cycle (Figure 19-8*a* and Figure 19-8*b*). Because all Carnot cycles have the same efficiency independent of the working substance, our result will be valid in general.

The efficiency of the Carnot cycle (Equation 19-2) is

$$\varepsilon = 1 - \frac{Q_c}{Q_h}$$

The heat Q_h is absorbed during the isothermal expansion from state 1 to state 2. The first law of thermodynamics is $\Delta E_{int} = Q_{in} + W_{on}$. For an isothermal expansion of an ideal gas $\Delta E_{int} = 0$. Applying the first law to the isothermal expansion from state 1 to state 2, we have $Q_h = Q_{in}$, so Q_h equals the work done by the gas:

$$Q_h = W_{by\ gas} = \int_{V_1}^{V_2} P\,dV = \int_{V_1}^{V_2} \frac{nRT_h}{V}dV = nRT_h \int_{V_1}^{V_2} \frac{dV}{V} = nRT_h \ln \frac{V_2}{V_1}$$

Similarly, the heat released to the cold reservoir equals the work done on the gas during the isothermal compression at temperature T_c from state 3 to state 4. This work has the same magnitude as that done by the gas if it expands from state 4 to state 3. The heat rejected is thus

$$Q_c = W_{on\ gas} = nRT_c \ln \frac{V_3}{V_4}$$

The ratio of these heats is

$$\frac{Q_c}{Q_h} = \frac{T_c \ln \frac{V_3}{V_4}}{T_h \ln \frac{V_2}{V_1}} \qquad \text{19-4}$$

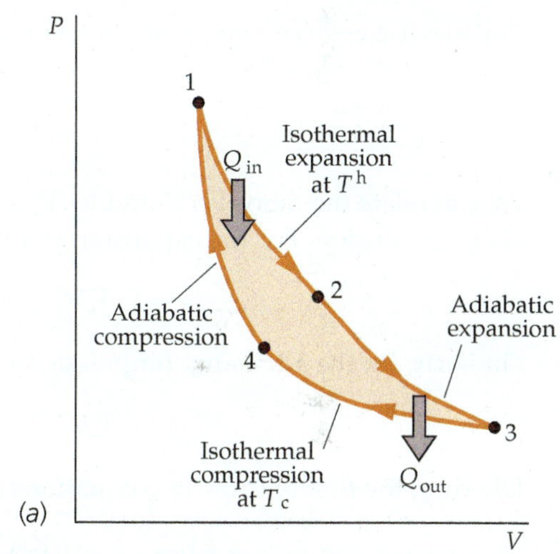

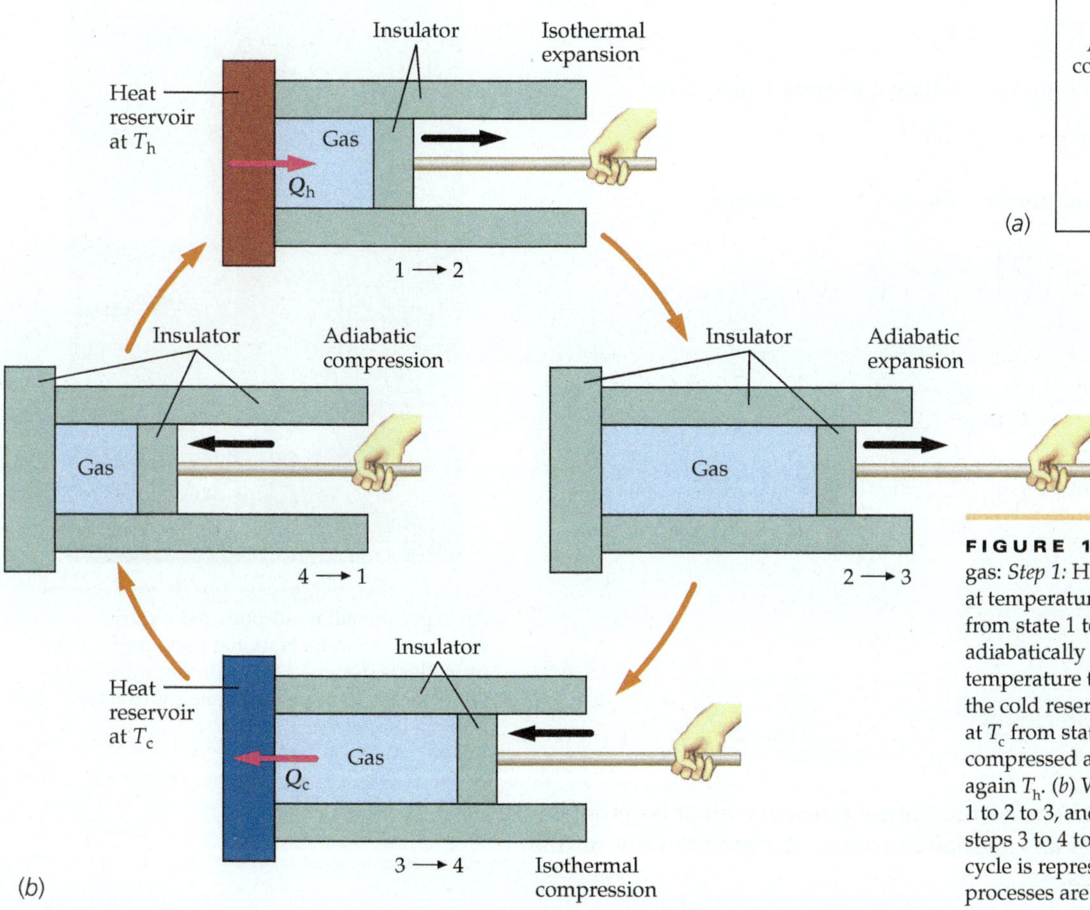

FIGURE 19-8 (*a*) Carnot cycle for an ideal gas: *Step 1:* Heat is absorbed from a hot reservoir at temperature T_h during an isothermal expansion from state 1 to state 2. *Step 2:* The gas expands adiabatically from state 2 to state 3, reducing its temperature to T_c. *Step 3:* The gas releases heat to the cold reservoir as it is compressed isothermally at T_c from state 3 to state 4. *Step 4:* The gas is compressed adiabatically until its temperature is again T_h. (*b*) Work is done by the gas during steps 1 to 2 to 3, and work is done on the gas during steps 3 to 4 to 1. The net work done during the cycle is represented by the shaded area. All processes are reversible. All steps are quasi-static.

Coal-fueled electric generating plant at Four Corners, New Mexico. *(Michael Collier/Stock, Boston.)*

Power plant at Wairakei, New Zealand, that converts geothermal energy into electricity. *(Jean-Pierre Horlin/The Image Bank.)*

We can relate the ratios V_2/V_1 and V_3/V_4 using Equation 18-37 for a quasi-static adiabatic expansion. For the expansion from state 2 to state 3, we have

$$T_h V_2^{\gamma-1} = T_c V_3^{\gamma-1}$$

Similarly, for the adiabatic compression from state 4 to state 1, we have

$$T_h V_1^{\gamma-1} = T_c V_4^{\gamma-1}$$

Dividing the first of these two equations by the second, we obtain

$$\left(\frac{V_2}{V_1}\right)^{\gamma-1} = \left(\frac{V_3}{V_4}\right)^{\gamma-1} \Rightarrow \frac{V_2}{V_1} = \frac{V_3}{V_4}$$

Thus, Equation 19-4 gives

$$\frac{Q_c}{Q_h} = \frac{T_c \ln \frac{V_2}{V_1}}{T_h \ln \frac{V_2}{V_1}} = \frac{T_c}{T_h} \qquad \text{19-5}$$

The Carnot efficiency ε_C is thus

$$\varepsilon_C = 1 - \frac{T_c}{T_h} \qquad \text{19-6}$$

CARNOT EFFICIENCY

An experimental wind-powered electric generator at Sandia National Laboratory. The propeller is designed for optimum transfer of wind energy to mechanical energy. *(Sandia National Laboratory.)*

Equation 19-6 demonstrates that because the Carnot efficiency must be independent of the working substance of any particular engine, it depends only on the temperatures of the two reservoirs.

Solar energy is focused and collected individually for these heliostats, which are being tested at Sandia National Laboratory, to produce electricity. *(Sandia National Laboratory.)*

Control rods are inserted into this nuclear reactor at Tihange, Belgium. *(Peter Miller/The Image Bank.)*

Example 19-4 Efficiency of a Steam Engine

A steam engine works between a hot reservoir at 100°C (373 K) and a cold reservoir at 0.0°C (273 K). (*a*) What is the maximum possible efficiency of this engine? (*b*) If the engine is run backwards as a refrigerator, what is its maximum coefficient of performance?

PICTURE The maximum efficiency is the Carnot efficiency given by Equation 19-6. To find the maximum COP, we use the definition of efficiency ($\varepsilon = W/Q_h$), the definition of COP ($\text{COP} = Q_c/W$), and, because it is a reversible cycle, $Q_c/Q_h = T_c/T_h$ (Equation 19-5).

SOLVE

(*a*) The maximum efficiency is the Carnot efficiency:

$$\varepsilon_{max} = \varepsilon_C = 1 - \frac{T_c}{T_h} = 1 - \frac{273\text{ K}}{373\text{ K}} = 0.268 = \boxed{26.8\%}$$

(*b*) 1. Write the expression for the COP if the engine is run in reverse for a single cycle:

$$\text{COP} = \frac{Q_c}{W}$$

2. The work is equal to $Q_h - Q_c$ (the heat absorbed from the high-temperature reservoir minus the heat released to the low-temperature reservoir):

$$\text{COP} = \frac{Q_c}{Q_h - Q_c} = \frac{1}{\dfrac{Q_h}{Q_c} - 1}$$

3. Substitute for Q_h/Q_c using $Q_c/Q_h = T_c/T_h$ (Equation 19-5) and solve for the COP:

$$\text{COP}_{max} = \frac{1}{\dfrac{T_h}{T_c} - 1} = \frac{1}{\dfrac{373\text{ K}}{273\text{ K}} - 1} = \boxed{2.73}$$

CHECK Solving the Part-(*a*) result for the ratio of temperatures, substituting into the Part-(*b*), step-3 result, and rearranging gives $\text{COP}_{max} = \varepsilon_C^{-1} - 1$. Substituting 0.268 (the Part-(*a*) result) for ε_C gives $\text{COP}_{max} = 2.73$ (the Part-(*b*) result).

TAKING IT FURTHER Even though an efficiency of 26.8 percent seems to be quite low, it is the greatest efficiency possible for any engine working between these two temperatures. Real engines operating between these temperatures will have lower efficiencies because of friction, heat leaks, and other irreversible processes. Real refrigerators will have lower coefficients of performance than 2.73. It can be shown that the coefficient of performance of a Carnot refrigerator is $T_c/\Delta T$, where $\Delta T = T_h - T_c$.

The Carnot efficiency gives us an upper limit on possible efficiencies, and is therefore useful to know. For example, we calculated in Example 19-4 that the Carnot efficiency is 26.8 percent. This means that, no matter how much we reduce friction and other irreversible losses, the best efficiency obtained between reservoirs at 373 K and 273 K is 26.8 percent. We would know, then, that an engine working between those two temperatures with an efficiency of 25 percent is a very good engine!

For an actual engine, the **work lost** is the work done by a reversible engine operating between the same two temperatures minus the work done by the actual engine, assuming both engines complete an integral number of cycles and both absorb the same amount of heat from the high-temperature reservoir. The ratio of the work done by the actual engine and the work done by a reversible engine operating between the same two temperatures is called the ***second law efficiency.***

Example 19-5 Work Lost by an Engine

An engine absorbs 200 J from a hot reservoir at 373 K, does 48.0 J of work, and releases 152 J to a cold reservoir at 273 K. What is the work lost due to irreversible processes in this engine?

PICTURE The work lost is the work done by a reversible engine operating between the same two temperatures minus the 18 J of work done by the engine, assuming both engines absorb the same amount of heat from the high-temperature reservoir.

SOLVE

1. The work lost is the maximum amount of work that could be done minus the work actually done:
$$W_{\text{lost}} = W_{\text{max}} - W$$
2. The maximum amount of work that could be done is the work done using a Carnot engine:
$$W_{\text{max}} = \varepsilon_C Q_h$$
3. The work lost is then:
$$W_{\text{lost}} = \varepsilon_C Q_h - W$$
4. The Carnot efficiency can be expressed in terms of the temperatures:
$$\varepsilon_C = 1 - \frac{T_c}{T_h}$$
5. Substituting for ε_C gives:
$$W_{\text{lost}} = \left(1 - \frac{T_c}{T_h}\right)Q_h - W = \left(1 - \frac{273\text{ K}}{373\text{ K}}\right)(200\text{ J}) - 48\text{ J}$$
$$= 53.6\text{ J} - 48.0\text{ J} = \boxed{5.6\text{ J}}$$

CHECK The Carnot efficiency for these two temperatures is 26.8%. The work done by the engine in this example is given as 48.0 J, and 48.0 J is 24% of 200 J. In addition, the 5.6 J of work lost is 2.4% of 200 J. Because 24% plus 2.4% equals 26.8%, our answer is plausible.

TAKING IT FURTHER The 5.6 J of energy in the answer is not "lost" to the universe—total energy is conserved. That 5.6 J of energy transferred to the cold reservoir by the nonideal engine of the problem is only lost in that it would have been converted into useful work if an ideal (reversible) engine had been used.

Example 19-6 Work Lost Between Heat Reservoirs

If 200 J of heat are released by a reservoir at 373 K and absorbed by a second reservoir at 273 K, how much work capability is "lost" in this process?

PICTURE No work is done during the transfer of the 200 J. Thus, the work lost is 100 percent of the work that would be done by a reversible engine operating between the same two reservoirs that absorbs 200 J from the high-temperature reservoir.

SOLVE

1. The work lost is the work done by a reversible engine minus the work done by the process described here. This process is the transfer of 200 J of heat from the high-temperature reservoir to the low-temperature reservoir, so the work done is zero:

$$W_{\text{lost}} = W_{\text{rev}} - W = W_{\text{rev}} - 0$$

2. The work done by a reversible engine operating between the same two reservoirs and the absorption of 200 J from the high-temperature reservoir is:

$$W_{\text{rev}} = \varepsilon Q_h = \left(1 - \frac{T_c}{T_h}\right)Q_h = \left(1 - \frac{273\text{ K}}{373\text{ K}}\right)(200\text{ J}) = 53.6\text{ J}$$

3. Calculate the work lost:

$$W_{\text{lost}} = W_{\text{rev}} = \boxed{53.6\text{ J}}$$

CHECK In Example 19-4, we calculated the efficiency of a reversible engine operating between 273 K and 373 K to be 26.8%. Our step-3 result is plausible because 53.6 J is 26.8% of the 200 J absorbed from the reservoir.

PRACTICE PROBLEM 19-3 A reversible engine works between heat reservoirs at 500 K and 300 K. (*a*) What is its efficiency? (*b*) If during each cycle the engine absorbs 200 kJ of heat from the hot reservoir, how much work does it do during each cycle?

PRACTICE PROBLEM 19-4 A real engine works between heat reservoirs at 500 K and 300 K. It absorbs 500 kJ of heat from the hot reservoir and does 150 kJ of work during each cycle. What is its efficiency?

THE THERMODYNAMIC OR ABSOLUTE TEMPERATURE SCALE

In Chapter 17, the ideal-gas temperature scale was defined in terms of the properties of gases that have low densities. Because the Carnot efficiency depends only on the temperatures of the two heat reservoirs, it can be used to define the ratio of the temperatures of the reservoirs independent of the properties of any substance. We *define* the ratio of the thermodynamic temperatures of the hot and cold reservoirs to be

$$\frac{T_c}{T_h} = \frac{Q_c}{Q_h} \qquad 19\text{-}7$$

DEFINITION OF THERMODYNAMIC TEMPERATURE

where Q_h is the energy absorbed from the hot reservoir and Q_c is the energy released to the cold reservoir by a Carnot engine operating in a cycle and working between the two reservoirs. Thus, to find the ratio of two reservoir temperatures, we set up a reversible engine operating between them and measure the energy transferred as heat to or from each reservoir during one cycle. The **thermodynamic temperature** is completely specified by Equation 19-7 *and* the choice of one fixed point. If the fixed point is defined to be 273.16 K for the triple point of water, then the thermodynamic temperature scale matches the ideal-gas temperature scale for the range of temperatures over which a gas thermometer can be used. Any temperature that reads zero at absolute zero is called an *absolute-temperature scale.*

* 19-4 HEAT PUMPS

A **heat pump** is a refrigerator with a different objective. Typically, the objective of a refrigerator is to cool an object or region of interest. The objective of a heat pump, however, is to heat an object or region of interest. For example, if you use a heat pump to heat your house, you transfer heat from the cold air outside the house to the warmer air inside it. Your objective is to heat the region inside your house. If work W is done on a heat pump to absorb heat Q_c from the cold reservoir and release heat Q_h to the hot reservoir, the coefficient of performance for a heat pump is defined as

$$\text{COP}_{\text{HP}} = \frac{Q_h}{W} \tag{19-8}$$

DEFINITION: COEFFICIENT OF PERFORMANCE (HEAT PUMP)

This coefficient of performance differs from that for the refrigerator, which is Q_c/W (Equation 19-3). Using $W = Q_h - Q_c$, this can be written

$$\text{COP}_{\text{HP}} = \frac{Q_h}{Q_h - Q_c} = \frac{1}{1 - \frac{Q_c}{Q_h}} \tag{19-9}$$

The maximum coefficient of performance is obtained using a Carnot heat pump. Then Q_c and Q_h are related by Equation 19-5. Substituting $Q_c/Q_h = T_c/T_h$ into Equation 19-9, we obtain for the maximum coefficient of performance

$$\text{COP}_{\text{HP max}} = \frac{1}{1 - \frac{T_c}{T_h}} = \frac{T_h}{T_h - T_c} = \frac{T_h}{\Delta T} \tag{19-10}$$

where ΔT is the difference in temperature between the hot and cold reservoirs. Real heat pumps have coefficients of performance less than the $\text{COP}_{\text{HP max}}$ because of friction, heat leaks, and other irreversible processes.

The two COPs are related. Using $Q_h = Q_c + W$, we can relate Equations 19-3 and 19-10:

$$\text{COP}_{\text{HP}} = \frac{Q_h}{W} = \frac{Q_c + W}{W} = 1 + \frac{Q_c}{W} = 1 + \text{COP} \tag{19-11}$$

where COP is the coefficient of performance of a refrigerator.

Example 19-7 An Ideal Heat Pump

An ideal heat pump is used to pump heat from the outside air at −5°C to the hot-air supply for the heating fan in a house, which is at 40°C. How much work is required to pump 1.0 kJ of heat into the house?

PICTURE Use Equation 19-11 with $\text{COP}_{\text{HP max}}$ calculated from Equation 19-10 for $T_c = 25°\text{C} = 268\ \text{K}$ and $\Delta T = 45\ \text{K}$.

SOLVE

1. Using the definition of COP_{HP} ($\text{COP}_{\text{HP}} = Q_h/W$), relate the work done to the heat released: $\text{COP}_{\text{HP}} = \frac{Q_h}{W}$
2. Relate the ideal or maximum COP_{HP} to the temperatures (Equation 19-10): $\text{COP}_{\text{HP}} = \text{COP}_{\text{HP max}} = \frac{T_h}{\Delta T}$
3. Solve for the work: $W = \frac{Q_h}{\text{COP}_{\text{HP}}} = Q_h \frac{\Delta T}{T_h} = (1.0\ \text{kJ})\frac{45\ \text{K}}{313\ \text{K}}$

   $W = \boxed{0.14\ \text{kJ}}$

CHECK Our step-3 expression for the work ensures the work has the same dimensions as heat. (The ratio $\Delta T/T_h$ is dimensionless.)

TAKING IT FURTHER The $\text{COP}_{\text{HP max}} = T_h/\Delta T = 7.0$. That is, the amount of heat released inside the house by the heat pump is 7 times larger than the amount of work done on the heat pump. (Only 0.14 kJ of work is needed to pump 1.0 kJ of heat into the hot-air supply in the house.)

19-5 IRREVERSIBILITY, DISORDER, AND ENTROPY

Many irreversible processes exist that cannot be described by the heat-engine or refrigerator statements of the second law, such as a glass falling to the floor and breaking or a balloon popping. However, all irreversible processes have one thing in common—the system plus its surroundings move toward a less ordered state.

Suppose a box of negligible mass that contains a gas of mass M at a temperature T is moving along a frictionless table with a velocity v_{cm} (Figure 19-9*a*). The total kinetic energy of the gas has two components: that associated with the movement of its center of mass $\frac{1}{2}Mv_{cm}^2$, and the kinetic energy of the motion of its molecules relative to its center of mass. The center-of-mass energy $\frac{1}{2}Mv_{cm}^2$ is ordered mechanical energy that could be converted entirely into work. (For example, if a weight were attached to the moving box by a string passing over a pulley, this energy could be used to lift the weight.) The relative energy is the internal thermal energy of the gas, which is related to its temperature T. It is random, nonordered energy that cannot be converted entirely into work.

Now, suppose that the box hits a fixed wall and stops (Figure 19-9*b*). This inelastic collision is clearly an irreversible process. The ordered mechanical energy of the gas $\frac{1}{2}Mv_{cm}^2$ is transformed into random internal energy, and the temperature of the gas increases. The gas still has the same total energy, but now all of that energy is associated with the random motion of the gas molecules relative to the center of mass, which is now at rest. Thus, the gas has become less ordered (more disordered), and has lost some of its ability to do work.

There is a thermodynamic function called **entropy** S that is a measure of the disorder of a system. Entropy S, like pressure P, volume V, temperature T, and internal energy E_{int}, is a function of the state of a system. As with potential energy and internal energy, it is the *change* in entropy that is important. The change in entropy dS of a system as it goes from one state to another is defined as

$$dS = \frac{dQ_{rev}}{T} \qquad \text{19-12}$$

DEFINITION: ENTROPY CHANGE

where dQ_{rev} is the heat absorbed by the system in a *reversible* process. If dQ_{rev} is negative, then the entropy change of the system has a negative value and the entropy of the system has decreased.

The term dQ_{rev} does not mean that a reversible heat transfer must take place in order for the entropy of a system to change. Indeed, there are many situations in which the entropy of a system changes when there is no transfer of heat whatsoever, for example, the box of gas colliding with the wall in Figure 19-9. Equation 19-12 simply gives us a method for *calculating* the entropy difference between two states of a system. Because entropy is a state function, the change in entropy when the

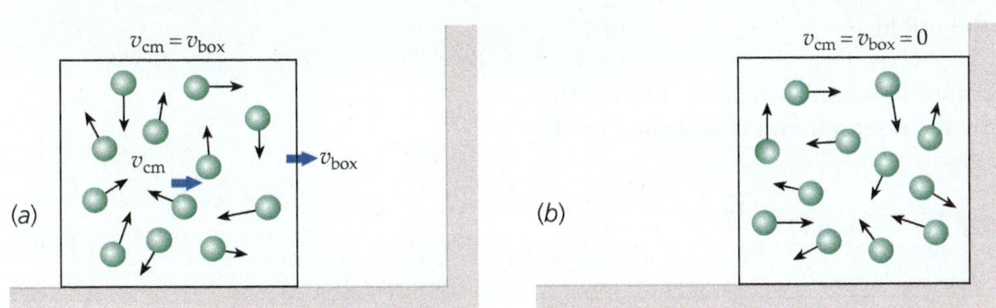

FIGURE 19-9 (*a*) A box of negligible mass contains a gas. The box and the center of mass of the gas move toward the wall at the same speed. (*b*) A short time after the box undergoes a perfectly inelastic collision with the wall, both the box and the center of mass of the gas are at rest, and the gas has a higher temperature.

system moves from an initial state to a final state depends only on the system's initial and final states, and not on the process by which the change occurs. That is, if S_1 is the entropy of the system when it is in state 1, and if S_2 is the entropy of the system when it is in state 2, then we calculate the difference in the entropy $S_2 - S_1$ by evaluating the integral $\int_1^2 dQ/T$ for *any* reversible path (process) that takes the system from state 1 to state 2.

ENTROPY OF AN IDEAL GAS

We can illustrate that dQ_{rev}/T is in fact the differential of a state function for an ideal gas (even though dQ_{rev} is not). Consider an arbitrary reversible quasi-static process in which a system consisting of an ideal gas absorbs an amount of heat dQ_{rev}. According to the first law, dQ_{rev} is related to the change in the internal energy dE_{int} of the gas and the work done on the gas ($dW_{on} = -P\,dV$) by

See
Math Tutorial ***for more information on***
Integrals

$$dE_{int} = dQ_{rev} + dW_{on} = dQ_{rev} - P\,dV$$

For an ideal gas, we can write dE_{int} in terms of the heat capacity, $dE_{int} = C_v\,dT$, and we can substitute nRT/V for P from the equation of state. Then

$$C_v\,dT = dQ_{rev} - nRT\frac{dV}{V} \qquad \text{19-13}$$

Equation 19-13 cannot be integrated directly unless we know how T depends on V and how C_v depends on T. This is just another way of saying that dQ_{rev} is not a differential of a state function Q_{rev}. But if we divide each term by T, we obtain

$$C_v\frac{dT}{T} = \frac{dQ_{rev}}{T} - nR\frac{dV}{V} \qquad \text{19-14}$$

Because C_v depends only on T, the term on the left can be integrated, as can the term on the right.* Thus, dQ_{rev}/T is the differential of a function, the entropy function S:

$$dS = \frac{dQ_{rev}}{T} = C_v\frac{dT}{T} + nR\frac{dV}{V} \qquad \text{19-15}$$

For simplicity, we will assume that C_v is constant. Integrating Equation 19-15 from state 1 to state 2, we obtain

$$\Delta S = \int\frac{dQ_{rev}}{T} = C_v\ln\frac{T_2}{T_1} + nR\ln\frac{V_2}{V_1} \qquad \text{19-16}$$

ENTROPY CHANGE FOR AN IDEAL GAS

Equation 19-16 gives the entropy change of an ideal gas that goes from an initial state of volume V_1 and temperature T_1 to a final state of volume V_2 and temperature T_2.

ENTROPY CHANGES FOR VARIOUS PROCESSES

ΔS for an isothermal expansion of an ideal gas If an ideal gas undergoes an isothermal expansion, then $T_2 = T_1$ and its entropy change is

$$\Delta S = \int\frac{dQ_{rev}}{T} = 0 + nR\ln\frac{V_2}{V_1} \qquad \text{19-17}$$

The entropy change of the gas is positive because V_2 is greater than V_1. During this process, an amount of heat Q_{rev} is released by the reservoir and is absorbed by the gas. This heat equals the work done by the gas:

$$Q_{rev} = W_{by} = \int_{V_1}^{V_2} P\,dV = nRT\int_{V_1}^{V_2}\frac{dV}{V} = nRT\ln\frac{V_2}{V_1} \qquad \text{19-18}$$

---

* Mathematically, the factor $1/T$ is called an integrating factor for Equation 19-13.

The entropy change of the gas is $+Q_{rev}/T$. Because the same amount of heat is released by the reservoir at temperature T, the entropy change of the reservoir is $-Q_{rev}/T$. The net entropy change of the gas plus the reservoir is zero. We will refer to the system under consideration plus its surroundings as the "universe." This example illustrates a general result:

During a reversible process, the entropy change of the universe is zero.

ΔS *for a free expansion of an ideal gas* During the free expansion of a gas discussed in Section 18-4, the gas is initially confined in one compartment of a container, which is connected by a stopcock to another compartment that is evacuated. The whole system has rigid walls and is thermally insulated from its surroundings so that no heat can be absorbed by, or released from, the system, and no work can be done on (or by) the system (Figure 19-10). When the stopcock is opened, the gas rushes into the evacuated chamber. The gas eventually reaches thermal equilibrium. Because no work is done on the gas and no heat is either absorbed or released by the gas, the final internal energy of the gas must equal its initial internal energy. If we assume that the gas is ideal, then its internal energy depends only on the temperature T, so its final temperature equals its initial temperature.

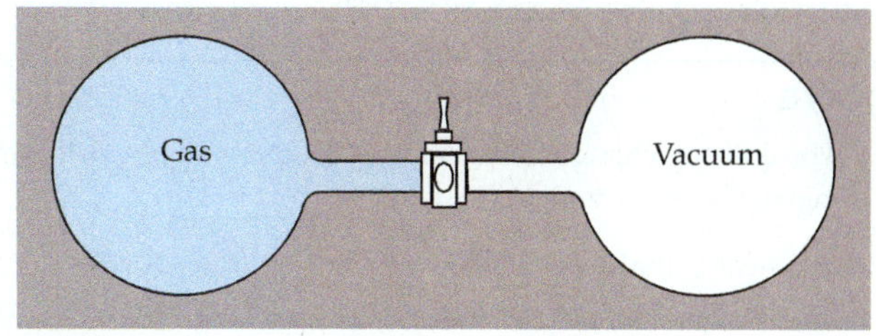

FIGURE 19-10 Adiabatic free expansion of a gas. When the stopcock is opened, the gas expands rapidly into the evacuated chamber. The initial and final internal energies of the gas are equal because no work is done on the gas during the expansion, because the entire system is insulated from its surroundings, and because the heat capacities of the chambers and valve are negligible.

We might think that there is no entropy change of the gas because there is no heat transfer. But this process is not reversible, so $\int dQ/T$ cannot be used to find the change in entropy of the gas. However, the initial and final states of the gas in the free expansion are the same as those of the gas in the isothermal expansion just discussed. *Because the change in the entropy of a system for any process depends only on the initial and final states of the system, the entropy change of the gas for the free expansion is the same as that for the isothermal expansion.* If V_1 is the initial volume of the gas and V_2 is its final volume, the entropy change of the gas is given by Equation 19-17, or

$$\Delta S_{gas} = nR\ln\frac{V_2}{V_1}$$

In this case, there is no change in the surroundings, so the entropy change of the gas is also the entropy change of the universe:

$$\Delta S_u = nR\ln\frac{V_2}{V_1} \qquad 19\text{-}19$$

Note that because V_2 is greater than V_1, the change in entropy of the universe for this irreversible process is positive; that is, the entropy of the universe increases. This is also a general result:

During an irreversible process, the entropy of the universe increases.

If the final volume during the free expansion were less than the initial volume, then the entropy of the universe would decrease—but this does not happen during free expansions. That is, a gas does not freely contract by itself into a smaller volume.* This leads us to yet another statement of the second law of thermodynamics:

For any process, the entropy of the universe never decreases.

* What is actually the case, is that the probability of a gas freely contracting into a smaller volume is minuscule (except when the gas contains only an extremely small number of molecules).

Example 19-8 Free Expansion of an Ideal Gas

Find the entropy change for the free expansion of 0.75 mol of an ideal gas from $V_1 = 1.5$ L to $V_2 = 3.0$ L.

PICTURE For a free expansion of an ideal gas the initial and final temperatures are the same. Thus, the entropy change ΔS for a free expansion from V_1 to V_2 is the same as ΔS for an isothermal process from V_1 to V_2. For the isothermal process $\Delta E_{\text{int}} = 0$, so $Q = W_{\text{by}}$. First, we calculate Q, then we set $\Delta S = Q/T$.

SOLVE

1. The entropy change is the same as for a reversible isothermal expansion from V_1 to V_2:

$$\Delta S = \Delta S_{\text{isothermal}} = \int_1^2 \frac{dQ_{\text{rev}}}{T} = \frac{1}{T}\int_1^2 dQ_{\text{rev}} = \frac{Q}{T}$$

2. The heat Q that would be absorbed by the gas during an isothermal expansion at temperature T equals the work done by the gas during the expansion:

$$Q = W_{\text{by}} = nRT \ln \frac{V_2}{V_1}$$

3. Substitute this value of Q to calculate ΔS:

$$\Delta S = \frac{Q}{T} = nR \ln \frac{V_2}{V_1} = (0.75\ \text{mol})(8.31\ \text{J/(mol}\cdot\text{K)}) \ln 2$$

$$= \boxed{4.3\ \text{J/K}}$$

CHECK The units in step 3 are such that moles cancel leaving joules per kelvin. Joules per kelvin are the correct units for entropy changes because, by definition, $\Delta S = \int dQ/T$.

CONCEPT CHECK 19-1

A living organism consists of highly organized matter. Does the growth of a living organism constitute a violation of the second law of thermodynamics? That is, during this process does the entropy of the universe increase or decrease?

ΔS for constant-pressure processes If a substance is heated from temperature T_1 to temperature T_2 at constant pressure, the heat absorbed dQ is related to its temperature change dT by

$$dQ = C_p\, dT$$

We can approximate reversible heat transfer if we have a large number of heat reservoirs with temperatures ranging from just slightly greater than T_1 to T_2 in very small steps. We could place the substance, with initial temperature T_1, in contact with the first reservoir at a temperature just slightly greater than T_1 and let the substance absorb a small amount of heat. Because the heat transfer from each reservoir is approximately isothermal, the process will be approximately reversible. We then place the substance in contact with the next reservoir at a slightly higher temperature, and so on, until the final temperature T_2 is reached. If heat dQ is absorbed reversibly, the entropy change of the substance is

$$dS = \frac{dQ}{T} = C_p \frac{dT}{T}$$

Integrating from T_1 to T_2, we obtain the total entropy change of the substance:

$$\Delta S = C_p \int_{T_1}^{T_2} \frac{dT}{T} = C_p \ln \frac{T_2}{T_1} \qquad \text{19-20}$$

This result gives the entropy change of a substance that is heated from T_1 to T_2 by any process, reversible or irreversible, as long as the final pressure equals the initial pressure and C_p is constant. It also gives the entropy change of a substance that is cooled. In the case of cooling, T_2 is less than T_1, and $\ln(T_2/T_1)$ is negative, giving a negative entropy change.

PRACTICE PROBLEM 19-5

Find the change in entropy of 1.00 kg of water that is heated at constant pressure from 0°C to 100°C.

PRACTICE PROBLEM 19-6

Derive Equation 19-20 directly from Equation 19-16.

Example 19-9 Entropy Changes During Heat Transfer

Suppose a 1.00-kg sample of water at temperature $T_1 = 30.0°C$ is added to a 2.00-kg sample of water at $T_2 = 90.0°C$ in an insulated calorimeter of negligible heat capacity at a constant pressure of 1.00 atm. (*a*) Find the change in entropy of the system. (*b*) Find the change in entropy of the universe.

PICTURE When the two samples of water are combined, they eventually come to a final equilibrium temperature, T_f, that can be found by setting the heat released equal to the heat absorbed. To calculate the entropy change of each sample of water, we consider a reversible isobaric (constant pressure) heating of the 1.00-kg sample of water from 30°C to T_f, and a reversible isobaric cooling of the 2.00-kg sample from 90°C to T_f using Equation 19-20. The entropy change of the system is the sum of the entropy changes of each part. The entropy change of the universe is the entropy change of the system plus the entropy change of its surroundings. To find the entropy change of the surroundings, assume a negligible amount of heat is absorbed or released by the calorimeter during the time it takes the water to reach its final temperature.

SOLVE

(*a*) 1. Calculate T_f by setting the heat released equal to the heat absorbed:

$$T_f = 70°C = 343\text{ K}$$

2. Use your result for T_f and Equation 19-20 to calculate ΔS_1 and ΔS_2:

$$\Delta S_1 = \int_1^f \frac{dQ_{rev}}{T} = \int_{T_1}^{T_f} \frac{C_p\,dT}{T} = C_p \int_{T_1}^{T_f} \frac{dT}{T} = C_p \ln\frac{T_f}{T_1} = m_1 c_p \ln\frac{T_f}{T_1}$$

$$= (1.00\text{ kg})(4.184\text{ kJ/kg}\cdot\text{K}) \ln\frac{343\text{ K}}{303\text{ K}} = 0.519\text{ kJ/K}$$

$$\Delta S_2 = (2.00\text{ kg})(4.184\text{ kJ/kg}\cdot\text{K}) \ln\frac{343\text{ K}}{363\text{ K}} = -0.474\text{ kJ/K}$$

3. Add ΔS_1 and ΔS_2 to find the total entropy change of the system:

$$\Delta S_{system} = \boxed{+0.045\text{ kJ/K}}$$

(*b*) 1. The calorimeter is insulated, so the surroundings are unchanged:

$$\Delta S_{surroundings} = 0$$

2. Add ΔS_{system} and $\Delta S_{surroundings}$ to find the entropy change of the universe:

$$\Delta S_u = \boxed{+0.045\text{ kJ/K}}$$

CHECK The Part-(*b*) result is a positive number, as expected. (The process is irreversible and the entropy change of the universe is never negative.)

***ΔS* for a perfectly inelastic collision** Because mechanical energy is converted into internal thermal energy during an inelastic collision, such a process is clearly irreversible. The entropy of the universe must therefore increase. Consider a block of mass m falling from a height h and making a perfectly inelastic collision with the ground. Let the block, ground, and atmosphere all be at the same temperature T, which is not significantly changed by the process. If we consider the block, ground, and atmosphere as our thermally isolated system, there is no heat absorbed or released by the system. The state of the system has been changed because its internal energy has been increased by an amount mgh. This change is the same as if we added heat $Q = mgh$ to the system at constant temperature T. To calculate the change in entropy of the system, we thus consider a reversible process in which heat $Q_{rev} = mgh$ is absorbed at a constant temperature T. According to Equation 19-12, the change in entropy of the system is then

$$\Delta S = \frac{Q_{rev}}{T} = \frac{mgh}{T}$$

This positive entropy change is also the entropy change of the universe.

***ΔS* for heat transfer from one reservoir to another** Heat transfer is also an irreversible process, and so we expect the entropy of the universe to increase when this occurs. Consider the simple case of heat Q transferred from a hot reservoir at a temperature T_h to a cold reservoir at a temperature T_c. The state of a heat reservoir is determined by its temperature and its internal energy only. The change in entropy of a heat reservoir due to a heat transfer is the same whether the heat transfer is reversible or not. If heat Q is absorbed by a reservoir at temperature T, then the entropy of the reservoir increases by Q/T, and if the heat Q is released by a reservoir at temperature T, then the entropy of the reservoir changes by $-Q/T$. In the case of heat transfer, the hot reservoir releases heat, so its entropy change is

$$\Delta S_h = -\frac{Q}{T_h}$$

The cold reservoir absorbs heat, so its entropy change is

$$\Delta S_c = +\frac{Q}{T_c}$$

The net entropy change of the universe is

$$\Delta S_u = \Delta S_c + \Delta S_h = \frac{Q}{T_c} - \frac{Q}{T_h} \qquad \text{19-21}$$

Note that, because heat transfers from a hot reservoir to a cold reservoir, the change in entropy of the universe is positive.

ΔS for a Carnot cycle Because a Carnot cycle is by definition reversible, the entropy change of the universe after a cycle must be zero. We demonstrate this by showing that the entropy change of the reservoirs in a Carnot engine is zero. (Because a Carnot engine works in a cycle, the entropy change of the engine itself is zero, so the entropy change of the universe is just the sum of the entropy changes of the reservoirs.) The entropy change of the hot reservoir is $\Delta S_h = -(Q_h/T_h)$, and the entropy change of the cold reservoir is $\Delta S_c = +(Q_c/T_c)$. The heats Q_h and Q_c are related to the temperatures T_h and T_c by the definition of thermodynamic temperature (Equation 19-7)

$$\frac{T_c}{T_h} = \frac{Q_c}{Q_h} \qquad \left(\text{or} \quad \frac{Q_h}{T_h} = \frac{Q_c}{T_c}\right)$$

The entropy change of the universe is thus

$$\Delta S_u = \Delta S_{engine} + \Delta S_h + \Delta S_c = 0 - \frac{Q_h}{T_h} + \frac{Q_c}{T_c} = 0$$

The entropy change of the universe is zero as expected.

Notice that we have ignored any entropy change associated with the energy transferred by work from the Carnot engine to its surroundings. If this work is used to raise a weight, or some other ordered process, then there is no entropy change. However, if this work is used to push a block across a tabletop or other surface where friction is involved, then there is an additional entropy increase associated with this work.

Example 19-10 Entropy Changes in a Carnot Cycle

During each cycle, a Carnot engine absorbs 100 J from a reservoir at 400 K, does work, and releases heat to a reservoir at 300 K. Compute the entropy change of each reservoir for each cycle, and show explicitly that the entropy change of the universe is zero for this reversible process.

PICTURE Because the engine works during a cycle, its entropy change is zero. We therefore compute the entropy change of each reservoir and add them to obtain the entropy change of the universe.

SOLVE

1. The entropy change of the universe equals the sum of the entropy changes of the reservoirs:

$$\Delta S_u = \Delta S_{400} + \Delta S_{300}$$

2. Calculate the entropy change of the hot reservoir:

$$\Delta S_{400} = -\frac{Q_h}{T_h} = -\frac{100\text{ J}}{400\text{ K}} = \boxed{-0.250\text{ J/K}}$$

3. The entropy change of the cold reservoir is Q_c divided by T_c, where $Q_c = Q_h - W$:

$$\Delta S_{300} = \frac{Q_c}{T_c} = \frac{Q_h - W}{T_c}$$

4. We use $W = \varepsilon_C Q_h$ (Equation 19-2) to relate W to Q_h. The efficiency is the Carnot efficiency (Equation 19-6):

$$W = \varepsilon Q_h, \quad \text{where} \quad \varepsilon = \varepsilon_C = 1 - (T_c/T_h),$$

$$\text{so} \quad W = \left(1 - \frac{T_c}{T_h}\right)Q_h$$

5. Calculate the entropy change of the cold reservoir:

$$\Delta S_{300} = \frac{Q_h - W}{T_c} = \frac{Q_h - Q_h\left(1 - \frac{T_c}{T_h}\right)}{T_c} = \frac{Q_h}{T_h}$$

$$= \frac{100\text{ J}}{400\text{ K}} = \boxed{0.250\text{ J/K}}$$

6. Substitute these results into step 1 to find the entropy change of the universe:

$$\Delta S_u = \Delta S_{400} + \Delta S_{300}$$

$$\Delta S_u = -0.250\text{ J/K} + 0.250\text{ J/K} = \boxed{0.000\text{ J/K}}$$

CHECK The entropy change of the universe is positive, as the second law of thermodynamics requires.

TAKING IT FURTHER Suppose that an ordinary, nonreversible engine removed 100 J from the hot reservoir. Because its efficiency must be less than that of a Carnot engine, it would do less work and release more heat to the cold reservoir. Then, the entropy increase of the cold reservoir would be greater than the entropy decrease of the hot reservoir, and the entropy change of the universe would be positive.

Example 19-11 The *ST* Plot

Conceptual

Because entropy is a state function, thermodynamic processes can be represented as *ST*, *SV*, or *SP* diagrams in addition to the *PV* diagrams we have used so far. Make a sketch of the Carnot cycle on an *ST* plot.

PICTURE The Carnot cycle consists of first a reversible isothermal expansion, then a reversible adiabatic expansion, followed by a reversible isothermal compression and then a reversible adiabatic compression. During the isothermal processes, heat is absorbed or released reversibly, so S increases or decreases , but T remains constant. During the adiabatic processes, the temperature changes, but because $\Delta Q_{rev} = 0$, S is constant.

SOLVE

1. During the reversible isothermal expansion (1 to 2 in Figure 19-11*a*), heat is absorbed reversibly, so S increases and T remains constant:
2. During the reversible adiabatic expansion (2 to 3 in Figure 19-11*b*), the temperature decreases while S remains constant as T decreases:
3. During the isothermal compression (3 to 4 in Figure 19-11*c*), heat is released reversibly, so S decreases and T remains constat:
4. During the reversible adiabatic compression (4 to 1 in Figure 19-11*d*), the temperature increases while S is constant as T increases:

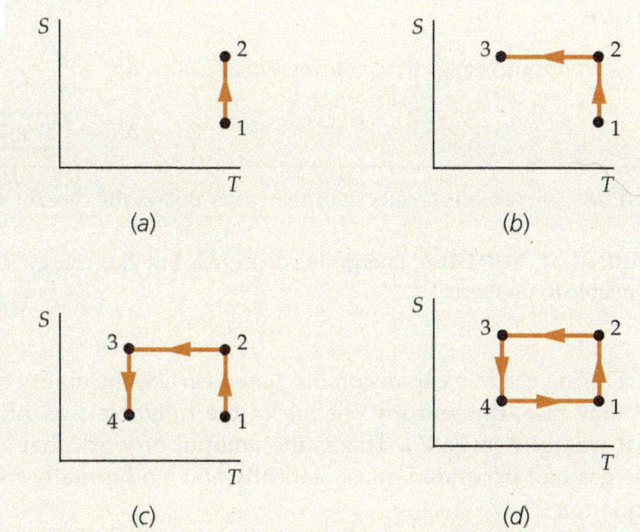

FIGURE 19-11 An *S*-versus-*T* plot of a Carnot cycle using an ideal gas.

CHECK The plot of S versus T is a closed curve, as expected. This is as expected for a complete cycle, because both S and T are state functions.

TAKING IT FURTHER The Carnot cycle is a rectangle if plotted on an *S*-versus-*T* diagram.

19-6 ENTROPY AND THE AVAILABILITY OF ENERGY

If an irreversible process occurs, energy is conserved, but some of the energy becomes unavailable to do work and is "wasted." Consider a block falling to the ground. The entropy change of the universe for this process is mgh/T. When the block was at a height h, its potential energy mgh could have been used to do useful work. But after the inelastic collision of the block with the ground, this energy is no longer fully available for doing useful work because it has become the disordered internal energy of the block and its surroundings.

The energy that has become unavailable (wasted) is equal to $mgh = T\,\Delta S_u$. This is a general result:

> During an irreversible process, energy equal to $T\,\Delta S_u$ becomes unavailable to do work, where T is the temperature of the coldest available reservoir.

For simplicity, we will call the energy that becomes unavailable to do work the "lost work":

$$W_{\text{lost}} = T\,\Delta S_u \qquad \text{19-22}$$

Example 19-12 A Sliding Box Revisited

Suppose that the box shown in Figure 19-9*a* and Figure 19-9*b* has a mass of 2.4 kg and slides with a speed of $v = 3.0$ m/s before crashing into a fixed wall and stopping. The temperature T of the box, table, and surroundings is 293 K and does not change appreciably as the box comes to rest. Find the entropy change of the universe.

PICTURE The initial mechanical energy of the box $\frac{1}{2}Mv^2$ is converted to the internal energy of the box–wall–surroundings system. The entropy change is equivalent to what would occur if the heat $Q = \frac{1}{2}Mv^2$ were absorbed by the box-wall system reversibly.

SOLVE

1. The entropy change of the universe is Q/T:

$$\Delta S_u = \frac{Q}{T} = \frac{\frac{1}{2}Mv^2}{T} = \frac{\frac{1}{2}(2.4\text{ kg})(3.0\text{ m/s})^2}{293\text{ K}}$$

$$\Delta S_u = \boxed{37\text{ mJ/K}}$$

CHECK The result is greater than zero, as is always the case for an irreversible process.

TAKING IT FURTHER Energy is conserved, but the energy $T\,\Delta S_u = \frac{1}{2}Mv^2$ is no longer available to do work.

During the free expansion discussed earlier, the ability to do work was also lost. In that case, the entropy change of the universe was $nR\ln(V_2/V_1)$, so the work lost was $nRT\ln(V_2/V_1)$. This is the amount of work that could have been done if the gas had expanded quasi-statically and isothermally from V_1 to V_2, as given by Equation 19-17.

If all the heat Q released by a hot reservoir is absorbed by a cold reservoir, the change in entropy of the universe is given by Equation 19-21, and the work lost is

$$W_{\text{lost}} = T_c\,\Delta S_u = T_c\left(\frac{Q}{T_c} - \frac{Q}{T_h}\right) = Q\left(1 - \frac{T_c}{T_h}\right)$$

We can see that this is just the work that could have been done by a Carnot engine running between these reservoirs, transferring heat Q from the hot reservoir and doing work $W = \varepsilon_C Q$, where $\varepsilon_C = 1 - T_c/T_h$.

19-7 ENTROPY AND PROBABILITY

Entropy, which is a measure of the disorder of a system, is related to probability. Essentially, a state of high order has a relatively low probability, whereas a state of low order has a relatively high probability. Thus, during an irreversible process, the universe moves from a state of relatively low probability to one of relatively high probability.

Let us consider a free expansion in which a gas expands from an initial volume V_1 to a final volume $V_2 = 2V_1$. The entropy change of the universe for this process is given by Equation 19-19:

$$\Delta S = nR \ln \frac{V_2}{V_1} = nR \ln 2$$

Why is this process irreversible? Why can the gas not spontaneously contract back into its original volume? Such a contraction would not violate the first law of thermodynamics, as there is no energy change involved. The reason that the gas does not compress to its original volume is merely that such a contraction is extremely *improbable*.

Example 19-13 The Probability of a Free Contraction

Suppose a gas consisting of only 10 molecules occupies a cube. What is the probability that all 10 molecules will be in the left half of the cube at a given instant?

PICTURE The chance that any one particular molecule will be in the left half of the container at any given instant is $\frac{1}{2}$. Using this information, we can calculate the probability that all 10 molecules will be in the left half at a given instant?

SOLVE

1. The chance that any one molecule is in the left half is the same as the chance that the same molecule is in the right half:

   The probability that any one particular molecule will be in the left half of the container at any given instant is $\frac{1}{2}$.

2. The chance that molecules 1 and 2 are both in the left half (at any given time) is the chance that molecule 1 is in the left half times the chance that molecule 2 is in the left half:

   The chance that any particular two molecules are both in the left half (at any given time) is $\frac{1}{2} \times \frac{1}{2} = \frac{1}{4}$.

   The chances are equal that molecules 1 and 2 are both in the left half, both in the right half, that molecule 1 is in the left half and molecule 2 is in the right half, or that molecule 2 is in the left half and molecule 1 is in the right half. The chance for any one of these options is $\frac{1}{4}$.

3. The chance that molecules 1, 2, and 3 are all in the left half at any given time is equal to the chance that molecules 1 and 2 are both in the left half times the chance that molecule 3 is in the left half:

   The chance that any particular three molecules are all in the left half (at any given instant) is $\frac{1}{2} \times \frac{1}{2} \times \frac{1}{2} = \left(\frac{1}{2}\right)^3 = \frac{1}{8}$.

4. Continuing this line of reasoning to determine the chance that all 10 molecules are in the left half gives:

   The chance that all 10 molecules are in the left half (at any given instant) is $\boxed{\left(\frac{1}{2}\right)^{10} = \frac{1}{1024}}$.

CHECK We intuitively know the chance that all 10 molecules are on the left side at a given instant is pretty small. It is the same as the chance that a tossed coin comes up heads 10 times in a row.

Although the probability of all 10 molecules in Example 19-13 being on one side of the container is small, we would not be completely surprised to see it occur. If we look at the gas once each second, we could expect to see it happen once in every 1024 seconds, or about once every 17 minutes. If we started with the 10 molecules randomly distributed and then found them all in the left half of the original volume, the entropy of the universe would have *decreased* by $nR \ln 2$. However, this decrease is extremely small, because the number of moles n corresponding to 10 molecules is only about 10^{-23}. Still, it would violate the entropy statement of the second law of thermodynamics, which says that for any process, the entropy of the universe never decreases. Therefore, if we wish to apply the second law of thermodynamics to microscopic systems such as a small number of molecules, we should consider the second law to be a statement of *probability*.

We can relate the probability of a gas spontaneously contracting into a smaller volume to the change in its entropy. If the original volume is V_1, the probability p of finding N molecules in a smaller volume V_2 is

$$p = \left(\frac{V_2}{V_1}\right)^N$$

Taking the natural logarithm of both sides of this equation, we obtain

$$\ln p = N \ln \frac{V_2}{V_1} = nN_A \ln \frac{V_2}{V_1} \qquad \text{19-23}$$

where n is the number of moles and N_A is Avogadro's number. The entropy change of the gas is

$$\Delta S = nR \ln \frac{V_2}{V_1} \qquad \text{19-24}$$

Substituting for $n \ln(V_1/V_2)$ in Equation 19-24, we see that

$$\Delta S = \frac{R}{N_A} \ln p = k \ln p \qquad \text{19-25}$$

where $k = R/N_a$ is Boltzmann's constant.

It may be disturbing to learn that irreversible processes, such as the spontaneous contraction of a gas or the spontaneous transfer of heat from a cold body to a hot body, are not impossible—they are just improbable. As we have just seen, there is a reasonable chance that an irreversible process will occur in a system consisting of a very small number of molecules; however, *thermodynamics itself is applicable only to macroscopic systems*, that is, to systems that have a very large number of molecules. Consider trying to measure the pressure of a gas consisting of only 10 molecules. The pressure would vary wildly depending on whether no molecule, 2 molecules, or 10 molecules were colliding with the wall of the container at the time of measurement. The macroscopic variables of pressure and temperature are not applicable to a microscopic system with only 10 molecules.

As we increase the number of molecules in a system, the chance of a process occurring in which the entropy of the universe decreases diminishes dramatically. For example, if we have 50 molecules in a container, the chance that they will all be in the left half of the container is $(\frac{1}{2})^{50} \approx 10^{-15}$. Thus, if we look at the gas once each second, we could expect to see all 50 molecules in the left half of the volume about once in every 10^{15} seconds or once in every 36 million years! For 1 mole (6×10^{23} molecules), the chance that all will end up in half of the volume is vanishingly small, essentially zero. For macroscopic systems, then, the probability of a process resulting in a decrease in the entropy of the universe is so extremely small that the distinction between improbable and impossible becomes blurred.

Physics Spotlight

The Perpetual Battle over Perpetual Motion

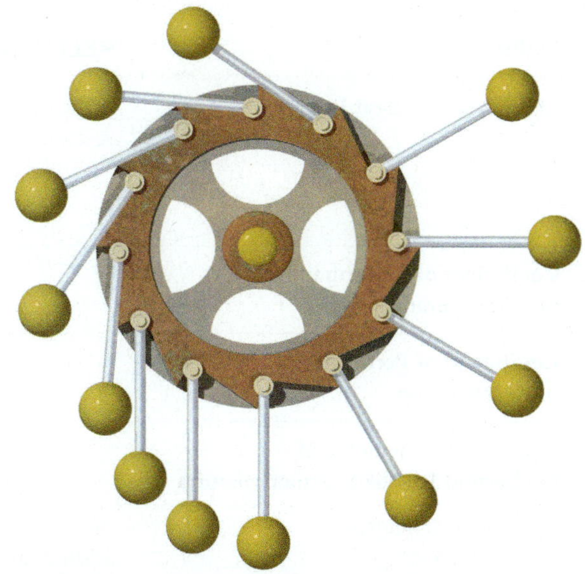

The wheel only appears to be unbalanced. However, note that there are more balls to the left of the axis than to the right of it. A calculation of the center of mass will show it is located directly below the axis. (Reducing the number of balls to four makes this calculation relatively simple.)

People dream of getting work done for free. Perpetual motion machines that do effective work with no energy, or are completely efficient, are a focus of those dreams. Perpetual motion cannot work. Physicists classify perpetual motion machines into two categories, depending on which law of thermodynamics they violate.

Category one machines violate the first law of thermodynamics—they claim to create energy out of nothing, or to create more energy than they use. The earliest known attempts at perpetual motion involved rotary motion and gravity. An overbalanced wheel had hinged rods that would supposedly cause the wheel to perpetually rotate toward one side. Another design, favored by Leonardo da Vinci, involved a water wheel that powered both a mill and a pump that pumped the water high enough that it would power the water wheel.* None of these attempts accounted for the energy needed to move the hinged rods relative to the larger wheel, or for the energy lost in turning the water wheel itself.

Later, buoyancy and rotary motion were part of popular attempts at perpetual motion.† Designs for buoyancy wheels and belts have included air pockets, chains, and connected hoses.‡ None of these designs has taken into account the work needed to fill the air pockets, nor the work needed to move the inner weights.

Category two machines violate the second law of thermodynamics. These machines do not claim to create energy. Instead, they are heat- or vapor-driven machines with impossible efficiency. One of the most famous machines was the Zeromotor, proposed by John Gamgee in 1880 to power ships' propellers.# The Zeromotor was a specially designed ammonia motor. Because ammonia boils at a temperature close to 0°C, liquid ammonia would be injected into a cylinder with a piston. The ammonia would expand and push the piston. The act of pushing the piston would cool the ammonia enough that it would condense, expand again, and go through the entire cycle without heat absorbed or released. This directly contradicts Carnot's work. Gamgee's Zeromotor, of course, did not work. Neither have any other engines based on the violation of the second law of thermodynamics. The vapor never condenses, and the heat engine never completes a single full cycle.

The French Academy of Science claimed that the pursuit of perpetual motion was a waste of time when it voted in 1775 to not consider patents for perpetual motion machines.° By 1856, submitting patents for perpetual motion machines to either the U.S. or the British Patent Office was not recommended.§ In the end, though, people still want free work. It is possible to find recent claims of engines that create more energy than they use.¶ It is also possible to find scientists upset by the latest unworkable patent** that slipped by the U.S. Patent and Trademark Office.†† The perpetual battle about perpetual motion rages on, but physicists are now able to explain why it is not possible.

---

* Leonardo3, "Pompe Meccaniche e a Moto Perpetuo," *Codex Atlanticus*, Milan: Leonardo3 srl, 2005.
† "Austin's Perpetual Motion," *Scientific American*, Mar. 27, 1847, Vol. 2, No. 27, 209.
‡ Diamond, David, "Gravity-Actuated Fluid Displacement Power Generator," United States Patent 3,934,964, Jan. 27, 1976.
# Park, Robert, *Voodoo Science*. Oxford: Oxford University Press, 2000, 129–130.
° Ward, James, *Naturalism and Agnosticism, Vol I.* London: Black, 1906.
§ "Patent Correspondence," *Scientific American*, Sep. 1856, Vol. 20, No. 1, 343.
¶ Wine, Byron, "Energy Information." https://byronwine.com as of July, 2006.
** Voss, David, "'New Physics' Finds a Haven at the Patent Office," *Science*, May 21, 1999, Vol. 284, No. 5418, 1252–1254.
†† Collins, G. P., "There's No Stopping Them," *Scientific American*, Oct. 22, 2002, 41.

Summary

The second law of thermodynamics is a fundamental law of nature.

TOPIC	RELEVANT EQUATIONS AND REMARKS	
1. Efficiency of a Heat Engine	If the engine absorbs Q_h from a hot reservoir, does work W, and releases heat Q_c to a cold reservoir, its efficiency is $$\varepsilon = \frac{W}{Q_h} = \frac{Q_h - Q_c}{Q_h} = 1 - \frac{Q_c}{Q_h}$$	19-2
2. Coefficient of Performance of a Refrigerator	$$\text{COP} = \frac{Q_c}{W}$$	19-3
3. Coefficient of Performance of a Heat Pump	$$\text{COP}_{HP} = \frac{Q_h}{W}$$	19-8
4. Equivalent Statements of the Second Law of Thermodynamics		
The Kelvin statement	No system can absorb heat from a single reservoir and convert it entirely into work without additional net changes in the system or its surroundings.	
The heat-engine statement	It is impossible for a heat engine working in a cycle to produce *only the effect* of absorbing heat from a single reservoir and performing an equivalent amount of work.	
The Clausius statement	A process whose only net result is to absorb heat from a cold reservoir and release the same amount of heat to a hot reservoir is impossible.	
The refrigerator statement	It is impossible for a refrigerator working in a cycle to produce *only the effect* of absorbing heat from a cold object and releasing the same amount of heat to a hot object.	
The entropy statement	The entropy of the universe (system plus surroundings) can never decrease.	
5. Conditions for a Reversible Process	1. No mechanical energy is transformed into internal thermal energy by friction, viscous forces, or other dissipative forces. 2. Energy transfer as heat can only occur between objects whose temperatures differ by an infinitesimal amount. 3. The process must be quasi-static so that the system is always in an equilibrium state (or infinitesimally near an equilibrium state).	
6. Carnot Engine	A Carnot engine is a reversible engine that works between two heat reservoirs. It operates in a Carnot cycle, which consists of:	
Carnot cycle	1. A quasi-static isothermal absorption of heat from a hot reservoir 2. A quasi-static adiabatic expansion to a lower temperature 3. A quasi-static isothermal release of heat to a cold reservoir at temperature T_c 4. A quasi-static adiabatic compression back to the original state	
Carnot efficiency	$$\varepsilon_C = 1 - \frac{Q_c}{Q_h} = 1 - \frac{T_c}{T_h}$$	19-6
7. Thermodynamic Temperature	The ratio of the thermodynamic temperatures of two reservoirs is defined to be the ratio of the heat released to the heat absorbed by Carnot engine running between the reservoirs: $$\frac{T_c}{T_h} = \frac{Q_c}{Q_h}$$ In addition, the triple point of water has a thermodynamic temperature of 273.16 K.	19-7

TOPIC	RELEVANT EQUATIONS AND REMARKS
8. **Entropy**	Entropy is a measure of the disorder of a system. The difference in entropy between two nearby states is given by $$dS = \frac{dQ_{rev}}{T} \quad (19\text{-}12)$$ where dQ_{rev} is the heat absorbed during a reversible process taking the system from one state to the other. The entropy change of a system can be positive or negative.
Entropy and loss of work capability	During an irreversible process, the entropy of the universe S_u increases and an amount of energy $$W_{lost} = T\,\Delta S_u \quad (19\text{-}22)$$ becomes unavailable for doing work.
Entropy and probability	Entropy is related to probability. A highly ordered system is one of low probability and low entropy. An isolated system moves toward a state of high probability, low order, and high entropy.

Answer to Concept Check

19-1 No. The development of a living organism comes at the expense of a great increase in disorder elsewhere. Much of this disorder can be traced back to the Sun, where nuclear reactions generate an increase in disorder, and thus an increase in entropy.

Answers to Practice Problems

19-1	(*a*) 52.5 J, (*b*) 97.5 J
19-2	250 kJ
19-3	(*a*) 40%, (*b*) 80 kJ
19-4	30%
19-5	$\Delta S = 1.31$ kJ/K

Problems

In a few problems, you are given more data than you actually need; in a few other problems, you are required to supply data from your general knowledge, outside sources, or informed estimate.

Interpret as significant all digits in numerical values that have trailing zeros and no decimal points.

- • Single-concept, single-step, relatively easy
- •• Intermediate-level, may require synthesis of concepts
- ••• Challenging
- SSM Solution is in the *Student Solutions Manual*
- Consecutive problems that are shaded are paired problems.

CONCEPTUAL PROBLEMS

1 • **ENGINEERING APPLICATION** Modern automobile gasoline engines have efficiencies of about 25 percent. About what percentage of the heat of combustion is not used for work but released as heat? (*a*) 25%, (*b*) 50%, (*c*) 75%, (*d*) 100%, (*e*) You cannot tell from the data given.

2 • If a heat engine does 100 kJ of work per cycle while releasing 400 kJ of heat, what is its efficiency? (*a*) 20%, (*b*) 25%, (*c*) 80%, (*d*) 400%, (*e*) You cannot tell from the data given.

3 • If the heat absorbed by a heat engine is 600 kJ per cycle, and it releases 480 kJ of heat each cycle, what is its efficiency? (*a*) 20%, (*b*) 80%, (*c*) 100%, (*d*) You cannot tell from the data given.

4 • Explain what distinguishes a refrigerator from a "heat pump."

5 • An air conditioner's COP is mathematically identical to that of a refrigerator, that is, $COP_{AC} = COP_{ref} = Q_c/W$. However a heat pump's COP is defined differently, as $COP_{hp} = Q_h/W$. Explain clearly *why* the two COPs are defined differently. *Hint: Think of the end use of the three different devices.* SSM

6 • Explain why you cannot cool your kitchen by leaving your refrigerator door open on a hot day. (Why does turning on a room air conditioner cool down the room, but opening a refrigerator door does not?)

7 • **ENGINEERING APPLICATION** Why do steam-power-plant designers try to increase the temperature of the steam as much as possible?

8 • To increase the efficiency of a Carnot engine, you should (*a*) decrease the temperature of the hot reservoir, (*b*) increase the temperature of the cold reservoir, (*c*) increase the temperature of the hot reservoir, (*d*) change the ratio of maximum volume to minimum volume.

9 •• Explain why the following statement is true: To increase the efficiency of a Carnot engine, you should make the difference between the two operating temperatures as large as possible; but to increase the efficiency of a Carnot cycle *refrigerator*, you should make the difference between the two operating temperatures as small as possible. SSM

10 •• A Carnot engine operates between a cold temperature reservoir of 27°C and a high temperature reservoir of 127°C. Its efficiency is (*a*) 21%, (*b*) 25%, (*c*) 75%, (*d*) 79%.

11 •• The Carnot engine in Problem 10 is run in reverse as a refrigerator. Its COP is (*a*) 0.33, (*b*) 1.3, (*c*) 3.0, (*d*) 4.7.

12 •• On a humid day, water vapor condenses on a cold surface. During condensation, the entropy of the water (*a*) increases, (*b*) remains constant, (*c*) decreases, (*d*) may decrease or remain unchanged. Explain your answer.

13 •• An ideal gas is taken reversibly from an initial state P_i, V_i, T_i to the final state P_f, V_f, T_f. Two possible paths are (A) an isothermal expansion followed by an adiabatic compression, and (B) an adiabatic compression followed by an isothermal expansion. For these two paths, (*a*) $\Delta E_{\text{int A}} > \Delta E_{\text{int B}}$, (*b*) $\Delta S_A > \Delta S_B$, (*c*) $\Delta S_A < \Delta S_B$, (*d*) none of the above.

14 •• Figure 19-12 shows a thermodynamic cycle for an ideal gas on an *ST* diagram. Identify this cycle and sketch it on a *PV* diagram.

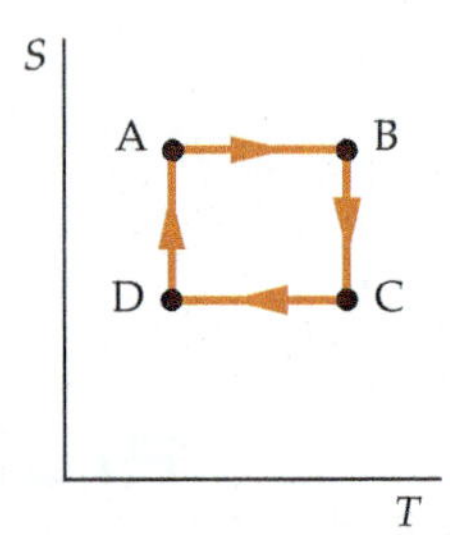

FIGURE 19-12 Problem 14, Problem 72

15 •• Figure 19-13 shows a thermodynamic cycle for an ideal gas on an *SV* diagram. Identify the type of engine represented by this diagram.

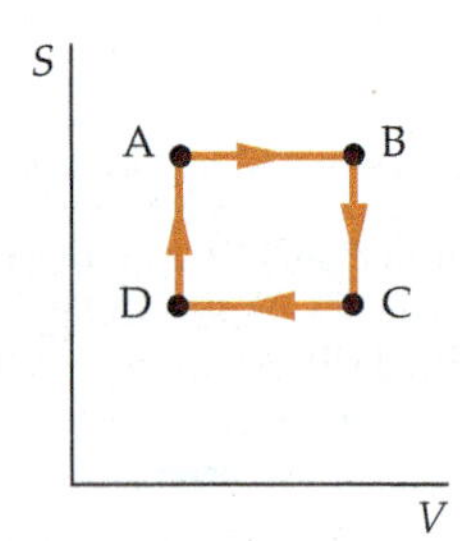

FIGURE 19-13 Problem 15

16 •• Sketch an *ST* diagram of the Otto cycle. (The Otto cycle is discussed in Section 19-1.)

17 •• Sketch an *SV* diagram of the Carnot cycle for an ideal gas. SSM

18 •• Sketch an *SV* diagram of the Otto cycle. (The Otto cycle is discussed in Section 19-1.)

19 •• Figure 19-14 shows a thermodynamic cycle for an ideal gas on an *SP* diagram. Make a sketch of this cycle on a *PV* diagram.

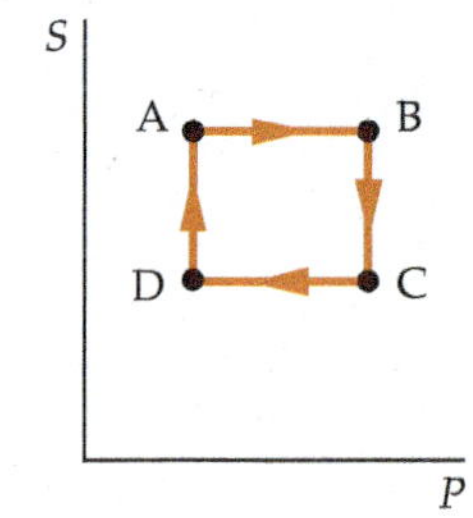

FIGURE 19-14 Problem 19

20 •• **CONTEXT-RICH** One afternoon, the mother of one of your friends walks into his room and finds a mess. She asks your friend how the room came to be in such a state, and your friend replies, "Well, it is the natural destiny of any closed system to degenerate toward greater and greater levels of entropy. That's all, Mom." Her reply is a sharp "Nevertheless, you'd better clean your room." Your friend retorts, "But that can't happen. It would violate the second law of thermodynamics." Critique your friend's response. Is his mother correct to ground him for not cleaning his room, or is cleaning the room really impossible?

ESTIMATION AND APPROXIMATION

21 • Estimate the change in COP of your electric food freezer when it is removed from your kitchen to its new location in your basement, which is 8°C cooler than your kitchen.

22 •• Estimate the probability that all the molecules in your bedroom are located in the (open) closet which accounts for about 10 percent of the total volume of the room.

23 •• Estimate the maximum efficiency of an automobile engine that has a compression ratio of 8.0:1.0. Assume the engine operates according to the Otto cycle and assume $\gamma = 1.4$. (The Otto cycle is discussed in Section 19-1.) SSM

24 •• **CONTEXT-RICH** You are working as an appliance salesperson during the summer. One day, your physics professor comes into your store to buy a new refrigerator. Wanting to buy the most efficient refrigerator possible, she asks you about the efficiencies of the available models. She decides to return the next day to buy the most efficient refrigerator. To make the sale, you need to provide her with the following estimates: (*a*) the highest COP possible for a household refrigerator, and (*b*) the highest rate possible for the heat to be released by the interior of the refrigerator if the refrigerator uses 600 W of electrical power.

25 •• The average temperature of the surface of the Sun is about 5400 K, and the average temperature of the surface of Earth is about 290 K. The solar constant (the intensity of sunlight reaching Earth's atmosphere) is about 1.37 kW/m². (*a*) Estimate the total power of the sunlight hitting Earth. (*b*) Estimate the net rate at which Earth's entropy is increasing due to this solar radiation. SSM

26 ••• A 1.0-L box contains *N* molecules of an ideal gas, and the positions of the molecules are observed 100 times per second. Calculate the average time it should take before we observe all *N* molecules in the left half of the box if *N* is equal to (*a*) 10, (*b*) 100, (*c*) 1000, and (*d*) 1.0 mole. (*e*) The best vacuums that have been created to date have pressures of about 10^{-12} torr. If a vacuum chamber has the same volume as the box, how long will a physicist have to wait before all of the gas molecules in the vacuum chamber occupy only the left half of it? Compare that to the expected lifetime of the universe, which is about 10^{10} years.

HEAT ENGINES AND REFRIGERATORS

27 • A heat engine with 20.0 percent efficiency does 0.100 kJ of work during each cycle. (*a*) How much heat is absorbed from the hot reservoir during each cycle? (*b*) How much heat is released to the cold reservoir during each cycle? SSM

28 • A heat engine absorbs 0.400 kJ of heat and does 0.120 kJ of work during each cycle. (*a*) What is the engine's efficiency? (*b*) How much heat is released to the cold reservoir during each cycle?

29 • A heat engine absorbs 100 J of heat from the hot reservoir and releases 60 J of heat to the cold reservoir during each cycle. (*a*) What is its efficiency? (*b*) If each cycle takes 0.50 s, find the power output of this engine.

30 • A refrigerator absorbs 5.0 kJ of heat from a cold reservoir and releases 8.0 kJ to a hot reservoir. (*a*) Find the coefficient of performance of the refrigerator. (*b*) The refrigerator is reversible. If it is run backward as a heat engine between the same two reservoirs, what is its efficiency?

31 •• The working substance of heat an engine is 1.00 mol of a monatomic ideal gas. The cycle begins at $P_1 = 1.00$ atm and $V_1 = 24.6$ L. The gas is heated at constant volume to $P_2 = 2.00$ atm. It then expands at constant pressure until it is 49.2 L. The gas is then cooled at constant volume until its pressure is again 1.00 atm. It is then compressed at constant pressure to its original state. All the steps are quasi-static and reversible. (*a*) Show this cycle on a *PV* diagram. For each step of the cycle, find the work done by the gas, the heat absorbed by the gas, and the change in the internal energy of the gas. (*b*) Find the efficiency of the cycle. SSM

32 •• The working substance of an engine is 1.00 mol of a diatomic ideal gas. The engine operates in a cycle consisting of three steps: (1) an adiabatic expansion from an initial volume of 10.0 L to a pressure of 1.00 atm and a volume of 20.0 L, (2) a compression at constant pressure to its original volume of 10.0 L, and (3) heating at constant volume to its original pressure. Find the efficiency of this cycle.

33 •• An engine using 1.00 mol of an ideal gas initially at a volume of 24.6 L and a temperature of 400 K performs a cycle consisting of four steps: (1) an isothermal expansion at a temperature of 400 K to twice its initial volume, (2) cooling at constant volume to a temperature of 300 K, (3) an isothermal compression to its original volume, and (4) heating at constant volume to its original temperature of 400 K. Assume that $C_v = 21.0$ J/K. Sketch the cycle on a *PV* diagram and find its efficiency.

34 •• Figure 19-15 shows the cycle followed by 1.00 mol of an ideal monatomic gas at an initial volume $V_1 = 25.0$ L. All the processes are quasi-static. Determine (*a*) the temperature of each numbered state of the cycle, (*b*) the heat transfer for each part of the cycle, and (*c*) the efficiency of the cycle.

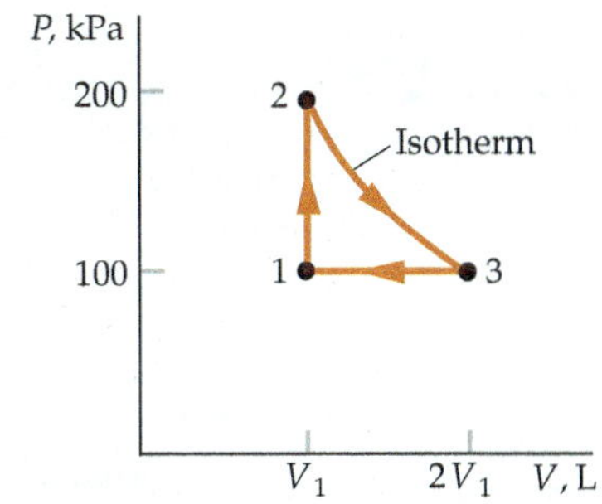

FIGURE 19-15 Problem 34

35 •• An ideal diatomic gas follows the cycle shown in Figure 19-16. The temperature of state 1 is 200 K. Determine (*a*) the temperatures of the other three numbered states of the cycle, and (*b*) the efficiency of the cycle.

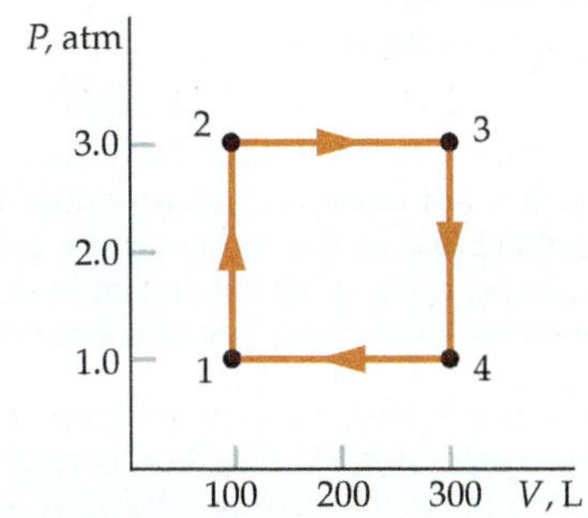

FIGURE 19-16 Problem 35

36 ••• **ENGINEERING APPLICATION** Recently, an old design for a heat engine, known as the *Stirling engine*, has been promoted as a means of producing power from solar energy. The cycle of a Stirling engine is as follows: (1) isothermal compression of the working gas, (2) heating of the gas at constant volume, (3) an isothermal expansion of the gas, and (4) cooling of the gas at constant volume. (*a*) Sketch *PV* and *ST* diagrams for the Stirling cycle. (*b*) Find the entropy change of the gas for each step of the cycle and show that the sum of these entropy changes is equal to zero.

37 ••• **BIOLOGICAL APPLICATION** "As far as we know, Nature has never evolved a heat engine"—Steven Vogel, *Life's Devices* (Princeton University Press, 1988). (*a*) Calculate the efficiency of a heat engine operating between body temperature (98.6°F) and a typical outdoor temperature (70°F), and compare this to the human body's efficiency for converting chemical energy into work (approximately 20 percent). Does this efficiency comparison contradict the second law of thermodynamics? (*b*) From the result of Part (*a*), and a general knowledge of the conditions under which most warm-blooded organisms exist, give a reason why no warm-blooded organisms have evolved heat engines to increase their internal energies.

38 ••• **ENGINEERING APPLICATION** The *diesel cycle* shown in Figure 19-17 approximates the behavior of a diesel engine. Process *ab* is an adiabatic compression, process *bc* is an expansion at constant pressure, process *cd* is an adiabatic expansion, and process *da* is cooling at constant volume. Find the efficiency of this cycle in terms of the volumes V_a, V_b, and V_c.

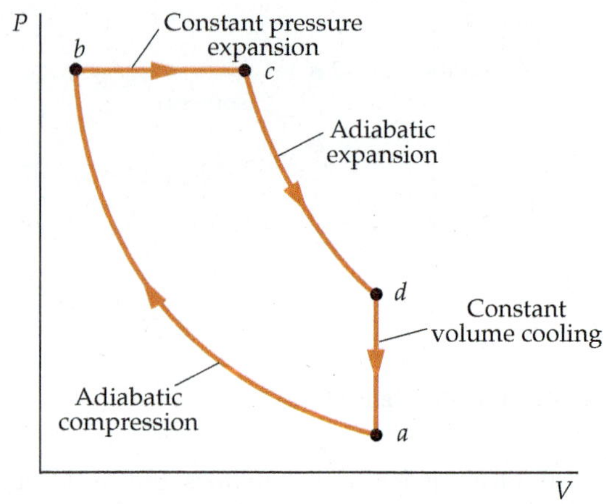

FIGURE 19-17 Problem 38

SECOND LAW OF THERMODYNAMICS

39 •• A refrigerator absorbs 500 J of heat from a cold reservoir and releases 800 J to a hot reservoir. Assume that the heat-engine statement of the second law of thermodynamics is false, and show how a perfect engine working with this refrigerator can violate the refrigerator statement of the second law of thermodynamics. SSM

40 •• If two curves that represent quasi-static adiabatic processes could intersect on a *PV* diagram, a cycle could be completed by an isothermal path between the two adiabatic curves shown in Figure 19-18. Show that such a cycle violates the second law of thermodynamics.

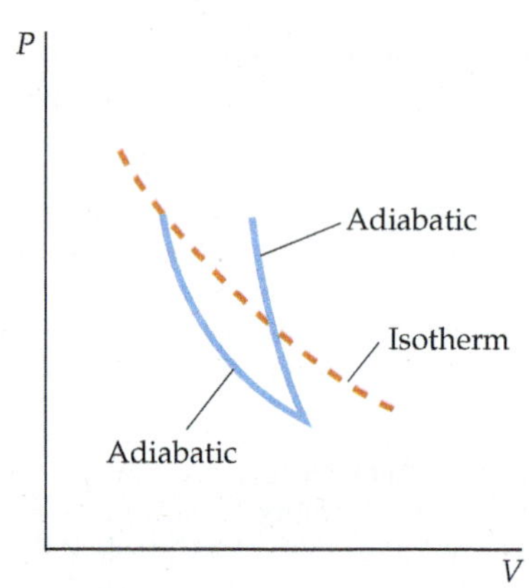

FIGURE 19-18 Problem 40

CARNOT CYCLES

41 • A Carnot engine works between two heat reservoirs at temperatures $T_h = 300$ K and $T_c = 200$ K. (*a*) What is its efficiency? (*b*) If it absorbs 100 J of heat from the hot reservoir during each cycle, how much work does it do each cycle? (*c*) How much heat does it release during each cycle? (*d*) What is the COP of this engine when it works as a refrigerator between the same two reservoirs? SSM

42 • An engine absorbs 250 J of heat per cycle from a reservoir at 300 K and releases 200 J of heat per cycle to a reservoir at 200 K. (*a*) What is its efficiency? (*b*) How much additional work per cycle could be done if the engine were reversible?

43 •• A reversible engine working between two reservoirs at temperatures T_h and T_c has an efficiency of 30 percent. Working as a heat engine, it releases 140 J per cycle of heat to the cold reservoir. A second engine working between the same two reservoirs also releases 140 J per cycle to the cold reservoir. Show that if the second engine has an efficiency greater than 30 percent, the two engines working together would violate the heat-engine statement of the second law.

44 •• A reversible engine working between two reservoirs at temperatures T_h and T_c has an efficiency of 20 percent. Working as a heat engine, it does 100 J of work per cycle. A second engine working between the same two reservoirs also does 100 J of work per cycle. Show that if the efficiency of the second engine is greater than 20 percent, the two engines working together would violate the refrigerator statement of the second law.

45 •• A Carnot engine works between two heat reservoirs as a refrigerator. During each cycle, 100 J of heat are absorbed from the cold reservoir and 150 J of heat are released to the hot reservoir. (*a*) What is the efficiency of the Carnot engine when it works as a heat engine between the same two reservoirs? (*b*) Show that no other engine working as a refrigerator between the same two reservoirs can have a COP greater than 2.00.

46 •• A Carnot engine works between two heat reservoirs at temperatures $T_h = 300$ K and $T_c = 77.0$ K. (*a*) What is its efficiency? (*b*) If it absorbs 100 J of heat from the hot reservoir during each cycle, how much work does it do? (*c*) How much heat does it release to the low-temperature reservoir during each cycle? (*d*) What is the coefficient of performance of this engine when it works as a refrigerator between these two reservoirs?

47 •• In the cycle shown in Figure 19-19, 1.00 mol of an ideal diatomic gas is initially at a pressure of 1.00 atm and a temperature of 0.0°C. The gas is heated at constant volume to $T_2 = 150$°C, and is then expanded adiabatically until its pressure is again 1.00 atm. It is then compressed at constant pressure back to its original state. Find (*a*) the temperature after the adiabatic expansion, (*b*) the heat absorbed or released by the system during each step, (*c*) the efficiency of this cycle, and (*d*) the efficiency of a Carnot cycle operating between the temperature extremes of this cycle. SSM

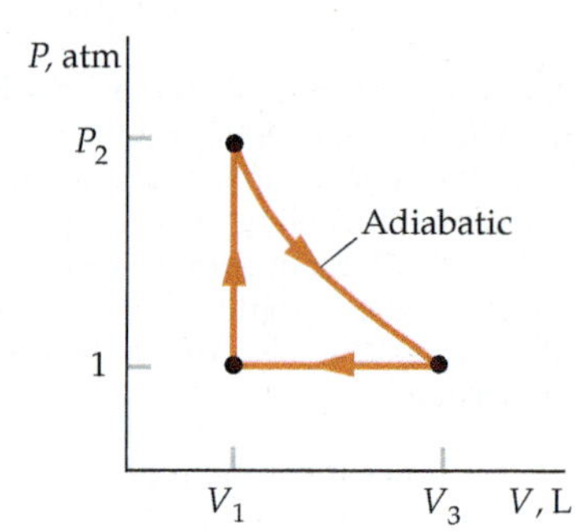

FIGURE 19-19 Problem 47

48 •• **Engineering Application, Context-Rich** You are a part of a team that is completing a mechanical engineering project. Your team built a steam engine that takes in superheated steam at 270°C and discharges condensed steam from its cylinder at 50.0°C. Your team has measured the engine's efficiency to be 30.0 percent. (*a*) How does this efficiency compare with the maximum possible efficiency for your engine? (*b*) If the useful power output of the engine is known to be 200 kW, how much heat does the engine release to its surroundings in 1.00 h?

* HEAT PUMPS

49 • **Engineering Application, Context-Rich** As an engineer, you are designing a heat pump that is capable of delivering heat at the rate of 20 kW to a house. The house is located where, in January, the average outside temperature is −10°C. The temperature of the air in the air handler inside the house is to be 40°C. (*a*) What is the maximum possible COP for a heat pump operating between these temperatures? (*b*) What must the minimum power of the electric motor driving the heat pump be? (*c*) In reality, the COP of the heat pump will be only 60 percent of the ideal value. What is the minimum power of the electric motor when the COP is 60 percent of the ideal value? SSM

50 • A refrigerator is rated at 370 W. (*a*) What is the maximum amount of heat it can absorb from the food compartment in 1.00 min if the food-compartment temperature of the refrigerator is 0.0°C and it releases heat into a room at 20.0°C? (*b*) If the COP of the refrigerator is 70 percent of that of a reversible refrigerator, how much heat can it absorb from the food compartment in 1.00 min under these conditions?

51 •• A refrigerator is rated at 370 W. (*a*) What is the maximum amount of heat it can absorb from the food compartment in 1.00 min if the temperature in the compartment is 0.0°C and it releases heat into a room at 35°C? (*b*) If the COP of the refrigerator is 70 percent of that of a reversible pump, how much heat can it absorb from the food compartment in 1.00 min? Is the COP for the refrigerator greater when the temperature of the room is 35°C or 20°C? Explain.

52 ••• **Context-Rich** You are installing a heat pump whose COP is half the COP of a reversible heat pump. You plan to use the pump on chilly winter nights to increase the air temperature in your bedroom. Your bedroom's dimensions are 5.00 m × 3.50 m × 2.50 m. The air temperature should increase from 63°F to 68°F. The outside temperature is 35°F, and the temperature at the air handler in the room is 112°F. If the pump's electric-power consumption is 750 W, how long will you have to wait for the room's air to warm if the specific heat of air is 1.005 kJ/(kg · °C)? Assume you have good window draperies and good wall insulation so that you can neglect the release of heat through windows, walls, ceilings, and floors. Also assume that the heat capacity of the floor, ceiling, walls, and furniture are negligible.

ENTROPY CHANGES

53 • You inadvertently leave a pan of water boiling on the hot stove. You return just in time to see the last drop converted into steam. The pan originally held 1.00 L of boiling water. What is the change in entropy of the water associated with its change of state from liquid to gas? SSM

54 • What is the change in entropy of 1.00 mol of liquid water at 0.0°C that freezes to ice at 0.0°C?

55 •• Consider the freezing of 50.0 g of water once it is placed in the freezer compartment of a refrigerator. Assume the walls of the freezer are maintained at −10°C. The water, initially liquid at 0.0°C, is frozen into ice and cooled to −10 °C. Show that even though the entropy of the water decreases, the net entropy of the universe increases.

56 • In this problem, 2.00 mol of an ideal gas at 400 K expand quasi-statically and isothermally from an initial volume of 40.0 L to a final volume of 80.0 L. (*a*) What is the entropy change of the gas? (*b*) What is the entropy change of the universe for this process?

57 •• A system completes a cycle consisting of six quasi-static steps, during which the total work done by the system is 100 J. During step 1 the system absorbs 300 J of heat from a reservoir at 300 K, during step 3 the system absorbs 200 J of heat from a reservoir at 400 K, and during step 5 it absorbs heat from a reservoir at temperature T_3. (During steps 2, 4 and 6 the system undergoes adiabatic processes in which the temperature of the system changes from one reservoir's temperature to that of the next.) (*a*) What is the entropy change of the system for the complete cycle? (*b*) If the cycle is reversible, what is the temperature T_3? SSM

58 •• In this problem, 2.00 mol of an ideal gas initially has a temperature of 400 K and a volume of 40.0 L. The gas undergoes a free adiabatic expansion to twice its initial volume. What are (*a*) the entropy change of the gas, and (*b*) the entropy change of the universe?

59 •• A 200-kg block of ice at 0.0°C is placed in a large lake. The temperature of the lake is just slightly higher than 0.0°C, and the ice melts very slowly. (*a*) What is the entropy change of the ice? (*b*) What is the entropy change of the lake? (*c*) What is the entropy change of the universe (the ice plus the lake)?

60 •• A 100-g piece of ice at 0.0°C is placed in an insulated calorimeter of negligible heat capacity containing 100 g of water at 100°C. (*a*) What is the final temperature of the water once thermal equilibrium is established? (*b*) Find the entropy change of the universe for this process.

61 •• A 1.00-kg block of copper at 100°C is placed in an insulated calorimeter of negligible heat capacity containing 4.00 L of liquid water at 0.0°C. Find the entropy change of (*a*) the copper block, (*b*) the water, and (*c*) the universe. SSM

62 •• If a 2.00-kg piece of lead at 100°C is dropped into a lake at 10°C, find the entropy change of the universe.

ENTROPY AND LOST WORK

63 •• A reservoir at 300 K absorbs 500 J of heat from a second reservoir at 400 K. (*a*) What is the change in entropy of the universe, and (*b*) how much work is lost during the process? SSM

64 •• In this problem, 1.00 mol of an ideal gas at 300 K undergoes a free adiabatic expansion from $V_1 = 12.3$ L to $V_2 = 24.6$ L. It is then compressed isothermally and reversibly back to its original state. (*a*) What is the entropy change of the universe for the complete cycle? (*b*) How much work is lost in this cycle? (*c*) Show that the work lost is $T\,\Delta S_u$.

GENERAL PROBLEMS

65 • A heat engine with an output of 200 W has an efficiency of 30 percent. It operates at 10.0 cycles/s. (*a*) How much work is done by the engine during each cycle? (*b*) How much heat is absorbed from the hot reservoir and how much is released to the cold reservoir during each cycle?

66 • During each cycle, a heat engine operating between two heat reservoirs absorbs 150 J from the reservoir at 100°C and releases 125 J to the reservoir at 20°C. (*a*) What is the efficiency of this engine? (*b*) What is the ratio of its efficiency to that of a Carnot engine working between the same reservoirs? (This ratio is called the *second law efficiency.*)

67 • An engine absorbs 200 kJ of heat per cycle from a reservoir at 500 K and releases heat to a reservoir at 200 K. The engine's efficiency is 85 percent of that of a Carnot engine working between the same reservoirs. (*a*) What is the efficiency of this engine? (*b*) How much work is done in each cycle? (*c*) How much heat is released to the low-temperature reservoir during each cycle? SSM

68 • Estimate the change in entropy of the universe associated with an Olympic diver diving into the water from the 10-m platform.

69 • To maintain the temperature inside a house at 20°C, the electric power consumption of the electric baseboard heaters is 30.0 kW on a day when the outside temperature is −7°C. At what rate does this house contribute to the increase in the entropy of the universe?

70 •• **ENGINEERING APPLICATION** Calvin Cliffs Nuclear Power Plant, located on the Hobbes River, generates 1.00 GW of power. In this plant, liquid sodium circulates between the reactor core and a heat exchanger located in the superheated steam that drives the turbine. Heat is absorbed by the liquid sodium in the core, and released by the liquid sodium (and into the superheated steam) in the heat exchanger. The temperature of the superheated steam is 500 K. Heat is released into the river, and the water in the river flows by at a temperature of 25°C. (*a*) What is the highest efficiency that this plant can have? (*b*) How much heat is released into the river every second? (*c*) How much heat must be released by the core to supply 1.00 GW of electrical power? (*d*) Assume that new environmental laws have been passed to preserve the unique wildlife of the river. Because of these laws, the plant is not allowed to heat the river by more than 0.50°C. What is the minimum flow rate that the water in the Hobbes River must have?

71 •• **ENGINEERING APPLICATION, CONTEXT-RICH** An inventor comes to you to explain his new invention. It is a novel heat engine using water vapor as the working substance. He claims that the water vapor absorbs heat at 100°C, does work at the rate of 125 W, and releases heat to the air at the rate of only 25.0 W, when the air temperature is 25°C. (*a*) Explain to him why he cannot be correct. (*b*) After careful analysis of the data in his prospectus folder, you decide he has made an error in the measurement of his exhausted-heat value. What is the minimum rate of exhausting heat that would make you consider believing him?

72 •• The cycle represented in Figure 19-12 (next to Problem 19-14) is for 1.00 mol of an ideal monatomic gas. The temperatures at points A and B are 300 and 750 K, respectively. What is the efficiency of the cyclic process ABCDA?

73 •• (*a*) Which of these two processes is more wasteful of available work? (1) A block moving with 0.50 J of kinetic energy being slowed to rest by sliding (kinetic) friction when the temperature of the environment is 300 K, or (2) A reservoir at 400 K releasing 1.00 kJ of heat to a reservoir at 300 K? Explain your choice. *Hint: How much of the 1.00 kJ of heat could be converted into work by an ideal cyclic process?* (*b*) What is the change in entropy of the universe for each process? SSM

74 •• Helium, a monatomic gas, is initially at a pressure of 16 atm, a volume of 1.0 L, and a temperature of 600 K. It is quasi-statically expanded at constant temperature until its volume is 4.0 L, and is then quasi-statically compressed at constant pressure until its volume and temperature are such that a quasi-static adiabatic compression will return the gas to its original state. (*a*) Sketch this cycle on a *PV* diagram. (*b*) Find the volume and temperature after the compression at constant pressure. (*c*) Find the work done during each step of the cycle. (*d*) Find the efficiency of the cycle.

75 •• A heat engine that does the work of blowing up a balloon at a pressure of 1.00 atm absorbs 4.00 kJ from a reservoir at 120°C. The volume of the balloon increases by 4.00 L, and heat is released to a reservoir at a temperature T_c, where $T_c < 120°C$. If the efficiency of the heat engine is 50 percent of the efficiency of a Carnot engine working between the same two reservoirs, find the temperature T_c. SSM

76 •• Show that the coefficient of performance of a Carnot engine run as a refrigerator is related to the efficiency of a Carnot engine operating between the same two temperatures by $\varepsilon_C \times COP_C = T_c/T_h$.

77 •• A freezer has a temperature $T_c = -23°C$. The air in the kitchen has a temperature $T_h = 27°C$. The freezer is not perfectly insulated, and heat leaks through the walls of the freezer at a rate of 50 W. Find the power of the motor that is needed to maintain the temperature in the freezer.

78 •• In a heat engine, 2.00 mol of a diatomic gas are taken through the cycle ABCA, as shown in Figure 19-20. (The *PV* diagram is not drawn to scale.) At A the pressure and temperature are 5.00 atm and 600 K. The volume at B is twice the volume at A. The segment BC is an adiabatic expansion, and the segment CA is an isothermal compression. (*a*) What is the volume of the gas at A? (*b*) What are the volume and temperature of the gas at B? (*c*) What is the temperature of the gas at C? (*d*) What is the volume of the gas at C? (*e*) How much work is done by the gas in each of the three segments of the cycle? (*f*) How much heat is absorbed or released by the gas in each segment of this cycle?

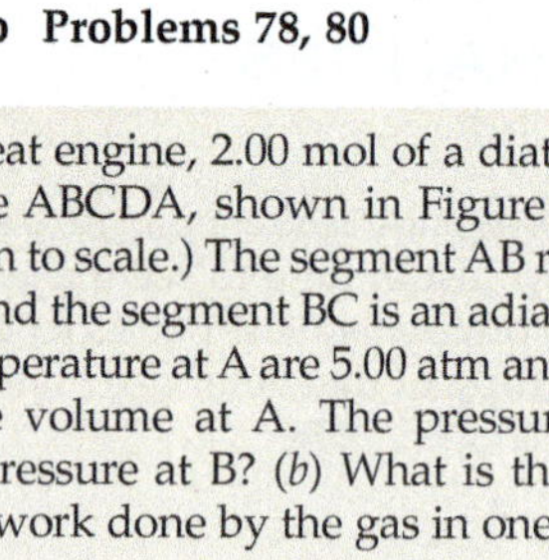

FIGURE 19-20 Problems 78, 80

79 •• In a heat engine, 2.00 mol of a diatomic gas are carried through the cycle ABCDA, shown in Figure 19-21. (The *PV* diagram is not drawn to scale.) The segment AB represents an isothermal expansion, and the segment BC is an adiabatic expansion. The pressure and temperature at A are 5.00 atm and 600 K. The volume at B is twice the volume at A. The pressure at D is 1.00 atm. (*a*) What is the pressure at B? (*b*) What is the temperature at C? (*c*) Find the total work done by the gas in one cycle. SSM

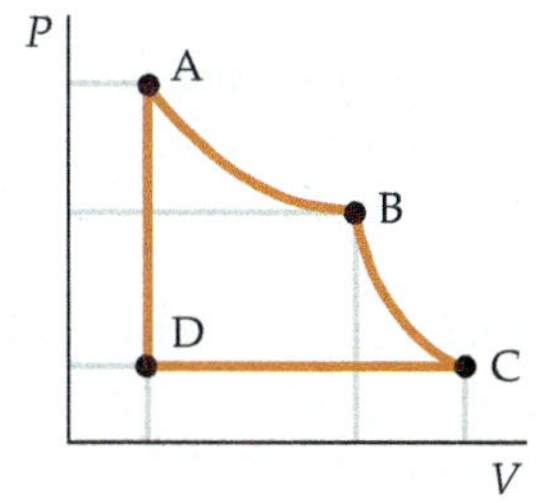

FIGURE 19-21 Problems 79, 81

80 •• In a heat engine, 2.00 mol of a monatomic gas are taken through the cycle ABCA, as shown in Figure 19-20. (The *PV* diagram is not drawn to scale.) At A the pressure and temperature are 5.00 atm and 600 K. The volume at B is twice the volume at A. The segment BC is an adiabatic expansion, and the segment CA is an isothermal compression. (*a*) What is the volume of the gas at A? (*b*) What are the volume and temperature of the gas at B? (*c*) What is the temperature of the gas at C? (*d*) What is the volume of the gas at C? (*e*) How much work is done by the gas in each of the three segments of the cycle? (*f*) How much heat is absorbed by the gas in each segment of the cycle?

81 •• In a heat engine, 2.00 mol of a monatomic gas are carried through the cycle ABCDA as shown in Figure 19-21. (The *PV* diagram is not drawn to scale.) The segment AB represents an isothermal expansion, and the segment BC is an adiabatic expansion. The pressure and temperature at A are 5.00 atm and 600 K. The volume at B is twice the volume at A. The pressure at D is 1.00 atm. (*a*) What is the pressure at B? (*b*) What is the temperature at C? (*c*) Find the total work done by the gas in one cycle.

82 •• Compare the efficiency of the Otto cycle to the efficiency of the Carnot cycle operating between the same maximum and minimum temperatures. (The Otto cycle is discussed in Section 19–1.)

83 ••• **ENGINEERING APPLICATION** A common practical cycle, often used in refrigeration, is the *Brayton cycle*, which involves (1) an adiabatic compression, (2) an isobaric (constant pressure) expansion, (3) an adiabatic expansion, and (4) an isobaric compression back to the original state. Assume the system begins the adiabatic compression at temperature T_1, and transitions to temperatures T_2, T_3, and T_4 after each leg of the cycle. (*a*) Sketch this cycle on a *PV* diagram. (*b*) Show that the efficiency of the overall cycle is given by $\varepsilon = 1 - (T_4 - T_1)/(T_3 - T_2)$. (*c*) Show that this efficiency, can be written as $\varepsilon = 1 - r^{(1-\gamma)/\gamma}$, where r is the pressure ratio $P_{\text{high}}/P_{\text{low}}$ (the ratio of the maximum and minimum pressures in the cycle). SSM

84 ••• **ENGINEERING APPLICATION** Suppose the Brayton cycle engine (see Problem 83) is run in reverse as a refrigerator in your kitchen. In this case, the cycle begins at temperature T_1 and expands at constant pressure until its temperature is T_4. Then the gas is adiabatically compressed until its temperature is T_3. And then it is compressed at constant pressure, until its temperature is T_2. Finally, it adiabatically expands until it returns to its initial state at temperature T_1. (*a*) Sketch this cycle on a *PV* diagram. (*b*) Show that the coefficient of performance, is

$$\text{COP}_{\text{B}} = \frac{(T_4 - T_1)}{(T_3 - T_2 - T_4 + T_1)}$$

(*c*) Suppose your "Brayton cycle refrigerator" is run as follows. The cylinder containing the refrigerant (a monatomic gas) has an initial volume and pressure of 60 mL and 1.0 atm. After the expansion at constant pressure, the volume and temperature are 75 mL and −25°C. The pressure ratio $r = P_{\text{high}}/P_{\text{low}}$ for the cycle is 5.0. What is the coefficient of performance for your refrigerator? (*d*) In order to absorb heat from the food compartment at the rate of 120 W, what is the rate at which electrical energy must be supplied to the motor of this refrigerator? (*e*) Assuming the refrigerator motor is actually running for only 4.0 h each day, how much does it add to your monthly electric bill. Assume 15 cents per kWh of electric energy and 30 days in a month.

85 •• Using $\Delta S = C_{\text{v}} \ln(T_2/T_1) - nR \ln(V_2/V_1)$ (Equation 19-16) for the entropy change of an ideal gas, show explicitly that the entropy change is zero for a quasi-static adiabatic expansion from state (V_1, T_1) to state (V_2, T_2).

86 ••• (*a*) Show that if the refrigerator statement of the second law of thermodynamics were not true, then the entropy of the universe could decrease. (*b*) Show that if the heat-engine statement of the second law were not true, then the entropy of the universe could decrease. (*c*) A third statement of the second law is that the entropy of the universe cannot decrease. Have you just proved that this statement is equivalent to the refrigerator and heat-engine statements?

87 ••• Suppose that two heat engines are connected in series, such that the heat released by the first engine is used as the heat absorbed by the second engine, as shown in Figure 19-22. The efficiencies of the engines are ε_1 and ε_2, respectively. Show that the net efficiency of the combination is given by $\varepsilon_{\text{net}} = \varepsilon_1 + \varepsilon_2 - \varepsilon_1\varepsilon_2$.

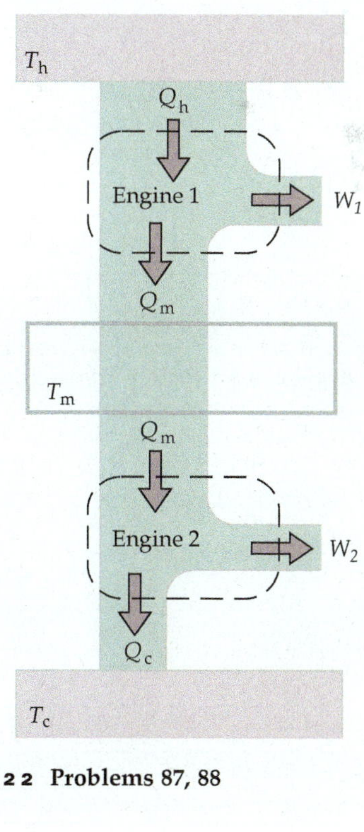

FIGURE 19-22 Problems 87, 88

88 ••• Suppose that two heat engines are connected in series, such that the heat released by the first engine is used as the heat absorbed by the second engine, as shown in Figure 19-22. Suppose that each engine is an ideal reversible heat engine. Engine 1 operates between temperatures T_h and T_m, and Engine 2 operates between T_m and T_c, where $T_h > T_m > T_c$. Show that the net efficiency of the combination is given by $\varepsilon_{net} = 1 - (T_c/T_h)$. (Note that this result means that two reversible heat engines operating "in series" are equivalent to one reversible heat engine operating between the hottest and coldest reservoirs.)

89 ••• The English mathematician and philospher Bertrand Russell (1872–1970) once said that if a million monkeys were given a million typewriters and typed away at random for a million years, they would produce all of Shakespeare's works. Let us limit ourselves to the following fragment of Shakespeare (*Julius Caesar* III:ii):

Friends, Romans, countrymen Lend me your ears.
I come to bury Caesar, not to praise him.
The evil that men do lives on after them,
The good is oft interred with the bones.
So let it be with Caesar.
The noble Brutus hath told you that Caesar was ambitious,
And, if so, it were a grievous fault,
And grievously hath Caesar answered it...

Even with this small fragment, it will take a lot longer than a million years! By what factor (roughly speaking) was Russell in error? Make any reasonable assumptions you want. (You can even assume that the monkeys are immortal.) SSM

CHAPTER 20

THE ALASKAN PIPELINE TRANSPORTS OIL THROUGH 800 MILES OF 48-IN.-DIAMETER STEEL PIPE. ZIGZAGS ARE PLACED IN THE PIPELINE TO ALLOW FOR THERMAL EXPANSION. (THE ZIGZAGS ALSO ALLOW FOR MOVEMENT DUE TO SEISMIC ACTIVITY.) THE PIPELINE WAS DESIGNED TO WITHSTAND TEMPERATURES RANGING FROM −60°F TO 145°F (THE TEMPERATURE OF THE PIPELINE WAS −60°F PRIOR TO THE START OF THE OIL FLOW.) *(Karen Kasmauski/CORBIS.)*

Thermal Properties and Processes

20-1 Thermal Expansion

20-2 The van der Waals Equation and Liquid–Vapor Isotherms

20-3 Phase Diagrams

20-4 The Transfer of Heat

When an object absorbs heat, various changes may occur in the physical properties of the object. For example, its temperature may increase, accompanied by an expansion or contraction of the object, or the object may melt or vaporize, during which its temperature remains constant.

The changes in objects related to temperature are concerns scientists and engineers in many industries must deal with. Civil engineers that design bridges and roads include expansion joints that allow for more subtle changes in road lengths that arise from changes in temperature. Other engineers create products to protect objects from extreme temperature changes. Materials are used to keep thermal energy within hot water heaters, ovens, and ships' turbines, as well as to protect automobile bodies and their occupants from heat from the car's exhaust system.

In this chapter, we examine some of the thermal properties of matter and some important processes involving heat.

? What was the change in length of a 720-ft-long section of the pipeline when the temperature changed from −60°F to 145°F? (See Example 20-2.)

20-1 THERMAL EXPANSION

When the temperature of an object increases, the object typically expands. Suppose that we have a long rod of length L at a temperature T. When the temperature of a solid changes by ΔT, the fractional change in length $\Delta L/L$ is proportional to ΔT:

$$\frac{\Delta L}{L} = \alpha\,\Delta T \qquad 20\text{-}1$$

where α, called the **coefficient of linear expansion,** is the ratio of the fractional change in length to the change in temperature:

$$\alpha = \frac{\Delta L/L}{\Delta T} \qquad 20\text{-}2$$

The SI units for the coefficient of linear expansion are reciprocal kelvins (1/K), which are the same as reciprocal Celsius degrees (1/°C). The value of α can vary with changes in pressure and temperature. Equation 20-2 gives the average value over the temperature interval ΔT with pressure held constant. The coefficient of linear expansion at a particular temperature T is found by taking the limit as ΔT approaches zero:

$$\alpha = \lim_{\Delta T \to 0} \frac{\Delta L/L}{\Delta T} = \frac{1}{L}\frac{dL}{dT} \qquad 20\text{-}3$$

DEFINITION: COEFFICIENT OF LINEAR EXPANSION

The accuracy obtained by using the average value of α over a wide temperature range is sufficient for most purposes.

For a liquid or a gas, the **coefficient of volume expansion** β is defined as the ratio of the fractional change in volume to the change in temperature (at constant pressure):

$$\beta = \lim_{\Delta T \to 0} \frac{\Delta V/V}{\Delta T} = \frac{1}{V}\frac{dV}{dT} \qquad 20\text{-}4$$

DEFINITION: COEFFICIENT OF VOLUME EXPANSION

Both α and β can vary with both pressure and temperature, but any variation with pressure is typically negligible. Average values for α and β for various substances are given in Table 20-1.

For a given material, $\beta = 3\alpha$. We can show this by considering a box of dimensions L_1, L_2, and L_3. Its volume at a temperature T is

$$V = L_1 L_2 L_3$$

The rate of change of the volume with respect to temperature is

$$\frac{\partial V}{\partial T} = L_1 L_2 \frac{\partial L_3}{dT} + L_1 \frac{\partial L_2}{dT} L_3 + \frac{\partial L_1}{dT} L_2 L_3$$

Dividing each side of the equation by the volume, we obtain

$$\beta = \frac{1}{V}\frac{\partial V}{\partial T} = \frac{1}{L_3}\frac{\partial L_3}{\partial T} + \frac{1}{L_2}\frac{\partial L_2}{\partial T} + \frac{1}{L_1}\frac{\partial L_1}{\partial T}$$

We can see that each term on the right side of the preceding equation equals α, and so we have

$$\beta = 3\alpha \qquad 20\text{-}5$$

Table 20-1 Approximate Values of the Coefficients of Thermal Expansion for Various Substances

(Log scale from 10^{-2} to 10^{-7})

Air	3.67×10^{-3}	β, K^{-1}
Acetone	1.5×10^{-3}	β, K^{-1}
Alcohol	1.1×10^{-3}	β, K^{-1}
Water (20°C)	0.207×10^{-3}	β, K^{-1}
Mercury	0.18×10^{-3}	β, K^{-1}
Ice	51×10^{-6}	α, K^{-1}
Aluminum	24×10^{-6}	α, K^{-1}
Brass	19×10^{-6}	α, K^{-1}
Copper	17×10^{-6}	α, K^{-1}
Steel	11×10^{-6}	α, K^{-1}
Glass (ordinary)	9×10^{-6}	α, K^{-1}
Graphite	7.9×10^{-6}	α, K^{-1}
Glass (Pyrex)	3.2×10^{-6}	α, K^{-1}
Diamond	1.2×10^{-6}	α, K^{-1}
Invar	1×10^{-6}	α, K^{-1}

In the derivation of Equation 20-5, we have assumed that the coefficient of linear expansion is independent of direction. (This assumption is approximately true for many materials, and it will be used for the calculations in this book.) A similar derivation shows that the coefficient of area expansion is twice that of linear expansion.

Example 20-1 Do Holes Expand? *Conceptual*

Suppose we have a steel object with a circular hole through it. If the temperature of the object increases, the metal expands. Does the diameter of the hole increase or decrease?

PICTURE The increase in size of any part of an object for a given temperature increase is proportional to the original size of that part of the object (in accord with Equation 20-2). For an object, consider a steel ruler that has a 1-cm-diameter hole in it, centered on the 3.5-cm mark.

SOLVE

1. For an object, we consider a steel ruler that has a 1-cm-diameter hole in it, centered on the 3.5-cm mark:

   If a steel ruler has a 1-cm-diameter hole in it that is centered on the 3.5-cm mark, the edge of the hole will touch both the 3-cm line and the 4-cm line

2. When the temperature of the ruler increases by a given amount, the ruler expands uniformly:

   The distance between the 3-cm line and the 4-cm line will increase.

3. The edge of the hole will remain touching the 3-cm and 4-cm marks as the ruler expands.

   If the distance between the 3-cm line and the 4-cm line increases, then **the diameter of the hole increases.**

CHECK If the hole were made by punching out at 1-cm hole, the punched out material would be a steel disk 1-cm in diameter. If the temperature of this disk were then increased by the same amount as the temperature of the ruler was increased, then the disk would fit the hole perfectly.

TAKING IT FURTHER An apparatus for demonstrating that a hole expands when heated is shown in Figure 20-1.

FIGURE 20-1 When the ball and ring are both at room temperature, the ball is too big to pass through the ring. The ring expands when heated, and when it is hot the ball, which remains at room temperature, is able to pass through the hole. *(Richard Megna/Fundamental Photographs.)*

Most materials expand when heated and contract when cooled. Water, however, is an important exception. Figure 20-2 shows the volume occupied by 1 g of water as a function of temperature. The minimum volume, and therefore the maximum density, is at 4.00°C. Thus, when water at 4.00°C is cooled, it expands rather than contracts. This property of water has important consequences for the ecology of

FIGURE 20-2 Volume of 1 g of water at atmospheric pressure versus temperature. The minimum volume, which corresponds to the maximum density, occurs at 4.0°C. At temperatures below 0.0°C, the curve shown is for supercooled water. (Supercooled water is water that is cooled below the normal freezing point without solidifying.)

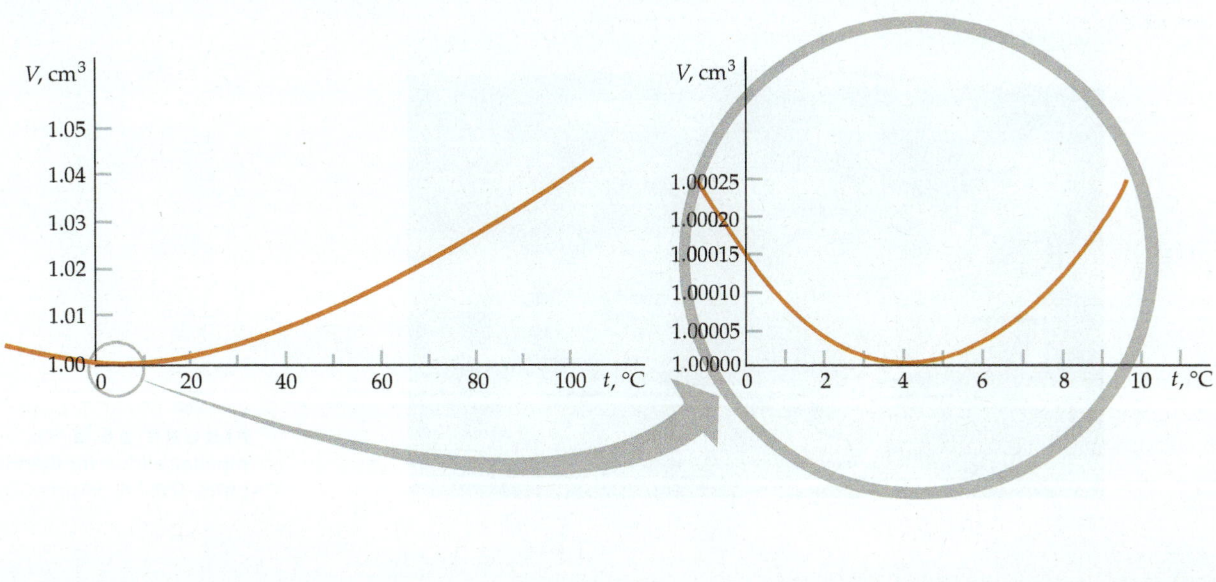

lakes. At temperatures above 4.00°C, the water in a lake becomes denser as it cools, and therefore sinks to the bottom. But as the water cools below 4.00°C, it becomes less dense and rises to the surface. This consequence is the reason that ice forms first on the surface of a lake. Water also expands when it freezes. Because ice is less dense than liquid water, it remains at the surface and acts as insulation for the water below. If water behaved like most substances and contracted when it froze, then ice would sink and expose more water at the surface that would then freeze. Lakes would fill with ice from the bottom up and would be much more likely to freeze completely in the winter.

Example 20-2 An Expanding Pipeline

A 720-ft-long straight section of the Alaskan pipeline was at a temperature of −60°F before it was filled with oil with a maximum temperature of 145°F. The pipeline is wrapped with insulation, so the oil and the steel pipe have the same temperature. (*a*) How much did the section expand when the temperature changed from −60°F to 145°F. (*b*) The sections of the pipeline that are above ground are 420 mi long. If the temperature of an entire 420-mi-long section increases from −60°F to 145°F, how much would it expand?

PICTURE Use $\alpha = 11 \times 10^{-6}\,\text{K}^{-1}$ from Table 20-1 and calculate ΔL from Equation 20-1.

SOLVE

(*a*) 1. The change in length for a given change in temperature is the product of α, L, and ΔT:

$$\Delta L = \alpha L\,\Delta T$$

2. The change in temperature is 205°F. Convert this change in Fahrenheit to a change in kelvins (by multiplying by 5/9):

$$\Delta T = \frac{5\text{ K}}{9°\text{F}}(205°\text{F}) = 114\text{ K}$$

3. Calculate the change in length:

$$\Delta L = \alpha L\,\Delta T = (11 \times 10^{-6}\,\text{K}^{-1})(720\text{ ft})(114\text{ K})$$
$$= \boxed{0.90\text{ ft} = 11\text{ in.}}$$

(*b*) The change in length is proportional to the length. Use this to calculate the change in length of the 420-mi section that is above ground:

$$\frac{\Delta L_2}{L_2} = \frac{\Delta L_1}{L_1} \Rightarrow \Delta L_2 = \frac{L_2}{L_1}\Delta L_1$$

$$\Delta L_2 = \frac{(420\text{ mi})(5280\text{ ft/mi})}{720\text{ ft}}(0.90\text{ ft}) = 2800\text{ ft} \approx \boxed{0.5\text{ mi}}$$

CHECK The 0.5-mi change in length is slightly more than one-tenth of one percent of the 420-mi length. That seems feasible for such a large temperature change and for such a great length.

TAKING IT FURTHER The ends of the sections of pipeline that are above ground do not move with temperature changes because the zigzagging (Figure 20-3) results in lateral movements that "absorb the expansion."

FIGURE 20-3 The zigzagging of the pipeline allows for thermal expansion of the pipes. *(Paul A. Souders/CORBIS.)*

FIGURE 20-4 Expansion joints, such as this one, allow bridges to expand with increases in temperature. *(Frank Siteman/Stock Boston, Inc./PictureQuest.)*

We can calculate the stress that would result in a 1000-m-long steel bridge without expansion joints (Figure 20-4) by using Young's modulus (Equation 12-1):

$$Y = \frac{\text{Stress}}{\text{Strain}} = \frac{F/A}{\Delta L/L}$$

Then

$$\frac{F}{A} = Y\frac{\Delta L}{L} = Y\alpha\,\Delta T$$

For $\Delta T = 30\,\text{K}$, $\Delta L/L = \alpha\,\Delta T = (11 \times 10^{-6}\,\text{K}^{-1})(30\,\text{K}) = 3.3 \times 10^{-4} = 0.33\,\text{m}/1000\,\text{m}$. Then using $Y = 2.0 \times 10^{11}\,\text{N/m}^2$ (from Table 12-1),

$$\frac{F}{A} = Y\frac{\Delta L}{L} = (2.0 \times 10^{11}\,\text{N/m}^2)\frac{0.33\text{ m}}{1000\text{ m}} = 6.6 \times 10^{7}\,\text{N/m}^2$$

This stress is about one-third of the breaking stress for steel under compression. A compression stress of this magnitude would cause a steel bridge to buckle and become permanently deformed.

Example 20-3 A Completely Filled Glass

While working in the laboratory, you fill a 1.000-L Pyrex glass flask to the brim with water at 10°C. You heat the flask, increasing the temperature of the water and flask to 30°C. How much water spills out of the flask?

PICTURE The water and the glass both expand when heated, but 1.000 L of water expands more than 1.000 L of glass, so some water spills out. We calculate the amount spilled by finding the changes in volume for $\Delta T = 20$ K using $\Delta V_{\text{water}} = \beta V_i\,\Delta T$ with $\beta = 0.207 \times 10^{-3}\,\text{K}^{-1}$ for water (from Table 20-1) and $\Delta V_{\text{glass}} = \beta V_i\,\Delta T = 3\alpha V\,\Delta T$ with $\alpha = 3.25 \times 10^{-6}\,\text{K}^{-1}$ for Pyrex glass, where $V_i = 1.000$ L. The difference in these volume changes equals the volume spilled.

SOLVE

1. The volume of water spilled V_{spill} is the difference in the changes in volume of the water and glass:
$$V_{\text{spill}} = \Delta V_{\text{water}} - \Delta V_{\text{glass}}$$
2. Find the increase in the volume of the water:
$$\Delta V_{\text{water}} = \beta_{\text{water}} V_i\,\Delta T$$
3. Find the increase in the volume of the glass flask:
$$\Delta V_{\text{glass}} = \beta_{\text{glass}} V_i\,\Delta T = 3\alpha_{\text{Pyrex}} V_i\,\Delta T$$
4. Subtract to find the amount of water spilled:
$$\begin{aligned} V_{\text{spill}} &= \Delta V_{\text{water}} - \Delta V_{\text{glass}} = \beta_{\text{water}} V_i\,\Delta T - \beta_{\text{glass}} V_i\,\Delta T \\ &= (\beta_{\text{water}} - \beta_{\text{glass}}) V_i\,\Delta T = (\beta_{\text{water}} - 3\alpha_{\text{Pyrex}}) V_i\,\Delta T \\ &= [0.207 \times 10^{-3}\,\text{K}^{-1} - 3(3.25 \times 10^{-6}\,\text{K}^{-1})](1.000\,\text{L})(20\,\text{K}) \\ &= 3.95 \times 10^{-3}\,\text{L} = \boxed{4.0\text{ mL}} \end{aligned}$$

CHECK The overflow of 4.0 mL represents only 0.4 percent of the initial volume of 1.000 L. It is feasible that this small amount would result from a 20 K temperature increase.

TAKING IT FURTHER The flask expands, making the space inside the flask larger, as if the flask were a piece of solid Pyrex glass.

Example 20-4 Breaking Copper *Context-Rich*

During a home plumbing project, you heat a length of copper pipe to 300°C. Then you clamp the pipe between two fixed points so that it cannot contract. If the breaking stress of copper is 230 MN/m^2, at what temperature will the bar break as it cools?

PICTURE As the copper pipe cools, the change in length ΔL that would occur if it was allowed to contract is offset by an equal stretching due to tensile stress in the bar. The stress F/A is related to the stretching ΔL by $Y = (F/A)/(\Delta L/L)$, where Young's modulus for copper is $Y = 110\ \text{GN/m}^2$ (from Table 12-1). The maximum allowable stretching occurs when F/A equals 230 MN/m^2. Thus, we find the temperature change that would produce this maximum contraction.

SOLVE

1. Calculate the change in length ΔL_1 that would occur if the pipe were allowed to contract as it cools:

$$\Delta L_1 = \alpha L\,\Delta T$$

2. A tensile stress F/A stretches the pipe by ΔL_2:

$$Y = \frac{F/A}{\Delta L_2/L} \quad \text{so} \quad \Delta L_2 = L\frac{F/A}{Y}$$

3. Substitute the step-1 and step-2 results into $\Delta L_1 + \Delta L_2 = 0$ and solve for ΔT with the stress equal to the breaking value:

$$\Delta L_1 + \Delta L_2 = 0$$

$$\alpha L\,\Delta T + L\frac{F/A}{Y} = 0$$

$$\text{so} \quad \Delta T = -\frac{F/A}{\alpha Y} = -\frac{230 \times 10^6\ \text{N/m}^2}{(17 \times 10^{-6}\ \text{K}^{-1})(110 \times 10^9\ \text{N/m}^2)}$$

$$= -123\ \text{K} = -123^\circ\text{C}$$

4. Add this result to the original temperature to find the final temperature at which the bar breaks:

$$T_f = T_1 + \Delta T = 300^\circ\text{C} - 123^\circ\text{C} = 177^\circ\text{C} = \boxed{180^\circ\text{C}}$$

CHECK If you look at copper plumbing you will see that the pipes are not rigidly clamped. In addition, in household plumbing the hot-water heater is almost always set so the water temperature does not exceed 60°C (140°F), and the cold water has to be at least 0°C or freezing occurs. Thus, changes in temperature greater than 60°C do not occur in home plumbing. It should not be too surprising that pipes designed for home plumbing might not hold up if expansion or contraction is not allowed for, and if temperature changes much greater than those expected in homes occur.

20-2 THE VAN DER WAALS EQUATION AND LIQUID–VAPOR ISOTHERMS

At ordinary pressures most gases behave like an ideal gas. However, this ideal behavior breaks down when the pressure is high enough or the temperature is low enough such that the density of the gas is high and the molecules are, on average, closer together. An equation of state called the **van der Waals equation** describes the behavior of many real gases over a wide range of pressures more accurately than does the ideal-gas equation of state ($PV = nRT$). The van der Waals equation for n moles of gas is

$$\left(P + \frac{an^2}{V^2}\right)(V - bn) = nRT \qquad \text{20-6}$$

THE VAN DER WAALS EQUATION OF STATE

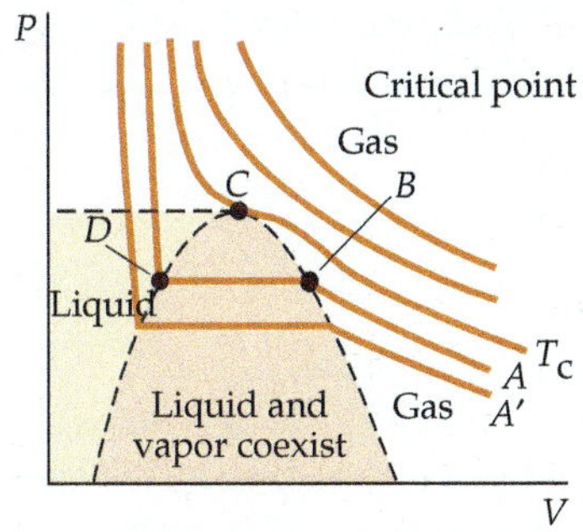

FIGURE 20-5 Isotherms on the *PV* diagram for a substance. For temperatures above the critical temperature T_c, the substance remains a gas at all pressures. Except for the region where the liquid and vapor coexist, these curves are described quite well by the van der Waals equation. The pressure for the horizontal portions of the curves in the shaded region is the vapor pressure, which is the pressure at which the vapor and liquid are in equilibrium. In the region shaded yellow, to the left of the region shaded pink, the substance is a liquid and is nearly incompressible.

The constant b in this equation arises because the gas molecules are not point particles, but objects that have a finite size; therefore, the volume available to each molecule is reduced. The magnitude of b is the volume of one mole of gas molecules. The term an^2/V^2 arises from the attraction of the gas molecules to each other. As a molecule approaches the wall of the container, it is pulled back by the molecules surrounding it with a force that is proportional to the density of those molecules n/V. Because the number of molecules that hit the wall in a given time is also proportional to the density of the molecules, the decrease in pressure due to the attraction of the molecules is proportional to the square of the density and therefore to n^2/V^2. The constant a depends on the gas and is small for inert gases, which have very weak chemical interactions. The terms bn and an^2/V^2 are both negligible when the volume V is large, so at low densities the van der Waals equation approaches the ideal-gas law. At high densities, the van der Waals equation provides a much better description of the behavior of real gases than does the ideal-gas law.

Figure 20-5 shows *PV* isothermal curves for a substance at various temperatures. Except for the region where the liquid and vapor coexist, these curves are described quite accurately by the van der Waals equation and can be used to determine the constants a and b. For example, the values of these constants that give the best fit to the experimental curves for nitrogen are $a = 1.370\ \text{L}^2\cdot\text{atm/mol}^2$ and $b = 38.7$ mL/mol. This volume of 38.7 mL/mole is about 0.2 percent of the volume of 22.4 L occupied by 1 mol of ideal gas under standard conditions. Because the molar mass of nitrogen is 28.02 g/mol, if 1 mol of nitrogen molecules were packed into a volume of 38.7 mL, then the density would be

$$\rho = \frac{M}{V} = \frac{28.0\ \text{g}}{38.7\ \text{mL}} = 0.724\ \text{g/mL} = 0.724\ \text{kg/L}$$

which is almost the same as the density of liquid nitrogen, 0.80 kg/L.

The value of the constant b can be used to estimate the size of a molecule. Because 1 mol (N_A molecules) of nitrogen has a volume of 38.7 cm³, the volume of one nitrogen molecule is

$$V = \frac{b}{N_A} = \frac{38.7\ \text{cm}^3/\text{mol}}{6.02 \times 10^{23}\ \text{molecules/mol}}$$
$$= 6.43 \times 10^{-23}\ \text{cm}^3/\text{molecule}$$

If we assume that each molecule occupies a cube of side d, we obtain

$$d^3 = 6.43 \times 10^{-23}\ \text{cm}^3$$

or

$$d = 4.0 \times 10^{-8}\ \text{cm} = 0.4\ \text{nm}$$

which is a plausible estimate for the "diameter" of a nitrogen molecule.

The values of the constants a and b that give the best fit to the experimental curves are listed in Table 20-2.

Table 20-2 The van der Waals *a* and *b* Coefficients for Several Gasses

	a ($\text{L}^2\cdot\text{atm/mol}^2$)	*b* (mL/mol)
He	0.0346	23.80
Ne	0.211	17.1
Ar	1.34	32.2
Kr	2.32	39.8
Xe	4.19	51.0
H_2	0.244	26.6
N_2	1.370	38.70
O_2	1.382	31.86
H_2O	5.46	30.5
CO_2	3.59	42.7

Example 20-5 Helium at High Density

A 20.0-L tank contains 300 mol of helium at a pressure of 400 atm. (*a*) What is the value of an^2/V^2, and what fraction of the pressure is it? (*b*) What is the value of bn, and what fraction of the volume of the container is it? (*c*) What is the temperature of the helium?

PICTURE To find the temperature use the van der Waals equation (Equation 20-6). The a and b coefficients for helium are found in Table 20-2.

SOLVE

(*a*) Calculate an^2/V^2 and compare it with 400 atm:

$$\frac{an^2}{V^2} = \frac{(0.0346\ \text{L}^2\cdot\text{atm/mol}^2)(300\ \text{mol})^2}{(20.0\ \text{L})^2}$$

$$= 7.785\ \text{atm} = \boxed{7.79\ \text{atm}}$$

(7.785 atm is about 2% of 400 atm)

(*b*) Calculate bn and compare it with 20 L:

$$bn = (0.0238\ \text{L/mol})(300\ \text{mol}) = \boxed{7.14\ \text{L}}$$

(7.14 L is about 36% of 20 L)

(*c*) 1. The van der Waals equation can be solved for the temperature:

$$\left(P + \frac{an^2}{V^2}\right)(V - bn) = nRT$$

2. Obtain the a and b coefficients for helium from Table 20-2:

$$a = 0.0346\ \text{L}^2\cdot\text{atm/mol}^2$$
$$b = 0.0238\ \text{L/mol}$$

3. Enter the given values and solve for the temperature. With the pressure in atmospheres and the volume in liters, we use $R = 0.082057\ \text{L}\cdot\text{atm/(mol}\cdot\text{K)}$:

$$T = \frac{\left(P + \frac{an^2}{V^2}\right)(V - bn)}{nR} = \frac{\left(400 + \frac{0.0346 \times 300^2}{20.0^2}\right)(20.0 - 0.0238 \times 300)}{300 \times 0.082057}$$

$$= \boxed{213\ \text{K}}$$

CHECK In the van der Waals equation the 2% correction to the pressure term [Part (*b*)] is dwarfed by the 36% correction to the volume term [Part (*a*)]. This is as expected. The correction to the pressure term is particularly small for helium because the helium atoms attract each other more weakly than do most other atoms.

At temperatures below T_c, the van der Waals equation describes those portions of the isotherms outside the shaded region in Figure 20-5, but not those portions inside the shaded region. Suppose we have a gas at a temperature below T_c that initially has a low pressure and a large volume. We begin to compress the gas while holding the temperature constant (isotherm *A* in the figure). At first the pressure

Cloud forming behind an aircraft as it breaks the sound barrier. As the aircraft moves through the air, an area of low pressure forms behind it. When the pressure of this air parcel falls below the vapor pressure of gaseous water, the water in the air condenses to form the cloud. Different atmospheric conditions cause the phenomenon to occur at different aircraft speeds. *(U.S. Department of Defense/Photo Researchers, Inc.)*

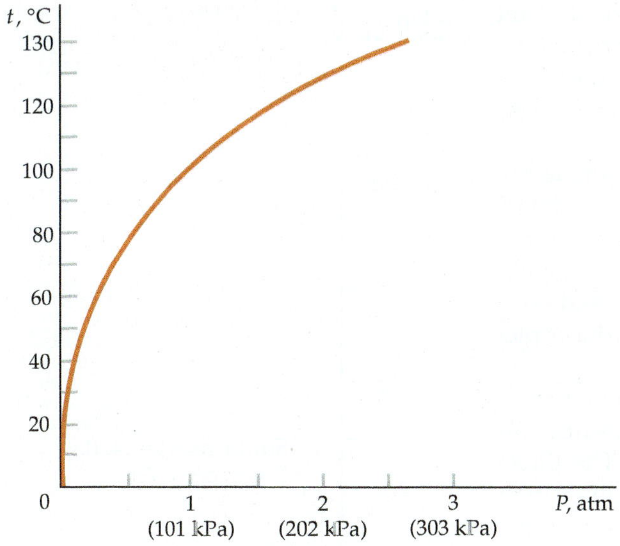

FIGURE 20-6 Boiling point of water versus pressure.

rises, but when we reach point B on the dashed curve, the pressure ceases to rise and the gas begins to liquefy at constant pressure. Along the horizontal line BD in the figure, the gas and liquid are in equilibrium. As we continue to compress the gas, more and more gas liquefies until point D on the dashed curve, at which point we have only liquid. Then, if we try to compress the substance further, the pressure rises sharply because a liquid is nearly incompressible.

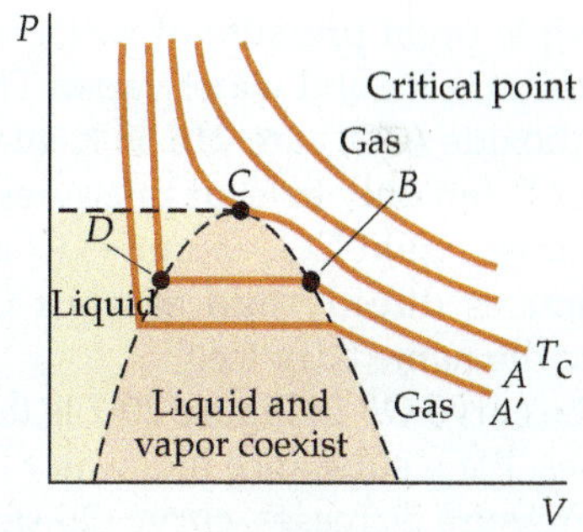

FIGURE 20-5 *(repeated)* Isotherms on the PV diagram for a substance.

Now consider injecting a liquid such as water into a sealed evacuated container. As some of the water evaporates, water-vapor molecules fill the previously empty space in the container. Some of these molecules will hit the liquid surface and rejoin the liquid water during a process called *condensation.* The rate of evaporation will initially be greater than the rate of condensation, but eventually equilibrium will be reached. The pressure at which a liquid is in equilibrium with its own vapor is called the **vapor pressure.** If we now heat the container slightly, the liquid boils, more liquid evaporates, and a new equilibrium is established at a higher vapor pressure. Vapor pressure thus depends on the temperature. We can see this from Figure 20-5. If we had started compressing the gas at a lower temperature, as with isotherm A' in Figure 20-5, the vapor pressure would be lower, as is indicated by the horizontal constant-pressure line for A' at a lower value of pressure. The temperature for which the vapor pressure for a substance equals 1 atm is the **normal boiling point** of that substance. For example, the temperature at which the vapor pressure of water is 1.00 atm is 373 K(= 100°C), so this temperature is the normal boiling point of water. At high altitudes, such as on the top of a mountain, the pressure is less than 1.00 atm, therefore, water boils at a temperature lower than 373 K. Figure 20-6 gives the vapor pressures of water at various temperatures.

At temperatures greater than the critical temperature T_c, a gas will not condense at any pressure. The critical temperature for water vapor is 647 K (374°C). The point at which the critical isotherm intersects the dashed curve in (point C) is called the **critical point.**

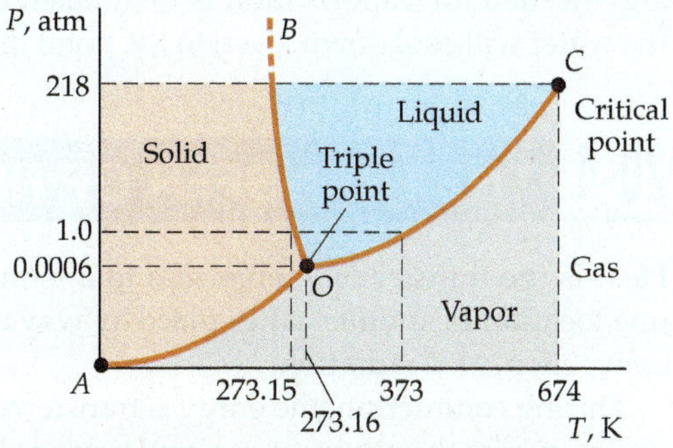

FIGURE 20-7 Phase diagram for water. The pressure and temperature scales are not linear but are compressed to show the points of interest. Curve OC is the curve of vapor pressure versus temperature. Curve OB is the melting curve, and curve OA is the sublimation curve.

20-3 PHASE DIAGRAMS

Figure 20-7 is a plot of pressure versus temperature at a constant volume for water. Such a plot is called a **phase diagram.** The portion of the diagram between points O and C shows vapor pressure versus temperature. As we continue to heat the container, the density of the liquid decreases and the density of the vapor increases. At point C on the diagram, these densities are equal. Point C is called the *critical point*. At this point and above it, there is no distinction between the liquid and the gas. Critical-point temperatures T_c for various substances are listed in Table 20-3. At temperatures greater than the critical temperature, a gas will not condense at any pressure.

If we now cool our container, some of the vapor condenses into a liquid as we move back down the curve *OC* in Figure 20-7 until the substance reaches point *O*. At this point, the liquid begins to solidify. Point *O* is the **triple point,** that one point at which the vapor, liquid, and solid phases of a substance can coexist in equilibrium. Every substance has a unique triple point at a specific temperature and pressure. The triple-point temperature for water is 273.16 K (0.01°C) and the triple-point pressure is 4.58 mmHg.

At temperatures and pressures below the triple point, the liquid cannot exist. The curve *OA* in the phase diagram of Figure 20-7 is the locus of pressures and temperatures for which the solid and vapor coexist in equilibrium. The direct change from a solid to a vapor is called **sublimation.** You can observe sublimation by putting a few loose ice cubes in the freezer compartment of a no-frost (self-defrosting) refrigerator. Over time, the ice cubes will shrink and eventually disappear due to sublimation. This happens because the atmospheric pressure is well above the triple-point pressure of water, and therefore, equilibrium is never established between the ice and water vapor. The triple-point temperature and pressure of carbon dioxide (CO_2) are 216.55 K and 3880 mmHg (5.1 atm), which means that liquid CO_2 can only exist at pressures above 5.1 atm. Thus, at ordinary atmospheric pressures, liquid CO_2 cannot exist at any temperature. When solid CO_2 "melts," it sublimates directly into gaseous CO_2 without going through the liquid phase, hence the name "dry ice."

The curve *OB* in Figure 20-7 is the melting curve separating the liquid and solid phases. For a substance like water for which the melting temperature decreases as the pressure increases, curve *OB* slopes upward to the left from the triple point, as in this figure. For most other substances, the melting temperature increases as the pressure increases. For such a substance, curve *OB* slopes upward to the right from the triple point.

For a molecule to escape (evaporate) from a substance in the liquid state, energy is required to break the intermolecular attractions at the liquid's surface. Vaporization cools the liquid left behind. If a pot of water is brought to a boil over a hot plate, this cooling effect keeps the temperature of the liquid constant at the boiling point. This is the reason that the boiling point of a substance can be used to calibrate thermometers. However, water can also be made to boil without adding heat by evacuating the air above it, thereby lowering the applied pressure. The energy needed for vaporization is then taken from the water left behind. As a result, the water will cool down, even to the point that ice forms on top of the boiling water!

Table 20-3 Critical Temperatures T_c for Various Substances

Substance	T_c, K
Water	647.4
Sulfur dioxide	430.9
Chlorine	417.12
Carbon dioxide	304.2
Nitric oxide	180.2
Oxygen	154.8
Argon	150.8
Neon	44.4
Hydrogen	33.3
Helium	5.3

20-4 THE TRANSFER OF HEAT

Heat is the transfer of energy due to a temperature difference. This transfer from one location to another takes place by way of three distinct processes: conduction, convection, and radiation.

During **conduction,** the enrgy is transferred by interactions among atoms or molecules, where the atoms or molecules are not themselves transported. For example, if one end of a solid bar is heated, the atoms in the heated end vibrate with greater energy than do those at the cooler end. The interaction of the more energetic atoms with the less energetic atoms causes this energy to be transported along the bar.*

* If the solid is a metal, the transfer of heat is made easier by *delocalized* electrons, which can move throughout the metal.

During **convection,** heat is transferred by direct transport of matter. For example, warm air in a region of a room expands, its density decreases, and the buoyant force on it due to the surrounding air causes it to rise. Energy is thus transported upward along with the molecules of warm air.

During **radiation,** the energy is transferred through space in the form of electromagnetic waves that move at the speed of light. Infrared waves, visible light waves, radio waves, television waves, and X rays are all forms of electromagnetic radiation that differ from one another in their wavelengths and frequencies.

During all mechanisms of heat transfer, the rate of cooling of a body is approximately proportional to the temperature difference between the body and its surroundings. This result is known as **Newton's law of cooling.**

During many real situations, all three mechanisms for energy transfer occur simultaneously, though one mechanism may be more dominant than the others. For example, an ordinary space heater uses both radiation and convection. If the heating element is quartz, then the main mechanism of transfer is radiation. If the heating element is metal (which does not radiate as efficiently as quartz), then convection is the main mechanism by which energy is transferred, with the heated air rising to be replaced by cooler air. Fans are often included in heaters to speed the convection process.

CONDUCTION

Figure 20-8*a* shows an insulated uniform solid bar of cross-sectional area A. If we keep one end of the bar at a high temperature and the other end at a low temperature, energy is conducted down the bar from the hot end to the cold end. In the steady state, the temperature varies linearly from the hot end to the cold end. The rate of change of the temperature along the bar dT/dx is called the **temperature gradient.***

Let dT be the temperature difference across a small segment of length dx (Figure 20-8*b*). If dQ is the amount of heat conducted through a cross section of the segment during some time dt, then the rate of conduction of heat dQ/dt is called the **thermal current *I*.** It has experimentally been found that the thermal current is proportional to the temperature gradient and to the cross-sectional area A:

$$I = \frac{dQ}{dt} = -kA\frac{dT}{dx} \qquad \text{20-7}$$

DEFINITION: THERMAL CURRENT

! Heat transports energy from a higher-temperature region to a lower-temperature region, so the thermal current is in the direction of decreasing temperature.

The proportionality constant k, called the *thermal conductivity,* depends on the composition of the bar.† The heat is transferred in the direction of decreasing temperature. That is, if the temperature increases with increasing x, then the heat transfer is in the negative x direction, and vice versa. In SI units, thermal current is

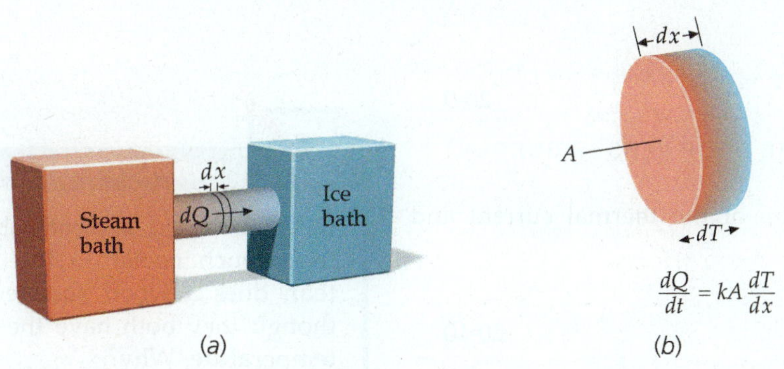

FIGURE 20-8 (*a*) An insulated conducting bar with its ends at two different temperatures. (*b*) A segment of the bar of length dx. The rate at which heat is conducted through a cross section of the segment is proportional to the cross-sectional area of the bar and the temperature drop dT across the segment, and it is inversely proportional to the length of the segment.

* The temperature gradient is actually a vector. The direction of this vector is the direction in which the temperature is increasing most rapidly, and the magnitude of this vector is the rate of change of temperature with respect to distance in this direction.

† Do not confuse the thermal conductivity with Boltzmann's constant, which is also designated by k.

Table 20-4 Thermal Conductivities k for Various Materials

Material	k, Btu·in./(h·ft^2·F°)	k, W/(m·K)
Silver	(2980)	(429)
Copper	(2780)	(401)
Gold	(2200)	(318)
Aluminum	(1644)	(237)
Iron	(558)	(80.4)
Steel	(319)	(46)
Lead	(20.4)	(35.3)
Concrete	(6–9)	(0.19–1.3)
Glass	(5–6)	(0.7–0.9)
Water at 27°C	(4.22)	(0.609)
Ice	(4.11)	(0.592)
Oak	(1.02)	(0.15)
White Pine	(0.78)	(0.11)
Air at 27°C	(0.18)	(0.026)

expressed in watts, and the thermal conductivity has units of W/(m·K).* In practical calculations in the United States, the thermal current is usually expressed in Btu per hour, the area is expressed in square feet, the length (or thickness) is expressed in inches, and the temperature is expressed in degrees Fahrenheit. The thermal conductivity is then given in Btu·in/(h·ft^2·°F). Table 20-4 gives the thermal conductivities of various materials.

If we solve Equation 20-7 for the temperature difference, we obtain

$$|\Delta T| = I\frac{|\Delta x|}{kA} \qquad 20\text{-}8$$

or

$$\Delta T = IR \qquad 20\text{-}9$$

TEMPERATURE DROP VERSUS CURRENT

where ΔT is the *temperature drop* in the direction of the thermal current and $|\Delta x|/(kA)$ is the **thermal resistance** R:

$$R = \frac{|\Delta x|}{kA} \qquad 20\text{-}10$$

DEFINITION: THERMAL RESISTANCE

CONCEPT CHECK 20-1

In a cool room, a metal tabletop feels much cooler to the touch than does a wood surface even though they both have the same temperature. Why?

* In some tables, the energy may be given in calories or kilocalories and the thickness in centimeters.

PRACTICE PROBLEM 20-1

Calculate the thermal resistance of an aluminum slab of cross-sectional area 15.0 cm^2 and thickness 2.00 cm.

PRACTICE PROBLEM 20-2

What thickness of a slab of silver would be required to give the same thermal resistance as a 1.00-cm-thick layer of dead air of the same area?

In many practical problems, we are interested in the rate of heat transfer through two or more conductors (or insulators) in series. For example, we may want to know the effect of adding insulating material of a certain thickness and thermal conductivity to the space between two layers of wallboard. Figure 20-9 shows two thermally conducting slabs of the same cross-sectional area, but of different materials and of different thicknesses. Let T_1 be the temperature on the warm side, T_2 be the temperature at the interface between the slabs, and T_3 be the temperature on the cool side. Under the conditions of steady-state heat transfer, the thermal current I through each of the slabs must be the same. This follows from energy conservation; for steady-state transfer, the rate at which energy enters any region must equal the rate at which it exits that region.

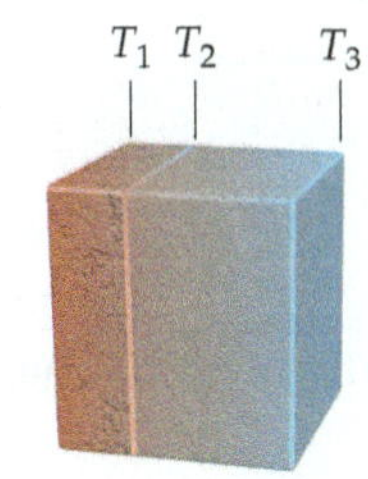

FIGURE 20-9 Two thermally conducting slabs of different materials in series. The equivalent thermal resistance of the slabs in series is the sum of their individual thermal resistances. The thermal current is the same through both slabs.

If R_1 and R_2 are the thermal resistances of the two slabs, we have by applying Equation 20-9 to each slab

$$T_1 - T_2 = IR_1$$

and

$$T_2 - T_3 = IR_2$$

Adding these equations gives

$$\Delta T = T_1 - T_3 = I(R_1 + R_2) = IR_{eq}$$

or

$$I = \frac{\Delta T}{R_{eq}} \qquad 20\text{-}11$$

where R_{eq} is the **equivalent resistance.** Thus, for thermal resistances in series, the equivalent resistance is the sum of the individual resistances:

$$R_{eq} = R_1 + R_2 + \cdots \qquad 20\text{-}12$$

THERMAL RESISTANCES IN SERIES

This result can be extended to any number of resistances in series. In Chapter 25, we will find that the same formula applies to electrical resistances in series.

This thermogram of a house shows the heat being radiated to its surroundings. *(Alfred Pasieka/Photo Researchers, Inc.)*

To calculate the rate at which energy is leaving a room by way of heat conduction, we need to know how much heat is released through the walls, the windows, the floor, and the ceiling. For this type of problem, in which there are several paths for heat transfer, the resistances are said to be in **parallel.** The temperature difference is the same for each path, but the thermal current is different. The total thermal current is the sum of the thermal currents through each of the parallel paths:

$$I_{\text{total}} = I_1 + I_2 + \cdots = \frac{\Delta T}{R_1} + \frac{\Delta T}{R_2} + \cdots = \Delta T\left(\frac{1}{R_1} + \frac{1}{R_2} + \cdots\right)$$

or

$$I_{\text{total}} = \frac{\Delta T}{R_{\text{eq}}} \qquad \text{20-13}$$

where the equivalent thermal resistance is given by

$$\frac{1}{R_{\text{eq}}} = \frac{1}{R_1} + \frac{1}{R_2} + \cdots \qquad \text{20-14}$$

THERMAL RESISTANCES IN PARALLEL

We will encounter this equation again in Chapter 25 when we study electric conduction through parallel resistances. Note that for both resistors in series (Equation 20-11) and resistors in parallel (Equation 20-13) I is proportional to ΔT, which is in agreement with Newton's law of cooling.

PROBLEM-SOLVING STRATEGY

Calculating Thermal Current

PICTURE Determine if the group of objects for which you are finding the total thermal current are in series or in parallel.

SOLVE

1. Using $R = |\Delta x|/(kA)$ (Equation 20-10), find the thermal resistance of each object.
2. If any objects are in series, use $R_{\text{eq}} = R_1 + R_2 + \cdots$ (Equation 20-12) to calculate their equivalent resistance.
3. If any objects are in parallel, use $\frac{1}{R_{\text{eq}}} = \frac{1}{R_1} + \frac{1}{R_2} + \cdots$ (Equation 20-14) to find their equivalent resistance.
4. Repeat steps 2 and 3 until you have calculated the equivalent resistance of the entire system of conducting objects.
5. Using $\Delta T = I_{\text{total}}R_{\text{eq}}$ (Equation 20-9), calculate the total thermal current.

CHECK For each parallel combination of objects, make sure the equivalent resistance is less than the resistance of the object with the least resistance. For each series combination of objects, make sure the equivalent resistance is greater than the resistance of the object with the greatest resistance.

Example 20-6 Two Metal Bars in Series

Two insulated metal bars, each of length 5.0 cm and rectangular cross section with sides 2.0 cm and 3.0 cm, are wedged between two walls, one held at 100°C and the other at 0.0°C (Figure 20-10). The bars are lead and silver. Find (*a*) the total thermal current through the two-bar combination, and (*b*) the temperature at the interface.

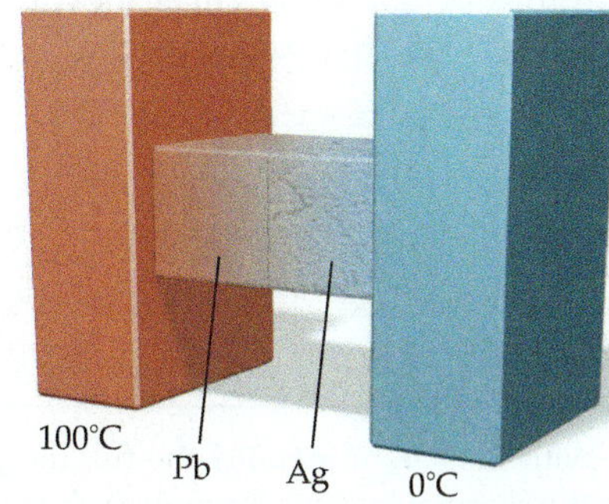

FIGURE 20-10 Two thermally conducting slabs of different materials in parallel.

PICTURE The bars are thermal resistors connected in series. (*a*) You can find the total thermal current from $I = \Delta T/R_{eq}$, where the equivalent resistance R_{eq} is the sum of the individual resistances. Using Equation 20-10 and the thermal conductivities given in Table 20-4, the individual resistances can be determined. (*b*) You can find the temperature at the interface by applying $I = \Delta T/R_1$, to the lead bar only, and solving for ΔT in terms of the value for I found in Part (*a*).

SOLVE

(*a*) 1. Use $\Delta T = IR$ (Equation 20-13) to relate the thermal current to the temperature difference:

$$I = \frac{\Delta T}{R}$$

2. Using $R = |\Delta x|/(kA)$ (Equation 20-10), write each thermal resistance in terms of the individual thermal conductivities and geometric parameters:

$$R_{Pb} = \frac{|\Delta x_{Pb}|}{k_{Pb}A_{Pb}} \qquad R_{Ag} = \frac{|\Delta x_{Ag}|}{k_{Ag}A_{Ag}}$$

$$R_{Pb} = \frac{0.050\text{ m}}{35.3\text{ W/(m}\cdot\text{K)} \times (0.020\text{ m} \times 0.030\text{ m})} = 2.36\text{ K/W}$$

$$R_{Ag} = \frac{0.050\text{ m}}{429\text{ W/(m}\cdot\text{K)} \times (0.020\text{ m} \times 0.030\text{ m})} = 0.194\text{ K/W}$$

3. Find R_{eq} using the formula for resistors in series:

$$R_{eq} = R_{Pb} + R_{Ag} = 2.36\text{ K/W} + 0.194\text{ K/W} = 2.55\text{ K/W}$$

4. Use $\Delta T = IR$ (Equation 20-13) to find the thermal current.

$$I = \frac{\Delta T}{R_{eq}} = \frac{100\text{ K}}{2.55\text{ K/W}} = \boxed{39.1\text{ W}}$$

(*b*) 1. Calculate the temperature difference across the lead bar using the current and thermal resistance found in Part (*a*).

$$\Delta T_{Pb} = IR_{Pb} = 39.1\text{ W} \times 2.36\text{ K/W} = 92.4\text{ K}$$

2. Use your result from the previous step to find the temperature at the interface.

$$T_{if} = 100°\text{C} - \Delta T_{Pb} = \boxed{8°\text{C}}$$

CHECK We check our Part-(*b*) result by calculating the temperature drop across the silver bar. That is, $\Delta T_{Ag} = IR_{Ag} = 39.1\text{ W} \times 0.194\text{ K/W} = 8°\text{C}$, which agrees with our Part-(*b*) result. Note that equivalent resistance (2.55 K/W) is greater than either of the individual resistances (2.36 K/W and 0.194 K/W).

Example 20-7 The Metal Bars in Parallel

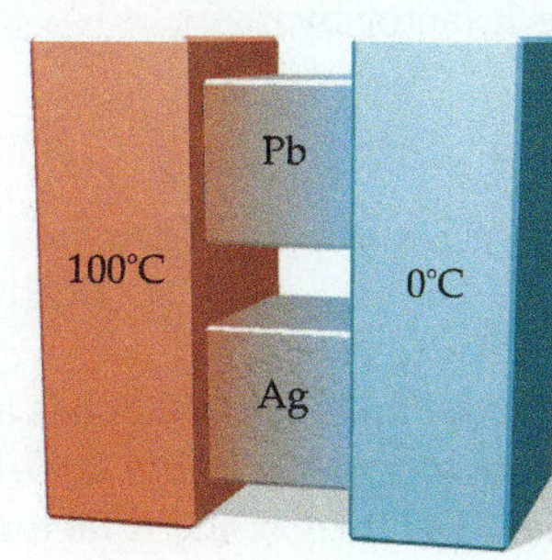

FIGURE 20-11

The metal bars in Example 20-6 are rearranged as shown in Figure 20-11. Find (*a*) the thermal current in each bar, (*b*) the total thermal current, and (*c*) the equivalent thermal resistance of the two-bar system.

PICTURE The current in each bar is found from $I = \Delta T/R$, where R is the thermal resistance of the bar (found in Example 20-6). The total current is the sum of the currents. The equivalent resistance can be found from Equation 20-14 or from $I_{total} = \Delta T/R_{eq}$.

SOLVE

(*a*) Calculate the thermal current for each bar:

$$I_{Pb} = \frac{\Delta T}{R_{Pb}} = \frac{100\text{ K}}{2.36\text{ K/W}} = \boxed{42.4\text{ W}}$$

$$I_{Ag} = \frac{\Delta T}{R_{Ag}} = \frac{100\text{ K}}{0.194\text{ K/W}} = \boxed{515\text{ W}}$$

(*b*) Add your results to find the total thermal current:

$$I_{\text{total}} = I_{\text{Pb}} + I_{\text{Ag}} = 42.4\text{ W} + 515\text{ W} = \boxed{557\text{ W}}$$

(*c*) 1. Use Equation 20-14 to calculate the equivalent resistance of the two bars in parallel:

$$\frac{1}{R_{\text{eq}}} = \frac{1}{R_{\text{Pb}}} + \frac{1}{R_{\text{Ag}}} = \frac{1}{2.36\text{ K/W}} + \frac{1}{0.194\text{ K/W}}$$

$$\text{so}\quad R_{\text{eq}} = \boxed{0.179\text{ K/W}}$$

2. Check your result using, $I_{\text{total}} = \Delta T/R_{\text{eq}}$:

$$I_{\text{total}} = \frac{\Delta T}{R_{\text{eq}}}$$

$$R_{\text{eq}} = \frac{\Delta T}{I_{\text{total}}} = \frac{100\text{ K}}{557\text{ W}} = \boxed{0.179\text{ K/W}}$$

CHECK With the bars in parallel the full temperature difference of 100 K is across each of them, so the thermal current through each is much higher than the thermal current was in either bar in Example 20-6, in which the bars were arranged in series so the temperature difference across either bar is considerably less than 100 K. In addition, in Example 20-7 the total current equals the sum of the currents in the bars, whereas in Example 20-6 the total current equals the current in either bar. Thus, that the total current (557 W) with the bars in parallel is more than 14 times larger than the total current (39.1 W) with the bars in series is a plausible result.

TAKING IT FURTHER Note that in Part (*b*), the equivalent resistance is less than either of the individual resistances. This is always the case for resistors connected in parallel.

In the building industry, the thermal resistance of a square foot of cross-sectional area of a material is called its **R factor**, R_{f}. Consider a 32-ft^2 sheet of insulating material with thickness Δx and R factor R_{f} of 7.2. That is, each square foot (Figure 20-12) has a thermal resistance of 7.2°F/(Btu/h). The 32 square feet are in parallel, so the net resistance R_{net} is calculated using Equation 20-14, giving

$$\frac{1}{R_{\text{net}}} = \frac{1}{R_1} + \frac{1}{R_2} + \cdots = \frac{1}{R_{\text{f}}} + \frac{1}{R_{\text{f}}} + \cdots = \frac{32}{R_{\text{f}}} \qquad \text{so} \qquad R_{\text{net}} = \frac{R_{\text{f}}}{32}$$

Thus, the total thermal resistance R in °F/(Btu/h) equals the R factor divided by the area A in square feet. That is,

$$R_{\text{net}} = \frac{R_{\text{f}}}{A}$$

Because the net (total) resistance R_{net} is related to the conductivity k by $R_{\text{net}} = |\Delta x|/(kA)$ (Equation 20-10), we can express the R factor by

$$R_{\text{f}} = R_{\text{net}}A = \frac{|\Delta x|}{k} \qquad \text{20-15}$$

DEFINITION: *R* FACTOR

where $|\Delta x|$ is the thickness in inches and k is the conductivity in Btu·in/(h·ft^2·°F). Table 20-5 lists R factors for several materials. In terms of the R factor, Equation 20-9 for the thermal current is

$$\Delta T = IR_{\text{net}} = \frac{I}{A}R_{\text{f}} \qquad \text{20-16}$$

For slabs of insulating material of the same area in series, R_{f} is replaced by the equivalent R factor $R_{\text{f eq}}$

$$R_{\text{f eq}} = R_{\text{f1}} + R_{\text{f2}} + \cdots$$

FIGURE 20-12 For a 1-in. thickness of this material, the $R_{\text{f}} = 7.2$. (*Courtesy of Eugene Mosca.*)

Table 20-5 *R* Factors for Various Building Materials

Material	Thickness, in.	R_f, h·ft²·F°/Btu
Building board		
Gypsum or plasterboard	0.375	0.32
Plywood (Douglas fir)	0.5	0.62
Plywood or wood panels	0.75	0.93
Particle board, medium density	1.0	1.06
Finish flooring materials		
Carpet and fibrous pad	1.0	2.08
Tile		0.5
Wood, hardwood finish	0.75	0.68
Roof insulation	1.0	2.8
Roofing		
Asphalt roll roofing		0.15
Asphalt shingles		0.44
Windows		
Single pane		0.9
Double pane		1.8

For parallel slabs, we calculate the thermal current through each slab and add all these currents to obtain the total current.

Example 20-8 Heat Loss Through a Roof

Context-Rich

You are helping your friend's family put new asphalt shingles on the roof of their winter cabin. The 60-ft × 20-ft roof is made of 1.0-in-thick pine board covered with asphalt shingles. There is room for 8.0 in of roof insulation, and your friend's family is wondering how much of a difference it would make to their energy bill if they were to install two inches of insulation. Knowing that you are studying physics, they ask for your opinion.

PICTURE To assess the situation, you first calculate the *R* factor for each layer of the roof. Because the layers are in series, the equivalent *R* factor is just the sum of the individual *R* factors. The aim is to calculate the equivalent *R* factor of the roof with and without the insulation. The *R* factors for asphalt shingles and for roof insulation are found in Table 20-5. The *R* factor for the pine board is calculated from its thermal conductivity, which is found in Table 20-4. Note that when you shingle a roof you have to overlap the shingles, so there are two layers of asphalt shingling on the roof.

SOLVE

1. For a series combination, the equivalent *R* factor is the sum of the individual *R* factors:
$$R_{\text{f eq}} = R_{\text{f pine}} + R_{\text{f asph}} + R_{\text{f insul}}$$

2. The *R* factor for the double layer of shingles is twice the *R* factor for one layer:
$$R_{\text{f asph}} = 2(0.44\ \text{h}\cdot\text{ft}^2\cdot{}^\circ\text{F/Btu}) = 0.88\ \text{h}\cdot\text{ft}^2\cdot{}^\circ\text{F/Btu}$$

3. The *R* factor for 2.0 in of roof insulation is twice that for 1.0 in:
$$R_{\text{f insul}} = 2(2.8\ \text{h}\cdot\text{ft}^2\cdot{}^\circ\text{F/Btu}) = 5.6\ \text{h}\cdot\text{ft}^2\cdot{}^\circ\text{F/Btu}$$

4. The *R* factor for 1.0-in-thick pine is obtained from the conductivity:
$$R_{\text{f pine}} = \frac{\Delta x_{\text{p}}}{k_{\text{p}}} = \frac{1.0\ \text{in}}{0.78\ \text{Btu}\cdot\text{in}/(\text{h}\cdot\text{ft}^2\cdot{}^\circ\text{F})} = 1.28\ \text{h}\cdot\text{ft}^2\cdot{}^\circ\text{F/Btu}$$

5. The equivalent *R* factor without the insulation is:
$$R'_{\text{f eq}} = R_{\text{f pine}} + R_{\text{f asph}} = 1.28\ \text{h}\cdot\text{ft}^2\cdot{}^\circ\text{F/Btu} + 0.88\ \text{h}\cdot\text{ft}^2\cdot{}^\circ\text{F/Btu}$$
$$= 2.16\ \text{h}\cdot\text{ft}^2\cdot{}^\circ\text{F/Btu} = 2.2\ \text{h}\cdot\text{ft}^2\cdot{}^\circ\text{F/Btu}$$

6. The equivalent R factor with insulation is:

$$R_{\text{f eq}} = R_{\text{f pine}} + R_{\text{f asph}} + R_{\text{f insul}} = R'_{\text{f eq}} + R_{\text{f insul}}$$
$$= 2.16\ \text{h}\cdot\text{ft}^2\cdot{}^\circ\text{F/Btu} + 5.6\ \text{h}\cdot\text{ft}^2\cdot{}^\circ\text{F/Btu}$$
$$= 7.76\ \text{h}\cdot\text{ft}^2\cdot{}^\circ\text{F/Btu}$$

7. One comparison of the two equivalent R factors is their ratio:

$$\frac{R'_{\text{f eq}}}{R_{\text{f eq}}} = \frac{2.16}{7.76} = 0.28$$

8. By adding the insulation, the heat loss rate per square foot is reduced by 72 percent. Is 72 percent of a large heat loss or a small heat loss? Using Equation 20-16 we calculate the thermal current I' through the entire roof:

$$\Delta T = IR_{\text{net}} = \frac{I}{A}R_{\text{f}}$$
$$I' = \frac{A}{R'_{\text{f eq}}}\Delta T = \frac{(60\ \text{ft})(20\ \text{ft})}{2.16\ \text{h}\cdot\text{ft}^2\cdot{}^\circ\text{F/Btu}}\Delta T = [556\ (\text{Btu/h})/{}^\circ\text{F}]\Delta T$$
$$= [5.6\times 10^2\ (\text{Btu/h})/{}^\circ\text{F}]\Delta T$$

9. To complete the calculation, we estimate that the temperature inside the cabin is maintained at 70°F and the temperature outside the cabin during the winter is typically 40°F colder:

$$I' = [556\ (\text{Btu/h})/{}^\circ\text{F}]\Delta T$$
$$= [556\ (\text{Btu/h})/{}^\circ\text{F}](40^\circ\text{F}) = 22\,200\ \text{Btu/h}$$

and $I = 0.28I' = 0.28(22\,200\ \text{Btu/h}) = 6200\ \text{Btu/h}$

so the reduction due to the insulation is

$$I' - I = 22\,200\ \text{Btu/h} - 6200\ \text{Btu/h} = 16\times 10^3\ \text{Btu/h}$$

10. Installing 2.0 in. of roof insulation reduces the heat loss through the roof by 16,000 Btu/h. The cabin is heated with propane, and the energy content of propane is 92,000 Btu/gal. Insulating the roof reduces consumption by approximately 4.2 gal of propane every 24 h of use.

Propane costs about $3.00/gal, so this amounts to a savings of approximately $12.60 per day, or $376 per month during the heating season. Your friend's family is impressed by the potential savings (and by the benefits of your physics knowledge). They decide to install the 2.0 in of roofing insulation.

CHECK It should come as no surprise that installing some insulation has a significant economic payback.

TAKING IT FURTHER These cost estimates do not include the cost of purchasing and installing the insulation.

PRACTICE PROBLEM 20-3 How much additional savings can possibly be had by adding even more insulation to the roof?

The thermal conductivity of air is very small compared with that of solid materials, which makes air a very good insulator. However, when there is a large air gap—say, between a storm window and the inside window—the insulating efficiency of air is greatly reduced because of convection. Whenever there is a temperature difference between different parts of the air space, convection currents act quickly to equalize the temperature, so the effective conductivity is greatly increased. For storm windows, air gaps of about 1 to 2 cm are optimal. Wider air gaps actually reduce the thermal resistance of a double-pane window due to convection.

The insulating properties of air are most effectively used when the air is trapped in small pockets that prevent convection from taking place. This is the principle underlying the excellent insulating properties of both goose down and Styrofoam.

If you touch the inside surface of a glass window when it is cold outside, you will observe that the surface is considerably colder than the inside air. The thermal resistance of windows is mainly due to thin films of insulating air that adhere to either side of the glass surface. The thickness of the glass has little effect on the overall thermal resistance. The air film on each side typically adds an R factor of about 0.45 per side. Thus, the R factor of a window with N separated glass layers is approximately $0.90N$, because of the two sides of each layer. Under windy conditions, the outside air film may be greatly decreased, leading to a smaller R factor for the window.

CONVECTION

Convection is the transfer of heat by the transport of the material medium itself. This thermal property is responsible for the great ocean currents as well as the global circulation of the atmosphere. In the simplest case, convection arises when a fluid (gas or liquid) is heated from below. The warm fluid then expands and rises as the cooler fluid sinks. The mathematical description of convection is very complex, because the flow depends on the temperature difference in different parts of the fluid, and this temperature difference is affected by the flow itself.

The heat transferred from an object to its surroundings by convection is approximately proportional to the area of the object and to the difference in temperature between the object and the surrounding fluid. It is possible to write an equation for the heat transferred by convection and to define a coefficient of convection, but the analyses of practical problems involving convection are quite complex and are not discussed here.

RADIATION

All objects emit and absorb electromagnetic radiation. When an object is in thermal equilibrium with its surroundings, it emits and absorbs radiation at the same rate. The rate at which an object radiates energy is proportional to both the surface area of the object and to the fourth power of its absolute temperature. This result, found empirically by Josef Stefan in 1879 and derived theoretically by Ludwig Boltzmann about five years later, is called the **Stefan–Boltzmann law:**

$$P_r = e\sigma AT^4 \tag{20-17}$$

STEFAN–BOLTZMANN LAW

where P_r is the power radiated, A is the surface area, σ is a universal constant called Stefan's constant, which has the value

$$\sigma = 5.6703 \times 10^{-8}\ \text{W}/(\text{m}^2 \cdot \text{K}^4) \tag{20-18}$$

and e is the **emissivity** of the radiating surface, a dimensionless quantity between 0 and 1 that is dependent upon the composition of the surface of the object.

When electromagnetic radiation falls on an opaque object, part of the radiation is reflected and part is absorbed. Light-colored objects reflect most visible radiation, whereas dark objects absorb most of it. The rate at which an object absorbs radiation is given by

$$P_a = e\sigma AT_0^4 \tag{20-19}$$

where T_0 is the temperature of the source of the radiation and e is the emissivity of the surface of the absorbing object.

Suppose an object at temperature T is surrounded by objects at temperature T_0. If the object emits radiant energy at a greater rate than the object absorbs it, then the object cools, while the object's surroundings absorb radiation from the object and becomes warmer. If the object absorbs radiant energy at a greater rate than the object emits it, then the object warms and its surroundings cool. The net power radiated by an object at temperature T in an environment at temperature T_0 is

$$P_{net} = e\sigma A(T^4 - T_0^4) \tag{20-20}$$

When an object is in thermal equilibrium with its surroundings, $T = T_0$, and the object emits and absorbs radiation at the same rate.

An object that absorbs all radiation incident upon it and has an emissivity equal to 1 is called a **blackbody.** A blackbody is also an ideal radiator. The concept of a blackbody is important because the characteristics of the radiation emitted by such an ideal object can be calculated theoretically. Materials such as black velvet come close to being ideal blackbodies. The best practical approximation of an ideal blackbody is a small hole leading into a cavity, such as a keyhole in a closet door (Figure 20-13). Radiation incident on the hole has little chance of being reflected out

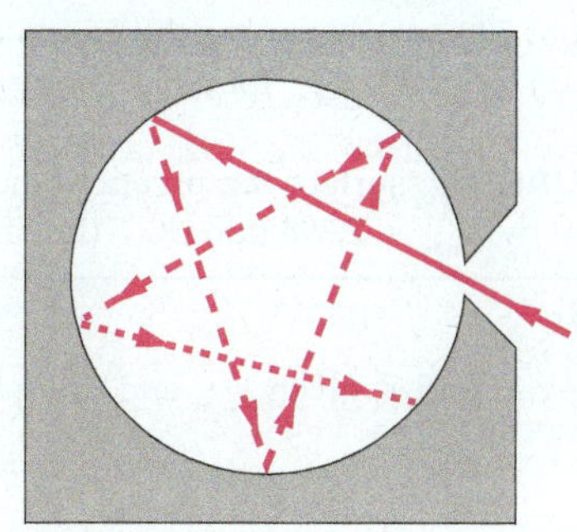

FIGURE 20-13 A hole in a cavity approximates an ideal blackbody. Radiation entering the cavity has little chance of leaving the cavity before it is completely absorbed. The radiation emitted through the hole (not shown) is therefore characteristic of the temperature of the walls of the cavity.

of the hole before the walls of the cavity absorb it. Thus, the radiation that is emitted out of the hole is characteristic of the temperature of the walls of the cavity.

The radiation emitted by an object at temperatures below approximately 600°C is not visible to the naked eye. Radiation of objects at room temperature is concentrated at wavelengths much longer than those of visible light. As an object is heated, the rate of energy emission increases, and the energy radiated extends to higher frequencies (and shorter wavelengths). Between about 600 and 700°C, enough of the radiated energy is in the visible spectrum for the object to glow a dull red. At higher temperatures, it may become bright red or even "white hot." Figure 20-14 shows the power radiated by a blackbody as a function of wavelength for three different temperatures. The wavelength at which the power is a maximum varies inversely with the temperature, a result known as Wien's displacement law:

$$\lambda_{max} = \frac{2.898 \text{ mm} \cdot \text{K}}{T} \qquad 20\text{-}21$$

WIEN'S DISPLACEMENT LAW

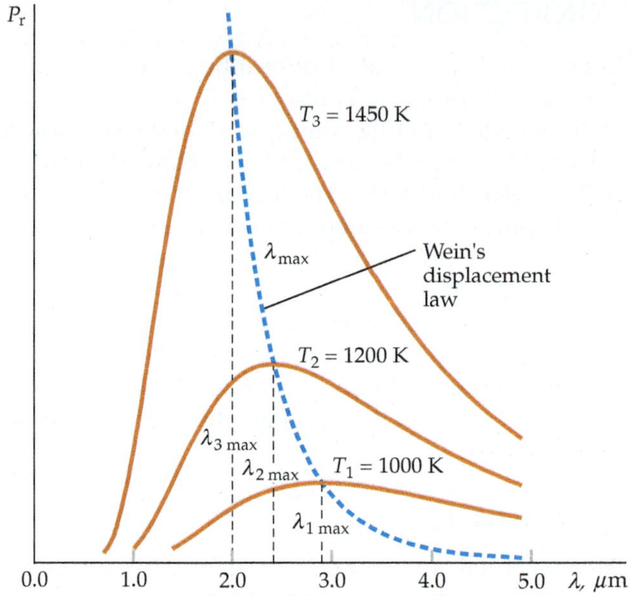

FIGURE 20-14 Radiated power versus wavelength for radiation emitted by a blackbody. The temperature of the emitting surface labels each plot on the graph. The wavelength λ_{max} at which the maximum power is emitted varies inversely with the absolute temperature T of the surface of the blackbody.

This law is used to determine the surface temperatures of stars by analyzing their radiation. It can also be used to map out the variation in temperature over different regions of the surface of an object. Such a map is called a thermograph. Thermographs can be used to detect cancer because cancerous tissue results in increased circulation, which produces a slight increase in skin temperature.

The spectral-distribution curves shown in Figure 20-14 played an important role in the history of physics. It was the discrepancy between theoretical calculations (using classic thermodynamics) that were used to generate blackbody spectral distributions and the actual experimental measurements of spectral distributions that led to Max Planck's first ideas about the quantization of energy in 1900.

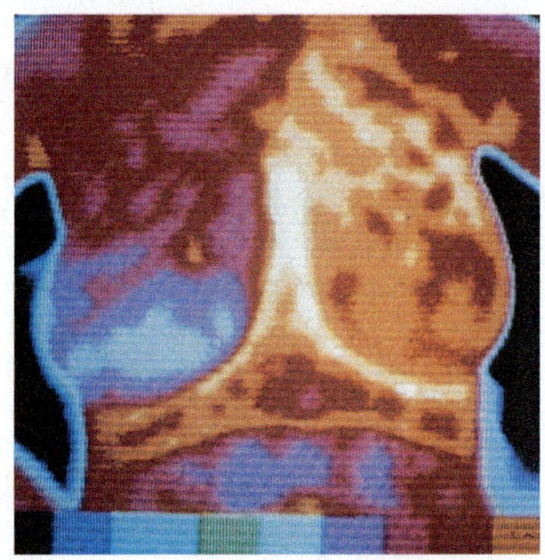

A thermograph was used to detect this cancerous tumor. *(Science Photo Library/ Photo Researchers, Inc.)*

Example 20-9 Radiation from the Sun

(*a*) The radiation emitted by the surface of the Sun emits maximum power at a wavelength of about 500 nm. Assuming the Sun to be a blackbody emitter, what is its surface temperature? (*b*) Calculate λ_{max} for a blackbody at room temperature, $T = 300$ K.

PICTURE The surface temperature and the wavelength at maximum emitted power are related by $\lambda_{max} = 2.898 \text{ mm} \cdot \text{K}/T$ (Equation 20-21).

SOLVE

(*a*) We can find T, given λ_{max} and using Wien's displacement law:

$$\lambda_{max} = \frac{2.898 \text{ mm} \cdot \text{K}}{T} \quad \text{so}$$

$$T = \frac{2.898 \text{ mm} \cdot \text{K}}{\lambda_{max}} = \frac{2.898 \text{ mm} \cdot \text{K}}{500 \text{ nm}} = \boxed{5800 \text{ K}}$$

(*b*) We can find λ_{max} from Wien's displacement law for $T = 300$ K:

$$\lambda_{max} = \frac{2.898 \text{ mm} \cdot \text{K}}{300 \text{ K}} = 9.66 \times 10^{-6} \text{ m} = \boxed{9.66 \ \mu\text{m}}$$

CHECK The Part-(*b*) result for λ_{max} is 19 times greater than 500 nm (the value of λ_{max} for the Sun), and the Part-(*a*) result of 5800 K for the surface temperature of the Sun is 19 times greater than 300 K, the surface temperature of the object in Part (*b*). Wien's law is that λ_{max} is inversely proportional to the temperature of the emitter, so the calculated results are as expected.

TAKING IT FURTHER The peak wavelength from the Sun is in the visible spectrum. The blackbody radiation spectrum describes the Sun's radiation spectrum fairly well, so the Sun is indeed a good example of a blackbody.

For $T = 300$ K, the spectrum peaks in the infrared at wavelengths much longer than the wavelengths visible to the eye. Surfaces that are not black to our eyes may act as blackbodies for infrared radiation and absorption. For example, it has been found experimentally that the skins of all human beings absorb all infrared radiation; hence, the emissivity of skin is 1.00 for this radiation process.

Example 20-10 Radiation from the Human Body

Try It Yourself

Calculate the net rate of heat loss in radiated energy for a naked person in a room at 20°C, assuming the person to be a blackbody with a surface area of 1.4 m^2 and a surface temperature of 33°C (= 306 K). (The surface temperature of the human body is slightly less than the internal temperature of 37°C because of the thermal resistance of the skin.)

PICTURE Use $P_{net} = e\sigma A(T^4 - T_0^4)$, with $e = 1$, $T = 306$ K, and $T_0 = 293$ K, to determine the difference between the emitted and absorbed power.

SOLVE

Cover the column to the right and try these on your own before looking at the answers.

Steps	Answer
Use $P_{net} = e\sigma A(T^4 - T_0^4)$, with $e = 1$, $T = 306$ K, and $T_0 = 293$ K.	$P_{net} = 111\text{ W} = \boxed{0.11\text{ kW}}$

CHECK A rate of 0.11 kW is equal to 2300 kcal/day. This is the correct order of magnitude.

TAKING IT FURTHER This large energy loss is approximately equal to the basal metabolic rate of about 120 W. We protect ourselves from this great loss of energy by wearing clothing, which, because of its low thermal conductivity, has a much lower outside temperature and therefore a much lower rate of thermal radiation.

When the temperature of an object T is not too different from the surrounding temperature T_0, a radiating object obeys Newton's law of cooling. We can see this by writing Equation 20-20 as

$$P_{net} = e\sigma A(T^4 - T_0^4) = e\sigma A(T^2 + T_0^2)(T^2 - T_0^2)$$
$$= e\sigma A(T^2 + T_0^2)(T + T_0)(T - T_0)$$

When $T - T_0$ is small, we can replace T by T_0 in the two sums with little change in the result. Then

$$P_{net} \approx e\sigma A(T_0^2 + T_0^2)(T_0 + T_0)(T - T_0) = 4e\sigma AT_0^3\,\Delta T$$

The net power radiated is approximately proportional to the temperature difference, in agreement with Newton's law of cooling. This result can also be obtained by using the differential approximation:

$$\Delta P_r \approx \left.\frac{dP_r}{dT}\right|_{T=T_0}(T - T_0)$$

where $P_r = e\sigma A(T^4 - T_0^4)$. For a small temperature difference $T - T_0$, we have

$$\Delta P_r = e\sigma A\,4T^3\Big|_{T=T_0}(T - T_0) = 4e\sigma AT_0^3\,\Delta T$$

Physics Spotlight

Urban Heat Islands: Hot Nights in the City

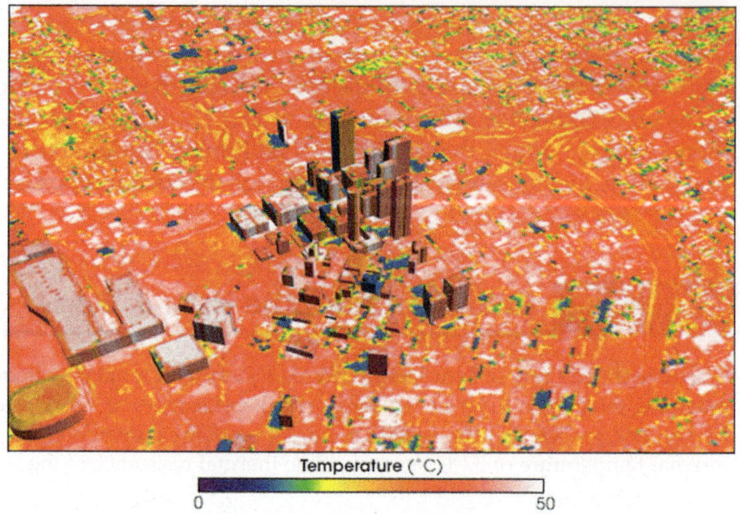

The sides of the buildings facing away from the Sun appear dark blue or black in this thermogram of Atlanta, Georgia. This is because these sides are cooler than the sides receiving direct sunlight. Particularly hot temperatures appear white. *(NASA/Goddard Space Flight Center Scientific Visualization Studio.)*

In 1820, Luke Howard published records that listed several years of temperatures from London and its suburbs for days and nights. His records showed that London was warmer than the surrounding suburban and rural areas, and that this difference was most pronounced at nighttime. He found that London was an average of 2.1°C warmer at night than the surrounding rural area.[*] In 2004, observations showed that in the summertime Phoenix could be up to 10°C warmer at night than its surrounding area.[†] London and Phoenix are urban heat islands (UHIs). In general, cities with paved streets and many buildings are warmer than the surrounding countryside.

One important factor in the formation of UHIs is the lack of trees and other plants. During the day, plants cool the area around them because of all the water they release, which has a high latent heat. In rural areas, and even in green areas within cities, much of the energy is used to overcome the latent heat of water, rather than in increasing the surface temperature.[‡] In addition, plants reflect back much of the infrared (heat) wavelengths from solar radiation, while asphalt, steel, glass, concrete, and aluminum absorb and retain these infrared wavelengths. Another factor is the geometry of a city. Urban canyons, formed by tall buildings on either side of the streets, reflect infrared wavelengths to other absorbent surfaces.[#] Open areas allow radiation of the heat away from the ground and nearby structures.

In addition to radiated heat, rainwater runoff in urban areas can be heated by conduction. In August of 2001, a rainstorm in Cedar Rapids, Iowa, increased the temperature of a stream by 10.5°C in one hour, which killed many fish.[°] The falling rain itself was cooler than the stream, but most of the water entering the stream in that first hour ran over hot pavement. Similar sudden temperature increases have been noted in urban streams in Minnesota, Wisconsin, Oregon,[§] and California.[¶]

In 1996, in order to support the Olympics, an unprecedented weather measurement effort took place near Atlanta, Georgia.[**] One interesting revelation from this effort was that the area directly downwind of the city got more precipitation, as UHI-related convection changed weather patterns.[††] It also has been found that Dallas, San Antonio,[‡‡] and even Saint Louis[##] have more precipitation directly downwind, for up to 64 km from the city center. Convection changes caused by UHIs are complex, but the precipitation effects are measurable.

Urban planners are implementing ways to cool UHIs.[°°] In Chicago, city hall now has a "green roof" with plants and reflective walkways. The temperature of the rooftop is monitored and compared to the nearby dark asphalt roof of the Cook County building. The temperature of the planted roof can be over 33°C cooler than the asphalt roof.[§§] Several cities encourage the planting of trees,[¶¶] and others encourage the use of reflective surfaces,[***] permeable pavement, and the use of green roofs.[†††] Urban heat islands are being mediated by cool technologies.

* Howard, Luke, *The Climate of London, Deduced from Meteorological Observations Made in the Metropolis and Various Places around It*, 2nd ed. London: Harvey and Darton, 1833.
† Fast, J. D, Torcolini, J. C., and Redman, R., "Pseudovertical Temperature Profiles and the Urban Heat Island Measured by a Temperature Datalogger Network in Phoenix, Arizona," *Journal of Applied Meteorology*, Jan. 2005, Vol. 44, No. 1, 3–13.
‡ Souch, C., and Grimmond, S., "Applied Climatology: Urban Climate," *Progress in Physical Geography*, Feb. 2006, Vol. 30, No. 2, 270–279.
\# Kusaka, H., and Kimura, F., "Thermal Effects of Urban Canyon Structure on the Nocturnal Heat Island: Numerical Experiment Using a Mesoscale Model Coupled with an Urban Canopy Model," *Journal of Applied Meteorology*, Dec. 2004, Vol. 43, No. 12, 1899–1910.
° Boshart, R., "Urban Trout Stream Still in Works—DNR Targeting McLoud Run in C. R., Despite Recent Fish Kill," *The Gazette*, Aug. 9, 2001, B1+.
§ Frazer, Lance, "Paving Paradise: The Perils of Impervious Surfaces," *Environmental Health Perspective*, Jul. 2005, Vol. 113, No. 7, 456–462.
¶ Fowler, B., and Rasmus, J., "Seaside Solution," *Civil Engineering*, Dec. 2005, 44–49.
** Skindrud, Erik, "Georgia on Their Minds," *Science News*, Jul. 13, 1996. http://www.sciencenews.org/pages/sn_arch/7_13_96/bob1.htm
†† Dixon, P. G., and Mote, T. L., "Patterns and Causes of Atlanta's Urban Heat Island-Initiated Precipitation," *Journal of Applied Meteorology*, Sept. 2003, Vol. 42, No. 9, 1273–1284.
‡‡ Shepard, J., Pierce, H., and Negri, A., "Rainfall Modification by Major Urban Areas: Observations from Spaceborne Rain Radar on the TRMM Satellite," *Journal of Applied Meteorology*, Jul. 2002, Vol. 41, No. 7, 689–701.
\#\# "Urban-related Weather Anomalies," *Science News*, Mar. 5, 1977, Vol. 111, No. 10, 152.
°° Wade, B., "Putting the Freeze on Heat Islands," *American City & County*, Feb. 2000, 30–40.
§§ "Monitoring the Rooftop Garden's Benefits," City of Chicago Department of Environment. http://chicagorooftops.notlong.com
¶¶ Duncan, H., "Trees, Please: More Cities Enacting Tree Ordinances, but Enforcement Is the Age-Old Problem," *Macon Telegraph*, July 6, 2006, A1.
*** "NASA Assesses Strategies to 'Turn Off the Heat' in New York City," *Engineered Systems*, April 2006, 79–80.
††† Hoffman, L., "Green Roof Storm Water Modeling," *BioCycle*, Feb. 2006, 38–40.

Summary

TOPIC	RELEVANT EQUATIONS AND REMARKS	
1. Thermal Expansion		
Coefficient of linear expansion	$\alpha = \frac{\Delta L/L}{\Delta T}$	20-2
Coefficient of volume expansion	$\beta = \frac{\Delta V/V}{\Delta T} = 3\alpha$	20-4, 20-5
2. The van der Waals Equation of State	The van der Waals equation of state describes the behavior of real gases over a wide range of temperatures and pressures, taking into account the space occupied by the gas molecules themselves and the attraction of the molecules to one another. $\left(P + \frac{an^2}{V^2}\right)(V - bn) = nRT$	20-6
3. Vapor Pressure	Vapor pressure is the pressure at which the liquid and gas phases of a substance are in equilibrium at a given temperature. The liquid boils at that temperature for which the external pressure equals the vapor pressure.	
4. The Triple Point	The triple point is the unique temperature and pressure at which the gas, liquid, and solid phases of a substance can coexist in equilibrium. At temperatures and pressures below the triple point, the liquid phase of a substance cannot exist.	
5. Heat Transfer	The three mechanisms by which energy is transferred due to a difference in temperature are radiation, conduction, and convection.	
Newton's law of cooling	For all mechanisms of heat transfer, if the temperature difference between the object and its surroundings is small, the rate of cooling of the object is approximately proportional to the temperature difference.	
6. Heat Conduction		
Current	The rate of conduction of heat is given by $I = \frac{dQ}{dt} = -kA\frac{dT}{dx}$ where I is the thermal current, k is the coefficient of thermal conductivity, and dT/dx is the temperature gradient.	20-7
Thermal resistance	$\Delta T = IR$	20-9
	where ΔT is the temperature decrease in the direction of the thermal current and R is the thermal resistance: $R = \frac{\lvert\Delta x\rvert}{kA}$	20-10
Equivalent resistance:		
series	$R_{eq} = R_1 + R_2 + \cdots$	20-12
parallel	$\frac{1}{R_{eq}} = \frac{1}{R_1} + \frac{1}{R_2} + \cdots$	20-14
R factor	The R factor is the thermal resistance in units of in·ft²·°F/(Btu/h) for a square foot of a slab of material $R_f = R_{net}A = \frac{\lvert\Delta x\rvert}{k}$	20-15
7. Thermal Radiation		
Rate of radiated power	$P_r = e\sigma AT^4$	20-17
	where $\sigma = 5.6703 \times 10^{-8}\ \text{W/m}^2\cdot\text{K}^4$ is Stefan's constant and e is the emissivity, which varies between 0 and 1 (depending on the composition of the surface of the object). Materials that are good heat absorbers are also good heat radiators.	

TOPIC	RELEVANT EQUATIONS AND REMARKS	
Net power radiated by an object at T to its environment at T_0	$P_{net} = e\sigma A(T^4 - T_0^4)$	20-20
Blackbody	A blackbody has an emissivity of 1. It is a perfect radiator, and it absorbs all radiation incident upon it.	
Wien's law	The power spectrum of electromagnetic energy radiated by a blackbody has a maximum at a wavelength λ_{max}, which varies inversely with the absolute temperature of the body: $\lambda_{max} = \frac{2.898 \text{ mm} \cdot \text{K}}{T}$	20-21

Answer to Concept Check

20-1 Wood is a poor conductor of heat and metal is a good conductor of heat. When your finger touches metal, the metal conducts heat away from your finger faster, and so the finger is cooled at a faster rate than if you touched wood.

Answers to Practice Problems

20-1 $0.0563 \text{ K/W} = 56.3 \text{ mK/W}$

20-2 $\Delta x = (1 \text{ cm})(429)/(0.026) = 16\,500 \text{ cm} = 165 \text{ m}$

20-3 The thermal current is 6200 Btu/h with 2.0 in. of insulation. Thus, the maximum additional savings is 6200 Btu/h, which would save spending an additional \$146 per month during the heating season.

Problems

In a few problems, you are given more data than you actually need; in a few other problems, you are required to supply data from your general knowledge, outside sources, or informed estimate.

Interpret as significant all digits in numerical values that have trailing zeros and no decimal points.

- • Single-concept, single-step, relatively easy
- •• Intermediate-level, may require synthesis of concepts
- ••• Challenging
- SSM Solution is in the *Student Solutions Manual*
- Consecutive problems that are shaded are paired problems.

CONCEPTUAL PROBLEMS

1 • Why does the mercury level of a thermometer first decrease slightly when the thermometer is first placed in warm water?

2 • A large sheet of metal has a hole cut in the middle of it. When the sheet is heated, the area of the hole will (*a*) not change, (*b*) always increase, (*c*) always decrease, (*d*) increase if the hole is not in the exact center of the sheet, (*e*) decrease only if the hole is in the exact center of the sheet.

3 • Why is it a bad idea to place a tightly sealed glass bottle that is completely full of water, into your kitchen freezer in order to make ice? SSM

4 • The windows of your physics laboratory are left open on a night when the temperature of the outside drops well below freezing. A steel ruler and a wooden ruler were left on the window sill, and when you arrive in the morning they are both very cold. The coefficient of linear expansion of wood is about $5 \times 10^6 \text{ K}^{-1}$. Which ruler should you use to make the most accurate length measurements? Explain your answer.

5 • **ENGINEERING APPLICATION** Bimetallic strips are used both for thermostats and electrical circuit breakers. A bimetallic strip consists of a pair of thin strips of metal that have different coefficients of linear expansion and are bonded together to form one doubly thick strip. Suppose a bimetallic strip is constructed out of one steel strip and one copper strip, and suppose the bimetallic strip is curled in the shape of a circular arc with the steel strip on the outside. If the temperature of the strip is decreased, will the strip straighten out or curl more tightly?

6 • Metal A has a coefficient of linear expansion that is three times the coefficient of linear expansion of metal B. How do their coefficients of volume expansion β compare? (a) $\beta_A = \beta_B$, (b) $\beta_A = 3\beta_B$, (c) $\beta_A = 6\beta_B$, (d) $\beta_A = 9\beta_B$, (e) You cannot tell from the data given.

7 • The summit of Mount Rainier is 14 410 ft above sea level. Mountaineers say that you cannot hard boil an egg at the summit. This statement is true because at the summit of Mount Rainier (*a*) the air temperature is too low to boil water, (*b*) the air pressure is too low for alcohol fuel to burn, (*c*) the temperature of boiling water is not hot enough to hard boil the egg, (*d*) the oxygen content of the air is too low to support combustion, (*e*) eggs always break in climbers' backpacks.

8 • Which gases in Table 20-3 cannot be condensed by applying pressure at 20°C? Explain your answer.

9 •• The phase diagram in Figure 20-15 can be interpreted to yield information on how the boiling and melting points of water change with altitude. (*a*) Explain how this information can be obtained. (*b*) How might this information affect cooking procedures in the mountains? SSM

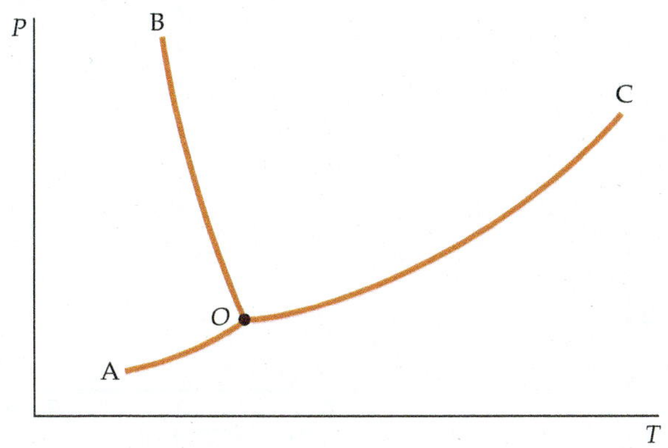

FIGURE 20-15 Problem 9

10 •• Sketch a phase diagram for carbon dioxide using information from Section 20-3.

11 •• Explain why the carbon dioxide on Mars is found in the solid state in the polar regions, even though the atmospheric pressure at the surface of Mars is only about 1 percent of the atmospheric pressure at the surface of Earth.

12 •• Explain why decreasing the temperature of your house at night in winter can save money on heating costs. Why does the cost of the fuel consumed to heat the house back to the daytime temperature in the morning not equal the savings realized by cooling it down in the evening and keeping it cool throughout the night?

13 •• Two solid cylinders made of materials A and B have the same lengths; their diameters are related by $d_A = 2d_B$. When the same temperature difference is maintained between the ends of the cylinders, they conduct heat at the same rate. Their thermal conductivities are therefore related by which of the following equations? (*a*) $k_A = k_B/4$, (*b*) $k_A = k_B/2$, (*c*) $k_A = k_B$, (*d*) $k_A = 2k_B$, (*e*) $k_A = 4k_B$ SSM

14 •• Two solid cylinders made of materials A and B have the same diameter; their lengths are related by $L_A = 2L_B$. When the same temperature difference is maintained between the ends of the cylinders, they conduct heat at the same rate. Their thermal conductivities are therefore related by which of the following equations? (*a*) $k_A = k_B/4$, (*b*) $k_A = k_B/2$, (*c*) $k_A = k_B$, (*d*) $k_A = 2k_B$, (*e*) $k_A = 4k_B$

15 •• If you feel the inside of a single-pane window during a very cold day, it is cold, even though the room temperature can be quite comfortable. Assuming the room temperature is 20.0°C and the outside temperature is 5.0°C, construct a plot of temperature versus position, starting from a point 5.0 m behind the window (inside the room) and ending at a point 5.0 m in front of the window. Explain the heat transfer mechanisms that occur along this path.

16 •• During the thermal retrofitting of many older homes in California, it was found that the 3.5-in-deep spaces between the wallboards and the outer sheathing were filled with just air (no insulation). Filling the spaces with insulating material certainly reduces heating and cooling costs; although, the insulating material is a better conductor of heat than air is. Explain why adding the insulation is a good idea.

ESTIMATION AND APPROXIMATION

17 •• You are using a cooking pot to boil water for a pasta dish. The recipe calls for at least 4.0 L of water to be used. You fill the pot with 4.0 L of room temperature water and note that this amount of water filled the pot to the brim. Knowing some physics, you are counting on the volume expansion of the steel pot to keep all of the water in the pot while the water is heated to a boil. Is your assumption correct? Explain. If your assumption is not correct, how much water runs over the sides of the pot due to the thermal expansion of the water? SSM

18 •• Liquid helium is stored in containers fitted with 7.00-cm-thick "superinsulation," consisting of numerous layers of very thin aluminized Mylar sheets. The rate of evaporation of liquid helium in a 200-L container is about 0.700 L per day when the container is stored at room temperature (20°C). The density of liquid helium is 0.125 kg/L and the latent heat of vaporization is 21.0 kJ/kg. Estimate the thermal conductivity of the superinsulation.

19 •• **BIOLOGICAL APPLICATION** Estimate the thermal conductivity of human skin. SSM

20 •• You are visiting Finland with a college friend and have met some Finnish friends. They talk you into taking part in a traditional Finnish after-sauna exercise, which consists of leaving the sauna, wearing only a bathing suit, and running out into the midwinter Finnish air. Estimate the rate at which you initially lose energy to the cold air. Compare this rate of initial energy loss to the resting metabolic rate of a typical human under normal temperature conditions. Explain the difference.

21 •• Estimate the rate of heat conduction through a 2.0-in-thick wooden door during a cold winter day in Minnesota. Include the brass doorknob. What is the ratio of the heat that escapes through the doorknob to the heat that escapes through the whole door? What is the total overall R factor for the door, including the knob? The thermal conductivity of brass is 85 W/(m · K).

22 •• Estimate the effective emissivity of Earth, given the following information. The solar constant, which is the intensity of radiation incident on Earth from the Sun, is about 1.37 kW/m^2. Seventy percent of this energy is absorbed by Earth, and Earth's average surface temperature is 288 K. (Assume that the effective area that is absorbing the light is πR^2, where R is Earth's radius, while the blackbody-emission area is $4\pi R^2$.)

23 •• Black holes are highly condensed remnants of stars. Some black holes, together with a normal star, form binary systems. In such systems the black hole and the normal star orbit about the center of mass of the system. One way black holes can be detected from Earth is by observing the frictional heating of the atmospheric gases from the normal star that fall into the black hole. These gases can reach temperatures greater than 1.0×10^6 K. Assuming that the falling gas can be modeled as a blackbody radiator, estimate λ_{max} for use in an astronomical detection of a black hole. (This region is in the X-ray region of the electromagnetic spectrum.)

24 ••• **ENGINEERING APPLICATION, CONTEXT-RICH** Your cabin in northern Michigan has walls that consist of pine logs that have average thicknesses of about 20 cm. You decide to finish the interior of the cabin to improve the look and to increase the insulation of the exterior walls. You choose to buy insulation with an R factor of 31 to cover the walls. In addition, you cover the insulation with 1.0-in.-thick gypsum wallboard. Assuming heat transfer is only due to conduction, estimate the ratio of thermal current through the walls during a cold winter night before the renovation to the thermal current through the walls following the renovation.

25 ••• **CONTEXT-RICH** You are in charge of transporting a liver from New York, New York, to Los Angeles, California, for a transplant surgery. The liver is kept cold in a styrofoam ice chest initially filled with 1.0 kg of ice. It is crucial that the liver temperature is never warmer than 5.0°C. Assuming the trip from the hospital in New York to the hospital in Los Angeles takes 7.0 h, estimate the R-value the styrofoam walls of the ice chest must have. SSM

THERMAL EXPANSION

26 •• You have inherited your grandfather's grandfather clock that was calibrated when the temperature of the room was 20°C. Assume that the pendulum consists of a thin brass rod of negligible mass with a compact heavy bob at its end. (*a*) During a hot day, when the temperature is 30°C, does the clock run fast or slow? Explain. (*b*) How much time does it gain or lose during this day?

27 •• **ENGINEERING APPLICATION** You need to fit a copper collar tightly around a steel shaft that has a diameter of 6.0000 cm at 20°C. The inside diameter of the collar at that temperature is 5.9800 cm. What temperature must the copper collar have so that it will just slip on the shaft, assuming the shaft itself remains at 20°C? SSM

28 •• **ENGINEERING APPLICATION** You have a copper collar and a steel shaft. At 20°C, the collar has an inside diameter of 5.9800 cm and the steel shaft has diameter of 6.0000 cm. The copper collar was heated. When its inside diameter exceeded 6.0000 cm it was slipped on the shaft. The collar fitted tightly on the shaft after they cooled to room temperature. Now, several years later, you need to remove the collar from the shaft. To do this you heat them both until you can just slip the collar off the shaft. What temperature must the collar have so that the collar will just slip off the shaft?

29 •• A container is filled to the brim with 1.4 L of mercury at 20°C. As the temperature of the container and mercury is increased to 60°C, a total of 7.5 mL of mercury spill over the brim of the container. Determine the linear expansion coefficient of the material that makes up the container.

30 •• A car with a 60.0-L steel gas tank is filled to the brim with 60.0 L of gasoline when the outside temperature is 10°C. The coefficient of volume expansion for gasoline at 20°C is $0.950 \times 10^{-3}\,K^{-1}$. How much gasoline spills out of the tank when the outside temperature increases to 25°C? Take the expansion of the steel tank into account.

31 ••• What is the tensile stress in the copper collar of Problem 27 when its temperature returns to 20°C?

THE VAN DER WAALS EQUATION, LIQUID–VAPOR ISOTHERMS, AND PHASE DIAGRAMS

32 • (*a*) Calculate the volume of 1.00 mol of an ideal gas at a temperature of 100°C and a pressure of 1.00 atm. (*b*) Calculate the temperature at which 1.00 mol of steam at a pressure of 1.00 atm has the volume calculated in part (*a*). Use $a = 0.550\,Pa \cdot m^6/mol^2$ and $b = 30.0\,cm^3/mol$.

33 •• Using Figure 20-16, find the following quantities. (*a*) The temperature at which water boils on a mountain where the atmospheric pressure is 70.0 kPa, (*b*) the temperature at which water boils in a container where the pressure inside the container is 0.500 atm, and (*c*) the pressure at which water boils at 115°C. SSM

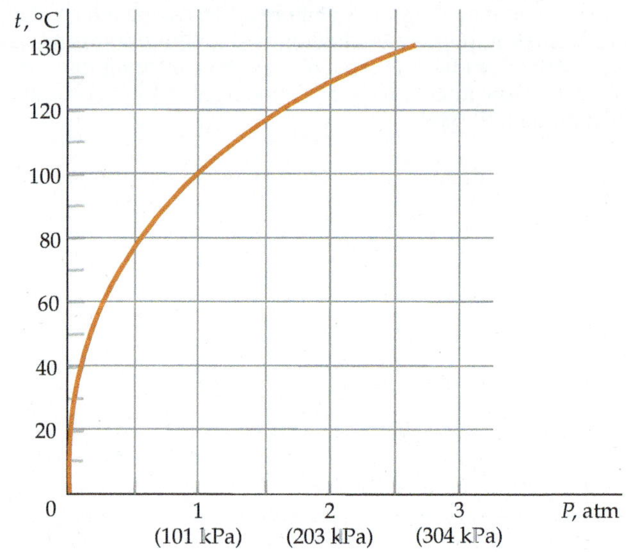

FIGURE 20-16 Problem 33

34 •• The van der Waals constants for helium are $a = 0.03412\,L^2 \cdot atm/mol^2$ and $b = 0.0237\,L/mol$. Use these data to find the volume in cubic centimeters occupied by one helium atom. Then, estimate the radius of the helium atom.

CONDUCTION

35 • A 20-ft × 30-ft slab of insulation has an R factor of 11. At what rate is heat conducted through the slab if the temperature on one side is a constant 68°F and the temperature on the other side is a constant 30°F? SSM

36 •• A copper cube and an aluminum cube, each with 3.00-cm-long edges, are arranged as shown in Figure 20-17. Find (*a*) the thermal resistance of each cube, (*b*) the thermal resistance of the two-cube combination, (*c*) the thermal current I, and (*d*) the temperature at the interface of the two cubes.

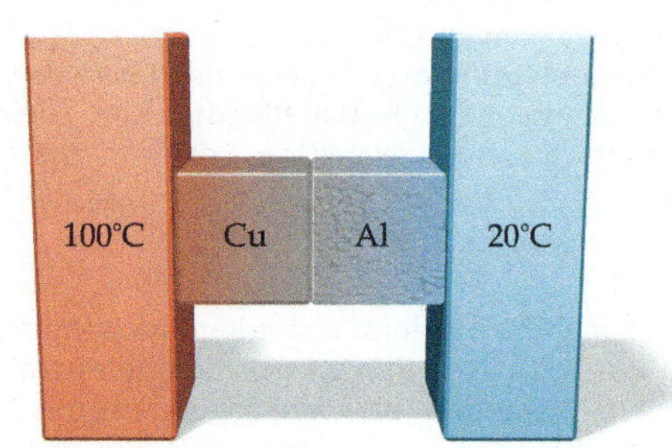

FIGURE 20-17 Problem 36

37 •• Two metal cubes, one copper and one aluminum, with 3.00-cm-long edges, are arranged in parallel, as shown in Figure 20-18. Find (*a*) the thermal current in each cube, (*b*) the total thermal current, and (*c*) the thermal resistance of the two-cube combination.

FIGURE 20-18 Problem 37

38 •• **Engineering Application** The cost of air conditioning a house is approximately proportional to the rate at which heat is absorbed by the house from its surroundings divided by the coefficient of performance (COP) of the air conditioner. Let us denote the temperature difference between the inside temperature and the outside temperature as ΔT. Assuming that the rate at which heat is absorbed by a house is proportional to ΔT and that the air conditioner is operating ideally, show that the cost of air conditioning is proportional to $(\Delta T)^2$ divided by the temperature inside the house.

39 •• A spherical shell of thermal conductivity k has inside radius r_1 and outside radius r_2 (Figure 20-19). The inside of the shell is held at a temperature T_1, and the outside of the shell is held at temperature T_2, with $T_1 < T_2$. In this problem, you are to show that the thermal current through the shell is given by

$$I = -\frac{4\pi k r_1 r_2}{r_2 - r_1}(T_2 - T_1)$$

where I is positive if heat is transferred in the $+r$ direction. Here is a suggested procedure for obtaining this result: (1) obtain an expression for the thermal current I through a thin spherical shell of radius r and thickness dr when there is a temperature difference dT across the thickness of the shell; (2) explain why the thermal current is the same through each such thin shell; (3) express the thermal current I through such a shell element in terms of the area $A = 4\pi r^2$, the thickness dr, and the temperature difference dT across the element; and (4) separate variables (solve for dT in terms of r and dr) and integrate.

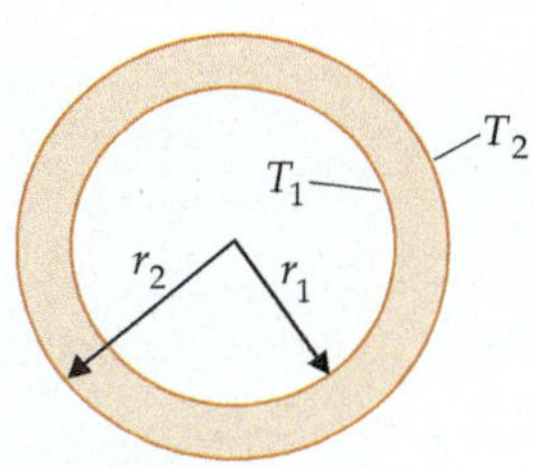

FIGURE 20-19 Problem 39

RADIATION

40 • **Biological Application** Calculate λ_{max} (the wavelength at which the emitted power is maximum) for a human skin. Assume the human skin is a blackbody emitter with a temperature of 33°C.

41 • The universe is filled with radiation that is believed to be left from the Big Bang. If the entire universe is considered to be a blackbody with a temperature equal to 2.3 K, what is the λ_{max} (the wavelength at which the power of the radiation is maximum) of this radiation? SSM

42 • What is the range of temperatures for star surfaces for which λ_{max} (the wavelength at which the power of the emitted radiation is maximum) is in the visible range?

43 • The heating wires of a 1.00-kW electric heater are red hot at a temperature of 900°C. Assuming that 100 percent of the heat released is due to radiation and that the wires act as blackbody emitters, what is the effective area of the radiating surface? (Assume a room temperature of 20°C.)

44 •• A blackened, solid copper sphere that has a radius equal to 4.0 cm hangs in an evacuated enclosure whose walls have a temperature of 20°C. If the sphere is initially at 0°C, find the initial rate at which its temperature changes, assuming that heat is transferred by radiation only. (Assume the sphere is a blackbody emitter.)

45 •• The surface temperature of the filament of an incandescent lamp is 1300°C. If the electric power input is doubled, what will the new temperature be? *Hint: Show that you can neglect the temperature of the surroundings.*

46 •• Liquid helium is stored at its boiling point (4.2 K) in a spherical can that is separated by an evacuated region of space from a surrounding shield that is maintained at the temperature of liquid nitrogen (77 K). If the can is 30 cm in diameter and is blackened on the outside so that it acts as a blackbody emitter, how much helium boils away per hour?

GENERAL PROBLEMS

47 • A steel tape is placed around Earth at the equator when the temperature is 0°C. What will the clearance between the tape and the ground (assumed to be uniform) be if the temperature of the tape increases to 30°C? Neglect the expansion of Earth.

48 •• Show that change in the density ρ of an isotropic material due to an increase in temperature ΔT is given by $\Delta\rho = -\beta\rho\,\Delta T$.

49 •• The solar constant is the power received from the Sun per unit area perpendicular to the Sun's rays at the mean distance of Earth from the Sun. Its value at the upper atmosphere of Earth is about 1.37 kW/m^2. Calculate the effective temperature of the Sun if it radiates like a blackbody. (The radius of the Sun is 6.96×10^8 m.) SSM

50 •• **Engineering Application** As part of your summer job as an engineering intern at an insulation manufacturer, you are asked to determine the R factor of insulating material. This particular material comes in $\frac{1}{2}$-in sheets. Using this material, you construct a hollow cube that has 12-in-long edges. You place a thermometer and a 100-W heater inside the box. After thermal equilibrium has been attained, the temperature inside the box is 90°C when the temperature outside the box is 20°C. Determine the R factor of the material.

51 • (*a*) From the definition of β, the coefficient of volume expansion (at constant pressure), show that $\beta = 1/T$ for an ideal gas. (*b*) The experimentally determined value of β for N_2 gas at 0°C is 0.003673 K^{-1}. By what percent does this measured value of β differ from the value obtained by modeling N_2 as an ideal gas?

52 •• A rod of length L_A, made from material A, is placed next to a rod of length L_B, made from material B. The rods remain in thermal equilibrium with each other. (*a*) Show that even though the lengths of each rod will change with changes in the ambient temperature, the difference between the two lengths will remain constant if the lengths L_A and L_B are chosen such that $L_A/L_B = \alpha_B/\alpha_A$, where α_A and α_B are the coefficients of linear expansion, respectively. (*b*) If material B is steel, material A is brass, and $L_A = 250$ cm at 0°C, what is the value of L_B?

53 •• On average, the temperature of Earth's crust increases 1.0°C for every increase in depth of 30 m. The average thermal conductivity of Earth's crust material is 0.74 J/(m · s · K). What is the heat loss of Earth per second due to conduction from the core? How does this heat loss compare with the average power received from the Sun? SSM

54 •• A copper-bottomed saucepan containing 0.800 L of boiling water boils dry in 10.0 min. Assuming that all the heat is transferred through the flat copper bottom, which has a diameter of 15.0 cm and a thickness of 3.00 mm, calculate the temperature of the outside of the copper bottom while some water is still in the pan.

55 •• **ENGINEERING APPLICATION** A cylindrical steel hot-water tank of cylindrical shape has an inside diameter of 0.550 m and inside height of 1.20 m. The tank is enclosed by a 5.00-cm-thick insulating layer of glass wool whose thermal conductivity is 0.0350 W/(m · K). The insulation is covered by a thin sheet-metal skin. The steel tank and the sheet-metal skin have thermal conductivities that are much greater than that of the glass wool. How much electrical power must be supplied to this tank in order to maintain the water temperature at 75.0°C when the external temperature is 1.0°C?

56 •• The diameter d of a tapered rod of length L is given by $d = d_0(1 + ax)$, where a is a constant and x is the distance from one end. If the thermal conductivity of the material is k, what is the thermal resistance of the rod?

57 ••• A solid disk of radius r and mass m is spinning about a frictionless axis through its center and perpendicular to the disk, with angular velocity ω_1 at temperature T_1. The temperature of the disk decreases to T_2. Express the angular velocity ω_2, rotational kinetic energy K_2, and angular momentum L_2 in terms of their values at the temperature T_1 and the linear expansion coefficient α of the disk.

58 ••• **SPREADSHEET** Write a **spreadsheet** program to graph the average temperature of the surface of Earth as a function of emissivity, using the results of Problem 22. How much does the emissivity have to change in order for the average temperature to increase by 1 K? This result can be thought of as a model for the effect of increasing concentrations of greenhouse gases such as methane and CO_2 in Earth's atmosphere.

59 ••• A small pond has a layer of ice 1.00 cm thick floating on its surface. (*a*) If the air temperature is −10°C on a day when there is a breeze, find the rate in centimeters per hour at which ice is added to the bottom of the layer. The density of ice is 0.917 g/cm³. (*b*) How long do you and your friends have to wait for a 20.0-cm layer to be built up so you can play hockey? SSM

60 ••• A blackened copper cube that has 1.00-cm-long edges is heated to a temperature of 300°C, and then is placed in a vacuum chamber whose walls are at a temperature of 0°C. In the vacuum chamber, the cube cools radiatively. (*a*) Show that the (absolute) temperature T of the cube follows the differential equation: $(dT/dt) = -(e\sigma A/C)(T^4 - T_0^4)$, where C is the heat capacity of the cube, A is its surface area, e the emissivity, and T_0 the temperature of the vacuum chamber. (*b*) Using Euler's method (Section 5.4 of Chapter 5), numerically solve the differential equation to find $T(t)$, and graph it. Assume $e = 1.00$. How long does it take the cube to cool to a temperature of 15°C? SSM

PART IV **ELECTRICITY AND MAGNETISM**

CHAPTER 21

COPPER IS A CONDUCTOR, A MATERIAL THAT HAS SPECIFIC PROPERTIES WE FIND USEFUL BECAUSE THESE PROPERTIES MAKE IT POSSIBLE TO TRANSPORT ELECTRICITY. *(Brooks R. Dillard/www.yuprocks.com.)*

The Electric Field I: Discrete Charge Distributions

? What is the total charge of all the electrons in a penny? (See Example 21-1.)

While just a century ago we had nothing more than a few electric lights, we are now extremely dependent on electricity in our daily lives. Yet, although the use of electricity has only recently become widespread, the study of electricity has a history reaching long before the first electric lamp glowed. Observations of electrical attraction can be traced back to the ancient Greeks, who noticed that after amber was rubbed, it attracted small objects such as straw or feathers. Indeed, the word *electric* comes from the Greek word for amber, *elektron*.

Today, the study and use of electricity continue. Electrical engineers improve existing electrical technologies, increasing performance and efficiency in devices such as hybrid cars and electric power plants. Electrostatic paints are used in the auto industry for engine parts and for car frames and bodies. This painting process creates a more durable coat than does liquid paint, and is easier on the environment because no solvents are used.

In this chapter, we begin our study of electricity with electrostatics, the study of charges at rest. After introducing the concept of charge, we briefly look at conductors and insulators and how conductors can be given a net charge. We then study Coulomb's law, which describes the force exerted by one charge on another. Next, we introduce the electric field and show how it can be visualized by electric field lines that indicate the magnitude and direction of the field, just as we visualized the velocity field of a flowing fluid using streamlines (Chapter 13). Finally, we discuss the behavior of point charges and dipoles in electric fields.

21-1 CHARGE

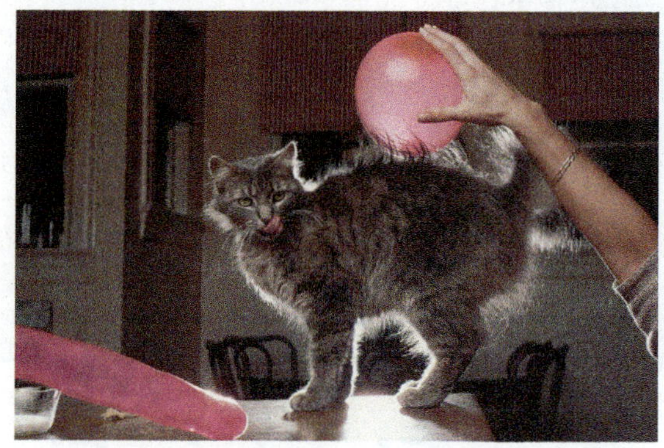

A cat and a balloon. *(Roger Ressmeyer/CORBIS.)*

Suppose we rub a hard-rubber rod with fur and then suspend the rod from a string so that it is free to rotate. Now we bring a second hard-rubber rod that has been rubbed with fur near it. The rods repel each other (Figure 21-1*a*). Two glass rods that have been rubbed with silk (Figure 21-1*b*) also repel each other. But, when we place a hard-rubber rod rubbed with fur near a glass rod rubbed with silk (Figure 21-1*c*) they attract each other.

Rubbing a rod causes the rod to become electrically charged. If we repeat the experiment with various materials, we find that all charged objects fall into one of just two groups—those like the hard-rubber rod rubbed with fur and those like the glass rod rubbed with silk. Objects from the same group repel each other, while objects from different groups attract each other. Benjamin Franklin explained this by proposing a model in which every object has a *normal* amount of electricity that can be transferred from one object to the other when two objects are in close contact, as when they are rubbed together. One object would have an excess charge and the other object would have a deficiency of charge, and the excess charge equals the deficiency of charge. Franklin described the resulting charges as positive (plus sign) or negative (minus sign). He also chose positive to be the charge acquired by a glass rod when it is rubbed with a piece of silk. The piece of silk then acquires a negative charge of equal magnitude during the procedure. Based on Franklin's convention, if hard rubber and fur are rubbed together, the hard rubber acquires a negative charge and the fur acquires a positive charge. Two objects that have the same sign (both + or both −) repel each other, and two objects that have oppositely

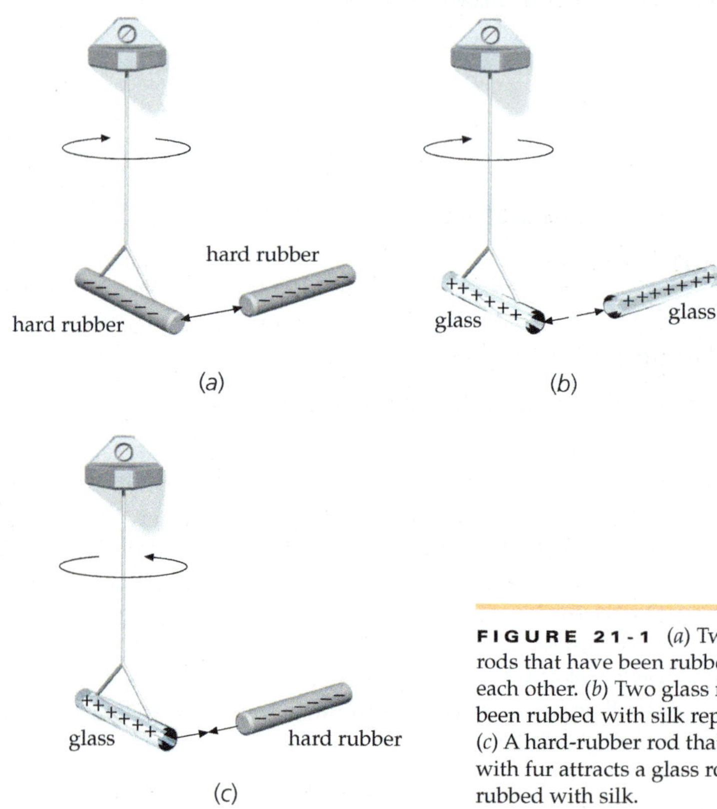

FIGURE 21-1 (*a*) Two hard-rubber rods that have been rubbed with fur repel each other. (*b*) Two glass rods that have been rubbed with silk repel each other. (*c*) A hard-rubber rod that has been rubbed with fur attracts a glass rod that has been rubbed with silk.

signed charges attract each other (Figure 21-1). An object that is neither positively nor negatively charged is said to be *electrically neutral.*

Today, we know that when glass is rubbed with silk, electrons are transferred from the glass to the silk. Because the silk is negatively charged (according to Franklin's convention, which we still use) electrons are said to have a negative charge. Table 21-1 is a short version of the **triboelectric series.** (In Greek *tribos* means "a rubbing.") The farther down the series a material is, the greater its affinity for electrons. If two of the materials are brought in contact, electrons are transferred from the material higher in the table to the one farther down the table. For example, if Teflon is rubbed with nylon, electrons are transferred from the nylon to the Teflon.

Table 21-1 The Triboelectric Series

+ Positive End of Series
Asbestos
Glass
Nylon
Wool
Lead
Silk
Aluminum
Paper
Cotton
Steel
Hard rubber
Nickel and copper
Brass and silver
Synthetic rubber
Orlon
Saran
Polyethylene
Teflon
Silicone rubber
− Negative End of Series

CHARGE QUANTIZATION

Matter consists of atoms that are neutral. Each atom has a tiny but massive nucleus that is composed of protons and neutrons. Protons are positively charged, whereas neutrons are neutral. The number of protons that an atom of a particular element has is the atomic number Z of that element. Surrounding the nucleus is an equal number of negatively charged electrons, leaving the atom with zero net charge. An electron is about 2000 times less massive than a proton, yet the charges of these two particles are exactly equal in magnitude. The charge of the proton is e and that of the electron is $-e$, where e is called the **fundamental unit of charge.** The charge of an electron or proton is an intrinsic property of the particle, just as mass and spin are intrinsic properties of these particles.

All observable charges occur in integral amounts of the fundamental unit of charge e; that is, *charge is quantized.* Any observable charge Q occurring in nature can be written $Q = \pm Ne$, where N is an integer.* For ordinary objects, however, N is usually very large and charge appears to be continuous, just as air appears to be continuous even though air consists of many discrete particles (molecules, atoms, and ions). To give an everyday example of N, charging a plastic rod by rubbing it with a piece of fur typically transfers 10^{10} or more electrons to the rod.

CHARGE CONSERVATION

When objects are rubbed together, one object is left with an excess of electrons and is therefore negatively charged; the other object is left with a deficit of electrons and is therefore positively charged. The net charge of the two objects remains constant; that is, *charge is conserved.* The **law of conservation of charge** is a fundamental law of nature. In certain interactions among elementary particles, charged particles such as electrons are created or annihilated. However, during these processes, equal amounts of positive and negative charge are produced or destroyed, so the net charge of the universe is unchanged.

The SI unit of charge is the coulomb, which is defined in terms of the unit of electric current, the ampere (A).† The **coulomb** (C) is the amount of charge flowing through a cross section of wire in one second when the current in the wire is one ampere. (The cross section of a solid object is the intersection of the object and a plane. Here we consider a plane that cuts across the wire.) The fundamental unit of electric charge e is related to the coulomb by

$$e = 1.602177 \times 10^{-19}\ \text{C} \approx 1.60 \times 10^{-19}\ \text{C} \qquad 21\text{-}1$$

FUNDAMENTAL UNIT OF CHARGE

* In the standard model of elementary particles, protons, neutrons, and some other elementary particles are made up of more fundamental particles called *quarks* that have charges of $\pm\frac{1}{3}e$ or $\pm\frac{2}{3}e$. Only combinations that result in a net charge of $\pm Ne$, where N is an integer, are observed.

† The ampere (A) is the unit of current used in everyday electrical work.

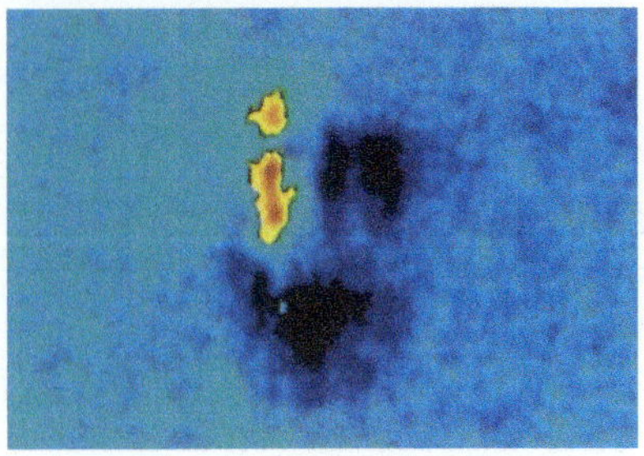

Charging by contact. A piece of plastic about 0.02 mm wide was charged by contact with a piece of nickel. Although the plastic carries a net positive charge, regions of negative charge (dark blue) as well as regions of positive charge (yellow) are indicated. The photograph was taken by sweeping a charged needle of width 10^{-7} m over the sample and recording the electrostatic force on the needle. *(Bruce Terris/IBM Almaden Research Center.)*

PRACTICE PROBLEM 21-1

A charge of magnitude 50 nC (1.0 nC = 10^{-9} C) can be produced in the laboratory by simply rubbing two objects together. How many electrons must be transferred to produce this charge?

Example 21-1 How Many in a Penny?

A copper penny* ($Z = 29$) has a mass of 3.10 grams. What is the total charge of all the electrons in the penny?

PICTURE The electrons have a total charge given by the number of electrons in the penny, N_e, multiplied by the charge of an electron, $-e$. The number of electrons in a copper atom is 29 (the atomic number of copper). So, the total charge of the electrons is 29 electrons multiplied by the number of copper atoms N_{at} in a penny. To find N_{at}, we use the fact that one mole of any substance has Avogadro's number ($N_A = 6.02 \times 10^{23}$) of particles (molecules, atoms, or ions), and the number of grams in one mole of copper is the molar mass M, which is 63.5 g/mol for copper.

SOLVE

1. The total charge Q is the number of electrons multiplied by the charge:
$$Q = N_e(-e)$$
2. The number of electrons is Z multiplied by the number of copper atoms N_{at}:
$$N_e = ZN_{at}$$
3. Compute the number of copper atoms in 3.10 g of copper:
$$N_{at} = (3.10\text{ g})\frac{6.02 \times 10^{23}\text{ atoms/mol}}{63.5\text{ g/mol}} = 2.94 \times 10^{22}\text{ atoms}$$
4. Compute the number of electrons N_e:
$$N_e = ZN_{at} = (29\text{ electrons/atom})(2.94 \times 10^{22}\text{ atoms}) = 8.53 \times 10^{23}\text{ electrons}$$
5. Use this value of N_e to find the total charge:
$$Q = N_e \times (-e) = (8.53 \times 10^{23}\text{ electrons})(-1.60 \times 10^{-19}\text{ C/electron}) = \boxed{-1.37 \times 10^5\text{ C}}$$

CHECK There are $29 \times (6.02 \times 10^{23})$ electrons in 63.5 g of copper, so in 3.10 g of copper there are $(3.10/63.5) \times 29 \times (6.02 \times 10^{23}) = 8.53 \times 10^{23}$ electrons—in agreement with our step-4 result.

PRACTICE PROBLEM 21-2 If one million electrons were given to each person in the United States (about 300 million people), what percentage of the number of electrons in a penny would this represent?

* The penny was composed of 100 percent copper from 1793 to 1837. In 1982, the composition changed from 95 percent copper and 5 percent zinc to 2.5 percent copper and 97.5 percent zinc.

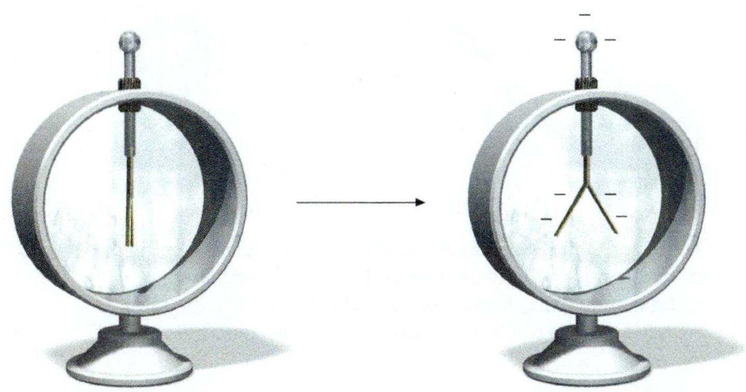

FIGURE 21-2 An electroscope. Two gold leaves are attached to a conducting post that has a conducting ball on top. The ball, post, and leaves are insulated from the container. When uncharged, the leaves hang together vertically. When the ball is touched by a negatively charged plastic rod, some of the negative charge from the rod is transferred to the ball and moves to the gold leaves, which then spread apart because of electrical repulsion between their negative charges. (Touching the ball with a positively charged glass rod would also cause the leaves to spread apart. In this case, the positively charged glass rod would remove electrons from the metal ball, leaving a net positive charge on the ball, rod, and leaves.)

21-2 CONDUCTORS AND INSULATORS

In many materials, such as copper and other metals, some of the electrons are free to move about the entire material. Such materials are called **conductors.** In other materials, such as wood or glass, all the electrons are bound to nearby atoms and none can move freely. These materials are called **insulators.**

In a single atom of copper, 29 electrons are bound to the nucleus by the electrostatic attraction between the negatively charged electrons and the positively charged nucleus. The outer (valence) electrons are more weakly bound to a nucleus than the inner (core) electrons. When a large number of copper atoms are combined in a piece of metallic copper, the strength of the attractions of electrons to a nucleus of an atom is reduced by their interactions with the electrons and nuclei of neighboring atoms. One or more of the valence electrons in each atom is no longer bound to the atom but is free to move about the whole piece of metal, much as an air molecule is free to move about in a room. The number of these free electrons depends on the particular metal, but it is typically about one per atom. (The free electrons are also referred to as conduction electrons or delocalized electrons.) An atom that has an electron removed or added, resulting in a net charge on the atom, is called an **ion.** In metallic copper, the copper ions are arranged in a regular array called a *lattice.* A conductor is neutral if for each lattice ion having a positive charge $+e$ there is a free electron having a negative charge $-e$. The net charge of the conductor can be changed by adding or removing electrons. A conductor that has a negative net charge has a surplus of free electrons, while a conductor that has a positive net charge has a deficit of free electrons.

CONCEPT CHECK 21-1

Two identical conducting spheres, one that has an initial charge $+Q$ and the other is initially uncharged, are brought into contact. (*a*) What is the new charge on each sphere? (*b*) While the spheres are in contact, a positively charged rod is moved close to one sphere, causing a redistribution of the charges on the two spheres so the charge on the sphere closest to the rod has a charge of $-Q$. What is the charge on the other sphere?

CHARGING BY INDUCTION

The conservation of charge is illustrated by a simple method of charging a conductor called **charging by induction,** as shown in Figure 21-3. Two uncharged metal spheres touch each other. When a positively charged rod (Figure 21-3*a*) is brought near one of the spheres, conduction electrons flow from one sphere to the other, toward the positively charged rod. The positively charged rod in Figure 21-3*a* attracts the negatively charged electrons, and the sphere nearest the rod acquires electrons from the sphere farther away. This leaves the near

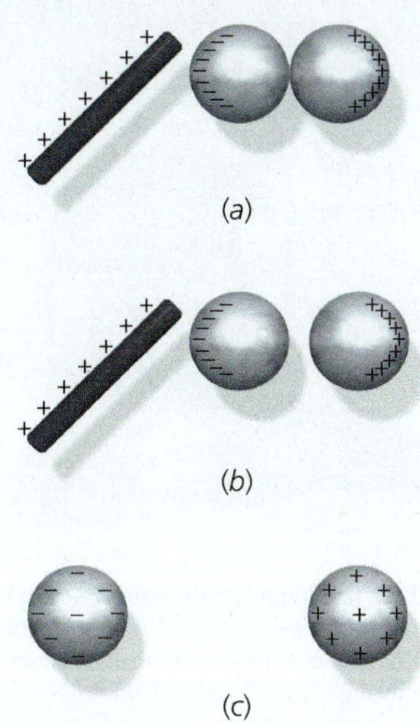

FIGURE 21-3 Charging by induction. (*a*) Neutral conductors in contact become oppositely charged when a charged rod attracts electrons to the left sphere. (*b*) If the spheres are separated before the rod is removed, they will retain their equal and opposite charges. (*c*) When the rod is removed and the spheres are far apart, the distribution of charge on each sphere approaches uniformity.

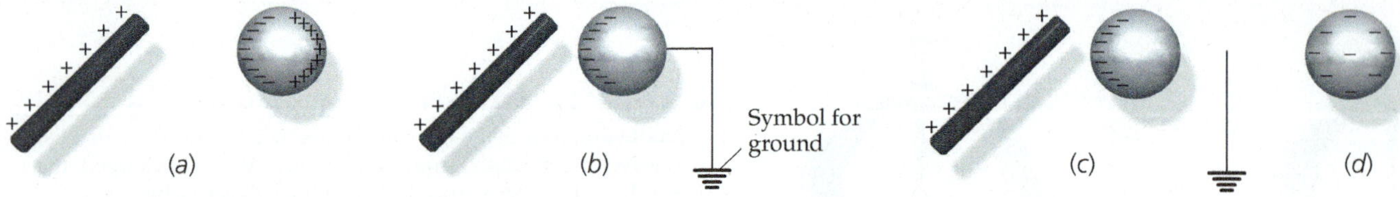

FIGURE 21-4 Induction via grounding. (*a*) The free charge on the single neutral conducting sphere is polarized by the positively charged rod, which attracts negative charges on the sphere. (*b*) When the conductor is grounded by connecting it with a wire to a very large conductor, such as Earth, electrons from the ground neutralize the positive charge on the far face. The conductor is then negatively charged. (*c*) The negative charge remains if the connection to the ground is broken before the rod is removed. (*d*) After the rod is removed, the sphere has a uniform negative charge.

sphere with a net negative charge and the far sphere with an equal net positive charge. A conductor that has *separated* equal and opposite charges is said to be **polarized.** If the spheres are separated before the rod is removed, they will be left with equal amounts of opposite charges (Figure 21-3*b*). A similar result would be obtained with a negatively charged rod, which would drive electrons from the near sphere to the far sphere.

For many purposes, Earth itself can be modeled as an infinitely large conductor that has an infinite supply of charged particles. If a conductor is electrically connected to Earth it is said to be **grounded.** Grounding a metal sphere is indicated schematically in Figure 21-4*b* by a connecting wire ending in parallel horizontal lines. Figure 21-4 demonstrates how we can induce a charge in a single conductor by transferring charge from Earth through a ground wire and then breaking the connection to the ground. (In practice, a person standing on the floor and touching the sphere with his hand provides an adequate ground for electrostatic demonstrations such the one described here.)

CONCEPT CHECK 21-2

Two identical conducting spheres are charged by induction and then separated by a large distance; sphere 1 has charge $+Q$ and sphere 2 has charge $-Q$. A third identical sphere is initially uncharged. If sphere 3 is touched to sphere 1 and separated, then touched to sphere 2 and separated, what is the final charge on each of the three spheres?

The lightning rod on this building is grounded so that it can conduct electrons from the ground to the positively charged clouds, thus neutralizing them. *(© Grant Heilman.)*

These fashionable ladies are wearing hats with metal chains that drag along the ground, which were supposed to protect them from lightning. *(Ann Roman Picture Library.)*

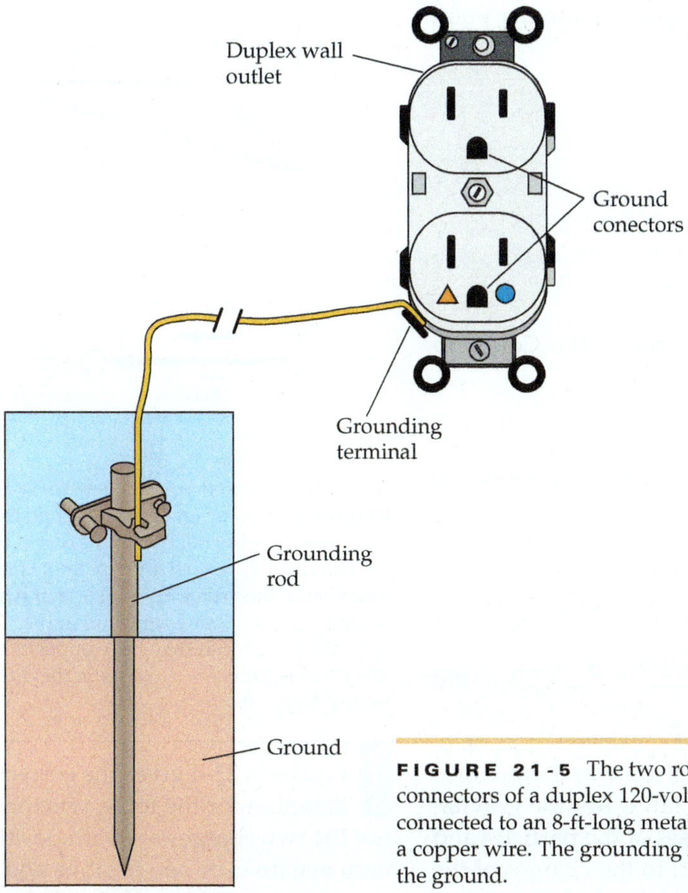

FIGURE 21-5 The two round ground connectors of a duplex 120-volt wall outlet are connected to an 8-ft-long metal grounding rod by a copper wire. The grounding rod is driven into the ground.

21-3 COULOMB'S LAW

Charles Coulomb (1736–1806) studied the force exerted by one charge on another using a torsion balance of his own invention.* In Coulomb's experiment, the charged spheres were much smaller than the distance between them so that the charges could be treated as point charges. Coulomb used the method of charging by induction to produce equally charged spheres and to vary the amount of charge on the spheres. For example, beginning with charge q_0 on each sphere, he could reduce the charge to $\frac{1}{2}q_0$ by temporarily grounding one sphere to discharge it, disconnecting it from ground, and then placing the two spheres in contact. The results of the experiments of Coulomb and others are summarized in **Coulomb's law:**

> The force exerted by one point charge on another acts along the line between the charges. It varies inversely as the square of the distance separating the charges and is proportional to the product of the charges. The force is repulsive if the charges have the same sign and attractive if the charges have opposite signs.
>
> COULOMB'S LAW

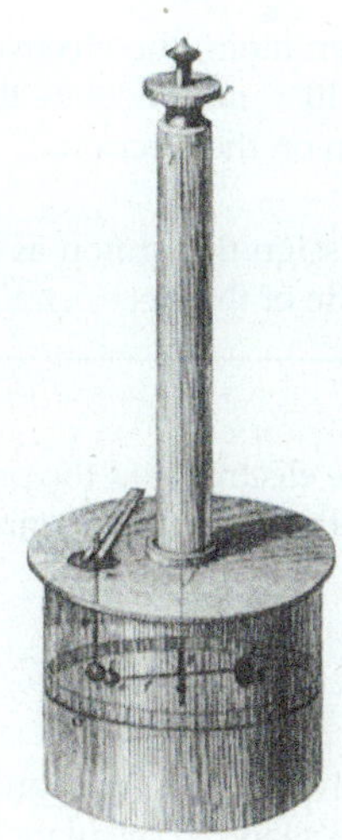

Coulomb's torsion balance. *(Bundy Library, Norwalk, CT.)*

* Coulomb's experimental apparatus was essentially the same as that described for the Cavendish experiment in Chapter 11, with the masses replaced by small charged spheres. For the magnitudes of charges easily transferred by rubbing, the gravitational attraction of the spheres is completely negligible compared with their electric attraction or repulsion.

The *magnitude* of the electric force exerted by a point charge q_1 on another point charge q_2 a distance r away is thus given by

$$F = \frac{k|q_1 q_2|}{r^2} \qquad \text{21-2}$$

COULOMB'S LAW FOR THE MAGNITUDE OF THE FORCE EXERTED BY q_1 ON q_2

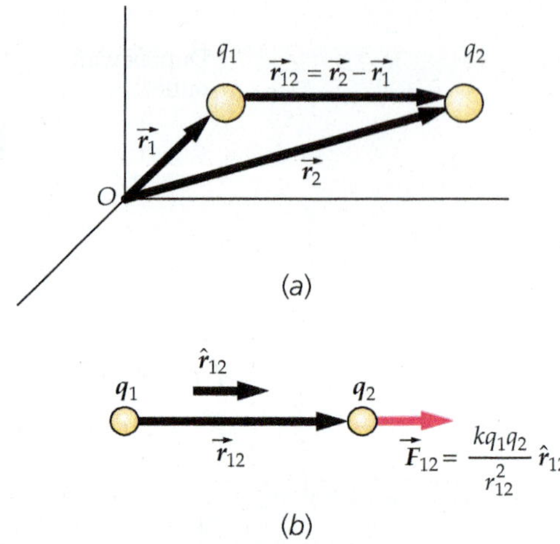

FIGURE 21-6 (*a*) Charge q_1 at position $\vec{r}_1$ and charge q_2 at $\vec{r}_2$ relative to the origin *O*. (*b*) The force $\vec{F}_{12}$ exerted by q_1 on q_2 is in the direction of the vector $\vec{r}_{12} = \vec{r}_2 - \vec{r}_1$ if both charges have the same sign, and in the opposite direction if they have opposite signs. The unit vector $\hat{r}_{12} = \vec{r}_{12}/r_{12}$ is in the direction from q_1 to q_2.

where k is an experimentally determined positive constant called the **Coulomb constant,** which has the value

$$k = 8.99 \times 10^9\ \text{N}\cdot\text{m}^2/\text{C}^2 \qquad \text{21-3}$$

If q_1 is at position $\vec{r}_1$ and q_2 is at $\vec{r}_2$ (Figure 21-6), the force $\vec{F}_{12}$ exerted by q_1 on q_2 is

$$\vec{F}_{12} = \frac{kq_1q_2}{r_{12}^2}\hat{r}_{12} \qquad \text{21-4}$$

COULOMB'S LAW (VECTOR FORM)

where $\vec{r}_{12} = \vec{r}_2 - \vec{r}_1$ is the vector pointing from q_1 to q_2, and $\hat{r}_{12} = \vec{r}_{12}/r_{12}$ is a unit vector in the same direction.

In accord with Newton's third law, the electrostatic force $\vec{F}_{21}$ exerted by q_2 on q_1 is the negative of $\vec{F}_{12}$. Note the similarity between Coulomb's law and Newton's law of gravity. (See Equation 11-3.) Both are inverse-square laws. But the gravitational force between two particles is proportional to the masses of the particles and is always attractive, whereas the electric force is proportional to the charges of the particles and is repulsive if the charges have the same sign and attractive if they have opposite signs.

! Equation 21-4 gives the correct direction for the force, whether or not the two charges are both positive, both negative, or one positive and one negative.

Example 21-2 Electric Force in Hydrogen

In a hydrogen atom, the electron is separated from the proton by an average distance of about 5.3×10^{-11} m. Calculate the magnitude of the electrostatic force of attraction exerted by the proton on the electron.

PICTURE Assign the proton as q_1 and the electron as q_2. Use Coulomb's law to determine the magnitude of the electrostatic force of attraction exerted by the proton on the electron.

SOLVE

1. Sketch the electron and the proton and label the sketch with the suitable symbols (Figure 21-7):

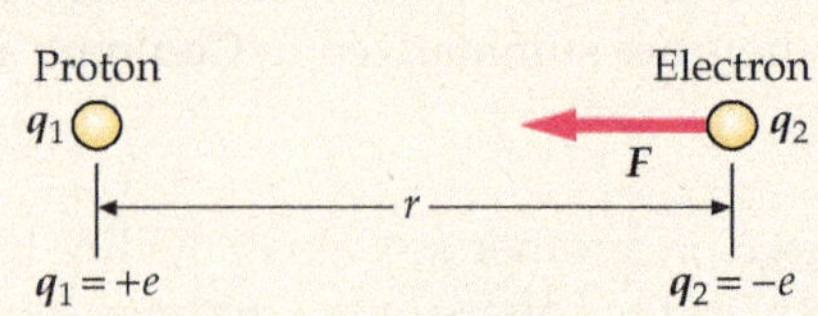

FIGURE 21-7

2. Use the given information and Equation 21-2 (Coulomb's law) to calculate the electrostatic force:

$$F = \frac{k|q_1q_2|}{r^2} = \frac{ke^2}{r^2} = \frac{(8.99 \times 10^9\ \text{N}\cdot\text{m}^2/\text{C}^2)(1.60 \times 10^{-19}\ \text{C})^2}{(5.3 \times 10^{-11}\ \text{m})^2}$$

$$= \boxed{8.2 \times 10^{-8}\ \text{N}}$$

CHECK The order of magnitude is plausible. The powers of ten in the numerator combined are $10^9 \times 10^{-38} = 10^{-29}$, the power of ten in the denominator is 10^{-22}, and $10^{-29}/10^{-22} = 10^{-7}$. In comparison, $8.2 \times 10^{-8} \approx 10^{-7}$.

TAKING IT FURTHER Compared with macroscopic interactions, this is a very small force. However, because the mass of the electron is only about 10^{-30} kg, this force produces an acceleration of $F/m \approx 9 \times 10^{22}\ \text{m/s}^2$. The proton is almost 2000 times more massive than the electron, so the acceleration of the proton is about $4 \times 10^{19}\ \text{m/s}^2$. To put these accelerations in perspective, the acceleration due to gravity g is a mere $10^1\ \text{m/s}^2$.

PRACTICE PROBLEM 21-3 Two point charges of 0.0500 μC each are separated by 10.0 cm. Find the magnitude of the force exerted by one point charge on the other.

Because the electrical force and the gravitational force between any two particles both vary inversely with the square of the distance between the particles, the ratio of these forces is independent of that distance. We can therefore compare the relative strengths of the electrical and gravitational forces for elementary particles such as the electron and proton.

Example 21-3 Ratio of Electric and Gravitational Forces

Compute the ratio of the electric force to the gravitational force exerted by a proton on an electron in a hydrogen atom.

PICTURE Use Coulomb's law and $q_1 = e$ and $q_2 = -e$ to find the electric force. Use Newton's law of gravity, the mass of the proton, $m_p = 1.67 \times 10^{-27}$ kg, and the mass of the electron, $m_e = 9.11 \times 10^{-31}$ kg, to find the gravitational force.

SOLVE

1. Express the magnitudes of the electric force F_e and the gravitational force F_g in terms of the charges, masses, separation distance r, and electrical and gravitational constants:

$$F_e = \frac{ke^2}{r^2} \qquad F_g = \frac{Gm_pm_e}{r^2}$$

2. Determine the ratio. Note that the separation distance r cancels:

$$\frac{F_e}{F_g} = \frac{ke^2}{Gm_pm_e}$$

3. Substitute numerical values:

$$\frac{F_e}{F_g} = \frac{(8.99 \times 10^9\ \text{N}\cdot\text{m}^2/\text{C}^2)(1.60 \times 10^{-19}\ \text{C})^2}{(6.67 \times 10^{-11}\ \text{N}\cdot\text{m}^2/\text{kg}^2)(1.67 \times 10^{-27}\ \text{kg})(9.11 \times 10^{-31}\ \text{kg})}$$

$$= \boxed{2.27 \times 10^{39}}$$

CHECK In the numerator of the fraction in step 3, the coulomb units cancel out. In the denominator of the fraction, the kilogram units cancel out. The result is that both numerator and denominator have units of $\text{N}\cdot\text{m}^2$. The fraction has no units, as expected for a ratio of two forces.

TAKING IT FURTHER The fact that the ratio (step 3) is so large reveals why the effects of gravity are not considered when discussing atomic or molecular interactions.

Although the gravitational force is incredibly weak compared with the electric force and plays essentially no role at the atomic level, it is the dominant force between large objects such as planets and stars. Because large objects contain almost equal numbers of positive and negative charges, the attractive and repulsive electrical forces cancel. The net force between astronomical objects is therefore essentially the force of gravitational attraction alone.

FORCE EXERTED BY A SYSTEM OF CHARGES

In a system of charges, each charge exerts a force, given by Equation 21-4, on every other charge. The net force on any charge is the vector sum of the individual forces exerted on that charge by all the other charges in the system. This result follows from the *principle of superposition* of forces.

***See* Math Tutorial *for more information on* Trigonometry**

Example 21-4 Electric Force on a Charge

Three point charges lie on the x axis; q_1 is at the origin, q_2 is at $x = 2.0$ m, and q_0 is at position x ($x > 2.0$ m). (*a*) Find the total electric force on q_0 due to q_1 and q_2 if $q_1 = +25$ nC, $q_2 = -10$ nC, $q_0 = +20$ nC, and $x = 3.5$ m. (*b*) Find an expression for the total electric force on q_0 due to q_1 and q_2 throughout the region $2.0 \text{ m} < x < \infty$.

PICTURE The total electric force on q_0 is the vector sum of the force $\vec{F}_{10}$ exerted by q_1 and the force $\vec{F}_{20}$ exerted by q_2. The individual forces are found using Coulomb's law and the principle of superposition. Note that $\hat{r}_{10} = \hat{r}_{20} = \hat{i}$ because both $\hat{r}_{10}$ and $\hat{r}_{20}$ are in the $+x$ direction.

SOLVE

(*a*) 1. Draw a sketch of the system of charges (Figure 21-8*a*). Identify the distances r_{10} and r_{20} on the graph:

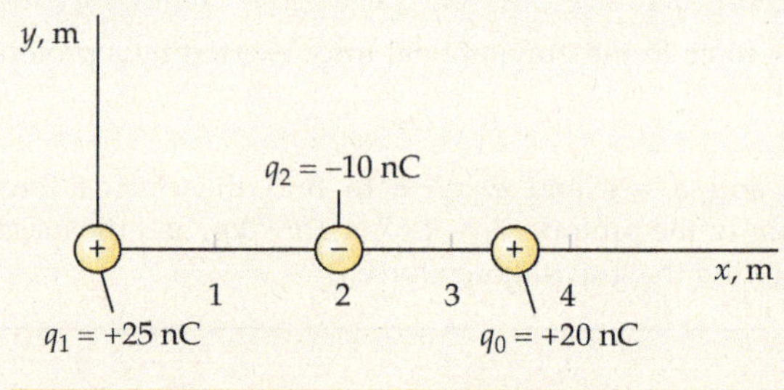

FIGURE 21-8*a*

2. Find the force exerted by q_1 on q_0. These charges have the same sign, so they repel. The force is in the $+x$ direction:

$$F_{10} = \frac{k|q_1 q_0|}{r_{10}^2}$$

$$\vec{F}_{10} = +F_{10}\hat{i} = +\frac{k|q_1 q_0|}{r_{10}^2}\hat{i} = \frac{(8.99 \times 10^9 \text{ N}\cdot\text{m}^2/\text{C}^2)(25 \times 10^{-9} \text{ C})(20 \times 10^{-9} \text{ C})}{(3.5 \text{ m})^2}\hat{i}$$

$$= (0.37 \times 10^{-6} \text{ N})\hat{i}$$

3. Find the force exerted by q_2 on q_0. These charges have opposite signs, so they attract. The force is in the $-x$ direction:

$$F_{20} = \frac{k|q_2 q_0|}{r_{20}^2}$$

$$\vec{F}_{20} = -F_{20}\hat{i} = -\frac{k|q_2 q_0|}{r_{20}^2}\hat{i} = -\frac{(8.99 \times 10^9 \text{ N}\cdot\text{m}^2/\text{C}^2)(10 \times 10^{-9} \text{ C})(20 \times 10^{-9} \text{ C})}{(1.5 \text{ m})^2}\hat{i}$$

$$= -(0.80 \times 10^{-6} \text{ N})\hat{i}$$

4. Combine your results to obtain the net force.

$$\vec{F}_{\text{net}} = \vec{F}_{10} + \vec{F}_{20} = \boxed{-(0.43 \times 10^{-6} \text{ N})\hat{i}}$$

(*b*) 1. Draw a sketch of the system of charges. Label the distances r_{10} and r_{20} (Figure 21-8*b*):

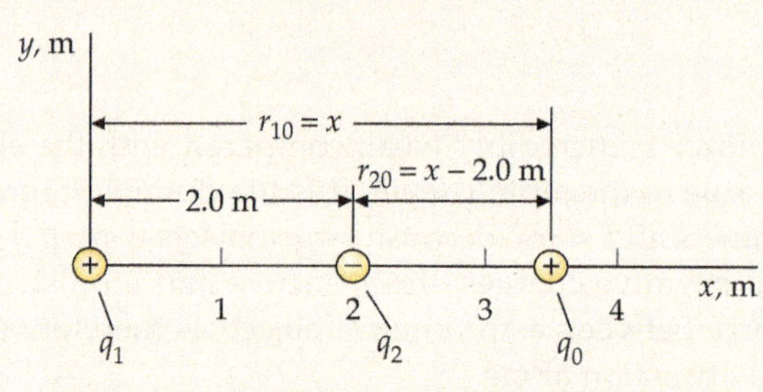

FIGURE 21-8*b*

2. Find an expression for the force on q_0 due to q_1.

$$\vec{F}_{10} = \frac{k|q_1q_0|}{x^2}\hat{i}$$

3. Find an expression for the force on q_0 due to q_2.

$$\vec{F}_{20} = -\frac{k|q_2q_0|}{(x - 2.0\text{ m})^2}\hat{i}$$

4. Combine your results to obtain an expression for the net force.

$$\vec{F}_{\text{net}} = \vec{F}_{10} + \vec{F}_{20} = \left(\frac{k|q_1q_0|}{x^2} - \frac{k|q_2q_0|}{(x - 2.0\text{ m})^2}\right)\hat{i}$$

CHECK In steps 2, 3, and 4 of Part (*b*), both forces approach zero as $x \to \infty$, as expected. In addition, the magnitude of the step-3 result approaches infinity as $x \to 2.0$ m, also as expected.

TAKING IT FURTHER The charge q_2 is located between charges q_1 and q_0, so you might think that the presence of q_2 will affect the force $\vec{F}_{10}$ exerted by q_1 on q_0. However, this is not the case. That is, the presence of q_2 does not effect the force $\vec{F}_{10}$ exerted by q_1 on q_0. (That this is so is called the principle of superposition.) Figure 21-9 shows the x component of the force on q_0 as a function of the position x of q_0 throughout the region $2.0\text{ m} < x < \infty$. Near q_2 the force due to q_2 dominates, and because opposite charges attract the force on q_2 is in the $-x$ direction. For $x \gg 2.0$ m the force is in the $+x$ direction. This is because for large x the distance between q_1 and q_2 is negligible so the force due to the two charges is almost the same as that for a single charge of +15 nC.

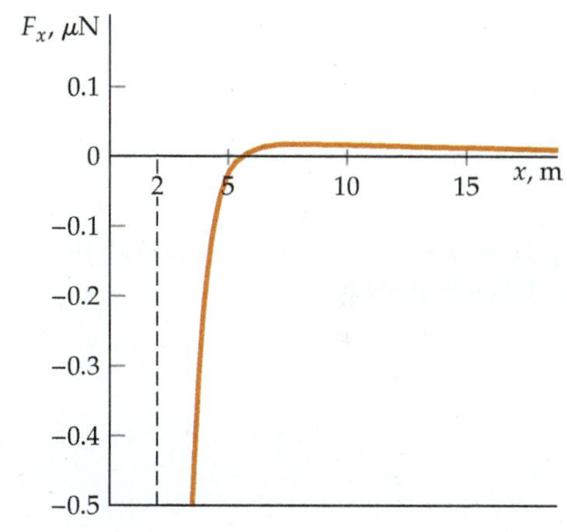

FIGURE 21-9

PRACTICE PROBLEM 21-4 If q_0 is at $x = 1.0$ m, find the total electric force acting on q_0.

For the charges in a system to remain stationary, there must be forces, other than the electric forces the charges exert on each other, acting on the charges so that the net force on each charge is zero. In the preceding example, and those that follow throughout the book, we assume that there are such forces so that all the charges remain stationary.

Example 21-5 Summing Forces in Two Dimensions

Charge q_1 = +25 nC is at the origin, charge q_2 = −15 nC is on the x axis at x = 2.0 m, and charge q_0 = +20 nC is at the point x = 2.0 m, y = 2.0 m as shown in Figure 21-10. Find the magnitude and direction of the resultant electric force on q_0.

PICTURE The resultant electric force is the vector sum of the individual forces exerted by each charge on q_0. We compute each force from Coulomb's law and write it in terms of its rectangular components.

SOLVE

1. Draw the coordinate axes showing the positions of the three charges. Show the resultant electric force $\vec{F}$ on charge q_0 as the vector sum of the forces $\vec{F}_{10}$ due to q_1 and $\vec{F}_{20}$ due to q_2 (Figure 21-10*a*):

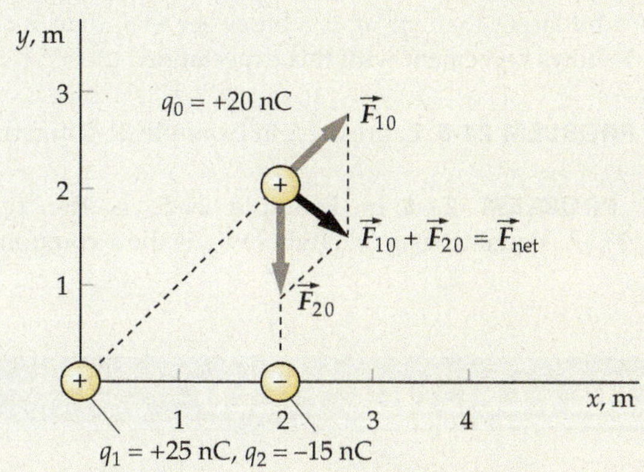

FIGURE 21-10*a*

2. The resultant force $\vec{F}$ on q_0 is the sum of the individual forces:

$$\vec{F} = \vec{F}_{10} + \vec{F}_{20}$$
$$\text{so} \quad \Sigma F_x = F_{10x} + F_{20x} \quad \text{and} \quad \Sigma F_y = F_{10y} + F_{20y}$$

3. The force $\vec{F}_{10}$ is directed away from the origin along the line from q_1 to q_0. Use $r_{10} = 2.0\sqrt{2}$ m for the distance between q_1 and q_0 to calculate its magnitude:

$$F_{10} = \frac{k|q_1q_0|}{r_{10}^2} = \frac{(8.99 \times 10^9\ \text{N}\cdot\text{m}^2/\text{C}^2)(25 \times 10^{-9}\ \text{C})(20 \times 10^{-9}\ \text{C})}{(2.0\sqrt{2}\ \text{m})^2}$$
$$= 5.62 \times 10^{-7}\ \text{N}$$

4. Because $\vec{F}_{10}$ makes an angle of 45° with the x and y axes, its x and y components are equal to each other:

$$F_{10x} = F_{10y} = F_{10}\cos 45° = (5.62 \times 10^{-7}\text{N})\cos 45°$$
$$= 3.97 \times 10^{-7}\ \text{N}$$

5. The force $\vec{F}_{20}$ exerted by q_2 on q_0 is attractive and in the $-y$ direction as shown in Figure 21-10*a*:

$$\vec{F}_{20} = -\frac{k|q_2q_0|}{r_{20}^2}\hat{j} = -\frac{(8.99 \times 10^9\ \text{N}\cdot\text{m}^2/\text{C}^2)(15 \times 10^{-9}\ \text{C})(20 \times 10^{-9}\ \text{C})}{(2.0\ \text{m})^2}\hat{j}$$
$$= -(6.74 \times 10^{-7}\text{N})\hat{j}$$

6. Calculate the components of the resultant force:

$$F_x = F_{10x} + F_{20x} = (3.97 \times 10^{-7}\ \text{N}) + 0 = 3.97 \times 10^{-7}\ \text{N}$$
$$F_y = F_{10y} + F_{20y} = (3.97 \times 10^{-7}\ \text{N}) + (-6.74 \times 10^{-7}\ \text{N})$$
$$F_y = -2.77 \times 10^{-7}\ \text{N}$$

7. Draw the resultant force (Figure 21-10*b*) along with its two components:

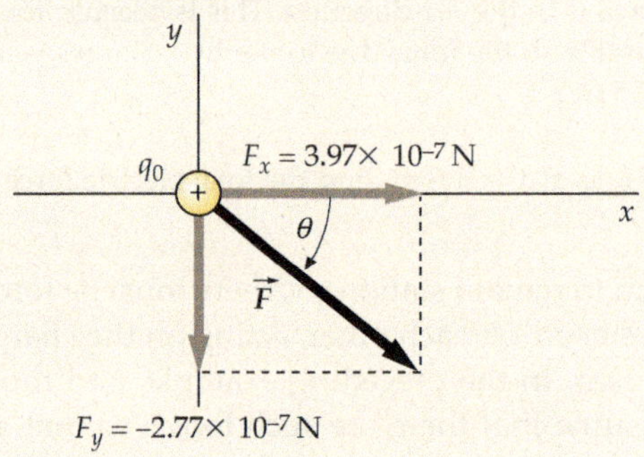

FIGURE 21-10*b*

8. The magnitude of the resultant force is found from its components:

$$F = \sqrt{F_x^2 + F_y^2} = \sqrt{(3.97 \times 10^{-7}\ \text{N})^2 + (-2.77 \times 10^{-7}\ \text{N})^2}$$
$$= 4.84 \times 10^{-7}\ \text{N} = \boxed{4.8 \times 10^{-7}\ \text{N}}$$

9. The resultant force points to the right and downward as shown in Figure 21-10*b*, making an angle θ with the x axis given by:

$$\tan\theta = \frac{F_y}{F_x} = \frac{-2.77}{3.97} = -0.698$$
$$\theta = \tan^{-1}(-0.698) = -34.9° = \boxed{-35°}$$

CHECK We expect the two forces to be approximately equal in magnitude because even though q_1 is a bit larger than $|q_2|$, q_2 is a bit closer to q_0 than is q_1. Comparing the results of steps 3 and 5 shows agreement with this expectation.

PRACTICE PROBLEM 21-5 Express $\hat{r}_{10}$ in Example 21-5 in terms of $\hat{i}$ and $\hat{j}$.

PRACTICE PROBLEM 21-6 In Example 21-5, is the x component of the force $\vec{F}_{10} = (kq_1q_0/r_{10}^2)\hat{r}_{10}$ equal to kq_1q_0/x_{10}^2 (where x_{10} is the x component of $\hat{r}_{10}$)?

21-4 THE ELECTRIC FIELD

The electric force exerted by one charge on another is an example of an action-at-a-distance force, similar to the gravitational force exerted by one mass on another. The idea of action at a distance presents a difficult conceptual challenge. What is the mechanism by which one particle can exert a force on another across the empty space between the particles? Suppose that a charged particle at some point is suddenly moved. Does the force exerted on the second particle some distance r away change instantaneously? To address the challenge of action at a distance, the concept of the **electric field** is introduced. One charge produces an electric field $\vec{E}$ everywhere in space, and this field exerts the force on the second charge. Thus, it is the *field* $\vec{E}$ at the location of the second charge that exerts the force on it, not the first charge itself (which is some distance away). Changes in the field propagate through space at the speed of light, c. Thus, if a charge is suddenly moved, the force it exerts on a second charge a distance r away does not change until a time r/c later.

Figure 21-11*a* shows a set of point charges q_1, q_2, and q_3 arbitrarily arranged in space. These charges produce an electric field $\vec{E}$ everywhere in space. If we place a small positive test charge q_0 at some point near the three charges, there will be a force exerted on q_0 due to the other charges. The net force on q_0 is the vector sum of the individual forces exerted on q_0 by the other charges in the system. Because each of these forces is proportional to q_0, the net force will be proportional to q_0. The electric field $\vec{E}$ at a point is this force divided by q_0:*

$$\vec{E} = \frac{\vec{F}}{q_0} \quad (q_0 \text{ is small}) \qquad 21\text{-}5$$

DEFINITION–ELECTRIC FIELD

The SI unit of the electric field is the newton per coulomb (N/C). In addition, the test charge q_0 will exert a force on each of the other point charges (Figure 21-11*b*). Because these forces on the other charges might cause some of the other charges to move, the charge q_0 must be so small that the forces it exerts on the other charges are negligible. Thus, the electric field at the location in question is actually defined by Equation 21-5, but in the limit that q_0 approaches zero. Table 21-2 lists the magnitudes of some of the electric fields found in nature.

The electric field describes the condition in space set up by the system of point charges. By moving a test charge q_0 from point to point, we can find $\vec{E}$ at all points in space (except at any point occupied by a charge q). The electric field $\vec{E}$ is thus a vector function of position. The force exerted on a test charge q_0 at any point is related to the electric field at that point by

$$\vec{F} = q_0\vec{E} \qquad 21\text{-}6$$

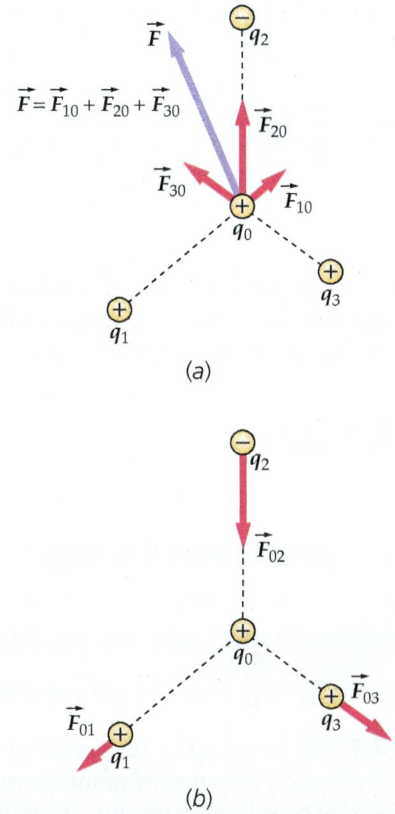

FIGURE 21-11 (*a*) A small test charge q_0 in the vicinity of a system of charges $q_1, q_2, q_3, \ldots$, experiences a resultant electric force $\vec{F}$ that is proportional to q_0. The ratio $\vec{F}/q_0$ is the electric field at that point. (*b*) The test charge q_0 also exerts a force on each of the surrounding charges, and each of these forces is proportional to q_0.

Table 21-2 Some Electric Fields in Nature

	E, N/C
In household wires	10^{-2}
In radio waves	10^{-1}
In the atmosphere	10^2
In sunlight	10^3
Under a thundercloud	10^4
In a lightning bolt	10^4
In an X-ray tube	10^6
At the electron in a hydrogen atom	5×10^{11}
At the surface of a uranium nucleus	2×10^{21}

PRACTICE PROBLEM 21-7

When a 5.0-nC test charge is placed at a certain point, it experiences a force of 2.0×10^{-4} N in the direction of increasing x. What is the electric field $\vec{E}$ at that point?

PRACTICE PROBLEM 21-8

What is the force on an electron placed at a point where the electric field is $\vec{E} = (4.0 \times 10^4\ \text{N/C})\hat{i}$?

The electric field due to a single point charge can be calculated from Coulomb's law. Consider a small, positive test charge q_0 at some point P a distance r_{iP} away from a charge q_i. The force on q_0 is

$$\vec{F}_{i0} = \frac{kq_iq_0}{r_{iP}^2}\hat{r}_{iP}$$

* This definition is similar to that for the gravitational field of Earth, which was defined in Section 4-3 as the force per unit mass exerted by Earth on an object.

The electric field at point P due to charge q_i (Figure 21-12) is thus

$$\vec{E}_{iP} = \frac{kq_i}{r_{iP}^2}\hat{r}_{iP} \qquad 21\text{-}7$$

COULOMB'S LAW FOR $\vec{E}$

where $\hat{r}_{iP}$ is the unit vector pointing from the **source point** i to the **field point** P.

The resultant electric field at P due to a distribution of point charges is found by summing the fields due to each charge separately:

$$\vec{E}_P = \sum_i \vec{E}_{iP} \qquad 21\text{-}8$$

ELECTRIC FIELD $\vec{E}$ DUE TO A SYSTEM OF POINT CHARGES

That is, electric fields follow the principle of superposition.

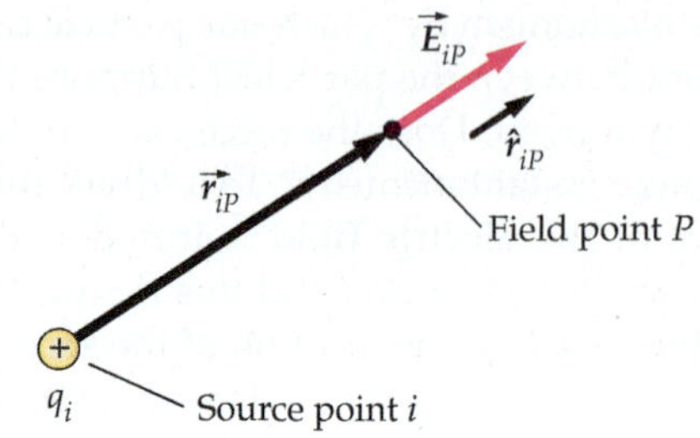

FIGURE 21-12 The electric field $\vec{E}$ at a field point P due to charge q_i at a source point i.

! Even though the expression for the electric field (Equation 21-7) does depend on the location of point P, it does *not* depend on the test charge q_0. That is, q_0 itself does not appear in Equation 21-7.

PROBLEM-SOLVING STRATEGY

Calculating the Resultant Electric Field

PICTURE To calculate the resultant electric field $\vec{E}_P$ at field point P due to a specified distribution of point charges, draw the charge configuration. Include coordinate axes and the field point on the drawing.

SOLVE

1. On the drawing label the distance r_{iP} from each charge to point P. Include an electric field vector $\vec{E}_{iP}$ for the electric field at P due to each point charge.
2. If the field point P and the point charges are not on a single line, then label the angle each individual electric field vector $\vec{E}_{iP}$ makes with one of the coordinate axes.
3. Calculate the component of each individual field vector $\vec{E}_{iP}$ along each axis and use these to calculate the components of the resultant electric field $\vec{E}_P$.

Example 21-6 Direction of Electric Field — *Conceptual*

A positive point charge $q_1 = +q$ and a negative point charge of $q_2 = -2q$ are located on the x axis at $x = a$ and $x = -a$, respectively, as shown in Figure 21-13. Consider the following regions on the x axis: region I ($x < -a$), region II ($-a < x < +a$), and region III ($x > a$). In which region, or regions, is there a point at which the resultant electric field is equal to zero?

PICTURE Let $\vec{E}_1$ and $\vec{E}_2$ be the electric fields due to q_1 and q_2, respectively. Because q_1 is positive, $\vec{E}_1$ points away from q_1 everywhere, and because q_2 is negative, $\vec{E}_2$ points toward from q_2 everywhere. The resultant electric field $\vec{E}$ is equal to the sum of the electric fields of the two charges ($\vec{E} = \vec{E}_1 + \vec{E}_2$). The resultant field is zero if $\vec{E}_1$ and $\vec{E}_2$ are equal in magnitude and oppositely directed. The magnitude of the electric field due to a point charge approaches infinity at points close to a point charge. In addition, at points far from the charge configuration, the electric field approaches the electric field of a point charge equal to $q_1 + q_2$ that is located at the center of charge. The electric field far from the charge configuration is that of a negative point charge because $q_1 + q_2$ is negative.

SOLVE

1. Sketch a figure showing the two charges, the x axis, and the electric fields due the charges at points on the x axis in each of regions I, II, and III. Label these points P_I, P_{II}, and P_{III}, respectively (Figure 21-13):

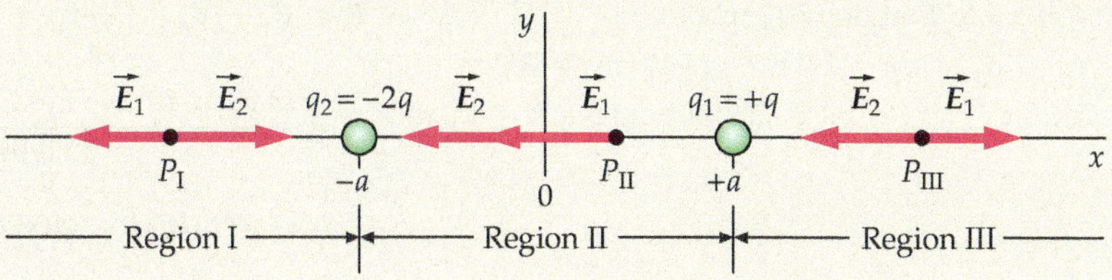

FIGURE 21-13

2. Check to see if the two electric field vectors can be equal in magnitude and opposite in direction anywhere in region I.

Throughout region I, the two electric field vectors are oppositely directed. However, each point in the region is closer to q_2 $(= -2q)$ than q_1 $(= +q)$, so E_2 is greater than E_1 at each point in the region. Thus, in region I there are no points where the electric field is equal to zero.

3. Check to see if the two electric field vectors can be equal in magnitude and opposite in direction anywhere in region II.

Throughout region II, the two electric field vectors are in the same direction at each point on the x axis. Thus, in region II there are no points where the electric field is equal to zero.

4. Check to see if the two electric field vectors can be equal in magnitude and opposite in direction anywhere in region III.

Throughout region III, the two electric field vectors are oppositely directed. At points very close to $x = a$, E_1 is greater than E_2 (because at points close to a point charge the magnitude of the electric field due to that charge approaches infinity). However, at points where $x \gg a$, E_2 is greater than E_1 (because at large distances from the two charges the field direction is determined by the sign of $q_1 + q_2$). Thus, there must be a point somewhere in region III where E_1 is equal to E_2. At that point the net electric field is zero.

CHECK The resultant electric field is zero at a point in region III, the region in which $\vec{E}_1$ and $\vec{E}_2$ are oppositely directed AND in which all points are farther from q_2, the charge with the larger magnitude, than from q_1. This result is as one would expect.

Example 21-7 Electric Field on a Line through Two Positive Point Charges

A positive point charge $q_1 = +8.0$ nC is on the x axis at $x = x_1 = -1.0$ m, and a second positive point charge $q_2 = +12$ nC is on the x axis at $x = x_2 = 3.0$ m. Find the net electric field (*a*) at point A on the x axis at $x = 6.0$ m, and (*b*) at point B on the x axis at $x = 2.0$ m.

PICTURE Let $\vec{E}_1$ and $\vec{E}_2$ be the electric fields due to q_1 and q_2, respectively. Because q_1 is positive, $\vec{E}_1$ points away from q_1 everywhere, and because q_2 is positive, $\vec{E}_2$ points away from q_2 everywhere. We calculate the resultant field using $\vec{E} = \vec{E}_1 + \vec{E}_2$.

SOLVE

(*a*) 1. Draw the charge configuration and place the field point A on the x axis at the appropriate place. Draw vectors representing the electric field at A due to each point charge. Repeat this procedure for field point B (Figure 21-14):

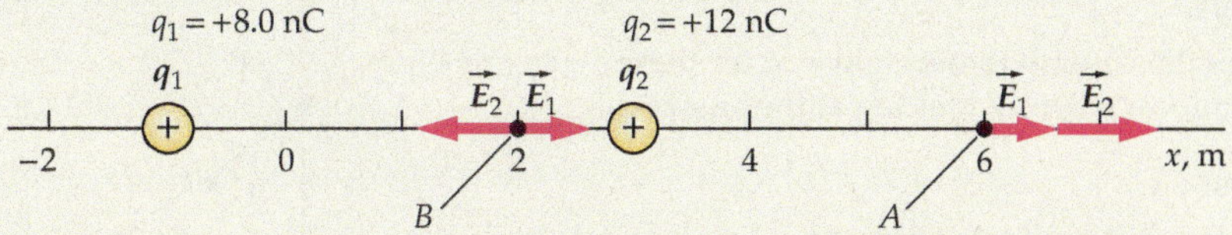

FIGURE 21-14 Because q_1 is a positive charge, $\vec{E}_1$ points away from q_1, at both point A and point B. Because q_2 is a positive charge, $\vec{E}_2$ points away from q_2 at both point A and point B.

2. Calculate $\vec{E}$ at point A, using $r_{1A} = |x_A - x_1| = 6.0\text{ m} - (-1.0\text{ m}) = 7.0\text{ m}$ and $r_{2A} = |x_A - x_2| = 6.0\text{ m} - (3.0\text{ m}) = 3.0\text{ m}$:

$$\vec{E} = \vec{E}_1 + \vec{E}_2 = \frac{kq_1}{r_{1A}^2}\hat{r}_{1A} + \frac{kq_2}{r_{2A}^2}\hat{r}_{2A} = \frac{kq_1}{(x_A - x_1)^2}\hat{i} + \frac{kq_2}{(x_A - x_2)^2}\hat{i}$$

$$= \frac{(8.99 \times 10^9\text{ N}\cdot\text{m}^2/\text{C}^2)(8.0 \times 10^{-9}\text{ C})}{(7.0\text{ m})^2}\hat{i} + \frac{(8.99 \times 10^9\text{ N}\cdot\text{m}^2/\text{C}^2)(12 \times 10^{-9}\text{ C})}{(3.0\text{ m})^2}\hat{i}$$

$$= (1.47\text{ N/C})\hat{i} + (12.0\text{ N/C})\hat{i} = \boxed{(13\text{ N/C})\hat{i}}$$

(*b*) Calculate $\vec{E}$ at point *B*, where $r_{1B} = |x_B - x_1| = 2.0\text{ m} - (-1.0\text{ m}) = 3.0\text{ m}$ and $r_{2B} = |x_B - x_2| = |2.0\text{ m} - (3.0\text{ m})| = 1.0\text{ m}$:

$$\vec{E} = \vec{E}_1 + \vec{E}_2 = \frac{kq_1}{r_{1B}^2}\hat{r}_{1B} + \frac{kq_2}{r_{2B}^2}\hat{r}_{2B} = \frac{kq_1}{(x_B - x_1)^2}\hat{i} + \frac{kq_2}{(x_B - x_2)^2}(-\hat{i})$$

$$= \frac{(8.99 \times 10^9\text{ N}\cdot\text{m}^2/\text{C}^2)(8.0 \times 10^{-9}\text{ C})}{(3.0\text{ m})^2}\hat{i} - \frac{(8.99 \times 10^9\text{ N}\cdot\text{m}^2/\text{C}^2)(12 \times 10^{-9}\text{ C})}{(1.0\text{ m})^2}\hat{i}$$

$$= (7.99\text{ N/C})\hat{i} - (108\text{ N/C})\hat{i} = \boxed{-(100\text{ N/C})\hat{i}}$$

CHECK The Part-(*b*) result is large and in the $-x$ direction. This result is expected because point *B* is close to q_2, and q_2 is a large positive charge (+12 nC) that produces electric field $\vec{E}_2$ in the $-x$ direction at *B*.

TAKING IT FURTHER The resultant electric field at source points close to $q_1 = +8.0$ nC is dominated by the field $\vec{E}_1$ due to q_1. There is one point between q_1 and q_2 where the resultant electric field is zero. A test charge placed at this point would experience no electric force. A sketch of E_x versus x for this charge configuration is shown in Figure 21-15.

PRACTICE PROBLEM 21-9 Regarding Example 21-7, find the point on the x axis where the electric field is zero.

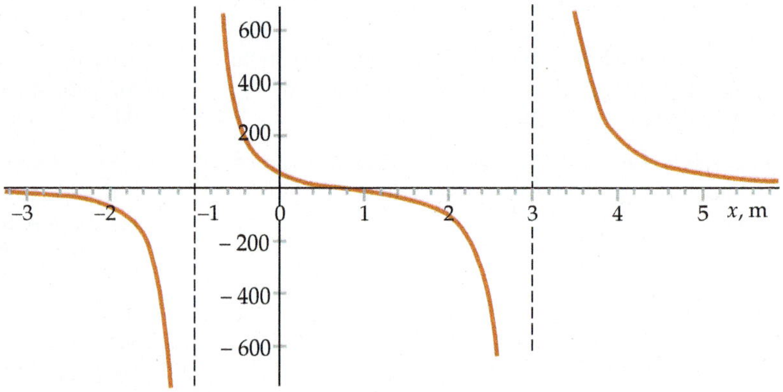

FIGURE 21-15

Example 21-8 Electric Field Due to Point Charges on the *x* Axis — *Try It Yourself*

A point charge $q_1 = +8.0$ nC is at the origin and a second point charge $q_2 = +12.0$ nC is on the x axis at $x = 4.0$ m. Find the electric field on the y axis at $y = 3.0$ m.

PICTURE As in Example 21-7, $\vec{E} = \vec{E}_1 + \vec{E}_2$. At points on the y axis, the electric field $\vec{E}_1$ due to charge q_1 is directed along the y axis, and the field $\vec{E}_2$ due to charge q_2 is in the second quadrant. To find the resultant field $\vec{E}$, we first find the x and y components of $\vec{E}$.

SOLVE

Cover the column to the right and try these on your own before looking at the answers.

Steps

1. Sketch the two charges and the field point. Include the coordinate axes. Draw the electric field due to each charge at the field point and label distances and angles appropriately (Figure 21-16*a*):

Answers

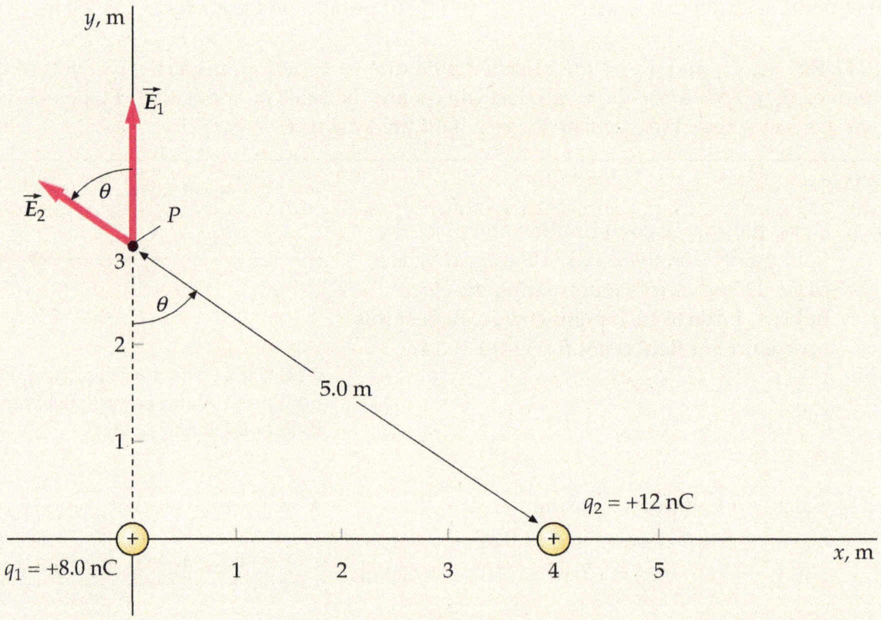

FIGURE 21-16*a*

2. Calculate the magnitude of the field $\vec{E}_1$ at (0, 3.0 m) due to q_1. Find the x and y components of $\vec{E}_1$.
$E_1 = kq_1/y^2 = 7.99$ N/C
$E_{1x} = 0,\ E_{1y} = E_1 = 7.99$ N/C

3. Calculate the magnitude of the field $\vec{E}_2$ at $(0, y)$ due to q_2.
$E_2 = 4.32$ N/C

4. Write the x and y components of $\vec{E}_2$ in terms of the angle θ.
$E_{2x} = -E_2 \sin\theta;\ E_{2y} = E_2 \cos\theta$

5. Compute $\sin\theta$ and $\cos\theta$.
$\sin\theta = 0.80;\ \cos\theta = 0.60$

6. Calculate E_{2x} and E_{2y}.
$E_{2x} = -3.46$ N/C; $E_{2y} = 2.59$ N/C

7. Sketch the components of the resultant field. Include both the vector $\vec{E}$ and the angle $\vec{E}$ makes with the x axis (Figure 21-16b):

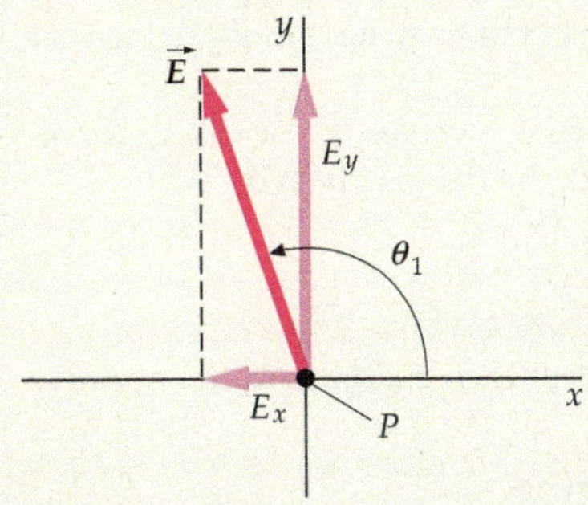

FIGURE 21-16b

8. Find the x and y components of the resultant field $\vec{E}$.
$E_x = E_{1x} + E_{2x} = -3.46$ N/C
$E_y = E_{1y} + E_{2y} = 10.6$ N/C

9. Calculate the magnitude of $\vec{E}$ from its components.
$E = \sqrt{E_x^2 + E_y^2} = 11.2\text{ N/C} = \boxed{11\text{ N/C}}$

10. Find the angle θ_1 made by $\vec{E}$ with the x axis.
$\theta_1 = \tan^{-1}\left(\frac{E_y}{E_x}\right) = \boxed{108^\circ}$

CHECK As expected, E is larger than either E_1 or E_2, but less than $E_1 + E_2$. (This result is expected because the angle between $\vec{E}_1$ and $\vec{E}_2$ is less than 90°.)

Example 21-9 Electric Field Due to Two Equal and Opposite Charges

A charge $+q$ is at $x = a$ and a second charge $-q$ is at $x = -a$ (Figure 21-17). (a) Find the electric field on the x axis at an arbitrary point $x > a$. (b) Find the limiting form of the electric field for $x \gg a$.

PICTURE We calculate the electric field at point P using the principle of superposition, $\vec{E}_P = \vec{E}_{1P} + \vec{E}_{2P}$. For $x > a$, the electric field $\vec{E}_+$ due to the positive charge is in the $+x$ direction and the electric field $\vec{E}_-$ due to the negative charge is in the $-x$ direction. The distances are $x - a$ to the positive charge and $x - (-a) = x + a$ to the negative charge.

SOLVE

(a) 1. Draw the charge configuration on a coordinate axis and label the distances from each charge to the field point (Figure 21-17):

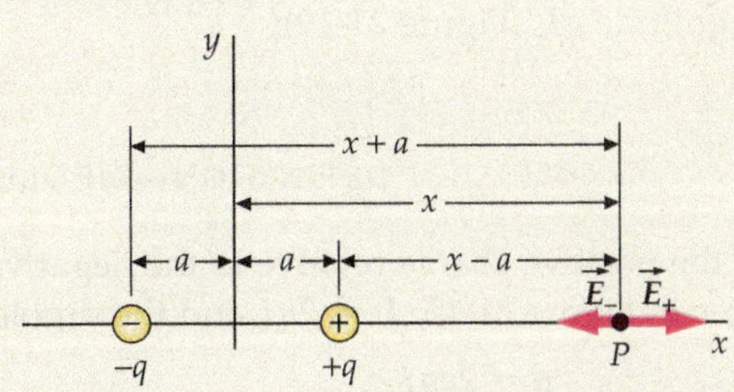

FIGURE 21-17

2. Calculate $\vec{E}$ due to the two charges for $x > a$: (*Note:* The equation on the right holds only for $x > a$.)

$$\vec{E} = \vec{E}_+ + \vec{E}_- = \frac{kq}{[x-a]^2}\hat{i} + \frac{kq}{[x-(-a)]^2}\left(-\hat{i}\right)$$
$$= kq\left[\frac{1}{(x-a)^2} - \frac{1}{(x+a)^2}\right]\hat{i}$$

3. Put the terms in square brackets under a common denominator and simplify:

$$\vec{E} = kq\left[\frac{(x+a)^2-(x-a)^2}{(x+a)^2(x-a)^2}\right]\hat{i} = \boxed{kq\frac{4ax}{(x^2-a^2)^2}\hat{i} \quad x > a}$$

(*b*) In the limit $x \gg a$, we can neglect a^2 compared with x^2 in the denominator:

$$\vec{E} = kq\frac{4ax}{(x^2-a^2)^2}\hat{i} \approx kq\frac{4ax}{x^4}\hat{i} = \boxed{\frac{4kqa}{x^3}\hat{i} \quad x \gg a}$$

CHECK Both boxed answers approach zero as x approaches infinity, which is as expected.

TAKING IT FURTHER Figure 21-18 shows E_x versus x for all x, for $q = 1.0$ nC and $a = 1.0$ m. For $|x| \gg a$ (far from the charges), the field is given by

$$\vec{E} = \frac{4kqa}{|x|^3}\hat{i} \qquad |x| \gg a$$

Between the charges, the contribution from each charge is in the negative direction. An expression for $\vec{E}$ is

$$\vec{E} = \frac{kq}{(x-a)^2}\hat{e}_+ + \frac{k(-q)}{(x+a)^2}\hat{e}_- \qquad -a < x < a$$

where $\hat{e}_+$ is a unit vector that points away from the point $x = a$ for all values of x (except $x = a$) and $\hat{e}_-$ is a unit vector that points away from the point $x = -a$ for all values of x (except $x = -a$). (Note that $\hat{e}_+ = \frac{x-a}{|x-a|}\hat{i}$ and $\hat{e}_- = \frac{x+a}{|x+a|}\hat{i}$.)

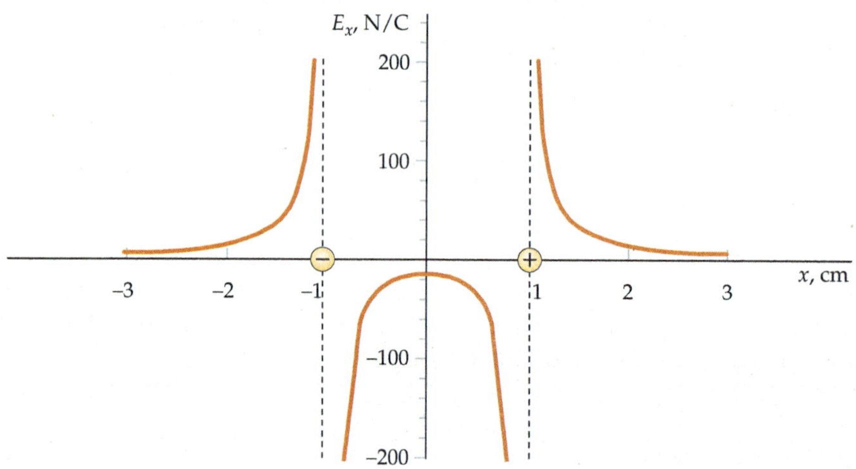

FIGURE 21-18 A plot of E_x versus x on the x axis for the charge distribution in Example 21-9.

ELECTRIC DIPOLES

A system of two equal and opposite charges q separated by a small distance L is called a **dipole.** Its strength and orientation are described by the **dipole moment** $\vec{p}$, which is a vector that points from the negative charge $-q$ toward the positive charge $+q$ and has the magnitude qL (Figure 21-19):

$$\vec{p} = q\vec{L} \qquad \text{21-9}$$

DEFINITION—DIPOLE MOMENT

where $\vec{L}$ is the position of the positive charge relative to the negative charge.

For the system of charges in Figure 21-17, $\vec{L} = 2a\hat{i}$ and the dipole moment is

$$\vec{p} = 2aq\hat{i}$$

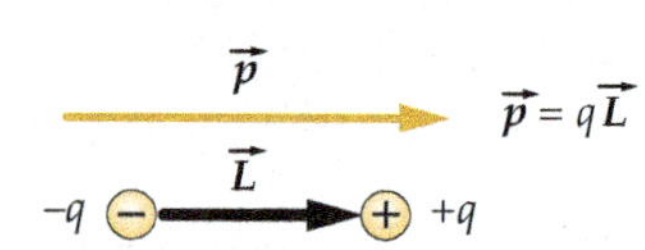

FIGURE 21-19 A dipole consists of a pair of equal and opposite charges. The dipole moment is $\vec{p} = q\vec{L}$, where q is the magnitude of one of the charges and $\vec{L}$ is the position of the negative charge relative to the positive charge.

In terms of the dipole moment $\vec{p}$, the electric field on the axis of the dipole at a point a great distance $|x|$ away is in the same direction as $\vec{p}$ and has magnitude

$$E = \frac{2kp}{|x|^3} \tag{21-10}$$

(see Example 21-9). At a point far from a dipole in any direction, the magnitude of the electric field is proportional to the magnitude of the dipole moment and decreases with the cube of the distance. If a system has a nonzero net charge, the electric field decreases as $1/r^2$ at large distances. In a system that has zero net charge, the electric field falls off more rapidly with distance. In the case of a dipole, the field falls off as $1/r^3$ in all directions.

21-5 ELECTRIC FIELD LINES

We can visualize the electric field by drawing a number of directed curved lines, called **electric field lines,** to indicate both the magnitude and the direction of the field. At any given point, the field vector $\vec{E}$ is tangent to the line through that point. (Electric field lines are also called *lines of force* because they show the direction of the electric force exerted on a positive test charge.) At points very near a positive point charge, the electric field $\vec{E}$ points directly away from the charge. Consequently, the electric field lines very near a positive charge also point directly away from the charge. Similarly, very near a negative point charge the electric field lines point directly toward the charge.

Figure 21-20 shows the electric field lines of a single positive point charge. The spacing of the lines is related to the strength of the electric field. As we move away from the charge, the field becomes weaker and the lines become farther apart. Consider an imaginary spherical surface of radius r that has its center at the charge. Its area is $4\pi r^2$. Thus, as r increases, the density of the field lines (the number of lines per unit area through a surface element normal to the field lines) decreases as $1/r^2$, the same rate of decrease as E. So, we adopt the convention of drawing a fixed number of lines from a point charge, the number being proportional to the charge q, and if we draw the lines equally spaced very near the point charge, the field strength is indicated by the density of the lines. The more closely spaced the lines, the stronger the electric field. The magnitude of the electric field is also called the **electric field strength.**

Figure 21-21 shows the electric field lines for two equal positive point charges q separated by a small distance. Near each point charge, the field is approximately that due to that charge alone. This is because the magnitude of the field of a single point charge is extremely large at points very close to the charge, and because the second charge is relatively far away. Consequently, the field lines near either charge are radial

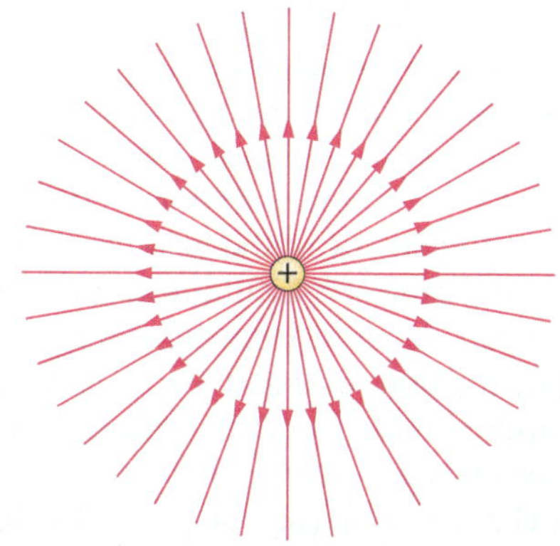

(a)

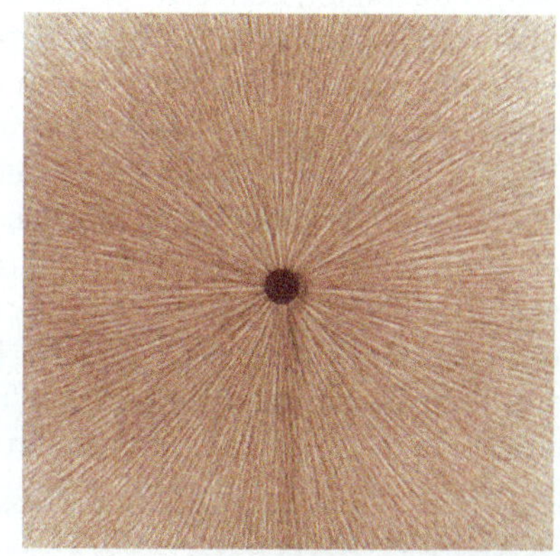

(b)

FIGURE 21-20 (*a*) Electric field lines of a single positive point charge. If the charge were negative, the arrows would be reversed. (*b*) The same electric field lines shown by bits of thread suspended in oil. The electric field of the charged object in the center induces opposite charges on the ends of each bit of thread, causing the threads to align themselves parallel to the field. *(Harold M. Waage.)*

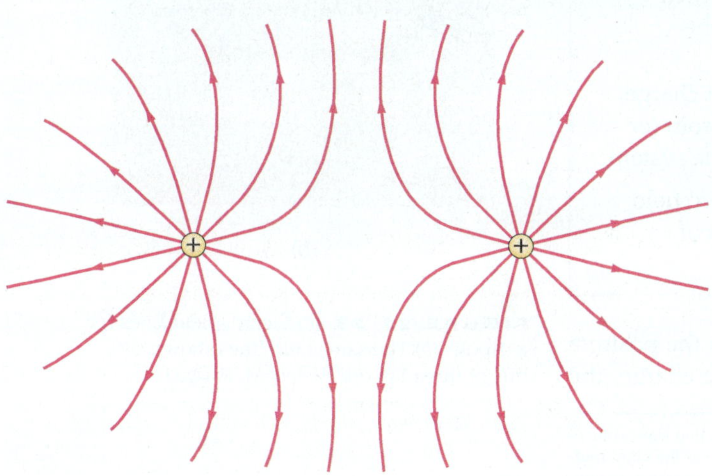

(a)

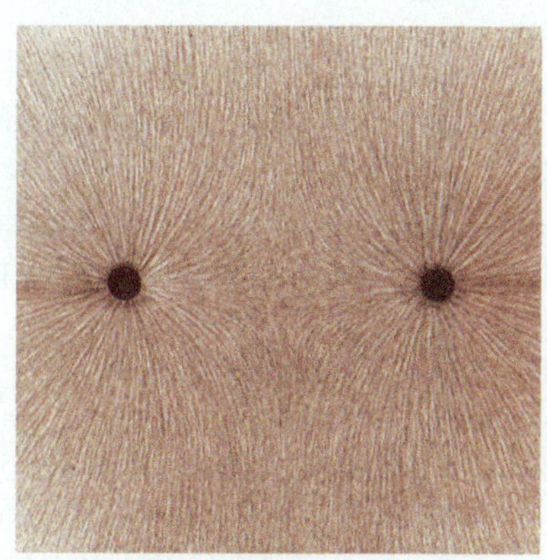

(b)

FIGURE 21-21 (*a*) Electric field lines due to two positive point charges. The arrows would be reversed if both charges were negative. (*b*) The same electric field lines shown by bits of thread in oil. *(Harold M. Waage.)*

and equally spaced. Because the charges are of equal magnitude, we draw an equal number of lines originating from each charge. At very large distances, the details of the charge configuration are not important and the electric field lines are indistinguishable from those of a point charge of magnitude $2q$ a very large distance away. (For example, if the two charges were 1 mm apart and we look at the field lines near a point 100 km away, the field lines would look like those of a single charge of magnitude $2q$ a distance 100 km away.) So at a large distance from the charges, the field is approximately the same as that due to a point charge $2q$ and the lines are approximately equally spaced. Looking at Figure 21-21, we see that the density of field lines in the region between the two charges is small compared to the density of lines in the region just to the left and just to the right of the charges. This indicates that the magnitude of the electric field is weaker in the region between the charges than it is in the region just to the right or left of the charges, where the lines are more closely spaced. This information can also be obtained by direct calculation of the field at points in these regions.

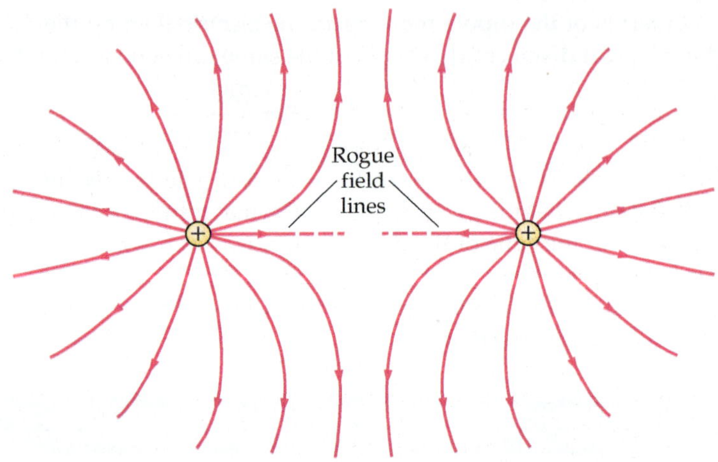

FIGURE 21-22 There are infinitely many field lines emanating from the two charges, two of which are rogue field lines. These rogue field lines terminate at the point midway between the two charges.

We can apply this reasoning to sketch the electric field lines for any system of point charges. Very near each charge, the field lines are equally spaced and emanate from or terminate on the charge radially, depending on the sign of the charge. Very far from all the charges, the detailed configuration of the system of charges is not important, so the field lines are like those of a single point charge having the net charge of the system. The rules for drawing electric field lines are summarized in the following Problem-Solving Strategy.

PROBLEM-SOLVING STRATEGY

Drawing Field Lines

PICTURE Electric field lines emanate from positive charges and terminate on negative charges.*

SOLVE

1. The lines emanating from (or terminating on) an isolated point charge are drawn uniformly spaced as they emanate (or terminate).
2. The number of lines emanating from a positive charge (or terminating on a negative charge) is proportional to the magnitude of the charge.
3. The density of the lines at any point (the number of lines per unit area through a surface element normal to the lines) is proportional to the magnitude of the field there.
4. At large distances from a system of charges that has a nonzero net charge, the field lines are equally spaced and radial, as if they emanated from (or terminated on) a single point charge equal to the total charge of the system.

CHECK Make sure that the field lines never intersect each other. (If two field lines intersected, that would indicate two directions for $\vec{E}$ at the point of intersection.)

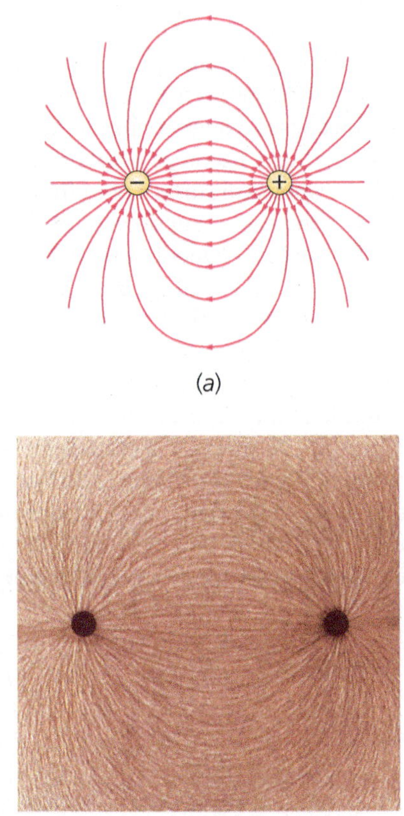

(a)

(b)

FIGURE 21-23 (*a*) Electric field lines for a dipole. (*b*) The same field lines shown by bits of thread in oil. *(Harold M. Waage.)*

Figure 21-23 shows the electric field lines due to a dipole. Very near the positive charge, the lines are directed radially outward. Very near the negative charge, the

* Rogue field lines are field lines that do not follow this rule. An example of a rogue field line is a line that leaves one of the positive charges in Figure 21-22 and is directed toward the other charge. This field line terminates at the point midway between the two charges-as does a corresponding field line emanating from the second positive charge in the figure. For these two charges there are infinitely many field lines, two of which are rogue field lines.

lines are directed radially inward. Because the charges have equal magnitudes, the number of lines that begin at the positive charge equals the number that end at the negative charge. In this case, the field is strong in the region between the charges, as indicated by the high density of field lines in this region.

Figure 21-24*a* shows the electric field lines for a negative charge $-q$ at a small distance from a positive charge $+2q$. Twice as many lines emanate from the positive charge as terminate on the negative charge. Thus, half the lines emanating from the positive charge $+2q$ terminate on the negative charge $-q$; the other half of the lines emanating from the positive charge continue on indefinitely. Very far from the charges (Figure 21-24*b*), the lines are approximately symmetrically spaced and point radially away from a single point, just as they would for a single positive point charge $+q$.

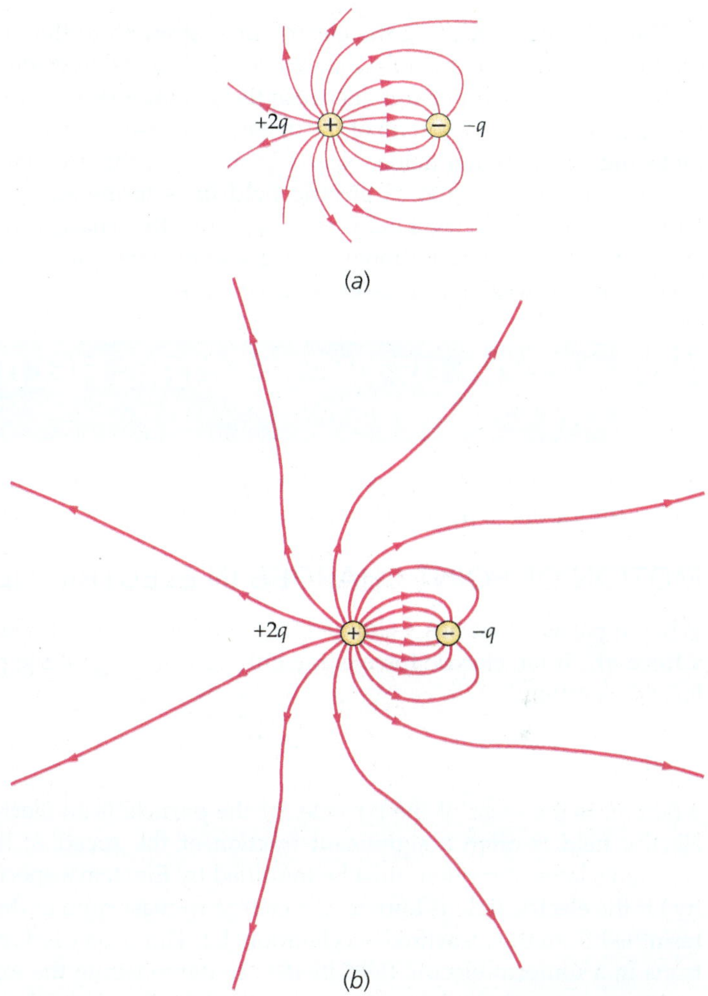

FIGURE 21-24 (*a*) Electric field lines for a point charge $+2q$ and a second point charge $-q$. (*b*) At great distances from the charges, the field lines approach those for a single point charge $+q$ located at the center of charge.

Example 21-10 Field Lines for Two Conducting Spheres *Conceptual*

The electric field lines for two conducting spheres are shown in Figure 21-25. What is the sign of the charge on each sphere, and what are the relative magnitudes of the charges on the spheres?

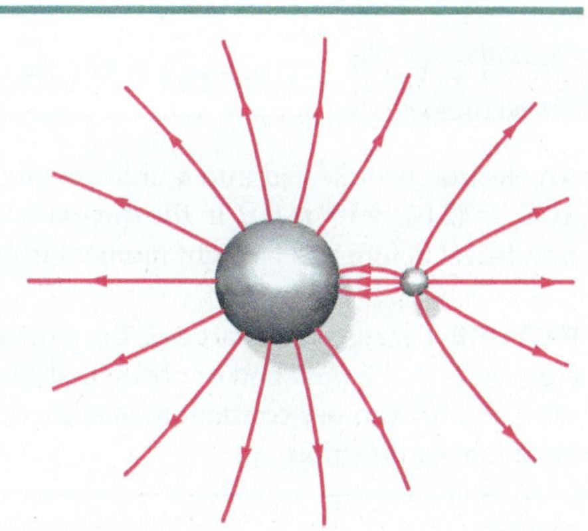

FIGURE 21-25

PICTURE The charge on an object is positive if more field lines emanate from it than terminate on it, and negative if more terminate on it than emanate from it. The ratio of the magnitudes of the charges equals the ratio of the net number of lines emanating from or terminating on the spheres.

SOLVE

1. By counting field lines, determine the net number of field lines emanating from the larger sphere: Because 11 electric field lines emanate from the larger sphere and 3 lines terminate on it, the net number of lines emanating from it is 8.
2. By counting field lines, determine the net number of field lines emanating from the smaller sphere: Because 8 electric field lines emanate from the smaller sphere and no lines terminate on it, the net number of lines emanating from it is 8.
3. Determine the sign of the charge on each sphere: Because both spheres have more field lines emanating from than terminating on them, **both spheres are positively charged.**
4. Determine the relative magnitudes of the charges on the two spheres: Because both spheres have the same net number of lines emanating from them, the **charges on them are equal in magnitude.**

The convention relating the electric field strength to the density of the electric field lines works only because the electric field varies inversely as the square of the distance from a point charge. Because the gravitational field of a point mass also varies inversely as the square of the distance, field-line drawings are also useful for picturing gravitational fields. Near a point mass, the gravitational field lines terminate on the mass just as electric field lines terminate on a negative charge. However, unlike electric field lines near a positive charge, there are no points in space from which gravitational field lines emanate. That is because the gravitational force between two masses is never repulsive.

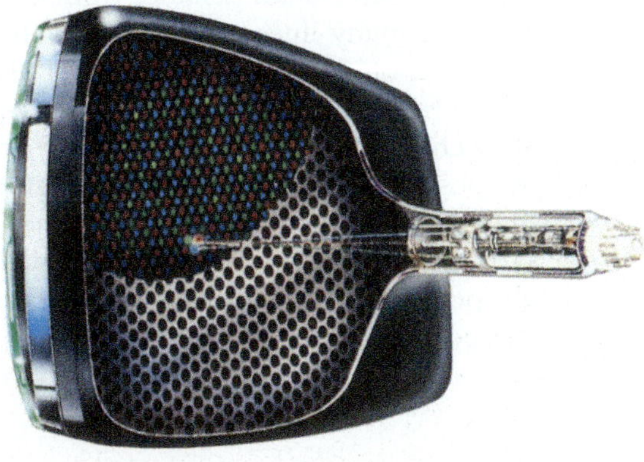

Schematic drawing of a cathode-ray-tube display used for color television. The beams of electrons from the electron gun on the right activate phosphors on the screen at the left, giving rise to bright spots whose colors depend on the relative intensity of each beam. Electric fields between deflection plates in the gun (or magnetic fields from coils surrounding the gun) deflect the beams. The beams sweep across the screen in a horizontal line, are deflected downward, then sweep across again. The entire screen is covered in this way 30 times per second. *(Courtesy of Hulon Forrester/Video Display Corporation, Tucker Georgia.)*

21-6 ACTION OF THE ELECTRIC FIELD ON CHARGES

A uniform electric field can exert a force on a single charged particle and can exert both a torque and a net force on an electric dipole.

MOTION OF POINT CHARGES IN ELECTRIC FIELDS

When a particle that has a charge q is placed in an electric field $\vec{E}$, it experiences a force $q\vec{E}$. If the electric force is the only force acting on the particle, the particle has acceleration

$$\vec{a} = \frac{\Sigma\vec{F}}{m} = \frac{q}{m}\vec{E}$$

where m is the mass of the particle. (If the particle is an electron, its speed in an electric field is often a significant fraction of the speed of light. In such cases, Newton's laws of motion must be modified by Einstein's special theory of relativity.) If the electric field is known, the charge-to-mass ratio of the particle can be determined from the measured acceleration. J. J. Thomson used the deflection of electrons in a uniform electric field in 1897 to demonstrate the existence of electrons and to measure their charge-to-mass ratio. Familiar examples of devices that rely on the motion of electrons in electric fields are oscilloscopes, computer monitors, and television sets that use cathode-ray-tube displays.

Example 21-11 Electron Moving Parallel to a Uniform Electric Field

An electron is projected into a uniform electric field $\vec{E} = (1000\text{ N/C})\hat{i}$ with an initial velocity $\vec{v}_0 = (2.00 \times 10^6\text{ m/s})\hat{i}$ in the direction of the field (Figure 21-26). How far does the electron travel before it is brought momentarily to rest?

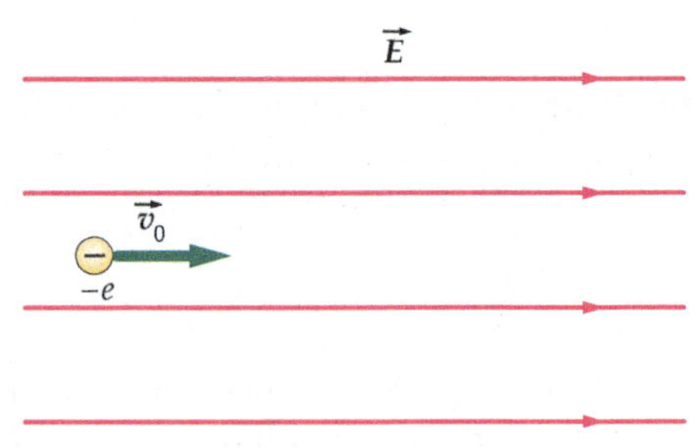

FIGURE 21-26

PICTURE Because the charge of the electron is negative, the force $\vec{F} = -e\vec{E}$ acting on the electron is in the direction opposite that of the field. Because $\vec{E}$ is constant, the force is constant and we can use constant acceleration formulas from Chapter 2. We choose the field to be in the $+x$ direction.

SOLVE

1. The displacement Δx is related to the initial and final velocities: $v_x^2 = v_{0x}^2 + 2a_x\Delta x$
2. The acceleration is obtained from Newton's second law: $a_x = \dfrac{F_x}{m} = \dfrac{-eE_x}{m}$
3. When $v_x = 0$, the displacement is:

$$\Delta x = \frac{v_x^2 - v_{0x}^2}{2a_x} = \frac{0 - v_{0x}^2}{2(-eE_x/m)} = \frac{mv_0^2}{2eE} = \frac{(9.11 \times 10^{-31}\text{ kg})(2.00 \times 10^6\text{ m/s})^2}{2(1.60 \times 10^{-19}\text{ C})(1000\text{ N/C})}$$

$$= 1.14 \times 10^{-2}\text{ m} = \boxed{1.14\text{ cm}}$$

CHECK The displacement Δx is positive, as is expected for something moving in the $+x$ direction.

Example 21-12 Electron Moving Perpendicular to a Uniform Electric Field

An electron enters a uniform electric field $\vec{E} = (-2.0\,\text{kN/C})\hat{j}$ with an initial velocity $\vec{v}_0 = (1.0 \times 10^6\,\text{m/s})\hat{i}$ perpendicular to the field (Figure 21-27). (*a*) Compare the gravitational force acting on the electron to the electric force acting on it. (*b*) By how much has the electron been deflected after it has traveled 1.0 cm in the x direction?

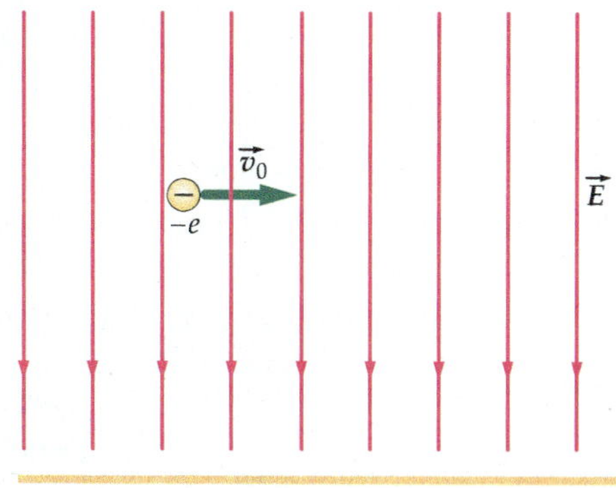

FIGURE 21-27

PICTURE (*a*) Calculate the ratio of the magnitude of the electric force $|q|E = eE$ to that of the gravitational force mg. (*b*) Because mg is, by comparison, negligible, the net force on the electron is equal to the vertically upward electric force. The electron thus moves with constant horizontal velocity v_x and is deflected upward by an amount $\Delta y = \frac{1}{2}at^2$, where t is the time to travel 1.0 cm in the x direction.

SOLVE

(*a*) 1. Calculate the ratio of the magnitude of the electric force, F_e, to the magnitude of the gravitational force, F_g:

$$\frac{F_e}{F_g} = \frac{eE}{mg} = \frac{(1.60 \times 10^{-19}\,\text{C})(2000\,\text{N/C})}{(9.11 \times 10^{-31}\,\text{kg})(9.81\,\text{N/kg})} = \boxed{3.6 \times 10^{13}}$$

(*b*) 1. Express the vertical deflection in terms of the acceleration a and time t:

$$\Delta y = \frac{1}{2}a_y t^2$$

2. Express the time required for the electron to travel a horizontal distance Δx with constant horizontal velocity v_0:

$$t = \frac{\Delta x}{v_0}$$

3. Use this result for t and eE/m for a_y to calculate Δy:

$$\Delta y = \frac{1}{2}\frac{eE}{m}\left(\frac{\Delta x}{v_0}\right)^2 = \frac{1}{2}\frac{(1.6 \times 10^{-19}\,\text{C})(2000\,\text{N/C})}{9.11 \times 10^{-31}\,\text{kg}}\left(\frac{0.010\,\text{m}}{10^6\,\text{m/s}}\right)^2$$

$$= \boxed{1.8\,\text{cm}}$$

CHECK The step-3 result is positive (upward), as is expected for an object accelerating upward that was initially moving horizontally.

TAKING IT FURTHER (*a*) As is usually the case, the electric force is huge compared with the gravitational force. Thus, it is not necessary to consider gravity when designing a cathode-ray tube, for example, or when calculating the deflection in the problem above. In fact, a television picture tube works equally well upside down and right side up, as if gravity were not even present. (*b*) The path of an electron moving in a uniform electric field is a parabola, the same as the path of a neutral particle moving in a uniform gravitational field.

Example 21-13 The Electric Field in an Ink-Jet Printer — *Context-Rich*

You have just finished printing out a long essay for your English professor, and you wonder about how the ink-jet printer knows where to place the ink. You search the Internet and find a picture (Figure 21-28) showing that the ink drops are given a charge and pass between a pair of oppositely charged metal plates that provide a uniform electric field in the region between the plates. Because you have been studying the electric field in physics class, you wonder if you can determine how large a field is used in this type of printer. You search further and find that the 40.0-μm-diameter ink drops have an initial velocity of 40.0 m/s, and that a drop that has a 2.00-nC charge is deflected upward a distance of 3.00 mm as the drop travels through the 1.00-cm-long region between the plates. Find the magnitude of the electric field. (Neglect any effects of gravity on the motion of the drops.)

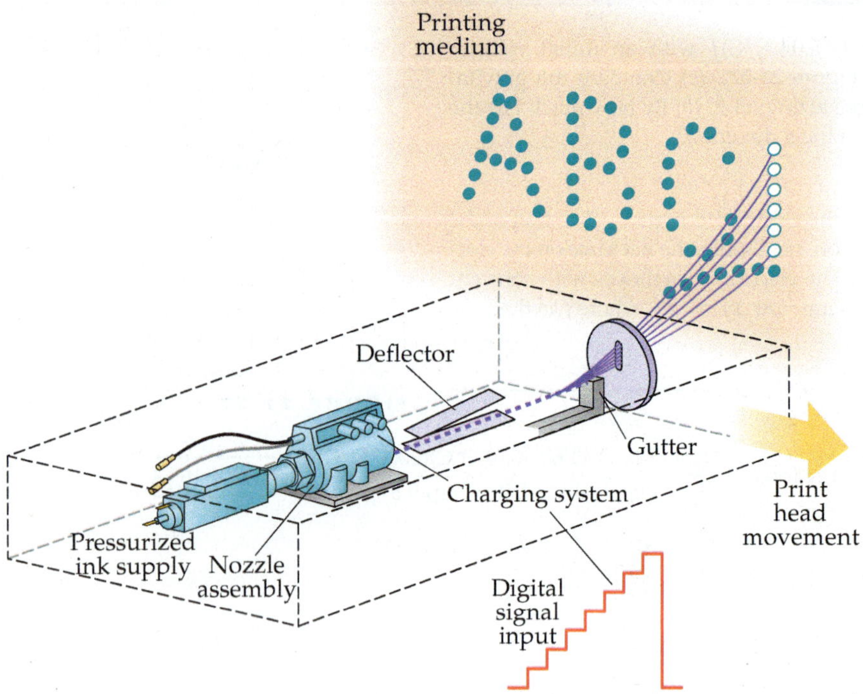

FIGURE 21-28 An ink-jet used for printing. The ink exits the nozzle in discrete droplets. Any droplet destined to form a dot on the image is given a charge. The deflector consists of a pair of oppositely charged plates. The greater the charge a drop receives, the higher the drop is deflected as it passes between the deflector plates. Drops that do not receive a charge are not deflected upward. These drops end up in the gutter, and the ink is returned to the ink reservoir. *(Courtesy of Videojet Systems International.)*

PICTURE The electric field $\vec{E}$ exerts a constant electric force $\vec{F}$ on the drop as it passes between the two plates, where $\vec{F} = q\vec{E}$. We are looking for E. We can get the force $\vec{F}$ by determining the mass and acceleration $\vec{F} = m\vec{a}$. The acceleration can be found from kinematics and mass can be found using the radius. Assume the density ρ of ink is 1000 kg/m^3 (the same as the density of water).

SOLVE

1. The electric field strength equals the force to charge ratio: $E = \dfrac{F}{q}$
2. The force, which is in the $+y$ direction (upward), equals the mass multiplied by the acceleration: $F = ma_y$
3. The vertical displacement is obtained using a constant-acceleration kinematic formula with $v_{0y} = 0$: $\Delta y = v_{0y}t + \frac{1}{2}a_y t^2 = 0 + \frac{1}{2}a_y t^2$
4. The time is how long it takes for the drop to travel the $\Delta x = 1.00$ cm at $v_0 = 40.0$ m/s: $\Delta x = v_{0x}t = v_0 t \quad \text{so} \quad t = \Delta x / v_0$
5. Solving for a_y gives: $a_y = \dfrac{2\Delta y}{t^2} = \dfrac{2\Delta y}{(\Delta x/v_0)^2} = \dfrac{2v_0^2\Delta y}{(\Delta x)^2}$
6. The mass equals the density multiplied by the volume: $m = \rho V = \rho\frac{4}{3}\pi r^3$
7. Solve for E:

$$E = \frac{F}{q} = \frac{ma}{q} = \frac{\rho\frac{4}{3}\pi r^3}{q}\frac{2v_0^2\Delta y}{(\Delta x)^2} = \frac{8\pi}{3}\frac{\rho r^3 v_0^2 \Delta y}{q(\Delta x)^2}$$

$$= \frac{8\pi}{3}\frac{(1000\text{ kg/m}^3)(20.0\times10^{-6}\text{ m})^3(40.0\text{ m/s})^2(3.00\times10^{-3}\text{ m})}{(2.00\times10^{-9}\text{ C})(0.0100\text{ m})^2} = \boxed{1.61\text{ kN/C}}$$

CHECK The units in last line of step 7 are kg·m/(C·s^2). The units work out because 1 N = 1 kg·m/s^2.

TAKING IT FURTHER The ink-jet in this example is called a multiple-deflection continuous ink-jet. It is used in some industrial printers. The low-cost ink-jet printers sold for home use do not use charged droplets deflected by an electric field.

DIPOLES IN ELECTRIC FIELDS

In Example 21-9 we found the electric field produced by a dipole, a system of two equal and opposite point charges that are close together. Here we consider the behavior of a dipole in an external electric field. Some molecules have permanent dipole moments due to a nonuniform distribution of charge within the molecule. Such molecules are called **polar molecules.** An example is HCl, which is essentially a positive hydrogen ion of charge $+e$ combined with a negative chloride ion of charge $-e$. The center of charge of the positive ion does not coincide with the center of charge for the negative ion, so the molecule has a permanent dipole moment. Another example is water (Figure 21-29).

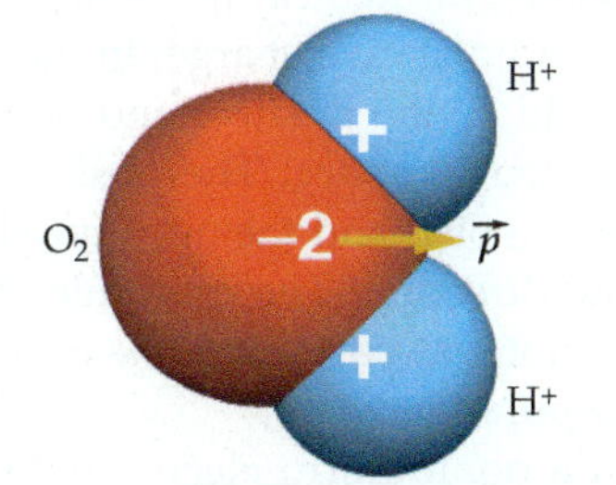

FIGURE 21-29 An H_2O molecule has a permanent dipole moment that points in the direction from the center of negative charge to the center of positive charge.

A uniform external electric field exerts no net force on a dipole, but it does exert a torque that tends to rotate the dipole so as to align it with the direction of the external field. We see in Figure 21-30 that the torque $\vec{\tau}$ calculated about the position of either charge has the magnitude $F_1 L \sin\theta = qEL \sin\theta = pE \sin\theta$.* The direction of the torque vector is into the paper such that it tends to rotate the dipole moment vector $\vec{p}$ so it aligns with the direction of $\vec{E}$. The torque can be expressed most concisely as the cross product:

$$\vec{\tau} = \vec{p} \times \vec{E} \qquad 21\text{-}11$$

If the dipole rotates through angle $d\theta$, the electric field does work:

$$dW = -\tau d\theta = -pE \sin\theta \, d\theta$$

(The minus sign arises because the torque opposes any increase in θ.) Setting the negative of this work value equal to the change in potential energy, we have

$$dU = -dW = +pE \sin\theta \, d\theta$$

Integrating, we obtain

$$U = -pE \cos\theta + U_0$$

If we choose the potential energy U to be zero when $\theta = 90°$, then $U_0 = 0$ and the potential energy of the dipole is

$$U = -pE \cos\theta = -\vec{p} \cdot \vec{E} \qquad 21\text{-}12$$

POTENTIAL ENERGY OF A DIPOLE IN AN ELECTRIC FIELD

Microwave ovens take advantage of the dipole moment of water molecules to cook food. Like other electromagnetic waves, microwaves have oscillating electric fields that exert torques on dipoles, torques that cause the water molecules to rotate with significant rotational kinetic energy. In this manner, energy is transferred from the microwave radiation to the water molecules at a high rate, accounting for the rapid cooking times that make microwave ovens so convenient.

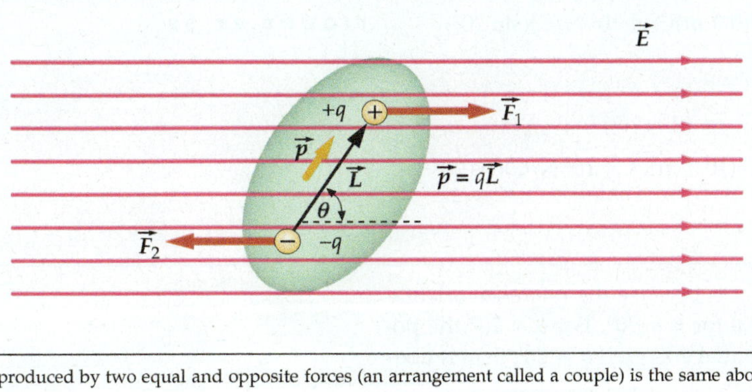

FIGURE 21-30 A dipole in a uniform electric field experiences equal and opposite forces that tend to rotate the dipole so that its dipole moment $\vec{p}$ is aligned with the electric field $\vec{E}$.

* The torque produced by two equal and opposite forces (an arrangement called a couple) is the same about any point in space.

Nonpolar molecules have no permanent dipole moment. However, all neutral molecules have equal amounts of positive and negative charge. In the presence of an external electric field $\vec{E}$, the positive and negative charge centers become separated in space. The positive charges are pushed in the direction of $\vec{E}$ and the negative charges are pushed in the opposite direction. The molecule thus acquires an induced dipole moment parallel to the external electric field and is said to be **polarized.**

In a nonuniform electric field, a dipole experiences a net force because the electric field has different magnitudes at the positive and negative charge centers. Figure 21-31 shows how a positive point charge polarizes a nonpolar molecule and then attracts it. A familiar example is the attraction that holds an electrostatically charged balloon against a wall. The nonuniform field produced by the charge on the balloon polarizes molecules in the wall and attracts them. An equal and opposite force is exerted by the wall molecules on the balloon.

The diameter of an atom or molecule is of the order of 10^{-12} m = 1 pm (one picometer). A convenient unit for the dipole moment of atoms and molecules is the fundamental charge e multiplied by the distance 1 pm. For example, the dipole moment of H_2O in these units has a magnitude of about 40 $e \cdot$pm.

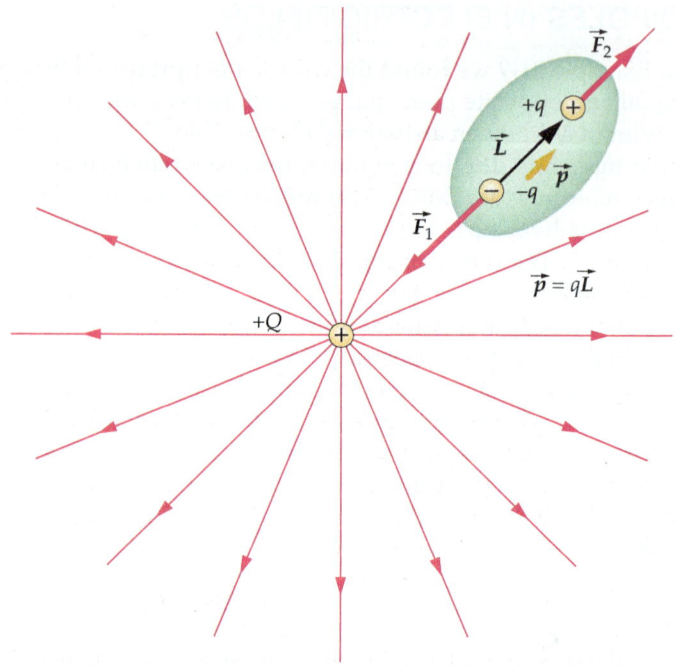

FIGURE 21-31 A nonpolar molecule in the nonuniform electric field of a positive point charge $+Q$. The point charge attracts the negative charges (the electrons) in the molecule and repels the positive charges (the protons). As a result the center of negative charge $-q$ is closer to $+Q$ than is the center of positive charge $+q$ and the induced dipole moment $\vec{p}$ is parallel to the field of the point charge. Because $-q$ is closer to $+Q$ than is $+q$, F_1 is greater than F_2 and the molecule is attracted to the point charge. In addition, if the point charge were negative, the induced dipole moment would be reversed, and the molecule would again be attracted to the point charge.

Example 21-14 Torque and Potential Energy

A polar molecule has a dipole moment of magnitude 20 $e \cdot$pm that makes an angle of 20° with a uniform electric field of magnitude 3.0×10^3 N/C (Figure 21-32). Find (*a*) the magnitude of the torque on the dipole, and (*b*) the potential energy of the system.

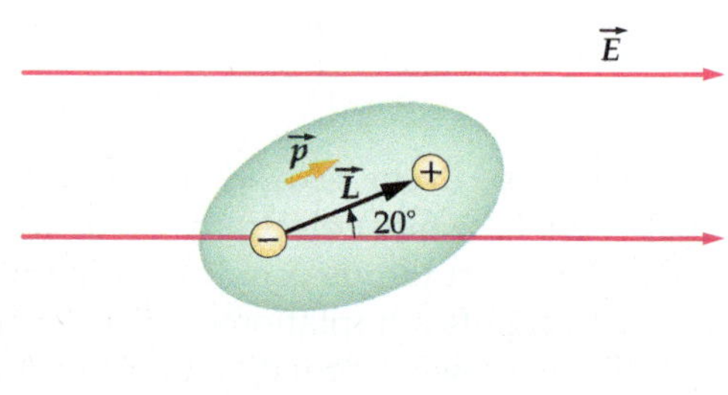

FIGURE 21-32

PICTURE The torque is found from $\vec{\tau} = \vec{p} \times \vec{E}$ and the potential energy is found from $U = -\vec{p} \cdot \vec{E}$.

SOLVE

1. Calculate the magnitude of the torque:

$$\tau = |\vec{p} \times \vec{E}| = pE\sin\theta = (20\,e\cdot\text{pm})(3 \times 10^3\ \text{N/C})(\sin 20°)$$
$$= (0.02)(1.6 \times 10^{-19}\ \text{C})(10^{-9}\ \text{m})(3 \times 10^3\ \text{N/C})(\sin 20°)$$
$$= \boxed{3.3 \times 10^{-27}\ \text{N}\cdot\text{m}}$$

2. Calculate the potential energy:

$$U = -\vec{p} \cdot \vec{E} = -pE\cos\theta$$
$$= -(0.02)(1.6 \times 10^{-19}\ \text{C})(10^{-9}\ \text{m})(3 \times 10^3\ \text{N/C})\cos 20°$$
$$= \boxed{-9.0 \times 10^{-27}\ \text{J}}$$

CHECK The sign of the potential energy is negative. That is because the reference orientation of the potential energy function $U = -\vec{p} \cdot \vec{E}$ is $U = 0$ for $\theta = 90°$. For $\theta = 20°$ the potential energy is less than zero. The system has more potential energy if $\theta = 20°$ than it does if $\theta = 90°$.

Physics Spotlight

Powder Coating—Industrial Static

Children around the world take advantage of triboelectric properties. The Ohio Art Company introduced the Etch A Sketch™ at about 1960.* Styrene beads provide a charge to very fine aluminum powder when shaken. The charged powder is attracted to the translucent screen of the toy. A stylus is then used to draw lines in the powder. The toy is based on the fact that the aluminum and screen attract each other with opposite charges.

A fine powder is attracted to the back of the screen by electrostatic. Turning the knobs results in the power being rubbed off by a small stylus. *(Courtesy of The Ohio Art Company.)*

Although charged powder can be a toy, it is serious business in many industries. Unprotected metal tends to corrode, so to prevent corrosion, metal parts of automobiles, metal appliances, and other metal objects are coated. In the past, coating involved paints, lacquers, varnishes, and enamels that were put on as liquids and dried. These liquids have disadvantages.† The solvents take a long time to dry or release unwanted volatile compounds. Surfaces at different angles can be coated unevenly. Liquid spray causes waste and cannot be easily recycled. Electrostatic powder coating reduces many of these problems.‡ This coating process was first used in the 1950s and is now popular with manufacturers adhering to environmental regulations by reducing the use of volatile chemicals.

Powder coating is applied by giving a charge to the item that will be coated.# To do this reliably, it is simplest if the object to be coated is conductive. Then very small (from 1 μm to 100 μm) particles° in a powder are given an opposite charge. The coating particles are strongly attracted to the object to be coated. Loose particles can be recycled and used again. Once the particles are on the object, the coating is then cured, either by increased temperature or by ultraviolet light. The curing process locks the molecules of the coating together, and the particles and the object lose their charges.

Coating particles are given a charge by either corona discharge or triboelectric charging.§ Corona discharge blows the particles through a plasma of electrons, giving them a negative charge. Triboelectric charging blows the particles through a tube that is made from a material on the opposite end of the triboelectric spectrum, often Teflon. The coating particles are given a positive charge from this rapid contact. The item to be coated is given a charge that depends on the coating method used. Depending on the coating and additives, coating charges range from 500 to 1000 μC/kg.¶ The curing process differs according to the coating materials and the coated item. The curing time can be anywhere from 1 to 30 minutes.**

Although powder coating is economical and environmentally friendly, it has difficulties. The abilities of the coating particles to hold a charge†† can vary with humidity, which must be precisely controlled.‡‡ If the electric field for corona discharge is too strong, the powder sprays too quickly toward the item to be coated, leaving a bare spot in the middle of a built-up ring, which gives an uneven "orange peel" finish.## Electrostatic powders can be child's play, but electrostatic powder coating is a complex, useful, and evolving process.

* Grandjean, A., "Tracing Device." *U.S. Patent No. 3,055,113*, Sept. 25, 1962.
† Matheson, R. D. "20th- to 21st-Century Technological Challenges in Soft Coatings." *Science*, Aug. 9, 2002, Vol. 297, No. 5583, pp. 976–979.
‡ Hammerton, D., and Buysens, K., "UV-Curable Powder Coatings: Benefits and Performance." *Paint and Coatings Industry*, Aug. 2000, p. 58.
# Zeren, S., and Renoux, D., "Powder Coatings Additives." *Paint and Coatings Industry*, Oct. 2002, p. 116.
° Hemphill, R., "Deposition of $BaTiO_3$ Nanoparticles by Electrostatic Spray Powder Charging." *Paint and Coatings Industry*, Apr. 2006, pp. 74–78.
§ Czyzak, S. J., and Williams, D. T., "Static Electrification of Solid Particles by Spraying." *Science*, Jul. 20, 1951, Vol 14, pp. 66–68.
¶ Zeren, S., and Renoux, D., op. cit.
** Hammerton, D., and Buysens, K., op. cit.
†† O'Konski, C. T., "The Exponential Decay Law in Spray De-electrification." *Science*, Oct. 5, 1951, Vol. 114, p. 368.
‡‡ Sharma, R., et al., "Effect of Ambient Relative Humidity and Surface in Modification on the Charge Decay Properties of Polymer Powders in Powder Coating." *IEEE Transactions on Industry Applications*, Jan./Feb. 2003, Vol. 39, No. 1, pp. 87–95.
Wostratzky, D., Lord, S., and Sitzmann, E. V., "Power!" *Paint and Coatings Industry*, Oct. 2000, p. 54.

Summary

1. Quantization and conservation are fundamental properties of electric charge.
2. Coulomb's law is the fundamental law of interaction between charges at rest.
3. The electric field describes the condition in space set up by a charge distribution.

TOPIC	RELEVANT EQUATIONS AND REMARKS	
1. Charge	There are two kinds of charge, positive and negative. Charges of like sign repel, those of opposite sign attract.	
Quantization	Charge is quantized—it always occurs in integer multiples of the fundamental charge unit e. The charge of the electron is $-e$ and that of the proton is $+e$.	
Magnitude	$e = 1.60 \times 10^{-19}$ C	
Conservation	Charge is conserved. When charged particles are created or annihilated, the total amount of charge carried by the created or annihilated particles is zero.	
2. Conductors and Insulators	In metals, about one electron per atom is delocalized (free to move about the entire material). In insulators, all the electrons are bound to nearby atoms.	
Ground	A very large conductor (such as Earth) that can supply or absorb a virtually unlimited amount of charge is called a ground.	
3. Charging by Induction	To charge a conductor by induction: connect a ground to the conductor, hold an external charge near the conductor (to attract or repel the conduction electrons), then disconnect the conductor from ground, and finally move the external charge away from the conductor.	
4. Coulomb's Law	The force exerted by point charge q_1 on point charge q_2 a distance r_{12} away is given by $\vec{F}_{12} = \frac{kq_1q_2}{r_{12}^2}\hat{r}_{12}$ where unit vector $\hat{r}_{12}$ points from q_1 toward q_2.	21-4
Coulomb constant	$k = 8.99 \times 10^9\ \text{N}\cdot\text{m}^2/\text{C}^2$	21-3
5. Electric Field	The electric field due to a system of charges at a point is defined as the net force $\vec{F}$, exerted by those charges on a very small positive test charge q_0, divided by q_0: $\vec{E} = \frac{\vec{F}}{q_0}$	21-5
Due to a point charge	$\vec{E}_{iP} = \frac{kq_i}{r_{iP}^2}\hat{r}_{iP}$	21-7
Due to a system of point charges	The electric field at P due to several charges is the vector sum of the fields at P due to the individual charges: $\vec{E}_P = \sum_i \vec{E}_{iP}$	21-8
6. Electric Field Lines	The electric field can be represented by electric field lines that emanate from positive charges and terminate on negative charges. The strength of the electric field is indicated by the density of the electric field lines.	
7. Dipole	A dipole is a system of two equal but opposite charges separated by a small distance.	
Dipole moment	$\vec{p} = q\vec{L}$ where $\vec{L}$ is the position of the positive charge relative to the negative charge.	21-9
Field due to dipole	The electric field strength far from a dipole is proportional to the magnitude of the dipole moment and decreases with the cube of the distance.	
Torque on a dipole	In a uniform electric field, the net force on a dipole is zero, but there is a torque that tends to align the dipole in the direction of the field. $\vec{\tau} = \vec{p} \times \vec{E}$	21-11

TOPIC	RELEVANT EQUATIONS AND REMARKS	
Potential energy of a dipole	$U = -\vec{p}\cdot\vec{E} + U_0$ where U_0 is usually taken to be zero.	21-12
8. **Polar and Nonpolar Molecules**	Polar molecules, such as H_2O and HCl, have permanent dipole moments because their centers of positive and negative charge do not coincide. They behave like simple dipoles in an electric field. Nonpolar molecules do not have permanent dipole moments, but they acquire induced dipole moments in the presence of an electric field.	

Answers to Concept Checks

21-1 (*a*) $+\frac{1}{2}Q$. Because the spheres are identical, they must share the total charge equally. (*b*) $+2Q$, which is necessary to satisfy the conservation of charge

21-2 $Q_1 = +Q/2$, $Q_2 = -Q/4$, and $Q_3 = -Q/4$

Answers to Practice Problems

21-1 $N = Q/e = (50 \times 10^{-9}\,\text{C})/(1.6 \times 10^{-19}\,\text{C}) = 3.1 \times 10^{11}$. Charge quantization cannot be detected in a charge of this size; even adding or subtracting a million electrons produces a negligibly small effect.

21-2 About 3.5×10^{-8} percent

21-3 2.25×10^{-3} N

21-4 $+(6.3\ \mu\text{N})\hat{i}$

21-5 $\hat{r}_{10} = (\hat{i} + \hat{j})/\sqrt{2}$

21-6 No, but suppose it were. Because the x component of $\vec{r}_{10}$ is less than the magnitude of $\vec{r}_{10}$, the denominator of kq_1q_0/x_{10}^2 is less than the denominator of kq_1q_0/r_{10}^2. This would imply that the x component of $\vec{F}_{10}$ is greater than the magnitude of $\vec{F}_{10}$, an impossibility because the component of a vector is never greater than the magnitude of the vector. Therefore, the x component of the force $\vec{F}_{10} = (kq_1q_0/r_{10}^2)\hat{r}_{10}$ is not necessarily equal to $F_{10x} = kq_1q_0/x_{10}^2$.

21-7 $\vec{E} = \vec{F}/q_0 = (4.0 \times 10^4\ \text{N/C})\hat{i}$

21-8 $\vec{F} = -(6.4 \times 10^{-15}\,\text{N})\hat{i}$

21-9 $x = 1.80$ m

Problems

In a few problems, you are given more data than you actually need; in a few other problems, you are required to supply data from your general knowledge, outside sources, or informed estimate.

Interpret as significant all digits in numerical values that have trailing zeros and no decimal points.

- • Single-concept, single-step, relatively easy
- •• Intermediate-level, may require synthesis of concepts
- ••• Challenging
- SSM Solution is in the *Student Solutions Manual*
- Consecutive problems that are shaded are paired problems.

CONCEPTUAL PROBLEMS

1 • Objects are composed of atoms which are composed of charged particles (protons and electrons); however, we rarely observe the effects of the electrostatic force. Explain why we do not observe these effects.

2 • A carbon atom can become a carbon *ion* if it has one or more of its electrons removed during a process called *ionization.* What is the net charge on a carbon atom that has had two of its electrons removed? (*a*) $+e$, (*b*) $-e$, (*c*) $+2e$, (*d*) $-2e$

3 •• You do a simple demonstration for your high school physics teacher in which you claim to disprove Coulomb's law. You first run a rubber comb through your dry hair, then use it to attract tiny neutral pieces of paper on the desk. You then say, "Coulomb's law states that for there to be electrostatic forces of attraction between two objects, both objects need to be charged. However, the paper was not charged. So according to Coulomb's law, there should be no electrostatic forces of attraction between them, yet there clearly was." You rest your case. (*a*) What is wrong with your assumptions? (*b*) Does attraction between the paper and the comb require that the net charge on the comb be negative? Explain your answer.

4 •• You have a positively charged insulating rod and two metal spheres on insulating stands. Give step-by-step directions of how the rod, without actually touching either sphere, can be used to give one of the spheres (*a*) a negative charge and (*b*) a positive charge.

5 •• (*a*) Two point particles that have charges of $+4q$ and $-3q$ are separated by distance d. Use field lines to draw a visualization of the electric field in the neighborhood of this system. (*b*) Draw the field lines at distances much greater than d from the charges.

6 •• A metal sphere is positively charged. Is it possible for the sphere to electrically attract another positively charged ball? Explain your answer.

7 •• A simple demonstration of electrostatic attraction can be done by dangling a small ball of crumpled aluminum foil on a string and bringing a charged rod near the ball. The ball initially will be attracted to the rod, but once they touch, the ball will be strongly repelled from it. Explain these observations.

8 •• Two positive point charges that are equal in magnitude are fixed in place, one at $x = 0.00$ m and the other at $x = 1.00$ m, on the x axis. A third positive point charge is placed at an equilibrium position. (*a*) Where is this equilibrium position? (*b*) Is the equilibrium position stable if the third particle is constrained to move parallel with the x axis? (*c*) What if it is constrained to move parallel with the y axis? Explain your answer.

9 •• Two neutral conducting spheres are in contact and are supported on a large wooden table by insulated stands. A positively charged rod is brought up close to the surface of one of the spheres on the side opposite its point of contact with the other sphere. (*a*) Describe the induced charges on the two conducting spheres and sketch the charge distributions on them. (*b*) The two spheres are separated and then the charged rod is removed. The spheres are then separated far apart. Sketch the charge distributions on the separated spheres.

10 •• Three point charges, $+q$, $+Q$, and $-Q$, are placed at the corners of an equilateral triangle as shown in Figure 21-33. No other charged objects are nearby. (*a*) What is the direction of the net force on charge $+q$ due to the other two charges? (*b*) What is the total electric force on the system of three charges? Explain.

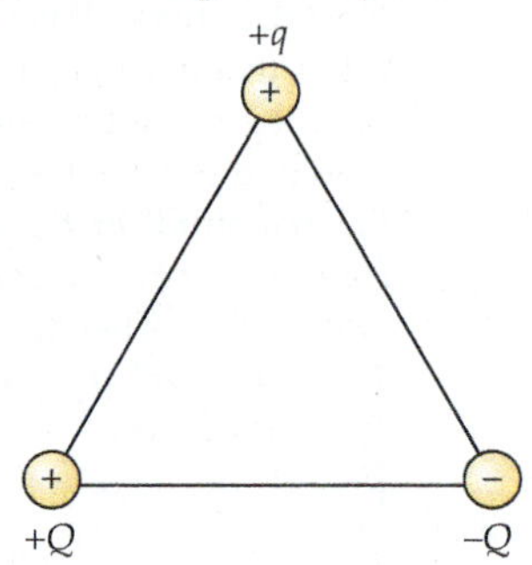

FIGURE 21-33 Problem 10

11 •• A positively charged particle is free to move in a region with a nonzero electric field $\vec{E}$. Which statement(s) must be true?

(*a*) The particle is accelerating in the direction perpendicular to $\vec{E}$.
(*b*) The particle is accelerating in the direction of $\vec{E}$.
(*c*) The particle is moving in the direction of $\vec{E}$.
(*d*) The particle could be momentarily at rest.
(*e*) The force on the particle is opposite the direction of $\vec{E}$.
(*f*) The particle is moving opposite the direction of $\vec{E}$.

12 •• Four charges are fixed in place at the corners of a square as shown in Figure 21-34. No other charges are nearby. Which of the following statements is true?

(*a*) $\vec{E}$ is zero at the midpoints of all four sides of the square.
(*b*) $\vec{E}$ is zero at the center of the square.
(*c*) $\vec{E}$ is zero midway between the top two charges and midway between the bottom two charges.

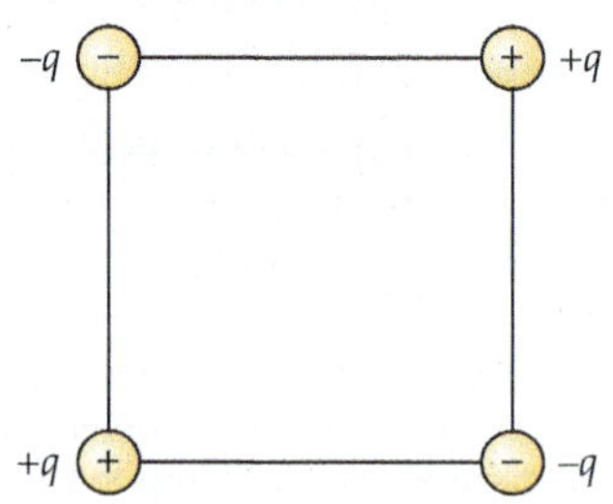

FIGURE 21-34 Problem 12

13 •• Two point particles that have charges of $+q$ and $-3q$ are separated by distance d. (*a*) Use field lines to sketch the electric field in the neighborhood of this system. (*b*) Draw the field lines at distances much greater than d from the charges. SSM

14 •• Three equal positive point charges (each charge $+q$) are fixed at the vertices of an equilateral triangle that has sides of length a. The origin is at the midpoint of one side the triangle, the center of the triangle on the x axis at $x = x_1$, and the vertex opposite the origin is on the x axis at $x = x_2$. (*a*) Express x_1 and x_2 in terms of a. (*b*) Write an expression for the electric field on the x axis a distance x from the origin on the interval $0\ x < x_2$. (*c*) Show that the expression you obtained in (*b*) gives the expected results for $x = 0$ and for $x = x_1$.

15 •• A molecule has a dipole moment given by $\vec{p}$. The molecule is momentarily at rest with $\vec{p}$ making an angle θ with a uniform electric field $\vec{E}$. Describe the subsequent motion of the dipole moment.

16 •• True or false:

(*a*) The electric field of a point charge always points away from the charge.
(*b*) The electric force on a charged particle in an electric field is always in the same direction as the field.
(*c*) Electric field lines never intersect.
(*d*) All molecules have dipole moments in the presence of an external electric field.

17 •• Two molecules have dipole moments of equal magnitude. The dipole moments are oriented in various configurations as shown in Figure 21-35. Determine the electric-field direction at each of the numbered locations. Explain your answers. SSM

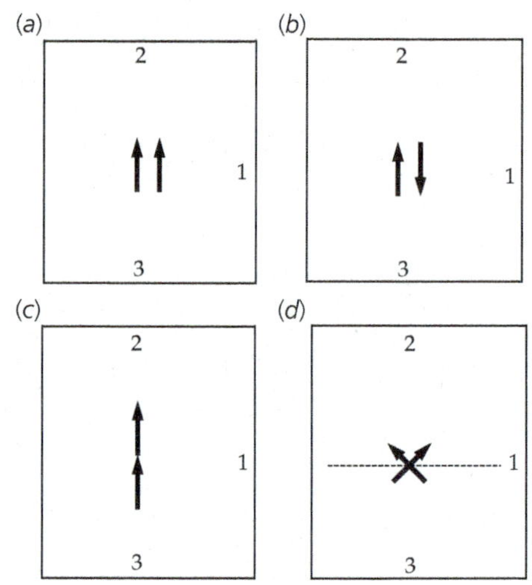

FIGURE 21-35 Problem 17

ESTIMATION AND APPROXIMATION

18 •• Estimate the force required to bind the two protons in the He nucleus together. *Hint: Model the protons as point charges. You will need to have an estimate of the distance between them.*

19 •• A popular classroom demonstration consists of rubbing a plastic rod with fur to give the rod charge, and then placing the rod near an empty soda can that is on its side (Figure 21-36). Explain why the can will roll toward the rod.

FIGURE 21-36 Problem 19

20 •• Sparks in air occur when ions in the air are accelerated to such a high speed by an electric field that when the ions impact on neutral gas molecules, the neutral molecules become ions. If the electric field strength is large enough, the ionized collision products are themselves accelerated and produce more ions on impact, and so forth. This avalanche of ions is what we call a spark. (*a*) Assume that an ion moves, on average, exactly one mean free path through the air before hitting a molecule. If the ion needs to acquire approximately 1.0 eV of kinetic energy in order to ionize a molecule, estimate the minimum strength of the electric field required at standard room pressure and temperature. Assume that the cross-sectional area of an air molecule is about 0.10 nm^2. (*b*) How does the strength of the electric field in Part (*a*) depend on temperature? (*c*) How does the strength of the electric field in Part (*a*) depend on pressure?

CHARGE

21 • A plastic rod is rubbed against a wool shirt, thereby acquiring a charge of $-0.80\ \mu C$. How many electrons are transferred from the wool shirt to the plastic rod?

22 • A charge equal to the charge of Avogadro's number of protons ($N_A = 6.02 \times 10^{23}$) is called a *faraday*. Calculate the number of coulombs in a faraday.

23 • What is the total charge of all of the protons in 1.00 kg of carbon? SSM

24 •• Suppose a cube of aluminum which is 1.00 cm on a side accumulates a net charge of +2.50 pC. (*a*) What percentage of the electrons originally in the cube was removed? (*b*) By what percentage has the mass of the cube decreased because of this removal?

25 •• During a process described by the *photoelectric effect*, ultraviolet light can be used to charge a piece of metal. (*a*) If such light is incident on a slab of conducting material and electrons are ejected with enough energy that they escape the surface of the metal, how long before the metal has a net charge of +1.50 nC if 1.00×10^6 electrons are ejected per second? (*b*) If 1.3 eV is needed to eject an electron from the surface, what is the power rating of the light beam? (Assume this process is 100% efficient.)

COULOMB'S LAW

26 • A point charge $q_1 = 4.0\ \mu C$ is at the origin and a point charge $q_2 = 6.0\ \mu C$ is on the x axis at $x = 3.0$ m. (*a*) Find the electric force on charge q_2. (*b*) Find the electric force on q_1. (*c*) How would your answers for Parts (*a*) and (*b*) differ if q_2 were $-6.0\ \mu C$?

27 • Three point charges are on the x axis: $q_1 = -6.0\ \mu C$ is at $x = -3.0$ m, $q_2 = 4.0\ \mu C$ is at the origin, and $q_3 = -6.0\ \mu C$ is at $x = 3.0$ m. Find the electric force on q_1. SSM

28 •• A 2.0-μC point charge and a 4.0-μC point charge are a distance L apart. Where should a third point charge be placed so that the electric force on that third charge is zero?

29 •• A -2.0-μC point charge and a 4.0-μC point charge are a distance L apart. Where should a third point charge be placed so that the electric force on that third charge is zero?

30 •• Three point charges, each of magnitude 3.00 nC, are at separate corners of a square of edge length 5.00 cm. The two point charges at opposite corners are positive, and the third point charge is negative. Find the force exerted by these point charges on a fourth point charge $q_4 = +3.00$ nC at the remaining corner.

31 •• A point charge of 5.00 μC is on the y axis at $y = 3.00$ cm, and a second point charge of $-5.00\ \mu C$ is on the y axis at $y = -3.00$ cm. Find the electric force on a point charge of 2.00 μC on the x axis at $x = 8.00$ cm.

32 •• A point particle that has a charge of $-2.5\ \mu C$ is located at the origin. A second point particle that has a charge of 6.0 μC is at $x = 1.0$ m, $y = 0.50$ m. A third point particle, an electron, is at a point that has coordinates (x, y). Find the values of x and y such that the electron is in equilibrium.

33 •• A point particle that has a charge of $-1.0\ \mu C$ is located at the origin; a second point particle that has a charge of 2.0 μC is located at $x = 0$, $y = 0.10$ m; and a third point particle that has a charge of 4.0 μC is located at $x = 0.20$ m, $y = 0$. Find the total electric force on each of the three point charges.

34 •• A point particle that has a charge of 5.00 μC is located at $x = 0$, $y = 0$ and a point particle that has a charge q is located at $x = 4.00$ cm, $y = 0$. The electric force on a point particle at $x = 8.00$ cm, $y = 0$ that has a charge of 2.00 μC is $-(19.7\ N)\hat{i}$. Determine the charge q.

35 ••• Five identical point charges, each having charge Q, are equally spaced on a semicircle of radius R as shown in Figure 21-37. Find the force (in terms of k, Q, and R) on a charge q located equidistant from the five other charges. SSM

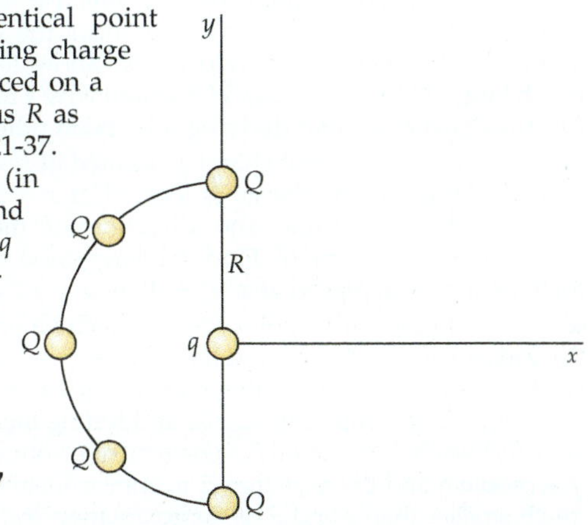

FIGURE 21-37 Problem 35

36 ••• The structure of the NH_3 molecule is approximately that of an equilateral tetrahedron, with three H^+ ions forming the base and an N^{3-} ion at the apex of the tetrahedron. The length of each side is 1.64×10^{-10} m. Calculate the electric force that acts on each ion.

THE ELECTRIC FIELD

37 • A point charge of 4.0 μC is at the origin. What are the magnitude and direction of the electric field on the x axis at (*a*) $x = 6.0$ m and (*b*) $x = -10$ m? (*c*) Sketch the function E_x versus x for both positive and negative values of x. (Remember that E_x is negative when $\vec{E}$ points in the $-x$ direction.) SSM

38 • Two point charges, each $+4.0\ \mu C$, are on the x axis; one point charge is at the origin and the other is at $x = 8.0$ m. Find the electric field on the x axis at (*a*) $x = -2.0$ m, (*b*) $x = 2.0$ m, (*c*) $x = 6.0$ m, and (*d*) $x = 10$ m. (*e*) At what point on the x axis is the electric field zero? (*f*) Sketch E_x versus x for $-3.0\ m < x < 11$ m.

39 • When a 2.0-nC point charge is placed at the origin, it experiences an electric force of 8.0×10^{-4} N in the $+y$ direction. (*a*) What is the electric field at the origin? (*b*) What would be the electric force on a -4.0-nC point charge placed at the origin? (*c*) If this force is due to the electric field of a point charge on the y axis at $y = 3.0$ cm, what is the value of that charge?

40 • The electric field near the surface of Earth points downward and has a magnitude of 150 N/C. (*a*) Compare magnitude of the upward electric force on an electron with the magnitude of gravitational force on the electron. (*b*) What charge should be placed on a Ping-Pong ball of mass 2.70 g so that the electric force balances the weight of the ball near Earth's surface?

41 •• Two point charges q_1 and q_2 both have a charge equal to +6.0 nC and are on the y axis at $y_1 = +3.0$ cm and $y_2 = -3.0$ cm, respectively. (*a*) What are the magnitude and direction of the electric field on the x axis at $x = 4.0$ cm? (*b*) What is the force exerted on a third charge $q_0 = 2.0$ nC when it is placed on the x axis at $x = 4.0$ cm? SSM

42 •• A point charge of $+5.0\ \mu C$ is located on the x axis at $x = -3.0$ cm, and a second point charge of $-8.0\ \mu C$ is located on the x axis at $x = +4.0$ cm. Where should a third charge of $+6.0\ \mu C$ be placed so that the electric field at the origin is zero?

43 •• A -5.0-μC point charge is located at $x = 4.0$ m, $y = -2.0$ m, and a 12-μC point charge is located at $x = 1.0$ m, $y = 2.0$ m. (*a*) Find the magnitude and direction of the electric field at $x = -1.0$ m, $y = 0$. (*b*) Calculate the magnitude and direction of the electric force on an electron that is placed at $x = -1.0$ m, $y = 0$.

44 •• Two equal positive charges q are on the y axis; one point charge is at $y = +a$ and the other is at $y = -a$. (*a*) Show that on the x axis the x component of the electric field is given by $E_x = 2kqx/(x^2 + a^2)^{3/2}$. (*b*) Show that near the origin, where x is much smaller than a, $E_x \approx 2kqx/a^3$. (*c*) Show that for values of x much larger than a, $E_x \approx 2kq/x^2$. Explain why a person might expect this result even without deriving it by taking the appropriate limit.

45 •• A 5.0-μC point charge is located at $x = 1.0$ m, $y = 3.0$ m, and a -4.0-μC point charge is located at $x = 2.0$ m, $y = -2.0$ m. (*a*) Find the magnitude and direction of the electric field at $x = -3.0$ m, $y = 1.0$ m. (*b*) Find the magnitude and direction of the force on a proton placed at $x = -3.0$ m, $y = 1.0$ m.

46 •• Two positive point charges, each having charge Q, are on the y axis—one at $y = +a$ and the other at $y = -a$. (*a*) Show that the electric field strength on the x axis is greatest at $x = a/\sqrt{2}$ and $x = -a/\sqrt{2}$ by computing $\partial E_x/\partial x$ and setting the derivative equal to zero. (*b*) Sketch the function E_x versus x using your results for Part (*a*) of this problem and the facts that E_x is approximately $2kqx/a^3$ when x is much smaller than a and E_x is approximately $2kq/x^2$ when x is much larger than a.

47 •• Two point particles, each having a charge q, sit on the base of an equilateral triangle that has sides of length L as shown in Figure 21-38. A third point particle that has a charge equal to $2q$ sits at the apex of the triangle. Where must a fourth point particle that has a charge equal to q be placed in order that the electric field at the center of the triangle be zero? (The center is in the plane of the triangle and equidistant from the three vertices.) SSM

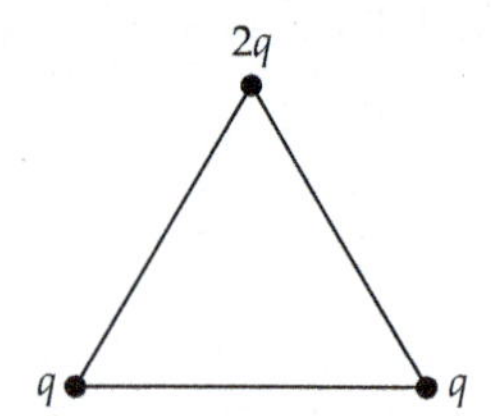

FIGURE 21-38 Problems 47 and 48

48 •• Two point particles, each having a charge equal to q, sit on the base of an equilateral triangle that has sides of length L as shown in Figure 21-38. A third point particle that has a charge equal to $2q$ sits at the apex of the triangle. A fourth point particle that has charge q' is placed at the midpoint of the baseline making the electric field at the center of the triangle equal to zero. What is the value of q'? (The center is in the plane of the triangle and equidistant from all three vertices.)

49 •• Two equal positive point charges $+q$ are on the y axis; one is at $y = +a$ and the other is at $y = -a$. The electric field at the origin is zero. A test charge q_0 placed at the origin will therefore be in equilibrium. (*a*) Discuss the stability of the equilibrium for a positive test charge by considering small displacements from equilibrium along the x axis and small displacements along the y axis. (*b*) Repeat Part (*a*) for a negative test charge. (*c*) Find the magnitude and sign of a charge q_0 that when placed at the origin results in a net force of zero on each of the three charges.

50 ••• Two positive point charges $+q$ are on the y axis at $y = +a$ and $y = -a$. A bead of mass m and charge $-q$ slides without friction along a taut thread that runs along the x axis. Let x be the position of the bead. (*a*) Show that for $x \ll a$, the bead experiences a linear restoring force (a force that is proportional to x and directed toward the equilibrium position at $x = 0$) and therefore undergoes simple harmonic motion. (*b*) Find the period of the motion.

POINT CHARGES IN ELECTRIC FIELDS

51 •• The acceleration of a particle in an electric field depends on q/m (the charge-to-mass ratio of the particle). (*a*) Compute q/m for an electron. (*b*) What are the magnitude and direction of the acceleration of an electron in a uniform electric field that has a magnitude of 100 N/C? (*c*) Compute the time it takes for an electron placed at rest in a uniform electric field that has a magnitude of 100 N/C to reach a speed 0.01c. (When the speed of an electron approaches the speed of light c, relativistic kinematics must be used to calculate its motion, but at speeds of 0.01c or less, nonrelativistic kinematics is sufficiently accurate for most purposes.) (*d*) How far does the electron travel in that time? SSM

52 • The acceleration of a particle in an electric field depends on the charge-to-mass ratio of the particle. (*a*) Compute q/m for a proton, and find its acceleration in a uniform electric field that has a magnitude of 100 N/C. (*b*) Find the time it takes for a proton initially at rest in such a field to reach a speed of 0.01c (where c is the speed of light). (When the speed of a proton approaches the speed of light c, relativistic kinematics must be used to calculate its motion, but at speeds of 0.01c or less, nonrelativistic kinematics is sufficiently accurate for most purposes.)

53 • An electron has an initial velocity of 2.00×10^6 m/s in the $+x$ direction. It enters a region that has a uniform electric field $\vec{E} = (300\ \text{N/C})\hat{j}$. (*a*) Find the acceleration of the electron. (*b*) How long does it take for the electron to travel 10.0 cm in the $+x$ direction in the region that has the field? (*c*) Through what angle, and in what direction, is the electron deflected while traveling the 10.0 cm in the x direction?

54 •• An electron is released from rest in a weak electric field given by $\vec{E} = -1.50 \times 10^{-10}\ \text{N/C}\hat{j}$. After the electron has traveled a vertical distance of 1.0 μm, what is its speed? (Do not neglect the gravitational force on the electron.)

55 •• A 2.00-g charged particle is released from rest in a region that has a uniform electric field $\vec{E} = (300\ \text{N/C})\hat{i}$. After traveling a distance of 0.500 m in this region, the particle has a kinetic energy of 0.120 J. Determine the charge of the particle.

56 •• A charged particle leaves the origin with a speed of 3.00×10^6 m/s at an angle of 35° above the x axis. A uniform electric field, given by $\vec{E} = -E_0\hat{j}$, exists throughout the region. Find E_0 such that the particle will cross the x axis at $x = 1.50$ cm if the particle is (*a*) an electron and (*b*) a proton.

57 •• An electron starts at the position shown in Figure 21-39 with an initial speed $v_0 = 5.00 \times 10^6$ m/s at 45° to the x axis. The electric field is in the $+y$ direction and has a magnitude of 3.50×10^3 N/C. The black lines in the figure are charged metal plates. On which plate and at what location will the electron strike? SSM

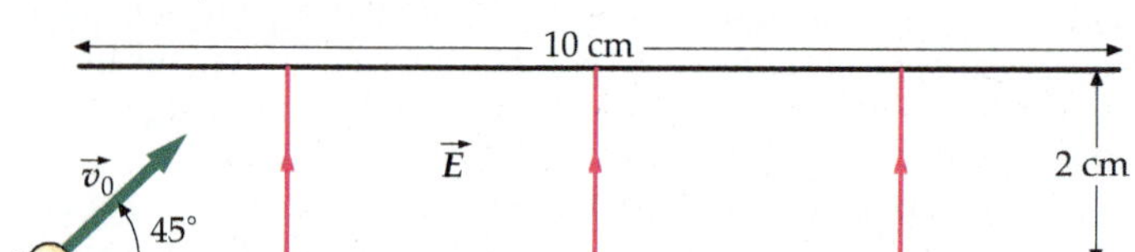

FIGURE 21-39 Problem 57

58 •• **ENGINEERING APPLICATION** An electron that has a kinetic energy equal to 2.00×10^{-16} J is moving to the right along the axis of a cathode-ray tube as shown in Figure 21-40. An electric field $\vec{E} = (2.00 \times 10^4\ \text{N/C})\hat{j}$ exists in the region between the deflection plates, and no electric field ($\vec{E} = 0$) exists outside this region. (*a*) How far is the electron from the axis of the tube when it exits the region between the plates? (*b*) At what angle is the electron moving, with respect to the axis, after exiting the region between the plates? (*c*) At what distance from the axis will the electron strike the fluorescent screen?

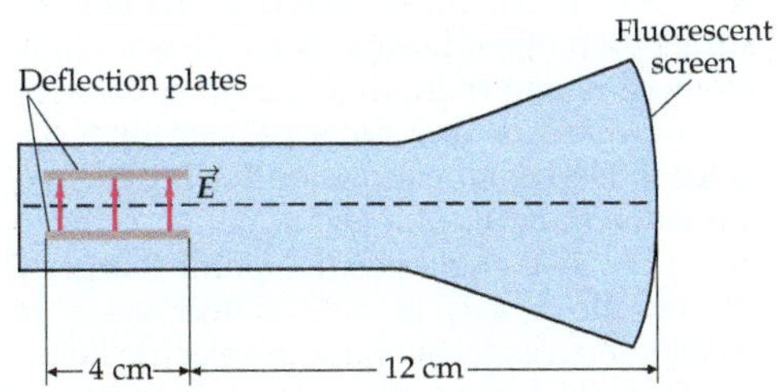

FIGURE 21-40 Problem 58

DIPOLES

59 • Two point charges, $q_1 = 2.0$ pC and $q_2 = -2.0$ pC, are separated by 4.0 μm. (*a*) What is the magnitude of the dipole moment of this pair of charges? (*b*) Sketch the pair and show the direction of the dipole moment.

60 • A dipole of moment 0.50 $e \cdot$nm is placed in a uniform electric field that has a magnitude of 4.0×10^4 N/C. What is the magnitude of the torque on the dipole when (*a*) the dipole is aligned with the electric field, (*b*) the dipole is transverse to (perpendicular to) the electric field, and (*c*) the direction of dipole makes an angle of 30° with the direction of electric field? (*d*) Defining the potential energy to be zero when the dipole is transverse to the electric field, find the potential energy of the dipole for the orientations specified in Parts (*a*) and (*c*).

GENERAL PROBLEMS

61 • Show that it is only possible to place one isolated proton in an ordinary empty coffee cup by considering the following situation. Assume the first proton is fixed at the bottom of the cup. Determine the distance directly above this proton where a second proton would be in equilibrium. Compare this distance to the depth of an ordinary coffee cup to complete the argument. SSM

62 •• Point charges of $-5.00\ \mu$C, $+3.00\ \mu$C, and $+5.00\ \mu$C are located on the x axis at $x = -1.00$ cm, $x = 0$, and $x = +1.00$ cm, respectively. Calculate the electric field on the x axis at $x = 3.00$ cm and at $x = 15.0$ cm. Are there any points on the x axis where the magnitude of the electric field is zero? If so, where are those points?

63 •• Point charges of $-5.00\ \mu$C and $+5.00\ \mu$C are located on the x axis at $x = -1.00$ cm and $x = +1.00$ cm, respectively. (*a*) Calculate the electric field strength at $x = 10.00$ cm. (*b*) Estimate the electric field strength at $x = 10.00$ cm by modeling the two charges as an electric dipole located at the origin and using $E = 2kp/|x|^3$ (Equation 21-10). Compare your result with the result obtained in Part (*a*), and explain the reason for the difference between the two results.

64 •• A fixed point charge of $+2q$ is connected by strings to point charges of $+q$ and $+4q$, as shown in Figure 21-41. Find the tensions T_1 and T_2.

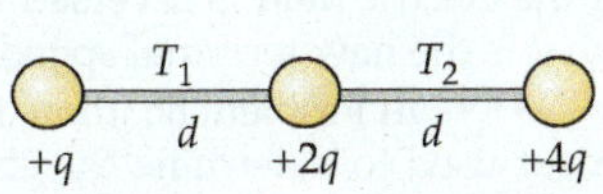

FIGURE 21-41 Problem 64

65 •• A positive charge Q is to be divided into two positive point charges q_1 and q_2. Show that, for a given separation D, the force exerted by one charge on the other is greatest if $q_1 = q_2 = \frac{1}{2}Q$. SSM

66 •• A point charge Q is located on the x axis at $x = 0$, and a point charge $4Q$ is located at $x = 12.0$ cm. The electric force on a point charge of $-2.00\ \mu$C is zero if that charge is placed at $x = 4.00$ cm, and is 126 N in the $+x$ direction if placed at $x = 8.00$ cm. Determine the charge Q.

67 •• Two point particles separated by 0.60 m have a total charge of 200 μC. (*a*) If the two particles repel each other with a force of 80 N, what is the charge on each of the two particles? (*b*) If the two particles attract each other with a force of 80 N, what are the charges on the two particles?

68 •• A point particle that has charge $+q$ and unknown mass m is released from rest in a region that has a uniform electric field $\vec{E}$ that is directed vertically downward. The particle hits the ground at a speed $v = 2\sqrt{gh}$, where h is the initial height of the particle. Find m in terms of E, q, and g.

69 •• A rigid 1.00-m-long rod is pivoted about its center (Figure 21-42). A charge $q_1 = 5.00 \times 10^{-7}$ C is placed on one end of the rod, and a charge $q_2 = -q_1$ is placed a distance $d = 10.0$ cm directly below it. (*a*) What is the force exerted by q_2 on q_1? (*b*) What is the torque (measured about the rotation axis) due to that force? (*c*) To counterbalance the attraction between the two charges, we hang a block 25.0 cm from the pivot as shown. What value should we choose for the mass m of the block? (*d*) We now move the block and hang it a distance of 25.0 cm from the balance point, on the same side of the balance as the charge. Keeping q_1 the same, and d the same, what value should we choose for q_2 to keep this apparatus in balance? SSM

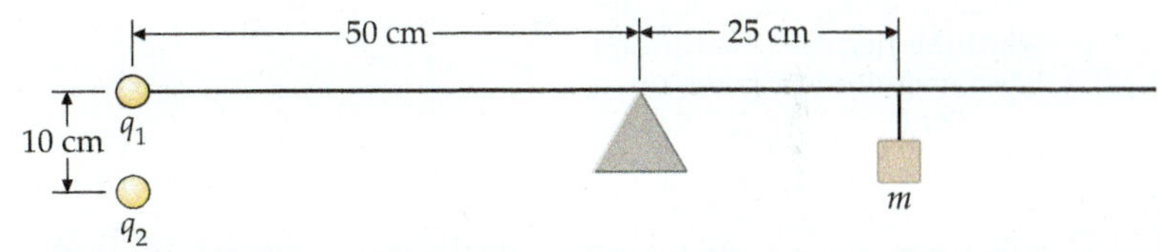

FIGURE 21-42 Problem 69

70 •• Two 3.0-μC point charges are located at $x = 0$, $y = 2.0$ m and at $x = 0$, $y = -2.0$ m. Two other point charges, each with charge Q, are located at $x = 4.0$ m, $y = 2.0$ m and at $x = 4.0$ m, $y = -2.0$ m (Figure 21-43). The electric field at $x = 0, y = 0$ due to the presence of the four charges is $(4.0 \times 10^3\ \text{N/C})\hat{\imath}$. Determine Q.

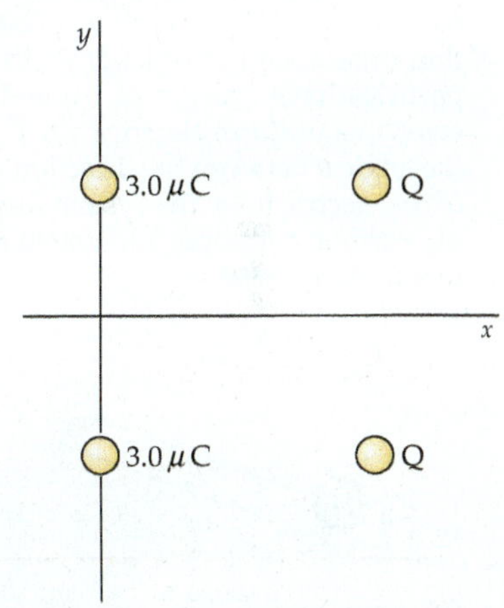

FIGURE 21-43 Problem 70

71 •• Two point charges have a total charge equal to 200 μC and are separated by 0.600 m. (*a*) Find the charge of each particle if the particles repel each other with a force of 120 N. (*b*) Find the force on each particle if the charge on each particle is 100 μC. SSM

72 •• Two point charges have a total charge equal to 200 μC and are separated by 0.600 m. (*a*) Find the charge of each particle if the particles attract each other with a force of 120 N. (*b*) Find the force on each particle if the charge on each particle is 100 μC. SSM

73 •• A point charge of $-3.00\ \mu$C is located at the origin; a point charge of 4.00 μC is located on the x axis at $x = 0.200$ m; a third point charge Q is located on the x axis at $x = 0.320$ m. The electric force on the 4.00-μC charge is 240 N in the $+x$ direction. (*a*) Determine the charge Q. (*b*) With this configuration of three charges, at what location(s) is the electric field zero?

74 •• Two point particles, each of mass m and charge q, are suspended from a common point by threads of length L. Each thread makes an angle θ with the vertical as shown in Figure 21-44. (*a*) Show that $q = 2L\sin\theta\sqrt{(mg/k)\tan\theta}$, where k is the Coulomb constant. (*b*) Find q if $m = 10.0$ g, $L = 50.0$ cm, and $\theta = 10.0°$.

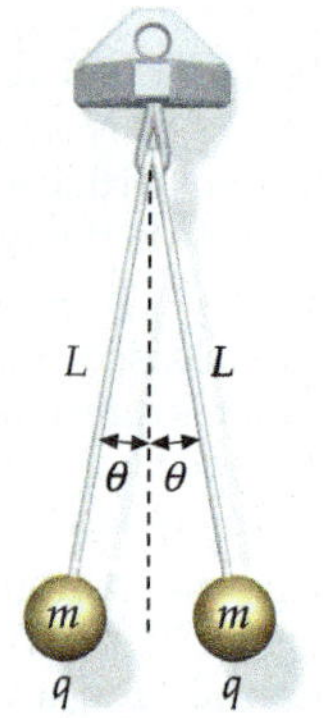

FIGURE 21-44 Problem 74

75 •• Suppose that in Problem 74 $L = 1.5$ m and $m = 0.010$ kg. (*a*) What is the angle that each string makes with the vertical if $q = 0.75\ \mu$C? (*b*) What is the angle that each string makes with the vertical if one particle has a charge of 0.50 μC and the other has a charge of 1.0 μC?

76 •• Four point charges of equal magnitude are arranged at the corners of a square of side L as shown in Figure 21-45. (*a*) Find the magnitude and direction of the force exerted on the charge in the lower left corner by the other three charges. (*b*) Show that the electric field at the midpoint of one of the sides of the square is directed along that side toward the negative charge and has a magnitude E given by

$$E = k\frac{8q}{L^2}\left(1 - \frac{1}{5\sqrt{5}}\right).$$

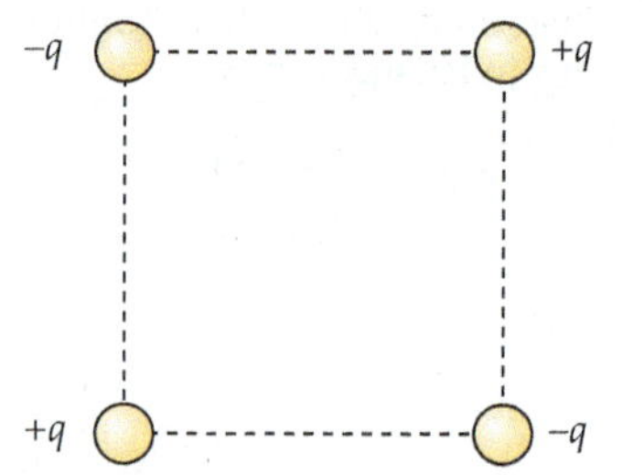

FIGURE 21-45 Problem 76

77 •• Figure 21-46 shows a dumbbell consisting of two identical small particles, each of mass m, attached to the ends of a thin (massless) rod of length a that is pivoted at its center. The particles have charges of $+q$ and $-q$, and the dumbbell is located in a uniform electric field $\vec{E}$. Show that for small values of the angle θ between the direction of the dipole and the direction of the electric field, the system displays a rotational form of simple harmonic motion, and obtain an expression for the period of that motion. SSM

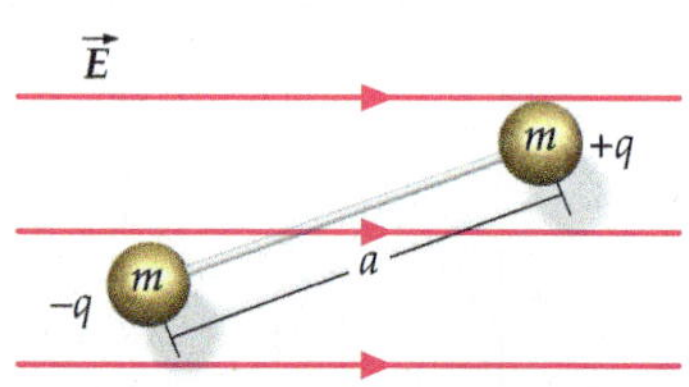

FIGURE 21-46 Problems 77 and 78

78 •• For the dumbbell in Problem 77, let $m = 0.0200$ kg, $a = 0.300$ m, and $\vec{E} = (600\text{ N/C})\hat{\imath}$. The dumbbell is initially at rest and makes an angle of 60° with the x axis. The dumbbell is then released, and when it is momentarily aligned with the electric field, its kinetic energy is 5.00×10^{-3} J. Determine the magnitude of q.

79 •• An electron (charge $-e$, mass m) and a positron (charge $+e$, mass m) revolve around their common center of mass under the influence of their attractive coulomb force. Find the speed v of each particle in terms of e, m, k, and their separation distance L. SSM

80 ••• A simple pendulum of length 1.0 m and mass 5.0×10^{-3} kg is placed in a uniform, electric field $\vec{E}$ that is directed vertically upward. The bob has a charge of $-8.0\ \mu$C. The period of the pendulum is 1.2 s. What are the magnitude and direction of $\vec{E}$?

81 ••• A point particle of mass m and charge q is constrained to move vertically inside a narrow, frictionless cylinder (Figure 21-47). At the bottom of the cylinder is a point charge Q having the same sign as q. (*a*) Show that the particle whose mass is m will be in equilibrium at a height $y_0 = (kqQ/mg)^{1/2}$. (*b*) Show that if the particle is displaced from its equilibrium position by a small amount and released, it will exhibit simple harmonic motion with angular frequency $\omega = (2g/y_0)^{1/2}$.

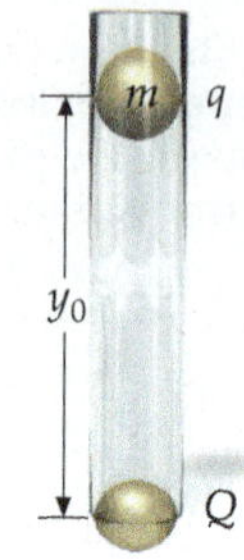

FIGURE 21-47 Problem 81

82 ••• Two neutral molecules on the x axis attract each other. Each molecule has a dipole moment $\vec{p}$, and these dipole moments are on the $+x$ direction and are separated by a distance d. Derive an expression for the force of attraction in terms of p and d.

83 ••• Two equal positive point charges Q are on the x axis at $x = \frac{1}{2}a$ and $x = -\frac{1}{2}a$. (*a*) Obtain an expression for the electric field on the y axis as a function of y. (*b*) A bead of mass m, which has a charge q, moves along the y axis on a thin frictionless taut thread. Find the electric force that acts on the bead as a function of y and determine the sign of q such that this force always points away from the origin. (*c*) The bead is initially at rest at the origin. If it is given a slight nudge in the $+y$ direction, how fast will the bead be traveling the instant the net force on it is a maximum? (Assume any effects due to gravity are negligible.)

84 ••• A gold nucleus is 100 fm (1 fm $= 10^{-15}$ m) from a proton, which initially is at rest. When the proton is released, it speeds away because of the repulsion that it experiences due to the charge on the gold nucleus. What is the proton's speed a large distance (assume to be infinity) from the gold nucleus? (Assume the gold nucleus remains stationary.)

85 ••• During a famous experiment in 1919, Ernest Rutherford shot doubly ionized helium nuclei (also known as alpha particles) at a gold foil. He discovered that virtually all of the mass of an atom resides in an extremely compact nucleus. Suppose that during such an experiment, an alpha particle far from the foil has a kinetic energy of 5.0 MeV. If the alpha particle is aimed directly at the gold nucleus, and the only force acting on it is the electric force of repulsion exerted on it by the gold nucleus, how close will it approach the gold nucleus before turning back? That is, what is the minimum center-to-center separation of the alpha particle and the gold nucleus? SSM

86 ••• During the Millikan experiment used to determine the charge on the electron, a charged polystyrene microsphere is released in still air in a known vertical electric field. The charged microsphere will accelerate in the direction of the net force until it reaches terminal speed. The charge on the microsphere is determined by measuring the terminal speed. During one such experiment, the microsphere has radius $r = 5.50 \times 10^7$ m, and the field has a magnitude $E = 6.00 \times 10^4$ N/C. The magnitude of the drag force on the sphere is given by $F_D = 6\pi\eta r v$, where v is the speed of the sphere and η is the viscosity of air ($\eta = 1.8 \times 10^{-5}\ \text{N}\cdot\text{s/m}^2$). Polystyrene has density 1.05×10^3 kg/m^3. (*a*) If the electric field is pointing down and the polystyrene microsphere is rising with a terminal speed of 1.16×10^{-4} m/s, what is the charge on the sphere? (*b*) How many excess electrons are on the sphere? (*c*) If the direction of the electric field is reversed but its magnitude remains the same, what is the new terminal speed?

87 ••• In Problem 86, there is a description of the Millikan experiment used to determine the charge on the electron. During the experiment, a switch is used to reverse the direction of the electric field without changing its magnitude, so that one can measure the terminal speed of the microsphere both as it is moving upward and as it is moving downward. Let v_u represent the terminal speed when the particle is moving up and v_d the terminal speed when moving down. (*a*) If we let $u = v_u + v_d$, show that $q = 3\pi\eta r u/E$, where q is the microsphere's net charge. For the purpose of determining q, what advantage does measuring both v_u and v_d have over measuring only one terminal speed? (*b*) Because charge is quantized, u can only change by steps of magnitude $N\Delta$, where N is an integer. Using the data from Problem 86, calculate Δ. SSM

CHAPTER 22

LIGHTNING IS AN ELECTRIC PHENOMENA. DURING A LIGHTNING STRIKE, CHARGES ARE TRANSFERRED BETWEEN THE CLOUDS AND THE GROUND. THE VISIBLE LIGHT GIVEN OFF COMES FROM AIR MOLECULES RETURNING TO LOWER ENERGY STATES. *(Photo Disc.)*

The Electric Field II: Continuous Charge Distributions

? How would you calculate the charge on the surface of Earth? (See Example 22-15.)

On a microscopic scale, charge is quantized. However, there are often situations in which many charges are so close together that the charge can be thought of as continuously distributed. We apply the concept of density to charge similarly to the way we use it to describe matter.

In addition to continuous charge distributions, we examine the importance of symmetry within the electric field. The mathematical findings of Carl Friedrich Gauss show that every electric field maintains symmetric properties. It is an understanding of charge distribution and symmetry within the electric field that aids scientists in a vast array of fields.

In this chapter, we show how Coulomb's law is used to calculate the electric field produced by various types of continuous charge distributions. We then introduce Gauss's law and use it to calculate the electric fields produced by charge distributions that have certain symmetries.

22-1 CALCULATING FROM COULOMB'S LAW

Figure 22-1 shows an element of charge $dq = \rho\, dV$ that is small enough to be considered a point charge. The element of charge dq is the amount of charge in volume element dV and ρ is the charge per unit volume. Coulomb's law states that the electric field $d\vec{E}$ at a field point P due to this element of charge is

$$d\vec{E} = dE_r\,\hat{r} = \frac{k\,dq}{r^2}\,\hat{r} \qquad \text{22-1}a$$

where $\hat{r}$ is a unit vector directed away from the charge element dq and toward point P, and dE_r (the component of $d\vec{E}$ in the direction of $\hat{r}$) is given by $k\,dq/r^2$.

The total field $\vec{E}$ at P is calculated by integrating this expression over the entire charge distribution. That is,

$$\vec{E} = \int d\vec{E} = \int \frac{k\hat{r}}{r^2}\,dq \qquad \text{22-1}b$$

ELECTRIC FIELD DUE TO A CONTINUOUS CHARGE DISTRIBUTION

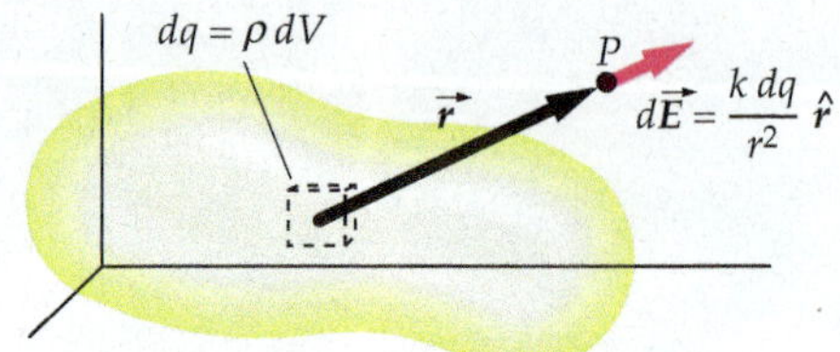

FIGURE 22-1 An element of charge dq produces a field $d\vec{E} = (k\,dq/r^2)\hat{r}$ at point P. The field at P is calculated by integrating Equation 22-1a over the entire charge distribution.

The use of a continuous charge density to describe a large number of discrete charges is similar to the use of a continuous mass density to describe air, which actually consists of a large number of discrete atoms and molecules. In both cases, it is usually easy to find a volume element ΔV that is large enough to contain a multitude of individual charge carriers and yet is small enough that replacing ΔV with a differential dV and using calculus introduces negligible error. If the charge is distributed over a surface or along a line, we use $dq = \sigma\, dA$ or $dq = \lambda\, dL$ and integrate over the surface or line. (In these cases σ and λ are the charge per unit area and charge per unit length, respectively.) The integration usually is done by expressing $\hat{r}$ in terms of its Cartesian components, and then integrating one component at a time.

! The x component of $\hat{r}$ is $\hat{r} \cdot \hat{i} = \cos\theta$, where θ is the angle between $\hat{r}$ and $\hat{i}$.* The y and z components of $\hat{r}$ are calculated in like manner.

PROBLEM-SOLVING STRATEGY

Calculating $\vec{E}$ Using Equations 22-1a and 22-1b

PICTURE Sketch the charge configuration along with a field point P (the point where $\vec{E}$ is to be calculated). In addition, the sketch should include an increment of charge dq at an arbitrary source point S.

SOLVE

1. Add coordinate axes to the sketch. The choice of axes should exploit any symmetry of the charge configuration. For example, if the charge is along a straight line, then select that line as one of the coordinate axes. Draw a second axis that passes through the field point P. In addition, include the coordinates of both P and S, the distance r between P and S, and the unit vector $\hat{r}$ directed away from S toward P.
2. To compute the electric field $\vec{E}$ using Equation 22-1b, we express $d\vec{E} = dE_r\,\hat{r}$ in component form. The x component of $d\vec{E}$ is $dE_x = dE_r\,\hat{r} \cdot \hat{i} = dE_r \cos\theta$, where θ is the angle between $\hat{r}$ and $\hat{i}$ (see Figure 22-2), and the y component of $d\vec{E}$ is $dE_y = dE_r\,\hat{r} \cdot \hat{j} = dE_r \sin\theta$.

* The componet of a vector in a given direction is equal to the scalar product of the vector with the unit vector in the given direction. Scalar products are discussed in Section 6-3.

3. Express $\vec{E}$ in Equation 22-1b in terms of its x and y components:

$$E_x = \int dE_x = \int dE_r \cos\theta = \int \frac{k\,dq}{r^2}\cos\theta$$

$$E_y = \int dE_y = \int dE_r \sin\theta = \int \frac{k\,dq}{r^2}\sin\theta$$

4. To calculate E_x, express dq as $\rho\,dV$ or $\sigma\,dA$ or $\lambda\,dL$ (whichever is appropriate) and integrate. To calculate E_y, follow a procedure similar to that used for calculating E_x.
5. Symmetry arguments are sometimes used to show that one or more components of $\vec{E}$ are equal to zero. (For example, a symmetry argument is used to show $E_y = 0$ in Example 22-5.)

CHECK If the charge distribution is confined to a finite region of space at points far from the charge distribution, the expression for the electric field will approach that of a point charge located at the center of charge. (If the charge configuration is sufficiently symmetric then the location of the center of charge can be obtained by inspection.)

See **Math Tutorial** *for more information on* **Trigonometry**

Example 22-1 Electric Field Due to a Line Charge of Finite Length

A thin rod of length L and charge Q is uniformly charged, so it has a linear charge density $\lambda = Q/L$. Find the electric field at point P, where P is an arbitrarily positioned point.

PICTURE Choose the x axis so the rod is on the x axis between points x_1 and x_2, and choose the y axis to be through the field point P. Let y_P be the radial distance of P from the x axis. To calculate the electric field $\vec{E}$ at P, we separately calculate E_x and E_y. Using Equations 22-1, first find the field increment $d\vec{E}$ at P due to an arbitrary increment dq of the charge distribution. Then integrate each component of $d\vec{E}$ over the entire charge distribution. (Because Q is distributed uniformly, the linear charge density λ equals Q/L.)

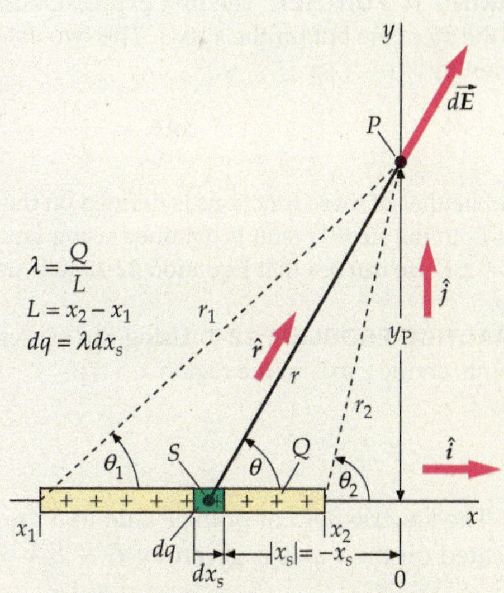

FIGURE 22-2 Geometry for the calculation of the electric field at field point P due to a uniformly charged rod.

SOLVE

1. Sketch the charge configuration and the field point P. Include the x and y axes with the x axis lying along the line of charge and the y axis passing through P. In addition, sketch an *arbitrary* increment of the line charge at point S (at $x = x_S$) that has a length dx_S and a charge dq, and the electric field at P due to dq. Sketch the electric field vector $d\vec{E}$ as if dq is positive (Figure 22-2):

2. $\vec{E} = E_x\hat{i} + E_y\hat{j}$. Find expressions for dE_x and dE_y in terms of dE_r and θ, where dE_r is the component of $d\vec{E}$ in the direction away from S toward P:

$$d\vec{E} = dE_r\,\hat{r}$$

so $dE_x = dE_r\,\hat{r}\cdot\hat{i} = dE_r\cos\theta$

$$dE_y = dE_r\,\hat{r}\cdot\hat{j} = dE_r\sin\theta$$

3. First we solve for E_x. Express dE_r using Equation 21-1a, where r is the distance from the source point S to the field point P. We see (Figure 22-2) that $\cos\theta = |x_S|/r = -x_S/r$. In addition, use $dq = \lambda\,dx_S$:

$$dE_r = \frac{k\,dq}{r^2} \text{ and } \cos\theta = \frac{-x_S}{r}$$

so

$$dE_x = \frac{k\,dq}{r^2}\cos\theta = \frac{k\cos\theta\,\lambda\,dx_S}{r^2}$$

4. Integrate the step-3 result:

$$dE_x = \int_{x_1}^{x_2}\frac{k\cos\theta\,\lambda\,dx_S}{r^2} = k\lambda\int_{x_1}^{x_2}\frac{\cos\theta\,dx_S}{r^2}$$

5. Next change the integration variable from x_S to θ. From Figure 22-2, find the relation between x_S and θ and between r and θ.

$$\tan\theta = \frac{y_P}{|x_S|} = \frac{y_P}{-x_S}, \quad \text{so} \quad x_S = -\frac{y_P}{\tan\theta} = -y_P\cot\theta$$

$$\sin\theta = \frac{y_P}{r}, \quad \text{so} \quad r = \frac{y_P}{\sin\theta}$$

6. Differentiate the step 5 result to obtain an expression for dx_S (the field point P remains fixed, so y_P is constant):

$$dx_S = -y_P\frac{d\cot\theta}{d\theta} = y_P\csc^2\theta\,d\theta$$

7. Substitute $y_P\csc^2\theta\,d\theta$ for dx_S and $y_P/\sin\theta$ for r in the integral in step 4 and simplify:

$$\int_{x_1}^{x_2}\frac{\cos\theta\,dx_S}{r^2} = \int_{\theta_1}^{\theta_2}\frac{\cos\theta\,y_P\csc^2\theta\,d\theta}{y_P^2/\sin^2\theta} = \frac{1}{y_P}\int_{\theta_1}^{\theta_2}\cos\theta\,d\theta \qquad (y_P \neq 0)$$

8. Evaluate the integral and solve for E_x:

$$E_x = k\lambda\frac{1}{y_P}\int_{\theta_1}^{\theta_2}\cos\theta\,d\theta = \frac{k\lambda}{y_P}(\sin\theta_2 - \sin\theta_1) = \frac{k\lambda}{y_P}\left(\frac{y_P}{r_2} - \frac{y_P}{r_1}\right)$$

$$= k\lambda\left(\frac{1}{r_2} - \frac{1}{r_1}\right) \qquad (r_1 > 0 \text{ and } r_2 > 0)$$

9. E_y can be found using a procedure that parallels the one in steps 3–7 for finding E_x (to find E_y, see Problem 22-21):

$$E_y = -\frac{k\lambda}{y_P}(\cos\theta_2 - \cos\theta_1) = -k\lambda\left(\frac{\cot\theta_2}{r_2} - \frac{\cot\theta_1}{r_1}\right) \qquad (y_P \neq 0)$$

and

$$E_y = 0 \qquad (y_P = 0)$$

10. Combine steps 8 and 9 to obtain and expression for the electric field at P:

$$\vec{E} = \boxed{E_x\hat{i} + E_y\hat{j}}$$

CHECK Consider the plane that is perpendicular to and bisecting the rod. At points on this plane, symmetry dictates that $\vec{E}$ points directly away from the center of the rod. That is, we expect that $E_x = 0$ throughout this plane. At all points on this plane $r_1 = r_2$. The step-8 result gives $E_x = 0$ if $r_1 = r_2$, as expected.

TAKING IT FURTHER The first expression for E_y in the step 9 result is valid everywhere in the xy plane but on the x axis. The two cotangent functions in the expression for E_y are given by

$$\cot\theta_1 = \frac{-x_1}{y_P} \quad \text{and} \quad \cot\theta_2 = \frac{-x_2}{y_P}$$

and neither of these functions is defined on the x axis (where $y_P = 0$). The second expression for E_y in the step-9 result is obtained using Equation 22-1*a*. By recognizing that on the x axis $\hat{r} = \pm\hat{i}$, we can see that Equation 22-1*a* tells us that $d\vec{E} = \pm dE\hat{i}$, which implies $E_y = 0$.

PRACTICE PROBLEM 22-1 Using the expression for E_x in step 8, show that $E_x > 0$ at all points on the x axis in the region $x > x_2$.

The electric field at point P due to a thin uniformly charged rod (see Figure 22-3) located on the z axis is given by $\vec{E} = E_z\hat{k} + E_R\hat{R}$, where

$$E_z = \frac{k\lambda}{R}(\sin\theta_2 - \sin\theta_1) = k\lambda\left(\frac{1}{r_2} - \frac{1}{r_1}\right) \qquad (r_1 \neq 0) \text{ and } (r_2 \neq 0) \qquad \text{22-2}a$$

$$E_R = -\frac{k\lambda}{R}(\cos\theta_2 - \cos\theta_1) = -k\lambda\left(\frac{\cot\theta_2}{r_2} - \frac{\cot\theta_1}{r_1}\right) \qquad (R \neq 0) \qquad \text{22-2}b$$

These equations are derived in Example 22-1. The expressions for E_z (Equation 22-2*a*) are undefined at the end points of the thin charged rod and the expressions for E_R (Equation 22-2*b*) are undefined at all points on the z axis (where $R = 0$). However, $E_R = 0$ at all points where $R = 0$.

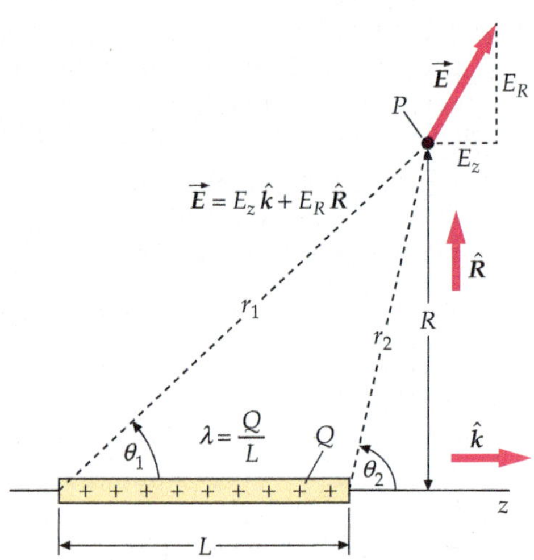

FIGURE 22-3 The electric field due to a uniformly charged thin rod.

Example 22-2 $\vec{E}$ of a Finite Line Charge and Far from the Charge

A charge Q is uniformly distributed along the z axis, from $z = -\frac{1}{2}L$ to $z = +\frac{1}{2}L$. Show that for large values of z the expression for the electric field of the line charge on the z axis approaches the expression for the electric field of a point charge Q at the origin.

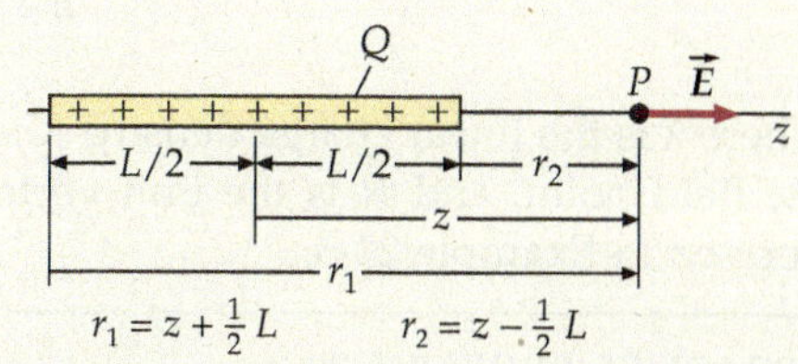

FIGURE 22-4 Geometry for the calculation of the electric field on the axis of a uniform line charge of length L, charge Q, and linear charge density $\lambda = Q/L$.

PICTURE Use Equation 22-2*a* to show that for large values of z the expression for the electric field of the line charge on the z axis approaches that of a point charge Q at the origin.

SOLVE

1. The electric field on the z axis has only a z component, given by Equation 22-2a: $$E_z = k\lambda\left(\frac{1}{r_2} - \frac{1}{r_1}\right)$$
2. Sketch the line charge. Include the z axis, the field point P, and r_1 and r_2 (Figure 22-4):
3. Substitute with $r_1 = z + \frac{1}{2}L$ and $r_2 = z - \frac{1}{2}L$ into the step 1 result and simplify: $$E_z = k\lambda\left(\frac{1}{z - \frac{1}{2}L} - \frac{1}{z + \frac{1}{2}L}\right) = \frac{kQ}{L}\frac{L}{z^2 - \left(\frac{1}{2}L\right)^2} = \frac{kQ}{z^2 - \left(\frac{1}{2}L\right)^2} \qquad \left(z > \tfrac{1}{2}L\right)$$
4. Find an approximate expression for E_z for $z \gg L$, which is done by neglecting $(\frac{1}{2}L)^2$ in comparison with z^2 in the step 3 result. $$\boxed{E_z \approx \frac{kQ}{z^2} \qquad (z \gg L)}$$

CHECK The approximate expression (step 4) falls off inversely as the square of z, the distance from the origin. This expression is the same as the expression for the electric field of a point charge Q located at the origin.

PRACTICE PROBLEM 22-2 The validity of the step 3 result is established for the region $L/2 > z > \infty$. Is the step 3 result also valid in the region $-L/2 < z < +L/2$? Explain your answer.

Example 22-3 $\vec{E}$ Due to an Infinite Line Charge

Find the electric field due to a uniformly charged line that extends to infinity in both directions and has linear charge density λ.

PICTURE A line charge is considered infinite if the distances between the ends of the line charge and the field points of interest are much much greater than the distances between any of the radial distances of the field points from the line charge. To calculate the electric field due to such a line charge we take the limit (see Figure 22-2) both as $x_1 \to -\infty$ and as $x_2 \to +\infty$. From the figure, we see that taking the limit as both $\theta_1 \to 0$ and as $\theta_2 \to \pi$ is needed. See Equations 22-2*a* and 22-2*b* for expressions for the electric field.

SOLVE

1. Choose the first expression for the electric field in each of Equations 22-2*a* and 22-2*b*: $$E_z = \frac{k\lambda}{R}(\sin\theta_2 - \sin\theta_1)$$ $$E_R = -\frac{k\lambda}{R}(\cos\theta_2 - \cos\theta_1)$$
2. Take the limit as both $\theta_1 \to 0$ and as $\theta_2 \to \pi$. $$E_z = \frac{k\lambda}{R}(\sin\pi - \sin 0) = \frac{k\lambda}{R}(0 - 0) = 0$$ $$E_R = -\frac{k\lambda}{R}(\cos\pi - \cos 0) = -\frac{k\lambda}{R}(-1 - 1) = 2\frac{k\lambda}{R}$$
3. Express the electric field in vector form: $$\vec{E} = E_z\hat{k} + E_R\hat{R} = 0\hat{k} + \frac{2k\lambda}{R}\hat{R} = \boxed{\frac{2k\lambda}{R}\hat{R}}$$

CHECK The electric field is in the radial direction as expected. We expected this due to the symmetry. (The line charge is uniformly distributed and extends to infinity in both directions.)

TAKING IT FURTHER The magnitude of the electric field decreases inversely with the radial distance from the line charge.

The electric field due to a uniformly charged line that extends to infinity in both directions is given by

$$\vec{E} = \frac{2k\lambda}{R}\hat{R} \qquad 22\text{-}3$$

where λ is the linear charge density, R is the radial distance from the line charge to the field point, and $\hat{R}$ is the unit vector in the radial direction. Equation 22-3 is derived in Example 22-3.

PRACTICE PROBLEM 22-3

Show that if k, λ, and R are in SI units then Equation 22-3 gives the electric field in newtons per coulomb.

It is customary to write the Coulomb constant k in terms of another constant, ϵ_0, called the **electric constant (permittivity of free space):**

$$k = \frac{1}{4\pi\epsilon_0} \qquad 22\text{-}4$$

Using this notation, Coulomb's law for $\vec{E}$ (Equation 21-7) is written

$$\vec{E} = k\frac{q}{r^2}\hat{r} = \frac{1}{4\pi\epsilon_0}\frac{q}{r^2}\hat{r} \qquad 22\text{-}5$$

and $\vec{E}$ for a uniformly charged infinite line (Equation 22-3) with linear charge density λ is written

$$\vec{E} = \frac{1}{2\pi\epsilon_0}\frac{\lambda}{R}\hat{R} \qquad 22\text{-}6$$

The value of ϵ_0 in SI units is

$$\epsilon_0 = \frac{1}{4\pi k} = 8.85 \times 10^{-12}\,\mathrm{C^2/(N \cdot m^2)} \qquad 22\text{-}7$$

Example 22-4 Approximating Equations 22-2*a* and 22-2*b* on the Symmetry Plane

A charge Q is uniformly distributed along the z axis, from $z = -\frac{1}{2}L$ to $z = +\frac{1}{2}L$. (*a*) Find an expression for the electric field on the $z = 0$ plane as a function of R, the radial distance of the field point from the z axis. (*b*) Show that for $R \gg L$, the expression found in Part (*a*) approaches that of a point charge at the origin of charge Q. (*c*) Show that for $R \ll L$, the expression found in Part (*a*) approaches that of an infinitely long line charge on the z axis with a uniform linear charge density $\lambda = Q/L$.

PICTURE The charge configuration is the same as that in Example 22-2, and the linear charge density is $\lambda = Q/L$. Sketch the line charge on the z axis and put the field point in the $z = 0$ plane. Then use Equations 22-2*a* and 22-2*b* to find the electric field expression for Part (*a*). The electric field due to a point charge decreases inversely with the square of the distance from the charge. Examine the Part-(*a*) result to see how it approaches that of a point charge at the origin for $R \gg L$. The electric field due to a uniform line charge of infinite length decreases inversely with the radial distance from the line (Equation 22-3). Examine the Part-(*a*) result to see how it approaches the expression for the electric field of a line charge of infinite length for $R \ll L$.

SOLVE

(*a*) 1. Choose the first expression for the electric field in each of Equations 22-2*a* and 22-2*b*:

$$E_z = \frac{k\lambda}{R}(\sin\theta_2 - \sin\theta_1)$$

$$E_R = -\frac{k\lambda}{R}(\cos\theta_2 - \cos\theta_1)$$

2. Sketch the charge configuration with the line charge on the z axis from $z = -\frac{1}{2}L$ to $z = +\frac{1}{2}L$. Show the field point P in the $z = 0$ plane a distance R from the origin (Figure 22-5):

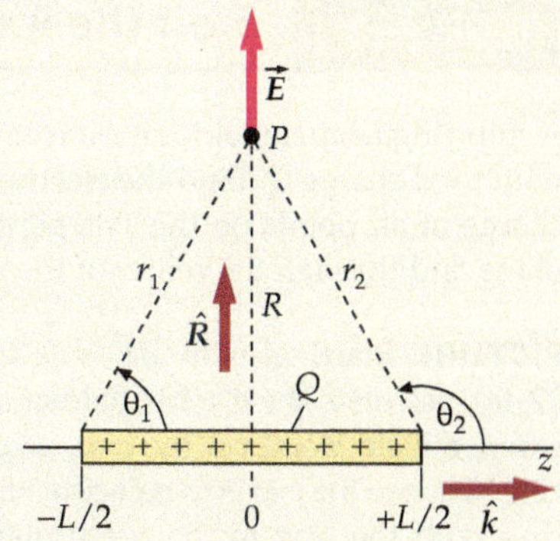

FIGURE 22-5

3. From the figure, we see that $\theta_2 + \theta_1 = \pi$, so $\sin\theta_2 = \sin(\pi - \theta_1) = \sin\theta_1$ and $\cos\theta_2 = \cos(\pi - \theta_1) = -\cos\theta_1$. Substitute into the step 1 results:

$$E_z = \frac{k\lambda}{R}(\sin\theta_1 - \sin\theta_1) = 0$$

$$E_R = -\frac{k\lambda}{R}(-\cos\theta_1 - \cos\theta_1) = \frac{2k\lambda}{R}\cos\theta_1$$

4. Express $\cos\theta_1$ in terms of R and L and substitute into the step-3 result:

$$\cos\theta_1 = \frac{\frac{1}{2}L}{\sqrt{R^2 + (\frac{1}{2}L)^2}}$$

so

$$E_R = \frac{2k\lambda}{R}\frac{\frac{1}{2}L}{\sqrt{R^2 + (\frac{1}{2}L)^2}} = \frac{k\lambda L}{R\sqrt{R^2 + (\frac{1}{2}L)^2}}$$

5. Express the electric field in vector form, and substitute Q for λL:

$$\vec{E} = E_z\hat{k} + E_R\hat{R} = 0\hat{k} + E_R\hat{R}$$

so $$\vec{E} = E_R\hat{R} = \boxed{\frac{kQ}{R\sqrt{R^2 + (\frac{1}{2}L)^2}}\hat{R}}$$

(*b*) 1. Examine the step-5 result. If $R \gg L$ then $R^2 + (\frac{1}{2}L)^2 \approx R^2$. Substitute R^2 for $R^2 + (\frac{1}{2}L)^2$:

$$\vec{E} \approx \frac{kQ}{R\sqrt{R^2}}\hat{R} = \frac{kQ}{R^2}\hat{R} \quad (R \gg L)$$

2. This (approximate) expression for the electric field decreases inversely with the square of the distance from the origin, just as it would for a point charge Q at the origin.

$$\vec{E} \approx \boxed{\frac{kQ}{R^2}\hat{R} \quad (R \gg L)}$$

(*c*) 1. Examine the Part (*a*), step-5 result. If $R \ll L$ then $R^2 + (\frac{1}{2}L)^2 \approx (\frac{1}{2}L)^2$. Substitute $(\frac{1}{2}L)^2$ for $R^2 + (\frac{1}{2}L)^2$. This (approximate) expression for the electric field falls off inversely with the radial distance from the line charge, just as the exact expression for an infinite line charge (Equation 22-3) would.

$$\vec{E} \approx \frac{k\lambda L}{R\sqrt{(\frac{1}{2}L)^2}}\hat{R} = \boxed{\frac{2k\lambda}{R}\hat{R}} \quad (R \ll L)$$

CHECK Parts (*b*) and (*c*) are themselves plausibility checks for the Part-(*a*) result. They reveal the validity of the Part-(*a*) result in two limiting cases, $R \gg L$ and $R \ll L$.

TAKING IT FURTHER Figure 22-6 shows the exact result for a line charge of length $L = 10$ cm and a linear charge density of $\lambda = 4.5$ nC/m. It also shows the limiting cases of an infinite line charge of the same charge density and a point charge $Q = \lambda L$.

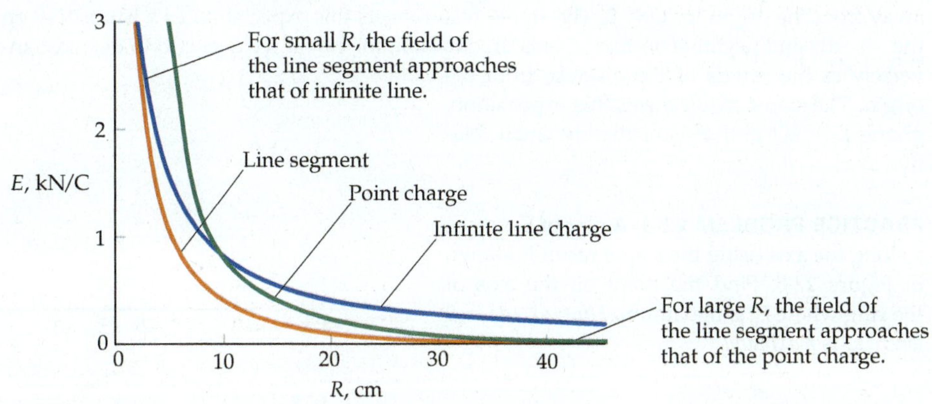

FIGURE 22-6 The magnitude of the electric field is plotted versus distance for a 10-cm-long line charge, a point charge, and an infinite line charge.

Example 22-5 $\vec{E}$ on the Axis of a Charged Ring

A thin ring (a circle) of radius a is uniformly charged with total charge Q. Find the electric field due to this charge at all points on the axis perpendicular to the plane and through the center of the ring.

PICTURE Starting with $d\vec{E} = (k\,dq/r^2)\hat{r}$ (Equation 22-1a), calculate the electric field at an arbitrarily positioned field point on the axis. Sketch the charged ring. Choose the z axis to coincide with the axis of the ring with the ring in the $z = 0$ plane. Label a field point P somewhere on the $+z$ axis, and place a source point S on the ring.

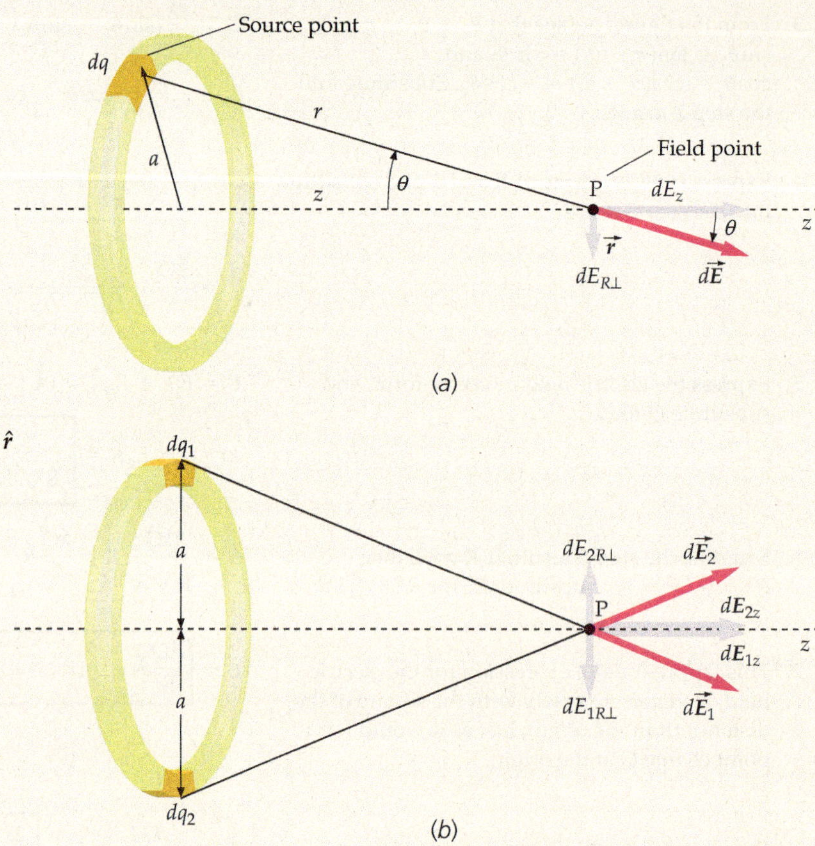

FIGURE 22-7 (a) A ring charge of radius a. The electric field at point P on the z axis due to the charge element dq shown has one component along the z axis and one perpendicular to the z axis. (b) For any charge element dq_1 there is an equal charge element dq_2 opposite it, and the electric field components perpendicular to the z axis sum to zero.

SOLVE

1. Write the equation (Equation 22-1a) giving the electric field due to an element of charge dq:
$$d\vec{E} = \frac{k\,dq}{r^2}\hat{r}$$

2. Sketch the ring (Figure 22-7a) and the axis (the z axis), and show the electric field vector at field point P due to an increment of charge dq at source point:

3. Sketch the ring (Figure 22-7b) and show the axial and radial components of $\vec{E}$ for identical charge elements on opposite sides of the ring. The radial components cancel in pairs, as can be seen, so the resultant field is axial:
$$E_R = 0$$

4. Express the z component of the electric field from the step-1 result:
$$dE_z = \frac{k\,dq}{r^2}\cos\theta = \frac{k\,dq}{r^2}\frac{z}{r} = \frac{k\,dq\,z}{r^3}$$

5. Integrate both sides of the step-4 result. Factor constant terms from the integral:
$$E_z = \int \frac{kz\,dq}{r^3} = \frac{kz}{r^3}\int dq = \frac{kz}{r^3}Q$$

6. Using the Pythagorean theorem gives $r = \sqrt{z^2 + a^2}$:
$$\vec{E} = E_z\hat{k} + E_R\hat{R} = E_z\hat{k} + 0 = \boxed{\frac{kQz}{(z^2 + a^2)^{3/2}}\hat{k}}$$

CHECK We expect the direction of the electric field at points on the z axis to be directed away from the origin for $Q > 0$. The step-6 result meets this expectation as z is positive on the $+z$ axis and negative on the $-z$ axis. In addition, for $z \gg a$ we expect E to decrease inversely as the square of the distance from the origin. The step-6 result meets this expectation, giving $E_z \approx kQ/z^2$ if a^2 is negligibly small relative to z^2.

PRACTICE PROBLEM 22-4 A plot of E_z versus z along the axis using the step-6 result is shown in Figure 22-8. Find the point on the axis of the ring where E_z is maximum. *Hint: $dE_z/dz = 0$ where E_z is maximum.*

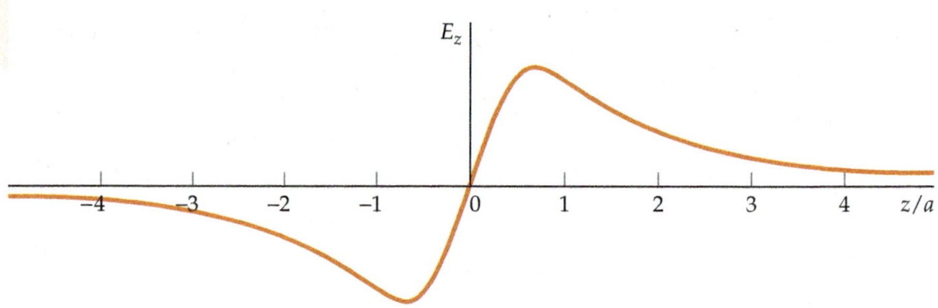

FIGURE 22-8

Example 22-6 $\vec{E}$ on the Axis of a Charged Ring — Conceptual

For the charged ring in Example 22-5, why is the magnitude of the electric field small near the origin, even though the origin is closer to the ring than any other points on the z axis (see Figure 22-9)?

PICTURE The key to solving this problem can be found in Figure 22-7*b*. Redraw this figure with the field point P on the z axis, but near the origin.

SOLVE

1. Redraw Figure 22-7*b* with the field point P near the origin:

2. The electric fields near the origin due to the two elements of charge (shown in Figure 22-9) are large but are of equal magnitude and nearly oppositely directed, so they nearly sum to zero. — Near the origin the resultant electric field is axial and small.

CHECK At the origin, the two electric fields are large, but are oppositely directed and so add to zero. Far from the origin ($|z| \gg a$), the two electric fields (Figure 22-7*b*) are in almost the same direction so they do not add to zero.

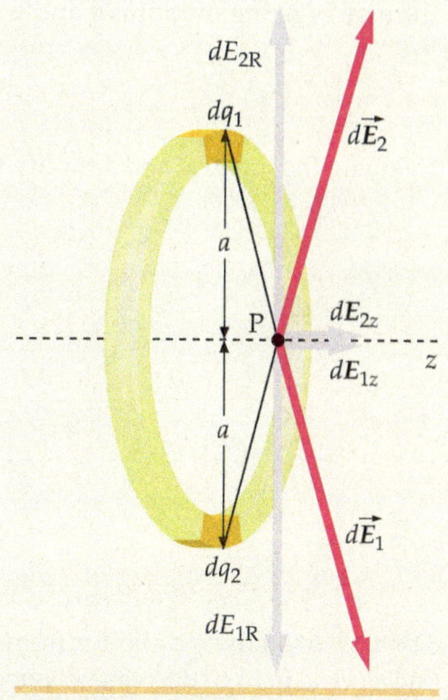

FIGURE 22-9

The electric field on the axis of a uniformly charged circular ring of radius a and charge Q is given by $\vec{E} = E_z\hat{k}$, where

$$E_z = \frac{kQz}{(z^2 + a^2)^{3/2}} \qquad 22\text{-}8$$

Equation 22-8 is derived in Example 22-5.

See **Math Tutorial** *for more information on* **Binomial Expansion**

Example 22-7 $\vec{E}$ on the Axis of a Charged Disk

Consider a uniformly charged thin disk of radius b and surface charge density σ. (*a*) Find the electric field at all points on the axis of the disk. (*b*) Show that for points on the axis and far from the disk, the electric field approaches that of a point charge at the origin with the same charge as the disk. (*c*) Show that for a uniformly charged disk of infinite radius, the electric field is uniform throughout the region on either side of the disk.

PICTURE We can calculate the field on the axis of the disk by treating the disk as a set of concentric, uniformly charged rings.

SOLVE

(*a*) 1. Calculate the field on the axis of the disk by treating the disk as a set of concentric rings of charge. The field of a single uniformly charged ring that has a charge Q and a radius a is shown in Equation 22-8:

$$\vec{E} = E_z\hat{k}, \text{ where } E_z = \frac{kQz}{(z^2 + a^2)^{3/2}}$$

2. Sketch the disk (Figure 22-10) and illustrate the electric field $d\vec{E}$ on its axis due to a single ring of charge dq, radius a, and width da:

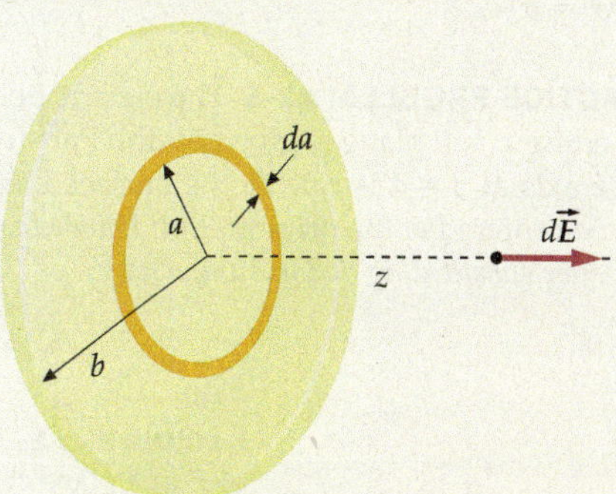

FIGURE 22-10 A uniform disk of charge can be treated as a set of ring charges, each of radius a.

3. Substitute dq for Q and dE_z for E_z in the step-1 result. Then integrate both sides to calculate the resultant field for the entire disk. The field point remains fixed, so z is constant:

$$dE_z = \frac{kz\,dq}{(z^2 + a^2)^{3/2}}$$

$$\text{so } E_z = \int \frac{kz\,dq}{(z^2 + a^2)^{3/2}} = kz\int \frac{dq}{(z^2 + a^2)^{3/2}}$$

4. To evaluate this integral we change integration variables from q to a. The charge $dq = \sigma\, dA$, where $dA = 2\pi a\, da$ is the area of a ring of radius a and width da:

$$dq = \sigma\, dA = \sigma 2\pi a da$$

$$\text{so}\quad E_z = \pi kz\sigma \int_0^b \frac{2ada}{(z^2 + a^2)^{3/2}} = \pi kz\sigma \int_{z^2+0^2}^{z^2+b^2} u^{-3/2}\, du$$

where $u = z^2 + a^2$, so $du = 2ada$

5. Evaluate the integral and simplify the result:

$$E_z = \pi kz\sigma \frac{u^{-1/2}}{-\frac{1}{2}}\bigg|_{z^2}^{z^2+b^2} = -2\pi kz\sigma\left(\frac{1}{\sqrt{z^2 + b^2}} - \frac{1}{\sqrt{z^2}}\right)$$

$$= \boxed{\text{sign}(z) \cdot 2\pi k\sigma\left(1 - \frac{1}{\sqrt{1 + \dfrac{b^2}{z^2}}}\right)}$$

where $\text{sign}(z) = z/|z|$. By definition*:

$$\text{sign}(z) = \begin{cases} +1 & z > 0 \\ 0 & z = 0 \\ -1 & z < 0 \end{cases}$$

(*b*) 1. For $z \gg b$ (on the $+z$ axis far from the disk) we expect the electric field to decrease inversely with z^2, like that of a point charge. To show this we use the binomial expansion:

The binomial expansion (to first order) is $(1 + x)^n \approx 1 + nx$ for $|x| \ll 1$.

2. Apply the binomial expansion to the rightmost term in the step-5 result:

$$\frac{1}{\sqrt{1 + \dfrac{b^2}{z^2}}} = \left(1 + \frac{b^2}{z^2}\right)^{-1/2} \approx 1 - \frac{1}{2}\frac{b^2}{z^2} \qquad z^2 \gg b^2$$

3. Substitute into the step-5 result and simplify. [For $z \gg b$, sign $(z) = 1$.] Thus, the approximate expression for the field for $z \gg b$ is the same as that of a point charge $Q = \sigma\pi b^2$ at the origin:

$$E_z \approx 2\pi k\sigma\left(1 - \left[1 - \frac{1}{2}\frac{b^2}{z^2}\right]\right) = 2\pi k\sigma\frac{1}{2}\frac{b^2}{z^2} = \boxed{\frac{kQ}{z^2}} \qquad z \gg b$$

where $Q = \sigma\pi b^2$.

(*c*) 1. Take the limit of the Part (*a*), step-5 result as $b \to \infty$. This result is an expression for E_z that is uniform, both in the region $z > 0$ and in the region $z < 0$:

$$E_z = \text{sign}(z) \cdot 2\pi k\sigma\left(1 - \frac{1}{\sqrt{1 + \infty}}\right) = \boxed{\text{sign}(z) \cdot 2\pi k\sigma}$$

CHECK We expect the electric field be in opposite directions on opposite sides of the disk. The Part (*a*), step-5 result meets this expectation.

TAKING IT FURTHER According to the Part-(*c*) result the electric field is discontinuous at $z = 0$ (Figure 22-11) where the field jumps from $-2\pi k\sigma\hat{i}$ to $+2\pi k\sigma\hat{i}$ as we cross the $z = 0$ plane. There is thus a discontinuity in E_z in the amount $4\pi k\sigma = \sigma/\epsilon_0$.

PRACTICE PROBLEM 22-5 The electric field due to a uniform surface charge on the entire $z = 0$ plane is given by the Part-(*c*) result. What fraction of the field on the z axis at $z = a$ is due to the surface charge within a circle that has a radius $r = 5a$ centered at the origin? *Hint: Divide the Part (a), step-5 result by the Part-(c) result after substituting 5a for b and a for z.*

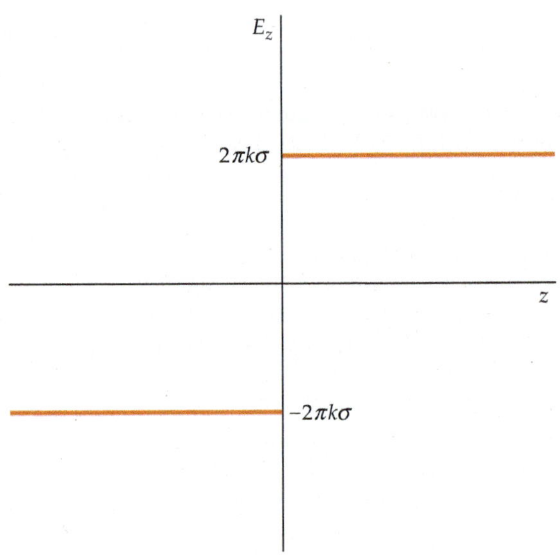

FIGURE 22-11 Graph showing the discontinuity of $\vec{E}$ at a plane charge. Can you see the similarity between this graph and the one in Figure 22-8?

* Both Excel and Mathematica use the definition of the sign function given here. Texas Instruments, however, uses a definition in which sign(0) returns ±1 instead of 0.

The answer to Practice Problem 22-5 depends not on a, but on the ratio $r/a = 5$. Eighty percent of the field at any distance a from a uniformly charged plane surface is due to the charge within a circle whose radius is equal to $5a$ multiplied by that distance.

The formula for the electric field on the axis of a uniformly charged circular disk, established in Example 22-7, is

$$E_z = \text{sign}(z) \cdot 2\pi k\sigma \left(1 - \frac{1}{\sqrt{1 + \frac{R^2}{z^2}}} \right) \qquad 22\text{-}9$$

ELECTRIC FIELD ON AXIS OF A UNIFORM DISK OF CHARGE

where sign(z) is defined in Part (a), step 5 of Example 22-7 and R is the radius of the disk. The field of a uniformly charged electric plane of charge can be obtained from Equation 22-9 by letting the ratio R/z go to infinity. Then

$$E_z = \text{sign}(z) \cdot 2\pi k\sigma = \text{sign}(z) \cdot \frac{\sigma}{2\epsilon_0} \qquad 22\text{-}10$$

ELECTRIC FIELD OF A UNIFORM PLANE OF CHARGE

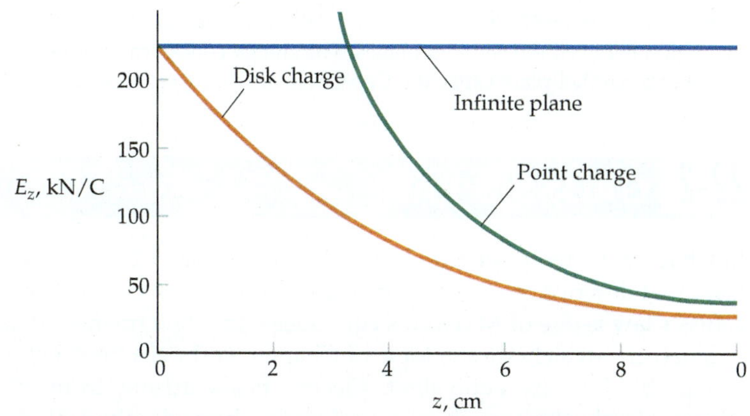

FIGURE 22-12 A disk and a point have equal charges, and an infinite plane and the disk have equal uniform surface-charge densities. Note that the field of the disk charge converges with the field of the point charge as z approaches infinity, and equals the field of the infinite plane charge as z approaches zero.

Figure 22-12 shows the electric fields of a point charge, a uniform disk of charge, and an infinite plane of charge as a function of position.

As we move along the z axis, the electric field jumps from $-2\pi k\sigma\hat{i}$ to $+2\pi k\sigma\hat{i}$ when we pass through the $z = 0$ plane (Figure 22-11). Thus, at $z = 0$ there is a discontinuity in E_z in the amount $4\pi k\sigma$.

Example 22-8 Electric Field Due to Two Infinite Planes

In Figure 22-13, an infinite plane of surface charge density $\sigma = +4.5\ \text{nC/m}^2$ lies in the $x = 0.00$ m plane, and a second infinite plane of surface charge density $\sigma = -4.50\ \text{nC/m}^2$ lies in the $x = 2.00$ m plane. Find the electric field at (a) $x = 1.80$ m and (b) $x = 5.00$ m.

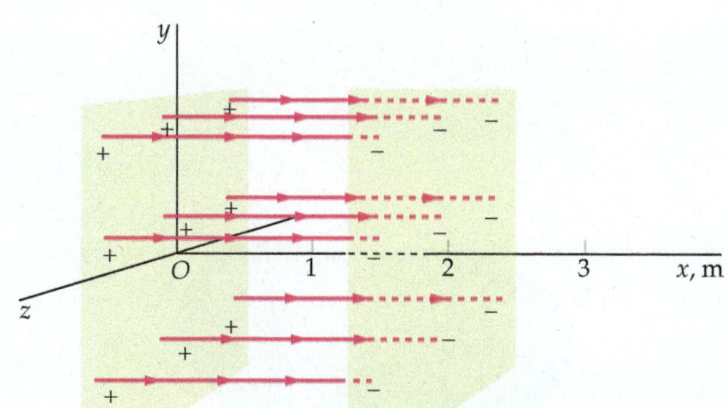

FIGURE 22-13

PICTURE Each charged plane produces a uniform electric field of magnitude $E = \sigma/(2\epsilon_0)$. We use superposition to find the resultant field. Between the planes the fields add, producing a net field of magnitude σ/ϵ_0 in the $+x$ direction. For $x > 2.00$ m and for $x < 0$, the two fields point in opposite directions and thus sum to zero.

SOLVE

(a) 1. Calculate the magnitude of the field E produced by each plane:

$$E = |\sigma|/(2\epsilon_0) = (4.50 \times 10^{-9}\ \text{N/C})/(2 \cdot 8.85 \times 10^{-12}) = 254\ \text{N/C}$$

2. At $x = 1.80$ m, between the planes, the field due to each plane points in the $+x$ direction:

$$E_{x\,\text{net}} = E_1 + E_2 = 254\ \text{N/C} + 254\ \text{N/C} = \boxed{508\ \text{N/C}}$$

(*b*) At $x = 5.00$ m, the fields due to the two planes are oppositely directed: $E_{x\,\text{net}} = E_1 - E_2 = \boxed{0.00\ \text{N/C}}$

CHECK Because the two planes have equal and opposite charge densities, electric field lines originate on the positive plane and terminate on the negative plane. $\vec{E}$ is equal to zero except between the planes.

TAKING IT FURTHER Note that $E_{x\,\text{net}} = 508$ N/C not just at $x = 1.8$ m but at any point in the region between the charged planes. The charge configuration described in this example is that of a parallel-plate capacitor. Capacitors are discussed in Chapter 24.

22-2 GAUSS'S LAW

In Chapter 21, the electric field is described visually by using electric field lines. Here that description is put in rigorous mathematical language called Gauss's law. Gauss's law is one of Maxwell's equations—the fundamental equations of electromagnetism—which are the topic of Chapter 30. In electrostatics, Gauss's law and Coulomb's law are equivalent. Electric fields arising from some symmetrical charge distributions, such as a uniformly charged spherical shell or uniformly charged infinite line, can be easily calculated using Gauss's law. In this section, we give an argument for the validity of Gauss's law based on the properties of electric field lines. A more rigorous derivation of Gauss's law is presented in Section 22-6.

A closed surface—like the surface of a soap bubble—is one that divides the universe into two distinct regions, the region enclosed by surface and the region outside the surface. Figure 22-14 shows a closed surface of arbitrary shape enclosing a dipole. The number of electric field lines beginning on the positive charge and penetrating the surface from the inside depends on where the surface is drawn, but any line penetrating the surface from the inside also penetrates it from the outside. To count the net number of lines out of any closed surface, count any penetration from the inside as +1, and any penetration from the outside as −1. Thus, for the surface shown (Figure 22-14), the net number of lines out of the surface is zero. For surfaces enclosing other types of charge distributions, such as that shown in Figure 22-15, *the net number of lines out of any surface enclosing the charges is proportional to the net charge enclosed by the surface.* This rule is a statement of Gauss's law.

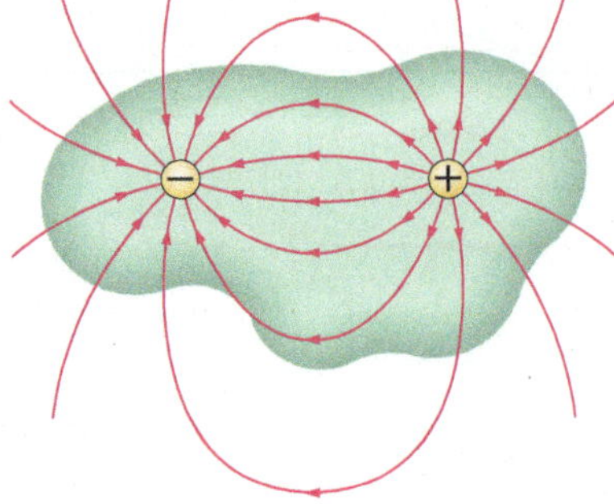

FIGURE 22-14 A surface of arbitrary shape enclosing an electric dipole. As long as the surface encloses both charges, the number of lines penetrating the surface from the inside is exactly equal to the number of lines penetrating the surface from the outside no matter where the surface is drawn.

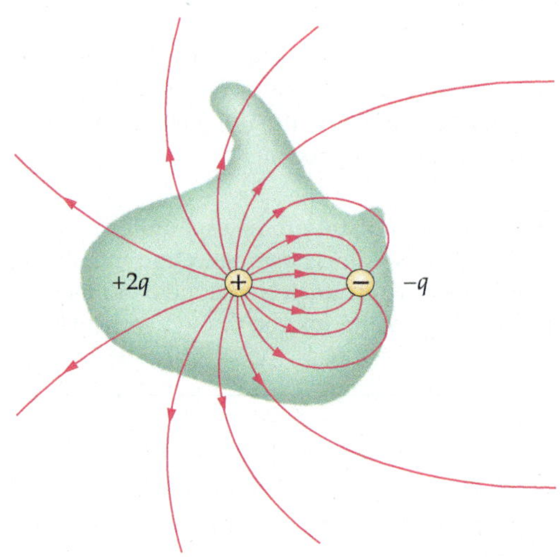

FIGURE 22-15 A surface of arbitrary shape enclosing the charges $+2q$ and $-q$. Either the field lines that end on $-q$ do not pass through the surface or they penetrate it from the inside the same number of times as from the outside. The net number that exit, the same as that for a single charge of $+q$, is equal to the net charge enclosed by the surface.

ELECTRIC FLUX

The mathematical quantity that corresponds to the number of field lines penetrating a surface is called the **electric flux** ϕ. For a surface perpendicular to $\vec{E}$ (Figure 22-16), the electric flux is the product of the magnitude of the field E and the area A:

$$\phi = EA$$

The units of electric flux are $N \cdot m^2/C$. Because E is proportional to the number of field lines per unit area, the flux is proportional to the number of field lines penetrating the surface.

In Figure 22-17, the surface of area A_2 is not perpendicular to the electric field $\vec{E}$. However, the number of lines that penetrate the surface of area A_2 is the same as the number that penetrate the surface of area A_1, which is normal (perpendicular) to $\vec{E}$. These areas are related by

$$A_2 \cos\theta = A_1 \qquad \text{22-11}$$

where θ is the angle between $\vec{E}$ and the unit vector $\hat{n}$ that is normal to the surface A_2, as shown in the figure. The electric flux through a surface is defined to be

$$\phi = \vec{E} \cdot \hat{n}A = EA\cos\theta = E_n A \qquad \text{22-12}$$

where $E_n = \vec{E} \cdot \hat{n}$ is the component of $\vec{E}$ normal to the surface.

Figure 22-18 shows a curved surface over which $\vec{E}$ may vary. If the area ΔA_i of the surface element that we choose is small enough, it can be modeled as a plane, and the variation of the electric field across the element can be neglected. The flux of the electric field through this element is

$$\Delta\phi_i = E_{ni}\,\Delta A_i = \vec{E}_i \cdot \hat{n}_i\,\Delta A_i$$

where $\hat{n}_i$ is the unit vector perpendicular to the surface element and $\vec{E}_i$ is the electric field on the surface element. If the surface is curved, the unit vectors for the different small surface elements will have different directions. The total flux through the surface is the sum of $\Delta\phi_i$ over all the elements making up the surface. In the limit, as the number of elements approaches infinity and the area of each element approaches zero, this sum becomes an integral. The general definition of electric flux is thus

$$\phi = \lim_{\Delta A_i \to 0} \sum_i \vec{E}_i \cdot \hat{n}_i\,\Delta A_i = \int_S \vec{E} \cdot \hat{n}\,dA \qquad \text{22-13}$$

DEFINITION—ELECTRIC FLUX

where the S stands for the surface we are integrating over.* The sign of the flux depends on the choice for the direction of the unit normal $\hat{n}$. By choosing $\hat{n}$ to be out of one side of a surface we are determining the sign of $\vec{E} \cdot \hat{n}$, and thus the sign of the flux through the surface.

On a *closed* surface we are interested in the electric flux through the surface, and by convention, we always choose the unit vector $\hat{n}$ to be out of the surface at each point. The integral over a closed surface is indicated by the symbol $\oint$. The total or net flux through a closed surface S is therefore written

$$\phi_{\text{net}} = \oint_S \vec{E} \cdot \hat{n}\,dA = \oint_S E_n\,dA \qquad \text{22-14}$$

* The flux of a vector field through a surface is a mathematical operation used to describe the flow rates of fluids and rates of heat transfers. In addition, it is used to relate electric fields with the charges that produce them.

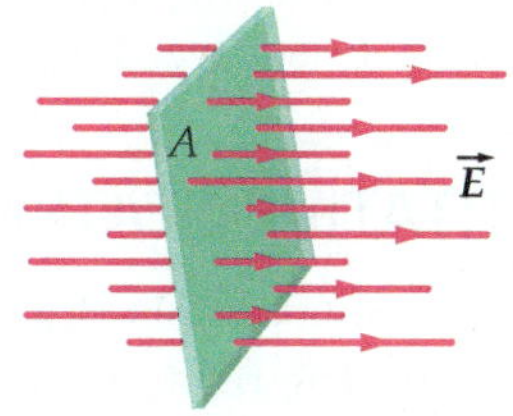

FIGURE 22-16 Electric field lines of a uniform field penetrating a surface of area A that is oriented perpendicular to the field. The product EA is the electric flux through the surface.

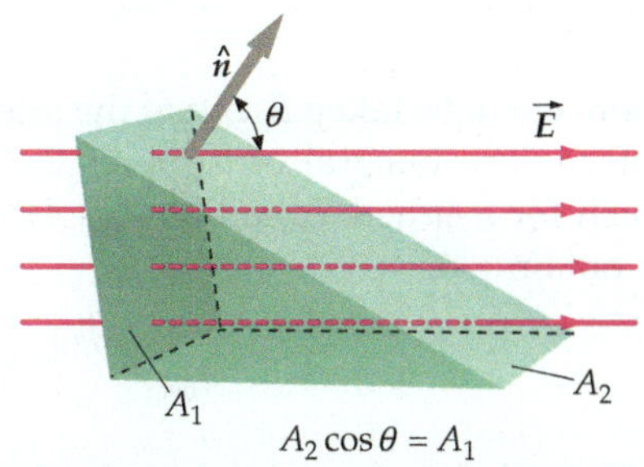

FIGURE 22-17 Electric field lines of a uniform electric field that is perpendicular to the surface of area A_1 but makes an angle θ with the unit vector $\hat{n}$ that is normal to the surface of area A_2. Where $\vec{E}$ is not perpendicular to the surface, the flux is E_nA, where $E_n = E\cos\theta$ is the component of $\vec{E}$ that is perpendicular to the surface. The flux through the surface of area A_2 is the same as that through the surface of area A_1.

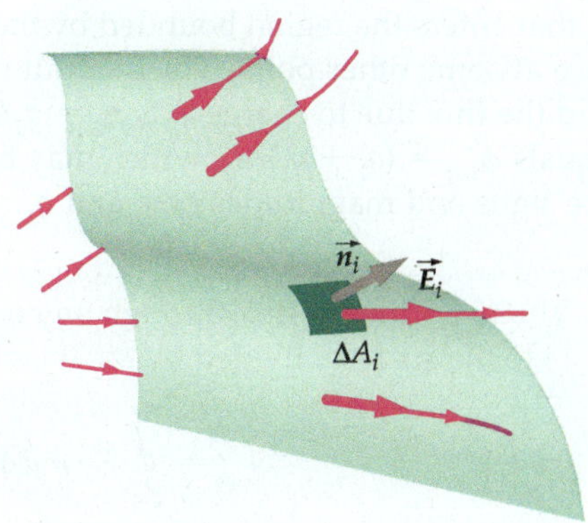

FIGURE 22-18 If E_n varies from place to place on a surface, either because the magnitude E varies or because the angle between $\vec{E}$ and $\hat{n}$ varies, the area of the surface is divided into small elements of area ΔA_i. The flux through the surface is computed by summing $\vec{E}_i \cdot \hat{n}_i\,\Delta A_i$ over all the area elements.

The net flux ϕ_{net} through the closed surface is positive or negative, depending on whether $\vec{E}$ is predominantly outward or inward at the surface. At points on the surface where $\vec{E}$ is inward, E_n is negative.

QUANTITATIVE STATEMENT OF GAUSS'S LAW

Figure 22-19 shows a spherical surface of radius R that has a point charge Q at its center. The electric field everywhere on this surface is normal to the surface and has the magnitude

$$E_n = \frac{kQ}{R^2}$$

The net flux of $\vec{E}$ out of this spherical surface is

$$\phi_{net} = \oint_S E_n\,dA = E_n \oint_S dA$$

where we have taken E_n out of the integral because it is constant everywhere on the surface. The integral of dA over the surface is just the total area of the surface, which for a sphere of radius R is $4\pi R^2$. Using this and substituting kQ/R^2 for E_n, we obtain

$$\phi_{net} = \frac{kQ}{R^2}4\pi R^2 = 4\pi kQ = Q/\epsilon_0 \qquad \text{22-15}$$

Thus, the net flux out of a spherical surface that has a point charge Q at its center is independent of the radius R of the sphere and is equal to Q divided by ϵ_0. This is consistent with our previous observation that the net number of lines through a closed surface is proportional to the net charge inside the surface. *This number of lines is the same for all closed surfaces surrounding the charge, independent of the shape of the surface.* Thus, the net flux out of *any surface* surrounding a point charge Q equals Q/ϵ_0.

We can extend this result to systems containing multiple charges. In Figure 22-20, the surface encloses two point charges, q_1 and q_2, and there is a third point charge q_3 outside the surface. Because the electric field at any point on the surface is the vector sum of the electric fields produced by each of the three charges, the net flux $\phi_{net} = \oint_S(\vec{E}_1 + \vec{E}_2 + \vec{E}_3)\cdot\hat{n}\,dA$ out of the surface is just the sum of the fluxes ($\phi_{net} = \Sigma\phi_i$, where $\phi_i = \oint_S \vec{E}_i\cdot\hat{n}\,dA$) due to the individual charges. The flux ϕ_3 (due to charge q_3 which is outside the surface) is zero because every field line from q_3 that enters the region bounded by the surface at one point leaves the region surface at some other point. The flux out of the surface due to charge q_1 is $\phi_1 = q_1/\epsilon_0$ and the flux due to charge q_2 is $\phi_2 = q_2/\epsilon_0$. The net flux out of the surface therefore equals $\phi_{net} = (q_1 + q_2)/\epsilon_0$, which may be positive, negative, or zero depending on the signs and magnitudes of q_1 and q_2.

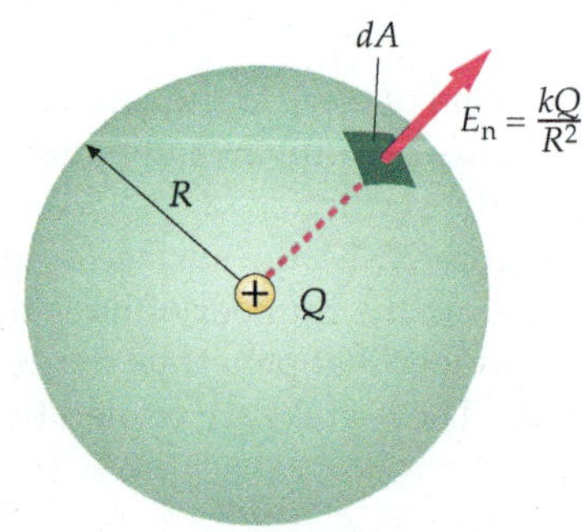

FIGURE 22-19 A spherical surface enclosing a point charge Q. The net flux is easily calculated for a spherical surface. It equals E_n multiplied by the surface area, or $E_n 4\pi R^2$.

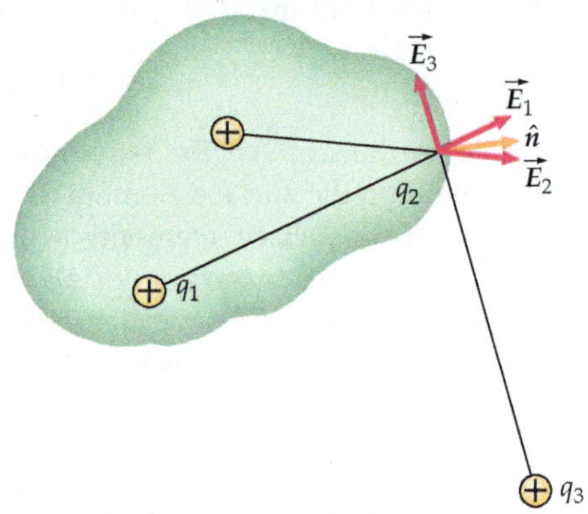

FIGURE 22-20 A surface enclosing point charges q_1 and q_2, but not q_3. The net flux out of this surface is $4\pi k(q_1 + q_2)$.

The net outward flux through any closed surface equals the net charge inside the surface divided by ϵ_0:

$$\phi_{net} = \oint_S \vec{E}\cdot\hat{n}\,dA = \oint_S E_n\,dA = \frac{Q_{inside}}{\epsilon_0} \qquad \text{22-16}$$

GAUSS'S LAW

This is **Gauss's law.** It reflects the fact that the electric field due to a single point charge varies inversely with the square of the distance from the charge. It was this property of the electric field that made it possible to draw a fixed number of electric field lines from a charge and have the density of lines be proportional to the field strength.

Gauss's law is valid for all surfaces and all charge distributions. For charge distributions that have high degrees of symmetry, it can be used to calculate the electric field, as we illustrate in the next section. For static charge distributions, Gauss's law and Coulomb's law are equivalent. However, Gauss's law is more general in that it is always valid whereas the validity of Coulomb's law is restricted to static charge distributions.

Example 22-9 Flux through a Piecewise-Continuous Closed Surface

An electric field is given by $\vec{E} = +(200\text{ N/C})\hat{k}$ throughout the region $z > 0$ and by $\vec{E} = -(200\text{ N/C})\hat{k}$ throughout the region $z < 0$. An imaginary soup-can-shaped surface that has a length equal to 20 cm and a radius R equal to 5.00 cm has its center at the origin and its axis along the z axis, so that one end is at $z = +10$ cm and the other is at $z = -10$ cm (Figure 22-21). (*a*) What is the net outward flux through the closed surface? (*b*) What is the net charge inside the closed surface?

PICTURE The closed surface described, which is piecewise continuous, consists of three pieces—two flat ends and a curved side. Separately calculate the flux of $\vec{E}$ out of each piece of this surface. To calculate the flux out of a piece draw the outward normal $\hat{n}$ at an arbitrarily chosen point on the piece and draw the vector $\vec{E}$ at the same point. If $E_n = \vec{E}\cdot\hat{n}$ is the same everywhere on the piece, then the outward flux through the piece is E_nA, where A is the area of the piece. The net outward flux through the entire closed surface is obtained by summing the fluxes through the individual pieces. The net outward flux is related to the charge inside by Gauss's law (Equation 22-16).

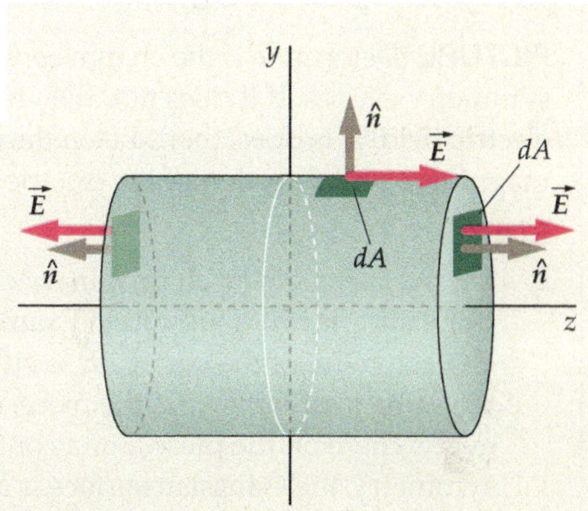

FIGURE 22-21

SOLVE

(*a*) 1. Sketch the soup-can-shaped surface. On each piece of the surface draw the outward normal $\hat{n}$ and the vector $\vec{E}$ (Figure 22-21):

2. Calculate the outward flux through the right end of the "can" (the piece of the surface at $z = +10$ cm). On this piece $\hat{n} = \hat{k}$:

$$\phi_{\text{right}} = \vec{E}_{\text{right}}\cdot\hat{n}_{\text{right}}A = \vec{E}_{\text{right}}\cdot\hat{k}\pi R^2 = +(200\text{ N/C})\hat{k}\cdot\hat{k}(\pi)(0.0500\text{ m})^2$$
$$= 1.57\text{ N}\cdot\text{m}^2/\text{C}$$

3. Calculate the outward flux through the left end of the "can" (the piece of the surface at $z = -10$ cm), where $\hat{n} = -\hat{k}$:

$$\phi_{\text{left}} = \vec{E}_{\text{left}}\cdot\hat{n}_{\text{left}}A = \vec{E}_{\text{left}}\cdot(-\hat{k})\pi R^2$$
$$= -(200\text{ N/C})\hat{k}\cdot(-\hat{k})(\pi)(0.0500\text{ m})^2$$
$$= 1.57\text{ N}\cdot\text{m}^2/\text{C}$$

4. Calculate the outward flux through the curved surface. On the curved surface $\hat{n}$ is in the radial direction, perpendicular to the z axis:

$$\phi_{\text{curved}} = \vec{E}_{\text{curved}}\cdot\hat{n}_{\text{curved}}A = 0$$

($\phi_{\text{curved}} = 0$ because $\vec{E}\cdot\hat{n} = 0$ everywhere on the curved piece.)

5. The net outward flux is the sum through all the individual surfaces:

$$\phi_{\text{net}} = \phi_{\text{right}} + \phi_{\text{left}} + \phi_{\text{curved}} = 1.57\text{ N}\cdot\text{m}^2/\text{C} + 1.57\text{ N}\cdot\text{m}^2/\text{C} + 0$$
$$= \boxed{3.14\text{ N}\cdot\text{m}^2/\text{C}}$$

(*b*) Gauss's law relates the charge inside to the net flux:

$$Q_{\text{inside}} = \epsilon_0\phi_{\text{net}} = (8.85\times10^{-12}\text{ C}^2/\text{N}\cdot\text{m}^2)(3.14\text{ N}\cdot\text{m}^2/\text{C})$$
$$= \boxed{2.78\times10^{-11}\text{ C} = 27.8\text{ pC}}$$

CHECK The flux through either end of the can does not depend on the length of the can. This result is expected for an electric field that does not vary with distance from the $z = 0$ plane.

TAKING IT FURTHER The net flux does not depend on the length of the can. Thus, the charge inside resides entirely on the $z = 0$ plane.

22-3 USING SYMMETRY TO CALCULATE WITH GAUSS'S LAW

Given a highly symmetrical charge distribution, the electric field can often be calculated more easily by using Gauss's law than it can by using Coulomb's law. There are three classes of symmetry to consider. A charge configuration has **cylindrical (or line) symmetry** if the charge density depends only on the distance from a line, **plane symmetry** if the charge density depends only on the distance from a plane, and **spherical (or point) symmetry** if the charge density depends only on the distance from a point.

PROBLEM-SOLVING STRATEGY

Calculating $\vec{E}$ Using Gauss's Law

PICTURE Determine if the charge configuration belongs to one of the three symmetry classes. If it does not, then try another method to calculate the electric field. If is does, then sketch the charge configuration and establish the magnitude and direction of the electric field $\vec{E}$ using symmetry considerations.

SOLVE

1. On the sketch draw an imaginary closed surface, called a **Gaussian surface** (for example, the soup can in Example 22-9). This surface is chosen so that on each piece of the surface $\vec{E}$ is either zero, normal to the surface with E_n the same everywhere on the piece, or parallel to the surface ($E_n = 0$) everywhere on the piece. For a configuration that has cylindrical (line) symmetry, the Gaussian surface is a cylinder coaxial with the symmetry line. For a configuration that has plane symmetry, the Gaussian surface is a cylinder bisected by the symmetry plane and with its symmetry axis normal to the symmetry plane. For a configuration that has spherical (point) symmetry, the Gaussian surface is a sphere centered on the symmetry point. On each piece of the Gaussian surface sketch an area element dA, an outward normal $\hat{n}$, and the electric field $\vec{E}$.
2. Closed cylindrical surfaces are piecewise continuous, with the surface divided into three pieces. Spherical surfaces consist of a single piece. The flux through each piece of a properly chosen Gaussian surface equals $E_n A$, where E_n is the component of $\vec{E}$ normal to the piece and A is the area of the piece. Add the fluxes to obtain the total outward flux through the closed surface.
3. Calculate the total charge inside the Gaussian surface.
4. Apply Gauss's law to relate E_n to the charges inside the closed surface and solve for E_n.

CONCEPT CHECK 22-1

Is the electric field $\vec{E}$ in Gauss's law only that part of the electric field due to the charges inside a surface, or is it the total electric field due to all the charges both inside and outside the surface?

Example 22-10 $\vec{E}$ Due to a Uniformly Charged Slab

A very large (infinite), uniformly charged slab of plastic of thickness $2a$ occupies the region between the $z = -a$ plane and the $z = +a$ plane. Find the electric field everywhere due this charge configuration. The charge per unit volume of the plastic is ρ.

PICTURE The charge configuration has plane symmetry, with the $z = 0$ plane as the symmetry plane. Use symmetry arguments to determine the direction of the electric field everywhere. Then, apply Gauss's law and solve for the electric field.

SOLVE

1. Use symmetry considerations to determine the direction of $\vec{E}$. Because the sheet is infinite, there is no preferred direction parallel to the sheet:

For $\rho > 0$, $\vec{E}$ points directly away from the $z = 0$ plane, and for $\rho < 0$, $\vec{E}$ points directly toward the $z = 0$ plane. On the $z = 0$ plane $\vec{E} = 0$.

2. Sketch the charge configuration that has a suitable Gaussian surface—a cylinder bisected by the symmetry plane (the $z = 0$ plane with its axis normal to the $z = 0$ plane). The cylinder extends from $-z$ to $+z$ (Figure 22-22):

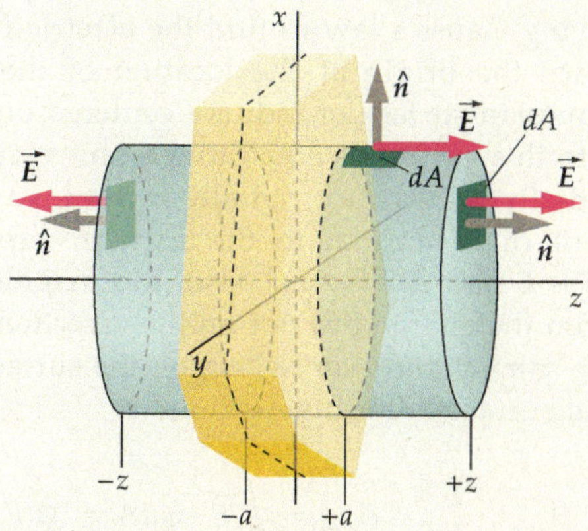

FIGURE 22-22 Gaussian surface for the calculation of $\vec{E}$ due to an infinite plane of charge. (Only the part of the plane that is inside the Gaussian surface is shown.) On the flat faces of this soup-can-shaped surface, $\vec{E}$ is perpendicular to the surface and constant in magnitude. On the curved surface $\vec{E}$ is parallel with the surface.

3. Write down Gauss's law (Equation 22-16):

$$\phi_{\text{net}} = \oint_S \vec{E}\cdot\hat{n}\,dA = \frac{Q_{\text{inside}}}{\epsilon_0}$$

4. The outward flux $\vec{E}$ through the surface is equal to the sum of the fluxes through each piece of the surface. Draw both $\hat{n}$ and $\vec{E}$ at an area element on each piece of the surface (Figure 22-22):

$$\phi_{\text{net}} = \phi_{\text{left end}} + \phi_{\text{right end}} + \phi_{\text{curved side}}$$

where $\phi_{\text{left end}} = \int_{\text{left end}} \vec{E}\cdot\hat{n}dA$

$$\phi_{\text{right end}} = \int_{\text{right end}} \vec{E}\cdot\hat{n}\,dA$$

$$\phi_{\text{curved side}} = \int_{\text{curved side}} \vec{E}\cdot\hat{n}\,dA$$

5. Because $\vec{E}\cdot\hat{n}$ is zero everywhere on the curved piece of the surface, the flux through the curved piece is zero:

$$\phi_{\text{curved side}} = 0$$

6. $\vec{E}$ is uniform on the right end of the surface, so $\vec{E}\cdot\hat{n} = E_n$ can be factored from the integral. Let A be the area of the end of right end of the surface:

$$\phi_{\text{right end}} = \int_{\text{right end}} \vec{E}\cdot\hat{n}\,dA = \int_{\text{right end}} E_n dA$$
$$= E_n \int_{\text{right end}} dA = E_n A$$

7. The two ends of the surface are the same distance from the symmetry plane (the $z = 0$ plane), so $\vec{E}$ on the left end is equal and opposite to $\vec{E}$ on the right end. The normals on the two ends are equal and opposite as well. Thus, $\vec{E}\cdot\hat{n} = E_n$ is the same on both ends. It follows that the flux out of both ends is the same as well:

$\vec{E}\cdot\hat{n} = E_n$ is the same on the two ends,

$$\therefore \phi_{\text{left end}} = \phi_{\text{right end}} = E_n A$$

8. Add the individual fluxes to get the net flux out of the surface:

$$\phi_{\text{net}} = \phi_{\text{left end}} + \phi_{\text{right end}} + \phi_{\text{curved side}} = E_n A + E_n A + 0 = 2E_n A$$

9. Solve for the charge inside the Gaussian surface. The volume of a cylinder is the cross-sectional area multiplied by the length. The cylinder has a length of $2z$.

$$Q_{\text{inside}} = \rho A 2a \qquad (z \geq a)$$
$$Q_{\text{inside}} = \rho A 2z \qquad (z \leq a)$$

10. Substitute the step-8 and step-9 results into $\phi_{\text{net}} = Q_{\text{inside}}/\epsilon_0$ (the step-3 result) and solve for E_n on the right end of the surface:

For $|z| \geq a$, $2E_n A = \rho A 2a/\epsilon_0$, so $E_n = \rho a/\epsilon_0$.
For $-a \leq z \leq a$, $2E_n A = \rho A 2|z|/\epsilon_0$, so $E_n = \rho|z|/\epsilon_0$.

11. Solve for $\vec{E}$ as a function of z. In the region $z < 0$, $\hat{n} = -\hat{k}$, so $E_z = -E_n$; this means $\vec{E}$ is in the $-z$ direction so E_z is negative:

$$\boxed{\vec{E} = E_z\hat{k} = \begin{cases} -(\rho a/\epsilon_0)\hat{k} & (z \leq -a) \\ (\rho z/\epsilon_0)\hat{k} & (-a \leq z \leq a) \\ +(\rho a/\epsilon_0)\hat{k} & (z \geq +a) \end{cases}}$$

or

$$\boxed{\vec{E} = E_z\hat{k} = \begin{cases} \text{sign}(z)\cdot(\rho a/\epsilon_0)\hat{k} & (|z| \geq a) \\ \text{sign}(z)\cdot(\rho|z|/\epsilon_0)\hat{k} & (|z| \leq a) \end{cases}}$$

CHECK The electric field has units of N/C. According to our step 11 results, $\rho a/\epsilon_0$ should have the same units. It does, as $\epsilon_0 = 8.85 \times 10^{-12}\ \text{C}^2/(\text{N}\cdot\text{m}^2)$, ρ has units of C/m^3, and a has units of m.

TAKING IT FURTHER Outside the slab the electric field is the same as that of the uniformly charged plane of Equation 22-10, with $\sigma = 2\rho a$. Figure 22-23 shows a graph of E_z versus z for the charged slab. Compare this graph with that of Figure 22-11 which shows a graph of E_z versus z for the charged plane. These graphs are readily compared if you recognize that $2\pi k = 1/(2\epsilon_0)$.

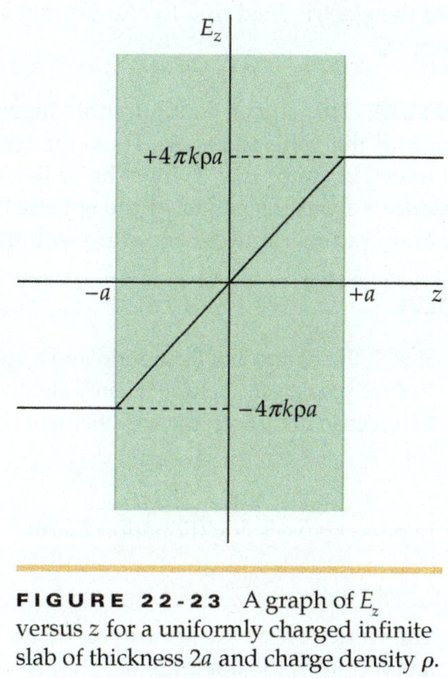

FIGURE 22-23 A graph of E_z versus z for a uniformly charged infinite slab of thickness $2a$ and charge density ρ.

We can use Gauss's law to derive Coulomb's law. This is accomplished by applying Gauss's law to find the electric field a distance r from a point charge q. Place the origin at the location of the point charge and choose a spherical Gaussian surface of radius r centered on the point charge. The outward normal $\hat{n}$ to this surface is equal to the unit vector $\hat{r}$. By symmetry, $\vec{E}$ is directed either radially outward or radially inward, so $\vec{E} = E_r\hat{r}$. It follows that E_n, the component of $\vec{E}$ normal to the surface, equals the radial component $\vec{E}$. That is, $E_n = \vec{E}\cdot\hat{n} = \vec{E}\cdot\hat{r} = E_r$. Also, the magnitude of $\vec{E}$ can depend on the distance from the charge but not on the direction from the charge. It follows that E_n has the same value everywhere on the surface. The net flux of $\vec{E}$ through the spherical surface of radius r is thus

$$\phi_{\text{net}} = \oint_S \vec{E}\cdot\hat{n}\,dA = \oint_S E_n\,dA = E_n\oint_S dA = E_r 4\pi r^2$$

where $\oint_S dA = 4\pi r^2$ (the area of the spherical surface). Because the total charge inside the surface is just the point charge q, Gauss's law gives

$$E_r 4\pi r^2 = \frac{q}{\epsilon_0}$$

Solving for E_r gives

$$E_r = \frac{1}{4\pi\epsilon_0}\frac{q}{r^2}$$

which is Coulomb's law. We have thus derived Coulomb's law from Gauss's law. Because for static charges Gauss's law can also be derived from Coulomb's law (see Section 22-6), we have shown that the two laws are equivalent (for static charges).

Example 22-11 $\vec{E}$ Due to a Thin Spherical Shell of Charge

Find the electric field due to a uniformly charged thin spherical shell of radius R and total charge Q.

PICTURE This charge configuration depends only on the distance from a single point—the center of the spherical shell. Thus, the configuration has spherical (point) symmetry. This symmetry dictates that $\vec{E}$ must be radial and have a magnitude that depends only on the distance r from the center of the spherical shell. A spherical Gaussian surface that has an arbitrary radius r and is concentric with the charge configuration is needed.

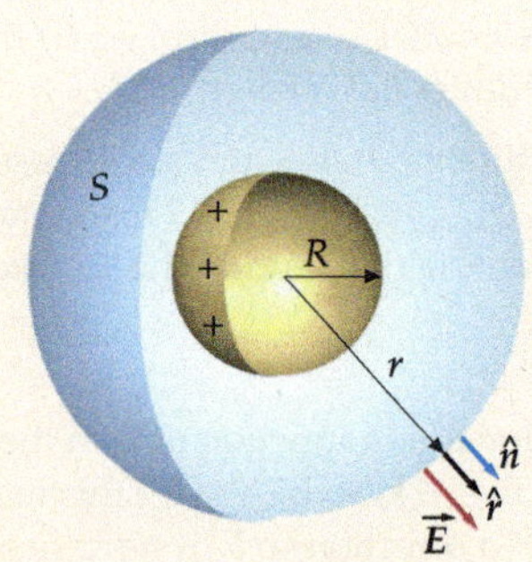

FIGURE 22-24 Spherical Gaussian surface of radius $r > R$ for the calculation of the electric field outside a uniformly charged thin spherical shell of radius R.

SOLVE

1. Sketch the charge configuration and a spherical Gaussian surface S of radius $r > R$. Include an area element dA, the normal $\hat{n}$, and the electric field $\vec{E}$ on the area element (Figure 22-24):

2. Express Gauss's law (Equation 22-16):
$$\phi_{\text{net}} = \oint_S E_n\,dA = \frac{Q_{\text{inside}}}{\epsilon_0}$$

3. The value of E_n is the same everywhere on S. Thus we can factor it from the integral:
$$E_n\oint_S dA = \frac{Q_{\text{inside}}}{\epsilon_0}$$

4. The integral of the area element over the surface S is just the area of the sphere. The area of the sphere is $4\pi r^2$:

$$E_n 4\pi r^2 = \frac{Q_{\text{inside}}}{\epsilon_0}$$

5. Due to the symmetry, $E_n = E_r$. Substitute E_r for E_n and solve for E_r:

$$E_r = \frac{1}{4\pi\epsilon_0}\frac{Q_{\text{inside}}}{r^2}$$

6. For $r > R$, $Q_{\text{inside}} = Q$. For $r < R$, $Q_{\text{inside}} = 0$:

$$\vec{E} = E_r\hat{r}, \quad \text{where} \quad \boxed{\begin{aligned} E_r &= \frac{1}{4\pi\epsilon_0}\frac{Q}{r^2} \qquad r > R \\ E_r &= 0 \qquad r < R \end{aligned}}$$

CHECK Outside the charged shell, the electric field is the same as that of a point charge Q at the shell's center. This result is expected for $r \gg R$.

TAKING IT FURTHER The step-6 result can also be obtained by direct integration of Coulomb's law, but that calculation is much more challenging.

Figure 22-25 shows E_r versus r for a spherical-shell charge distribution. Again, note that the electric field is discontinuous at $r = R$, where the surface charge density is $\sigma = Q/(4\pi R^2)$. Just outside the shell, the electric field is $E_r = Q/(4\pi\epsilon_0 R^2) = \sigma/\epsilon_0$, because $\sigma = Q/4\pi R^2$. Because the field just inside the shell is zero, the electric field is discontinuous at $r = R$ by the amount σ/ϵ_0.

The electric field of a uniformly charged thin spherical shell is given by $\vec{E} = E_r\hat{r}$, where

$$E_r = \frac{1}{4\pi\epsilon_0}\frac{Q}{r^2} \qquad r > R \qquad \text{22-17}a$$

$$E_r = 0 \qquad r < R \qquad \text{22-17}b$$

(a)

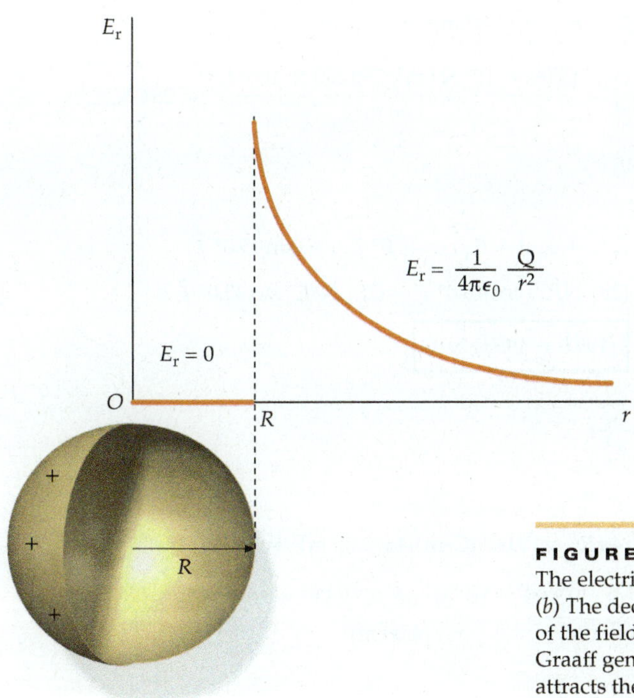

(b)

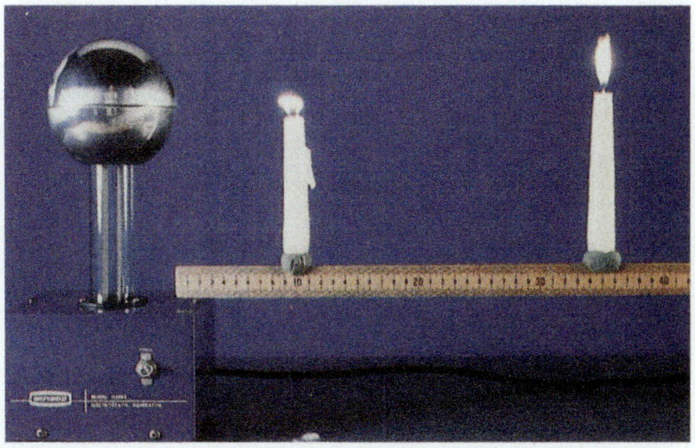

FIGURE 22-25 (a) A plot of E_r versus r for a thin spherical shell charge distribution. The electric field is discontinuous at $r = R$, where there is a surface charge of density σ. (b) The decrease in E_r over distance due to a charged spherical shell is evident by the effect of the field on the flames of two candles. The spherical shell at the left (part of a Van de Graaff generator, a device that is discussed in Chapter 23) has a large negative charge that attracts the positive ions in the nearby candle flame. The flame at right, which is much farther away, is not noticeably affected. *(Runk/Schoenberger from Grant Heilmann.)*

Example 22-12 Electric Field Due to a Point Charge and a Charged Spherical Shell

A spherical shell of radius $R = 3.00$ m has its center at the origin and has a surface charge density of $\sigma = 3.00\ \text{nC/m}^2$. A point charge $q = 250$ nC is on the y axis at $y = 2.00$ m. Find the electric field on the x axis at (*a*) $x = 2.00$ m and (*b*) $x = 4.00$ m.

PICTURE We separately find the field due to the point charge and that due to the spherical shell and sum the field vectors in accord with the principle of superposition. For Part (*a*), the field point is inside the shell, so the field is due only to the point charge (Figure 22-26*a*). For Part (*b*) , the field point is outside the shell, so the field due to the shell can be calculated as if the charge were a point charge at the origin. We then add the fields due to the two point charges (Figure 22-26*b*).

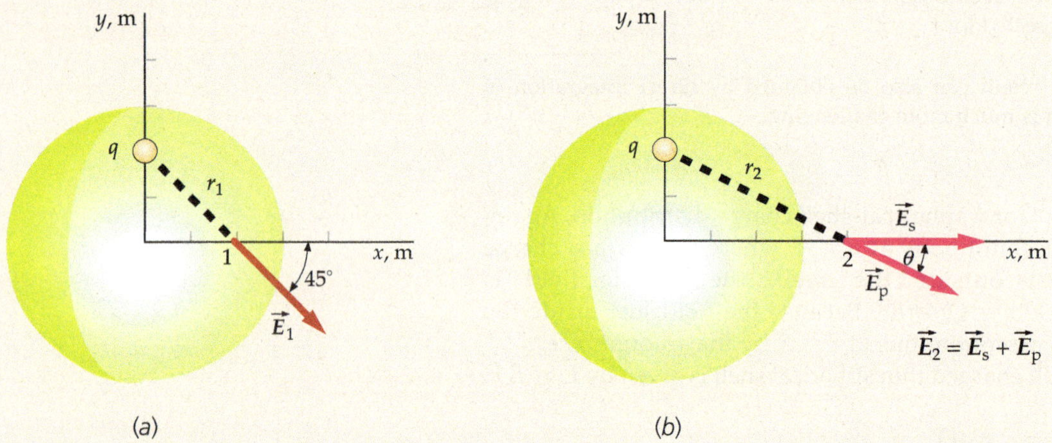

FIGURE 22-26

SOLVE

(*a*) 1. Inside the shell, $\vec{E}_1$ is due only to the point charge:

$$\vec{E}_1 = \frac{kq}{r_1^2}\hat{r}_1$$

2. Calculate the square of the distance r_1:

$$r_1^2 = (2.00\ \text{m})^2 + (2.00\ \text{m})^2 = 8.00\ \text{m}^2$$

3. Use r_1 to calculate the magnitude of the field:

$$E_1 = \frac{kq}{r_1^2} = \frac{(8.99 \times 10^9\ \text{N}\cdot\text{m}^2/\text{C}^2)(250 \times 10^{-9}\ \text{C})}{8.00\ \text{m}^2} = 281\ \text{N/C}$$

4. From Figure 22-26*a*, we can see that the field makes an angle of 45° with the x axis:

$$\theta_1 = 45.0°$$

5. Express $\vec{E}_1$ in terms of its components:

$$\vec{E}_1 = E_{1x}\hat{i} + E_{1y}\hat{j} = E_1 \cos 45.0°\ \hat{i} - E_1 \sin 45.0°\ \hat{j}$$
$$= (281\ \text{N/C}) \cos 45.0°\ \hat{i} - (281\ \text{N/C}) \sin 45.0°\ \hat{j}$$
$$= \boxed{(199\hat{i} - 199\hat{j})\ \text{N/C}}$$

(*b*) 1. Outside of its perimeter, the field of the shell can be calculated as if the shell were a point charge at the origin, and the field due to the shell $\vec{E}_S$ is therefore along the x axis:

$$\vec{E}_S = \frac{kQ}{x_2^2}\hat{i}$$

2. Calculate the total charge Q on the shell:

$$Q = \sigma 4\pi R^2 = (3.00\ \text{nC/m}^2)4\pi(3.00\ \text{m})^2 = 339\ \text{nC}$$

3. Use Q to calculate the field due to the shell:

$$E_S = \frac{kQ}{x_2^2} = \frac{(8.99 \times 10^9\ \text{N}\cdot\text{m}^2/\text{C}^2)(339 \times 10^{-9}\text{C})}{(4.00\ \text{m})^2} = 190\ \text{N/C}$$

4. The field due to the point charge is:

$$\vec{E}_P = \frac{kq}{r_2^2}\hat{r}_2$$

5. Calculate the square of the distance from the point charge q on the y axis to the field point at $x = 4.00$ m:

$$r_2^2 = (2.00\text{ m})^2 + (4.00\text{ m})^2 = 20.0\text{ m}^2$$

6. Calculate the magnitude of the field due to the point charge:

$$E_\text{P} = \frac{kq}{r_2^2} = \frac{(8.99 \times 10^9\text{ N}\cdot\text{m}^2/\text{C}^2)(250 \times 10^{-9}\text{ C})}{20.0\text{ m}^2} = 112\text{ N/C}$$

7. This field makes an angle θ with the x axis, where:

$$\tan\theta = \frac{2.00\text{ m}}{4.00\text{ m}} = 0.500 \Rightarrow \theta = \tan^{-1}0.500 = 26.6°$$

8. The x and y components of the net electric field are thus:

$$E_x = E_{\text{P}x} + E_{\text{S}x} = E_\text{P}\cos\theta + E_\text{S}$$
$$= (112\text{ N/C})\cos 26.6° + 190\text{ N/C} = 290\text{ N/C}$$
$$E_y = E_{\text{P}y} + E_{\text{S}y} = -E_\text{P}\sin\theta + 0$$
$$= -(112\text{ N/C})\sin 26.6° = -50.0\text{ N/C}$$
$$\vec{E}_2 = \boxed{(290\hat{i} - 50.0\hat{j})\text{ N/C}}$$

CHECK The Part (*b*), step 8 result is qualitatively in agreement with Figure 22-26*b*. That is, E_x is positive, E_y is negative, and $|E_y| < E_x$.

TAKING IT FURTHER Specifying the x, y, and z components of a vector completely specifies the vector. In these cases, the z component is zero.

$\vec{E}$ DUE TO A UNIFORMLY CHARGED SPHERE

Example 22-13 $\vec{E}$ Due to a Uniformly Charged Solid Sphere

Find the electric field everywhere for a uniformly charged solid sphere that has a radius R and a total charge Q that is uniformly distributed throughout the volume of the sphere that has a charge density $\rho = Q/V$, where $V = \frac{4}{3}\pi R^3$ is the volume of the sphere.

PICTURE The charge configuration has spherical symmetry. By symmetry, the electric field must be radial. We choose a spherical Gaussian surface of radius r (Figure 22-27*a* and Figure 22-27*b*). On the Gaussian surface, E_n is the same everywhere, and $E_n = E_r$. Gauss's law thus relates E_r to the total charge inside the Gaussian surface.

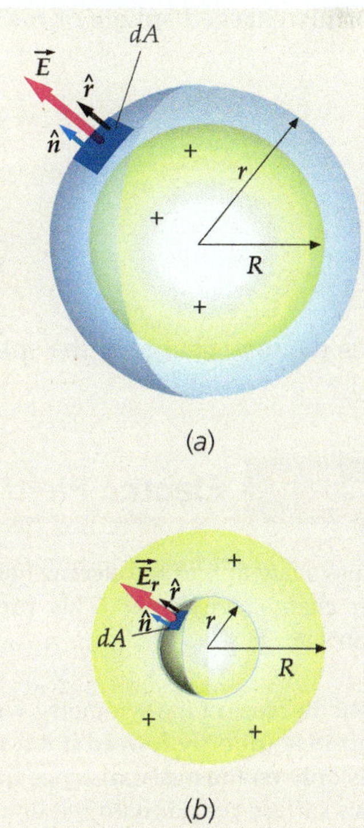

FIGURE 22-27

SOLVE

1. Draw a charged sphere of radius R and draw a spherical Gaussian surface with radius r (Figure 22-27*a* for $r > R$ and Figure 22-27*b* for $r < R$):

2. Relate the flux through the Gaussian surface to the electric field E_r on it. At every point on this surface $\hat{n} = \hat{r}$ and E_r has the same value:

$$\phi_\text{net} = \vec{E}\cdot\hat{n}A = \vec{E}\cdot\hat{r}A = E_r 4\pi r^2$$

(The surface area of a sphere of radius r is $4\pi r^2$.)

3. Apply Gauss's law to relate the field to the total charge inside the surface:

$$E_r 4\pi r^2 = \frac{Q_\text{inside}}{\epsilon_0}$$

4. Find Q_inside for all values of r. The charge density $\rho = Q/V$, where $V = \frac{4}{3}\pi R^3$:

For $r \geq R$, $Q_\text{inside} = Q$

For $r \leq R$, $Q_\text{inside} = \rho V'$,

where $V' = \frac{4}{3}\pi r^3$

so

$$Q_\text{inside} = \frac{Q}{V}V' = \frac{Q}{\frac{4}{3}\pi R^3}\frac{4}{3}\pi r^3 = Q\frac{r^3}{R^3}$$

5. Substitute into the step 3 result and solve for $\vec{E}$:

$\vec{E} = E_r\hat{r}$, where

$$E_r = \frac{1}{4\pi\epsilon_0}\frac{Q}{r^2} \qquad r \geq R$$

$$E_r = \frac{1}{4\pi\epsilon_0}\frac{Q}{r^2}\frac{r^3}{R^3} = \frac{1}{4\pi\epsilon_0}\frac{Q}{R^3}r \qquad r \leq R$$

CHECK At the center of the charged sphere the electric field is zero, as symmetry suggests. For $r > R$, the field is identical to the field of a point charge Q at the center of the sphere, also as expected.

TAKING IT FURTHER Figure 22-28 shows E_r versus r for the charge distribution in this example. Inside the sphere of charge, E_r increases with r. Note that E_r is continuous at $r = R$. A uniformly charged sphere is sometimes used as a model to describe the electric field of an atomic nucleus.

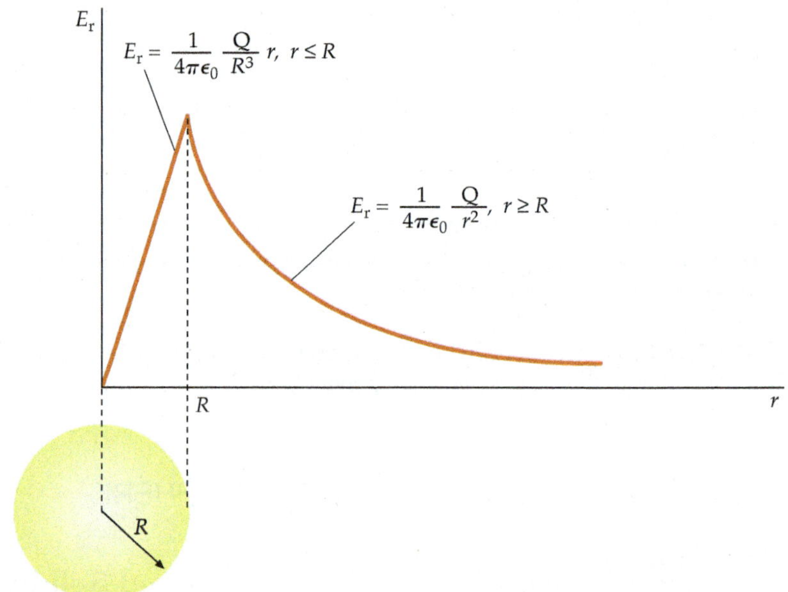

FIGURE 22-28

We see from Example 22-13 that the electric field a distance r from the center of a uniformly charged sphere of radius R is given by $\vec{E} = E_r\hat{r}$, where

$$E_r = \frac{1}{4\pi\epsilon_0}\frac{Q}{r^2} \qquad r \geq R \qquad \text{22-18}a$$

$$E_r = \frac{1}{4\pi\epsilon_0}\frac{Q}{R^3}r \qquad r \leq R \qquad \text{22-18}b$$

and Q is the total charge of the sphere.

Example 22-14 Electric Field Due to Infinite Line Charge

Use Gauss's law to find the electric field everywhere due to an infinitely long line charge of uniform charge density λ. (This problem was already solved in Example 22-3 using Coulomb's law.)

PICTURE Because of the symmetry, we know the electric field is directed away from the line if λ is positive (directly toward it if λ is negative), and we know the magnitude of the field depends only on the radial distance from the line charge. We therefore choose a cylindrical Gaussian surface coaxial with the line charge. We calculate the outward flux of $\vec{E}$ through each piece of the surface, and, using Gauss's law, relate the net outward flux of $\vec{E}$ to the charge inside the cylinder.

SOLVE

1. Sketch the wire and a coaxial cylindrical Gaussian surface (Figure 22-29) that has a length L and a radius R. The closed surface consists of three pieces: the two flat ends and the curved side. At a randomly chosen point on each piece, draw an area element and the vectors $\vec{E}$ and $\hat{n}$. Because of the symmetry, we know that the direction of $\vec{E}$ is radial (either toward or away from the line charge), and we know that the magnitude E depends only on the distance from the line charge.

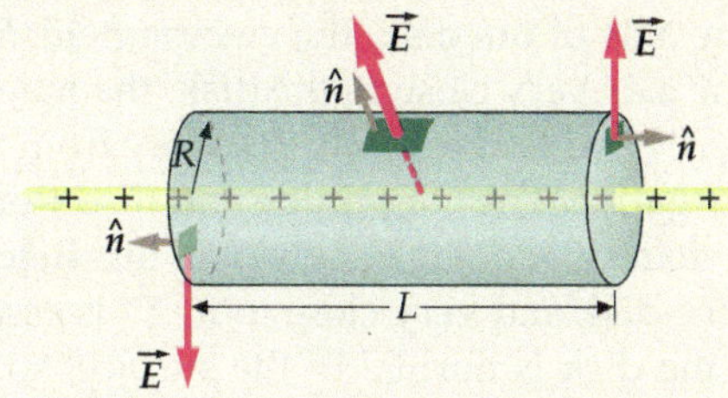

FIGURE 22-29

2. Calculate the outward flux through the curved piece of the Gaussian surface. At each point on the curved piece $\hat{R} = \hat{n}$, where $\hat{R}$ is the unit vector in the radial direction.

$$\phi_{\text{curved}} = \vec{E}\cdot\hat{n}A_{\text{curved}} = \vec{E}\cdot\hat{R}A_{\text{curved}} = E_R 2\pi RL$$

3. Calculate the outward flux through each of the flat ends of the Gaussian surface. On these pieces the direction of $\hat{n}$ is parallel with the line charge (and thus perpendicular to $\vec{E}$):

$$\phi_{\text{left}} = \vec{E}\cdot\hat{n}A_{\text{left}} = 0$$
$$\phi_{\text{right}} = \vec{E}\cdot\hat{n}A_{\text{right}} = 0$$

4. Apply Gauss's law to relate the field to the total charge inside the surface Q_{inside}. The net flux out of the Gaussian surface is the sum of the fluxes out of the three pieces of the surface, and Q_{inside} is the charge on a length L of the line charge:

$$\phi_{\text{net}} = \frac{Q_{\text{inside}}}{\epsilon_0}$$

$$E_R 2\pi RL = \frac{\lambda L}{\epsilon_0} \quad \text{so} \quad \vec{E} = E_R\hat{R}, \text{ where } \boxed{E_R = \frac{1}{2\pi\epsilon_0}\frac{\lambda}{R}}$$

CHECK Because $1/(2\pi\epsilon_0) = 2k$, the step-4 result can also be written $2k\lambda/R$. This is the same expression for E_R that was obtained by using Coulomb's law (see Example 22-3).

In the calculation of $\vec{E}$ for a line charge (Example 22-14), we needed to assume that the field point was very far from the ends of the line charge so that E_n would be constant everywhere on the cylindrical Gaussian surface. If we are near the end of a finite line charge, we cannot assume that $\vec{E}$ is perpendicular to the curved surface of the cylinder, or that E_n is constant everywhere on it, so we cannot use Gauss's law to calculate the electric field.

It is important to realize that although Gauss's law holds for any closed surface and any charge distribution, it is particularly useful for calculating the electric fields of charge distributions that have cylindrical, spherical, or plane symmetry. It is also particularly useful doing calculations involving conductors in electrostatic equilibrium, as we shall see in Section 22-5.

22-4 DISCONTINUITY OF

We have seen that the electric field for an infinite plane of charge and a thin spherical shell of charge is discontinuous by the amount σ/ϵ_0 at a surface having charge density σ. We now show that this is a general result for the component of the electric field that is perpendicular to a surface having a charge density of σ.

Figure 22-30 shows an arbitrary surface having a surface charge density σ. The surface is arbitrary in that it is arbitrarily curved, although it does not have any sharp folds where the normal direction is ambiguous, and σ may vary continuously on the surface from place to place. We consider electric field $\vec{E}$ in the vicinity of a point P on the surface as the superposition of electric field $\vec{E}_{\text{disk}}$, due just to the charge on a small disk centered at point P, and the electric field $\vec{E}'$ due to all other charges in the universe. Thus,

$$\vec{E} = \vec{E}_{\text{disk}} + \vec{E}' \qquad 22\text{-}19$$

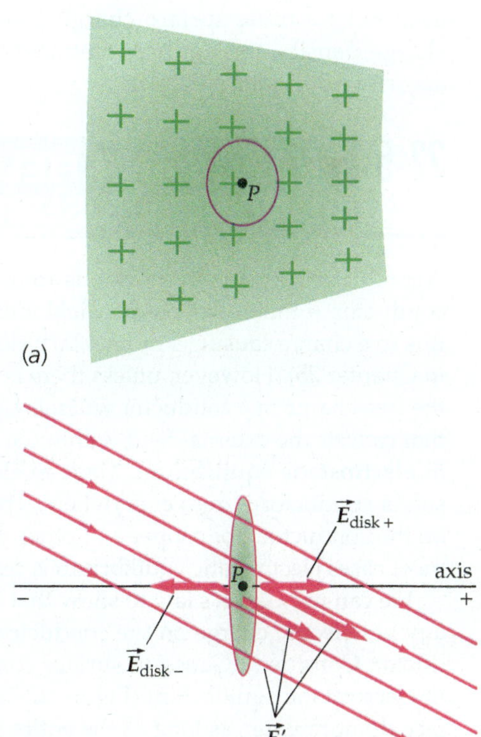

FIGURE 22-30 (*a*) A surface having surface charge. (*b*) The electric field $\vec{E}_{\text{disk}}$ due to the charge on a circular disk, plus the electric field $\vec{E}'$ due to all other charges.

The disk is small enough that it may be considered both flat and uniformly charged. On the axis of the disk, the electric field $\vec{E}_{\text{disk}}$ is given by Equation 22-9. At points on the axis very close to the disk, the magnitude of this field is given by $E_{\text{disk}} = |\sigma|/(2\epsilon_0)$. The direction of $\vec{E}_{\text{disk}}$ is away from the disk if σ is positive, and toward it if σ is negative. The magnitude and direction of the electric field $\vec{E}\,'$ are unknown. In the vicinity of point P, however, this field is continuous. Thus, at points on the axis of the disk and very close to it, $\vec{E}\,'$ is essentially uniform.

The axis of the disk is normal to the surface, so vector components along this axis can be referred to as normal components. The normal components of the vectors in Equation 22-19 are related by $E_n = E_{\text{disk}\,n} + E'_n$. If we refer to one side of the surface as the + side, and the other side as the − side, then $E_{n+} = \dfrac{\sigma}{2\epsilon_0} + E'_{n+}$ and $E_{n-} = -\dfrac{\sigma}{2\epsilon_0} + E'_{n+}$. Thus, E_n changes discontinuously from one side of the surface to the other. That is,

$$\Delta E_n = E_{n+} - E_{n-} = \frac{\sigma}{2\epsilon_0} - \left(-\frac{\sigma}{2\epsilon_0}\right) = \frac{\sigma}{\epsilon_0} \qquad \text{22-20}$$

DISCONTINUITY OF E_n AT A SURFACE CHARGE

where we have made use of the fact that near the disk $E'_{n+} = E'_{n-}$ (because $\vec{E}\,'$ is continuous and uniform).

Note that the discontinuity of E_n occurs at a finite disk of charge, an infinite plane of charge (see Figure 22-12), and a thin spherical shell of charge (see Figure 22-25). However, it does not occur at the perimeter of a solid sphere of charge (see Figure 22-28). The electric field is discontinuous at any location with an infinite volume charge density. These include locations that each have a finite point charge, locations that each have a finite line charge density, and locations that each have a finite surface charge density. At all locations with a finite surface charge density, the normal component of the electric field is discontinuous—in accord with Equation 22-20.

22-5 CHARGE AND FIELD AT CONDUCTOR SURFACES

A conductor contains an enormous amount of charge that can move freely within the conductor. If there is an electric field within a conductor, there will be a net force on this free charge causing a momentary electric current (electric currents are discussed in Chapter 25). However, unless there is a source of energy to maintain this current, the free charge in a conductor will merely redistribute itself to create an electric field that cancels the external field within the conductor. The conductor is then said to be in **electrostatic equilibrium.** Thus, in electrostatic equilibrium, the electric field inside a conductor is zero everywhere. The time taken to reach equilibrium depends on the conductor. For copper and other metal conductors, the time is so small that in most cases electrostatic equilibrium is reached in a few nanoseconds.*

We can use Gauss's law to show that for a conductor in electrostatic equilibrium, any net electric charge on the conductor resides entirely on the surface of the conductor. Consider a Gaussian surface completely inside the material of a conductor in electrostatic equilibrium (Figure 22-31). The size and shape of the Gaussian surface do not matter, as long as the entire surface is embedded within the material of the conductor. The electric field is zero everywhere on the Gaussian surface because the surface is completely within the conductor, where the field is everywhere zero. The net flux of the electric field through the surface must therefore be zero, and, by Gauss's law, the net charge inside the surface must be zero. Thus, there can be no

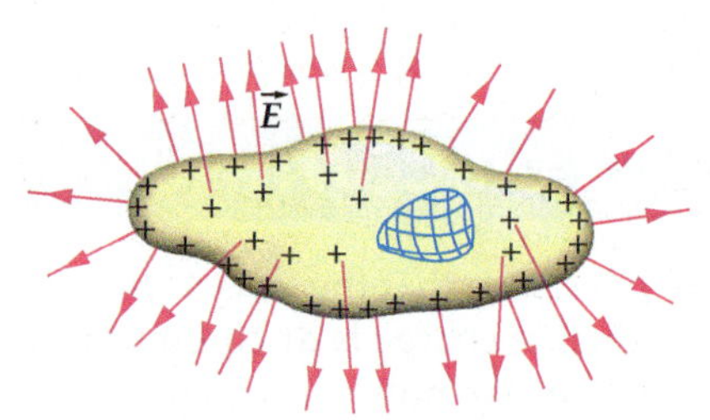

FIGURE 22-31 A Gaussian surface completely within the material of a conductor. Because the electric field is zero inside the material a conductor in electrostatic equilibrium, the net flux through this surface must also be zero. Therefore, the net charge density ρ must be zero everywhere within the material of a conductor.

* At very low temperatures some metals become superconducting. In a superconductor, a current is sustained for a much longer time, even without an energy source. Superconducting metals are discussed in Chapters 27 and 38.

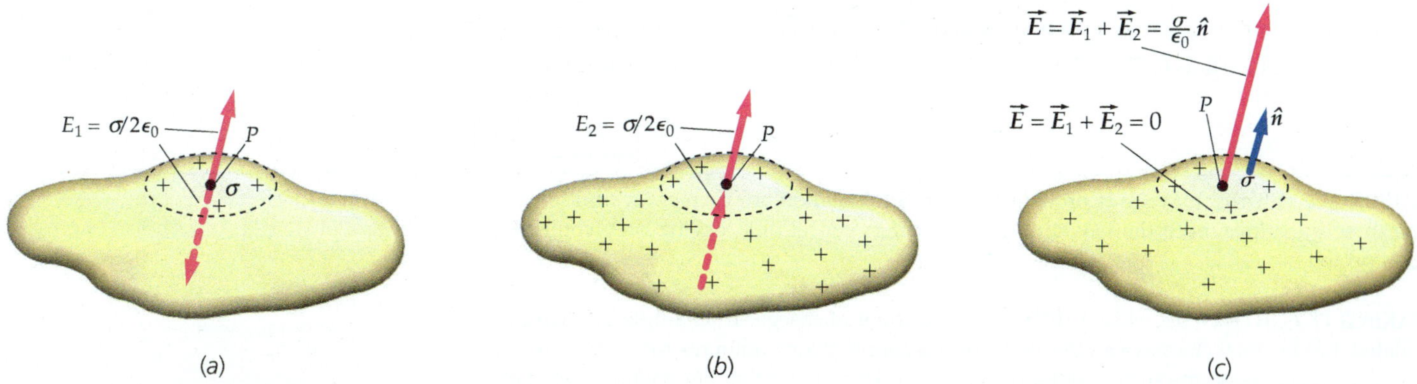

FIGURE 22-32 An arbitrarily shaped conductor having a charge on its surface. (*a*) The charge in the vicinity of point P near the surface looks like a small uniformly charged circular disk centered at P, giving an electric field of magnitude $\sigma/(2\epsilon_0)$ pointing away from the surface both inside and outside the surface. Inside the conductor, this field points away from point P in the opposite direction. (*b*) Because the net field inside the conductor is zero, the rest of the charges in the universe must produce a field of magnitude $\sigma/(2\epsilon_0)$ in the outward direction. The field due to these charges is the same just inside the surface as it is just outside the surface. (*c*) Inside the surface, the fields shown in (*a*) and (*b*) cancel, but outside they add to give $E_n = \sigma/\epsilon_0$.

net charge inside any surface lying completely within the material of the conductor. Therefore, if a conductor has a net charge, it must reside on the conductor's surface. At the surface of a conductor in electrostatic equilibrium, $\vec{E}$ must be perpendicular to the surface. (If the electric field did have a tangential component at the surface, the free charge would be accelerated tangential to the surface until electrostatic equilibrium was reestablished.)

Because E_n is discontinuous at any charged surface by the amount σ/ϵ_0, and because $\vec{E}$ is zero inside the material of a conductor, the field just outside the surface of a conductor is given by

$$E_n = \frac{\sigma}{\epsilon_0} \qquad \text{22-21}$$

E_n JUST OUTSIDE THE SURFACE OF A CONDUCTOR

This result is exactly twice the field produced by a uniform disk of surface charge density σ. We can understand this result from Figure 22-32. The charge on the conductor consists of two parts: (1) the charge near point P and (2) all the rest of the charge. The charge near point P looks like a small, uniformly charged circular disk centered at P that produces a field near P of magnitude $\sigma/(2\epsilon_0)$ just inside and just outside the conductor. The rest of the charges in the universe must produce a field of magnitude $\sigma/(2\epsilon_0)$ that exactly cancels the field inside the conductor. This field due to the rest of the charge in the universe adds to the field due to the small charged disk just outside the conductor to give a total field of σ/ϵ_0.

Example 22-15 The Charge of Earth — *Context-Rich*

While watching a science show on the atmosphere, you find out that on average the electric field of Earth is about 100 N/C directed vertically downward. Given that you have been studying electric fields in your physics class, you wonder if you can determine what the total charge on Earth's surface is.

PICTURE Earth is a conductor, so any charge it has resides on the surface of Earth. The surface charge density σ is related to the normal component of the electric field E_n by Equation 22-21. The total charge Q equals the charge density σ multiplied by the surface area A.

SOLVE

1. The surface charge density σ is related to the normal component of the electric field E_n by Equation 22-21:
   $$E_n = \frac{\sigma}{\epsilon_0}$$
2. On the surface of Earth, $\hat{n}$ is upward and $\vec{E}$ is downward so E_n is negative:
   $$E_n = \vec{E}\cdot\hat{n} = E\cos 180° = -E = -100\text{ N/C}$$
3. The charge Q is the charge per unit area multiplied by the area. Combine this with the step-1 and -2 results to obtain an expression for Q:
   $$Q = \sigma A = \epsilon_0 E_n A = -\epsilon_0 E A$$

4. The surface area of a sphere of radius r is given by $A = 4\pi r^2$: $Q = -\epsilon_0 EA = -\epsilon_0 E4\pi R_E^2 = -4\pi\epsilon_0 ER_E^2$

5. The radius of Earth is 6.37×10^6 m:

$$Q = -4\pi\epsilon_0 ER_E^2$$
$$= -4\pi(8.85 \times 10^{-12}\,\mathrm{C^2/N\cdot m^2})(100\,\mathrm{N/C})(6.37 \times 10^6\,\mathrm{m})^2$$
$$= \boxed{-4.51 \times 10^5\,\mathrm{C}}$$

CHECK We will check to see if units in the step 5 calculation are correct. In multiplying the three quantities, both the newtons and the meters cancel out, leaving only coulombs as expected.

TAKING IT FURTHER Is -4.53×10^5 C a large amount of charge? In Example 21-1 we calculated that the total charge of all the electrons in a copper penny amounts to -1.37×10^5 C, so the total charge on the surface of Earth is only 3.3 times larger than the total charge of all the electrons in a single copper penny.

Figure 22-33 shows a positive point charge q at the center of a spherical cavity inside a spherical conductor. Because the net charge must be zero within any Gaussian surface drawn within the material of the conductor, there must be a negative charge $-q$ induced on the surface of the cavity. In Figure 22-34, the point charge has been moved so that it is no longer at the center of the cavity. The field lines in the cavity are altered, and the surface charge density of the induced negative charge on the inner surface is no longer uniform. However, the positive surface charge density on the outside surface is not disturbed—it is still uniform—because it is electrically shielded from the cavity by the conducting material. The electric field of the point charge q and that of the surface charge $-q$ on the inner surface of the cavity superpose to produce an electric field that is exactly zero everywhere outside the cavity. This is obviously true if the point charge is at the center of the cavity, but it is true even if the point charge is somewhere else in the cavity. In addition, the surface charge on the outer surface of the conductor produces an electric field that is exactly zero everywhere inside the outer surface of the conductor. Furthermore, these statements are valid even if both the outer surface and the inner surface of the conductor are nonspherical.

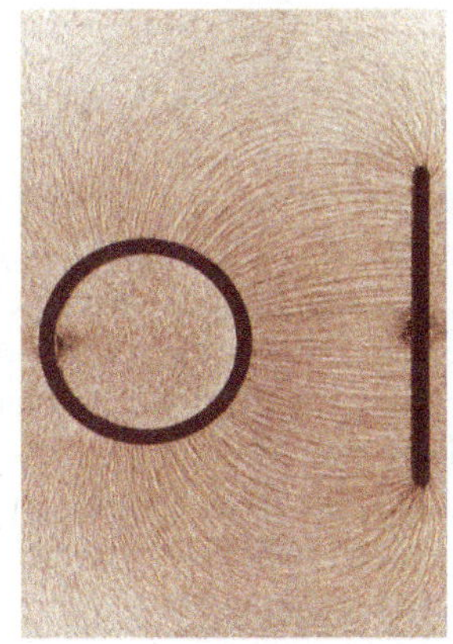

Electric field lines for an oppositely charged cylinder and plate, shown by bits of fine thread suspended in oil. Note that the field lines are normal to the surfaces of the conductors and that there are no lines inside the cylinder. The region inside the cylinder is electrically shielded from the region outside the cylinder. *(Harold M. Waage.)*

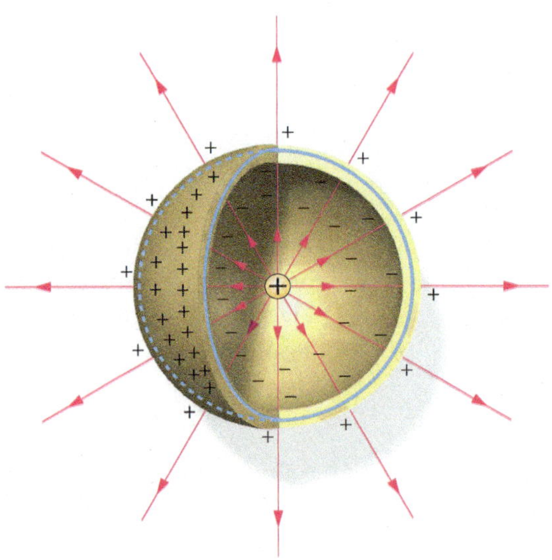

FIGURE 22-33 A point charge q in the cavity at the center of a thick spherical conducting shell. Because the net charge within the Gaussian surface (indicated in blue) must be zero, we know a surface charge $-q$ is induced on the inner surface of the shell, and because the conductor is neutral, an equal but opposite charge $+q$ is induced on the outer surface. Electric field lines begin on the point charge and end on the inner surface. Field lines begin again on the outer surface.

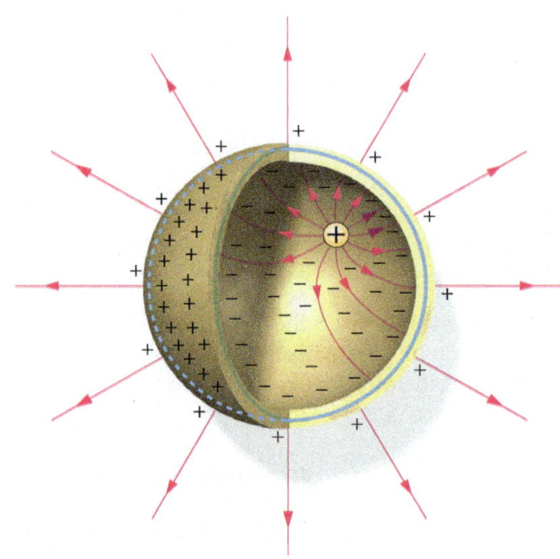

FIGURE 22-34 The same conductor as in Figure 22-33 with the point charge moved away from the center of the sphere. The charge on the outer surface and the electric field lines outside the sphere are not affected.

* 22-6 THE EQUIVALENCE OF GAUSS'S LAW AND COULOMB'S LAW IN ELECTROSTATICS

Gauss's law can be derived mathematically from Coulomb's law for the electrostatic case using the concept of the **solid angle.** Consider an area element ΔA on a spherical surface. The solid angle $\Delta\Omega$ subtended by ΔA at the center of the sphere is defined to be

$$\Delta\Omega = \frac{\Delta A}{r^2}$$

where r is the radius of the sphere. Because ΔA and r^2 both have dimensions of length squared, the solid angle is dimensionless. The SI unit of the solid angle is the **steradian** (sr). Because the total area of a sphere is $4\pi r^2$, the total solid angle subtended by a spherical surface is

$$\frac{4\pi r^2}{r^2} = 4\pi \text{ steradians}$$

There is a close analogy between the solid angle and the ordinary plane angle $\Delta\theta$, which is defined to be the ratio of an element of arc length of a circle Δs to the radius of the circle:

$$\Delta\theta = \frac{\Delta s}{r} \text{ radians}$$

The total plane angle subtended by a circle is 2π radians.

In Figure 22-35, the area element ΔA is not perpendicular to the radial lines from point O. The unit vector $\hat{n}$ normal to the area element makes an angle θ with the radial unit vector $\hat{r}$. In this case, the solid angle subtended by ΔA at point O is

$$\Delta\Omega = \frac{\Delta A\, \hat{n} \cdot \hat{r}}{r^2} = \frac{\Delta A \cos\theta}{r^2} \qquad 22\text{-}22$$

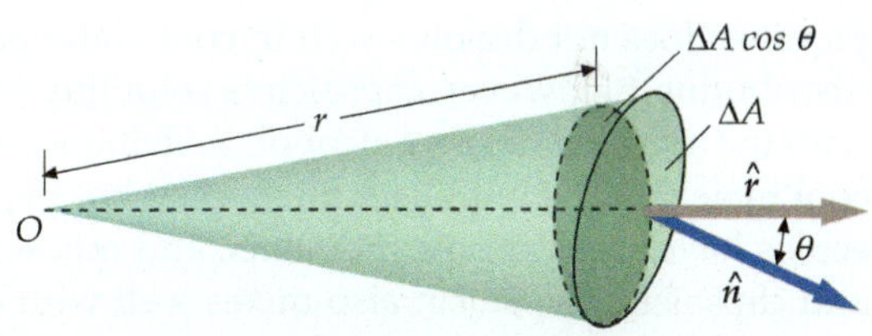

FIGURE 22-35 An area element ΔA whose normal is not parallel to the radial line from O to the center of the element. The solid angle subtended by this element at O is defined to be $(\Delta A \cos\theta)/r^2$.

The solid angle $\Delta\Omega$ is the same as that subtended by the corresponding area element of a spherical surface of any radius.

Figure 22-36 shows a point charge q surrounded by a surface of arbitrary shape. To calculate the flux of $\vec{E}$ through this surface, we want to find $\vec{E} \cdot \hat{n}\Delta A$ for each element of area on the surface and sum over the entire surface. The electric field at the area element shown is given by

$$\vec{E} = \frac{kq}{r^2}\hat{r}$$

so the flux through the element is

$$\Delta\phi = \vec{E} \cdot \hat{n}\Delta A = \frac{kq}{r^2}\hat{r} \cdot \hat{n}\Delta A = kq\, \Delta\Omega$$

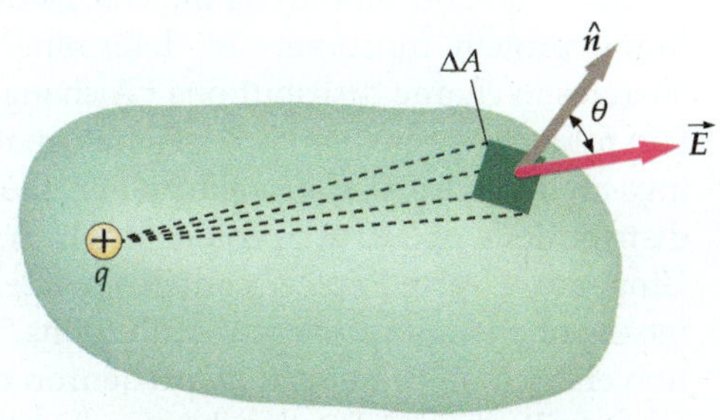

FIGURE 22-36 A point charge enclosed by an arbitrary surface. The flux through an area element ΔA is proportional to the solid angle subtended by the area element at the charge. The net flux through the surface, found by summing over all the area elements, is proportional to the total solid angle 4π at the charge, which is independent of the shape of the surface.

The sum of the fluxes through the entire surface is kq multiplied by the total solid angle subtended by the closed surface, which is 4π steradians:

$$\phi_{\text{net}} = \oint_S \vec{E} \cdot \hat{n}\, dA = kq \oint d\Omega = kq4\pi = 4\pi kq = \frac{q}{\epsilon_0} \qquad 22\text{-}23$$

which is Gauss's law.

Physics Spotlight

Charge Distribution—Hot and Cold

Electrical dipole moment, or *polarity*, affects the solubility of substances. Because water has such a strong electric dipole moment, it works very well as a solvent for other molecules that have both weak and strong dipole moments and ions. On the other hand, molecules without dipole moments, or molecules that are so big that they have large regions without dipole moments, do not dissolve well in water. Some oils, for example, do not have dipole moments and are immiscible with water.

Charge distributions that molecules can have also control whether substances that are not strictly classified as oils dissolve well in water. Anyone who has ever bitten into a very spicy-hot pepper and then taken a large drink of water can testify that water does not wash away the sensation of pain. Capsaicin, the active chemical in spicy-hot peppers such as habañeros, serranos, and piquins, does not dissolve well in cold water because of its charge distribution.* However, capsaicin's solubility in water is increased with the addition of ethyl alcohol, as demonstrated by people who cool their mouths with beer after spicy-hot peppers. Alcohol molecules have weak dipole moments, and mix well with both water and capsaicin. Capsaicin also mixes well with oils, some starches, and proteins. In many cultures, rice or meat, rather than alcohol, is used to dissolve capsaicin.

The molecules of the active ingredient in these spicy hot peppers do not dissolve in water because they do not have electric dipole moments. *(Stockbyte Platinum/Getty Images.)*

The sensation of pain that people who eat peppers feel is also due to the charge distributions in molecules. The protein TRPV1 is a neuron receptor in humans that signals how hot—temperature wise—something is. This protein has a charge distribution that is changed by a temperature above 43°C. Proteins change their shapes (fold and unfold) as the charge distribution changes across the molecule.† Many protein functions are determined by folding and unfolding caused by changes in charge distributions.‡ A change in charge distribution on a TRPV1 protein folds the protein and passes information about how hot a human's surroundings are to neurons. Capsaicin creates the same changes as heat does in the charge distributions of TRPV1 proteins,# which is why people perceive peppers as hot. Ginger, a "warm" spice, contains gingerols, which trigger similar receptors by means of changing charge distributions.° Menthol causes similar charge distribution changes in proteins that are neuron receptors in humans and signal how cold surroundings are.§ This is why people perceive mint as cool.

Changes in charge distributions of proteins can cause textural changes in proteins. The salting of caviar, for example, changes the charge distribution of proteins inside the fish eggs. As the proteins unfold, they thicken the formerly thin fluid inside the egg to a creamy texture.¶

---

* Turgut, C., Newby, B., and Cutright, T., "Determination of Optimal Water Solubility of Capsaicin for Its Usage as a Non-Toxic Antifoulant." *Environmental Science Pollution Research International*, Jan.–Feb. 2004, Vol. 11, No. 1, pp. 7–10.

† Suydam, I. T., et al., "Electric Fields at the Active Site of an Enzyme: Direct Comparison of Experiment with Theory." *Science*, Jul. 14, 2006, Vol. 313, No. 5784, pp. 200–204.

‡ Honig, B., and Nicholls, A., "Classical Electrostatics in Biology and Chemisty." *Science*, May 26, 1995, Vol. 268, p. 1144.

\# Montell, C., "Thermosensation: Hot Findings Make TRPNs Very Cool." *Current Biology*, Jun. 17, 2003, Vol. 13, No. 12, pp. R476–R478.

° Dedov, V. N., et al., "Gingerols: A Novel Class of Vanilloid Receptor (VR1) Agonists." *British Journal of Pharmacology*, 2002, Vol. 137, pp. 793–798.

§ Montell, C., op. cit.

¶ Sternin, V., and Dorè, I, *Caviar: The Resource Book*. Moscow: Cultura, 1993, in McGee, H., *On Food and Cooking: The Science and Lore of the Kitchen*. New York: Scribner, 2004.

Summary

1. Gauss's law is a fundamental law of physics that is equivalent to Coulomb's law for static charges.

2. For highly symmetric charge distributions, Gauss's law can be used to calculate the electric field.

TOPIC	RELEVANT EQUATIONS AND REMARKS	
1. Electric Field for a Continuous Charge Distribution	$\vec{E} = \int d\vec{E} = \int \frac{k\hat{r}}{r^2}\,dq$ (Coulomb's law) where $dq = \rho\,dV$ for a charge distributed throughout a volume, $dq = \sigma\,dA$ for a charge distributed on a surface, and $dq = \lambda\,dL$ for a charge distributed along a line.	22-1*b*
2. Electric Flux	$\phi = \lim_{\Delta A_i \to 0} \sum_i \vec{E}_i \cdot \hat{n}_i \Delta A_i = \int_S \vec{E} \cdot \hat{n}\,dA$	22-13
3. Gauss's Law	$\phi_{\text{net}} = \oint_S \vec{E} \cdot \hat{n}\,dA = \oint_S E_n\,dA = \frac{Q_{\text{inside}}}{\epsilon_0}$ The net outward electric flux through a closed surface equals the net charge within the surface divided by ϵ_0.	22-16
4. Coulomb Constant k and Electric Constant (Permittivity of Free Space) ϵ_0	$k = \frac{1}{4\pi\epsilon_0} = 8.99 \times 10^9\ \text{N}\cdot\text{m}^2/\text{C}^2$ $\epsilon_0 = \frac{1}{4\pi k} = 8.85 \times 10^{-12}\ \text{C}^2/(\text{N}\cdot\text{m}^2)$	22-7
5. Coulomb's Law and Gauss's Law	$\vec{E} = \frac{1}{4\pi\epsilon_0}\frac{q}{r^2}\hat{r}$	22-5
	$\phi_{\text{net}} = \oint_S E_n\,dA = \frac{Q_{\text{inside}}}{\epsilon_0}$	22-16
6. Discontinuity of E_n	At a surface having a surface charge density σ, the component of the electric field normal to the surface is discontinuous by σ/ϵ_0. $E_{n+} - E_{n-} = \frac{\sigma}{\epsilon_0}$	22-20
7. Charge on a Conductor	In electrostatic equilibrium, the charge density is zero throughout the material of the conductor. All excess or deficit charge resides on the surfaces of the conductor.	
8. $\vec{E}$ Just Outside a Conductor	The resultant electric field just outside the surface of a conductor is normal to the surface and has the magnitude σ/ϵ_0, where σ is the local surface charge density on the conductor: $E_n = \frac{\sigma}{\epsilon_0}$	22-21
9. Electric Fields for Selected Uniform Charge Distributions		
Of a line charge of infinite length	$E_R = 2k\frac{\lambda}{R} = \frac{1}{2\pi\epsilon_0}\frac{\lambda}{R}$	22-6
On the axis of a charged ring	$E_z = \frac{kQz}{(z^2 + a^2)^{3/2}}$	22-8
On the axis of a charged disk	$E_z = \text{sign}(z)\cdot\frac{\sigma}{2\epsilon_0}\left[1 - \left(1 + \frac{R^2}{z^2}\right)^{-1}\right]$	22-9

TOPIC	RELEVANT EQUATIONS AND REMARKS	
Of a charged infinite plane	$E_z = \text{sign}(z) \cdot \dfrac{\sigma}{2\epsilon_0}$	22-10
Of a charged thin spherical shell	$E_r = \dfrac{1}{4\pi\epsilon_0}\dfrac{Q}{r^2} \qquad r > R$	22-17*a*
	$E_r = 0 \qquad r < R$	22-17*b*

Answers to Concept Checks

22-1 The $\vec{E}$ in Gauss's law is the electric field due to all charges. However, the flux of the electric field due to all the charges outside the surface equals zero, so the flux of the electric field due to all charges equals the flux of the field due to the charges inside the surface alone.

Answers to Practice Problems

22-1 $E_x = k\lambda\left(\dfrac{1}{r_2} - \dfrac{1}{r_1}\right)$. For $x > x_2$, $r_2 < r_1$ so $\dfrac{1}{r_2} > \dfrac{1}{r_1}$ which means $E_x > 0$.

22-2 No. Symmetry dictates that E_z is zero at $z = 0$ whereas the equation in step 3 gives a negative value for E_z at $z = 0$. These contradictory results cannot both be valid.

22-3 The SI units for k, λ, and R are $\text{N} \cdot \text{m}^2/\text{C}^2$, C/m, and m, respectively. It follows that $k\lambda/R$ has units of $(\text{N} \cdot \text{m}^2/\text{C}^2)(\text{C/m})(1/\text{m}) = \text{N/C}$.

22-4 $z = a/\sqrt{2}$

22-5 80%

Problems

In a few problems, you are given more data than you actually need; in a few other problems, you are required to supply data from your general knowledge, outside sources, or informed estimate.

Interpret as significant all digits in numerical values that have trailing zeros and no decimal points.

- • Single-concept, single-step, relatively easy
- •• Intermediate-level, may require synthesis of concepts
- ••• Challenging
- SSM Solution is in the *Student Solutions Manual*
- Consecutive problems that are shaded are paired problems.

CONCEPTUAL PROBLEMS

1 • Figure 22-37 shows an L-shaped object that has sides which are equal in length. Positive charge is distributed uniformly along the length of the object. What is the direction of the electric field along the dashed 45° line? Explain your answer. SSM

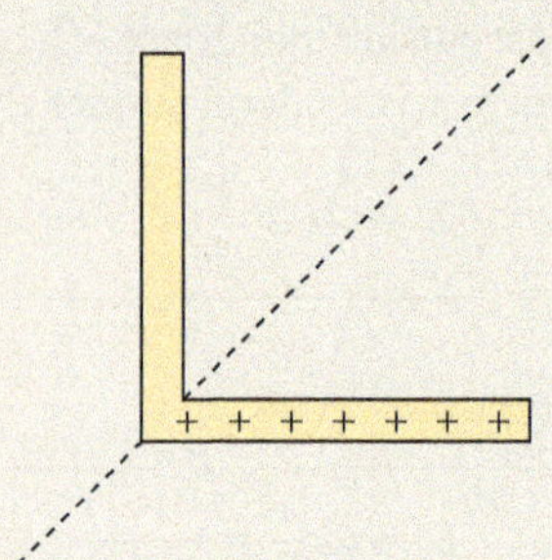

FIGURE 22-37 Problem 1

2 • Positive charge is distributed uniformly along the entire length of the x axis, and negative charge is distributed uniformly along the entire length of the y axis. The charge per unit length on the two axes is identical, except for the sign. Determine the direction of the electric field at points on the lines defined by $y = x$ and $y = -x$. Explain your answer.

3 • True or false:

(*a*) The electric field due to a hollow uniformly charged thin spherical shell is zero at all points inside the shell.
(*b*) In electrostatic equilibrium, the electric field everywhere inside the material of a conductor must be zero.
(*c*) If the net charge on a conductor is zero, the charge density must be zero at every point on the surface of the conductor.

4 • If the electric flux through a closed surface is zero, must the electric field be zero everywhere on that surface? If not, give a specific example. From the given information can the net charge inside the surface be determined? If so, what is it?

5 • True or false:

(*a*) Gauss's law holds only for symmetric charge distributions.
(*b*) The result that $E = 0$ everywhere inside the material of a conductor under electrostatic conditions can be derived from Gauss's law.

6 •• A single point charge q is located at the center of both an imaginary cube and an imaginary sphere. How does the electric flux through the surface of the cube compare to that through the surface of the sphere? Explain your answer.

7 •• An electric dipole is completely inside a closed imaginary surface and there are no other charges. True or false:

(*a*) The electric field is zero everywhere on the surface.
(*b*) The electric field is normal to the surface everywhere on the surface.
(*c*) The electric flux through the surface is zero.
(*d*) The electric flux through the surface could be positive or negative.
(*e*) The electric flux through a portion of the surface might not be zero. SSM

8 •• Explain why the electric field strength increases linearly with r, rather than decreases inversely with r^2, between the center and the surface of a uniformly charged solid sphere.

9 •• Suppose that the total charge on the conducting spherical shell in Figure 22-38 is zero. The negative point charge at the center has a magnitude given by Q. What is the direction of the electric field in the following regions: (*a*) $r < R_1$, (*b*) $R_2 > r > R_1$, (*c*) and $r > R_2$? Explain your answer. SSM

10 •• The conducting shell in Figure 22-38 is grounded, and the negative point charge at the center has a magnitude given by Q. Which of the following statements is correct?

(*a*) The charge on the inner surface of the shell is $+Q$ and the charge on the outer surface is $-Q$.
(*b*) The charge on the inner surface of the shell is $+Q$ and the charge on the outer surface is zero.
(*c*) The charge on both surfaces of the shell is $+Q$.
(*d*) The charge on both surfaces of the shell is zero.

11 •• The conducting shell in Figure 22-38 is grounded, and the negative point charge at the center has a magnitude given by Q. What is the direction of the electric field in the following regions: (*a*) $r < R_1$, (*b*) $R_2 > r > R_1$, (*c*) and $r > R_2$? Explain your answers.

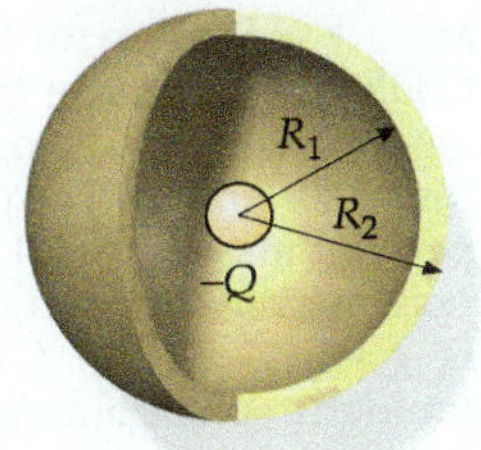

FIGURE 22-38 Problems 9, 10, and 11

ESTIMATION AND APPROXIMATION

12 •• In the chapter, the expression for the electric field due to a uniformly charged disk (on its axis), was derived. At any location on the axis, the field magnitude is given by $|E| = 2\pi k\sigma\left[1 - \left(1 + \frac{R^2}{z^2}\right)^{-1/2}\right]$. At large distances ($|z| \gg R$), it was shown that this equation approaches $E \approx kQ/z^2$. Very near the disk ($|z| \ll R$), the field strength is approximately that of an infinite plane of charge or $|E| \approx 2\pi k\sigma$. Suppose you have a disk of radius 2.5 cm that has a uniform surface charge density of 3.6 μC/m^2. Use both the exact and approximate expression from those given above to find the electric field strength on the axis at distances of (*a*) 0.010 cm, (*b*) 0.040 cm, and (*c*) 5.0 m. Compare the two values in each case and comment on how well the approximations work in their region of validity.

CALCULATING $\vec{E}$ FROM COULOMB'S LAW

13 • A uniform line charge that has a linear charge density λ equal to 3.5 nC/m is on the x axis between $x = 0$ and $x = 5.0$ m. (*a*) What is its total charge? Find the electric field on the x axis at (*b*) $x = 6.0$ m, (*c*) $x = 9.0$ m, and (*d*) $x = 250$ m. (*e*) Estimate the electric field at $x = 250$ m, using the approximation that the charge is a point charge on the x axis at $x = 2.5$ m, and compare your result with the result calculated in Part (*d*). (To do this, you will need to assume that the values given in this problem statement are valid to more than two significant figures.) Is your approximate result greater or smaller than the exact result? Explain your answer. SSM

14 • Two infinite nonconducting sheets of charge are parallel to each other, with sheet A in the $x = -2.0$ m plane and sheet B in the $x = +2.0$ m plane. Find the electric field in the region $x < -2.0$ m, in the region $x > +2.0$ m, and between the sheets for the following situations. (*a*) When each sheet has a uniform surface charge density equal to $+3.0\ \mu$C/m^2 and (*b*) when sheet A has a uniform surface charge density equal to $+3.0\ \mu$C/m^2 and sheet B has a uniform surface charge density equal to $-3.0\ \mu$C/m^2. (c) Sketch the electric field line pattern for each case.

15 • A charge of 2.75 μC is uniformly distributed on a ring of radius 8.5 cm. Find the electric field strength on the axis at distances of (*a*) 1.2 cm, (*b*) 3.6 cm, and (*c*) 4.0 m from the center of the ring. (*d*) Find the field strength at 4.0 m using the approximation that the ring is a point charge at the origin, and compare your results for Parts (c) and (d). Is your approximate result a good one? Explain your answer.

16 • A nonconducting disk of radius R lies in the $z = 0$ plane with its center at the origin. The disk has a uniform surface charge density σ. Find the value of z for which $E_z = \sigma/(4\epsilon_0)$. Note that at this distance, the magnitude of the electric field strength is half the electric field strength at points on the x axis that are very close to the disk.

17 • A ring that has radius a lies in the $z = 0$ plane with its center at the origin. The ring is uniformly charged and has a total charge Q. Find E_z on the z axis at (*a*) $z = 0.2a$, (*b*) $z = 0.5a$, (*c*) $z = 0.7a$, (*d*) $z = a$, and (*e*) $z = 2a$. (*f*) Use your results to plot E_z versus z for both positive and negative values of z. (Assume that these distances are exact.) SSM

18 • A nonconducting disk of radius a lies in the $z = 0$ plane with its center at the origin. The disk is uniformly charged and has a total charge Q. Find E_z on the z axis at (*a*) $z = 0.2a$, (*b*) $z = 0.5a$, (*c*) $z = 0.7a$, (*d*) $z = a$, and (*e*) $z = 2a$. (*f*) Use your results to plot E_z versus z for both positive and negative values of z. (Assume that these distances are exact.)

19 •• SPREADSHEET (*a*) Using a **spreadsheet** program or graphing calculator, make a graph of the electric field strength on the axis of a disk that has a radius $a = 30.0$ cm and a surface charge density $\sigma = 0.500$ nC/m^2. (*b*) Compare your results to the results based on the approximation $E = 2\pi k\sigma$ (the formula for the electric field strength of a uniformly charged infinite sheet). At what distance does the solution based on approximation differ from the exact solution by 10.0 percent?

20 •• (*a*) Show that the electric field strength E on the axis of a ring charge of radius a has maximum values at $z = \pm a/\sqrt{2}$. (*b*) Sketch the field strength E versus z for both positive and negative values of z. (*c*) Determine the maximum value of E.

21 •• A line charge that has a uniform linear charge density λ lies along the x axis from $x = x_1$ to $x = x_2$ where $x_1 < x_2$. Show that the x component of the electric field at a point on the y axis is given by $E_x = \frac{k\lambda}{y}(\cos\theta_2 - \cos\theta_1)$ where $\theta_1 = \tan^{-1}(x_1/y)$, $\theta_2 = \tan^{-1}(x_2/y)$, and $y \neq 0$.

22 •• A ring of radius a has a charge distribution on it that varies as $\lambda(\theta) = \lambda_0 \sin\theta$, as shown in Figure 22-39. (*a*) What is the direction of the electric field at the center of the ring? (*b*) What is the magnitude of the field at the center of the ring?

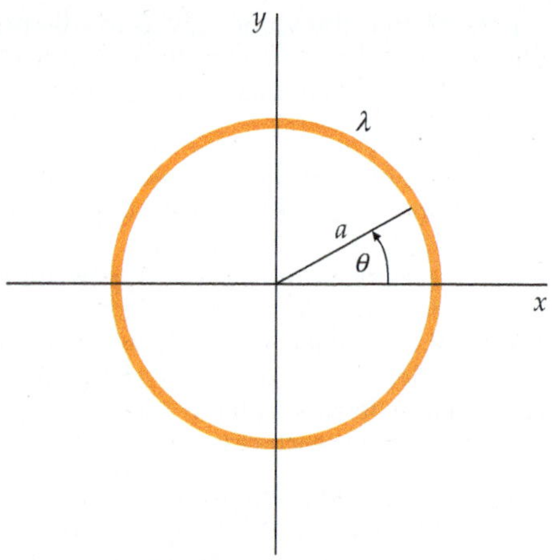

FIGURE 22-39 Problem 22

23 •• A line of charge that has uniform linear charge density λ lies on the x axis from $x = 0$ to $x = a$. Show that the y component of the electric field at a point on the y axis is given by $E_y = \frac{k\lambda}{y}\frac{a}{\sqrt{y^2 + a^2}}, y \neq 0.$

24 ••• Calculate the electric field a distance z from a uniformly charged infinite flat nonconducting sheet by modeling the sheet as a continuum of infinite straight lines of charge.

25 •• Calculate the electric field a distance z from a uniformly charged infinite flat nonconducting sheet by modeling the sheet as a continuum of infinite circular rings of charge. SSM

26 ••• A thin hemispherical shell of radius R has a uniform surface charge σ. Find the electric field at the center of the base of the hemispherical shell.

GAUSS'S LAW

27 • A square that has 10-cm-long edges is centered on the x axis in a region where there exists a uniform electric field given by $\vec{E} = (2.00\ \text{kN/C})\hat{i}$. (*a*) What is the electric flux of this electric field through the surface of a square if the normal to the surface is in the $+x$ direction? (*b*) What is the electric flux through the same square surface if the normal to the surface makes a 60° angle with the y axis and an angle of 90° with the z axis?

28 • A single point charge ($q = +2.00\ \mu\text{C}$) is fixed at the origin. An imaginary spherical surface of radius 3.00 m is centered on the x axis at $x = 5.00$ m. (*a*) Sketch electric field lines for this charge (in two dimensions) assuming twelve equally spaced field lines in the xy plane leave the charge location, with one of the lines in the $+x$ direction. Do any lines enter the spherical surface? If so, how many? (*b*) Do any lines leave the spherical surface? If so, how many? (*c*) Counting the lines that enter as negative and the ones that leave as positive, what is the net number of field lines that penetrate the spherical surface? (*d*) What is the net electric flux through this spherical surface?

29 • An electric field is given by $\vec{E} = \text{sign}(x) \cdot (300\ \text{N/C})\hat{i}$, where sign (x) equals -1 if $x < 0$, 0 if $x = 0$, and $+1$ if $x > 0$. A cylinder of length 20 cm and radius 4.0 cm has its center at the origin and its axis along the x axis such that one end is at $x = +10$ cm and the other is at $x = -10$ cm. (*a*) What is the electric flux through each end? (*b*) What is the electric flux through the curved surface of the cylinder? (*c*) What is the electric flux through the entire closed surface? (*d*) What is the net charge inside the cylinder? SSM

30 • Careful measurement of the electric field at the surface of a black box indicates that the net outward electric flux through the surface of the box is $6.0\ \text{kN} \cdot \text{m}^2/\text{C}$. (*a*) What is the net charge inside the box? (*b*) If the net outward electric flux through the surface of the box were zero, could you conclude that there were no charges inside the box? Explain your answer.

31 • A point charge ($q = +2.00\ \mu\text{C}$) is at the center of an imaginary sphere that has a radius equal to 0.500 m. (*a*) Find the surface area of the sphere. (*b*) Find the magnitude of the electric field at all points on the surface of the sphere. (*c*) What is the flux of the electric field through the surface of the sphere? (*d*) Would your answer to Part (*c*) change if the point charge were moved so that it was inside the sphere but not at its center? (*e*) What is the flux of the electric field through the surface of an imaginary cube that has 1.00-m-long edges and encloses the sphere?

32 • What is the electric flux through one side of a cube that has a single point charge of $-3.00\ \mu\text{C}$ placed at its center? *Hint: You do not need to integrate any equations to get the answer.*

33 • A single point charge is placed at the center of an imaginary cube that has 20-cm-long edges. The electric flux out of one of the cube's sides is $-1.50\ \text{kN} \cdot \text{m}^2/\text{C}$. How much charge is at the center? SSM

34 •• Because the formulas for Newton's law of gravity and for Coulomb's law have the same inverse-square dependence on distance, a formula analogous to the formula for Gauss's law can be found for gravity. The gravitational field $\vec{g}$ at a location is the force per unit mass on a test mass m_0 placed at that location. (Then, for a point mass m at the origin, the gravitational field g at some position $\hat{r}$ is $\vec{g} = -(Gm/r^2)\hat{r}$.) Compute the flux of the gravitational field through a spherical surface of radius R centered at the origin, and verify that the gravitational analog of Gauss's law is $\phi_{\text{net}} = -4\pi G m_{\text{inside}}$.

35 •• An imaginary right circular cone (Figure 22-40) that has a base angle θ and a base radius R is in charge free region that has a uniform electric field $\vec{E}$ (field lines are vertical and parallel to the cone's axis). What is the ratio of the number of field lines per unit area penetrating the base to the number of field lines per unit area penetrating the conical surface of the cone? Use Gauss's law in your answer. (The field lines in the figure are only a representative sample.)

36 •• In the atmosphere and at an altitude of 250 m, you measure the electric field to be 150 N/C directed downward, and you measure the electric field to be 170 N/C directed downward at an altitude of 400 m. Calculate the volume charge density of the atmosphere in the region between altitudes of 250 m and 400 m, assuming it to be uniform. (You may neglet the curvature of Earth. Why?)

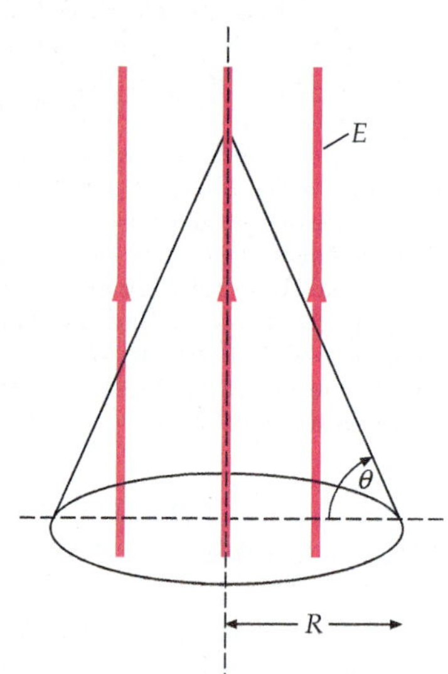

FIGURE 22-40 Problem 35

GAUSS'S LAW APPLICATIONS IN SPHERICAL SYMMETRY SITUATIONS

37 • A thin nonconducting spherical shell of radius R_1 has a total charge q_1 that is uniformly distributed on its surface. A second, larger thin nonconducting spherical shell of radius R_2 that is coaxial with the first has a charge q_2 that is uniformly distributed on its surface. (*a*) Use Gauss's law to obtain expressions for the electric field in each of the three regions: $r < R_1$, $R_1 < r < R_2$, and $r > R_2$. (*b*) What should the ratio of the charges q_1/q_2 and the relative signs for q_1 and q_2 be for the electric field to be zero throughout the region $r > R_2$? (*c*) Sketch the electric field lines for the situation in Part (*b*) when q_1 is positive.

38 • A nonconducting thin spherical shell of radius 6.00 cm has a uniform surface charge density of 9.00 nC/m^2. (*a*) What is the total charge on the shell? Find the electric field at the following distances from the sphere's center: (*b*) 2.00 cm, (*c*) 5.90 cm, (*d*) 6.10 cm, and (*e*) 10.0 cm.

39 •• A nonconducting sphere of radius 6.00 cm has a uniform volume charge density of 450 nC/m^3. (*a*) What is the total charge on the sphere? Find the electric field at the following distances from the sphere's center: (*b*) 2.00 cm, (*c*) 5.90 cm, (*d*) 6.10 cm, and (*e*) 10.0 cm. SSM

40 •• Consider the solid conducting sphere and the concentric conducting spherical shell in Figure 22-41. The spherical shell has a charge $-7Q$. The solid sphere has a charge $+2Q$. (*a*) How much charge is on the outer surface and how much charge is on the inner surface of the spherical shell? (*b*) Suppose a metal wire is now connected between the solid sphere and the shell. After electrostatic equilibrium is reestablished, how much charge is on the solid sphere and on each surface of the spherical shell? Does the electric field at the surface of the solid sphere change when the wire is connected? If so, in what way? (*c*) Suppose we return to the conditions in Part (*a*), with $+2Q$ on the solid sphere and $-7Q$ on the spherical shell. We next connect the solid sphere to ground with a metal wire, and then disconnect it. Then how much total charge is on the solid sphere and on each surface of the spherical shell?

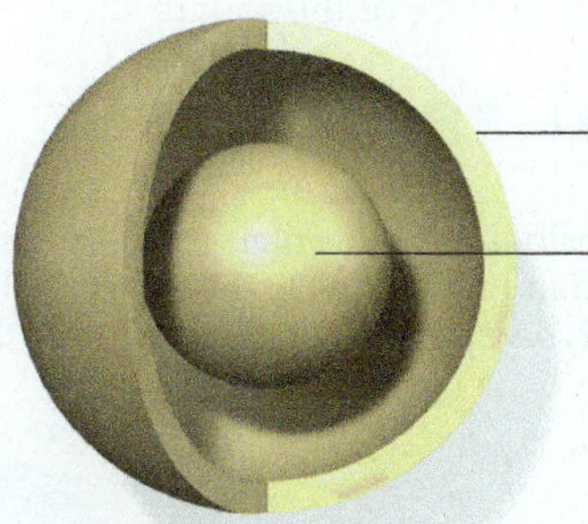

FIGURE 22-41
Problem 40

41 •• A nonconducting solid sphere of radius 10.0 cm has a uniform volume charge density. The magnitude of the electric field at 20.0 cm from the sphere's center is 1.88×10^3 N/C. (*a*) What is the sphere's volume charge density? (*b*) Find the magnitude of the electric field at a distance of 5.00 cm from the sphere's center.

42 •• A nonconducting solid sphere of radius R has a volume charge density that is proportional to the distance from the center. That is, $\rho = Ar$ for $r \leq R$, where A is a constant. (*a*) Find the total charge on the sphere. (*b*) Find the expressions for the electric field inside the sphere ($r < R$) and outside the sphere ($r > R$). (*c*) Sketch the magnitude of the electric field as a function of the distance r from the sphere's center.

43 •• A sphere of radius R has volume charge density $\rho = B/r$ for $r < R$, where B is a constant and $\rho = 0$ for $r > R$. (*a*) Find the total charge on the sphere. (*b*) Find the expressions for the electric field inside and outside the charge distribution. (*c*) Sketch the magnitude of the electric field as a function of the distance r from the sphere's center. SSM

44 •• A sphere of radius R has volume charge density $\rho = C/r^2$ for $r < R$, where C is a constant and $\rho = 0$ for $r > R$. (*a*) Find the total charge on the sphere. (*b*) Find the expressions for the electric field inside and outside the charge distribution. (*c*) Sketch the magnitude of the electric field as a function of the distance r from the sphere's center.

45 ••• A nonconducting spherical shell of inner radius R_1 and outer radius R_2 has a uniform volume charge density ρ. (*a*) Find the total charge on the shell. (*b*) Find expressions for the electric field everywhere.

GAUSS'S LAW APPLICATIONS IN CYLINDRICAL SYMMETRY SITUATIONS

46 • **CONTEXT-RICH, ENGINEERING APPLICATION** For your senior project, you are designing a Geiger tube for detecting radiation in the nuclear physics laboratory. This instrument will consist of a long metal cylindrical tube that has a long straight metal wire running down its central axis. The diameter of the wire will be 0.500 mm and the inside diameter of the tube will be 4.00 cm. The tube is to be filled with a dilute gas in which an electrical discharge (breakdown of the gas) occurs when the electric field reaches 5.50×10^6 N/C. Determine the maximum linear charge density on the wire if breakdown of the gas is not to happen. Assume that the tube and the wire are infinitely long.

47 ••• In Problem 46, suppose ionizing radiation produces an ion and an electron at a distance of 1.50 cm from the long axis of the central wire of the Geiger tube. Suppose that the central wire is positively charged and has a linear charge density equal to 76.5 pC/m. (*a*) In this case, what will be the electron's speed as it impacts the wire? (*b*) How will the electron's speed compare to the ion's final speed when it impacts the outside cylinder? Explain your answer.

48 •• Show that the electric field due to an infinitely long, uniformly charged thin cylindrical shell of radius a having a surface charge density σ is given by the following expressions: $E = 0$ for $0 \leq R < a$ and $E_R = \sigma a/(\epsilon_0 R)$ for $R > a$.

49 • A thin cylindrical shell of length 200 m and radius 6.00 cm has a uniform surface charge density of 9.00 nC/m^2. (*a*) What is the total charge on the shell? Find the electric field at the following radial distances from the long axis of the cylinder: (*b*) 2.00 cm, (*c*) 5.90 cm, (*d*) 6.10 cm, and (*e*) 10.0 cm. (Use the results of Problem 48.)

50 •• An infinitely long nonconducting solid cylinder of radius a has a uniform volume charge density of ρ_0. Show that the electric field is given by the following expressions: $E_R = \rho_0 R/(2\epsilon_0)$ for $0 \leq R < a$ and $E_R = \rho_0 a^2/(2\epsilon_0 R)$ for $R > a$, where R is the distance from the long axis of the cylinder.

51 •• A solid cylinder of length 200 m and radius 6.00 cm has a uniform volume charge density of 300 nC/m^3. (*a*) What is the total charge of the cylinder? Use the formulas given in Problem 50 to calculate the electric field at a point equidistant from the ends at the following radial distances from the cylindrical axis: (*b*) 2.00 cm, (*c*) 5.90 cm, (*d*) 6.10 cm, and (*e*) 10.0 cm. SSM

52 •• Consider two infinitely long, coaxial thin cylindrical shells. The inner shell has a radius a_1 and has a uniform surface charge density of σ_1, and the outer shell has a radius a_2 and has a uniform surface charge density of σ_2. (*a*) Use Gauss's law to find expressions for the electric field in the three regions: $0 \leq R < a_1$, $a_1 < R < a_2$, and $R > a_2$, where R is the distance from the axis. (*b*) What is the ratio of the surface charge densities σ_2/σ_1 and their relative signs if the electric field is to be zero everywhere outside the largest cylinder? (*c*) For the case in Part (*b*), what would be the electric field between the shells? (*d*) Sketch the electric field lines for the situation in Part (*b*) if σ_1 is positive.

53 •• Figure 22-42 shows a portion of an infinitely long, concentric cable in cross section. The inner conductor has a linear charge density of 6.00 nC/m and the outer conductor has no net charge. (*a*) Find the electric field for all values of R, where R is the perpendicular distance from the common axis of the cylindrical system. (*b*) What are the surface charge densities on the inside and the outside surfaces of the outer conductor?

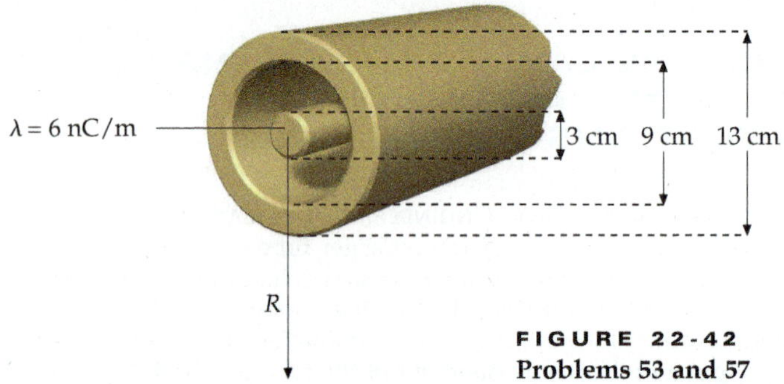

FIGURE 22-42
Problems 53 and 57

54 •• An infinitely long nonconducting solid cylinder of radius a has a nonuniform volume charge density. This density varies linearly with R, the perpendicular distance from its axis, according to $\rho(R) = \beta R$, where β is a constant. (*a*) Show that the linear charge density of the cylinder is given by $\lambda = 2\pi\beta a^3/3$. (*b*) Find expressions for the electric field for $R < a$ and $R > a$.

55 •• An infinitely long nonconducting solid cylinder of radius a has a nonuniform volume charge density. This density varies with R, the perpendicular distance from its axis, according to $\rho(R) = bR^2$, where b is a constant. (*a*) Show that the linear charge density of the cylinder is given by $\lambda = \pi b a^4/2$. (*b*) Find expressions for the electric field for $R < a$ and $R > a$. SSM

56 ••• An infinitely long, nonconducting cylindrical shell of inner radius a_1 and outer radius a_2 has a uniform volume charge density ρ. Find expressions for the electric field everywhere.

57 ••• The inner cylinder of Figure 22-42 is made of nonconducting material and has a volume charge distribution given by $\rho(R) = C/R$, where $C = 200\ \text{nC/m}^2$. The outer cylinder is metallic, and both cylinders are infinitely long. (*a*) Find the charge per unit length (that is, the linear charge density) on the inner cylinder. (*b*) Calculate the electric field for all values of R. SSM

ELECTRIC CHARGE AND FIELD AT CONDUCTOR SURFACES

58 • An uncharged penny is in a region that has a uniform electric field of magnitude 1.60 kN/C directed perpendicular to its faces. (*a*) Find the charge density on each face of the penny, assuming the faces are planes. (*b*) If the radius of the penny is 1.00 cm, find the total charge on one face.

59 • A thin metal slab has a net charge of zero and has square faces that have 12-cm-long sides. It is in a region that has a uniform electric field that is perpendicular to its faces. The total charge induced on one of the faces is 1.2 nC. What is the magnitude of the electric field?

60 • A charge of −6.00 nC is uniformly distributed on a thin square sheet of nonconducting material of edge length 20.0 cm. (*a*) What is the surface charge density of the sheet? (*b*) What are the magnitude and direction of the electric field next to the sheet and proximate to the center of the sheet?

61 • A conducting spherical shell that has zero net charge has an inner radius R_1 and an outer radius R_2. A positive point charge q is placed at the center of the shell. (*a*) Use Gauss's law and the properties of conductors in electrostatic equilibrium to find the electric field in the three regions: $0 \leq r < R_1$, $R_1 < r < R_2$, and $r > R_2$, where r is the distance from the center. (*b*) Draw the electric field lines in all three regions. (*c*) Find the charge density on the inner surface ($r = R_1$) and on the outer surface ($r = R_2$) of the shell.

62 •• The electric field just above the surface of Earth has been measured to typically be 150 N/C pointing downward. (*a*) What is the sign of the net charge on Earth's surface under typical conditions? (*b*) What is the total charge on Earth's surface implied by this measurement?

63 •• A positive point charge of 2.5 μC is at the center of a conducting spherical shell that has a net charge of zero, an inner radius equal to 60 cm, and an outer radius equal to 90 cm. (*a*) Find the charge densities on the inner and outer surfaces of the shell and the total charge on each surface. (*b*) Find the electric field everywhere. (*c*) Repeat Part (*a*) and Part (*b*) with a net charge of +3.5 μC placed on the shell. SSM

64 •• If the magnitude of an electric field in air is as great as 3.0×10^6 N/C, the air becomes ionized and begins to conduct electricity. This phenomenon is called *dielectric breakdown*. A charge of 18 μC is to be placed on a conducting sphere. What is the minimum radius of a sphere that can hold this charge without breakdown?

65 •• A thin square conducting sheet that has 5.00-m-long edges has a net charge of 80.0 μC. The square is in the $x = 0$ plane and is centered at the origin. (Assume the charge on each surface is uniformly distributed.) (*a*) Find the charge density on each side of the sheet and find the electric field on the x axis in the region $|x| \ll 5.00$ m. (*b*) A thin but infinite nonconducting sheet that has a uniform charge density of 2.00 μC/m² is now placed in the $x = -2.50$ m plane. Find the electric field on the x axis on each side of the square sheet in the region $|x| \ll 2.50$ m. Find the charge density on each surface of the square sheet. SSM

GENERAL PROBLEMS

66 •• Consider the concentric metal sphere and spherical shells that are shown in Figure 22-43. The innermost is a solid sphere that has a radius R_1. A spherical shell surrounds the sphere and has an

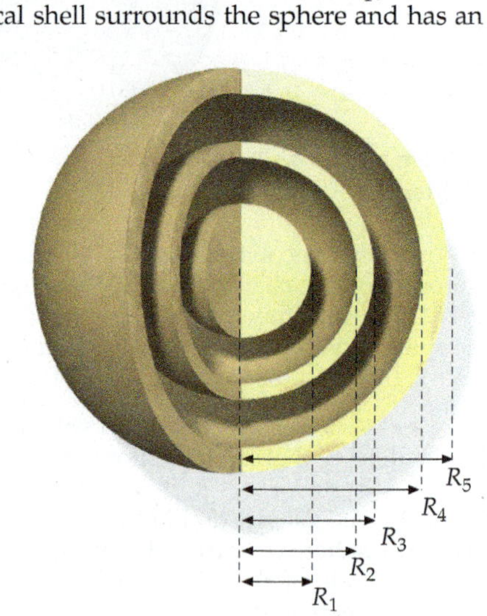

FIGURE 22-43
Problem 66

inner radius R_2 and an outer radius R_3. The sphere and the shell are both surrounded by a second spherical shell that has an inner radius R_4 and an outer radius R_5. None of the three objects initially have a net charge. Then, a negative charge $-Q_0$ is placed on the inner sphere and a positive charge $+Q_0$ is placed on the outermost shell. (*a*) After the charges have reached equilibrium, what will be the direction of the electric field between the inner sphere and the middle shell? (*b*) What will be the charge on the inner surface of the middle shell? (*c*) What will be the charge on the outer surface of the middle shell? (*d*) What will be the charge on the inner surface of the outermost shell? (*e*) What will be the charge on the outer surface of the outermost shell? (*f*) Plot E as a function of r for all values of r.

67 •• A large, flat, nonconducting, nonuniformly charged surface lies in the $x = 0$ plane. At the origin, the surface charge density is $+3.10\ \mu\text{C/m}^2$. A small distance away from the surface on the positive x axis, the x component of the electric field is 4.65×10^5 N/C. What is E_x a small distance away from the surface on the negative x axis? SSM

68 •• An infinitely long line charge that has a uniform linear charge density equal to $-1.50\ \mu\text{C/m}$ lies parallel to the y axis at $x = -2.00$ m. A positive point charge that has a magnitude equal to $1.30\ \mu\text{C}$ is located at $x = 1.00$ m, $y = 2.00$ m. Find the electric field at $x = 2.00$ m, $y = 1.50$ m.

69 •• A thin, nonconducting, uniformly charged spherical shell of radius R (Figure 22-44*a*) has a total positive charge of Q. A small circular plug is removed from the surface. (*a*) What are the magnitude and direction of the electric field at the center of the hole? (*b*) The plug is now put back in the hole (Figure 22-44*b*). Using the result of Part (*a*), find the electric force acting on the plug. (*c*) Using the magnitude of the force, calculate the "electrostatic pressure" (force/unit area) that tends to expand the sphere. SSM

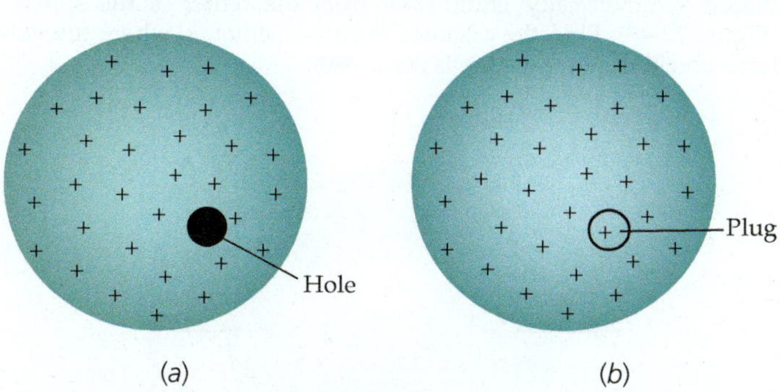

FIGURE 22-44 Problem 69

70 •• An infinite thin sheet in the $y = 0$ plane has a uniform surface charge density $\sigma_1 = +65\ \text{nC/m}^2$. A second infinite thin sheet has a uniform charge density $\sigma_2 = +45\ \text{nC/m}^2$ and intersects the $y = 0$ plane at the z axis and makes an angle of 30° with the xz plane, as shown in Figure 22-45. Find the electric field at (*a*) $x = 6.0$ m, $y = 2.0$ m and (*b*) $x = 6.0$ m, $y = 5.0$ m.

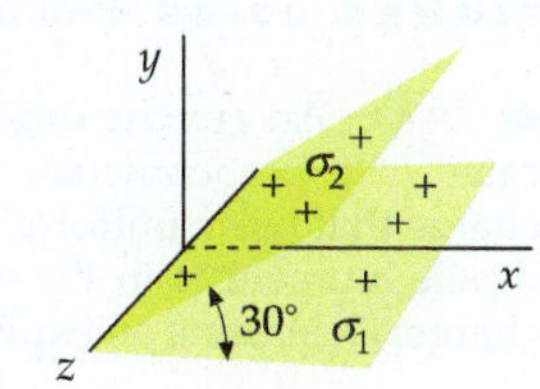

FIGURE 22-45
Problem 70

71 ••• Two identical square parallel metal plates each have an area of 500 cm². They are separated by 1.50 cm. They are both initially uncharged. Now a charge of +1.50 nC is transferred from the plate on the left to the plate on the right and the charges then establish electrostatic equilibrium. (Neglect edge effects.) (*a*) What is the electric field between the plates at a distance of 0.25 cm from the plate on the right? (*b*) What is the electric field between the plates a distance of 1.00 cm from the plate on the left? (*c*) What is the electric field just to the left of the plate on the left? (*d*) What is the electric field just to the right of the plate on the right? SSM

72 •• Two infinite nonconducting uniformly charged planes lie parallel to each other and to the yz plane. One is at $x = -2.00$ m and has a surface charge density of $-3.50\ \mu\text{C/m}^2$. The other is at $x = 2.00$ m and has a surface charge density of $6.00\ \mu\text{C/m}^2$. Find the electric field in the region: (*a*) $x < -2.00$ m (*b*) $-2.00\ \text{m} < x < 2.00$ m, and (*c*) $x > 2.00$ m.

73 ••• A quantum-mechanical treatment of the hydrogen atom shows that the electron in the atom can be treated as a smeared-out distribution of negative charge of the form $\rho(r) = -\rho_0 e^{-2r/a}$. Here r represents the distance from the center of the nucleus and a represents the *first Bohr radius*, which has a numerical value of 0.0529 nm. Recall that the nucleus of a hydrogen atom consists of just one proton and treat this proton as a positive point charge. (*a*) Calculate ρ_0, using the fact that the atom is neutral. (*b*) Calculate the electric field at any distance r from the nucleus. SSM

74 •• A uniformly charged ring has a radius a, lies in a horizontal plane, and has a negative charge given by $-Q$. A small particle of mass m has a positive charge given by q. The small particle is located on the axis of the ring. (*a*) What is the minimum value of q/m such that the particle will be in equilibrium under the action of gravity and the electrostatic force? (*b*) If q/m is twice the value calculated in Part (*a*), where will the particle be when it is in equilibrium? Express your answer in terms of a.

75 •• A long, thin, nonconducting plastic rod is bent into a circular loop that has a radius a. Between the ends of the rod a short gap of length ℓ, where $\ell \ll a$, remains. A positive charge of magnitude Q is evenly distributed on the loop. (*a*) What is the direction of the electric field at the center of the loop? Explain your answer. (*b*) What is the magnitude of the electric field at the center of the loop?

76 •• A nonconducting solid sphere that is 1.20 m in diameter and has its center on the x axis at $x = 4.00$ m has a uniform volume charge of density of $+5.00\ \mu\text{C/m}^3$. Concentric with the sphere is a thin nonconducting spherical shell that has a diameter of 2.40 m and a uniform surface charge density of $-1.50\ \mu\text{C/m}^2$. Calculate the magnitude and direction of the electric field at (*a*) $x = 4.50$ m, $y = 0$, (*b*) $x = 4.00$ m, $y = 1.10$ m, and (*c*) $x = 2.00$ m, $y = 3.00$ m.

77 •• An infinite nonconducting plane sheet of charge that has a surface charge density $+3.00\ \mu\text{C/m}^2$ lies in the $y = -0.600$ m plane. A second infinite nonconducting plane sheet of charge that has a surface charge density of $-2.00\ \mu\text{C/m}^2$ lies in the $x = 1.00$ m plane. Lastly, a nonconducting thin spherical shell that has a radius of 1.00 m and its center in the $z = 0$ plane at the intersection of the two charged planes has a surface charge density of $-3.00\ \mu\text{C/m}^2$. Find the magnitude and direction of the electric field on the x axis at (*a*) $x = 0.400$ m and (*b*) $x = 2.50$ m.

78 •• An infinite nonconducting plane sheet lies in the $x = 2.00$ m plane and has a uniform surface charge density of $+2.00\ \mu\text{C/m}^2$. An infinite nonconducting line charge of uniform linear charge density $4.00\ \mu\text{C/m}$ passes through the origin at an angle of 45.0° with the x axis in the xy plane. A solid nonconducting sphere of volume charge density $-6.00\ \mu\text{C/m}^3$ and radius 0.800 m is centered on the x axis at $x = 1.00$ m. Calculate the magnitude and direction of the electric field in the $z = 0$ plane at $x = 1.50$ m, $y = 0.50$ m.

79 •• A uniformly charged, infinitely long line of negative charge has a linear charge density of $-\lambda$ and is located on the z axis. A small positively charged particle that has a mass m and a charge q is in a circular orbit of radius R in the xy plane centered on the line of charge. (*a*) Derive an expression for the speed of the particle. (*b*) Obtain an expression for the period of the particle's orbit. SSM

80 •• A stationary ring of radius a lies in the yz plane and has a uniform positive charge Q. A small particle that has mass m and a negative charge $-q$ is located at the center of the ring. (*a*) Show that if $x \ll a$, the electric field along the axis of the ring is proportional to x. (*b*) Find the force on the particle as a function of x. (*c*) Show that if the particle is given a small displacement in the $+x$ direction, it will perform simple harmonic motion. (*d*) What is the frequency of that motion?

81 •• The charges Q and q of Problem 80 are $+5.00\ \mu$C and $-5.00\ \mu$C, respectively, and the radius of the ring is 8.00 cm. When the particle is given a small displacement in the x direction, it oscillates about its equilibrium position at a frequency of 3.34 Hz. (*a*) What is the particle's mass? (*b*) What is the frequency if the radius of the ring is doubled to 16.0 cm and all other parameters remain unchanged? SSM

82 •• If the radius of the ring in Problem 80 is doubled while keeping the linear charge density on the ring the same, does the frequency of oscillation of the particle change? If so, by what factor does it change?

83 ••• A uniformly charged nonconducting solid sphere of radius R has its center at the origin and has a volume charge density of ρ. (*a*) Show that at a point within the sphere a distance r from the center $\vec{E} = \frac{\rho}{3\epsilon_0} r\hat{r}$. (*b*) Material is removed from the sphere leaving a spherical cavity that has a radius $b = R/2$ and its center at $x = b$ on the x axis (Figure 22-46). Calculate the electric field at points 1 and 2 shown in Figure 22-46. *Hint: Model the sphere-with-cavity as two uniform spheres of equal positive and negative charge densities.*

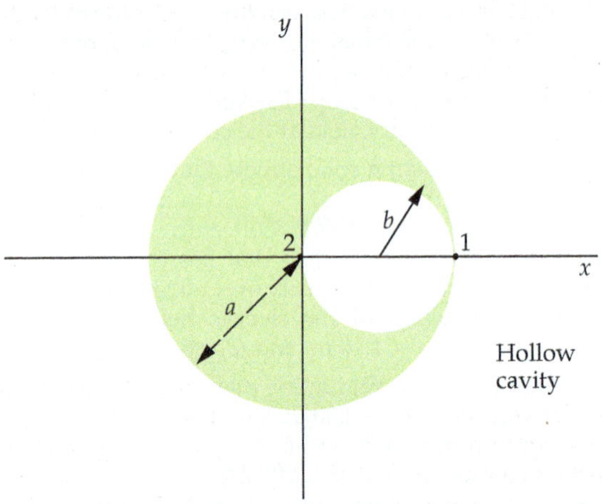

FIGURE 22-46 Problems 83 and 85

84 ••• Show that the electric field throughout the cavity of Problem 83*b* is uniform and is given by $\vec{E} = \frac{\rho}{3\epsilon_0} b\hat{i}$.

85 ••• The cavity in Problem 83*b* is now filled with a uniformly charged nonconducting material with a total charge of Q. Calculate the new values of the electric field at points 1 and 2 shown in Figure 22-46.

86 ••• A *small* Gaussian surface in the shape of a cube has faces parallel to the xy, xz, and yz planes (Figure 22-47) and is in a region in which the electric field is parallel to the x axis. (*a*) Using the differential approximation, show that the net electric flux of the electric field out of the Gaussian surface is given by $\phi_{\text{net}} \approx \frac{\partial E_x}{\partial x}\Delta V$, where ΔV is the volume enclosed by the Gaussian surface. (*b*) Using Gauss's law and the results of Part (*a*) show that $\frac{\partial E_x}{\partial x} = \frac{\rho}{\epsilon_0}$, where ρ is the volume charge density inside the cube. (This equation is the one-dimensional version of the point form of Gauss's law.)

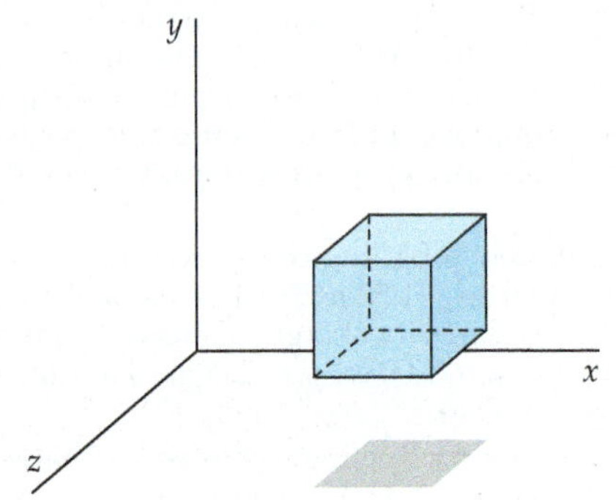

FIGURE 22-47 Problem 86

87 ••• Consider a simple but surprisingly accurate model for the hydrogen molecule: two positive point charges, each having charge $+e$, are placed inside a uniformly charged sphere of radius R, which has a charge equal to $-2e$. The two point charges are placed symmetrically, equidistant from the center of the sphere (Figure 22-48). Find the distance from the center, a, where the net force on either point charge is zero. SSM

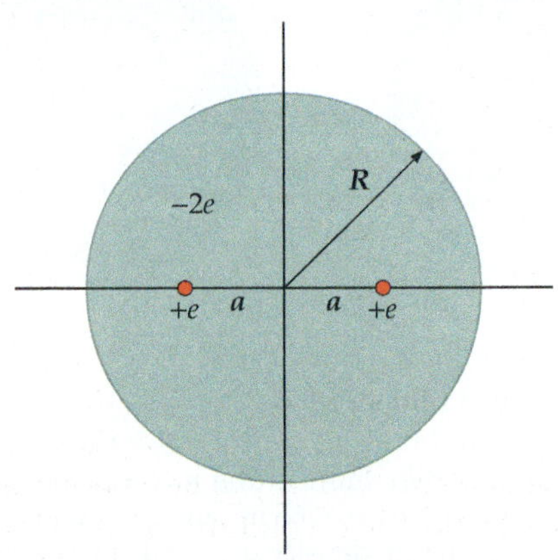

FIGURE 22-48 Problem 87

88 ••• An electric dipole that has a dipole moment of $\vec{p}$ is located at a perpendicular distance R from an infinitely long line charge that has a uniform linear charge density λ. Assume that the dipole moment is in the same direction as the field of the line of charge. Determine an expression for the electric force on the dipole.

CHAPTER 23

THIS GIRL HAS BEEN INCREASED TO A HIGH ELECTRIC POTENTIAL THROUGH CONTACT WITH THE DOME OF A VAN DE GRAAFF GENERATOR. SHE IS STANDING ON A PLATFORM THAT ELECTRICALLY INSULATES HER FROM THE FLOOR, SO SHE ACCUMULATES CHARGE FROM THE VAN DE GRAAFF. HER HAIR STANDS UP BECAUSE THE CHARGES ON HER HAIR STRANDS HAVE THE SAME SIGN, AND LIKE CHARGES REPEL EACH OTHER. *(Courtesy of the U.S. Department of Energy.)*

Electric Potential

? Did you know that the maximum potential that the dome of a Van de Graaff generator can be increased to is determined by the radius of the dome? (See Example 23-14.)

Gravitational potential energy, introduced in Chapter 7, is a powerful conceptual and computational aid. In this chapter, we introduce electrical potential energy, and like gravitational potential energy, we will find electrical potential energy to be a powerful conceptual and computational aid. In addition, we continue to develop the concept of the electric field. Electric fields are discussed in Chapters 21 and 22, and we continue this discussion in this chapter by introducing the electric potential—a scalar field that is directly related to the electric field. Because it is a scalar field, it is easier to use during calculations than the electric field (a vector) in many circumstances. Also, the potential is usually much easier to measure—using a voltmeter—than is the electric field. Both electric potential energy and the electrical potential field will be essential tools in the analysis of capacitance, resistance, and electric circuits—topics that are developed in Chapters 24 and 25.

In this chapter, we will establish the relationship between the electric field and electric potential and calculate the electric potential of various continuous charge distributions. We will also calculate the electrical potential energy of a system of point charges and a system of charged conductors.

23-1 POTENTIAL DIFFERENCE

The electrostatic force exerted by a point charge on another point charge is directed along the line joining the charges and varies inversely with the square of their separation distance. This same dependence can be seen when analyzing the gravitational force between two masses. Like the gravitational force, the electric force is conservative, so there is a potential energy function U associated with it. If the point of application of a conservative force $\vec{F}$ undergoes a displacement $d\vec{\ell}$, the change in the potential energy function dU associated with this displacement is given by

$$dU = -\vec{F} \cdot d\vec{\ell}$$

If the conservative force is exerted by electrostatic field $\vec{E}$ on point charge q, then the force is given by

$$\vec{F} = q\vec{E}$$

and if the point charge q undergoes a displacement $d\vec{\ell}$, the corresponding change in the electrostatic potential energy is given by

$$dU = -q\vec{E} \cdot d\vec{\ell} \qquad \text{23-1}$$

In Section 21-4, we revealed that the electrostatic force $\vec{F}$ on a test charge q_0 is proportional to q_0, and this relationship led to defining a quantity (the force per unit charge at the location of the test charge) called the electric field $\vec{E}$. There is an analogous situation here. The potential energy change associated with the displacement of a test charge q_0 that undergoes a displacement $d\vec{\ell}$ is given by $dU = -q_0\vec{E} \cdot d\vec{\ell}$. Thus, the potential energy change is proportional to the test charge. This relation suggests that we define a quantity—the potential energy change per unit charge—called the **potential difference** dV:

$$dV = \frac{dU}{q_0} = -\vec{E} \cdot d\vec{\ell} \qquad \text{23-2}a$$

DEFINITION—POTENTIAL DIFFERENCE

For a finite displacement from point a to point b, the change in potential is

$$\Delta V = V_b - V_a = \frac{\Delta U}{q_0} = -\int_a^b \vec{E} \cdot d\vec{\ell} \qquad \text{23-2}b$$

DEFINITION—FINITE POTENTIAL DIFFERENCE

The potential difference $V_b - V_a$ is the negative of the work per unit charge done by the electric field on a test charge when the test charge moves from point a to point b (along *any* path). During this calculation, the positions of any and all other charges remain fixed. (Recall that a test charge is a point charge whose magnitude is so small that it exerts only negligible forces on any and all other charges. For convenience, test charges are invariably considered to be positive charges.)

The function V is called the **electric potential;** it is often referred to as the **potential.** Like the electric field, the potential V is a function of position. Unlike the electric field, V is a scalar function, whereas $\vec{E}$ is a vector function. As with potential energy U, only *differences* in the potential V are physically significant. We are free to choose the potential to be zero at any convenient point, just as we are when dealing with potential energy. For convenience, the electric potential and the potential energy of a test charge are chosen to be zero at the same reference point. Under this restriction they are related by

$$U = q_0 V \qquad \text{23-3}$$

RELATION BETWEEN POTENTIAL ENERGY AND POTENTIAL

CONTINUITY OF V

In Chapter 22, we saw that the electric field is discontinuous by σ/ϵ_0 at points where there is a surface charge density σ. The potential function, on the other hand, is continuous everywhere, except at points where the electric field is infinite (points occupied by a point charge or a line charge). We can see this result from the definition of potential. Consider a region occupied by an electric field $\vec{E}$. The difference in potential between two nearby points separated by displacement $d\vec{\ell}$ is related to the electric field by $dV = -\vec{E}\cdot d\vec{\ell}$ (Equation 23-2*a*). The dot product can be expressed as $E_{\parallel}\,d\ell$, where $E_{\parallel}$ is the component of $\vec{E}$ in the direction of $d\vec{\ell}$ and $d\ell$ is the magnitude of $d\vec{\ell}$. Substituting into Equation 23-2*a* gives $dV = -E_{\parallel}\,d\ell$. If $\vec{E}$ is finite at each of the two points and along the line segment of infinitesimal length $d\ell$ joining them, then dV is infinitesimal. Thus, the potential function V is continuous at any point not occupied by a point charge or a line charge.

UNITS

Because electric potential is the potential energy per unit charge, the SI unit for potential and potential difference is the joule per coulomb, called the **volt** (V):

$$1\ \text{V} = 1\ \text{J/C} \qquad 23\text{-}4$$

The potential difference between two points (measured in volts) is commonly referred to as the **voltage** between the two points. In a 12-V car battery, the positive terminal has a potential 12 V higher than the negative terminal. If we attach an external circuit to the battery and one coulomb of charge is transferred from the positive terminal through the circuit to the negative terminal, the potential energy of the charge decreases by $Q\,\Delta V = (1\ \text{C})(12\ \text{V}) = 12\ \text{J}$.

We can see from Equation 23-2 that the dimensions of potential are also those of the product of electric field and distance. Thus, the unit of the electric field is equal to one volt per meter:

$$1\ \text{N/C} = 1\ \text{V/m} \qquad 23\text{-}5$$

so we may think of the magnitude of the electric field E as either a force per unit charge or as a rate of change of potential (V) with respect to distance in a given direction. In atomic and nuclear physics, we often have particles that have charges of magnitude e, such as electrons and protons, moving through potential differences of several thousands or even millions of volts. Because energy has dimensions of electric charge multiplied by electric potential, a unit of energy is defined as the product of the fundamental charge unit e and a volt. This particularly useful unit is called an **electron volt** (eV). Energies used in atomic and molecular physics are typically a few eV, making the electron volt a convenient-sized unit for atomic and molecular processes. The conversion between electron volts and joules is obtained by expressing the fundamental charge unit in coulombs:

$$1\ \text{eV} = 1.60 \times 10^{-19}\ \text{C}\cdot\text{V} = 1.60 \times 10^{-19}\ \text{J} \qquad 23\text{-}6$$

THE ELECTRON VOLT

For example, an electron moving from the negative terminal to the positive terminal of a 12-V car battery loses 12 eV of potential energy.

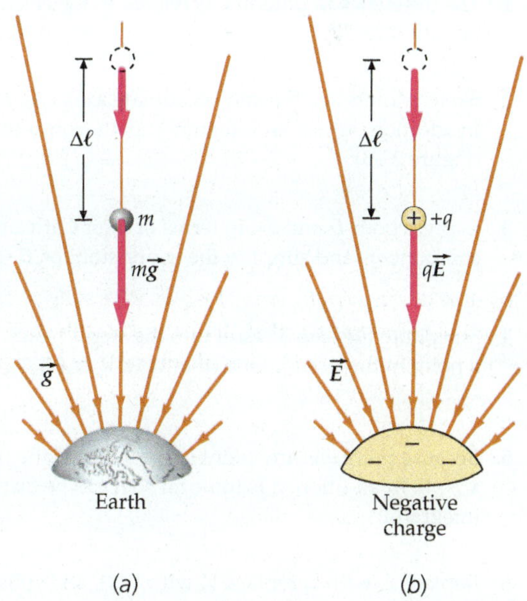

FIGURE 23-1 (*a*) The work done by the gravitational field $\vec{g}$ on a mass m is equal to the decrease in the gravitational potential energy. (*b*) The work done by the electric field $\vec{E}$ on a charge q is equal to the decrease in the electric potential energy.

POTENTIAL AND ELECTRIC FIELDS

If we place a positive charge q_0 in an electric field $\vec{E}$ and release it, it accelerates in the direction of $\vec{E}$. As the kinetic energy of the charge increases, its potential energy decreases. The charge therefore accelerates toward a region where its electric potential energy is less, just as a mass in a gravitational field accelerates toward a region where its gravitational potential energy is less (Figure 23-1).

The electric potential energy U is related to the electric potential V by $U = qV$, so *for a positive charge* a region where the charge has lower potential energy U is also a region of lower electric potential V. In summary, a positive charge accelerates in the direction of $\vec{E}$ (Figure 23-2) toward a region of lower electric potential V. Thus,

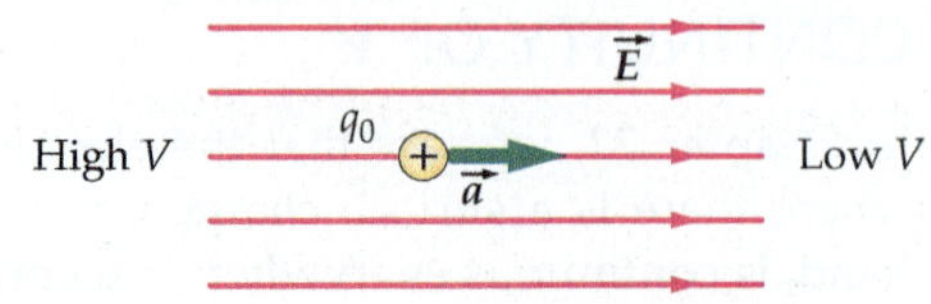

FIGURE 23-2 The electric field points in the direction in which the potential decreases most rapidly. If a positive test charge q_0 is in an electric field, it accelerates in the direction of the field. If it is released from rest, its kinetic energy increases and its potential energy decreases.

> The electric field $\vec{E}$ points in the direction in which the potential V decreases most rapidly.

PRACTICE PROBLEM 23-1

If you place a negative charge in an electric field, would the negative charge accelerate in the direction of increasing or decreasing potential?

Example 23-1 Find V for Uniform $\vec{E}$

A uniform electrostatic field points in the $+x$ direction and has a magnitude of $E = 10$ N/C $= 10$ V/m. Find the potential as a function of x, assuming that $V = 0$ at $x = 0$.

PICTURE We can solve for V by using $V_b - V_a = -\int_a^b \vec{E}\cdot d\vec{\ell}$ (Equation 23-2b). Let a be a point in the $x = 0$ plane (where $V = 0$) and let b be an arbitrary positioned point. Express both $\vec{E}$ and $d\vec{\ell}$ in terms of their Cartesian components and then compute the integral.

SOLVE

1. The difference in potential is related to the electric field by Equation 23-2b:
$$V_b - V_a = -\int_a^b \vec{E}\cdot d\vec{\ell}$$

2. Sketch points a and b and coordinate axes x, y, and z. In addition, sketch an integration path from a to b (Figure 23-3):

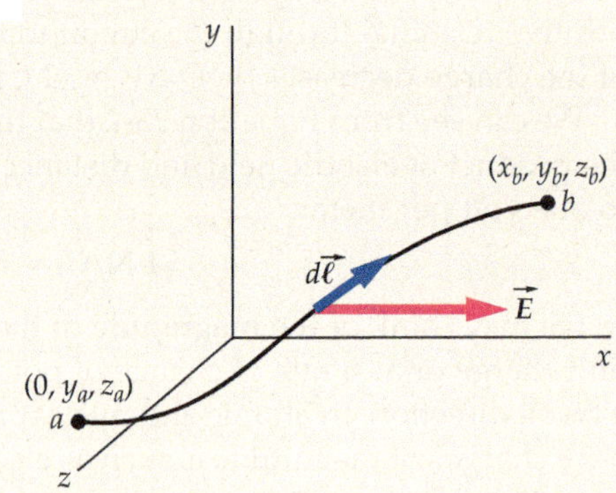

FIGURE 23-3

3. Express both $\vec{E}$ and $d\vec{\ell}$ in terms of their Cartesian components and simplify the expression for $\vec{E}\cdot d\vec{\ell}$:
$$\vec{E}\cdot d\vec{\ell} = E\hat{i}\cdot(dx\,\hat{i} + dy\,\hat{j} + dz\,\hat{k}) = E\,dx$$

4. Substitute the step 3 result into the step 1 result. Let point a be a point in the $x = 0$ plane (that way, $V_a = 0$):
$$V_b - V_a = -\int_{x_a}^{x_b} E\,dx$$

5. Because point a is any point in the $x = 0$ plane, $V_a = 0$ and $x_a = 0$. In addition, E is uniform so it can be factored from the integrand:
$$V_b - 0 = -E\int_0^{x_b} dx \quad \text{so} \quad V_b = -Ex_b$$

6. Replace x_b with x, replace V_b with $V(x)$, and substitute 10 V/m for E:
$$V(x) = -Ex = \boxed{-(10\text{ V/m})x}$$

CHECK The step 6 result is equal to zero if $x = 0$, which is in agreement with the assumption that $V = 0$ at $x = 0$ in the problem statement.

PRACTICE PROBLEM 23-2 Repeat this example for the electric field $\vec{E} = (10\text{ V/m}^2)x\hat{i}$.

In Example 23-1, point a—the point where the value of the potential is specified—is called the **reference point** for the potential function V. The potential at a field point b is obtained by calculating $V - 0 = -\int_a^b \vec{E}\cdot d\vec{\ell}$, where the potential at a is taken to be zero. The integral is to be evaluated along any path from a to b.

We now show how to calculate the potential for a number of different charge distributions.

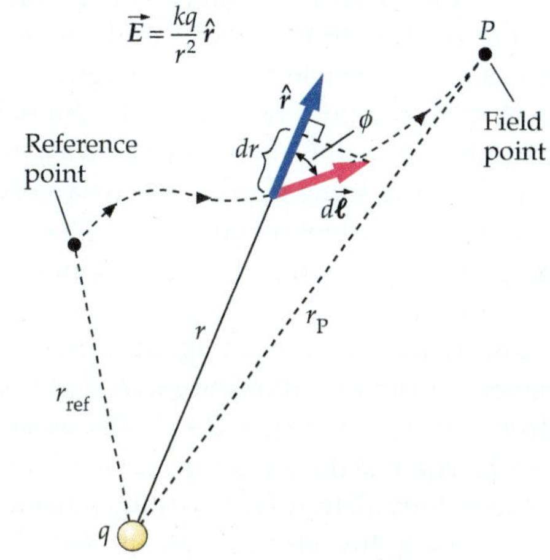

FIGURE 23-4 The change in r is dr. It is the component of $d\vec{\ell}$ in the direction of $\hat{r}$. It can be seen from the figure that $|d\vec{\ell}|\cos\phi = dr$. Because $\hat{r}\cdot d\vec{\ell} = |d\vec{\ell}|\cos\phi$, it follows that $dr = \hat{r}\cdot d\vec{\ell}$.

23-2 POTENTIAL DUE TO A SYSTEM OF POINT CHARGES

The electric potential a distance r from a point charge q at the origin can be calculated using $V_P - V_{ref} = -\int_{ref}^{P}\vec{E}\cdot d\vec{\ell}$ (Equation 23-2b), where at the reference point the potential equals V_{ref}, and P is an arbitrary field point (Figure 23-4). The electric field due to the point charge is given by

$$\vec{E} = \frac{kq}{r^2}\hat{r}$$

Substituting for $\vec{E}$ in the integral gives

$$V_P - V_{ref} = -\int_{ref}^{P}\vec{E}\cdot d\vec{\ell} = -\int_{ref}^{P}\frac{kq}{r^2}\hat{r}\cdot d\vec{\ell} = -\int_{r_{ref}}^{r_P}\frac{kq}{r^2}\,dr$$

where $dr = \hat{r}\cdot d\vec{\ell}$ (see Figure 23-4) is the change in the distance r associated with the displacement $d\vec{\ell}$. Setting V_{ref} equal to zero and integrating along a path from an arbitrary reference point to an arbitrary field point gives

$$V_P - 0 = -\int_{ref}^{P}\vec{E}\cdot d\vec{\ell} = -kq\int_{r_{ref}}^{r_P}\frac{1}{r^2}\,dr = \frac{kq}{r_P} - \frac{kq}{r_{ref}}$$

or

$$V = \frac{kq}{r} - \frac{kq}{r_{ref}} \qquad 23\text{-}7$$

POTENTIAL DUE TO A POINT CHARGE

where we have replaced r_P (the distance to the field point P) with r and replaced V_P with V. We are free to choose the location of the reference point, so we choose it to give the potential the simplest algebraic form. Choosing the reference point infinitely far from the point charge (so $r_{ref} \rightarrow \infty$) accomplishes this. Thus,

$$V = \frac{kq}{r} \qquad 23\text{-}8$$

COULOMB POTENTIAL

The potential given by Equation 23-8 is called the **Coulomb potential.** It is positive or negative depending on whether q is positive or negative.

The potential energy U of a point charge q' placed a distance r from the point charge q is

$$U = q'V = q'\frac{kq}{r} = \frac{kq'q}{r} \qquad 23\text{-}9$$

ELECTROSTATIC POTENTIAL ENERGY OF A TWO-CHARGE SYSTEM

See **Math Tutorial** *for more information on*

Integrals

This is the electric potential energy of the two-charge system relative to $U = 0$ at infinite separation. If we release a point particle that has charge q' from rest at a distance r_0 from q (and hold q fixed), the point particle that has charge q' will be accelerated away from q (assuming that q has the same sign as q'). At a very great distance from q, the potential energy of the particle that has charge q' approaches zero so its kinetic energy approaches kqq'/r_0.

The work an external agent must do to move a test charge q_0 from rest at infinity to rest at point P, a distance r from q, is kq_0q/r (Figure 23-5). The work per unit charge is kq/r, which is the electric potential V at point P relative to the potential an infinite distance from P.

Choosing the electrostatic potential energy of two point charges to be zero at an infinite separation is analogous to the choice we made in Chapter 11 when we chose the gravitational potential energy of two point masses to be zero at an infinite separation. If two charges (or two masses) are at infinite separation, we think of them as not interacting. That the potential energy is zero if the particles are not interacting has a certain aesthetic appeal.

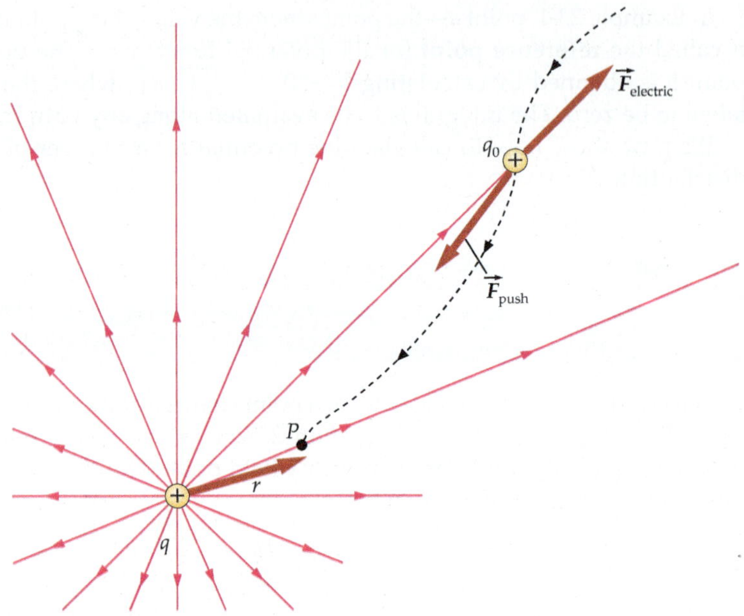

FIGURE 23-5 The work required to bring a test charge q_0 from rest at infinity to rest at point P is kq_0q/r, where r is the distance from P to the positive point charge q. The work per unit charge is thus kq/r, which is the electric potential at point P relative to zero potential at infinity. If the test charge is released from point P, the electric field does work kq_0q/r on the charge as the charge accelerates to infinity.

Example 23-2 Potential Energy of a Hydrogen Atom

(*a*) What is the electric potential at a distance $r_0 = 0.529 \times 10^{-10}$ m from a proton? This is the average distance between the proton and the electron in a hydrogen atom. (*b*) What is the electric potential energy of the electron and the proton at this separation?

PICTURE The electric potential due to the charge of the proton and the potential energy of two point charges are given by Equations 23-8 and 23-9.

SOLVE

(*a*) Use $V = kq/r$ to calculate the potential V due to the charge on the proton at $r = r_0$. For a proton, $q = e$:

$$V = \frac{kq}{r_0} = \frac{ke}{r_0} = \frac{(8.99 \times 10^9\ \mathrm{N \cdot m^2/C^2})(1.6 \times 10^{-19}\ \mathrm{C})}{0.529 \times 10^{-10}\ \mathrm{m}}$$

$$= 27.2\ \mathrm{N \cdot m/C} = \boxed{27.2\ \mathrm{V}}$$

(*b*) Use $U = q'V$, with $q' = -e$ to calculate the potential energy:

$$U = q'V = (-e)(27.2\ \mathrm{V}) = \boxed{-27.2\ \mathrm{eV}}$$

CHECK By examining the units in the equation $V = kq/r$, we can see that the units work out to N·m/C. Because 1 N·m = 1 J and 1 J/C = 1 V, we have 1 N·m/C = 1 J/C = 1 V.

TAKING IT FURTHER If the electron were at rest at this distance from the proton, it would take a minimum of 27.2 eV to remove it from the atom. However, the electron has kinetic energy equal to 13.6 eV, so its total energy in the atom is 13.6 eV − 27.2 eV = −13.6 eV. The minimum energy needed to remove the electron from a hydrogen atom is thus 13.6 eV. This energy is called the *ionization energy*.

PRACTICE PROBLEM 23-3 What is the potential energy of the two point charges in Example 23-2 in SI units?

Example 23-3 Potential Energy of Nuclear-Fission Products

During nuclear fission, a uranium-235 nucleus captures a neutron to form an unstable uranium-236 nucleus. The unstable nucleus then splits apart into two lighter nuclei (Figure 23-6). In addition, two or three neutrons are released. Sometimes the two fission products are a barium nucleus (charge 56*e*) and a krypton nucleus (charge 36*e*). Assume that immediately after the split these nuclei are positive point charges separated by $r = 14.6 \times 10^{-15}$ m (14.6×10^{-15} m is the sum of the radii of the barium and krypton nuclei). Calculate the potential energy of this two-charge system in electron volts.

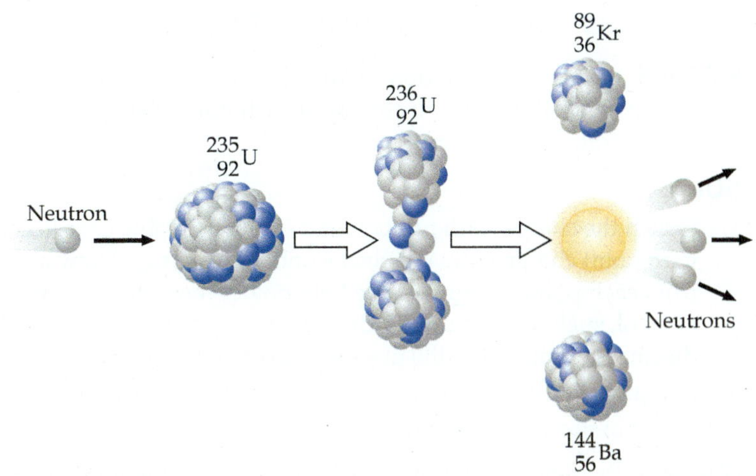

FIGURE 23-6 A uranium-235 nucleus absorbs a neutron and fissions into a barium nucleus and a krypton nucleus.

PICTURE The potential energy for two point charges separated by a distance r is $U = kq_1q_2/r$. To find this energy in electron volts, we calculate the potential due to one of the charges kq_1/r, in volts, and multiply this quantity by the other charge expressed as a multiple of e.

SOLVE

1. Equation 23-9 gives the potential energy of the two charges: $U = q_2\dfrac{kq_1}{r}$

2. Substitute the given values and factor out e:

$$U = e\frac{36 \cdot 56ke}{r}$$

$$= e\frac{36 \cdot 56 \cdot (8.99 \times 10^9\ \text{N}\cdot\text{m}^2/\text{C}^2)(1.60 \times 10^{-19}\ \text{C})}{14.6 \times 10^{-15}\ \text{m}}$$

$$= e(199 \times 10^6\ \text{V}) = \boxed{199\ \text{MeV}}$$

CHECK The potential energy of a proton and an electron in a hydrogen atom, calculated in Example 23-2, is seven orders of magnitude smaller than the potential energy calculated in this example. Because we expect energies for nuclear processes to be much larger than energies for atomic processes, this result is as expected.

TAKING IT FURTHER After the fission, the two nuclei fly apart because of their electrostatic repulsion. Their potential energy of 199 MeV is converted into kinetic energy. Upon colliding with surrounding atoms, this kinetic energy is distributed as thermal energy. During a chain reaction, one or more of the released neutrons produces a fission of another uranium nucleus. The average energy released during chain reactions of this type is about 200 MeV per nucleus, as calculated in this example.

The potential at a field point due to the presence of several point charges is the sum of the potentials due to each of these charges separately. (This result follows from the superposition principle for the electric field.) The potential due to a system of point charges q_i is thus given by

$$V = \sum_i \frac{kq_i}{r_i} \qquad 23\text{-}10$$

POTENTIAL DUE TO A SYSTEM OF POINT CHARGES

where the sum is over all the charges, and r_i is the distance from the ith charge to the field point at which the potential is to be found. Using this formula, the reference point (where $V = 0$) is at infinity and the distance between any two point charges in the system is finite.

PROBLEM-SOLVING STRATEGY

Calculating *V* Using Equation 23-10

PICTURE We can use Equation 23-10 to calculate the potential at a field point due to any collection of point charges if each point charge is a finite distance from every other point charge.

SOLVE

1. Sketch the charge configuration and include suitable coordinate axes. Label each point charge with a distinct symbol, such as q_1. Draw a straight line from each point charge q_i to the field point P and label it with a suitable symbol, such as r_{iP}. A careful drawing can be very helpful in relating the distances of interest to the distances given in the problem statement.
2. Use the formula $V = \Sigma\, kq_i/r_{iP}$ (Equation 23-10) to calculate the potential at P due to the presence of the point charges.

CHECK If the field point is arbitrarily chosen, take the limit as the field point goes to infinity. In that limit, the potential must approach zero.

Example 23-4 Potential Due to Two Point Charges

Two +5.0-nC point charges are on the x axis, one at the origin and the other at $x = 8.0$ cm. Find the potential at (*a*) point P_1 on the x axis at $x = 4.0$ cm and (*b*) point P_2 on the y axis at $y = 6.0$ cm. The reference point (where $V = 0$) is at infinity.

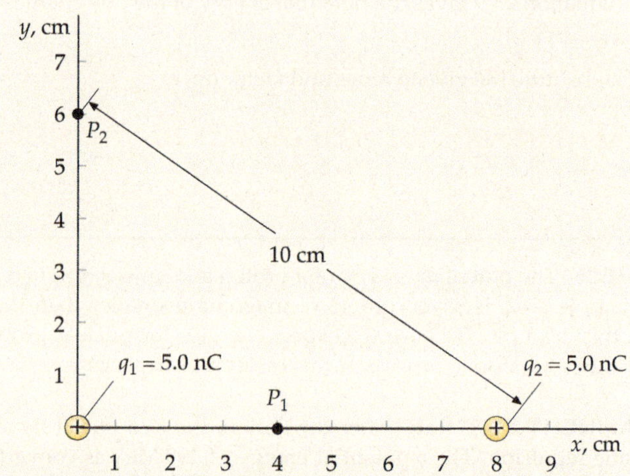

FIGURE 23-7

PICTURE The two positive point charges on the x axis are shown in Figure 23-7, and the potential is to be found at points P_1 and P_2.

SOLVE

(*a*) 1. Use Equation 23-10 to write V as a function of the distances r_1 and r_2 to the charges:

$$V = \sum_i \frac{kq_i}{r_i} = \frac{kq_1}{r_1} + \frac{kq_2}{r_2}$$

2. Point P_1 is 4.0 cm from each charge, and the charges are equal:

$$r_1 = r_2 = r = 0.040 \text{ m}$$

$$q_1 = q_2 = q = 5.0 \times 10^{-9}\text{ C}$$

3. Use these to find the potential at point P_1:

$$V = \frac{kq}{r} + \frac{kq}{r} = \frac{2kq}{r}$$

$$= \frac{2 \times (8.99 \times 10^9 \text{ N}\cdot\text{m}^2/\text{C}^2)(5.0 \times 10^{-9}\text{ C})}{0.040 \text{ m}} = 2247 \text{ V} = \boxed{2.2 \text{ kV}}$$

(*b*) Point P_2 is 6.0 cm from one charge and 10 cm from the other. Use these to find the potential at point P_2:

$$V = \frac{(8.99 \times 10^9 \text{ N}\cdot\text{m}^2/\text{C}^2)(5.0 \times 10^{-9}\text{ C})}{0.060 \text{ m}} + \frac{(8.99 \times 10^9 \text{ N}\cdot\text{m}^2/\text{C}^2)(5.0 \times 10^{-9}\text{ C})}{0.10 \text{ m}}$$

$$= 749 \text{ V} + 450 \text{ V} = \boxed{1.2 \text{ kV}}$$

CHECK The calculated potentials are both positive. The potential at a field point is the work per unit charge to bring a test charge from a reference point (where the potential is zero) to the field point. For the potential function used here, the reference point is at infinity. A positive test charge at any location would be repelled by both q_1 and q_2. Thus, an external agent would have to do work on the test charge to bring the test charge from rest at the reference point at infinity to rest at any field point. Thus, we expect the potential at any field point to be positive.

TAKING IT FURTHER Note that in Part (*a*), the electric field is zero at the point midway between the charges but the potential is not. An external agent must do positive work to bring a test charge to this point from a long distance away, because the electric field is zero only at the final position.

Example 23-5 Potential throughout the *x* Axis

A point charge q_1 is at the origin, and a second point charge q_2 is on the x axis at $x = a$. Using Equation 23-10, find an expression for the potential everywhere on the x axis as a function of x.

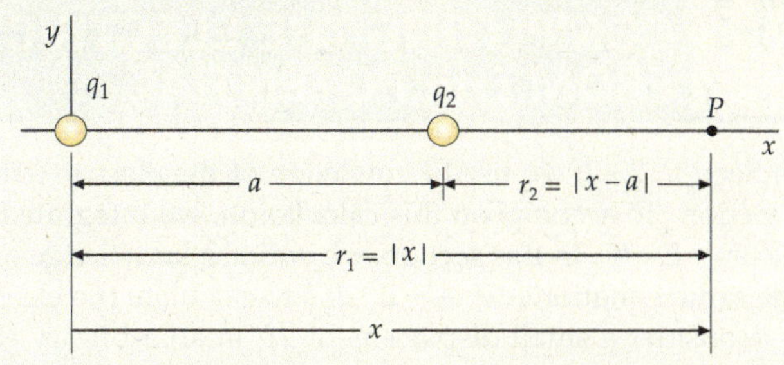

FIGURE 23-8

PICTURE The total potential at a field point is the sum of the potentials due to each charge separately.

SOLVE

1. Sketch the x axis and place the two charges on it. Let r_1 be the distance from q_1 to an arbitrary field point P at position x on the x axis, that is, $r_1 = |x|$. Let r_2 be the distance r_2 from q_2 to P, that is, $r_2 = |x - a|$ (Figure 23-8):

2. Write the potential as a function of the distances to the two charges:
$$V = \frac{kq_1}{r_1} + \frac{kq_2}{r_2}$$
$$= \boxed{\frac{kq_1}{|x|} + \frac{kq_2}{|x - a|} \qquad x \neq 0, \quad x \neq a}$$

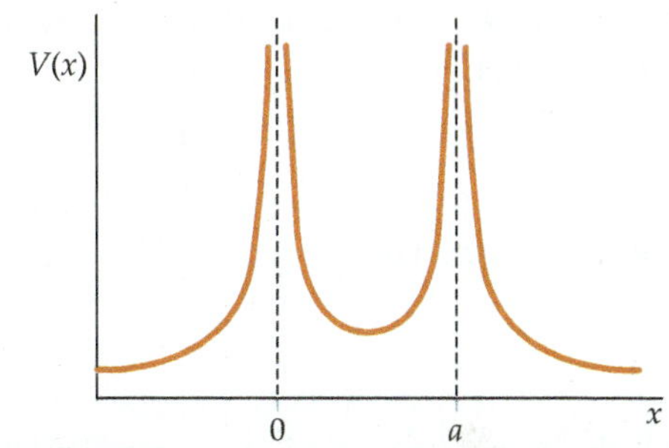

FIGURE 23-9

CHECK Note that $V \to \infty$ both as $x \to 0$ and as $x \to a$, and $V \to 0$ both as $x \to -\infty$ and as $x \to +\infty$, as one would expect.

TAKING IT FURTHER Figure 23-9 shows V versus x on the x axis for $q_1 = q_2 > 0$.

Example 23-6 Potential Due to an Electric Dipole

An electric dipole consists of a positive point charge $+q$ on the x axis at $x = +\ell/2$ and a negative point charge $-q$ on the x axis at $x = -\ell/2$. Find the potential on the x axis for $x \gg +\ell/2$ in terms of the dipole moment $\vec{p} = q\ell\hat{i}$.

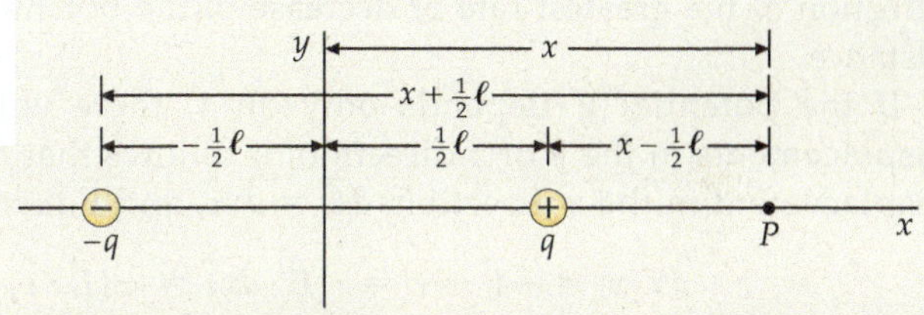

FIGURE 23-10

PICTURE The potential at a field point is the sum of the potentials for each charge.

SOLVE

1. Sketch the x axis and place the two charges on it. For $x > \ell/2$, the distance from the field point P to the positive charge is $x - \frac{1}{2}\ell$ and the distance from the field point to the negative charge is $x + \frac{1}{2}\ell$ (Figure 23-10).

2. For $x > \ell/2$, the potential due to the two charges is
$$V = \frac{kq}{x - (\ell/2)} + \frac{k(-q)}{x + (\ell/2)}$$
$$= \frac{kq\ell}{x^2 - (\ell^2/4)} \qquad x > \frac{\ell}{2}$$

3. The magnitude of $\vec{p}$ is $p = q\ell$. For $x \gg \ell/2$, we can neglect $\ell^2/4$ compared with x^2 in the denominator.
$$V \approx \boxed{\frac{kq\ell}{x^2} = \frac{kp}{x^2} \qquad x \gg \ell}$$

CHECK A dipole has a total charge of zero, so we expect that far from the dipole, the potential would decrease with increasing distance from the dipole more rapidly than it would for a configuration that has a nonzero net charge. The step 3 result is that the potential decreases inversely with the square of the distance. Far from a configuration with a net charge the potential decreases inversely with distance from the configuration, which is less rapidly than inversely with the square of the distance.

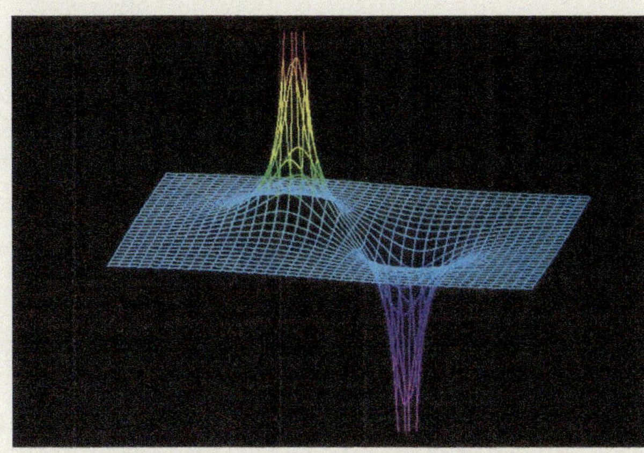

The electrostatic potential in a plane containing both point charges of an electric dipole. The potential due to each point charge is proportional to the charge and inversely proportional to the distance from the charge. (*© 1990 Richard Menga/Fundamental Photographs.*)

23-3 COMPUTING THE ELECTRIC FIELD FROM THE POTENTIAL

In Section 23-2, we used knowledge of the electric field to calculate the potential function. To accomplish this calculation, we integrated both sides of the equation $dV = -\vec{E}\cdot d\vec{\ell}$. In this section, we will use knowledge of the potential function and the same equation ($dV = -\vec{E}\cdot d\vec{\ell}$) to calculate the electric field.

Consider a small displacement $d\vec{\ell}$ in an arbitrary electrostatic field $\vec{E}$. The associated change in potential is given by $dV = -\vec{E}\cdot d\vec{\ell}$. If the displacement $d\vec{\ell}$ is perpendicular to $\vec{E}$, then $dV = 0$ (the potential does not change). For a given $|d\vec{\ell}|$, the maximum increase in V occurs when the displacement $d\vec{\ell}$ is in the opposite direction as $\vec{E}$. To solve for $\vec{E}$, we first solve for the component of $\vec{E}$ in the direction of $d\vec{\ell}$. That is,

$$dV = -\vec{E}\cdot d\vec{\ell} = -E\cos\theta\, d\ell = E_{\text{tan}} d\ell \qquad \text{23-11}$$

where $E_{\text{tan}} = E\cos\theta$ (the tangential component of $\vec{E}$) is the component of $\vec{E}$ in the direction of $d\vec{\ell}$. Then

$$E_{\text{tan}} = -\frac{dV}{d\ell} \qquad \text{23-12}$$

If the displacement $d\vec{\ell}$ is perpendicular to the electric field, then $dV = 0$ (the potential does not change). For a given $d\vec{\ell}$, the maximum increase in V occurs when the displacement $d\vec{\ell}$ is in the same direction as $-\vec{E}$. A vector that points in the direction of the greatest change in a scalar function and that has a magnitude equal to the derivative of that function with respect to the distance in that direction is called the **gradient** of the function. Thus, the electric field $\vec{E}$ is the negative gradient of the potential V. That is, the direction of the electric field is the same as the direction of the greatest rate of decrease of the potential function with respect to distance.

If the potential V depends only on x, there will be no change in V for displacements in the y or z direction; it follows that E_y and E_z equal zero. For a displacement in the x direction, $d\vec{\ell} = dx\hat{i}$, and Equation 23-11 becomes

$$dV(x) = -\vec{E}\cdot d\vec{\ell} = -\vec{E}\cdot dx\hat{i} = -(\vec{E}\cdot\hat{i})\,dx = -E_x\,dx$$

Then

$$E_x = -\frac{dV(x)}{dx} \qquad \text{23-13}$$

For a spherically symmetric charge distribution centered at the origin, the potential can be a function only of the radial coordinate r. Displacements perpendicular to the radial direction give no change in $V(r)$, so the electric field must be radial. A displacement in the radial direction is written $d\vec{\ell} = dr\hat{r}$. Equation 23-11 is then

$$dV(r) = -\vec{E}\cdot d\vec{\ell} = -\vec{E}\cdot dr\hat{r} = -E_r\,dr$$

and

$$E_r = -\frac{dV(r)}{dr} \qquad \text{23-14}$$

If we know either the potential or the electric field throughout some region of space, we can use one to calculate the other. The potential is often easier to calculate because it is a scalar function, whereas the electric field is a vector function. Note that we cannot calculate $\vec{E}$ if we know the potential V at just a single point—we must know V over a region of space to compute the derivative necessary to obtain $\vec{E}$ throughout that region. If we only know V along a curve or on a surface, then we can only calculate the component of $\vec{E}$ tangent to the curve or surface.

CONCEPT CHECK 23-1

In what direction can you move relative to an electric field so that the electric potential does not change?

CONCEPT CHECK 23-2

In what direction can you move relative to an electric field so that the electric potential increases at the greatest rate?

Example 23-7 $\vec{E}$ for a Potential that Varies with x

Find the electric field for the electric potential function V given by $V = 100\text{ V} - (25\text{ V/m})x$.

PICTURE This potential function depends only on x. Use $E_x = -dV/dx$ (Equation 23-13) and solve for E_x. Because the potential does not vary with y or z, $E_y = E_z = 0$.

SOLVE

The electric field is found from $E_x = -dV/dx$ (Equation 23-13) using $V = 100\text{ V} - (25\text{ V/m})x$:

$$E_x = -\frac{dV}{dx} \text{ and } E_y = E_z = 0 \quad \text{so} \quad \vec{E} = \boxed{+(25\text{ V/m})\hat{i}}$$

CHECK The potential decreases as x increases. Note that the electric field is in the $+x$ direction, the direction of decreasing potential, as expected.

TAKING IT FURTHER This electric field is uniform and in the $+x$ direction. Note that the constant 100 V in the expression for $V(x)$ has no effect on the electric field. The electric field does not depend on the choice of zero for the potential function.

PRACTICE PROBLEM 23-4 (*a*) At what points does V equal zero in this example? (*b*) Write the potential function corresponding to the same electric field with $V = 0$ everywhere on the $x = 0$ plane.

GENERAL RELATION BETWEEN $\vec{E}$ AND V

In vector notation, the gradient of V is written as either $\overrightarrow{grad}V$ or $\vec{\nabla}V$. Then

$$\vec{E} = -\vec{\nabla}V \qquad 23\text{-}15$$

In general, the potential function can depend on x, y, and z. The Cartesian components of the electric field are related to the partial derivatives of the potential with respect to x, y, or z. For example, the x component of the electric field is given by

$$E_x = -\frac{\partial V}{\partial x} \qquad 23\text{-}16a$$

Similarly, the y and z components of the electric field are related to the potential by

$$E_y = -\frac{\partial V}{\partial y} \qquad 23\text{-}16b$$

and

$$E_z = -\frac{\partial V}{\partial z} \qquad 23\text{-}16c$$

Thus, Equation 23-15 in Cartesian coordinates is written

$$\vec{E} = -\vec{\nabla}V = -\left(\frac{\partial V}{\partial x}\hat{i} + \frac{\partial V}{\partial y}\hat{j} + \frac{\partial V}{\partial z}\hat{k}\right) \qquad 23\text{-}17$$

23-4 CALCULATIONS OF V FOR CONTINUOUS CHARGE DISTRIBUTIONS

The potential due to a continuous distribution of charge can be calculated by choosing an element of charge dq, which we treat as a point charge, and invoking superposition, changing the sum in $V = \Sigma kq_i/r_i$ (Equation 23-10) to an integral:

$$V = \int \frac{k\,dq}{r} \qquad 23\text{-}18$$

POTENTIAL DUE TO A CONTINUOUS CHARGE DISTRIBUTION

This equation assumes that $V = 0$ at an infinite distance from the charges, so we cannot use it for any charge distributions of infinite extent, as is the case for artificial charge distributions like an infinite line charge or an infinite plane charge.

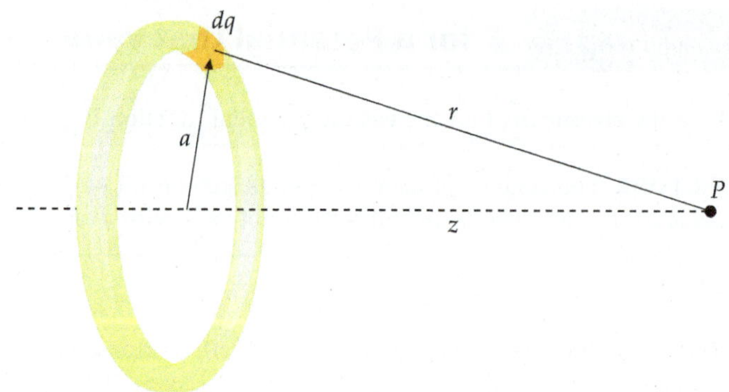

FIGURE 23-11 Geometry for the calculation of the electric potential at a point on the axis of a charged ring of radius a.

V ON THE AXIS OF A CHARGED RING

Figure 23-11 shows a uniformly charged ring of radius a and charge Q in the $z = 0$ plane and centered at the origin. The distance from an element of charge dq to the field point P on the axis of the ring is $r = \sqrt{z^2 + a^2}$. Because this distance is the same for all elements of charge on the ring, we can remove this term from the integral in Equation 23-18. The potential at point P due to the ring is thus

$$V = \int \frac{k\,dq}{r} = \frac{k}{r}\int dq = \frac{kQ}{r}$$

or

$$V = \frac{kQ}{\sqrt{z^2 + a^2}} \qquad 23\text{-}19$$

POTENTIAL ON THE AXIS OF A CHARGED RING

Note that when $|z|$ is much greater than a, the potential approaches $kQ/|z|$, the same as the potential due to a point charge Q at the origin.

Example 23-8 A Ring and a Particle — *Try It Yourself*

A ring of radius 4.0 cm is in the $z = 0$ plane and has its center at the origin. The ring has a uniform charge of 8.0 nC. A small particle that has a mass equal to 6.0 mg (6.0×10^{-6} kg) and a charge equal to 5.0 nC is placed on the z axis at $z = 3.0$ cm and released. Find the speed of the particle when it is a great distance from the ring. Assume effects due to gravity are negligible.

PICTURE The particle is repelled by the ring. As the particle moves along the z axis, its potential energy decreases and its kinetic energy increases. Use conservation of mechanical energy to find the kinetic energy of the particle when it is far from the ring. The final speed is found from the final kinetic energy.

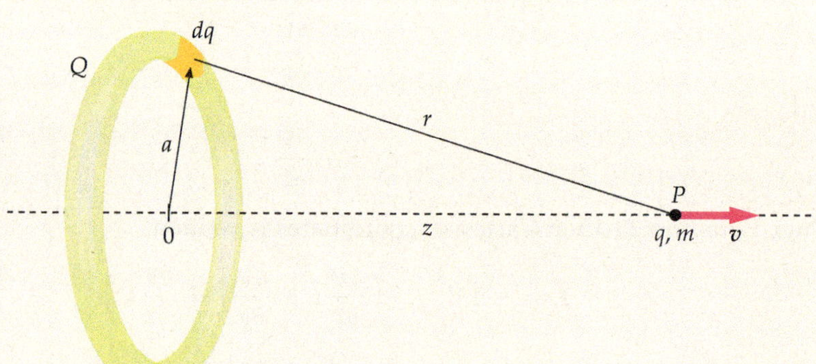

FIGURE 23-12

SOLVE

Cover the column to the right and try these on your own before looking at the answers.

Steps	Answers
1. Sketch the ring, the particle, and the z axis. Label the sketch appropriately (Figure 23-12).	
2. Write down the relation between the kinetic energy and the speed.	$K = \frac{1}{2}mv^2$
3. Use $U = qV$, with V given by $V = kQ/\sqrt{z^2 + a^2}$ (Equation 23-19), to obtain an expression for the potential energy V as a function of the distance z of the point charge from the center of the ring.	$U = qV = \dfrac{kqQ}{\sqrt{z^2 + a^2}}$

4. Use conservation of mechanical energy to relate the energy at $z_i = 0.030$ m to the energy as $z_f \rightarrow \infty$. Solve for the speed as z_f approaches infinity.

$$U_f + K_f = U_i + K_i$$

$$\frac{kqQ}{\sqrt{z_f^2 + a^2}} + \frac{1}{2}mv_f^2 = \frac{kqQ}{\sqrt{z_i^2 + a^2}} + \frac{1}{2}mv_i^2$$

$$\text{so} \quad v_f^2 = \frac{2kqQ}{m\sqrt{z_i^2 + a^2}} = 2.40\ \text{m}^2/\text{s}^2$$

$$v_f = \boxed{1.6\ \text{m/s}}$$

CHECK In step 3, we found that $v_f^2 = 2.40\ \text{m}^2/\text{s}^2$, a positive number. If our result was that v_f^2 had equaled a negative number it would be a clear indication that a mistake had been made.

PRACTICE PROBLEM 23-5 What is the potential energy of the particle when it is at $z = 9.0$ cm?

V ON THE AXIS OF A UNIFORMLY CHARGED DISK

We can use our result for the potential on the axis of a ring charge to calculate the potential on the axis of a uniformly charged disk.

Example 23-9 Find V for a Charged Disk

Find the potential on the axis of a disk of radius R that carries a total charge Q distributed uniformly on its surface.

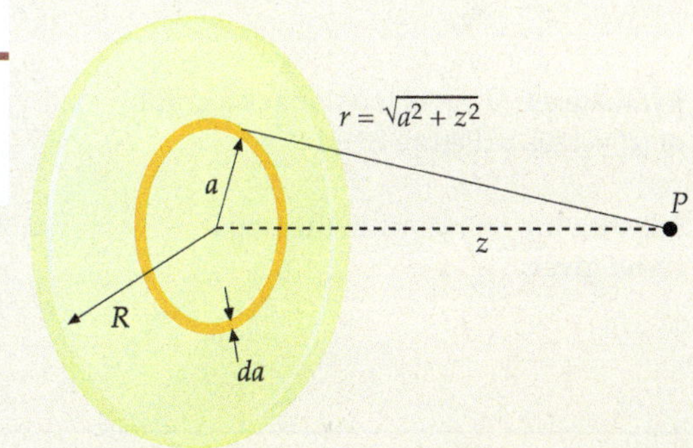

FIGURE 23-13

PICTURE We take the axis of the disk to be the z axis, and we treat the disk as a set of ring charges. The ring of radius a and thickness da in Figure 23-13 has an area of $2\pi a\, da$. The charge of the ring is $dq = \sigma\, dA = \sigma 2\pi a\, da$, where $\sigma = Q/(\pi R^2)$ is the surface charge density. The potential at point P due to the charge on this ring is given by $k\, dq/(z^2 + a^2)^{1/2}$ (Equation 23-19). We then integrate from $a = 0$ to $a = R$ to find the total potential due to the charge on the disk.

SOLVE

1. Write the potential dV at point P due to the charged ring of radius a:

$$dV = \frac{k\, dq}{(z^2 + a^2)^{1/2}} = \frac{k\sigma 2\pi a\, da}{(z^2 + a^2)^{1/2}}$$

2. Integrate from $a = 0$ to $a = R$:

$$V = \int_0^R \frac{k\sigma 2\pi a\, da}{(z^2 + a^2)^{1/2}} = k\sigma\pi \int_0^R (z^2 + a^2)^{-1/2} 2a\, da$$

3. The integral is of the form $\int u^n\, du$, with $u = z^2 + a^2$, $du = 2z\, dz$, and $n = -\frac{1}{2}$. When $a = 0$, $u = z^2 + 0^2$ and when $a = R$, $u = z^2 + R^2$:

$$V = k\sigma\pi \int_{z^2+0^2}^{z^2+R^2} u^{-1/2}\, du = k\sigma\pi \left.\frac{u^{1/2}}{\frac{1}{2}}\right|_{z^2}^{z^2+R^2}$$

$$= 2k\sigma\pi\left(\sqrt{z^2 + R^2} - \sqrt{z^2}\right)$$

4. Rearranging this result to find V gives

$$V = \boxed{2\pi k\sigma|z|\left(\sqrt{1 + \frac{R^2}{z^2}} - 1\right)}$$

CHECK For $|z| \gg R$, the potential function V should approach the potential function of a point charge Q at the origin. That is, we expect that for large $|z|$, $V \approx kQ/|z|$. To approximate our result for $|z| \gg R$, we use the binomial expansion:

$$\left(1 - \frac{R^2}{z^2}\right)^{1/2} = 1 + \frac{1}{2}\frac{R^2}{z^2} + \cdots$$

Then

$$V = 2\pi k\sigma|z|\left[\left(1 + \frac{1}{2}\frac{R^2}{z^2} + \cdots\right) - 1\right] \approx \frac{k(\sigma\pi R^2)}{|z|} = \frac{kQ}{|z|}$$

From Example 23-9, we see that the potential on the axis of a uniformly charged disk in the $z = 0$ plane is

$$V = 2\pi k\sigma|z|\left(\sqrt{1 + \frac{R^2}{z^2}} - 1\right) \qquad 23\text{-}20$$

POTENTIAL ON THE AXIS OF A UNIFORMLY CHARGED DISK

Example 23-10 Find $\vec{E}$ Given V

Calculate the electric field on the axis of a uniformly charged disk that has a charge q and a radius R using the potential function given in Equation 23-20.

PICTURE Using $E_z = -dV/dz$, we can evaluate E_z by direct differentiation. We cannot evaluate either E_x or E_y by direct differentiation because we do not know how V varies in those directions. However, the symmetry of the charge distribution dictates that *on* the x axis, $E_x = E_y = 0$.

SOLVE

1. Write Equation 23-20 for the potential on the axis of a uniformly charged disk:

$$V = 2\pi k\sigma|z|\left(\sqrt{1 + \frac{R^2}{z^2}} - 1\right) = 2\pi k\sigma[(z^2 + R^2)^{1/2} - |z|]$$

2. Compute $-dV/dz$ to find E_z:

$$E_z = -\frac{dV}{dz} = -2\pi k\sigma\left[\frac{1}{2}(z^2 + R^2)^{-1/2}2z - \frac{d|z|}{dz}\right]$$

3. Evaluate $d|z|/dz$. It is the slope of a graph of $|z|$ versus z (Figure 23-14):*

$$\frac{d|z|}{dz} = \text{sign}(z) = \begin{cases} +1 & z > 0 \\ 0 & z = 0 \\ -1 & z < 0 \end{cases}$$

4. Substituting for $d|z|/dz$ in the step 2 result gives:

$$E_z = -2\pi k\sigma\left(\frac{z}{\sqrt{z^2 + R^2}} - \text{sign}(z)\right)$$

$$= \boxed{2\pi k\sigma\left(\text{sign}(z) - \frac{z}{\sqrt{z^2 + R^2}}\right)}$$

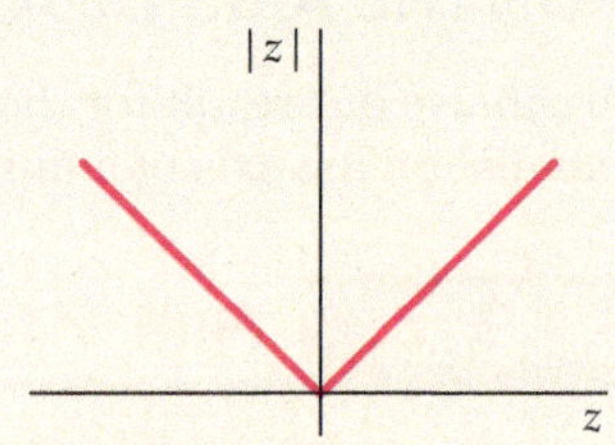

FIGURE 23-14 A plot of $|z|$ versus z.

CHECK By factoring $|z|$ from the radical in the step 4 result, we obtain

$$E_z = -2\pi k\sigma\left(\frac{z}{|z|\sqrt{1 + (a^2/z^2)}} - \text{sign}(z)\right) = \text{sign}(z)\cdot 2\pi k\sigma\left(1 - \frac{1}{\sqrt{1 + (a^2/z^2)}}\right)$$

where we have used $z/|z| = \text{sign}(z)$. This expression for E_z has the same form as the expression for E_z found in Equation 22-9.

TAKING IT FURTHER The step 3 result ($d|z|/dz = \text{sign}\, z$) defines $d|z|/dz$ to equal zero at $z = 0$. In like manner, using $d|z|/dz = \text{sign}\, z$ in the Check defines $z/|z|$ to equal zero at $z = 0$. It is common practice to define the value of a function at a point where it is not continuous to equal the average of the values of the function on either side of the discontinuity. That is what we have done here with $d|z|/dz$ and with $z/|z|$.

PRACTICE PROBLEM 23-6 Using the expression for the potential V on the axis of a uniformly charged ring of radius R (Equation 23-20), compute $-dV/dz$ on the axis and obtain an expression for E_z on the axis. Show that this expression has the same form as that shown in Equation 22-8.

V DUE TO AN INFINITE PLANE OF CHARGE

If we let R become very large, our uniformly charged disk approaches an infinite plane. As R approaches infinity, the potential function $V = 2\pi k\sigma|z|(\sqrt{1 + (R^2/z^2)} - 1)$ (Equation 23-20) approaches infinity. However, we obtained

* See Taking It Further at the end of this example.

Equation 23-20 from Equation 23-18, which assumes that $V = 0$ at infinity. We have a contradiction—Equation 23-20 is not a valid potential function for a uniformly charge disk of infinite radius. For charge distributions that extend to infinity, we cannot choose $V = 0$ at a point at an infinite distance from the charges. Instead, we first find the electric field $\vec{E}$ (by direct integration or from Gauss's law) and then calculate the potential function V from its defining relation $dV = -\vec{E}\cdot d\vec{\ell}$. For an infinite plane of uniform charge of density σ in the $x = 0$ plane, the electric field in the region $x > 0$ is given by Equation 22-10:

$$\vec{E} = \frac{\sigma}{2\epsilon_0}\hat{i} = 2\pi k\sigma\hat{i} \qquad x > 0$$

The potential increment dV for an arbitrary displacement increment $d\vec{\ell} = dx\hat{i} + dy\hat{j} + dy\hat{k}$ is then

$$dV = -\vec{E}\cdot d\vec{\ell} = -(2\pi k\sigma\hat{i})\cdot(dx\hat{i} + dy\hat{j} + dz\hat{k}) = -2\pi k\sigma\, dx \qquad x > 0$$

Integrating both sides of this equation, we obtain

$$V = -2\pi k\sigma x + V_0 \qquad x > 0$$

where the arbitrary integration constant V_0 is the potential at $x = 0$. Note that the value of this potential function decreases with distance from the plane and approaches $-\infty$ as x approaches $+\infty$.

For negative x, the electric field is

$$\vec{E} = -2\pi k\sigma\hat{i} \qquad x > 0$$

so

$$dV = -\vec{E}\cdot d\vec{\ell} = +2\pi k\sigma\, dx \qquad x > 0$$

and the potential is

$$V = V_0 + 2\pi k\sigma x = V_0 - 2\pi k\sigma|x| \qquad x > 0$$

Figure 23-15 is a plot of this potential function. The potential again decreases with distance from the charged plane and approaches $-\infty$ as x approaches $-\infty$. For either positive or negative x, the potential V can be written

$$V = V_0 - 2\pi k\sigma|x| \qquad \text{23-21}$$

POTENTIAL NEAR AN INFINITE PLANE OF CHARGE

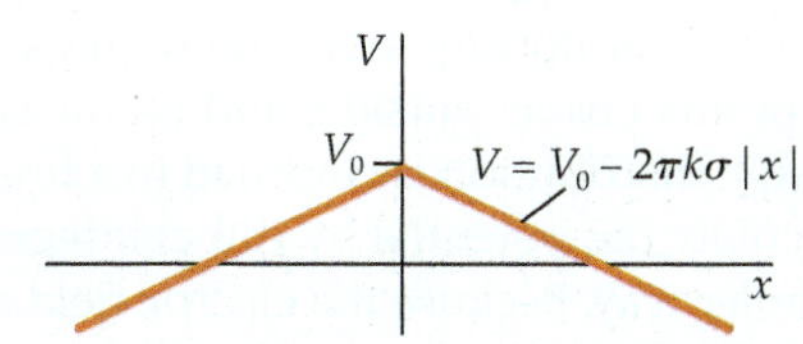

FIGURE 23-15 Plot of V versus x for an infinite plane of charge in the $x = 0$ plane. Note that the potential is continuous at $x = 0$ even though $E_x = -dV/dx$ is not continuous there. The reference point where $V = V_0$ is at the origin.

Example 23-11 A Sheet of Charge and a Point Charge

An infinite flat sheet that has a uniform charge density σ lies in the $x = 0$ plane, and a point charge q is on the x axis at $x = a$ (Figure 23-16). Find the potential at some point P a distance r from the point charge.

PICTURE We can use the principle of superposition. The total potential V is the sum of the individual potentials due to the plane and the point charge. We must add an arbitrary constant in our expression for V, which is determined by our choice of the reference point, where $V = 0$. We are free to choose the location of the reference point to be anywhere but at $x = \pm\infty$ or at $x = a$ on the x axis. For this calculation, we choose $V = 0$ at the origin.

SOLVE

1. Sketch the charge configuration. Include the coordinate axes and a field point at (x, y, z):

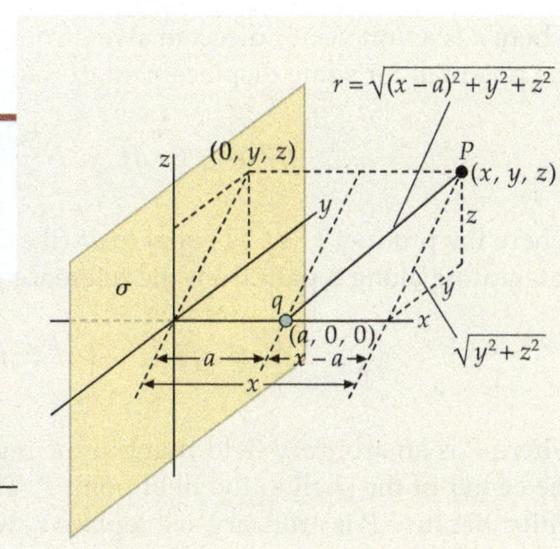

FIGURE 23-16

2. The potential due to the charged plane is given by $V_{plane} = V_0 - 2\pi k\sigma|x|$ (Equation 23-21) and the potential due to a point charge is given by $V_{point} = kq/r - kq/r_{ref}$ (Equation 23-7) where r is the distance from the point charge to the field point. The total potential is the sum of the two potentials:

$$V = V_{plane} + V_{point} = -2\pi k\sigma|x| + \frac{kq}{r} + C$$

where the constant $C\ (= V_0 - kq/r_{ref})$ is chosen to set the potential at the reference point to zero.

3. The distance r from the point charge at $(a, 0, 0)$ to the field point at (x, y, z) is $\sqrt{(x-a)^2 + y^2 + z^2}$:

$$V = -2\pi k\sigma|x| + \frac{kq}{\sqrt{(x-a)^2 + y^2 + z^2}} + C$$

4. We choose to let $V = 0$ at the origin. To do that, set $V = 0$ at $x = y = z = 0$ and solve for the constant C:

$$0 = 0 + \frac{kq}{a} + C \quad \text{so} \quad C = -\frac{kq}{a}$$

5. Substitute $-kq/a$ for C in the step 3 result:

$$V = -2\pi k\sigma|x| + \frac{kq}{\sqrt{(x-a)^2 + y^2 + z^2}} - \frac{kq}{a}$$

$$= \boxed{-2\pi k\sigma|x| + kq\left(\frac{1}{r} - \frac{1}{a}\right)}$$

CHECK The step 5 result is what you would expect by superposing the potential for a uniformly charged plane and a point charge.

TAKING IT FURTHER The answer is not unique. We could have specified the potential at any point, other than at $x = a$ or at $x = \pm\infty$.

V INSIDE AND OUTSIDE A SPHERICAL SHELL OF CHARGE

Here, we find the potential due to a thin spherical shell that has a radius R and a charge Q uniformly distributed on its surface. We are interested in the potential at all points inside, outside, and on the shell. Unlike the infinite plane of charge, this charge distribution is confined to a finite region of space, so, in principle, we could calculate the potential by direct integration of Equation 23-18. However, there is a simpler way. Because the electric field for this charge distribution is easily obtained from Gauss's law, we will calculate the potential from the known electric field using $dV = -\vec{E}\cdot d\vec{\ell}$.

Outside the spherical shell, the electric field is radial and is the same as if all the charge Q were a point charge at the origin:

$$\vec{E} = \frac{kQ}{r^2}\hat{r}$$

where $\hat{r}$ is a unit vector directed away from the center of the sphere. The change in the potential for some displacement $d\vec{\ell}$ outside the shell is then

$$dV = -\vec{E}\cdot d\vec{\ell} = -\frac{kQ}{r^2}\hat{r}\cdot d\vec{\ell} = -\frac{kQ}{r^2}\,dr$$

where the product $\hat{r}\cdot d\vec{\ell}$ is equal to dr (the component of $d\vec{\ell}$ in the direction of $\hat{r}$). Integrating along a path from the reference point at infinity, we obtain

$$V_P = -\int_{\infty}^{\vec{r}_P} \vec{E}\cdot d\vec{\ell} = -\int_{\infty}^{r_P} \frac{kQ}{r^2}\,dr = -kQ\int_{\infty}^{r_P} r^{-2}\,dr = \frac{kQ}{r_P}$$

where P is an arbitrary field point in the region $r \geq R$, and r_P is the distance from the center of the shell to the field point P. The potential is chosen to be zero at infinity. Because P is arbitrary, we replace r_P with r to obtain

$$V = \frac{kQ}{r} \qquad r \geq R$$

Inside the spherical shell, the electric field is zero everywhere. Again integrating from the reference point at infinity, we obtain

$$V_P = -\int_{\infty}^{\vec{r}_P} \vec{E}\cdot d\vec{r} = -\int_{\infty}^{R} \frac{kQ}{r^2}\,dr - \int_{R}^{r_P} (0)\,dr = \frac{kQ}{R}$$

where P is an arbitrary field point in the region $r < R$, and r_P is the distance from the center of the shell to the field point P. The potential at all points inside the shell is kQ/R, where R is the radius of the shell. Inside the shell V is the same everywhere. The potential at any point inside the shell is the work per unit charge to bring a test charge from infinity to the shell. No additional work is required to bring it from the shell to any point inside the shell. Thus,

$$V = \begin{cases} \dfrac{kQ}{r} & (r \geq R) \\ \dfrac{kQ}{R} & (r \leq R) \end{cases} \qquad \text{23-22}$$

POTENTIAL DUE TO A THIN SPHERICAL SHELL

This potential function is plotted in Figure 23-17.

A region of zero electric field merely implies that the potential field is uniform throughout the region. Consider a spherical shell that has a small hole through it so that we can move a test charge in and out of the shell. If we move the test charge from an infinite distance to the shell, the work per charge we must do is kQ/R. Inside the shell there is no electric field, so it takes no work to move the test charge around inside the shell. The total amount of work per unit charge it takes to bring the test charge from infinity to any point inside the shell is just the work per charge it takes to bring the test charge up to the shell radius R, which is kQ/R. The potential is therefore kQ/R everywhere inside the shell.

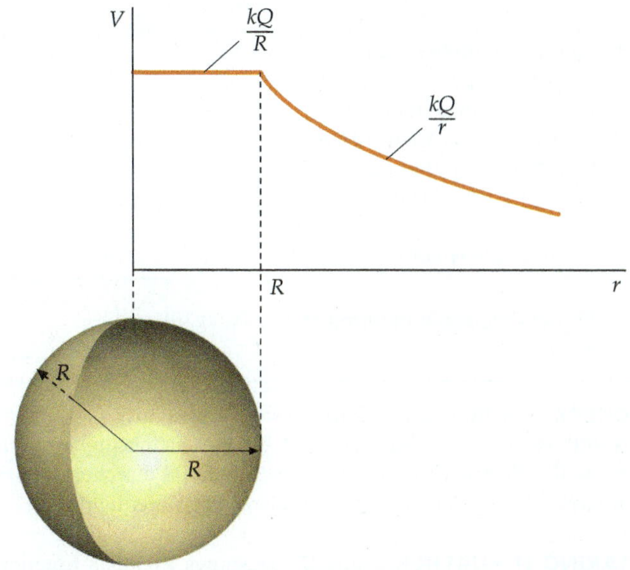

FIGURE 23-17 Electric potential of a uniformly charged thin spherical shell that has a radius R as a function of the distance r from the center of the shell. Inside the shell, the potential has the constant value kQ/R. Outside the shell, the potential is the same as that due to a point charge Q at the center of the sphere.

! A common mistake is to think that the potential must be zero inside a spherical shell because the electric field is zero throughout that region.

PRACTICE PROBLEM 23-7

What is the potential of a spherical shell of radius 10.0 cm carrying a charge of 6.00 μC?

Example 23-12 Find V for a Uniformly Charged Sphere — *Try It Yourself*

In one model, a proton is considered to be a uniformly charged solid sphere that has a radius R and a charge Q. The electric field inside the sphere is given by $E_r = k\dfrac{Q}{R^3}r$ (Equation 22-18*b*). Find the potential V both inside and outside the sphere.

PICTURE Outside the sphere, the charge looks like a point charge, so the potential is given by $V = kQ/r$. Inside the sphere, V can be found by integrating $dV = -\vec{E}\cdot d\vec{\ell}$, where the electric field inside the sphere given by $\vec{E} = (kQr/R^3)\hat{r}$ (Equation 22-18*b*).

SOLVE

Cover the column to the right and try these on your own before looking at the answers.

Steps	Answers
1. Outside the sphere, the electric field is the same as that of a point charge. If we set the potential equal to zero at infinity, the potential there is also the same as that of a point charge.	$V(r) = \boxed{\dfrac{kQ}{r} \quad r \geq R}$
2. For $r \leq R$, find dV from $dV = -\vec{E}\cdot d\vec{\ell}$, where the electric field inside the sphere given by $\vec{E} = (kQr/R^3)\hat{r}$ (Equation 22-18*b*).	$dV = -\vec{E}\cdot d\vec{\ell} = -\dfrac{kQr}{R^3}\hat{r}\cdot d\vec{\ell} = -\dfrac{kQr}{R^3}dr$

3. Find the definite integral using the expression in step 2. Find the change in potential from infinity to an arbitrary field point P in the region $r_P < R$, where r_P is the distance of point P from the center of the sphere.

$$V_P = -\int_\infty^{r_P} E_r\,dr = -\int_\infty^{R} \frac{kQ}{r^2}\,dr - \int_R^{r_P} \frac{kQ}{R^3}r\,dr$$

$$= \frac{kQ}{R} - \frac{kQ}{2R^3}(r_P^2 - R^2) = \frac{kQ}{2R}\left(3 - \frac{r_P^2}{R^2}\right)$$

4. Express the result in terms of $r = r_P$:

$$V(r) = \boxed{\frac{kQ}{2R}\left(3 - \frac{r^2}{R^2}\right) \quad r \le R}$$

CHECK Substituting $r = R$ in the step 4 result gives $V = kQ/R$ as required by the step-1 result. At $r = 0$, $V = 3kQ/2R = 1.5\,kQ/R$, which is greater than kQ/R, as it should be, because the electric field is in the positive radial direction for $r < R$. (An electrostatic field always points in the direction of decreasing potential.)

TAKING IT FURTHER Figure 23-18 shows $V(r)$ as a function of r. Note that both $V(r)$ and $E_r = -dV/dr$ are continuous everywhere.

PRACTICE PROBLEM 23-8 Find the potential function if the reference point where $V = 0$ is at $r = R$ (instead of at $r = \infty$).

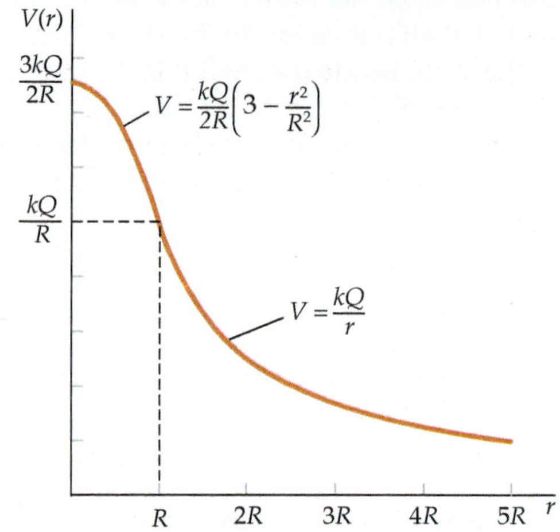

FIGURE 23-18

V DUE TO AN INFINITE LINE CHARGE

We will now calculate the potential due to a uniformly charged infinite line. Let the charge per unit length be λ. Like the infinite plane of charge, this charge distribution is not confined to a finite region of space, so, in principle, we cannot calculate the potential by direct integration of $dV = k\,dq/r$ (Equation 23-18). Instead, we find the potential by integrating the electric field directly. The electric field of a uniformly charged infinite line is given by $\vec{E} = (2k\lambda/R)\,\hat{R}$ (Equation 22-3), where λ is the linear charge density and R is the radial distance from the line. The change in potential for an arbitrary displacement $d\vec{\ell}$ is given by

$$dV = -\vec{E}\cdot d\vec{\ell} = -\frac{2k\lambda}{R}\hat{R}\cdot d\vec{\ell}$$

where $\hat{R}$ is in the radial direction. The product $\hat{R}\cdot d\vec{\ell} = dR$ (the component of $d\vec{\ell}$ in the direction of $\hat{R}$), so $dV = -(2k\lambda/R)\,dR$. Integrating from an arbitrary reference point to an arbitrary field point P (Figure 23-19) gives

$$V_P - V_{\text{ref}} = -2k\lambda\int_{R_{\text{ref}}}^{R_P} \frac{dR}{R} = -2k\lambda \ln\frac{R_P}{R_{\text{ref}}}$$

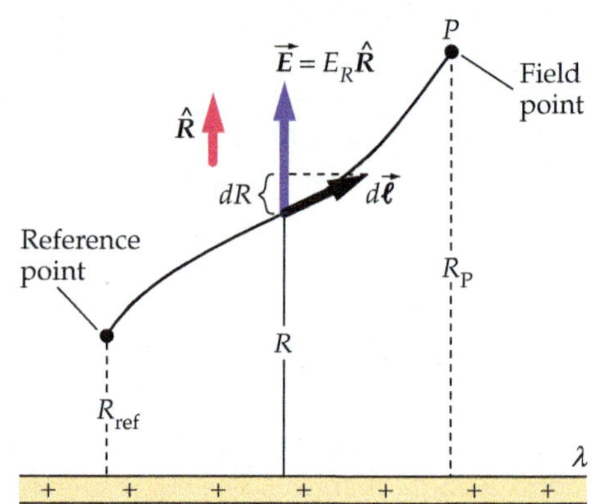

FIGURE 23-19

where R_P and R_{ref} are the radial distances of the field and reference points, respectively, from the line charge. For convenience, we choose the potential to equal zero at the reference point ($V_{\text{ref}} = 0$). We cannot choose R_{ref} to be zero because $\ln(0) = -\infty$, and we cannot choose R_{ref} to be infinity because $\ln(\infty) = +\infty$. However, any other choice in the interval $0 < R_{\text{ref}} < \infty$ is acceptable, and the potential function is given by

$$V = 2k\lambda \ln\frac{R_{\text{ref}}}{R} \qquad \text{23-23}$$

POTENTIAL DUE TO A UNIFORM LINE CHARGE OF INFINITE LENGTH

We do not encounter charge distributions in nature that actually extend to infinity. However, such distributions make excellent models for some real-world situations. An example is the potential of a 500-m-long, nearly straight, high-voltage transmission power line (but not close to either end).

23-5 EQUIPOTENTIAL SURFACES

Because there is no electric field inside the material of a conductor that is in static equilibrium, the value of the potential is the same throughout the region occupied by the conducting material. That is, the conductor is a three-dimensional **equipotential region** and the surface of a conductor is an **equipotential surface.**

The potential V has the same value everywhere on an equipotential surface. If a test charge on an equipotential surface is given a small displacement $d\vec{\ell}$ parallel to the surface, $dV = -\vec{E}\cdot d\vec{\ell} = 0$. Because $\vec{E}\cdot d\vec{\ell}$ is zero for any $d\vec{\ell}$ parallel to the surface, $\vec{E}$ must either be zero or be perpendicular to any and every $d\vec{\ell}$ that is parallel to the surface. The only way $\vec{E}$ can be perpendicular to every $d\vec{\ell}$ parallel to the surface is for $\vec{E}$ to be normal to the surface. Therefore, we conclude that electric field lines are normal to any equipotential surfaces they intersect. Figures 23-20 and 23-21 show equipotential surfaces near a spherical conductor and a nonspherical conductor. Note that anywhere a field line meets or penetrates an equipotential surface, shown in gray, the field line is normal to the equipotential surface. If we go from one equipotential surface to a neighboring equipotential surface by undergoing a displacement $d\vec{\ell}$ along a field line in the direction of the field, the potential changes by $dV = -\vec{E}\cdot d\vec{\ell} = -E d\ell$. It follows that equipotential surfaces that have a fixed potential difference between them are more closely spaced where the electric field strength E is greater.

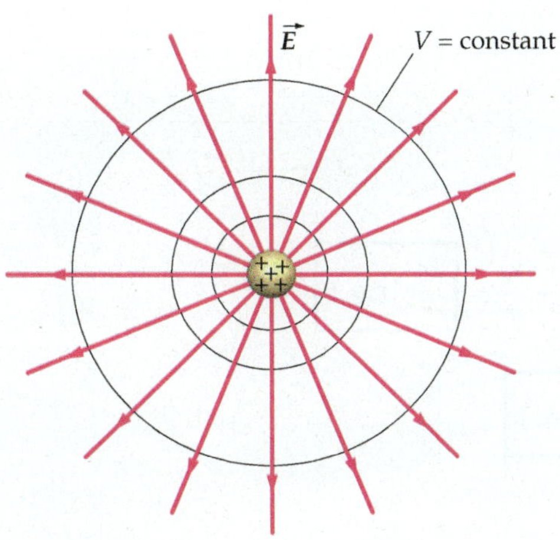

FIGURE 23-20 Equipotential surfaces and electric field lines outside a uniformly charged spherical conductor. The equipotential surfaces are spherical and the field lines are radial. The field lines are normal to the equipotential surfaces.

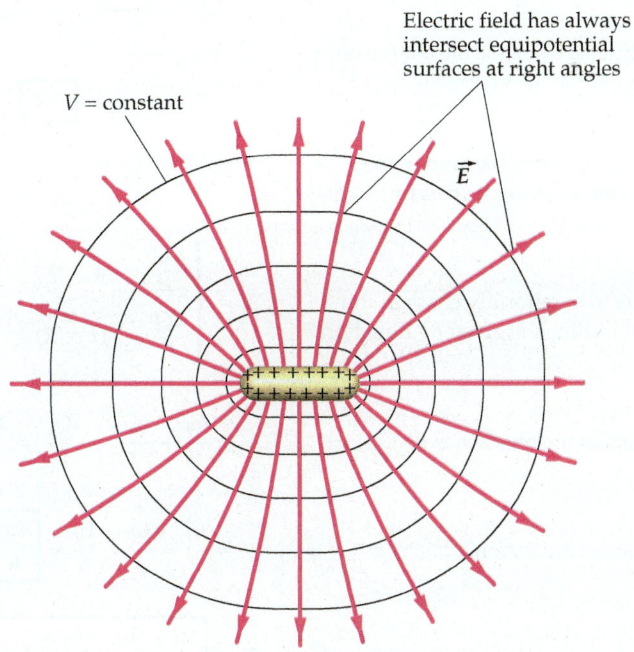

FIGURE 23-21 Equipotential surfaces and electric field lines outside a nonspherical conductor.

Example 23-13 A Hollow Spherical Shell

A hollow, uncharged spherical conducting shell has an inner radius a and an outer radius b. A positive point charge $+q$ is located at the center of the shell. (*a*) Find the charge on each surface of the conductor. (*b*) Find the potential $V(r)$ everywhere, assuming that $V = 0$ at $r = \infty$.

PICTURE (*a*) The charge distribution is spherical, so applying Gauss's law should be a good method for finding the charges on the inner and outer surfaces of the shell. (*b*) Sum the individual potentials for the individual charges to obtain the resultant potential. The potential for a point charge and for a uniform thin spherical shell of charge have already been established (Equations 23-8 and 23-22).

SOLVE

(*a*) 1. The charge inside a closed surface is proportional to the outward flux of $\vec{E}$ through the surface:

$$\phi_{\text{net}} = 4\pi k Q_{\text{inside}}$$

where $\phi_{\text{net}} = \oint_S E_n\, dA$

2. Sketch the point charge and the spherical shell. On a conducting object, charge can reside on its surfaces but not within the conducting material. Label the charge on each surface of the shell. Include a Gaussian surface completely inside the conducting material and enclosing the inner surface (Figure 23-22):

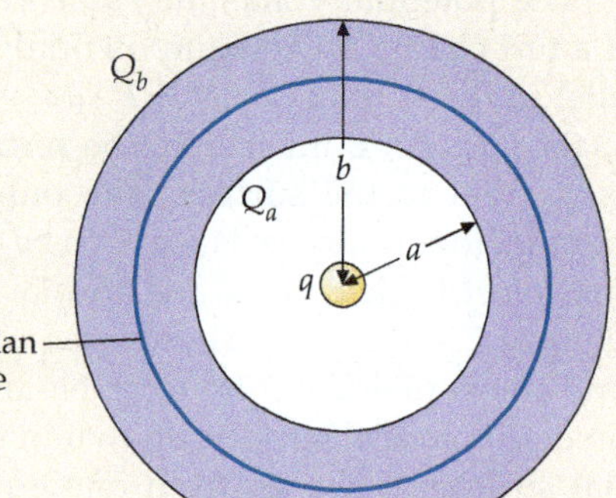

FIGURE 23-22

3. Apply Gauss's law (the step 1 result) to the Gaussian surface and solve for the charge on the inner surface of the shell:

$$E_n = 0 \Rightarrow Q_{\text{inside}} = q + Q_a = 0$$

so $\boxed{Q_a = -q}$

4. The shell is neutral, so solve for the charge on its outer surface:

$$Q_a + Q_b = 0$$

so $Q_b = -Q_a = \boxed{+q}$

(*b*) 1. The potential at any point is the sum of the potentials due to the individual charges:

$$V = V_q + V_{Q_a} + V_{Q_b}$$

2. The potential due to a uniformly charged thin spherical shell of radius R is given by Equation 23-22:

$$V = \begin{cases} \dfrac{kQ}{r} & (r \geq R) \\[2ex] \dfrac{kQ}{R} & (r \leq R) \end{cases}$$

3. Add the potentials in the region $r \geq b$:

$$V = \frac{kq}{r} + \frac{kQ_a}{r} + \frac{kQ_b}{r} = \frac{kq}{r} - \frac{kq}{r} + \frac{kq}{r} = \boxed{\frac{kq}{r} \quad r \geq b}$$

4. Add the potentials in the region $a \leq r \leq b$:

$$V = \frac{kq}{r} - \frac{kq}{r} + \frac{kq}{b} = \boxed{\frac{kq}{b} \quad a \leq r \leq b}$$

5. Add the potentials in the region $0 < r \leq a$:

$$V = \boxed{\frac{kq}{r} - \frac{kq}{a} + \frac{kq}{b} \quad 0 < r \leq a}$$

CHECK All potential functions must be continuous. Thus, we expect the Part-(*b*), step-3 and step-4 results to be equal at $r = b$, and the Part-(*b*), step-4 and step-5 results to be equal at $r = a$. This expectation is realized by the results obtained. At $r = b$ the step-3 and -4 results both equal kq/b. The same is true of the step-4 and -5 results at $r = a$.

TAKING IT FURTHER Each of the individual potential functions in step 1 of Part (*b*) has its zero-potential reference point at $r = \infty$. Thus, the sum of these functions also has its zero-potential reference point at $r = \infty$. The potential arrived at in the example can be obtained by directly evaluating $-\int_{\infty}^{P} \vec{E} \cdot d\vec{\ell} = -\int_{\infty}^{r_P} E_r\, dr$. Yet a third way to obtain the potential is by evaluating the indefinite integral $-\int E_r\, dr$ in each region to find the integration constants by matching the potential functions at the boundaries. Matching the potential functions at the boundaries is valid because the potential must be a continuous function.

Figure 23-23 shows the electric potential as a function of the distance from the center of the cavity. Inside the conducting material, where $a \le r \le b$, the potential has the constant value kq/b. Outside the shell, the potential is the same as that of a point charge q at the center of the shell. Note that $V(r)$ is continuous everywhere. The electric field is discontinuous at the conductor surfaces, as reflected in the discontinuous slope of $V(r)$ at $r = a$ and $r = b$.

Two conductors that are separated in space will typically not be at the same potential. The potential difference between such conductors depends on their geometrical shapes, their separation in space, and the net charge on each. When two conductors touch, the charge on the conductors redistributes itself so that electrostatic equilibrium is established and the electric field is zero inside both conductors. While touching, the two conductors can be considered to be a single conductor with a single potential. If we put a spherical charged conductor in contact with a second spherical conductor that is uncharged, charge will flow between them until both conductors are at the same potential. If the spherical conductors are identical, after touching they share the original charge equally. If the identical spherical conductors are now separated, each has half the original charge.

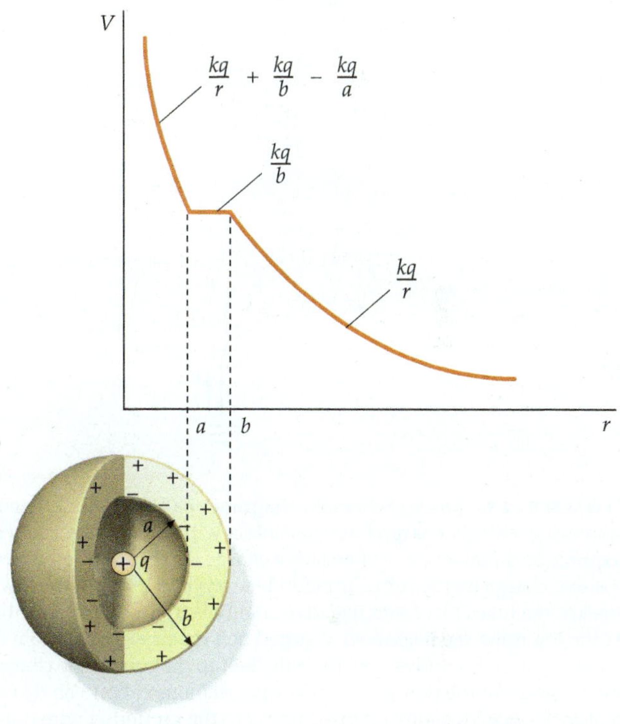

FIGURE 23-23

THE VAN DE GRAAFF GENERATOR

In Figure 23-24, a small conductor carrying a positive charge q is inside the cavity of a larger conductor. In equilibrium, the electric field is zero inside the conducting material of both conductors. The electric field lines that begin on the positive charge q must terminate on the inner surface of the large conductor. This must occur no matter what the charge may be on the outside surface of the large conductor. Regardless of the charge on the large conductor, the small conductor in the cavity is at a greater potential because the electric field lines go from this conductor to the larger conductor. If the conductors are now connected, say, with a fine conducting wire, *all* the charge originally on the smaller conductor will flow to the larger conductor. When the connection is broken, there is no charge on the small conductor in the cavity, and there are no field lines between the conductors. The positive charge transferred from the smaller conductor resides completely on the outside surface of the larger conductor. If we put more positive charge on the small conductor in the cavity and again connect the conductors with a fine wire, all of the charge on the inner conductor will again flow to the outer conductor. The procedure can be repeated indefinitely. This method is used to produce large potentials in a

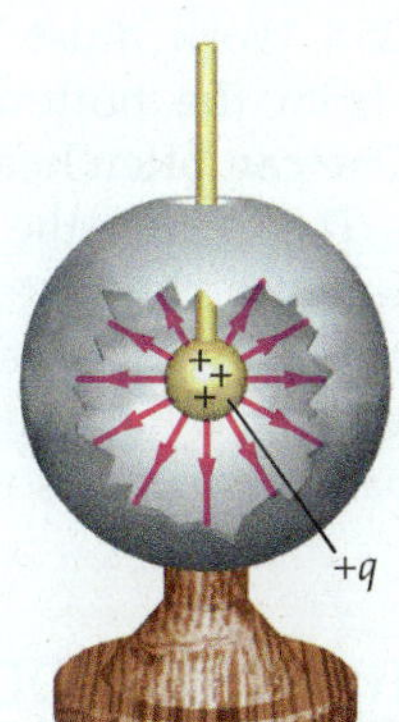

FIGURE 23-24 Small conductor that has a positive charge q inside a larger hollow conductor.

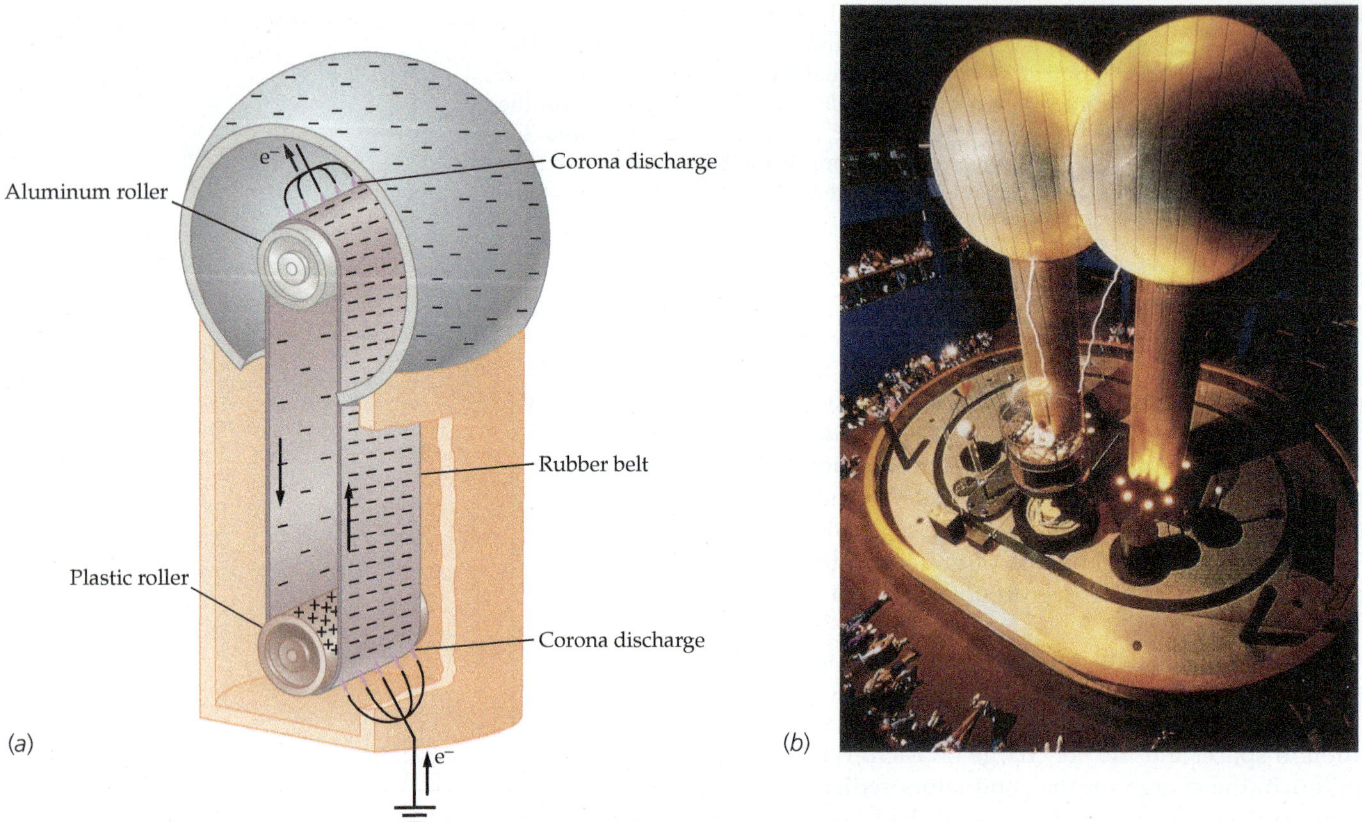

FIGURE 23-25 (*a*) Schematic diagram of a Van de Graaff generator. The lower roller becomes positively charged due to contact with the moving belt. (The inner surface of the belt acquires an equal amount of negative charge that is distributed over a larger area.) The dense positive charge on the roller attracts electrons to the tips of the lower comb where dielectric breakdown takes place and negative charge is transported to the belt via corona discharge. At the top roller the negatively charged belt repels electrons from the tips of the comb and negative charge is transferred from the belt to the comb. The charge is then transferred to the outer surface of the dome. (*b*) These large demonstration Van de Graaff generators in the Boston Science Museum are discharging to the grounded wire cage housing the operator. *((b) © Karen R. Preuss.)*

device called the *Van de Graaff generator*, in which the charge is brought to the inner surface of a larger spherical conductor by a continuous charged belt (Figure 23-25). Work must be done by the motor driving the belt to bring the charge from the bottom to the top of the belt, where the potential is very high. One can often hear the motor speed decrease as the sphere accumulates charge. The greater the charge on the outer conductor, the greater the potential of the outer conductor and the greater the electric field just outside its outer surface. A Van de Graaff accelerator is a device that uses the intense electric field produced by a Van de Graaff generator to accelerate ions and charged subatomic particles such as protons.

DIELECTRIC BREAKDOWN

Many nonconducting materials become ionized in very high electric fields and become conductors. This phenomenon, called **dielectric breakdown,** occurs in air at an electric field strength of $E_{max} \approx 3 \times 10^6$ V/m = 3 MN/C. In air, some of the existing ions are accelerated to greater kinetic energies before they collide with neighboring molecules. Dielectric breakdown occurs when these ions are accelerated to kinetic energies sufficient to result in a growth in ion concentration due to the collisions with neighboring molecules. The maximum potential that can be obtained in a Van de Graaff generator is limited by the dielectric breakdown of the air. Van de Graaff generators can achieve much higher potentials in a controlled atmosphere than they can in air at atmospheric pressure. Sulfur hexafluoride gas at several atmospheres of pressure is used when optimal performance is desired. The magnitude of the electric field for which dielectric breakdown occurs in a material is called the **dielectric strength** of that material. The dielectric strength of air is about 3 MV/m. The discharge through the conducting air resulting from dielectric breakdown is called **arc discharge.** The electric shock you receive when you touch a metal doorknob after walking across a rug on a dry day is a familiar example of arc discharge. These breakdowns occur more often on dry days because moist air can conduct the charge away before the breakdown condition is reached. Lightning is an example of arc discharge on a large scale.

Example 23-14 Dielectric Breakdown for a Charged Sphere

A spherical conductor has a radius of 30 cm (~1.0 ft). (*a*) What is the maximum charge that can be placed on the sphere before dielectric breakdown of the surrounding air occurs? (*b*) What is the maximum potential of the sphere?

PICTURE (*a*) We find the maximum charge by relating the charge to the electric field and setting the field equal to the dielectric strength of air, E_{max}. (*b*) The maximum potential is then found from the maximum charge calculated in Part (*a*).

SOLVE

(*a*) 1. The electric field at the surface of a conductor is proportional to the charge density σ on the surface of the conductor (Equation 22-21):

$$E = \frac{\sigma}{\epsilon_0} = 4\pi k\sigma$$

2. Set this field equal to E_{max}:

$$E_{max} = 4\pi k\sigma_{max}$$

3. The maximum charge Q_{max} is found from σ_{max}:

$$\sigma_{max} = \frac{\text{charge}}{\text{area}} = \frac{Q_{max}}{4\pi R^2}$$

4. Solving for Q_{max} gives:

$$Q_{max} = 4\pi R^2\sigma_{max} = 4\pi R^2\frac{E_{max}}{4\pi k} = \frac{R^2E_{max}}{k}$$

$$= \frac{(0.30\text{ m})^2(3 \times 10^6\text{ N/C})}{(8.99 \times 10^9\text{ N}\cdot\text{m}^2/\text{C}^2)} = \boxed{3 \times 10^{-5}\text{ C}}$$

(*b*) Use the expression for the maximum charge to calculate the maximum potential of the sphere:

$$V_{max} = \frac{kQ_{max}}{R} = \frac{k}{R}\left(\frac{R^2E_{max}}{k}\right) = RE_{max}$$

$$= (0.30\text{ m})(3 \times 10^6\text{ N/C}) = \boxed{9 \times 10^5\text{ V}}$$

CHECK Small van de Graaff generators are commonly used in hair-raising demonstrations that must achieve a high potential. Our Part-(*b*) result is certainly a high potential.

TAKING IT FURTHER The values calculated are for a Van de Graaff generator that has a 2.0-ft-diameter dome. For safety reasons, most classroom Van de Graaff generator domes have a diameter of 1.0 ft or less.

PRACTICE PROBLEM 23-9 Calculate the maximum charge and maximum potential of a Van de Graaff generator that has a 1.0-ft-diameter dome.

Example 23-15 Two Charged Spherical Conductors

Two uncharged spherical conductors of radius $R_1 = 6.0$ cm and $R_2 = 2.0$ cm (Figure 23-26) and separated by a distance much greater than 6.0 cm are connected by a long, very thin conducting wire. A total charge $Q = +80$ nC is placed on one of the spheres and the system is allowed to reach electrostatic equilibrium. (*a*) What is the charge on each sphere? (*b*) What is the electric field strength at the surface of each sphere? (*c*) What is the electric potential of each sphere? (Assume that the charge on the connecting wire is negligible.)

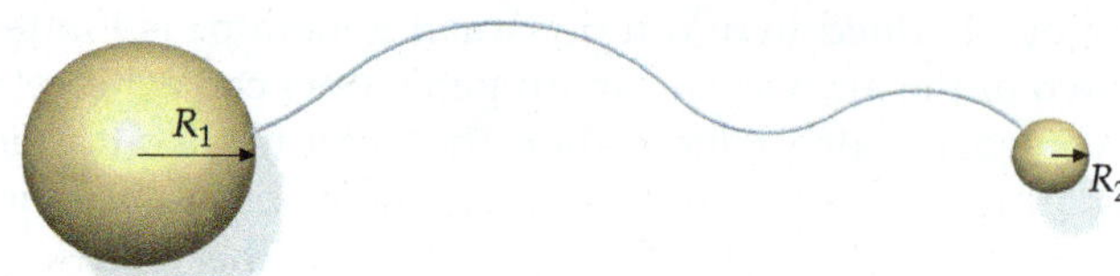

FIGURE 23-26

PICTURE The total charge will be distributed with Q_1 on sphere 1 and Q_2 on sphere 2 so that the spheres will be at the same potential. We can use $V = kQ/R$ for the potential of each sphere.

SOLVE

(*a*) 1. Conservation of charge gives us one relation between the charges Q_1 and Q_2:

$$Q_1 + Q_2 = Q$$

2. Equating the potential of the spheres gives us a second relation for the charges Q_1 and Q_2:

$$\frac{kQ_1}{R_1} = \frac{kQ_2}{R_2} \quad \Rightarrow \quad Q_2 = \frac{R_2}{R_1}Q_1$$

3. Combine the results from steps 1 and 2 and solve for Q_1 and Q_2:

$$Q_1 + \frac{R_1}{R_2}Q_1 = Q \quad \text{so}$$

$$Q_1 = \frac{R_1}{R_1 + R_2}Q = \frac{6.0\text{ cm}}{8.0\text{ cm}}(80\text{ nC}) = \boxed{60\text{ nC}}$$

$$Q_2 = Q - Q_1 = \boxed{20\text{ nC}}$$

(*b*) Use these results to calculate the electric field strengths at the surface of the spheres:

$$E_1 = \frac{kQ_1}{R_1^2} = \frac{(8.99 \times 10^9\text{ N}\cdot\text{m}^2/\text{C}^2)(60 \times 10^{-9}\text{ C})}{(0.060\text{ m})^2}$$

$$= \boxed{150\text{ kN/C}}$$

$$E_2 = \frac{kQ_2}{R_2^2} = \frac{(8.99 \times 10^9\text{ N}\cdot\text{m}^2/\text{C}^2)(20 \times 10^{-9}\text{ C})}{(0.020\text{ m})^2}$$

$$= \boxed{450\text{ kN/C}}$$

(*c*) Calculate the common potential from kQ/R for either sphere:

$$V_1 = \frac{kQ_1}{R_1} = \frac{(8.99 \times 10^9\text{ N}\cdot\text{m}^2/\text{C}^2)(60 \times 10^{-9}\text{ C})}{0.060\text{ m}}$$

$$= \boxed{9.0\text{ kV}}$$

CHECK If we use sphere 2 to calculate V, we obtain $V_2 = kQ_2/R_2 = (8.99 \times 10^9\text{ N}\cdot\text{m}^2/\text{C}^2)(20 \times 10^{-9}\text{ C})/0.020\text{ m} = 9.0 \times 10^3$ V. An additional check is available, because the electric field strength at the surface of each sphere is proportional to its charge density. The radius of sphere 1 is three times the radius of sphere 2, so its surface area is nine times the surface area of sphere 2. And because sphere 1 has three times the charge, its charge density is one-third the charge density of sphere 2. Therefore, the electric field strength at the surface of sphere 1 should be one-third of the electric field strength at the surface of sphere 2, which is what we found in Part (*b*).

TAKING IT FURTHER The presence of the long, very thin wire connecting the spheres makes the result of this example only approximate because the potential function $V = kQ/r$ is valid for the region outside an isolated conducting sphere. With the wire in place the spheres cannot accurately be modeled as isolated spheres.

When a charge is placed on a conductor of nonspherical shape, like that in Figure 23-27*a*, the surface of the conductor will be an equipotential surface, but the surface charge density and the electric field just outside the conductor will vary from point to point. Near a point where the radius of curvature is small, such as point A in the figure, the surface charge density and electric field will be large, whereas near a point where the radius of curvature is large, such as point B in the figure, the field and surface charge density will be small. We can understand this qualitatively by considering the ends of the conductor to be spheres of different radii. Let σ be the surface charge density.

The potential of a sphere of radius R is

$$V = \frac{kq}{R} = \frac{1}{4\pi\epsilon_0}\frac{Q}{R} \qquad \text{23-24}$$

Because the area of a sphere is $4\pi R^2$, the charge on a sphere is related to the charge density by $Q = 4\pi R^2\sigma$. Substituting this expression for Q into Equation 23-24 we have

$$V = \frac{1}{4\pi\epsilon_0}\frac{4\pi R^2\sigma}{R} = \frac{R\sigma}{\epsilon_0}$$

Solving for σ, we obtain

$$\sigma = \frac{\epsilon_0 V}{R} \qquad \text{23-25}$$

Because both *spheres* are at the same potential, the sphere that has the smaller radius must have the greater surface charge density. And because $E = \sigma/\epsilon_0$ at the surface of a conductor, the electric field strength is greatest at points on the conductor where the radius of curvature is least.

For an arbitrarily shaped conductor, the potential at which dielectric breakdown occurs depends on the smallest radius of curvature of any part of the conductor. If the conductor has sharp points of very small radius of curvature, dielectric breakdown will occur at relatively low potentials. In the Van de Graaff generator (see Figure 23-25*a*), the charge is transferred onto the belt by sharp-edged conductors near the bottom of the belt. The charge is removed from the belt by sharp-edged conductors near the top of the belt.

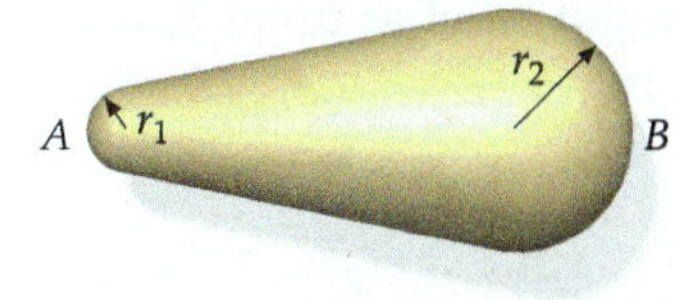

(*a*)

(*b*)

FIGURE 23-27 (*a*) A nonspherical conductor. If a charge is placed on such a conductor, it will produce an electric field that is stronger near point *A*, where the radius of curvature is small, than near point *B*, where the radius of curvature is large. (*b*) Electric field lines near a nonspherical conductor and plate that have equal and opposite charges. The lines are shown by small bits of thread suspended in oil. Note that the electric field is strongest near points of small radius of curvature, such as at the ends of the plate and at the pointed left side of the conductor. The equipotential surfaces are more closely spaced where the field strength is greater. *((b) Harold M. Waage.)*

23-6 ELECTROSTATIC POTENTIAL ENERGY

Objects that repel each other have more potential energy if they are close together, and objects that attract each other have more potential energy if they are far apart. Suppose there is a point charge q_1 at point 1. To bring a second point charge q_2 from rest at infinity to rest at point 2, a distance r_{12} from point 1, requires that we do work:

$$W_2 = q_2V_2 = q_2\frac{kq_1}{r_{12}} = \frac{kq_2q_1}{r_{12}}$$

where V_2 is the potential at point 2 due to the presence of charge q_1. (It follows that the potential energy of these two point charges is the negative of this work value.)

$$V_2 = \frac{kq_1}{r_{12}}$$

The potential at point 3, a distance r_{13} from q_1 and a distance r_{23} from q_2, is given by

$$V_3 = \frac{kq_1}{r_{13}} + \frac{kq_2}{r_{23}}$$

so to bring in an additional point charge q_3 from rest at infinity to rest at point 3 requires that we do the additional work:

$$W_3 = q_3 V_3 = \frac{kq_3 q_1}{r_{13}} + \frac{kq_3 q_2}{r_{23}}$$

The total work required to assemble the three charges is the **electrostatic potential energy** U of the system of three point charges:

$$U = \frac{kq_2 q_1}{r_{12}} + \frac{kq_3 q_1}{r_{13}} + \frac{kq_3 q_2}{r_{23}} \qquad \text{23-26}$$

This quantity of work is independent of the order in which the charges are brought to their final positions. In general,

The electrostatic potential energy of a system of point charges is the work needed to bring the charges from an infinite separation to their final positions.

ELECTROSTATIC POTENTIAL ENERGY OF A SYSTEM

The first two terms on the right-hand side of Equation 23-26 can be written

$$\frac{kq_2 q_1}{r_{12}} + \frac{kq_3 q_1}{r_{13}} = q_1\left(\frac{kq_2}{r_{12}} + \frac{kq_3}{r_{13}}\right) = q_1 V_1$$

where V_1 is the potential at the location of q_1 due to charges q_2 and q_3. Similarly, the second and third terms represent the charge q_3 multiplied by the potential due to charges q_1 and q_2, and the first and third terms equal the charge q_2 multiplied by the potential due to charges q_1 and q_2. We can thus rewrite Equation 23-26 as

$$\begin{aligned} U &= \frac{1}{2}U + \frac{1}{2}U \\ &= \frac{1}{2}\left(\frac{kq_2 q_1}{r_{12}} + \frac{kq_3 q_1}{r_{13}} + \frac{kq_3 q_2}{r_{23}}\right) + \frac{1}{2}\left(\frac{kq_2 q_1}{r_{12}} + \frac{kq_3 q_1}{r_{13}} + \frac{kq_3 q_2}{r_{23}}\right) \\ &= \frac{1}{2}q_1\left(\frac{kq_2}{r_{12}} + \frac{kq_3}{r_{13}}\right) + \frac{1}{2}q_2\left(\frac{kq_3}{r_{23}} + \frac{kq_1}{r_{12}}\right) + \frac{1}{2}q_3\left(\frac{kq_1}{r_{13}} + \frac{kq_2}{r_{23}}\right) \\ &= \frac{1}{2}(q_1 V_1 + q_2 V_2 + q_3 V_3) \end{aligned}$$

The electrostatic potential energy U of a system of n point charges is thus

$$U = \frac{1}{2}\sum_{i=1}^{n} q_i V_i \qquad \text{23-27}$$

ELECTROSTATIC POTENTIAL ENERGY OF A SYSTEM OF POINT CHARGES

where V_i is the potential at the location of the ith charge due to the presence of all the other charges in the system.

Equation 23-27 can also describe the electrostatic potential energy of a continuous charge distribution. Consider a spherical conductor of radius R. When the sphere carries a charge q, its potential relative to $V = 0$ at infinity is

$$V = \frac{kq}{R}$$

The work we must do to bring an additional amount of charge dq from infinity to the conductor is $V\,dq$. This work equals the increase in the potential energy of the conductor:

$$dU = V\,dq = \frac{kq}{R}\,dq$$

The total potential energy U is the integral of dU as q increases from zero to its final value Q. Integrating, we obtain

$$U = \frac{k}{R}\int_0^Q q\,dq = \frac{kQ^2}{2R} = \frac{1}{2}QV \qquad \text{23-28}$$

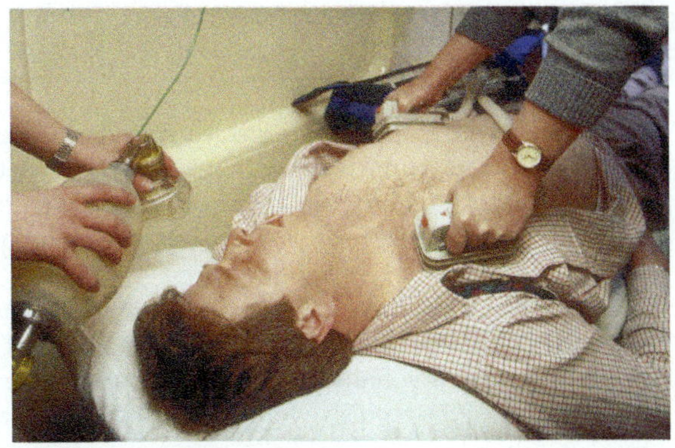

Hearts reach a state called ventricular fibrillation in about two-thirds of people that experience cardiac arrest. In this state, a heart quivers, spasms chaotically, and does not pump. To take a heart out of this state, a significant current is passed through the heart. Then the pacemaker cells in the heart can again establish a regular heartbeat. An external defibrillator applies a large voltage across the chest. *(© Steve Allen/The Image Bank/Getty Images.)*

where $V = kQ/R$ is the potential on the surface of the fully charged sphere. We can interpret Equation 23-28 as $U = Q \times \frac{1}{2}V$ where $\frac{1}{2}V$ is the average potential of the spherical conductor during the charging process. During the charging process, bringing the first element of charge in from infinity to the uncharged sphere requires no work because the charge being brought in is not being repelled by the charge already on the sphere. As the charge on the sphere accumulates, bringing in each additional element of charge to the sphere requires additional work; when the sphere is almost fully charged, bringing the last element of charge in against the repulsive force of the charge on the sphere requires the most work.* The average potential of the sphere during the charging process is one-half its final potential V, so the total work required to bring in the entire charge Q equals $\frac{1}{2}QV$.

Alternatively, if we set $V_i = V$ and $Q = \sum_i q_i$, Equation 23-27 becomes Equation 23-28. We can think of the charge on the uniformly charged spherical shell as a collection of infinitesimal point charges—all at the same potential V. Thus, Equation 23-27 leads directly Equation 23-28.

Although we derived Equation 23-28 for a spherical conductor, it holds for any conductor. The potential of any conductor is proportional to its charge q, so we can write $V = \alpha q$, where α is a proportionality constant. The work needed to bring an additional charge dq from infinity to the conductor is $V\,dq = \alpha q\,dq$, so the total work needed to put a charge Q on the conductor is $\frac{1}{2}\alpha Q^2 = \frac{1}{2}QV$. If we have a set of n conductors with the ith conductor at potential V_i and carrying a charge Q_i, the electrostatic potential energy is

$$U = \frac{1}{2}\sum_{i=1}^{n} Q_i V_i \qquad \text{23-29}$$

ELECTROSTATIC POTENTIAL ENERGY OF A SYSTEM OF CONDUCTORS

* We are assuming that each element of charge is the same size.

Example 23-16 Work Required to Move Point Charges

Four identical positive point charges, each having charge q, are initially at rest at infinite separation. (*a*) Calculate the total work required to move the point charges to the four corners of the square of edge length a by separately calculating the work required to sequentially move each charge to its final position. (*b*) Show that Equation 23-27 gives the total work.

PICTURE Move the charges to the corners of the square sequentially. No work is needed to move the first charge to a corner because the potential at the corner is zero when the other three charges are at infinity. As each additional charge is moved to a corner, work must be done because of the repulsive forces of the previous placed charges.

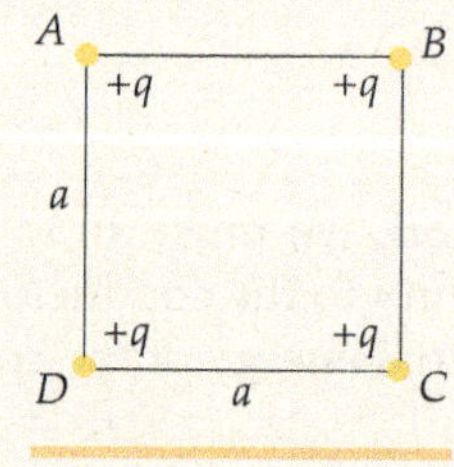

FIGURE 23-28

SOLVE

(*a*) 1. Sketch the square and label the corners A, B, C, and D (Figure 23-28):

2. Place the first charge at point A. To accomplish this step, the work W_A that is needed is zero:

$$W_A = 0 \qquad x > 0$$

3. Bring the second charge to point B. The work required is $W_B = qV_A$, where V_A is the potential at point B due to the first charge at point A a distance a from it:

$$W_B = qV_A = q\left(\frac{kq}{a}\right) = \frac{kq^2}{a}$$

4. $W_C = qV_C$, where V_C is the potential at point C due to q at point A, a distance $\sqrt{2}a$ away, and q at point B, a distance a from it:

$$W_C = qV_C = q\left(\frac{kq}{a} + \frac{kq}{\sqrt{2}a}\right) = \left(1 + \frac{1}{\sqrt{2}}\right)\frac{kq^2}{a}$$

5. Similar considerations give W_D, the work needed to bring the fourth charge to point D:

$$W_D = qV_D = q\left(\frac{kq}{a} + \frac{kq}{\sqrt{2}a} + \frac{kq}{a}\right) = \left(2 + \frac{1}{\sqrt{2}}\right)\frac{kq^2}{a}$$

6. Summing the individual contributions gives the total work required to assemble the four charges:

$$W_{\text{total}} = W_A + W_B + W_C + W_D = \boxed{(4 + \sqrt{2})\frac{kq^2}{a}}$$

(*b*) 1. Calculate W_{total} from Equation 23-27. Use V_D from step 5 of Part (*a*) for the potential V_i at the location of each charge:

$$W_{\text{total}} = U = \frac{1}{2}\sum_{i=1}^{4} q_i V_i$$

where $V_1 = V_2 = V_3 = V_4 = V_D$ and $q_1 = q_2 = q_3 = q_4 = q$

2. The potential at the location of each charge is V_D. Substitute V_D and q for V_i and q_i, respectively, and solve for W_{total}:

$$W_{\text{total}} = \frac{1}{2}\sum_{i=1}^{4} q_i V_i = \frac{1}{2}\sum_{i=1}^{4} qV_D = \frac{1}{2}qV_D\sum_{i=1}^{4} 1$$

$$= \frac{1}{2}qV_D 4 = 2q\left(2 + \frac{1}{\sqrt{2}}\right)\frac{kq}{a}$$

$$= \boxed{(4 + \sqrt{2})\frac{kq^2}{a}}$$

CHECK Parts (*a*) and (*b*) have identical results.

TAKING IT FURTHER W_{total} equals the total electrostatic energy of the charge distribution. It is the work an external agent must do to assemble the configuration, beginning with the four charges at infinite separation.

PRACTICE PROBLEM 23-10 (*a*) How much additional work is required to bring a fifth positive charge q from infinity to the center of the square? (*b*) What is the total work required to assemble the five-charge system?

Physics Spotlight

Lightning—Fields of Attraction

Scientists have observed and analyzed lightning ~~for~~ more than 100 years. In recent years, high-speed digital recording,[*] low-light television cameras,[†] and satellites that have synchronized clocks[‡] have given atmospheric scientists new information about the events that occur with a bolt of cloud-to-ground, or CG, lightning.

A bolt of lightning strikes near an airport terminal. *(Tom Fox/Dallas Morning News/Corbis.)*

Thunderstorm clouds have layers of positive and negative charges and act like enormous, very powerful dipoles. CG lightning is usually negative charge from the lower part of a cloud that travels to the ground by ionization of the air. This charge is often "stepped" through the air with several pauses on the order of milliseconds. The visible bolt is a return positive *stroke* following the ionized path back up from the ground. Most flashes are 3 to 10 strokes back and forth between cloud and ground, several milliseconds apart. The strokes follow the initial path, as it already consists of ionized and heated air, and usually transfer a total negative charge of 20-35 C.[#] CG lightning strikes carrying negative charge to ground have been recorded with more than 1 million volts of potential difference.[°]

Some extremely powerful CG lightning strokes that carry positive charge to ground have transferred total positive charges of up to 400 C,[§] and have been recorded with over 10 million volts of potential. In large supercell thunderstorms that have hail and tornadoes,[¶,**] a majority of CG lightning strokes carry positive charge from the tops of the clouds to the ground rather than carry negative charge from the lower middle of the clouds to the ground. This lightning is associated with strong bursts of energy radiated near the start of the lightning and with brief bursts of light many kilometers above the cloud tops shortly after the strokes.[††]

But very strong bursts of radiated energy have been observed microseconds before less powerful negative lightning.[‡‡,##,°°,§§] Some bursts of energy repeat, detectable by satellite for up to nearly an hour after a lightning strike. Although the bursts last less than a millisecond, they are so energetic that they are associated with radio noise detected as far away as the opposite hemisphere of the planet.[¶¶]

Because high-energy electromagnetic radiation has repeatedly been measured in association with lightning, scientists are coming up with new models about how lightning is formed. One possible model involves "runaway breakdown." Because electric fields associated with thunderstorms are very large, it might be possible for a stray electron or ion to be accelerated by the electric field of a thunderstorm to approach the speed of light.[***] At that speed, the electron would be so energetic that colliding with molecules in the cloud would not halt it, even as it ionized those molecules. The ions could then be further accelerated by the electric field within the storm to produce a shower, or burst, of energy. Many scientists feel that runaway breakdown explains the formation of CG lightning in clouds that have measured electric fields ten times lower than the required potential needed to overcome air's insulating ability.[†††]

Because the technology to detect and to time the energy bursts in relation to lightning flashes is recent, scientists are now developing ways to confirm or refute these new models. The study of lightning is a very attractive field with great potential.

---

[*] Wang, D., et al., "Observed Leader and Return-Stroke Propagation Characteristics in the Bottom 400 m of a Rocket-Triggered Lightning Channel." *Journal of Geophysical Research*, Jun. 27, 1999, Vol. 104, No. D12, pp. 14,369–14,376.

[†] Lyons, W. A., et al., "Upward Electrical Discharges from Thunderstorm Tops." *Bulletin of the American Meteorological Society*, Apr. 2003, pp. 445–454.

[‡] Gurevich, A. V., and Zybin, K. P., "Runaway Breakdown and the Mysteries of Lightning." *Physics Today*, May 2005, pp. 37–43.

[#] Uman, M. A., *Lightning*. New York: Dover, 1984.

[°] Uman, M. A., op. cit.

[§] Rakov, V. A., "A Review of Positive and Bipolar Lightning Discharges." *Bulletin of the American Meteorological Society*, Jun. 2003, pp. 767–776.

[¶] Lang, T. J., et al., "The Severe Thunderstorm Electrification and Precipitation Study." *Bulletin of the American Meteorological Society*, Aug. 2004, pp. 1107–1125

[**] Wiens, K. C., "The 29 June 2000 Supercell Observed During STEPS. Part II: Lightning and Charge Structure."[Need journal name, volume, pages.]

[††] Lyons, W. A., et al., op. cit.

[‡‡] Dwyer, J. H., et al., "X-Ray Bursts Associated Leader Steps in Cloud-to-Ground Lightning." *Geophysical Research Letters*, Vol. 32, Letter 01803, 2005.

[##] Dwyer, J. R., "A Ground Level Gamma-Ray Burst Observed in Association with Rocket-Triggered Lightning." *Geophysical Research Letters*, Vol. 31, Letter 05119, 2004.

[°°] Greenfield, M. B., et al., "Near-Ground Detection of Atmospheric γ Rays Associated with Lightning." *Journal of Applied Physics*, Feb. 1, 2003, Vol. 93, No. 3, pp. 1839–1844.

[§§] Gurevich, A. V., and Zybin, K. P., op. cit.

[¶¶] Inan, U., "Gamma Rays Made on Earth." *Science*, Feb. 18, 2005, Vol. 307, No. 5712, pp. 1054–1055.

[***] Inan, U., op. cit.

[†††] Schrope, M., "The Bolt Catchers." *Nature*, Sept. 19, 2004, Vol. 431, pp. 120–121.

Summary

1. Electric potential at a location, which is defined as the electric potential energy per unit charge that a test charge would have at that location, is an important derived physical concept that is related to the electric field.
2. Because potential is a scalar quantity, it is often easier to calculate than the electric field. Once V is known, $\vec{E}$ can be calculated from V.

TOPIC	RELEVANT EQUATIONS AND REMARKS	
1. Potential Difference	The potential difference $V_b - V_a$ is defined as the negative of the work per unit charge done by the electric field on a test charge as it moves from point a to point b: $\Delta V = V_b - V_a = \frac{\Delta U}{q_0} = -\int_a^b \vec{E}\cdot d\vec{\ell}$	23-2*b*
Potential difference for infinitesimal displacements	$dV = -\vec{E}\cdot d\vec{\ell}$	23-2*a*
2. Electric Potential		
Potential due to a point charge	$V = \frac{kq}{r} - \frac{kq}{r_{ref}}$ $(V = 0 \text{ if } r = r_{ref})$	23-7
Coulomb potential	$V = \frac{kq}{r}$ $(V = 0 \text{ if } r = \infty)$	23-8
Potential due to a system of point charges	$V = \sum_i \frac{kq_i}{r_i}$ $(V = 0 \text{ if } r_i = \infty, i = 1, 2, \ldots)$	23-10
Potential due to a continuous charge distribution	$V = \int \frac{k\,dq}{r}$ $(V = 0 \text{ if } r = \infty)$ where dq is an increment of charge and r is the distance from the increment to the field point. This expression can be used only if the charge distribution is contained in a finite volume so that the potential can be chosen to be zero at infinity.	23-18
Continuity of electric potential	The potential function V is continuous everywhere in space.	
3. Computing the Electric Field from the Potential	The electric field points in the direction of the most rapid decrease in the potential.	
	The change in potential when a test charge undergoes a displacement $d\vec{\ell}$ is given by $E_{tan} = -\frac{dV}{d\ell}$	23-12
Gradient	A vector that points in the direction of the greatest rate of change in a scalar function and that has a magnitude equal to the derivative of that function, with respect to the distance in that direction, is called the gradient of the function. $\vec{E}$ is the negative gradient of V.	
Potential a function of x alone	$E_x = -\frac{dV(x)}{dx}$	23-13
Potential a function of r alone	$E_r = -\frac{dV(r)}{dr}$	23-14
4. *General Relation between $\vec{E}$ and V	$\vec{E} = -\vec{\nabla}V = -\left(\frac{\partial V}{\partial x}\hat{i} + \frac{\partial V}{\partial y}\hat{j} + \frac{\partial V}{\partial z}\hat{k}\right)$ or $V_b - V_a = -\int_a^b \vec{E}\cdot d\vec{\ell}$	23-17

TOPIC	RELEVANT EQUATIONS AND REMARKS	
5. Units		
V and ΔV	The SI unit of potential and potential difference is the volt (V): $1\ \text{V} = 1\ \text{J/C}$	23-4
Electric field	$1\ \text{N/C} = 1\ \text{V/m}$	23-5
Electron volt	The electron volt (eV) is the change in potential energy of a particle of charge e as it moves from a to b, where $V_b - V_a = 1$ volt: $1\ \text{eV} = 1.60 \times 10^{-19}\ \text{C}\cdot\text{V} = 1.60 \times 10^{-19}\ \text{J}$	23-6
6. Potential Energy of Two Point Charges	$U = q_0 V = \dfrac{kq_0 q}{r} \quad (U = 0 \text{ if } r = \infty)$	23-9
7. Potential Functions		
On the axis of a uniformly charged ring	$V = \dfrac{kQ}{\sqrt{z^2 + a^2}} \quad (V = 0 \text{ if } \lvert z\rvert = \infty)$	23-19
On the axis of a uniformly charged disk	$V = 2\pi k\sigma\lvert z\rvert\left(\sqrt{1 + \dfrac{R^2}{z^2}} - 1\right) \quad (V = 0 \text{ if } \lvert z\rvert = \infty)$	23-20
For an infinite plane of charge	$V = V_0 - 2\pi k\sigma\lvert x\rvert \quad (V = V_0 \text{ if } x = 0)$	23-21
For a spherical shell of charge	$V = \begin{cases} \dfrac{kQ}{r} & r \geq R \\ \dfrac{kQ}{R} & r \leq R \end{cases} \quad (V = 0 \text{ if } r = \infty)$	23-22
For an infinite line charge	$V = 2k\lambda \ln \dfrac{R_{ref}}{R} \quad (V = 0 \text{ if } r = R_{ref})$	23-23
8. Charge on a Nonspherical Conductor	On a conductor of arbitrary shape, the surface charge density σ is greatest at points where the radius of curvature is smallest.	
9. Dielectric Breakdown	The amount of charge that can be placed on a conductor is limited by the fact that molecules of the surrounding medium undergo dielectric breakdown at very high electric fields, causing the medium to become a conductor.	
Dielectric strength	The dielectric strength is the magnitude of the electric field at which dielectric breakdown occurs. The dielectric strength of dry air is $E_{max} \approx 3 \times 10^6\ \text{V/m} = 3\ \text{MV/m}$	
10. Electrostatic Potential Energy	The electrostatic potential energy of a system of point charges is the work needed to bring the charges from an infinite separation to their final positions.	
Of point charges	$U = \dfrac{1}{2}\sum_{i=1}^{n} q_i V_i$	23-27
Of a conductor with charge Q at potential V	$U = \tfrac{1}{2}QV$	23-28
Of a system of conductors	$U = \dfrac{1}{2}\sum_{i=1}^{n} Q_i V_i$	23-29

Answers to Concept Checks

23-1 The change in potential is zero if you move in a direction perpendicular to the direction of $\vec{E}$.

23-2 The potential increases at the greatest rate with respect to distance if you move in the direction opposite to the direction of $\vec{E}$.

Answers to Practice Problems

23-1 Increasing potential

23-2 $V(x) = -(5\ \text{V/m}^2)x^2$

23-3 -4.35×10^{-18} J

23-4 (*a*) the $x = 4.0$ m plane, (*b*) $V = -(25\ \text{V/m})x$

23-5 3.7×10^{-6} J

23-6 $V = \text{sign}(z) \cdot 2\pi k\sigma \left(1 - \dfrac{1}{\sqrt{1 + (R^2/z^2)}}\right)$

23-7 $5.39 \times 10^5\ V = 539$ kV

23-8 $V(r) = kQ/r - kQ/R$ for $r \geq R$;

$V(r) = \frac{1}{2}(kQ/R)(1 - r^2/R^2)$ for $r \leq R$

23-9 7.5×10^{-6} C, 4.4×10^5 V

23-10 (*a*) $4\sqrt{2}kq^2/a$, (*b*) $(4 + 5\sqrt{2})kq^2/a$

Problems

In a few problems, you are given more data than you actually need; in a few other problems, you are required to supply data from your general knowledge, outside sources, or informed estimate.

Interpret as significant all digits in numerical values that have trailing zeros and no decimal points.

- • Single-concept, single-step, relatively easy
- •• Intermediate-level, may require synthesis of concepts
- ••• Challenging
- SSM Solution is in the *Student Solutions Manual*
- Consecutive problems that are shaded are paired problems.

CONCEPTUAL PROBLEMS

1 • A proton is moved to the left in a uniform electric field that points to the right. Is the proton moving in the direction of increasing or decreasing electric potential? Is the electrostatic potential energy of the proton increasing or decreasing? SSM

2 • An electron is moved to the left in a uniform electric field that points to the right. Is the electron moving in the direction of increasing or decreasing electric potential? Is the electrostatic potential energy of the electron increasing or decreasing?

3 • If the electric potential is uniform throughout a region of space, what can be said about the electric field in that region?

4 • If V is known at only a single point in space, can $\vec{E}$ be found at that point? Explain your answer.

5 •• Figure 23-29 shows a point particle that has a positive charge $+Q$ and a metal sphere that has a charge $-Q$. Sketch the electric field lines and equipotential surfaces for this system of charges. SSM

FIGURE 23-29
Problem 5

6 •• Figure 23-30 shows a point particle that has a negative charge $-Q$ and a metal sphere that has a charge $+Q$. Sketch the electric field lines and equipotential surfaces for this system of charges.

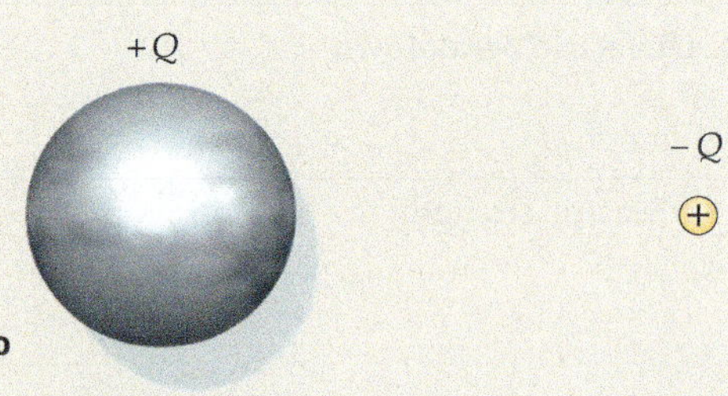

FIGURE 23-30
Problem 6

7 •• Sketch the electric field lines and equipotential surfaces for the region surrounding the charged conductor shown in Figure 23-31, assuming that the conductor has a net positive charge.

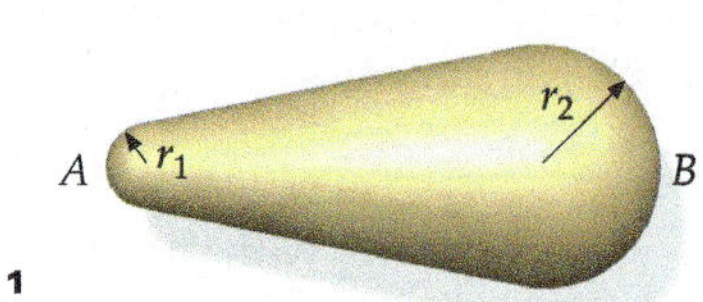

FIGURE 23-31
Problem 7

8 •• Two equal positive point charges are separated by a finite distance. Sketch the electric field lines and the equipotential surfaces for this system.

9 •• Two point charges are fixed on the x axis. (*a*) Each has a positive charge q. One is at $x = -a$ and the other is at $x = +a$. At the origin, which of the following is true?

(1) $\vec{E} = 0$ and $V = 0$
(2) $\vec{E} = 0$ and $V = 2kq/a$
(3) $\vec{E} = (2kq/a^2)\hat{i}$ and $V = 0$
(4) $\vec{E} = (2kq/a^2)\hat{i}$ and $V = 2kq/a$
(5) None of the above

(*b*) One point charge has a positive charge $+q$ and the other has a negative charge $-q$. The positive point charge is at $x = -a$ and the negative point charge is at $x = +a$. At the origin, which of the following is true?

(1) $\vec{E} = 0$ and $V = 0$
(2) $\vec{E} = 0$ and $V = 2kq/a$
(3) $\vec{E} = (2kq/a^2)\hat{i}$ and $V = 0$
(4) $\vec{E} = (2kq/a^2)\hat{i}$ and $V = 2kq/a$
(5) None of the above

10 •• The electrostatic potential (in volts) is given by $V(x, y, z) = 4.00|x| + V_0$, where V_0 is a constant, and x is in meters. (*a*) Sketch the electric field for this potential. (*b*) Which of the following charge distributions is most likely responsible for this potential: (1) A negatively charged flat sheet in the $x = 0$ plane, (2) a point charge at the origin, (3) a positively charged flat sheet in the $x = 0$ plane, or (4) a uniformly charged sphere centered at the origin? Explain your answer.

11 •• The electric potential is the same everywhere on the surface of a conductor. Does this mean that the surface charge density is also the same everywhere on the surface? Explain your answer. SSM

12 •• Three identical positive point charges are located at the vertices of an equilateral triangle. If the length of each side of the triangle shrinks to one-fourth of its original length, by what factor does the electrostatic potential energy of this system change? (The electrostatic potential energy approaches zero if the length of each side of the triangle approaches infinity.)

ESTIMATION AND APPROXIMATION

13 • Estimate the maximum potential difference between a thundercloud and Earth, given that the electrical breakdown of air occurs at fields of roughly 3.0×10^6 V/m. SSM

14 • The specifications for the gap width of typical automotive spark plugs is approximately equal to the thickness of the cardboard used for matchbook covers. Because of the high compression of the air–gas mixture in the cylinder, the dielectric strength of the mixture is roughly 2.0×10^7 V/m. Estimate the maximum potential difference across the spark gap during operating conditions.

15 • The radius of a proton is approximately 1.0×10^{-15} m. Suppose two protons having equal and opposite momenta undergo a head-on collision. Estimate the minimum kinetic energy (in MeV) required by each proton to allow the protons to overcome electrostatic repulsion and collide. *Hint: The rest energy of a proton is 938 MeV. If the kinetic energies of the protons are much less than this rest energy, then a nonrelativistic calculation is justified.*

16 • When you touch a friend after walking across a rug on a dry day, you typically draw a spark of about 2.0 mm. Estimate the potential difference between you and your friend just before the spark.

17 • Estimate the maximum surface charge density that can exist at the end of a sharp lightning rod so that no dielectric breakdown of air occurs.

18 •• The electric field strength near the surface of Earth is about 300 V/m. (*a*) Estimate the magnitude of the charge density on the surface of Earth. (*b*) Estimate the total charge on Earth. (*c*) What is value of the electric potential at Earth's surface? (Assume the potential is zero at infinity.) (*d*) If all Earth's electrostatic potential energy could be harnessed and converted to electric energy at reasonable efficiency, how long could it be used to run the consumer households in the United States? Assume the average American household consumes about 500 kW · h of electric energy per month.

ELECTROSTATIC POTENTIAL DIFFERENCE, ELECTROSTATIC ENERGY, AND ELECTRIC FIELD

19 • A point particle has a charge equal to $+2.00$ μC and is fixed at the origin. (*a*) What is the electric potential V at a point 4.00 m from the origin assuming that $V = 0$ at infinity? (*b*) How much work must be done to bring a second point particle that has a charge of $+3.00$ μC from infinity to a distance of 4.00 m from the $+2.00$-μC charge?

20 •• The facing surfaces of two large parallel conducting plates separated by 10.0 cm have uniform surface charge densities that are equal in magnitude but opposite in sign. The difference in potential between the plates is 500 V. (*a*) Is the positive plate or the negative plate at the higher potential? (*b*) What is the magnitude of the electric field between the plates? (*c*) An electron is released from rest next to the negatively charged surface. Find the work done by the electric field on the electron as the electron moves from the release point to the positive plate. Express your answer in both electron volts and joules. (*d*) What is the change in potential energy of the electron when it moves from the release point to the positive plate? (*e*) What is its kinetic energy when it reaches the positive plate?

21 •• A uniform electric field has a magnitude 2.00 kV/m and points in the $+x$ direction. (*a*) What is the electric potential difference between the $x = 0.00$ m plane and the $x = 4.00$ m plane? A point particle that has a charge of $+3.00$ μC is released from rest at the origin. (*b*) What is the change in the electric potential energy of the particle as it travels from the $x = 0.00$ m plane to the $x = 4.00$ m plane? (*c*) What is the kinetic energy of the particle when it arrives at the $x = 4.00$ m plane? (*d*) Find the expression for the electric potential $V(x)$ if its value is chosen to be zero at $x = 0$.

22 •• In a potassium chloride unit, the distance between the potassium ion (K^+) and the chloride ion (Cl^-) is 2.80×10^{-10} m. (*a*) Calculate the energy (in eV) required to separate the two ions to an infinite distance apart. (Model the two ions as two point particles initially at rest.) (*b*) If twice the energy determined in Part (*a*) is actually supplied, what is the total amount of kinetic energy that the two ions have when they were an infinite distance apart?

23 •• Protons are released from rest in a Van de Graaff accelerator system. The protons initially are located where the electric potential has a value of 5.00 MV and then they travel through a vacuum to a region where the potential is zero. (*a*) Find the final speed of these protons. (*b*) Find the accelerating electric field strength if the potential changed *uniformly* over a distance of 2.00 m. SSM

24 •• The picture tube of a television set was, until recently, invariably a cathode-ray tube. In a typical cathode-ray tube, an electron "gun" arrangement is used to accelerate electrons from rest to the screen. The electrons are accelerated through a potential difference of 30.0 kV. (*a*) Which region is at a higher electric potential, the screen or the electron's starting location? Explain your answer. (*b*) What is the kinetic energy (in both eV and J) of an electron as it reaches the screen?

25 ••• (*a*) A positively charged particle is on a trajectory to collide head-on with a massive positively charged nucleus that is initially at rest. The particle initially has kinetic energy K_i. In addition, the particle is initially far from the nucleus. Derive an expression for the distance of closest approach. Your expression should be in terms of the initial kinetic energy K_i of the particle, the charge ze on the particle, and the charge Ze on the nucleus, where both z and Z are integers. (*b*) Find the numerical value for the distance of closest approach between a 5.00-MeV α particle and a stationary gold nucleus and between a 9.00-MeV α particle and a stationary gold nucleus. (The values 5.00 MeV and 9.00 MeV are the initial kinetic energies of the alpha particles. Neglect the motion of the gold nucleus following the collisions.) (*c*) The radius of the gold nucleus is about 7×10^{-15} m. If α particles approach the nucleus closer than 7×10^{-15} m, they experience the strong nuclear force in addition to the electric force of repulsion. In the early twentieth century, before the strong nuclear force was known, Ernest Rutherford bombarded gold nuclei with α particles that had kinetic energies of about 5 MeV. Would you expect this experiment to reveal the existence of this strong nuclear force? Explain your answer.

POTENTIAL DUE TO A SYSTEM OF POINT CHARGES

Note: **In all the problems in this section, assume that the electric potential is zero at distances far from all charges unless otherwise stated.**

26 • Four point charges, each having a magnitude of 2.00 μC, are fixed at the corners of a square whose edges are 4.00 m long. Find the electric potential at the center of the square if (*a*) all the charges are positive, (*b*) three of the charges are positive and one charge is negative, and (*c*) two charges are positive and two charges are negative. (Assume the potential is zero very far from all charges.)

27 • Three point charges are fixed at locations on the x axis: q_1 is at $x = 0.00$ m, q_2 is at $x = 3.00$ m, and q_3 is at $x = 6.00$ m. Find the electric potential at the point on the y axis at $y = 3.00$ m if (*a*) $q_1 = q_2 = q_3 = +2.00\ \mu$C, (*b*) $q_1 = q_2 = +2.00\ \mu$C and $q_3 = -2.00\ \mu$C, and (*c*) $q_1 = q_3 = +2.00\ \mu$C and $q_2 = -2.00\ \mu$C. (Assume the potential is zero very far from all charges.) SSM

28 • Points *A*, *B*, and *C* are fixed at the vertices of an equilateral triangle whose edges are 3.00 m long. A point particle that has a charge of +2.00 μC is fixed at each of vertices *A* and *B*. (*a*) What is the electric potential at point *C*? (Assume the potential is zero very far from all charges.) (*b*) How much work is required to move a point particle having a charge of +5.00 μC from a distance of infinity to point *C*? (*c*) How much additional work is required to move the +5.00-μC point particle from point *C* to the midpoint of side *AB*?

29 •• Three identical point particles that have charge q are at the vertices of an equilateral triangle that is circumscribed by a circle of radius a that lies in the $z = 0$ plane and is centered at the origin. The values of q and a are +3.00 μC and 60.0 cm, respectively. (Assume the potential is zero very far from all charges.) (*a*) What is the electric potential at the origin? (*b*) What is the electric potential at the point on the z axis at $z = a$? (*c*) How would your answers to Parts (*a*) and (*b*) change if the charges were still on the circle but one is no longer at a vertex of the triangle? Explain your answer.

30 •• Two point charges q and q' are separated by a distance a. At a point $a/3$ from q and along the line joining the two charges the potential is zero. (Assume the potential is zero very far from all charges.) (*a*) Which of the following statements is true?
(1) The charges have the same sign.
(2) The charges have opposite signs.
(3) The relative signs of the charges cannot be determined by using the data given.

(*b*) Which of the following statements is true?
(1) $|q| > |q'|$.
(2) $|q| < |q'|$.
(3) $|q| = |q'|$.
(4) The relative magnitudes of the charges cannot be determined by using the data given.

(*c*) Find the ratio q/q'.

31 •• Two identical positively charged point particles are fixed on the x axis at $x = +a$ and $x = -a$. (*a*) Write an expression for the electric potential $V(x)$ as a function of x for all points on the x axis. (*b*) Sketch $V(x)$ versus x for all points on the x axis. SSM

32 •• A point charge of $+3e$ is at the origin and a second point charge of $-2e$ is on the x axis at $x = a$. (*a*) Sketch the potential function $V(x)$ versus x for all points on the x axis. (*b*) At what point or points, if any, is $V = 0$ on the x axis? (*c*) At what point or points, if any, on the x axis is the electric field zero? Are these locations the same locations found in Part (*b*) ? Explain your answer. (*d*) How much work is needed to bring a third charge $+e$ to the point $x = \frac{1}{2}a$ on the x axis?

33 ••• A dipole consists of equal but opposite point charges $+q$ and $-q$. It is located so that its center is at the origin, and its axis is aligned with the z axis (Figure 23-32). The distance between the charges is L. Let $\vec{r}$ be the vector from the origin to an arbitrary field point and θ be the angle that $\vec{r}$ makes with the $+z$ direction. (*a*) Show that at large distances from the dipole (i.e., for $r \gg L$), the dipole's electric potential is given by $V(r, \theta) \approx k\vec{p} \cdot \hat{r}/r^2 = kp\cos\theta/r^2$, where $\vec{p}$ is the dipole moment of the dipole and θ is the angle between $\vec{r}$ and $\vec{p}$. (*b*) At what points in the region $r \gg L$, other than at infinity, is the electric potential zero? SSM

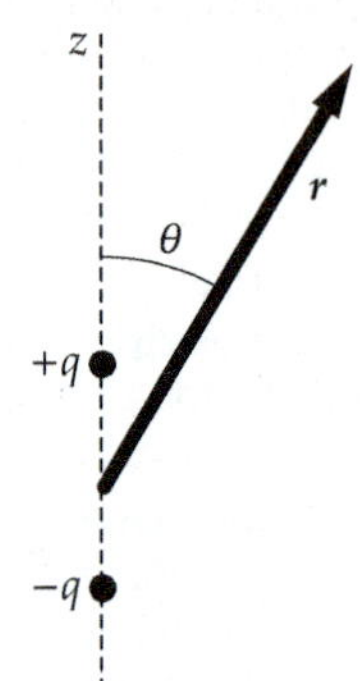

FIGURE 23-32
Problem 33

34 ••• A charge configuration consists of three point charges located on the z axis (Figure 23-33). One has a charge equal to $-2q$, and is located at the origin. The other two each have a charge equal to $+q$, one is located at $z = +L$ and the other is located at $z = -L$. This charge configuration can be modeled as two dipoles: one centered at $z = +L/2$ and with a dipole moment in the $+z$ direction, the other centered at $z = -L/2$ and with a dipole moment in the $-z$ direction. Each of these dipoles has a dipole moment that has a magnitude equal to qL. Two dipoles arranged in this fashion form a *linear electric quadrupole*. (There are other geometrical arrangements of dipoles that create quadrupoles but they are not linear.) (*a*) Using the result from Problem 33, show that at large distances from the quadrupole (i.e., for $r \gg L$), the electric potential is given by $V_{quad}(r, \theta) = 2kB\cos^2\theta/r^3$, where $B = qL^2$. (B is the magnitude of the quadrupole moment of the charge configuration.) (*b*) Show that on the positive z axis, this potential gives an electric field (for $z \gg L$) of $\vec{E} = (6kB/z^4)\hat{k}$. (*c*) Show you get the result of Part (*b*) by adding the electric fields from the three point charges.

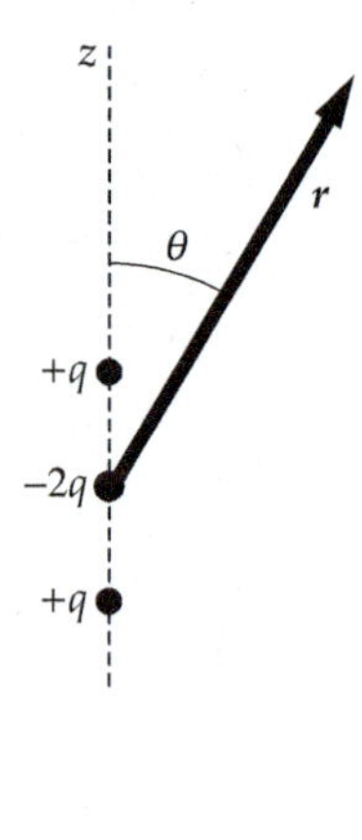

FIGURE 23-33
Problem 34

COMPUTING THE ELECTRIC FIELD FROM THE POTENTIAL

35 • A uniform electric field is in the $-x$ direction. Points a and b are on the x axis, with a at $x = 2.00$ m and b at $x = 6.00$ m. (*a*) Is the potential difference $V_b - V_a$ positive or negative? (*b*) If $|V_b - V_a|$ is 100 kV, what is the magnitude of the electric field?

36 • An electric field is given by the expression $\vec{E} = bx^3\hat{i}$, where $b = 2.00\ \text{kV/m}^4$. Find the potential difference between the point at $x = 1.00$ m and the point $x = 2.00$ m. Which of these points is at the higher potential?

37 •• The electric field on the x axis due to a point charge fixed at the origin is given by $\vec{E} = (b/x^2)\hat{i}$, where $b = 6.00\ \text{kV}\cdot\text{m}$ and $x \neq 0$. (*a*) Find the magnitude and sign of the point charge. (*b*) Find the potential difference between the points on the x axis at $x = 1.00$ m and $x = 2.00$ m. Which of these points is at the higher potential?

38 •• The electric potential due to a particular charge distribution is measured at many points along the x axis. A plot of the data is shown in Figure 23-34. At what location (or locations) is the x component of the electric field equal to zero? At this location (or these locations) is the potential also equal to zero? Explain your answer.

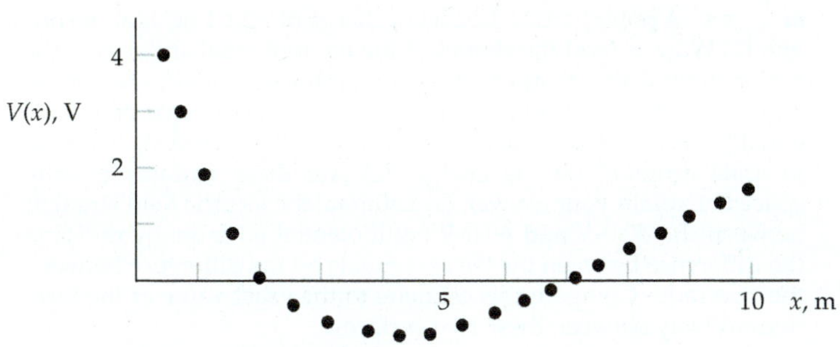

FIGURE 23-34 Problem 38

39 •• Three identical point charges, each with a charge equal to q, lie in the xy plane. Two of the charges are on the y axis at $y = -a$ and $y = +a$, and the third charge is on the x axis at $x = a$. (*a*) Find the potential as a function of position along the x axis. (*b*) Use the Part-(*a*) result to obtain an expression for $E_x(x)$, the x component of the electric field as a function of x. Check your answers to Parts (*a*) and (*b*) at the origin and as x approaches ∞ to see if they yield the expected results.

CALCULATIONS OF *V* FOR CONTINUOUS CHARGE DISTRIBUTIONS

40 • A charge of $+10.0\ \mu\text{C}$ is uniformly distributed on a thin spherical shell of radius 12.0 cm. (Assume the potential is zero very far from all charges.) (*a*) What is the magnitude of the electric field just outside and just inside the shell? (*b*) What is the magnitude of the electric potential just outside and just inside the shell? (*c*) What is the electric potential at the center of the shell? (*d*) What is the magnitude of the electric field at the center of the shell?

41 • An infinite line charge of linear charge density $+1.50\ \mu\text{C/m}$ lies on the z axis. Find the electric potential at distances from the line charge of (*a*) 2.00 m, (*b*) 4.00 m, and (*c*) 12.0 m. Assume that we choose $V = 0$ at a distance of 2.50 m from the line of charge. SSM

42 • (*a*) Find the maximum net charge that can be placed on a spherical conductor of radius 16 cm before dielectric breakdown of the air occurs. (*b*) What is the electric potential of the sphere when it has this maximum charge? (Assume the potential is zero very far from all charges.)

43 • Find the maximum surface charge density σ_{max} that can exist on the surface of any conductor before dielectric breakdown of the air occurs.

44 •• A conducting spherical shell of inner radius b and outer radius c is concentric with a small metal sphere of radius $a < b$. The metal sphere has a positive charge Q. The total charge on the conducting spherical shell is $-Q$. (Assume the potential is zero very far from all charges.) (*a*) What is the electric potential of the spherical shell? (*b*) What is the electric potential of the metal sphere?

45 •• Two coaxial conducting cylindrical shells have equal and opposite charges. The inner shell has charge $+q$ and an outer radius a, and the outer shell has charge $-q$ and an inner radius b. The length of each cylindrical shell is L, and L is very long compared with b. Find the potential difference $V_a - V_b$ between the shells. SSM

46 •• Positive charge is placed on two conducting spheres that are very far apart and connected by a long, very thin conducting wire. The radius of the smaller sphere is 5.00 cm and the radius of the larger sphere is 12.0 cm. The electric field strength at the surface of the larger sphere is 200 kV/m. Estimate the surface charge density on each sphere.

47 •• Two concentric conducting spherical shells have equal and opposite charges. The inner shell has outer radius a and charge $+q$; the outer shell has inner radius b and charge $-q$. Find the potential difference $V_a - V_b$ between the shells.

48 •• The electric potential at the surface of a uniformly charged sphere is 450 V. At a point outside the sphere at a (radial) distance of 20.0 cm from its surface, the electric potential is 150 V. (The potential is zero very far from the sphere.) What is the radius of the sphere, and what is the charge of the sphere?

49 •• Consider two infinite parallel thin sheets of charge, one in the $x = 0$ plane and the other in the $x = a$ plane. The potential is zero at the origin. (*a*) Find the electric potential everywhere in space if the planes have equal positive charge densities $+\sigma$. (*b*) Find the electric potential everywhere in space if the sheet in the $x = 0$ plane has a charge density $+\sigma$ and the sheet in the $x = a$ plane has a charge density $-\sigma$.

50 ••• The expression for the potential along the axis of a thin uniformly charged disk is given by $V = 2\pi k\sigma|z|\left(\sqrt{1 + \frac{R^2}{z^2}} - 1\right)$ (Equation 23-20), where R and σ are the radius and the charge per unit area of the disk, respectively. Show that this expression reduces to $V = kQ/|z|$ for $|z| \gg R$, where $Q = \sigma\pi R^2$ is the total charge on the disk. Explain why this result is expected. *Hint: Use the binomial theorem to expand the radical.*

51 •• A rod of length L has a total charge Q uniformly distributed along its length. The rod lies along the y axis with its center at the origin. (*a*) Find an expression for the electric potential as a function of position along the x axis. (*b*) Show that the result obtained in Part (*a*) reduces to $V = kQ/|x|$ for $|x| \gg L$. Explain why this result is expected. SSM

52 •• A rod of length L has a charge Q uniformly distributed along its length. The rod lies along the y axis with one end at the origin. (*a*) Find an expression for the electric potential as a function of position along the x axis. (*b*) Show that the result obtained in Part (*a*) reduces to $V = kQ/|x|$ for $|x| \gg L$. Explain why this result is expected.

53 •• A disk of radius R has a surface charge distribution given by $\sigma = \sigma_0 r^2/R^2$ where σ_0 is a constant and r is the distance from the center of the disk. (*a*) Find the total charge on the disk. (*b*) Find an

expression for the electric potential at a distance z from the center of the disk on the axis that passes through the disk's center and is perpendicular to its plane. SSM

54 ••• A disk of radius R has a surface charge distribution given by $\sigma = \sigma_0 R/r$ where σ_0 is a constant and r is the distance from the center of the disk. (*a*) Find the total charge on the disk. (*b*) Find an expression for the electric potential at a distance x from the center of the disk on the axis that passes through the disk's center and is perpendicular to its plane.

55 •• A rod of length L has a total charge Q uniformly distributed along its length. The rod lies along the x axis with its center at the origin. (*a*) What is the electric potential as a function of position along the x axis for $x > L/2$? (*b*) Show that for $x \gg L/2$, your result reduces to that due to a point charge Q.

56 ••• A circle of radius a is removed from the center of a uniformly charged thin circular disk of radius b and charge per unit area σ. (*a*) Find an expression for the potential on the x axis a distance x from the center of the disk. (*b*) Show that for $x \gg b$ the electric potential on the axis of the uniformly charged disk with cutout approaches kQ/x, where $Q = \sigma\pi(b^2 - a^2)$ is the total charge on the disk.

57 ••• The expression for the electric potential inside a uniformly charged solid sphere is given by $V(r) = \frac{kQ}{2R}\left(3 - \frac{r^2}{R^2}\right)$, where R is the radius of the sphere and r is the distance from the center. This expression was obtained in Example 23-12 by first finding the electric field. In this problem, you derive the same expression by modeling the sphere as a nested collection of thin spherical shells, and then adding the potentials of these shells at a field point inside the sphere. The potential dV that is a distance r from the center of a uniformly charged thin spherical shell that has a radius r' and a charge dQ is given by $dV = k\,dQ/r$ for $r \geq r'$ and $dV = k\,dQ/r'$ for $r \leq r'$ (Equation 23-22). Consider a sphere of radius R containing a charge Q that is uniformly distributed and you want to find V at some point inside the sphere (i.e., for $r < R$). (*a*) Find an expression for the charge dQ on a spherical shell of radius r' and thickness dr'. (*b*) Find an expression for the potential dV at r due to the charge on a shell of radius r' and thickness dr', where $r \leq r' \leq R$. (*c*) Integrate your expression in Part (*b*) from $r' = r$ to $r' = R$ to find the potential at r due to all the charge in the region farther than r from the center of the sphere. (*d*) Find an expression for the potential dV at r due to the charge in a shell of radius r' and thickness dr', where $r' \leq r$. (*e*) Integrate your expression in Part (*d*) from $r' = 0$ to $r' = r$ to find the potential at r due to all the charge in the region closer than r to the center of the sphere. (*f*) Find the total potential V at r by adding your Part-(*c*) and Part-(*e*) results.

58 •• Calculate the electric potential at the point a distance $R/2$ from the center of a uniformly charged thin spherical shell of radius R and charge Q. (Assume the potential is zero far from the shell.)

59 •• A circle of radius a is removed from the center of a uniformly charged thin circular disk of radius b. Show that the potential at a point on the central axis of the disk a distance z from its geometrical center is given by $V(z) = 2\pi k\sigma(\sqrt{z^2 + b^2} + \sqrt{z^2 + a^2})$, where σ is the charge density of the disk. SSM

EQUIPOTENTIAL SURFACES

60 • An infinite flat sheet of charge has a uniform surface charge density equal to 3.50 μC/m^2. How far apart are the equipotential surfaces whose potentials differ by 100 V?

61 •• Consider two parallel uniformly charged infinite planes that are equal but oppositely charged. (*a*) What is (are) the shape(s) of the equipotential surfaces in the region between them? Explain your answer. (*b*) What is (are) the shape(s) of the equipotential surfaces in the regions not between them? Explain your answer. SSM

62 •• A Geiger tube consists of two elements, a long metal cylindrical shell and a long straight metal wire running down its central axis. Model the tube as if both the wire and cylinder are infinitely long. The central wire is positively charged and the outer cylinder is negatively charged. The potential difference between the wire and the cylinder is 1.00 kV. (*a*) What is the direction of the electric field inside the tube? (*b*) Which element is at a higher electric potential? (*c*) What is (are) the shape(s) of the equipotential surfaces inside the tube? (*d*) Consider two equipotential surfaces described in Part (*c*). Suppose they differ in electric potential by 10 V. Do two such equipotential surfaces near the central wire have the same spacing as they would near the outer cylinder? If not, where in the tube are the equipotential surfaces that are more widely spaced? Explain your answer.

63 •• Suppose the cylinder in the Geiger tube in Problem 62 has an inside diameter of 4.00 cm and the wire has a diameter of 0.500 mm. The cylinder is grounded so its potential is equal to zero. (*a*) What is the radius of the equipotential surface that has a potential equal to 500 V? Is this surface closer to the wire or to the cylinder? (*b*) How far apart are the equipotential surfaces that have potentials of 200 and 225 V? (*c*) Compare your result in Part (*b*) to the distance between the two surfaces that have potentials of 700 and 725 V, respectively. What does this comparison tell you about the electric field strength as a function of the distance from the central wire? SSM

64 •• A point particle that has a charge of +11.1 nC is at the origin. (*a*) What is (are) the shapes of the equipotential surfaces in the region around this charge? (*b*) Assuming the potential to be zero at $r = \infty$, calculate the radii of the five surfaces that have potentials equal to 20.0 V, 40.0 V, 60.0 V, 80.0 V, and 100.0 V, and sketch them to scale centered on the charge. (*c*) Are these surfaces equally spaced? Explain your answer. (*d*) Estimate the electric field strength between the 40.0-V and 60.0-V equipotential surfaces by dividing the difference between the two potentials by the difference between the two radii. Compare this estimate to the exact value at the location midway between these two surfaces.

ELECTROSTATIC POTENTIAL ENERGY

65 • Three point charges are on the x axis: q_1 is at the origin, q_2 is at $x = +3.00$ m, and q_3 is at $x = +6.00$ m. Find the electrostatic potential energy of this system of charges for the following charge values: (*a*) $q_1 = q_2 = q_3 = +2.00\ \mu$C; (*b*) $q_1 = q_2 = +2.00\ \mu$C and $q_3 = -2.00\ \mu$C; and (*c*) $q_1 = q_3 = +2.00\ \mu$C and $q_2 = -2.00\ \mu$C. (Assume the potential energy is zero when the charges are very far from each other.)

66 • Point charges q_1, q_2, and q_3 are fixed at the vertices of an equilateral triangle whose sides are 2.50 m long. Find the electrostatic potential energy of this system of charges for the following charge values: (*a*) $q_1 = q_2 = q_3 = +4.20\ \mu$C, (*b*) $q_1 = q_2 = +4.20\ \mu$C and $q_3 = -4.20\ \mu$C; and (*c*) $q_1 = q_2 = -4.20\ \mu$C and $q_3 = +4.20\ \mu$C. (Assume the potential energy is zero when the charges are very far from each other.)

67 •• (*a*) How much charge is on the surface of an isolated spherical conductor that has a 10.0-cm radius and is charged to 2.00 kV? (*b*) What is the electrostatic potential energy of this conductor? (Assume the potential is zero far from the sphere.) SSM

68 ••• Four point charges, each having a charge with a magnitude of 2.00 μC, are at the corners of a square whose sides are 4.00 m long. Find the electrostatic potential energy of this system under the following conditions: (*a*) all of the charges are negative, (*b*) three of the charges are positive and one of the charges is negative, (*c*) the charges at two adjacent corners are positive and the other two charges are negative, and (*d*) the charges at two opposite corners are positive and the other two charges are negative. (Assume the potential energy is zero when the point charges are very far from each other.)

69 •• Four point charges are fixed at the corners of a square centered at the origin. The length of each side of the square is $2a$. The charges are located as follows: $+q$ is at $(-a, +a)$, $+2q$ is at $(+a, +a)$, $-3q$ is at $(+a, -a)$, and $+6q$ is at $(-a, -a)$. A fifth particle that has a mass m and a charge $+q$ is placed at the origin and released from rest. Find its speed when it is a very far from the origin. SSM

70 •• Consider two point particles that have charge $+e$, are at rest, and are separated by 1.50×10^{-15} m. (*a*) How much work was required to bring them together from a very large separation distance? (*b*) If they are released, how much kinetic energy will each have when they are separated by twice their separation at release? (*c*) The mass of each particle is 1.00 u (1.00 amu). What speed will each have when they are very far from each other?

71 ••• Consider an electron and a proton that are initially at rest and are separated by 2.00 nm. Neglecting any motion of the much more massive proton, what is the minimum (*a*) kinetic energy and (*b*) speed with which the electron must be projected at so it reaches a point a distance of 12.0 nm from the proton? Assume the electron's velocity is directed radially away from the proton. (*c*) How far will the electron travel away from the proton if it has twice that initial kinetic energy?

GENERAL PROBLEMS

72 • A positive point charge equal to 4.80×10^{-19} C is separated from a negative point charge of the same magnitude by 6.40×10^{-10} m. What is the electric potential at a point 9.20×10^{-10} m from each of the two charges?

73 • Two positive point charges each have a charge of $+q$ and are fixed on the y axis at $y = +a$ and $y = -a$. (*a*) Find the electric potential at any point on the x axis. (*b*) Use your result in Part (*a*) to find the electric field at any point on the x axis. SSM

74 • If a conducting sphere is to be charged to a potential of 10.0 kV, what is the smallest possible radius of the sphere so that the electric field near the surface of the sphere will not exceed the dielectric strength of air?

75 •• **SPREADSHEET** Two infinitely long parallel wires have a uniform charge per unit length λ and $-\lambda$, respectively. The wires are parallel with the z axis. The positively charged wire intersects the x axis at $x = -a$, and the negatively charged wire intersects the x axis at $x = +a$. (*a*) Choose the origin as the reference point where the potential is zero, and express the potential at an arbitrary point (x, y) in the xy plane in terms of x, y, λ, and a. Use this expression to solve for the potential everywhere on the y axis. (*b*) Using $a = 5.00$ cm and $\lambda = 5.00$ nC/m, obtain the equation for the equipotential surface in the xy plane that passes through the point $x = \frac{1}{4}a$, $y = 0$. (*c*) Use a spreadsheet program to plot the equipotential surface found in Part (*b*). SSM

76 •• The equipotential curve graphed in Problem 75 should be a circle. (*a*) Show mathematically that it is a circle. (*b*) The equipotential circle in the xy plane is the intersection of a three-dimensional equipotential surface and the xy plane. Describe the three-dimensional surface using one or two sentences.

77 ••• The hydrogen atom in its ground state can be modeled as a positive point charge of magnitude $+e$ (the proton) surrounded by a negative charge distribution that has a charge density (the electron) that varies with the distance from the center of the proton r as $\rho(r) = -\rho_0 e^{-2r/a}$ (a result obtained from quantum mechanics), where $a = 0.523$ nm is the most probable distance of the electron from the proton. (*a*) Calculate the value of ρ_0 needed for the hydrogen atom to be neutral. (*b*) Calculate the electrostatic potential (relative to infinity) of this system as a function of the distance r from the proton.

78 •• Charge is supplied to the metal dome of a Van de Graaff generator by the belt at the rate of 200 μC/s when the potential difference between the belt and the dome is 1.25 MV. The dome transfers charge to the atmosphere at the same rate, so the 1.25 MV potential difference is maintained. What minimum power is needed to drive the moving belt and maintain the 1.25 MV potential difference?

79 •• A positive point charge $+Q$ is located on the x axis at $x = -a$. (*a*) How much work is required to bring an identical point charge from infinity to the point on the x axis at $x = +a$? (*b*) With the two identical point charges in place at $x = -a$ and $x = +a$, how much work is required to bring a third point charge $-Q$ from infinity to the origin? (*c*) How much work is required to move the charge $-Q$ from the origin to the point on the x axis at $x = 2a$ along the semicircular path shown (Figure 23-35)?

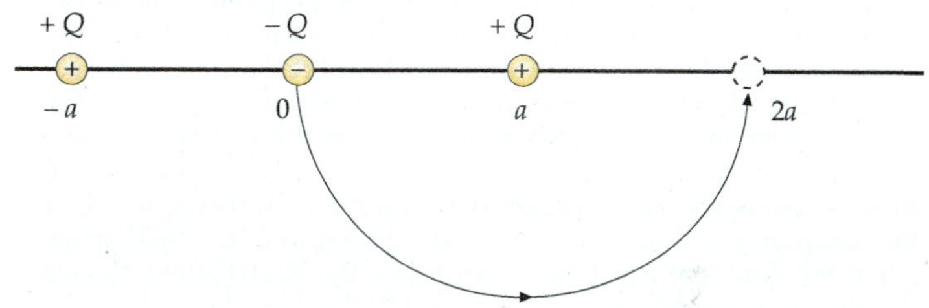

FIGURE 23-35 Problem 79

80 •• A charge of +2.00 nC is uniformly distributed on a ring of radius 10.0 cm that lies in the $x = 0$ plane and is centered at the origin. A point charge of +1.00 nC is initially located on the x axis at $x = 50.0$ cm. Find the work required to move the point charge to the origin.

81 •• Two metal spheres each have a radius of 10.0 cm. The centers of the two spheres are 50.0 cm apart. The spheres are initially neutral, but a charge Q is transferred from one sphere to the other, creating a potential difference between the spheres of 100 V. A proton is released from rest at the surface of the positively charged sphere and travels to the negatively charged sphere. (*a*) What is the proton's kinetic energy just as it strikes the negatively charged sphere? (*b*) At what speed does it strike the sphere?

82 • **SPREADSHEET** (*a*) Using a spreadsheet program, graph $V(z)$ versus z for a uniformly charged ring in the $z = 0$ plane and centered at the origin. The potential on the z axis is given by $V(z) = kQ/\sqrt{a^2 + z^2}$ (Equation 23-19). (*b*) Use your graph to estimate the points on the z axis where the electric field strength is greatest.

83 •• A spherical conductor of radius R_1 is charged to 20 kV. When it is connected by a long, very thin conducting wire to a second conducting sphere far away, its potential drops to 12 kV. What is the radius of the second sphere?

84 •• A metal sphere centered at the origin has a surface charge density that has a magnitude of 24.6 nC/m^2 and a radius less than 2.00 m. A distance of 2.00 m from the origin, the electric potential is 500 V and the electric field strength is 250 V/m. (Assume the potential is zero very far from the sphere.) (*a*) What is the radius of the metal sphere? (*b*) What is the sign of the charge on the sphere? Explain your answer.

85 •• Along the central axis of a uniformly charged disk, at a point 0.60 m from the center of the disk, the potential is 80 V and the magnitude of the electric field is 80 V/m. At a distance of 1.5 m, the potential is 40 V and the magnitude of the electric field is 23.5 V/m. (Assume the potential is zero very far from the sphere.) Find the total charge on the disk.

86 •• A radioactive ^{210}Po nucleus emits an α particle that has a charge $+2e$. When the α particle is a large distance from the nucleus, it has a kinetic energy of 5.30 MeV. Assume that the α particle had negligible kinetic energy as it left the surface of the nucleus. The "daughter" (or residual) nucleus ^{206}Pb has a charge $+82e$. Determine the radius of the ^{206}Pb nucleus. (Neglect the radius of the α particle and assume the ^{206}Pb nucleus remains at rest.)

87 ••• (*a*) Configuration *A* consists of two point particles; one particle has a charge of $+q$ and is on the x axis at $x = +d$ and the other particle has a charge of $-q$ and is at $x = -d$ (Figure 23-36*a*). Assuming the potential is zero at large distances from these charged particles, show that the potential is also zero everywhere on the $x = 0$ plane. (*b*) Configuration *B* consists of a flat metal plate of infinite extent and a point particle located a distance d from the plate (Figure 23-36*b*). The point particle has a charge equal to $+q$ and the plate is grounded. (Grounding the plate forces its potential to equal zero.) Choose the line perpendicular to the plate and through the point charge as the x axis, and choose the origin at the surface of the plate nearest the particle. (These choices put the particle on the x axis at $x = +d$.) For configuration *B*, the electric potential is zero both at all points in the half-space $x \geq 0$ that are very far from the particle and at all points on the $x = 0$ plane—just as was the case for configuration *A*. A theorem, called the *uniqueness theorem*, implies that throughout the half-space $x \geq 0$ the potential function V—and thus the electric field $\vec{E}$—for the two configurations are identical. Using this result, obtain the electric field $\vec{E}$ at every point in the $x = 0$ plane in configuration *B*. (The uniqueness theorem tells us that in configuration *B* the electric field at each point in the $x = 0$ plane is the same as it is in configuration *A*.) Use this result to find the surface charge density σ at each point in the conducting plane (in configuration *B*). SSM

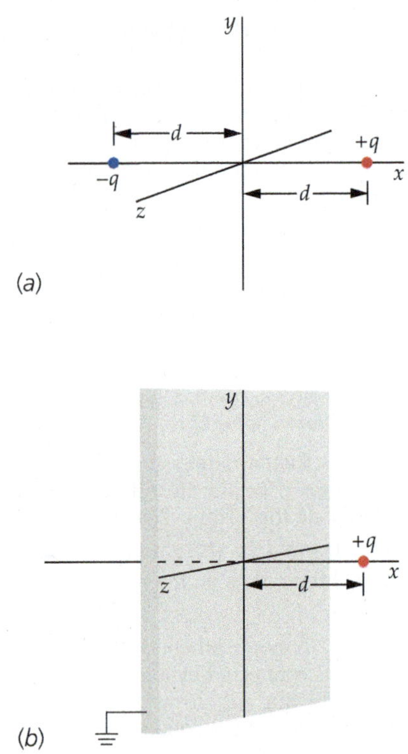

FIGURE 23-36
Problem 87

88 ••• A particle that has a mass m and a positive charge q is constrained to move along the x axis. At $x = -L$ and $x = L$ are two ring charges of radius L (Figure 23-37). Each ring is centered on the x axis and lies in a plane perpendicular to it. Each ring has a total positive charge Q uniformly distributed on it. (*a*) Obtain an expression for the potential $V(x)$ on the x axis due to the charge on the rings. (*b*) Show that $V(x)$ has a minimum at $x = 0$. (*c*) Show that for $|x| \ll L$, the potential approaches the form $V(x) = V(0) + \alpha x^2$. (*d*) Use the result of Part (*c*) to derive an expression for the angular frequency of oscillation of the mass m if it is displaced slightly from the origin and released. (Assume the potential equals zero at points far from the rings.)

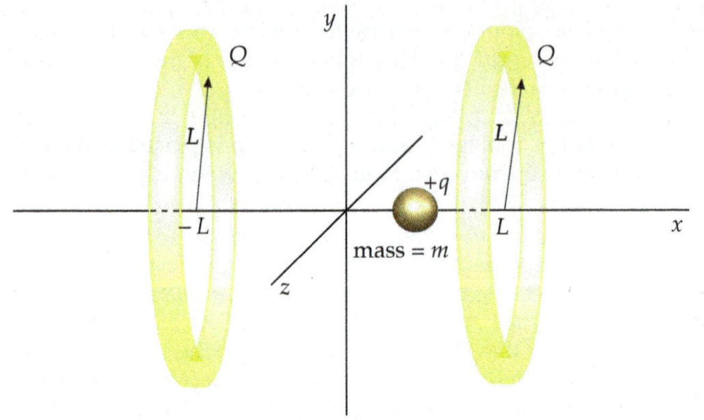

FIGURE 23-37 Problem 88

89 ••• Three concentric conducting thin spherical shells have radii a, b, and c so that $a < b < c$. Initially, the inner shell is uncharged, the middle shell has a positive charge $+Q$, and the outer shell has a charge $-Q$. (Assume the potential equals zero at points far from the shells.) (*a*) Find the electric potential of each of the three shells. (*b*) If the inner and outer shells are now connected by a conducting wire that is insulated as it passes through a small hole in the middle shell, what is the electric potential of each of the three shells, and what is the final charge on each shell?

90 ••• Consider two concentric spherical thin metal shells of radii a and b, where $b \geq a$. The outer shell has a charge Q, but the inner shell is grounded. This means that the potential on the inner shell is the same as the potential at points far from the shells. Find the charge on the inner shell.

91 ••• Show that the total work needed to assemble a uniformly charged sphere that has a total charge of Q and radius R is given by $3Q^2/(20\pi\epsilon_0 R)$. Energy conservation tells us that this result is the same as the resulting electrostatic potential energy of the sphere. *Hint: Let ρ be the charge density of the sphere that has charge Q and radius R. Calculate the work dW to bring in charge dq from infinity to the surface of a uniformly charged sphere of radius r ($r < R$) and charge density ρ. (No additional work is required to smear dq throughout a spherical shell of radius r, thickness dr, and charge density ρ. Why?)* SSM

92 ••• (*a*) Use the result of Problem 91 to calculate the *classical electron radius*, the radius of a uniform sphere that has a charge $-e$ and an electrostatic potential energy equal to the rest energy of the electron (5.11×10^5 eV). Comment on the shortcomings of this model for the electron. (*b*) Repeat the calculation in Part (*a*) for a proton using its rest energy of 938 MeV. Experiments indicate the proton has an approximate radius of about 1.2×10^{-15} m. Is your result close to this value?

93 ••• (*a*) Consider a uniformly charged sphere that has radius R and charge Q and is composed of an incompressible fluid, such as water. If the sphere fissions (splits) into two halves of equal volume and equal charge, and if these halves stabilize into uniformly charged spheres, what is the radius R' of each? (*b*) Using the expression for potential energy shown in Problem 91, calculate the change in the total electrostatic potential energy of the charged fluid. Assume that the spheres are separated by a large distance. SSM

94 ••• Problem 93 can be modified to be used as a very simple model for nuclear fission. When a ^{235}U nucleus absorbs a neutron, it can fission into the fragments ^{140}Xe, ^{94}Sr, and 2 neutrons. The ^{235}U has 92 protons, while ^{140}Xe has 54 protons and ^{94}Sr has 38 protons. Estimate the energy released during this fission process (in MeV), assuming that the mass density of the nucleus is constant and has a value of 4×10^{17} kg/m^3.

CHAPTER 24

THE ENERGY FOR THE ELECTRONIC FLASH OF THIS CAMERA HAS BEEN TRANSFERRED FROM A BATTERY TO A CAPACITOR. *(PhotoDisc/Getty Images.)*

Capacitance

? How do you determine how much energy can be stored in a capacitor? (See Example 24-3.)

How many people do you know that *don't* have a digital camera, a cell phone, a cell-phone/digital-camera combination, or any of a myriad of additional portable electronic devices? Virtually all portable electronic devices contain one or more capacitors, and life without portable electronic devices seems unthinkable today. We live in an on-the-run age yet still manage to stay in communication with people important to us using cell phones, to enjoy music using mp3 players, and even to check and send email using PDA (personal digital assistant) devices.

In the previous chapters, we discussed the relation of electric fields to charges and how the relation of charges translates into electric potential energy. Here we show that potential energy can be stored and released using the concept of capacitance.

In this chapter, we will discuss circuits containing batteries and capacitors. In the next few chapters, the concepts of electric potential and capacitance will be further developed as they relate to circuits containing resistors, inductors, and other devices.

24-1 CAPACITANCE

The potential V of a single isolated conductor due to the charge Q on it is proportional to Q and depends on the size and shape of the conductor. Typically, the larger the surface area of a conductor, the more charge it can carry for a given potential. For example, if the potential is chosen to be zero at infinity, the potential of a spherical conductor having a radius R and a charge Q is

$$V = \frac{kQ}{R}$$

(The equation for an isolated sphere $V = kQ/R$ (Equation 23-22) is established in Chapter 23.) The ratio Q/V of the charge to the potential of an isolated conductor is called its **self-capacitance *C*.** A **capacitor** is a device consisting of two conductors, one that has a charge Q and the other that has a charge $-Q$. The ratio of charge Q to the potential difference V between the two conductors is called the **capacitance** of the capacitor.

Capacitors are used in large numbers in common electronic devices such as television sets. Some capacitors are used to store energy, but most are used to filter unwanted electrical frequencies. (© *Tom Pantages Images.*)

$$C = \frac{Q}{V} \qquad \text{24-1}$$

DEFINITION—CAPACITANCE

Capacitance is a measure of the capacity to store charge for a given potential difference. Because the potential difference is proportional to the charge, this ratio does not depend on either Q or V, but only on the sizes, shapes, and relative positions of the conductors. The self-capacitance of an isolated spherical conductor is

$$C = \frac{Q}{V} = \frac{Q}{kQ/R} = \frac{R}{k} = 4\pi\epsilon_0 R \qquad \text{24-2}$$

The SI unit of capacitance is the coulomb per volt, which is called a **farad** (F) after the great English experimentalist Michael Faraday:

$$1\text{ F} = 1\text{ C/V} \qquad \text{24-3}$$

The farad is a rather large unit, so submultiples such as the microfarad ($1\ \mu\text{F} = 10^{-6}$ F) or the picofarad ($1\text{ pF} = 10^{-12}$ F) are more commonly used. Because capacitance is in farads and R is in meters, we can see from Equation 24-2 that the SI unit for the electric constant (the permittivity of empty space), ϵ_0, can also be written as a farad per meter:

$$\epsilon_0 = 8.85 \times 10^{-12}\text{ F/m} = 8.85\text{ pF/m} \qquad \text{24-4}$$

ELECTRIC CONSTANT

PRACTICE PROBLEM 24-1

Find the radius of a spherical conductor that has a capacitance of 1.0 F.

The farad is indeed a very large unit.

CONCEPT CHECK 24-1

A sphere of capacitance C_1 carries a charge of 20 μC. If the charge is increased to 60 μC, what is the new capacitance C_2?

CAPACITORS

A capacitor is usually charged by transferring a charge Q from one conductor to the other conductor, which leaves one of the conductors having a charge $+Q$ and the other conductor having a charge $-Q$. The capacitance of the device is defined to be Q/V, where Q is the magnitude of the charge on either conductor and V is the magnitude of the potential difference between the conductors. To calculate the capacitance, we place equal and opposite charges on the conductors and then find the potential difference V by first finding the electric field $\vec{E}$ due to the charges and then calculating V from $\vec{E}$.

When we speak of the charge on a capacitor, we mean the magnitude of the charge on either conductor. The use of V rather than ΔV for the magnitude of the potential difference between the plates is standard and simplifies many of the equations relating to capacitance.

The first capacitor was the Leyden jar (Figure 24-1), a glass container lined with metal on its outside and base and either filled with water or lined with foil on its inside. It was invented at the University of Leyden in the Netherlands by eighteenth-century experimenters who, while studying the effects of electric charges on people and animals, got the idea of trying to store a large amount of charge in a bottle of water. An experimenter held up a jar of water in one hand while charge was conducted to the water by a chain from a static electric generator. When the experimenter reached over to lift the chain out of the water with his other hand, he was knocked unconscious. Benjamin Franklin realized that the device for storing charge did not have to be jar shaped and used foil-covered window glass, called Franklin panes. Using several of these panes connected in parallel, Franklin stored a large charge and attempted to kill a turkey with them. Instead, he knocked himself out. Franklin later wrote, "I tried to kill a turkey but nearly succeeded in killing a goose."

FIGURE 24-1 Leyden jar with bells. The bell on the pole through the stopper is connected to a conductor on the inside surface of the jar. The second bell is connected to the conductor on the outside surface of the jar. The system is energized by connecting a battery between the two bells for a short time. After the battery is removed the conducting ball swings from one bell to the other, transferring charge a little bit at a time. *(Courtesy of Bernhard Thomas.)*

PARALLEL-PLATE CAPACITORS

A common capacitor is the **parallel-plate capacitor,** which uses two parallel conducting plates. In practice, the plates are often thin metallic foils that are separated and insulated from one another by a thin plastic film. This "sandwich" is then rolled up, which allows for a large surface area in a relatively small space. Let A be the area of the surface (the area of that side of each plate that faces the other plate), and let d be the separation distance, which is very small compared to the length and width of the plates. We place a charge $+Q$ on one plate and $-Q$ on the other plate. These charges attract each other and become uniformly distributed on the inside surfaces of the plates. Because the plates are very close together, the electric field between them is uniform and has a magnitude of $E = \sigma/\epsilon_0$. [That the electric field strength just outside the surface of a conductor is given by $E = \sigma/\epsilon_0$ (Equation 22-21) is established in Chapter 22.] Because $\vec{E}$ is uniform between the plates (Figure 24-2), the potential difference between the plates equals the field strength E multiplied by the plate separation d:

$$V = Ed = \frac{\sigma}{\epsilon_0}d = Qd/(\epsilon_0 A) \qquad 24\text{-}5$$

where we have substituted Q/A for σ. The capacitance of the parallel-plate capacitor is thus

$$C = \frac{Q}{V} = \frac{Q}{Qd/(\epsilon_0 A)} = \frac{\epsilon_0 A}{d} \qquad 24\text{-}6$$

CAPACITANCE OF A PARALLEL-PLATE CAPACITOR

Note that because V is proportional to Q, the capacitance does not depend on either Q or V. For a parallel-plate capacitor, the capacitance is proportional to the area of the plates and is inversely proportional to the gap width (separation distance). In general, capacitance depends on the size, shape, and geometrical arrangement of the conductors. Capacitance also depends on the properties of the insulating medium between the conductors as we shall see in Section 24-4.

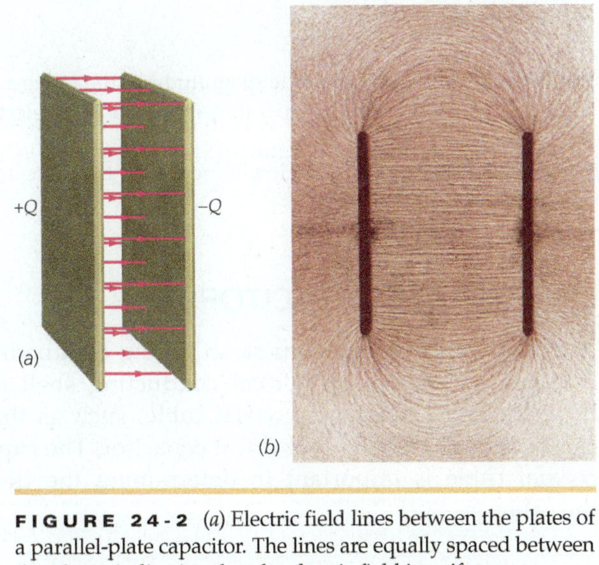

FIGURE 24-2 (*a*) Electric field lines between the plates of a parallel-plate capacitor. The lines are equally spaced between the plates, indicating that the electric field is uniform. (*b*) Electric field lines in a parallel-plate capacitor shown by small bits of thread suspended in oil. *(Harold M. Waage.)*

PROBLEM-SOLVING STRATEGY

Calculating Capacitance

PICTURE Make a sketch of the capacitor that has a charge of $+Q$ on one conductor and a charge of $-Q$ on the other conductor.

SOLVE

1. Determine the electric field $\vec{E}$, usually by using Gauss's law.
2. Determine the magnitude of the potential difference V between the two conductors by integrating $dV = -\vec{E} \cdot d\vec{\ell}$ (Equation 23-2*a*).
3. The capacitance is equal to $C = Q/V$.

CHECK Check that the result depends only on the electric constant* and on geometrical factors such as lengths and areas.

Example 24-1 The Capacitance of a Parallel-Plate Capacitor

A parallel-plate capacitor has square metallic plates of edge length 10 cm separated by 1.0 mm. (*a*) Calculate the capacitance of this device. (*b*) As this capacitor is charged to 12 V, how much charge is transferred from one plate to another?

PICTURE The capacitance C is determined by the area and the separation of the plates. Once C is found, the charge for a given voltage V is found from the definition of capacitance $C = Q/V$.

SOLVE

(*a*) We find the capacitance using $C = \epsilon_0 A/d$ (Equation 24-6):

$$C = \frac{\epsilon_0 A}{d} = \frac{(8.85\ \text{pF/m})(0.10\ \text{m})^2}{0.0010\ \text{m}} = 88.5\ \text{pF} = \boxed{89\ \text{pF}}$$

(*b*) The charge transferred is found from $Q = CV$ (the definition of capacitance):

$$Q = CV = (88.5\ \text{pF})(12\ \text{V}) = 1.06 \times 10^{-9}\ \text{C} = \boxed{1.1\ \text{nC}}$$

CHECK The Part (*b*) expression has units of farads multiplied by volts. Because $1\ \text{F} = 1\ \text{C/V}$ (Equation 24-3), the product of farads and volts equals coulombs, which is the appropriate unit for charge.

TAKING IT FURTHER Q is the magnitude of the charge on each plate of the capacitor. A charge of 1.1 nC corresponds to a transfer of 6.6×10^9 electrons from one plate to the other.

PRACTICE PROBLEM 24-2 How large would the plate area have to be for the capacitance to equal 1.0 F?

CYLINDRICAL CAPACITORS

A cylindrical capacitor consists of a long conducting cylinder of radius R_1 and a larger, concentric cylindrical conducting shell of radius R_2. The cylinders have the same length. A coaxial cable, such as that used for cable television, can be thought of as a cylindrical capacitor. The capacitance per unit length of a coaxial cable is important in determining the transmission characteristics of the cable.

* Capacitance also depends on a property of any nonconducting material placed between the conductors. This dependence is introduced in Section 24-4.

Example 24-2 An Expression for the Capacitance of a Cylindrical Capacitor

Find an expression for the capacitance of a cylindrical capacitor consisting of two conductors, each of length L. One conductor is a cylinder of radius R_1 and the second conductor is a coaxial cylindrical shell of inner radius R_2, where $R_1 < R_2 \ll L$ as shown in Figure 24-3.

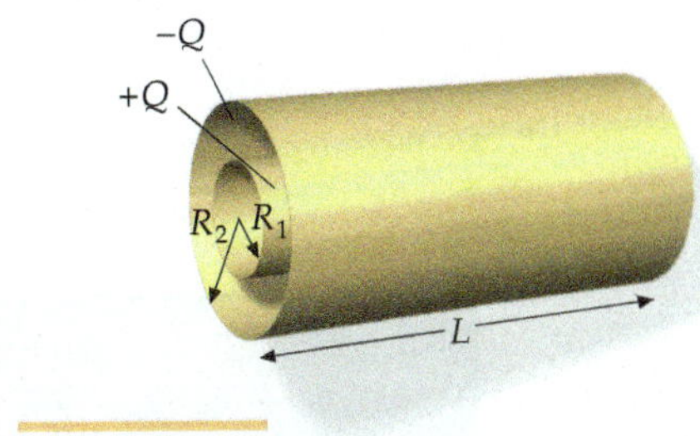

FIGURE 24-3

PICTURE We place charge $+Q$ on the inner conductor and charge $-Q$ on the outer conductor and calculate the potential difference $V = V_{R_2} - V_{R_1}$ from the electric field between the conductors, which is found from Gauss's law. Because the electric field is not uniform (it depends on the distance R from the axis) we must integrate $\vec{E}$ to find the potential difference.

SOLVE

1. The capacitance is defined as the ratio Q/V:

$$C = Q/V$$

2. V is related to the electric field:

$$dV = -\vec{E}\cdot d\vec{\ell}$$

3. To find E_R we choose a soup-can-shaped Gaussian surface of radius R and length ℓ, where $R_1 < R < R_2$ and $\ell \ll L$. The entire Gaussian surface is located far from the ends of the coaxial conductors (Figure 24-4):

FIGURE 24-4

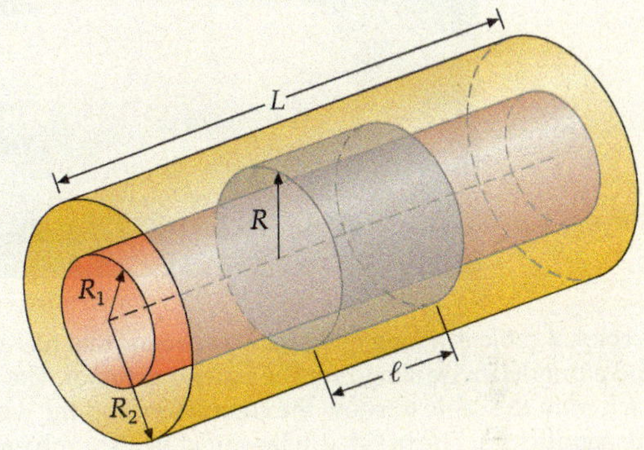

4. On the Gaussian surface, $\vec{E}$ is either zero or is in the radial direction. Thus, there is no flux of $\vec{E}$ through either of the two flat ends of the can. The area of the curved side of the can is $2\pi R\ell$, and on this side $E_n = E_R$, so Gauss's law gives:

$$\phi_{\text{net}} = \oint_S E_n\,dA = \frac{Q_{\text{inside}}}{\epsilon_0}$$

where

$$\oint_S E_n\,dA = \int_{\text{left end}} E_n\,dA + \int_{\text{curved side}} E_n\,dA + \int_{\text{right end}} E_n\,dA$$

$$= 0 + \int_{\text{curved side}} E_R\,dA + 0 = E_R\int_{\text{curved side}} dA = E_R 2\pi R\ell$$

5. Substituting for $\oint_S E_n\,dA$ in the previous step gives:

$$E_R 2\pi R\ell = \frac{Q_{\text{inside}}}{\epsilon_0}$$

6. Assuming the charge per unit length on the inner shell is uniformly distributed, find Q_{inside}:

$$\frac{Q_{\text{inside}}}{Q} = \frac{\ell}{L} \quad \text{so} \quad Q_{\text{inside}} = \frac{\ell}{L}Q$$

7. Substitute for Q_{inside} in the step 5 result and solve for E_R:

$$E_R 2\pi R\ell = \frac{1}{\epsilon_0}\frac{\ell}{L}Q \quad \text{so} \quad E_R = \frac{Q}{2\pi L\epsilon_0 R}$$

8. Integrate to find $V = |V_{R_2} - V_{R_1}|$:

$$V_{R_2} - V_{R_1} = \int_{V_{R_1}}^{V_{R_2}} dV = -\int_{R_1}^{R_2} E_R\,dR = -\frac{Q}{2\pi L\epsilon_0}\int_{R_1}^{R_2}\frac{dR}{R} = -\frac{Q}{2\pi L\epsilon_0}\ln\frac{R_2}{R_1}$$

$$\text{so} \quad V = |V_{R_2} - V_{R_1}| = \frac{Q}{2\pi L\epsilon_0}\ln\frac{R_2}{R_1}$$

9. Rearrange this result to find $C = Q/V$:

$$C = \frac{Q}{V} = \boxed{\frac{2\pi\epsilon_0 L}{\ln(R_2/R_1)}}$$

CHECK The step 9 result is dimensionally correct. Capacitances always have the dimension of ϵ_0 multiplied by length.

TAKING IT FURTHER The capacitance of a cylindrical capacitor is proportional to the length of the cylinders. In addition, we were able to factor E_R from the integrand in step 4 because the symmetry reveals that E_R is the same everywhere on the curved side of the can.

CONCEPT CHECK 24-2

How is the capacitance affected if the potential across a cylindrical capacitor is increased from 20 V to 80 V?

From Example 24-2, we see that the capacitance of a cylindrical capacitor is given by

$$C = \frac{2\pi\epsilon_0 L}{\ln(R_2/R_1)} \qquad 24\text{-}7$$

CAPACITANCE OF A CYLINDRICAL CAPACITOR

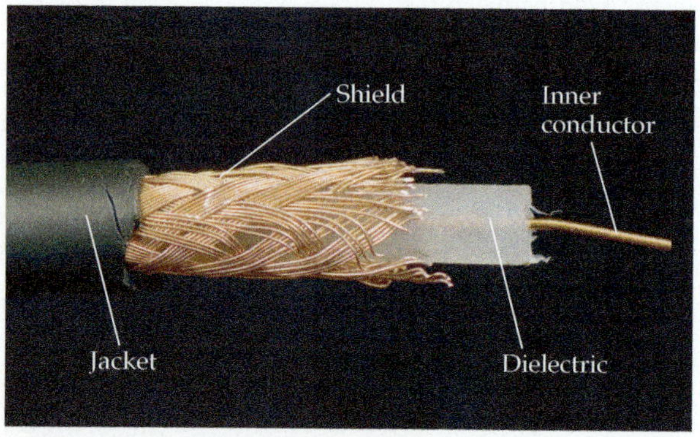

A coaxial cable is a long cylindrical capacitor that has a solid wire for the inner conductor and a braided-wire shield for the outer conductor. The outer rubber coating has been peeled back from the cable to show the conductors and the white plastic insulator that separates the conductors. The braided-wire shield blocks ambient electric fields from reaching the inner conductor, which carries the information of interest (such as the video and audio signal for a television show). *(John Perry Fish.)*

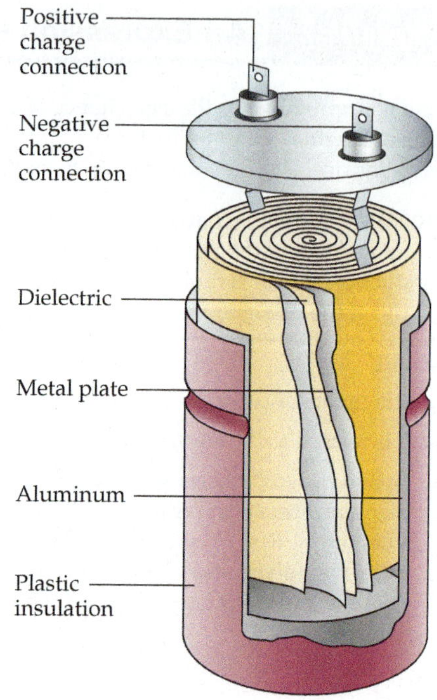

Cutaway of an electrolytic capacitor. The dielectric is an insulator.

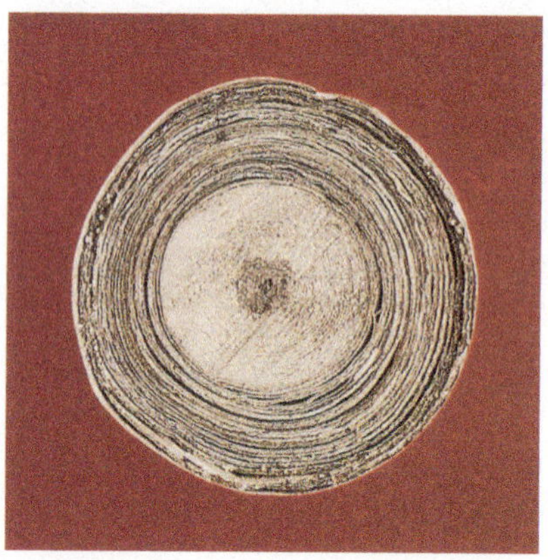

Cross section of a foil-wound capacitor. *(© Bruce Iverson.)*

A variable air-gap capacitor such as those used in the tuning circuits of old radios. The semicircular plates rotate through the fixed plates, which changes the amount of surface area between the plates, and hence the capacitance. *(Loren Winters/Visuals Unlimited.)*

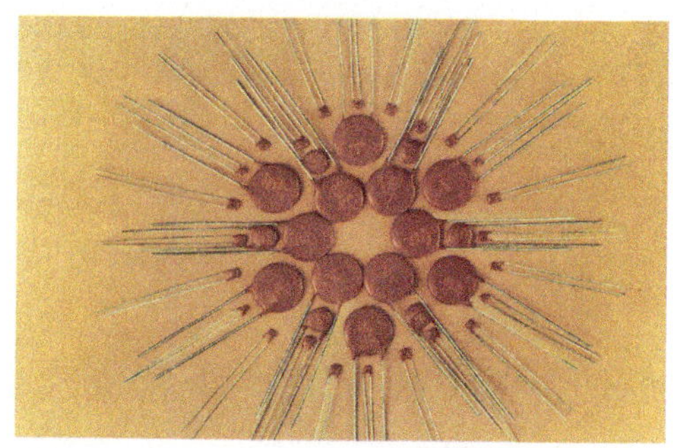

Ceramic capacitors for use in electronic circuits. *(Courtesy Tusonix, Tucson, AZ.)*

24-2 THE STORAGE OF ELECTRICAL ENERGY

When a capacitor is being charged, electrons are transferred from the positively charged conductor to the negatively charged conductor. This leaves the positively charged conductor with an electron deficit and the negatively charged conductor with an electron surplus. Alternatively, transferring positive charges from the negatively charged conductor to the positively charged conductor could also charge a capacitor. Either way, work must be done to charge a capacitor, and at least some of this work is stored as electrostatic potential energy.

We start with two uncharged conductors that do not touch each other. Let q be the positive charge that has been transferred during the initial stages of the charging process. The potential difference then is $V = q/C$. If a small amount of additional positive charge dq is now transferred from the negative conductor to the positive conductor through a potential increase of V (Figure 24-5), the electrical potential energy of the charge, and thus the capacitor, is increased by

$$dU = V\,dq = \frac{q}{C}\,dq$$

The total increase in potential energy U is the integral of dU as q increases from zero to its final value Q (Figure 24-6):

$$U = \int dU = \int_0^Q \frac{q}{C}\,dq = \frac{1}{C}\int_0^Q q\,dq = \frac{1}{2}\frac{Q^2}{C}$$

This potential energy is the energy stored in the capacitor. Using the definition of capacitance ($C = Q/V$), we can express this energy in terms of either Q and V, C and V, or Q and C:

$$U = \frac{1}{2}\frac{Q^2}{C} = \frac{1}{2}QV = \frac{1}{2}CV^2 \qquad 24\text{-}8$$

ENERGY STORED IN A CAPACITOR

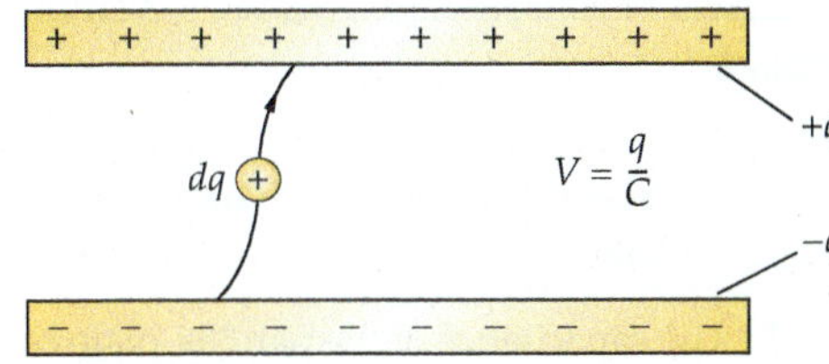

FIGURE 24-5 When a small amount of positive charge dq is moved from the negative conductor to the positive conductor, its potential energy is increased by $dU = V\,dq$, where V is the potential difference between the conductors.

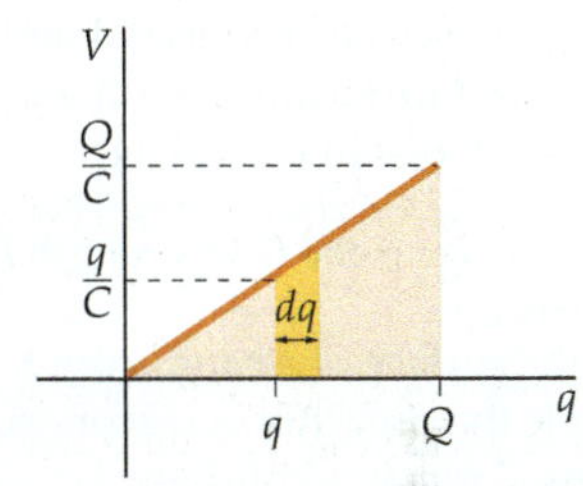

FIGURE 24-6 The work needed to charge a capacitor is the integral of $V\,dq$ from the original charge of $q = 0$ to the final charge of $q = Q$. This work is equal to area under the curve. That is, the work equals the area of the triangle of height Q/C and width Q.

PRACTICE PROBLEM 24-3

A 185-μF capacitor is charged to 200 V. How much energy is stored in the capacitor?

PRACTICE PROBLEM 24-4

Obtain the expression for the electrostatic energy stored in a capacitor (Equation 24-8) from $U = \frac{1}{2}\sum_{i=1}^{n} Q_i V_i$ (Equation 23-29), using $Q_1 = -Q$, $Q_2 = +Q$, $n = 2$, and $V_2 = V_1 + V$.

Suppose we charge a capacitor by connecting it to a battery. The potential difference V when the capacitor is fully energized with charge $+Q$ on one conductor and charge $-Q$ on the other is just the potential difference between the terminals of the battery before they were connected to the capacitor. The total work done *by the battery* in charging the capacitor is QV, which is twice the amount of energy stored in the capacitor. The additional work done by the battery is either dissipated as thermal energy in the battery and in the connecting wires* or radiated as electromagnetic waves.†

Example 24-3 Charging a Parallel-Plate Capacitor with a Battery

A parallel-plate capacitor that has square plates which are 14 cm on a side and are separated by 2.0 mm is connected to a battery and charged to 12 V. (*a*) What is the charge on the capacitor? (*b*) How much energy is stored in the capacitor? (*c*) The battery is then disconnected from the capacitor and the plates are pulled apart until the plate separation is increased to 3.5 mm. How much does the stored energy change as the plate separation is increased from 2.0 mm to 3.5 mm?

PICTURE (*a*) The charge on the capacitor can be calculated from the capacitance. (*b*) The energy stored in the capacitor can be calculated if we know both the charge and the capacitance. (*c*) The charge remains constant as the plates are separated because the capacitor is not connected to the battery during separation. The energy change is found by using the charge and new potential difference to calculate the new energy, from which we subtract the original energy.

* We show in Section 25-6 that if the capacitor is connected to an ideal battery by wires of some resistance R, half the energy supplied by the battery in charging the capacitor is dissipated as thermal energy in the wires.

† We show in Section 30-3 that under certain circumstances the circuit will act as a broadcast antenna and a significant portion of the work will be broadcast as electromagnetic radiation.

SOLVE

(*a*) 1. The charge Q on the capacitor equals the product of C_0 and V_0, where C_0 is the capacitance and $V_0 = 12$ V is the battery voltage:

$$Q = C_0V_0$$

2. The capacitance of the parallel-plate capacitor is given by Equation 24-6:

$$C_0 = \frac{\epsilon_0 A}{d_0}$$

3. Substitute for C_0 and calculate Q:

$$Q = C_0V_0 = \frac{\epsilon_0 A}{d_0}V_0 = \frac{(8.85\ \text{pF/m})(0.14\ \text{m})^2}{0.0020\ \text{m}}(12\ \text{V}) = 1.04\ \text{nC}$$

$$= \boxed{1.0\ \text{nC}}$$

(*b*) Calculate the energy stored:

$$U_0 = \tfrac{1}{2}QV_0 = \tfrac{1}{2}(1.04\ \text{nC})(12\ \text{V}) = 6.24\ \text{nJ} = \boxed{6.2\ \text{nJ}}$$

(*c*) 1. The battery is disconnected and the plate separation is increased to 3.5 mm. The change in energy is proportional to the change in the voltage:

$$\Delta U = U - U_0 = \tfrac{1}{2}QV - \tfrac{1}{2}QV_0 = \tfrac{1}{2}Q(V - V_0)$$

2. The voltage is the field strength E multiplied by the separation distance d:

$$V = Ed \quad \text{and} \quad V_0 = E_0d_0$$

3. At the surface of a conductor, $E = \sigma/\epsilon_0$ (Equation 22-21). While the capacitor is disconnected, σ remains constant. Thus, E remains constant:

$$E = E_0$$

4. Combining the last two steps establishes that V is proportional to d:

$$E = \frac{V}{d} = \frac{V_0}{d_0} \quad \text{so} \quad V = \frac{d}{d_0}V_0$$

5. Substitute for V in the equation of Part (*c*), step 1 by using the result from Part (*c*), step 4. Solve for ΔU, by using the value for U_0 from Part (*b*):

$$\Delta U = \frac{1}{2}Q\left(\frac{d}{d_0}V_0 - V_0\right) = \left(\frac{d}{d_0} - 1\right)\left(\frac{1}{2}QV_0\right) = \left(\frac{d}{d_0} - 1\right)U_0$$

$$= \left(\frac{3.5\ \text{mm}}{2.0\ \text{mm}} - 1\right)(6.24\ \text{nJ}) = \boxed{4.7\ \text{nJ}}$$

CHECK An increase in potential energy with an increase in plate separation is expected. The plates are oppositely charged, so they attract each other. Thus, it takes work to separate them farther. This work done on the plates results in an increase in the potential energy of the system.

TAKING IT FURTHER An application of the dependence of capacitance on separation distance is shown in Figure 24-7.

PRACTICE PROBLEM 24-5 Find the final voltage V between the capacitor plates.

PRACTICE PROBLEM 24-6 (*a*) Find the initial capacitance C_0 in this example when separation of the plates is 2.0 mm. (*b*) Find the final capacitance C when separation of the plates is 3.5 mm.

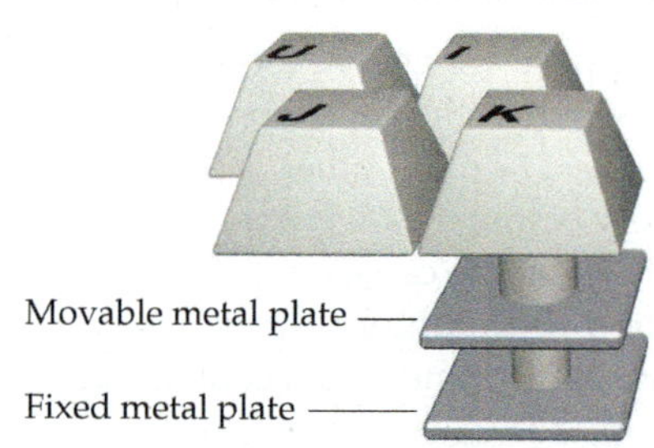

FIGURE 24-7 Capacitance switching in computer keyboards. A metal plate attached to each key acts as the top plate of a capacitor. Depressing the key decreases the separation between the top and bottom plates and increases the capacitance, which triggers the electronic circuitry of the computer to acknowledge the keystroke.

It is instructive to work Part (*c*) of Example 24-3 in another way. The oppositely charged plates of a capacitor exert attractive forces on one another. Work must be done on the plates to overcome the forces to increase the plate separation. Assume that the lower plate is held fixed and the upper plate is moved. The force on the upper plate is the charge $+Q$ on the plate multiplied by the electric field $\vec{E}'$ *due to the charge* $-Q$ *on the lower plate*. This field is half the total field $\vec{E}$ between the plates because the charge on the upper plate and the charge on the lower plate contribute equally to the electric field in the region between the plates. When the potential difference is 12 V and the separation is 2.0 mm, the total field strength between the plates is

$$E = \frac{V}{d} = \frac{12\text{ V}}{2.0\text{ mm}} = 6.0\text{ V/mm} = 6.0\text{ kV/m}$$

The magnitude of the force exerted on the upper plate by the bottom plate is thus

$$F = QE' = Q(\tfrac{1}{2}E) = (1.04\text{ nC})(3.0\text{ kV/m}) = 3.1\ \mu\text{N}$$

The work that must be done to move the upper plate a distance of $\Delta d = 1.5$ mm is then

$$W = F\,\Delta d = (3.1\ \mu\text{N})(1.5\text{ mm}) = 4.7\text{ nJ}$$

This value is the same as the number of joules calculated in Part (*c*) of Example 24-3. This work is equal to the increase in the potential energy.

ELECTROSTATIC FIELD ENERGY

During the process of charging a capacitor, an electric field is produced between the plates. The work required to charge the capacitor can be thought of as the work required to establish the electric field. That is, we can think of the energy stored in a capacitor as energy stored in the electric field, called **electrostatic field energy.**

Consider a parallel-plate capacitor. We can relate the energy stored in the capacitor to the electric field strength E between the plates. The potential difference between the plates is related to the electric field by $V = Ed$, where d is the plate separation distance. The capacitance is given by $C = \epsilon_0 A/d$ (Equation 24-6). The energy stored (Equation 24-8) is

$$U = \frac{1}{2}CV^2 = \frac{1}{2}\left(\frac{\epsilon_0 A}{d}\right)(Ed)^2 = \frac{1}{2}\epsilon_0 E^2(Ad)$$

The quantity Ad is the volume of the space between the plates of the capacitor. This volume is the volume of the region containing the electric field. The energy per unit volume is called the **energy density** u_e. The energy density in an electric field of strength E is thus

$$u_e = \frac{\text{energy}}{\text{volume}} = \frac{1}{2}\epsilon_0 E^2 \qquad 24\text{-}9$$

ENERGY DENSITY OF AN ELECTROSTATIC FIELD

Thus, the energy per unit volume of the electrostatic field is proportional to the square of the electric field strength. *Although we obtained Equation 24-9 by considering the electric field between the plates of a parallel-plate capacitor, the result applies to any electric field.* Whenever there is an electric field in space, the electrostatic energy per unit volume is given by Equation 24-9.

PRACTICE PROBLEM 24-7

(*a*) Calculate the energy density u_e for Example 24-3 when the plate separation is 2.0 mm.
(*b*) Show that the increase in energy in Example 24-3 equal to u_e multiplied by the increase in volume ($A\,\Delta d$) of the region between the plates.

We can illustrate the generality of Equation 24-9 by calculating the electrostatic field energy of a spherical conductor that has a radius R and a charge Q. The self-capacitance of a spherical conductor is given by $C = R/k$ (Equation 24-2) and the electrostatic potential energy is given by $U = \frac{1}{2}Q^2/C$ (Equation 24-8). Thus, for a spherical conductor:

$$U = \frac{1}{2}\frac{Q^2}{C} = \frac{1}{2}\frac{Q^2}{R/k} = \frac{kQ^2}{2R} \qquad \text{24-10}$$

We now obtain the same result by considering the energy density of an electric field given by Equation 24-9. When the conductor has a charge Q, the electric field is radial and is given by

$$E_r = 0 \qquad r < R \text{ (inside the conductor)}$$

$$E_r = \frac{kQ}{r^2} \qquad r > R \text{ (outside the conductor)}$$

Because the electric field is spherically symmetric, we choose a spherical shell for our volume element. If the radius of the shell is r and its thickness is dr, the volume is $d\mathcal{V} = 4\pi r^2\,dr$ (Figure 24-8). The energy dU in this volume element is

$$dU = u_e\,d\mathcal{V} = \frac{1}{2}\epsilon_0 E^2 4\pi r^2\,dr$$

$$= \frac{1}{2}\epsilon_0\left(\frac{kQ}{r^2}\right)^2(4\pi r^2\,dr) = \frac{1}{2}(4\pi\epsilon_0 k^2)Q^2\frac{dr}{r^2} = \frac{1}{2}kQ^2\frac{dr}{r^2}$$

where we have used $4\pi\epsilon_0 = 1/k$. Because the electric field is zero for $r < R$, we obtain the total energy in the electric field by integrating from $r = R$ to $r = \infty$:

$$U = \int u_e\,d\mathcal{V} = \frac{1}{2}kQ^2\int_R^\infty r^{-2}\,dr = \frac{1}{2}k\frac{Q^2}{R} = \frac{1}{2}\frac{Q^2}{C} \qquad \text{24-11}$$

which is the same as Equation 24-8.

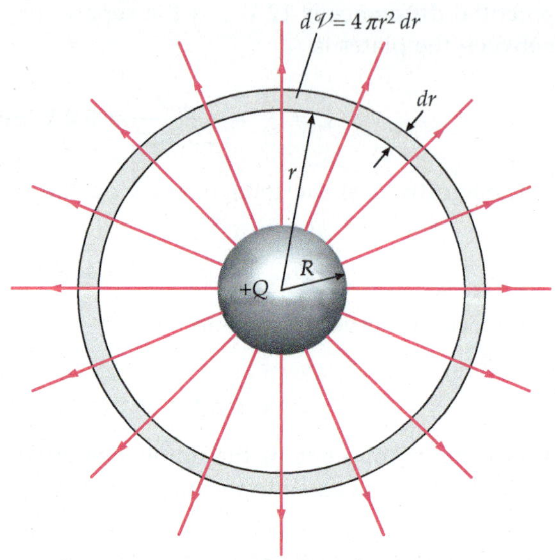

FIGURE 24-8 Geometry for the calculation of the electrostatic energy of a spherical conductor that has a charge Q. The volume of the space between r and $r + dr$ is $d\mathcal{V} = 4\pi r^2\,dr$. The electrostatic field energy in this volume element is $u_e\,d\mathcal{V}$, where $u_e = \frac{1}{2}\epsilon_0 E^2$ is the energy density.

24-3 CAPACITORS, BATTERIES, AND CIRCUITS

Next, we examine what happens when an initially uncharged capacitor is connected to the terminals of a battery. The potential difference between the two terminals of a battery is called its **terminal voltage.** The terminals of a battery (Figure 24-9) are connected to dissimilar conductors called *electrodes,* and within the battery the electrodes are separated by a conducting liquid or paste called an *electrolyte.* Because of chemical reactions in the battery, charge is transferred from one electrode to the other. This leaves one electrode of the battery (the anode) positively charged and the other electrode (the cathode) negatively charged; this charge separation is maintained by chemical reactions within the battery. Within the battery, there is an electric field directed away from the positive electrode and toward the negative electrode.* When the plates of an uncharged capacitor are connected to the terminals of the battery, the negative electrode shares its negative charge with the plate connected to it and the positive battery terminal shares its positive charge with the plate connected to it. This charge sharing momentarily reduces the amount of charge on each of the battery electrodes, and thus decreases the potential difference between the electrodes. This decrease in terminal voltage triggers the chemical reactions within the battery, and charge is transferred from one electrode to the other electrode in an effort to maintain the terminal voltage at its initial level, which is called the **open-circuit terminal voltage.** These chemical reactions cease when the battery has transferred sufficient charge from one capacitor plate to the other capacitor plate to increase the potential difference between the capacitor plates to the open-circuit terminal voltage of the battery.

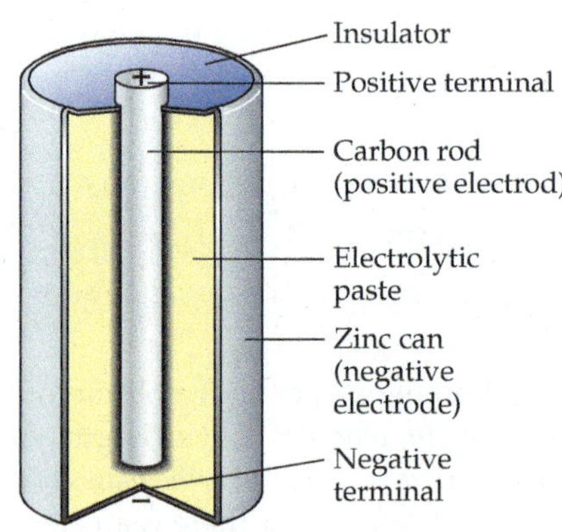

FIGURE 24-9 A carbon-zinc cell.

* This electric field from the positive to the negative terminal exists outside as well as within the battery.

A battery is a "charge pump." When we connect the plates of an uncharged capacitor to the terminals of a battery (Figure 24-10), the terminal voltage of the battery drops. This results in the battery pumping charge from one plate to the other plate until the open-circuit terminal voltage is again reached.

In electric circuit diagrams, the symbol representing a battery is $\dashv\vdash^{+}$, where the longer vertical line represents the positive terminal and the shorter vertical line represents the negative terminal. The symbol representing a capacitor is $\dashv\vdash$.

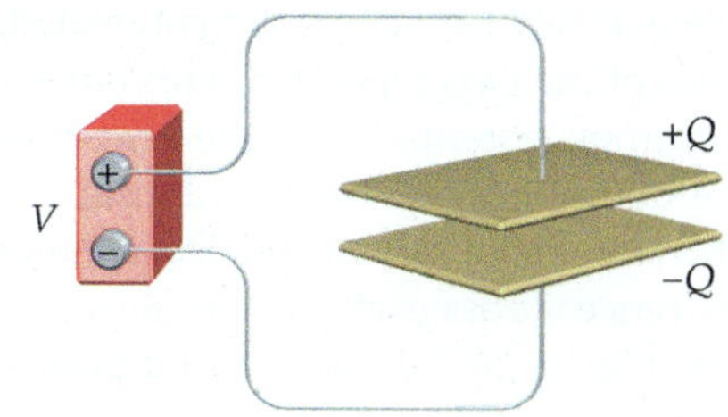

FIGURE 24-10 When the conductors of an uncharged capacitor are connected to the terminals of a battery, the battery "pumps" charge from one conductor to the other until the potential difference between the conductors equals the open circuit potential difference between the battery terminals.* The amount of charge transferred through the battery is $Q = CV$.

> **PRACTICE PROBLEM 24-8**
> A 6.0-μF capacitor, initially uncharged, is connected to the terminals of a 9.0-V battery. What total amount of charge then flows through the battery as the capacitor is charged to the open-circuit terminal voltage of the battery?

COMBINATIONS OF CAPACITORS

Example 24-4 Capacitors Connected in Parallel

A circuit consists of a 6.0-μF capacitor, a 12.0-μF capacitor, a 12.0-V battery, and a switch, connected as shown in Figure 24-11. The switch is initially open and the capacitors are initially uncharged. The switch is then closed and the capacitors charge. When the capacitors are fully charged and the open-circuit terminal voltage of the battery is restored, (*a*) what is the potential of each conductor in the circuit? (Choose the negative battery terminal to be the zero-potential reference point.) (*b*) What is the charge on each capacitor plate? (*c*) What total charge passed through the battery?

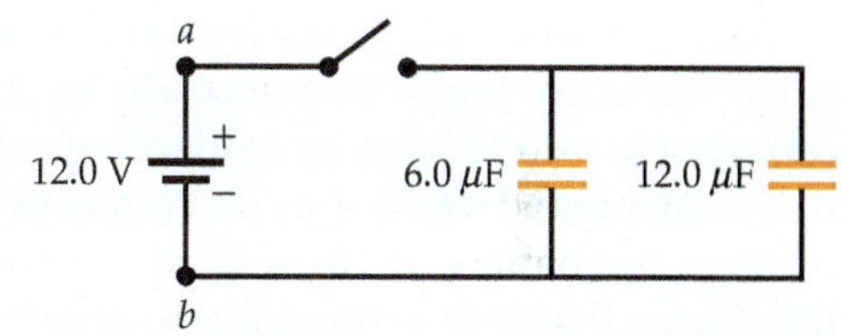

FIGURE 24-11

PICTURE The potential is the same throughout a conductor in electrostatic equilibrium. Thus, after the charges stop flowing, all of the conductors connected to each other by a conducting wire are at the same potential. The charge Q on a capacitor is related to the potential difference V across the capacitor by $Q = CV$. In addition, the charges on the plates of a single capacitor are equal in magnitude but opposite in sign.

SOLVE

(*a*) Use a red marker to color the positive (+) battery terminal and all the conductors connected to it (Figure 24-12), and use a blue marker to color the negative (−) battery terminal and all the conductors connected to it:

All points colored red are at potential

$\boxed{V_a = 12\text{ V}}$

All points colored blue are at potential

$\boxed{V_b = 0}$

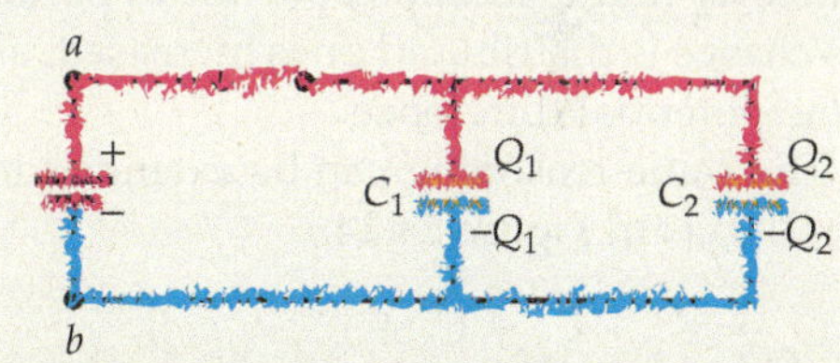

FIGURE 24-12

(*b*) Use $Q = CV$ to find the magnitude of the charge on the plates. (The plate of a capacitor at the higher potential has a positive charge):

$Q_1 = C_1V = (6.0\ \mu\text{F})(12.0\text{ V}) = \boxed{72\ \mu\text{C}}$

$Q_2 = C_2V = (12.0\ \mu\text{F})(12.0\text{ V}) = \boxed{144\ \mu\text{C}}$

(*c*) The plates become charged because the battery acts as a charge pump:

$Q = Q_1 + Q_2 = \boxed{216\ \mu\text{C}}$

CHECK The charge on the 12.0-μF capacitor is two times the charge on the 6.0-μF capacitor when the voltage across each is 12.0 V. This result is as expected. The capacitance of a capacitor is a measure of its capacity to store charge for a given voltage.

TAKING IT FURTHER The equivalent capacitance of the two-capacitor combination is Q/V, where Q is the charge passing through the battery and V is the open-circuit terminal voltage of the battery. For this example $C_{eq} = (216\ \mu\text{C})/(12.0\text{ V}) = 18.0\ \mu\text{F}$.

* We will discuss batteries more fully in Chapter 25. Here, all we need to know is that a battery is a device that stores energy, supplies electrical energy, and pumps charge in an effort to restore the potential difference between its terminals to the open circuit terminal voltage V.

When two capacitors are connected, as shown in Figure 24-13, so that the upper plates of the two capacitors are connected by a conducting wire and are therefore at a common potential, and the lower plates are also connected together and are at a common potential, just like the capacitors in Example 24-4, the capacitors are said to be **connected in parallel.** Devices connected in parallel share a common potential difference across each device *due solely to the way they are connected.*

In Figure 24-13, assume that points a and b are connected to a battery or some other device that maintains a potential difference $V = V_a - V_b$ between the plates of each capacitor. If the capacitances are C_1 and C_2, the charges Q_1 and Q_2 stored on the plates are given by

$$Q_1 = C_1 V$$

and

$$Q_2 = C_2 V$$

The total charge stored is

$$Q = Q_1 + Q_2 = C_1 V + C_2 V = (C_1 + C_2)V$$

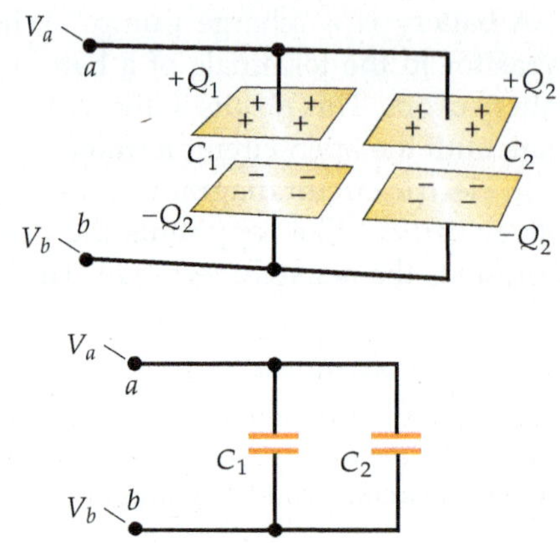

FIGURE 24-13 Two capacitors in parallel. The upper plates are connected to each other by a conductor and are therefore at a common potential V_a; the lower plates are similarly connected together and therefore at a common potential V_b.

A combination of capacitors in a circuit can sometimes be substituted with a single capacitor that is operationally equivalent to the combination. The substitute capacitor is said to have an **equivalent capacitance.** That is, if a combination of initially uncharged capacitors is connected to a battery, the charge Q that flows through the battery as the capacitor combination becomes charged is the same as the charge that flows through the same battery if connected to a single uncharged capacitor of equivalent capacitance. Therefore, the equivalent capacitance of two capacitors in parallel is the ratio of the charge $Q_1 + Q_2$ to the potential difference:

$$C_{eq} = \frac{Q}{V} = \frac{Q_1 + Q_2}{V} = \frac{Q_1}{V} + \frac{Q_2}{V} = C_1 + C_2 \qquad \text{24-12}$$

Thus, for two capacitors in parallel, C_{eq} is the sum of the individual capacitances. When we add a second capacitor in parallel with the first capacitor, the area that the charge is distributed on is increased, allowing more charge to be stored for the same potential difference.

The same reasoning can be extended to three or more capacitors connected in parallel, as in Figure 24-14:

$$C_{eq} = C_1 + C_2 + C_3 + \ldots \qquad \text{24-13}$$

EQUIVALENT CAPACITANCE FOR CAPACITORS IN PARALLEL

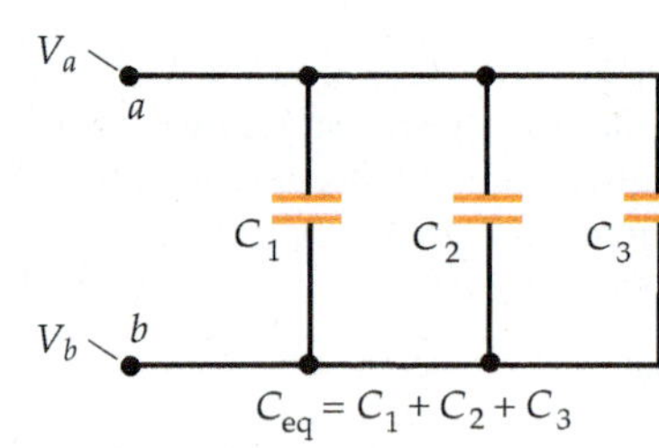

FIGURE 24-14 Three capacitors in parallel. The effect of adding an additional capacitor in parallel with a combination of capacitors is to increase the equivalent capacitance.

Example 24-5 Capacitors Connected in Series

A circuit consists of a 6.0-μF capacitor, a 12-μF capacitor, a 12-V battery, and a switch, connected as shown in Figure 24-15. Initially, the switch is open and the capacitors are uncharged. The switch is then closed and the capacitors charge. When the open-circuit terminal voltage is restored to the battery, the capacitors are fully charged. (*a*) What is the potential of each conductor in the circuit? (Choose the negative battery terminal to be the zero-potential reference point.) If the potential of a conductor is not known, represent its potential symbolically. (*b*) What is the charge on each capacitor plate? (*c*) What total charge passed through the battery?

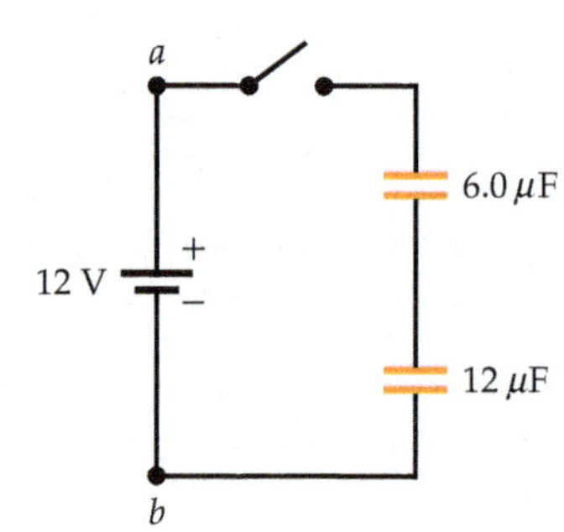

FIGURE 24-15

PICTURE The potential is the same throughout a conductor in electrostatic equilibrium. After the charges stop moving, all of the conductors connected by a conducting wire are at the same potential. The charge on a capacitor is related to the potential difference across the capacitor by $Q = CV$. Charge does not travel across the gap between the plates of a capacitor.

SOLVE

(*a*) Use a red marker to color the positive (+) battery terminal and all conductors connected to it, use a blue marker to color the negative (−) battery terminal and all the conductors connected to it, and use a green marker to color all other mutually connected conductors (Figure 24-16):

All points colored red are at potential $\boxed{V_a = 12\text{ V}}$

All points colored blue are at potential $\boxed{V_b = 0}$

All points colored green are at the yet unknown potential $\boxed{V_m}$

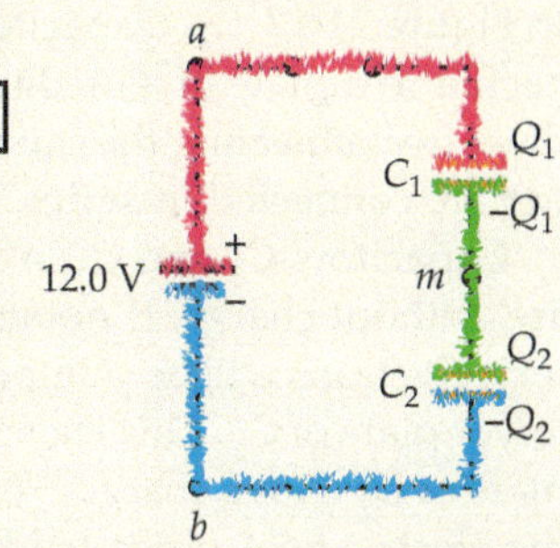

FIGURE 24-16

(*b*) 1. Express the potential difference across each capacitor in terms of the Part-(*a*) results:

$V_1 = V_a - V_m$ and $V_2 = V_m - V_b$

2. Use $Q = CV$ to relate the charge on each capacitor to the potential difference:

$Q_1 = C_1V_1 = C_1(V_a - V_m)$

and

$Q_2 = C_2V_2 = C_2(V_m - V_b)$

3. Eliminating V_m gives:

$$\left.\begin{aligned} V_a - V_m &= \frac{Q_1}{C_1} \\ V_m - V_b &= \frac{Q_2}{C_2} \end{aligned}\right\} \Rightarrow V_a - V_b = \frac{Q_1}{C_1} + \frac{Q_2}{C_2}$$

4. During charging, there is no charge transferred either to or from the green region in Figure 24-16, so its net charge remains zero:

$(-Q_1) + Q_2 = 0$ so $Q_1 = Q_2$

5. Let $Q = Q_1 = Q_2$. Substitute Q for both Q_1 and Q_2 and solve for Q:

$$V_a - V_b = \frac{Q}{C_1} + \frac{Q}{C_2} \quad \text{so} \quad Q = \frac{V_a - V_b}{\dfrac{1}{C_1} + \dfrac{1}{C_2}} = \frac{12\text{ V} - 0}{\dfrac{1}{6.0\ \mu\text{F}} + \dfrac{1}{12\ \mu\text{F}}} = 48\ \mu\text{C}$$

$Q_1 = Q_2 = \boxed{48\ \mu\text{C}}$

(*c*) All the charge passing through the battery ends up on the upper plate of C_1:

$Q_1 = Q = \boxed{48\ \mu\text{C}}$

CHECK The potential difference across a capacitor is equal to Q/C. Thus the potential difference across the 6.0-μF and 12-μF capacitors are $(48\ \mu\text{C})/(6.0\ \mu\text{F}) = 8.0\text{ V}$ and $(48\ \mu\text{C})/(12\ \mu\text{F}) = 4.0\text{ V}$, respectively. The sum of these potential differences is $8.0\text{ V} + 4.0\text{ V} = 12.0\text{ V}$, which is as expected because the battery is a 12-V battery.

TAKING IT FURTHER The equivalent capacitance of the two-capacitor combination is Q/V, where Q is the charge passing through the battery and V is the open-circuit terminal voltage of the battery. For this example $C_{eq} = (48\ \mu\text{C})/(12\text{ V}) = 4.0\ \mu\text{F}$.

PRACTICE PROBLEM 24-9 Find the potential V_m on the conductors colored green in Figure 24-16.

CONCEPT CHECK 24-3

During the charging of the capacitors in Example 24-5, did the net charge within the battery increase, decrease, or remain the same?

Consider the circuit shown in Figure 24-15. If we start at point *b* and follow a path that goes once around the circuit in the clockwise direction, the potential goes up by 12 V as we step across the battery, drops by 4 V as we step across the 6-μF capacitor, drops by an additional 8 V as we step across the 12-μF capacitor, and remains the same on the path back to point *b* (thus completing the trip around the circuit). The changes in potential (+12 V, −4 V, and −8 V) add to zero, which is not a special circumstance. The changes in potential around any closed path always sum to zero. Adding the changes in potential around a circuit loop and setting the sum equal to zero is a very useful tool for analyzing electric circuits. Known as **Kirchhoff's loop rule,** it is a consequence of the fact that the potential difference between any two points does not depend on the path taken from one of the points to the other.

The changes in potential around any closed path always sum to zero.

KIRCHHOFF'S LOOP RULE

A **junction** is a point in a wire where the wire divides into two or more wires. In Figure 24-17, two capacitors are connected so that a plate of one capacitor is connected to a plate of a second capacitor by a wire containing no junctions, just like the wire connecting the capacitors in Example 24-5. Devices connected in this manner are connected in **series.**

Capacitors C_1 and C_2 in Figure 24-17 are connected in series and initially they are without charge. If points a and b are then connected to the terminals of a battery, electrons will be pumped through the battery from the upper plate of C_1 to the lower plate of C_2. This leaves the upper plate of C_1 with a charge $+Q$ and the lower plate of C_2 with a charge $-Q$. When a charge $+Q$ appears on the upper plate of C_1, the electric field produced by that charge induces an equal negative charge, $-Q$, on the lower plate of C_1. The charge on the lower plate of C_1 comes from electrons drawn from the upper plate of C_2. Thus, there will be a charge equal to $+Q$ on the upper plate of C_2. The potential difference across C_1 is

$$V_1 = \frac{Q}{C_1}$$

Similarly, the potential difference across the second capacitor is

$$V_2 = \frac{Q}{C_2}$$

The potential difference across the two capacitors in series is the sum of these potential differences:

$$V = V_a - V_b = V_1 + V_2 = \frac{Q}{C_1} + \frac{Q}{C_2} = Q\left(\frac{1}{C_1} + \frac{1}{C_2}\right) \qquad 24\text{-}14$$

The equivalent capacitance of the two capacitors is

$$C_{eq} = \frac{Q}{V} \qquad 24\text{-}15$$

where Q is the charge that passes through the battery during the charging process. Substituting Q/C_{eq} for V in Equation 24-14, and then dividing both sides by Q, gives

$$\frac{1}{C_{eq}} = \frac{1}{C_1} + \frac{1}{C_2} \qquad 24\text{-}16$$

Equation 24-16 can be generalized to three or more capacitors connected in series:

$$\frac{1}{C_{eq}} = \frac{1}{C_1} + \frac{1}{C_2} + \frac{1}{C_3} + \cdots \qquad 24\text{-}17$$

EQUIVALENT CAPACITANCE FOR EQUALLY CHARGED CAPACITORS IN SERIES

PRACTICE PROBLEM 24-10

Two capacitors have capacitances of 20 μF and 30 μF. Find the equivalent capacitance if the capacitors are connected (*a*) in parallel and (*b*) in series.

Note that in Practice Problem 24-10, the equivalent capacitance of the two capacitors in series is less than the capacitance of either capacitor. Adding a capacitor in series increases $1/C_{eq}$, which means the equivalent capacitance C_{eq} decreases. When we add a second capacitor in series, we decrease the equivalent capacitance of the combination. The plate separation is, in effect, increased, so it requires a greater potential difference to store the same charge.

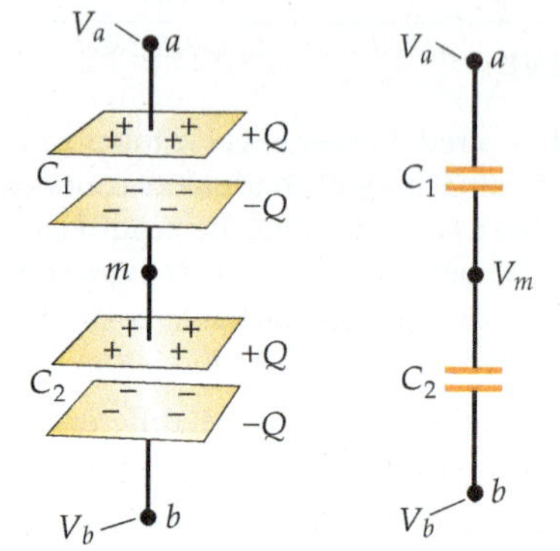

FIGURE 24-17 The total charge on the two interconnected capacitor plates equals zero. The potential difference across the pair equals the sum of the potential differences across the individual capacitors. The two capacitors are connected in series.

A capacitor bank for storing energy to be used by the pulsed Nova laser at Lawrence Livermore Laboratories. The laser is used in fusion studies. (*Lawrence Livermore National Laboratory.*)

! Equation 24-10 is valid if the capacitors are connected in series AND if the total charge on each pair of capacitor plates connected by an otherwise isolated conductor is zero.

Example 24-6 Using the Equivalence Formula

A 6.0-μF capacitor and a 12-μF capacitor, each initially uncharged, are connected in series across a 12-V battery. Using the equivalence formula for capacitors in series, find the charge on each capacitor and the potential difference across each.

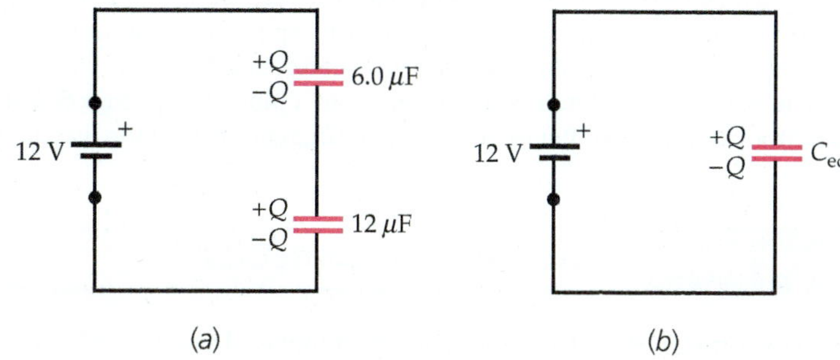

FIGURE 24-18

PICTURE Figure 24-18*a* shows the circuit in this example and Figure 24-18*b* shows an equivalent capacitor that has the same charge $Q = C_{eq}V$. After finding the charge, we can find the potential drop across each capacitor.

SOLVE

1. The charge on each capacitor equals the charge on the equivalent capacitor:
$$Q = C_{eq}V$$
2. The equivalent capacitance of the series combination is found from:
$$\frac{1}{C_{eq}} = \frac{1}{C_1} + \frac{1}{C_2} = \frac{1}{6.0\,\mu\text{F}} + \frac{1}{12\,\mu\text{F}} = \frac{3}{12\,\mu\text{F}}$$
$$C_{eq} = 4.0\,\mu\text{F}$$
3. Use this value to find the charge Q. This is the charge that passed through the battery. It is also the charge on each capacitor:
$$Q = C_{eq}V = (4.0\,\mu\text{F})(12\text{ V}) = \boxed{48\,\mu\text{C}}$$
4. Use the result for Q to find the potential across the 6.0-μF capacitor:
$$V_1 = \frac{Q}{C_1} = \frac{48\,\mu\text{C}}{6.0\,\mu\text{F}} = \boxed{8.0\text{ V}}$$
5. Again, use the result for Q to find the potential across the 12-μF capacitor:
$$V_2 = \frac{Q}{C_2} = \frac{48\,\mu\text{C}}{12\,\mu\text{F}} = \boxed{4.0\text{ V}}$$

CHECK The sum of these potential differences is 12 V, as required.

TAKING IT FURTHER The results are the same as those obtained in Example 24-5.

Example 24-7 Series, Parallel, or Neither *Conceptual*

Consider the capacitors shown in Figure 24-19*a*. (*a*) Identify all parallel combinations of capacitors. (*b*) Identify all series combinations of capacitors.

PICTURE Capacitors connected in parallel share a common potential difference across each capacitor due solely to the way they are connected. The potential along a conducing path remains constant. Use markers to color each conducting path a unique color. Two capacitors are connected in series if a plate of one capacitor is connected to a plate of a second capacitor by a conducting wire that contains no junctions.

SOLVE

(*a*) 1. Use markers to give each electric potential unique color (Figure 24-19*b*). The potential can change only at a circuit device such as a capacitor or a battery.

2. Capacitors connected in parallel share a common potential difference across each capacitor due solely to the way they are connected.

Capacitors 4 and 7 are the only two capacitors that are connected in parallel with each other.

(*b*) If two capacitors are connected so that a plate of one capacitor is connected to a plate of a second capacitor by a conducting wire containing no junctions then the capacitors are connected in series.

Capacitors 8 and 9 are the only two capacitors that are connected in series with each other.

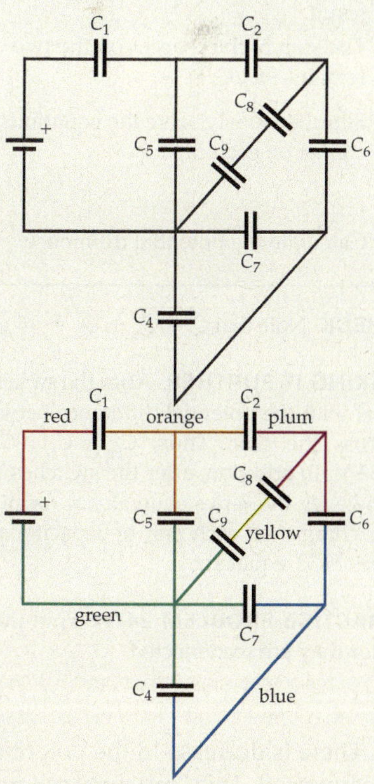

FIGURE 24-19

TAKING IT FURTHER Capacitors 1 and 2 are not in series. We make this conclusion because there is a junction connecting the wire that connects capacitors 1 and 2. Capacitors 2 and 5 are not in parallel even though one plate of each capacitor has an orange-colored wire attached to it. However, there is plum-colored wire connected to one of the plates of capacitor 2 and there is not a plum-colored wire connected to a plate of capacitor 5. Thus, we know the two capacitors do not share the same potential difference due to the way they are wired.

Example 24-8 Capacitors Reconnected

The two capacitors in Example 24-6 are removed from the battery and carefully disconnected from each other so that the charge on each of the plates is not changed (Figure 24-20*a*). They are then reconnected, positive plate to positive plate and negative plate to negative plate (Figure 24-20*b*), in a circuit containing open switches S_1 and S_2. Find the potential difference across the capacitors and the charge on each capacitor after the switches are closed and the charges have stopped flowing.

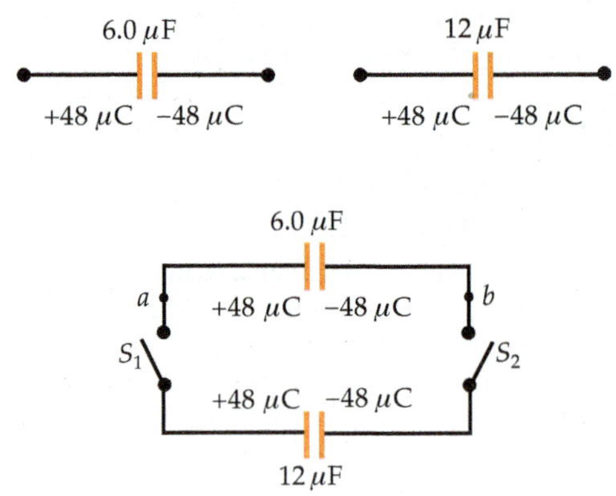

FIGURE 24-20

PICTURE Just after the two capacitors are disconnected from the battery, they have equal charges of 48 μC. After switches S_1 and S_2 in the new circuit are closed the voltages across the capacitors are the same. Use the definition of capacitance and conservation of charge to solve for the charge on each capacitor. Once these charges are known, use them to solve for the voltage.

SOLVE

1. Draw and label the circuit after both switches have been closed. Let C_1 and C_2 be 6.0 μF and 12 μF, respectively (Figure 24-21).

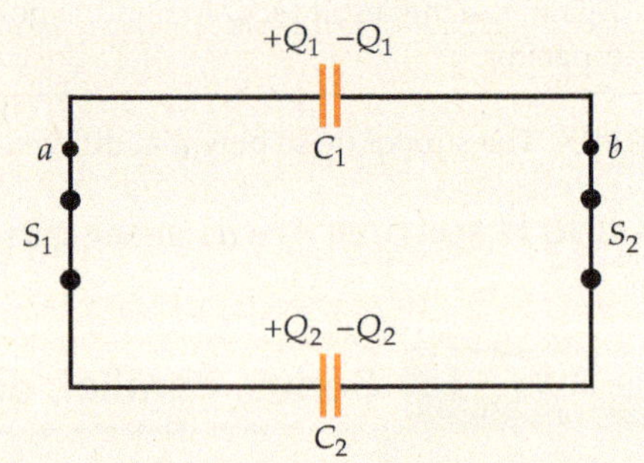

FIGURE 24-21

2. The wiring is such that after the switches are closed the voltage is the same across each capacitor. $V = V_1 = V_2$

3. For each capacitor, $V = Q/C$. Substitute this into the step 2 result. $\dfrac{Q_1}{C_1} = \dfrac{Q_2}{C_2}$

4. The sum of the charges on the two capacitor plates on the left remains 96 μC. $Q_1 + Q_2 = 96\ \mu\text{C}$

5. Simultaneously solve the equations in steps 3 and 4 for the charge on each capacitor. $Q_1 = \boxed{32\ \mu\text{C}}$ $Q_2 = \boxed{64\ \mu\text{C}}$

6. Calculate the potential difference. $V = \dfrac{Q_1}{C_1} = \boxed{5.3\ \text{V}}$

CHECK Note that $Q = Q_1 + Q_2 = 96\ \mu\text{C}$ and that $Q_2/C_2 = 5.33\ \text{V}$, as required for consistency.

TAKING IT FURTHER After the switches are closed, the two capacitors are connected in parallel with the potential difference between point *a* and point *b* being the potential difference across the pair. Thus, $C_{eq} = C_1 + C_2 = 18\ \mu\text{F}$, $Q = Q_1 + Q_2 = 96\ \mu\text{C}$, and $V = Q/C_{eq} = 5.33\ \text{V}$. In addition, after the switches are closed, the two capacitors are connected in series. However, the series equivalence formula (Equation 24-17) is NOT valid because the sum of the charges on each pair of capacitor plates connected by a single, otherwise isolated, wire does NOT equal zero.

PRACTICE PROBLEM 24-11 Find the potential energy stored in the capacitors before and after they are reconnected.

There is decrease in the potential energy stored in the capacitors when they are reconnected. This "missing" potential energy is either dissipated as thermal energy in the wires or radiated away.

! Charge IS conserved when capacitors are reconnected, but while energy IS conserved, electrical potential energy is NOT conserved.

Example 24-9 Capacitors in Series and in Parallel

Three capacitors are connected as shown in Figure 24-22. (*a*) Find the equivalent capacitance of the three-capacitor combination. (*b*) The capacitors are initially uncharged. The combination is then connected to a 6.0-V battery. Find the potential difference across each capacitor and the charge on each capacitor after the battery is connected and the charges have stopped flowing.

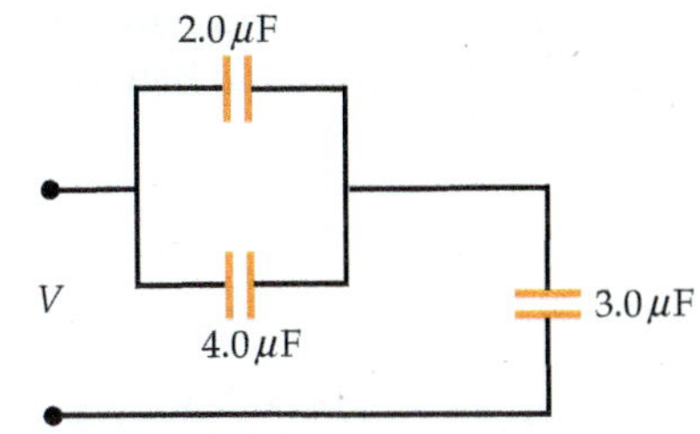

FIGURE 24-22

PICTURE The 2.0-μF capacitor and the 4.0-μF capacitor are connected in parallel, and the parallel combination is connected in series with the 3.0-μF capacitor. We first find the equivalent capacitance of the parallel combination (Figure 24-23*a*), then we combine this equivalent capacitance with the 3.0-μF capacitor to reach a final equivalent capacitance (Figure 24-23*b*). The charge on the positive plate of the 3.0-μF capacitor is the charge passing through the battery $Q = C_{eq}V$ as shown in Figure 24-23*a*.

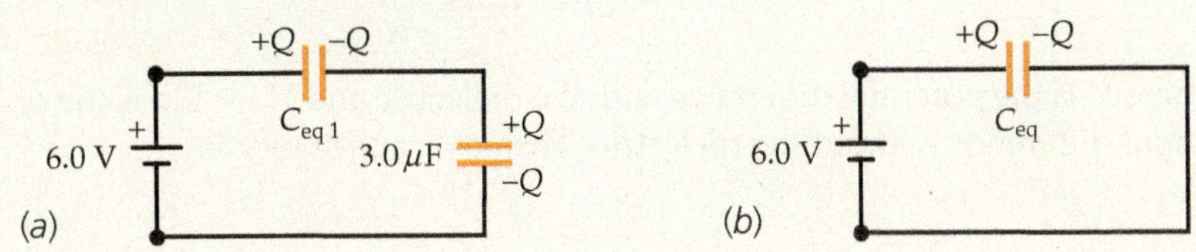

FIGURE 24-23

SOLVE

(*a*) 1. The equivalent capacitance of the two capacitors in parallel is the sum of the capacitances:

$$C_{eq\,1} = C_1 + C_2 = 2.0\ \mu\text{F} + 4.0\ \mu\text{F} = 6.0\ \mu\text{F}$$

2. Find the equivalent capacitance of the 6.0-μF capacitor in series with the 3.0-μF capacitor:

$$\frac{1}{C_{eq}} = \frac{1}{C_{eq\,1}} + \frac{1}{C_3} = \frac{1}{6.0\ \mu\text{F}} + \frac{1}{3.0\ \mu\text{F}} = \frac{1}{2.0\ \mu\text{F}}$$

$$C_{eq} = \boxed{2.0\ \mu\text{F}}$$

(*b*) 1. Calculate the charge Q that passed through the battery during charging. This is the charge on the 3.0-μF capacitor:

$$Q = C_{eq}V = (2.0\ \mu\text{F})(6.0\ \text{V}) = \boxed{12\ \mu\text{C}}$$

2. The potential drop across the 3.0-μF capacitor is Q/C_3:

$$V_3 = \frac{Q_3}{C_3} = \frac{Q}{C_3} = \frac{12\ \mu\text{C}}{3.0\ \mu\text{F}} = \boxed{4.0\ \text{V}}$$

3. The potential drop across the parallel combination V_{24} is $Q/C_{eq\,1}$:

$$V_{24} = \frac{Q}{C_{eq\,1}} = \frac{12\ \mu\text{C}}{6.0\ \mu\text{F}} = \boxed{2.0\ \text{V}}$$

4. The charge on each of the parallel capacitors is found from $Q_i = C_iV_{24}$, where $V_{24} = 2.0$ V:

$$Q_2 = C_2V_{24} = (2.0\ \mu\text{F})(2.0\ \text{V}) = \boxed{4.0\ \mu\text{C}}$$

$$Q_4 = C_4V_{24} = (4.0\ \mu\text{F})(2.0\ \text{V}) = \boxed{8.0\ \mu\text{C}}$$

CHECK The voltage drop across the parallel combination (2.0 V) plus the voltage drop across the 3.0-μF capacitor (4.0 V) equals the voltage of the battery. Also, the sum of the charges on the parallel capacitors (4.0 μC + 8.0 μC) equals the total charge (12 μC) on the 3.0-μF capacitor.

PRACTICE PROBLEM 24-12 Find the energy stored in each capacitor.

24-4 DIELECTRICS

A nonconducting material (for example, air, glass, paper, or wood) is called a **dielectric.** When the space between the two conductors of a capacitor is occupied by a dielectric, the capacitance is increased by a factor that is characteristic of the dielectric, a fact discovered experimentally by Michael Faraday. The reason for this increase is that the electric field between the plates of a capacitor is weakened by the dielectric. Thus, for a given charge on the plates, the potential difference V is reduced and the capacitance (Q/V) is increased.

Consider an isolated charged capacitor without a dielectric between its plates. A dielectric slab is then inserted between the plates, completely filling the space between

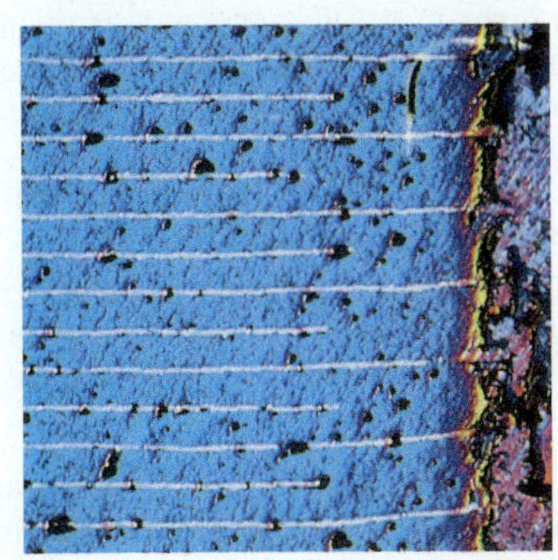

A cut section of a multilayer capacitor that has a blue ceramic dielectric. The white lines are the edges of the conducting plates.(© *Manfred Kage/Peter Arnold, Inc.)*

the plates. If the electric field strength is E_0 before the dielectric slab is inserted, after the dielectric slab is inserted between the plates the electric field strength is

$$E = \frac{E_0}{\kappa} \tag{24-18}$$

ELECTRIC FIELD INSIDE A DIELECTRIC

where κ (kappa) is called the **dielectric constant** of the inserted material. For a parallel-plate capacitor of separation d, the potential difference V between the plates is

$$V = Ed = \frac{E_0 d}{\kappa} = \frac{V_0}{\kappa}$$

where V is the potential difference with the dielectric and $V_0 = E_0 d$ is the original potential difference without the dielectric. The new capacitance is

$$C = \frac{Q}{V} = \frac{Q}{V_0/\kappa} = \kappa\frac{Q}{V_0}$$

or

$$C = \kappa C_0 \tag{24-19}$$

EFFECT OF A DIELECTRIC ON CAPACITANCE

where $C_0 = Q/V_0$ is the capacitance without the dielectric. The capacitance of a parallel-plate capacitor filled with a dielectric of constant κ is thus

$$C = \frac{\kappa\epsilon_0 A}{d} = \frac{\epsilon A}{d} \tag{24-20}$$

where

$$\epsilon = \kappa\epsilon_0 \tag{24-21}$$

The parameter ϵ is called the **permittivity** of the dielectric.

In the preceding discussion, the capacitor was electrically isolated (not part of a circuit), so we assumed that the charge on its plates did not change as the dielectric was inserted. This is the case if the capacitor is charged and then removed from the charging source (the battery) before the insertion of the dielectric. If the dielectric is inserted while the battery remains connected, the battery pumps additional charge to maintain the original potential difference. The total charge on the plates is then $Q = \kappa Q_0$. In either case, the capacitance (Q/V) is increased by the factor κ.

PRACTICE PROBLEM 24-13

The 89-pF capacitor of Example 24-1 is filled with a dielectric of constant $\kappa = 2.0$. (*a*) Find the new capacitance. (*b*) Find the charge on the capacitor when the dielectric is in place and the capacitor is attached to a 12-V battery.

PRACTICE PROBLEM 24-14

The capacitor in the previous problem is charged to 12 V without the dielectric and is then disconnected from the battery. The dielectric of constant $\kappa = 2.0$ is then inserted. Find the new values for (*a*) the charge Q, (*b*) the voltage V, and (*c*) the capacitance C.

Dielectrics not only increase the capacitance of a capacitor, they also provide a means for keeping parallel conducting plates apart and they increase the potential difference at which dielectric breakdown occurs.* Consider a parallel-plate capacitor made from two sheets of metal foil that are separated by a thin plastic sheet.

* Recall from Chapter 23 that for electric fields greater than about 3×10^6 V/m, air breaks down; that is, it becomes ionized and begins to conduct.

The plastic sheet allows the metal sheets to be very close together without actually being in electrical contact, and because the dielectric strength of plastic is greater than that of air, a greater potential difference can be attained before dielectric breakdown occurs. Table 24-1 lists the dielectric constants and dielectric strengths of some dielectrics. Note that for air $\kappa \approx 1$; so, for most situations we do not need to distinguish between air and a vacuum.

Table 24-1 Dielectric Constants and Dielectric Strengths of Various Materials

Material	Dielectric Constant κ	Dielectric Strength, kV/mm
Air	1.00059	3
Bakelite	4.9	24
Gasoline	2.0 (70°F)	
Glass (Pyrex)	5.6	14
Mica	5.4	10–100
Neoprene	6.9	12
Paper	3.7	16
Paraffin	2.1–2.5	10
Plexiglas	3.4	40
Polystyrene	2.55	24
Porcelain	7	5.7
Strontium titanate	240	8
Transformer oil	2.24	12

Example 24-10 Using a Dielectric in a Parallel-Plate Capacitor

A parallel-plate capacitor has square plates of edge length 10 cm and a separation of $d = 4.0$ mm. A dielectric slab of constant $\kappa = 2.0$ has dimensions 10 cm × 10 cm × 4.0 mm. (*a*) What is the capacitance without the dielectric? (*b*) What is the capacitance if the dielectric slab fills the space between the plates? (*c*) What is the capacitance if a dielectric slab that has dimensions 10 cm × 10 cm × 3.0 mm is inserted into the 4.0-mm gap?

PICTURE The capacitance without the dielectric, C_0, is found from the area and spacing of the plates (Figure 24-24*a*). When the capacitor is filled with a slab of dielectric constant κ (Figure 24-24*b*), the capacitance is $C = \kappa C_0$ (Equation 24-19). If the dielectric only partially fills the capacitor (Figure 24-24*c*), we isolate the capacitor and calculate the potential difference V with a given charge Q_0, then apply the definition of capacitance, $C = Q/V$.

FIGURE 24-24

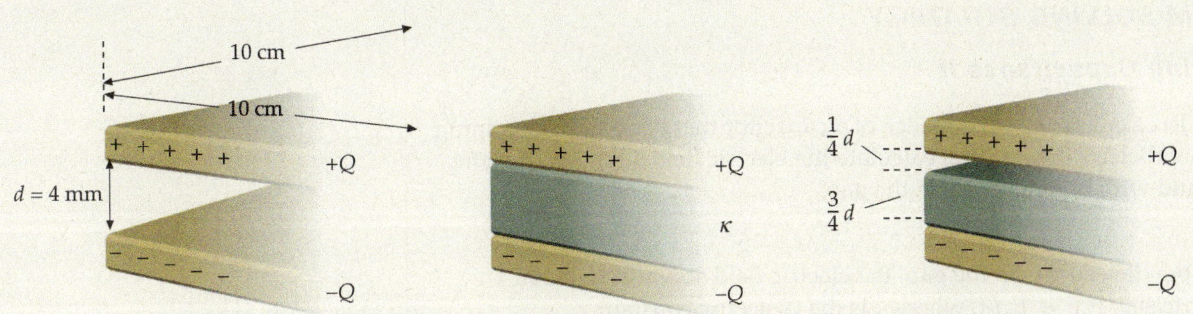

SOLVE

(*a*) If there is no dielectric, the capacitance C_0 is given by Equation 24-6:

$$C_0 = \frac{\epsilon_0 A}{d} = \frac{(8.85\ \text{pF/m})(0.10\ \text{m})^2}{0.0040\ \text{m}} = 22.1\ \text{pF} = \boxed{22\ \text{pF}}$$

(*b*) When the capacitor is filled with a material that has dielectric constant κ, its capacitance C is increased by the factor κ:

$$C = \kappa C_0 = (2.0)(22.1\ \text{pF}) = 44.2\ \text{pF} = \boxed{44\ \text{pF}}$$

(*c*) 1. We keep the capacitor electrically isolated, so the charge remains constant when dielectric slabs are inserted or removed. The capacitance is related to the charge Q_0 and the new potential difference V:

$$C = \frac{Q_0}{V}$$

2. When the 3.0-mm-thick slab is in place, the potential difference V across the entire gap is the potential difference across the empty portion of the gap plus the potential difference across dielectric slab:

$$V = V_{\text{gap}} + V_{\text{slab}} = E_{\text{gap}}(\tfrac{1}{4}d) + E_{\text{slab}}(\tfrac{3}{4}d)$$

3. The field strength E_{gap} in the empty gap is σ_0/ϵ_0, where $\sigma_0 = Q_0/A$. This is the same as the field strength E_0 when no dielectric is between the plates:

$$E_{\text{gap}} = E_0 = \frac{\sigma_0}{\epsilon_0} = \frac{Q_0}{\epsilon_0 A}$$

4. The field in the dielectric slab is reduced by the factor κ^{-1}:

$$E_{\text{slab}} = \frac{E_0}{\kappa}$$

5. Substituting the results of the previous two steps into the step 2 result yields V in terms of κ. Note that the potential difference when no dielectric is between the plates is $V_0 = E_0 d$:

$$V = E_0 d_{\text{gap}} + E_{\text{slab}} d_{\text{slab}} = E_0\left(\frac{1}{4}d\right) + \frac{E_0}{\kappa}\left(\frac{3}{4}d\right)$$

$$= E_0 d\left(\frac{1}{4} + \frac{3}{4\kappa}\right) = V_0\left(\frac{\kappa + 3}{4\kappa}\right)$$

6. Using $C = Q_0/V$, we find the new capacitance in terms of the original capacitance, $C_0 = Q_0/V_0$:

$$C = \frac{Q_0}{V} = \frac{Q_0}{V_0 \dfrac{\kappa + 3}{4\kappa}} = \frac{Q_0}{V_0}\left(\frac{4\kappa}{\kappa + 3}\right) = C_0\left(\frac{4\kappa}{\kappa + 3}\right)$$

$$= (22.1\ \text{pF})\left(\frac{4 \cdot 2.0}{2.0 + 3}\right) = \boxed{35\ \text{pF}}$$

CHECK The absence of a dielectric corresponds to $\kappa = 1$. Substituting 1 for κ in the final step in Part (*c*) would result $C = C_0$, as expected. Suppose that the dielectric slab were a conducting slab instead of a dielectric. In a conductor, $E = 0$; so, according to $E = E_0/\kappa$ (Equation 24-18), κ for a conductor would equal infinity. As κ approaches infinity, the quantity $4\kappa/(\kappa + 3)$ approaches 4, so the result for the final step in Part (*c*) approaches $4C_0$. A conducting slab simply reduces the plate separation by the thickness of the slab. The plate separation when the conducting slab is in place would be $\frac{1}{4}d$. According to Equation 24-20 ($C = \kappa\epsilon_0 A/d$), C should be $4C_0$, as it is for very large values of κ.

TAKING IT FURTHER The results of this example are independent of the vertical position of the dielectric (or conducting) slab in the space between the plates.

PROBLEM-SOLVING STRATEGY

Calculating Capacitance II

PICTURE To calculate the capacitance of a capacitor that has a gap containing two or more dielectric slabs, first calculate the electric field strength E_0 using charge Q and with no dielectrics in the gap.

SOLVE

1. When the dielectric is in the gap, the electric field strength within a dielectric slab is $E = E_0/\kappa$, where κ is the dielectric constant.

2. Use $\vec{E}$ within a dielectric slab to calculate the voltage V_{slab} across the slab. The voltage V across the entire gap is the sum of the voltages across the individual slabs in the gap plus the sum of the voltages across any empty regions of the gap.
3. Then, calculate C using $C = Q/V$.

CHECK Evaluate your expression for C by setting κ equal to 1. Then compare your result with the expression for C_0 (the capacitance without a dielectric present).

Example 24-11 A Homemade Capacitor

Context-Rich

When studying capacitors in physics class, your professor claims that you could build a parallel-plate capacitor from waxed paper and aluminum foil. You decide to build one about the size of a piece of notebook paper. Before testing its charge-storing power on your gullible roommate, you decide to calculate the amount of charge the capacitor will store when connected to a 9.0-V battery.

PICTURE We want charge, which we can get from the definition $C = Q/V$ if we know the capacitance. We can get the capacitance from the parallel-plate capacitor formula $C = \epsilon_0 A/d$. We will need to either measure or estimate the thickness of the waxed paper.

SOLVE

1. The charge on a capacitor is related to the voltage and the capacitance by the definition of capacitance:

$$Q = CV$$

2. The capacitance is obtained from the parallel-plate capacitance formula:

$$C = \frac{\kappa\epsilon_0 A}{d}$$

3. Substituting for C and solving for Q give:

$$Q = CV = \frac{\kappa\epsilon_0 VA}{d}$$

4. A sheet of notebook paper is approximately 8.5 by 11 in:

$$A = 8.5\text{ in} \times 11\text{ in} = 93.5\text{ in}^2 = 0.0603\text{ m}^2$$

5. We assume a sheet of wax paper is the same thickness as a sheet of the paper your physics textbook is made of. Measure the thickness of 300 sheets of paper in a book (from page 1 through page 600):

The 300 sheets of paper are 2.0 cm (0.020 m) thick. So, the thickness of a single sheet of paper is $0.020\text{ m}/300 = 66.7\ \mu\text{m}$.

6. Using the step 3 result, solve for the charge. Assume the dielectric constant of wax paper is 2.3 (the same as that of paraffin):

$$Q = \frac{\kappa\epsilon_0 AV}{d} = \frac{2.3\,(8.85\text{ pF/m})(0.0603\text{ m}^2)(9.0\text{ V})}{66.7 \times 10^{-6}\text{ m}}$$
$$= 1.66 \times 10^5\text{ pC} = \boxed{0.17\ \mu\text{C}}$$

CHECK A farad is a coulomb per volt, so the units do reduce to coulombs.

ENERGY STORED IN THE PRESENCE OF A DIELECTRIC

The energy stored in a parallel-plate capacitor that has a dielectric is

$$U = \tfrac{1}{2}QV = \tfrac{1}{2}CV^2$$

We can express the capacitance C in terms of the area and separation of the plates, and the voltage difference V in terms of the electric field and plate separation, to obtain

$$U = \frac{1}{2}CV^2 = \frac{1}{2}\left(\frac{\epsilon A}{d}\right)(Ed)^2 = \frac{1}{2}\epsilon E^2(Ad)$$

The quantity Ad is the volume of the region where there is an electric field. (This is the region between the two plates.) The energy per unit volume is thus

$$u_e = \tfrac{1}{2}\epsilon E^2 = \tfrac{1}{2}\kappa\epsilon_0 E^2 \qquad \text{24-22}$$

Part of this energy is the energy associated with the electric field (Equation 24-9) and the rest is the energy associated with mechanical stress associated with the polarization of the dielectric (discussed in Section 24-5).

Example 24-12 Inserting the Dielectric—Battery Disconnected

A parallel combination of two parallel-plate air-gap capacitors, each having a capacitance of 2.00 μF, are connected in parallel across a 12.0-V battery. The battery is disconnected from the parallel combination, and then a slab that has a dielectric constant $\kappa = 2.50$ is inserted between the plates of one of the capacitors, completely filling the gap. Before the dielectric slab is inserted, find (*a*) the charge on and energy stored in each capacitor, and (*b*) the total energy stored in the capacitors. After the dielectric is inserted, find (*c*) the potential difference across each capacitor, (*d*) the charge on each capacitor, and (*e*) the total energy stored in the capacitors.

PICTURE The capacitors are connected in parallel, so the voltage across each is the same. The charge Q and total energy U can be found for each capacitor from its capacitance C and voltage V. After the capacitors are removed from the battery, the total charge on the pair remains the same. When the dielectric is inserted into one of the capacitors, its capacitance changes. The potential across the parallel combination can be found from the total charge and the equivalent capacitance.

SOLVE

(*a*) The charge on each capacitor is found from its capacitance C and voltage $V = 12.0$ V:

$$Q = CV = (2.00\ \mu\text{F})(12.0\ \text{V}) = \boxed{24.0\ \mu\text{C}}$$

(*b*) 1. The energy stored in each capacitor is found from its charge Q and its voltage V:

$$U = \tfrac{1}{2}QV = \tfrac{1}{2}(24.0\ \mu\text{C})(12.0\ \text{V}) = 144\ \mu\text{J}$$

2. The total potential energy is twice that stored in each capacitor:

$$U_{\text{total}} = 2U = \boxed{288\ \mu\text{J}}$$

(*c*) 1. The potential across the parallel combination is related to the total charge Q_{total} and the equivalent capacitance C_{eq}:

$$V = \frac{Q_{\text{total}}}{C_{\text{eq}}}$$

2. The capacitance of the capacitor that has the dielectric is increased by the factor κ. The equivalent capacitance is the sum of the capacitances:

$$C_{\text{eq}} = C_1 + C_2 = C_1 + \kappa C_2 = (2.00\ \mu\text{F}) + 2.50(2.00\ \mu\text{F})$$
$$= 2.00\ \mu\text{F} + 5.00\ \mu\text{F} = 7.00\ \mu\text{F}$$

3. The total charge remains 48.0 μC. Substitute for Q_{total} and C_{eq} to calculate V:

$$V = \frac{Q_{\text{total}}}{C_{\text{eq}}} = \frac{48.0\ \mu\text{C}}{7.00\ \mu\text{F}} = \boxed{6.86\ \text{V}}$$

(*d*) The charge on each capacitor is again obtained from $Q = CV$:

$$Q_1 = (2.00\ \mu\text{F})(6.86\ \text{V}) = \boxed{13.7\ \mu\text{C}}$$
$$Q_2 = (5.00\ \mu\text{F})(6.86\ \text{V}) = \boxed{34.3\ \mu\text{C}}$$

(*e*) The potential energy stored in each capacitor is found from its new charge and new voltage:

$$U = U_1 + U_2 = \tfrac{1}{2}Q_1V + \tfrac{1}{2}Q_2V = \tfrac{1}{2}(Q_1 + Q_2)V$$
$$= \tfrac{1}{2}(13.7\ \mu\text{C} + 34.3\ \mu\text{C})(6.86\ \text{V}) = \boxed{165\ \mu\text{J}}$$

CHECK When the dielectric is inserted into one of the capacitors, the electric field is weakened so the potential difference across it is lowered. Because the two capacitors are connected in parallel, charge must flow from the other capacitor so that the potential difference is the same across both capacitors. Note that the capacitor that has the dielectric has the greater charge, and that when the charges calculated for each capacitor in Part (*d*) are added, $Q_1 + Q_2 = 13.7\ \mu\text{C} + 34.3\ \mu\text{C} = 48.0\ \mu\text{C}$, the result is the same as the original net charge.

TAKING IT FURTHER The total energy of 165 μJ is 123 μJ less than the original energy of 288 μJ. When the dielectric is inserted, it is attracted by the charges on the plates, so it must be restrained from accelerating into the gap. During this process $-123\ \mu$J of work ($165\ \mu\text{J} + 123\ \mu\text{J} = 288\ \mu\text{J}$) is done on the dielectric by the forces of restraint. To remove the dielectric from the gap, $+123\ \mu$J must be done on it, and this work is stored as potential energy in the capacitors.

Example 24-13 Running Out of Fuel

Context-Rich

You are flying from New Zealand to Hawaii when the electronic components of the fuel gauge on the instrument panel of the small plane you are in start to combust. Your companion becomes very concerned and asks you to try to find a solution for the problem. The gauge consists of an air-gap cylindrical capacitor in the fuel tank (Figure 24-25). The axis of this capacitor is vertical, and fuel fills the gap up to the level of the fuel in the tank. Can you find a way make the gauge work? You had noticed the tank was half full when the gauge broke. In addition, a handheld multimeter capable of measuring capacitance (Figure 24-26) is on board.

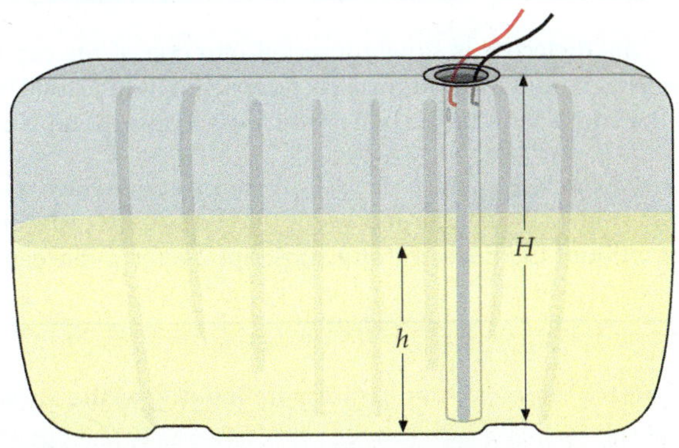

FIGURE 24-25 The cylinder is one of the plates of the capacitor. A coaxial rod is the other plate. The top of the cylinder is to be capped with the two wires feeding through the cap. This way no fuel can exit the tank through the top of the cylinder. One of the wires is attached to the cylinder and the second wire is attached to the rod. There needs to be a small hole though the side of the cylinder—near the top. This hole prevents pressure from building up as the tank is filled.

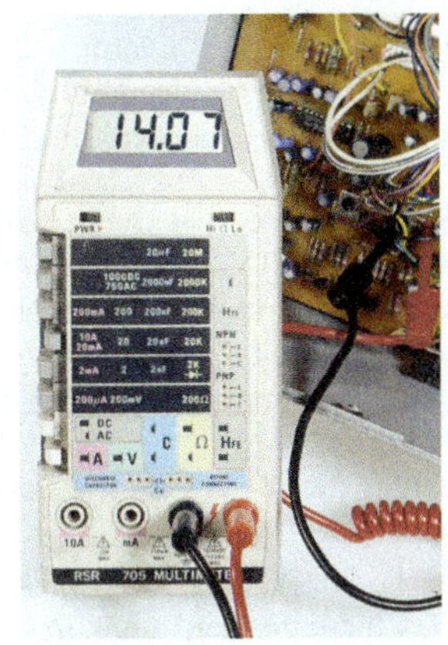

FIGURE 24-26 *(Paul Silverman/ Fundamental Photographs.)*

PICTURE The cylindrical capacitor can be modeled as a parallel combination of two capacitors, with the submerged portion as one of the capacitors and the portion above the fuel as the other. The ratio of the length of the submerged portion to the entire length is the desired reading.

SOLVE

1. Disconnect the two wires from the fuel tank at the instrument panel, connect them to the multimeter, and measure the capacitance so you know the reading $C_{1/2}$ when the tank is half full:

$$C = C_{1/2}$$

2. Model the capacitor as a parallel combination of two capacitors, one submerged and one not, and make a schematic diagram of the combination. Label the capacitances C_1 and C_2, where C_2 is the capacitance of the submerged portion.

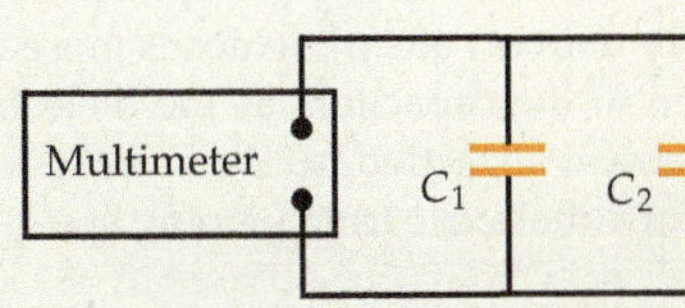

3. The capacitance of a cylindrical capacitor is proportional to its length. Let H be the height of the tank (and the length of the capacitor) and let h be the height of the fuel. The capacitance of the capacitor is C_0 when the tank is empty:

$$C_1 = \frac{H-h}{H}C_0 \quad \text{and} \quad C_2 = \frac{h}{H}\kappa C_0$$

4. The equivalent capacitance C is the sum of the capacitances:

$$C = C_1 + C_2 = \frac{H-h}{H}C_0 + \frac{h}{H}\kappa C_0 = \left[1 + (\kappa - 1)\frac{h}{H}\right]C_0$$

5. Look up the dielectric constant of gasoline in Table 24-1. (Lucky you had your physics book with you):

$$\kappa = 2.0$$

$$C = \left[1 + (2.0 - 1)\frac{h}{H}\right]C_0 = \left[1 + 1.0\frac{h}{H}\right]C_0$$

6. Just before the gauge burned the tank was half full. Set $C = C_{1/2}$ and $h = h/H$ and solve for C_0:

$$C_{1/2} = \left[1 + 1.0\frac{1}{2}\right]C_0 \Rightarrow C_0 = \frac{2}{3}C_{1/2}$$

7. Substitute for C_0 in the step 4 result, and then solve for h/H. You now have a formula to convert the readings of the meter C into the fraction of the fuel remaining:

$$C = \left[1 + 1.0\frac{h}{H}\right]\frac{2}{3}C_{1/2} \quad \text{so} \quad \boxed{\frac{h}{H} = \frac{3}{2}\frac{C}{C_{1/2}} - 1}$$

CHECK Substituting $C_{1/2}$ for C in our step 7 result gives $h/H = \frac{1}{2}$, as expected. In addition, substituting zero for h, C_0 for C, and solving for C_0 gives $C_0 = \frac{2}{3}C_{1/2}$, which is the expression obtained for C_0 in step 6.

TAKING IT FURTHER Because fuel tanks are not of uniform height, this fuel gauge will not be very accurate. This is the case with many automotive fuel gauges.

Example 24-14 Inserting the Dielectric—Battery Connected

Try It Yourself

For the circuit of Example 24-12, the dielectric is slowly inserted into one of the capacitors while the battery remains connected. Find (*a*) the charge on each capacitor, (*b*) the total energy stored in the capacitors, and (*c*) the work done by the battery during the insertion process.

PICTURE Because the battery is still connected, the potential difference across the capacitors remains 12.0 V. This condition determines the charge and energy stored in each capacitor. Let subscript 1 refer to the capacitor without the dielectric and subscript 2 refer to the capacitor with the dielectric.

SOLVE

Cover the column to the right and try these on your own before looking at the answers.

Steps	Answers
(*a*) Calculate the charge on each capacitor from $Q = CV$ using the result that $C_1 = 2.00\ \mu\text{F}$ and $C_2 = 5.00\ \mu\text{F}$ as found in Example 24-12.	$Q_1 = C_1V = \boxed{24.0\ \mu\text{C}}$ $Q_2 = C_2V = \boxed{60.0\ \mu\text{C}}$
(*b*) 1. Calculate the energy stored in each capacitor from $U = \frac{1}{2}CV^2$. (Check your results by using $U = \frac{1}{2}QV$.)	$U_1 = 144\ \mu\text{J}$ $U_2 = 360\ \mu\text{J}$
2. Add your results for U_1 and U_2 to obtain the final energy.	$U_{\text{total}} = \boxed{504\ \mu\text{J}}$
(*c*) The work done by the battery during the insertion process is the battery voltage multiplied by the charge passing through the battery. This charge is the increase of the charge on C_2.	$W = V\Delta Q = (12.0\ \text{V})(60.0\ \mu\text{C} - 24.0\ \mu\text{C}) = \boxed{432\ \mu\text{J}}$

CHECK The total energy of the two capacitors is larger when the dielectric is in place by $504\ \mu\text{J} - 288\ \mu\text{J} = 216\ \mu\text{J}$ than when the dielectric is not in place. This result is expected because during the insertion the battery delivers $432\ \mu\text{J}$, which is more than enough to account for the increase in the energy stored in the capacitors as the dielectric is inserted. (The dielectric is pulled in by forces of electrical attraction, so work must be done on the dielectric by the restraining forces to prevent the dielectric from gaining speed during the insertion.)

CONCEPT CHECK 24-4

Does the work done by the restraining forces to prevent the dielectric from gaining speed during insertion have a positive value or a negative value?

24-5 MOLECULAR VIEW OF A DIELECTRIC

A dielectric weakens the electric field between the plates of a capacitor. This happens because the polarized molecules of the dielectric produce an electric field within the dielectric in a direction opposite to the field produced by the charges on the plates. The electric field produced by the dielectric is due to the electric dipole moments of the molecules of the dielectric.

Although atoms and molecules are neutral, they are affected by electric fields because they contain positive and negative charges that can individually respond to external fields. We can think of an atom as a very small, positively charged nucleus surrounded by a negatively charged electron cloud. In some atoms and molecules, the charge configuration is sufficiently symmetric so that the "center of negative charge" coincides with the center of positive charge. An atom or molecule that has this symmetry has zero dipole moment and is said to be nonpolar. In the presence of an external electric field, however, the positive and negative charges experience forces in opposite directions, so the positive and negative charges then separate until the attractive force they exert on each other balances the forces due to the external electric field (Figure 24-27). The molecule is then said to be polarized and it behaves like the electric dipole it is.

In some molecules (for example, HCl and H_2O), the centers of positive and negative charge do not coincide, even in the absence of an external electric field. As we noted in Chapter 21, these polar molecules have a permanent electric dipole moment.

When a dielectric is placed in the field of a charged capacitor, its molecules are polarized in such a way that there is a net dipole moment parallel to the field. If the molecules are polar, their dipole moments, originally oriented at random, tend to become aligned due to the torque exerted by the field.* If the molecules are nonpolar, the field induces dipole moments that are parallel to the field. In either case, the molecules in the dielectric are polarized in the direction of the external field (Figure 24-28).

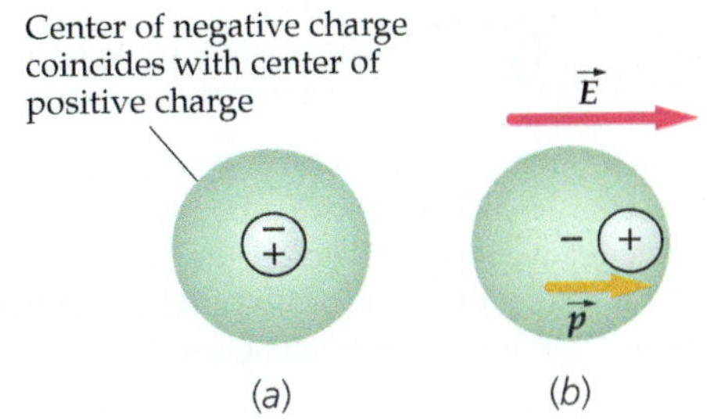

FIGURE 24-27 Schematic diagrams of the charge distributions of an atom or nonpolar molecule. (*a*) In the absence of an external electric field, the center of positive charge coincides with the center of negative charge. (*b*) In the presence of an external electric field, the centers of positive and negative charge are displaced, producing an induced dipole moment in the direction of the external field.

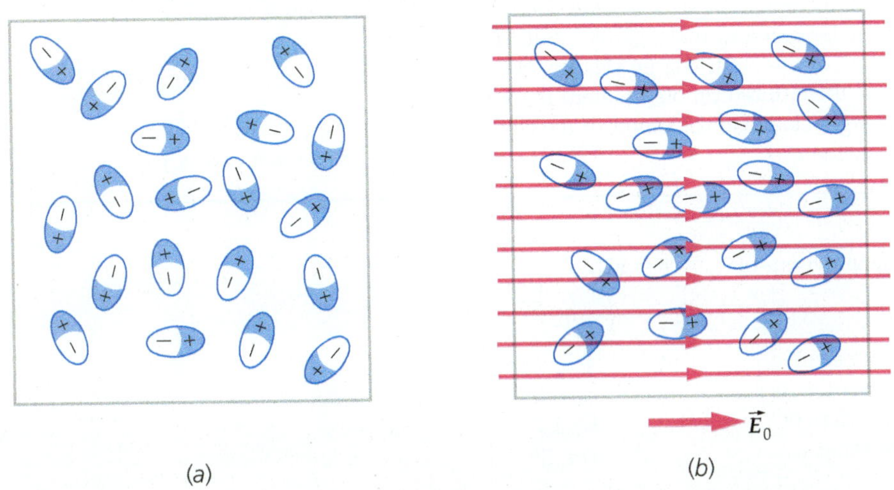

FIGURE 24-28 (*a*) The randomly oriented electric dipoles of a polar dielectric in the absence of an external electric field. (*b*) In the presence of an external electric field, the dipoles are partially aligned parallel to the field.

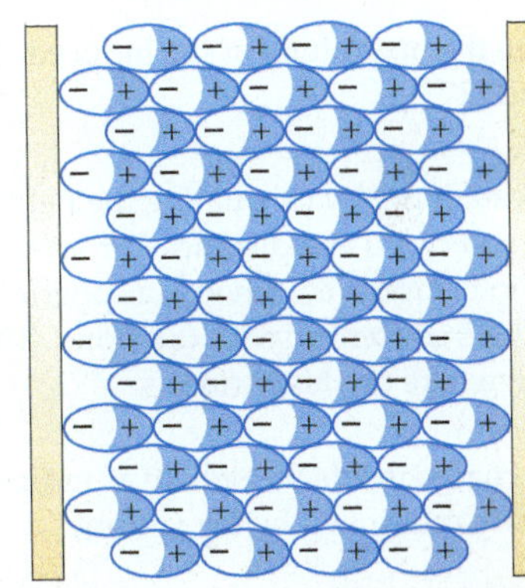

FIGURE 24-29 When a dielectric is placed between the plates of a capacitor, the electric field of the capacitor polarizes the molecules of the dielectric. The result is a bound charge on the surface of the dielectric that produces its own electric field; this field opposes the external field. The field of the bound surface charges thus weakens the electric field within the dielectric.

The net effect of the polarization of a homogeneous dielectric in a parallel-plate capacitor is the creation of surface charges on the dielectric faces near the plates, as shown in Figure 24-29. The surface charge on the dielectric is called a **bound charge,** because the surface charge is bound to the surface molecules of the dielectric and cannot move about like the free charge on the conducting capacitor plates. This bound charge produces an electric field opposite in direction to the electric field produced by the free charge on the conductors. Thus, the net electric field between the plates is reduced, as illustrated in Figure 24-30.

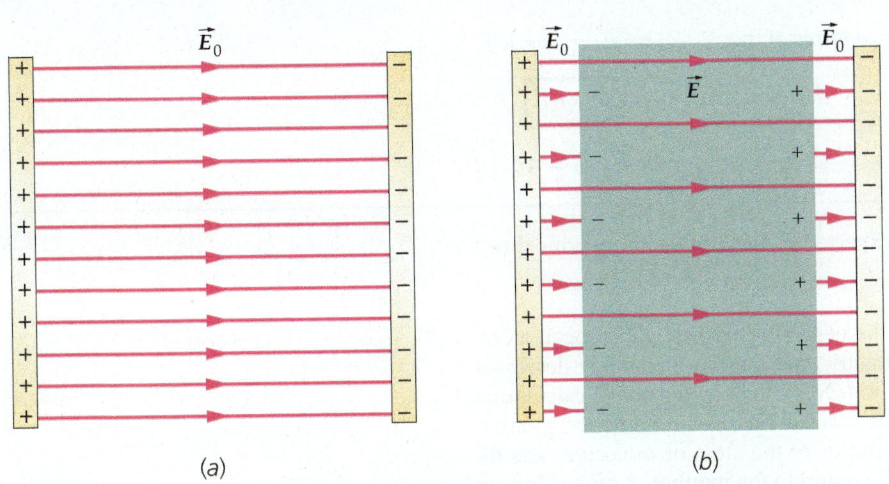

FIGURE 24-30 The electric field between the plates of a capacitor that has (*a*) no dielectric and (*b*) a dielectric. The surface charge on the dielectric weakens the original field between the plates.

* The degree of alignment depends on the external field and on the temperature. It is approximately proportional to $pE/(kT)$, where pE is the maximum energy of a dipole in a field E, and kT is the characteristic thermal energy.

Example 24-15 Induced Dipole Moment—Hydrogen Atom

A hydrogen atom consists of a proton of charge $+e$ and an electron of charge $-e$. The charge distribution of the atom is spherically symmetric, so the atom is nonpolar. Consider a model in which the hydrogen atom consists of a positive point charge $+e$ at the center of a uniformly charged spherical cloud of radius R and total charge $-e$. Show that when such an atom is placed in a uniform external electric field $\vec{E}$, the induced dipole moment is proportional to $\vec{E}$; that is, $\vec{p} = \alpha\vec{E}$, where α is called the *polarizability*.

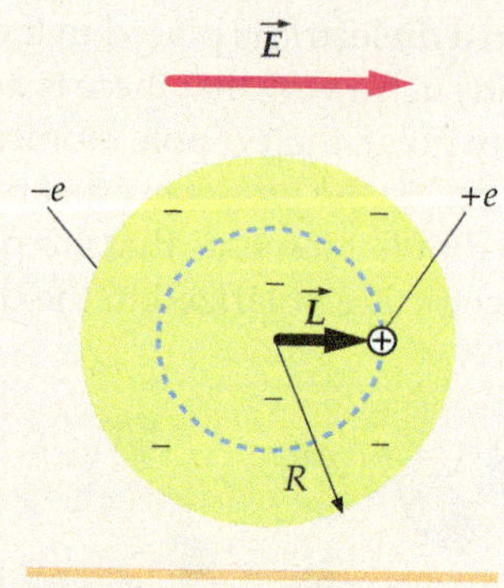

FIGURE 24-31

PICTURE In the external field, the center of the uniform negative cloud is displaced from the positive charge by an amount L so that the force exerted by the field $e\vec{E}$ on the positive point charge is balanced by the force on it exerted by the negative cloud $e\vec{E}'$, where $\vec{E}'$ is the field due to the cloud at the location of the point charge (Figure 24-31). We use Gauss's law to find E', and then we calculate the induced dipole moment $\vec{p} = q\vec{L}$, where $q = e$ and $\vec{L}$ is the position of the positive charge relative to the center of the cloud. The dipole moment, defined as $q\vec{L}$, is discussed in Section 21-4.

SOLVE

1. Write the magnitude of the induced dipole moment in terms of e and L:

$$p = eL$$

2. We can find L by calculating the field E'_n due to the negatively charged cloud at a distance L from the center. We use Gauss's law to compute E'_n. Choose a spherical Gaussian surface of radius L concentric with the cloud. Then E'_n is the same everywhere on this surface:

$$\phi_{\text{net}} = \oint E_n\, dA = \frac{Q_{\text{inside}}}{\epsilon_0}$$

$$E'_n = \frac{Q_{\text{inside}}}{4\pi\epsilon_0 L^2}$$

3. The charge inside the sphere of radius L equals the charge density multiplied by the volume:

$$Q_{\text{inside}} = \rho\tfrac{4}{3}\pi L^3 = \frac{-e}{\tfrac{4}{3}\pi R^3}\tfrac{4}{3}\pi L^3 = -e\frac{L^3}{R^3}$$

4. Substitute this value of Q_{inside} to calculate E'_n:

$$E'_n = \frac{Q_{\text{inside}}}{4\pi\epsilon_0 L^2} = \frac{-eL^3/R^3}{4\pi\epsilon_0 L^2} = -\frac{e}{4\pi\epsilon_0 R^3}L$$

5. Solve for L:

$$L = -\frac{4\pi\epsilon_0 R^3}{e}E'_n$$

6. E'_n is negative because $\vec{E}'$ points inward on the Gaussian surface. At the positive charge, $\vec{E}'$ points to the left. Because E' is equal to E, we conclude that $E'_n = -E$:

$$E'_n = -E \quad \text{so} \quad L = \frac{4\pi\epsilon_0 R^3}{e}E$$

7. Substitute these results for L and E'_n to express p in terms of the external field strength E:

$$p = eL = 4\pi\epsilon_0 R^3 E$$

so $\boxed{\vec{p} = \alpha\vec{E}}$

where $\alpha = 4\pi\epsilon_0 R^3$

CHECK We expected α to be positive because we expected $\vec{p}$ and $\vec{E}$ to be in the same direction. Our step 7 result met this expectation.

TAKING IT FURTHER The charge distribution of the negative charge in a hydrogen atom, obtained from quantum theory, is spherically symmetric, but the charge density decreases exponentially with distance rather than being uniform. Nevertheless, the above calculation shows that the dipole moment is proportional to the external field $p = \alpha E$, and the polarizability α is of the order of $4\pi\epsilon_0 R^3$, where R is the radius of the atom or molecule. The dielectric constant κ can be related to the polarizability α and to the number of molecules per unit volume.

MAGNITUDE OF THE BOUND CHARGE

The bound charge density σ_b on the surfaces of the dielectric is related to the dielectric constant κ and to the free charge density σ_f on the surfaces of the plates. Consider a dielectric slab between the plates of a parallel-plate capacitor, as shown in Figure 24-32. If the dielectric is a thin slab between plates that are close together, the electric field inside the dielectric slab due to the bound charge densities, $+\sigma_b$ on the right and $-\sigma_b$ on the left, is just the field due to two infinite-plane charge densities. Thus, the field E_b has the magnitude

$$E_b = \frac{\sigma_b}{\epsilon_0}$$

This field is directed to the left and subtracts from the electric field E_0 due to the free charge density on the capacitor plates, which has the magnitude

$$E_0 = \frac{\sigma_f}{\epsilon_0}$$

The strength of the net field $E = E_0/\kappa$ is the difference between these magnitudes:

$$E = E_0 - E_b = \frac{E_0}{\kappa}$$

or

$$E_b = \left(1 - \frac{1}{\kappa}\right)E_0$$

Writing σ_b/ϵ_0 for E_b and σ_f/ϵ_0 for E_0, we obtain

$$\sigma_b = \left(1 - \frac{1}{\kappa}\right)\sigma_f \qquad \text{24-23}$$

The bound charge density σ_b is always less than or equal to the free charge density σ_f on the capacitor plates, and it is zero if $\kappa = 1$, which is the case when there is no dielectric. For a conducting slab, $\kappa = \infty$ and $\sigma_b = \sigma_f$.

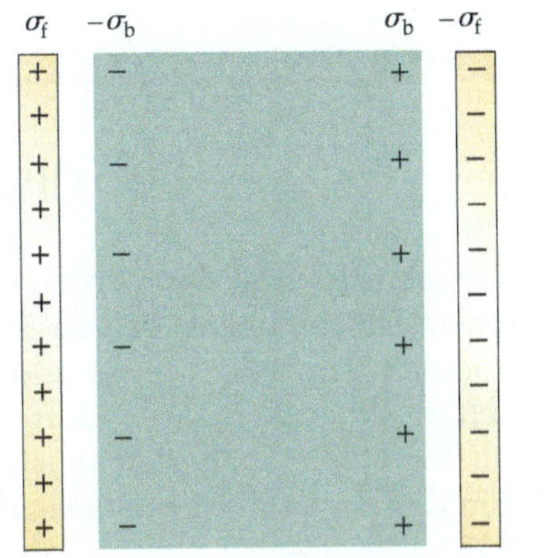

FIGURE 24-32 A parallel-plate capacitor with a dielectric slab between the plates. If the plates are closely spaced, each of the surface charges can be considered as an infinite plane charge. The electric field due to the free charge on the plates is directed to the right and has a magnitude $E_0 = \sigma_f/\epsilon_0$. That due to the bound charge is directed to the left and has a magnitude $E_b = \sigma_b/\epsilon_0$.

CONCEPT CHECK 24-5

Does the capacitance always increase when a dielectric is inserted into the gap of a capacitor? Explain your answer.

*THE PIEZOELECTRIC AND PYROELECTRIC EFFECTS

In certain crystals that have polar molecules (for example, quartz, tourmaline, and topaz), a mechanical stress applied to the crystal produces polarization of the molecules. This is known as the **piezoelectric effect.** The polarization of the stressed crystal causes a potential difference across the crystal, which can be used to produce an electric current. Piezoelectric crystals are used in transducers (for example, microphones, phonograph pickups, and vibration-sensing devices) to convert mechanical strain into electrical signals. The converse piezoelectric effect, in which a voltage applied to such a crystal induces mechanical strain (deformation), is used in headphones and many other devices. Because the natural frequency of vibration of quartz is in the range of radio frequencies, and because its resonance curve is very sharp,* quartz is used extensively to stabilize radio-frequency oscillators and to make accurate clocks.

Many crystals that exhibit the piezoelectric effect also exhibit the **pyroelectric effect,** which is the generation of a large electric field within the crystal when the temperature of the crystal is increased. Pyroelectric crystals are sometimes used to accelerate charged particles to such high speeds that X rays, and even nuclear fusion, result when the charged particles impact a target material.

* Resonance in AC circuits, which will be discussed in Chapter 29, is analogous to mechanical resonance, which was discussed in Chapter 14.

Physics Spotlight

Changes in Capacitors—Charging Ahead

In 1746, shortly after the Leyden jar was publicized, 180 soldiers demonstrated the power of a large Leyden jar to the French court. They joined hands in a circle and waited to be connected to the Leyden jar. When a single shock from the jar passed through the circle, all the soldiers simultaneously jumped and shouted.[*,†] Some Leyden jars have since been measured with a capacitance of 2.5 nF at 10 kV.

Capacitors have come a long way since then. One change (of many) that was made during the nineteenth century was the addition of mineral oil to capacitors as a dielectric. However, oil-filled condensers, as they were known, are fire hazards when heated. In 1929, the Swann chemical company produced polychlorinated biphenyls, or PCBs, for use as dielectrics in industrial capacitors.[‡] PCBs are resistant to burning and do not easily react with other substances. They also have dielectric constants slightly larger than mineral oil's. Unfortunately, PCBs proved to be carcinogenic, and they are extremely toxic when partially burned.[#] In 1979, PCB manufacture was banned in the United States, and the use of PCBs as dielectrics for capacitors[°] was discontinued. (Numerous older capacitors using PCBs are still in service as of 2006.[§]) The ban on PCBs in new capacitors caused researchers to try to develop more efficient capacitors. (In this context, efficient usually means with greater capacitance per unit mass.)

Capacitor come in many different sizes and shapes as well as several different types. Circuit designers choose the size, shape and type to suit the requirements for specific circumstances. *(Maynard & Bouchard/Scientifica/Visuals Unlimited.)*

Several types of very efficient capacitors are available. Many capacitors now take advantage of the large dielectric coefficients of specialized ceramics,[¶] plastic films, and polymer gels. But the most efficient capacitors are electrical double-layer capacitors (EDLCs). EDLCs are composed of electrodes that are made of porous carbon deposited on either side of an *electrolyte separator*. The layers are tightly wound and placed into a container. The carbon and the electrolyte separator are so thin that the distance between the layers of carbon is molecules thick.[**] The capacitors are called double layered because each layer of electrolyte has two layers of charge.

Owing to the porous nature of the carbon, each layer has a very large surface area for the carbon to be in contact with the electrolyte—from 400 up to 2000 m^2/g. This large surface area combined with the very thin electrolyte layer yields a large capacitance. Because the electrolyte layers are very thin, most electrical double-layer capacitors have low breakdown voltages. An EDLC the size of a D-cell battery weighs 60 grams, has a capacitance of 350 farads, and is rated at 2.5 volts.[††] Because of the low breakdown voltage, EDLCs are rarely used individually. A package of six of the D-cell–sized capacitors in series has an equivalent capacitance of 58 farads with a rating of 15 volts.[‡‡]

EDLCs are already incorporated into cell phones, cameras, and automobiles. For frequently used rechargeable items, EDLCs may soon be inexpensive and powerful enough to be used instead of batteries.

* Dray, P., *Stealing God's Thunder: Benjamin Franklin's Lightning Rod and the Invention of America.* New York: Random House, 2005, pp. 45–46.
† Cohen, I. B., *Benjamin Franklin's Science.* Cambridge: Harvard University Press, 1990, pp. 4–37.
‡ *History of PCB Manufacturing in Anniston.* 2000. Solutia http://www.solutia.com/pages/anniston/pcbhistory.asp As of Sept. 2006.
\# Lloyd, R. J. W., et al., *Current Intelligence Bulletin 7—Polychlorinated Biphenyls (PCBs).* Washington, D.C.: Centers for Disease Control, Nov. 3, 1975. http://www.cdc.gov/niosh/78127_7.html As of Sept. 2006.
° *EPA Bans PCB Manufacture; Phases Out Uses.* United States Environmental Protection Agency, Apr. 19, 1979. http://www.epa.gov/history/topics/pcbs/01.htm As of Sept. 2006.
§ *Brookhaven National Laboratory Reduces Mercury and PCBs.* United States Environmental Protection Agency, http://www.epa.gov/epaoswer/hazwaste/minimize/brookhav.htm As of Sept. 2006.
¶ Chen, L., et al., "Migration and Redistribution of Oxygen Vacancy in Barium Titanate Ceramics." *Applied Physics Letters,* Aug. 14, 2006, Vol. 89, No. 7, Letter 071916.
** Prophet, G., "Supercaps for Supercaches." *Electronic Design News,* Jan. 9, 2003, pp. 53–58.
†† Blankenship, S., "It Looks Like a Battery, but It's an Ultracapacitor." *Power Engineering,* May 2004, pp. 64–65.
‡‡ Everett, M., "Ultracapacitors Turn Malibus into Mercedes." *Machine Design,* Dec. 8, 2005, pp. 82–88.

Summary

1. Capacitance is an important defined quantity that relates charge to potential difference.
2. Two devices connected in *parallel* share a common potential difference across each device *due solely to the way they are connected.*
3. Two devices connected in *series* are connected by a conducting path *that contains no junctions.*
4. The changes in potential around any closed path *always* sum to zero. This is known as Kirchhoff's loop rule

TOPIC	RELEVANT EQUATIONS AND REMARKS	
1. Capacitor	A capacitor is a device for storing charge and energy. It consists of two conductors that are insulated from each other and carry equal and opposite charges.	
2. Capacitance	Definition of capacitance $$C = \frac{Q}{V}$$	24-1
Single conductor	Q is the conductor's total charge, V is the conductor's potential relative to its surroundings.	
Capacitor	Q is the magnitude of the charge on either conductor, V is the magnitude of the potential difference between the conductors.	
Of an isolated spherical conductor	$$C = 4\pi\epsilon_0 R$$	24-2
Of a parallel-plate capacitor	$$C = \frac{\epsilon_0 A}{d}$$	24-6
Of a cylindrical capacitor	$$C = \frac{2\pi\epsilon_0 L}{\ln(R_2/R_1)}$$	24-7
Energy stored in a capacitor	$$U = \frac{1}{2}QV = \frac{1}{2}\frac{Q^2}{C} = \frac{1}{2}CV^2$$	24-8
Energy density of an electric field	$$u_e = \tfrac{1}{2}\epsilon_0 E^2$$	24-9
3. Equivalent Capacitance		
Parallel capacitors	When devices are connected in parallel, the voltage drop is the same across each. $$C_{eq} = C_1 + C_2 + C_3 + \dots$$	24-13
Series capacitors	When capacitors are in series, the voltage drops add. If the total charge on each connected pair of plates is zero, then: $$\frac{1}{C_{eq}} = \frac{1}{C_1} + \frac{1}{C_2} + \frac{1}{C_3} + \dots$$	24-17
4. Dielectrics		
Macroscopic behavior	A nonconducting material is called a dielectric. When a dielectric is inserted between the plates of a charged, electrically isolated capacitor, the electric field between the plates is weakened and the capacitance is thereby increased by the factor κ, which is the dielectric constant.	
Microscopic view	The electric field in the dielectric of a capacitor is weakened because the molecular dipole moments (either preexisting or induced) tend to align with the applied field and thereby produce a second electric field inside the dielectric that opposes the applied field. The aligned dipole moment of the dielectric is proportional to the applied field.	

TOPIC	RELEVANT EQUATIONS AND REMARKS	
Electric field inside	$E = \frac{E_0}{\kappa}$	24-18
Effect on capacitance	$C = \kappa C_0$	24-19
Permittivity ϵ	$\epsilon = \kappa \epsilon_0$	24-21
Uses of a dielectric	1. Increases capacitance 2. Increases dielectric strength 3. Physically separates conductors	
***5. Piezoelectric Effect**	In certain crystals a mechanical stress changes the polarization of the material, which results in a voltage across the crystal. Conversely, an applied voltage induces mechanical strain (deformation) in the crystal.	
***Pyroelectric Effect**	In certain crystals an increase in temperature changes the polarization of the material, which results in a voltage across the crystal.	

Answers to Concept Checks

24-1 $C_2 = C_1$. The capacitance does not depend on the charge. If the charge is tripled, the potential of the sphere will be tripled and the ratio Q/V, which depends only on the radius of the sphere, remains unchanged.

24-2 The capacitance of any capacitor does not depend on the potential. To increase V you must increase the charge Q, and vice versa. The ratio Q/V depends only on the geometry of the capacitor and on the nature of any dielectric material separating the plates.

24-3 The net charge remains the same. Like a water pump transfers water, a battery transfers charge. The amount of water in a water pump does not change and the amount of charge in a battery does not change.

24-4 A negative value.

24-5 Yes. Capacitance is defined as $C = Q/V$. So for an isolated charged capacitor, a capacitor for which Q is constant, the capacitance C is inversely proportional to the voltage V. When a dielectric is inserted into an isolated capacitor the bound surface charges induced on the dielectric result in a reduced electric-field strength within the dielectric. The voltage is directly proportional to the electric-field strength, so a reduced electric-field strength means a reduced voltage and an increased capacitance.

Answers to Practice Problems

24-1 9.0×10^9 m, which is about 1400 times the radius of Earth. (The farad is indeed a very large unit.)

24-2 $A = 1.1 \times 10^8\ \text{m}^2$, which corresponds to a square 11 km on a side

24-3 3.7 J

24-4 $U = \frac{1}{2}\sum_{i=1}^{n} Q_i V_i = \frac{1}{2}Q_1V_1 + \frac{1}{2}Q_2V_2$

$= \frac{1}{2}(-Q)V_1 + \frac{1}{2}(+Q)(V_1 + V) = \frac{1}{2}QV$

24-5 21 V

24-6 (*a*) $C_0 = 87$ pF, (*b*) $C = 50$ pF

24-7 (*a*) $u_e = \frac{1}{2}\epsilon_0 E^2 = 160\ \mu\text{J/m}^3$,
(*b*) $\Delta\text{vol} = A\,\Delta d = 2.9 \times 10^{-5}\ \text{m}^3$, $u_e \Delta\text{vol} = 4.7$ nJ, in agreement with Example 24-3

24-8 54 μC

24-9 4.0 V

24-10 (*a*) 50 μF, (*b*) 12 μF

24-11 $U_i = q_2/(2C_1) + q_2/(2C_2)$, where $q = 48\ \mu$C. Thus, $U_i = 288\ \mu$J. $U_f = Q_1^2/(2C_1) + Q_2^2/(2C_2) = 256\ \mu$J.

24-12 $U_2 = 4.0\ \mu$J, $U_3 = 24\ \mu$J, $U_4 = 8.0\ \mu$J. Note that $U_2 + U_3 + U_4 = 36\ \mu\text{J} = \frac{1}{2}QV = \frac{1}{2}Q^2/C_{eq} = \frac{1}{2}C_{eq}V^2$.

24-13 (*a*) 0.18 nF, (*b*) 2.1 nC

24-14 (*a*) $Q = 1.1$ nC (which is unchanged), (*b*) $V = 6.0$ V, (*c*) $C = 180$ pF

Problems

In a few problems, you are given more data than you actually need; in a few other problems, you are required to supply data from your general knowledge, outside sources, or informed estimate.

Interpret as significant all digits in numerical values that have trailing zeros and no decimal points.

- • Single-concept, single-step, relatively easy
- •• Intermediate-level, may require synthesis of concepts
- ••• Challenging
- SSM Solution is in the *Student Solutions Manual*
- Consecutive problems that are shaded are paired problems.

CONCEPTUAL PROBLEMS

1 • If the voltage across a parallel-plate capacitor is doubled, its capacitance (*a*) doubles, (*b*) drops by half, (*c*) remains the same.

2 • If the charge on an isolated spherical conductor is doubled, its self-capacitance (*a*) doubles, (*b*) drops by half, (*c*) remains the same.

3 • True or false: The electrostatic energy density is uniformly distributed in the region between the conductors of a cylindrical capacitor.

4 • If the distance between the plates of a charged and isolated parallel-plate capacitor is doubled, what is the ratio of the final stored energy to the initial stored energy?

5 • A parallel-plate capacitor is connected to a battery. The space between the two plates is empty. If the separation between the capacitor plates is tripled while the capacitor remains connected to the battery, what is the ratio of the final stored energy to the initial stored energy? SSM

6 • If the capacitor of Problem 5 is disconnected from the battery before the separation between the plates is tripled, what is the ratio of the final stored energy to the initial stored energy?

7 • True or false:
(*a*) The equivalent capacitance of two capacitors in parallel is always greater than the larger of the two capacitance values.
(*b*) The equivalent capacitance of two capacitors in series is always less than the least of the two capacitance values if the charges on the two plates that are connected by an otherwise isolated conductor sum to zero.

8 • Two uncharged capacitors have capacitances C_0 and $2C_0$, respectively, and are connected in series. This series combination is then connected across the terminals a battery. Which of the following is true?
(*a*) The capacitor $2C_0$ has twice the charge of the other capacitor.
(*b*) The voltage across each capacitor is the same.
(*c*) The energy stored by each capacitor is the same.
(*d*) The equivalent capacitance is $3C_0$.
(*e*) The equivalent capacitance is $2C_0/3$.

9 • A dielectric is inserted between the plates of a parallel-plate capacitor, completely filling the region between the plates. Air initially filled the region between the two plates. The capacitor was connected to a battery during the entire process. True or false:
(*a*) The capacitance value of the capacitor increases as the dielectric is inserted between the plates.
(*b*) The charge on the capacitor plates decreases as the dielectric is inserted between the plates.
(*c*) The electric field between the plates does not change as the dielectric is inserted between the plates.
(*d*) The energy storage of the capacitor decreases as the dielectric is inserted between the plates. SSM

10 •• Capacitors A and B (Figure 24-33) have identical plate areas and gap separations. The space between the plates of each capacitor is half-filled with a dielectric as shown. Which has the larger capacitance, capacitor A or capacitor B? Explain your answer.

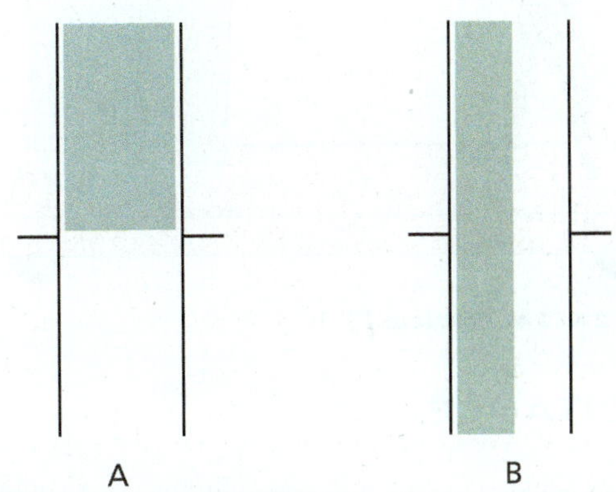

FIGURE 24-33 Problem 10

11 •• (*a*) Two identical capacitors are connected in parallel. This combination is then connected across the terminals of a battery. How does the total energy stored in the parallel combination of the two capacitors compare to the total energy stored if just one of the capacitors were connected across the terminals of the same battery? (*b*) Two identical capacitors that have been discharged are connected in series. This combination is then connected across the terminals of a battery. How does the total energy stored in the series combination of the two capacitors compare to the total energy stored if just one of the capacitors were connected across the terminals of the same battery? SSM

12 •• Two identical capacitors that have been discharged are connected in series across the terminals of a 100-V battery. When only one of the capacitors is connected across the terminals of the battery, the energy stored is U_0. What is the total energy stored in the two capacitors when the series combination is connected to the battery? (*a*) $4U_0$, (*b*) $2U_0$, (*c*) U_0, (*d*) $U_0/2$, (*e*) $U_0/4$

ESTIMATION AND APPROXIMATION

13 •• Disconnect the coaxial cable from a television or other device and estimate the diameter of the inner conductor and the diameter of the shield. Assume a plausible value (see Table 24-1) for the dielectric constant of the dielectric separating the two conductors and estimate the capacitance per unit length of the cable. SSM

14 •• **ENGINEERING APPLICATION, CONTEXT-RICH** You are part of an engineering research team that is designing a pulsed nitrogen laser. To create the high energy densities needed to operate such a laser, the electrical discharge from a high-voltage capacitor is used. Typically, the energy requirement per pulse (i.e., per discharge) is 100 J. Estimate the capacitance required if the discharge is to create a spark across a gap of about 1.0 cm. Assume that the dielectric breakdown of nitrogen is the same as the value for normal air.

15 •• Estimate the capacitance of the Leyden jar shown in Figure 24-34. The figure of a man is one-tenth the height of an average man. SSM

FIGURE 24-34 Problem 15

CAPACITANCE

16 • An isolated conducting sphere that has a 10.0-cm radius has an electric potential of 2.00 kV (the potential far from the sphere is zero). (*a*) How much charge is on the sphere? (*b*) What is the self-capacitance of the sphere? (*c*) By how much does the self-capacitance change if the sphere's electric potential is increased to 6.00 kV?

17 • The charge on one plate of a capacitor is $+30.0\ \mu C$ and the charge on the other plate is $-30.0\ \mu C$. The potential difference between the plates is 400 V. What is the capacitance of the capacitor?

18 •• Two isolated conducting spheres of equal radius R have charges $+Q$ and $-Q$, respectively. Their centers are separated by a distance d that is large compared to their radius. Estimate the capacitance of this unusual capacitor.

THE STORAGE OF ELECTRICAL ENERGY

19 • (*a*) The potential difference between the plates of a $3.00\text{-}\mu F$ capacitor is 100 V. How much energy is stored in the capacitor? (*b*) How much additional energy is required to increase the potential difference between the plates from 100 V to 200 V? SSM

20 • The charges on the plates of a $10\text{-}\mu F$ capacitor are $\pm 4.0\ \mu C$. (*a*) How much energy is stored in the capacitor? (*b*) If charge is transferred until the charges on the plates are equal to $\pm 2.0\ \mu C$, how much stored energy remains?

21 • (*a*) Find the energy stored in a 20.0-nF capacitor when the charges on the plates are $\pm 5.00\ \mu C$. (*b*) How much additional energy is stored if charges are increased from $\pm 5.00\ \mu C$ to $\pm 10.0\ \mu C$?

22 • What is the maximum electric energy density in a region containing dry air at standard conditions?

23 •• An air-gap parallel-plate capacitor that has a plate area of $2.00\ m^2$ and a separation of 1.00 mm is charged to 100 V. (*a*) What is the electric field between the plates? (*b*) What is the electric energy density between the plates? (*c*) Find the total energy by multiplying your answer from Part (*b*) by the volume between the plates. (*d*) Determine the capacitance of this arrangement. (*e*) Calculate the total energy from $U = \frac{1}{2}CV^2$, and compare your answer with your result from Part (*c*). SSM

24 •• A solid metal sphere has radius of 10.0 cm and a concentric metal spherical shell has an inside radius of 10.5 cm. The solid sphere has a charge 5.00 nC. (*a*) Estimate the energy stored in the electric field in the region between the spheres. *Hint: You can treat the spheres essentially as parallel flat slabs separated by 0.5 cm.* (*b*) Estimate the capacitance of this two-sphere system. (*c*) Estimate the total energy stored in the electric field from $\frac{1}{2}Q^2/C$ and compare it to your answer in Part (*a*).

25 •• A parallel-plate capacitor has plates of area $500\ cm^2$ and is connected across the terminals of a battery. After some time has passed, the capacitor is disconnected from the battery. When the plates are then moved 0.40 cm farther apart, the charge on each plate remains constant but the potential difference between the plates increases by 100 V. (*a*) What is the magnitude of the charge on each plate? (*b*) Do you expect the energy stored in the capacitor to increase, decrease, or remain constant as the plates are moved this way? Explain your answer. (*c*) Support your answer to Part (*b*) by determining the change in stored energy in the capacitor due to the movement of the plates.

COMBINATIONS OF CAPACITORS

26 • (*a*) How many $1.00\text{-}\mu F$ capacitors connected in parallel would it take to store a total charge of 1.00 mC if the potential difference across each capacitor is 10.0 V? Diagram the parallel combination. (*b*) What would be the potential difference across this parallel combination? (*c*) If the capacitors in Part (*a*) are discharged, connected in series, and then energized until the potential difference across each is equal to 10.0 V, find the charge on each capacitor and the potential difference across the connection.

27 • A $3.00\text{-}\mu F$ capacitor and a $6.00\text{-}\mu F$ capacitor are discharged and then connected in series, and the series combination is then connected in parallel with an $8.00\text{-}\mu F$ capacitor. Diagram this combination. What is the equivalent capacitance of this combination?

28 • Three capacitors are connected in a triangle as shown in Figure 24-35. Find an expression for the equivalent capacitance between points a and c in terms of the three capacitance values.

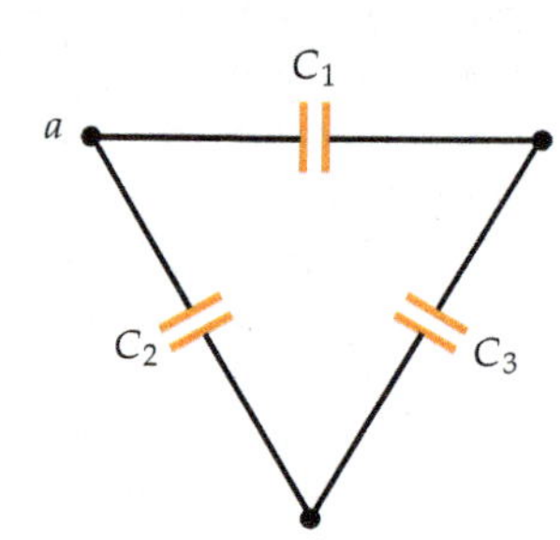

FIGURE 24-35 Problem 28

29 •• A 10.0-μF capacitor and a 20.0-μF capacitor are connected in parallel across the terminals of a 6.00-V battery. (*a*) What is the equivalent capacitance of this combination? (*b*) What is the potential difference across each capacitor? (*c*) Find the charge on each capacitor. (*d*) Find the energy stored in each capacitor.

30 •• A 10.0-μF capacitor and a 20.0-μF capacitor are discharged and then connected in series. The series combination is then connected across the terminals of a 6.00-V battery. (*a*) What is the equivalence capacitance of this combination? (*b*) Find the charge on each capacitor. (*c*) Find the potential difference across each capacitor. (*d*) Find the energy stored in each capacitor.

31 •• Three identical capacitors are connected so that their maximum equivalent capacitance, which is 15.0 μF, is obtained. (*a*) Determine how the capacitors are connected and diagram the combination. (*b*) There are three additional ways to connect all three capacitors. Diagram these three ways and determine the equivalent capacitances for each arrangement.

32 •• For the circuit shown in Figure 24-36, the capacitors were each discharged before being connected to the voltage source. Find (*a*) the equivalent capacitance of the combination, (*b*) the charge stored on the positively charged plate of each capacitor, (*c*) the voltage across each capacitor, and (*d*) the energy stored in each capacitor.

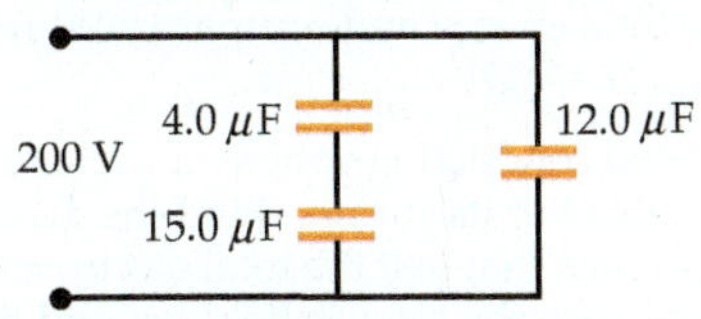

FIGURE 24-36 Problem 32

33 •• (*a*) Show that the equivalent capacitance of two capacitors in series can be written

$$C_{eq} = \frac{C_1C_2}{C_1 + C_2}$$

(*b*) Using only this formula and some algebra, show that C_{eq} must always be less than C_1 and C_2, and hence must be less than the smaller of the two values. (*c*) Show that the equivalent capacitance of three capacitors in series can be written

$$C_{eq} = \frac{C_1C_2C_3}{C_1C_2 + C_2C_3 + C_1C_3}$$

(*d*) Using only this formula and some algebra, show that C_{eq} must always be less than each of C_1, C_2, and C_3, and hence must be less than the least of the three values.

34 •• For the circuit shown in Figure 24-37 find (*a*) the equivalent capacitance between the terminals, (*b*) the charge stored on the positively charged plate of each capacitor, (*c*) the voltage across each capacitor, and (*d*) the total stored energy.

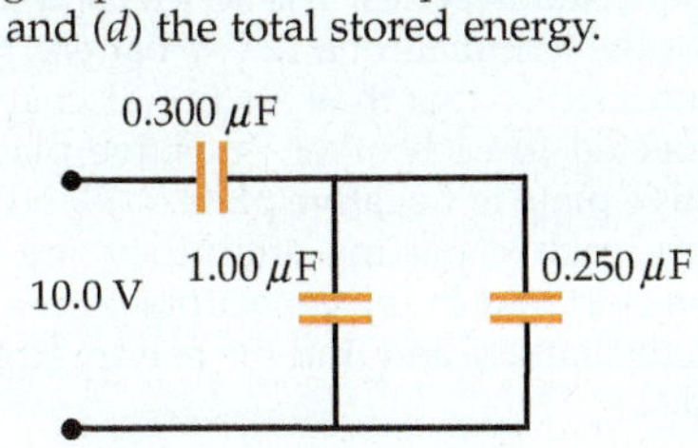

FIGURE 24-37 Problem 34

35 •• Five identical capacitors of capacitance C_0 are connected in a so-called bridge network, as shown in Figure 24-38. (*a*) What is the equivalent capacitance between points *a* and *b*? (*b*) Find the equivalent capacitance between points *a* and *b* if the capacitor at the center is replaced by a capacitor that has a capacitance of $10C_0$.

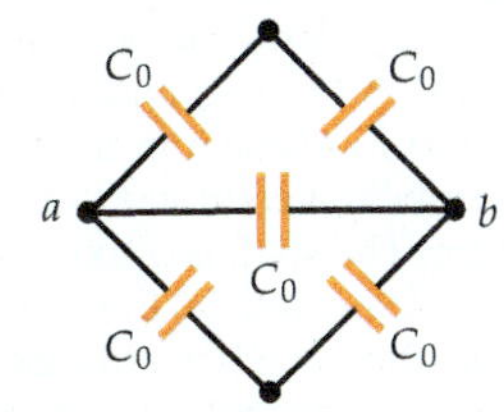

FIGURE 24-38 Problem 35

36 •• You and your laboratory team have been given a project by your electrical engineering professor. Your team must design a network of capacitors that has an equivalent capacitance of 2.00 μF and breakdown voltage of 400 V. The restriction is that your team must use only 2.00-μF capacitors that have individual breakdown voltages of 100 V. Diagram the combination.

37 •• Find the different equivalent capacitances that can be obtained by using two or three of the following capacitors: a 1.00-μF capacitor, a 2.00-μF capacitor, and a 4.00-μF capacitor.

38 ••• What is the equivalent capacitance (in terms of *C*, which is the capacitance of one of the capacitors) of the infinite ladder of capacitors shown in Figure 24-39?

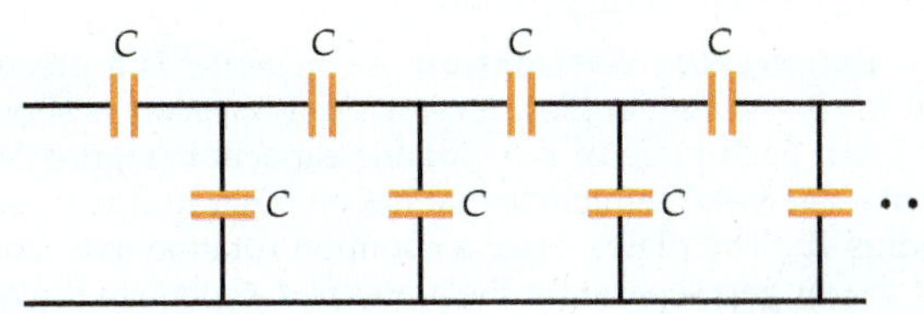

FIGURE 24-39 Problem 38

PARALLEL-PLATE CAPACITORS

39 • A parallel-plate capacitor has a capacitance of 2.00 μF and a plate separation of 1.60 mm. (*a*) What is the maximum potential difference between the plates, so that dielectric breakdown of the air between the plates does not occur? (*b*) How much charge is stored at this potential difference?

40 • An electric field of 2.00×10^4 V/m exists between the circular plates of a parallel-plate capacitor that has a plate separation of 2.00 mm. (*a*) What is the potential difference across the capacitor plates? (*b*) What plate radius is required if the positively charged plate is to have a charge of 10.0 μC?

41 •• A parallel-plate, air-gap capacitor has a capacitance of 0.14 μF. The plates are 0.50 mm apart. (*a*) What is the area of each plate? (*b*) What is the potential difference between the plates if the positively charged plate has a charge of 3.2 μC? (*c*) What is the stored energy? (*d*) What is the maximum energy this capacitor can store before dielectric breakdown of the air between the plates occurs?

42 •• Design a 0.100-μF parallel-plate capacitor that has air between its plates and that can be charged to a maximum potential difference of 1000 V before dielectric breakdown occurs. (*a*) What is the minimum possible separation between the plates? (*b*) What minimum area must each plate of the capacitor have?

CYLINDRICAL CAPACITORS

43 • In preparation for an experiment that you will do in your introductory nuclear physics lab, you are shown the inside of a Geiger tube. You measure the radius and the length of the central wire of the Geiger tube to be 0.200 mm and 12.0 cm, respectively. The outer surface of the tube is a conducing cylindrical shell that has an inner radius of 1.50 cm. The shell is coaxial with the wire and has the same length (12.0 cm). Calculate (*a*) the capacitance of your tube, assuming that the gas in the tube has a dielectric constant of 1.00, and (*b*) the value of the linear charge density on the wire when the potential difference between the wire and shell is of 1.20 kV.

44 •• A cylindrical capacitor consists of a long wire that has a radius R_1, a length L, and a charge $+Q$. The wire is enclosed by a coaxial outer cylindrical shell that has an inner radius R_2, length L, and charge $-Q$. (*a*) Find expressions for the electric field and energy density as a function of the distance R from the axis. (*b*) How much energy resides in a region between the conductors that has a radius R, a thickness dR, and a volume $2\pi rL\,dR$? (*c*) Integrate your expression from Part (*b*) to find the total energy stored in the capacitor. Compare your result with that obtained by using the formula $U = Q^2/(2C)$ in conjunction with the known expression for the capacitance of a cylindrical capacitor.

45 ••• Three concentric, thin, long conducting cylindrical shells have radii of 2.00 mm, 5.00 mm, and 8.00 mm. The space between the shells is filled with air. The innermost and outermost shells are connected at one end by a conducting wire. Find the capacitance per unit length of this configuration.

46 ••• **ENGINEERING APPLICATION** A *goniometer* is a precise instrument for measuring angles. A *capacitive goniometer* is shown in Figure 24-40*a*. Each plate of the variable capacitor (Figure 24-40*b*) consists of a flat metal semicircle that has an inner radius R_1 and an outer radius R_2. The plates share a common rotation axis, and the width of the air gap separating the plates is d. Calculate the capacitance as a function of the angle θ and the parameters given.

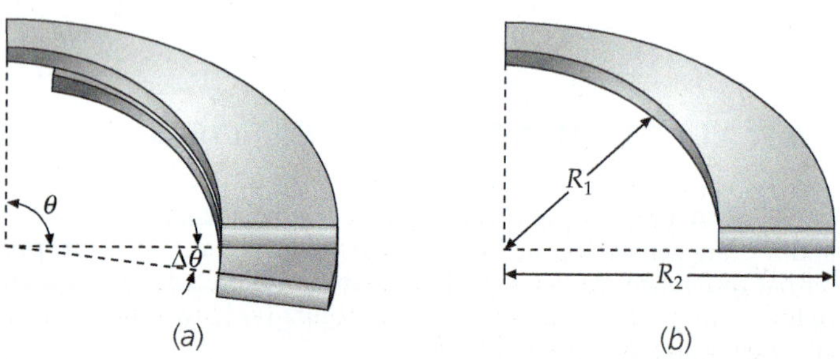

FIGURE 24-40 Problem 46

47 ••• **ENGINEERING APPLICATION** A *capacitive pressure gauge* is shown in Figure 24-41. Each plate has an area A. The plates are separated by a material that has a dielectric constant κ, a thickness d, and a Young's modulus Y. If a pressure increase of ΔP is applied to the plates, derive an expression for the change in capacitance.

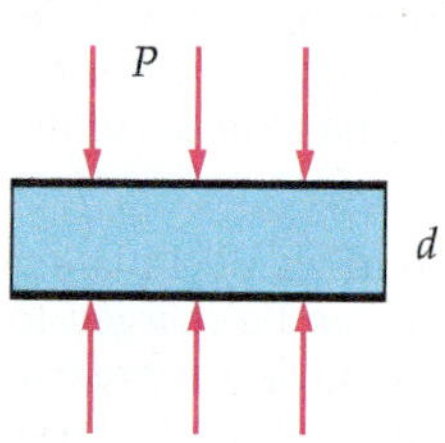

FIGURE 24-41 Problem 47

SPHERICAL CAPACITORS

48 •• Model Earth as a conducting sphere. (*a*) What is its self-capacitance? (*b*) Assume the magnitude of the electric field at Earth's surface is 150 V/m. What charge density does this correspond to? Express this value in fundamental charge units e per square centimeter.

49 •• A spherical capacitor consists of a thin spherical shell that has a radius R_1 and a thin, concentric spherical shell that has a radius R_2, where $R_2 > R_1$. (*a*) Show that the capacitance is given by $C = 4\pi\epsilon_0 R_1R_2/(R_2 - R_1)$. (*b*) Show that when the radii of the shells are nearly equal, the capacitance is approximately given by the expression for the capacitance of a parallel-plate capacitor, $C = \epsilon_0 A/d$, where A is the area of the sphere and $d = R_2 - R_1$. SSM

50 •• A spherical capacitor is composed of an inner sphere which has a radius R_1 and a charge $+Q$ and an outer concentric spherical thin shell which has a radius R_2 and a charge $-Q$. (*a*) Find the electric field and the energy density as a function of r, where r is the distance from the center of the sphere, for $0 \le r < \infty$. (*b*) Calculate the energy associated with the electrostatic field in a spherical shell between the conductors that has a radius r, a thickness dr, and a volume $4\pi r^2\,dr$. (*c*) Integrate your expression from Part (*b*) to find the total energy and compare your result with the result obtained using $U = \frac{1}{2}QV$.

51 ••• An isolated conducting sphere of radius R has a charge Q distributed uniformly over its surface. Find the distance R' from the center of the sphere such that half the total electrostatic energy of the system is associated with the electric field beyond that distance.

DISCONNECTED AND RECONNECTED CAPACITORS

52 •• A 2.00-μF capacitor is energized to a potential difference of 12.0 V. The wires connecting the capacitor to the battery are then disconnected from the battery and connected across a second, initially uncharged capacitor. The potential difference across the 2.00-μF capacitor then drops to 4.00 V. What is the capacitance of the second capacitor?

53 •• A 100-pF capacitor and a 400-pF capacitor are both charged to 2.00 kV. They are then disconnected from the voltage source and are connected together, positive plate to negative plate and negative plate to positive plate. (*a*) Find the resulting potential difference across each capacitor. (*b*) Find the energy dissipated when the connections are made. SSM

54 •• Two capacitors, one that has a capacitance of 4.00 μF and one that has a capacitance of 12.0 μF, are first discharged and then are connected in series. The series combination is then connected across the terminals of a 12.0-V battery. Next, they are carefully disconnected so that they are not discharged and they are then reconnected to each other—positive plate to positive plate and negative plate to negative plate. (*a*) Find the potential difference across each capacitor after they are reconnected. (*b*) Find the energy stored in the capacitors before they are disconnected from the battery, and find the energy stored after they are reconnected.

55 •• A 1.2-μF capacitor is charged to 30 V. After charging, the capacitor is disconnected from the voltage source and is connected across the terminals of a second capacitor that had previously been discharged. The final voltage across the 1.2-μF capacitor is 10 V. (*a*) What is the capacitance of the second capacitor? (*b*) How much energy was dissipated when the connection was made?

56 •• A 12-μF capacitor and a capacitor of unknown capacitance are both charged to 2.00 kV. After charging, the two capacitors are disconnected from the voltage source. The capacitors are then connected to each other—positive plate to negative plate and negative plate to positive plate. The final voltage across the terminals of the 12-μF capacitor is 1.00 kV. (*a*) What is the capacitance of the second capacitor? (*b*) How much energy was dissipated when the connection was made?

57 •• Two capacitors, one that has a capacitance of 4.00 μF and one that has a capacitance of 12.0 μF, are connected in parallel. The parallel combination is then connected across the terminals of a 12.0-V battery. Next, they are carefully disconnected so that they are not discharged. They are then reconnected to each other— the positive plate of each capacitor connected to the negative plate of the other. (*a*) Find the potential difference across each capacitor after they are reconnected. (*b*) Find the energy stored in the capacitors before they are disconnected from the battery, and find the energy stored after they are reconnected.

58 •• A 20-pF capacitor is charged to 3.0 kV and then removed from the battery and connected to an uncharged 50-pF capacitor. (*a*) What is the new charge on each capacitor? (*b*) Find the energy stored in the 20-pF capacitor before it is disconnected from the battery, and the energy stored in the two capacitors after they are connected to each other. Does the stored energy increase or decrease when the two capacitors are connected to each other?

59 •• Capacitors 1, 2, and 3 have capacitances equal to 2.00 μF, 4.00 μF, and 6.00 μF, respectively. The capacitors are connected in parallel, and the parallel combination is connected across the terminals of a 200-V source. The capacitors are then disconnected from both the voltage source and each other, and are connected to three switches as shown in Figure 24-42. (*a*) What is the potential difference across each capacitor when switches S_1 and S_2 are closed but switch S_3 remains open? (*b*) After switch S_3 is closed, what is the final charge on the leftmost plate of each capacitor? (*c*) Give the final potential difference across each capacitor after switch S_3 is closed. SSM

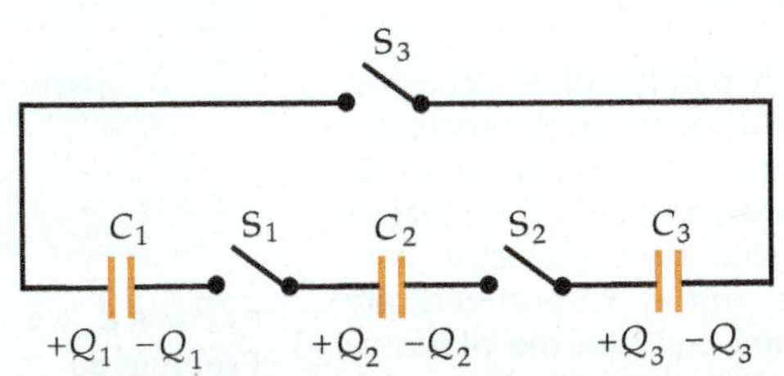

FIGURE 24-42 Problem 59

60 •• A capacitor has a capacitance C and a charge Q on its positively charged plate. A student connects one terminal of the capacitor to a terminal of an identical capacitor whose plates are electrically neutral. When the remaining two terminals are connected, charge flows until electrostatic equilibrium is reestablished and both capacitors have charge $Q/2$ on them. Compare the total energy initially stored in the one capacitor to the total energy stored in the two capacitors in which electrostatic equilibrium is reestablished. If there is less energy afterward, where do you think the missing energy went? *Hint: Wires that transport charge can heat up, which is called Joule heating and is discussed in detail in Chapter 25.*

DIELECTRICS

61 • ENGINEERING APPLICATION, CONTEXT-RICH You are a laboratory assistant in a physics department that has budget problems. Your supervisor wants to construct inexpensive parallel-plate capacitors for use in introductory laboratory experiments. The design uses polyethylene, which has a dielectric constant of 2.30, between two sheets of aluminum foil. The area of each sheet of foil is 400 cm^2 and the thickness of the polyethylene is 0.300 mm. Find the capacitance of this arrangement.

62 •• The radius and the length of the central wire in a Geiger tube are 0.200 mm and 12.0 cm, respectively. The outer surface of the tube is a conducing cylindrical shell that has an inner radius of 1.50 cm. The shell is coaxial with the wire and has the same length (12.0 cm). The tube is filled with a gas that has a dielectric constant of 1.08 and a dielectric strength of 2.00×10^6 V/m. (*a*) What is the maximum potential difference that can be maintained between the wire and shell? (*b*) What is the maximum charge per unit length on the wire?

63 •• ENGINEERING APPLICATION, CONTEXT-RICH You are a materials science engineer and your group has fabricated a new dielectric, that has a dielectric constant of 24 and a dielectric strength of 4.0×10^7 V/m. Suppose you want to use this material to construct a 0.10-μF parallel plate capacitor that can withstand a potential difference of 2.0 kV. (*a*) What is the minimum plate separation required to do this? (*b*) What is the area of each plate at this separation?

64 •• A parallel-plate capacitor has plates separated by a distance d. The capacitance of this capacitor is C_0 when no dielectric is in the space between the plates. However, the space between the plates is completely filled by two different dielectrics. One dielectric has a thickness $\frac{1}{4}d$ and a dielectric constant κ_1, and the other dielectric has a thickness $\frac{3}{4}d$ and a dielectric constant κ_2. Find the capacitance of this capacitor.

65 •• Two capacitors each have two conducting plates of surface area A and an air gap of width d. They are connected in parallel, as shown in Figure 24-43, and each has a charge Q on the positively charged plate. A slab that has a width d, an area A, and a dielectric constant κ is inserted between the plates of *one* of the capacitors. Calculate the new charge Q' on the positively charged plate of that capacitor after electrostatic equilibrium is reestablished.

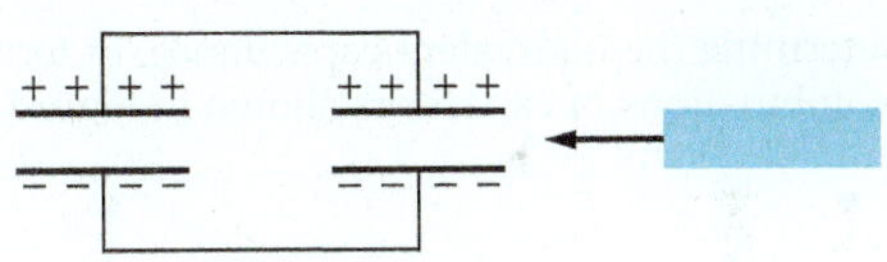

FIGURE 24-43 Problem 65

66 •• A parallel-plate capacitor has a plate separation d and has a capacitance equal to C_0 when there is only empty space in the space between the plates. A slab of thickness t, where $t < d$, that has a dielectric constant κ is placed in the space between the plates—completely covering one of the plates. What is the capacitance with the slab inserted?

67 •• BIOLOGICAL APPLICATION The membrane of the axon of a nerve cell can be modeled as a thin cylindrical shell of radius 1.00×10^{-5} m, having a length of 10.0 cm and a thickness of 10.0 nm. The membrane has a positive charge on one side and a negative charge on the other, and the membrane acts as a parallel-plate capacitor of area $2\pi rL$ and separation d. Assume the membrane is filled with a material whose dielectric constant is 3.00. (*a*) Find the capacitance of the membrane. If the potential difference across the membrane is 70.0 mV, find (*b*) the charge on the positively charged side of the membrane, and (*c*) the electric field strength in the membrane.

68 •• The space between the plates of a capacitor that is connected across the terminals of a battery is filled with a dielectric material. Determine the dielectric constant of the material if the induced bound charge per unit area on it is (*a*) 80 percent of the free charge per unit area on the plates, (*b*) 20 percent of the free charge per unit area on the plates, and (*c*) 98 percent of the free charge per unit area on the plates.

69 •• The positively charged plate of a parallel-plate capacitor has a charge equal to Q. When the space between the plates is evacuated of air, the electric field strength between the plates is 2.5×10^5 V/m. When the space is filled with a certain dielectric material, the field strength between the plates is reduced to 1.2×10^5 V/m. (*a*) What is the dielectric constant of the material? (*b*) If $Q = 10$ nC, what is the area of the plates? (*c*) What is the total induced bound charge on either face of the dielectric material?

70 •• Find the capacitance of the parallel-plate capacitor shown in Figure 24-44.

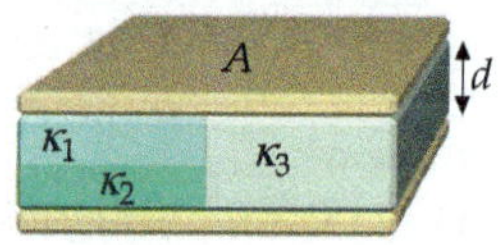

FIGURE 24-44 Problem 70

GENERAL PROBLEMS

71 • You are given four identical capacitors and a 100-V battery. When only one of the capacitors is connected to the battery the energy stored is U_0. Combine the four capacitors in such a way that the total energy stored in all four capacitors is U_0. Describe the combination and explain your answer.

72 • Three capacitors have capacitances of 2.00 μF, 4.00 μF, and 8.00 μF. Find the equivalent capacitance if (*a*) the capacitors are connected in parallel and (*b*) the capacitors are connected in series.

73 • A 1.00-μF capacitor is connected in parallel with a 2.00-μF capacitor, and this combination is connected in series with a 6.00-μF capacitor. What is the equivalent capacitance of this combination?

74 • The voltage across a parallel-plate capacitor that has a plate separation equal to 0.500 mm is 1.20 kV. The capacitor is disconnected from the voltage source and the separation between the plates is increased until the energy stored in the capacitor has been doubled. Determine the final separation between the plates.

75 •• Determine the equivalent capacitance, in terms of C_0, of each of the combinations of capacitors shown in Figure 24-45.

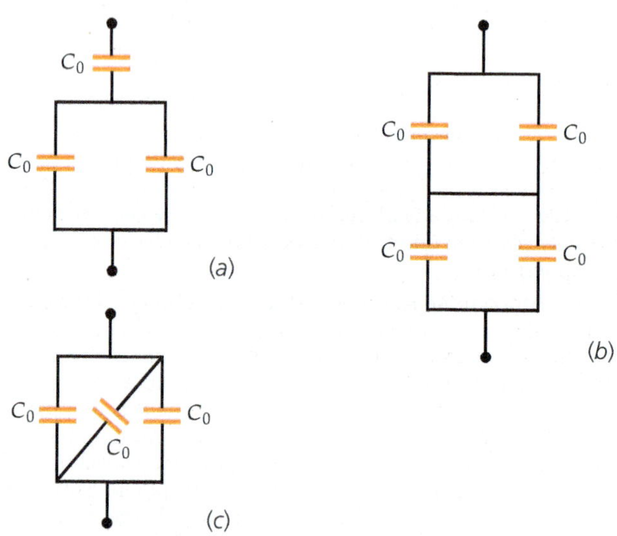

FIGURE 24-45 Problem 75

76 •• Figure 24-46 shows four capacitors connected in the arrangement known as a capacitance bridge. The capacitors are initially uncharged. What must the relation between the four capacitances be so that the potential difference between points *c* and *d* remains zero when a voltage *V* is applied between points *a* and *b*?

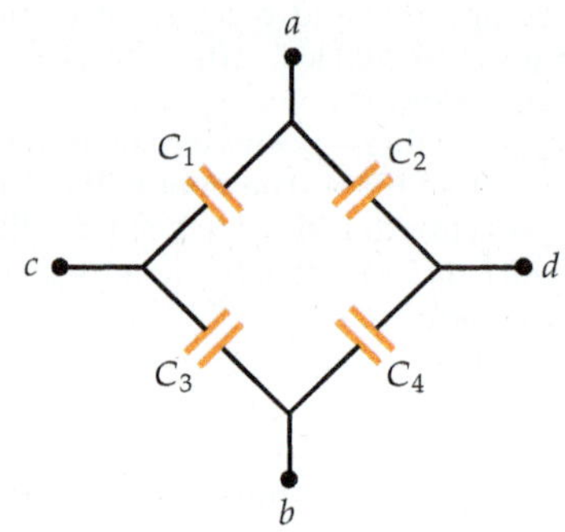

FIGURE 24-46 Problem 76

77 •• The plates of a parallel-plate capacitor are separated by distance *d*, and each plate has area *A*. The capacitor is charged to a potential difference *V* and then disconnected from the voltage source. The plates are then pulled apart until the separation is 3*d*. Find (*a*) the new capacitance, (*b*) the new potential difference, and (*c*) the new stored energy. (*d*) How much work was required to change the plate separation from *d* to 3*d*?

78 •• A parallel-plate capacitor has capacitance C_0 when there is no dielectric in the space between the plates. The space between the plates is then filled with a material that has a dielectric constant of κ. When a second capacitor of capacitance C' is connected in series with the first, the capacitance of the series combination is C_0. Find C' in terms of C_0.

79 •• A parallel combination of two identical 2.00-μF parallel-plate capacitors (no dielectric is in the space between the plates) is connected to a 100-V battery. The battery is then removed and the separation between the plates of one of the capacitors is doubled. Find the charge on the positively charged plate of each of the capacitors. SSM

80 •• A parallel-plate capacitor that has no dielectric in the space between the plates has a capacitance C_0 and a plate separation *d*. Two dielectric slabs that have dielectric constants of κ_1 and κ_2, respectively, are then inserted between the plates as shown in Figure 24-47. Each slab has a thickness $\frac{1}{2}d$ and has area *A*, the same area as each capacitor plate. When the charge on the positively charged capacitor plate is *Q*, find (*a*) the electric field in each dielectric and (*b*) the potential difference between the plates. (*c*) Show that the capacitance of the system after the slabs are inserted is given by $[2\kappa_1\kappa_2/(\kappa_1 + \kappa_2)]C_0$. (*d*) Show that $[2\kappa_1\kappa_2/(\kappa_1 + \kappa_2)]C_0$ is the equivalent capacitance of a series combination of two capacitors, each having plates of area *A* and a gap width equal to $d/2$. The space between the plates of one is filled with a material that has a dielectric constant equal to κ_1 and the space between the plates of the other is filled with a material that has a dielectric constant equal to κ_2.

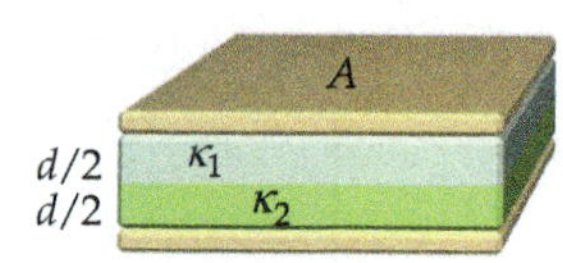

FIGURE 24-47 Problem 80

81 •• The plates of a parallel-plate capacitor are separated by distance d_0, and each plate has area *A*. A metal slab of thickness *d* and area *A* is inserted between the plates in such a way that the slab is parallel with the capacitor plates. (*a*) Show that the new capacitance is given by $\epsilon_0 A/(d_0 - d)$, regardless of the distance between the metal slab and the positively charged plate. (*b*) Show that this arrangement can be modeled as a capacitor that has plate separation *a* in series with a capacitor of plate separation *b*, where $a + b + d = d_0$.

82 •• A parallel-plate capacitor that has plate area A is filled with two dielectrics of equal size, as shown in Figure 24-48. (*a*) Show that this system can be modeled as two capacitors that are connected in parallel and each have an area $\frac{1}{2}A$. (*b*) Show that the capacitance is given by $\frac{1}{2}(\kappa_1 + \kappa_2)C_0$, where C_0 is the capacitance if there were no dielectric materials in the space between the plates.

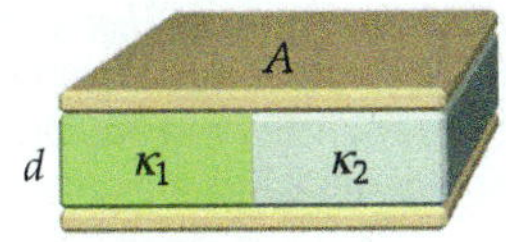

FIGURE 24-48 Problem 82

83 •• A parallel-plate capacitor that has no dielectric in the space between the plates has a plate area A and a gap width x. A charge Q is on the positively charged plate. (*a*) Find the stored electrostatic energy as a function of x. (*b*) Find the increase in energy dU due to an increase in plate separation dx from $dU = (dU/dx)\,dx$. (*c*) If F is the force exerted by one plate on the other, the work needed to move one plate a distance dx is $F\,dx = dU$. Show that $F = Q_2/(2\epsilon_0 A)$. (*d*) Show that the force in Part (*c*) equals $\frac{1}{2}EQ$, where Q is the charge on one plate and E is the electric field between the plates. Give a conceptual explanation for the factor $\frac{1}{2}$ in this result.

84 •• A rectangular parallel-plate capacitor that has a length a and a width b has a dielectric that has a width b partially inserted a distance x between the plates, as shown in Figure 24-49. (*a*) Find the capacitance as a function of x. Neglect edge effects. (*b*) Show that your answer gives the expected results for $x = 0$ and $x = a$.

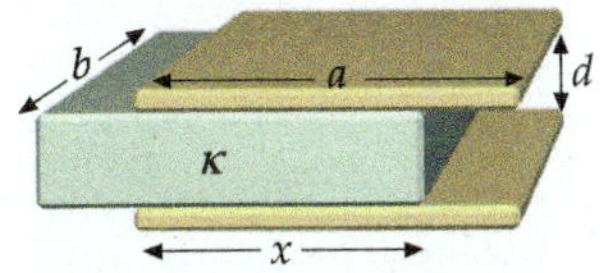

FIGURE 24-49 Problems 84 and 85

85 ••• An electrically isolated capacitor that has a charge Q on its positively charged plate is partly filled with a dielectric substance as shown in Figure 24-49. The capacitor consists of two rectangular plates that have edge lengths a and b and are separated by distance d. The dielectric is inserted into the gap a distance x. (*a*) What is the energy stored in the capacitor? *Hint: The capacitor can be modeled as two capacitors connected in parallel.* (*b*) Because the energy of the capacitor decreases as x increases, the electric field must be doing work on the dielectric, meaning that there must be an electric force pulling it in. Calculate this force by examining how the stored energy varies with x. (*c*) Express the force in terms of the capacitance and potential difference V between the plates. (*d*) From where does this force originate? SSM

86 ••• A spherical capacitor consists of a solid conducting sphere that has a radius a and a charge $+Q$ and a concentric conducting spherical shell that has an inner radius b and a charge $-Q$. The space between the two is filled with two different dielectric materials of dielectric constants κ_1 and κ_2. The boundary between the two dielectrics occurs a distance $\frac{1}{2}(a + b)$ from the center. (*a*) Calculate the electric field in the regions $a < r < \frac{1}{2}(a + b)$ and $\frac{1}{2}(a + b) < r < b$. (*b*) Integrate the expression $dV = -\vec{E} \cdot d\vec{\ell}$ to obtain the potential difference, V, between the two conductors. (*c*) Use $C = Q/V$ to obtain an expression for the capacitance of this system. (*d*) Show that your answer from Part (*c*) simplifies to the expected one if κ_1 equals κ_2.

87 ••• A capacitance balance is shown in Figure 24-50. The balance has a weight attached on one side and a capacitor that has a variable gap width on the other side. Assume the upper plate of the capacitor has negligible mass. When the capacitor potential difference between the plates is V_0, the attractive force between the plates balances the weight of the hanging mass. (*a*) Is the balance stable? That is, if we balance it out, and then move the plates a little closer together, will they snap shut or move back to the equilibrium point? (*b*) Calculate the value of V_0 required to balance an object of mass M, assuming the plates are separated by distance d_0 and have area A. *Hint: A useful relation is that the force between the plates is equal to the derivative of the stored electrostatic energy with respect to the plate separation.*

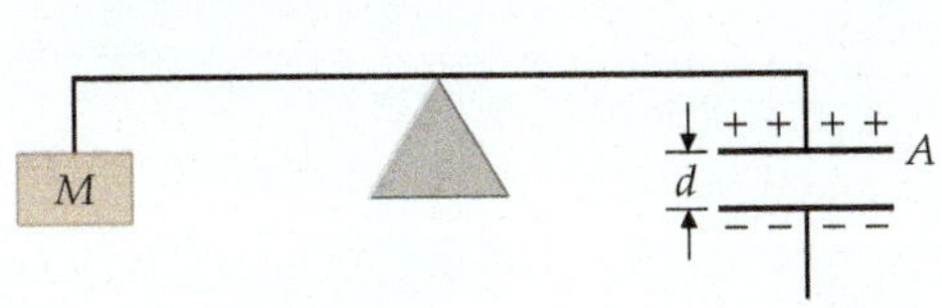

FIGURE 24-50 Problem 87

88 ••• **ENGINEERING APPLICATION, CONTEXT-RICH** You are an intern at an engineering company that makes capacitors used for energy storage in pulsed lasers. Your manager asks your team to construct a parallel-plate, air-gap capacitor that will store 100 kJ of energy. (*a*) What minimum volume is required between the plates of the capacitor? (*b*) Suppose you have developed a dielectric that has a dielectric strength of 3.00×10^8 V/m and a dielectric constant of 5.00. What volume of this dielectric, between the plates of the capacitor, is required for it to be able to store 100 kJ of energy?

89 ••• Consider two parallel-plate capacitors, C_1 and C_2, that are connected in parallel. The capacitors are identical except that C_2 has a dielectric inserted between its plates. A 200 V battery is connected across the combination until electrostatic equilibrium is established, and then the battery is disconnected. (*a*) What is the charge on each capacitor? (*b*) What is the total stored energy of the capacitors? (*c*) The dielectric is removed from C_2. What is the final stored energy of the capacitors? (*d*) What is the final voltage across the two capacitors?

90 ••• A capacitor is constructed of two coaxial conducting thin cylindrical shells of radii a and b ($b > a$), which have a length $L \gg b$. A charge of $+Q$ is on the inner cylinder, and a charge of $-Q$ is on the outer cylinder. The region between the two cylinders is filled with a material that has a dielectric constant κ. (*a*) Find the potential difference between the cylinders. (*b*) Find the density of the free charge σ_f on the inner cylinder and the outer cylinder. (*c*) Find the bound charge density σ_b on the inner cylindrical surface of the dielectric and on the outer cylindrical surface of the dielectric. (*d*) Find the total stored energy. (*e*) If the dielectric will move without friction, how much mechanical work is required to remove the dielectric cylindrical shell?

91 ••• Before switch S is closed, as shown in Figure 24-51, the voltage across the terminals of the switch is 120 V and the voltage across the capacitor labeled C_1 is 40.0 V. The capacitance of C_1 is 0.200 μF. The total energy stored in the two capacitors is 1.44 mJ. After closing the switch, the voltage across each capacitor is 80.0 V, and the energy stored by the two capacitors has dropped to 960 μJ. Determine the capacitance of C_2 and the charge on that capacitor before the switch was closed.

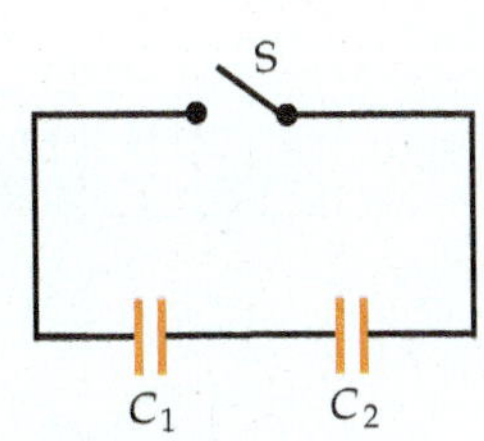

FIGURE 24-51 Problem 91

92 ••• An air-filled parallel-plate capacitor that has gap width d has plates that each have an area A. The capacitor is charged to a potential difference V and is then removed from the voltage source. A dielectric slab that has a dielectric constant of 2.00, a thickness d, and an area $\frac{1}{2}A$ is then inserted, as shown in Figure 24-52. Let σ_1 be the free charge density at the conductor–dielectric surface, and

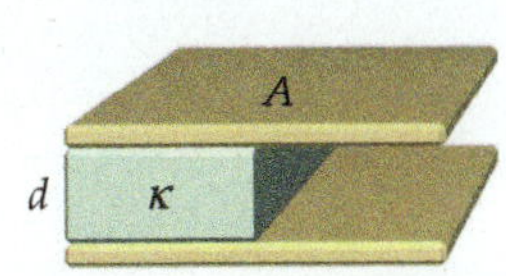

FIGURE 24-52 Problem 92

let σ_2 be the free charge density at the conductor–air surface. (*a*) Explain why the electric field must have the same value inside the dielectric as in the free space between the plates. (*b*) Show that $\sigma_1 = 2\sigma_2$. (*c*) Show that the final capacitance (after the slab is inserted) is 1.50 times the capacitance when the capacitor is filled with air. (*d*) Show that the final potential difference is $\frac{2}{3}V$. (*e*) Show that energy stored after the slab is inserted is only two-thirds of the energy stored before insertion.

93 ••• A capacitor has rectangular plates of length a and width b. The top plate is inclined at a small angle, as shown in Figure 24-53. The plate separation varies from y_0 at the left to $2y_0$ at the right, where y_0 is much less than a or b. Calculate the capacitance of this arrangement. *Hint: Break the problem up into a parallel combination. Choose strips of width dx and length b to approximate small (differential) capacitors (each having a value of dC). Each will have a plate area of b dx and separation distance* $y_0 + (y_0/a)x$. *Then argue that these differential capacitors are connected in parallel.*

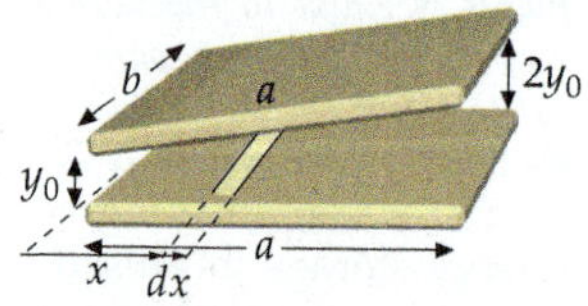

FIGURE 24-53
Problem 93

94 ••• Not all dielectrics that separate the plates of a capacitor are rigid. For example, the membrane of a nerve axon is a lipid bilayer that has a finite compressibility. Consider a parallel-plate capacitor whose plate separation is maintained by a material that has a dielectric constant of 3.00, a dielectric strength of 40.0 kV/mm, and a Young's modulus for compressive stress of 5.00×10^6 N/m^2. When the potential difference between the capacitor plates is zero, the thickness of the dielectric is equal to 0.200 mm and the capacitance of the capacitor is given by C_0. (*a*) Derive an expression for the capacitance as a function of the potential difference between the capacitor plates. (*b*) What is the maximum value of this potential difference? (Assume that the dielectric constant and the dielectric strength do *not* change under compression.)

CHAPTER 25

UNDERSTANDING DIRECT-CURRENT CIRCUITS CAN HELP YOU PERFORM POTENTIALLY DANGEROUS TASKS LIKE JUMP-STARTING A VEHICLE. *(© Tom Stewart/CORBIS.)*

Electric Current and Direct-Current Circuits

? When jump-starting your car using a second car, which terminal of the battery of your car should be connected to the positive terminal of the battery of the second car? (See Example 25-15.)

When we turn on a light, we connect the wire filament in the lightbulb across a potential difference that causes charge to flow through the wire, which is similar to the way a pressure difference in a garden hose causes water to flow through the hose. The flow of charge constitutes an electric current. We usually think of currents as being in conducting wires, but the electron beam in a cathode-ray tube (CRT) video monitor and a beam of charged ions from a particle accelerator also constitute electric currents.

In Chapter 25, we look at direct-current (dc) circuits, which are circuits where the direction of the current in a circuit element does not vary with time. Direct currents can be produced by batteries connected to resistors and capacitors. In Chapter 29, we discuss alternating-current (ac) circuits, in which the direction of the current changes.

25-1 CURRENT AND THE MOTION OF CHARGES

When a switch is thrown to turn on a circuit, a very small amount of charge accumulates along the surfaces of the wires and other conducting elements of the circuit, and these surface charges produce electric fields that drive the motion of charges through the conducting materials of the circuit. In the circuits we consider here, the time required for these small surface charges to be established is very short. The time for steady-state flow to be established depends on the size and the conductivity of the elements in the circuit, but the time is instantaneous as far as our perceptions are concerned. In steady state, charge no longer continues to accumulate at points along the circuit and the current is steady. (For the circuits in this chapter containing capacitors and resistors, the current may increase or decrease slowly, but appreciable changes occur only over a period that is much longer than the time needed to reach the steady state.)

Electric current is the rate of flow of charge through a surface—typically a cross-sectional surface of a conducting wire. Figure 25-1 shows a segment of a wire that is carrying a current (charges are moving). If ΔQ is the charge that flows through the cross-sectional area A in time Δt, the current I is

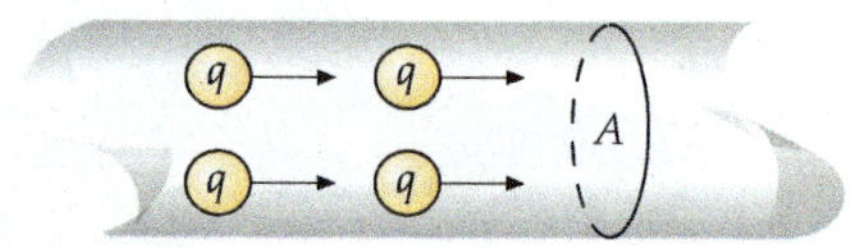

FIGURE 25-1 A segment of a current-carrying wire. If ΔQ is the amount of charge that flows through the cross-sectional area A in time Δt, the current through A is $I = \Delta Q/\Delta t$ in the limit that Δt approaches zero.

$$I = \frac{\Delta Q}{\Delta t} \qquad 25\text{-}1$$

in the limit that Δt approaches zero. The SI unit of current is the **ampere** (A)*:

$$1\text{ A} = 1\text{ C/s} \qquad 25\text{-}2$$

Mobile charges can be negatively charged or positively charged. In addition, a direction along the wire is designated as the positive direction. By convention, the sign of the current is positive if the current is due either to positive charges moving in the positive direction or to negative charges moving in the negative direction. However, the current is negative if it is due either to positive charges moving in the negative direction or to negative charges moving in the positive direction. This convention was established before it was known that the mobile charge carriers in metals were free electrons. Thus, in a current-carrying metallic wire, the free electrons move in the negative direction when the current is positive, and vice versa.

In a metal wire, the motion of negatively charged free electrons is quite complex. When there is no electric field in the wire, the free electrons move in random directions with relatively large speeds of the order of 10^6 m/s.† In addition, the electrons collide repeatedly with the lattice ions in the wire. Because the velocity vectors of the electrons are randomly oriented, the *average* velocity is zero. When an electric field is applied, the field exerts a force $-e\vec{E}$ on each free electron, giving it a change in velocity in the direction opposite the field. However, any additional kinetic energy acquired is quickly dissipated by collisions with the lattice ions in the wire. During the time between collisions with the lattice ions, the free electrons, on average, acquire an additional velocity in the direction opposite to the field. The net result of this repeated acceleration and dissipation of energy is that the electrons drift along the wire with a small average velocity, directed opposite to the electric-field direction, called the **drift velocity.** The **drift speed** is the magnitude of the drift velocity.

* The ampere is operationally defined (see Chapter 26) in terms of the magnetic force that current-carrying wires exert on one another. The coulomb is then defined as the ampere-second.

† The average kinetic energy of the free electrons in a metal is quite large, even at very low temperatures. These electrons do not have the classical Maxwell–Boltzmann energy distribution and do not obey the classical equipartition theorem. We discuss the energy distribution of these electrons and calculate their average speed in Chapter 38.

The motion of the free electrons in a metal is similar to the motion of the molecules of a gas, such as air. In still air at room temperature, the gas molecules move with large speeds (about 500 m/s) due to their thermal energy, but their average velocity is zero. When there is a breeze, the air molecules have a small average velocity or drift velocity in the direction of the breeze superimposed on their random high-speed motions. Similarly, when there is no applied electric field, the average velocity of all the free electrons in a metal is zero, but when there is an applied electric field, the average velocity is not zero due to the small drift velocities of the free electrons.

Let n be the number of mobile charged particles (charge carriers) per unit volume in a conducting wire of cross-sectional area A. We call n the **number density** of the charge carriers. Assume that each particle carries a charge q and moves in the positive direction with a drift speed v_d. During time Δt, all the particles in the volume $Av_d\,\Delta t$, shown in Figure 25-2 as a shaded region, pass through the area element. The number of particles in this volume is $nAv_d\,\Delta t$, and the total free charge in the volume is

$$\Delta Q = qnAv_d\,\Delta t$$

The current is thus

$$I = \frac{\Delta Q}{\Delta t} = qnAv_d \qquad 25\text{-}3$$

RELATION BETWEEN CURRENT AND DRIFT SPEED

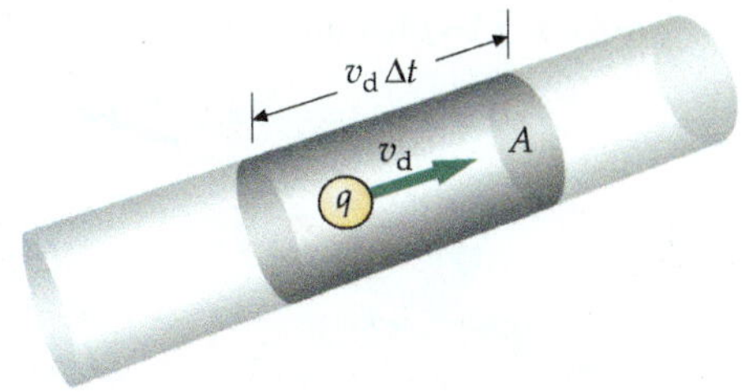

FIGURE 25-2 During time Δt, all the free charges initially in the shaded volume pass through A. If there are n charge carriers per unit volume, each having charge q, the total free charge in this volume is $\Delta Q = qnAv_d\,\Delta t$, where v_d is the drift speed of the charge carriers.

Equation 25-3 can be used to find the current due to the flow of any species of charged particle. If the current is the result of the motion of more than one species of mobile charge, as it sometimes is in ionic solutions such as salt water, then the total current is the sum of the currents for each of the individual species of mobile charges.

The number density of charge carriers in a conductor can be measured by the Hall effect, which is discussed in Chapter 26. The result is that, in most metals, there is approximately one free electron per atom.

The current per unit area is qnv_d, which is gotten by dividing both sides of Equation 25-3 by the area A. The **current density** vector, $\vec{J}$, is specified by

$$\vec{J} = qn\vec{v}_d \qquad 25\text{-}4$$

DEFINITION—CURRENT DENSITY

The **current** through a surface S is defined as the flux of the current density vector $\vec{J}$ through the surface. That is,

$$I = \int_S \vec{J}\cdot d\vec{A} = \int_S \vec{J}\cdot\hat{n}\,dA \qquad 25\text{-}5$$

DEFINITION—CURRENT

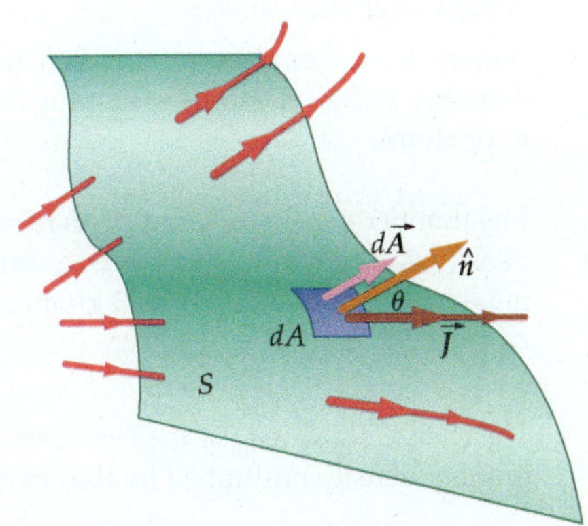

FIGURE 25-3 Current density $\vec{J}$ is a vector field that we can visualize by drawing field lines. The red directed lines are field lines of the current density field. These field lines serve as streamlines for the flow of charge. The current I (through S) is the flux of $\vec{J}$ through surface S.

where $d\vec{A}$ is an element of area for the surface and $\hat{n}$ is the unit vector normal to the surface S in the direction of $d\vec{A}$ (see Figure 25-3). If $\vec{J}$ is uniform and if the surface is flat, which means that $\hat{n}$ would be uniform, then the flux can be expressed

$$I = \int_S \vec{J}\cdot d\vec{A} = \vec{J}\cdot\vec{A} = \vec{J}\cdot\hat{n}A = JA\cos\theta$$

where $\vec{A}$ is the area of the surface and θ is the angle between $\vec{J}$ and $\hat{n}$. The sign of the current I is the same as the sign of $\cos\theta$. If $\theta < 90°$ I is positive, and if $\theta > 90°$

then I is negative (Figure 25-4). The black arrow with the plus sign next to each wire in the figure indicates the choice for the direction of $\hat{n}$ on the cross-sectional surfaces of the wire.

> **!** For a given current density $\vec{J}$ and surface S, the sign of the current I is determined by the choice of the direction of $\hat{n}$.

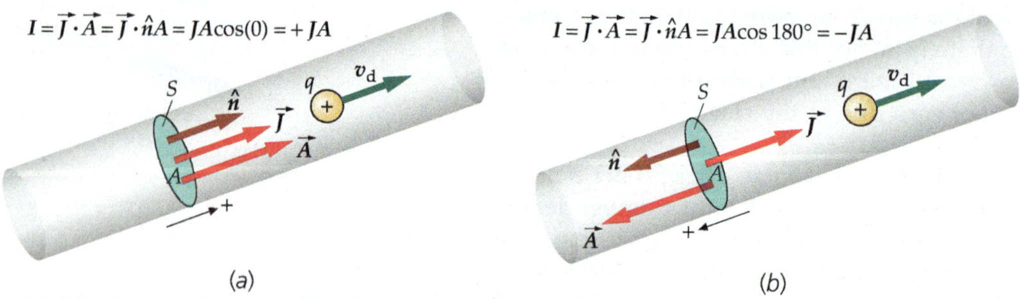

FIGURE 25-4 The flat surface S is perpendicular to the current density vector $\vec{J}$. The area vector $\vec{A}$ for the surface S is defined to be in the same direction as normal $\hat{n}$ for the surface. However, there are two choices for the direction of $\hat{n}$. (*a*) The current I through surface S is positive if the direction of $\hat{n}$ is chosen so that $\hat{n}$ and $\vec{J}$ are in the same direction. (*b*) The current I through surface S is negative if the direction of $\hat{n}$ is chosen so that $\hat{n}$ and $\vec{J}$ are in opposite directions.

Example 25-1 Finding the Drift Speed

The wire used for student laboratory experiments is typically made of copper and has a radius equal to 0.815 mm. (*a*) Estimate the total charge of the free electrons in each meter of such a wire carrying a current that has a magnitude equal to 1.0 A. Assume one free electron per atom. (*b*) Calculate the drift speed of the free electrons.

PICTURE Equation 25-3 relates the drift speed to the number density of charge carriers, which approximately equals the number density n_a of copper atoms. We can find n_a from the mass density and molar mass of copper and Avogadro's number.

SOLVE

(*a*) 1. The drift speed is related to the current and number density of charge carriers:

$$I = nqv_dA$$

2. If there is one free electron per atom, the number density n of free electrons equals the number density n_a of atoms:

$$n = n_a$$

3. The number density of atoms n_a is related to the mass density ρ_m, Avogadro's number N_A, and the molar mass M. For copper, $\rho_m = 8.93\ \text{g/cm}^3$ and $M = 63.5\ \text{g/mol}$:

$$n_a = \frac{\rho_m N_A}{M} = \frac{(8.93\ \text{g/cm}^3)(6.02 \times 10^{23}\ \text{atoms/mol})}{63.5\ \text{g/mol}}$$
$$= 8.47 \times 10^{22}\ \text{atoms/cm}^3 = 8.47 \times 10^{28}\ \text{atoms/m}^3$$

4. The charge density ρ_{fe} of the free electrons equals the number density multiplied by the charge:

$$\rho_{fe} = -en = -(1.60 \times 10^{-19}\ \text{C})(8.47 \times 10^{28}\ \text{m}^{-3}) = -1.36 \times 10^{10}\ \text{C/m}^3$$

5. The charge is the charge density multiplied by the volume:

$$Q = \rho_{fe}AL = -enAL \quad \text{so}$$
$$Q/L = -enA = (-1.36 \times 10^{10}\ \text{C/m}^3)\,\pi(8.15 \times 10^{-4}\ \text{m})^2$$
$$= -2.83 \times 10^4\ \text{C/m} = \boxed{-2.8 \times 10^4\ \text{C/m}}$$

(*b*) Substituting numerical values in Equation 25-3 yields v_d (The current is negative because Equation 25-3 is valid only for charges moving in the positive direction.):

$$v_d = \frac{I}{nqA} = \frac{I}{-neA} = \frac{I}{Q/L}$$
$$= \frac{-1.0\ \text{C/s}}{(-2.83 \times 10^4\ \text{C/m})} = \boxed{3.5 \times 10^{-2}\ \text{mm/s}}$$

CHECK Because there is 28 000 coulombs of mobile charge per meter of wire [Part (*a*), step 5], only a small drift speed is expected for a current of one coulomb per second. The Part-(*b*) result is in agreement with this expectation.

PRACTICE PROBLEM 25-1 How long would it take for an electron to drift from your car battery to the starter motor, a distance of about 1 m, if its drift speed is 3.5×10^{-5} m/s?

The drift speed of the mobile electrons in the wire in Example 25-1 is only a few hundredths of a millimeter per second. If electrons drift along wires at such low speeds, why does an electric light on the ceiling come on instantly when the wall switch is turned on? A comparison with water in a hose may prove useful. If you attach an empty 100-ft-long hose to a water faucet and turn on the water, it typically takes several seconds for the water to travel the length of the hose to the nozzle. However, if the hose is already full of water when the faucet is opened, the water emerges from the nozzle almost instantaneously. Because of the water pressure at the faucet, the segment of water near the faucet pushes on the water immediately next to it, which pushes on the next segment of water, and so on, until the last segment of water is pushed out the nozzle. This pressure wave moves down the hose at the speed of sound in water, and the water quickly reaches a steady flow rate.

Unlike a water hose, a metal wire is never empty. That is, there are always a very large number of conduction electrons throughout the metal wire. Thus, electrons start moving along the entire length of the wire (including the wire inside the lightbulb) almost immediately after the light switch is turned on. The transport of a significant amount of electrons in a wire is accomplished not by a few electrons moving rapidly down the wire, but by a very large number of electrons slowly drifting down the wire. Surface charges are established on the wires, and these surface charges produce an electric field. It is the electric field produced by these surface charges that drives the conduction electrons through the wire.

Example 25-2 Finding the Number Density

In a certain particle accelerator, a current of 0.50 mA is carried by a 5.0-MeV proton beam that has a radius equal to 1.5 mm. (*a*) Find the number density of protons in the beam. (*b*) If the beam hits a target, how many protons hit the target in 1.0 s?

PICTURE To find the number density, we use the relation $I = qnAv$ (Equation 25-3), where v is the drift speed of the charge carriers. (The drift speed is the magnitude of the average velocity.) We can find v from the energy. (In a 5.0-MeV beam, each particle in the beam has a kinetic energy equal to 5.0 MeV.) The amount of charge Q that hits the target in time Δt is $I\,\Delta t$, and the number N of protons that hits the target is Q divided by the charge per proton.

SOLVE

(*a*) 1. The number density is related to the current, the charge, the cross-sectional area, and the speed:

$$I = qnAv$$

2. We find the speed of the protons from their kinetic energy:

$$K = \tfrac{1}{2}mv^2 = 5.0\text{ MeV}$$

3. Use $m = 1.67 \times 10^{-27}$ kg for the mass of a proton, and solve for the speed:

$$v = \sqrt{\frac{2K}{m}} = \sqrt{\frac{(2)(5.0 \times 10^6\text{ eV})}{1.67 \times 10^{-27}\text{ kg}} \times \frac{1.60 \times 10^{-19}\text{ J}}{1\text{ eV}}}$$

$$= 3.09 \times 10^7\text{ m/s} = \boxed{3.1 \times 10^7\text{ m/s}}$$

4. Substitute to calculate n:

$$n = \frac{I}{qAv} = \frac{0.50 \times 10^{-3}\ \text{A}}{(1.60 \times 10^{-19}\ \text{C/proton})\ \pi(1.5 \times 10^{-3}\ \text{m})^2\ (3.10 \times 10^{7}\ \text{m/s})}$$

$$= 1.43 \times 10^{13}\ \text{protons/m}^3 = \boxed{1.4 \times 10^{13}\ \text{protons/m}^3}$$

(*b*) 1. The number of protons N that hit the target in 1.0 s is related to the total charge ΔQ that hits in 1.0 s and the proton charge q:

$$\Delta Q = Nq$$

2. The charge ΔQ that strikes the target in time Δt is the current multiplied by the time:

$$\Delta Q = I\,\Delta t$$

3. The number of protons is then:

$$N = \frac{\Delta Q}{q} = \frac{I\,\Delta t}{q} = \frac{(0.50 \times 10^{-3}\ \text{A})(1.0\ \text{s})}{1.60 \times 10^{-19}\ \text{C/proton}}$$

$$= 3.13 \times 10^{15}\ \text{protons} = \boxed{3.1 \times 10^{15}\ \text{protons}}$$

CHECK The number of protons N hitting the target in time Δt is the number in the volume $Av\,\Delta t$. Thus, $N = nAv\,\Delta t$. Substituting $n = I/(qAv)$ then gives $N = nAv\,\Delta t = [I/(qAv)](Av)\,\Delta t = I\,\Delta t/q = \Delta Q/q$, which is the expression for N that we used in Part (*b*).

TAKING IT FURTHER We were able to use the classical expression for kinetic energy in step 2 of Part (*a*) without taking relativity into consideration, because the proton kinetic energy of 5.0 MeV is much less than the proton rest energy (about 931 MeV). The speed found, 3.1×10^7 m/s, is about one-tenth the speed of light.

PRACTICE PROBLEM 25-2 Using the number density found in Part (*a*), how many protons are there in a volume of 1.0 mm³ of the space containing the beam?

25-2 RESISTANCE AND OHM'S LAW

Current in a conductor is driven by an electric field $\vec{E}$ inside the conductor that exerts a force $q\vec{E}$ on the free charges. (In electrostatic equilibrium, the electric field must be zero inside a conductor, but when there is a current in a conductor, the conductor is no longer in electrostatic equilibrium.) The free charges drifts down the conductor, driven by forces exerted on the charges by the electric field. In a metal, the free charges are negatively charged, so the free charges are driven in a direction opposite to the direction of the electric field $\vec{E}$. If the only forces on the free charges were the electric forces, then the free charges would gain speed indefinitely. However, this does not happen because the free electrons interact with the lattice of ions that make up the metal, and the interaction forces oppose the drifting motion of the free electrons.

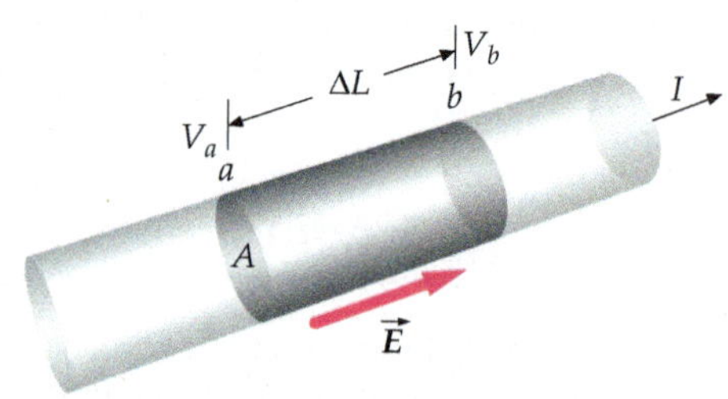

FIGURE 25-5 A segment of wire that has a current I. The potential drop $V_a - V_b$ is related to the electric field by $V_a - V_b = E\,\Delta L$.

Figure 25-5 shows a wire segment that has a length ΔL, a cross-sectional area A, and a current I. Because electric fields point in the direction of decreasing potential, the potential at point a is greater than the potential at point b. If we model the current as the flow of positive charge carriers, these positive charge carriers drift in the direction of decreasing potential. Assuming the electric field $\vec{E}$ to be uniform throughout the segment, the **potential drop** V between points a and b is

$$V = V_a - V_b = E\,\Delta L \qquad 25\text{-}6$$

The ratio of the potential drop in the direction of the current* to the current is called the **resistance** of the segment,

$$R = \frac{V}{I} \qquad 25\text{-}7$$

DEFINITION—RESISTANCE

where the *direction of the current* refers to the direction of the current density vector. The SI unit of resistance, the volt per ampere, is called an **ohm** (Ω):

$$1\Omega = 1\ \text{V/A} \qquad 25\text{-}8$$

* Because current is a scalar it does not have a direction.

For many materials, the resistance of a sample of the material does not depend on either the potential drop or the current. Such materials, which include most metals, are called **ohmic materials.** For many ohmic materials resistance remains essentially constant over a wide range of conditions. In that case the potential drop across a segment of the material is proportional to the current in the material. Equation 25-7 is typically written:

$$V = IR \qquad 25\text{-}9$$

OHM'S LAW

The relation $V = IR$ is commonly referred to as Ohm's law, even when the resistance R varies with the current I.

Figure 25-6 shows the potential difference V versus the current I for two conductors. For one conductor (Figure 25-6*a*), the relation is linear, but for the other conductor (Figure 25-6*b*), the relation is not linear. Ohm's law is not a fundamental law of nature, like Newton's laws or the laws of thermodynamics, but rather is an empirical description of a property shared by many materials under specified conditions. As we shall see, the resistance of a conductor does vary with the temperature of the conductor.

See
Math Tutorial *for more information on*
Direct and Inverse Proportions

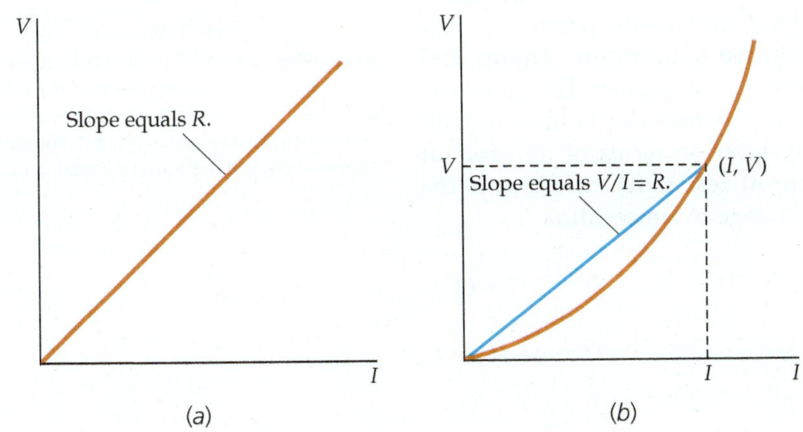

FIGURE 25-6 Plots of V versus I. (*a*) The potential drop is proportional to the current in accord with Ohm's law. The resistance $R = V/I$, equal to the slope of the line, is independent of I as indicated by the constant slope of the line. (*b*) The potential drop is not proportional to the current. The resistance $R = V/I$, equal to the slope of the chord connecting the origin with the point (I, V), increases with increasing I.

PRACTICE PROBLEM 25-3

A wire of resistance 3.0 Ω carries a current of 1.5 A. What is the potential drop across the wire?

The resistance R of a conducting wire is found to be proportional to the length L of the wire and inversely proportional to its cross-sectional area A:

$$R = \rho \frac{L}{A} \qquad 25\text{-}10$$

where the proportionality constant ρ is called the **resistivity** of the conducting material.* The unit of resistivity is the ohm-meter ($\Omega \cdot m$). Note that Equation 25-9 and Equation 25-10 for electrical conduction and electrical resistance are of the same form as Equation 20-9 ($\Delta T = IR$) and Equation 20-10 [$R = \Delta|x|/(kA)$] for thermal conduction and thermal resistance. For the equations that describe current, the potential difference V is substituted for the temperature difference ΔT and $1/\rho$ is substituted for the thermal conductivity k. (In fact, $1/\rho$ is called the *electrical conductivity*.†) Ohm was led to his law by the similarity between the conduction of electricity and the conduction of heat.

* The symbol ρ used here for the resistivity was used in previous chapters for volume charge density. Care must be taken to identify what quantity ρ refers to from context.

† The unit of conductance is the siemens (S), $1\ S = 1\ \Omega^{-1}$, and the units for conductivity are siemens per meter (S/m).

PRACTICE PROBLEM 25-4

A Nichrome wire ($\rho = 110 \times 10^{-8}\ \Omega \cdot \text{m}$) has a radius of 0.65 mm. What length of wire is needed to obtain a resistance of 2.0 Ω?

For a segment of wire that has a length L, a cross-sectional area A, a current I, and a resistance R, the voltage drop V across the length of the segment is related to the current I in the segment by

$$V = IR = I\rho\frac{L}{A}$$

The voltage drop V and the electric field strength E are related by $V = EL$. Substituting EL for V and J for I/A gives

$$EL = \rho J L$$

Dividing both sides by L and expressing E and J as vectors, we obtain

$$\vec{E} = \rho\vec{J} \qquad \text{25-11}$$

Equation 25-11 is an alternative version of Ohm's law. It states that the current density vector $\vec{J}$ at a point in a current-carrying conductor is equal to the reciprocal of the resistivity multiplied by the electric field vector $\vec{E}$ at the same point.

The resistivity of any given metal depends on the temperature. Figure 25-7 shows the temperature dependence of the resistivity of copper. This graph is nearly a straight line, which means that the resistivity varies nearly linearly with temperature.* In tables, the resistivity is usually given in terms of its value at 20°C, ρ_{20}, along with the **temperature coefficient of resistivity,** α, which is the ratio of the fractional change in resistivity to the change in temperature:

$$\alpha = \frac{(\rho - \rho_0)/\rho_0}{T - T_0} \qquad \text{25-12}$$

where ρ_0 is the resistivity at temperature T_0 and ρ is the resistivity at temperature T.

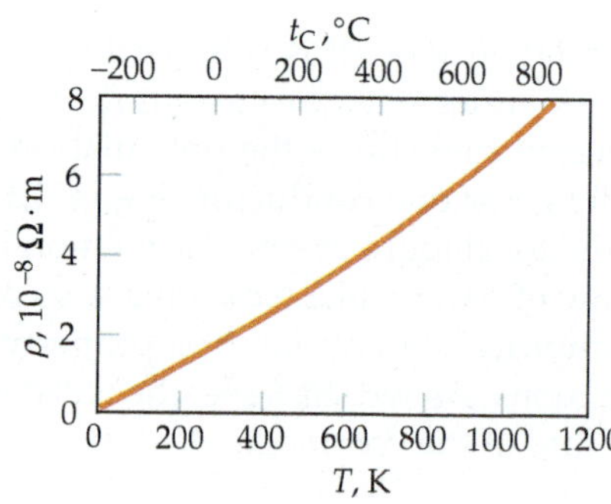

FIGURE 25-7 Plot of resistivity ρ versus temperature for copper. Because the Celsius and absolute temperatures differ only in the choice of zero, the resistivity has the same slope whether it is plotted against t_C or T.

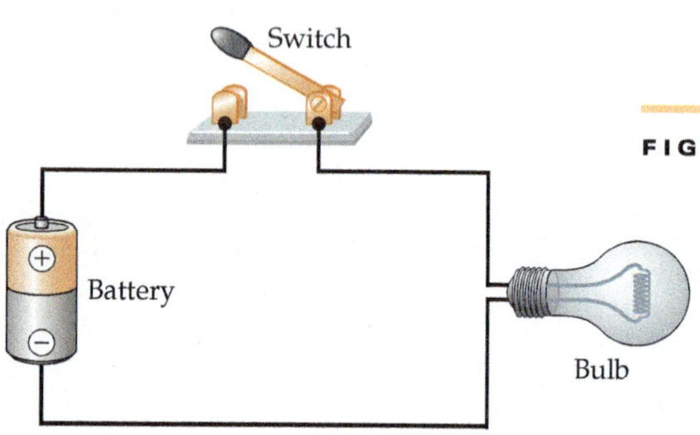

FIGURE 25-8

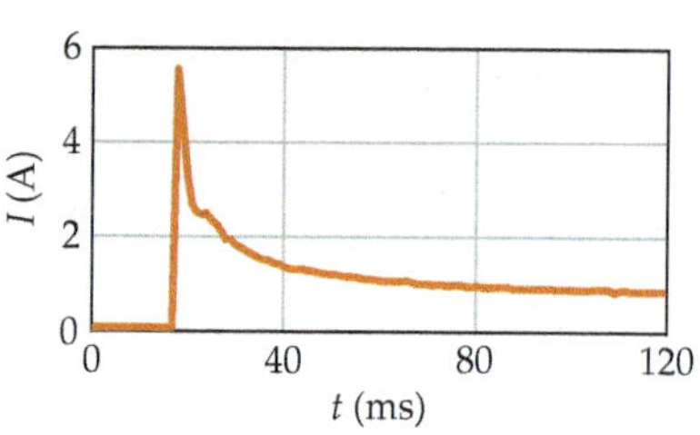

FIGURE 25-9 The current in the tungsten filament of an incandescent light bulb peaks when the bulb is first connected to the battery, but during the next 100 ms or so the current drops to its steady-state value of about 0.75 A. This occurs because the resistance of the filament increases with increasing temperature.

CONCEPT CHECK 25-1

The filament in the lightbulb shown in Figure 25-8 is a thin tungsten wire, and Figure 25-9 is a plot of the current in the filament as a function of time. Note that the current increases rapidly when the switch is closed, and then decreases until the current is maintained at a constant value. (*a*) Why does the current initially get larger than the constant value? (*b*) Why does the current remain constant following the initial surge?

* There is a breakdown in this linearity for all metals at very low temperatures that is not shown in Figure 25-7.

Table 25-1 Resistivities and Temperature Coefficients

Material	Resistivity ρ at 20°C, $\Omega \cdot m$	Temperature Coefficient α at 20°C, K^{-1}
Conducting Elements		
Aluminum	2.8×10^{-8}	3.9×10^{-3}
Copper	1.7×10^{-8}	3.93×10^{-3}
Iron	10×10^{-8}	5.0×10^{-3}
Lead	22×10^{-8}	4.3×10^{-3}
Mercury	96×10^{-8}	0.89×10^{-3}
Platinum	100×10^{-8}	3.927×10^{-3}
Silver	1.6×10^{-8}	3.8×10^{-3}
Tungsten	5.5×10^{-8}	4.5×10^{-3}
Carbon	3500×10^{-8}	-0.5×10^{-3}
Conducting alloys		
Brass	$\sim 8 \times 10^{-8}$	2×10^{-3}
Constantin (60% Cu, 40% Ni)	$\sim 44 \times 10^{-8}$	0.002×10^{-3}
Manganin ($\sim$84% Cu, $\sim$12% Mn, $\sim$4% Ni)	44×10^{-8}	0.000×10^{-3}
Nichrome	100×10^{-8}	0.4×10^{-3}
Semiconductors		
Germanium	0.45	-4.8×10^{-2}
Silicon	640	-7.5×10^{-2}
Insulators		
Neoprene	$\sim 10^{9}$	
Polystyrene	$\sim 10^{8}$	
Porcelain	$\sim 10^{11}$	
Wood	$10^{8} - 10^{14}$	
Glass	$10^{10} - 10^{14}$	
Hard rubber	$10^{13} - 10^{16}$	
Amber	5×10^{14}	
Sulfur	1×10^{15}	
Teflon	1×10^{14}	
Body material		
Blood	1.5	
Fat	25	

Table 25-2 Wire Diameters and Cross-Sectional Areas for Commonly Used Copper Wires

AWG* Gauge Number	Diameter† at 20°C, mm	Area, mm^2
4	5.189	21.15
6	4.115	13.30
8	3.264	8.366
10	2.588	5.261
12	2.053	3.309
14	1.628	2.081
16	1.291	1.309
18	1.024	0.8235
20	0.8118	0.5176
22	0.6438	0.3255

* American wire gauge.

† The diameter d is related to the gauge number n by $d = 0.127 \times 92^{[(36-n)/39]}$.

Table 25-1 gives the resistivity and the temperature coefficient at 20°C for various materials. Note the tremendous range of resistivity values for various materials at 20°C. The classical theory of conduction predicts that the resistivities of metals decrease with increases in temperature, which is one of several reasons that the classical theory of conduction has been discredited. However, the increases in resistivities of metals with increases in temperature are consistent with the quantum-mechanical theory of conduction. Both the classical and the quantum-mechanical theories of conduction are presented in Chapter 38.

Electrical wires are manufactured in standard sizes. The diameter of the circular cross section is indicated by a *gauge number*—larger numbers correspond to smaller diameters—as can be seen in Table 25-2.

Example 25-3 Resistance per Unit Length

Calculate the resistance per unit length of a 14-gauge copper wire.

PICTURE To calculate the resistance per unit length of the 14-gauge wire, you will need to find the resistivity of copper from Table 25-1 and the cross-sectional area of the copper wire from Table 25-2.

SOLVE

1. From Equation 25-10, the resistance per unit length equals the resistivity divided by unit area:
$$R = \rho\frac{L}{A} \quad \text{so} \quad \frac{R}{L} = \frac{\rho}{A}$$
2. Find the resistivity of copper from Table 25-1 and the cross-sectional area of the copper wire from Table 25-2:
$$\rho = 1.7 \times 10^{-8}\ \Omega\cdot\text{m}$$
$$A = 2.08\ \text{mm}^2$$
3. Use these values to find R/L:
$$\frac{R}{L} = \frac{\rho}{A} = \frac{1.7 \times 10^{-8}\ \Omega\cdot\text{m}}{2.08 \times 10^{-6}\ \text{m}^2} = \boxed{8.2 \times 10^{-3}\ \Omega/\text{m}}$$

CHECK 14-gauge copper wire is commonly used for household lighting circuits. The resistance of a 100-W, 120-V lightbulb filament is 144 Ω and the resistance of 100 m of 14-gauge copper wire is 0.82 Ω, so the resistance of the 14-gauge copper wire is negligible compared to the resistance of the lightbulb filament as expected.

Carbon, which has a relatively high resistivity, is used in resistors found in electronic equipment. Resistors are often marked with colored stripes that indicate their resistance value. The code for interpreting these colors is given in Table 25-3.

Table 25-3 The Color Code for Resistors and Other Devices

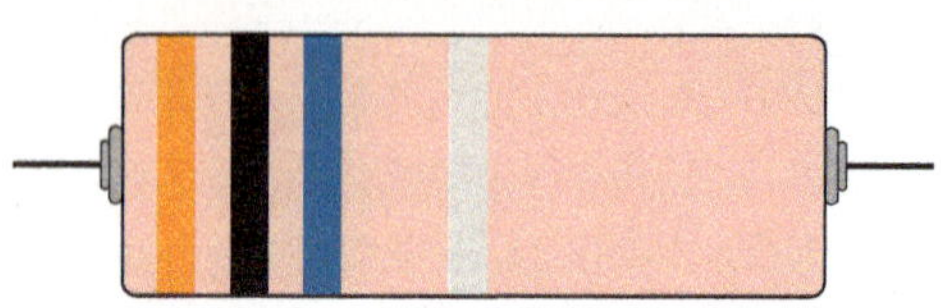

Colors	Numeral	Tolerance	
Black	= 0	Brown	= 1%
Brown	= 1	Red	= 2%
Red	= 2	Gold	= 5%
Orange	= 3	Silver	= 10%
Yellow	= 4	None	= 20%
Green	= 5		
Blue	= 6		
Violet	= 7		
Gray	= 8		
White	= 9		

The color bands consist of a group of three or four equally spaced bands that represent the value of the resistance in ohms, plus an additional tolerance band that is separate from the group. The value bands are read starting with the band closest to the end of the resistor. If there are three value bands, the first two bands represent a number between 1 and 99 the third band represents the number of zeros that follow. For the resistor shown, the colors of the first three bands are, respectively, orange, black, and blue. Thus, the number is 30 000 000 and the resistance value is 30 MΩ. (If a green band had been inserted between the black and blue bands, the resistance value would have been 305 MΩ.) The band separated from the others is the tolerance band. If the tolerance band is silver, as shown here, the tolerance is 10 percent. Ten percent of 30 is 3, so the resistance value is (30 ± 3) MΩ.

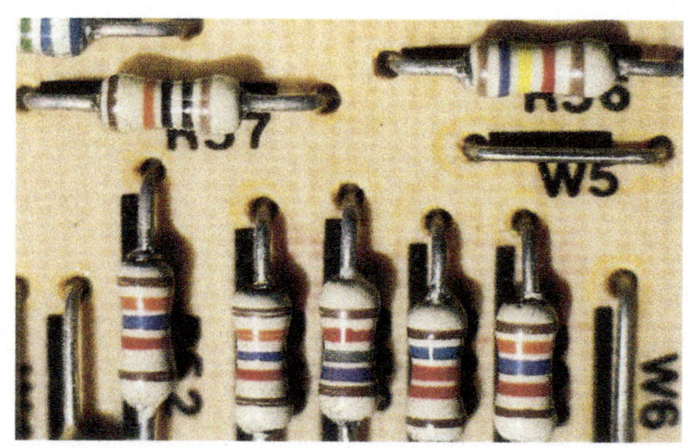

Color-coded carbon resistors on a circuit board. *(© Chris Rogers/The Stock Market.)*

PRACTICE PROBLEM 25-5

What are the values of the resistance and tolerance of the resistor shown on the lower left in the photo?

Example 25-4 The Electric Field That Drives the Current

Find the electric field strength E in the 14-gauge copper wire of Example 25-3 when the wire has a current equal to 1.3 A.

PICTURE We find the electric field strength as the potential drop for a given length of wire, $E = V/L$. The potential drop is found using Ohm's law, $V = IR$, and the resistance per length is given in Example 25-3.

SOLVE

1. The electric field strength equals the potential drop per unit length: $E = \dfrac{V}{L}$
2. Write Ohm's law for the potential drop: $V = IR$
3. Substitute this expression into the equation for E: $E = \dfrac{V}{L} = \dfrac{IR}{L} = I\dfrac{R}{L}$
4. Substitute the value of R/L found in Example 25-3 to calculate E: $E = I\dfrac{R}{L} = (1.3\text{ A})(8.2 \times 10^{-3}\ \Omega/\text{m}) = \boxed{0.011\text{ V/m}}$

CHECK An electric field of 0.011 V/m means that the potential drop for a 100-m length of the wire is 1.1 V. This result seems acceptable for a 120-V household circuit. However, a 13-A current would mean an 11-V drop, which is much less acceptable. (It is unacceptable because many devices do not function properly if the potential difference applied across their terminals is significantly less than the full 120 volts.)

25-3 ENERGY IN ELECTRIC CIRCUITS

When there is an electric field in a conductor, the free electrons gain kinetic energy due to the work done on the free electrons by the field. However, steady state is soon achieved as the kinetic energy gain is continuously dissipated into the thermal energy of the conductor by interactions between the free electrons and the lattice ions of the conductor. This mechanism for increasing the thermal energy of a conductor is called **Joule heating.**

Consider the segment of wire of length L and cross-sectional area A shown in Figure 25-10*a*. The wire carries a steady current that we will model as positive free charge in the wire that is moving to the right. Consider the free charge Q initially in the segment. During time Δt, this free charge undergoes a small displacement to the right (Figure 25-10*b*). This displacement is equivalent to an amount of charge ΔQ (Figure 25-10*c*) being moved from its left end, where it had potential energy $\Delta Q\, V_a$, to its right end, where it has potential energy $\Delta Q\, V_b$. The net change in the potential energy of Q is thus

$$\Delta U = \Delta Q(V_b - V_a)$$

Because $V_a > V_b$, this represents a net loss in the potential energy. The potential energy lost is then

$$-\Delta U = \Delta Q\, V$$

where $V = V_a - V_b$ is the *potential drop* across the segment in the direction of the current. The rate of potential energy loss is

$$-\frac{\Delta U}{\Delta t} = \frac{\Delta Q}{\Delta t} V$$

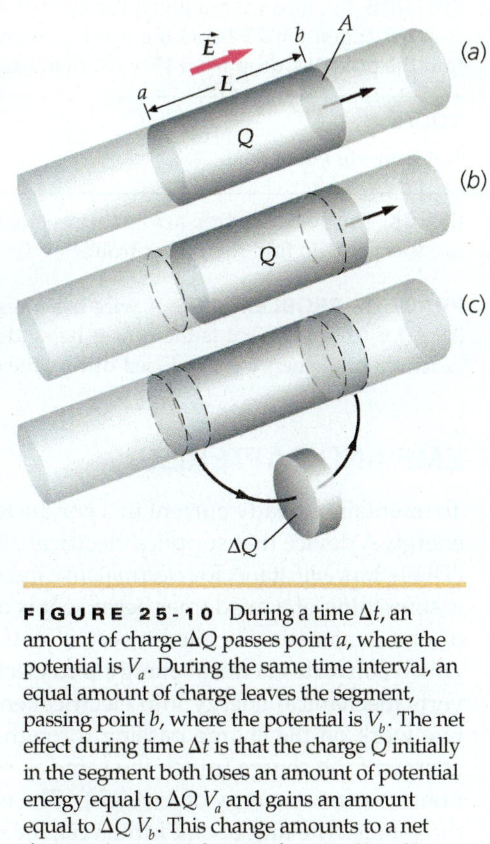

FIGURE 25-10 During a time Δt, an amount of charge ΔQ passes point a, where the potential is V_a. During the same time interval, an equal amount of charge leaves the segment, passing point b, where the potential is V_b. The net effect during time Δt is that the charge Q initially in the segment both loses an amount of potential energy equal to $\Delta Q\, V_a$ and gains an amount equal to $\Delta Q\, V_b$. This change amounts to a net decrease in potential energy because $V_a > V_b$.

Taking the limit as Δt approaches zero gives

$$-\frac{dU}{dt} = \frac{dQ}{dt}V = IV$$

where $I = dQ/dt$ is the current. The rate of potential energy loss is the power P delivered to the conducting segment, and it is equal to the rate of dissipation of electrical potential energy in the segment:

$$P = IV \qquad 25\text{-}13$$

RATE OF POTENTIAL ENERGY LOSS

If V is in volts and I is in amperes, the power is in watts. The power loss is the product IV, where V is the decrease in potential energy per unit charge, and I is the rate at which the charge flows past a cross section of the segment. Equation 25-13 applies to any device in a circuit. The rate at which potential energy is delivered to the device is the product of the potential drop across the device in the direction of the current, and the current through the device. In a conductor (a resistor is a conductor), the potential energy is dissipated as thermal energy. Using $V = IR$, or $I = V/R$, we can write Equation 25-13 in other convenient forms

$$P = IV = I^2R = \frac{V^2}{R} \qquad 25\text{-}14$$

POWER DELIVERED TO A RESISTOR

Example 25-5 Power Delivered to a Resistor

A 12.0-Ω resistor has a current equal to 3.00 A. Find the power delivered to this resistor.

PICTURE Because we are given the current and the resistance, but not the potential drop, $P = I^2R$ (Equation 25-14) is the most convenient equation to use. Alternatively, we could find the potential drop from $V = IR$, then use $P = IV$.

SOLVE

1. Compute I^2R: $P = I^2R = (3.00\text{ A})^2(12.0\ \Omega) = \boxed{108\text{ W}}$

CHECK The potential drop across the resistor is $V = IR = (3.00\text{ A})(12.0\ \Omega) = 36.0\text{ V}$. We can use this value to find the power from $P = IV = (3.00\text{ A})(36.0\text{ V}) = 108\text{ W}$.

PRACTICE PROBLEM 25-6 A wire has a resistance equal to 5.0 Ω and a current equal to 3.0 A for 6.0 s. (*a*) What is the power being delivered to the wire during the 6.0 s? (*b*) How much thermal energy is produced during the 6.0 s?

EMF AND BATTERIES

To maintain a steady current in a conductor, we need a constant supply of electrical energy. A device that supplies electrical energy to a circuit is called a **source of emf.** (The letters *emf* stand for *electromotive force* a term that is now rarely used. The term is something of a misnomer because it is definitely not a force. In addition, a source of emf is sometimes called a seat of emf.) Examples of emf sources are a battery, which converts chemical energy into electrical energy, and a generator, which converts mechanical energy into electrical energy. A source of emf does nonconservative work on the charge passing through it, increasing or decreasing the potential energy of the charge (much the same as your lifting a weight increases the gravitational potential energy of a weight). The work per unit charge is called the **emf** $\mathcal{E}$ of the source. The unit of emf is the volt, the same unit as potential difference. An **ideal battery** is a source of emf that maintains a constant potential difference between its two terminals, independent of the current through the battery. The potential difference between the terminals of an ideal battery is equal in magnitude to the emf of the battery.

Figure 25-11 shows a simple circuit consisting of a resistance R connected to an ideal battery. The resistance is indicated by the symbol -ΛΛΛ-. The straight lines indicate connecting wires of negligible resistance. The source of emf ideally maintains a constant potential difference equal to emf $\mathcal{E}$ between points a and b, with point a being at the higher potential. There is negligible potential difference between points a and c and between points d and b, because the connecting wire is assumed to have negligible resistance. The potential drop from point c to d is therefore equal in magnitude to the emf $\mathcal{E}$, and the current I through the resistor is given by $I = \mathcal{E}/R$. The direction of the current in this circuit is clockwise, as shown in the figure.

Note that *inside* the source of emf, the charge flows from a region where its potential energy is low to a region where its potential is high, so the charge gains electric potential energy.* When charge ΔQ flows through the ideal source of emf $\mathcal{E}$, its potential energy is increased by the amount $\Delta Q\mathcal{E}$. The charge then flows through the resistor, where this potential energy is dissipated as thermal energy. The rate at which energy is supplied by the source of emf is the power output of the source:

The electric ray has two large electric organs on each side of its head, where current passes from the lower to the upper surface of the body. These organs are composed of columns, with each column consisting of one hundred forty to half a million gelatinous plates. In saltwater fish, these batteries are connected in parallel, whereas in freshwater fish the batteries are connected in series, transmitting discharges of higher voltage. Fresh water has a higher resistivity than salt water, so to be effective a higher voltage is required. It is with such a battery that an average electric ray can electrocute a fish, delivering 50 A at 50 V. *(Stephen Frink/CORBIS.)*

$$P = \frac{(\Delta Q)\mathcal{E}}{\Delta t} = I\mathcal{E} \qquad 25\text{-}15$$

POWER SUPPLIED BY AN IDEAL EMF SOURCE

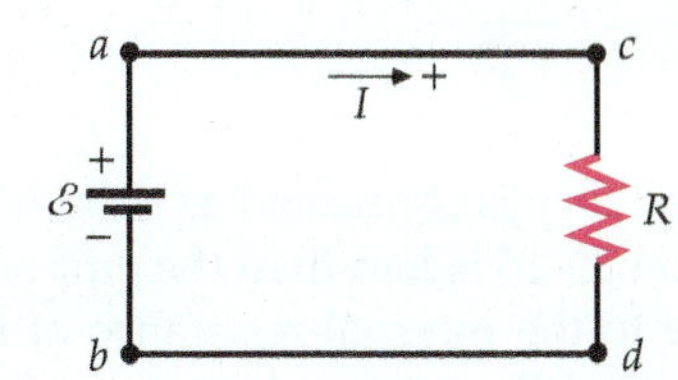

FIGURE 25-11 A simple circuit consisting of an ideal battery of emf $\mathcal{E}$, a resistance R, and connecting wires that are assumed to be of negligible resistance.

In the simple circuit of Figure 25-11, the power output by the ideal source of emf equals the power delivered to the resistor.

The battery in Figure 25-11 can be thought of as a charge pump that pumps the charge from a region where its potential energy is low to a region where its potential energy high. Figure 25-12 shows a mechanical analog of the simple electric circuit just discussed.

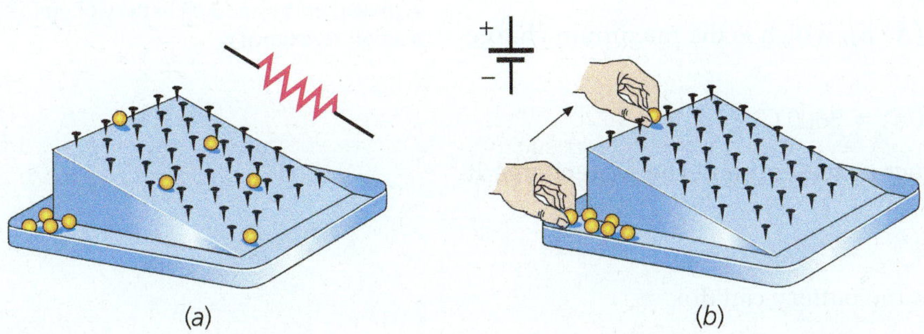

FIGURE 25-12 A mechanical analog of a simple circuit consisting of a resistance and source of emf. (*a*) The marbles start at some height h above the bottom and are accelerated between collisions with the nails by the gravitational field. The nails are analogous to the lattice ions in the resistor. During the collisions, the marbles transfer the kinetic energy they obtained between collisions to the nails. Because of the many collisions, the marbles have only a small, approximately constant, drift velocity toward the bottom. (*b*) When the marbles reach the bottom, a child picks them up, lifts them to their original height h, and starts them again. The child, who does work mgh on each marble of mass m, is analogous to the source of emf. The energy source in this case is the internal chemical energy of the child.

* When a battery is being charged (by a generator or by another battery), within the battery the charge flows from a region where its potential energy is high to a region where its potential energy is low, thus losing electric potential energy. The energy lost is converted to chemical energy and stored in the battery being charged.

In a **real battery,** the potential difference across the battery terminals, called the **terminal voltage,** is not simply equal to the emf of the battery. Consider a circuit consisting of a real battery and a variable resistor. If the current is varied by varying the resistance R and the terminal voltage V of the battery is measured, the terminal voltage is found to decrease as the current increases (Figure 25-13), as if there were a resistor internal to the battery.

Thus, we can consider a real battery to consist of an ideal source of emf $\mathcal{E}$ and a resistor with resistance r, called the **internal resistance** of the battery.

The circuit diagram for a nonideal battery and resistor is shown in Figure 25-14. If the current in the circuit is I, the potential at point a is related to the potential at point b by

$$V_a = V_b + \mathcal{E} - Ir$$

The terminal voltage is thus

$$V_a - V_b = \mathcal{E} - Ir \qquad 25\text{-}16$$

The terminal voltage of the battery decreases linearly with current, as we saw in Figure 25-13. The potential drop across the resistor R is IR and is equal to the terminal voltage:

$$IR = V_a - V_b = \mathcal{E} - Ir$$

Solving for the current I, we obtain

$$I = \frac{\mathcal{E}}{R + r} \qquad 25\text{-}17$$

If a battery is connected as shown in Figure 25-14, the terminal voltage given by Equation 25-16 is less than the emf of the battery because of the decrease in potential due to the internal resistance of the battery. Real batteries, such as a good car battery, usually have an internal resistance of the order of a few hundredths of an ohm, so the terminal voltage is nearly equal to the emf unless the current is very large. One sign of a bad battery is an unusually high internal resistance. If you suspect that your car battery is bad, checking the terminal voltage with a voltmeter, which draws very little current, is not always sufficient. You need to check the terminal voltage while current is being drawn from the battery, such as while you are trying to start your car. Then the terminal voltage may drop considerably, indicating a high internal resistance and a bad battery.

Batteries are often rated in ampere-hours (A · h), which is the maximum charge that batteries can deliver:

$$1\ \text{A}\cdot\text{h} = (1\ \text{C/s})(3600\ \text{s}) = 3600\ \text{C}$$

The energy stored in the battery is the product of the emf and the total charge it can deliver.

$$E_{\text{stored}} = Q\mathcal{E} \qquad 25\text{-}18$$

The stored energy is the amount of work that the battery can do.

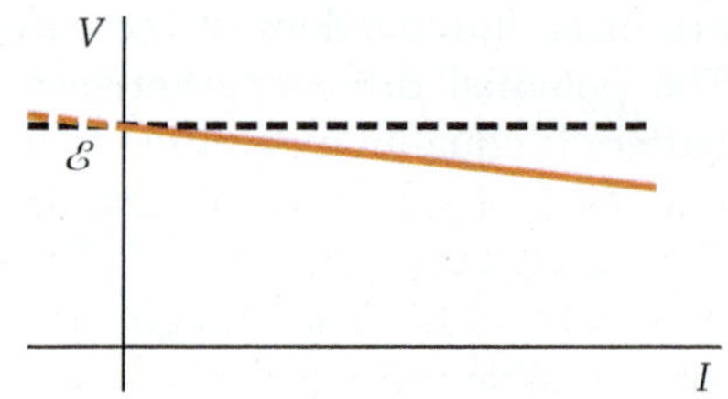

FIGURE 25-13 Terminal voltage V versus I for a real battery. The dashed line shows the terminal voltage of an ideal battery, which has the same magnitude as $\mathcal{E}$.

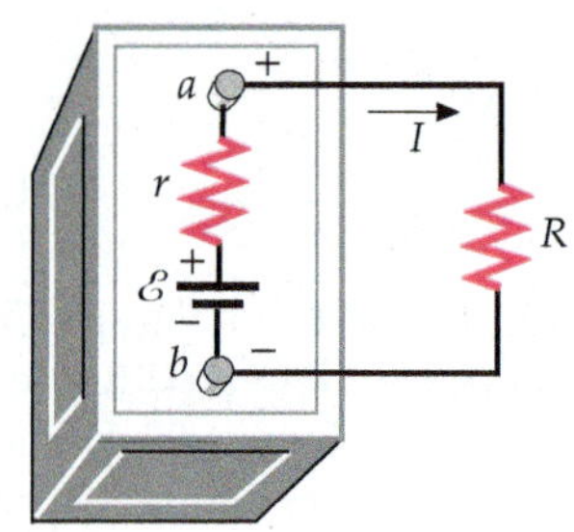

FIGURE 25-14 A real battery can be represented by an ideal battery of emf $\mathcal{E}$ and a small resistance r.

Example 25-6 Terminal Voltage, Power, and Stored Energy

An 11.0-Ω resistor is connected across a battery of emf 6.00 V and internal resistance 1.00 Ω. Find (*a*) the current, (*b*) the terminal voltage of the battery, (*c*) the power supplied by the chemical reactions within the battery, (*d*) the power delivered to the external resistor, and (*e*) the power delivered to the battery's internal resistance. (*f*) If the battery is rated at 150 A · h, how much energy does the battery store?

PICTURE The circuit diagram is the same as the circuit diagram shown in Figure 25-14. We find the current from $I = \mathcal{E}/(R + r)$ (Equation 25-17) and then use it to find the terminal voltage and power delivered to the resistors.

SOLVE

(*a*) Equation 25-17 gives the current: $I = \frac{\mathcal{E}}{R + r} = \frac{6.00\text{ V}}{11.0\ \Omega + 1.00\ \Omega} = \boxed{0.500\text{ A}}$

(*b*) Use the current to calculate the terminal voltage of the battery: $V_a - V_b = \mathcal{E} - Ir = 6.00\text{ V} - (0.500\text{ A})(1.00\ \Omega) = \boxed{5.50\text{ V}}$

(*c*) The power supplied by the chemical reactions within the battery equals $\mathcal{E}I$: $P = \mathcal{E}I = (6.00\text{ V})(0.500\text{ A}) = \boxed{3.00\text{ W}}$

(*d*) The power delivered to the external resistance equals I^2R (Equation 25-14): $I^2R = (0.500\text{ A})^2(11.0\ \Omega) = \boxed{2.75\text{ W}}$

(*e*) The power delivered to the internal resistance is I^2r: $I^2r = (0.500\text{ A})^2(1.00\ \Omega) = \boxed{0.250\text{ W}}$

(*f*) The energy stored is the emf of the battery multiplied by the total charge the battery can deliver: $W = Q\mathcal{E} = \left(150\text{ A}\cdot\text{h} \times \frac{3600\text{ C}}{\text{A}\cdot\text{h}}\right)(6.00\text{ V}) = \boxed{3.24\text{ MJ}}$

CHECK Of the 3.00 W of power supplied by the chemical reactions of the battery, 2.75 W is delivered to the external resistor and 0.250 W is dissipated in the battery due the internal resistance of the battery.

TAKING IT FURTHER The value of the internal resistance of the battery in this example is larger than that of most batteries. This value was chosen to simplify the calculations. In other examples, we may simply assume that the internal resistance of the battery is negligible.

Example 25-7 Maximum Power Delivered

For a battery that has an emf equal to $\mathcal{E}$ and internal resistance equal to r, what value of external resistance R should be placed across the terminals to obtain the maximum power delivered to the resistor?

PICTURE The circuit diagram is shown in Figure 25-14. The power delivered to the resistor is I^2R (Equation 25-14), where $I = \mathcal{E}/(R + r)$ (Equation 25-17). To find the value of R that results in the maximum power being delivered to the resistor, we set dP/dR equal to zero and solve for R.

SOLVE

1. Use $I = \mathcal{E}/(R + r)$ (Equation 25-17) to eliminate I from $P = I^2R$ so that P is written as a function of R and the constants $\mathcal{E}$ and r: $P = \frac{\mathcal{E}^2R}{(R + r)^2}$
2. Calculate the derivative dP/dR (We use the quotient rule.): $\frac{dP}{dR} = \frac{(R + r)^2\mathcal{E}^2 - 2\mathcal{E}^2R(R + r)}{(R + r)^4} = \frac{\mathcal{E}^2(r - R)}{(R + r)^3}$
3. Solve for the value of R for which dP/dR equals zero: $\boxed{R = r}$

CHECK For $R = 0$, the current is maximum but $P = 0$, so no power is delivered to the external resistor when $R = 0$. To take the limit of P as $R \to \infty$ we factor R from the denominator to obtain

$$P = \frac{\mathcal{E}^2R}{(R + r)^2} = \frac{\mathcal{E}^2}{R(1 + r/R)^2}$$

From this result we can see that as $R \to \infty$, $P \to 0$. This means that P must be maximum for R in the range $0 < R < \infty$, so $R = r$ is a plausible result.

TAKING IT FURTHER The maximum value of P occurs when $R = r$, that is, when the load resistance equals the internal resistance. A similar result holds for alternating-current circuits. Choosing $R = r$ to maximize the power delivered to the load is known as *impedance matching*. A graph of P versus R is shown in Figure 25-15.

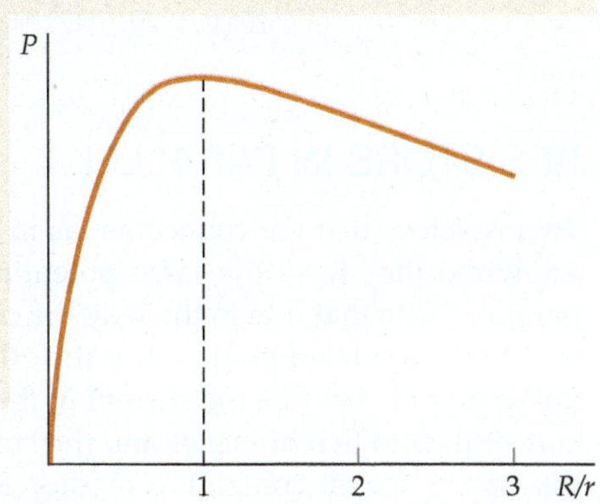

FIGURE 25-15 The power delivered to the external resistor is maximum if $R = r$.

25-4 COMBINATIONS OF RESISTORS

The analysis of a circuit can often be simplified by replacing a combination of two or more resistors with a single equivalent resistor that has the same current and potential drop as the combination of resistors. The replacement of a combination of resistors by an equivalent resistor is similar to the replacement of a combination of capacitors by an equivalent capacitor, discussed in Chapter 24.

RESISTORS IN SERIES

When two or more resistors are connected like R_1 and R_2 in Figure 25-16 so that due to the way they are connected the current in each resistor is the same, the resistors are said to be connected in series. The potential drop across R_1 is IR_1 and the potential drop across R_2 is IR_2, where I is the current in each resistor. The potential drop across the two resistors is the sum of the potential drops across the individual resistors:

$$V = IR_1 + IR_2 = I(R_1 + R_2) \qquad 25\text{-}19$$

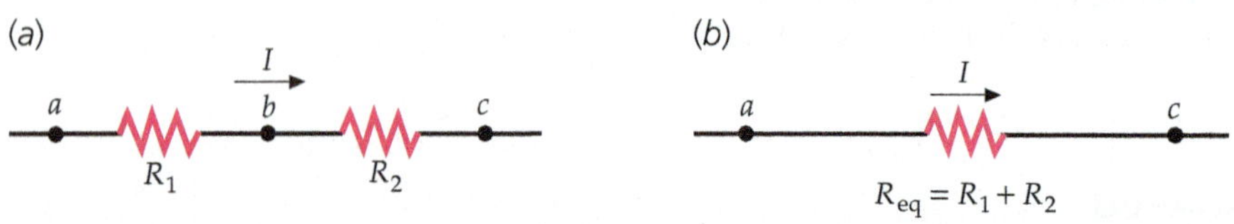

FIGURE 25-16 (*a*) Two resistors connected in series that carry the same current. (*b*) The resistors in Figure 25-16*a* can be replaced by a single equivalent resistance $R_{eq} = R_1 + R_2$ that gives the same total potential drop when carryng the same current as in Figure 25-16*a*.

The single equivalent resistance R_{eq} that yields the same total potential drop V when carrying the same current I is found by setting V equal to IR_{eq} (Figure 25-16*b*). Then R_{eq} is given by

$$R_{eq} = R_1 + R_2$$

When there are more than two resistors connected in series, the equivalent resistance is

$$R_{eq} + R_1 + R_2 + R_3 + \ldots \qquad 25\text{-}20$$

EQUIVALENT RESISTANCE FOR RESISTORS IN SERIES

RESISTORS IN PARALLEL

Two resistors that are connected, as in Figure 25-17*a*, so that due to the way they are wired they have the same potential difference across them, are connected in parallel. Note that due to the way the circuit is wired, one terminal of each resistor is at the potential of point *a*, and the other terminal of each resistor is at the potential of point *b*. Let I be the current in the wire leading to point *a*. At point *a*, the circuit splits into two branches and the current I divides into two parts—current I_1 in the upper branch containing resistor R_1 and current I_2 in the lower branch containing R_2. The *branch currents* I_1 and I_2 sum to the current I in the wire leading into point *a*:

$$I = I_1 + I_2 \qquad 25\text{-}21$$

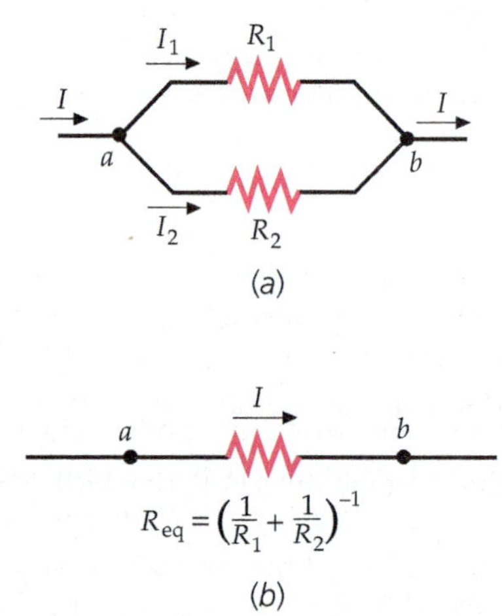

FIGURE 25-17 (*a*) Two resistors are in parallel when they are connected together at both ends so that the potential drop is the same across each. (*b*) The two resistors in Figure 25-17*a* can be replaced by an equivalent resistance R_{eq} that is related to R_1 and R_2 by $1/R_{eq} = 1/R_1 + 1/R_2$.

At point b the branch currents recombine so the current in the wire following point b is also equal to $I = I_1 + I_2$. The potential drop V across either resistor, $V = V_a - V_b$, is related to the branch currents by

$$V = I_1R_1 \text{ and } V = I_2R_2 \qquad 25\text{-}22$$

The equivalent resistance for parallel resistors is the resistance R_{eq} for which the same total current I requires the same potential drop V (Figure 25-17b):

$$V = IR_{eq} \qquad 25\text{-}23$$

Solving Equations 25-22 and 25-23 for I, I_1 and I_2 and substituting into $I = I_1 + I_2$ (Equation 25-21), we have

$$\frac{V}{R_{eq}} = \frac{V}{R_1} + \frac{V}{R_2} = V\left(\frac{1}{R_1} + \frac{1}{R_2}\right) \qquad 25\text{-}24$$

Dividing both sides by V gives

$$\frac{1}{R_{eq}} = \frac{1}{R_1} + \frac{1}{R_2}$$

which can be solved for the equivalent resistance, R_{eq}, for two resistors in parallel. This result can be generalized for parallel combinations such as that shown in Figure 25-18, in which three or more resistors are connected in parallel:

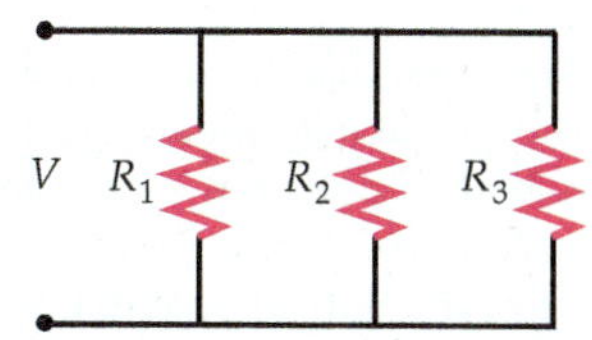

FIGURE 25-18 Three resistors in parallel.

$$\frac{1}{R_{eq}} = \frac{1}{R_1} + \frac{1}{R_2} + \frac{1}{R_3} + \cdots \qquad 25\text{-}25$$

EQUIVALENT RESISTANCE FOR RESISTORS IN PARALLEL

! The equivalent resistance of a parallel combination of resistors is less than the resistance of any single resistor in the combination.

PRACTICE PROBLEM 25-7

A 2.0-Ω resistor and a 4.0-Ω resistor are connected (a) in series and (b) in parallel. Find the equivalent resistances for both combinations.

The equivalent resistance of a parallel combination of resistors is less than the resistance of any single resistor in the combination. From Equation 25-25, we see that

$$\frac{1}{R_{eq}} > \frac{1}{R_i}$$

where R_i is the resistance of any single resistor in the combination. Multiplying both sides of this inequality by the product $R_{eq}R_i$, we obtain

$$R_i > R_{eq}$$

Resistors are, in reality, conductors. (They do not conduct as well as the wires connecting them in a circuit, but they are conductors nevertheless.) Adding more resistors in parallel means adding more conducting paths for charges to flow along. The creation of additional parallel paths lowers the equivalent resistance of the combination.

Example 25-8 Identifying Series and Parallel Combinations — Conceptual

The circuit shown in Figure 25-19 has a battery and six resistors. (a) Which resistors, if any, are connected in series? (b) Which resistors, if any, are connected in parallel?

PICTURE Resistors are connected in series if the current through each of them is the same due to the way they are connected. Resistors are connected in parallel if the potential difference (voltage) across each of them is the same due to the way they are connected.

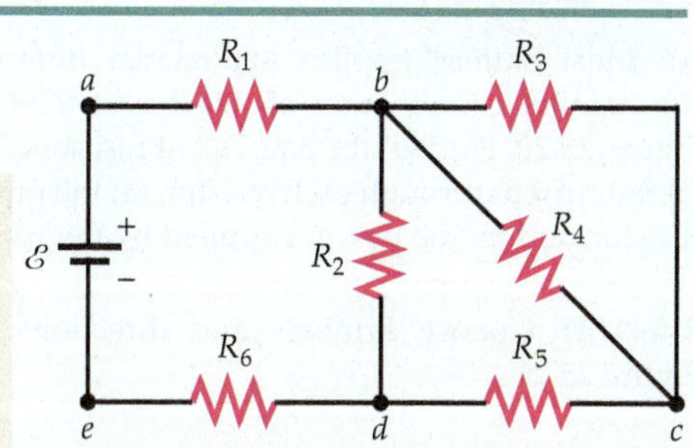

FIGURE 25-19

SOLVE

(*a*) In a circuit, the current only changes at junctions (points *b*, *c*, and *d*):

Resistors 1 and 6 are connected in series.

(*b*) 1. The potential along any path does not change except in batteries, resistors, or capacitors. Let V_a, V_b, V_c, V_d, and V_e be the potentials at points *a*, *b*, *c*, *d*, and *e*, respectively. Construct a table that lists the potential of both terminals of each resistor:

Resistor	V_a	V_b	V_c	V_d	V_e
1	X	X			
2		X		X	
3		X	X		
4		X	X		
5			X	X	
6				X	X

2. The table reveals that a terminal of resistor 3 and a terminal of resistor 4 are both at potential V_b, and the other terminals of the same resistors are both at potential V_c:

Resistors 3 and 4 are connected in parallel.

TAKING IT FURTHER Resistor 5 is in series with the parallel combination consisting of resistors 3 and 4. Resistor 2 is in parallel with the combination consisting of resistors 3, 4, and 5. In addition, resistor 6, the battery, resistor 1, and the combination of resistors 2, 3, 4, and 5 are in series.

PROBLEM-SOLVING STRATEGY

Problems Involving Series and/or Parallel Combinations of Resistors

PICTURE If no circuit diagram is provided, draw one.

SOLVE

1. Identify each series and/or parallel combination of resistors and calculate its equivalent resistance.
2. Redraw the circuit so that each series or parallel combination of resistors is replaced by a single resistor of equivalent resistance.
3. Repeat steps 1 and 2 until there are no more series or parallel combinations. (At this point the circuit should contain only a single resistor.) Apply $V = IR$ and calculate the current.
4. Return to the previous drawing and calculate the voltage across and/or the current in each resistor in that drawing.
5. Repeat step 4 until you calculate all currents and/or voltages of interest.

CHECK Calculate the power delivered to each resistor (using $P = IV$ or its equivalent) and calculate the power supplied by the chemical reactions in each battery using $P = I\mathcal{E}$. Then check to see that the total power being delivered equals the total power being supplied.

Example 25-9 Resistors in Parallel

An ideal battery applies a potential difference of 12 V across the parallel combination of 4.0-Ω and 6.0-Ω resistors shown in Figure 25-20. Find (*a*) the equivalent resistance, (*b*) the total current, (*c*) the current through each resistor, (*d*) the power delivered to each resistor, and (*e*) the power supplied by the battery.

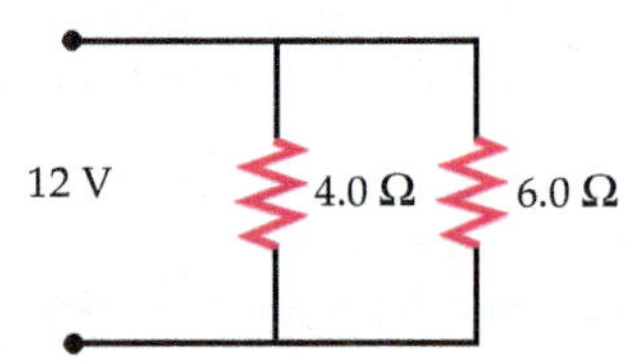

FIGURE 25-20

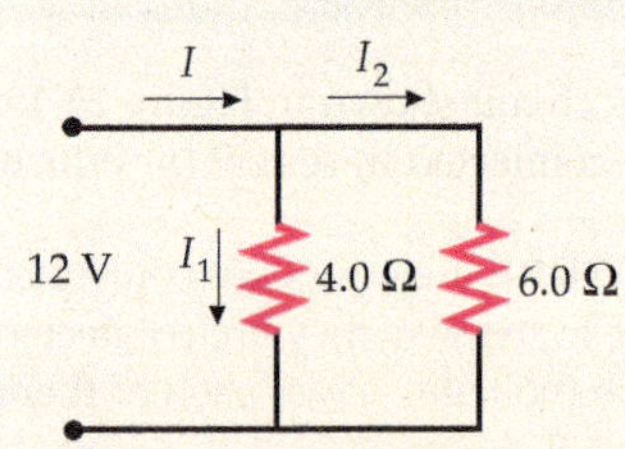

FIGURE 25-21

PICTURE Choose symbols and directions for the currents in Figure 25-21.

SOLVE

(*a*) Calculate the equivalent resistance:

$$\frac{1}{R_{eq}} = \frac{1}{4.0\ \Omega} + \frac{1}{6.0\ \Omega} = \frac{3.0}{12.0\ \Omega} + \frac{2.0}{12.0\ \Omega} = \frac{5.0}{12.0\ \Omega}$$

$$R_{eq} = \frac{12.0\ \Omega}{5.0} = \boxed{2.4\ \Omega}$$

(*b*) The total current is the potential drop divided by the equivalent resistance:

$$I = \frac{V}{R_{eq}} = \frac{12\ V}{2.4\ \Omega} = \boxed{5.0\ \text{A}}$$

(*c*) We obtain the current through each resistor using Equation 25-22 and using the potential drop across the parallel combination, 12 V:

$$V = IR$$

$$I_1 = \frac{12\ V}{4.0\ \Omega} = \boxed{3.0\ \text{A}}$$

$$I_2 = \frac{12\ V}{6.0\ \Omega} = \boxed{2.0\ \text{A}}$$

(*d*) Use $P = VI$ together with $V = IR$ to find the power delivered to each resistor:

$$P = VI = (IR)R = I^2R$$

$$P_1 = I_1^2R = (3.0\ \text{A})^2(4.0\ \Omega) = \boxed{36\ \text{W}}$$

$$P_2 = I_2^2R = (2.0\ \text{A})^2(6.0\ \Omega) = \boxed{24\ \text{W}}$$

(*e*) Use $P = \mathcal{E}I$ to find the power supplied by the battery:

$$P = \mathcal{E}I = (12\ \text{V})(5.0\ \text{A}) = \boxed{60\ \text{W}}$$

CHECK The power supplied by the battery equals the total power delivered to the two resistors $P = 60\ \text{W} = 36\ \text{W} + 24\ \text{W}$. In Part (*d*), we could just as well have calculated the power delivered to each resistor from $P_1 = VI_1 = (12\ \text{V})(3.0\ \text{A}) = 36\ \text{W}$ and $P_2 = VI_2 = (12\ \text{V})(2.0\ \text{A}) = 24\ \text{W}$.

TAKING IT FURTHER The ratio of the currents in the two parallel resistors equals the inverse ratio of the resistances. This result follows from $I_1R_1 = I_2R_2$ (Equation 25-22). Rearranging gives

$$\frac{I_1}{I_2} = \frac{R_2}{R_1} \text{ (two parallel resistors)} \qquad 25\text{-}26$$

Example 25-10 Resistors in Series

Try It Yourself

A 4.0-Ω resistor and a 6.0-Ω resistor are connected in series to a battery that has an emf equal to 12.0 V and has a negligible internal resistance. Find (*a*) the equivalent resistance of the two resistors, (*b*) the current in the circuit, (*c*) the potential drop across each resistor, (*d*) the power delivered to each resistor, and (*e*) the total power delivered to the resistors.

SOLVE

Cover the column to the right and try these on your own before looking at the answers.

Steps	**Answers**
(*a*) 1. Draw a circuit diagram (Figure 25-22):	
2. Calculate R_{eq} for the two series resistors:	$R_{eq} = \boxed{10.0\ \Omega}$
(*b*) Use $V = IR_{eq}$ to find the current through the battery:	$I = \boxed{1.2\ \text{A}}$
(*c*) Use Ohm's law to find the potential drop across each resistor:	$V_4 = \boxed{4.8\ \text{V}}$ $V_6 = \boxed{7.2\ \text{V}}$
(*d*) Find the power delivered to each resistor using $P = I^2R$:	$P_4 = \boxed{5.8\ \text{W}}$ $P_6 = \boxed{8.6\ \text{W}}$
(*e*) Add your results from Part (*d*) to find the total power:	$P = \boxed{14.4\ \text{W}}$

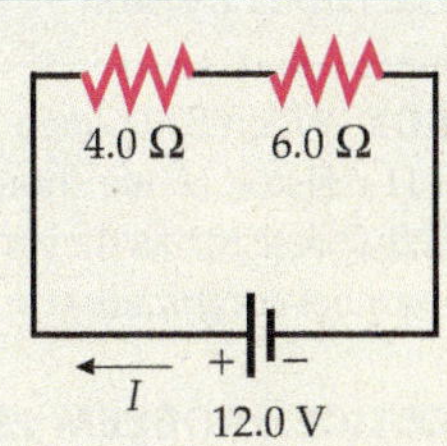

FIGURE 25-22

CHECK The current through the battery in this example is 1.2 A, but in the corresponding parallel circuit with the same resistors (Example 25-9) the current in the battery is 5.0 A. The current in a circuit is expected to be less when the resistors are connected in series.

Example 25-11 Series and Parallel Combinations

Try It Yourself

Consider the circuit in Figure 25-23. When switch S_1 is open and switch S_2 is closed, find (*a*) the equivalent resistance of the circuit, (*b*) the current in the source of emf, (*c*) the potential drop across each resistor, and (*d*) the current in each resistor. (*e*) If switch S_1 is now closed, find the current in the 2.0-Ω resistor. (*f*) If switch S_2 is now opened (while switch S_1 remains closed), find the potential drops across the 6.0-Ω resistor and across switch S_2.

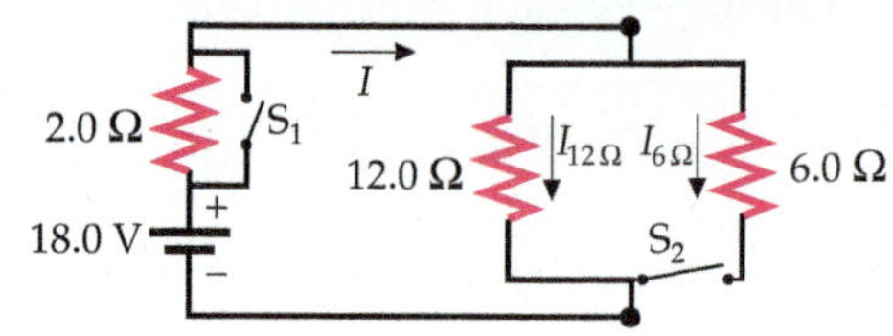

FIGURE 25-23

PICTURE (*a*) To find the equivalent resistance of the circuit, first replace the two parallel resistors by their equivalent resistance. Ohm's law can then be used to find the current and potential drops. For Part (*b*) and Part (*c*), use Ohm's law.

SOLVE

Cover the column to the right and try these on your own before looking at the answers.

Steps	Answers
(*a*) 1. Find the equivalent resistance of the 6.0- and 12.0-Ω parallel combination.	$R_{eq} = 4.0\ \Omega$
2. Combine your result in step 1 with the 2.0-Ω resistor in series to find the total equivalent resistance of the circuit.	$R'_{eq} = \boxed{6.0\ \Omega}$
(*b*) Find the current using Ohm's law. This is the current both in the battery and in the 2-Ω resistor.	$I_{2\Omega} = \boxed{3.0\ \text{A}}$
(*c*) 1. Find the potential drop across the 2.0-Ω resistor from $V_{2\Omega} = IR$.	$V_{2\Omega} = \boxed{6.0\ \text{V}}$
2. Find the potential drop across each resistor in the parallel combination using $V_p = IR_{eq}$, where V_p is the potential drop across the parallel combination.	$V_{6\Omega} = V_{12\Omega} = \boxed{12.0\ \text{V}}$
(*d*) Find the current in the 6.0-Ω and 12.0-Ω resistors from $I = V_p/R$.	$I_{6\Omega} = \boxed{2.0\ \text{A}}$ $I_{12\Omega} = \boxed{1.0\ \text{A}}$
(*e*) When S_1 is closed, the potential drop across the 2.0-Ω resistor is zero. Using $V_{2\Omega} = IR$, calculate the current through the 2.0-Ω resistor.	$I_{2\Omega} = \boxed{0}$
(*f*) When S_2 is open, the current through the 6.0-Ω resistor is zero. Using $V_{6\Omega} = IR$, calculate the potential drop across the 6.0-Ω resistor. The potential drop across the 6.0-Ω resistor plus the potential drop across switch S_2 equals the potential drop across the 12.0-Ω resistor.	$V_{6\Omega} = \boxed{0}$ $V_{S_2} = V_{12\Omega} = \boxed{18\ \text{V}}$

CHECK When S_1 is open and S_2 is closed, the current in the 6.0-Ω resistor is twice that in the 12.0-Ω resistor, as we should expect. In addition, these two currents sum to give the current in the 2.0-Ω resistor, as they must. Finally, note that the potential drops across the 2.0-Ω resistor and the parallel combination sum to the emf of the battery; $V_{2\Omega} + V_p = 6.0\ \text{V} + 12.0\ \text{V} = 18.0\ \text{V}$.

PRACTICE PROBLEM 25-8 Repeat Part (*a*) through Part (*d*) of this example, but with the 6.0-Ω resistor having been replaced by a wire of negligible resistance.

Example 25-12 Combinations of Combinations

Try It Yourself

Find the equivalent resistance of the combination of resistors shown in Figure 25-24.

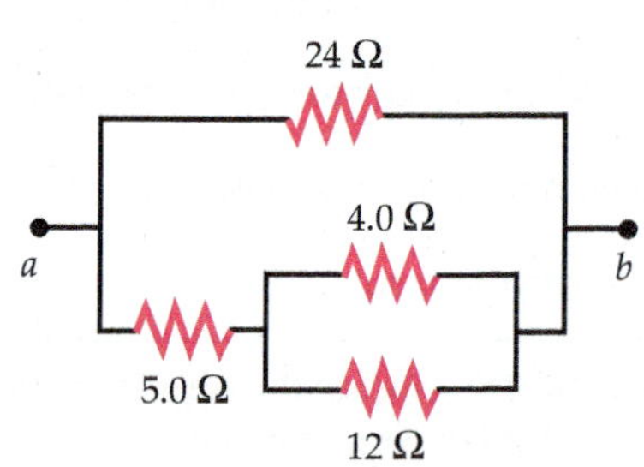

FIGURE 25-24

PICTURE You can analyze this complicated combination step by step. First, find the equivalent resistance R_{eq} of the 4.0-Ω and 12-Ω parallel combination. Next, find the equivalent resistance R'_{eq} of the series combination of the 5.0-Ω resistor and R_{eq}. Finally, find the equivalent resistance R''_{eq} of the parallel combination of the 24-Ω resistor and R'_{eq}.

SOLVE

Cover the column to the right and try these on your own before looking at the answers.

Steps	Answers
1. Find the equivalent resistance R_{eq} of the 4.0-Ω and 12-Ω resistors in parallel.	$R_{eq} = 3.0\ \Omega$
2. Find the equivalent resistance R'_{eq} of R_{eq} in series with the 5.0-Ω resistor.	$R'_{eq} = 8.0\ \Omega$
3. Find the equivalent resistance of R'_{eq} in parallel with the 24-Ω resistor.	$R''_{eq} = \boxed{6.0\ \Omega}$

CHECK As expected for parallel combinations, the step-1 and step-3 results are less than the resistance of either of the two resistors in parallel. In addition, the step-2 result is greater than either of the two resistors in series, as expected for series combinations.

Example 25-13 Blowing the Fuse

Context-Rich

You are making a snack for you and some friends to help you get ready for a full night of studying. You decide that coffee, toast, and popcorn would be a good start. You turn on the toaster, place some popcorn in the microwave, and turn on the microwave. Because your apartment is in an older building, you know that the fuse may blow when you turn on too many appliances. Should you start the coffeemaker? You look on the appliances and find that the toaster has a rating of 900 W, the microwave has a rating of 1200 W, and the coffeemaker has a rating of 600 W. You know that your apartment uses 20-A fuses.

PICTURE We can assume that household circuits are wired in parallel, because plugging in one device does not affect others that are in the circuit. Household voltage in the United States is 120 V. (We can neglect the fact that it is not dc.) If we can determine the current through each device, we can add up the total current in the circuit and see how it compares to the fuse current.

SOLVE

1. The power delivered to a device is the current multiplied by the voltage. That is, $P = IV$. Solve for the current for each device:

$$I_{toaster} = \frac{P_{toaster}}{V} = \frac{900\text{ W}}{120\text{ V}} = 7.5\text{ A}$$

$$I_{m\text{-}wave} = \frac{P_{m\text{-}wave}}{V} = \frac{1200\text{ W}}{120\text{ V}} = 10.0\text{ A}$$

$$I_{c\text{-}maker} = \frac{P_{c\text{-}maker}}{V} = \frac{600\text{ W}}{120\text{ V}} = 5.0\text{ A}$$

2. The current through the fuse is the sum of these currents:

$$I_{fuse} = 22.5\text{ A}$$

3. A current this large is above the 20-A rating of the fuse:

Your guests will have to wait on the coffee. (boxed)

CHECK The maximum possible power that can be delivered by a 120-V circuit that has a 20-A fuse is $P_{max} = I_{max}V = (20\text{ A})(120\text{ V}) = 2400\text{ W}$. The total power needed to run the three appliances simultaneously is 900 W + 1200 W + 600 W = 2700 W, which is 300 W more than the maximum that the circuit can deliver.

TAKING IT FURTHER We have assumed that the apartment has only one circuit, and thus only one fuse. Typically, there are several circuits, each fused separately. The coffeemaker can be plugged into an outlet that is on a different circuit than the outlet for the toaster and microwave without a fuse blowing.

25-5 KIRCHHOFF'S RULES

There are many circuits, such as the circuit shown in Figure 25-25, that cannot be analyzed by merely replacing combinations of resistors by an equivalent resistance. The two resistors R_1 and R_2 in this circuit appear as though they might be in parallel, but they are not. The potential drop is not the same across both resistors because of the presence of the emf source $\mathcal{E}_2$ in series with R_2. Nor are R_1 and R_2 in series, because the wire connecting them has a branch point—they do not have the same current due to the way they are connected.

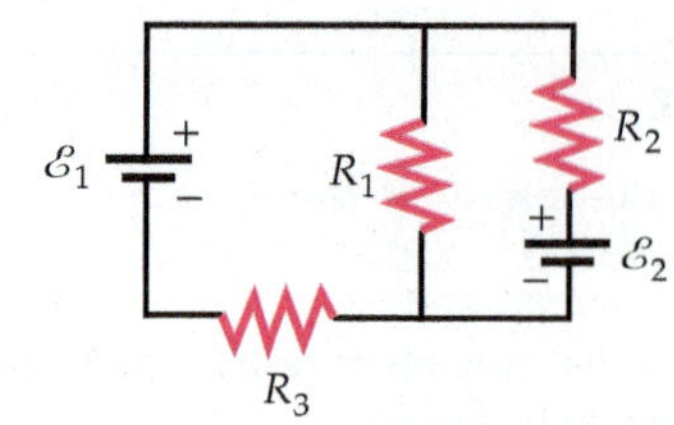

FIGURE 25-25 An example of a circuit that cannot be analyzed by replacing combinations of resistors in series or parallel with their equivalent resistances. The potential drops across R_1 and R_2 are not equal because of the emf source $\mathcal{E}_2$, so these resistors are not in parallel. (Parallel resistors would be connected together at both ends.) The resistors do not have the same current, so they are not in series.

Two rules, called **Kirchhoff's rules,** apply to this circuit and to any other circuit:

1. When any closed loop is traversed, the algebraic sum of the changes in potential around the loop must equal zero.
2. At any junction (branch point) in a circuit where the current can divide, the sum of the currents into the junction must equal the sum of the currents out of the junction.

KIRCHHOFF'S RULES

Kirchhoff's first rule, called the **loop rule,** was introduced in Chapter 24. This rule follows directly from the presence of a conservative field $\vec{E}$.* To say $\vec{E}$ is conservative means that

$$\oint_C \vec{E} \cdot d\vec{r} = 0 \qquad 25\text{-}27$$

where the integral is taken around any closed curve C. Changes in potential ΔV and $\vec{E}$ are related by $\Delta V = V_b - V_a = -\int_a^b \vec{E} \cdot d\vec{r}$. Thus, Equation 25-27 implies that the sum of the changes in potential (the sum of the ΔVs) around any closed path equals zero.

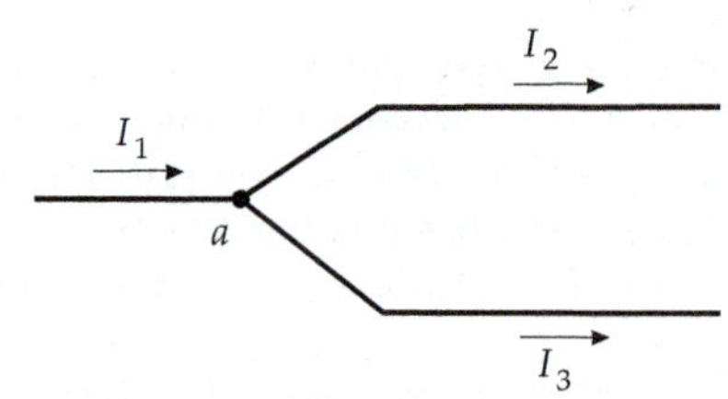

FIGURE 25-26 Illustration of Kirchhoff's junction rule. The current I_1 into point a equals the sum $I_2 + I_3$ of the currents out of point a.

Kirchhoff's second rule, called the **junction rule,** follows from the conservation of charge. Figure 25-26 shows the junction of three wires carrying currents I_1, I_2, and I_3. Because charge does not originate or accumulate at this point, the conservation of charge implies the junction rule, which for this case gives

$$I_1 = I_2 + I_3 \qquad 25\text{-}28$$

Charges do accumulate on the surfaces of conductors. However, it would require a very large surface area, such as the surface area of some capacitor plates, to accumulate a significant amount of charge. The surface areas of the conductors that are used in common circuits are much too small to accumulate a significant amount of charge.

SINGLE-LOOP CIRCUITS

As an example of using Kirchhoff's loop rule, consider the circuit shown in Figure 25-27, which contains two batteries that have internal resistances r_1 and r_2 and three external resistors. We wish to find the current in terms of the emfs and resistances.

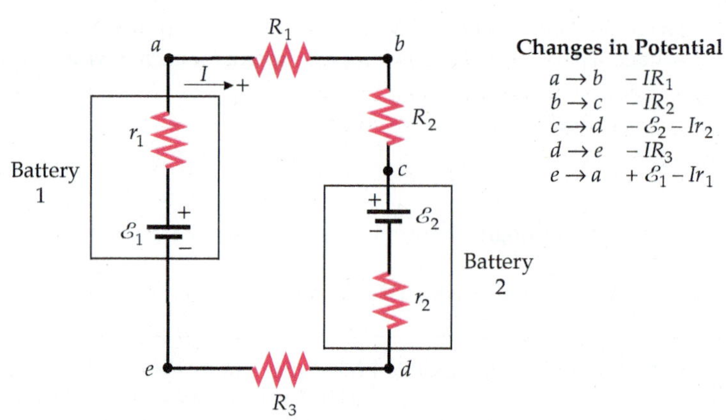

FIGURE 25-27 Circuit containing two batteries and three external resistors.

* There is also such a thing as a nonconservative electric field that is discussed in Chapter 28. The resultant electric field is the superposition of the conservative electric field and the nonconservative electric field. Kirchhoff's loop rule applies only to the conservative part of the electric field.

We choose clockwise as positive, as indicated by the arrow with the plus sign next to it in Figure 25-27. We then apply Kirchhoff's loop rule as we traverse the circuit in the positive direction, beginning at point a. Note that we encounter a potential drop as we traverse the source of emf between points c and d and we encounter a potential rise as we traverse the source of emf between e and a. Assuming that I is positive, we encounter a potential drop as we traverse each resistor. Beginning at point a, we obtain from Kirchhoff's loop rule

$$(V_b - V_a) + (V_c - V_b) + (V_d - V_c) + (V_e - V_d) + (V_a - V_e) = 0$$

Expressing the changes in potential in terms of the current, the emfs, and the resistances gives

$$(-IR_1) + (-IR_2) + (-\mathcal{E}_2 - Ir_2) + (-IR_3) + (\mathcal{E}_1 - Ir_1) = 0$$

Solving for the current I, we obtain

$$I = \frac{\mathcal{E}_1 - \mathcal{E}_2}{R_1 + R_2 + R_3 + r_1 + r_2} \qquad \text{25-29}$$

If $\mathcal{E}_2$ is greater than $\mathcal{E}_1$, we get a negative value for the current I, indicating that the current is in the negative (counterclockwise) direction.

For this example, we assume that $\mathcal{E}_1$ is greater than $\mathcal{E}_2$, so the current is positive. In addition, we model the current as positive charge carriers that move clockwise around the circuit. (The actual charge carriers are negatively charged electrons traveling counterclockwise.) Then, charge flows through battery 2 from the high-potential end to the low-potential end. Therefore, a positive charge ΔQ moving through battery 2 from point c to point d loses potential energy $\Delta Q\mathcal{E}_2$ (in addition to any energy dissipated within the battery due to the internal resistance r_2). If battery 2 is a rechargeable battery, much of this lost potential energy is stored in the battery as chemical energy, which means that battery 2 is *charging.*

The analysis of a circuit is usually simplified if we define the potential to equal zero at a convenient point in the circuit. Then we calculate the potential at the other points relative to it. Because only potential differences are important, any point in a circuit can be chosen to have zero potential. In many circuits, however, one point is connected to a rod that is driven into the ground. Such a point is said to be grounded or put to earth, and the potential is defined to be zero at that point. However, in an automobile the negative terminal of the battery is connected to the engine block by a heavy cable (called a grounding cable), and the point where the cable is connected to the engine block is referred to as ground. In the following example, we choose point e in the figure to be at zero potential. This is indicated by the ground symbol ⏚ at point e.

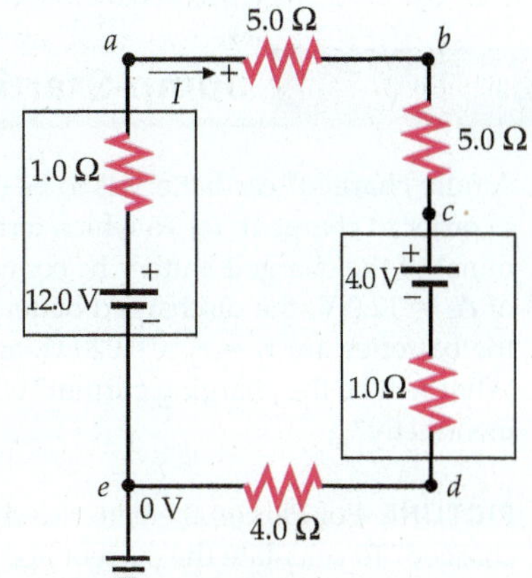

FIGURE 25-28

Example 25-14 Finding the Potential

Suppose the elements in the circuit in Figure 25-28 have the values $\mathcal{E}_1 = 12.0$ V, $\mathcal{E}_2 = 4.0$ V, $r_1 = r_2 = 1.0\ \Omega$, $R_1 = R_2 = 5.0\ \Omega$, and $R_3 = 4.0\ \Omega$. (*a*) Find the potentials at points a through e in the figure, assuming that the potential at point e is zero. (*b*) Discuss the energy transfers in the circuit.

PICTURE To find the potential differences, we first need to find the current I in the circuit. The potential drop across each resistor is equal to the product IR. To discuss the energy transfers, we calculate the power delivered to or supplied by each element using Equations 25-14 and 25-15.

SOLVE

(*a*) 1. The current I in the circuit is found using Equation 25-29:

$$I = \frac{12.0\ \text{V} - 4.0\ \text{V}}{5.0\ \Omega + 5.0\ \Omega + 4.0\ \Omega + 1.0\ \Omega + 1.0\ \Omega} = \frac{8.0\ \text{V}}{16\ \Omega} = 0.50\ \text{A}$$

2. We now find the potential at each labeled point in the circuit:

$$V_a = V_e + \mathcal{E}_1 - Ir_1 = 0 + 12.0\ \text{V} - (0.50\ \text{A})(1.0\ \Omega) = \boxed{11.5\ \text{V}}$$

$$V_b = V_a - IR_1 = 11.5\ \text{V} - (0.50\ \text{A})(5.0\ \Omega) = \boxed{9.0\ \text{V}}$$

$$V_c = V_b - IR_2 = 9.0\ \text{V} - (0.50\ \text{A})(5.0\ \Omega) = \boxed{6.5\ \text{V}}$$

$$V_d = V_c - \mathcal{E}_2 - Ir_2 = 6.5\ \text{V} - 4.0\ \text{V} - (0.50\ \text{A})(1.0\ \Omega) = \boxed{2.0\ \text{V}}$$

$$V_e = V_d + IR_3 = 2.0\ \text{V} - (0.50\ \text{A})(4.0\ \Omega) = \boxed{0.0\ \text{V}}$$

(*b*) 1. First, calculate the power supplied by the chemical reactions in the emf source that has emf $\mathcal{E}_1$:

$$P_{\mathcal{E}_1} = \mathcal{E}_1 I = (12.0\ \text{V})(0.50\ \text{A}) = 6.0\ \text{W}$$

2. Part of this power is delivered to the resistors, both inside and outside the batteries:

$$P_R = I^2R_1 + I^2R_2 + I^2R_3 + I^2r_1 + I^2r_2$$
$$= (0.50\ \text{A})^2(5.0\ \Omega + 5.0\ \Omega + 4.0\ \Omega + 1.0\ \Omega + 1.0\ \Omega) = 4.0\ \text{W}$$

3. The remaining 2.0 W of power goes into charging battery 2:

$$P_{\mathcal{E}_2} = \mathcal{E}_2 I = (4.0\ \text{V})(0.50\ \text{A}) = 2.0\ \text{W}$$

4. The rate at which potential energy is being delivered in the circuit is

$$P = P_R + P_{\mathcal{E}_2} = 6.0\ \text{W}$$

CHECK The rate at which the 12-V battery converts chemical energy to electrical potential energy (6.0 W) is equal to the rate at which the 4.0-V battery converts electrical potential energy to chemical energy (2.0 W) plus the rate at which potential energy is dissipated (4.0 W).

Note that the terminal voltage of the battery that is being charged in Example 25-14 is $V_c - V_d = 4.5$ V, which is greater than the emf of the battery. If the same 4.0-V battery were used to deliver 0.50 A to an external circuit, its terminal voltage would be 3.5 V (again assuming that its internal resistance is 1.0 Ω). If the internal resistance is very small, the terminal voltage of a battery is nearly equal to its emf, whether the battery is delivering energy to an external circuit or is being charged. Some real batteries, such as those used in automobiles, are nearly reversible and can easily be recharged. Other types of batteries are not reversible. If you attempt to recharge one of those batteries by driving current through it from its positive to its negative terminal, virtually all of the energy will be dissipated into thermal energy rather than being transformed into the chemical energy of the battery.

Example 25-15 Jump-Starting a Car

A fully charged* car battery is to be connected by jumper cables to a discharged car battery in order to charge it. (*a*) To which terminal of the discharged battery should the positive terminal of the charged battery be connected? (*b*) Assume that the charged battery has an emf of $\mathcal{E}_1 = 12.0$ V, the discharged battery has an emf of $\mathcal{E}_2 = 11.0$ V, the internal resistances of the batteries are $r_1 = r_2 = 0.020\ \Omega$, and the resistance of the jumper cables is $R = 0.010\ \Omega$. What will be the charging current? (*c*) What will be the current if the batteries are connected incorrectly?

PICTURE For Part (*a*) the batteries should be connected so the initially discharged battery is charged. To calculate the current apply Kirchhoff's loop rule.

* Batteries do not store charge. A *fully charged* battery is one with a maximum amount of stored chemical energy.

SOLVE

(*a*) To charge the discharged battery, we connect positive terminal to positive terminal and negative terminal to negative terminal, to drive current through the discharged battery from the positive terminal to the negative terminal (Figure 25-29):

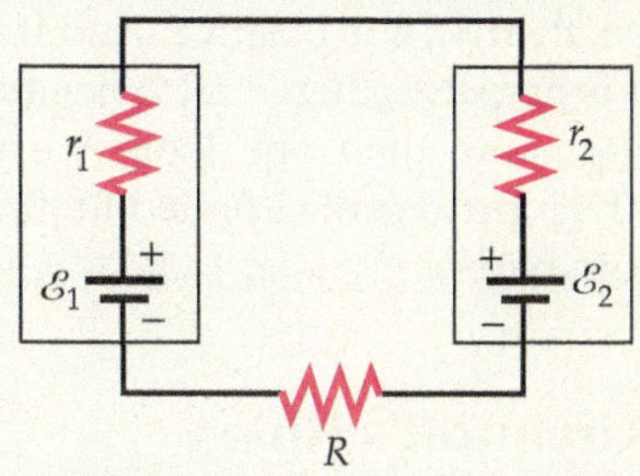

FIGURE 25-29

(*b*) Use Kirchhoff's loop rule to find the charging current:

$$\mathcal{E}_1 - Ir_1 - Ir_2 - \mathcal{E}_2 - IR = 0$$

so

$$I = \frac{\mathcal{E}_1 - \mathcal{E}_2}{R + r_1 + r_2} = \frac{12.0\text{ V} - 11.0\text{ V}}{0.050\ \Omega} = \boxed{20\text{ A}}$$

(*c*) When the batteries are connected incorrectly, positive terminals to negative terminals, the emfs add:

$$\mathcal{E}_1 - Ir_1 + \mathcal{E}_2 - Ir_2 - IR = 0$$

so

$$I = \frac{\mathcal{E}_1 + \mathcal{E}_2}{R + r_1 + r_2} = \frac{12.0\text{ V} + 11.0\text{ V}}{0.050\ \Omega} = \boxed{460\text{ A}}$$

CHECK If the batteries are connected incorrectly, as shown in Figure 25-30, the current is very large and the batteries could explode—producing a shower of boiling battery acid.

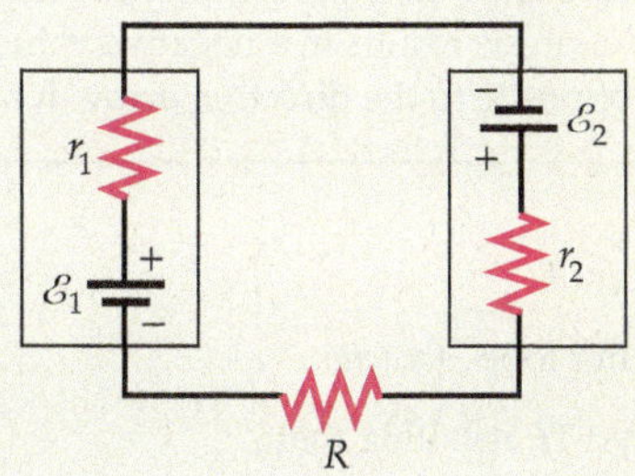

FIGURE 25-30 Two batteries connected incorrectly—dangerous!

MULTILOOP *CIRCUITS*

In multiloop circuits, often the direction of the current in one or more branches of the circuit are not obvious. Fortunately, Kirchhoff's rules do not require that we know these directions initially. In fact, the opposite is true. Kirchhoff's rules enable us to determine the directions of the currents. To accomplish this, for each branch of the circuit we arbitrarily assign a positive direction along the branch, and we indicate this assignment by placing a corresponding arrow on the circuit diagram (Figure 25-31). If the current density in the branch is in this positive direction, then when we solve for this current we will get a positive value. However, if the current density is opposite to the assigned positive direction, when we solve for the current we will get a negative value. In a resistor, an electric field within the resistor causes the current, and the current is in the same direction as the electric field. Because the electric field always points in the direction of decreasing potential, we know that in a resistor the direction of the current is also the direction of decreasing potential. Therefore, any time we traverse a resistor in the direction of the current, the change in potential is negative, and vice versa. Here is the rule:

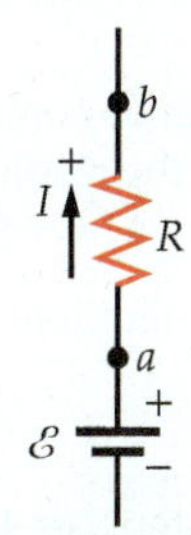

FIGURE 25-31 It is not known whether the current I has a positive or a negative value. Whether it is positive or negative, $V_b - V_a = -IR$. If the current is in the positive direction, then I is positive and $-IR$ is negative. If the current is in the negative direction, however, then I is negative and $-IR$ is positive.

> For each branch of a circuit, we draw an arrow to indicate the positive direction for that branch. Then, if we traverse a resistor in the direction of the arrow, the change in potential ΔV is equal to $-IR$ (and if we traverse a resistor in the direction opposite the direction of the arrow, ΔV is equal to $+IR$).
>
> SIGN RULE FOR THE CHANGE IN POTENTIAL ACROSS A RESISTOR

If we traverse a resistor in the positive direction, and if I is positive, then $-IR$ is negative. This is as expected, because in a resistor, the current is always in the direction of decreasing potential. However, if we traverse a resistor in the positive direction, and if I is negative, then $-IR$ is positive. Similarly, if we traverse a resistor in the negative direction, and if I is positive, then $+IR$ is positive. And if we traverse a resistor in the negative direction and if I is negative, then $+IR$ is negative.

To analyze circuits containing more than one loop, we need to use both of Kirchhoff's rules, with Kirchhoff's junction rule applied to junctions (points where the current splits into two or more parts).

Example 25-16 Applying Kirchhoff's Rules

(*a*) Find the current in each branch of the circuit shown in Figure 25-32. (*b*) Find the energy dissipated in the 4.0-Ω resistor in 3.0 s.

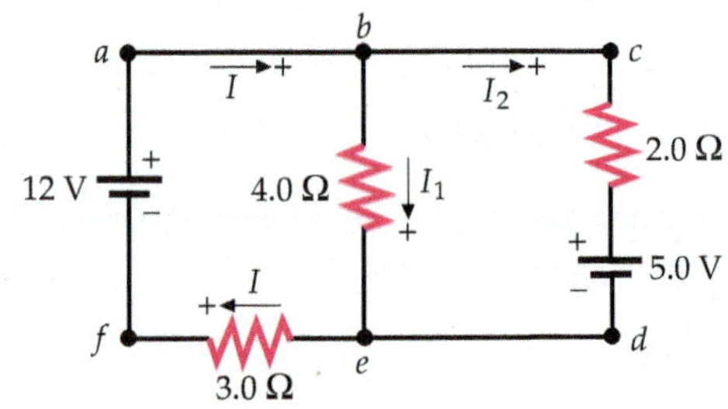

FIGURE 25-32

PICTURE There are three branch currents, I, I_1, and I_2, to be determined, so we need three equations. One equation comes from applying the junction rule to point *b*. (We can also apply the junction rule to point *e*, the only other junction in the circuit, but it gives exactly the same information.) The other two relations are obtained by applying the loop rule. There are three loops in the circuit: the two interior loops, *abefa* and *bcdeb*, and the exterior loop, *abcdefa*. We can use any two of these loops—the third will give redundant information. There is at least one direction arrow on each branch in Figure 25-32. Each direction arrow indicates the positive direction for that branch. If our analysis results in a negative value for a branch current, then that current is in the direction opposite to the direction arrow for that branch.

SOLVE

(*a*) 1. Apply the junction rule to point *b*:

$$I = I_1 + I_2$$

2. Apply the loop rule to the outer loop, *abcdefa*:

$$-(2.0\ \Omega)I_2 - 5.0\ \text{V} - (3.0\ \Omega)(I_1 + I_2) + 12\ \text{V} = 0$$

3. Divide the above equation by 1 Ω, recalling that (1 V)/(1 Ω) = 1 A, then simplify:

$$7.0\ \text{A} - 3.0I_1 - 5.0I_2 = 0$$

4. For the third condition, apply the loop rule to the loop on the right, *bcdeb*:

$$-(2.0\ \Omega)I_2 - 5.0\ \text{V} + (4.0\ \Omega)I_1 = 0 \quad \text{so} \quad -5.0\ \text{A} + 4.0I_1 - 2.0I_2 = 0$$

5. The results for steps 3 and 4 can be combined to solve for I_1 and I_2. To do so, first multiply the result for step 3 by 2, and then multiply the result for step 4 by -5:

$$14\ \text{A} - 6.0I_1 - 10I_2 = 0$$
$$25\ \text{A} - 20I_1 + 10I_2 = 0$$

6. Add the equations in step 5 to eliminate I_2, then solve for I_1;

$$39\ \text{A} - 26I_1 = 0$$
$$I_1 = \frac{39\ \text{A}}{26} = \boxed{1.5\ \text{A}}$$

7. Substitute I_1 in the results for step 3 or 4 to solve for I_2:

$$7.0\ \text{A} - 3.0(1.5\ \text{A}) - 5.0I_2 = 0$$
$$I_2 = \frac{2.5\ \text{A}}{5.0} = \boxed{0.50\ \text{A}}$$

8. Finally, I_1 and I_2 determine I using the equation in step 1:

$$I = I_1 + I_2 = 1.5\ \text{A} + 0.50\ \text{A} = \boxed{2.0\ \text{A}}$$

(*b*) 1. The power delivered to the 4.0-Ω resistor is found using $P = I_1^2R$:

$$P = I_1^2R = (1.5\ \text{A})^2(4.0\ \Omega) = 9.0\ \text{W}$$

2. The total energy dissipated in the 4.0-Ω resistor during time Δt is $W = P\,\Delta t$. In this case, $\Delta t = 3.0$ s:

$$W = P\,\Delta t = (9.0\ \text{W})(3.0\ \text{s}) = \boxed{27\ \text{J}}$$

CHECK In Figure 25-33, we have chosen the potential to be zero at point f, and we have labeled the currents and the potentials at the other points. Note that $V_b - V_e = 6.0$ V and $V_e - V_f = 6.0$ V. Applying the loop rule to the loop on the left gives $+12\ \text{V} - 6.0\ \text{V} - 6.0\ \text{V} = 0$.

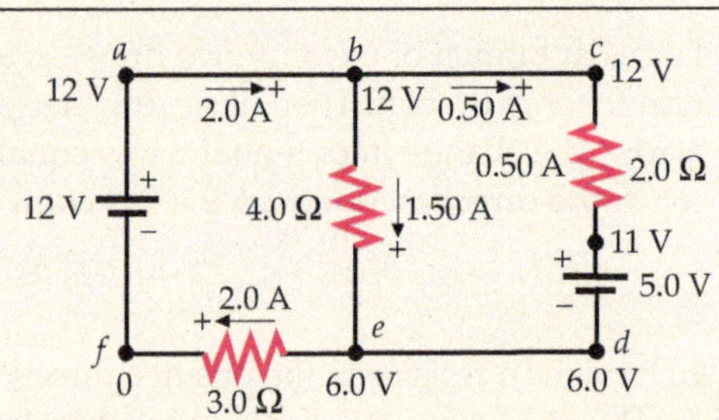

FIGURE 25-33

TAKING IT FURTHER Applying the loop rule to the loop on the left, *abefa*, gives $12\ \text{V} - (4.0\ \Omega)I_1 - (3.0\ \Omega)(I_1 + I_2) = 0$, or $12\ \text{A} - 7.0I_1 - 3.0I_2 = 0$. Note that this is just the result for step 3 minus the result for step 4 and hence contains no new information, as expected.

PRACTICE PROBLEM 25-9 Find I_1 for the case in which the 3.0-Ω resistor approaches (*a*) zero resistance and (*b*) infinite resistance.

Example 25-16 illustrates the general methods for the analysis of multiloop circuits. These methods are listed in the following problem-solving strategy.

PROBLEM-SOLVING STRATEGY

Method for Analyzing Multiloop Circuits

PICTURE Draw a sketch of the circuit.

SOLVE

1. Replace any series or parallel resistor combinations or capacitor combinations with their equivalent values.
2. Repeat step 1 as many times as possible.
3. Next, assign a positive direction for each branch of the circuit and indicate this direction with an arrow. Label the current in each branch. Add a plus sign and a minus sign to indicate the high-potential terminal and low-potential terminal of each source of emf.
4. Apply the junction rule to all but one of the junctions.
5. Apply the loop rule to the different loops until the total number of independent equations equals the total number of unknowns. When traversing a resistor in the positive direction, the change in potential equals $-IR$. When traversing a battery from the negative terminal to the positive terminal, the change in potential equals $\mathcal{E} \pm Ir$.
6. Solve the equations to obtain the values of the unknowns.

CHECK Check your results by assigning a potential of zero to one point in the circuit and use the values of the currents found to determine the potentials at other points in the circuit.

Example 25-17 A Three-Branch Circuit

(*a*) Find the current in each branch of the circuit shown in Figure 25-34. (*b*) Assign $V = 0$ to point *c* and then find the potential at each other point *a* through *f*.

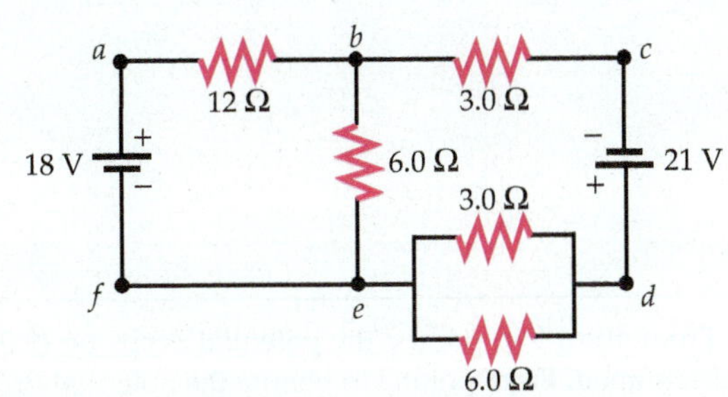

FIGURE 25-34

PICTURE First, replace the two parallel resistors by an equivalent resistance. Second, assign a positive direction to each branch and indicate each choice with an arrow. Third, place a plus sign and a minus sign at the high-potential and low-potential terminals of each battery. Assign a symbol for the current in each branch. These branch currents can then be found by applying the junction rule at either junction *b* or junction *e* and by applying the loop rule twice.

SOLVE

(*a*) 1. Find the equivalent resistance of the 3.0-Ω and 6.0-Ω parallel resistors:

$R_{eq} = 2.0\ \Omega$

2. Redraw the circuit (Figure 25-35) with the 2.0-Ω resistor in place of the parallel combination. Place an arrow on each branch indicating your direction-sign assignments. Let I be the current in branch with the 18-V battery, let I_1 be the current in the 6.0-Ω resistor, and let I_2 be the current in branch with the 21-V battery:

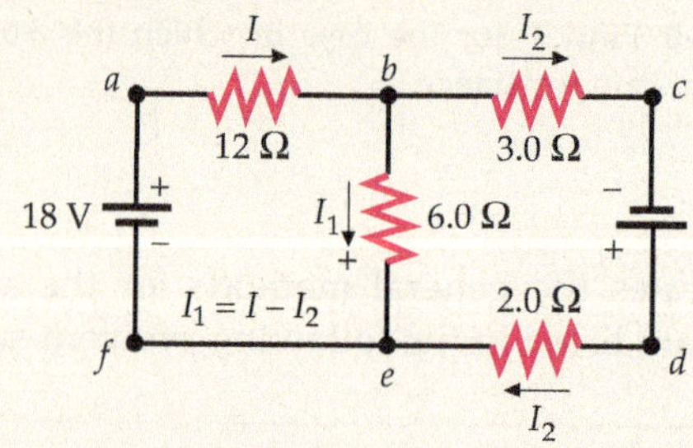

FIGURE 25-35

3. Apply the junction rule at point *b*:

$I = I_1 + I_2$

4. Apply Kirchhoff's loop rule to loop *abefa* to obtain an equation involving I and I_1:

$18\ \text{V} - (12\ \Omega)I - (6.0\ \Omega)I_1 = 0$

5. Simplify the equation from step 4 (dividing both sides by 6.0 Ω):

$3.0\ \text{A} - 2.0I - 1.0I_1 = 0$

6. Apply Kirchhoff's loop rule to loop *bcdeb*:

$-(3.0\ \Omega)I_2 + 21\ \text{V} - (2.0\ \Omega)I_2 + (6.0\ \Omega)I_1 = 0$

7. Simplify the equation in step 6 (dividing both sides by 1.0 Ω):

$21\ \text{A} + 6.0I_1 - 5.0I_2 = 0$

8. Solve the simultaneous equations (from steps 3, 5, and 7) for I, I_1, and I_2. One way to do this calculation is first substitute $I_1 + I_2$ for I in the step 5 equation to obtain $3.0\ \text{A} - 3.0I_1 - 2.0I_2 = 0$. This equation and the step 7 equation constitute two equations and two unknowns. Solve for the currents:

$\boxed{I_1 = -1.0\ \text{A}}$ $\boxed{I_2 = 3.0\ \text{A}}$ $\boxed{I = 2.0\ \text{A}}$

9. Use $V = I_2R_{eq}$ to find the potential drop across the parallel combination consisting of the 3.0-Ω and 6.0-Ω resistors:

$V = 6.0\ \text{V}$

10. Use the result of step 9 and Ohm's law to find the current in each of the parallel resistors;

$\boxed{I_{3\,\Omega} = 2.0\ \text{A}}$ $\boxed{I_{6\,\Omega} = 1.0\ \text{A}}$

(*b*) Redraw Figure 25-35 showing the direction and value of the current in each branch of the circuit (Figure 25-36). Begin with $V = 0$ at point *c* and calculate the potential at points *d*, *e*, *f*, *a*, and *b*:

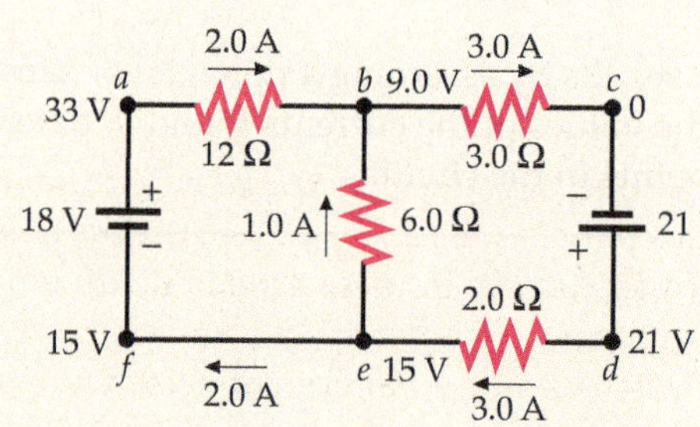

FIGURE 25-36

$V_d = V_c + 21\ \text{V} = 0 + 21\ \text{V} = \boxed{21\ \text{V}}$

$V_e = V_d - (3.0\ \text{A})(2.0\ \Omega) = 21\ \text{V} - 6.0\ \text{V} = \boxed{15\ \text{V}}$

$V_f = V_e = \boxed{15\ \text{V}}$

$V_a = V_f + 18\ \text{V} = 15\ \text{V} + 18\ \text{V} = \boxed{33\ \text{V}}$

$V_b = V_a - (2.0\ \text{A})(12.0\ \Omega) = 33\ \text{V} - 24\ \text{V} = \boxed{9\ \text{V}}$

CHECK From point *b* to point *c* the potential drops by (3.0 A)(3.0 Ω) = 9.0 V, which gives $V_c = 0$, as assumed. From point *e* to point *b* the potential drops by (1.0 A)(6.0 Ω) = 6.0 V, so $V_b = V_e - 6.0\ \text{V} = 15\ \text{V} - 6.0\ \text{V} = 9\ \text{V}$.

AMMETERS, VOLTMETERS, AND OHMMETERS

The devices that measure current, potential difference, and resistance are called **ammeters, voltmeters,** and **ohmmeters,** respectively. Often, all three of these meters are included in a single *multimeter* that can be switched from one use to another. You might use a voltmeter to measure the terminal voltage of your car battery and an ohmmeter to measure the resistance of some electrical device at home (for example, a toaster or lightbulb) when you suspect a short circuit or a broken wire.

To measure the current in a resistor in a simple circuit, we place an ammeter in series with the resistor, as shown in Figure 25-37, so that the current is the same in the ammeter and the resistor. Because the ammeter has a very low (but finite) resistance, the current in the circuit decreases very slightly when the ammeter is inserted. Ideally, the ammeter should have a negligibly small resistance so that the current to be measured is only negligibly affected.

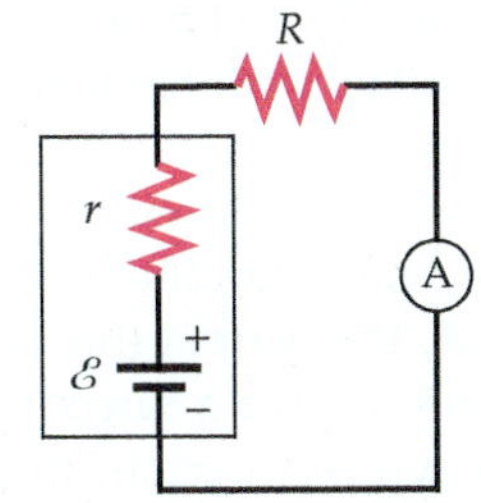

FIGURE 25-37 To measure the current in a resistor *R*, an ammeter A (circled) is placed in series with the resistor so that it carries the same current as the resistor.

The potential difference across a resistor is measured by placing a voltmeter across the resistor (in parallel with the resistor), as shown in Figure 25-38, so that the potential drop is the same across both the voltmeter and the resistor. The voltmeter reduces the resistance between points *a* and *b*, thus increasing the total current in the circuit and changing the potential drop across the resistor. A voltmeter should have an extremely high resistance so that its effect on the current in the circuit is negligible.

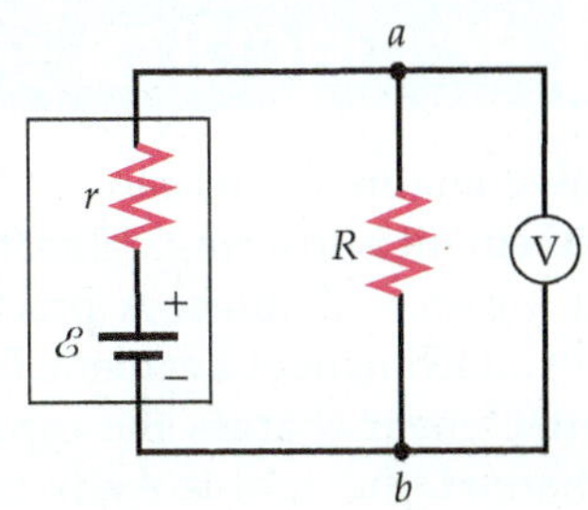

FIGURE 25-38 To measure the potential drop across a resistor, a voltmeter V (circled) is placed in parallel with the resistor so that the potential difference across the voltmeter is equal to the potential difference across the resistor.

The principal component of many commonly used ammeters and voltmeters is a **galvanometer,** a device that detects small currents passing through it. The galvanometer is designed so that the scale reading is proportional to the current passing through. The type of galvanometer used in many student laboratories consists of a coil of wire in the magnetic field of a permanent magnet. When there is current in the coil, the magnetic field exerts a torque on the coil, which causes the coil to rotate. A pointer attached to the coil indicates the reading on a scale. The coil itself contributes some resistance when the galvanometer is placed within a circuit.

To construct an ammeter from a galvanometer, we place a small resistor called a **shunt resistor** in *parallel* with the galvanometer. The resistance of the shunt resistor is usually much smaller than the resistance of the galvanometer so that the majority of the current is carried by the shunt resistor. The equivalent resistance of the ammeter is then approximately equal to the resistance of the shunt resistor, which is much smaller than the internal resistance of the galvanometer alone. To construct a voltmeter, we place a resistor that has a large resistance in *series* with the galvanometer so that the equivalent resistance of the voltmeter is much larger than the resistance of the galvanometer coil alone. Figure 25-39 illustrates the construction of an ammeter and voltmeter from a galvanometer. The resistance of the galvanometer R_g is shown separately in these schematic drawings, but it is actually part of the galvanometer.

A simple ohmmeter consists of a battery connected in series with a galvanometer and a resistor, as shown in Figure 25-40*a*.

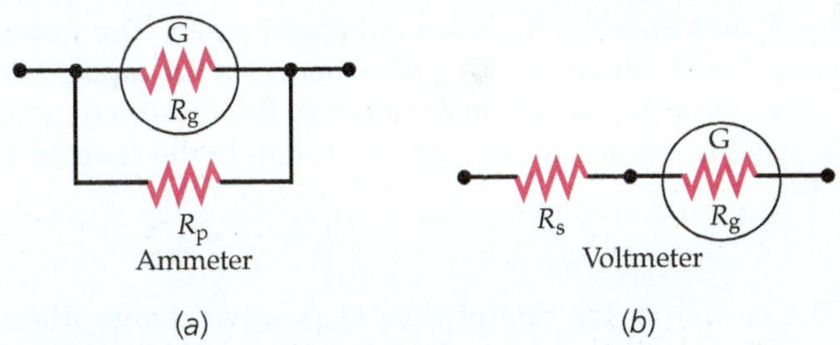

FIGURE 25-39 (*a*) An ammeter consists of a galvanometer G (circled) whose resistance is R_g and a small parallel resistance R_p. (*b*) A voltmeter consists of a galvanometer G (circled) and a large series resistance R_s.

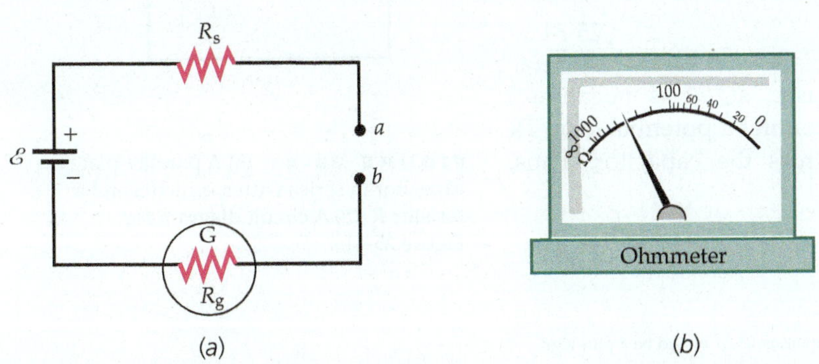

FIGURE 25-40 (*a*) An ohmmeter consists of a battery connected in series with a galvanometer and a resistor R_s, which is chosen so that the galvanometer gives full-scale deflection when points *a* and *b* are shorted. (*b*) When a resistor *R* is connected between terminals *a* and *b*, the galvanometer needle deflects by an amount that depends on the value of *R*. The galvanometer scale is calibrated to give a readout in ohms.

The resistance R_s is chosen so that when the terminals *a* and *b* are shorted (put in electrical contact with negligible resistance between them), the current through the galvanometer gives a full-scale deflection. Thus, a full-scale deflection indicates no resistance between terminals *a* and *b*. A zero deflection indicates an infinite resistance between the terminals. When the terminals are connected across an unknown resistance *R*, the current through the galvanometer depends on *R*, so the scale can be calibrated to give a direct reading of *R*, as shown in Figure 25-40*b*. Because an ohmmeter sends a current through the resistance to be measured, some caution must be exercised when using this instrument. For example, you would not want to try to measure the resistance of a sensitive galvanometer with an ohmmeter, because the current provided by the battery in the ohmmeter could possibly damage the galvanometer.

25-6 *RC* CIRCUITS

A circuit containing a resistor and a capacitor is called an ***RC* circuit.** The current in an *RC* circuit is in a single direction, as in all dc circuits, but the magnitude of the current varies with time. A practical example of an *RC* circuit is the circuit in the flash attachment of a camera. Before a flash photograph is taken, a battery in the flash attachment charges the capacitor through a resistor. When the capacitor is fully charged, the light is ready to flash. When the picture is taken, the capacitor discharges through the flashbulb. The battery then recharges the capacitor, and a short time later the flash is ready for another picture. Using Kirchhoff's rules, we can obtain equations for the charge *Q* and the current *I* as functions of time for both the charging and discharging of a capacitor through a resistor.

DISCHARGING A CAPACITOR

Figure 25-41 shows a capacitor that has an initial charge $+Q_0$ on the upper plate and an initial charge $-Q_0$ on the lower plate. The capacitor is connected to a resistor *R* and a switch S, which is initially open. The potential difference across the capacitor is initially $V_0 = Q_0/C$, where *C* is the capacitance.

We close the switch at time $t = 0$. Because there is now a potential difference across the resistor, there is also a current in the resistor. This initial current is

$$I_0 = \frac{V_0}{R} = \frac{Q_0}{RC} \qquad 25\text{-}30$$

The current is the rate of flow of positive charge from the positive plate of the capacitor to the negative plate through the resistor. (We model the charge carriers as being positively charged, when actually they are negatively charged electrons.) As time passes, the charge on the capacitor decreases. If we choose the positive direction to be clockwise, then the current equals the rate of decrease of the charge. If *Q* is the charge on the upper plate of the capacitor at time *t*, the current is related to *Q* by

$$I = -\frac{dQ}{dt} \qquad 25\text{-}31$$

(The minus sign is needed because while *Q* decreases, dQ/dt is negative.*) Traversing the circuit in the clockwise direction, we encounter a potential drop *IR* across the resistor and a potential increase Q/C across the capacitor. Thus, Kirchhoff's loop rule gives

$$\frac{Q}{C} - IR = 0 \qquad 25\text{-}32$$

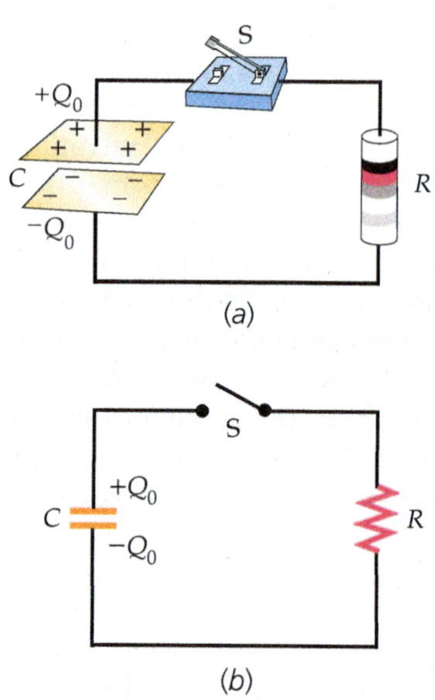

FIGURE 25-41 (*a*) A parallel-plate capacitor in series with a switch S and a resistor *R*. (*b*) A circuit diagram for Figure 25-41*a*.

* If the positive direction were chosen to be counterclockwise, then the sign in Equation 25-31 would be a plus sign.

where Q and I, both functions of time, are related by Equation 25-31. Substituting $-dQ/dt$ for I in Equation 25-32, we have

$$\frac{Q}{C} + R\frac{dQ}{dt} = 0$$

or

$$\frac{dQ}{dt} = -\frac{1}{RC}Q \qquad 25\text{-}33$$

To solve this equation, we first separate the variables Q and t by multiplying both sides of the equation by dt/Q, and then we integrate both sides. Multiplying both sides by dt/Q, we obtain

$$\frac{dQ}{Q} = -\frac{1}{RC}dt \qquad 25\text{-}34$$

The variables Q and t are now in separate terms. Integrating from Q_0 at $t = 0$ to Q' at time t' gives

$$\int_{Q_0}^{Q'} \frac{dQ}{Q} = -\frac{1}{RC}\int_0^{t'} dt$$

so

$$\ln\frac{Q'}{Q_0} = -\frac{t'}{RC}$$

Because t' is arbitrary, we can replace t' with t, and then $Q' = Q(t)$. Solving for $Q(t)$ gives

$$Q(t) = Q_0 e^{-t/(RC)} = Q_0 e^{-t/\tau} \qquad 25\text{-}35$$

where τ, called the **time constant,** is the time it takes for the charge to decrease by a factor of e^{-1}:

$$\tau = RC \qquad 25\text{-}36$$

DEFINITION–TIME CONSTANT

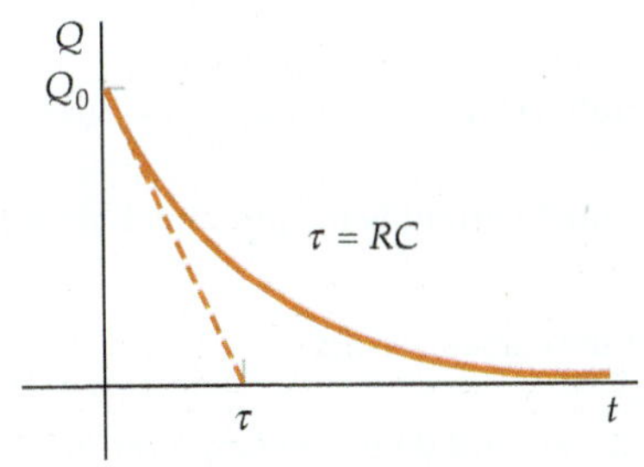

FIGURE 25-42 Plot of the charge on the capacitor versus time for the circuit shown in Figure 25-41. The switch is closed at time $t = 0$. The time constant $\tau = RC$ is the time it takes for the charge to decrease by a factor of e^{-1}. (The time constant is also the time it would take the capacitor to discharge fully if its discharge rate remains constant, as indicated by the dashed line.)

Figure 25-42 shows the charge on the capacitor in the circuit of Figure 25-41 as a function of time. After a time $t = \tau$, the charge is $Q = e^{-1}Q_0 = 0.37Q_0$. After a time $t = 2\tau$, the charge is $Q = e^{-2}Q_0 = 0.135Q_0$, and so forth. After a time equal to several time constants, the charge Q is negligible. This type of decrease, which is called an **exponential decrease,** is very common in nature. It occurs whenever the rate at which a quantity decreases is proportional to the quantity itself.*

The decrease in the charge on a capacitor can be likened to the decrease in the amount of water in a bucket that has vertical sides and a small hole in the bottom. (The rate at which the water flows out of the hole is proportional to the difference in pressure of the water on either side of the hole, which is in turn proportional to the amount of water still in the bucket.)

The current is obtained by differentiating Equation 25-35

$$I = -\frac{dQ}{dt} = \frac{Q_0}{RC}e^{-t/(RC)}$$

Substituting, using Equation 25-30, we obtain

$$I = I_0 e^{-t/\tau} \qquad 25\text{-}37$$

where $I_0 = V_0/R = Q_0/(RC)$ is the initial current. The current as a function of time is shown in Figure 25-43. As with the charge, the current decreases exponentially with time constant $\tau = RC$.

See **Math Tutorial** *for more information on* **Exponential Functions**

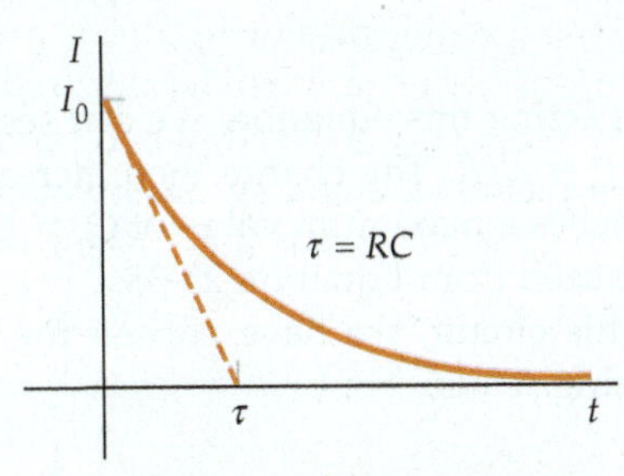

FIGURE 25-43 Plot of the current versus time for the circuit in Figure 25-41. The curve has the same shape as that in Figure 25-42. If the rate of decrease of the current remains constant, the current would reach zero after one time constant, as indicated by the dashed line.

* We encountered exponential decreases in Chapter 14 when we studied the damped oscillator.

Example 25-18 Discharging a Capacitor

A 4.0-μF capacitor is charged to 24 V and then connected across a 200-Ω resistor. Find (*a*) the initial charge on the capacitor, (*b*) the initial current through the 200-Ω resistor, (*c*) the time constant, and (*d*) the charge on the capacitor after 4.0 ms.

PICTURE The circuit diagram is the same as the circuit diagram shown in Figure 25-41.

SOLVE

(*a*) The initial charge is related to the capacitance and voltage:

$$Q_0 = CV_0 = (4.0\ \mu\text{F})(24\ \text{V}) = \boxed{96\ \mu\text{C}}$$

(*b*) The initial current is the initial voltage divided by the resistance:

$$I_0 = \frac{V_0}{R} = \frac{24\ \text{V}}{200\ \Omega} = \boxed{0.12\ \text{A}}$$

(*c*) The time constant is RC:

$$\tau = RC = (200\ \Omega)(4.0\ \mu\text{F}) = 800\ \mu\text{s} = \boxed{0.80\ \text{ms}}$$

(*d*) Substitute $t = 4.0$ ms into Equation 25-35 to find the charge on the capacitor at that time:

$$Q = Q_0 e^{-t/\tau} = (96\ \mu\text{C})e^{-(4.0\ \text{ms})/(0.80\ \text{ms})}$$

$$= (96\ \mu\text{C})e^{-5} = \boxed{0.65\ \mu\text{C}}$$

CHECK At the initial current of $I_0 = 0.12$ A it would take $Q_0/I_0 = 96\ \mu\text{C}/0.12\ \text{A} = 0.80$ ms for the capacitor to fully discharge. Because the current decreases exponentially during discharge, it is not surprising that it takes 4.0 ms for the charge to lose 99.3% of its initial charge.

TAKING IT FURTHER After five time constants, Q is less than 1 percent of its initial value.

PRACTICE PROBLEM 25-10 Find the current through the 200-Ω resistor at $t = 4.0$ ms.

CHARGING A CAPACITOR

Figure 25-44*a* shows a circuit for charging a capacitor. The capacitor is initially uncharged. The switch S, originally open, is closed at time $t = 0$. Charge immediately begins to flow through the battery (Figure 25-44*b*). If the charge on the rightmost plate of the capacitor at time t is Q, the current in the circuit is I, and clockwise is positive, then Kirchhoff's loop rule gives

$$\mathcal{E} - IR - \frac{Q}{C} = 0 \qquad \text{25-38}$$

By inspecting this equation, we can see that when Q is zero (at $t = 0$) the current is $I = I_0 = \mathcal{E}/R$. The charge then increases and the current decreases. The charge approaches a maximum value of $Q_f = C\mathcal{E}$ as the current I approaches zero, as can also be seen from Equation 25-38.

In this circuit, we have chosen the positive direction so if I is positive Q is increasing. Thus,

$$I = +\frac{dQ}{dt}$$

Substituting dQ/dt for I in Equation 25-38 gives

$$\mathcal{E} - R\frac{dQ}{dt} - \frac{Q}{C} = 0 \qquad \text{25-39}$$

Equation 25-39 can be solved in the same way that we solved Equation 25-33.

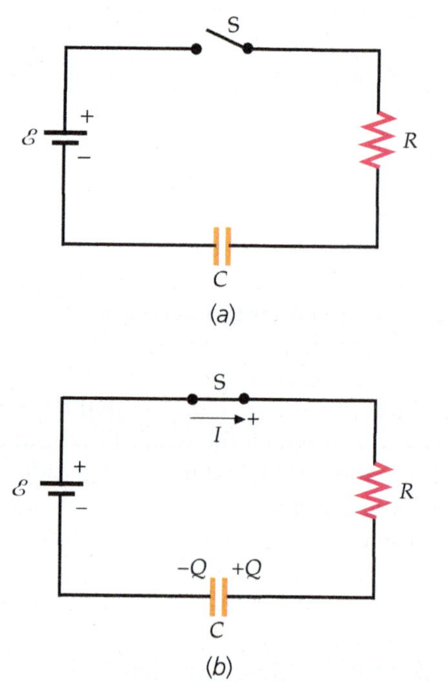

FIGURE 25-44 (*a*) A circuit for charging a capacitor to a potential difference $\mathcal{E}$. (*b*) After the switch is closed, there is both a current through and a potential drop across the resistor and both a charge on and a potential drop across the capacitor.

The details are left as a problem (see Problem 101). The result is

$$Q = C\mathcal{E}[1 - e^{-t/(RC)}] = Q_f(1 - e^{-t/\tau}) \qquad \text{25-40}$$

where $Q_f = C\mathcal{E}$ is the final charge. The current is obtained from $I = dQ/dt$:

$$I = \frac{dQ}{dt} = C\mathcal{E}\left[-\frac{-1}{RC}e^{-t/(RC)}\right] = \frac{\mathcal{E}}{R}e^{-t/(RC)}$$

or

$$I = \frac{\mathcal{E}}{R}e^{-t/(RC)} = I_0 e^{-t/\tau} \qquad \text{25-41}$$

where the initial current is $I_0 = \mathcal{E}/R$.

Figure 25-45 and Figure 25-46 show the charge and the current as functions of time.

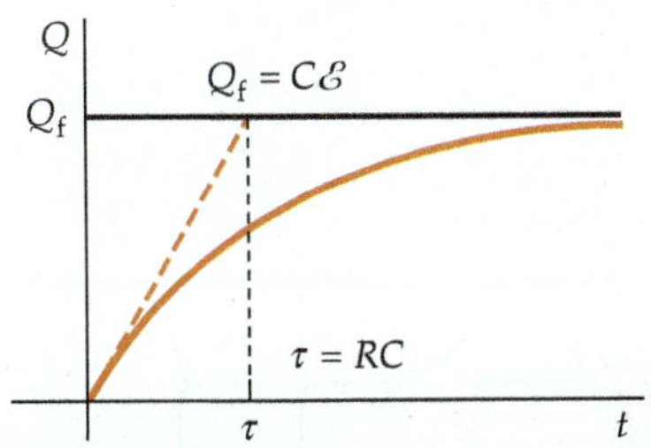

FIGURE 25-45 Plot of the charge on the capacitor versus time for the charging circuit of Figure 25-44 after the switch is closed (at $t = 0$). After a time $t = \tau = RC$, the charge on the capacitor is $0.63C\mathcal{E}$, where $C\mathcal{E}$ is its final charge. If the charging rate were constant, the capacitor would be fully charged after a time $t = \tau$.

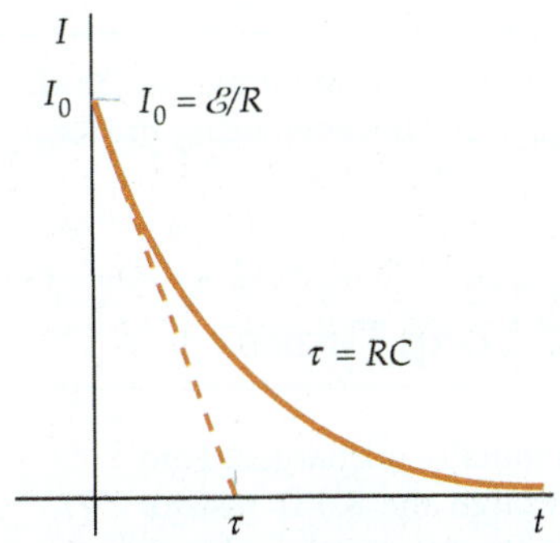

FIGURE 25-46 Plot of the current versus time for the charging circuit of Figure 25-44. The current is initially $\mathcal{E}/R$, and the current decreases exponentially with time.

PRACTICE PROBLEM 25-11

Show that Equation 25-40 does indeed satisfy Equation 25-39 by substituting the expressions for Q and dQ/dt into Equation 25-39.

PRACTICE PROBLEM 25-12

What fraction of the maximum charge is on the charging capacitor after a time $t = 2\tau$?

Example 25-19 Charging a Capacitor *Try It Yourself*

A 6.0-V battery that has a negligible internal resistance is used to charge a 2.0-μF capacitor through a 100-Ω resistor. Find (*a*) the initial current, (*b*) the final charge on the capacitor, (*c*) the time required for the charge to reach 90 percent of its final value, and (*d*) the charge when the current is half its initial value.

PICTURE The charge initially is zero so the voltage across the resistor is equal to the emf of the battery. Apply Ohm's law to the resistor and solve for the current. After a long time, the current is zero so the voltage across the capacitor is equal to the emf of the battery. Apply the definition of capacitance and solve for the charge. Use Equation 25-40 to relate the charge to the time, and use Kirchhoff's loop rule to relate the charge to the current.

SOLVE

Cover the column to the right and try these on your own before looking at the answers.

Steps	Answers
(*a*) Find the initial current from $I_0 = \mathcal{E}/R$.	$I_0 = 0.060\ \text{A} = \boxed{60\ \text{mA}}$
(*b*) Find the final charge from $Q = C\mathcal{E}$.	$Q_f = \boxed{12\ \mu\text{C}}$
(*c*) Set $Q = 0.90Q_f$ in Equation 25-40 and solve for t. (First solve for $e^{t/\tau}$, then take the natural log of both sides and solve for t.)	$t = 2.3\tau = \boxed{0.46\ \text{ms}}$
(*d*) 1. Apply Kirchhoff's loop rule to the circuit using Figure 25-44*b*.	$\mathcal{E} - IR - \dfrac{Q}{C} = 0$
2. Set $I = I_0/2$ and solve for Q.	$Q = \dfrac{Q_f}{2} = \boxed{6.0\ \mu\text{C}}$

CHECK The answer to Part (*d*) can be obtained by first solving for t using Equation 25-41, then substituting that time into Equation 25-40 and solving for Q. However, using the loop rule is certainly the more direct approach.

Example 25-20 Finding Values at Short and Long Times

The 6.0-μF capacitor in the circuit shown in Figure 25-47 is initially uncharged. Find the current through the 4.0-Ω resistor and the current through the 8.0-Ω resistor (*a*) immediately after the switch is closed and (*b*) a long time after the switch is closed. (*c*) Find the charge on the capacitor a long time after the switch is closed.

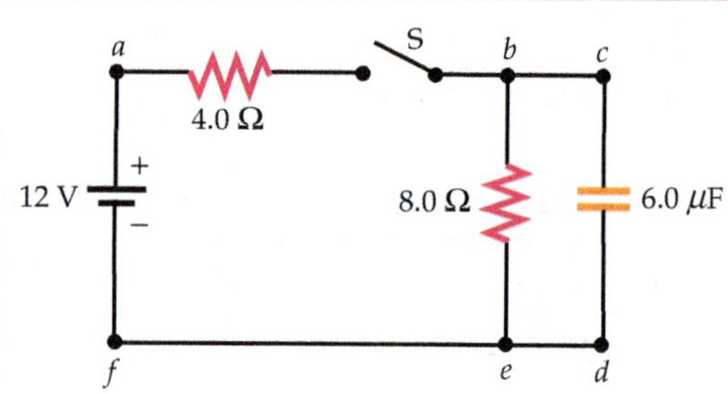

FIGURE 25-47

PICTURE Because the capacitor is initially uncharged (and the 4.0-Ω resistor limits the current through the battery), the initial potential difference across the capacitor is zero. The capacitor and the 8.0-Ω resistor are connected in parallel, and the difference in potential across each is the same. Thus, the initial potential difference across the 8.0-Ω resistor is also zero. The positive direction for the branch that has the battery is up the page, and the positive direction for the other two branches is down the page. Let Q be the charge on the upper plate of the capacitor.

SOLVE

(*a*) The charge on the capacitor initially is zero. Apply the loop rule to the outer loop and solve for the current through the 4.0-Ω resistor. Apply the loop rule to the loop containing the 8.0-Ω resistor and the capacitor and solve for the current through the 8.0-Ω resistor.

$$12\ \text{V} - (4.0\ \Omega)I_{4\Omega} - \frac{0}{C} = 0$$

$$I_{4\Omega} = \boxed{3.0\ \text{A}}$$

$$I_{8\Omega}(8\ \Omega) - \frac{0}{C} = 0$$

$$I_{8\Omega} = \boxed{0}$$

(*b*) After a long time, the capacitor is fully charged (no more charge flows onto its plates) and the current through both resistors is the same. Apply the loop rule to the loop on the left and solve for the current:

$$12\ \text{V} - (4.0\ \Omega)I_f - (8.0\ \Omega)I_f = 0$$

$$I_f = \boxed{1.0\ \text{A}}$$

(*c*) The potential difference across the 8.0-Ω resistor and the capacitor are equal. Use this to solve for Q_f:

$$I_f(8.0\ \Omega) = \frac{Q_f}{C}$$

$$Q_f = (1.0\ \text{A})(8.0\ \Omega)(6.0\ \mu\text{F}) = \boxed{48\ \mu\text{C}}$$

CHECK The analysis of this circuit at the extreme times when the capacitor is either uncharged or fully charged is simple. When the capacitor is uncharged, it acts like a good conductor (a short circuit) between points c and d; that is, the circuit is the same as the one shown in Figure 25-48*a* where we have replaced the capacitor by a wire of zero resistance. When the capacitor is fully charged, it acts like an open switch, as shown in Figure 25-48*b*.

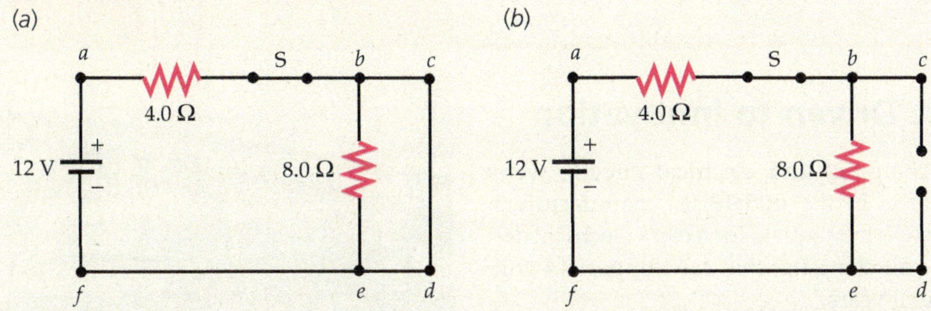

FIGURE 25-48

ENERGY CONSERVATION IN CHARGING A CAPACITOR

During the charging process, a total charge $Q_f = \mathcal{E}C$ flows through the battery. The battery therefore does work

$$W = Q_f\mathcal{E} = C\mathcal{E}^2$$

Half of this work is accounted for by the energy stored in the capacitor (see Equation 24-8):

$$U = \tfrac{1}{2}Q_f\mathcal{E}$$

We now show that the other half of work done by the battery is dissipated as thermal energy by the resistance of the circuit. The rate at which energy is dissipated by the resistance R is

$$\frac{dW_R}{dt} = I^2R$$

where $I = (\mathcal{E}/R)e^{-t/(RC)}$ (Equation 25-41). Substituting for I we have

$$\frac{dW_R}{dt} = \left(\frac{\mathcal{E}}{R}e^{-t/(RC)}\right)^2 R = \frac{\mathcal{E}^2}{R}e^{-2t/(RC)}$$

We find the total energy dissipated by integrating from $t = 0$ to $t = \infty$:

$$W_R = \int_0^\infty \frac{\mathcal{E}^2}{R}e^{-2t/(RC)}\,dt = \frac{\mathcal{E}^2}{R}\int_0^\infty e^{-at}\,dt$$

where $a = 2/(RC)$. Thus,

$$W_R = \frac{\mathcal{E}^2}{R}\frac{e^{-at}}{-a}\bigg|_0^\infty = -\frac{\mathcal{E}^2}{Ra}(0 - 1) = \frac{\mathcal{E}^2}{R}\frac{1}{a} = \frac{\mathcal{E}^2}{R}\frac{RC}{2}$$

The total amount of Joule heating is thus

$$W_R = \frac{1}{2}\mathcal{E}^2C = \frac{1}{2}Q_f\mathcal{E}$$

where $Q_f = \mathcal{E}C$. This result is independent of the resistance R. Thus, when a capacitor is charged through a resistor by a constant source of emf, half the energy provided by the source of emf is stored in the capacitor and half goes into thermal energy. This thermal energy includes the energy that is dissipated by the internal resistance of the source of emf.

Physics Spotlight

Vehicle Electrical Systems: Driven to Innovation

(Graham Harrison/Alamy.)

During the 1930s, 6-volt batteries (7-volt charging) and electrical circuits were standard for automobiles in the United States. In the mid-1950s, manufacturers around the world recognized that this battery was inadequate for a car's electrical demands, and changed to 12-volt batteries (14-volt charging) that can support 14-volt electrical systems.[*] The changeover took several years.[†]

In the mid-1960s, demands on a car's electrical system included the starter, ignition, lights, radio, and air-conditioning in luxury cars.[‡] Today, the electrical system and electronics of a car[#] can include crash sensors, automatic braking systems, seat motors, power steering, power braking, windshield washers with intermittent timing, video-based entertainment systems, engine controllers, cruise control, and window motors. Some high-end luxury cars put further demands on their electrical systems with electronic throttle control, radar to detect the distance to objects,[°] electronic stability and suspension control, and heated seats.[§] The power requirement for an automobile is 1.5 to 2.0 kW today, and is expected to increase to 3.0 to 3.5 kW or more in the near future.[¶] The electrical system and electronics of a car now account for more than 20 percent of the manufacturing cost of the average car.[**]

Because demands on the electrical system of cars are expected to increase still further,[††] many people have suggested that it would be a good idea to upgrade the electrical system of cars to a 36-volt battery and a 42-volt system. (Because power is the product of voltage and current, this means that when the voltage increased, the current would decrease in order to supply the same power.) Many people were excited about the prospect of using smaller wires and lighter wiring harnesses to carry power to all the electrical devices.[‡‡] In addition, the higher voltage would mean smaller and lighter starter motors and alternators.

But switching over to a 42-volt system is proving to be more difficult than expected. Although a concept car was built with a 42-volt system, it was made from nonstandard customized parts.[##] In a 14-V system, a connection that has vibrated loose will not arc persistently over a gap that is about a millimeter. In a 42-V system, the same loose connection will arc, which creates the danger of an electrical fire.[°°] At 42 V, more expensive electrical connectors are required. By mid-2005, several manufacturers had confirmed that they were not interested in using 42-volt systems within the next few years.[§§,¶¶] A consortium of researchers continues to look at issues involving 42-volt automotive systems.[***] When the economics of a switch to the 42-volt systems make sense, then 42-volt electrical systems will be put into mass-production automobiles.

* Ribbens, W. B., *Understanding Automotive Electronics, 6th ed.* New York: Newnes (Elsevier), 2003.
† Corbett, B., "No Flick of the Switch." *Ward's Auto World,* April 2001, Vol. 37, No. 4, p. 50.
‡ Ribbens, W. B., op. cit.
\# *Automotive Electronics Handbook,* R. Jurgen, ed., New York: McGraw-Hill, 1995.
° Allen, R., "New Technologies Make Roads Safer… One Smart Car at a Time." *Electronic Design,* Jun. 29, 2006, pp. 41–44.
§ "The 2007 S600 Sedan," Mercedes-Benz, http://www.mbusa.com/models/features/specs/overview.do?modelCode=S600V&class=07_S As of Sept. 2006.
¶ Masrur, M. A., Monroe, J., Patel, R., and Garg, V. K., "42-volt Electrical Power System for Military Vehicles—Comparison with Commercial Automotive Systems," *Vehicular Technology Conference,* 2002. Proceedings, VTC 2002-Fall, 2002 IEEE 56th, Vol. 3, pp. 1846–1850.
** Marsh, D., "LIN Simplifies and Standardizes In-Vehicle Networks." *Electronic Design News,* Apr. 8, 2005, pp. 29+.
†† Huber, P. W., and Mills, M. P., "The End of the M. E.?" *Mechanical Engineering,* May 2005, pp. 26–29.
‡‡ Truett, R., "42-Volt Systems Boost Fuel Economy Efforts." *Automotive News,* Oct. 21, 2001, Vol. 77, No. 6008, p. 6i.
\## "No-Compromise Mild Hybrid Car Engine Has a Promising Future." *Asia-Pacific Engineer,* Jun. 1, 2003. http://www.engineerlive.com/asiapacific-engineer/automotive-design/1603/nocompromise-mild-hybrid-car-engine-has-a-promising-future.thtml As of Sept. 2006.
°° Moran, T., "42-Volt Challenges: Arcs and Sparks." *Automotive News,* Mar. 12, 2001, Vol. 75, No. 5920, p. 8.
§§ Kelly, K., "DC Dumps 42-Volts." *Ward's AutoWorld,* Jun. 2004, p. 9.
¶¶ Crain, K., "Let's Step Back, Rethink Technology." *Automotive News,* Jan. 3, 2005, Vol. 79, No. 6128, p. 12.
*** MIT/Industry Consortium on Advanced Automotive Electrical/Electronic Components and Systems. "Consortium Research Units." http://lees-web.mit.edu/public/Public%20Documents/Research_Units_and_Deliverables.pdf As of Sept. 2006.

Summary

1. Ohm's law is an empirical law that holds only for certain materials.
2. Current, resistance, and emf are important *defined* quantities.
3. Kirchhoff's rules follow from the conservation of charge and the conservative nature of the electric field.

TOPIC	RELEVANT EQUATIONS AND REMARKS	
1. Electric Current	Electric current is the rate of flow of electric charge through a cross-sectional area. $$I = \frac{\Delta Q}{\Delta t}$$ in the limit that Δt approaches zero.	25-1
Drift velocity	In a conducting wire, electric current is the result of the slow drift of negatively charged electrons that are accelerated by an electric field in the wire and then collide with the lattice ions. Typical drift speeds of electrons in wires are of the order of a few millimeters per second. For mobile charges moving in the positive direction, $$I = qnAv_{\mathrm{d}}$$ where $q = -e$, n is the number density of free electrons, A is the cross-sectional area, and v_{d} is the drift speed.	25-3
Current density	The current density $\vec{J}$ is related to the drift velocity by $$\vec{J} = qn\vec{v}_{\mathrm{d}}$$ The current I through a cross-sectional surface is the flux of the current density through the surface.	25-4
2. Resistance		
Definition of resistance	$$R = \frac{V}{I}$$	25-7
Resistivity, ρ	$$R = \rho\frac{L}{A}$$	25-10
Temperature coefficient of resistivity, α	$$\alpha = \frac{(\rho - \rho_0)/\rho_0}{T - T_0}$$	25-12
3. Ohm's Law	For ohmic materials, the resistance does not depend on either the current or the potential drop: $$V = IR,\ R \text{ constant}$$	25-9
4. Power		
Supplied to a device or segment	$$P = IV$$	25-13
Delivered to a resistor	$$P = IV = I^2R = \frac{V^2}{R}$$	25-14
5. Emf		
Source of emf	A device that supplies electrical energy to a circuit.	
Power supplied by an ideal emf source	$$P = I\mathcal{E}$$	25-15
6. Battery		
Ideal	An ideal battery is a source of emf that maintains a constant potential difference between its two terminals, independent of the current through the battery.	
Real	A real battery can be considered as an ideal battery in series with a small resistance, called its internal resistance.	
Terminal voltage	$$V_a - V_b = \mathcal{E} - Ir$$ where in the battery the positive direction is the direction of increasing potential.	25-16

TOPIC	RELEVANT EQUATIONS AND REMARKS	
Total energy stored	$E_{\text{stored}} = Q\mathcal{E}$	25-18
7. Equivalent Resistance		
Resistors in series	$R_{\text{eq}} = R_1 + R_2 + R_3 + \ldots$	25-20
Resistors in parallel	$\frac{1}{R_{\text{eq}}} = \frac{1}{R_1} + \frac{1}{R_1} + \frac{1}{R_3} + \ldots$	25-25
8. Kirchhoff's Rules	1. When any closed loop is traversed, the algebraic sum of the changes in potential around the loop must equal zero. 2. At any junction (branch point) in a circuit where the current can divide, the sum of the currents into the junction must equal the sum of the currents out of the junction.	
9. Measuring Devices		
Ammeter	An ammeter is a very low resistance device that is placed in series with a circuit element to measure the current in the element.	
Voltmeter	A voltmeter is a very high resistance device that is placed in parallel with a circuit element to measure the potential difference across the element.	
Ohmmeter	An ohmmeter is a device containing a battery connected in series with a galvanometer and a resistor that is used to measure the resistance of a circuit element placed across its terminals.	
10. Discharging a Capacitor		
Charge on the capacitor	$Q(t) = Q_0 e^{-t/(RC)} = Q_0 e^{-t/\tau}$	25-35
Current in the circuit	$I = -\frac{dQ}{dt} = \frac{V_0}{R} e^{-t/(RC)} = I_0 e^{-t/\tau}$	25-37
Time constant	$\tau = RC$	25-36
11. Charging a Capacitor		
Charge on the capacitor	$Q = C\mathcal{E}[1 - e^{-t/(RC)}] = Q_f(1 - e^{-t/\tau})$	25-40
Current in the circuit	$I = \frac{\mathcal{E}}{R} e^{-t/(RC)} = I_0 e^{-t/\tau}$	25-41

Answers to Concept Checks

25-1 (*a*) The current is larger just after the switch is thrown because the bulb filament is a metal and is relatively cool, so its resistance is less than it will be when the bulb has been running for a while. Lower resistance means higher current. (*b*) Energy from the battery initially is delivered to the filament at a greater rate than the relatively cool filament releases heat. After a while, energy from the battery is delivered to the filament at the same rate that the now hot filament releases heat. Under these conditions the filament's temperature, and thus its resistance, remains constant.

Answers to Practice Problems

25-1 7.9 h

25-2 14 000

25-3 4.5 V

25-4 2.4 m

25-5 The color bands are colored, from top to bottom, brown, orange, blue, red, and brown. The value of the resistance is 13.6 kΩ and the tolerance is 1%.

25-6 (*a*) 45 W, (*b*) 270 J

25-7	(*a*) 6.0 Ω, (*b*) 1.3 Ω
25-8	(*a*) $R'_{eq} = 2.0\ \Omega$; (*b*) $I = 9.0$ A; (*c*) $V_2 = 18$ V, $V_0 = 0$, $V_{12} = 0$; (*d*) $I_2 = 9.0$ A, $I_0 = 9.0$ A, $I_{12} = 0$
25-9	(*a*) 3.0 A (*b*) 0.83 A
25-10	0.81 mA
25-12	0.86

Problems

In a few problems, you are given more data than you actually need; in a few other problems, you are required to supply data from your general knowledge, outside sources, or informed estimate.

Interpret as significant all digits in numerical values that have trailing zeros and no decimal points.

- • Single-concept, single-step, relatively easy
- •• Intermediate-level, may require synthesis of concepts
- ••• Challenging
- SSM Solution is in the *Student Solutions Manual*
- Consecutive problems that are shaded are paired problems.

CONCEPTUAL PROBLEMS

1 • In our study of electrostatics, we concluded that no electric field exists within the material of a conductor in electrostatic equilibrium. Why can we discuss electric fields within the material of conductors in this chapter?

2 • Figure 25-12 shows a mechanical analog of a simple electric circuit. Devise another mechanical analog in which the current is represented by a flow of water instead of marbles. In the water circuit, what would be analogous to the battery? What would be analogous to the wire? What would be analogous to the resistor?

3 • Wires A and B are both made of copper. The wires are connected in series, so we know they carry the same current. However, the diameter of wire A is twice the diameter of wire B. Which wire has the higher number density of charge carriers (number per unit charge)? (*a*) A, (*b*) B, (*c*) They have the same number density of charge carriers.

4 • The diameters of copper wires A and B are equal. The current carried by wire A is twice the current carried by wire B. In which wire do the charge carriers have the higher drift speed? (*a*) A, (*b*) B, (*c*) They have the same drift speed.

5 • Wire A and wire B are identical copper wires. The current carried by wire A is twice the current carried by wire B. Which wire has the higher current density? (*a*) A, (*b*) B, (*c*) They have the same current density. (*d*) none of the above

6 •• Consider a metal wire that has each end connected to a different terminal of the same battery. Your friend argues that no matter how long the wire is, the drift speed of the charge carriers in the wire is the same. Evaluate your friend's claim.

7 • In a resistor, the direction of the current must always be in the "downhill" direction, that is, in the direction of decreasing electric potential. Is it also the case that in a battery, the direction of the current must always be the "downhill"? Explain your answer.

8 • Discuss the distinction between an emf and a potential difference.

9 • Wire A and wire B are made of the same material and have the same length. The diameter of wire A is twice the diameter of wire B. If the resistance of wire B is R, then what is the resistance of wire A? (Neglet any effects that temperature may have on resistance.) (*a*) R, (*b*) $2R$, (*c*) $R/2$, (*d*) $4R$, (*e*) $R/4$

10 • Two cylindrical copper wires have the same mass. Wire A is twice as long as wire B. (Neglet any effects that temperature may have on resistance.) Their resistances are related by (*a*) $R_A = 8R_B$, (*b*) $R_A = 4R_B$, (*c*) $R_A = 2R_B$, (*d*) $R_A = R_B$

11 • If the current in a resistor is I, the power delivered to the resistor is P. If the current in the resistor is increased to $3I$, what is the power then delivered to the resistor? (Assume the resistance of the resistor does not change.) (*a*) P, (*b*) $3P$, (*c*) $P/3$, (*d*) $9P$, (*e*) $P/9$

12 • If the potential drop across a resistor is V, the power delivered to the resistor is P. If the potential drop is increased to $2V$, what is the power delivered to the resistor then equal to? (*a*) P, (*b*) $2P$, (*c*) $4P$, (*d*) $P/2$, (*e*) $P/4$

13 • A heater consists of a variable resistor (a resistor whose resistance can be varied) connected across an ideal voltage supply. (An ideal voltage supply is one that has a constant emf and a negligible internal resistance.) To increase the heat output, should you decrease the resistance or increase the resistance? Explain your answer. SSM

14 • One resistor has a resistance R_1 and another resistor has a resistance R_2. The resistors are connected in parallel. If $R_1 \gg R_2$, the equivalent resistance of the combination is approximately (*a*) R_1, (*b*) R_2, (*c*) 0, (*d*) infinity.

15 • One resistor has a resistance R_1 and another resistor has a resistance R_2. The resistors are connected in series. If $R_1 \gg R_2$, the equivalent resistance of the combination is approximately (*a*) R_1, (*b*) R_2, (*c*) 0, (*d*) infinity.

16 • A parallel combination consisting of resistors A and B is connected across the terminals of a battery. The resistor A has twice the resistance of resistor B. If the current carried by resistor A is I, then what is the current carried by resistor B? (*a*) I, (*b*) $2I$, (*c*) $I/2$, (*d*) $4I$, (*e*) $I/4$

17 • A series combination consisting of resistors A and B is connected across the terminals of a battery. The resistor A has twice the resistance of resistor B. If the current carried by resistor A is I, then what is the current carried by resistor B? (*a*) I, (*b*) $2I$, (*c*) $I/2$, (*d*) $4I$, (*e*) $I/4$

18 • Kirchhoff's junction rule is considered to be a consequence of (*a*) conservation of charge, (*b*) conservation of energy, (*c*) Newton's laws, (*d*) Coulomb's law, (*e*) quantization of charge.

19 • True or false:

(*a*) An ideal voltmeter has a zero internal resistance.
(*b*) An ideal ammeter has a zero internal resistance.
(*c*) An ideal voltage source has a zero internal resistance.

20 • Before you and your classmates run an experiment, your professor lectures about safety. She reminds you that to measure the voltage across a resistor you connect a voltmeter in parallel with the resistor, and to measure the current in a resistor you connect an ammeter in series with the resistor. She also states that connecting a voltmeter in series with a resistor will not measure the voltage across the resistor, but also cannot do any damage to the circuit or the instrument. In addition, connecting an ammeter in parallel with a resistor will not measure the current in the resistor, but could cause significant damage to the circuit and the instrument. Explain why connecting a voltmeter in series with a resistor causes no damage while connecting an ammeter in parallel with a resistor can cause significant damage.

21 • The capacitor C in Figure 25-49 is initially uncharged. Just after the switch S is closed, (*a*) the voltage across C equals $\mathcal{E}$, (*b*) the voltage across R equals $\mathcal{E}$, (*c*) the current in the circuit is zero, (*d*) both (*a*) and (*c*) are correct.

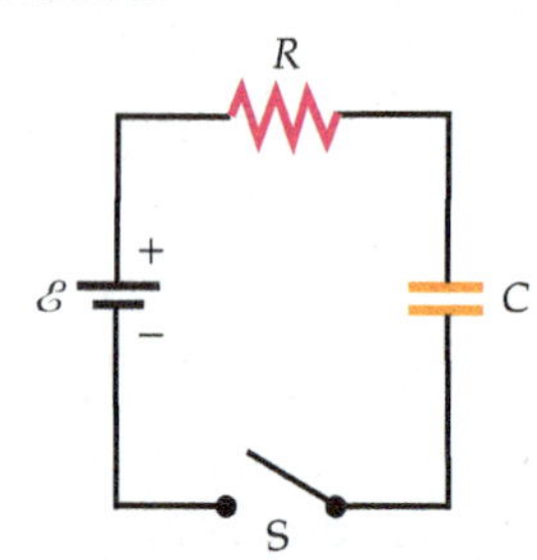

FIGURE 25-49 Problems 21 and 24

22 •• A capacitor is discharging through a resistor. If it takes a time T for the charge on a capacitor to drop to half its initial value, how long (in terms of T) does it take for the stored energy to drop to half its initial value?

23 •• In Figure 25-50, the values of the resistances are related as follows: $R_2 = R_3 = 2R_1$. If power P is delivered to R_1, what is the power delivered to R_2 and R_3? SSM

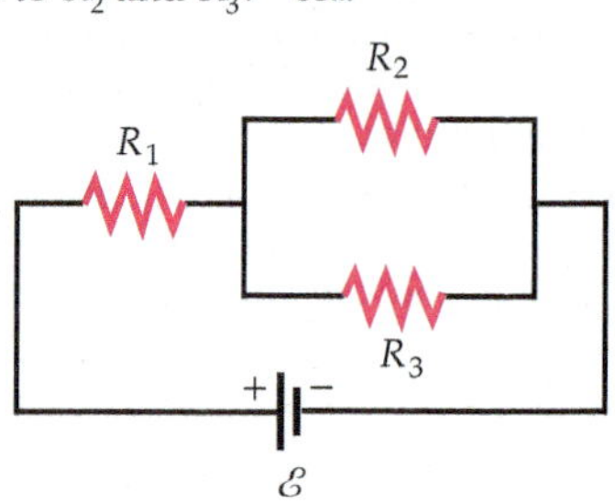

FIGURE 25-50 Problem 23

24 •• The capacitor in Figure 25-49 is initially uncharged. The switch S is closed and remains closed for a very long time. During this time, (*a*) the energy supplied by the battery is $\frac{1}{2}C\mathcal{E}^2$, (*b*) the energy dissipated in the resistor is $\frac{1}{2}C\mathcal{E}^2$, (*c*) the energy in the resistor is dissipated at a constant rate, (*d*) the total charge passing through the resistor is $\frac{1}{2}C\mathcal{E}$.

ESTIMATION AND APPROXIMATION

25 •• It is not a good idea to stick the ends of a metal paper clip into the two rectangular slots of a household electrical wall outlet in the United States. Explain why by estimating the current that a paper clip would carry until either the fuse blows or the breaker trips.

26 •• (*a*) Estimate the resistance of an automobile jumper cable. (*b*) Look up the current required to start a typical car. At that current, what is the potential drop that occurs across the jumper cable? (*c*) How much power is dissipated in the jumper cable when it carries that current?

27 •• **ENGINEERING APPLICATION, CONTEXT-RICH** Your manager wants you to design a new superinsulated water heater for the residential market. A coil of Nichrome wire is to be used as the heating element. Estimate the length of wire required. *Hint: You will need to determine the size of a typical water heater and a reasonable time period for creating hot water.*

28 •• A compact fluorescent lightbulb costs about \$6.00 and has a typical lifetime of 10 000 h. The bulb uses 20 W of power but produces illumination equivalent to that of a 75-W incandescent bulb. An incandescent bulb costs about \$1.50 and has a typical lifetime of 1000 h. Your family wonders whether it should buy fluorescent lightbulbs. Estimate the amount of money your household would save each year by using compact fluorescent lightbulbs instead of the incandescent bulbs.

29 •• **CONTEXT-RICH** The wires in a house must be large enough in diameter so that they do not get hot enough to start a fire. While working for a building contractor during the summer, you are involved in remodeling a house. The local building code states that the Joule heating of the wire used in houses should not exceed 2.0 W/m. Estimate the maximum gauge of the copper wire that you can use during the rewiring of the house with 20-A circuits.

30 •• A laser diode used in making a laser pointer is a highly nonlinear circuit element. Its behavior is as follows. For any voltage drop across it that is less than about 2.30 V, it behaves as if it has an infinite internal resistance, but for voltages higher than 2.30 V it has a very low internal resistance—effectively zero. (*a*) A laser pointer is made by putting two 1.55-V watch batteries in series across the laser diode. If the batteries each have an internal resistance between 1.00 Ω and 1.50 Ω, estimate the current in the laser diode. (*b*) About half of the power delivered to the laser diode goes into radiant energy. Using this fact, estimate the power of the laser beam, and compare this value to typical quoted values of about 3.00 mW. (*c*) If the batteries each have a capacity of 20.0 mA·h (i.e., they can deliver a constant current of 20.0 mA for approximately 1 hour before discharging), estimate how long one can continuously operate the laser pointer before replacing the batteries.

CURRENT, CURRENT DENSITY, DRIFT SPEED, AND THE MOTION OF CHARGES

31 • A 10-gauge copper wire carries a current equal to 20 A. Assuming copper has one free electron per atom, calculate the drift speed of the free electrons in the wire. SSM

32 •• A thin nonconducting ring that has a radius a and a linear charge density λ rotates with angular speed ω about an axis through its center and perpendicular to the plane of the ring. Find the current of the ring.

33 •• A length of 10-gauge copper wire and a length of 14-gauge copper wire are welded together end to end. The wires carry a current of 15 A. (*a*) If there is one free electron for each copper atom in each wire, find the drift speed of the electrons in each wire. (*b*) What is the ratio of the magnitude of the current density in the length of 10-gauge wire to the magnitude of the current density in the length of 14-gauge wire? SSM

34 •• An accelerator produces a beam of protons with a circular cross section that is 2.0 mm in diameter and has a current of 1.0 mA. The current density is uniformly distributed throughout the beam. The kinetic energy of each proton is 20 MeV. The beam strikes a metal target and is absorbed by the target. (*a*) What is the number density of the protons in the beam? (*b*) How many protons strike the target each minute? (*c*) What is the magnitude of the current density in this beam?

35 •• In one of the colliding beams of a planned proton *super-collider,* the protons are moving at nearly the speed of light and the beam current is 5.00 mA. The current density is uniformly distributed throughout the beam. (*a*) How many protons are there per meter of length of the beam? (*b*) If the cross-sectional area of the beam is $1.00 \times 10^{-6}\ \text{m}^2$, what is the number density of protons? (*c*) What is the magnitude of the current density in this beam? SSM

36 •• **CONTEXT-RICH** The *solar wind* consists of protons from the Sun moving toward Earth (the wind actually consists of about 95% protons). The number density of protons at a distance from the Sun equal to the orbital radius of Earth is about 7.0 protons per cubic centimeter. Your research team monitors a satellite that is in orbit around the Sun at a distance from the Sun equal to Earth's orbital radius. You are in charge of the satellite's *mass spectrometer,* an instrument used to measure the composition and intensity of the solar wind. The aperture of your spectrometer is a circle of radius 25 cm. The rate of collection of protons by the spectrometer is such that they constitute a measured current of 85 nA. What is the speed of the protons in the solar wind? (Assume the protons enter the aperture at normal incidence.)

37 •• A gold wire has a 0.10-mm-diameter cross section. Opposite ends of this wire are connected to the terminals of a 1.5-V battery. If the length of the wire is 7.5 cm, how much time, on average, is required for electrons leaving the negative terminal of the battery to reach the positive terminal? Assume the resistivity of gold is $2.44 \times 10^{-8}\ \Omega \cdot \text{m}$.

RESISTANCE, RESISTIVITY, AND OHM'S LAW

***Note:* In this section, assume the resistors are ohmic (constant resistance) unless stated otherwise.**

38 • A 10-m-long wire has a resistance equal to 0.20 Ω and carries a current equal to 5.0 A. (*a*) What is the potential difference across the entire length of the wire? (*b*) What is the electric-field strength in the wire?

39 • A potential difference of 100 V across the terminals of a resistor produces a current of 3.00 A in the resistor. (*a*) What is the resistance of the resistor? (*b*) What is the current in the resistor when the potential difference is only 25.0 V? (Assume the resistance of the resistor remains constant.) SSM

40 • A block of carbon is 3.0 cm long and has a square cross-section whose sides are 0.50 cm long. A potential difference of 8.4 V is maintained across its length. (*a*) What is the resistance of the block? (*b*) What is the current in this resistor?

41 • An extension cord consists of a pair of 30-m-long 16-gauge copper wires. What is the potential difference that must be applied across one of the wires if it is to carry a current of 5.0 A? SSM

42 • (*a*) How long is a 14-gauge copper wire that has a resistance of 12.0 Ω? (*b*) How much current will it carry if a 120-V potential difference is applied across its length?

43 • A cylinder of glass is 1.00 cm long and has a resistivity of $1.01 \times 10^{12}\ \Omega \cdot \text{m}$. What length of copper wire that has the same cross-sectional area will have the same resistance as the glass cylinder?

44 •• **ENGINEERING APPLICATION** While remodeling your garage, you need to temporarily splice, end to end, an 80-m-long copper wire that is 1.00 mm in diameter with a 49-m-long aluminum wire that has the same diameter. The maximum current in the wires is 2.00 A. (*a*) Find the potential drop across each wire of this system when the current is 2.00 A. (*b*) Find the electric field in each wire when the current is 2.00 A.

45 •• A 1.00-m-long wire has a resistance equal to 0.300 Ω. A second wire made of identical material has a length of 2.00 m and a mass equal to the mass of the first wire. What is the resistance of the second wire? SSM

46 •• A 10-gauge copper wire can safely carry currents up to 30.0 A. (*a*) What is the resistance of a 100-m length of the wire? (*b*) What is the electric field in the wire when the current is 30.0 A? (*c*) How long does it take for an electron to travel 100 m in the wire when the current is 30.0 A?

47 •• A cube of copper has edges that are 2.00 cm long. If copper in the cube is drawn to form a length of 14-gauge wire, what will the resistance of the length of wire be? Assume the density of the copper does not change.

48 ••• Find an expression for the resistance between the ends of the half-ring shown in Figure 25-51. The resistivity of the material constituting the half-ring is ρ. *Hint: Model the half-ring as a parallel combination of a large number of thin half-rings. Assume the current is uniformly distributed on a cross section of the half-ring.*

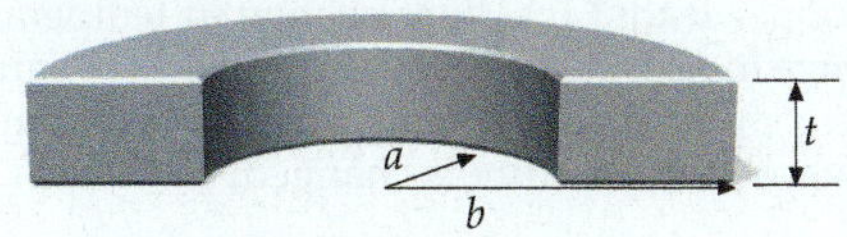

FIGURE 25-51 Problem 48

49 ••• Consider a wire of length L in the shape of a truncated cone. The radius of the wire varies with distance x from the narrow end according to $r = a + [(b - a)/L]x$, where $0 < x < L$. Derive an expression for the resistance of this wire in terms of its length L, radius a, radius b, and resistivity ρ. *Hint: Model the wire as a series combination of a large number of thin disks. Assume the current is uniformly distributed on a cross section of the cone.* SSM

50 ••• The space between two metallic concentric spherical shells is filled with a material that has a resistivity of $3.50 \times 10^{-5}\ \Omega \cdot \text{m}$. If the inner metal shell has an outer radius of 1.50 cm and the outer metal shell has an inner radius of 5.00 cm, what is the resistance between the conductors? *Hint: Model the material as a series combination of a large number of thin spherical shells.*

51 ••• The space between two metallic coaxial cylinders that have the same length L is completely filled with a nonmetallic material having a resistivity ρ. The inner metal shell has an outer radius a and the outer metal shell has an inner radius b. (*a*) What is the resistance between the two cylinders? *Hint: Model the material as a series combination of a large number of thin cylindrical shells.* (*b*) Find the current between the two metallic cylinders if $\rho = 30.0\ \Omega \cdot \text{m}$, $a = 1.50$ cm, $b = 2.50$ cm, $L = 50.0$ cm, and a potential difference of 10.0 V is maintained between the two cylinders.

TEMPERATURE DEPENDENCE OF RESISTANCE

52 • A tungsten rod is 50 cm long and has a square cross section that has 1.0-mm-long edges. (*a*) What is its resistance at 20°C? (*b*) What is its resistance at 40°C?

53 • At what temperature will the resistance of a copper wire be 10 percent greater than its resistance at 20°C? SSM

54 •• **ENGINEERING APPLICATION** You have a toaster that uses a Nichrome wire as a heating element. You need to determine the temperature of the Nichrome wire under operating conditions. First, you measure the resistance of the heating element at 20°C and find it to be 80.0 Ω. Then you measure the current immediately after you plug the toaster into a wall outlet—before the temperature of the Nichrome wire increases significantly. You find this startup current to be 8.70 A. When the heating element reaches its operating temperature, you measure the current to be 7.00 A. Use your data to determine the maximum operating temperature of the heating element.

55 •• **ENGINEERING APPLICATION** Your electric space heater has a Nichrome heating element that has a resistance of 8.00 Ω at 20.0°C. When 120 V is applied, the electric current heats the Nichrome wire to 1000°C. (*a*) What is the initial current in the heating element at 20.0°C? (*b*) What is the resistance of the heating element at 1000°C? (*c*) What is the operating power of this heater?

56 •• A 10.0-Ω Nichrome resistor is wired into an electronic circuit using copper leads (wires) that have diameters equal to 0.600 mm. The copper leads have a total length of 50.0 cm. (*a*) What additional resistance is due to the copper leads? (*b*) What percentage error in the total resistance is produced by neglecting the resistance of the copper leads? (*c*) What change in temperature would produce a change in resistance of the Nichrome wire equal to the resistance of the copper leads? Assume that the Nichrome section is the only one whose temperature is changed.

57 ••• A wire that has a cross-sectional area A, a length L_1, a resistivity ρ_1, and a temperature coefficient α_1 is connected end to end to a second wire that has the same cross-sectional area, a length L_2, a resistivity ρ_2, and a temperature coefficient α_2, so that the wires carry the same current. (*a*) Show that if $\rho_1 L_1 \alpha_1 + \rho_2 L_2 \alpha_2 = 0$, then the total resistance is independent of temperature for small temperature changes. (*b*) If one wire is made of carbon and the other wire is made of copper, find the ratio of their lengths for which the total resistance is approximately independent of temperature. SSM

58 ••• The resistivity of tungsten increases approximately linearly with temperature from 56.0 nΩ · m at 293 K to 1.10 μΩ · m at 3500 K. A lightbulb is powered by a 100-V dc power supply. Under those operating conditions the temperature of the tungsten filament is 2500 K, the length of the filament is equal to 5.00 cm, and the power delivered to the filament is 40 W. Estimate (*a*) the resistance of the filament and (*b*) the diameter of the filament.

59 ••• A 5.00-V lightbulb used in an electronics class has a carbon filament that has a length of 3.00 cm and a diameter of 40.0 μm. At temperatures between 500 K and 700 K, the resistivity of the carbon used in making small lightbulb filaments is about $3.00 \times 10^{-5}\ \Omega \cdot \text{m}$. (*a*) Assuming that the bulb is a perfect blackbody radiator, calculate the temperature of the filament under operating conditions. (*b*) One concern about carbon filament bulbs, unlike tungsten filament bulbs, is that the resistivity of carbon decreases with increasing temperature. Explain why this decrease in resistivity is a concern.

ENERGY IN ELECTRIC CIRCUITS

60 • A 1.00-kW heater is designed to operate at 240 V. (*a*) What is the heater's resistance and what is the current in the wires that supply power to the heater? (*b*) What is the power delivered to the heater if it operates at 120 V? Assume that its resistance remains the same.

61 • A battery has an emf of 12 V. How much work does it do in 5.0 s if it delivers a current of 3.0 A?

62 • An automotive battery has an emf of 12.0 V. When supplying power to the starter motor, the current in the battery is 20.0 A and the terminal voltage of the battery is 11.4 V. What is the internal resistance of the battery?

63 • (*a*) How much power is delivered by the battery in Problem 62 due to the chemical reactions within the battery when the current in the battery is 20 A? (*b*) How much of this power is delivered to the starter when the current in the battery is 20 A? (*c*) By how much does the chemical energy of the battery decrease if the current in the starter is 20 A for 7.0 s? (*d*) How much energy is dissipated in the battery during those 7.0 seconds? SSM

64 • A battery that has an emf of 6.0 V and an internal resistance of 0.30 Ω is connected to a variable resistor with resistance R. Find the current and power delivered by the battery when R is (*a*) 0, (*b*) 5.0 Ω, (*c*) 10 Ω, and (*d*) infinite.

65 •• **ENGINEERING APPLICATION, CONTEXT-RICH** A 12.0-V automobile battery that has a negligible internal resistance can deliver a total charge of 160 A · h. (*a*) What is the amount of energy stored in the battery? (*b*) After studying all night for a calculus test, you try to drive to class to take the test. However, you find that the car's battery is "dead" because you had left the headlights on! Assuming the battery was able to produce current at a constant rate until it died, how long were your lights on? Assume the pair of headlights together operates at a power of 150 W.

66 •• **ENGINEERING APPLICATION** The measured current in a circuit in your uncle's house is 12.5 A. In this circuit, the only appliance that is on is a space heater that is being used to heat the bathroom. A pair of 12-gauge copper wires carries the current from the supply panel in your basement to the wall outlet in the bathroom, a distance of 30.0 m. You measure the voltage at the supply panel to be exactly 120 V. What is the voltage at the wall outlet in the bathroom that the space heater is connected to?

67 •• **ENGINEERING APPLICATION** A lightweight electric car is powered by a series combination of ten 12.0-V batteries, each having negligible internal resistance. Each battery can deliver a charge of 160 A · h before needing to be recharged. At a speed of 80.0 km/h, the average force due to air drag and rolling friction is 1.20 kN. (*a*) What must be the minimum power delivered by the electric motor if the car is to travel at a speed of 80.0 km/h? (*b*) What is the

total charge, in coulombs, that can be delivered by the series combination of ten batteries before recharging is required? (*c*) What is the total electrical energy delivered by the ten batteries before recharging? (*d*) How far can the car travel (at 80.0 km/h) before the batteries must be recharged? (*e*) What is the cost per kilometer if the cost of recharging the batteries is 9.00 cents per kilowatt-hour? SSM

68 ••• A 100-W heater is designed to operate with an applied voltage of 120 V. (*a*) What is the heater's resistance, and what current does the heater carry? (*b*) Show that if the potential difference V across the heater changes by a small amount ΔV, the power P changes by a small amount ΔP, where $\Delta P/P \approx 2\,\Delta V/V$. *Hint: Approximate the changes by modeling them as differentials, and assume the resistance is constant.* (*c*) Using the Part-(*b*) result, find the approximate power delivered to the heater if the potential difference is decreased to 115 V. Compare your result to the exact answer.

COMBINATIONS OF RESISTORS

69 • If the potential drop from point *a* to point *b* (Figure 25-52) is 12.0 V, find the current in each resistor. SSM

70 • If the potential drop between point *a* and point *b* (Figure 25-53) is 12.0 V, find the current in each resistor.

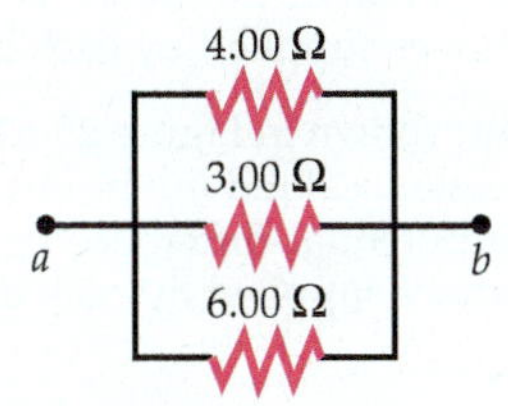

FIGURE 25-52 Problem 69

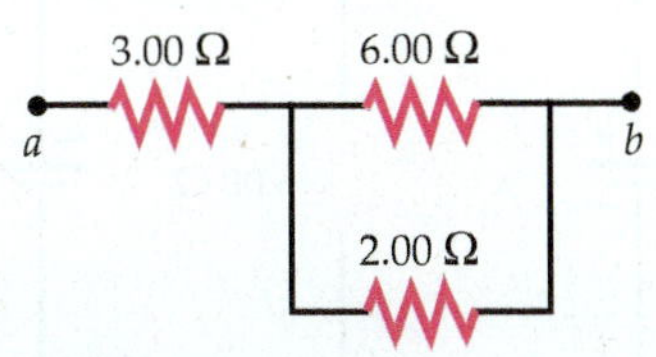

FIGURE 25-53 Problem 70

71 • (*a*) Show that the equivalent resistance between point *a* and point *b* in Figure 25-54 is *R*. (*b*) How would adding a fifth resistor that has resistance *R* between point *c* and point *d* affect the equivalent resistance between point *a* and point *b*?

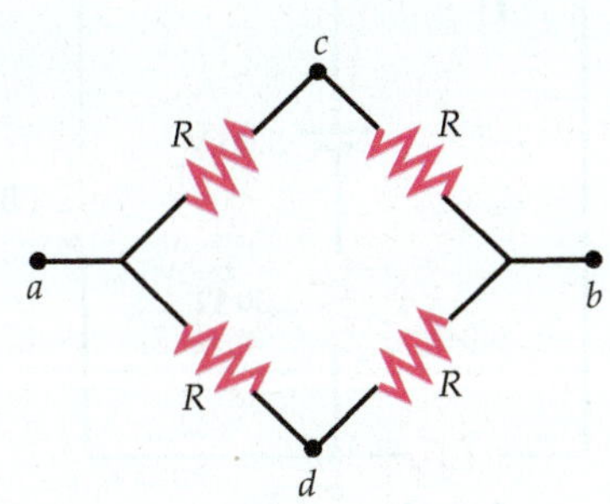

FIGURE 25-54 Problem 71

72 •• The battery in Figure 25-55 has negligible internal resistance. Find (*a*) the current in each resistor and (*b*) the power delivered by the battery.

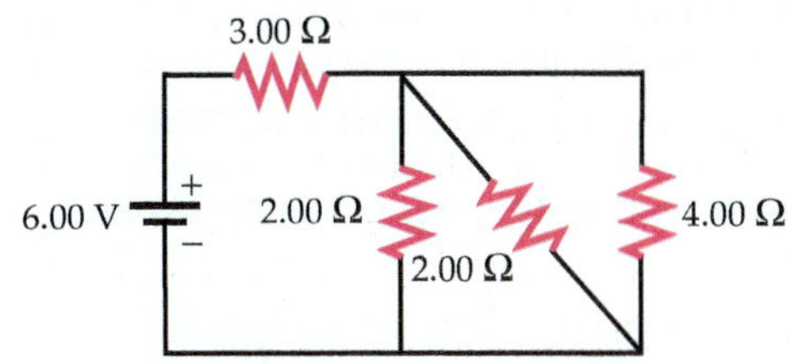

FIGURE 25-55 Problem 72

73 •• A 5.00-V power supply has an internal resistance of 50.0 Ω. What is the smallest resistor that can be put in series with the power supply so that the voltage drop across the resistor is larger than 4.50 V? SSM

74 •• **ENGINEERING APPLICATION** You have been handed an unknown battery. Using your multimeter, you determine that when a 5.00-Ω resistor is connected across the battery's terminals, the current in the battery is 0.500 A. When this resistor is replaced by an 11.0-Ω resistor, the current drops to 0.250 A. From those data, find (*a*) the emf and (*b*) internal resistance of the battery.

75 •• (*a*) Find the equivalent resistance between point *a* and point *b* in Figure 25-56. (*b*) If the potential drop between point *a* and point *b* is 12.0 V, find the current in each resistor.

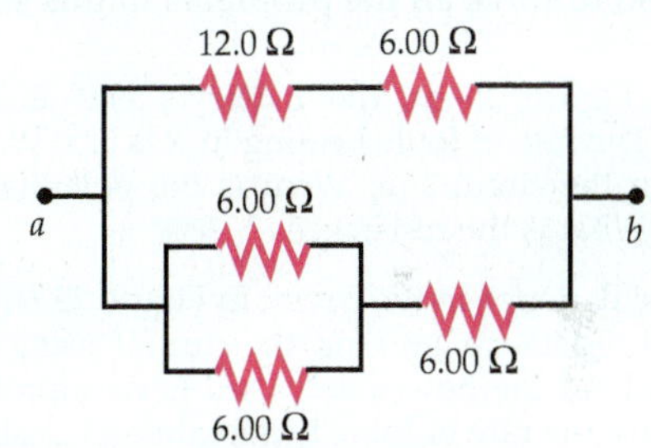

FIGURE 25-56 Problem 75

76 •• (*a*) Find the equivalent resistance between point *a* and point *b* in Figure 25-57. (*b*) If the potential drop between point *a* and point *b* is 12.0 V, find the current in each resistor.

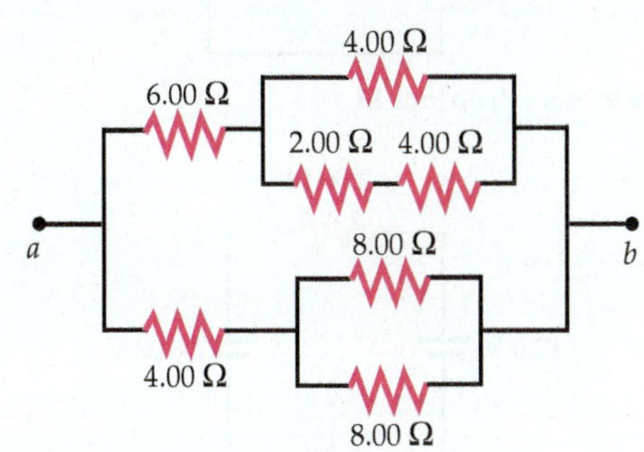

FIGURE 25-57 Problem 76

77 •• A length of wire has a resistance of 120 Ω. The wire is cut into pieces that have the same length, and then the wires are connected in parallel. The resistance of the parallel arrangement is 1.88 Ω. Find the number of pieces into which the wire was cut. SSM

78 •• A parallel combination of an 8.00-Ω resistor and a resistor of unknown resistance is connected in series with a 16.0-Ω resistor and an ideal battery. The circuit is disassembled and the three resistors are then connected in series with each other and the same battery. In both arrangements, the current through the 8.00-Ω resistor is the same. What is the resistance of the unknown resistor?

79 •• For the network shown in Figure 25-58, let R_{ab} denote the equivalent resistance between terminals *a* and *b*. Find (*a*) R_3, so that $R_{ab} = R_1$; (*b*) R_2, so that $R_{ab} = R_3$; and (*c*) R_1, so that $R_{ab} = R_1$.

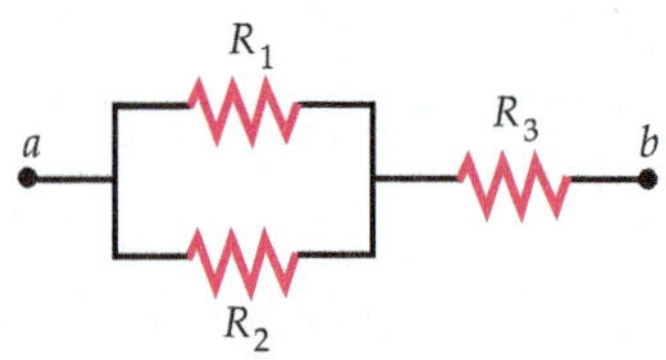

FIGURE 25-58 Problems 79 and 80

80 •• Check your results for Problem 79 using the following specific values: (*a*) $R_1 = 4.00\ \Omega$, $R_2 = 6.00\ \Omega$; (*b*) $R_1 = 4.00\ \Omega$, $R_3 = 3.00\ \Omega$; and (*c*) $R_2 = 6.00\ \Omega$, $R_3 = 3.00\ \Omega$.

KIRCHHOFF'S RULES

Note: **While simpler circuits in this section can be solved using the concepts of equivalent parallel and series resistor combinations, the intent is to gain practice using Kirchhoff's rules. Use them to solve all the problems in this section.**

81 • In Figure 25-59, the battery's emf is 6.00 V and *R* is 0.500 Ω. The rate of Joule heating in *R* is 8.00 W. (*a*) What is the current in the circuit? (*b*) What is the potential difference across *R*? (*c*) What is the resistance *r*? SSM

82 • The batteries in the circuit in Figure 25-60 have negligible internal resistance. (*a*) Find the current using Kirchhoff's loop rule. (*b*) Find the power delivered to or supplied by each battery. (*c*) Find the rate of Joule heating in each resistor.

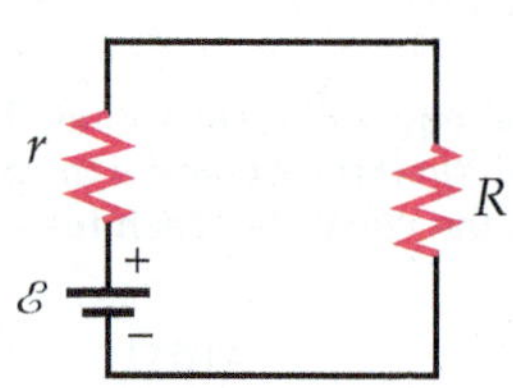

FIGURE 25-59 Problem 81

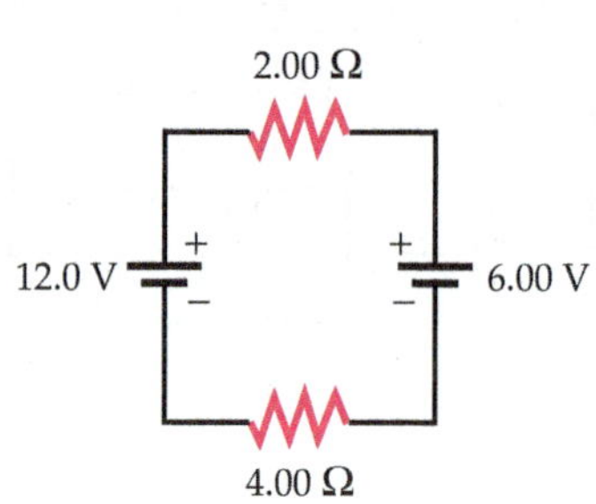

FIGURE 25-60 Problem 82

83 •• **ENGINEERING APPLICATION** An old car battery that has an emf of $\mathcal{E}_1 = 11.4$ V and an internal resistance of 50.0 mΩ is connected to a 2.00-Ω resistor. In an attempt to recharge the battery, you connect a second battery that has an emf of $\mathcal{E}_2 = 12.6$ V and an internal resistance of 10.0 mΩ in parallel with the first battery and the resistor with a pair of jumper cables. (*a*) Draw a diagram of the circuit. (*b*) Find the current in each branch of the circuit. (*c*) Find the power supplied by the second battery and discuss where that power is delivered. Assume that the emfs and internal resistances of both batteries remain constant.

84 •• In the circuit in Figure 25-61, the reading of the ammeter is the same when both switches are open and when both switches are closed. What is the unknown resistance *R*?

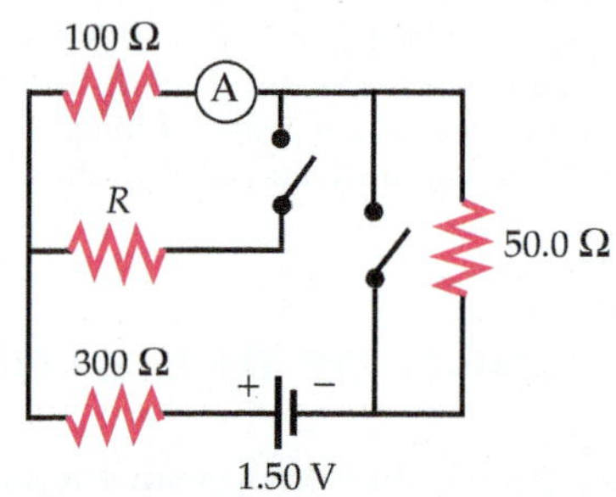

FIGURE 25-61 Problem 84

85 •• In the circuit shown in Figure 25-62, the batteries have negligible internal resistance. Find (*a*) the current in each branch of the circuit, (*b*) the potential difference between point *a* and point *b*, and (*c*) the power supplied by each battery. SSM

86 •• In the circuit shown in Figure 25-63, the batteries have negligible internal resistance. Find (*a*) the current in each branch of the circuit, (*b*) the potential difference between point *a* and point *b*, and (*c*) the power supplied by each battery.

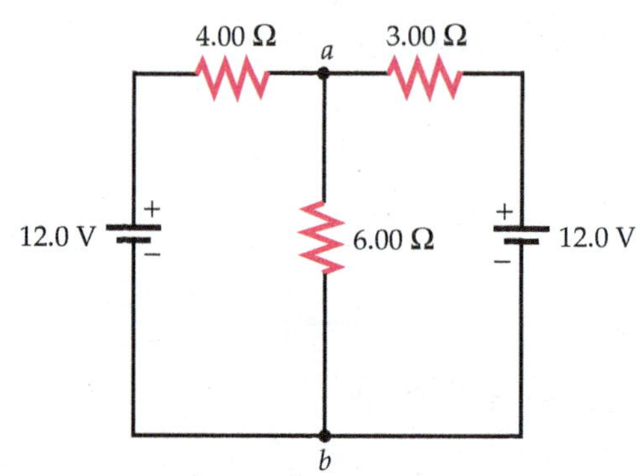

FIGURE 25-62 Problem 85

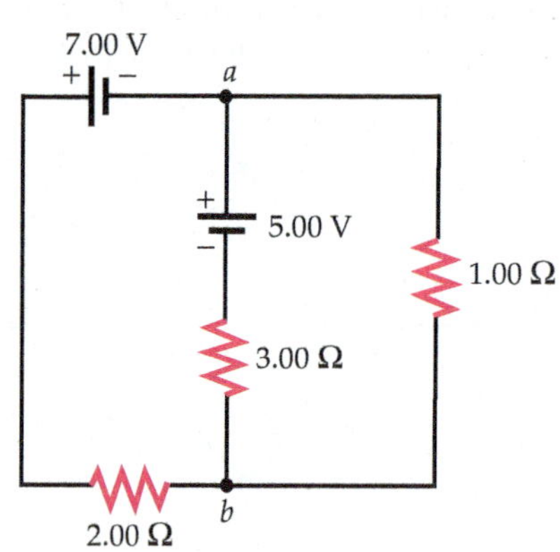

FIGURE 25-63 Problem 86

87 ••• Two identical batteries, each having an emf $\mathcal{E}$ and an internal resistance r, can be connected across a resistance R with the batteries connected either in series or in parallel. In each situation, determine explicitly whether the power supplied to R is greater when R is less than r or when R is greater than r.

88 •• **ENGINEERING APPLICATION** The circuit fragment shown in Figure 25-64 is called a *voltage divider.* (*a*) If R_{load} is not attached, show that $V_{out} = VR_2/(R_1 + R_2)$. (*b*) If $R_1 = R_2 = 10\ \text{k}\Omega$, what is the smallest value of R_{load} that can be used so that V_{out} drops by less than 10 percent from its unloaded value? (V_{out} is measured with respect to ground.)

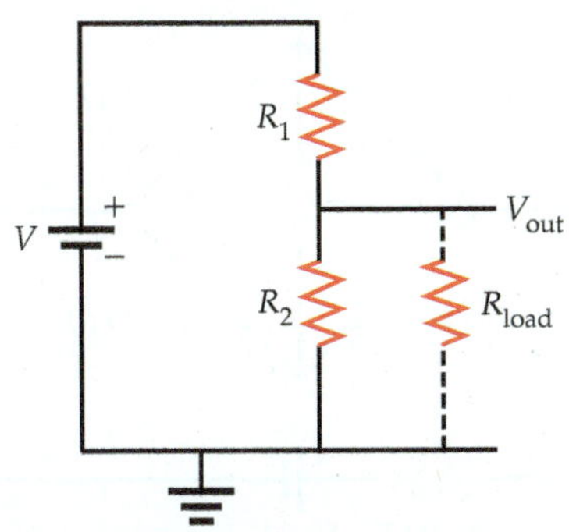

FIGURE 25-64 Problem 88

89 ••• For the circuit shown in Figure 25-65, find the potential difference between point a and point b. SSM

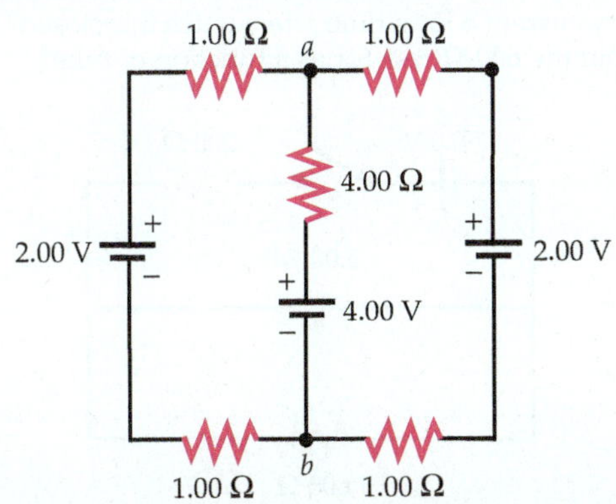

FIGURE 25-65 Problem 89

90 ••• For the circuit shown in Figure 25-66, find (*a*) the current in each resistor, (*b*) the power supplied by each source of emf, and (*c*) the power delivered to each resistor.

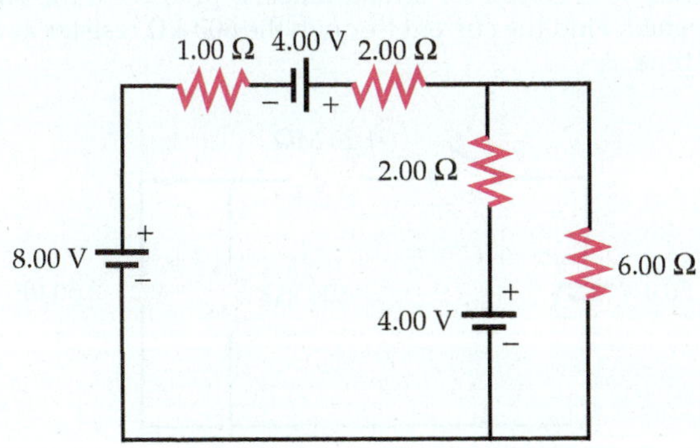

FIGURE 25-66 Problem 90

AMMETERS AND VOLTMETERS

91 •• The voltmeter shown in Figure 25-67 can be modeled as an ideal voltmeter (a voltmeter that has an infinite internal resistance) in parallel with a 10.0-MΩ resistor. Calculate the reading on the voltmeter when (*a*) $R = 1.00\ \text{k}\Omega$, (*b*) $R = 10.0\ \text{k}\Omega$, (*c*) $R = 1.00\ \text{M}\Omega$, (*d*) $R = 10.0\ \text{M}\Omega$, and (*e*) $R = 100\ \text{M}\Omega$. (*f*) What is the largest value of R possible if the measured voltage is to be within 10 percent of the *true* voltage (i.e., the voltage drop across R without the voltmeter in place)? SSM

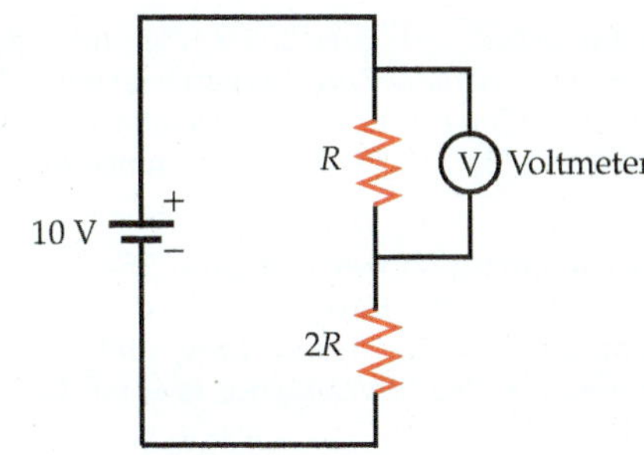

FIGURE 25-67 Problem 91

92 •• You are given a D'Arsonval galvanometer that will deflect full scale if a current of 50.0 μA runs through the galvanometer. At that current, there is a voltage drop of 0.250 V across the meter. What is the internal resistance of the galvanometer?

93 •• You are given a D'Arsonval galvanometer that will deflect full scale if a current of 50.0 μA runs through the galvanometer. At that current, there is a voltage drop of 0.250 V across the meter. You wish to use the galvanometer to construct an ammeter that can measure currents up to 100 mA. Show that this can be done by placing a resistor in parallel with the meter, and find the value of its resistance.

94 •• You are given a D'Arsonval galvanometer that will deflect full scale if a current of 50.0 μA runs through the galvanometer. At that current, there is a voltage drop of 0.250 V across the meter. You wish to use the galvanometer to construct a voltmeter that can measure potential differences up to 10.0 V. Show that this can be done by placing a large resistance in series with the meter movement, and find the resistance needed.

RC CIRCUITS

95 • For the circuit shown in Figure 25-68, $C = 6.00\ \mu\text{F}$, $\mathcal{E} = 100$ V, and $R = 500\ \Omega$. After having been at contact a for a long time, the switch throw is rotated to contact b. (*a*) What is the charge on the upper plate of the capacitor just as the switch throw is moved to contact b? (*b*) What is the current just after the switch throw is rotated to contact b? (*c*) What is the time constant of this circuit? (*d*) How much charge is on the upper plate of the capacitor 6.00 ms after the switch throw is rotated to contact b?

96 • At $t = 0$ the switch throw in Figure 25-68 is rotated to contact b after having been at contact a for a long time. (*a*) How much energy is stored in the capacitor at $t = 0$? (*b*) For $t > 0$, find the energy stored in the capacitor as a function of time. (*c*) Sketch a plot of the energy stored in the capacitor versus time t.

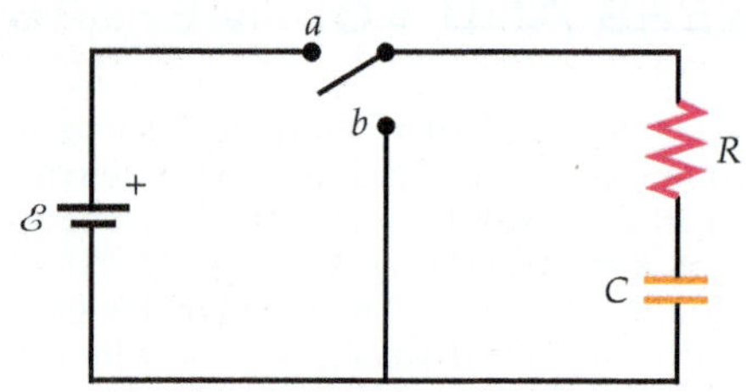

FIGURE 25-68 Problems 95, 96, and 98

97 •• In the circuit in Figure 25-69, the emf equals 50.0 V and the capacitance equals 2.00 μF. Switch S is opened after having been closed for a long time, and 4.00 s later the voltage drop across the resistor is 20.0 V. Find the resistance of the resistor. SSM

98 •• For the circuit shown in Figure 25-68, $C = 0.120\ \mu$F and $\mathcal{E} = 100$ V. The switch throw is rotated to contact b after having been at contact a for a long time, and 4.00 s later the potential difference across the capacitor is equal to $\frac{1}{2}\mathcal{E}$. What is the value of R?

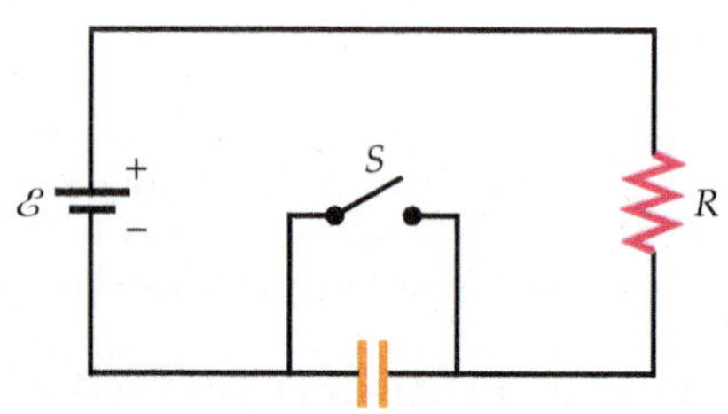

FIGURE 25-69 Problems 97 and 99

99 •• In the circuit in Figure 25-69, the emf equals 6.00 V and has negligible internal resistance. The capacitance equals 1.50 μF and the resistance equals 2.00 MΩ. Switch S has been closed for a long time. Switch S is opened. After a time interval equal to one time constant of the circuit has elapsed, find (*a*) the charge on the capacitor plate on the right, (*b*) the rate at which the charge is increasing, (*c*) the current, (*d*) the power supplied by the battery, (*e*) the power delivered to the resistor, and (*f*) the rate at which the energy stored in the capacitor is increasing.

100 •• A constant charge of 1.00 mC is on the positively charged plate of the 5.00-μF capacitor shown in Figure 25-70. Find (*a*) the battery current and (*b*) the resistances R_1, R_2, and R_3.

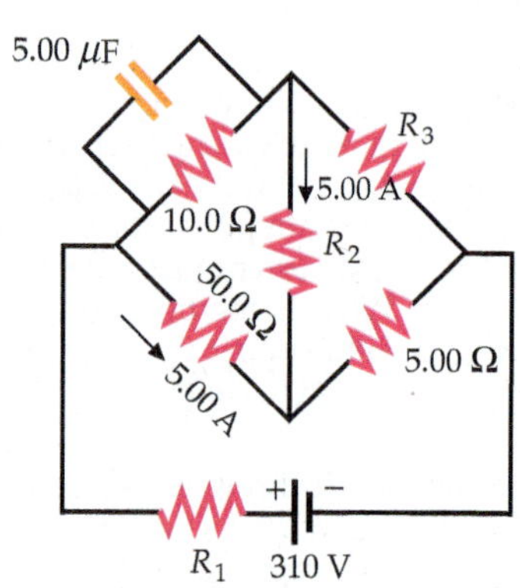

FIGURE 25-70 Problem 100

101 •• Show that Equation 25-39 can be rearranged and written as $\dfrac{dQ}{\mathcal{E}C - Q} = \dfrac{dt}{RC}$. Integrate this equation to derive the solution given by Equation 25-40.

102 •• Switch S, shown in Figure 25-71, is closed after having been open for a long time. (*a*) What is the initial value of the battery current just after switch S is closed? (*b*) What is the battery current a long time after switch S is closed? (*c*) What are the charges on the plates of the capacitors a long time after switch S is closed? (*d*) Switch S is reopened. What are the charges on the plates of the capacitors a long time after switch S is reopened?

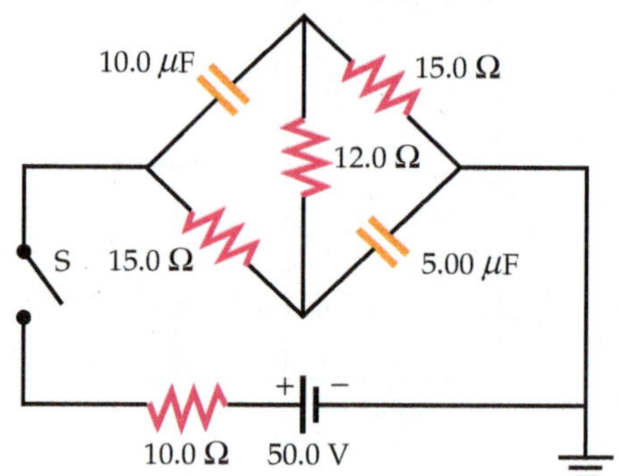

FIGURE 25-71 Problem 102

103 ••• For the circuit shown in Figure 25-72, switch S has been open for a long time. At time $t = 0$ the switch is then closed. (*a*) What is the battery current just after switch S is closed? (*b*) What is the battery current a long time after switch S is closed? (*c*) What is the current in the 600-Ω resistor as a function of time?

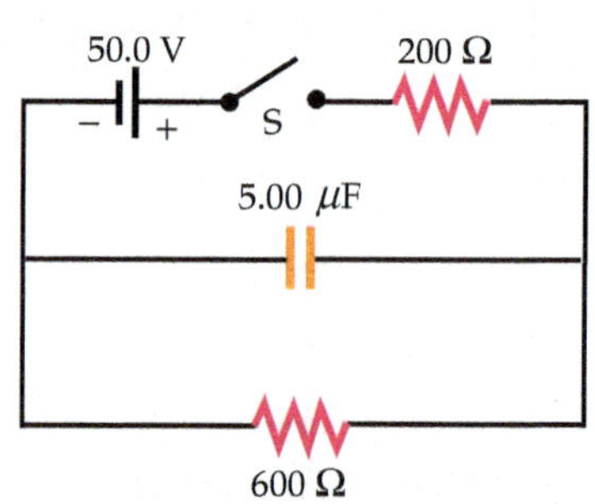

FIGURE 25-72 Problem 103

104 ••• For the circuit shown in Figure 25-73, switch S has been open for a long time. At time $t = 0$ the switch is then closed. (*a*) What is the battery current just after switch S is closed? (*b*) What is the battery current a long time after switch S is closed? (*c*) The switch has been closed for a long time. At time $t = 0$ the switch is then opened. Find the current through the 600-kΩ resistor as a function of time.

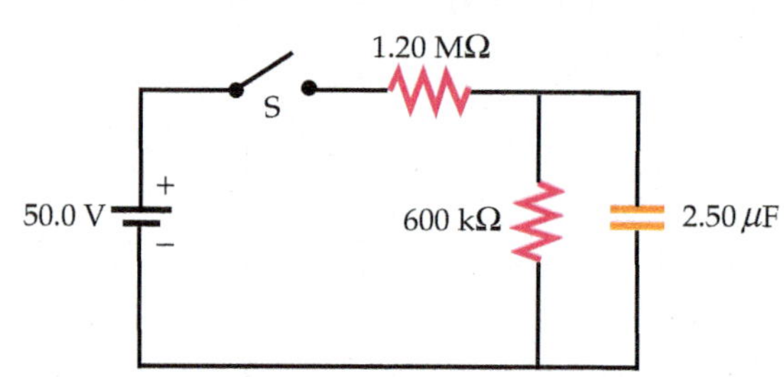

FIGURE 25-73 Problem 104

105 ••• In the circuit shown in Figure 25-74, the capacitor has a capacitance of 2.50 μF and the resistor has a resistance of 0.500 MΩ. Before the switch is closed, the potential drop across the capacitor is 12.0 V, as shown. Switch S is closed at $t = 0$. (*a*) What is the current immediately after switch S is closed? (*b*) At what time t is the voltage across the capacitor 24.0 V? SSM

106 ••• Repeat Problem 105 if the initial polarity of the capacitor is opposite to that shown in Figure 25-74.

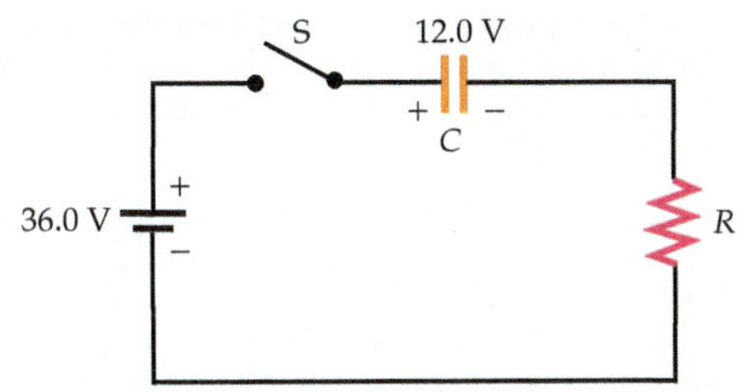

FIGURE 25-74 Problems 105 and 106

GENERAL PROBLEMS

107 •• In Figure 25-75, $R_1 = 4.00\ \Omega$, $R_2 = 6.00\ \Omega$, $R_3 = 12.0\ \Omega$, and the battery emf is 12.0 V. Denote the currents through the resistors as I_1, I_2, and I_3, respectively. (*a*) Decide which of the following inequalities holds for the circuit. Explain your answer conceptually. (1) $I_1 > I_2 > I_3$, (2) $I_2 = I_3$, (3) $I_3 > I_2$, (4) None of the above (*b*) To verify that your answer to Part (*a*) is correct, calculate all three currents. SSM

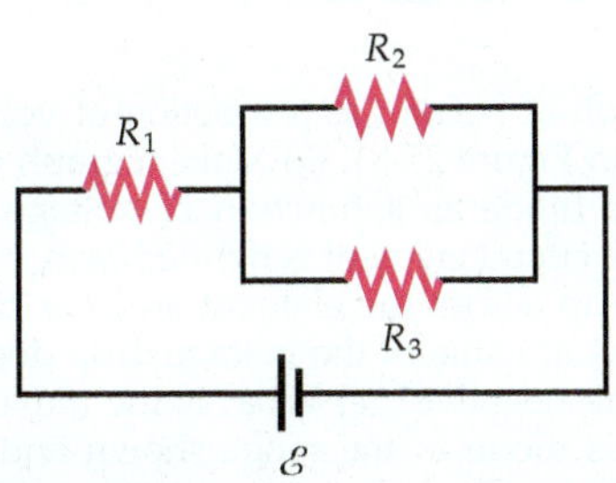

FIGURE 25-75 Problem 107

108 •• A 120 V, 25.0-W lightbulb is connected in series with a 120 V, 100-W lightbulb and a potential difference of 120 V is placed across the combination. Assume the bulbs have constant resistance. (*a*) Which bulb should be brighter under those conditions? Explain your answer conceptually. *Hint: What does the phrase "25.0-W lightbulb" mean? That is, under what conditions is 25 W of power delivered to the bulb?* (*b*) Determine the power delivered to each bulb under those conditions. Do your results support your answer to Part (*a*)?

109 •• The circuit shown in Figure 25-76 is a *Wheatstone bridge,* and the variable resistor is being used as a slide-wire potentiometer. The resistance R_0 is known. This "bridge" is used to determine an unknown resistance R_x. The resistances R_1 and R_2 comprise a wire 1.00 m long. Point *a* is a sliding contact that is moved along the wire to vary the resistances. Resistance R_1 is proportional to the distance from the left end of the wire (labeled 0.00 cm) to point *a*, and R_2 is proportional to the distance from point *a* to the right end of the wire (labeled 100 cm). The sum of R_1 and R_2 remains constant. When points *a* and *b* are at the same potential, there is no current in the galvanometer and the bridge is said to be balanced. (Because the galvanometer is used to detect the absence of a current, it is called a *null detector.*) If the fixed resistance $R_0 = 200\ \Omega$, find the unknown resistance R_x if (*a*) the bridge balances at the 18.0-cm mark, (*b*) the bridge balances at the 60.0-cm mark, and (*c*) the bridge balances at the 95.0-cm mark.

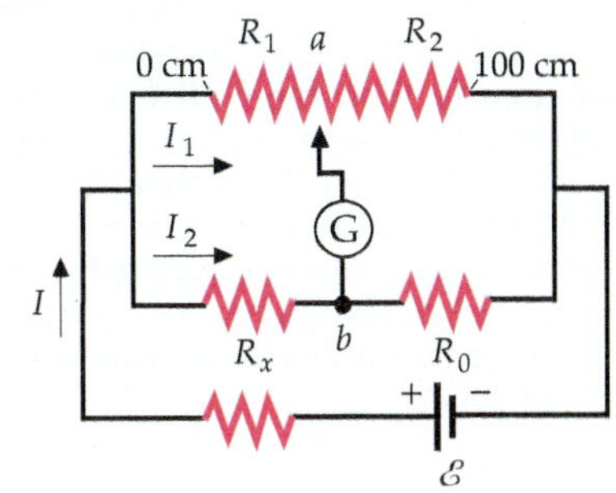

FIGURE 25-76 Problems 109 and 110

110 •• For the Wheatstone bridge in Problem 109, suppose the bridge balances at the 98.0-cm mark. (*a*) What is the unknown resistance? (*b*) What is the percentage error in the measured value of R_x if there is an error of 2.00 mm in the location of the balance point? (*c*) To what value should R_0 be changed to so that the balance point for the unknown resistor will be nearer the 50.0-cm mark? (*d*) If the balance point is at the 50.0-cm mark, what is the percentage error in the measured value of R_x if there is an error of 2.00 mm in the location of the balance point?

111 •• You are running an experiment that uses an accelerator that produces a 3.50-μA proton beam. Each proton in the beam has 60.0 MeV of kinetic energy. The protons impinge on, and come to rest inside, a 50.0-g copper target within a vacuum chamber. You are concerned that the target will get too hot and melt the solder on some connecting wires that are crucial to the experiment. (*a*) Determine the number of protons that strike the target per second. (*b*) Find the amount of energy delivered to the target each second. (*c*) Determine how much time elapses before the target temperature increases to 300°C. (Neglect any heat released by the target.) SSM

112 •• The belt of a Van de Graaff generator carries a surface charge density of 5.00 mC/m^2. The belt is 0.500 m wide and moves at 20.0 m/s. (*a*) What current does the belt carry? (*b*) If the potential of the dome of the generator is 100 kV above ground, what is the minimum power of the motor needed to drive the belt?

113 •• **ENGINEERING APPLICATION** Large conventional electromagnets use water cooling to prevent excessive heating of the magnet coils. A large laboratory electromagnet has a current equal to 100 A when a voltage of 240 V is applied to the terminals of the energizing coils. To cool the coils, water at an initial temperature of 15°C is circulated around the coils. How many liters of water must circulate by the coils each second if the temperature of the coils is not to exceed 50°C?

114 ••• (*a*) Give support to the assertion that a leaky capacitor (one for which the resistance of the dielectric is finite) can be modeled as a capacitor that has an infinite resistance in parallel with a resistor. (*b*) Show that the time constant for discharging the capacitor is given by $\tau = \kappa \epsilon_0 \rho$. (For simplicity, assume the capacitor is a parallel-plate variety filled completely with a leaky dielectric.) (*c*) Mica has a dielectric constant equal to about 5.0 and a resistivity equal to about $9.0 \times 10^{13}\ \Omega \cdot \text{m}$. Calculate the time it takes for the charge of a mica-filled capacitor to decrease to 10 percent of its initial value.

115 ••• **ENGINEERING APPLICATION** Figure 25-77 shows the basis of the sweep circuit used in an oscilloscope. Switch S is an electronic switch that closes whenever the potential across the switch increases to a value V_c and opens when the potential across the switch decreases to 0.200 V. The emf $\mathcal{E}$, which is much greater than V_c,

charges the capacitor C through a resistor R_1. The resistor R_2 represents the small but finite resistance of the electronic switch. In a typical circuit, $\mathcal{E} = 800$ V, $V_c = 4.20$ V, $R_2 = 1.00$ mΩ, $R_1 = 0.500$ MΩ, and $C = 20.0$ nF. (*a*) What is the time constant for charging of the capacitor C? (*b*) Show that as the potential across switch S increases from 0.200 V to 4.20 *V*, the potential across the capacitor increases almost linearly with time. *Hint: Use the approximation* $e^x \approx 1 + x$, *for* $|x| \ll 1$. *(This approximation of* e^x *can be derived using the differential approximation.)* (*c*) What should the value of R_1 be changed to so that the capacitor charges from 0.200 V to 4.20 V in 0.100 s? (*d*) How much time elapses during the discharge of the capacitor when switch S closes? (*e*) At what average rate is energy delivered to the resistor R_1 during charging and to the switch resistance R_2 during discharge?

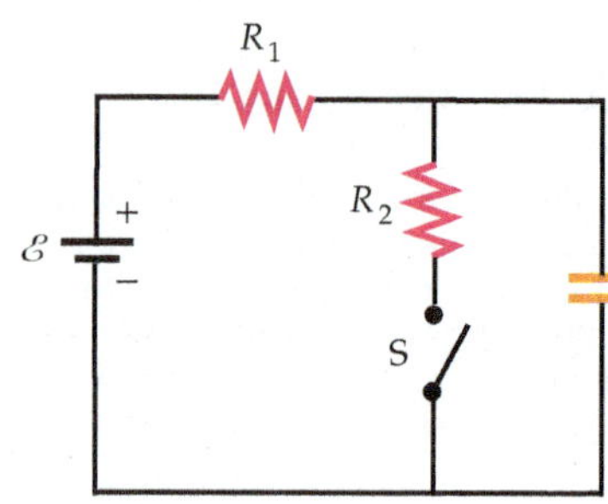

FIGURE 25-77 Problem 115

116 ••• In the circuit shown in Figure 25-78, $R_1 = 2.00$ MΩ, $R_2 = 5.00$ MΩ, and $C = 1.00$ μF. The capacitor is initially without charge on either plate. At $t = 0$, switch S is closed, and at $t = 2.00$ s switch S is opened. (*a*) Sketch a graph of the voltage across C and the current in R_2 between $t = 0$ and $t = 10.0$ s. (*b*) Find the voltage across the capacitor at $t = 2.00$ s and at $t = 8.00$ s.

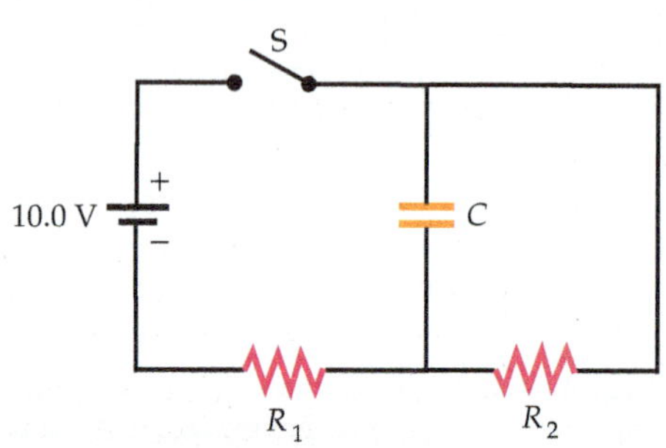

FIGURE 25-78 Problem 116

117 ••• Two batteries that have emfs $\mathcal{E}_1$ and $\mathcal{E}_2$ and internal resistances r_1 and r_2 are connected in parallel. Prove that if a resistor of resistance R is connected in parallel with combination, the optimal load resistance (the value of R at which maximum power is delivered) is given by $R = r_1r_2/(r_1 + r_2)$.

118 ••• Capacitors C_1 and C_2 are connected to a resistor of resistance R and an ideal battery that has $\mathcal{E}$ as shown in Figure 25-79. The throw of switch S is initially at contact *a* and both capacitors are without charge. The throw is then rotated to contact *b* and left there for a long time. Finally, at time $t = 0$, the throw is returned to contact *a*. (*a*) Quantitatively compare the total energy stored in the two capacitors at $t = 0$ and a long time later. (*b*) Find the current

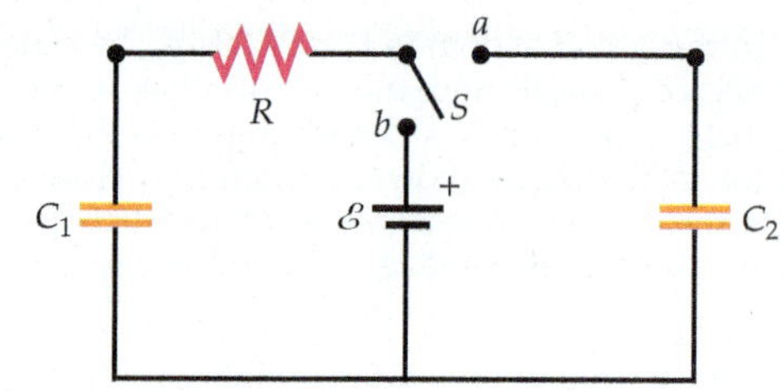

FIGURE 25-79 Problem 118

through R as a function of t for $t > 0$. (*c*) Find the energy delivered to the resistor as a function of t for $t > 0$. (*d*) Find the total energy dissipated in the resistor after $t = 0$ and compare it with the loss of stored energy found in Part (*a*).

119 ••• (*a*) Calculate the equivalent resistance (in terms of R, the resistance of each individual resistor) between points *a* and *b* for the infinite ladder of resistors shown in Figure 25-80 assuming the resistors are identical. That is, assuming $R = R_1 = R_2$. (*b*) Repeat Part (*a*) but do not assume that $R_1 = R_2$ and express your answer in terms of R_1 and R_2. (*c*) Check your results by showing that your result from Part (*b*) agrees with your result from Part (*a*) if you substitute R for both R_1 and R_2.

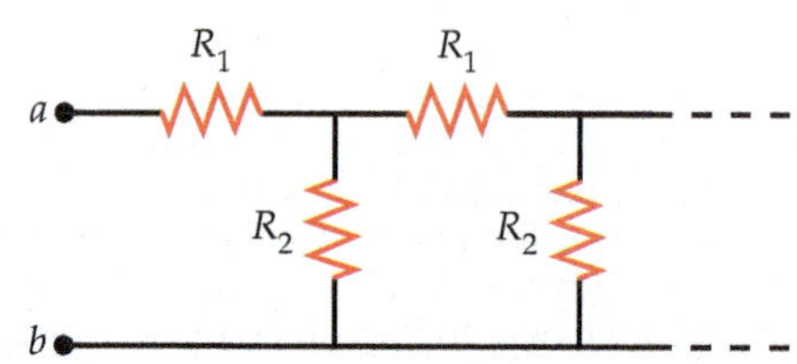

FIGURE 25-80 Problem 119

120 ••• A graph of current as a function of voltage for an Esaki diode is shown in Figure 25-81. (*a*) Make a graph of the differential resistance of the diode as a function of voltage. The differential resistance R_d of a circuit element is defined as $R_d = dV/dI$, where V is the voltage drop across the element and I is the current in the element. (*b*) At what value of the voltage drop does the differential resistance become negative? (*c*) What is the maximum differential resistance for this diode in the range shown and at what voltage does it occur? (*d*) Are there any places in the voltage range shown where the diode exhibits a differential resistance equal to zero? If so, under what value(s) of the voltage does this (do these) occur?

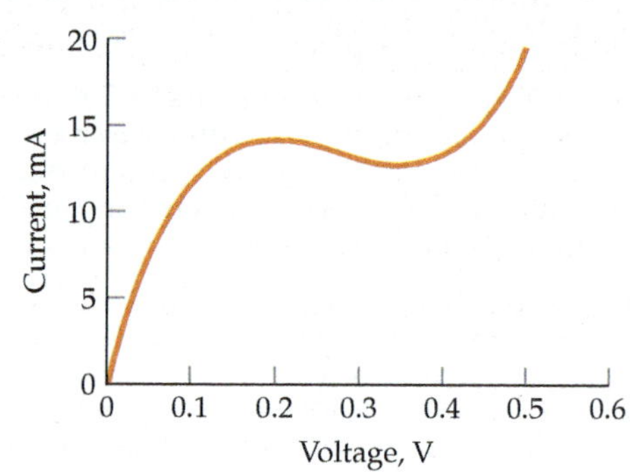

FIGURE 25-81 Problem 120

CHAPTER 26

THE AURORA BOREALIS APPEARS WHEN THE SOLAR WIND, CHARGED PARTICLES PRODUCED BY NUCLEAR FUSION REACTIONS IN THE SUN, BECOMES TRAPPED BY EARTH'S MAGNETIC FIELD. *(Atlas Photo Bank/Photo Researchers, Inc.)*

The Magnetic Field

? How does Earth's magnetic field act on subatomic particles? (See Example 26-1.)

More than 2000 years ago, the Greeks were aware that a certain type of stone (now called magnetite) attracts pieces of iron, and written references exist which describe the use of magnets for navigation dating from the twelfth century.

In 1269, Pierre de Maricourt discovered that a needle laid at various positions on a spherical natural magnet orients itself along lines that pass through points at opposite ends of the sphere. He called these points the *poles of the magnet*. Subsequently, many experimenters noted that every magnet of any shape has two poles, called the north and the south pole, where the force exerted by the magnet is strongest. It was also noted that the *like poles* of two magnets repel each other and the *unlike poles* of two magnets attract each other.

In 1600, William Gilbert discovered that Earth is a natural magnet and has magnetic poles near the north and south geographic poles. Because the north pole of a compass needle points toward the south pole of a given magnet, what we call the north pole of Earth is actually a south magnetic pole, as illustrated in Figure 26-1. Thus, the north and south poles of a magnet are sometimes referred to as the north-seeking and south-seeking poles, respectively.

Although electric charges and magnetic poles are similar in many respects, there is an important difference: Magnetic poles always occur in pairs. When a magnet is broken in half, equal and opposite poles appear at either side of the break point. The result is two magnets, each with a north and a south pole.

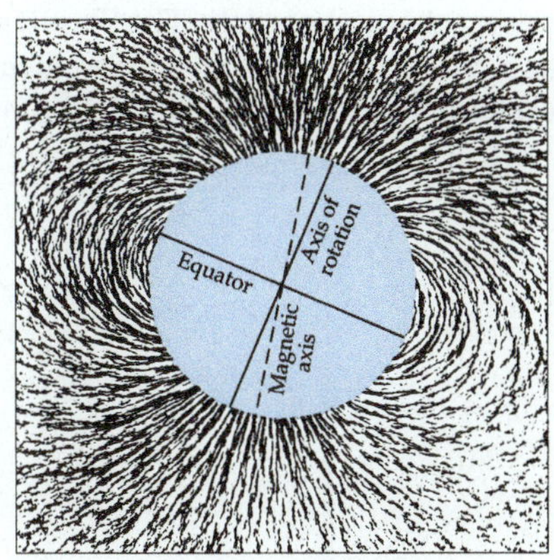

FIGURE 26-1 Magnetic field lines of Earth depicted by iron filings around a uniformly magnetized sphere. The field lines exit from the north magnetic pole, which is near the south geographic pole, and enter the south magnetic pole, which is near the north geographic pole.

There has long been speculation about the existence of an isolated magnetic pole, and in recent years considerable experimental effort has been made to find such an object. Thus far, there is no conclusive evidence that an isolated magnetic pole exists.

In this chapter, we consider the effects of a given magnetic field on moving charges and on wires carrying currents. The sources of magnetic fields are discussed in the next chapter.

26-1 THE FORCE EXERTED BY A MAGNETIC FIELD

The existence of a magnetic field $\vec{B}$ at some point in space can be demonstrated using a compass needle. If there is a magnetic field, the needle will align itself in the direction of the field.*

It has been experimentally observed that, when a particle that has charge q and velocity $\vec{v}$ is in a region with a magnetic field $\vec{B}$, a force acts on the particle that is proportional to q, to v, to B, and to the sine of the angle between the directions of $\vec{v}$ and $\vec{B}$. Surprisingly, the force is perpendicular to both the velocity and the magnetic field. These experimental results can be summarized as follows: When a particle that has a charge q and a velocity $\vec{v}$ is in a region with a magnetic field $\vec{B}$, the magnetic force $\vec{F}$ on the particle is

$$\vec{F} = q\vec{v} \times \vec{B} \qquad 26\text{-}1$$

MAGNETIC FORCE ON A MOVING CHARGED PARTICLE

Because $\vec{F}$ is perpendicular to both $\vec{v}$ and $\vec{B}$, $\vec{F}$ is perpendicular to the plane defined by these two vectors. The direction of $\vec{v} \times \vec{B}$ is given by the right-hand rule as $\vec{v}$ is rotated into $\vec{B}$, as illustrated in Figure 26-2. If q is positive, then $\vec{F}$ is in the same direction as $\vec{v} \times \vec{B}$.

Examples of the direction of the forces exerted on moving charged particles when the magnetic field vector $\vec{B}$ is in the vertical direction are shown in Figure 26-3.

The direction of any particular magnetic field $\vec{B}$ can be found experimentally by measuring $\vec{F}$ and $\vec{v}$ for several velocities in different directions and then applying Equation 26-1.

Equation 26-1 defines the **magnetic field** $\vec{B}$ in terms of the force exerted on a moving charged particle. The SI unit of magnetic field is the **tesla** (T). A particle that has a charge of one coulomb and is moving with a velocity of one meter per second perpendicular to a magnetic field of one tesla experiences a force of one newton:

$$1\,\text{T} = 1\frac{\text{N}}{\text{C}\cdot\text{m/s}} = 1\,\text{N/(A}\cdot\text{m)} \qquad 26\text{-}2$$

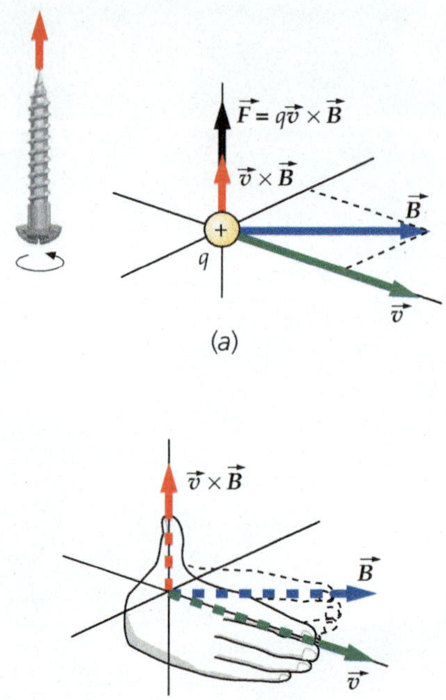

FIGURE 26-2 Right-hand rule for determining the direction of a force exerted on a charged particle moving in a magnetic field. If q is positive, then $\vec{F}$, is in the same direction as $\vec{v} \times \vec{B}$. (*a*) The vector product $\vec{v} \times \vec{B}$ is perpendicular to both $\vec{v}$ and $\vec{B}$ and is in the direction of the advance of a right-hand-threaded screw if turned in the same direction as to rotate $\vec{v}$ into $\vec{B}$. (*b*) If the fingers of the right hand are in the direction of $\vec{v}$ so that they can be curled toward $\vec{B}$, the thumb points in the direction of $\vec{v} \times \vec{B}$.

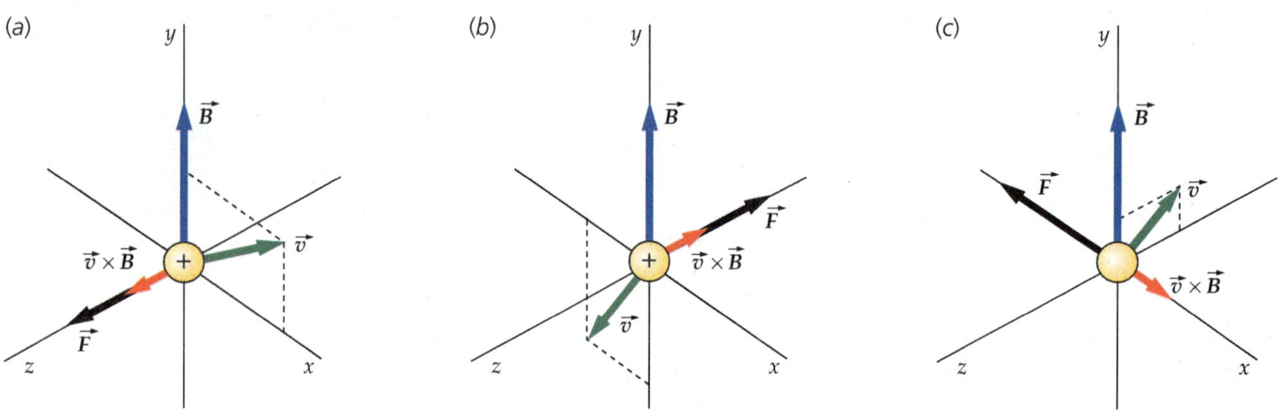

FIGURE 26-3 Part (*a*) and (*b*) show the direction of the magnetic force on a positively charged particle moving with velocity $\vec{v}$ in a magnetic field $\vec{B}$. In Concept Check 26-2 (see page 889), you are asked to find the sign of the charge on the particle shown in part (*c*) of this figure.

* Compass needles are suspended so they remain horizontal. This results in the compass needle aligning itself in the horizontal component of the magnetic field. A compass needle with an unrestricted suspension would align itself in the magnetic field.

CONCEPT CHECK 26-1

The direction of any magnetic field $\vec{B}$ is specified as the direction that the north pole of a compass needle points toward when the needle is aligned in the field. Suppose that the direction of the magnetic field $\vec{B}$ were instead specified as the direction pointed toward by the south pole of a compass needle aligned in the field. Would the right-hand rule shown in Figure 26-2 then give the direction of the magnetic force on the moving positive charge, or would a left-hand rule be required? Explain your answer.

CONCEPT CHECK 26-2

The particle in Figure 26-3(*c*) (*a*) is positively charged, (*b*) is negatively charged, (*c*) could be either positively or negatively charged. Explain your answer.

Like the farad, the tesla is a large unit. The magnetic field strength of Earth has a magnitude somewhat less than 10^{-4} T on Earth's surface. The magnetic field strengths near powerful permanent magnets are about 0.1 T to 0.5 T, and powerful laboratory and industrial electromagnets produce fields of 1 T to 2 T. Fields greater than 10 T are extremely difficult to produce because the resulting magnetic forces will either tear the magnets apart or crush the magnets. A commonly used unit, derived from the cgs system, is the **gauss** (G), which is related to the tesla as follows:

$$1\ \text{G} = 10^{-4}\ \text{T} \qquad 26\text{-}3$$

DEFINITION—GAUSS

Because magnetic fields are often given in gauss, which is not an SI unit, you need to remember to convert from gauss to teslas when making calculations.

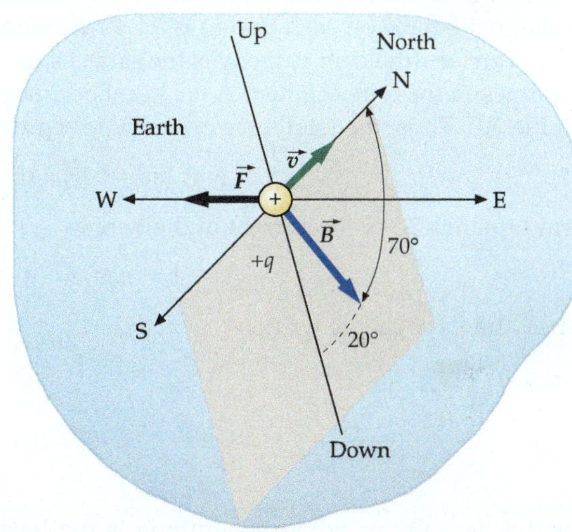

FIGURE 26-4

Example 26-1 Force on a Proton Going North

The magnetic field strength of Earth is measured at a point on the surface to have a magnitude of about 0.6 G and is tilted downward in the northern hemisphere, making an angle of about 70° with the horizontal, as shown in Figure 26-4. (Earth's magnetic field varies from place to place. These data are approximately correct for the central United States.) A proton ($q = +e$) is moving horizontally in the northward direction with speed $v = 1.0 \times 10^7$ m/s. Calculate the magnetic force on the proton (*a*) using $F = qvB\sin\theta$ and (*b*) by first expressing $\vec{v}$ and $\vec{B}$ in terms of the unit vectors $\hat{i}$, $\hat{j}$, and $\hat{k}$, and then computing $\vec{F} = q\vec{v} \times \vec{B}$.

PICTURE Let the x and y directions be to the east and to the north, respectively, and let the z direction be vertically upward (Figure 26-5). The velocity vector is then in the $+y$ direction.

SOLVE

(*a*) Calculate $F = qvB\sin\theta$ using $\theta = 70°$. From Figure 26-4, we see that the direction of the force is westward.

$$F = qvB\sin 70° = (1.6 \times 10^{-19}\ \text{C})(10 \times 10^6\ \text{m/s})(0.6 \times 10^{-4}\ \text{T})(0.94) = \boxed{9.0 \times 10^{-17}\ \text{N}}$$

(*b*) 1. The magnetic force is the vector product of $q\vec{v}$ and $\vec{B}$:

$$\vec{F} = q\vec{v} \times \vec{B}$$

2. Express $\vec{v}$ and $\vec{B}$ in terms of their components:

$$\vec{v} = v_y\hat{j}$$
$$\vec{B} = B_y\hat{j} + B_z\hat{k}$$

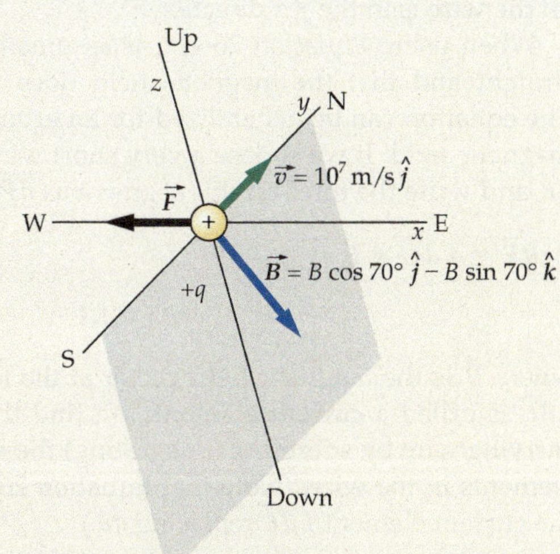

FIGURE 26-5

3. Write $\vec{F} = q\vec{v} \times \vec{B}$ in terms of these components:
$$\vec{F} = q\vec{v} \times \vec{B} = q(v_y\hat{j}) \times (B_y\hat{j} + B_z\hat{k})$$
$$= qv_yB_y(\hat{j} \times \hat{j}) + qv_yB_z(\hat{j} \times \hat{k}) = qv_yB_z\hat{i}$$

4. Evaluate $\vec{F}$:
$$\vec{F} = qv(-B\sin\theta)\hat{i}$$
$$= -(1.6 \times 10^{-19}\text{C})(10^7\text{ m/s})(0.6 \times 10^{-4}\text{ T})\sin 70°\,\hat{i}$$
$$= \boxed{-9.0 \times 10^{-17}\text{ N}\hat{i}}$$

CHECK The Part-(*a*) result is equal to the magnitude of the Part-(*b*) result.

TAKING IT FURTHER Note that the direction of $\hat{i}$ is to the east, so the force is directed to the west as shown in Figure 26-5.

PRACTICE PROBLEM 26-1 Find the force on a proton moving with velocity $\vec{v} = 4 \times 10^6\text{ m/s}\,\hat{i}$ in a magnetic field $\vec{B} = 2.0\text{ T}\hat{k}$.

When a current-carrying wire is in a region that has a magnetic field, there is a force on the wire that is equal to the sum of the magnetic forces on the individual charge carriers in the wire. Figure 26-6 shows a short segment of wire that has cross-sectional area A, length L, and current I. If the wire is in a magnetic field $\vec{B}$, the magnetic force on each charge is $q\vec{v}_d \times \vec{B}$, where $\vec{v}_d$ is the drift velocity of the charge carriers (the drift velocity is the same as the average velocity). The number of charges in the wire segment is the number n per unit volume multiplied by the volume AL. Thus, the total force on the wire segment is

$$\vec{F} = (q\vec{v}_d \times \vec{B})nAL$$

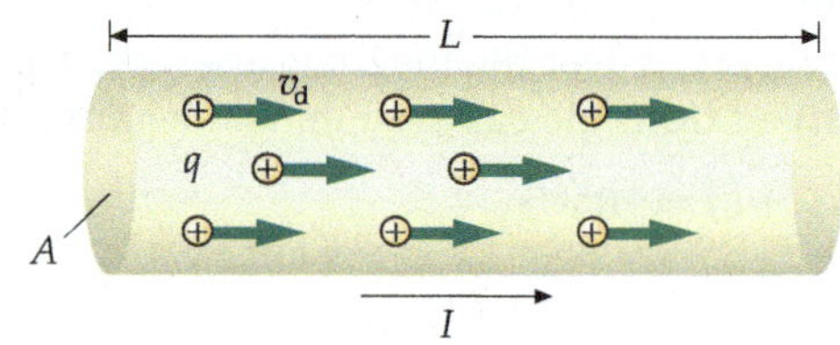

FIGURE 26-6 Wire segment that has a length L and carries a current I. If the wire is in a magnetic field $\vec{B}$, there will be a force on each charge carrier resulting in a force on the wire.

From Equation 25-3, the current in the wire is

$$I = nqv_dA$$

Hence, the force can be written

$$\vec{F} = I\vec{L} \times \vec{B} \qquad 26\text{-}4$$

MAGNETIC FORCE ON A STRAIGHT SEGMENT OF CURRENT-CARRYING WIRE

where $\vec{L}$ is a vector whose magnitude is the length of the segment and whose direction is the same as that of the current.* For the current in the $+x$ direction (Figure 26-7) and the magnetic field vector at the segment in the xy plane, the force on the wire is in the $+z$ direction.

When using Equation 26-4, it is assumed that the wire segment is straight and that the magnetic field does not vary over its length. The equation can be generalized for an arbitrarily shaped wire in any magnetic field. If we choose a very short wire segment that has length $d\vec{\ell}$ and write the force on this segment as $d\vec{F}$, we have

$$d\vec{F} = I\,d\vec{\ell} \times \vec{B} \qquad 26\text{-}5$$

MAGNETIC FORCE ON A CURRENT ELEMENT

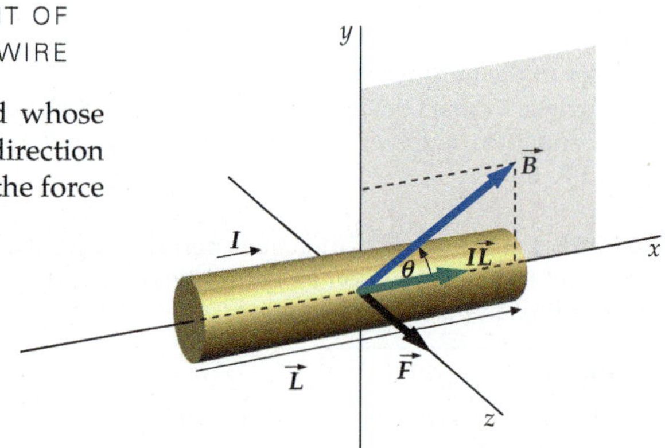

FIGURE 26-7 Magnetic force on a current-carrying segment of wire in a magnetic field. The current is in the x direction, and the magnetic field is in the xy plane and makes an angle θ with the $+x$ direction. The force $\vec{F}$ is in the $+z$ direction, perpendicular to both $\vec{B}$ and $\vec{L}$, and has magnitude $ILB\sin\theta$.

where $\vec{B}$ is the magnetic field vector at the location of the segment. The quantity $I\,d\vec{\ell}$ is called a **current element.** We find the total magnetic force on a current-carrying wire by summing (integrating) the magnetic forces due to all the current elements in the wire. (Note that Equation 26-5 is the same as Equation 26-1 with the current element $I\,d\vec{\ell}$ replacing $q\vec{v}$.)

* By the direction of the current we mean the direction of the current-density vector $\vec{J}$.

Just as the electric field $\vec{E}$ can be represented by electric field lines, the magnetic field $\vec{B}$ can be represented by **magnetic field lines.** In both cases, the direction of the field is indicated by the direction of the field lines and the magnitude of the field is indicated by the density (number per unit area) of the lines on surface perpendicular to the lines. There are, however, two important differences between electric field lines and magnetic field lines:

1. Electric field lines are in the direction of the electric force on a positive charge, but the magnetic field lines are perpendicular to the magnetic force on a moving charge.
2. Electric field lines begin on positive charges and end on negative charges; magnetic field lines neither begin nor end.

Figure 26-8 shows the magnetic field lines both inside and outside a bar magnet.

> ! Do not think the field lines for the magnetic field of a magnet begin on magnetic south poles and end on magnetic north poles. In reality, they neither begin nor end. Instead they enter the magnet at one end and exit the magnet at the other end.

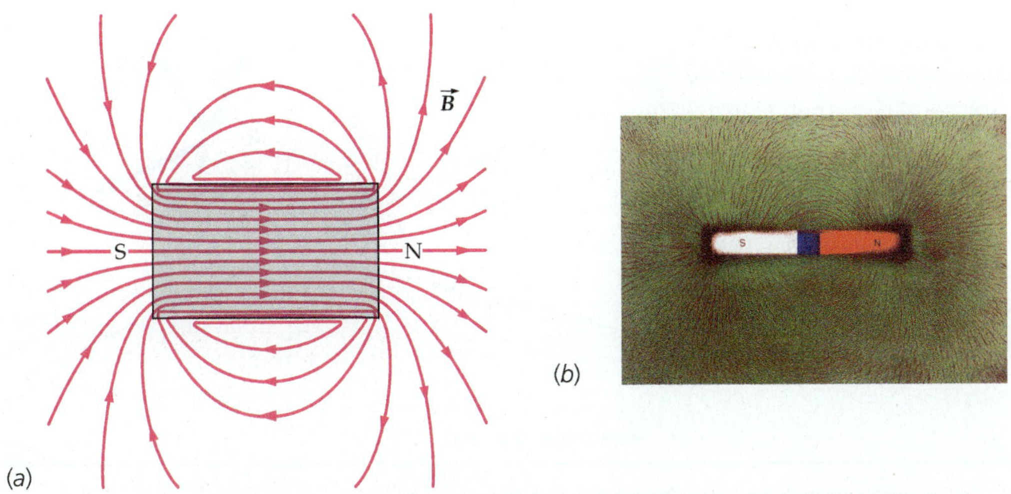

FIGURE 26-8 (*a*) Magnetic field lines inside and outside a bar magnet. The lines emerge from the north pole and enter the south pole, but they have no beginning or end. Instead, they form closed loops. (*b*) Magnetic field lines outside a bar magnet as indicated by iron filings. *(© 1995 Tom Pantages.)*

Example 26-2 Force on a Straight Wire

A 3.0-mm-long segment of wire carries a current of 3.0 A in the $+x$ direction. It lies in a magnetic field of magnitude 0.020 T that is in the xy plane and makes an angle of 30° with the $+x$ direction, as shown in Figure 26-9. What is the magnetic force exerted on the wire segment?

FIGURE 26-9

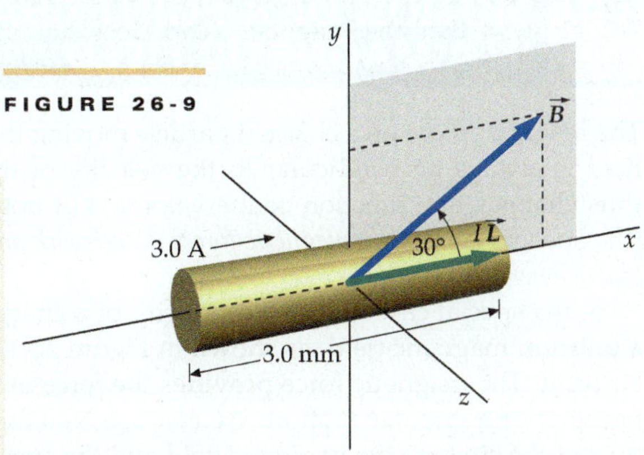

PICTURE The magnetic force is in the direction of $\vec{L} \times \vec{B}$, which we see from Figure 26-9 is in the $+z$ direction.

SOLVE

1. The magnetic force is given by Equation 26-4:

$$\vec{F} = I\vec{L} \times \vec{B} = ILB \sin 30^\circ\, \hat{k}$$
$$= (3.0\text{ A})(0.0030\text{ m})(0.020\text{ T})(\sin 30^\circ)\hat{k}$$
$$= \boxed{9.0 \times 10^{-5}\,\text{N}\hat{k}}$$

CHECK The force is perpendicular to the wire, as expected.

Example 26-3 Force on a Bent Wire

A wire bent into a semicircular loop of radius R lies in the xy plane. It carries a current I from point a to point b, as shown in Figure 26-10. Throughout the region there is a uniform magnetic field $\vec{B} = B\hat{k}$ that is perpendicular to the plane of the loop. Find the magnetic force acting on the semicircular loop section of the wire.

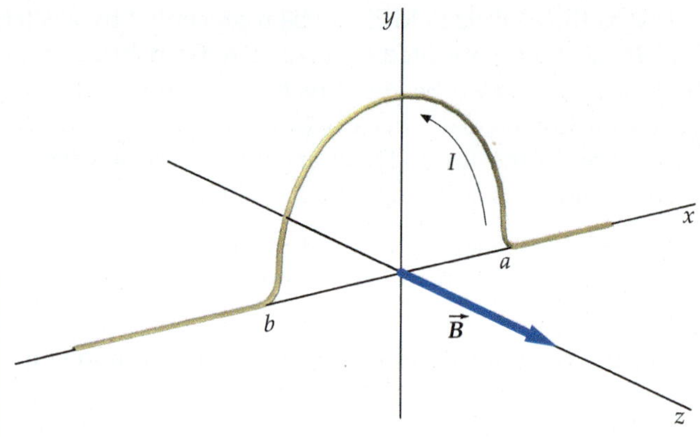

FIGURE 26-10

PICTURE The magnetic force $d\vec{F}$ is exerted on a segment of the semicircular wire that lies in the xy plane, as shown in Figure 26-11. We find the total magnetic force by expressing the x and y components of $d\vec{F}$ in terms of θ and integrating them separately from $\theta = 0$ to $\theta = \pi$.

SOLVE

1. Write the force $d\vec{F}$ on a current element $I\,d\vec{\ell}$.

$$d\vec{F} = I\,d\vec{\ell} \times \vec{B}$$

2. Express $d\vec{\ell}$ in terms of the unit vectors $\hat{i}$ and $\hat{j}$:

$$d\vec{\ell} = -d\ell\,\sin\theta\hat{i} + d\ell\,\cos\theta\hat{j}$$

3. Compute $I\,d\vec{\ell}$ using $d\ell = R\,d\theta$ and $\vec{B} = B\hat{k}$:

$$\begin{aligned} d\vec{F} &= I\,d\vec{\ell} \times \vec{B} \\ &= I(-R\sin\theta\,d\theta\hat{i} + R\cos\theta\,d\theta\hat{j}) \times B\hat{k} \\ &= IRB\sin\theta\,d\theta\hat{j} + IRB\cos\theta\,d\theta\hat{i} \end{aligned}$$

4. Integrate each component of $d\vec{F}$ from $\theta = 0$ to $\theta = \pi$.

$$\begin{aligned} \vec{F} &= \int d\vec{F} = IRB\hat{i}\int_0^{\pi}\cos\theta\,d\theta + IRB\hat{j}\int_0^{\pi}\sin\theta\,d\theta \\ &= IRB\hat{i}(0) + IRB\hat{j}(2) = \boxed{2IRB\hat{j}} \end{aligned}$$

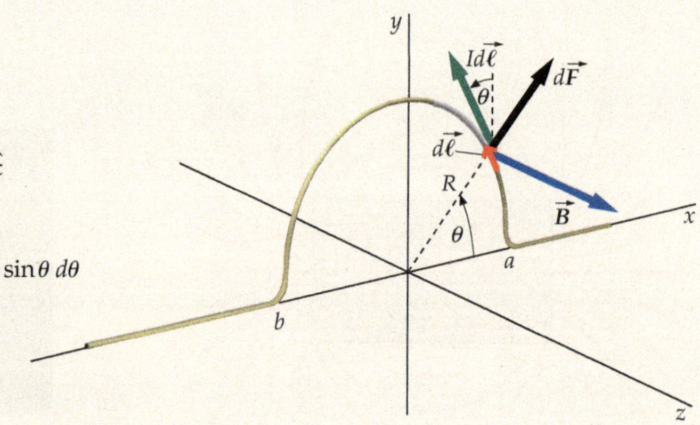

FIGURE 26-11

CHECK The result that the x component of $\vec{F}$ is zero can be seen from symmetry. For the right half of the loop, $d\vec{F}$ tilts to the right; for the left half of the loop, $d\vec{F}$ tilts to the left.

TAKING IT FURTHER The net force on the semicircular wire is the same as if the semicircle were replaced by a straight-line segment of length $2R$ connecting points a and b. (This is a general result that is derived in Problem 26.)

26-2 MOTION OF A POINT CHARGE IN A MAGNETIC FIELD

The magnetic force on a charged particle moving through a region with a magnetic field is always perpendicular to the velocity of the particle. The magnetic force thus changes the direction of the velocity but not the magnitude of the velocity (the speed). Therefore, *magnetic forces do no work on particles and do not change their kinetic energy*.

In the special case where the velocity of a charged particle is perpendicular to a uniform magnetic field, as shown in Figure 26-12, the particle moves in a circular orbit. The magnetic force provides the force in the centripetal direction that is necessary for circular motion. We can use Newton's second law to relate the radius of the circle to the magnetic field and the speed of the particle. If the velocity is $\vec{v}$, the magnetic force on a particle that has charge q is given by $\vec{F} = q\vec{v} \times \vec{B}$. The magnitude of the net force is equal to qvB, because $\vec{v}$ and $\vec{B}$ are perpendicular.

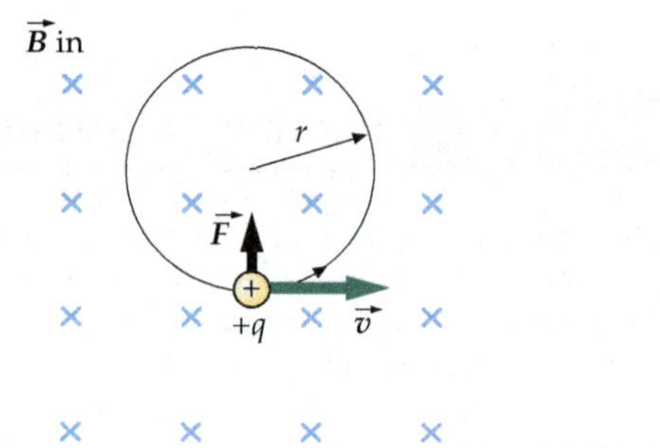

FIGURE 26-12 Charged particle moving in a plane perpendicular to a uniform magnetic field. The magnetic field is into the page as indicated by the crosses. (Each cross represents the tail feathers of an arrow. A field out of the plane of the page would be indicated by dots, each dot representing the point of an arrow.) The magnetic force is perpendicular to the velocity of the particle, causing it to move in a circular orbit.

Newton's second law gives

$$F = ma$$

$$qvB = m\frac{v^2}{r}$$

or

$$r = \frac{mv}{qB} \qquad 26\text{-}6$$

where m is the mass of the particle.

The period of the circular motion is the time it takes the particle to travel once around the circumference of the circle. The period is related to the speed by

$$T = \frac{2\pi r}{v}$$

Substituting $mv/(qB)$ for r (Equation 26-6), we obtain the period of the particle's circular orbit, which is called the **cyclotron period:**

$$T = \frac{2\pi(mv/qB)}{v} = \frac{2\pi m}{qB} \qquad 26\text{-}7$$

CYCLOTRON PERIOD

The frequency of the circular motion, called the **cyclotron frequency**, is the reciprocal of the period:

$$f = \frac{1}{T} = \frac{qB}{2\pi m} \quad \text{so} \quad \omega = 2\pi f = \frac{q}{m}B \qquad 26\text{-}8$$

CYCLOTRON FREQUENCY

Note that the period and the frequency given by Equations 26-7 and 26-8 depend on the charge-to-mass ratio q/m, but the period and the frequency are independent of the velocity v or the radius r. Two important applications of the circular motion of charged particles in a uniform magnetic field, the mass spectrometer and the cyclotron, are discussed later in this section.

(a)

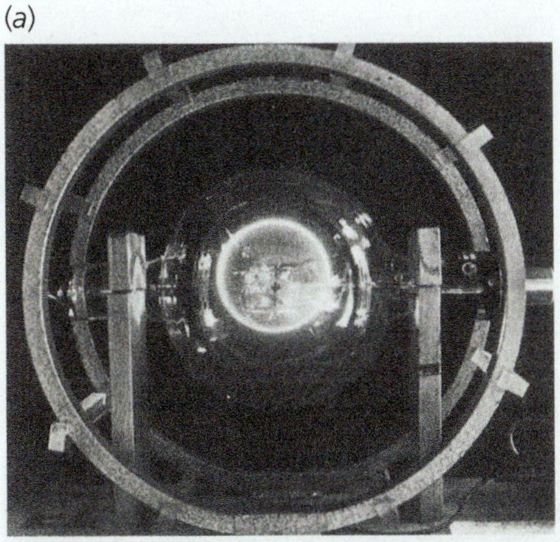

(b)

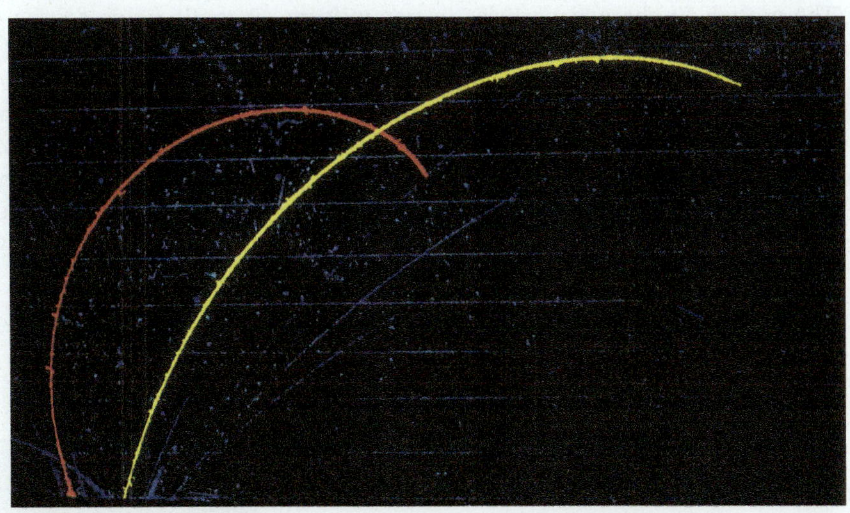

(*a*) Circular path of electrons moving in the magnetic field produced by the current in two large coils. The electrons ionize the dilute gas in the tube, causing it to give off a glow that indicates the path of the beam. (*b*) False-color photograph showing tracks of a 1.6-MeV proton (red) and a 7-MeVα particle (yellow) in a cloud chamber. The radius of curvature is proportional to the momentum and inversely proportional to the charge of the particle. For these energies, the momentum of the α particle, which has twice the charge of the proton, is about four times that of the proton and so its radius of curvature is greater. *((a) Larry Langrill. (b) © Lawrence Berkeley Laboratory/Science Photo Library.)*

Example 26-4 Cyclotron Period

A proton has a mass equal to 1.67×10^{-27} kg, has a charge equal to 1.60×10^{-19} C, and moves in a circle of radius $r = 21.0$ cm perpendicular to a magnetic field equal to 4000 G. Find (*a*) the speed of the proton and (*b*) the period of the motion.

PICTURE Apply Newton's second law to find the speed, and use distance equals speed multiplied by time to find the period.

SOLVE

(*a*) 1. Apply Newton's second law ($F = ma$):

$$F = ma \quad \Rightarrow \quad qvB = m\frac{v^2}{r}$$

2. Solve for the speed:

$$v = \frac{rqB}{m} = \frac{(0.210\ \text{m})(1.60 \times 10^{-19}\ \text{C})(0.400\ \text{T})}{1.67 \times 10^{-27}\ \text{kg}}$$

$$= \boxed{8.05 \times 10^6\ \text{m/s} = 0.0268c}$$

(*b*) Use distance equals speed multiplied by time and solve for the period:

$$2\pi r = vT$$

so

$$T = \frac{2\pi r}{v} = \frac{2\pi(0.210\ \text{m})}{(8.05 \times 10^6\ \text{m/s})} = 1.64 \times 10^{-7}\ \text{s} = \boxed{164\ \text{ns}}$$

TAKING IT FURTHER The radius of the circular orbit is proportional to the speed, but the period of the orbit is independent of both the speed and radius.

Suppose that a charged particle is in a region that has a uniform magnetic field and is moving with a velocity that is not perpendicular to $\vec{B}$. There is no magnetic force component, and thus no acceleration component, parallel to $\vec{B}$, so the component of the velocity that is parallel to $\vec{B}$ remains constant. The magnetic force on the particle is perpendicular to $\vec{B}$, so the change in motion of the particle due to this force is the same as that just discussed. The path of the particle is thus a helix, as shown in Figure 26-13.

The motion of charged particles in nonuniform magnetic fields can be quite complex. Figure 26-14 shows a *magnetic bottle,* an interesting magnetic field configuration in which the field is weak at the center and strong at both ends. A detailed analysis of the motion of a charged particle in such a field shows that the particle spirals around the field lines and becomes trapped, oscillating back and forth between points P_1 and P_2 in the figure. Such magnetic field configurations are used to confine dense beams of charged particles, called plasmas, in nuclear fusion research. A similar phenomenon is the oscillation of ions back and forth between Earth's magnetic poles in the Van Allen belts (Figure 26-15).

(*a*)

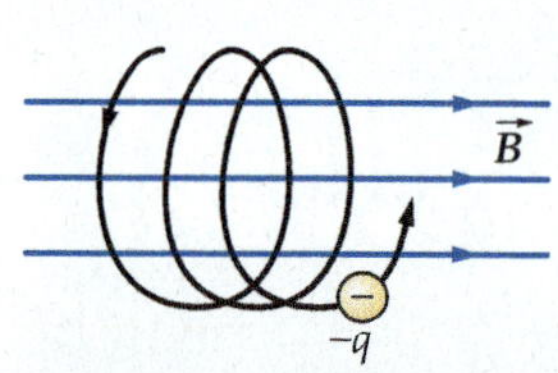

(*b*)

FIGURE 26-13 (*a*) When a particle has a velocity component parallel to a magnetic field as well as a velocity component perpendicular to the magnetic field the particle moves in a helical path around the field lines. (*b*) Cloud-chamber photograph of the helical path of an electron moving in a magnetic field. The path of the electron is made visible by the condensation of water droplets in the cloud chamber. *(Carl E. Nielson.)*

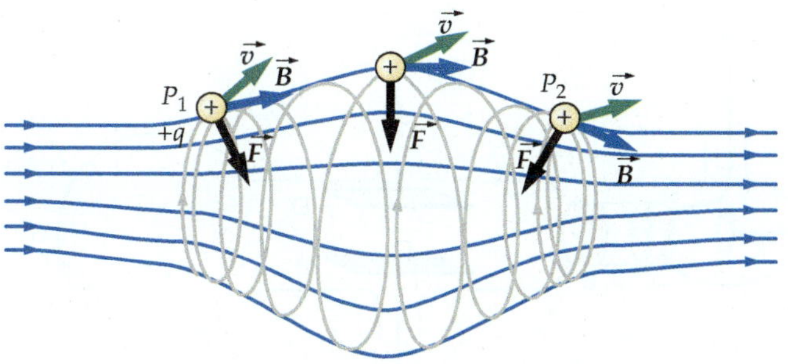

FIGURE 26-14 Magnetic bottle. When a charged particle moves in such a field, which is strong at both ends and weak in the middle, the particle becomes trapped and moves back and forth, spiraling around the field lines.

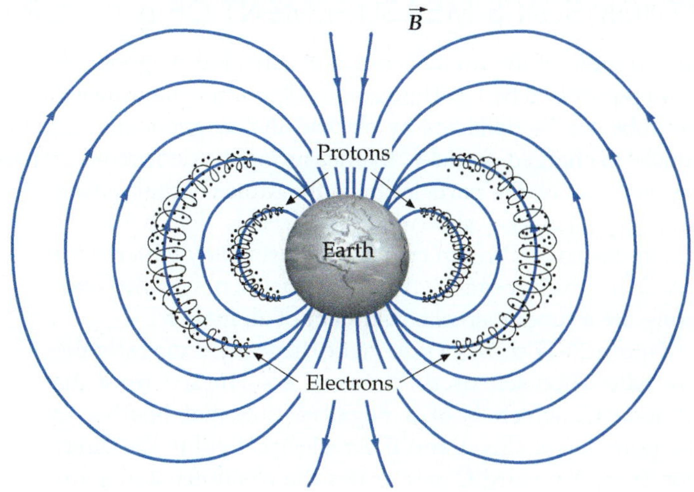

FIGURE 26-15 Van Allen belts. Protons (inner belts) and electrons (outer belts) are trapped in Earth's magnetic field and spiral around the field lines between the north and south poles.

*THE VELOCITY SELECTOR

The magnetic force on a charged particle moving in a uniform magnetic field can be balanced by an electric force if the magnitudes and directions of the magnetic field and the electric field are properly chosen. Because the electric force is in the direction of the electric field (for particles with positive charge) and the magnetic force is perpendicular to the magnetic field, the electric and magnetic fields in the region through which the particle is moving must be perpendicular to each other if the forces are to balance. Such a region is said to have **crossed fields.**

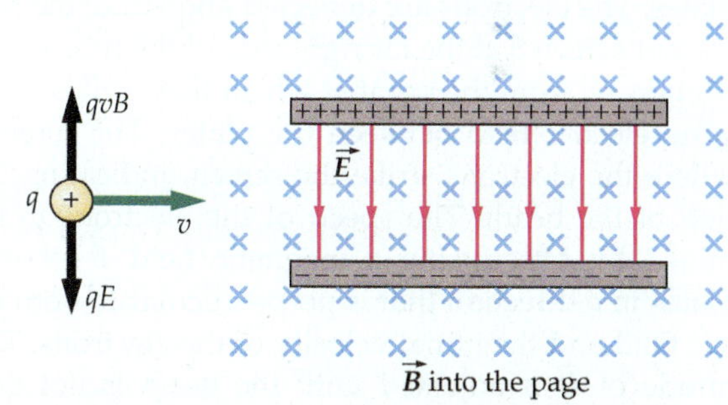

FIGURE 26-16 Crossed electric and magnetic fields. When a particle that has a positive charge moves to the right, the particle experiences a downward electric force and an upward magnetic force. These forces balance if the speed of the particle is related to the field strengths by $vB = E$.

Figure 26-16 shows a region of space between the plates of a capacitor where there is an electric field and a perpendicular magnetic field (produced by a magnet that has one pole on each side of this sheet of paper). Consider a particle that has charge q entering this space from the left. The net force on the particle is

$$\vec{F} = q\vec{E} + q\vec{v} \times \vec{B}$$

If q is positive, the electric force of magnitude qE is down the page and the magnetic force of magnitude qvB is up the page. If the charge is negative, the direction of each of these forces is reversed. The two forces balance if $qE = qvB$ that is, if

$$v = \frac{E}{B} \qquad 26\text{-}9$$

For given magnitudes of the electric and magnetic fields, the forces balance only for particles that have the exact speed given by Equation 26-9. Any particle that has this speed, regardless of its mass or charge, will traverse the space undeflected. A particle that has a greater speed will be deflected toward the direction of the magnetic force, and a particle that has a lesser speed will be deflected in the direction of the electric force. This arrangement of fields is often used as a **velocity selector,** which is a device that allows only particles with the speed specified by Equation 26-9 to pass.

PRACTICE PROBLEM 26-2

A proton is moving in the $+x$ direction in a region of crossed fields where $\vec{E} = 2.00 \times 10^5$ N/C $\hat{k}$ and $\vec{B} = 0.300$ T $\hat{j}$. (*a*) What is the speed of the proton if it is not deflected? (*b*) If the proton moves with twice this speed, in which direction will it be deflected?

*THOMSON'S MEASUREMENT OF q/m FOR ELECTRONS

An example of the use of crossed electric and magnetic fields is the famous experiment performed by J. J. Thomson in 1897 where he showed that the rays of a cathode-ray tube can be deflected by electric and magnetic fields, indicating that they must consist of charged particles. By measuring the deflections of these particles, Thomson showed that all the particles have the same charge-to-mass ratio q/m. He also showed that particles that have this charge-to-mass ratio can be obtained using any material for a source, which means that these particles, now called electrons, are a fundamental constituent of all matter.

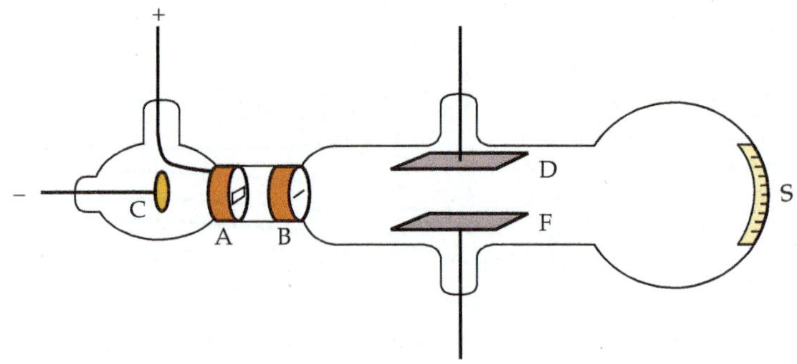

FIGURE 26-17 Thomson's tube for measuring q/m for the particles of cathode rays (electrons). Electrons from the cathode C pass through the slits at A and B and strike a phosphorescent screen S. The beam can be deflected by an electric field between plates D and F or by a magnetic field (not shown).

Figure 26-17 shows a schematic diagram of the cathode-ray tube Thomson used. Electrons are emitted from the cathode C, which is at a negative potential relative to the potential at slits A and B. An electric field in the direction from A toward C accelerates the electrons, and some of the electrons pass through slits A and B into a field-free region. The electrons then enter the electric field between the capacitor plates D and F that is perpendicular to the velocity of the electrons. This field accelerates the electrons vertically for the short time that they are between the plates. The electrons are deflected and strike the phosphorescent screen S at the far right side of the tube at some deflection Δy from the point at which they strike when there is no electric field between the plates. The screen glows where the electrons strike the screen, indicating the location of the beam. The speed of the electrons v_0 is determined by introducing a magnetic field $\vec{B}$ between the plates in a direction that is perpendicular to both the electric field and the initial velocity of the electrons. The magnitude of $\vec{B}$ is adjusted until the beam is not deflected. The speed is then found from Equation 26-9.

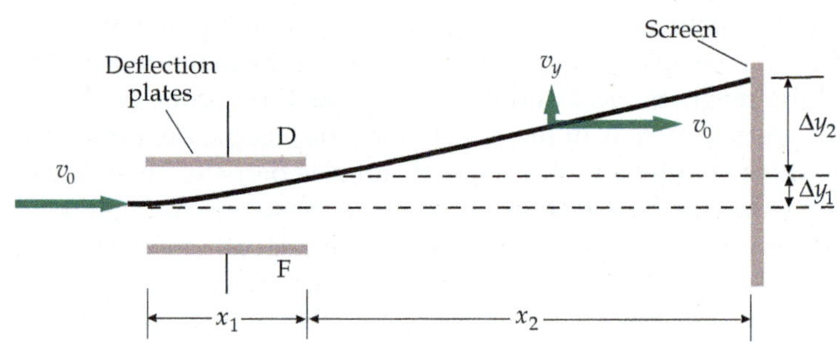

FIGURE 26-18 The total deflection of the beam in the J. J. Thomson experiments consists of the deflection Δy_1 while the electrons are between the plates plus the deflection Δy_2 that occurs in the field-free region between the plates and the screen.

With the magnetic field turned off, the beam is deflected by an amount Δy, which consists of two parts: the deflection Δy_1, which occurs while the electrons are between the plates, and the deflection Δy_2, which occurs after the electrons leave the region between the plates (Figure 26-18).

Let x_1 be the horizontal distance across the deflection plates D and F. If the electron is moving horizontally with speed v_0 when it enters the region between the plates, the time spent between the plates is $t_1 = x_1/v_0$, and the vertical velocity when it leaves the plates is

$$v_y = a_y t_1 = \frac{qE_y}{m} t_1 = \frac{qE_y}{m}\frac{x_1}{v_0}$$

where E_y is the upward component of the electric field between the plates. The deflection in this region is

$$\Delta y_1 = \frac{1}{2} a_y t_1^2 = \frac{1}{2}\frac{qE_y}{m}\left(\frac{x_1}{v_0}\right)^2$$

The electron then travels an additional horizontal distance x_2 in the field-free region from the deflection plates to the screen. Because the velocity of the electron is constant in this region, the time to reach the screen is $t_2 = x_2/v_0$, and the additional vertical deflection is

$$\Delta y_2 = v_y t_2 = \frac{qE_y}{m}\frac{x_1}{v_0}\frac{x_2}{v_0}$$

The total deflection at the screen is therefore

$$\Delta y = \Delta y_1 + \Delta y_2 = \frac{1}{2}\frac{qE_y}{mv_0^2}x_1^2 + \frac{qE_y}{mv_0^2}x_1x_2 \qquad 26\text{-}10$$

The measured deflection Δy can be used to determine the charge-to-mass ratio, q/m, from Equation 26-10.

Example 26-5 Electron Beam Deflection

Electrons pass undeflected through the plates of Thomson's apparatus when the electric field is 3000 V/m and there is a crossed magnetic field of 0.140 mT. If the plates are 4.00 cm long and the ends of the plates are 30.0 cm from the screen, find the deflection on the screen when the magnetic field is turned off.

PICTURE The mass and charge of the electron are known: $m = 9.11 \times 10^{-31}$ kg and $q = -e = -1.60 \times 10^{-19}$ C. The speed of the electron can be found from the ratio of the magnetic and electric fields.

SOLVE

1. The total deflection of the electron is given by Equation 26-10:

$$\Delta y = \Delta y_1 + \Delta y_2 = \frac{1}{2}\frac{qE_y}{mv_0^2}x_1^2 + \frac{qE_y}{mv_0^2}x_1x_2$$

2. The speed v_0 equals E/B:

$$v_0 = \frac{E}{B} = \frac{3000\ \text{V/m}}{1.40 \times 10^{-4}\ \text{T}} = 2.14 \times 10^7\ \text{m/s}$$

3. Substitute the value for v_0 determined in step 2, the given value of E, and the known values for m and q into Equation 26-10 to find Δy:

$$\Delta y_1 = \frac{1}{2}\frac{(-1.60 \times 10^{-19}\ \text{C})(-3000\ \text{V/m})}{(9.11 \times 10^{-31}\ \text{kg})(2.14 \times 10^7\ \text{m/s})^2}(0.0400\ \text{m})^2$$
$$= 9.20 \times 10^{-4}\ \text{m}$$

$$\Delta y_2 = \frac{(-1.60 \times 10^{-19}\ \text{C})(-3000\ \text{V/m})}{(9.11 \times 10^{-31}\ \text{kg})(2.14 \times 10^7\ \text{m/s})^2}(0.0400\ \text{m})(0.300\ \text{m})$$
$$= 1.38 \times 10^{-2}\ \text{m}$$

$$\Delta y = \Delta y_1 + \Delta y_2$$
$$= 9.20 \times 10^{-4}\ \text{m} + 1.38 \times 10^{-2}\ \text{m}$$
$$= 0.92\ \text{mm} + 13.8\ \text{mm} = \boxed{14.7\ \text{mm}}$$

CHECK As expected, Δy_2 is an order of magnitude greater than Δy_1. This was expected because the distance from the plates to the screen is an order of magnitude greater than the length of the plates.

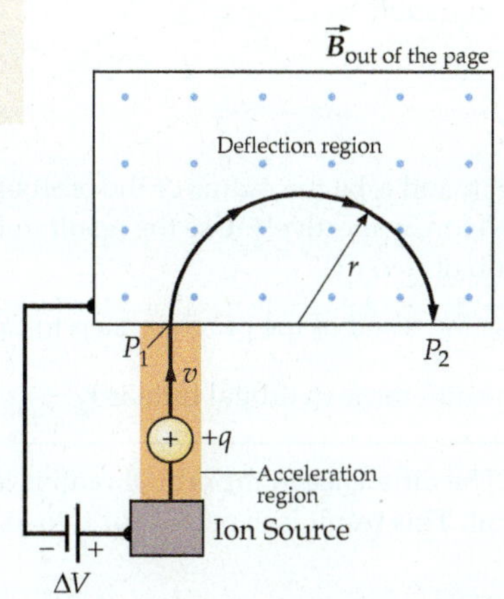

FIGURE 26-19 Schematic drawing of a mass spectrometer. Positive ions from an ion source are accelerated through a potential difference ΔV and enter a uniform magnetic field at P_1. The magnetic field is out of the plane of the page as indicated by the dots. The ions are bent into a circular arc and emerge at P_2. The radius r of the circle varies with the mass of the ion.

*THE MASS SPECTROMETER

The **mass spectrometer,** first designed by Francis William Aston in 1919, was developed as a means of measuring the masses of isotopes. Such measurements are important in determining both the presence of isotopes and their abundance in nature. On Earth, for example, naturally occurring magnesium has been found to consist of 78.7 percent ^{24}Mg, 10.1 percent ^{25}Mg, and 11.2 percent ^{26}Mg. These isotopes have masses in the approximate ratio 24:25:26.

Figure 26-19 shows a simple schematic drawing of a mass spectrometer. Positive ions are formed by bombarding atoms with X rays or a beam of electrons. (Electrons are knocked out of the atoms by the X rays or bombarding electrons to form positive ions.) The ions are accelerated by an electric field and enter a uniform magnetic field. If the positive ions start from rest and move through a potential difference ΔV, the ions' kinetic energy when they enter the magnetic field equals their loss in potential energy, $q|\Delta V|$:

$$\tfrac{1}{2}mv^2 = q|\Delta V| \qquad 26\text{-}11$$

The ions move in a semicircle of radius r given by Equation 26-6, $r = mv/qB$, and strike a photographic plate at point P_2, a distance $2r$ from the point P_1 where the ions entered the magnetic field.

The speed v can be eliminated from Equations 26-6 and 26-11 to find m/q in terms of the known quantities V, B, and r. We first solve Equation 26-6 for v and square each term, which gives

$$v^2 = \frac{r^2q^2B^2}{m^2}$$

Substituting this expression for v^2 into Equation 26-11, we obtain

$$\frac{1}{2}m\left(\frac{r^2q^2B^2}{m^2}\right) = q|\Delta V|$$

Simplifying this equation and solving for m/q, we obtain

$$\frac{m}{q} = \frac{B^2r^2}{2|\Delta V|} \qquad \text{26-12}$$

In Aston's original mass spectrometer, mass differences could be measured to a precision of about 1 part in 10,000. The precision has been improved by introducing a velocity selector between the ion source and the magnet, which increases the degree of accuracy with which the speeds of the incoming ions can be determined.

Example 26-6 Separating Isotopes of Nickel

A ^{58}Ni ion that has a charge equal to $+e$ and a mass equal to 9.62×10^{-26} kg is accelerated through a potential drop of 3.00 kV and deflected in a magnetic field of 0.120 T. (*a*) Find the radius of curvature of the orbit of the ion. (*b*) Find the difference in the radii of curvature of ^{58}Ni ions and ^{60}Ni ions. (Assume that the mass ratio is 58:60.)

PICTURE The radius of curvature r can be found using Equation 26-12. Using the mass dependence of r, we can find the radius of curvature for the orbit of the ^{60}Ni ions from the radius of curvature for the orbit of the ^{58}Ni ions, and then take the difference between the two radii.

SOLVE

(*a*) Solve Equation 26-12 for r:

$$r = \sqrt{\frac{2m|\Delta V|}{qB^2}} = \left[\frac{2(9.62 \times 10^{-26}\text{ kg})(3000\text{ V})}{(1.60 \times 10^{-19}\text{ C})(0.120\text{ T})^2}\right]^{1/2} = \boxed{0.501\text{ m}}$$

(*b*) 1. Let r_1 and r_2 be the radius of the orbit of the ^{58}Ni ion and the ^{60}Ni ion, respectively. Use the result in Part (*a*) to find the ratio of r_2 to r_1:

$$\frac{r_2}{r_1} = \sqrt{\frac{m_2}{m_1}} = \sqrt{\frac{60}{58}} = 1.017$$

2. Use the result of the previous step to calculate r_2 for ^{60}Ni:

$$r_2 = 1.017r_1 = (1.017)(0.501\text{ m}) = 0.510\text{ m}$$

3. The difference in orbital radii is $r_2 - r_1$:

$$r_2 - r_1 = 0.510\text{ m} - 0.501\text{ m} = \boxed{9\text{ mm}}$$

CHECK The difference in the orbital radii is less than 2 percent of the radius of curvature of either orbit. This result is expected for two ions whose masses differ by less than 4 percent.

THE CYCLOTRON

The cyclotron was invented by E. O. Lawrence and M. S. Livingston in 1934 to accelerate particles, such as protons or deuterons, to large kinetic energies.* The high-energy particles are used to bombard atomic nuclei, causing nuclear reactions that

* A deuteron is the nucleus of heavy hydrogen, ^{2}H, which consists of a proton and neutron tightly bound together.

are then studied to obtain information about nuclei. High-energy protons and deuterons are also used to produce radioactive materials and for medical purposes.

Figure 26-20 is a schematic drawing of a cyclotron. The particles move in two semicircular metal containers called *dees* (because they are the shape of the letter "D"). The dees are housed in a vacuum chamber that is in a region with a uniform magnetic field provided by an electromagnet. The region in which the particles move must be evacuated so that the particles will not be scattered in collisions with air molecules. A potential difference ΔV, which alternates in time with a period T, is maintained between the dees. The period is chosen to be the cyclotron period $T = 2\pi m/(qB)$ (Equation 26-7). The potential difference creates an electric field across the gap between the dees. At the same time, there is no electric field within each dee because the metal dees act as shields.

Positively charged particles are initially injected into dee$_1$ with a small velocity from an ion source S near the center of the dees. They move in a semicircle in dee$_1$ and arrive at the gap between dee$_1$ and dee$_2$ after a time $\frac{1}{2}T$. The potential is adjusted so that dee$_1$ is at a higher potential than dee$_2$ when the particles arrive at the gap between them. Each particle is therefore accelerated across the gap by the electric field and gains kinetic energy equal to $q\Delta V$.

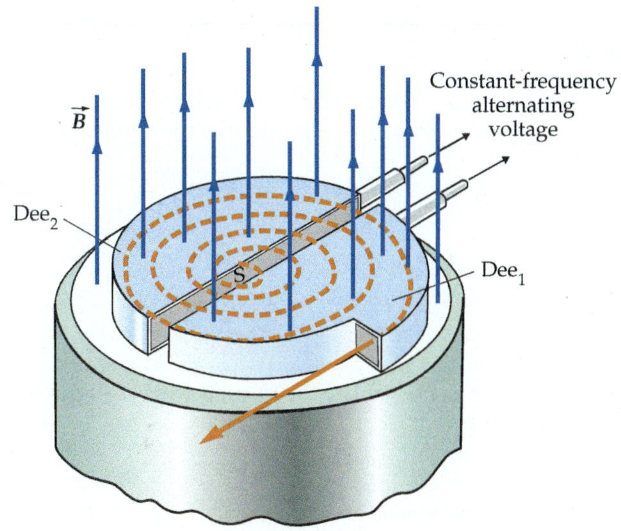

FIGURE 26-20 Schematic drawing of a cyclotron. The upper-pole face of the magnet has been omitted. Charged particles, such as protons, are accelerated from a source S at the center by the potential difference across the gap between the dees. When the charged particles arrive at the gap again the potential difference has changed sign so they are again accelerated across the gap and move in a larger circle. The potential difference across the gap alternates with the cyclotron frequency of the particle, which is independent of the radius of the circle.

Because the particle now has more kinetic energy, the particle moves in a semicircle of larger radius in dee$_2$. It arrives at the gap again after a time $\frac{1}{2}T$, because the period is independent of the particle's speed. By this time, the potential difference between the dees has been reversed so that dee$_2$ is now at the higher potential. Once more the particle is accelerated across the gap and gains additional kinetic energy equal to $q\Delta V$. Each time the particle arrives at the gap, it is accelerated and gains kinetic energy equal to $q\Delta V$. Thus, the particle moves in larger and larger semicircular orbits until it eventually leaves the magnetic field. In the typical cyclotron, each particle may make 50 to 100 revolutions and exit with energies of up to several hundred megaelectron volts.

The kinetic energy of a particle leaving a cyclotron can be calculated by setting r in Equation 26-6 equal to the maximum radius of the dees and solving the equation for v:

$$r = \frac{mv}{qB} \quad \Rightarrow \quad v = \frac{qBr}{m}$$

Then

$$K = \frac{1}{2}mv^2 = \frac{1}{2}\left(\frac{q^2B^2}{m}\right)r^2 \qquad \text{26-13}$$

Example 26-7 Energy of Accelerated Proton

A cyclotron for accelerating protons has a magnetic field of 0.150 T and a maximum radius of 0.500 m. (*a*) What is the cyclotron frequency? (*b*) What is the kinetic energy of the protons when they emerge?

PICTURE Apply Newton's second law ($F = ma$) with $F = |q\vec{v} \times \vec{B}|$. Use $v = r\omega$ and solve for the frequency and the speed.

SOLVE

(*a*) 1. Apply $F = ma$, where F is the magnetic force and a is the centripetal acceleration. Substitute ωr for v and solve for ω:

$$F = ma$$
$$qvB = m\frac{v^2}{r}$$
$$q\omega rB = m\frac{\omega^2 r^2}{r}$$
$$\omega = \frac{qB}{m} = \frac{(1.60 \times 10^{-19}\ \text{C})(0.150\ \text{T})}{1.67 \times 10^{-27}\ \text{kg}}$$
$$= 1.44 \times 10^7\ \text{rad/s}$$

2. Use $2\pi f = \omega$ to calculate the frequency in cycles per second (hertz):

$$f = \frac{\omega}{2\pi} = \frac{1.44 \times 10^7\ \text{rad/s}}{2\pi\ \text{rad}}$$
$$= 2.29 \times 10^6\ \text{Hz} = \boxed{2.29\ \text{MHz}}$$

(*b*) 1. Calculate the kinetic energy:

$$K = \frac{1}{2}mv^2 = \frac{1}{2}m\omega^2 r^2$$
$$= \frac{1}{2}(1.67 \times 10^{-27}\ \text{kg})(1.44 \times 10^7\ \text{rad/s})^2(0.500\ \text{m})^2$$
$$= 4.33 \times 10^{-14}\ \text{J}$$

2. The energies of protons and other elementary particles are usually expressed in electron volts. Use $1\ \text{eV} = 1.60 \times 10^{-19}$ J to convert to eV:

$$K = 4.33 \times 10^{-14}\ \text{J} \times \frac{1\ \text{eV}}{1.60 \times 10^{-19}\ \text{J}} = \boxed{271\ \text{keV}}$$

CHECK The exit speed of the proton is $v = r\omega = (0.500\ \text{m})(1.44 \times 10^7\ \text{rad/s}) = 7.20 \times 10^6$ m/s. The speed of light is 3.00×10^8 m/s. Our calculated value of 1.44×10^7 rad/s for the angular frequency is plausible because it is a high speed that is less than ten percent of the speed of light.

26-3 TORQUES ON CURRENT LOOPS AND MAGNETS

A current-carrying loop experiences no net force in a uniform magnetic field, but it does experience a net torque. The orientation of the loop can be described conveniently by a unit vector $\hat{\boldsymbol{n}}$ that is normal to the plane of the loop, as illustrated in Figure 26-21. If the fingers of the right hand curl around the loop in the direction of the current, the thumb points in the direction of $\hat{\boldsymbol{n}}$.

Figure 26-22 shows the forces exerted by a uniform magnetic field on a rectangular current-carrying loop whose vector $\hat{\boldsymbol{n}}$ makes an angle θ with the direction of the magnetic field $\vec{B}$. The net force on the loop is zero. The forces $\vec{F}_1$ and $\vec{F}_2$ have the magnitude

$$F_1 = F_2 = IaB$$

The forces form a couple, so the torque they exert is the same about any point. Point P in Figure 26-22 is a convenient point about which to compute the torque. The magnitude of the torque is

$$\tau = F_2 b \sin\theta = IaBb \sin\theta = IAB \sin\theta$$

where $A = ab$ is the area of the loop. For a loop that has N turns, the torque has the magnitude

$$\tau = NIAB \sin\theta$$

This torque tends to twist the loop so that $\hat{\boldsymbol{n}}$ is in the same direction as $\vec{B}$.

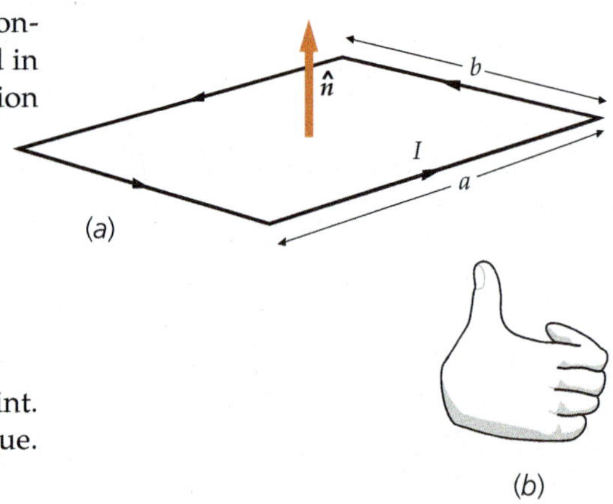

FIGURE 26-21 (*a*) The orientation of a current loop is described by the unit vector $\hat{\boldsymbol{n}}$ perpendicular to the plane of the loop. (*b*) Right-hand rule for determining the direction of $\hat{\boldsymbol{n}}$. If the fingers of the right hand curl around the loop in the direction of the current, the thumb points in the direction of $\hat{\boldsymbol{n}}$.

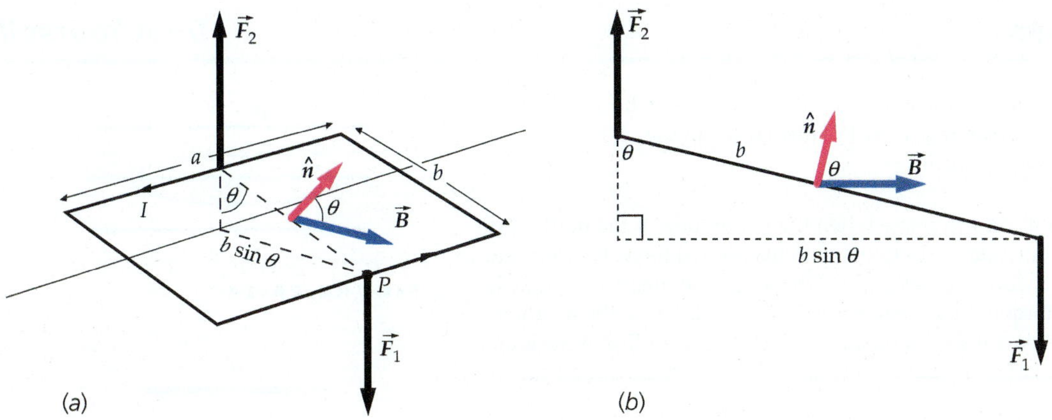

FIGURE 26-22 (*a*) Rectangular current loop whose unit normal $\hat{n}$ makes an angle θ with a uniform magnetic field $\vec{B}$. (*b*) An edge-on view of the current loop. The torque on the loop has magnitude $IAB\sin\theta$ and is in the direction such that $\hat{n}$ tends to rotate so as to align itself with $\vec{B}$.

The torque can be written conveniently in terms of the **magnetic dipole moment** $\vec{\mu}$ (also referred to simply as the **magnetic moment**) of the current loop, which is defined as

$$\vec{\mu} = NIA\hat{n} \tag{26-14}$$

MAGNETIC DIPOLE MOMENT OF A CURRENT LOOP

The SI unit of magnetic moment is the ampere-square meter ($A\cdot m^2$). In terms of the magnetic dipole moment, the torque on the current loop is given by

$$\vec{\tau} = \vec{\mu} \times \vec{B} \tag{26-15}$$

TORQUE ON A CURRENT LOOP

Equation 26-15, which we have derived for a rectangular loop, holds in general for a loop of any shape that lies in a single plane. The torque on any such loop is the vector product of the magnetic moment $\vec{\mu}$ of the loop and the magnetic field $\vec{B}$, where the magnetic moment (Figure 26-23) is defined as a vector that has a magnitude equal to NIA and has the same direction as $\hat{n}$. Comparing Equation 26-15 with Equation 21-11 ($\vec{\tau} = \vec{p} \times \vec{E}$) for the torque on an electric dipole, we see that the expression for the torque on a magnetic dipole in a magnetic field has the same form as that for the torque on an electric dipole in an electric field.

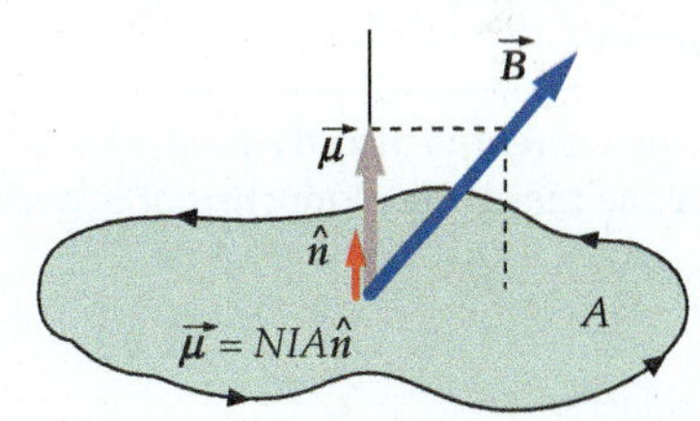

FIGURE 26-23 A flat current loop of arbitrary shape is described by its magnetic moment $\vec{\mu} = NIA\hat{n}$. In a magnetic field $\vec{B}$, the loop experiences a torque $\vec{\mu} \times \vec{B}$.

Example 26-8 Torque on a Current Loop

A circular loop has a radius equal to 2.00 cm, has 10 turns of wire, and carries a current equal to 3.00 A. The axis of the loop makes an angle of 30.0° with a magnetic field of 8000 G. Find the magnitude of the torque on the loop.

PICTURE The torque on a current loop is given by $\vec{\tau} = \vec{\mu} \times \vec{B}$ (Equation 26-15) where $\vec{\mu} = NIA\hat{n}$ (Equation 26-14).

SOLVE

The magnitude of the torque is given by Equation 26-15:

$$\tau = |\vec{\mu} \times \vec{B}| = \mu B\sin\theta = NIAB\sin\theta$$
$$= (10.0)(3.00\ \text{A})\pi(0.0200\ \text{m})^2(0.800\ \text{T})\sin 30.0°$$
$$= \boxed{1.51 \times 10^{-2}\ \text{N}\cdot\text{m}}$$

CHECK From $\vec{F} = I\vec{L} \times \vec{B}$ (Equation 26-4) we can see that the SI unit for magnetic field (the tesla) must have units of N/(A·m). With this in mind, one can see by inspection that the units for the right-hand side of the equation in the solution work out to N·m, which are SI units for torque.

Example 26-9 Tilting a Loop

Try It Yourself

A circular wire loop that has a radius R, a mass m, and a current I lies on a horizontal surface (Figure 26-24). There is a horizontal magnetic field $\vec{B}$. How large can the current I be before one edge of the loop will lift off the surface?

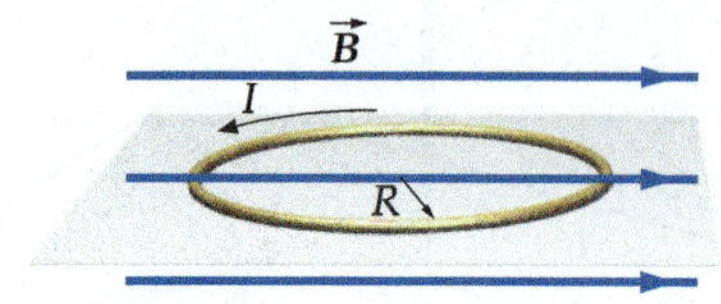

FIGURE 26-24

PICTURE The loop (Figure 26-25) will start to rotate when the magnitude of the net torque on the loop is greater than zero. To eliminate the torque due to the normal force, we calculate torques about the point of contact between the surface and the loop. The magnetic torque is given by $\vec{\tau} = \vec{\mu} \times \vec{B}$. The magnetic torque is the same about any point because the magnetic torque consists of couples. The lever arm for the gravitational torque is the radius of the loop.

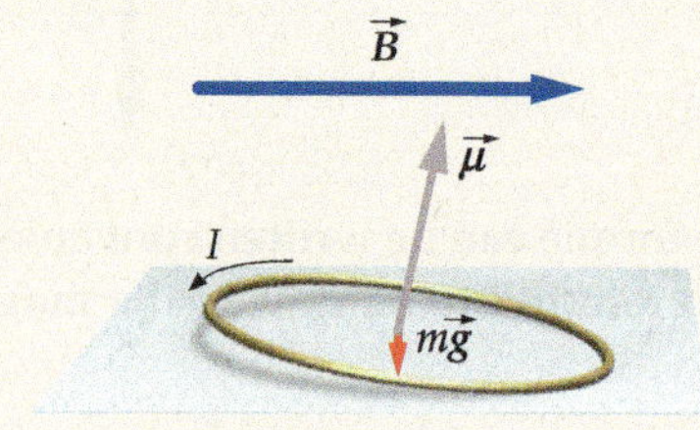

FIGURE 26-25

SOLVE

Cover the column to the right and try these on your own before looking at the answers.

Steps	Answers
1. Find the magnitude of the magnetic torque acting on the loop.	$\tau_{m} = \mu B \sin(90°) = I\pi R^2 B$
2. Find the magnitude of the gravitational torque exerted on the loop.	$\tau_{g} = mgR$
3. Equate the magnitudes of the torques and solve for the current I.	$I = \boxed{\dfrac{mg}{\pi RB}}$

CHECK The current is directly proportional to the mass for constant B, which makes sense. The larger the mass, the more current is needed to start to rotate the ring.

POTENTIAL ENERGY OF A MAGNETIC DIPOLE IN A MAGNETIC FIELD

When a torque is exerted on a rotating object, work is done. When a magnetic dipole is rotated through an angle $d\theta$, the work done is

$$dW = -\tau\, d\theta = -\mu B \sin\theta\, d\theta$$

where θ is the angle between $\vec{\mu}$ and $\vec{B}$. The minus sign arises because the magnetic torque tends to decrease θ. Setting this work equal to the decrease in potential energy U, we have

$$dU = -dW = +\mu B \sin\theta\, d\theta$$

Integrating, we obtain

$$U = -\mu B \cos\theta + U_0$$

We choose the potential energy to be zero when $\theta = 90°$. Then $U_0 = 0$ and the potential energy of the dipole is given by

$$U = -\mu B \cos\theta = -\vec{\mu} \cdot \vec{B} \qquad \text{26-16}$$

POTENTIAL ENERGY OF A MAGNETIC DIPOLE

Equation 26-16 gives the potential energy of a magnetic dipole at an angle θ to the direction of a magnetic field.

Example 26-10 Torque on a Coil

A square 12-turn coil has an edge-length equal to 40.0 cm and carries a current of 3.00 A. It lies in the $z = 0$ plane, as shown in a uniform magnetic field $\vec{B} = 0.300\ \text{T}\ \hat{i} + 0.400\ \text{T}\ \hat{k}$. The current is counterclockwise when viewed from a point on the positive z axis. Find (*a*) the magnetic moment of the coil and (*b*) the torque exerted on the coil. (*c*) Find the potential energy of the coil.

PICTURE From Figure 26-26, we see that the magnetic moment of the loop is in the $+z$ direction.

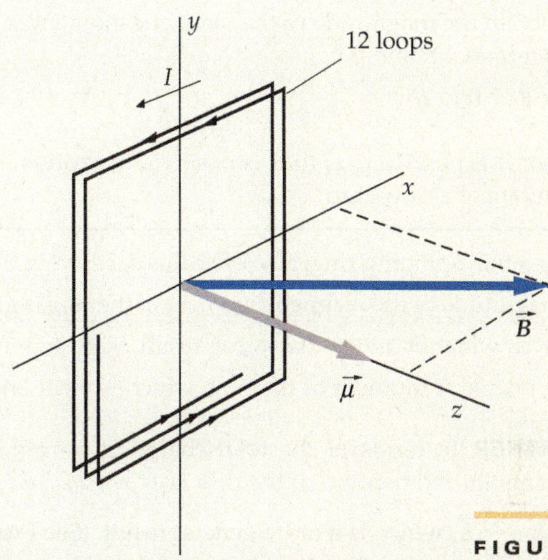

FIGURE 26-26

SOLVE

(a) Calculate the magnetic moment of the loop:

$$\vec{\mu} = NIA\hat{k} = (12)(3.00\ \text{A})(0.400\ \text{m})^2\hat{k}$$
$$= \boxed{5.76\ \text{A}\cdot\text{m}^2\ \hat{k}}$$

(b) The torque on the current loop is given by Equation 26-15:

$$\vec{\tau} = \vec{\mu} \times \vec{B}$$
$$= (5.76\ \text{A}\cdot\text{m}^2\ \hat{k}) \times (0.300\ \text{T}\ \hat{i} + 0.400\ \text{T}\ \hat{k})$$
$$= \boxed{1.73\ \text{N}\cdot\text{m}\ \hat{j}}$$

(c) The potential energy is the negative dot product of $\vec{\mu}$ and $\vec{B}$:

$$U = -\vec{\mu}\cdot\vec{B}$$
$$= -(5.76\ \text{A}\cdot\text{m}^2\hat{k})\cdot(0.300\ \text{T}\ \hat{i} + 0.400\ \text{T}\ \hat{k})$$
$$= \boxed{-2.30\ \text{J}}$$

CHECK The torque in the Part-(*b*) result is perpendicular to both the magnetic moment vector and the magnetic field vector, as is expected for a vector product.

PRACTICE PROBLEM 26-3 The potential energy of a current-carrying coil in a uniform magnetic field $\vec{B}$ is equal to zero when the angle between the magnetic dipole moment of the coil $\vec{\mu}$ and the magnetic field is 90°. Calculate the potential energy of the system if the coil is oriented so $\vec{B}$ and $\vec{\mu}$ are (*a*) in the same direction and (*b*) in opposite directions.

When a permanent magnet, such as a compass needle or a bar magnet, is placed in a region where there is a magnetic field $\vec{B}$, the field exerts a torque on the magnet that tends to rotate the magnet so that it lines up with the field. (This effect also occurs with previously unmagnetized iron filings, which become magnetized in the presence of a field $\vec{B}$.) The bar magnet is characterized by a magnetic moment $\vec{\mu}$, a vector that points in the same direction as an arrow drawn from the south pole of the magnet to the north pole of the magnet. A short bar magnet thus behaves like a current loop.

Example 26-11 $\vec{\mu}$ of a Spinning Disk

A thin nonconducting disk that has a mass m, a radius a, and a uniform surface charge per unit area σ spins with angular velocity $\vec{\omega}$ about an axis through the center of the disk and perpendicular to the plane of the disk. Find the magnetic moment of the spinning disk.

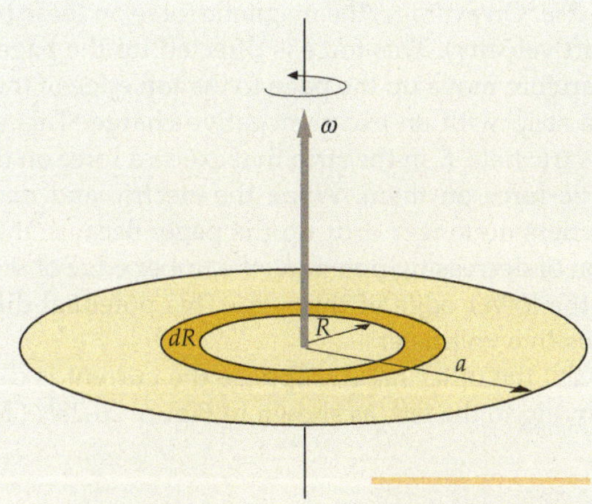

FIGURE 26-27

PICTURE We find the magnetic moment of a circular element that has a radius R and a width dR and integrate (Figure 26-27). The charge on the element is $dq = \sigma\, dA = \sigma 2\pi R\, dR$. If the charge is positive, the magnetic moment is in the direction of $\vec{\omega}$, so we need only calculate its magnitude.

SOLVE

1. The magnitude of the magnetic moment of the strip shown is the current multiplied by the area of the loop:

$$d\mu = A\,dI = \pi R^2\,dI$$

2. The current in the strip is the total charge dq on the strip divided by the period T. During one period the charge dq passes by a point not rotating with the strip. The period is equal to the reciprocal of the frequency f of rotation $1/T = f = \omega/(2\pi)$:

$$dI = \frac{dq}{T} = \frac{\omega}{2\pi}dq = \frac{\omega}{2\pi}\sigma\,dA$$
$$= \frac{\omega}{2\pi}\sigma 2\pi R\,dR = \sigma\omega R\,dR$$

3. Substitute to obtain the magnitude of the magnetic moment of the strip $d\mu$ in terms of R and dR:

$$d\mu = \pi R^2\,dI = \pi R^2\sigma\omega R\,dR = \pi\sigma\omega R^3\,dR$$

4. Integrate from $R = 0$ to $R = a$:

$$\mu = \int_0^a \pi\sigma\omega R^3\,dR = \frac{1}{4}\pi\sigma\omega a^4$$

5. Use the fact that $\vec{\mu}$ is parallel to $\vec{\omega}$ (if σ is positive) to express the magnetic moment as a vector:

$$\vec{\mu} = \boxed{\tfrac{1}{4}\pi\sigma a^4\vec{\omega}}$$

CHECK Consider a thin spinning ring, also of radius a, carrying the same charge, $Q = \sigma\pi a^2$, as the disk. The magnitude of the magnetic moment of the ring is given by $\mu = IA = \frac{Q}{T}\pi a^2 = \frac{\sigma\pi a^2}{2\pi/\omega}\pi a^2 = \frac{1}{2}\pi\sigma a^4\omega$, which is twice the step-5 result. The step-5 result is smaller than the magnitude of the magnetic moment of the ring, which is what one would expect.

TAKING IT FURTHER In terms of the total charge $Q = \sigma\pi a^2$, the magnetic moment is $\vec{\mu} = \frac{1}{4}Qa^2\vec{\omega}$. The angular momentum of the disk is $\vec{L} = (\frac{1}{2}ma^2)\vec{\omega}$, so the magnetic moment can be written $\vec{\mu} = \frac{Q}{2m}\vec{L}$, which is a more general result. (See Problem 57.)

26-4 THE HALL EFFECT

As we have seen, charges moving in a region where there is a magnetic field each experience a force perpendicular to their motion. When these charges are traveling in a conducting wire, they will be pushed to one side of the wire. This results in a separation of charge in the wire—a phenomenon called the **Hall effect.** This phenomenon allows us to determine the sign of the charge on the charge carriers and the number of charge carriers per unit volume n in a conductor. The Hall effect also provides a convenient method for measuring magnetic fields.

Figure 26-28 shows two conducting strips; each conducting strip carries a current I to the right because the left sides of the strips are connected to the positive terminal of a battery and the right sides are connected to the negative terminal. A magnetic field $\vec{B}$ is directed into the paper. Let us suppose the current in the strip is due to positively charged particles moving to the right, as shown in Figure 26-28*a*. On average, the magnetic force on these particles is $q\vec{v}_d \times \vec{B}$ (where $\vec{v}_d$ is the drift velocity). This force is directed up the page. The positively charged particles therefore move up the page to the top edge of the strip, leaving the bottom edge of the strip with an excess negative charge. This separation of charge produces an electric field $\vec{E}$ in the strip that exerts a force on the particles that opposes the magnetic force on them. When the electric and magnetic forces balance, the charge carriers no longer drift up the page. Because the electric field points in the direction of decreasing potential, the upper edge of the strip is at a higher potential than is the lower edge of the strip. This potential difference can be measured using a sensitive voltmeter.

On the other hand, suppose the current is due to negatively charged particles moving to the left, as shown in Figure 26-28*b*. (The negatively charged particles in the strip must move to the left because the current, as before, is to the right.) The magnetic force $q\vec{v}_d \times \vec{B}$ is again up the page, because the signs of both q and the direction of $\vec{v}_d$ have been reversed. Again the carriers are forced to the upper edge of the strip, but the upper edge of the strip now carries a negative charge (because the charge carriers are negative) and the lower edge of the strip now carries a positive charge.

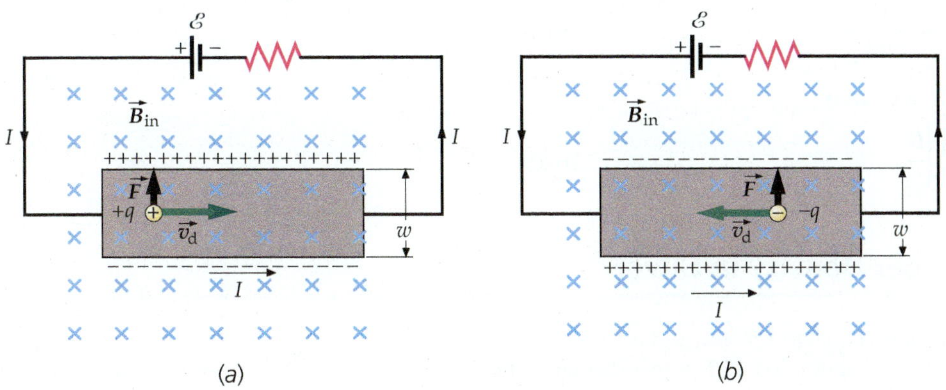

FIGURE 26-28 The Hall effect. The magnetic field is directed into the plane of the page as indicated by the crosses. The magnetic force on a charged particle is upward for a current to the right whether the current is due to (*a*) positive particles moving to the right or (*b*) negative particles moving to the left.

A measurement of the sign of the potential difference between the upper and lower parts of the strip tells us the sign of the charge carriers. In semiconductors, the charge carriers may be negative electrons or positive "holes." A measurement of the sign of the potential difference tells us which are dominant for a particular semiconductor. For a metal strip, we find that the upper edge of the strip in Figure 26-28*b* is at a lower potential than is the lower edge of the strip—which means that the upper part must carry a negative charge. Thus, Figure 26-28*b* is the correct illustration of the current in a metal strip. It was a measurement like this which led to the discovery that the charge carriers in metals are negatively charged.

The potential difference between the top of the strip and the bottom of the strip is called the **Hall voltage**. We can calculate the magnitude of the Hall voltage in terms of the drift velocity. The magnitude of the magnetic force on the charge carriers in the strip is qv_dB. This magnetic force is balanced by the electrostatic force of magnitude qE_H, where E_H is the electric field due to the charge separation. Thus, we have $E_H = v_dB$. If the width of the strip is w, the potential difference is E_Hw. The Hall voltage is therefore

$$V_H = E_Hw = v_dBw \qquad 26\text{-}17$$

PRACTICE PROBLEM 26-4

A conducting strip of width $w = 2.0$ cm is placed in a magnetic field of 0.80 T. The Hall voltage is measured to be 0.64 μV. Calculate the drift velocity of the electrons.

Because the drift velocity for ordinary currents is very small, we can see from Equation 26-17 that the Hall voltage is very small for ordinary-sized strips and magnetic fields. From measurements of the Hall voltage for a strip of a given size, we can determine the number of charge carriers per unit volume in the strip.

The magnitude of the current is given by Equation 26-3:

$$|I| = |q|nv_{\text{d}}A$$

where A is the cross-sectional area of the strip. For a strip of width w and thickness t, the cross-sectional area is $A = wt$. Because the charge carriers are electrons, the quantity $|q|$ is the charge on one electron e. The number density of charge carriers n is thus given by

$$n = \frac{|I|}{A|q|v_{\text{d}}} = \frac{|I|}{wtev_{\text{d}}} \qquad \text{26-18}$$

Substituting V_{H}/B for $v_{\text{d}}w$ (Equation 26-17), we have

$$n = \frac{|I|B}{teV_{\text{H}}} \qquad \text{26-19}$$

Example 26-12 Charge Carrier Number Density in Silver

A silver slab has a thickness equal to 1.00 mm, a width equal to 1.50 cm, and a current equal to 2.50 A in a region where there is a magnetic field of magnitude 1.25 T perpendicular to the slab. The Hall voltage is measured to be 0.334 μV. (*a*) Calculate the number density of the charge carriers. (*b*) Calculate the number density of atoms in silver, which has a mass density of $\rho = 10.5\ \text{g/cm}^3$ and a molar mass of $M = 107.9$ g/mol, and compare the number density of atoms in silver with the Part-(*a*) result.

PICTURE We can use Equation 26-19 to find the number density of charge carriers. The number density of atoms can be obtained from knowledge of the density and the molar mass.

SOLVE

(*a*) Substitute numerical values into Equation 26-19 to find n:

$$n = \frac{|I|B}{teV_{\text{H}}} = \frac{(2.50\ \text{A})(1.25\ \text{T})}{(1.00 \times 10^{-3}\ \text{m})(1.60 \times 10^{-19}\ \text{C})(3.34 \times 10^{-7}\ \text{V})}$$

$$= \boxed{5.85 \times 10^{28}\ \text{electrons/m}^3}$$

(*b*) 1. The number of atoms per unit volume is $\rho N_{\text{A}}/M$:

$$n_{\text{a}} = \rho\frac{N_{\text{A}}}{M} = (10.5\ \text{g/cm}^3)\frac{6.02 \times 10^{23}\ \text{atoms/mol}}{107.9\ \text{g/mol}}$$

$$= \boxed{5.86 \times 10^{22}\ \text{atoms/cm}^3 = 5.86 \times 10^{28}\ \text{atoms/m}^3}$$

2. Compare the Part-(*b*), step-1 result with the Part-(*a*) result:

These results indicate that the number of charge carriers in silver is very nearly one per atom.

CHECK We should expect the number density of charge carriers and the number density of atoms in a metal to be the same order of magnitude. Our results validate that expectation.

The Hall voltage provides a convenient method for measuring magnetic fields. If we rearrange Equation 26-19, we can write for the Hall voltage

$$V_{\text{H}} = \frac{|I|}{nte}B \qquad \text{26-20}$$

A given strip can be calibrated by measuring the Hall voltage for a given current in a known magnetic field. The strip can then be used to measure an unknown magnetic field B by measuring the Hall voltage for a given current.

*THE QUANTUM HALL EFFECTS

According to Equation 26-20, the Hall voltage should increase linearly with magnetic field strength B for a given current in a given slab. In 1980, while studying the Hall effect in semiconductors at very low temperatures and very large magnetic fields, Klaus von Klitzing discovered that a plot of V_H versus B resulted in a series of plateaus, as shown in Figure 26-29, rather than a straight line. That is, the Hall voltage is quantized. For the discovery of the integer quantum Hall effect, von Klitzing won the Nobel Prize in Physics in 1985.

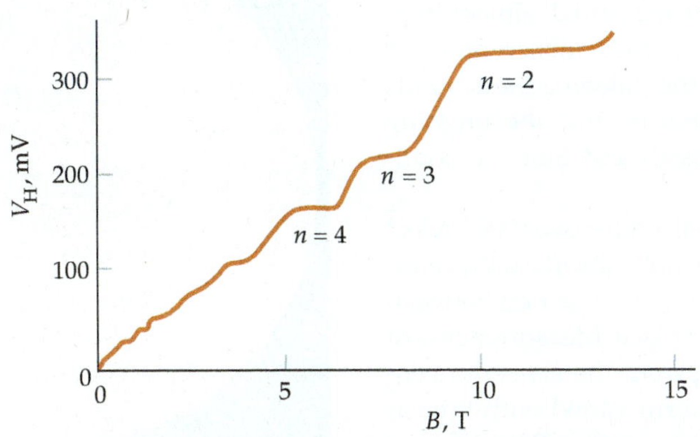

FIGURE 26-29 A plot of the Hall voltage versus applied magnetic field shows plateaus, indicating that the Hall voltage is quantized. The data were taken at a temperature of 1.39 K with the current I held fixed at 25.52 μA.

In the theory of the integer quantum Hall effect, the Hall resistance, defined as $R_H = V_H/I$, can take on only the values

$$R_H = \frac{V_H}{I} = \frac{R_K}{n} \qquad n = 1, 2, 3, \ldots \tag{26-21}$$

where n is an integer, and R_K, called the **von Klitzing constant,** is related to the fundamental electric charge e and Planck's constant h by

$$R_K = \frac{h}{e^2} \tag{26-22}$$

Because the von Klitzing constant can be measured to an accuracy of a few parts per billion, the quantum Hall effect is now used to define a standard of resistance. As of January 1990, the **ohm** is defined in terms of the conventional value* of the von Klitzing constant R_{K-90}, which has the value

$$R_{K-90} = 25\,812.8076\ \Omega \text{ (exact)} \tag{26-23}$$

In 1982, it was observed that under certain special conditions the Hall resistance is given by Equation 26-22 but with the integer n replaced by a series of rational fractions. This is called the *fractional quantum Hall effect.* For the discovery and explanation of the fractional quantum Hall effect, American professors Laughlin, Störmer, and Tsui won the Nobel Prize in Physics in 1998.

* The value of R_{K-90} differs only slightly from that of R_K. The currently used value of the von Klitzing constant is $R_K = (25\,812.807\,572 \pm 0.000\,095)\ \Omega$.

Physics Spotlight

Earth and the Sun—Magnetic Changes

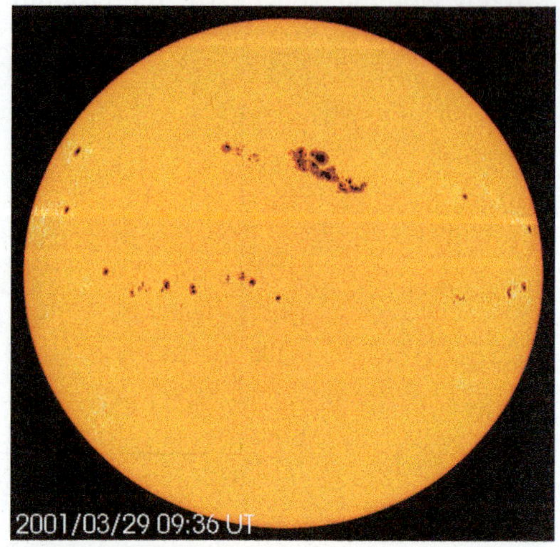

Sunspots are regions where the magnetic field strength is very high. They are darker than the surrounding surface because the temperature in the sunspot is cooler than the temperature of the surrounding area. *(SOHO/NASA.)*

The magnetic fields of the Sun and Earth have been measured almost constantly in recent years by satellite and ground-based magnetic observatories.[*] Geologists and physicists have collaborated to study the paleomagnetic fields of both Earth[†] and the Sun.[‡] The paleomagnetic studies and the ongoing observations show that the magnetic fields of the Earth and Sun are continuously changing.

Earth's magnetic field has been used as a navigational aid for over 900 years.[#] Navigators were soon aware that magnetic north does not coincide with celestial north, and that the magnetic declination (the difference in direction between magnetic north and celestial north) varied from place to place. Measurements of magnetic declination taken in the same places dating from the sixteenth century[°] showed that the apparent location of magnetic north varied with time at the same place.[§] These measurements are the first evidence that Earth's magnetic field is dynamic.

In the 1960s, drill cores showed many layers of magnetic reversals in volcanic rocks.[¶] It became clear that Earth's magnetic field reverses around every 200 000 years, but there have been durations of over six million years during which there were no geomagnetic reversals. Immediately surrounding the reversal, the record shows that the field strength decreases, reverses, and then increases over a period of a few thousand years.[**] The last geomagnetic reversal was 700 000 years ago. Lately, Earth's magnetic field strength has been decreasing. From 1840 to the present, Earth's magnetic field has decreased by 15 nT/y,[††] which is a decrease of 3% per century, and reconstruction of data from ships' logs shows a decrease of about 2 nT/y from 1590 to 1840.

In the early twentieth century, G. E. Hale noted that sunspots, which had been observed for hundreds of years, had magnetic fields. He demonstrated that during a 22-year sunspot cycle, the Sun's magnetic field gradually decreased, reversed, increased, and returned back to the original configuration.[‡‡] Sunspots themselves have been measured with a magnetic field strength in excess of 200 mT.[##] Recent observation has shown that sunspots are magnetically powered vortices in the Sun. Although the surface of the Sun has an apparent average field of 0.10 mT in regions without sunspots, small areas of such regions have magnetic strengths varying from below 20 mT up to 100 mT.[°°]

The solar wind, which consists of charged sub-atomic particles ejected from the Sun at around 400 km/s,[§§] carries a magnetic field. Satellite data show that the interplanetary magnetic field is complex and dynamic.[¶¶,**] Near Earth, the strength of the interplanetary magnetic field varies between 1 and 37 nT. Sometimes, the Sun ejects a large burst of charged particles. When a large burst arrives at Earth's magnetic field, it causes a magnetic storm that can block radio communications and cause widespread power blackouts. The *Voyager 1* spacecraft was more than 94 AU from the Sun when it measured the strength of the interplanetary magnetic field as 0.03 nT.[††,‡‡] The solar wind still carries a measurable magnetic field well beyond the orbit of Pluto.

* "Geomagnetic Frequently Asked Questions." United States National Geophysical Data Center, National Oceanic and Atmospheric Administration. http://www.ngdc.noaa.gov/seg/geomag/faqgeom.shtml As of Sept., 2006.

† Yamazaki, T., and Oda, H., "Orbital Influence on Earth's Magnetic Field: 100,000-Year Periodicity in Inclination." *Science*, Mar. 29, 2002, Vol. 294 pp. 2435–2437.

‡ Solanki, S. K., et al., "11,000 Year Sunspot Number Reconstruction." *IBGP Pages/World Data Center for Paleoclimatology Data Contribution Series #2005-015.* 2005. ftp://ftp.ncdc.noaa.gov/pub/data/paleo/climate-forcing/solar_variability/solanki2004-ssn.txt As of Sept., 2006.

\# Hellemans, A., and Bunch, B., *The Timetables of Science*. New York: Simon and Schuster, 1988. p. 75.

° Kono, M., "Ships' Logs and Archeomagnetism." *Science*, May 12, 2006, Vol. 312, pp. 865–66.

§ Hermanus Magnetic Observatory, "Detailed History." http://www.hmo.ac.za/detailed-history.html As of Sept., 2006.

¶ Dunn, J. R., et al., "Paleomagnetic Study of a Reversal of the Earth's Magnetic Field." *Science*, May 21, 1971, Vol. 172, pp. 840–844.

** Merrill, R. T., and McFadden, P. L., "Geomagnetic Polarity Transitions." *Reviews of Geophysics*, May 1999, Vol. 37, No. 2, pp. 201–226.

†† Gubbins, D., Jones, A. L., and Finlay, C., "Fall in Earth's Magnetic Field is Erratic." *Science*, May 12, 2006, Vol 312, pp. 900–902.

‡‡ Abbot, C. G., "Sun-Spots and Weather." *Science*, Dec. 8, 1933, Vol. 78, pp. 518–519.

\#\# Liang, H.-F., Zhao, H.-J., and Xiang, F.-Y., "Vector Magnetic Field Measurement of NOAA AR 10197." *Chinese Journal of Astronomy and Astrophysics*, Aug. 2006, Vol. 6, No. 4, pp. 470–476.

°° Lin, H., and Rimmele, T., "The Granular Magnetic Fields of the Quiet Sun." *The Astrophysical Journal*, Mar. 20, 1999, Vol. 514, Pt. 1, pp. 448–455.

§§ Hathaway, D., "The Solar Wind." *Solar Physics*, Marshall Space Flight Center, NASA http://solarscience.msfc. nasa.gov/SolarWind.shtml. Jun. 1, 2006, As of Oct., 2006.

¶¶ Smith, E. J., et al., "The Sun and Heliosphere at Solar Maximum." *Science*, Nov. 14, 2003, Vol. 302, pp. 1165–1168.

*** Arnold, N., and Lyons, A., "Granta MIST: Meeting Report." *Astronomy and Geophysics*, Aug. 2006, Vol. 46, pp. 4.18–4.21.

††† Gurnett, D. A., and Kurth, W. S., "Electron Plasma Oscillations Upstream of the Solar Wind Termination Shock." *Science*, Sept. 23, 2005, Vol. 309, pp. 2025–2027.

‡‡‡ Burlaga, L. F., et al., "Crossing the Termination Shock into the Helosheath: Magnetic Fields." *Science*, Sept. 23, 2005, Vol. 309, pp. 2027–2029.

Summary

1. The magnetic field describes the condition in space in which moving charges experience a force perpendicular to their velocity.
2. The magnetic force is part of the electromagnetic interaction, one of the three known fundamental interactions in nature.
3. The magnitude and direction of a magnetic field $\vec{B}$ are defined by the formula $\vec{F} = q\vec{v} \times \vec{B}$, where $\vec{F}$ is the force exerted on a particle with charge q moving with velocity $\vec{v}$.

TOPIC	RELEVANT EQUATIONS AND REMARKS	
1. Magnetic Force		
On a moving charge	$\vec{F} = q\vec{v} \times \vec{B}$	26-1
On a current element	$d\vec{F} = I\,d\vec{\ell} \times \vec{B}$	26-5
Unit of the magnetic field	The SI unit of magnetic fields is the tesla (T). A commonly used unit is the gauss (G), which is related to the tesla by $1\text{ G} = 10^{-4}\text{ T}$	26-3
2. Motion of Point Charges	A particle of mass m and charge q moving with speed v in a plane perpendicular to a uniform magnetic field moves in a circular orbit. The period and frequency of the circular motion are independent of the radius of the orbit and of the speed of the particle.	
Newton's second law	$qvB = m\frac{v^2}{r}$	26-6
Cyclotron period	$T = \frac{2\pi m}{qB}$	26-7
Cyclotron frequency	$f = \frac{1}{T} = \frac{qB}{2\pi m}$	26-8
*Velocity selector	A velocity selector consists of crossed electric and magnetic fields so that the electric and magnetic forces balance for a particle moving with speed v. $v = \frac{E}{B}$	26-9
*Thomson's measurement of q/m	The deflection of a charged particle in an electric field depends on the speed of the particle and is proportional to the charge-to-mass ratio q/m of the particle. J. J. Thomson used crossed electric and magnetic fields to measure the speed of cathode rays and then measured q/m for these particles by deflecting them in an electric field. He showed that all cathode rays consist of particles which all have the same charge-to-mass ratio. These particles are now called electrons.	
*Mass spectrometer	The mass-to-charge ratio of an ion of known speed can be determined by measuring the radius of the circular path taken by the ion in a known magnetic field.	
3. Current Loops		
Magnetic dipole moment	$\vec{\mu} = NIA\hat{n}$	26-14
Torque	$\vec{\tau} = \vec{\mu} \times \vec{B}$	26-15
Potential energy of a magnetic dipole	$U = -\vec{\mu} \cdot \vec{B}$	26-16
Net force	The net force on a current loop in a *uniform* magnetic field is zero.	

TOPIC	RELEVANT EQUATIONS AND REMARKS	
4. The Hall Effect	When a conducting strip carrying a current is placed in a magnetic field, the magnetic force on the charge carriers causes a separation of charge called the Hall effect. This results in a voltage V_H, called the Hall voltage. The sign of the charge carriers can be determined from a measurement of the sign of the Hall voltage, and the number of carriers per unit volume can be determined from the magnitude of V_H.	
Hall voltage	$V_H = E_H w = v_d B w = \frac{\|I\|}{nte} B$	26-17, 26-20
*Quantum Hall effects	Measurements at very low temperatures in very large magnetic fields indicate that the Hall resistance $R_H = V_H/I$ is quantized and can take on only the values given by $R_H = \frac{V_H}{I} = \frac{R_K}{n} \quad n = 1, 2, 3, \ldots$	26-21
*Conventional von Klitzing constant (definition of ohm)	$R_{K-90} = 25\,812.8076\ \Omega$ (exact)	26-23

Answers to Concept Checks

26-1 A left-hand rule is one way to answer the question. The definition for the direction of $\vec{B}$ is a convention. If the definition for the direction of $\vec{B}$ were changed as described in the question statement, a correct force law could be written $\vec{F} = q\vec{v} \otimes \vec{B}$, where the symbol $\otimes$ denotes the same operation as the symbol $\times$, except the product denoted by $\otimes$ requires using the left-hand rule instead of the right-hand rule. Alternatively, the force law could be revised to $\vec{F} = \vec{B} \times q\vec{v}$ and then you could stay with the right-hand rule.

26-2 (*b*) Negatively charged. The force $\vec{F}$ and the vector $\vec{v} \times \vec{B}$ are in opposite directions only if the particle is negatively charged. This is consistent with the relation $\vec{F} = q\vec{v} \times \vec{B}$.

Answers to Practice Problems

26-1 $-1.3 \times 10^{-12}\ \text{N}\ \hat{j}$

26-2 (*a*) 667 km/s, (*b*) in the $-z$ direction

26-3 (*a*) -2.88 J. Note that this potential energy is lower than the potential energy calculated in the example. (The potential energy is lowest when $\vec{\mu}$ and $\vec{B}$ are in the same direction.) (*b*) $+2.88$ J

26-4 4.0×10^{-5} m/s

Problems

In a few problems, you are given more data than you actually need; in a few other problems, you are required to supply data from your general knowledge, outside sources, or informed estimate.

Interpret as significant all digits in numerical values that have trailing zeros and no decimal points.

- • Single-concept, single-step, relatively easy
- •• Intermediate-level, may require synthesis of concepts
- ••• Challenging
- SSM Solution is in the *Student Solutions Manual*
- Consecutive problems that are shaded are paired problems.

CONCEPTUAL PROBLEMS

1 • When the axis of a cathode-ray tube is horizontal in a region in which there is a magnetic field that is directed vertically upward, the electrons emitted from the cathode follow one of the dashed paths to the face of the tube in Figure 26-30. The correct path is (*a*) 1, (*b*) 2, (*c*) 3, (*d*) 4, (*e*) 5. SSM

2 •• We define the direction of the electric field to be the same as the direction of the force on a positive test charge. Why then do we *not* define the direction of the magnetic field to be the same as the direction of the magnetic force on a moving positive test charge?

3 • A *flicker bulb* is a lightbulb that has a long, thin flexible filament. It is meant to be plugged into an ac outlet that delivers current at a frequency of 60 Hz. There is a small permanent magnet

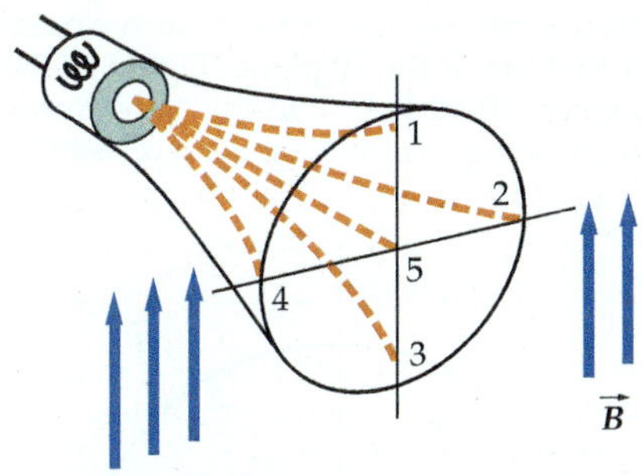

FIGURE 26-30 Problem 1

inside the bulb. When the bulb is plugged in the filament oscillates back and forth. At what frequency does it oscillate? Explain your answer. SSM

4 • In a cyclotron, the potential difference between the dees oscillates with a period given by $T = 2\pi m/(qB)$. Show that the expression to the right of the equal sign has units of seconds if q, B, and m have units of coulombs, teslas, and kilograms, respectively.

5 • A ^{7}Li nucleus has a charge equal to $+3e$ and a mass that is equal to the mass of seven protons. A ^{7}Li nucleus and a proton are both moving perpendicular to a uniform magnetic field $\vec{B}$. The magnitude of the momentum of the proton is equal to the magnitude of the momentum of the nucleus. The path of the proton has a radius of curvature equal to R_p and the path of the ^{7}Li nucleus has a radius of curvature equal to R_{Li}. The ratio R_p/R_{Li} is closest to (*a*) 3/1, (*b*) 1/3, (*c*) 1/7, (*d*) 7/1, (*e*) 3/7, (*f*) 7/3.

6 • An electron moving in the $+x$ direction enters a region that has a uniform magnetic field in the $+y$ direction. When the electron enters this region, it will (*a*) be deflected toward the $+y$ direction, (*b*) be deflected toward the $-y$ direction, (*c*) be deflected toward the $+z$ direction, (*d*) be deflected toward the $-z$ direction, (*e*) continue undeflected in the $+x$ direction.

7 • In a velocity selector, the speed of the undeflected charged particle is given by the ratio of the magnitude of the electric field to the magnitude of the magnetic field. Show that E/B in fact does have the units of m/s if E and B are in units of volts per meter and teslas, respectively. SSM

8 • How are the properties of magnetic field lines similar to the properties of electric field lines? How are they different?

9 • True or false:

(*a*) The magnetic moment of a bar magnet points from its north pole to its south pole.
(*b*) Inside the material of a bar magnet, the magnetic field due to the bar magnet points from the magnet's south pole toward its north pole.
(*c*) If a current loop simultaneously has its current doubled and its area cut in half, then the magnitude of its magnetic moment remains the same.
(*d*) The maximum torque on a current loop placed in a magnetic field occurs when the plane of the loop is perpendicular to the direction of the magnetic field.

10 •• Show that the von Klitzing constant, h/e^2, gives the SI unit for resistance (the ohm) if h and e are in units of joule-seconds and coulombs, respectively.

11 ••• The theory of relativity states that no law of physics can be described using the absolute velocity of an object, which is in fact impossible to define due to a lack of an absolute reference frame. Instead, the behavior of interacting objects can only be described by the relative velocities between the objects. New physical insights result from this idea. For example, in Figure 26-31 a magnet moving at high speed relative to some observer passes by an electron that is at rest relative to the same observer. Explain why you are sure that a force must be acting on the electron. In what direction will the force point at the instant the north pole of the magnet passes directly underneath the electron? Explain your answer.

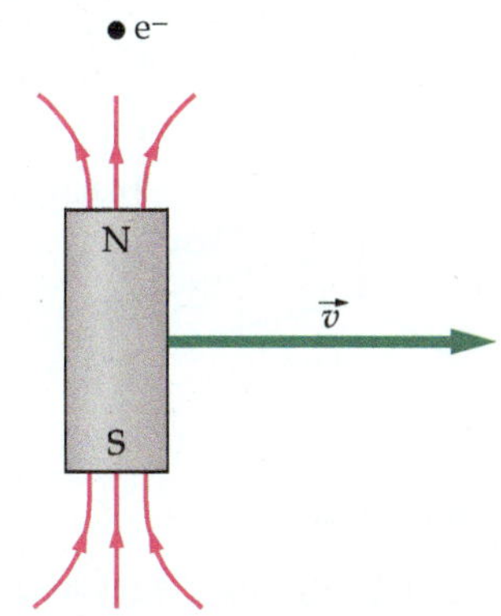

FIGURE 26-31 Problem 11

ESTIMATION AND APPROXIMATION

12 • Estimate the maximum magnetic force per meter that Earth's magnetic field could exert on a current-carrying wire in a 20-A circuit in your house.

13 •• CONTEXT-RICH Your friend wants to be a magician and intends to use Earth's magnetic field to suspend a current-carrying wire above the stage. He asks you to estimate the minimum current needed to suspend the wire just above Earth's surface at the equator (where Earth's magnetic field is horizontal). Assume the wire has a linear mass density of 10 g/m. Would you advise him to proceed with his plans for this act?

THE FORCE EXERTED BY A MAGNETIC FIELD

14 • Find the magnetic force on a proton moving in the $+x$ direction at a speed of 0.446 Mm/s in a uniform magnetic field of 1.75 T in the $+z$ direction.

15 • A point particle has a charge equal to -3.64 nC and a velocity equal to 2.75×10^3 m/s $\hat{i}$. Find the force on the charge if the magnetic field is (*a*) 0.38 T $\hat{j}$, (*b*) 0.75 T $\hat{i}$ + 0.75 T $\hat{j}$, (*c*) 0.65 T $\hat{i}$, and (*d*) 0.75 T $\hat{i}$ + 0.75 T $\hat{k}$.

16 • A uniform magnetic field equal to 1.48 T $\hat{k}$ is in the $+z$ direction. Find the force exerted by the field on a proton if the velocity of the proton is (*a*) 2.7 km/s $\hat{i}$, (*b*) 3.7 km/s $\hat{j}$, (*c*) 6.8 km/s $\hat{k}$, and (*d*) 4.0 km/s $\hat{i}$ + 3.0 km/s $\hat{j}$.

17 • A straight wire segment that is 2.0 m long makes an angle of 30° with a uniform 0.37-T magnetic field. Find the magnitude of the force on the wire if the wire has a current of 2.6 A.

18 • A straight segment of a current-carrying wire has a current element $I\vec{L}$, where $I = 2.7$ A and $\vec{L} = 3.0$ cm $\hat{i}$ + 4.0 cm $\hat{j}$. The segment is in a region with a uniform magnetic field given by 1.3 T $\hat{i}$. Find the force on the segment of wire.

19 • What is the force on an electron that has a velocity equal to 2.0×10^6 m/s $\hat{i} - 3.0 \times 10^6$ m/s $\hat{j}$ when it is in a region with a magnetic field given by 0.80 T $\hat{i}$ + 0.60 T $\hat{j}$ − 0.40 T $\hat{k}$?

20 •• The section of wire shown in Figure 26-32 carries a current equal to 1.8 A from a to b. The segment is in a region that has a magnetic field whose value is $1.2\ \mathrm{T}\ \hat{k}$. Find the total force on the wire and show that the total force is the same as if the wire were in the form of a straight wire directly from a to b and carrying the same current.

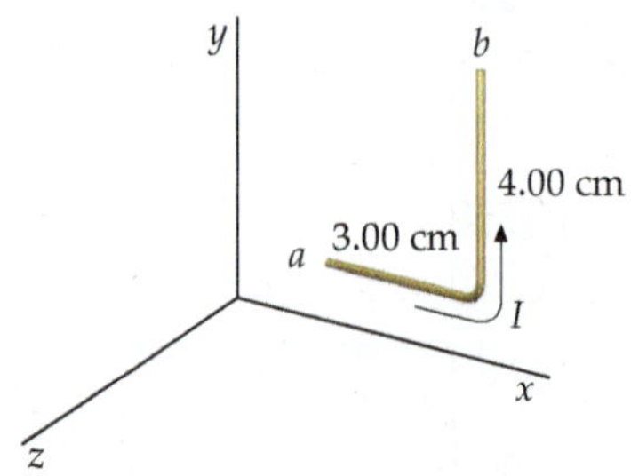

FIGURE 26-32 Problem 20

21 •• A straight, stiff, horizontal 25-cm-long wire that has a mass equal to 50 g is connected to a source of emf by light, flexible leads. A magnetic field of 1.33 T is horizontal and perpendicular to the wire. Find the current necessary to "float" the wire, that is, when the wire is released from rest it remains at rest.

22 •• **ENGINEERING APPLICATION** In your physics laboratory class, you have constructed a simple *gaussmeter* for measuring the horizontal component of magnetic fields. The setup consists of a stiff 50-cm wire that hangs vertically from a conducting pivot so that its free end makes contact with a pool of mercury in a dish below (Figure 26-33). The mercury provides an electrical contact without constraining the movement of the wire. The wire has a mass of 5.0 g and conducts a current downward. (*a*) What is the equilibrium angular displacement of the wire from vertical if the horizontal component of the magnetic field is 0.040 T and the current is 0.20 A? (*b*) What is the sensitivity of this gaussmeter? That is, what is the ratio of the output to the input (in radians per tesla)?

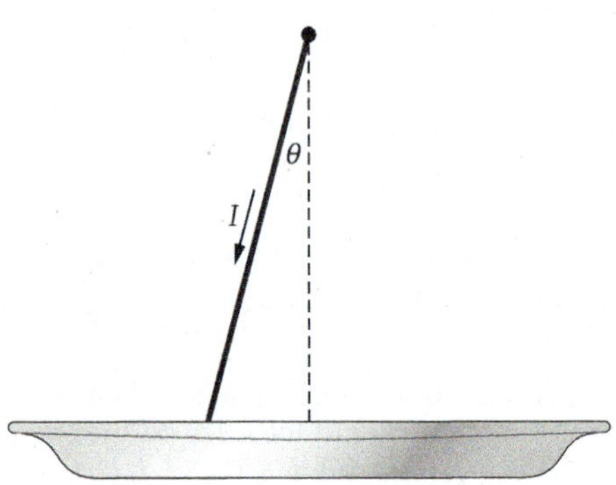

FIGURE 26-33 Problem 22

23 •• A 10-cm-long straight wire is parallel with the x axis and carries a current of 2.0 A in the $+x$ direction. The force on this wire due to the presence of a magnetic field $\vec{B}$ is $3.0\ \mathrm{N}\ \hat{j} + 2.0\ \mathrm{N}\ \hat{k}$. If this wire is rotated so that it is parallel with the y axis with the current in the $+y$ direction, the force on the wire becomes $-3.0\ \mathrm{N}\ \hat{j} - 2.0\ \mathrm{N}\ \hat{k}$. Determine the magnetic field $\vec{B}$. SSM

24 •• A 10-cm-long straight wire is parallel with the z axis and carries a current of 4.0 A in the $+z$ direction. The force on this wire due to a uniform magnetic field $\vec{B}$ is $-0.20\ \mathrm{N}\ \hat{i} + 0.20\ \mathrm{N}\ \hat{j}$. If this wire is rotated so that it is parallel with the x axis with the current in the $+x$ direction, the force on the wire becomes $0.20\ \hat{k}\ \mathrm{N}$. Find $\vec{B}$.

25 •• A current-carrying wire is bent into a closed semicircular loop of radius R that lies in the xy plane (Figure 26-34). The wire is in a uniform magnetic field that is in the $+z$ direction, as shown. Verify that the force acting on the loop is zero. SSM

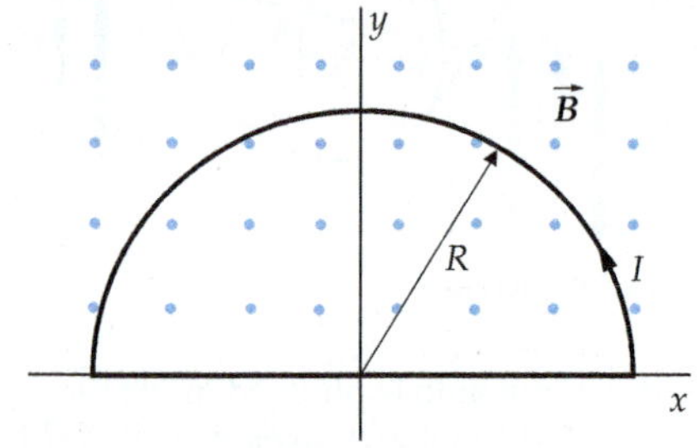

FIGURE 26-34 Problem 25

26 ••• A wire bent in some arbitrary shape carries a current I. The wire is in a region with a uniform magnetic field $\vec{B}$. Show that the total force on the part of the wire from some arbitrary point on the wire (designated as a) to some other arbitrary point on the wire (designated as b) is $\vec{F} = I\vec{L} \times \vec{B}$, where $\vec{L}$ is the vector from point a to point b. In other words, show that the force on an arbitrary section of the bent wire is the same as the force on a straight section wire carrying the same current and connecting the two endpoints of the arbitrary section.

MOTION OF A POINT CHARGE IN A MAGNETIC FIELD

27 • A proton moves in a 65-cm-radius circular orbit that is perpendicular to a uniform magnetic field of magnitude 0.75 T. (*a*) What is the orbital period for the motion? (*b*) What is the speed of the proton? (*c*) What is the kinetic energy of the proton? SSM

28 • A 4.5-keV electron (an electron that has a kinetic energy equal to 4.5 keV) moves in a circular orbit that is perpendicular to a magnetic field of 0.325 T. (*a*) Find the radius of the orbit. (*b*) Find the frequency and period of the orbital motion.

29 •• A proton, deuteron, and an alpha particle in a region with a uniform magnetic field each follow circular paths that have the same radius. The deuteron has a charge that is equal to the charge a proton has, and an alpha particle has a charge that is equal to twice the charge a proton has. Assume that $m_\alpha = 2m_\mathrm{d} = 4m_\mathrm{p}$. Compare (*a*) their speeds, (*b*) their kinetic energies, and (*c*) the magnitudes of their angular momenta about the centers of the orbits.

30 •• A particle has a charge q, a mass m, a linear momentum of magnitude p, and a kinetic energy K. The particle moves in a circular orbit of radius R perpendicular to a uniform magnetic field $\vec{B}$. Show that (*a*) $p = BqR$ and (*b*) $K = \frac{1}{2}B^2q^2R^2/m$.

31 •• A beam of particles with velocity $\vec{v}$ enters a region that has a uniform magnetic field $\vec{B}$ in the $+x$ direction. Show that when the x component of the displacement of one of the particles is $2\pi(m/qB)v\cos\theta$, where θ is the angle between $\vec{v}$ and $\vec{B}$, the velocity of the particle is in the same direction as it was when the particle entered the field. SSM

32 •• A proton that has a speed equal to 1.00×10^6 m/s enters a region with a uniform magnetic field that has a magnitude of 0.800 T and points into the page, as shown in Figure 26-35. The proton enters the region at an angle $\theta = 60°$. Find the exit angle ϕ and the distance d.

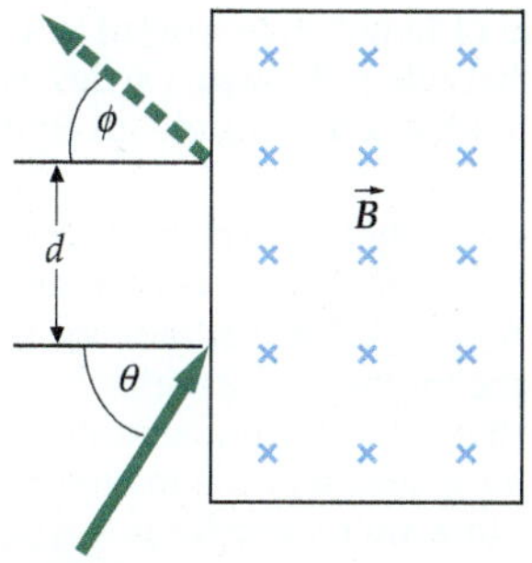

FIGURE 26-35
Problems 32 and 33

33 •• Suppose that in Figure 26-35, the magnetic field has a magnitude of 60 mT, the distance d is 40 cm, and θ is 24°. Find the speed v at which a particle enters the region and the exit angle ϕ if the particle is (*a*) a proton and (*b*) a deuteron. Assume that $m_d = 2m_p$. SSM

34 •• The galactic magnetic field in some region of interstellar space has a magnitude of 1.00×10^{-9} T. A particle of interstellar dust has a mass of 10.0 μg and a total charge of 0.300 nC. How many years does it take for the particle to complete a revolution of the circular orbit caused by its interaction with the magnetic field?

APPLICATIONS OF THE MAGNETIC FORCE ACTING ON CHARGED PARTICLES

35 • A velocity selector has a magnetic field that has a magnitude equal to 0.28 T and is perpendicular to an electric field that has a magnitude equal to 0.46 MV/m. (*a*) What must the speed of a particle be for that particle to pass through the velocity selector undeflected? What kinetic energy must (*b*) protons and (*c*) electrons have in order to pass through the velocity selector undeflected? SSM

36 •• A beam of protons is moving in the $+x$ direction with a speed of 12.4 km/s through a region in which the electric field is perpendicular to the magnetic field. The beam is not deflected in this region. (*a*) If the magnetic field has a magnitude of 0.85 T and points in the $+y$ direction, find the magnitude and direction of the electric field. (*b*) Would electrons that have the same velocity as the protons be deflected by these fields? If so, in what direction would they be deflected? If not, why not?

37 •• The plates of a Thomson q/m apparatus are 6.00 cm long and are separated by 1.20 cm. The end of the plates is 30.0 cm from the tube screen. The kinetic energy of the electrons is 2.80 keV. If a potential difference of 25.0 V is applied across the deflection plates, by how much will the point where the beam of electrons strikes the screen be displaced?

38 •• Chlorine has two stable isotopes, ^{35}Cl and ^{37}Cl. Chlorine gas which consists of singly ionized ions is to be separated into its isotopic components using a mass spectrometer. The magnetic field strength in the spectrometer is 1.2 T. What is the minimum value of the potential difference through which these ions must be accelerated so that the separation between them, after they complete their semicircular path, is 1.4 cm?

39 •• In a mass spectrometer, a singly ionized ^{24}Mg ion has a mass equal to 3.983×10^{-26} kg and is accelerated through a 2.50-kV potential difference. It then enters a region where it is deflected by a magnetic field of 557 G. (*a*) Find the radius of curvature of the ion's orbit. (*b*) What is the difference in the orbital radii of the ^{26}Mg and ^{24}Mg ions? Assume that their mass ratio is 26:24. SSM

40 •• A beam of singly ionized ^{6}Li and ^{7}Li ions passes through a velocity selector and enters a region of uniform magnetic field with a velocity that is perpendicular to the direction of the field. If the diameter of the orbit of the ^{6}Li ions is 15 cm, what is the diameter of the orbit for ^{7}Li ions? Assume their mass ratio is 7:6.

41 •• Using Example 26-6, determine the time required for a ^{58}Ni ion and a ^{60}Ni ion to complete the semicircular path.

42 •• Before entering a mass spectrometer, ions pass through a velocity selector consisting of parallel plates that are separated by 2.0 mm and have a potential difference of 160 V. The magnetic field strength is 0.42 T in the region between the plates. The magnetic field strength in the mass spectrometer is 1.2 T. Find (*a*) the speed of the ions entering the mass spectrometer and (*b*) the difference in the diameters of the orbits of singly ionized ^{238}U and ^{235}U. The mass of a ^{235}U ion is 3.903×10^{-25} kg.

43 •• A cyclotron for accelerating protons has a magnetic field strength of 1.4 T and a radius of 0.70 m. (*a*) What is the cyclotron's frequency? (*b*) Find the kinetic energy of the protons when they emerge. (*c*) How will your answers change if deuterons are used instead of protons? SSM

44 •• A certain cyclotron that has a magnetic field whose magnitude is 1.8 T is designed to accelerate protons to a kinetic energy of 25 MeV. (*a*) What is the cyclotron frequency for this cyclotron? (*b*) What must the minimum radius of the magnet be to achieve this energy? (*c*) If the alternating potential difference applied to the dees has a maximum value of 50 kV, how many revolutions must the protons make before emerging with kinetic energies of 25 MeV?

45 •• Show that for a given cyclotron the cyclotron frequency for accelerating deuterons is the same as the frequency for accelerating alpha particles and is half the frequency for accelerating protons in the same magnetic field. The deuteron has a charge that is equal to the charge a proton has, and an alpha particle has a charge that is equal to twice the charge a proton has. Assume that $m_\alpha = 2m_d = 4m_p$.

46 ••• Show that the radius of the orbit of a charged particle in a cyclotron is proportional to the square root of the number of orbits completed.

TORQUES ON CURRENT LOOPS, MAGNETS, AND MAGNETIC MOMENTS

47 • A small circular coil consisting of 20 turns of wire lies in a region with a uniform magnetic field whose magnitude is 0.50 T. The arrangement is such that the normal to the plane of the coil makes an angle of 60° with the direction of the magnetic field. The radius of the coil is 4.0 cm, and the wire carries a current of 3.0 A. (*a*) What is the magnitude of the magnetic moment of the coil? (*b*) What is the magnitude of the torque exerted on the coil? SSM

48 • What is the maximum torque on a 400-turn circular coil of radius 0.75 cm that carries a current of 1.6 mA and is in a region with a uniform magnetic field of 0.25 T?

49 • A current-carrying wire is in the shape of a square of edge length 6.0 cm. The square lies in the $z = 0$ plane. The wire carries a current of 2.5 A. What is the magnitude of the torque on the wire if it is in a region with a uniform magnetic field of magnitude 0.30 T that points in the (*a*) $+z$ direction and (*b*) $+x$ direction? SSM

50 • A current-carrying wire is in the shape of an equilateral triangle of edge length 8.0 cm. The triangle lies in the $z = 0$ plane. The wire carries a current of 2.5 A. What is the magnitude of the torque on the wire if it is in a region with a uniform magnetic field of magnitude 0.30 T that points in the (*a*) $+z$ direction and (*b*) $+x$ direction?

51 •• A rigid wire is in the shape of a square of edge length L. The square has mass m and the wire carries current I. The square lies on a flat horizontal surface in a region where there is a magnetic field of magnitude B that is parallel to two edges of the square. What is the minimum value of B so that one edge of the square will lift off the surface?

52 •• A rectangular current-carrying 50-turn coil, as shown in Figure 26-36, is pivoted about the z axis. (*a*) If the wires in the $z = 0$ plane make an angle $\theta = 37°$ with the y axis, what angle does the magnetic moment of the coil make with the unit vector $\hat{i}$? (*b*) Write an expression for $\hat{n}$ in terms of the unit vectors $\hat{i}$ and $\hat{j}$, where $\hat{n}$ is a unit vector in the direction of the magnetic moment. (*c*) What is the magnetic moment of the coil? (*d*) Find the torque on the coil when there is a uniform magnetic field $\vec{B} = 1.5\text{ T}\,\hat{j}$ in the region occupied by the coil. (*e*) Find the potential energy of the coil in this field. (The potential energy is zero when $\theta = 0$.)

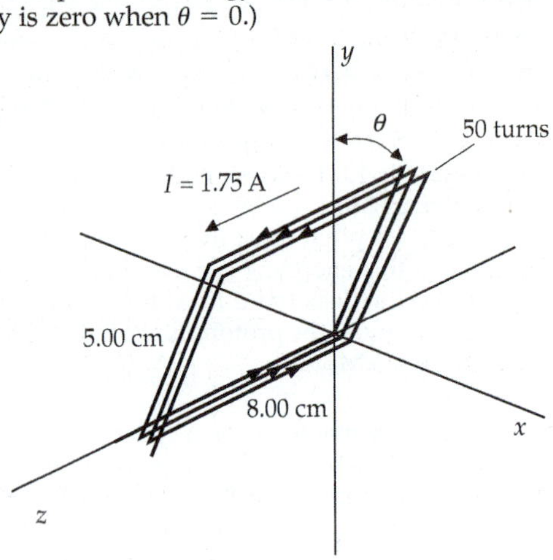

FIGURE 26-36
Problems 52 and 53

53 •• For the coil in Problem 52 the magnetic field is now $\vec{B} = 2.0\text{ T}\,\hat{j}$. Find the torque exerted on the coil when $\hat{n}$ is equal to (*a*) $\hat{i}$, (*b*) $\hat{j}$, (*c*) $-\hat{j}$, and (*d*) $(\hat{i} + \hat{j})/\sqrt{2}$. SSM

54 •• A small bar magnet has a length equal to 6.8 cm and its magnetic moment is aligned with a uniform magnetic field of magnitude 0.040 T. The bar magnet is then rotated through an angle of 60° about an axis perpendicular to its length. The observed torque on the bar magnet has a magnitude of 0.10 N·m. (*a*) Find the magnetic moment of the magnet. (*b*) Find the potential energy of the magnet.

55 •• A wire loop consists of two semicircles connected by straight segments (Figure 26-37). The inner and outer radii are 0.30 m and 0.50 m, respectively. A current of 1.5 A is in this wire and the current in the outer semicircle is in the clockwise direction. What is the magnetic moment of this current loop?

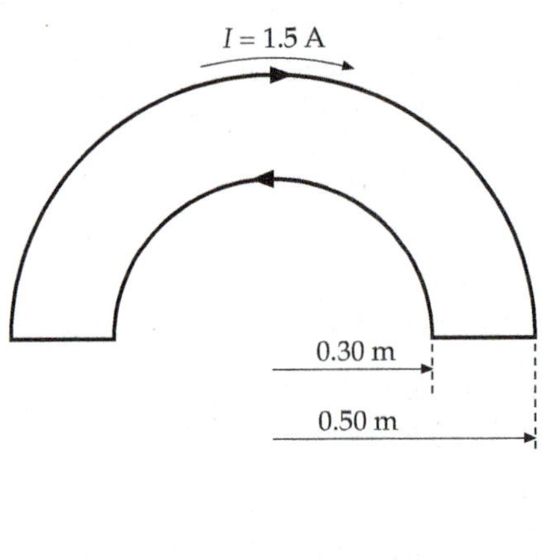

FIGURE 26-37
Problem 55

56 •• A wire of length L is wound into a circular coil that has N turns. Show that when the wire carries a current I, the magnetic moment of the coil has a magnitude given by $IL^2/(4\pi N)$.

57 •• A particle that has a charge q and a mass m moves with angular velocity ω in a circular path of radius r. (*a*) Show that the average current created by this moving particle is $\omega q/(2\pi)$ and that the magnetic moment of its orbit has a magnitude of $\frac{1}{2}q\omega r^2$. (*b*) Show that the angular momentum of this particle has the magnitude of $mr^2\omega$ and that the magnetic moment and angular momentum vectors are related by $\vec{\mu} = \frac{1}{2}(q/m)\,\vec{L}$, where $\vec{L}$ is the angular momentum about the center of the circle. SSM

58 ••• A uniformly charged nonconducting cylindrical shell (Figure 26-38) has length L, inner and outer radii R_i and R_o, respectively, a charge density ρ, and an angular velocity ω about its axis. Derive an expression for the magnetic moment of the cylinder.

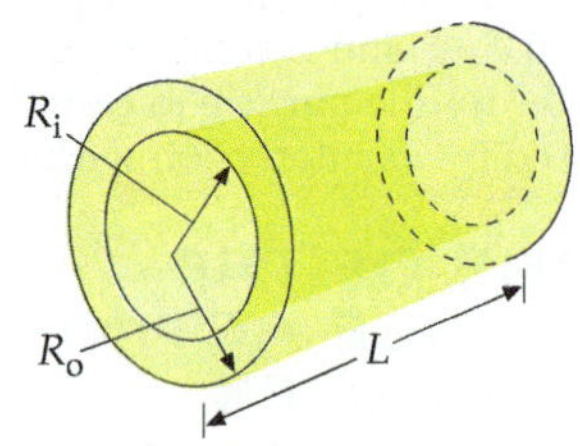

FIGURE 26-38 **Problem 58**

59 ••• A uniform nonconducting thin rod of mass m and length L has a uniform charge per unit length λ and rotates with angular speed ω about an axis through one end and perpendicular to the rod. (*a*) Consider a small segment of the rod of length dx and charge $dq = \lambda\,dr$ at a distance r from the pivot (Figure 26-39). Show that the average current created by this moving segment is $\omega\,dq/(2\pi)$ and show that the magnetic moment of this segment is $\frac{1}{2}\lambda\omega r^2\,dx$. (*b*) Use this to show that the magnitude of the magnetic moment of the rod is $\frac{1}{6}\lambda\omega L^3$. (*c*) Show that the magnetic moment $\vec{\mu}$ and angular momentum $\vec{L}$ are related by $\vec{\mu} = \frac{1}{2}(Q/m)\vec{L}$, where Q is the total charge on the rod. SSM

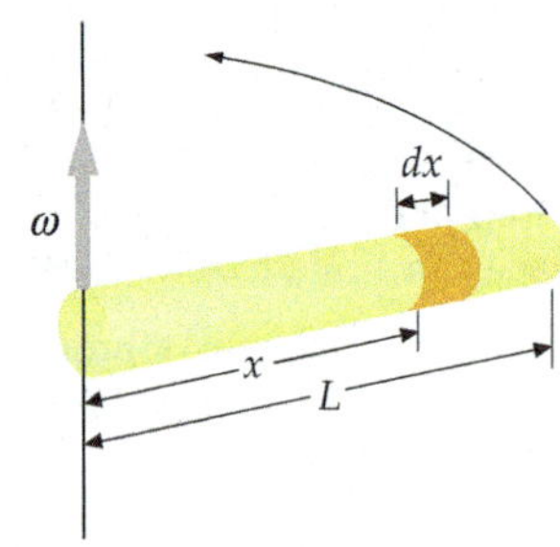

FIGURE 26-39
Problem 59

60 ••• A nonuniform, nonconducting thin disk of mass m, radius R, and total charge Q has a charge per unit area σ that varies as $\sigma_0 r/R$ and a mass per unit area σ_m that is given by $(m/Q)\sigma$. The disk rotates with angular speed ω about its central axis. (*a*) Show that the magnetic moment of the disk has a magnitude $\frac{1}{5}\pi\omega\sigma_0 R^4$, which can be alternatively rewritten as $\frac{3}{10}\omega Q R^2$. (*b*) Show that the magnetic moment $\vec{\mu}$ and angular momentum $\vec{L}$ are related by $\vec{\mu} = \frac{1}{2}(Q/m)\vec{L}$.

61 ••• A spherical shell of radius R carries a constant surface charge density σ. The shell rotates about its diameter with angular speed ω. Find the magnitude of the magnetic moment of the rotating spherical shell. SSM

62 ••• A uniform, solid, uniformly charged sphere of radius R has a volume charge density ρ. The sphere rotates about an axis through its center with angular speed ω. Find the magnitude of the magnetic moment of the rotating sphere.

63 ••• A uniform, thin, uniformly charged disk of mass m, radius R, and uniform surface charge density σ rotates with angular speed ω about an axis through its center and perpendicular to the disk (Figure 26-40). The disk is in a region with a uniform magnetic field $\vec{B}$ that makes an angle θ with the rotation axis. Calculate (*a*) the magnitude of the torque exerted on the disk by the magnetic field and (*b*) the precession frequency of the disk in the magnetic field.

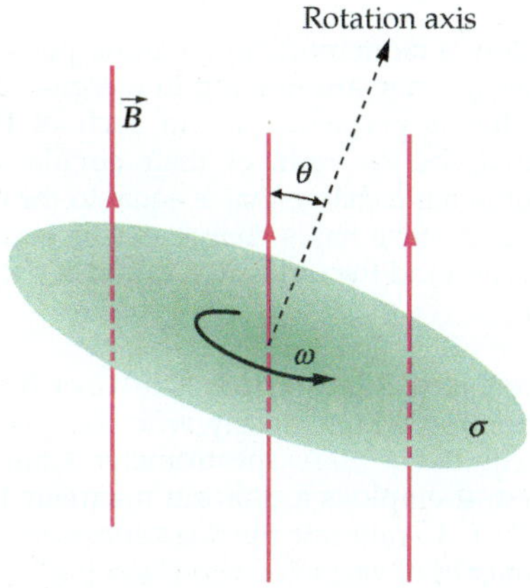

FIGURE 26-40 Problem 63

THE HALL EFFECT

64 • A metal strip that is 2.00 cm wide and 0.100 cm thick carries a current of 20.0 A in a region with a uniform magnetic field of 2.00 T, as shown in Figure 26-41. The Hall voltage is measured to be 4.27 μV. (*a*) Calculate the drift speed of the free electrons in the strip. (*b*) Find the number density of the free electrons in the strip. (*c*) Is point *a* or point *b* at the higher potential? Explain your answer.

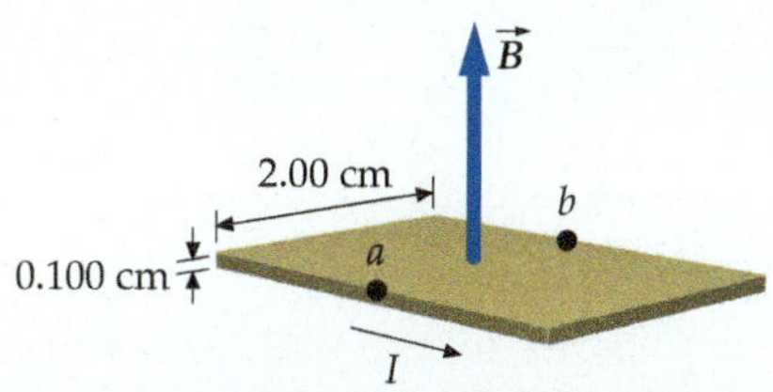

FIGURE 26-41 Problems 64 and 65

65 •• The number density of free electrons in copper is 8.47×10^{22} electrons per cubic centimeter. If the metal strip in Figure 26-41 is copper and the current is 10.0 A, find (*a*) the drift speed v_d and (*b*) the potential difference $V_a - V_b$. Assume that the magnetic field strength is 2.00 T. SSM

66 •• **ENGINEERING APPLICATION** A copper strip has 8.47×10^{22} free electrons per cubic centimeter, is 2.00 cm wide, is 0.100 cm thick, and is used to measure the magnitudes of unknown magnetic fields that are perpendicular to it. Find the magnitude B when the current is 20.0 A and the Hall voltage is (*a*) 2.00 μV, (*b*) 5.25 μV, and (*c*) 8.00 μV.

67 •• **BIOLOGICAL APPLICATION** Because blood contains ions, moving blood in the presence of a magnetic field develops a Hall voltage across the diameter of an artery. A large artery that has a diameter of 0.85 cm can have blood flowing through it with a maximum speed of 0.60 m/s. If a section of the artery is in a magnetic field of 0.20 T, what is the maximum potential difference across the diameter of the artery?

68 •• The Hall coefficient R_H is a property of conducting material (just as resistivity is). It is defined as $R_H = E_y/(J_x B_z)$, where J_x is the x component of the current density in the material, B_z is the z component of the magnetic field, and E_y is the y component of the resulting Hall electric field. Show that the Hall coefficient is equal to $1/(nq)$, where q is the charge of the charge carriers ($-e$ if they are electrons). (The Hall coefficients of monovalent metals, such as copper, silver, and sodium, are therefore negative.)

69 •• Aluminum has a density of 2.7×10^3 kg/m^3 and a molar mass of 27 g/mol. The Hall coefficient of aluminum is $R = -0.30 \times 10^{-10}$ m^3/C. (See Problem 68 for the definition of R.) What is the number of conduction electrons per aluminum atom? SSM

GENERAL PROBLEMS

70 • A long wire parallel to the x axis carries a current of 6.50 A in the $+x$ direction. The wire occupies a region that has a uniform magnetic field $\vec{B} = 1.35\text{ T}\,\hat{j}$. Find the magnetic force per unit length on the wire.

71 • An alpha particle (charge $+2e$) travels in a circular path of radius 0.50 m in a region with a magnetic field whose magnitude is 0.10 T. Find (*a*) the period, (*b*) the speed, and (*c*) the kinetic energy (in electron volts) of the alpha particle. (The mass of an alpha particle is 6.65×10^{-27} kg.)

72 •• The pole strength q_m of a bar magnet is defined by $\vec{\mu} = q_m\vec{\ell}$, where $\vec{\mu}$ is the magnetic moment of the magnet and $\vec{\ell}$ is the position of the north-pole end of the magnet relative to the south-pole end. Show that the torque exerted on a bar magnet in a uniform magnetic field $\vec{B}$ is the same as if a force $+q_m\vec{B}$ is exerted on the north pole of the magnet and a force $-q_m\vec{B}$ is exerted on the south pole.

73 •• A particle of mass m and charge q enters a region where there is a uniform magnetic field $\vec{B}$ parallel with the x axis. The initial velocity of the particle is $\vec{v} = v_{0x}\hat{i} + v_{0y}\hat{j}$, so the particle moves in a helix. (*a*) Show that the radius of the helix is $r = mv_{0y}/qB$. (*b*) Show that the particle takes a time $\Delta t = 2\pi m/qB$ to complete each turn of the helix. (*c*) What is the x component of the displacement of the particle during the time given in Part (*b*)? SSM

74 •• A metal cross-bar of mass m rides on a parallel pair of long horizontal conducting rails separated by a distance L and connected to a device that supplies constant current I to the circuit, as shown in Figure 26-42. The circuit is in a region with a uniform magnetic field $\vec{B}$ whose direction is vertically downward. There is no friction and the bar starts from rest at $t = 0$. (*a*) In which direction will the bar start to move? (*b*) Show that at time t the bar has a speed of $(BIL/m)t$.

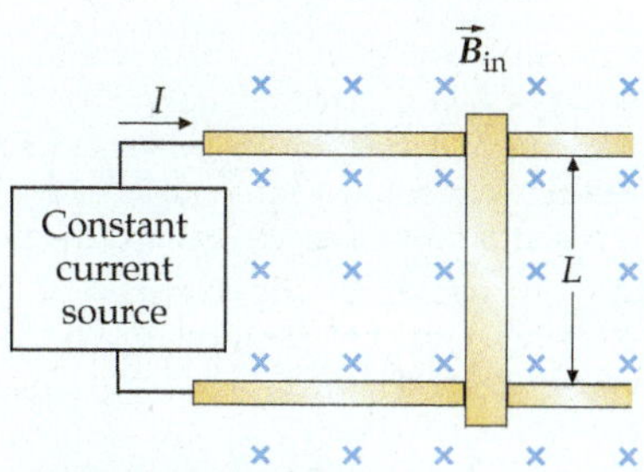

FIGURE 26-42 Problems 74 and 75

75 •• Assume that the rails Problem 74 are frictionless but tilted upward so that they make an angle θ with the horizontal, and with the current source attached to the low end of the rails. The magnetic field is still directed vertically downward. (*a*) What minimum value of B is needed to keep the bar from sliding down the rails? (*b*) What is the acceleration of the bar if B is twice the value found in Part (*a*)? SSM

76 •• A long, narrow bar magnet that has magnetic moment $\vec{\mu}$ parallel to its long axis is suspended at its center as a frictionless compass needle. When placed in region with a horizontal magnetic field $\vec{B}$, the needle lines up with the field. If it is displaced by a small angle θ, show that the needle will oscillate about its equilibrium position with frequency $f = \frac{1}{2\pi}\sqrt{\frac{\mu B}{I}}$, where I is the moment of inertia of the needle about the point of suspension.

77 •• A straight 20-m-long conducting wire is parallel to the y axis and is moving in the $+x$ direction with a speed of 20 m/s in a region that has a magnetic field given by 0.50 T $\hat{k}$. (*a*) A magnetic force acting on the conduction electrons leaves one end negatively charged due to an electron surplus and the other end positively charged due to an electron deficit. This charge separation process continues until the electric field due to the accumulated positive and negative charges exerts forces on the remaining conduction electrons that exactly balance the magnetic forces acting on them. Find the magnitude and direction of this electric field in the steady-state situation. (*b*) Which end of the wire is positively charged and which end is negatively charged? (*c*) Suppose the moving wire is 2.0 m long. What is the potential difference between its two ends due to this electric field?

78 ••• A circular loop of wire that has a mass m and a constant current I is in a region with a uniform magnetic field. It is initially in equilibrium and its magnetic moment is aligned with the magnetic field. The loop is given a small angular displacement about an axis through its center and perpendicular to the magnetic field and then released. What is the period of the subsequent motion? (Assume that the only torque exerted on the loop is due to the magnetic field and that there are no other forces acting on the loop.)

79 ••• A small bar magnet has a magnetic moment $\vec{\mu}$ that makes an angle θ with the x axis. The magnet is in a region that has a *nonuniform* magnetic field given by $\vec{B} = B_x(x)\hat{i} + B_y(y)\hat{j}$. Using $F_x = -\partial U/\partial x$, $F_y = -\partial U/\partial y$, and $F_z = -\partial U/\partial z$, show that there is a net magnetic force on the magnet that is given by

$$\vec{F} = \mu_x \frac{\partial B_x}{\partial x}\hat{i} + \mu_y \frac{\partial B_y}{\partial y}\hat{j}$$

80 •• A proton, a deuteron, and an alpha particle all have the same kinetic energy. They are moving in a region with a uniform magnetic field that is perpendicular to each of their velocities. Let R_p, R_d, and R_α be the radii of their circular orbits, respectively. The deuteron has a charge that is equal to the charge a proton has, and an alpha particle has a charge that is equal to twice the charge a proton has. Find the ratios R_d/R_p and R_α/R_p. Assume that $m_\alpha = 2m_d = 4m_p$.

81 ••• **ENGINEERING APPLICATION, CONTEXT-RICH** Your forensic chemistry group, working closely with local law enforcement agencies, has acquired a mass spectrometer similar to that discussed in the text. It employs a uniform magnetic field that has a magnitude of 0.75 T. To calibrate the mass spectrometer, you decide to measure the masses of various carbon isotopes by measuring the position of impact of the various singly ionized carbon ions that have entered the spectrometer with a kinetic energy of 25 keV. A wire chamber with position sensitivity of 0.50 mm is part of the apparatus. What will be the limit on its mass resolution (in kg) for ions in this mass range, that is, those whose mass is on the order of that of a carbon atom?

CHAPTER 27

THESE COILS AT THE KETTERING MAGNETICS LABORATORY AT OAKLAND UNIVERSITY ARE CALLED HELMHOLTZ COILS. THEY ARE USED TO CANCEL EARTH'S MAGNETIC FIELD AND TO PROVIDE A UNIFORM MAGNETIC FIELD IN A SMALL REGION OF SPACE FOR STUDYING THE MAGNETIC PROPERTIES OF MATTER. *(Bob Williamson, Oakland University, Rochester, Michigan.)*

Sources of the Magnetic Field

? Do you know how to calculate the magnitude of the magnetic field of a current-carrying coil? (See Example 27-2.)

As we discussed in Chapter 26, an awareness of the power of permanent magnets has been around since the year 1000. However, the study of magnets as they relate to electricity did not occur until 1819 when a Danish physicist, Hans Christian Oersted, discovered that a compass needle is deflected by an electric current. Just a month after Oersted's discovery, Jean-Baptiste Biot and Félix Savart announced the results of their measurements of the torque on a magnet near a long, current-carrying wire and they analyzed these results in terms of the magnetic field produced by each element of the current. André-Marie Ampère did additional experiments and showed that current elements also experience a force in the presence of a magnetic field and that two current elements exert forces on each other.

In this chapter, we begin by considering the magnetic field produced by a single moving charge and by the moving charges in a current element. We then calculate the magnetic fields produced by some common current configurations, such as a straight-wire segment; a long, straight wire; a current loop; and a solenoid. Next we discuss Ampère's law. Finally, we consider the magnetic properties of matter.

27-1 THE MAGNETIC FIELD OF MOVING POINT CHARGES

When a point charge q moves with velocity $\vec{v}$, the moving point charge produces a magnetic field $\vec{B}$ in space, given by*

$$\vec{B} = \frac{\mu_0}{4\pi}\frac{q\vec{v}\times\hat{r}}{r^2} \qquad 27\text{-}1$$

MAGNETIC FIELD OF A MOVING POINT CHARGE

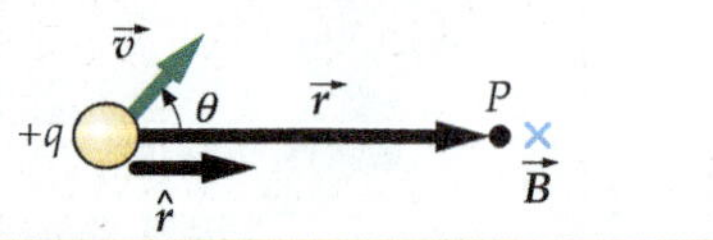

FIGURE 27-1 A positive point charge q moving with velocity $\vec{v}$ produces a magnetic field $\vec{B}$ at a field point P. The magnetic field at P is in the direction of $\vec{v}\times\hat{r}$, where $\vec{v}$ is the velocity of the point charge and $\hat{r}$ is the unit vector pointing from the charge to the field point. The field varies inversely as the square of the distance from the charge to the field point and is proportional to the sine of the angle between $\vec{v}$ and $\hat{r}$. (The blue $\times$ at the field point indicates that the direction of the field is into the page.)

where $\hat{r}$ is a unit vector (see Figure 27-1) that points to the field point P from the charge q moving with velocity $\vec{v}$, and μ_0 is a constant of proportionality called the **magnetic constant (permeability of free space),**† which, by definition, has the exact value

$$\mu_0 = 4\pi\times10^{-7}\,\text{T}\cdot\text{m/A} = 4\pi\times10^{-7}\,\text{N/A}^2 \qquad 27\text{-}2$$

The units of μ_0 are such that B is in teslas when q is in coulombs, v is in meters per second, and r is in meters. The unit N/A^2 comes from the fact that $1\text{ T} = 1\text{ N}\cdot\text{s/(C}\cdot\text{m)}$. The constant $1/(4\pi)$ is arbitrarily included in Equation 27-1 so that the factor 4π does not appear in Ampère's law (Equation 27-16), which we will study in Section 27-4.

Example 27-1 Magnetic Field of a Moving Point Charge

A point particle that has charge $q = 4.5\ \mu\text{C}$ is moving with velocity $\vec{v} = 3.0\text{ m/s}\ \hat{i}$ along the line $y = 3.0$ m in the $z = 0$ plane. Find the magnetic field at the origin produced by this charge when the charge is at the point $x = -4.0$ m, $y = 3.0$ m, as shown in Figure 27-2.

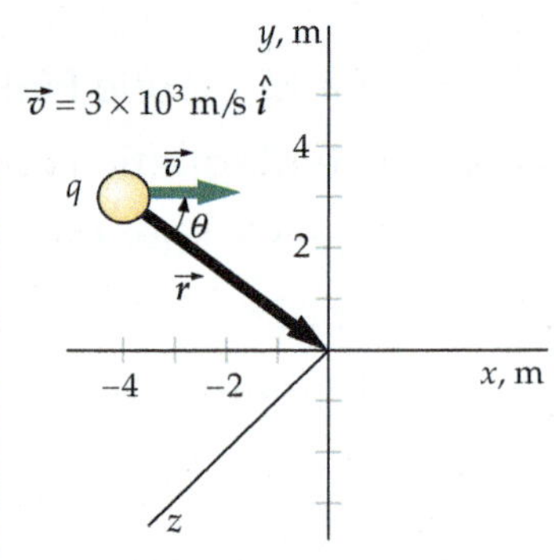

FIGURE 27-2

PICTURE The magnetic field associated with a moving charged particle is given by Equation 27-1.

SOLVE

1. The magnetic field is given by Equation 27-1:

$$\vec{B} = \frac{\mu_0}{4\pi}\frac{q\vec{v}\times\hat{r}}{r^2} \quad\text{where}\quad \vec{v} = v\hat{i}$$

2. Find $\hat{r}$ and r from Figure 27-2 and write $\hat{r}$ in terms of $\hat{i}$ and $\hat{j}$:

$$\vec{r} = 4.0\text{ m}\ \hat{i} - 3.0\text{ m}\ \hat{j}$$

$$r = \sqrt{4.0^2 + 3.0^2}\text{ m} = 5.0\text{ m}$$

$$\hat{r} = \frac{\vec{r}}{r} = \frac{4.0\text{ m}\ \hat{i} - 3.0\text{ m}\ \hat{j}}{5.0\text{ m}} = 0.80\,\hat{i} - 0.60\,\hat{j}$$

3. Substitute the above results in Equation 27-1:

$$\vec{B} = \frac{\mu_0}{4\pi}\frac{q\vec{v}\times\hat{r}}{r^2} = \frac{\mu_0}{4\pi}\frac{q(v\hat{i})\times(0.80\hat{i} - 0.60\hat{j})}{r^2} = \frac{\mu_0}{4\pi}\frac{q(-0.60v\hat{k})}{r^2}$$

$$= -(10^{-7}\,\text{T}\cdot\text{m/A})\frac{(4.5\times10^{-6}\text{ C})(0.60)(3.0\text{ m/s})}{(5.0\text{ m})^2}\hat{k}$$

$$= \boxed{-3.2\times10^{-14}\text{ T}\ \hat{k}}$$

CHECK It is also possible to obtain $\vec{B}$ without finding an explicit expression for the unit vector $\hat{r}$. From Figure 27-2 we note that $\vec{v}\times\hat{r}$ is in the $-z$ direction. In addition, the magnitude of $\vec{v}\times\hat{r}$ is $v\sin\theta$, where $\sin\theta = (3.0\text{ m})/(5.0\text{ m}) = 0.60$. Combining these results, we have $\vec{v}\times\hat{r} = v\sin\theta(-\hat{k}) = -v(0.60)\hat{k}$, in agreement with our result in line 1 of step 3.

PRACTICE PROBLEM 27-1 At the same instant, find the magnetic field on the y axis both at $y = 3.0$ m and at $y = 6.0$ m.

---

* This expression is used only for speeds much much less than the speed of light.

† Some care must be taken not to confuse the constant μ_0 with the magnitude of the magnetic moment vector $\vec{\mu}$.

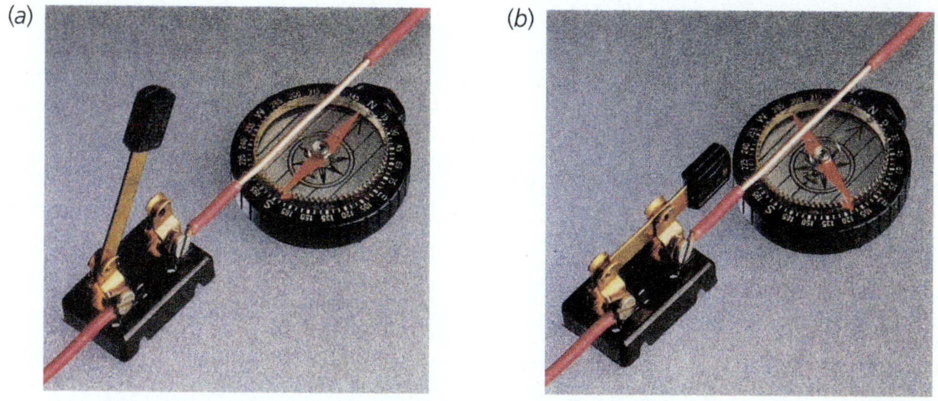

Oersted's experiment. (*a*) When no current exists in the wire, the compass needle points north. (*b*) When the wire carries a current, the compass needle is deflected in the direction of the resultant magnetic field. The current in the wire is directed upward, from left to right. The insulation has been stripped from the wire to improve the contrast of the photograph. *(© 1990 Richard Menga/Fundamental Photographs.)*

27-2 THE MAGNETIC FIELD OF CURRENTS: THE BIOT–SAVART LAW

In the previous chapter, we extended our discussion of forces on point charges to forces on current elements by substituting $q\vec{v}$ with the current element $I\,d\vec{\ell}$. We do the same for the magnetic field produced by a current element. The magnetic field $d\vec{B}$ produced by a current element $I\,d\vec{\ell}$ is given by Equation 27-1, where $q\vec{v}$ is replaced by $I\,d\vec{\ell}$:

$$d\vec{B} = \frac{\mu_0}{4\pi}\frac{I\,d\vec{\ell} \times \hat{r}}{r^2} \qquad 27\text{-}3$$

BIOT–SAVART LAW

Equation 27-3, known as the **Biot–Savart law,** was also deduced by Ampère. The Biot–Savart law and Equation 27-1 are analogous to Coulomb's law for the electric field of a point charge. The source of the magnetic field is a moving charge $q\vec{v}$ or a current element $I\,d\vec{\ell}$, just as the charge q is the source of the electrostatic field. The magnetic field decreases with the square of the distance from the moving charge or current element, just as the electric field decreases with the square of the distance from a point charge. However, the directional aspects of the electric and magnetic fields are quite different. Whereas the electric field points in the radial direction $\hat{r}$ from the point charge to the field point (for a positive charge), the magnetic field is perpendicular to both $\hat{r}$ and $\vec{v}$, in the case of a point charge, or $\hat{r}$ and $d\vec{\ell}$ in the case of a current element. At a point along the line of a current element, such as point P_2 in Figure 27-3, the magnetic field due to that current element is zero. (Equation 27-3 gives $d\vec{B} = 0$ if $d\vec{\ell}$ and $\hat{r}$ are either parallel or antiparallel.)

The magnetic field due to the total current in a circuit can be calculated by using the Biot–Savart law to find the field due to each current element, and then summing (integrating) over all the current elements in the circuit. This calculation is challenging for all but the simplest circuit geometries.

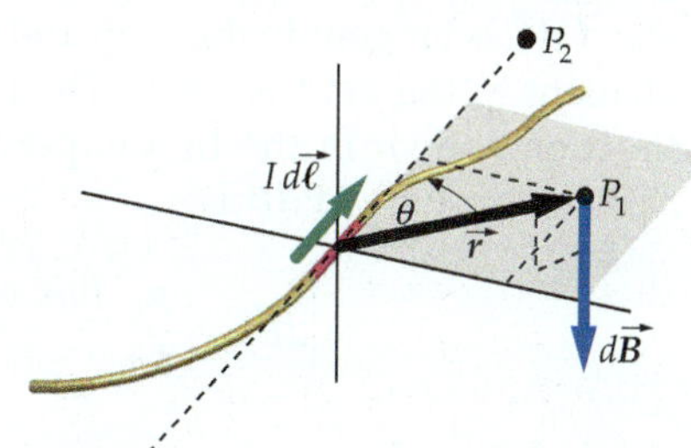

FIGURE 27-3 The current element $I\,d\vec{\ell}$ produces a magnetic field $d\vec{B}$ at point P_1 that is in the direction of $I\,d\vec{\ell} \times \hat{r}$, and thus perpendicular to both $d\vec{\ell}$ and $\hat{r}$. The current element produces no magnetic field at point P_2, which is along the line of $d\vec{\ell}$.

$\vec{B}$ DUE TO A CURRENT LOOP

Figure 27-4 shows a current element $I\,d\vec{\ell}$ of a current loop of radius R and the unit vector $\hat{r}$ that is directed from the element to the center of the loop. The magnetic field at the center of the loop due to this element is directed along the axis of the loop, and its magnitude, obtained by taking the magnitude of both sides of Equation 27-3, is given by

$$dB = \frac{\mu_0}{4\pi}\frac{I\,d\ell\,\sin\theta}{R^2} \qquad 27\text{-}4$$

where θ is the angle between $d\vec{\ell}$ and $\hat{r}$, which is 90° for each current element, so $\sin\theta = 1$. The resultant magnetic field due to all the current elements in the loop is found by integrating Equation 27-4 over all the current elements in the loop.

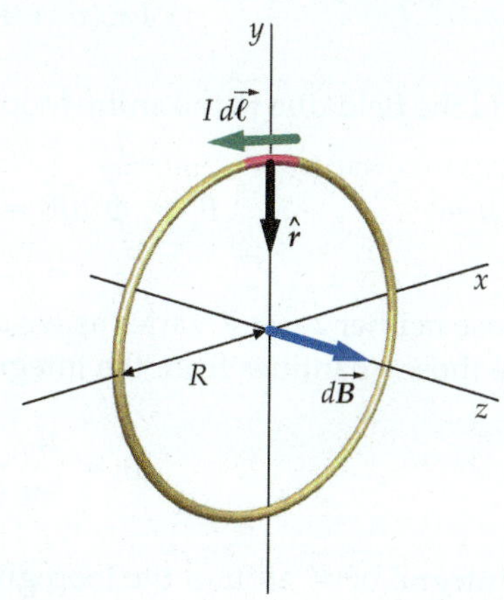

FIGURE 27-4 Current element for calculating the magnetic field at the center of a circular current loop. Each element produces a magnetic field that is directed along the axis of the loop.

Because I and R are the same for all elements, we obtain

$$B = \frac{\mu_0}{4\pi}\frac{I}{R^2}\oint d\ell$$

The integral of $d\ell$ around the complete loop gives the total length $2\pi R$, the circumference of the loop. The magnetic field due to the entire loop is thus

$$B = \frac{\mu_0}{4\pi}\frac{I}{R^2}2\pi R = \frac{\mu_0 I}{2R} \qquad 27\text{-}5$$

B AT THE CENTER OF A CURRENT LOOP

PRACTICE PROBLEM 27-2

Find the current in a circular loop of radius 8.0 cm that will produce a magnetic field of 0.20 mT at the center of the loop.

Figure 27-5 shows the geometry for calculating the magnetic field at a point on the axis of a circular current loop a distance z from the circular loop's center. We first consider the current element at the top of the loop. Here, as everywhere on the loop, $I\,d\vec{\ell}$ is tangent to the loop and perpendicular to the vector $\vec{r}$ from the current element to the field point P. The magnetic field $d\vec{B}$ due to this element is in the direction shown in the figure, perpendicular to $\hat{r}$ and also perpendicular to $I\,d\vec{\ell}$. The magnitude of $d\vec{B}$ is

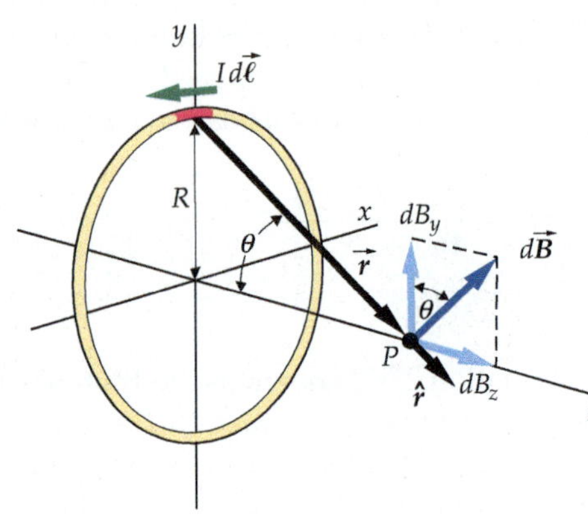

FIGURE 27-5 Geometry for calculating the magnetic field at a point on the axis of a circular current loop.

$$|d\vec{B}| = \frac{\mu_0}{4\pi}\frac{I|d\hat{\ell}\times\hat{r}|}{r^2} = \frac{\mu_0}{4\pi}\frac{I\,d\ell}{(z^2+R^2)}$$

where we have used the facts that $r^2 = z^2 + R^2$ and that $d\vec{\ell}$ and $\hat{r}$ are perpendicular, so $|d\vec{\ell}\times\hat{r}| = d\ell$.

When we sum around all the current elements in the loop, the components of $d\vec{B}$ perpendicular to the axis of the loop, such as dB_y in Figure 27-5, sum to zero, which leave only the components dB_z that are parallel to the axis. We thus compute only the z component of the field. From Figure 27-5, we have

$$dB_z = dB\sin\theta = \left(\frac{\mu_0}{4\pi}\frac{I\,d\ell}{(z^2+R^2)}\right)\left(\frac{R}{\sqrt{z^2+R^2}}\right) = \frac{\mu_0}{4\pi}\frac{IR\,d\ell}{(z^2+R^2)^{3/2}}$$

To find the field due to the entire loop of current, we integrate dB_z around the loop:

$$B_z = \oint dB_z = \oint \frac{\mu_0}{4\pi}\frac{IR}{(z^2+R^2)^{3/2}}d\ell$$

Because neither z nor R varies as we sum over the elements in the loop, we can remove those quantities from the integral. Then,

$$B_z = \frac{\mu_0}{4\pi}\frac{IR}{(z^2+R^2)^{3/2}}\oint d\ell$$

The integral of $d\ell$ around the loop gives $2\pi R$. Thus,

$$B_z = \frac{\mu_0}{4\pi}\frac{IR}{(z^2+R^2)^{3/2}}2\pi R = \frac{\mu_0}{4\pi}\frac{2\pi R^2 I}{(z^2+R^2)^{3/2}} \qquad 27\text{-}6$$

B ON THE AXIS OF A CURRENT LOOP

PRACTICE PROBLEM 27-3
Show that Equation 27-6 reduces to $B_z = \frac{1}{2}\mu_0 I/R$ (Equation 27-5) at the center of the loop.

At great distances from the loop, $|z|$ is much greater than R, so $(z^2 + R^2)^{3/2} \approx (z^2)^{3/2} = |z|^3$. Then,

$$B_z = \frac{\mu_0}{4\pi}\frac{2I\pi R^2}{|z|^3}$$

or

$$B_z = \frac{\mu_0}{4\pi}\frac{2\mu}{|z|^3} \qquad 27\text{-}7$$

MAGNETIC-DIPOLE FIELD ON THE AXIS OF THE DIPOLE

where $\mu = I\pi R^2$ is the magnitude of the magnetic moment of the loop. Note the similarity of this expression and the electric field on the axis of an electric dipole whose electric dipole moment has magnitude p (Equation 21-10):

$$E_z = \frac{1}{4\pi\epsilon_0}\frac{2p}{|z|^3}$$

Although it has not been demonstrated, our result that a current loop produces a magnetic dipole field far away holds in general for any point whether it is on the axis of the loop or off of the axis of the loop. Thus, a current loop behaves as a magnetic dipole because it experiences a torque $\vec{\mu} \times \vec{B}$ when placed in an external magnetic field (as was shown in Chapter 26) and it also produces a magnetic dipole field at field points far from the current loop. Figure 27-6 shows the magnetic field lines for a current loop.

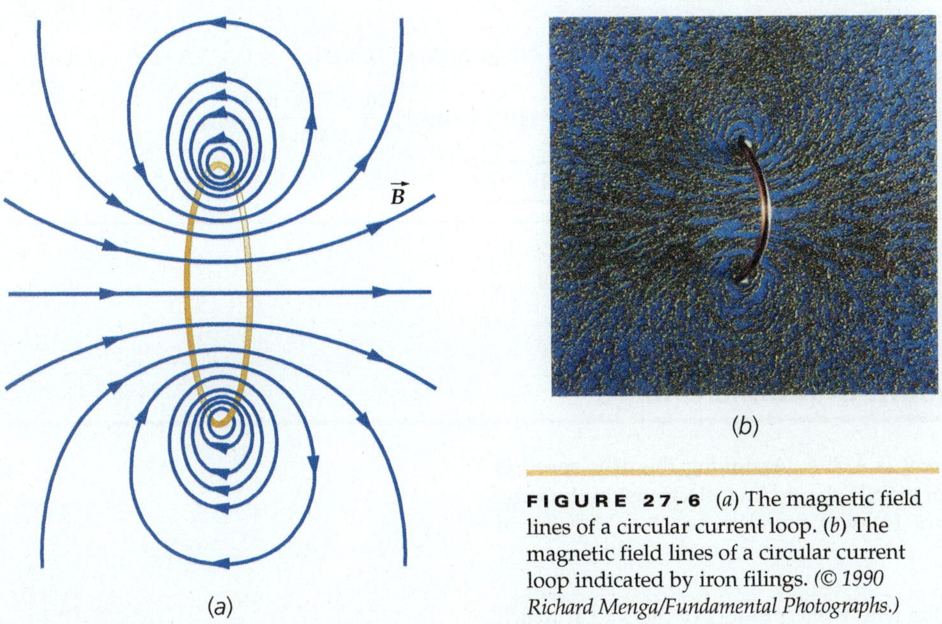

FIGURE 27-6 (*a*) The magnetic field lines of a circular current loop. (*b*) The magnetic field lines of a circular current loop indicated by iron filings. (*© 1990 Richard Menga/Fundamental Photographs.*)

Example 27-2 Find $\vec{B}$ on Axis of Coil

A circular coil has a radius equal to 5.00 cm, has 12 turns, lies in the $z = 0$ plane, and is centered at the origin. It carries a current of 4.00 A, and the magnetic moment of the coil is in the $+z$ direction. Using Equation 27-6, find the magnetic field on the z axis at (*a*) $z = 0$, (*b*) $z = 15.0$ cm, and (*c*) $z = 3.00$ m. (*d*) Using Equation 27-7, find the magnetic field on the z axis at $z = 3.00$ m.

PICTURE The magnetic field due to a loop that has N turns is N times that due to a single turn. (*a*) At $z = 0$, $B = \frac{1}{2}\mu_0 N/R$ (from Equation 27-5). Equation 27-6 gives the magnetic field on the axis due to the current in a single turn. Far from the loop, as in Part (*c*), the field can be found using Equation 27-7. In that case, because we have N loops, the magnetic moment is $\mu = NIA$, where $A = \pi R^2$.

SOLVE

(*a*) B_z at the center is N times that given by Equation 27-5 for a single turn of the coil:

$$B_z = \frac{\mu_0 NI}{2R}$$
$$= (4\pi \times 10^{-7}\,\mathrm{T \cdot m/A})\frac{(12)(4.00\,\mathrm{A})}{2(0.0500\,\mathrm{m})}$$
$$= \boxed{6.03 \times 10^{-4}\,\mathrm{T}}$$

(*b*) B_z on the axis is N times that given by Equation 27-6:

$$B_z = \frac{\mu_0}{4\pi}\frac{2\pi R^2 NI}{(z^2 + R^2)^{3/2}}$$
$$= (10^{-7}\,\mathrm{T \cdot m/A})\frac{2\pi(0.0500\,\mathrm{m})^2(12)(4.00\,\mathrm{A})}{\left[(0.1500\,\mathrm{m})^2 + (0.0500\,\mathrm{m})^2\right]^{3/2}}$$
$$= \boxed{1.91 \times 10^{-5}\,\mathrm{T}}$$

(*c*) Use Equation 27-6 again:

$$B_z = \frac{\mu_0}{4\pi}\frac{2\pi R^2 NI}{(z^2 + R^2)^{3/2}}$$
$$= (10^{-7}\,\mathrm{T \cdot m/A})\frac{2\pi(0.0500\,\mathrm{m})^2(12)(4.00\,\mathrm{A})}{\left[(3.00\,\mathrm{m})^2 + (0.0500\,\mathrm{m})^2\right]^{3/2}}$$
$$= \boxed{2.79 \times 10^{-9}\,\mathrm{T}}$$

(*d*) 1. Because 3.00 m is much greater than the radius $R = 0.0500$ m, we can use Equation 27-7 for the magnetic field far from the loop:

$$B_z = \frac{\mu_0}{4\pi}\frac{2\mu}{|z|^3}$$

2. The magnitude of the magnetic moment of the loop is N/A:

$$\mu = NI\pi R^2 = (12)(4.00\,\mathrm{A})\,\pi(0.0500\,\mathrm{m})^2 = 0.377\,\mathrm{A \cdot m^2}$$

3. Substitute $\mu = 0.377\,\mathrm{A \cdot m^2}$ and $z = 3.00$ m into the expression for B_z in step 1:

$$B_z = \frac{\mu_0}{4\pi}\frac{2\mu}{|z|^3} = (10^{-7}\,\mathrm{T \cdot m/A})\frac{2(0.377\,\mathrm{A \cdot m^2})}{(3.00\,\mathrm{m})^3}$$
$$= \boxed{2.79 \times 10^{-9}\,\mathrm{T}}$$

CHECK In Part (*d*) $z = 60R$, so we were able to use an approximation that is valid for $z \gg R$. The result differs from the exact value, calculated in Part (*c*), by less than one part in 279.

Example 27-3 Calculating the Amount of Mobile Charge

In the coil described in Example 27-2, the current is 4.00 A. Assuming the drift speed is 1.40×10^{-4} m/s, find the number of coulombs of mobile charge (free electrons) in the wire. (The drift speed for a wire carrying a current of 1 A was found to be 3.5×10^{-5} m/s in Example 25-1.)

PICTURE The amount of moving charge Q in the wire is the product of the rate at which charge enters one end of the wire and the time it takes the charge to travel the length of the wire. The rate at which charge enters one end of the wire is the current I, and the time for the charge to travel the length L of the wire is L/v_d, where v_d is the drift speed.

SOLVE

1. The amount of moving charge is the product of the current and the time for a charge carrier to travel the length of the wire:

$$Q = I\,\Delta t$$

2. The drift speed is the length of the wire divided by the time:

$$v_d = \frac{L}{\Delta t}$$

3. The length L is the number of turns multiplied by the length per turn. Also, we solve the step-2 result for the time:

$$L = N2\pi R = (12)2\pi(0.0500\text{ m}) = 3.77\text{ m}$$

and

$$\Delta t = \frac{L}{v_d} = \frac{3.77\text{ m}}{1.40 \times 10^{-4}\text{ m/s}} = 2.69 \times 10^4\text{ s}$$

4. Solve the step-1 result for the amount of moving charge in the wire:

$$Q = I\,\Delta t = (4.00\text{ A})(2.69 \times 10^4\text{ s}) = \boxed{1.08 \times 10^5\text{ C}}$$

CHECK There is approximately one conduction electron for each atom in a metal. If the wire is made of copper (with an molar mass equal to 63.5 g/mol), it is plausible that 3.77 m of the wire would have a mass approximately equal to 63.5 g. Thus, we estimate about one mole of copper exists in the wire. That means the number of conduction electrons in the wire is approximately equal to Avogadro's number. The total charge Q carried by these electrons is the number of electrons multiplied by the charge per electron. That is, $Q = -N_A e = -(6.02 \times 10^{23})(1.60 \times 10^{-19}\text{ C}) = -0.965 \times 10^5\text{ C}$. The magnitude of this result is very close to our step-4 result.

TAKING IT FURTHER The current consists of about 10^5 C of moving charges. In comparison to the amount of charge stored in an ordinary capacitor, this amount is an enormous amount of charge.

Example 27-4 Torque on a Bar Magnet

Try It Yourself

A small bar magnet that has a magnetic moment whose magnitude is 0.0300 A · m² is placed at the center of the coil of Example 27-2 so that its magnetic moment vector lies in the $x = 0$ plane and makes an angle of 30° with the $+z$ direction as shown. Neglecting any variation in $\vec{B}$ over the region occupied by the magnet, find the torque on the magnet.

PICTURE The torque on a magnetic moment is given by $\vec{\tau} = \vec{\mu} \times \vec{B}$. $\vec{B}$ is in the $+z$ direction, so you can use the right-hand rule to show that $\vec{\mu} \times \vec{B}$ is in the $+x$ direction (Figure 27-7).

SOLVE

Cover the column to the right and try these on your own before looking at the answers.

Steps	Answers
1. Compute the magnitude of the torque from $\vec{\tau} = \vec{\mu} \times \vec{B}$.	$\tau = 9.04 \times 10^{-6}\text{ N}\cdot\text{m}$
2. Indicate the direction using a unit vector.	$\vec{\tau} = \boxed{(9.04 \times 10^{-6}\text{ N}\cdot\text{m})\hat{i}}$

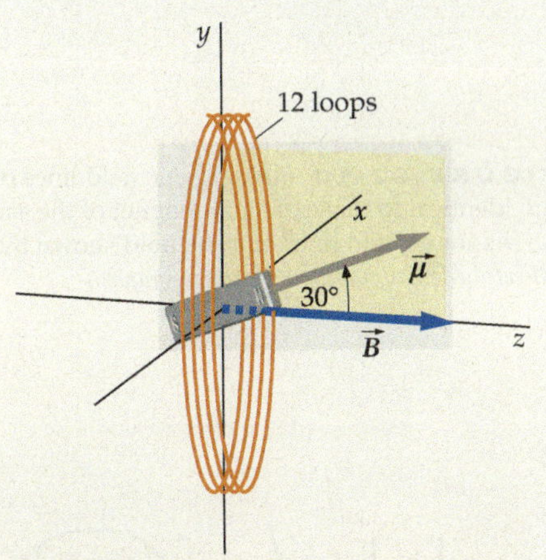

FIGURE 27-7

CHECK We expect the torque to tend to align the magnetic moment with the magnetic field. Thus, a torque vector in the $+x$ direction is as expected.

$\vec{B}$ DUE TO A CURRENT IN A SOLENOID

A **solenoid** is a conducting wire wound into a helix of closely spaced turns, as illustrated in Figure 27-8. A solenoid is used to produce a strong, uniform magnetic field in the region surrounded by its loops. The solenoid's role in magnetism is analogous to that of the parallel-plate capacitor, which produces a strong, uniform electric field between its plates.

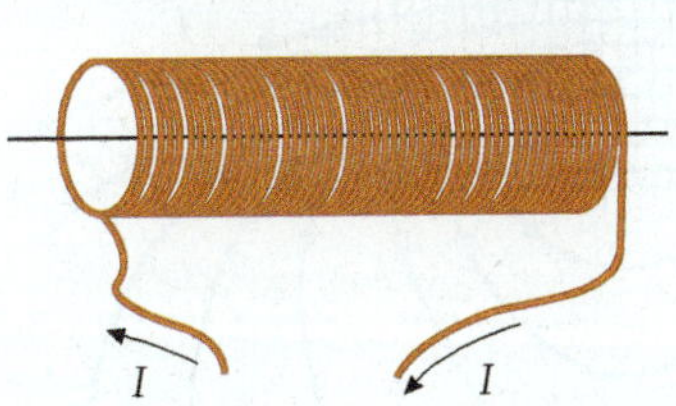

FIGURE 27-8 A tightly wound solenoid can be considered as a set of circular current loops placed side by side that carry the same current. The solenoid produces a uniform magnetic field inside the loops and distant from the ends of the solenoid.

The magnetic field of a solenoid is essentially that of a set of N identical current loops placed side by side. Figure 27-9 shows the magnetic field lines for two such loops.

Figure 27-10*a* shows the magnetic field lines for a tightly wound solenoid. Inside the solenoid and away from the ends, the field lines are approximately parallel to the axis and are closely and uniformly spaced, indicating a strong, uniform magnetic field. Outside the solenoid (above and below it) the lines are much less dense. In addition, the field lines become farther apart as you move away from either end of the solenoid. Comparing this figure with Figure 27-10*b*, we see that the magnetic field of a solenoid, both inside and outside the solenoid, are virtually identical to the magnetic field of a bar magnet of the same size and shape as the solenoid. In Figure 27-10*c*, iron filings align with the field of a current-carrying solenoid.

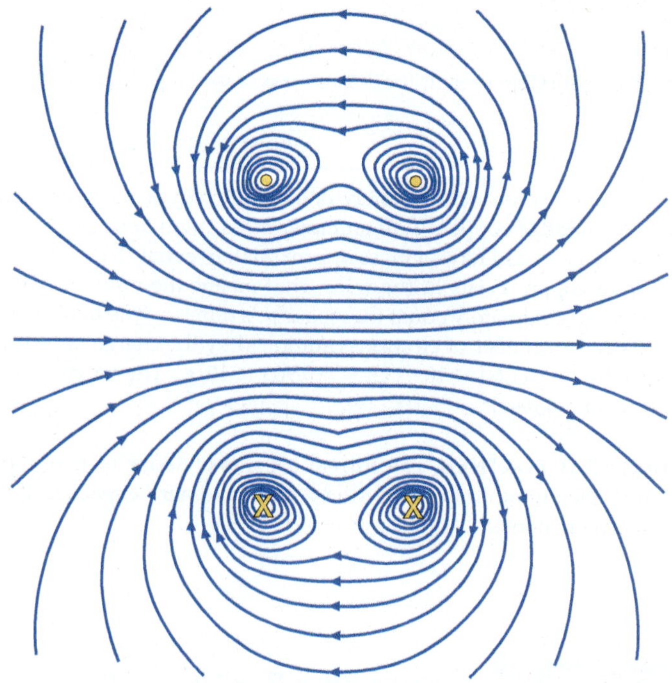

FIGURE 27-9 Magnetic field lines due to two identical coaxial loops carrying the same current. The points where the loops intersect the plane of the page are each marked by an × where the current enters the page and by a dot • where the current emerges from the page. In the region between the loops near the axis the magnetic fields of the individual loops superpose, so the resultant field is strong and surprisingly uniform. The region where the field is uniform is greatest if the planes of the two loops are separated by a distance equal to the radius of the loops.

FIGURE 27-10 (*a*) Magnetic field lines of a solenoid. The lines are identical to those of a bar magnet of the same shape, as in (*b*). (*c*) Magnetic field lines of a solenoid shown by iron filings. *(© 1990 Richard Menga/Fundamental Photographs.)*

(*a*)

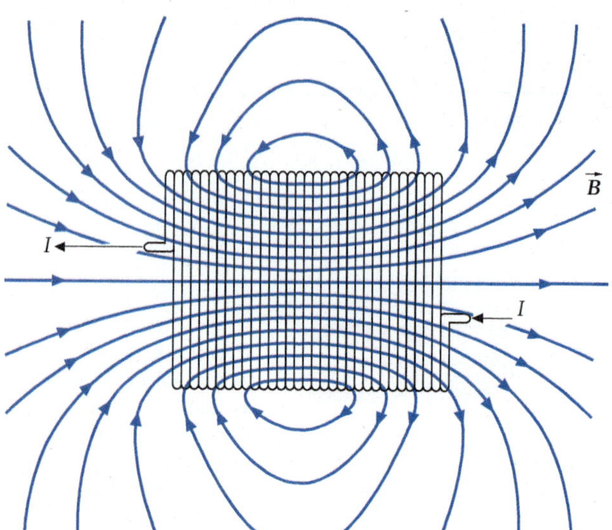

(*b*)

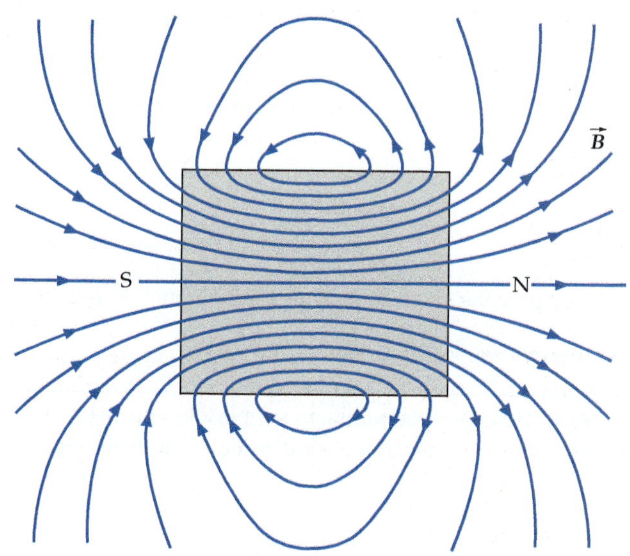

(*c*)

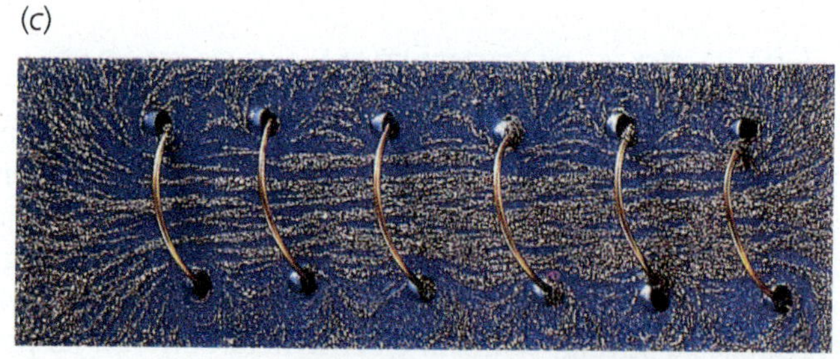

Example 27-5 The Field of a Long, Tightly Wound Solenoid *Conceptual*

The previous paragraph asserts that magnetic field inside and far from the ends of a long, tightly wound, current-carrying solenoid is uniform and parallel with the axis of the solenoid, and that the magnetic field is zero outside the solenoid. Validate this assertion by modeling the solenoid as a tightly packed stack of current-carrying rings, and by using the field line drawing of a single current-carrying ring (Figure 27-11).

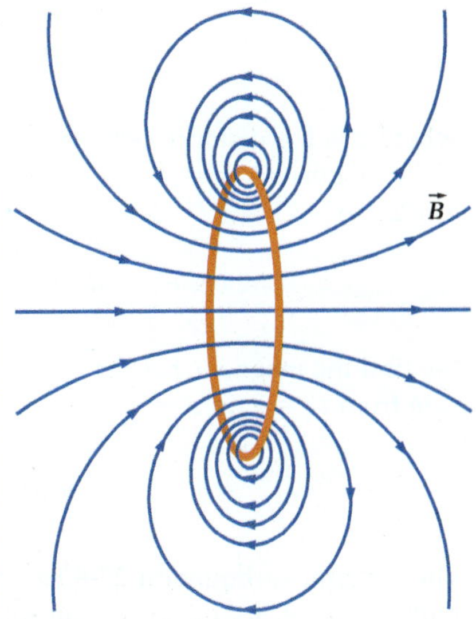

FIGURE 27-11

PICTURE Figure 27-12 shows three equally spaced current-carrying rings representing three loops of a long, tightly wound solenoid. At each of points A, B, and C, where A is just inside loop 2, B is just outside loop 2, and C is at the center of loop 2, sketch the three magnetic field vectors due to the three loops shown. Use the field line drawing of a single current-carrying ring (Figure 27-11) to obtain the directions and relative magnitudes of the three fields. Using your sketch, present an argument that the resultant magnetic fields at points A and B are equal in magnitude and are parallel with the axis of the solenoid. Using your sketch, present an argument that the resultant magnetic field at point C is zero.

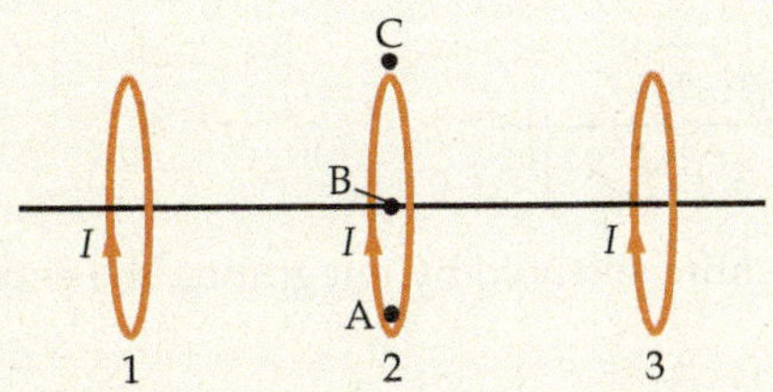

FIGURE 27-12

SOLVE

1. At point A, sketch the magnetic-field vectors $\vec{B}_1$, $\vec{B}_2$, and $\vec{B}_3$ due to the currents in loops 1, 2, and 3, respectively (Figure 27-13). Use Figure 27-11 for guidance:

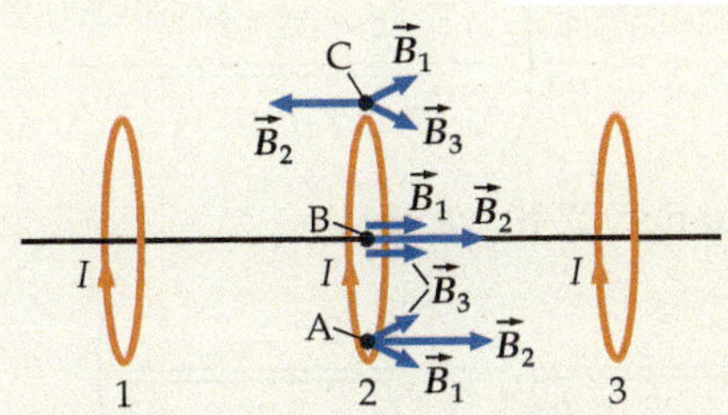

FIGURE 27-13

2. The magnitude of the magnetic field is greater where the field lines are closer together. An examination of the field lines in Figure 27-11 reveals that magnitude of the field $\vec{B}_2$ (due to loop 2) at point A is greater than it is at point B:

The magnitude of the field $\vec{B}_2$ (due to loop 2) is greater at point A than the magnitude of $\vec{B}_2$ is at point B. However, because $\vec{B}_1$, $\vec{B}_2$ and $\vec{B}_3$ are in the same direction, it is feasible that the resultant field ($\vec{B}_1 + \vec{B}_2 + \vec{B}_3$) at point B has the same magnitude as the resultant magnetic field at A.

3. An examination of the field lines in Figure 27-11 reveals that in the plane of the current-carrying ring, at points outside the ring the magnetic field is in the opposite direction than it is at points inside the ring:

At point C, the direction of $\vec{B}_2$ is to the left and the direction of $\vec{B}_1 + \vec{B}_3$ is to the right. In addition, additional loops in the solenoid that are near loops 1 and 3 will produce additional magnetic fields at C whose vector sum is to the right. It is feasible, therefore, that the resultant magnetic field at C is zero.

TAKING IT FURTHER The arguments presented in this example only hold for those sections of the solenoid located far from either end of the solenoid. Suppose loop 2 in Figure 27-13 is not near the center of a long solenoid, but is the last loop at the right end of the solenoid. Then, loop 3 would be absent from the picture, and the three vectors labeled $\vec{B}_3$ would be absent as well.

Consider a solenoid that has a length L, consists of N turns, and carries a current I. We choose the axis of the solenoid to be the z axis, with the left end at $z = z_1$ and the right end at $z = z_2$, as shown in Figure 27-14. We will calculate the magnetic field at field point P on the z axis a distance z from the origin. The figure shows an element of the solenoid of length dz' at a distance z' from the origin. If $n = N/L$ is the number of turns per unit length, there are $n\,dz'$ turns of wire in this element,

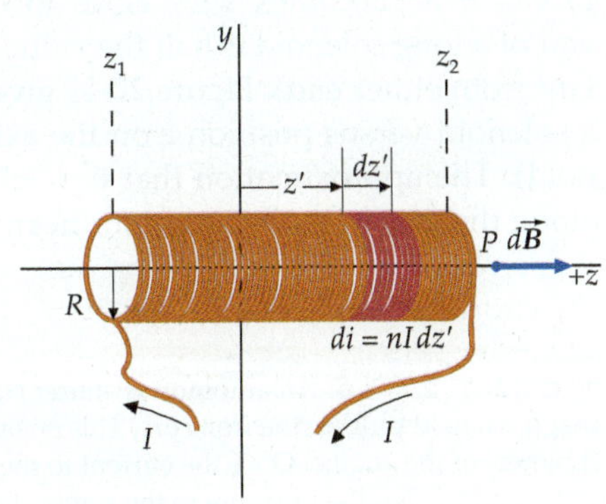

FIGURE 27-14 Geometry for calculating the magnetic field inside a solenoid on its axis. The number of turns in the element dz' is $n\,dz'$, where $n = N/L$ is the number of turns per unit length. The element dz' is treated as a current loop carrying a current $di = nI\,dz'$.

with each turn carrying a current I. The element is thus equivalent to a single loop carrying a current $di = nI\,dz'$. The magnetic field at a point on the z axis due to a loop at the origin carrying current di is given by Equation 27-6:

$$dB_z = \tfrac{1}{2}\mu_0 \frac{R^2\,di}{(z^2 + R^2)^{3/2}}$$

where z is the distance between the loop and the field point. For a loop at $z = z'$ carrying current $di = nI\,dz'$, the distance between the loop and the field point P is $z - z'$, so

$$dB_z = \tfrac{1}{2}\mu_0 \frac{R^2 nI\,dz'}{\left[(z - z')^2 + R^2\right]^{3/2}}$$

We find the magnetic field at P due to the entire solenoid by integrating the expression from $z' = z_1$ to $z' = z_2$:

$$B_z = \tfrac{1}{2}\mu_0 nIR^2 \int_{z_1}^{z_2} \frac{dz'}{\left[(z - z')^2 + R^2\right]^{3/2}} \qquad \text{27-8}$$

The integral in Equation 27-8 can be evaluated using trigonometric substitution with $z - z' = R\tan\theta$. Also, the integral can be looked up in standard tables of integrals. The integral's value is

$$\int_{z_1}^{z_2} \frac{dz'}{\left[(z - z')^2 + R^2\right]^{3/2}} = \frac{1}{R^2}\left(\frac{z - z_1}{\sqrt{(z - z_1)^2 + R^2}} - \frac{z - z_2}{\sqrt{(z - z_2)^2 + R^2}}\right)$$

Substituting this into Equation 27-8, we obtain

$$B_z(z) = \tfrac{1}{2}\mu_0 nI\left(\frac{z - z_1}{\sqrt{(z - z_1)^2 + R^2}} - \frac{z - z_2}{\sqrt{(z - z_2)^2 + R^2}}\right) \qquad \text{27-9}$$

B_z ON THE AXIS OF A SOLENOID

A solenoid is called a long solenoid if its length L is much greater than its radius R. Inside and far from the ends of a long solenoid, the fraction on the left in the parentheses approaches $+1$ and the fraction on the right approaches -1. This means the expression in the parentheses tends toward $+2$. Thus, in the region inside and far from either end of the solenoid the magnetic field is given by

$$B_z = \mu_0 nI \qquad \text{27-10}$$

B_z INSIDE A LONG SOLENOID

To evaluate B_z at the right end of the solenoid we use Equation 27-9 with $z = z_2$. This gives $B_z(z_2) = \frac{1}{2}\mu_0 nI\,L/\sqrt{L^2 + R^2}$, where $L = z_2 - z_1$. Then, if $L \gg R$, the ratio $L/\sqrt{L^2 + R^2}$ becomes very close to one, so $B_z(z_2) \approx \frac{1}{2}\mu_0 nI$. Thus, B_z at either end of a long solenoid is half the value of B at points deep inside the solenoid (distant from either end). Figure 27-15 gives a plot of the magnetic field on the axis of a solenoid versus position z on the axis (with the origin at the center of the solenoid). The approximation that the field is uniform (independent of the position) along the axis is good, except for near the ends.

See
Math Tutorial *for more information on*
Integrals

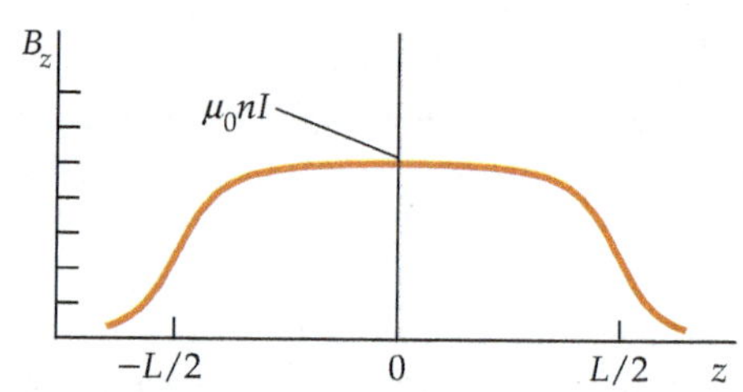

FIGURE 27-15 Graph of the magnetic field on the axis inside a solenoid versus the position z on the axis. The field inside the solenoid is nearly constant except near the ends. The length L of the solenoid is ten times longer than the radius.

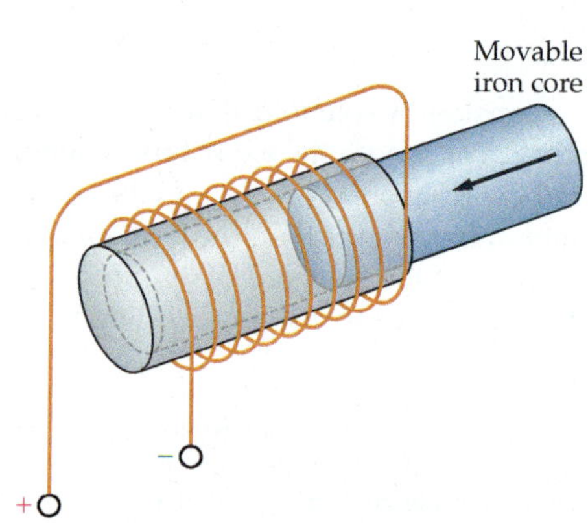

FIGURE 27-16 An automotive starter solenoid. When the solenoid is energized, its magnetic field pulls in the iron core. This engages gears that connect the starter motor to the flywheel of the engine. Once the current to the solenoid is interrupted, a spring disengages the gears and pushes the iron core to the right.

Example 27-6 $\vec{B}$ at Center of a Solenoid

Find the magnetic field at the center of a solenoid that has a length equal to 20.0 cm, a radius equal to 1.40 cm, 600 turns, and a current equal to 4.00 A.

PICTURE To find B on the axis of the solenoid we apply Equation 27-9 with the origin at the center of the solenoid.

SOLVE

1. We will calculate the field exactly, using Equation 27-9:

$$B_z(z) = \tfrac{1}{2}\mu_0 nI\left(\frac{z - z_1}{\sqrt{(z - z_1)^2 + R^2}} - \frac{z - z_2}{\sqrt{(z - z_2)^2 + R^2}}\right)$$

2. To find the magnetic field at the middle of the solenoid, we choose the middle of the solenoid as the origin. Then we set $z = 0$, $z_1 = -\frac{1}{2}L$, and $z_2 = \frac{1}{2}L$:

$$B_z(0) = \tfrac{1}{2}\mu_0 nI\left(\frac{0 - (-\frac{1}{2}L)}{\sqrt{[0 - (-\frac{1}{2}L)]^2 + R^2}} - \frac{0 - (\frac{1}{2}L)}{\sqrt{[(0 - (\frac{1}{2}L))^2] + R^2}}\right)$$

$$= \tfrac{1}{2}\mu_0 nI\frac{L}{\sqrt{\frac{1}{4}L^2 + R^2}} = \mu_0 nI\frac{L}{\sqrt{L^2 + 4R^2}}$$

3. Substituting in the given information, we have:

$$\frac{L}{\sqrt{L^2 + 4R^2}} = \frac{20.0\ \text{cm}}{\sqrt{(20.0\ \text{cm})^2 + 4(1.40\ \text{cm})^2}} = 0.990$$

$$B_z(0) = 0.990\,\mu_0 nI$$

$$= 0.990(4\pi \times 10^{-7}\ \text{T}\cdot\text{m/A})\frac{600}{0.200\ \text{m}}(4.00\ \text{A})$$

$$= \boxed{1.50 \times 10^{-2}\ \text{T}}$$

CHECK Note that the approximation obtained using Equation 27-10 amounts to replacing 0.990 by 1 in step 3. Doing so gives a result that differs from the step-3 result by only one percent. This result is expected for a solenoid whose length to radius ratio is 20 cm/1.4 cm ≈ 14.

PRACTICE PROBLEM 27-4 Calculate B on the axis a distance halfway between the center and one end of the solenoid. Compare this result with the step-3 result.

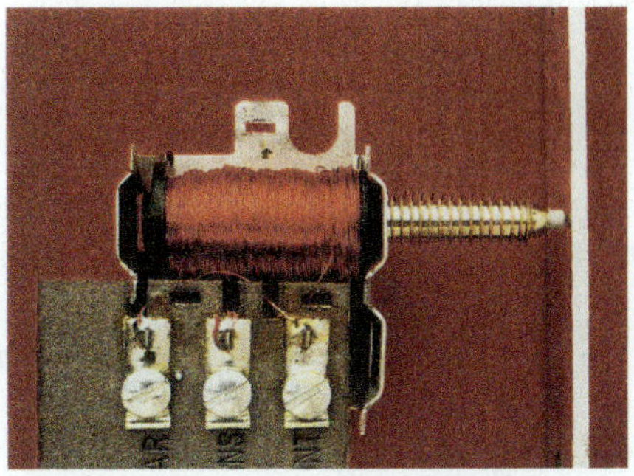

A cross section of a doorbell. When the solenoid is energized, its magnetic field pulls on the plunger, causing it to strike the bell (not shown). The spring returns the plunger to its normal position. (© *Bruce Iverson.*)

$\vec{B}$ DUE TO A CURRENT IN A STRAIGHT WIRE

Figure 27-17 shows the geometry for calculating the magnetic field $\vec{B}$ at a point P due to the current in the straight-wire segment shown. We choose R to be the perpendicular distance from the wire to point P, and we choose the x axis to be along the wire with $x = 0$ at the projection of P onto the x axis.

A typical current element $I\,d\vec{\ell}$ at a distance x from the origin is shown. The vector $\vec{r}$ points from the element to the field point P. The direction of the magnetic field at P due to this element is the direction of $I\,d\vec{\ell} \times \hat{r}$, which is out of the paper. Note that at P, the magnetic fields due to all the current elements of the wire are in

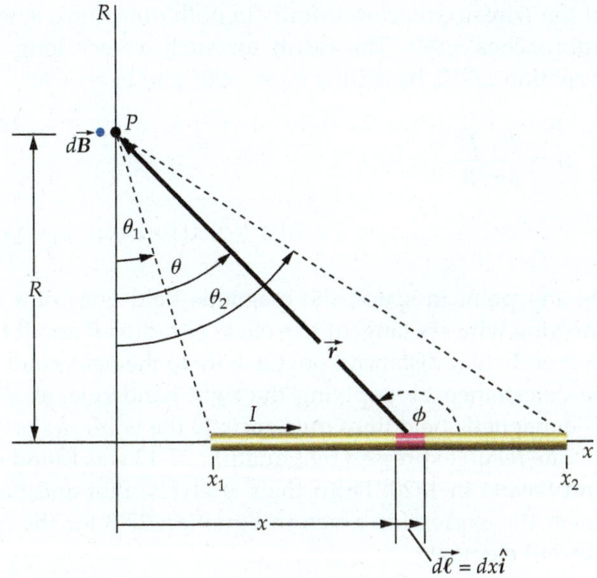

FIGURE 27-17 Geometry for calculating the magnetic field at point P due to a straight current segment. Each element of the segment contributes to the total magnetic field at point P, which is directed out of the paper. The result is expressed in terms of the angles θ_1 and θ_2.

this same direction. Thus, we need to compute only the magnitude of the field. The field due to the current element shown has the magnitude (Equation 27-3)

$$dB = \frac{\mu_0}{4\pi}\frac{Idx}{r^2}\sin\phi$$

It is more convenient to write this in terms of θ rather than ϕ:

$$dB = \frac{\mu_0}{4\pi}\frac{I\,dx}{r^2}\cos\theta \qquad \text{27-11}$$

To sum over all the current elements, we need to relate the variables θ, r, and x. It turns out to be easiest to express x and r in terms of θ. We have

$$x = R\tan\theta$$

Then, taking the differential of each side with R as a constant gives

$$dx = R\sec^2\theta\,d\theta = R\frac{r^2}{R^2}d\theta = \frac{r^2}{R}d\theta$$

where we have used $\sec\theta = r/R$. Substituting this expression for dx into Equation 27-11, we obtain

$$dB = \frac{\mu_0}{4\pi}\frac{I}{r^2}\frac{r^2\,d\theta}{R}\cos\theta = \frac{\mu_0}{4\pi}\frac{I}{R}\cos\theta\,d\theta$$

We sum over these elements by integrating from $\theta = \theta_1$ to $\theta = \theta_2$, where θ_1 and θ_2 are shown in Figure 27-17. This calculation gives

$$B = \int_{\theta_1}^{\theta_2}\frac{\mu_0}{4\pi}\frac{I}{R}\cos\theta\,d\theta = \frac{\mu_0}{4\pi}\frac{I}{R}\int_{\theta_1}^{\theta_2}\cos\theta\,d\theta$$

Evaluating the integral, we obtain

$$B = \frac{\mu_0}{4\pi}\frac{I}{R}(\sin\theta_2 - \sin\theta_1) \qquad \text{27-12}$$

B DUE TO A STRAIGHT-WIRE SEGMENT

This result gives the magnetic field due to any straight, current-carrying wire segment in terms of the perpendicular distance R and θ_1 and θ_2, which are the angles subtended at the field point by the ends of the wire. If the length of the wire approaches infinity in both directions, θ_2 approaches $+90°$ and θ_1 approaches $-90°$. The result for such a very long wire is obtained from Equation 27-12, by setting $\theta_1 = -90°$ and $\theta_2 = +90°$:

$$B = \frac{\mu_0}{4\pi}\frac{2I}{R} \qquad \text{27-13}$$

B DUE TO AN INFINITELY LONG, STRAIGHT WIRE

At any point in space, the magnetic field lines of a long, straight, current-carrying wire are tangent to a circle of radius R about the wire, where R is the perpendicular distance from the wire to the field point. The direction of $\vec{B}$ can be determined by applying the right-hand rule, as shown in Figure 27-18*a*. The magnetic field lines thus encircle the wire, as shown in Figure 27-18*b*.

The result expressed by Equation 27-13 was found experimentally by Biot and Savart in 1820. From their analysis, Biot and Savart were able to discover the expression given in Equation 27-3 for the magnetic field due to a current element.

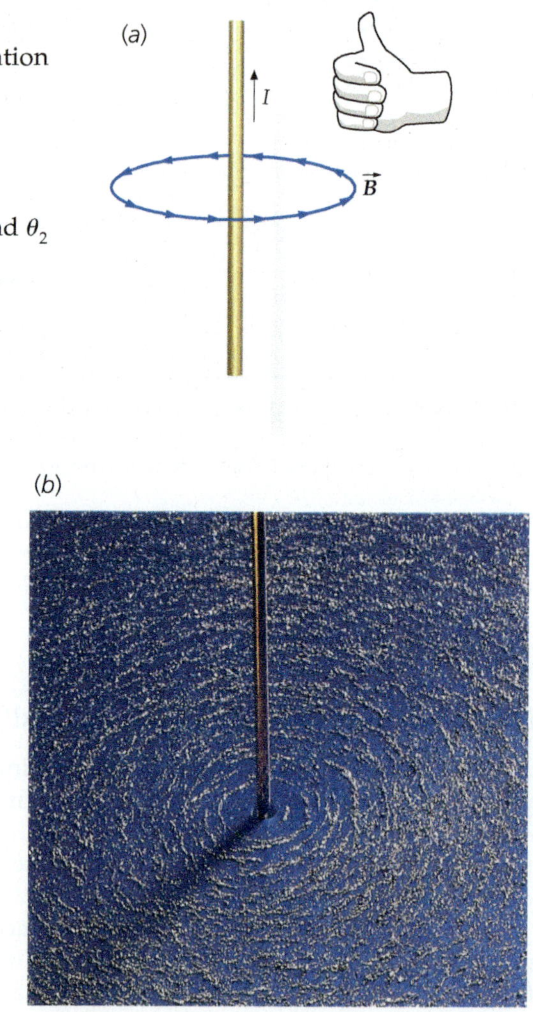

FIGURE 27-18 (*a*) Right-hand rule for determining the direction of the magnetic field due to a long, straight, current-carrying wire. The magnetic field lines encircle the wire in the direction of the fingers of the right hand when the thumb points in the direction of the current. (*b*) Magnetic field lines due to a long wire, which is indicated by iron filings. *(© 1990 Richard Menga/Fundamental Photographs.)*

Example 27-7 $\vec{B}$ at Center of Square Current Loop

Find the magnetic field at the center of a square current loop that has an edge length L, which is equal to 50 cm, and carries a current equal to 1.5 A.

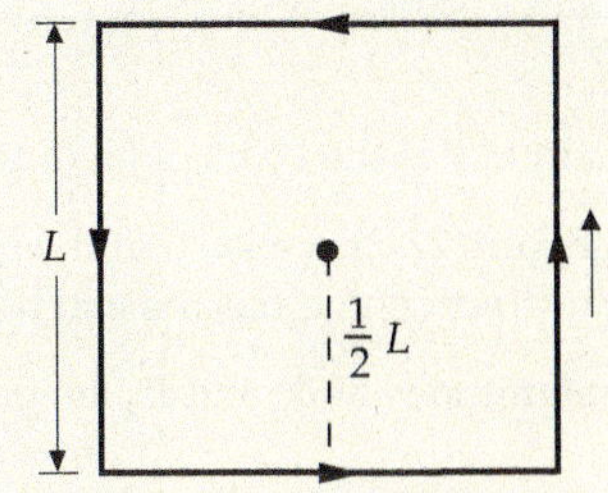

FIGURE 27-19

PICTURE The magnetic field at the center of the loop is the sum of the contributions from each of the four sides of the loop. From Figure 27-19, we can see that each side of the loop produces a field of equal magnitude pointing out of the page. Thus, we use Equation 27-12 for a given side, then multiply by 4 for the total field.

SOLVE

1. The magnitude of the resultant field is 4 times the magnitude of the field B_s due to a single side:

$$B = 4B_s$$

2. Calculate the magnitude of the magnetic field B_s due to a single side of the loop. Note from the figure that $R = \frac{1}{2}L$ and $\theta_1 = -45°$ and $\theta_2 = +45°$:

$$B_s = \frac{\mu_0}{4\pi}\frac{I}{R}(\sin\theta_2 - \sin\theta_1) = \frac{\mu_0}{4\pi}\frac{I}{\frac{1}{2}L}[\sin(+45°) - \sin(-45°)]$$

$$= (10^{-7}\,\text{T}\cdot\text{m/A})\frac{1.5\text{ A}}{0.25\text{ m}}2\sin 45° = 8.5 \times 10^{-7}\,\text{T}$$

3. Multiply this value by 4 to find the magnitude of the resultant field:

$$B = 4B_s = 4(8.5 \times 10^{-7}\,\text{T}) = \boxed{3.4 \times 10^{-6}\,\text{T}}$$

CHECK Practice Problem 27-5 serves as a check.

PRACTICE PROBLEM 27-5 Compare the magnetic field at the center of a circular current loop of radius R with the magnetic field at the center of a square current loop of side $L = 2R$ carrying the same current. Which is larger?

PRACTICE PROBLEM 27-6 Find the distance from a long, straight wire carrying a current of 12 A to a point where the magnetic field due to the current in the wire has a magnitude equal to 60 μT.

A current gun used to measure electric current. The jaws of the current gun clamp around a current-carrying wire without touching the wire. The magnetic field produced by the wire is measured with a Hall-effect device mounted in the current gun. The Hall-effect device puts out a voltage proportional to the magnetic field, which in turn is proportional to the current in the wire. *(Courtesy of F. W. Bell.)*

Example 27-8 $\vec{B}$ Due to Two Parallel Wires

A long, straight wire carries a current of 1.7 A in the $+z$ direction and lies along the line $x = -3.0$ cm, $y = 0$. A second such wire carries a current of 1.7 A in the $+z$ direction and lies along the line $x = +3.0$ cm, $y = 0$, as shown in Figure 27-20. Find the magnetic field at a point P on the y axis at $y = 6.0$ cm.

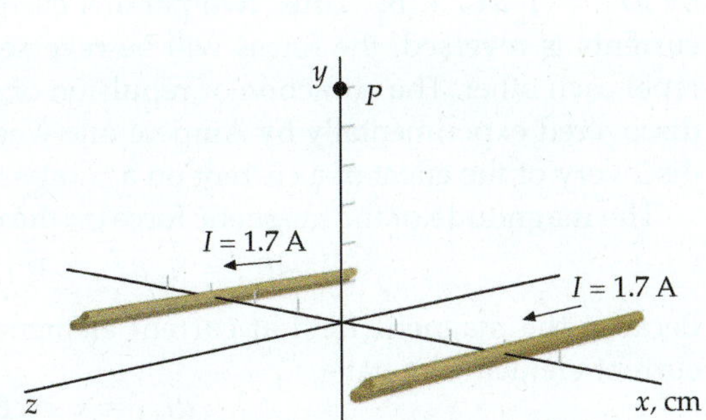

FIGURE 27-20

PICTURE The magnetic field at point P is the vector sum of the field $\vec{B}_L$ due to the wire on the left in Figure 27-21, and the field $\vec{B}_R$ due to the wire on the right. Because each wire carries the same current and each wire is the same distance from point P, the magnitudes B_L and B_R are equal. $\vec{B}_L$ is perpendicular to the radius from the left wire to point P, and $\vec{B}_R$ is perpendicular to the radius from the right wire to the point P.

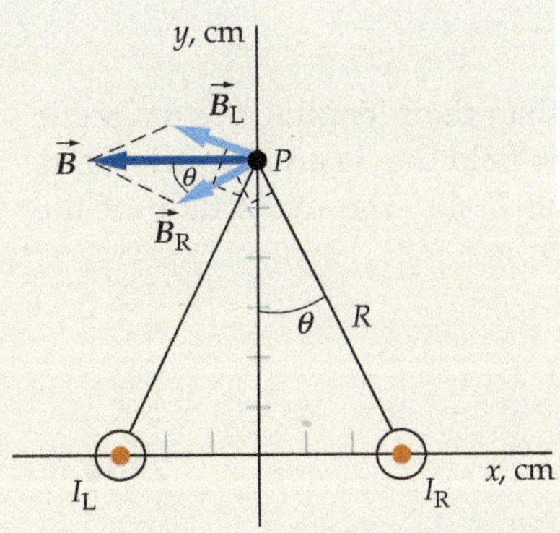

FIGURE 27-21

SOLVE

1. The field at P is the vector sum of the fields $\vec{B}_L$ and $\vec{B}_R$: $\vec{B} = \vec{B}_L + \vec{B}_R$

2. From Figure 27-21 we see that the resultant magnetic field is in the $-x$ direction and has the magnitude $2B_L \cos\theta$. $\vec{B} = -2B_L \cos\theta\, \hat{i}$

3. The magnitudes of $\vec{B}_L$ and $\vec{B}_R$ are given by Equation 27-13: $B_L = B_R = \dfrac{\mu_0}{4\pi}\dfrac{2I}{R}$

4. R is the radial distance from each wire to the point P. We find R from the figure and substitute R into the expression for B_L and B_R: $R = \sqrt{(3.0\text{ cm})^2 + (6.0\text{ cm})^2} = 6.7\text{ cm}$

so

$$B_L = B_R = (10^{-7}\,\text{T}\cdot\text{m/A})\frac{2(1.7\text{ A})}{0.067\text{ m}} = 5.07 \times 10^{-6}\,\text{T}$$

5. We obtain $\cos\theta$ from the figure: $\cos\theta = \dfrac{6.0\text{ cm}}{R} = \dfrac{6.0\text{ cm}}{6.7\text{ cm}} = 0.894$

6. Substitute the values of $\cos\theta$ and B_L into the equation in step 2 for $\vec{B}$: $\vec{B} = -2(5.07 \times 10^{-6}\,\text{T})(0.894)\hat{i} = \boxed{-9.1 \times 10^{-6}\,\text{T}\,\hat{i}}$

CHECK The magnitude of the step-6 result is less than twice the step-4 result, which is expected because the vectors being added are not parallel.

PRACTICE PROBLEM 27-7 Find $\vec{B}$ at the origin.

PRACTICE PROBLEM 27-8 Find $\vec{B}$ at the origin assuming that the direction of the current is reversed in the wire along the line $x = 3.0$ cm, $y = 0$.

MAGNETIC FORCE BETWEEN PARALLEL WIRES

We can use Equation 27-13 for the magnetic field due to a long, straight, current-carrying wire and $d\vec{F} = I\,d\vec{\ell} \times \vec{B}$ (Equation 26-5) for the force exerted by a magnetic field on a segment of a current-carrying wire to find the magnetic force exerted by one long, straight, current-carrying wire on another. Figure 27-22 shows two long parallel wires carrying currents in the same direction. We consider the force on a segment $d\vec{\ell}_2$ carrying current I_2, as shown. The magnetic field $\vec{B}_1$ at this segment due to current I_1 is perpendicular to the segment $d\vec{\ell}_2$, as shown. This is true for all current elements along wire 2. That the force, given by $d\vec{F}_{12} = I_2\,d\vec{\ell}_2 \times \vec{B}_1$, on current element $I_2\,d\vec{\ell}_2$ is directed toward wire 1 can be revealed by applying the right hand rule. Similarly, a current segment $I_1\,d\vec{\ell}_1$ will experience a magnetic force directed toward current I_2, due to a magnetic field $\vec{B}_2$ arising from current I_2, given by $d\vec{F}_{21} = I_1\,d\vec{\ell}_1 \times \vec{B}_2$. Thus, two parallel currents attract each other. If one of the currents is reversed, the forces will be reversed, so two antiparallel currents will repel each other. The attraction or repulsion of parallel or antiparallel currents was discovered experimentally by Ampère one week after he heard of Oersted's discovery of the effect of a current on a compass needle.

The magnitude of the magnetic force on the current element $I_2\,d\vec{\ell}_2$ is

$$dF_{12} = |I_2\,d\vec{\ell}_2 \times \vec{B}_1|$$

Because the magnetic field at current element $I_2\,d\vec{\ell}_2$ is perpendicular to the current element, we have

$$dF_{12} = I_2\,d\ell_2 B_1$$

If the distance R between the wires is much less than their lengths, the magnetic field at $I_2\,d\vec{\ell}_2$ due to current I_1 will approximate the field due to an infinitely long, current-carrying wire, which is given by Equation 27-13. The magnitude of the force on the segment $I_2\,d\vec{\ell}_2$ is therefore

$$dF_{12} = I_2\,d\ell_2\frac{\mu_0 I_1}{2\pi R}$$

and the force per unit length is

$$\frac{dF_{12}}{d\ell_2} = \frac{\mu_0}{2\pi}\frac{I_1 I_2}{R} \qquad 27\text{-}14$$

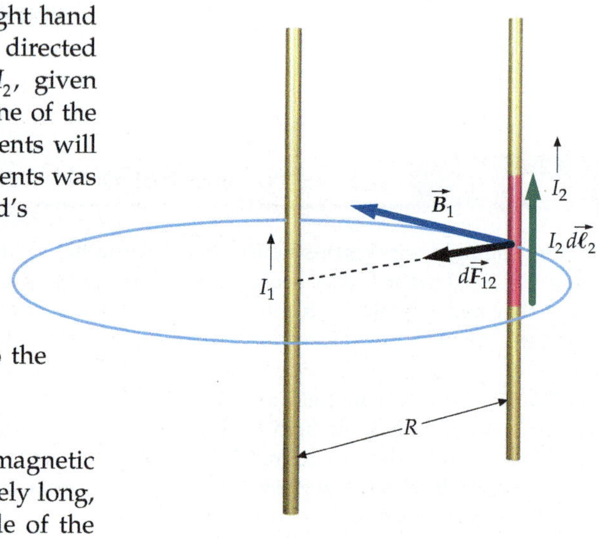

FIGURE 27-22 Two long, straight, wires carrying parallel currents. The magnetic field $\vec{B}_1$ due to current I_1 is perpendicular to current I_2. The force on current I_2 is toward current I_1. There is an equal and opposite force exerted by current I_2 on I_1. The current-carrying wires thus attract each other.

In Chapter 21, the coulomb was defined in terms of the ampere, but the definition of the ampere was deferred. The ampere is defined as follows:

> The ampere is that constant current which, if maintained in two straight, parallel conductors of infinite length and of negligible circular cross sections placed one meter apart in a vacuum, would produce a force between the conductors equal to 2×10^{-7} newton per meter of length.
>
> DEFINITION—AMPERE

That definition of the ampere allows the unit of current (and therefore the unit of electric charge) to be determined by a mechanical measurement. In practice, currents much closer together than 1 m are used so that the force can be measured more accurately.

Figure 27-23 shows a **current balance,** which is a device that can be used to calibrate an ammeter from the definition of the ampere. The upper conductor, directly above the lower conductor, is pivoted on knife-edge contacts and is balanced so that the wires (or conducting rods) are a small distance apart. The conductors are connected in series to carry the same current but in opposite directions so that the currents will repel each other. Weights are placed on the upper conductor until it balances again at the original separation. The force of repulsion is thus determined by measuring the total weight required to balance the upper conductor.

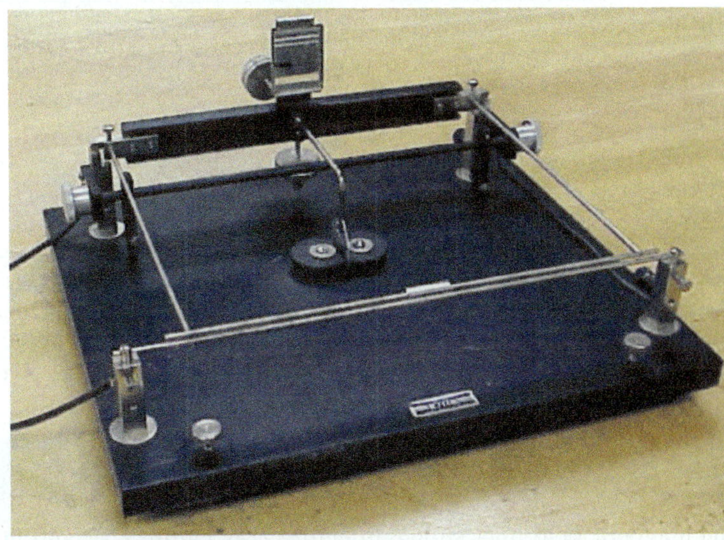

(a)

FIGURE 27-23 (*a*) A picture of a current balance used in a general physics lab. (*b*) A schematic diagram of a current balance. The two parallel rods in front carry equal but oppositely directed currents and therefore repel each other. The force of repulsion is balanced by weights placed on the upper rod, which is part of a rectangle that is balanced on knife edges at the back. The mirror on top is used to reflect a beam of laser light to accurately determine the position of the upper rod. *(Photo by Gene Mosca.)*

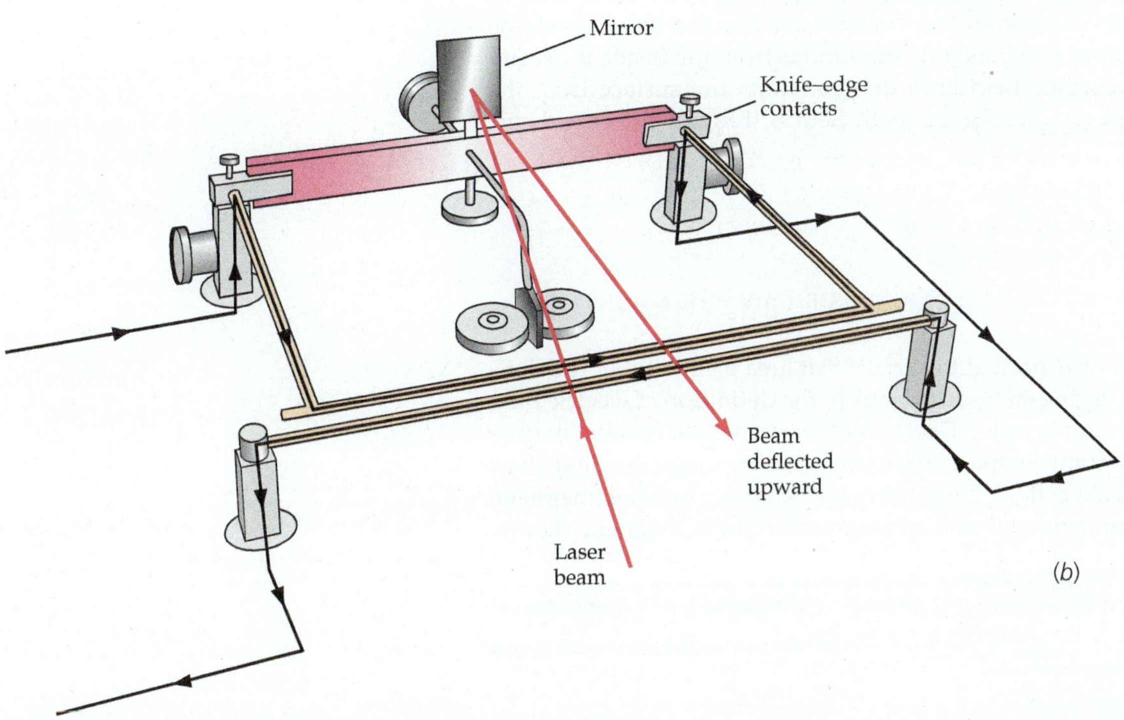

(b)

Example 27-9 Balancing the Magnetic Force

Try It Yourself

Two straight 50.0-cm-long rods have central axes that are 1.50 mm apart in a current balance, and carry currents of 15.0 A each in opposite directions. What mass must be placed on the upper rod to balance the magnetic force of repulsion?

PICTURE Equation 27-14 gives the magnitude of the magnetic force per unit length exerted by the lower rod on the upper rod. Find this force for a rod of length L and set it equal to the weight mg.

SOLVE

Cover the column to the right and try these on your own before looking at the answers.

Steps	Answers
1. Set the weight mg equal to the magnetic force of repulsion of the rods.	$mg = \frac{\mu_0}{2\pi}\frac{I_1 I_2}{R}L$
2. Solve for the mass m.	$m = 1.53 \times 10^{-3}\text{ kg} = \boxed{1.53\text{ g}}$

TAKING IT FURTHER Because only 1.53 g is required to balance the system, we see that the magnetic force between two straight current-carrying wires is relatively small, even for currents as large as 15.0 A separated by only 1.50 mm.

27-3 GAUSS'S LAW FOR MAGNETISM

The magnetic field lines shown in Figure 27-6, Figure 27-9, and Figure 27-10 differ from electric field lines because the lines of $\vec{B}$ form closed curves, whereas lines of $\vec{E}$ begin and end on electric charges. The magnetic equivalent of an electric charge is a magnetic pole, such as appears to be at the ends of a bar magnet. Magnetic field lines appear to diverge from the north-pole end of a bar magnet (Figure 27-10*b*) and appear to converge on the south-pole end. Inside the magnet, however, the magnetic field lines neither diverge from a point near the north-pole end, nor do they converge on a point near the south-pole end. Instead, the magnetic field lines pass through the bar magnet from the south-pole end to the north-pole end, as shown in Figure 27-10*b*. If a Gaussian surface encloses one end of a bar magnet, the number of magnetic field lines that penetrate the surface from the inside is exactly equal to the number of magnetic field lines that penetrate the surface from the outside. That is, the net flux $\phi_{\text{m net}}$ of the magnetic field $\vec{B}$ through any closed surface S is always zero.*

$$\phi_{\text{m net}} = \oint_S \vec{B}\cdot\hat{n}\,dA = \oint_S B_{\text{n}}\,dA = 0 \qquad 27\text{-}15$$

GAUSS'S LAW FOR MAGNETISM

where B_{n} is the component of $\vec{B}$ normal to surface S at area element dA. The definition of the magnetic flux ϕ_{m} is exactly analogous to the definition of electric flux, with $\vec{B}$ replacing $\vec{E}$. This result is called *Gauss's law for magnetism*. It is the mathematical statement that no point in space exists from which magnetic field lines diverge, or to which magnetic field lines converge. That is, isolated magnetic poles do not exist.† The fundamental unit of magnetism is the magnetic dipole.

* Recall that the net flux of the electric field is a measure of the net number of field lines that leave a closed surface and is equal to $Q_{\text{inside}}/\epsilon_0$.

† The existence of magnetic monopoles is a subject of great debate, and the search for magnetic monopoles remains active. To date, however, none have been discovered.

Figure 27-24 compares the field lines of $\vec{B}$ for a magnetic dipole with the field lines of $\vec{E}$ for an electric dipole. Note that far from the dipoles the field lines are identical. But inside the dipole, the field lines of $\vec{E}$ are directed opposite to the field lines of $\vec{B}$. The field lines of $\vec{E}$ diverge from the positive charge and converge to the negative charge, whereas the field lines of $\vec{B}$ are continuous loops.

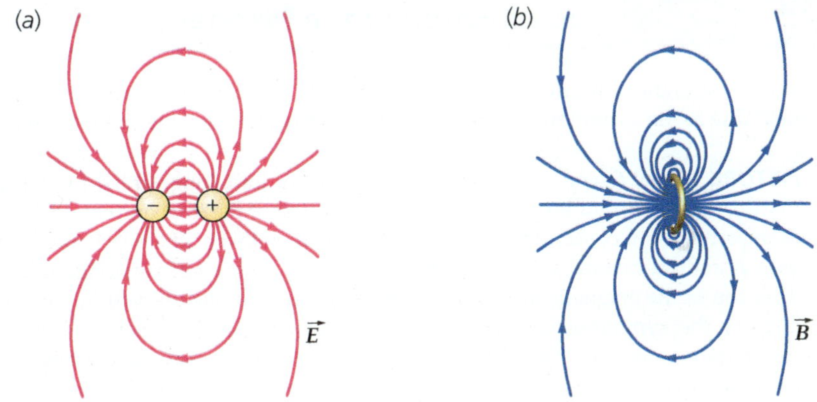

FIGURE 27-24 (*a*) Electric field lines of an electric dipole. (*b*) Magnetic field lines of a magnetic dipole. Far from the dipoles, the field lines are identical. In the region between the charges in (*a*), the electric field lines are approximately opposite the direction of the dipole moment, whereas inside the loop in (*b*), the magnetic field lines are approximately parallel to the direction of the dipole moment.

27-4 AMPÈRE'S LAW

In Chapter 22, we found that for highly symmetric charge distributions we could calculate the electric field more easily using Gauss's law than Coulomb's law. A similar situation exists in magnetism. Ampère's law relates the tangential component B_t of the magnetic field summed (integrated) around a closed curve C to the current I_C that passes through any surface bounded by C. This law can be used to obtain an expression for the magnetic field in situations that have a high degree of symmetry. In mathematical form, **Ampère's law** is

$$\oint_C B_t\, d\ell = \oint_C \vec{B}\cdot d\vec{\ell} = \mu_0 I_C \quad \text{C is any closed curve} \qquad 27\text{-}16$$

AMPÈRE'S LAW

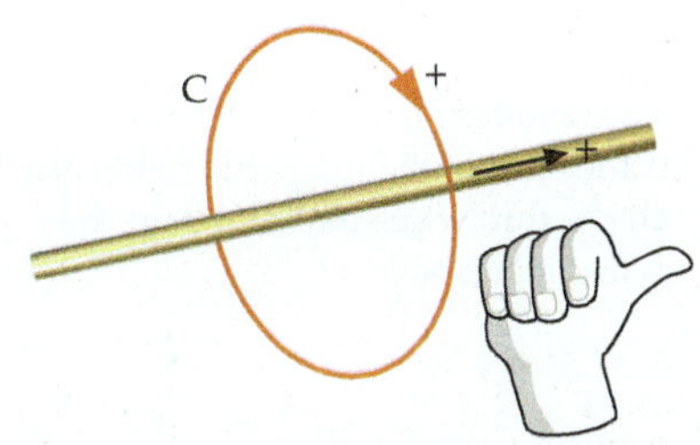

FIGURE 27-25 The positive direction for the path integral for Ampère's law is related to the positive direction for the current passing through the surface by a right-hand rule.

where I_C is the net current that penetrates any surface S bounded by the curve C. The positive tangential direction for the path integral along C is related to the choice for the positive direction for the current I_C through S by the right-hand rule shown in Figure 27-25. Ampère's law holds *as long as the currents are steady and continuous*. This means the current does not change in time and that charge is not accumulating anywhere. Ampère's law is useful in calculating the magnetic field $\vec{B}$ in situations that have a high degree of symmetry so that the line integral $\oint_C \vec{B}\cdot d\vec{\ell}$ can be written as $B\oint_C d\ell$ (the product of B and some distance). The integral $\oint_C \vec{B}\cdot d\vec{\ell}$ is called a **circulation integral.** More specifically, $\oint_C \vec{B}\cdot d\vec{\ell}$ is called the circulation of $\vec{B}$ around curve C. Ampère's law and Gauss's law are both of considerable theoretical importance, and both laws hold whether there is symmetry or there is no symmetry. If there is no symmetry, neither law is very useful in calculating electric or magnetic fields.

> ! Ampère's law holds as long as the currents are steady and continuous.

The simplest application of Ampère's law is to find the magnetic field due to the current in an infinitely long, straight wire. Figure 27-26 shows a circular curve C around a long wire with its center at the wire. We know the direction of the magnetic field due to each current element is tangent to this circle from the Biot–Savart law. Assuming that the magnetic field $\vec{B}$ is tangent to this circle, is in the same direction as $d\vec{\ell}$, and has the same magnitude B at any point on the circle, Ampère's law ($\oint_C B_t\, d\ell = \mu_0 I_C$) then gives

$$B\oint_C d\ell = \mu_0 I_C$$

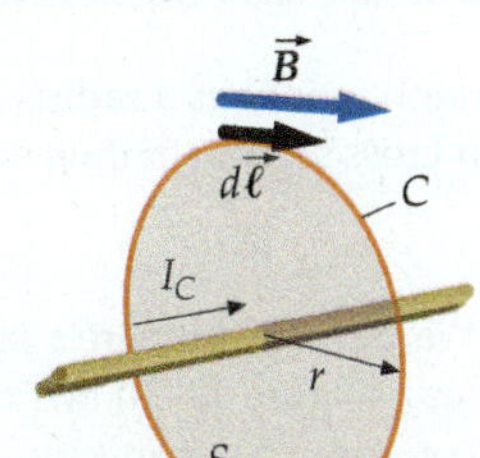

FIGURE 27-26 Geometry for calculating the magnetic field of a long, straight, current-carrying wire using Ampère's law. The magnetic field is tangent to the circle and the magnitude of the magnetic field is the same everywhere in the circle.

where $B = B_t$. We can factor B out of the integral because B has the same value everywhere on the circle. The integral of $d\ell$ around the circle equals $2\pi R$ (the circumference of the circle). The current I_C is the current I in the wire. We thus obtain $B2\pi R = \mu_0 I$, or

$$B = \frac{\mu_0 I}{2\pi R}$$

which is Equation 27-13.

Example 27-10 The Direction of the Magnetic Field — Conceptual

A long, straight cylindrical shell carries a current. Show that the direction of the magnetic field due to the current in the shell is tangent to a circle coaxial with the shell (Figure 27-27).

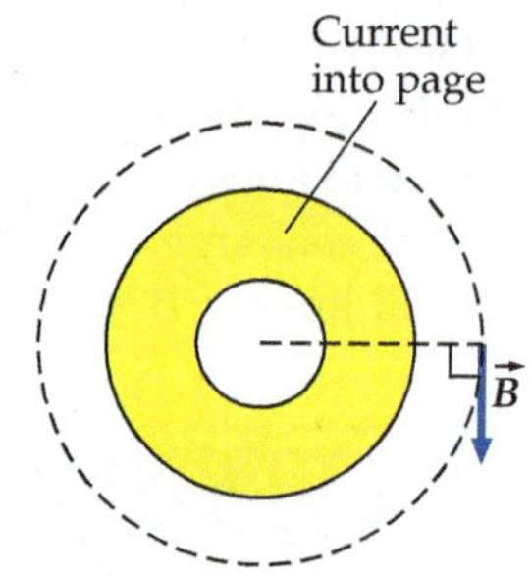

FIGURE 27-27

PICTURE Model the cylindrical shell as a bundle of thin, long, straight wires, each parallel to the central axis and each carrying a small fraction of the total current. Pick a field point P at an arbitrary location. Divide the shell in half with an imaginary plane that contains both P and the central axis of the shell. Using the right-hand rule (Figure 27-25), find the direction of the magnetic field at P due to the current in one of the thin wires in the model. Identify the symmetric counterpart to the thin wire in the other half of the shell. The counterpart is the wire equidistant from the plane and opposite the initially identified wire. Find the direction of the magnetic field at P due to the current in the counterpart thin wire. The direction of the magnetic field at P is midway between the directions of the magnetic fields due to currents in the thin wire and its counterpart.

SOLVE

1. Choose a field point P. Use the right-hand rule (Figure 27-25) to find the directions of the magnetic fields at P due to the current in a thin wire and the current in its symmetric counterpart. Draw a sketch of these two wires and their magnetic fields at the field point (Figure 27-28). Also show the sum of the two magnetic fields:

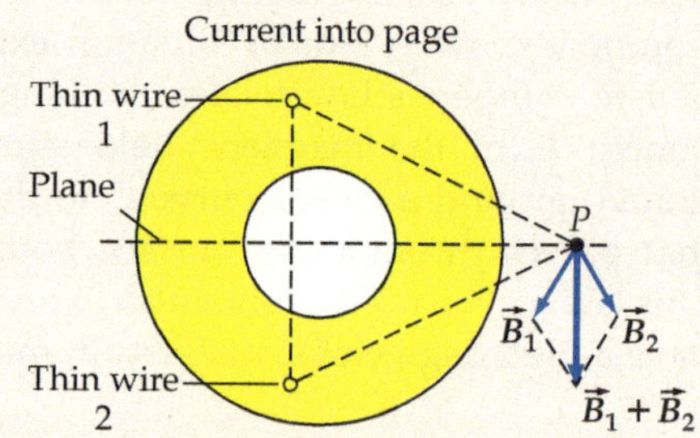

FIGURE 27-28

2. The resultant magnetic field at P is the sum of the magnetic fields due to all the thin wires that make up the cylindrical shell:

   The resultant magnetic field at P is in the same direction as the sum $\vec{B}_1 + \vec{B}_2$. This is because the sum of the magnetic fields due to the currents in any thin wire and its symmetric counterpart will point in the same direction as $\vec{B}_1 + \vec{B}_2$.

3. If the field point P is inside the shell, the magnetic field at P due to the currents in the thin wires to the right of P (Figure 27-29) will point in the opposite direction of those to the left of P (Figure 27-28):

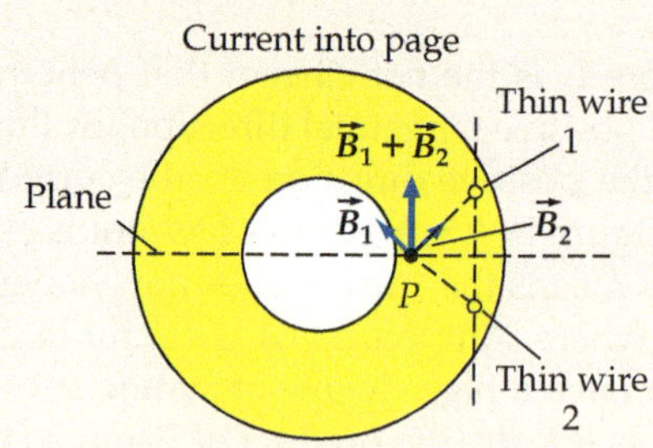

FIGURE 27-29

Example 27-11 $\vec{B}$ Inside and Outside a Wire

A long, straight wire has a radius R and carries a current I that is uniformly distributed over the circular cross section of the wire. Find the magnetic field both outside the wire and inside the wire.

PICTURE We can use Ampère's law to calculate $\vec{B}$ because of the high degree of symmetry. At a distance r (Figure 27-30), we know that $\vec{B}$ is tangent to the circle of radius r about the wire and $\vec{B}$ is constant in magnitude everywhere on the circle. The expression for the current through the surface S that is bounded by C depends on whether r is less than or greater than the radius of the wire R.

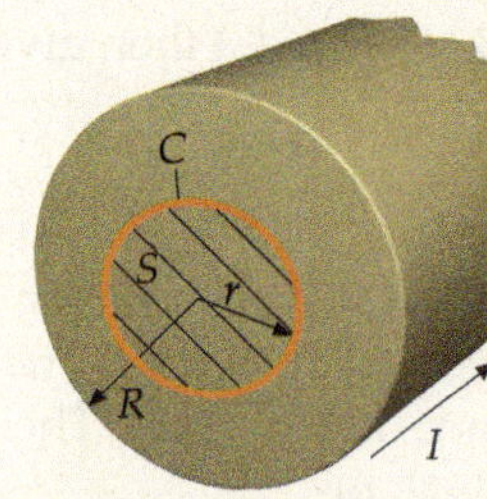

FIGURE 27-30

SOLVE

1. Ampère's law is used to relate the circulation of $\vec{B}$ around curve C to the current passing through the surface S bounded by C:

$$\oint_C \vec{B}\cdot d\vec{\ell} = \mu_0 I_C$$

2. Evaluate the circulation of $\vec{B}$ around a circle of radius r that is coaxial with the wire:

$$\oint_C \vec{B}\cdot d\vec{\ell} = B\oint_C d\ell = B2\pi r$$

3. Substitute into Ampère's law and solve for B:

$$B2\pi r = \mu_0 I_C$$

so

$$B = \frac{\mu_0 I_C}{2\pi r}$$

4. Outside the wire, $r > R$, and the total current passes through the surface bounded by C:

$$I_C = I$$

or

$$B = \boxed{\frac{\mu_0}{2\pi}\frac{I}{r} \qquad r \geq R}$$

5. Inside the wire, $r < R$. Assume that the current is distributed uniformly to solve for I_C. Solve for B:

$$\frac{I_C}{\pi r^2} = \frac{I}{\pi R^2}$$

or

$$\left(I_C = \frac{r^2}{R^2} I\right)$$

so

$$B = \frac{\mu_0}{2\pi}\frac{I_C}{r} = \frac{\mu_0}{2\pi}\frac{(r^2/R^2)I}{r} = \boxed{\frac{\mu_0 I}{2\pi R^2} r \leq R}$$

CHECK The step 4 and step 5 results give the same expression for B for $r = R$, as expected.

TAKING IT FURTHER Inside the wire, the field increases with distance from the center of the wire. Figure 27-31 shows the graph of B versus r for this example.

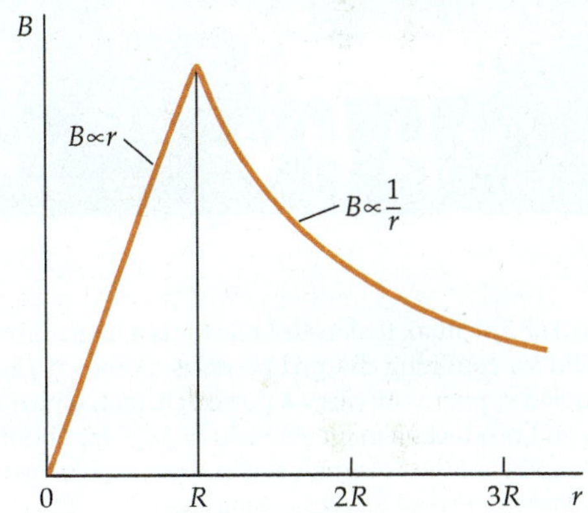

FIGURE 27-31

We see from Example 27-11 that the magnetic field due to a current uniformly distributed over a wire of radius R is given by

$$B = \begin{cases} \dfrac{\mu_0 I}{2\pi R^2} r & r \leq R \\ \dfrac{\mu_0}{2\pi}\dfrac{I}{r} & r \geq R \end{cases} \qquad \text{27-17}$$

B FOR A LONG STRAIGHT WIRE

For the next application of Ampère's law, we calculate the magnetic field of a tightly wound **toroid,** which consists of loops of wire wound around a doughnut-shaped form, as shown in Figure 27-32. There are N turns of wire, each carrying a current I. To calculate B, we evaluate the line integral $\oint_C \vec{B} \cdot d\vec{\ell}$ around a circle of radius r that is coaxial with the toroid and is inside its loops. By symmetry, $\vec{B}$ is tangent to this circle and constant in magnitude at every point on the circle. Then,

$$\oint_C \vec{B} \cdot d\vec{\ell} = B2\pi r = \mu_0 I_C$$

Let a and b be the inner and outer radii of the toroid, respectively. The total current through the surface S bounded by a circle of radius r for $a < r < b$ is $I_C = NI$. Ampère's law then gives

$$\oint_C \vec{B} \cdot d\vec{\ell} = \mu_0 I_C \qquad \text{or} \qquad (B2\pi r = \mu_0 NI)$$

or

$$B = \frac{\mu_0 NI}{2\pi r} \qquad a < r < b \qquad \text{27-18}$$

B INSIDE A TIGHTLY WOUND TOROID

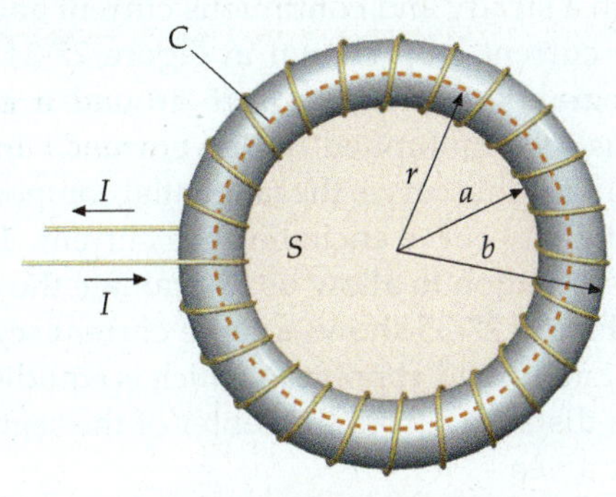

FIGURE 27-32 A toroid consists of loops of wire wound around an imaginary doughnut-shaped form. The magnetic field at any distance r can be found by applying Ampère's law to the circle of radius r. The surface S is bounded by curve C. The wire penetrates S once for each turn.

If r is less than a, there is no current through the surface S. If r is greater than b, the total current through S is zero because for each turn of the wire the current penetrates the surface S twice (Figure 27-33), once going into the page and once coming out of the page. Thus, the magnetic field is zero for both the region $r < a$ and the region $r > b$:

$$B = 0, \qquad r < a \text{ or } r > b$$

The magnetic field intensity inside the toroid is not uniform but decreases with increasing r. However, if $b - a$ (the diameter of the loops of the coil) is much less than $2b$, the variation in r from $r = a$ to $r = b$ is small, so the variation of B inside the loops is also small.

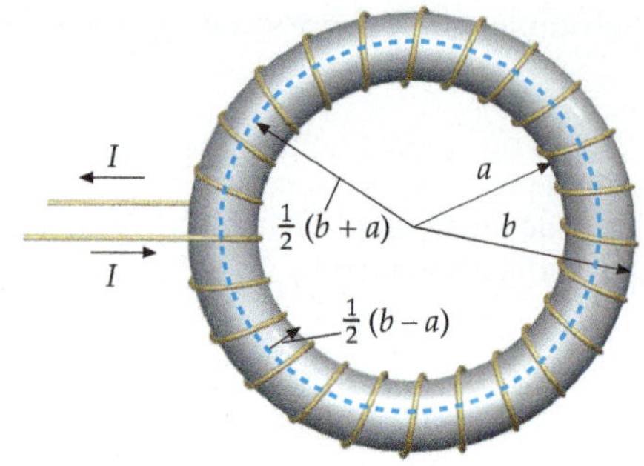

FIGURE 27-33 The toroid has mean radius $r = \frac{1}{2}(b + a)$, where a and b are the inner and outer radii of the toroid. Each turn of the wire is a circle of radius $\frac{1}{2}(b - a)$.

(a)

(b)

(*a*) The Tokamak fusion-test reactor is a large toroid that produces a magnetic field for confining charged particles. Coils containing over 10 km of water-cooled copper wire carry a pulsed current, which has a peak value of 73,000 A and produces a magnetic field of 5.2 T for about 3 s. (*b*) Inspection of the assembly of the Tokamak reactor from inside the toroid. *(Courtesy of Princeton University Plasma Physics Laboratory.)*

LIMITATIONS OF AMPÈRE'S LAW

Ampère's law is useful for calculating the magnetic field only when there is both a steady and continuous current and a high degree of symmetry. Consider the current loop shown in Figure 27-34. According to Ampère's law, the line integral $\oint_C \vec{B} \cdot d\vec{\ell} = \oint_C B_t\, d\ell$ around a curve, such as curve C in the figure, equals μ_0 multiplied by the current I in the loop. Although Ampère's law is valid for this curve, the tangential component of magnetic field B_t is not constant along any curve encircling the current. Thus, there is not enough symmetry in this situation to allow us to evaluate the integral $\oint_C B_t\, d\ell$ and solve for B_t.

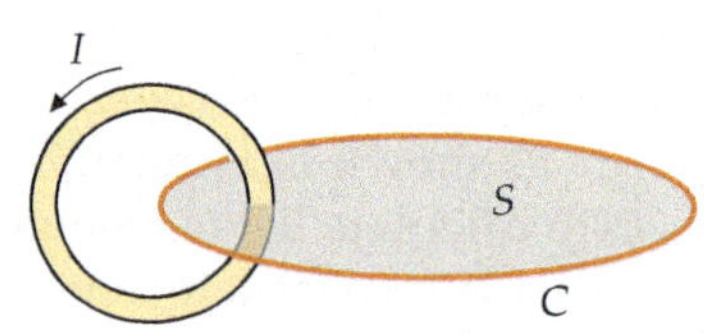

FIGURE 27-34 Ampère's law holds for the curve C encircling the current in the circular loop, but it is not useful for finding B_t because B_t cannot be factored out of the circulation integral.

Figure 27-35 shows a finite current segment of length ℓ. We wish to find the magnetic field at point P, which is equidistant from the ends of the segment and at a distance r from the center of the segment. A direct application of Ampère's law gives

$$B = \frac{\mu_0}{2\pi}\frac{I}{r}$$

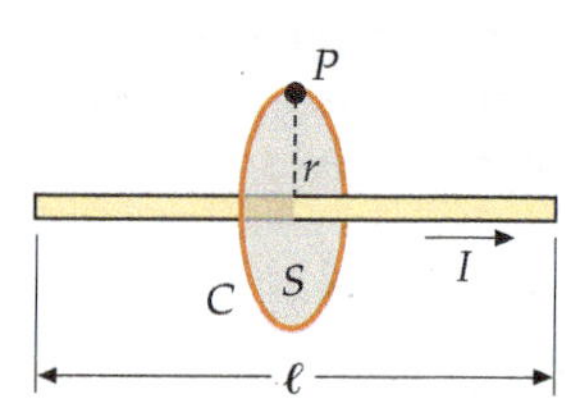

FIGURE 27-35 The application of Ampère's law to find the magnetic field on the bisector of a finite current segment gives an incorrect result.

This result is the same as for an infinitely long wire, because the same symmetry arguments apply. It does not agree with the result obtained from the Biot–Savart law, which depends on the length of the current segment and which agrees with experiment. If the current segment is just one part of a continuous circuit carrying a current, as shown in Figure 27-36, Ampère's law for curve C is valid, but it cannot be used to find the magnetic field at point P because there is insufficient symmetry.

In Figure 27-37, the current in the segment arises from a small spherical conductor that has an initial charge $+Q$ at the left of the segment and another small spherical conductor at the right that has a charge $-Q$. When they are connected, a current $I = -dQ/dt$ exists in the segment for a short time, until the spheres are uncharged. For this case, we *do* have the symmetry needed to assume that $\vec{B}$ is tangential to the curve and $\vec{B}$ is constant in magnitude along the curve. For a situation like this, in which the current is discontinuous in space, Ampère's law is not valid. In Chapter 30, we will see how Maxwell was able to modify Ampère's law so that it holds for all currents. When Maxwell's generalized form of Ampère's law is used to calculate the magnetic field for a current segment, such as the current segment shown in Figure 27-37, the result agrees with the result found from the Biot–Savart law.

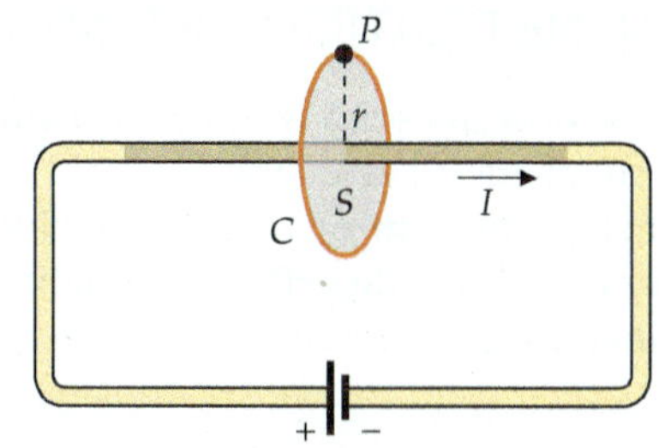

FIGURE 27-36 If the current segment in Figure 27-34 is part of a complete circuit, Ampère's law for the curve C is valid, but there is not enough symmetry to use Ampère's law to find the magnetic field at point P.

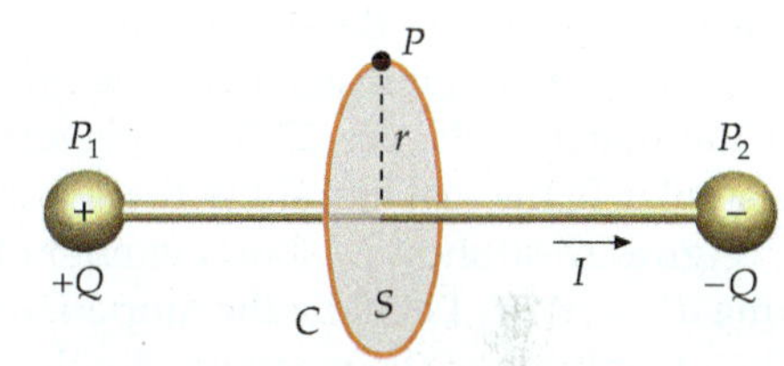

FIGURE 27-37 If the current segment in Figure 27-35 is due to a momentary flow of charge from a small conductor on the left to a small conductor on the right, there is enough symmetry to use Ampère's law to compute the magnetic field at P, but Ampère's law is not valid because the current is not steady.

27-5 MAGNETISM IN MATTER

Atoms have magnetic dipole moments due to the motion of their electrons and due to the intrinsic magnetic dipole moment associated with the spin of the electrons. Unlike the situation with electric dipoles, the alignment of magnetic dipoles parallel to an external magnetic field tends to *increase* the field. We can see this difference by comparing the electric field lines of an electric dipole with the magnetic field lines of a magnetic dipole, such as a small current loop, as was shown in Figure 27-24. Far from the dipoles, the field lines are identical. However, between the charges of the electric dipole, the electric field lines are opposite the direction of the dipole moment, whereas inside the current loop, the magnetic field lines are parallel to the magnetic dipole moment. Thus, inside a magnetically polarized material, the magnetic dipoles create a magnetic field that is parallel to the magnetic dipole moment vectors.

Materials fall into three categories—**paramagnetic, ferromagnetic,** and **diamagnetic**—according to the behavior of their magnetic moments in an external magnetic field. Paramagnetism arises from the partial alignment of the electron spins (in metals) or from the atomic or molecular magnetic moments by an applied magnetic field in the direction of the field. In paramagnetic materials, the magnetic dipoles do not interact strongly with each other and are normally randomly oriented. In the presence of an applied magnetic field, the dipoles are partially aligned in the direction of the field, thereby increasing the field. However, in external magnetic fields of ordinary strength at ordinary temperatures, only a very small fraction of the atoms are aligned because thermal motion tends to randomize their orientation. The increase in the total magnetic field is therefore very small. Ferromagnetism is much more complicated. Because of a strong interaction between neighboring magnetic dipoles, a high degree of alignment occurs even in weak external magnetic fields, which causes a very large increase in the total field. Even when there is no external magnetic field, a ferromagnetic material may have its magnetic dipoles aligned, as in permanent magnets. Diamagnetism arises from the orbital magnetic dipole moments induced by an applied magnetic field. These magnetic moments are opposite the direction of the applied magnetic field, thereby decreasing the field. This effect actually occurs in all materials; however, because the induced magnetic moments are very small compared to the permanent magnetic moments, diamagnetism is often masked by paramagnetic or ferromagnetic effects. Diamagnetism is thus observed only in materials whose atoms have no permanent magnetic moments.

MAGNETIZATION AND MAGNETIC SUSCEPTIBILITY

When some material is placed in a strong magnetic field, such as that of a solenoid, the magnetic field of the solenoid tends to align the magnetic dipole moments (either permanent or induced) inside the material and the material is said to be magnetized. We describe a magnetized material by its **magnetization** $\vec{M}$, which is defined as the net magnetic dipole moment per unit volume of the material:

$$\vec{M} = \frac{d\vec{\mu}}{dV} \qquad 27\text{-}19$$

Long before we had any understanding of atomic or molecular structure, Ampère proposed a model of magnetism in which the magnetization of materials is due to microscopic current loops inside the magnetized material. We now know that these current loops are a classical model for the orbital motion and spin of the electrons in atoms. Consider a cylinder of magnetized material. Figure 27-38 shows atomic current loops in the cylinder aligned with their magnetic moments along the axis of the cylinder. Because of cancellation of neighboring current loops, the net current at any point inside the material is zero, leaving a net current on the surface of the material (Figure 27-39). This surface current, called an **amperian current,** is similar to the real current in the windings of the solenoid.

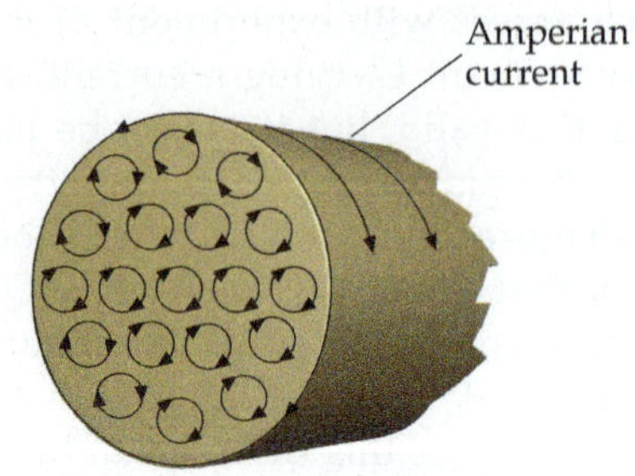

FIGURE 27-38 A model of atomic current loops in which all the atomic dipoles are parallel to the axis of the cylinder. The net current at any point inside the material is zero due to cancellation of neighboring atoms. The result is a surface current similar to that of a solenoid.

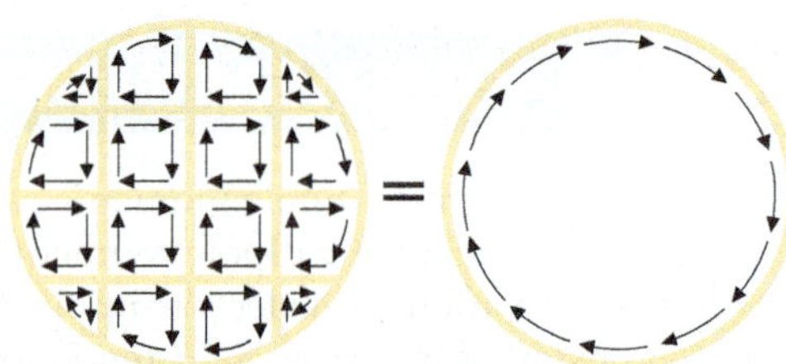

FIGURE 27-39 The currents in the adjacent current loops in the interior of a uniformly magnetized material cancel, leaving only a surface current. Cancellation occurs at every interior point independent of the shape of the very small loops.

Figure 27-40 shows a short cylinder of cross-sectional area A, length $d\ell$, and volume $dV = A\,d\ell$. Let di be the amperian current on the curved surface of the disk. The magnitude of the magnetic dipole moment of the disk is the same as that of a current loop that has an area A and carries a current di:

$$d\mu = A\,di$$

The magnitude of the magnetization of the disk is the magnetic moment per unit volume:

$$M = \frac{d\mu}{dV} = \frac{A\,di}{A\,d\ell} = \frac{di}{d\ell} \qquad 27\text{-}20$$

Thus, the magnitude of the magnetization vector is the amperian current per unit length along the surface of the magnetized material. We see from this result that the SI units for magnetization M are amperes per meter.

Consider a cylinder that has a uniform magnetization $\vec{M}$ parallel to its axis. The effect of the magnetization is the same as if the cylinder carried a surface current per unit length of magnitude M. This current is similar to the current carried by a tightly wound solenoid. For a solenoid, the current per unit length is nI, where n is the number of turns per unit length and I is the current in each turn. The magnitude of the magnetic field B_m inside the cylinder and far from its ends is thus given by $B = \mu_0 nI$ (Equation 27-10) for a solenoid with nI replaced by M:

$$B_m = \mu_0 M \qquad 27\text{-}21$$

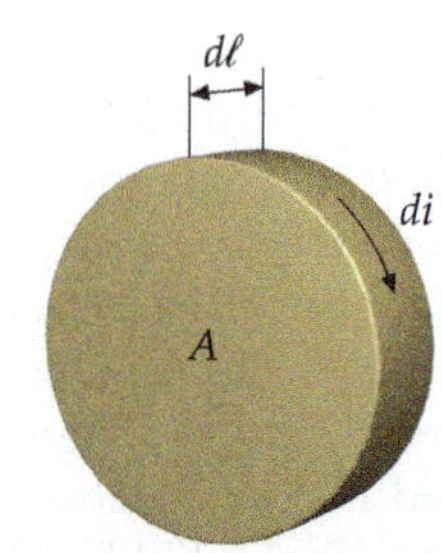

FIGURE 27-40 A disk element for relating the magnetization M to the surface current per unit length $di/d\ell$.

Suppose we place a cylinder of magnetic material inside a long solenoid that has n turns per unit length and carries a current I. The applied field of the solenoid $\vec{B}_{app}$ ($B_{app} = \mu_0 nI$) magnetizes the material so that it has a magnetization $\vec{M}$. The resultant magnetic field at a point inside the solenoid and far from its ends due to the current in the solenoid plus the magnetized material is

$$\vec{B} = \vec{B}_{app} + \mu_0\vec{M} \qquad 27\text{-}22$$

For paramagnetic and ferromagnetic materials, $\vec{M}$ is in the same direction as $\vec{B}_{app}$; for diamagnetic materials, $\vec{M}$ is opposite to $\vec{B}_{app}$. For paramagnetic and diamagnetic materials, the magnetization is found to be proportional to the applied magnetic field that produces the alignment of the magnetic dipoles in the material. We can thus write

$$\vec{M} = \chi_m \frac{\vec{B}_{app}}{\mu_0} \qquad 27\text{-}23$$

where the proportionality constant χ_m is a dimensionless number called the **magnetic susceptibility.** Equation 27-22 is then

$$\vec{B} = \vec{B}_{app} + \mu_0\vec{M} = \vec{B}_{app}(1 + \chi_m) = K_m\vec{B}_{app} \tag{27-24}$$

where

$$K_m = 1 + \chi_m \tag{27-25}$$

is called the **relative permeability** of the material. For paramagnetic materials, χ_m is a small positive number that depends on temperature. For diamagnetic materials (other than superconductors), it is a small negative constant independent of temperature. Table 27-1 lists the magnetic susceptibility of various paramagnetic and diamagnetic materials. We see that the magnetic susceptibility for the solids listed is of the order of 10^{-5}, and $K_m \approx 1$.

The magnetization of ferromagnetic materials, which we discuss shortly, is much more complicated. The relative permeability K_m defined as the ratio B/B_{app} is not constant and has maximum values ranging from 5000 to 100,000. In the case of permanent magnets, K_m is not even defined because such materials exhibit magnetization even in the absence of an applied field.

Table 27-1 Magnetic Susceptibility of Various Materials at

Material	χ_m
Aluminum	2.3×10^{-5}
Bismuth	-1.66×10^{-5}
Copper	-0.98×10^{-5}
Diamond	-2.2×10^{-5}
Gold	-3.6×10^{-5}
Magnesium	1.2×10^{-5}
Mercury	-3.2×10^{-5}
Silver	-2.6×10^{-5}
Sodium	-0.24×10^{-5}
Titanium	7.06×10^{-5}
Tungsten	6.8×10^{-5}
Hydrogen (1 atm)	-9.9×10^{-9}
Carbon dioxide (1 atm)	-2.3×10^{-9}
Nitrogen (1 atm)	-5.0×10^{-9}
Oxygen (1 atm)	2090×10^{-9}

ATOMIC MAGNETIC MOMENTS

The magnetization of a paramagnetic or ferromagnetic material can be related to the permanent magnetic moments of the individual atoms or electrons of the material. The orbital magnetic moment of an atomic electron can be derived semiclassically, even though it is quantum mechanical in origin. Consider a particle of mass m and charge q moving with speed v in a circle of radius r, as shown in Figure 27-41. The magnitude of the angular momentum of the particle about the center of the circle is

$$L = mvr \tag{27-26}$$

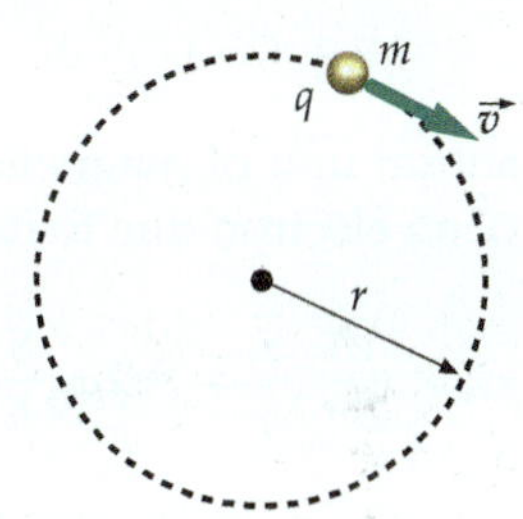

FIGURE 27-41 A particle of charge q and mass m moving with speed v in a circle of radius r. The angular momentum is directed into the paper and has a magnitude mvr. The magnetic moment is directed into the paper (if q is positive) and has a magnitude $\frac{1}{2}qvr$.

The magnitude of the magnetic moment is the product of the current and the area of the circle:

$$\mu = IA = I\pi r^2$$

If T is the time for the charge to complete one revolution, the current (the charge passing a point per unit time) is q/T. Because the period T is the distance $2\pi r$ divided by the velocity v, the current is

$$I = \frac{q}{T} = \frac{qv}{2\pi r}$$

The magnetic moment is then

$$\mu = IA = \frac{qv}{2\pi r}\pi r^2 = \tfrac{1}{2}qvr \tag{27-27}$$

Using $vr = L/m$ from Equation 27-26, we have for the magnetic moment

$$\mu = \frac{q}{2m}L$$

If the charge q is positive, the angular momentum and magnetic moment are in the same direction. We can therefore write

$$\vec{\mu} = \frac{q}{2m}\vec{L} \tag{27-28}$$

CLASSICAL RELATION BETWEEN MAGNETIC MOMENT AND ANGULAR MOMENTUM

Equation 27-28 is the general classical relation between magnetic moment and angular momentum. It also holds in the quantum theory of the atom for orbital angular momentum, but the equation does not hold for the intrinsic spin angular momentum of the electron. For electron spin, the magnetic moment is twice that predicted by the equation.* The extra factor of 2, which is accounted for by quantum theory, has no analog in classical mechanics.

Because angular momentum is quantized, the magnetic moment of an atom is also quantized. The quantum of angular momentum is $\hbar = h/(2\pi)$, where h is Planck's constant, so we express the magnetic moment in terms of $\vec{L}/\hbar$:

$$\vec{\mu} = \frac{q\hbar}{2m}\frac{\vec{L}}{\hbar}$$

For an electron, $m = m_e$ and $q = -e$, so the magnetic moment of the electron due to its orbital motion is

$$\vec{\mu}_\ell = -\frac{e\hbar}{2m_e}\frac{\vec{L}}{\hbar} = -\mu_B\frac{\vec{L}}{\hbar} \qquad \text{27-29}$$

MAGNETIC MOMENT DUE TO THE ORBITAL MOTION OF AN ELECTRON

where

$$\mu_B = \frac{e\hbar}{2m_e} = 9.27 \times 10^{-24}\ \text{A}\cdot\text{m}^2 = 9.27 \times 10^{-24}\ \text{J/T}$$
$$= 5.79 \times 10^{-5}\ \text{eV/T} \qquad \text{27-30}$$

BOHR MAGNETON

is the quantum unit of magnetic moment called a **Bohr magneton.** The magnetic moment of an electron due to its intrinsic spin angular momentum $\vec{S}$ is

$$\vec{\mu}_s = -2 \times \frac{e\hbar}{2m_e}\frac{\vec{S}}{\hbar} = -2\mu_B\frac{\vec{S}}{\hbar} \qquad \text{27-31}$$

MAGNETIC MOMENT DUE TO ELECTRON SPIN

Although the calculation of the magnetic moment of any atom is a complicated problem in quantum theory, the result for all electrons, according to both theory and experiment, is that the magnetic moment is of the order of a few Bohr magnetons. In addition, any atom that has an angular momentum equal to zero has a net magnetic moment equal to zero. (This topic is further discussed in Chapter 36.)

If all the atoms in a sample of material have their magnetic moments aligned, the magnetic moment per unit volume of the sample is the product of the number of atoms per unit volume n and the magnetic moment μ of each atom. For this extreme case, the **saturation magnetization** M_s is

$$M_s = n\mu \qquad \text{27-32}$$

The number of atoms per unit volume can be found from the molar mass M, the density ρ of the material, and Avogadro's number N_A:

$$n = \frac{N_A\,(\text{atoms/mol})}{M\,(\text{kg/mol})}\rho(\text{kg/m}^3) \qquad \text{27-33}$$

* This result and the phenomenon of electron spin was predicted in 1927 by Paul Dirac, who combined special relativity and quantum mechanics into a relativistic wave equation called the Dirac equation. Precise measurements indicate that the magnetic moment of the electron due to its spin is 2.00232 times that predicted by Equation 27-28. The fact that the intrinsic magnetic moment of the electron is approximately twice what we would expect makes it clear that the simple model of the electron as a spinning ball is not to be taken literally.

Example 27-12 Saturation Magnetization for Iron

Find the saturation magnetization of a sample of iron and find the magnetic field it produces deep within the sample. Assume that each iron atom has a magnetic moment of 1 Bohr magneton.

PICTURE We find the number of atoms per unit volume from the density of iron, $\rho = 7.87 \times 10^3$ kg/m^3, and its molar mass $M = 55.8 \times 10^{-3}$ kg/mol.

SOLVE

1. The saturation magnetization is the product of the number of atoms per unit volume and the magnetic moment of each molecule:

$$M_s = n\mu$$

2. Calculate the number of atoms per unit volume from Avogadro's number, the molar mass, and the density:

$$n = \frac{N_A}{M}\rho = \frac{6.02 \times 10^{23}\text{ atoms/mol}}{55.8 \times 10^{-3}\text{kg/mol}}(7.87 \times 10^3\text{ kg/m}^3)$$
$$= 8.49 \times 10^{28}\text{ atoms/m}^3$$

3. Substitute this result and $\mu = 1$ Bohr magneton to calculate the saturation magnetization:

$$M_s = n\mu$$
$$= (8.49 \times 10^{28}\text{ atoms/m}^3)(9.27 \times 10^{-24}\text{ A}\cdot\text{m}^2)$$
$$= \boxed{7.88 \times 10^5\text{ A/m}}$$

4. The magnetic field on the axis inside and far from the ends of a long iron cylinder resulting from this maximum magnetization is given by $B = \mu_0 M_s$:

$$B = \mu_0 M_s$$
$$= (4\pi \times 10^{-7}\text{T}\cdot\text{A})(7.86 \times 10^5\text{ A/m})$$
$$= \boxed{0.990\text{ T} \approx 1\text{ T}}$$

CHECK The step-4 result of $B \approx 1$ T is a large magnetic field. This result is as expected for the saturation magnetic field inside a ferromagnetic material.

TAKING IT FURTHER The measured saturation magnetic field of annealed iron is about 2.16 T, indicating that the magnetic moment of an iron atom is slightly greater than 2 Bohr magnetons. This magnetic moment is due mainly to the spins of two unpaired electrons in the iron atom.

*PARAMAGNETISM

Paramagnetism occurs in materials whose atoms have permanent magnetic moments that interact with each other only very weakly, resulting in a very small, positive magnetic susceptibility χ_m. When there is no external magnetic field, these magnetic moments are randomly oriented. In the presence of an external magnetic field, the magnetic moments tend to line up parallel to the field, but this is counteracted by the tendency for the magnetic moments to be randomly oriented due to thermal motion. The degree to which the moments line up with the field depends on the strength of the field and on the temperature. This degree of alignment usually is small because the energy of a magnetic moment in an external magnetic field is typically much smaller than the thermal energy of an atom of the material, which is of the order of kT, where k is Boltzmann's constant and T is the absolute temperature.

Liquid oxygen, which is paramagnetic, is attracted by the magnetic field of a permanent magnet. A net force is exerted on the magnetic dipoles because the magnetic field is not uniform. *(J. F. Allen, St. Andrews University, Scotland.)*

The potential energy of a magnetic dipole of moment $\vec{\mu}$ in an external magnetic field $\vec{B}$ is given by Equation 27-16:

$$U = -\mu B\cos\theta = -\vec{\mu}\cdot\vec{B}$$

The potential energy U_{min} when the moment and the field are parallel ($\theta = 0$) is thus lower than the potential energy U_{max} when the moment and the field are antiparallel ($\theta = 180°$) by $2\mu B$. For a typical atomic magnetic moment of 1 Bohr magneton and a typical strong magnetic field of 1 T, the difference in these potential energies is

$$\Delta U = 2\mu_B B = 2(5.79 \times 10^{-5}\text{ eV/T})(1\text{ T}) = 1.16 \times 10^{-4}\text{ eV}$$

At a normal temperature of $T = 300$ K, the typical thermal energy kT is

$$kT = (8.62 \times 10^{-5}\ \text{eV/K})(300\ \text{K}) = 2.59 \times 10^{-2}\ \text{eV}$$

which is more than 200 times greater than $2\mu_B B$. Thus, even in a very strong magnetic field of 1 T, most of the magnetic moments will be randomly oriented because of thermal motions (unless the temperature is very low).

Figure 27-42 shows a plot of the magnetization M versus an applied external magnetic field B_{app} at a given temperature. In very strong applied fields, nearly all the magnetic moments are aligned with the field and $M \approx M_s$. (For magnetic fields attainable in the laboratory, this can occur only for very low temperatures.) When $B_{app} = 0$, $M = 0$, indicating that the orientation of the moments is completely random. In weak fields, the magnetization is approximately proportional to the applied field, as indicated by the orange dashed line in the figure. In this region, the magnetization is given by

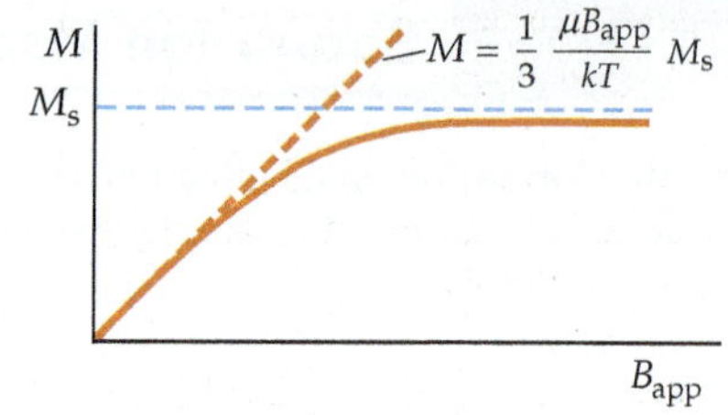

FIGURE 27-42 Plot of magnetization M versus an applied magnetic field B_{app}. In very strong fields, the magnetization approaches the saturation value M_s. This can be achieved only at very low temperatures. In weak fields, the magnetization is approximately proportional to B_{app}, a result known as Curie's law.

$$M = \frac{1}{3}\frac{\mu B_{app}}{kT}M_s \qquad 27\text{-}34$$

CURIE'S LAW

Note that $\mu B_{app}/(kT)$ is the ratio of the maximum energy of a dipole in the magnetic field to the characteristic thermal energy. The result that the magnetization varies inversely with the absolute temperature was discovered experimentally by Pierre Curie and is known as **Curie's law**.

Example 27-13 Applying Curie's Law

If $\mu = \mu_B$, at what temperature will the magnetization be 1.00 percent of the saturation magnetization in an applied magnetic field of 1.00 T?

PICTURE Using Equation 27-34, solve for the temperature when M/M_s equals 0.0100.

SOLVE

1. Curie's law relates M, T, M_s, and B_{app}:

$$M = \frac{1}{3}\frac{\mu B_{app}}{kT}M_s$$

2. Solve for T using $\mu = \mu_B$ and $M/M_s = 0.0100$:

$$T = \frac{\mu_B B_{app}}{3k}\frac{M_s}{M} = \frac{(5.79 \times 10^{-5}\ \text{eV/T})(1.00\ \text{T})}{3(8.62 \times 10^{-5}\ \text{eV/K})}100$$

$$= \boxed{22.4\ \text{K}}$$

CHECK The step-2 result is greater than absolute zero, as expected.

TAKING IT FURTHER From this example, we see that even in a strong applied magnetic field of 1.00 T, the magnetization is less than 1.00 percent of saturation at temperatures above 22.4 K.

PRACTICE PROBLEM 27-9 If $\mu = \mu_B$, what fraction of the saturation magnetization is M at 300 K for an external magnetic field of 1.5 T?

*FERROMAGNETISM

Ferromagnetism occurs in pure iron, cobalt, and nickel as well as in alloys of those metals with each other. It also occurs in gadolinium, dysprosium, and a few compounds. Ferromagnetism arises from a strong interaction between the electrons in a partially full band in a metal or between the localized electrons that form magnetic moments on neighboring atoms. This interaction, called the **exchange interaction**, lowers the energy of a pair of electrons with parallel spins.

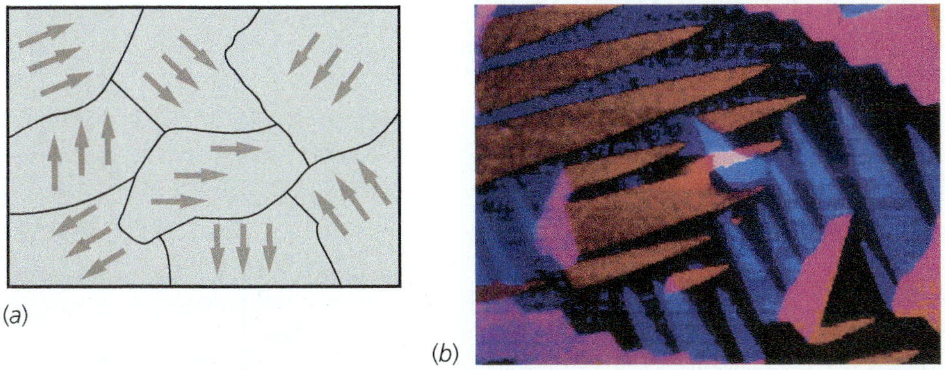

FIGURE 27-43 (*a*) Schematic illustration of ferromagnetic domains. Within a domain, the magnetic dipoles are aligned, but the direction of alignment varies from domain to domain so that the net magnetic moment is zero. A small external magnetic field may cause the enlargement of those domains that are aligned parallel to the field (at the expense of the neighboring domains), or it may cause the alignment within a domain to rotate. In either case, the result is a net magnetic moment parallel to the field. (*b*) Magnetic domains on the surface of a 97 percent Fe–3 percent Si crystal observed using a scanning electron microscope with polarization analysis. The four colors indicate four possible domain orientations. *(Robert J. Celotta, National Institute of Standards and Technology.)*

A Canadian quarter that is attracted by a magnet. Canadian coins often contain significant amounts of nickel, which is ferromagnetic. *(Photo by Gene Mosca.)*

A chunk of magnetite (lodestone) attracts the needle of a compass. *(© Paul Silverman/Fundamental Photographs.)*

Ferromagnetic materials have very large positive values of magnetic susceptibility χ_m (as measured under conditions described, which follow). In samples of these substances, a small external magnetic field can produce a very large degree of alignment of the atomic magnetic dipole moments. In some cases, the alignment can persist even when the external magnetizing field is removed. This alignment persists because the magnetic dipole moments exert strong forces on their neighbors so that over a small region of space the moments are aligned with each other even when there is no external field. The region of space over which the magnetic dipole moments are aligned is called a **magnetic domain.** The size of a domain is usually microscopic. Within the domain, all the permanent atomic magnetic moments are aligned, but the direction of alignment varies from domain to domain so that the net magnetic moment of a macroscopic piece of ferromagnetic material is zero in the normal state. Figure 27-43 illustrates this situation. The dipole forces that produce this alignment are predicted by quantum theory and cannot be explained with classical physics. At temperatures above a critical temperature, called the **Curie temperature,** thermal agitation is great enough to break up this alignment and ferromagnetic materials become paramagnetic.

When an external magnetic field is applied, the boundaries of the domains may shift or the direction of alignment within a domain may change so that there is a net macroscopic magnetic moment in the direction of the applied field. Because the degree of alignment is large for even a small external field, the magnetic field produced in the material by the dipoles is often much greater than the external field.

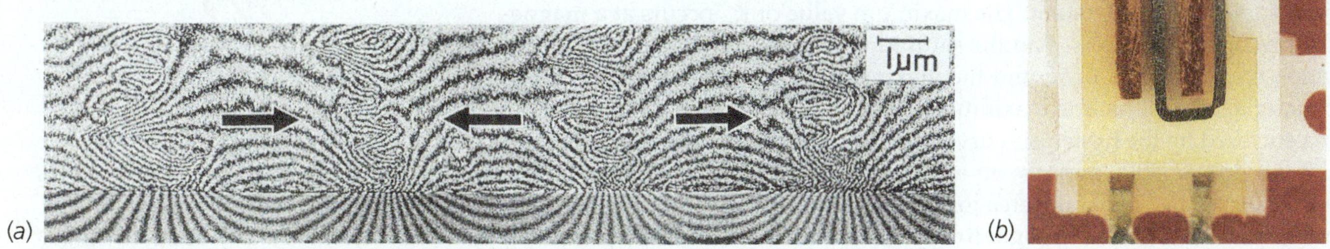

(*a*) Magnetic field lines on a cobalt magnetic recording tape. The solid arrows indicate the encoded magnetic bits. (*b*) Cross section of a magnetic tape recording head. Current from an audio amplifier is sent to wires around a magnetic core in the recording head where it produces a magnetic field. When the tape passes over a gap in the core of the recording head, the fringing magnetic field encodes information on the tape. *((a) Akira Tonomura, Hitachi Advanced Research Library, Hatomaya, Japan; (b) © Bruce Iverson.)*

Let us consider what happens when we magnetize a long iron rod by placing it inside a solenoid and gradually increase the current in the solenoid windings. We assume that the rod and the solenoid are long enough to permit us to neglect end effects. Because the induced magnetic moments are in the same direction as the applied field, $\vec{B}_{app}$ and $\vec{M}$ are in the same direction. Then,

$$B = B_{app} + \mu_0 M = \mu_0 nI + \mu_0 M \qquad 27\text{-}35$$

In ferromagnetic materials, the magnetic field $\mu_0 M$ due to the magnetic moments is often greater than the magnetizing field B_{app} by a factor of several thousand.

Figure 27-44 shows a plot of B versus the magnetizing field B_{app}. As the current is gradually increased from zero, B increases from zero along the part of the curve from the origin O to point P_1. The flattening of this curve near point P_1 indicates that the magnetization M is approaching its saturation value M_s, at which all the atomic magnetic moments are aligned. Above saturation, B increases only because the magnetizing field $B_{app} = \mu_0 nI$ increases. When B_{app} is gradually decreased from point P_1, there is not a corresponding decrease in the magnetization. The shift of the domains in a ferromagnetic material is not completely reversible, and some magnetization remains even when B_{app} is reduced to zero, as indicated in the figure. This effect is called **hysteresis,** from the Greek word *hysteros* meaning later or behind, and the curve in Figure 27-44 is called a **hysteresis curve.** The value of the magnetic field at point P_4 when B_{app} is zero is called the **remnant field** B_{rem}. At that point, the iron rod is a permanent magnet. If the current in the solenoid is now reversed so that B_{app} is in the opposite direction, the magnetic field B is gradually brought to zero at point c. The remaining part of the hysteresis curve is obtained by further increasing the current in the opposite direction until point P_2 is reached, which corresponds to saturation in the opposite direction, and then decreasing the current to zero at point P_3 and increasing it again in its original direction.

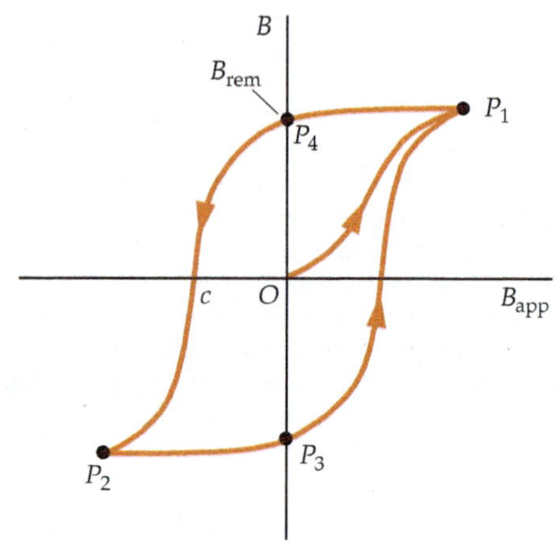

FIGURE 27-44 Plot of B versus the applied magnetizing field B_{app}. The outer curve is called a hysteresis curve. The field B_{rem} is called the remnant field. It remains when the applied field returns to zero.

Because the magnetization M depends on the previous history of the material, and because it can have a large value even when the applied field is zero, it is not simply related to the applied field B_{app}. However, if we confined ourselves to that part of the magnetization curve from the origin to point P_1 in Figure 27-44, $\vec{B}_{app}$ and $\vec{M}$ are parallel and M is zero when B_{app} is zero. We can then define the magnetic susceptibility as in Equation 27-23,

$$M = \chi_m \frac{B_{app}}{\mu_0}$$

and

$$B = B_{app} + \mu_0 M = B_{app}(1 + \chi_m) = K_m \mu_0 nI = \mu nI \qquad 27\text{-}36$$

where

$$\mu = (1 + \chi_m)\mu_0 = K_m \mu_0 \qquad 27\text{-}37$$

is called the **permeability** of the material. [For paramagnetic and diamagnetic materials, χ_m is much less than 1 so the permeability μ and the magnetic constant (permeability of empty space) μ_0 are very nearly equal.]

Because B does not vary linearly with B_{app}, as can be seen from Figure 27-44, the relative permeability is not constant. The maximum value of K_m occurs at a magnetization that is considerably less than the saturation magnetization. Table 27-2 lists the saturation magnetic field $\mu_0 M_s$ and the maximum values of K_m for some ferromagnetic materials. Note that the maximum values of K_m are much greater than 1.

The area enclosed by the hysteresis curve is proportional to the energy dissipated as heat in the irreversible process of magnetizing and demagnetizing. If the hysteresis effect is small, so that the area inside the curve is small, indicating a small energy loss, the material is called **magnetically soft.** Soft iron (chemically pure iron) is an example. The hysteresis curve for a magnetically soft material is shown in Figure 27-45. Here the remnant field B_{rem} is nearly zero, and the energy loss per cycle is small. Magnetically soft materials are used for transformer cores to allow the magnetic field B to change without incurring large energy losses as the field changes.

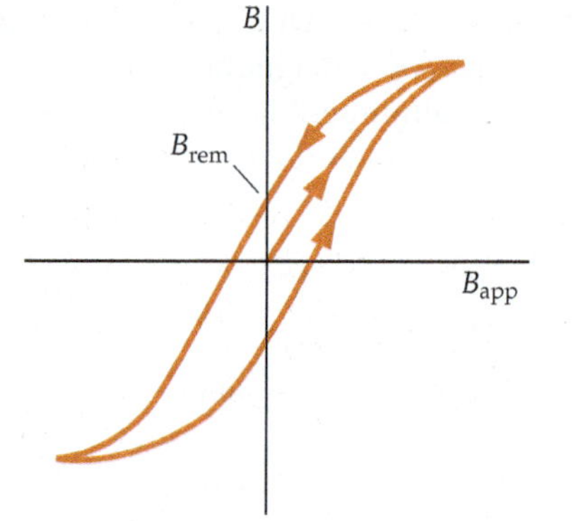

FIGURE 27-45 Hysteresis curve for a magnetically soft material. The remnant field is very small compared with the remnant field for a magnetically hard material such as that shown in Figure 27-44.

Table 27-2 Maximum Values of and for Some Ferromagnetic Materials

Material	$\mu_0 M_s$, T	K_m
Iron (annealed)	2.16	5500
Iron–silicon (96 percent Fe, 4 percent Si)	1.95	7000
Permalloy (55 percent Fe, 45 percent Ni)	1.60	25,000
Mu-metal (77 percent Ni, 16 percent Fe, 5 percent Cu, 2 percent Cr)	0.65	100,000

On the other hand, a large remnant field is desirable in a permanent magnet. **Magnetically hard** materials, such as carbon steel, the alloy Alnico 5, and the rare earths samarium and neodymium (samarium–cobalt and neodymium–iron–boron) are used for permanent magnets.

(*a*) An extremely high-capacity, hard-disk drive for magnetic storage of information, capable of storing over 250 gigabytes of information. (*b*) A magnetic test pattern on a hard disk, magnified 2400 times. The light and dark regions correspond to oppositely directed magnetic fields. The smooth region just outside the pattern is a region of the disk that has been erased just prior to writing. *((a) © 2003 Western Digital Corporation. All rights reserved. (b) Tom Chang/ IBM Storage Systems Division, San Jose, CA.)*

Example 27-14 Solenoid with Iron Core

A long solenoid has 12 turns per centimeter and a core of soft iron. When the current is 0.500 A, the magnetic field inside the iron core is 1.36 T. Find (*a*) the applied field B_{app}, (*b*) the relative permeability K_m, and (*c*) the magnetization M.

PICTURE The applied field is just that of a long solenoid given by $B_{app} = \mu_0 nI$. Because the total magnetic field is given, we can find the relative permeability from its definition ($K_m = B/B_{app}$) and we can find M from $B = B_{app} + \mu_0 M$.

SOLVE

(*a*) The applied field is given by Equation 27-10:

$$B_{app} = \mu_0 nI = (4\pi \times 10^{-7}\ \text{T}\cdot\text{m/A})(1200\ \text{m}^{-1})(0.500\ \text{A}) = \boxed{7.54 \times 10^{-4}\ \text{T}}$$

(*b*) The relative permeability is the ratio of B to B_{app}:

$$K_m = \frac{B}{B_{app}} = \frac{1.36\ \text{T}}{7.54 \times 10^{-4}\ \text{T}} = \boxed{1.80 \times 10^3}$$

(*c*) The magnetization M is found from Equation 27-35:

$$\mu_0 M = B - B_{\text{app}}$$

$$= 1.36\text{ T} - 7.54 \times 10^{-4}\text{ T} \approx B = 1.36\text{ T}$$

$$M = \frac{B}{\mu_0} = \frac{1.36\text{ T}}{4\pi \times 10^{-7}\text{ T}\cdot\text{m/A}} = \boxed{1.08 \times 10^6\text{ A/m}}$$

CHECK Table 27-2 gives 5500 for the maximum value for K_m. Our Part-(*b*) result is less than this maximum value, as expected.

TAKING IT FURTHER The applied magnetic field of 7.54×10^{-4} T is a negligible fraction of the total field of 1.36 T.

*DIAMAGNETISM

Diamagnetic materials are those materials that have very small negative values of magnetic susceptibility χ_m. Diamagnetism was discovered by Michael Faraday in 1845 when Faraday found that a piece of bismuth is repelled by either pole of a magnet, indicating that the external field of the magnet induces a magnetic moment in bismuth in the direction opposite the field.

We can understand this effect qualitatively from Figure 27-46, which shows two positive charges moving in circular orbits with the same speed but in opposite directions. Their magnetic moments are in opposite directions and therefore cancel.* In the presence of an external magnetic field $\vec{B}$ directed into the paper, the charges experience an extra force $q\vec{v} \times \vec{B}$, which is along the radial direction. For the charge on the left, this extra force is inward, increasing the centripetal force. If the charge is to remain in the same circular orbit, it must speed up so that mv^2/r equals the total centripetal force.† Its magnetic moment, which is out of the page, is thus increased. For the charge on the right, the additional force is outward, so the particle must slow down to maintain its circular orbit. Its magnetic moment, which is into the page, is decreased. In each case, the *change* in the magnetic moment of the charges is in the direction out of the page, opposite that of the external applied field. Because the permanent magnetic moments of the two charges are equal and oppositely directed they add to zero, leaving only the induced magnetic moments, which are both in the direction opposite to the direction of the applied magnetic field.

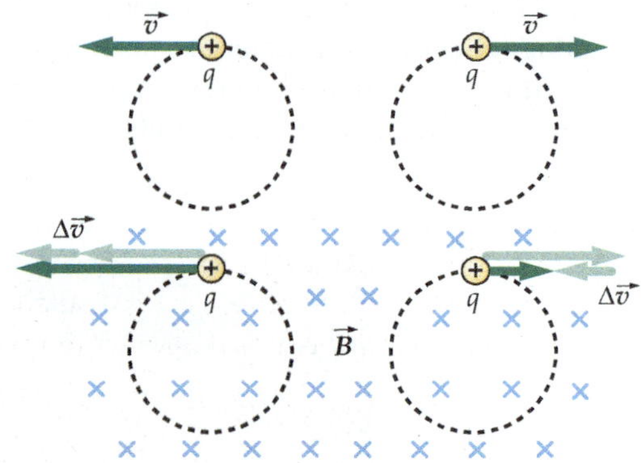

FIGURE 27-46 (*a*) A positive charge moving counterclockwise in a circle has its magnetic moment directed out of the paper. When an external magnetic field directed into the page is turned on, the magnetic force increases the centripetal force so the speed of the particle must increase. The change in the magnetic moment is out of the page. (*b*) A positive charge moving clockwise in a circle has its magnetic moment directed into the page. When an external magnetic field directed into the page is turned on, the magnetic force decreases the centripetal force so the speed of the particle must decrease. As in (*a*), the change in the magnetic moment is directed out of the page.

A material is diamagnetic if its atoms have zero net angular momentum and therefore no permanent magnetic moment. (The net angular momentum of an atom depends on the electronic structure of the atom, which is presented in Chapter 36.) The induced magnetic moments that cause diamagnetism have magnitudes of the order of 10^{-5} Bohr magnetons. Because this is much smaller than the permanent magnetic moments of the atoms of paramagnetic or ferromagnetic materials, the diamagnetic effect in these atoms is masked by the alignment of their permanent magnetic moments. However, because this alignment decreases with temperature, all materials are theoretically diamagnetic at sufficiently high temperatures.

When a superconductor is placed in an external magnetic field, electric currents are induced on the superconductor's surface so that the net magnetic field in the superconductor is zero. Consider a superconducting rod inside a solenoid of n turns per unit length. When the solenoid is connected to a source of emf so that it carries a current I, the magnetic field due to the solenoid is $\mu_0 nI$. A surface current of $-nI$ per unit length is induced on the superconducting rod that cancels out the field due to the solenoid so that the net field inside the superconductor is zero. From Equation 27-24,

$$\vec{B} = \vec{B}_{\text{app}}(1 + \chi_m) = 0$$

so

$$\chi_m = -1$$

A superconductor is thus a perfect diamagnet with a magnetic susceptibility of -1.

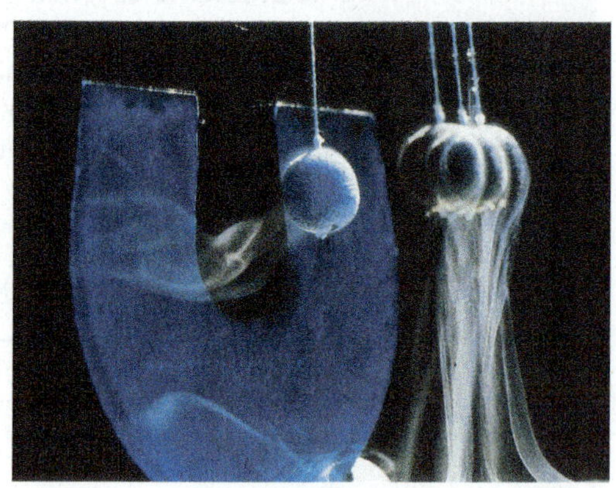

A superconductor is a perfect diamagnet. Here the superconducting pendulum bob is repelled by the permanent magnet. (*© Bill Pierce/ Time Magazine, Inc.*)

* It is simpler to consider positive charges even though negatively charged electrons provide the magnetic moments in matter.

† The electron speeds up because of an electric field induced by the changing magnetic field, an effect called induction, which we discuss in Chapter 28.

Physics Spotlight

Solenoids at Work

Why solenoids? Unlike gears, solenoids do not depend on friction for transferring motion, which means that solenoid-based movement is less likely to wear out machine parts. Solenoid valves, switches, and actuators are all based on the same principle—a central core in the solenoid is moved when current exists in the solenoid coils. Solenoid valves to control the flow of liquids and gases are the most popular mechanical use of solenoids. Some solenoid valves are opened directly by the motion of solenoid cores. When the solenoids turn off, springs return the valves to the off position.[*] Other solenoid valves, known as pilot-operated valves, use solenoid cores as switches for pistons that have very large ports, or use the motion of solenoid cores to open small pilot ports, which cause enough of a pressure differential in the main fluid line to open the main valve port.[†]

The world's largest superconducting solenoid magnet first reached its full field strength of 4 T in December 2006. Weighing in at over 10,000 tonnes, the magnet is built around a 6-m diameter, 13-m long superconducting solenoid. The solenoid, which is at CERN, will be used as part of a muon detector. *(CERN.)*

Because the cost of replacing a small valve is large due to lost time on a manufacturing line, manufacturers often choose solenoid valves in these lines.[‡] Some solenoid valves have been rated at several million cycles.[#] Solenoid valves also have been designed for use in very challenging environments. Solenoid valves can operate in corrosive[°,§] areas and in areas that have explosive atmospheres.[¶] Landscaping and irrigation applications[**] require solenoids that can operate outdoors. Solenoids are increasingly used for complex automated manufacturing processes.[††]

Because of their high reliability, long duty life, and low power draw compared to strictly mechanical systems, solenoids are used in robotics, aviation, and automotive applications. In robotics, solenoids often control air valves. In automotive applications, some solenoids control fluid pressure in the transmission, while others control automatic door locks.

One shortcoming of solenoids is that they are liable to overheating if they are overpowered[‡‡] or if the solenoid continually receives the power that is required to originally turn the solenoid on.[##] Overheating can fuse solenoid coils, shut down manufacturing lines, keep engines from starting, and even start fires. Because of these problems, designers are very careful to fit the solenoid to the use and the circuit.

Not all solenoids are used in mechanical applications. Some of the most powerful solenoids on Earth are used to provide large uniform magnetic fields for particle physics experiments. Many of these solenoids use cryogenically cooled superconductors to reach their full strength without overheating. A 5-T superconducting solenoid is set up at the Deutsches Elektronen-Synchrotron, or DESY. This magnet produces a magnetic field of 5.25 T when it carries currents of up to 1000 A. For the superconductors in the coil to work most efficiently, as well as to prevent overheating, the DESY solenoid must be cooled to 4.4 K.[°°] In Cessy, France, the largest superconducting solenoid magnet on Earth, the Compact Muon Solenoid, is scheduled to begin operating in November 2007.[§§] The coils of the solenoid contain 1947 km of superconducting niobium/titanium strands, and the solenoid has an internal diameter of 6.0 m. When it is cooled to 4.5 K, a current of more than 56 kA produces a magnetic field of 4 T.[¶¶] Whether very large solenoids for particle physics or miniature solenoids for chemical plants, the reliability and predictability of solenoids are prized assets.

* Hargraves, D., "Solenoid Valves: Operation, Selection, and Application." *Air Conditioning, Heating, & Refrigeration News*, Apr. 5, 1999, pp. 26–28.

† Zdobinski, D., Mudd, W., and Byrne, G., "Understanding Applications, Uses, Key to Solenoid Valve Selection." *Plant Engineering*, Jun. 2006, pp. 65–68.

‡ Heney, P. J., "Wide Variety of Solenoid Valves Available to Designers." *Hydraulics and Pneumatics*, Sept. 1998, Vol. 51, No. 9, pp. 51–56.

\# "Updated Solenoid Survives 20 Million Cycles." *Machine Design*, Aug. 23, 2001, p. 54.

° "Direct-Acting Solenoid Valves." *Design News*, Jun. 5, 2006, pp. 83–84.

§ "Solenoid Valve Handles Acids." *Manufacturing Chemist*, Jul. 1996, Vol. 67, No. 7, p. 51.

¶ "Solenoid Valve for Hazardous Areas." *Offshore*, Nov. 1998, Vol. 58, No. 11, p. 216.

** Mentzer, T., "Control Gets 'Smart'." *Landscape Management*, Jan. 2000, Vol. 39, No. 1, pp. 38+.

†† Mervartova, K., Martinez Calatayud, J., and Catala Icardo, M., "A Fully Automated Assembly Using Solenoid Valves for the Photodegradation and Chemiluminometric Determination of the Herbicide Chlorsulfuron." *Analytical Letters*, Jan. 2005, Vol. 38, No. 1, pp. 179–194.

‡‡ Zdobinski, D., Mudd, W., and Byrne, G., op. cit.

\#\# Nakhe, S. V., "Smart Solenoid Driver Reduces Power Loss." *Electronic Design*, Oct. 13, 2005, Vol. 53, No. 22, pp. 62–64.

°° Gadwinkel, E., et al., "Cryogenics for a 5 Tesla Superconducting Solenoid with Large Aperture at DESY." *CP170, Advances in Cryogenic Engineering: Transactions of the Cryogenic Engineering Conference—CEC, Vol. 49, AIP Conference Proceedings*, 2004, Vol. 710, Issue 1, pp. 719–725.

§§ Science Daily, "World's Largest Superconducting Solenoid Magnet Reaches Full Field." *Science Daily*, Sept. 26, 2006. http://www.sciencedaily.com/releases/2006/09/060925075001.htm As of Oct. 2006.

¶¶ Blau, B., and Pauss, F., "Superconducting Magnet: ETH Zürich and Superconductor Manufacture for CMS." *CMS Info, CERN*, Apr. 2003, http://cmsinfo.cern.ch/outreach/CMSdocuments/MagnetBrochure/MagnetBrochure.pdf As of Oct. 2006.

Summary

1. Magnetic fields arise from moving charges, and therefore from currents.
2. The Biot–Savart law describes the magnetic field produced by a current element.
3. Ampère's law relates the line integral of the magnetic field along some closed curve to the current that passes through any surface bounded by the curve.
4. The magnetization vector $\vec{M}$ describes the magnetic dipole moment per unit volume of matter.
5. The classical relation $\vec{\mu} = [q/(2m)]\vec{L}$ is derived from the definitions of angular momentum and magnetic moment.
6. The Bohr magneton is a convenient unit for atomic and nuclear magnetic moments.

TOPIC	RELEVANT EQUATIONS AND REMARKS	
1. Magnetic Field $\vec{B}$		
Due to a moving point charge	$\vec{B} = \frac{\mu_0}{4\pi}\frac{q\vec{v} \times \hat{r}}{r^2}$	27-1
	where $\hat{r}$ is a unit vector that points to the field point P from the charge q moving with velocity $\vec{v}$, and μ_0 is a constant of proportionality called the magnetic constant (the permeability of empty space):	
	$\mu_0 = 4\pi \times 10^{-7}\,\mathrm{T \cdot m/A} = 4\pi \times 10^{-7}\,\mathrm{N/A^2}$	27-2
Due to a current element (Biot–Savart law)	$d\vec{B} = \frac{\mu_0}{4\pi}\frac{I\,d\vec{\ell} \times \hat{r}}{r^2}$	27-3
On the axis of a current loop	$B_z = \frac{\mu_0}{4\pi}\frac{2\pi R^2 I}{(z^2 + R^2)^{3/2}}$	27-6
Inside a long solenoid, far from the ends	$B_z = \mu_0 n I$	27-10
	where n is the number of turns per unit length.	
Due to a straight-wire segment	$B = \frac{\mu_0}{4\pi}\frac{I}{R}(\sin\theta_2 - \sin\theta_1)$	27-12
	where R is the perpendicular distance to the wire and θ_1 and θ_2 are the angles subtended at the field point by the ends of the wire.	
Due to a long, straight wire	Use Equation 27-12 with $\theta_2 = 90°$ and $\theta_2 = -90°$, or derive using Ampere's law.	
	The direction of $\vec{B}$ is such that the magnetic field lines of $\vec{B}$ encircle the wire in the direction of the fingers of the right hand if the thumb points in the direction of the current.	
Inside the loops of a tightly wound toroid	$B = \frac{\mu_0}{2\pi}\frac{NI}{r}$	27-18
2. Magnetic Field Lines	Magnetic lines neither begin nor end. Either they form closed loops or they continue indefinitely.	
3. Gauss's Law for Magnetism	$\phi_{\mathrm{m\,net}} = \oint_S \vec{B} \cdot \hat{n}\,dA = \oint_S B_n\,dA = 0$	27-15
4. Magnetic Poles	Magnetic poles always occur in north–south pairs. Isolated magnetic poles have not been found.	
5. Ampère's Law	$\oint_C \vec{B} \cdot d\vec{\ell} = \oint_C B_t\,d\ell = \mu_0 I_C$	27-16
	where C is any closed curve.	

TOPIC	RELEVANT EQUATIONS AND REMARKS	
Validity of Ampère's law	Ampère's law is valid only if the currents are steady and continuous. It can be used to derive expressions for the magnetic field for situations with a high degree of symmetry, such as a long, straight, current-carrying wire or a long, tightly wound solenoid.	
6. Magnetism in Matter	Matter can be classified as paramagnetic, ferromagnetic, or diamagnetic.	
Magnetization	A magnetized material is described by its magnetization vector $\vec{M}$, which is defined as the magnetic dipole moment per unit volume of the material: $$\vec{M} = \frac{d\vec{\mu}}{dV}$$	27-19
	The magnetic field due to a uniformly magnetized cylinder is the same as if the cylinder carried a current per unit length of magnitude M on its surface. This current, which is due to the intrinsic motion of the atomic charges in the cylinder, is called an amperian current.	
7. $\vec{B}$ in Magnetic Materials	$$\vec{B} = \vec{B}_{app} + \mu_0 \vec{M}$$	27-22
Magnetic susceptibility χ_m	$$\vec{M} = \chi_m \frac{\vec{B}_{app}}{\mu_0}$$	27-23
	For paramagnetic materials, χ_m is a small positive number that depends on temperature. For diamagnetic materials (other than superconductors), it is a small negative constant independent of temperature. For superconductors, $\chi_m = -1$. For ferromagnetic materials, the magnetization depends not only on the magnetizing current but also on the past history of the material.	
Relative permeability	$$\vec{B} = K_m \vec{B}_{app}$$	27-24
	where $$K_m = 1 + \chi_m$$	27-25
8. Atomic Magnetic Moments	$$\vec{\mu} = \frac{q}{2m}\vec{L}$$	27-28
	where $\vec{L}$ is the orbital angular momentum of the particle.	
Bohr magneton	$$\mu_B = \frac{e\hbar}{2m_e} = 9.27 \times 10^{-24}\ \mathrm{A \cdot m^2}$$ $$= 9.27 \times 10^{-24}\ \mathrm{J/T} = 5.79 \times 10^{-5}\ \mathrm{eV/T}$$	27-30
Due to the orbital motion of an electron	$$\vec{\mu}_\ell = -\mu_B \frac{\vec{L}}{\hbar}$$	27-29
Due to electron spin	$$\vec{\mu}_s = -2\mu_B \frac{\vec{S}}{\hbar}$$	27-31
***9. Paramagnetism**	Paramagnetic materials have permanent atomic magnetic moments that have random directions in the absence of an applied magnetic field. In an applied field these dipoles are aligned with the field to some degree, producing a small contribution to the total field that adds to the applied field. The degree of alignment is small except in very strong fields and at very low temperatures. At ordinary temperatures, thermal motion tends to maintain the random directions of the magnetic moments.	
Curie's law	In weak fields, the magnetization is approximately proportional to the applied field and inversely proportional to the absolute temperature. $$M = \frac{1}{3}\frac{\mu B_{app}}{kT} M_s$$	27-34

TOPIC	RELEVANT EQUATIONS AND REMARKS
***10. Ferromagnetism**	Ferromagnetic materials have small regions of space called magnetic domains in which all the permanent atomic magnetic moments are aligned. When the material is unmagnetized, the direction of alignment in one domain is independent of that in another domain so that no net magnetic field is produced. When the material is magnetized, the domains of a ferromagnetic material are aligned, producing a very strong contribution to the magnetic field. This alignment can persist in magnetically hard materials, even when the external field is removed, thus leading to permanent magnets.
***11. Diamagnetism**	Diamagnetic materials are those materials in which the magnetic moments of all electrons in each atom cancel, leaving each atom with a zero magnetic moment in the absence of an external field. In an applied magnetic field, a very small magnetic moment is induced that tends to weaken the field. This effect is independent of temperature. Superconductors are diamagnetic with a magnetic susceptibility equal to -1.

Answers to Practice Problems

27-1 $\vec{B} = 0$, $\vec{B} = 3.2 \times 10^{-14}\ \mathrm{T}\ \hat{k}$

27-2 25 A

27-4 1.48×10^{-2} T. This is about 2 percent less than the step-3 result.

27-5 B at the center is larger for the circle.

27-6 $R = 4.0$ cm

27-7 0

27-8 $\vec{B} = 2.3 \times 10^{-5}\ \mathrm{T}\ \hat{j}$

27-9 $M/M_s = 1.12 \times 10^{-3}$

Problems

In a few problems, you are given more data than you actually need; in a few other problems, you are required to supply data from your general knowledge, outside sources, or informed estimate.

Interpret as significant all digits in numerical values that have trailing zeros and no decimal points.

- • Single-concept, single-step, relatively easy
- •• Intermediate-level, may require synthesis of concepts
- ••• Challenging
- SSM Solution is in the *Student Solutions Manual*
- Consecutive problems that are shaded are paired problems.

CONCEPTUAL PROBLEMS

1 • Sketch the field lines for the electric dipole and the magnetic dipole shown in Figure 27-47. How do the field lines differ in appearance close to the center of each dipole?

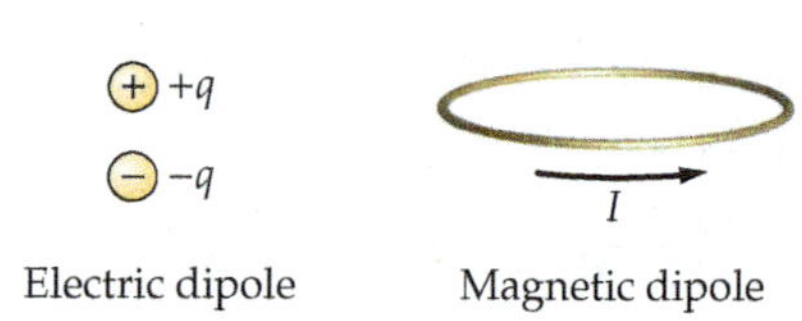

FIGURE 27-47 Problem 1

2 • Two wires lie in the plane of the page and carry equal currents in opposite directions, as shown in Figure 27-48. At a point midway between the wires, the magnetic field is (*a*) zero, (*b*) into the page, (*c*) out of the page, (*d*) toward the top or bottom of the page, (*e*) toward one of the two wires.

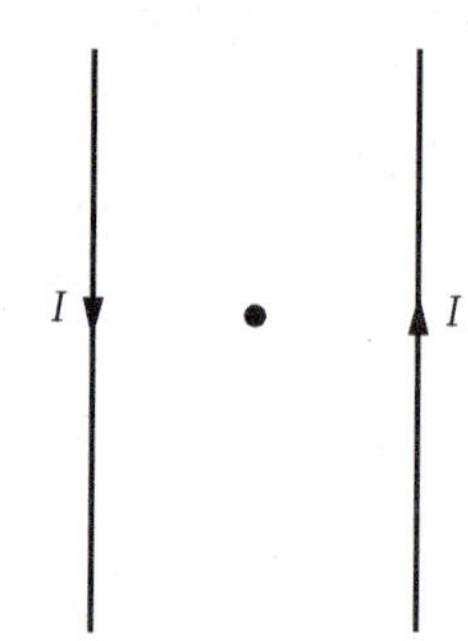

FIGURE 27-48 Problem 2

3 • Parallel wires 1 and 2 carry currents I_1 and I_2, respectively, where $I_2 = 2I_1$. The two currents are in the same direction. The magnitudes of the magnetic force by current 1 on wire 2 and by current 2 on wire 1 are F_{12} and F_{21}, respectively. These magnitudes are related by (*a*) $F_{12} = F_{21}$, (*b*) $F_{21} = 2F_{12}$, (*c*) $2F_{21} = F_{12}$, (*d*) $F_{21} = 4F_{12}$, (*e*) $4F_{21} = F_{12}$.

4 • Make a field-line sketch of the magnetic field due to the currents in the pair of identical coaxial coils shown in Figure 27-49. Consider two cases: (*a*) the currents in the coils have the same magnitude and have the same direction and (*b*) the currents in the coils have the same magnitude and have the opposite directions.

FIGURE 27-49 Problem 4

5 • Discuss the differences and similarities between Gauss's law for magnetism and Gauss's law for electricity. SSM

6 • Explain how you would modify Gauss's law if scientists discovered that single, isolated magnetic poles actually existed.

7 • You are facing directly into one end of a long solenoid and the magnetic field inside of the solenoid points away from you. From your perspective, is the direction of the current in the solenoid coils clockwise or counterclockwise? Explain your answer. SSM

8 • Opposite ends of a helical metal spring are connected to the terminals of a battery. Do the spacings between the coils of the spring tend to increase, decrease, or remain the same when the battery is connected? Explain your answer.

9 • The current density is constant and uniform in a long, straight wire that has a circular cross section. True or false:

(*a*) The magnitude of the magnetic field produced by the wire is greatest at the surface of the wire.

(*b*) The magnetic field strength in the region surrounding the wire varies inversely with the square of the distance from the wire's central axis.

(*c*) The magnetic field is zero at all points on the wire's central axis.

(*d*) The magnitude of the magnetic field inside the wire increases linearly with the distance from the wire's central axis.

10 • If the magnetic susceptibility of a material is positive, (*a*) paramagnetic effects or ferromagnetic effects must be greater than diamagnetic effects, (*b*) diamagnetic effects must be greater than paramagnetic effects, (*c*) diamagnetic effects must be greater than ferromagnetic effects, (*d*) ferromagnetic effects must be greater than paramagnetic effects, (*e*) paramagnetic effects must be greater than ferromagnetic effects.

11 • Of the four gases listed in Table 27-1, which are diamagnetic and which are paramagnetic? SSM

12 • When a current is passed through the wire in Figure 27-50, will the wire tend to bunch up or will it tend to form a circle? Explain your answer.

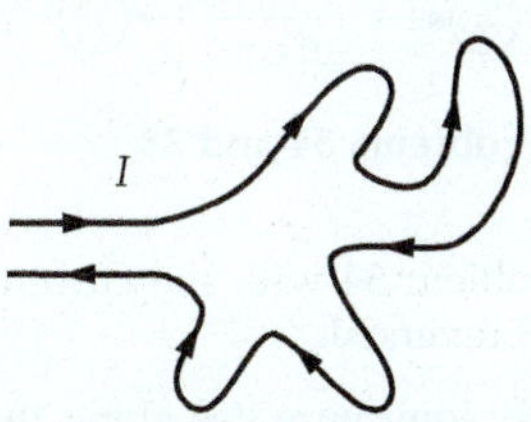

FIGURE 27-50 Problem 12

THE MAGNETIC FIELD OF MOVING POINT CHARGES

13 • At time $t = 0$, a particle has a charge of $12\,\mu$C, is located in the $z = 0$ plane at $x = 0, y = 2.0$ m, and has a velocity equal to $30\text{ m/s}\,\hat{i}$. Find the magnetic field in the $z = 0$ plane at (*a*) the origin, (*b*) $x = 0, y = 1.0$ m, (*c*) $x = 0, y = 3.0$ m, and (*d*) $x = 0, y = 4.0$ m. SSM

14 • At time $t = 0$, a particle has a charge of $12\,\mu$C, is located in the $z = 0$ plane at $x = 0, y = 2.0$ m, and has a velocity equal to $30\text{ m/s}\,\hat{i}$. Find the magnetic field in the $z = 0$ plane at (*a*) $x = 1.0$ m, $y = 3.0$ m, (*b*) $x = 2.0$ m, $y = 2.0$ m, and (*c*) $x = 2.0$ m, $y = 3.0$ m.

15 • A proton has a velocity of $1.0 \times 10^2\text{ m/s}\,\hat{i} + 2.0 \times 10^2\text{ m/s}\,\hat{j}$ and is located in the $z = 0$ plane at $x = 3.0$ m, $y = 4.0$ m at some time $t = T$. Find the magnetic field in the $z = 0$ plane at (*a*) $x = 2.0$ m, $y = 2.0$ m, (*b*) $x = 6.0$ m, $y = 4.0$ m, and (*c*) $x = 3.0$ m, $y = 6.0$ m.

16 •• In a pre–quantum-mechanical model of the hydrogen atom, an electron orbits a proton at a radius of 5.29×10^{-11} m. According to this model, what is the magnitude of the magnetic field at the proton due to the orbital motion of the electron? Neglect any motion of the proton.

17 •• Two equal point charges q are, at some instant, located at (0, 0, 0) and at (0, b, 0). They are both moving with speed v in the $+x$ direction (assume $v \ll c$). Find the ratio of the magnitude of the magnetic force to the magnitude of the electric force on each charge.

THE MAGNETIC FIELD USING THE BIOT–SAVART LAW

18 • A small current element at the origin has a length of 2.0 mm and carries a current of 2.0 A in the $+z$ direction. Find the magnetic field due to the current element (*a*) on the x axis at $x = 3.0$ m, (*b*) on the x axis at $x = -6.0$ m, (*c*) on the z axis at $z = 3.0$ m, and (*d*) on the y axis at $y = 3.0$ m.

19 • A small current element at the origin has a length of 2.0 mm and carries a current of 2.0 A in the $+z$ direction. Find the magnitude and direction of the magnetic field due to this current element at the point (0, 3.0 m, 4.0 m). SSM

20 • A small current element at the origin has a length of 2.0 mm and carries a current of 2.0 A in the $+z$ direction. Find the magnitude of the magnetic field due to this current element and indicate its direction on a diagram at (*a*) $x = 2.0$ m, $y = 4.0$ m, $z = 0$ and (*b*) $x = 2.0$ m, $y = 0$, $z = 4.0$ m.

THE MAGNETIC FIELD DUE TO CURRENT LOOPS AND COILS

21 • A single conducting loop has a radius equal to 3.0 cm and carries a current equal to 2.6 A. What is the magnitude of the magnetic field on the line through the center of the loop and perpendicular to the plane of the loop (*a*) at the center of the loop, (*b*) 1.0 cm from the center, (*c*) 2.0 cm from the center, and (*d*) 35 cm from the center?

22 ••• **SPREADSHEET** A pair of identical coils, each having a radius of 30 cm, are separated by a distance equal to their radii, that is, 30 cm. Called *Helmholtz coils*, they are coaxial and carry equal currents in directions such that their axial fields are in the same direction. A feature of Helmholtz coils is that the resultant magnetic field in the region between the coils is very uniform. Assume the current in each is 15 A and there are 250 turns for each coil. Using a spreadsheet program, calculate and graph the magnetic field as a function of z, the distance from the center of the coils along the common axis, for $-30\text{ cm} < z < +30\text{ cm}$. Over what range of z does the field vary by less than 20%?

23 ••• A pair of Helmholtz coils that have radii R have their axes along the z axis (see Problem 22). One coil is in the $z = -\frac{1}{2}R$ plane and the second coil is in the $z = +\frac{1}{2}R$ plane. Show that on the z axis at $z = 0$ $dB_z/dz = 0$, $d^2B_z/dz^2 = 0$, and $d^3B_z/dz^3 = 0$. (*Note:* These results show that the magnitude and direction of the magnetic field in the region to either side of the midpoint are approximately equal to the magnitude and direction of the magnetic field at the midpoint.)

24 ••• **ENGINEERING APPLICATION** *Anti-Helmholtz* coils are used in many physics applications, such as laser cooling and trapping, where a field with a uniform gradient is desired. These coils have the same construction as a Helmholtz coil, except that the currents have opposite directions, so that the axial fields are in opposite directions, and the coil separation is $\sqrt{3}R$ rather than R. Graph the magnetic field as a function of z, the axial distance from the center of the coils, for an anti-Helmholtz coil using the same parameters as in Problem 22. Over what interval of the z axis is dB_z/dz within one percent of its value at the midpoint between the coils?

THE MAGNETIC FIELD DUE TO STRAIGHT-LINE CURRENTS

Problems 25 to 30 refer to Figure 27-51 which shows two long, straight wires in the xy plane and parallel to the x axis. One wire is at $y = -6.0$ cm and the other wire is at $y = +6.0$ cm. The current in each wire is 20 A.

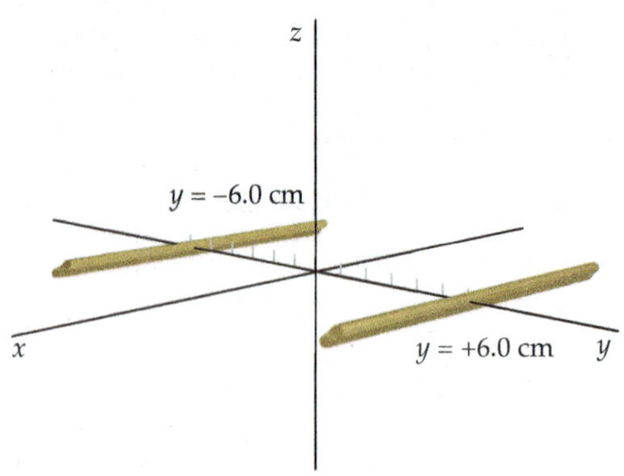

FIGURE 27-51 Problems 25–30

25 •• If the currents are both in the $-x$ direction, find the magnetic field at the following points on the y axis: (*a*) $y = -3.0$ cm, (*b*) $y = 0$, (*c*) $y = +3.0$ cm, and (*d*) $y = +9.0$ cm. SSM

26 •• **SPREADSHEET** Using a spreadsheet program or graphing calculator, graph B_z versus y when both currents are in the $-x$ direction.

27 •• The current in the wire at $y = -6.0$ cm is in the $-x$ direction and the current in the wire at $y = +6.0$ cm is in the $+x$ direction. Find the magnetic field at the following points on the y axis: (*a*) $y = -3.0$ cm, (*b*) $y = 0$, (*c*) $y = +3.0$ cm, and (*d*) $y = +9.0$ cm.

28 •• **SPREADSHEET** The current in the wire at $y = -6.0$ cm is in the $+x$ direction and the current in the wire at $y = +6.0$ cm is in the $-x$ direction. Using a spreadsheet program or graphing calculator, graph B_z versus y.

29 • Find the magnetic field on the z axis at $z = +8.0$ cm if (*a*) the currents are both in the $-x$ direction and (*b*) the current in the wire at $y = -6.0$ cm is in the $-x$ direction and the current in the wire at $y = +6.0$ cm is in the $+x$ direction.

30 • Find the magnitude of the force per unit length exerted by one wire on the other.

31 • Two long, straight parallel wires 8.6 cm apart carry equal currents. The wires repel each other with a force per unit length of 3.6 nN/m. (*a*) Are the currents parallel or antiparallel? Explain your answer. (*b*) Determine the current in each wire.

32 •• The current in the wire shown in Figure 27-52 is 8.0 A. Find the magnetic field at point P.

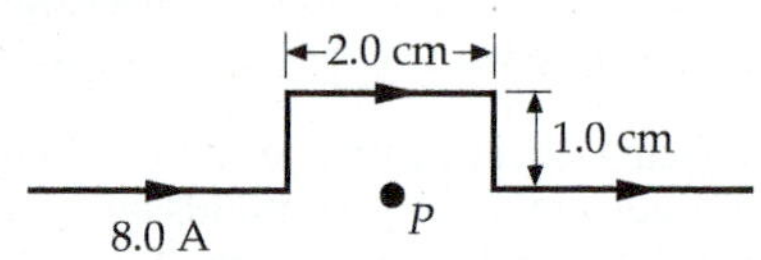

FIGURE 27-52 Problem 32

33 •• As a student technician, you are preparing a lecture demonstration on "magnetic suspension." You have a 16-cm-long straight, rigid wire that will be suspended by flexible conductive lightweight leads above a long, straight wire. Currents that are equal but are in opposite directions will be established in the two wires so the 16-cm wire "floats" a distance h above the long wire with no tension in its suspension leads. If the mass of the 16-cm wire is 14 g and if h (the distance between the central axes of the two wires) is 1.5 mm, what should their common current be? SSM

34 •• Three long, parallel straight wires pass through the vertices of an equilateral triangle that has sides equal to 10 cm, as shown in Figure 27-53. A dot indicates that the direction of the current is out of the page and a cross indicates that the direction of the current is into the page. If each current is 15 A, find (*a*) the magnetic field at the location of the upper wire due to the currents in the two lower wires and (*b*) the force per unit length on the upper wire.

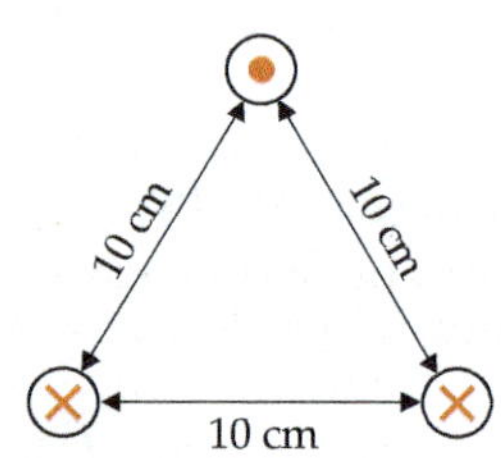

FIGURE 27-53 Problems 34 and 35

35 •• Rework Problem 34 with the current in the lower right corner of Figure 27-53 reversed.

36 •• An infinitely long wire lies along the x axis and carries current I in the $+x$ direction. A second infinitely long wire lies along the y axis and carries current I in the $+y$ direction. At what points in the $z = 0$ plane is the resultant magnetic field zero?

37 •• An infinitely long wire lies along the z axis and carries a current of 20 A in the $+z$ direction. A second infinitely long wire is parallel to the z axis and intersects the x axis at $x = 10.0$ cm.

(*a*) Find the current in the second wire if the magnetic field is zero at (2.0 cm, 0, 0). (*b*) What is the magnetic field at (5.0 cm, 0, 0)? SSM

38 •• Three long parallel wires are at the corners of a square, as shown in Figure 27-54. The wires each carry a current I. Find the magnetic field at the unoccupied corner of the square when (*a*) all the currents are into the page, (*b*) I_1 and I_3 are into the page and I_2 is out, and (*c*) I_1 and I_2 are into the page and I_3 is out. Your answers should be in terms of I and L.

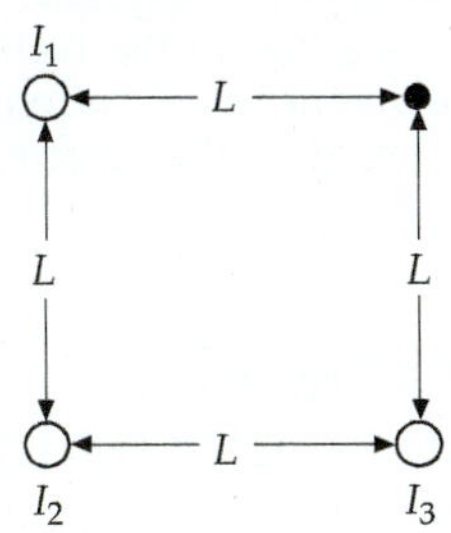

FIGURE 27-54 Problem 38

39 •• Four long, straight parallel wires each carry current I. In a plane perpendicular to the wires, the wires are at the corners of a square of side length a. Find the magnitude of the force per unit length on one of the wires if (*a*) all the currents are in the same direction and (*b*) the currents in the wires at adjacent corners are oppositely directed. SSM

40 •• Five long, straight, current-carrying wires are parallel to the z axis, and each carries a current I in the $+z$ direction. The wires each are a distance R from the z axis. Two of the wires intersect the x axis, one at $x = R$ and the other at $x = -R$. Another wire intersects the y axis at $y = R$. One of the remaining wires intersects the $z = 0$ plane at the point $(R/\sqrt{2}, R/\sqrt{2})$ and the last remaining wire intersects the $z = 0$ plane at the point $(-R/\sqrt{2}, R/\sqrt{2})$. Find the magnetic field on the z axis.

MAGNETIC FIELD DUE TO A CURRENT-CARRYING SOLENOID

41 •• A solenoid that has length 30 cm, radius 1.2 cm, and 300 turns carries a current of 2.6 A. Find the magnitude of the magnetic field on the axis of the solenoid (*a*) at the center of the solenoid and (*b*) at one end of the solenoid. SSM

42 • A solenoid is 2.7 m long, has a radius of 0.85 cm, and has 600 turns. It carries a current I of 2.5 A. What is the magnitude of the magnetic field B inside the solenoid and far from either end?

43 •• A solenoid has n turns per unit length, has a radius R, and carries a current I. Its axis coincides with the z axis—one end at $z = -\frac{1}{2}\ell$ and the other end at $z = +\frac{1}{2}\ell$. Show that the magnitude of the magnetic field at a point on the z axis is given by $B = \frac{1}{2}\mu_0 nI(\cos\theta_1 - \cos\theta_2)$, where the angles are related to the geometry by $\cos\theta_1 = (z + \frac{1}{2}\ell)/\sqrt{(z + \frac{1}{2}\ell)^2 + R^2}$ and $\cos\theta_2 = (z - \frac{1}{2}\ell)/\sqrt{(z - \frac{1}{2}\ell)^2 + R^2}$.

44 ••• In Problem 43, an expression for the magnitude of the magnetic field along the axis of a solenoid is given. For $z \gg \ell$ and $z \gg R$, the angles θ_1 and θ_2 are very small, so the small-angle approximations $\cos\theta \approx 1 - \frac{1}{2}\theta^2$ and $\sin\theta \approx \tan\theta \approx \theta$ are highly accurate. (*a*) Draw a diagram and use it to show that, for these conditions, the angles can be approximated as $\theta_1 \approx R/(z + \frac{1}{2}\ell)$ and $\theta_2 \approx R/(z - \frac{1}{2}\ell)$. (*b*) Using these approximations, show that the magnetic field at points on the z axis where $z \gg \ell$ can be written as $B = \frac{\mu_0}{4\pi}\left(\frac{q_m}{r_2^2} - \frac{q_m}{r_1^2}\right)$ where $r_2 = z - \frac{1}{2}\ell$ is the distance to the near end of the solenoid, $r_1 = z + \frac{1}{2}\ell$ is the distance to the far end, and the quantity q_m is defined by $q_m = nI\pi R^2 = \mu/\ell$, where $\mu = NI\pi R^2$ is the magnitude of the magnetic moment of the solenoid.

USING AMPÈRE'S LAW

45 • A long, straight, thin-walled cylindrical shell of radius R carries a current I parallel to the central axis of the shell. Find the magnetic field (including direction) both inside and outside the shell. SSM

46 • In Figure 27-55, one current is 8.0 A into the page, the other current is 8.0 A out of the page, and each curve is a circular path. (*a*) Find $\oint_C \vec{B} \cdot d\vec{\ell}$ for each path, assuming that each integral is to be evaluated in the counterclockwise direction. (*b*) Which path, if any, can be used to find the combined magnetic field of the currents?

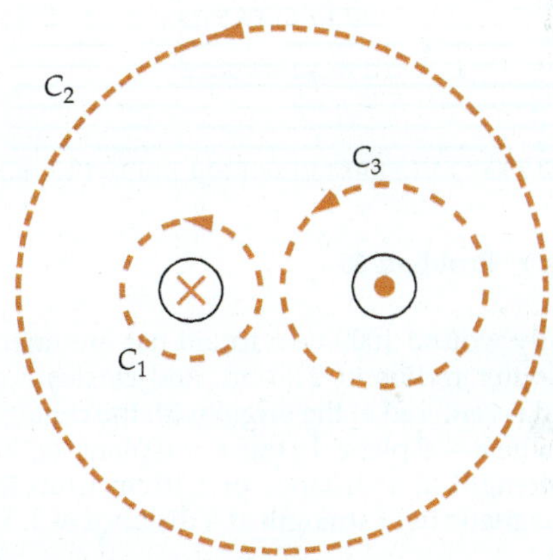

FIGURE 27-55 Problem 46

47 •• Show that a uniform magnetic field that has no fringing field, such as that shown in Figure 27-56, is impossible because it violates Ampère's law. Do this calculation by applying Ampère's law to the rectangular curve shown by the dashed lines. SSM

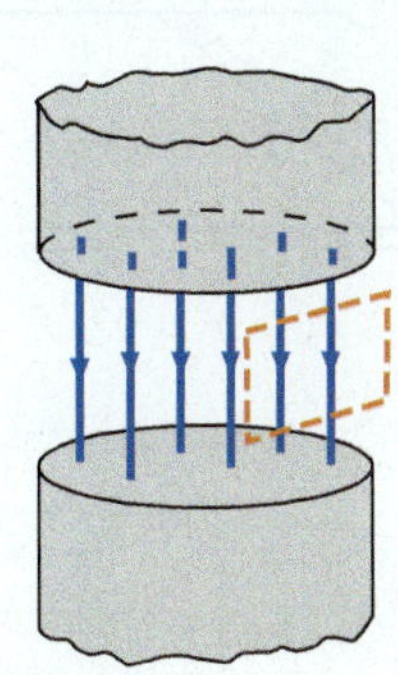

FIGURE 27-56 Problem 47

48 •• **SPREADSHEET** A coaxial cable consists of a solid conducting cylinder that has a radius equal to 1.00 mm and a conducting cylindrical shell that has an inner radius equal to 2.00 mm and an outer radius equal to 3.00 mm. The solid cylinder carries a current of 15.0 A parallel to the central axis. The cylindrical shell carries a current of 15.0 A in the opposite direction. Assume that the current densities are uniformly distributed in both conductors. (*a*) Using a spreadsheet program or graphing calculator, graph the magnitude of the magnetic field as a function of the radial distance r from the central axis for $0 < R < 3.00$ mm. (*b*) What is the magnitude of the field for $R > 3.00$ mm?

49 •• A long cylindrical shell has an inner radius a and an outer radius b and carries a current I parallel to the central axis. Assume that within the material of the shell the current density is uniformly distributed. Find an expression for the magnitude of the magnetic field for (*a*) $0 < R < a$, (*b*) $a < R < b$, and (*c*) $R > b$. SSM

50 •• Figure 27-57 shows a solenoid that has n turns per unit length and carries a current I. Apply Ampère's law to the rectangular curve shown in the figure to derive an expression for the magnitude of the magnetic field. Assume that inside the solenoid the magnetic field is uniform and parallel with the central axis, and that outside the solenoid there is no magnetic field.

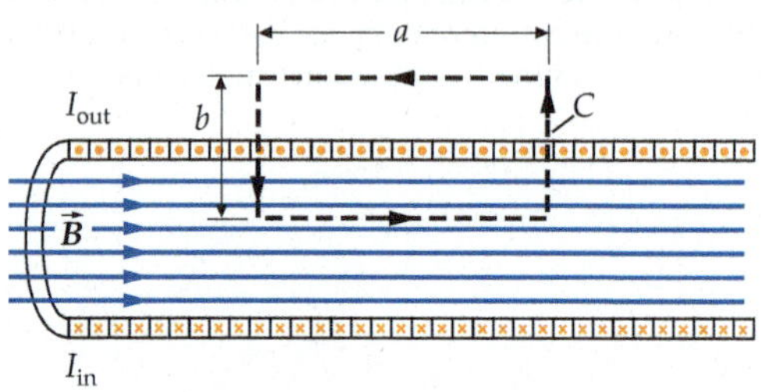

FIGURE 27-57 Problem 50

51 •• A tightly wound 1000-turn toroid has an inner radius of 1.00 cm and an outer radius of 2.00 cm, and carries a current of 1.50 A. The toroid is centered at the origin with the centers of the individual turns in the $z = 0$ plane. In the $z = 0$ plane: (*a*) What is the magnetic field strength at a distance of 1.10 cm from the origin? (*b*) What is the magnetic field strength at a distance of 1.50 cm from the origin? SSM

52 ••• A thin conducting sheet in the $z = 0$ plane carries current in the $-x$ direction (Figure 27-58*a*). The sheet extends indefinitely in all directions and the current is uniformly distributed throughout the sheet. To find the direction of the magnetic field at point P consider the field due only to the currents I_1 and I_2 in the two narrow strips shown. The strips are identical, so $I_1 = I_2$. (*a*) What is the direction of the magnetic field at point P due to just I_1 and I_2? Explain your answer using a sketch. (*b*) What is the direction of the magnetic field at point P due to the *entire* sheet? Explain your answer. (*c*) What is the direction of the field at a point to the right of point P (where $y \neq 0$)? Explain your answer. (*d*) What is the direction of the field at a point below the sheet (where $z < 0$)? Explain your answer using a sketch. (*e*) Apply Ampère's law to the rectangular curve (Figure 27-58*b*) to show that the magnetic field strength at point P is given by $B = \frac{1}{2}\mu_0\lambda$, where $\lambda = dI/dy$ is the current per unit length along the y axis.

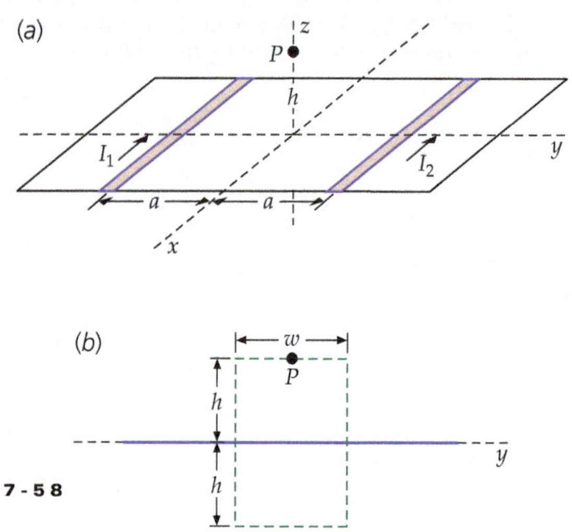

FIGURE 27-58 Problem 52

MAGNETIZATION AND MAGNETIC SUSCEPTIBILITY

53 • A tightly wound solenoid is 20.0 cm long, has 400 turns, and carries a current of 4.00 A so that its axial field is in the $+z$ direction. Find B and B_{app} at the center when (*a*) there is no core in the solenoid and (*b*) there is a soft iron core that has a magnetization of 1.2×10^6 A/m. SSM

54 • A long tungsten-core solenoid carries a current. (*a*) If the core is removed while the current is held constant, does the magnetic field strength in the region inside the solenoid decrease or increase? (*b*) By what percentage does the magnetic field strength in the region inside the solenoid decrease or increase?

55 • As a liquid fills the interior volume of a solenoid that carries a constant current, the magnetic field inside the solenoid *decreases* by 0.0040 percent. Determine the magnetic susceptibility of the liquid.

56 • A long thin solenoid carrying a current of 10 A has 50 turns per centimeter of length. What is the magnetic field strength in the region occupied by the interior of the solenoid when the interior is (*a*) a vacuum, (*b*) filled with aluminum, and (*c*) filled with silver?

57 •• A cylinder of iron, initially unmagnetized, is cooled to 4.00 K. What is the magnetization of the cylinder at that temperature due to the influence of Earth's magnetic field of 0.300 G? Assume a magnetic moment of 2.00 Bohr magnetons per atom. SSM

58 •• A cylinder of silver at a temperature of 77 K has a magnetization equal to 0.075% of its saturation magnetization. Assume a magnetic moment of one Bohr magneton per atom. The density of silver is 1.05×10^4 kg/m^3. (*a*) What value of applied magnetic field parallel to the central axis of the cylinder is required to reach this magnetization? (*b*) What is the magnetic field strength at the center of the cylinder?

59 •• During a solid-state physics lab, you are handed a cylindrically shaped sample of unknown magnetic material. You and your lab partners place the sample in a long solenoid that has n turns per unit length and a current I. The values for magnetic field B

within the material versus nI are given below. Use these values to plot B versus B_{app} and K_m versus nI, where B_{app} is the field due to the current I and K_m is the relative permeability of the sample.

nI, A/m	0	50	100	150	200	500	1000	10 000
B, T	0	0.04	0.67	1.00	1.2	1.4	1.6	1.7

ATOMIC MAGNETIC MOMENTS

60 •• Nickel has a density of 8.70 g/cm^3 and a molar mass of 58.7 g/mol. Nickel's saturation magnetization is 0.610 T. Calculate the magnetic moment of a nickel atom in Bohr magnetons.

61 •• Repeat Problem 60 for cobalt, which has a density of 8.90 g/cm^3, a molar mass of 58.9 g/mol, and a saturation magnetization of 1.79 T.

*PARAMAGNETISM

62 • Show that Curie's law predicts that the magnetic susceptibility of a paramagnetic substance is given by $\chi_m = \mu_0 M_s/(3kT)$.

63 •• In a simple model of paramagnetism, we can consider that some fraction f of the atoms have their magnetic moments aligned with the external magnetic field and that the rest of the atoms are randomly oriented and therefore do not contribute to the magnetic field. (*a*) Use this model and Curie's law to show that at temperature T and external magnetic field B, the fraction of aligned atoms f is given by $\mu B/(3kT)$. (*b*) Calculate this fraction for a sample temperature of 300 K and an external field of 1.00 T. Assume that μ has a value of 1.00 Bohr magneton.

64 •• Assume that the magnetic moment of an aluminum atom is 1.00 Bohr magneton. The density of aluminum is 2.70 g/cm^3 and its molar mass is 27.0 g/mol. (*a*) Calculate the value of the saturation magnetization and the saturation magnetic field for aluminum. (*b*) Use the result of Problem 62 to calculate the magnetic susceptibility at 300 K. (*c*) Explain why the result for Part (*b*) is larger than the value listed in Table 27-1.

65 •• A toroid has N turns, carries a current I, has a mean radius R, and has a cross-sectional radius r, where $r \ll R$ (Figure 27-59). When the toroid is filled with material, it is called a *Rowland ring*. Find B_{app} and B in such a ring, assuming a magnetization that is everywhere parallel to $\vec{B}_{app}$. SSM

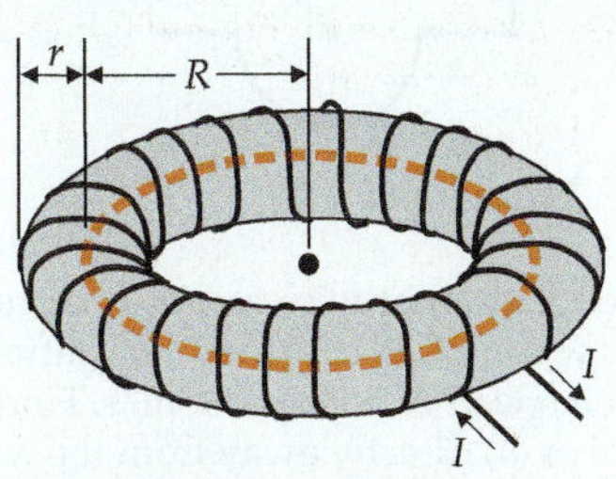

FIGURE 27-59 Problems 65 and 73

66 •• A toroid is filled with liquid oxygen that has a magnetic susceptibility of 4.00×10^{-3}. The toroid has 2000 turns and carries a current of 15.0 A. Its mean radius is 20.0 cm, and the radius of its cross section is 8.00 mm. (*a*) What is the magnetization? (*b*) What is the magnetic field? (*c*) What is the percentage change in the magnetic field produced by the liquid oxygen?

67 •• The centers of the turns of a toroid form a circle with a radius of 14.0 cm. The cross-sectional area of each turn is 3.00 cm^2. It is wound with 5278 turns of fine wire, and the wire carries a current of 4.00 A. The core is filled with a paramagnetic material of magnetic susceptibility 2.90×10^{-4}. (*a*) What is the magnitude of the magnetic field within the substance? (*b*) What is the magnitude of the magnetization? (*c*) What would the magnitude of the magnetic field be if there were no paramagnetic core present?

*FERROMAGNETISM

68 • For annealed iron, the relative permeability K_m has its maximum value of approximately 5500 at $B_{app} = 1.57 \times 10^{-4}$ T. Find the magnitude of the magnetization and magnetic field in annealed iron when K_m is maximum.

69 •• The saturation magnetization for annealed iron occurs when $B_{app} = 0.201$ T. Find the permeability and the relative permeability of annealed iron at saturation. (See Table 27-2.) SSM

70 •• The *coercive force* (which is a misnomer because it is really a magnetic field value) is defined as the applied magnetic field needed to bring the magnetic field back to zero along the hysteresis curve (which is point *c* in Figure 27-44). For a certain permanent bar magnet, the coercive force is known to be 5.53×10^{-2} T. The bar magnet is to be demagnetized by placing it inside a 15.0-cm-long solenoid that has 600 turns. What minimum current is needed in the solenoid to demagnetize the magnet?

71 •• A long thin solenoid has 50 turns/cm and carries a current of 2.00 A. The solenoid is filled with iron and the magnetic field is measured to be 1.72 T. (*a*) Neglecting end effects, what is the magnitude of the applied magnetic field? (*b*) What is the magnetization? (*c*) What is the relative permeability?

72 •• When the current in Problem 71 is 0.200 A, the magnetic field is measured to be 1.58 T. (*a*) Neglecting end effects, what is the applied magnetic field? (*b*) What is the magnetization? (*c*) What is the relative permeability?

73 •• A toroid has N turns, carries a current I, has a mean radius R, and has a cross-sectional radius r, where $r \ll R$ (Figure 27-59). The core of the toroid is filled with iron. When the current is 10.0 A, the magnetic field in the region where the iron is has a magnitude of 1.80 T. (*a*) What is the magnetization? (*b*) Find the values for the relative permeability, the permeability, and the magnetic susceptibility for this iron sample. SSM

74 • The centers of the turns of a toroid form a circle with a radius of 14.0 cm. The cross-sectional area of each turn is 3.00 cm^2. It is wound with 5278 turns of fine wire, and the wire carries a current of 0.200 A. The core is filled with soft iron, which has a relative permeability of 500. What is the magnetic field strength in the core?

75 ••• A long, straight wire that has a radius of 1.00 mm is coated with an insulating ferromagnetic material that has a thickness of 3.00 mm and a relative magnetic permeability of 400. The coated wire is in air and the wire itself is nonmagnetic. The wire carries a current of 40.0 A. (*a*) Find the magnetic field in the region occupied by the inside of the wire as a function of the perpendicular distance, r, from the central axis of the wire. (*b*) Find the magnetic field in the region occupied by the inside of the ferromagnetic

material as a function of the perpendicular distance, r, from the central axis of the wire. (*c*) Find the magnetic field in the region surrounding the wire and coating as a function of the perpendicular distance, r, from the central axis of the wire. (*d*) What must the magnitudes and directions of the amperian currents be on the surfaces of the ferromagnetic material to account for the magnetic fields observed?

GENERAL PROBLEMS

76 • Find the magnetic field at point P in Figure 27-60.

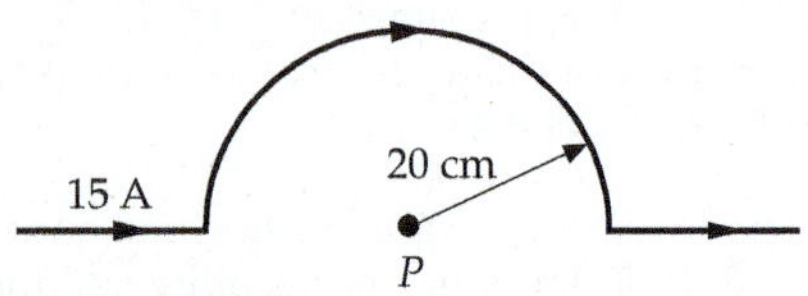

FIGURE 27-60 Problem 76

77 • Using Figure 27-61, find the magnetic field (in terms of the parameters given in the figure) at point P, the common center of the two arcs. SSM

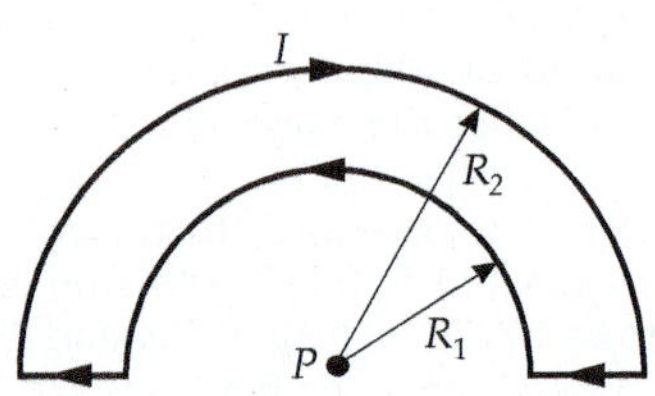

FIGURE 27-61 Problem 77

78 •• A wire of length ℓ is wound into a circular coil of N turns and carries a current I. Show that the magnetic field strength in the region occupied by the center of the coil is given by $\mu_0 \pi N^2 I/\ell$.

79 •• A very long wire carrying a current I is bent into the shape shown in Figure 27-62. Find the magnetic field at point P.

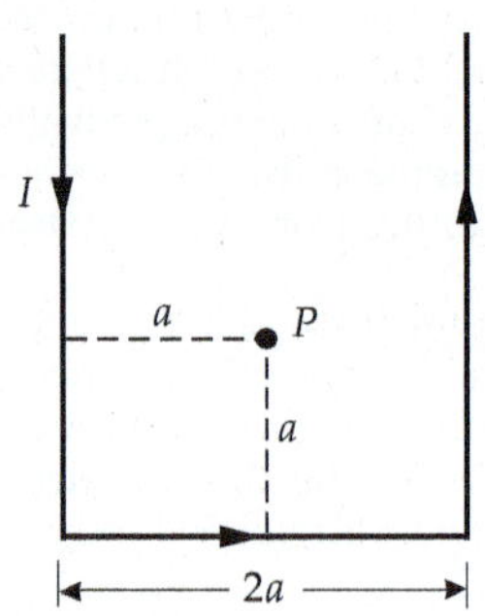

FIGURE 27-62 Problem 79

80 •• A power cable carrying 50 A is 2.0 m below Earth's surface, but the cable's direction and precise position are unknown. Explain how you could locate the cable using a compass. Assume that you are at the equator, where Earth's magnetic field is horizontal and 0.700 G due north.

81 •• A long, straight wire carries a current of 20.0 A, as shown in Figure 27-63. A rectangular coil that has two sides parallel to the straight wire has sides that are 5.00 cm long and 10.0 cm long. The side nearest to the wire is 2.00 cm from the wire. The coil carries a current of 5.00 A. (*a*) Find the force on each segment of the rectangular coil due to the current in the long, straight wire. (*b*) What is the net force on the coil? SSM

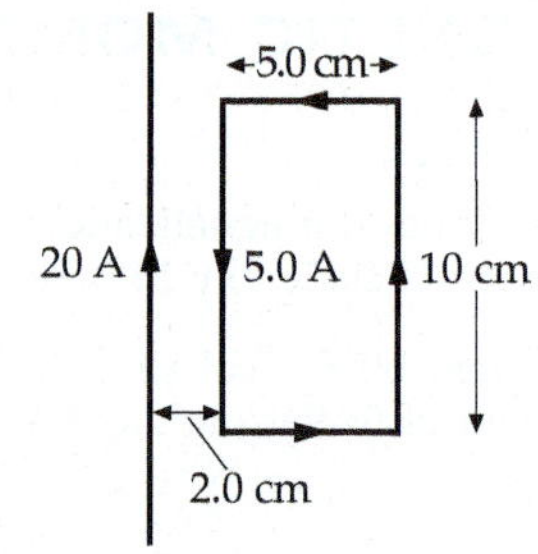

FIGURE 27-63 Problem 81

82 •• The closed loop shown in Figure 27-64 carries a current of 8.0 A in the counterclockwise direction. The radius of the outer arc is 0.60 m and that of the inner arc is 0.40 m. Find the magnetic field at point P.

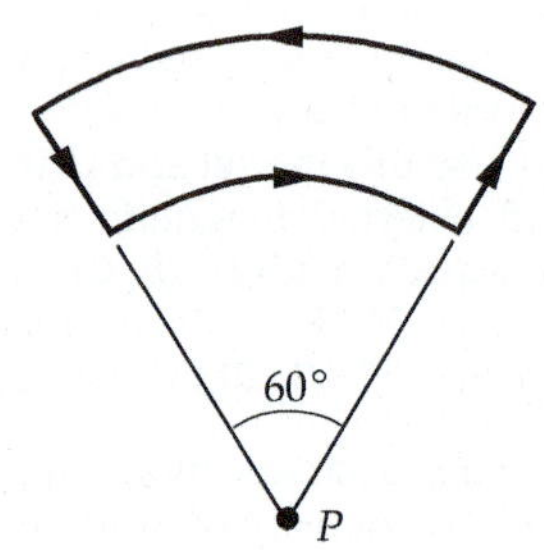

FIGURE 27-64 Problem 82

83 •• A closed circuit consists of two semicircles of radii 40 cm and 20 cm that are connected by straight segments, as shown in Figure 27-65. A current of 3.0 A exists in this circuit and has a clockwise direction. Find the magnetic field at point P.

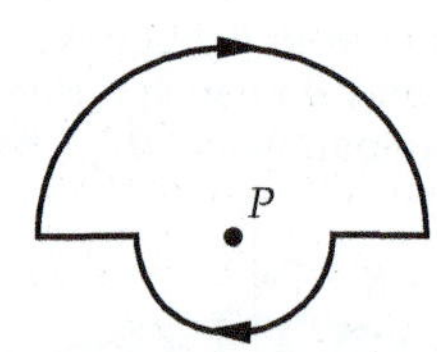

FIGURE 27-65 Problem 83

84 •• A very long, straight wire carries a current of 20.0 A. An electron outside the wire is 1.00 cm from the central axis of the wire and is moving with a speed of 5.00×10^6 m/s. Find the force on the electron when it moves (*a*) directly away from the wire, (*b*) parallel to the wire in the direction of the current, and (*c*) perpendicular to the central axis of wire and tangent to a circle that is coaxial with the wire.

85 •• SPREADSHEET A current of 5.00 A is uniformly distributed over the cross section of a long, straight wire of radius $R_0 = 2.55$ mm. Using a spreadsheet program, graph the magnetic field strength as a function of R (the distance from the central axis of the wire) for $0 \leq R \leq 10R_0$.

86 •• A 50-turn coil of radius 10.0 cm carries a current of 4.00 A and a concentric 20-turn coil of radius 0.500 cm carries a current of 1.00 A. The planes of the two coils are perpendicular. Find the magnitude of the torque exerted by the large coil on the small coil. (Neglect any variation in magnetic field due to the current in the large coil over the region occupied by the small coil.)

87 •• The magnetic needle of a compass is a uniform rod with a length of 3.00 cm, a radius of 0.850 mm, and a density of $7.96 \times 10^3\ \mathrm{kg/m^3}$. The needle is free to rotate in a horizontal plane, where the horizontal component of Earth's magnetic field is 0.600 G. When disturbed slightly, the compass executes simple harmonic motion about its midpoint with a frequency of 1.40 Hz. (*a*) What is the magnetic dipole moment of the needle? (*b*) What is the magnetization of the needle? (*c*) What is the amperian current on the surface of the needle?

88 •• A relatively inexpensive ammeter, called a *tangent galvanometer*, can be made using Earth's magnetic field. A plane circular coil that has N turns and a radius R is oriented so the magnetic field B_c it produces in the center of the coil is either east or west. A compass is placed at the center of the coil. When there is no current in the coil, assume the compass needle points due north. When there is a current in the coil (I), the compass needle points in the direction of the resultant magnetic field at an angle θ to the north. Show that the current I is related to θ and to the horizontal component of Earth's magnetic field B_e by $I = \dfrac{2RB_e}{\mu_0 N}\tan\theta$.

89 •• Earth's magnetic field is about 0.600 G at the magnetic poles, and is pointed vertically downward at the magnetic pole in the northern hemisphere. If the magnetic field were due to an electric current circulating in a loop at the radius of the inner iron core of Earth (approximately 1300 km), (*a*) what would be the magnitude of the current required? (*b*) What direction would this current have–the same as Earth's spin, or opposite? Explain your answer.

90 •• A long, narrow bar magnet has its magnetic moment $\vec{\mu}$ parallel to its long axis and is suspended at its center—in essence becoming a frictionless compass needle. When the magnet is placed in a magnetic field $\vec{B}$, it lines up with the field. If it is displaced by a small angle and released, show that the magnet will oscillate about its equilibrium position with frequency given by $\dfrac{1}{2\pi}\sqrt{\dfrac{\mu B}{I}}$, where I is the moment of inertia about the point of suspension.

91 •• An infinitely long, straight wire is bent, as shown in Figure 27-66. The circular portion has a radius of 10.0 cm and its center is a distance r from the straight part. Find the value of r such that the magnetic field at the center of the circular portion is zero.

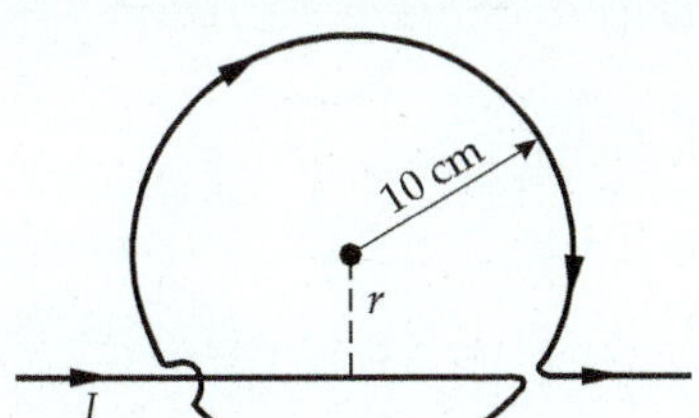

FIGURE 27-66 Problem 91

92 •• (*a*) Find the magnetic field strength at point P on the perpendicular bisector of a wire segment carrying current I, as shown in Figure 27-67. (*b*) Use your result from Part (*a*) to find the magnetic field strength at the center of a regular polygon of N sides. (*c*) Show that when N is very large, your result approaches that for the magnetic field strength at the center of a circle.

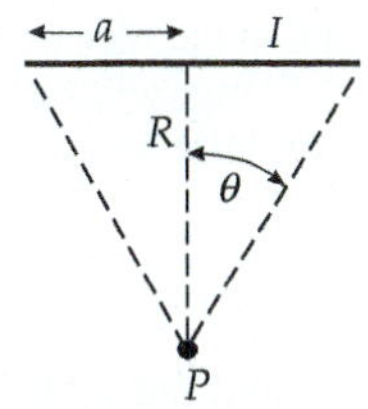

FIGURE 27-67 Problem 92

93 •• The current in a long cylindrical conductor of radius 10 cm varies with distance from the axis of the cylinder according to the relation $I(r) = (50\ \mathrm{A/m})r$. Find the magnetic field at the following perpendicular distances from the wire's central axis: (*a*) 5.0 cm, (*b*) 10 cm, and (*c*) 20 cm.

94 •• Figure 27-68 shows a square loop that has 20-cm-long sides and is in the $z = 0$ plane with its center at the origin. The loop carries a current of 5.0 A. An infinitely long wire that is parallel to the x axis and carries a current of 10 A intersects the z axis at $z = 10$ cm. The directions of the currents are shown in the figure. (*a*) Find the net torque on the loop. (*b*) Find the net force on the loop.

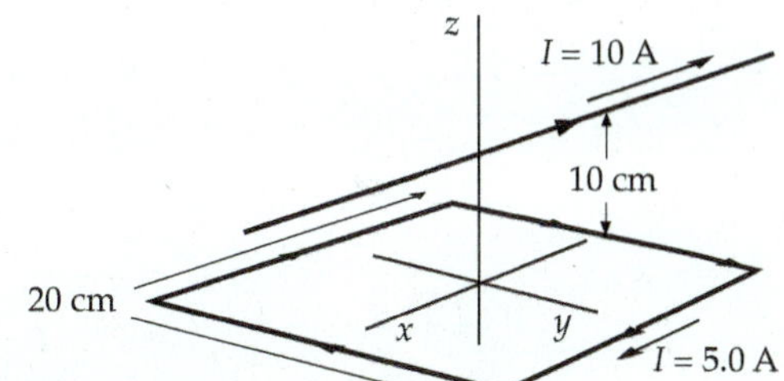

FIGURE 27-68 Problem 94

95 •• A current balance is constructed in the following way: A straight 10.0-cm-long section of wire is placed on top of the pan of an electronic balance (Figure 27-69). This section of wire is connected in series with a power supply and a long straight horizontal section of wire that is parallel to it and positioned directly above it. The distance between the central axes of the two wires is 2.00 cm. The power supply provides a current in the wires. When the power supply is switched on, the reading on the balance increases by 5.00 mg. What is the current in the wire? SSM

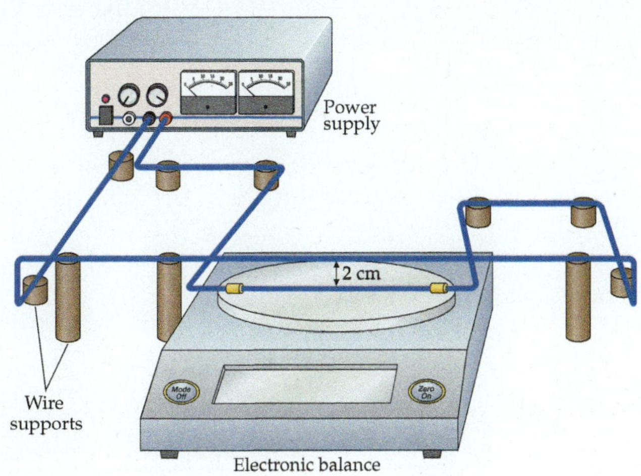

FIGURE 27-69 Problem 95

96 •• Consider the current balance of Problem 95. If the sensitivity of the balance is 0.100 mg, what is the minimum current detectable using this current balance?

97 ••• A nonconducting disk has radius R, carries a uniform surface charge density σ, and rotates with angular speed ω. (*a*) Consider an annular strip that has a radius r, a width dr, and a charge dq. Show that the current (dI) produced by this rotating strip is given by $\omega\sigma r\, dr$. (*b*) Use your result from Part (*a*) to show that the magnetic field strength at the center of the disk is given by the expression $\frac{1}{2}\mu_0\sigma\omega R$. (*c*) Use your result from Part (*a*) to find an expression for the magnetic field strength at a point on the central axis of the disk a distance z from its center. SSM

98 ••• A square loop that has sides of length ℓ lies in the $z = 0$ plane with its center at the origin. The loop carries a current I. (*a*) Derive an expression for the magnetic field strength at any point on the z axis. (*b*) Show that for z much larger than ℓ, your result from Part (*a*) becomes $B \approx \mu\mu_0/(2\pi z^3)$, where μ is the magnitude of the magnetic moment of the loop.

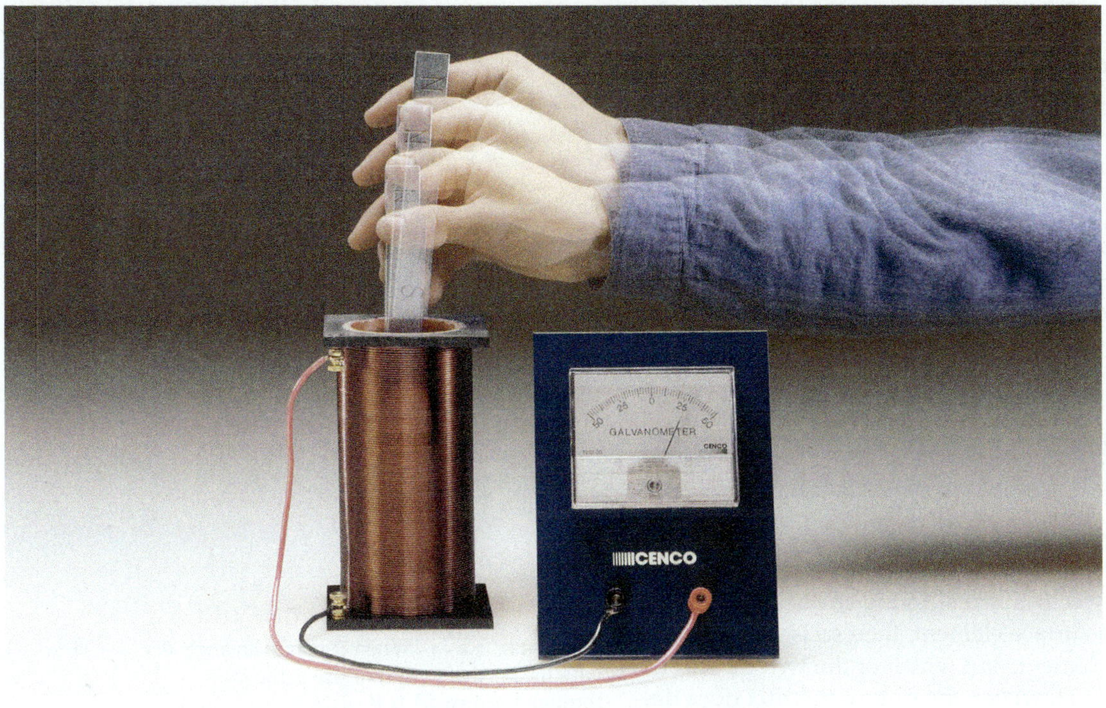

CHAPTER 28

DEMONSTRATION OF INDUCED EMF. WHEN THE MAGNET IS MOVING TOWARD OR AWAY FROM THE COIL, AN EMF IS INDUCED IN THE COIL, AS SHOWN BY THE GALVANOMETER'S DEFLECTION. NO DEFLECTION IS OBSERVED WHEN THE MAGNET IS STATIONARY. *(Richard Megna/ Fundamental Photographs.)*

Magnetic Induction

? How do you calculate the magnitude of an induced emf in a coil? (See Example 28-2.)

In the early 1830s, Michael Faraday in England and Joseph Henry in the United States independently discovered that in a *changing* magnetic field a changing magnetic flux through a surface bounded by a closed stationary loop of wire induces a current in the wire. The emfs and currents caused by such changing magnetic fluxes are called **induced emfs** and **induced currents.** The process itself is referred to as **induction.** Faraday and Henry also discovered that in a *static* magnetic field a changing magnetic flux through a surface bounded by a moving loop of wire induces an emf in the wire. An emf caused by the motion of a conductor in a region with a magnetic field is called a **motional emf.**

When you pull the plug of an electric cord from its socket, you sometimes observe a small spark. Before the cord is disconnected, the cord carries a current that produces a magnetic field encircling the current. When the cord is disconnected, the current abruptly ceases and the magnetic field encircling the cord collapses.

This changing magnetic field induces an emf that tends to maintain the original current, resulting in a spark at the points of the disconnect. Once the magnetic field collapses to zero it is no longer changing, and the induced emf is zero.

This chapter will explore the various methods of magnetic induction, all of which can be summarized by a single relation known as Faraday's law. Faraday's law relates the induced emf in a circuit to the rate of change in magnetic flux through the circuit. (The magnetic flux through the circuit *refers to the flux of the magnetic field through a surface bounded by the circuit.)*

28-1 MAGNETIC FLUX

The flux of any vector field through a surface is calculated in the same way as the flux of an electric field through a surface (Section 22-2). Let dA be an element of area on the surface S, and let $\hat{n}$ be a unit vector normal to the surface element of area dA (Figure 28-1). If $\hat{n}$ is a normal to a surface element, then so is $-\hat{n}$, so there are two directions normal to any surface element, and which of the two directions is selected for the direction of $\hat{n}$ is optional. However, the sign of the flux does depend on the choice for the direction of $\hat{n}$. The magnetic flux ϕ_m through S is

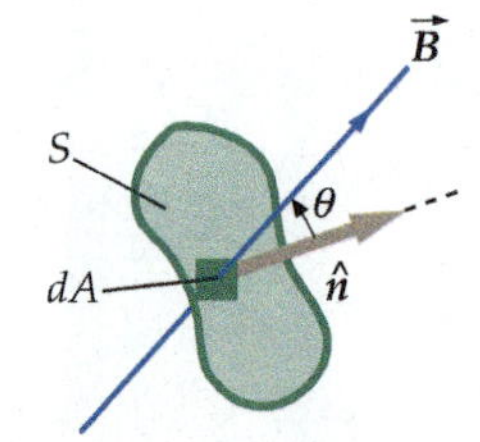

FIGURE 28-1 When $\vec{B}$ makes an angle θ with the normal to the surface S, the flux through the area dA is $\vec{B}\cdot\hat{n}\,dA = B\,dA\cos\theta$, where dA is the area of the surface element.

$$\phi_m = \int_S \vec{B}\cdot\hat{n}\,dA = \int_S B_n\,dA \qquad 28\text{-}1$$

MAGNETIC FLUX

The unit of magnetic flux is that of magnetic field strength multiplied by area, namely, the tesla-meter-squared, which is called a **weber** (Wb):

$$1\text{ Wb} = 1\text{ T}\cdot\text{m}^2 \qquad 28\text{-}2$$

Because B is proportional to the number of field lines per unit area, the magnetic flux is proportional to the number of field lines through an element of area.

> **PRACTICE PROBLEM 28-1**
> Show that one weber per second is one volt.

If the surface is flat and has an area A, and if $\vec{B}$ is uniform (has the same magnitude and direction) everywhere on the surface, the magnetic flux through the surface is

$$\phi_m = \vec{B}\cdot\hat{n}A = BA\cos\theta = B_nA \qquad 28\text{-}3$$

where θ is the angle between the direction of $\vec{B}$ and the direction of $\hat{n}$. We shall refer to the direction of $\hat{n}$ as the positive normal direction. We are often interested in the flux through a surface bounded by a coil that has several turns of wire. If the coil has N turns, the flux through the surface is N multiplied by the flux through each turn (Figure 28-2). That is,

$$\phi_m = NBA\cos\theta \qquad 28\text{-}4$$

where A is the area of the flat surface bounded by a single turn. (*Note:* Only a closed curve can actually bound a surface. A single turn of a multiturn coil is not closed, so a single turn cannot actually bound a surface. However, if a coil is tightly wound a single turn is almost closed, and A is the area of the flat surface that it almost bounds.)

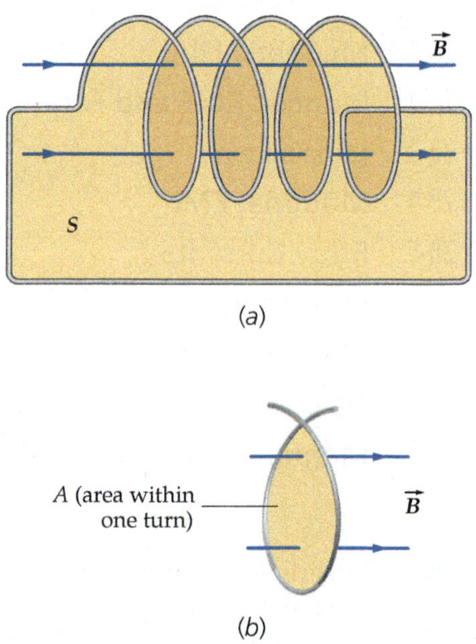

FIGURE 28-2 (*a*) The flux through the surface S bounded by a coil that has N turns is proportional to the number of field lines penetrating the surface. The coil shown has 4 turns. For the two field lines shown, each line penetrates the surface S four times, once for each turn, so the flux through S is four times greater than the flux through the surface "bounded" by a single turn of the coil. The coil shown is not tightly wound so the surface S can be seen better. (*b*) The area A of the flat surface is (almost) bounded by a single turn.

Example 28-1 Flux through a Solenoid

Find the magnetic flux through a solenoid that is 40 cm long, has a radius of 2.5 cm, has 600 turns, and carries a current of 7.5 A.

PICTURE The magnetic field $\vec{B}$ inside this long solenoid is uniform and parallel with the axis of the solenoid. (We are neglecting the end effects.) It is therefore perpendicular to the plane of each loop of the solenoid. Therefore, to find the flux we need to find B inside the solenoid and then multiply B by NA.

SOLVE

1. The magnetic flux is the product of the number of turns, the magnetic field strength, and the area bounded by one turn (Equation 28-4):
$$\phi_m = NBA$$
2. The magnetic field inside the solenoid is given by $B = \mu_0 nI$ (Equation 27-10), where $n = N/\ell$ is the number of turns per unit length:
$$\phi_m = N\mu_0 nIA = N\mu_0 \frac{N}{\ell} IA = \frac{\mu_0 N^2 IA}{\ell}$$
3. Express the area A in terms of its radius:
$$A = \pi r^2$$
4. Substitute the given values to calculate the flux:
$$\phi_m = \frac{\mu_0 N^2 I \pi r^2}{\ell}$$
$$= \frac{(4\pi \times 10^{-7}\ \text{T}\cdot\text{m/A})(600)^2(7.5\ \text{A})\pi(0.025\ \text{m})^2}{0.40\ \text{m}}$$
$$= \boxed{1.67 \times 10^{-2}\ \text{Wb}}$$

CHECK The units in line 2 of step 4 work out to $\text{T}\cdot\text{m}^2$, and a weber is defined as $1\ \text{T}\cdot\text{m}^2$. These are the correct units for magnetic flux.

TAKING IT FURTHER Note that because $\phi_m = NBA$ and B is proportional to the number of turns N, ϕ_m is proportional to N^2.

28-2 INDUCED EMF AND FARADAY'S LAW

Experiments by Faraday, Henry, and others showed that if the magnetic flux through a surface bounded by a wire (a conducting path) changes, an emf equal in magnitude to the rate of change of the flux is induced in the wire. We usually detect the emf by observing a current in the conductor, but the emf around the boundary of the surface exists even if the conducting path does not exist or is incomplete (not closed) and no current exists. In previous chapters, we considered emfs that were localized in a specific part of a circuit, such as between the terminals of a battery. However, induced emfs can be distributed throughout a circuit.

The magnetic flux ϕ_m through a flat surface of area A in a uniform magnetic field $\vec{B}$ is given by $\phi_m = BA\cos\theta$ (Equation 28-3), where θ is the angle between $\vec{B}$ and the normal to the surface. The flux can be changed by increasing or decreasing B, by increasing or decreasing A, or by changing the angle θ. If the magnetic field is due to a permanent magnet, the magnitude of the magnetic field can be increased or decreased by moving a permanent magnet toward or away from the surface. If the magnetic field is due to a current in a circuit, the magnitude of the magnetic field can be increased or decreased by increasing or decreasing the current. The flux through the surface can also be changed by varying the angle θ. To vary θ, we can change either the orientation of the surface or the direction of the magnetic field. In each case, if along the perimeter of the surface there is a conducting path, such as a metal wire, an emf $\mathcal{E}$ is induced along the path that is equal in magnitude to the rate of change of the magnetic flux through the surface. That is,

$$\mathcal{E} = -\frac{d\phi_m}{dt} \qquad 28\text{-}5$$

FARADAY'S LAW

This result is known as **Faraday's law.** The minus sign in Faraday's law has to do with the direction of the induced emf (clockwise or counterclockwise), which is addressed later in this section.

Figure 28-3 shows a single stationary loop of wire in a magnetic field. The flux through the loop is changing because the magnetic field strength on surface S is increasing, so an emf is induced in the loop. Because emf is the work done per unit charge, we know there must be forces exerted on the mobile charges that are doing work on these charges. Magnetic forces can do no work; therefore, we cannot attribute the emf to the work done by magnetic forces. It is electric forces associated with a nonconservative electric field $\vec{E}_{nc}$ doing the work on the mobile charges. The line integral of this electric field around a complete circuit equals the work done per unit charge, which is equal to the induced emf in the circuit.

The electric fields that we studied in earlier chapters resulted from static electric charges. Such electric fields are conservative, meaning that their circulation about any closed path C is zero. (The circulation of a vector field $\vec{E}$ about a closed path C is defined as $\oint_C \vec{E}\cdot d\vec{\ell}$.) However, the electric field associated with a changing magnetic field is nonconservative. Its circulation about C is equal to the induced emf in the loop of wire. The circulation of the electric field is equal to the negative of the rate of change of the magnetic flux through any surface S bounded by C:

$$\mathcal{E} = \oint_C \vec{E}_{nc}\cdot d\vec{\ell} = -\frac{d}{dt}\int_S \vec{B}\cdot\hat{n}\,dA = -\frac{d\phi_m}{dt} \qquad 28\text{-}6$$

INDUCED EMF FOR A STATIONARY CIRCUIT IN A CHANGING MAGNETIC FIELD

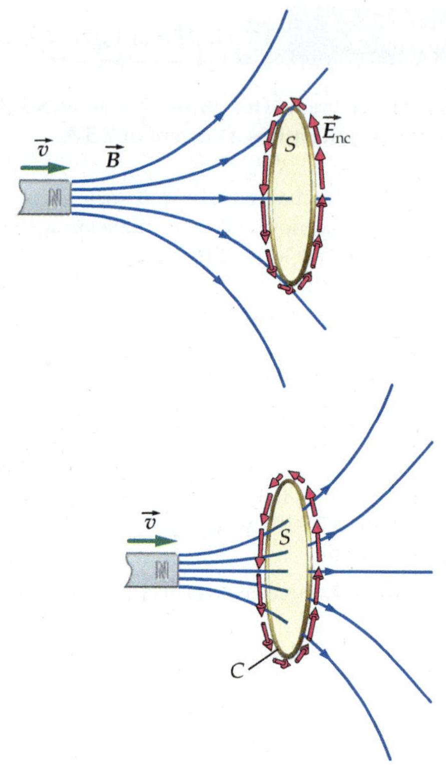

FIGURE 28-3 If the magnetic flux through the stationary wire loop is changing, an emf is induced in the loop. The emf is distributed throughout the loop, which is due to a nonconservative electric field $\vec{E}_{nc}$ tangent to the wire. The closed path C is within the material of the conducting loop.

Example 28-2 Induced EMF in a Circular Coil I

A uniform magnetic field makes an angle of 30.0° with the axis of a circular coil that has 300 turns and a radius equal to 4.00 cm. The magnitude of the magnetic field increases at a rate of 85.0 T/s while its direction remains fixed. Find the magnitude of the induced emf in the coil.

PICTURE The induced emf equals the number of turns N multiplied by the rate of change of the flux through a single turn. Because $\vec{B}$ is uniform, the flux through each turn is simply $\phi_m = BA\cos\theta$, where $A = \pi r^2$ is the area of the circle bounded by one turn of the coil.

SOLVE

1. The magnitude of the induced emf is given by Faraday's law: $\mathcal{E} = -\frac{d\phi_m}{dt}$

2. For a uniform field, the flux is: $\phi_m = N\vec{B}\cdot\hat{n}A = NBA\cos\theta$

3. Substitute this expression for ϕ_m and calculate $\mathcal{E}$:

$$\mathcal{E} = -\frac{d\phi_m}{dt} = -\frac{d}{dt}(NBA\cos\theta) = -N\pi r^2\cos\theta\frac{dB}{dt}$$
$$= (300)\pi(0.0400\text{ m})^2\cos 30.0°(85.0\text{ T/s}) = -111\text{ V}$$
$$\mathcal{E} = \boxed{111\text{ V}}$$

CHECK Line 2 of step 3 has units of $\text{T}\cdot\text{m}^2/\text{s}$, where $1\text{ T}\cdot\text{m}^2/\text{s} = 1\text{ Wb/s} = 1$ volt. [Use the formula $\vec{F} = q\vec{v}\times\vec{B}$ as a reminder that $1\text{ N} = 1\text{ C}\cdot\text{m}\cdot\text{T/s}$, so $1\text{ T} = 1\text{ N}\cdot\text{s/(C}\cdot\text{m)}$.]

PRACTICE PROBLEM 28-2 If the resistance of the coil is 200 Ω, what is the induced current?

Example 28-3 Induced EMF in a Circular Coil II — *Try It Yourself*

An 80.0-turn coil that has a radius equal to 5.00 cm and a resistance equal to 30.0 Ω sits in a region that has a uniform magnetic field normal to the plane of the coil. At what rate must the magnitude of the magnetic field change to produce a current of 4.00 A in the coil?

PICTURE The number of turns, multiplied by the rate of change of the magnetic flux through a surface bounded by a single turn, is equal to the negative of the induced emf using Faraday's law. The emf in the coil equals IR.

SOLVE

Cover the column to the right and try these on your own before looking at the answers.

Steps	Answers
1. Write the magnetic flux in terms of B, N, and the radius r, and solve for B.	$\phi_m = NBA = NB\pi r^2$ $B = \dfrac{\phi_m}{N\pi r^2}$
2. Take the time derivative of B.	$\dfrac{dB}{dt} = \dfrac{1}{N\pi r^2}\dfrac{d\phi_m}{dt}$
3. Use Faraday's law to relate the rate of change of the flux to the emf.	$\mathcal{E} = -\dfrac{d\phi_m}{dt}$
4. Calculate the magnitude of the emf in the coil from the current and resistance of the coil.	$\lvert\mathcal{E}\rvert = IR = 120\text{ V}$
5. Substitute numerical values of $\mathcal{E}$, N, and r to calculate $\lvert dB/dt\rvert$.	$\left\lvert\dfrac{dB}{dt}\right\rvert = \dfrac{1}{N\pi r^2}\lvert\mathcal{E}\rvert = \boxed{191\text{ T/s}}$

A sign convention allows us to use the minus sign in Faraday's law to find the direction of the induced emf. According to this convention, the positive tangential direction along the integration path C is related to the direction of the unit normal $\hat{n}$ on the surface S bounded by C by a right-hand rule (Figure 28-4). By placing your right thumb in the direction of $\hat{n}$, the fingers of your hand curl in the positive tangential direction on C. If $d\phi_m/dt$ is positive, then in accord with Faraday's law (Equation 28-6), $\mathcal{E}$ is in the negative tangential direction. (The direction of $\mathcal{E}$ can also be determined via Lenz's law, which is discussed in Section 28-3.)

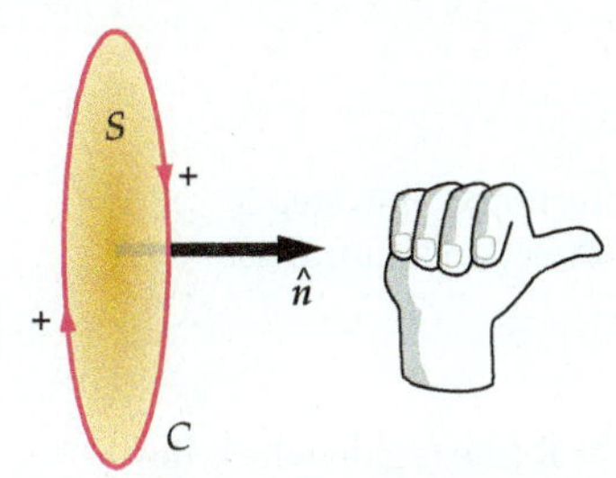

FIGURE 28-4 By placing your right thumb in the direction of $\hat{n}$ on the surface S, the fingers of your hand curl in the positive tangential direction on C.

Example 28-4 Induced Nonconservative Electric Field

A magnetic field $\vec{B}$ is perpendicular to the plane of the page. $\vec{B}$ is uniform throughout a circular region that has a radius R, as shown in Figure 28-5. Outside this region, B equals zero. The direction of $\vec{B}$ remains fixed and the rate of change of B is dB/dt. What are the magnitude and direction of the induced electric field in the plane of the page (*a*) a distance $r < R$ from the center of the circular region and (*b*) a distance $r > R$ from the center, where $B = 0$?

PICTURE The magnetic field $\vec{B}$ is into the page and uniform over a circular region of radius R, as shown in Figure 28-6. As B increases or decreases, the magnetic flux through a surface bounded by closed curve C also changes, and an emf $\mathcal{E} = \oint_C \vec{E}\cdot d\vec{\ell}$ is induced around C. The induced electric field is found by applying $\oint_C \vec{E}\cdot d\vec{\ell} = -d\phi_m/dt$ (Equation 28-6). To take advantage of the system's symmetry, we

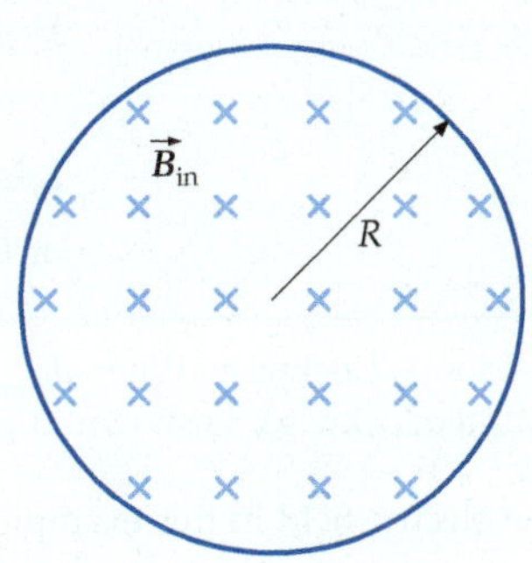

FIGURE 28-5

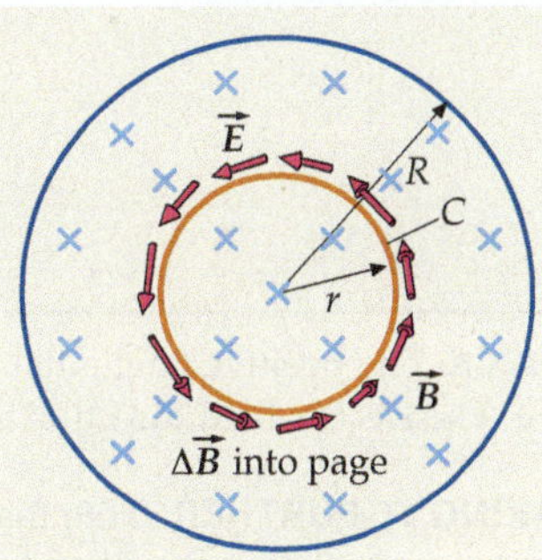

FIGURE 28-6

choose C to be a circular curve of radius r and then evaluate the line integral. By symmetry, $\vec{E}$ is tangent to circle C and has the same magnitude at any point on the circle. We will assign into the page as the direction of $\hat{n}$. The sign convention then tells us that the positive tangential direction is clockwise. We then calculate the magnetic flux ϕ_m, take its time derivative, and solve for E_t (the tangential component of $\vec{E}$).

SOLVE

(*a*) 1. The $\vec{E}$ and $\vec{B}$ fields are related by Equation 28-6:

$$\oint_C \vec{E}\cdot d\vec{\ell} = -\frac{d\phi_m}{dt}$$

where

$$\phi_m = \int_S \vec{B}\cdot\hat{n}\,dA$$

2. E_t (the tangential component of $\vec{E}$) is found from the line integral for a circle of radius $r < R$. $\vec{E}$ is tangent to the circle and has the same magnitude at all points on the circle:

$$\oint_C \vec{E}\cdot d\vec{\ell} = \oint_C E_t\,d\ell = E_t\oint_C d\ell = E_t 2\pi r$$

3. For $r < R$, $\vec{B}$ is uniform on the flat surface S bounded by the circle C. We choose into the page as the direction of $\hat{n}$. Because $\vec{B}$ is also into the page, the flux through S is simply BA:

$$\phi_m = \int_S \vec{B}\cdot\hat{n}\,dA = \int_S B_n\,dA = B_n\int_S dA$$
$$= BA = B\pi r^2$$

4. Calculate the time derivative of ϕ_m:

$$\frac{d\phi_m}{dt} = \frac{d}{dt}(B\pi r^2) = \frac{dB}{dt}\pi r^2$$

5. Substitute the step 2 and step 4 results into the step 1 result and solve for E_t:

$$E_t 2\pi r = -\frac{dB}{dt}\pi r^2$$

so

$$\boxed{E_t = -\frac{r}{2}\frac{dB}{dt} \quad r < R}$$

6. For the choice of the direction of $\hat{n}$ in step 3, the positive tangential direction is clockwise:

E_t is negative, so the direction of $\vec{E}$ is $\boxed{\text{counterclockwise}}$.

(*b*) 1. For a circle of radius $r > R$ (the region where the magnetic field is zero), the line integral is the same as before:

$$\oint_C \vec{E}\cdot d\vec{\ell} = E_t 2\pi r$$

2. Because $B = 0$ for $r > R$, the magnetic flux through S is $B\pi R^2$:

$$\phi_m = B\pi R^2$$

3. Apply Faraday's law to find E_t:

$$E_t 2\pi r = -\frac{dB}{dt}\pi R^2$$

so

$$\boxed{E_t = -\frac{R^2}{2r}\frac{dB}{dt} \quad r > R}$$

E_t is negative, so the direction of $\vec{E}$ is $\boxed{\text{counterclockwise}}$.

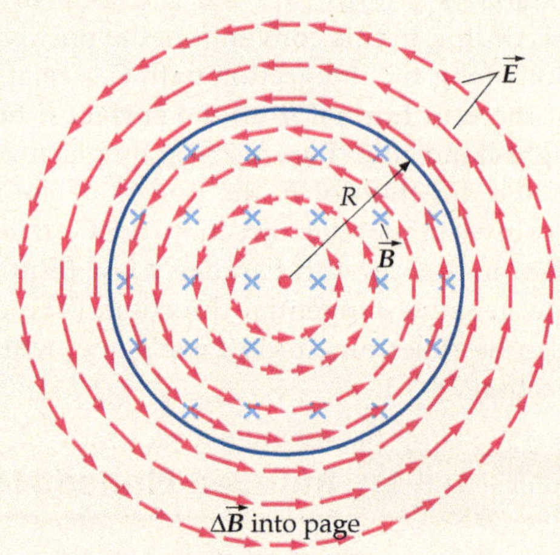

FIGURE 28-7 The magnetic field is into the page and increasing in magnitude. The induced electric field is counterclockwise.

CHECK The positive tangential direction is clockwise. When $d\phi_m/dt$ is positive, E_t is negative. The electric field direction is counterclockwise, as shown in Figure 28-7.

TAKING IT FURTHER Note that the electric field in this example is produced by a changing magnetic field rather than by electric charges. Note also that $\vec{E}$, and thus an emf, exists along any closed curve bounding the area through which the magnetic flux is changing, whether there is a wire or circuit along the curve or there is not.

! Note also that $\vec{E}$, and thus the emf, exists along any closed curve bounding the area through which the magnetic flux is changing, whether there is a wire or circuit along the curve or there is not.

28-3 LENZ'S LAW

The minus sign in Faraday's law has to do with the direction of the induced emf. This can be obtained by applying the sign convention described in the previous section, or by applying a general physical principle known as **Lenz's law:**

> The induced emf is in such a direction as to oppose, or tend to oppose, the change that produces it.
>
> LENZ'S LAW

Note that Lenz's law does not specify just what kind of change causes the induced emf and current. The statement of Lenz's law is purposely left vague to cover a variety of conditions, which we will now illustrate.

Figure 28-8 shows a bar magnet moving toward a conducting loop. It is the motion of the bar magnet to the right that induces an emf and current in the loop. Lenz's law tells us that this induced emf and current must be in a direction to oppose the motion of the bar magnet. That is, the current induced in the loop produces a magnetic field of its own, and this magnetic field must exert a force to the left on the approaching bar magnet. Figure 28-9 shows the induced magnetic moment of the current loop when the magnet is moving toward it. The loop acts like a small magnet with its north pole to the left and its south pole to the right. Because like poles repel, the induced magnetic moment of the loop repels the bar magnet; that is, it opposes its motion toward the loop. This result means the direction of the induced current in the loop must be as shown in Figure 28-9.

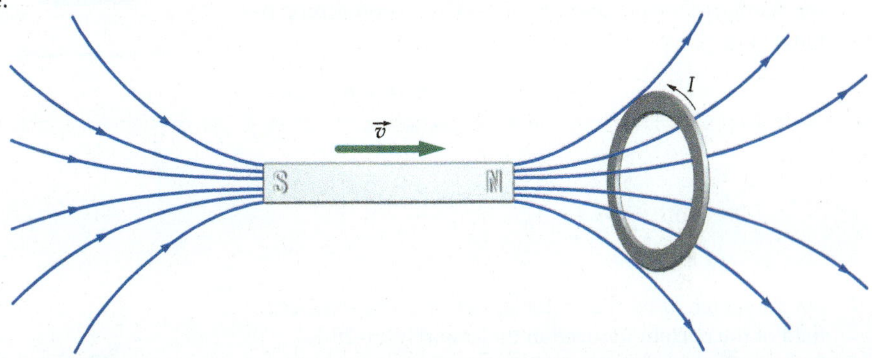

FIGURE 28-8 When the bar magnet is moving to the right, toward the loop, the emf induced in the loop produces an induced current in the direction shown. The magnetic field due to this induced current in the loop produces a magnetic field that exerts a force on the bar magnet opposing its motion to the right.

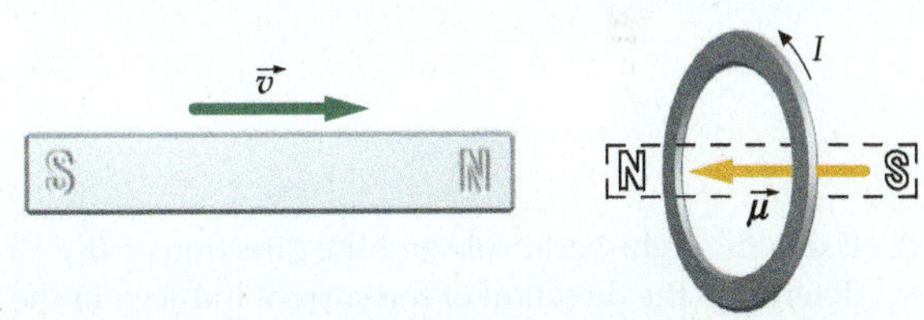

FIGURE 28-9 The magnetic moment of the loop $\vec{\mu}$ (shown in outline as if it were a bar magnet) due to the induced current is such as to oppose the motion of the bar magnet. The bar magnet is moving toward the loop, so the induced magnetic moment repels the bar magnet.

Suppose the induced current in the loop shown in Figure 28-9 was opposite to the direction shown. Then there would be a magnetic force toward the right on the approaching bar magnet, causing the bar magnet to gain speed. This gain in speed would cause an increase in the induced current, which in turn would cause the force on the bar magnet to increase, and so on. This result is too good to be true. Any time we nudge a bar magnet toward a conducting loop it would move toward the loop with ever increasing speed and with no significant effort on our part. Were this situation to occur, it would be a violation of energy conservation. The reality, however, is that energy is conserved, and Lenz's law is consistent with this reality.

An alternative statement of Lenz's law in terms of magnetic flux is frequently of use. This statement is

> When a magnetic flux through a surface changes, the magnetic field due to any induced current produces a flux of its own—through the same surface and opposite in sign to the initial change in flux.
>
> ALTERNATIVE STATEMENT OF LENZ'S LAW

For an example which shows how this alternative statement is applied, see Example 28-5.

Example 28-5 Lenz's Law and Induced Current

Using the alternative statement of Lenz's law, find the direction of the induced current in the loop shown in Figure 28-8.

PICTURE Use the alternative statement of Lenz's law to determine the direction of the magnetic field due to the current induced in the loop. When a magnetic flux through a surface changes, the magnetic field due to any induced current produces a flux of its own—through the same surface and opposite in sign to the initial change in flux. Then use a right-hand rule to determine the direction of the induced current.

SOLVE

1. Draw a sketch of the loop bounding the flat surface *S* (Figure 28-10). On surface *S* draw the vector $\Delta\vec{B}_1$, which is the change in the magnetic field $\vec{B}_1$ of the approaching bar magnet on *S*:

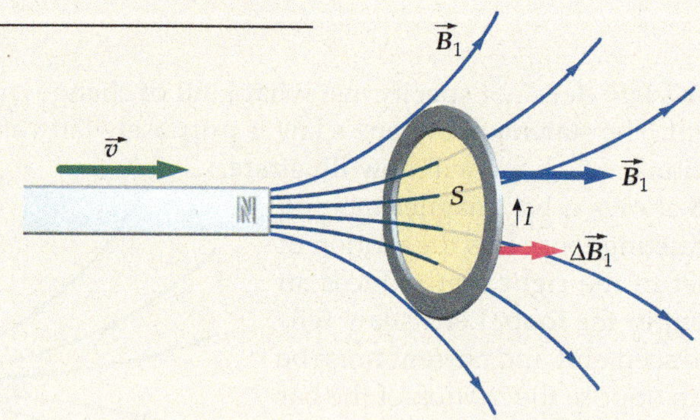

FIGURE 28-10

2. On the sketch draw the vector $\vec{B}_2$, which is the magnetic field of the current induced in the loop (Figure 28-11). Because $\vec{B}_2$ was initially zero, $\vec{B}_2$ is in the same direction as $\Delta\vec{B}_2$. Use the alternative statement of Lenz's law to determine the direction of $\vec{B}_2$. $\vec{B}_2$ and $\Delta\vec{B}_1$ must penetrate *S* in opposite directions for the change in the flux of $\vec{B}_2$ to be opposite in sign to the change in flux of $\vec{B}_1$:

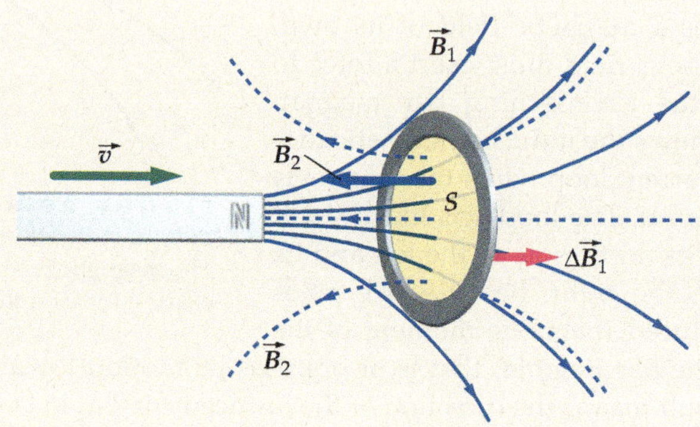

FIGURE 28-11

3. Using the right-hand rule and the direction of $\vec{B}_2$, determine the direction of the current induced in the loop (Figure 28-12):

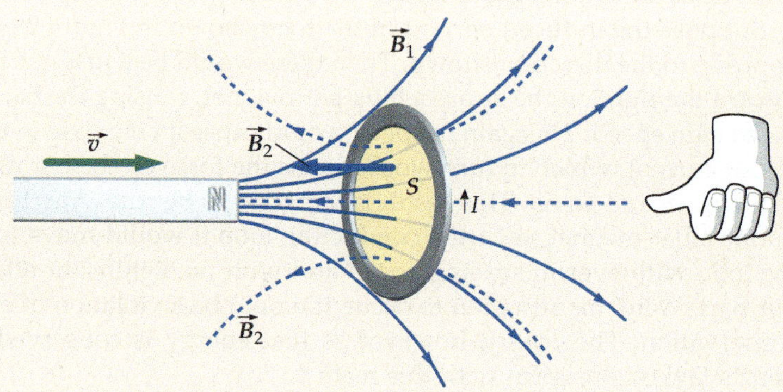

FIGURE 28-12

CHECK The step-3 result gives the same direction as was obtained on page 965 using the initial statement of Lenz's law.

CONCEPT CHECK 28-1

Using the alternative statement of Lenz's law, find the direction of the induced current in the loop shown in Figure 28-8 if the magnet is moving to the left (away from the loop).

In Figure 28-13, the bar magnet is at rest and the loop is moving away from the magnet. The induced current and magnetic moment are shown in the figure. In this case, the bar magnet attracts the loop, thus opposing the motion of the loop as required by Lenz's law.

In Figure 28-14, when the current in circuit 1 is changing, there is a changing flux through circuit 2. Suppose that the switch S in circuit 1 is initially open so that there is no current in the circuit (Figure 28-14*a*). When we close the switch (Figure 28-14*b*), the current in circuit 1 does not instantaneously reach its steady value $\mathcal{E}_1/R_1$ but takes some time to change from zero to that value. During the time the current is increasing, the flux through circuit 2 is changing and a current is induced in circuit 2 in the direction shown. When the current in circuit 1 reaches its steady value, the flux through circuit 2 is no longer changing, so there is no longer an induced current in circuit 2. An induced current in circuit 2 in the opposite direction appears briefly when the switch in circuit 1 is opened (Figure 28-14*c*) and the current in circuit 1 is decreasing to zero. It is important to understand that there is an induced emf *only while the flux is changing*. The emf does not depend on the magnitude of the flux itself, but only on its rate of change. If there is a large steady flux through a circuit, there is no induced emf.

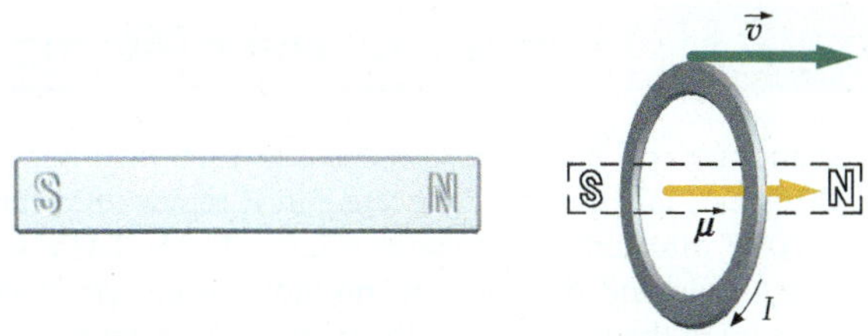

FIGURE 28-13 When the loop is moving away from the stationary bar magnet, the bar magnet attracts the magnetic moment of the loop, again opposing the relative motion.

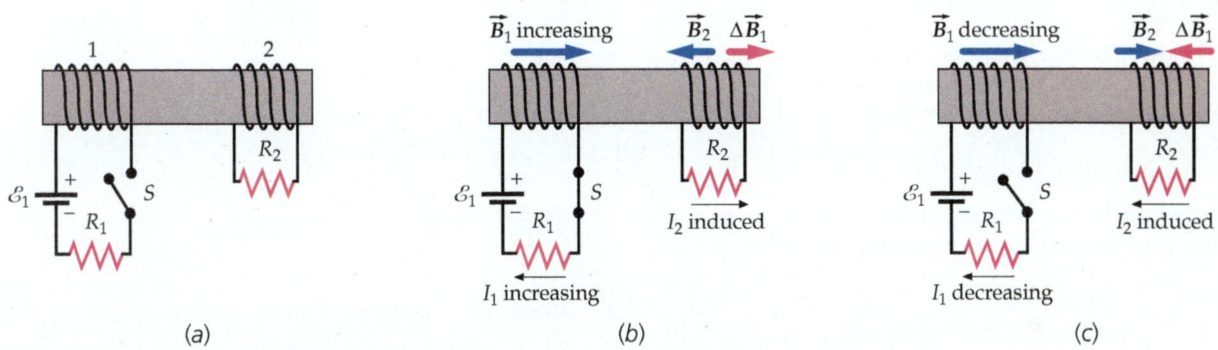

FIGURE 28-14 (*a*) Two adjacent circuits. (*b*) Just after the switch is closed, I_1 is increasing in the direction shown. The changing flux through circuit 2 induces the current I_2. The flux through circuit 2 due to I_2 opposes the change in flux due to I_1. (*c*) As the switch is opened, I_1 decreases and the flux through circuit 2 changes. The induced current I_2 then tends to maintain the flux through circuit 2.

For our next example, we consider the single isolated circuit shown in Figure 28-15. If there is a current in the circuit, there is a magnetic flux through the coil due to its own current. If the current is changing, the flux through the coil is changing and there is an induced emf in the circuit while the flux is changing. This *self-induced emf* opposes the change in the current. It is therefore called a **back emf.** Because of this self-induced emf, the current in a circuit cannot jump instantaneously from zero to some finite value or from some finite value to zero. Henry first noticed this effect when he experimented with a circuit consisting of many turns of a wire like that in Figure 28-15. This arrangement gives a large flux through the circuit for even a small current. Joseph Henry noticed a spark across the switch when he tried to break the circuit. Such a spark is due to the large induced emf that occurs when the current varies rapidly, as during the opening of the switch. In this case, the induced emf is directed so as to maintain the original current. The large induced emf produces a large potential difference across the switch as it is opened. The electric field between the contacts of the switch is large enough to produce dielectric breakdown in the surrounding air. When dielectric breakdown occurs, the air conducts electric current in the form of a spark.

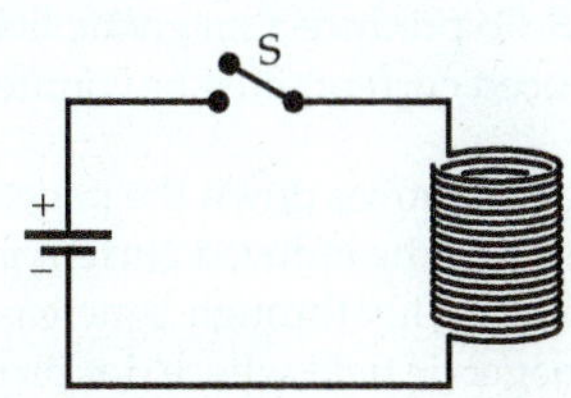

FIGURE 28-15 The coil that has many turns of wire yields a large flux for a given current in the circuit. Thus, when the current changes, there is a large emf induced in the coil opposing the change.

Example 28-6 Lenz's Law and a Moving Coil

A rectangular coil has $N = 80$ turns and each turn has a width $a = 20.0$ cm and a length $b = 30.0$ cm. Half the coil is located in a region that has a magnetic field of magnitude $B = 0.800$ T directed into the page (Figure 28-16). The resistance R of the coil is 30.0 Ω. Find the magnitude and direction of the induced current if the coil is moved with a speed of 2.00 m/s (*a*) to the right, (*b*) up the page, and (*c*) down the page.

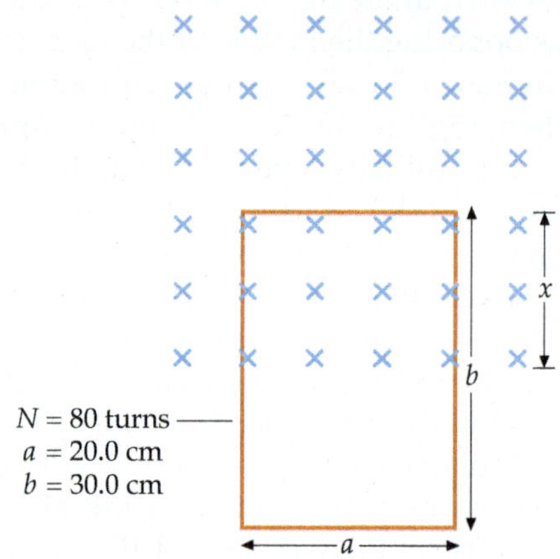

FIGURE 28-16

PICTURE The induced current equals the induced emf divided by the resistance. We can calculate the emf induced in the circuit as the coil moves by calculating the rate of change of the flux through the coil. The direction of the induced current is found from Lenz's law.

SOLVE

(*a*) 1. The induced current equals the emf divided by the resistance:

$$I = \frac{\mathcal{E}}{R}$$

2. The induced emf and the magnetic flux are related by Faraday's law:

$$\mathcal{E} = -\frac{d\phi_m}{dt}$$

3. The flux through the surface bounded by the coil is N multiplied by the flux through each turn of the coil. We choose into the page as the direction of $\hat{n}$. The flux through the surface S bounded by a single turn is Bax:

$$\phi_m = N\vec{B}\cdot\hat{n}A$$
$$= N[Bax - (0)a(b - x)] = NBax$$

4. When the coil is moving to the right (or to the left), x does not change and the flux does not change (until the coil leaves the region of the magnetic field). The current is therefore zero:

$$\mathcal{E} = -\frac{d\phi_m}{dt} = 0$$

so

$$I = \boxed{0}$$

(*b*) 1. Compute the rate of change of the flux when the coil is moving up the page. In this case x is increasing, so dx/dt is positive:

$$\frac{d\phi_m}{dt} = \frac{d}{dt}(NBax) = NBa\frac{dx}{dt}$$

2. The derivative dx/dt is equal to the speed of the coil.

$$|I| = \frac{|\mathcal{E}|}{R} = \frac{NBa|dx/dt|}{R} = \frac{(80)(0.800\text{ T})(0.200\text{ m})(2.00\text{ m/s})}{30.0\ \Omega}$$
$$= 0.853\text{ A}$$

3. As the coil moves up the page, the flux of $\vec{B}$ through S is increasing. The induced current must produce a magnetic field whose flux through S decreases as x increases. That would be a magnetic field whose dot product with $\hat{n}$ is negative. Such a magnetic field is directed out of the page on S. To produce a magnetic field in this direction the induced current must be counterclockwise:

$$\boxed{I = 0.853\text{ A, counterclockwise}}$$

(*c*) As the coil moves down the page, the flux of $\vec{B}$ through S is decreasing. The induced current must produce a magnetic field whose flux through S increases as x decreases. That would be a magnetic field whose dot product with $\hat{n}$ is positive. Such a magnetic field is directed into the page on S. To produce a magnetic field in this direction the induced current must be clockwise:

$$\boxed{I = 0.853\text{ A, clockwise}}$$

CHECK In Part (*b*) the motion up the page causes the current to be induced, so the direction of the induced current must result in a force opposing the upward motion. Applying $\vec{F} = I\vec{L} \times \vec{B}$ (Equation 26-4) to the upper section of the loop gives a force down the page if the current in the loop is counterclockwise. This is in agreement with our Part-(*b*) result.

TAKING IT FURTHER In this example the magnetic field is static, so no nonconservative electric field exists. Thus, the emf is not the work done by a nonconservative electric field. The cause of this emf is examined in the next section.

28-4 MOTIONAL EMF

The emf induced in a conductor due to its motion in a region in which there exists a magnetic field is called motional emf. More generally,

> Motional emf is any emf induced by the motion of a conductor in a region in which there exists a magnetic field.
>
> DEFINITION—MOTIONAL EMF

Example 28-7 Total Charge through a Flipped Coil

A small coil of N turns has its plane perpendicular to a uniform static magnetic field $\vec{B}$, as shown in Figure 28-17. The coil is connected to a current integrator (C.I.), which is a device used to measure the total charge passing through the coil. Find the charge passing through the coil if the coil is rotated through 180° about the axis shown.

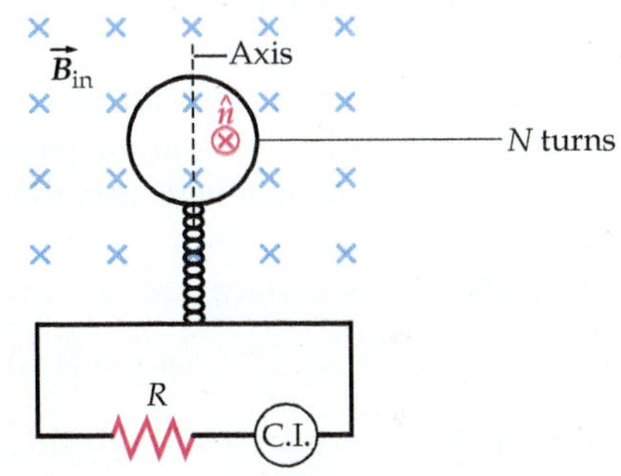

FIGURE 28-17

PICTURE When the coil in Figure 28-17 is rotated, the magnetic flux through the coil changes, causing an induced emf $\mathcal{E}$. The emf in turn causes a current $I = \mathcal{E}/R$, where R is the total resistance of the circuit. Because $I = dq/dt$, we can find the charge Q passing through the integrator by integrating I; that is, $Q = \int I\,dt$.

SOLVE

1. The increment of charge dq equals the current I multiplied by the increment of time dt:

$$dq = I\,dt$$

2. The emf $\mathcal{E}$ is related to I by Ohm's law:

$$\mathcal{E} = RI$$

so

$$\mathcal{E}\,dt = RI\,dt$$

3. The emf is related to the flux ϕ_m by Faraday's law:

$$\mathcal{E} = -\frac{d\phi_m}{dt}$$

or

$$\mathcal{E}\,dt = -d\phi_m$$

4. Substitute $-d\phi_m$ for $\mathcal{E}\,dt$ and dq for $I\,dt$ in the step 2 result and solve for dq:

$$-d\phi_m = R\,dq$$

so

$$dq = -\frac{1}{R}\,d\phi_m$$

5. Integrate to find the total charge Q:

$$Q = \int_0^Q dq = -\frac{1}{R}\int_{\phi_{mi}}^{\phi_{mf}} d\phi_m = -\frac{1}{R}(\phi_{mf} - \phi_{mi}) = -\frac{\Delta\phi_m}{R}$$

6. The flux through the coil is $\phi_m = n\vec{B}\cdot\hat{n}A$, where $\hat{n}$ is the normal to the flat surface bounded by the coil (Figure 28-18). The normal initially is directed into the page. When the coil rotates, so does the surface and its normal. Find the change in ϕ_m when the coil rotates 180°:

$$\Delta\phi_m = \phi_{mf} - \phi_{mi} = N\vec{B}\cdot\hat{n}_f A - N\vec{B}\cdot\hat{n}_i A$$
$$= NA(\vec{B}\cdot\hat{n}_f - \vec{B}\cdot\hat{n}_i) = NA[(-B) - (+B)] = -2NBA$$

FIGURE 28-18

7. Combining the results from the previous two steps yields Q:

$$\boxed{Q = \frac{2NBA}{R}}$$

TAKING IT FURTHER Note that the charge Q does not depend on whether or not the coil is rotated slowly or quickly—all that matters is the change in the magnetic flux through the coil. A coil used in this way is called a *flip coil*. It is used to measure magnetic fields. For example, if the current integrator (C.I.) measures a total charge Q passing through the coil when it is flipped, the magnetic field strength is given by $B = RQ/(2NA)$, which can be obtained directly from the step-7 result.

PRACTICE PROBLEM 28-3 A flip coil has 40 turns, a radius of 3.00 cm, and a resistance of 16.0 Ω, and the plane of the coil is initially perpendicular to a static, uniform 0.500-T magnetic field. If the coil is rotated 90° about an axis perpendicular to the magnetic field, how much charge passes through the coil?

Figure 28-19 shows a thin conducting rod sliding to the right along conducting rails that are connected by a resistor. A uniform magnetic field $\vec{B}$ is directed into the page.

Consider the magnetic flux through the flat surface S bounded by the circuit. Let the normal $\hat{n}$ to the surface be into the page. As the rod moves to the right, the surface S increases, as does the magnetic flux through S. Thus, an emf is induced in the circuit. Let ℓ be the separation of the rails and x be the distance from the left end of the rails to the rod. The area of surface S is then ℓx, and the magnetic flux through S is

$$\phi_{\mathrm{m}} = \vec{B} \cdot \hat{n} A = B_n A = B\ell x$$

Taking the time derivative of both sides gives

$$\frac{d\phi_{\mathrm{m}}}{dt} = B\ell \frac{dx}{dt} = B\ell v$$

where $v = dx/dt$ is the speed of the rod. The emf induced in this circuit is therefore

$$\mathcal{E} = -\frac{d\phi_{\mathrm{m}}}{dt} = -B\ell v$$

where the negative sign tells us that the emf is in the negative tangential direction. Put your right thumb in the direction of $\hat{n}$ (into the page) and your fingers will curl in the positive tangential direction (clockwise). Thus, the induced emf is counterclockwise.

We can check this result (the direction of the induced emf) using Lenz's law. It is the motion of the rod to the right that produces the induced current, so the magnetic force on this rod due to the induced current must be to the left. The magnetic force on a current-carrying conductor is given by $I\vec{L} \times \vec{B}$ (Equation 26-4), where $\vec{L}$ is in the direction of the current. If $\vec{L}$ is up the page and $\vec{B}$ is into the page, the force is to the left, which affirms our previous result (that the induced emf is counterclockwise). If the rod is given some initial velocity $\vec{v}$ to the right and is then released, the force due to the induced current slows the rod until it stops. To maintain the motion of the rod, an external force pushing the rod to the right must be maintained.

A second check on the direction of the induced emf and current is implemented by considering the direction of the magnetic force on the charge carriers moving to the right with the rod. The charge carriers move rightward with the same velocity $\vec{v}$ as the rod, so the charge carriers experience a magnetic force $\vec{F} = q\vec{v} \times \vec{B}$. If q is positive that force is upward, which means the induced emf is counterclockwise.

$$|\mathcal{E}| = B\ell v \qquad 28\text{-}7$$

MAGNITUDE OF EMF FOR A ROD MOVING PERPENDICULAR TO BOTH THE LENGTH OF THE ROD AND $\vec{B}$

(If the magnetic field is not normal to the plane of the circuit, the B in Equation 28-7 should be replaced with the component of B normal to the plane of the circuit.)

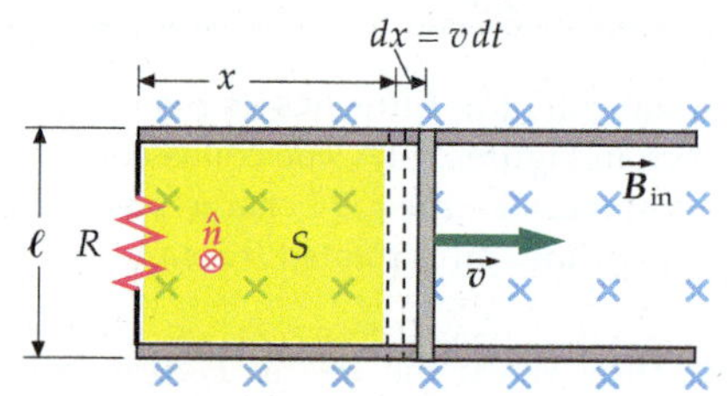

FIGURE 28-19 A conducting rod sliding on conducting rails in a magnetic field. As the rod moves to the right, the area of the surface S increases, so the magnetic flux through S into the paper increases. An emf of magnitude $B\ell v$ is induced in the circuit, inducing a counterclockwise current that produces flux through the surface S directed out of the paper opposing the change in flux due to the motion of the rod.

Figure 28-20 shows a positive charge carrier in a conducting rod that is moving at constant speed through a uniform magnetic field directed into the paper. Because the charge carrier is moving horizontally with the rod, there is an upward magnetic force on the charge carrier of magnitude qvB. Responding to that force, the charge carriers in the rod move upward, producing a net positive charge at the top of the rod and leaving a net negative charge at the bottom of the rod. The charge carriers continue to move upward until the electric field $\vec{E}_{\parallel}$ produced by the separated charges exerts a downward force of magnitude $qE_{\parallel}$ on the separated charges, which balances the upward magnetic force qvB. In equilibrium, the magnitude of this electric field in the rod is

$$E_{\parallel} = vB$$

The direction of this electric field is parallel to the rod, directed downward. The associated potential difference across the length ℓ of the rod is

$$\Delta V = E_{\parallel}\ell = vB\ell$$

with the potential being higher at the top. That is, when there is no current through the rod, the potential difference across the length of the rod equals $vB\ell$ (the motional emf). When there is a current I through the rod, the potential difference is

$$\Delta V = vB\ell - Ir \qquad 28\text{-}8$$

where r is the resistance of the rod.

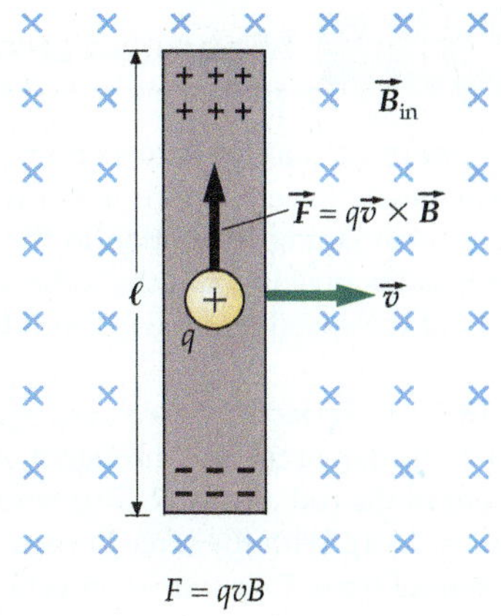

FIGURE 28-20 A positive charge carrier in a conducting rod that is moving through a magnetic field experiences a magnetic force that has an upward component. Some of these charge carriers move to the top of the rod, leaving the bottom of the rod negative. The charge separation produces a downward electric field of magnitude $E_{\parallel} = vB$ in the rod. Thus, the potential at the top of the rod is greater than the potential at the bottom of the rod by $E_{\parallel}\ell = vB\ell$.

PRACTICE PROBLEM 28-4

A rod 40 cm long moves at 12 m/s in a plane perpendicular to a magnetic field of 0.30 T. The rod's velocity is perpendicular to its length. Find the emf induced in the rod.

Example 28-8 A U-Shaped Conductor and a Sliding Rod — *Try It Yourself*

Using Figure 28-19, let $B = 0.600$ T, $v = 8.00$ m/s, $\ell = 15.0$ cm, and $R = 25.0\ \Omega$; assume that the resistances of the rod and the rails are negligible. Find (*a*) the induced emf in the circuit, (*b*) the current in the circuit, (*c*) the force needed to move the rod with constant velocity, and (*d*) the power dissipated in the resistor.

SOLVE

Cover the column to the right and try these on your own before looking at the answers.

Steps	Answers
1. Calculate the induced emf from Equation 28-7.	$\mathcal{E} = Bv\ell =$ 0.720 V
2. Find the current from Ohm's law.	$I = \frac{\mathcal{E}}{R} =$ 28.8 mA
3. The force needed to move the rod with constant velocity is equal and opposite to the force exerted by the magnetic field on the rod, which has the magnitude $IB\ell$ (Equation 26-4). Calculate the magnitude of this force.	$F = IB\ell =$ 2.59 mN
4. Find the power dissipated in the resistor.	$P = I^2R =$ 20.7 mW

CHECK Using $P = Fv$, we confirm that the power is 20.7 mW.

TAKING IT FURTHER The potential at the top of the rod is greater than the potential at the bottom of the rod by the emf.

Example 28-9 Magnetic Drag

A rod that has a mass m slides on frictionless conducting rails in a region that has a static uniform magnetic field $\vec{B}$ directed into the page (Figure 28-21). An external agent is pushing the rod, maintaining its motion to the right at constant speed v_0. At time $t = 0$, the agent abruptly stops pushing and the rod continues forward while being slowed by the magnetic force. Find the speed v of the rod as a function of time.

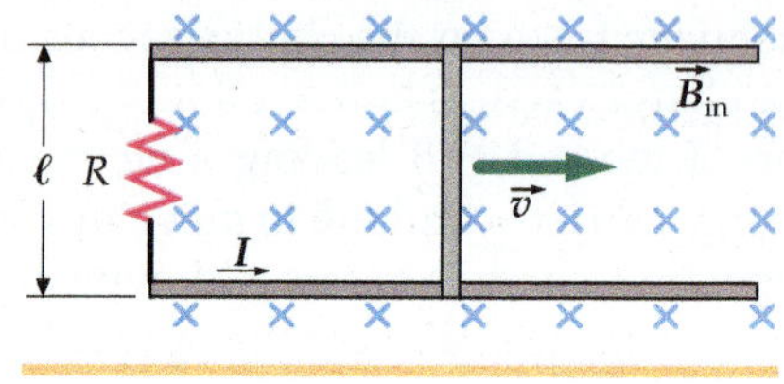

FIGURE 28-21

PICTURE The speed of the rod changes because a magnetic force acts on the induced current. The motion of the rod through a magnetic field induces an emf $\mathcal{E} = B\ell v$ and, therefore, a current in the rod, $I = \mathcal{E}/R$. This result causes a magnetic force to act on the rod, $F = I\ell B$ (Equation 26-4). With the force known, we apply Newton's second law to find the speed as a function of time. Take the $+x$ direction as being to the right.

SOLVE

1. Apply Newton's second law to the rod:

$$F_x = ma_x = m\frac{dv}{dt}$$

2. The force exerted on the rod is the magnetic force (Equation 26-4), which is proportional to the current and in the $-x$ direction, as shown in Figure 28-21:

$$F_x = -I\ell B$$

3. The current equals the motional emf divided by the resistance of the rod:

$$I = \frac{\mathcal{E}}{R} = \frac{B\ell v}{R}$$

4. Combining these results, we find the x component of the magnetic force exerted on the rod:

$$F_x = -IB\ell = -\frac{B\ell v}{R}B\ell = -\frac{B^2\ell^2 v}{R}$$

5. Newton's second law then gives:

$$-\frac{B^2\ell^2 v}{R} = m\frac{dv}{dt}$$

6. Separate the variables, then integrate the velocity from v_0 to v_f and the time from 0 to t_f:

$$\frac{dv}{v} = -\frac{B^2\ell^2}{mR}dt$$

$$\int_{v_0}^{v_f}\frac{dv}{v} = -\frac{B^2\ell^2}{mR}\int_0^{t_f} dt$$

$$\ln\frac{v_f}{v_0} = -\frac{B^2\ell^2}{mR}t_f$$

7. Let $v = v_f$ and $t = t_f$, then solve for v:

$$v = \boxed{v_0 e^{-t/\tau} \quad \text{where } \tau = \frac{mR}{B^2\ell^2}}$$

CHECK The kinetic energy of the rod is transformed into thermal energy in the resistor. To conserve energy the kinetic energy of the rod must decrease, which means its speed must decrease. The step 7 result is in agreement with the conservation of energy.

TAKING IT FURTHER If the force were constant, the rod's speed would decrease linearly with time. However, because the force is proportional to the rod's speed, as found in step 4, the force is large initially but the force decreases as the speed decreases. In principle, the rod never stops moving. Even so, the rod travels only a finite distance.

GENERATORS AND MOTORS

Most electrical energy used today is produced by electric generators in the form of alternating current (ac). A simple **generator** of alternating current is a coil rotating in a uniform magnetic field as shown in Figure 28-22. The ends of the coil are connected to rings called slip rings that rotate with the coil. Electrical contact is made with the coil by stationary graphite brushes in contact with the rings.

When the normal to the plane of the coil $\hat{n}$ makes an angle θ with a uniform magnetic field $\vec{B}$, as shown in the figure, the magnetic flux through the coil is

$$\phi_m = NBA\cos\theta \qquad 28\text{-}9$$

where N is the number of turns in the coil and A is the area of the flat surface bounded by the coil. When the coil is mechanically rotated, the flux through it will change, and an emf will be induced in the coil according to Faraday's law. If the initial angle between $\hat{n}$ and $\vec{B}$ is zero, then the angle at some later time t is given by

$$\theta = \omega t$$

where ω is the angular frequency of rotation. Substituting this expression for θ into Equation 28-9, we obtain

$$\phi_m = NBA\cos\omega t = NBA\cos 2\pi f t$$

The emf in the coil will then be

$$\mathcal{E} = -\frac{d\phi_m}{dt} = -NBA\frac{d}{dt}\cos\omega t = \omega NBA\sin\omega t \qquad 28\text{-}10$$

This can be written

$$\mathcal{E} = \mathcal{E}_{max}\sin\omega t$$

where

$$\mathcal{E}_{max} = \omega NBA$$

is the maximum value of the emf. We can thus produce a sinusoidal emf in a coil by rotating it with constant frequency in a magnetic field. In this source of emf, the mechanical energy of the rotating coil is converted into electric energy. The mechanical energy usually comes from a waterfall or a steam turbine. Although practical generators are considerably more complicated, they work on the same principle that an alternating emf is produced in a coil rotating in a magnetic field, and they are designed so that the emf produced is sinusoidal.

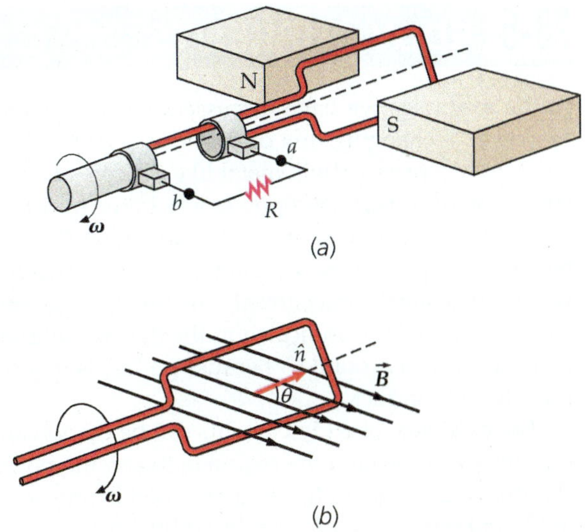

FIGURE 28-22 (a) An ac generator. A coil rotating with constant angular speed ω in a magnetic field $\vec{B}$ generates a sinusoidal emf. Energy from a waterfall or a steam turbine is used to rotate the coil to produce electrical energy. The emf is supplied to an external circuit by the brushes in contact with the rings. (b) At this instant, the normal to the plane of the coil makes an angle θ with the magnetic field and the flux is equal to $BA\sin\theta$.

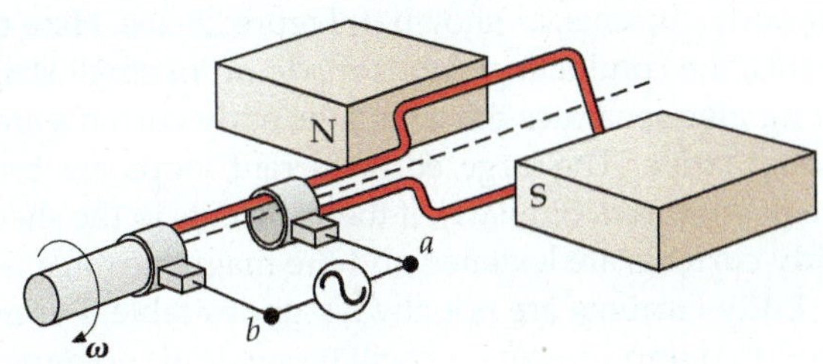

FIGURE 28-23 When alternating current is supplied to the coil of Figure 28-22, the coil becomes a motor. As the coil rotates, a back emf is generated, limiting the current.

The same coil in a magnetic field that can be used to generate an alternating emf can also be used as an ac **motor.** Instead of mechanically rotating the coil to generate an emf, we apply an alternating current to the coil from another ac generator as shown in Figure 28-23. (In circuit diagrams, an ac generator is represented by the symbol ⊙.) A current loop in a magnetic field experiences a torque that tends to rotate the loop such that its magnetic moment $\vec{\mu}$ points in the direction of $\vec{B}$ and the plane of the loop is perpendicular to $\vec{B}$. If direct current were supplied to the coil in Figure 28-23, the torque on the coil would change directions when the coil rotates past its equilibrium position, which is when the plane of the coil is vertical in the figure. The coil would then oscillate about its equilibrium position, eventually coming to rest there with its plane vertical. However, if the direction of the current is reversed just as the coil passes the vertical position, the torque does not change direction but continues to rotate the coil in the same direction. As the coil rotates in the magnetic field, a back emf is generated that tends to oppose the current. When the motor is first turned on, there is no back emf and the current is very large, being limited only by the resistance in the circuit. As the motor begins to rotate, the back emf increases and the current decreases.

PRACTICE PROBLEM 28-5

A 250-turn coil has an area per turn of 3.0 cm^2. If it rotates at 60 rev/s in a 0.40-T magnetic field at 60 Hz, what is the maximum emf in the coil?

CONCEPT CHECK 28-2

When a generator delivers electric energy to a circuit, where does the energy come from?

28-5 EDDY CURRENTS

In the examples we have discussed, currents were induced in thin wires or rods. However, a changing flux often induces circulating currents, which are called *eddy currents*, in a piece of bulk metal like the core of a transformer. The heat produced by such current constitutes a power loss in the transformer. Consider a conducting slab between the pole faces of an electromagnet (Figure 28-24). If the magnetic field $\vec{B}$ between the pole faces is changing with time (as it will if the current in the magnet windings is alternating current), the flux through any closed loop in the slab, such as through the curve C indicated in the figure, will change. Consequently, there will be an induced emf around C. Because path C is in a conductor, the emf will drive currents in the conductor.

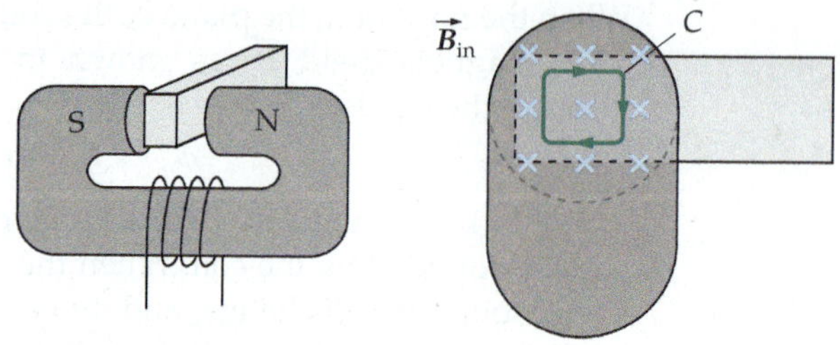

FIGURE 28-24 Eddy currents. When the magnetic field through a metal slab is changing, an emf is induced in any closed loop in the metal, such as loop C. The induced emfs drive currents, which are called eddy currents.

The existence of eddy currents can be demonstrated by pulling a copper or aluminum sheet through the region between the poles of a strong permanent magnet (Figure 28-25). Part of the area enclosed by curve C in the figure is in the magnetic field, and part of the area enclosed by curve C is outside the magnetic field. As the sheet is pulled to the right, the flux through this curve decreases (assuming that into the paper is the positive normal direction). A clockwise emf is induced around this curve. This emf drives a current that is directed upward in the region between the pole faces, and the magnetic field exerts a force on this current to the left opposing motion of the sheet. You can feel this drag force on the sheet if you pull a conducting sheet rapidly through a region that has a strong magnetic field.

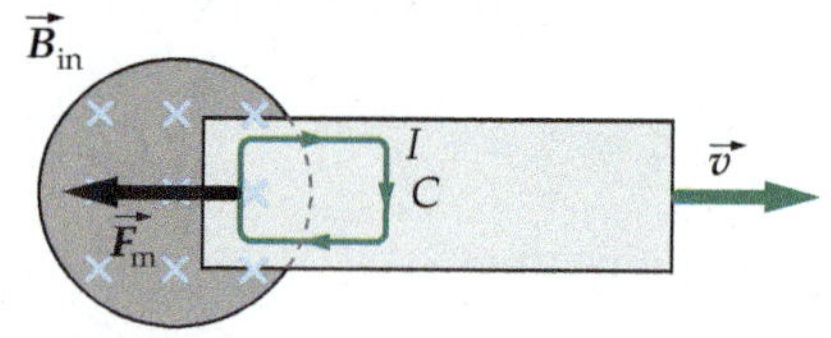

FIGURE 28-25 Demonstration of eddy currents. When the metal sheet is pulled to the right, there is a magnetic force to the left on the induced current opposing the motion.

Eddy currents are frequently undesirable because power is lost due to Joule heating by the current, and this dissipated energy must be transferred to the environment. The power loss can be reduced by increasing the resistance of the possible paths for the eddy currents, as shown in Figure 28-26*a*. Here the conducting slab is laminated; that is, the conducting slab is made up of small strips glued together. Because insulating glue separates the strips, the eddy currents are essentially confined to the individual strips. The large eddy-current loops are broken up, and the power loss is greatly reduced. Similarly, if there are cuts in the sheet, as shown in Figure 28-26*b*, the eddy currents are lessened and the magnetic force is greatly reduced.

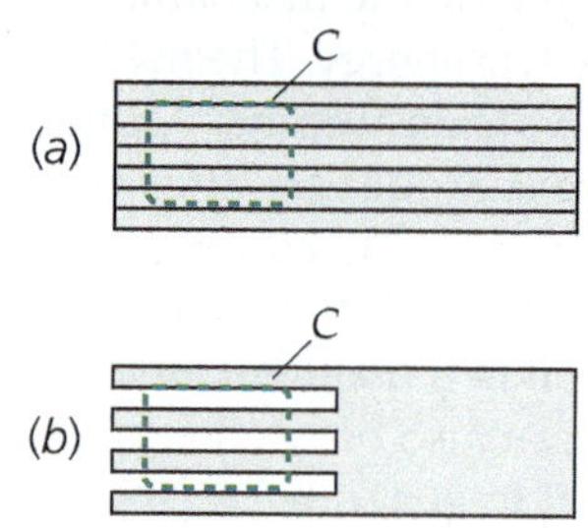

FIGURE 28-26 Disrupting the conduction paths in the metal slab can reduce the eddy current. (*a*) If the slab is constructed from strips of metal glued together, the insulating glue between the slabs increases the resistance of the closed loop C. (*b*) Slots cut into the metal slab also reduce the eddy current.

Eddy currents are not always undesirable. For example, eddy currents are often used to damp unwanted oscillations. With no damping present, sensitive mechanical balance scales that are used to measure small masses might oscillate back and forth around their equilibrium reading many times. Such scales are usually designed so that a small sheet of aluminum (or some other metal) moves between the poles of a permanent magnet as the scales oscillate. The resulting eddy currents dampen the oscillations so that equilibrium is quickly reached. Eddy currents also play a role in the magnetic braking systems of some rapid transit cars. A large electromagnet is positioned in the vehicle over the rails. If the magnet is energized by a current in its windings, eddy currents are induced in the rails by the motion of the magnet and the magnetic forces provide a drag force on the magnet that slows the car.

28-6 INDUCTANCE

SELF-INDUCTANCE

Consider a coil carrying a current I. The current in the coil produces a magnetic field $\vec{B}$ that varies from point to point, but at each point in space the value of B is proportional to I. The magnetic flux of $\vec{B}$ through the coil is therefore also proportional to I:

$$\phi_m = LI \qquad 28\text{-}11$$

DEFINITION—SELF-INDUCTANCE

where L, the proportionality constant, is called the self-inductance of the coil. The self-inductance depends on the geometric shape of the coil. The SI unit of inductance is the **henry** (H). From Equation 28-11, we can see that the unit of inductance equals the unit of flux divided by the unit of current:

$$1\text{ H} = 1\text{ Wb/A} = 1\text{ T}\cdot\text{m}^2/\text{A}$$

In principle, the self-inductance of any coil or circuit can be calculated by assuming a current I, calculating $\vec{B}$ at every point on a surface bounded by the coil, calculating the flux ϕ_m, and using $L = \phi_m/I$. In actual practice, the calculation is often very challenging. However, the self-inductance of a long, tightly wound solenoid can be calculated directly. The magnetic flux through a long, thin solenoid is given by NBA, where $B = \mu_0 nI$, N is the number of turns, n is the number of turns per unit length, I is the current and A is the area per turn. Thus, the magnetic flux through the coil is

$$\phi_m = NBA = \mu_0 N(\mu_0 nI)A = \frac{\mu_0 N^2 IA}{\ell} = \mu_0 n^2 IA\ell \qquad 28\text{-}12$$

where ℓ is the length of the solenoid. As expected, the flux is proportional to the current. The proportionality constant is the self-inductance L:

$$L = \frac{\phi_m}{I} = \mu_0 n^2 A\ell \qquad 28\text{-}13$$

SELF-INDUCTANCE OF A LONG SOLENOID

The self-inductance of a long solenoid is proportional to the square of the number of turns per unit length n and to the volume $A\ell$. Thus, like capacitance, self-inductance depends only on geometric factors.* From the dimensions of Equation 28-13, we can see that μ_0 can be expressed in henrys per meter:

$$\mu_0 = 4\pi \times 10^{-7}\text{ H/m}$$

Example 28-10 Self-Inductance of a Solenoid

Find the self-inductance of a solenoid of length 10.0 cm, area 5.00 cm^2, and 100 turns.

PICTURE We can calculate the self-inductance in henrys from Equation 28-13.

SOLVE

1. L is given by Equation 28-13: $L = \mu_0 n^2 A\ell$
2. Convert the given quantities to SI units:
   $\ell = 10.0\text{ cm} = 0.100\text{ m}$
   $A = 5.00\text{ cm}^2 = 5.00 \times 10^{-4}\text{ m}^2$
   $n = N/\ell = (100\text{ turns})/(0.100\text{ m}) = 1000\text{ turns/m}$
   $\mu_0 = 4\pi \times 10^{-7}\text{ H/m}$
3. Substitute the given quantities:
   $L = \mu_0 n^2 A\ell$
   $= (4\pi \times 10^{-7}\text{ H/m})(1000\text{ turns/m})^2(5.00 \times 10^{-4}\text{ m}^2)(0.100\text{ m})$
   $= \boxed{6.28 \times 10^{-5}\text{ H}}$

CHECK The inductance of a solenoid that does not have a soft iron core is expected to be a small fraction of a henry. That is the case for the solenoid in this example.

* If the inductor has a material in the core, the self-inductance also depends on the properties of the material.

When the current in a circuit is changing, the magnetic flux due to the current is also changing, so an emf is induced in the circuit. Because the self-inductance L of a circuit is constant, the rate of change of the flux is related to the rate of change of the current by

$$\frac{d\phi_{\mathrm{m}}}{dt} = \frac{d(LI)}{dt} = L\frac{dI}{dt}$$

According to Faraday's law, we have

$$\mathcal{E} = -\frac{d\phi_{\mathrm{m}}}{dt} = -L\frac{dI}{dt} \qquad \text{28-14}$$

SELF-INDUCED EMF

Thus, the self-induced emf is proportional to the rate of change of the current. Because of the negative sign in Equation 28-14, the self-induced emf is often called a back emf. A coil or solenoid that has enough turns to have a significant self-inductance is called an **inductor.** In circuits, it is denoted by the symbol . Typically, the self-inductance of the rest of the circuit is negligible in comparison to the self-inductance of a coil or solenoid. The potential difference across an inductor is given by

$$\Delta V = \mathcal{E} - Ir = -L\frac{dI}{dt} - Ir \qquad \text{28-15}$$

POTENTIAL DIFFERENCE ACROSS AN INDUCTOR

where r is the internal resistance of the inductor.* For an ideal inductor, $r = 0$.

PRACTICE PROBLEM 28-6

At what rate must the current in the solenoid of Example 28-10 change in order to induce a back emf of 20.0 V?

MUTUAL INDUCTANCE

When two or more circuits are close to each other, as in Figure 28-27, the magnetic flux through one circuit depends not only on the current in that circuit but also on the current in the nearby circuits. Let I_1 be the current in circuit 1, on the left in Figure 28-27, and let I_2 be the current in circuit 2, on the right. The magnetic field $\vec{B}$ at surface S_2 is the superposition of $\vec{B}_1$ due to I_1 and $\vec{B}_2$ due to I_2, where B_1 is proportional to I_1 and B_2 is proportional to I_2. We can therefore write the flux of $\vec{B}_1$ through circuit 2, $\phi_{\mathrm{m}12}$, as

$$\phi_{\mathrm{m}12} = M_{12}I_1 \qquad \text{28-16}a$$

DEFINITION—MUTUAL INDUCTANCE

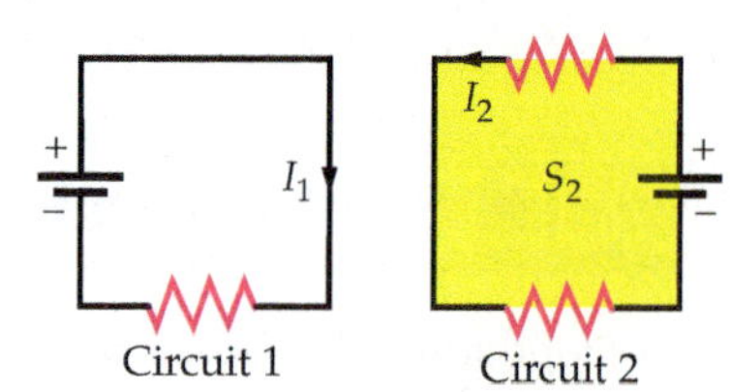

FIGURE 28-27 Two adjacent circuits. The magnetic field on S_2 is partly due to current I_1 and partly due to current I_2. The flux through S_2 is the sum of two terms, one proportional to I_1 and the other to I_2.

where M_{12} is called the **mutual inductance** of the two circuits. The mutual inductance depends on the geometrical arrangement of the two circuits. For instance, if the circuits are far apart, the flux of $\vec{B}_1$ through circuit 2 will be small and the mutual inductance will be small. (The net flux $\phi_{\mathrm{m}2}$ of $\vec{B} = \vec{B}_1 + \vec{B}_2$ through circuit 2 is given by $\phi_{\mathrm{m}2} = \phi_{\mathrm{m}12} + \phi_{\mathrm{m}22}$.) An equation similar to Equation 28-16a can be written for the flux of $\vec{B}_2$ through circuit 1:

$$\phi_{\mathrm{m}21} = M_{21}I_2 \qquad \text{28-16}b$$

We can calculate the mutual inductance for two tightly wound coaxial solenoids like the solenoids shown in Figure 28-28. Let ℓ be the length of both solenoids, and let the inner solenoid have N_1 turns and radius r_1 and the outer solenoid have N_2 turns and radius r_2. We will first calculate the mutual inductance M_{12} by assuming that the inner solenoid carries a current I_1 and finding the magnetic flux $\phi_{\mathrm{m}2}$ due to this current through the outer solenoid.

* If the inductor has an iron core, the internal resistance includes properties of the core.

The magnetic field $\vec{B}_1$ due to the current in the inner solenoid is uniform in the space inside the inner solenoid and has magnitude

$$B_1 = \mu_0(N_1/\ell)I_1 = \mu_0 n_1 I_1 \quad r < r_1 \qquad 28\text{-}17$$

and outside the inner solenoid this magnetic field B_1 is essentially zero. The flux of $\vec{B}_1$ through the outer solenoid is therefore

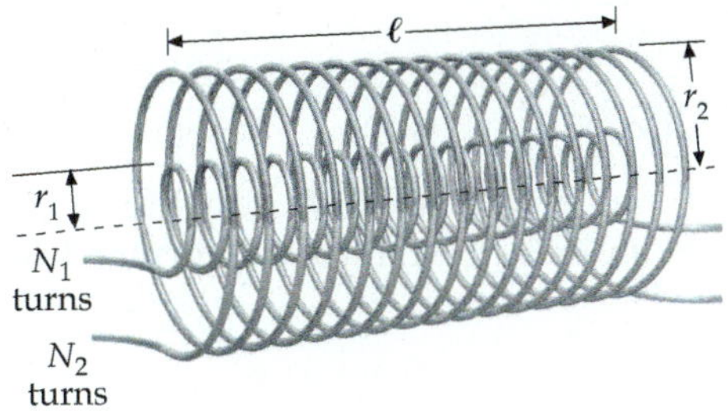

(a)

$$\phi_{m2} = N_2 B_1(\pi r_1^2) = n_2 \ell B_1(\pi r_1^2) = \mu_0 n_2 n_1 \ell(\pi r_1^2) I_1$$

Note that the area used to compute the flux through the outer solenoid is not the area of the surface bounded by a loop of that solenoid, πr_2^2, but rather is the area of the surface bounded by a loop of the inner solenoid, πr_1^2. This is because the magnetic field due to the inner solenoid is zero outside the inner solenoid. The mutual inductance M_{12} is thus

$$M_{12} = \frac{\phi_{m12}}{I_1} = \mu_0 n_2 n_1 \ell \pi r_1^2 \qquad 28\text{-}18$$

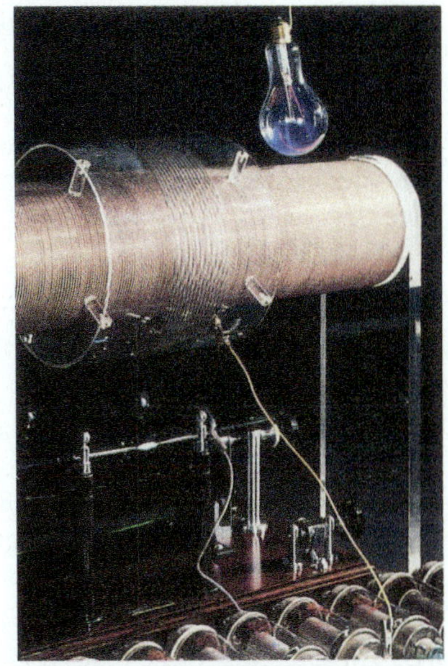
(b)

FIGURE 28-28 (*a*) A long, narrow solenoid inside a second solenoid of the same length. A current in either solenoid produces a magnetic flux in the other. (*b*) A tesla coil illustrating the geometry of the wires in Figure 28-28*a*. Such a device functions as a transformer.* Here, low-voltage alternating current in the outer winding is transformed into a higher-voltage alternating current in the inner winding. The emf induced in the inner coil by the field of the charging current in the outer coil is high enough to light the bulb above the coils. *((b) © Michael Holford, Collection of the Science Museum, London.)*

PRACTICE PROBLEM 28-7

Calculate the mutual inductance M_{21} of the coaxial solenoids of Figure 28-28 by finding the flux through the inner solenoid due to a current I_2 in the outer solenoid.

Note the result of Practice Problem 28-7 reveals that $M_{12} = M_{21}$. It can be shown that this is a general result. Therefore in the future we will drop the subscripts for mutual inductance and simply write M.

28-7 MAGNETIC ENERGY

An inductor stores magnetic energy, just as a capacitor stores electrical energy. Consider the circuit shown in Figure 28-29, which consists of an ideal inductor that has an inductance L and a resistor that has a resistance R in series with an ideal battery that has emf $\mathcal{E}_0$ and a switch S. We assume that R and L are the resistance and inductance of the entire circuit. The switch is initially open, so no current exists in the circuit. A short time after the switch is closed, there are a current I in the circuit, a potential difference $-IR$ across the resistor, and a potential difference $-L\,dI/dt$ across the inductor. (For an ideal inductor, the difference in potential across the inductor equals the back emf, which was given in Equation 28-14.) Applying Kirchhoff's loop rule to this circuit gives

$$\mathcal{E}_0 - IR - L\frac{dI}{dt} = 0 \qquad 28\text{-}19$$

If we multiply each term by the current I and rearrange, we obtain

$$\mathcal{E}_0 I = I^2 R + LI\frac{dI}{dt} \qquad 28\text{-}20$$

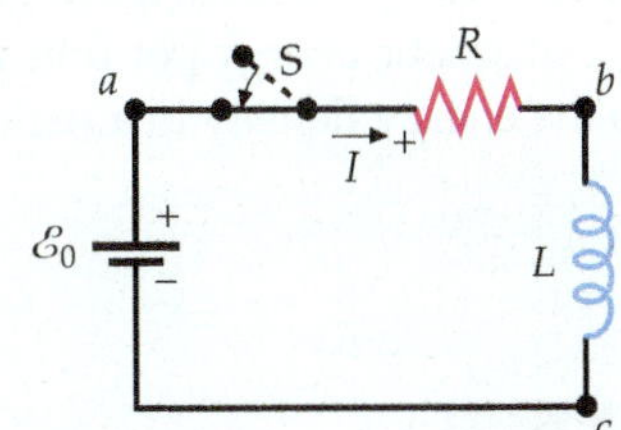

FIGURE 28-29 Just after the switch S is closed in this circuit, the current begins to increase and a back emf of magnitude $L\,dI/dt$ is induced in the inductor. The potential drop across the resistor IR plus the potential drop across the inductor ($L\,dI/dt$) equals the emf of the battery ($\mathcal{E}_0$).

The term $\mathcal{E}_0 I$ is the rate at which electrical potential energy is supplied by the battery. The term I^2R is the rate at which potential energy is delivered to the resistor. (It is also the rate at which potential energy is dissipated by the resistance in the circuit.) The term $LI\,dI/dt$ is the rate at which potential energy is delivered to the inductor, so if U_m is the energy stored in the inductor, then

$$\frac{dU_m}{dt} = LI\frac{dI}{dt}$$

which implies

$$dU_m = LI\,dI$$

* The transformer is discussed in Chapter 29.

Integrating this equation, we obtain

$$U_m = \frac{1}{2}LI^2 + C$$

where C is a constant of integration. To evaluate C, we set U_m equal to zero when I is equal to zero. The energy stored in an inductor carrying a current I is thus given by

$$U_m = \frac{1}{2}LI^2 \qquad \text{28-21}$$

ENERGY STORED IN AN INDUCTOR

When a current is produced in an inductor, a magnetic field is created in the region in and around the inductor coil. We can think of the energy stored in an inductor as energy stored in this magnetic field. For the special case of a long, thin solenoid, the magnetic field strength is zero except for the region inside the inductor, where it is given by

$$B = \mu_0 nI$$

The self-inductance of a long, thin solenoid is given by Equation 28-13:

$$L = \mu_0 n^2 A\ell$$

where A is the cross-sectional area and ℓ is the length. Substituting $B/(\mu_0 n)$ for I and $\mu_0 n^2 A\ell$ for L in Equation 28-21, we obtain

$$U_m = \frac{1}{2}LI^2 = \frac{1}{2}\mu_0 n^2 A\ell\left(\frac{B}{\mu_0 n}\right)^2 = \frac{B^2}{2\mu_0}A\ell$$

The quantity $A\ell$ is the volume of the space within the solenoid containing the magnetic field. The energy per unit volume is the **magnetic energy density** u_m:

$$u_m = \frac{B^2}{2\mu_0} \qquad \text{28-22}$$

MAGNETIC ENERGY DENSITY

Although we derived this expression by considering the special case of the magnetic field in a long solenoid, it is a general result. Whenever a magnetic field exists in space, the magnetic energy per unit volume is given by Equation 28-22. Note the similarity to the energy density in a region where there is an electric field (Equation 24-9):

$$u_e = \frac{1}{2}\epsilon_0 E^2$$

Example 28-11 Electromagnetic Energy Density

A certain region of space has a uniform magnetic field of 0.0200 T and a uniform electric field of 2.50×10^6 N/C. Find (*a*) the total electromagnetic energy density in the region, and (*b*) the energy in a cubical box of edge length $\ell = 12.0$ cm.

PICTURE The total energy density u is the sum of the electrical and magnetic energy densities, $u = u_e + u_m$. The energy in a volume $\mathcal{V}$ is given by $U = u\mathcal{V}$.

SOLVE

(*a*) 1. Calculate the electrical energy density:

$$u_e = \frac{1}{2}\epsilon_0 E^2$$

$$= \frac{1}{2}(8.85 \times 10^{-12}\ \text{C}^2/\text{N}\cdot\text{m}^2)(2.50 \times 10^6\ \text{N/C})^2$$

$$= 27.7\ \text{J/m}^3$$

2. Calculate the magnetic energy density: $u_m = \frac{B^2}{2\mu_0} = \frac{(0.0200\text{ T})^2}{2(4\pi \times 10^{-7}\text{ N/A}^2)} = 159\text{ J/m}^3$

3. The total energy density is the sum of the above two contributions: $u = u_e + u_m = 27.7\text{ J/m}^3 + 159\text{ J/m}^3 = \boxed{187\text{ J/m}^3}$

(*b*) The total energy in the box is $U = u\mathcal{V}$, where $\mathcal{V} = \ell^3$ is the volume of the box: $U = u\mathcal{V} = u\ell^3 = (187\text{ J/m}^3)(0.120\text{ m})^3 = \boxed{0.323\text{ J}}$

*28-8 CIRCUITS

A circuit containing a resistor and an inductor, such as that shown in Figure 28-29, is called an ***RL* circuit.** Because all circuits have resistance and self-inductance at room temperature, the analysis of an *RL* circuit can be applied to some extent to all circuits.*

For the circuit shown in Figure 28-29, application of Kirchhoff's loop rule gave us

$$\mathcal{E}_0 - IR - L\frac{dI}{dt} = 0 \qquad \text{28-19}$$

Let us look at some general features of this equation. First, the sum $IR + L\,dI/dt$ equals the emf of the battery, which is constant. Immediately after we close the switch in the circuit, the current is still zero, so IR is zero, and $L\,dI/dt$ equals the emf of the battery, $\mathcal{E}_0$. Setting $I = 0$ in Equation 28-19, we get

$$\left.\frac{dI}{dt}\right|_{I=0} = \frac{\mathcal{E}_0}{L} \qquad \text{28-23}$$

As the current increases, IR increases and dI/dt decreases. Note that the current cannot abruptly jump from zero to some final value as it would if the inductance L were zero. When the inductance L is greater than zero dI/dt is finite, and therefore the current must be continuous in time. After a short time, the current has reached a positive value I, and the rate of change of the current is

$$\frac{dI}{dt} = \frac{\mathcal{E}_0 - IR}{L}$$

At this time the current is still increasing, but its rate of increase is less than it was at $t = 0$. The final value I_f of the current can be obtained by setting dI/dt equal to zero in Equation 28-19:

$$I_f = \frac{\mathcal{E}_0}{R} \qquad \text{28-24}$$

Figure 28-30 shows the current in this circuit as a function of time. This figure is the same as that for the charge on a capacitor as a function of time when the capacitor is being charged in an *RC* circuit (Figure 25-45).

Equation 28-19 is of the same form as Equation 25-38 for the charging of a capacitor—and can be solved in the same way (by separating variables and integrating). The result is

$$I = \frac{\mathcal{E}_0}{R}(1 - e^{-(R/L)t}) = I_f(1 - e^{-t/\tau}) \qquad \text{28-25}$$

where $I_f = \mathcal{E}_0/R$ is the current as $t \to \infty$, and

$$\tau = \frac{L}{R} \qquad \text{28-26}$$

is the **time constant** of the circuit. The larger the self-inductance L or the smaller the resistance R, the longer it takes for the current to reach any specified fraction of its final current I_f.

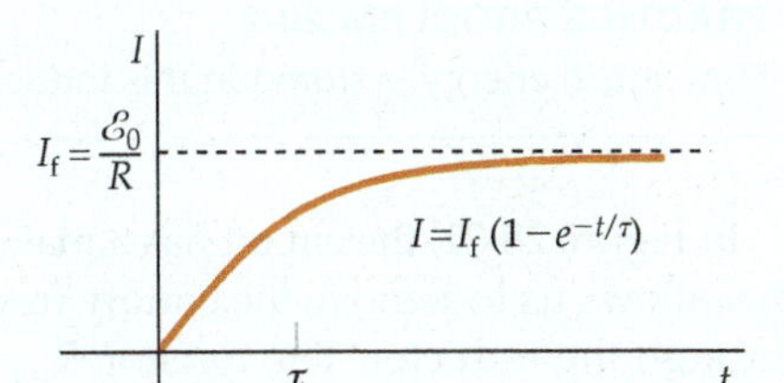

FIGURE 28-30 Current versus time in an *RL* circuit. At a time $t = \tau = L/R$, the current is at 63 percent of its maximum value $\mathcal{E}_0/R$.

* All circuits also have some capacitance between parts of the circuits at different potentials. We will consider the effects of capacitance in Chapter 29 when we study ac circuits. Here we will neglect capacitance to simplify the analysis and to focus on the effects of inductance.

Example 28-12 Energizing a Coil

A coil that has a self-inductance equal to 5.00 mH and a resistance equal to 15.0 Ω is placed across the terminals of a 12.0-V battery that has a negligible internal resistance. (*a*) What is the final current? (*b*) What is the time constant? (*c*) How many time constants does it take for the current to reach 99.0 percent of its final value?

PICTURE The final current is the current when $dI/dt = 0$. The current as a function of time is given by Equation 28-25, $I = I_f(1 - e^{-t/\tau})$, where $\tau = L/R$.

SOLVE

(*a*) Using Equation 28-19, set dI/dt equal to zero to find the final current, I_f:

$$\mathcal{E}_0 - IR - L\frac{dI}{dt} = 0$$

$$\mathcal{E}_0 - I_f R - 0 = 0$$

$$I_f = \frac{\mathcal{E}_0}{R} = \frac{12.0\ \text{V}}{15.0\ \Omega} = \boxed{0.800\ \text{A}}$$

(*b*) Calculate the time constant τ.

$$\tau = \frac{L}{R} = \frac{5.00 \times 10^{-3}\text{H}}{15.0\ \Omega} = \boxed{333\ \mu\text{s}}$$

(*c*) Use Equation 28-25 and calculate the time t for $I = 0.990 I_f$:

$$I = I_f(1 - e^{-t/\tau}),$$

so

$$e^{-t/\tau} = \left(1 - \frac{I}{I_f}\right)$$

Taking the logarithm of both sides gives

$$-\frac{t}{\tau} = \ln\left(1 - \frac{I}{I_f}\right)$$

Thus,

$$t = -\tau \ln\left(1 - \frac{I}{I_f}\right) = -\tau \ln(1 - 0.990)$$

$$= -\tau \ln(0.010) = \tau \ln 100 = \boxed{4.61\tau}$$

CHECK In five time constants, the current is within one percent of its final value. This is consistent with the results of Example 25-18 where we found that after five time constants the charge on a discharging capacitor was less that one percent of its initial charge.

PRACTICE PROBLEM 28-8

How much energy is stored in this inductor when the final current has been attained?

In Figure 28-31, the circuit has a make-before-break switch (shown in Figure 28-32) that allows us to remove the battery from the circuit without interrupting the current through the inductor. The resistor R_1 protects the battery so that the battery is not shorted when the switch is thrown. If the switch throw is in position *e*, the battery, the inductor, and the two resistors are connected in series and the current builds up in the circuit as just discussed, except that the total resistance is now $R_1 + R$ and the final current is $\mathcal{E}_0/(R + R_1)$. Suppose that the throw has been in position *e* for a long time, so that the current remains at its final value, which we will call I_0. At time $t = 0$, we rapidly move the throw from position *e* to position *f*. With the throw at *f*, the current is zero in the branch with the battery and R_1. We now have a closed single-loop circuit (loop *abcdfa*) that has a resistor and an inductor carrying an initial current I_0. Applying Kirchoff's loop rule to this circuit gives

$$-IR - L\frac{dI}{dt} = 0$$

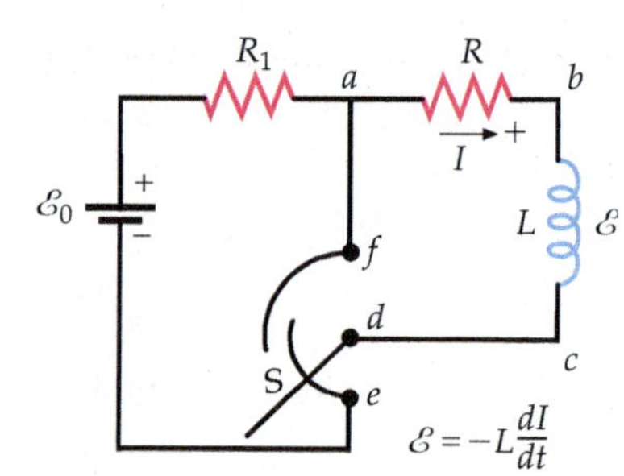

FIGURE 28-31 An *RL* circuit that has a make-before-break switch so that the battery can be removed from the circuit without interrupting the current through the inductor. The current in the inductor reaches its steady-state value with the switch pole in position *e*. The pole is then rapidly moved to position *f*.

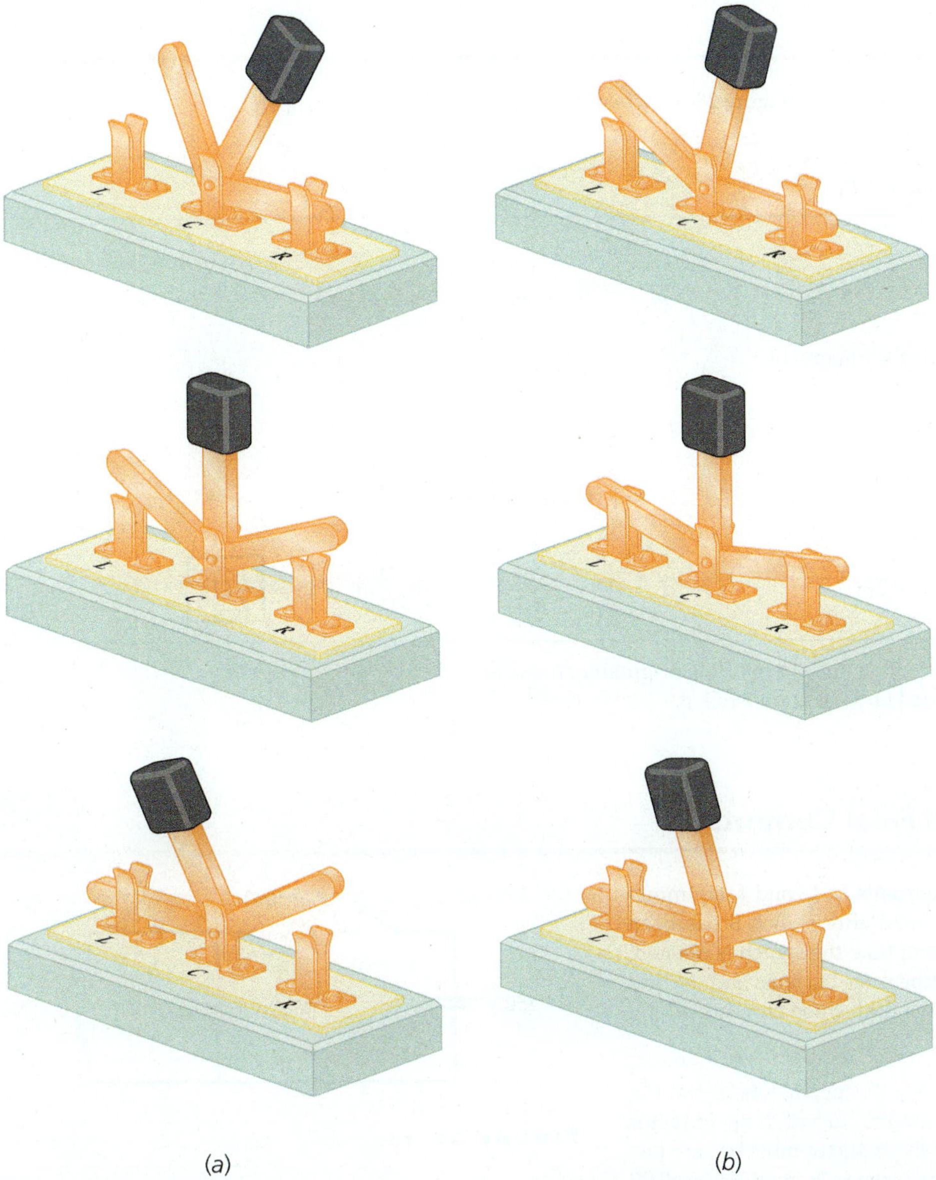

FIGURE 28-32 (*a*) The standard single-pole, double-throw switch is a break-before-make switch. That is, it breaks the first contact before making the second contact. (*b*) In a make-before-break, single-pole, double-throw switch, the throw makes the second contact before breaking the first contact. With the throw in the middle position, the throw is simultaneously in electrical contact with contacts *L* and *R*.

Rearranging this equation to separate the variables *I* and *t* gives

$$\frac{dI}{I} = -\frac{R}{L}\,dt \qquad \text{28-27}$$

(Equation 28-27 is of the same form as Equation 25-34 for the discharge of a capacitor.) Integrating and then solving for *I* gives

$$I = I_0 e^{-t/\tau} \qquad \text{28-28}$$

where $\tau = L/R$ is the time constant. Figure 28-33 shows the current as a function of time.

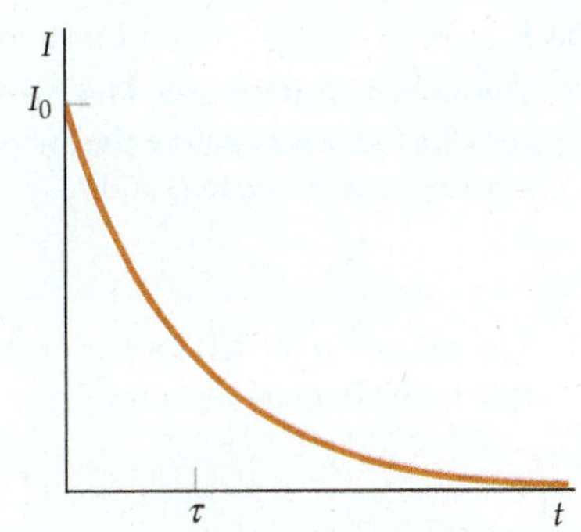

FIGURE 28-33 Current versus time for the circuit in Figure 28-31. The current decreases exponentially with time.

PRACTICE PROBLEM 28-9

What is the time constant of a closed single-loop circuit that has a resistance equal to 85 Ω and an inductance equal to 6.0 mH?

Example 28-13 Energy Dissipated

Find the total energy dissipated in the resistor R, as shown in Figure 28-31, when the current in the circuit decreases from its initial value of I_0 to 0.

PICTURE The rate of energy dissipation is equal to I^2R.

SOLVE

1. The rate of energy dissipation is I^2R: $\dfrac{dU}{dt} = I^2R$

2. The total energy U dissipated in the resistor is the integral of $P\,dt$ from $t = 0$ to $t = \infty$: $U = \int_0^\infty I^2R\,dt$

3. The current I is given by Equation 28-28: $I = I_0e^{-(R/L)t}$

4. Substitute this current into the integral: $U = \int_0^\infty I^2R\,dt = \int_0^\infty I_0^2e^{-2(R/L)t}R\,dt = I_0^2R\int_0^\infty e^{-2(R/L)t}\,dt$

5. The integration can be done by substituting $x = 2Rt/L$: $U = I_0^2R\dfrac{e^{-2(R/L)t}}{-2(R/L)}\bigg|_0^\infty = I_0^2R\dfrac{-L}{2R}(0-1) = \boxed{\dfrac{1}{2}LI_0^2}$

CHECK The total amount of energy dissipated equals the energy $\frac{1}{2}LI_0^2$ originally stored in the inductor. (The energy stored in an inductor is $\frac{1}{2}LI^2$ (Equation 28-21.)

Example 28-14 Initial Currents and Final Currents

For the circuit shown in Figure 28-34, find the currents I_1, I_2, and I_3 (*a*) immediately after switch S is closed and (*b*) a long time after switch S has been closed. After the switch has been closed for a long time the switch is opened. Immediately after the switch is opened (*c*) find the three currents and (*d*) find the potential drop across the 20-Ω resistor. (*e*) Find all three currents a long time after switch S was opened.

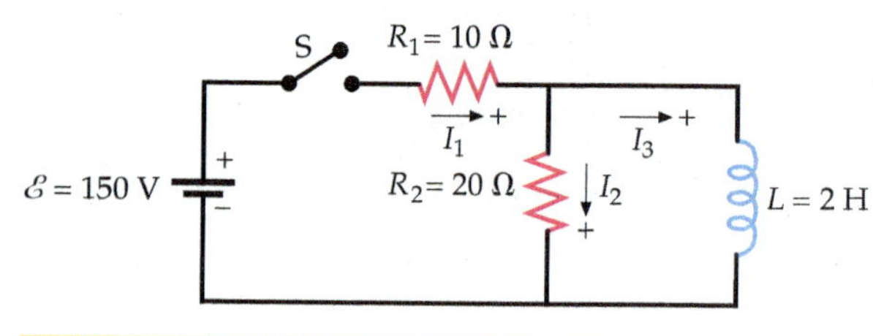

FIGURE 28-34

PICTURE (*a*) We simplify our calculations by using the fact that the current in an inductor cannot change abruptly. Thus, because the current in the inductor is zero before the switch is closed, the current in the inductor must be zero just after the switch is closed. (*b*) When the current reaches its final value dI/dt equals zero, so there is no potential drop across the inductor. The inductor thus acts like a short circuit; that is, the inductor acts like a wire with zero resistance. (*c*) Immediately after the switch is opened, the current in the inductor is the same as it was just before the switch was opened. (*d*) A long time after the switch is opened, all the currents must be zero.

SOLVE

(*a*) 1. The switch is just closed. The current through the inductor is zero, just as it was before the switch was closed. Apply the junction rule to relate I_1 and I_2:

$I_3 = \boxed{0}$

$I_1 = I_2 + I_3$

so

$I_1 = I_2$

2. The current in the left loop is obtained by applying the loop rule to the loop on the left:

$\mathcal{E} - I_1R_1 - I_1R_2 = 0$

so

$I_1 = I_2 = \dfrac{\mathcal{E}}{R_1 + R_2} = \dfrac{150\text{ V}}{10\ \Omega + 20\ \Omega} = \boxed{5.0\text{ A}}$

(*b*) 1. After a long time, the currents are steady and the inductor acts like a short circuit, so the potential drop across R_2 is zero. Apply the loop rule to the loop on the right and solve for I_2:

$-L\dfrac{dI_3}{dt} + I_2R_2 = 0$

$0 + I_2R_2 = 0 \quad\Rightarrow\quad I_2 = \boxed{0}$

2. Apply the loop rule to the loop on the left and solve for I_1:

$$\mathcal{E} - I_1R_1 - I_2R_2 = 0$$
$$\mathcal{E} - I_1R_1 - 0 = 0$$

so

$$I_1 = \frac{\mathcal{E}}{R_1} = \frac{150\text{ V}}{10\ \Omega} = \boxed{15\text{ A}}$$

3. Apply the junction rule and solve for I_3:

$$I_1 = I_2 + I_3$$
$$15\text{ A} = 0 + I_3$$

so

$$I_3 = \boxed{15\text{ A}}$$

(*c*) When the switch is reopened, I_1 *"instantaneously"* becomes zero. The current I_3 in the inductor changes continuously, so at that "instant" $I_3 = 15$ A. Apply the junction rule and solve for I_2:

$$I_3 = \boxed{15\text{ A}}$$
$$I_1 = I_2 + I_3$$

so

$$I_2 = I_1 - I_3 = 0 - 15\text{ A} = \boxed{-15\text{ A}}$$

(*d*) Apply Ohm's law to find the potential drop across R_2:

$$V = I_2R_2 = (15\text{ A})(20\ \Omega) = \boxed{300\text{ V}}$$

(*e*) A long time after the switch is opened, all the currents must equal zero.

$$I_1 = I_2 = I_3 = \boxed{0}$$

TAKING IT FURTHER Were you surprised to find the potential drop across R_2 in Part (*d*) to be larger than the emf of the battery? This potential drop is equal to the emf of the inductor.

PRACTICE PROBLEM 28-10 Suppose $R_2 = 200\ \Omega$ and the switch has been closed for a long time. What is the potential drop across R_2 immediately after the switch is then opened?

28-9 MAGNETIC PROPERTIES OF SUPERCONDUCTORS

A superconductor has a resistivity equal to zero below a critical temperature T_c, which varies from material to material. In the presence of a magnetic field $\vec{B}$, the critical temperature is lower than the critical temperature is when there is no magnetic field. As the magnetic field increases, the critical temperature decreases. If the magnetic field strength is greater than some critical field strength B_c, superconductivity does not exist at any temperature.

*MEISSNER EFFECT

As a superconductor in a region that has a magnetic field is cooled below its critical temperature, the magnetic field in the region within the superconducting material becomes zero (Figure 28-35). This effect was discovered by Walter Meissner and Robert Ochsenfeld in 1933 and is now known as the **Meissner effect.** The magnetic field becomes zero because superconducting currents induced on the surface of the superconductor produce a

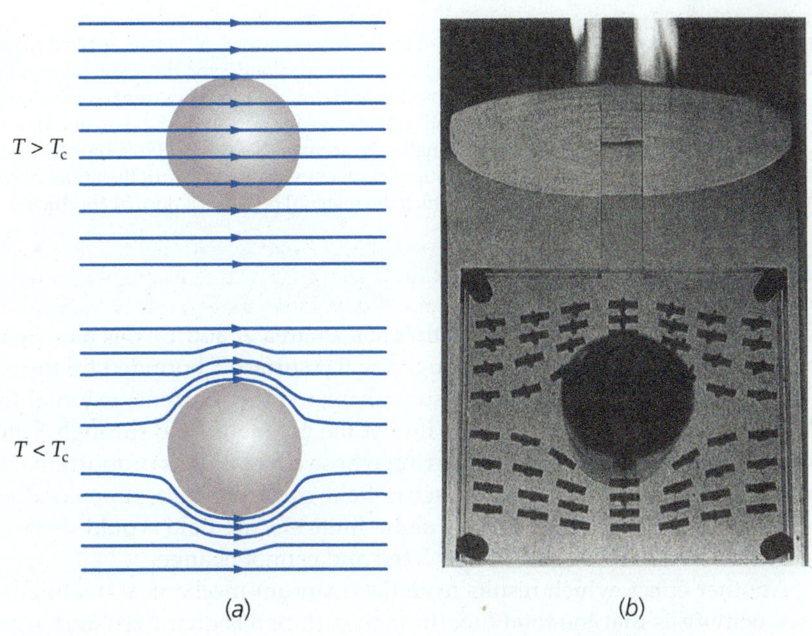

FIGURE 28-35 (*a*) The Meissner effect in a superconducting solid sphere cooled in a constant applied magnetic field. As the temperature drops below the critical temperature T_c, the magnetic field inside the sphere becomes zero. (*b*) Demonstration of the Meissner effect. A superconducting tin cylinder is situated with its axis perpendicular to a horizontal magnetic field. The directions of the field lines are indicated by weakly magnetized compass needles mounted in a Lucite sandwich so that they are free to turn. *(A. Leitner/Rensselaer Polytechnic Institute.)*

second magnetic field that cancels out the applied one. The magnetic levitation (see the photo) results from the repulsion between the permanent magnet producing the applied field and the magnetic field produced by the currents induced in the superconductor.

The disk is a superconductor. The magnetic levitation results from the repulsion between the permanent magnet producing the applied field and the magnetic field produced by the currents induced in the superconductor. *(© Palmer/Kane, Inc./CORBIS.)*

Only certain superconductors, called **type I superconductors,** exhibit the complete Meissner effect. Figure 28-36*a* shows a plot of the magnetization M multiplied by μ_0 versus the applied magnetic field B_{app} for a type I superconductor. For a magnetic field less than the critical field strength B_c, the magnetic field $\mu_0 M$ induced in the superconductor is equal and opposite to the applied magnetic field. The values of B_c for type I superconductors are always too small for such materials to be useful in the coils of a superconducting magnet.

Other materials, known as **type II superconductors,** have a magnetization curve similar to that in Figure 28-36*b*. Such materials are usually alloys or metals that have large resistivities in the normal state. Type II superconductors exhibit the electrical properties of superconductors, except for the Meissner effect, up to the critical field B_{c2}, which may be several hundred times the typical values of critical fields for type I superconductors. For example, the alloy Nb_3Ge has a critical field $B_{c2} = 34$ T. Such materials can be used for high-field superconducting magnets. Below the critical field B_{c1}, the behavior of a type II superconductor is the same as that of a type I superconductor.

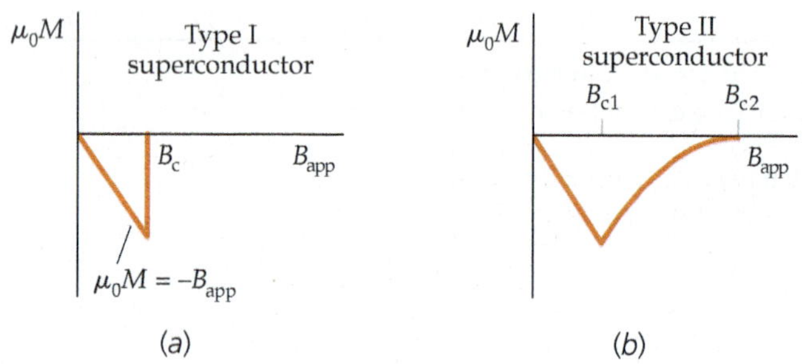

FIGURE 28-36 Plots of μ_0 multiplied by the magnetization M versus applied magnetic field for type I and type II superconductors. (*a*) In a type I superconductor, the resultant magnetic field is zero below a critical applied field B_c because the field due to induced currents on the surface of the superconductor exactly cancels the applied field. Above the critical field, the material is a normal conductor and the magnetization is too small to be seen on this scale. (*b*) In a type II superconductor, the magnetic field starts to penetrate the superconductor at a field B_{c1}, but the material remains superconducting up to a field B_{c2}, after which the material becomes a normal conductor.

*FLUX QUANTIZATION

Consider a superconducting ring that has an area A and carries a current. A magnetic flux $\phi_m = B_n A$ can exist through the flat surface S bounded by the ring due to the current in the ring and due also perhaps to other currents external to the ring. According to Equation 28-6, if the flux of the magnetic field through S changes, an electric field will be induced in the ring whose circulation is proportional to the rate of change of the flux. But no electric field can exist in a superconducting ring because the ring has no resistance and a finite electric field would drive an infinite current. The flux through S is thus fixed and cannot change.

Another effect, which results from the quantum-mechanical treatment of superconductivity, is that the total flux through surface S is quantized and is given by

$$\phi_m = n\frac{h}{2e} \qquad n = 1, 2, 3, \ldots \qquad 28\text{-}29$$

The smallest unit of flux, called a **fluxon,** is

$$\phi_0 = \frac{h}{2e} = 2.0678 \times 10^{-15}\ \text{T}\cdot\text{m}^2 \qquad 28\text{-}30$$

Physics Spotlight

The Promise of Superconductors

In 1986, a pair of IBM researchers tested a metal oxide—a ceramic—and found it was superconductive at 23 K.[*] Researchers worldwide then began testing different ceramics for superconductivity. By early 1987, a high-temperature superconductor (HTS) ceramic had been found. It is superconducting at 90 K—warm enough to be cooled by liquid nitrogen rather than liquid helium.[†] Superconducting ceramics were also found that could carry very large currents. The popular press assumed that room-temperature superconductors would be discovered. Books written in the late 1980s discussed the possibilities of superconductor-assisted levitating trains, superconductor-enabled computers, transfer of power over long distances without large losses due to resistance, and even superconductor-assisted satellite lasers.[‡, #]

The researcher is filling tubes with high-temperature superconducting powder in order to make wire. *(Courtesy of Department of Energy.)*

Unfortunately, room-temperature superconductors have not been reliably observed. Furthermore, high-temperature ceramic superconductors are difficult to work with.[°] They are brittle and cannot be connected easily to wires, so several ways of depositing the superconducting ceramics on other surfaces had to be invented. In addition, if the boundaries between the tiny ceramic grains are not properly oriented or if the layers are too thick, the ceramic is not superconductive.[§]

These difficulties have, however, slowly been overcome. HTS are now used in a growing number of applications. Superconducting quantum interference detectors, or SQUIDs, use interruptions[¶] in superconductivity to detect extremely small amounts of energy. They are used in extraordinarily sensitive metal detectors,[**] light detectors,[††] and even to detect magnetic fields in the nervous systems of newborns.[‡‡] HTS have been tested in short lengths of electrical cable that have been nitrogen-cooled and carry large currents[##] and in fine superconducting wires.[°°]

Superconductors become resistive conductors when they carry large currents, which can be advantageous for long-distance power distribution systems. When short circuits occur in electrical circuits, the current rapidly increases unless the circuit is protected by a fuse or circuit breaker. Without the protections, the large currents can damage equipment and cause fires. Superconducting current limiters are being developed[§§] to protect electrical distribution networks from those excessively large currents.[¶¶]

In 2001, Japanese researchers discovered that magnesium diboride, MgB_2, is superconductive at 39 K, much warmer than any other metallic superconductor. Unlike other metallic superconductors, it can be cooled by liquid neon instead of the more costly liquid helium. Because MgB_2 is a metallic alloy, it is easily made into wire.[***] Pure MgB_2 does develop resistance at a lower critical current than other metallic superconductors, so it is not currently used for high-current applications.[†††] Researchers are studying the additions of small amounts of other elements to improve the characteristics of MgB_2.[‡‡‡]

* Yamazaki, S., "Superconducting Ceramics," *United States Patent* 7,1112,556 *B1*. September 26, 2006.
† Chu, C. W., "Superconductivity Above 90 K." *Proceedings of the National Academy of Sciences*, Jul. 1987, Vol. 84, pp. 4681–4682.
‡ Asimov, I., *How Did We Find Out About Superconductivity?* New York: Walker and Company, 1988, pp. 57–62.
# Lampton, C. E., *Superconductors*. Hillside, New Jersey: Enslow, 1989, pp. 7–8, 53–69.
° Pool, R., "Superconductors' Material Problems." *Science*, Apr. 1, 1988, Vol. 240, No. 4848, pp. 25–27.
§ Service, R. F., "YBCO Confronts Life in the Slow Lane." *Science*, Feb. 1, 2002, Vol. 295, p. 787.
¶ Irwin, K. D., "Seeing with Superconductors." *Scientific American*, Nov. 2006, pp. 86–94.
** Bick, M., et al., "A SQUID-Based Metal Detector-Comparison to Coil and X-Ray Systems." *Superconducting Science and Technology*, Jan. 18, 2005, Vol. 18, pp. 346–351.
†† "Color Video Streaming from Space" *Machine Design*, May 25, 2006, p. 40.
‡‡ Draganova, R., et al., "Sound Frequency Change Detection in Fetuses and Newborns, a Magnetoencephalographic Study." *Neuroimage*, Nov. 1, 2005, Vol. 28, No. 2, pp. 354–361.
Malozemoff, A. P., Mannhart, J., and Scalapino, D., "High-Temperature Cuprate Superconductors Get to Work." *Physics Today*, April 2005, pp. 41–47.
°° Kang, S., "High-Performance, High-T_C Superconducting Wires." *Science*, Mar. 31, 2006, Vol. 311, pp. 1911–1914.
§§ Malozemoff, A. P., Mannhart, J., and Scalapino, D., op. cit.
¶¶ Meerovich, V., and Sokolovsky, V., "Experimental Study of a Transformer with Superconducting Elements for Fault Current Limitation and Energy Redistribution." *Cryogenics*, Aug. 2005, Vol. 45, No. 8, pp. 572–577.
*** Service, R., "MgB_2 Trades Performance for a Shot at the Real World." *Science*, Feb. 1, 2002, Vol. 295, pp. 786–788.
††† Canfield, P., and Bud'ko, S., "Low-Temperature Superconductivity Is Warming Up." *Scientific American*, Apr. 2005, pp. 80–87.
‡‡‡ Senkowicz, B. J., et al., "Atmospheric Conditions and Their Effect on Ball-Milled Magnesium Diboride." *Superconductor Science and Technology*, Oct. 2006, Vol. 19, pp. 1173–1177.

Summary

1. Faraday's law and Lenz's law are fundamental laws of physics.
2. Self-inductance is a property of a circuit element that relates the flux through the element to the current.

TOPIC	RELEVANT EQUATIONS AND REMARKS	
1. Magnetic Flux ϕ_m		
General definition	$\phi_m = \int_S \vec{B} \cdot \hat{n}\, dA$	28-1
Uniform field, flat surface bounded by coil of N turns	$\phi_m = NBA \cos\theta$ where A is the area of the flat surface bounded by a single turn.	28-4
Units	$1\text{ Wb} = 1\text{ T}\cdot\text{m}^2$	28-2
Due to current in a circuit	$\phi_m = LI$	28-11
Due to current in two circuits	$\phi_{m1} = L_1 I_1 + MI_2$ $\phi_{m2} = L_2 I_2 + MI_1$	28-16
2. EMF		
Faraday's law (includes both induction and motional emf)	$\mathcal{E} = -\frac{d\phi_m}{dt}$	28-5
Induction (time-varying magnetic field, C stationary)	$\mathcal{E} = \oint_C \vec{E} \cdot d\vec{\ell}$	28-6
Rod moving perpendicular to both its length and $\vec{B}$	$\lvert\mathcal{E}\rvert = vB\ell$	28-7
Self-induced (back emf)	$\mathcal{E} = -L\frac{dI}{dt}$	28-14
3. Faraday's Law	$\mathcal{E} = -\frac{d\phi_m}{dt}$	28-5
4. Lenz's Law	The induced emf and induced current are in such a direction as to oppose, or tend to oppose, the change that produces them.	
Alternative statement	When a magnetic flux through a surface changes, the magnetic field due to any induced current produces a flux of its own—through the same surface and opposite in sign to the change in flux.	
5. Inductance		
Self-inductance	$L = \frac{\phi_m}{I}$	28-11
Self-inductance of a solenoid	$L = \mu_0 n^2 A\ell$	28-13

TOPIC	RELEVANT EQUATIONS AND REMARKS	
Mutual inductance	$M = \frac{\phi_{m21}}{I_1} = \frac{\phi_{m12}}{I_2}$	28-18
Units and constants	$1\ \text{H} = 1\ \text{Wb/A} = 1\ \text{T}\cdot\text{m}^2/\text{A}$ $\mu_0 = 4\pi \times 10^{-7}\ \text{H/m}$	
6. Magnetic Energy		
Energy stored in an inductor	$U_m = \frac{1}{2}LI^2$	28-21
Energy density in a magnetic field	$u_m = \frac{B^2}{2\mu_0}$	28-22
***7. *RL* Circuits**		
Potential difference across an inductor	$\Delta V = \mathcal{E} - Ir = -L\frac{dI}{dt} - Ir$ where r is the internal resistance of the inductor. For an ideal inductor $r = 0$.	28-15
Energizing an inductor with a battery	In a single-loop circuit consisting of a resistor that has a resistance R, an inductor that has a self-inductance L, and a battery that has an emf $\mathcal{E}_0$, the current does not reach its maximum value I_f instantaneously, but rather takes some time to build up. If the current is initially zero, its value at some later time t is given by $I = \frac{\mathcal{E}_0}{R}(1 - e^{-t/\tau}) = I_f(1 - e^{-t/\tau})$	28-25
Time constant τ	$\tau = \frac{L}{R}$	28-26
De-energizing an inductor	In a single-loop circuit consisting of a resistor that has a resistance R and an inductor that has a self-inductance L, the current does not drop to zero through a resistor instantaneously, but rather takes some time to decrease. If the current is initially I_0, its value at some later time t is given by $I = I_0 e^{-t/\tau}$	28-28

Answers to Concept Checks

28-1 Opposite to the direction shown in Figure 28-12.

28-2 The external agent turning the coil does work on the coil. The energy comes from the external agent.

Answers to Practice Problems

28-2	0.555 A
28-3	3.53 mC
28-4	1.4 V
28-5	11 V
28-6	3.18×10^5 A/s
28-7	$M_{12} = \mu_0 n_2 n_1 \ell \pi r_1^2$
28-8	$U_m = \frac{1}{2}LI_f^2 = 1.60 \times 10^{-3}$ J
28-9	71 μs
28-10	3.0 kV

Problems

In a few problems, you are given more data than you actually need; in a few other problems, you are required to supply data from your general knowledge, outside sources, or informed estimate.

Interpret as significant all digits in numerical values that have trailing zeros and no decimal points.

- • Single-concept, single-step, relatively easy
- •• Intermediate-level, may require synthesis of concepts
- ••• Challenging
- SSM Solution is in the *Student Solutions Manual*
- Consecutive problems that are shaded are paired problems.

CONCEPTUAL PROBLEMS

1 • (*a*) The magnetic equator is a line on the surface of Earth on which Earth's magnetic field is horizontal. At the magnetic equator, how would you orient a flat sheet of paper so as to create the maximum magnitude of magnetic flux through it? (*b*) How about the minimum magnitude of magnetic flux? SSM

2 • At one of Earth's magnetic poles, how would you orient a flat sheet of paper so as to create the maximum magnitude of magnetic flux through it?

3 • Show that the following combination of SI units is equivalent to the volt: $T \cdot m^2/s$ SSM

4 • Show that the following combination of SI units is equivalent to the ohm: $Wb/(A \cdot s)$

5 • A current is induced in a conducting loop that lies in a horizontal plane, and the induced current is clockwise when viewed from above. Which of the following statements could be true? (*a*) A constant magnetic field is directed vertically downward. (*b*) A constant magnetic field is directed vertically upward. (*c*) A magnetic field whose magnitude is increasing is directed vertically downward. (*d*) A magnetic field whose magnitude is decreasing is directed vertically downward. (*e*) A magnetic field whose magnitude is decreasing is directed vertically upward. SSM

6 • Give the direction of the induced current in the circuit, shown on the right in Figure 28-37, when the resistance in the circuit on the left is suddenly (*a*) increased and (*b*) decreased. Explain your answer.

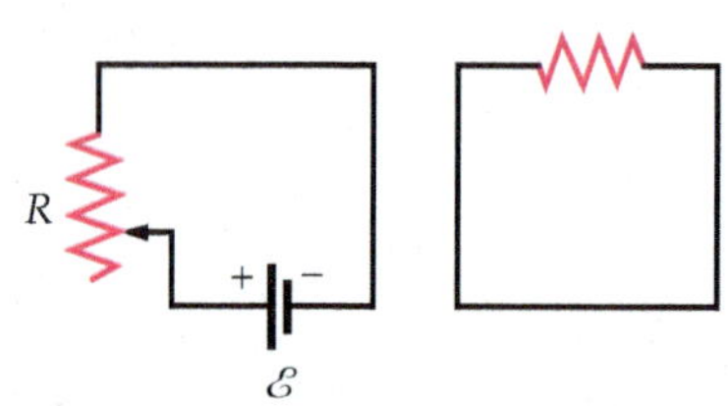

FIGURE 28-37 Problem 6

7 • The planes of the two circular loops in Figure 28-38 are parallel. As viewed from the left, a counterclockwise current exists in loop A. If the magnitude of the current in loop A is increasing, what is the direction of the current induced in loop B? Do the loops attract or repel each other? Explain your answer. SSM

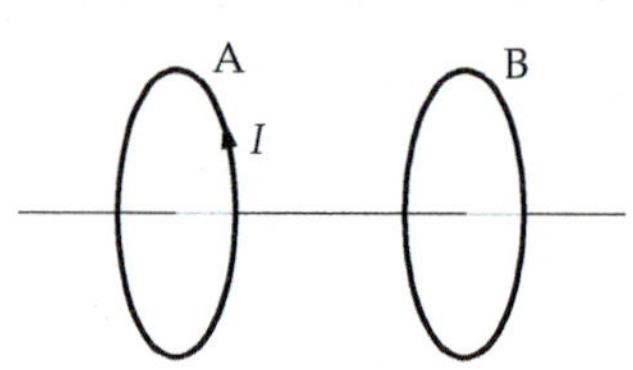

FIGURE 28-38 Problem 7

8 • A bar magnet moves with constant velocity along the axis of a loop, as shown in Figure 28-39. (*a*) Make a graph of the magnetic flux through the loop as a function of time. Indicate on the graph when the magnet is halfway through the loop by designating this time t_1. Choose the direction of the normal to the flat surface bounded by the loop to be to the right. (*b*) Make a graph of the induced current in the loop as a function of time. Choose the positive direction for the current to be clockwise as viewed from the left.

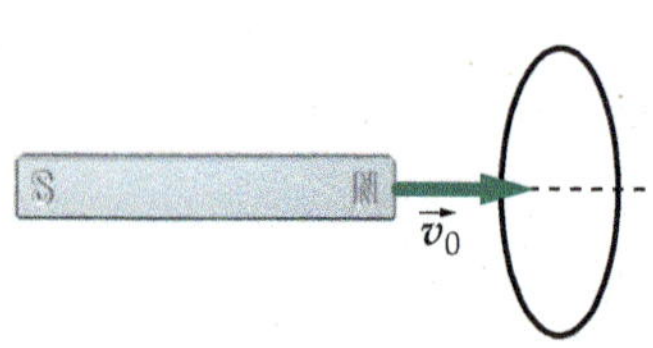

FIGURE 28-39 Problem 8

9 • A bar magnet is mounted on the end of a coiled spring and is oscillating in simple harmonic motion along the axis of a loop, as shown in Figure 28-40. The magnet is in its equilibrium position when its midpoint is in the plane of the loop. (*a*) Make a graph of the magnetic flux through the loop as a function of time. Indicate when the magnet is halfway through the loop by designating these times t_1 and t_2. (*b*) Make a graph of the induced current in the loop as a function of time, choosing the current to be positive when it is clockwise as viewed from above.

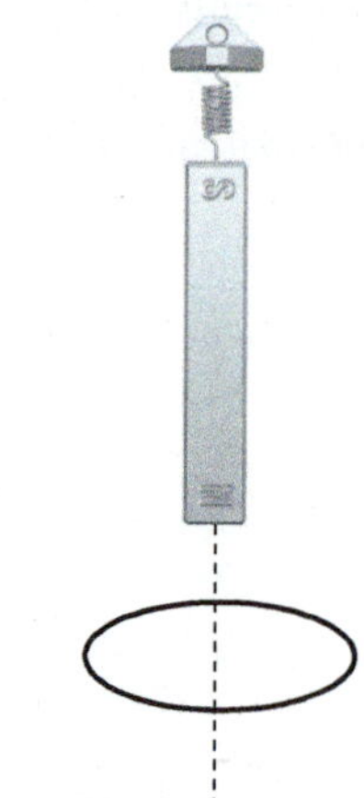

FIGURE 28-40 Problem 9

10 • A pendulum is fabricated from a thin, flat piece of aluminum. At the bottom of its arc, it passes between the poles of a strong permanent magnet. In Figure 28-41*a*, the metal sheet is continuous, whereas in Figure 28-41*b*, there are slots in the sheet. When released from the same angle, the pendulum that has slots swings back and forth many times, but the pendulum that does not have slots stops swinging after no more than one complete oscillation. Explain why.

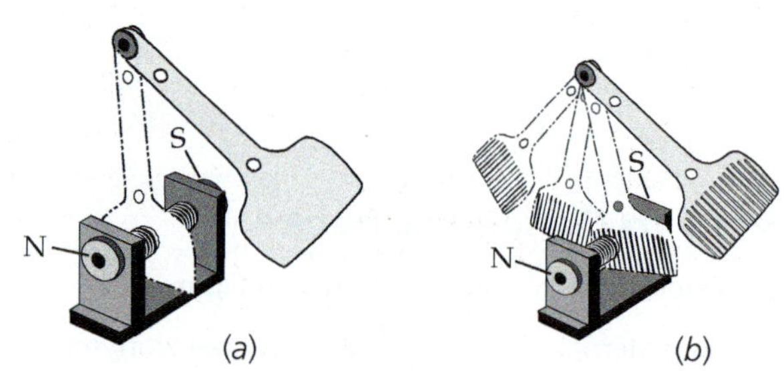

FIGURE 28-41 Problem 10 *(Courtesy of PASCO Scientific Co.)*

11 • A bar magnet is dropped inside a long, vertical tube. If the tube is made of metal, the magnet quickly approaches a terminal speed, but if the tube is made of cardboard, the magnet falls with constant acceleration. Explain why the magnet falls differently in the metal tube than it does in the cardboard tube.

12 • A small square wire loop lies in the plane of this page, and a constant magnetic field is directed into the page. The loop is moving to the right, which is the $+x$ direction. Find the direction of the induced current, if any, in the loop if (*a*) the magnetic field is uniform, (*b*) the magnetic field strength increases as x increases, and (*c*) the magnetic field strength decreases as x increases.

13 • If the current in an inductor doubles, the energy stored in the inductor will (*a*) remain the same, (*b*) double, (*c*) quadruple, (*d*) halve.

14 • Two solenoids are equal in length and radius, and the cores of both are identical cylinders of iron. However, solenoid A has three times the number of turns per unit length as solenoid B. (*a*) Which solenoid has the larger self-inductance? (*b*) What is the ratio of the self-inductance of solenoid A to the self-inductance of solenoid B?

15 • True or false:

(*a*) The induced emf in a circuit is equal to the negative of the magnetic flux through the circuit.
(*b*) There can be a nonzero induced emf at an instant when the flux through the circuit is equal to zero.
(*c*) The self-inductance of a solenoid is proportional to the rate of change of the current in the solenoid.
(*d*) The magnetic energy density at some point in space is proportional to the square of the magnitude of the magnetic field at that point.
(*e*) The inductance of a solenoid is proportional to the current in it. SSM

ESTIMATION AND APPROXIMATION

16 • CONTEXT-RICH Your baseball teammates, having just studied this chapter, are concerned about generating enough voltage to shock them while swinging aluminum bats at fastballs. Estimate the maximum possible motional emf measured between the ends of an aluminum baseball bat during a swing. Do you think your team should switch to wooden bats to avoid electrocution?

17 • Compare the energy density stored in Earth's electric field near its surface to that stored in Earth's magnetic field near its surface.

18 •• A physics teacher does the following emf demonstration. She has two students hold a long wire connected to a voltmeter. The wire is held slack, so that it sags with a large arc in it. When she says, "Start," the students begin rotating the wire as if they were playing jump rope. The students stand 3.0 m apart, and the sag in the wire is about 1.5 m. The motional emf from the "jump rope" is then measured on the voltmeter. (*a*) Estimate a reasonable value for the maximum angular speed at which the students can rotate the wire. (*b*) From this, estimate the maximum motional emf in the wire. *Hint: What field is involved in creating the induced emf?*

19 •• (*a*) Estimate the maximum possible motional emf between the wingtips of a typical commercial airliner in flight. (*b*) Estimate the magnitude of the electric field between the wingtips.

MAGNETIC FLUX

20 • A uniform magnetic field of magnitude 0.200 T is in the $+x$ direction. A square coil that has 5.00-cm-long sides has a single turn and makes an angle θ with the z axis, as shown in Figure 28-42. Find the magnetic flux through the coil when θ is (*a*) 0°, (*b*) 30°, (*c*) 60°, and (*d*) 90°.

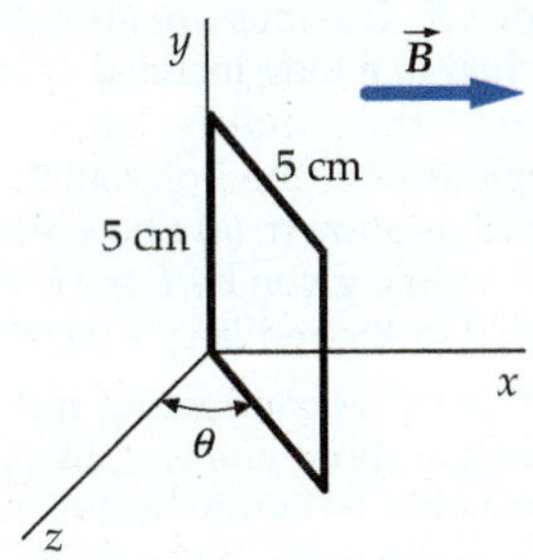

FIGURE 28-42 Problem 20

21 • A circular coil has 25 turns and a radius of 5.0 cm. It is at the equator, where Earth's magnetic field is 0.70 G, north. The axis of the coil is the line that passes through the center of the coil and is perpendicular to the plane of the coil. Find the magnetic flux through the coil when the axis of the coil is (*a*) vertical, (*b*) horizontal with the axis pointing north, (*c*) horizontal with the axis pointing east, and (*d*) horizontal with the axis making an angle of 30° with north. SSM

22 • A magnetic field of 1.2 T is perpendicular to the plane of a 14-turn square coil with sides 5.0 cm long. (*a*) Find the magnetic flux through the coil. (*b*) Find the magnetic flux through the coil if the magnetic field makes an angle of 60° with the normal to the plane of the coil.

23 • A uniform magnetic field $\vec{B}$ is perpendicular to the base of a hemisphere of radius R. Calculate the magnetic flux (in terms of B and R) through the spherical surface of the hemisphere.

24 • Find the magnetic flux through a 400-turn solenoid that has a length equal to 25.0 cm, has a radius equal to 1.00 cm, and carries a current of 3.00 A.

25 • Find the magnetic flux through a 800-turn solenoid that has a length equal to 30.0 cm, has a radius equal to 1.00 cm, and carries a current of 2.00 A.

26 •• A circular coil has 15.0 turns, has a radius of 4.00 cm, and is in a uniform magnetic field of 4.00 kG in the $+x$ direction. Find the flux through the coil when the unit normal to the plane of the coil is (*a*) $\hat{i}$, (*b*) $\hat{j}$, (*c*) $(\hat{i} + \hat{j})/\sqrt{2}$, (*d*) $\hat{k}$, and (*e*) $0.60\hat{i} + 0.80\hat{j}$.

27 •• A long solenoid has n turns per unit length, has a radius R_1, and carries a current I. A circular coil with radius R_2 and with N total turns is coaxial with the solenoid and equidistant from its ends. (*a*) Find the magnetic flux through the coil if $R_2 > R_1$. (*b*) Find the magnetic flux through the coil if $R_2 < R_1$. SSM

28 ••• (*a*) Compute the magnetic flux through the rectangular loop shown in Figure 28-43. (*b*) Evaluate your answer for $a = 5.0$ cm, $b = 10$ cm, $d = 2.0$ cm, and $I = 20$ A.

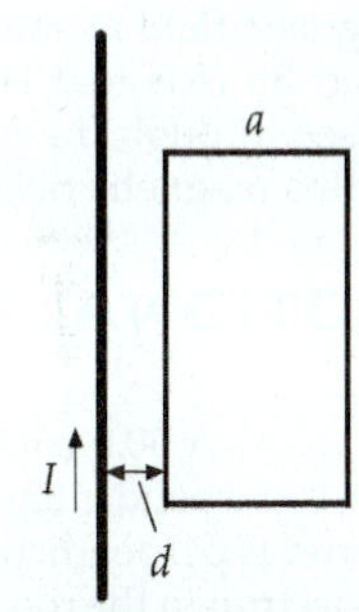

FIGURE 28-43 Problem 28

29 ••• A long, cylindrical conductor with a radius R and a length L carries a current I. Find the magnetic flux per unit length through the area indicated in Figure 28-44.

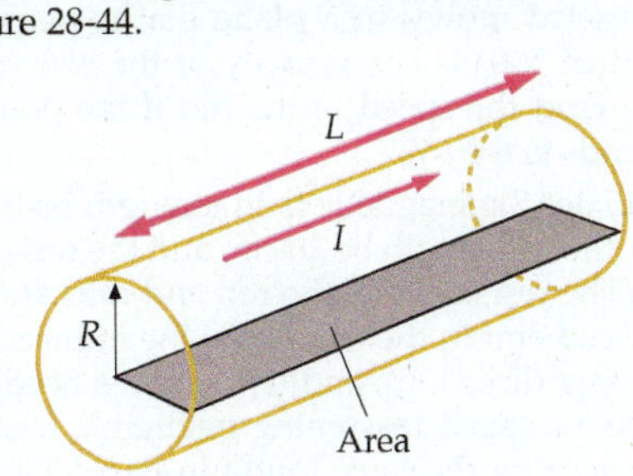

FIGURE 28-44 Problem 29

INDUCED EMF AND FARADAY'S LAW

30 • The flux through a loop is given by $\phi_m = (0.10t^2 - 0.40t)$, where ϕ_m is in webers and t is in seconds. (*a*) Find the induced emf as a function of time. (*b*) Find both ϕ_m and $\mathcal{E}$ at $t = 0$, $t = 2.0$ s, $t = 4.0$ s, and $t = 6.0$ s.

31 • The flux through a loop is given by $\phi_m = (0.10t^2 - 0.40t)$, where ϕ_m is in webers and t is in seconds. (*a*) Sketch graphs of magnetic flux and induced emf as a function of time. (*b*) At what time(s) is the flux minimum? What is the induced emf at that (those) time(s)? (*c*) At what time(s) is the flux zero? What is (are) the induced emf(s) at those time(s)?

32 • A solenoid that has a length equal to 25.0 cm, a radius equal to 0.800 cm, and 400 turns is in a region where a magnetic field of 600 G exists and makes an angle of 50° with the axis of the solenoid. (*a*) Find the magnetic flux through the solenoid. (*b*) Find the magnitude of the average emf induced in the solenoid if the magnetic field is reduced to zero in 1.40 s.

33 •• A 100-turn circular coil has a diameter of 2.00 cm and a resistance of 50.0 Ω, and the two ends of the coil are connected together. The plane of the coil is perpendicular to a uniform magnetic field of magnitude 1.00 T. The direction of the field is reversed. (*a*) Find the total charge that passes through a cross section of the wire. If the reversal takes 0.100 s, find (*b*) the average current and (*c*) the average emf during the reversal. SSM

34 •• At the equator, a 1000-turn coil that has a cross-sectional area of 300 cm^2 and a resistance of 15.0 Ω is aligned so that its plane is perpendicular to Earth's magnetic field of 0.700 G. (*a*) If the coil is flipped over in 0.350 s, what is the average induced current in it during the 0.350 s? (*b*) How much charge flows through a cross section of the coil wire during the 0.350 s?

35 •• **ENGINEERING APPLICATION** A *current integrator* measures the current as a function of time and integrates (adds) the current to find the total charge passing through it. (Because $I = dq/dt$, the integrator calculates the integral of the current or $Q = \int I\,dt$.) A circular coil that has 300 turns and a radius equal to 5.00 cm is connected to such an instrument. The total resistance of the circuit is 20.0 Ω. The plane of the coil is originally aligned perpendicular to Earth's magnetic field at some point. When the coil is rotated through 90° about an axis that is in the plane of the coil a charge of 9.40 μC passes through the current integrator. Calculate the magnitude of Earth's magnetic field at that point.

MOTIONAL EMF

36 • A 30.0-cm-long rod moves steady at 8.00 m/s in a plane that is perpendicular to a magnetic field of 500 G. The velocity of the rod is perpendicular to its length. Find (*a*) the magnetic force on an electron in the rod, (*b*) the electrostatic field in the rod, and (*c*) the potential difference between the ends of the rod.

37 • A 30.0-cm-long rod moves in a plane that is perpendicular to a magnetic field of 500 G. The velocity of the rod is perpendicular to its length. Find the speed of the rod if the potential difference between the ends is 6.00 V.

38 •• In Figure 28-45, let the magnetic field strength be 0.80 T, the rod speed be 10 m/s, the rod length be 20 cm, and the resistance of the resistor be 2.0 Ω. (The resistance of the rod and rails are negligible.) Find (*a*) the induced emf in the circuit, (*b*) the induced current in the circuit (including direction), and (*c*) the force needed to move the rod with constant speed (assuming negligible friction). Find (*d*) the power delivered by the force found in Part (*c*) and (*e*) the rate of Joule heating in the resistor.

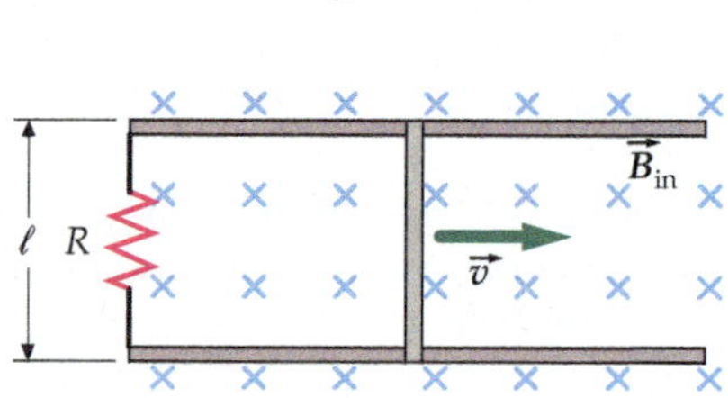

FIGURE 28-45 Problem 38

39 •• A 10-cm by 5.0-cm rectangular loop (Figure 28-46) that has a resistance equal to 2.5 Ω moves at a constant speed of 2.4 cm/s through a region that has a uniform 1.7-T magnetic field directed out of the page as shown. The front of the loop enters the field region at time $t = 0$. (*a*) Graph the flux through the loop as a function of time. (*b*) Graph the induced emf and the current in the loop as functions of time. Neglect any self-inductance of the loop and construct your graphs to include the interval $0 \le t \le 16$ s.

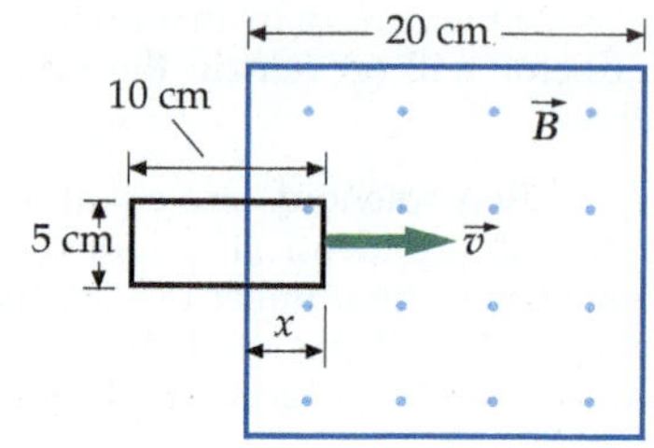

FIGURE 28-46 Problem 39

40 •• A uniform 1.2-T magnetic field is in the $+z$ direction. A conducting rod of length 15 cm lies parallel to the y axis and oscillates in the x direction with displacement given by $x = (2.0\text{ cm})\cos(120\pi t)$, where 120π has units of rad/s. (*a*) Find an expression for potential difference between the ends of the rod as a function of time? (*b*) What is the maximum potential difference between the ends the rod?

41 •• In Figure 28-47, the rod has a mass m, and a resistance R. The rails are horizontal, frictionless, and have negligible resistances. The distance between the rails is ℓ. An ideal battery that has an emf $\mathcal{E}$ is connected between points a and b so that the current in the rod is downward. The rod is released from rest at $t = 0$. (*a*) Derive an expression for the force on the rod as a function of the speed. (*b*) Show that the speed of the rod approaches a terminal speed and find an expression for the terminal speed. (*c*) What is the current when the rod is moving at its terminal speed? SSM

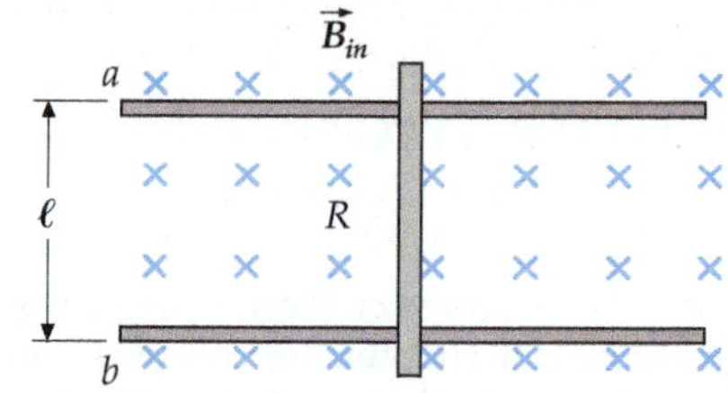

FIGURE 28-47 Problem 41

42 • A uniform magnetic field is established perpendicular to the plane of a loop that has a radius equal to 5.00 cm and a resistance equal to 0.400 Ω. The magnitude of the field is increasing at a rate of 40.0 mT/s. Find (*a*) the magnitude of the induced emf in the loop, (*b*) the induced current in the loop, and (*c*) the rate of Joule heating in the loop.

43 •• In Figure 28-48, a conducting rod that has a mass m and a negligible resistance is free to slide without friction along two parallel frictionless rails that have negligible resistances separated by a distance ℓ and connected by a resistance R. The rails are attached to a long inclined plane that makes an angle θ with the horizontal. There is a magnetic field directed upward as shown. (*a*) Show that there is a retarding force directed up the incline given by $F = (B^2\ell^2 v\cos^2\theta)/R$. (*b*) Show that the terminal speed of the rod is $v_t = mgR\sin\theta/(B^2\ell^2\cos^2\theta)$.

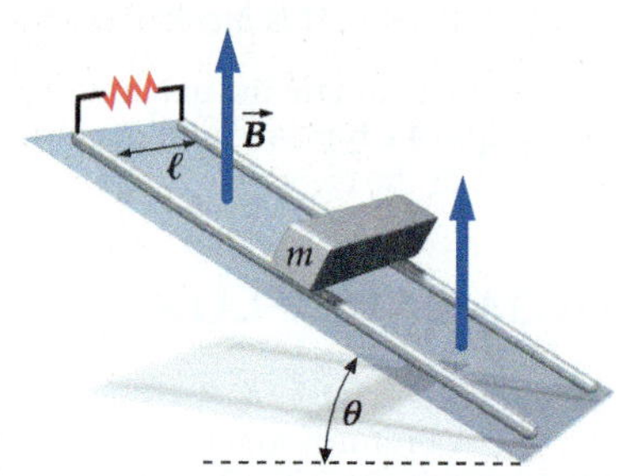

FIGURE 28-48 Problem 43

44 ••• A conducting rod of length ℓ rotates at constant angular speed ω about one end, in a plane perpendicular to a uniform magnetic field B (Figure 28-49). (*a*) Show that the potential difference between the ends of the rod is $\frac{1}{2}B\omega\ell^2$. (*b*) Let the angle θ between the

rotating rod and the dashed line be given by $\theta = \omega t$. Show that the area of the pie-shaped region swept out by the rod during time t is $\frac{1}{2}\ell^2\theta$. (*d*) Compute the flux ϕ_m through that area, and apply $\mathcal{E} = -d\phi_m/dt$ (Faraday's law) to show that the motional emf is given by $\frac{1}{2}B\,\omega\ell^2$.

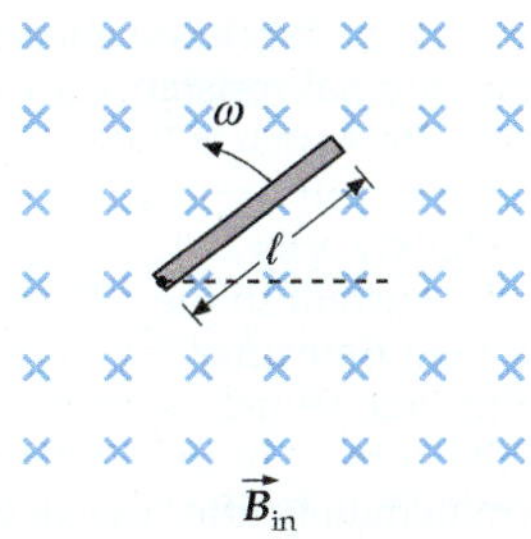

FIGURE 28-49 Problem 44

GENERATORS AND MOTORS

45 • A 2.00-cm by 1.50-cm rectangular coil has 300 turns and rotates in a region that has magnetic field of 0.400 T. (*a*) What is the maximum emf generated when the coil rotates at 60 rev/s? (*b*) What must its angular speed be to generate a maximum emf of 110 V? SSM

46 • The coil of Problem 45 rotates at 60 rev/s in a magnetic field. If the maximum emf generated by the coil is 24 V, what is the magnitude of the magnetic field?

INDUCTANCE

47 • When the current in an 8.00-H coil is equal to 3.00 A and is increasing at 200 A/s, find (*a*) the magnetic flux through the coil and (*b*) the induced emf in the coil.

48 •• A 300-turn solenoid has a radius equal to 2.00 cm and a length equal to 25.0 cm; a 1000-turn solenoid has a radius equal to 5.00 cm and is also 25.0 cm long. The two solenoids are coaxial, with one nested completely inside the other. What is their mutual inductance?

49 •• An insulated wire that has a resistance of 18.0 Ω/m and a length of 9.00 m will be used to construct a resistor. First, the wire is bent in half and then the doubled wire is wound on a cylindrical form (Figure 28-50) to create a 25.0-cm-long helix that has a diameter equal to 2.00 cm. Find both the resistance and the inductance of this wire-wound resistor. SSM

FIGURE 28-50 Problem 49

50 •• You are given a length ℓ of wire that has radius a and are told to wind it into an inductor in the shape of a helix that has a circular cross section of radius r. The windings are to be as close together as possible without overlapping. Show that the self-inductance of this inductor is $L = \frac{1}{4}\mu_0 r\ell/a$.

51 • Using the result of Problem 50, calculate the self-inductance of an inductor wound from 10 cm of wire that has a diameter of 1.0 mm into a coil that has a radius of 0.25 cm.

52 ••• In Figure 28-51, circuit 2 has a total resistance of 300 Ω. After switch *S* is closed, the current in Circuit 1 increases—reaching a value of 5.00 A after a long time. A charge of 200 μC passes through the galvanometer in Circuit 2 during the time that the current in Circuit 1 is increasing. What is the mutual inductance between the two coils?

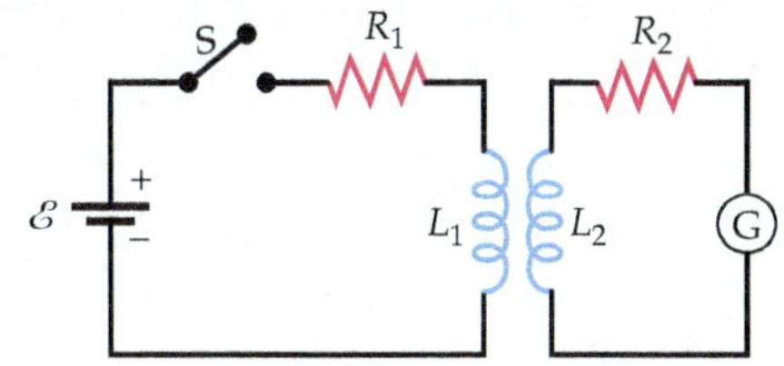

FIGURE 28-51 Problem 52

53 ••• Show that the inductance of a toroid of rectangular cross section, as shown in Figure 28-52, is given by $L = \dfrac{\mu_0 N^2 H \ln(b/a)}{2\pi}$, where N is the total number of turns, a is the inside radius, b is the outside radius, and H is the height of the toroid. SSM

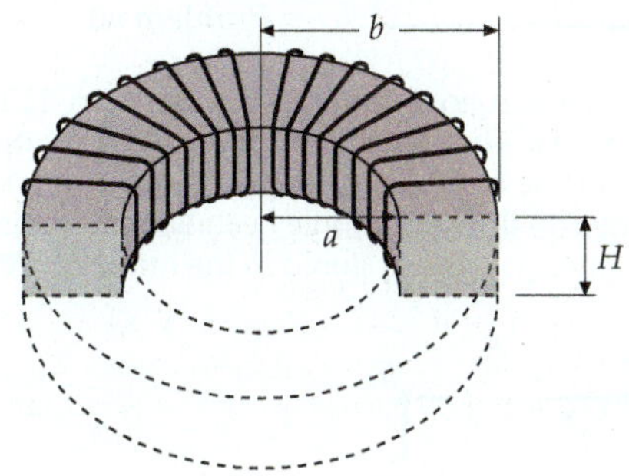

FIGURE 28-52 Problem 53

MAGNETIC ENERGY

54 • A coil that has a self-inductance of 2.00 H and a resistance of 12.0 Ω is connected to an ideal 24.0-V battery. (*a*) What is the steady-state current? (*b*) How much energy is stored in the inductor when the steady-state current is established?

55 • In a plane electromagnetic wave, the magnitudes of the electric fields and magnetic fields are related by $E = cB$, where $c = 1/\sqrt{\epsilon_0\mu_0}$ is the speed of light. Show that when $E = cB$ the electric and the magnetic energy densities are equal. SSM

56 •• A 2000-turn solenoid has a cross-sectional area equal to 4.0 cm² and a length equal to 30 cm. The solenoid carries a current of 4.0 A. (*a*) Calculate the magnetic energy stored in the solenoid using $U = \frac{1}{2}LI^2$, where $L = \mu_0 n^2 A\ell$. (*b*) Divide your answer in Part (*a*) by the volume of the region inside the solenoid to find the magnetic energy per unit volume in the solenoid. (*c*) Check your Part-(*b*) result by computing the magnetic energy density from $u_m = B^2/(2\mu_0)$ where $B = \mu_0 nI$.

57 •• A long cylindrical wire has a radius equal to 2.0 cm and carries a current of 80 A uniformly distributed over its cross-sectional area. Find the magnetic energy per unit length within the wire.

58 •• A toroid that has a mean radius equal to 25.0 cm and circular loops with radii equal to 2.00 cm is wound with a superconducting wire. The wire has a length equal to 1000 m and carries a current of 400 A. (*a*) What is the number of turns of the wire? (*b*) What is the magnetic field strength and magnetic energy density at the mean radius? (*c*) Estimate the total energy stored in this toroid by assuming that the energy density is uniformly distributed in the region inside the toroid.

*RL CIRCUITS

59 • A circuit consists of a coil that has a resistance equal to 8.00 Ω and a self-inductance equal to 4.00 mH, an open switch, and an ideal 100-V battery—all connected in series. At $t = 0$ the switch is closed. Find the current and its rate of change at times (*a*) $t = 0$, (*b*) $t = 0.100$ ms, (*c*) $t = 0.500$ ms, and (*d*) $t = 1.00$ ms. SSM

60 • In the circuit shown in Figure 28-53, the throw of the make-before-break switch has been at contact *a* for a long time and the current in the 1.00 mH coil is equal to 2.00 A. At $t = 0$ the throw is quickly moved to contact *b*. The total resistance $R + r$ of the coil and the resistor is 10.0 Ω. Find the current when (*a*) $t = 0.500$ ms and (*b*) $t = 10.0$ ms.

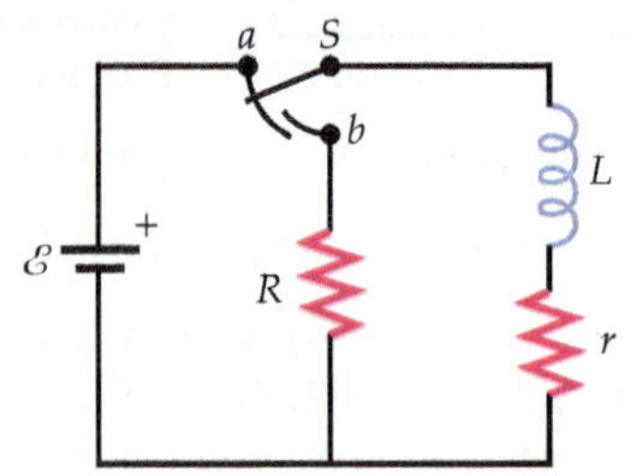

FIGURE 28-53 Problem 60

61 •• In the circuit shown in Figure 28-54, let $\mathcal{E}_0 = 12.0$ V, $R = 3.00$ Ω, and $L = 0.600$ H. The switch, which was initially open, is closed at time $t = 0$. At time $t = 0.500$ s, find (*a*) the rate at which the battery supplies energy, (*b*) the rate of Joule heating in the resistor, and (*c*) the rate at which energy is being stored in the inductor. SSM

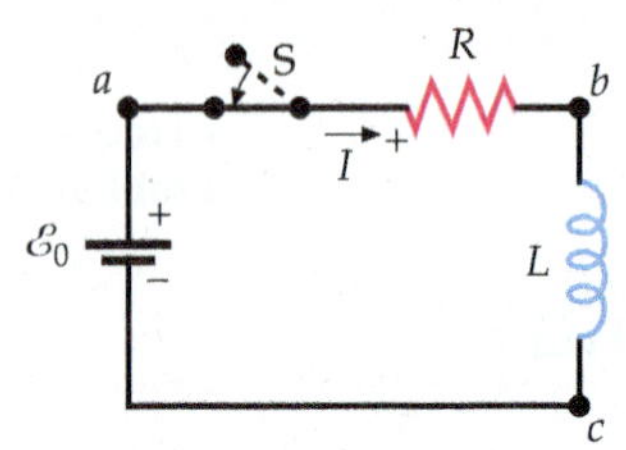

FIGURE 28-54 Problems 61, 62 and 69

62 •• How many time constants must elapse before the current in an *RL* circuit (Figure 28-54) that is initially zero reaches (*a*) 90 percent, (*b*) 99 percent, and (*c*) 99.9 percent of its steady-state value?

63 •• A circuit consists of a 4.00-mH coil, a 150-Ω resistor, a 12.0-V ideal battery, and an open switch—all connected in series. After the switch is closed: (*a*) What is the initial rate of increase of the current? (*b*) What is the rate of increase of the current when the current is equal to half its steady-state value? (*c*) What is the steady-state value of the current? (*d*) How long does it take for the current to reach 99 percent of its steady state value? SSM

64 •• A circuit consists of a large electromagnet that has an inductance of 50.0 H and a resistance of 8.00 Ω, a dc 250-V power source, and an open switch—all connected in series. How long after the switch is closed is the current equal to (*a*) 10 A and (*b*) 30 A.

65 •• **SPREADSHEET** Given the circuit shown in Figure 28-55, assume that the inductor has negligible internal resistance and that the switch S has been closed for a long time so that a steady current exists in the inductor. (*a*) Find the battery current, the current in the 100 Ω resistor, and the current in the inductor. (*b*) Find the potential drop across the inductor immediately after the switch S is opened. (*c*) Using a spreadsheet program, make graphs of the current in the inductor and the potential drop across the inductor as functions of time, for the period during which the switch is open. SSM

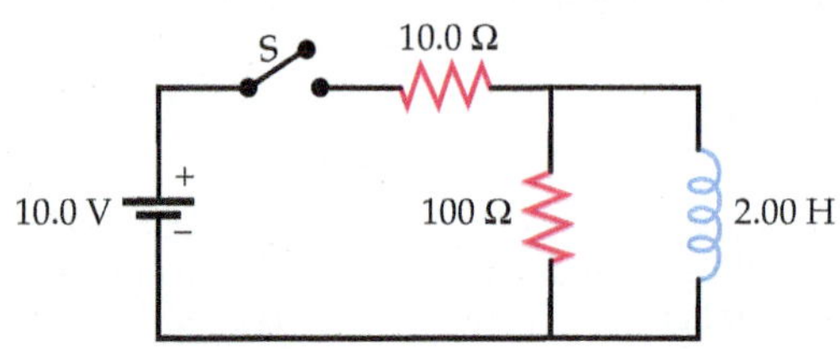

FIGURE 28-55 Problem 65

66 •• Given the circuit shown in Figure 28-56, the inductor has negligible internal resistance and the switch *S* has been open for a long time. The switch is then closed. (*a*) Find the current in the battery, the current in the 100-Ω resistor, and the current in the inductor immediately after the switch is closed. (*b*) Find the current in the battery, the current in the 100-Ω resistor, and the current in the inductor a long time after the switch is closed. After being closed for a long time the switch is now opened. (*c*) Find the current in the battery, the current in the 100-Ω resistor, and the current in the inductor immediately after the switch is opened. (*d*) Find the current in the battery, the current in the 100-Ω resistor, and the current in the inductor after the switch is opened for a long time.

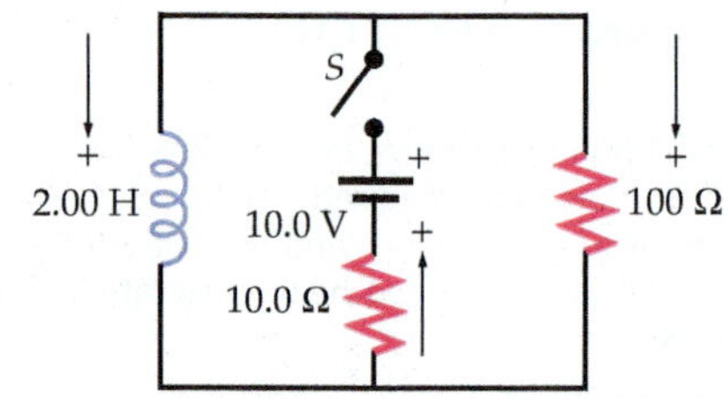

FIGURE 28-56 Problem 66

67 •• An inductor, two resistors, a make-before-break switch, and a battery are connected as shown in Figure 28-57. The switch throw has been at contact *e* for a long time and the current in the inductor is 2.5 A. Then, at $t = 0$, the throw is quickly moved to contact *f*. During the next 45 ms the current in the inductor drops to 1.5 A. (*a*) What is the time constant for this circuit? (*b*) If the resistance *R* is equal to 0.40 Ω, what is the value of the inductance *L*?

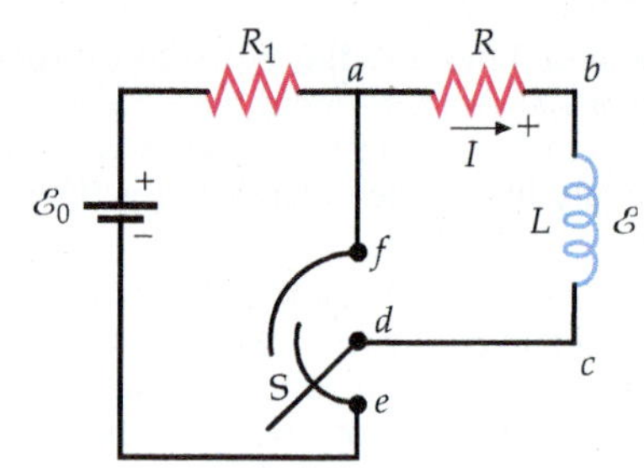

FIGURE 28-57 Problem 67

68 •• A circuit consists of a coil that has a self-inductance equal to 5.00 mH and an internal resistance equal to 15.0 Ω, an ideal 12.0-V battery, and an open switch—all connected in series (Figure 28-58). At $t = 0$ the switch is closed. Find the time when the rate at which energy is dissipated in the coil equals the rate at which magnetic energy is stored in the coil.

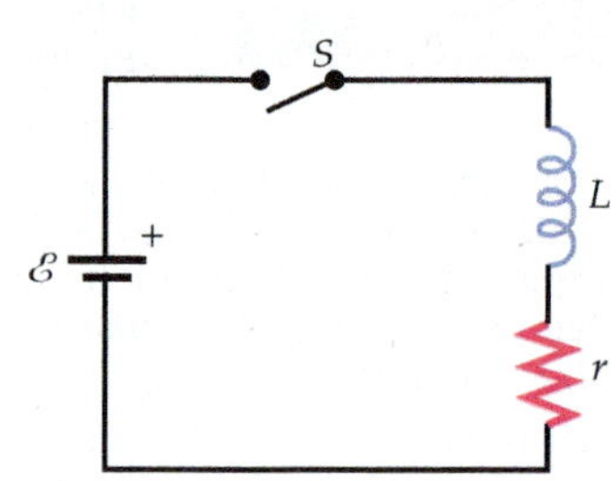

FIGURE 28-58 Problem 68

69 ••• In the circuit shown in Figure 28-54, let $\mathcal{E}_0 = 12.0$ V, $R = 3.00$ Ω, and $L = 0.600$ H. The switch is closed at time $t = 0$. During the time from $t = 0$ to $t = L/R$, find (*a*) the amount of energy supplied by the battery, (*b*) the amount of energy dissipated in the resistor, and (*c*) the amount of energy delivered to the inductor. *Hint: Find the energy transfer rates as functions of time and integrate.* SSM

GENERAL PROBLEMS

70 • A 100-turn coil has a radius of 4.00 cm and a resistance of 25.0 Ω. (*a*) The coil is in a uniform magnetic field that is perpendicular to the plane of the coil. What rate of change of the magnetic field strength will induce a current of 4.00 A in the coil? (*b*) What rate of change of the magnetic field strength is required if the magnetic field makes an angle of 20° with the normal to the plane of the coil?

71 •• **ENGINEERING APPLICATION** Figure 28-59 shows a schematic drawing of an *ac generator*. The basic generator consists of a rectangular loop of dimensions *a* and *b* and has *N* turns connected to *slip rings*. The loop rotates (driven by a gasoline engine) at an angular speed of ω in a uniform magnetic field $\vec{B}$. (*a*) Show that the induced potential difference between the two slip rings is given by $\mathcal{E} = NBab\omega \sin\omega t$. (*b*) If $a = 2.00$ cm, $b = 4.00$ cm, $N = 250$, and $B = 0.200$ T, at what angular frequency ω must the coil rotate to generate an emf whose maximum value is 100 V? SSM

72 •• **ENGINEERING APPLICATION** Prior to 1960, magnetic field strengths were usually measured by a *rotating coil gaussmeter*. The device uses a small multiturn coil rotating at a high speed on an axis perpendicular to the magnetic field. The coil is connected to an ac voltmeter by means of slip rings, like those shown in Figure 28-59. In one specific design, the rotating coil has 400 turns and an area of 1.40 cm^2. The coil rotates at 180 rev/min. If the magnetic field strength is 0.450 T, find the maximum induced emf in the coil and the orientation of the normal to the plane of the coil relative to the field for which the maximum induced emf occurs.

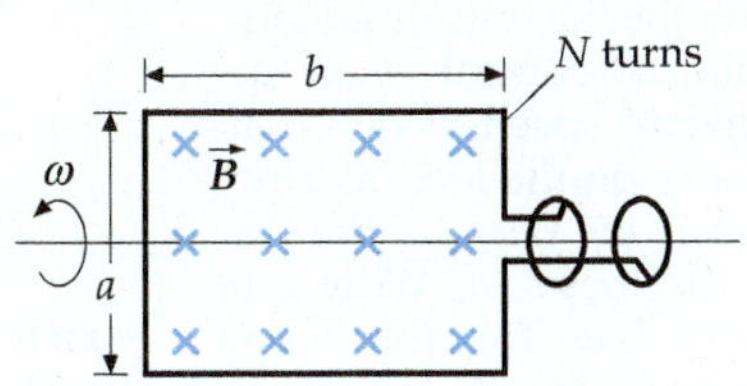

FIGURE 28-59 Problems 71 and 72

73 •• Show that the equivalent self-inductance for two inductors that have self-inductances L_1 and L_2 and are connected in series is given by $L_{eq} = L_1 + L_2$ if there is no flux linkage between the two inductors. (Saying there is no flux linkage between them is equivalent to saying that the mutual inductance between them is zero.)

74 •• Show that the equivalent self-inductance for two in ductors that have self-inductances L_1 and L_2 and are connected in parallel is given by $\frac{1}{L_{eq}} = \frac{1}{L_1} + \frac{1}{L_2}$ if there is no flux linkage between the two inductors. (Saying there is no flux linkage between them is equivalent to saying that the mutual inductance between them is equal to zero.)

75 •• A circuit consists of a 12-V battery, a switch, and a lightbulb—all connected in series. It is known that the lightbulb requires a minimum current of 0.10 A in order to produce a visible glow. In the circuit, that particular bulb draws 2.0 W when the switch has been closed for a long time. Now, an inductor is put in series with the bulb and the rest of the circuit. If the lightbulb begins to glow 3.5 ms after the switch is closed, how large is the self-inductance of the inductor? Ignore any heating time of the filament and assume the glow is observed as soon as the current in the filament reaches the 0.10-A threshold.

76 •• Your friend decides to generate electrical power by rotating a 100000-turn coil of wire around an axis in the plane of the coil and through its center. The coil is perpendicular to Earth's magnetic field in a region where the field strength is equal to 0.300 G. The loops of the coil have a radius of 25.0 cm, and the coil has negligible resistance. (*a*) If your friend turns the coil at a rate of 150 rev/s, what peak current will exist in a 1500-Ω resistor that is connected across the terminals of the coil? (*b*) The average of the square of the current will equal half of the square of the peak current. What will be the average power delivered to the resistor? Is this an economical way to generate power? *Hint: Energy has to be expended to keep the coil rotating.*

77 •• Figure 28-60*a* shows an experiment designed to measure the acceleration due to gravity. A large plastic tube is encircled by a wire, which is arranged in single loops separated by a distance of 10 cm. A strong magnet is dropped through the top of the loop. As the magnet falls through each loop the voltage rises; then the voltage rapidly falls through zero to a large negative value and then returns to zero. The shape of the voltage signal is shown in Figure 28-60*b*. (*a*) Explain the basic physics behind the generation of this voltage pulse. (*b*) Explain why the tube cannot be made of a conductive material. (*c*) Qualitatively explain the *shape* of the voltage signal in Figure 28-60*b*. (*d*) The times at which the voltage crosses zero as the magnet falls through each loop in succession are given in the table on the next page. Use these data to calculate a value for *g*. SSM

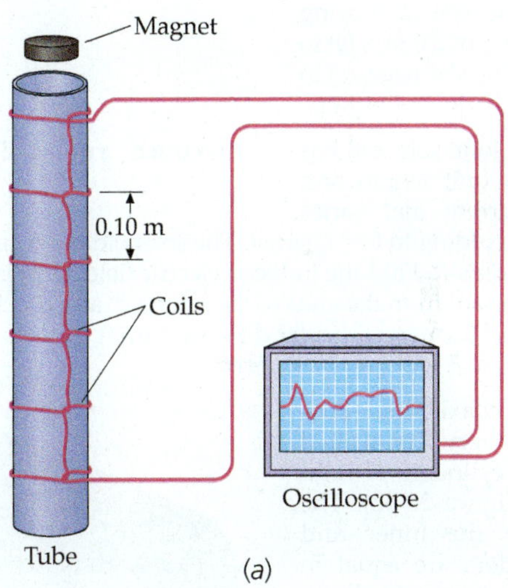

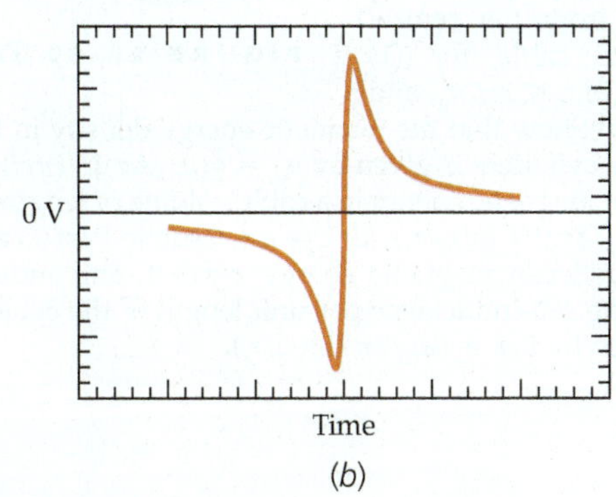

FIGURE 28-60 Problem 77

Loop Number	Zero Crossing Time (s)
1	0.011189
2	0.063133
3	0.10874
4	0.14703
5	0.18052
6	0.21025
7	0.23851
8	0.26363
9	0.28853
10	0.31144
11	0.33494
12	0.35476
13	0.37592
14	0.39107

78 •• The rectangular coil shown in Figure 28-61 has 80 turns, is 25 cm wide, is 30 cm long, and is located in a magnetic field of 0.14 T directed out of the page, as shown. Only half of the coil is in the region of the magnetic field. The resistance of the coil is 24 Ω. Find the magnitude and the direction of the induced current if the coil is moving with a velocity of 2.0 m/s (*a*) to the right, (*b*) up the page, (*c*) to the left, and (*d*) down the page.

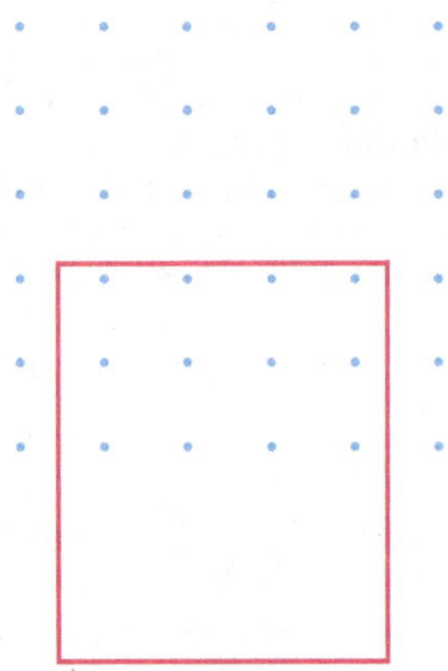

FIGURE 28-61 Problem 78

79 •• A long solenoid has n turns per unit length and carries a current that varies with time according to $I = I_0 \sin \omega t$. The solenoid has a circular cross section of radius R. Find the induced electric field, at points near the plane equidistant from the ends of the solenoid, as a function of both the time t and the perpendicular distance r from the axis of the solenoid for (*a*) $r < R$ and (*b*) $r > R$. SSM

80 ••• A coaxial cable consists of two very thin-walled conducting cylinders of radii r_1 and r_2 (Figure 28-62). The currents in the inner and outer cylinders are equal in magnitude but opposite in direction. (*a*) Use Ampère's law to find the magnetic field as a function of the perpendicular distance r from the central axis of the cable for (1) $0 < r < r_1$, (2) $r_1 < r < r_2$, and (3) $r > r_2$. (*b*) Show that the magnetic energy density in the region between the cylinders is given by $u_m = \frac{1}{2}(\mu_0/4\pi)I^2/(\pi r^2)$. (*c*) Show that the total magnetic energy in a cable volume of length ℓ is given by $U_m = (\mu_0/4\pi)I^2\ell \ln(r_2/r_1)$. (*d*) Use the result in Part (*c*) and the relationship between magnetic energy, current, and inductance to show that the self-inductance per unit length of the cable arrangement is given by $L/\ell = (\mu_0/2\pi)\ln(r_2/r_1)$.

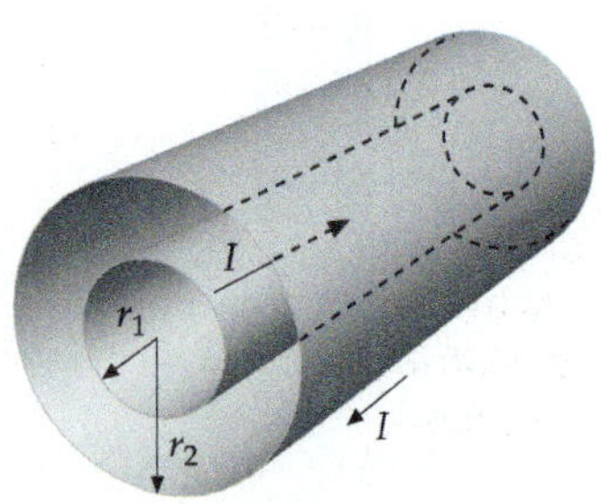

FIGURE 28-62 Problem 80

81 ••• A coaxial cable consists of two very thin-walled conducting cylinders of radii r_1 and r_2 (Figure 28-63). The currents in the inner and outer cylinders are equal in magnitude but opposite in direction. Compute the flux through a rectangular area of sides ℓ and $r_2 - r_1$ between the conductors shown in Figure 28-63. Use the relationship between flux and current ($\phi_m = LI$) to show that the self-inductance per unit length of the cable is given by $L/\ell = (\mu_0/2\pi)\ln(r_2/r_1)$.

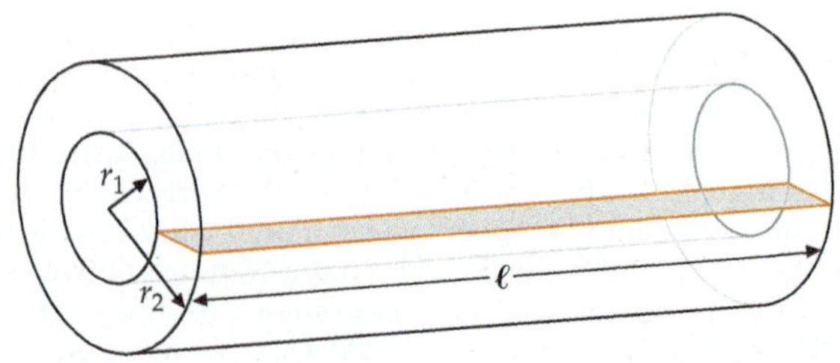

FIGURE 28-63 Problem 81

82 ••• **SPREADSHEET** Figure 28-64 shows a rectangular loop of wire that is 0.300 m wide, is 1.50 m long, and lies in the vertical plane which is perpendicular to a region that has a uniform magnetic field. The magnitude of the uniform magnetic field is 0.400 T and the direction of the magnetic field is into the page. The portion of the loop not in the magnetic field is 0.100 m long. The resistance of the loop is 0.200 Ω and its mass is 50.0 g. The loop is released from rest at $t = 0$. (*a*) What are the magnitude and direction of the induced current when the loop has a downward speed v? (*b*) What is the force that acts on the loop as a result of the current? (*c*) What is the net force acting on the loop? (*d*) Write out Newton's second law for the loop. (*e*) Obtain an expression for the speed of the loop as a function of time. (*f*) Integrate the expression obtained in Part (*e*) to find the distance the loop falls as a function of time. (*g*) Using a spreadsheet program, make a graph of the position of the loop as a function of time (letting $t = 0$ at the start) for values of y between 0 m and 1.40 m (i.e., when the loop leaves the magnetic field). (*h*) At what time does the loop completely leave the field region? Compare this to the time it would have taken if there were no field.

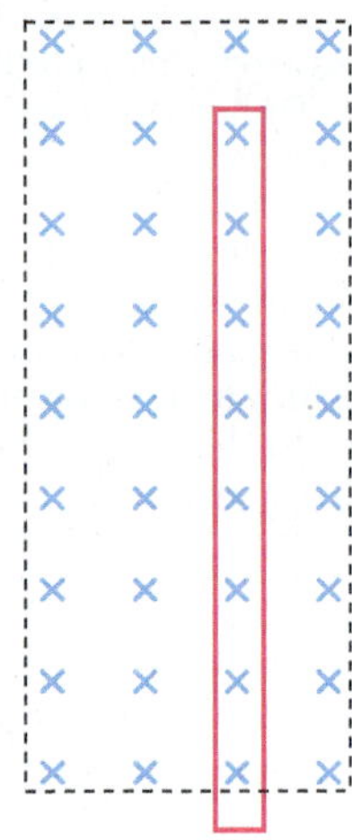

FIGURE 28-64 Problem 82

83 ••• A coil that has N turns and an area A is suspended from the ceiling by a wire that provides a linear restoring torque that has a torsion constant κ. The two ends of the coil are connected to each other, the coil has resistance R, and the moment of inertia of the coil is I. The plane of the coil is vertical and parallel to a uniform horizontal magnetic field $\vec{B}$ when the wire is not twisted (i.e., $\theta = 0$). The coil is twisted about a vertical axis through its center by a small angle θ_0 and released. The coil then undergoes damped harmonic oscillation. Show that its angle with its equilibrium position will vary with time according to $\theta(t) = \theta_0 e^{-t/2\tau} \cos \omega' t$, where $\tau = RI/(NBA)^2$, $\omega = \sqrt{\kappa/I}$ and $\omega' = \omega_0\sqrt{1 - (2\omega_0\tau)^{-2}}$.

CHAPTER 29

THIS LISTENER DIALS IN HER FAVORITE RADIO STATION. THIS CHANGES THE RESONANT FREQUENCY OF AN OSCILLATING ELECTRIC CIRCUIT WITHIN THE TUNER, SO ONLY THE STATION SHE SELECTS IS AMPLIFIED. *(© Roger Ressmeyer/Corbis.)*

? What component of the circuit is modified as she turns the dial? (See Example 29-11.)

Alternating-Current Circuits

More than 99 percent of the electrical energy used today is produced by electrical generators in the form of alternating current, which has a great advantage over direct current. Electrical energy can be distributed over large regions at very high voltages and low currents to reduce energy losses due to Joule heating. With alternating current, electrical energy can then be transformed, with almost no energy loss, to lower and safer voltages and correspondingly higher currents for local distribution and use.* The transformer that accomplishes these changes in potential difference and current works on the basis of magnetic induction. In North America, power is delivered by a sinusoidal current that has a frequency equal to 60 Hz. Devices such as radios, television sets, and microwave ovens detect or generate alternating currents of much higher frequencies.

Alternating current is produced by motional emf or by magnetic induction in an ac generator, which is designed to provide a sinusoidal emf.

* High voltage direct current is sometimes used to transmit electrical power between one point and another distant point. However, alternating current is always used to transmit power from one point to two or more distant points.

In this chapter, we will see that when the generator output is sinusoidal, the current in an inductor, a capacitor, or a resistor is also sinusoidal, although it is generally not in phase with the generator's emf. When the emf and current are both sinusoidal, their maximum values are proportional. The study of sinusoidal currents is particularly important because even currents that are not sinusoidal can be analyzed in terms of sinusoidal components using Fourier analysis.

29-1 ALTERNATING CURRENT IN A RESISTOR

Figure 29-1 shows a simple **ac generator.** An analysis of such a generator is presented in Chapter 28. The emf of such a generator is given by the equation immediately following Equation 28-10:

$$\mathcal{E} = \mathcal{E}_{\text{peak}} \cos \omega t \qquad 29\text{-}1$$

where ω is the angular speed of the coil. (Equation 28-10 has the emf proportional to $\sin \omega t$ rather than $\cos \omega t$. The distinction between the two is the choice of when $t = 0$.) If the N-turn coil has area A, and if the magnetic field is uniform and has magnitude B, the peak emf is given by ωNBA. Although practical generators are considerably more complicated, they all produce a sinusoidal emf either by induction or by motional emf. In circuit diagrams, an ac generator is represented by the symbol ⊙.

Figure 29-2 shows a simple ac circuit that consists of an ideal generator and a resistor. (A generator is ideal if its internal resistance, its self-inductance, and its capacitance are negligible.) The voltage drop across the resistor V_R is equal to the emf $\mathcal{E}$ of the generator. If the generator produces an emf given by Equation 29-1, we have

$$V_R = V_{R\,\text{peak}} \cos \omega t$$

Applying Ohm's law, we have

$$V_R = IR \qquad 29\text{-}2$$

Thus,

$$V_{R\,\text{peak}} \cos \omega t = IR \qquad 29\text{-}3$$

so the current in the resistor is

$$I = \frac{V_{R\,\text{peak}}}{R} \cos \omega t = I_{\text{peak}} \cos \omega t \qquad 29\text{-}4$$

where

$$I_{\text{peak}} = \frac{V_{R\,\text{peak}}}{R} \qquad 29\text{-}5$$

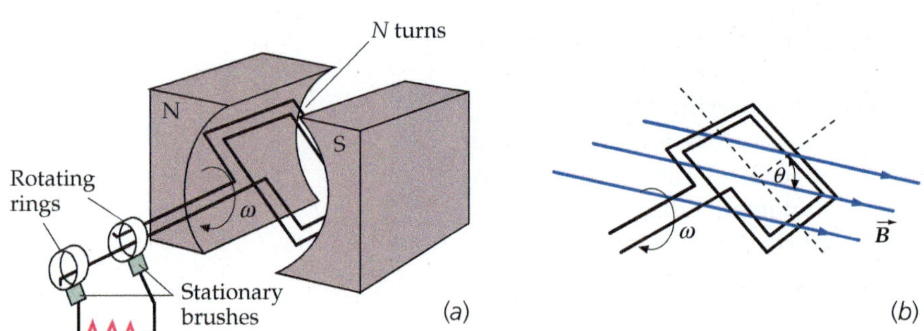

FIGURE 29-1 (*a*) An ac generator. A coil rotating with constant angular frequency ω in a static magnetic field $\vec{B}$ generates a sinusoidal emf. Energy from falling water or from a steam turbine is used to rotate the coil to produce electrical energy. The emf is supplied to an external circuit by the brushes that are in contact with the rings. (*b*) At this instant, the normal to the plane of the coil makes an angle θ with the magnetic field, and the flux through the each turn of the coil is $BA \cos \theta$.

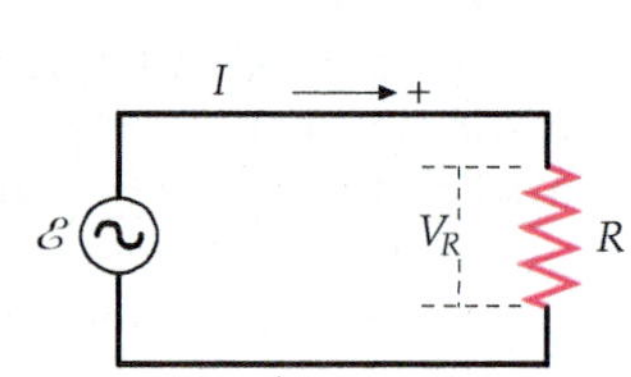

FIGURE 29-2 An ac generator in series with a resistor R.

(a)

(b)

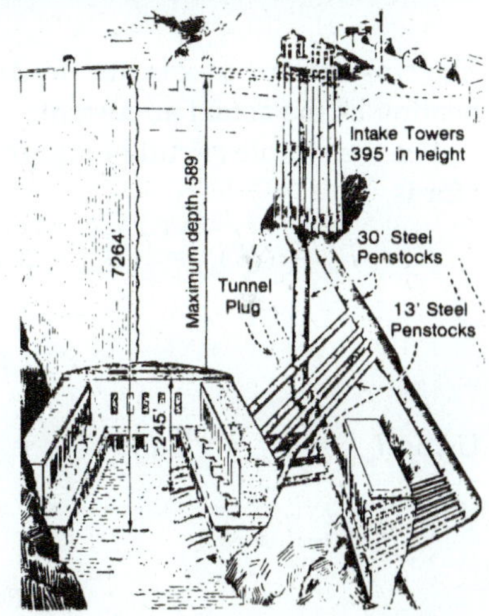

(c)

(*a*) The mechanical energy of falling water drives turbines (*b*) for the generation of electricity. (*c*) Schematic drawing of the Hoover Dam showing the intake towers and pipes (penstocks) that carry the water to the generators below. ((*a*) *Courtesy of U.S. Department of the Interior, Department of Reclamation.* (*b*) *© Lee Langum/ Photo Researchers, Inc.*)

Note that the current through the resistor is in phase with the potential drop across the resistor, as shown in Figure 29-3.

The power delivered to the resistor varies with time. Its instantaneous value is

$$P = I^2R = (I_{\text{peak}} \cos \omega t)^2 R = I_{\text{peak}}^2 R \cos^2 \omega t \qquad 29\text{-}6$$

Figure 29-4 shows the power as a function of time. The power varies from zero to its peak value I_{peak}^2, as shown. We are usually interested in the average power over one or more complete cycles:

$$P_{\text{av}} = (I^2R)_{\text{av}} = I_{\text{peak}}^2 R(\cos^2 \omega t)_{\text{av}}$$

The average value of $\cos^2 \omega t$ over one or more complete periods is $\frac{1}{2}$. This result can be seen from the identity $\cos^2 \omega t + \sin^2 \omega t = 1$. A plot of $\sin^2 \omega t$ looks the same as a plot of $\cos^2 \omega t$ except that the plot is shifted by 90°. Both have the same average value over one or more complete periods, and because their sum is 1, the average value of each must be $\frac{1}{2}$. The average power dissipated in the resistor is thus

$$P_{\text{av}} = (I^2R)_{\text{av}} = \frac{1}{2} I_{\text{peak}}^2 R \qquad 29\text{-}7$$

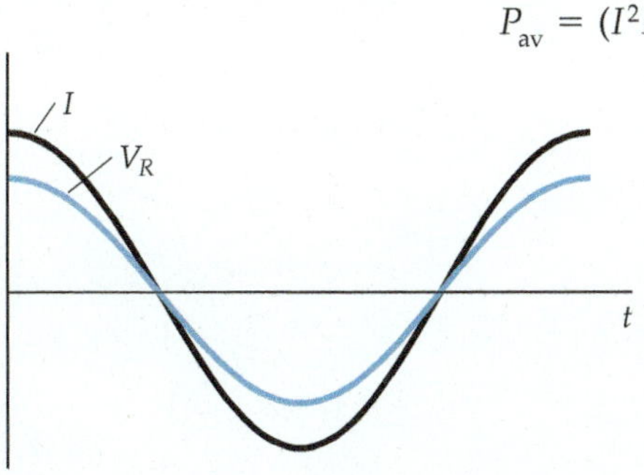

FIGURE 29-3 The voltage drop across a resistor is in phase with the current.

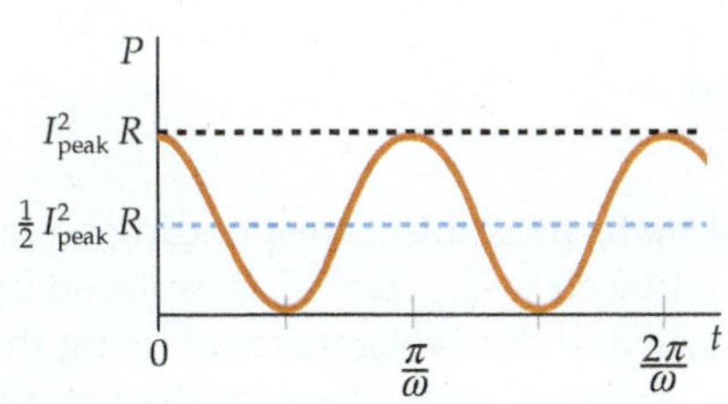

FIGURE 29-4 Plot of the power delivered to the resistor shown in Figure 29-2 versus time. The power varies from zero to a peak value $I_{\text{peak}}^2 R$. The average power is half the peak power.

ROOT-MEAN-SQUARE VALUES

Most ac ammeters and voltmeters are designed to measure the **root-mean-square (rms) values** of current and potential difference. The rms value I_{rms} of a current is defined by

$$I_{\text{rms}} = \sqrt{(I^2)_{\text{av}}} \qquad 29\text{-}8$$

DEFINITION—RMS CURRENT

For a sinusoidal current, the average value of I^2 is

$$(I^2)_{av} = \left[(I_{peak}\cos\omega t)^2\right]_{av} = \frac{1}{2}I^2_{peak}$$

Substituting $\frac{1}{2}I^2_{peak}$ for $(I^2)_{av}$ in Equation 29-8, we obtain

$$I_{rms} = \frac{1}{\sqrt{2}}I_{peak} \approx 0.707I_{peak} \qquad 29\text{-}9$$

RMS VALUE RELATED TO PEAK VALUE

The rms value of *any quantity* that varies sinusoidally with time is equal to the peak value of that quantity divided by $\sqrt{2}$.

Substituting $(I_{rms})^2$ for $\frac{1}{2}I^2_{peak}$ in Equation 29-7, we obtain for the average power delivered to the resistor

$$P_{av} = (I_{rms})^2R \qquad 29\text{-}10$$

The rms current equals the steady dc current that would produce the same Joule heating as the actual ac current.

For the simple circuit in Figure 29-2, the average power delivered by the generator is

$$P_{av} = (\mathcal{E}I)_{av} = \left[\left(\mathcal{E}_{peak}\cos\omega t\right)\left(I_{peak}\cos\omega t\right)\right]_{av} = \mathcal{E}_{peak}I_{peak}(\cos^2\omega t)_{av}$$

or

$$P_{av} = \tfrac{1}{2}\mathcal{E}_{peak}I_{peak}$$

Using $I_{rms} = I_{peak}/\sqrt{2}$ and $\mathcal{E}_{rms} = \mathcal{E}_{peak}/\sqrt{2}$, this can be written

$$P_{av} = \mathcal{E}_{rms}I_{rms} \qquad 29\text{-}11$$

AVERAGE POWER DELIVERED BY A GENERATOR

> ! The rms current equals the steady dc current that would produce the same Joule heating as the actual ac current.

The rms current is related to the rms potential drop in the same way that the peak current is related to the peak potential drop. We can see this by dividing each side of Equation 29-5 by $\sqrt{2}$ and substituting for I_{peak} and $V_{R\,rms}$ using $I_{rms} = I_{peak}/\sqrt{2}$ and $V_{R\,rms} = V_{R\,peak}/\sqrt{2}$.

$$I_{rms} = \frac{V_{R\,rms}}{R} \qquad 29\text{-}12$$

Equations 29-10, 29-11, and 29-12 are of the same form as the corresponding equations for direct-current circuits; however, I is replaced by I_{rms} and V_R is replaced by $V_{R\,rms}$. We can therefore calculate the power input and the heat generated using the same equations that we used for direct current, if we use rms values for the current and potential drop.

PRACTICE PROBLEM 29-1

The sinusoidal potential drop across a 12-Ω resistor has a peak value of 48 V. Find (*a*) the rms current, (*b*) the average power, and (*c*) the maximum power delivered to the resistor.

The ac power supplied to domestic wall outlets and light fixtures in the United States has an rms potential difference of 120 V at a frequency of 60 Hz. This potential difference is maintained, independent of the current. If you plug a 1600-W space heater into a wall outlet it will draw a current of

$$I_{rms} = \frac{P_{av}}{V_{rms}} = \frac{1600\text{ W}}{120\text{ V}} = 13.3\text{ A}$$

All appliances plugged into the outlets of a single 120-V circuit are connected in parallel. If you plug a 500-W toaster into another outlet of the same circuit, it will draw a current of 500 W/120 V = 4.17 A, and the total current through the parallel combination will be 17.5 A. Typical household wall outlets are rated at 15 A and are part of a circuit using wires rated at either 15 A or 20 A, with each circuit having several outlets. A total current greater than the rated current for the wiring is likely to overheat the wiring and is a fire hazard. Each circuit is therefore equipped with a circuit breaker (or a fuse in older houses) that trips (or blows) when the total current exceeds the 15-A or 20-A rating.

High-power domestic appliances, such as electric clothes dryers, kitchen ranges, and hot water heaters, typically require power delivered at 240 V rms. For a given power requirement, only half as much current is required at 240 V as at 120 V, but 240 V is more likely to deliver a fatal shock or to start a fire than 120 V.

Example 29-1 Sawtooth Waveform

Find (*a*) the average current and (*b*) the rms current for the sawtooth waveform shown in Figure 29-5. In the region $0 < t < T$, the current is given by $I = (I_0/T)t$.

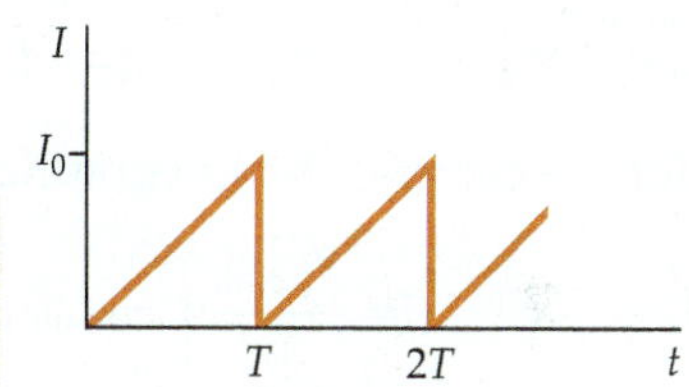

FIGURE 29-5

PICTURE The average of any quantity over a time interval T is the integral of the quantity over the interval divided by T. We use this to find both the average current, I_{av}, and the average of the square of the current, $(I^2)_{av}$.

SOLVE

(*a*) Calculate I_{av} by integrating I from $t = 0$ to $t = T$ and dividing by T:

$$I_{av} = \frac{1}{T}\int_0^T I\,dt = \frac{1}{T}\int_0^T \frac{I_0}{T}t\,dt = \frac{I_0}{T^2}\frac{T^2}{2} = \boxed{\frac{1}{2}I_0}$$

(*b*) 1. Find $(I^2)_{av}$ by integrating I^2:

$$(I^2)_{av} = \frac{1}{T}\int_0^T I^2\,dt = \frac{1}{T}\left(\frac{I_0}{T}\right)^2\int_0^T t^2\,dt = \frac{I_0^2}{T^3}\frac{T^3}{3} = \frac{1}{3}I_0^2$$

2. The rms current is the square root of $(I^2)_{av}$:

$$I_{rms} = \sqrt{(I^2)_{av}} = \boxed{\frac{I_0}{\sqrt{3}}}$$

CHECK Both the average current and the rms current are less than I_0, as expected.

29-2 ALTERNATING-CURRENT CIRCUITS

Alternating current behaves differently than direct current in inductors and capacitors. When a capacitor becomes fully charged in a dc circuit, the capacitor blocks the current; that is, the capacitor acts like an open circuit. However, if the current alternates, charge continually flows onto the plates and off of the plates of the capacitor. We will see that at high frequencies, a capacitor hardly impedes the current at all. That is, the capacitor acts like a short circuit. Conversely, an induction coil that has a low internal resistance is essentially a short circuit for direct current; however, when the current is changing, a back emf is generated in an inductor that is proportional to dI/dt. At high frequencies, the back emf is large and the inductor acts like an open circuit.

INDUCTORS IN ALTERNATING-CURRENT CIRCUITS

Figure 29-6 shows an induction coil in series with an ac generator. When the current changes in the inductor, a back emf equal to $L\,dI/dt$ is generated due to the changing flux. Usually this back emf is much greater than the Ir drop due to the resistance r of the coil, so we normally neglect the resistance of the coil.

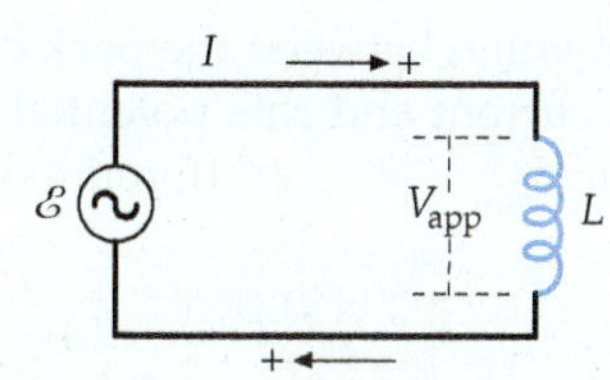

FIGURE 29-6 An ac generator in series with an inductor L. The arrow indicates the positive direction along the wire. Note that for a positive value of dI/dt, the potential drop V_L across the inductor is positive.

The potential drop across the inductor V_L is then given by

$$V_L = L\frac{dI}{dt} \qquad 29\text{-}13$$

POTENTIAL DROP ACROSS AN IDEAL INDUCTOR

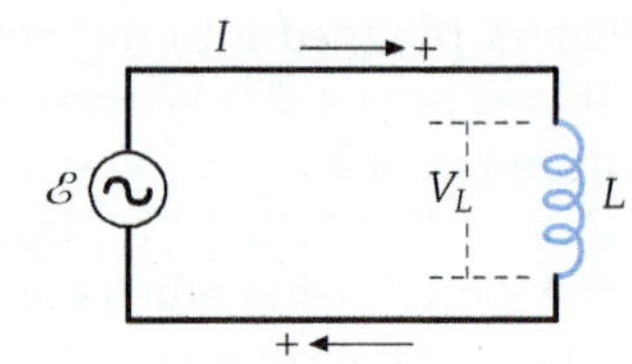

FIGURE 29-6 *(repeated)* An ac generator in series with an inductor L. The arrow indicates the positive direction along the wire. Note that for a positive value of dI/dt, the potential drop V_L across the inductor is positive.

In this circuit, the potential drop V_L across the inductor equals the emf $\mathcal{E}$ of the generator. That is,

$$V_L = \mathcal{E} = \mathcal{E}_{\text{max}} \cos \omega t = V_{L\text{ peak}} \cos \omega t$$

where $V_{L\text{ peak}} = \mathcal{E}_{\text{peak}}$. Substituting for V_L in Equation 29-13 gives

$$V_{L\text{ peak}} \cos \omega t = L\frac{dI}{dt} \qquad 29\text{-}14$$

Rearranging, we obtain

$$dI = \frac{V_{L\text{ peak}}}{L} \cos \omega t\, dt \qquad 29\text{-}15$$

We solve for the current I by integrating both sides of the equation:

$$I = \frac{V_{L\text{ peak}}}{L}\int \cos \omega t\, dt = \frac{V_{L\text{ peak}}}{\omega L} \sin \omega t + C \qquad 29\text{-}16$$

where the constant of integration C is the dc component of the current. Setting the dc component of the current to be zero, we have

$$I = \frac{V_{L\text{ peak}}}{\omega L} \sin \omega t = I_{\text{peak}} \sin \omega t \qquad 29\text{-}17$$

where

$$I_{\text{peak}} = \frac{V_{L\text{ peak}}}{\omega L} \qquad 29\text{-}18$$

The potential drop $V_L = V_{L\text{ peak}} \cos \omega t$ across the inductor is 90° out of phase with the current $I = I_{\text{peak}} \sin \omega t$. From Figure 29-7, which shows I and V_L as functions of time, we can see that the peak value of the potential drop occurs $\frac{1}{4}T$ earlier in time than the corresponding peak value of the current, where T is the period. The potential drop across an inductor is said to *lead the current by* 90°. We can also conceptually understand this result. The potential drop across the inductor its equal to the emf induced in it. When I is zero but decreasing, dI/dt is at its minimum, which is negative, so the emf induced in the inductor is at its maximum. One-quarter cycle later, I is maximum. At this time, dI/dt is zero, so V_L is zero. Using the trigonometric identity $\sin\theta = \cos(\theta - \frac{\pi}{2})$, Equation 29-17 for the current can be written

$$I = I_{\text{peak}} \cos(\omega t - \tfrac{\pi}{2}) \qquad 29\text{-}19$$

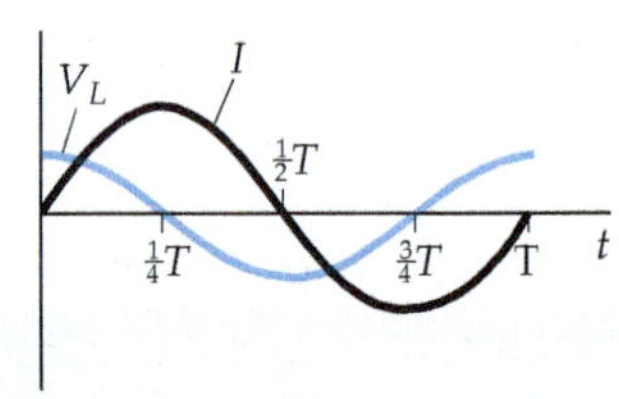

FIGURE 29-7 Current and potential drop across the inductor shown in Figure 29-6 as functions of time. The maximum potential drop occurs one-fourth period before the maximum current. Thus, the potential drop is said to lead the current by one-fourth period or 90°.

The relation between the peak current and the peak potential drop (or between the rms current and rms potential drop) for an inductor can be written in a form similar to $I_{\text{rms}} = V_{R\text{ rms}}/R$ (Equation 29-12). From Equation 29-18, we have

$$I_{\text{peak}} = \frac{V_{L\text{ peak}}}{\omega L} = \frac{V_{L\text{ peak}}}{X_L} \qquad 29\text{-}20$$

where

$$X_L = \omega L \qquad 29\text{-}21$$

DEFINITION—INDUCTIVE REACTANCE

is called the **inductive reactance.** Because $I_{rms} = I_{peak}/\sqrt{2}$ and $V_{L\,rms} = V_{L\,peak}/\sqrt{2}$, the rms current is given by

$$I_{rms} = \frac{V_{L\,rms}}{X_L} \tag{29-22}$$

Like resistance, inductive reactance has units of ohms. As we can see from Equation 29-22, the larger the reactance for a given rms potential drop, the smaller the rms current. Unlike resistance, the inductive reactance depends on the frequency—the greater the frequency, the greater the reactance.

The *instantaneous* power delivered to the inductor is

$$P = V_L I = (V_{L\,peak}\cos\omega t)(I_{peak}\sin\omega t) = V_{L\,peak}I_{peak}\cos\omega t\sin\omega t$$

The *average* power delivered to the inductor is zero. We can see this by using the trigonometric identity

$$2\cos\omega t\sin\omega t = \sin 2\omega t$$

The value of $\sin 2\omega t$ oscillates twice during each cycle of the current, and so is negative as often as it is positive. Thus, when averaged over an integral number of cycles, no energy is delivered to an inductor. (If the internal resistance r of the inductor is not negligible, then the average power delivered is equal to $(I_{rms})^2 r$.)

Example 29-2 Inductive Reactance

The potential drop across a 40.0-mH inductor is sinusoidal and has an rms potential drop of 120 V. Find the inductive reactance and the rms current when the frequency is (*a*) 60.0 Hz and (*b*) 2000 Hz.

PICTURE We calculate the inductive reactance at each frequency and use Equation 29-20 to find the peak current.

SOLVE

(*a*) 1. The peak current equals the rms potential drop divided by the inductive reactance:

$$I_{rms} = \frac{V_{L\,rms}}{X_L}$$

2. Compute the inductive reactance at 60.0 Hz:

$$X_{L1} = \omega_1 L = 2\pi f_1 L = (2\pi)(60.0\text{ Hz})(40.0\times10^{-3}\text{ H}) = \boxed{15.1\ \Omega}$$

3. Use this value of X_L to compute the rms current at 60.0 Hz:

$$I_{1\,rms} = \frac{120\text{ V}}{15.1\ \Omega} = \boxed{7.95\text{ A}}$$

(*b*) 1. Compute the inductive reactance at 2000 Hz:

$$X_{L2} = \omega_2 L = 2\pi f_2 L = (2\pi)(2000\text{ Hz})(40.0\times10^{-3}\text{ H}) = \boxed{503\ \Omega}$$

2. Use this value of X_L to compute the rms current at 2000 Hz:

$$I_{2\,rms} = \frac{120\text{ V}}{503\ \Omega} = \boxed{0.239\text{ A}}$$

CHECK The rms current at 2000 Hz is about 3 percent of the rms current at 60.0 Hz. This result is expected because we expect the inductor to behave more and more like an open circuit as the frequency increases.

CAPACITORS IN ALTERNATING-CURRENT CIRCUITS

When a capacitor is connected across the terminals of an ac generator (Figure 29-8), the voltage drop across the capacitor is

$$V_C = \frac{Q}{C} \qquad 29\text{-}23$$

where Q is the charge on the upper plate of the capacitor.

In this circuit, the potential drop V_C across the capacitor equals the emf $\mathcal{E}$ of the generator. That is,

$$V_C = \mathcal{E}_{\text{peak}} \cos \omega t = V_{C\,\text{peak}} \cos \omega t$$

where $V_{C\,\text{peak}} = \mathcal{E}_{\text{peak}}$. Substituting for V_C in Equation 29-23 and solving for Q gives

$$Q = V_C C = V_{C\,\text{peak}} C \cos \omega t = Q_{\text{peak}} \cos \omega t$$

The current is

$$I = \frac{dQ}{dt} = -\omega Q_{\text{peak}} \sin \omega t = -I_{\text{peak}} \sin \omega t$$

where

$$I_{\text{peak}} = \omega Q_{\text{peak}} \qquad 29\text{-}24$$

Using the trigonometric identity $\sin\theta = -\cos(\theta + \frac{\pi}{2})$, where $\theta = \omega t$, we obtain

$$I = -\omega Q_{\text{peak}} \sin \omega t = I_{\text{peak}} \cos(\omega t + \tfrac{\pi}{2}) \qquad 29\text{-}25$$

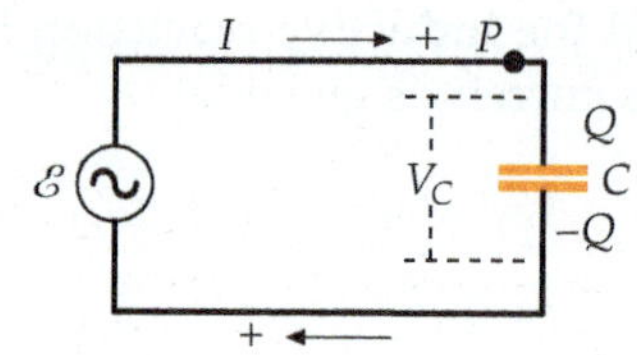

FIGURE 29-8 An ac generator in series with a capacitor C. The positive direction along the circuit is such that when the current is positive the charge Q on the upper capacitor plate is increasing, so the current is related to the charge by $I = +dQ/dt$.

The voltage drop V_C across the capacitor is in phase with the charge Q (Equation 29-23), so as with the inductor, the voltage drop across the capacitor is 90° out of phase with the current in the circuit. From Figure 29-9, we see that the maximum value of the potential drop occurs 90° or one-fourth period later in time than the maximum value of the current. Thus, *the potential drop across a capacitor lags the current by* 90°. We can also understand this result in another way. The charge Q is proportional to the potential drop V_C, so the maximum value of $dQ/dt = I$ occurs when the charge Q, and therefore V_C, is zero. As the charge on the capacitor plate increases the current decreases, until one-fourth period later the charge Q, and therefore V_C, is a maximum and the current is zero. The current then becomes negative as the charge Q decreases.

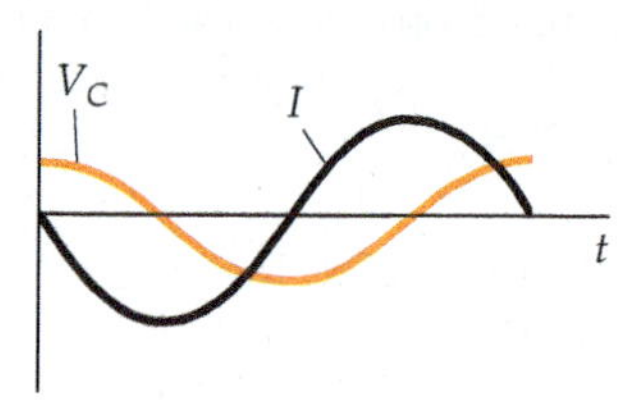

FIGURE 29-9 Current and potential drop across the capacitor shown in Figure 29-8 versus time. The maximum potential drop occurs one-fourth period after the maximum current. Thus, the potential drop is said to lag the current by 90°.

We can relate the current to the potential drop in a form similar to $I_{\text{rms}} = V_{R\,\text{rms}}/R$ (Equation 29-5) for a resistor. From Equation 29-24, we have

$$I_{\text{peak}} = \omega Q_{\text{peak}} = \omega C V_{C\,\text{peak}} = \frac{V_{C\,\text{peak}}}{1/(\omega C)} = \frac{V_{C\,\text{peak}}}{X_C}$$

and, similarly,

$$I_{\text{rms}} = \frac{V_{C\,\text{rms}}}{X_C} \qquad 29\text{-}26$$

where

$$X_C = \frac{1}{\omega C} \qquad 29\text{-}27$$

DEFINITION—CAPACITIVE REACTANCE

is called the **capacitive reactance** of the circuit. Like resistance and inductive reactance, capacitive reactance has units of ohms and, like inductive reactance, capacitive reactance depends on the frequency of the current. In this case, the greater the frequency, the smaller the reactance. The average power delivered to a capacitor in an ac circuit is zero, as it is for an inductor. This is so because the potential drop is proportional to $\cos\omega t$ and the current is proportional to $\sin\omega t$, and $(\cos\omega t\ \sin\omega t)_{av} = 0$. Thus, like inductors with no resistance, capacitors dissipate no energy.

Because charge cannot pass across the space between the plates of a capacitor, it may seem strange that there is a continuing alternating current in the circuit shown in Figure 29-8. Suppose we choose the time to be zero at the instant that the voltage drop V_C across the capacitor is both zero and increasing. (At this same instant, the charge Q on the upper plate of the capacitor is also both zero and increasing.) As V_C then increases, positive charge flows off the lower plate and onto the upper plate, and Q reaches its maximum value Q_{peak} a quarter period later. After Q reaches its maximum value Q continues to change, reaching zero at the half-period point, $-Q_{peak}$ at the three-quarter-period point, and zero (again) at the completion of the cycle at the full-period point. The charge Q_{peak} flows past point P (see Figure 29-8) on the wire each quarter period. If we double the frequency, we halve the period. Thus, if we double the frequency we halve the time for the charge Q_{peak} to flow past point P on the wire, so we have doubled the current amplitude I_{peak}. Hence, the greater the frequency, the less the capacitor impedes the flow of charge.

Example 29-3 Capacitive Reactance

A 20.0-μF capacitor is placed across an ac generator that applies a potential drop which has an amplitude (peak value) of 100 V. Find the capacitive reactance and the current amplitude when the frequency is (*a*) 60.0 Hz and (*b*) 6000 Hz.

PICTURE The capacitive reactance is $X_C = 1/(\omega C)$ and the peak current is $I_{peak} = V_{C\,peak}/X_C$.

SOLVE

(*a*) Calculate the capacitive reactance at 60.0 Hz and use this value to find the peak current at 60.0 Hz:

$$X_{C1} = \frac{1}{\omega_1 C} = \frac{1}{2\pi f_1 C}$$

$$= \frac{1}{2\pi(60.0\ \text{Hz})(20.0 \times 10^{-6}\ \text{F})} = \boxed{133\ \Omega}$$

$$I_{1\,peak} = \frac{V_{C\,peak}}{X_{C1}} = \frac{100\ \text{V}}{133\ \Omega} = \boxed{0.752\ \text{A}}$$

(*b*) Calculate the capacitive reactance at 6000 Hz and use this value to find the peak current at 6000 Hz:

$$X_{C2} = \frac{1}{\omega_2 C} = \frac{1}{2\pi f_2 C}$$

$$= \frac{1}{2\pi(6000\ \text{Hz})(20.0 \times 10^{-6}\ \text{F})} = \boxed{1.33\ \Omega}$$

$$I_{2\,peak} = \frac{V_{C\,peak}}{X_{C2}} = \frac{100\ \text{V}}{1.33\ \Omega} = \boxed{75.2\ \text{A}}$$

CHECK The current at 60.0 Hz is about 1 percent of the current at 6000 Hz. This result is expected because we expect the capacitor to be more like an open circuit at lower frequencies.

TAKING IT FURTHER Note that the capacitive reactance is inversely proportional to the frequency, so increasing the frequency by two orders of magnitude decreases the reactance by two orders of magnitude. The current is directly proportional to the frequency, as expected.

* 29-3 THE TRANSFORMER

A transformer is a device used to increase or decrease the voltage in a circuit without an appreciable loss of power. Figure 29-10 shows a simple transformer consisting of two wire coils around a common iron core. The coil carrying the input power is called the primary, and the other coil is called the secondary. Either coil of a transformer can be used for the **primary** or **secondary.** The transformer operates on the principle that an alternating current in one circuit induces an alternating emf in a nearby circuit due to the mutual inductance of the two circuits. The iron core increases the magnetic field for a given current and guides its direction so that the flux linkage between the coils approaches 100%. (At 100% flux linkage, all the magnetic field lines through one coil also go through the other coil.) If no power were lost, the product of the potential drop across and the current in the secondary windings would equal the product of the potential drop across and the current in the primary windings. Thus, if the potential difference across the secondary coil is larger than the potential difference across the primary coil, the current in the secondary coil is smaller than the current in the primary coil, and vice versa. Power losses arise because of Joule heating, both in the two coils and in the iron core currents,* and from hysteresis in the iron core. We will neglect those losses and consider an ideal transformer of 100 percent efficiency, for which all of the power supplied to the primary coil appears in the secondary coil. Actual power distribution transformers often have efficiencies of 98 percent or more.

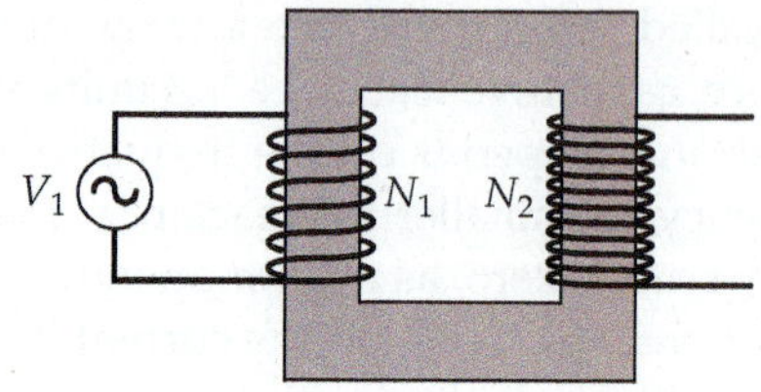

FIGURE 29-10 A transformer that has N_1 turns in the primary and N_2 turns in the secondary.

Consider a transformer with a potential drop V_1 across the primary coil of N_1 turns; the secondary coil of N_2 turns is an open circuit. Because of the iron core, there is a large flux through each coil even when the magnetizing current I_m in the primary circuit is very small. (The magnetizing current is the current in the primary when the secondary circuit is open.) We can ignore the resistances of the coils, which are negligible in comparison with their inductive reactances. The primary circuit is then a simple circuit consisting of an ac generator and a pure inductance, like that discussed in Section 29-2. The magnetizing current in the primary coil and the voltage drop across the primary coil are out of phase by 90°, and the average power dissipated in the primary coil is zero. If ϕ_{turn} is the magnetic flux per turn of the primary coil, the potential drop across the primary coil is equal to the back emf, so

$$V_1 = N_1 \frac{d\phi_{turn}}{dt} \qquad 29\text{-}28$$

If there is no flux leakage out of the iron core, the flux through each turn is the same for both coils. Thus, the total flux through the secondary coil is $N_2\phi_{turn}$, and the potential difference across the secondary coil is

$$V_2 = N_2 \frac{d\phi_{turn}}{dt} \qquad 29\text{-}29$$

Comparing Equations 29-28 and 29-29, we can see that

$$V_2 = \frac{N_2}{N_1} V_1 \qquad 29\text{-}30$$

If N_2 is greater than N_1, the potential difference across the secondary coil is greater than the potential drop across the primary coil, and the transformer is called a *step-up transformer*. If N_2 is less than N_1, the potential difference across the secondary coil is less than the potential drop across the primary coil, and the transformer is called a *step-down transformer*.

(*a*) A power box that has a transformer for stepping down voltage for distribution to homes. (*b*) A suburban power substation where transformers step down voltage from high-voltage transmission lines. (*c*) A 9-volt ac plug-in transformer. *((a) Yaov/Phototake. (b) Daniel S. Brody/Stock Boston. (c) Ramòn Rivera Moret.)*

* The induced currents, called eddy currents, can be greatly reduced by using a core of laminated metal to break up current paths.

When we put a resistance R, called a *load resistance*, across the secondary coil, there will then be a current I_2 in the secondary circuit that is in phase with the potential drop V_2 across the resistance. This current sets up an additional flux ϕ'_{turn} through each turn that is proportional to N_2I_2. This flux opposes the original flux set up by the original magnetizing current I_{m} in the primary. However, the potential drop across the primary coil is determined by the generator emf, which is unaffected by the secondary circuit. According to Equation 29-29, the flux in the iron core must change at the original rate; that is, the total flux in the iron core must be the same as when no load exists across the secondary. The primary coil thus draws an additional current I_1 to maintain the original flux ϕ_{turn}. The flux through each turn produced by this additional current is proportional to N_1I_1. Because this flux equals $-\phi'_{\text{turn}}$, the additional current I_1 in the primary is related to the current I_2 in the secondary by

$$N_1I_1 = -N_2I_2 \qquad \text{29-31}$$

The currents are 180° out of phase and produce counteracting fluxes. Because I_2 is in phase with V_2, the additional current I_1 is in phase with the potential drop across the primary circuit. The power input from the generator is $V_{1\,\text{rms}}I_{1\,\text{rms}}$, and the power output is $V_{2\,\text{rms}}I_{2\,\text{rms}}$. (The magnetizing current does not contribute to the power input because it is 90° out of phase with the generator voltage.) If there are no losses,

$$V_{1\,\text{rms}}I_{1\,\text{rms}} = V_{2\,\text{rms}}I_{2\,\text{rms}} \qquad \text{29-32}$$

In most cases, the additional current in the primary I_1 is much greater than the original magnetizing current I_{m} that is drawn from the generator when no load exists. This can be demonstrated by putting a lightbulb in series with the primary coil. The lightbulb is much brighter when there is a load across the secondary circuit than when the secondary circuit is open. If I_{m} can be neglected, Equation 29-32 relates the total currents in the primary and secondary circuits.

Example 29-4 Doorbell Transformer

A doorbell requires 0.40 A rms of alternating current at 6.0 V rms. It is connected to a transformer whose primary has 2000 turns and is connected to a 120-V-rms ac line. (*a*) How many turns should there be in the secondary? (*b*) What is the current in the primary?

PICTURE We can find the number of turns from the turns ratio, which equals the voltage ratio. The primary current can be found by equating the power out to the power in.

SOLVE

(*a*) The turns ratio can be obtained from Equation 29-30. Solve for the number of turns in the secondary, N_2:

$$\frac{N_2}{N_1} = \frac{V_2}{V_1}$$

so

$$N_2 = \frac{V_{2\,\text{rms}}}{V_{1\,\text{rms}}}N_1 = \frac{6.0\text{ V}}{120\text{ V}}2000\text{ turns} = \boxed{100\text{ turns}}$$

(*b*) Because we are assuming 100 percent efficiency in power transmission, the input and output currents are related by Equation 29-32. Solve for the current in the primary, I_1:

$$V_{2\,\text{rms}}I_{2\,\text{rms}} = V_{1\,\text{rms}}I_{1\,\text{rms}}$$

so

$$I_{1\,\text{rms}} = \frac{V_{2\,\text{rms}}}{V_{1\,\text{rms}}}I_{2\,\text{rms}} = \frac{6.0\text{ V}}{120\text{ V}}(0.40\text{ A}) = \boxed{0.020\text{ A}}$$

CHECK To step-down the voltage requires fewer turns in the secondary than in the primary. In addition, a transformer that steps down the voltage steps up the current. Our results reflect both of these attributes.

An important use of transformers is in the transmission and distribution of electrical power. To minimize the I^2R heat loss (Joule heating) in transmission lines, it is economical to use a high voltage and a low current. On the other hand, safety and other considerations require that power be delivered to consumers at lower voltages and therefore with higher currents. Suppose, for example, that each person in a city with a population of 50000 uses consumes electrical energy at a rate of 1.2 kW. (The per capita rate of consumption of electrical energy in the United States is actually somewhat higher than this value.) At 120 V, the current required for each person would be

$$I = \frac{1200\text{ W}}{120\text{ V}} = 10\text{ A}$$

The total current for 50000 people would then be 500000 A. The transport of such a current from a power-plant generator to a city many kilometers away would require conductors of enormous thickness, and the I^2R power loss would be substantial. Rather than transmit the power at 120 V, step-up transformers are used at power plants to step up the voltage to values as great as 600000 V. For this voltage, the current needed is only

$$I = \frac{120\text{ V}}{600\,000\text{ V}}(500\,000\text{ A}) = 100\text{ A}$$

To reduce the voltage to a safer level for transport within a city, power substations are located just outside the city to step down the voltage to a safer value, such as 10000 V. Transformers in boxes attached to the power poles outside each house again step down the voltage to 120 V (or 240 V) for distribution to the house. Because of the ease of stepping the voltage up or down with transformers, alternating current rather than direct current is in common use.

Example 29-5 Transmission Losses

A transmission line has a resistance of 0.020 Ω/km. Calculate the power loss due to Joule heating if 200 kW of power is transmitted from a power generator to a city 10 km away at (*a*) 240 V rms and (*b*) 4.4 kV rms.

PICTURE First, note that the total resistance of 10 km of wire is $R = (0.020\ \Omega/\text{km})(10\text{ km}) = 0.20\ \Omega$. In each case, begin by finding the current needed to transmit 200 kW using $P = IV$, then find the power loss using $(I_{\text{rms}})^2R$. In the solution, the voltages and currents are rms values and the power is the average power.

SOLVE

(*a*) 1. Find the current needed to transmit 200 kW of power at 240 V:
$$I = \frac{P}{V} = \frac{200\text{ kW}}{240\text{ V}} = 833\text{ A}$$

2. Calculate the power loss:
$$I^2R = (833\text{ A})^2(0.20\ \Omega) = \boxed{1.4 \times 10^2\text{ kW}}$$

(*b*) 1. Now, find the current needed to transmit 200 kW of power at 4.4 kV:
$$I = \frac{P}{V} = \frac{200\text{ kW}}{4.4\text{ kV}} = 45.4\text{ A}$$

2. Calculate the power loss:
$$I^2R = (45.4\text{ A})^2(0.20\ \Omega) = \boxed{0.41\text{ kW}}$$

CHECK The power loss at 4.4 kV is less than one percent of the power loss at 240 V. This result is consistent with the reason for stepping up the voltage for transmission.

TAKING IT FURTHER Note that with a transmission voltage of 240 V, almost 70 percent of the power is wasted through heat loss. In addition, there is an *IR* (voltage) drop across the transmission line of 167 V, so the power is delivered at only 73 V. However, with transmission at 4.4 kV, only about 0.2 percent of the power is lost during transmission and there is an *IR* drop across the transmission line of only 9 V, so the power is delivered with only a 0.2 percent voltage drop.

* 29-4 *LC* AND *RLC* CIRCUITS WITHOUT A GENERATOR

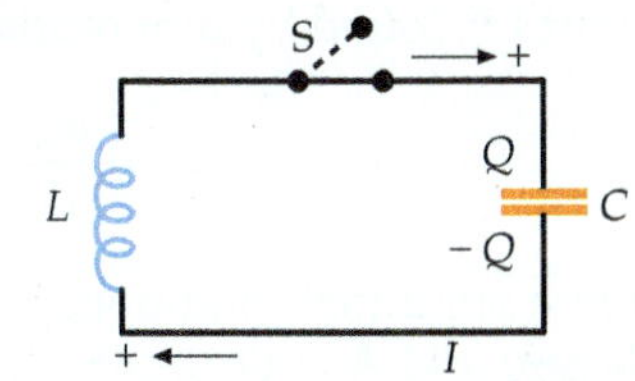

FIGURE 29-11 An *LC* circuit. When the switch is closed, the initially charged capacitor discharges through the inductor, producing a back emf.

Figure 29-11 shows a simple circuit that has inductance and capacitance but has no resistance. Such a circuit is called an ***LC* circuit.** We assume that the upper capacitor plate carries an initial positive charge Q_0 and that the switch is initially open. After the switch is closed at $t = 0$, the charge begins to flow through the inductor. Let Q be the charge on the upper plate of the capacitor and let the positive direction around the circuit be clockwise, as shown. Then,

$$I = +\frac{dQ}{dt}$$

Applying Kirchhoff's loop rule to the circuit, we have

$$L\frac{dI}{dt} + \frac{Q}{C} = 0 \qquad 29\text{-}33$$

Substituting dQ/dt for I gives

$$L\frac{d^2Q}{dt^2} + \frac{1}{C}Q = 0 \qquad 29\text{-}34$$

This equation is of the same form as Equation 14-2 for the acceleration of a mass on a spring:

$$m\frac{d^2x}{dt^2} + kx = 0$$

The behavior of an *LC* circuit is thus analogous to that of a mass on a spring—L analogous to the mass m, Q analogous to the position x, and $1/C$ analogous to the spring constant k. Also, the current I is analogous to the velocity v, because $\omega = dx/dt$ and $I = dQ/dt$. In mechanics, the mass of an object describes the inertia of the object. The greater the mass, the more opposition there is to change the velocity of the object. Similarly, the inductance L can be thought of as the inertia of an ac circuit. The greater the inductance, the more opposition there is to changes in the current I.

If we divide each term in Equation 29-34 by L and rearrange, we obtain

$$\frac{d^2Q}{dt^2} = -\frac{1}{LC}Q \qquad 29\text{-}35$$

which is analogous to

$$\frac{d^2x}{dt^2} = -\frac{k}{m}x \qquad 29\text{-}36$$

In Chapter 14, we found that we could write the solution of Equation 29-36 for simple harmonic motion in the form

$$x = A\cos(\omega t - \delta)$$

where $\omega = \sqrt{k/m}$ is the angular frequency, A is the displacement amplitude, and δ is the phase constant, which depends on the initial conditions. The solution to Equation 29-35 is thus

$$Q = A\cos(\omega t - \delta)$$

with

$$\omega = \frac{1}{\sqrt{LC}} \qquad 29\text{-}37$$

The current I is found by differentiating:

$$I = \frac{dQ}{dt} = -\omega A \sin(\omega t - \delta)$$

If we choose our initial conditions to be $Q = Q_{peak}$ and $I = 0$ at $t = 0$, the phase constant δ is zero and $A = Q_{peak}$. Our solutions are then

$$Q = Q_{peak} \cos \omega t \qquad 29\text{-}38$$

and

$$I = -\omega Q_{peak} \sin \omega t = -I_{peak} \sin \omega t \qquad 29\text{-}39$$

where $I_{peak} = \omega Q_{peak}$.

Figure 29-12 shows graphs of Q and I versus time. The charge oscillates between the values $+Q_{peak}$ and $-Q_{peak}$ with angular frequency $\omega = 1/\sqrt{LC}$. The current oscillates between $+\omega Q_{peak}$ and $-\omega Q_{peak}$ with the same frequency. Also, the charge lags behind the current by 90° (see Problem 29-33). The current is maximum when the charge is zero and the current is zero when the charge is maximum.

In our study of the oscillations of a mass on a spring, we found that the total energy is constant, and that the total energy oscillates between potential energy and kinetic energy. We also have two kinds of energy in the LC circuit—electric energy and magnetic energy. The electric energy stored in the capacitor is

$$U_e = \frac{1}{2}QV_C = \frac{1}{2}\frac{Q^2}{C}$$

Substituting $Q_{peak} \cos \omega t$ for Q, we have for the electric energy

$$U_e = \frac{1}{2}\frac{Q_{peak}^2}{C}\cos^2 \omega t \qquad 29\text{-}40$$

The electric energy oscillates between its maximum value $Q_{peak}^2/(2C)$ and zero at an angular frequency of 2ω (see Problem 29-33). The magnetic energy stored in the inductor is

$$U_m = \frac{1}{2}LI^2 \qquad 29\text{-}41$$

Substituting $I = -\omega Q_{peak} \sin \omega t$ (Equation 29-39), we get

$$U_m = \frac{1}{2}L\omega^2 Q_{peak}^2 \sin^2 \omega t = \frac{1}{2}\frac{Q_{peak}^2}{C}\sin^2 \omega t \qquad 29\text{-}42$$

where we have used $\omega^2 = 1/LC$ (Equation 29-37). The magnetic energy also oscillates between its maximum value of $Q_{peak}^2/2C$ and zero at an angular frequency of 2ω. The sum of the electrostatic energy and the magnetic energy is the total energy, which is constant in time:

$$U_{total} = U_e + U_m = \frac{1}{2}\frac{Q_{peak}^2}{C}\cos^2 \omega t + \frac{1}{2}\frac{Q_{peak}^2}{C}\sin^2 \omega t = \frac{1}{2}\frac{Q_{peak}^2}{C}$$

This sum equals the energy initially stored on the capacitor.

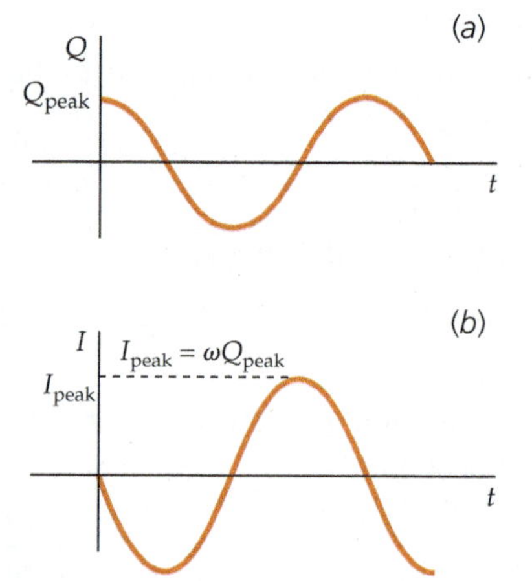

FIGURE 29-12 Graphs of (*a*) Q versus t and (*b*) I versus t for the LC circuit shown in Figure 29-11.

Example 29-6 *LC* Oscillator

A 2.0-μF capacitor is charged to 20 V and the capacitor is then connected across a 6.0-μH inductor. (*a*) What is the frequency of oscillation? (*b*) What is the peak value of the current?

PICTURE In (*b*), the current is maximum when dQ/dt is maximum, so the current amplitude is ωQ_{peak}. Also, $Q = Q_{peak}$ when $V = V_{peak}$, where V is the voltage across the capacitor.

SOLVE

(*a*) The frequency of oscillation depends only on the values of the capacitance and the inductance:

$$f = \frac{\omega}{2\pi} = \frac{1}{2\pi\sqrt{LC}} = \frac{1}{2\pi\sqrt{(6.0 \times 10^{-6}\,\text{H})(2.0 \times 10^{-6}\,\text{F})}}$$

$$= \boxed{4.6 \times 10^4\ \text{Hz}}$$

(*b*) 1. The peak value of the current is related to the peak value of the charge:

$$I_{\text{peak}} = \omega Q_{\text{peak}} = \frac{Q_{\text{peak}}}{\sqrt{LC}}$$

2. The peak charge on the capacitor is related to the peak potential drop across the capacitor:

$$Q_{\text{peak}} = CV_{\text{peak}}$$

3. Substitute CV_{peak} for Q_{peak} and calculate I_{peak}:

$$I_{\text{peak}} = \frac{CV_{\text{peak}}}{\sqrt{LC}} = \frac{V_{\text{peak}}}{\sqrt{L/C}}$$

$$= \frac{(20\ \text{V})}{\sqrt{(6.0\ \mu\text{H})/(2.0\ \mu\text{F})}} = \boxed{12\ \text{A}}$$

PRACTICE PROBLEM 29-2 A 5.0-μF capacitor is charged and is then discharged through an inductor. What should the value of the inductance be so that the current oscillates with frequency 8.0 kHz?

If we include a resistor in series with the capacitor and the inductor, as in Figure 29-13, we have an ***RLC* circuit.** Kirchhoff's loop rule gives

$$L\frac{dI}{dt} + IR + \frac{Q}{C} = 0 \qquad \text{29-43}a$$

or

$$L\frac{d^2Q}{dt^2} + R\frac{dQ}{dt} + \frac{1}{C}Q = 0 \qquad \text{29-43}b$$

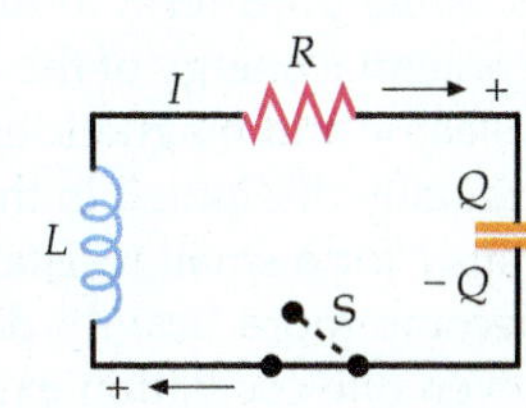

FIGURE 29-13 An *RLC* circuit.

where we have used $I = dQ/dt$ as before. Equations 29-43*a* and 29-43*b* are analogous to the equation for a damped harmonic oscillator (see Equation 14-38):

$$m\frac{d^2x}{dt^2} + b\frac{dx}{dt} + kx = 0$$

The first term, $L\,dI/dt = L\,d^2Q/dt^2$, is analogous to the mass multiplied by the acceleration, $m\,dv/dt = m\,d^2x/dt^2$; the second term, $IR = R\,dQ/dt$, is analogous to the damping term, $bv = b\,dx/dt$; and the third term, Q/C, is analogous to the restoring force kx. In the oscillation of a mass on a spring, the damping constant b leads to a dissipation of energy. In an *RLC* circuit, the resistance R is analogous to the damping constant b and leads to a dissipation of energy.

If the resistance is small, the charge and the current oscillate with (angular) frequency* that is very nearly equal to $\omega_0 = 1/\sqrt{LC}$, which is called the **natural frequency** of the circuit, but the oscillations are damped. We can understand this qualitatively from energy considerations. If we multiply each term in Equation 29-43*a* by the current I, we obtain

$$LI\frac{dI}{dt} + I^2R + I\frac{Q}{C} = 0 \qquad \text{29-44}$$

* As in Chapter 14 when we discussed mechanical oscillations, we usually omit the word *angular* when the omission will not cause confusion.

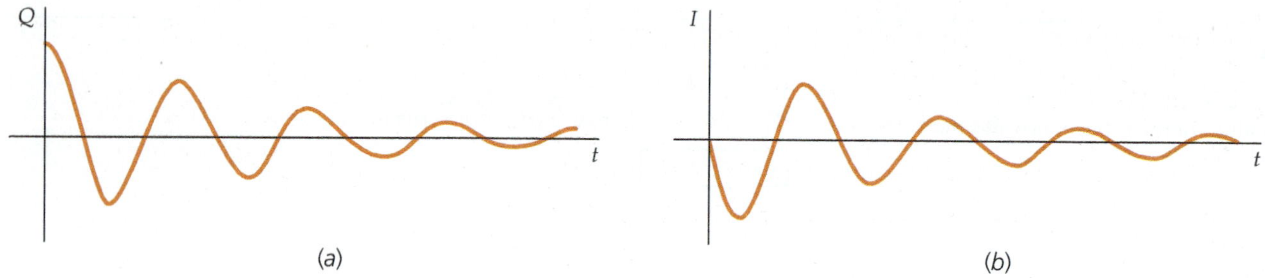

FIGURE 29-14 Graphs of (*a*) Q versus t and (*b*) I versus t for the RLC circuit shown in Figure 29-13 when the value of R is small enough so that the oscillations are underdamped.

The magnetic energy in the inductor is given by $\frac{1}{2}LI^2$ (see Equation 28-21). Note that

$$\frac{d\left(\frac{1}{2}LI^2\right)}{dt} = LI\frac{dI}{dt}$$

where $LI\,dI/dt$ is the first term in Equation 29-44. If $LI\,dI/dt$ is positive, it equals the rate at which electrical potential energy is transformed into magnetic energy. If $LI\,dI/dt$ is negative, it equals the rate at which magnetic energy is transformed back into electrical potential energy. Note that $LI\,dI/dt$ is positive or negative depending on whether I and dI/dt have the same sign or different signs. The second term in Equation 29-44 is I^2R, the rate at which electrical potential energy is dissipated in the resistor. I^2R is never negative. Note that

$$\frac{d\left(\frac{1}{2}Q^2/C\right)}{dt} = \frac{Q}{C}\frac{dQ}{dt} = I\frac{Q}{C}$$

where IQ/C is the third term in Equation 29-44. This result is the rate of change of the electric potential energy of the capacitor, which may be positive or negative. The sum of the electric and magnetic energies is not constant for this circuit because energy is continually dissipated in the resistor. Figure 29-14 shows graphs of Q versus t and I versus t for a small resistance R in an RLC circuit. If we increase R, the oscillations become more heavily damped until a critical value of R is reached for which not even one oscillation exists. Figure 29-15 shows a graph of Q versus t in an RLC circuit when the value of R is greater than the critical damping value.

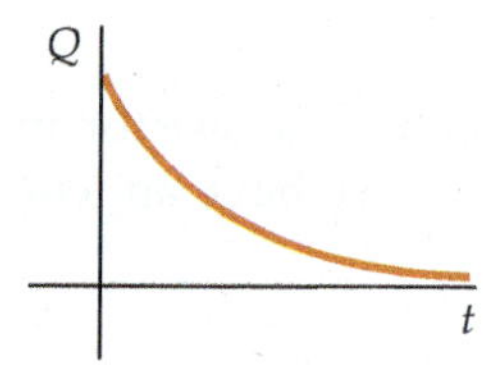

FIGURE 29-15 A graph of Q versus t for the RLC circuit shown in Figure 29-13 when the value of R is so large that the oscillations are overdamped.

* 29-5 PHASORS

Until this point, the circuits considered contained an ideal ac generator and only a single passive element (for example, resistor, inductor, or capacitor). In those circuits, the potential drop across the passive element equaled the emf of the generator. In circuits that contain an ideal ac generator and two or more additional elements connected in series, the sum of the potential drops across the elements at a given instant is equal to the generator emf at that instant; this is the same as with dc circuits. However, in series ac circuits the potential drops typically are not in phase, so the sum of their rms values does not equal the rms value of the generator emf.

Two-dimensional vectors, which are called **phasors,** can represent the phase relations between the current and the potential drops across resistors, capacitors, or inductors. In Figure 29-16, the potential drop across a resistor V_R is represented by a vector $\vec{V}_R$ that has magnitude $I_{peak}R$ and makes an angle θ with the x axis. This potential drop is in phase with the current. The current in a steady-state ac circuit varies with time, as

$$I = I_{peak}\cos\theta = I_{peak}\cos(\omega t - \delta) \qquad \text{29-45}$$

where ω is the angular frequency and δ is some phase constant. The potential drop across a resistor is then given by

$$V_R = IR = I_{peak}R\cos(\omega t - \delta) \qquad \text{29-46}$$

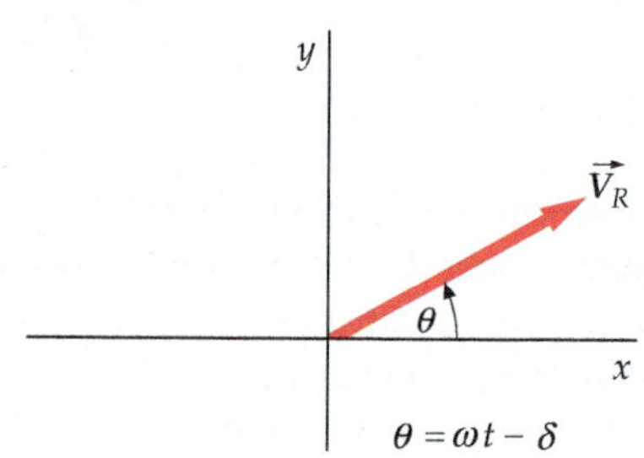

FIGURE 29-16 The potential drop across a resistor can be represented by a vector $\vec{V}_R$, which is called a phasor, that has magnitude $I_{peak}R$ and makes an angle $\theta = \omega t - \delta$ with the x axis. The phasor rotates with an angular frequency ω. The potential drop $V_R = IR$ is the x component of $\vec{V}_R$.

The potential drop across a resistor is thus equal to the x component of the phasor vector $\vec{V}_R$, which rotates counterclockwise with an angular frequency ω. The current I may be written as the x component of a phasor $\vec{I}$ having the same direction as $\vec{V}_R$.

When several components are connected together in a series combination, their potential drops add. When several components are connected in parallel, their currents add. Unfortunately, adding sines or cosines of different amplitudes and phases algebraically is awkward. It is much easier to do this by vector addition.*

Let us look at how phasors are used. Any ac current or any potential drop is written in the form $A\cos(\omega t - \delta)$, which in turn is treated as A_x, the x component of a phasor that makes an angle $(\omega t - \delta)$ with the $+x$ direction. Instead of adding two potential drops or currents algebraically, as $A\cos(\omega t - \delta_1) + B\cos(\omega t - \delta_2)$, we represent the quantities as phasors $\vec{A}$ and $\vec{B}$ and find the phasor sum $\vec{C} = \vec{A} + \vec{B}$ geometrically. The resultant potential drop or current is then the x component of the resultant phasor, $C_x = A_x + B_x$. The geometric representation conveniently shows the relative amplitudes and phases of the phasors.

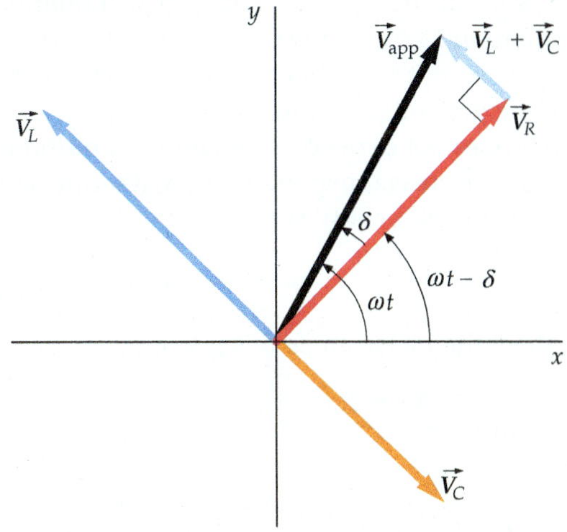

FIGURE 29-17 Phasor representations of the potential drops V_R, V_L, and V_C. Each vector rotates in the counterclockwise direction with an angular frequency ω. At any instant, the potential drop across an element equals the x component of the corresponding phasor, and the potential drop V_{app} across the *RLC*-series combination, which equals the sum of the potential drops, equals the x component of the vector sum $\vec{V}_R + \vec{V}_L + \vec{V}_C$.

Consider an ac circuit that contains an inductor L, a capacitor C, and a resistor R connected in series. They all carry the same current, which is represented as the x component of the current phasor $\vec{I}$. The potential drop across the resistor V_R is represented by a phasor $\vec{V}_R$ that has magnitude $I_{peak}R$ and is in phase with the current phasor $\vec{I}$. The potential drop across the inductor V_L is represented by a phasor $\vec{V}_L$ that has magnitude $I_{peak}X_L$ and leads the current phasor $\vec{I}$ by 90°. Similarly, the potential drop across the capacitor V_C is represented by a phasor $\vec{V}_C$ that has magnitude $I_{peak}X_C$ and lags $\vec{I}$ by 90°. Figure 29-17 shows the phasors $\vec{V}_R$, $\vec{V}_L$, $\vec{V}_C$, and $\vec{V}_{app}$, where the x component of V_{app} is the potential drop across the series combination. The phasors all rotate counterclockwise with an angular frequency ω. At any instant in time, the instantaneous value of the potential drop across any of these elements equals the x component of the corresponding phasor.

* 29-6 DRIVEN *RLC* CIRCUITS

SERIES *RLC* CIRCUIT

Figure 29-18 shows a series *RLC* circuit being sinusoidally driven by an ac generator. If the potential drop applied by the generator to the series *RLC* combination is $V_{app} = V_{app\ peak}\cos\omega t$, applying Kirchhoff's loop rule gives

$$V_{app\ peak}\cos\omega t - L\frac{dI}{dt} - IR - \frac{Q}{C} = 0$$

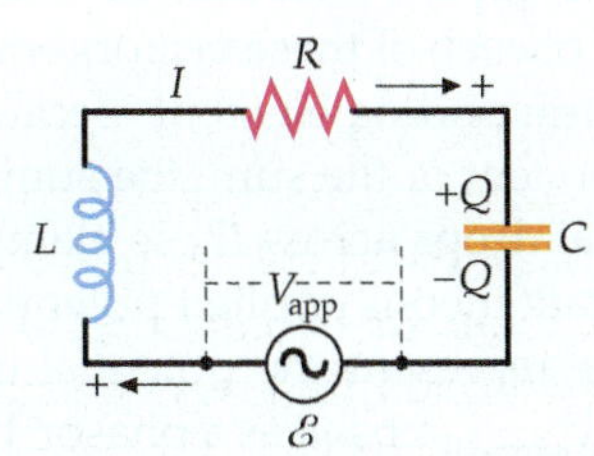

FIGURE 29-18 A series *RLC* circuit with an ac generator.

Using $I = dQ/dt$ and rearranging, we obtain

$$L\frac{d^2Q}{dt^2} + R\frac{dQ}{dt} + \frac{1}{C}Q = V_{app\ peak}\cos\omega t \qquad 29\text{-}47$$

This equation is analogous to Equation 14-53 for the forced oscillation of a mass on a spring:

$$m\frac{d^2x}{dt^2} + b\frac{dx}{dt} + m\omega_0^2 x = F_0\cos\omega t$$

(In Equation 14-53, the force constant k was written in terms of the mass m and the natural angular frequency ω_0 using $k = m\omega_0^2$. The capacitance in Equation 29-47 could be similarly written in terms of L and the natural angular frequency using $1/C = L\omega_0^2$.)

* It is also easier to do using complex numbers.

We will discuss the solution of Equation 29-47 qualitatively as we did with Equation 14-53 for the forced oscillator. The current in the circuit consists of a transient current that depends on the initial conditions (for example, the initial phase of the generator and the initial charge on the capacitor) and a steady-state current that does not depend on the initial conditions. We will ignore the transient current, which decreases exponentially with time and is eventually negligible, and concentrate on the steady-state current. The steady-state current obtained by solving Equation 29-47 is

$$I = I_{\text{peak}} \cos(\omega t - \delta) \qquad \text{29-48}$$

where the phase angle δ is given by

$$\tan \delta = \frac{X_L - X_C}{R} \qquad \text{29-49}$$

PHASE CONSTANT FOR A SERIES *RLC* CIRCUIT

The peak current is

$$I_{\text{peak}} = \frac{V_{\text{app peak}}}{\sqrt{R^2 + (X_L - X_C)^2}} = \frac{V_{\text{app peak}}}{Z} \qquad \text{29-50}$$

PEAK CURRENT IN A SERIES *RLC* CIRCUIT

where

$$Z = \sqrt{R^2 + (X_L - X_C)^2} \qquad \text{29-51}$$

IMPEDANCE OF A SERIES *RLC* CIRCUIT

The quantity $X_L - X_C$ is called the **total reactance,** and Z is called the **impedance.** Combining these results, we have

$$I = \frac{V_{\text{app peak}}}{Z} \cos(\omega t - \delta) \qquad \text{29-52}$$

Equation 29-52 can also be obtained from a simple diagram using vectors called phasors. Figure 29-19 shows the phasors representing the potential drops across the resistance, the inductance, and the capacitance. The x component of each of these vectors equals the instantaneous potential drop across the corresponding element. Because the sum of the x components equals the x component of the sum, the sum of the x components equals the sum of the potential drops across these elements, which by Kirchhoff's loop rule equals the instantaneous applied potential drop.

If we represent the potential drop applied across the series combination $V_{\text{app}} = V_{\text{app peak}} \cos \omega t$ as a phasor $\vec{V}_{\text{app}}$ that has the magnitude $V_{\text{app peak}}$, we have

$$\vec{V}_{\text{app}} = \vec{V}_R + \vec{V}_L + \vec{V}_C \qquad \text{29-53}$$

In terms of the magnitudes,

$$V_{\text{app peak}} = |\vec{V}_R + \vec{V}_L + \vec{V}_C| = \sqrt{V_{R\,\text{peak}}^2 + (V_{L\,\text{peak}} - V_{C\,\text{peak}})^2}$$

But $V_R = I_{\text{peak}}R$, $V_L = I_{\text{peak}}X_L$, and $V_C = I_{\text{peak}}X_C$. Thus,

$$V_{\text{app peak}} = I_{\text{peak}}\sqrt{R^2 + (X_L - X_C)^2} = I_{\text{peak}}Z$$

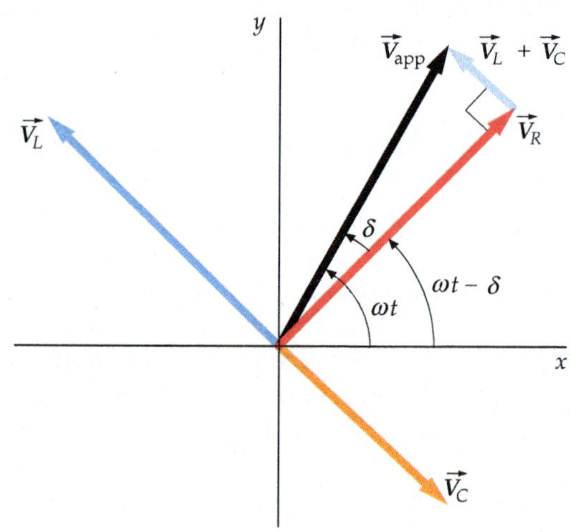

FIGURE 29-19 Phase relations among potential drops in a series *RLC* circuit. The potential drop across the resistor is in phase with the current. The potential drop across the inductor V_L leads the current by 90°. The potential drop across the capacitor lags the current by 90°. The sum of the vectors representing the potential drops gives a vector at an angle δ with the current representing the applied emf. For the case shown here, V_L is greater than V_C, and the current lags the applied potential drop by δ.

The phasor $\vec{V}_{app}$ makes an angle δ with $\vec{V}_R$, as shown in Figure 29-19. From the figure, we can see that

$$\tan\delta = \frac{|\vec{V}_L + \vec{V}_C|}{|\vec{V}_R|} = \frac{I_{peak}X_L - I_{peak}X_C}{I_{peak}R} = \frac{X_L - X_C}{R}$$

in agreement with Equation 29-49. Because $\vec{V}_{app}$ makes an angle ωt with the x axis, $\vec{V}_R$ makes an angle $\omega t - \delta$ with the x axis. This applied potential drop is in phase with the current, which is therefore given by

$$I = I_{peak}\cos(\omega t - \delta) = \frac{V_{app\ peak}}{Z}\cos(\omega t - \delta)$$

This is Equation 29-52. The relation between the impedance Z, the resistance R, and the total reactance $X_L - X_C$ is best remembered by using the right triangle shown in Figure 29-20.

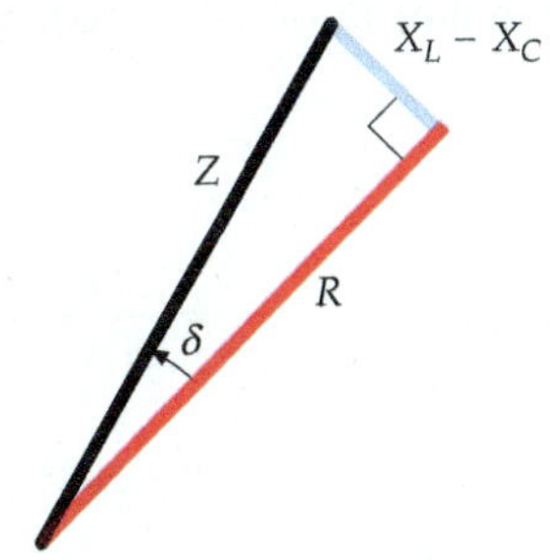

FIGURE 29-20 A right triangle relating capacitive and inductive reactance, resistance, impedance, and the phase angle in an *RLC* circuit.

RESONANCE

When X_L and X_C are equal, the total reactance is zero, and the impedance Z has its smallest value R. Then I_{peak} has its greatest value and the phase angle δ is zero, which means that the current is in phase with the applied potential drop. Let ω_{res} be the value of ω for which X_L and X_C are equal. It is obtained from

$$X_L = X_C$$

$$\omega_{res}L = \frac{1}{\omega_{res}C}$$

or

$$\omega_{res} = \frac{1}{\sqrt{LC}}$$

which equals the natural frequency ω_0. When the frequency of the applied potential drop equals the natural frequency ω_0, the impedance is smallest, I_{peak} is greatest, and the circuit is said to be at **resonance.** The natural frequency ω_0 is therefore also called the **resonance frequency.** This resonance condition in a driven *RLC* circuit is similar to that in a driven simple harmonic oscillator.

Because neither an inductor nor a capacitor dissipates energy, the average power delivered to a series *RLC* circuit is the average power supplied to the resistor. The instantaneous power supplied to the resistor is

$$P = I^2R = [I_{peak}\cos(\omega t - \delta)]^2R$$

Averaging over one or more cycles and using $(\cos^2\theta)_{av} = \frac{1}{2}$, we obtain for the average power

$$P_{av} = \frac{1}{2}I_{peak}^2R = (I_{rms})^2R \qquad 29\text{-}54$$

Using $R/Z = \cos\delta$ from Figure 29-20 and $I_{peak} = V_{app\ peak}/Z$, this can be written

$$P_{av} = \frac{1}{2}V_{app\ peak}I_{peak}\cos\delta = V_{app\ rms}I_{rms}\cos\delta \qquad 29\text{-}55$$

The quantity $\cos\delta$ is called the **power factor** of the *RLC* circuit. At resonance, δ is zero, and the power factor is 1.

The power can also be expressed as a function of the angular frequency ω. Using $I_{rms} = V_{app\ rms}/Z$, Equation 29-54 becomes

$$P_{av} = (I_{rms})^2 = (V_{app\ rms})^2\frac{R}{Z^2}$$

From the definition of impedance Z, we have

$$Z^2 = (X_L - X_C)^2 + R^2 = \left(\omega L - \frac{1}{\omega C}\right)^2 + R^2$$
$$= \frac{L^2}{\omega^2}\left(\omega^2 - \frac{1}{LC}\right)^2 + R^2$$
$$= \frac{L^2}{\omega^2}(\omega^2 - \omega_0^2)^2 + R^2$$

where we have used $\omega_0 = 1/\sqrt{LC}$. Using this expression for Z^2, we obtain the average power as a function of ω:

$$P_{av} = \frac{(V_{app\ rms})^2 R\omega^2}{L^2(\omega^2 - \omega_0^2)^2 + \omega^2 R^2} \qquad \text{29-56}$$

Figure 29-21 shows the average power supplied by the generator to the series combination as a function of generator frequency for two different values of the resistance R. These curves, called **resonance curves,** are the same as the power-versus-frequency curves for a driven damped oscillator (see Section 14-5). The average power is greatest when the generator frequency equals the resonance frequency. When the resistance is small, the resonance curve is narrow; when the resistance is large, the resonance curve is broad. A resonance curve can be characterized by the **resonance width** $\Delta\omega$. As shown in Figure 29-21, the resonance width is the frequency difference between the two points on the curve where the power is half its maximum value. When the width is small compared with the resonance frequency, the resonance is sharp; that is, the resonance curve is narrow.

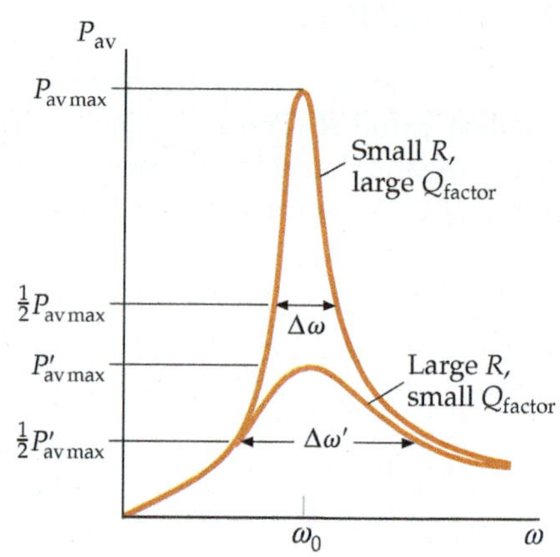

FIGURE 29-21 Plot of average power versus frequency for a series RLC circuit. The power is maximum when the frequency of the generator ω equals the natural frequency of the circuit $\omega_0 = 1/\sqrt{LC}$. If the resistance is small, the Q factor is large and the resonance is sharp. The resonance width $\Delta\omega$ of the curves is measured between points where the power is half its maximum value.

The Q factor for a mechanical oscillator is defined as $Q_{factor} = \omega_0 m/b$ (Equations 14-42 and 14-45), where m is the mass and b is the damping constant. We then saw that for weakly-damped oscillator $Q_{factor} = 2\pi E/|\Delta E|$, where E is the total energy of the system at the beginning of a cycle and ΔE is the energy dissipated during the cycle. The **Q factor** for an RLC circuit can be defined in a similar way. Because L is analogous to the mass m and R is analogous to the damping constant b, the Q factor for an RLC circuit that has a small resistance is given by

$$Q_{factor} = 2\pi\left(\frac{E}{|\Delta E|}\right)_{cycle} = \frac{\omega_0 L}{R} \qquad \text{29-57}$$

When the resonance curve is reasonably narrow (that is, when Q is greater than about 2 or 3), the Q factor can be approximated by

$$Q_{factor} = \frac{\omega_0}{\Delta\omega} = \frac{f_0}{\Delta f} \qquad \text{29-58}$$

Q FACTOR FOR AN *RLC* CIRCUIT

Resonance circuits are used in radio receivers, where the resonance frequency of the circuit is varied either by varying the capacitance or the inductance. Resonance occurs when the natural frequency of the circuit equals one of the frequencies of the radio waves picked up at the antenna. At resonance, a relatively large current exists in the antenna circuit. If the Q factor of the circuit is sufficiently high, currents due to other station frequencies off resonance will be negligible compared with those currents due to the station frequency to which the circuit is tuned.

Example 29-7 Driven Series *RLC* Circuit

A series *RLC* combination that has $L = 2.0\text{ H}$, $C = 2.0\ \mu\text{F}$, and $R = 20\ \Omega$ is driven by an ideal generator that has a peak emf of 100 V and a frequency which can be varied. Find (*a*) the resonance frequency f_0, (*b*) the Q factor, (*c*) the width of the resonance Δf, and (*d*) the current amplitude at resonance.

PICTURE The resonance frequency is found from $\omega_0 = 1/\sqrt{LC}$ and the Q factor is found from $Q_{\text{factor}} = \omega_0 L/R$ (Equation 29-57).

SOLVE

(*a*) The resonance frequency is $f_0 = \omega_0/2\pi$:

$$f_0 = \frac{\omega_0}{2\pi} = \frac{1}{2\pi\sqrt{LC}}$$

$$= \frac{1}{2\pi\sqrt{(2.0\text{ H})(2.0\times10^{-6}\text{ F})}} = \boxed{80\text{ Hz}}$$

(*b*) Use this result to calculate Q_{factor}:

$$Q_{\text{factor}} = \frac{\omega_0 L}{R} = \frac{2\pi(80\text{ Hz})(2.0\text{ H})}{20\ \Omega} = \boxed{50}$$

(*c*) Use the value of Q_{factor} to find the width of the resonance Δf, where $Q_{\text{factor}} = f_0/\Delta f$:

$$\Delta f = \frac{f_0}{Q_{\text{factor}}} = \frac{80\text{ Hz}}{50} = \boxed{1.6\text{ Hz}}$$

(*d*) At resonance, the impedance is equal to R and I_{peak} is $V_{\text{app peak}}/R$:

$$I_{\text{peak}} = \frac{V_{\text{app peak}}}{R} = \frac{\mathcal{E}_{\text{peak}}}{R} = \frac{100\text{ V}}{20\ \Omega} = \boxed{5.0\text{ A}}$$

CHECK At resonance the inductive and capacitances each equal $X_L = \omega_0 L = 2\pi(80\text{ Hz})(2.0\text{ H}) = 1.0\text{ k}\Omega$. The resistance is given as $R = 20\ \Omega$. Because the resistance is much smaller than the inductive reactance at resonance, we expect the Q factor to be high and the resonance to be sharp. The Part-(*b*) and -(*c*) results meet this expectation.

TAKING IT FURTHER The width of 1.6 Hz is about 2.0 percent of the resonance frequency of 80 Hz, so the resonance peak is quite sharp.

Example 29-8 Driven Series *RLC* Circuit Current, Phase, and Power

Try It Yourself

If the generator in Example 29-7 has a frequency of 60 Hz, find (*a*) the peak current, (*b*) the phase constant δ, (*c*) the power factor, and (*d*) the average power delivered.

PICTURE The peak current is the peak applied potential drop divided by the total impedance of the series combination. The phase angle δ is found from $\tan\delta = (X_L - X_C)/R$. You can use either Equation 29-54 or Equation 29-55 to find the average power delivered.

SOLVE

Cover the column to the right and try these on your own before looking at the answers.

Steps	**Answers**
(*a*) 1. Write the peak current in terms of $V_{\text{app peak}}$ and the impedance.	$I_{\text{peak}} = \frac{V_{\text{app peak}}}{Z} = \frac{\mathcal{E}_{\text{peak}}}{Z}$
2. Calculate the capacitive and inductive reactances and the total reactance.	$X_C = 1326\ \Omega$, $X_L = 754\ \Omega$ so $X_L - X_C = -572\ \Omega$
3. Calculate the total impedance Z.	$Z = 573\ \Omega$
4. Use the results of steps 2 and 3 to calculate I_{peak}.	$I_{\text{peak}} = \boxed{0.17\text{ A}}$
(*b*) Use the results of Part (*a*), steps 2 and 3 to calculate δ.	$\delta = \tan^{-1}\frac{X_L - X_C}{R} = \boxed{-88.0^\circ}$

(*c*) Use your value of δ to compute the power factor. $\cos\delta = \boxed{0.035}$

(*d*) Calculate the average power delivered from Equation 29-54. $P_{av} = \frac{1}{2}I_{peak}^2 R = \boxed{0.29\text{ W}}$

CHECK To check our result for the average power using the power factor found in Part (*c*), we have $P_{av} = \frac{1}{2}V_{app\,peak}I_{peak}\cos\delta = \frac{1}{2}\mathcal{E}_{peak}I_{peak}\cos\delta = 0.29$ W. This result is in agreement with our result for Part (*d*).

TAKING IT FURTHER The frequency of 60 Hz is well below the resonance frequency of 80 Hz. (Recall that the width of the resonance peak, calculated in Example 29-7, is only 1.6 Hz.) As a result, the total reactance is much greater in magnitude than the resistance. This result is always the case far from resonance. Similarly, a peak current of 0.17 A is much less than the peak current at resonance, which was found to be 5.0 A in Example 29-7. Finally, we see from Figure 29-19 that a negative phase angle δ means that the current leads the applied potential drop.

CONCEPT CHECK 29-1

A circuit consists of an ideal constant-frequency generator, a resistor, a capacitor and an inductor with a moveable soft-iron core—all connected in series. You notice that if you nudge the soft-iron core a bit deeper into the induction coil, the rms current increases slightly. Prior to the nudge, the resonance frequency of the circuit was (*a*) below the generator frequency, (*b*) equal to the generator frequency, (*c*) above the generator frequency.

Example 29-9 Driven Series *RLC* Circuit at Resonance — *Try It Yourself*

Find the peak potential drop across the resistor, the inductor, and the capacitor at resonance for the circuit in Example 29-7.

PICTURE The peak potential drop across the resistor is I_{peak} multiplied by R. Similarly, the peak potential drop across the inductor or capacitor is I_{peak} multiplied by the appropriate reactance. We found that at resonance $I_{peak} = 5.0$ A and $f_0 = 80$ Hz in Example 29-7.

SOLVE

Cover the column to the right and try these on your own before looking at the answers.

Steps	Answers
1. Calculate $V_{R\,peak} = I_{peak}R$.	$V_{R\,peak} = I_{peak}R = \boxed{100\text{ V}}$
2. Express $V_{L\,peak}$ in terms of I_{peak} and X_L.	$V_{L\,peak} = I_{peak}X_L = I_{peak}\omega_0 L = \boxed{5.0\text{ kV}}$
3. Express $V_{C\,peak}$ in terms of I_{peak} and X_C.	$V_{C\,peak} = I_{peak}X_C = \dfrac{I_{peak}}{\omega_0 C} = \boxed{5.0\text{ kV}}$

CHECK The inductive and capacitive reactances are equal, as we would expect at resonance. (We solved for the resonance frequency by setting them equal.)

TAKING IT FURTHER The phasor diagram for the potential drops across the resistor, capacitor, and inductor is shown in Figure 29-22. The peak potential drop across the resistor is a relatively safe 100 V, equal to the peak emf of the generator. However, the peak potential drops across the inductor and the capacitor are a dangerously high 5.0 kV. These potential drops are 180° out of phase. At resonance, the potential drop across the inductor at any instant is the negative of the potential drop across the capacitor, so they always sum to zero, leaving the potential drop across the resistor equal to the emf in the circuit.

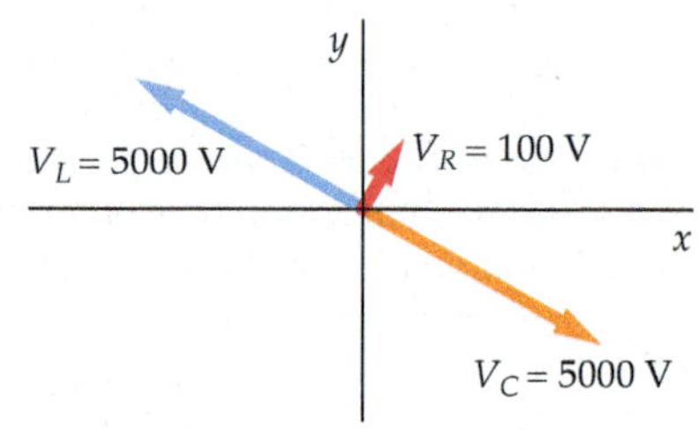

FIGURE 29-22

Example 29-10 *RC* Low-Pass Filter

A resistor R and capacitor C are in series with an ideal generator, as shown in Figure 29-23. The generator applies a potential difference across the RC combination given by $V_{app} = \sqrt{2}\,V_{app\,rms}\cos\omega t$. Find the rms potential difference across the capacitor $V_{out\,rms}$ as a function of frequency ω.

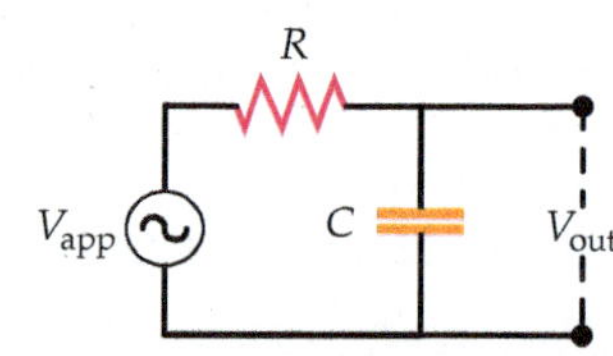

FIGURE 29-23 The peak output voltage decreases as frequency increases.

PICTURE The rms potential difference across the capacitor is the product of the rms current and the capacitive reactance. The rms current is found from the potential difference applied by the generator and the impedance of the series RC combination.

SOLVE

1. The potential drop across the capacitor is I_{rms} multiplied by X_C: $V_{out\ rms} = I_{rms}X_C$

2. The rms current depends on the applied rms potential difference and the impedance: $I_{rms} = \dfrac{V_{app\ rms}}{Z}$

3. In this circuit, only R and X_C contribute to the total impedance: $Z = \sqrt{R^2 + X_C^2}$

4. Substitute those expressions and $X_C = 1/(\omega C)$ to find the output rms potential difference:

$$V_{out\ rms} = I_{rms}X_C = \frac{V_{app\ rms}X_C}{\sqrt{R^2 + X_C^2}} = \frac{V_{app\ rms}}{\sqrt{1 + \dfrac{R^2}{X_C^2}}} = \boxed{\frac{V_{app\ rms}}{\sqrt{1 + (\omega RC)^2}}}$$

CHECK The dimensions of our step 4 result are correct. The dimension of ω is 1/T and the dimension of RC is T, so the product ωRC is dimensionless.

TAKING IT FURTHER This circuit is called an *RC low-pass filter*, because it transmits low frequencies with greater amplitude than high frequencies. In fact, the output potential difference equals the potential difference applied by the generator in the limit that $\omega \to 0$, but also approaches zero for $\omega \to \infty$, as shown in the graph of the ratio of output potential difference to applied potential difference in Figure 29-24.

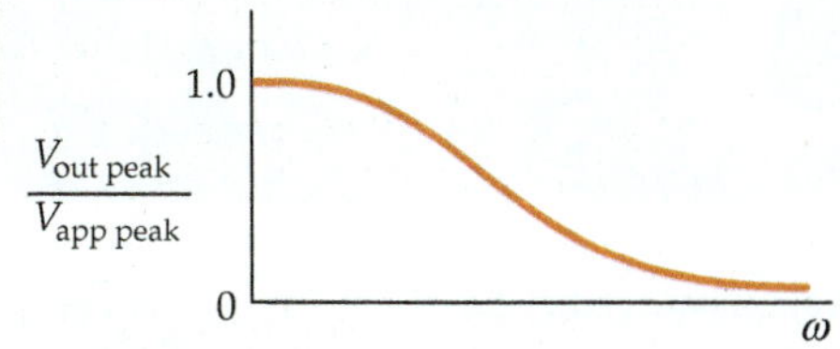

FIGURE 29-24

PRACTICE PROBLEM 29-3 Find the output potential difference for this circuit if the capacitor is replaced by an inductor L.

Example 29-11 An FM Tuner

Context-Rich

You have been tinkering with building a radio tuner using your new knowledge of physics. You know that the FM dial gives its frequencies in megahertz, and you would like to determine what percentage of change in an inductor would allow you to tune for the whole FM range. You decide to start at midrange and determine a percent increase and decrease needed for inductance. A variable inductor is usually an iron-core solenoid, and the inductance is increased by further inserting the core. The FM dial goes from 88 MHz to 108 MHz.

PICTURE We can relate inductance to the resonant frequency with $\omega = 2\pi f$ and $\omega = 1\sqrt{LC}$. Then, if we find the percent change in frequency, we can determine the percent change in inductance. The capacitance C does not vary.

SOLVE

1. The resonant angular frequency ω is related to the inductance L: $\omega = 1/\sqrt{LC}$
and
$\omega = 2\pi f$
so
$$f = \frac{1}{2\pi\sqrt{LC}}$$

2. Solving for L gives: $L = a/f^2$
where
$a = (4\pi^2 C)^{-1}$

3. Express the fractional change in L in terms of the frequencies. When L is maximum, f is minimum and vice versa. The middle frequency f_{mid} is halfway between the maximum and minimum frequency and L_{mid} is the inductance when $f = f_{mid}$:

$$\frac{\Delta L}{L} = \frac{L_{max} - L_{min}}{L_{mid}} = \frac{a/f_{max}^2 - a/f_{min}^2}{a/f_{mid}^2}$$
$$= f_{mid}^2\left(\frac{1}{f_{max}^2} - \frac{1}{f_{min}^2}\right) = 98^2\left(\frac{1}{108^2} - \frac{1}{88^2}\right)$$
$$= -0.417$$

4. The minus sign is not relevant, except as an indication that when the inductance increases the resonant frequency decreases. Express the step 3 result as a percentage: The inductance varies by about $\boxed{\text{42 percent}}$.

A shipboard radio, circa 1920. Exposed at the operator's left are the inductance coils and capacitor plates of the tuning circuit. *(© George H. Clark Radioana Collection-Archive Center, National Museum of American History.)*

PARALLEL RLC CIRCUIT

Figure 29-25 shows a resistor R, a capacitor C, and an inductor L connected in parallel across an ac generator. The total current I from the generator divides into three currents: the current I_R in the resistor, the current I_C in the capacitor, and the current I_L in the inductor. The instantaneous potential drop V_{app} is the same across each element. The current in the resistor is in phase with the potential drop and the phasor $\vec{I}_R$ has magnitude V_{peak}/R. Because the potential drop across an inductor *leads* the current in the inductor by 90°, I_L *lags* the potential drop by 90°, and the phasor $\vec{I}_L$ has magnitude V_{peak}/X_L. Similarly, I_C leads the potential drop by 90° and the phasor $\vec{I}_C$ has magnitude V_{peak}/X_C. These currents are represented by phasors in Figure 29-26. The total current I is the x component of the vector sum of the individual current phasors as shown in the figure. The magnitude of the total current phasor is

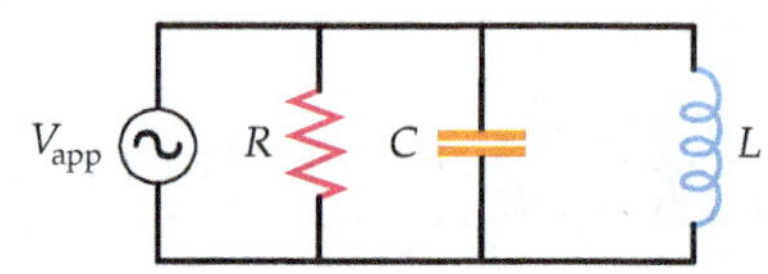

FIGURE 29-25 A parallel *RLC* circuit.

$$I = \sqrt{I_R^2 + (I_L - I_C)^2} = \sqrt{\left(\frac{V_{peak}}{R}\right)^2 + \left(\frac{V_{peak}}{X_L} - \frac{V_{peak}}{X_C}\right)^2} = \frac{V_{peak}}{Z} \qquad 29\text{-}59$$

where the total impedance Z is related to the resistance and the capacitive and inductive reactances by

$$\frac{1}{Z^2} = \frac{1}{R^2} + \left(\frac{1}{X_L} - \frac{1}{X_C}\right)^2 \qquad 29\text{-}60$$

At resonance, the currents in the inductor and capacitor are 180° out of phase, so the total current is a minimum and is just the current in the resistor. We see from Equation 29-59 that this occurs if Z is maximum, so $1/Z$ is minimum. Then, we see from Equation 29-60 that if $X_L = X_C$, $1/Z$ has its minimum value $1/R$. Equating X_L with X_C and solving for ω obtains the resonant frequency, which equals the natural frequency $\omega_0 = 1/\sqrt{LC}$.

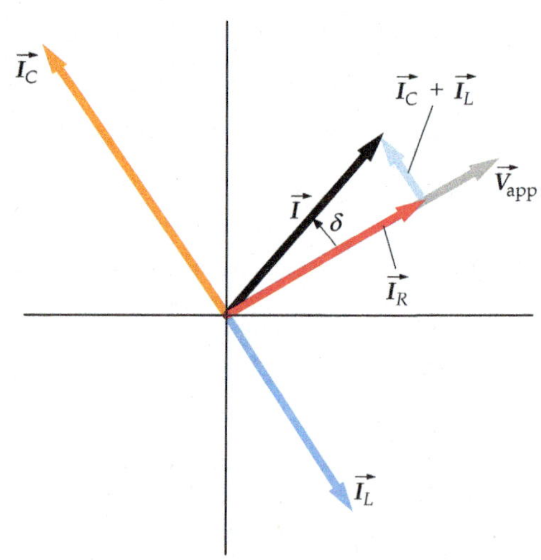

FIGURE 29-26 A phasor diagram for the currents in the parallel *RLC* circuit shown in Figure 29-25. The potential drop is the same across each element. The current in the resistor is in phase with the potential drop. The current in the capacitor leads the potential drop by 90° and the current in the inductor lags the potential drop by 90°. The phase difference δ between the total current and the potential drop depends on the relative magnitudes of the currents, which depend on the values of the resistance and of the capacitive and inductive reactances.

Physics Spotlight

The Electric Grid: Power to the People

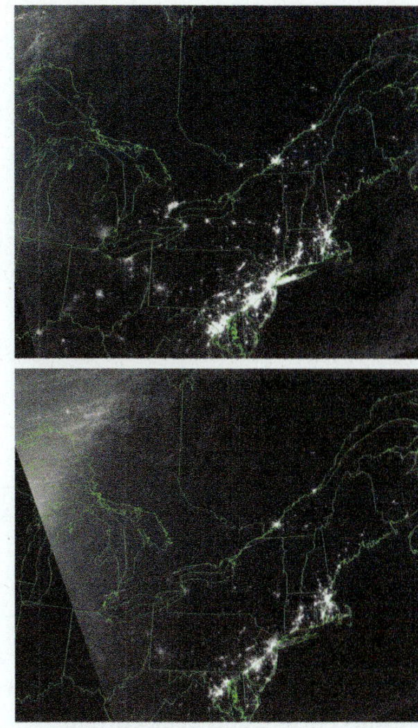

The pair of satellite images shows how a power failure affected many American and Canadian cities during the evening of Thursday, August 14, 2003. The top image was taken 20 hours before the blackout while the bottom image was taken 7 hours after the blackout. *(Courtesy of Chris Elridge/U.S. Airforce.)*

All around the world, people depend on electrical distribution systems, or grids, for reliable electric power. Generators, substations that have transformers, and high-voltage transmission lines are all necessary to efficiently move electrical energy from one place to another.[*] As of 2002, over 150 000 miles of high-voltage ac transmission lines and more than 10 000 transmission substations were part of the grid in the United States alone.[†] Grids are growing in complexity[‡] all over the world.[#, °] Unfortunately, as grids grow, so do the number of possible points of failure.

Most electrical grid failures are small scale, caused by local weather, by equipment failure,[§] or even by animals.[¶] But these failures give clues to the causes of large-scale power outages. Power surges within transformers and lines are the primary cause of local outages. Damage is prevented from propagating, or cascading, by relay switches that close off the line and act as surge suppressors for the system as a whole. Rarely, local outages are caused when there is much more demand than local generators can supply.

Sometimes the same mechanisms meant to prevent damage from a local outage can cause the outage to cascade. On November 9, 1965, a relay switch tripped at a hydroelectric plant in southern Ontario. The current from that line was shunted to five other transmission lines, which caused the relay switches for those lines to trip. Because of this dramatic reduction in load, the generators sped up, which meant that the power they provided was out of phase with power from other providers.[**] Over the course of a few minutes, relay after relay tripped, and many generators were shut down as they were isolated from their loads. Within four seconds, relays had tripped throughout the northeastern United States. Within a few minutes, generators were taken off-line, and over 30 million people were left without electricity for several hours.

That blackout prompted the formation of the National Electric Reliability Council.[††] Measures put in place for coordination of electric loads[‡‡, ##] have prevented many large blackouts, but blackouts have still happened. In July 1977, a lightning strike to transmission lines in New York tripped relays, and because of the slow response of the system operator,[°°] New York City was left without power for three days.[§§] On August 14, 2003, an unfortunate combination of high demand, a transmission line shorting against an untrimmed tree, and inadequate communications led to a blackout in the northeast United States and Canada that left 50 million people without power, some for days.[¶¶]

To prevent future outages, technical improvements to the grid are being actively sought. One improvement is software able to monitor and control portions of the grid with speed and flexibility.[***] Other improvements may include higher capacity transmission lines, improved transformers, and more responsive maintenance programs.[†††, ‡‡‡]

[*] *The Electricity Delivery System.* United States Department of Energy, Office of Electricity Delivery and Energy Reliability, Feb. 2006. http://www.energetics.com/gridworks/pdfs/factsheet.pdf As of Nov. 2006.

[†] Ibid.

[‡] Harris, J. L., et al., "Peak Demand and Energy Projection Bandwidths 2005–2014 Regional and National." National Energy Reliability Council, Sept. 14, 2005. ftp://www.nerc.com/pub/sys/all_updl/docs/pubs/Final_NERC_2005-2014_REGIONAL_BANDWIDTH_REPORT.pdf As of Nov. 2006.

[#] "Towards National Power Grid." [sic] Power Grid Corporation of India Limited. http://www.powergridindia.com/pgnew/01-0001-003.asp As of Nov. 2006.

[°] Chow, J., Kopp, R., and Portney, P., "Energy Resources and Global Development." *Science*, Nov. 28, 2003, Vol. 302, pp. 1528–1531.

[§] Chowdhury, A., et al., "MAPP Bulk Transmission System Outage Report." Mid-Continent Area Power Pool, Jun. 2001. http://www.mapp.org/assets/pdf/BTOR19_1.PDF As of Nov. 2006.

[¶] Orso, J., "Bangor Hit with Power Outage." *La Crosse Tribune*, Jul. 16, 2006.

[**] U.S. Federal Power Commission, "Northeast Power Failure: November 9 and 10, 1965." Washington, DC: U.S. Government Printing Office. At http://blackout.gmu.edu/archive/pdf/fpc_65.pdf As of Nov. 2006.

[††] Central Maine Power Company, "The Great Northeast Blackout of 1965." http://www.cmpco.com/about/system/blackout.html As of Nov. 2006.

[‡‡] California Independent System Operator, "Load Reduction Programs." California Independent System Operator Procedure E-502, Mar. 15, 2005. http://www.caiso.com/docs/2000/06/15/2000061511113359621.pdf As of Nov. 2006.

[##] "Emergency Manual Load Shedding." California Independent System Operator Procedure E-502, Feb. 17, 2006. http://www.caiso.com/docs/1998/12/02/1998120218100812000.pdf As of Nov. 2006.

[°°] Boffey, P. M., "Investigators Agree N. Y. Blackout of 1977 Could Have Been Avoided." *Science*, Sept. 15, 1978, Vol. 201, No. 4360, pp. 994–998.

[§§] Metz, W. D., " New York Blackout: Weak Links Tie Con Ed to Neighboring Utilities." *Science*, Jul. 29, 1977, Vol. 197, No. 4302, pp. 441–442.

[¶¶] U.S.–Canada Power System Outage Task Force, "Final Report on the August 14, 2003 Blackout in the United States and Canada: Causes and Recommendations." ftp://www.nerc.com/pub/sys/all_updl/docs/blackout/ch1-3.pdf As of Nov. 2006.

[***] Brown, E., "Creating Stability in a World of Unstable Electricity Distribution." *Logos*, Argonne National Laboratories, Spring 2004, Vol. 22, No. 1. At http://www.anl.gov/Media_Center/logos22-1/electricity.htm As of Nov. 2006.

[†††] Office of Electric Transmission and Distribution, "GridWorks Multi-Year Plan." United States Department of Energy. http://www.oe.energy.gov/DocumentsandMedia/multiyear-plan_final.pdf As of Nov. 2006.

[‡‡‡] U.S.–Canada Power System Outage Task Force, "The August 14, 2003 Blackout One Year Later: Actions Taken in the United States and Canada to Reduce Blackout Risk." Natural Resources Canada and the U.S. Department of Energy, Aug. 13, 2004. ftp://www.nerc.com/pub/sys/all_updl/docs/blackout/Blackout-OneYearLater(PRINT).pdf As of Nov. 2006.

Summary

1. Reactance is a frequency-dependent property of capacitors and inductors that is analogous to the resistance of a resistor.
2. Impedance is a frequency-dependent property of an ac circuit or circuit loop that is analogous to the resistance in a dc circuit.
3. Phasors are two-dimensional vectors that allow us to picture the phase relations in a circuit.
4. Resonance occurs when the frequency of the generator equals the natural frequency of the oscillating circuit.

TOPIC	RELEVANT EQUATIONS AND REMARKS	
1. Alternating-Current Generators	An ac generator is a device that transforms mechanical energy into electrical energy. This transformation can be accomplished by using the mechanical energy to either rotate a conducting coil in a magnetic field or to rotate a magnet in a conducting coil.	
Emf generated	$\mathcal{E} = \mathcal{E}_{\text{peak}} \cos(\omega t + \delta)$	29-1
2. Current		
Rms current	$I_{\text{rms}} = \sqrt{(I^2)_{\text{av}}}$	29-8
Rms current and peak current	$I_{\text{rms}} = \frac{I}{\sqrt{2}} I_{\text{peak}}$	29-9
For a resistor	$I_{\text{rms}} = \frac{V_{R\,\text{rms}}}{R}$ potential drop and current in phase	29-12
For an inductor	$I_{\text{rms}} = \frac{V_{L\,\text{rms}}}{\omega L} = \frac{V_{L\,\text{rms}}}{X_L}$ potential drop leads current by 90°	29-22
For a capacitor	$I_{\text{rms}} = \frac{V_{C\,\text{rms}}}{1/\omega C} = \frac{V_{C\,\text{rms}}}{X_C}$ potential drop lags current by 90°	29-26
3. Reactance		
Inductive reactance	$X_L = \omega L$	29-21
Capacitive reactance	$X_C = \frac{1}{\omega C}$	29-27
4. Average Power		
To a resistor	$P_{\text{av}} = V_{R\,\text{rms}} I_{\text{rms}} = (I_{\text{rms}})^2 R$	29-10, 29-12
To an inductor or to a capacitor	$P_{\text{av}} = 0$	
5. *Transformers	A transformer is a device used to increase or decrease the voltage in a circuit without an appreciable loss in power. For a transformer with N_1 turns in the primary and N_2 turns in the secondary, the potential difference across the secondary coil is related to the potential drop across the primary coil by $V_2 = \frac{N_2}{N_1} V_1$	29-30

TOPIC	RELEVANT EQUATIONS AND REMARKS	
	If there are no power losses, $V_{1\,\text{rms}}I_{1\,\text{rms}} = V_{2\,\text{rms}}I_{2\,\text{rms}}$	29-32
6. *LC and RLC Series Circuits	If a capacitor is discharged through an inductor, the charge and the voltage on the capacitor oscillate with angular frequency $\omega = \frac{1}{\sqrt{LC}}$	29-37
	The current in the inductor oscillates with the same frequency, but it is out of phase with the charge by 90°. The energy oscillates between electric energy in the capacitor and magnetic energy in the inductor. If the circuit also has resistance, the oscillations are damped because energy is dissipated in the resistor.	
7. *Phasors	Phasors are two-dimensional vectors that represent the current $\vec{I}$, the potential drop across a resistor $\vec{V}_R$, the potential drop across a capacitor $\vec{V}_C$, and the potential drop across an inductor $\vec{V}_L$ in an ac circuit. These phasors rotate in the counterclockwise direction with an angular velocity that is equal to the angular frequency ω of the current. $\vec{V}_R$ is in phase with the current, $\vec{V}_L$ leads the current by 90°, and $\vec{V}_C$ lags the current by 90°. The x component of each phasor equals the magnitude of the current or the corresponding potential drop at any instant.	
8. *Driven Series RLC Circuit		
Applied potential drop	$V_{\text{app}} = V_{\text{app peak}} \cos \omega t$	
Current	$I = \frac{V_{\text{app peak}}}{Z} \cos(\omega t - \delta)$	29-52
Impedance Z	$Z = \sqrt{R^2 + (X_L - X_C)^2}$	29-51
Phase angle δ	$\tan \delta = \frac{X_L - X_C}{R}$	29-49
Average power	$P_{\text{av}} = (I_{\text{rms}})^2 R = V_{\text{app rms}} I_{\text{rms}} \cos \delta = \frac{(V_{\text{app rms}})^2 R \omega^2}{L^2(\omega^2 - \omega_0^2)^2 + \omega^2 R^2}$	29-54, 29-55, 29-56
Power factor	The quantity $\cos \delta$ in Equation 29-55 is called the power factor of the RLC circuit. At resonance, δ is zero, the power factor is 1, and $P_{\text{av}} = V_{\text{app rms}} I_{\text{rms}}$	
Resonance	When the rms current is maximum, the circuit is said to be at resonance. The conditions for resonance are $Z = \sqrt{R^2 + (X_L - X_C)^2} = R$ $X_L = X_C$, so $\omega = \omega_0 = \frac{1}{\sqrt{LC}}$ and $\delta = 0$	
9. *Q Factor	The sharpness of the resonance curve is described by the Q factor: $Q_{\text{factor}} = \frac{\omega_0 L}{R}$	29-57
	When the resonance curve is reasonably narrow, the Q factor can be approximated by $Q_{\text{factor}} = \frac{\omega_0}{\Delta \omega} = \frac{f_0}{\Delta f}$	29-58

Answers to Concept Check

29-1 (*c*)

Answers to Practice Problems

29-1 (*a*) 2.8 A, (*b*) 96 W, (*c*) 1.9×10^2 W

29-2 79 μH

29-3 $V_{\text{out rms}} = V_{\text{in rms}}/\sqrt{1 + (R/L)^2/\omega^2}$. This circuit is a *high-pass filter*.

Problems

In a few problems, you are given more data than you actually need; in a few other problems, you are required to supply data from your general knowledge, outside sources, or informed estimate.

Interpret as significant all digits in numerical values that have trailing zeros and no decimal points.

- • Single-concept, single-step, relatively easy
- •• Intermediate-level, may require synthesis of concepts
- ••• Challenging
- SSM Solution is in the *Student Solutions Manual*
- Consecutive problems that are shaded are paired problems.

CONCEPTUAL PROBLEMS

1 • A coil in an ac generator rotates at 60 Hz. How much time elapses between successive peak emf values of the coil?

2 • If the rms voltage in an ac circuit is doubled, the peak voltage is (*a*) doubled, (*b*) halved, (*c*) increased by a factor of $\sqrt{2}$, (*d*) not changed.

3 • If the frequency in the circuit shown in Figure 29-27 is doubled, the inductance of the inductor will (*a*) double, (*b*) not change, (*c*) halve, (*d*) quadruple. SSM

4 • If the frequency in the circuit shown in Figure 29-27 is doubled, the inductive reactance of the inductor will (*a*) double, (*b*) not change, (*c*) halve, (*d*) quadruple.

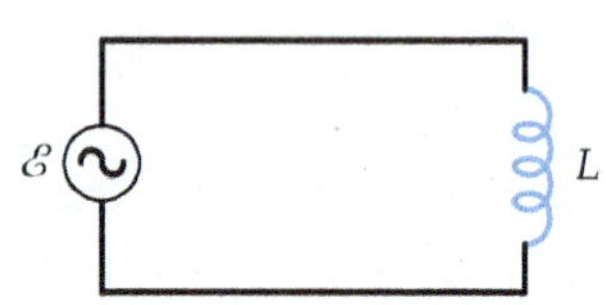

FIGURE 29-27 Problems 3 and 4

5 • If the frequency in the circuit in Figure 29-28 is doubled, the capacitive reactance of the circuit will (*a*) double, (*b*) not change, (*c*) halve, (*d*) quadruple.

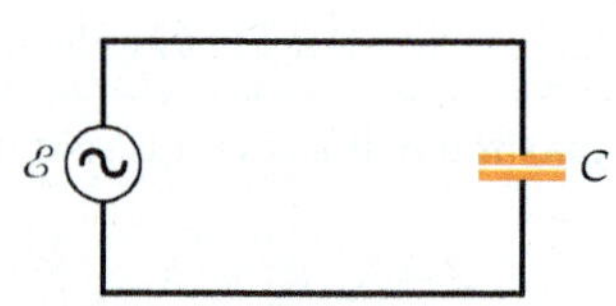

FIGURE 29-28 Problem 5

6 • (*a*) In a circuit consisting solely of an ac generator and an ideal inductor, are there any time intervals when the inductor receives energy from the generator? If so, when? Explain your answer. (*b*) Are there any time intervals when the inductor supplies energy back to the generator? If so when? Explain your answer.

7 • (*a*) In a circuit consisting of a generator and a capacitor, are there any time intervals when the capacitor receives energy from the generator? If so, when? Explain your answer. (*b*) Are there any time intervals when the capacitor supplies power to the generator? If so, when? Explain your answer. SSM

8 • (*a*) Show that the SI unit of inductance multiplied by the SI unit of capacitance is equivalent to seconds squared. (*b*) Show that the SI unit of inductance divided by the SI unit of resistance is equivalent to seconds.

9 • Suppose you increase the rotation rate of the coil in the generator shown in the simple ac circuit in Figure 29-29. Then the rms current (*a*) increases, (*b*) does not change, (*c*) may increase or decrease depending on the magnitude of the original frequency, (*d*) may increase or decrease depending on the magnitude of the resistance, (*e*) decreases. SSM

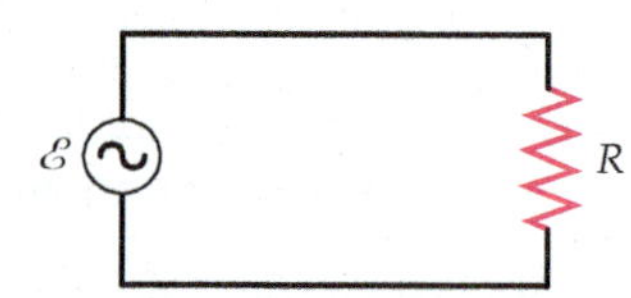

FIGURE 29-29 Problem 9

10 • If the inductance value is tripled in a circuit consisting solely of a variable inductor and a variable capacitor, how would you have to change the capacitance so that the natural frequency of the circuit is unchanged? (*a*) Triple the capacitance. (*b*) Decrease the capacitance to one-third of its original value. (*c*) You should not change the capacitance. (*d*) You cannot determine how to change the capacitance from the data given.

11 • Consider a circuit consisting solely of an ideal inductor and an ideal capacitor. How does the maximum energy stored in the capacitor compare to the maximum value stored in the inductor? (*a*) They are the same and each is equal to the total energy stored in the circuit. (*b*) They are the same and each is equal to half of the total energy stored in the circuit. (*c*) The maximum energy stored in the capacitor is larger than the maximum energy stored in the inductor. (*d*) The maximum energy stored in the inductor is larger than the maximum energy stored in the capacitor. (*e*) You cannot compare the maximum energies based on the data given because the ratio of the maximum energies depends on the actual capacitance and inductance values. SSM

12 • True or false:

(*a*) A driven series *RLC* circuit that has a high *Q* factor has a narrow resonance curve.

(*b*) A circuit consists solely of a resistor, an inductor, and a capacitor, all connected in series. If the resistance of the resistor is doubled, the natural frequency of the circuit remains the same.

(*c*) At resonance, the impedance of a driven series *RLC* combination equals the resistance *R*.

(*d*) At resonance, the current in a driven series *RLC* circuit is in phase with the voltage applied to the combination.

13 • True or false:

(*a*) Near resonance, the power factor of a driven series *RLC* circuit is close to zero.

(*b*) The power factor of a driven series *RLC* circuit does not depend on the value of the resistance.
(*c*) The resonance frequency of a driven series *RLC* circuit does not depend on the value of the resistance.
(*d*) At resonance, the peak current of a driven series *RLC* circuit does not depend on the capacitance or the inductance.
(*e*) For frequencies below the resonant frequency, the capacitive reactance of a driven series *RLC* circuit is larger than the inductive reactance.
(*f*) For frequencies below the resonant frequency of a driven series *RLC* circuit, the phase of the current leads (is ahead of) the phase of the applied voltage.

14 • You may have noticed that sometimes two radio stations can be heard when your receiver is tuned to a specific frequency. This situation often occurs when you are driving and are between two cities. Explain how this situation can occur.

15 • True or false:
(*a*) At frequencies much higher than or much lower than the resonant frequency of a driven series *RLC* circuit, the power factor is close to zero.
(*b*) The larger the resonance width of a driven series *RLC* circuit is, the larger the *Q* factor for the circuit is.
(*c*) The larger the resistance of a driven series *RLC* circuit is, the larger the resonance width for the circuit is.

16 • An ideal transformer has N_1 turns on its primary and N_2 turns on its secondary. The average power delivered to a load resistance *R* connected across the secondary is P_2 when the primary rms voltage is V_1. The rms current in the primary windings can then be expressed as (*a*) P_2/V_1, (*b*) $(N_1/N_2)(P_2/V_1)$, (*c*) $(N_2/N_1)(P_2/V_1)$, (*d*) $(N_2/N_1)2(P_2/V_1)$.

17 • True or false:
(*a*) A transformer is used to change frequency.
(*b*) A transformer is used to change voltage.
(*c*) If a transformer steps up the current, it must step down the voltage.
(*d*) A step-up transformer steps down the current.
(*e*) The standard household wall-outlet voltage in Europe is 220 V, about twice that used in the United States. If a European traveler wants her hair dryer to work properly in the United States, she should use a transformer that has more windings in its secondary coil than in its primary coil.
(*f*) The standard household wall-outlet voltage in Europe is 220 V, about twice that used in the United States. If an American traveler wants his electric razor to work properly in Europe, he should use a transformer that steps up the current. SSM

ESTIMATION AND APPROXIMATION

18 •• **ENGINEERING APPLICATION** The impedances of motors, transformers, and electromagnets include both resistance and inductive reactance. Suppose that the phase of the current to a large industrial plant lags the phase of the applied voltage by 25° when the plant is under full operation and using 2.3 MW of power. The power is supplied to the plant from a substation 4.5 km from the plant; the 60 Hz rms line voltage at the plant is 40 kV. The resistance of the transmission line from the substation to the plant is 5.2 Ω. The cost per kilowatt-hour to the company that owns the plant is \$0.14, and the plant pays only for the actual energy used. (*a*) Estimate the resistance and inductive reactance of the plant's total load. (*b*) Estimate the rms current in the power lines and the rms voltage at the substation. (*c*) How much power is lost in transmission? (*d*) Suppose that the phase at which the current lags the phase of the applied voltage is reduced to 18° by adding a bank of capacitors in series with the load. How much money would be saved by the electric utility during one month of operation, assuming the plant operates at full capacity for 16 h each day? (*e*) What must be the capacitance of this bank of capacitors to achieve this change in phase angle?

ALTERNATING CURRENT IN RESISTORS, INDUCTORS, AND CAPACITORS

19 • A 100-W lightbulb is screwed into a standard 120-V-rms. socket. Find (*a*) the rms current, (*b*) the peak current, and (*c*) the peak power. SSM

20 • A circuit breaker is rated for a current of 15 A rms at a voltage of 120 V rms. (*a*) What is the largest value of peak current that the breaker can carry? (*b*) What is the maximum average power that can be supplied by this circuit?

21 • What is the reactance of a 1.00-μH inductor at (*a*) 60 Hz, (*b*) 600 Hz, and (*c*) 6.00 kHz? SSM

22 • An inductor has a reactance of 100 Ω at 80 Hz. (*a*) What is its inductance? (*b*) What is its reactance at 160 Hz?

23 • At what frequency would the reactance of a 10-μF capacitor equal the reactance of a 1.0-mH inductor?

24 • What is the reactance of a 1.00-nF capacitor at (*a*) 60.0 Hz, (*b*) 6.00 kHz, and (*c*) 6.00 MHz?

25 • A 20-Hz ac generator that produces a peak emf of 10 V is connected to a 20-μF capacitor. Find (*a*) the peak current and (*b*) the rms current. SSM

26 • At what frequency is the reactance of a 10-μF capacitor (*a*) 1.00 Ω, (*b*) 100 Ω, and (*c*) 10.0 mΩ?

27 •• A circuit consists of two ideal ac generators and a 25-Ω resistor, all connected in series. The potential difference across the terminals of one of the generators is given by $V_1 = (5.0\text{ V})\cos(\omega t - \alpha)$, and the potential difference across the terminals of the other generator is given by $V_2 = (5.0\text{ V})\cos(\omega t + \alpha)$, where $\alpha = \pi/6$. (*a*) Use Kirchhoff's loop rule and a trigonometric identity to find the peak current in the circuit. (*b*) Use a phasor diagram to find the peak current in the circuit. (*c*) Find the current in the resistor if $\alpha = \pi/4$ and the amplitude of V_2 is increased from 5.0 V to 7.0 V.

*UNDRIVEN CIRCUITS CONTAINING CAPACITORS, RESISTORS, AND INDUCTORS

28 • (*a*) Show that $1/\sqrt{LC}$ has units of inverse seconds by substituting SI units for inductance and capacitance into the expression. (*b*) Show that $\omega_0 L/R$ (the expression for the *Q* factor) is dimensionless by substituting SI units for angular frequency, inductance, and resistance into the expression.

29 • (*a*) What is the period of oscillation of an *LC* circuit consisting of a 2.0-mH coil and a 20-μF capacitor? (*b*) A circuit that oscillates consists solely of an 80-μF capacitor and a variable ideal inductor. What inductance is needed in order to tune this circuit to oscillate at 60 Hz? SSM

30 • An *LC* circuit has capacitance C_0 and inductance *L*. A second *LC* circuit has capacitance $\frac{1}{2}C_0$ and inductance 2*L*, and a third *LC* circuit has capacitance $2C_0$ and inductance $\frac{1}{2}L$. (*a*) Show that each circuit oscillates with the same frequency. (*b*) In which circuit would the peak current be greatest if the peak voltage across the capacitor in each circuit were the same?

31 •• A 5.0-μF capacitor is charged to 30 V and is then connected across an ideal 10-mH inductor. (*a*) How much energy is stored in the system? (*b*) What is the frequency of oscillation of the circuit? (*c*) What is the peak current in the circuit?

32 •• A coil with internal resistance can be modeled as a resistor and an ideal inductor in series. Assume that the coil has an internal resistance of 1.00 Ω and an inductance of 400 mH. A 2.00-μF capacitor is charged to 24.0 V and is then connected across coil. (*a*) What is the initial voltage across the coil? (*b*) How much energy is dissipated in the circuit before the oscillations die out? (*c*) What is the frequency of oscillation of the circuit? (Assume the internal resistance is sufficiently small that it has no impact on the frequency of the circuit.) (*d*) What is the quality factor of the circuit?

33 •• An inductor and a capacitor are connected, as shown in Figure 29-30. Initially, the switch is open, and the left plate of the capacitor has charge Q_0. The switch is then closed. (*a*) Plot both Q versus t and I versus t on the same graph, and explain how it can be seen from these two plots that the current leads the charge by 90°. (*b*) The expressions for the charge and for the current are given by Equations 29-38 and 29-39, respectively. Use trigonometry and algebra to show that the current leads the charge by 90°. SSM

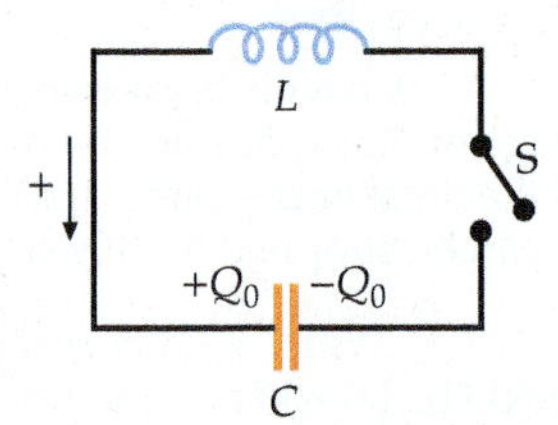

FIGURE 29-30 Problem 33

DRIVEN *RL* CIRCUITS

34 •• A circuit consists of a resistor, an ideal 1.4-H inductor, and an ideal 60-Hz generator, all connected in series. The rms voltage across the resistor is 30 V and the rms voltage across the inductor is 40 V. (*a*) What is the resistance of the resistor? (*b*) What is the peak emf of the generator?

35 •• A coil that has a resistance of 80.0 Ω has an impedance of 200 Ω when driven at a frequency of 1.00 kHz. What is the inductance of the coil? SSM

36 •• **Engineering Application** A two-conductor transmission line simultaneously carries a superposition of two voltage signals, so the potential difference between the two conductors is given by $V = V_1 + V_2$, where $V_1 = (10.0\text{ V})\cos(\omega_1 t)$ and $V_2 = (10.0\text{ V})\cos(\omega_2 t)$, where $\omega_1 = 100$ rad/s and $\omega_2 = 10\,000$ rad/s. A 1.00-H inductor and a 1.00-kΩ shunt resistor are inserted into the transmission line as shown in Figure 29-31. (Assume that the output is connected to a load that draws only an insignificant amount of current.) (*a*) What is the voltage (V_{out}) at the output of the transmission line? (*b*) What is the ratio of the low-frequency amplitude to the high-frequency amplitude at the output?

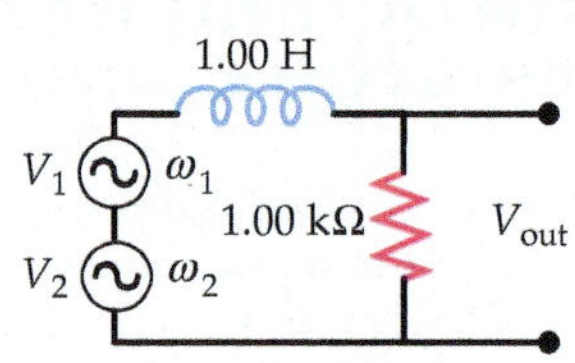

FIGURE 29-31 Problem 36

37 •• A coil is connected to a 120-V-rms, 60-Hz line. The average power supplied to the coil is 60 W, and the rms current is 1.5 A. Find (*a*) the power factor, (*b*) the resistance of the coil, and (*c*) the inductance of the coil. (*d*) Does the current lag or lead the voltage? Explain your answer. (*e*) Support your answer to Part (*d*) by determining the phase angle.

38 •• A 36-mH inductor that has a resistance of 40 Ω is connected to an ideal ac voltage source whose output is given by $\mathcal{E} = (345\text{ V})\cos(150\pi t)$, where t is in seconds. Determine (*a*) the peak current in the circuit, (*b*) the peak and rms voltages across the inductor, (*c*) the average power dissipation, and (*d*) the peak and average magnetic energy stored in the inductor.

39 •• A coil that has a resistance R and an inductance L has a power factor equal to 0.866 when driven at a frequency of 60 Hz. What is the coil's power factor if it is driven at 240 Hz? SSM

40 •• A resistor and an inductor are connected in parallel across an ideal ac voltage source whose output is given by $\mathcal{E} = \mathcal{E}_{max}\cos\omega t$ as shown in Figure 29-32. Show that (*a*) the current in the resistor is given by $I_R = (\mathcal{E}_{peak}/R)\cos\omega t$, (*b*) the current in the inductor is given by $I_L = (\mathcal{E}_{peak}/X_L)\cos(\omega t - 90°)$, and (*c*) the current in the voltage source is given by $I = I_R + I_L = I_{peak}\cos(\omega t - \delta)$, where $I_{peak} = \mathcal{E}_{peak}/Z$.

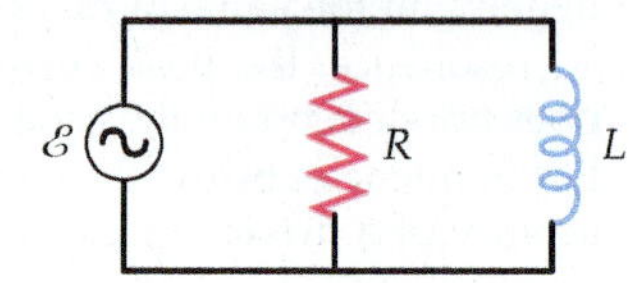

FIGURE 29-32 Problem 40

41 •• Figure 29-33 shows a load resistor that has a resistance of $R_L = 20.0$ Ω connected to a high-pass filter consisting of an inductor that has an inductance $L = 3.20$-mH inductor and a resistor that has resistance $R = 4.00$ Ω. The output of the ideal ac generator is given by $\mathcal{E} = (100\text{ V})\cos(2\pi ft)$. Find the rms currents in all three branches of the circuit if the driving frequency is (*a*) 500 Hz and (*b*) 2000 Hz. Find the fraction of the total average power supplied by the ac generator that is delivered to the load resistor if the frequency is (*c*) 500 Hz and (*d*) 2000 Hz. SSM

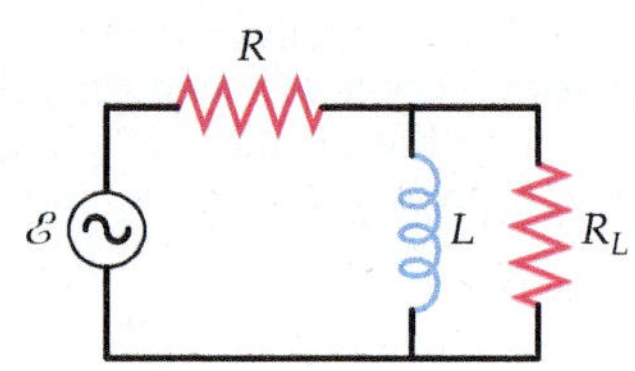

FIGURE 29-33 Problem 41

42 •• An ideal ac voltage source whose emf $\mathcal{E}_1$ is given by $(20\text{ V})\cos(2\pi ft)$ and an ideal battery whose emf $\mathcal{E}_2$ is 16 V are connected to a combination of two resistors and an inductor (Figure 29-34), where $R_1 = 10$ Ω, $R_2 = 8.0$ Ω, and $L = 6.0$ mH. Find the average power delivered to each resistor if the driving frequency is (*a*) 100 Hz, (*b*) 200 Hz, and (*c*) 800 Hz.

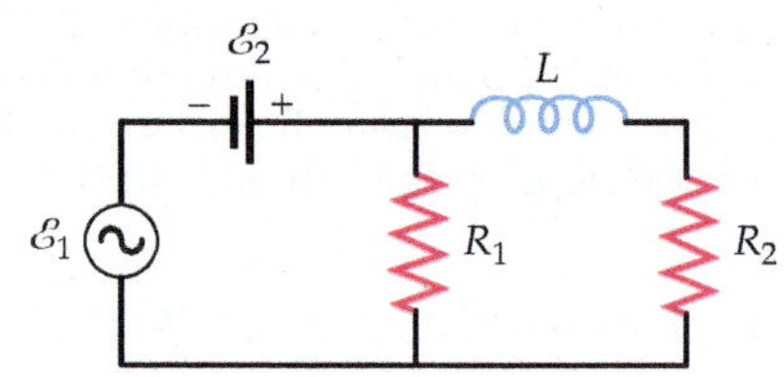

FIGURE 29-34 Problem 42

43 •• An ac circuit contains a resistor and an ideal inductor connected in series. The rms voltage drop across the series combination is 100 V and the rms voltage drop across the inductor alone is 80 V. What is the rms voltage drop across the resistor?

FILTERS AND RECTIFIERS

44 •• **Engineering Application** The circuit shown in Figure 29-35 is called an *RC high-pass filter* because it transmits input voltage signals that have high frequencies with greater amplitude than it transmits input voltage signals that have low frequencies. If the input voltage is given by $V_{in} = V_{in\,peak}\cos\omega t$, show that the output voltage is $V_{out} = V_H\cos(\omega t - \delta)$ where $V_H = V_{in\,peak}/\sqrt{1 + (\omega RC)^{-2}}$.

(Assume that the output is connected to a load that draws only an insignificant amount of current.) Show that this result justifies calling this circuit a high-pass filter.

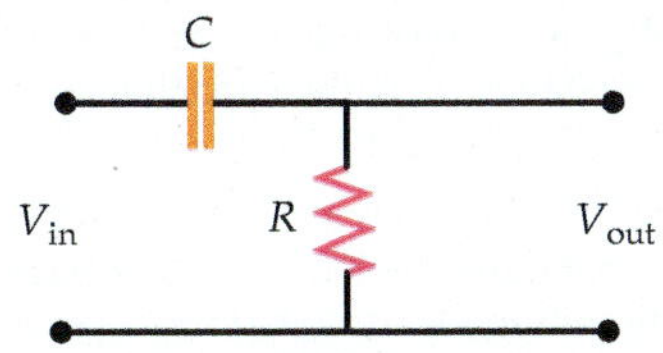

FIGURE 29-35 Problem 44

45 •• (*a*) Find an expression for the phase constant δ in Problem 44 in terms of ω, R, and C. (*b*) What is the value of δ in the limit that $\omega \to 0$? (*c*) What is the value of δ in the limit that $\omega \to \infty$? (*d*) Explain your answers to Parts (*b*) and (*c*).

46 •• **SPREADSHEET** Assume that in Problem 44, $R = 20\ \text{k}\Omega$ and $C = 15\ \text{nF}$. (*a*) At what frequency is $V_H = \frac{1}{\sqrt{2}} V_{\text{in peak}}$? That particular frequency is known as the 3-dB frequency, or f_{3dB}, for the circuit. (*b*) Using a spreadsheet program, make a graph of $\log_{10}(V_H)$ versus $\log_{10}(f)$, where f is the frequency. Make sure that the scale extends from at least 10% of the 3-dB frequency to ten times the 3-dB frequency. (*c*) Make a graph of δ versus $\log_{10}(f)$ for the same range of frequencies as in Part (*b*). What is the value of the phase constant when the frequency is equal to the 3-dB frequency?

47 •• A slowly varying voltage signal $V(t)$ is applied to the input of the high-pass filter of Problem 44. Slowly varying means that during one time constant (equal to RC) there is no significant change in the voltage signal. Show that under those conditions the output voltage is proportional to the time derivative of $V(t)$. This situation is known as a *differentiation circuit*. SSM

48 •• We can describe the output from the high-pass filter from Problem 44 using a decibel scale: $\beta = (20\ \text{dB})\log_{10}(V_H/V_{\text{in peak}})$, where β is the output in decibels. Show that for $V_H = \frac{1}{\sqrt{2}} V_{\text{in peak}}$, $\beta = 3.0\ \text{dB}$. The frequency at which $V_H = \frac{1}{\sqrt{2}} V_{\text{in peak}}$ is known as f_{3dB} (the 3-dB frequency). Show that for $f \ll f_{3dB}$, the output β drops by 6 dB if the frequency f is halved.

49 •• Show that the average power dissipated in the resistor of the high-pass filter of Problem 44 is given by

$$P_{av} = \frac{V_{\text{in peak}}^2}{2R[1 + (\omega RC)^{-2}]}.$$ SSM

50 •• One application of the high-pass filter of Problem 44 is as a noise filter for electronic circuits (a filter that blocks out low-frequency noise). Using a resistance value of 20 kΩ, find a value for the capacitance for the high-pass filter that attenuates a 60-Hz input voltage signal by a factor of 10, that is, so $V_H = \frac{1}{10} V_{\text{in peak}}$.

51 •• **ENGINEERING APPLICATION** The circuit shown in Figure 29-36 is an example of a low-pass filter. (Assume that the output is connected to a load that draws only an insignificant amount of current.) (*a*) If the input voltage is given by $V_{in} = V_{\text{in peak}} \cos \omega t$, show that the output voltage is $V_{out} = V_L \cos(\omega t - \delta)$ where $V_L = V_{\text{in peak}}/\sqrt{1 + (\omega RC)^2}$. (*b*) Discuss the trend of the output voltage in the limiting cases $\omega \to 0$ and $\omega \to \infty$. SSM

52 •• (*a*) Find an expression for the phase angle δ for the low-pass filter of Problem 51 in terms of ω, R, and C. (*b*) Find the value of δ in the limit that $\omega \to 0$ and in the limit that $\omega \to \infty$. Explain your answer.

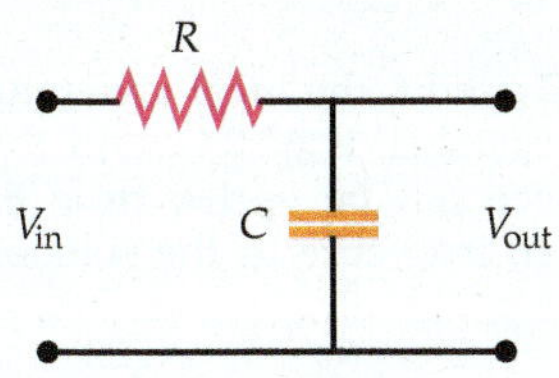

FIGURE 29-36 Problems 51 and 52

53 •• **SPREADSHEET** Using a spreadsheet program, make a graph of V_L versus input frequency f and a graph of phase angle δ versus input frequency for the low-pass filter of Problems 51 and 52. Use a resistance value of 10 kΩ and a capacitance value of 5.0 nF.

54 ••• A rapidly varying voltage signal $V(t)$ is applied to the input of the low-pass filter of Problem 51. Rapidly varying means that during one time constant (equal to RC) there are significant changes in the voltage signal. Show that under those conditions the output voltage is proportional to the integral of $V(t)$ with respect to time. This situation is known as an *integration circuit*.

55 ••• **ENGINEERING APPLICATION** The circuit shown in Figure 29-37 is a *trap filter*. (Assume that the output is connected to a load that draws only an insignificant amount of current.) (*a*) Show that the *trap filter* acts to reject signals in a band of frequencies centered at $\omega = 1/\sqrt{LC}$. (*b*) How does the width of the frequency band rejected depend on the resistance R? SSM

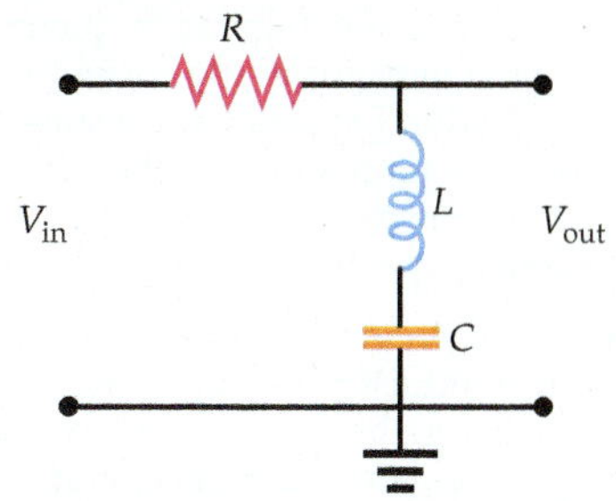

FIGURE 29-37 Problem 55

56 ••• **ENGINEERING APPLICATION** A half-wave rectifier for transforming an ac voltage into a dc voltage is shown in Figure 29-38. The diode in the figure can be thought of as a one-way valve for current. It allows current to pass in the forward direction (the direction of the arrowhead) only when V_{in} is at a higher electric potential than V_{out} by 0.60 V (i.e., whenever $V_{in} - V_{out} \geq +0.60\ \text{V}$). The resistance of the diode is effectively infinite when $V_{in} - V_{out}$ is less than +0.60 V. Plot two cycles of both input and output voltages as a function of time, on the same graph, assuming the input voltage is given by $V_{in} = V_{\text{in peak}} \cos \omega t$.

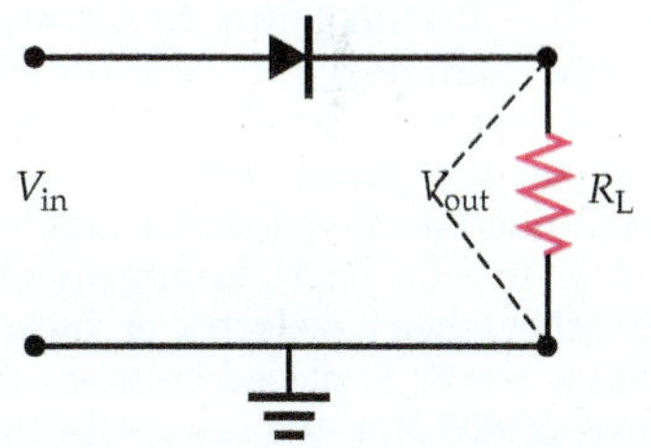

FIGURE 29-38 Problem 56

57 ••• **ENGINEERING APPLICATION** The output of the rectifier of Problem 56 can be further filtered by putting its output through a low-pass filter as shown in Figure 29-39*a*. The resulting output is a dc voltage with a small ac component (ripple) shown in Figure 29-39*b*. If the input frequency is 60 Hz and the load resistance is 1.00 kΩ, find the value for the capacitance so that the output voltage varies by less than 50 percent of the mean value over one cycle.

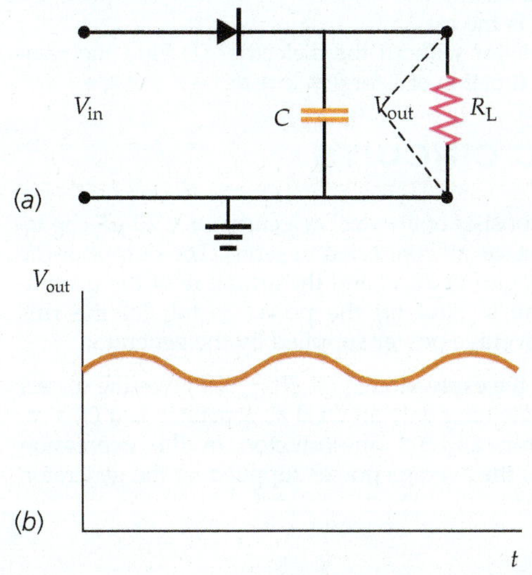

FIGURE 29-39 Problem 57

DRIVEN *LC* CIRCUITS

58 •• The generator voltage in Figure 29-40 is given by $\mathcal{E} = (100\ \text{V})\cos(2\pi f t)$. (*a*) For each branch, what is the peak current and what is the phase of the current relative to the phase of the generator voltage? (*b*) At the resonance frequency there is no current in the generator. What is the angular frequency at resonance? (*c*) At the resonance frequency, find the current in the inductor and the current in the capacitor. Express your results as functions of time. (*d*) Draw a phasor diagram showing the phasors for the applied voltage, the generator current, the capacitor current, and the inductor current for the case where the frequency is higher than the resonance frequency.

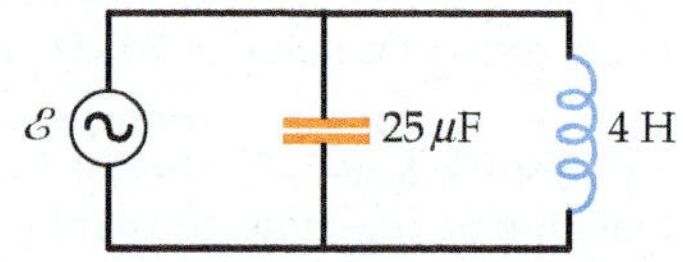

FIGURE 29-40 Problem 58

59 •• A circuit consists of an ideal ac generator, a capacitor, and an ideal inductor, all connected in series. The charge on the capacitor is given by $Q = (15\ \mu\text{C})\cos(\omega t + \frac{\pi}{4})$, where $\omega = 1250$ rad/s. (*a*) Find the current in the circuit as a function of time. (*b*) Find the capacitance if the inductance is 28 mH. (*c*) Write expressions for the electrical energy U_e, the magnetic energy U_m, and the total energy U as functions of time.

60 ••• **ENGINEERING APPLICATION** One method for determining the compressibility of a dielectric material uses a driven *LC* circuit that has a parallel-plate capacitor. The dielectric is inserted between the plates and the change in resonance frequency is determined as the capacitor plates are subjected to a compressive stress. In one such arrangement, the resonance frequency is 120 MHz when a dielectric of thickness 0.100 cm and dielectric constant $\kappa = 6.80$ is placed between the plates. Under a compressive stress of 800 atm, the resonance frequency decreases to 116 MHz. Find the Young's modulus of the dielectric material. (Assume that the dielectric constant does not change with pressure.)

61 ••• Figure 29-41 shows an inductor in series with a parallel-plate capacitor. The capacitor has a width w of 20 cm and a gap of 2.0 mm. A dielectric that has a dielectric constant of 4.8 can be slid in and out of the gap. The inductor has an inductance of 2.0 mH. When half the dielectric is between the capacitor plates (when $x = \frac{1}{2}w$), the resonant frequency of this combination is 90 MHz. (*a*) What is the capacitance of the capacitor without the dielectric? (*b*) Find the resonance frequency as a function of x for $0 \le x \le w$.

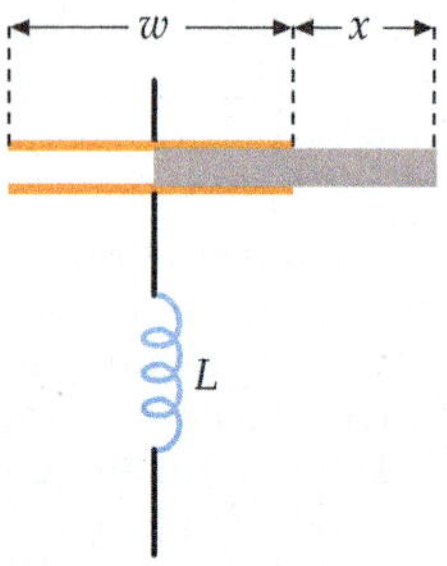

FIGURE 29-41 Problem 61

DRIVEN *RLC* CIRCUITS

62 • A circuit consists of an ideal ac generator, a 20-μF capacitor, and an 80-Ω resistor, all connected in series. The output of the generator has a peak emf of 20 V, and the armature of the generator rotates at 400 rad/s. Find (*a*) the power factor, (*b*) the rms current, and (*c*) the average power supplied by the generator.

63 •• Show that the expression $P_{av} = R\mathcal{E}_{rms}^2/Z^2$ gives the correct result for a circuit containing only an ideal ac generator and (*a*) a resistor, (*b*) a capacitor, and (*c*) an inductor. In the expression $P_{av} = R\mathcal{E}_{rms}^2/Z^2$, P_{av} is the average power supplied by the generator, $\mathcal{E}_{rms}$ is the root-mean-square value of the emf of the generator, R is the resistance, C is the capacitance, and L is the inductance. [In Part (*a*), $C = L = 0$; in Part (*b*), $R = L = 0$; and in Part (*c*), $R = C = 0$.] SSM

64 •• A series *RLC* circuit that has an inductance of 10 mH, a capacitance of 2.0 μF, and a resistance of 5.0 Ω is driven by an ideal ac voltage source that has a peak emf of 100 V. Find (*a*) the resonant frequency and (*b*) the root-mean-square current at resonance. When the frequency is 8000 rad/s, find (*c*) the capacitive and inductive reactances, (*d*) the impedance, (*e*) the root-mean-square current, and (*f*) the phase angle.

65 •• Find (*a*) the Q factor and (*b*) the resonance width (in hertz) for the circuit in Problem 64. (*c*) What is the power factor when $\omega = 8000$ rad/s? SSM

66 •• **ENGINEERING APPLICATION** FM radio stations typically operate at frequencies separated by 0.20 MHz. Thus, when your radio is tuned to a station operating at a frequency of 100.1 MHz, the resonance width of the receiver circuit should be much smaller than 0.20 MHz, so that you do not receive signal from stations operating at adjacent frequencies. Assume your receiving circuit has a resonance width of 0.050 MHz. When tuned in to that particular station, what is the Q factor of your circuit?

67 •• A coil is connected to a 60-Hz ac generator with a peak emf equal to 100 V. At this frequency, the coil has an impedance of 10 Ω and a reactance of 8.0 Ω. (*a*) What is the peak current in the coil? (*b*) What is the phase angle between the current and the applied voltage? (*c*) A capacitor is put in series with the coil and the generator. What capacitance is required so that the current is in phase with the generator emf? (*d*) What is the peak voltage measured across the capacitor?

68 •• An ideal 0.25-H inductor and a capacitor are connected in series with an ideal 60-Hz generator. A digital voltmeter is used to measure the rms voltages across the inductor and capacitor independently. The voltmeter reading across the capacitor is 75 V and that across the inductor is 50 V. (*a*) Find the capacitance and the rms current in the circuit. (*b*) What is the rms voltage across the series combination of the capacitor and the inductor?

69 •• In the circuit shown in Figure 29-42 the ideal generator produces an rms voltage of 115 V when operated at 60 Hz. What is the rms voltage between points (*a*) *A* and *B*, (*b*) *B* and *C*, (*c*) *C* and *D*, (*d*) *A* and *C*, and (*e*) *B* and *D*? SSM

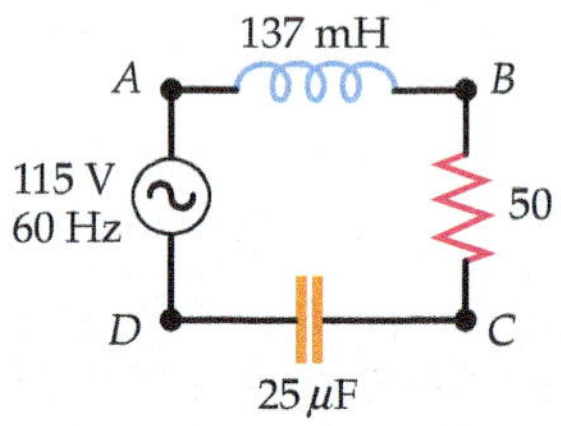

FIGURE 29-42 Problem 69

70 •• When an *RLC* series circuit is connected to a 120-V-rms, 60-Hz line, the rms current in the circuit is 11 A and the current leads the line voltage by 45°. (*a*) Find the average power supplied to the circuit. (*b*) What is the resistance in the circuit? (*c*) If the inductance in the circuit is 50 mH, find the capacitance in the circuit. (*d*) Without changing the inductance, by how much should you change the capacitance to make the power factor equal to 1? (*e*) Without changing the capacitance, by how much should you change the inductance to make the power factor equal to 1?

71 •• **SPREADSHEET** Plot the circuit impedance versus the angular frequency for each of the following circuits: (*a*) a driven series *LR* circuit, (*b*) a driven series *RC* circuit, and (*c*) a driven series *RLC* circuit.

72 •• In a driven series *RLC* circuit, the ideal generator has a peak emf equal to 200 V, the resistance is 60.0 Ω, and the capacitance is 8.00 μF. The inductance can be varied from 8.00 mH to 40.0 mH by the insertion of an iron core in the solenoid. The angular frequency of the generator is 2500 rad/s. If the capacitor voltage is not to exceed 150 V, find (*a*) the peak current and (*b*) the range of inductances that is safe to use.

73 •• A certain electrical device draws an rms current of 10 A at an average power of 720 W when connected to a 120-V-rms, 60-Hz power line. (*a*) What is the impedance of the device? (*b*) What series combination of resistance and reactance would have the same impedance as this device? (*c*) If the current leads the emf, is the reactance inductive or capacitive?

74 •• A method for measuring inductance is to connect the inductor in series with a known capacitance, a known resistance, an ac ammeter, and a variable-frequency signal generator. The frequency of the signal generator is varied and the emf is kept constant until the current is maximum. (*a*) If the capacitance is 10 μF, the peak emf is 10 V, the resistance is 100 Ω, and the rms current in the circuit is maximum when the driving frequency is 5000 rad/s, what is the value of the inductance? (*b*) What is the maximum rms current?

75 •• A resistor and a capacitor are connected in parallel across an ac generator (Figure 29-43) that has an emf given by $\mathcal{E} = \mathcal{E}_{peak} \cos \omega t$. (*a*) Show that the current in the resistor is given by $I_R = (\mathcal{E}_{peak}/R) \cos \omega t$. (*b*) Show that the current in the capacitor branch is given by $I_C = (\mathcal{E}_{peak}/X_C) \cos(\omega t + 90°)$.(*c*) Show that the current in the generator is given by $I = I_{peak} \cos(\omega t + \delta)$, where $\tan \delta = R/X_C$ and $I_{peak} = \mathcal{E}_{peak}/Z$.

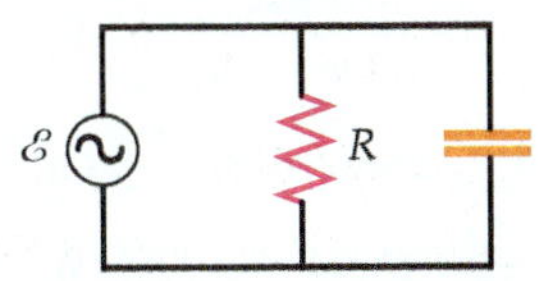

FIGURE 29-43 Problem 75

76 ••• Figure 29-44 shows a plot of average power P_{av} versus generator frequency ω for a series *RLC* circuit driven by an ac generator. The average power P_{av} is given by Equation 29-56. The full width at half-maximum, $\Delta\omega$, is the width of the resonance curve between the two points, where P_{av} is one-half its maximum value. Show that for a sharply peaked resonance, $\Delta\omega \approx R/L$ and that $Q_{factor} \approx \omega_0/\Delta\omega$ (Equation 29-58). *Hint: The half-power points occur when the denominator of Equation 29-56 is equal to twice the value it has at resonance; that is, when* $L^2(\omega^2 - \omega_0^2)^2 + \omega^2R^2 \approx +\omega_0^2R^2$. *Let* ω_1 *and* ω_2 *be the solutions of this equation. Then, show that* $\Delta\omega = \omega_2 - \omega_1 \approx R/L$.

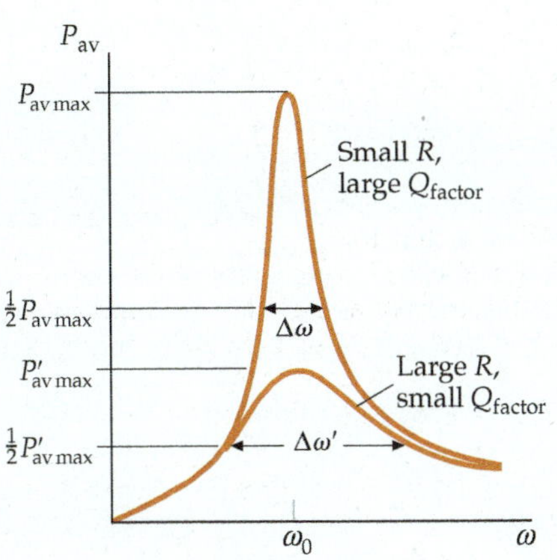

FIGURE 29-44 Problem 76

77 ••• Show by direct substitution that $L\frac{d^2Q}{dt^2} + R\frac{dQ}{dt} + \frac{1}{C}Q = 0$ (Equation 29-43*b*) is satisfied by $Q = Q_0 e^{-t/\tau} \cos \omega' t$, where $\tau = 2L/R$, $\omega' = \sqrt{1/(LC) - 1/\tau^2}$, and Q_0 is the charge on the capacitor at $t = 0$.

78 ••• **ENGINEERING APPLICATION** One method for measuring the magnetic susceptibility of a sample uses an *LC* circuit consisting of an air-core solenoid and a capacitor. The resonant frequency of the circuit without the sample is determined and then measured again with the sample inserted in the solenoid. Suppose you have a solenoid that is 4.00 cm long, is 3.00 mm in diameter, and has 400 turns of fine wire. You have a sample that is inserted in the solenoid and completely fills the air space. Neglect end effects. (*a*) Calculate the inductance of the empty solenoid. (*b*) What value for the capacitance of the capacitor should you choose so that the resonance frequency of the circuit without a sample is exactly 6.0000 MHz? (*c*) When a sample is inserted in the solenoid, you determine that the resonance frequency drops to 5.9989 MHz. Use your data to determine the sample's susceptibility.

*THE TRANSFORMER

79 • A rms voltage of 24 V is required for a device whose impedance is 12 Ω. (*a*) What should the turns ratio of a transformer be, so that the device can be operated from a 120-V-rms line? (*b*) Suppose the transformer is accidentally connected in reverse with the secondary winding across the 120-V-rms line and the 12-Ω load across the primary. How much rms current will then be in the primary winding? SSM

80 • A transformer has 400 turns in the primary and 8 turns in the secondary. (*a*) Is this a step-up or a step-down transformer? (*b*) If the primary is connected to a 120-V-rms voltage source, what is the open-circuit rms voltage across the secondary? (*c*) If the primary rms current is 0.100 A, what is the secondary rms current, assuming negligible magnetization current and no power loss?

81 • The primary of a step-down transformer has 250 turns and is connected to a 120-V-rms line. The secondary is to supply 20 A rms at 9.0 V rms. Find (*a*) the rms current in the primary and (*b*) the number of turns in the secondary, assuming 100 percent efficiency.

82 •• An audio oscillator (ac source) that has an internal resistance of 2000 Ω and an open-circuit rms output voltage of 12.0 V is to be used to drive a loudspeaker coil that has a resistance of 8.00 Ω. (*a*) What should be the ratio of primary to secondary turns of a transformer so that maximum average power is transferred to the speaker? (*b*) Suppose a second identical speaker is connected in parallel with the first speaker. How much average power is then supplied to the two speakers combined?

83 • The distribution circuit of a residential power line is operated at 2000 V rms. This voltage must be reduced to 240 V rms for use within residences. If the secondary side of the transformer has 400 turns, how many turns are in the primary?

GENERAL PROBLEMS

84 •• A resistor that has a resistance *R* carries a current given by (5.0 A) $\sin 2\pi ft$ + (7.0 A) $\sin 4\pi ft$, where $f = 60$ Hz. (*a*) What is the rms current in the resistor? (*b*) If $R = 12\ \Omega$, what is the average power delivered to the resistor? (*c*) What is the rms voltage across the resistor?

85 •• Figure 29-45 shows the voltage versus time for a *square-wave* voltage source. If $V_0 = 12$ V, (*a*) what is the rms voltage of this source? (*b*) If this alternating waveform is rectified by eliminating the negative voltages, so that only the positive voltages remain, what is the new rms voltage? SSM

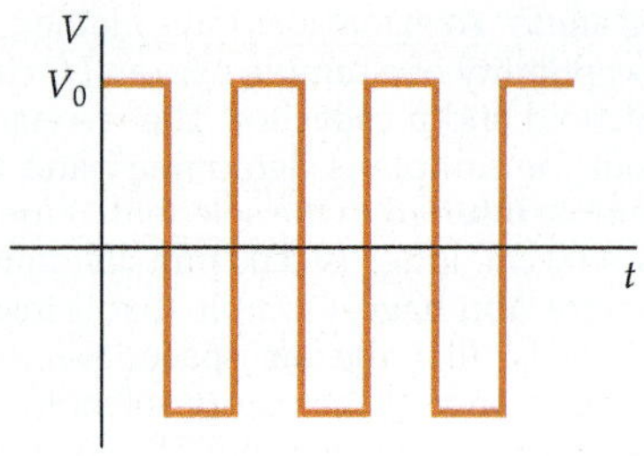

FIGURE 29-45 Problem 85

86 •• What are the average values and rms values of current for the two current waveforms shown in Figure 29-46?

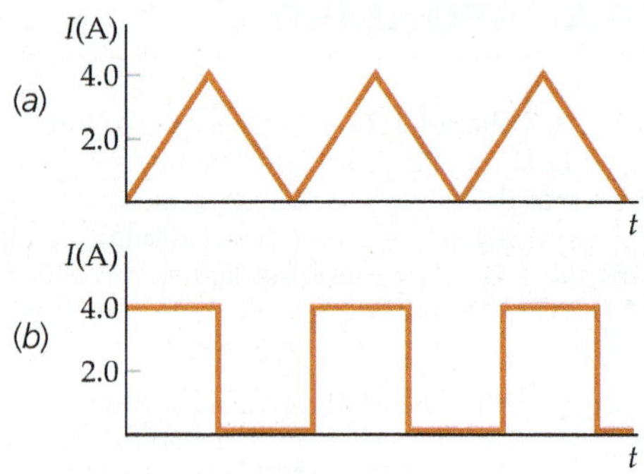

FIGURE 29-46 Problem 86

87 •• In the circuit shown in Figure 29-47, $\mathcal{E}_1 = (20\text{ V})\cos 2\pi ft$, where $f = 180$ Hz; $\mathcal{E}_2 = 18$ V; and $R = 36\ \Omega$. Find the maximum, minimum, average, and rms values of the current in the resistor.

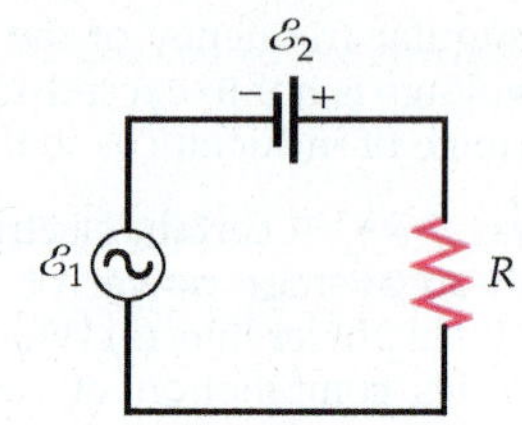

FIGURE 29-47 Problems 87, 88, and 89

88 •• Repeat Problem 87 if the resistor is replaced by a 2.0-μF capacitor.

89 ••• A circuit consists of an ac generator, a capacitor and an ideal inductor—all connected in series. The emf of the generator is given by $\mathcal{E}_{\text{peak}} \cos \omega t$. (*a*) Show that the charge on the capacitor obeys the equation $L\dfrac{d^2Q}{dt^2} + \dfrac{Q}{C} = \mathcal{E}_{\text{peak}} \cos \omega t$. (*b*) Show by direct substitution that this equation is satisfied by $Q = Q_{\text{peak}} \cos \omega t$ where $Q_{\text{peak}} = -\dfrac{\mathcal{E}_{\text{peak}}}{L(\omega^2 - \omega_0^2)}$. (*c*) Show that the current can be written as $I = I_{\text{peak}} \cos(\omega t - \delta)$, where $I_{\text{peak}} = \dfrac{\omega \mathcal{E}_{\text{peak}}}{L|\omega^2 - \omega_0^2|} = \dfrac{\mathcal{E}_{\text{peak}}}{|X_L - X_C|}$, $\delta = -90°$ for $\omega < \omega_0$, and $\delta = 90°$ for $\omega > \omega_0$, where ω_0 is the resonance frequency. SSM

CHAPTER 30

SET ACROSS THE DESERT NEAR SOCORRO, NEW MEXICO, THE NATIONAL RADIO ASTRONOMY OBSERVATORY'S VERY LARGE ARRAY IS A SYSTEM OF 27 RADIO ANTENNAS SET IN A Y-SHAPED CONFIGURATION. BECAUSE THE INFORMATION GATHERED FROM THE ARRAY IS COMBINED ELECTRONICALLY, THE INSTRUMENT HAS A RESOLUTION THAT IS 22 MILES WIDE. *(NRAO/AUI.)*

Maxwell's Equations and Electromagnetic Waves

? Did you ever wonder whether a radio antenna generates a wave equally in all directions?

(See Example 30-5.)

Maxwell's equations, first proposed by the great Scottish physicist James Clerk Maxwell, relate the electric and magnetic field vectors $\vec{E}$ and $\vec{B}$ and their sources, which are electric charges and currents. These equations summarize the experimental laws of electricity and magnetism—the laws of Coulomb, Gauss, Biot–Savart, Ampère, and Faraday. These experimental laws hold in general except for Ampère's law, which applies only to steady continuous currents.

In this chapter, we will see how Maxwell was able to generalize Ampère's law with the invention of the displacement current (Section 30-1). Maxwell was then able to show that the generalized laws of electricity and magnetism imply the existence of electromagnetic waves.

30-1 MAXWELL'S DISPLACEMENT CURRENT

Maxwell's equations play a role in classical electromagnetism analogous to the role of Newton's laws in classical mechanics. In principle, all problems in classical electricity and magnetism can be solved using Maxwell's equations, just as all problems in classical mechanics can be solved by using Newton's laws. Maxwell's equations are considerably more complicated than Newton's laws, however, and their application to most problems involves mathematics beyond the scope of this book. Nevertheless, Maxwell's equations are of great theoretical importance. For example, Maxwell showed that these equations can be combined to yield a wave equation for the electric and magnetic field vectors $\vec{E}$ and $\vec{B}$. Such **electromagnetic waves** are generated by accelerating charges (for example, the charges in an alternating current in an antenna). Electromagnetic waves were first produced in the laboratory by Heinrich Hertz in 1887. Maxwell showed that his equations predicted the speed of electromagnetic waves in free space to be

$$c = \frac{1}{\sqrt{\mu_0 \epsilon_0}} \qquad 30\text{-}1$$

THE SPEED OF ELECTROMAGNETIC WAVES

where ϵ_0, the electric constant, is the constant appearing in Coulomb's and Gauss's laws and μ_0, the magnetic constant, is the constant appearing in the Biot–Savart law and Ampère's law. Maxwell noticed with great excitement the coincidence that the measure for the speed of light equaled $1/\sqrt{\mu_0\epsilon_0}$, and Maxwell correctly surmised that light itself is an electromagnetic wave. Today, the value of c is defined as $2.997\,924\,58 \times 10^8$ m/s, the value of μ_0 is defined as $4\pi \times 10^{-7}$ N/A^2, and the value of ϵ_0 is defined by Equation 30-1.

Ampère's law (Equation 27-16) relates the line integral of the magnetic field around some closed curve C to the current that passes through any surface bounded by that curve:

$$\oint_C \vec{B}\cdot d\vec{\ell} = \mu_0 I_S \quad \text{for any closed curve } C \qquad 30\text{-}2$$

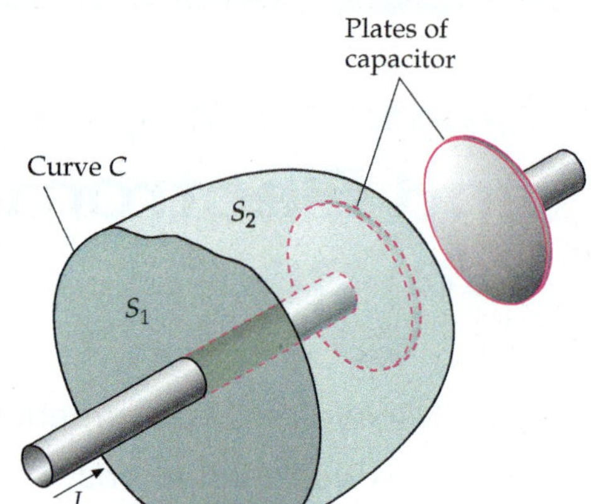

FIGURE 30-1 Two surfaces S_1 and S_2 bounded by the same curve C. The current I passes through surface S_1 but not through surface S_2. Ampère's law, which relates the line integral of the magnetic field around the curve C to the total current passing through any surface bounded by C, is not valid when the current is not continuous, as when it stops at the capacitor plate here.

Maxwell recognized a flaw in Ampère's law. Figure 30-1 shows two different surfaces, S_1 and S_2, bounded by the same curve C, which encircles a current carrying wire that is connected to a capacitor plate. The current through surface S_1 is I, but no current exists through surface S_2 because the charge stops on the capacitor plate. Thus, ambiguity exists in the phrase "the current through any surface bounded by the curve." Such a problem arises when the current is not continuous.

Maxwell showed that the law can be generalized to include all situations if the current I in the equation is replaced by the sum of the current I and another term I_d, called **Maxwell's displacement current,** defined as

$$I_d = \epsilon_0 \frac{d\phi_e}{dt} \qquad 30\text{-}3$$

DEFINITION—DISPLACEMENT CURRENT

where ϕ_e is the flux of the electric field through the same surface bounded by the curve C. The generalized form of Ampère's law is then

$$\oint_C \vec{B}\cdot d\vec{\ell} = \mu_0(I + I_d) = \mu_0 I + \mu_0\epsilon_0 \frac{d\phi_e}{dt} \qquad 30\text{-}4$$

GENERALIZED FORM OF AMPÈRE'S LAW

We can understand this generalization by considering Figure 30-1 again. Let us call the sum $I + I_d$* the generalized current. According to the argument just stated, the same generalized current must cross any surface bounded by the curve C. The surfaces S_1 and S_2 together form a single closed surface. Thus, the sum of the generalized currents into the region enclosed by the two surfaces S_1 and S_2, is equal to the sum of the generalized currents out of the region. If a net current I into the enclosed region exists, an equal net displacement current I_d out of the enclosed region must exist. In the enclosed region in the figure, a net current I into the region exists that increases the charge Q_{inside} within the region:

$$I = \frac{dQ_{\text{inside}}}{dt}$$

The net flux of the electric field out of the enclosed region is related to the charge enclosed by Gauss's law:

$$\phi_{\text{e net}} = \oint_S E_n \, dA = \frac{1}{\epsilon_0} Q_{\text{inside}}$$

Solving for the charge gives

$$Q_{\text{inside}} = \epsilon_0 \phi_{\text{e net}}$$

and taking the derivative of each side gives

$$\frac{dQ_{\text{inside}}}{dt} = \epsilon_0 \frac{d\phi_{\text{e net}}}{dt}$$

The rate of increase of the enclosed charge is thus proportional to the rate of increase of the net flux of the electric field out of the region:

$$\frac{dQ_{\text{inside}}}{dt} = \epsilon_0 \frac{d\phi_{\text{e net}}}{dt} = I_d$$

Thus, the net current into the volume equals the net displacement current out of the volume. The generalized current is thus continuous, and this is *always* the case.

It is interesting to compare Equation 30-4 to Equation 28-6:

$$\mathcal{E} = \oint_C \vec{E} \cdot d\vec{\ell} = -\frac{d\phi_m}{dt} = -\int_S \frac{\partial B_n}{\partial t} dA \qquad \text{30-5}$$

FARADAY'S LAW

which in this chapter will be referred to as Faraday's law. (Equation 30-5 is a restricted form of Faraday's law, a form that does include emfs associated with time-varying magnetic fields, but does not include emfs associated with moving conductors.) According to Faraday's law, a changing magnetic flux produces an electric field whose line integral around a closed curve is proportional to the rate of change of magnetic flux through any surface bounded by the curve. Maxwell's modification of Ampère's law shows that a changing electric flux produces a magnetic field whose line integral around a curve is proportional to the rate of change of the electric flux. We thus have the interesting reciprocal result that a changing magnetic field produces an electric field (Faraday's law) and a changing electric field produces a magnetic field (generalized form of Ampère's law). Note, no magnetic analog of a current I exists. This is consistent with the observation that the magnetic monopole, the magnetic analog of an electric charge, does not exist.†

---

* In more advanced treatments, the generalized current is taken as the sum of a conduction current and a displacement current, where the conduction current is attributed to the motion of free (delocalized) charge carriers, and the displacement current is what is referred to in this book as the displacement current and a term associated with the motion of bound (localized) charge carriers.

† The question of the existence of magnetic monopoles has theoretical importance. Numerous attempts to observe magnetic monopoles have been made but to date no one has been unambiguously successful.

Example 30-1 Calculating Displacement Current

A parallel-plate capacitor has closely spaced circular plates of radius R. The current I in the wires connected to the plates is 2.5 A, as shown in Figure 30-2. Compute the displacement current I_d through surface S passing between the plates by directly computing the rate of change of the flux of $\vec{E}$ through surface S.

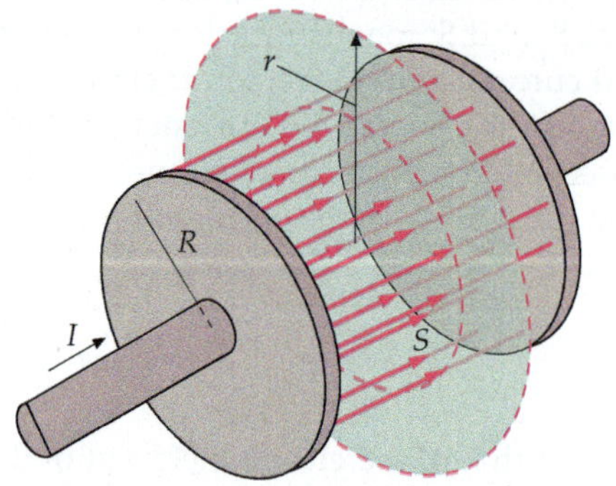

FIGURE 30-2 The surface S between the capacitor plates is penetrated by electric field lines. The charge Q on the positively charged plate is increasing at 2.5 C/s = 2.5 A. The distance between the plates is not drawn to scale. The plates are much closer together than the plates shown in the figure.

PICTURE The displacement current is $I_d = \epsilon_0\, d\phi_e/dt$, where ϕ_e is the electric flux through the surface between the plates. Because the parallel plates are closely spaced, the electric field in the region between the plates is uniform and perpendicular to the plates. Outside the region between the plates the electric field is negligible. Thus, the electric flux is simply $\phi_e = EA$, where E is the electric field between the plates and A is the plate area.

SOLVE

1. The displacement current is found by taking the time derivative of the electric flux: $$I_d = \epsilon_0 \frac{d\phi_e}{dt}$$
2. The flux equals the electric field magnitude multiplied by the plate area: $$\phi_e = EA$$
3. The electric field is proportional to the charge density on the plates, which we treat as uniformly distributed: $$E = \frac{\sigma}{\epsilon_0} = \frac{Q/A}{\epsilon_0}$$
4. Substitute these results to calculate I_d: $$I_d = \epsilon_0 \frac{d(EA)}{dt} = \epsilon_0 A \frac{dE}{dt} = \epsilon_0 A \frac{d}{dt}\left(\frac{Q}{A\epsilon_0}\right)$$ $$= \frac{dQ}{dt} = \boxed{2.5\text{ A}}$$

CHECK The step-4 result is equal to the current in the wires, as expected.

Example 30-2 Calculating $\vec{B}$ from Displacement Current

The circular plates in Example 30-1 have a radius of $R = 3.0$ cm. Find the magnetic field strength B at a point between the plates a distance $r = 2.0$ cm from the axis through the centers of the plates when the current into the positive plate is 2.5 A.

PICTURE We find B from the generalized form of Ampère's law (Equation 30-4). We chose a circular path C of radius $r = 2.0$ cm about the centerline joining the plates, as shown in Figure 30-3. We then calculate the displacement current through the surface S bounded by C. By symmetry, $\vec{B}$ is tangent to C and has the same magnitude everywhere on C.

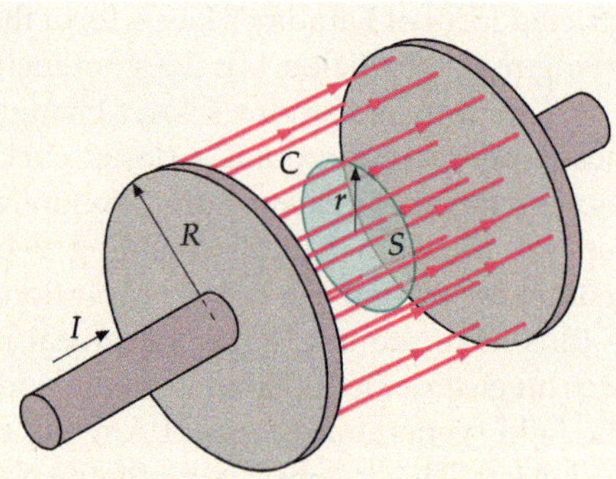

FIGURE 30-3 The distance between the plates is not drawn to scale. The plates are much closer together than they appear.

SOLVE

1. We find B from the generalized form of Ampère's law: $$\oint_C \vec{B}\cdot d\vec{\ell} = \mu_0(I + I_d)$$ where $$I_d = \epsilon_0 \frac{d\phi_e}{dt}$$
2. The line integral is B multiplied by the circumference of the circle: $$\oint_C \vec{B}\cdot d\vec{\ell} = B\cdot 2\pi r$$

3. Because no charges are moving through the surface S, $I = 0$. Thus, the generalized current through S is just the displacement current:

$$\oint_C \vec{B}\cdot d\vec{\ell} = \mu_0 I + \mu_0\epsilon_0\frac{d\phi_e}{dt}$$

$$B\cdot 2\pi r = 0 + \mu_0\epsilon_0\frac{d\phi_e}{dt}$$

4. The electric flux through S equals the product of the uniform field strength E and the area A of the flat surface S bounded by the curve C, and E is equal to σ/ϵ_0:

$$\phi_e = AE = \pi r^2 E = \pi r^2\frac{\sigma}{\epsilon_0}$$

$$= \pi r^2\frac{Q}{\epsilon_0\pi R^2} = \frac{Qr^2}{\epsilon_0 R^2}$$

5. Substitute these results into step 3 and solve for B:

$$B\cdot 2\pi r = \mu_0\epsilon_0\frac{d}{dt}\left(\frac{Qr^2}{\epsilon_0 R^2}\right) = \mu_0\frac{r^2}{R^2}\frac{dQ}{dt}$$

$$B = \frac{\mu_0}{2\pi}\frac{r}{R^2}\frac{dQ}{dt} = \frac{\mu_0}{2\pi}\frac{r}{R^2}I$$

$$= (2\times 10^{-7}\,\text{T}\cdot\text{m/A})\frac{0.02\text{ m}}{(0.03\text{ m})^2}(2.5\text{ A})$$

$$= \boxed{1.11\times 10^{-5}\,\text{T}}$$

30-2 MAXWELL'S EQUATIONS

Maxwell's equations are

$$\oint_S E_n\,dA = \frac{1}{\epsilon_0}Q_{\text{inside}} \qquad \text{30-6}a$$

GAUSS'S LAW

$$\oint_S B_n\,dA = 0 \qquad \text{30-6}b$$

GAUSS'S LAW FOR MAGNETISM

$$\oint_C \vec{E}\cdot d\vec{\ell} = -\frac{d}{dt}\int_S B_n\,dA = -\int_S\frac{\partial B_n}{\partial t}dA \qquad \text{30-6}c$$

FARADAY'S LAW

$$\oint_C \vec{B}\cdot d\vec{\ell} = \mu_0(I + I_d),\text{ where } I_d = \epsilon_0\int_S\frac{\partial E_n}{\partial t}dA \qquad \text{30-6}d$$

AMPÈRE'S LAW

MAXWELL'S EQUATIONS*

Gauss's law (Equation 30-6*a*) states that the flux of the electric field through any closed surface equals $1/\epsilon_0$ multiplied by the net charge inside the surface. As discussed in Chapter 22, Gauss's law implies that the electric field $\vec{E}$ due to a point charge varies inversely as the square of the distance from the charge. This law describes how electric field lines diverge from a positive charge and converge on a negative charge. Its experimental basis is Coulomb's law.

* In all four of Maxwell's equations, the integration paths C and the integration surfaces S are at rest and the integrations take place at an instant in time.

Gauss's law for magnetism (Equation 30-6*b*) states that the flux of the magnetic field $\vec{B}$ through *any* closed surface is zero. This equation describes the experimental observation that magnetic field lines do not diverge from any point in space or converge to any point in space; that is, it implies that isolated magnetic poles do not exist.

Faraday's law (Equation 30-6*c*) states that the line integral of the electric field $\vec{E}$ around any closed curve C equals the negative of the rate of change of the flux of the magnetic field $\vec{B}$ through any surface S bounded by curve C. (S is not a closed surface, so the magnetic flux through S is not necessarily zero.) Faraday's law describes how electric field lines encircle any area through which the magnetic flux is changing, and it relates the electric field vector $\vec{E}$ to the rate of change of the magnetic field vector $\vec{B}$.

Ampère's law modified to include Maxwell's displacement current (Equation 30-6*d*) states that the line integral of the magnetic field $\vec{B}$ around any closed curve C equals μ_0 multiplied by the sum of the current I through any surface S bounded by the curve and the displacement current I_d through the same surface. This law describes how the magnetic field lines encircle an area through which a current or a displacement current is passing.

In Section 30-3, we show how wave equations for both the electric field $\vec{E}$ and the magnetic field $\vec{B}$ can be derived from Maxwell's equations.

30-3 THE WAVE EQUATION FOR ELECTROMAGNETIC WAVES

In Section 15-1, we saw that waves on a string obey a partial differential equation called the **wave equation:**

$$\frac{\partial^2 y(x,t)}{\partial x^2} = \frac{1}{v^2}\frac{\partial^2 y(x,t)}{\partial t^2} \qquad \text{30-7}$$

where $y(x, t)$ is the wave function, which for string waves is the displacement of the string. The velocity of the wave is given by $v = \sqrt{F_T/\mu}$, where F_T is the tension and μ is the linear mass density. The general solution to this equation is

$$y(x,t) = y_1(x - vt) + y_2(x + vt)$$

where y_1 and y_2 are functions of $x - vt$ and $x + vt$, respectively. The general solution functions can be expressed as a superposition of harmonic wave functions of the form

$$y(x,t) = y_0 \sin(kx - \omega t) \qquad \text{and} \qquad y(x,t) = y_0 \sin(kx + \omega t)$$

where $k = 2\pi/\lambda$ is the wave number and $\omega = 2\pi f$ is the angular frequency.

Maxwell's equations imply that $\vec{E}$ and $\vec{B}$ obey wave equations similar to Equation 30-7. We consider only empty space (space in which no charges or currents exist) and we assume that the electric and magnetic fields $\vec{E}$ and $\vec{B}$ are functions of time and one space coordinate only, which we will take to be the x coordinate. Such a wave is called a **plane wave,** because $\vec{E}$ and $\vec{B}$ are uniform throughout any plane perpendicular to the x axis. For a plane electromagnetic wave traveling parallel to the x axis, the x components of the fields are zero, so the vectors $\vec{E}$ and $\vec{B}$ are perpendicular to the x axis and each obeys the wave equation:

$$\frac{\partial^2 \vec{E}}{\partial x^2} = \frac{1}{c^2}\frac{\partial^2 \vec{E}}{\partial t^2} \qquad \text{30-8a}$$

WAVE EQUATION FOR $\vec{E}$

$$\frac{\partial^2 \vec{B}}{\partial x^2} = \frac{1}{c^2}\frac{\partial^2 \vec{B}}{\partial t^2} \qquad \text{30-8}b$$

WAVE EQUATION FOR $\vec{B}$

where $c = 1/\sqrt{\mu_0\epsilon_0}$ is the speed of the waves. (*Note:* Dimensional analysis helps in remembering these equations. For each equation, the numerators on both sides are the same and the denominators on both sides have the dimension of length squared.)

DERIVATION OF THE WAVE EQUATION

We can relate the space derivative of one of the field vectors to the time derivative of the other field vector by applying Faraday's law (Equation 30-6*c*) and the modified version of Ampère's law (Equation 30-6*d*) to appropriately chosen curves in space. We first relate the space derivative of E_y to the time derivative of B_z by applying Equation 30-6*c* (Faraday's law) to the rectangular curve of sides Δx and Δy lying in the xy plane (Figure 30-4). The circulation of $\vec{E}$ around C, for small Δx and Δy, is given by

$$\oint_C \vec{E}\cdot d\vec{\ell} = E_y(x_2)\,\Delta y - E_y(x_1)\,\Delta y = [E_y(x_2) - E_y(x_1)]\,\Delta y$$

where $E_y(x_1)$ is the value of E_y at $x = x_1$ and $E_y(x_2)$ is the value of E_y at $x = x_2$. The contributions of the type $E_x\Delta x$ from the top and bottom of this curve are zero because $E_x = 0$. Because Δx is very small (compared to the wavelength), we can approximate the difference in E_y on the left and right sides of this curve (at x_1 and at x_2) by

$$E_y(x_2) - E_y(x_1) = \Delta E_y \approx \frac{\partial E_y}{\partial x}\Delta x$$

Then

$$\oint_C \vec{E}\cdot d\vec{\ell} \approx \frac{\partial E_y}{\partial x}\Delta x\,\Delta y$$

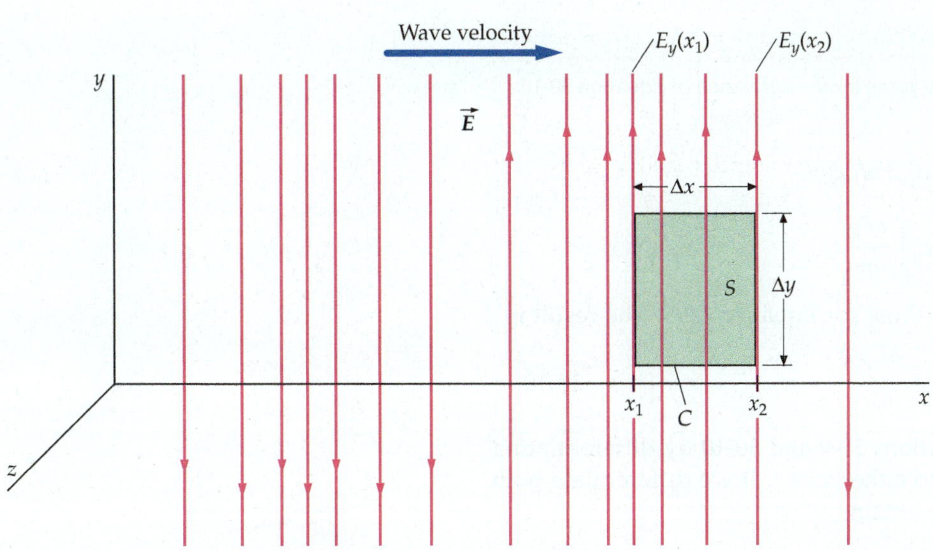

FIGURE 30-4 A rectangular curve in the xy plane for the derivation of Equation 30-9.

Faraday's law is

$$\oint_C \vec{E} \cdot d\vec{\ell} = -\int_S \frac{\partial B_n}{\partial t} dA$$

The time derivative of the flux of $\partial B_n/\partial t$ through the rectangular surface bounded by this curve is given by

$$\int_S B_n \, dA \approx \frac{\partial B_z}{\partial t} \Delta x \, \Delta y$$

Faraday's law then gives

$$\frac{\partial E_y}{\partial x} \Delta x \, \Delta y = -\frac{\partial B_z}{\partial t} \Delta x \, \Delta y$$

or

$$\frac{\partial E_y}{\partial x} = -\frac{\partial B_z}{\partial t} \qquad \text{30-9}$$

Equation 30-9 implies that if there is a component of the electric field E_y that depends on x, there must be a component of the magnetic field B_z that depends on time or, conversely, if there is a component of the magnetic field B_z that depends on time, there must be a component of the electric field E_y that depends on x. We can get a similar equation relating the space derivative of the magnetic field B_z to the time derivative of the electric field E_y by applying Ampère's law (Equation 30-6*d*) to the curve of sides Δx and Δz in the xz plane shown in Figure 30-5.

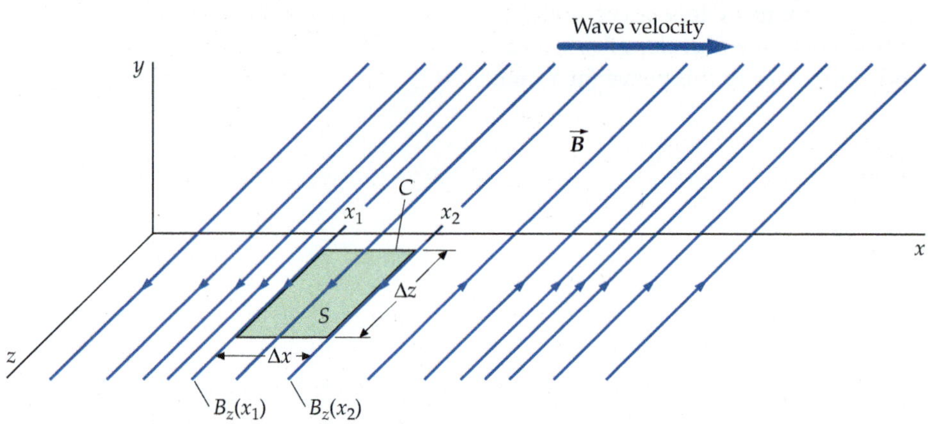

FIGURE 30-5 A rectangular curve in the $y = 0$ plane for the derivation of Equation 30-10.

For the case of no currents ($I = 0$), Equation 30-6*d* is

$$\oint_C \vec{B} \cdot d\vec{\ell} = \mu_0 \epsilon_0 \int_S \frac{\partial E_n}{\partial t} dA$$

The details of this calculation are similar to those for Equation 30-9. The result is

$$\frac{\partial B_z}{\partial x} = -\mu_0 \epsilon_0 \frac{\partial E_y}{\partial t} \qquad \text{30-10}$$

We can eliminate either B_z or E_y from Equations 30-9 and 30-10 by differentiating both sides of either equation with respect to either x or t. If we differentiate both sides of Equation 30-9 with respect to x, we obtain

$$\frac{\partial}{\partial x}\left(\frac{\partial E_y}{\partial x}\right) = -\frac{\partial}{\partial x}\left(\frac{\partial B_z}{\partial t}\right)$$

Interchanging the order of the time and space derivatives on the term to the right of the equal sign gives

$$\frac{\partial^2 E_y}{\partial x^2} = -\frac{\partial}{\partial t}\left(\frac{\partial B_z}{\partial x}\right)$$

Using Equation 30-10, we substitute for $\partial B_z/\partial x$ to obtain

$$\frac{\partial^2 E_y}{\partial x^2} = -\frac{\partial}{\partial t}\left(-\mu_0\epsilon_0\frac{\partial E_y}{\partial t}\right)$$

which yields the wave equation

$$\frac{\partial^2 E_y}{\partial x^2} = \mu_0\epsilon_0\frac{\partial^2 E_y}{\partial t^2} \qquad \text{30-11}$$

Comparing Equation 30-11 with Equation 30-7, we see that E_y obeys a wave equation for waves with speed $v = 1/\sqrt{\mu_0\epsilon_0}$, which is Equation 30-1.

If we had instead chosen to eliminate E_y from Equations 30-9 and 30-10 (by differentiating Equation 30-9 with respect to t, for example), we would have obtained an equation identical to Equation 30-11 except with B_z replacing E_y. We can thus see that both the electric field E_y and the magnetic field B_z obey a wave equation for waves traveling with the velocity $1/\sqrt{\mu_0\epsilon_0}$. By substituting the measured values for μ_0 and ϵ_0, Maxwell showed that the value of $1/\sqrt{\mu_0\epsilon_0}$ is equal to the measured value for speed of light.

By following the same line of reasoning as used above, and applying Equation 30-6*c* (Faraday's law) to the curve in the *xz* plane (Figure 30-5), we would obtain

$$\frac{\partial E_z}{\partial x} = \frac{\partial B_y}{\partial t} \qquad \text{30-12}$$

Similarly, the application of Equation 30-6*d* to the curve in the *xy* plane (Figure 30-4) gives

$$\frac{\partial B_y}{\partial x} = \mu_0\epsilon_0\frac{\partial E_z}{\partial t} \qquad \text{30-13}$$

We can use these results to show that, for a wave propagating in the x direction, the components E_z and B_y also obey the wave equation.

To show that the magnetic field B_z is in phase with the electric field E_y, consider the harmonic wave function of the form

$$E_y = E_0 \sin(kx - \omega t) \qquad \text{30-14}$$

If we substitute this solution into Equation 30-9, we have

$$\frac{\partial B_z}{\partial t} = -\frac{\partial E_y}{\partial x} = -kE_0\cos(kx - \omega t)$$

To solve for B_z, we take the integral of $\partial B_z/\partial t$ with respect to time. Doing so yields

$$B_z = \int \frac{\partial B_z}{\partial t}\,dt = \frac{k}{\omega}E_0 \sin(kx - \omega t) + f(x) \qquad \text{30-15}$$

where $f(x)$ is an arbitrary function of x.

PRACTICE PROBLEM 30-1

Verify Equation 30-15 by showing that $\frac{\partial}{\partial t}\left[\frac{k}{\omega}E_0\sin(kx - \omega t) + f(x)\right]$ is equal to $-kE_0\cos(kx - \omega t)$.

We next substitute the solution (Equation 30-14) into Equation 30-10 and obtain

$$\frac{\partial B_z}{\partial x} = -\mu_0\epsilon_0\frac{\partial E_y}{\partial t} = \omega\mu_0\epsilon_0 E_0\cos(kx - \omega t)$$

Solving for B_z gives

$$B_z = \int \frac{\partial B_z}{\partial x}\,dx = \frac{\omega\mu_0\epsilon_0}{k}E_0\sin(kx - \omega t) + g(t) \qquad \text{30-16}$$

where $g(t)$ is an arbitrary function of time. Equating the right sides of Equations 30-15 and 30-16 gives

$$\frac{k}{\omega}E_0\sin(kx - \omega t) + f(x) = \frac{\omega\mu_0\epsilon_0}{k}E_0\sin(kx - \omega t) + g(t)$$

Substituting c for ω/k and $1/c^2$ for $\mu_0\epsilon_0$ gives

$$\frac{1}{c}E_0\sin(kx - \omega t) + f(x) = \frac{1}{c}E_0\sin(kx - \omega t) + g(t)$$

which implies $f(x) = g(t)$ for all values of x and t. These remain equal only if $f(x) = g(t) =$ constant (independent of both x and t). Thus, Equation 30-15 becomes

$$B_z = \frac{k}{\omega}E_0\sin(kx - \omega t) + \text{constant} = B_0\sin(kx - \omega t) \qquad \text{30-17}$$

where $B_0 = (k/\omega)E_0 = (1/c)E_0$. The integration constant was dropped because it plays no part in the wave. It merely allows for the presence of a static uniform magnetic field. Because the electric and magnetic fields oscillate in phase and have the same frequency, we have the general result that the magnitude of the electric field is c multiplied by the magnitude of the magnetic field for an electromagnetic wave:

$$E = cB \qquad \text{30-18}$$

The direction of propagation of an electromagnetic wave is always the direction of the vector product $\vec{E} \times \vec{B}$. For the wave described in the preceding discussion, the electric and magnetic fields are given by $\vec{E} = E_0\sin(kx - \omega t)\hat{j}$ and $\vec{B} = B_0\sin(kx - \omega t)\hat{k}$. Thus,

$$\vec{E} \times \vec{B} = [E_0\sin(kx - \omega t)\hat{j}] \times [B_0\sin(kx - \omega t)\hat{k}] = E_0B_0\sin^2(kx - \omega t)\hat{i}$$

The term on the right is a vector in the $+x$ direction, so we have verified that $\vec{E} \times \vec{B}$ is in the direction of propagation for this electromagnetic wave.

We see that Maxwell's equations imply wave equations 30-8*a* and 30-8*b* for the electric and magnetic fields; and that if E_y varies harmonically, as in Equation 30-14, the magnetic field B_z is in phase with E_y and has an amplitude related to the amplitude of E_y by $B_z = E_y/c$. The electric and magnetic fields are perpendicular to each other and to the direction of the wave propagation.

> ! The direction of propagation of an electromagnetic wave is always the direction of the vector product $\vec{E} \times \vec{B}$.

Example 30-3 $\vec{B}(y, t)$ for a Linearly Polarized Plane Wave

The expression for the electric field of a certain electromagnetic wave is given by $\vec{E}(y, t) = E_0\sin(ky + \omega t)\hat{k}$. (*a*) What is the direction of propagation of the wave? (*b*) What is the corresponding expression for the magnetic field on the wave?

PICTURE The argument of the sine function gives the direction of propagation. $\vec{B}$ is perpendicular to both $\vec{E}$ and to the direction of propagation. $\vec{B}$ and $\vec{E}$ are in phase and $\vec{E} \times \vec{B}$ is in the direction of propagation.

SOLVE

(*a*) The argument of the sine function $(ky + \omega t)$ tells us the direction of propagation:

The direction of propagation is the $-y$ direction, which is the direction of $-\hat{j}$.

(*b*) 1. $\vec{B}$ is in phase with $\vec{E}$ and is perpendicular to both $\vec{E}$ and the direction of propagation $-\hat{j}$. (That is, $\vec{B}$ is perpendicular to both $\hat{j}$ and $\hat{k}$.) This result means:

Either $\vec{B}(y,t) = +B_0 \sin(ky + \omega t)\hat{i}$ or

$\vec{B}(y,t) = -B_0 \sin(ky + \omega t)\hat{i}$

2. $\vec{E} \times \vec{B}$ is in the direction of propagation, $-\hat{j}$. Assume $\vec{B}(x,t) = +B_0 \sin(ky + \omega t)\hat{i}$ and calculate the product $\vec{E} \times \vec{B}$:

$$\begin{aligned}\vec{E} \times \vec{B} &= E_0 \sin(ky + \omega t)\hat{k} \times B_0 \sin(ky + \omega t)\hat{i}\\ &= E_0B_0 \sin^2(ky + \omega t)(\hat{k} \times \hat{i})\\ &= E_0B_0 \sin^2(ky + \omega t)\hat{j}\end{aligned}$$

3. The step-2 result contradicts the reality that the direction of propagation is the $-y$ direction. Take the cross product $\vec{E} \times \vec{B}$ with the other choice for the expression for the magnetic field:

$$\begin{aligned}\vec{E} \times \vec{B} &= E_0 \sin(ky + \omega t)\hat{k} \times (-B_0) \sin(ky + \omega t)\hat{i}\\ &= E_0(-B_0) \sin^2(ky + \omega t)(\hat{k} \times \hat{i})\\ &= -E_0B_0 \sin^2(ky + \omega t)\,\hat{j}\end{aligned}$$

4. The step-3 result is in the direction of propagation. The correct expression for the magnetic field is:

$$\boxed{\vec{B}(y,t) = -B_0 \sin(ky + \omega t)\hat{i}}$$

where $B_0 = E_0/c$ (Equation 30-18).

CHECK The step-4 result is perpendicular to both $\vec{E}$ and to the direction of propagation, as expected.

Example 30-4 $\vec{B}$ (*x*, *t*) for a Circular Polarized Plane Wave

The expression for the electric field of a certain electromagnetic wave is given by $\vec{E}(x,t) = E_0 \sin(kx - \omega t)\hat{j} + E_0\cos(kx - \omega t)\hat{k}$. (*a*) Find the corresponding magnetic field of the same wave. (*b*) Compute $\vec{E} \cdot \vec{B}$ and $\vec{E} \times \vec{B}$.

PICTURE We can solve this example by using the principle of superposition. The given electric field is the superposition of two fields, the one given in Equation 30-14 and the other given by $E_0 \cos(kx - \omega t)\hat{k}$.

SOLVE

(*a*) 1. From the arguments of the trigonometric functions we can see that the direction of propagation is the $+x$ direction:

The wave is traveling in the $+x$ direction.

2. The given electric field can be considered as the superposition of $\vec{E}_1 = E_0 \sin(kx - \omega t)\hat{j}$ and $\vec{E}_2 = E_0 \cos(kx - \omega t)\hat{k}$. Find the magnetic fields $\vec{B}_1$ and $\vec{B}_2$ associated with these electric fields, respectively. Use the procedure followed in Example 30-3:

For $\vec{E}_1 = E_0 \sin(kx - \omega t)\hat{j}$, $\vec{B}_1 = B_0 \sin(kx - \omega t)\hat{k}$ where $B_0 = E_0/c$ (Equation 30-18),

and

For $\vec{E}_2 = E_0 \cos(kx - \omega t)\hat{k}$, $\vec{B}_2 = -B_0 \cos(kx - \omega t)\hat{j}$ where $B_0 = E_0/c$.

3. The superposition of magnetic fields gives the resultant magnetic field:

$$\boxed{\begin{aligned}\vec{B}(x,t) &= \vec{B}_1 + \vec{B}_2\\ &= B_0 \sin(kx - \omega t)\hat{k} - B_0 \cos(kx - \omega t)\hat{j}\\ &\text{where}\\ &B_0 = E_0/c\end{aligned}}$$

(*b*) 1. Let $\theta = kx - \omega t$ to simplify the notation and calculate $\vec{E} \cdot \vec{B}$:

$$\begin{aligned}\vec{E} \cdot \vec{B} &= (E_0 \sin\theta\,\hat{j} + E_0 \cos\theta\,\hat{k}) \cdot (B_0 \sin\theta\,\hat{k} - B_0 \cos\theta\,\hat{j})\\ &= E_0B_0 \sin^2\theta\,\hat{j} \cdot \hat{k} - E_0B_0 \sin\theta \cos\theta\,\hat{j} \cdot \hat{j}\\ &\quad + E_0B_0 \cos\theta \sin\theta\,\hat{k} \cdot \hat{k} - E_0B_0 \cos^2\theta\,\hat{k} \cdot \hat{j}\\ &= 0 - E_0B_0 \sin\theta \cos\theta + E_0B_0 \cos\theta \sin\theta - 0 = \boxed{0}\end{aligned}$$

2. Calculate $\vec{E} \times \vec{B}$:

$$\begin{aligned}\vec{E} \times \vec{B} &= (E_0 \sin\theta\, \hat{j} + E_0 \cos\theta\, \hat{k}) \times (-B_0 \cos\theta\, \hat{j} + B_0 \sin\theta\, \hat{k}) \\ &= -E_0B_0 \sin\theta \cos\theta\, (\hat{j} \times \hat{j}) + E_0B_0 \sin^2\theta\, (\hat{j} \times \hat{k}) \\ &\quad - E_0B_0 \cos^2\theta\, (\hat{k} \times \hat{j}) + E_0B_0 \cos\theta \sin\theta\, (\hat{k} \times \hat{k}) \\ &= 0 + E_0B_0 \sin^2\theta\, \hat{i} + E_0B_0 \cos^2\theta\, \hat{i} + 0 = \boxed{E_0B_0\hat{i}}\end{aligned}$$

CHECK The Part (*b*), step 1 result verifies that $\vec{E}$ and $\vec{B}$ are perpendicular to one another and the the Part (*b*), step 2 result verifies that the $+x$ direction is the direction of propagation.

TAKING IT FURTHER This type of electromagnetic wave is said to be *circularly polarized*. At a fixed value of x, both $\vec{E}$ and $\vec{B}$ rotate in a circle with angular frequency ω.

PRACTICE PROBLEM 30-2 Calculate $\vec{E} \cdot \vec{E}$ and $\vec{B} \cdot \vec{B}$. Note that the fields $\vec{E}$ and $\vec{B}$ are constant in magnitude.

30-4 ELECTROMAGNETIC RADIATION

Figure 30-6 shows the electric and magnetic field vectors of an electromagnetic wave. The electric and magnetic fields are perpendicular to each other and perpendicular to the direction of propagation of the wave. Electromagnetic waves are thus transverse waves. The electric and magnetic fields are in phase and, at each point in space and at each instant in time, their magnitudes are related by

$$E = cB \qquad 30\text{-}18$$

where $c = 1/\sqrt{\mu_0\epsilon_0}$ is the speed of the wave. The direction of propagation of an electromagnetic wave is the direction of the cross product $\vec{E} \times \vec{B}$.

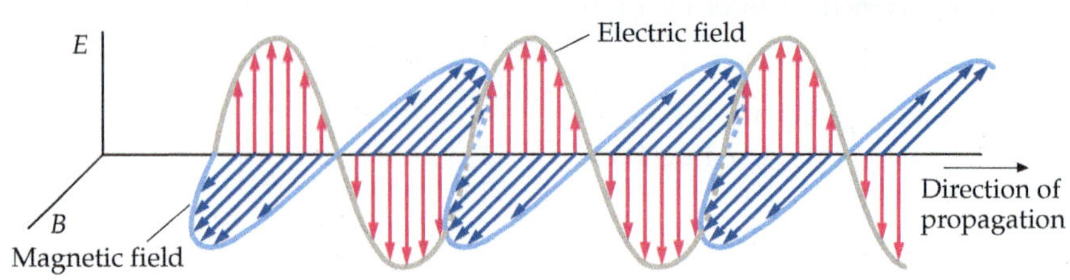

FIGURE 30-6 The electric and magnetic field vectors in an electromagnetic wave. The fields are in phase, perpendicular to each other, and perpendicular to the direction of propagation of the wave.

THE ELECTROMAGNETIC SPECTRUM

The various types of electromagnetic waves (for example, radio waves and gamma rays) differ only in wavelength and frequency, which are related according to the equation $f\lambda = c$. Table 30-1 lists the **electromagnetic spectrum** and the names usually associated with the various frequency and wavelength ranges. These ranges are often not well defined and sometimes overlap. For example, electromagnetic waves that have wavelengths of approximately 0.1 nm are usually called X rays, but if the electromagnetic waves originate from nuclear radioactivity, they are called gamma rays.

The human eye is sensitive to electromagnetic radiation that has wavelengths between 400 and 780 nm,* which is the range called **visible light.** The shortest wavelengths of visible light are those of violet light and the longest wavelengths are those of red light. Electromagnetic waves that have wavelengths shorter than 400 nm but longer than 10 nm (the longest wavelength in the X-rays region) are called **ultraviolet rays. Infrared waves** have wavelengths longer than 780 nm but

* Light whose wavelength is between 700 and 780 nm can only be seen under special circumstances that include the light intensity being very high.

Table 30-1 The Electromagnetic Spectrum

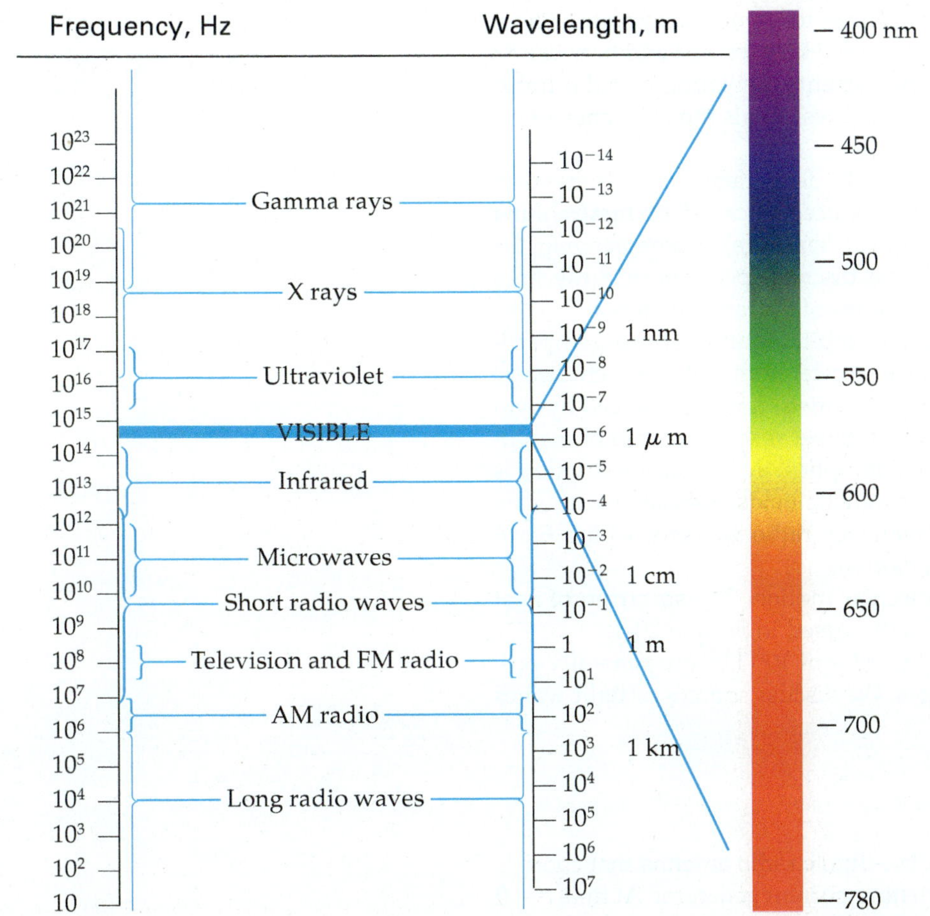

shorter than 100 μm. Heat radiation emitted by objects at temperatures in the range of room temperature is in the infrared region of the electromagnetic spectrum. There are no limits on the wavelengths of electromagnetic radiation; that is, all wavelengths (or frequencies) are theoretically possible.

The differences in wavelengths of the various kinds of electromagnetic waves have important physical consequences. As you know, the behavior of waves depends strongly on the relative sizes of the wavelengths and the objects or apertures (openings) the waves encounter. Because the wavelengths of visible light are in the rather narrow range between 400 and 780 nm, they are much smaller than most obstacles. Thus, the ray approximation (introduced in Section 15-4) is often valid. The wavelength and frequency are also important in determining the kinds of interactions between electromagnetic waves and matter. X rays, for example, have very short wavelengths and high frequencies. They easily penetrate many materials that are opaque to lower-frequency light waves, which are absorbed by the materials. Microwaves have wavelengths between 1 mm and 30 cm. Wavelengths in that range are used to heat food in microwave ovens. The main mechanism of this heating is that molecules which have large dipole moments align themselves in the electric field of the radiation. This electric field flips its direction at twice the frequency of the radiation, so the polar molecules must rotate rapidly to keep up with the alternating electric field. These rapidly rotating molecules bump into surrounding molecules—causing them to heat up. Bluetooth and other wireless local-area-network protocols use wavelengths in the microwave region.

PRODUCTION OF ELECTROMAGNETIC WAVES

Electromagnetic waves are produced when free charges accelerate or when electrons bound to atoms and molecules make transitions to lower energy states. Radio waves, which have frequencies between 550 and 1600 kHz for AM and between 88 and 108 MHz for FM, are produced by electric currents oscillating in radio transmission antennas. The frequency of the emitted waves equals the frequency of oscillation of the charges.

A continuous spectrum of X rays is produced by the deceleration of electrons when they crash into a metal target. The radiation produced is called **bremsstrahlung** (which means "braking radiation" in the German language). Accompanying the broad, continuous bremsstrahlung spectrum is a discrete spectrum of X-ray lines produced by transitions of core electrons in the atoms of the target material.

Synchrotron radiation arises from the circular orbital motion of charged particles (usually electrons or positrons) in nuclear accelerators called synchrotrons. Originally considered a nuisance by accelerator scientists, X rays produced by synchrotrons are now used as a medical diagnostic tool because of the ease of manipulating the beams with reflection and diffraction optics. Synchrotron radiation is also emitted by charged particles trapped in magnetic fields associated with stars and galaxies. It is believed that most low-frequency radio waves reaching Earth from outer space originate as synchrotron radiation.

Heat is radiated by thermally excited molecular motion. The spectrum of heat radiation is the blackbody radiation spectrum discussed in Section 20-4.

Light waves, which have frequencies of the order of 10^{14} Hz, are generally produced by transitions of bound atomic charges. We discuss sources of light waves in Chapter 31.

ELECTRIC DIPOLE RADIATION

Figure 30-7 is a schematic drawing of an electric-dipole radio antenna that consists of two conducting rods connected to an alternating-current generator. At time $t = 0$ (Figure 30-7*a*), the ends of the rods are charged, and an electric field parallel to the rod exists near the rod. A magnetic field also exists, which is not shown, encircling the rods due to the current in the rods. The fluctuations in these fields move out away from the rods with the speed of light. After one-fourth period, at $t = T/4$ (Figure 30-7*b*), the rods are uncharged, and the electric field near the rod is zero. At $t = T/2$ (Figure 30-7*c*), the rods are again charged, but the charges are opposite those at $t = 0$. The electric and magnetic fields at a great distance from the antenna are quite different from the fields near the antenna. Far from the antenna, the electric and magnetic fields oscillate in phase with simple harmonic motion, perpendicular to each other and to the direction of propagation of the wave. Figure 30-8 shows the electric and magnetic fields far from an electric dipole antenna.

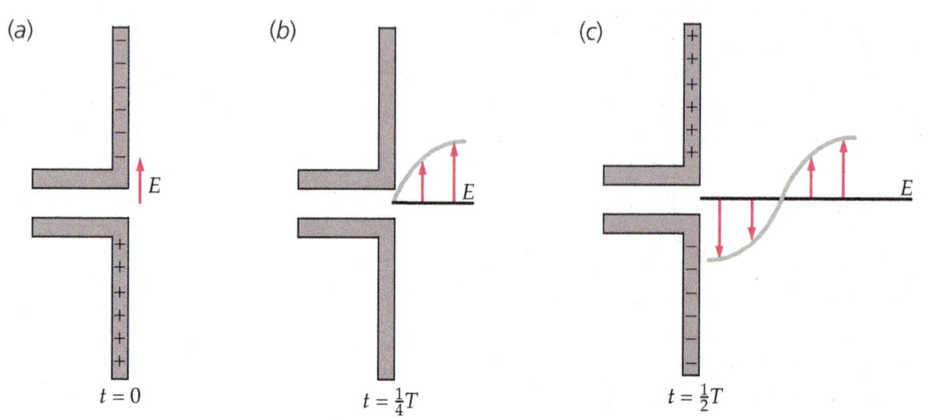

FIGURE 30-7 An electric dipole radio antenna for radiating electromagnetic waves. Alternating current is supplied to the antenna by a generator (not shown). The fluctuations in the electric field due to the fluctuations in the charges in the antenna propagate outward at the speed of light. There is also a fluctuating magnetic field (not shown) perpendicular to the paper due to the current in the antenna.

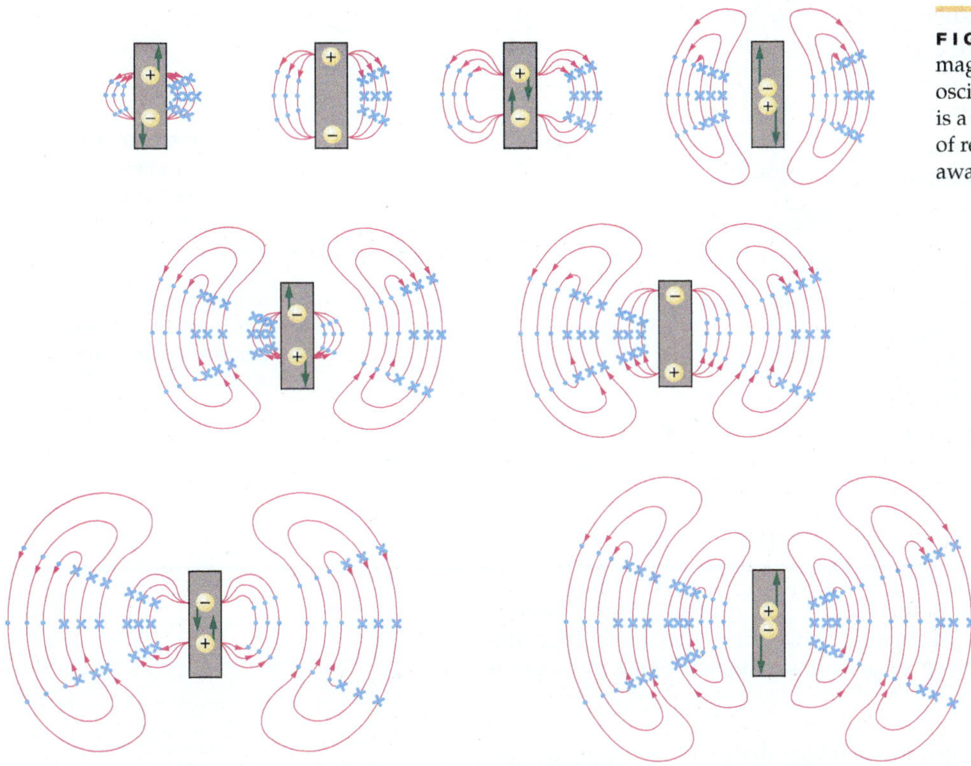

FIGURE 30-8 Electric field lines (in red) and magnetic field lines (in blue) produced by an oscillating electric dipole. Each magnetic field line is a circle with the long axis of the dipole as its axis of revolution. The cross product $\vec{E} \times \vec{B}$ is directed away from the dipole at all points.

Electromagnetic waves of radio or television frequencies can be detected by an electric dipole antenna placed parallel to the electric field of the incoming wave, so that it induces an alternating current in the antenna (Figure 30-9). These electromagnetic waves can also be detected by a loop antenna placed perpendicular to the magnetic field, so that the changing magnetic flux through the loop induces a current in the loop (Figure 30-10). Electromagnetic waves of frequencies in the visible light range are detected by the eye or by photographic film, both of which are mainly sensitive to the electric field.

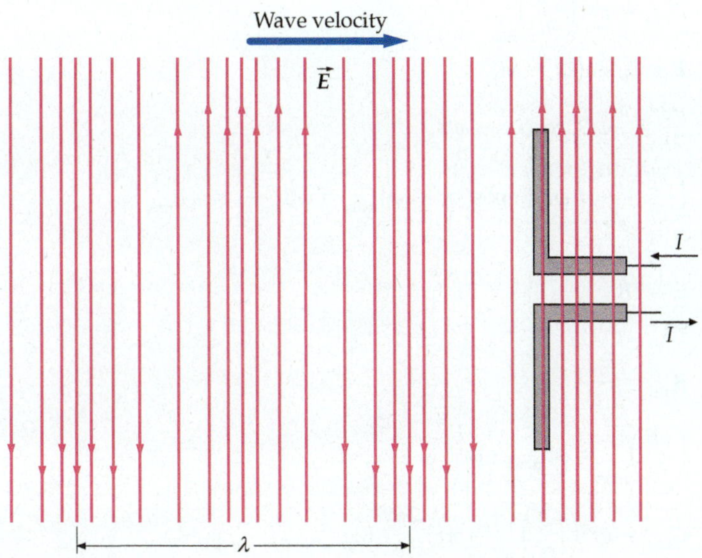

FIGURE 30-9 An electric dipole antenna for detecting electromagnetic waves. The alternating electric field of the incoming wave produces an alternating current in the antenna. The magnetic field lines (not shown) are perpendicular to the plane of the page.

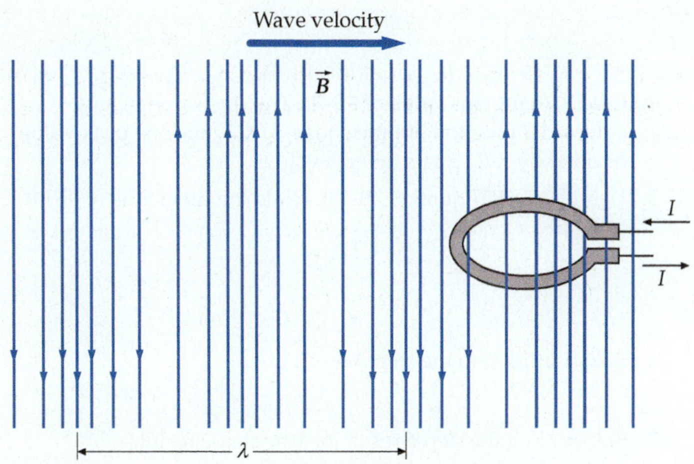

FIGURE 30-10 Loop antenna for detecting electromagnetic radiation. The alternating magnetic flux through the loop due to the magnetic field of the radiation induces an alternating current in the loop. The electric field lines (not shown) are perpendicular to the plane of the page.

The radiation from a dipole antenna, such as that shown in Figure 30-7, is called electric dipole radiation. Many electromagnetic waves exhibit the characteristics of electric dipole radiation. An important feature of this type of radiation is that the intensity of the electromagnetic waves radiated by a dipole antenna is zero along the axis of the antenna and maximum in the radial direction (away from the axis). If the dipole is in the y direction with its center at the origin, as in Figure 30-11, the intensity is zero along the y axis and maximum in the xz plane. In the direction of a line making an angle θ with the y axis, the intensity is proportional to $\sin^2\theta$.

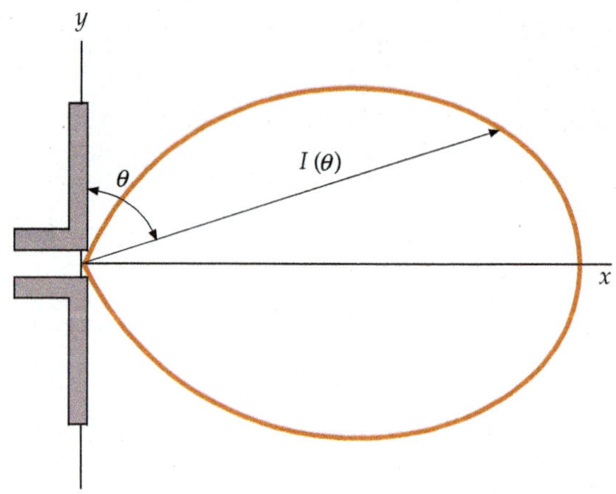

FIGURE 30-11 Polar plot of the intensity of electromagnetic radiation from an electric dipole antenna versus angle. The intensity $I(\theta)$ is proportional to the length of the arrow. The intensity is maximum perpendicular to the antenna (at $\theta = 90°$) and minimum along the antenna at $\theta = 0$ or $\theta = 180°$.

Example 30-5 EMF Induced in a Loop Antenna

A loop antenna consisting of a single 10.0-cm radius loop of wire is used to detect electromagnetic waves for which $E_{rms} = 0.150$ V/m. If the plane of the loop is perpendicular to the magnetic field, find the rms emf induced in the loop when the frequency of the plane wave is (*a*) 600 kHz and (*b*) 60.0 MHz.

PICTURE The induced emf in the wire is related to the rate of change of the magnetic flux through the loop by Faraday's law (Equation 30-5). Using Equation 30-18, we can obtain the rms value of the magnetic field from the given rms value of the electric field.

SOLVE

(*a*) 1. Faraday's law relates the emf to the rate of change of the magnetic flux through the flat stationary surface bounded by the loop:

$$\mathcal{E} = -\frac{d\phi_m}{dt}$$

2. The wavelength of a 600-kHz wave traveling at speed c is $\lambda = c/f = 500$ m. Over the flat surface bounded by the 10-cm radius loop $\vec{B}$ is quite uniform.

$$\phi_m = BA = \pi r^2 B \quad \text{so} \quad \mathcal{E} = -\frac{d\phi_m}{dt} = -\pi r^2 \frac{\partial B}{\partial t}$$

and

$$\mathcal{E}_{rms} = \pi r^2 \left(\frac{\partial B}{\partial t}\right)_{rms}$$

3. Compute $\partial B_{rms}/\partial t$ from a sinusoidal B:

$$B = B_0 \sin(kx - \omega t)$$

$$\frac{\partial B}{\partial t} = -\omega B_0 \cos(kx - \omega t)$$

4. Calculate the rms value of $\partial B/\partial t$. The rms value of any sinusoidal function of time equals $1/\sqrt{2}$, and the peak value divided by $\sqrt{2}$ equals the rms value:

$$\left(\frac{\partial B}{\partial t}\right)_{rms} = \omega B_0[-\cos(kx - \omega t)]_{rms} = \omega B_0 \frac{1}{\sqrt{2}} = \omega B_{rms}$$

5. Using $E = cB$ (Equation 30-18), relate the rms value of $\partial B/\partial t$ to E_{rms}:

$$E = cB$$

so

$$B_{rms} = \frac{E_{rms}}{c}$$

6. Substituting into the step-3 result gives:

$$\left(\frac{\partial B}{\partial t}\right)_{rms} = \omega B_{rms} = \omega \frac{E_{rms}}{c} = \frac{2\pi f}{c} E_{rms}$$

7. Substituting the step-6 result into the step-2 result, calculate $\mathcal{E}_{rms}$ at $f = 600$ kHz:

$$\mathcal{E}_{rms} = \pi r^2 \left(\frac{\partial B}{\partial t}\right)_{rms} = \pi r^2 \frac{2\pi f}{c} E_{rms}$$

$$= \pi(0.100\text{ m})^2 \frac{2\pi(6.00 \times 10^5\text{ Hz})}{3.00 \times 10^8\text{ m/s}}(0.150\text{ V/m})$$

$$= \boxed{5.92 \times 10^{-5}\text{ V} = 59.2\ \mu\text{V}}$$

(*b*) The induced emf is proportional to the frequency [Part (*a*), step 4], so at 60 MHz it will be 100 times greater than at 600 kHz:

$\mathcal{E}_{\text{rms}} = (100)(5.92 \times 10^{-5}\ \text{V}) = 0.00592\ \text{V}$

$= \boxed{5.92\ \text{mV}}$

CHECK Step 7 of Part (*a*) shows that $\mathcal{E}_{\text{rms}}$ increases with increases in frequency, E_{rms}, and area. These results are all expected.

TAKING IT FURTHER For Part (*b*) the frequency is 60.0 MHz, so $\lambda = c/f = 5.00$ m. $\vec{B}$ is not as uniform over the surface bounded by the 10-cm radius loop when $\lambda = 5.00$ m as it is when $\lambda = 500$ m, as in Part (*a*). However, $\vec{B}$ on the surface when $\lambda = 5.00$ m is sufficiently uniform that the Part-(*b*) result is sufficiently accurate for most purposes.

ENERGY AND MOMENTUM IN AN ELECTROMAGNETIC WAVE

Like other waves, electromagnetic waves carry energy and momentum. The energy carried is described by the intensity, which is the average power per unit area incident on a surface perpendicular to the direction of propagation. The momentum per unit time per unit area carried by an electromagnetic wave is called the **radiation pressure.**

Intensity Consider an electromagnetic wave traveling toward the right and a cylindrical region that has a length L, a cross-sectional area A, and its central axis from left to right. The average amount of electromagnetic energy U_{av} within this region equals $u_{\text{av}}\mathcal{V}$ where u_{av} is the average energy density and $\mathcal{V} = LA$ is the volume of the cylindrical region. In the time it takes the electromagnetic wave to travel the distance L, all of the electromagnetic energy equal to $u_{\text{av}}LA$ passes through the right end of the region. The time Δt for the wave to travel the distance L is L/c, so the power P_{av} (the energy per unit time) passing out the right end of the region is

$$P_{\text{av}} = U_{\text{av}}/\Delta t = u_{\text{av}}LA/(L/c) = u_{\text{av}}Ac$$

and the intensity I (the average power per unit area) is

$$I = P_{\text{av}}/A = u_{\text{av}}c$$

The total energy density in the wave u is the sum of the electric and magnetic energy densities. The electric energy density u_{e} (Equation 24-9) and magnetic energy density u_{m} (Equation 28-22) are given by

$$u_{\text{e}} = \frac{1}{2}\epsilon_0 E^2 \quad \text{and} \quad u_{\text{m}} = \frac{B^2}{2\mu_0}$$

At each point in a region where there is an electromagnetic wave in free space, E equals cB, so we can express the magnetic energy density in terms of the electric field:

$$u_{\text{m}} = \frac{B^2}{2\mu_0} = \frac{(E/c)^2}{2\mu_0} = \frac{E^2}{2\mu_0 c^2} = \frac{1}{2}\epsilon_0 E^2$$

where we have used $\mu_0\epsilon_0 = 1/c^2$. Thus, the electric and magnetic energy densities are equal. Using $E = cB$, we may express the total energy density in several useful ways:

$$u = u_{\text{e}} + u_{\text{m}} = \epsilon_0 E^2 = \frac{B^2}{\mu_0} = \frac{EB}{\mu_0 c} \qquad \text{30-19}$$

ENERGY DENSITY IN AN ELECTROMAGNETIC WAVE

To compute the average energy density, we replace the instantaneous fields E and B by their rms values $E_{rms} = \frac{1}{\sqrt{2}}E_0$ and $B_{rms} = \frac{1}{\sqrt{2}}B_0$, where E_0 and B_0 are the maximum values of the fields. The intensity is then

$$I = u_{av}c = \frac{E_{rms}B_{rms}}{\mu_0} = \frac{1}{2}\frac{E_0B_0}{\mu_0} = |\vec{S}|_{av} \qquad 30\text{-}20$$

INTENSITY OF AN ELECTROMAGNETIC WAVE

where the vector

$$\vec{S} = \frac{\vec{E} \times \vec{B}}{\mu_0} \qquad 30\text{-}21$$

DEFINITION—POYNTING VECTOR

is called the **Poynting vector** after its discoverer, John Poynting. The average magnitude of $\vec{S}$ is the intensity of the wave, and the direction of $\vec{S}$ is the direction of propagation of the wave.

Radiation pressure We now show by a simple example that an electromagnetic wave carries momentum. Consider a plane wave traveling in the $+x$ direction that is incident on a stationary charge, as shown in Figure 30-12. Let $\vec{E}$ be in the $+y$ direction and $\vec{B}$ in the $+z$ direction, and neglect the time dependence of the fields. The particle experiences a force $q\vec{E}$ in the $+y$ direction and is thus accelerated by the electric field. At any time t, the velocity in the $+y$ direction is

$$v_y = a_y t = \frac{qE}{m}t$$

After a short time t_1, the charge has acquired kinetic energy equal to

$$K = \frac{1}{2}mv_y^2 = \frac{1}{2}\frac{mq^2E^2t_1^2}{m^2} = \frac{1}{2}\frac{q^2E^2}{m}t_1^2 \qquad 30\text{-}22$$

When the charge is moving in the y direction, it experiences a magnetic force

$$\vec{F}_m = q\vec{v} \times \vec{B} = qv_y\,\hat{j} \times B\,\hat{k} = qv_yB\,\hat{i} = \frac{q^2EB}{m}t\,\hat{i}$$

Note that this force is in the direction of propagation of the wave. Using $dp_x = F_x\,dt$, we find for the momentum p_x transferred by the wave to the particle in time t_1:

$$p_x = \int_0^{t_1} F_x\,dt = \int_0^{t_1}\frac{q^2EB}{m}t\,dt = \frac{1}{2}\frac{q^2EB}{m}t_1^2$$

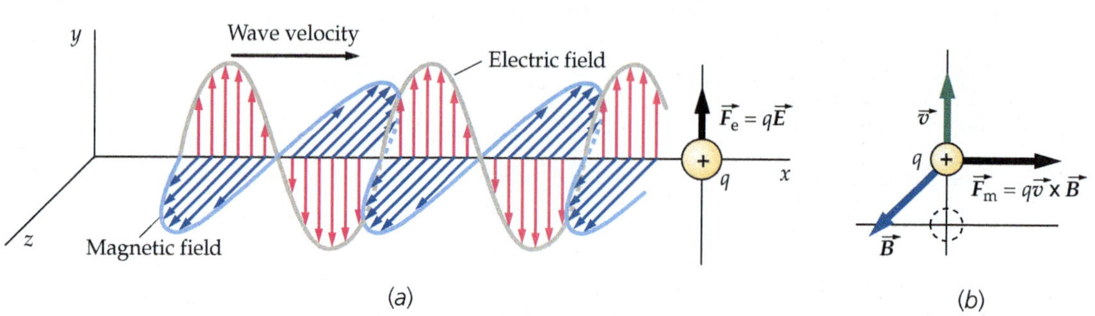

FIGURE 30-12 An electromagnetic wave incident on a point charge that is initially at rest on the x axis. (a) The electric force $q\vec{E}$ accelerates the charge in the $+y$ direction. (b) When the velocity $\vec{v}$ of the charge is in the $+y$ direction, the magnetic force $q\vec{v} \times \vec{B}$ accelerates the charge in the direction of propagation (the $+x$ direction) of the wave.

If we use $B = E/c$, this becomes

$$p_x = \frac{1}{c}\left(\frac{1}{2}\frac{q^2E^2}{m}t_1^2\right) \qquad 30\text{-}23$$

Comparing Equations 30-22 and 30-23, we see that the momentum acquired by the charge in the direction of the wave is $1/c$ multiplied by the energy. Although our simple calculation was not rigorous, the results are correct. The magnitude of the momentum carried by an electromagnetic wave is $1/c$ multiplied by the energy carried by the wave:

$$p = \frac{U}{c} \qquad 30\text{-}24$$

MOMENTUM AND ENERGY IN AN ELECTROMAGNETIC WAVE

Because the intensity is the energy per unit area per unit time, the intensity divided by c is the momentum carried by the wave per unit area per unit time. The momentum carried per unit time is a force. The intensity divided by c is thus a force per unit area, which is a pressure. This pressure is the radiation pressure P_r:

$$P_r = \frac{I}{c} \qquad 30\text{-}25$$

RADIATION PRESSURE AND INTENSITY

We can relate the radiation pressure to the electric or magnetic fields by using Equation 30-20 to relate I to E and B, and Equation 30-18 to eliminate either E or B:

$$P_r = \frac{I}{c} = \frac{E_0B_0}{2\mu_0c} = \frac{E_{rms}B_{rms}}{\mu_0c} = \frac{E_0^2}{2\mu_0c^2} = \frac{B_0^2}{2\mu_0} \qquad 30\text{-}26$$

RADIATION PRESSURE IN TERMS OF E AND B

Consider an electromagnetic wave incident normally on some surface. If the surface absorbs energy U from the electromagnetic wave, it also absorbs momentum p given by Equation 30-24, and the pressure exerted on the surface equals the radiation pressure. If the wave is reflected, the momentum transferred is $2p$ because the wave now carries momentum in the opposite direction. The pressure exerted on the surface by the wave is then twice that given by Equation 30-26.

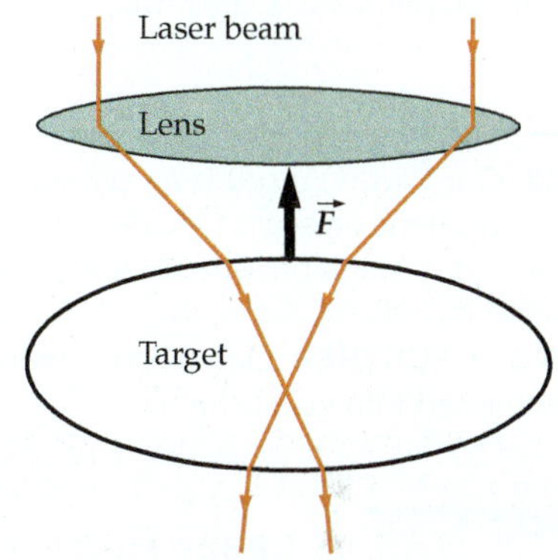

"Laser tweezers" make use of the momentum carried by electromagnetic waves to manipulate targets on a molecular scale. The two rays shown are refracted as they pass through a transparent target, such as a biological cell or, on an even smaller scale, a tiny transparent bead attached to a large molecule within a cell. At each refraction, the rays are bent downward, which increases the downward component of momentum of the rays. The target thus exerts a downward force on the laser beams, and the laser beams exert an upward force on the target, which pulls the target toward the laser source. The force is typically of the order of piconewtons. Laser tweezers have been used to accomplish such astonishing feats as stretching out coiled molecules of DNA.

Example 30-6 Radiation Pressure 3.0 m from a Lightbulb

A lightbulb emits spherical electromagnetic waves uniformly in all directions. Find (*a*) the intensity, (*b*) the radiation pressure, and (*c*) the electric and magnetic field magnitudes at a distance of 3.0 m from the lightbulb, assuming that 50 W of electromagnetic radiation is emitted.

PICTURE At a distance r from the lightbulb, the energy is spread uniformly over the surface of a sphere of radius r—an area equal to $4\pi r^2$. The intensity is the power divided by the area. The radiation pressure can then be found from $P_r = I/c$.

SOLVE

(*a*) 1. Divide the power output by the area to find the intensity: $I = \dfrac{50\text{ W}}{4\pi r^2}$

2. Substitute $r = 3.0$ m: $I = \dfrac{50\text{ W}}{4\pi(3.0\text{ m})^2} = \boxed{0.44\text{ W/m}^2}$

(*b*) The radiation pressure is the intensity divided by the speed of light:

$$P_r = \frac{I}{c} = \frac{0.44\ \text{W/m}^2}{3.00 \times 10^8\ \text{m/s}} = \boxed{1.5 \times 10^{-9}\ \text{Pa}}$$

(*c*) 1. B_0 is related to P_r by Equation 30-26:

$$B_0 = \sqrt{2\mu_0 P_r}$$
$$= [2(4\pi \times 10^{-7}\ \text{T}\cdot\text{m/A})(1.5 \times 10^{-9}\ \text{Pa})]^{1/2}$$
$$= 6.1 \times 10^{-8}\ \text{T}$$

2. The maximum value of the electric field E_0 is c multiplied by B_0:

$$E_0 = cB_0 = (3.00 \times 10^8\ \text{m/s})(6.1 \times 10^{-8}\ \text{T})$$
$$= 18\ \text{V/m}$$

3. The electric and magnetic field magnitudes at that point are of the form:

$$\boxed{E = E_0 \sin\omega t \quad \text{and} \quad B = B_0 \sin\omega t \\ \text{with } E_0 = 18\ \text{V/m} \\ \text{and } B_0 = 6.1 \times 10^{-8}\ \text{T}}$$

CHECK Our Part-(*b*) result is a very small pressure. (It is fourteen orders of magnitude less than atmospheric pressure.) We do not perceive any pressure on us by the light from a lightbulb, so a very small pressure is as expected.

TAKING IT FURTHER Only about 2 percent of the power consumed by incandescent bulbs is transformed into visible light.

Example 30-7 A Laser Rocket

Context-Rich

You are stranded in space a distance of 20 m from your spaceship. You carry a 1.0-kW laser. If your total mass, including your space suit and laser, is 95 kg, how long will it take you to reach the spaceship if you point the laser beam directly away from the ship?

PICTURE The laser emits light, which carries with it momentum. By momentum conservation, you are given an equal and opposite momentum toward the spaceship. The momentum carried by light is $p = U/c$, where U is the energy of the light. If the power of the laser is $P = dU/dt$, then the rate of change of momentum produced by the laser is $dp/dt = (dU/dt)/c = P/c$. This force is the force exerted on you, which is constant.

SOLVE

1. The time taken is related to the distance and the acceleration. We assume that you are initially at rest relative to the spaceship:

$$x = \frac{1}{2}at^2 \quad t = \sqrt{\frac{2x}{a}}$$

2. Your acceleration is the force divided by your mass, and the force is the power divided by c:

$$a = \frac{F}{m} = \frac{P/c}{m} = \frac{P}{mc}$$

3. Use $x = \frac{1}{2}at^2$ to calculate the time t:

$$t = \sqrt{\frac{2x}{a}} = \sqrt{\frac{2xmc}{P}}$$
$$= \sqrt{\frac{2(20\ \text{m})(95\ \text{kg})(3.00 \times 10^8\ \text{m/s})}{1000\ \text{W}}}$$
$$= 3.38 \times 10^4\ \text{s} = \boxed{9.4\ \text{h}}$$

CHECK You expect the time to be long because you know from experience that the pressure from a lightbulb is very small. The step-3 result is as expected.

TAKING IT FURTHER Note that the acceleration is extremely small—only about $10^{-9}\,g$. Your speed when you reach the spaceship would be $v = at = 1.2$ mm/s, which is practically imperceptible.

PRACTICE PROBLEM 30-3 How long would it take you to reach the spaceship if you took off one of your shoelaces and threw it as fast as you could in the direction opposite the ship? (To answer this, you must first estimate the mass of the shoelace and the maximum speed that you can throw the shoelace.) Compare this time with the step-3 result.

Physics Spotlight

Wireless: Sharing the Spectrum

Marconi's transmitter. This transmitter was used in 1901 during the first transatlantic radio broadcast. *(Cornwall, UK, to Newfoundland, USA.)*

One day in March 1998,[*] the remote heart monitors at Baylor University Medical Center and Dallas Methodist Hospital stopped working. WFFA, a Dallas television station, was testing its new digital broadcast system on its licensed frequency. The heart monitors, longtime unlicensed low-power users of the same frequency, were overwhelmed by the test. No patients were harmed, and the station stopped further testing until the hospitals could replace their monitors with ones that used different frequencies.[†] In 2000, the Wireless Medical Telemetry Service established a set of licensed frequencies for medical monitoring devices.[‡]

When Guglielmo Marconi transmitted signals on his wireless telegraph in 1896, he used a spark-gap transmitter.[#] The sparks produced electromagnetic radiation at frequencies over a range of 5 or more orders of magnitude (from a few kilohertz up to 2 GHz). When more than one wireless telegraph was transmitting in the same area, they had to take turns and follow rules. One reckless operator could wreck communications for an entire area.[°]

In 1903, the International Telegraph Union began to study the problems of radiotelegraphy. In 1906, the first Radio Telegraph Convention, signed in Berlin, assigned the frequency of 500 kHz to maritime distress signals.[§] Ships were directed to use less than 1 kW of power, unless they were over 300 km from the nearest shore station.[¶] (These communications were still broadband, but the transmitted power spectrum peaked at 500 kHz.) The first practical alternative to spark-gap transmitters was a continuous wave circuit invented by Edwin Armstrong in 1912.[**] Also in 1912, the International Radiotelegraph Convention issued the first table of frequency allocations,[††] but spark-gap transmitters were still plentiful and could overwhelm local and regional communications.[‡‡]

Radio transmissions became concentrated around narrowing frequency bands. In 1927, national bodies were established to coordinate use of the electromagnetic spectrum.[##] By 1934, the international body was renamed the International Telecommunication Union.[°°] The Federal Communications Commission regulates the radio frequency portions of the spectrum in the United States.[§§] Since then, the ITU and the FCC have cooperated in international frequency allocation with other regulatory bodies around the world.

As new services have been added, changes and accommodations have been made to allocations of the frequency spectrum. These changes are not always made globally. For instance, in the United States, cell phone service frequencies are at 850 and 1900 MHz. In many other countries, cell phone bands are at 900 and 1800 MHz.[¶¶]

No matter what the power, devices that have the potential of emitting electromagnetic interference have to be certified that they do not disturb the spectrum beyond their frequency allocation outside of a small radius.[***] Many applications share frequency bands that do not have exclusive licenses granted to any one application. For instance, microwave ovens, wireless computer devices, and some cordless telephones all operate at frequencies close to 2.4 GHz.[†††] Low-power applications that are not the licensed users of a frequency sometimes need frequencies of their own, as demonstrated by the new medical telemetry bands. Occasionally, wireless receivers in an area pick up broadband interference caused by the intermittent sparking of a short circuit. In essence, the malfunctioning electrical equipment has become an unlicensed spark-gap transmitter.

[*] "Wireless Medical Telemetry—Electromagnetic Interference." *United States Food and Drug Administration Center for Devices and Radiological Health*, Sept. 1, 2002. http://www.fda.gov/cdrh/emc/wmt-emi.html#1 As of Nov. 2006.
[†] McClain, J. P., "Time to Upgrade." *American Society for Healthcare Engineering*. www.ashe.org/ashe/wmts/pdfs/timetoupgrade.pdf As of Nov. 2006.
[‡] Federal Communications Commission. *FCC-00211*. Washington, DC: United States Federal Communications Commission, Jun. 12, 2000. http://www.fcc.gov/Bureaus/Engineering_Technology/Orders/2000/fcc00211.doc As of Nov. 2006.
[#] Thomson, E., "The Field of Experimental Research." *Science*, Aug. 25, 1899, Vol. X, No. 243, pp. 236–245.
[°] Pitts, A., "Backgrounder: What Is Amateur Radio?" *American Radio Relay League*, Oct. 4, 2004. http://www.arrl.org/pio/bwhatis.html As of Nov. 2006.
[§] "ARRL Granted Experimental License for 500 kHz Research by Radio Amateurs." *American Radio Relay League*, Sep. 15, 2006. http://www.arrl.org/news/stories/2006/09/15/104/ As of Nov. 2006.
[¶] "Service Regulations Affixed to the International Wireless Convention." *United States Early Radio History*. http://earlyradiohistory.us/1906conv.htm#SR As of Nov. 2006.
[**] Lewis, T., *Empire of the Air.* New York: HarperCollins, 1991, pp. 70–74.
[††] "History." *International Telecommunication Union*, Nov. 15, 2004. http://www.itu.int/aboutitu/overview/history.html As of Nov. 2006.
[‡‡] Lapin, G. D., "Lessons Learned about Frequency Sharing in the Amateur Radio Service." *American Radio Relay League*. http://www.arrl.org/tis/info/HTML/plc/files/ Lessons%20Learned%20About%20Frequency%20Sharing%20in%20the%20Amateur%20Radio%20Service%20Rev%202.ppt As of Nov. 2006.
[##] *Radio Act of 1927.* United States Public Law 632. Feb. 23, 1927. Available at http://showcase.netins.net/web/akline/ pdf/1927act.pdf As of Nov. 2006.
[°°] "History." *International Telecommunication Union*, Nov. 15, 2004. http://www.itu.int/aboutitu/overview/history.html As of Nov. 2006.
[§§] "About the FCC." *United States Federal Communications Commission*, Sept. 26, 2006. http://www.fcc.gov/aboutus.html As of Nov. 2006.
[¶¶] Luna, N. "Globetrotting with Cell Phones Tricky but Not Impossible." *The Orange County Register*, May 4, 2005.
[***] "Rule 47 CFR Part 15." *United States Federal Register*, Washington, DC: Aug. 14, 2006. http://www.fcc.gov/oet/info/rules/part15/part15-8-14-06.pdf As of Nov. 2006.
[†††] Lowe, M., "Muting Microwaves." *Appliance Design*, Jan. 2006, Vol. 54, No. 1, pp. 74–75.

Summary

1. Maxwell's equations summarize the fundamental laws of physics that govern electricity and magnetism.
2. Electromagnetic waves include light, radio waves, television waves, X rays, gamma rays, microwaves, and others.

TOPIC	RELEVANT EQUATIONS AND REMARKS	
1. Maxwell's Displacement Current	Ampère's law can be generalized to apply to currents that are not steady (and not continuous) if the current I is replaced by $I + I_d$, where I_d is Maxwell's displacement current: $I_d = \epsilon_0 \frac{d\phi_e}{dt}$	30-3
Generalized form of Ampère's law	$\oint_C \vec{B} \cdot d\vec{\ell} = \mu_0(I + I_d) = \mu_0 I + \mu_0\epsilon_0 \frac{d\phi_e}{dt}$	30-4
2. Maxwell's Equations	The laws of electricity and magnetism are summarized by Maxwell's equations.	
Gauss's law	$\oint_S E_n\, dA = \frac{1}{\epsilon_0} Q_{\text{inside}}$	30-6*a*
Gauss's law for magnetism (isolated magnetic poles do not exist)	$\oint_S B_n\, dA = 0$	30-6*b*
Faraday's law (form that does not include motional emf)	$\oint_C \vec{E} \cdot d\vec{\ell} = -\frac{d}{dt}\int_S B_n\, dA = -\int_S \frac{\partial B_n}{\partial t} dA$	30-6*c*
Ampère's law modified	$\oint_C \vec{B} \cdot d\vec{\ell} = \mu_0(I + I_d) = \mu_0 I + \mu_0\epsilon_0 \int_S \frac{\partial E_n}{\partial t} dA$	30-6*d*
3. The Wave Equation for Electromagnetic Waves	Maxwell's equations imply that the electric and magnetic field vectors in free space obey a wave equation. $\frac{\partial^2 \vec{E}}{\partial x^2} = \frac{1}{c^2}\frac{\partial^2 \vec{E}}{\partial t^2}$	30-8*a*
	$\frac{\partial^2 \vec{B}}{\partial x^2} = \frac{1}{c^2}\frac{\partial^2 \vec{B}}{\partial t^2}$	30-8*b*
4. Electromagnetic Waves	In an electromagnetic wave, the electric and magnetic field vectors are perpendicular to each other and to the direction of propagation. Their magnitudes are related by $E = cB$	30-18
	The vector product $\vec{E} \times \vec{B}$ is in the direction of propagation.	
Wave speed	$c = \frac{1}{\sqrt{\mu_0\epsilon_0}} = 3.00 \times 10^8$ m/s	30-1
Electromagnetic spectrum	The various types of electromagnetic waves—light, radio waves, X rays, gamma rays, microwaves, and others—differ only in wavelength and frequency. The human eye is sensitive to the range from about 400 nm to 780 nm.	
Electric dipole radiation	Electromagnetic waves are produced when free electric charges accelerate. Oscillating charges in an electric dipole antenna radiate electromagnetic waves with an intensity that is greatest in directions perpendicular to the antenna. There is no radiated intensity along the long axis of the antenna. Perpendicular to the antenna and far away from it, the electric field of the electromagnetic wave is parallel to the antenna.	
Energy density in an electromagnetic wave	$u = u_e + u_m = \epsilon_0 E^2 = \frac{B^2}{\mu_0} = \frac{EB}{\mu_0 c}$	30-19

TOPIC	RELEVANT EQUATIONS AND REMARKS	
Intensity of an electromagnetic wave	$I = u_{av}c = \frac{E_{rms}B_{rms}}{\mu_0} = \frac{1}{2}\frac{E_0B_0}{\mu_0} = \lvert\vec{S}\rvert_{av}$	30-20
Poynting vector	$\vec{S} = \frac{\vec{E}\times\vec{B}}{\mu_0}$	30-21
Momentum and energy in an electromagnetic wave	$p = \frac{U}{c}$	30-24
Radiation pressure and intensity	$P_r = \frac{I}{c}$	30-25

Answers to Practice Problems

30-2 $\vec{E}\cdot\vec{E} = E_0^2$ and $\vec{B}\cdot\vec{B} = B_0^2$

30-3 About 5 h for a 10-g shoelace thrown at 10 m/s. Light beam propulsion takes almost twice as long as shoelace propulsion.

Problems

In a few problems, you are given more data than you actually need; in a few other problems, you are required to supply data from your general knowledge, outside sources, or informed estimate.

Interpret as significant all digits in numerical values that have trailing zeros and no decimal points.

- • Single-concept, single-step, relatively easy
- •• Intermediate-level, may require synthesis of concepts
- ••• Challenging
- SSM Solution is in the *Student Solutions Manual*
- Consecutive problems that are shaded are paired problems.

CONCEPTUAL PROBLEMS

1 • True or false:
(*a*) The displacement current has different units than the conduction current.
(*b*) Displacement current only exists if the electric field in the region is changing with time.
(*c*) In an oscillating *LC* circuit, no displacement current exists between the capacitor plates when the capacitor is momentarily fully charged.
(*d*) In an oscillating *LC* circuit, no displacement current exists between the capacitor plates when the capacitor is momentarily uncharged. SSM

2 • Using SI units, show that $\epsilon_0\, d\phi_e/dt$ has units of current.

3 • True or false:
(*a*) Maxwell's equations apply only to electric and magnetic fields that are constant over time.
(*b*) The electromagnetic wave equation can be derived from Maxwell's equations.
(*c*) Electromagnetic waves are transverse waves.
(*d*) The electric and magnetic fields of an electromagnetic wave in free space are in phase. SSM

4 • Theorists have speculated about the existence of *magnetic monopoles*, and several experimental searches for such monopoles have occurred. Suppose magnetic monopoles were found and that the magnetic field at a distance r from a monopole of strength q_m is given by $B = (\mu_0/4\pi)q_m/r^2$. Modify the Gauss's law for magnetism equation to be consistent with such a discovery.

5 • (*a*) For each of the following pairs of electromagnetic waves, which has the higher frequency: (1) visible light or X rays, (2) green light or red light, (3) infrared waves or red light. (*b*) For each of the following pairs of electromagnetic waves, which has the longer wavelength: (1) visible light or microwaves, (2) green light or ultraviolet light, (3) gamma rays or ultraviolet light.

6 • The detection of radio waves can be accomplished with either an electric dipole antenna or a loop antenna. True or false:
(*a*) The electric dipole antenna works according to the Faraday's law.
(*b*) If a linearly polarized radio wave is approaching you head on such that its electric field oscillates vertically, to best detect this wave the normal to a loop antenna's plane should be oriented so that it points either right or left.
(*c*) If a linearly polarized radio wave is approaching you such that its electric field oscillates in a horizontal plane, to best detect this wave using a dipole antenna the antenna should be oriented vertically.

7 • A transmitter emits electromagnetic waves using an electric dipole antenna oriented vertically. (*a*) A receiver to detect the waves also uses an electric dipole antenna that is one mile from the transmitting antenna and at the same altitude. How should the receiver's electric dipole antenna be oriented for optimum signal reception? (*b*) A receiver to detect these waves uses a loop antenna that is one mile from the transmitting antenna and at the same altitude. How should the loop antenna be oriented for optimum signal reception?

8 • Show that the expression $(\vec{E}\times\vec{B})/\mu_0$ for the Poynting vector $\vec{S}$ (Equation 30-21) has units of watts per square meter (the SI units for electromagnetic wave intensity).

9 • If a red light beam, a green light beam, and a violet light beam, all traveling in empty space, have the same intensity, which light beam carries more momentum? (*a*) the red light beam, (*b*) the green light beam, (*c*) the violet light beam, (*d*) They all have the same momentum. (*e*) You cannot determine which beam carries the most momentum from the data given. SSM

10 • If a red light plane wave, a green light plane wave, and a violet light plane wave, all traveling in empty space, have the same intensity, which wave has the largest peak electric field? (*a*) the red light wave, (*b*) the green light wave, (*c*) the violet light wave, (*d*) They all have the same peak electric field. (*e*) You cannot determine the largest peak electric field from the data given.

11 • Two sinusoidal plane electromagnetic waves are identical except that wave A has a peak electric field that is three times the peak electric field of wave B. How do their intensities compare? (*a*) $I_A = \frac{1}{3}I_B$, (*b*) $I_A = \frac{1}{9}I_B$, (*c*) $I_A = 3I_B$, (*d*) $I_A = 9I_B$, (*e*) You cannot determine how their intensities compare from the data given.

ESTIMATION AND APPROXIMATION

12 •• ENGINEERING APPLICATION In *laser cooling and trapping*, the forces associated with radiation pressure are used to slow down atoms from thermal speeds of hundreds of meters per second at room temperature to speeds of a few meters per second or slower. An isolated atom will absorb only radiation of specific frequencies. If the frequency of the laser-beam radiation is tuned so that the target atoms will absorb the radiation, then the radiation is absorbed during a process called *resonant absorption*. The cross-sectional area of the atom for resonant absorption is approximately equal to λ^2, where λ is the wavelength of the laser light. (*a*) Estimate the acceleration of a rubidium atom (molar mass 85 g/mol) in a laser beam whose wavelength is 780 nm and intensity is 10 W/m^2. (*b*) About how long would it take such a light beam to slow a rubidium atom in a gas at room temperature (300 K) to near-zero speed?

13 •• ENGINEERING APPLICATION One of the first successful satellites launched by the United States in the 1950s was essentially a large spherical (aluminized) Mylar balloon from which radio signals were reflected. After several orbits around Earth, scientists noticed that the orbit itself was changing with time. They eventually determined that radiation pressure from sunlight was causing the orbit of this object to change—a phenomenon not taken into account in planning the mission. Estimate the ratio of the radiation-pressure force by the sunlight on the satellite to the gravitational force by Earth's gravity on the satellite. SSM

14 •• Some science fiction writers have described solar sails that could propel interstellar spaceships. Imagine a giant sail on a spacecraft subjected to radiation pressure from our Sun. (*a*) Explain why this arrangement works better if the sail is highly reflective rather than highly absorptive. (*b*) If the sail is assumed highly reflective, show that the force exerted by the sunlight on the spacecraft's sail is given by $P_S A/(4\pi r^2 c)$ where P_S is the power output of the Sun (3.8×10^{26} W), A is the surface area of the sail, r is the distance from the Sun, and c is the speed of light. (Assume the area of the sail is much larger than the area of the spacecraft so that all the force is due to radiation pressure on the sail only.) (*c*) Using a reasonable value for A, compare the force on the spacecraft due to the radiation pressure and the force on the spacecraft due to the gravitational force of the Sun on the spacecraft. Does the result imply that such a system will work? Explain your answer.

MAXWELL'S DISPLACEMENT CURRENT

15 • A parallel-plate capacitor has circular plates and no dielectric between the plates. Each plate has a radius equal to 2.3 cm and the plates are separated by 1.1 mm. Charge is flowing onto the upper plate (and off of the lower plate) at a rate of 5.0 A. (*a*) Find the rate of change of the electric field strength in the region between the plates. (*b*) Compute the displacement current in the region between the plates and show that it equals 5.0 A. SSM

16 • In a region of space, the electric field varies with time as (0.050 N/C) $\sin \omega t$, where $\omega = 2000$ rad/s. Find the peak displacement current through a surface that is perpendicular to the electric field and has an area equal to 1.00 m^2.

17 •• For Problem 15, show that the magnetic field strength between the plates a distance r from the axis through the centers of both plates is given by $B = (1.9 \times 10^{-3}\ \text{T/m})r$.

18 •• The capacitors referred to in this problem have only empty space between the plates. (*a*) Show that a parallel-plate capacitor has a displacement current in the region between its plates that is given by $I_d = C\,dV/dt$, where C is the capacitance and V is the potential difference between the plates. (*b*) A 5.00–nF parallel-plate capacitor is connected to an ideal ac generator so the potential difference between the plates is given by $V = V_0 \cos \omega t$, where $V_0 = 3.00$ V and $\omega = 500\pi$ rad/s. Find the displacement current in the region between the plates as a function of time.

19 •• There is a current of 10 A in a resistor that is connected in series with a parallel-plate capacitor. The plates of the capacitor have an area of 0.50 m^2, and no dielectric exists between the plates. (*a*) What is the displacement current between the plates? (*b*) What is the rate of change of the electric field strength between the plates? (*c*) Find the value of the line integral $\oint_C \vec{B} \cdot d\vec{\ell}$, where the integration path C is a 10-cm-radius circle that lies in a plane that is parallel with the plates and is completely within the region between them. SSM

20 ••• Demonstrate the validity of the generalized form of Ampère's law (Equation 30-4) by showing that it gives the same result as the Biot–Savart law (Equation 27-3) in a specified situation. Figure 30-13 shows two momentarily equal but opposite point charges ($+Q$ and $-Q$) on the x axis at $x = -a$ and $x = +a$, respectively. At the same instant there is a current I in the wire connecting them, as shown. Point P is on the y axis at $y = R$. (*a*) Use the Biot–Savart law to show that the magnitude of the magnetic field at point P is given by $B = \dfrac{\mu_0 I a}{2\pi R}\dfrac{1}{\sqrt{R^2 + a^2}}$. (*b*) Now consider a circular strip of radius r and width dr in the $x = 0$ plane that has its center at the origin. Show that the flux of the electric field through this strip is given by $E_x\,dA = \dfrac{Q}{\epsilon_0 (r^2 + a^2)^{3/2}}\pi r\,dr$. (*c*) Use the result from Part (*b*) to show that the total electric flux ϕ_e through a circular surface S of radius R is given by $\phi_e = \dfrac{Q}{\epsilon_0}\left(1 - \dfrac{a}{\sqrt{a^2 + R^2}}\right)$. (*d*) Find the displacement current I_d through S, and show that $I + I_d = I\dfrac{a}{\sqrt{a^2 + R^2}}$. (*e*) Finally, show that the generalized form of Ampère's law (Equation 30-4) gives the same result for the magnetic field as found in Part (*a*).

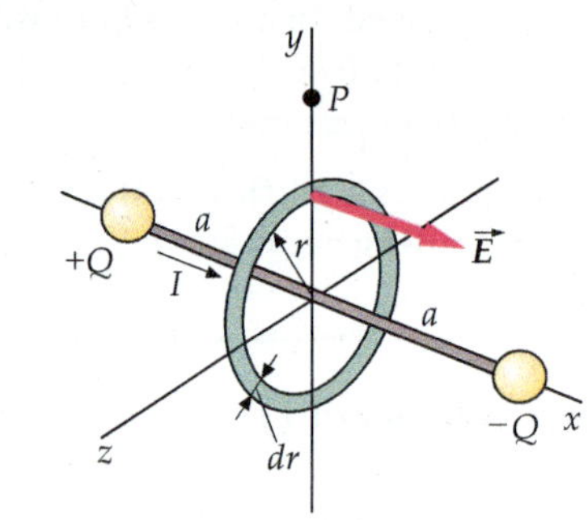

FIGURE 30-13 Problem 20

MAXWELL'S EQUATIONS AND THE ELECTROMAGNETIC SPECTRUM

21 • The color of the dominant light from the Sun is in the yellow-green region of the visible spectrum. Estimate the wavelength and frequency of the dominant light emitted by our Sun. *Hint: See Table 30-1.*

22 • (*a*) What is the frequency of microwave radiation that has a 3.00-cm-long wavelength? (*b*) Using Table 30-1, estimate the ratio of the shortest wavelength of green light to the shortest wavelength of red light.

23 • (*a*) What is the frequency of an X ray that has a 0.100-nm-long wavelength? (*b*) The human eye is sensitive to light that has a wavelength equal to 550 nm. What are the color and frequency of this light? Comment on how this answer compares to your answer for Problem 21.

ELECTRIC DIPOLE RADIATION

Note: **All of the problems in this section are based on the following information. Refer to Figure 30-11. It can be shown that the intensity of radiation from a radiating electric dipole at a field point far from the antenna is proportional to $\sin^2\theta/r^2$, where θ is the angle between the electric dipole moment vector $\vec{p}$ and the position vector $\vec{r}$ of the field point relative to the center of the antenna. The pattern of radiation from this type of antenna is independent of the azimuthal angle, that is, you can rotate the pattern about the antenna axis and it does not change shape.**

24 •• Suppose a radiating electric dipole lies along the z axis. Let I_1 be the intensity of the radiation at a distance of 10 m and at angle of 90°. Find the intensity (in terms of I_1) at (*a*) a distance of 30 m and an angle of 90°, (*b*) a distance of 10 m and an angle of 45°, and (*c*) a distance of 20 m and an angle of 30°.

25 •• (*a*) For the situation described in Problem 24, at what angle is the intensity at a distance of 5.0 m equal to I_1? (*b*) At what distance is the intensity equal to I_1 when $\theta = 45°$?

26 •• ENGINEERING APPLICATION, CONTEXT-RICH You and your engineering crew are in charge of setting up a wireless telephone network for a village in a mountainous region. The transmitting antenna of one station is an electric dipole antenna located atop a mountain 2.00 km above sea level. There is a nearby mountain that is 4.00 km from the antenna and is also 2.00 km above sea level. At that location, one member of the crew measures the intensity of the signal to be $4.00 \times 10^{-12}\ \text{W/m}^2$. What should be the intensity of the signal at the village that is located at sea level and 1.50 km from the transmitter?

27 ••• ENGINEERING APPLICATION A radio station that uses a vertical electric dipole antenna broadcasts at a frequency of 1.20 MHz and has a total power output of 500 kW. Calculate the intensity of the signal at a horizontal distance of 120 km from the station. SSM

28 ••• ENGINEERING APPLICATION Regulations require that licensed radio stations have limits on their broadcast power so as to avoid interference with signals from distant stations. You are in charge of checking compliance with the law. At a distance of 30.0 km from a radio station that broadcasts from a single vertical electric dipole antenna at a frequency of 800 kHz, the intensity of the electromagnetic wave is $2.00 \times 10^{-13}\ \text{W/m}^2$. What is the total power radiated by the station?

29 ••• ENGINEERING APPLICATION A small private plane approaching an airport is flying at an altitude of 2.50 km above sea level. As a flight controller at the airport, you know your system uses a vertical electric dipole antenna to transmit 100 W at 24.0 MHz. What is the intensity of the signal at the plane's receiving antenna when the plane is 4.00 km from the airport? Assume the airport is at sea level.

ENERGY AND MOMENTUM IN AN ELECTROMAGNETIC WAVE

30 • An electromagnetic wave has an intensity of $100\ \text{W/m}^2$. Find its (*a*) rms electric field strength, and (*b*) rms magnetic field strength.

31 • The amplitude of an electromagnetic wave's electric field is 400 V/m. Find the wave's (*a*) rms electric field strength, (*b*) rms magnetic field strength, (*c*) intensity, and (*d*) radiation pressure (P_r). SSM

32 • The rms value of an electromagnetic wave's electric field strength is 400 V/m. Find the wave's (*a*) rms magnetic field strength, (*b*) average energy density, and (*c*) intensity.

33 •• (*a*) An electromagnetic wave that has an intensity equal to $200\ \text{W/m}^2$ is normal to a black 20 cm by 30 cm rectangular card that absorbs 100 percent of the wave. Find the force exerted on the card by the radiation. (*b*) Find the force exerted by the same wave if the card reflects 100 percent of the wave.

34 •• Find the force exerted by the electromagnetic wave on the card in Part (*b*) of Problem 33 if both the incident and the reflected waves are at angles of 30° to the normal.

35 • (*a*) For a given distance from a radiating electric dipole, at what angle (expressed as θ and measured from the dipole axis) is the intensity equal to 50 percent of the maximum intensity? (*b*) At what angle θ is the intensity equal to 1 percent of the maximum intensity? SSM

36 •• A laser pulse has an energy of 20.0 J and a beam radius of 2.00 mm. The pulse duration is 10.0 ns and the energy density is uniformly distributed within the pulse. (*a*) What is the spatial length of the pulse? (*b*) What is the energy density within the pulse? (*c*) Find the rms values of the electric and magnetic fields in the pulse.

37 •• An electromagnetic plane wave has an electric field that is parallel to the y axis, and has a Poynting vector given by $\vec{S}(x, t) = (100\ \text{W/m}^2)\cos^2(kx - \omega t)\hat{i}$, where x is in meters, $k = 10.0$ rad/m, $\omega = 3.00 \times 10^9$ rad/s, and t is in seconds. (*a*) What is the direction of propagation of the wave? (*b*) Find the wavelength and frequency of the wave. (*c*) Find the electric and magnetic fields of the wave as functions of x and t. SSM

38 •• A parallel-plate capacitor is being charged. The capacitor consists of a pair of identical circular parallel plates that each have a radius b and a separation distance d. (*a*) Show that the displacement current in the capacitor gap has the same value as the conduction current in the capacitor leads. (*b*) What is the direction of the Poynting vector in the region between the capacitor plates? (*c*) Find an expression for the Poynting vector in this region and show that its flux into the region between the plates is equal to the rate of change of the energy stored in the capacitor.

39 •• A pulsed laser fires a 1000-MW pulse that has a 200-ns duration at a small object that has a mass equal to 10.0 mg and is suspended by a fine fiber that is 4.00 cm long. If the radiation is completely absorbed by the object, what is the maximum angle of deflection of this pendulum? (Think of the system as a ballistic pendulum and assume the small object was hanging vertically before the radiation hit it.) SSM

40 •• The mirrors used in a particular type of laser are 99.99% reflecting. (*a*) If the laser has an average output power of 15 W, what is the average power of the radiation incident on one of the mirrors? (*b*) What is the force due to radiation pressure on one of the mirrors?

41 •• (*a*) Estimate the force on Earth due to the pressure of the radiation on Earth by the Sun, and compare this force to the gravitational force of the Sun on Earth. (At Earth's orbit, the intensity of sunlight is $1.37\ \text{kW/m}^2$.) (*b*) Repeat Part (*a*) for Mars which is at an average distance of 2.28×10^8 km from the Sun and has a radius of 3.40×10^3 km. (*c*) Which planet has the larger ratio of radiation pressure to gravitational attraction? SSM

THE WAVE EQUATION FOR ELECTROMAGNETIC WAVES

42 • Show by direct substitution that Equation 30-8*a* is satisfied by the wave function $E_y = E_0 \sin(kx - \omega t) = E_0 \sin k(x - ct)$ where $c = \omega/k$.

43 • Use the values of μ_0 and ϵ_0 in SI units to compute $1/\sqrt{\epsilon_0\mu_0}$ and show that it is equal to 3.00×10^8 m/s.

44 •• (*a*) Use Maxwell's equations to show for a plane wave, in which $\vec{E}$ and $\vec{B}$ are independent of y and z, that $\dfrac{\partial E_z}{\partial x} = \dfrac{\partial B_y}{\partial t}$ and $\dfrac{\partial B_y}{\partial x} = \mu_0\epsilon_0\dfrac{\partial E_z}{\partial t}$. (*b*) Show that E_z and B_y also satisfy the wave equation.

45 •• Show that any function of the form $y(x, t) = f(x - vt)$ or $y(x, t) = g(x + vt)$ satisfies the wave equation (Equation 30-7). SSM

GENERAL PROBLEMS

46 • An electromagnetic wave has a frequency of 100 MHz and is traveling in a vacuum. The magnetic field is given by $\vec{B}(z, t) = (1.00 \times 10^{-8}\ \text{T}) \cos(kz - \omega t)\hat{i}$. (*a*) Find the wavelength and the direction of propagation of this wave. (*b*) Find the electric field vector $\vec{E}(z, t)$. (*c*) Determine the Poynting vector, and use it to find the intensity of the wave.

47 •• **ENGINEERING APPLICATION** A circular loop of wire can be used to detect electromagnetic waves. Suppose the signal strength from a 100-MHz FM radio station 100 km distant is 4.0 μW/m^2, and suppose the signal is vertically polarized. What is the maximum rms voltage induced in your antenna, assuming your antenna is a 10.0-cm-radius loop? SSM

48 •• **ENGINEERING APPLICATION** The electric field strength from a radio station some distance from the electric dipole transmitting antenna is given by $(1.00 \times 10^{-4}\ \text{N/C}) \cos[(1.00 \times 10^6\ \text{rad/s})t]$. (*a*) What peak voltage is picked up on a 50.0-cm-long wire oriented parallel with the electric field direction? (*b*) What is the maximum voltage that can be induced by this electromagnetic wave in a conducting loop of radius 20.0 cm, and what orientation of the loop does this require?

49 ••• A parallel-plate capacitor has circular plates of radius a that are separated by a distance d. In the gap between the two plates is a thin straight wire of resistance R that connects the centers of the two plates. A time-varying voltage given by $V_0 \sin \omega t$ is applied across the plates. (*a*) What is the current drawn by the capacitor? (*b*) What is the magnetic field as a function of the radial distance r from the centerline within the capacitor plates? (*c*) What is the phase angle between the current drawn by the capacitor and the applied voltage?

50 •• A 20-kW beam of electromagnetic radiation is normal to a surface that reflects 50 percent of the radiation. What is the force exerted by the radiation on the surface?

51 •• The electric fields of two harmonic electromagnetic waves of angular frequency ω_1 and ω_2 are given by $\vec{E}_1 = E_{10} \cos(k_1x - \omega_1 t)\hat{j}$ and by $\vec{E}_2 = E_{20} \cos(k_2x - \omega_2 t + \delta)\hat{j}$. For the resultant of these two waves, find (*a*) the instantaneous Poynting vector and (*b*) the time-averaged Poynting vector. (*c*) Repeat Parts (*a*) and (*b*) if the direction of propagation of the second wave is reversed so that $\vec{E}_2 = E_{20} \cos(k_2x + \omega_2 t + \delta)\hat{j}$. SSM

52 •• Show that $\dfrac{\partial B_z}{\partial x} = -\mu_0\epsilon_0\dfrac{\partial E_y}{\partial t}$ (Equation 30-10) follows from $\oint_C \vec{B} \cdot d\vec{\ell} = \mu_0\epsilon_0 \int_S \dfrac{\partial E_n}{\partial t}\, dA$ (Equation 30-6*d* and where $I = 0$) by integrating along a suitable curve C and over a suitable surface S in a manner that parallels the derivation of Equation 30-9.

53 •• For your backpacking excursions, you have purchased a radio capable of detecting a signal as weak as 1.00×10^{-14} W/m^2. This radio has a 2000-turn coil antenna that has a radius of 1.00 cm wound on an iron core that increases the magnetic field by a factor of 200. The broadcast frequency of the radio station is 1400 kHz. (*a*) What is the peak magnetic field strength of an electromagnetic wave of this minimum intensity? (*b*) What is the peak emf that it is capable of inducing in the antenna? (*c*) What would be the peak emf induced in a straight 2.00-m-long metal wire oriented parallel to the direction of the electric field?

54 •• The intensity of the sunlight striking Earth's upper atmosphere is 1.37 kW/m^2. (*a*) Find the rms values of the magnetic and electric fields of this light. (*b*) Find the average power output of the Sun. (*c*) Find the intensity and the radiation pressure at the surface of the Sun.

55 ••• A conductor in the shape of a long solid cylinder that has a length L, a radius a, and a resistivity ρ carries a steady current I that is uniformly distributed over its cross section. (*a*) Use Ohm's law to relate the electric field E in the conductor to I, ρ, and a. (*b*) Find the magnetic field $\vec{B}$ just outside the conductor. (*c*) Use the results from Part (*a*) and Part (*b*) to compute the Poynting vector $\vec{S} = (\vec{E} \times \vec{B})/\mu_0$ at $r = a$ (the edge of the conductor). In what direction is $\vec{S}$? (*d*) Find the flux $\oint S_n\, dA$ through the surface of the cylinder, and use the flux to show that the rate of energy flow into the conductor equals I^2R, where R is the resistance of the cylinder. SSM

56 ••• A long solenoid that has n turns per unit length carries a current that increases linearly with time. The solenoid has radius R, length L, and the current I in the windings is given by $I = at$. (*a*) Find the induced electric field at a distance $r < R$ from the central axis of the solenoid. (*b*) Find the magnitude and direction of the Poynting vector $\vec{S}$ at $r = R$ (just inside the solenoid windings). (*c*) Calculate the flux $\oint S_n\, dA$ into the region inside the solenoid, and show that the flux equals the rate of increase of the magnetic energy inside the solenoid.

57 ••• Small particles are blown out of the solar system by the radiation pressure of sunlight. Assume that each particle is spherical, has a radius r, has a density of 1.00 g/cm^3, and absorbs all the radiation in a cross-sectional area of πr^2. Assume the particles are located at some distance d from the Sun, which has a total power output of 3.83×10^{26} W. (*a*) What is the critical value for the radius r of the particle for which the radiation force of repulsion just balances the gravitational force of attraction to the Sun? (*b*) Do particles that have radii larger than the critical value get ejected from the solar system, or is it only particles that have radii smaller than the critical value that get ejected? Explain your answer.

58 ••• When an electromagnetic wave at normal incidence on a perfectly conducting surface is reflected, the electric field of the reflected wave at the reflecting surface is equal and opposite to the electric field of the incident wave at the reflecting surface. (*a*) Explain why this assertion is valid. (*b*) Show that the superposition of incident and reflected waves results in a standing wave. (*c*) Are the magnetic fields of the incident waves and reflected waves at the reflecting surface equal and opposite as well? Explain your answer.

59 ••• An intense point source of light radiates 1.00 MW isotropically (uniformly in all directions). The source is located 1.00 m above an infinite, perfectly reflecting plane. Determine the force that the radiation pressure exerts on the plane. SSM

PART V LIGHT

CHAPTER 31

LIGHT IS TRANSMITTED BY TOTAL INTERNAL REFLECTION THROUGH TINY GLASS FIBERS. *(© James L. Amos/Corbis.)*

? How large must the angle of incidence of the light on the wall of the tube be so that no light escapes? (See Example 31-4.)

Properties of Light

The human eye is sensitive to electromagnetic radiation with wavelengths from approximately 400 nm to 700 nm.* The shortest wavelengths in the visible spectrum correspond to violet light and the longest to red light. The perceived colors of light are the result of the physiological and psychological responses of the eyes and brain to the different frequencies of visible light. Although the correspondence between perceived color and frequency is quite good, there are many interesting deviations. For example, a mixture of red light and green light is perceived by the eyes and brain as yellow—even in the absence of light in the yellow region of the spectrum.

* Wavelengths as short as 380 nm and as long as 780 nm can be seen by some individuals.

In this chapter, we study how light is produced; how its speed is measured; and how light is scattered, reflected, refracted, and polarized.

31-1 THE SPEED OF LIGHT

Prior to the seventeenth century the speed of visible light was thought by many people to be infinite, and an effort to measure the speed of visible light was made by Galileo. He and a partner stood on hilltops about three kilometers apart, each with a lantern and a shutter to cover it. Galileo proposed to measure the time it took for visible light to travel back and forth between the experimenters. First, one would uncover his lantern, and when the other saw the light he would uncover his lantern. The time between the first partner's uncovering his lantern and his seeing the light from the other lantern would be the time it took for light to travel back and forth between the experimenters. Though this method is sound in principle, the speed of light is so great that the time interval to be measured is much smaller than variations in human response time, so Galileo was unable to obtain a value for the speed of light.

The first indication of the true magnitude of the speed of light came from astronomical observations of the period of Io, one of the moons of Jupiter. This period is determined by measuring the time between eclipses of Io. An eclipse occurs when Io enters the region behind Jupiter where no direct sunlight reaches. The eclipse period is about 42.5 h, but measurements made when Earth is moving away from Jupiter along path *ABC* in Figure 31-1 give a greater time for this period than do measurements made when Earth is moving toward Jupiter along path *CDA* in the figure. Because these measurements differ from the average value by only about 15 s, the discrepancies were difficult to measure accurately. In 1675, the astronomer Ole Römer attributed these discrepancies to the fact that the speed of light is finite, and that during the 42.5 h between eclipses of Jupiter's moon, the distance between Earth and Jupiter changes, making the path for the light longer or shorter. Römer devised the following method for measuring the cumulative effect of these discrepancies. Jupiter is moving much more slowly than Earth, so we can neglect its motion. When Earth is at point *A*, nearest to Jupiter, the distance between Earth and Jupiter is changing negligibly. The period of Io's eclipse is measured, providing the time between the beginnings of successive eclipses. Based on this measurement, the number of eclipses during 6 months is computed, and the time when an eclipse should begin a half-year later when Earth is at point C is predicted. When Earth is actually at point C, the observed beginning of the eclipse is about 16.6 min later than predicted. This is the time it takes light to travel a distance equal to the diameter of Earth's orbit. This calculation neglects the distance traveled by Jupiter toward Earth. However, because the orbital speed of Jupiter is so much slower than that of Earth, the distance Jupiter moves toward (or away from) Earth during the 6 months is much less than the diameter of Earth's orbit.

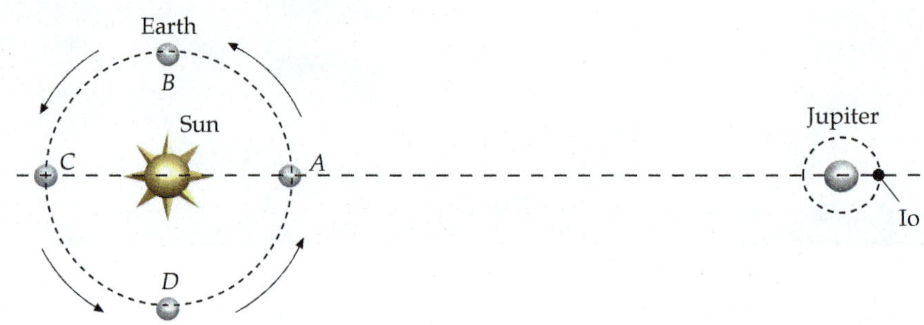

FIGURE 31-1 Römer's method of measuring the speed of light. The time between eclipses of Jupiter's moon Io appears to be greater when Earth is moving along path *ABC* than when Earth is moving along path *CDA*. The difference is due to the time it takes light to travel the distance traveled by Earth along the line of sight during one period of Io.

PRACTICE PROBLEM 31-1

Calculate (*a*) the distance traveled by Earth between successive eclipses of Io and (*b*) the speed of light, given that the time between successive eclipses is 15 s longer than average when Earth is moving directly away from Jupiter.

The French physicist Armand Fizeau made the first nonastronomical measurement of the speed of visible light in 1849. On a hill in Paris, Fizeau placed a light source and a system of lenses arranged so that the light reflected from a semitransparent mirror was focused on a gap in a toothed wheel, as shown in Figure 31-2.

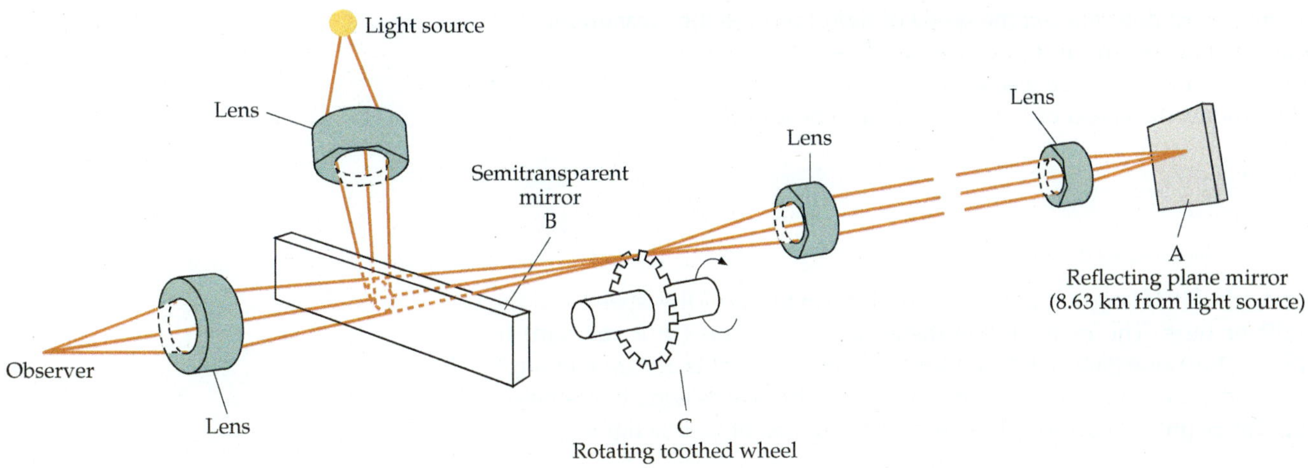

FIGURE 31-2 Fizeau's method of measuring the speed of light. Light from the source is reflected by mirror B and is transmitted through a gap in the toothed wheel to mirror A. The speed of light is determined by measuring the angular speed of the wheel that will permit the reflected light to pass through the next gap in the toothed wheel so that an image of the source is observed.

On a distant hill (about 8.63 km away) Fizeau placed a mirror to reflect the light back, to be viewed by an observer as shown. The toothed wheel was rotated, and the speed of rotation was varied. At low speeds of rotation, no light was visible because the light that passed through a gap in the rotating wheel and was reflected back by the mirror was obstructed by the next tooth of the wheel. The speed of rotation was then increased. The light suddenly became visible when the rotation speed was such that the reflected light passed through the next gap in the wheel. The time for the wheel to rotate through the angle between successive gaps equals the time for the light to make the round trip to the distant mirror and back.

Fizeau's method was improved upon by Jean Foucault, who replaced the toothed wheel with a rotating mirror, as shown in Figure 31-3. Light that strikes the rotating mirror is reflected toward a distant fixed mirror, where it is reflected back toward the rotating mirror. The light is then reflected toward the telescope by the rotating mirror. During the time taken for the light to travel from the rotating mirror to the distant fixed mirror and back, the mirror rotates through a small angle θ. By measuring the angle θ, the time for the light to travel to the distant mirror and back is determined. In approximately 1850, Foucault measured the speed of light in air and in water, and he showed that the speed of light in water is less than the speed of light in air. Using essentially the same method, the American physicist Albert Michelson made more precise measurements of the speed of light in approximately 1880. A half-century later, Michelson made even more precise measurements of the speed of light, using an octagonal rotating mirror (Figure 31-4). In these measurements, the mirror rotates through one-eighth of a turn during the time it takes for the light to travel to the fixed mirror and back. The rotation rate is varied until another face of the mirror is in the right position for the reflected light to enter the telescope.

FIGURE 31-3 Simplified drawing of Foucault's method of measuring the speed of light.

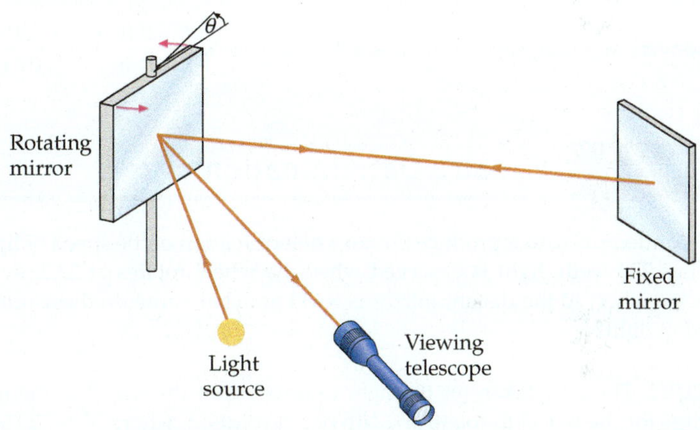

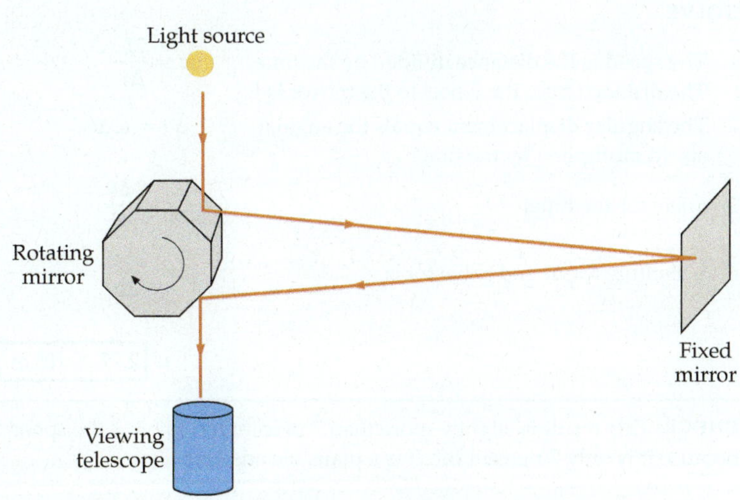

FIGURE 31-4 Simplified drawing of Michelson's method of measuring the speed of light at Mt. Wilson in the late 1920s.

Another method of determining the speed of light involves the measurement of the electrical constants ϵ_0 and μ_0 to determine c from $c = 1/\sqrt{\epsilon_0\mu_0}$.

The various methods we have discussed for measuring the speed of light are all in agreement. Today, the speed of light is defined to be exactly

$$c = 299\,792\,458 \text{ m/s} \qquad 31\text{-}1$$

DEFINITION—SPEED OF LIGHT

and the standard unit of length, the meter, is defined in terms of this speed and the standard unit of time. The meter is the distance light travels (in a vacuum) in $1/299\,792\,458$ s. The value 3.00×10^8 m/s for the speed of light is accurate enough for nearly all calculations in this book. The speed of radio waves and all other electromagnetic waves (in a vacuum) is the same as the speed of visible light.

Example 31-1 The Speed of Light

What is the speed of light in feet per nanosecond?

PICTURE This is an exercise in unit conversions. There are ~30 cm = 0.30 m in 1.0 ft.

SOLVE

1. Convert m/s to ft/ns: $c = 3.0 \times 10^8 \text{ m/s} \times \left(\frac{1.0 \text{ ft}}{0.30 \text{ m}}\right) \times \left(\frac{1.0 \text{ s}}{10^9 \text{ ns}}\right) = \boxed{1.0 \text{ ft/ns}}$

Example 31-2 Fizeau's Determination of *c*

You are attempting to reproduce Fizeau's determination of the speed of light. Using a wheel that has 720 teeth, light is observed when the wheel rotates at 22.3 rev/s. If the distance from the wheel to the distant mirror is 8.63 km, what value do these numbers give for the speed of light?

PICTURE The time taken for the light to travel from the wheel to the mirror and back is the time for the wheel to rotate one Nth of a revolution, where $N = 720$ is the total number of teeth.

SOLVE

1. The speed is the distance divided by the time. The distance from the wheel to the mirror is L: $c = \frac{2L}{\Delta t}$
2. The angular displacement equals the angular speed multiplied by the time: $\Delta\theta = \omega\,\Delta t$
3. Solve for the time: $\Delta t = \frac{\Delta\theta}{\omega}$
4. Substitute for Δt and solve for c: $c = \frac{2L\omega}{\Delta\theta} = \frac{2(8.63 \times 10^3 \text{ m})(22.3 \text{ rev/s})}{\frac{1}{720}\text{ rev}}$

$$= \boxed{2.77 \times 10^8 \text{ m/s}}$$

CHECK This result is slightly more than 7 percent too low for the speed of light. However, because it is only 7 percent off, it is a plausible answer.

PRACTICE PROBLEM 31-2 Space travelers on the moon use electromagnetic waves to communicate with the space control center on Earth. Use $c = 3.00 \times 10^8$ m/s to calculate the time delay for their signal to reach Earth, which is 3.84×10^8 m away.

CONCEPT CHECK 31-1

The step-4 result to Example 31-2 is 7 percent too low. What might account for this discrepancy between this measured value and the known value for the speed of light? An error in counting the number of teeth, in measuring the angular speed, or in measuring the distance to the mirror are possible sources of error, but they are unlikely sources of error. There is a more likely source of error. Can you find it?

Large distances are often given in terms of the distance traveled by light in a given time. For example, the distance to the Sun is 8.33 light-minutes, written 8.33 $c \cdot$ min. A light-year is the distance light travels in one year. We can easily find a conversion factor between light-years and meters. The number of seconds in one year is

$$1\text{ y} = 1\text{ y} \times \frac{365.24\text{ d}}{1\text{ y}} \times \frac{24\text{ h}}{1\text{ d}} \times \frac{3600\text{ s}}{1\text{ h}} = 3.156 \times 10^7\text{ s}$$

(*Note:* There are approximately π multiplied by 10^7 seconds per year, which is the mechanism by which some individuals remember the approximate value of the conversion.) The number of meters in one light-year is thus

$$1\ c \cdot y = (2.998 \times 10^8\text{ m/s})(3.156 \times 10^7\text{ s}) = 9.46 \times 10^{15}\text{ m} \qquad 31\text{-}2$$

31-2 THE PROPAGATION OF LIGHT

The propagation of light is governed by the wave equation discussed in Chapter 30. But long before Maxwell's theory of electromagnetic waves, the propagation of light and other waves was described empirically by two interesting and very different principles attributed to the Dutch physicist Christian Huygens (1629–1695) and the French mathematician Pierre de Fermat (1601–1665).

HUYGENS'S CONSTRUCTION

Figure 31-5 shows a portion of a spherical wavefront emanating from a point source. The wavefront is the locus of points of constant phase. If the radius of the wavefront is r at time t, its radius at time $t + \Delta t$ is $r + c\,\Delta t$, where c is the speed of the wave. However, if a part of the wave is blocked by some obstacle or if the wave passes through a different medium, as in Figure 31-6, the determination of the new wavefront position at time $t + \Delta t$ is much more difficult. The propagation of any wavefront through space can be described using a geometric construction invented by Huygens in approximately 1678, which is now known as **Huygens's construction** or **Huygens's principle:**

> Each point on a primary wavefront serves as the source of spherical secondary wavelets that advance at the wavespeed for the propagating medium. The primary wavefront at some later time is the envelope of these wavelets.
>
> HUYGENS'S CONSTRUCTION

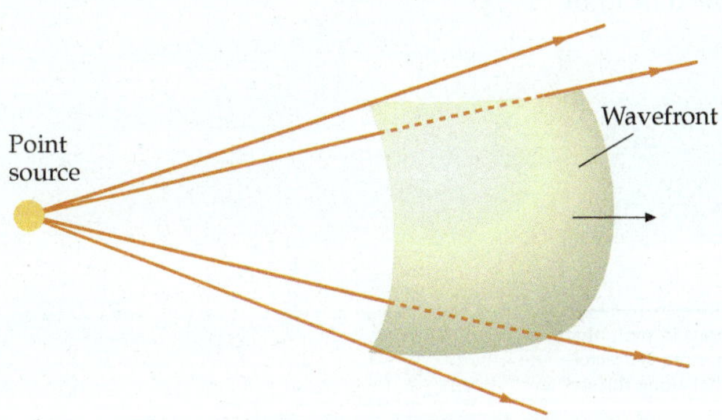

FIGURE 31-5 Spherical wavefront from a point source.

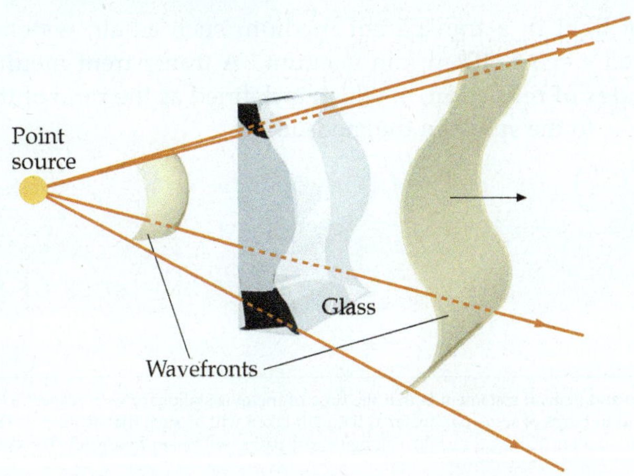

FIGURE 31-6 Wavefront from a point source before and after passing through a piece of glass of varied thickness.

Figure 31-7 shows the application of Huygens's construction to the propagation of a plane wave and the propagation of a spherical wave. Of course, if each point on a wavefront were really a point source, there would be waves in the backward direction as well. Huygens ignored those back waves.

Huygens's construction was later modified by Augustin Fresnel, so that the new wavefront was calculated from the old wavefront by superposition of the wavelets considering their relative amplitudes and phases. Kirchhoff later showed that the Huygens–Fresnel construction was a consequence of the wave equation (Equation 30-8*a*), thus putting it on a firm mathematical basis. Kirchhoff showed that the intensity of each wavelet depends on the angle and is zero at 180° (the backward direction).

We will use Huygens's construction to derive the laws of reflection and refraction in Section 31-5. In Chapter 33, we apply Huygens's construction with Fresnel's modification to calculate the diffraction pattern of a single slit. Because the wavelength of light is so small, we can often use the ray approximation to describe its propagation.

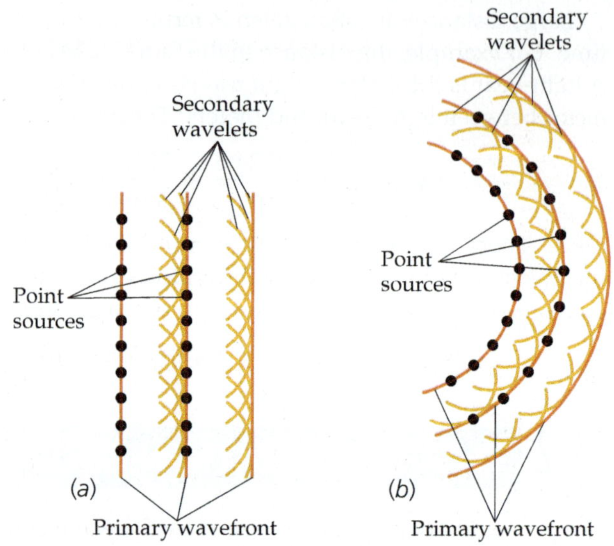

FIGURE 31-7 Huygens's construction for the propagation from left to right of a primary wavefront of (*a*) a plane wave and (*b*) an outgoing spherical, or circular, wave.

FERMAT'S PRINCIPLE

The propagation of light can also be described by Fermat's principle:

> The path taken by light traveling from one point to another is such that the time of travel is a minimum. That is, light travels along the path of least time.*
>
> FERMAT'S PRINCIPLE

The path of least time is not necessarily the path of shortest distance. For example, suppose you are a lifeguard at one end of a pool and a person needs immediate assistance at the far end of the pool. You could get to the person by swimming the entire length of the pool, but you would get to the person more quickly if you ran along the perimeter of the pool and entered the water only when you are near the person.

In Section 31-5, we will use Fermat's principle to derive the laws of reflection and refraction.

31-3 REFLECTION AND REFRACTION

The speed of light in a transparent medium such as air, water, or glass is less than the speed $c = 3 \times 10^8$ m/s in vacuum.† A transparent medium is characterized by an **index of refraction,** n, which is defined as the ratio of the speed of light in a vacuum, c, to the speed in the medium, v:

$$n = \frac{c}{v} \qquad 31\text{-}3$$

DEFINITION—INDEX OF REFRACTION

* A more complete and general statement is that the time of travel is stationary with respect to variations in path; that is, if t is expressed in terms of some parameter x, the path taken will be such that $dt/dx = 0$. The important characteristic of a stationary path is that the time taken along nearby paths will be approximately the same as that along the stationary path.

† It is not the case that the wave speed is never greater than c. In certain materials it is greater than c. However, this does not mean that information can travel at speeds greater than c.

For water, $n = 1.33$, whereas for glass n ranges from approximately 1.50 to 1.66, depending on the type of glass. Diamond has a very high index of refraction—approximately 2.4. The index of refraction of air is approximately 1.0003, so for most purposes we can assume that the speed of light in air is the same as the speed of light in vacuum.

When a beam of light strikes a boundary surface separating two different media, such as an air–glass interface, part of the light energy is reflected and part of the light energy enters the second medium. If the incident light is not perpendicular to the surface, then the transmitted beam is not parallel to the incident beam. The change in direction of the transmitted ray is called **refraction.** Figure 31-8 shows a light ray striking a smooth air–glass interface. The angle θ_1 between the incident ray and the normal (the line perpendicular to the surface) is called the **angle of incidence,** and the plane containing the incident ray and the normal is called the **plane of incidence.** The reflected ray lies in the plane of incidence and makes an angle θ_1' with the normal that is equal to the angle of incidence as shown in the figure:

$$\theta_1' = \theta_1 \qquad \text{31-4}$$

LAW OF REFLECTION

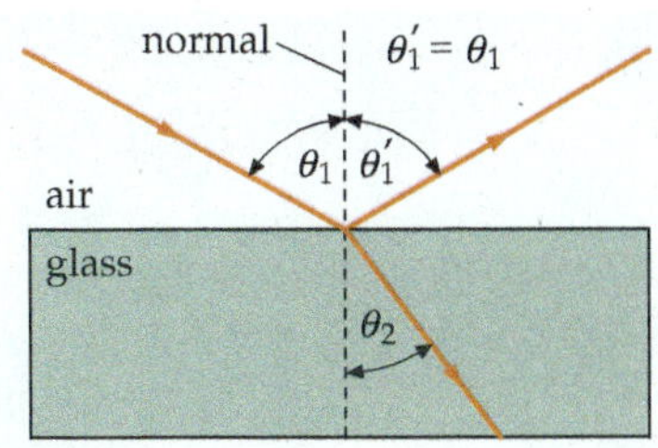

FIGURE 31-8 The angle of reflection θ_1' equals the angle of incidence θ_1. The angle of refraction θ_2 is less than the angle of incidence if the light speed in the second medium is less than that in the incident medium.

See **Math Tutorial** *for more information on* **Trigonometry**

This result is known as the **law of reflection.** The law of reflection holds for any type of wave. Figure 31-9 illustrates the law of reflection for rays of light and for wavefronts of ultrasonic waves.

The ray that enters the glass in Figure 31-8 is called the *refracted ray,* and the angle θ_2 is called the angle of refraction. When a wave crosses a boundary where the wave speed is reduced, as in the case of light entering glass from air, the angle of refraction is less than the angle of incidence θ_1, as shown in Figure 31-8; that is, the refracted ray is bent toward the normal. If, on the other hand, the light beam originates in the glass and is refracted into the air, then the refracted ray is bent away from the normal.

The angle of refraction θ_2 depends on the angle of incidence and on the relative speed of light waves in the two mediums. If v_1 is the wave speed in the incident medium and v_2 is the wave speed in the transmission medium, the angles of incidence and refraction are related by

$$\frac{1}{v_1}\sin\theta_1 = \frac{1}{v_2}\sin\theta_2 \qquad \text{31-5}a$$

Equation 31-5*a* holds for the refraction of any kind of wave incident on a boundary separating two media.

The indices of refraction of the two media are n_1 and n_2. Combining Equations 31-3 and 31-5*a* gives

$$n_1 \sin\theta_1 = n_2 \sin\theta_2 \qquad \text{31-5}b$$

SNELL'S LAW OF REFRACTION

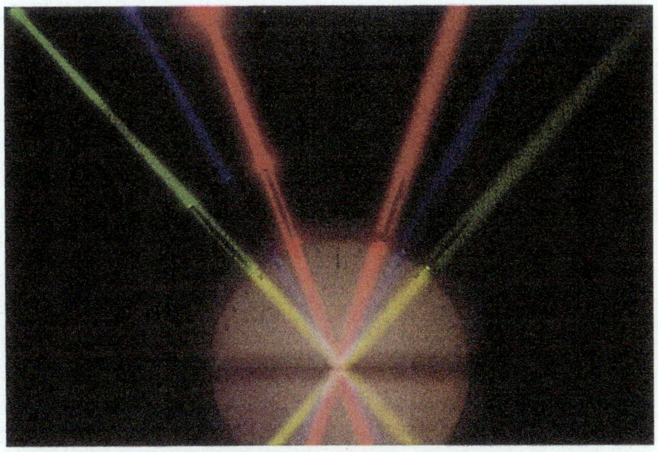

(*a*)

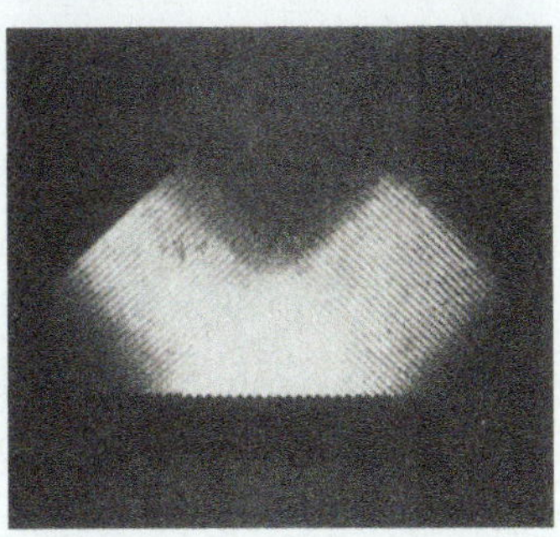

(*b*)

FIGURE 31-9 (*a*) Light rays reflecting from an air–glass interface showing equal angles of incidence and reflection. (*b*) Ultrasonic plane waves in water reflecting from a steel plate. (*(a) Ken Kay/Fundamental Photographs. (b) Courtesy Battelle-Northwest Laboratories.)*

Reflection and refraction of a beam of light incident on a glass slab. *(Richard Megna/Fundamental Photographs.)*

This result was discovered experimentally in 1621 by the Dutch scientist Willebrord Snell and is known as **Snell's law** or the **law of refraction.** It was independently discovered a few years later by the French mathematician and philosopher René Descartes.

PHYSICAL MECHANISMS FOR REFLECTION AND REFRACTION

The physical mechanism of the reflection and refraction of light can be understood in terms of the absorption and reradiation of the light by the atoms in the reflecting or refracting medium. When light traveling in air strikes a glass surface, the atoms in the glass absorb the light and reradiate it at the same frequency in all directions. The waves radiated backward by the glass atoms interfere constructively at an angle equal to the angle of incidence to produce the reflected wave.

The transmitted wave is the result of the interference of the incident wave and the wave produced by the absorption and reradiation of light energy by the atoms in the medium. For light entering glass from air, a phase lag exists between the reradiated wave and the incident wave. Therefore, a phase lag also exists between the resultant wave and the incident wave. This phase lag means that the position of a wave crest of the transmitted wave is retarded relative to the position of a wave crest of the incident wave in the medium. As a result, a transmitted wave crest does not travel as far in a given time as the original incident wave crest; that is, the wave speed of the transmitted wave is less than that of the incident wave. The index of refraction is therefore greater than 1. The frequency of the light in the second medium is the same as the frequency of the incident light—the atoms absorb and reradiate the light at the same frequency—but the wave speed is different, so the wavelength of the transmitted light is different from that of the incident light. If λ is the wavelength of light in a vacuum, then $\lambda f = c$, and if λ_n is the wavelength in a medium that has an index of refraction n in which the wave has speed v, then $\lambda_n f = v$. Combining these two relations gives $\lambda/\lambda_n = c/v$, or

$$\lambda_n = \frac{\lambda}{c/v} = \frac{\lambda}{n} \qquad \text{31-6}$$

SPECULAR REFLECTION AND DIFFUSE REFLECTION

Figure 31-10*a* shows a bundle of light rays from a point source P that are reflected from a flat surface. After reflection, the rays diverge exactly as if they came from a point P' behind the surface. (The point P' is called the *image point*. We will study the formation of images by reflecting and refracting surfaces in the next chapter.) When the rays enter the eye, they cannot be distinguished from rays actually diverging from a source at P'.

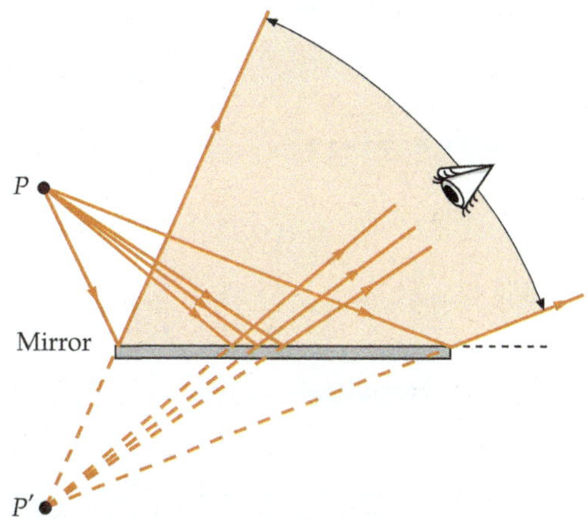

(a)

(b)

FIGURE 31-10 (*a*) Specular reflection from a smooth surface. (*b*) Specular reflection of trees from water. *(Macduff Everton/Corbis.)*

Reflection from a smooth surface is called **specular reflection.** It differs from **diffuse reflection,** which is illustrated in Figure 31-11. Here, because the surface is rough, the rays from a point reflect in random directions and do not diverge from any point, so no image exists. The reflection of light from the page of this book is diffuse reflection. The glass used in picture frames is sometimes ground slightly to give diffuse reflection and thereby cut down on glare from the light used to illuminate the picture. Diffuse reflection from the surface of a road allows you to see the road when you are driving at night because some of the light from your headlights reflects back toward you. In wet weather the reflection is mostly specular; therefore, little light is reflected back toward you, which makes the road difficult to see.

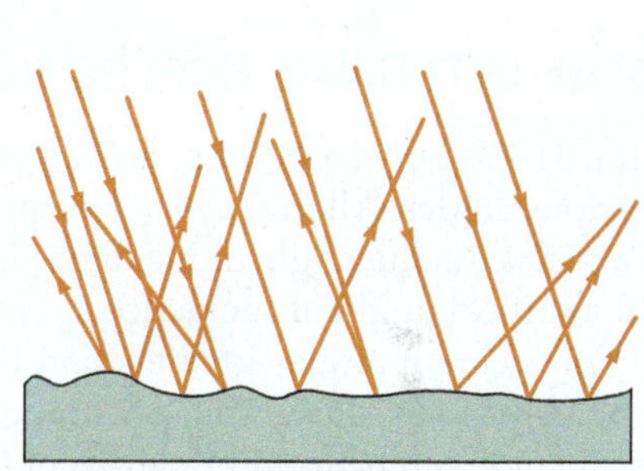

(a)

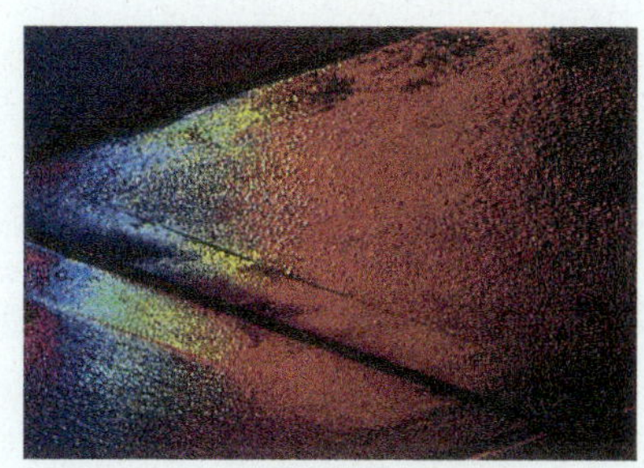

(b)

FIGURE 31-11 (*a*) Diffuse reflection from a rough surface. (*b*) Diffuse reflection of colored lights from a sidewalk. *((b) Pete Saloutos/The Stock Market.)*

RELATIVE INTENSITY OF REFLECTED AND TRANSMITTED LIGHT

The fraction of light energy reflected at a boundary, such as an air–glass interface, depends in a complicated way on the angle of incidence, the orientation of the electric field vector associated with the wave, and the indices of refraction of the two media. For the special case of normal incidence ($\theta_1 = \theta_1' = 0$), the reflected intensity can be shown to be

$$I = \left(\frac{n_1 - n_2}{n_1 + n_2}\right)^2 I_0 \qquad 31\text{-}7$$

where I_0 is the incident intensity and n_1 and n_2 are the indices of refraction of the two media.* For a typical case of reflection from a clean air–glass interface for which $n_1 = 1$ and $n_2 = 1.5$, Equation 31-7 gives $I = I_0/25$. Only about 4 percent of the energy is reflected; the remainder of the energy is transmitted.

* An equation for waves on a string that is similar to Equation 31-7 is presented in Section 4 of Chapter 15.

Example 31-3 Refraction from Air to Water

Light traveling in air enters water with an angle of incidence of 45.0°. If the index of refraction of water is 1.33, what is the angle of refraction?

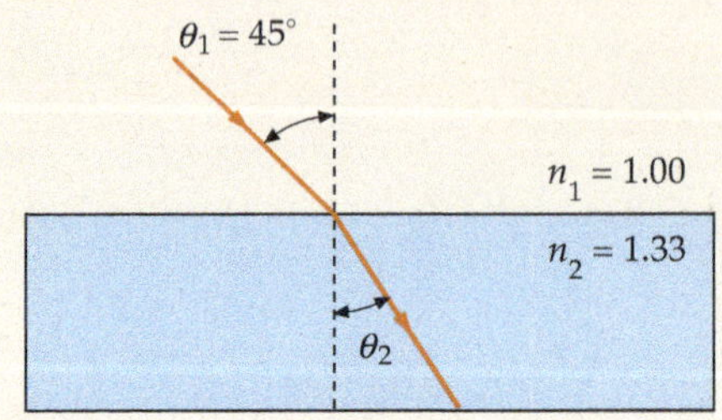

FIGURE 31-12

PICTURE The angle of refraction is found using Snell's law of refraction. Let subscripts 1 and 2 refer to the air and water, respectively. Then $n_1 = 1.00$, $\theta_1 = 45.0°$, $n_2 = 1.33$, and θ_2 is the angle of refraction (Figure 31-12).

SOLVE

1. Use Snell's law of refraction to solve for $\sin\theta_2$, the sine of the angle of refraction:

$$n_1 \sin\theta_1 = n_2 \sin\theta_2$$

so

$$\sin\theta_2 = \frac{n_1}{n_2}\sin\theta_1$$

2. Find the angle whose sine is 0.532:

$$\theta_2 = \sin^{-1}\left(\frac{n_1}{n_2}\sin\theta_1\right) = \sin^{-1}\left(\frac{1.00}{1.33}\sin 45.0°\right)$$

$$= \sin^{-1}(0.532) = \boxed{32.1°}$$

CHECK When entering a medium in which light travels more slowly, light bends toward the normal; so we expect θ_2 to be less than θ_1. The step-2 result meets this expectation.

TAKING IT FURTHER Note that the light is bent toward the normal as the light travels into the medium with the larger index of refraction.

TOTAL INTERNAL REFLECTION

Figure 31-13 shows a point source in glass and rays striking the glass–air interface at various angles. All the rays not perpendicular to the interface are bent away from the normal. As the angle of incidence is increased, the angle of refraction increases until a critical angle of incidence θ_c is reached for which the angle of refraction is 90°. For incident angles greater than the critical angle, no refracted ray exists. All the energy is reflected. This phenomenon is called **total internal reflection.** The critical angle can be found in terms of the indices of refraction of the two media by solving Equation 31-5*b* ($n_1 \sin\theta_1 = n_2 \sin\theta_2$) for $\sin\theta_1$ and setting θ_2 equal to 90°.

(a)

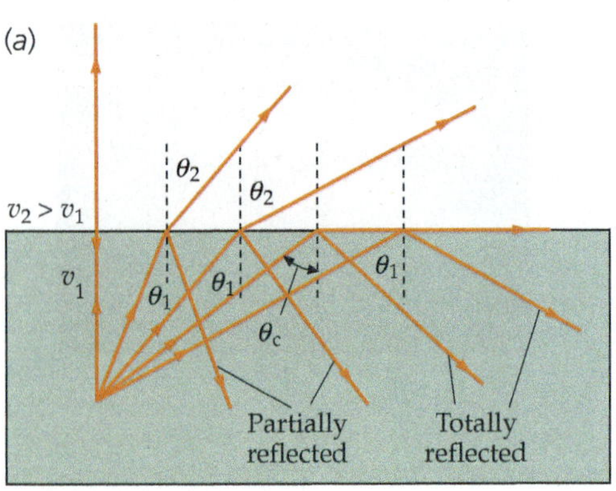

(b)

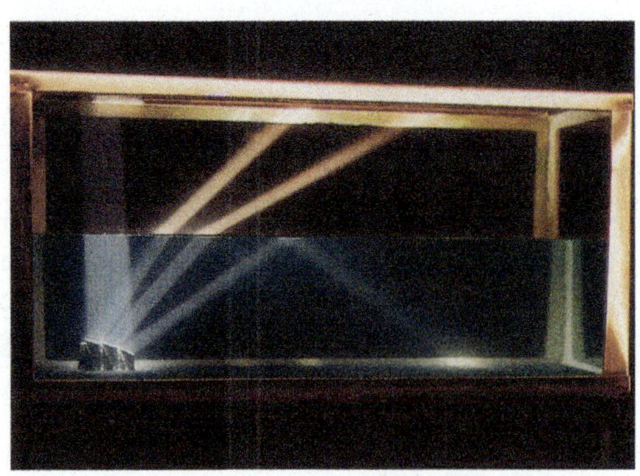

FIGURE 31-13 (*a*) Total internal reflection. As the angle of incidence is increased, the angle of refraction is increased until, at a critical angle of incidence θ_c, the angle of refraction is 90°. For angles of incidence greater than the critical angle, there is no refracted ray. (*b*) A photograph of refraction and total internal reflection from a water–air interface. (*Ken Kay/Fundamental Photographs.*)

That is,

$$\sin\theta_c = \frac{n_2}{n_1}\sin 90° = \frac{n_2}{n_1} \qquad \text{31-8}$$

CRITICAL ANGLE FOR TOTAL INTERNAL REFLECTION

Note that total internal reflection occurs only when the incident light is in the medium that has the higher index of refraction. Mathematically, if n_2 is greater than n_1, Snell's law of refraction cannot be satisfied because there is no angle whose sine is greater than 1.

Example 31-4 Total Internal Reflection

Try It Yourself

A particular glass has an index of refraction of $n = 1.50$. What is the critical angle for total internal reflection for light leaving the glass and entering air, for which $n = 1.00$?

PICTURE Apply the law of refraction (Equation 31-5*b*) with the angle of refraction equal to 90°.

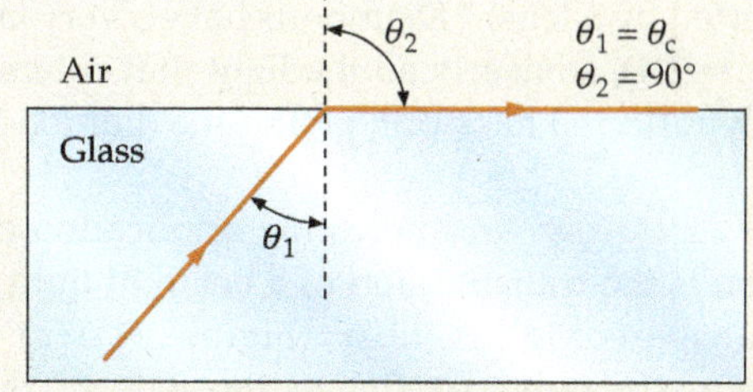

FIGURE 31-14

SOLVE

Cover the column to the right and try these on your own before looking at the answers.

Steps	Answers
1. Make a diagram (Figure 31-14) showing the incident and refracted rays. For the critical angle, the angle of refraction is 90°.	
2. Apply the law of refraction (Equation 31-5*b*). The critical angle is the angle of incidence.	$\theta_c = \boxed{41.8°}$

CHECK Figure 31-13*b* shows that the critical angle for a water-air boundary is a bit larger than 45°. Because the index of refraction for glass is somewhat larger than the index for water, we expect a critical angle for a glass-water boundary to be a bit less than 45°. An angle of 41.8° meets this expectation.

Example 31-5 How Deep Are You?

Context-Rich

You are spending time at the pool. While under the water, you look up and notice that you see objects above water level in a circle of light of radius approximately 2.0 m, and the rest of your vision is the color of the sides of the pool. How deep are you below the surface of the water?

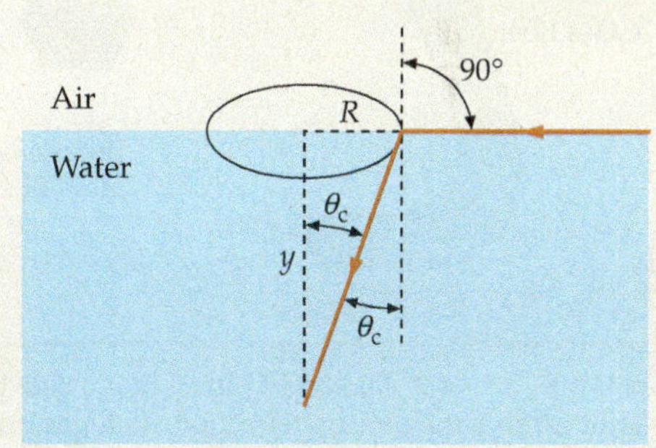

FIGURE 31-15

PICTURE We can determine the depth of the pool from the radius of the light and the angle at which the light is entering your eye from the edge of the circle. At the edge of the circle the light is entering the water at 90.0°, so the angle of refraction at the air–water surface is the critical angle for total internal reflection at the water–air surface. From Figure 31-15, we see that the depth y is related to this angle and the radius of the circle R by $\tan\theta_c = R/y$. The critical angle is found from Equation 31-8 with $n_2 = 1.00$ and $n_1 = 1.33$.

SOLVE

1. The depth y is related to the radius of the circle R and the critical angle θ_c:
$$\tan\theta_c = R/y$$

2. Solve for the depth y:
$$y = \frac{R}{\tan\theta_c}$$

3. Find the critical angle for total internal reflection at a water–air surface:
$$\sin\theta_c = \frac{n_2}{n_1} = \frac{1.00}{1.33} = 0.752$$
$$\theta_c = 48.8°$$

4. Solve for the depth y:
$$y = \frac{R}{\tan\theta_c} = \frac{2.0\text{ m}}{\tan 48.8°} = \boxed{1.7\text{ m}}$$

CHECK The step-4 result seems like a plausible value. Most swimming pools are at least that deep.

Figure 31-16*a* shows light reaching one of the short sides of a 45–45–90° glass prism at normal incidence. If the index of refraction of the glass is 1.5, the critical angle for total internal reflection is 41.8°, as we calculated in Example 31-4. Because the angle of incidence of the ray on the glass–air interface is 45°, the light will be totally reflected and will exit perpendicular to the other face of the prism, as shown. In Figure 31-16*b*, the light is incident perpendicular to the hypotenuse of the prism and is totally reflected twice so that it emerges at 180° to its original direction. Prisms are used to change the directions of light rays. In binoculars, two prisms are used on each side. These prisms reflect the light, thus shortening the required length, and reinvert the image (first inverted by a lens).* Diamonds have a very large index of refraction ($n \approx 2.4$), so nearly all the light that enters a diamond is eventually reflected back out, giving the diamond its sparkle.

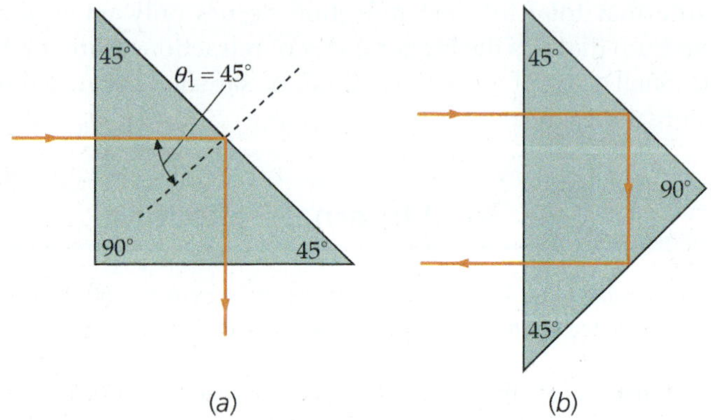

FIGURE 31-16 (*a*) Light entering through one of the short sides of a 45–45–90° glass prism is totally reflected. (*b*) Light entering through the long side of the prism is totally reflected twice.

Fiber optics An interesting application of total internal reflection is the transmission of a beam of light down a long, narrow, transparent glass fiber (Figure 31-17*a*). If the beam begins approximately parallel to the axis of the fiber, it will strike the walls of the fiber at angles greater than the critical angle (if the bends in the fiber are not too sharp) and no light energy will be lost through the walls of the fiber. A bundle of such fibers can be used for imaging, as illustrated in Figure 31-17*b*.

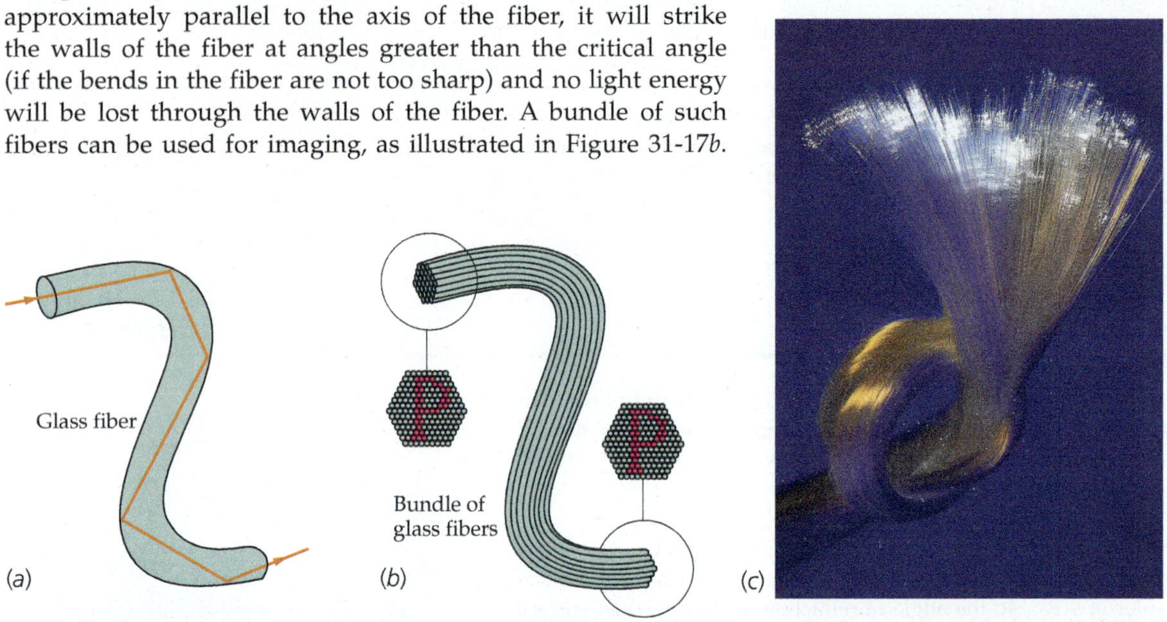

FIGURE 31-17 (*a*) A light pipe. Light inside the pipe is always incident at an angle greater than the critical angle, so no light escapes the pipe by refraction. (*b*) Light from the object is transported by a bundle of glass fibers to form an image of the object at the other end of the pipe. (*c*) Light emerging from a bundle of glass fibers. *((c) Ted Horowitz/The Stock Market.)*

Fiber optics has many applications in medicine and in communications. In medicine, light is transmitted along tiny fibers to visually probe various internal organs without surgery. In communications, the rate at which information can be transmitted is related to the signal frequency. A transmission system using light of frequencies of the order of 10^{14} Hz can transmit information at a much greater rate than one using radio waves, which have frequencies of the order of 10^6 Hz. In telecommunication systems, a single glass fiber that is the thickness of a human hair can transmit audio or video information equivalent to 32 000 voices speaking simultaneously.

* The image produced by the objective lens of a telescope is discussed in Section 32-4.

MIRAGES

When the index of refraction of a medium changes gradually, the refraction is continuous, leading to a gradual bending of the light. An interesting example of this is the formation of a mirage. On a hot and sunny day, the surface of exposed rocks, pavement, and sand often gets very hot. In this case there is often a layer of air near the ground that is warmer, and therefore less dense, than the air just above it. The speed of any light wave is slightly greater in this less dense layer, so a light beam passing from the cooler layer into the warmer layer is bent. Figure 31-18*a* shows the light from a tree when all the surrounding air is at the same temperature. The wavefronts are spherical, and the rays are straight lines. In Figure 31-18*b*, the air near the ground is warmer, resulting in the wavefronts traveling faster there. The portions of the wavefronts near the hot ground get ahead of the higher portions, creating a nonspherical wavefront and causing a curving of the rays. Thus, the ray shown initially heading for the ground is bent upward. As a result, the viewer sees an image of the tree looking as if it were reflected off a water surface on the ground. When driving on a hot sunny day, you may have noticed apparent wet spots on the highway ahead that disappear as you approach them. These mirages are due to the refraction of light from the sky by a layer of air that has been heated due to its proximity to the hot pavement.

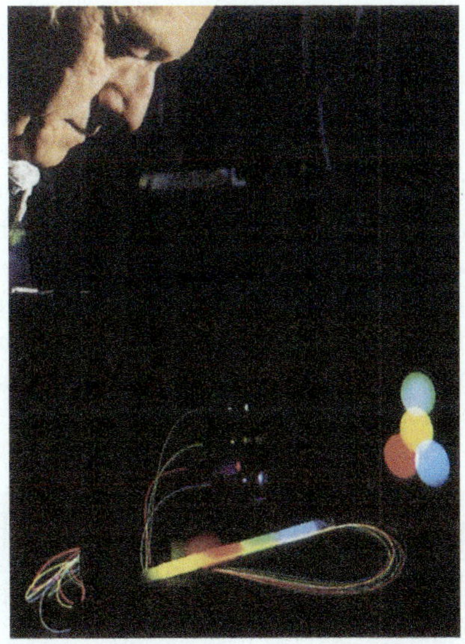
(*a*)

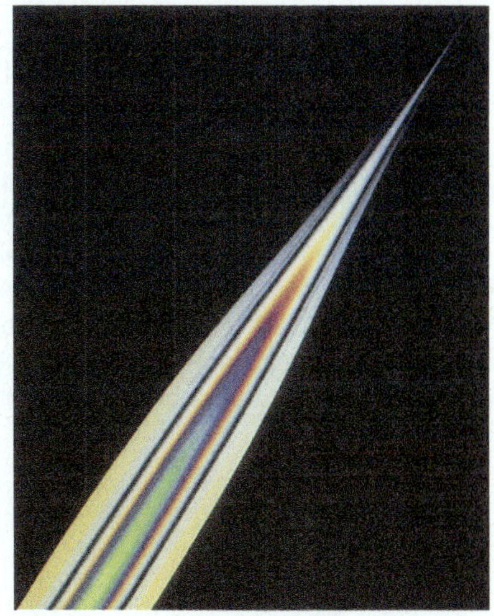
(*b*)

(*a*) In this demonstration at the Naval Research Laboratory, a combination of laser sources generates different colors that excite adjacent fiber sensor elements, leading to a separation of the information as indicated by the separation of the colors. (*b*) The tip of a light guide preform is softened by heat and drawn into a long, tiny fiber. The colors in the preform indicate a layered structure of differing compositions, which is retained in the fiber. *((a) Dan Boyd/Courtesy Naval Research Laboratory. (b) Courtesy AT&T Archives.)*

(*a*)

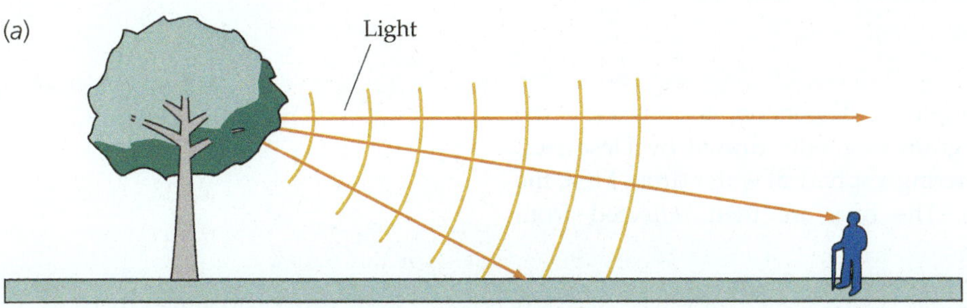

(*b*)

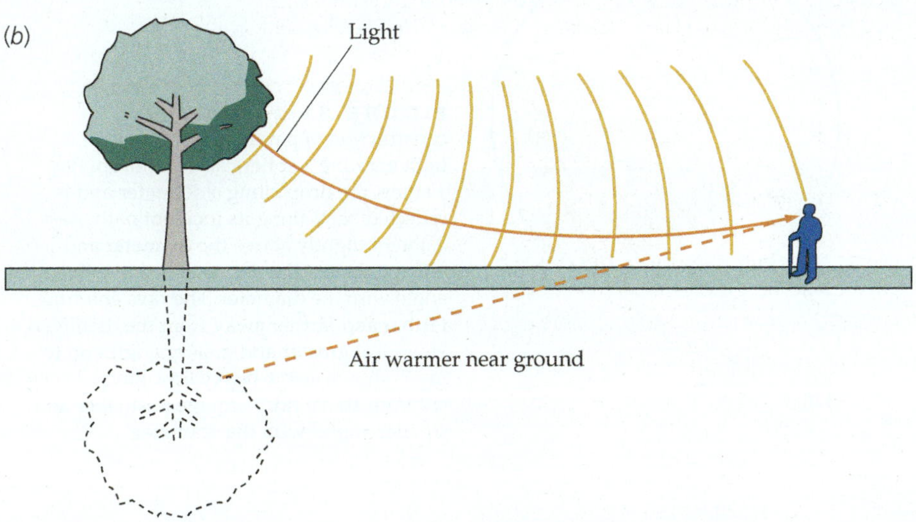

(*c*)

FIGURE 31-18 A mirage. (*a*) When the air is at a uniform temperature, the wavefronts of the light from the tree are spherical. (*b*) When the air near the ground is warmer, the wavefronts are not spherical and the light from the tree is continuously refracted into a curved path. (*c*) Apparent reflections of motorcycles on a hot road. *(Robert Greenler.)*

DISPERSION

The index of refraction of a material has a slight dependence on wavelength. For many materials, n decreases slightly as the wavelength increases, as shown in Figure 31-19. The dependence of the index of refraction on wavelength (and therefore on frequency) is called **dispersion.** When a beam of white light is incident at some angle on the surface of a glass prism, the angle of refraction (which is measured relative to the normal) for the shorter wavelengths is slightly smaller than the angle of refraction for the longer wavelengths. The light of shorter wavelength (toward the violet end of the spectrum) is therefore bent more toward the normal than that of longer wavelength. The beam of white light is thus spread out or dispersed into its component colors or wavelengths (Figure 31-20).

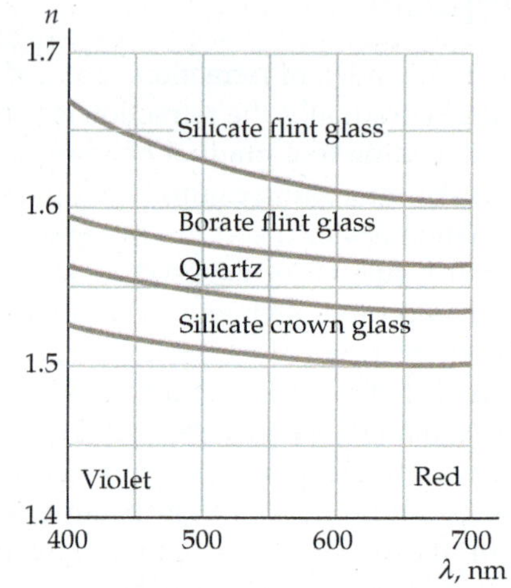

FIGURE 31-19 Index of refraction versus wavelength for various materials.

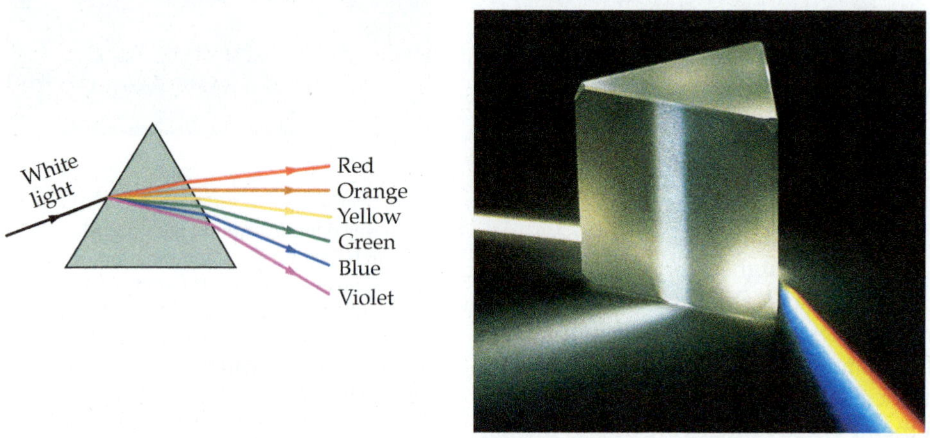

FIGURE 31-20 A beam of white light incident on a glass prism is dispersed into its component colors. The index of refraction decreases as the wavelength increases so that the longer wavelengths (red) are bent less than the shorter wavelengths (blue). *(David Parker/Science Photo Library/Photo Researchers.)*

Rainbows The rainbow is a familiar example of dispersion, in this case the dispersion of sunlight. Figure 31-21 is a diagram originally drawn by Descartes, showing parallel rays of light from the Sun entering a spherical water drop. First, the rays are refracted as they enter the drop. The rays are then reflected from the water–air interface on the other side of the drop and finally are refracted again as they leave the drop.

From Figure 31-21, we can see that the angle made by the emerging rays and the diameter (along ray 1) reaches a maximum around ray 7 and then decreases. The concentration of rays emerging at approximately the maximum angle gives rise to the rainbow. By construction, using the law of refraction, Descartes showed that the maximum angle is about 42°. To observe a rainbow, we must therefore look at the water drops at an angle of 42° relative to the line back to the Sun, as shown in Figure 31-22. The angular radius of the rainbow is therefore 42°.

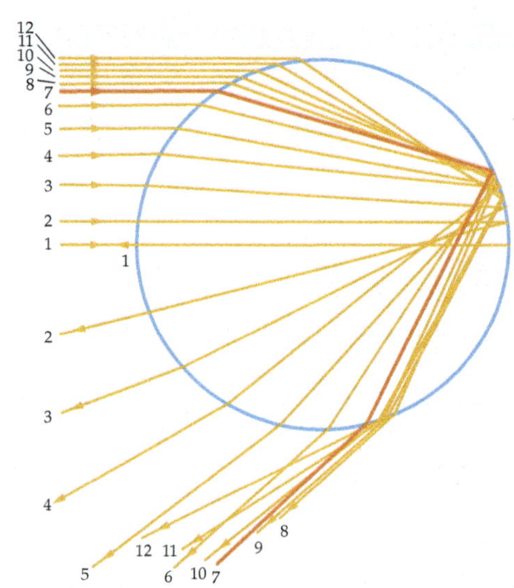

FIGURE 31-21 Descartes's construction of parallel rays of light entering a spherical water drop. Ray 1 enters the drop along a diameter and is reflected back along its incident path. Ray 2 enters slightly above the diameter and emerges below the diameter at a small angle with the diameter. The rays entering farther and farther away from the diameter emerge at greater and greater angles up to ray 7, shown as the heavy line. Rays entering above ray 7 emerge at smaller and smaller angles with the diameter.

The separation of the colors in the rainbow results from the fact that the index of refraction of water depends slightly on the wavelength of light. The angular radius of the rainbow will therefore depend slightly on the wavelength of the light. The observed rainbow is made up of light rays from many different droplets of water (Figure 31-23). The color seen at a particular angular radius corresponds to the wavelength of light that allows the light to reach the eye from the droplets at that angular radius. Because n_{water} is smaller for red light than for blue light, the red part of the rainbow is at a slightly greater angular radius than the blue part of the rainbow, so red is at the outer side of the rainbow.

When a light ray strikes a surface separating water and air, part of the light is reflected and part of the light is refracted. A secondary rainbow results from the light rays that are reflected twice within a droplet (Figure 31-24). The secondary rainbow has an angular radius of 51°, and its color sequence is the reverse of that of the primary rainbow; that is, the violet is on the outside in the secondary rainbow. Because of the small fraction of light reflected from a water–air interface, the secondary rainbow is considerably fainter than the primary rainbow.

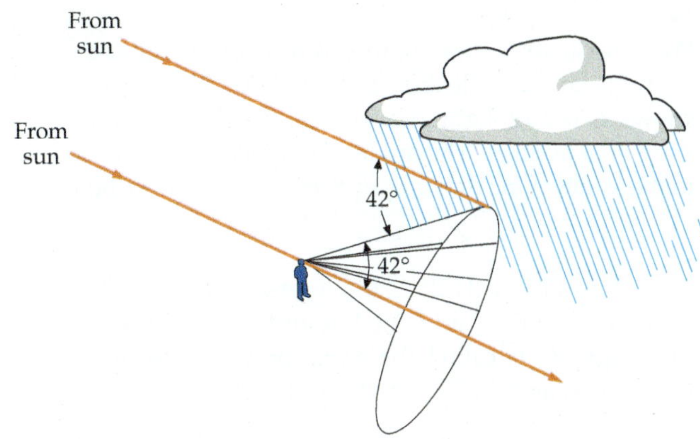

FIGURE 31-22 A rainbow is viewed at an angle of 42° from the line to the Sun, as predicted by Descartes's construction, as shown in Figure 31-21.

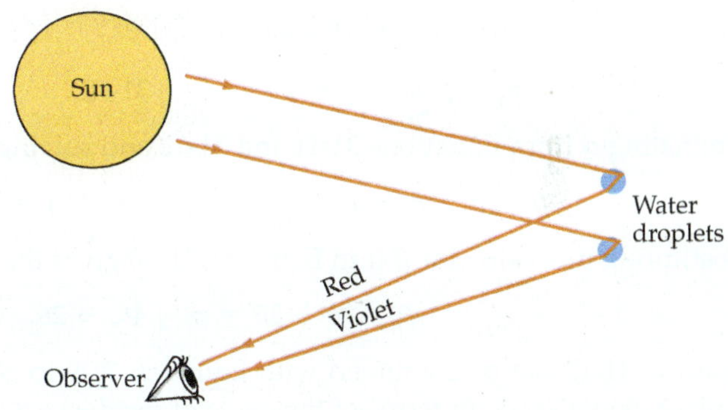

FIGURE 31-23 The rainbow results from light rays from many different water droplets.

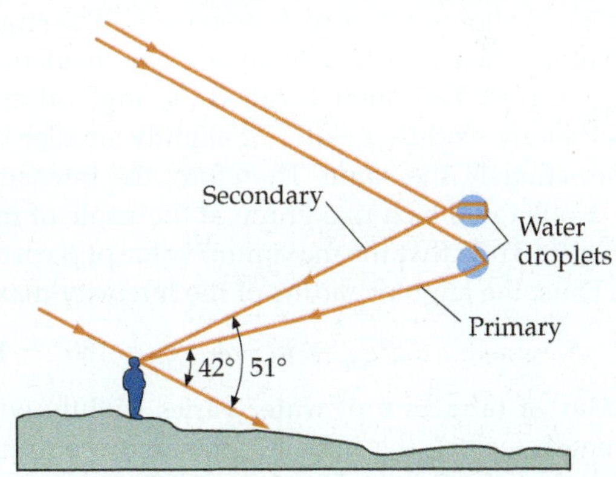

FIGURE 31-24 The secondary rainbow results from light rays that are reflected twice within a water droplet.

(*a*) This 22° halo around the Sun results from refraction by hexagonal ice crystals that are randomly oriented in the upper atmosphere. (*b*) When the ice crystals are not randomly oriented but are falling with their flat bases horizontal, only parts of the halo to the left and to the right of the Sun, called *sun dogs*, are seen. *((a) Robert Greenler. (b) Giovanni DeAmici, NSF, Lawrence Berkeley Laboratory.)*

*Calculating the angular radius of the rainbow

We can calculate the angular radius of the rainbow from the laws of reflection and refraction. Figure 31-25 shows a ray of light incident on a spherical water droplet at point A. The angle of refraction θ_2 is related to the angle of incidence θ_1 by Snell's law of refraction:

$$n_{\text{air}} \sin\theta_1 = n_{\text{water}} \sin\theta_2 \qquad 31\text{-}9$$

Point P in Figure 31-25 is the intersection of the line of the incident ray and the line of the emerging ray. The angle ϕ_d is called the angle of deviation of the ray, and ϕ_d and 2β form a straight angle. Thus,

$$\phi_d + 2\beta = \pi \qquad 31\text{-}10$$

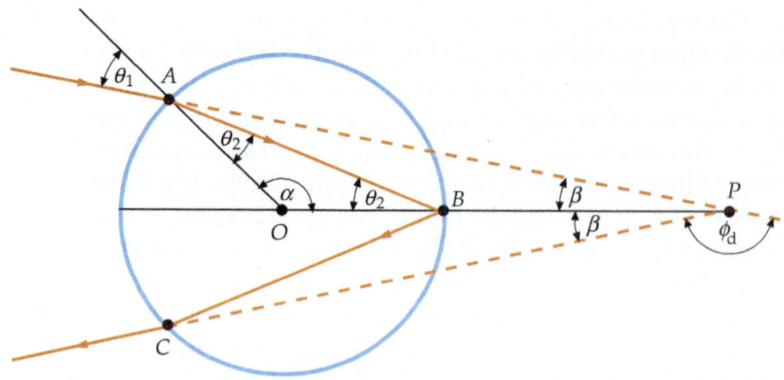

FIGURE 31-25 Light ray incident on a spherical water drop. The refracted ray strikes the back of the water droplet at point B. It makes an angle θ_2 with the radial line OB and is reflected at an equal angle. The ray is refracted again at point C, where it leaves the droplet.

We wish to relate the angle of deviation ϕ_d to the angle of incidence θ_1. From the triangle AOB, we have

$$2\theta_2 + \alpha = \pi \qquad 31\text{-}11$$

Similarly, from the triangle AOP, we have

$$\theta_1 + \beta + \alpha = \pi \qquad 31\text{-}12$$

Eliminating α from Equations 31-11 and 31-12 and solving for β gives

$$\beta = \pi - \theta_1 - \alpha = \pi - \theta_1 - (\pi - 2\theta_2) = 2\theta_2 - \theta_1$$

Substituting this value for β into Equation 31-10 gives the angle of deviation:

$$\phi_d = \pi - 2\beta = \pi - 4\theta_2 + 2\theta_1 \qquad 31\text{-}13$$

Equation 31-13 can be combined with Equation 31-9 to eliminate θ_2 and give the angle of deviation ϕ_d in terms of the angle of incidence θ_1:

$$\phi_d = \pi + 2\theta_1 - 4\sin^{-1}\left(\frac{n_{\text{air}}}{n_{\text{water}}}\sin\theta_1\right) \qquad 31\text{-}14$$

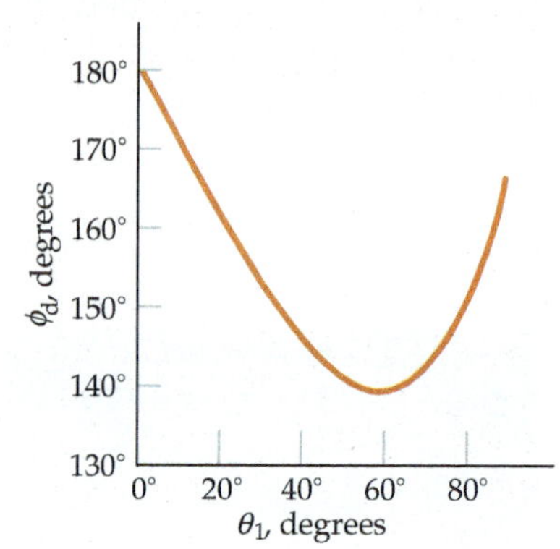

FIGURE 31-26 Plot of the angle of deviation ϕ_d as a function of incident angle θ_1. The angle of deviation has its minimum value of 138° when the angle of incidence is 60°. Because $d\phi_d/d\theta_1 = 0$ at minimum deviation, the deviation of rays with incident angles slightly less or slightly greater than 60° will be approximately the same.

Figure 31-26 shows a plot of ϕ_d versus θ_1. The angle of deviation ϕ_d has its minimum value when $\theta_1 \approx 60°$. At an angle of incidence of 60°, the angle of deviation is $\phi_{d,\,\text{min}} = 138°$. This angle is called the **angle of minimum deviation.** At incident angles that are slightly greater or slightly smaller than 60°, the angle of deviation is approximately the same. Therefore, the intensity of the light reflected by the water droplet will be a maximum at the angle of minimum deviation. We can see from Figure 31-25 that the maximum value of β corresponds to the minimum value of ϕ_d. Thus, the angular radius of the intensity maximum, given by $2\beta_{\text{max}}$, is

$$2\beta_{\text{max}} = \pi - \phi_{d,\,\text{min}} = 180° - 138° = 42° \qquad 31\text{-}15$$

The index of refraction of water varies slightly with wavelength. Thus, for each wavelength (color), the intensity maximum occurs at an angular radius slightly different than that of neighboring wavelengths.

31-4 POLARIZATION

In an electromagnetic wave, the direction of the electric field is perpendicular to the direction of propagation of the wave. If the electric field remains parallel to a line perpendicular to the direction of propagation, the wave is said to be **linearly polarized.** A wave produced by an electric dipole antenna is polarized with the electric field vector at any field point remaining in the plane containing the field point and the antenna axis. Waves produced by numerous sources are usually not polarized. An incandescent light source, for example, contains millions of atoms acting independently. The electric field for such a wave can be resolved into x and y components that vary randomly, because there is no correlation between the individual atoms producing the light.

The polarization of electromagnetic waves can be demonstrated with microwaves, which have wavelengths on the order of centimeters. In a typical microwave generator, polarized waves are radiated by an electric dipole antenna. In Figure 31-27, the electric dipole antenna is vertical, so the electric field vector $\vec{E}$ of the horizontally radiated waves is also vertical. An absorber can be made of a screen of parallel straight wires. When the wires are vertical, as in Figure 31-27*a*, the electric field parallel to the wires sets up currents in the wires and energy is absorbed. When the wires are horizontal and therefore perpendicular to $\vec{E}$, as in Figure 31-27*b*, no currents are set up and the waves are transmitted.

There are four phenomena that produce polarized electromagnetic waves from unpolarized waves: (1) absorption, (2) reflection, (3) scattering, and (4) birefringence (also called double refraction), each of which is examined in the upcoming sections.

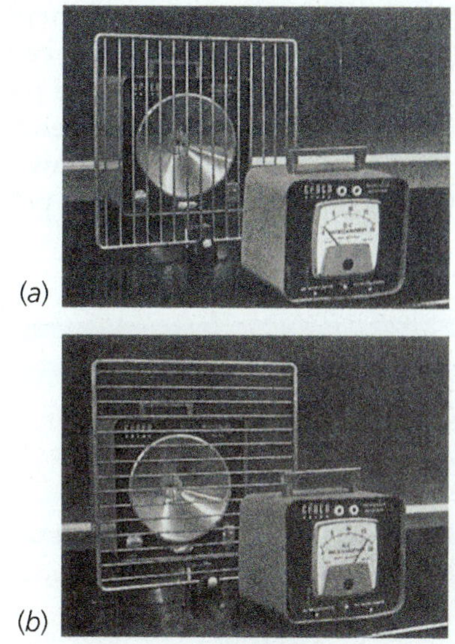

FIGURE 31-27 Demonstration showing the polarization of microwaves. The electric field of the microwaves is vertical, parallel to the vertical dipole antenna. (*a*) When the metal wires of the absorber are vertical, electric currents are set up in the wires and energy is absorbed, as indicated by the low reading on the microwave detector. (*b*) When the wires are horizontal, no currents are set up and the microwaves are transmitted, as indicated by the high reading on the detector. *(Larry Langrill.)*

POLARIZATION BY ABSORPTION

Several naturally occurring crystals, when cut into appropriate shapes, absorb and transmit light differently depending on the polarization of the light. These crystals can be used to produce linearly polarized light. In 1938, E. H. Land invented a simple commercial polarizing sheet called Polaroid. This material contains long-chain hydrocarbon molecules that are aligned when the sheet is stretched in one direction during the manufacturing process. These chains become conducting at optical frequencies when the sheet is dipped in a solution containing iodine. When light is incident with its electric field vector parallel to the chains, electric currents are set up along the chains, and the light energy is absorbed, just as the microwaves are absorbed by the wires in Figure 31-27. If the electric field is perpendicular to the chains, the light is transmitted. The direction perpendicular to the chains is called the **transmission axis.** We will make the simplifying assumption that all the light is transmitted when the electric field is parallel to the transmission axis and all the light is absorbed when it is perpendicular to the transmission axis. (In reality, polarizing sheets absorb some of the light, even when the electric field is parallel to the transmission axis.)

Consider an unpolarized light beam incident on a polarizing sheet with its transmission axis along the x direction, as shown in Figure 31-28. The beam is incident on a second polarizing sheet, the analyzer, whose transmission axis makes an angle θ with the x axis. If E is the electric field amplitude of the incident beam, the component parallel with the transmission axis is $E_{\parallel} = E\cos\theta$, and the component perpendicular to the transmission axis is $E_{\perp} = E\sin\theta$. The sheet absorbs $E_{\perp}$ and transmits $E_{\parallel}$, so the transmitted beam has an electric field amplitude of $E_{\parallel} = E\cos\theta$ and is linearly polarized in the direction of the transmission axis. Because the intensity of light is proportional to the square of the magnitude of the electric field amplitude, the intensity I of light transmitted by the sheet is given by

$$I = I_0\cos^2\theta \qquad \text{31-16}$$

LAW OF MALUS

where I_0 is the intensity of the incident beam. If we have an incident beam of unpolarized light of intensity I_0 incident on a polarizing sheet, the direction of the incident electric field varies from location to location on the sheet, and at each location it fluctuates in time. At each location, the average value of $\cos^2\theta$ is one-half, so applying Equation 31-16 gives $I = I_0(\cos^2\theta)_{\text{av}} = \frac{1}{2}I_0$, where I is the intensity of the transmitted beam.

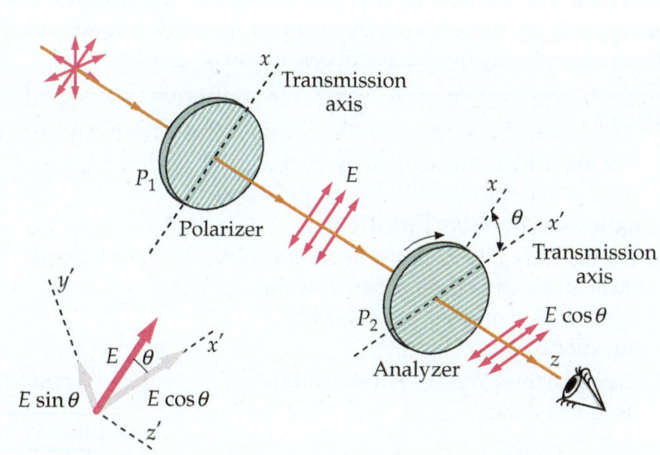

FIGURE 31-28 A vertically polarized beam is incident on a polarizing sheet with its transmission axis x' making an angle θ with the vertical. Only the component $E\cos\theta$ is transmitted through the second sheet, and the transmitted beam is linearly polarized in the direction of the transmission axis x'. If the intensity between the sheets is I_0, the intensity transmitted by the second sheet is $I_0\cos^2\theta$.

When two polarizing elements are placed in succession in a beam of unpolarized light, the first polarizing element is called the **polarizer** and the second polarizing element is called the **analyzer.** If the polarizer and the analyzer are crossed, that is, if their transmission axes are perpendicular to each other, no light gets through. Equation 31-16 is known as the **law of Malus** after its discoverer, E. L. Malus (1775–1812). It applies to any two polarizing elements whose transmission axes make an angle θ with each other.

(*a*) Crossed polarizers block out all of the light. (*b*) In a liquid crystal display, the crystal is between crossed polarizers. Light incident on the crystal is transmitted because the crystal rotates the direction of polarization of the light 90°. The light is reflected back out through the crystal by a mirror behind the crystal, and a uniform background is seen. When a voltage is applied across a small segment of the crystal, the polarization is not rotated, so no light is transmitted and the segment appears black. *((a) Fundamental Photographs. (b) 1990 PAR/NYC, Inc./Photo by Elizabeth Algieri.)*

Example 31-6 Intensity Transmitted

Unpolarized light of intensity 3.0 W/m² is incident on two polarizing sheets whose transmission axes make an angle of 60° (Figure 31-29). What is the intensity of light transmitted by the second sheet?

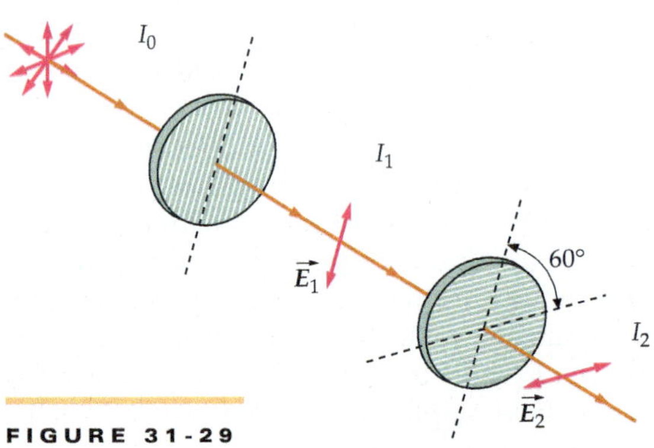

FIGURE 31-29

PICTURE The incident light is unpolarized, so the intensity transmitted by the first polarizing sheet is half the incident intensity. The second sheet further reduces the intensity by a factor of $\cos^2\theta$, with $\theta = 60°$.

SOLVE

1. The intensity I_1 transmitted by the first sheet is half the intensity I_0 of unpolarized light incident on the first sheet: $I_1 = \frac{1}{2}I_0$
2. The intensity I_2 transmitted by the second sheet is related to the intensity I_1 of the light incident on the second sheet by Equation 31-16: $I_2 = I_1\cos^2\theta$
3. Combine these results and substitute the given data: $I_2 = \frac{1}{2}I_0\cos^2 60° = \frac{1}{2}(3.0\text{ W/m}^2)(0.500)^2 = \boxed{0.38\text{ W/m}^2}$

CHECK The first polarizer cuts the intensity in half, so we should expect the intensity transmitted through both sheets to be less than half the incident intensity of 3.0 W/m². The step-3 result meets this expectation.

TAKING IT FURTHER Note that the second sheet rotates the plane of polarization by 60°.

POLARIZATION BY REFLECTION

When unpolarized light is reflected from a plane surface boundary between two transparent media, such as air and glass or air and water, the reflected light is partially polarized. The degree of polarization depends on the angle of incidence and on the ratio of the wave speeds in the two media. For a certain angle of incidence called the polarizing angle θ_p, the reflected light is completely polarized. At the polarizing angle, the reflected and refracted rays are perpendicular to each other. David Brewster (1781–1868), a Scottish scientist and an inventor of numerous instruments (including the kaleidoscope), discovered this experimentally in 1812. The polarizing angle is also referred to as the Brewster angle.

Figure 31-30 shows light incident at the polarizing angle θ_p for which the reflected light is completely polarized. The electric field of the incident light can be resolved into components parallel and perpendicular to the plane of incidence. The reflected light is linearly polarized with its electric field perpendicular to the plane of incidence. We can relate the polarizing angle to the indices of refraction of the media using Snell's law (the law of refraction). If n_1 is the index of refraction of the first medium and n_2 is the index of refraction of the second medium, the law of refraction gives

$$n_1 \sin\theta_p = n_2 \sin\theta_2$$

where θ_2 is the angle of refraction. From Figure 31-30, we can see that the sum of the angle of reflection and the angle of refraction is 90°. Because the angle of reflection equals the angle of incidence, we have

$$\theta_2 = 90° - \theta_p$$

Then

$$n_1 \sin\theta_p = n_2 \sin(90° - \theta_p) = n_2 \cos\theta_p$$

or

$$\tan\theta_p = \frac{n_2}{n_1} \qquad 31\text{-}17$$

POLARIZING ANGLE

Although the reflected light is completely polarized for this angle of incidence, the transmitted light is only partially polarized (because only a small fraction of the incident light is reflected). If the incident light itself is polarized with the electric field in the plane of incidence, no reflected light exists when the angle of incidence is θ_p. We can qualitatively understand this result by using Figure 31-31. If we consider the charges in the atoms next to the surface of the second medium to be driven, by the electric field of the refracted light, to oscillate parallel to the direction of electric field, no reflected ray can exist because for an electric dipole antenna, no energy is radiated along the line of oscillation. (Each of the oscillating atoms is a small electric dipole antenna.)

Because of the polarization of reflected light, sunglasses that contain a polarizing sheet can be very effective in cutting out glare. If light is reflected from a horizontal surface, such as a lake surface or snow on the ground, the electric field of the reflected light will be predominantly horizontal. Polarizing sunglasses with a vertical transmission axis will then reduce glare by absorbing much of the reflected light. If you have polarizing sunglasses, you can observe this effect by looking through the glasses at reflected light and then rotating the glasses 90°; much more of the light will be transmitted.

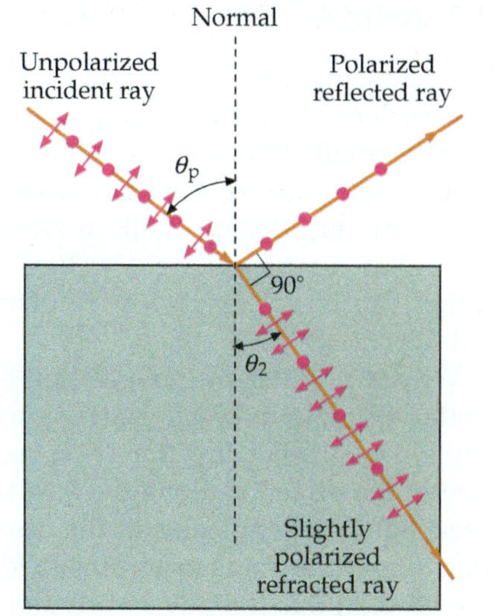

FIGURE 31-30 Polarization by reflection. The incident wave is unpolarized and has components of the electric field parallel to the plane of incidence (arrows) and components perpendicular to that plane (dots). For incidence at the polarizing angle, the reflected wave is completely polarized, with its electric field perpendicular to the plane of incidence.

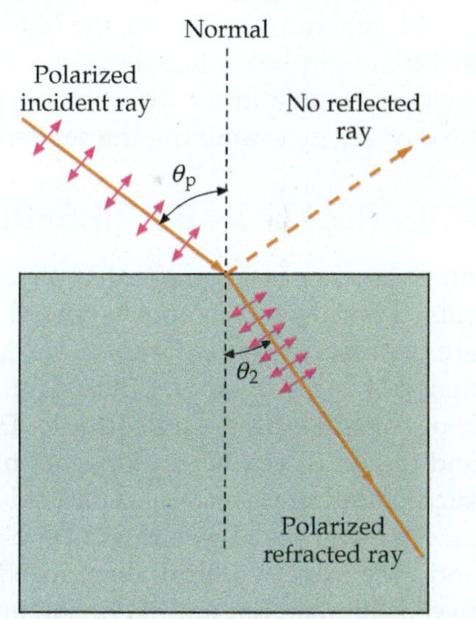

FIGURE 31-31 Polarized light incident at the polarizing angle. When the incident light is polarized with $\vec{E}$ in the plane of incidence, there is no reflected ray.

POLARIZATION BY SCATTERING

The phenomenon of absorption and reradiation is called **scattering.** Scattering can be demonstrated by passing a light beam through a container of water to which a small amount of powdered milk has been added. The milk particles scatter the light, making the light beam visible. Similarly, laser beams can be made visible by introducing chalk or smoke particles into the air to scatter the light. A familiar example of light scattering is that from air molecules, which tend to scatter short wavelengths more than long wavelengths, thereby giving the sky its blue color.

We can understand polarization by scattering if we think of the charges in a scattering atom as electric dipole antennas that radiate waves that have maximum intensities in directions perpendicular to the antenna axes and intensities of zero in the direction along the antenna axis. The electric field vector of the scattered light perpendicular to the direction of propagation is in the plane of the long axis of the antenna and the field point. Figure 31-32 shows a beam of unpolarized light that initially travels along the z axis, striking a particle at the origin. The electric field in the light beam has components in both the x and y directions perpendicular to the direction of motion of the light beam. These fields set up oscillations of the charges within the molecule in the $z = 0$ plane, and no oscillation exists along the z direction. These oscillations can be thought of as a superposition of an oscillation along the x axis and another along the y axis, and each of these oscillations produce dipole radiation. Thus, the oscillation along the x axis produces no radiation along the x axis, which means the light radiated along the x axis is produced only by the oscillation along the y axis. It follows that the light radiated along the x axis is polarized with its electric field parallel with the y axis. There is nothing special about the choice of axes for this discussion, so the result can be generalized. That is, the light scattered in a direction perpendicular to the incident light beam is polarized with its electric field perpendicular to both the incident beam and the direction of propagation of the scattered light. This can be seen easily by examining the scattered light with a piece of polarizing sheet.

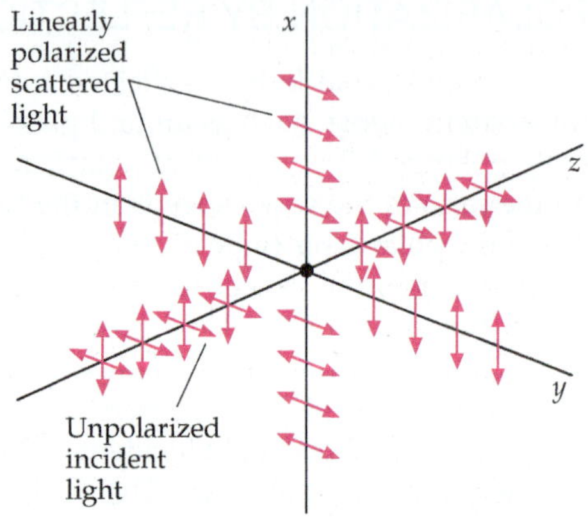

FIGURE 31-32 Polarization by scattering. Unpolarized light propagating in the $+z$ direction is incident on a scattering center at the origin. The light scattered in the $z = 0$ plane along the $\pm x$ direction is polarized parallel with the y axis (and the light scattered in the $\pm y$ direction is polarized parallel with the x axis).

POLARIZATION BY BIREFRINGENCE

Birefringence is a complicated phenomenon that occurs in calcite and other noncubic crystals and in some stressed plastics, such as cellophane. Most materials are **isotropic,** that is, the speed of light passing through the material is independent of the polarization of the light. Because of their microscopic structure, birefringent materials are **anisotropic.** The speed of light depends on the polarization and on the direction of propagation of the light. When a light ray is incident on such materials, it may be separated into two rays called the *ordinary ray* and the *extraordinary ray*. These rays are polarized in mutually perpendicular directions, and they travel with different speeds. Depending on the relative orientation between the material and the incident light beam, the two rays may also travel in different directions.

There is one particular direction in a birefringent material in which both rays propagate with the same speed. This direction is called the **optic axis** of the material. (The optic axis is actually a *direction* rather than a line in the material.) Nothing unusual happens when light travels in the direction of the optic axis. However, when light is incident at an angle to the optic axis, as shown in Figure 31-33, the

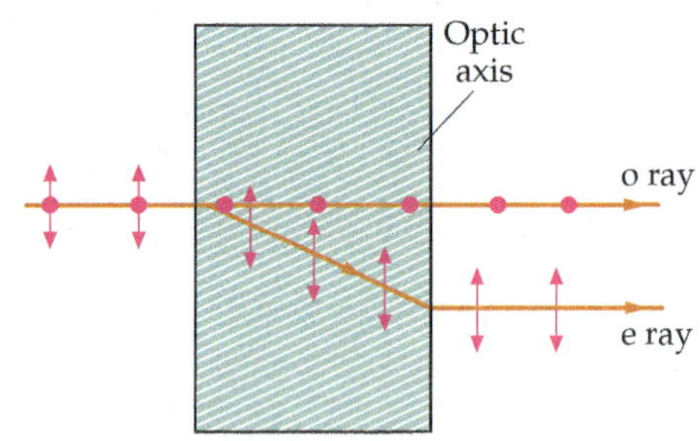

(a)

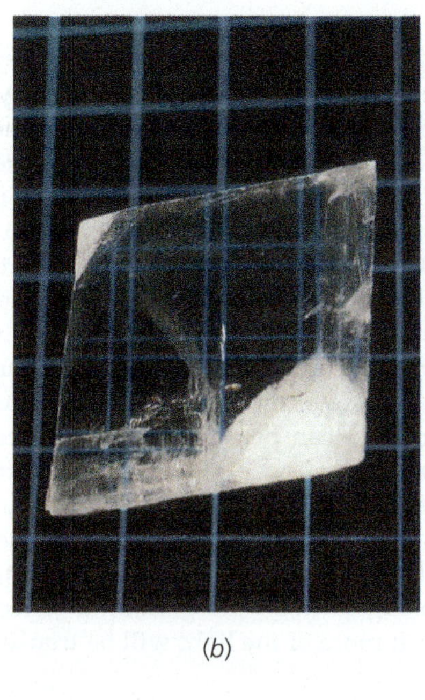

(b)

FIGURE 31-33 (*a*) A narrow beam of light incident on a birefringent crystal such as calcite is split into two beams, called the ordinary ray (o ray) and the extraordinary ray (e ray), that have mutually perpendicular polarizations. If the crystal is rotated, the extraordinary ray rotates in space. (*b*) A double image of the cross-hatching is produced by this birefringent crystal of calcium carbonate. *(Paul Silverman Photographs.)*

rays travel in different directions and emerge separated in space. If the material is rotated, the extraordinary ray (the e ray in the figure) revolves in space around the ordinary ray (o ray).

If light is incident on a birefringent plate perpendicular to its crystal face and perpendicular to the optic axis, the two rays travel in the same direction but at different speeds. The number of wavelengths in the two rays in the plate is different because the wavelengths ($\lambda = v/f$) of the rays differ. The rays emerge with a phase difference that depends on the thickness of the plate and on the wavelength of the incident light. In a **quarter-wave plate,** the thickness is such that a 90° phase difference exists between the waves of a particular wavelength when they emerge. In a **half-wave plate,** the rays emerge with a phase difference of 180°.

Suppose that the incident light is linearly polarized so that the electric field vector is at 45° to the optic axis, as illustrated in Figure 31-34. The ordinary and extraordinary rays start out in phase and have equal amplitudes. With a quarter-wave plate, the waves emerge with a phase difference of 90°, so the resultant electric field has components $E_x = E_0 \sin \omega t$ and $E_y = E_0 \sin(\omega t + 90°) = E_0 \cos \omega t$. The electric field vector thus rotates in a circle and has constant magnitude. Such a wave is said to be **circularly polarized.**

With a half-wave plate, the waves that emerge have a phase difference of 180°, so the resultant electric field is linearly polarized with components $E_x = E_0 \sin \omega t$ and $E_y = E_0 \sin(\omega t + 180°) = -E_0 \sin \omega t$. The net effect is that the direction of polarization of the wave is rotated by 90° relative to that of the incident light, as shown in Figure 31-35.

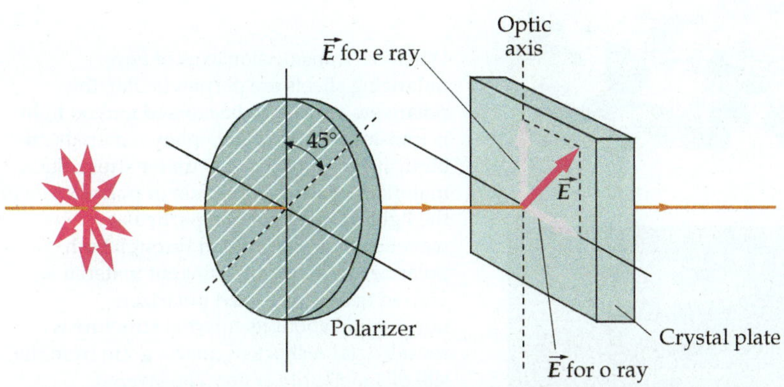

FIGURE 31-34 Polarized light emerging from the polarizer is incident on a birefringent crystal so that the electric field vector makes a 45° angle with the optic axis, which is perpendicular to the light beam. The ordinary and extraordinary rays travel in the same direction but at different speeds. The polarization of the emerging light depends on the thickness of the crystal and the wavelength of the light.

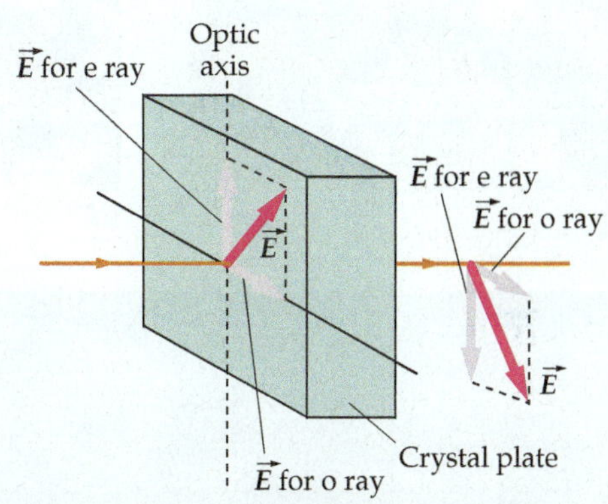

FIGURE 31-35 If the birefringent crystal in Figure 31-34 is a half-wave plate, and if the electric field vector of the incident light makes an angle of 45° with the optic axis, then the direction of polarization of the emerging light is rotated by 90°.

Interesting and beautiful patterns can be observed by placing birefringent materials, such as cellophane or stressed plastic, between two polarizing sheets with their transmission axes perpendicular to each other. Ordinarily, no light is transmitted through crossed polarizing sheets. However, if we place a birefringent material between the crossed polarizing sheets, the material acts as a half-wave plate for light of a certain color depending on the material's thickness. The direction of polarization is rotated and some light gets through both sheets. Various glasses and plastics become birefringent when under stress. The stress patterns can be observed when the material is placed between crossed polarizing sheets.

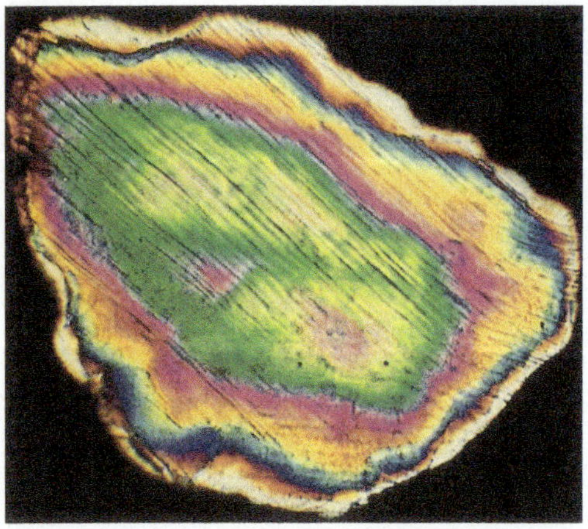

(*a*)

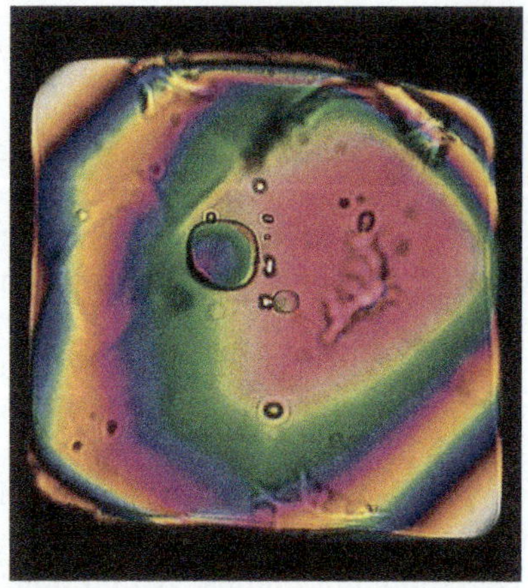

(*b*)

(*c*)

(*d*)

(*e*)

When the transmission axes of two polarizing sheets are perpendicular, the polarizers are said to be crossed and no light is transmitted. However, many materials are birefringent or become so under stress. Such materials rotate the direction of polarization of the light so that light of a particular wavelength is transmitted through both polarizers. When a birefringent material is viewed between crossed polarizers, information about its internal structure is revealed. (*a*) A shocked quartz grain from the site of a meteorite crater. The layered structure, evidenced by the parallel lines, arises from the shock of the impact of the meteor. (*b*) A grain of quartz typically found in silicic volcanic rocks. No shock lines are seen. (*c*) Thin sections of an ice core from the Antarctic ice sheet reveal bubbles of trapped CO_2, which appear amber-colored. The sample was taken from a depth of 194 m, corresponding to air trapped 1600 years ago, whereas the sample in (*d*) is from a depth of 56 m, corresponding to air trapped 450 years ago. Ice core measurements have replaced the less reliable technique of analyzing carbon in tree rings to compare current atmospheric CO_2 levels with those of the recent past. (*e*) Robert Mark of the Princeton School of Architecture examines the stress patterns in a plastic model of the nave structure of Chartres Cathedral. *((a, b) Glen A. Izett, US Geological Survey. (c, d) Dr. Anthony J Gow/Cold Regions Research and Engineering Laboratory, Hanover New Hampshire. (e) Sepp Seitz/Woodfin Camp and Associates.)*

31-5 DERIVATION OF THE LAWS OF REFLECTION AND REFRACTION

The laws of reflection and refraction can be derived from either Huygens's principle or Fermat's principle.

HUYGENS'S CONSTRUCTION

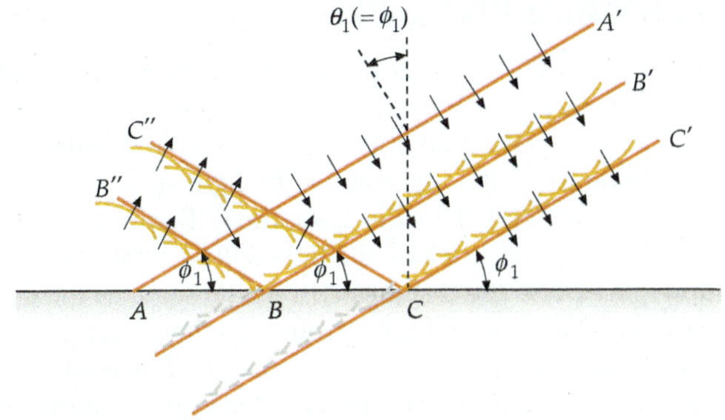

FIGURE 31-36 Plane wave reflected at a plane mirror. The angle θ_1 between the incident ray and the normal to the mirror is the angle of incidence. It is equal to the angle ϕ_1 between the incident wavefront and the mirror.

Reflection Figure 31-36 shows a plane wavefront AA' striking a mirror at point A. As can be seen from the figure, the angle ϕ_1 between the wavefront and the mirror is the same as the angle of incidence θ_1, which is the angle between the normal to the mirror and the rays (which are perpendicular to the wavefronts). According to Huygens's construction, each point on a given wavefront can be considered a point source of secondary wavelets. The position of the wavefront after a time t is found by constructing wavelets of radius ct with their centers on the wavefront AA'. Wavelets that have not yet reached the mirror form the portion of the new wavefront BB'. Wavelets that have already reached the mirror are reflected and form the portion of the new wavefront $B''B$. By a similar construction, the wavefront $C''C$ is obtained from the Huygens's wavelets originating on the wavefront $B''B$. Figure 31-37 is an enlargement of a portion of Figure 31-36 showing AP, which is part of the initial position of the wavefront. During the time t, the wavelet from point P reaches the mirror at point B, and the wavelet from point A reaches point B''. The reflected wavefront $B''B$ makes an angle ϕ_1' with the mirror that is equal to the angle of reflection θ_1' between the reflected ray and the normal to the mirror. The triangles $AB''B$ and APB are both right triangles that have a common side AB and equal sides $AB'' = BP = ct$. Hence, these triangles are congruent, and the angles ϕ_1 and ϕ_1' are equal, implying that the angle of reflection θ_1' equals the angle of incidence θ_1.

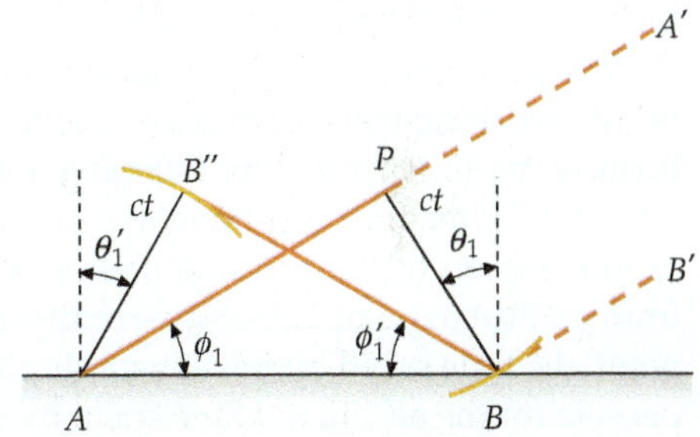

FIGURE 31-37 Geometry of Huygens's construction for the calculation of the law of reflection. The wavefront AP initially strikes the mirror at point A. After a time t, the Huygens wavelet from P strikes the mirror at point B, and the Huygens wavelet from point A reaches point B''.

Refraction Figure 31-38 shows a plane wave incident on an air–glass interface. We apply Huygens's construction to find the wavefront in the transmitted wave. Line AP indicates a portion of the wavefront in medium 1 that strikes the glass surface at an angle ϕ_1. In time t, the wavelet from P travels the distance v_1t and reaches the point B on the line AB separating the two media, while the wavelet from point A travels a shorter distance v_2t into the second medium. The new wavefront BB' is not parallel to the original wavefront AP because the speeds v_1 and v_2 are different. From the triangle APB,

$$\sin\phi_1 = \frac{v_1t}{AB}$$

or

$$AB = \frac{v_1t}{\sin\phi_1} = \frac{v_1t}{\sin\theta_1}$$

using the fact that the angle ϕ_1 equals the angle of incidence θ_1. Similarly, from triangle $AB'B$,

$$\sin\phi_2 = \frac{v_2t}{AB}$$

or

$$AB = \frac{v_2t}{\sin\phi_2} = \frac{v_2t}{\sin\theta_2}$$

where $\theta_2 = \phi_2$ is the angle of refraction. Equating the reciprocals of the two values for AB, we obtain

$$\frac{1}{v_1}\sin\theta_1 = \frac{1}{v_2}\sin\theta_2 \qquad \text{31-18}$$

Multiplying both sides by c, and then substituting n_1 for c/v_1 and n_2 for c/v_2, we obtain $n_1\sin\theta_1 = n_2\sin\theta_2$, which is Snell's law.

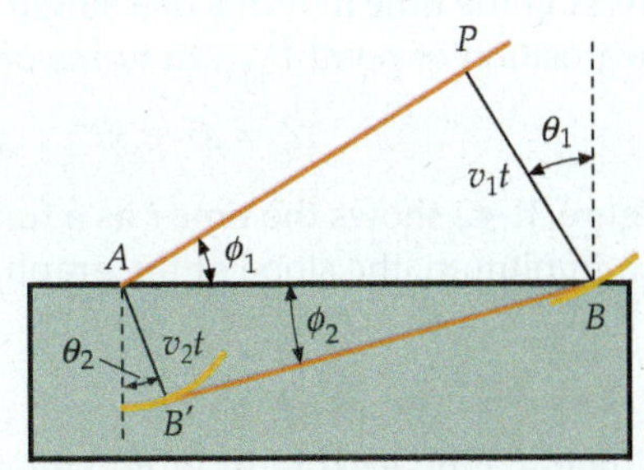

FIGURE 31-38 Application of Huygens's principle to the refraction of plane waves at the surface separating a medium in which the wave speed is v_1 from a medium in which the wave speed v_2 is less than v_1. The angle of refraction θ_2 is less than the angle of incidence θ_1.

FERMAT'S PRINCIPLE

Reflection Figure 31-39 shows two paths in which light leaves point A, strikes the plane surface, which we can consider to be a mirror, and travels to point B. The problem for the application of Fermat's principle to reflection can be stated as follows: At what point P in the figure must the light strike the mirror so that it will travel from point A to point B in the least time? Because the light is traveling in the same medium for this problem, the time will be minimum when the distance is minimum. In Figure 31-39 the distance APB is the same as the distance $A'PB$, where point A' lies along the perpendicular from A to the mirror and is equidistant behind the mirror. As we vary point P, the distance $A'PB$ is least when the points A', P, and B lie on a straight line. We can see from the figure that this occurs when the angle of incidence equals the angle of reflection.

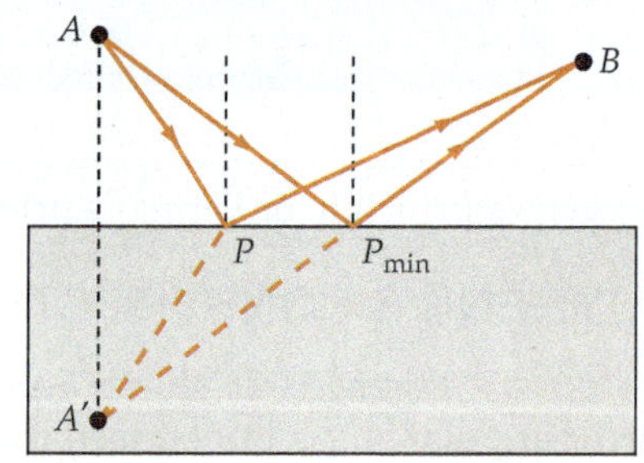

FIGURE 31-39 Geometry for deriving the law of reflection from Fermat's principle. The time it takes for the light to travel from point A to the surface and then on to point B is a minimum for the light striking the surface at point P_{min}.

Refraction The derivation of Snell's law of refraction from Fermat's principle is slightly more complicated. Figure 31-40 shows several possible paths for light traveling from point A in air to point B in glass. Point P_1 is on the straight line between A and B, but this path is not the one for the shortest travel time because light travels at a slower speed in the glass. If we move slightly to the right of P_1, the total path length is longer, but the distance traveled in the slower medium is less than for the path through P_1. It is not apparent from the figure which path is the path of least time. However, it is not surprising that a path slightly to the right of the straight-line path takes less time because the time gained by traveling a shorter distance in the glass more than compensates for the time lost traveling a longer distance in the air. As we move the point of intersection of the possible path to the right of point P_1, the total time of travel from point A to point B decreases until we reach a minimum at point P_{min}. Beyond this point, the time saved by traveling a shorter distance in the glass is not enough to compensate for the additional time spent by traveling the greater distance in the air.

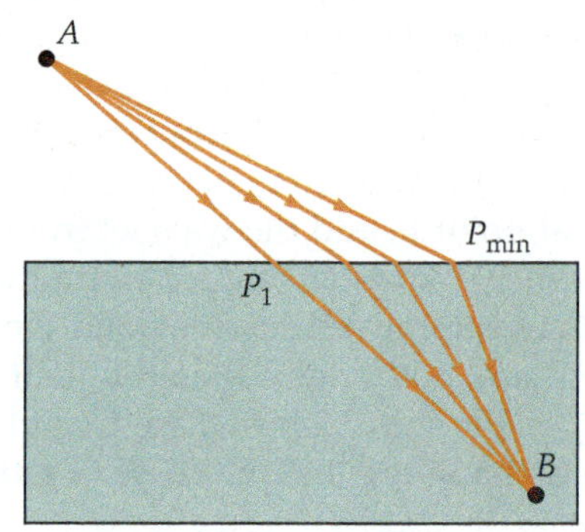

FIGURE 31-40 Geometry for deriving Snell's law from Fermat's principle. The point P_{min} is the point at which light must strike the glass in order that the travel time from point A to point B is a minimum.

Figure 31-41 shows the geometry for finding the path of least time. If L_1 is the distance traveled in medium 1 that has an index of refraction n_1, and L_2 is the distance traveled in medium 2 that has an index of refraction n_2, the time for light to travel the total path AB is

$$t = \frac{L_1}{v_1} + \frac{L_2}{v_2} = \frac{L_1}{c/n_1} + \frac{L_2}{c/n_2} = \frac{n_1L_1}{c} + \frac{n_2L_2}{c} \qquad \text{31-19}$$

We wish to find the point P_{min} for which this time is a minimum. We do this by expressing the time in terms of a single parameter x, as shown in the figure, indicating the position of point P_{min}. In terms of the distance x,

$$L_1^2 = a^2 + x^2 \quad \text{and} \quad L_2^2 = b^2 + (d - x)^2 \qquad \text{31-20}$$

Figure 31-42 shows the time t as a function of x. At the value of x for which the time is a minimum, the slope of the graph of t versus x is zero:

$$\frac{dt}{dx} = 0$$

Differentiating each term in Equation 31-19 with respect to x and setting the result equal to zero, we obtain

$$\frac{dt}{dx} = \frac{1}{c}\left(n_1\frac{dL_1}{dx} + n_2\frac{dL_2}{dx}\right) = 0 \qquad \text{31-21}$$

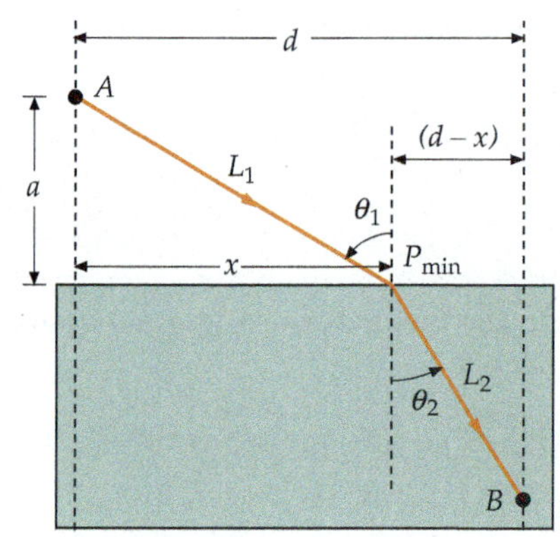

FIGURE 31-41 Geometry for calculating the minimum time in the derivation of Snell's law from Fermat's principle.

We can compute these derivatives from Equations 31-20. We have

$$2L_1\frac{dL_1}{dx} = 2x \quad \text{or} \quad \frac{dL_1}{dx} = \frac{x}{L_1}$$

where x/L_1 is just $\sin\theta_1$, and where θ_1 is the angle of incidence. Thus,

$$\frac{dL_1}{dx} = \sin\theta_1 \qquad \text{31-22}$$

Similarly,

$$2L_2\frac{dL_2}{dx} = 2(d - x)(-1)$$

or

$$\frac{dL_2}{dx} = -\frac{d - x}{L_2} = -\sin\theta_2 \qquad 31\text{-}23$$

where θ_2 is the angle of refraction. From Equation 31-21,

$$n_1\frac{dL_1}{dx} + n_2\frac{dL_2}{dx} = 0 \qquad 31\text{-}24$$

Substituting the results of Equations 31-22 and 31-23 for dL_1/dx and dL_2/dx gives

$$n_1 \sin\theta_1 + n_2(-\sin\theta_2) = 0$$

or

$$n_1 \sin\theta_1 = n_2 \sin\theta_2$$

which is Snell's law.

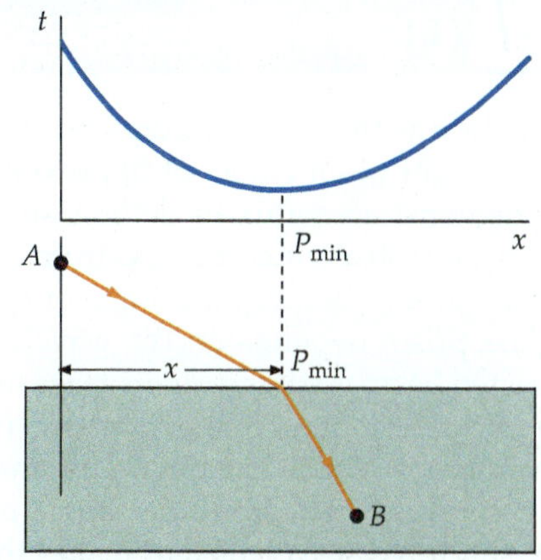

FIGURE 31-42 Graph of the time it takes for light to travel from point A to point B versus x, measured along the refracting surface. The time is a minimum at the point at which the angles of incidence and refraction obey Snell's law.

31-6 WAVE–PARTICLE DUALITY

The wave nature of light was first demonstrated by Thomas Young, who in 1801 observed the interference pattern of two coherent light sources produced by illuminating a pair of narrow, parallel slits with a single source. (Young's experiment is presented in Section 3 of Chapter 33.) The wave theory of light culminated in 1860 with Maxwell's prediction of electromagnetic waves. The particle nature of light was first proposed by Albert Einstein in 1905 in his explanation of the photoelectric effect.* A particle of light called a **photon** has energy E that is related to the frequency f and wavelength λ of the light wave by the Einstein equation

$$E = hf = \frac{hc}{\lambda} \qquad 31\text{-}25$$

EINSTEIN'S EQUATION FOR PHOTON ENERGY

where c is the speed of light and h is Planck's constant:

$$h = 6.626 \times 10^{-34}\,\text{J}\cdot\text{s} = 4.136 \times 10^{-15}\,\text{eV}\cdot\text{s}$$

Because energies are often given in electron volts and wavelengths are given in nanometers, it is convenient to express the combination hc in eV · nm. We have

$$hc = (4.1357 \times 10^{-15}\,\text{eV}\cdot\text{s})(2.9979 \times 10^{8}\,\text{m/s}) = 1.2398 \times 10^{-6}\,\text{eV}\cdot\text{m}$$

or

$$hc = 1240\,\text{eV}\cdot\text{nm} \qquad 31\text{-}26$$

The propagation of light is governed by its wave properties, whereas the exchange of energy between light and matter is governed by its particle properties. This wave–particle duality is a general property of nature. For example, the propagation of electrons (and other so-called particles) is also governed by wave properties, whereas the exchange of energy between the electrons and other particles is governed by particle properties.

* The photoelectric effect is discussed in Chapter 34.

31-7 LIGHT SPECTRA

Newton was the first to recognize that white light is a mixture of light of all colors of approximately equal intensity. He demonstrated this by letting sunlight fall on a glass prism and observing the spectrum of the refracted light (Figure 31-43). Because the angle of refraction produced by a glass prism depends on the wavelength of the light, the refracted beam is spread out in space into its component colors or wavelengths, like a rainbow. Figure 31-44 shows a spectroscope, which is a device for analyzing the spectra of light sources. Light from the source passes through a narrow slit, traverses a lens that makes the beam parallel, and falls on a glass prism, where it is refracted twice (once as it enters the glass and again as it leaves the glass). The refracted beam is viewed with a telescope, which is mounted on a rotating platform so that the angle of the refracted beam, which depends on the wavelength of the light, can be measured. The spectrum of the light source can thus be analyzed in terms of its component wavelengths. The spectrum of sunlight has a continuous range of wavelengths and is therefore called a **continuous spectrum.** The light emitted by the atoms in low-pressure gases, such as mercury atoms in a fluorescent lamp, contains only a discrete set of wavelengths. Each wavelength emitted by the source produces a separate image of the collimating slit in the spectroscope. Such a spectrum is called a **line spectrum.** The continuous visible spectrum and the line spectra from several elements are shown in the photograph.

FIGURE 31-43 Newton demonstrating the spectrum of sunlight with a glass prism. *(Corbis/Bettmann.)*

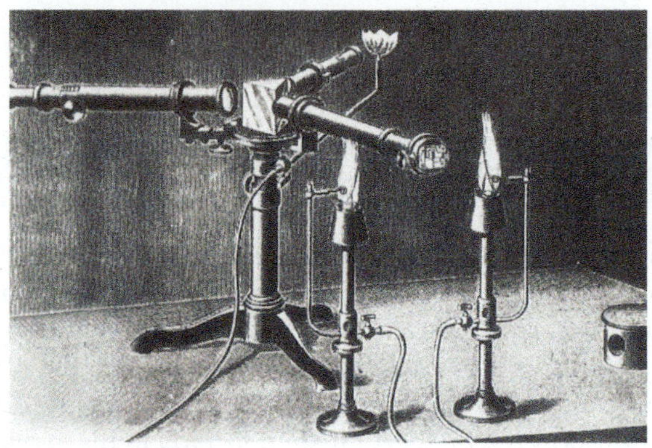

FIGURE 31-44 A late nineteenth-century spectroscope belonging to Gustav Kirchhoff. Modern student spectroscopes usually have the same general design. *(Corbis/Bettmann.)*

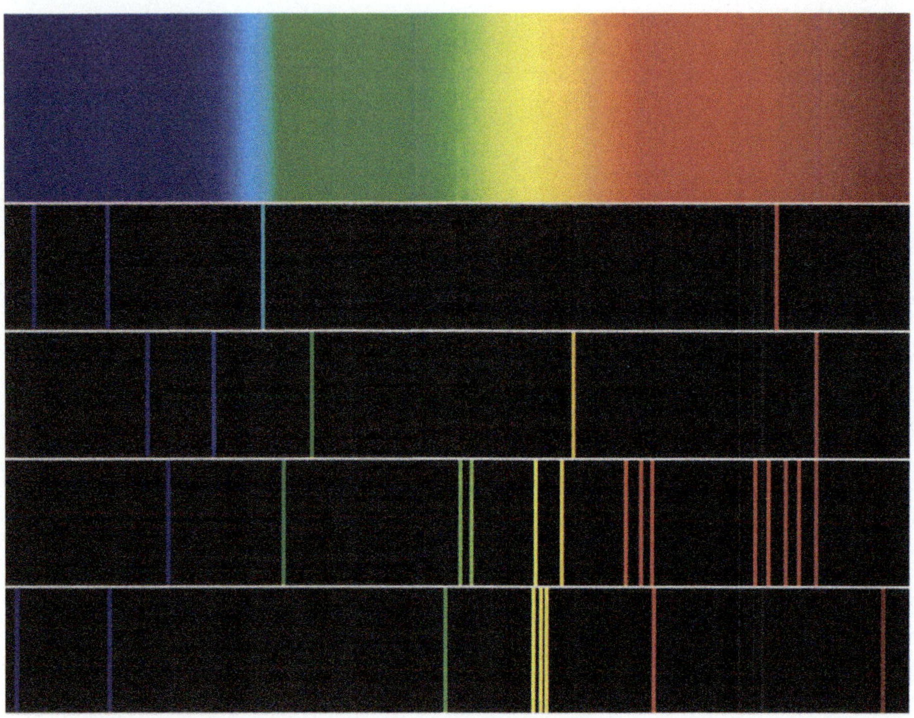

The continuous visible spectrum (top) and the line spectra of (from top to bottom) hydrogen, helium, barium, and mercury. *(Adapted from Eastman Kodak and Wabash Instrument Corporation.)*

*31-8 SOURCES OF LIGHT

LINE SPECTRA

The most common sources of visible light are transitions of the valence electrons in atoms. Normally, an atom is in its ground state where its electrons are at their lowest allowed energy levels consistent with the exclusion principle. (The exclusion principle, which was first proposed by Wolfgang Pauli in 1925 to explain the electronic structure of atoms, states that no two electrons in an atom can have the same quantum state.) The lowest energy electrons are tightly bound to the nucleus, forming a stable core of electrons. The one or two electrons in the highest energy states are much less tightly bound to the nucleus and are relatively easily excited to vacant higher energy states. These outer electrons are responsible for the energy changes in the atom that result in the emission or absorption of visible light.

When an atom collides with another atom or with a free electron, or when the atom absorbs electromagnetic energy, the valence electrons can be excited to higher energy states. After a time of approximately 10 ns (1 ns = 10^{-9} s), the valence electrons spontaneously make transitions to lower energy states with the emission of a photon. This process, called **spontaneous emission,** is random; the photons emitted from two different atoms are not correlated. The emitted light is thus incoherent. By conservation of energy, the energy of an emitted photon is the energy difference $|\Delta E|$ between the initial state and the final state of the atom. The frequency of the light wave is related to the energy by the Einstein equation, $|\Delta E| = hf$ (Equation 31-25). The wavelength of the emitted light is then

$$\lambda = \frac{c}{f} = \frac{hc}{hf} = \frac{hc}{|\Delta E|} \qquad \text{31-27}$$

The photon energies corresponding to shortest wavelengths (400 nm) and longest (700 nm) wavelengths in the visible spectrum are

$$E_{400\text{ nm}} = \frac{hc}{\lambda} = \frac{1240\text{ eV}\cdot\text{nm}}{400\text{ nm}} = 3.10\text{ eV} \qquad \text{31-28}a$$

and

$$E_{700\text{ nm}} = \frac{hc}{\lambda} = \frac{1240\text{ eV}\cdot\text{nm}}{700\text{ nm}} = 1.77\text{ eV} \qquad \text{31-28}b$$

Because the energy levels in atoms form a discrete set, the emission spectrum of light from single atoms or from atoms in low-pressure gases consists of a set of sharp discrete lines that are characteristic of the element. These narrow lines are broadened somewhat by Doppler shifts, due to the motion of the atom relative to the observer and by collisions with other atoms; generally, however, if the gas density is low enough, the lines are narrow and well separated from one another. The study of the line spectra of hydrogen and other atoms led to the first understanding of the energy levels of atoms.

Continuous spectra When atoms are close together and interact strongly, as in liquids and solids, the energy levels of the individual atoms are spread out into energy bands, resulting in essentially continuous bands of energy levels. When the bands overlap, as they often do, the result is a continuous spectrum of possible energies and a continuous emission spectrum. In an incandescent material such as a hot metal filament, electrons are randomly accelerated by frequent collisions, resulting in a broad spectrum of thermal radiation. The rate at which an object radiates thermal energy is proportional to the fourth power of its absolute temperature.*

* This is known as the Stefan–Boltzmann law. This property and other properties of thermal radiation, such as Wien's displacement law, are discussed more fully in Section 20-4.

The radiation emitted by an object at temperatures below approximately 600°C is concentrated in the infrared and is not visible. As an object is heated, the energy radiated extends to shorter and shorter wavelengths. Between approximately 600°C and 700°C, enough of the radiated energy is in the visible spectrum for the object to glow a dull red. At higher and higher temperatures, the object becomes bright red and then white. For a given temperature, the wavelength λ_{peak} at which the emitted power is a maximum varies inversely with the temperature, a result known as Wien's displacement law. The surface of the Sun at $T = 6000$ K emits a continuous spectrum of approximately constant intensity over the visible range of wavelengths.

ABSORPTION, SCATTERING, SPONTANEOUS EMISSION, AND STIMULATED EMISSION

When radiation is emitted, an atom (the words atom and molecule are used interchangeably in this section) makes a transition from an excited state to a state of lower energy; when radiation is absorbed, an atom makes a transition from a lower state to a higher state. When atoms are irradiated with a continuous spectrum of radiation, the transmitted spectrum shows dark lines corresponding to the absorption of light at discrete wavelengths. The absorption spectra of atoms were the first line spectra observed. Because atoms at normal temperatures are in either their ground states or low-energy excited states, only transitions from a ground state (or a near ground state) to a more highly excited state are observed. Thus, absorption spectra usually have far fewer lines than do emission spectra.

Figure 31-45 illustrates several interesting phenomena that can occur when a photon is incident on an atom. In Figure 31-45*a*, the energy of the incoming photon is too small to excite the atom to an excited state, so the atom remains in its ground state and the photon is said to be scattered. Because the incoming and outgoing or scattered photons have the same energy, the scattering is said to be elastic. If the wavelength of the incident light is large compared with the size of the atom, the scattering can be described in terms of classical electromagnetic theory and is called **Rayleigh scattering** after Lord Rayleigh, who worked out the theory in 1871. The probability of Rayleigh scattering varies as $1/\lambda^4$. This means that blue light is scattered much more readily than red light, which accounts for the bluish color of the sky. The removal of blue light by Rayleigh scattering also accounts for some of the reddish color of the light transmitted through the atmosphere at sunrise and sunset.

Inelastic scattering, also called **Raman scattering,** occurs when an incident photon that has just the right amount of energy is absorbed and the atom undergoes a transition to a more energetic state. Then, the atom emits a photon as it undergoes a transition to a less energetic state, whose energy differs from that of the initial state. If the energy of the scattered photon hf' is less than that of the incident photon hf (Figure 31-45*b*), it is called **Stokes Raman scattering.** If the energy of the scattered photon is greater than that of the incident photon (Figure 31-45*c*), it is called **anti-Stokes Raman scattering.**

In Figure 31-45*d*, the energy of the incident photon is just equal to the difference in energy between the initial state and a more energetic state. The atom absorbs the photon and makes a transition to the more excited state in a process called **resonance absorption.**

In Figure 31-45*e*, an atom in an excited state spontaneously undergoes a transition to a less energetic state, in a process called **spontaneous emission.** Often an atom in an excited state undergoes transitions to one or more intermediate states as it returns to the ground state. A common example occurs when an atom is excited by ultraviolet light and emits visible light as it returns to its ground state by multiple transitions. This process, often called **fluorescence,** occurs in a thin film lining on the inside of the glass tubes of fluorescent lightbulbs. Because the lifetime of a typical excited atomic energy state is of the order of 10 ns, this process appears to occur instantaneously. However, some excited states have much longer lifetimes—of the

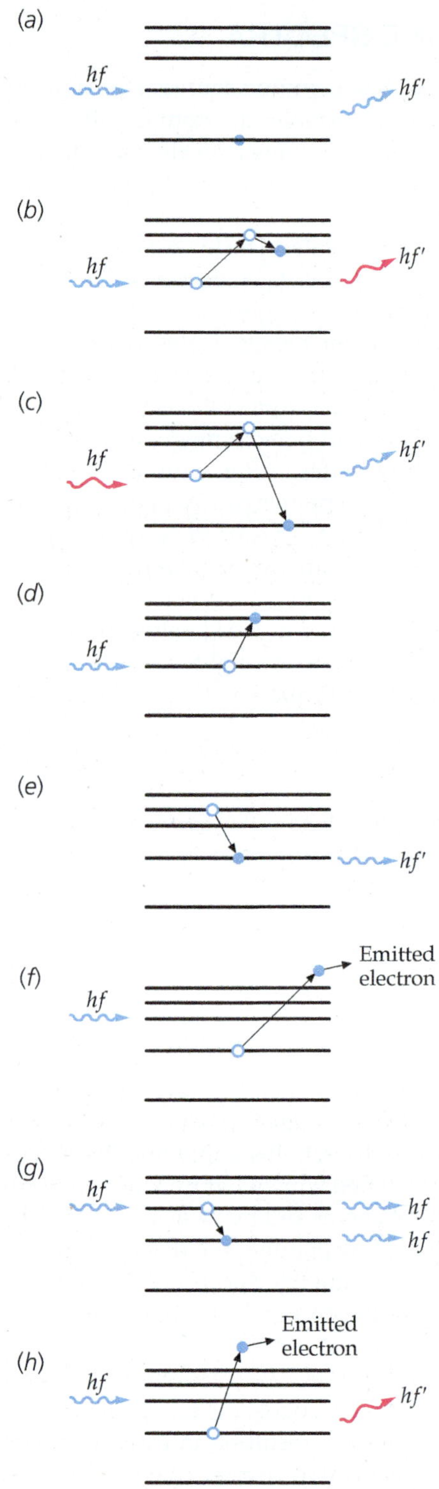

FIGURE 31-45 Photon–atom and photon–molecule interactions. (*a*) Elastic scattering. (*b*) Stokes Raman scattering. (*c*) Anti-Stokes Raman scattering. (*d*) Resonance absorption. (*e*) Spontaneous emission. (*f*) Photoelectric effect. (*g*) Stimulated emission. (*h*) Compton scattering.

order of milliseconds or occasionally seconds or even minutes. Such a state is called a **metastable state.** Materials that have very long-lived metastable states and emit light long after the original excitation are called **phosphorescent materials.**

Figure 31-45*f* illustrates the photoelectric effect, in which the absorption of the photon ionizes the atom by causing the emission of an electron. Figure 31-45*g* illustrates **stimulated emission.** This process occurs if the atom is initially in an excited state of energy E_H, and the energy of the incident photon is equal to $E_H - E_L$, where E_H and E_L are the energies of higher and lower energy states, respectively. In this case, the oscillating electromagnetic field associated with the incident photon can stimulate the excited atom, which then emits a photon in the same direction as the incident photon and in phase with it. The photons from the stimulated atoms can stimulate the emission of additional photons propagating in the same direction with the same phase. This process amplifies the initially emitted photon, yielding a beam of light originating from different atoms that is coherent. As a result, interference of the light from a large number of atoms can easily be observed.

Figure 31-45*h* illustrates **Compton scattering,** which occurs if the energy of the incident photon is much greater than the ionization energy. Note that in Compton scattering, a photon is absorbed and a photon is emitted, whereas in the photoelectric effect, a photon is absorbed with none emitted.

(*a*)

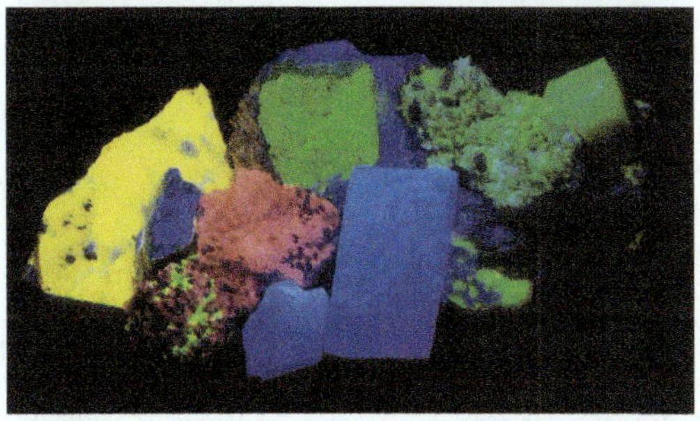
(*b*)

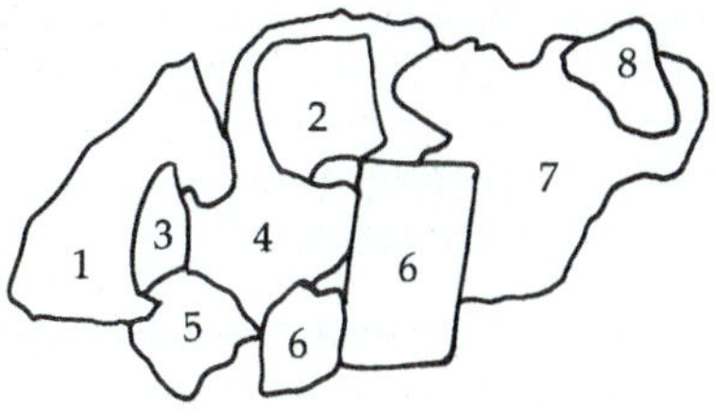

(*c*)

A collection of minerals in (*a*) daylight and in (*b*) ultraviolet light (sometimes called *black light*). Identified by number in the schematic (*c*), they are 1, powerllite; 2, willemite; 3, scheelite; 4, calcite; 5, calcite and willemite composite; 6, optical calcite; 7, willemite; and 8, opal. The change in color is due to the minerals fluorescing under the ultraviolet light. In optical calcite, both fluorescence and phosphorescence occur. *(Paul Silverman/Fundamental Photographs.)*

Example 31-7 Resonant Absorption and Emission

The energy level E_1 of the first excited state of a potassium atom is 1.62 eV above the energy level E_0 of the ground state. The energy levels E_2 and E_3 of the second and third excited states of a potassium atom are 2.61 eV and 3.07 eV, respectively, above the ground state energy E_0. (*a*) What is the longest wavelength of radiation that can be absorbed by a potassium atom in its ground state? Calculate the wavelength of the emitted photon when the atom makes a transition from (*b*) the third excited state (E_3) to the ground state and from (*c*) the third excited state (E_3) to the second excited state (E_2).

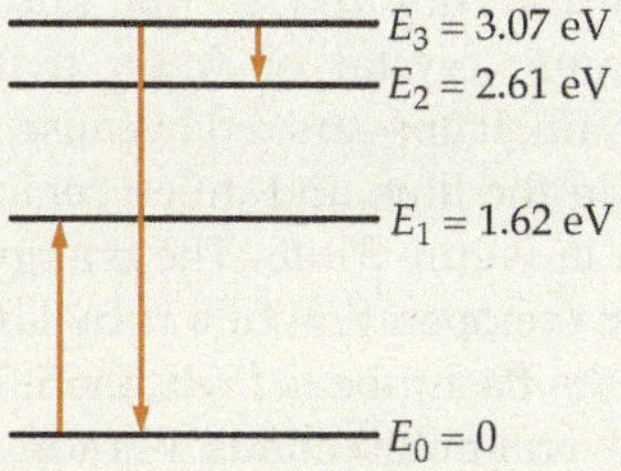

FIGURE 31-46

PICTURE The ground state and the first three excited energy levels are shown in Figure 31-46. (*a*) Because the wavelength is related to the energy of a photon by $\lambda = hc/\Delta E$, longer wavelengths correspond to smaller energy differences, and the smallest energy difference for a transition originating at the ground state is from the ground state to the first excited state. (*b*) The wavelengths of the photons emitted when the atom transitions to lower energy states are related to the energy differences by $\lambda = hc/|\Delta E|$.

SOLVE

(*a*) Calculate the wavelength of radiation absorbed in a transition from the ground state to the first excited state:

$$\lambda = \frac{hc}{\Delta E} = \frac{hc}{E_1 - E_0} = \frac{1240\ \text{eV}\cdot\text{nm}}{1.62\ \text{eV} - 0} = \boxed{765\ \text{nm}}$$

(*b*) For the transition from E_3 to the ground state, the photon energy is $E_3 - E_0 = E_3$. Calculate the wavelength of radiation emitted in this transition:

$$\lambda = \frac{hc}{|\Delta E|} = \frac{hc}{E_3 - E_0} = \frac{1240\ \text{eV}\cdot\text{nm}}{3.07\ \text{eV} - 0} = \boxed{404\ \text{nm}}$$

(*c*) For the transition from E_3 to E_2, the photon energy is $E_3 - E_2$. Calculate the wavelength of radiation emitted in this transition:

$$\lambda = \frac{hc}{|\Delta E|} = \frac{hc}{E_3 - E_2} = \frac{1240\ \text{eV}\cdot\text{nm}}{3.07\ \text{eV} - 2.61\ \text{eV}} = \boxed{2.70\ \mu\text{m}}$$

CHECK The Part-(*b*) result is smaller than the Part-(*a*) result. This result is expected because the more energy a photon has, the shorter the wavelength.

TAKING IT FURTHER The wavelength of radiation emitted in the transition from E_1 to the ground state E_0 is 765 nm, the same as the wavelength for radiation absorbed in the transition from the ground state to E_1. This transition and the transition from E_3 to the ground state both result in photons in the visible spectrum. The photon emitted during the transition from E_3 to E_2 is in the infrared region of the electromagnetic spectrum.

LASERS

The *laser* (*l*ight *a*mplification by *s*timulated *e*mission of *r*adiation) is a device that produces a strong beam of coherent photons by stimulated emission. Consider a system consisting of atoms that have a ground state of energy E_0 and an excited metastable state of energy E_1. If the atoms are irradiated by photons of energy $E_1 - E_0$, those atoms in the ground state can absorb a photon and make the transition to state E_1, whereas those atoms already in the excited state may be stimulated to decay back to the ground state. The relative probabilities of absorption and stimulated emission, first worked out by Einstein, are equal. Ordinarily, nearly all the atoms of the system at normal temperature will initially be in the ground state, so absorption will be the main effect. To produce more stimulated-emission transitions than absorption transitions, we must arrange to have more atoms in the excited state than in the ground state. This condition, called *population inversion*, can be achieved by a method called optical pumping in which atoms are *pumped* up to levels of energy greater than E_1 by the absorption of an intense auxiliary radiation. The atoms then decay down to state E_1 either by spontaneous emission or by nonradiative transitions, such as those transitions due to collisions.

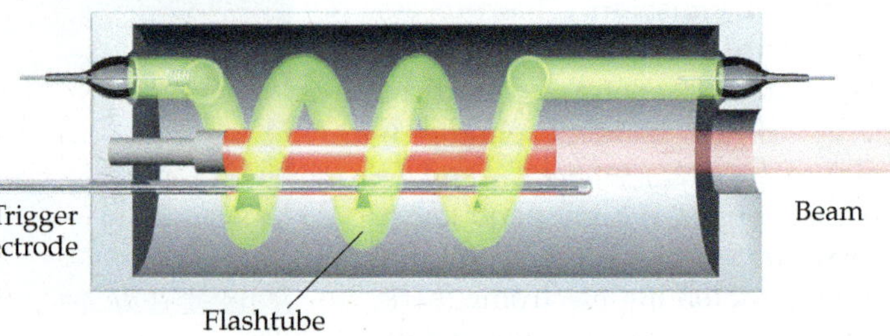

FIGURE 31-47 Schematic diagram of the first ruby laser.

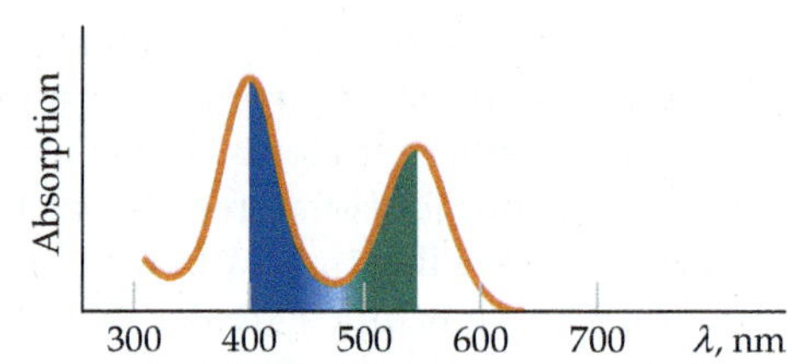

FIGURE 31-48 Absorption versus wavelength for Cr^{3+} in ruby. Ruby appears red because of the strong absorption of green and blue light by the chromium ions.

Figure 31-47 shows a schematic diagram of the first laser, a ruby laser built by Theodore Maiman in 1960. The laser consists of a ruby rod a few centimeters long surrounded by a helical gaseous flashtube that emits a broad spectrum of light. The ends of the ruby rod are flat and perpendicular to the axis of the rod. Ruby is a transparent crystal of Al_2O_3 that has a small amount (about 0.05 percent) of chromium. It appears red because the chromium ions (Cr^{3+}) have strong absorption bands in the blue and green regions of the visible spectrum, as shown in Figure 31-48. The energy levels of chromium—important for the operation of a ruby laser–are shown in Figure 31-49. When the flashtube is fired, there is an intense burst of light that lasts several milliseconds. Photon absorption excites many of the chromium ions to the bands of energy levels indicated by the shading in Figure 31-49. The excited chromium ions then rapidly drop down to a closely spaced pair of metastable states labeled E_1 in the figure. The metastable states are approximately 1.79 eV

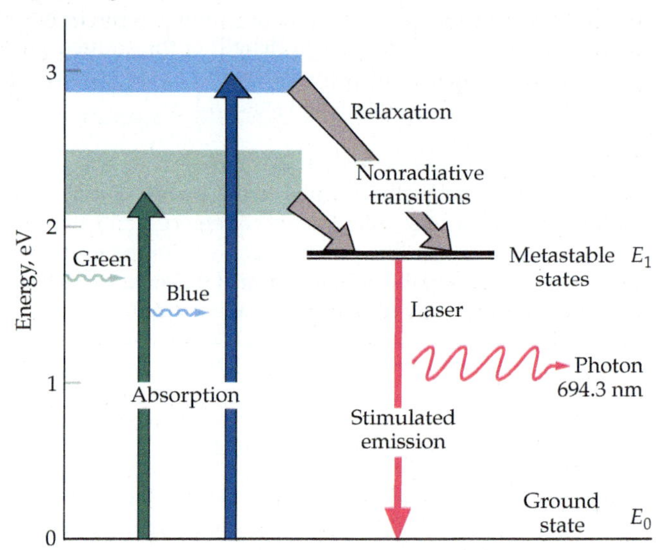

FIGURE 31-49 Energy levels in a ruby laser. To make the population of the metastable states greater than that of the ground state, the ruby crystal is subjected to intense radiation that contains energy in the green and blue wavelengths. This excites atoms from the ground state to the bands of energy levels indicated by the shading, from which the atoms decay to the metastable states by nonradiative transitions. Then, by stimulated emission, the atoms undergo the transition from the metastable states to the ground state.

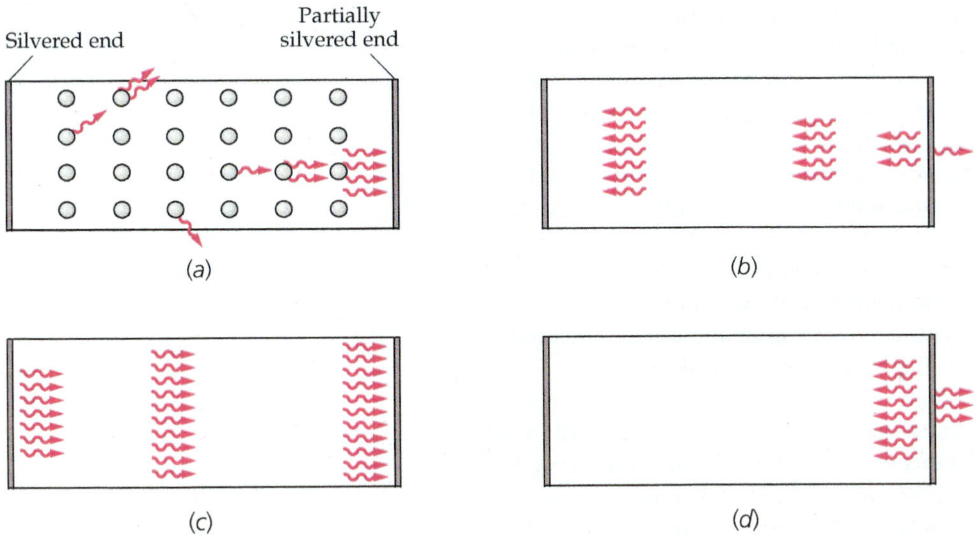

FIGURE 31-50 Buildup of a photon beam in a laser. (*a*) When irradiated, some atoms spontaneously emit photons, some of which travel to the right and stimulate other atoms to emit photons parallel to the axis of the crystal. (*b*) Of the four photons that strike the right face, one is transmitted and three are reflected. As the reflected photons traverse the lasing material, they stimulate other atoms to emit photons, and the beam builds up. By the time the beam reaches the right face again (*c*), it comprises many photons. (*d*) Some of the photons are transmitted, and the rest of the photons are reflected.

above the ground state. The expected lifetime for a chromium ion to remain in one of the metastable states is about 5 ms, after which the chromium ion spontaneously emits a photon and decays to the ground state. A millisecond is a long time for an atomic process. Consequently, if the flash is intense enough, the number of chromium ions populating the two metastable states will exceed the population of chromium ions in the ground state. It follows that during the time the flashtube is firing, the populations of ions in the ground state and in the metastable states are inverted. When the chromium ions in the state E_1 decay to the ground state by spontaneous emission, they emit photons of energy 1.79 eV and wavelength 694.3 nm. The photons have just the right energy to stimulate chromium ions in the metastable states to emit photons of the same energy (and wavelength) as they undergo the transition to the ground state. The photons also have just the right energy to stimulate chromium ions in the ground state to absorb a photon as they undergo the transition to one of the metastable states. These processes are competing processes, but the stimulated emission process dominates as long as the population of chromium ions in the metastable states exceeds the population in the ground state.

In the ruby laser, one end of the crystal is fully silvered, so it is 100 percent* reflecting; the other end of the crystal, called the output coupler, is partially silvered, leaving it about 85 percent reflecting. When photons traveling parallel to the axis of the crystal strike the silvered ends, all are reflected from the back face and 85 percent are reflected from the front face, with 15 percent of the photons escaping through the partially silvered front face. During each pass through the crystal, the photons stimulate more and more atoms so that an intense beam is emitted from the partially silvered end (Figure 31-50). Because the duration of each flash of the flashtube is between two and three seconds, the laser beam is produced in pulses lasting a few milliseconds. Modern ruby lasers generate intense light beams with energies ranging from 50 J to 100 J per pulse. The beam can have a diameter as small as 1 mm and an angular divergence as small as 0.25 mrad to about 7 mrad.

Population inversion is achieved somewhat differently in the continuous helium–neon laser. The energy levels of helium and neon that are important for the operation of the laser are shown in Figure 31-51. Helium has an excited energy state $E_{1\,\mathrm{He}}$ that is 20.61 eV above its ground state. Helium atoms are excited to state $E_{1\,\mathrm{He}}$ by an electric discharge. Neon has an excited state $E_{2\,\mathrm{Ne}}$ that is 20.66 eV above its ground state. This is just 0.05 eV above the first excited state of helium. The neon atoms are excited to state $E_{2\,\mathrm{Ne}}$ by collisions with excited helium atoms. The kinetic energy of the helium atoms provides the extra 0.05 eV of energy needed to excite the neon atoms. Another excited state of neon $E_{1\,\mathrm{Ne}}$ exists that is 18.70 eV above its ground state and 1.96 eV below state $E_{2\,\mathrm{Ne}}$. Because state $E_{1\,\mathrm{Ne}}$ is normally unoccupied, population

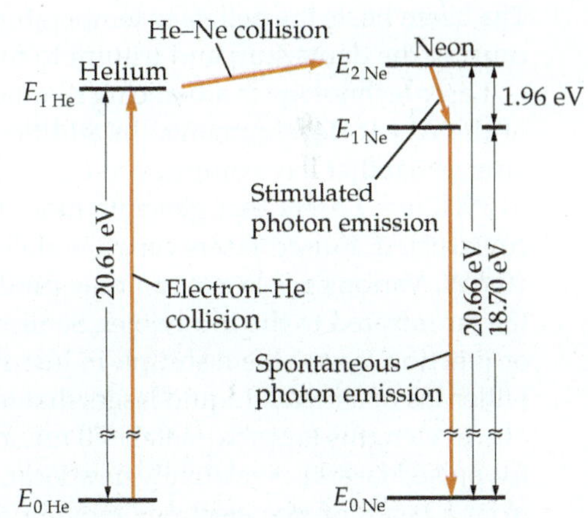

FIGURE 31-51 Energy levels of helium and neon that are important for the helium–neon laser. The helium atoms are excited by electrical discharge to an energy state 20.61 eV above the ground state. They collide with neon atoms, exciting some neon atoms to an energy state 20.66 eV above the ground state. Population inversion is thus achieved between this level and one 1.96 eV below it. The spontaneous emission of photons of energy 1.96 eV stimulates other atoms in the upper state to emit photons of energy 1.96 eV.

* In actuality, the light is only 99.7 percent reflected by a "fully silvered" end. In addition, the reflective coating consists of multiple dielectric layers, not silver.

inversion between states $E_{2\,\text{Ne}}$ and $E_{1\,\text{Ne}}$ is obtained immediately. The stimulated emission that occurs between these states results in photons of energy 1.96 eV and wavelength 632.8 nm, which produces a bright red light. After stimulated emission, the neon atoms in state $E_{1\,\text{Ne}}$ decay to the ground state $E_{0\,\text{Ne}}$ by spontaneous emission.

Note that stimulated emission involves transitions between two excited states of the neon atom in the helium–neon laser, whereas stimulated emission involves transitions between an excited state and the ground state of the chromium ion in the ruby laser. For stimulated emission between an excited state and a ground state, population inversion is difficult to achieve because more than half the atoms in the ground state must be excited. However, for stimulated emission between two excited states, population inversion is easily achieved because the state after stimulated emission is not the ground state but an excited state that is normally unpopulated.

Figure 31-52 shows a schematic diagram of a helium–neon laser commonly used for physics demonstrations. The helium–neon laser consists of a gas tube that contains 15 percent helium gas and 85 percent neon gas. A totally reflecting flat mirror is mounted at one end of the gas tube and a 99 percent reflecting concave mirror is placed at the other end of the gas tube. The concave mirror focuses parallel light at the flat mirror and also acts as a lens that transmits part of the light, so that the light emerges as a parallel beam.

A laser beam is coherent, very narrow, and intense. Its coherence makes the laser beam useful in the production of holograms, which we discuss in Chapter 33. The precise direction and small angular spread of the laser beam make it useful as a surgical tool for destroying cancer cells or reattaching a detached retina. Lasers are also used by surveyors for precise alignment over large distances. Distances can be accurately measured by reflecting a laser pulse from a mirror and measuring the time the pulse takes to travel to the mirror and back. The distance to the moon has been measured to within a few centimeters using an array of mirrors placed on the moon for that purpose. Laser beams are also used in fusion research. An intense laser pulse is focused on tiny pellets of deuterium–tritium in a combustion chamber. The beam heats the pellets to temperatures of the order of 10^8 K in a very short time, causing the deuterium and tritium to fuse and release energy.

Laser technology is advancing so quickly that it is possible to mention only a few of the recent developments. In addition to the ruby laser, many other solid-state lasers exist that have output wavelengths which range from approximately 170 nm to 3900 nm. Lasers that generate more than 1 kW of continuous power have been constructed. Pulsed lasers can now deliver nanosecond pulses of power exceeding 10^{14} W. Various gas lasers can now produce beams of wavelengths that range from the far infrared to the ultraviolet. Semiconductor lasers (also known as diode lasers or junction lasers) have shrunk in just 10 years from the size of a pinhead to mere billionths of a meter. Liquid lasers that use chemical dyes can be tuned over a range of wavelengths (approximately 70 nm for continuous lasers and more than 170 nm for pulsed lasers). A relatively new laser, the free-electron laser, extracts light energy from a beam of free electrons moving through a spatially varying magnetic field.

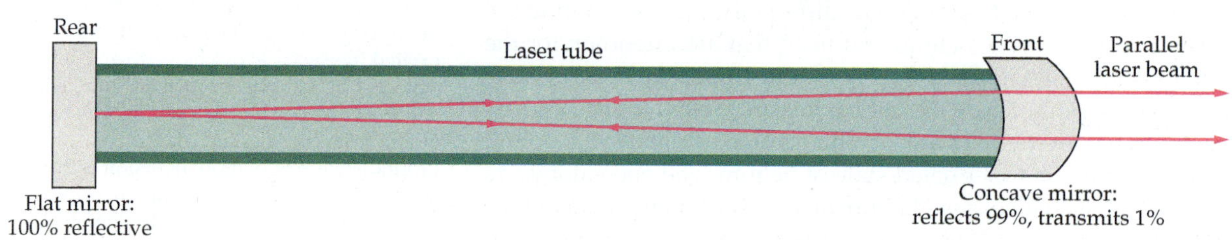

FIGURE 31-52 Schematic drawing of a helium–neon laser. The use of a concave mirror rather than a second plane mirror makes the alignment of the mirrors less critical than it is for the ruby laser. The concave mirror on the right also serves as a lens that focuses the emitted light into a parallel beam.

The free-electron laser has the potential to have very high power and large efficiency and can be tuned over a large range of wavelengths. There appears to be no limit to the variety and uses of modern lasers.

(*a*)

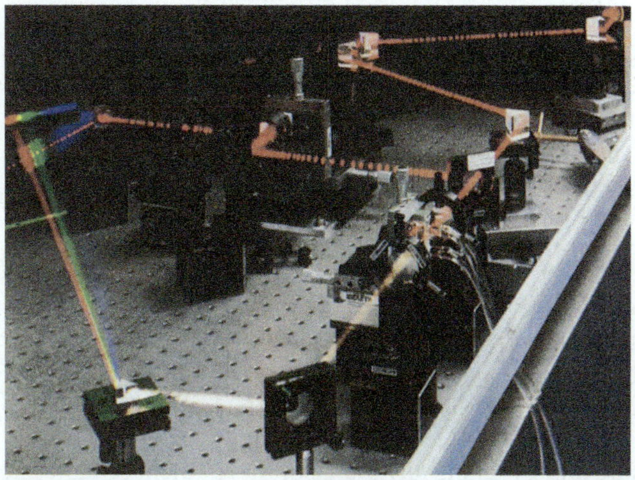

(*b*)

(*c*)

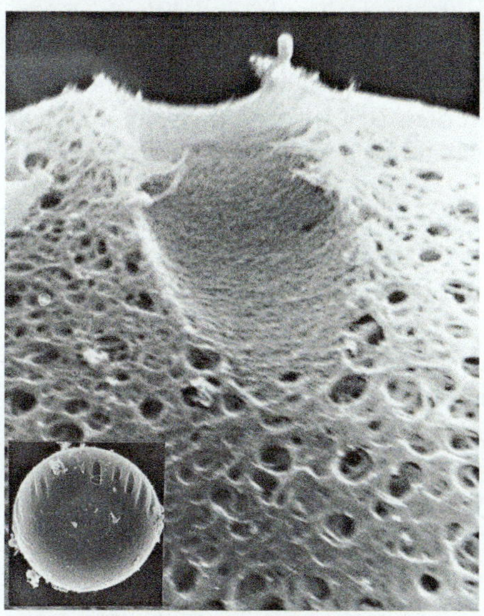

(*d*)

(*e*)

(*a*) Beams from a krypton laser and an argon laser, split into their component wavelengths. In these gas lasers, krypton and argon atoms have been stripped of multiple electrons, forming positive ions. The light-emitting energy transitions occur when excited electrons in the ions decay from one upper energy level to another. Here, several energy transitions are occurring at once, each corresponding to emitted light of a different wavelength. (*b*) A femtosecond pulsed laser. By a technique known as *modelocking*, different excited modes within a laser's cavity can be made to interfere with one another and create a series of ultrashort pulses, which are picoseconds long, that correspond to the time it takes light to bounce back and forth once within the cavity. Ultrashort pulses have been used as probes to study the behavior of atoms and molecules during chemical reactions. (*c*) A carbon dioxide laser takes just 2 minutes to cut out a steel saw blade. (*d*) A groove etched in the zona pellucida (protective outer covering) of a mouse egg by a *laser scissor* facilitates implantation. This technique has already been applied in human fertility therapies. Several effects contribute to the ability of the finely focused laser to cut on such a delicate scale—photon absorption may heat the target, break molecular bonds, or drive chemical reactions. (*e*) The so-called nanolasers shown are semiconductor disks mere microns in diameter and fractions of a micron in width. These tiny lasers work like their larger counterparts. Exploiting quantum effects that prevail on the microscopic scale, nanolasers promise great efficiency and they are being explored as ultrafast, low-energy switching devices. *((a, c) Chuck O'rear/West Light. (b) Courtesy of Ahmed H. Zewail. (d) Michael W. Berns/Scientific American. (e) David Scharf.)*

Physics Spotlight

Optical Tweezers and Vortices: Light at Work

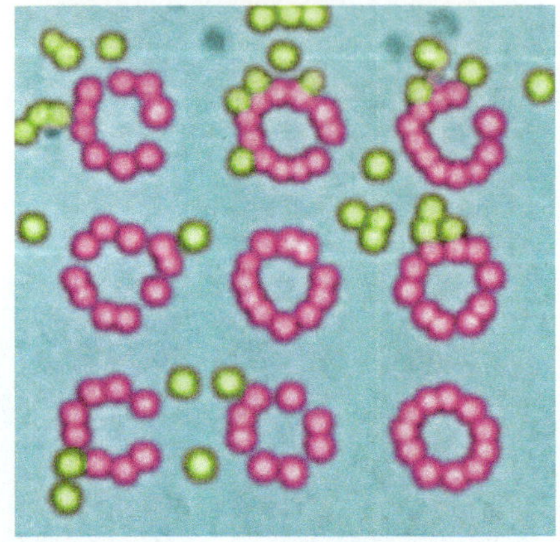

Silicon spheres in water are caught in a three by three array of optical vortices. The optical vortices trap the spheres and exert torques on them. *(Courtesy David G. Grier, from E. Curtis, B. A. Koss and D. G. Grier, "Dynamic holographic optical tweezers," Optics Communications 207, 169-175 (2002).)*

Light pressure has been used to measure the force exerted by biological molecules,[*] to unfold and refold proteins,[†] and even to assist in the trapping and study of atoms.[‡] Using the radiation pressure of light to hold microscopic particles in place is called *optical trapping*. Some optical traps, often called *optical tweezers*, can move and manipulate particles.

In the 1970s, a research group led by Arthur Ashkin at Bell Laboratories used radiation pressure of light to levitate droplets of water from 1 to 40 micrometers in diameter.[#] After many years of experimentation, this group demonstrated that a single laser could control the position of a virus in a solution on a microscope slide.[°] Molecular biologists and microbiologists quickly began to use optical tweezers in their studies.

Often, optical trapping is done using lasers that transmit light which has a wavelength near 1000 nanometers,[§] because many biological materials are relatively transparent to such near-infrared wavelengths. The liquid used for holding biological specimens absorbs scattered light near this wavelength.[¶,**] (This means that the trapped object is less likely to be cooked by the light.) Other wavelengths of light may be used, depending on the items to be trapped. The force used in optical traps for examining biological molecules is a few piconewtons.[††]

Optical trapping works both by light pressure and by taking advantage of the gradient of light intensity of a tightly focused laser beam. If a beam of light shines through a small translucent spherical object within the beam, the light will refract. The average of the pressure of refraction of an intense beam will act to keep the object centered in the beam. The more tightly the beam of light is focused, the closer a particle is trapped along the beam near the focus by the strength of the gradient of light intensity.[‡‡,##] This allows for positional control of an object in three dimensions. During the study of biological molecules, a molecule is usually attached to a polystyrene sphere that can be anywhere from 100 nm to 2 μm in diameter. Moving the sphere allows the molecule to be stretched, folded, and brought into focus with the help of optical tweezers. Much larger objects, such as entire cells, can also be moved by optical tweezers.[°°]

Specialized digital lenses can give laser light carefully calculated twists. The twisted light is called an *optical vortex*. Although they have many other potential uses, optical vortices can be used as specialized optical tweezers that have angular momentum.[§§] Different twists have different amounts of angular momentum and can be used to spin and rotate particles. Optical vortices have even been used to spin particles into one another to combine particles.

Physicists at the University of Chicago have come up with a method of generating hundreds of different optical tweezers from the same beam by passing laser beams through a digitally controlled lens.[¶¶] These tweezers can include optical vortices that exert different amounts of torque on particles. The holographic optical tweezers (HOT) method of creating optical vortices has been patented for manipulating particles, and for pumping, mixing, and sorting fluids and objects on a microscopic scale.[***,†††] Makers of miniature machines are enthusiastic about this technology because, unlike delicate tiny machines, light does not wear out.[‡‡‡]

* Mehta, A. D., et al., "Single-Molecule Biomechanics with Optical Methods." *Science*, Mar. 12, 1999, Vol. 283, No. 5408, pp. 1689–1695.

† Cecconi, C., et al., "Direct Observation of the Three-State Folding of a Single Protein Molecule." *Science*, Sept. 23, 2005, Vol. 309, No. 5743, pp. 2057–2060.

‡ Nagel, B., "Presentation Speech for 1997 Nobel Prize in Physics." *Les Prix Nobel. The Nobel Prizes 1997*, Ed. Tore Frängsmyr [Nobel Foundation], Stockholm, 1998, at http://nobelprize.org/nobel_prizes/physics/laureates/1997/presentation-speech.html As of Dec. 2006.

\# Ashkin, A., and Dziedzic, J., M., "Optical Levitation of Liquid Drops by Radiation Pressure." *Science*, Mar. 21, 1975, Vol. 187, No. 4181, pp. 1073–1075.

° Ashkin, A., and Dziedzic, J., M., "Optical Trapping and Manipulation of Viruses and Bacteria." *Science*, Mar. 20, 1987, Vol. 235, No. 4795, pp. 1517–1520.

§ Mohanty, S. K., Dasgupta, R., and Gupta, P. K., "Three-Dimensional Orientation of Microscopic Objects Using Combined Elliptical and Point Optical Tweezers." *Applied Physics B*, Dec. 2005, Vol. 81, No. 8, pp. 1063–1066.

¶ Molloy, J. E., and Padgett, M. E., "Lights, Action: Optical Tweezers." *Contemporary Physics*, Jul./Aug. 2002, Vol. 43, No. 4, pp. 241–258.

** Block, S. M., "Construction of Optical Tweezers." *Cells: A Laboratory Manual. Vol. 2, Sect. 7*, Eds. D. L. Spector, R. D. Goldman, and L. A. Leinwand. Cold Spring Harbor: Cold Spring Harbor Laboratory Press, 1998. At http://www.cshlpress.com/chap_cells.tpl#intro As of Dec. 2006.

†† Mehta, A. D., et al., op. cit.

‡‡ Block, S. M., op. cit.

\#\# Molloy, J. E., and Padgett, M. J., op. cit.

°° Pool, R., "Trapping with Optical Tweezers." *Science*, Aug. 26, 1988, Vol. 241, No. 4869, p. 1042.

§§ Dholakia, K., Spalding, G., and MacDonald, M., "Optical Tweezers: The Next Generation." *Physics World*, Oct. 2002, Vol. 15, No. 9, pp. 31–35.

¶¶ Curtis, J. E., Koss, B. A., and Grier, D. G., "Dynamic Holographic Optical Tweezers." *Optics Communications*, 2002, Vol. 207, pp. 169–175

*** Curtis, J. E., Koss, B. A., and Grier, D. G., *Use of Multiple Optical Vortices for Pumping, Mixing and Sorting*. U.S. Patent 6,858,833 B2. Feb. 22, 2005.

††† Curtis, J. E., Koss, B. A., and Grier, D. G., *Multiple Optical Vortices for Manipulating Particles* U.S. Patent 6,995,351 B2. Feb. 7, 2006.

‡‡‡ "Motors of Light May Resolve Dilemma of How to Power MEMS." *Small Times*, April 4, 2004. At http://www.smalltimes.com/document_display.cfm?document_id=5796 As of May 2006.

Summary

TOPIC	RELEVANT EQUATIONS AND REMARKS	
1. Speed of Light	The SI unit of length, the meter, is defined so that the speed of light in vacuum is exactly $c = 299\,792\,458 \text{ m/s}$	31-1
v in a transparent medium	$v = \frac{c}{n}$ where n is the index of refraction.	31-3
2. Reflection and Refraction	When light is incident on a surface separating two media in which the speed of light differs, part of the light energy is transmitted and part of the light energy is reflected.	
Law of reflection	The reflected ray lies in the plane of incidence and makes an angle θ_1' with the normal that is equal to the angle of incidence. $\theta_1' = \theta_1$	31-4
Reflected intensity, normal incidence	$I = \left(\frac{n_1 - n_2}{n_1 + n_2}\right)^2 I_0$	31-7
Index of refraction	$n = \frac{c}{v}$	31-3
Law of refraction (Snell's law)	$n_1 \sin\theta_1 = n_2 \sin\theta_2$	31-5*b*
Total internal reflection	When light is traveling in a medium that has an index of refraction n_1 and is incident on the boundary of a second medium that has a lower index of refraction $n_2 < n_1$, the light is totally reflected if the angle of incidence is greater than the critical angle θ_c given by	
Critical angle	$n_1 \sin\theta_c = n_2 \sin 90° \qquad n_1 > n_2$	31-8
Dispersion	The speed of light in a medium, and therefore the index of refraction of that medium, depends on the wavelength of light. Because of dispersion, a beam of white light incident on a refracting prism is dispersed into its component colors. Similarly, the reflection and refraction of sunlight by raindrops produce a rainbow.	
3. Polarization	Transverse waves can be polarized. The four phenomena that produce polarized electromagnetic waves from unpolarized waves are (1) absorption, (2) scattering, (3) reflection, and (4) birefringence.	
Malus's law	When two polarizers have their transmission axes at an angle θ, the intensity transmitted by the second polarizer is reduced by the factor $\cos^2\theta$: $I = I_0 \cos^2\theta$	31-16
4. Huygens's Construction	Each point on a primary wavefront serves as the source of spherical secondary wavelets that advance with a speed and frequency equal to that of the primary wave. The primary wavefront at some later time is the envelope of these wavelets.	
5. Wave-Particle Duality	Light propagates like a wave, but interacts with matter like a particle.	
Photon energy	$E = hf = \frac{hc}{\lambda}$	31-25
Planck's constant	$h = 6.626 \times 10^{-34} \text{ J}\cdot\text{s} = 4.136 \times 10^{-15} \text{ eV}\cdot\text{s}$	
hc	$hc = 1240 \text{ eV}\cdot\text{nm}$	31-26

TOPIC	RELEVANT EQUATIONS AND REMARKS
6. **Emission of Light**	Light is emitted when a valence electron makes a transition from an excited state to a state of lower energy.
Line spectra	Atoms in dilute gases emit a discrete set of wavelengths called a line spectrum. The photon energy $E = hf = hc/\lambda$ equals the difference in energy of the initial and final states of the atom.
Continuous spectra	Atoms in high-density gases, liquids, or solids have continuous bands of energy levels, so they emit a continuous spectrum of light. Thermal radiation is visible if the temperature of the emitting object is above approximately 600°C.
Spontaneous emission	An atom in an excited state will spontaneously make a transition to a lower state with the emission of a photon. This process is random, with a characteristic lifetime of about 10^{-8} s. The photons from two or more atoms are not correlated, so the light is incoherent.
Stimulated emission	Stimulated emission occurs if an atom is initially in an excited state and a photon of energy equal to the energy difference between that state and a lower state is incident on the atom. The oscillating electromagnetic field of the incident photon stimulates the excited atom to emit another photon in the same direction and in phase with the incident photon. The emitted light is coherent with the incident light.
7. **Visible Light**	The human eye is sensitive to electromagnetic radiation that has wavelengths from approximately 400 nm (violet) to 700 nm (red). The photon energies range from approximately 1.8 eV to 3.1 eV. A uniform distribution of wavelengths, such as the wavelengths emitted by the Sun, appears white to our eyes.
8. **Lasers**	A laser produces an intense, coherent, and narrow beam of photons as the result of stimulated emission. The operation of a laser depends on population inversion, in which there are more atoms in an excited state than in the ground state or a lower state.

Answers to Concept Check

31-1 There are 720 teeth, but there are also 720 gaps, so the width of a tooth is less than $\frac{1}{720}$ of the circumference of the wheel. Consequently, the wheel actually has to rotate through less than $\frac{1}{720}$ rev for the light from the distant mirror to be again observed.

Answers to Practice Problems

31-1 (*a*) 4.57×10^6 km (*b*) 3.05×10^8 m/s

31-2 1.28 s each way

Problems

In a few problems, you are given more data than you actually need; in a few other problems, you are required to supply data from your general knowledge, outside sources, or informed estimate.

Interpret as significant all digits in numerical values that have trailing zeros and no decimal points.

- • Single-concept, single-step, relatively easy
- •• Intermediate-level, may require synthesis of concepts
- ••• Challenging
- SSM Solution is in the *Student Solutions Manual*
- Consecutive problems that are shaded are paired problems.

CONCEPTUAL PROBLEMS

1 • A ray of light reflects from a plane mirror. The angle between the incoming ray and the reflected ray is 70°. What is the angle of reflection? (*a*) 70°, (*b*) 140°, (*c*) 35°, (*d*) Not enough information is given to determine the reflection angle. SSM

2 • A ray of light in air is incident on the surface of a piece of glass. The angle between the normal to the surface and the incident ray is 40°, and the angle between the normal and the refracted ray is 28°. What is the angle between the incident ray and the refracted ray? (*a*) 12°, (*b*) 28°, (*c*) 40°, (*d*) 68°

3 • **Engineering Application** During a physics experiment, you are measuring refractive indices of different transparent materials using a red helium–neon laser beam. For a given angle of incidence, the beam has an angle of refraction equal to 28° in material A, and an angle of refraction equal to 26° in material B. Which material has the larger index of refraction? (*a*) A, (*b*) B, (*c*) The indices of refraction are the same. (*d*) You cannot determine the relative magnitudes of the indices of refraction from the data given.

4 • A ray of light passes from air into water, striking the surface of the water at an angle of incidence of 45°. Which, if any, of the following four quantities changes as the light enters the water: (*a*) wavelength, (*b*) frequency, (*c*) speed of propagation, (*d*) direction of propagation, (*e*) none of the above?

5 • Earth's atmosphere decreases in density as the altitude increases. As a consequence, the index of refraction of the atmosphere also decreases as altitude increases. Explain how one can see the Sun when it is below the horizon. (The horizon is the extension of a plane that is tangent to the Earth's surface.) Why does the setting Sun appear flattened?

6 • A physics student playing pocket billiards wants to strike her cue ball so that it hits a cushion and then hits the eight ball squarely. She chooses several points on the cushion and then measures the distances from each point to the cue ball and to the eight ball. She aims at the point for which the sum of these distances is least. (*a*) Will her cue ball hit the eight ball? (*b*) How is her method related to Fermat's principle? Neglect any effects due to ball rotation.

7 • A swimmer at point *S* in Figure 31-53 develops a leg cramp while swimming near the shore of a calm lake and calls for help. A lifeguard at point *L* hears the call. The lifeguard can run 9.0 m/s and swim 3.0 m/s. She knows physics and chooses a path that will take the least time to reach the swimmer. Which of the paths shown in the figure does the lifeguard take? SSM

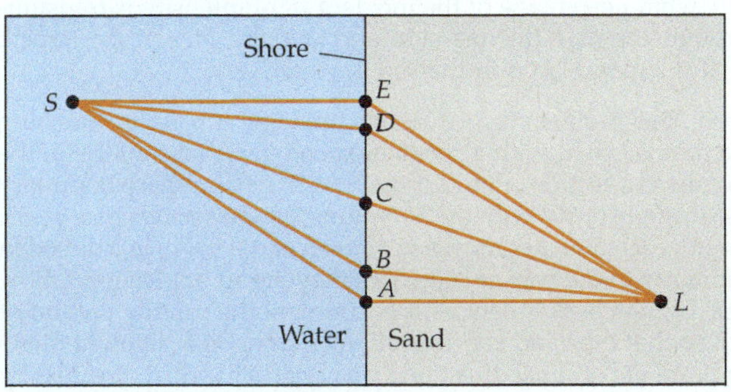

FIGURE 31-53 Problem 7

8 • Material A has a higher index of refraction than material B. Which material has the larger critical angle for total internal reflection when the material is in air? (*a*) A, (*b*) B, (*c*) The angles are the same. (*d*) You cannot compare the angles based on the data given.

9 • **Biological Application** A human eye perceives color using a structure which is called a *cone* that is located on the retina. Three types of molecules compose these cones and each type of molecule absorbs either red, green, or blue light by resonance absorption. Use this fact to explain why the color of an object that appears blue in air appears blue underwater, in spite of the fact that the wavelength of the light is shortened in accordance with Equation 31-6. SSM

10 • Let θ be the angle between the transmission axes of two polarizing sheets. Unpolarized light of intensity I is incident on the first sheet. What is the intensity of the light transmitted through both sheets? (*a*) $I\cos^2\theta$, (*b*) $(I\cos^2\theta)/2$, (*c*) $(I\cos^2\theta)/4$, (*d*) $I\cos\theta$, (*e*) $(I\cos\theta)/4$, (*f*) none of the above

11 •• Draw a diagram to explain how Polaroid sunglasses reduce glare from sunlight reflected from a smooth horizontal surface, such as the surface found on a pool of water. Your diagram should clearly indicate the direction of polarization of the light as it propagates from the Sun to the reflecting surface and then through the sunglasses into the eye. SSM

12 • **Biological Application** Why is it far less dangerous to stand in front of an intense beam of red light than in front of a very low-intensity beam of gamma rays?

13 • Three energy states of an atom are A, B, and C. State B is 2.0 eV above state A and state C is 3.00 eV above state B. Which atomic transition results in the emission of the shortest wavelength of light? (*a*) B → A, (*b*) C → B, (*c*) C → A, (*d*) A → C

14 • In Problem 13, if the atom is initially in state A, which transition results in the emission of the longest wavelength of light? (*a*) A → B, (*b*) B → C, (*c*) A → C, (*d*) B → A

15 • What role does the helium play in a helium–neon laser? SSM

16 • When a beam of visible white light that passes through a gas of atomic hydrogen at room temperature is viewed with a spectroscope, dark lines are observed at the wavelengths of the hydrogen atom emission series. The atoms that participate in the resonance absorption then emit light of same wavelength as they return to the ground state. Explain why the observed spectrum nevertheless exhibits pronounced dark lines.

17 • Which of the following types of light would have the highest energy photons? (*a*) red, (*b*) infrared, (*c*) blue, (*d*) ultraviolet SSM

ESTIMATION AND APPROXIMATION

18 • Estimate the time required for light to make the round trip during Galileo's experiment to measure the speed of light. Compare the time of the round trip to typical human response times. How accurate do you think this experiment is?

19 • Estimate the time delay in receiving a light on your retina when you are wearing eyeglasses compared to when you are not wearing your eyeglasses.

20 •• **Biological Application** Estimate the number of photons that enter your eye if you look for a tenth of a second at the Sun. What energy is absorbed by your eye during that time, assuming that all the photons are absorbed? The total power output of the Sun is 4.2×10^2 W.

21 •• Römer was observing the eclipses of Jupiter's moon Io with the hope that they would serve as a highly accurate clock that would be independent of longitude. (Prior to GPS, such a clock was needed for accurate navigation.) Io eclipses (enters the umbra of Jupiter's shadow) every 42.5 h. Assuming an eclipse of Io is observed on Earth on June 1 at midnight when Earth is at location *A*

(as shown in Figure 31-54), predict the expected time of observation of an eclipse one-quarter of a year later when Earth is at location *B*, assuming (*a*) the speed of light is infinite and (*b*) the speed of light is 2.998×10^8 m/s.

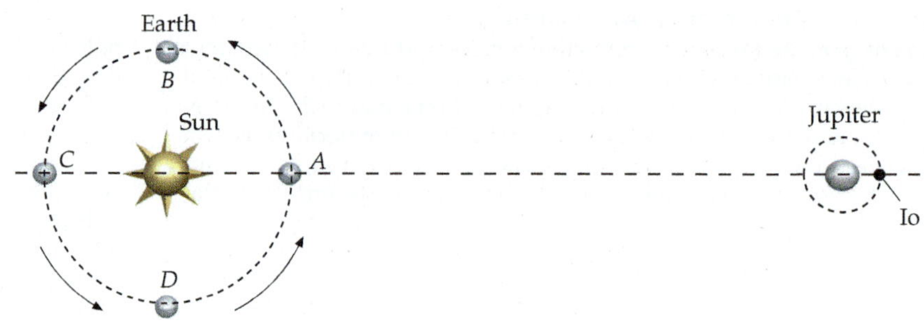

FIGURE 31-54 Problem 21

22 •• If the angle of incidence is small enough, the small angle approximation $\sin\theta \approx \theta$ may be used to simplify Snell's law of refraction. Determine the maximum value of the angle that would make the value for the angle differ by no more than one percent from the value for the sine of the angle. (This approximation will be used in connection with image formation by spherical surfaces in Chapter 32.)

THE SPEED OF LIGHT

23 • Mission Control sends a brief wake-up call to astronauts in a spaceship that is far from Earth. At a time 5.0 s after the call is sent, Mission Control can hear the groans of the astronauts. How far from Earth is the spaceship? (*a*) 7.5×10^8 m, (*b*) 15×10^8 m, (*c*) 30×10^8 m, (*d*) 45×10^8 m, (*e*) The spaceship is on the moon.

24 •• **ENGINEERING APPLICATION** The distance from a point on the surface of Earth to a point on the surface of the moon is measured by aiming a laser light beam at a reflector on the surface of the moon and measuring the time required for the light to make a round trip. The uncertainty in the measured distance Δx is related to the uncertainty in the measured time Δt by $\Delta x = \frac{1}{2}c\,\Delta t$. If the time intervals can be measured to ± 1.00 ns, (*a*) find the uncertainty of the distance. (*b*) Estimate the percentage uncertainty in the distance.

25 •• Ole Römer discovered the finiteness of the speed of light by observing Jupiter's moons. Approximately how sensitive would the timing apparatus need to be in order to detect a shift in the predicted time of the moon's eclipses that occur when the moon happens to be at perigee (3.63×10^5 km) and those that occur when the moon is at apogee (4.06×10^5 km)? Assume that an instrument should be able to measure to at least one-tenth the magnitude of the effect it is to measure. SSM

REFLECTION AND REFRACTION

26 • Calculate the fraction of light energy reflected from an air–water interface at normal incidence.

27 • A ray of light is incident on one of two mirrors that are set at right angles to each other. The plane of incidence is perpendicular to both mirrors. Show that after reflecting from each mirror, the ray will emerge traveling in the direction opposite to the incident direction, regardless of the angle of incidence.

28 •• **SPREADSHEET** (*a*) A ray of light in air is incident on an air–water interface. Using a spreadsheet or graphing program, plot the angle of refraction as a function of the angle of incidence from 0° to 90°. (*b*) Repeat Part (*a*), but for a ray of light in water that is incident on a water–air interface. [For Part (*b*), there is no reflected ray for angles of incidence that are greater than the critical angle.]

29 •• The red light from a helium–neon laser has a wavelength of 632.8 nm in air. Find the (*a*) speed, (*b*) wavelength, and (*c*) frequency of helium–neon laser light in air, water, and glass. (The glass has an index of refraction equal to 1.50.)

30 •• The index of refraction for silicate flint glass is 1.66 for violet light that has a wavelength in air equal to 400 nm and 1.61 for red light that has a wavelength in air equal to 700 nm. A ray of 700-nm-wavelength red light and a ray of 400-nm-wavelength violet light both have angles of refraction equal to 30° upon entering the glass from air. (*a*) Which is greater, the angle of incidence of the ray of red light or the angle of incidence of the ray of violet light? Explain your answer. (*b*) What is the difference between the angles of incidence of the two rays?

31 •• A slab of glass that has an index of refraction of 1.50 is submerged in water that has an index of refraction of 1.33. Light in the water is incident on the glass. Find the angle of refraction if the angle of incidence is (*a*) 60°, (*b*) 45°, and (*c*) 30°. SSM

32 •• Repeat Problem 31 for a beam of light initially in the glass that is incident on the glass–water interface at the same angles.

33 •• A beam of light in air strikes a glass slab at normal incidence. The glass slab has an index of refraction of 1.50. (*a*) Approximately what percentage of the incident light intensity is transmitted through the slab (in one side and out the other)? (*b*) Repeat Part (*a*) if the glass slab is immersed in water.

34 •• This problem is a refraction analogy. A band is marching down a football field with a constant speed v_1. About midfield, the band comes to a section of muddy ground that has a sharp boundary making an angle of 30° with the 50-yd line, as shown in Figure 31-55. In the mud, each marcher moves at a speed equal to $\frac{1}{2}v_1$ in a direction perpendicular to the row of marchers they are in. (*a*) Diagram how each line of marchers is bent as it encounters the muddy section of the field so that the band is eventually marching in a different direction. Indicate the original direction by a ray and the final direction by a second ray. (*b*) Find the angles between these rays and the line normal to the boundary. Is their direction of motion "bent" toward the normal or away from it? Explain your answer in terms of refraction.

35 •• In Figure 31-56, light is initially in a medium that has an

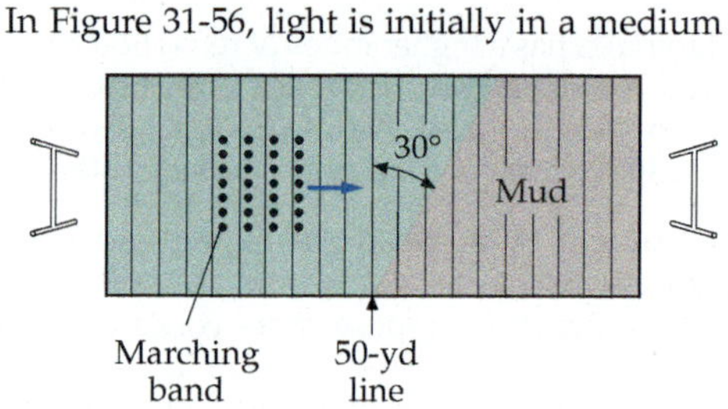

FIGURE 31-55 Problem 34

index of refraction n_1. It is incident at angle θ_1 on the surface of a liquid that has an index of refraction n_2. The light passes through the layer of liquid and enters glass that has an index of refraction n_3. If θ_3 is the angle of refraction in the glass, show that $n_1 \sin\theta_1 = n_3 \sin\theta_3$. That is, show that the second medium can be neglected when finding the angle of refraction in the third medium. SSM

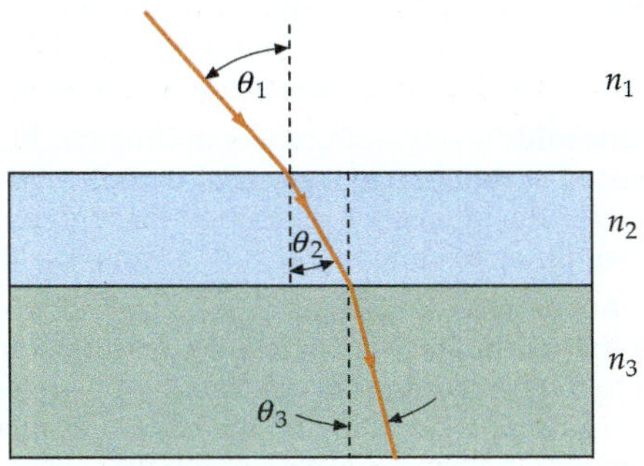

FIGURE 31-56 Problem 35

36 •• On a safari, you are spearfishing while wading in a river. You observe a fish gliding by you. If your line of sight to the fish is 64.0° below the horizontal in air, and assuming the spear follows a straight-line path through the air and water after it is released, determine the angle below the horizontal that you should aim your spear gun in order to catch dinner. Assume the spear gun barrel is 1.50 m above the water surface, the fish is 1.20 m below the surface, and the spear travels in a straight line all the way to the fish.

37 ••• You are standing on the edge of a swimming pool and looking directly across at the opposite side. You notice that the bottom edge of the opposite side of the pool appears to be at an angle of 28° below the horizontal. However, when you sit on the pool edge, the bottom edge of the opposite side of the pool appears to be at an angle of only 14° below the horizontal. Use these observations to determine the width and depth of the pool. *Hint: You will need to estimate the height of your eyes above the surface of the water when standing and sitting.*

38 ••• Figure 31-57 shows a beam of light incident on a glass plate of thickness d and index of refraction n. (*a*) Find the angle of incidence so that the separation b between the ray reflected from the top surface and the ray reflected from the bottom surface and exiting the top surface is a maximum. (*b*) What is this angle of incidence if the index of refraction of the glass is 1.60? (*c*) What is the separation of the two beams if the thickness of the glass plate is 4.0 cm?

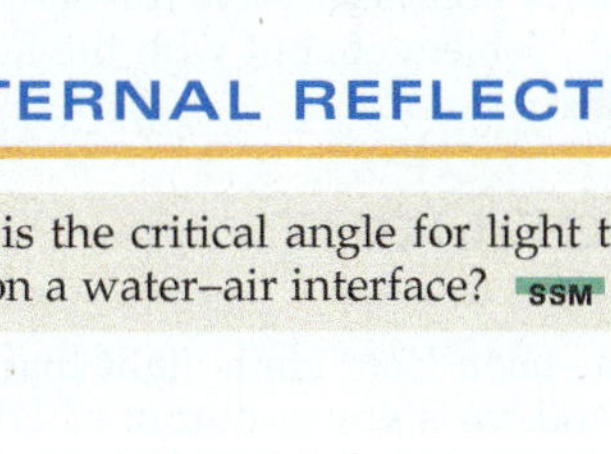

FIGURE 31-57 Problems 38 and 48

TOTAL INTERNAL REFLECTION

39 • What is the critical angle for light traveling in water that is incident on a water–air interface? SSM

40 • A glass surface ($n = 1.50$) has a layer of water ($n = 1.33$) on it. Light in the glass is incident on the glass–water interface. Find the critical angle for total internal reflection.

41 • A point source of light is located 5.0 m below the surface of a large pool of water. Find the area of the largest circle on the pool's surface through which light coming directly from the source can emerge.

42 •• Light traveling in air strikes the largest face of an isosceles-right-triangle prism at normal incidence. What is the speed of light in this prism if the prism is just barely able to produce total internal reflection?

43 •• A point source of light is located at the bottom of a steel tank, and an opaque circular card of radius 6.00 cm is placed horizontally over it. A transparent fluid is gently added to the tank so that the card floats on the fluid surface with its center directly above the light source. No light is seen by an observer above the surface until the fluid is 5.00 cm deep. What is the index of refraction of the fluid?

44 •• **ENGINEERING APPLICATION** An optical fiber allows rays of light to propagate long distances by using total internal reflection. Optical fibers are used extensively in medicine and in digital communications. As shown in Figure 31-58 the fiber consists of a core material that has an index of refraction n_2 and radius b surrounded by a cladding material that has an index of refraction $n_3 < n_2$. The *numerical aperture* of the fiber is defined as $\sin\theta_1$, where θ_1 is the angle of incidence of a ray of light that impinges on the center of the end of the fiber and then reflects off the core-cladding interface just at the critical angle. Using the figure as a guide, show that the numerical aperture is given by $\sin\theta_1 = \sqrt{n_2^2 - n_3^2}$ assuming the ray is initially in air. *Hint: Use of the Pythagorean theorem may be required.*

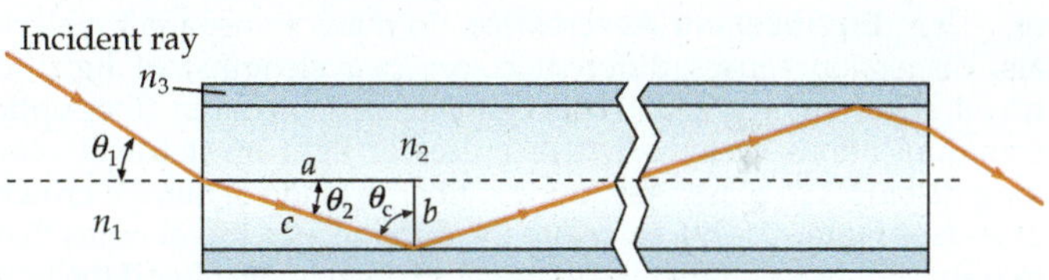

FIGURE 31-58 Problems 44, 45, and 46

45 •• **ENGINEERING APPLICATION** Find the maximum angle of incidence θ_1 of a ray that would propagate through an optical fiber that has a core index of refraction of 1.492, a core radius of 50.00 μm, and a cladding index of 1.489. See Problem 44. SSM

46 •• **ENGINEERING APPLICATION** Calculate the difference in time needed for two pulses of light to travel down 15.0 km of the fiber that is described in Problem 44. Assume that one pulse enters the fiber at normal incidence and the second pulse enters the fiber at the maximum angle of incidence calculated in Problem 45. In fiber optics, this effect is known as *modal dispersion*.

47 ••• Investigate how a thin film of water on a glass surface affects the critical angle for total reflection. Use $n = 1.50$ for glass and $n = 1.33$ for water. (*a*) What is the critical angle for total internal reflection at the glass–water interface? (*b*) Does a range of incident angles exist such that the angles are greater than θ_c for glass-to-air refraction and for which the light rays will leave the glass, travel through the water, and then pass into the air?

48 •• A laser beam is incident on a plate of glass that is 3.0 cm thick (Figure 31-57). The glass has an index of refraction of 1.5 and the angle of incidence is 40°. The top and bottom surfaces of the glass are parallel. What is the distance b between the beam formed by reflection off the top surface of the glass and the beam reflected off the bottom surface of the glass.

DISPERSION

49 • A beam of light strikes the plane surface of silicate flint glass at an angle of incidence of 45°. The index of refraction of the glass varies with wavelength (see Figure 31-59). How much smaller is the angle of refraction for violet light of wavelength 400 nm than the angle of refraction for red light of wavelength 700 nm?

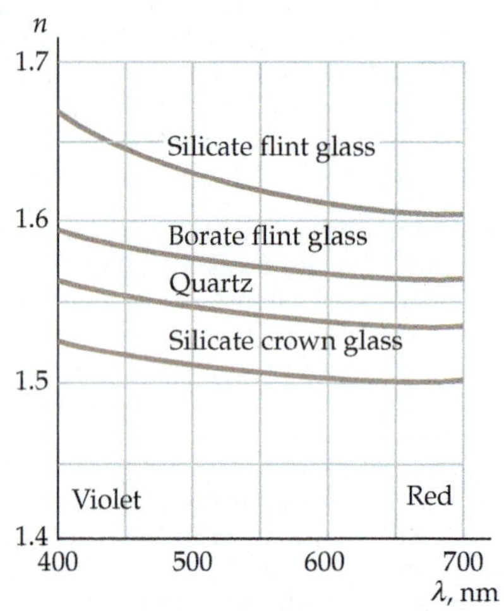

FIGURE 31-59 Problems 49, 50, 73, and 77

50 •• **ENGINEERING APPLICATION** In many transparent materials, dispersion causes different colors (wavelengths) of light to travel at different speeds. This can cause problems in fiber-optic communications systems where pulses of light must travel very long distances in glass. Assuming a fiber is made of silicate crown glass (see Figure 31-59), calculate the difference in travel times that two short pulses of light take to travel 15.0 km in the fiber if the first pulse has a wavelength of 700 nm and the second pulse has a wavelength of 500 nm.

POLARIZATION

51 • What is the polarizing angle for light in air that is incident on (*a*) water ($n = 1.33$) and (*b*) glass ($n = 1.50$)? SSM

52 • Light that is horizontally polarized is incident on a polarizing sheet. It is observed that only 15 percent of the intensity of the incident light is transmitted through the sheet. What angle does the transmission axis of the sheet make with the horizontal?

53 • Two polarizing sheets have their transmission axes crossed so that no light gets through. A third sheet is inserted between the first two so that its transmission axis makes an angle θ with the transmission axis of the first sheet. Unpolarized light of intensity I_0 is incident on the first sheet. Find the intensity of the light transmitted through all three sheets if (*a*) $\theta = 45°$ and (*b*) $\theta = 30°$.

54 • A horizontal 5.0 mW laser beam that is vertically polarized is incident on a polarizing sheet that is oriented with its transmission axis vertical. Behind the first sheet is a second sheet that is oriented so that its transmission axis makes an angle of 27° with respect to vertical. What is the power of the beam transmitted through the second sheet?

55 •• The polarizing angle for light in air that is incident on a certain substance is 60°. (*a*) What is the angle of refraction of light incident at this angle? (*b*) What is the index of refraction of this substance? SSM

56 •• Two polarizing sheets have their transmission axes crossed so that no light is transmitted. A third sheet is inserted so that its transmission axis makes an angle θ with the transmission axis of the first sheet. (*a*) Derive an expression for the intensity of the transmitted light as a function of θ. (*b*) Show that the intensity transmitted through all three sheets is maximum when $\theta = 45°$.

57 •• If the middle polarizing sheet in Problem 56 is rotating at an angular speed ω about an axis parallel with the light beam, find an expression for the intensity transmitted through all three sheets as a function of time. Assume that $\theta = 0°$ at time $t = 0$.

58 •• **SPREADSHEET** A stack of $N + 1$ ideal polarizing sheets is arranged so that each sheet is rotated by an angle of $\pi/(2N)$ rad with respect to the preceding sheet. A linearly polarized light wave of intensity I_0 is incident normally on the stack. The incident light is polarized along the transmission axis of the first sheet and is therefore perpendicular to the transmission axis of the last sheet in the stack. (*a*) Show that the intensity of the light transmitted through the entire stack is given by $I_0 \cos^{2N}[\pi/(2N)]$. (*b*) Using a spreadsheet or graphing program, plot the transmitted intensity as a function of N for values of N from 2 to 100. (*c*) What is the direction of polarization of the transmitted beam in each case?

59 •• **SPREADSHEET, ENGINEERING APPLICATION** The device described in Problem 58 could serve as a *polarization rotator*, which changes the linear plane of polarization from one direction to another. The efficiency of such a device is measured by taking the ratio of the output intensity at the desired polarization to the input intensity. The result of Problem 58 suggests that the highest efficiency is achieved by using a large value for the number N. A small amount of intensity is lost regardless of the input polarization when using a real polarizer. For each polarizer, assume the transmitted intensity is 98 percent of the amount predicted by the law of Malus and use a **spreadsheet** or graphing program to determine the optimum number of sheets you should use to rotate the polarization 90°. SSM

60 •• Show mathematically that a linearly polarized wave can be thought of as a superposition of a right and a left circularly polarized wave.

61 •• Suppose that the middle sheet in Problem 53 is replaced by two polarizing sheets. If the angles between the transmission axes of the second, third and fourth sheets in the stack make angles of 30°, 60°, and 90°, respectively, with the transmission axis of the first sheet, (*a*) what is the intensity of the transmitted light? (*b*) How does this intensity compare with the intensity obtained in Part (*a*) of Problem 53?

62 •• Show that the electric field of a circularly polarized wave propagating parallel to the x axis can be expressed by $\vec{E} = E_0 \sin(kx + \omega t)\hat{j} + E_0 \cos(kx + \omega t)\hat{k}$.

63 •• A circularly polarized wave is said to be *right circularly polarized* if the electric and magnetic fields rotate clockwise when viewed along the direction of propagation and *left circularly polarized* if the fields rotate counterclockwise. (*a*) What is the sense of the circular polarization for the wave described by the expression in Problem 62? (*b*) What would be the expression for the electric field of a circularly polarized wave traveling in the same direction as the wave in Problem 60, but with the fields rotating in the opposite sense? SSM

SOURCES OF LIGHT

64 • A helium–neon laser emits light that has a wavelength equal to 632.8 nm and has a power output of 4.00 mW. How many photons are emitted per second by this laser?

65 •• The first excited state of an atom of a gas is 2.85 eV above the ground state. (*a*) What is the maximum wavelength of radiation for resonance absorption by atoms of the gas that are in the ground state? (*b*) If the gas is irradiated with monochromatic light that has a wavelength of 320 nm, what is the wavelength of the Raman scattered light?

66 •• A gas is irradiated with monochromatic ultraviolet light that has a wavelength of 368 nm. Scattered light that has a wavelength equal to 368 nm is observed, and scattered light that has a wavelength of 658 nm is also observed. Assuming that the gas atoms were in their ground state prior to irradiation, find the energy difference between the ground state and the excited state obtained by the irradiation.

67 •• Sodium has excited states 2.11 eV, 3.20 eV, and 4.35 eV above the ground state. Assume that the atoms of the gas are all in the ground state prior to irradiation. (*a*) What is the maximum wavelength of radiation that will result in resonance fluorescence? What is the wavelength of the fluorescent radiation? (*b*) What wavelength will result in excitation of the state 4.35 eV above the ground state? If that state is excited, what are the possible wavelengths of resonance fluorescence that might be observed? SSM

68 •• Singly ionized helium is a hydrogen-like atom that has a nuclear charge of $+2e$. Its energy levels are given by $E_n = -4E_0/n^2$, where $n = 1, 2, \ldots$ and $E_0 = 13.6$ eV. If a beam of visible white light is sent through a gas of singly ionized helium, at what wavelengths will dark lines be found in the spectrum of the transmitted radiation? (Assume that the ions of the gas are all in the state with energy E_1 prior to irradiation.)

69 • A pulse from a ruby laser has an average power of 10 MW and lasts 1.5 ns. (*a*) What is the total energy of the pulse? (*b*) How many photons are emitted in the pulse? SSM

GENERAL PROBLEMS

70 • A beam of red light that has a wavelength of 700 nm in air travels in water. (*a*) What is the wavelength in water? (*b*) Does a swimmer underwater observe the same color or a different color for this light?

71 • The critical angle for total internal reflection for a substance is 48°. What is the polarizing angle for the substance? SSM

72 •• Show that when a flat mirror is rotated through an angle θ about an axis in the plane of the mirror, a reflected beam of light (from a fixed incident beam) that is perpendicular to the rotation axis is rotated through 2θ.

73 •• Use Figure 31-59 to calculate the critical angles for light initially in silicate flint glass that is incident on a glass–air interface if the light is (*a*) violet light of wavelength 400 nm and (*b*) red light of wavelength 700 nm. SSM

74 •• Light is incident on a slab of transparent material at an angle θ_1, as shown in Figure 31-60. The slab has a thickness t and an index of refraction n. Show that $n = \sin(\theta_1)/\sin[\tan^{-1}(d/t)]$, where d is the distance shown in the figure.

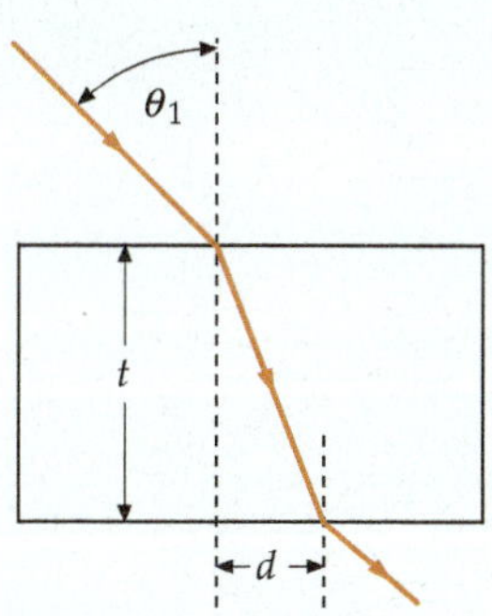

FIGURE 31-60 Problem 74

75 •• A ray of light begins at the point (−2.00 m, 2.00 m, 0.00 m), strikes a mirror in the $y = 0$ plane at some point $(x, 0, 0)$, and reflects through the point (2.00 m, 6.00 m, 0.00 m). (*a*) Find the value of x that makes the total distance traveled by the ray a minimum. (*b*) What is the angle of incidence on the reflecting plane? (*c*) What is the angle of reflection?

76 •• **ENGINEERING APPLICATION** To produce a polarized laser beam, a plate of transparent material (Figure 31-61) is placed in the laser cavity and oriented so the light strikes it at the polarizing angle. Such a plate is called a Brewster window. Show that if θ_{P1} is the polarizing angle for the n_1 to n_2 interface, then θ_{P2} is the polarizing angle for the n_2 to n_1 interface.

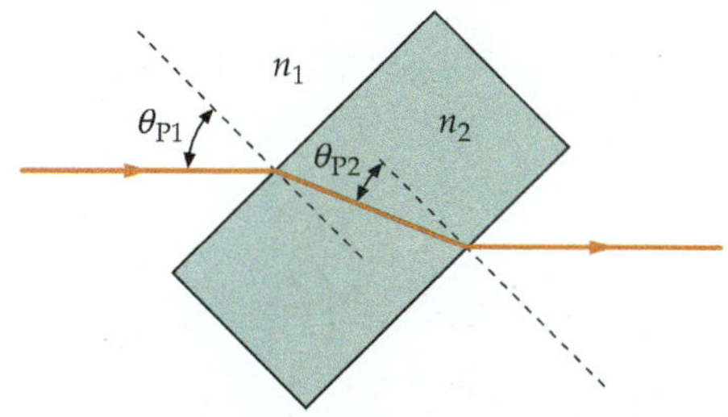

FIGURE 31-61 Problem 76

77 •• From the data provided in Figure 31-59, calculate the polarization angle for an air–glass interface, using light of wavelength 550 nm in each of the four types of glass shown. SSM

78 •• A light ray passes through a prism with an apex angle of α, as shown in Figure 31-62. The ray and the bisector of the apex angle intersect at right angles. Show that the *angle of deviation* δ is related to the apex angle and the index of refraction of the prism material by $\sin\left[\frac{1}{2}(\alpha + \delta)\right] = n \sin\left(\frac{1}{2}\alpha\right)$.

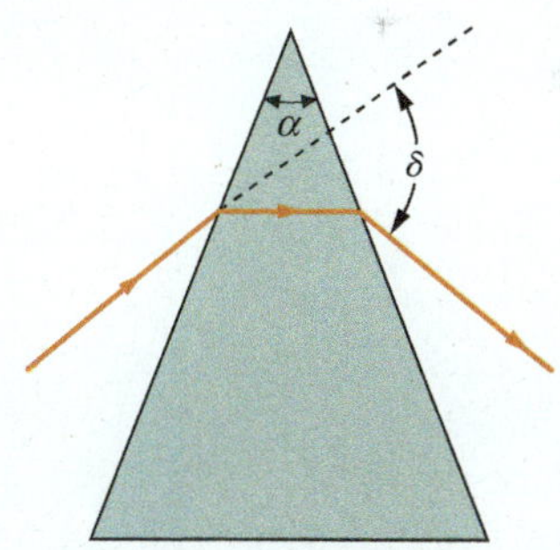

FIGURE 31-62 Problems 78 and 84

79 •• (*a*) For light rays inside a transparent medium that is surrounded by a vacuum, show that the polarizing angle and the critical angle for total internal reflection satisfy $\tan\theta_p = \sin\theta_c$. (*b*) Which angle is larger, the polarizing angle or the critical angle for total internal reflection? SSM

80 •• Light in air is incident on the surface of a transparent substance at an angle of 58° with the normal. The reflected and refracted rays are observed to be mutually perpendicular. (*a*) What is the index of refraction of the transparent substance? (*b*) What is the critical angle for total internal reflection in this substance?

81 •• A light ray in dense flint glass that has an index of refraction of 1.655 is incident on the glass surface. An unknown liquid condenses on the surface of the glass. Total internal reflection on the glass–liquid interface occurs for a minimum angle of incidence on the glass–liquid interface of 53.7°. (*a*) What is the refractive index of

the unknown liquid? (*b*) If the liquid is removed, what is the minimum angle of incidence for total internal reflection? (*c*) For the angle of incidence found in Part (*b*), what is the angle of refraction of the ray into the liquid film? Does a ray emerge from the liquid film into the air above? Assume the glass and liquid have parallel planar surfaces.

82 ••• (*a*) Show that for normally incident light, the intensity transmitted through a glass slab that has an index of refraction of n and is surrounded by air is approximately given by $I_T = I_0[4n/(n+1)^2]^2$. (*b*) Use the Part-(*a*) result to find the ratio of the transmitted intensity to the incident intensity through N parallel slabs of glass for light of normal incidence. (*c*) How many slabs of a glass that has an index of refraction of 1.5 are required to reduce the intensity to 10 percent of the incident intensity?

83 ••• Equation 31-14 gives the relation between the angle of deviation ϕ_d of a light ray incident on a spherical drop of water in terms of the incident angle θ_1 and the index of refraction of water. (*a*) Assume that $n_{air} = 1$ and derive an expression for $d\phi_d/d\theta_1$. *Hint: If* $y = \sin^{-1}x$, *then* $dy/dx = (1 - x^2)^{-1/2}$. (*b*) Use the result to show that the angle of incidence for minimum deviation θ_{1m} is given by $\cos\theta_{1m} = \sqrt{\tfrac{1}{3}(n^2 - 1)}$. (*c*) The index of refraction for a certain red light in water is 1.3318 and the index of refraction for a certain blue light in water is 1.3435. Use the result of Part (*a*) to find the angular separation of these colors in the primary rainbow.

84 ••• Show that the angle of deviation δ is a minimum if the angle of incidence is such that the ray and the bisector of the apex angle α (Figure 31-62) intersect at right angles.

CHAPTER 32

THIS IMAGE IS OF A FEMALE MOSQUITO AND WAS PRODUCED BY A MICROSCOPE. *(Nuridsany & Perennou/Photo Researchers.)*

Optical Images

? How do you determine the magnifying power of a compound microscope? (See Example 32-15.)

Because the wavelength of light is very small compared with most obstacles and openings, diffraction—the bending of waves around corners—is often negligible, and the ray approximation, in which waves are considered to propagate in straight lines, accurately describes observations.

In this chapter, we apply the laws of reflection and refraction to the formation of images by mirrors and lenses.

32-1 MIRRORS

PLANE MIRRORS

Figure 32-1 shows a bundle of light rays emanating from a point source P and reflected from a plane mirror. After reflection, the rays diverge exactly as if they came from a point P' behind the plane of the mirror. The point P' is called the **image** of the **object** P. When the reflected rays enter the eye, they cannot be distinguished from rays diverging from a source at P' when no mirror is present. This image at P' is called a **virtual image,** virtual because the light does not really emanate from it.

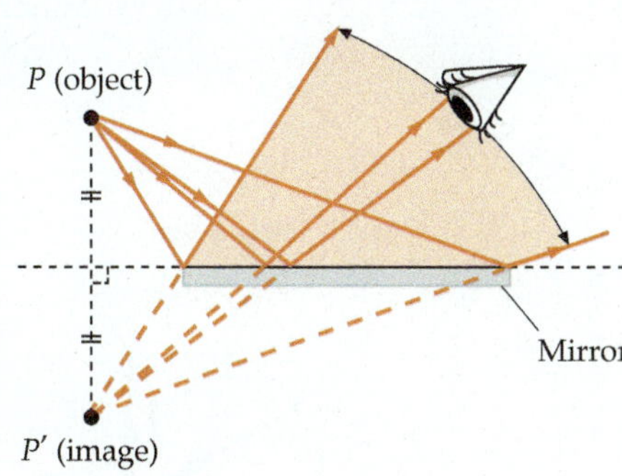

FIGURE 32-1 Image formed by a plane mirror. The rays from point P that strike the mirror and enter the eye appear to come from the image point P' behind the mirror. The image can be seen by an eye located anywhere in the shaded region.

The plane of the mirror is the perpendicular bisector of the line from the object point P to the image point P' as shown. The image can be seen by an eye located anywhere in the shaded region indicated, in which a straight line from the image to the eye passes through the mirror. The object need not be directly in front of the mirror. As long as the object is not located behind the plane of the mirror, some location exists at which the eye can be positioned so as to view the image.

If you stand in front of a mirror and hold up your right hand, the image you see is neither magnified nor reduced, but it looks like a left hand (Figure 32-2). This right-to-left reversal is a result of **depth inversion**—the hand is transformed from a right hand to a left hand because the front and the back of the hand are reversed by the mirror. Depth inversion is also illustrated in Figure 32-3. Figure 32-4 shows the image of a simple rectangular coordinate system. The mirror transforms a right-handed coordinate system, for which $\hat{i} \times \hat{j} = \hat{k}$, into a left-handed coordinate system, for which $\hat{i} \times \hat{j} = -\hat{k}$.

Figure 32-5 shows an arrow of height y standing parallel to a plane mirror a distance s from the mirror. We can locate the image of the tip of the arrowhead (and of any other point on the arrow) by drawing two rays. One ray, drawn perpendicular to the mirror, hits the mirror at point A and is reflected back onto itself. The other ray, making an angle θ with the normal to the mirror, is reflected, making an equal angle θ with the x axis. The extension of these two rays back behind the mirror locates the image of the arrowhead, as shown by the dashed lines in the figure. We can see from this figure that the image is the same distance behind the mirror as the object is in front of the mirror, and that the image is upright (the image of the arrow points in the same direction as the object) and is the same size as the object.

The formation of multiple images by two flat mirrors whose planes make an angle with each other is illustrated in Figure 32-6. We frequently see multiple images like this in clothing stores that provide mirrors for this purpose. The light from source point P that is reflected from mirror 1 strikes mirror 2 just as if it came from the image point P_1'. The image P_1' is the object for mirror 2. Its image is behind the plane of mirror 2 at point P_{12}''. This image will be formed as long as the image point P_1' is in front of the plane of mirror 2. The image at point P_2' is due to rays from P that reflect directly from mirror 2. Because P_2' is behind the plane of mirror 1, it cannot serve as an object point for an additonal image in mirror 1. The image at point P_2' cannot serve as an object for mirror 1 because the geometry dictates that none of the rays from P that reflect directly from mirror 2 can then strike mirror 1.

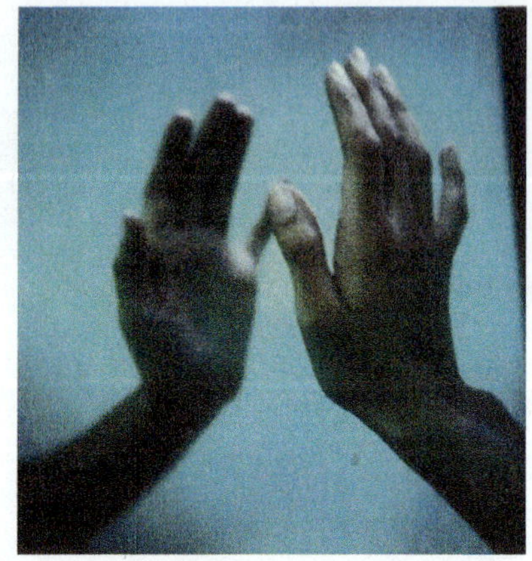

FIGURE 32-2 The image of a right hand in a plane mirror is transformed to a left hand. This right-to-left reversal is a result of depth inversion. *(Demitrios Zangos.)*

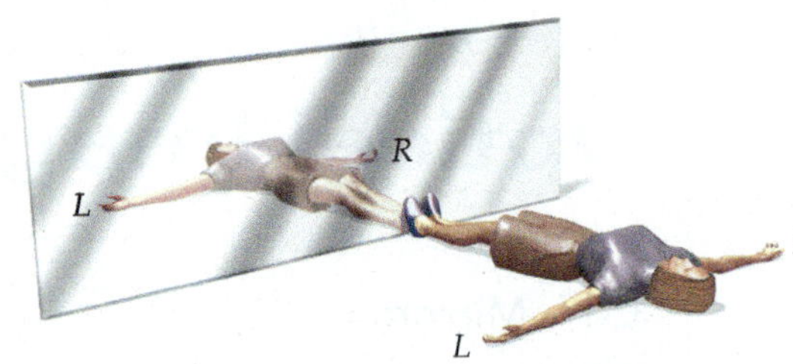

FIGURE 32-3 A person lying down with her feet against the mirror. The image is depth inverted.

Do not make the mistake of thinking that a mirror is an inversion in the left to right direction. Instead, it is an inversion in the front to back direction.

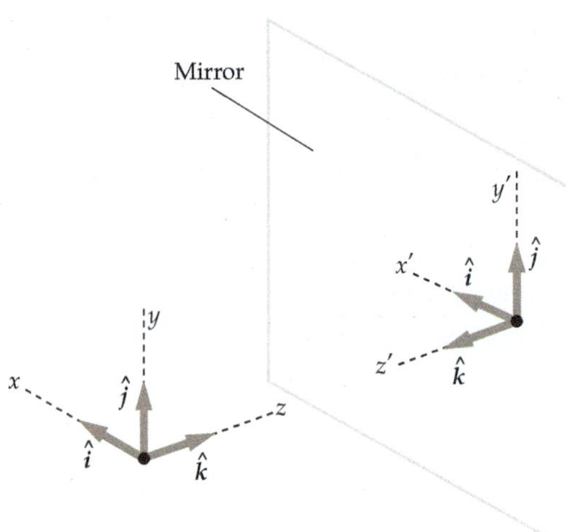

FIGURE 32-4 Image of a rectangular coordinate system in a plane mirror. The arrow along the z axis is reversed in the image. The image of the original right-handed coordinate system, for which $\hat{i} \times \hat{j} = \hat{k}$, is a left-handed coordinate system, for which $\hat{i} \times \hat{j} = -\hat{k}$.

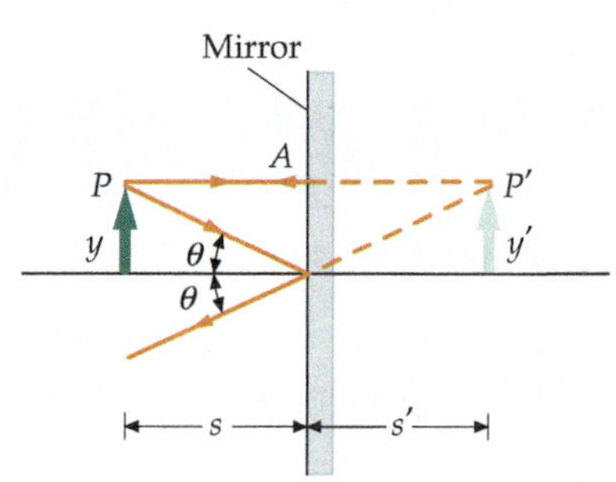

FIGURE 32-5 Ray diagram for locating the image of an arrow in a plane mirror.

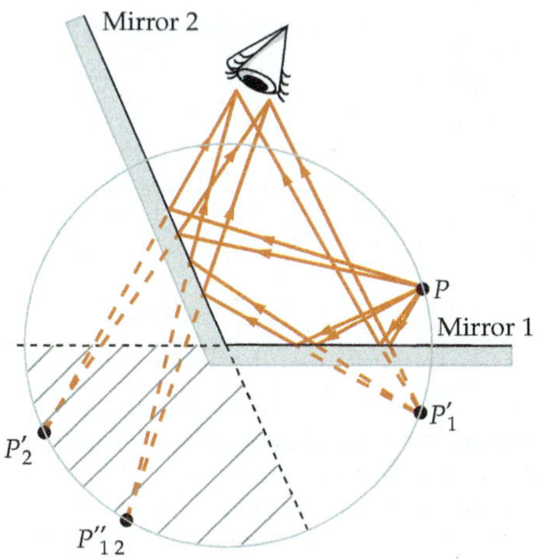

FIGURE 32-6 Images formed by two plane mirrors. P'_1 is the image of the object P in mirror 1, and P'_2 is the image of the object in mirror 2. Point P''_{12} is the image of P'_1 in mirror 2, which is seen when light rays from the object reflect first from mirror 1 and then from mirror 2. The image P'_2 does not have an image in mirror 1 because it is behind that mirror.

An alternative way of stating this is that because P'_2 is behind the plane of mirror 1, the image at P'_2 cannot serve as an object for mirror 1. The number of images formed by two mirrors depends on the angle between the mirrors and the position of the object.

Suppose your friend Ben is standing at point P (Figure 32-6) and is wearing a sweatshirt with BEN printed on it. In addition, suppose Ben is facing the intersection of the two mirrors and waving his right hand. Also, suppose that you are standing at the location of the eye. You can see an image of Ben at all three image locations. For the images at P'_1 and P'_2, Ben is waving his left hand and the printing on his sweatshirt appears as ИƎB. However, for the image at P''_{12} Ben is waving his right hand and the printing appears as BEN. For the image at P''_{12} depth inversion occurs twice, once for each reflection, so the result is as if no depth inversion occurs.

Figure 32-7 illustrates the fact that a horizontal ray reflected from two mutually perpendicular vertical mirrors is reflected back along a parallel path no matter what angle the ray makes with the mirrors. If three mirrors are mutually perpendicular to each other, like the sides of an inside corner of a box, any ray incident on any one of the mirrors from any direction is reflected back on a path parallel to the incident ray. A set of three mirrors arranged in this manner is called a corner-cube reflector. An array of corner-cube reflectors was placed on the moon in the Sea of Tranquility by the *Apollo 11* astronauts in 1969. A laser beam from Earth that is directed at the mirrors is reflected back to the same place on Earth. Such a beam has been used to measure the distance from the laser to the mirrors to within a few centimeters by measuring the time it takes for the light to travel to the mirrors and return.

SPHERICAL MIRRORS

Figure 32-8 shows a bundle of rays from a point source P on the axis of a concave spherical mirror reflecting from the mirror and converging at point P'. (A concave mirror is shaped like a cave when you look into it.) The rays then diverge from point P', just as if there were an object at the point P'. This image is called a **real image,** because light really does emanate from the image point. The image can be seen by an eye at the left of the image looking into the mirror. It can also be observed on a small viewing screen* or on a small piece of photographic film placed at the image point.

* A viewing screen must produce either diffuse reflection or diffuse transmission of the light. Ground glass is commonly used for this purpose. The screen must be relatively small, so that some of the light from the source reaches the mirror without being blocked by the screen.

CONCEPT CHECK 32-1

Show that a source point and all consequent image points formed by two flat mirrors are equidistant from the intersection of the planes of the two mirrors. (The circle shown in Figure 32-6 is equidistant from such an intersection.)

CONCEPT CHECK 32-2

Which of the images of himself can Ben see?

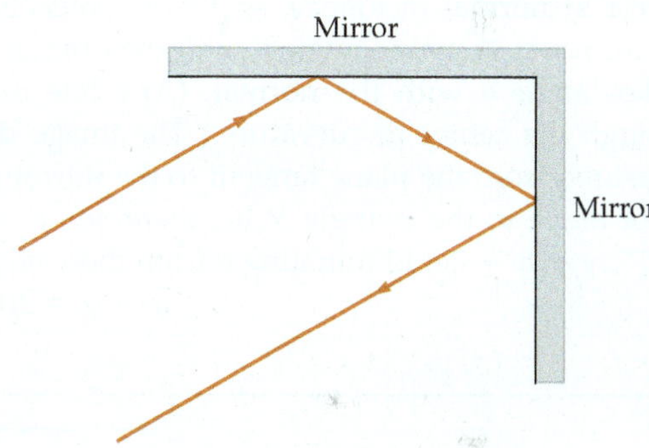

FIGURE 32-7 A ray striking one of two perpendicular plane mirrors is reflected from the second mirror in the direction opposite the original direction for any angle of incidence. The plane of the rays is perpendicular to both mirrors.

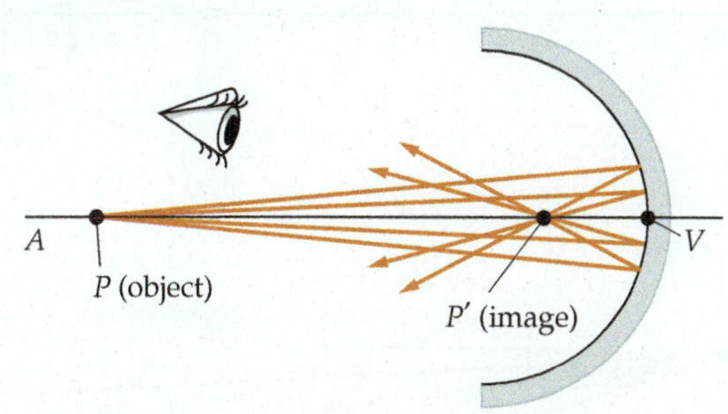

FIGURE 32-8 Rays from a point object P on the axis AV of a concave spherical mirror form an image at P'. The image is sharp if the rays strike the mirror near the axis and if the rays are almost parallel with the axis.

A virtual image, such as that formed by a plane mirror as discussed in the previous section, cannot be observed on a screen at the image point because no light from the object point exists at the image point. Despite this distinction between real and virtual images, the eye makes no distinction between them. The light rays diverging from a real image and those appearing to diverge from a virtual image are the same to the eye.

From Figure 32-9, we can see that only rays that strike the spherical mirror at points near the axis (line AV) are reflected through the image point. Rays that are almost parallel with the axis and are near to the axis are called **paraxial rays.** Rays that strike the mirror at points far from the axis upon reflection pass near the image point, but not through it. Such rays cause the image to appear blurred, an effect called **spherical aberration.** The image can be sharpened by blocking off all but the central part of the mirror, so that rays far from the axis do not strike it. The image is then sharper, but its brightness is reduced because less light is reflected to the image point.

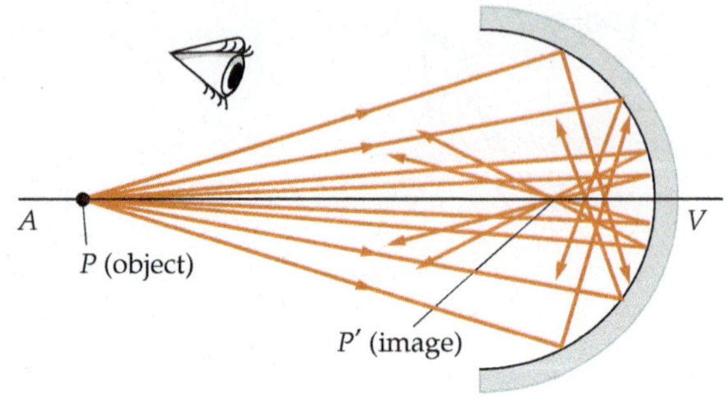

FIGURE 32-9 Spherical aberration of a mirror. Nonparaxial rays that strike the mirror at points far from the axis AV are not reflected through the image point P' formed by the paraxial rays. The nonparaxial rays blur the image.

In order to describe image formation, we wish to obtain an equation relating the position of the image point to the position of the object point. To do this, we draw two rays (Figure 32-10*a*) from an arbitrarily positioned object point P. One ray passes through point C, the center of curvature of the mirror, and the other ray strikes point A, an arbitrarily positioned point on the mirror. The image point P' is where these two rays intersect after reflecting off the mirror. Using the law of reflection, we obtain the location of P'. The ray passing through point C strikes the mirror at normal incidence, so the ray reflects back upon itself. The ray striking the mirror at A makes angle θ with the normal, so, as shown, the reflected ray also makes angle θ with the normal. (Any line normal to a spherical surface passes through the center of curvature.) The image distance s' and object distance s are measured from the plane tangent to the mirror at its vertex V. The angle β is an exterior angle to the triangle PAC, therefore, $\beta = \alpha + \theta$. Similarly, from the triangle PAP', $\gamma = \alpha + 2\theta$. Eliminating θ from these equations gives

$$\alpha + \gamma = 2\beta$$

See **Math Tutorial** *for more information on* **Geometry and Trigonometry**

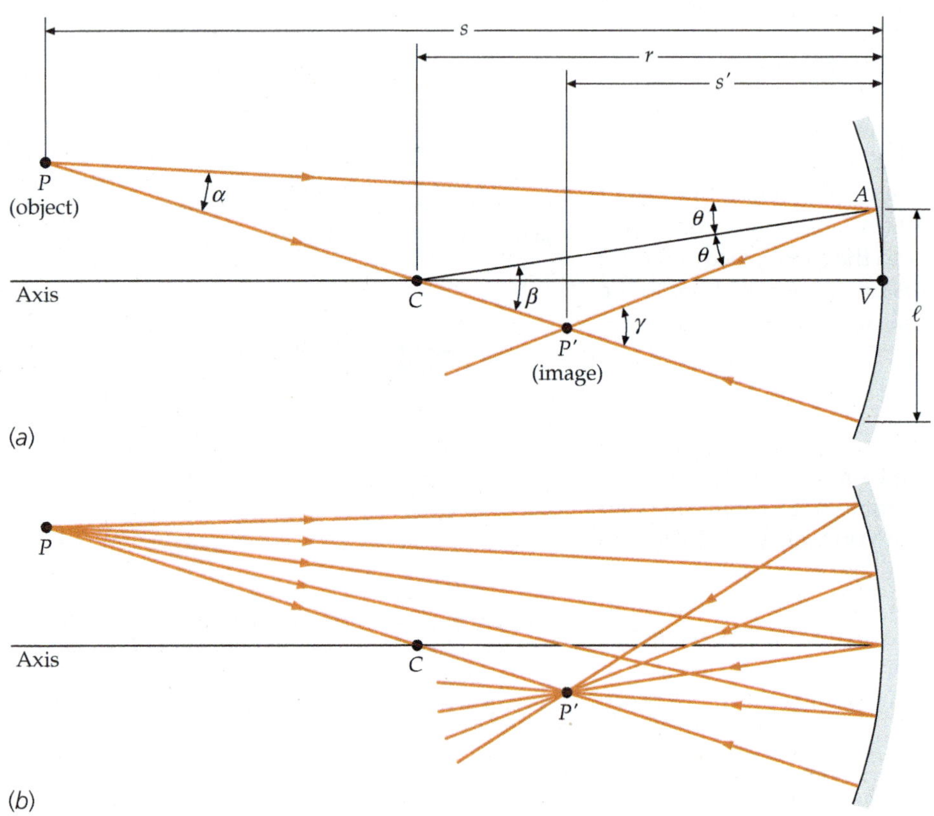

FIGURE 32-10 (a) Geometry for calculating the image distance s' from the object distance s and the radius of curvature r. The angle β is an exterior angle to the triangle PAC; therefore, $\beta = \alpha + \theta$. Similarly, from the triangle PAP', $\gamma = \alpha + 2\theta$. Eliminating θ from these equations gives $2\beta = \alpha + \gamma$. Equation 32-1 follows directly, if we assume the following small-angle approximations: $\alpha \approx \ell/s$, $\beta \approx \ell/r$, and $\gamma \approx \ell/s'$. (b) All paraxial rays from object point P pass through image point P' after reflecting off the mirror.

By assuming all rays are paraxial, we can substitute using the small-angle approximations: $\alpha \approx \ell/s$, $\beta \approx \ell/r$, and $\gamma \approx \ell/s'$. Equation 32-1 follows directly:

$$\frac{1}{s} + \frac{1}{s'} = \frac{2}{r} \qquad \text{32-1}$$

This equation relates the object and image distances with the radius of curvature. The striking thing about this equation is that it has absolutely no information about the location of point A. Therefore, the equation is valid for *any* choice for the location of point A, as long as point A is on the surface of the mirror and all rays are paraxial. That is, as shown in Figure 32-10*b*, *all* paraxial rays emanating from an object point will, upon reflection, pass through a *single* image point.

Equation 32-1 specifies the image position in terms of its distance from the mirror. We now specify the image position in terms of its distance from the axis. We first draw a single ray (Figure 32-11) that reflects off the mirror at its vertex. The two right triangles formed are similar. Corresponding sides of similar triangles are equal, so

$$\frac{y'}{y} = -\frac{s'}{s} \qquad \text{32-2}$$

The minus sign takes into account that y'/y is negative as P and P' are on opposite sides of the axis. Thus, if y is positive, y' is negative and if y is negative, y' is positive.

PRACTICE PROBLEM 32-1

For the image point and object point shown in Figure 32-11, show that

$$\frac{y'}{y} = -\frac{r/2}{s - (r/2)}$$

Hint: Solve Equation 32-1 for s' and substitute your result into Equation 32-2.

When the object distance is large compared with the radius of curvature of the mirror, the term $1/s$ in Equation 32-1 is much smaller than $2/r$ and can be neglected. That is, as $s \to \infty$, $s' \to \frac{1}{2}r$, where s' is the image distance. This distance is called the **focal length** f of the mirror, and the plane on which parallel rays incident on the mirror are focused is called the **focal plane.** The intersection of the axis with the focal plane is called the **focal point** F, as illustrated in Figure 32-12*a*. (Again, only paraxial rays are focused at a single point.)

$$f = \tfrac{1}{2}r \qquad \text{32-3}$$

FOCAL LENGTH FOR A MIRROR

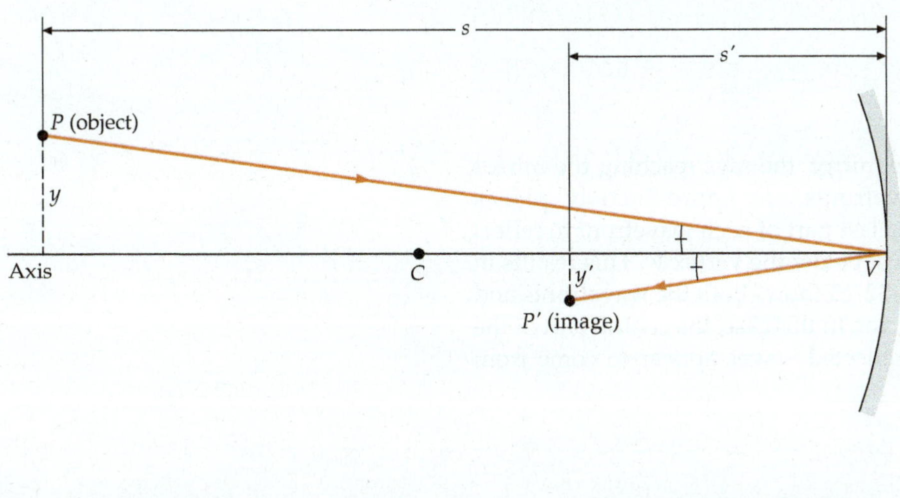

FIGURE 32-11 Geometry for calculating the position y' of the image point with respect to its distance from the axis.

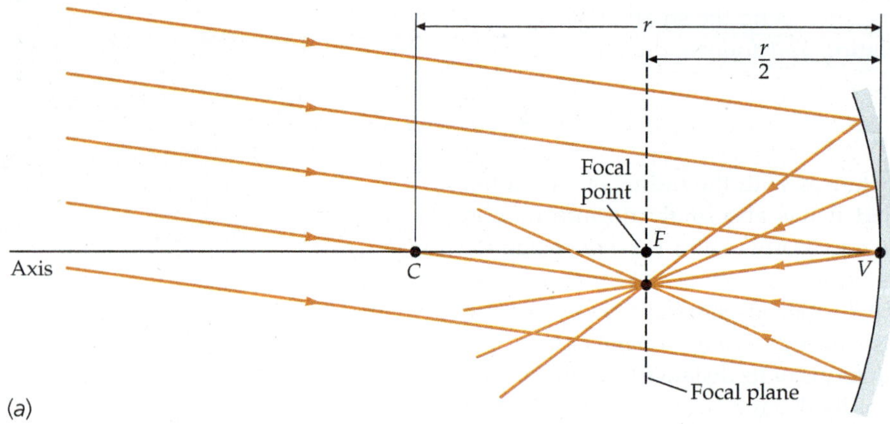

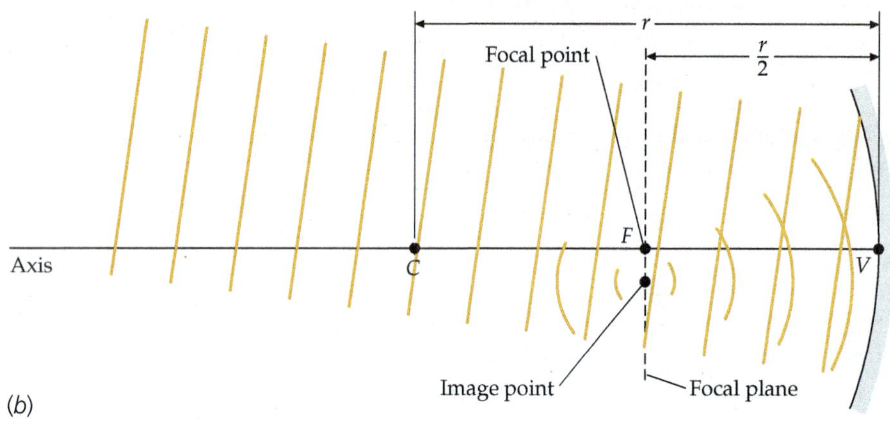

FIGURE 32-12 (*a*) Parallel rays strike a concave mirror and are reflected to a point on the focal plane a distance $\frac{1}{2}r$ to the left of the mirror. (*b*) The incoming wavefronts are planes; upon reflection, they become spherical wavefronts that converge to, and then diverge from, the image point.

PRACTICE PROBLEM 32-2

Show that solving Equation 32-1 for s' gives

$$s' = \frac{r}{2 - \frac{r}{s}}$$

Then show that as $s \rightarrow \infty$, $s' \rightarrow \frac{1}{2}r$.

The focal length of a spherical mirror is half the radius of curvature. In terms of the focal length f, Equation 32-1 is

$$\frac{1}{s} + \frac{1}{s'} = \frac{1}{f} \qquad 32\text{-}4$$

MIRROR EQUATION

Equation 32-4 is called the **mirror equation.**

When an object point is very far from the mirror, the rays reaching the mirror are approximately parallel, and the wavefronts are approximately planes (Figure 32-12*b*). In Figure 32-12*b*, note that the last part of each wavefront to reflect from the concave mirror surface is the part just below the vertex V. This results in a spherical wavefront upon reflection. Figure 32-13 shows both the wavefronts and the rays for plane waves striking a convex mirror. In this case, the central part of the wavefront strikes the mirror first, and the reflected waves appear to come from the focal point behind the mirror.

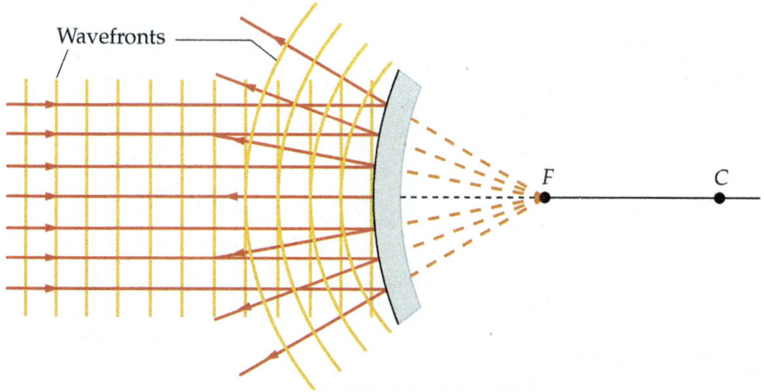

FIGURE 32-13 Reflection of plane waves from a convex mirror. The outgoing wavefronts are spherical, as if emanating from the focal point F behind the mirror. The rays are normal to the wavefronts and appear to diverge from F.

Figure 32-14 illustrates a property of waves called **reversibility.** If we reverse the direction of a reflected ray, the law of reflection assures us that the reflected ray will be along the original incoming ray, but in the opposite direction. (Reversibility also holds for refracted rays, which are discussed in later sections.) Thus, if we have a real image of an object formed by a reflecting (or refracting) surface, we can place an object at the image point and a new, real image will be formed at the position of the original object.

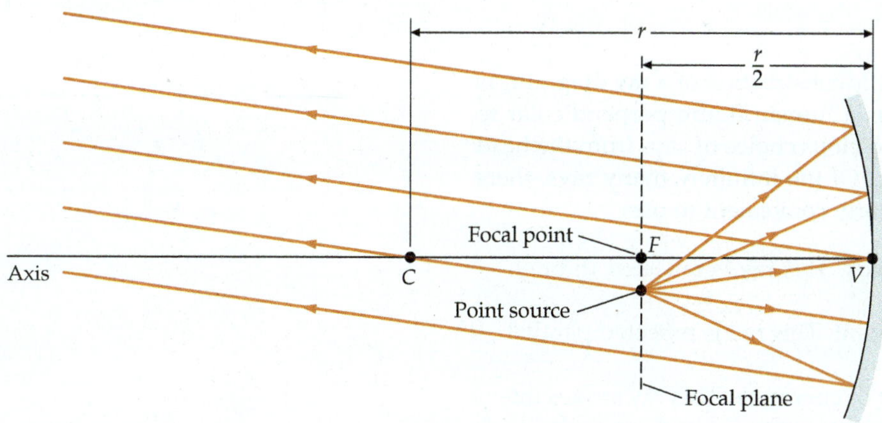

FIGURE 32-14 Reversibility. Rays diverging from a point source on the focal plane of a concave mirror are reflected from the mirror as parallel rays. The rays follow the same paths as in Figure 32-12*a* but in the reverse direction.

Example 32-1 Image in a Concave Mirror

A point source is 12 cm from a concave mirror and 3.0 cm above the axis of the mirror. The radius of curvature of the mirror is 6.0 cm. Find (*a*) the focal length of the mirror and (*b*) the image distance. (*c*) Find the position of the image relative to the axis.

PICTURE The focal length of a spherical mirror is half the radius of curvature. Once the focal length is known, the image distance can be found using the mirror equation (Equation 32-4), and the distance of the image from the axis can be found using Equation 32-2. The image distance from the mirror is the distance from the plane tangent to the mirror at its vertex.

SOLVE

(*a*) The focal length is half the radius of curvature:

$$f = \tfrac{1}{2}r = \tfrac{1}{2}(6.0\text{ cm}) = \boxed{3.0\text{ cm}}$$

(*b*) 1. Use the mirror equation to find a relation for the image distance s':

$$\frac{1}{s} + \frac{1}{s'} = \frac{1}{f}$$

or

$$\frac{1}{12\text{ cm}} + \frac{1}{s'} = \frac{1}{3.0\text{ cm}}$$

2. Solve for s':

$$\frac{1}{s'} = \frac{4}{12\text{ cm}} - \frac{1}{12\text{ cm}} = \frac{3}{12\text{ cm}}$$

$$s' = \boxed{4.0\text{ cm}}$$

(*c*) 1. Use Equation 32-2 to find the distance y' of image from the axis:

$$\frac{y'}{y} = -\frac{s'}{s}$$

2. Solve for y':

$$y' = -\frac{s'}{s}y = -\frac{4.0\text{ cm}}{12\text{ cm}}(3.0\text{ cm})$$

$$= \boxed{-1.0\text{ cm}}$$

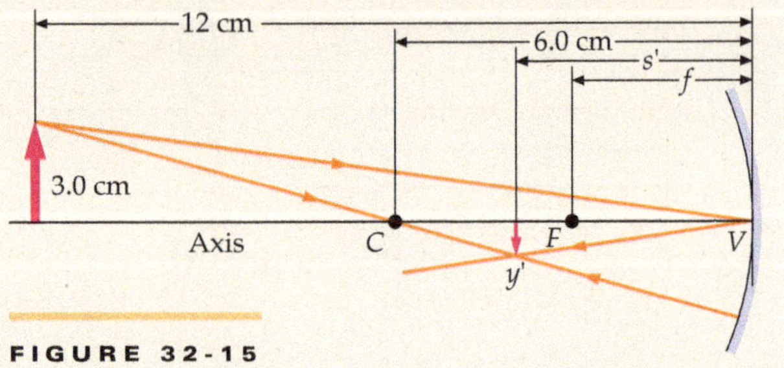

FIGURE 32-15

CHECK In Figure 32-15, two rays from the tip of the arrow have been drawn to locate the corresponding point on the image. By choosing to draw the ray that passes through C and the ray that reflects off the mirror at *V*, we have chosen to draw two rays that are easy to trace. From this figure, we can see that the results of the solution are very plausible.

PRACTICE PROBLEM 32-3 A concave mirror has a focal length of 4.0 cm. (*a*) What is the mirror's radius of curvature? (*b*) Find the image distance for an object 2.0 cm from the mirror.

CONCEPT CHECK 32-3

What is the radius of curvature of a plane mirror?

RAY DIAGRAMS FOR MIRRORS

A useful method to locate images is by geometric construction of a **ray diagram,** as illustrated in Figure 32-16, where the object is a human figure perpendicular to the axis a distance *s* from the mirror. By the judicious choice of rays from the head of the figure, we can quickly locate the image. Of the infinitely many rays, there are three rays, **principal rays,** that are particularly convenient to use:

1. The **parallel ray,** drawn parallel to the axis. This ray is reflected through the focal point.
2. The **focal ray,** drawn through the focal point. This ray is reflected parallel to the axis.
3. The **radial ray,** drawn through the center of curvature. This ray strikes the mirror perpendicular to its surface and is thus reflected back on itself.

PRINCIPAL RAYS FOR A MIRROR

These rays are shown in Figure 32-16. The intersection of any two paraxial rays locates the image point of the head. The three principal rays are easier to draw than any of the other rays. Typically, you draw two of the principal rays to locate the image, and then draw the third principal ray as a check to verify the result. Ray diagrams are best drawn with the mirror replaced by a straight line that extends as far as necessary to intercept the rays, as shown in Figure 32-17. (Note that the image in this case is real, inverted, and smaller than the object.)

When the object is between the mirror and its focal point, the rays from an object point reflected from the mirror do not converge but appear to diverge from a point behind the mirror,

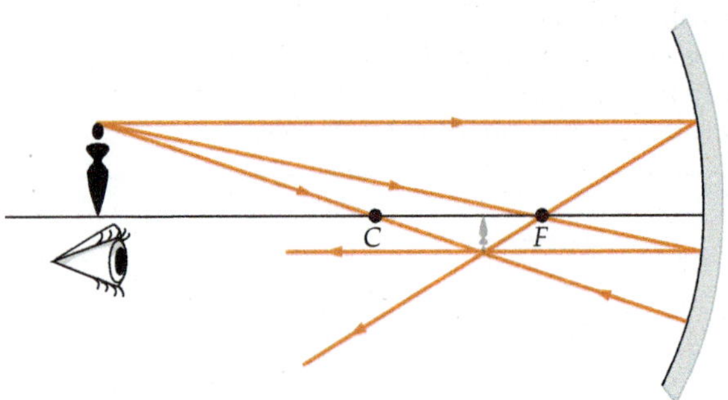

FIGURE 32-16 Ray diagram for the location of the image by geometric construction.

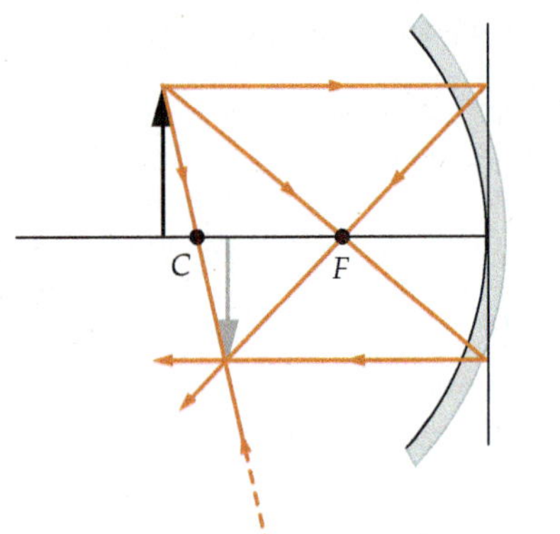

FIGURE 32-17 Ray diagrams are easier to construct if the curved surface is replaced by a plane tangent to the surface at the vertex.

as illustrated in Figure 32-18. In this case, the image is virtual and upright (*upright* meaning not inverted relative to the object). For an object between the mirror and the focal point, s is less than $r/2$, so the image distance s' calculated from Equation 32-1 turns out to be negative. We can apply Equations 32-1, 32-2, 32-3, and 32-4 to this case and to convex mirrors if we adopt a convenient sign convention. Whether the mirror is convex or concave, real images can be formed only in front of the mirror, that is, on the same side of the mirror as the reflected light (and the object). Virtual images are formed behind the mirror where there is no actual light from the object. Our sign convention is as follows:

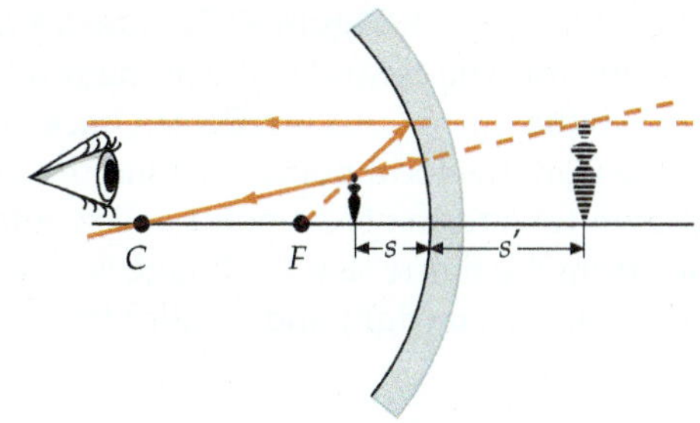

FIGURE 32-18 A virtual image is formed by a concave mirror when the object is inside the focal point. Here the image is located by the radial ray, which is reflected back on itself, and the focal ray, which is reflected parallel to the axis. The two reflected rays appear to diverge from an image point behind the mirror. This image point is found by constructing extensions to the reflected rays.

1. s is positive if the object is on the incident-light side of the mirror.
2. s' is positive if the image is on the reflected-light side of the mirror.
3. r (and thus f) is positive if the mirror is concave so the center of curvature is on the reflected-light side of the mirror.

SIGN CONVENTIONS FOR REFLECTION

The incident-light side and the reflected-light side are, of course, the same side of the mirror. The parameters s, s', r, and f are all positive if a real object* is in front of a concave mirror that forms a real image. A parameter is negative if it does not meet the stated condition for being positive.

The ratio of the image height to the object height is defined as the **lateral magnification** m of the image. From Figure 32-19 and Equation 32-2, we see that the lateral magnification is

$$m = \frac{y'}{y} = -\frac{s'}{s} \qquad 32\text{-}5$$

LATERAL MAGNIFICATION

A negative magnification, which occurs when both s and s' are positive, indicates that the image is inverted.

For plane mirrors, the radius of curvature is infinite. The focal length given by Equation 32-3 is then also infinite. Equation 32-4 then gives $s' = -s$, indicating that the image is behind the mirror at a distance equal to the object distance. The magnification given by Equation 32-5 is then $+1$, indicating that the image is upright and the same size as the object.

Although the preceding equations, coupled with our sign conventions, are relatively easy to use, we often need to know only the approximate location and magnification of the image and whether it is real or virtual and upright or inverted. This knowledge is usually easiest to obtain by constructing a ray diagram. It is always a good idea to use both this graphical method and the algebraic method to locate an image, so that one method serves as a check on the results of the other.

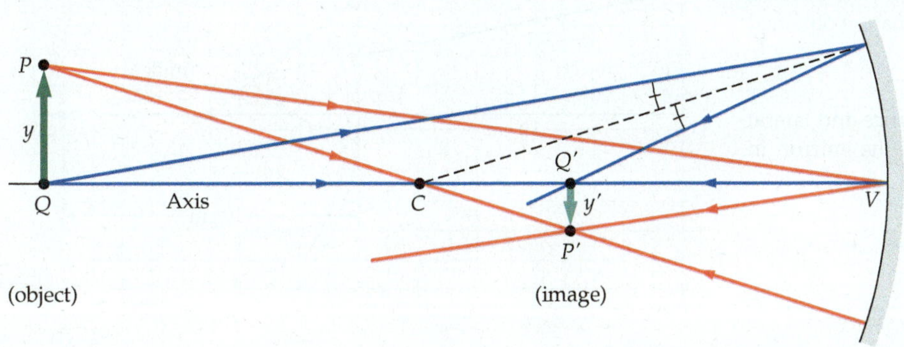

FIGURE 32-19 Geometry for showing the lateral magnification. Rays from the top of the object at P, upon reflection, intersect at P'; and rays from the bottom of the object at Q intersect at Q', where points P and P' have vertical positions y and y', respectively. The lateral magnification m is given by the ratio y'/y. In accord with Equation 32-2, $y'/y = -s'/s$. The minus sign results from the fact that y'/y is negative when s and s' are both positive. A negative m means the image is inverted.

* An object is real if it is on the same side of the mirror as the incident light.

Convex mirrors Figure 32-20 shows a ray diagram for an object in front of a convex mirror. The central ray heading toward the center of curvature C is perpendicular to the mirror and is reflected back on itself. The parallel ray is reflected as if it came from the focal point F behind the mirror. The focal ray (not shown) would be drawn toward the focal point and would be reflected parallel to the axis. We can see from the figure that the image is behind the mirror and is therefore virtual. The image is also upright and smaller than the object.

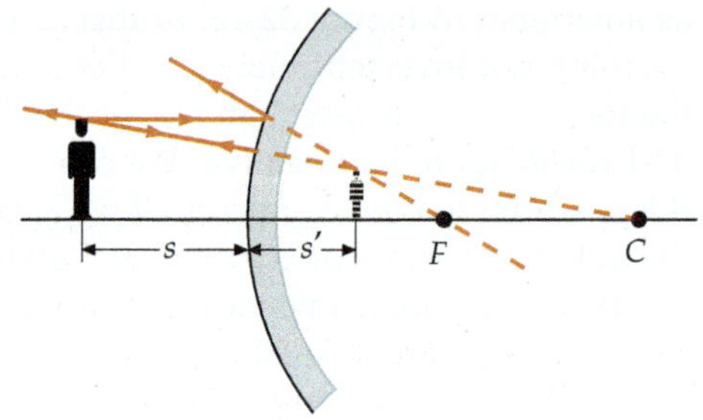

FIGURE 32-20 Ray diagram for an object in front of a convex mirror.

Example 32-2 Image in a Convex Mirror

An object that is 2.0 cm high is 10 cm from a convex mirror that has a radius of curvature equal to 10 cm. (*a*) Locate the image and (*b*) find the height of the image.

PICTURE The ray diagram for this problem is the same as shown in Figure 32-20. From the figure, we see that the image is upright, virtual, and smaller than the object. To find the exact location and height of the image, we use the mirror equation, with $s = 10$ cm and $r = -10$ cm.

SOLVE

(*a*) 1. The image distance s' is related to the object distance s and the focal length f by the mirror equation: $\frac{1}{s} + \frac{1}{s'} = \frac{1}{f}$

2. Calculate the focal length of the mirror: $f = \frac{1}{2}r = \frac{1}{2}(-10\text{ cm}) = -5.0\text{ cm}$

3. Substitute $s = 10$ cm and $f = -5.0$ cm into the mirror equation to find the image distance: $\frac{1}{10\text{ cm}} + \frac{1}{s'} = \frac{1}{-5.0\text{ cm}}$

4. Solve for s': $s' = \boxed{-3.3\text{ cm}}$

(*b*) 1. The height of the image is m multiplied by the height of the object: $y' = my$

2. Calculate the magnification m: $m = -\frac{s'}{s} = -\frac{-3.3\text{ cm}}{10\text{ cm}} = +0.33$

3. Use m to find the height of the image: $y' = my = (0.33)(2.0\text{ cm}) = \boxed{0.67\text{ cm}}$

CHECK The image distance is negative, indicating a virtual image behind the mirror. The magnification is positive and less than one, indicating that the image is upright and smaller than the object. The results are in agreement with the information obtained from the ray diagram (Figure 32-21).

PRACTICE PROBLEM 32-4 Find the image distance and lateral magnification for an object 5.0 cm away from the mirror in Example 32-2.

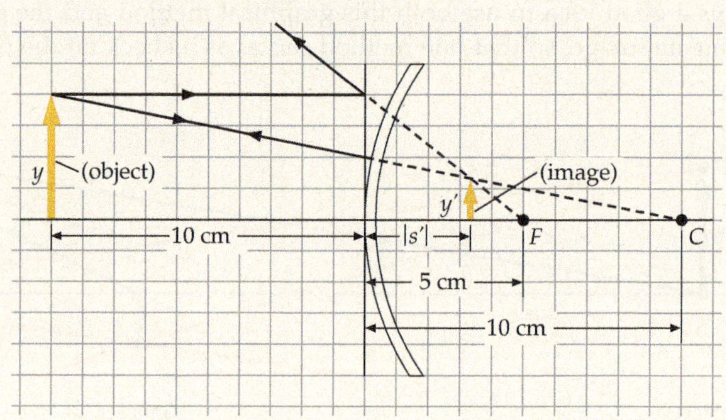

FIGURE 32-21

Example 32-3 Determining the Range — *Context-Rich*

You have a part-time job at Pleasant Hills Golf Course. The fairway of the 16th hole is horizontal for the first 50 yd and then goes down a not-too-steep hill (Figure 32-22), so the people on the tee cannot see the party in front. To prevent people from driving off the tee into the party in front, a convex mirror is mounted on a pole, enabling golfers on the tee to see whether or not the party in front is out of range.* Your boss says that a range finder that works by triangulation could be placed facing the mirror, so the golfers could measure how far the image of the party in front is behind the mirror. Then the golfers could be given a chart telling them how far the next party is from the tee. Your boss knows you are taking a physics course, so he asks you to calculate the distance of the image behind the mirror if the next party is 250 yd from the tee. The radius of curvature of the mirror has a magnitude of 20.0 yd.

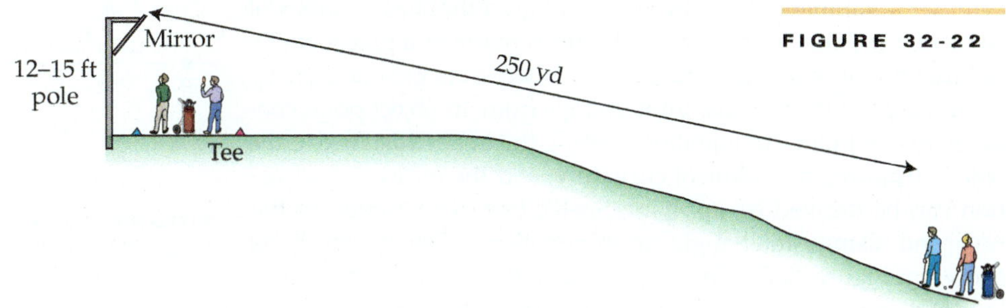

FIGURE 32-22

PICTURE The image distance is related to the object distance by the mirror formula, and the focal length of the lens is half the radius of curvature.

SOLVE

1. Use the mirror equation. For a convex mirror, the radius of curvature is negative:

$$\frac{1}{s} + \frac{1}{s'} = \frac{1}{f}$$

and

$$f = \frac{2}{r}$$

so

$$\frac{1}{250\text{ yd}} + \frac{1}{s'} = \frac{2}{-20.0\text{ yd}}$$

2. The image is 9.62 yd behind the mirror:

$$s' = \boxed{-9.62\text{ yd}}$$

CHECK That the step-2 result is negative is as expected. That is, it was expected that the image would be behind the mirror.

PRACTICE PROBLEM 32-5 What is the distance to the party in front if the image is 9.75 yd behind the mirror?

(a)

(b)

(*a*) A convex mirror resting on paper that has equally spaced parallel stripes. Note the large number of lines imaged in a small space and the reduction in size and distortion in shape of the image. (*b*) A convex mirror is used for security in a store. *(Richard Megna/Fundamental Photographs.)*

* The mirror was not replaced after it was knocked down during a storm a couple of years ago.

32-2 LENSES

IMAGES FORMED BY REFRACTION

One end of a long transparent cylinder is machined and polished to form a convex spherical surface. Figure 32-23 illustrates the formation of an image by refraction at such a surface. Suppose the cylinder is submerged in a transparent liquid that has an index of refraction n_1, and suppose the cylinder is made of a plastic material that has an index of refraction n_2, where n_2 is greater than n_1. Again, only in the paraxial limit do rays from an object point converge to one point. An equation relating the image distance to the object distance, the radius of curvature, and the indices of refraction can be derived by applying Snell's law of refraction to the rays and using small-angle approximations. The geometry is shown in Figure 32-24. The angles θ_1 and θ_2 are related by Snell's law of refraction: $n_1 \sin\theta_1 = n_2 \sin\theta_2$. Using the small-angle approximation ($\sin\theta = \theta$), Snell's law becomes $n_1\theta_1 = n_2\theta_2$. From triangle ACP', we have $\beta = \theta_2 + \gamma = (n_1/n_2)\theta_1 + \gamma$; and from triangle PAC, we have $\theta_1 = \alpha + \beta$. Eliminating θ_1 from these two equations gives $n_1\alpha + n_2\gamma = (n_2 - n_1)\beta$. Substituting for α, β and γ and using the small-angle approximations $\alpha \approx \ell/s, \beta \approx \ell/r$, and $\gamma = \ell/s'$ gives

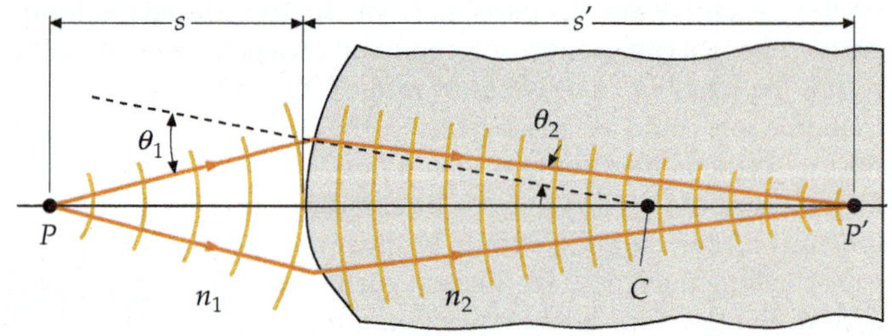

FIGURE 32-23 Image formed by refraction at a spherical surface between two media where the waves move slower in the second medium.

$$\frac{n_1}{s} + \frac{n_2}{s'} = \frac{n_2 - n_1}{r} \qquad \text{32-6}$$

REFRACTION AT A SINGLE SURFACE

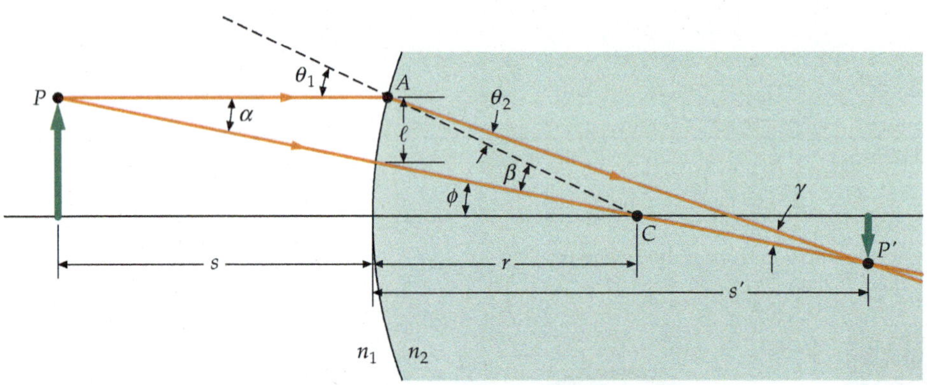

FIGURE 32-24 Geometry for relating the image position to the object position for refraction at a single spherical surface.

In refraction, real images are formed in back of the surface, which we will call the refracted-light side, whereas virtual images occur on the incident-light side, in front of the surface. The sign conventions we use for refraction are similar to those for reflection:

1. s is positive for objects on the incident-light side of the surface.
2. s' is positive for images on the refracted-light side of the surface.
3. r is positive if the center of curvature is on the refracted-light side.

SIGN CONVENTIONS FOR REFRACTION*

* The sign convention of choice for advanced work on optical design is the Cartesian sign convention. It can readily be found on the Internet.

We see that parameters s, s', and r are all positive if a real object is in front of a convex refracting surface that forms a real image. A parameter is negative if it does not meet the stated condition for being positive.

Example 32-4 Magnification by a Refracting Surface *Try It Yourself*

Derive an expression for the magnification $m = y'/y$ of an image formed by a spherical refracting surface.

PICTURE The magnification is the ratio of y' to y. Using Figure 32-19 and Figure 32-24 as guides, draw a ray diagram suitable for this derivation. The heights are related to the tangents of the angles θ_1 and θ_2, as shown in Figure 32-25. The angles are related by Snell's law. For paraxial rays, you can use the approximations $\tan\theta \approx \sin\theta \approx \theta$, and $\cos\theta \approx 1$.

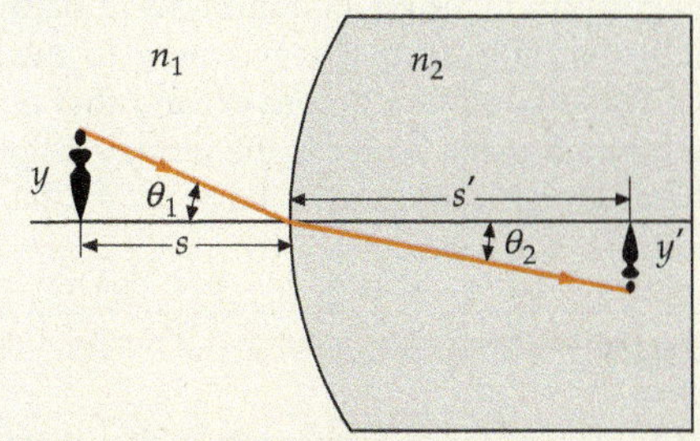

FIGURE 32-25

SOLVE

Cover the column to the right and try these on your own before looking at the answers.

Steps:	Answers
1. Using Figure 32-19 and Figure 32-24 as guides, draw a ray diagram suitable for this derivation. The drawing should include an object, a real image, a refracting surface, and an axis. Then draw an incident ray from the top of the object to the intersection of the axis with the refracting surface, and draw the refracted ray to the corresponding image point (Figure 32-25).	
2. Write expressions for $\tan\theta_1$ and $\tan\theta_2$ in terms of the heights y and $-y'$ and the object and image distances s and s'. (Because y' is negative, use $-y'$, so that $\tan\theta_2$ is positive.)	$\tan\theta_1 = \frac{y}{s};\quad \tan\theta_2 = \frac{-y'}{s'}$
3. Apply the small-angle approximation $\tan\theta \approx \theta$ to your expressions.	$\theta_1 = \frac{y}{s};\quad \theta_2 = \frac{-y'}{s'}$
4. Write Snell's law of refraction relating the angles θ_1 and θ_2 using the small-angle approximation $\sin\theta \approx \theta$.	$n_1 \sin\theta_1 = n_2 \sin\theta_2$ $n_1\theta_1 = n_2\theta_2$
5. Substitute the expressions for θ_1 and θ_2 found in step 3.	$n_1\left(\frac{y}{s}\right) = n_2\left(\frac{-y'}{s'}\right)$
6. Solve for the magnification $m = y'/y$.	$\boxed{m = \frac{y'}{y} = -\frac{n_1 s'}{n_2 s}}$

CHECK The step-6 result for the lateral magnification is dimensionless, as expected.

We see from Example 32-4 that the magnification due to refraction at a spherical surface is

$$m = \frac{y'}{y} = -\frac{n_1 s'}{n_2 s} \tag{32-7}$$

MAGNIFICATION FOR A REFRACTING BOUNDARY

Example 32-5 Image Seen from a Goldfish Bowl

Goldie the goldfish is in a 15.0-cm-radius spherical bowl of water that has an index of refraction of 1.33. Fluffy the cat is sitting on the table with her nose 10.0 cm from the surface of the bowl (Figure 32-26). The light from Fluffy's nose is refracted by the air–water boundary to form an image. Find (*a*) the image distance and (*b*) the magnification of the image of Fluffy's nose. Neglect any effect of the bowl's thin glass wall.

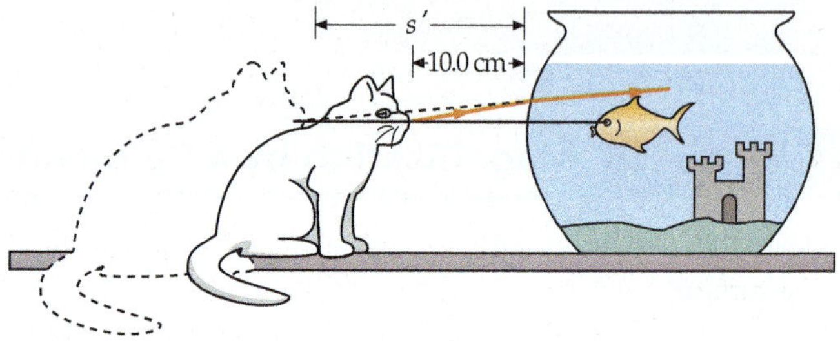

FIGURE 32-26 Goldie sees Fluffy's image farther from the bowl than Fluffy actually is.

PICTURE We find the image distance s' using Equation 32-6 and the magnification using Equation 32-7. Because we are interested in light that goes from Fluffy's nose to the bowl, it follows that the air–water boundary is convex, and that air is the incident-light side of boundary and water is the refracted-light side of boundary. With these identifications, we have $n_1 = 1.00$, $n_2 = 1.33$, $s = +10.0$ cm, and $r = +15.0$ cm.

SOLVE

(*a*) 1. The equation relating the object distance to the image distance is Equation 32-6:

$$\frac{n_1}{s} + \frac{n_2}{s'} = \frac{n_2 - n_1}{r}$$

2. Identify and assign signs to the parameters in the previous step:

$n_1 = 1.00$, $n_2 = 1.33$, $s = +10.0$ cm, and $r = +15.0$ cm

3. Substitute numerical values and solve for s':

$$\frac{1.00}{10.0\text{ cm}} + \frac{1.33}{s'} = \frac{1.33 - 1.00}{15.0\text{ cm}}$$

so

$$s' = \boxed{-17.1\text{ cm}}$$

(*b*) Substitute numerical values into Equation 32-7 to find the magnification m:

$$m = -\frac{n_1 s'}{n_2 s} = -\frac{(1.00)(-17.1\text{ cm})}{(1.33)(10.0\text{ cm})} = \boxed{1.29}$$

CHECK Because s' is negative, the image is virtual; that is, the image is on the side of the refracting surface opposite the refracted light, as shown in Figure 32-26. The fish, Goldie, would see Fluffy to be slightly farther away ($|s'| > s$) than she actually is, and larger ($|m| > 1$) than she actually is. That m is positive indicates the image is upright.

PRACTICE PROBLEM 32-6 If Goldie is 7.5 cm from the side of the bowl nearest Fluffy, find (*a*) the location and (*b*) the magnification of Goldie's image, as seen by Fluffy.

PRACTICE PROBLEM 32-7 The bowl is replaced by an aquarium that has flat sides and Goldie is 7.5 cm from the side where Fluffy is. Use Equation 32-6 to find the location of the image of Goldie that Fluffy sees.

Example 32-6 Image Seen from an Overhead Branch

During the summer months, Goldie the fish spends much of her time in a small pond in her owner's backyard. While enjoying a rest at the bottom of the 1.00-m-deep pond, Goldie is being watched by Fluffy the cat, who is perched on a tree limb 3.00 m above the surface of the pond. How far below the surface is the image of the fish that Fluffy sees? (The index of refraction of water is 1.33.)

PICTURE The surface of the pond is a spherical refracting surface that has a radius of curvature equal to the radius of Earth. (Neglecting the curvature of the surface of Earth, we use $r \approx \infty$.) Thus, Equation 32-6 applies. Because the light reaching Fluffy originates in the water, use $n_1 = 1.33$ and $n_2 = 1.00$.

SOLVE

1. Draw a picture of the situation. Label the object distance and the indices of refraction of the media. Goldie is the object (Figure 32-27):

2. Using Equation 32-6, relate the image position s' to the other relevant parameters:

$$\frac{n_1}{s} + \frac{n_2}{s'} = \frac{n_2 - n_1}{r}$$

3. The refracting surface is virtually flat. Using $r = \infty$, solve for s':

$$s' = -\frac{n_2}{n_1}s$$

4. Using the given values $n_1 = 1.33$, $n_2 = 1.00$ and $s = 1.00$ m, substitute to obtain s':

$$s' = -\frac{1}{1.33}(1.00\text{ m}) = -0.752\text{ m}$$

That the image is negative means that the image is on the side of the surface opposite the refracted light. That is, it is 0.752 m below the surface.

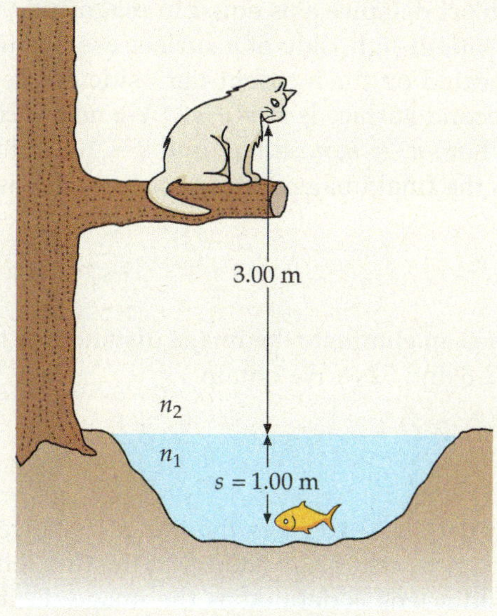

FIGURE 32-27

CHECK The image is between the location of the fish and the surface of the water. This result is expected. Recall that if you dip an oar into the water at an angle, the part of the oar that is underwater appears above where you know it has to be.

TAKING IT FURTHER (1) This image can be seen at the calculated position only when the object is viewed from directly overhead, or nearly so. From that observation point the rays are paraxial, a condition necessary for Equation 32-6 to be valid. If Fluffy is standing on the edge of the pond, the rays will not satisfy the paraxial approximation and Equation 32-6 will not correctly predict the location of the image. (2) The distance (n_2/n_1) multiplied by s is called the apparent depth of the submerged object. If $n_2 = 1$, the apparent depth equals s/n_1.

CONCEPT CHECK 32-4

Draw a ray diagram for the image of Goldie, as described in Example 32-6. That is, draw several rays diverging from an object point P on Goldie, and show that after refracting the rays appear to diverge from an image point P' somewhat above the object point.

THIN LENSES

An important application of Equation 32-6 for refraction at a single surface is finding the position of the image formed by a lens. This determination is done by considering the refraction at each surface of the lens separately to derive an equation relating the image distance to the object distance, the radius of curvature of each surface of the lens, and the index of refraction of the lens material.

We will consider a thin lens that has an index of refraction n and air on both sides. Let the radii of curvature of the surfaces of the lens be r_1 and r_2. If an object is at a distance s from the first surface (and therefore from the lens), the distance s_1' of the image due to refraction at the first surface can be found using Equation 32-6:

$$\frac{n_{\text{air}}}{s} + \frac{n}{s_1'} = \frac{n - n_{\text{air}}}{r_1} \qquad 32\text{-}8$$

The light refracted at the first surface is again refracted at the second surface. Figure 32-28 shows the case when the image distance s_1' for the first surface is negative, indicating a virtual image to the left of the surface. Rays in the glass refracted from the first surface diverge as if they came from the image point P_1'. The rays strike the second surface at the same angles as if there were an object at image point P_1'. The image for the first surface therefore becomes

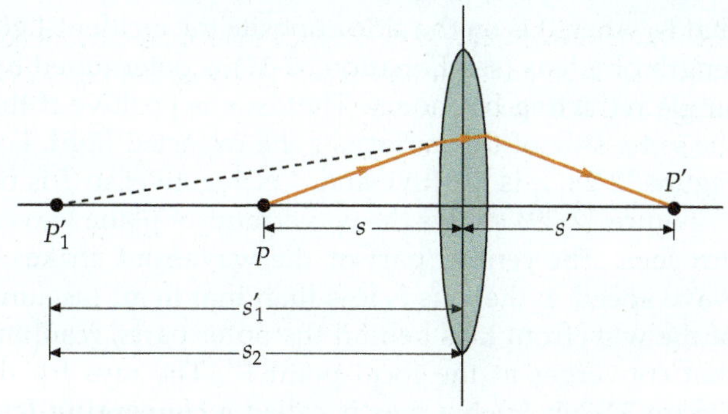

FIGURE 32-28 Refraction occurs at both surfaces of a lens. Here, the refraction at the first surface leads to a virtual image at P_1'. The rays strike the second surface as if they came from P_1'. Image distances are negative when the image is on the incident-light side of the surface, whereas object distances are positive for objects located on that side. Thus, $s_2(\approx -s_1')$ is the object distance for the second surface of the lens.

the object for the second surface. Because the lens is of negligible thickness, the object distance s_2 is equal in magnitude to s_1'. Object distances for objects on the incident-light side of a surface are positive, whereas image distances for images located on the incident-light side are negative. Thus, the object distance for the second surface is $s_2' = -s_1'$.* We now write Equation 32-6 for the second surface, where $n_1 = n$, $n_2 = n_{air}$, and $s = -s_1'$. The image distance for the second surface is the final image distance s' for the lens:

$$\frac{n}{-s_1'} + \frac{n_{air}}{s'} = \frac{n_{air} - n}{r_2} \qquad 32\text{-}9$$

We can eliminate the image distance for the first surface $-s_1'$ by adding Equations 32-8 and 32-9. We obtain

$$\frac{1}{s} + \frac{1}{s'} = \left(\frac{n}{n_{air}} - 1\right)\left(\frac{1}{r_1} - \frac{1}{r_2}\right) \qquad 32\text{-}10$$

Equation 32-10 gives the image distance s' in terms of the object distance s and the properties r_1, r_2, and n of the thin lens. As with mirrors, the focal length f of a thin lens is defined as the image distance when the object distance is infinite. Setting s equal to infinity and writing f for the image distance s', we obtain

$$\frac{1}{f} = \left(\frac{n}{n_{air}} - 1\right)\left(\frac{1}{r_1} - \frac{1}{r_2}\right) \qquad 32\text{-}11$$

LENS-MAKER'S EQUATION

Equation 32-12 is called the **lens-maker's** equation. Substituting $1/f$ for the right side of Equation 32-10, we obtain

$$\frac{1}{s} + \frac{1}{s'} = \frac{1}{f} \qquad 32\text{-}12$$

THIN-LENS EQUATION

This **thin-lens equation** is the same as the mirror equation (Equation 32-4). Recall, however, that the sign conventions for refraction are somewhat different from the sign conventions for reflection. For refraction, the image distance s' is positive when the image is on the refracted-light side of the refracting surface(s), that is, when it is on the side opposite the incident-light side. The sign of the focal length of a lens (see Equation 32-11) is determined by the sign convention for a single refracting boundary. That is, r is positive if the center of curvature is on the same side of the surface as the refracted light. For a lens like that shown in Figure 32-28, r_1 is positive and r_2 is negative, so f is positive.

Figure 32-29*a* shows the wavefronts of plane waves incident on a double convex lens. The central part of the wavefront strikes the lens first. Because the wave speed in the lens is less than that in air (assuming $n > 1$), the central part of the wavefront lags behind the outer parts, resulting in a spherical wavefront that converges at the focal point F'. The rays for this situation are shown in Figure 32-29*b*. Such a lens is called a **converging lens.** Because its focal length as calculated from Equation 32-2 is positive, it is also called a **positive lens.**

* If s_1' were positive, the rays would be converging as they strike the second surface. The object for the second surface would then be a virtual object located to the right of the second surface. This object would be a virtual object. Again, $s_2 = -s_1'$.

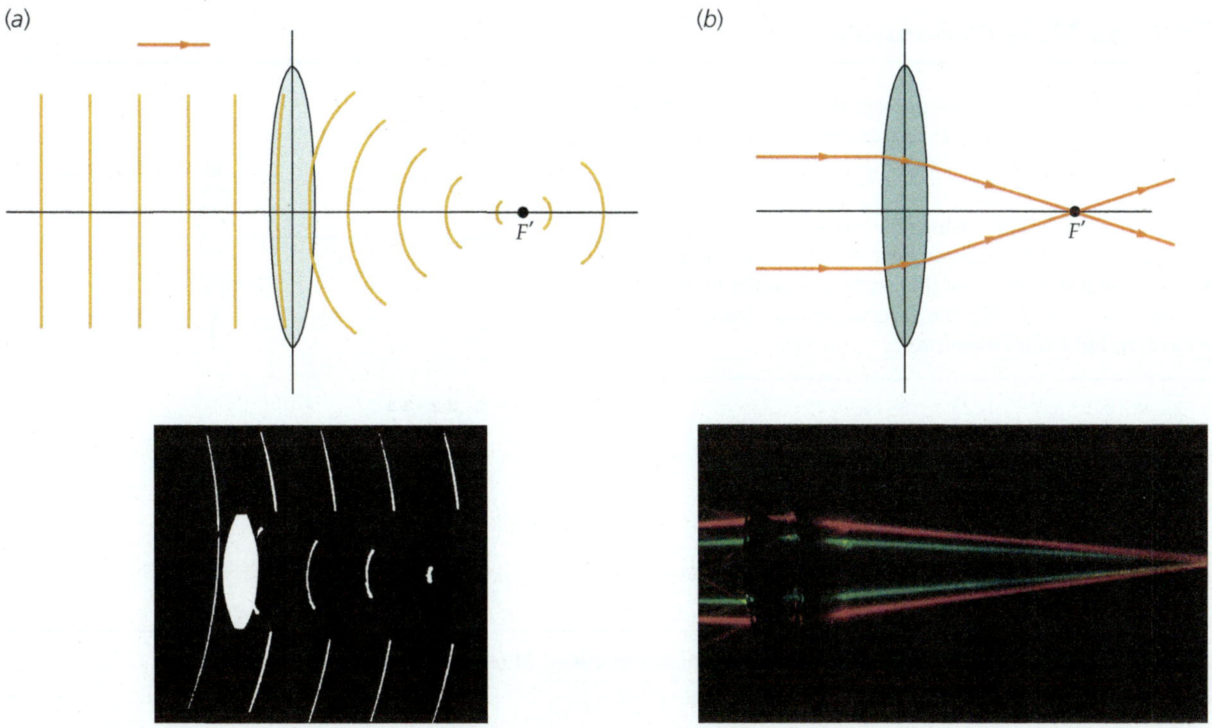

FIGURE 32-29 (*a*) *Top:* Wavefronts for plane waves striking a converging lens. The central part of the wavefront is retarded more by the lens than the outer part, resulting in a spherical wave that converges at the focal point F'. *Bottom:* Wavefronts passing through a lens, shown by a photographic technique called *light-in-flight recording* that uses a pulsed laser to make a hologram of the wavefronts of light. (*b*) *Top:* Rays for plane waves striking a converging lens. The rays are bent at each surface and converge at the focal point. *Bottom:* A photograph of rays focused by a converging lens. *((a) Nils Abramson, (b) Fundamental Photographs.)*

Any lens that is thicker in the middle than at the edges is a converging lens (providing that the index of refraction of the lens is greater than that of the surrounding medium). Figure 32-30 shows the wavefronts and rays for plane waves incident on a double concave lens. In this case, the outer part of the wavefronts lag behind the central parts, resulting in outgoing spherical waves that diverge from a focal point on the incident-light side of the lens. The focal length of this lens is negative. Any lens that is thinner in the middle than at the edges is a **diverging,** or **negative,** lens.

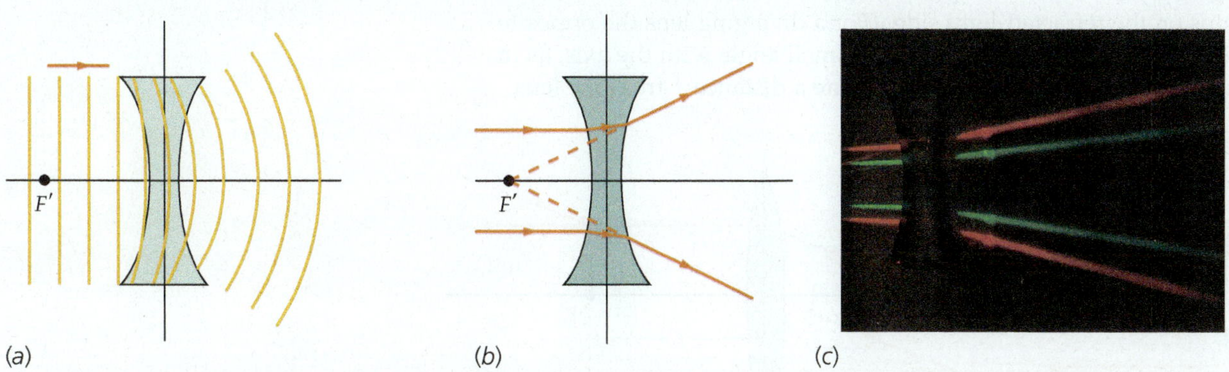

FIGURE 32-30 (*a*) Wavefronts for plane waves striking a diverging lens. Here, the outer part of the wavefront is retarded more than the central part, resulting in a spherical wave that diverges as it moves out, as if it came from the focal point F' to the left of the lens. (*b*) Rays for plane waves striking the same diverging lens. The rays are bent outward and diverge, as if they came from the focal point F'. (*c*) A photograph of rays passing through a diverging lens. *(Fundamental Photographs.)*

Example 32-7 The Lens-Maker's Formula

A double convex, thin glass lens that has an index of refraction $n = 1.50$ has radii of curvature whose magnitudes are 10 cm and 15 cm, as shown in Figure 32-31. Find its focal length in air.

PICTURE We can find the focal length using the lens-maker's equation (Equation 32-11). Here, light is incident on the surface that has the smaller radius of curvature. The center of curvature C_1 of this surface is on the refracted-light side of the lens; thus, $r_1 = +10$ cm. For the second surface, the center of curvature C_2 is on the incident-light side; therefore, $r_2 = -15$ cm.

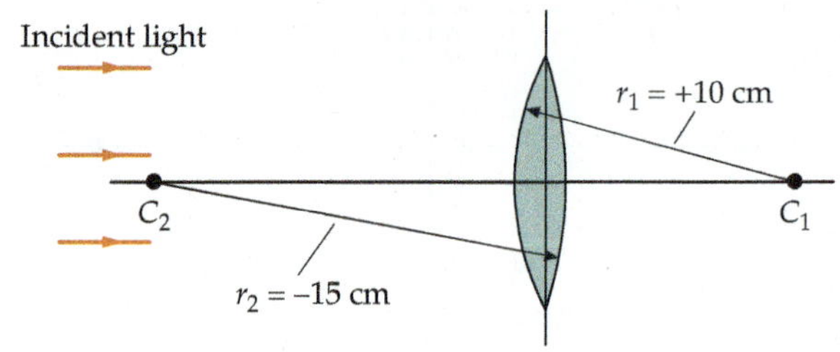

FIGURE 32-31

SOLVE

Numerical substitution in Equation 32-11 yields the focal length f:

$$\frac{1}{f} = \left(\frac{n}{n_{\text{air}}} - 1\right)\left(\frac{1}{r_1} - \frac{1}{r_2}\right)$$

$$= \left(\frac{1.50}{1.00} - 1\right)\left(\frac{1}{10\text{ cm}} - \frac{1}{-15\text{ cm}}\right) = 0.50\left(\frac{5.0}{30\text{ cm}}\right)$$

$$\therefore f = \boxed{12\text{ cm}}$$

CHECK The calculated focal length is positive, as expected. The lens is thicker in the middle than at the edges, so the focal length is expected to be positive.

PRACTICE PROBLEM 32-8 A double convex thin lens has an index of refraction $n = 1.6$ and its surfaces have radii of curvature of equal magnitude. If the focal length of the lens is 15 cm, what is the magnitude of the radii of curvature of the surfaces?

PRACTICE PROBLEM 32-9 Show that if you reverse the direction of the incoming light for the lens shown in Example 32-7, so that the incoming light is incident on the surface that has the greater radius of curvature, you get the same result for the focal length.

If parallel light strikes the lens of Example 32-7 from the left, it is focused at a point 12 cm to the right of the lens; whereas if parallel light strikes the lens from the right, it is focused at 12 cm to the left of the lens. Both of these points are focal points of the lens. Using the reversibility property of light rays, we can see that light diverging from a focal point and striking a lens will leave the lens as a parallel beam, as shown in Figure 32-32. Incident rays parallel to the axis emerge directed either toward or away from the **first focal point** F. Incident rays directed either toward or away from the **second focal point** F' emerge parallel with the axis. For a converging lens, the first focal point is on the incident-light side and the second focal point is on the refracted-light side. (For a diverging lens the opposite is true.) If parallel light is incident on the lens at a small angle with the axis, as in Figure 32-33, it is focused at a point in the **focal plane** a distance f from the lens.

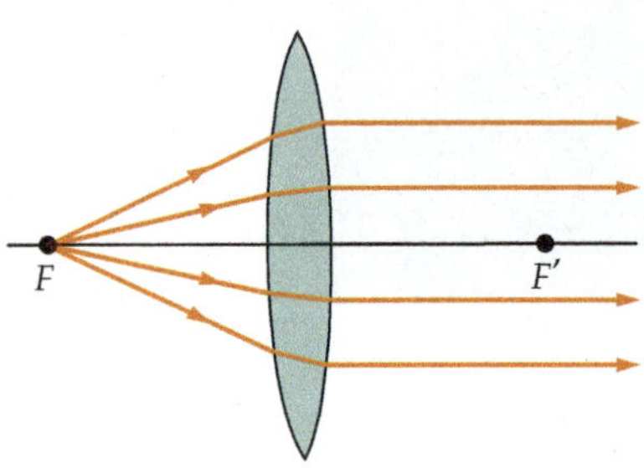

FIGURE 32-32 Light rays diverging from the focal point of a positive lens emerge parallel to the axis.

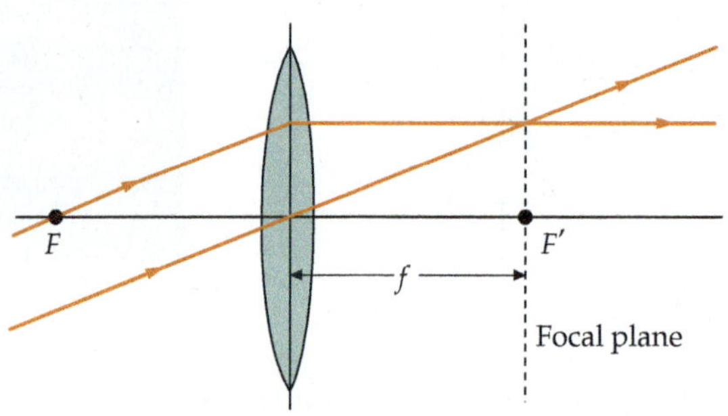

FIGURE 32-33 Parallel rays incident on the lens at an angle to its axis are focused at a point in the focal plane of the lens.

The reciprocal of the focal length is called the **power of a lens.** When the focal length is expressed in meters, the power is given in reciprocal meters, called **diopters** (D):

$$P = \frac{1}{f} \qquad 32\text{-}13$$

The power of a lens measures its ability to focus parallel light at a short distance from the lens. The shorter the focal length, the greater the power. For example, a lens that has a focal length of 25 cm = 0.25 m has a power of 4.0 D. A lens that has a focal length of 10 cm = 0.10 m has a power of 10 D. Because the focal length of a diverging lens is negative, its power is negative.

Example 32-8 Power of a Lens

The lens shown in Figure 32-34 has an index of refraction of 1.50 and radii of curvature whose magnitudes are 10.0 cm and 13.0 cm. Find (*a*) its focal length in air and (*b*) its power.

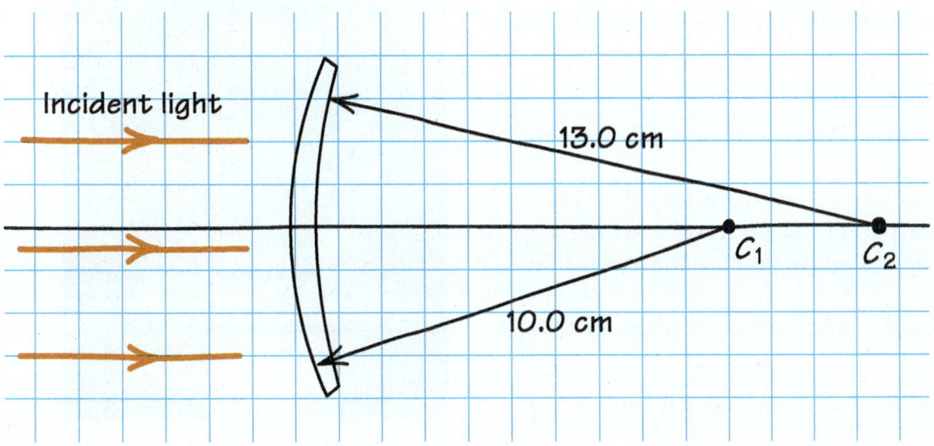

FIGURE 32-34

PICTURE For the orientation of the lens relative to the incident light shown in Figure 32-34, the radius of curvature of the first surface is $r_1 = +10.0$ cm and that of the second surface is $r_2 = +13.0$ cm.

SOLVE

(*a*) Calculate f from the lens-maker's equation using the given value of n and the values of r_1 and r_2 for the orientation shown:

$$\frac{1}{f} = \left(\frac{n}{n_{\text{air}}} - 1\right)\left(\frac{1}{r_1} - \frac{1}{r_2}\right)$$

$$= \left(\frac{1.50}{1.00} - 1\right)\left(\frac{1}{10.0\text{ cm}} - \frac{1}{13.0\text{ cm}}\right)$$

$$\therefore f = \boxed{87\text{ cm}}$$

(*b*) The power is the reciprocal of the focal length expressed in meters:

$$P = \frac{1}{f} = \frac{1}{0.867\text{ m}} = \boxed{1.2\text{ D}}$$

CHECK The values for the focal length and the power are positive. That is what is expected for a lens that is thicker in the middle than at the edges.

TAKING IT FURTHER We obtain the same result no matter which surface the light strikes first.

During laboratory experiments involving lenses, it is usually much easier to measure the focal length than to measure the radii of curvature of the surfaces.

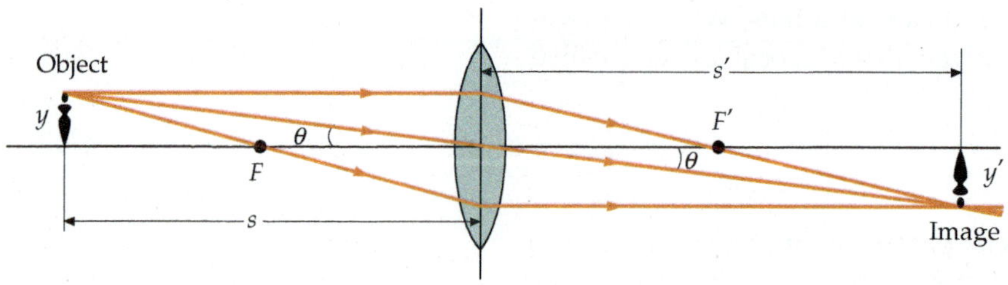

FIGURE 32-35 Ray diagram for a thin converging lens. We draw the rays as if all the bending of light takes place at the central plane. The ray through the center is undeflected because the lens surfaces there are parallel and close together.

RAY DIAGRAMS FOR LENSES

As with images formed by mirrors, it is convenient to locate the images of lenses by graphical methods. Figure 32-35 illustrates the graphical method for a thin converging lens. In the thin lens approximation, we consider the rays to bend at the plane through the center of the lens and perpendicular to the optic axis. The three principal rays are as follows:

1. The **parallel ray,** drawn parallel to the axis. The emerging ray is directed toward the second focal point of the lens.
2. The **central ray,** drawn through the center (the vertex) of the lens. This ray is undeflected. (The faces of the lens are parallel at the center, so the ray emerges in the same direction but displaced slightly. Because the lens is thin, the displacement is negligible.)
3. The **focal ray,** drawn through the first focal point.* This ray emerges parallel to the axis.

PRINCIPAL RAYS FOR A THIN LENS

These three rays converge to the image point, as shown in Figure 32-35. In this case, the image is real and inverted. From the figure, we have $\tan\theta = y/s = -y'/s'$. The lateral magnification is then

$$m = \frac{y'}{y} = -\frac{s'}{s} \qquad \text{32-14}$$

This expression is the same as the expression for mirrors. Again, a negative magnification indicates that the image is inverted. The ray diagram for a diverging lens is shown in Figure 32-36.

The weight and bulk of a large-diameter lens can be reduced by constructing the lens from annular segments at different angles so that light from a point is refracted by the segments into a parallel beam. Such an arrangement is called a Fresnel lens. Several Fresnel lenses are used in this lighthouse to produce intense parallel beams of light from a source at the focal point of the lenses. The illuminated surface of an overhead projector is a Fresnel lens. *(Bohdan Hrynewych/Stock Boston.)*

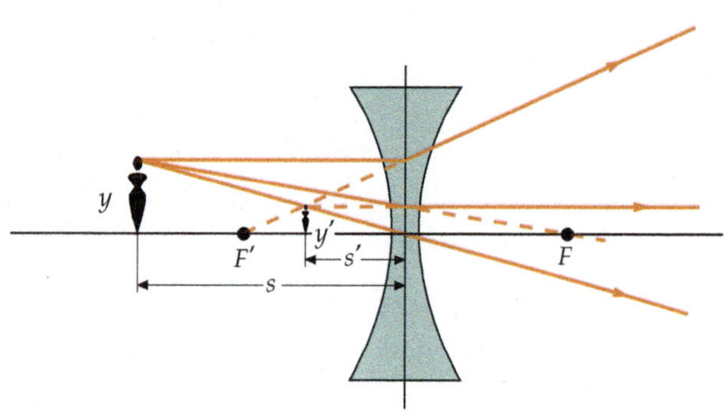

FIGURE 32-36 Ray diagram for a diverging lens. The parallel ray is bent away from the axis, as if it came from the second focal point F'. The ray toward the first focal point F emerges parallel to the axis. For a diverging lens the first focal point F is on the refracted-light side of the lens.

* The focal ray is drawn toward the first focal point F for a diverging lens.

Example 32-9 Image Formed by a Lens

An object that is 1.2 cm high is placed 4.0 cm from a double convex lens that has a focal length of 12 cm. Locate the image both graphically and algebraically, state whether the image is real or virtual, and find its height. Place an eye on the figure positioned and oriented so as to view the image.

PICTURE Locate the image by graphical methods. That means by drawing the three principal rays. The eye is positioned and oriented so the light from the image enters the eye.

SOLVE

1. Draw the parallel ray. This ray leaves the object parallel to the axis, then is bent by the lens to pass through the second focal point, F' (Figure 32-37):

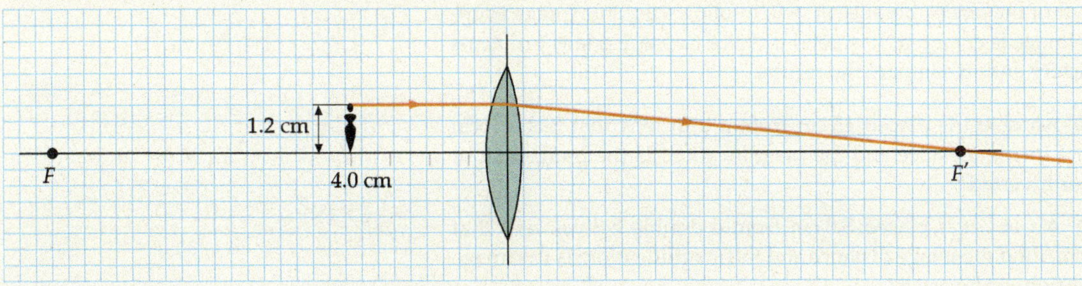

FIGURE 32-37

2. Draw the central ray, which passes undeflected through the center of the lens. Because the two rays are diverging on the refracted-light side, we extend them back to the incident-light side to find the image (Figure 32-38 here):

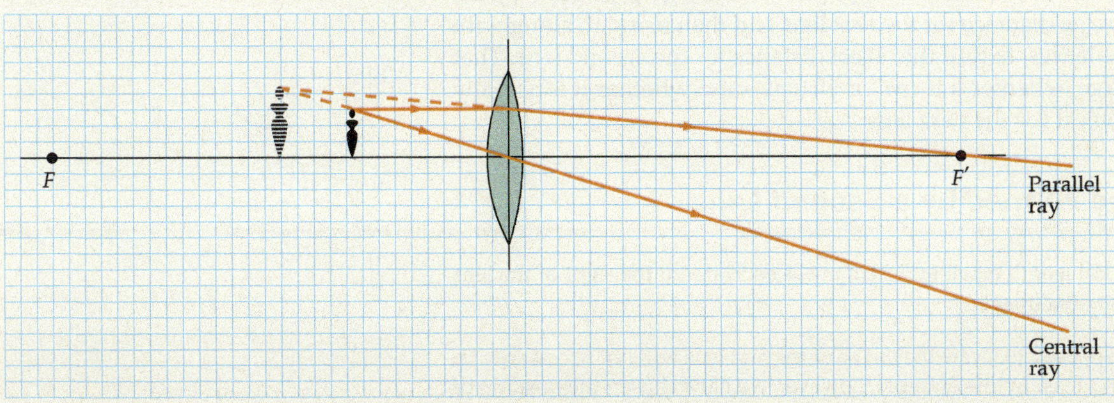

FIGURE 32-38

3. As a check, we also draw the focal ray. This ray leaves the object on a line passing through the first focal point, then emerges parallel to the axis. Note that the image is virtual, upright, and enlarged (Figure 32-39):

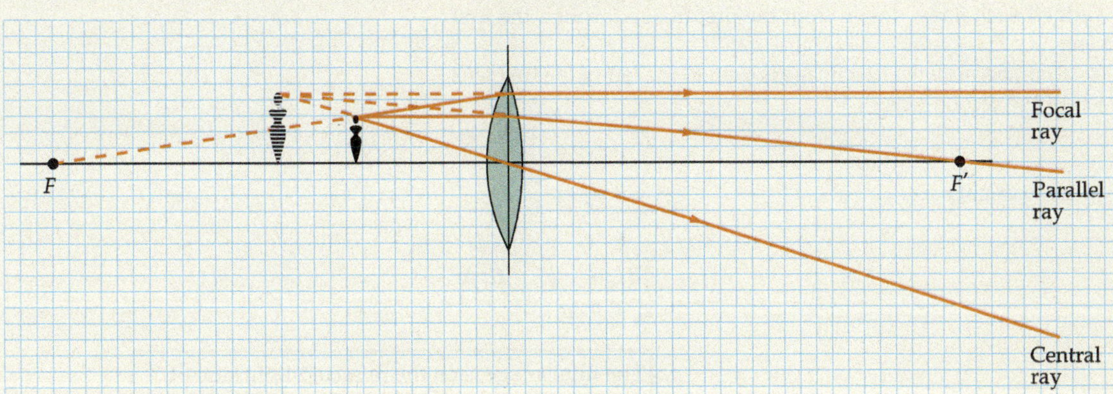

FIGURE 32-39

4. The eye must be positioned so the light from the image enters the eye.

5. We now verify the results of the ray diagram algebraically. First, find the image distance using Equation 32-12:

$$\frac{1}{4.0\text{ cm}} + \frac{1}{s'} = \frac{1}{12\text{ cm}}$$

$$\frac{1}{s'} = \frac{1}{12\text{ cm}} - \frac{1}{4.0\text{ cm}} = -\frac{1}{6.0\text{ cm}}$$

$$s' = -6.0\text{ cm}$$

6. The height of the image is found from the height of the object and the magnification:

$$h' = mh$$

7. The magnification m is given by Equation 32-14:

$$m = -\frac{s'}{s} = -\frac{-6.0\text{ cm}}{4.0\text{ cm}} = \boxed{+1.5}$$

8. Using this result, we find the height of the image, h':

$$h' = mh = (1.5)(1.2\text{ cm}) = \boxed{1.8\text{ cm}}$$

CHECK Note the agreement between the algebraic and ray diagram results. Algebraically, we find that the image is 6.0 cm from the lens on the incident-light side (because $s' < 0$); that is, the image is 2.0 cm to the left of the object. Because $m > 0$, it follows that the image is upright, and because $m > 1$, the image is enlarged. It is good practice to process lens problems both graphically and algebraically and to compare the results.

PRACTICE PROBLEM 32-10 An object is placed 15 cm from a thin lens that has a focal length equal to 10 cm. Find the image distance and the magnification. Is the image real or virtual? Is the image upright or inverted?

PRACTICE PROBLEM 32-11 An object is placed 5.0 cm from a double convex lens that has a focal length equal to 10 cm. Find the image distance and the magnification. Is the image real or virtual? Is the image upright or inverted?

COMBINATIONS OF LENSES

If we have two or more thin lenses, we can find the final image produced by the system by finding the image distance for the first lens and then using it, along with the distance between the lenses, to find the object distance for the second lens. That is, we consider each image, whether it is real or virtual—and whether it is actually formed or not—as the object for the next lens.

Example 32-10 Image Formed by a Second Lens

A second lens that has a focal length equal to +6 cm is placed 12 cm to the right of the lens in Example 32-9. Locate the final image.

PICTURE The principal rays used to locate the image of the first lens will not necessarily be principal rays for the second lens. In this example, however, we have chosen the position of the second lens (Figure 32-40*a*) so that the parallel ray for the first lens turns out to be the central ray for the second lens. Also, the focal ray for the first lens emerges parallel to the axis and is therefore the parallel ray for the second lens. If additional principal rays for the second lens are needed, we simply draw them from the image formed by the first lens. For example, in Figure 32-40*b* we added such a ray, drawn from the first image through the first focal point F_2 of the second lens.

Algebraically we use $s_2 = 18$ cm, because the first image is 6 cm to the left of the first lens and therefore 18 cm to the left of the second lens.

SOLVE

1. Use $s_2 = 18$ cm and $f = 6$ cm to calculate s_2':

$$\frac{1}{s_2} + \frac{1}{s_2'} = \frac{1}{f_2}$$

$$\frac{1}{18\text{ cm}} + \frac{1}{s_2'} = \frac{1}{6\text{ cm}}$$

$$s_2' = 9\text{ cm}$$

The final image is on the refracted-light side of the second lens and is 9 cm from the second lens.

2.

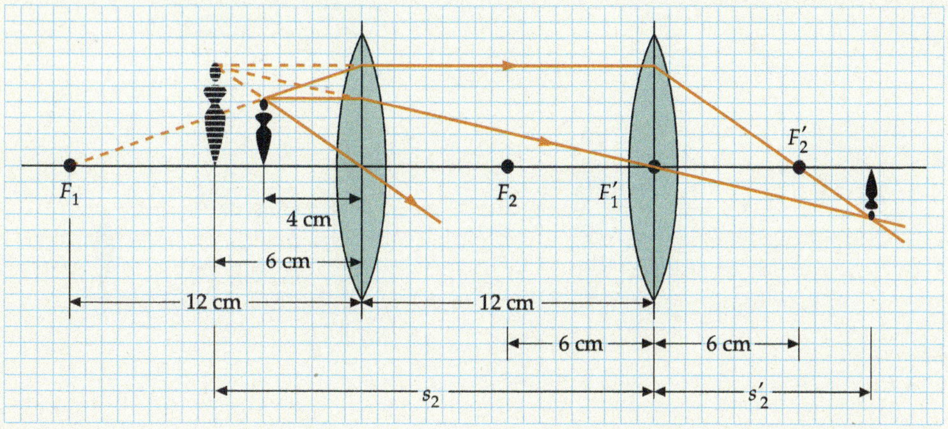

FIGURE 32-40*a*

CHECK The ray diagram in step 2 checks the calculation in step 1. In addition, to check the ray diagram in step 2, we draw the focal ray for the second lens.

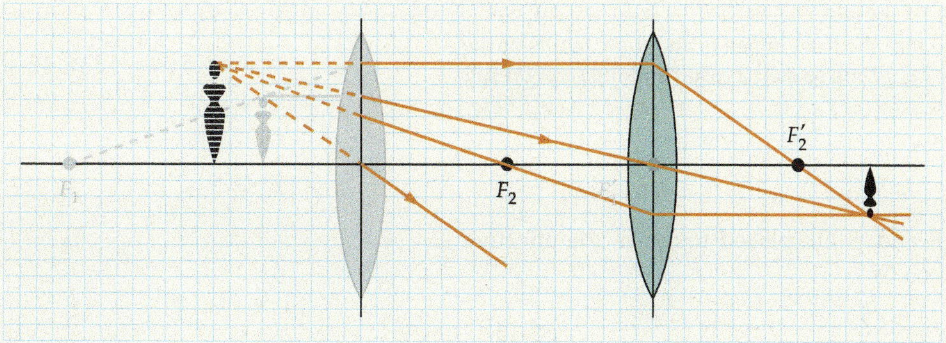

FIGURE 32-40*b*

Example 32-11 A Combination of Two Lenses — *Try It Yourself*

Two lenses, each having a focal length equal to 10 cm, are 15 cm apart. Find the final image location for an object 15 cm from one of the lenses and 30 cm from the other.

PICTURE Use a ray diagram to find the location of the image formed by lens 1. When these rays strike lens 2 they are further refracted, leading to the final image. More accurate results are obtained algebraically using the thin-lens equation for both lens 1 and lens 2.

SOLVE

Cover the column to the right and try these on your own before looking at the answers.

Steps

1. Draw the (*a*) parallel, (*b*) central, and (*c*) focal rays for lens 1 (Figure 32-41). If lens 2 did not alter these rays, they would form the image I_1.

Answers

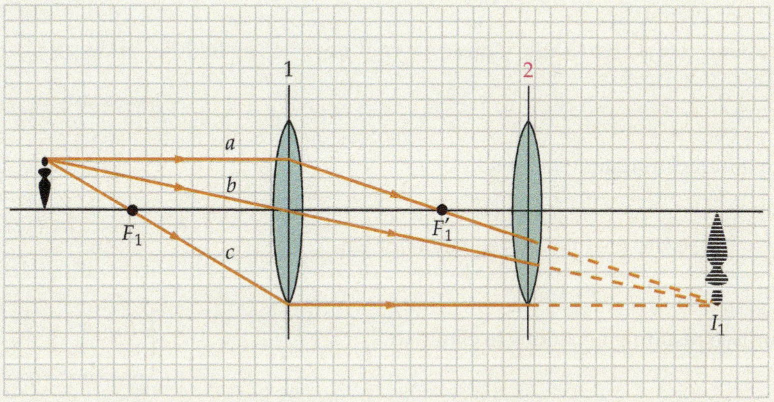

FIGURE 32-41

2. To locate the final image, add three principal rays (d, e, and f) for lens 2. The intersection of these rays gives the location of image I_2 (Figure 32-42).

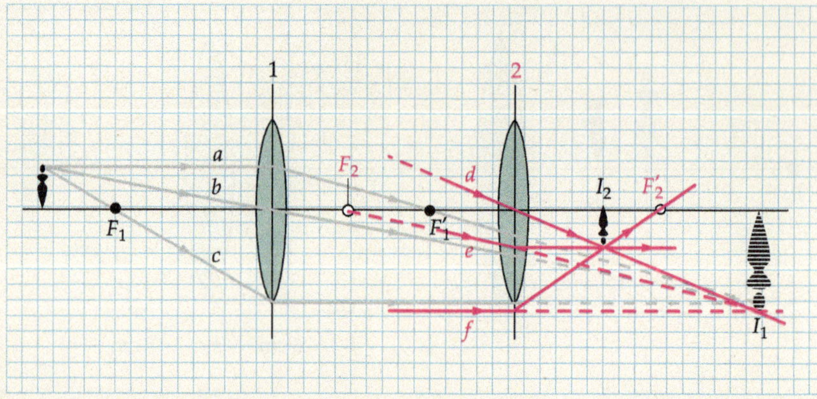

FIGURE 32-42

3. To proceed algebraically, use the thin-lens equation to find the image distance s_1' produced by lens 1. $s_1' = 30$ cm

4. For lens 2, the image, I_1 is 15 cm from the lens on the refracted-light side; hence, $s_2 = -15$ cm. Use this to find the final image distance s_2'. $s_2' = \boxed{6 \text{ cm}}$

CHECK From the step-2 ray diagram we see that the final image is about six tenths of the focal length of lens 2. The focal length of lens 2 is 10 cm, so step-2 result and the step-4 result are in agreement.

COMPOUND LENSES

When two thin lenses of focal lengths f_1 and f_2 are placed together, the effective focal length of the combination f_{eff} is given by

$$\frac{1}{f_{\text{eff}}} = \frac{1}{f_1} + \frac{1}{f_2} \qquad \text{32-15}$$

as is shown in the following example (Example 32-12). The power of two lenses in contact is given by

$$P_{\text{eff}} = P_1 + P_2 \qquad \text{32-16}$$

Example 32-12 Two Lenses in Contact *Try It Yourself*

For two lenses very close together, derive the relation $\frac{1}{f_{\text{eff}}} = \frac{1}{f_1} + \frac{1}{f_2}$.

PICTURE Apply the thin-lens equation to each lens using the fact that the distance between the lenses is negligible, so the object distance for the second lens is the negative of the image distance for the first lens.

SOLVE

Cover the column to the right and try these on your own before looking at the answers.

Steps	Answers
1. Write the thin-lens equation for lens 1.	$\frac{1}{s_1} + \frac{1}{s_1'} = \frac{1}{f_1}$
2. Using $s_2 = -s_1'$, write the thin-lens equation for lens 2.	$\frac{1}{-s_1'} + \frac{1}{s_2'} = \frac{1}{f_2}$
3. Add your two resulting equations to eliminate s_1'.	$\boxed{\frac{1}{s_1} + \frac{1}{s_2'} = \frac{1}{f_1} + \frac{1}{f_2} = \frac{1}{f_{\text{eff}}}}$

* 32-3 ABERRATIONS

When all the rays from a point object are not focused at a single image point, the resultant blurring of the image is called **aberration.** Figure 32-43 shows rays from a point source on the axis traversing a thin lens with spherical surfaces. Rays that strike the lens far from the axis are bent much more than are the rays near the axis, with the result that not all the rays are focused at a single point. Instead, the image appears as a circular disk at point C, where the diameter is minimum. This type of aberration in a lens is called **spherical aberration;** it is the same as the spherical aberration of mirrors discussed in Section 32-1. Similar but more complicated aberrations called *coma* and *astigmatism* occur when objects are off axis. The aberration in the shape of the image of an extended object that occurs, because the magnification depends on the distance of the object point from the axis, is called **distortion.** We will not discuss these aberrations further, except to point out that they do not arise from any defect in the lens or mirror but instead result from the application of the laws of refraction and reflection to spherical surfaces. These aberrations are not evident in our simple equations, because we used small-angle approximations in the derivation of the equations.

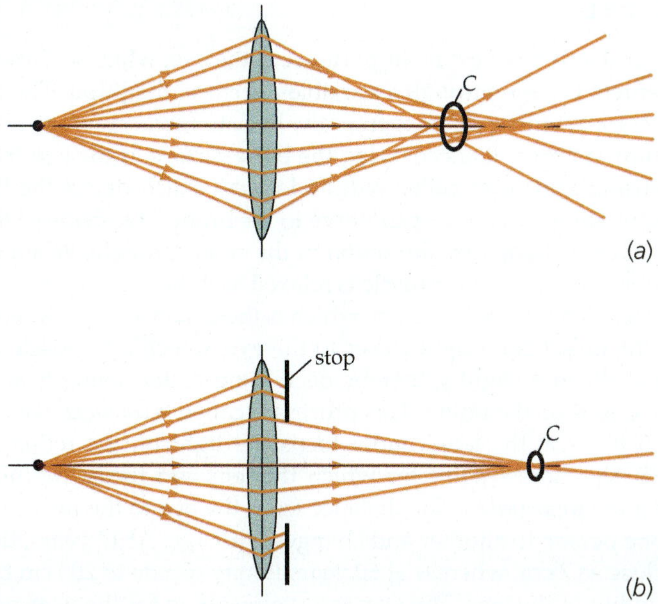

FIGURE 32-43 Spherical aberration in a lens. (*a*) Rays from a point object on the axis are not focused at a point. (*b*) Spherical aberration can be reduced by using a stop to block off the outer parts of the lens, but this also reduces the amount of light reaching the image.

Some aberrations can be eliminated or partially corrected by using nonspherical surfaces for mirrors or lenses, but nonspherical surfaces are usually much more difficult and costly to produce than spherical surfaces. One example of a nonspherical reflecting surface is the parabolic mirror illustrated in Figure 32-44. Rays that are parallel to the axis of a parabolic surface are reflected and focused at a common point, no matter how far the rays are from the axis. Parabolic reflecting surfaces are sometimes used in large astronomical telescopes, which need a large reflecting surface to gather as much light as possible to make the image as intense as possible (reflecting telescopes are described in the upcoming optional Section 32-4). Satellite dishes use parabolic surfaces to focus microwaves from communications satellites. A parabolic surface can also be used in a searchlight to produce a parallel beam of light from a small source placed at the focal point of the surface.

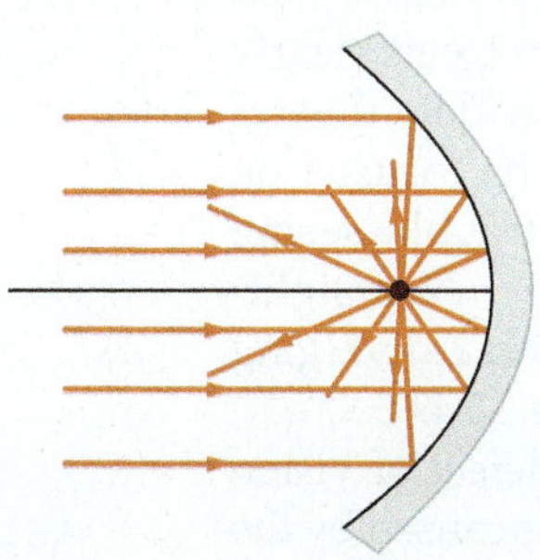

FIGURE 32-44 A parabolic mirror focuses all rays parallel to the axis to a single point with no spherical aberration.

An important aberration found with lenses but not found with mirrors is **chromatic aberration,** which is due to variations in the index of refraction with wavelength. From Equation 32-11, we can see that the focal length of a lens depends on its index of refraction and is therefore different for different wavelengths. Because n is slightly greater for blue light than for red light, the focal length for blue light will be shorter than the focal length for red light.

Chromatic aberration and other aberrations can be partially corrected by using combinations of lenses instead of a single lens. For example, a positive lens and a negative lens of greater focal length can be used together to produce a converging lens system that has much less chromatic aberration than a single lens of the same focal length. A high quality camera lens typically contains six elements to correct the various aberrations that are present.

* 32-4 OPTICAL INSTRUMENTS

*THE EYE

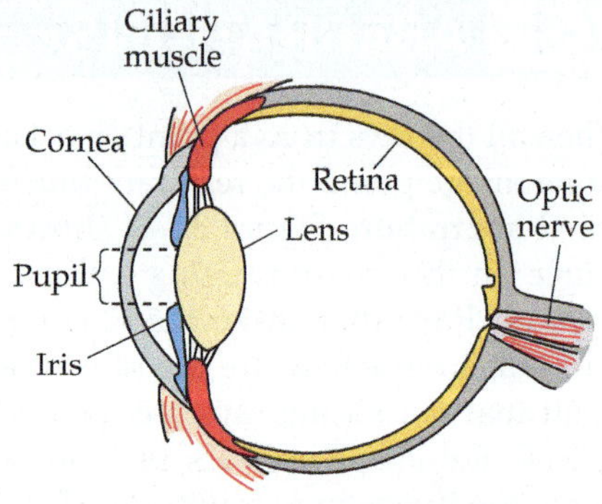

FIGURE 32-45 The human eye. The amount of light entering the eye is controlled by the iris, which regulates the size of the pupil. The lens thickness is controlled by the ciliary muscle. The cornea and lens together converge the light to focus the image on the retina, which contains approximately 125 million receptors, called rods and cones, and approximately 1 million optic-nerve fibers.

The optical system of prime importance is the eye, which is shown in Figure 32-45. Light enters the eye through a variable aperture, the pupil. The light is converged by the cornea, with assistance from the lens, to form an image on the retina, which has a film of nerve fibers covering the back surface of the eye. The retina contains tiny sensing structures called *rods* and *cones,* which detect the light and transmit the information along the optic nerve to the brain. The shape of the crystalline lens can be altered slightly by the action of the ciliary muscle. When the eye is focused on an object far away, the muscle is relaxed and the cornea-lens system has its maximum focal length, about 2.5 cm, which is the distance from the cornea to the retina. When the object is brought closer to the eye, the ciliary muscle increases the curvature of the lens slightly, thereby decreasing its focal length, so that the image is again focused on the retina. This process is called *accommodation*. If the object is too close to the eye, the lens cannot focus the light on the retina and the image is blurred. The closest point for which the lens can focus the image on the retina is called the **near point.** The distance from the eye to the near point varies greatly from one person to another and changes with age. At 10 years, the near point may be as close as 7 cm, whereas at 60 years it may recede to 200 cm because of the loss of flexibility of the lens. The standard value taken for the near point is 25 cm.

If the eye underconverges, resulting in the images being focused behind the retina, the person is said to be farsighted. A farsighted person can see distant objects clearly where little convergence is required, but has trouble seeing close objects clearly. Farsightedness is corrected with a converging (positive) lens (Figure 32-46).

On the other hand, the eye of a nearsighted person overconverges and focuses light from distant objects in front of the retina. A nearsighted person can see nearby objects clearly (objects for which the widely diverging incident rays can be focused on the retina), but has trouble seeing distant objects clearly. Nearsightedness is corrected with a diverging (negative) lens (Figure 32-47).

Another common defect of vision is astigmatism, which is caused by the cornea being not quite spherical but having a different curvature in one plane than in another. This results in a blurring of the image of a point object into a short line. Astigmatism is corrected by glasses using lenses of cylindrical rather than spherical shape.

The *apparent size* of an object is determined by the actual size of the image on the retina. The larger the image on the retina, the larger the apparent size and the greater the number of rods and cones that are illuminated by the light from the object. From Figure 32-48, we see that the size of the image on the retina is greater

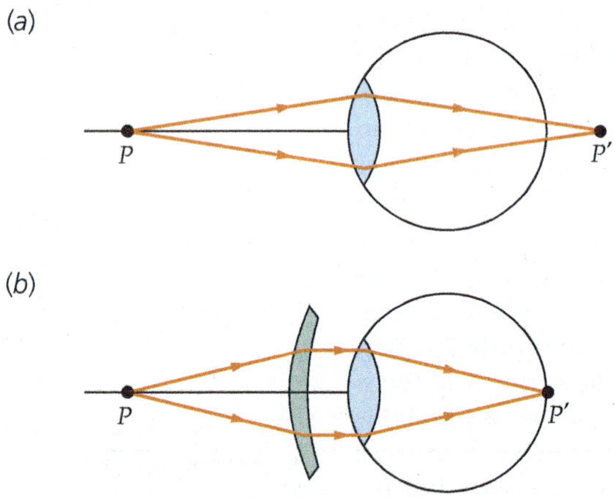

FIGURE 32-46 (*a*) A farsighted eye focuses rays from a nearby object to a point behind the retina. (*b*) A converging lens corrects this defect by bringing the image onto the retina. These diagrams, and those following, are drawn as if all the focusing of the eye is done at the cornea; in fact, the lens and cornea system act more like a spherical refracting surface than a thin lens.

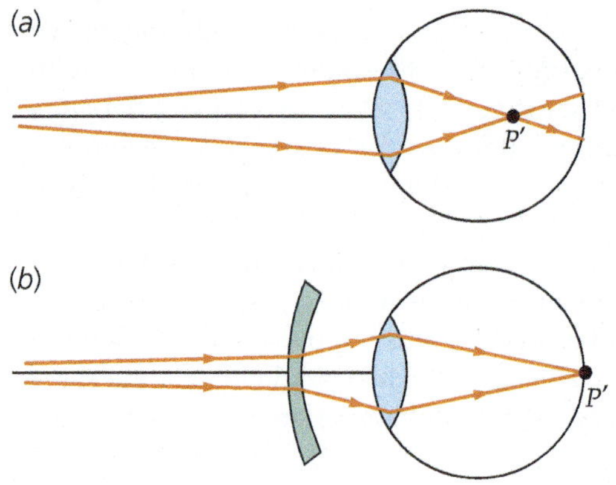

FIGURE 32-47 (*a*) A nearsighted eye focuses rays from a distant object to a point in front of the retina. (*b*) A diverging lens corrects this defect.

when the object is close than it is when the object is far away. The apparent size of an object is thus greater when the object is closer to the eye. The image size is proportional to the angle θ subtended by the object at the eye. For Figure 32-48,

$$\phi = \frac{y'}{2.5\text{ cm}} \quad \text{and} \quad \theta \approx \frac{y}{s} \qquad 32\text{-}17$$

for small angles. Applying the law of refraction gives $n_{air} \sin\theta = n \sin\phi$, where $n_{air} = 1.00$ and n is the refractive index inside the eye. For small angles this becomes

$$\theta \approx n\phi \qquad 32\text{-}18$$

Combining Equations 32-17 and 32-18 gives

$$\frac{y}{s} \approx n\frac{y'}{2.5\text{ cm}} \quad \text{or} \quad y' \approx \frac{2.5\text{ cm}}{n}\frac{y}{s} \qquad 32\text{-}19$$

The size of the image on the retina is proportional to the size of the object and inversely proportional to the distance from the cornea to the object. Because the near point is the closest point to the eye for which a sharp image can be formed on the retina, the distance from the cornea to the near point is called the *distance of most distinct vision.*

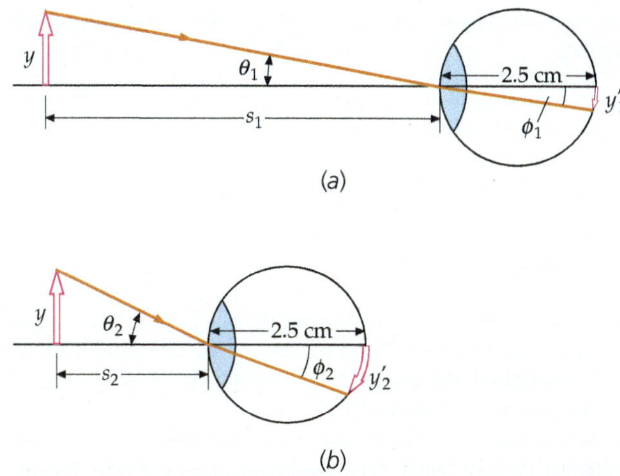

FIGURE 32-48 (*a*) A distant object of height *y* looks small because the image on the retina is small. (*b*) When the same object is closer, it looks larger because the image on the retina is larger.

Example 32-13 Reading Glasses

The near-point distance of a person's eye is 75 cm. Using a reading glasses lens placed a negligible distance from the eye, the near-point distance of the lens-eye system is 25 cm. That is, the image on the retina is blurry if an object is placed less than 25 cm in front of the lens. (*a*) What power is the reading glasses lens and (*b*) what is the lateral magnification of the image formed by the lens? (*c*) Which produces the bigger image on the retina, (1) the object at the near point of, and viewed by, the unaided eye or (2) the object at the near point of the lens-eye system and viewed through the lens that is immediately in front of the eye?

PICTURE A near-point distance of the lens-eye system of 25 cm means the lens forms a virtual image 75 cm in front of the lens if an object is placed 25 cm in front of the lens. Figure 32-49*a* shows a diagram of an object 25 cm from a converging lens that produces a virtual, upright image at $s' = -75$ cm. Figure 32-49*b* shows the image on the retina formed by the focusing power of the eye.

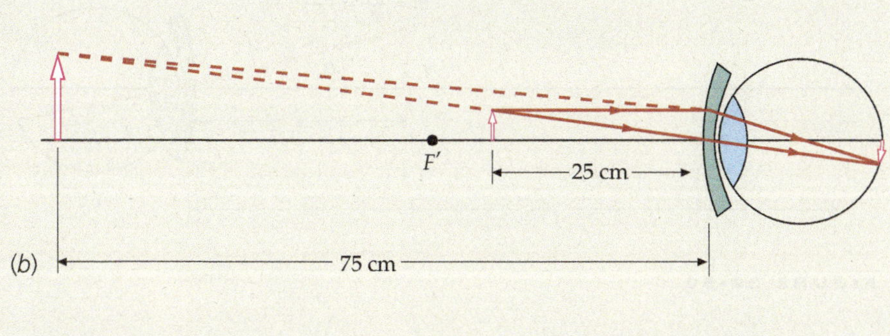

FIGURE 32-49

SOLVE

(*a*) Use the thin-lens equation, $s = 25$ cm, and $s' = -75$ cm to calculate the power, $1/f$:

$$\frac{1}{f} = \frac{1}{s} + \frac{1}{s'} = \frac{1}{25\text{ cm}} + \frac{1}{-75\text{ cm}}$$

$$= \frac{2}{75\text{ cm}} = \frac{2}{0.75\text{ m}} = \boxed{2.7\text{ D}}$$

(*b*) Using $m = -s'/s$, find m:

$$m = -\frac{s'}{s} = -\frac{-75\text{ cm}}{25\text{ cm}} = \boxed{3.0}$$

(*c*) In both cases, the rays entering the eye appear to diverge from points on an image 75 cm in front of the eye. However, with the lens in place, the image 75 cm in front of the eye is larger than the object by a factor of 3:

$\boxed{\text{Option 2}}$

TAKING IT FURTHER (1) If your near point is 75 cm, you are farsighted. To read a book you must hold it at least 75 cm from your eye to be able to focus on the print. The image of the print on your retina is then very small. The reading glasses lens produces an image also 75 cm from your eye, and this image is three times larger than the actual print. Thus, looking through the lens, the image of the print on the retina is larger by a factor of 3. (2) In this example, the distance from the lens to the eye was negligible. The results are slightly different if this distance is not negligible, which is factored into the calculations.

PRACTICE PROBLEM 32-12 Calculate the power of the eye for which the near point is 75 cm and the cornea-retina distance is 2.5 cm, and calculate the combined power of the lens and eye when they are in contact. Compare this result with the power of a lens for which $s' = 2.5$ cm, when $s = 25$ cm.

*THE SIMPLE MAGNIFIER

We saw in Example 32-13 that the *apparent* size of an object can be increased by using a converging lens placed next to the eye. A converging lens is called a **simple magnifier** if it is placed next to the eye and if the object is placed closer to the lens than its focal length, as was the case for the lens in Example 32-13. In that example, the lens formed a virtual image at the near point of the eye, the same location that the object must be placed for best viewing by the unaided eye. So, with the lens in place, the magnitude of the image distance $|s'|$ was greater than the object distance s, so the image seen by the eye is magnified by $m = |s'|/s$. If the actual height of the object is y, then the height y' of the image formed by the lens is my. To the eye, this image subtends an angle θ (Figure 32-50) given approximately by

$$\theta = \frac{my}{|s'|} = m\frac{y}{|s'|} = \frac{|s'|}{s}\frac{y}{|s'|} = \frac{y}{s}$$

which is the *very same angle* the object would subtend if the lens were removed while the object and the eye are left in place. That is, the apparent size of the image seen by the eye through the lens is the same as the apparent size of the object that would be seen by the eye were the lens removed (assuming the eye could focus at that distance). Thus, the apparent size of the object seen through the lens is inversely proportional to the distance from the object to the eye with the lens in place. The smaller s is, the larger the subtended angle θ and the larger the apparent size of the object.

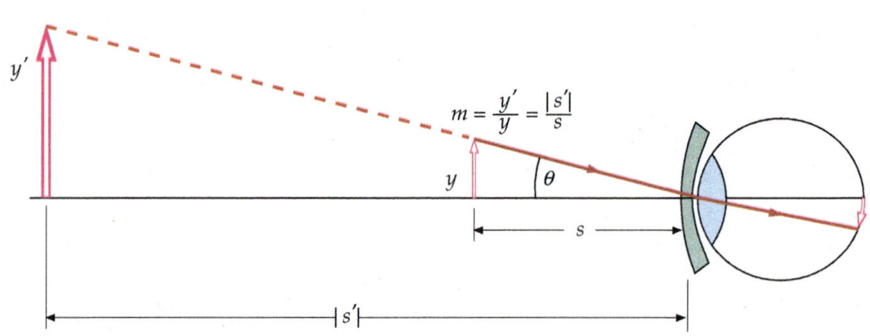

FIGURE 32-50

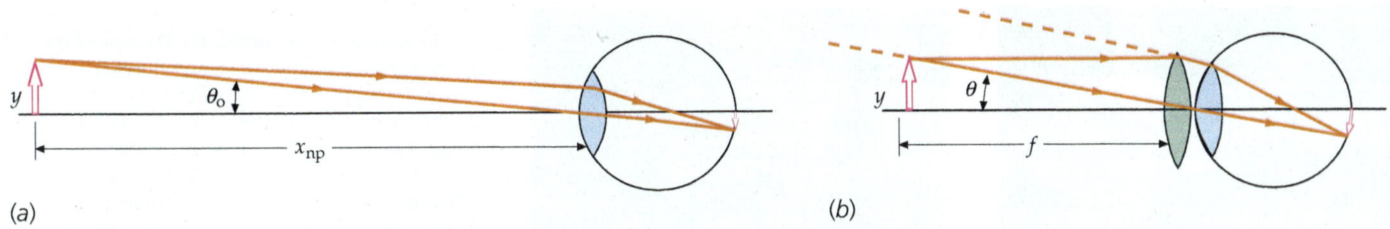

FIGURE 32-51 (*a*) An object at the near point subtends an angle θ_o at the naked eye. (*b*) When the object is at the focal point of the converging lens, the rays emerge from the lens parallel and enter the eye as if they came from an object a very large distance away. The image can thus be viewed at infinity by the relaxed eye. When f is less than the near-point distance, the converging lens allows the object to be brought closer to the eye. This increases the angle subtended by the object to θ, thereby increasing the size of the image on the retina.

In Figure 32-51*a*, a small object of height y is at the near point of the eye at a distance x_{np}. The angle subtended, θ_o, is given approximately by

$$\theta_o = \frac{y}{x_{np}}$$

In Figure 32-51*b*, a converging lens of focal length f that is smaller than x_{np} is placed a negligible distance in front of the eye, and the object is placed in the focal plane of the lens. The rays emerge from the lens parallel, indicating that the image is located an infinite distance in front of the lens. The parallel rays are focused by the relaxed eye on the retina. The angle subtended by this image is equal to the angle subtended by the object (assuming that the lens is a negligible distance from the eye). The angle subtended by the object is approximately

$$\theta = \frac{y}{f}$$

The ratio θ/θ_o is called the *angular magnification* or *magnifying power* M of the lens:

$$M = \frac{\theta}{\theta_o} = \frac{x_{np}}{f} \qquad 32\text{-}20$$

Simple magnifiers are used as eyepieces (called oculars) in microscopes and telescopes to view a real image formed by another lens or lens system. To correct aberrations, combinations of lenses that result in a short positive focal length may be used in place of a single lens, but the principle of the simple magnifier is the same.

Example 32-14 Angular Magnification of a Simple Magnifier

Try It Yourself

A person who has a near point of 25 cm uses a 40-D lens as a simple magnifier. What angular magnification is obtained?

PICTURE The angular magnification is found from the focal length f (Equation 32-20), which is the reciprocal of the power.

SOLVE

Cover the column to the right and try these on your own before looking at the answers.

Steps	Answers
1. Calculate the focal length of the lens using $P = 1/f$ (Equation 32-13).	$f = 2.5$ cm
2. Use your result from step 1 and incorporate the result into Equation 32-20 to calculate the angular magnification.	$M = \boxed{10}$

TAKING IT FURTHER Looking through the lens, the object appears 10 times larger because it can be placed at 2.5 cm rather than at 25 cm from the eye, thus increasing the size of the image on the retina tenfold.

PRACTICE PROBLEM 32-13 What is the angular magnification in this example if the near point of the person is 30 cm rather than 25 cm?

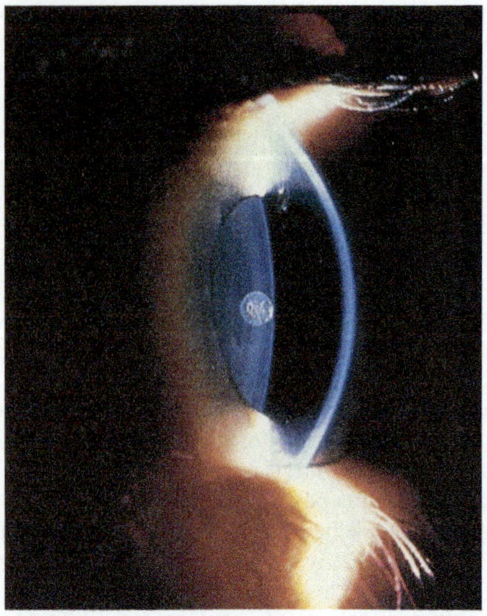

(a)

(b)

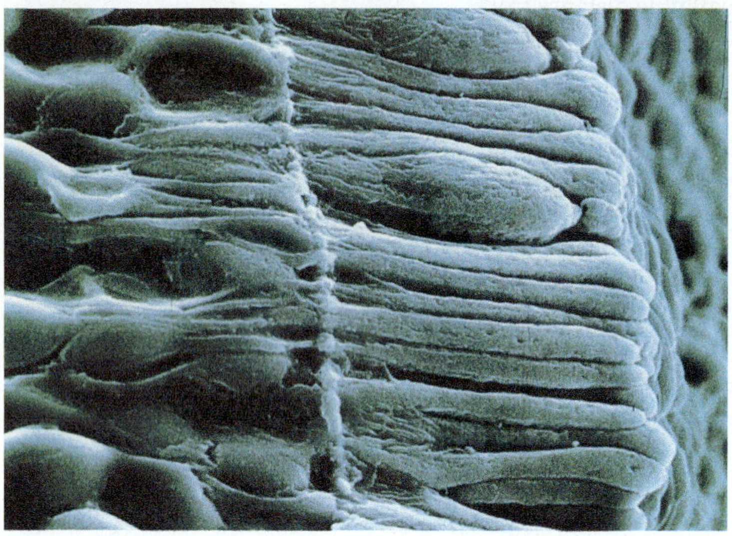

(c)

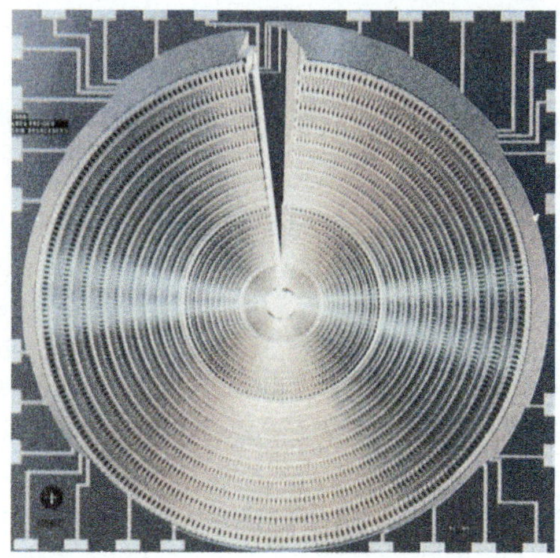

(d)

(*a*) The human eye in profile. (*b*) The lens of the eye is kept in place by the ciliary muscle (shown here in the upper left), which rings the lens. When the ciliary muscle contracts, the lens tends to bulge. The greater lens curvature enables the eye to focus on nearby objects. (*c*) Some of the 120 million rods and 7 million cones in the eye, magnified approximately 5000 times. The rods (the more slender of the two) are more sensitive in dim light, whereas the cones are more sensitive to color. The rods and cones form the bottom layer of the retina and are covered by nerve cells, blood vessels, and supporting cells. Most of the light entering the eye is reflected or absorbed before reaching the rods and cones. The light that does reach them triggers electrical impulses along nerve fibers that ultimately reach the brain. (*d*) A neural net used in the vision system of certain robots. Loosely modeled on the human eye, it contains 1920 sensors. *((a), (b) and (c) Lennart Nilsson, (d) Courtesy IMEC and University of Pennsylvania Department of Electrical Engineering.)*

*THE COMPOUND MICROSCOPE

The compound microscope (Figure 32-52) is used to look at very small objects at short distances. In its simplest form, it consists of two converging lenses. The lens nearest the object, called the **objective,** forms a real image of the object. This image is enlarged and inverted. The lens nearest the eye, called the **eyepiece** or **ocular,** is used as a simple magnifier to view the image formed by the objective. The eyepiece is placed so that the image formed by the objective falls at the first focal point of the eyepiece. The light from each point on the object thus emerges from the eyepiece as a parallel beam, as if it were coming from a point a great distance in front of the eye. (This is commonly called *viewing the image at infinity.*)

The distance between the second focal point of the objective and the first focal point of the eyepiece is called the **tube length** L. The tube length is fixed at 16 cm. The object is placed just outside the first focal point of the objective so that an enlarged image is formed at the first focal point of the eyepiece a distance $L + f_o$ from the objective, where f_o is the focal length of the objective. From Figure 32-52, $\tan\beta = y/f_o = -y'/L$. The lateral magnification of the objective is therefore

$$m_o = \frac{y'}{y} = -\frac{L}{f_o} \tag{32-21}$$

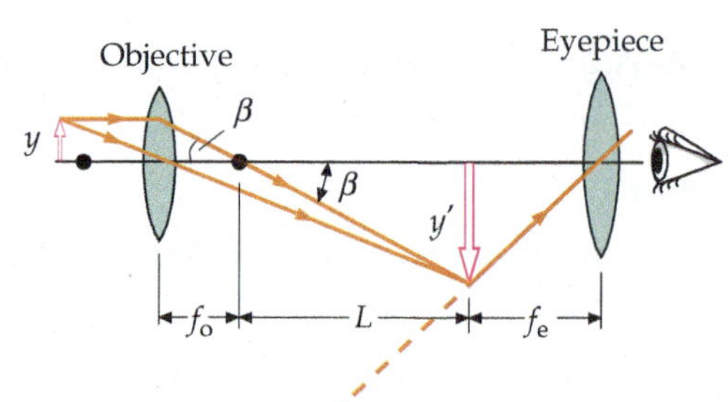

FIGURE 32-52 Schematic diagram of a compound microscope consisting of two positive lenses, the objective of focal length f_o and the eyepiece of focal length f_e. The real image of the object formed by the objective is viewed by the eyepiece, which acts as a simple magnifier. The final image is at infinity.

The angular magnification of the eyepiece (from Equation 32-20) is

$$M_e = \frac{x_{np}}{f_e}$$

where x_{np} is the near-point distance of the viewer and f_e is the focal length of the eyepiece. The magnifying power of the compound microscope is the product of the lateral magnification of the objective and the angular magnification of the eyepiece:

$$M = m_o M_e = -\frac{L}{f_o}\frac{x_{np}}{f_e} \qquad 32\text{-}22$$

MAGNIFYING POWER OF A MICROSCOPE

Example 32-15 The Compound Microscope

A microscope has an objective lens that has a focal length equal to 1.2 cm and an eyepiece that has a focal length equal to 2.0 cm. These lenses are separated by 20.0 cm. (*a*) Find the magnifying power if the near point of the viewer is 25.0 cm. (*b*) Where should the object be placed if the final image is to be viewed at infinity?

PICTURE To calculate the magnifying power, we use Equation 32-22. To calculate the object distance for the objective, we use the lens equation.

SOLVE

(*a*) 1. The magnifying power is given by Equation 32-22:

$$M = -\frac{L}{f_o}\frac{x_{np}}{f_e}$$

2. The tube length L is the distance between the lenses minus the focal distances:

$$L = 20.0\text{ cm} - 2.0\text{ cm} - 1.2\text{ cm} = 16.8\text{ cm}$$

3. Substitute this value for L and the given values of x_{np}, f_o, and f_e to calculate M:

$$M = -\frac{L}{f_o}\frac{x_{np}}{f_e} = -\frac{16.8\text{ cm}}{1.2\text{ cm}}\frac{25.0\text{ cm}}{2.0\text{ cm}} = \boxed{-180}$$

(*b*) 1. Calculate the object distance s in terms of the image distance for the objective s' and the focal length f_o:

$$\frac{1}{s} + \frac{1}{s'} = \frac{1}{f_o}$$

2. From Figure 32-52, the image distance for the image of the objective is $f_o + L$:

$$s' = f_o + L = 1.2\text{ cm} + 16.8\text{ cm} = 18.0\text{ cm}$$

3. Substitute to calculate s:

$$\frac{1}{s} + \frac{1}{18.0\text{ cm}} = \frac{1}{1.2\text{ cm}}$$

$$s = \boxed{1.3\text{ cm}}$$

CHECK The magnifying power is very large, which is the purpose of a compound microscope.

TAKING IT FURTHER The object should thus be placed at 1.3 cm from the objective or 0.1 cm outside its first focal point.

*THE TELESCOPE

A telescope is used to view objects that are far away and are often large. The telescope works by creating a real image of the object that is much closer than the object. The astronomical telescope, illustrated schematically in Figure 32-53, consists of two converging lenses—an objective lens that forms a real, inverted image and an eyepiece that is used as a simple magnifier to view that image. Because the object is very far away, the image of the objective lies in the focal plane of the objective, and the image distance equals the focal length f_o. The image formed by the objective is much smaller than the object (because the object distance is

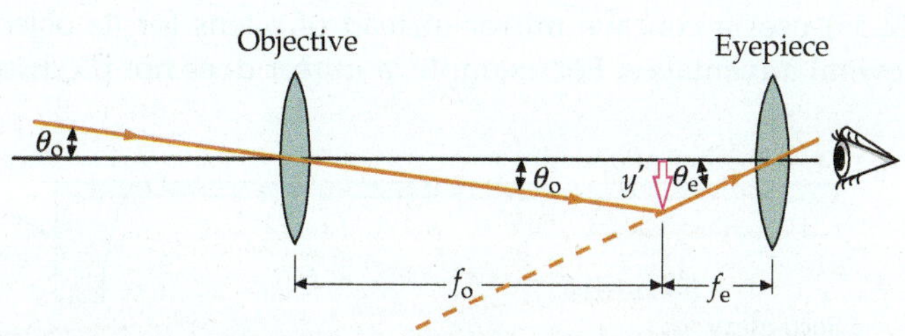

FIGURE 32-53 Schematic diagram of an astronomical telescope. The objective lens forms a real, inverted image of a distant object near its second focal point, which coincides with the first focal point of the eyepiece. The eyepiece serves as a simple magnifier to view the image.

much larger than the focal length of the objective). For example, if we are looking at the moon, the image of the moon formed by the objective is much smaller than the moon itself. The function of the objective is not to magnify the object, but to produce an image that is close enough to us so it can be magnified by the eyepiece acting as a simple magnifier. The eyepiece is placed a distance f_e from the image, where f_e is the focal length of the eyepiece, so the final image can be viewed at infinity. Because this image is at the second focal plane of the objective and at the first focal plane of the eyepiece, the objective lens and the eyepiece must be separated by the sum of the focal lengths of the objective lens and eyepiece, $f_o + f_e$.

The magnifying power of the telescope is the angular magnification θ_e/θ_o, where θ_e is the angle subtended by the virtual image produced by the eyepiece as viewed through the eyepiece, and θ_o is the angle subtended by the object when it is viewed directly by the unaided eye. The angle θ_o is the same as that subtended by the object at the objective shown in Figure 32-53. (The distance from a distant object, such as the moon, to the objective is essentially the same as the distance to the eye.) From this figure, we can see that

$$\tan\theta_o = \frac{y}{s} = -\frac{y'}{f_o} \approx \theta_o$$

where we have used the small-angle approximation $\tan\theta \approx \theta$. The angle θ_e in the figure is that subtended by the image at infinity formed by the eyepiece:

$$\tan\theta_e = \frac{y'}{f_e} \approx \theta_e$$

Because y' is negative, θ_e is negative, indicating that the image is inverted. The magnifying power of the telescope is then

$$M = \frac{\theta_e}{\theta_o} = -\frac{f_o}{f_e} \qquad \text{32-23}$$

MAGNIFYING POWER OF A TELESCOPE

From Equation 32-23, we can see that a large magnifying power is obtained with an objective of large focal length and an eyepiece of short focal length.

PRACTICE PROBLEM 32-14

The world's largest refracting telescope is at the Yerkes Observatory of the University of Chicago at Williams Bay, Wisconsin. The telescope's objective has a diameter of 1.02 m and a focal length of 19.5 m. The focal length of the eyepiece is 10.0 cm. What is its magnifying power?

The main consideration with an astronomical telescope is not its magnifying power but its light-gathering power, which depends on the size of the objective. The larger the objective, the brighter the image. Very large lenses without aberrations are difficult to produce. In addition, there are mechanical problems in supporting very large, heavy lenses by their edges. A reflecting telescope (Figure 32-54 and Figure 32-55) uses a concave mirror instead of a lens for its objective. The mirror offers several advantages. For example, a mirror does not produce chromatic aberration.

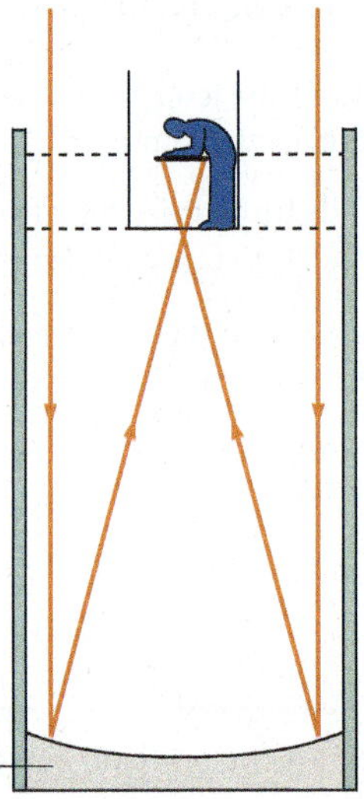

FIGURE 32-54 A reflecting telescope uses a concave mirror instead of a lens for its objective. Because the viewer compartment blocks off some of the incoming light, the arrangement shown here is used only in telescopes with very large objective mirrors.

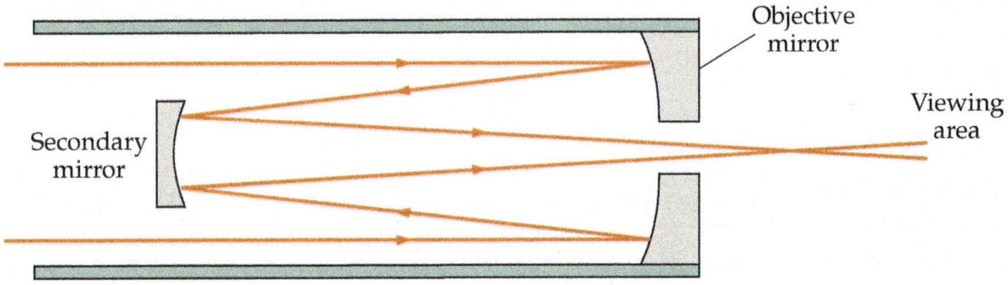

FIGURE 32-55 This reflecting telescope has a secondary mirror to redirect the light through a small hole in the objective mirror, thus providing more room for auxiliary instruments in the viewing area.

In addition, mechanical support is much simpler, because the mirror weighs far less than a lens of equivalent optical quality, and the mirror can be supported over its entire back surface. In modern Earth-based telescopes, the objective mirror consists of several dozen adaptive mirror segments that can be adjusted individually to correct for minute variations in gravitational stress when the telescope is tilted, and to compensate for thermal expansions and contractions and other changes caused by climatic conditions. In addition, they can adjust to nullify the distortions produced by atmospheric fluctuations.

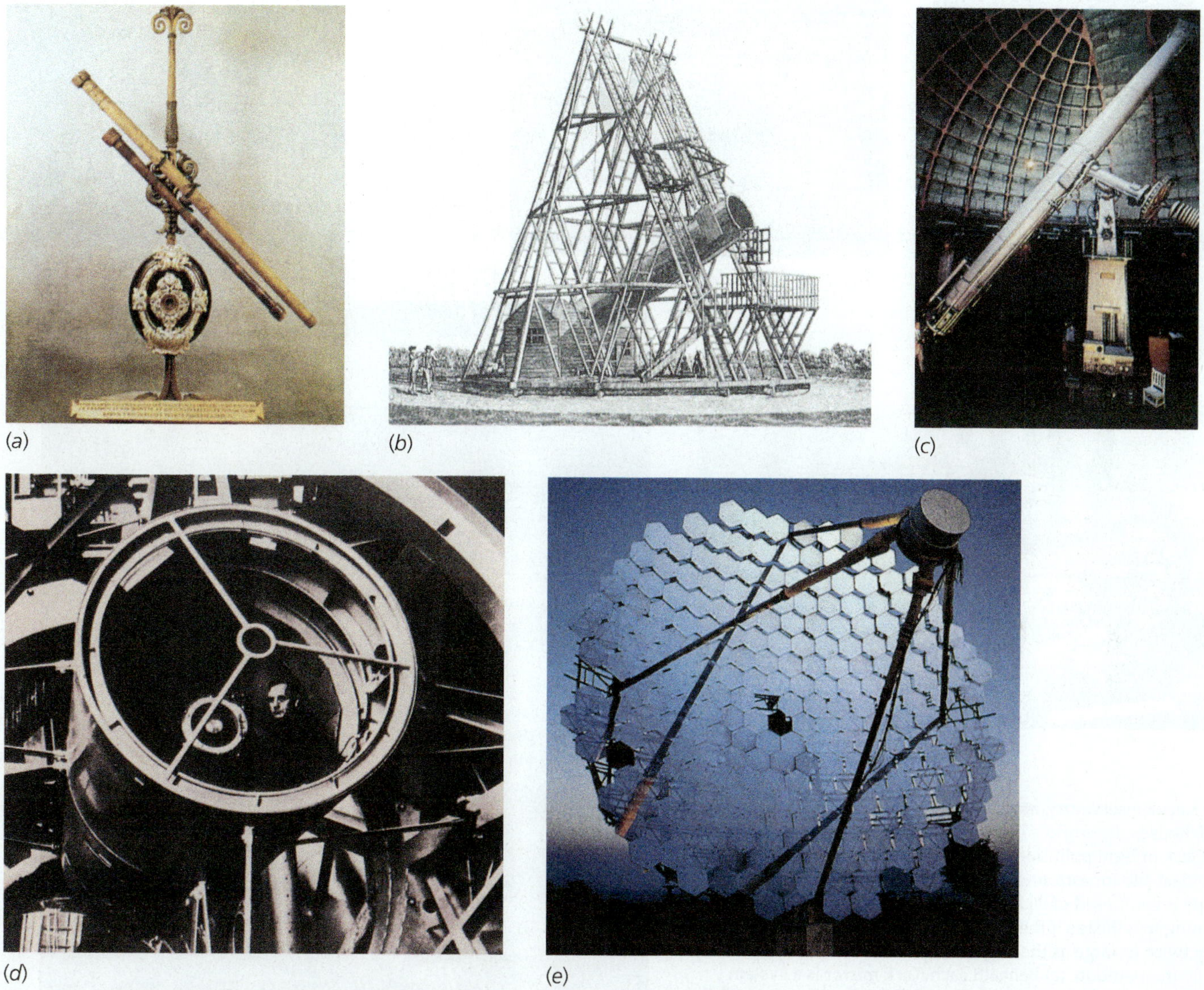

Astronomy at optical wavelengths began with Galileo approximately 400 years ago. In the twentieth century, astronomers began to explore the electromagnetic spectrum at other wavelengths; beginning with radio astronomy in the 1940s, satellite-based X-ray astronomy in the early 1960s, and more recently, ultraviolet, infrared, and gamma-ray astronomy. (*a*) Galileo's seventeenth-century telescope, with which he discovered mountains on the moon, sunspots, Saturn's rings, and the bands and moons of Jupiter. (*b*) An engraving of the reflector telescope built in the 1780s and used by the great astronomer Friedrich Wilhelm Herschel, who was the first to observe galaxies outside our own. (*c*) Because it is difficult to make large, flaw-free lenses, refractor telescopes like this 91.4-cm telescope at Lick Observatory have been superseded in light-gathering power by reflector telescopes. (*d*) The great astronomer Edwin Powell Hubble, who discovered the apparent expansion of the universe, is shown seated in the observer's cage of the 5.08-m Hale reflecting telescope, which is large enough for the observer to sit at the prime focus itself. (*e*) This 10-m optical reflector at the Whipple Observatory in southern Arizona is the largest instrument designed exclusively for use in gamma-ray astronomy. High-energy gamma rays of unknown origin strike the upper atmosphere and create cascades of particles. Among these particles are high-energy electrons that emit Cerenkov radiation observable from the ground. According to one hypothesis, high-energy gamma rays are emitted as matter is accelerated toward ultradense rotating stars called pulsars. *((a) Scala/Art Resource, (b) Royal Astronomical Society Library, (c) Lick Observatory, courtesy of the University of California Regents, (d) California Institute of Technology, (e) Gary Ladd.)*

(*a*)

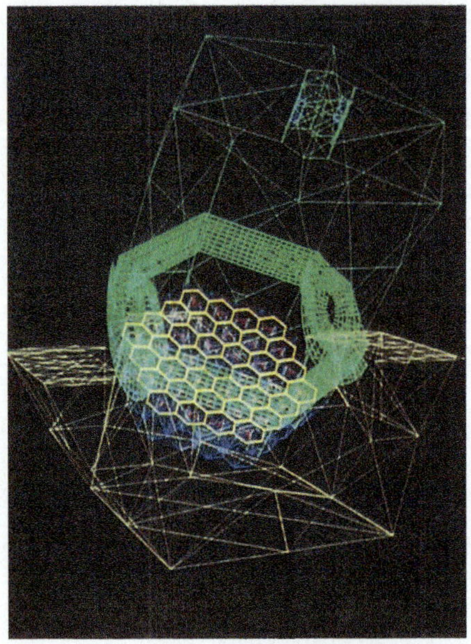

(*b*)

(*c*)

(*a*) The Keck Observatory, atop the inactive volcano of Mauna Kea, Hawaii, houses the world's largest optical telescope. The clear, dry air and lack of light pollution make the remote heights of Mauna Kea an ideal site for astronomical observations. (*b*) The Keck telescope is composed of 36 hexagonal mirror segments performing together as if they were a single mirror 10 m wide—roughly twice as large as the largest single-mirror telescope presently in operation. (*c*) Beneath each Keck mirror is a system of computer-controlled sensors and motor-driven actuators that can continuously vary the mirror's shape. These variations, which are sensitive to within 100 nm, enable the system to correct for variations in the alignments of the segments due to minute variations in gravitational stress when the telescope is tilted and to compensate for thermal expansions and contractions and fluctuations caused by gusts of wind on the mountaintop. *(California Association for Research in Astronomy.)*

The Hubble Space Telescope is high above the atmospheric turbulence that limits the ability of ground-based telescopes to resolve images at optical wavelengths. *(NASA.)*

Physics Spotlight

Eye Surgery: New Lenses for Old

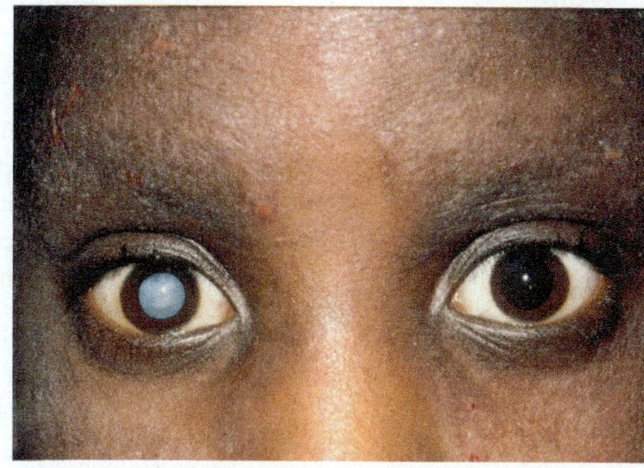

Unfortunately, this person has a cataract affecting the entire lens of her right eye. The presence of the cataract is associated with the severe eczema that can be seen on her forehead. *(Western Ophthalmic Hospital/Photo Researchers.)*

The first documented eye surgery took place over 2000 years ago and was performed by a surgeon in India known today as Susruta.[*, †] The lens of the human eye is susceptible to clouding with age and exposure to ultraviolet light or with exposure to some diseases or chemicals. This opacity is called a cataract. In the *Susruta Samhita*, the surgeon wrote of diagnosing and removing cataract-clouded lenses.

In 2005, more than two million cataract surgeries were performed in the United States.[‡] Most involved removing the lens and implanting an artificial lens inside the eye. Although permanent *i*ntra*o*cular *l*enses (IOLs) were pioneered in 1951,[#] they did not gain widespread acceptance until the 1980s. Before then, patients had to wear strong external corrective lenses (eyeglasses or contact lenses after the 1960s) after cataract surgery.

Cataract patients who have had IOLs implanted have had to wear external corrective lenses, because the IOLs are fixed focus lenses. However, several different IOLs that have variable focus have been recently introduced.[°, §] These accommodative, multifocal lenses are flexed and focused by the eye muscles. Some patients have had such good results with accommodative IOLs that they do not need supplemental glasses.

Many people without cataracts wear external corrective lenses. In the United States alone, over 150 million people spend at least $15 billion on external corrective lenses each year.[¶] In addition, nearly a million get refractive surgery each year that promises to reduce their need for external corrective lenses.[**] Refractive surgery reshapes the cornea to minimize refractive errors. It was pioneered in the 1930s, but did not become popular until techniques that use lasers were developed in the late 1980s.[††] Refractive surgery can be done by using chemicals or mechanical abrasion, but it is most commonly done by using lasers that vaporize minute pieces of the cornea. Some refractive surgeries involve cutting a thin flap in the outer layers of the cornea, shaping the stroma, the thickest layer of the cornea underneath, and then replacing the flap. Other surgeries just involve shaping the outer surface of the cornea.[‡‡]

Refractive surgeries work best on patients who have medium to low refractive error. In those patients, however, the results are promising. Up to 72% of the patients who have low to moderate nearsightedness ended up with postsurgical uncorrected vision that is considered 20/20, or normal.[##] Recently, customized wavefront-guided laser surgery has been used to minimize refractive errors that are not fixed by earlier techniques of eye surgery.[°°] A light is bounced off of the retina through many locations on the cornea. The returning wavefront is mapped, and very tiny refractive errors are located for elimination. With custom wavefront-guided laser surgery, up to 89% of patients ended up with uncorrected 20/20 vision.

But even with the best surgeons, there can be complications with refractive surgeries.[§§, ¶¶] The most common is dry eyes, but in up to 2% of patients vision can actually be worse after surgery.[***] Complications are more likely in people who have large refractive errors, and refractive surgery is not recommended for them. But those people may also eventually be able to not wear external corrective lenses. Recently, accommodative IOLs have been tested on people who have high refractive errors. Many were able to have near-normal vision after surgery without additional corrective lenses.[†††]

* Raju, V. K., "Susruta of Ancient India." *Indian Journal of Ophthalmology*, Feb. 2003, Vol. 51, No. 2, pp. 119–122. http://www.ijo.in/article.asp?issn=0301-4738;year=2003;volume=51;issue=2;spage=119;epage=122;aulast=Raju As of Nov. 2006.

† Hellemans, A., and Bunch, B., *The Timetables of Science*. New York: Simon & Schuster, 1988, p. 28.

‡ American Academy of Ophthalmology, "Industry News." *Academy Express*, Vol. 5, No. 14, http://www.aao.org/news/academy_express/20060405.cfm#asc As of Nov. 2006.

\# Apple, D. J., "Sir Harold Ridley." *Journal of Cataract and Refractive Surgery*, Mar. 2004, Vol. 30, No. 3, pp. 47–52.

° Cummings, J. S., et al., "Clinical Evaluation of the Crystalens AT-45 Accommodating Intraocular Lens: Results of the U.S. Food and Drug Administration Clinical Trial." *Journal of Cataract and Refractive Surgery*, May 2006, Vol. 32, No. 5, pp. 812–825.

§ Charters, L., "Dual-Optic IOL Effective Answer to Presbyopia." *Ophthalmology Times*, Feb. 1, 2006, pp. 20–21.

¶ Rados, C., "A Focus on Vision." *FDA Consumer*, Jul.–Aug. 2006, pp. 10–17.

** American Academy of Ophthalmology, "Industry News." *Academy Express*, Vol. 5, No. 14, http://www.aao.org/news/academy_express/20060405.cfm#asc As of Nov. 2006.

†† Kornmehl, E., "The Start of Something Big." *Ophthalmology Times*, Nov. 1, 2006, Vol. 31, No. 21, p. 24.

‡‡ Sakimoto, T., Rosenblatt, M., and Azar, D., "Laser Eye Surgery for Refractive Errors." *The Lancet*, Apr. 29, 2006, Vol. 367, No. 9520, pp. 1432–1447.

\#\# Sakimoto, T., Rosenblatt, M., and Azar, D., op. cit.

°° Mackenzie, D., "Coming Soon: 'Wavefront Eye Surgery'?" *Science*, Mar. 14, 2003, Vol. 299, No. 5613, p. 1655.

§§ Potter, J., "Do What's Right When Refractive Surgery Goes Wrong." *Review of Optometry*, Oct. 15, 2006, pp. 52–62.

¶¶ Guttman, C., "DLK a Lifelong Risk in Post-LASIK eyes." *Ophthalmology Times*, Nov. 1, 2006, Vol. 31, No. 21, pp. 1+.

*** Sakimoto, T., Rosenblatt, M., and Azar, D., op. cit.

††† Charters, L., "Accommodating IOL Improves Vision for High Refractive Errors in Analysis." *Ophthalmology Times*, Jun. 1, 2006, Vol. 31, No. 11, p. 43.

Summary

TOPIC	RELEVANT EQUATIONS AND REMARKS	
1. Virtual and Real Images and Objects		
Images	An image is *real* if actual light rays converge to each image point. This can occur on the reflected-light side of a mirror or on the refracted-light side of a thin lens or refracting surface. An image is *virtual* if only the extensions of the actual light rays converge to each image point. This can occur behind a mirror or on the incident-light side of a lens or refracting surface.	
Objects	A real object is either a physical object or a real image. An object is *real* if actual light rays diverge from each object point. This can occur only on the incident-light side of a mirror, lens, or refracting surface. An object is *virtual* if only extensions of actual light rays diverge from each object point. This can occur only behind a mirror or on the refracted-light side of a lens or refracting surface.	
2. Spherical Mirrors		
Focal length	The focal length is the image distance when the object is at infinity, so the incident light is parallel to the axis.	
Mirror equation (for locating an image)	$\frac{1}{s} + \frac{1}{s'} = \frac{1}{f}$	32-4
	where $f = \frac{r}{2}$	32-3
Lateral magnification	$m = \frac{y'}{y} = -\frac{s'}{s}$	32-5
Ray diagrams	Images can be located by a ray diagram using any two paraxial rays. The parallel, focal, and radial rays are the easiest to draw: 1. The parallel ray, drawn parallel to the axis, is reflected through the focal point. 2. The focal ray, drawn through the focal point, is reflected parallel to the axis. 3. The radial ray, drawn through the center of curvature, is reflected back on itself.	
Sign conventions for reflection	1. s is positive if the object is on the incident-light side of the mirror. 2. s' is positive if the image is on the reflected-light side of the mirror. 3. r and f are positive if the mirror is concave so the center of curvature is on the reflected-light side of the mirror.	
3. Images Formed by Refraction		
Refraction at a single surface	$\frac{n_1}{s} + \frac{n_2}{s'} = \frac{n_2 - n_1}{r}$	32-6
	where n_1 is the index of refraction of the medium on the incident-light side of the surface.	
Magnification	$m = \frac{y'}{y} = -\frac{n_1 s'}{n_2 s}$	32-7
Sign conventions for refraction	1. s is positive for objects on the incident-light side of the surface.	
	2. s' is positive for images on the refracted-light side of the surface.	
	3. r is positive if the center of curvature is on the refracted-light side of the surface.	

TOPIC	RELEVANT EQUATIONS AND REMARKS	
4. Thin Lenses		
Focal length (lens-maker's equation)	$\frac{1}{f} = \left(\frac{n}{n_{\text{air}}} - 1\right)\left(\frac{1}{r_1} - \frac{1}{r_2}\right)$ A positive lens ($f > 0$) is a converging lens. A negative lens ($f < 0$) is a diverging lens.	32-11
First and second focal points	Incident rays parallel to the axis emerge directed either toward or away from the *first focal point* F'. Incident rays directed either toward or away from the *second focal point* F emerge parallel with the axis.	
Power	$P = \frac{1}{f}$	32-13
Thin-lens equation (for locating image)	$\frac{1}{s} + \frac{1}{s'} = \frac{1}{f}$	32-12
Magnification	$m = \frac{y'}{y} = -\frac{s'}{s}$	32-14
Ray diagrams	Images can be located by a ray diagram using any two paraxial rays. The parallel, central, and focal rays are the easiest to draw: 1. The parallel ray, drawn parallel to the axis, emerges directed toward (or away from) the second focal point. 2. The central ray, drawn through the center of the lens, is not deflected. 3. The focal ray, drawn through (or toward) the first focal point, emerges parallel to the axis.	
Sign conventions for lenses	The sign conventions are the same as for refraction at a spherical surface.	
5. *Aberrations	Blurring of the image of a single object point is called aberration. Spherical aberration results from the fact that a spherical surface focuses only paraxial rays (those that travel close to the axis) to a single point. Nonparaxial rays are focused at nearby points depending on the angle made with and distance from the axis. Spherical aberration can be reduced by blocking the rays farthest from the axis. This, of course, reduces the amount of light reaching the image. Chromatic aberration, which occurs with lenses but not mirrors, results from the variation in the index of refraction with wavelength. Lens aberrations are most commonly reduced by using a series of lens elements.	
6. *The Eye	The cornea–lens system of the eye focuses light on the retina, where it is sensed by the rods and cones that send information along the optic nerve to the brain. When the eye is relaxed, the focal length of the cornea-lens system is about 2.5 cm, which is the distance to the retina. When objects are brought near the eye, the lens changes shape to decrease the overall focal length so that the image remains focused on the retina. The closest distance for which the image can be focused on the retina is called the near point, typically about 25 cm. The apparent size of an object depends on the size of the image on the retina. The closer the object, the larger the image on the retina and therefore the larger the apparent size of the object.	
7. *The Simple Magnifier	A simple magnifier consists of a lens with a positive focal length that is smaller than the near point.	
Magnifying power (angular magnification)	$M = \frac{\theta}{\theta_o} = \frac{x_{\text{np}}}{f}$	32-20
8. *The Compound Microscope	The compound microscope is used to look at very small objects that are nearby. It consists of two converging lenses (or lens systems), an objective and an eyepiece. The object to be viewed is placed just outside the focal point of the objective, which forms an enlarged image of the object at the focal plane of the eyepiece. The eyepiece acts as a simple magnifier to view the final image.	

TOPIC	RELEVANT EQUATIONS AND REMARKS	
Magnifying power (angular magnification)	$M = m_o M_e = -\frac{L}{f_o}\frac{x_{np}}{f_e}$ where L is the tube length, the distance between the second focal point of the objective and the first focal point of the eyepiece.	32-22
9. *The Telescope	The telescope is used to view objects that are far away. The objective of the telescope forms a real image of the object that is much smaller than the object but much closer. The eyepiece is then used as a simple magnifier to view the image. A reflecting telescope uses a mirror for its objective.	
Magnifying power (angular magnification)	$M = \frac{\theta_e}{\theta_o} = -\frac{f_o}{f_e}$	32-23

Answers to Concept Checks

32-2 Ben can see only the image at P_1'.

32-3 Infinity

32-4

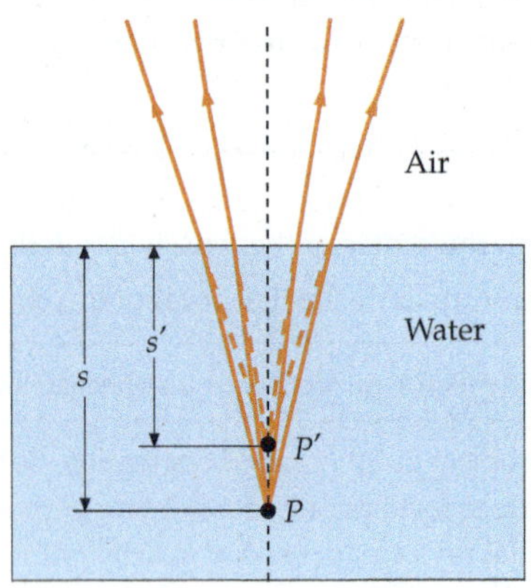

FIGURE 32-56 Ray diagram for the image of an object in water as viewed from directly overhead. The depth of the image is less than the depth of the object.

Answers to Practice Problems

32-3	(*a*) 8.0 cm (*b*) $s' = -4.0$ cm
32-4	$s' = -2.5$ cm, $m = +0.50$; the image is upright, virtual, and reduced in size.
32-5	390 yd
32-6	(*a*) $s' = -6.44$ cm and (*b*) $m = 1.14$. Fluffy sees Goldie to be 1.1 cm closer and 14 percent larger than she actually is.
32-7	The image is 5.6 cm from the near side of the aquarium.
32-8	18 cm
32-10	$s' = 30$ cm, $m = -2.0$; real, inverted
32-11	$s' = -10$ cm, $m = 2.0$; virtual, upright
32-12	$P_{eye} = 41.33$ D; $P_c = 41.33$ D $+ 2.67$ D $= 44$ D; the two powers are equal
32-13	$M = 12$
32-14	$M = -195$

Problems

In a few problems, you are given more data than you actually need; in a few other problems, you are required to supply data from your general knowledge, outside sources, or informed estimate.

Interpret as significant all digits in numerical values that have trailing zeros and no decimal points.

- • Single-concept, single-step, relatively easy
- •• Intermediate-level, may require synthesis of concepts
- ••• Challenging
- SSM Solution is in the *Student Solutions Manual*
- Consecutive problems that are shaded are paired problems.

CONCEPTUAL PROBLEMS

1 • Can a virtual image be photographed? If so, give an example. If not, explain why.

2 • Suppose the x, y, and z axes of a 3-D coordinate system are painted different colors. One photograph is taken of the coordinate system and another is taken of its image in a plane mirror. Is it possible to tell that one of the photographs is of a mirror image? Or could both photographs be of the real coordinate system from different angles?

3 • True or false

(*a*) The virtual image formed by a concave mirror is always smaller than the object.
(*b*) A concave mirror always forms a virtual image.
(*c*) A convex mirror never forms a real image of a real object.
(*d*) A concave mirror never forms an enlarged real image of an object. SSM

4 • An ant is crawling along the axis of a concave mirror that has radius of curvature R. At what object distances, if any, will the mirror produce (*a*) an upright image, (*b*) a virtual image, (*c*) an image smaller than the object, and (*d*) an image larger than the object?

5 • An ant is crawling along the axis of a convex mirror that has radius of curvature R. At what object distances, if any, will the mirror produce (*a*) an upright image, (*b*) a virtual image, (*c*) an image smaller than the object, and (*d*) an image larger than the object? SSM

6 •• Convex mirrors are often used for rearview mirrors on cars and trucks to give a wide-angle view. "Warning, objects are closer than they appear" is written below the mirrors. Yet, according to a ray diagram, the image distance for distant objects is much shorter than the object distance. Why then do they appear more distant?

7 • As an ant on the axis of a concave mirror crawls from a great distance to the focal point of a concave mirror, the image of the ant moves (*a*) from a great distance toward the focal point and is always real, (*b*) from the focal point to a great distance from the mirror and is always real, (*c*) from the focal point to the center of curvature of the mirror and is always real, (*d*) from the focal point to a great distance from the mirror and changes from a real image to a virtual image.

8 • A kingfisher bird that is perched on a branch a few feet above the water is viewed by a scuba diver submerged beneath the surface of the water directly below the bird. Does the bird appear to the diver to be closer to or farther from the surface than the actual bird? Explain your answer using a ray diagram.

9 • An object is placed on the axis of a diverging lens whose focal length has a magnitude of 10 cm. The distance from the object to the lens is 40 cm. The image is (*a*) real, inverted, and diminished; (*b*) real, inverted, and enlarged; (*c*) virtual, inverted, and diminished; (*d*) virtual, upright, and diminished; (*e*) virtual, upright, and enlarged.

10 • If an object is placed between the focal point of a converging lens and the optical center of the lens, the image is (*a*) real, inverted, and enlarged; (*b*) virtual, upright, and diminished; (*c*) virtual, upright, and enlarged; (*d*) real, inverted, and diminished.

11 • A converging lens is made of glass that has an index of refraction of 1.6. When the lens is in air, its focal length is 30 cm. When the lens is immersed in water, its focal length (*a*) is greater than 30 cm, (*b*) is between zero and 30 cm, (*c*) is equal to 30 cm, (*d*) has a negative value.

12 • True or false:

(*a*) A virtual image cannot be displayed on a screen.
(*b*) A negative image distance implies that the image is virtual.
(*c*) All rays parallel to the axis of a spherical mirror are reflected through a single point.
(*d*) A diverging lens cannot form a real image from a real object.
(*e*) The image distance for a converging lens is always positive.

13 • **Biological Application** Both the human eye and the digital camera work by forming real images on light-sensitive surfaces. The eye forms a real image on the retina and the camera forms a real image on a CCD array. Explain the difference between the ways in which these two systems *accommodate*. That is, the difference between how an eye adjusts and how a camera adjusts (or can be adjusted) to form a focused image for objects at both large and short distances from the camera.

14 • **Biological Application** If an object is 25 cm in front of the naked eye of a farsighted person, an image (*a*) would be formed behind the retina if it were not for the fact that that the light is blocked (by the back of the eyeball) and the corrective contact lens should be convex, (*b*) would be formed behind the retina if it were not for the fact that that the light is blocked (by the back of the eyeball) and the corrective contact lens should be concave, (*c*) is formed in front of the retina and the corrective contact lens should be convex, (*d*) is formed in front of the retina and the corrective contact lens should be concave.

15 •• Explain the following statement: A microscope is an object magnifier, but a telescope is an angle magnifier. *Hint: Take a look at the ray diagram for each magnifier and use it to explain the difference in adjectives.*

ESTIMATION AND APPROXIMATION

16 • Estimate the location and size of the image of your face when you hold a shiny new tablespoon a foot in front of your face and with the convex side toward you.

17 • Estimate the focal length of the "mirror" produced by the surface of the water in the reflection pool in front of the Lincoln Memorial on a still night.

18 •• Estimate the maximum value that could be obtained for the magnifying power of a simple magnifier, using Equation 32-20. *Hint: Think about the smallest focal length lens that could be made from glass and still be used as a magnifier.*

PLANE MIRRORS

19 • The image of the object point P in Figure 32-57 is viewed by an eye, as shown. Draw two rays from the object point that reflect from the mirror and enter the eye. If the object point and the mirror are fixed in their locations, indicate the range of locations where the eye can be positioned and still see the image of the object point. SSM

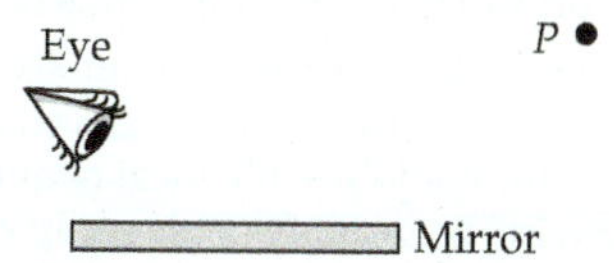

FIGURE 32-57 Problem 19

20 • You are 1.62 m tall and want to be able to see your full image in a vertical plane mirror. (*a*) What is the minimum height of the mirror that will meet your needs? (*b*) How far above the floor should the bottom of mirror in (*a*) be placed, assuming that the top of your head is 14 cm above your eye level? Use a ray diagram to explain your answer.

21 •• (*a*) Two plane mirrors make an angle of 90°. The light from a point object that is arbitrarily positioned in front of the mirrors produces images at three locations. For each image location, draw two rays from the object that, after one or two reflections, appear to come from the image location. (*b*) Two plane mirrors make an angle of 60° with each other. Draw a sketch to show the location of all the images formed of an object on the bisector of the angle between the mirrors. (*c*) Repeat Part (*b*) for an angle of 120°. SSM

22 •• Show that the mirror equation (Equation 32-4 where $f = r/2$) yields the correct image distance and magnification for a plane mirror.

23 •• When two plane mirrors are parallel, such as on opposite walls in a barber shop, multiple images arise because each image in one mirror serves as an object for the other mirror. An object is placed between parallel mirrors separated by 30 cm. The object is 10 cm in front of the left mirror and 20 cm in front of the right mirror. (*a*) Find the distance from the left mirror to the first four images in that mirror. (*b*) Find the distance from the right mirror to the first four images in that mirror. (*c*) Explain why each more distant image becomes fainter and fainter.

SPHERICAL MIRRORS

24 •• A concave mirror has a radius of curvature equal to 24 cm. Use ray diagrams to locate the image, if it exists, for an object near the axis at distances of (*a*) 55 cm, (*b*) 24 cm, (*c*) 12 cm, and (*d*) 8.0 cm from the mirror. For each case, state whether the image is real or virtual; upright or inverted; and enlarged, reduced, or the same size as the object.

25 • (*a*) Use the mirror equation (Equation 32-4 where $f = r/2$) to calculate the image distances for the object distances and mirror of Problem 24. (*b*) Calculate the magnification for each given object distance. SSM

26 • A convex mirror has a radius of curvature that has a magnitude equal to 24 cm. Use ray diagrams to locate the image, if it exists, for an object near the axis at distances of (*a*) 55 cm, (*b*) 24 cm, (*c*) 12 cm, (*d*) 8.0 cm, and (*e*) 1.0 cm from the mirror. For each case, state whether the image is real or virtual; upright or inverted; and enlarged, reduced, or the same size as the object.

27 • (*a*) Use the mirror equation (Equation 32-4 where $f = r/2$) to calculate the image distances for the object distances and mirror of Problem 26. (*b*) Calculate the magnification for each given object distance.

28 •• Use the mirror equation (Equation 32-4 where $f = r/2$) to prove that a convex mirror cannot form a real image of a real object, no matter where the object is placed.

29 • A dentist wants a small mirror that will produce an upright image that has a magnification of 5.5 when the mirror is located 2.1 cm from a tooth. (*a*) Should the mirror be concave or convex? (*b*) What should the radius of curvature of the mirror be? SSM

30 •• **CONTEXT-RICH** Convex mirrors are used in many stores to provide a wide angle of surveillance for a reasonable mirror size. Your summer job is at a local convenience store that uses the mirror shown in Figure 32-58. This setup allows you (or the clerk) to survey the entire store when you are 5.0 m from the mirror. The mirror has a radius of curvature equal to 1.2 m. Assume all rays are paraxial. (*a*) If a customer is 10 m from the mirror, how far from the mirror is his image? (*b*) Is the image in front of or behind the mirror? (*c*) If the customer is 2.0 m tall, how tall is his image?

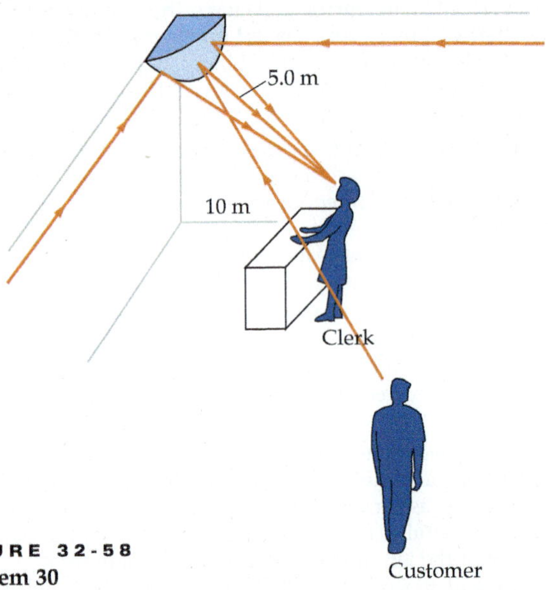

FIGURE 32-58
Problem 30

31 •• A certain telescope uses a concave spherical mirror that has a radius equal to 8.0 m. Find the location and diameter of the image of the moon formed by this mirror. The moon has a diameter of 3.5×10^6 m and is 3.8×10^8 m from Earth.

32 •• A piece of a thin spherical shell that has a radius of curvature of 100 cm is silvered on both sides. The concave side of the piece forms a real image 75 cm from the piece. The piece is then turned around so that its convex side faces the object. The piece is moved so that the image is now 35 cm from the piece on the concave side. (*a*) How far was the piece moved? (*b*) Was it moved toward the object or away from the object?

33 •• Two light rays parallel to the optic axis of a concave mirror strike that mirror as shown in Figure 32-59. The mirror has a radius of curvature equal to 5.0 m. They then strike a small spherical mirror that is 2.0 m from the large mirror. The light rays finally meet at the vertex of the large mirror. (*Note:* The small mirror is shown as planar, so as not to give away the answer, but it is *not* actually planar.) (*a*) What is the radius of curvature of the small mirror? (*b*) Is that mirror convex or concave? Explain your answer.

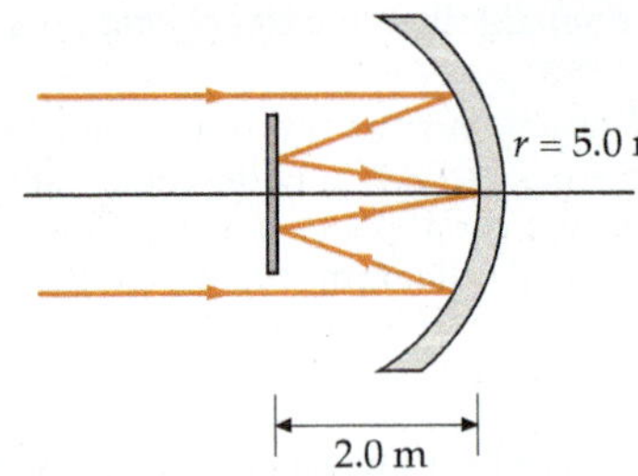

FIGURE 32-59
Problem 33

IMAGES FORMED BY REFRACTION

34 •• A very long 1.75-cm-diameter glass rod has one end ground and polished to a convex spherical surface that has a 7.20-cm radius. The glass material has an index of refraction of 1.68. (*a*) A point object in air is on the axis of the rod and 30.0 cm from the spherical surface. Find the location of the image and state whether the image is real or virtual. (*b*) Repeat Part (*a*) for a point object in air, on the axis, and 5.00 cm from the spherical surface. Draw a ray diagram for each case.

35 • A fish is 10 cm from the front surface of a spherical fish bowl of radius 20 cm. (*a*) How far behind the surface of the bowl does the fish appear to someone viewing the fish from in front of the bowl? (*b*) By what distance does the fish's apparent location change (relative to the front surface of the bowl) when it swims away to 30 cm from the front surface? SSM

36 •• A very long 1.75-cm-diameter glass rod has one end ground and polished to a concave spherical surface that has a 7.20-cm radius. The glass material has an index of refraction of 1.68. A point object in air is on the axis of the rod and 15.0 cm from the spherical surface. Find the location of the image and state whether the image is real or virtual. Draw a ray diagram.

37 •• Repeat Problem 34 for when the glass rod and the object are immersed in water, and (*a*) the object is 6.00 cm from the spherical surface, and (*b*) the object is 12.0 cm from the spherical surface. SSM

38 •• Repeat Problem 36 for when the glass rod and the object are immersed in water and the object is 20 cm from the spherical surface.

39 •• A rod that is 96.0 cm long is made of glass that has an index of refraction equal to 1.60. The rod has its ends ground to convex spherical surfaces that have radii equal to 8.00 cm and 16.0 cm. An object is in air on the long axis of the rod 20.0 cm from the end that has the 8.00-cm radius. (*a*) Find the image distance due to refraction at the 8.00-cm-radius surface. (*b*) Find the position of the final image due to refraction at both surfaces. (*c*) Is the final image real or virtual?

40 •• Repeat Problem 39 for an object in air on the axis of the glass rod 20.0 cm from the end that has the 16.0-cm radius.

THIN LENSES AND THE LENS-MAKER'S EQUATION

41 • A double concave lens that has an index of refraction equal to 1.45 has radii whose magnitudes are equal to 30.0 cm and 25.0 cm. An object is located 80.0 cm to the left of the lens. Find (*a*) the focal length of the lens, (*b*) the location of the image, and (*c*) the magnification of the image. (*d*) Is the image real or virtual? Is the image upright or inverted? SSM

42 • The following thin lenses are made of glass that has an index of refraction equal to 1.60. Make a sketch of each lens and find each focal length in air: (*a*) $r_1 = 20.0$ cm and $r_2 = 10.0$ cm, (*b*) $r_1 = 10.0$ cm and $r_2 = 20.0$ cm, and (*c*) $r_1 = -10.0$ cm and $r_2 = -20.0$ cm.

43 • The following four thin lenses are made of glass that has an index of refraction of 1.5. The radii given are *magnitudes*. Make a sketch of each lens and find each focal length in air: (*a*) double-convex that has radii of curvature equal to 15 cm and 26 cm, (*b*) plano-convex that has a radius of curvature equal to 15 cm, (*c*) double concave that has radii of curvature equal to 15 cm, and (*d*) plano-concave that has a radius of curvature equal to 26 cm.

44 • Find the focal length of a glass lens that has an index of refraction equal to 1.62, a concave surface that has a radius of curvature of magnitude 100 cm, and a convex surface that has a radius of curvature of magnitude 40.0 cm.

45 •• (*a*) An object that is 3.00 cm high is placed 25.0 cm in front of a thin lens that has a power equal to 10.0 D. Draw a ray diagram to find the position and the size of the image and check your results using the thin-lens equation. (*b*) Repeat Part (*a*) if the object is placed 20.0 cm in front of the lens. (*c*) Repeat Part (*a*) for an object placed 20.0 cm in front of a thin lens that has a power equal to −10.0 D. SSM

46 •• The lens-maker's equation has three design parameters. They consist of the index of refraction of the lens and the radii of curvature for its two surfaces. Thus, there are many ways to design a lens that has a particular focal length in air. Use the lens-maker's equation to design three different thin converging lenses, each having a focal length of 27.0 cm and each made from glass that has an index of refraction of 1.60. Sketch each of your designs.

47 •• Repeat Problem 46, but for a diverging lens that has a focal length in air of the same magnitude.

48 •• (*a*) What is meant by a negative object distance? Describe a specific situation in which a negative object distance can occur. (*b*) Find the image distance and the magnification for a thin lens in air when the object distance is −20 cm and the lens is a converging lens that has a focal length of 20 cm. Describe the image—is it virtual or real, upright or inverted? (*c*) Repeat Part (*b*) if the object distance is, instead, −10 cm, and the lens is diverging and has a focal length (magnitude) of 30 cm.

49 •• Two converging lenses, each having a focal length equal to 10 cm, are separated by 35 cm. An object is 20 cm to the left of the first lens. (*a*) Find the position of the final image using both a ray diagram and the thin-lens equation. (*b*) Is the final image real or virtual? Is the final image upright or inverted? (*c*) What is the overall lateral magnification? SSM

50 •• Repeat Problem 49, but with the second lens replaced by a diverging lens that has a focal length equal to −15 cm.

51 •• (*a*) Show that to obtain a magnification of magnitude $|m|$ using a converging thin lens of focal length f, the object distance must be equal to $(1 + |m|^{-1})f$. (*b*) You want to use a digital camera which has a lens whose focal length is 50.0 mm to take a picture of a person 1.75 m tall. How far from the camera lens should you have that person stand so that the image size on the light-receiving sensors of your camera is 24.0 mm?

52 •• **SPREADSHEET** A converging lens has a focal length of 12.0 cm. (*a*) Using a spreadsheet program or graphing calculator, plot the image distance as a function of the object distance, for object distances ranging from $1.10f$ to $10.0f$, where f is the focal length. (*b*) On the same graph used in Part (*a*), but using a different y axis, plot the magnification of the lens as a function of the object distance. (*c*) What type of image is produced for this range of object distances—real or virtual, upright or inverted? (*d*) Discuss the significance of any asymptotic limits your graphs have.

53 •• **SPREADSHEET** A converging lens has a focal length of 12.0 cm. (*a*) Using a spreadsheet program or graphing calculator, plot the image distance as a function of the object distance, for object distances ranging from $0.010f$ to $0.90f$, where f is the focal length. (*b*) On the same graph used in Part (*a*), but using a different y axis, plot the magnification of the lens as a function of the object distance. (*c*) What type of image is produced for this range of object distances—real or virtual, upright or inverted? (*d*) Discuss the significance of any asymptotic limits your graphs have.

54 •• An object is 15.0 cm in front of a converging lens that has a focal length equal to 15.0 cm. A second converging lens that also has a focal length equal to 15.0 cm is located 20.0 cm in back of the first. (*a*) Find the location of the final image and describe its properties (for example, real and inverted) and (*b*) draw a ray diagram to corroborate your answers to Part (*a*).

55 •• An object is 15.0 cm in front of a converging lens that has a focal length equal to 15.0 cm. A diverging lens that has a focal length whose magnitude is equal to 15.0 cm is located 20.0 cm in back of the first. (*a*) Find the location of the final image and describe its properties (for example, real and inverted) and (*b*) draw a ray diagram to corroborate your answers to Part (*a*). SSM

56 ••• In a convenient form of the thin-lens equation used by Newton, the object and image distances x and x' are measured from the focal points F and F', and not from the center of the lens. (*a*) Indicate x and x' on a sketch of a lens and show that if $x = s - f$ and $x' = s' - f$, the thin-lens equation (Equation 32-12) can be rewritten as $xx' = f^2$. (*b*) Show that the lateral magnification is given by $m = -x'/f = -f/x$.

57 ••• In *Bessel's method* for finding the focal length f of a lens, an object and a screen are separated by distance L, where $L > 4f$. It is then possible to place the lens at either of two locations, both between the object and the screen, so that there is an image of the object on the screen, in one case magnified and in the other case reduced. Show that if the distance between those two lens locations is D, then the focal length is given by $f = \frac{1}{4}(L^2 - D^2)/L$. *Hint: Refer to Figure 32-60.* SSM

58 ••• **ENGINEERING APPLICATION, CONTEXT-RICH** You are working for an optician during the summer. The optician needs to measure an unknown focal length and you suggest using *Bessel's method* (see Problem 57), which you used during a physics lab. You set the object-to-image distance at 1.70 m. The lens position is adjusted to get a sharp image on the screen. A second sharp image is found when the lens is moved a distance of 72 cm from its first location. (*a*) Sketch the ray diagram for the two locations. (*b*) Find the focal length of the lens using Bessel's method. (*c*) What are the two locations of the lens with respect to the object? (*d*) What are the magnifications of the images when the lens is in the two locations?

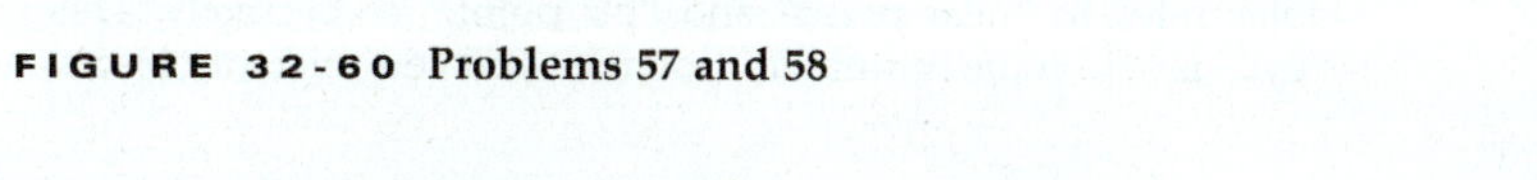

FIGURE 32-60 Problems 57 and 58

59 ••• An object is 17.5 cm to the left of a lens that has a focal length of +8.50 cm. A second lens, which has a focal length of −30.0 cm, is 5.00 cm to the right of the first lens. (*a*) Find the distance between the object and the final image formed by the second lens. (*b*) What is the overall magnification? (*c*) Is the final image real or virtual? Is the final image upright or inverted?

*ABERRATIONS

60 • Chromatic aberration is a common defect of (*a*) concave and convex lenses, (*b*) concave lenses only, (*c*) concave and convex mirrors, (*d*) all lenses and mirrors.

61 • **ENGINEERING APPLICATION** Discuss some of the reasons why most telescopes that are used by astronomers are reflecting rather than refracting telescopes.

62 • A symmetric double-convex lens has radii of curvature equal to 10.0 cm. It is made from glass that has an index of refraction equal to 1.530 for blue light and equal to 1.470 for red light. Find the focal length of this lens for (*a*) red light and (*b*) blue light.

*THE EYE

63 • **BIOLOGICAL APPLICATION** Find the change in the focal length of the eye when an object originally at 3.0 m is brought to 30 cm from the eye.

64 • **BIOLOGICAL APPLICATION** A farsighted person requires lenses that have powers equal to 1.75 D to read comfortably from a book that is 25.0 cm from his eyes. What is that person's near point without the lenses?

65 • **BIOLOGICAL APPLICATION** If two point objects close together are to be seen as two distinct objects, the images must fall on the retina on two different cones that are not adjacent. That is, there must be an unactivated cone between them. The separation of the cones is about 1.00 μm. Model the eye as a uniform 2.50-cm-diameter sphere that has a refractive index of 1.34. (*a*) What is the smallest angle the two points can subtend? (See Figure 32-61.) (*b*) How close together can two points be if they are 20.0 m from the eye? SSM

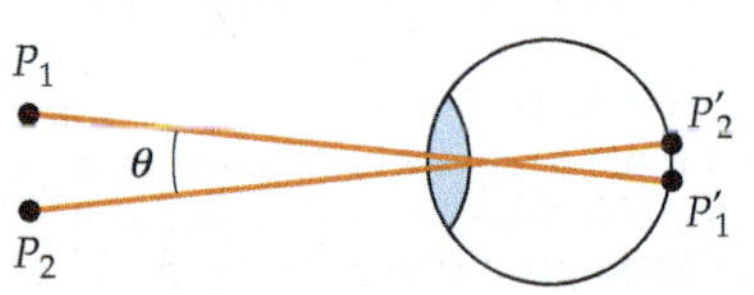

FIGURE 32-61 Problem 65

66 • **BIOLOGICAL APPLICATION** Suppose the eye were designed like a camera that has a lens of fixed focal length equal to 2.50 cm that could move toward or away from the retina, has air on both sides of the lens, and has no cornea. Approximately how far would the lens have to move to focus the image of an object 25.0 cm from the eye onto the retina? *Hint: Find the distance from the retina to the image behind it for an object at* 25.0 cm.

Note: **Problems 67 through 69 refer to the model of the eye shown in Figure 32-62.**

67 •• **BIOLOGICAL APPLICATION** A simple model for the eye is a lens that has a variable power P located a fixed distance d in front of a screen, with the space between the lens and the screen filled by air. This "eye" can focus for all values of object distance s such that $x_{np} \le s \le x_{fp}$ where the subscriptson the variables refer to "near point" and "far point," respectively. This "eye" is said to be normal if it can focus on very distant objects. (*a*) Show that for a normal "eye" of this type, the required minimum value of P is given by $P_{min} = 1/d$. (*b*) Show that the maximum value of P is given by $P_{max} = 1/x_{np} + 1/d$. (*c*) The difference between the maximum and minimum powers, symbolized by A, is defined as $A = P_{max} - P_{min}$ and is called the *accommodation*. Find the minimum power and accommodation for this model eye that has a screen distance of 2.50 cm, a far point distance of infinity, and a near point distance of 25.0 cm. SSM

68 •• **BIOLOGICAL APPLICATION** (This problem refers to the model eye described in Problem 67.) In an eye that exhibits nearsightedness, the eye cannot focus on distant objects. (*a*) Show that for a nearsighted model eye capable of focusing out to a maximum distance x_{fp}, the minimum value of the power P is greater than that of a normal eye (that has a far point at infinity) and is given by $P_{min} = 1/x_{fp} + 1/d$. (*b*) To correct for nearsightedness, a contact lens may be placed directly in front of the lens of the model eye. What power contact lens would be needed to correct the vision of a nearsighted model eye that has a far-point distance of 50.0 cm?

69 •• **BIOLOGICAL APPLICATION** (This problem refers to the model eye described in Problem 67.) In an eye that exhibits farsightedness, the eye may be able to focus on distant objects but cannot focus on close objects. (*a*) Show that for a farsighted model eye capable of focusing only as close as a distance x'_{np}, the maximum value of the power P is given by $P_{max} = 1/x'_{np} + 1/d$. (*b*) Show that, compared to a model eye capable of focusing as close as a distance x_{np} (where $x_{np} < x'_{np}$), the maximum power of the farsighted lens is too small by $1/x_{np} - 1/x'_{np}$. (*c*) What power contact lens would be needed to correct the vision of a farsighted model eye, with $x'_{np} = 150$ cm, so that the eye may focus on objects as close as 15 cm?

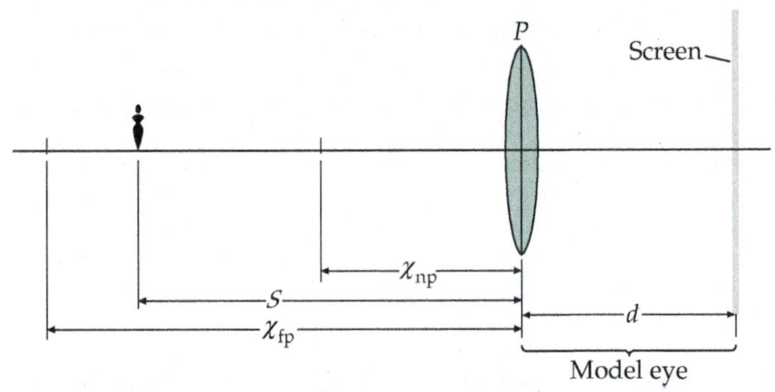

FIGURE 32-62 Problems 67, 68, and 69

70 •• **BIOLOGICAL APPLICATION** A person who has a near point of 80 cm needs to read from a computer screen that is only 45 cm from her eye. (*a*) Find the focal length of the lenses in reading glasses that will produce an image of the screen at a distance of 80 cm from her eye. (*b*) What is the power of the lenses?

71 •• **BIOLOGICAL APPLICATION** A nearsighted person cannot focus clearly on objects that are more distant than 2.25 m from her eye. What power lenses are required for her to see distant objects clearly?

72 •• **BIOLOGICAL APPLICATION** Because the index of refraction of the lens of the eye is not very different from that of the surrounding material, most of the refraction takes place at the cornea, where the index changes abruptly from 1.00 (air) to approximately 1.38. (*a*) Modeling the cornea, aqueous humor, lens and vitreous humor as a transparent homogeneous solid sphere that has an index of refraction of 1.38, calculate the sphere's radius if it focuses parallel light on the retina a distance 2.50 cm away. (*b*) Do you expect your result to be larger or smaller than the actual radius of the cornea? Explain your answer.

73 •• **Biological Application** The near point of a certain person's eyes is 80 cm. Reading glasses are prescribed so that he can read a book at 25 cm from his eye. The glasses are 2.0 cm from the eye. What diopter lenses should be used in the glasses?

74 ••• **Biological Application** At age 45, a person is fitted for reading glasses that have a power equal to 2.10 D in order to read at 25 cm. By the time she reaches the age of 55, she discovers herself holding her newspaper at a distance of 40 cm in order to see it clearly with her glasses on. (*a*) Where was her near point at age 45? (*b*) Where is her near point at age 55? (*c*) What power is now required for the lenses of her reading glasses so that she can again read at 25 cm? Assume the glasses are placed 2.2 cm from her eyes.

*THE SIMPLE MAGNIFIER

75 • What is the magnifying power of a lens that has a focal length equal to 7.0 cm when the image is viewed at infinity by a person whose near point is at 35 cm? SSM

76 •• A lens that has a focal length equal to 6.0 cm is used as a simple magnifier by one person whose near point is 25 cm and by another person whose near point is 40 cm. (*a*) What is the effective magnifying power of the lens for each person? (*b*) Compare the sizes of the images on the retinas when each person looks at the same object with the magnifier.

77 •• In your botany class, you examine a leaf using a convex 12-D lens as a simple magnifier. What is the angular magnification of the leaf if the image formed by the lens is (*a*) at infinity and (*b*) at 25 cm? SSM

*THE MICROSCOPE

78 •• Your laboratory microscope objective has a focal length of 17.0 mm. It forms an image of a tiny specimen at 16.0 cm from its second focal point. (*a*) How far from the objective is the specimen located? (*b*) What is the magnifying power for you if your near point distance is 25.0 cm and the focal length of the eyepiece is 51.0 mm?

79 •• A microscope has an objective that has a focal length equal to 8.5 mm. The eyepiece provides an angular magnification of 10 for a person whose near point distance is 25 cm. The tube length is 16 cm. (*a*) What is the lateral magnification of the objective? (*b*) What is the magnifying power of the microscope? SSM

80 •• A crude, symmetric handheld microscope consists of two 20-D lenses fastened at the ends of a tube 30 cm long. (*a*) What is the tube length of this microscope? (*b*) What is the lateral magnification of the objective? (*c*) What is the magnifying power of the microscope? (*d*) How far from the objective should the object be placed?

81 •• A compound microscope has an objective lens that has a power of 45 D and an eyepiece that has a power of 80 D. The lenses are separated by 28 cm. Assuming that the final image formed by the microscope is 25 cm from the eye, what is the magnifying power?

82 ••• A microscope has a magnifying power of 600. The eyepiece has an angular magnification of 15.0. The objective lens is 22.0 cm from the eyepiece. Calculate (*a*) the focal length of the eyepiece, (*b*) the location of the object so that it is in focus for a normal relaxed eye, and (*c*) the focal length of the objective lens.

*THE TELESCOPE

83 • You have a simple telescope that has an objective which has a focal length of 100 cm and an eyepiece which has a focal length of 5.00 cm. You are using it to look at the moon, which subtends an angle of about 9.00 mrad. (*a*) What is the diameter of the image formed by the objective? (*b*) What angle is subtended by the image formed at infinity by the eyepiece? (*c*) What is the magnifying power of your telescope? SSM

84 • The objective lens of the refracting telescope at the Yerkes Observatory has a focal length of 19.5 m. The moon subtends an angle of about 9.00 mrad. When the telescope is used to look at the moon, what is the diameter of the image of the moon formed by the objective?

85 •• The 200-in (5.10-m) diameter mirror of the reflecting telescope at Mt. Palomar has a focal length of 16.8 m. (*a*) By what factor is the light-gathering power increased over the 40.0-in (1.02-m) diameter refracting lens of the Yerkes Observatory telescope? (*b*) If the focal length of the eyepiece is 1.25 cm, what is the magnifying power of the 200-in telescope?

86 •• An astronomical telescope has a magnifying power of 7.0. The two lenses are 32 cm apart. Find the focal length of each lens.

87 •• A disadvantage of the astronomical telescope for terrestrial use (for example, at a football game) is that the image is inverted. A Galilean telescope uses a converging lens as its objective, but a diverging lens as its eyepiece. The image formed by the objective is at the second focal point of the eyepiece (the focal point on the refracted-light side of the eyepiece), so that the final image is virtual, upright, and at infinity. (*a*) Show that the magnifying power is given by $M = -f_o/f_e$, where f_o is the focal length of the objective and f_e is that of the eyepiece (which is negative). (*b*) Draw a ray diagram to show that the final image is indeed virtual, upright, and at infinity. SSM

88 •• A Galilean telescope (see Problem 87) is designed so that the final image is at the near point, which is 25 cm (rather than at infinity). The focal length of the objective is 100 cm and the focal length of the eyepiece is −5.0 cm. (*a*) If the object distance is 30.0 m, where is the image of the objective? (*b*) What is the object distance for the eyepiece so that the final image is at the near point? (*c*) How far apart are the lenses? (*d*) If the object height is 1.5 m, what is the height of the final image? What is the angular magnification?

89 ••• If you look into the wrong end of a telescope, that is, into the objective, you will see distant objects reduced in angular size. For a refracting telescope that has an objective with a focal length equal to 2.25 m and an eyepiece with a focal length equal to 1.50 cm, by what factor is the angular size of the object changed?

GENERAL PROBLEMS

90 • To focus a camera, the distance between the lens and the image-sensing surface is varied. A wide-angle lens of a camera has a focal length of 28 mm. By how much must the lens move to change from focusing on an object at infinity to an object at a distance of 5.00 m in front of the camera?

91 • A converging lens that has a focal length equal to 10 cm is used to obtain an image that is twice the height of the object. Find the object and image distances if (*a*) the image is to be upright and (*b*) the image is to be inverted. Draw a ray diagram for each case.

92 •• You are given two converging lenses that have focal lengths of 75 mm and 25 mm. (*a*) Show how the lenses should be arranged to form a telescope. State which lens to use as the objective, which lens to use as the eyepiece, how far apart to place the lenses, and what angular magnification you expect. (*b*) Draw a ray diagram to show how rays from a distant object are refracted by the two lenses.

93 •• (*a*) Show how the same two lenses in Problem 92 should be arranged to form a compound microscope that has a tube length of 160 mm. State which lens to use as the objective, which lens to use as the eyepiece, how far apart to place the lenses, and what

overall magnification you expect to get, assuming the user has a near point of 25 cm. (*b*) Draw a ray diagram to show how rays from a close object are refracted by the lenses. SSM

94 •• **CONTEXT-RICH** On a vacation, you are scuba diving and using a diving mask that has a face plate that bulges outward with a radius of curvature of 0.50 m. As a result, a convex spherical surface exists between the water and the air in the mask. A fish swims by you 2.5 m in front of your mask. (*a*) How far in front of the mask does the fish appear to be? (*b*) What is the lateral magnification of the image of the fish?

95 •• A 35-mm digital camera has a rectangular array of CCDs (light sensors) that is 24 mm by 36 mm. It is used to take a picture of a person 175 cm tall so that the image just fills the height (24 mm) of the CCD array. How far should the person stand from the camera if the focal length of the lens is 50 mm? SSM

96 •• A 35-mm camera that has interchangeable lenses is used to take a picture of a hawk that has a wingspan of 2.0 m. The hawk is 30 m away. What would be the ideal focal length of the lens used so that the image of the wings just fills the width of the light-sensitive area of the camera, which is 36 mm?

97 •• An object is placed 12.0 cm in front of a lens that has a focal length equal to 10.0 cm. A second lens that has a focal length equal to 12.5 cm is placed 20.0 cm in back of the first lens. (*a*) Find the position of the final image. (*b*) What is the magnification of the image? (*c*) Sketch a ray diagram showing the final image.

98 •• (*a*) Show that if f_a is the focal length of a thin lens in air, its focal length in water is given by $f_w = -(n_w/n_a)(n - n_a)/(n - n_w)f_a$, where n_w is the index of refraction of water, n is that of the lens material and n_a is that of air. (*b*) Calculate the focal length in air and in water of a double concave lens that has an index of refraction of 1.50 and radii of magnitudes 30 cm and 35 cm.

99 •• While parked in your car, you see a jogger in your rear view mirror, which is convex and has a radius of curvature whose magnitude is equal to 2.00 m. The jogger is 5.00 m from the mirror and is approaching at 3.50 m/s. How fast is the image of the jogger moving relative to the mirror?

100 •• A 2.00-cm-thick layer of water ($n = 1.33$) floats on top of a 4.00-cm-thick layer of carbon tetrachloride ($n = 1.46$) in a tank. How far below the top surface of the water does the bottom of the tank appear, according to an observer looking down from above at normal incidence?

101 ••• An object is 15.0 cm in front of a thin converging lens that has a focal length equal to 10.0 cm. A concave mirror that has a radius equal to 10.0 cm is 25.0 cm in back of the lens. (*a*) Find the position of the final image formed by the mirror–lens combination. (*b*) Is the image real or virtual? Is the image upright or inverted? (*c*) On a diagram, show where your eye must be to see this image. SSM

102 ••• When a bright light source is placed 30 cm in front of a lens, there is an upright image 7.5 cm from the lens. There is also a faint inverted image 6.0 cm from the lens on the incident-light side due to reflection from the front surface of the lens. When the lens is turned around, this weaker, inverted image is 10 cm in front of the lens. Find the index of refraction of the lens.

103 ••• A concave mirror that has a radius of curvature equal to 50.0 cm is oriented with its axis vertical. The mirror is filled with water that has an index of refraction equal to 1.33 and a maximum depth of 1.00 cm. At what height above the vertex of the mirror must an object be placed so that its image is at the same position as the object?

104 ••• The concave side of a lens has a radius of curvature that has a magnitude equal to 17.0 cm, and the convex side of the lens has a radius of curvature that has a magnitude equal to 8.00 cm. The focal length of the lens in air is 27.5 cm. When the lens is placed in a liquid that has an unknown index of refraction, the focal length increases to 109 cm. What is the index of refraction of the liquid?

105 ••• A solid glass ball of radius 10.0 cm has an index of refraction equal to 1.500. The right half of the ball is silvered so that it acts as a concave mirror (Figure 32-63). Find the position of the final image formed for an object located at (*a*) 40.0 cm and (*b*) 30.0 cm to the left of the center of the ball.

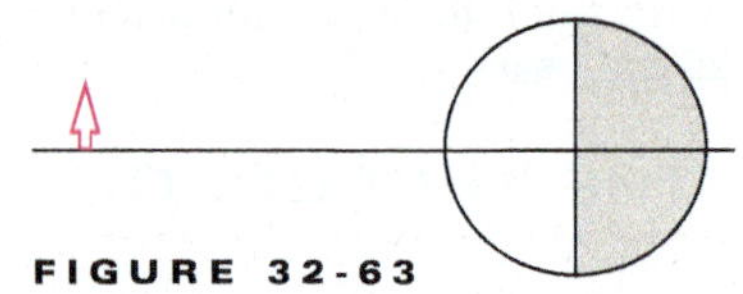

FIGURE 32-63
Problem 105

106 ••• (*a*) Show that a small change dn in the index of refraction of a lens material produces a small change in the focal length df given approximately by $df/f = -dn/(n - n_{air})$. (*b*) Use this result to estimate the focal length of a thin lens for blue light, for which $n = 1.530$, if the focal length for red light, for which $n = 1.470$, is 20.0 cm.

107 ••• The lateral magnification of a spherical mirror or a thin lens is given by $m = -s'/s$. Show that for objects of small horizontal extent, the longitudinal magnification is approximately $-m^2$. *Hint: Show that* $ds'/ds = -s'^2/s^2$. SSM

CHAPTER 33

WHITE LIGHT IS REFLECTED OFF A SOAP BUBBLE. WHEN LIGHT OF ONE WAVELENGTH IS INCIDENT ON A THIN SOAP-AND-WATER FILM, LIGHT IS REFLECTED FROM BOTH THE FRONT AND THE BACK SURFACES OF THE FILM. IF THE ORDER OF MAGNITUDE OF THE THICKNESS OF THE FILM IS ONE WAVELENGTH OF THE LIGHT, THE TWO REFLECTED LIGHT WAVES INTERFERE. IF THE TWO REFLECTED WAVES ARE 180° OUT OF PHASE, THE REFLECTED WAVES INTERFERE DESTRUCTIVELY, SO THE NET RESULT IS THAT NO LIGHT IS REFLECTED. IF WHITE LIGHT, WHICH CONTAINS A CONTINUUM OF WAVELENGTHS, IS INCIDENT ON THE THIN FILM, THEN THE REFLECTED WAVES WILL INTERFERE DESTRUCTIVELY ONLY FOR CERTAIN WAVELENGTHS, AND FOR OTHER WAVELENGTHS THEY WILL INTERFERE CONSTRUCTIVELY. THIS PROCESS PRODUCES THE COLORED FRINGES THAT YOU SEE IN THE SOAP BUBBLE. *(Aaron Haupt/ Photo Researchers.)*

Interference and Diffraction

? Have you ever wondered if the phenomenon that produces the bands that you see in the light reflected off a soap bubble has any practical applications? (See Example 33-2.)

Interference and diffraction are the important phenomena that distinguish waves from particles.* Interference is the formation of a lasting intensity pattern by two or more waves that superpose in space. Diffraction is the bending of waves around corners that occurs when a portion of a wavefront is cut off by a barrier or obstacle.

In this chapter, we will see how the pattern of the resulting wave can be calculated by treating each point on the original wavefront as a point source, according to Huygens's principle, and calculating the interference pattern resulting from these sources.

* Before you study this chapter, you may wish to review Chapter 15 and Chapter 16, where the general topics of interference and diffraction of waves are first discussed.

33-1 PHASE DIFFERENCE AND COHERENCE

When two harmonic sinusoidal waves of the same frequency and wavelength but of different phase combine, the resultant wave is a harmonic wave whose amplitude depends on the phase difference. If the phase difference is zero, or an integer multiplied by 360°, the waves are in phase and interfere constructively. The resultant amplitude equals the sum of the two individual amplitudes, and the intensity (which is proportional to the square of the amplitude) is maximum. (If the amplitudes are equal and the waves are in phase, the intensity is four times that of either individual wave.) If the phase difference is 180° or any odd integer multiplied by 180°, the waves are out of phase and interfere destructively. The resultant amplitude is then the difference between the two individual amplitudes, and the intensity is a minimum. (If the amplitudes are equal and the waves are 180° out of phase, the intensity is zero.)

A phase difference between two waves is often the result of a difference in path lengths. When a light wave reflects from a thin transparent film, such as a soap bubble, the reflected light is a superposition of the light reflected from the front surface of the film and the light reflected from the back surface of the film. The additional distance traveled by the light reflected from the back surface is called the path-length difference between the two reflected waves. A path-length difference of one wavelength produces a phase difference of 360°, which is equivalent to no phase difference at all. A path-length difference of one-half wavelength produces a 180° phase difference. In general, a path-length difference of Δr contributes a phase difference δ given by

$$\delta = \frac{\Delta r}{\lambda}2\pi = \frac{\Delta r}{\lambda}360° \qquad 33\text{-}1$$

PHASE DIFFERENCE DUE TO A PATH-LENGTH DIFFERENCE

Example 33-1 Phase Difference

(*a*) What is the minimum path-length difference that will produce a phase difference of 180° for light of wavelength 800 nm? (*b*) What phase difference will that path-length difference produce in light of wavelength 700 nm?

PICTURE The phase difference is to 360° as the path-length difference is to the wavelength.

SOLVE

(*a*) The phase difference δ is to 360° as the path-length difference Δr is to the wavelength λ. We know that $\lambda = 800$ nm and $\delta = 180°$:

$$\frac{\delta}{360°} = \frac{\Delta r}{\lambda}$$

$$\Delta r = \frac{\delta}{360°}\lambda = \frac{180°}{360°}(800\text{ nm}) = \boxed{400\text{ nm}}$$

(*b*) Set $\lambda = 700$ nm, $\Delta r = 400$ nm, and solve for δ:

$$\delta = \frac{\Delta r}{\lambda}360° = \frac{400\text{ nm}}{700\text{ nm}}360° = \boxed{206° = 3.59\text{ rad}}$$

CHECK The Part-(*b*) result is somewhat larger than 180°. This result is expected because 400 nm is longer than half of the 700-nm wavelength.

Another cause of phase difference is the 180° phase change a wave sometimes undergoes upon reflection from a surface. This phase change is analogous to the inversion of a pulse on a string when it reflects from a point where the density suddenly increases, such as when a light string is attached to a heavier string or rope. The inversion of the reflected pulse is equivalent to a phase change of 180° for a sinusoidal wave (which can be thought of as a series of pulses). When light traveling in air strikes the surface of a medium in which light travels more slowly, such as glass or water, there is a 180° phase change in the reflected light.

When light is traveling in the liquid wall of a soap bubble, there is no phase change in the light reflected from the surface between the liquid and the air. This situation is analogous to the reflection without inversion of a pulse on a heavy string at a point where the heavy string is attached to a lighter string.

> If light traveling in one medium strikes the surface of a medium in which light travels more slowly, there is a 180° phase change in the reflected light.
>
> PHASE DIFFERENCE DUE TO REFLECTION

As we saw in Chapter 16, interference of waves is observed when two or more coherent waves overlap. Interference of overlapping waves from two sources is not observed unless the sources are coherent. Because the light from each source is usually the result of millions of atoms radiating independently, the phase difference between the waves from such sources fluctuates randomly many times per second, so two light sources are usually not coherent. Coherence in optics is often achieved by splitting the light beam from a single source into two or more beams that can then be combined to produce an interference pattern. The light beam can be split by reflecting the light from the two surfaces of a thin film (Section 33-2), by diffracting the beam through two small openings or slits in an opaque barrier (Section 33-3), or by using a single point source and its image in a plane mirror for the two sources (Section 33-3). Today, lasers are the most important sources of coherent light in the laboratory.

Light from an ideal monochromatic source is a sinusoidal wave of infinite duration, and light from certain lasers approaches this ideal. However, light from conventional *monochromatic* sources, such as gas discharge tubes designed for this purpose, consists of packets of sinusoidal light that are only a few million wavelengths long. The light from such a source consists of many such packets, each approximately the same length. The packets have essentially the same wavelength, but the packets differ in phase in a random manner. The length of the individual packets is called the **coherence length** of the light, and the time it takes one of the packets to pass a point in space is the **coherence time.** The light emitted by a gas discharge tube designed to produce monochromatic light has a coherence length of only a few millimeters. By comparison, some highly stable lasers produce light that has a coherence length many kilometers long.

33-2 INTERFERENCE IN THIN FILMS

You have probably noticed the colored bands in a soap bubble or in the film on the surface of oily water. These bands are due to the interference of light reflected from the top and bottom surfaces of the film. The different colors arise because of variations in the thickness of the film, causing interference for different wavelengths at different points.

When waves traveling in a medium cross a surface where the wave speed changes, part of the wave is reflected and part is transmitted. In addition, the reflected wave undergoes a 180° phase shift upon reflection if the transmitted wave travels at a slower speed than do the incident and reflected waves. (This 180° phase shift is established for waves on a string in Section 15-4 of Chapter 15.) The reflected wave does not undergo a phase shift upon reflection if the transmitted wave travels at a faster speed than do the incident and reflected waves.

Consider a thin film of water (such as a small section of a soap bubble) of uniform thickness viewed at small angles with the normal, as shown in Figure 33-1. Part of the light is reflected from the upper air–water interface where it undergoes a 180° phase change. Most of the light enters the film and part of it is reflected by the bottom water–air interface. There is no phase change in this reflected light. If the light is nearly perpendicular to the surfaces, both the light reflected from the top surface and the light reflected from the bottom surface can enter the eye. The path-length

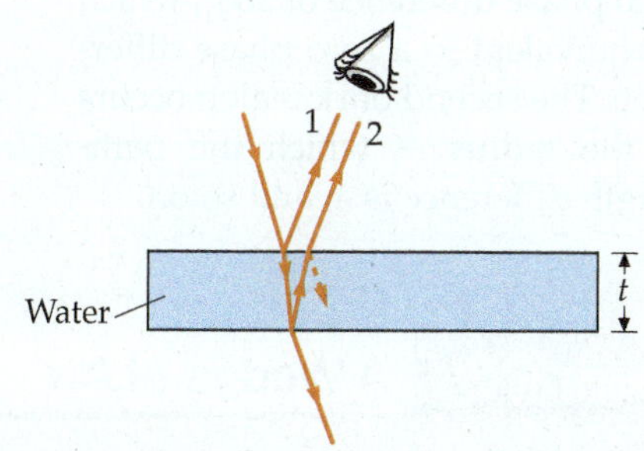

FIGURE 33-1 Light rays reflected from the top and bottom surfaces of a thin film are coherent because both rays come from the same source. If the light is incident almost normally, the two reflected rays will be very close to each other and will produce interference.

difference between these two rays is $2t$, where t is the thickness of the film. This path-length difference produces a phase difference of $(2t/\lambda')360°$, where $\lambda' = \lambda/n$ is the wavelength of the light in the film, n is the index of refraction of the film, and λ is the wavelength of the light in vacuum. The total phase difference between the two rays is thus 180° plus the phase difference due to the path-length difference. Destructive interference occurs when the path-length difference $2t$ is zero or a whole number of wavelengths λ' (in the film). Constructive interference occurs when the path-length difference is an odd number of half-wavelengths.

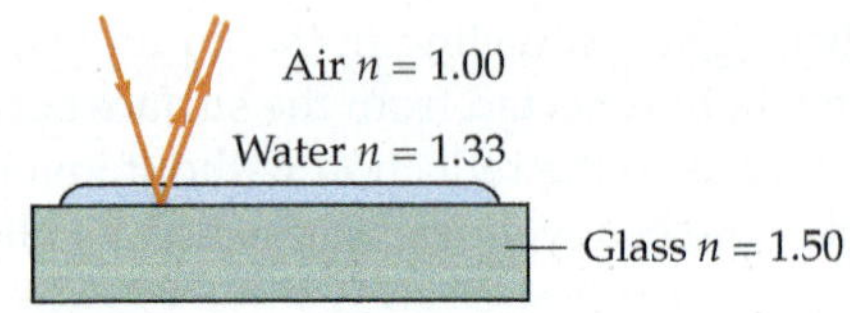

FIGURE 33-2 The interference of light reflected from a thin film of water resting on a glass surface. In this case, both rays undergo a change in phase of 180° upon reflection.

When a thin film of water lies on a glass surface, as in Figure 33-2, the ray that reflects from the lower water–glass interface also undergoes a 180° phase change, because the index of refraction of glass (approximately 1.50) is greater than that of water (approximately 1.33). Thus, both the rays shown in the figure have undergone a 180° phase change upon reflection. The phase difference δ between these rays is due solely to the path-length difference and is given by $\delta = (2t/\lambda')360°$.

When a thin film of varying thickness is viewed with monochromatic light, such as the yellow light from a sodium lamp, alternating bright and dark bands or lines, called **interference fringes,** are observed. The distance between a bright fringe and a dark fringe is that distance over which the film's thickness t changes so that the path-length difference $2t$ changes by $\lambda'/2$. Figure 33-3*a* shows the interference pattern observed when light is reflected from an air film between a spherical glass surface and a plane glass surface in contact. These circular interference fringes are known as **Newton's rings.** Typical rays reflected at the top and bottom of the air film are shown in Figure 33-3*b*. Near the point of contact of the surfaces, where the path-length difference between the ray reflected from the upper glass–air interface and the ray reflected from the lower air–glass interface is approximately zero (it is small compared with the wavelength of light) the interference is destructive because of the 180° phase shift of the ray reflected from the lower air–glass interface. This central region in Figure 33-3*a* is therefore dark. The first bright fringe occurs at the radius at which the path-length difference is $\lambda/2$, which contributes a phase difference of 180°. This adds to the phase shift due to reflection to produce a total phase difference of 360°, which is equivalent to a zero phase difference. The second dark region occurs at the radius at which the path-length difference is λ, and so on.

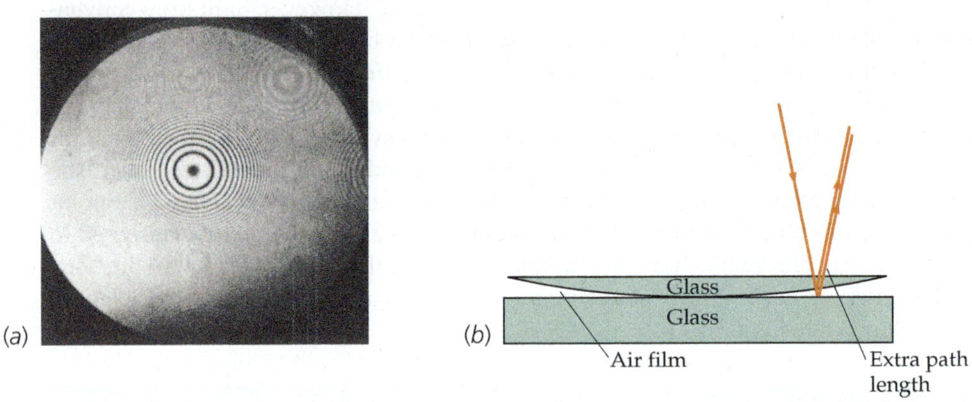

FIGURE 33-3 (*a*) Newton's rings observed when light is reflected from a thin film of air between a plane glass surface and a spherical glass surface. At the center, the thickness of the air film is negligible and the interference is destructive because of the 180° phase change of one of the rays upon reflection. (*b*) Glass surfaces for the observation of Newton's rings shown in Figure 33-3*a*. The thin film in this case is the film of air between the glass surfaces. *(Courtesy of Bausch & Lomb.)*

Example 33-2 A Wedge of Air

A wedge-shaped film of air is made by placing a small slip of paper between the edges of two flat pieces of glass, as shown in Figure 33-4. Light of wavelength 500 nm is incident normally on the glass, and interference fringes are observed by reflection. If the angle θ made by the plates is 3.0×10^{-4} rad (0.017°), how many dark interference fringes per centimeter are observed?

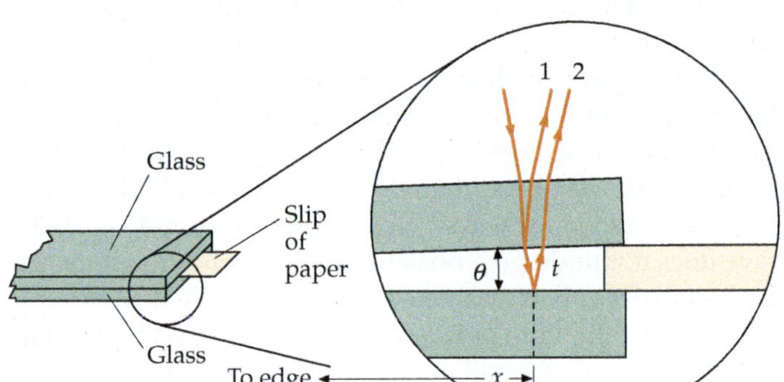

FIGURE 33-4 The angle θ, which is less than 0.02°, is exaggerated. The incoming and outgoing rays are virtually perpendicular to all air–glass interfaces.

PICTURE We find the number of fringes per centimeter by finding the horizontal distance x to the mth fringe and solving for m/x. Because the ray reflected from the bottom plate undergoes a 180° phase shift, the point of contact (where the path-length difference is zero) will be dark. The mth dark fringe after the contact point occurs when $2t = m\lambda'$, where $\lambda' = \lambda$ is the wavelength in the air film, and t is the plate separation at x, as shown in Figure 33-4. Because the angle θ is small, we can use the small-angle approximation $\theta \approx \tan\theta = t/x$.

SOLVE

1. The mth dark fringe from the contact point occurs when the path-length difference $2t$ equals m wavelengths:
$$2t = m\lambda' = m\lambda$$
$$m = \frac{2t}{\lambda}$$
2. The thickness t is related to the angle θ:
$$\theta = \frac{t}{x}$$
3. Substitute $t = x\theta$ into the equation for m:
$$m = \frac{2x\theta}{\lambda}$$
4. Calculate m/x:
$$\frac{m}{x} = \frac{2\theta}{\lambda} = \frac{2(3.0 \times 10^{-4})}{5.0 \times 10^{-7}\ \text{m}} = 1200\ \text{m}^{-1} = \boxed{12\ \text{cm}^{-1}}$$

CHECK The expression for the number of dark fringes per unit length in step 4 shows that the number per centimeter would decrease if light of a longer wavelength is used. This result is as expected.

TAKING IT FURTHER We observe 12 dark fringes per centimeter. In practice, the number of fringes per centimeter, which is easy to count, can be used to determine the angle. Note that if the angle of the wedge is increased, the fringes become more closely spaced.

PRACTICE PROBLEM 33-1 How many dark fringes per centimeter are observed if light of wavelength 650 nm is used?

Figure 33-5*a* shows interference fringes produced by a wedge-shaped air film between two flat glass plates, as in Example 33-2. Plates that produce straight fringes, such as those in Figure 33-5*a*, are said to be **optically flat.** To be optically flat, a surface must be flat to within a small fraction of a wavelength. A similar wedge-shaped air film formed by two ordinary glass plates yields the irregular fringe pattern in Figure 33-5*b*, which indicates that these plates are not optically flat.

One application of interference effects in thin films is in nonreflecting lenses, which are made by coating the surface of a lens with a thin film of a material that has an index of refraction equal to approximately 1.38, which is between the index of refraction of glass and that of air. The intensities of the light reflected from the top and bottom surfaces of the film are approximately equal, and because the reflected rays undergo a 180° phase change at both surfaces there is no phase difference due to reflection between the two rays. The thickness of the film is chosen to be $\frac{1}{4}\lambda' = \frac{1}{4}\lambda n$, where λ is the wavelength, in vacuum, that is in the middle of the visible spectrum, so that there is a phase change of 180° due to the path-length difference of $\lambda'/2$ for light of normal incidence. Reflection from the coated surface is thus minimized, which means that transmission through the surface is maximized.

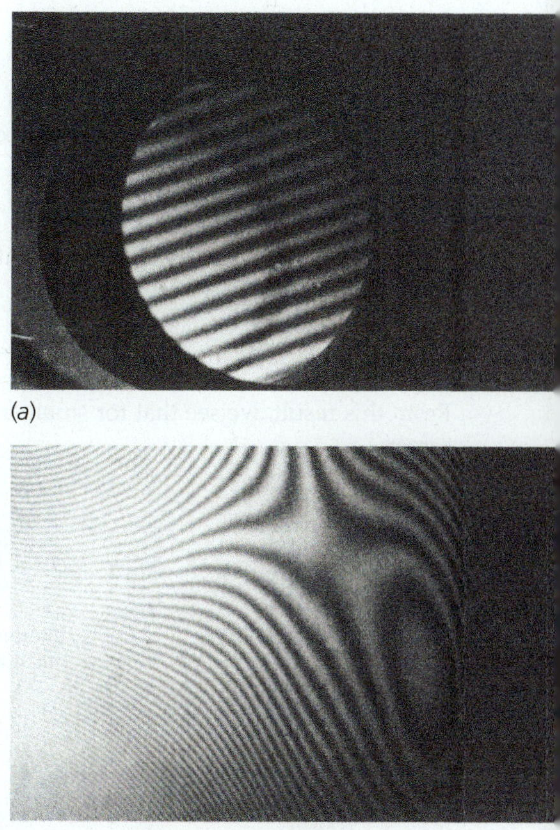

(*a*)

(*b*)

FIGURE 33-5 (*a*) Straight-line fringes from a wedge-shaped film of air, like that shown in Figure 33-4. The straightness of the fringes indicates that the glass plates are optically flat. (*b*) Fringes from a wedge-shaped film of air between glass plates that are not optically flat. *(Courtesy T. A. Wiggins.)*

33-3 TWO-SLIT INTERFERENCE PATTERN

Interference patterns of light from two or more sources can be observed only if the sources are coherent. The interference in thin films discussed previously can be observed because the two beams come from the same light source but are separated by reflection. In Thomas Young's famous 1801 experiment, in which he demonstrated the wave nature of light, two coherent light sources are produced by illuminating two very narrow parallel slits using a single light source. We saw in Chapter 15 that when a wave encounters a barrier that has a very small opening, the opening acts as a point source of waves (Figure 33-6).

During Young's experiment, diffraction causes each slit to act as a line source (which is equivalent to a point source in two dimensions). The interference pattern is observed on a screen far from the slits (Figure 33-7*a*). At very large distances from the slits, the lines from the two slits to some point P on the screen are approximately parallel, and the path-length difference is approximately $d\sin\theta$, where d is the separation of the slits, as shown in Figure 33-7*b*. When the path-length difference is equal to an integral number of wavelengths, the interference is constructive. We thus have interference maxima at an angle θ_m given by

$$d\sin\theta_m = m\lambda \qquad m = 0, 1, 2, \ldots \tag{33-2}$$

TWO-SLIT INTERFERENCE MAXIMA

FIGURE 33-6 Plane water waves in a ripple tank encountering a barrier that has a small opening. The waves to the right of the barrier are circular waves that are concentric about the opening, just as if there were a point source at the opening. *(From PSSC Physics, 2nd Edition, 1965. D. C. Heath & Co. and Education Development Center, Newton MA.)*

where m is called the **order number.** The interference minima occur at

$$d\sin\theta_m = \left(m - \tfrac{1}{2}\right)\lambda \qquad m = 1, 2, 3, \ldots \tag{33-3}$$

TWO-SLIT INTERFERENCE MINIMA

The phase difference δ at a point P is related to the path-length difference $d\sin\theta$ by

$$\delta = \frac{\Delta r}{\lambda}2\pi = \frac{d\sin\theta}{\lambda}2\pi \tag{33-4}$$

We can relate the distance y_m measured along the screen from the central point to the mth bright fringe (see Figure 33-7*b*) to the distance L from the slits to the screen:

$$\tan\theta_m = \frac{y_m}{L}$$

For small angles, $\tan\theta \approx \sin\theta$. Substituting y_m/L for $\sin\theta_m$ in Equation 33-2 and solving for y_m gives

$$y_m = m\frac{\lambda L}{d} \tag{33-5}$$

From this result, we see that for small angles the fringes are equally spaced on the screen.

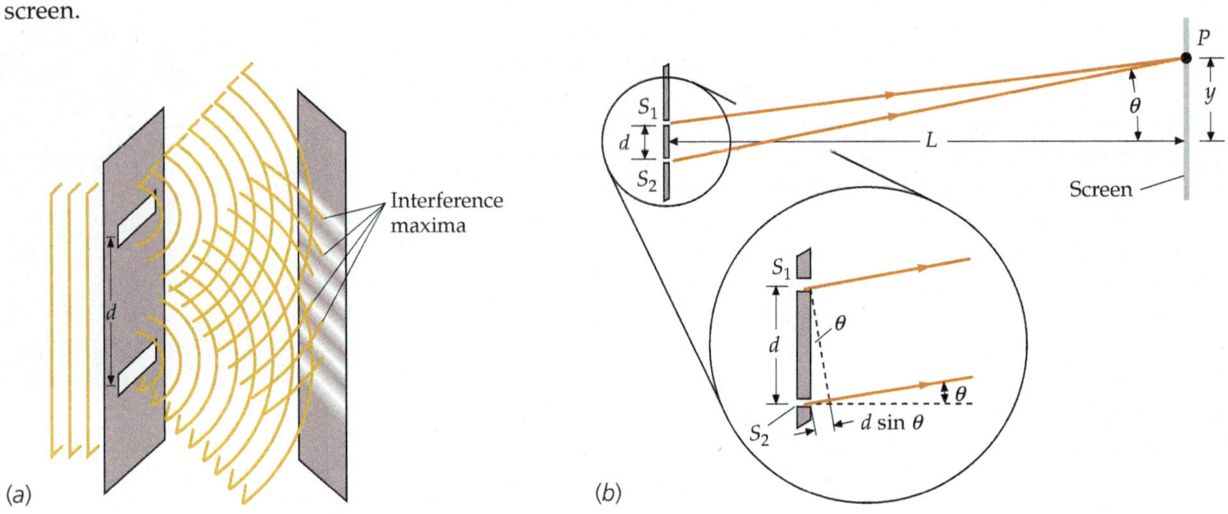

FIGURE 33-7 (*a*) Two slits act as coherent sources of light for the observation of interference in Young's experiment. Cylindrical waves from the slits overlap and produce an interference pattern on a screen. (*b*) Geometry for relating the distance y measured along the screen to L and θ. When the screen is very far away compared with the slit separation, the rays from the slits to a point on the screen are approximately parallel, and the path-length difference between the two rays is $d\sin\theta$.

Example 33-3 Fringe Spacing from Slit Spacing — *Try It Yourself*

Two narrow slits separated by 1.50 mm are illuminated by yellow light from a sodium lamp that has a wavelength equal to 589 nm. Find the spacing of the bright fringes observed on a screen 3.00 m away.

PICTURE The distance y_m measured along the screen to the mth bright fringe is given by Equation 33-5, where $L = 3.00$ m, $d = 1.50$ mm, and $\lambda = 589$ nm.

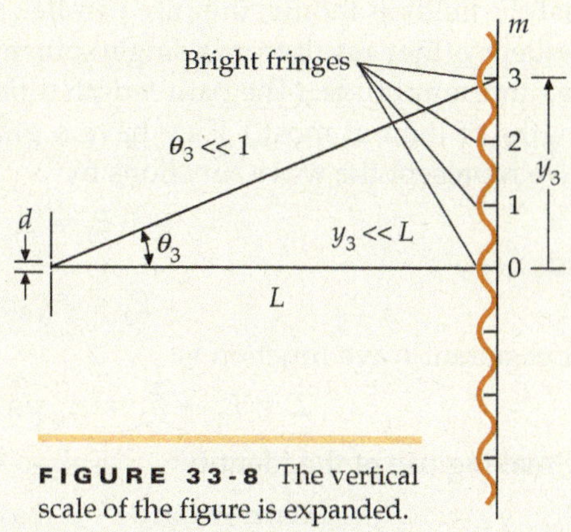

FIGURE 33-8 The vertical scale of the figure is expanded.

SOLVE

Cover the column to the right and try these on your own before looking at the answers.

Steps	Answers
1. Make a sketch of the situation (Figure 33-8).	
2. The fringe spacing is the distance between the mth bright fringe and the $(m + 1)$th bright fringe. Using the sketch, obtain an expression for the spacing between fringes.	fringe spacing $= y_{m+1} - y_m$
3. Apply Equation 33-5 to the mth and $(m + 1)$th fringe.	$y_m = m\dfrac{\lambda L}{d}$ and $y_{m+1} = (m + 1)\dfrac{\lambda L}{d}$
4. Substitute into the step-2 result and simplify.	$y_{m+1} - y_m = \dfrac{\lambda L}{d}$
5. Substitute into the step-4 result and solve for the fringe spacing.	fringe spacing $=$ $\boxed{1.18 \text{ mm}}$

TAKING IT FURTHER The fringes are uniformly spaced only to the degree that the small-angle approximation is valid, that is, to the degree that $\lambda/d \ll 1$. In this example, $\lambda/d = (589 \text{ nm})/(1.50 \text{ mm}) \approx 0.0004$.

See **Math Tutorial** *for more information on* **Trigonometry**

Example 33-4 How Many Fringes? — *Conceptual*

Two narrow slits are illuminated by monochromatic light. If the distance between the slits is equal to 2.75 wavelengths, what is the maximum number of bright fringes that can be seen on a screen? (*a*) 1, (*b*) 2, (*c*) 3, (*d*) 4, (*e*) 5, (*f*) 6 or more

PICTURE A bright fringe (constructive interference) exists at points on the screen for which the distance to the two slits differs by an integer multiplied by the wavelength. However, the maximum difference in distance possible is equal to the distance between the two slits.

SOLVE

1. Find the maximum difference in distance from points on the screen to the two slits:	At all points on the screen, the difference in distance from the two slits is 2.75 wavelengths or less.
2. A bright fringe (constructive interference) exists at points on the screen for which the distance to the two slits differs by an integer multiplied by the wavelength:	Bright fringes exist on the screen at places where the difference in distance to the slits is 2 wavelengths, 1 wavelength, or zero wavelengths.
3. Count up the bright fringes. There is the central maximum and two on either side of the central maximum:	$\boxed{(e)\ 5}$

CONCEPT CHECK 33-1

What is the maximum number of dark fringes that can be seen on a screen?

CALCULATION OF INTENSITY

To calculate the intensity of the light on the screen at a general point P, we need to add two harmonic wave functions that differ in phase.* The wave functions for electromagnetic waves are the electric field vectors. Let E_1 be the electric field at some point P on the screen due to the waves from slit 1, and let E_2 be the electric field at that point due to waves from slit 2. Because the angles of interest are small, we can treat the fields as though they are parallel. Both electric fields oscillate with the same frequency (they result from a single source that illuminates both slits) and they have the same amplitude. (The path-length difference is only of the order of a few wavelengths of light at most.) They have a phase difference δ given by Equation 33-4. If we represent the wave functions by

$$E_1 = A_0 \sin \omega t$$

and

$$E_2 = A_0 \sin(\omega t + \delta)$$

the resultant wave function is

$$E = E_1 + E_2 = A_0 \sin \omega t + A_0 \sin(\omega t + \delta) \qquad \text{33-6}$$

By making use of the identity

$$\sin \alpha + \sin \beta = 2 \cos \tfrac{1}{2}(\alpha - \beta) \sin \tfrac{1}{2}(\alpha + \beta)$$

the resultant wave function is given by

$$E = \left[2A_0 \cos \tfrac{1}{2}\delta\right] \sin\left(\omega t + \tfrac{1}{2}\delta\right) \qquad \text{33-7}$$

The amplitude of the resultant wave is thus $2A_0 \cos \frac{1}{2}\delta$. It has its maximum value of $2A_0$ when the waves are in phase and is zero when they are 180° out of phase. Because the intensity is proportional to the square of the amplitude, the intensity at any point P is

$$I = 4I_0 \cos^2 \tfrac{1}{2}\delta \qquad \text{33-8}$$

INTENSITY IN TERMS OF PHASE DIFFERENCE

where I_0 is the intensity of the light reaching the screen from either slit separately. The phase angle δ is related to the position on the screen by $\delta = (d \sin\theta/\lambda)2\pi$ (Equation 33-4).

Figure 33-9*a* shows the intensity pattern as seen on a screen. A graph of the intensity as a function of $\sin\theta$ is shown in Figure 33-9*b*. For small θ, this graph is equivalent to a plot of intensity versus y (because $y = L \tan\theta \approx L \sin\theta$). The intensity I_0 is the intensity from each slit separately. The dashed line in Figure 33-9*b* shows the average intensity $2I_0$, which is the result of averaging over a distance containing many interference maxima and minima. This is the intensity that would arise from the two sources if they acted independently without interference, that is, if they were not coherent. Then, the phase difference between the two sources would fluctuate randomly, so that only the average intensity would be observed.

Figure 33-10 shows another method of producing the two-slit interference pattern, an arrangement known as **Lloyd's mirror.** A monochromatic horizontal line source is placed at a distance $\frac{1}{2}d$ above the plane of a mirror. Light striking the screen directly from the source interferes with the light that is reflected from the mirror. The reflected light can be considered to come from the virtual image of the line source formed by the mirror. Because of the 180° change in phase upon reflection at the mirror, the interference pattern is that of two coherent line sources that differ in phase by 180°. The pattern is the same as that shown in Figure 33-9 for two slits, except that the maxima and minima are interchanged. Constructive interference occurs at points for which the path-length difference is a half-wavelength or any odd number of half-wavelengths. At those points, the 180° phase difference due to the path-length difference combines with the 180° phase difference of the sources to produce constructive interference.

(*a*)

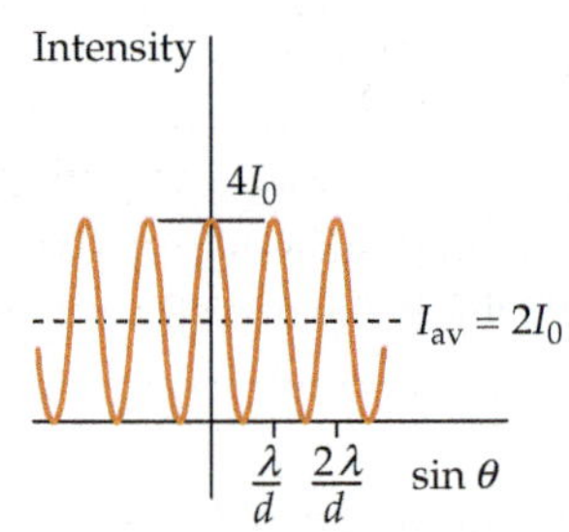

(*b*)

FIGURE 33-9 (*a*) The interference pattern observed on a screen far away from the two slits shown in Figure 33-7. (*b*) Plot of intensity versus $\sin\theta$. The maximum intensity is $4I_0$, where I_0 is the intensity due to each slit separately. The average intensity (dashed line) is $2I_0$. *(Courtesy of Michael Cagnet.)*

* We did this in Chapter 16 where we first discussed the superposition of two waves.

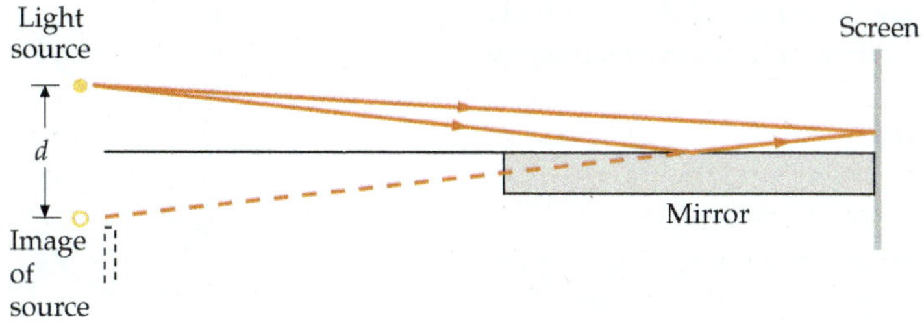

FIGURE 33-10 Lloyd's mirror for producing a two-slit interference pattern. The two sources (the source and its virtual image) are coherent and are 180° out of phase.

> **PRACTICE PROBLEM 33-2**
>
> A point source of light ($\lambda = 589$ nm) is placed 0.40 mm above the surface of a glass mirror. Interference fringes are observed on a screen 6.0 m away, and the interference is between the light reflected from the front surface of the glass and the light traveling from the source directly to the screen. Find the spacing of the fringes on the screen.

The physics of Lloyd's mirror was used in the early days of radio astronomy to determine the location of distant radio sources on the celestial sphere. A radio-wave receiver was placed on a cliff overlooking the sea, and the surface of the sea served as the mirror.

33-4 DIFFRACTION PATTERN OF A SINGLE SLIT

In our discussion of the interference patterns produced by two or more slits, we assumed that the slits were very narrow so that we could consider the slits to be line sources of cylindrical waves, which in our two-dimensional diagrams are point sources of circular waves. We could therefore assume that the value of the intensity due to one slit acting alone was the same (I_0) at any point P on the screen, independent of the angle θ made between the ray to point P and the normal line between the slit and the screen. When the slit is not narrow, the intensity on a screen far away is not independent of angle but decreases as the angle increases. Consider a slit of width a. Figure 33-11 shows the intensity pattern on a screen far away from the slit of width a as a function of $\sin\theta$. We can see that the intensity is maximum in the forward direction ($\sin\theta = 0$) and decreases to zero at an angle that depends on the slit width a and the wavelength λ.

Most of the light intensity is concentrated in the broad **central diffraction maximum,** although there are minor secondary maxima bands on either side of the central maximum. The first zeroes in the intensity occur at angles specified by

$$\sin\theta_1 = \lambda/a \qquad 33\text{-}9$$

Note that for a given wavelength λ, Equation 33-9 describes how variations in the slit width result in variations in the angular width of the central maximum. If we *increase* the slit width a, the angle θ_1 at which the intensity first becomes zero *decreases*, giving a more narrow central diffraction maximum. Conversely, if we *decrease* the slit width, the angle of the first zero *increases*, giving a wider central diffraction maximum. When a is smaller than λ, then $\sin\theta_1$ would have to exceed 1

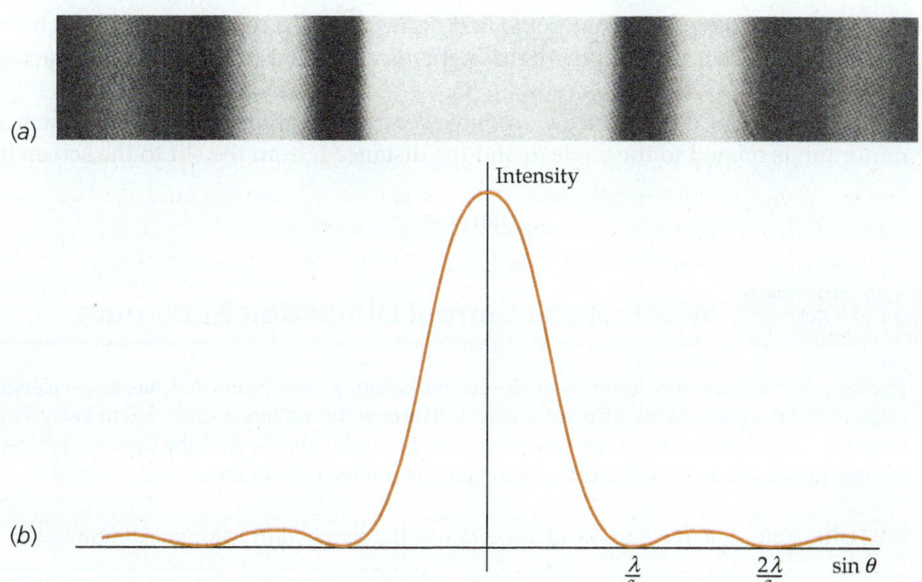

FIGURE 33-11 (*a*) Diffraction pattern of a single slit as observed on a screen far away. (*b*) Plot of intensity versus $\sin\theta$ for the pattern in Figure 33-11*a*. *(Courtesy of Michael Cagnet.)*

to satisfy Equation 33-9. Thus, for a less than λ, there are no points of zero intensity in the pattern, and the slit acts as a line source (a point source in two dimensions) radiating light energy essentially equal in all directions.

Multiplying both sides of Equation 33-9 by $a/2$ gives

$$\tfrac{1}{2}a\sin\theta_1 = \tfrac{1}{2}\lambda \qquad \text{33-10}$$

The quantity $\frac{1}{2}a\sin\theta_1$ is the path-length difference between a light ray leaving the middle of the upper half of the slit and one leaving the middle of the lower half of the slit. We see that the first diffraction *minimum* occurs when these two rays are 180° out of phase, that is, when their path-length difference equals a half-wavelength. We can understand this result by considering each point on a wavefront to be a point source of light in accordance with Huygens's principle. In Figure 33-12, we have placed a line of dots on the wavefront at the slit to represent these point sources schematically. Suppose, for example, that we have 100 such dots and that we look at an angle θ_1 for which $a\sin\theta_1 = \lambda$. Let us consider the slit to be divided into two halves, with sources 1 through 50 in the upper half and sources 51 through 100 in the lower half. When the path-length difference between the middle of the upper half and the middle of the lower half of the slit equals a half-wavelength, the path-length difference between source 1 (the first source in the upper half) and source 51 (the first source in the lower half) is also $\frac{1}{2}\lambda$. The waves from those two sources will be out of phase by 180° and will thus cancel. Similarly, waves from the second source in each region (source 2 and source 52) will cancel. Continuing this argument, we can see that the waves from each pair of sources separated by $a/2$ will cancel. Thus, there will be no light energy at that angle. We can extend this argument to the second and third minima in the diffraction pattern of Figure 33-11. At an angle θ_2 where $a\sin\theta_2 = 2\lambda$, we can divide the slit into four regions, two regions for the top half and two regions for the bottom half. Using this same argument, the light intensity from the top half is zero because of the cancellation of pairs of sources; similarly, the light intensity from the bottom half is zero. The general expression for the points of zero intensity in the diffraction pattern of a single slit is thus

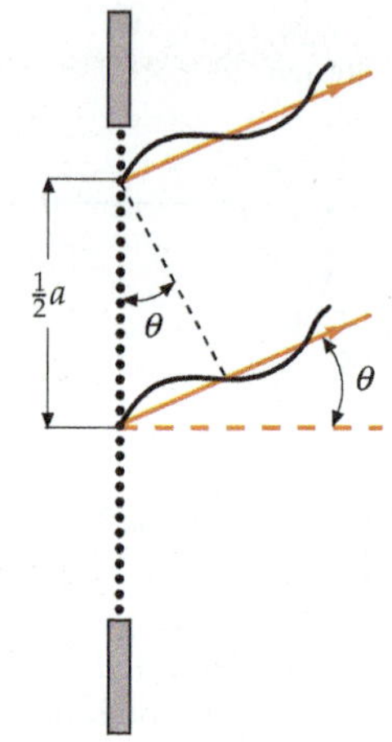

FIGURE 33-12 A single slit is represented by a large number of point sources of equal amplitude. At the first diffraction minimum of a single slit, the waves from each point source in the upper half of the slit are 180° out of phase with the wave from the point source a distance $a/2$ lower in the slit. Thus, the interference from each such pair of point sources is destructive.

$$a\sin\theta_m = m\lambda \qquad m = 1, 2, 3, \ldots \qquad \text{33-11}$$

POINTS OF ZERO INTENSITY FOR A SINGLE-SLIT DIFFRACTION PATTERN

Usually, we are just interested in the first occurrence of a minimum in the light intensity because nearly all of the light energy is contained in the central diffraction maximum.

In Figure 33-13, the distance y_1 from the central maximum to the first diffraction minimum is related to the angle θ_1 and the distance L from the slit to the screen by

$$\tan\theta_1 = \frac{y_1}{L}$$

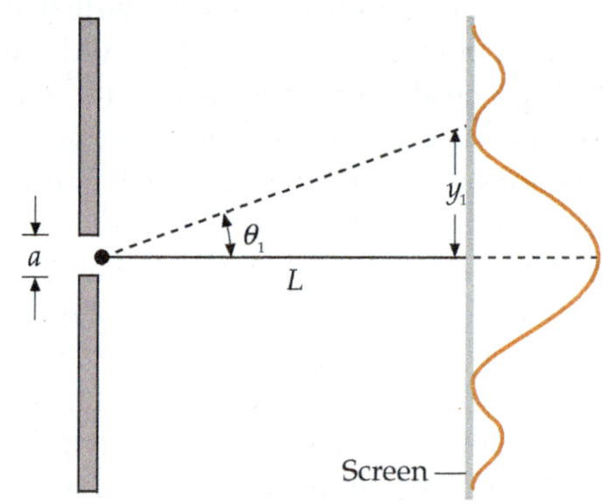

FIGURE 33-13 The distance y_1 measured along the screen from the central maximum to the first diffraction minimum is related to the angle θ_1 by $\tan\theta_1 = y_1/L$, where L is the distance to the screen.

Example 33-5 Width of the Central Diffraction Maximum

During a lecture demonstration of single-slit diffraction, a laser beam that has a wavelength equal to 700 nm passes through a vertical slit 0.20 mm wide and hits a screen 6.0 m away. Find the width of the central diffraction maximum on the screen; that is, find the distance between the first minimum on the left and the first minimum on the right of the central maximum.

PICTURE Referring to Figure 33-13, the width of the central diffraction maximum is $2y_1$.

SOLVE

1. The half-width of the central maximum y_1 is related to the angle θ_1 by: $\qquad \tan\theta_1 = \dfrac{y_1}{L}$

2. The angle θ_1 is related to the slit width a by Equation 33-11:

$$\sin\theta_1 = \lambda/a$$

3. Solve the step-2 result for θ_1, substitute into the step-1 result, and solve for $2y_1$:

$$2y_1 = 2L\tan\theta_1 = 2L\tan\left(\sin^{-1}\frac{\lambda}{a}\right)$$

$$= 2(6.0\text{ m})\tan\left(\sin^{-1}\frac{700\times10^{-9}\text{ m}}{0.00020\text{ m}}\right)$$

$$= 4.2\times10^{-2}\text{ m} = \boxed{4.2\text{ cm}}$$

CHECK Because $\sin\theta_1 = \lambda/a = (700\text{ nm})/(0.20\text{ mm}) = 0.0035$, we can use the small-angle approximation to evaluate $2y_1$. In this approximation, $\sin\theta_1 = \tan\theta_1$, so $\lambda/a = y_1/L$ and $2y_1 = 2L\lambda/a = 2(6.0\text{ m})(700\text{ nm})/(0.20\text{ mm}) = 4.2\text{ cm}$. (This approximate value is in agreement with the exact value to within 0.0006 percent.)

INTERFERENCE–DIFFRACTION PATTERN OF TWO SLITS

When there are two or more slits, the intensity pattern on a screen far away is a combination of the single-slit diffraction pattern of the individual slits and the multiple-slit interference pattern we have studied. Figure 33-14 shows the intensity pattern on a screen far from two slits whose separation d is $10a$, where a is the width of each slit. The pattern is the same as the two-slit pattern that has very narrow slits (Figure 33-11) except that it is modulated by the single-slit diffraction pattern; that is, the intensity due to each slit separately is now not constant but decreases with angle, as shown in Figure 33-14*b*.

Note that the central diffraction maximum in Figure 33-14 has 19 interference maxima—the central interference maximum and 9 maxima on either side. The tenth interference maximum on either side of the central one is at the angle θ_{10}, given by $\sin\theta_{10} = 10\lambda/d = \lambda/a$, because $d = 10a$. This coincides with the position of the first diffraction minimum, so this interference maximum is not seen. At these points, the light from the two slits would be in phase and would interfere constructively, but there is no light coming from either slit because the points are at diffraction minima of each slit. In general, we can see that if $m = d/a$, the mth interference maximum will fall at the first diffraction minimum. Because the mth fringe is not seen, there will be $m - 1$ fringes on each side of the central fringe for a total of N fringes in the central maximum, where N is given by

$$N = 2(m-1) + 1 = 2m - 1 \qquad 33\text{-}12$$

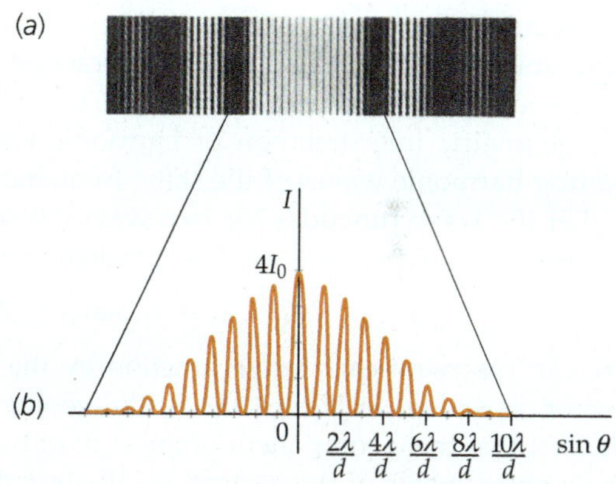

FIGURE 33-14 (*a*) Interference–diffraction pattern for two slits whose separation d is equal to 10 times their width a. The tenth interference maximum on either side of the central interference maximum is missing because it falls at the first diffraction minimum. (*b*) Plot of intensity versus $\sin\theta$ for the central band of the pattern in Figure 33-14*a*. *(Courtesy of Michael Cagnet.)*

Example 33-6 Interference and Diffraction

Two slits that each have a width $a = 0.015$ mm are separated by a distance $d = 0.060$ mm and are illuminated by light of wavelength $\lambda = 650$ nm. How many bright fringes are seen in the central diffraction maximum?

PICTURE We need to find the value of m for which the mth interference maximum coincides with the first diffraction minimum. Then there will be $N = 2m - 1$ fringes in the central maximum.

SOLVE

1. Relate the angle θ_1 of the first diffraction minimum to the slit width a:

$$\sin\theta_1 = \frac{\lambda}{a} \quad \text{(first diffraction minimum)}$$

2. Relate the angle θ_m of the mth interference maxima to the slit separation d:

$$\sin\theta_m = \frac{m\lambda}{d} \quad (m\text{th interference maxima})$$

3. Set the angles equal and solve for m:

$$\frac{m\lambda}{d} = \frac{\lambda}{a}$$

$$m = \frac{d}{a} = \frac{0.060\text{ mm}}{0.015\text{ mm}} = 4.0$$

4. The first diffraction minimum coincides with the fourth bright fringe. Therefore, there are 3 bright fringes visible on either side of the central diffraction maximum. These 6 maxima, plus the central interference maximum, combine for a total of 7 bright fringes in the central diffraction maximum:

$$N = \boxed{7\text{ bright fringes}}$$

* 33-5 USING PHASORS TO ADD HARMONIC WAVES

To calculate the interference pattern produced by three, four, or more coherent light sources and to calculate the diffraction pattern of a single slit, we need to combine several harmonic waves of the same frequency that differ in phase. A simple geometric interpretation of harmonic wave functions leads to a method of adding harmonic waves of the same frequency by geometric construction.

Let the wave functions for two waves at some point be $E_1 = A_1 \sin\alpha$ and $E_2 = A_2 \sin(\alpha + \delta)$, where $\alpha = \omega t$. Our problem is then to find the sum:

$$E_1 + E_2 = A_1 \sin\alpha + A_2 \sin(\alpha + \delta)$$

We can represent each wave function by the y component of a two-dimensional vector, as shown in Figure 33-15. The geometric method of addition is based on the fact that the y component of the sum of two or more vectors equals the sum of the y components of the vectors, as illustrated in the figure. The wave function E_1 is represented by the y component of the vector $\vec{A}_1$. As the time continues on, this vector rotates in the xy plane with angular frequency ω. The vector $\vec{A}_1$ is called a **phasor.** (We encountered phasors in our study of ac circuits in Section 29-5.) The wave function E_2 is the y component of a phasor of magnitude A_2 that makes an angle $\alpha + \delta$ with the x axis. By the laws of vector addition, the sum of the y components of the individual phasors equals the y component of the resultant phasor $\vec{A}$, as shown in Figure 33-15. The y component of the resultant phasor, $A \sin(\alpha + \delta')$, is a harmonic wave function that is the sum of the two original wave functions. That is,

$$A_1 \sin\alpha + A_2 \sin(\alpha + \delta) = A \sin(\alpha + \delta') \qquad 33\text{-}13$$

where A (the amplitude of the resultant wave) and δ' (the phase of the resultant wave relative to the phase of the first wave) are found by adding the phasors representing the waves. As time varies, α varies. The phasors representing the two wave functions and the resultant phasor representing the resultant wave function rotate in space, but their relative positions do not change because they all rotate with the same angular velocity ω.

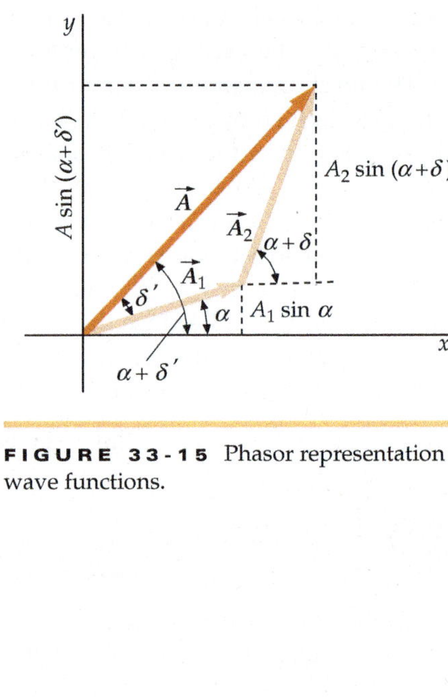

FIGURE 33-15 Phasor representation of wave functions.

Example 33-7 Wave Superposition Using Phasors — *Try It Yourself*

Use the phasor method of addition to derive $E = \left[2A_0 \cos\frac{1}{2}\delta\right] \sin\left(\omega t + \frac{1}{2}\delta\right)$ (Equation 33-7) for the superposition of two waves of the same amplitude.

PICTURE Represent the waves $y_1 = A_0 \sin\alpha$ and $y_2 = A_0 \sin(\alpha + \delta)$ by vectors (phasors) of length A_0 making an angle δ with one another. The resultant wave $y_r = A \sin(\alpha + \delta')$ is represented by the sum of these vectors, which form an isosceles triangle, as shown in Figure 33-16.

SOLVE

Cover the column to the right and try these on your own before looking at the answers.

Steps	Answers
1. Relate δ and δ' using the theorem: "An external angle to a triangle is equal to the sum of the two non-adjacent internal angles."	$\delta' + \delta' = \delta$
2. Solve for δ'.	$\delta' = \frac{1}{2}\delta$
3. Write $\cos\delta'$ in terms of A and A_0.	$\cos\delta' = \dfrac{\frac{1}{2}A}{A_0}$
4. Solve for A in terms of δ.	$A = 2A_0\cos\delta' = 2A_0\cos\frac{1}{2}\delta$
5. Use your results for A and δ' to write the resultant wave function.	$y_r = A\sin(\alpha + \delta') = \boxed{[2A_0\cos\frac{1}{2}\delta]\sin(\alpha + \frac{1}{2}\delta)}$

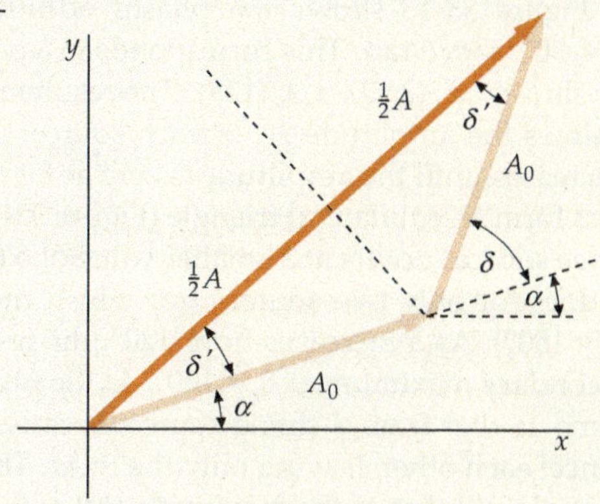

FIGURE 33-16

CHECK The step-5 result is identical to Equation 33-7 (see Problem statement).

PRACTICE PROBLEM 33-3 Find the amplitude and phase constant of the resultant wave function produced by the superposition of the two waves $E_1 = (4.0 \text{ V/m})\sin(\omega t)$ and $E_2 = (3.0 \text{ V/m})\sin(\omega t + 90°)$.

*THE INTERFERENCE PATTERN OF THREE OR MORE EQUALLY SPACED SOURCES

We can apply the phasor method of addition to calculate the interference pattern of three or more coherent sources that are equally spaced and in phase. We are most interested in the location of the interference maxima and minima. Figure 33-17 illustrates the case of three such sources. The geometry is the same as for two sources. At a great distance from the sources, the rays from the sources to a point P on the screen are approximately parallel. The path-length difference between the first and second source is then $d\sin\theta$, as before, and the path-length difference between the first and third source is $2d\sin\theta$. The wave at point P is the sum of the three waves. Let $\alpha = \omega t$ be the phase of the first wave at point P. We thus have the problem of adding three waves of the form

$$E_1 = A_0\sin\alpha$$
$$E_2 = A_0\sin(\alpha + \delta) \qquad 33\text{-}14$$
$$E_3 = A_0\sin(\alpha + 2\delta)$$

where

$$\delta = \frac{2\pi}{\lambda}d\sin\theta \approx \frac{2\pi}{\lambda}\frac{yd}{L} \qquad 33\text{-}15$$

as in the two-slit problem.

At $\theta = 0$, $\delta = 0$, so all the waves are in phase. The amplitude of the resultant wave is 3 times that of each individual wave and the intensity is 9 times that due to each source acting separately. As the angle θ increases from $\theta = 0$, the phase angle δ increases and the intensity decreases. The position $\theta = 0$ is thus a position of maximum intensity.

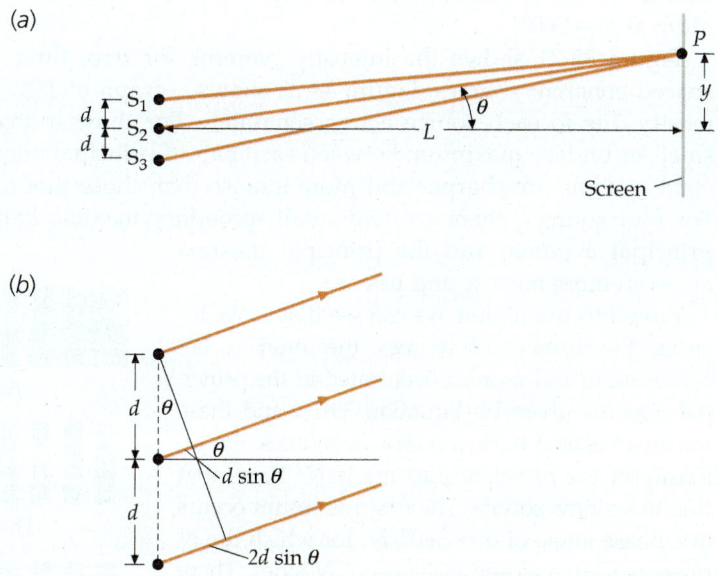

FIGURE 33-17 Geometry for calculating the intensity pattern far away from three equally spaced, coherent sources that are in phase.

Figure 33-18 shows the phasor addition of three waves for a phase angle $\delta = 30° = \pi/6$ rad. This corresponds to a point P on the screen for which θ is given by $\sin\theta = \lambda\delta/(2\pi d) = \lambda/(12d)$. The resultant amplitude A is considerably less than 3 times the amplitude A_0 of each source. As δ increases, the resultant amplitude decreases until the amplitude is zero at $\delta = 120°$. For this value of δ, the three phasors form an equilateral triangle (Figure 33-19). This first interference minimum for three sources occurs at a smaller value of δ (and therefore at a smaller angle θ) than it does for only two sources (for which the first interference minimum occurs at $\delta = 180°$). As δ increases from 120°, the resultant amplitude increases, reaching a secondary maximum at $\delta = 180°$. At the phase angle $\delta = 180°$, the amplitude is the same as that from a single source, because the waves from the first two sources cancel each other, leaving only the third. The intensity of the secondary maximum is one-ninth that of the maximum at $\theta = 0$. As δ increases beyond 180°, the amplitude again decreases and is zero at $\delta = 180° + 60° = 240°$. For δ greater than 240°, the amplitude increases and is again 3 times that of each source when $\delta = 360°$. This phase angle corresponds to a path-length difference of 1 wavelength for the waves from the first two sources and 2 wavelengths for the waves from the first and third sources. Hence, the three waves are in phase at this point. The largest maxima, called the principal maxima, are at the same positions as for just two sources, which are those points corresponding to the angles θ given by

$$d \sin\theta_m = m\lambda \quad m = 0, 1, 2, \ldots \qquad 33\text{-}16$$

These maxima are stronger and narrower than those for two sources. They occur at points for which the path-length difference between adjacent sources is zero or an integral number of wavelengths.

These results can be generalized to more than three sources. For four coherent sources that are equally spaced and in phase, the principal interference maxima are again given by Equation 33-16, but the maxima are even more intense, they are narrower, and there are two small secondary maxima between each pair of principal maxima. At $\theta = 0$, the intensity is 16 times that due to a single source. The first interference minimum occurs when δ is 90°, as can be seen from the phasor diagram of Figure 33-20. The first secondary maximum is near $\delta = 132°$. The intensity of the secondary maximum is about one-fourteenth that of the central maximum. There is another minimum at $\delta = 180°$, another secondary maximum near $\delta = 228°$, and another minimum at $\delta = 270°$ before the next principal maximum at $\delta = 360°$.

Figure 33-21 shows the intensity patterns for two, three, and four equally spaced coherent sources. Figure 33-22 shows a graph of I/I_0, where I_0 is the intensity due to each source acting separately. For three sources, there is a very small secondary maximum between each pair of principal maxima, and the principal maxima are sharper and more intense than those due to just two sources. For four sources, there are two small secondary maxima between each pair of principal maxima, and the principal maxima are even more narrow and intense.

From this discussion, we can see that as we increase the number of sources, the intensity becomes more and more concentrated in the principal maxima given by Equation 33-16, and these maxima become narrower. For N sources, the intensity of the principal maxima is N^2 times that due to a single source. The first minimum occurs at a phase angle of $\delta = 360°/N$, for which the N phasors form a closed polygon of N sides. There are $N - 2$ secondary maxima between each pair of principal maxima. These secondary maxima are very weak compared with the principal maxima.

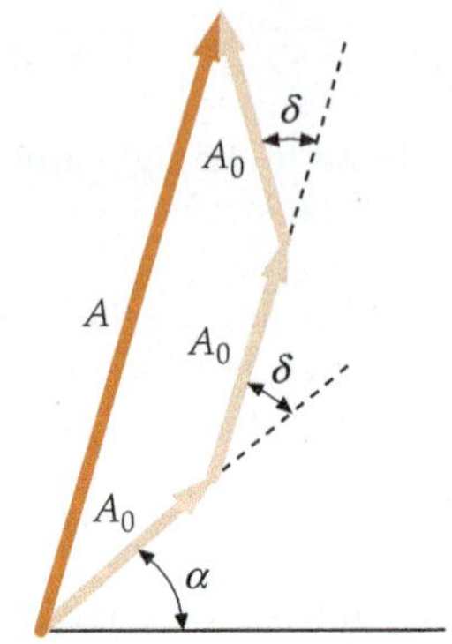

FIGURE 33-18 Phasor diagram for determining the resultant amplitude A due to three waves, each of amplitude A_0, that have phase differences of δ and 2δ due to path-length differences of $d \sin\theta$ and $2d \sin\theta$. The angle $\alpha = \omega t$ varies with time, but this does not affect the calculation of A.

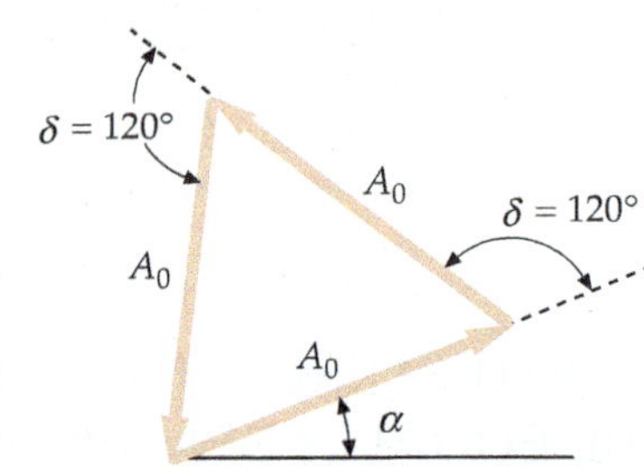

FIGURE 33-19 The resultant amplitude for the waves from three sources is zero when δ is 120°. This interference minimum occurs at a smaller angle θ than does the first minimum for two sources, which occurs when δ is 180°.

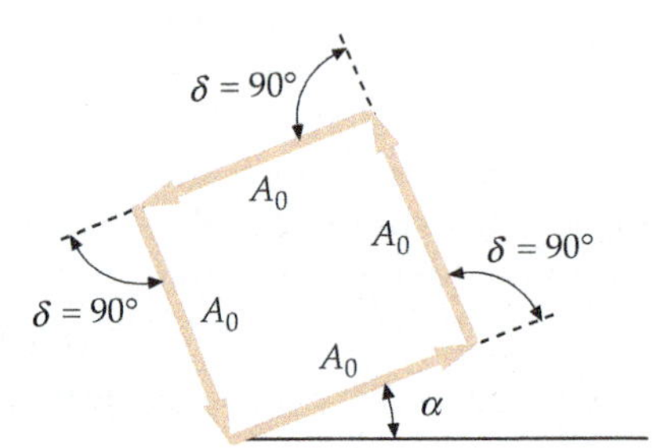

FIGURE 33-20 Phasor diagram for the first minimum for four coherent sources that are equally spaced and in phase. The amplitude is zero when the phase difference of the waves from adjacent sources is 90°.

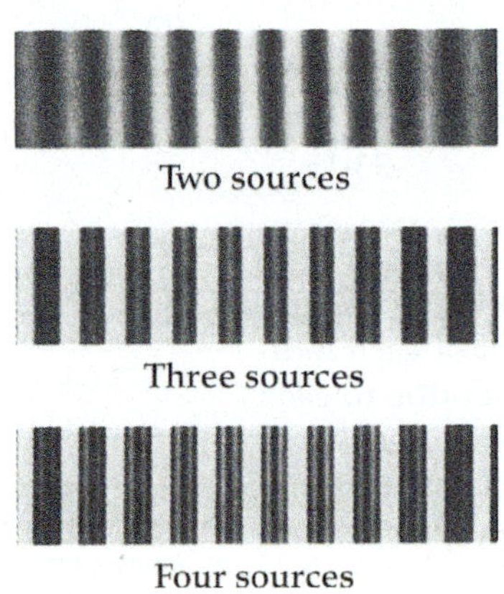

FIGURE 33-21 Intensity patterns for two, three, and four coherent sources that are equally spaced and in phase. There is a secondary maximum between each pair of principal maxima for three sources, and two secondary maxima between each pair of principal maxima for four sources. *(Courtesy of Michael Cagnet.)*

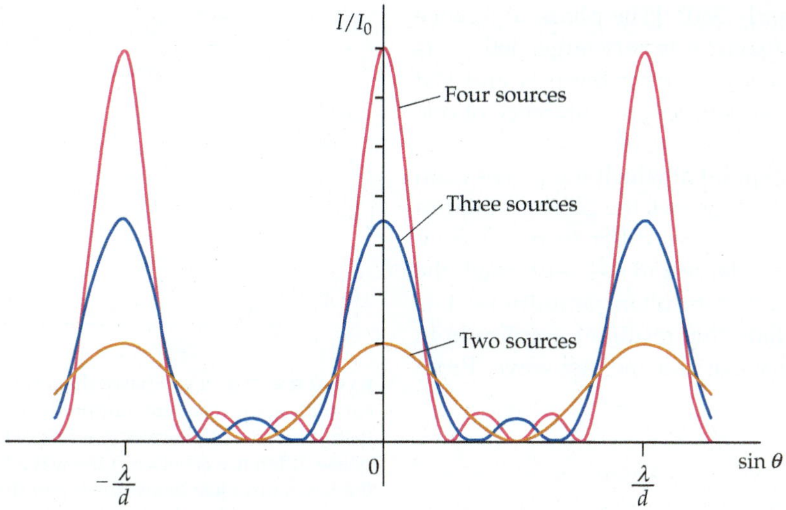

FIGURE 33-22 Plot of relative intensity versus sin θ for two, three, and four coherent sources that are equally spaced and in phase.

As the number of sources is increased, the principal maxima become sharper and more intense, and the intensities of the secondary maxima become negligible compared to those of the principal maxima.

*CALCULATING THE SINGLE-SLIT DIFFRACTION PATTERN

We now use the phasor method for the addition of harmonic waves to calculate the intensity pattern shown in Figure 33-11. We assume that the slit of width a is divided into N equal intervals and that there is a point source of waves at the midpoint of each interval (Figure 33-23). If d is the distance between two adjacent sources and a is the width of the opening, we have $d = a/N$. Because the screen on which we are calculating the intensity is far from the sources, the rays from the sources to a point P on the screen are approximately parallel. The path-length difference between any two adjacent sources is $\delta \sin\theta$, and the phase difference δ is related to the path-length difference by

$$\delta = \frac{d \sin\theta}{\lambda} 2\pi$$

If A_0 is the amplitude due to a single source, the amplitude at the central maximum, where $\theta = 0$ and all the waves are in phase, is $A_{max} = NA_0$ (Figure 33-24).

We can find the amplitude at some other point at an angle θ by using the phasor method for the addition of harmonic waves. As in the addition of two, three, or four waves, the intensity is zero at any point where the phasors representing the waves form a closed polygon. In this case, the polygon has N sides (Figure 33-25). At the first minimum, the wave from the first source just below the top of the opening and the wave from the source just below the middle of the opening are 180° out of phase. In this case, the waves from the source near the top of the opening differ

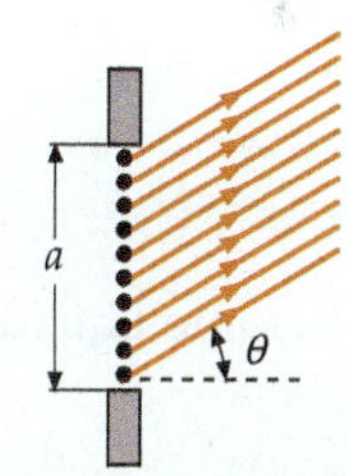

FIGURE 33-23 Diagram for calculating the diffraction pattern far away from a narrow slit. The slit width a is assumed to contain a large number of in-phase, equally spaced point sources separated by a distance d. The rays from the sources to a point far away are approximately parallel. The path-length difference for the waves from adjacent sources is $d \sin\theta$.

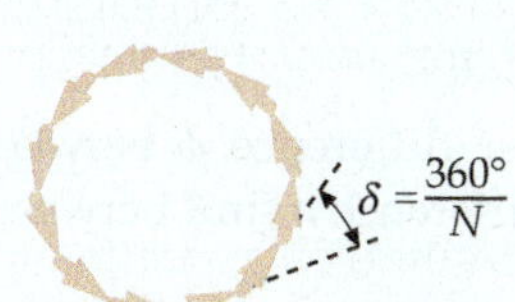

FIGURE 33-25 Phasor diagram for calculating the first minimum in the single-slit diffraction pattern. When the waves from the N sources completely cancel, the N phasors form a closed polygon. The phase difference between the waves from adjacent sources is then $\delta = 360°/N$. When N is very large, the waves from the first and last sources are approximately in phase.

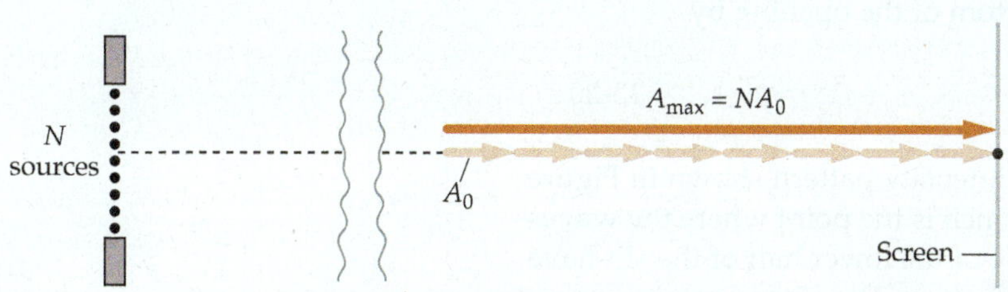

FIGURE 33-24 A single slit is represented by N sources, each of amplitude A_0. At the central maximum point, where $\theta = 0$, the waves from the sources add in phase, giving a resultant amplitude $A_{max} = NA_0$.

from those from the bottom of the opening by nearly 360°. [The phase difference is, in fact, $360° - (360°/N)$.] Thus, if the number of sources is very large, $360°/N$ is negligible and we get complete cancellation if the waves from the first and last sources are out of phase by 360°, corresponding to a path-length difference of one wavelength, in agreement with Equation 33-11.

We will now calculate the amplitude at a general point at which the waves from two adjacent sources differ in phase by δ. Figure 33-26 shows the phasor diagram for the addition of N waves, where the subsequent waves differ in phase from the first wave by $\delta, 2\delta, \ldots, (N-1)\delta$. When N is very large and δ is very small, the phasor diagram approximates the arc of a circle. The resultant amplitude A is the length of the chord of this arc. We will calculate this resultant amplitude in terms of the phase difference ϕ between the first wave and the last wave. From Figure 33-26, we have

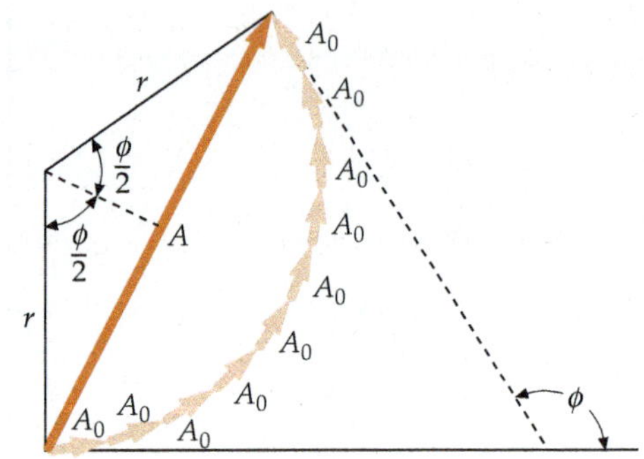

FIGURE 33-26 Phasor diagram for calculating the resultant amplitude due to the waves from N sources in terms of the phase difference ϕ between the wave from the first source just below the top of the slit and the wave from the last source just above the bottom of the slit. When N is very large, the resultant amplitude A is the chord of a circular arc of length $NA_0 = A_{max}$.

$$\sin \tfrac{1}{2}\phi = \frac{A/2}{r}$$

or

$$A = 2r \sin \tfrac{1}{2}\phi \qquad 33\text{-}17$$

where r is the radius of the arc. Because the length of the arc is $A_{max} = NA_0$ and the angle subtended is ϕ, we have

$$\phi = \frac{A_{max}}{r} \qquad 33\text{-}18$$

or

$$r = \frac{A_{max}}{\phi}$$

Substituting this into Equation 33-17 gives

$$A = \frac{2A_{max}}{\phi} \sin \tfrac{1}{2}\phi = A_{max} \frac{\sin \frac{1}{2}\phi}{\frac{1}{2}\phi}$$

Because the amplitude at the center of the central maximum ($\theta = 0$) is A_{max}, the ratio of the intensity at any other point to that at the center of the central maximum is

$$\frac{I}{I_0} = \frac{A^2}{A_{max}^2} = \left(\frac{\sin \frac{1}{2}\phi}{\frac{1}{2}\phi}\right)^2$$

or

$$I = I_0 \left(\frac{\sin \frac{1}{2}\phi}{\frac{1}{2}\phi}\right)^2 \qquad 33\text{-}19$$

INTENSITY FOR A SINGLE-SLIT DIFFRACTION PATTERN

The phase difference ϕ between the first and last waves is related to the path-length difference $a \sin\theta$ between the top and bottom of the opening by

$$\phi = \frac{a \sin\theta}{\lambda} 2\pi \qquad 33\text{-}20$$

Equation 33-19 and Equation 33-20 describe the intensity pattern shown in Figure 33-11. The first minimum occurs at $a \sin\theta = \lambda$, which is the point where the waves from the middle of the upper half and the middle of the lower half of the slit have a path-length difference of $\lambda/2$ and are 180° out of phase. The second minimum occurs at $a \sin\theta = 2\lambda$, where the waves from the upper half of the upper half of the slit and those from the lower half of the upper half of the slit have a path-length difference of $\lambda/2$ and are 180° out of phase.

There is a secondary maximum approximately midway between the first and second minima at $a \sin\theta \approx \frac{3}{2}\lambda$. Figure 33-27 shows the phasor diagram for determining the approximate intensity of this secondary maximum. The phase difference ϕ between the first and last waves is approximately $2\pi + \pi$. The phasors thus complete $1\frac{1}{2}$ circles. The resultant amplitude is the diameter of a circle that has a circumference which is two-thirds the total length A_{max}. If $C = \frac{2}{3}A_{max}$ is the circumference, the diameter A is

$$A = \frac{C}{\pi} = \frac{\frac{2}{3}A_{max}}{\pi} = \frac{2}{3\pi}A_{max}$$

and

$$A^2 = \frac{4}{9\pi^2}A_{max}^2$$

The intensity at this point is

$$I = \frac{4}{9\pi^2}I_0 = \frac{1}{22.2}I_0 \qquad 33\text{-}21$$

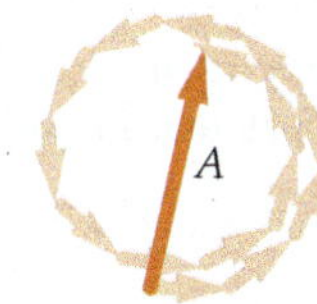

FIGURE 33-27 Phasor diagram for calculating the approximate amplitude of the first secondary maximum of the single-slit diffraction pattern. The secondary maximum occurs near the midpoint between the first and second minima when the N phasors complete $1\frac{1}{2}$ circles.

*CALCULATING THE INTERFERENCE–DIFFRACTION PATTERN OF MULTIPLE SLITS

The intensity of the two-slit interference–diffraction pattern can be calculated from the two-slit pattern (Equation 33-8) where the intensity of each slit (I_0 in that equation) is replaced by the diffraction pattern intensity due to each slit, I, given by Equation 33-19. The intensity for the two-slit interference–diffraction pattern is thus

$$I = 4I_0\left(\frac{\sin\frac{1}{2}\phi}{\frac{1}{2}\phi}\right)^2 \cos^2\tfrac{1}{2}\delta \qquad 33\text{-}22$$

INTERFERENCE–DIFFRACTION INTENSITY FOR TWO SLITS

where ϕ is the difference in phase between rays from the top and bottom of each slit, which is related to the width of each slit by

$$\phi = \frac{a\sin\theta}{\lambda}2\pi$$

and δ is the difference in phase between rays from the centers of two adjacent slits, which is related to the slit separation by

$$\delta = \frac{d\sin\theta}{\lambda}2\pi$$

In Equation 33-22, the intensity I_0 is the intensity at $\theta = 0$ due to one slit alone.

Example 33-8 Five-Slit Interference–Diffraction Pattern

Find the interference–diffraction intensity pattern for five equally spaced slits, where a is the width of each slit and d is the distance between adjacent slits.

PICTURE First, find the interference intensity pattern for the five slits, assuming no angular variations in the intensity due to diffraction. To do this, first construct a phasor diagram to find the amplitude of the resultant wave in an arbitrary direction θ. Intensity is proportional to the square of the amplitude. Next, correct for the variation of intensity with θ by using the single-slit diffraction pattern intensity relation (Equation 33-20 and Equation 33-20).

SOLVE

1. The diffraction pattern intensity I' due to a slit of width a is given by Equation 33-19 and Equation 33-20:

$$I' = I_0\left(\frac{\sin\frac{1}{2}\phi}{\frac{1}{2}\phi}\right)^2$$

where

$$\phi = \frac{2\pi}{\lambda}a\sin\theta$$

2. The interference pattern intensity I is proportional to the square of the amplitude A of the superposition of the wave functions for the light from the five slits:

$$I \propto A^2$$

where

$$A\sin(\alpha+\delta') = A_0\sin\alpha + A_0\sin(\alpha+\delta) + A_0\sin(\alpha+2\delta) + A_0\sin(\alpha+3\delta) + A_0\sin(\alpha+4\delta)$$

and where $\alpha = \omega t$ and $\delta = \frac{d\sin\theta}{\lambda}2\pi$

3. To solve for A, we construct a phasor diagram (Figure 33-28). The amplitude A equals the sum of the projections of the individual phasors onto the resultant phasor:

$$\delta' = \beta + \delta$$

so

$$\beta = \delta' - \delta = 2\delta - \delta = \delta$$

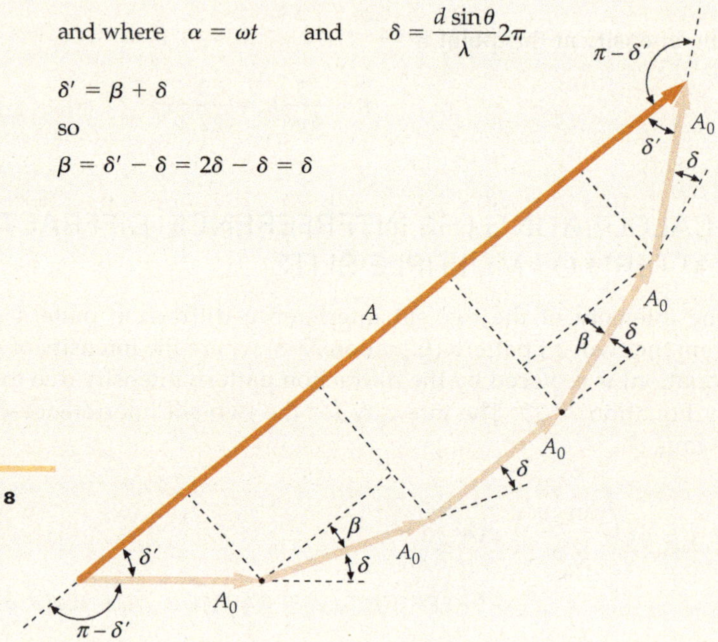

FIGURE 33-28

4. To find δ', we add the exterior angles. The sum of the exterior angles equals 2π (if you walk the perimeter of any polygon you rotate through the sum of the exterior angles, and you rotate through 2π radians):

$$2(\pi - \delta') + 4\delta = 2\pi \quad \Rightarrow \quad \delta' = 2\delta$$

5. Solve for A from the figure:

$$A = 2A_0\cos\delta' + 2A_0\cos\beta + A_0$$

6. Substitute for δ' using the step-4 result, and substitute for β using the relation $\beta = \delta$. (That β and δ are equal follows from the theorem "If two parallel lines are cut by a transversal, the interior and exterior angles on the same side of the transversal are equal.")

$$A = A_0(2\cos 2\delta + 2\cos\delta + 1)$$

7. Square both sides to relate the intensities. Recall, I' is the intensity from a single slit, and A_0 is the amplitude from a single slit:

$$A^2 = A_0^2(2\cos 2\delta + 2\cos\delta + 1)^2$$

so

$$I = I'(2\cos 2\delta + 2\cos\delta + 1)^2$$

8. Substitute for I' using the step-1 result:

$$\boxed{I = I_0\left(\frac{\sin\frac{1}{2}\phi}{\frac{1}{2}\phi}\right)^2(2\cos 2\delta + 2\cos\delta + 1)^2}$$

where $\phi = \frac{a\sin\theta}{\lambda}2\pi$ and $\delta = \frac{d\sin\theta}{\lambda}2\pi$

CHECK If $\theta = 0$, both $\phi = 0$ and $\delta = 0$. So, for $\theta = 0$, step 5 becomes $A = 5A_0$ and step 8 becomes $I = 5^2 I_0 = 25I_0$ as expected.

33-6 FRAUNHOFER AND FRESNEL DIFFRACTION

Diffraction patterns, like the single-slit pattern shown in Figure 33-11, that are observed at points for which the rays from an aperture or an obstacle are nearly parallel are called **Fraunhofer diffraction patterns.** Fraunhofer patterns can be observed at great distances from the obstacle or the aperture so that the rays reaching any point are approximately parallel, or they can be observed using a lens to focus parallel rays on a viewing screen placed in the focal plane of the lens.

The diffraction pattern observed near an aperture or an obstacle is called a **Fresnel diffraction pattern.** Because the rays from an aperture or an obstacle close to a screen cannot be considered parallel, Fresnel diffraction is much more difficult to analyze. Figure 33-29 illustrates the difference between the Fresnel and the Fraunhofer patterns for a single slit.*

Figure 33-30*a* shows the Fresnel diffraction pattern of an opaque disk. Note the bright spot at the center of the pattern caused by the constructive interference of the light waves diffracted from the edge of the disk. This pattern is of some historical interest. In an attempt to discredit Augustin Fresnel's wave theory of light, Siméon Poisson pointed out that it predicted a bright spot at the center of the shadow, which he assumed was a ridiculous contradiction of fact. However, Fresnel immediately demonstrated experimentally that such a spot does, in fact, exist. This demonstration convinced many doubters of the validity of the wave theory of light. The Fresnel diffraction pattern of a circular aperture is shown in Figure 33-30*b*. Comparing this with the pattern of the opaque disk in Figure 33-30*a*, we can see that the two patterns are complements of each other.

Figure 33-31*a* shows the Fresnel diffraction pattern of a straightedge illuminated by light from a point source. A graph of the intensity versus distance (measured along a line perpendicular to the edge) is shown in Figure 33-31*b*. The light intensity does not fall abruptly to zero in the geometric shadow, but it decreases rapidly and is negligible within a few wavelengths of the edge. The Fresnel diffraction pattern

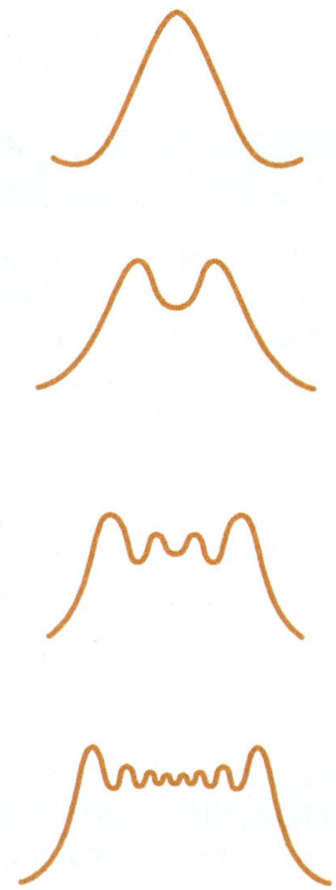

FIGURE 33-29 Diffraction patterns for a single slit at various screen distances.

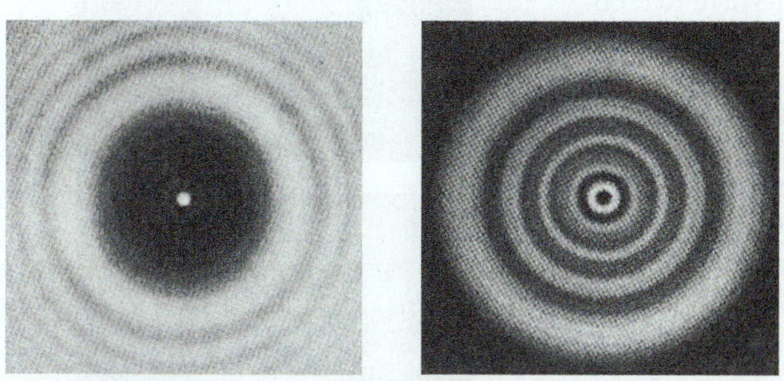

FIGURE 33-30 (*a*) The Fresnel diffraction pattern of an opaque disk. At the center of the shadow, the light waves diffracted from the edge of the disk are in phase and produce a bright spot called the *Poisson spot.* (*b*) The Fresnel diffraction pattern of a circular aperture. Compare this with Figure 33-30*a*. *((a) and (b) M. Cagnet, M. Fraçon, J. C. Thrierr, Atlas of Optical Phenomena.)*

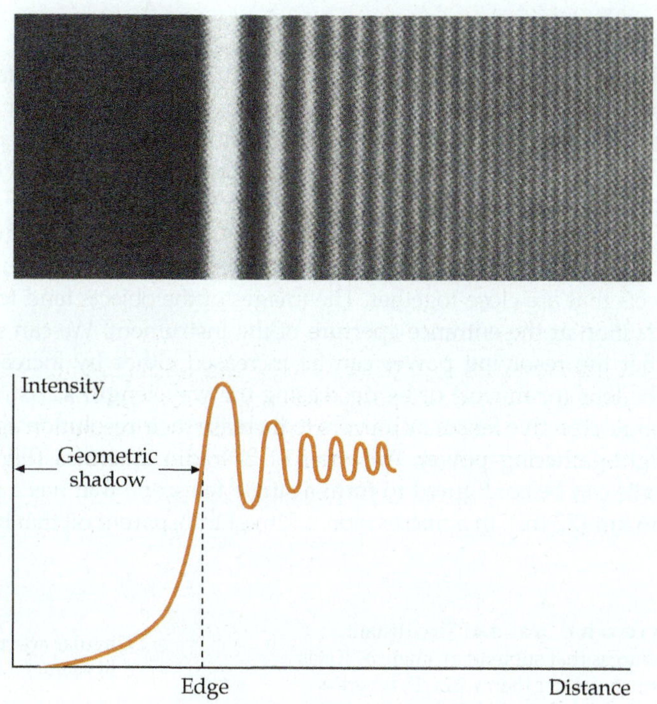

FIGURE 33-31 (*a*) The Fresnel diffraction of a straightedge. (*b*) A graph of intensity versus distance along a line perpendicular to the edge. *(Courtesy Battelle-Northwest Laboratories.)*

* See Richard E. Haskel, "A Simple Experiment on Fresnel Diffraction," *American Journal of Physics* 38 (1970): 1039.

of a rectangular aperture is shown in Figure 33-32. These patterns cannot be seen using extended light sources like an ordinary lightbulb, because the dark fringes of the pattern produced by light from one point on the source overlap the bright fringes of the pattern produced by light from another point.

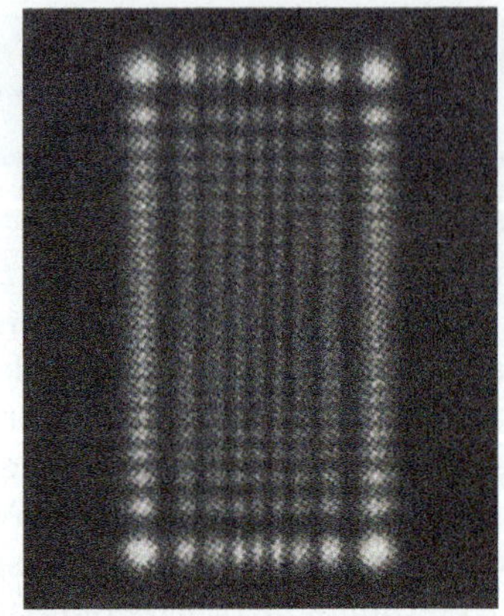

FIGURE 33-32 The Fresnel diffraction pattern of a rectangular aperture. *(Courtesy of Michael Cagnet.)*

33-7 DIFFRACTION AND RESOLUTION

Diffraction due to a circular aperture has important implications for the resolution of many optical instruments. Figure 33-33 shows the Fraunhofer diffraction pattern of a circular aperture. The angle θ subtended by the first diffraction minimum is related to the wavelength and the diameter of the opening D by

$$\sin\theta = 1.22\frac{\lambda}{D} \qquad 33\text{-}23$$

Equation 33-23 is similar to Equation 33-9 except for the factor 1.22, which arises from the mathematical analysis, and is similar to the equation for a single slit but more complicated because of the circular geometry. In many applications, the angle θ is small, so $\sin\theta$ can be replaced by θ. The first diffraction minimum is then at an angle θ given by

$$\theta \approx 1.22\frac{\lambda}{D} \qquad 33\text{-}24$$

FIGURE 33-33 The Fraunhofer diffraction pattern of a circular aperture. *(Courtesy of Michael Cagnet.)*

Figure 33-34 shows two point sources that subtend an angle α at a circular aperture far from the sources. The intensities of the Fraunhofer diffraction pattern are also indicated in this figure. If α is much greater than $1.22\lambda/D$, the sources will be seen as two sources. However, as α is decreased, the overlap of the diffraction patterns increases, and it becomes difficult to distinguish the two sources from one source. At the critical angular separation, α_c, given by

$$\alpha_c = 1.22\frac{\lambda}{D} \qquad 33\text{-}25$$

the first minimum of the diffraction pattern of one source falls on the central maximum of the other source. These objects are said to be just resolved by **Rayleigh's criterion for resolution.** Figure 33-35 shows the diffraction patterns for two sources when α is greater than the critical angle for resolution and when α is just equal to the critical angle for resolution.

Equation 33-25 has many applications. The *resolving power* of an optical instrument, such as a microscope or telescope, is the ability of the instrument to resolve two objects that are close together. The images of the objects tend to overlap because of diffraction at the entrance aperture of the instrument. We can see from Equation 33-25 that the resolving power can be increased either by increasing the diameter D of the lens (or mirror) or by decreasing the wavelength λ. Astronomical telescopes use large objective lenses or mirrors to increase their resolution as well as to increase their light-gathering power. An array of 27 radio antennas (Figure 33-36) mounted on rails can be configured to form a single telescope that has a resolution distance D of 36 km (22 mi). In a microscope, a film of transparent oil that has an index of refraction

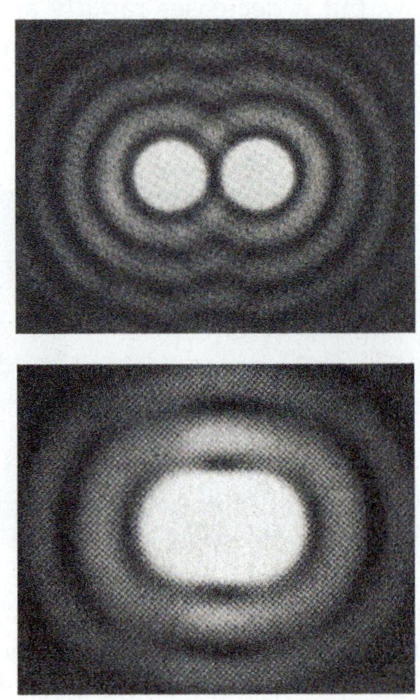

(a)

(b)

FIGURE 33-35 The diffraction patterns for a circular aperture and two incoherent point sources when (*a*) α is a factor of 2 or so greater than $\alpha_c = 1.22\lambda/\text{D}$ and (*b*) when α is equal to the limit of resolution, $\alpha_c = 1.22\lambda/\text{D}$. *((a) and (b) Courtesy of Michael Cagnet.)*

FIGURE 33-34 Two distant sources that subtend an angle α. If α is much greater than $1.22\lambda/D$, where λ is the wavelength of light and D is the diameter of the aperture, the diffraction patterns have little overlap and the sources are easily seen as two distinct sources. If α is not much greater than $1.22\lambda/D$, the overlap of the diffraction patterns makes it difficult to distinguish two sources from one.

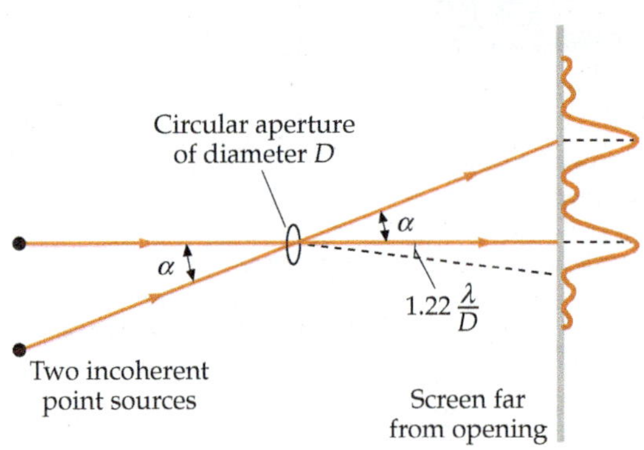

of approximately 1.55 is sometimes used under the objective to decrease the wavelength of the light ($\lambda' = \lambda/n$). The wavelength can be reduced further by using ultraviolet light and photographic film; however, ordinary glass is opaque to ultraviolet light, so the lenses in an ultraviolet microscope must be made from quartz or fluorite. To obtain very high resolutions, electron microscopes are used—microscopes that use electrons rather than light. The wavelengths of electrons vary inversely with the square root of their kinetic energy and can be made as small as desired.*

FIGURE 33-36 The very large array (VLA) of radio antennas is located near Socorro, New Mexico. The 25-m-diameter antennas are mounted on rails, which can be arranged in several configurations, and can be extended over a diameter of 36 km. The data from the antennas are combined electronically, so the array is really a single high-resolution telescope. *(Courtesy of National Radio Astronomy Observatory/Associated Universities, Inc./National Science Foundation. Photographer: Kelly Gatlin. Digital composite: Patricia Smiley.)*

Example 33-9 Physics in the Library

Context-Rich

While studying in the library, you lean back in your chair and ponder the small holes you notice in the ceiling tiles. You notice that the holes are approximately 5.0 mm apart. You can clearly see the holes directly above you, about 2 m up, but the tiles far away do not appear to have the holes. You wonder if the reason you cannot see the distant holes is because they are not within the criteria for resolution established by Rayleigh. Is this a feasible explanation for the disappearance of the holes? You notice the holes disappear about 20 m from you.

PICTURE We will need to make assumptions about the situation. If we use Equation 33-25, we will need to know the wavelength of light and the aperture diameter. Assuming our pupil is the aperture, we can assume approximately 5.0 mm for the diameter. (This is the value used in this physics textbook.) The wavelength of the light is probably about 500 nm or so.

SOLVE

1. The angular limit for resolution by the eye depends on the ratio of the wavelength and the diameter of the pupil: $\theta_c \approx 1.22\dfrac{\lambda}{D}$
2. The angle subtended by two holes depends on their separation distance d and their distance L from your eye: $\theta \approx \dfrac{d}{L}$
3. Equating the two angles and putting in the numbers give: $\dfrac{d}{L} \approx 1.22\dfrac{\lambda}{D}$

   $\dfrac{5.0\text{ mm}}{L} \approx 1.22\dfrac{500\text{ nm}}{5.0\text{ mm}}$
4. Solving for L gives: $L = 41\text{ m}$
5. By a factor of 2, 41 m is too large. However, you are suspect of the value given for the pupil diameter in your physics textbook. You know the pupil is smaller when the light is bright, and the library ceiling is very bright and colored white. An online search for eye pupil diameter soon yields the information you need. The pupil diameter ranges from 2 to 3 mm up to 7 mm:

   Sucess. If the pupil diameter is 2.5 mm, the value of L is 20 m.

It is instructive to compare the limitation on resolution of the eye due to diffraction, as seen in Example 33-9, with the limitation on resolution due to the separation of the receptors (cones) on the retina. To be seen as two distinct objects, the images of the objects must fall on the retina on two nonadjacent cones. (See Problem 65 in Chapter 32.) Because the retina is about 2.5 cm from the cornea, the distance y on the retina corresponding to an angular separation of 1.5×10^{-4} rad is found from

$$\alpha_c = 1.5 \times 10^{-4}\text{ rad} = \frac{y}{2.5\text{ cm}}$$

or

$$y = 3.8 \times 10^{-4}\text{ cm} = 3.8 \times 10^{-6}\text{ m} = 3.8\ \mu\text{m}$$

The actual separation of the cones in the fovea centralis, where the cones are the most tightly packed, is about 1 μm. Outside this region, they are about 3 μm to 5 μm apart.

CONCEPT CHECK 33-2

True or False:
Fraunhofer diffraction is a limiting case of Fresnel diffraction.

* The wave properties of electrons are discussed in Chapter 34.

* 33-8 DIFFRACTION GRATINGS

A widely used tool for measuring the wavelength of light is the **diffraction grating,** which consists of a large number of equally spaced lines or slits on a flat surface. Such a grating can be made by cutting parallel, equally spaced grooves on a glass or metal plate with a precision ruling machine. With a reflection grating, light is reflected from the ridges between the lines or grooves. Phonograph records and compact disks exhibit some of the properties of reflection gratings. In a transmission grating, the light passes through the clear gaps between the rulings. Inexpensive, optically produced plastic gratings that have 10 000 or more slits per centimeter are common items in teaching laboratories. The spacing of the slits in a grating that has 10 000 slits per centimeter is $d = (1\text{ cm})/10\,000\text{ slits} = 10^{-4}\text{ cm}$ slit.

Compact disks act as reflection gratings. *(Kevin R. Morris/Corbis.)*

Consider a plane wave of monochromatic light that is incident normally on a transmission grating (Figure 33-37). Assume that the width of each slit is very small so that it produces a widely diffracted beam. The interference pattern produced on a screen a large distance from the grating is due to a large number of coherent, equally spaced light sources. Suppose we have N slits that have separation d between adjacent slits. At $\theta = 0$, the light from each slit is in phase with that from all the other slits, so the amplitude of the wave is NA_0, where A_0 is the amplitude from each slit, and the intensity is N^2I_0, where I_0 is the intensity due to a single slit alone. At an angle θ_1, where $d\sin\theta_1 = \lambda_1$, the path-length difference between any two successive slits is λ_1, so again the light from each slit is in phase with that from all the other slits and the intensity is N^2I_0. The interference maxima are thus at angles θ given by

$$d\sin\theta_m = m\lambda \qquad m = 0, 1, 2, \ldots \qquad \text{33-26}$$

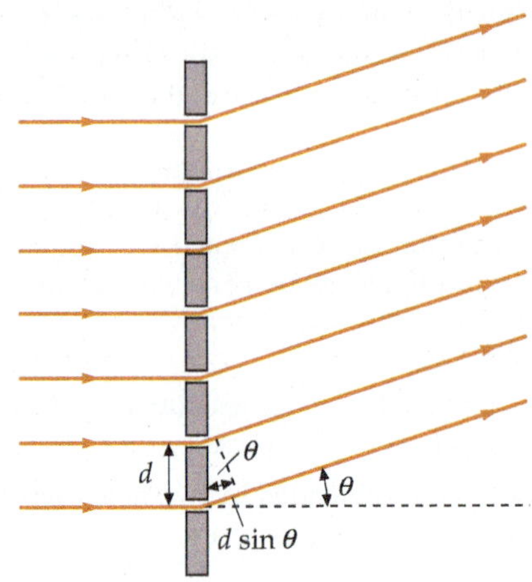

FIGURE 33-37 Light incident normally on a diffraction grating. At an angle θ, the path-length difference between rays from adjacent slits is $d\sin\theta$.

The positions of the interference maxima do not depend on the number of sources, but the more sources there are, the sharper (narrower) and more intense the maxima will be.

To see that the interference maxima will be sharper when there are many slits, consider the case of N illuminated slits, where N is large ($N \gg 1$). The distance from the first slit to the Nth slit is $(N-1)d \approx Nd$. When the path-length difference for the light from the first slit and that from the Nth slit is λ, the resulting intensity will be zero because the light from any two slits separated by $\frac{1}{2}Nd$ interferes destructively. (We saw this in our discussion of single-slit diffraction in Section 33-4.) Because the first and Nth slits are separated by approximately Nd, the intensity will be zero at angle θ_{min} given by

$$Nd\sin\theta_{\text{min}} = \lambda$$

so

$$\theta_{\text{min}} \approx \sin\theta_{\text{min}} = \frac{\lambda}{Nd}$$

The angular width of the interference maximum, which is equal to $2\theta_{\text{min}}$, is thus inversely proportional to N. Therefore, the greater the number of illuminated slits N, the sharper the maximum. Because the intensity in the maximum is proportional to N^2I_0, the intensity in the maximum multiplied by the width of the maximum is proportional to NI_0. The intensity multiplied by the width is a measure of power per unit length in the maximum.

Figure 33-38*a* shows a student spectroscope that uses a diffraction grating to analyze light. In student laboratories, the light source is typically a glass tube containing atoms of a gas (for example, helium or sodium vapor) that are excited by a bombardment of electrons accelerated by high voltage across the tube. The light emitted by such a source contains only certain wavelengths that are characteristic of the atoms in the source. Light from the source passes through a narrow collimating slit and is made parallel by a lens. Parallel light from the lens is incident on the grating. Instead of falling on a screen a large distance away, the parallel light from the grating is focused by a telescope and viewed by the eye. The telescope is

(a)

(b)

mounted on a rotating platform that has been calibrated so that the angle θ can be measured. In the forward direction ($\theta = 0$), the central maximum for all wavelengths is seen. If light of a particular wavelength λ is emitted by the source, the first interference maximum is seen at the angle θ given by $d \sin\theta_m = m\lambda$ (Equation 33-26) with $m = 1$. Each wavelength emitted by the source produces a separate image of the collimating slit in the spectroscope called a **spectral line.** The set of lines corresponding to $m = 1$ is called the **first-order spectrum.** The **second-order spectrum** corresponds to $m = 2$ for each wavelength. Higher orders may be seen, providing the angle θ given by $d \sin\theta_m = m\lambda$ is less than 90°. Depending on the wavelengths, the orders may be mixed; that is, the third-order line for one wavelength may occur at a smaller value of θ than does the second-order line for another wavelength. If the spacing of the slits in the grating is known, the wavelengths emitted by the source can be determined by measuring the angles.

FIGURE 33-38 *(a)* A typical student spectroscope. Light from a collimating slit near the source is made parallel by a lens and falls on a grating. The diffracted light is viewed with a telescope at an angle that can be accurately measured. *(b)* Aerial view of the very large array (VLA) radio telescope in New Mexico. Radio signals from distant galaxies add constructively when Equation 33-26 is satisfied, where d is the distance between two adjacent telescopes. *((a) Clarence Bennett/ Oakland University, Rochester, Michigan. (b) NRAO/AUI/Science Photo Library/ Photo Researchers.)*

Example 33-10 Sodium D Lines

Sodium light is incident on a diffraction grating with 12 000 lines per centimeter. At what angles will the two yellow lines (called the sodium D lines) of wavelengths 589.00 nm and 589.59 nm be seen in the first order?

PICTURE Apply $d \sin\theta_m = m\lambda$ to each wavelength, with $m = 1$ and $d = (1/12\,000)$ cm.

SOLVE

1. The angle θ_m is given by $d \sin\theta_m = m\lambda$ with $m = 1$: $\sin\theta_1 = \dfrac{\lambda}{d}$

2. Calculate θ_1 for $\lambda = 589.00$ nm: $\theta_1 = \sin^{-1}\left[\dfrac{589.00 \times 10^{-9}\text{ m}}{(1/12\,000)\text{ cm}} \times \left(\dfrac{100\text{ cm}}{1\text{ m}}\right)\right] = \boxed{44.98°}$

3. Repeat the calculation for $\lambda = 589.59$ nm: $\theta_1 = \sin^{-1}\left[\dfrac{589.59 \times 10^{-9}\text{ m}}{(1/12\,000)\text{ cm}} \times \left(\dfrac{100\text{ cm}}{1\text{ m}}\right)\right] = \boxed{45.03°}$

CHECK The first-order intensity maximum for the longer wavelength appears at the larger angle, as expected.

PRACTICE PROBLEM 33-4 Find the angles for the first-order intensity maxima of the two yellow lines if the grating has 15 000 lines per centimeter.

An important feature of a spectroscope is its ability to resolve spectral lines of two nearly equal wavelengths λ_1 and λ_2. For example, the two prominent yellow lines in the spectrum of sodium have wavelengths 589.00 and 589.59 nm. These can be seen as two separate wavelengths if their interference maxima do not overlap. According to Rayleigh's criterion for resolution, these wavelengths are resolved if the angular separation of their interference maxima is greater than the angular separation between an interference maximum and the first interference minimum on either side of it. The **resolving power** of a diffraction grating is defined to be $\lambda/|\Delta\lambda|$, where $|\Delta\lambda|$ is the smallest observable difference between two nearby wavelengths, each approximately equal to λ, that may be resolved. The resolving power is proportional to the number of slits illuminated because the more slits illuminated, the sharper the interference maxima. The resolving power R can be shown to be

$$R = \frac{\lambda}{|\Delta\lambda|} = mN \qquad 33\text{-}27$$

where N is the number of illuminated slits and m is the order number (see Problem 78). We can see from Equation 33-27 that to resolve the two yellow lines in the first order ($m = 1$) of the sodium spectrum the resolving power must be

$$R = 1 \times \frac{589.00\text{ nm}}{589.59\text{ nm} - 589.00\text{ nm}} = 998$$

Thus, to resolve the two yellow sodium lines in the first order, we need a grating containing 998 or more slits in the area illuminated by the light.

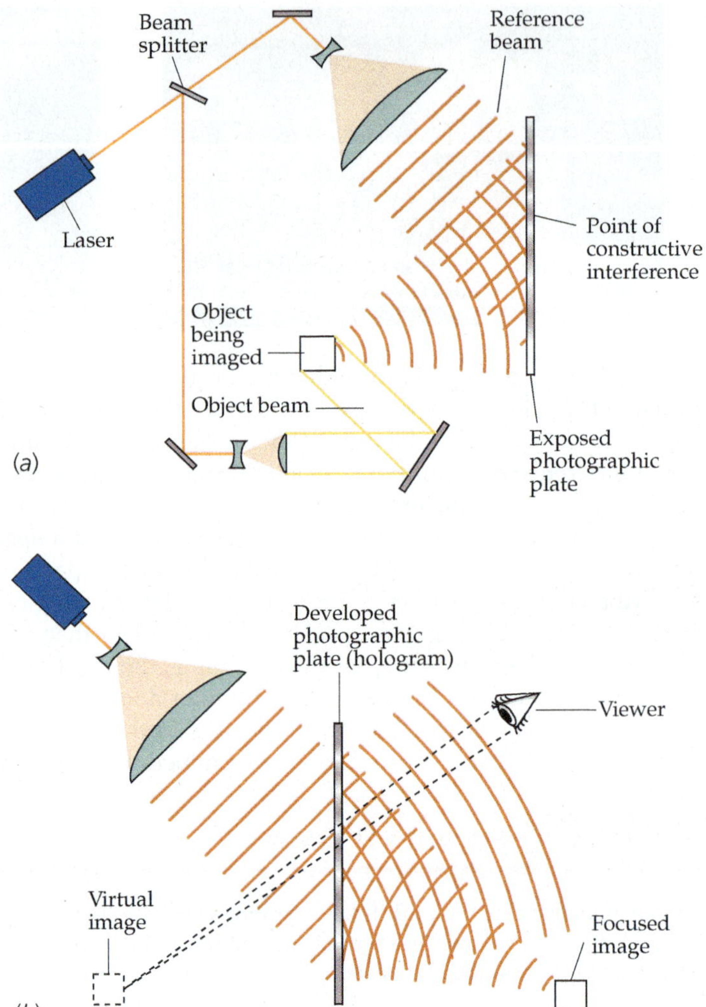

FIGURE 33-39 (*a*) The production of a hologram. The interference pattern produced by the reference beam and object beam is recorded on a photographic film. (*b*) When the film is developed and illuminated by coherent laser light, a three-dimensional image is seen.

*HOLOGRAMS

An interesting application of diffraction gratings is the production of a three-dimensional photograph called a **hologram** (Figure 33-39). In an ordinary photograph, the intensity of reflected light from an object is focused on a light-sensitive surface. As a result, a two-dimensional image is recorded. In a hologram, a beam from a laser is split into two beams, a reference beam and an object beam. The object beam reflects from the object to be photographed, and the interference pattern between it and the reference beam is recorded on a transparent film coated with a photosensitive emulsion. This can be done because the laser beam is coherent so that the relative phase difference between the reference beam and the object beam can be kept constant during the exposure. The film can be used to produce a holographic image after the emulsion is developed (chemically processed). The interference fringes on the film act as a diffraction grating. When the developed film is illuminated with a laser, a three-dimensional holographic image of the object is produced.

Holograms that you see on credit cards or postage stamps, called rainbow holograms, are more complicated. A horizontal strip of the original hologram is used to make a second hologram. The three-dimensional image can be seen as the viewer moves from side to side, but if viewed using monochromatic light, the image disappears when the viewer's eyes move above or below the slit image. When viewed using white light, the image is seen in different colors as the viewer moves in the vertical direction.

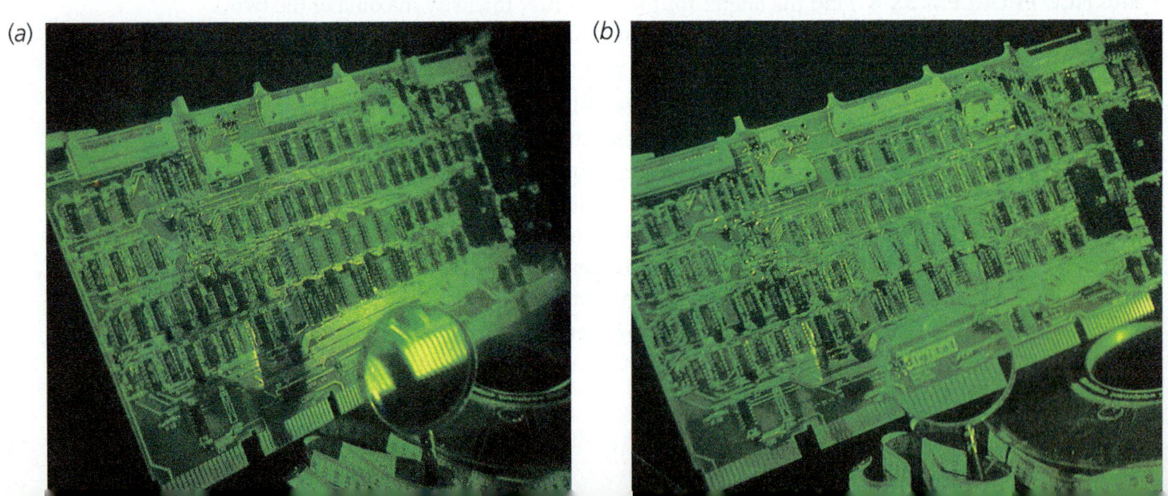

A hologram viewed from two different angles. Note that different parts of the circuit board appear behind the front magnifying lens. (*© 1981 by Ronald R. Erickson, Hologram by Nicklaus Phillips, 1978, for Digital Equipment Corporation.)*

Physics Spotlight

Holograms: Guided Interference

The machine is using heat to emboss holograms on plastic cards (perhaps credit cards). Such holograms serve as both a security feature and aesthetic enrichment. *(Pascal Goetgheluck/Photo Researchers.)*

Holography was invented by Dennis Gabor in 1948 when he tried to improve the resolution of electron microscopy.[*] He reconstructed wavefronts using interference on the photographic plate to make a picture that contained phase information as well as intensity information. He named this type of imagery holography, after the Greek words for "whole" and "writing," for he felt that including the phase information gave a complete picture.[†]

It was extremely difficult to create those first few holograms, and they did not achieve the desired resolution. He used mercury vapor lamps as a light source. The light was highly monochromatic, but incoherent. (The phase of the light fluctuated randomly.) A decade or so later, after the laser was invented, the use of coherent laser light made holography practical for many purposes.

Embossed holograms are frequently used because they are inexpensive. Embossed holograms are made by hot-stamping a metallized plastic film[‡] with a die that is a negative copy of the extremely shallow (around 0.3–0.5 micron deep) interference lines present in a hologram.[#] The plastic film is then a duplicate of the very tiny interference lines in the original hologram. When light shines through the film and reflects from the metallic backing, the holographic image is reconstructed. Almost all embossed holograms are rainbow holograms—able to be viewed without a laser. Creating the master of a rainbow hologram is a complex process involving multiple exposures at precise angles.[°]

Embossed holograms are highly visible, easy to recognize, and difficult to forge.[§] Because they can take the place of paper labels or be added to paper or plastic, they are used on credit cards, pharmaceutical packaging, currency, and traveler's checks as a quick method for authentication.[¶,**]

In January 1999, the Ford Motor Company used a series of digital holograms to create a 10 foot by 4 foot hologram of a concept car. The holograms were printed directly from computer design data.[††] Digital holography is now used to help physicians visualize the results of either computed tomography scans or magnetic resonance scans.[‡‡] The output from a series of MR or CT slices is collected, digitally processed, and then printed onto a single hologram, which can be viewed on a portable viewer. The resulting hologram allows surgeons to prepare for difficult surgeries[##] and may also have biomedical and industrial engineering applications.[°°] Digital holography is beginning to be used in holographic video applications.[§§]

Holograms have also been used as substitutes for traditional lenses. Holographic optical elements allow smaller and more compact displays to be built. Heads-up displays for airplane pilots are created using holographic optical elements.[¶¶] An extremely compact system that uses digitally calculated holograms as the optical element has been tested for use as a cell-phone based projector.[***] The use of holograms as optical elements and in optical data storage depends on advances in materials that are lightweight, tough, and have the desired optical properties.[†††]

Recently, holograms have been used to measure the electrostatic potential[‡‡‡] and magnetic fields[###] of very small objects. They have also been used to create higher resolution optics for X-ray lenses.[°°°] More than fifty years after holograms were invented, they are used to improve the resolution of microscopic images.

* Gabor, D., "Nobel Lecture." *Nobel Prize Lectures, 1971*, Dec. 11, 1971, at http://nobelprize.org/nobel_prizes/physics/laureates/1971/gabor-lecture.pdf As of Nov. 2006.
† Scanlon, L., "The Whole Picture." *Technology Review*, Dec. 2002/Jan. 2003, Vol. 105, No. 10, p. 88.
‡ Ruschmann, H. W., "Apparatus for Embossing Holograms on Metallized Thermoplastic Films." *United States Patent 4,547,141*, Oct. 15, 1985.
# Abraham, N. C., "Optical Data Storage Disc." *United States Patent 5,452,282*, Sept. 19, 1995.
° Benton, S., Houde-Walter, W., and Mingace, Jr., H., "Methods of Making Holographic Images." *United States Patent 4,415,225.*, Nov. 15, 1983.
§ Cross, L., "Brand Security." *Graphic Arts Monthly*, Jan. 2006, Vol. 78, No. 1, pp. 32–33.
¶ "MasterCard Renews Hologram Contract." *American Banker*, Mar. 3, 2003, Vol. 168, No. 44, p. 18.
** Miller, H. I., "Fear and Pharmaceutical Failure." *The Washington Times*, Oct. 5, 2006, p. A16.
†† Mahoney, D. P., "Ford Drives Holography Development." *Computer Graphics World*, Feb. 1999, Vol. 22, No. 2, pp. 12–13.
‡‡ Samudhram, A., "Digital Holography Opens New Frontiers." *New Straits Times (Malaysia)*, Nov. 23, 2000, p. 2W.
Penrod, S., "3D Imaging Assisting Surgeons in Separation Surgery." *Local News*, KSL, Salt Lake City, Aug. 7, 2006. At http://www.ksl.com/?nid=148&sid=408002 As of Dec. 2006.
°° Liu, C., Yan, C., and Gao, S., "Digital Holographic Method for Tomography Reconstruction." *Applied Physics Letters*, Feb. 9, 2004, Vol. 84, No. 6, pp. 1010–1012.
§§ Freedman, D. H., "Holograms in Motion." *Technology Review*, Nov. 2002, Vol. 105, No. 9, pp. 48–55.
¶¶ Stevens, T., "Holograms: More than Pretty Pictures." *Industry Week*, Oct. 4, 1993, Vol. 242, No. 19, pp. 34–46.
*** Buckley, E., "Miniature Projectors Based on LBO Technology." *SID Mobile Displays Conference*, San Diego: Oct. 3–5. 2006. At http://www.lightblueoptics.com/images/news/SID_Mobile_Displays_2006.pdf As of Nov. 2006.
††† Huang, G. T., "Holographic Memory." *Technology Review*, Sept. 2005, Vol. 108, No. 9, pp. 64–67.
‡‡‡ Chou, L.-J., Chang, M.-T., and Chueh, Y.-L., "Electron Holography for Improved Measurement of Microfields in Nanoelectrode Assemblies." *Applied Physics Letters*, Jul. 10, 2006, Vol. 89, No. 2, Letter 023112, 3 pp.
Nepijko, S., and Wiesendanger, R., "Studies of Magnetic Properties of Small Particles by Electron Holography." *Applied Physics A, Materials Science and Processing*, 1997, Vol. 65, No. 4/5, pp. 361–366.
°°° "Solak, H. H., David, C., and Gobrecht, J., "Fabrication of High-Resolution Zone Plates with Wideband Extreme-Ultraviolet Holography." *Applied Physics Letters*, Oct. 4, 2004, Vol. 85, No. 14, pp. 2700–2702.

Summary

TOPIC	RELEVANT EQUATIONS AND REMARKS	
1. Interference	Two superposing light waves interfere if their phase difference remains constant for a time long enough to observe. They interfere constructively if their phase difference is zero or an integer multiplied by 360°. They interfere destructively if their phase difference is 180° or an odd integer multiplied by 180°.	
Phase difference due to a path-length difference	$\delta = \frac{\Delta r}{\lambda} 2\pi$	33-1
Phase difference due to reflection	A phase difference of 180° is introduced when a light wave is reflected from a boundary between two media for which the wave speed is greater on the incident-wave side of the boundary.	
Thin films	The interference of light waves reflected from the front and back surfaces of a thin film produces interference fringes, commonly observed in soap films or oil films. The difference in phase between the two reflected waves results from the path-length difference of twice the thickness of the film plus any phase change due to reflection of one or both of the rays.	
Two slits	The path-length difference at an angle θ on a screen far away from two narrow slits separated by a distance d is $d \sin\theta$. If the intensity due to each slit separately is I_0, the intensity at points of constructive interference is $4I_0$, and the intensity at points of destructive interference is zero.	
Two slit interference maxima (sources in phase)	$d \sin\theta_m = m\lambda \qquad m = 0, 1, 2, \ldots$	33-2
Two slit interference maxima (sources 180° out of phase)	$d \sin\theta_m = (m - \frac{1}{2})\lambda \qquad m = 1, 2, 3, \ldots$	33-3
2. Diffraction	Diffraction occurs whenever a portion of a wavefront is limited by an obstacle or an aperture. The intensity of light at any point in space can be computed using Huygens's construction by taking each point on the wavefront to be a point source and computing the resulting interference pattern.	
Fraunhofer patterns	Fraunhofer patterns are observed at great distances from the obstacle or aperture so that the rays reaching any point are approximately parallel, or they can be observed using a lens to focus parallel rays on a viewing screen placed in the focal plane of the lens.	
Fresnel patterns	Fresnel patterns are observed at points not necessarily far from the source.	
Single slit	When light is incident on a single slit of width a, the intensity pattern on a screen far away shows a broad central diffraction maximum that decreases to zero at an angle θ_1 given by	
	$\sin\theta_1 = \frac{\lambda}{a}$	33-9
	The width of the central maximum is inversely proportional to the width of the slit. The zeros in the single-slit diffraction pattern occur at angles given by	
	$a \sin\theta_m = m\lambda \qquad m = 1, 2, 3, \ldots$	33-11
	The maxima on either side of the central maximum have intensities that are much smaller than the intensity of the central maximum.	
Two slits	The interference-diffraction pattern of two slits is the two-slit interference pattern modulated by the single-slit diffraction pattern.	
Resolution of two sources	When light from two point sources that are close together passes through an aperture, the diffraction patterns of the sources may overlap. If the overlap is too great, the two sources cannot be resolved as two separate sources. When the central diffraction maximum of one source falls at the diffraction minimum of the other source, the two sources are said to be just resolved by Rayleigh's criterion for resolution. For a circular aperture of diameter D, the critical angular separation of two sources for resolution by Rayleigh's criterion is given by	
Rayleigh's criterion	$\alpha_c = 1.22\frac{\lambda}{D}$	33-25

TOPIC	RELEVANT EQUATIONS AND REMARKS
*Gratings	A diffraction grating consisting of a large number of equally spaced lines or slits is used to measure the wavelength of light emitted by a source. The positions of the mth-order interference maxima from a grating are at angles given by $d \sin\theta_m = m\lambda \quad m = 0, 1, 2, \ldots$ (33-26). The resolving power of a grating is $R = \frac{\lambda}{\lvert\Delta\lambda\rvert} = mN$ (33-27), where N is the number of slits of the grating that are illuminated and m is the order number.
3. *Phasors	Two or more harmonic waves can be added by representing each wave as the y component of a two-dimensional vector called a phasor. The phase difference between two harmonic waves is represented as the angle between the phasors.

Answers to Concept Checks

33-1	6
33-2	True. Fresnel diffraction is the name that describes the observations when the observing screen is any distance from the source of the diffraction. Fraunhofer diffraction is the name that describes the observations in the limit that the observing screen is far from the source of the diffraction.

Answers to Practice Problems

33-1	$9.2\ \text{cm}^{-1}$
33-2	4.4 mm
33-3	$A = 5.0\ \text{V/m}, \delta = 37°$
33-4	62.07° and 62.18°

Problems

In a few problems, you are given more data than you actually need; in a few other problems, you are required to supply data from your general knowledge, outside sources, or informed estimate.

Interpret as significant all digits in numerical values that have trailing zeros and no decimal points.

- • Single-concept, single-step, relatively easy
- •• Intermediate-level, may require synthesis of concepts
- ••• Challenging
- SSM Solution is in the *Student Solutions Manual*
- Consecutive problems that are shaded are paired problems.

CONCEPTUAL PROBLEMS

1 • A phase difference due to a path-length difference is observed for monochromatic visible light. Which phase difference requires the least (minimum) path-length difference? (*a*) 90°,(*b*) 180°, (*c*) 270°,(*d*) The answer depends on the wavelength of the light.

2 • Which of the following pairs of light sources are coherent: (*a*) two candles, (*b*) one point source and its image in a plane mirror, (*c*) two pinholes uniformly illuminated by the same point source, (*d*) two headlights of a car, (*e*) two images of a point source due to reflection from the front and back surfaces of a soap film?

3 • The spacing between Newton's rings decreases rapidly as the diameter of the rings increases. Explain why this result occurs. SSM

4 • If the angle of a wedge-shaped air film such as the angle in Example 33-2 is too large, fringes are not observed. Why?

5 • Why must a film that is used to observe interference colors be thin?

6 • A loop of wire is dipped in soapy water and held up so that the soap film is vertical. (*a*) Viewed by reflection using white light, the top of the film appears black. Explain why. (*b*) Below the black region are colored bands. Is the first band red or violet?

7 • A two-slit interference pattern is formed using monochromatic laser light that has a wavelength of 640 nm. At the second maximum from the central maximum, what is the path-length difference between the light coming from each of the slits? (*a*) 640 nm, (*b*) 320 nm, (*c*) 960 nm, (*d*) 1280 nm SSM

8 • A two-slit interference pattern is formed using monochromatic laser light that has a wavelength of 640 nm. At the first minimum from the central maximum, what is the path-length difference between the light coming from each of the slits? (*a*) 640 nm, (*b*) 320 nm, (*c*) 960 nm, (*d*) 1280 nm

9 • A two-slit interference pattern is formed using monochromatic laser light that has a wavelength of 450 nm. What happens to the distance between the first maximum and the central maximum as the two slits are moved closer together? (*a*) The distance increases. (*b*) The distance decreases. (*c*) The distance remains the same.

10 • A two-slit interference pattern is formed using two different monochromatic lasers, one green and one red. Which color light has its first maximum closer to the central maximum? (*a*) green, (*b*) red, (*c*) Both maxima are at the same location.

11 • A single-slit diffraction pattern is formed using monochromatic laser light that has a wavelength of 450 nm. What happens to the distance between the first maximum and the central maximum as the slit is made narrower? (*a*) The distance increases. (*b*) The distance decreases. (*c*) The distance remains the same.

12 • Equation 33-2, which is $d \sin\theta_m = m\lambda$, and Equation 33-11, which is $a \sin\theta_m = m\lambda$, are sometimes confused. For each equation, define the symbols and explain the equation's application.

13 • When a diffraction grating is illuminated by white light, the first-order maximum of green light (*a*) is closer to the central maximum than the first-order maximum of red light, (*b*) is closer to the central maximum than the first-order maximum of blue light, (*c*) overlaps the second-order maximum of red light, (*d*) overlaps the second-order maximum of blue light.

14 • A double-slit interference experiment is set up in a chamber that can be evacuated. Using light from a helium–neon laser, an interference pattern is observed when the chamber is open to air. As the chamber is evacuated, one will note that (*a*) the interference fringes remain fixed, (*b*) the interference fringes move closer together, (*c*) the interference fringes move farther apart, (*d*) the interference fringes disappear completely.

15 • True or false:

(*a*) When waves interfere destructively, the energy is converted into heat.
(*b*) Interference patterns are observed only if the relative phases of the waves that superpose remain constant.
(*c*) In the Fraunhofer diffraction pattern for a single slit, the narrower the slit, the wider the central maximum of the diffraction pattern.
(*d*) A circular aperture can produce both a Fraunhofer diffraction pattern and a Fresnel diffraction pattern.
(*e*) The ability to resolve two point sources depends on the wavelength of the light. SSM

16 • You observe two very closely spaced sources of white light through a circular opening using various filters. Which color filter is most likely to prevent your resolving the images on your retinas as coming from two distinct sources? (*a*) red, (*b*) yellow, (*c*) green, (*d*) blue, (*e*) The filter choice is irrelevant.

17 •• Explain why the ability to distinguish the two headlights of an oncoming car at a given distance is easier for a human eye at night than during the daytime. Assume the headlights of the oncoming car are on both during both daytime and nighttime hours.

ESTIMATION AND APPROXIMATION

18 • It is claimed that the Great Wall of China is the only human-made object that can be seen from space using no equipment. Check to see if this claim is true, based on the resolving power of the human eye. Assume the observers are in a low-Earth orbit that has an altitude of about 250 km.

19 •• (*a*) Estimate how close an approaching car at night on a flat, straight part of a highway must be before its headlights can be distinguished from the single headlight of a motorcycle. (*b*) Estimate how far ahead of you a car is if its two red taillights merge to look as if they were one. SSM

20 •• A small loudspeaker is located at a large distance to the east from you. The loudspeaker is driven by a sinusoidal current whose frequency can be varied. Estimate the lowest frequency for which one of your ears would receive the sound waves exactly out of phase with that received by your other ear when you are facing north.

21 •• Estimate the maximum distance at which a binary star system could be resolvable by the human eye. Assume the two stars are about fifty times farther apart than Earth and the Sun are. Neglect any atmospheric effects. (A test similar to this "eye test" was used in ancient Rome to test for eyesight acuity before entering the army. A person who had normal eyesight could just barely resolve two well-known stars that appear close in the sky. Anyone who could not tell there were two stars failed the test.) SSM

PHASE DIFFERENCE AND COHERENCE

22 • Light that has a wavelength of 500 nm is incident normally on a film of water 1.00 μm thick. (*a*) What is the wavelength of the light in the water? (*b*) How many wavelengths are contained in the distance $2t$, where t is the thickness of the film? (*c*) The film has air on both sides. What is the phase difference between the wave reflected from the front surface and the wave reflected from the back surface in the region where the two reflected waves superpose?

23 •• Two coherent microwave sources both produce waves that each have a wavelength equal to 1.50 cm. The sources are located in the $z = 0$ plane, one at $x = 0$, $y = 15.0$ cm and the other at $x = 3.00$ cm, $y = 14.0$ cm. If the sources are in phase, find the difference in phase between the two waves for a receiver located at the origin. SSM

INTERFERENCE IN THIN FILMS

24 • A wedge-shaped film of air is made by placing a small slip of paper between the edges of two flat plates of glass. Light that has a wavelength of 700 nm is incident normally on the glass plates, and interference fringes are observed by reflection. (*a*) Is the first fringe near the point of contact of the plates dark or bright? Why? (*b*) If there are five dark fringes per centimeter, what is the angle of the wedge?

25 •• The diameters of fine fibers can be accurately measured using interference patterns. Two optically flat pieces of glass that each have a length L are arranged with the fiber between them, as shown in Figure 33-40. The setup is illuminated by monochromatic light, and the resulting interference fringes are observed. Suppose that L is 20.0 cm and that yellow sodium light (590 nm) is used for illumination. If 19 bright fringes are seen along this 20.0-cm distance, what are the limits on the diameter of the fiber? *Hint: The nineteenth fringe might not be right at the end, but you do not see a twentieth fringe at all.* SSM

26 •• Light that has a wavelength equal to 600 nm is used to illuminate two glass plates at normal incidence. The plates are 22 cm in length, touch at one end, and are separated at the other end by a wire that has a radius equal to 0.025 mm. How many bright fringes appear along the total length of the plates?

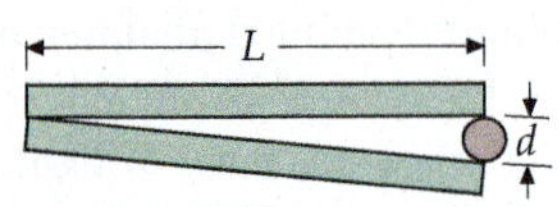

FIGURE 33-40 Problem 25

27 •• A thin film having an index of refraction of 1.50 is surrounded by air. It is illuminated normally by white light. Analysis of the reflected light shows that the wavelengths 360, 450, and 602 nm are the only missing wavelengths in or near the visible portion of the spectrum. That is, for those wavelengths, there is destructive interference. (*a*) What is the thickness of the film? (*b*) What visible wavelengths are brightest in the reflected interference pattern? (*c*) If this film were resting on glass that has an index of refraction of 1.60, what wavelengths in the visible spectrum would be missing from the reflected light?

28 •• A drop of oil (refractive index of 1.22) floats on water (refractive index of 1.33). When reflected light is observed from above, as shown in Figure 33-41, what is the thickness of the drop at

the point where the second red fringe, counting from the edge of the drop, is observed? Assume red light has a wavelength of 650 nm.

FIGURE 33-41 Problem 28

29 •• A film of oil that has an index of refraction of 1.45 rests on an optically flat piece of glass that has an index of refraction of 1.60. When illuminated by white light at normal incidence, light of wavelengths 690 nm and 460 nm is predominant in the reflected light. Determine the thickness of the oil film. SSM

30 •• A film of oil that has an index of refraction equal to 1.45 floats on water. When illuminated by white light at normal incidence, light of wavelengths 700 nm and 500 nm is predominant in the reflected light. Determine the thickness of the oil film.

NEWTON'S RINGS

31 •• A Newton's ring apparatus consists of a plano-convex glass lens that has a radius of curvature R and rests on a flat glass plate, as shown in Figure 33-42. The thin film is air of variable thickness. The apparatus is illuminated from above by light from a sodium lamp that has a wavelength equal to 590 nm. The pattern is viewed by reflected light. (*a*) Show that for a thickness t the condition for a bright (constructive) interference ring is $2t = \left(m + \frac{1}{2}\right)\lambda$, where $m = 0, 1, 2, \ldots$. (*b*) Show that for $t \ll R$, the radius r of a fringe is related to t by $r = \sqrt{2tR}$. (*c*) For a radius of curvature of 10.0 m and a lens diameter of 4.00 cm, how many bright fringes would you see in the reflected light? (*d*) What would be the diameter of the sixth bright fringe? (*e*) If the glass used in the apparatus has an index of refraction $n = 1.50$ and water replaces the air between the two pieces of glass, explain qualitatively the changes that will take place in the bright-fringe pattern. SSM

32 •• A plano-convex glass lens of radius of curvature 2.00 m rests on an optically flat glass plate. The arrangement is illuminated from above using monochromatic light that has a 520-nm wavelength. The indices of refraction of the lens and plate are 1.60. Determine the radii of the first and second bright fringe from the center in the reflected light.

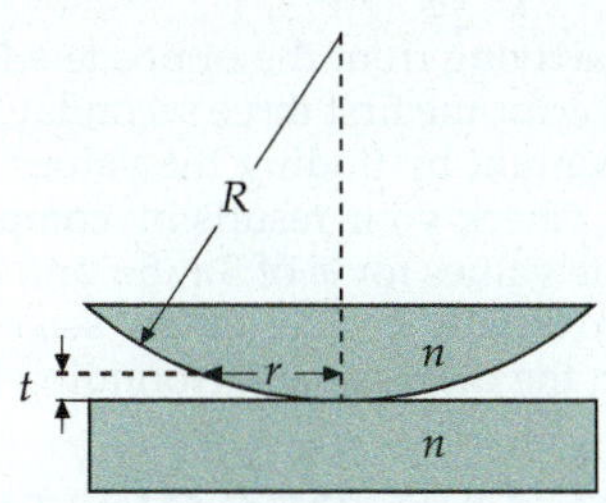

FIGURE 33-42 Problem 31

33 ••• Suppose that before the lens of Problem 32 is placed on the plate, a film of oil that has a refractive index equal to 1.82 is deposited on the plate. What will then be the radii of the first and second bright fringes?

TWO-SLIT INTERFERENCE PATTERNS

34 • Two narrow slits separated by 1.00 mm are illuminated by light that has a 600-nm wavelength, and the interference pattern is viewed on a screen 2.00 m away. Calculate the number of bright fringes per centimeter on the screen in the region near the center fringe.

35 • Using a conventional two-slit apparatus and light that has a 589-nm wavelength, 28 bright fringes per centimeter are observed near the center of a screen 3.00 m away. What is the slit separation? SSM

36 • Light that has a 633-nm wavelength and is from a helium–neon laser is shone normal to a plane having two slits. The first interference maximum is 82 cm from the central maximum on a screen 12 m away. (*a*) Find the separation of the slits. (*b*) How many interference maxima is it, in principle, possible to observe?

37 •• Two narrow slits are separated by a distance d. Their interference pattern is to be observed on a screen a large distance L away. (*a*) Calculate the spacing between successive maxima near the center fringe for light that has a 500-nm wavelength, when L is 1.00 m and d is 1.00 cm. (*b*) Would you expect to be able to observe the interference of light on the screen for this situation? (*c*) How close together should the slits be placed for the maxima to be separated by 1.00 mm for this wavelength and screen distance?

38 •• Light is incident at an angle ϕ with the normal to a vertical plane that has two slits of separation d (Figure 33-43). Show that the interference maxima are located at angles θ_m given by $\sin\theta_m + \sin\phi = m\lambda/d$.

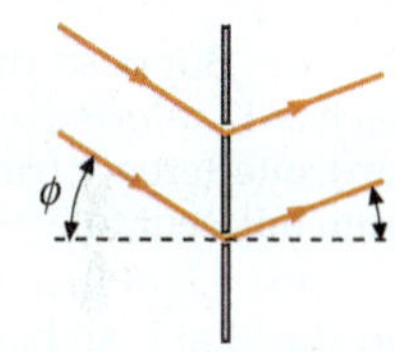

FIGURE 33-43 Problems 38 and 39

39 •• White light falls at an angle of 30° to the normal of a plane that has a pair of slits separated by 2.50 μm. What visible wavelengths give a bright interference maximum in the transmitted light in the direction normal to the plane? (See Problem 38.) SSM

40 •• Two small loudspeakers are separated by a distance of 5.0 cm, as shown in Figure 33-44. The speakers are driven in phase with a sine wave signal of frequency 10 kHz. A small microphone is placed a distance 1.00 m away from the speakers on the axis running through the middle of the two speakers, and the microphone is then moved perpendicular to the axis. Where does the microphone record the first minimum and the first maximum of the interference pattern from the speakers? The speed of sound in air is 343 m/s.

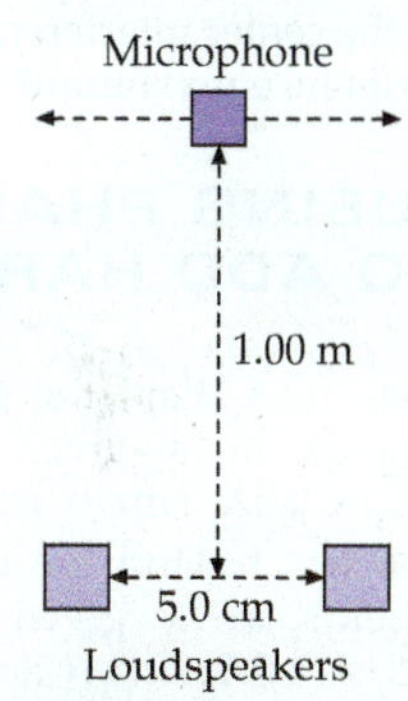

FIGURE 33-44 Problem 40

DIFFRACTION PATTERN OF A SINGLE SLIT

41 • Light that has a 600-nm wavelength is incident on a long narrow slit. Find the angle of the first diffraction minimum if the width of the slit is (*a*) 1.0 mm, (*b*) 0.10 mm, and (*c*) 0.010 mm.

42 • Plane microwaves are incident on a thin metal sheet that has a long, narrow slit of width 5.0 cm in it. The microwave radiation strikes the sheet at normal incidence. The first diffraction minimum is observed at $\theta = 37°$. What is the wavelength of the microwaves?

43 ••• Measuring the distance to the moon (lunar ranging) is routinely done by firing short-pulse lasers and measuring the time it takes for the pulses to reflect back from the moon. A pulse is fired from Earth. To send the pulse out, the pulse is expanded so that it fills the aperture of a 6.00-in-diameter telescope. Assuming the only thing spreading the beam out is diffraction and that the light wavelength is 500 nm, how large will the beam be when it reaches the moon, 3.82×10^5 km away? SSM

INTERFERENCE–DIFFRACTION PATTERN OF TWO SLITS

44 • How many interference maxima will be contained in the central diffraction maximum in the interference–diffraction pattern of two slits if the separation of the slits is exactly 5 times their width? How many will there be if the slit separation is an integral multiple of the slit width (that is, $d = na$ for any value of n)?

45 •• A two-slit Fraunhofer interference–diffraction pattern is observed using light that has a wavelength equal to 500 nm. The slits have a separation of 0.100 mm and an unknown width. (*a*) Find the width if the fifth interference maximum is at the same angle as the first diffraction minimum. (*b*) For that case, how many bright interference fringes will be seen in the central diffraction maximum? SSM

46 •• A two-slit Fraunhofer interference–diffraction pattern is observed using light that has a wavelength equal to 700 nm. The slits have widths of 0.010 mm and are separated by 0.20 mm. How many bright fringes will be seen in the central diffraction maximum?

47 •• Suppose that the *central* diffraction maximum for two slits has 17 interference fringes for some wavelength of light. How many interference fringes would you expect in the diffraction maximum adjacent to one side of the central diffraction maximum?

48 •• Light that has a wavelength equal to 550 nm illuminates two slits that both have widths equal to 0.030 mm and separations equal to 0.15 mm. (*a*) How many interference maxima fall within the full width of the central diffraction maximum? (*b*) What is the ratio of the intensity of the third interference maximum to one side of the center interference maximum to the intensity of the center interference maximum?

*USING PHASORS TO ADD HARMONIC WAVES

49 • Find the resultant of the two waves whose electric fields at a given location vary with time as follows: $\vec{E}_1 = 2.0A_0 \sin\omega t\,\hat{i}$ and $\vec{E}_2 = 3.0A_0 \sin\left(\omega t + \frac{3}{2}\pi\right)\hat{i}$. SSM

50 • Find the resultant of the two waves whose electric fields at a given location vary with time as follows: $\vec{E}_1 = 4.0A_0 \sin\omega t\,\hat{i}$ and $\vec{E}_2 = 3.0A_0 \sin\left(\omega t + \frac{1}{6}\pi\right)\hat{i}$.

51 •• Monochromatic light is incident on a sheet with a long narrow slit (Figure 33-45). Let I_0 be the intensity at the central maximum of the diffraction pattern on a distant screen, and let I be the intensity at the second intensity maximum from the central intensity maximum. The distance from this second intensity maximum to the far edge of the slit is longer than the distance from the second intensity maximum to the near edge of the slit by approximately 2.5 wavelengths. What is the ratio of I to I_0?

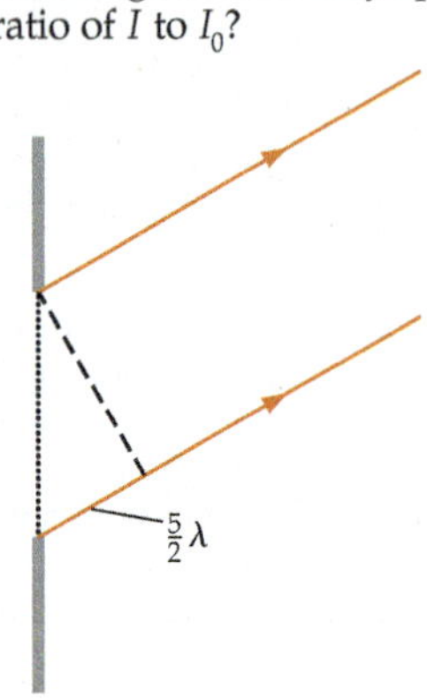

FIGURE 33-45 Problem 51

52 •• Monochromatic light is incident on a sheet that has three long narrow parallel equally spaced slits a distance d apart. (*a*) Show that the positions of the interference minima on a screen a large distance L away from the sheet that has the three equally spaced slits (spacing d, with $d \gg \lambda$) are given approximately by $y_m = m\lambda L/3d$, where $m = 1, 2, 4, 5, 7, 8, 10, \ldots,$ that is, m is *not* a multiple of 3. (*b*) For a screen distance of 1.00 m, a light wavelength of 500 nm, and a source spacing of 0.100 mm, calculate the width of the principal interference maxima (the distance between successive minima) for three sources.

53 •• Monochromatic light is incident on a sheet that has four long narrow parallel equally spaced slits a distance d apart. (*a*) Show that the positions of the interference minima on a screen a large distance L away from the sheet that has the four equally spaced slits (spacing d, with $d \gg \lambda$) are given approximately by $y_m = m\lambda L/4d$, where $m = 1, 2, 3, 5, 6, 7, 9, 10, \ldots,$ that is, m is *not* a multiple of 4. (*b*) For a screen distance of 2.00 m, a light wavelength of 600 nm, and a source spacing of 0.100 mm, calculate the width of the principal interference maxima (the distance between successive minima) for four sources. Compare the width with that for two sources with the same spacing. SSM

54 •• Light that has a wavelength equal to 480 nm falls normally on four slits. Each slit is 2.00 μm wide and the center-to-center separation between it and the next slit is 6.00 μm. (*a*) Find the angular width of the central intensity maximum of the single-slit diffraction pattern on a distant screen. (*b*) Find the angular position of all interference intensity maxima that lie inside the central diffraction maximum. (*c*) Find the angular width of the central interference maximum. That is, find the angle between the first interference intensity minima on either side of the central interference maximum. (*d*) Sketch the relative intensity as a function of the sine of the angle.

55 ••• Three slits, each separated from its neighbor by 60.0 μm, are illuminated at the central intensity maximum by a coherent light source that has a wavelength equal to 550 nm. The slits are extremely narrow. A screen is located 2.50 m from the slits. The intensity is 50.0 mW/m^2. Consider a location 1.72 cm from the central maximum. (*a*) Draw a phasor diagram suitable for the addition of the three harmonic waves at that location. (*b*) From the phasor diagram, calculate the intensity of light at that location. SSM

56 ••• In single-slit Fraunhofer diffraction, the intensity pattern (Figure 33-11) consists of a broad central maximum with a sequence of secondary maxima to either side of the central maximum. The intensity is given by $I = I_0\left(\dfrac{\sin\frac{1}{2}\phi}{\frac{1}{2}\phi}\right)^2$, where ϕ is the phase difference between the wavelets arriving from the opposite edges of the slits. Calculate the values of ϕ for the first three seconday maxima to one side of the central maximum by finding the values of ϕ for which $dI/d\phi$ is equal to zero. Check your results by comparing your answers with approximate values for ϕ of 3π, 5π and 7π. (That these values for ϕ are approximatly correct at the secondary intensity maxima is discussed in the discussion surrounding Figure 33-27.)

DIFFRACTION AND RESOLUTION

57 • Light that has a wavelength equal to 700 nm is incident on a pinhole of diameter 0.100 mm. (*a*) What is the angle between the central maximum and the first diffraction minimum for a Fraunhofer diffraction pattern? (*b*) What is the distance between the central maximum and the first diffraction minimum on a screen 8.00 m away? SSM

58 • Two sources of light that both have wavelengths equal to 700 nm are 10.0 m away from the pinhole of Problem 57. How far apart must the sources be for their diffraction patterns to be resolved by Rayleigh's criterion?

59 • Two sources of light that both have wavelengths equal to 700 nm are separated by a horizontal distance x. They are 5.00 m from a vertical slit of width 0.500 mm. What is the smallest value of x for which the diffraction pattern of the sources can be resolved by Rayleigh's criterion?

60 •• The ceiling of your lecture hall is probably covered with acoustic tile, which has small holes separated by about 6.0 mm. (*a*) Using light that has a wavelength of 500 nm, how far could you be from this tile and still resolve the holes? Assume the diameter of the pupil of each of your eyes is about 5.0 mm. (*b*) Could you resolve the holes better using red light or violet light? Explain your answer.

61 •• The telescope on Mount Palomar has a diameter of 200 in. Suppose a double star were 4.00 light-years away. Under ideal conditions, what must be the minimum separation of the two stars for their images to be resolved using light that has a wavelength equal to 550 nm? SSM

62 •• The star Mizar in Ursa Major is a binary system of stars that have nearly equal magnitudes. The angular separation between the two stars is 14 seconds of arc. What is the minimum diameter of the pupil that allows resolution of the two stars using light that has a wavelength equal to 550 nm?

*DIFFRACTION GRATINGS

63 • A diffraction grating that has 2000 slits per centimeter is used to measure the wavelengths emitted by hydrogen gas. (*a*) At what angles in the first-order spectrum would you expect to find the two violet lines that have wavelengths of 434 nm and 410 nm? (*b*) What are the angles if the grating has 15000 slits per centimeter? SSM

64 • Using a diffraction grating that has 2000 slits per centimeter, two lines in the first-order hydrogen spectrum are found at angles of 9.72×10^{-2} rad and 1.32×10^{-1} rad. What are the wavelengths of the lines?

65 • The colors of many butterfly wings and beetle carapaces are due to the effects of diffraction. The *Morpho* butterfly has structural elements on its wings that effectively act as a diffraction grating that has an 880-nm spacing. At what angle will the first diffraction maximum occur for normally incident light diffracted by the butterfly's wings? Assume the light is blue and has a wavelength of 440 nm.

66 •• A diffraction grating that has 2000 slits per centimeter is used to analyze the spectrum of mercury. (*a*) Find the angular separation in the first-order spectrum of the two lines that have wavelengths equal to 579 nm and 577 nm. (*b*) How wide must the beam on the grating be for the lines to be resolved?

67 •• A diffraction grating that has 4800 lines per centimeter is illuminated at normal incidence using white light (wavelength range of 400 nm to 700 nm). How many orders of complete spectra can one observe in the transmitted light? Do any of these orders overlap? If so, describe the overlapping regions. SSM

68 •• A square diffraction grating that has an area of 25.0 cm^2 has a resolution of 22000 in the fourth order. At what angle should you look to see a wavelength of 510 nm in the fourth order?

69 •• Sodium light that has a wavelength equal to 589 nm falls normally on a 2.00-cm-square diffraction grating ruled with 4000 lines per centimeter. The Fraunhofer diffraction pattern is projected onto a screen a distance of 1.50 m from the grating by a 1.50-m-focal-length lens that is placed immediately in front of the grating. Find (*a*) the distances of the first and second order intensity maxima from the central intensity maximum, (*b*) the width of the central maximum, and (*c*) the resolution in the first order. (Assume the entire grating is illuminated.)

70 •• The spectrum of neon is exceptionally rich in the visible region. Among the many lines are two lines at wavelengths of 519.313 nm and 519.322 nm. If light from a neon discharge tube is normally incident on a transmission grating that has 8400 lines per centimeter and the spectrum is observed in second order, what must be the width of the grating that is illuminated, so that the two lines can be resolved?

71 •• Mercury has several stable isotopes, among them ^{198}Hg and ^{202}Hg. The strong spectral line of mercury, at about 546.07 nm, is a composite of spectral lines from the various mercury isotopes. The wavelengths of the line for ^{198}Hg and ^{202}Hg are 546.07532 nm and 546.07355 nm, respectively. What must be the resolving power of a grating capable of resolving the two isotopic lines in the third-order spectrum? If the grating is illuminated over a 2.00-cm-wide region, what must be the number of lines per centimeter of the grating? SSM

72 ••• A diffraction grating has n lines per unit length. Show that the angular separation ($\Delta\theta$) of two lines of wavelengths λ and $\lambda + \Delta\lambda$ is approximately $\Delta\theta = \Delta\lambda/\sqrt{\frac{1}{n^2m^2} - \lambda^2}$, where m is the order number.

73 ••• For a diffraction grating in which all the surfaces are normal to the incident radiation, most of the energy goes into the zeroth order, which is useless from a spectroscopic point of view, because in zeroth order all the wavelengths are at 0°. Therefore, modern reflection gratings have shaped, or *blazed*, grooves, as shown in Figure 33-46. This shifts the specular reflection, which contains most of the energy, from the zeroth order to some higher order. (*a*) Calculate the *blaze angle* ϕ_m in terms of the groove separation d, the wavelength λ, and the order number m in which specular reflection is to occur for $m = 1, 2, \ldots$. (*b*) Calculate the proper blaze angle for the specular reflection to occur in the second order for light of wavelength 450 nm incident on a grating with 10000 lines per centimeter. SSM

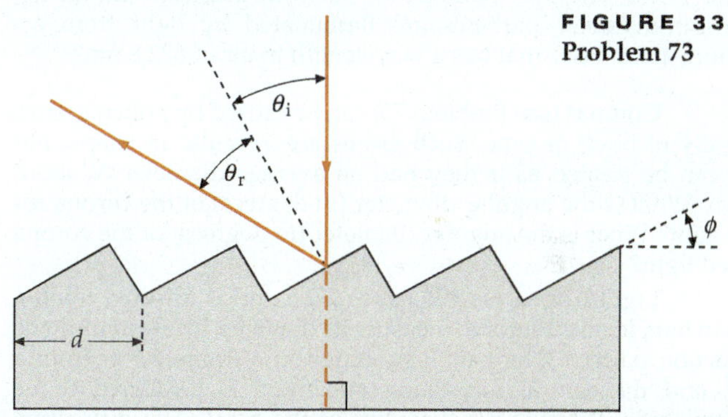

FIGURE 33-46 Problem 73

74 ••• In this problem, you will derive the relation $R = \lambda/|\Delta\lambda| = mN$ (Equation 33-27) for the resolving power of a diffraction grating having N slits separated by a distance d. To do this, you will calculate the angular separation between the intensity maximum and intensity minimum for some wavelength λ and set it equal to the angular separation of the mth-order maximum for two nearby wavelengths. (*a*) First show that the phase difference ϕ waves between the waves from two adjacent slits is given by $\phi = \frac{2\pi d}{\lambda}\sin\theta$. (*b*) Next, differentiate that expression to show that a small change in angle $d\theta$ results in a change in phase of $d\phi$ given by $d\phi = \frac{2\pi d}{\lambda}\cos\theta\, d\theta$. (*c*) Then for N slits, the angular separation between an interference maximum and an interference minimum corresponds to a phase change of $d\phi = 2\pi/N$. Use that to show that the angular separation $d\theta$ between the intensity maximum and intensity minimum for some wavelength λ is given by $d\theta = \frac{\lambda}{Nd\cos\theta}$. (*d*) Next, use the fact that the angle of the mth-order interference maximum for wavelength λ is specified by $d\sin\theta = m\lambda$ (Equation 33-26). Compute the differential of each side of the equation to show that angular separation of the mth-order

maximum for two nearly equal wavelengths differing by $d\lambda$ is given by $d\theta = \dfrac{m\,d\lambda}{d\cos\theta}$. (*e*) According to Rayleigh's criterion, two wavelengths will be resolved in the *m*th order if the angular separation of the wavelengths, given by the Part-(*d*) result, equals the angular separation of the interference maximum and the interference minimum given by the Part-(*c*) result. Use this to arrive at $R = \lambda/|\Delta\lambda| = mN$ (Equation 33-27) for the resolving power of a grating.

GENERAL PROBLEMS

75 • Naturally occuring coronas (brightly colored rings) are sometimes seen around the moon or the Sun when viewed through a thin cloud. (Warning: When viewing a Sun corona, be sure that the entire Sun is blocked by the edge of a building, a tree, or a traffic pole to safeguard your eyes.) These coronas are due to diffraction of light by small water droplets in the cloud. A typical angular diameter for a coronal ring is about 10°. From this, estimate the size of the water droplets in the cloud. Assume that the water droplets can be modeled as opaque disks that have the same radius as the droplet, and that the Fraunhofer diffraction pattern from an opaque disk is the same as the pattern from an aperture of the same diameter. (This last statement is known as *Babinet's principle*.) SSM

76 • An artificial corona (see Problem 75) can be made by placing a suspension of polystyrene microspheres in water. Polystyrene microspheres are small, uniform spheres that are made of plastic and have indices of refraction equal to 1.59. Assuming that the water has an index of refraction equal to 1.33, what is the angular diameter of such an artificial corona if 5.00-μm-diameter particles are illuminated by light from a helium–neon laser that has a wavelength in air of 632.8 nm?

77 • Coronas (see Problem 75) can be caused by pollen grains, typically of birch or pine. Such grains are irregular in shape, but they can be treated as if they had an average diameter of about 25 μm. What is the angular diameter (in degrees) of the corona for blue light? What is the angular diameter (in degrees) of the corona for red light?

78 • Light from a He–Ne laser (632.8 nm) is directed upon a human hair, in an attempt to measure its diameter by examining the diffraction pattern. The hair is mounted in a frame 7.5 m from a wall, and the central diffraction maximum is measured to be 14.6 cm wide. What is the diameter of the hair? (The diffraction pattern of a hair that has a diameter d is the same as the diffraction pattern of a single slit that has a width $a = d$. See Babinet's principle discussed in Problem 75.)

79 • A long, narrow horizontal slit lies 1.00 μm above a plane mirror, which is in the horizontal plane. The interference pattern produced by the slit and its image is viewed on a screen 1.00 m from the slit. The wavelength of the light is 600 nm. (*a*) Find the distance from the mirror to the first maximum. (*b*) How many dark bands per centimeter are seen on the screen? SSM

80 • A radio telescope is situated at the edge of a lake. The telescope is looking at light from a radio galaxy that is just rising over the horizon. If the height of the antenna is 20 m above the surface of the lake, at what angle above the horizon will the radio galaxy be when the telescope is centered in the first intensity interference maximum of the radio waves? Assume the wavelength of the radio waves is 20 cm. *Hint: The interference is caused by the light reflecting off the lake and remember that this reflection will result in a 180° phase shift.*

81 • The diameter of the radio telescope at Arecibo, Puerto Rico, is 300 m. What is the smallest angular separation of two objects that this telescope can detect when it is tuned to detect microwaves of 3.2-cm wavelength?

82 •• A thin layer of a transparent material that has an index of refraction of 1.30 is used as a nonreflective coating on the surface of glass that has an index of refraction of 1.50. What should the minimum thickness of the material be for the material to be nonreflecting for light that has a wavelength of 600 nm?

83 •• A *Fabry–Perot interferometer* (Figure 33-47) consists of two parallel, half-silvered mirrors that face each other and are separated by a small distance a. A half-silvered mirror is one that transmits 50% of the incident intensity and reflects 50% of the incident intensity. Show that when light is incident on the interferometer at an angle of incidence θ, the transmitted light will have maximum intensity when $2a = m\lambda\cos\theta$. SSM

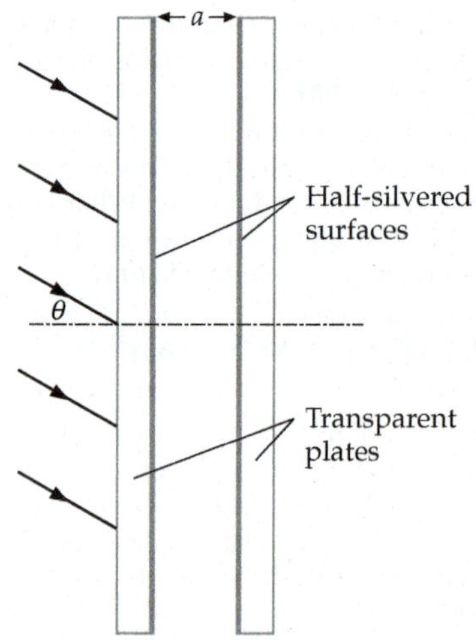

FIGURE 33-47 Problem 83

84 •• A mica sheet that is 1.20 μm thick is suspended in air. In reflected light, there are gaps in the visible spectrum at 421, 474, 542, and 633 nm. Find the index of refraction of the mica sheet.

85 •• A camera lens is made of glass that has an index of refraction of 1.60. This lens is coated with a magnesium fluoride film (index of refraction equal to 1.38) to enhance its light transmission. The purpose of this film is to produce zero reflection for light that has a wavelength of 540 nm. Treat the lens surface as a flat plane and the film as a uniformly thick flat film. (*a*) What minimum thickness of this film will accomplish its objective? (*b*) Would there be destructive interference for any other visible wavelengths? (*c*) By what factor would the reflection for light of 400 nm wavelength be reduced by the presence of this film? Neglect the variation in the reflected light amplitudes from the two surfaces. SSM

86 •• In a pinhole camera, the image is fuzzy because of geometry (rays arrive at the film after passing through different parts of the pinhole) and because of diffraction. As the pinhole is made smaller, the fuzziness due to geometry is reduced, but the fuzziness due to diffraction is increased. The optimum size of the pinhole for the sharpest possible image occurs when the spread due to diffraction equals the spread due to the geometric effects of the pinhole. Estimate the optimum size of the pinhole if the distance from the pinhole to the film is 10.0 cm and the wavelength of the light is 550 nm.

87 •• The Impressionist painter Georges Seurat used a technique called *pointillism*, in which his paintings are composed of small, closely spaced dots of pure color, each about 2.0 mm in diameter. The illusion of the colors blending together smoothly is produced in the eyes of the viewer by diffraction effects. Calculate the minimum viewing distance for this effect to work properly. Use the wavelength of visible light that requires the *greatest* distance between dots, so that you are sure the effect will work for *all* visible wavelengths. Assume the pupil of the eye has a diameter of 3.0 mm. SSM

PART VI **MODERN PHYSICS: Quantum Mechanics, Relativity, and the Structure of Matter**

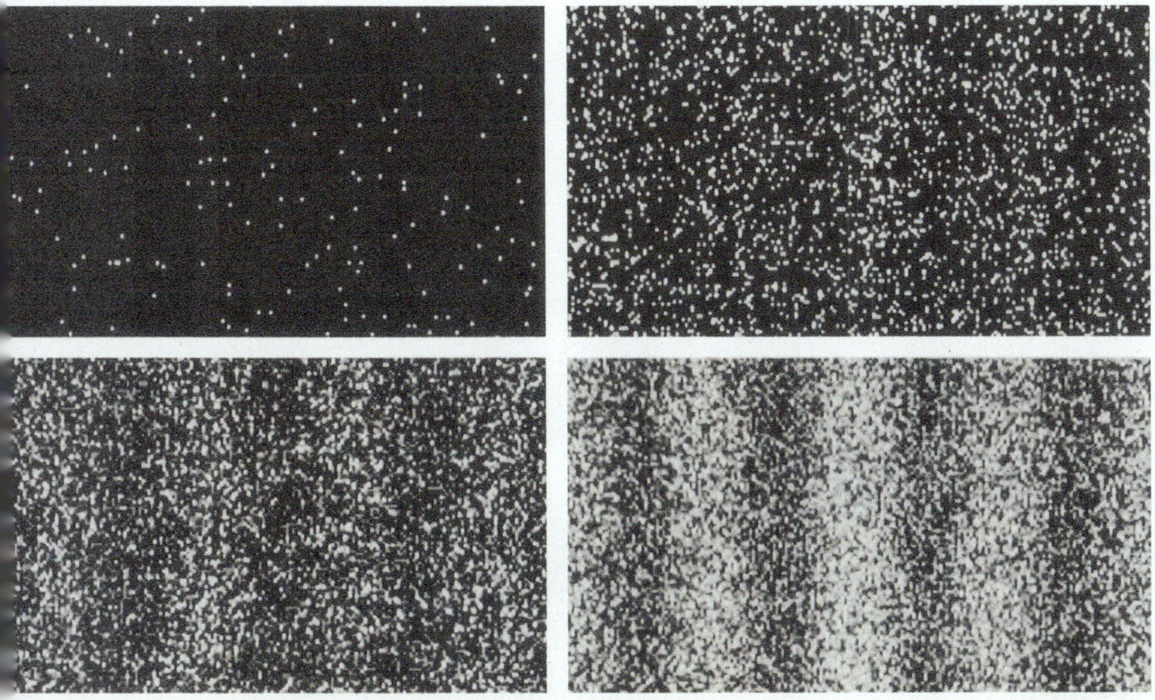

CHAPTER 34

ELECTRON INTERFERENCE PATTERN PRODUCED BY ELECTRONS INCIDENT ON A BARRIER CONTAINING TWO SLITS: (*A*) 10 ELECTRONS, (*B*) 100 ELECTRONS, (*C*) 3000 ELECTRONS, AND (*D*) 70,000 ELECTRONS. THE MAXIMA AND MINIMA DEMONSTRATE THE WAVE NATURE OF THE ELECTRON AS IT TRAVERSES THE SLITS. INDIVIDUAL DOTS ON THE SCREEN INDICATE THE PARTICLE NATURE OF THE ELECTRON AS IT EXCHANGES ENERGY WITH THE DETECTOR. THE PATTERN IS THE SAME WHETHER ELECTRONS OR PHOTONS (PARTICLES OF LIGHT) ARE USED. *(Courtesy of Akira Tononmura, Advanced Research Laboratory, Hitachi, Ltd.)*

Wave–Particle Duality and Quantum Physics

? How do you calculate the wavelength of an electron? (See Example 34-4.)

At the beginning of the twentieth century, it was thought that sound, light, and other electromagnetic radiation (such as radio waves) were waves, and electrons, protons, atoms, and similar units were particles. The first 30 years of that century revealed startling developments in theoretical and experimental physics, such as the finding that light actually exchanges energy in discrete lumps or quanta, just like particles, and the finding that an electron exhibits diffraction and interference as it propagates through space, just like a wave. The fact that light exchanges energy like a particle implies that light energy is not continuous but is *quantized*. Similarly, the wave nature of the electron, along with the fact that the standing wave condition requires a discrete set of frequencies, implies that the energy of an electron in a confined region of space is not continuous, but is quantized to a discrete set of values.

In this chapter, we begin by discussing some basic properties of light and electrons, examining their wave and particle characteristics. We then consider some of the detailed properties of matter waves, showing, in particular, how standing waves imply the quantization of energy. Finally, we discuss some of the important features of the theory of quantum physics, which was developed in the 1920s and which has been extremely successful in describing nature. Quantum physics is now the basis of our understanding of both atomic and subatomic systems and systems that have very low temperatures.

34-1 WAVES AND PARTICLES

We have seen that the propagation of waves through space is quite different from the propagation of particles. Waves bend around corners (diffraction) and interfere with one another, producing interference patterns. If a wave encounters a small aperture, the wave spreads out on the other side as if the aperture were a point source. The propagation of particles is quite unlike the propagation of waves. Particles travel in straight lines until they collide with something, after which the particles again travel in straight lines. If two particle beams meet in space, they never produce an interference pattern.

Particles and waves also exchange energy differently. Particles exchange energy in collisions that occur at specific points in space and in time. The energy of waves, on the other hand, is spread out in space and deposited continuously as the wave fronts interact with matter.

Sometimes the propagation of a wave cannot be distinguished from the propagation of a beam of particles. If the wavelength λ is very small compared to distances from the edges of objects, diffraction effects are negligible and the wave travels in straight lines. Also, interference maxima and minima are so close together in space as to be unobservable. The result is that the wave interacts with a detector, like a beam of numerous small particles each exchanging a small amount of energy; the exchange cannot distinguish particles from waves. For example, you do not observe the individual air molecules bouncing off your face if the wind blows on it. Instead, the interaction of billions of particles appears to be continuous, like that of a wave.

34-2 LIGHT: FROM NEWTON TO MAXWELL

The question of whether light consists of a beam of particles or waves in motion is one of the most interesting in the history of science (see Chapter 31). Isaac Newton used a particle theory of light to explain the laws of reflection and refraction; however, for refraction, Newton needed to assume that light travels faster in water or glass than in air, an assumption later shown to be incorrect. The chief early proponents of the wave theory were Robert Hooke and Christian Huygens, who explained refraction by assuming that light travels more slowly in glass or water than it does in air.* Newton favored the theory that light consists of particles and does not consist of waves because, in his time, light was believed to travel through a medium only in straight lines—diffraction had not yet been observed.

Because of Newton's great reputation and authority, his particle theory of light was accepted for more than a century. Then, in 1801, Thomas Young demonstrated the wave nature of light in a famous experiment in which two coherent light sources are produced by illuminating a pair of narrow, parallel slits with a single source (Figure 34-1*a*).† In Chapter 33, we saw that when light encounters a small opening, the opening acts as a point source of waves (Figure 33-7). In Young's experiment, each slit acts as a line source, which is equivalent to a point source in two dimensions.‡ The interference pattern is observed on a screen placed behind the slits. Interference maxima occur at angles so that the path difference is an integral number of wavelengths. Similarly, interference minima occur if the path difference is one-half the wavelength or any odd number of half wavelengths. Figure 34-1*b* shows the intensity pattern as seen on the screen. Remember that if two coherent waves of equal intensity I_0 meet in space, the result can be a wave of intensity $4I_0$ (constructive interference), an intensity of zero (destructive interference), or a wave of intensity between zero and $4I_0$, depending on the phase difference between the waves at the observation point. Young's experiment and many other experiments demonstrate that light propagates like a wave.

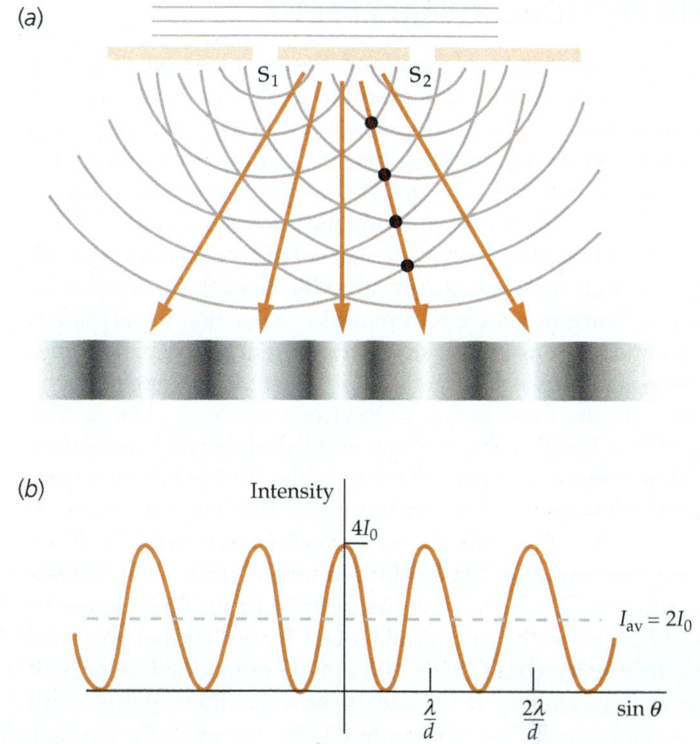

FIGURE 34-1 (*a*) Two slits act as coherent sources of light for the observation of interference in Young's experiment. Cylindrical waves from the slits overlap and produce an interference pattern on a screen far away. (*b*) The intensity pattern produced in Figure 34-1*a*. The intensity is maximum at points where the path difference is an even number of half wavelengths, and the intensity is zero where the path difference is an odd number of half wavelengths.

In the early nineteenth century, the French physicist Augustin Fresnel (1788–1827) performed extensive experiments on interference and diffraction and put the wave theory on a rigorous mathematical basis. Fresnel showed that the observed straight-line propagation of light is a result of the very short wavelengths of visible light.

The classical wave theory of light culminated in 1860 when James Clerk Maxwell published his mathematical theory of electromagnetism. This theory yielded a wave equation that predicted the existence of electromagnetic waves that propagate with a speed that can be calculated from the laws of electricity and magnetism.# The fact that the result of this calculation was $c \approx 3 \times 10^8$ m/s, the same as the speed of light, suggested to Maxwell that light is an electromagnetic wave. The eye is sensitive to electromagnetic waves that have wavelengths in the range from approximately 400 nm (1 nm $= 10^{-9}$ m) to approximately 700 nm. This range is called *visible light*. Other electromagnetic waves (for example, microwave waves, radio waves, and X rays) differ from visible light waves only in wavelength and in frequency.

34-3 THE PARTICLE NATURE OF LIGHT: PHOTONS

The diffraction of light and the existence of an interference pattern in the two-slit experiment give clear evidence that light has wave properties. However, early in the twentieth century, it was found that light energy comes in discrete amounts.

* See Section 5 of Chapter 31.
† See Section 3 of Chapter 33.
‡ See Section 4 of Chapter 33.
# See Section 3 of Chapter 30

THE PHOTOELECTRIC EFFECT

The quantum nature of light and the quantization of energy were suggested by Albert Einstein in 1905 in his explanation of the photoelectric effect. Einstein's work marked the beginning of quantum theory, and for his work, Einstein received the Nobel Prize in Physics. Figure 34-2 shows a schematic diagram of the basic apparatus for studying the photoelectric effect. Light of a single frequency enters an evacuated chamber and falls on a clean metal surface C (C for cathode), causing electrons to be emitted. Some of these electrons strike the second metal plate A (A for anode), constituting an electric current between the plates. Plate A is negatively charged, so the electrons are repelled by it, with only the most energetic electrons reaching plate A. The maximum kinetic energy of the emitted electrons is measured by slowly increasing the voltage until the current becomes zero. Experiments give the surprising result that the maximum kinetic energy of the emitted electrons is *independent of the intensity* of the incident light. Classically, we would expect that increasing the rate at which light energy falls on the metal surface would increase the energy absorbed by individual electrons and, therefore, would increase the maximum kinetic energy of the electrons emitted. Experiments show that this classical result does not happen. The maximum kinetic energy of the emitted electrons is the same for a given wavelength of incident light, no matter how intense the light is. Einstein demonstrated that this experimental result can be explained if light energy is quantized in small bundles called **photons.** The energy E of each photon is given by

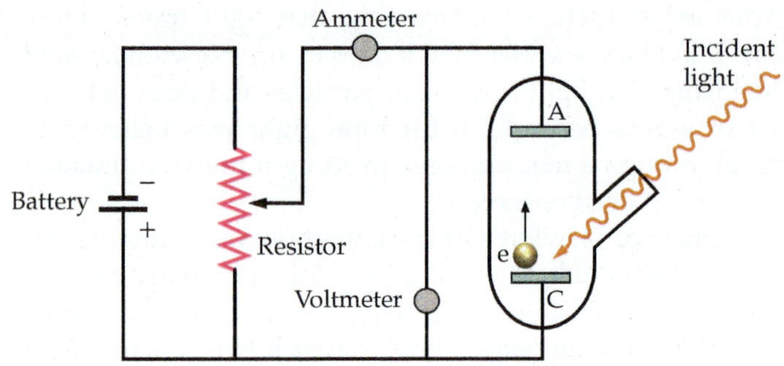

FIGURE 34-2 A schematic drawing of the apparatus for studying the photoelectric effect. Light of a single frequency enters an evacuated chamber and strikes the cathode C, which then ejects electrons (electron in figure is not drawn to scale). The current in the ammeter measures the number of these electrons that reach the anode A per unit time. The anode is made electrically negative with respect to the cathode to repel the electrons. Only those electrons that have enough initial kinetic energy to overcome the repulsion can reach the anode. The voltage between the two plates is slowly increased until the current becomes zero, which happens when even the most energetic electrons do not make it to plate A.

$$E = hf = \frac{hc}{\lambda} \tag{34-1}$$

EINSTEIN EQUATION FOR PHOTON ENERGY

where f is the frequency, and h is a constant now known as **Planck's constant.*** The measured value of this constant is

$$h = 6.626 \times 10^{-34}\ \text{J}\cdot\text{s} = 4.136 \times 10^{-15}\ \text{eV}\cdot\text{s} \tag{34-2}$$

PLANCK'S CONSTANT

Equation 34-1 is sometimes called the **Einstein equation.**

A light beam consists of a beam of particles—photons—each having energy hf. The intensity (power per unit area) of a monochromatic light beam is the number of photons per unit area per unit of time, multiplied by the energy per photon. The interaction of the light beam with the metal surface consists of collisions between photons and electrons. During each of these collisions, the photon gives all its energy to an electron and the photon no longer exists. The electron is emitted from the surface after it receives the energy from a single photon. If the intensity of light is increased, more photons fall on the surface per unit time, and more electrons are emitted per unit time. However, each photon still has the same energy hf, so the energy absorbed by each electron is unchanged.

* In 1900, the German physicist Max Planck introduced this constant to explain discrepancies between the theoretical curves and experimental data on the spectrum of blackbody radiation. Planck also assumed that the radiation was emitted and absorbed by a blackbody in quanta of energy hf, but he considered his assumption to be just a computational device rather than a fundamental property of electromagnetic radiation. (Blackbody radiation is discussed in Chapter 20.)

If ϕ is the minimum energy necessary to remove an electron from a metal surface, the maximum kinetic energy of the electrons emitted is given by

$$K_{max} = \left(\tfrac{1}{2}mv^2\right)_{max} = hf - \phi \qquad 34\text{-}3$$

EINSTEIN'S PHOTOELECTRIC EQUATION

where f is the frequency of the photons. The quantity ϕ, called the **work function,** is a characteristic of the particular metal. (Some electrons will have kinetic energies less than $hf - \phi$, because of the loss of energy from traveling through the metal.)

According to Einstein's photoelectric equation, a plot of K_{max} versus frequency f should be a straight line that has the slope h. This was a bold prediction, because, at the time, no evidence existed that Planck's constant had any application outside of blackbody radiation. In addition, there was no experimental data on K_{max} versus frequency f, because no one before had even suspected that the frequency of the light was related to K_{max}. This prediction was difficult to verify experimentally, but careful experiments by R. A. Millikan approximately 10 years later showed that Einstein's equation was correct. Figure 34-3 shows a plot of Millikan's data.

Photons that have frequencies less than a **threshold frequency** f_t, and therefore have wavelengths greater than a **threshold wavelength** $\lambda_t = c/f_t$, do not have enough energy to eject an electron from a particular metal. The threshold frequency and the corresponding threshold wavelength can be related to the work function ϕ by setting the maximum kinetic energy of the electrons equal to zero in Equation 34-3. Then

$$\phi = hf_t = \frac{hc}{\lambda_t} \qquad 34\text{-}4$$

Work functions for metals are typically a few electron volts. Because wavelengths are usually given in nanometers and energies in electron volts, it is useful to have the value of hc in electron volt–nanometers:

$$hc = (4.1357 \times 10^{-15}\ \text{eV}\cdot\text{s})(2.9979 \times 10^{8}\ \text{m/s}) = 1.240 \times 10^{-6}\ \text{eV}\cdot\text{m}$$

or

$$hc = 1240\ \text{eV}\cdot\text{nm} \qquad 34\text{-}5$$

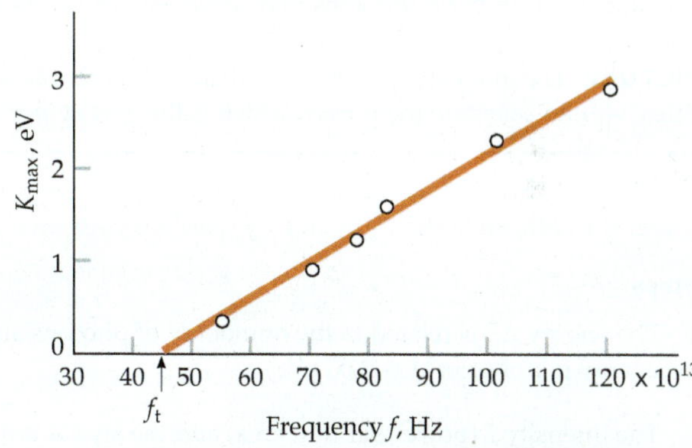

FIGURE 34-3 Millikan's data for the maximum kinetic energy K_{max} versus frequency f for the photoelectric effect. The data fall on a straight line that has a slope h, as predicted by Einstein approximately a decade before the experiment was performed.

Example 34-1 Photon Energies for Visible Light

Calculate the photon energies for light that has a wavelength equal to 400 nm (violet) and light that has a wavelength equal to 700 nm (red). (The wavelengths of 400 nm and 700 nm are the approximate wavelengths for the two extremes of the visible light spectrum.)

PICTURE Photon energies are related to photon frequencies and wavelengths by $E = hf = hc/\lambda$ (Equation 34-1).

SOLVE

1. The energy is related to the wavelength by Equation 34-1: $E = hf = \dfrac{hc}{\lambda}$

2. For $\lambda = 400$ nm, the energy is $E = \dfrac{hc}{\lambda} = \dfrac{1240\ \text{eV}\cdot\text{nm}}{400\ \text{nm}} = \boxed{3.10\ \text{eV}}$

3. For $\lambda = 700$ nm, the energy is $E = \dfrac{hc}{\lambda} = \dfrac{1240\ \text{eV}\cdot\text{nm}}{700\ \text{nm}} = \boxed{1.77\ \text{eV}}$

CHECK The shorter the wavelength of light, the greater the energy and 3.10 eV for 400 nm is greater than 1.77 eV for 700 nm.

TAKING IT FURTHER We can see from these calculations that visible light has photons that have energies which range from approximately 1.8 eV to 3.1 eV. X rays, which have much shorter wavelengths, have photons that have energies of the order of keV. Gamma rays emitted by nuclei have even shorter wavelengths and photons that have energies of the order of MeV.

PRACTICE PROBLEM 34-1 Find the energy of a photon corresponding to electromagnetic radiation in the FM radio band of wavelength 3.00 m.

PRACTICE PROBLEM 34-2 Find the wavelength of a photon whose energy is (*a*) 0.100 eV, (*b*) 1.00 keV, and (*c*) 1.00 MeV.

Example 34-2 The Number of Photons per Second in Sunlight — *Try It Yourself*

The intensity of sunlight at Earth's surface is approximately 1400 W/m^2. Assuming the average photon energy is 2.00 eV (corresponding to a wavelength of approximately 600 nm), calculate the number of photons that strike an area of 1.00 cm^2 each second.

PICTURE The intensity (power per unit area) is given, as is the area. From these given quantities, we can calculate the power, which is the energy per unit time.

SOLVE

Cover the column to the right and try these on your own before looking at the answers.

Steps	Answers
1. The energy ΔE is related to the number N of photons and the energy per photon $hf = 2.00$ eV:	$\Delta E = Nhf$
2. The intensity I (power per unit area) and the area A are given, so we can find the power:	$I = \frac{P}{A}$
3. Knowing the power (energy per unit time) and the time, we can find the energy:	$\Delta E = P\Delta t$
4. Combine the results from steps 1–3 and solve for N (take care to get the units to cancel):	$N = \boxed{4.38 \times 10^{17}}$

CHECK This is an enormous number of photons. However, in everyday situations we do not notice that the energy of sunlight arrives in discrete amounts. Thus, an enormous number is expected.

PRACTICE PROBLEM 34-3 Calculate the photon density (in photons per cubic centimeter) of the sunlight in Example 34-2. The number arriving on an area of 1.00 cm^2 in one second is the number in a column whose cross section is 1.00 cm^2 and whose height is the distance light travels in one second.

COMPTON SCATTERING

The first use of the photon concept was to explain the results of photoelectric-effect experiments. In the photoelectric effect, all the energy of the photon is transferred to an electron. However, in Compton scattering only some of the energy of the photon is transferred to an electron. The photon concept was also used by Arthur H. Compton to explain the results of his measurements of the scattering of X rays by free electrons in 1923. According to classical theory, if an electromagnetic wave of frequency f_i is incident on material containing free charges, the charges will oscillate with this frequency and reradiate electromagnetic waves of the same frequency.

Compton considered these reradiated waves as scattered photons, and he pointed out that if the scattering process were a collision between a photon and an electron (Figure 34-4), the electron would recoil and thus absorb energy. The scattered photon would then have less energy, and therefore a lower frequency and longer wavelength, than the incident photon.

According to classical electromagnetic theory (see Section 30-3), the energy and momentum of an electromagnetic wave are related by

$$E = pc \tag{34-6}$$

The momentum of a photon is thus related to its wavelength λ by $p = E/c = hf/c = h/\lambda$.

$$p = \frac{h}{\lambda} \tag{34-7}$$

MOMENTUM OF A PHOTON

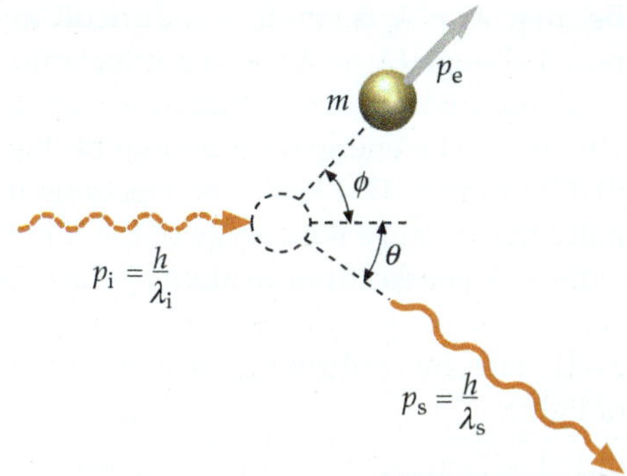

FIGURE 34-4 The scattering of light by an electron is considered as a collision of a photon of momentum h/λ_i and a stationary electron. The scattered photon has less energy and therefore has a longer wavelength than does the incident photon.

Compton applied the laws of conservation of momentum and energy to the collision of a photon and an electron to calculate the momentum p_s and thus the wavelength $\lambda_s = h/p_s$ of the scattered photon (see Figure 34-4). Applying conservation of momentum to the collision gives

$$\vec{p}_i = \vec{p}_s + \vec{p}_e \tag{34-8}$$

where $\vec{p}_i$ is the momentum of the incident photon and $\vec{p}_e$ is the momentum of the electron after the collision. The initial momentum of the electron is zero. Rearranging Equation 34-8, we have $\vec{p}_e = \vec{p}_i - \vec{p}_s$. Taking the dot product of each side with itself gives

$$p_e^2 = p_i^2 + p_s^2 - 2p_ip_s\cos\theta \tag{34-9}$$

where θ is the angle the direction of motion of the scattered photon makes with the direction of motion of the incident photon. Because the kinetic energy of the electron after the collision can be a significant fraction of the rest energy of an electron, the relativistic expression relating the total energy E of the electron to its momentum is used (see Chapter R). This expression (Equation R-17) is

$$E = \sqrt{p_e^2c^2 + (m_ec^2)^2}$$

where m_e is the mass of the electron. Applying conservation of energy to the collision gives

$$p_ic + m_ec^2 = p_sc + \sqrt{p_e^2c^2 + (m_ec^2)^2} \tag{34-10}$$

where pc (Equation 34-6) has been used to express the energies of the photons. Eliminating p_e^2 from Equations 34-9 and 34-10 gives

$$\frac{1}{p_s} - \frac{1}{p_i} = \frac{1}{m_ec}(1 - \cos\theta)$$

and substituting for p_i and p_s, using Equation 34-7, gives

$$\lambda_s - \lambda_i = \frac{h}{m_ec}(1 - \cos\theta) \tag{34-11}$$

COMPTON EQUATION

The increase in wavelength is independent of the wavelength λ_i of the incident photon. The quantity $h/(m_ec)$ has dimensions of length and is called the *Compton wavelength* λ_C. Its value is

$$\lambda_C = \frac{h}{m_ec} = \frac{hc}{m_ec^2} = \frac{1240\text{ eV}\cdot\text{nm}}{5.110 \times 10^5\text{ eV}} = 2.426 \times 10^{-12}\text{ m} = 2.426\text{ pm} \tag{34-12}$$

Because $\lambda_s - \lambda_i$ is small, it is difficult to observe unless λ_i is so small that the fractional change $(\lambda_s - \lambda_i)/\lambda_i$ is appreciable.

Compton used X rays that have wavelengths equal to 71.1 pm (1 pm $= 10^{-12}$ m $= 10^{-3}$ nm). The energy of a photon of this wavelength is $E = hc/\lambda = (1240\ \text{eV}\cdot\text{nm})/(0.0711\ \text{nm}) = 17.4$ keV. The electrons in the experiment can be considered essentially free because the energy of the X rays is much greater than the binding energies of the valence electrons in atoms (which are of the order of a few eV). Compton's measurements of $\lambda_s - \lambda_i$ as a function of scattering angle θ agreed with Equation 34-11, thereby confirming the correctness of the photon concept (the particle nature of light).

Example 34-3 Finding the Increase in Wavelength

An X-ray photon of wavelength 6.00 pm makes a head-on collision with an electron, so that the scattered photon goes in a direction opposite to that of the incident photon. The electron is initially at rest. (*a*) How much longer is the wavelength of the scattered photon than the wavelength of the incident photon? (*b*) What is the kinetic energy of the recoiling electron?

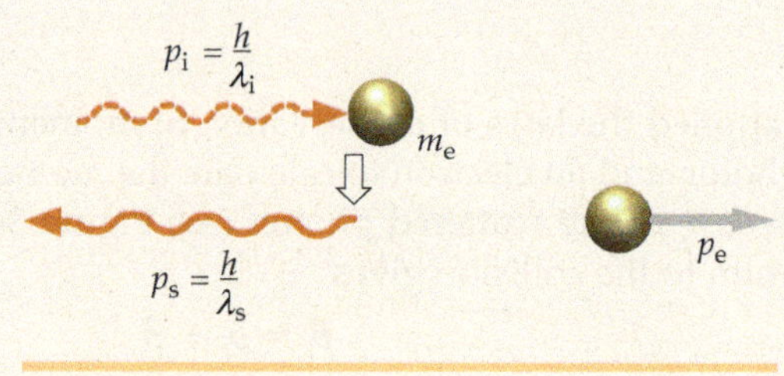

FIGURE 34-5

PICTURE We can calculate the increase in wavelength, and thus the new wavelength, from the Compton equation (Equation 34-11). We then use the new wavelength to find the energy of the scattered photon and then to find the kinetic energy of the recoiling electron from conservation of energy (Figure 34-5).

SOLVE

(*a*) Use Equation 34-11 to calculate the increase in wavelength:

$$\Delta\lambda = \lambda_s - \lambda_i = \frac{h}{m_e c}(1 - \cos\theta)$$

$$= (2.43\ \text{pm})(1 - \cos 180°) = \boxed{4.86\ \text{pm}}$$

(*b*) 1. The kinetic energy of the recoiling electron equals the energy of the incident photon E_i minus the energy of the scattered photon E_s:

$$K_e = E_i - E_s = hf_i - hf_s = \frac{hc}{\lambda_i} - \frac{hc}{\lambda_s}$$

2. Calculate λ_s from the given wavelength of the incident photon and the change found in Part (*a*):

$$\lambda_s = \lambda_i + \Delta\lambda = 6.00\ \text{pm} + 4.86\ \text{pm}$$
$$= 10.86\ \text{pm}$$

3. Substitute the values of λ_i and λ_s into the Part (*b*), step-1 result to find the energy of the recoiling electron:

$$K_e = \frac{hc}{\lambda_i} - \frac{hc}{\lambda_s}$$
$$= \frac{1240\ \text{eV}\cdot\text{nm}}{6.00\ \text{pm}} - \frac{1240\ \text{eV}\cdot\text{nm}}{10.86\ \text{pm}}$$
$$= \frac{1.240\ \text{keV}\cdot\text{nm}}{6.00 \times 10^{-3}\ \text{nm}} - \frac{1.240\ \text{keV}\cdot\text{nm}}{10.86 \times 10^{-3}\ \text{nm}}$$
$$= 207\ \text{keV} - 114\ \text{keV} = \boxed{93\ \text{keV}}$$

TAKING IT FURTHER The kinetic energy of the scattered electron is 93 keV and the rest energy of an electron is 511 keV, so the kinetic energy is 18 percent of the rest energy. Thus, the nonrelativistic formula for the kinetic energy ($\frac{1}{2}m_e v^2$) is not valid.

PRACTICE PROBLEM 34-4 What is the speed of the scattered electron given by the nonrelativistic formula for the kinetic energy ($\frac{1}{2}m_e v^2$)?

34-4 ENERGY QUANTIZATION IN ATOMS

Ordinary white light has a continuous spectrum; that is, it contains *all* the wavelengths in the visible spectrum. But if atoms in a gas at low pressure are excited by an electric discharge, they emit light of specific wavelengths that are characteristic of the element or the compound. Because the energy of a photon is related to its wavelength by $E = hf = hc/\lambda$, a discrete set of wavelengths implies a discrete set of energies. Conservation of energy then implies that if an atom absorbs a photon, its internal energy increases by a discrete amount, an amount equal to the energy of the photon. (It also implies that if an atom emits a photon, its internal energy decreases by a discrete amount that is equal to the energy of the photon.) In 1913, this led Niels Bohr to postulate that the internal energy of an atom can have only a discrete set of values. That is, the internal energy of an atom is quantized. If an excited atom radiates light of frequency f, the atom makes a transition from one allowed level to another level that has less energy by $|\Delta E| = hf$. Bohr was able to construct a semiclassical model of the hydrogen atom that had a discrete set of energy levels consistent with the observed spectrum of emitted light.* However, the *reason* for the quantization of energy levels in atoms and other systems remained a mystery until the wave nature of electrons was discovered a decade later.

34-5 ELECTRONS AND MATTER WAVES

In 1897, J. J. Thomson showed that the rays of a cathode-ray tube (Figure 34-6) can be deflected by electric and magnetic fields and therefore must consist of electrically charged particles. By measuring the deflections of these particles, Thomson showed that all the particles have the same charge-to-mass ratio q/m. He also showed that particles with this charge-to-mass ratio can be obtained using any material for the cathode, which means that these particles, now called electrons, are a fundamental constituent of all matter.

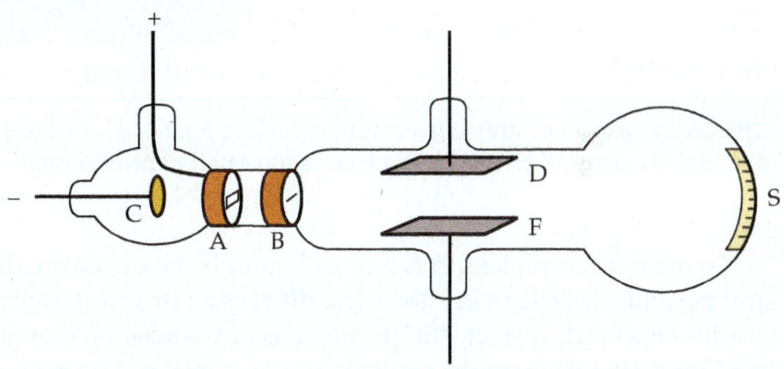

FIGURE 34-6 Schematic diagram of the cathode-ray tube Thomson used to measure q/m for the particles that comprise cathode rays (electrons). Electrons from the cathode C pass through the slits at A and B and strike a phosphorescent screen S. The beam can be deflected by an electric field between plates D and F or by a magnetic field (not shown).

THE DE BROGLIE HYPOTHESIS

Because light seems to have both wave and particle properties, it is natural to ask whether matter (for example, electrons and protons) might also have both wave and particle characteristics. In 1924, a French physics student, Louis de Broglie, suggested this idea in his doctoral dissertation. de Broglie's work was highly speculative, because no evidence existed at that time of any wave aspects of matter.

For the wavelength of electron waves, de Broglie chose

$$\lambda = \frac{h}{p} \qquad \text{34-13}$$

DE BROGLIE RELATION FOR THE WAVELENGTH OF ELECTRON WAVES

where p is the momentum of the electron. Note that this is the same as Equation 34-7 for a photon. For the frequency of electron waves, de Broglie chose the Einstein equation relating the frequency and energy of a photon.

$$f = \frac{E}{h} \qquad \text{34-14}$$

DE BROGLIE RELATION FOR THE FREQUENCY OF ELECTRON WAVES

* The Bohr model is reviewed in Chapter 36.

These equations are thought to apply to all matter. However, for macroscopic objects, the wavelengths calculated from Equation 34-13 are so small that it is impossible to observe the usual wave properties of interference or diffraction. Even a dust particle that has a mass as small as 1 μg is much too massive for any wave characteristics to be noticed, as we see in the following example.

Example 34-4 The de Broglie Wavelength — *Try It Yourself*

Find the de Broglie wavelength of a 1.00×10^{-6} g particle moving with a speed of 1.00×10^{-6} m/s.

PICTURE The wavelength λ and the momentum p of a particle are related by $\lambda = h/p$.

SOLVE

Cover the column to the right and try this on your own before looking at the answers.

Steps	Answers
Write the definition of the de Broglie wavelength and substitute the given data.	$\lambda = \frac{h}{p} = \frac{h}{mv} = \frac{6.63 \times 10^{-34}\ \text{J} \cdot \text{s}}{(1.00 \times 10^{-9}\ \text{kg})(1.00 \times 10^{-6}\ \text{m/s})} = \boxed{6.63 \times 10^{-19}\ \text{m}}$

CHECK As expected, this wavelength, which is four or five orders of magnitude smaller than the diameter of an atomic nucleus, is too small to be observed.

Because the wavelength found in Example 34-4 is so small, much smaller than any possible apertures or obstacles, diffraction or interference of such waves cannot be observed. In fact, the propagation of waves of very small wavelengths is indistinguishable from the propagation of particles. The momentum of the particle in Example 34-4 is only 10^{-15} kg · m/s. A macroscopic particle that has a greater momentum would have an even smaller de Broglie wavelength. We therefore do not observe the wave properties of such macroscopic objects as baseballs and billiard balls.

PRACTICE PROBLEM 34-5

Find the de Broglie wavelength of a baseball of mass 0.17 kg moving at 100 km/h.

The situation is different for low-energy electrons and other subatomic particles. Consider a particle with kinetic energy K. Its momentum is found from

$$K = \frac{p^2}{2m}$$

or

$$p = \sqrt{2mK}$$

Its wavelength is then

$$\lambda = \frac{h}{p} = \frac{h}{\sqrt{2mK}}$$

If we multiply the numerator and the denominator by c, we obtain

$$\lambda = \frac{hc}{\sqrt{2mc^2K}} = \frac{1240\ \text{eV} \cdot \text{nm}}{\sqrt{2mc^2K}} \qquad 34\text{-}15$$

WAVELENGTH ASSOCIATED WITH A PARTICLE OF MASS *m*

where we have used $hc = 1240$ eV · nm. For electrons, $mc^2 = 0.5110$ MeV. Then,

$$\lambda = \frac{1240 \text{ eV} \cdot \text{nm}}{\sqrt{2mc^2K}} = \frac{1240 \text{ eV} \cdot \text{nm}}{\sqrt{2(0.5110 \times 10^6 \text{ eV})K}}$$

or

$$\lambda = \frac{1.226}{\sqrt{K}} \text{ nm} \qquad (K \text{ in electron volts}) \qquad 34\text{-}16$$

ELECTRON WAVELENGTH

Equation 34-15 and Equation 34-16 do not hold for relativistic particles whose kinetic energies are a significant fraction of their rest energies mc^2. (Rest energies are discussed in Chapter 7 and in Chapter R.)

PRACTICE PROBLEM 34-6

Find the wavelength of an electron whose kinetic energy is 10.0 eV.

ELECTRON INTERFERENCE AND DIFFRACTION

The observation of diffraction and interference of electron waves would provide the crucial test of the existence of wave properties of electrons. This observation was first seen serendipitously in 1927 by C. J. Davisson and L. H. Germer as they were studying electron scattering from a nickel target at the Bell Telephone Laboratories. After heating the target to remove an oxide coating that had accumulated during an accidental break in the vacuum system, they found that the scattered electron intensity as a function of the scattering angle showed maxima and minima. Their target had crystallized, and they had observed electron diffraction by accident. Davisson and Germer then prepared a target consisting of a single crystal of nickel and investigated this phenomenon extensively. Figure 34-7*a* illustrates their experiment. Electrons from an electron gun are directed at a crystal and detected at some angle ϕ that can be varied. Figure 34-7*b* shows a typical pattern observed. There is a strong scattering maximum at an angle of 50°. The angle for maximum scattering of waves from a crystal depends on the wavelength of the waves and the spacing of the atoms in the crystal. Using the known spacing of atoms in their crystal, Davisson and Germer calculated the wavelength that could

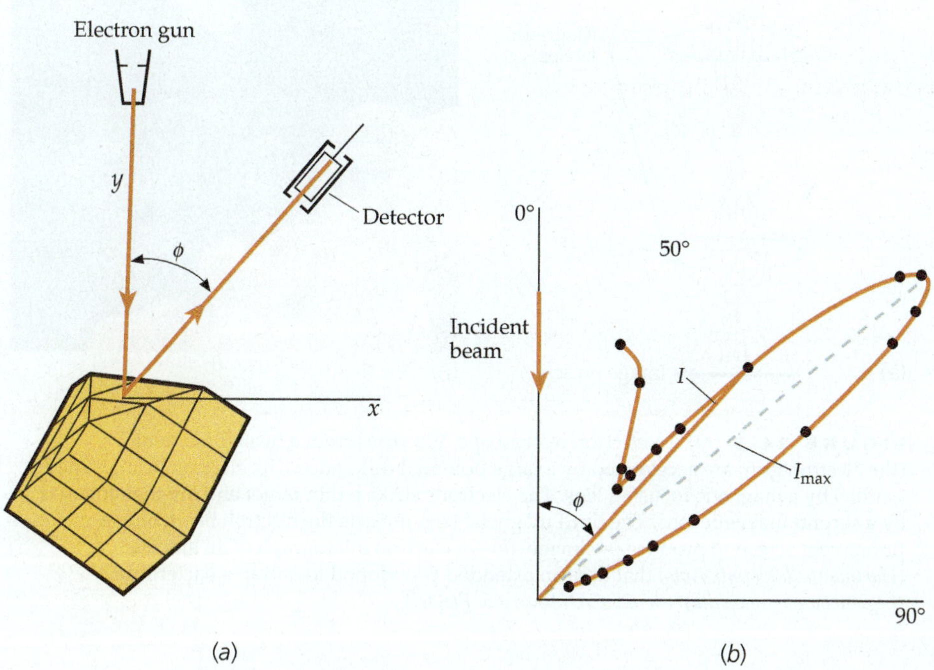

FIGURE 34-7 The Davisson–Germer experiment. (*a*) Electrons are scattered from a nickel crystal into a detector. (*b*) A polar plot of the intensity I of scattered electrons versus scattering angle. The maximum intensity I_{max} is at the angle predicted by the diffraction of waves of wavelength λ given by the de Broglie formula.

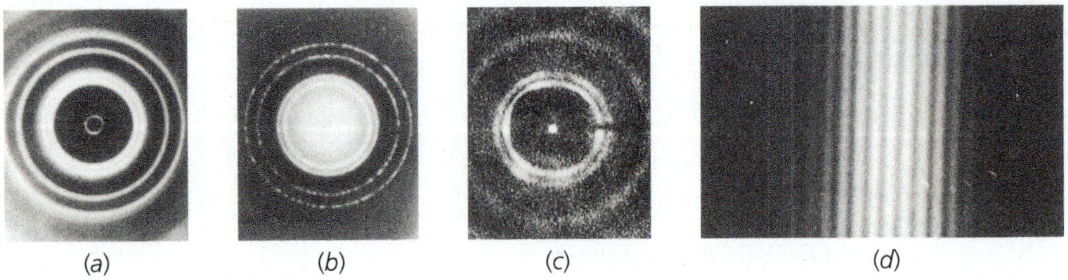

FIGURE 34-8 (*a*) The diffraction pattern produced by X rays of wavelength 0.071 nm on an aluminum foil target. (*b*) The diffraction pattern produced by 600-eV electrons (λ = 0.050 nm) on an aluminum foil target. (*c*) The diffraction of 0.0568 eV neutrons (λ = 0.12 nm) incident on a copper foil. (*d*) A two-slit electron diffraction–interference pattern. (*(a) and (b) PSSC Physics, 2nd ed., 1965. D.C. Heath & Co., and Education Development Center, Inc., Newton, MA, (c) C.G. Shull, (d) Claus Jönsson.*)

produce such a maximum and found that it agreed with the de Broglie equation (Equation 34-16) for the electron energy they were using. By varying the energy of the incident electrons, they could vary the electron wavelengths and produce maxima and minima at different locations in the diffraction patterns. In all cases, the measured wavelengths agreed with de Broglie's hypothesis.

Another demonstration of the wave nature of electrons was provided in the same year by G. P. Thomson (son of J. J. Thomson) who observed electron diffraction in the transmission of electrons through thin metal foils. A metal foil consists of tiny, randomly oriented crystals. The diffraction pattern resulting from such a foil is a set of concentric circles. Figure 34-8*a* and Figure 34-8*b* show the diffraction pattern observed using X rays and electrons on an aluminum foil target. Figure 34-8*c* shows the diffraction patterns of neutrons on a copper foil target. Note the similarity of the patterns. The diffraction of hydrogen and helium atoms was observed in 1930. In all cases, the measured wavelengths agree with the de Broglie predictions. Figure 34-8*d* shows a diffraction pattern produced by electrons incident on two narrow slits. This experiment is equivalent to Young's famous double-slit experiment with light. The pattern is identical to the pattern observed with photons of the same wavelength. (Compare with Figure 34-1.)

Shortly after the wave properties of the electron were demonstrated, it was suggested that electrons rather than light might be used to *see* small objects. As discussed in Chapter 33, reflected waves or transmitted waves can resolve details of objects only if the details are larger than the wavelength of the reflected wave. Beams of electrons, which can be focused by electric and magnetic fields, can have very small wavelengths—much shorter than visible light. Today, the electron microscope (Figure 34-9) is an important research tool used to visualize specimens at scales far smaller than those possible with a light microscope.

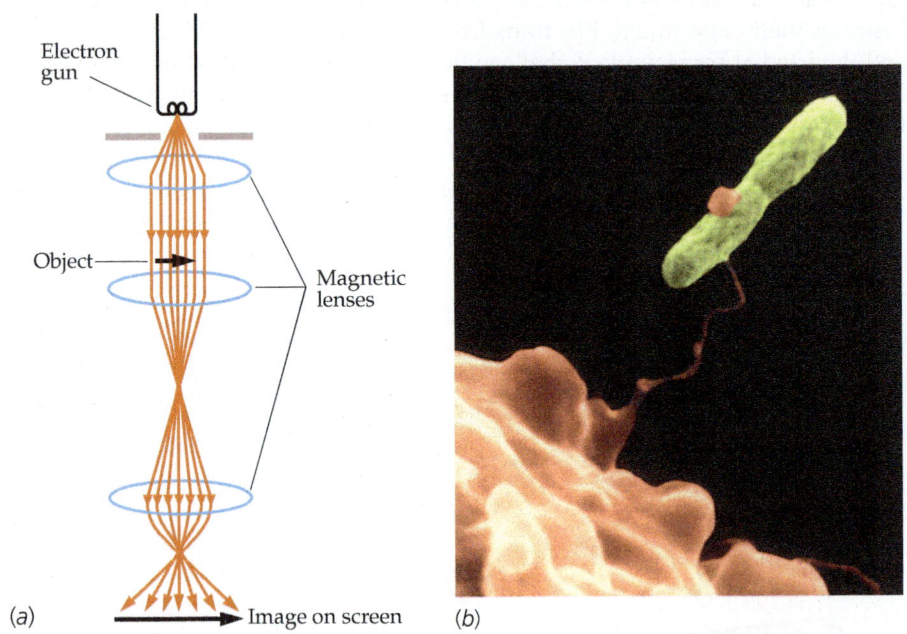

FIGURE 34-9 (*a*) An electron microscope. Electrons from a heated filament (the electron gun) are accelerated by a large potential difference. The electron beam is made parallel by a magnetic focusing lens. The electrons strike a thin target and are then focused by a second magnetic lens. The third magnetic lens projects the electron beam onto a fluorescent screen to produce the image. (*b*) An electron micrograph of an amoeba (*Hartmannella vermiformis*) that uses an extended pseudopod to entrap a bacterium (*Legionella pneumophila*). (*(b) CDC/Dr. Barry S. Fields.*)

STANDING WAVES AND ENERGY QUANTIZATION

Given that electrons have wavelike properties, it should be possible to produce standing electron waves. If energy is associated with the frequency of a standing wave, as in $E = hf$ (Equation 34-14), then standing waves imply quantized energies.

The idea that the discrete energy states in atoms could be explained by standing waves led to the development of a detailed mathematical theory known as quantum theory, quantum mechanics, or wave mechanics by Erwin Schrödinger and others in 1926. In this theory, the electron is described by a wave function ψ that obeys a wave equation called the **Schrödinger equation.** The form of the Schrödinger equation of a particular system depends on the forces acting on the particle, which are described by the potential energy functions associated with those forces. In Chapter 35, we discuss this equation, which is somewhat similar to the classical wave equations for sound or for light. Schrödinger solved the standing wave problem for the hydrogen atom, the simple harmonic oscillator, and other systems of interest. He found that the allowed frequencies, combined with $E = hf$, resulted in the set of energy levels found experimentally for the hydrogen atom, thereby demonstrating that quantum theory provides a general method of finding the quantized energy levels for a given system. Quantum theory is the basis for our modern understanding of the world—from the inner workings of the atomic nucleus to the radiation spectra of distant galaxies.

! Do not think energy is always quantized. It is not unless the system is bound. The energy of a system consisting of a proton and an electron is quantized only if the electron is bound to the proton—as it is in the hydrogen atom. If the electron is not bound to the proton, then the energy of the system is not quantized.

34-6 THE INTERPRETATION OF THE WAVE FUNCTION

The wave function for waves on a string is the string displacement y. The wave function for sound waves can be either the displacement of the air molecules s, or the pressure P. The wave function for electromagnetic waves is the electric field $\vec{E}$ and the magnetic field $\vec{B}$. What is the wave function for electron waves? The symbol we use for this wave function is ψ (the Greek letter psi). When Schrödinger published his wave equation, neither he nor anyone else knew just how to interpret the wave function ψ. We can get a hint about how to interpret ψ by considering the quantization of light waves. For classical waves, such as sound or light, the energy per unit volume in the wave is proportional to the square of the wave function. Because the energy of a light wave is quantized, the energy per unit volume is proportional to the number of photons per unit volume. We might therefore expect the square of the photon's wave function to be proportional to the number of photons per unit volume in a light wave. But suppose we have a very low-energy source of light that emits just one photon at a time. In any unit volume, there is either one photon or none. The square of the wave function must then describe the *probability* of finding a photon in some unit volume.

The Schrödinger equation describes a single particle. The square of the wave function for a particle must then describe the *probability density*, which is the probability per unit volume, of finding the particle at a location. The probability of finding the particle in some volume element must also be proportional to the size of the volume element dV. Thus, in one dimension, the probability of finding a particle in a region of length dx at the position x is $\psi^2(x)\,dx$. If we call this probability $P(x)\,dx$, where $P(x)$ is the **probability density,** we have

$$P(x) = \psi^2(x) \tag{34-17}$$

PROBABILITY DENSITY

Generally the wave function depends on time as well as position, and is written $\psi(x,t)$. However, for standing waves, the probability density is independent of time. Because we will be concerned mostly with standing waves in this chapter, we omit the time dependence of the wave function and write it $\psi(x)$ or just ψ.

The probability of finding the particle either in the region between x_1 and $x_1 + dx$ or in the region between x_2 and $x_2 + dx$ is the sum of the separate probabilities $P(x_1)\,dx + P(x_2)\,dx$. If we have a particle at all, the probability of finding the particle somewhere must be 1. Then the sum of the probabilities over all the possible values of x must equal 1. That is,

$$\int_{-\infty}^{\infty} \psi^2\,dx = 1 \qquad 34\text{-}18$$

NORMALIZATION CONDITION

Equation 34-18 is called the **normalization condition.** If ψ is to satisfy the normalization condition, it must approach zero as $|x|$ approaches infinity. This condition places restrictions on the possible solutions of the Schrödinger equation. There are mathematical solutions to the Schrödinger equation that do not approach zero as $|x|$ approaches infinity. However, these solutions are not acceptable as wave functions.

Example 34-5 Probability Calculation for a Classical Particle

It is known that a classical point particle moves back and forth with constant speed between two walls at $x = 0$ and $x = 8.0$ cm (Figure 34-10). No additional information about the location of the particle is known. (*a*) What is the probability density $P(x)$? (*b*) What is the probability of finding the particle at the point where x equals exactly 2 cm? (*c*) What is the probability of finding the particle between $x = 3.0$ cm and $x = 3.4$ cm?

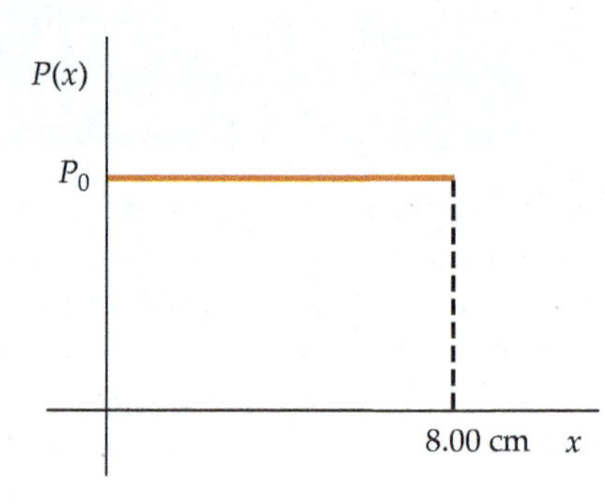

FIGURE 34-10 The probability function $P(x)$.

PICTURE We do not know the initial position of the particle. Because the particle moves with constant speed, it is equally likely to be anywhere in the region $0 < x < 8.0$ cm. The probability density $P(x)$ is therefore independent of x, for $0 < x < 8.0$ cm, and zero outside of this range. We can find $P(x)$, for $0 < x < 8.0$ cm, by normalization, that is, by requiring that the probability that the particle is somewhere between $x = 0$ and $x = 8.0$ cm is 1.

SOLVE

(*a*) 1. The probability density $P(x)$ is uniform between the walls and zero elsewhere:

$$P(x) = \begin{cases} 0 & x < 0 \\ P_0 & 0 < x < 8.0\text{ cm} \\ 0 & x > 8.0\text{ cm} \end{cases}$$

2. Apply the normalization condition:

$$\int_{-\infty}^{+\infty} P(x)\,dx = \int_{-\infty}^{0} P(x)\,dx + \int_{0}^{8.0\text{ cm}} P(x)\,dx + \int_{8.0\text{ cm}}^{\infty} P(x)\,dx$$

$$= 0 + \int_{0}^{8.0\text{ cm}} P_0\,dx + 0 = P_0\,(8.0\text{ cm}) = 1$$

3. Solve for P_0:

$$\boxed{P_0 = \frac{1}{8.0\text{ cm}}}$$

(*b*) On the interval $0 < x < 8.0$ cm, the probability of finding the particle in some range Δx is proportional to $P_0\Delta x = \Delta x/(8\text{ cm})$. The probability of finding the particle at the point $x = 2$ cm is zero because Δx is zero (no range exists). Alternatively, because an infinite number of points exists between $x = 0$ and $x = 8$ cm, and the particle is equally likely to be at any point, the chance that the particle will be at any one particular point must be zero.

The probability of finding the particle at the point where x equals exactly 2 cm is 0.

(*c*) Because the probability density is uniform, the probability of a particle being in some range Δx in the region $0 < x < 8.0$ cm is $P_0\Delta x$. The probability of the particle being in the region $3.0\text{ cm} < x < 3.4$ cm is thus:

$$P_0\Delta x = \left(\frac{1}{8.0\text{ cm}}\right)0.4\text{ cm} = \boxed{0.05}$$

CHECK The length of the interval $3.0\text{ cm} < x < 3.4$ cm is 0.4 cm, which is 5 percent of $L = 8.0$ cm. Because the particle is moving at constant speed v, we expect it to be in the interval $3.0\text{ cm} < x < 3.4$ cm during 5 percent of the time, provided the total time is much longer than the time L/v (the time required for the particle to travel 8.0 cm). Our Part-(*c*) result meets this expectation.

34-7 WAVE–PARTICLE DUALITY

We have seen that light, which we ordinarily think of as a wave, exhibits particle properties when it interacts with matter, as in the photoelectric effect or in Compton scattering. Electrons, which we usually think of as particles, exhibit the wave properties of interference and diffraction when they pass near the edges of obstacles. All carriers of momentum and energy (for example, electrons, atoms, or photons) exhibit both wave and particle characteristics. It might be tempting to say that an electron, for example, is both a wave and a particle, but what does this mean? In classical physics, the concepts of waves and particles are mutually exclusive. A **classical particle** behaves like a piece of shot; it can be localized and scattered, it exchanges energy suddenly at a point in space, and it obeys the laws of conservation of energy and momentum in collisions. It does *not* exhibit interference or diffraction. A **classical wave,** on the other hand, behaves like a sound or light wave; it exhibits diffraction and interference, and its energy is spread out continuously in space and time. A classical wave and a classical particle are mutually exclusive. Nothing can be both a classical particle and a classical wave at the same time.

After Thomas Young observed the two-slit interference pattern by using light in 1801, light was thought to be a classical wave. On the other hand, the electrons discovered by J. J. Thomson were thought to be classical particles. We now know that these classical concepts of waves and particles do not adequately describe the complete behavior of any phenomenon.

Everything propagates like a wave and exchanges energy like a particle.

Often the concepts of the classical particle and the classical wave give the same results. If the wavelength is very small, diffraction effects are negligible, so the waves travel in straight lines like classical particles. Also, interference is not seen for waves of very short wavelength, because the interference fringes are too closely spaced to be observed. It then makes no difference which concept we use. If diffraction is negligible, we can think of light as a wave propagating along rays, as in geometrical optics, or as a beam of photon particles. Similarly, we can think of an electron as a wave propagating in straight lines along rays or, more commonly, as a particle.

We can also use either the wave or particle concept to describe exchanges of energy if we have a large number of particles and we are interested only in the average values of energy and momentum exchanges.

THE TWO-SLIT EXPERIMENT REVISITED

The wave–particle duality of nature is illustrated by the analysis of the experiment in which a single electron is incident on a barrier that has two slits. The analysis is virtually the same whether we use an electron or a photon (light). To describe the propagation of an electron, we must use wave theory. Let us assume the source is a point source, such as a needle point, so we have spherical waves spreading out from the source. After passing through the two slits, the wavefronts spread out—as if each slit were a source of wavefronts. The wave function ψ at a point on a screen or film far from the slits depends on the difference in path lengths from the source to the point, one path through one slit, and the other path through the other slit. At points on the screen for which the difference in path lengths is either zero or an integral number of wavelengths, the amplitude of the wave function is a maximum. Because the probability of detecting the electron is proportional to ψ^2, the electron is very likely to arrive near these points. At points for which the path difference is an odd number of half wavelengths, the wave function ψ is zero, so the electron is very unlikely to arrive near these points. The chapter opening photos show the interference pattern produced by 10 electrons, 100 electrons, 3000 electrons, and 70,000 electrons. Note that, although the electron propagates through the slits like a wave, the electron interacts with the screen at a single point—like a particle.

THE UNCERTAINTY PRINCIPLE

An important principle consistent with the wave–particle duality of nature is the **uncertainty principle.** It states that, in principle, it is impossible to simultaneously measure both the position and the momentum of a particle with unlimited precision. A common way to measure the position of an object is to look at the object by using light. If we do this, we scatter light from the object and determine the position by the direction of the scattered light. If we use light of wavelength λ, we can measure the position x only to an uncertainty Δx of the order of λ because of diffraction effects.

$$\Delta x \sim \lambda$$

To reduce the uncertainty in position, we therefore use light of very short wavelength, perhaps even X rays. In principle, there is no limit to the accuracy of such a position measurement, because there is no limit on how small the wavelength λ can be.

We can determine the momentum p_x of the object if we know the mass and can determine its velocity. The momentum of the object can be found by measuring the object's position at two nearby times and computing its velocity. If we use light of wavelength λ, the photons carry momentum h/λ. If these photons are scattered by the object we are looking at, the scattering changes the momentum of the object in an uncontrollable way. Each photon carries momentum h/λ, so the uncertainty in the momentum Δp_x of the object is of the order of h/λ:

$$\Delta p_x \sim \frac{h}{\lambda}$$

If the wavelength of the radiation is small, the momentum of each photon will be large and the momentum measurement will have a large uncertainty. Reducing the intensity of light cannot eliminate this uncertainty; such a reduction merely reduces the number of photons in the beam. To see the object, we must scatter at least one photon. Therefore, the uncertainty in the momentum measurement of the object will be large if λ is small, and the uncertainty in the position measurement of the object will be large if λ is large.

Of course, we could always look at the objects by scattering electrons instead of photons, but the same difficulty remains. If we use low-momentum electrons to reduce the uncertainty in the momentum measurement, we have a large uncertainty in the position measurement because of diffraction of the electrons. The relation between the wavelength and momentum $\lambda = h/p_x$ is the same for electrons as it is for photons.

The product of the intrinsic uncertainties in position and momentum is

$$\Delta x \, \Delta p_x \sim \lambda \times \frac{h}{\lambda} = h$$

This relation between the uncertainties in position and momentum is called the uncertainty principle. If we define precisely what we mean by uncertainties in measurement, we can give a precise statement of the uncertainty principle. If Δx and Δp are defined to be the standard deviations in the measurements of position and momentum, it can be shown that their product must be greater than or equal to $\hbar/2$.

$$\Delta x \, \Delta p_x \geq \tfrac{1}{2}\hbar \qquad 34\text{-}19$$

where $\hbar = h/2\pi$.*

Equation 34-19 provides a statement of the uncertainty principle first given by Werner Heisenberg in 1927. In practice, the experimental uncertainties are usually much greater than the intrinsic lower limit that results from wave–particle duality.

34-8 A PARTICLE IN A BOX

We can illustrate many of the important features of quantum physics without solving the Schrödinger equation by considering a simple problem of a particle of mass m confined to a one-dimensional box of length L, like the particle in Example 34-5. This can be considered a crude description of an electron that is constrained to be within an atom or a proton that is constrained to be within a nucleus. If a classical particle bounces back and forth between the walls of the box, the particle's energy and momentum can have any values. However, according to quantum theory, the particle is described by a wave function ψ, whose square describes the probability of finding the particle in some region. Because we are assuming that the particle is indeed inside the box, the wave function must be zero everywhere outside the box. If the box is between $x = 0$ and $x = L$, we have

$$\psi = 0 \qquad \text{for } x \leq 0 \text{ and for } x \geq L$$

In particular, if we assume the wave function to be continuous everywhere, it must be zero at the end points of the box $x = 0$ and $x = L$. This is the same condition as the condition for standing waves on a string fixed at $x = 0$ and $x = L$, and the results are the same. The allowed wavelengths for a particle in the box are those where the length L equals an integral number of half wavelengths (Figure 34-11).

$$L = n\frac{\lambda_n}{2} \qquad n = 1, 2, 3, \ldots \qquad 34\text{-}20$$

STANDING-WAVE CONDITION FOR A PARTICLE IN A BOX OF LENGTH L

The total energy E of the particle is its kinetic energy

$$E = \frac{1}{2}mv^2 = \frac{p^2}{2m}$$

Substituting the de Broglie relation $p_n = h/\lambda_n$,

$$E_n = \frac{p_n^2}{2m} = \frac{(h/\lambda_n)^2}{2m} = \frac{h^2}{2m\lambda_n^2}$$

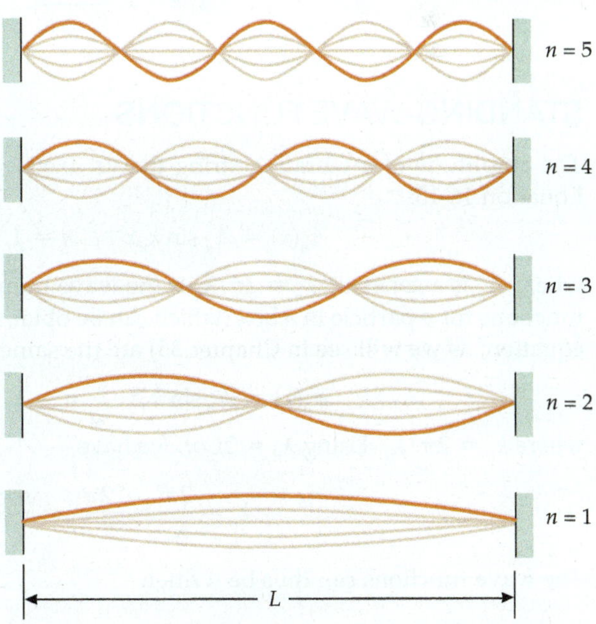

FIGURE 34-11 Standing waves on a string fixed at both ends. The standing-wave condition is the same as for standing electron waves in a box.

* The combination $h/2\pi$ occurs so often it is given a special symbol, somewhat analogous to giving the special symbol ω for $2\pi f$, which occurs often in oscillations.

Then the standing-wave condition $\lambda_n = 2L/n$ gives the allowed energies.

$$E_n = n^2\frac{h^2}{8mL^2} = n^2E_1 \qquad 34\text{-}21$$

ALLOWED ENERGIES FOR A PARTICLE IN A BOX

where

$$E_1 = \frac{h^2}{8mL^2} \qquad 34\text{-}22$$

GROUND-STATE ENERGY FOR A PARTICLE IN A BOX

is the energy of the lowest state, which is the ground state.

The condition $\psi = 0$ at $x = 0$ and $x = L$ is called a **boundary condition.** Boundary conditions in quantum theory lead to energy quantization. Figure 34-12 shows the energy-level diagram for a particle in a box. Note that the lowest energy is not zero. This result is a general feature of quantum theory. If a particle is confined to some region of space, the particle has a minimum kinetic energy, called the **zero-point energy** that is greater than zero. The smaller the region of space the particle is confined to, the greater its zero-point energy. In Equation 34-22, this is indicated by the fact that E_1 varies as $1/L^2$.

If an electron is confined (bound to an atom) in some energy state E_i, the electron can make a transition to another energy state E_f by the emission of a photon if E_f is less than E_i. (If E_f is greater than E_i, the system absorbs a photon.) The transition from state 3 to the ground state is indicated in Figure 34-12 by the vertical arrow. The frequency of the emitted photon is found from the conservation of energy*

$$hf = E_i - E_f \qquad 34\text{-}23$$

The wavelength of the photon is then

$$\lambda = \frac{c}{f} = \frac{hc}{E_i - E_f} \qquad 34\text{-}24$$

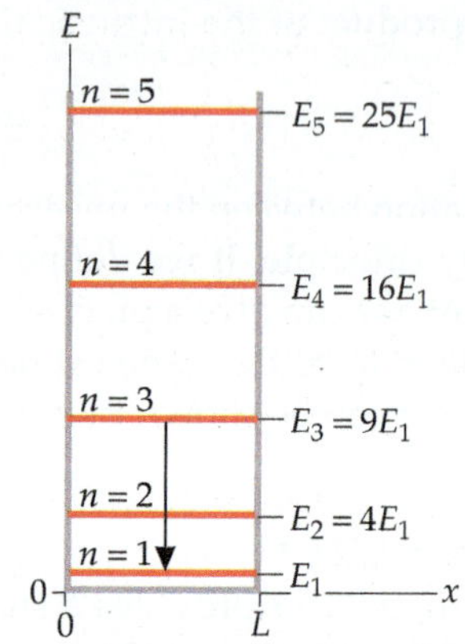

FIGURE 34-12 Energy-level diagram for a particle in a box. Classically, a particle can have any energy value. Quantum mechanically, only those energy values given by Equation 34-21 are allowed. A transition between the state $n = 3$ and the ground state $n = 1$ is indicated by the vertical arrow.

STANDING-WAVE FUNCTIONS

The amplitude of a vibrating string that is fixed at $x = 0$ and $x = L$ is given by Equation 16-15:

$$A_n(x) = A_n \sin k_n x \qquad n = 1, 2, 3, \ldots$$

where A_n is a constant, $k_n = 2\pi/\lambda_n$ is the wave number, and $\lambda_n = 2L/n$. The wave functions for a particle in a box (which can be obtained by solving the Schrödinger equation, as we will see in Chapter 35) are the same:

$$\psi_n(x) = A_n \sin k_n x \qquad n = 1, 2, 3, \ldots$$

where $k_n = 2\pi/\lambda_n$. Using $\lambda_n = 2L/n$, we have

$$k_n = \frac{2\pi}{\lambda_n} = \frac{2\pi}{2L/n} = \frac{n\pi}{L}$$

The wave functions can thus be written

$$\psi_n(x) = A_n \sin\left(n\pi\frac{x}{L}\right)$$

* This equation was first proposed by Niels Bohr in his semiclassical model of the hydrogen atom in 1913, about 10 years before de Broglie's suggestion that electrons have wave properties. The Bohr model is presented in Chapter 36.

The constant A_n is determined by the normalization condition (Equation 34-18):

$$\int_{-\infty}^{\infty} \psi^2\, dx = \int_0^L A_n^2 \sin^2\left(n\pi \frac{x}{L}\right) dx = 1$$

Note that we need integrate only from $x = 0$ to $x = L$ because $\psi(x)$ is zero everywhere else. The result of evaluating the integral and solving for A_n is

$$A_n = \sqrt{\frac{2}{L}}$$

which is independent of n. The normalized standing-wave functions for a particle in a box are thus

$$\psi_n(x) = \sqrt{\frac{2}{L}} \sin\left(n\pi \frac{x}{L}\right) \qquad n = 1, 2, 3, \ldots \qquad \text{34-25}$$

STANDING-WAVE FUNCTIONS FOR A PARTICLE IN A BOX

The standing-wave functions for $n = 1$, $n = 2$, and $n = 3$ are shown in Figure 34-13.

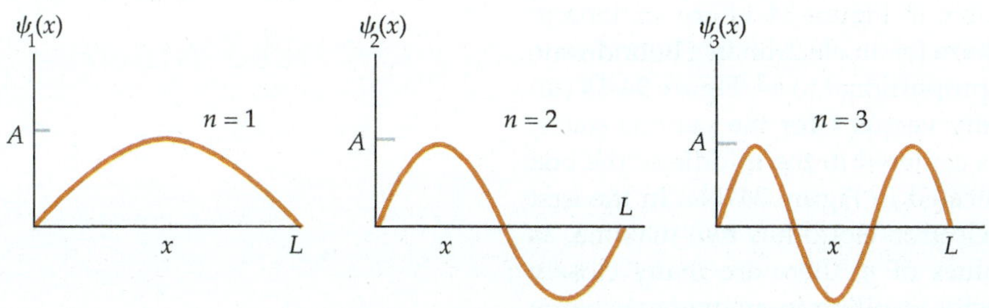

FIGURE 34-13 Standing-wave functions for $n = 1$, $n = 2$, and $n = 3$.

The number n is called a **quantum number.** It characterizes the wave function for a particular state and for the energy of that state. In our one-dimensional problem, a quantum number arises from the boundary condition on the wave function that it must be zero at $x = 0$ and $x = L$. In three-dimensional problems, three quantum numbers arise, one associated with a boundary condition in each dimension.

Figure 34-14 shows plots of ψ^2 for the ground state $n = 1$, the first excited state $n = 2$, the second excited state $n = 3$, and the state $n = 10$. In the ground state, the particle is most likely to be found near the center of the box, as indicated by

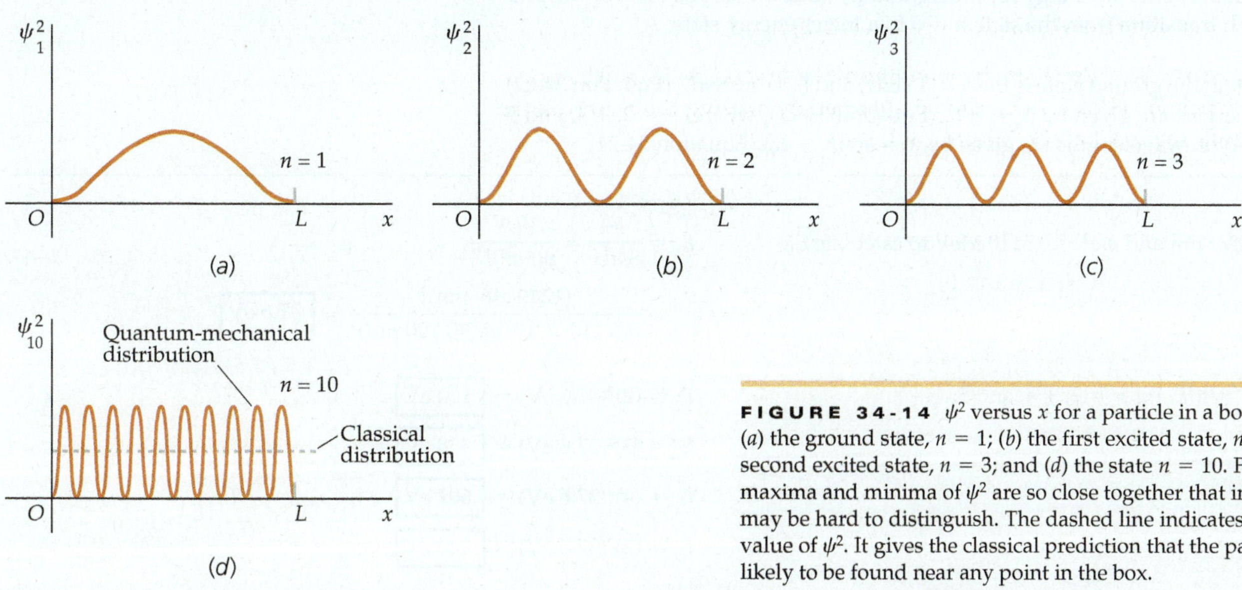

FIGURE 34-14 ψ^2 versus x for a particle in a box of length L for (*a*) the ground state, $n = 1$; (*b*) the first excited state, $n = 2$; (*c*) the second excited state, $n = 3$; and (*d*) the state $n = 10$. For $n = 10$, the maxima and minima of ψ^2 are so close together that individual maxima may be hard to distinguish. The dashed line indicates the average value of ψ^2. It gives the classical prediction that the particle is equally likely to be found near any point in the box.

the maximum value of ψ^2 at $x = L/2$. In the first excited state, the particle is least likely to be found near the center of the box because ψ^2 is small near $x = L/2$. For very large values of n, the maxima and minima of ψ^2 are very close together, as illustrated for $n = 10$. The average value of ψ^2 is indicated in this figure by the dashed line. For very large values of n, the maxima and minima are so closely spaced that ψ^2 cannot be distinguished from its average value. The fact that $(\psi^2)_{av}$ is constant across the whole box means that the particle is equally likely to be found anywhere in the box—the same as in the classical result. This is an example of **Bohr's correspondence principle:**

> In the limit of very large quantum numbers, the classical calculation and the quantum calculation must yield the same results.
>
> BOHR'S CORRESPONDENCE PRINCIPLE

The region of very large quantum numbers is also the region of very large energies. For large energies, the percentage change in energy between adjacent quantum states is very small, so energy quantization is not important (see Problem 71).

We are so accustomed to thinking of the electron as a classical particle that we tend to think of an electron in a box as a particle bouncing back and forth between the walls. But the probability distributions shown in Figure 34-14 are stationary; that is, they do not depend on time. A better picture for an electron in a bound state is a cloud of charge that has the charge density proportional to ψ^2. Figure 34-14 can then be thought of as plots of the charge density versus x for the various states. In the ground state, $n = 1$, the electron cloud is centered in the middle of the box and is spread out over most of the box, as indicated in Figure 34-14*a*. In the first excited state, $n = 2$, the charge density of the electron cloud has two maxima, as indicated in Figure 34-14*b*. For very large values of n, there are many closely spaced maxima and minima in the charge density resulting in an average charge density that is approximately uniform throughout the box. This electron-cloud picture of an electron is very useful in understanding the structure of atoms and molecules. However, it should be noted that whenever an electron is observed to interact with matter or radiation, it is always observed as a whole unit charge.

Example 34-6 Photon Emission by a Particle in a Box

An electron is in a one-dimensional box of length 0.100 nm. (*a*) Find the ground-state energy. (*b*) Find the energies of the four lowest-energy states that have energies above the ground-state energy, and then sketch an energy-level diagram. (*c*) Find the wavelength of the photon emitted for each transition from the state $n = 3$ to a lower-energy state.

PICTURE For Part (*a*) the ground state is the $n = 1$ state, and $E_1 = h^2/8mL^2$ (Equation 34-22). For Part (*b*), the energies are given by $E_n = n^2E_1$ (Equation 34-21), where, $n = 2, 3, 4$, and 5. For Part (*c*), the photon wavelengths are given by $\lambda = hc/(E_i - E_f)$ (Equation 34-24).

SOLVE

(*a*) Use $hc = 1240\ \text{eV}\cdot\text{nm}$ and $mc^2 = 0.5110\ \text{MeV}$ to calculate E_1:

$$E_1 = \frac{h^2}{8mL^2} = \frac{(hc)^2}{8(mc^2)L^2}$$

$$= \frac{(1240\ \text{eV}\cdot\text{nm})^2}{8(5.110\times 10^5\ \text{eV})(0.100\ \text{nm})^2} = \boxed{37.6\ \text{eV}}$$

(*b*) 1. Calculate $E_n = n^2E_1$ for $n = 2, 3, 4$, and 5:

$$E_2 = (2)^2(37.6\ \text{eV}) = \boxed{150\ \text{eV}}$$

$$E_3 = (3)^2(37.6\ \text{eV}) = \boxed{338\ \text{eV}}$$

$$E_4 = (4)^2(37.6\ \text{eV}) = \boxed{602\ \text{eV}}$$

$$E_5 = (5)^2(37.6\ \text{eV}) = \boxed{940\ \text{eV}}$$

2. Sketch an energy-level diagram using the values for the five energy states (Figure 34-15).

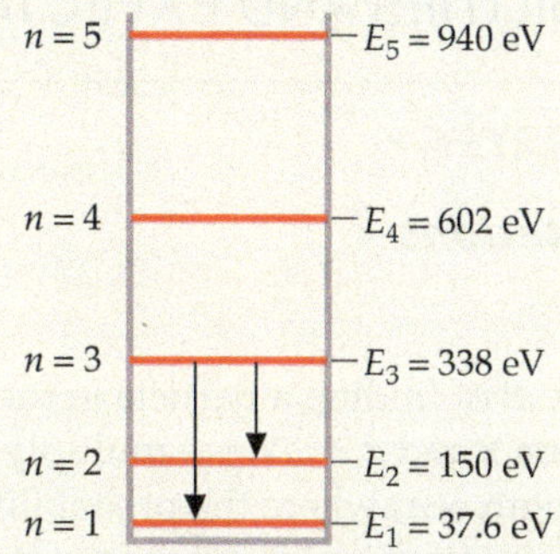

FIGURE 34-15

(c) 1. Use the energies found in Part (b) to calculate the wavelength for a transition from state 3 to state 2:

$$\lambda = \frac{hc}{E_3 - E_2} = \frac{1240\ \text{eV}\cdot\text{nm}}{338\ \text{eV} - 150\ \text{eV}} = \boxed{6.60\ \text{nm}}$$

2. Then use the energies in Part (a) and Part (b) to calculate the wavelength for a transition from state 3 to state 1:

$$\lambda = \frac{hc}{E_3 - E_1} = \frac{1240\ \text{eV}\cdot\text{nm}}{338\ \text{eV} - 37.6\ \text{eV}} = \boxed{4.13\ \text{nm}}$$

CHECK The wavelength of the photon emitted during the transition from the $n = 3$ to the $n = 1$ state is shorter than the wavelength of the photon emitted during the transition from the $n = 3$ to the $n = 2$ state. This result is expected—the greater the energy of the photon the shorter its wavelength.

TAKING IT FURTHER The energy-level diagram is shown in Figure 34-15. The transitions from $n = 3$ to $n = 2$ and from $n = 3$ to $n = 1$ are indicated by the vertical arrows. The ground-state energy of 37.6 eV is on the same order of magnitude as the kinetic energy of the electron in the ground state of the hydrogen atom, which is 13.6 eV. In the hydrogen atom, the electron has potential energy of −27.2 eV in the ground state, giving it a total ground-state energy (potential energy plus kinetic energy) of −13.6 eV.

PRACTICE PROBLEM 34-7 Calculate the wavelength of the photon emitted if the electron in the box makes a transition from $n = 4$ to $n = 3$.

34-9 EXPECTATION VALUES

The solution of a classical mechanics problem is typically specified by giving the position of a particle as a function of time. But the wave nature of matter prevents us from doing this for microscopic systems. The most that we can know is the relative probability of measuring a certain value of the position x. If we measure the position for a large number of identical systems, we get a range of values corresponding to the probability distribution. The average value of x obtained from such measurements is called the **expectation value** and is written $\langle x\rangle$. The expectation value of x is the same as the average value of x that we would expect to obtain from a measurement of the positions of a large number of particles that have the same wave function $\psi(x)$.

Because $\psi^2(x)\,dx$ is the probability of finding a particle in the region dx, the expectation value of x is

$$\langle x\rangle = \int_{-\infty}^{+\infty} x\psi^2(x)\,dx \qquad 34\text{-}26$$

EXPECTATION VALUE OF x DEFINED

The expectation value of any function $F(x)$ is given by

$$\langle F(x)\rangle = \int_{-\infty}^{+\infty} F(x)\psi^2(x)\,dx \qquad 34\text{-}27$$

EXPECTATION VALUE OF $F(x)$ DEFINED

See Math Tutorial for more information on Integrals

CALCULATING PROBABILITIES AND EXPECTATION VALUES

PROBLEM-SOLVING STRATEGY

Probabilities and Expectations

SOLVE

1. To calculate the probability P of finding a particle in the region of infinitesimal length between x and $x + dx$, we multiply the length dx by the probability per unit length at x, where the probability per unit length (called the probability density function) is given by ψ^2.
2. To calculate the probability P of finding a particle in the region $x_1 < x < x_2$, we, in principle, divide the region into an infinite number of regions of infinitesimal length dx, calculate the probability P of finding the particle in each infinitesimal length, and then sum the probabilities. That is, we evaluate the integral $\int_{x_1}^{x_2} \psi^2\, dx$.
3. To calculate the expected value of a function $F(x)$, we evaluate the integral $\int_{-\infty}^{+\infty} F(x)\psi^2(x)\, dx$. The result of this calculation is called the expected value of $F(x)$.

The problem of a particle in a box allows us to illustrate the calculation of the probability of finding the particle in various regions of the box and the expectation values for various energy states. We give two examples, using the wave functions given by Equation 34-25.

Example 34-7 The Probability of the Particle Being Found in a Specified Region of a Box

A particle in a one-dimensional box of length L is in the ground state. Find the probability of finding the particle (*a*) in the region that has a length $\Delta x = 0.01L$ and is centered at $x = \frac{1}{2}L$ and (*b*) in the region $0 < x < \frac{1}{4}L$.

PICTURE The probability P of finding the particle in some infinitesimal range dx is $\psi^2\, dx$. For a particle in the nth state, the wave function is given by $\psi_n = \sqrt{2/L}\,\sin(n\pi x/L)$ (Equation34-25). For a particle in the ground state, $n = 1$; and ψ_1^2 is illustrated in Figure 34-14. The probability of finding x in some region is just the area under this curve for the region. For Part (*a*) , the region is $\Delta x = 0.01L$, centered at $x = L/2$, and the area under the ψ_1^2 versus x curve is shown in Figure 34-16*a*. This area is $\sim\psi_1^2\Delta x$. For Part (*b*) , the region is $0 < x < L/4$, and the area under the curve is shown in Figure 34-16*b*. To calculate this area, we must integrate ψ_1^2 from $x = 0$ to $x = L/4$.

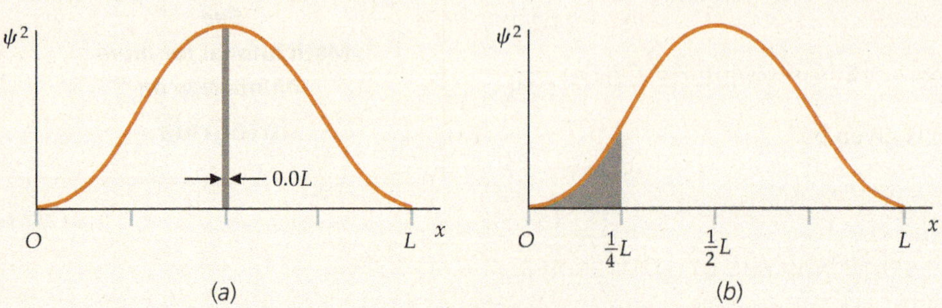

FIGURE 34-16

SOLVE

(*a*) 1. The probability of finding the particle is the area under the curve shown in Figure 34-16*a*. To calculate this area, we need to calculate the height of curve at $x = \frac{1}{2}L$:

$$\psi(x) = \psi_1(x) = \sqrt{\frac{2}{L}}\sin\left(\pi\frac{x}{L}\right)$$

so

$$\psi^2(\tfrac{1}{2}L) = \frac{2}{L}\sin^2\frac{\pi}{2} = \frac{2}{L}$$

2. The area is the height multiplied by the width, and the width is $\Delta x = 0.01L$:

$$P = \psi^2(\tfrac{1}{2}L)\Delta x = \frac{2}{L} \times 0.01L = \boxed{0.02}$$

(*b*) 1. The probability of finding the particle is the area under the curve shown in Figure 34-16*b*. To calculate this area, we need to integrate from $x = 0$ to $x = L/4$:

$$P = \int_0^{L/4} \psi^2(x)\,dx = \int_0^{L/4} \frac{2}{L}\sin^2\frac{\pi x}{L}\,dx$$

2. The integral can be evaluated a number of ways. If a table of integrals is used, a change in the integration variable in required. Changing the integration variable to $\theta = \pi x/L$ gives:

$$P = \frac{2}{\pi}\int_0^{\pi/4} \sin^2\theta\,d\theta$$

3. The integral can be found in tables:

$$\int_0^{\pi/4} \sin^2\theta\,d\theta = \left(\frac{\theta}{2} - \frac{\sin 2\theta}{4}\right)\bigg|_0^{\pi/4} = \left(\frac{\pi}{8} - \frac{1}{4}\right)$$

4. Use the result from Part (*b*), step 3 to calculate the probability:

$$P = \frac{2}{\pi}\left(\frac{\pi}{8} - \frac{1}{4}\right) = \boxed{0.091}$$

CHECK If ψ_1^2 were uniformly distributed on the interval $0 < x < L$, the step-4 result would be 0.25. However, instead of ψ_1^2 being uniformly distributed, it is relatively small on the interval $0 < x < \frac{1}{4}L$, so a step-4 result that is less than 0.25 is expected.

TAKING IT FURTHER An integral was not necessary for Part (*a*) because the area of interest could be well approximated by a rectangle of height ψ^2 and width Δx. The chance of finding the particle in the region $\Delta x = 0.01L$ at $x = \frac{1}{2}L$ is approximately 2 percent. The chance of finding the particle in the region $0 < x < \frac{1}{4}L$ is about 9.1 percent.

CONCEPT CHECK 34-1

A fair six-sided die has the number 1 printed on four faces and the number 6 printed on the other two faces. What is the probability that a 1 comes up when the die is thrown? *Hint: The probability that a specific value comes up for one throw is the fraction of the throws that that value comes up after a large number of throws.*

Example 34-8 Calculating Expectation Values

Find (*a*) $\langle x\rangle$ and (*b*) $\langle x^2\rangle$ for a particle in its ground state in a box of length L.

PICTURE We use $\langle F(x)\rangle = \int F(x)\psi^2(x)\,dx$, with $\psi_n(x) = \sqrt{\frac{2}{L}}\sin\frac{n\pi x}{L}$.

SOLVE

(*a*) 1. Write $\langle x\rangle$ using the ground-state wave function given by Equation 34-25, with $n = 1$:

$$\langle x\rangle = \int_{-\infty}^{+\infty} x\psi^2(x)\,dx = \frac{2}{L}\int_0^L x\sin^2\left(\frac{\pi x}{L}\right)dx$$

2. To evaluate this integral by using a table of integrals, first change the integration variable to $\theta = \pi x/L$:

$$\langle x\rangle = \frac{2}{L}\left(\frac{L}{\pi}\right)^2\int_0^{\pi} \theta\sin^2\theta\,d\theta$$

$$= \frac{2L}{\pi^2}\int_0^{\pi} \theta\sin^2\theta\,d\theta$$

3. The table of integrals gives:

$$\int_0^{\pi} \theta\sin^2\theta\,d\theta = \left[\frac{\theta^2}{4} - \frac{\theta\sin 2\theta}{4} - \frac{\cos 2\theta}{8}\right]_0^{\pi} = \frac{\pi^2}{4}$$

4. Substitute this value into the expression in step 2:

$$\langle x\rangle = \frac{2L}{\pi^2}\int_0^{\pi} \theta\sin^2\theta\,d\theta = \frac{2L}{\pi^2}\frac{\pi^2}{4} = \boxed{\frac{L}{2}}$$

(*b*) 1. Repeat step 1 and step 2 of Part (*a*) for $\langle x^2 \rangle$:

$$\langle x^2 \rangle = \int_{-\infty}^{+\infty} x^2\psi^2(x)\,dx = \int_0^L x^2 \frac{2}{L}\sin^2(\pi x/L)\,dx$$

$$= \frac{2}{L}\left(\frac{L}{\pi}\right)^3 \int_0^{\pi} \theta^2 \sin^2\theta\,d\theta = \frac{2L^2}{\pi^3}\int_0^{\pi} \theta^2 \sin^2\theta\,d\theta$$

2. Evaluating the integral using a table of integrals gives:

$$\int_0^{\pi} \theta^2 \sin^2\theta\,d\theta = \left[\frac{\theta^3}{6} - \left(\frac{\theta^2}{4} - \frac{1}{8}\right)\sin 2\theta - \frac{\theta\cos 2\theta}{4}\right]\Bigg|_0^{\pi}$$

$$= \frac{\pi^3}{6} - \frac{\pi}{4}$$

3. Substitute this value into the expression in step 1 of Part (*b*):

$$\langle x^2 \rangle = \frac{2L^2}{\pi^3}\left(\frac{\pi^3}{6} - \frac{\pi}{4}\right) = \left(\frac{1}{3} - \frac{1}{2\pi^2}\right)L^2 = \boxed{0.283L^2}$$

CHECK The expectation value of x is $L/2$, as we would expect, because the probability distribution is symmetric about the midpoint of the box.

TAKING IT FURTHER Note that $\langle x^2 \rangle$ is greater than $\langle x \rangle^2$.

CONCEPT CHECK 34-2

A fair six-sided die has the number 1 printed on four faces and the number 6 printed on the other two faces. Let N be the number that comes up when the die is thrown. What is the expectation value of N? What is the expectation value of N^2? *Hint: The expectation value of a quantity is the average value of that quantity after a large number of throws.*

34-10 ENERGY QUANTIZATION IN OTHER SYSTEMS

The quantized energies of a system are generally determined by solving the Schrödinger equation for that system. The form of the Schrödinger equation depends on the potential energy of the particle. The potential energy for a one-dimensional box from $x = 0$ to $x = L$ is shown in Figure 34-17. This potential energy function is called an **infinite square-well potential,** and it is described mathematically by

$$U(x) = \begin{cases} \infty & x < 0 \\ 0 & 0 < x < L \\ \infty & x > L \end{cases} \qquad \text{34-28}$$

The particle moves freely inside the box, so the potential energy is uniform. For convenience, we choose the value of this potential energy to be zero. Outside the box the potential energy is infinite, so the particle cannot exist outside the box no matter what its energy. We did not need to solve the Schrödinger equation for this potential because the wave functions and quantized frequencies are the same as for a string fixed at both ends, which we studied in Chapter 16. Although this problem seems artificial, actually it is useful for some physical problems, such as a neutron that is constrained to a nucleus that has a large number of protons and neutrons.

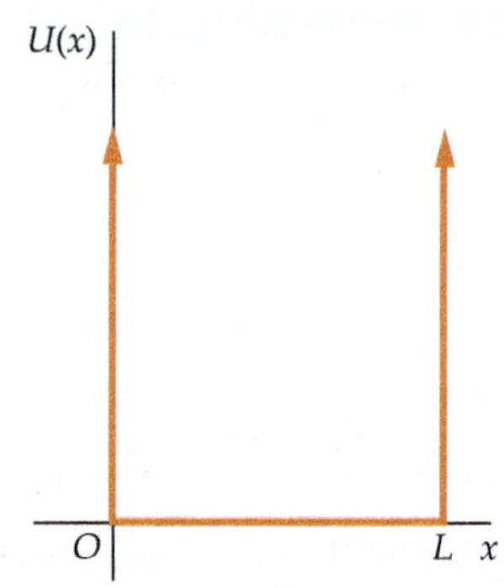

FIGURE 34-17 The infinite square-well potential energy. For $x < 0$ and $x > L$, the potential energy $U(x)$ is infinite. The particle is confined to the region in the well ($0 < x < L$).

THE HARMONIC OSCILLATOR

More realistic than the particle in a box is the harmonic oscillator, which applies to an object of mass m on a spring that has a force constant k or to any system undergoing small oscillations about a stable equilibrium. Figure 34-18 shows the potential energy function

$$U(x) = \tfrac{1}{2}kx^2 = \tfrac{1}{2}m\omega_0^2 x^2$$

where $\omega_0 = \sqrt{k/m}$ is the natural frequency of the oscillator. Classically, the object oscillates between $x = +A$ and $x = -A$. Its total energy is $E = \frac{1}{2}m\omega_0^2 A^2$, which can have any nonnegative value, including zero.

In quantum theory, the particle is represented by the wave function $\psi(x)$, which is determined by solving the Schrödinger equation for this potential. Normalizable wave functions $\psi_n(x)$ occur only for discrete values of the energy E_n given by

$$E_n = (n + \tfrac{1}{2})hf_0 \qquad n = 0, 1, 2, 3, \ldots \qquad \text{34-29}$$

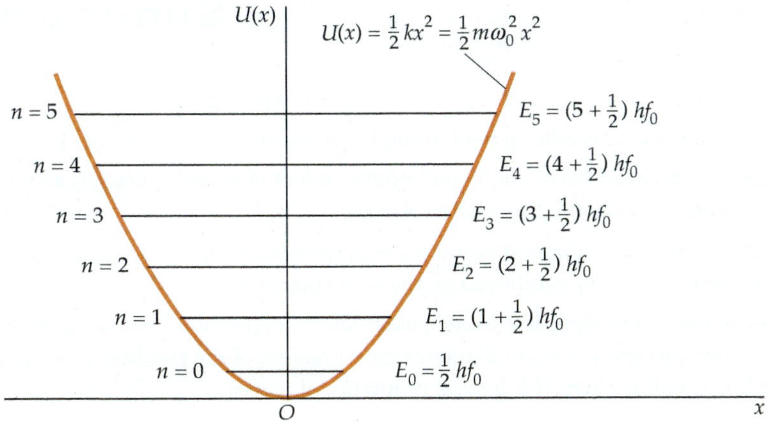

FIGURE 34-18 Harmonic oscillator potential energy function. The allowed energy levels are indicated by the equally spaced horizontal lines. Also, $\omega_0 = 2\pi f_0$.

where $f_0 = \omega_0/2\pi$ is the classical frequency of the oscillator. Note that the energy levels of a harmonic oscillator are evenly spaced with separation hf_0, as indicated in Figure 34-18. Compare this with the uneven spacing of the energy levels for the particle in a box, as shown in Figure 34-12. If a harmonic oscillator makes a transition from energy level n to the next lowest energy level $n - 1$, the frequency f of the photon emitted is given by $hf = E_i - E_f$ (Equation 34-23). Applying this equation gives

$$hf = E_n - E_{n-1} = (n + \tfrac{1}{2})hf_0 - (n - 1 + \tfrac{1}{2})hf_0 = hf_0$$

The frequency f of the emitted photon is therefore equal to the classical frequency f_0 of the oscillator.

THE HYDROGEN ATOM

In the hydrogen atom, an electron is bound to a proton by the electrostatic force of attraction (discussed in Chapter 21). This force varies inversely as the square of the separation distance (exactly like the gravitational attraction of Earth and the Sun). The potential energy of the electron–proton system therefore varies inversely with separation distance (Equation 23-9). As in the case of gravitational potential energy, the potential energy of the electron–proton system is chosen to be zero if the electron is an infinite distance from the proton. Then for all finite distances, the potential energy is negative. Like the case of an object orbiting Earth, the electron–proton system is a bound system if its total energy is negative. Like the energies of a particle in a box and of a harmonic oscillator, the energies are described by a quantum number n. As we will see in Chapter 36, the allowed energies of the hydrogen atom are given by

$$E_n = -\frac{13.6 \text{ eV}}{n^2} \qquad n = 1, 2, 3, \ldots \qquad \text{34-30}$$

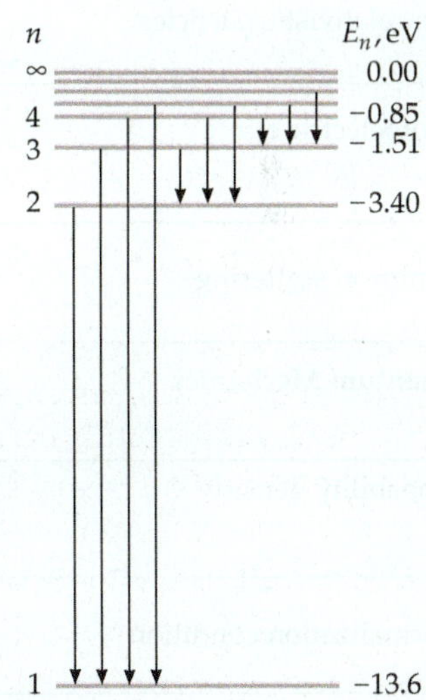

FIGURE 34-19 Energy-level diagram for the hydrogen atom. The energy of the ground state is −13.6 eV. As n approaches infinity, the energy approaches 0, which is the highest energy state for which an electron is bound to the nucleus.

The lowest energy corresponds to $n = 1$. The ground-state energy is thus −13.6 eV. The energy of the first excited state is $-(13.6 \text{ eV})/2^2 = -3.40$ eV. Figure 34-19 shows the energy-level diagram for the hydrogen atom. The vertical arrows indicate transitions from a higher state to a lower state that accompany the emission of electromagnetic radiation. Only those transitions ending at the first excited state ($n = 2$) involve energy differences in the range of visible light of 1.77 eV to 3.10 eV, as calculated in Example 34-1.

Other atoms are more complicated than the hydrogen atom, but their energy levels are similar in many ways to those of hydrogen. Their ground-state energies are of the order of −1 eV to −10 eV, and many transitions involve energies corresponding to photons in the visible range.

! Do not think an electron orbits a proton in a classical orbit like Earth's orbit around the Sun. It doesn't.

Summary

1. All phenomena propagate like waves and interact like particles.
2. The quantum of light is called a photon and has energy $E = hf$, where h is Planck's constant.
3. The relation between wavelength and momentum of electrons, photons, and other particles is given by the de Broglie relation $\lambda = h/p$.
4. Energy quantization in bound systems arises from standing-wave conditions, which are equivalent to boundary conditions on the wave function.
5. The uncertainty principle is a fundamental law of nature that places theoretical restrictions on the precision of a simultaneous measurement of the position and momentum of a particle. It follows from the general properties of waves.

TOPIC	RELEVANT EQUATIONS AND REMARKS	
1. Constants and Values		
Planck's constant	$h = 6.626 \times 10^{-34}\ \text{J}\cdot\text{s} = 4.136 \times 10^{-15}\ \text{eV}\cdot\text{s}$	34-2
hc	$hc = 1240\ \text{eV}\cdot\text{nm}$	34-5
2. The Particle Nature of Light: Photons	Energy is quantized.	
Photon energy and momentum	$E = hf$ and $E = pc$	34-1 and 34-6
3. Frequency–Wavelength (Energy–Momentum) Relations		
Photons and material particles (de Broglie relations)	$E = hf$ and $p = \dfrac{h}{\lambda}$	34-14 and 34-13
Nonrelativistic particles	$K = \dfrac{p^2}{2m}$ so $\lambda = \dfrac{hc}{\sqrt{2mc^2K}}$	34-15
Photoelectric effect	$K_{\text{max}} = \left(\tfrac{1}{2}mv^2\right)_{\text{max}} = hf - \phi$ where ϕ is the work function of the cathode.	34-3
Compton scattering	$\lambda_s - \lambda_i = \dfrac{h}{m_e c}(1 - \cos\theta) = \lambda_C(1 - \cos\theta) = 2.426\ \text{pm}\,(1 - \cos\theta)$	34-11
4. Quantum Mechanics	The state of a particle, such as an electron, is described by its wave function ψ, which is the solution of the Schrödinger wave equation.	
Probability density	The probability of finding the particle in some region of space dx is given by $P(x) = \psi^2(x)\,dx$	34-17
Normalization condition	$\int_{-\infty}^{\infty} \psi^2\,dx = 1$	34-18
Quantum number	The wave function for a particular energy state is characterized by a quantum number n. In three dimensions there are three quantum numbers—one associated with a boundary condition in each dimension.	
Expectation value	The expectation value of x is the same as the average value of x that we would expect to obtain from a measurement of the positions of a large number of particles with the same wave function $\psi(x)$. $\langle x\rangle = \int_{-\infty}^{+\infty} x\psi^2(x)\,dx$ (34-26); $\langle F(x)\rangle = \int_{-\infty}^{+\infty} F(x)\psi^2(x)\,dx$ (34-27)	34-26, 34-27

TOPIC	RELEVANT EQUATIONS AND REMARKS
5. **Wave–Particle Duality**	Photons, electrons, neutrons, and all other carriers of momentum and energy exhibit both wave and particle properties. Everything propagates like a classical wave, exhibiting diffraction and interference, but exchanges energy in discrete lumps like a classical particle. Because the wavelength of macroscopic objects is so small, diffraction and interference are not observed. Also, if a macroscopic amount of energy is exchanged, so many quanta are involved that the particle nature of the energy is not evident.
6. **Uncertainty Principle**	The wave–particle duality of nature leads to the uncertainty principle, which states that the product of the uncertainty in a measurement of position and the uncertainty in a measurement of momentum must be greater than or equal to $\frac{1}{2}\hbar$, where $\hbar$ is Planck's constant divided by 2π. $\Delta x\, \Delta p_x \geq \frac{1}{2}\hbar$ 34-19

Answers to Concept Checks

34-1 2/3

34-2 $\langle N \rangle = 8/3$ $\langle N^2 \rangle = 38/3$

Answers to Practice Problems

34-1 4.13×10^{-7} eV

34-2 (*a*) 12.4 μm, (*b*) 1.24 nm, (*c*) 1.24 pm

34-3 $1.46 \times 10^{7}\ \text{cm}^{-3}$

34-4 $0.6c$

34-5 1.4×10^{-34} m

34-6 0.388 nm. From this result, we see that a 10-eV electron has a de Broglie wavelength of about 0.4 nm. This quantity is of the same order of magnitude as the size of the atom and the spacing of atoms in a crystal.

34-7 4.70 nm

Problems

In a few problems, you are given more data than you actually need; in a few other problems, you are required to supply data from your general knowledge, outside sources, or informed estimate.

Interpret as significant all digits in numerical values that have trailing zeros and no decimal points.

- • Single-concept, single-step, relatively easy
- •• Intermediate-level, may require synthesis of concepts
- ••• Challenging
- SSM Solution is in the *Student Solutions Manual*
- Consecutive problems that are shaded are paired problems.

CONCEPTUAL PROBLEMS

1 • The quantized character of electromagnetic radiation is observed by (*a*) the Young double-slit experiment, (*b*) diffraction of light by a small aperture, (*c*) the photoelectric effect, (*d*) the J. J. Thomson cathode-ray experiment. SSM

2 •• Two monochromatic light sources, A and B, emit the same number of photons per second. The wavelength of A is $\lambda_A = 400$ nm and the wavelength of B is $\lambda_B = 600$ nm. The power radiated by source B (*a*) is equal to the power of source A, (*b*) is less than the power of source A, (*c*) is greater than the power of source A, (*d*) cannot be compared to the power from source A using the available data.

3 • The work function of a surface is ϕ. The threshold wavelength for emission of photoelectrons from the surface is equal to (*a*) hc/ϕ, (*b*) ϕ/hf, (*c*) hf/ϕ, (*d*) none of the above. SSM

4 •• When light of wavelength λ_1 is incident on a certain photoelectric cathode, no electrons are emitted, no matter how intense the incident light is. Yet, when light of wavelength $\lambda_2 < \lambda_1$ is incident, electrons are emitted, even when the incident light has low intensity. Explain this observation.

5 • True or false: (*a*) The wavelength of an electron's matter wave varies inversely with the momentum of the electron. (*b*) Electrons can undergo diffraction. (*c*) Neutrons can undergo diffraction.

6 • If the wavelength of an electron is equal to the wavelength of a proton, then (*a*) the speed of the proton is greater than the speed of the electron, (*b*) the speeds of the proton and the electron are equal, (*c*) the speed of the proton is less than the speed of the electron, (*d*) the energy of the proton is greater than the energy of the electron, (*e*) both (*a*) and (*d*) are correct.

7 • A proton and an electron have equal kinetic energies. It follows that the wavelength of the proton is (*a*) greater than the wavelength of the electron, (*b*) equal to the wavelength of the electron, (*c*) less than the wavelength of the electron.

8 • The parameter x represents the position of a particle. Can the expectation value of x ever have a value such that the probability density function $P(x)$ is zero? Give a specific example.

9 •• It was once believed that if two identical experiments are done on identical systems under the same conditions, the results must be identical. Explain how this statement can be modified so that it is consistent with quantum physics.

10 •• A six-sided die has the numeral 1 painted on three sides and the numeral 2 painted on the other three sides. (*a*) What is the probability of a 1 coming up when the die is thrown? (*b*) What is the expectation value of the numeral that comes up when the die is thrown? (*c*) What is the expectation value of the cube of the numeral that comes up when the die is thrown?

ESTIMATION AND APPROXIMATION

11 •• During an advanced physics lab, students measure the Compton wavelength, λ_C. The students obtain the following wavelength shifts $\lambda_s - \lambda_i$ as a function of scattering angle θ.

θ	45°	75°	90°	135°	180°
$\lambda_s - \lambda_i$	0.647 pm	1.67 pm	2.45 pm	3.98 pm	4.95 pm

Use their data to estimate the value for the Compton wavelength. Compare this number with the accepted value. SSM

12 •• **SPREADSHEET** Students in a physics lab are trying to determine the value of Planck's constant h, using a photoelectric apparatus similar to the one shown in Figure 34-2. The students are using a helium–neon laser that has a tunable wavelength as the light source. The data that the students obtain for the maximum electron kinetic energies are

λ	544 nm	594 nm	604 mn	612 nm	633 mn
K_{max}	0.360 eV	0.199 eV	0.156 eV	0.117 eV	0.062 eV

(*a*) Using a spreadsheet program or graphing calculator, plot K_{max} versus light frequency. (*b*) Use the graph to estimate the value of Planck's constant. (*Note:* You may wish to use a feature of your spreadsheet program or graphing calculator to obtain the best straight-line fit to the data.) (*c*) Compare your result with the accepted value for Planck's constant.

13 •• **SPREADSHEET** The cathode that was used by the students in the experiment described in Problem 12 is constructed from one of the following metals:

Metal	Tungsten	Silver	Potassium	Cesium
Work function	4.58 eV	2.4 eV	2.1 eV	1.9 eV

Determine which metal composes the cathode by using the same data given in Problem 12. (*a*) Using a spreadsheet program or graphing calculator, plot K_{max} versus frequency. (*b*) Use the graph to estimate the value of the work function based on the students' data. (*Note:* You may wish to use a feature of your spreadsheet program or graphing calculator to obtain the best straight-line fit to the data.) (*c*) Which metal was most likely used for the cathode in their experiment?

THE PARTICLE NATURE OF LIGHT: PHOTONS

14 • Find the photon energy in electron volts for light of wavelength (*a*) 450 nm, (*b*) 550 nm, and (*c*) 650 nm.

15 • Find the photon energy in electron volts for an electromagnetic wave of frequency (*a*) 100 MHz in the FM radio band and (*b*) 900 kHz in the AM radio band.

16 • What are the frequencies of photons that have the following energies: (*a*) 1.00 eV, (*b*) 1.00 keV, and (*c*) 1.00 MeV?

17 • Find the photon energy in electron volts if the wavelength is (*a*) 0.100 nm (about 1 atomic diameter) and (*b*) 1.00 fm (1 fm $= 10^{-15}$ m, about 1 nuclear diameter).

18 •• The wavelength of red light emitted by a 3.00-mW helium–neon laser is 633 nm. If the diameter of the laser beam is 1.00 mm, what is the density of photons in the beam? Assume that the intensity is uniformly distributed across the beam.

19 • **ENGINEERING APPLICATION** Lasers used in a telecommunications network typically produce light that has a wavelength near 1.55 μm. How many photons per second are being transmitted if such a laser has an output power of 2.50 mW? SSM

THE PHOTOELECTRIC EFFECT

20 • The work function for tungsten is 4.58 eV. (*a*) Find the threshold frequency and wavelength for the photoelectric effect to occur when monochromatic electromagnetic radiation is incident on the surface of a sample of tungsten. Find the maximum kinetic energy of the electrons if the wavelength of the incident light is (*b*) 200 nm and (*c*) 250 nm.

21 • When monochromatic ultraviolet light that has a wavelength equal to 300 nm is incident on a sample of potassium, the emitted electrons have maximum kinetic energy of 2.03 eV. (*a*) What is the energy of an incident photon? (*b*) What is the work function for potassium? (*c*) What would be the maximum kinetic energy of the electrons if the incident electromagnetic radiation had a wavelength of 430 nm? (*d*) What is the maximum wavelength of incident electromagnetic radiation that will result in the photoelectric emission of electrons by a sample of potassium?

22 • The maximum wavelength of electromagnetic radiation that will result in the photoelectric emission of electrons from a sample of silver is 262 nm. (*a*) Find the work function for silver. (*b*) Find the maximum kinetic energy of the electrons if the incident radiation has a wavelength of 175 nm.

23 • The work function for cesium is 1.90 eV. (*a*) Find the minimum frequency and maximum wavelength of electromagnetic radiation that will result in the photoelectric emission of electrons from a sample of cesium. Find the maximum kinetic energy of the electrons if the wavelength of the incident radiation is (*b*) 250 nm and (*c*) 350 nm.

24 •• When a surface is illuminated with electromagnetic radiation of wavelength 780 nm, the maximum kinetic energy of the emitted electrons is 0.37 eV. What is the maximum kinetic energy if the surface is illuminated using radiation of wavelength 410 nm?

COMPTON SCATTERING

25 • Find the shift in wavelength of photons scattered by free stationary electrons at $\theta = 60°$. (Assume that the electrons are initially moving with negligible speed and are virtually free of (unattached to) any atoms or molecules.)

26 • When photons are scattered by electrons in a carbon sample, the shift in wavelength is 0.33 pm. Find the scattering angle. (Assume that the electrons are initially moving with negligible speed and are virtually free of (unattached to) any atoms or molecules.)

27 • The photons in a monochromatic beam are scattered by electrons. The wavelength of the photons that are scattered at an angle of 135° with the direction of the incident photon beam is 2.3 percent less than the wavelength of the incident photons. What is the wavelength of the incident photons?

28 • Compton used photons of wavelength 0.0711 nm. (*a*) What is the energy of one of those photons? (*b*) What is the wavelength of the photons scattered in the direction opposite to the direction of the incident photons? (*c*) What is the energy of the photon scattered in that direction?

29 • For the photons used by Compton (see Problem 28), find the momentum of the incident photon and the momentum of the photon scattered in the direction opposite to the direction of the incident photons. Use the conservation of momentum to find the momentum of the recoil electron in this case.

30 •• A beam of photons that have a wavelength equal to 6.00 pm is scattered by electrons initially at rest. A photon in the beam is scattered in a direction perpendicular to the direction of the incident beam. (*a*) What is the change in wavelength of the photon? (*b*) What is the kinetic energy of the electron?

ELECTRONS AND MATTER WAVES

31 • An electron is moving at 2.5×10^5 m/s. Find the electron's wavelength.

32 • An electron has a wavelength of 200 nm. Find (*a*) the magnitude of its momentum and (*b*) its kinetic energy.

33 •• An electron, a proton, and an alpha particle each have a kinetic energy of 150 keV. Find (*a*) the magnitudes of their momenta and (*b*) their de Broglie wavelengths.

34 • A neutron in a reactor has a kinetic energy of approximately 0.020 eV. Calculate the wavelength of the neutron.

35 • Find the wavelength of a proton that has a kinetic energy of 2.00 MeV.

36 • What is the kinetic energy of a proton whose wavelength is (*a*) 1.00 nm and (*b*) 1.00 fm?

37 • The kinetic energy of the electrons in the electron beam in a run of Davisson and Germer's experiment was 54 eV. Calculate the wavelength of the electrons in the beam.

38 • The distance between Li^+ and Cl^- ions in a LiCl crystal is 0.257 nm. Find the energy of electrons that have a wavelength equal to that spacing.

39 • An electron microscope uses electrons that have energies equal to 70 keV. Find the wavelength of the electrons. SSM

40 • What is the wavelength of a neutron that has a speed of 1.00×10^6 m/s?

A PARTICLE IN A BOX

41 •• (*a*) Find the energy of the ground state ($n = 1$) and the first two excited states of a neutron in a one-dimensional box of length $L = 1.00 \times 10^{-15}$ m = 1.00 fm (about the diameter of an atomic nucleus). Make an energy-level diagram for the system. Calculate the wavelength of electromagnetic radiation emitted when the neutron makes a transition from (*b*) $n = 2$ to $n = 1$, (*c*) $n = 3$ to $n = 2$, and (*d*) $n = 3$ to $n = 1$.

42 •• (*a*) Find the energy of the ground state ($n = 1$) and the first two excited states of a neutron in a one-dimensional box of length 0.200 nm (about the diameter of a H_2 molecule). Calculate the wavelength of electromagnetic radiation emitted when the neutron makes a transition from (*b*) $n = 2$ to $n = 1$, (*c*) $n = 3$ to $n = 2$, and (*d*) $n = 3$ to $n = 1$.

CALCULATING PROBABILITIES AND EXPECTATION VALUES

43 •• A particle is in the ground state of a one-dimensional box that has length L. (The box has one end at the origin and the other end on the positive x axis.) Determine the probability of finding the particle in the interval of length $\Delta x = 0.002L$ and centered at (*a*) $\frac{1}{4}x = L$, (*b*) $x = \frac{1}{2}L$, and (*c*) $x = \frac{3}{4}L$. (Because Δx is very small you need not do any integration.)

44 •• A particle is in the second excited state ($n = 3$) of a one-dimensional box that has length L. (The box has one end at the origin and the other end on the positive x axis.) Determine the probability of finding the particle in the interval of length $\Delta x = 0.002L$ and centered at (*a*) $x = \frac{1}{3}L$, (*b*) $x = \frac{1}{2}L$, and (*c*) $x = \frac{2}{3}L$. (Because Δx is very small you need not do any integration.)

45 •• A particle is in the first excited ($n = 2$) state of a one-dimensional box that has length L. (The box has one end at the origin and the other end on the positive x axis.) Find (*a*) $\langle x \rangle$ and (*b*) $\langle x^2 \rangle$.

46 •• A particle in a one-dimensional box that has length L is in the first excited state ($n = 2$). (The box has one end at the origin and the other end on the positive x axis.) (*a*) Sketch $\psi^2(x)$ versus x for this state. (*b*) What is the expectation value $\langle x \rangle$ for this state? (*c*) What is the probability of finding the particle in some small region dx centered at $x = L/2$? (*d*) Are your answers for Part (*b*) and Part (*c*) contradictory? If not, explain why your answers are not contradictory.

47 •• A particle of mass m has a wave function given by $\psi(x) = Ae^{-|x|/a}$, where A and a are positive constants. (*a*) Find the normalization constant A. (*b*) Calculate the probability of finding the particle in the region $-a \le x \le a$.

48 •• A one-dimensional box is on the x axis in the region of $0 \le x \le L$. A particle in this box is in its ground state. Calculate the probability that the particle will be found in the region (*a*) $0 < x < \frac{1}{2}L$, (*b*) $0 < x < \frac{1}{3}L$, and (*c*) $0 < x < \frac{3}{4}L$.

49 •• A one-dimensional box is on the x axis in the region of $0 \le x \le L$. A particle in this box is in its first excited state. Calculate the probability that the particle will be found in the region (*a*) $0 < x < \frac{1}{2}L$, (*b*) $0 < x < \frac{1}{3}L$, and (*c*) $0 < x < \frac{3}{4}L$.

50 •• The classical probability distribution function for a particle in a one-dimensional box on the x axis in the region of $0 < x < L$ is given by $P(x) = 1/L$. Use this expression to show that $\langle x \rangle = \frac{1}{2}L$ and $\langle x^2 \rangle = \frac{1}{3}L^2$ for a classical particle in the box.

51 •• A one-dimensional box is on the x axis in the region of $0 \le x \le L$. (*a*) The wave functions for a particle in the box are given by

$$\psi_n(x) = \sqrt{\frac{2}{L}} \sin \frac{n\pi x}{L} \qquad n = 1, 2, 3, \ldots$$

For a particle in the nth state, show that $\langle x \rangle = \frac{1}{2}L$ and $\langle x^2 \rangle = L^2/3 - L^2/(2n^2\pi^2)$. (*b*) Compare these expressions for $\langle x \rangle$ and $\langle x^2 \rangle$, for $n \gg 1$, with the expressions for $\langle x \rangle$ and $\langle x^2 \rangle$ for the classical distribution of Problem 50.

52 •• **SPREADSHEET** (*a*) Use a spreadsheet program or graphing calculator to plot $\langle x^2 \rangle$ as a function of the quantum number n for the particle in the box described in Problem 48 and for values of n from 1 to 100. Assume $L = 1.00$ m for your graph. Refer to Problem 51. (*b*) Comment on the significance of any asymptotic limits that your graph shows.

53 •• The wave functions for a particle of mass m in a one-dimensional box of length L *centered at the origin* (so that the ends are at $x = \pm\frac{1}{2}L$) are given by

$$\psi(x) = \sqrt{\frac{2}{L}} \cos \frac{n\pi x}{L} \qquad n = 1, 3, 5, 7, \ldots$$

and

$$\psi(x) = \sqrt{\frac{2}{L}} \sin \frac{n\pi x}{L} \qquad n = 2, 4, 6, 8, \ldots$$

Calculate $\langle x \rangle$ and $\langle x^2 \rangle$ for the ground state ($n = 1$).

54 •• Calculate $\langle x \rangle$ and $\langle x^2 \rangle$ for the first excited state ($n = 2$) of the box described in Problem 53.

GENERAL PROBLEMS

55 • Photons in a uniform 4.00-cm-diameter light beam have wavelengths equal to 400 nm and the beam has an intensity of 100 W/m^2. (*a*) What is the energy of each photon in the beam? (*b*) How much energy strikes an area of 1.00 cm^2 perpendicular to the beam in 1.00 s? (*c*) How many photons strike this area in 1.00 s? SSM

56 • A 1-μg particle is moving with a speed of approximately 1 mm/s in a one-dimensional box that has a length equal to 1 cm. Calculate the approximate value of the quantum number n of the state occupied by the particle.

57 • (*a*) For the particle and box of Problem 56, find Δx and Δp_x, assuming that these uncertainties are given by $\Delta x/L = 0.01$ percent and $\Delta p_x/p_x = 0.01$ percent. (*b*) What is $(\Delta x\,\Delta p_x)/\hbar$?

58 • In 1987, a laser at Los Alamos National Laboratory produced a flash that lasted 1×10^{-12} s and had a power of 5×10^{15} W. Estimate the number of emitted photons, assuming they all had wavelengths equal to 400 nm.

59 • **ENGINEERING APPLICATION** You cannot "see" anything smaller than the wavelength of the wave used to make the observation. What is the minimum energy of an electron needed in an electron microscope to "see" an atom that has a diameter of about 0.1 nm?

60 • A common flea that has a mass of 0.008 g can jump vertically as high as 20 cm. Estimate the wavelength for the flea immediately after takeoff.

61 •• **BIOLOGICAL APPLICATION** A 100-W source radiates light of wavelength 600 nm uniformly in all directions. An eye that has been adapted to the dark has a 7-mm-diameter pupil and can detect the light if at least 20 photons per second enter the pupil. How far from the source can the light be detected under these rather extreme conditions? SSM

62 •• **BIOLOGICAL APPLICATION** The diameter of the pupil of an eye under room-light conditions is approximately 5 mm. Find the intensity of light that has a wavelength equal to 600 nm so that 1 photon per second passes through the pupil.

63 •• A 100-W incandescent lightbulb radiates 2.6 W of visible light uniformly in all directions. (*a*) Find the intensity of the light from the bulb at a distance of 1.5 m. (*b*) If the average wavelength of the visible light is 650 nm, and counting only those photons in the visible spectrum, find the number of photons per second that strike a surface that has an area equal to 1.0 cm^2, is oriented so that the line to the bulb is perpendicular to the surface, and is a distance of 1.5 m from the bulb.

64 •• When light of wavelength λ_1 is incident on the cathode of a photoelectric tube, the maximum kinetic energy of the emitted electrons is 1.8 eV. If the wavelength is reduced to $\frac{1}{2}\lambda_1$, the maximum kinetic energy of the emitted electrons is 5.5 eV. Find the work function ϕ of the cathode material.

65 •• An incident photon of energy E_i undergoes Compton scattering at an angle of θ. Show that the energy E_s of the scattered photon is given by

$$E_s = \frac{E_i}{1 + (E_i/m_e c^2)(1 - \cos\theta)}$$

66 •• A particle is confined to a one-dimensional box. While the particle makes a transition from the state n to the state $n - 1$, radiation of 114.8 nm is emitted. While the particle makes the transition from the state $n - 1$ to the state $n - 2$, radiation of wavelength 147 nm is emitted. The ground-state energy of the particle is 1.2 eV. Determine n.

67 •• The Pauli exclusion principle states that no more than one electron may occupy a particular quantum state at a time. Electrons intrinsically occupy two spin states. Therefore, if we wish to model an atom as a collection of electrons trapped in a one-dimensional box, no more than two electrons in the box can have the same value of the quantum number n. Calculate the energy that the most energetic electron(s) would have for the uranium atom that has an atomic number 92. Assume the box has a length of 0.050 nm and the electrons are in the lowest possible energy states. How does this energy compare to the rest energy of the electron? SSM

68 •• A beam of electrons that each have the same kinetic energy illuminates a pair of slits separated by a distance 54 nm. The beam forms bright and dark fringes on a screen located a distance 1.5 m beyond the two slits. The arrangement is otherwise identical to that used in the optical two-slit interference experiment described in Chapter 33 and in Figure 33-7 and the fringes have the appearance shown in Figure 34-8*d*. The bright fringes are found to be separated by a distance of 0.68 mm. What is the kinetic energy of the electrons in the beam?

69 •• When a surface is illuminated by light of wavelength λ, the maximum kinetic energy of the emitted electrons is 1.20 eV. If the wavelength $\lambda' = 0.800\lambda$ is used, the maximum kinetic energy increases to 1.76 eV. For wavelength $\lambda' = 0.600\lambda$, the maximum kinetic energy of the emitted electrons is 2.676 eV. Determine the work function of the surface and the wavelength λ.

70 •• A simple pendulum has a length equal to 1.0 m and has a bob that has a mass equal to 0.30 kg. The energy of this oscillator is quantized, and the allowed values of the energy are given by $E_n = \left(n + \frac{1}{2}\right)hf_0$, where n is an integer and f_0 is the frequency of the pendulum. (*a*) Find n if the angular amplitude is 1.0°. (*b*) Find n such that E_{n+1} exceeds E_n by 0.010 percent.

71 •• (*a*) Show that for large n, the fractional difference in energy between state n and state $n + 1$ for a particle in a one-dimensional box is given approximately by

$$(E_{n+1} - E_n)/E_n \approx 2/n$$

(*b*) What is the approximate percentage energy difference between the states $n_1 = 1000$ and $n_2 = 1001$? (*c*) Comment on how this result is related to Bohr's correspondence principle. SSM

72 •• A mode-locked, titanium–sapphire laser has a wavelength of 850 nm and produces 100 million pulses of light each second. Each pulse has a duration of 125 femtoseconds (1 fs = 10^{-15} s) and consists of 5×10^9 photons. What is the average power produced by the laser?

73 •• This problem estimates the time lag in the photoelectric effect that is expected classically but not observed. Let the intensity of the incident radiation falling on an atom be 0.010 W/m^2. (*a*) If the area presented by an atom is 0.010 nm^2, find the energy per second falling on an atom. (*b*) If the work function is 2.0 eV, how long would it take for this much energy to fall on the atom if the radiation energy was distributed uniformly rather than in compact packets (photons)?

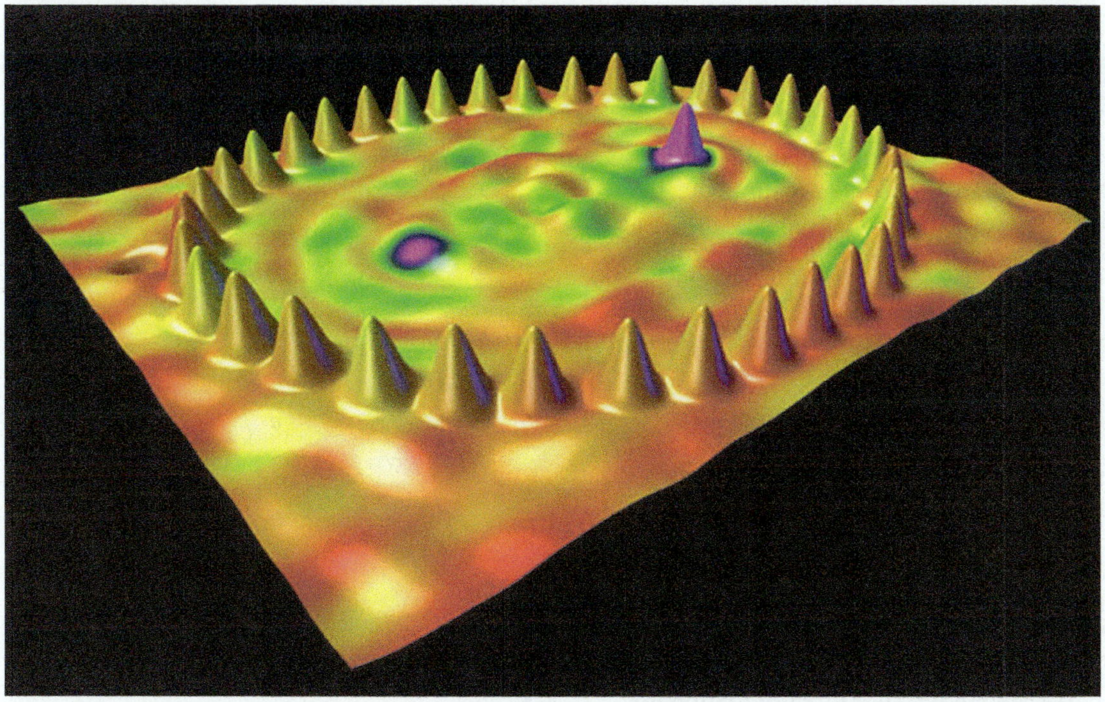

CHAPTER 35

A QUANTUM MIRAGE. THE SCANNING TUNNELING MICROSCOPE (STM) ALLOWS ONE TO PUSH INDIVIDUAL ATOMS AROUND ON A SURFACE AND TO IMAGE THEM. ESPECIALLY INTRIGUING ARE IMAGES OF QUANTUM CORRALS, WHICH ARE CIRCULAR OR ELLIPTICAL ARRANGEMENTS ON A SURFACE INSIDE OF WHICH THE WAVES CORRESPONDING TO ELECTRONS NEAR THE SUBSTRATE SURFACE CAN BE REVEALED. THIS IMAGE COMES FROM IBM, WHERE PHYSICISTS PLACED THIRTY-SIX COBALT ATOMS IN AN ELLIPTICAL "STONEHENGE" PATTERN ON A COPPER SURFACE. AN EXTRA MAGNETIC COBALT ATOM WAS PLACED AT ONE OF THE TWO FOCI OF THE ELLIPSE, CAUSING VISIBLE INTERACTIONS WITH THE SURFACE ELECTRON WAVES. BUT THE WAVES ALSO SEEM TO BE INTERACTING WITH A PHANTOM COBALT ATOM AT THE OTHER FOCUS, AN ATOM THAT IS NOT REALLY THERE. *(Courtesy of IBM and the IBM Almadin Laboratories.)*

Applications of the Schrödinger Equation

In Chapter 34, we found that electrons and other particles have wave properties and are described by wave functions in the form $\Psi(x, t)$. We also mentioned that the wave function is a solution of the Schrödinger equation, and we discussed some solutions qualitatively without reference to the equation itself. In particular, we showed how the standing-wave conditions lead to quantization of energy for a particle confined to a one-dimensional box.

In this chapter we continue our discussion of the material introduced in Chapter 34. We discuss the Schrödinger equation and apply the equation to the particle-in-the-box problem and to several other situations in which a particle is confined to a region of space to illustrate how boundary conditions lead to energy quantization. We then show how the Schrödinger equation leads to barrier penetration and discuss the extension of the Schrödinger equation to more than one dimension and to more than one particle.

? Could the phantom cobalt atom described above be caused by reflections of waves from the corral of cobalt atoms? (See Section 35-4).

35-1 THE SCHRÖDINGER EQUATION

Like the classical wave equation (Equation 15-10*b*), the Schrödinger equation is a partial differential equation in space and time. Like Newton's laws of motion, the Schrödinger equation cannot be derived. Its validity, like that of Newton's laws, lies in its agreement with experiment. In one dimension, the Schrödinger equation is*

$$-\frac{\hbar^2}{2m}\frac{\partial^2\Psi(x,t)}{\partial x^2} + U\Psi(x,t) = i\hbar\frac{\partial\Psi(x,t)}{\partial t} \qquad \text{35-1}$$

TIME-DEPENDENT SCHRÖDINGER EQUATION

where U is the potential energy function and $\Psi(x, t)$ is a wave function. Equation 35-1 is called the **time-dependent Schrödinger equation.** Unlike the classical wave equation, it relates the *second* space derivative of the wave function to the *first* time derivative of the wave function, and it contains the imaginary number $i = \sqrt{-1}$. The wave functions that are solutions of this equation are not necessarily real. $\Psi(x, t)$ is not a measurable function like the classical wave functions for sound or electromagnetic waves. The probability of finding a particle in some region of space dx certainly has a real value, though. We can modify slightly the equation for probability density given in Chapter 34 (Equation 34-17) to determine the probability of finding a particle in some region dx

See Math Tutorial for more information on Complex Numbers

$$P(x,t)\,dx = |\Psi(x,t)|^2\,dx = \Psi^*\Psi\,dx \qquad \text{35-2}$$

where Ψ^*, the complex conjugate of Ψ, is identical to Ψ, except that $-i$ is substituted for i wherever i appears in the expression for Ψ.†

In classical mechanics, the standing-wave solutions to the wave equation (Equation 16-16) are of great interest and value. Not surprisingly, standing-wave solutions to the Schrödinger wave equation are also of great interest and value. The wave function for the standing-wave motion of a uniform taut string is $A\sin(kx)\cos(\omega t + \delta)$, which is representative of all standing waves. A standing wave function can always be expressed as a function of position multiplied by a function of time, where the function of time is one that varies sinusoidally with time. Standing-wave solutions to the one-dimensional Schrödinger wave equation are thus expressed

$$\Psi(x,t) = \psi(x)e^{-i\omega t} \qquad \text{35-3}$$

where $e^{-i\omega t} = \cos(\omega t) - i\sin(\omega t)$. The right side of Equation 35-1 is then

$$i\hbar\frac{\partial\Psi(x,t)}{\partial t} = i\hbar(-i\omega)\psi(x)e^{-i\omega t} = \hbar\omega\psi(x)e^{-i\omega t} = E\psi(x)e^{-i\omega t}$$

where $E = \hbar\omega$ is the energy of the particle.

The Schrödinger wave equation has standing-wave solutions only if the potential energy function U depends on position x alone. Substituting $\psi(x)e^{-i\omega t}$ into Equation 35-1 and canceling the common factor $e^{-i\omega t}$, we obtain an equation for $\psi(x)$, called the **time-independent Schrödinger equation:**

$$-\frac{\hbar^2}{2m}\frac{d^2\psi(x)}{dx^2} + U(x)\psi(x) = E\psi(x) \qquad \text{35-4}$$

TIME-INDEPENDENT SCHRÖDINGER EQUATION

* Although we simply state the Schrödinger equation, Schrödinger himself had a vast knowledge of classical wave theory that led him to this equation.

† Every complex number can be written in the form $z = a + bi$, where a and b are real numbers and $i = \sqrt{-1}$. The complex conjugate of z is $z^* = a - bi$, so $z^*z = (a + bi)(a - bi) = a^2 + b^2 = |z|^2$.

where we have written U as $U(x)$ to emphasize that while U may depend on position, U does not depend on time. The function $U(x)$ represents the interaction between the environment and the particle being observed. Different environments require different expressions for the potential energy function U in the Schrödinger equation.

The calculation of the allowed energy levels in a system involves only the time-independent Schrödinger equation, whereas finding the probabilities of transition between these levels requires the solution of the time-dependent equation. In this book, we will be concerned only with the time-independent Schrödinger equation (Equation 35-4).

The solution of Equation 35-4 depends on the form of the potential energy function $U(x)$. When $U(x)$ is such that the particle is confined to some region of space, only certain discrete energies E_n give solutions ψ_n that can satisfy the normalization condition (Equation 34-18):

$$\int_{-\infty}^{\infty} |\psi_n|^2 dx = 1$$

The complete time-dependent wave functions are then given, from Equation 35-3, by

$$\Psi_n(x, t) = \psi_n(x)e^{-i\omega_n t} = \psi_n(x)e^{-i(E_n/\hbar)t} \qquad 35\text{-}5$$

A PARTICLE IN AN INFINITE SQUARE-WELL POTENTIAL

We will illustrate the use of the time-independent Schrödinger equation by solving it for the problem of a particle in a box. The potential energy for a one-dimensional box from $x = 0$ to $x = L$ is shown in Figure 35-1. It is called an **infinite square-well potential** and is described mathematically by

$$U(x) = \begin{cases} \infty & x < 0 \\ 0 & 0 < x < L \\ \infty & x > L \end{cases} \qquad 35\text{-}6$$

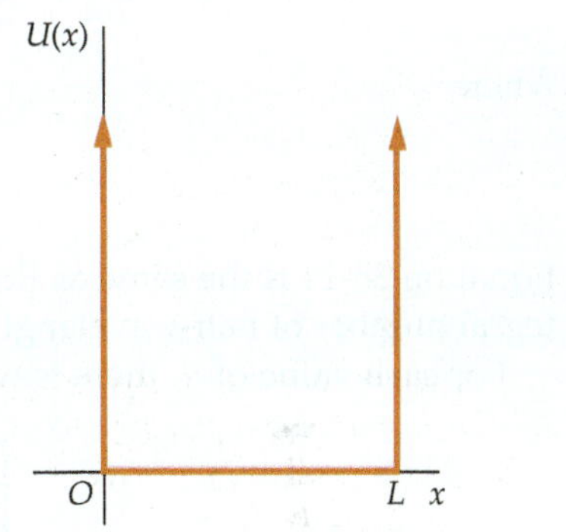

FIGURE 35-1 The infinite square-well potential energy function. For both $x < 0$ and $x > L$, the potential energy $U(x)$ is infinite. The particle is confined to the region in the well $0 < x < L$.

Inside the box, the potential energy is zero, whereas outside the box it is infinite. Because we require the particle to be in the box, we have $\psi(x) = 0$ everywhere outside the box. We then need to solve the Schrödinger equation inside the box for wave functions $\psi(x)$ that must be zero at $x = 0$ and at $x = L$.

Inside the box $U(x) = 0$, so the Schrödinger equation is written

$$-\frac{\hbar^2}{2m}\frac{d^2\psi(x)}{dx^2} = E\psi(x)$$

or

$$\frac{d^2\psi(x)}{dx^2} + k^2\psi(x) = 0 \qquad 35\text{-}7$$

where

$$k^2 = \frac{2mE}{\hbar^2} \qquad 35\text{-}8$$

The general solution of Equation 35-7 can be written as

$$\psi(x) = A\sin kx + B\cos kx \qquad 35\text{-}9$$

where A and B are constants. At $x = 0$, we have

$$\psi(0) = A\sin(0) + B\cos(0) = 0 + B$$

The boundary condition $\psi(x) = 0$ at $x = 0$ thus gives $B = 0$, and Equation 35-9 becomes

$$\psi(x) = A \sin kx \qquad 35\text{-}10$$

The wave function is thus a sine wave where the wavelength λ is related to the wave number k in the usual way, $\lambda = 2\pi/k$. The boundary condition $\psi(x) = 0$ at $x = L$ restricts the possible values of k and therefore the values of the wavelength λ, and (from Equation 35-8) the energy $E = \frac{1}{2}\hbar^2k^2/m$. We have

$$\psi(L) = A \sin kL = 0 \qquad 35\text{-}11$$

This condition is satisfied if kL is π or any integer multiplied by π, that is, if k is restricted to the values k_n given by

$$k_n = n\frac{\pi}{L} \qquad n = 1, 2, 3, \ldots \qquad 35\text{-}12$$

The condition (Equation 35-11) is also satisfied for $n = 0$. The function $\psi(x) = A \sin 0 = 0$ for all values of x, in the interval $0 < x < L$, is also a solution to the wave equation. However, if the wave function has a value of zero everywhere inside the box, then the box is empty. Furthermore, the wave function cannot be normalized and cannot be a wave function for a particle. Substituting $n\pi/L$ for k_n into Equation 35-8 and solving for E gives us the allowed energy values:

$$E_n = \frac{\hbar^2 k_n^2}{2m} = \frac{\hbar^2}{2m}\left(n\frac{\pi}{L}\right)^2 = n^2\left(\frac{h^2}{8mL^2}\right) = n^2 E_1 \qquad 35\text{-}13$$

where

$$E_1 = \frac{h^2}{8mL^2} \qquad 35\text{-}14$$

Equation 35-14 is the same as Equation 34-22, which we obtained by fitting an integral number of half-wavelengths into the box.

For each value of n, there is wave function $\psi_n(x)$ given by

$$\psi_n(x) = \begin{cases} 0 & x < 0 \\ A_n \sin\dfrac{n\pi x}{L} & 0 < x < L \\ 0 & x > L \end{cases} \qquad 35\text{-}15$$

which is the same as Equation 34-25, where the constant $A_n = \sqrt{2/L}$ is determined by normalization.*

CONCEPT CHECK 35-1

The function below is not an acceptable wave function. What makes the function unacceptable?

$$\psi_n(x) = A_n \sin\frac{n\pi x}{L} \qquad -\infty < x < \infty$$

35-2 A PARTICLE IN A FINITE SQUARE WELL

The quantization of energy that we found for a particle in an infinite square well is a result that follows from the general solution of the Schrödinger equation for any particle confined to some region of space. We will illustrate this by considering the qualitative behavior of the wave function for a slightly more general potential energy function, the finite square well, which is shown in Figure 35-2. This potential energy function is described mathematically by

$$U(x) = \begin{cases} U_0 & x < 0 \\ 0 & 0 < x < L \\ U_0 & x > L \end{cases} \qquad 35\text{-}16$$

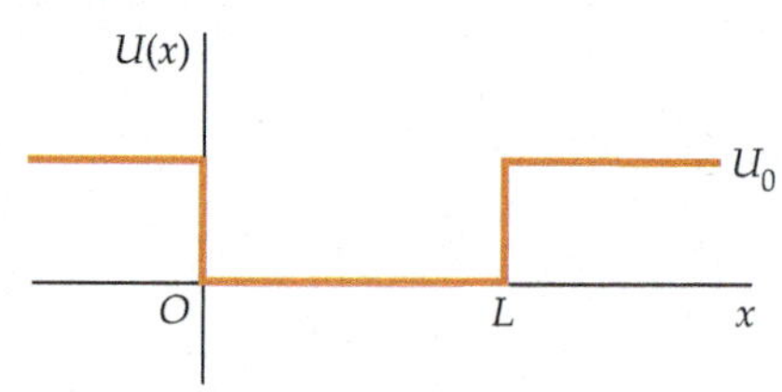

FIGURE 35-2 The finite square-well potential energy function.

* See Equation 34-18.

This potential energy function is discontinuous at $x = 0$ and $x = L$, but it is finite everywhere. The solutions of the Schrödinger equation for this type of potential energy function depend on whether the total energy E is greater or less than U_0. We will not discuss the case of $E > U_0$, except to remark that in that case the particle is not confined and any value of the energy is allowed. That is, there is no energy quantization when $E > U_0$. Here, we assume that $0 \leq E < U_0$.

Inside the well, $U(x) = 0$, and the time-independent Schrödinger equation is the same as for the infinite well (Equation 35-7):

$$-\frac{\hbar^2}{2m}\frac{d^2\psi(x)}{dx^2} = E\psi(x) \qquad 0 \leq x \leq L$$

or

$$\frac{d^2\psi(x)}{dx^2} + k^2\psi(x) = 0$$

where $k^2 = 2mE/\hbar^2$. The general solution is of the form

$$\psi(x) = A\sin kx + B\cos kx$$

In this case, $\psi(x)$ is not required to be zero at $x = 0$ (the particle is not required to be inside the box), so B is not zero. Outside the well, the time-independent Schrödinger equation is

$$-\frac{\hbar^2}{2m}\frac{d^2\psi(x)}{dx^2} + U_0\psi(x) = E\psi(x) \qquad x < 0 \text{ and } x > L$$

or

$$\frac{d^2\psi(x)}{dx^2} - \alpha^2\psi(x) = 0 \qquad \text{35-17}$$

where

$$\alpha^2 = \frac{2m}{\hbar^2}(U_0 - E) \qquad U_0 > E \qquad \text{35-18}$$

The wave functions and allowed energies for the particle can be found by solving Equation 35-17 for $\psi(x)$ outside the well and then requiring that both $\psi(x)$ and $d\psi(x)/dx$ be continuous at the boundaries $x = 0$ and $x = L$. The solution of Equation 35-17 is not difficult [in the region $x > L$, it is of the form $\psi(x) = Ce^{-\alpha x}$ and in the region $x < 0$ it is of the form $\psi(x) = Ce^{+\alpha x}$], but applying the boundary conditions involves much tedious algebra and is not important for our purpose. The important feature of Equation 35-17 is that $d^2\psi/dx^2$ has the same sign as ψ. Thus, if ψ is positive, $d^2\psi/dx^2$ is also positive and the wave function curves away from the axis as x approaches either $+\infty$ or $-\infty$, as shown in Figure 35-3*a*. Similarly, if ψ is negative, $d^2\psi/dx^2$ is negative and ψ again curves away from the axis as x approaches either $+\infty$ or $-\infty$, as shown in Figure 35-3*b*. This behavior is very different from the behavior inside the well, where ψ and $d^2\psi/dx^2$ have opposite signs so that ψ always curves toward the axis like a sine or cosine function. Because of this behavior outside the well, for most values of the energy E in Equation 35-17, $\psi(x)$ becomes infinite as x approaches $\pm\infty$; that is, most wave functions $\psi(x)$ are not well behaved outside the well. Though they satisfy the Schrödinger equation, such functions are not proper wave functions because they cannot be normalized. The solutions of the Schrödinger equation are well behaved (that is, they approach 0 as $|x|$ becomes very large) only for certain values of the energy. These energy values are the allowed energies for the finite square well.

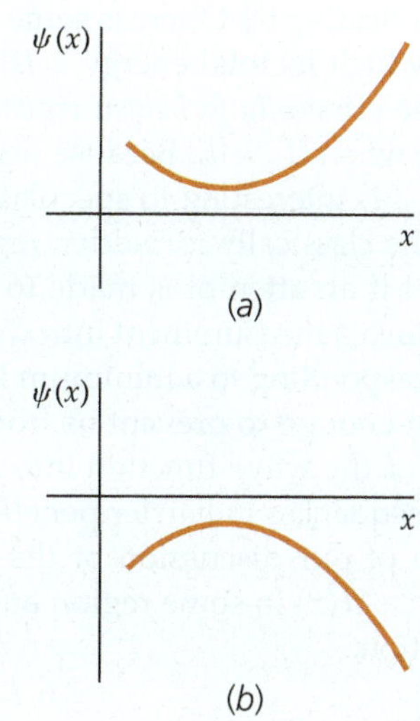

FIGURE 35-3 (*a*) A function ψ that has both a positive value and a positive concavity throughout the region shown. (Concavity is the sign of $d\psi/dx$.) (*b*) A function ψ that has both a negative value and a negative concavity throughout the region shown.

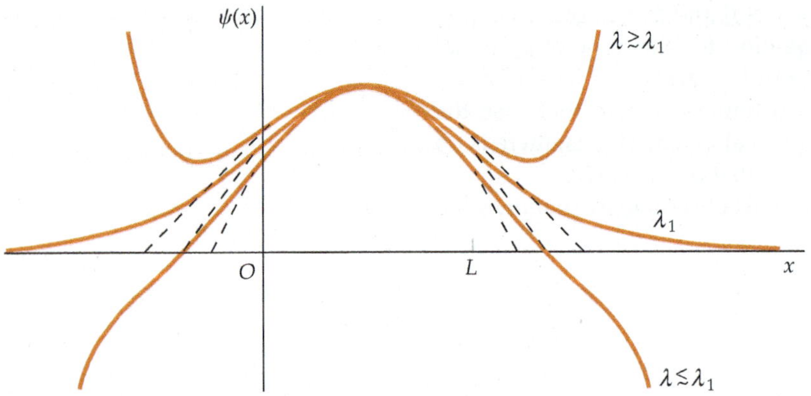

FIGURE 35-4 Functions satisfying the Schrödinger equation that have a wavelength λ that is almost equal to the wavelength λ_1, which is the wavelength that corresponds to the ground-state energy $E_1 = \hbar^2/2m\lambda_1^2$ in the finite well. If λ is slightly greater than λ_1, the function approaches plus infinity as $|x|$ approaches infinity, like the function in Figure 35-3*a*. At the critical wavelength λ_1, the function and its slope approach zero together as $|x|$ approaches infinity. If λ is slightly less than λ_1, the function crosses the x axis while the slope is still negative. The slope then becomes more negative because its rate of change $d^2\psi/dx^2$ is now negative. This function approaches negative infinity as $|x|$ approaches infinity.

Figure 35-4 shows a well-behaved wave function, a wave function that has a wavelength λ_1 inside the well that corresponds to the ground-state energy. The behavior of the wave functions corresponding to nearby wavelengths and energies is also shown. Figure 35-5 shows the wave functions and probability distributions for the ground state and first two excited states. From this figure, we can see that the wavelengths inside the well are slightly longer than the corresponding wavelengths for the infinite well (Figure 34-14), so the corresponding energies are slightly less than those for the infinite well. Another feature of the finite-well problem is that there are only a finite number of allowed energies. For very small values of U_0, there is only one allowed energy.

Note that the wave function penetrates beyond the edges of the well at $x = L$ and $x = 0$, indicating that there is some small probability of finding the particle in the region in which its total energy E is less than its potential energy U_0. This region is called the *classically forbidden region* because the kinetic energy, $E - U_0$, would be negative when $U_0 > E$. Because negative kinetic energy has no meaning in classical physics, it is interesting to speculate on the result of an attempt to observe the particle in the classically forbidden region. It can be shown from the uncertainty principle that if an attempt is made to localize the particle in the classically forbidden region, such a measurement introduces an uncertainty in the momentum of the particle corresponding to a minimum kinetic energy that is greater than $U_0 - E$. This is just great enough to prevent us from measuring a negative kinetic energy. The penetration of the wave function into a classically forbidden region does have important consequences in barrier penetration, which will be discussed in Section 35-4.

Much of our discussion of the finite-well problem applies to any problem in which $E > U(x)$ in some region and $E < U(x)$ outside that region, as we see in the next section.

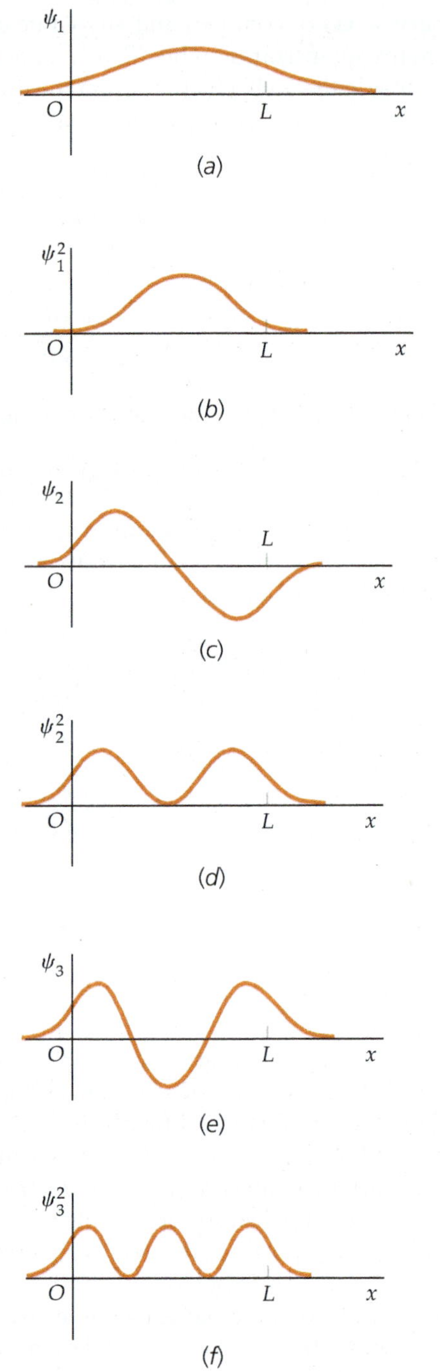

FIGURE 35-5 Graphs of the wave functions $\psi_n(x)$ and probability distributions $\psi^2(x)$ for $n = 1$, $n = 2$, and $n = 3$ for the finite square well. Compare these graphs with those of Figure 34-14 for the infinite square well, where the wave functions are zero at $x = 0$ and $x = L$. The wavelengths here are slightly longer than the corresponding wavelengths for the infinite well, so the allowed energies are somewhat smaller.

35-3 THE HARMONIC OSCILLATOR

The potential energy for a particle that has mass m and is attached to a spring that has force constant k is

$$U(x) = \tfrac{1}{2}kx^2 = \tfrac{1}{2}m\omega_0^2 x^2 \qquad 35\text{-}19$$

where $\omega_0 = \sqrt{k/m}$ is the natural frequency of the oscillator. Classically, the object oscillates between $x = +A$ and $x = -A$. The object's total energy is $E = \frac{1}{2}m\omega_0^2 A^2$, which can have any positive value or zero.

This potential energy function, shown in Figure 35-6, applies to virtually any system undergoing small oscillations about a position of stable equilibrium. For example, it could apply to the oscillations of the atoms of a diatomic molecule (such as H_2 or HCl) in which the atoms are oscillating about their equilibrium positions. Between the classical turning points $(-A < x < A)$, the total energy is greater than the potential energy, and the Schrödinger equation can be written

$$\frac{d^2\psi(x)}{dx^2} = -k^2\psi(x) \qquad -A < x < A \qquad \text{35-20}$$

where $k^2 = (2m/\hbar^2)[E - U(x)]$ now depends on x. The solutions of this equation are no longer simple sine or cosine functions because the wave number $k = 2\pi/\lambda$ now varies with x; but because $d^2\psi/dx^2$ and ψ have opposite signs throughout the region $-A < x < A$, ψ will always curve toward the axis and the solutions will oscillate.

Outside the classical turning points $(|x| > A)$, the potential energy is greater than the total energy and the Schrödinger equation is similar to Equation 35-17:

$$\frac{d^2\psi(x)}{dx^2} - \alpha^2\psi(x) = 0 \qquad |x| > A \qquad \text{35-21}$$

except that here $\alpha^2 = (2m/\hbar^2)[U(x) - E] > 0$, where $U(x) > E$, depends on x. For $|x| > A$, $d^2\psi/dx^2$ and ψ have the same sign, so ψ will curve away from the axis and there will be only certain values of E for which solutions exist that approach zero as x approaches infinity.

For the harmonic oscillator potential energy function, the Schrödinger equation is

$$-\frac{\hbar^2}{2m}\frac{d^2\psi(x)}{dx^2} + \tfrac{1}{2}m\omega_0^2 x^2\psi(x) = E\psi(x) \qquad \text{35-22}$$

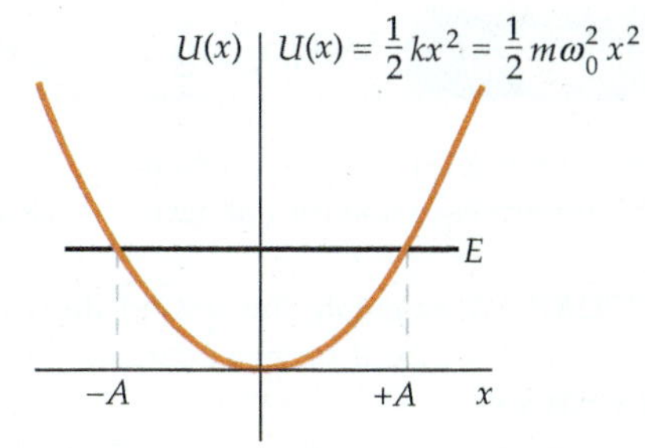

FIGURE 35-6 Harmonic oscillator potential.

WAVE FUNCTIONS AND ENERGY LEVELS

Rather than pursue a general solution to the Schrödinger equation for this system, we simply present the solution for the ground state and the first excited state.

The ground-state wave function $\psi_0(x)$ is found to be a Gaussian function centered at the origin:

$$\psi_0(x) = A_0 e^{-ax^2} \qquad \text{35-23}$$

where A_0 and a are positive constants. This wave function and the wave function for the first excited state are shown in Figure 35-7.

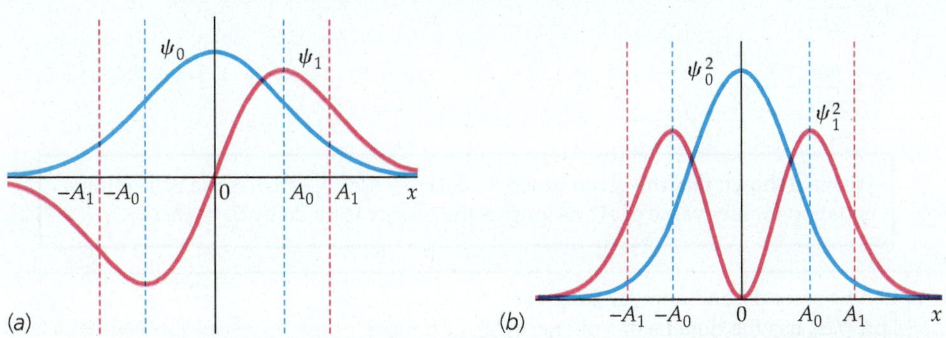

FIGURE 35-7 (*a*) The ground-state ψ_0 and first excited state ψ_1 wave functions for the harmonic oscillator potential. Classically, the motion of a harmonic oscillator with the ground-state energy E_0 would be restricted to the region $-A_0 \leq x \leq +A_0$ and the motion of a harmonic oscillator with the first excited state energy E_1 would be restricted to the region $-A_1 \leq x \leq +A_1$. (*b*) The ground-state ψ_0^2 and first excited state ψ_1^2 probability density functions for the harmonic oscillator potential.

Example 35-1 Verifying the Ground-State Wave Function

Verify that $\psi_0(x) = A_0 e^{-ax^2}$, where A_0 and a are positive constants, is a solution of the Schrödinger equation for the harmonic oscillator.

PICTURE We calculate the second derivative of ψ_0 with respect to x and substitute into Equation 35-22. Because this expression is the ground-state wave function, we write E_0 for the energy E.

SOLVE

1. Compute $d\psi_0/dx$:

$$\frac{d\psi_0(x)}{dx} = \frac{d}{dx}(A_0 e^{-ax^2}) = -2axA_0 e^{-ax^2}$$

2. Compute $d^2\psi_0/dx^2$:

$$\frac{d^2\psi_0(x)}{dx^2} = -2aA_0 e^{-ax^2} + 4a^2x^2 A_0 e^{-ax^2}$$
$$= (4a^2x^2 - 2a)A_0 e^{-ax^2}$$

3. Substitute into the Schrödinger equation (Equation 35-22):

$$-\frac{\hbar^2}{2m}\frac{d^2\psi(x)}{dx^2} + \frac{1}{2}m\omega_0^2 x^2\psi(x) = E\psi(x)$$
$$-\frac{\hbar^2}{2m}(4a^2x^2 - 2a)A_0 e^{-ax^2} + \frac{1}{2}m\omega_0^2 x^2 A_0 e^{-ax^2} = E_0 A_0 e^{-ax^2}$$

4. Cancel the common factor $A_0 e^{-ax^2}$ and show the result in standard polynomial form:

$$-\frac{\hbar^2}{2m}(4a^2x^2 - 2a) + \frac{1}{2}m\omega_0^2 x^2 = E_0$$

so

$$\left(\frac{1}{2}m\omega_0^2 - \frac{2\hbar^2a^2}{m}\right)x^2 + \left(\frac{\hbar^2 a}{m} - E_0\right) = 0$$

5. The equation in step 4 must hold for all x. Set $x = 0$ and solve for E_0:

$$0 + \left(\frac{\hbar^2 a}{m} - E_0\right) = 0$$

so

$$E_0 = \frac{\hbar^2 a}{m}$$

6. Substitute this result for E_0 into the equation in step 4 and simplify:

$$\left(-\frac{2\hbar^2a^2}{m} + \frac{1}{2}m\omega_0^2\right)x^2 + 0 = 0$$

7. It follows that the coefficient of x^2 must equal zero:

$$-\frac{2\hbar^2a^2}{m} + \frac{1}{2}m\omega_0^2 = 0$$

8. Solve for a:

$$a = \frac{m\omega_0}{2\hbar}$$

9. Substitute this result into the equation for E_0 in step 5:

$$E_0 = \frac{\hbar^2 a}{m} = \frac{1}{2}\hbar\omega_0$$

> We have shown that the given function, $\psi_0(x) = A_0 e^{-ax^2}$, satisfies the Schrödinger equation for any value of A_0, as long as the energy is given by $E_0 = \frac{1}{2}\hbar\omega_0$.

CHECK Planck's constant has units of joules multiplied by seconds, and angular frequency has units of reciprocal seconds, so the step-9 expression $\frac{1}{2}\hbar\omega_0$ has the dimensions of energy, as expected.

TAKING IT FURTHER The step-4 equation is a polynomial that is equal to zero. A theorem that would have simplified the solution is "If a polynomial is equal to zero over a continuous range of values of x, then each of the polynomial coefficients is equal to zero. For example, if $Ax^3 + Bx^2 + Cx + D = 0$ on the interval $1 < x < 2$, then $A = B = C = D = 0$."

We see from this example that the ground-state energy is given by

$$E_0 = \frac{\hbar^2 a}{m} = \frac{1}{2}\hbar\omega_0 \qquad 35\text{-}24$$

The first excited state has a node in the center of the potential well, just as with the particle in a box.* The wave function $\psi_1(x)$ is

$$\psi_1(x) = A_1 x e^{-ax^2} \qquad 35\text{-}25$$

where $a = \frac{1}{2}m\omega_0/\hbar$, as in Example 35-1. This function is also shown in Figure 35-7. Substituting $\psi_1(x)$ into the Schrödinger equation, as was done for $\psi_0(x)$ in Example 35-1, yields the energy of the first excited state,

$$E_1 = \tfrac{3}{2}\hbar\omega_0$$

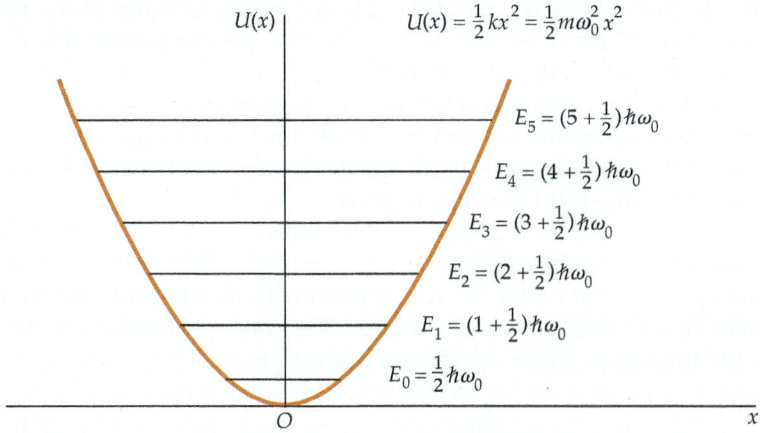

FIGURE 35-8 Energy levels in the harmonic oscillator potential.

In general, the energy of the nth excited state of the harmonic oscillator is

$$E_n = \left(n + \tfrac{1}{2}\right)\hbar\omega_0 \qquad n = 0, 1, 2, \ldots \qquad 35\text{-}26$$

as indicated in Figure 35-8. The fact that the energy levels are evenly spaced by the amount $\hbar\omega_0$ is a peculiarity of the harmonic oscillator potential. As we saw in Chapter 34, the energy levels for a particle in a box, or for the hydrogen atom, are not evenly spaced. The precise spacing of energy levels is closely tied to the particular form of the potential energy function.

35-4 REFLECTION AND TRANSMISSION OF ELECTRON WAVES: BARRIER PENETRATION

In Sections 35-2 and 35-3, we were concerned with bound-state problems in which the potential energy is larger than the total energy for large values of $|x|$. In this section, we consider some simple examples of unbound states for which E is greater than $U(x)$. For these problems, $d^2\psi/dx^2$ and ψ have opposite signs, so $\psi(x)$ curves toward the axis and does not become infinite as x approaches either $+\infty$ or $-\infty$.

STEP POTENTIAL

Consider a particle of energy E moving in a region in which the potential energy is the step function

$$U(x) = \begin{cases} 0 & x < 0 \\ U_0 & x > 0 \end{cases}$$

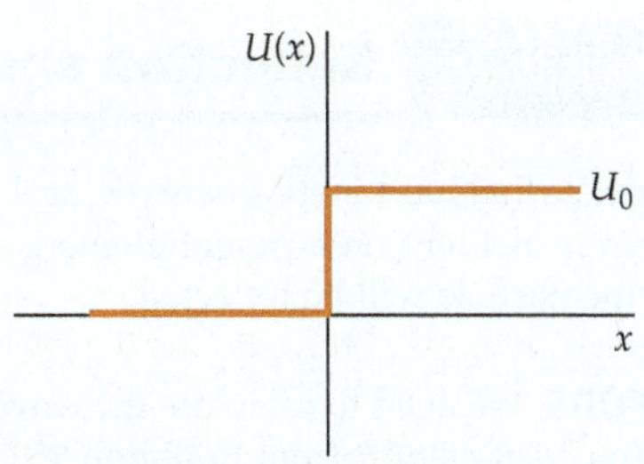

FIGURE 35-9 Step potential. A classical particle that is incident from the left and has total energy $E > U_0$ is always transmitted. The change in potential energy at $x = 0$ merely provides an impulsive force that reduces the speed of the particle. A wave incident from the left is partially transmitted and partially reflected because the wavelength changes abruptly at $x = 0$.

as shown in Figure 35-9. We are interested in what happens when a particle moving from left to right encounters the step.

The classical answer is simple. To the left of the step, the particle moves with a speed $v = \sqrt{2E/m}$. At $x = 0$, an impulsive force acts on the particle. If the initial energy E is less than U_0, the particle will be turned around and will then move to the left at its original speed; that is, the particle will be reflected by the step.

* Each higher-energy state has one additional node in the wave function.

If E is greater than U_0, the particle will continue to move to the right but with reduced speed given by $v = \sqrt{2(E - U_0)/m}$. We can picture this classical problem as a ball rolling along a level surface and coming to a steep hill of height h given by $mgh = U_0$. If the initial kinetic energy of the ball is less than mgh, the ball will roll part way up the hill and then back down and to the left along the lower surface at its original speed. If E is greater than mgh, the ball will roll up the hill and proceed to the right at a lesser speed.

The quantum-mechanical result is similar when E is less than U_0. Figure 35-10 shows the wave function for the case $E < U_0$. The wave function does not go to zero at $x = 0$ but rather decays exponentially, like the wave function for the bound state in a finite square-well problem. The wave penetrates slightly into the classically forbidden region $x > 0$, but it is eventually completely reflected. This problem is somewhat similar to that of total internal reflection in optics.

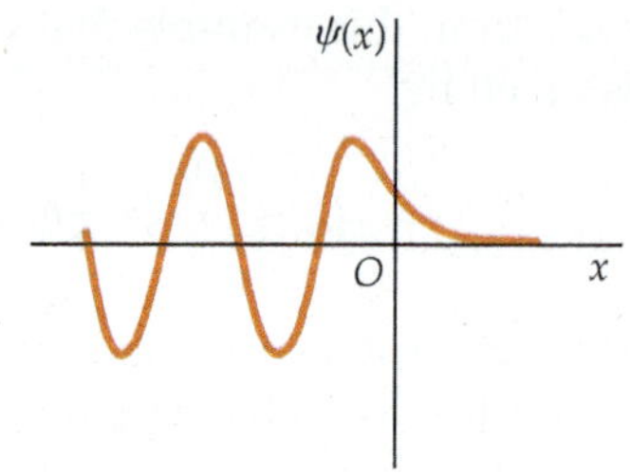

FIGURE 35-10 When the total energy E is less than U_0, the wave function penetrates slightly into the region $x > 0$. However, the probability of reflection for this case is 1, so no energy is transmitted.

For $E > U_0$, the quantum-mechanical result differs markedly from the classical result. At $x = 0$, the wavelength changes abruptly from $\lambda_1 = h/p_1 = h/\sqrt{2mE}$ to $\lambda_2 = h/p_2 = h/\sqrt{2m(E - U_0)}$. We know from our study of waves that when the wavelength changes suddenly, part of the wave is reflected and part of the wave is transmitted. Because the motion of an electron (or other particle) is governed by a wave equation, the electron sometimes will be transmitted and sometimes will be reflected. The probabilities of reflection and transmission can be calculated by solving the Schrödinger equation in each region of space and comparing the amplitudes of the transmitted waves and reflected waves with the amplitudes of the incident wave. These calculations and their results are similar to finding the fraction of light reflected from an air–glass interface. If R is the probability of reflection, called the reflection coefficient, this calculation gives

$$R = \frac{(k_1 - k_2)^2}{(k_1 + k_2)^2} \qquad 35\text{-}27$$

where k_1 is the wave number for the incident wave and k_2 is the wave number for the transmitted wave. This result is the same as the result in optics for the reflection of light at normal incidence from the boundary between two media having different indexes of refraction n (Equation 31-17). The probability of transmission T, called the **transmission coefficient,** can be calculated from the reflection coefficient, because the probability of transmission plus the probability of reflection must equal 1:

$$T + R = 1 \qquad 35\text{-}28$$

Example 35-2 Reflection and Transmission at a Step Barrier

A particle that has kinetic energy E_0 and is traveling in a region in which the potential energy is zero is incident on a potential-energy barrier of height $U_0 = 0.20E_0$. Find the probability that the particle will be reflected.

PICTURE We need to calculate the wave numbers k_1 and k_2 and use them to calculate the reflection coefficient R from Equation 35-27. The wave numbers are related to the momentum by the de Broglie relation $p = h/\lambda$ (Equation 34-13), where $k = 2\pi/\lambda$. Combining these two equations gives $p = \hbar k$. Thus, the kinetic energy K is related to the wave number by $K = \frac{1}{2}p^2/m = \frac{1}{2}\hbar^2k^2/m$.

SOLVE

1. The probability of reflection is the reflection coefficient: $R = \dfrac{(k_2 - k_1)^2}{(k_1 + k_2)^2}$

2. Calculate k_1 from the initial kinetic energy E_0: $E_0 = \dfrac{\hbar^2 k_1^2}{2m}$

$k_1 = \sqrt{2mE_0/\hbar^2}$

3. Relate the final kinetic energy K_2 to the initial kinetic energy E_0 and the potential energy U_0 in the region $x > 0$:

$$K_2 = E_0 - U_0 = E_0 - 0.2E_0 = 0.8E_0$$

4. Relate k_2 to the final kinetic energy K_2 and solve for k_2:

$$K_2 = \frac{\hbar k_2^2}{2m}$$

so

$$k_2 = \sqrt{2mK_2/\hbar^2} = \sqrt{2m(0.8E_0)/\hbar^2}$$
$$= \sqrt{0.80}\sqrt{2mE_0/\hbar^2}$$

5. Substitute these values into Equation 35-27 to calculate R:

$$R = \frac{(k_1 - k_2)^2}{(k_1 + k_2)^2} = \left(\frac{1 - \sqrt{0.80}}{1 + \sqrt{0.80}}\right)^2 = \boxed{0.0031}$$

CHECK Classically, the particle would not be reflected by such a low barrier. The step-5 result gives a probability of 0.31 percent that the particle will be reflected. Such a low probability approaches being in agreement with our classical expectations.

TAKING IT FURTHER The probability of reflection is only 0.31 percent. This probability is small because the barrier height reduces the kinetic energy by only 20 percent. Because k is proportional to the square root of the kinetic energy, the wave number and therefore the wavelength is changed by only 10 percent.

PRACTICE PROBLEM 35-1 Express the index of refraction n of light in terms of the wave number k and the frequency ω, and show that the expression $(n_1 - n_2)^2/(n_1 + n_2)^2$ (Equation 31-7) for the reflection coefficient of light at normal incidence is the same as Equation 35-27. *Hint: Express the index of refraction n of light in terms of the wave number k and the angular frequency ω.*

In quantum mechanics, a localized particle is represented by a wave packet, which has a maximum at the most probable position of the particle. Figure 35-11 shows a wave packet representing a particle of energy E incident on a step potential of height U_0, which is less than E. After the encounter, there are two wave packets. The relative heights of the transmitted packet and reflected packet indicate the relative probabilities of transmission and reflection. For the situation shown here, E is much greater than U_0, and the probability of transmission is much greater than that of reflection.

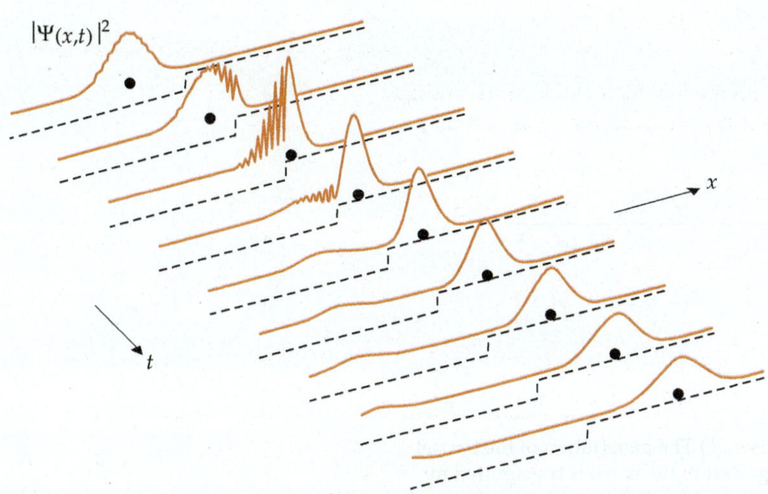

FIGURE 35-11 Time development of a one-dimensional wave packet representing a particle incident on a step potential for $E > U_0$. The position of a classical particle is indicated by the dot. Note that part of the packet is transmitted and part is reflected.

BARRIER PENETRATION

Figure 35-12*a* shows a rectangular potential-energy barrier of height U_0 and width *a* given by

$$U(x) = \begin{cases} 0 & x < 0 \\ U_0 & 0 < x < a \\ 0 & x > a \end{cases}$$

We consider a particle of energy *E*, which is slightly less than U_0, that is incident on the barrier from the left. Classically, the particle would always be reflected. However, a wave incident from the left does not decrease immediately to zero at the barrier, but it will instead decay exponentially in the classically forbidden region $0 < x < a$. On reaching the far wall of the barrier ($x = a$), the wave function must join smoothly to a sinusoidal wave function to the right of the barrier, as shown in Figure 35-12*b* This implies that there is some probability of the particle (which is represented by the wave function) being found on the far side of the barrier even though, classically, it should never pass through the barrier. For the case in which the quantity αa [where $\alpha^2 = 2m(U_0 - E)/\hbar^2$] is much greater than 1 the transmission coefficient *T* is equal to $e^{-2\alpha a}$:

! Do not think it might be possible to detect a particle in the classically forbidden region. It cannot. A proof shows this claim to be a consequence of the uncertainty principle.

$$T = e^{-2\alpha a} \qquad \alpha a \gg 1 \qquad \text{35-29}$$

TRANSMISSION THROUGH A BARRIER

The probability of penetration of the barrier thus decreases exponentially with both the barrier thickness *a* and the square root of the relative barrier height $(U_0 - E)$. This phenomenon, the penetrating of a classically forbidden region, is called **quantum tunneling.**

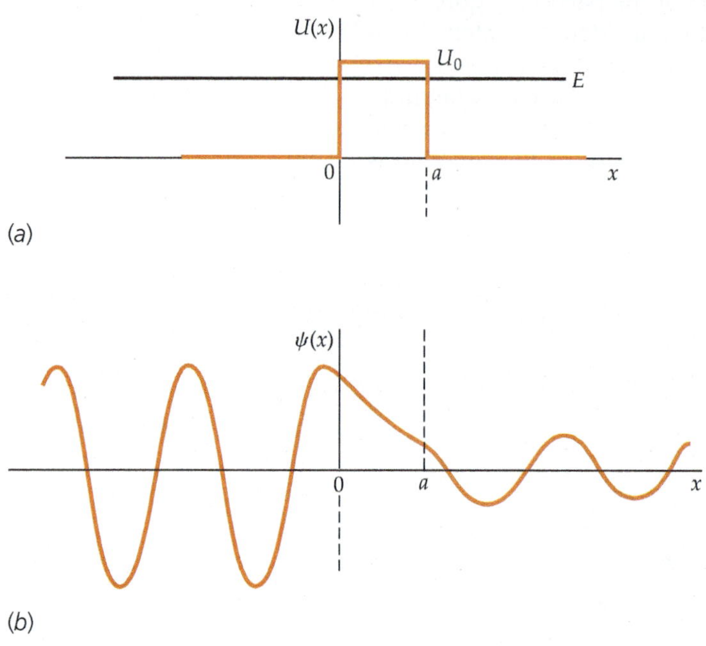

FIGURE 35-12 (*a*) A rectangular potential-energy barrier. (*b*) The penetration of the barrier by a wave that has a total energy less than the barrier energy. Part of the wave is transmitted by the barrier even though, classically, the particle cannot enter the region $0 < x < a$ in which the potential energy is greater than the total energy. To the left of the barrier, there is both an incident wave and a reflected wave. These waves form a resultant wave so that ψ is a superposition of a standing wave and a traveling wave (traveling toward the barrier). Only the transmitted wave exists in the region $x > a$, and it is traveling away from the barrier.

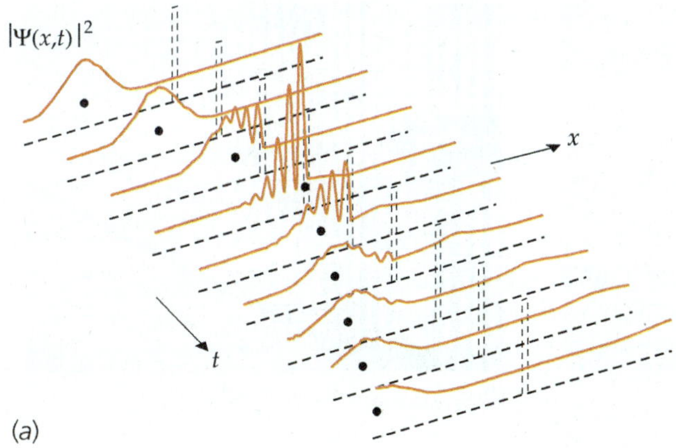

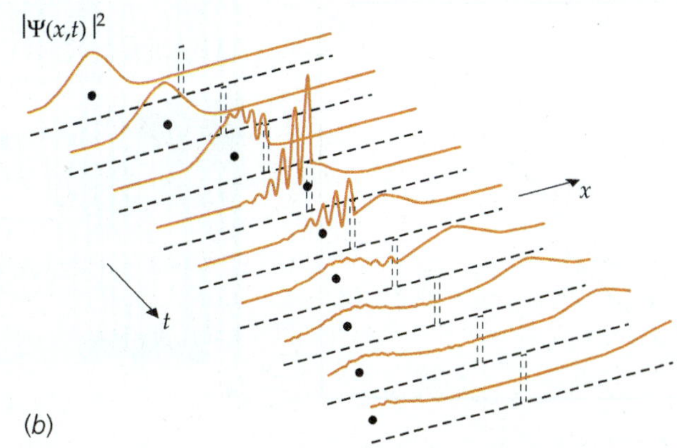

FIGURE 35-13 Barrier penetration. (*a*) The same particle incident on a barrier of height much greater than the energy of the particle. A very small part of the packet tunnels through the barrier. In both drawings, the position of a classical particle is indicated by a dot. (*b*) A wave packet representing a particle incident on a barrier of height just slightly greater than the energy of the particle. For this particular choice of energies, the probability of transmission is approximately equal to the probability of reflection, as indicated by the relative sizes of the transmitted and reflected packets.

Figure 35-13*a* shows a wave packet incident on a potential-energy barrier of height U_0 that is considerably greater than the energy of the particle. The probability of penetration is very small, as indicated by the relative sizes of the reflected and transmitted packets. In Figure 35-13*b*, the barrier is just slightly greater than the energy of the particle. In this case, the probability of penetration is about the same as the probability of reflection. Figure 35-14 shows a particle incident on two potential-energy barriers of height just slightly greater than the energy of the particle.

As we have mentioned, the penetration of a barrier is not unique to quantum mechanics. When light is totally reflected from a glass–air interface, the light wave can penetrate the air barrier if a second piece of glass is brought within a few

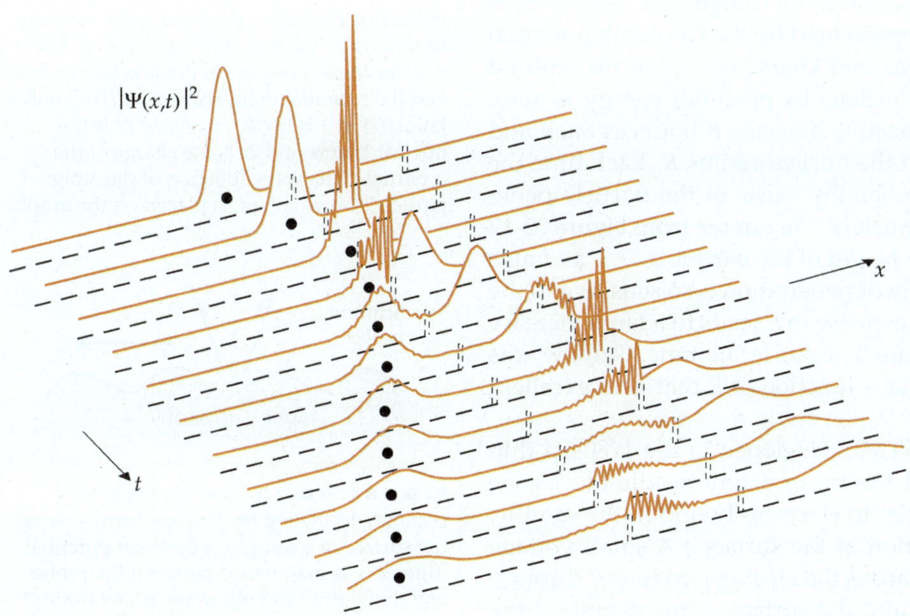

FIGURE 35-14 A wave packet representing a particle incident on two barriers. At each encounter, part of the packet is transmitted and part reflected, resulting in part of the packet being trapped between the barriers for some time.

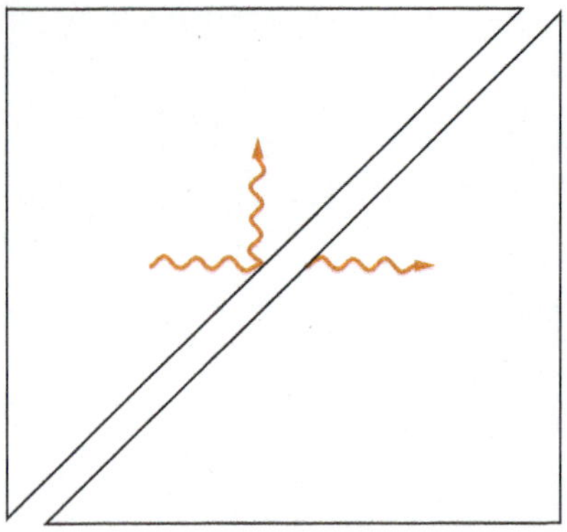

FIGURE 35-15 The penetration of an optical barrier. If the second prism is close enough to the first, part of the wave penetrates the air barrier even when the angle of incidence in the first prism is greater than the critical angle.

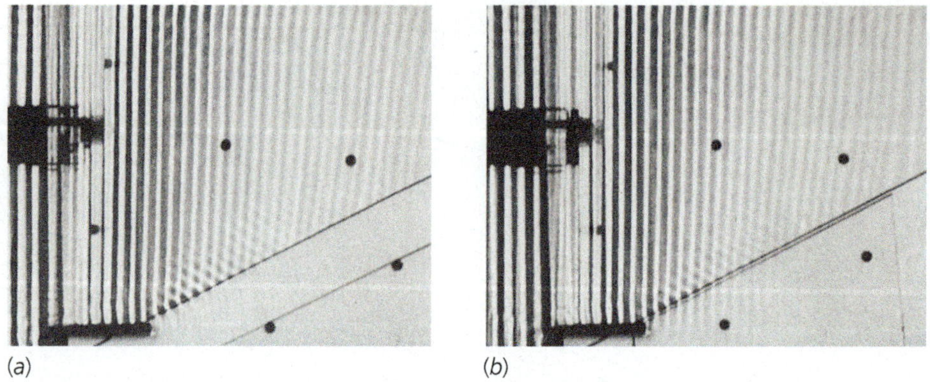

FIGURE 35-16 The penetration of a barrier by water waves in a ripple tank. In Figure 35-16*a*, the waves are totally reflected from a gap of deeper water. When the gap is very narrow, as in Figure 35-16*b*, a transmitted wave appears. The dark circles are spacers that are used to support the prisms from below. *(Education Development Center.)*

wavelengths of the first. This effect can be demonstrated with a laser beam and two 45° prisms (Figure 35-15). Similarly, water waves in a ripple tank can penetrate a gap of deep water (Figure 35-16).

The theory of barrier penetration was used by George Gamow in 1928 to explain the enormous variation in the half-lives for α decay of radioactive nuclei. (Alpha particles are emitted from atoms during radioactive decay and consist of two protons and two neutrons tightly bound together.) In general, the smaller the energy of the emitted α particle, the longer the half-life of the particle is. The energies of α particles from natural radioactive sources range from approximately 4 MeV to 7 MeV, whereas the half-lives range from approximately 10^{-5} seconds to 10^{10} years. Gamow represented a radioactive nucleus by a potential well of finite depth containing an α particle, as shown in Figure 35-17. Without knowing very much about the nuclear force that is exerted by the nucleus on the particle, Gamow represented it by a square well. Just outside the well, the α particle that has a charge of $+2e$ is repelled by the nucleus that has a charge $+Ze$, where Ze is the remaining nuclear charge. This force is represented by the Coulomb potential energy $+k(2e)(Ze)/r$. The energy E is the measured kinetic energy of the emitted α particle, because when it is far from the nucleus its potential energy is zero. After the α particle is formed from the radioactive nucleus, it bounces back and forth inside the nucleus, hitting the barrier at the nuclear radius R. Each time the α particle strikes the barrier, some small probability exists of the particle penetrating the barrier and appearing outside the nucleus. We can see from Figure 35-17 that a small increase in E reduces the relative height of the barrier $U - E$ and also the barrier's thickness. Because the probability of penetration is so sensitive to the barrier thickness and relative height, a small increase in E leads to a large increase in the probability of transmission and therefore to a shorter lifetime. Gamow was able to derive an expression for the half-life as a function of E that is in excellent agreement with experimental results.

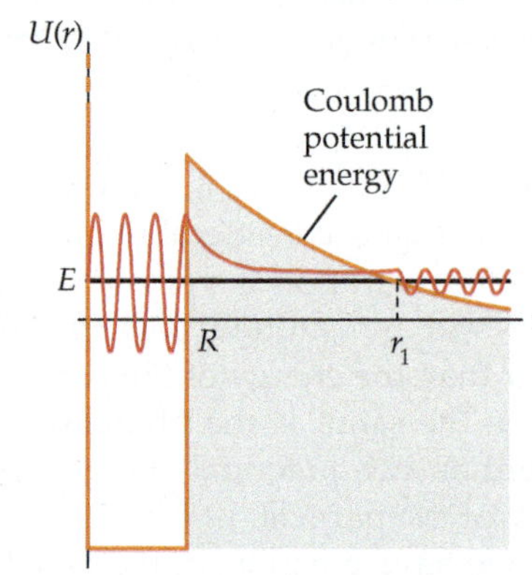

FIGURE 35-17 Model of a potential energy function for an α particle of a radioactive nucleus. The strong attractive nuclear force when r is approximately equal to the nuclear radius R can be approximately described by the potential well shown. The nuclear force is negligible outside the nucleus, and the potential there is given by Coulomb's law, $U(r) = +k(2e)(Ze)/r$, where Ze is the nuclear charge and $2e$ is the charge of the α particle. The wave function of the alpha particle, shown in red, is placed on the graph.

In the **scanning tunneling microscope (STM),** developed in the 1980s, a thin space between the surface of a sample and the tip of a tiny needle-like probe (Figure 35-18) acts as a potential-energy barrier to electrons bound in the sample. (The height of the barrier is the work function of the surface.) A small voltage applied between the probe and the sample causes the electrons to *tunnel* through the vacuum separating the tip of the probe and the surface of the sample if the surfaces are close enough together. The tunneling current is extremely sensitive to the size of the gap between the probe and sample. A constant tunneling current is maintained as the probe scans (travels along) the surface by a feedback

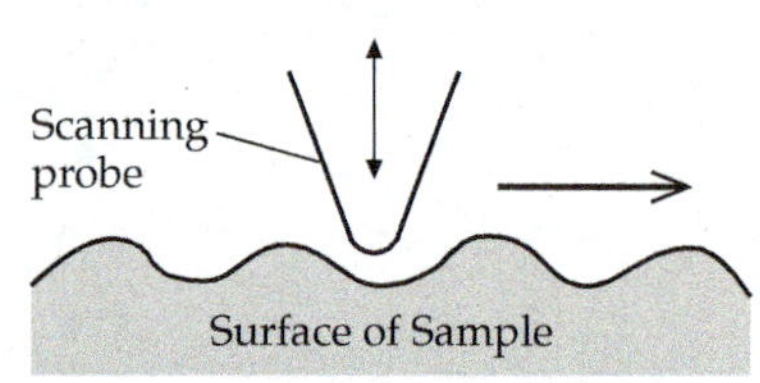

FIGURE 35-18 The tiny probe of a scanning tunneling microscope travels along the surface of a sample. A constant potential difference is maintained between the probe and the sample and electrons tunnel through the potential-energy barrier at the surface. A feedback mechanism maintains a constant tunneling current by moving the probe up and down as it travels along the surface.

mechanism that moves the probe up and down (farther from or closer to the surface). The surface of the sample is mapped out by the tracking of the motions of the probe. In this way, the surface features of the sample can be measured with a resolution of the order of the size of an atom.

35-5 THE SCHRÖDINGER EQUATION IN THREE DIMENSIONS

The one-dimensional time-independent Schrödinger equation is easily extended to three dimensions. In rectangular coordinates, it is

$$-\frac{\hbar^2}{2m}\left(\frac{\partial^2\psi}{\partial x^2}+\frac{\partial^2\psi}{\partial y^2}+\frac{\partial^2\psi}{\partial z^2}\right)+U\psi=E\psi \qquad \text{35-30}$$

where the wave function ψ and the potential energy U are generally functions of all three coordinates, x, y, and z. To illustrate some of the features of problems in three dimensions, we consider a particle in a three-dimensional infinite square well given by $U(x, y, z) = 0$ for $0 < x < L$, $0 < y < L$, and $0 < z < L$. Outside this cubical region, $U(x, y, z) = \infty$. For this problem, the wave function must be zero at the edges of the well.

There are standard methods in partial differential equations for solving Equation 35-30. We can guess the form of the solution from our knowledge of probability. For a one-dimensional box along the x axis, we have found the probability that a particle is in the region between x and $x + dx$ to be $A_1^2 \sin^2(k_1x)\,dx$ (from Equation 35-10), where A_1 is a normalization constant and $k_1 = n\pi/L$ is the wave number. Similarly, for a box along the y axis, the probability of a particle being in a region between y and $y + dy$ is $A_2^2\sin^2(k_2y)dy$. The probability of two independent events occurring is the product of the probabilities of each event occurring.* So the probability of a particle being in region between x and $x + dx$ *and* in region between y and $y + dy$ is $A_1^2\sin^2(k_1x)\,dx\,A_2^2\sin^2(k_2y)\,dy = A_1^2\sin^2(k_1x)\,A_2^2\sin^2(k_2y)\,dx\,dy$. The probability of a particle being in the region between x and $x + dx$, y and $y + dy$, and z and $z + dz$ is $\psi^2(x, y, z)\,dx\,dy\,dz$, where $\psi(x, y, z)$ is the solution of Equation 35-30. This solution is of the form

$$\psi(x, y, z) = A\sin(k_1x)\sin(k_2y)\sin(k_3z) \qquad \text{35-31}$$

where the constant A is determined by normalization. Inserting this solution into Equation 35-30, we obtain for the energy

$$E = \frac{\hbar^2}{2m}(k_1^2 + k_2^2 + k_3^2)$$

which is equivalent to $E = \frac{1}{2}(p_x^2 + p_y^2 + p_z^2)/m$, where $p_x = \hbar k_1$, $p_y = \hbar k_2$, and $p_z = \hbar k_3$. The wave function (Equation 35-31) will be zero at $x = L$ if $k_1 = n_1\pi/L$, where n_1 is an integer. Similarly, the wave function will be zero at $y = L$ if $k_2 = n_2\pi/L$, and the wave function will be zero at $z = L$ if $k_3 = n_3\pi/L$. (It is also zero at $x = 0$, $y = 0$, and $z = 0$.) The energy is thus quantized to the values

$$E_{n_1n_2n_3} = \frac{\hbar^2\pi^2}{2mL^2}(n_1^2 + n_2^2 + n_3^2) = E_1(n_1^2 + n_2^2 + n_3^2) \qquad \text{35-32}$$

where n_1, n_2, and n_3 are positive integers and $E_1 = \hbar^2\pi^2/(2mL^2)$ is the ground-state energy of the one-dimensional well. Note that the energy and wave function are characterized by three quantum numbers, each arising from the boundary conditions for one of the coordinates x, y, and z.

* For example, if you throw two dice, the probability of the first die coming up 6 is 1/6 and the probability of the second die coming up an odd number is 1/2. The probability of the first die coming up 6 *and* the second die coming up an odd number is (1/6)(1/2) = 1/12.

The lowest energy state (the ground state) for the cubical well occurs when $n_1 = n_2 = n_3 = 1$ and has the value

$$E_{111} = \frac{3\hbar^2\pi^2}{2mL^2} = 3E_1$$

The first excited energy level can be obtained in three different ways: $n_1 = 2$, $n_2 = n_3 = 1$; $n_2 = 2$, $n_1 = n_3 = 1$; or $n_3 = 2$, $n_1 = n_2 = 1$. Each has a different wave function. For example, the wave function for $n_1 = 2$ and $n_2 = n_3 = 1$ is

$$\psi_{211} = A \sin\frac{2\pi x}{L} \sin\frac{\pi y}{L} \sin\frac{\pi z}{L} \qquad \text{35-33}$$

There are thus three different quantum states as described by the three different wave functions corresponding to the same energy level. An energy level with which more than one wave function is associated is said to be **degenerate.** In this case, there is threefold degeneracy. Degeneracy is related to the spatial symmetry of the system. If, for example, we consider a noncubic well, where $U = 0$ for $0 < x < L_1$, $0 < y < L_2$, and $0 < z < L_3$, the boundary conditions at the edges would lead to the quantum conditions $k_1L_1 = n_1\pi$, $k_2L_2 = n_2\pi$, and $k_3L_3 = n_3\pi$, and the total energy would be

$$E_{n_1n_2n_3} = \frac{\hbar^2\pi^2}{2m}\left(\frac{n_1^2}{L_1^2} + \frac{n_2^2}{L_2^2} + \frac{n_3^2}{L_3^2}\right) \qquad \text{35-34}$$

These energy levels are not degenerate if L_1, L_2, and L_3 are all different. Figure 35-19 shows the energy levels for the ground state and first two excited states for an infinite cubic well in which the excited states are degenerate and for a noncubic infinite well in which L_1, L_2, and L_3 are all slightly different so that the excited levels are slightly split apart and the degeneracy is removed. The ground state is the state where the quantum numbers n_1, n_2, and n_3 all equal 1. None of the three quantum numbers can be zero. If any one of n_1, n_2, and n_3 were zero, the corresponding wave number k would also equal zero and the corresponding wave function (Equation 35-31) would equal zero for all values of x, y, and z.

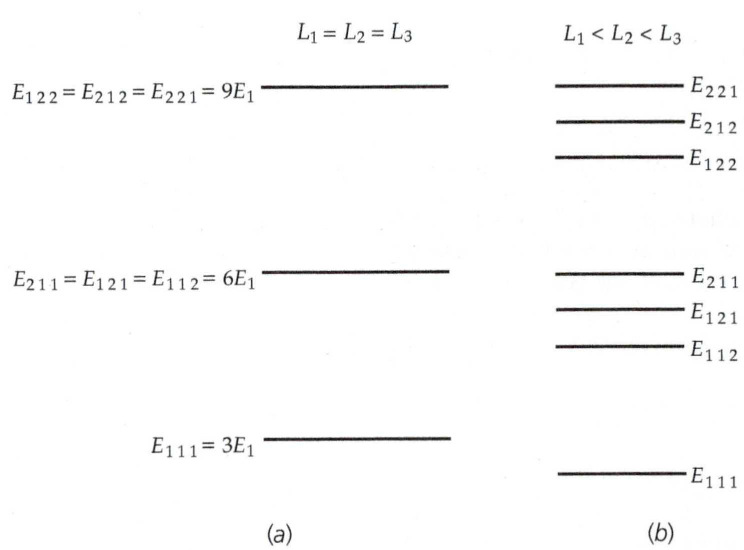

FIGURE 35-19 Energy-level diagrams for (*a*) a cubic infinite well and (*b*) a noncubic infinite well. In Figure 35-19*a* the energy levels are degenerate; that is, there are two or more wave functions having the same energy. The degeneracy is removed when the symmetry of the potential is removed, as in Figure 35-19*b*.

Example 35-3 Energy Levels for a Particle in a Three-Dimensional Box

A particle is in a three-dimensional box where $L_3 = L_2 = 2L_1$. Give the quantum numbers n_1, n_2, and n_3 that correspond to the state(s) in each of the seven lowest energy levels of this box.

PICTURE We can use Equation 35-34 to write the energy in terms of L_1 and the quantum numbers n_1, n_2, and n_3. Then we can find by inspection the values of the quantum numbers that give the lowest energies.

SOLVE

1. The energy of a level is given by Equation 35-34:

$$E_{n_1n_2n_3} = \frac{\hbar^2\pi^2}{2m}\left(\frac{n_1^2}{L_1^2} + \frac{n_2^2}{L_2^2} + \frac{n_3^2}{L_3^2}\right)$$

2. Factor out $1/L_1^2$:

$$E_{n_1n_2n_3} = \frac{\hbar^2\pi^2}{2m}\left(\frac{n_1^2}{L_1^2} + \frac{n_2^2}{4L_1^2} + \frac{n_3^2}{4L_1^2}\right) = \frac{\hbar^2\pi^2}{8mL_1^2}(4n_1^2 + n_2^2 + n_3^2)$$

3. The lowest energy is E_{111}:

$$E_{111} = E_1(4\cdot 1^2 + 1^2 + 1^2) = \boxed{6E_1} \quad \text{(1st)}$$

where $E_1 = \hbar^2\pi^2/8mL_1^2$.

4. The energy increases the least when we increase n_2 or n_3. Trying various values of the quantum numbers:

$$E_{121} = E_{112} = E_1(4\cdot 1^2 + 2^2 + 1^2) = \boxed{9E_1} \quad \text{(2nd)}$$

$$E_{122} = E_1(4\cdot 1^2 + 2^2 + 2^2) = \boxed{12E_1} \quad \text{(3rd)}$$

$$E_{131} = E_{113} = E_1(4\cdot 1^2 + 3^2 + 1^2) = \boxed{14E_1} \quad \text{(4th)}$$

$$E_{132} = E_{123} = E_1(4\cdot 1^2 + 3^2 + 2^2) = \boxed{17E_1} \quad \text{(5th)}$$

$$E_{211} = E_1(4\cdot 2^2 + 1^2 + 1^2) = \boxed{18E_1} \quad \text{(6th)}$$

$$\left.\begin{aligned} E_{221} = E_{212} &= E_1(4\cdot 2^2 + 2^2 + 1^2) = \boxed{21E_1} \\ E_{141} = E_{114} &= E_1(4\cdot 1^2 + 4^2 + 1^2) = \boxed{21E_1} \end{aligned}\right\} \text{(7th)}$$

CHECK Because two of the lengths are equal, degenerate energy levels are expected. Our results meet this expectation.

TAKING IT FURTHER Energies E_{221} and E_{212} are exactly equal because L_2 and L_3 are exactly equal. However, energies E_{221} and E_{141} are exactly equal because L_1 is exactly half of L_2.

PRACTICE PROBLEM 35-2 Find the quantum numbers and energies of the next two energy levels in step 4.

Example 35-4 Wave Functions for a Particle in a Three-Dimensional Box

Try It Yourself

Write the degenerate wave functions for the fourth and fifth excited states (the 5th and 6th levels) of the results in step 4 of Example 35-3.

PICTURE Use $\psi(x, y, z) = A\sin(k_1x)\sin(k_2y)\sin(k_3z)$ (a generalized version of Equation 35-31) with $k_i = n_i\pi/L_i$.

SOLVE

Cover the column to the right and try these on your own before looking at the answers.

Steps	Answers
Write the wave functions corresponding to the energies E_{131} and E_{113}	$\psi_{131} = A \sin\frac{\pi x}{L_1} \sin\frac{3\pi y}{2L_1} \sin\frac{\pi z}{2L_1}$ $\psi_{113} = A \sin\frac{\pi x}{L_1} \sin\frac{\pi y}{2L_1} \sin\frac{3\pi z}{2L_1}$

35-6 THE SCHRÖDINGER EQUATION FOR TWO IDENTICAL PARTICLES

Our discussion of quantum mechanics has thus far been limited to situations in which a single particle moves in some force field characterized by a potential energy function U. The most important physical problem of this type is the hydrogen atom, in which a single electron moves in the Coulomb potential of a proton. This problem is actually a two-body problem, because the proton also moves in the field of the electron. However, the motion of the much more massive proton requires only a very small correction to the energy of the atom that is easily made in both classical and quantum mechanics. When we consider more complicated problems, such as the helium atom, we must apply quantum mechanics to two or more electrons moving in an external field. Such problems are complicated, not only by the interaction of the electrons with each other, but also by the fact that the electrons are identical.

The interaction of two electrons with each other is electromagnetic and is essentially the same as the classical interaction of two charged particles. The Schrödinger equation for an atom that has two or more electrons cannot be solved exactly, so approximation methods must be used. This situation is not very different from the situation in which a classical expression describes three or more particles. However, the complications arising from the identity of electrons are purely quantum mechanical and have no classical counterpart. They are due to the fact that it is impossible to keep track of which electron is which. Classically, identical particles can be identified by their positions, which in principle can be determined with unlimited accuracy. This is impossible quantum mechanically because of the uncertainty principle. Figure 35-20 offers a schematic illustration of the problem.

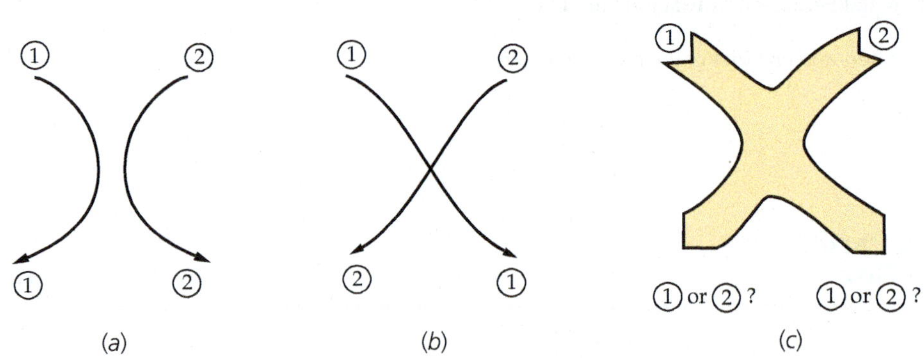

FIGURE 35-20 (*a*, *b*) Two possible classical electron paths. If electrons were classical particles, they could be distinguished by the paths followed. (*c*) However, because of the quantum-mechanical wave properties of electrons, the paths are spread out, as indicated by the shaded region. It is impossible to distinguish which electron is which after they separate.

The indistinguishability of identical particles has important consequences. For instance, consider the very simple case of two identical, noninteracting particles in a one-dimensional infinite square well. The time-independent Schrödinger equation for two particles, each of mass m, is

$$-\frac{\hbar^2}{2m}\frac{\partial^2\psi(x_1, x_2)}{\partial x_1^2} - \frac{\hbar^2}{2m}\frac{\partial^2\psi(x_1, x_2)}{\partial x_2^2} + U\psi(x_1, x_2) = E\psi(x_1, x_2) \qquad \text{35-35}$$

where x_1 and x_2 are the coordinates of the two particles. If the particles interact, the potential energy U is described by terms that have both x_1 and x_2 and cannot be separated into separate terms having only x_1 or x_2. For example, the electrostatic repulsion of two electrons in one dimension is represented by the potential energy function $ke^2/|x_2 - x_1|$. However, if the particles do not interact (as we are assuming here), we can write $U = U_1(x_1) + U_2(x_2)$. For the infinite square well, we need only solve the Schrödinger equation inside the well where $U = 0$ and require that the wave function be zero at the walls of the well. With $U = 0$, Equation 35-35 looks just like the expression for a particle in a two-dimensional well (Equation 35-30, but no z terms exist and y is replaced by x_2).

Solutions of this equation can be written in the form*

$$\psi_{nm} = \psi_n(x_1)\psi_m(x_2) \qquad \text{35-36}$$

where ψ_n and ψ_m are the single-particle wave functions for a particle in an infinite well and n and m are the quantum numbers of particles 1 and 2, respectively. For example, for $n = 1$ and $m = 2$, the wave function is

$$\psi_{12} = A \sin\frac{\pi x_1}{L} \sin\frac{2\pi x_2}{L} \qquad \text{35-37}$$

The probability of finding particle 1 in the region between $x = x_1$ and $x = x_1 + dx_1$ and particle 2 in the region between $x = x_2$ and $x = x_2 + dx_2$ is $\psi_{nm}^2(x_1, x_2)\,dx_1\,dx_2$, which is just the product of the separate probabilities $\psi_n^2(x_1)\,dx_1$ and $\psi_m^2(x_2)\,dx_2$. However, even though we have labeled the particles 1 and 2, we cannot distinguish which is between x_1 and $x_1 + dx_1$ and which is between x_2 and $x_2 + dx_2$ if they are identical. The mathematical descriptions of identical particles must be the same if we interchange the labels. The probability density $\psi^2(x_1, x_2)$ must therefore be the same as $\psi^2(x_2, x_1)$:

$$\psi^2(x_2, x_1) = \psi^2(x_1, x_2) \qquad \text{35-38}$$

Equation 35-38 is satisfied if $\psi(x_2, x_1)$ is either **symmetric** or **antisymmetric** on the exchange of particles—that is, if either

$$\psi(x_2, x_1) = \psi(x_1, x_2) \quad \text{symmetric} \qquad \text{35-39}$$

or

$$\psi(x_2, x_1) = -\psi(x_1, x_2) \quad \text{antisymmetric} \qquad \text{35-40}$$

Note that the wave functions given by Equations 35-36 and 35-37 are neither symmetric nor antisymmetric. If we interchange x_1 and x_2 in these wave functions, we get a different wave function, which implies that the particles can be distinguished.

---

* Again, this result can be obtained by solving Equation 35-35, but it also can be understood in terms of our knowledge of probability. The probability of electron 1 being between $x = x_1$ and $x = x_1 + dx_1$ and electron 2 being between $x = x_2$ and $x = x_2 + dx_2$ is the product of the individual probabilities.

We can find symmetric and antisymmetric wave functions that are solutions of the Schrödinger equation by adding or subtracting ψ_{nm} and ψ_{mn}. Adding them, we obtain

$$\psi_S = A'[\psi_n(x_1)\psi_m(x_2) + \psi_n(x_2)\psi_m(x_1)] \quad \text{symmetric} \qquad 35\text{-}41$$

and subtracting them, we obtain

$$\psi_A = A'[\psi_n(x_1)\psi_m(x_2) - \psi_n(x_2)\psi_m(x_1)] \quad \text{antisymmetric} \qquad 35\text{-}42$$

For example, the symmetric and antisymmetric wave functions for the first excited state of two identical particles in an infinite square well would be

$$\psi_S = A'\left(\sin\frac{\pi x_1}{L}\sin\frac{2\pi x_2}{L} + \sin\frac{\pi x_2}{L}\sin\frac{2\pi x_1}{L}\right) \qquad 35\text{-}43$$

and

$$\psi_A = A'\left(\sin\frac{\pi x_1}{L}\sin\frac{2\pi x_2}{L} - \sin\frac{\pi x_2}{L}\sin\frac{2\pi x_1}{L}\right) \qquad 35\text{-}44$$

There is an important difference between antisymmetric and symmetric wave functions. If $n = m$, the antisymmetric wave function is identically zero for all values of x_1 and x_2, whereas the symmetric wave function is not. Thus, if the wave function describing two identical particles is antisymmetric, the quantum numbers n and m of two particles cannot be the same. The idea that no two electrons in an atom can occupy the same quantum state, and thus have the same quantum numbers, was first stated by Wolfgang Pauli in 1925. This idea was soon generalized to include systems other than atoms and particles other than electrons. For example, no two protons of a nucleus can occupy the same quantum state, and no two neutrons of a nucleus can occupy the same quantum state. Electrons, protons, neutrons, neutrinos, and quarks all have a spin quantum number s equal to one-half, and all particles that have half-integer spin are called fermions. The two allowed values for the secondary spin quantum number m_s are plus and minus one half. Pauli's idea is called the **Pauli exclusion principle:**

> No two identical fermions can simultaneously occupy the same quantum state.
>
> PAULI EXCLUSION PRINCIPLE

The wave function for two or more identical fermions must be an antisymmetric wave function. Other particles (for example, α particles, deuterons, photons, and mesons) have integer spin and symmetric wave functions. These particles are called **bosons.**

A wave function that is a solution to the multiparticle time-independent wave equation (Equation 35-35) is called a spatial state. A bound system that contains fermions has either one or two identical fermions in each occupied spatial state. However, for a bound system that contain bosons there is no limit to the number of identical bosons in each spatial state.

Summary

1. The Schrödinger equation is a partial differential equation that relates the second space derivative of a wave function to its first time derivative. Wave functions that describe physical situations are solutions of this differential equation.
2. Because a wave function must satisfy the normalization condition, it must be well behaved; this means, among other things, that it must approach zero as $|x|$ approaches infinity. For bound systems such as a particle in a box, a simple harmonic oscillator, or an electron in an atom, this requirement leads to energy quantization.
3. The well-behaved wave functions for bound systems describe standing waves.

TOPIC	RELEVANT EQUATIONS AND REMARKS	
1. Time-Independent Schrödinger Equation	$-\frac{\hbar^2}{2m}\frac{d^2\psi(x)}{dx^2} + U(x)\psi(x) = E\psi(x)$	35-4
Allowable solutions	In addition to satisfying the Schrödinger equation, a wave function $\psi(x)$ must be continuous and must have a continuous first derivative $d\psi/dx$.* Because the probability of finding an electron somewhere must be 1, the wave function must obey the normalization condition $\int_{-\infty}^{\infty} \lvert\psi\rvert^2\,dx = 1$ This condition implies the boundary condition that ψ must approach 0 as $\lvert x\rvert$ approaches ∞. Such boundary conditions lead to the quantization of energy.	
2. Confined Particles	When the total energy E is greater than the potential energy $U(x)$ in some region (the classically allowed region) and less than $U(x)$ outside that region, the wave function ψ oscillates within the classically allowed region and $\lvert\psi\rvert$ decreases exponentially outside that region. The wave function approaches zero as $\lvert x\rvert$ approaches ∞ only for certain values of the total energy E. The energy is thus quantized.	
In a finite square well	In a finite well of height U_0, there are only a finite number of allowed energies, and these are slightly less than the corresponding energies in an infinite well.	
In the simple harmonic oscillator	For the oscillator with potential energy function $U(x) = \frac{1}{2}m\omega_0^2x^2$, the allowed energies are equally spaced and given by $E_n = \left(n + \frac{1}{2}\right)\hbar\omega_0 \quad n = 0, 1, 2, \ldots$	35-26
	The ground-state wave function is given by $\psi_0(x) = A_0e^{-ax^2}$	35-23
	where A_0 is the normalization constant and $a = \frac{1}{2}m\omega_0/\hbar$.	
3. Reflection and Barrier Penetration	When the potential changes abruptly over a small distance, a particle may be reflected even though $E > U(x)$. A particle may penetrate a region in which $E < U(x)$. Reflection and penetration of matter waves are similar to those for other kinds of waves.	
4. The Schrödinger Equation in Three Dimensions	The wave function for a particle in a three-dimensional box can be written $\psi(x, y, z) = \psi_1(x)\psi_2(y)\psi_3(z)$ where ψ_1, ψ_2, and ψ_3 are wave functions for a one-dimensional box.	
Degeneracy	When more than one wave function is associated with the same energy level, the energy level is said to be degenerate. Energy-level degeneracy occurs because of spatial symmetry.	

* An exception to this claim is for the infinite well potential (where U is equal to zero inside the well and to infinity outside the well). For this potential function $d\psi/dx$ is not continuous at the boundary of the well (see Figure 34-17).

TOPIC	RELEVANT EQUATIONS AND REMARKS
5. The Schrödinger Equation for Two Identical Particles	A wave function that describes two identical particles must be either symmetric or antisymmetric when the coordinates of the particles are exchanged. Fermions (which include electrons, protons, and neutrons) are described by antisymmetric wave functions and obey the Pauli exclusion principle, which states that no two identical particles can simultaneously have the same values for their quantum number. Bosons (which include α particles, deuterons, photons, and mesons) have symmetric wave functions and do not obey the Pauli exclusion principle.

Answers to Concept Checks

35-1 The wave function cannot be normalized.

Answers to Practice Problems

35-2 $E_{133} = 22E_1, E_{142} = E_{124} = E_{222} = 24E_1$

Problems

In a few problems, you are given more data than you actually need; in a few other problems, you are required to supply data from your general knowledge, outside sources, or informed estimate.

Interpret as significant all digits in numerical values that have trailing zeros and no decimal points.

- • Single-concept, single-step, relatively easy
- •• Intermediate-level, may require synthesis of concepts
- ••• Challenging
- SSM Solution is in the *Student Solutions Manual*
- Consecutive problems that are shaded are paired problems.

CONCEPTUAL PROBLEMS

1 • Sketch (*a*) the wave function and (*b*) the probability density function for the $n = 5$ state of the finite square-well potential.

2 • Sketch (*a*) the wave function and (*b*) the probability density function for the $n = 4$ state of the finite square-well potential.

THE SCHRÖDINGER EQUATION

3 •• Show that if $\psi_1(x)$ and $\psi_2(x)$ are each solutions to the time-independent Schrödinger equation (Equation 35-4), then $\psi_3(x) = \psi_1(x) + \psi_2(x)$ is also a solution. This result, known as the superposition principle, applies to the solutions of all linear equations.

THE HARMONIC OSCILLATOR

4 •• The harmonic oscillator problem may be used to describe the vibrations of molecules. For example, the hydrogen molecule H_2 is found to have equally spaced vibrational energy levels separated by 8.7×10^{-20} J. What value of the force constant of the spring would be needed to get this energy spacing, assuming that half the molecule can be modeled as a hydrogen atom attached to one end of a spring that has its other end fixed? *Hint: The spacing for the energy levels of this half-molecule would be half of the spacing for the energy levels of the complete molecule. In addition, the force constant of a spring is inversely proportional to its relaxed length, so if half of the spring has force constant k, the entire spring has a force constant that is equal to* $\frac{1}{2}k$.

5 •• Use the procedure of Example 35-1 to verify that the energy of the first excited state of the harmonic oscillator is $E_1 = \frac{3}{2}\hbar\omega_0$. (*Note:* Rather than solve for a again, use the step-8 result $a = \frac{1}{2}m\omega_0/\hbar$ obtained in Example 35-1). SSM

6 •• Show that the expectation value $\langle x\rangle = \int_{-\infty}^{\infty} x|\psi|^2\,dx$ is zero for both the ground state and the first excited state of the harmonic oscillator.

7 •• Verify that the normalization constant A_0 in the ground-state harmonic-oscillator wave function $\psi_0(x) = A_0e^{-ax^2}$ (Equation 35-23) is given by $A_0 = (2m\omega_0/h)^{1/4}$.

8 •• Using the result of Problem 7, show that for the ground state of the harmonic oscillator $\langle x^2\rangle = \int x^2|\psi|^2\,dx = \hbar/(2m\omega_0) = 1/(4a)$. Use this result to show that the average potential energy equals half the total energy.

9 •• The quantity $\sqrt{\langle x^2\rangle - \langle x\rangle^2}$ is a measure of the average spread in the location of a particle. (*a*) Consider an electron trapped in a harmonic oscillator potential. Its lowest energy level is found to be 2.1×10^{-4} eV. Calculate $\sqrt{\langle x^2\rangle - \langle x\rangle^2}$ for this electron. (See Problems 6 and 8.) (*b*) Now consider an electron trapped in an infinite square-well potential. If the width of the well is equal to $\sqrt{\langle x^2\rangle - \langle x\rangle^2}$, what would be the lowest energy level for this electron?

10 ••• Classically, the average kinetic energy of the harmonic oscillator equals the average potential energy. Assume that this result is also true for the quantum-mechanical harmonic oscillator, and use this result, along with the result of Problem 8, to determine the expectation value of p_x^2 (where $p_x = mv_x$) for the ground state of the one-dimensional harmonic oscillator.

11 ••• We know that for the classical harmonic oscillator, $p_{x\,\mathrm{av}} = 0$. It can be shown that for the quantum-mechanical harmonic oscillator, $\langle p_x\rangle = 0$. Use the results of Problems 6, 8, and 10 to determine the uncertainty product $\Delta x\,\Delta p_x$ for the ground state of the harmonic oscillator. The uncertainties are defined by $(\Delta x)^2 = \langle(x - \langle x\rangle)^2\rangle$ and $(\Delta p_x)^2 = \langle(p_x - \langle p_x\rangle)^2\rangle$.

REFLECTION AND TRANSMISSION OF ELECTRON WAVES: BARRIER PENETRATION

12 •• A particle of energy E approaches a step barrier of height U_0. What should be the ratio E/U_0 so that the reflection coefficient is $\frac{1}{2}$?

13 •• **SPREADSHEET** A particle that has mass m is traveling in the direction of increasing x. The potential energy of the particle is equal to zero everywhere in the region $x < 0$ and is equal to U_0 everywhere in the region $x > 0$, where $U_0 > 0$. (*a*) Show that if the total energy is $E = \alpha U_0$, where $\alpha \geq 1$, then the wave number k_2 in the region $x > 0$ is given by $k_2 = k_1\sqrt{(\alpha - 1)/\alpha}$, where k_1 is the wave number in the region $x < 0$. (b) Using a spreadsheet program or graphing calculator, graph the reflection coefficient R and the transmission coefficient T as functions of α, for $1 \leq \alpha \leq 5$. SSM

14 •• Suppose that the potential energy in Problem 13 is equal to zero everywhere in the region $x < 0$ and is equal to $-U_0$ everywhere in the region $x > 0$, where $U_0 > 0$. The wave number for the incident particle is again k_1, and the total energy is $2U_0$. (*a*) What is the wave number for the particle in the region where $x > 0$? (*b*) What is the reflection coefficient R? (*c*) What is the transmission coefficient T? (*d*) If each of one million particles that are in the region $x < 0$ are traveling with wave number k_1 in the direction of increasing x and are incident upon the potential energy drop at $x = 0$, how many of these particles are expected to continue along in the direction of increasing x? How does this compare with the classical prediction?

15 •• A 10-eV electron (an electron with a kinetic energy of 10 eV) is incident on a potential-energy barrier that has a height equal to 25 eV and a width equal to 1.0 nm. (*a*) Use Equation 35-29 to calculate the order of magnitude of the probability that the electron will tunnel through the barrier. (*b*) Repeat your calculation for a width of 0.10 nm. SSM

16 •• Use Equation 35-29 to calculate the order of magnitude of the probability that a proton will tunnel out of a nucleus in one collision with the nuclear barrier if the proton has an energy 6.0 MeV below the top of the potential-energy barrier and the barrier thickness is 1.2×10^{-15} m.

17 ••• To understand how a small change in α-particle energy can dramatically change the probability of the α particle tunneling from a nucleus, consider an α particle emitted by a uranium nucleus ($Z = 92$). (*a*) Referring to Figure 35-17, calculate the center-to-center distance of closest approach r_1 that α particles that have kinetic energies of 4.0 MeV and 7.0 MeV could make to the uranium nucleus. (*b*) Use the result from Part (*a*) to calculate the relative transmission coefficient $e^{-2\alpha a}$ for the same α particles. (*Note:* The actual half-lives of uranium nuclei vary over nine orders of magnitude. Your calculation will show a smaller range than this; however, to find half-life, you must also include the frequency with which the α particle strikes the barrier.)

THE SCHRÖDINGER EQUATION IN THREE DIMENSIONS

18 •• (*a*) A particle is confined to a three-dimensional box that has sides L_1, $L_2 = 2L_1$, and $L_3 = 3L_1$. Give the quantum numbers n_1, n_2 and n_3 that correspond to the ten lowest-energy quantum states of this box. *Hint: A spreadsheet can be helpful.* (*b*) What quantum numbers, if any, correspond to degenerate energy levels? (*c*) Give a wave function for the fifth excited state. (There are only five states that have energy levels below the energy level of the fifth excited state.)

19 •• (*a*) A particle is confined to a three-dimensional box that has sides L_1, $L_2 = 2L_1$, and $L_3 = 4L_1$. Give the quantum numbers n_1, n_2 and n_3 that correspond to the ten lowest-energy quantum states of this box. *Hint: A spreadsheet can be helpful.* (*b*) What combinations of these quantum numbers, if any, correspond to degenerate energy levels? (*c*) Give the wave function for the fourth excited energy state. (There are only four states that have energy levels below the energy level of the fourth excited state.)

20 • A particle moves in a potential well given by $U(x, y, z) = 0$ for $-L/2 < x < L/2, 0 < y < L$, and $0 < z < L$; $U = \infty$ outside these ranges. (*a*) Write an expression for the ground-state wave function for the particle. (*b*) How do the allowed energies compare with those for a well having $U = 0$ for $0 < x < L$, rather than for $-L/2 < x < L/2$? Explain your answer.

21 •• A particle is constrained to the two-dimensional region defined by $0 \leq x \leq L$ and $0 \leq y \leq L$ and moves freely throughout that region. (*a*) Find the wave functions that meet these conditions and are solutions of the Schrödinger equation. (*b*) Find the energies that correspond to the wave functions in Part (*a*). (*c*) Find the quantum numbers of the two lowest states that have the same energy (that are degenerate). (*d*) Find the quantum numbers of the three lowest states that have the same energy.

THE SCHRÖDINGER EQUATION FOR TWO IDENTICAL PARTICLES

22 • Show that the two-particle wave function $\psi_{12} = A\sin(\pi x_1/L)\sin(2\pi x_2/L)$, $0 < x_1 < L$ and $0 < x_2 < L$, (Equation 35-37), is a solution of

$$-\frac{\hbar^2}{2m}\frac{\partial^2\psi(x_1, x_2)}{\partial x_1^2} - \frac{\hbar^2}{2m}\frac{\partial^2\psi(x_1, x_2)}{\partial x_2^2} + U\psi(x_1, x_2) = E\psi(x_1, x_2)$$

(Equation 35-35), if $U(x_1, x_2) = 0$, and find the energy of the state represented by this wave function.

23 • What is the ground-state energy of ten noninteracting bosons in a one-dimensional box of length L?

24 •• What is the ground-state energy of seven identical noninteracting fermions in a one-dimensional box of length L? (Because the quantum number associated with spin can have two values, each spatial state can be occupied by two fermions.)

ORTHOGONALITY OF WAVE FUNCTIONS

The integral of two functions over some space interval is somewhat analogous to the dot product of two vectors. If this integral is zero, the functions are said to be orthogonal, which is analogous to two vectors being perpendicular. The following problems illustrate the general principle that any two wave functions corresponding to different energy levels in the same potential are orthogonal. A general hint for all these problems is that the integral $\int_{x_1}^{x_2} f(x)\,dx$ *is equal to zero if* x_1 *is equal to* $-x_2$ *and if* $f(x)$ *is equal to* $-f(-x)$.

25 •• Show that the ground-state and the first excited state wave functions of the harmonic oscillator are orthogonal; that is, show that $\int_{-\infty}^{\infty} \psi_0(x)\psi_1(x)\,dx = 0$.

26 •• The wave function for the state $n = 2$ of the harmonic oscillator is $\psi_2(x) = A_2\left(2ax^2 - \frac{1}{2}\right)e^{-ax^2}$, where A_2 is the normalization constant and a is a positive constant. Show that the wave functions for the states $n = 1$ and $n = 2$ of the harmonic oscillator are orthogonal.

27 •• For the wave functions

$$\psi_n(x) = A\sin(n\pi x/L) \qquad n = 1, 2, 3, \ldots$$

corresponding to a particle in an infinite square-well potential from 0 to L, show that $\int_0^L \psi_m(x)\psi_n(x)\,dx = 0$ for all positive integers m and n, where $m \neq n$; that is, show that the wave functions are orthogonal.

GENERAL PROBLEMS

28 •• Consider a particle in an infinite one-dimensional box that has a length L and is centered at the origin. (*a*) What are the values of $\psi_1(0)$ and $\psi_2(0)$? (*b*) What are the values of $\langle x\rangle$ for the $n = 1$ and $n = 2$ states? (*c*) Evaluate $\langle x^2\rangle$ for the $n = 1$ and $n = 2$ states.

29 •• Eight identical noninteracting fermions are confined to an infinite two-dimensional square box of side length L. Determine the energies of the three lowest energy states. (See Problem 22.) SSM

30 •• A particle is confined to a two-dimensional box defined by the following boundary conditions: $U(x, y) = 0$ for $-L/2 \leq x \leq L/2$ and $-3L/2 \leq y \leq 3L/2$, and $U(x, y) = \infty$ outside these ranges. (*a*) Determine the energies of the three lowest energy states. Are any of these states degenerate? (*b*) Identify the quantum numbers of the two lowest energy degenerate states and determine the energy of these states.

31 ••• The classical probability distribution function for a particle in an infinite one-dimensional well of length L is $P = 1/L$. (See Example 34-5.) (*a*) Show that the classical expectation value of x^2 for a particle in an infinite one-dimensional well of length L that is centered at the origin is $L^2/12$. (*b*) Find the quantum expectation value of x^2 for the nth state of a particle in the one-dimensional box and show that it approaches the classical limit $L^2/12$ as n approaches infinity.

32 •• Show that Equations 35-27 and 35-28 imply that the transmission coefficient for particles of energy E incident on a step barrier $U_0 < E$ is given by

$$T = \frac{4k_1k_2}{(k_1 + k_2)^2} = \frac{4r}{(1 + r)^2}$$

where $r = k_2/k_1$.

33 •• (*a*) Show that for the case of a particle of energy E incident on a step barrier $U_0 < E$, the wave numbers k_1 and k_2 are related by

$$\frac{k_2}{k_1} = r = \sqrt{1 - \frac{U_0}{E}}$$

(*b*) Use this result to show that $R = (1 - r)^2/(1 + r)^2$.

34 •• **SPREADSHEET** (*a*) Using a spreadsheet program or graphing calculator and the results of Problem 32 and Problem 33, graph the transmission coefficient T and reflection coefficient R as a function of incident energy E for values of E ranging from $E = U_0$ to $E = 10.0U_0$. (*b*) What limiting values do your graphs indicate?

35 ••• The wave function for the state $n = 2$ of the harmonic oscillator is $\psi_2(x) = A_2(2ax^2 - \frac{1}{2})e^{-ax^2}$, where $a = \frac{1}{2}m\omega_0/\hbar$. Determine the normalization constant A_2.

36 ••• Consider the time-independent, one-dimensional Schrödinger equation when the potential function is symmetric about the origin, that is, when $U(x)$ is even.* (*a*) Show that if $\psi(x)$ is a solution of the Schrödinger equation and has energy E, then $\psi(-x)$ is also a solution that has the same energy E. Furthermore, $\psi(x)$ and $\psi(-x)$ can differ by only a multiplicative constant. (*b*) Write $\psi(x) = C\psi(-x)$, and show that $C = \pm 1$. Note that $C = +1$ means that $\psi(x)$ is an even function of x, and $C = -1$ means that $\psi(x)$ is an odd function of x.

37 ••• In this problem, you will derive the ground-state energy of the harmonic oscillator using the precise form of the uncertainty principle, $\Delta x\,\Delta p_x \geq \hbar/2$, where Δx and Δp_x are defined to be the standard deviations $(\Delta x)^2 = \langle(x - \langle x\rangle)^2\rangle$ and $(\Delta p_x)^2 = \langle(p_x - \langle p_x\rangle)^2\rangle$.

Proceed as follows:

1. Write the total classical energy in terms of the position x and momentum p_x using $U(x) = \frac{1}{2}m\omega_0^2x^2$ and $K = \frac{1}{2}p_x^2/m$.
2. Show that $(\Delta x)^2 = \langle(x - \langle x\rangle)^2\rangle = \langle x^2\rangle - \langle x\rangle^2$ and $(\Delta p_x)^2 = \langle(p_x - \langle p_x\rangle)^2\rangle = \langle p_x^2\rangle - \langle p_x\rangle^2$. *Hint: See Equations 17-34a and 17-34b.*
3. Use the symmetry of the potential energy function to argue that $\langle x\rangle$ and $\langle p_x\rangle$ must be zero, so that $(\Delta x)^2 = \langle x^2\rangle$ and $(\Delta p_x)^2 = \langle p_x^2\rangle$.
4. Assume that $\Delta p_x \Delta x = \hbar/2$ to eliminate $\langle p_x^2\rangle$ from the average energy $\langle E\rangle = \langle\frac{1}{2}p_x^2/m + \frac{1}{2}m\omega_0^2x^2\rangle = \frac{1}{2}\langle p_x^2\rangle/m + \frac{1}{2}m\omega_0^2\langle x^2\rangle$ and write $\langle E\rangle$ as $\langle E\rangle = \hbar^2/(8mZ) + \frac{1}{2}m\omega^2 Z$, where $Z = \langle x^2\rangle$.
5. Set $d\langle E\rangle/dZ = 0$ to find the value of Z for which $\langle E\rangle$ is a minimum.
6. Show that the minimum energy is given by $\langle E\rangle_{\text{min}} = +\frac{1}{2}\hbar\omega_0$. SSM

38 ••• A particle that has mass m and is near Earth's surface, at which $z = 0$, can be described by the potential energy function $U = mgz$ in the region $z > 0$, and by $U = \infty$ in the region $z < 0$. Sketch a graph of $U(z)$ versus z. For some positive value of total energy E, indicate the classically allowed region on the graph and plot the classical kinetic energy versus z on the graph. The Schrödinger equation for this problem is quite difficult to solve. Using arguments similar to those in Section 35-2 about the concavity of the wave function as given by the Schrödinger equation, sketch the shape of the wave function for the ground state and for the first two excited states.

* A function $f(x)$ is even if $f(x) = f(-x)$ for all x, and a function $f(x)$ is odd if $f(x) = -f(-x)$ for all x.

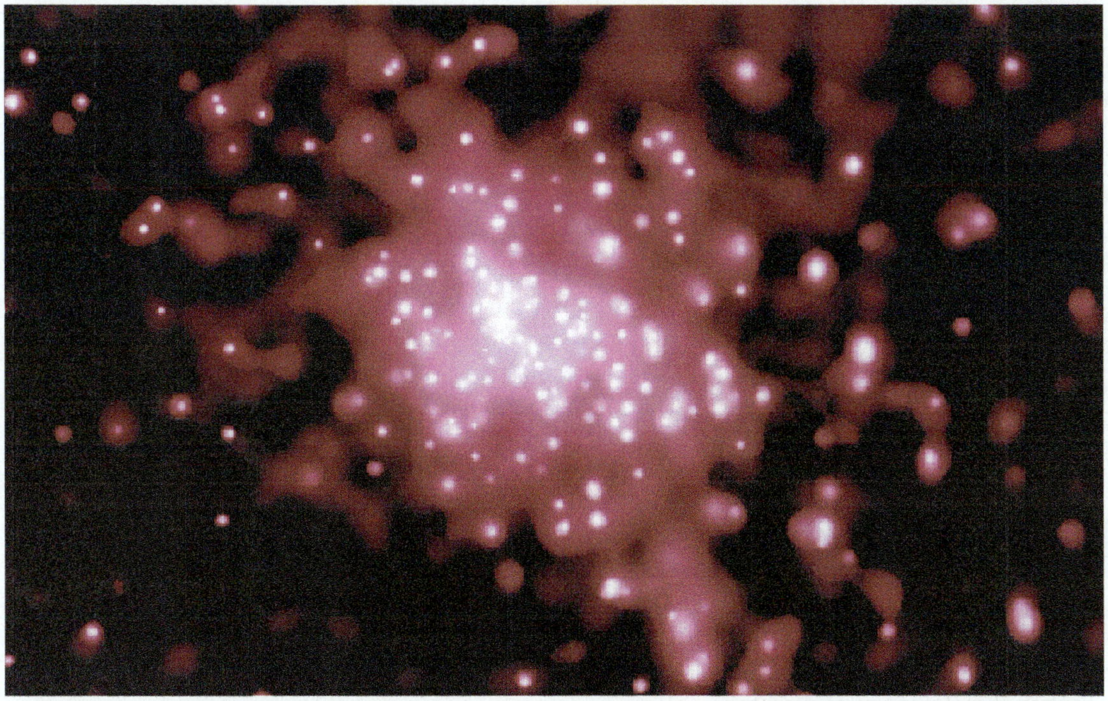

CHAPTER 36

AT A DISTANCE OF 6000 LIGHTYEARS FROM EARTH, THE STAR CLUSTER RCW 38 IS A RELATIVELY CLOSE STAR-FORMING REGION. THIS IMAGE COVERS AN AREA ABOUT 5 LIGHTYEARS ACROSS AND CONTAINS THOUSANDS OF HOT, VERY YOUNG STARS FORMED LESS THAN A MILLION YEARS AGO. X RAYS FROM THE HOT UPPER ATMOSPHERES OF 190 OF THESE STARS WERE DETECTED BY CHANDRA, AN X-RAY OBSERVATORY ORBITING EARTH. THE MECHANISMS GENERATING THESE X RAYS ARE NOT KNOWN. ON EARTH, X-RAY MACHINES PRODUCE X RAYS BY BOMBARDING A TARGET WITH HIGH-ENERGY ELECTRONS. THE ATOMIC NUMBER OF THE ATOMS THAT MAKE UP THE TARGET CAN BE DETERMINED BY ANALYZING THE RESULTING X-RAY SPECTRA. *(NASA/CXC/CfA/S. Wolk et al.)*

? How is the atomic number obtained from the spectral analysis? (See Example 36-8.)

Atoms

One hundred eleven chemical elements have been discovered, and several additional chemical elements recently have been reported but not authenticated. Each element is characterized by an atom that has a number of protons Z and an equal number of electrons. The number of protons Z is called the **atomic number.** The atom that has the fewest protons is called hydrogen (H) and has $Z = 1$. A helium (He) atom has two protons ($Z = 2$), a lithium (Li) atom has three protons ($Z = 3$), and so forth. Nearly all the mass of an atom is concentrated in its tiny nucleus, which is made up of protons and neutrons. An atom's nuclear radius is approximately 1 fm to 10 fm (1 fm $= 10^{-15}$ m) and the radius of an atom is approximately 0.1 nm = 100 000 fm.

The chemical properties and physical properties of an element are determined by the number and arrangement of the electrons in an atom of the element. Because each proton has a positive charge $+e$, the nucleus has a total positive charge $+Ze$. The electrons are negatively charged ($-e$), so they are attracted to the nucleus and repelled by each other. Because electrons and protons have equal but opposite charges and an atom has equal numbers of electrons and protons, atoms are electrically neutral. Atoms that lose or gain one or more electrons are then electrically charged and are called *ions*.

We will begin our study of atoms by discussing the Bohr model, a semiclassical model developed by Niels Bohr in 1913 to explain the electromagnetic spectra produced by hydrogen atoms. Although this pre-quantum mechanics model has many shortcomings, it provides a useful framework for the discussion of atomic phenomena. After discussing the Bohr model, we will then apply our knowledge of quantum mechanics from Chapter 35 to give a much more productive model of the hydrogen atom. We will then discuss the structure of other atoms and the periodic table of the elements. Finally, we will discuss both optical and X-ray spectra.

36-1 THE ATOM

ATOMIC SPECTRA

By the beginning of the twentieth century, a large body of data had been collected on the emission of light by atoms in a gas when the atoms are excited by an electric discharge. Viewed through a spectroscope that has a narrow-slit aperture, light that has been emitted by atoms of a particular element appears as a discrete set of lines of different colors or wavelengths. The spacing and relative intensities of the lines are characteristic of the element. The wavelengths of these spectral lines could be accurately determined, and much effort went into finding regularities in the spectra. Figure 36-1 shows line spectra for hydrogen and for mercury.

In 1885, Johann Balmer determined that the wavelengths of the lines in the visible spectrum of hydrogen can be represented by the formula

$$\lambda = (364.6\text{ nm})\frac{m^2}{m^2 - 4} \qquad m = 3, 4, 5, \ldots \qquad \text{36-1}$$

Balmer suggested that this expression might be a special case of a more general expression that would be applicable to the spectra of other elements. Such an expression, found by Johannes R. Rydberg and Walter Ritz and known as the **Rydberg–Ritz formula,** gives the reciprocal wavelength as

$$\frac{1}{\lambda} = R\left(\frac{1}{n_2^2} - \frac{1}{n_1^2}\right) \qquad \text{36-2}$$

(a)

(b)

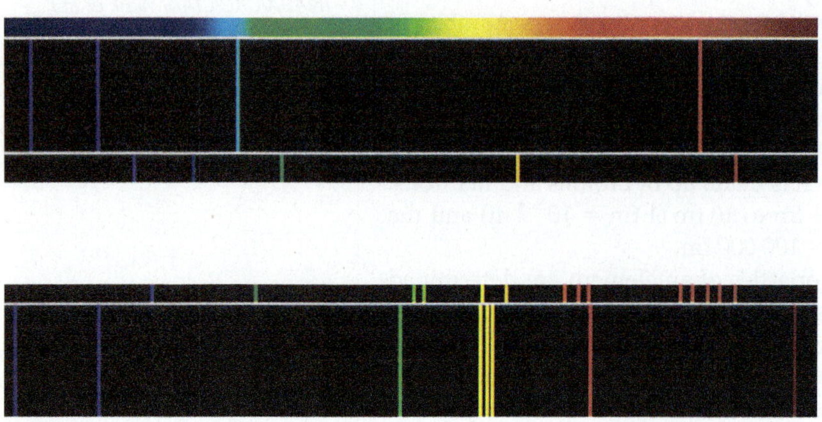

FIGURE 36-1 (*a*) Line spectrum of hydrogen and (*b*) line spectrum of mercury. *((a) and (b) adapted from Eastern Kodak and Wabash Instrument Corporation.)*

where n_1 and n_2 are integers, $n_1 > n_2$, and R is the Rydberg constant. The Rydberg constant does vary slightly, and in a regular way, from element to element. For hydrogen, R has the value

$$R_H = 1.097776 \times 10^7 \text{ m}^{-1}$$

RYDBERG CONSTANT FOR HYDROGEN

The Rydberg–Ritz formula gives the wavelengths for all the lines in the spectra of hydrogen, as well as alkali elements such as lithium and sodium. The hydrogen Balmer series given by Equation 36-1 is also given by Equation 36-2, with $R = R_H$, $n_2 = 2$, and $n_1 = m$.

Many attempts were made to construct a model of the atom that would yield these formulas for an atom's radiation spectrum. The most popular model, created by J. J. Thomson, considered various arrangements of electrons embedded in some kind of fluid that had most of the mass of the atom and had enough positive charge to make the atom electrically neutral. Thomson's model, called the "plum pudding" model, is illustrated in Figure 36-2. Because classical electromagnetic theory predicted that a charge oscillating with frequency f would radiate electromagnetic energy of that frequency, Thomson searched for configurations that were stable and that had normal modes of vibration of frequencies equal to those of the spectrum of the atom. A difficulty of this model and all other models was that, according to classical physics, electric forces alone cannot produce stable equilibrium. Thomson was unsuccessful in finding a model that predicted the observed frequencies for any atom.

The Thomson model was essentially ruled out by a set of experiments by H. W. Geiger and E. Marsden, under the supervision of E. Rutherford in approximately 1911, in which alpha particles from radioactive radium were scattered by atoms in a gold foil. Rutherford showed that the number of alpha particles scattered at large angles could not be accounted for by an atom in which the positive charge was distributed throughout the atom (known to be about 0.1 nm in diameter). Instead, the results suggested that the positive charge and most of the mass of an atom is concentrated in a very small region, now called the nucleus, which has a diameter of the order of 10^{-6} nm = 1 fm.

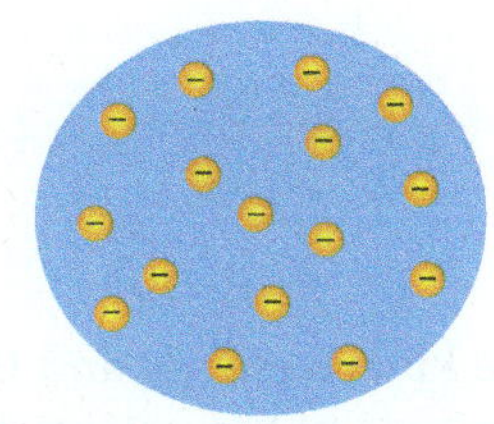

FIGURE 36-2 J. J. Thomson's plum pudding model of the atom. In this model the electrons, which have a negative charge, are embedded in a fluid of positive charge. For a given configuration in such a system, the resonance frequencies of oscillations of the electrons can be calculated. According to classical theory, the atom should radiate light of frequency equal to the frequency of oscillation of the electrons. Thomson could not find any configuration that would give frequencies in agreement with the measured frequencies of the spectrum of any atom.

36-2 THE BOHR MODEL OF THE HYDROGEN ATOM

Niels Bohr, working in the Rutherford laboratory in 1912, proposed a model of the hydrogen atom that extended the work of Planck, Einstein, and Rutherford and successfully predicted the observed spectra. According to Bohr's model, the electron of the hydrogen atom moves in either a circular or an elliptical orbit around the positive nucleus according to Coulomb's law and classical mechanics like the planets orbit the Sun. For simplicity, Bohr chose a circular orbit, as shown in Figure 36-3.

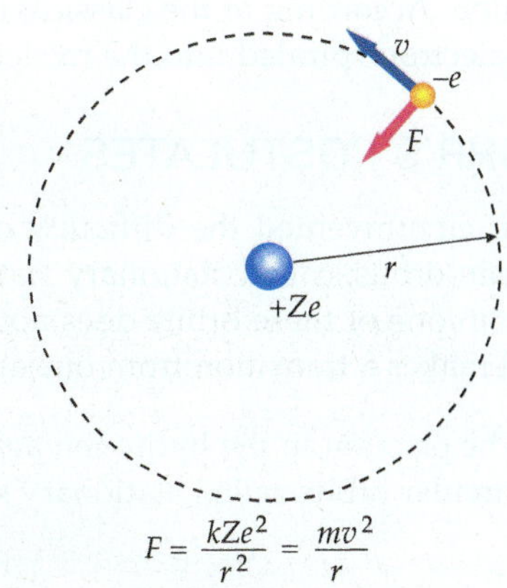

FIGURE 36-3 Electron of charge $-e$ traveling in a circular orbit of radius r around the nuclear charge $+Ze$. The attractive electrical force kZe^2/r^2 keeps the electron in its orbit.

ENERGY FOR A CIRCULAR ORBIT

Consider an electron of charge $-e$ moving in a circular orbit of radius r about a positive charge $+Ze$ such as the nucleus of a hydrogen atom ($Z = 1$) or of a singly ionized helium atom ($Z = 2$). The total energy of the electron can be related to the radius of the orbit. The potential energy of the electron of charge $-e$ at a distance r from a positive charge Ze is

$$U = \frac{kq_1q_2}{r} = \frac{k(Ze)(-e)}{r} = -\frac{kZe^2}{r} \qquad 36\text{-}3$$

where k is the Coulomb constant. The kinetic energy K can be obtained as a function of r by using Newton's second law, $F_{net} = ma$. Setting the Coulomb attractive force equal to the mass multiplied by the centripetal acceleration gives

$$\frac{kZe^2}{r^2} = m\frac{v^2}{r} \qquad 36\text{-}4a$$

Multiplying both sides by $r/2$ gives

$$K = \frac{1}{2}mv^2 = \frac{1}{2}\frac{kZe^2}{r} \qquad 36\text{-}4b$$

Thus, the kinetic energy and the potential energy vary inversely with r. Note that the magnitude of the potential energy is twice that of the kinetic energy:

$$U = -2K \qquad 36\text{-}5$$

Equation 36-5 a general result for particles orbiting under the influence of forces that vary inversely with the square of the distance from a fixed point. [It also holds for circular orbits in a gravitational field (see Example 11-6 in Section 11-3)]. The total energy is the sum of the kinetic energy and the potential energy:

$$E = K + U = \frac{1}{2}\frac{kZe^2}{r} - \frac{kZe^2}{r}$$

or

$$E = -\frac{1}{2}\frac{kZe^2}{r} \qquad 36\text{-}6$$

ENERGY IN A CIRCULAR ORBIT FOR A $1/r^2$ FORCE

Although mechanical stability is achieved because the Coulomb attractive force provides the centripetal force necessary for the electron to remain in orbit, classical *electromagnetic* theory says that such an atom would be electrically unstable. The atom would be unstable because the electron must accelerate when moving in a circle and therefore radiate electromagnetic energy of frequency equal to that of its motion. According to the classical theory, such an atom would quickly collapse as the electron spiraled into the nucleus and radiated away the energy.

BOHR'S POSTULATES

Bohr circumvented the difficulty of the collapsing atom by *postulating* that only certain orbits, called stationary states, are allowed and that an atom with an electron in one of these orbits does not radiate. An atom radiates only when the electron makes a transition from one allowed orbit (stationary state) to another.

The electron in the hydrogen atom can move only in certain nonradiating, circular orbits called stationary states.

BOHR'S FIRST POSTULATE—NONRADIATING ORBITS

Bohr's second postulate relates the frequency of radiation to the energies of the stationary states. If E_i and E_f are the initial and final energies of the atom, the frequency of the emitted radiation during a transition is given by

$$f = \frac{E_i - E_f}{h} \qquad 36\text{-}7$$

BOHR'S SECOND POSTULATE—PHOTON FREQUENCY FROM ENERGY CONSERVATION where h is Planck's constant. This postulate is equivalent to the assumption of conservation of energy when a photon of energy hf is emitted. Combining Equation 36-6 and Equation 36-7, we obtain for the frequency

$$f = \frac{E_i - E_f}{h} = \frac{1}{2}\frac{kZe^2}{h}\left(\frac{1}{r_f} - \frac{1}{r_i}\right) \qquad \text{36-8}$$

where r_i and r_f are the radii of the initial and final orbits.

To obtain the frequencies implied by the Rydberg–Ritz formula, $f = c/\lambda = cR(1/n_2^2 - 1/n_1^2)$, it is evident that the radii of stable orbits must be proportional to the squares of integers. Bohr searched for a quantum condition for the radii of the stable orbits that would yield this result. After much trial and error, Bohr found that he could obtain the desired result if he postulated that the magnitude of the angular momentum of the electron in a stable orbit equals an integer multiplied by $\hbar$. Because the magnitude of the angular momentum of a circular orbit is just mvr, this postulate is

$$mv_n r_n = n\hbar \qquad n = 1, 2, 3, \ldots \qquad \text{36-9}$$

BOHR'S THIRD POSTULATE—QUANTIZED ANGULAR MOMENTUM

where $\hbar = h/2\pi = 1.055 \times 10^{-34}\,\text{J}\cdot\text{s} = 6.582 \times 10^{-16}\,\text{eV}\cdot\text{s}$.

Equation 36-9 relates the speed v_n to the radius r_n of the orbit that has angular momentum $n\hbar$. Equation 36-4*a* gives us another equation relating the speed to the radius:

$$\frac{kZe^2}{r_n^2} = m\frac{v_n^2}{r_n}$$

or

$$v_n^2 = \frac{kZe^2}{mr_n} \qquad \text{36-10}$$

We can determine r_n by first solving for v_n in Equation 36-9. Squaring the result then gives

$$v_n^2 = n^2\frac{\hbar^2}{m^2 r_n^2}$$

Equating this expression for v_n^2 with the expression given by Equation 36-10, we get

$$n^2\frac{\hbar^2}{m^2 r_n^2} = \frac{kZe^2}{mr_n}$$

Solving for r_n, we obtain

$$r_n = n^2\frac{\hbar^2}{mkZe^2} = n^2\frac{a_0}{Z} \qquad \text{36-11}$$

RADII OF THE BOHR ORBITS

where a_0 is called the **first Bohr radius.** According to the Bohr model, a_0 is the orbital radius of the electron in a hydrogen atom that has $n = 1$.

$$a_0 = \frac{\hbar^2}{mke^2} = \frac{\epsilon_0 h^2}{\pi me^2} = 0.0529\text{ nm} \qquad \text{36-12}$$

FIRST BOHR RADIUS

Substituting the expressions for r_n in Equation 36-11 into Equation 36-8 for the frequency gives

$$f = \frac{1}{2}\frac{kZe^2}{h}\left(\frac{1}{r_f} - \frac{1}{r_i}\right) = Z^2\frac{mk^2e^4}{4\pi\hbar^3}\left(\frac{1}{n_f^2} - \frac{1}{n_i^2}\right) \qquad \text{36-13}$$

If we compare this expression where $Z = 1$ and $f = c/\lambda$ with the empirical Rydberg–Ritz formula (Equation 36-2), we obtain for the Rydberg constant

$$R = \frac{mk^2e^4}{4\pi c\hbar^3} = \frac{me^4}{8\epsilon_0^2ch^3} \qquad \text{36-14}$$

Using the values of m, e, c, k, and $\hbar$ known in 1913, Bohr calculated R and found his result to agree (within the limits of the uncertainties of the constants) with the value obtained from spectroscopy.

Example 36-1 Standing-Wave Condition Implies Quantization of Angular Momentum

For waves in a circle, the standing-wave condition is that there is an integral number of wavelengths in the circumference. That is, $n\lambda_n = 2\pi r_n$, where $n = 1, 2, 3$, and so on. Show that this condition for electron waves implies quantization of angular momentum.

PICTURE The wavelength and the momentum are related by the de Broglie relation $p = h/\lambda$ (Equation 34-13). Using this relation and the standing wave condition $n\lambda_n = 2\pi r_n$ to relate the momentum to the radius.

SOLVE

1. Write the standing-wave condition: $n\lambda_n = 2\pi r_n$

2. Use the de Broglie relation (Equation 34-13) to relate the momentum p to λ_n: $p = \dfrac{h}{\lambda_n} = \dfrac{nh}{2\pi r_n} = n\dfrac{\hbar}{r_n}$

3. Solve for pr_n. The angular momentum of an electron in a circular orbit is $mvr_n = pr_n$, where $p = mv$: $pr_n = \boxed{mvr_n = n\hbar}$

CHECK The step-3 result is the check. It is what the problem statement asks us to show.

ENERGY LEVELS

The total mechanical energy of the electron in the hydrogen atom is related to the radius of the circular orbit by Equation 36-6. If we substitute the quantized values of r as given by Equation 36-11, we obtain

$$E_n = -\frac{1}{2}\frac{kZe^2}{r_n} = -\frac{1}{2}\frac{kZ^2e^2}{n^2a_0} = -\frac{1}{2}\frac{mk^2Z^2e^4}{n^2\hbar^2}$$

or

$$E_n = -Z^2\frac{E_0}{n^2} \qquad \text{36-15}$$

ENERGY LEVELS IN THE HYDROGEN ATOM

where

$$E_0 = \frac{mk^2e^4}{2\hbar^2} = \frac{1}{2}\frac{ke^2}{a_0} = 13.6\text{ eV} \qquad \text{36-16}$$

The energies E_n corresponding to $Z = 1$ are the quantized allowed energies for the hydrogen atom.

Transitions between these allowed energies result in the emission or absorption of a photon whose frequency is given by $f = (E_\text{i} - E_\text{f})/h$, and whose wavelength is

$$\lambda = \frac{c}{f} = \frac{hc}{E_\text{i} - E_\text{f}} \qquad 36\text{-}17$$

As we found in Chapter 34, it is convenient to have the value of hc in electron-volt nanometers:

$$hc = 1240 \text{ eV} \cdot \text{nm} \qquad 36\text{-}18$$

Because the energies are quantized, the frequencies and wavelengths of the radiation emitted by the hydrogen atom are quantized in agreement with the observed line spectrum.

Figure 36-4 shows the energy-level diagram for hydrogen. The energy of the hydrogen atom in the ground state is $E_1 = -13.6$ eV. As n approaches infinity the energy approaches zero. The process of removing an electron from an atom is called **ionization,** and the minimum amount of energy required to remove the electron is the **ionization energy.** The ionization energy of the ground-state hydrogen atom, which is also its **binding energy,** is 13.6 eV. A few transitions from a higher state to a lower state are indicated in Figure 36-4. When Bohr published his model of the hydrogen atom, the Balmer series (corresponding to $n_\text{f} = 2$ and $n_\text{i} = 3, 4, 5, \ldots$) and the Paschen series (corresponding to $n_\text{f} = 3$ and $n_\text{i} = 4, 5, 6, \ldots$) were known. In 1916, T. Lyman found the series corresponding to $n_\text{f} = 1$. F. Brackett in 1922 and H. A. Pfund in 1924 found the series corresponding to $n_\text{f} = 4$ and $n_\text{f} = 5$, respectively. Only the Balmer series lies in the visible portion of the electromagnetic spectrum.

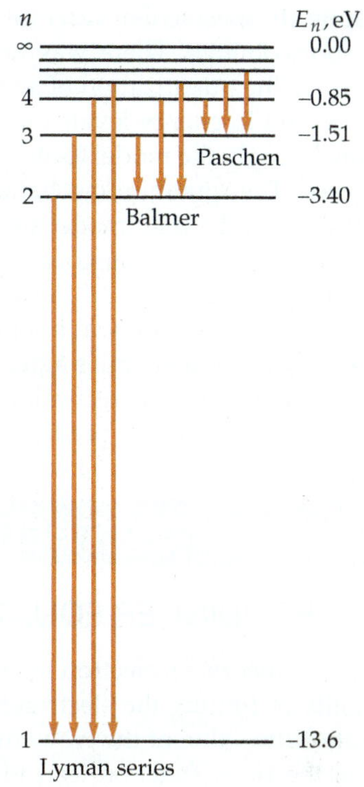

FIGURE 36-4 Energy-level diagram for hydrogen showing the first few transitions in each of the Lyman, Balmer, and Paschen series. The energies of the levels are given by Equation 36-15.

! Do not think the λ in Equation 36-17 is the wavelength of the electron. It is not. It is the wavelength of the emitted or absorbed photon.

Example 36-2 Longest Wavelength in the Lyman Series

Find (*a*) the energy and (*b*) the wavelength of the spectral line that has the longest wavelength in the Lyman series.

PICTURE For the Lyman series $n_\text{f} = 1$. From Figure 36-4, we can see that the Lyman series corresponds to transitions ending at the ground-state energy, $E_\text{f} = E_1 = -13.6$ eV. Because the photon wavelength λ varies inversely with energy, the transition that has the longest wavelength is the transition that has the lowest energy, which is from the first excited state $n = 2$ to the ground state $n = 1$.

SOLVE

1. The energy of the photon is the difference in the energies of the initial and final atomic states:

$$E_\text{photon} = \Delta E_\text{atom} = E_\text{i} - E_\text{f}$$
$$= E_2 - E_1 = \frac{-13.6 \text{ eV}}{2^2} - \frac{-13.6 \text{ eV}}{1^2}$$
$$= -3.40 \text{ eV} + 13.6 \text{ eV} = \boxed{10.2 \text{ eV}}$$

2. The wavelength of the photon is

$$\lambda = \frac{hc}{E_2 - E_1} = \frac{1240 \text{ eV} \cdot \text{nm}}{10.2 \text{ eV}} = \boxed{122 \text{ nm}}$$

CHECK The step-1 result of 10.2 eV is less than 13.6 eV (the binding energy of ground-state hydrogen). This result is expected.

TAKING IT FURTHER This photon has a wavelength that corresponds to the ultraviolet region of the electromagnetic spectrum. Because all the other lines in the Lyman series have even greater energies and shorter wavelengths, the Lyman series is completely in the ultraviolet region.

PRACTICE PROBLEM 36-1 Find the shortest wavelength for a line in the Lyman series.

Despite its spectacular successes, the Bohr model of the hydrogen atom had many shortcomings. There was no justification for the postulates of stationary states or for the quantization of angular momentum other than the fact that these postulates led to energy levels that agreed with spectroscopic data. Furthermore, attempts to apply the model to atoms that have more electrons and protons had little success. The quantum-mechanical model resolves these difficulties. The stationary states of the Bohr model are replaced by the standing-wave solutions of the Schrödinger equation analogous to the standing electron waves for a particle in a box discussed in Chapter 34 and Chapter 35. Energy quantization is a direct consequence of the standing-wave solutions of the Schrödinger equation. For hydrogen, these quantized energies agree with those obtained from the Bohr model and with experiment. The quantization of angular momentum that had to be postulated in the Bohr model is predicted by the quantum theory.

36-3 QUANTUM THEORY OF ATOMS

THE SCHRÖDINGER EQUATION IN SPHERICAL COORDINATES

In quantum theory, an electron in an atom is described by its wave function ψ. The probability of finding the electron in some volume dV of space equals the square of the absolute value of the wave function $|\psi|^2$ multiplied by dV. Boundary conditions on the wave function lead to the quantization of the wavelengths and frequencies and thereby to the quantization of the electron energy.

Consider a single electron of mass m moving in three dimensions in a region in which the potential energy is U. The time-independent Schrödinger equation for such a particle is given by Equation 35-30:

$$-\frac{\hbar^2}{2m}\left(\frac{\partial^2\psi}{\partial x^2}+\frac{\partial^2\psi}{\partial y^2}+\frac{\partial^2\psi}{\partial z^2}\right)+U(x,y,z)\psi=E\psi \qquad \text{36-19}$$

For a single isolated atom, the potential energy U depends only on the radial distance $r=\sqrt{x^2+y^2+z^2}$ of the electron from the center of the nucleus. The problem is then most conveniently treated using the spherical coordinates r, θ, and ϕ, which are related to the rectangular coordinates x, y, and z by

$$z = r\cos\theta$$
$$x = r\sin\theta\cos\phi \qquad \text{36-20}$$
$$y = r\sin\theta\sin\phi$$

These relations are shown in Figure 36-5. The transformation of Equation 36-19 from rectangular to spherical coordinates is a straightforward, but tedious, calculation, which we will omit. The result of this transformation can be found in Problem 42. We will discuss qualitatively some of the interesting features of the wave functions that satisfy this equation.

The transformed version of Equation 36-19 can be solved using the technique called separation of variables. This is accomplished by expressing the wave function $\psi(r,\theta,\phi)$ as a product of three functions, each of which is a function of only one of the three spherical coordinates:

$$\psi(r,\theta,\phi)=R(r)f(\theta)g(\phi) \qquad \text{36-21}$$

where R is a function that depends only on the radial coordinate r, f is a function that depends only on the polar coordinate θ, and g is a function that depends only on the azimuth coordinate ϕ. When this form of $\psi(r,\theta,\phi)$ is substituted into the Schrödinger equation, the Schrödinger equation can be transformed into three ordinary differential equations, one for $R(r)$, one for $f(\theta)$, and one for $g(\phi)$. The potential energy $U(r)$ appears only in the equation for $R(r)$, which is called the **radial equation.**

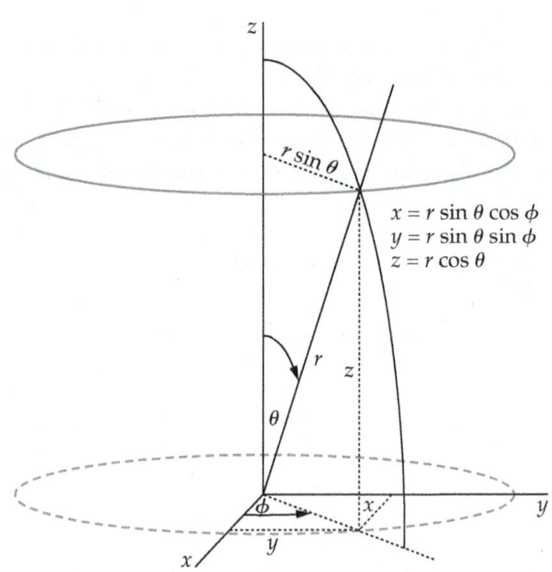

FIGURE 36-5 Geometric relations between spherical coordinates and rectangular coordinates.

Because the potential energy depends only on the coordinate r, the potential energy has no effect on the solutions of the equations for $f(\theta)$ and $g(\phi)$, and therefore has no effect on the angular dependence of the wave function $\psi(r, \theta, \phi)$. These solutions are applicable to *any* problem in which the potential energy depends only on r.

QUANTUM NUMBERS IN SPHERICAL COORDINATES

In three dimensions, the requirement that the wave function be continuous and normalizable introduces three quantum numbers, one associated with each spatial dimension. In spherical coordinates, the quantum number associated with r is labeled n, that associated with θ is labeled ℓ, and that associated with ϕ is labeled m_ℓ.* The quantum numbers n_1, n_2, and n_3 that we found in Chapter 35 for a particle in a three-dimensional square well in rectangular coordinates x, y, and z were independent of one another, but the quantum numbers associated with wave functions in spherical coordinates are dependent on each other. The possible values of these quantum numbers are

$$n = 1, 2, 3, \ldots$$
$$\ell = 0, 1, 2, 3, \ldots, n-1$$
$$m_\ell = -\ell, -\ell+1, -\ell+2, \ldots, 0, \ldots, \ell-2, \ell-1, \ell \qquad \text{36-22}$$

QUANTUM NUMBERS IN SPHERICAL COORDINATES

That is, n can be any positive integer; ℓ can be 0 or any positive integer up to $n-1$; and m_ℓ can have $2\ell+1$ possible values, ranging from $-\ell$ to $+\ell$ in integral steps.

The number n is called the **principal quantum number.** It is associated with the dependence of the wave function on the distance r and therefore with the probability of finding the electron at various distances from the nucleus. The quantum numbers ℓ and m_ℓ are associated with the orbital angular momentum of the electron and with the angular dependence of the electron wave function. The quantum number ℓ is called the **orbital quantum number.** The magnitude L of the orbital angular momentum $\vec{L}$ is related to the orbital quantum number ℓ by

$$L = \sqrt{\ell(\ell+1)}\hbar \qquad \text{36-23}$$

The quantum number m_ℓ is called the **magnetic quantum number.** It is related to the component of the orbital angular momentum along some direction in space. All spatial directions are equivalent for an isolated atom, but placing the atom in a magnetic field results in the direction of the magnetic field being separated out from the other directions. The convention is that the $+z$ direction is chosen for the magnetic-field direction. Then the z component of the orbital angular momentum of the electron is given by the quantum condition

$$L_z = m_\ell \hbar \qquad \text{36-24}$$

This quantum condition arises from the boundary condition on the azimuth coordinate ϕ that the probability of finding the electron at some arbitrary angle ϕ_1 must be the same as that of finding the electron at angle $\phi_2 = \phi_1 + 2\pi$, because these two values of ϕ represent the same point in space.

If we measure the angular momentum of the electron in units of $\hbar$, we see that the magnitude of the orbital angular momentum is quantized to the value $\sqrt{\ell(\ell+1)}$ units, and that its component along any direction can have only the $2\ell+1$ values ranging from $-\ell$ to $+\ell$ units. Figure 36-6 shows a vector-model diagram illustrating the possible orientations of the angular-momentum vector for $\ell = 2$. Note that only specific values of θ are allowed; that is, the directions in space are quantized.

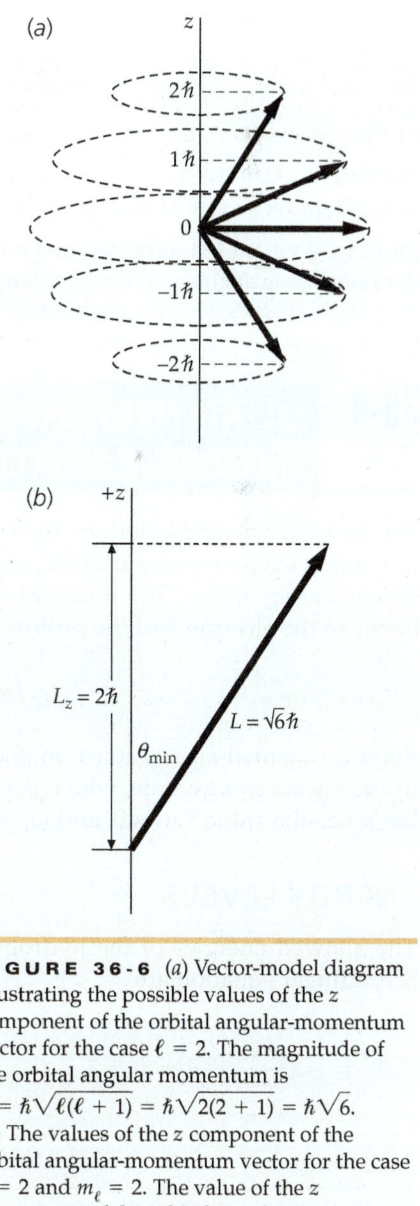

FIGURE 36-6 (*a*) Vector-model diagram illustrating the possible values of the z component of the orbital angular-momentum vector for the case $\ell = 2$. The magnitude of the orbital angular momentum is $L = \hbar\sqrt{\ell(\ell+1)} = \hbar\sqrt{2(2+1)} = \hbar\sqrt{6}$. (*b*) The values of the z component of the orbital angular-momentum vector for the case $\ell = 2$ and $m_\ell = 2$. The value of the z component of the orbital angular momentum is $L_z = 2\hbar$.

* For simplicity, m_ℓ is sometimes written as m.

Example 36-3 The Directions of the Angular Momentum

If the orbital angular momentum is characterized by the quantum number $\ell = 2$, what are the possible values of L_z, and what is the smallest possible angle between $\vec{L}$ and the direction of increasing z?

PICTURE The possible orientations of $\vec{L}$ and the z axis are shown in Figure 36-6. The direction of increasing z is the direction of the external magnetic field in the vicinity of the atom.

SOLVE

1. Write the possible values of L_z:

$$\boxed{L_z = m_\ell \hbar \quad \text{where} \quad m_\ell = -2, -1, 0, 1, 2}$$

2. Express the angle θ between $\vec{L}$ and the $+z$ direction in terms of L and L_z:

$$\cos\theta = \frac{L_z}{L} = \frac{m_\ell \hbar}{\sqrt{\ell(\ell+1)}\hbar} = \frac{m_\ell}{\sqrt{\ell(\ell+1)}}$$

3. The smallest angle occurs when $\ell = 2$ and $m_\ell = 2$:

$$\cos\theta_{\text{min}} = \frac{2}{\sqrt{2(2+1)}} = \frac{2}{\sqrt{6}} = 0.816$$

$$\theta_{\text{min}} = \boxed{35.3^\circ}$$

CHECK The angle in Figure 36-6*b* looks to be between about 30° and 40°, so the step-3 result of 35.3° is plausible.

TAKING IT FURTHER We note the somewhat strange result that the orbital angular-momentum vector cannot be parallel to the z axis.

PRACTICE PROBLEM 36-2 An atom is in a region that has a magnetic field. An electron in the atom has an angular momentum characterized by the quantum number $\ell = 4$. What are the possible values of m_ℓ for this electron?

36-4 QUANTUM THEORY OF THE HYDROGEN ATOM

We can treat the simplest atom, the hydrogen atom, as a stationary nucleus (a proton) and a single moving particle, an electron, which has linear momentum p and kinetic energy $p^2/2m$. The potential energy $U(r)$ due to the electrostatic attraction between the electron and the proton* is

$$U(r) = \frac{kZe^2}{r} \qquad \text{36-25}$$

For this potential-energy function, the Schrödinger equation can be solved exactly. In the lowest energy state, which is the ground state, the principal quantum number n has the value 1, ℓ is 0, and m_ℓ is 0.

ENERGY LEVELS

The allowed energies of the hydrogen atom that result from the solution of the Schrödinger equation are

$$E_n = -Z^2 \frac{E_0}{n^2} \qquad n = 1, 2, 3, \ldots \qquad \text{36-26}$$

ENERGY LEVELS FOR HYDROGEN

* We include the factor Z, which is 1 for hydrogen, so that we can apply our results to other one-electron "atoms," such as singly ionized helium He^+, for which $Z = 2$.

where

$$E_0 = \frac{mk^2e^4}{2\hbar^2} = 13.6 \text{ eV} \qquad 36\text{-}27$$

These energies are the same as those obtained using the Bohr model. Note that the energies E_n are negative, indicating that the electron is bound to the nucleus (thus the term *bound state*), and that the energies depend only on the principal quantum number n. The fact that the energy does not depend on the orbital quantum number ℓ is a peculiarity of the inverse-square force and holds only for an inverse r potential such as Equation 36-25. For atoms having multiple electrons, the interaction of the electrons leads to a dependence of the energy on ℓ. In general, the lower the value of ℓ, the lower the energy for such atoms. Because there is usually no preferred direction in space, the energy for any atom does not ordinarily depend on the magnetic quantum number m_ℓ, which is related to the z component of the angular momentum. However, the energy does depend on m_ℓ if the atom is in a magnetic field.

Figure 36-7 shows an energy-level diagram for hydrogen. This diagram is similar to Figure 36-4, except that the states which have the same value of n but different values of ℓ are shown separately. These states (called *terms*) are referred to by giving the value of n along with a code letter: s for $\ell = 0$, p for $\ell = 1$, d for $\ell = 2$, and f for $\ell = 3$.* (Lowercase letters s, p, d, f, and so on, are used to specify the orbital angular momentum of an individual electron, whereas uppercase letters S, P, D, F, and so on, are used to identify the orbital angular momentum for the entire multielectron atom. For a single-electron atom, like hydrogen, either uppercase or lowercase letters will suffice.) When an atom makes a transition from one allowed energy state to another, electromagnetic radiation in the form of a photon is emitted or absorbed. Such transitions result in spectral lines that are characteristic of the atom. The transitions obey the **selection rules:**

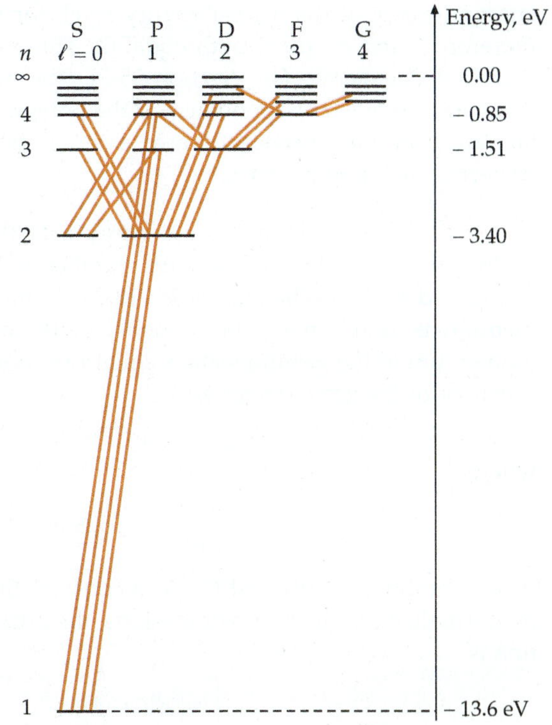

FIGURE 36-7 Energy-level diagram for hydrogen. The diagonal lines show transitions that involve emission or absorption of radiation and obey the selection rule $\Delta\ell = \pm 1$. States that have the same value of n but with different values of ℓ have the same energy $-E_0/n^2$, where $E_0 = 13.6$ eV as in the Bohr model.

$$\Delta m_\ell = -1, 0, \text{ or } +1 \qquad 36\text{-}28$$
$$\Delta\ell = -1 \text{ or } +1$$

These selection rules are related to the conservation of angular momentum and to the fact that the photon itself has an intrinsic angular momentum that has a maximum angular-momentum component of $1\hbar$ in any direction. The wavelengths of the light emitted by hydrogen (and by other atoms) are related to the energy levels by

$$hf = \frac{hc}{\lambda} = E_\text{i} - E_\text{f} \qquad 36\text{-}29$$

where E_i and E_f are the energies of the initial and final states.

WAVE FUNCTIONS AND PROBABILITY DENSITIES

The solutions of the Schrödinger equation in spherical coordinates are characterized by the quantum numbers n, ℓ, and m_ℓ, and are written $\psi_{n\ell m_\ell}$. The principal quantum number n can take on any of the values 1, 2, 3, ... In addition, for each value of n, ℓ can take on any of the values $0, 1, \ldots, n-1$, and for each value of ℓ, m_ℓ can take on any of the values $-\ell, -\ell+1, -\ell+2, \ldots, +\ell$. Thus, for any given value of n, there are n possible values of ℓ, and for any given value of ℓ, there are $2\ell + 1$ possible values of m_ℓ. For hydrogen, the energy depends only on n, so there are generally many different wave functions that correspond to the same

* These code letters for the values of ℓ are remnants of spectroscopists' descriptions of various spectral lines as *sharp*, *principal*, *diffuse*, and *fundamental*. For values greater than 3, the letters follow alphabetically, starting with g for $\ell = 4$.

energy (except at the lowest energy level, for which $n = 1$ and therefore ℓ and m_ℓ must both equal 0). These energy levels are therefore degenerate (see Section 35-5). The origins of this degeneracy are the $1/r$ dependence of the potential energy and the fact that, in the absence of any external fields, there is no preferred direction in space.*

The ground state In the lowest energy state, the ground state of hydrogen, the principal quantum number n has the value 1, ℓ is 0, and m_ℓ is 0. The energy is –13.6 eV, and the angular momentum is zero. (In the Bohr model of the atom the angular momentum in the ground state is equal to $\hbar$, not zero.) The wave function for the ground state is

$$\psi_{100} = C_{100}e^{-Zr/a_0} \qquad 36\text{-}30$$

where

$$a_0 = \frac{\hbar^2}{mke^2} = 0.0529 \text{ nm}$$

is the first Bohr radius and C_{100} is a constant that is determined by normalization. In three dimensions, the normalization condition is

$$\int |\psi|^2 dV = 1$$

where dV is a volume element and the integration is performed over all space. In spherical coordinates, the volume element (Figure 36-8) is

$$dV = (r \sin\theta\, d\phi)(r\, d\theta)dr = r^2 \sin\theta\, d\theta\, d\phi\, dr$$

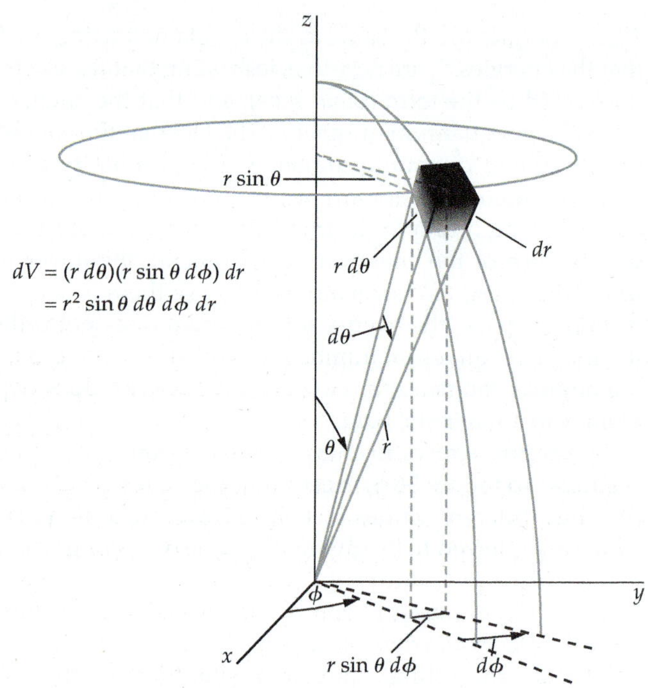

FIGURE 36-8 Volume element in spherical coordinates.

We integrate over all space by integrating over ϕ from $\phi = 0$ to $\phi = 2\pi$, over θ from $\theta = 0$ to $\theta = \pi$, and over r from $r = 0$ to $r = \infty$. The normalization condition is thus

$$\int |\psi|^2\, dV = \int_0^\infty \left[\int_0^\pi \left(\int_0^{2\pi} |\psi|^2 r^2 \sin\theta\, d\phi\right) d\theta\right] dr$$

$$= \int_0^\infty \left[\int_0^\pi \left(\int_0^{2\pi} C_{100}^2 e^{-2Zr/a_0} r^2 \sin\theta\, d\phi\right) d\theta\right] dr = 1$$

Because there is no θ or ϕ dependence in ψ_{100}, the triple integral can be factored into the product of three integrals. This gives

$$\int |\psi|^2\, dV = \left(\int_0^{2\pi} d\phi\right)\left(\int_0^\pi \sin\theta\, d\theta\right)\left(\int_0^\infty C_{100}^2 e^{-2Zr/a_0} r^2\, dr\right)$$

$$= 2\pi \cdot 2 \cdot C_{100}^2 \left(\int_0^\infty e^{-2Zr/a_0} r^2\, dr\right) = 1$$

The remaining integral is of the form $\int_0^\infty x^n e^{-ax}\, dx$, where n a positive integer and $a > 0$. Using successive integration-by-parts operations† yields the result

$$\int_0^\infty x^n e^{-ax}\, dx = \frac{n!}{a^{n+1}}$$

so

$$\int_0^\infty r^2 e^{-2Zr/a_0}\, dr = \frac{a_0^3}{4Z^3}$$

See **Math Tutorial** *for more information on* **Integrals**

* If spin, relativistic effects, the spin of the nucleus, and quantum electrodynamics are considered, the degeneracy is broken.

† This integral can also be looked up in a table of integrals.

then

$$4\pi C_{100}^2\left(\frac{a_0^3}{4Z^3}\right) = 1$$

so

$$C_{100} = \frac{1}{\sqrt{\pi}}\left(\frac{Z}{a_0}\right)^{3/2} \qquad \text{36-31}$$

The normalized ground-state wave function is thus

$$\psi_{100} = \frac{1}{\sqrt{\pi}}\left(\frac{Z}{a_0}\right)^{3/2} e^{-Zr/a_0} \qquad \text{36-32}$$

The probability of finding the electron in a volume dV is $|\psi|^2\,dV$. The probability density $|\psi|^2$ is shown in Figure 36-9. Note that this probability density is spherically symmetric; that is, the probability density depends only on r, and is independent of θ or ϕ. The probability density is maximum at the origin.

We are more often interested in the probability of finding the electron at some radial distance r between r and $r + dr$. This radial probability $P(r)\,dr$ is the probability density $|\psi|^2$ multiplied by the volume of the spherical shell of thickness dr, which is $dV = 4\pi r^2\,dr$. The probability of finding the electron in the range from r to $r + dr$ is thus $P(r)\,dr = |\psi|^2\,4\pi r^2\,dr$, and the **radial probability density** is

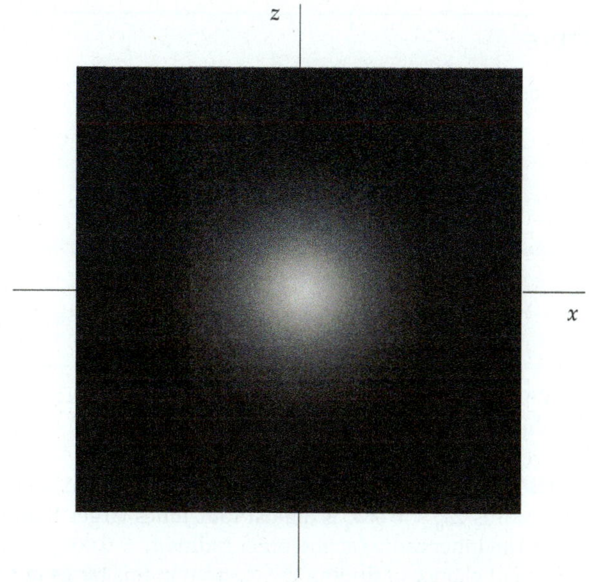

FIGURE 36-9 Computer-generated picture of the probability density $|\psi|^2$ for the ground state of hydrogen. The quantity $-e|\psi|^2$ can be thought of as the electron charge density in the atom. The density is spherically symmetric, is greatest at the origin, and decreases exponentially with r.

$$P(r) = 4\pi r^2|\psi|^2 \qquad \text{36-33}$$

RADIAL PROBABILITY DENSITY

For the hydrogen atom in the ground state, the radial probability density is

$$P(r) = 4\pi r^2|\psi|^2 = 4\pi C_{100}^2 r^2 e^{-2Zr/a_0} = 4\left(\frac{Z}{a_0}\right)^3 r^2 e^{-2Zr/a_0} \qquad \text{36-34}$$

Figure 36-10 shows the radial probability density $P(r)$ as a function of r. The maximum value of $P(r)$ occurs at $r = a_0/Z$, which for $Z = 1$ is the first Bohr radius. In contrast to the Bohr model, in which the electron stays in a well-defined orbit at $r = a_0$, we see that it is possible for the electron to be found at any distance from the nucleus. However, the most probable distance is a_0 (assuming $Z = 1$), and the chance of finding the electron at a much different distance is small. It is often useful to think of the electron in an atom as a charged cloud of charge density $\rho = -e|\psi|^2$, but we should remember that when it interacts with matter, an electron is always observed as a single charge.

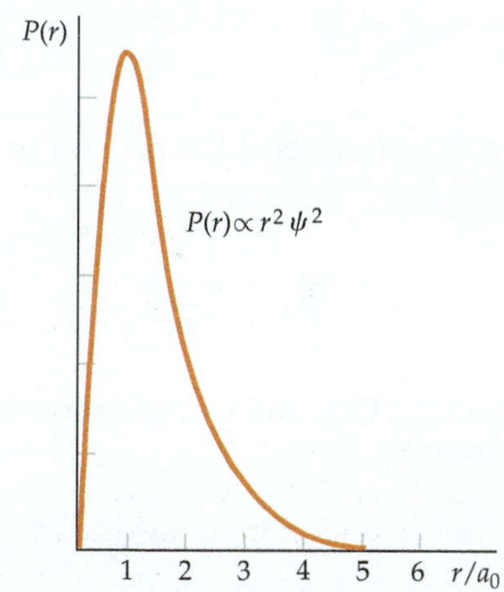

FIGURE 36-10 Radial probability density $P(r)$ versus r/a_0 for the ground state of the hydrogen atom. $P(r)$ is proportional to $r^2\psi^2$. The value of r for which $P(r)$ is maximum is the most probable distance $r = a_0$.

Example 36-4 Probability That the Electron Is in a Thin Spherical Shell

Consider a hydrogen atom that is in the ground state. Estimate the probability of finding the electron in a thin spherical shell of inner radius r and outer radius $r + \Delta r$, where $\Delta r = 0.06a_0$ at (*a*) $r = a_0$ and (*b*) $r = 2a_0$.

PICTURE Because Δr is so small compared to r, the variation in the radial probability density $P(r)$ in the shell can be neglected. The probability of finding the electron in some small range Δr is then $P(r)\,\Delta r$.

SOLVE

1. Substitute $Z = 1$ and $r = a_0$ into Equation 36-34:

$$P(r)\Delta r = \left[4\left(\frac{1}{a_0}\right)^3 r^2 e^{-2r/a_0}\right]\Delta r$$

$$P(a_0)(0.06a_0) = \left[4\left(\frac{1}{a_0}\right)^3 a_0^2 e^{-2}\right](0.06a_0) = \boxed{0.0325}$$

2. Substitute $Z = 1$ and $r = 2a_0$ into Equation 36-34:

$$P(r)\Delta r = \left[4\left(\frac{1}{a_0}\right)^3 r^2 e^{-2r/a_0}\right]\Delta r$$

$$P(2a_0)(0.06a_0) = \left[4\left(\frac{1}{a_0}\right)^3 4a_0^2 e^{-4}\right](0.06a_0) = \boxed{0.0176}$$

CHECK The probability of finding the electron between $r = a_0$ and $r = a_0 + 0.06a_0$ is larger than the probability of finding the particle between $r = 2a_0$ and $r = 2a_0 + 0.06a_0$, as expected.

TAKING IT FURTHER The volume of the spherical shell that has an inner radius $2a_0$ and outer radius $2a_0 + 0.06a_0$ is almost four times larger than the volume of the spherical shell that has an inner radius a_0 and outer radius $a_0 + 0.06a_0$. In spite of this there is approximately a 3 percent chance of finding the electron in this range at $r = a_0$, but at $r = 2a_0$ the chance is slightly less than 2 percent.

The first excited state In the first excited state of a hydrogen atom, n is equal to 2 and ℓ can equal either 0 or 1. For $\ell = 0$, $m_\ell = 0$, and we again have a spherically symmetric wave function, this time given by

$$\psi_{200} = C_{200}\left(2 - \frac{Zr}{a_0}\right)e^{-Zr/(2a_0)} \qquad \text{36-35}$$

For $\ell = 1$, m_ℓ can be $+1$, 0, or -1. The corresponding wave functions are

$$\psi_{210} = C_{210}\frac{Zr}{a_0}e^{-Zr/(2a_0)}\cos\theta \qquad \text{36-36}$$

$$\psi_{21\pm1} = C_{211}\frac{Zr}{a_0}e^{-Zr/(2a_0)}\sin\theta\, e^{\pm i\phi} \qquad \text{36-37}$$

where C_{200}, C_{210}, and C_{211} are normalization constants. The probability densities are given by

$$\psi_{200}^2 = C_{200}^2\left(2 - \frac{Zr}{a_0}\right)^2 e^{-Zr/a_0} \qquad \text{36-38}$$

$$\psi_{210}^2 = C_{210}^2\left(\frac{Zr}{a_0}\right)^2 e^{-Zr/a_0}\cos^2\theta \qquad \text{36-39}$$

$$|\psi_{21\pm1}|^2 = C_{211}^2\left(\frac{Zr}{a_0}\right)^2 e^{-Zr/a_0}\sin^2\theta \qquad \text{36-40}$$

The wave functions and probability densities for $\ell \neq 0$ are not spherically symmetric, but instead depend on the angle θ. The probability densities do not depend on ϕ. Figure 36-11 shows the probability density $|\psi|^2$ for $n = 2$, $\ell = 0$, and $m_\ell = 0$ (Figure 36-11*a*); for $n = 2$, $\ell = 1$, and $m_\ell = 0$ (Figure 36-11*b*); and for $n = 2$, $\ell = 1$, and $m_\ell = \pm1$ (Figure 36-11*c*). An important feature of these plots is that the electron cloud is spherically symmetric for $\ell = 0$ and is not spherically symmetric for $\ell \neq 0$. These angular distributions of the electron charge density depend only on the values of ℓ and m_ℓ and not on the radial part of the wave function. Similar charge distributions for the valence electrons of more complicated atoms play an important role in the chemistry of molecular bonding. (Electrons in the outermost shell are called **valence electrons.**)

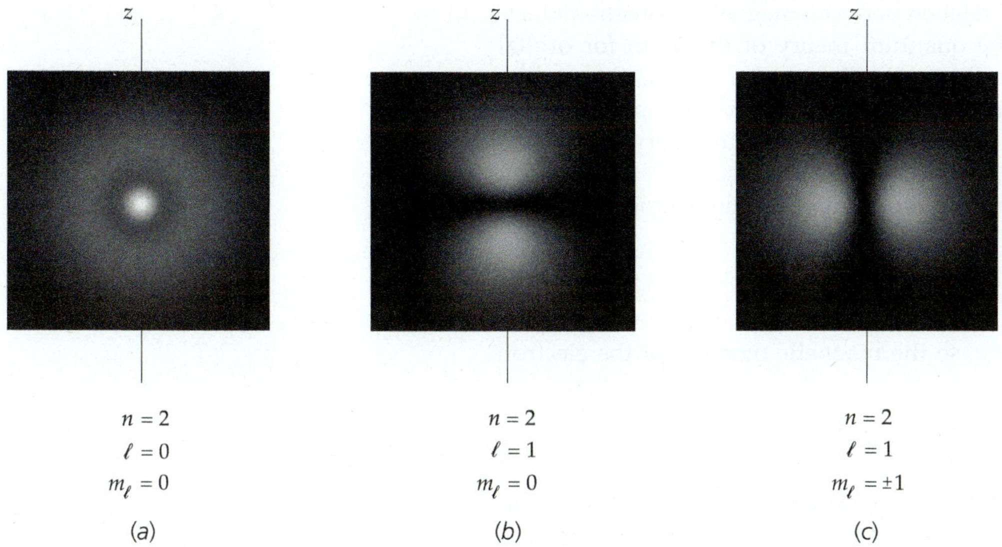

FIGURE 36-11 Computer-generated picture of the probability densities $|\psi|^2$ for the electron in the $n = 2$ states of hydrogen. All three images represent figures of revolution about the z axis. (*a*) For $\ell = 0$, $|\psi|^2$ is spherically symmetric. (*b*) For $\ell = 1$ and $m_\ell = 0$, $|\psi|^2$ is proportional to $\cos^2\theta$. (*c*) For $\ell = 1$ and $m_\ell = +1$ or -1, $|\psi|^2$ is proportional to $\sin^2\theta$.

Figure 36-12 shows the probability density for finding the electron at a distance r from the nucleus for $n = 2$, when $\ell = 1$ and when $\ell = 0$. We can see from the figure that this radial probability density depends on ℓ as well as on n.

For $n = 1$, we found that the most likely distance between the electron and the nucleus is a_0, which is the first Bohr radius, whereas for $n = 2$ and $\ell = 1$, the most probable distance between the electron and the nucleus is $4a_0$. These are the orbital radii for the first and second Bohr orbits (Equation 36-11). For $n = 3$ (and $\ell = 2$),* the most likely distance between the electron and nucleus is $9a_0$, which is the radius of the third Bohr orbit.

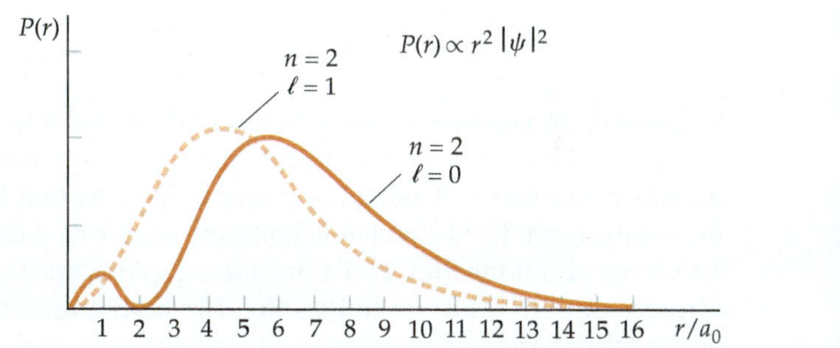

FIGURE 36-12 Radial probability density $P(r)$ versus r/a_0 for the $n = 2$ states of hydrogen. For $\ell = 1$, $P(r)$ is maximum at the Bohr value $r_2 = 2^2a_0$. For $\ell = 0$, there is a maximum near this value and a much smaller maximum near the origin.

36-5 THE SPIN–ORBIT EFFECT AND FINE STRUCTURE

The orbital magnetic moment of an electron in an atom can be derived semiclassically, even though it is quantum mechanical in origin.† Consider a particle of mass m and charge q moving with speed v in a circle of radius r. The magnitude of the orbital angular momentum of the particle is $L = mvr$, and the magnitude of the magnetic moment is the product of the current and the area of the circle $\mu = IA = I\pi r^2$. If T is the time for the charge to complete one revolution, the current (charge passing a point per unit time) is q/T. Because the period T is the distance $2\pi r$ divided by the velocity v, the current is $I = q/T = qv/(2\pi r)$. The magnetic moment is then

$$\mu = IA = \frac{qv}{2\pi r}\pi r^2 = \frac{1}{2}qvr = \frac{q}{2m}L$$

where we have substituted L/m for vr. If the charge q is positive, the orbital angular momentum and orbital magnetic moment vectors are in the same direction. We can therefore write

$$\vec{\mu} = \frac{q}{2m}\vec{L} \qquad 36\text{-}41$$

* The correspondence with the Bohr model is closest for the maximum value of ℓ, which is $n - 1$.

† This topic was first presented in Section 27-5.

Equation 36-41 is the general classical relation between magnetic moment and angular momentum. It also holds in the quantum theory of the atom for orbital angular momentum, but not for the intrinsic spin angular momentum of the electron. For electron spin, the magnetic moment is twice that predicted by Equation 36-41.* The extra factor of 2 is a result from quantum theory that has no analog in classical mechanics.

The quantum of angular momentum is $\hbar$, so we express the magnetic moment in terms of $\vec{L}/\hbar$:

$$\vec{\mu} = \frac{q\hbar}{2m}\frac{\vec{L}}{\hbar}$$

For an electron, $m = m_e$ and $q = -e$, so the magnetic moment of the electron due to its orbital motion is

$$\vec{\mu}_\ell = -\frac{e\hbar}{2m_e}\frac{\vec{L}}{\hbar} = -\mu_B\frac{\vec{L}}{\hbar}$$

where $\mu_B = e\hbar/(2m_e) = 5.79 \times 10^{-5}$ eV/T is the quantum unit of magnetic moment called a Bohr magneton. The magnetic moment of an electron due to its intrinsic spin angular momentum $\vec{S}$ is

$$\vec{\mu}_S = -2\frac{e\hbar}{2m_e}\frac{\vec{S}}{\hbar} = -2\mu_B\frac{\vec{S}}{\hbar}$$

In general, an electron in an atom has both orbital angular momentum characterized by the quantum number ℓ and spin angular momentum characterized by the quantum number s. Analogous classical systems that have two kinds of angular momentum are Earth, which is spinning about its axis of rotation in addition to revolving about the Sun, and a precessing gyroscope that has angular momentum of precession in addition to its spin. The total angular momentum $\vec{J}$ is the sum of the orbital angular momentum $\vec{L}$ and the spin angular momentum $\vec{S}$, where

$$\vec{J} = \vec{L} + \vec{S} \qquad 36\text{-}42$$

Classically $\vec{J}$ is an important quantity because the resultant torque on a system equals the rate of change of the total angular momentum, and in the case of only central forces, the total angular momentum is conserved. For a classical system, the direction of the total angular momentum $\vec{J}$ is without restrictions and the magnitude of $\vec{J}$ can take on any value between $J_{max} = L + S$ and $J_{min} = |L - S|$. In quantum mechanics, however, the directions of both $\vec{L}$ and $\vec{S}$ are more restricted and the magnitudes L and S are both quantized. Furthermore, like $\vec{L}$ and $\vec{S}$, the direction of the total angular momentum $\vec{J}$ is restricted and the magnitude of $\vec{J}$ is quantized. For an electron that has an orbital angular momentum characterized by the quantum number ℓ and spin $s = \frac{1}{2}$, the total angular-momentum magnitude J is equal to $\sqrt{j(j+1)}\hbar$, where the quantum number j is given by

$$j = +\tfrac{1}{2} \quad \text{if} \quad \ell = 0$$

and either

$$j = \ell + \tfrac{1}{2} \quad \text{or} \quad j = \ell - \tfrac{1}{2} \quad \text{if} \quad \ell > 0 \qquad 36\text{-}43$$

Figure 36-13 is a vector model illustrating the two possible combinations $j = \frac{3}{2}$ and $j = \frac{1}{2}$ for the case of $\ell = 1$. The lengths of the vectors are proportional to $\sqrt{\ell(\ell+1)}\hbar$, $\sqrt{s(s+1)}\hbar$, and $\sqrt{j(j+1)}\hbar$. The spin angular momentum and the orbital angular momentum are said to be *parallel* when $j = \ell + s$ and *antiparallel* when $j = \ell - s$.

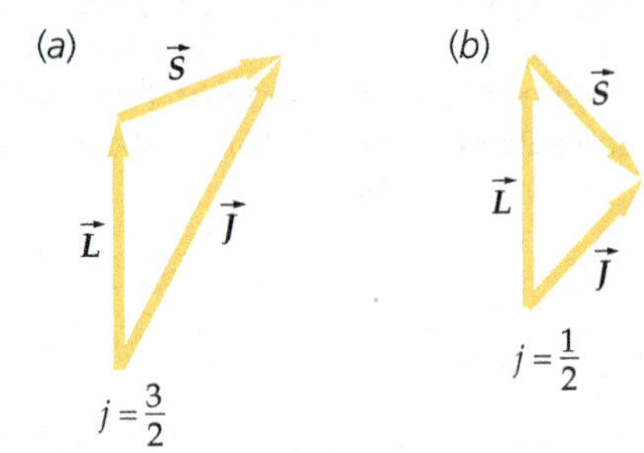

FIGURE 36-13 Vector diagrams illustrating the addition of orbital angular momentum and spin angular momentum for the case $\ell = 1$ and $s = \frac{1}{2}$. There are two possible values of the quantum number for the total angular momentum: $j = \ell + s = \frac{3}{2}$ and $j = \ell - s = \frac{1}{2}$.

* This result and the phenomenon of electron spin itself, was predicted in 1927 by Paul Dirac, who combined special relativity and quantum mechanics into a relativistic wave equation called the *Dirac equation.* Precise measurements indicate that the magnetic moment of the electron due to its spin is 2.00232 times that predicted by Equation 36-42. The fact that the intrinsic magnetic moment of the electron is approximately twice what we would expect makes it clear that the simple model of the electron as a spinning ball is not to be taken literally.

Atomic states that have the same n and ℓ values but with different j values have slightly different energies because of the interaction of the spin of the electron with its orbital motion. This effect is called the **spin–orbit effect.** The resulting splitting of spectral lines is called **fine-structure splitting.**

In the notation $n\ell_j$, the ground state of the hydrogen atom is written $1s_{1/2}$, where the 1 indicates that $n = 1$, the s indicates that $\ell = 0$, and the 1/2 indicates that $j = \frac{1}{2}$. The $n = 2$ states can have either $\ell = 0$ or $\ell = 1$, and the $\ell = 1$ state can have either $j = \frac{3}{2}$ or $j = \frac{1}{2}$. These states are thus denoted by $2s_{1/2}$, $2p_{3/2}$, and $2p_{1/2}$. Because of the spin–orbit effect, the $2p_{3/2}$ and $2p_{1/2}$ states have slightly different energies resulting in the fine-structure splitting of the transitions $2p_{3/2} \rightarrow 2s_{1/2}$ and $2p_{1/2} \rightarrow 2s_{1/2}$.

We can understand the spin–orbit effect qualitatively from a simple Bohr-model picture, as shown in Figure 36-14. In this figure, the electron moves in a circular orbit around a fixed proton. In Figure 36-14*a*, the orbital angular momentum $\vec{L}$ is up. In an inertial reference frame in which the electron is momentarily at rest (see Figure 36-14*b*), the proton is moving at right angles to the line connecting the proton and the electron. The moving proton produces a magnetic field $\vec{B}$ at the position of the electron. The direction of $\vec{B}$ is up, parallel to $\vec{L}$. The energy of the electron depends on its spin because of the magnetic moment $\vec{\mu}_s$ associated with the electron's spin. The energy is lowest when $\vec{\mu}_s$ is parallel to $\vec{B}$ and the energy is highest when it is antiparallel. This energy is given by (Equation 36-16).

$$U = -\vec{\mu}_s \cdot \vec{B} = -\mu_{s_z} B \approx -\mu_B B \qquad \text{36-44*}$$

Because $\vec{\mu}_s$ is directed opposite to its spin (because the electron has a negative charge), the energy is lowest when the spin $\vec{S}$ is antiparallel to $\vec{B}$ and thus to $\vec{L}$. The energy of the $2p_{1/2}$ state in hydrogen, in which $\vec{L}$ and $\vec{S}$ are antiparallel (Figure 36-15), is therefore slightly lower than that of the $2p_{3/2}$ state, in which $\vec{L}$ and $\vec{S}$ are parallel.

* Transferring the energy of the dipole to the frame of the proton gives a factor of 2, which is included in this result.

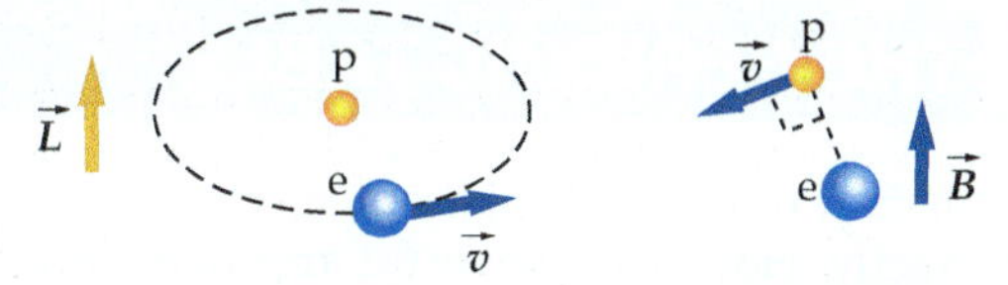

FIGURE 36-14 (*a*) An electron moving about a proton in a circular orbit in the horizontal plane with angular momentum $\vec{L}$ up. (*b*) In an inertial reference frame in which the electron is momentarily at rest there is, at the location of the electron, a magnetic field $\vec{B}$ due to the motion of the proton that is also directed up. When the electron spin $\vec{S}$ is parallel to $\vec{L}$, its magnetic moment $\vec{\mu}_s$ is antiparallel to $\vec{L}$ and $\vec{B}$, so the spin–orbit energy is at its greatest.

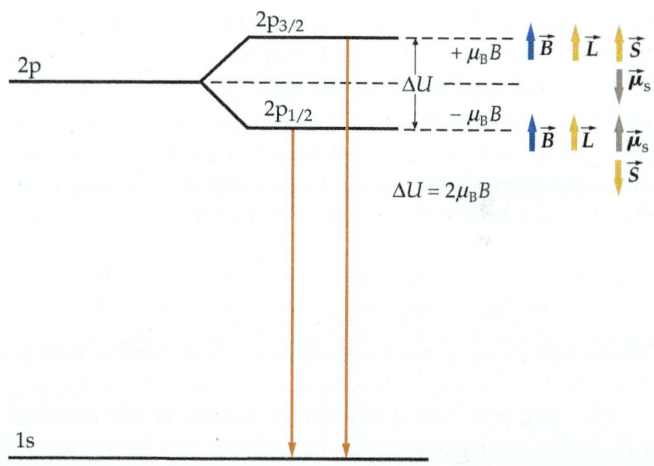

FIGURE 36-15 Fine-structure energy-level diagram. On the left, the levels in the absence of a magnetic field are shown. The effect of the field is shown on the right. Because of the spin-orbit interaction, the magnetic field splits the 2p level into two energy levels, with the $j = \frac{3}{2}$ level having slightly greater energy than the $j = \frac{1}{2}$ level. The spectral line due to the transition $2p \rightarrow 1s$ is therefore split into two lines of slightly different wavelengths.

Example 36-5 Determining *B* by Fine-Structure Splitting

As a consequence of fine-structure splitting, the energies of the $2p_{3/2}$ and $2p_{1/2}$ levels in hydrogen differ by 4.5×10^{-5} eV. If the 2p electron sees an internal magnetic field of magnitude B, the spin–orbit energy splitting will be of the order of $\Delta E = 2\mu_B B$, where μ_B is the Bohr magneton. From this, estimate the magnetic field that the 2p electron in hydrogen experiences.

PICTURE Use the equation $\Delta E = 2\mu_B B$ along with the given value of the energy difference and known value of μ_B.

SOLVE

1. Write the spin–orbit energy splitting in terms of the magnetic moment:

$$\Delta E = 2\mu_B B$$

where

$$\Delta E = 4.5 \times 10^{-5}\ \text{eV}$$

2. Solve for the magnetic field B:

$$B = \frac{\Delta E}{2\mu_B} = \frac{4.5 \times 10^{-5}\ \text{eV}}{2(5.79 \times 10^{-5}\ \text{eV/T})} = \boxed{0.389\ \text{T}}$$

36-6 THE PERIODIC TABLE

For atoms that have more than one electron, the Schrödinger equation cannot be solved exactly. However, powerful approximation methods allow us to calculate the energy levels of the atoms and wave functions of the electrons to a high degree of accuracy. As a first approximation, the Z electrons in an atom are assumed to be noninteracting. The Schrödinger equation can then be solved, and the resulting wave functions used to calculate the interaction of the electrons, which in turn can be used to better approximate the wave functions. Because the spin of an electron can have two possible components along an axis, there is an additional quantum number m_s, which can have the possible values $+\frac{1}{2}$ or $-\frac{1}{2}$. The state of each electron is thus described by the four quantum numbers n, ℓ, m_ℓ, and m_s, and such states are called **stationary states.** The energy of the electron is determined mainly by the principal quantum number n (which is related to the radial dependence of the wave function) and by the orbital angular-momentum quantum number ℓ. Generally, the lower the values of n, the lower the energy; and for a given value of n, the lower the value of ℓ, the lower the energy. The dependence of the energy on ℓ is due to the interaction of the electrons in the atom with each other. In hydrogen, of course, there is only one electron, and the energy is independent of ℓ. The specification of n, ℓ, m_ℓ, and m_s for each electron in an atom is called the **electron configuration.** Customarily, ℓ is specified according to the same code used to label the states of the hydrogen atom rather than by its numerical value. The code is

	s	p	d	f	g	h
ℓ value	0	1	2	3	4	5

The n values are sometimes referred to as shells, which are identified by another letter code: $n = 1$ denotes the K shell*; $n = 2$, the L shell; and so on.

The electron configuration of atoms is constrained by the Pauli exclusion principle—no two electrons in an atom can be in the same quantum state. That is, no two electrons can have the same set of values for the quantum numbers n, ℓ, m_ℓ, and m_s. Using the exclusion principle and the restrictions on the quantum numbers discussed in the previous sections (n is a positive integer, ℓ is an integer that ranges from 0 to $n - 1$, m_ℓ can have $2\ell + 1$ values from $-\ell$ to ℓ in integral steps, and m_s can be either $+\frac{1}{2}$ or $-\frac{1}{2}$), we can understand much of the structure of the periodic table.

We have already discussed the lightest element, hydrogen, which has just one electron. In the ground (lowest energy) state, the electron has $n = 1$ and $\ell = 0$, with $m_\ell = 0$ and $m_s = +\frac{1}{2}$ or $-\frac{1}{2}$. We call this a 1s electron. The 1 signifies that $n = 1$, and the s signifies that $\ell = 0$.

Electrons of atoms whose atomic numbers are greater than 1 will have states that will give the lowest total energy consistent with the Pauli exclusion principle.

CONCEPT CHECK 36-1

The following table lists candidates for the quantum numbers of an electron in an atom. Which of these candidates are not found in nature?

	n	ℓ	m_ℓ	m_s
(*a*)	2	2	-1	$+\frac{1}{2}$
(*b*)	3	2	-1	$+\frac{1}{2}$
(*c*)	2	-1	-1	$-\frac{1}{2}$
(*d*)	3	0	1	$+\frac{1}{2}$
(*e*)	3	1	1	$+\frac{1}{2}$

HELIUM ($Z = 2$)

The next element after hydrogen in the periodic table is helium ($Z = 2$); a helium atom has two electrons. In its ground state, both electrons are in the K shell, where $n = 1$, $\ell = 0$, and $m_\ell = 0$; one electron has $m_s = +\frac{1}{2}$ and the other has $m_s = -\frac{1}{2}$. This configuration is lower in energy than any other two-electron configuration. The resultant spin of the two electrons is zero. Because the orbital angular momentum is also zero, the total angular momentum is zero. The electron configuration for helium is written $1s^2$. The 1 signifies that $n = 1$, the s signifies that $\ell = 0$, and the superscript 2 signifies that there are two electrons in this state. Because ℓ

* The designation of the $n = 1$ shell as K is usually found when dealing with X-ray levels where the final shell in an inner electron transition is labeled as K, L, M, and so on.

can be only 0 for $n = 1$, these two electrons fill the K ($n = 1$) shell. The energy required to remove the most loosely bound electron from an atom in the ground state is called the **first ionization energy.** For a helium atom, the first ionization energy is 24.6 eV, which is relatively large. Helium is therefore basically inert.

Example 36-6 Electron Interaction Energy in Helium

(a) Use the measured first ionization energy to calculate the energy of interaction of the two electrons in the ground state of the helium atom. (b) Use your result to estimate the average separation of the two electrons.

PICTURE The energy of one electron in the ground state of helium is given by $E_n = -Z^2E_0/n^2$ (Equation 36-26), where $n = 1$ and $Z = 2$. If the electrons did not interact, the energy of the second electron would also be E_1, the same as that of the first electron. Thus, for an atom that has noninteracting electrons, the ionization energy of the first electron removed would be $|E_1|$ and the ground-state energy would be $E_{non} = 2E_1$. This is represented by the lowest level in Figure 36-16. Because of the interaction energy, the ground-state energy is greater than $2E_1$, which is represented by the higher level labeled E_g in the figure. The measured first ionization energy of helium is 24.6 eV. When we add $E_{ion} = 24.6$ eV to ionize He, we obtain ionized helium, written He^+, which has just one electron and therefore energy E_1.

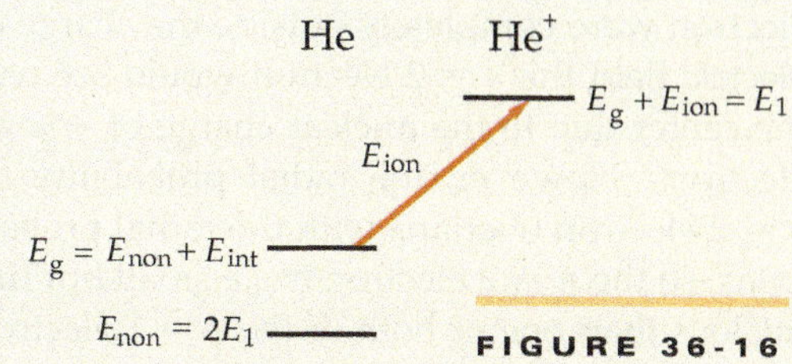

FIGURE 36-16

SOLVE

(a) 1. The sum of the energy of interaction E_{int} and the energy of two noninteracting electrons E_{non} equals the ground-state energy of helium:

$$E_{int} + E_{non} = E_g$$

2. Solve for E_{int} and substitute $E_{non} = 2E_1$:

$$E_{int} = E_g - E_{non} = E_g - 2E_1$$

3. Use Equation 36-26 to calculate the energy E_1 of one electron in the ground state:

$$E_n = -Z^2\frac{E_0}{n^2}$$

so

$$E_1 = -(2)^2\frac{13.6\text{ eV}}{1^2} = -54.4\text{ eV}$$

4. Substitute this value for E_1:

$$E_{int} = E_g - 2E_1 = E_g - 2(-54.4\text{ eV}) = E_g + 108.8\text{ eV}$$

5. The sum of the ground-state energy of He E_g and the ionization energy E_{ion} equals the ground-state energy of He^+, which is E_1:

$$E_g + E_{ion} = E_1 = -54.4\text{ eV}$$

6. Substitute $E_{ion} = 24.6$ eV to calculate E_g:

$$E_g = E_1 - E_{ion} = -54.4\text{ eV} - 24.6\text{ eV} = -79.0\text{ eV}$$

7. Substitute this result for E_g to obtain E_{int}:

$$E_{int} = E_g + 108.8\text{ eV} = -79.0\text{ eV} + 108.8\text{ eV} = \boxed{29.8\text{ eV}}$$

(b) 1. The energy of interaction of two electrons separated by distance r_s is the potential energy:

$$U = +\frac{ke^2}{r_s}$$

2. Set U equal to 29.8 eV, and solve for r_s. It is convenient to express r in terms of a_0, the radius of the first Bohr orbit in hydrogen, and to use $E_0 = ke^2/(2a_0) = 13.6$ eV (Equation 36-16):

$$r_s = \frac{ke^2}{U} = \frac{ke^2}{a_0}\frac{a_0}{U} = 2\frac{ke^2}{2a_0}\frac{a_0}{U} = 2\frac{E_0}{U}a_0$$

$$= 2\frac{13.6\text{ eV}}{29.8\text{ eV}}a_0 = \boxed{0.913a_0}$$

CHECK This separation is approximately the size of the diameter d_1 of the first Bohr orbit for an electron in helium, which is $d_1 = 2r_1 = 2a_0/Z = a_0$.

LITHIUM ($Z = 3$)

The next element, lithium, has an atom that has three electrons. Because the K shell ($n = 1$) of a ground-state lithium atom is completely filled with two electrons, the third electron occupies a higher energy shell. The next lowest energy shell after $n = 1$ is the $n = 2$ or L shell. This $n = 2$ electron has a greater probability of being much farther from the nucleus than the two $n = 1$ electrons. It is most likely to be found at a radius near that of the second Bohr orbit, which is four times the radius of the first Bohr orbit.

The nuclear charge is partially screened from the $n = 2$ electron by the two $n = 1$ electrons. Recall that the electric field outside a spherically symmetric charge density is the same as if all the charge were at the center of the sphere. If the $n = 2$ electron were completely outside the charge cloud of the two $n = 1$ electrons, the electric field the $n = 2$ electron would see would be that of a single charge $+e$ at the center due to the nuclear charge of $+3e$ and the charge $-2e$ of the two $n = 1$ electrons. However, the radial probability distribution (Equation 36-33) of the $n = 2$ electron overlaps with the radial probability distributions of the $n = 1$ electrons, so the $n = 2$ electron has a small but finite probability of being closer to the nucleus than one or both of the $n = 1$ electrons. Because of this, the effective nuclear charge $Z'e$ seen by the $n = 2$ electron is somewhat greater than $+1e$. Consequently, the energy of the $n = 2$ electron at a distance r from a point charge $+Z'e$ is given by Equation 36-6, where the nuclear charge $+Z$ replaced by $+Z'$.

$$E = -\frac{1}{2}\frac{kZ'e^2}{r} \qquad \text{36-45}$$

The greater the overlap of the radial probability distributions of a higher energy electron with lower energy electrons, the greater is the effective nuclear charge $Z'e$ seen by the higher energy electron and the lower is the energy of the higher energy electron. Because the overlap is greater for ℓ values closer to zero (see Figure 36-12), the energy of the $n = 2$ electron in lithium is lower for the s state ($\ell = 0$) than for the p state ($\ell = 1$). The electron configuration of a lithium atom in the ground state is therefore $1s^22s$. The first ionization energy of a lithium atom is only 5.39 eV. Because its $n = 2$ electron is so loosely bound to the atom, lithium is very active chemically. It behaves like a one-electron atom, similar to hydrogen.

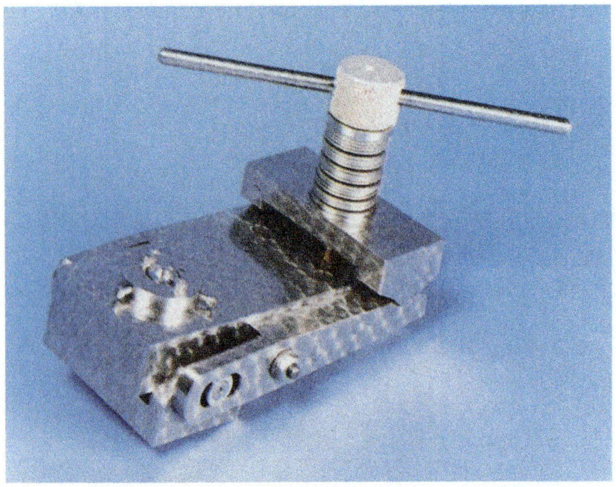

(a)

(b)

(*a*) A diamond anvil cell, in which the facets of two diamonds (approximately 1 mm^2 each) are used to compress a sample substance, subjecting it to very high pressure. (*b*) Samarium monosulfide (SmS) is normally a black, dull-looking semiconductor. When it is subjected to pressure above 7000 atm, an electron from the 4f state moves into the 5d state. The resulting compound glitters like gold and behaves like a metal. *((a) and (b) A. Jayaraman/AT&T Bell Labs.)*

Example 36-7 Effective Nuclear Charge for an Electron of a Lithium Atom

Suppose the radial probability distribution of the $n = 2$ electron in the lithium atom in the ground state did not overlap the probability distribution of the two $n = 1$ electrons; the nuclear charge would be shielded by the two $n = 1$ electrons and the effective nuclear charge would be $Z'e$, where $Z' = 1$. Then the energy of the $n = 2$ electron would be $-(13.6\text{ eV})/2^2 = -3.4\text{ eV}$. However, the first ionization energy of lithium is 5.39 eV, not 3.4 eV. Use this fact to calculate the effective nuclear charge Z' seen by the $n = 2$ electron in lithium.

PICTURE Because the $n = 2$ electron is in the $n = 2$ shell, we will take $r = 4a_0$ for its average distance from the nucleus. We can then calculate Z' from Equation 36-45. Because r is given in terms of a_0, it will be convenient to use the fact that $E_0 = ke^2/(2a_0) = 13.6\text{ eV}$ (Equation 36-16).

SOLVE

1. Equation 36-45 relates the energy of the $n = 2$ electron to its average distance r from the nucleus and the effective nuclear charge Z':

$$E = -\frac{1}{2}\frac{kZ'e^2}{r}$$

2. Substitute the given values $r = 4a_0$ and $E = -5.39\text{ eV}$:

$$-5.39\text{ eV} = -\frac{1}{2}\frac{kZ'e^2}{4a_0}$$

3. Use $ke^2/(2a_0) = E_0 = 13.6\text{ eV}$ and solve for Z':

$$-5.39\text{ eV} = -\frac{Z'}{4}\frac{ke^2}{2a_0} = -\frac{Z'}{4}(13.6\text{ eV})$$

so

$$Z' = 4\frac{5.39\text{ eV}}{13.6\text{ eV}} = \boxed{1.59}$$

CHECK We expected Z' to be larger than one and certainly less than 3. Our step-3 result meets these expectations.

TAKING IT FURTHER This calculation is interesting but not very rigorous. We essentially used the radius ($r = 4a_0$) for the circular orbit from the semiclassical Bohr model and the measured first ionization energy to calculate the effective nuclear charge seen by the $n = 2$ electron. We know, of course, that this $n = 2$ electron does not move in a circular orbit of constant radius, but is better represented by a probability density $|\psi|^2$ that overlaps the probability distributions of the $n = 1$ electrons.

BERYLLIUM ($Z = 4$)

The energy of the beryllium atom is a minimum if both $n = 2$ electrons are in the 2s state. There can be two electrons that have $n = 2$, $\ell = 0$, and $m_\ell = 0$ because of the two possible values for the spin quantum number m_s. The configuration of a beryllium atom is thus $1s^22s^2$.

BORON TO NEON ($Z = 5$ TO $Z = 10$)

If the 2s subshell of a ground-state boron atom is filled, the fifth electron must be in the next available (lowest energy) subshell, which is the 2p subshell, where $n = 2$ and $\ell = 1$. Because there are three possible values of m_ℓ (+1, 0, and −1) and two values of m_s for each value of m_ℓ, there can be six electrons in this subshell.

Hydrogen

Carbon

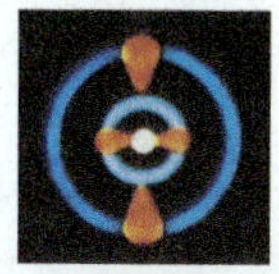
Silicon

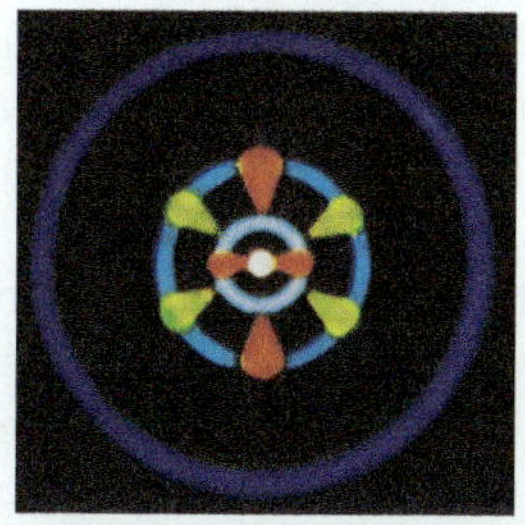
Iron

Silver

Europium

A schematic depiction of the electron configurations in atoms. The spherically symmetric s states can have 2 electrons and are shown as white and blue. The dumbbell-shaped p states can have up to 6 electrons and are shown as orange. The d states can have up to 10 electrons and are shown as yellow-green. The f states can have up to 14 electrons and are shown as purple. *(David Parker/Photo Researchers.)*

The electron configuration for a boron atom is $1s^2 2s^2 2p$. The electron configurations for carbon atoms ($Z = 6$) to neon atoms ($Z = 10$) differ from that for boron atoms only in the number of electrons in the 2p subshell. The first ionization energy increases with Z for these elements, reaching the value of 21.6 eV for the last element in the group, neon. A neon atom has the maximum number of electrons allowed in the $n = 2$ shell and its electron configuration is $1s^2 2s^2 2p^6$. Because of its very large first ionization energy, neon, like helium, basically is chemically inert. The atom whose atomic number is one less than neon's atomic number is fluorine, which has an unoccupied electron state in the 2p subshell; that is, a fluorine atom can have one more electron in its 2p subshell. Fluorine readily combines with elements such as lithium that have one electron in its highest energy shell (that is, an electron in an unfilled highest energy shell of an atom in the ground state). Lithium, for example, will donate its single valence electron to the fluorine atom to make an F^- ion and a Li^+ ion. These ions then bond together to form lithium fluoride.

SODIUM TO ARGON ($Z = 11$ TO $Z = 18$)

The eleventh electron of a ground-state sodium atom must be the $n = 3$ shell. Because this electron is shielded from the nucleus by $n = 2$ and $n = 1$ electrons, it is weakly bound in the sodium ($Z = 11$) atom. The first ionization energy of sodium is only 5.14 eV. Sodium atoms therefore combine readily with atoms such as fluorine. With $n = 3$, the value of ℓ can be 0, 1, or 2. A 3s electron has a lower energy than a 3p or 3d electron because its probability density overlap with the probability densities of $n = 2$ and $n = 1$ electrons is greatest. This energy difference between subshells of the same n value becomes greater as the number of electrons increases. The electron configuration of a sodium atom is $1s^2 2s^2 2p^6 3s^1$. For elements whose atoms have larger values of Z, the 3s subshell and then the 3p subshell are occupied. These two subshells can accommodate $2 + 6 = 8$ electrons. The configuration of an argon ($Z = 18$) atom is $1s^2 2s^2 2p^6 3s^2 3p^6$. One might expect the nineteenth electron of potassium would occupy the third subshell (the d subshell where $\ell = 2$), but the overlap effect is now so strong that the energy of the nineteenth electron is lower in the 4s subshell than in the 3d subshell. There is thus another large energy difference between the eighteenth and nineteenth electrons of a potassium atom, and so an argon atom, with its full 3p subshell, is basically stable and inert.

ELEMENTS WITH Z > 18

The nineteenth electron in a potassium ($Z = 19$) atom and the twentieth electron in a calcium ($Z = 20$) atom occupy the 4s subshell rather than the 3d subshell. The electron configurations of the atoms of the next ten elements, scandium ($Z = 21$) through zinc ($Z = 30$), differ only in the number of electrons in the 3d shell, except for a chromium ($Z = 24$) atom and a copper ($Z = 29$) atom, each of which has only one 4s electron. These ten elements are called **transition elements.**

Figure 36-17 shows a plot of the first ionization energies versus Z for $Z = 1$ to $Z = 60$. The peaks in first ionization energy at $Z = 2$, 10, 18, 36, and 54 mark a filled shell or subshell. Table 36-1 gives the ground-state electron configurations of atoms up to atomic number 111.

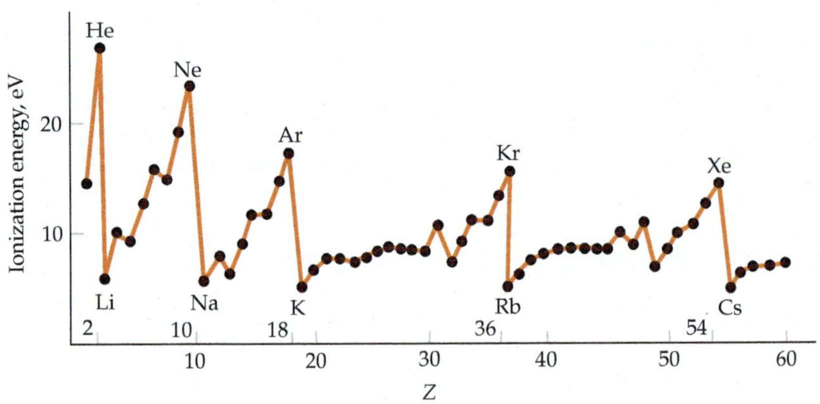

FIGURE 36-17 First ionization energies versus Z for $Z = 1$ to $Z = 60$. The first ionization energy increases with Z until a shell is filled at $Z = 2$, 10, 18, 36, and 54. An atom that has a filled shell and a single valence electron, such as sodium ($Z = 11$), has a very low ionization energy because the valence electron is shielded by the core electrons.

Table 36-1 Electron Configurations of the Atoms in Their Ground States
For some of the rare-earth elements and the heavy elements the configurations are not firmly established.

		Shell (n):	K (1)	L (2)		M (3)			N (4)				O (5)				P (6)			Q (7)
Z	Element	Subshell (ℓ):	s (0)	s (0)	p (1)	s (0)	p (1)	d (2)	s (0)	p (1)	d (2)	f (3)	s (0)	p (1)	d (2)	f (3)	s (0)	p (1)	d (2)	s (1)
1	H	hydrogen	1																	
2	He	helium	2																	
3	Li	lithium	2	1																
4	Be	beryllium	2	2																
5	B	boron	2	2	1															
6	C	carbon	2	2	2															
7	N	nitrogen	2	2	3															
8	O	oxygen	2	2	4															
9	F	fluorine	2	2	5															
10	Ne	neon	2	2	6															
11	Na	sodium	2	2	6	1														
12	Mg	magnesium	2	2	6	2														
13	Al	aluminum	2	2	6	2	1													
14	Si	silicon	2	2	6	2	2													
15	P	phosphorus	2	2	6	2	3													
16	S	sulfur	2	2	6	2	4													
17	Cl	chlorine	2	2	6	2	5													
18	Ar	argon	2	2	6	2	6													
19	K	potassium	2	2	6	2	6	.	1											
20	Ca	calcium	2	2	6	2	6	.	2											
21	Sc	scandium	2	2	6	2	6	1	2											
22	Ti	titanium	2	2	6	2	6	2	2											
23	V	vanadium	2	2	6	2	6	3	2											
24	Cr	chromium	2	2	6	2	6	5	1											
25	Mn	manganese	2	2	6	2	6	5	2											
26	Fe	iron	2	2	6	2	6	6	2											
27	Co	cobalt	2	2	6	2	6	7	2											
28	Ni	nickel	2	2	6	2	6	8	2											
29	Cu	copper	2	2	6	2	6	10	1											
30	Zn	zinc	2	2	6	2	6	10	2											
31	Ga	gallium	2	2	6	2	6	10	2	1										
32	Ge	germanium	2	2	6	2	6	10	2	2										
33	As	arsenic	2	2	6	2	6	10	2	3										
34	Se	selenium	2	2	6	2	6	10	2	4										
35	Br	bromine	2	2	6	2	6	10	2	5										
36	Kr	krypton	2	2	6	2	6	10	2	6										
37	Rb	rubidium	2	2	6	2	6	10	2	6	.	.	1							
38	Sr	strontium	2	2	6	2	6	10	2	6	.	.	2							

Continued on next page

Table 36-1 Continued

		Shell (n):	K (1)	L (2)		M (3)			N (4)				O (5)				P (6)			Q (7)
Z	Element	Subshell (ℓ):	s (0)	s (0)	p (1)	s (0)	p (1)	d (2)	s (0)	p (1)	d (2)	f (3)	s (0)	p (1)	d (2)	f (3)	s (0)	p (1)	d (2)	s (1)
39	Y	yttrium	2	2	6	2	6	10	2	6	1	.	2							
40	Zr	zirconium	2	2	6	2	6	10	2	6	2	.	2							
41	Nb	niobium	2	2	6	2	6	10	2	6	4	.	1							
42	Mo	molybdenum	2	2	6	2	6	10	2	6	5	.	1							
43	Tc	technetium	2	2	6	2	6	10	2	6	6	.	1							
44	Ru	ruthenium	2	2	6	2	6	10	2	6	7	.	1							
45	Rh	rhodium	2	2	6	2	6	10	2	6	8	.	1							
46	Pd	palladium	2	2	6	2	6	10	2	6	10	.	.							
47	Ag	silver	2	2	6	2	6	10	2	6	10	.	1							
48	Cd	cadmium	2	2	6	2	6	10	2	6	10	.	2							
49	In	indium	2	2	6	2	6	10	2	6	10	.	2	1						
50	Sn	tin	2	2	6	2	6	10	2	6	10	.	2	2						
51	Sb	antimony	2	2	6	2	6	10	2	6	10	.	2	3						
52	Te	tellurium	2	2	6	2	6	10	2	6	10	.	2	4						
53	I	iodine	2	2	6	2	6	10	2	6	10	.	2	5						
54	Xe	xenon	2	2	6	2	6	10	2	6	10	.	2	6						
55	Cs	cesium	2	2	6	2	6	10	2	6	10	.	2	6	.	.	1			
56	Ba	barium	2	2	6	2	6	10	2	6	10	.	2	6	.	.	2			
57	La	lanthanum	2	2	6	2	6	10	2	6	10	.	2	6	1	.	2			
58	Ce	cerium	2	2	6	2	6	10	2	6	10	1	2	6	1	.	2			
59	Pr	praseodymium	2	2	6	2	6	10	2	6	10	3	2	6	.	.	2			
60	Nd	neodymium	2	2	6	2	6	10	2	6	10	4	2	6	.	.	2			
61	Pm	promethium	2	2	6	2	6	10	2	6	10	5	2	6	.	.	2			
62	Sm	samarium	2	2	6	2	6	10	2	6	10	6	2	6	.	.	2			
63	Eu	europium	2	2	6	2	6	10	2	6	10	7	2	6	.	.	2			
64	Gd	gadolinium	2	2	6	2	6	10	2	6	10	7	2	6	1	.	2			
65	Tb	terbium	2	2	6	2	6	10	2	6	10	9	2	6	.	.	2			
66	Dy	dysprosium	2	2	6	2	6	10	2	6	10	10	2	6	.	.	2			
67	Ho	holmium	2	2	6	2	6	10	2	6	10	11	2	6	.	.	2			
68	Er	erbium	2	2	6	2	6	10	2	6	10	12	2	6	.	.	2			
69	Tm	thulium	2	2	6	2	6	10	2	6	10	13	2	6	.	.	2			
70	Yb	ytterbium	2	2	6	2	6	10	2	6	10	14	2	6	.	.	2			
71	Lu	lutetium	2	2	6	2	6	10	2	6	10	14	2	6	1	.	2			
72	Hf	hafnium	2	2	6	2	6	10	2	6	10	14	2	6	2	.	2			
73	Ta	tantalum	2	2	6	2	6	10	2	6	10	14	2	6	3	.	2			
74	W	tungsten (wolfram)	2	2	6	2	6	10	2	6	10	14	2	6	4	.	2			
75	Re	rhenium	2	2	6	2	6	10	2	6	10	14	2	6	5	.	2			
76	Os	osmium	2	2	6	2	6	10	2	6	10	14	2	6	6	.	2			
77	Ir	iridium	2	2	6	2	6	10	2	6	10	14	2	6	7	.	2			
78	Pt	platinum	2	2	6	2	6	10	2	6	10	14	2	6	9	.	1			
79	Au	gold	2	2	6	2	6	10	2	6	10	14	2	6	10	.	1			

Continued on next page

Table 36-1 Continued

		Shell (n):	K (1)	L (2)		M (3)			N (4)				O (5)				P (6)			Q (7)
Z	Element	Subshell (ℓ):	s (0)	s (0)	p (1)	s (0)	p (1)	d (2)	s (0)	p (1)	d (2)	f (3)	s (0)	p (1)	d (2)	f (3)	s (0)	p (1)	d (2)	s (1)
80	Hg	mercury	2	2	6	2	6	10	2	6	10	14	2	6	10	.	2			
81	Tl	thallium	2	2	6	2	6	10	2	6	10	14	2	6	10	.	2	1		
82	Pb	lead	2	2	6	2	6	10	2	6	10	14	2	6	10	.	2	2		
83	Bi	bismuth	2	2	6	2	6	10	2	6	10	14	2	6	10	.	2	3		
84	Po	polonium	2	2	6	2	6	10	2	6	10	14	2	6	10	.	2	4		
85	At	astatine	2	2	6	2	6	10	2	6	10	14	2	6	10	.	2	5		
86	Rn	radon	2	2	6	2	6	10	2	6	10	14	2	6	10	.	2	6		
87	Fr	francium	2	2	6	2	6	10	2	6	10	14	2	6	10	.	2	6	.	1
88	Ra	radium	2	2	6	2	6	10	2	6	10	14	2	6	10	.	2	6	.	2
89	Ac	actinium	2	2	6	2	6	10	2	6	10	14	2	6	10	.	2	6	1	2
90	Th	thorium	2	2	6	2	6	10	2	6	10	14	2	6	10	.	2	6	2	2
91	Pa	protactinium	2	2	6	2	6	10	2	6	10	14	2	6	10	2	2	6	1	2
92	U	uranium	2	2	6	2	6	10	2	6	10	14	2	6	10	3	2	6	1	2
93	Np	neptunium	2	2	6	2	6	10	2	6	10	14	2	6	10	4	2	6	1	2
94	Pu	plutonium	2	2	6	2	6	10	2	6	10	14	2	6	10	6	2	6	.	2
95	Am	americium	2	2	6	2	6	10	2	6	10	14	2	6	10	7	2	6	.	2
96	Cm	curium	2	2	6	2	6	10	2	6	10	14	2	6	10	7	2	6	1	2
97	Bk	berkelium	2	2	6	2	6	10	2	6	10	14	2	6	10	9	2	6	.	2
98	Cf	californium	2	2	6	2	6	10	2	6	10	14	2	6	10	10	2	6	.	2
99	Es	einsteinium	2	2	6	2	6	10	2	6	10	14	2	6	10	11	2	6	.	2
100	Fm	fermium	2	2	6	2	6	10	2	6	10	14	2	6	10	12	2	6	.	2
101	Md	mendelevium	2	2	6	2	6	10	2	6	10	14	2	6	10	13	2	6	.	2
102	No	nobelium	2	2	6	2	6	10	2	6	10	14	2	6	10	14	2	6	.	2
103	Lr	lawrencium	2	2	6	2	6	10	2	6	10	14	2	6	10	14	2	6	1	2
104	Rf	rutherfordium	2	2	6	2	6	10	2	6	10	14	2	6	10	14	2	6	2	2
105	Db	dubnium	2	2	6	2	6	10	2	6	10	14	2	6	10	14	2	6	3	2
106	Sg	seaborgium	2	2	6	2	6	10	2	6	10	14	2	6	10	14	2	6	4	2
107	Bh	bohrium	2	2	6	2	6	10	2	6	10	14	2	6	10	14	2	6	5	2
108	Hs	hassium	2	2	6	2	6	10	2	6	10	14	2	6	10	14	2	6	6	2
109	Mt	meitnerium	2	2	6	2	6	10	2	6	10	14	2	6	10	14	2	6	7	2
110	Ds	darmstadtium	2	2	6	2	6	10	2	6	10	14	2	6	10	14	2	6	9	1
111	Rg	roentgenium	2	2	6	2	6	10	2	6	10	14	2	6	10	14	2	6	10	1

36-7 OPTICAL SPECTRA AND X-RAY SPECTRA

When an atom is in an excited state (when it is in an energy state above the ground state), it makes transitions to lower energy states, and in doing so emits electromagnetic radiation. The wavelength of the electromagnetic radiation emitted is related to the initial and final states by the Bohr formula (Equation 36-17), $\lambda = hc/(E_i - E_f)$, where E_i and E_f are the initial and final energies and h is Planck's constant. The atom can be excited to a higher energy state by bombarding the atom with a beam of electrons, as in a spectral tube that has a high voltage across it. Because the excited energy states of an atom form a discrete (rather than continuous) set, only certain wavelengths are emitted. These wavelengths of the emitted radiation constitute the emission spectrum of the atom.

OPTICAL SPECTRA

To understand atomic spectra we need to understand the excited states of the atom. The situation for an atom that has many electrons is, in general, much more complicated than that of a hydrogen atom that has just one electron. An excited state of the atom may involve a change in the state occupied by any one of the electrons, or even two or more electrons. Fortunately, in most cases, an excited state of an atom involves the excitation of just one of the electrons in the atom. The energies of excitation of the valence electrons of an atom are of the order of a few electron volts. Transitions involving these electrons result in photons in or near the visible or **optical spectrum.** (Recall that the energies of visible photons range from approximately 1.5 eV to 3 eV.) The excitation energies can often be calculated from a simple model in which the atom is pictured as a single electron plus a stable core consisting of the nucleus plus the other electrons. This model works particularly well for the alkali metals: Li, Na, K, Rb, and Cs. These elements are in the first column of the periodic table. The optical spectra of these elements are similar to the optical spectra of hydrogen.

Figure 36-18 shows an energy-level diagram for the optical transitions of a sodium atom, whose electrons form a neon core plus one electron. Because the total spin angular momentum of the core adds up to zero, the spin of each state of the sodium atom is $\frac{1}{2}$ (the spin of the valence electron). Because of the spin–orbit effect, the atomic states for which $J = L - \frac{1}{2}$ have a slightly different energy than those for which $J = L + \frac{1}{2}$ (except for states with $L = 0$). Each state (except for the $L = 0$ states) is therefore split into two states, called a doublet. The doublet splitting is very small and not evident on the energy scale of this diagram. The usual spectroscopic notation is that the states of these atoms are labeled with a superscript given by $2S + 1$, followed by a letter denoting the orbital angular momentum, followed by a subscript denoting the total angular momentum J. For states that have a total spin angular momentum $S = \frac{1}{2}$ the superscript is 2, indicating the state is a doublet. Thus, ${}^2P_{3/2}$, read as "doublet P three halves," denotes a state in which $L = 1$ and $J = \frac{3}{2}$. (The $L = 0$, or S, states are customarily labeled as if they were doublets even though they are not.) For a sodium atom in the first excited state, the electron is excited from the 3s level to the 3p level, which is approximately 2.1 eV above the ground state. The energy difference between the $P_{3/2}$ and $P_{1/2}$ states due to the spin–orbit effect is about 0.002 eV. Transitions from these states to the ground state give the familiar sodium yellow doublet:

$$3p({}^2P_{1/2}) \rightarrow 3s({}^2S_{1/2}) \qquad \lambda = 589.6\text{ nm}$$
$$3p({}^2P_{3/2}) \rightarrow 3s({}^2S_{1/2}) \qquad \lambda = 589.0\text{ nm}$$

The energy levels and spectra of other alkali metal atoms are similar to those for sodium. The optical spectrum for atoms such as helium, beryllium, and magnesium that have two valence electrons is considerably more complex because of the interaction of the two electrons.

A neon sign outside a Chinatown restaurant in Paris. Neon atoms in the tube are excited by an electron current passing through the tube. The excited neon atoms emit light in the visible range as they decay toward their ground states. The colors of neon signs result from the characteristic red-orange spectrum of neon plus the color of the glass tube itself. *(Robert Landau/Westlight.)*

X-RAY SPECTRA

X rays are usually produced in the laboratory by bombarding a target element with a high-energy beam of electrons in an X-ray tube. The result (Figure 36-19) consists of a continuous spectrum that depends only on the energy of the bombarding electrons and a line spectrum that is characteristic of the target element. The characteristic spectrum results from excitation of the core electrons in the target element.

The energy needed to excite a core electron—for example, an electron in the $n = 1$ state (K shell)—is much greater than the energy required to excite a valence

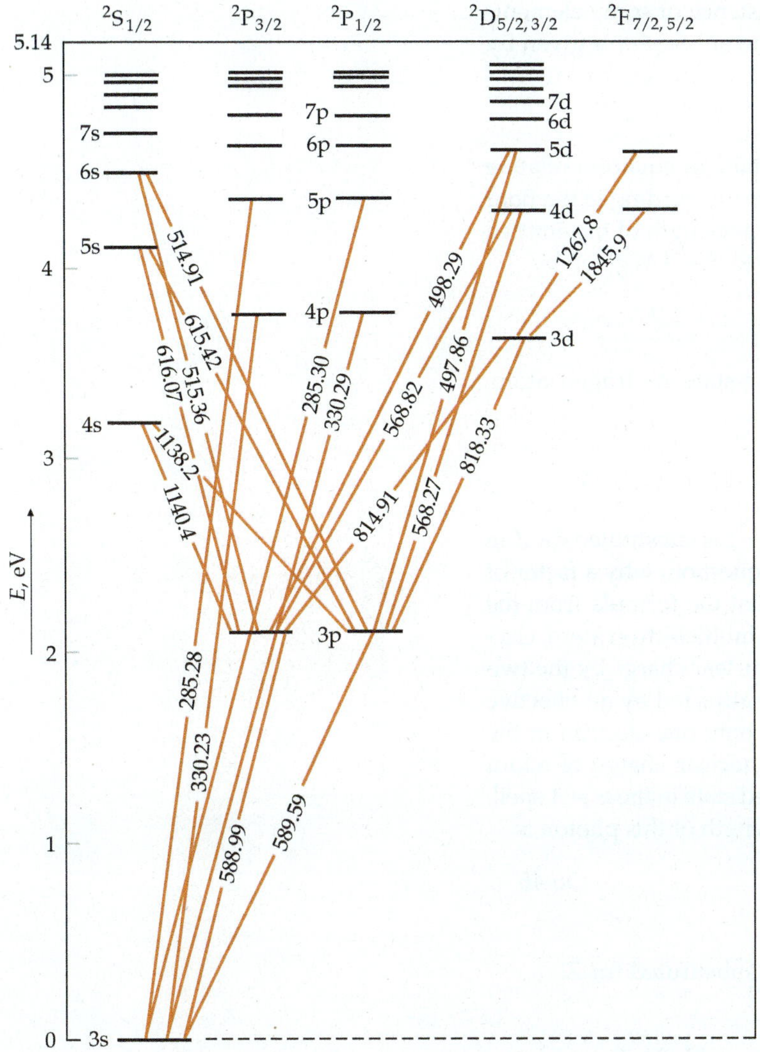

FIGURE 36-18 Energy-level diagram for sodium. The diagonal lines show observed optical transitions, where wavelengths are given in nanometers. The energy of the ground state has been chosen as the zero point for the scale on the left.

electron. A core electron cannot be excited to any of the filled states (for example, the $n = 2$ states in an atom with $Z \geq 10$) because of the exclusion principle. The energy required to excite a core electron to an unoccupied state is typically of the order of several thousand electron volts. If an electron is knocked out of the $n = 1$ shell (the K shell), there is a vacancy left in that shell. The vacancy can be filled if an electron in a higher energy shell makes a transition into the K shell. The photons emitted by electrons making such transitions also have energies of the order of several thousand electron volts and produce the sharp peaks in the X-ray spectrum, as shown in Figure 36-19. The K_α spectral line arises from transitions from the $n = 2$ shell (the L shell) to the $n = 1$ shell (the K shell). The K_β spectral line arises from transitions from the $n = 3$ shell to the $n = 1$ shell. These and other lines arising from transitions ending at the $n = 1$ shell make up the K series of the characteristic X-ray spectrum of the target element. Similarly, a second series, the L series, is produced by transitions from higher energy states to a vacated place in the $n = 2$ (L) shell. The letters K, L, M, and so on, designate the final shell of the electron making the transition and the series α, β, and so on, designates the number of shells above the final shell for the initial state of the electron.

In 1913, the English physicist Henry Moseley measured the wavelengths of the characteristic K_α X-ray spectra for approximately 40 elements. Using this data, Moseley showed that a plot of $\lambda^{-1/2}$ versus the order in which the elements appeared in the periodic table resulted in a straight line (with a few gaps and a few outliers). From his data, Moseley was able to accurately determine the atomic

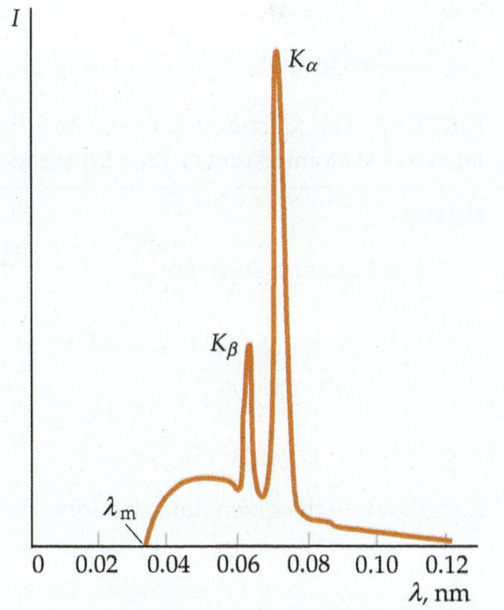

FIGURE 36-19 X-ray spectrum of molybdenum. The sharp peaks labeled K_α and K_β are characteristic of the element. The cutoff wavelength λ_m is independent of the target element and is related to the voltage V of the X-ray tube by $\lambda_m = hc/eV$.

number Z for each known element, and to predict the existence of some elements that were later discovered. The equation of the straight line of his plot is given by

$$\frac{1}{\sqrt{\lambda_{K_\alpha}}} = a(Z - 1)$$

The work of Bohr and Moseley can be combined to obtain an equation relating the wavelength of the emitted photon and the atomic number. According to the Bohr model of a single-electron atom (see Equation 36-13), the wavelength of the emitted photon when the electron makes the transition from $n = 2$ to $n = 1$ is given by

$$\frac{1}{\lambda} = Z^2 \frac{E_0}{hc}\left(1 - \frac{1}{2^2}\right)$$

where $E_0 = 13.6$ eV is the binding energy of the ground-state hydrogen atom. Taking the square root of both sides gives

$$\frac{1}{\sqrt{\lambda_{K_\alpha}}} = \left[\frac{E_0}{hc}\left(1 - \frac{1}{2^2}\right)\right]^{1/2} Z$$

Moseley's equation and this equation are in agreement if $Z - 1$ is substituted for Z in Bohr's equation and if $a = 3E_0/(4hc)$. This result raises the question, why a factor of $Z - 1$ instead of a factor of Z? Part of the explanation is that the formula from the Bohr theory ignores the shielding of the nuclear charge. In a multielectron atom, electrons in the $n = 2$ states are electrically shielded from the nuclear charge by the two electrons in the $n = 1$ state, so the $n = 2$ state electrons are attracted by an effective nuclear charge of about $(Z - 2)e$. However, when there is only one electron in the K shell, the $n = 2$ electrons are attracted by an effective nuclear charge of about $(Z - 1)e$. When an electron from state n drops into the vacated state in the $n = 1$ shell, a photon of energy $E_n - E_1$ is emitted. For $n = 2$, the wavelength of this photon is

$$\lambda_{K_\alpha} = \frac{hc}{(Z - 1)^2 E_0\left(1 - \frac{1}{2^2}\right)} \qquad \text{36-46}$$

which is obtained from the previous equation with $Z - 1$ substituted for Z.

Example 36-8 Identifying the Element from the K_α X-Ray Line

The wavelength of the K_α X-ray line for a certain element is $\lambda = 0.0721$ nm. What is the element?

PICTURE The K_α line corresponds to a transition from $n = 2$ to $n = 1$. The wavelength is related to the atomic number Z by Equation 36-46.

SOLVE

1. Solve Equation 36-46 for $(Z - 1)^2$:

$$\lambda_{K_\alpha} = \frac{hc}{(Z - 1)^2 E_0\left(1 - \frac{1}{2^2}\right)}$$

so

$$(Z - 1)^2 = \frac{4hc}{3\lambda_{K_\alpha} E_0}$$

2. Substitute the given data and solve for Z:

$$(Z - 1)^2 = \frac{4(1240\text{ eV}\cdot\text{nm})}{3(0.0721\text{ nm})(13.6\text{ eV})} = 1686$$

so

$$Z = 1 + \sqrt{1686} = 42.06$$

3. Because Z is an integer, we round to the nearest integer:

$$Z = 42$$

The element is molybdenum.

CHECK The naturally occurring atom that has the largest atomic number is uranium, which has an atomic number $Z = 92$. That our step-3 result is greater than 0 and less than 93 is as expected.

Summary

1. The Bohr model is important because it was the first model to succeed at explaining the discrete optical spectrum of atoms in terms of the quantization of energy. It has been superceded by the quantum-mechanical model.
2. The quantum theory of atoms results from the application of the Schrödinger equation to a bound system consisting of nucleus of charge $+Ze$ and Z electrons of charge $-e$.
3. For the hydrogen atom, an atom that consists of one proton and one electron, the time-independent Schrödinger equation can be solved exactly to obtain the wave functions ψ, which depend on the quantum numbers n, ℓ, m_ℓ, and m_s.
4. The electron configuration of atoms is governed by the Pauli exclusion principle—no two electrons in an atom can have the same set of values for the quantum numbers n, ℓ, m_ℓ, and m_s. Using the exclusion principle and the restrictions on the quantum numbers, we can understand much of the structure of the periodic table.

TOPIC	RELEVANT EQUATIONS AND REMARKS	
1. The Bohr Model of the Hydrogen Atom		
Postulates for the hydrogen atom		
Nonradiating orbits	The idea that an electron moves in a circular nonradiating orbit around the proton.	
Photon frequency from energy conservation	$f = \dfrac{E_i - E_f}{h}$	36-7
Quantized angular momentum	$L_n = mv_n r_n = n\hbar \qquad n = 1, 2, 3, \ldots$	36-9
First Bohr radius	$a_0 = \dfrac{\hbar^2}{mke^2} = 0.0529\text{ nm}$	36-12
Radii of the Bohr orbits	$r_n = n^2 \dfrac{a_0}{Z}$	36-11
Energy levels in hydrogen-like atoms	$E_n = -Z^2 \dfrac{E_0}{n^2}$	36-15
	where $E_0 = -\dfrac{mk^2e^4}{2\hbar^2} = \dfrac{1}{2}\dfrac{ke^2}{a_0} = 13.6\text{ eV}$	36-16
Wavelengths emitted by the hydrogen atom	$\lambda = \dfrac{c}{f} = \dfrac{hc}{E_i - E_f} = \dfrac{1240\text{ eV}\cdot\text{nm}}{E_i - E_f}$	36-17, 36-18
2. Quantum Theory of Atoms	The electron is described by a wave function ψ that is a solution of the Schrödinger equation. Energy quantization arises from standing-wave conditions. ψ is described by the principal, orbital, and magnetic quantum numbers n, ℓ, and m_ℓ, and the spin quantum number $m_s = \pm\frac{1}{2}$.	
Time-independent Schrödinger equation	$-\dfrac{\hbar^2}{2m}\left(\dfrac{\partial^2\psi}{\partial x^2} + \dfrac{\partial^2\psi}{\partial y^2} + \dfrac{\partial^2\psi}{\partial z^2}\right) + U\psi = E\psi$	36-19
For an isolated atom, the solutions can be written as products of functions of r, θ, and ϕ separately	$\psi(r, \theta, \phi) = R(r)f(\theta)g(\phi)$	36-21
Quantum numbers in spherical coordinates		
Principal quantum number	$n = 1, 2, 3, \ldots$	
Orbital quantum number	$\ell = 0, 1, 2, 3, \ldots, n - 1$	
Magnetic quantum number	$m_\ell = -\ell, (-\ell + 1), \ldots, 0, \ldots, (\ell + 1), \ell$	36-22
Orbital angular momentum	$L = \sqrt{\ell(\ell + 1)}\hbar$	36-23

TOPIC	RELEVANT EQUATIONS AND REMARKS
z component of orbital angular momentum	$L_z = m_\ell \hbar$ 36-24
3. Quantum Theory of the Hydrogen Atom	
Energy levels for hydrogen-like atoms (same as for the Bohr model)	$E_n = -Z^2 \frac{E_0}{n^2} \quad n = 1, 2, 3, \ldots$ 36-26 where $E_0 = -\frac{mk^2e^4}{2\hbar^2} = 13.6\text{ eV}$ 36-27
Wavelengths emitted by the hydrogen atom (same as for Bohr model)	$\lambda = \frac{c}{f} = \frac{hc}{E_i - E_f} = \frac{1240\text{ eV}\cdot\text{nm}}{E_i - E_f}$ 36-17, 36-18
Wave functions	
The ground state	$\psi_{100} = C_{100}e^{-Zr/a_0} = \frac{1}{\sqrt{\pi}}\left(\frac{Z}{a_0}\right)^{3/2} e^{-Zr/a_0}$ 36-30, 36-32
The first excited state	$\psi_{200} = C_{200}\left(2 - \frac{Zr}{a_0}\right)e^{-Zr/2a_0}$ 36-35 $\psi_{210} = C_{210}\frac{Zr}{a_0}e^{-Zr/2a_0}\cos\theta$ 36-36 $\psi_{21\pm1} = C_{21\pm1}\frac{Zr}{a_0}e^{-Zr/2a_0}\sin\theta\, e^{\pm i\phi}$ 36-37
Probability densities	For $\ell = 0$, $\lvert\psi\rvert^2$ is spherically symmetric. For $\ell > 0$, $\lvert\psi\rvert^2$ depends on the angle θ.
Radial probability density	$P(r) = 4\pi r^2 \lvert\psi\rvert^2$ 36-33 The radial probability density is maximum at the distances corresponding roughly to the Bohr orbits.
4. The Spin–Orbit Effect and Fine Structure	The total angular momentum of an electron in an atom is a combination of the orbital angular momentum and spin angular momentum. It is characterized by the quantum number j, which can be either $\lvert\ell - \frac{1}{2}\rvert$ or $\ell + \frac{1}{2}$. Because of the interaction of the orbital and spin magnetic moments, the state $j = \lvert\ell - \frac{1}{2}\rvert$ has lower energy than the state $j = \ell + \frac{1}{2}$, for $\ell > 0$. This small splitting of the energy states gives rise to a small splitting of the spectral lines called fine structure.
5. The Periodic Table	An atom of an element has Z electrons, where Z is the atomic number of the element. For an atom in the ground state, the electrons are in those states that will give the lowest energy consistent with the Pauli exclusion principle. The state of an atom is described by its electron configuration, which gives the values of n and ℓ for each electron. The ℓ values are specified by a code: s p d f g h ℓ value 0 1 2 3 4 5
Pauli exclusion principle	No two electrons in an atom can have the same set of values for the quantum numbers n, ℓ, m_ℓ, and m_s.
6. Atomic Spectra	Atomic spectra include optical spectra and X-ray spectra. Optical spectra result from transitions between energy levels of a single valence electron moving in the field of the nucleus and core electrons of the atom. Characteristic X-ray spectra result from the excitation of a core electron and the subsequent filling of the vacancy by other electrons in the atom.
Selection rules	Transitions between energy states with the emission of a photon are governed by the following selection rules $\Delta m_\ell = 0 \quad \text{or} \quad \Delta m_\ell = \pm 1$ 36-28 $\Delta\ell = \pm 1$

Answers to Concept Checks

36-1 (*a*), (*c*), and (*d*)

Answers to Practice Problems

36-1 91.2 nm

36-2 $-4, -3, -2, -1, 0, 1, 2, 3, 4$

Problems

In a few problems, you are given more data than you actually need; in a few other problems, you are required to supply data from your general knowledge, outside sources, or informed estimate.

Interpret as significant all digits in numerical values that have trailing zeros and no decimal points.

- • Single-concept, single-step, relatively easy
- •• Intermediate-level, may require synthesis of concepts
- ••• Challenging
- SSM Solution is in the *Student Solutions Manual*
- Consecutive problems that are shaded are paired problems.

CONCEPTUAL PROBLEMS

1 • For the hydrogen atom, as n increases, does the spacing of adjacent energy levels on an energy-level diagram increase or decrease? SSM

2 • The energy of the ground state of doubly ionized lithium ($Z = 3$) is ______, where $E_0 = 13.6$ eV. (*a*) $-9E_0$, (*b*) $-3E_0$, (*c*) $-E_0/3$, (*d*) $-E_0/9$

3 • Bohr's quantum condition on electron orbits requires (*a*) that the orbital angular momentum of the electron about the hydrogen nucleus equals $n\hbar$, where n is an integer, (*b*) that no more than one electron occupy a given state, (*c*) that the electrons spiral into the nucleus while radiating electromagnetic waves, (*d*) that the energies of an electron in a hydrogen atom be equal to nE_0, where E_0 is a constant and n is an integer, (*e*) none of the above.

4 • According to the Bohr model, if an electron moves to a larger orbit, does the electron's total energy increase or decrease? Does the electron's kinetic energy increase or decrease?

5 • According to the Bohr model, the kinetic energy of the electron in the ground state of hydrogen is E_0, where $E_0 = 13.6$ eV. The kinetic energy of the electron in the state $n = 2$ is (*a*) $4E_0$, (*b*) $2E_0$, (*c*) $E_0/2$, (*d*) $E_0/4$.

6 • According to the Bohr model, the radius of the $n = 1$ orbit in the hydrogen atom is $a_0 = 0.053$ nm. What is the radius of the $n = 5$ orbit? (*a*) $25a_0$, (*b*) $5a_0$, (*c*) a_0, (*d*) $a_0/5$, (*e*) $a_0/25$

7 • For the principal quantum number $n = 4$, how many different values can the orbital quantum number ℓ have? (*a*) 4, (*b*) 3, (*c*) 7, (*d*) 16, (*e*) 25 SSM

8 • For the principal quantum number $n = 4$, how many different combinations of ℓ and m_ℓ can occur? (*a*) 4, (*b*) 3, (*c*) 7, (*d*) 16, (*e*) 25

9 •• Why is the energy of the 3s state considerably lower than the energy of the 3p state for sodium, whereas in hydrogen 3s and 3p states have essentially the same energy? SSM

10 • The d state of an electron configuration corresponds to (*a*) $n = 2$, (*b*) $\ell = 3$, (*c*) $\ell = 2$, (*d*) $n = 3$, (*e*) $\ell = 0$.

11 •• Why are three quantum numbers inadequate to describe the states of the electrons in atoms that have more than one electron?

12 •• Group the following six atoms—potassium, calcium, titanium, chromium, manganese, and copper—according to their ground-state electron configurations for the $n = 4$ shell.

13 • What element has the electron configuration (*a*) $1s^22s^22p^63s^23p^3$ and (*b*) $1s^22s^22p^63s^23p^63d^54s^1$?

14 • For the principal quantum number $n = 3$, what are the possible combinations of the quantum numbers ℓ and m_ℓ?

15 • An electron in the L shell means that the electron is represented by (*a*) $\ell = 0$, (*b*) $\ell = 1$, (*c*) $n = 1$, (*d*) $n = 2$, or (*e*) $m_\ell = 2$.

16 •• The Bohr model and the quantum-mechanical model of the hydrogen atom give the same results for the energy levels. Discuss the advantages and disadvantages of each model.

17 •• The Sommerfeld–Hosser displacement theorem states that the optical spectrum of any atom is very similar to the spectrum of the singly charged positive ion of the element immediately following it in the periodic table. Discuss why this theorem is accurate.

18 • Using the triplet of numbers (n, ℓ, m_ℓ) to represent an electron that has the principal quantum number n, orbital quantum number ℓ, and magnetic quantum number m_ℓ, which of the following transitions is allowed? (*a*) $(5, 2, 2) \rightarrow (3, 1, -2)$, (*b*) $(2, 1, 0) \rightarrow (3, 0, 0)$, (*c*) $(4, 3, -2) \rightarrow (3, 2, -1)$, (*d*) $(1, 0, 0) \rightarrow (2, 1, -1)$, (*e*) $(2, 1, 0) \rightarrow (3, 1, 0)$

19 •• The Ritz combination principle states that for any atom, one can find different spectral lines λ_1, λ_2, λ_3, and λ_4, so that $1/\lambda_1 + 1/\lambda_2 = 1/\lambda_3 + 1/\lambda_4$. Show why this is true using an energy-level diagram. SSM

ESTIMATION AND APPROXIMATION

20 •• (*a*) We can define a thermal de Broglie wavelength λ_T for an atom in a gas at temperature T as being the de Broglie wavelength for an atom moving at the rms speed appropriate to that temperature. (The average kinetic energy of an atom is equal to $\frac{3}{2}kT$, where k is the Boltzmann constant. Use this value to calculate the rms speed of the atoms.) Show that $\lambda_T = h/\sqrt{3mkT}$ where m is the mass of the atom. (*b*) Cooled atoms can form a Bose *condensate* (a new state of matter) when their thermal de Broglie wavelength becomes larger than the average interatomic spacing. From this criterion, estimate the temperature needed to create a Bose condensate in a gas of ^{85}Rb atoms whose number density is 10^{12} atoms/cm^3.

21 •• In laser cooling and trapping, a beam of atoms traveling in one direction is slowed by interaction with an intense laser beam in the opposite direction. The photons scatter off the atoms by resonance absorption, a process by which the incident photon is

absorbed by the atom, and a short time later a photon of equal energy is emitted in a random direction. The net result of a single such scattering event is a transfer of momentum to the atom in a direction opposite to the motion of the atom, followed by a second transfer of momentum to the atom in a random direction. Thus, during photon absorption the atom loses speed, but during photon emission the change in speed of the atom is, on average, zero (because the directions of the emitted photons are random). An analogy often made to this process is that of slowing down a bowling ball by bouncing ping-pong balls off of it. (*a*) Given that the typical photon energy used in these experiments is about 1 eV, and that the typical kinetic energy of an atom in the beam is the typical kinetic energy of the atoms in a gas that has a temperature of about 500 K (a typical temperature for an oven that produces an atomic beam), estimate the number of photon–atom collisions that are required to bring an atom to rest. (The average kinetic energy of an atom is equal to $\frac{3}{2}kT$, where k is the Boltzmann constant and T is the temperature. Use this to estimate the speed of the atoms.) (*b*) Compare the Part-(*a*) result with the number of ping-pong ball–bowling ball collisions that are required to bring the bowling ball to rest. (Assume the typical speed of the incident ping-pong balls are all equal to the initial speed of the bowling ball.) (*c*) ^{85}Rb is a type of atom often used during cooling experiments. The wavelength of the light resonant with the cooling transition of the atoms is $\lambda = 780.24$ nm. Estimate the number of photons needed to slow down an ^{85}Rb atom from a typical thermal velocity of 300 m/s to a stop. SSM

THE BOHR MODEL OF THE HYDROGEN ATOM

22 • The first Bohr radius is given by $a_0 = \hbar^2/(mke^2) = 0.0529$ nm (Equation 36-12). Use the known values of the constants in the equation to show that a_0 is equal to 0.0529 nm.

23 • The longest wavelength in the Lyman series for the hydrogen atom was calculated in Example 36-2. Find the wavelengths for the transitions (*a*) $n_i = 3$ to $n_f = 1$ and (*b*) $n_i = 4$ to $n_f = 1$.

24 • Find the photon energies for the three longest wavelengths in the Balmer series for the hydrogen atom, and calculate the three wavelengths.

25 •• Find the photon energy and wavelength for the series limit (shortest wavelength) in the Paschen series ($n_f = 3$) for the hydrogen atom. (*b*) Calculate the wavelength for the three longest wavelengths in Paschen series.

26 •• (*a*) Find the photon energy and wavelength for the series limit (shortest wavelength) in the Brackett series ($n_f = 4$) for the hydrogen atom. (*b*) Calculate the wavelength for the three longest wavelengths in Brackett series.

27 ••• In the center-of-mass reference frame of a hydrogen atom, the electron and nucleus have momenta that have equal magnitudes p and opposite directions. (*a*) Using the Bohr model, show that the total kinetic energy of the electron and nucleus can be written $K = p^2/(2\mu)$, where $\mu = m_e M/(M + m_e)$ is called the reduced mass, m_e is the mass of the electron, and M is the mass of the nucleus. (*b*) For the equations for the Bohr model of the atom, the motion of the nucleus can be taken into account by replacing the mass of the electron with the reduced mass. Use Equation 36-14 to calculate the Rydberg constant for a hydrogen atom that has a nucleus of mass $M = m_p$. Find the approximate value of the Rydberg constant by letting M go to infinity in the reduced mass formula. To how many figures does this approximate value agree with the actual value? (*c*) Find the percentage correction for the ground-state energy of the hydrogen atom by using the reduced mass in Equation 36-16.

Note: In general, the reduced mass for a two-body problem with masses m_1 and m_2 is given by

$$\mu = \frac{m_1 m_2}{m_1 + m_2}$$ SSM

28 •• The Pickering series of the spectrum of He^+ (singly ionized helium) consists of spectral lines due to transitions to the $n = 4$ state of He^+. Every other line of the Pickering series is very close to a spectral line in the Balmer series for hydrogen transitions to $n = 2$. (*a*) Show that this statement is accurate. (*b*) Calculate the wavelength of the photon during a transition from the $n = 6$ level to the $n = 4$ level of He^+, and show that it corresponds to one of the Balmer series lines.

QUANTUM NUMBERS IN SPHERICAL COORDINATES

29 • For an electron in an atom that has an orbital quantum number $\ell = 1$, find (*a*) the magnitude of the angular momentum L and (*b*) the possible values of the magnetic quantum number m_ℓ. (*c*) Draw a vector diagram to scale showing the possible orientations of $\vec{L}$ relative to the $+z$ direction.

30 • For an electron in an atom that has an orbital quantum number $\ell = 3$, find (*a*) the magnitude of the angular momentum L and (*b*) the possible values of m_ℓ. (*c*) Draw a vector diagram to scale showing the possible orientations of $\vec{L}$ relative to the $+z$ direction.

31 • An electron in an atom has principal quantum number $n = 3$. (*a*) What are the possible values of ℓ? (*b*) What are the possible combinations of ℓ and m_ℓ? (*c*) Using the fact that there are two quantum states for each combination of ℓ and m_ℓ because of electron spin, find the total number of electron states for $n = 3$.

32 • In an atom, find the total number of electron states that have (*a*) $n = 4$ and (*b*) $n = 2$. (See Problem 31.)

33 •• Find the minimum value of the angle θ between $\vec{L}$ and the $+z$ direction for an electron in an atom that has (*a*) $\ell = 1$, (*b*) $\ell = 4$, and (*c*) $\ell = 50$. SSM

34 •• What are the possible values of n and m_ℓ for an electron in an atom that has (*a*) $\ell = 3$, (*b*) $\ell = 4$, and (*c*) $\ell = 0$?

35 •• For an electron in an atom that is in an $\ell = 2$ state, find (*a*) the magnitude of the angular momentum squared L^2, (*b*) the maximum value of L_z^2, and (*c*) the smallest value of $L_x^2 + L_y^2$.

QUANTUM THEORY OF THE HYDROGEN ATOM

36 • For the ground state of the hydrogen atom, find the values of (*a*) $\psi(r)$ at $r = a_0$, (*b*) $\psi^2(r)$ at $r = a_0$, and (*c*) the radial probability density $P(r)$ at $r = a_0$. Give your answers in terms of a_0.

37 • (*a*) If electron spin is not included, how many different wave functions are there corresponding to the first excited energy level $n = 2$ for a hydrogen atom? (*b*) Specify the quantum numbers for each of these wave functions. SSM

38 •• For the ground state of the hydrogen atom, calculate the probability of finding the electron in the region between r and $r + \Delta r$, where $\Delta r = 0.03a_0$ and (*a*) $r = a_0$ and (*b*) $r = 2a_0$.

39 •• The value of the constant C_{200} in the equation

$$\psi_{200} = C_{200}\left(2 - \frac{Zr}{a_0}\right)e^{-Zr/(2a_0)}$$

(Equation 36-35) is given by

$$C_{200} = \frac{1}{4\sqrt{2\pi}}\left(\frac{Z}{a_0}\right)^{3/2}$$

Find the values of (*a*) $\psi(r)$ at $r = a_0$, (*b*) $\psi^2(r)$ at $r = a_0$, and (*c*) the radial probability density $P(r)$ at $r = a_0$ for the state $n = 2$, $\ell = 0$, and $m_\ell = 0$ of a hydrogen atom. Give your answers in terms of a_0.

40 ••• Show that the radial probability density for the $n = 2$, $\ell = 1$, and $m_\ell = 0$ state of a one-electron atom can be written as $P(r) = A\cos^2\theta\, r^4 e^{-Zr/a_0}$, where A is a constant.

41 ••• Calculate the probability of finding the electron in the region between r and $r + \Delta r$, where $\Delta r = 0.02a_0$ and (*a*) $r = a_0$ and (*b*) $r = 2a_0$ for the state $n = 2$, $\ell = 0$, and $m_\ell = 0$ in hydrogen. (See Problem 39 for the value of C_{200}.)

42 •• Show that the ground-state hydrogen atom wave function $\psi_{100} = \pi^{-1/2}(Z/a_0)^{3/2}e^{-Zr/a_0}$ (Equation 36-32) is a solution to Schrödinger's equation in spherical coordinates:

$$\frac{-\hbar}{2mr^2}\left\{\frac{\partial}{\partial r}\left(r^2\frac{\partial\psi}{\partial r}\right) + \left[\frac{1}{\sin\theta}\frac{\partial}{\partial\theta}\left(\sin\theta\frac{\partial\psi}{\partial\theta}\right) + \frac{1}{\sin^2\theta}\frac{\partial^2\psi}{\partial\phi^2}\right]\right\} + U(r)\psi = E\psi$$

where $U(r) = kZe^2/r$ (Equation 36-25).

43 •• Show by dimensional analysis that the expression for the hydrogen atom ground-state energy given by $E_0 = \frac{1}{2}mk^2e^4/\hbar^2$ (Equation 36-27) has the dimensions of energy.

44 •• By dimensional analysis, show that the expression for the first Bohr radius given by $a_0 = \hbar^2/(mke^2)$ (Equation 36-12) has the dimensions of length.

45 •• The radial probability distribution function for a one-electron atom in its ground state can be written $P(r) = Cr^2e^{-2Zr/a_0}$, where C is a constant. Show that $P(r)$ has its maximum value at $r = a_0/Z$.

46 ••• Show that the number of states in the hydrogen atom for a given n is $2n^2$.

47 ••• Calculate the probability that the electron in the ground state of a hydrogen atom is in the region $0 < r < a_0$.

THE SPIN-ORBIT EFFECT AND FINE STRUCTURE

48 • The potential energy of a magnetic moment in an external magnetic field is given by $U = -\vec{\mu}\cdot\vec{B}$. (*a*) Calculate the difference in energy between the two possible orientations of an electron in a magnetic field $\vec{B} = 1.50\,\text{T}\hat{k}$. (*b*) If the electrons are bombarded with photons of energy equal to that energy difference, "spin flip" transitions can be induced. Find the wavelength of the photons needed for such transitions. This phenomenon is called *electron spin resonance*.

49 • The total angular momentum of a hydrogen atom in a certain excited state has the quantum number $j = \frac{1}{2}$. What can you say about the value of the orbital angular-momentum quantum number ℓ?

50 • A hydrogen atom is in the state $n = 3$, $\ell = 2$. What are the possible values of j?

51 • Using a scaled vector diagram, show how the orbital angular momentum $\vec{L}$ combines with the spin angular momentum $\vec{S}$ to produce the two possible values of total angular momentum $\vec{J}$ for the $\ell = 3$ state of the hydrogen atom.

THE PERIODIC TABLE

52 • The total number of states of a hydrogen atom that has principal quantum number $n = 4$ is (*a*) 4, (*b*) 16, (*c*) 32, (*d*) 36, (*e*) 48.

53 • How many of the eight electrons in an oxygen atom in the ground state are in a p state? (*a*) 0, (*b*) 2, (*c*) 4, (*d*) 6, (*e*) 8

54 • Write the ground-state electron configuration of (*a*) an atom of carbon and (*b*) an atom of oxygen.

55 • Give the possible values of the z component of the orbital angular momentum of (*a*) a d electron and (*b*) an f electron.

OPTICAL SPECTRA AND X-RAY SPECTRA

56 • The optical spectra of atoms that have two electrons in the same highest energy shell are similar, but they are quite different from the spectra of atoms that have just one electron in the highest energy shell because of the interaction of the two electrons. Group the elements according to similar spectra: lithium, beryllium, sodium, magnesium, potassium, calcium, chromium, nickel, cesium, and barium.

57 • Write down the possible electron configurations for the first excited state of (*a*) a hydrogen atom, (*b*) a sodium atom, and (*c*) a helium atom.

58 • Indicate which of the following atoms should have optical spectra similar to a hydrogen atom and which of the following atoms should have optical spectra similar to a helium atom: Li, Ca, Ti, Rb, Hg, Ag, Cd, Ba, Fr, and Ra.

59 • (*a*) Calculate the next two longest wavelengths in the K series (after the K_α line) of molybdenum. (*b*) What is the wavelength of the shortest wavelength in this series?

60 • The wavelength of the K_α line for a certain element is 0.3368 nm. What is the element?

61 • Calculate the wavelength of the K_α line in (*a*) a magnesium ($Z = 12$) atom and (*b*) a copper ($Z = 29$) atom.

GENERAL PROBLEMS

62 • What is the energy of the shortest wavelength photon emitted by the hydrogen atom?

63 • The wavelength of a spectral line of hydrogen is 97.254 nm. Identify the transition that results in this line, assuming that the transition is to the ground state.

64 •• The wavelength of a spectral line of hydrogen is 1093.8 nm. Identify the transition that results in this line.

65 •• Spectral lines of the following wavelengths are emitted by a singly ionized helium atom: 164 nm, 230.6 nm, and 541 nm. Identify the transitions that result in those spectral lines.

66 •• The combination of physical constants $\alpha = e^2k/\hbar c$, where k is the Coulomb constant, is known as the *fine-structure constant*. It appears in numerous relations in atomic physics. (*a*) Show that α is dimensionless. (*b*) Show that in the Bohr model of the hydrogen atom $v_n = c\alpha/n$, where v_n is the speed of the electron in the state of quantum number n.

67 •• The wavelengths of the photons emitted by a potassium atom corresponding to transitions from the $4P_{3/2}$ and $4P_{1/2}$ states to the ground state are 766.41 nm and 769.90 nm. (*a*) Calculate the energies of the photons in electron volts. (*b*) The difference in the energies of the photons equals the difference in energy ΔE between the $4P_{3/2}$ and $4P_{1/2}$ states in potassium. Calculate ΔE. (*c*) Estimate the magnetic field that the 4p electron in potassium experiences.

68 •• To observe the characteristic K lines of the X-ray spectrum, one of the $n = 1$ electrons must be ejected from the atom. This is generally accomplished by bombarding the target material with electrons of sufficient energy to eject this tightly bound electron. What is the minimum energy required to observe the K lines of (*a*) a tungsten atom, (*b*) a molybdenum atom, and (*c*) a copper atom?

69 •• We are often interested in finding the quantity ke^2/r in electron volts when r is given in nanometers. Show that $ke^2 = 1.44\ \text{eV}\cdot\text{nm}$. SSM

70 •• The *positron* is a particle that has the same mass as the electron and carries a charge equal to $+e$. *Positronium* is a bound state of an electron–positron combination. (*a*) Calculate the energies of the five lowest energy states of positronium using the reduced mass, as given in Problem 27. (*b*) Do transitions between any of the levels found in Part (*a*) fall in the visible range of wavelengths? If so, which transitions are they?

71 • In 1947, Lamb and Retherford showed that there is a very small energy difference between the $2S_{1/2}$ and the $2P_{1/2}$ states of the hydrogen atom. They measured this difference essentially by causing transitions between the two states using very long wavelength electromagnetic radiation. The energy difference (the Lamb shift) is 4.372×10^{-6} eV and is explained by quantum electrodynamics as being due to fluctuations in the energy level of the vacuum. (*a*) What is the frequency of a photon whose energy is equal to the Lamb shift energy? (*b*) What is the wavelength of that photon? In what spectral region does it belong?

72 • A Rydberg atom is one in which an electron is in a *very* high excited state ($n \approx 40$ or higher). Such atoms are useful for experiments that probe the transition from quantum-mechanical behavior to classical. Furthermore, these excited states have extremely long lifetimes (i.e., the electron will stay in this high excited state for a very long time). A hydrogen atom is in the $n = 45$ state. (*a*) What is the ionization energy of the atom when it is in that state? (*b*) What is the energy level separation (in electron volts) between that state and the $n = 44$ state? (*c*) What is the wavelength of a photon resonant with the transition between these two states? (*d*) What is the radius of the atom when it is in the $n = 45$ state?

73 •• The deuteron, the nucleus of deuterium (heavy hydrogen), was first recognized from the spectrum of hydrogen. The deuteron has a mass that is approximately twice the mass of the proton. (*a*) Calculate the Rydberg constant for hydrogen and for deuterium using the reduced mass as given in Problem 27. (*b*) Using the result obtained in Part (*a*), determine the difference between the longest wavelength Balmer line of hydrogen (protium) and the longest wavelength Balmer line of deuterium.

74 •• The muonium atom is a hydrogen atom that has the electron replaced by a μ^- particle. The μ^- has a mass 207 times as great as the electron. (*a*) Calculate the energies of the five lowest energy levels of muonium using the reduced mass as given in Problem 27. (*b*) Do transitions between any of the levels found in Part (*a*) fall in the visible range of wavelengths (for example, between $\lambda = 700$ nm and $\lambda = 400$ nm)? If so, which transitions are they?

75 •• The triton, a nucleus consisting of a proton and two neutrons, is unstable and has a half-life of approximately 12 years. Tritium is an atom consisting of an electron and a triton. (*a*) Calculate the Rydberg constant of tritium using the reduced mass as given in Problem 27. (*b*) Determine the difference between the longest wavelength of the Balmer lines of tritium and the longest wavelength of the Balmer lines of deuterium (see Problem 73). In addition, (*c*) determine the difference between the longest wavelength of the Balmer lines of tritium and the longest wavelength of the Balmer lines of hydrogen (protium).

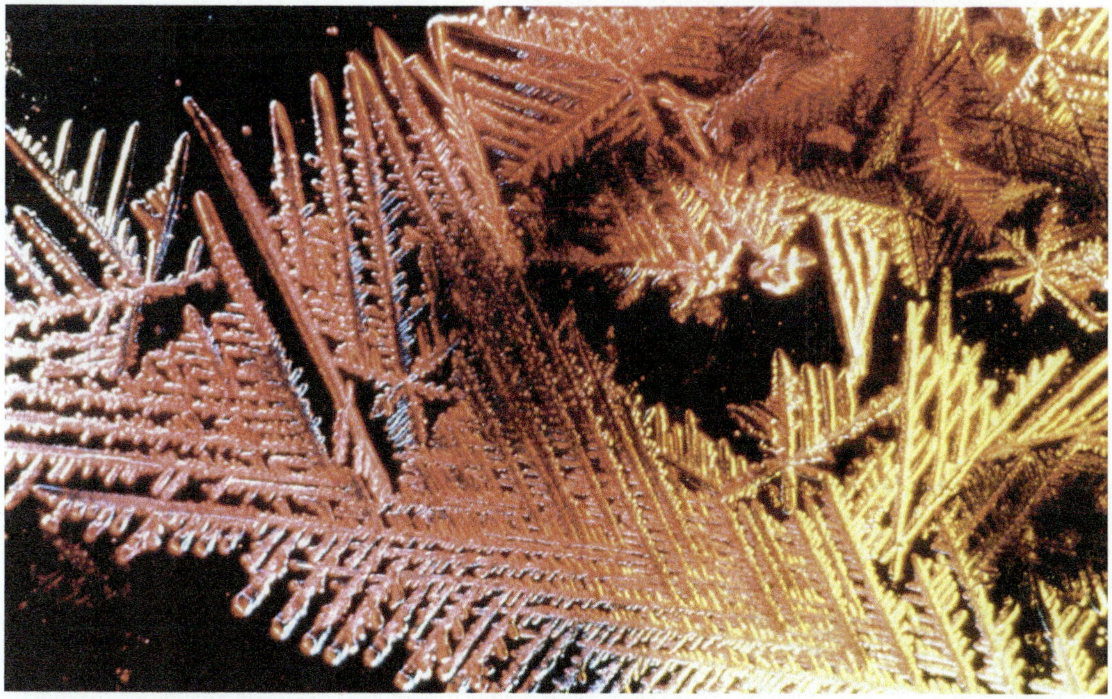

CHAPTER 37

A MICROGRAPH OF SODIUM FLUORIDE CRYSTALS. SODIUM FLUORIDE IS OFTEN ADDED TO PUBLIC WATER SUPPLIES AS A TOOTH-DECAY PREVENTATIVE. *(National Institutes of Health/Photo Researchers.)*

?
How much energy is needed to form sodium fluoride? (See Example 37-1).

Molecules

Most atoms bond together to form molecules or solids. Molecules may exist as separate entities, as in gaseous O_2 or N_2, or they may bond together to form liquids or solids. A molecule is the smallest constituent of a substance that retains its chemical properties.

In this chapter, we use our understanding of quantum mechanics to discuss bonding and the energy levels and spectra of diatomic molecules. Much of our discussion will be qualitative because, as in atomic physics, the quantum-mechanical calculations are very difficult.

37-1 BONDING

Consider a hydrogen molecule (H_2). We can think of H_2 either as two H atoms joined together or as a quantum-mechanical system of two protons and two electrons. The latter picture is more useful in this case because neither of the electrons in the H_2 molecule is confined to the region surrounding either one of the two protons. Instead, each electron is equally shared by both protons. For more complicated molecules, however, an intermediate picture is useful. For example, the fluorine molecule F_2 consists of 18 protons and 18 electrons, but only two of the electrons take part in the bonding. We therefore can consider this molecule as two F^+ ions and two electrons that belong to the molecule as a whole. The molecular wave functions for the bonding electrons are called **molecular orbitals.** In many cases, these molecular wave functions can be constructed from combinations of the atomic wave functions with which we are familiar.

The two principal types of bonds responsible for the formation of solids and molecules are the ionic bond and the covalent bond. Other types of bonds that are important in the bonding of liquids and solids are van der Waals bonds, metallic bonds, and hydrogen bonds. In many cases, bonding is a mixture of these mechanisms.

THE IONIC BOND

The simplest type of bond is the **ionic bond,** which is found in salts such as sodium chloride (NaCl). The sodium atom has one 3s electron outside a stable ten-electron core. The first ionization energy of sodium is the energy needed to remove the 3s electron from an isolated sodium atom. This energy is just 5.14 eV (see Figure 36-18). The removal of this electron results in an isolated positive ion that has its $n = 1$ and $n = 2$ electron shells filled. A chlorine atom has 17 electrons, and so is one electron short of having its first three shells filled. A measure of the energy released when an isolated atom gains one electron is called its **electron affinity;** a chlorine atom releases 3.62 eV of energy when it acquires an electron to form a Cl^- ion. Thus, the chlorine atom is said to have an electron affinity of -3.62 eV. The acquisition of one electron by chlorine results in a negative ion that has a filled outer electron shell. Thus, the formation of a Na^+ ion and a Cl^- ion by the donation of one electron of sodium to chlorine requires only 5.14 eV $-$ 3.62 eV $=$ 1.52 eV at infinite separation. The electrostatic potential energy U_e of the two ions when they are a distance r apart is $-ke^2/r$. When the separation of the ions is less than approximately 0.95 nm, the negative potential energy of attraction is of greater magnitude than the 1.52 eV of energy needed to create the ions. Thus, at separation distances less than 0.95 nm, it is energetically favorable (the total energy of the system is reduced) for the sodium atom to donate an electron to the chlorine atom to form NaCl.

Because the electrostatic attraction increases as the ions get closer together, it might seem that equilibrium could not exist. However, when the separation of the ions is very small, there is a strong repulsion that can be described by quantum mechanics and the exclusion principle. This repulsion is also responsible for the repulsion of the atoms in all molecules and ions (except H_2).* We can understand it qualitatively as follows. When the ions are very far apart, the probability distribution for a core electron in one of the ions does not overlap the probability distribution of any electron in the other ion. We can distinguish the electrons by the ion to which they belong. This means that electrons in the two ions can have the same quantum numbers because they occupy different regions of space. However, as the distance between the ions decreases, the probability distributions of the core electrons begin to overlap; that is, the electrons in the two ions begin to occupy the same region of space. Some of these electrons must go into higher energy quantum states as described by the exclusion principle.† But energy is required to shift the electrons into higher energy quantum states. This increase in energy when the ions are pushed closer together is equivalent to the repulsion energy of the ions. It is not a sudden process. The energy states of the electrons change gradually as the ions are brought together. A sketch of the potential energy $U(r)$ of the Na^+ and Cl^- ions versus separation distance r is shown in Figure 37-1. The energy is lowest at an

---

* In H_2, the repulsion is simply that of the two positively charged protons.

† Recall from our discussion in Chapter 35 that the exclusion principle is related to the fact that the wave function for two identical electrons is antisymmetric on the exchange of the electrons and that an antisymmetric wave function for two electrons with the same quantum numbers is zero if the space coordinates of the electrons are the same.

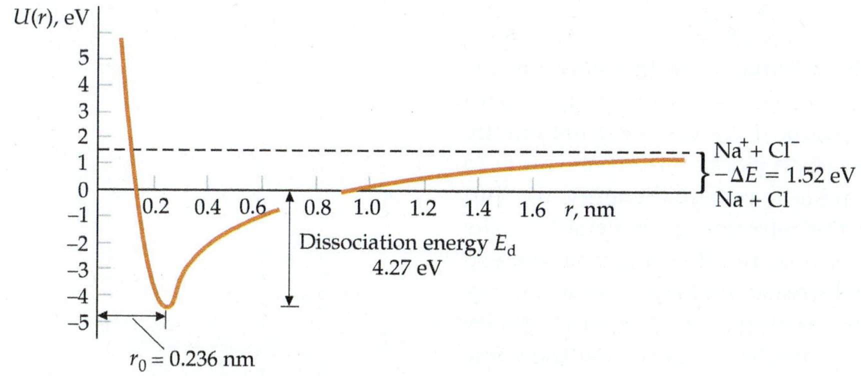

FIGURE 37-1 Potential energy for Na^+ and Cl^- ions as a function of separation distance r. The energy at infinite separation is chosen to be 1.52 eV, corresponding to the energy $-\Delta E$ needed to form the ions from atoms. The minimum energy is at the equilibrium separation r_0 = 0.236 nm for the ions.

equilibrium separation r_0 of approximately 0.236 nm. At smaller separations, the energy increases steeply. The energy required to separate the ions and form sodium and chlorine atoms is called the **dissociation energy** E_d, which is approximately 4.27 eV for NaCl.

The equilibrium separation distance of the gaseous NaCl, which can be obtained by evaporating solid NaCl, is 0.236 nm. Normally, NaCl exists as a solid in a cubic crystal structure, where the Na^+ and Cl^- ions are at the alternate corners of a cube. The separation of the Na^+ and Cl^- ions in a crystal is approximately 0.28 nm, which is somewhat larger than the 0.236 nm separation for the gaseous NaCl. Because of the presence of neighboring ions of opposite charge, the electrostatic energy per ion pair is lower when the ions are in a crystal.

Example 37-1 The Energy of Sodium Fluoride

The electron affinity of fluorine is −3.40 eV and the equilibrium separation of sodium fluoride (NaF) is 0.193 nm. (*a*) How much energy is needed to form Na^+ and F^- ions from sodium and fluorine atoms? (*b*) What is the electrostatic potential energy of the Na^+ and F^- ions at their equilibrium separation? (*c*) The dissociation energy of NaF is 5.38 eV. What is the energy due to repulsion of the ions at the equilibrium separation?

PICTURE (*a*) The energy ΔE needed to form Na^+ and F^- ions from the sodium and fluorine atoms is the sum of the first ionization energy of sodium (5.14 eV) and the electron affinity of fluorine. (*b*) The electrostatic potential energy, where $U = 0$ at infinity, is $U_e = -ke^2/r$. (*c*) If we choose the potential energy at infinity to be ΔE, the total potential energy is $U_{tot} = U_e + \Delta E + U_{rep}$, where U_{rep} is the energy of repulsion, which is found by setting the dissociation energy equal to $-U_{tot}$.

SOLVE

(*a*) Calculate the energy needed to form Na^+ and F^- ions from the sodium and fluorine atoms (see the Picture section):

$$\Delta E = 5.14\text{ eV} - 3.40\text{ eV} = \boxed{1.74\text{ eV}}$$

(*b*) Calculate the electrostatic potential energy at the equilibrium separation of r = 0.193 nm:

$$U_e = -\frac{ke^2}{r}$$

$$= -\frac{(8.99\times10^9\text{ N}\cdot\text{m}^2/\text{C}^2)(1.60\times10^{-19}\text{ C})^2}{1.93\times10^{-10}\text{ m}}$$

$$= -1.19\times10^{-18}\text{ J} = \boxed{-7.45\text{ eV}}$$

(*c*) The dissociation energy equals the negative of the total potential energy:

$$E_d = -U_{tot} = -(U_e + \Delta E + U_{rep})$$

so

$$U_{rep} = -(E_d + \Delta E + U_e)$$

$$= -(5.38\text{ eV} + 1.74\text{ eV} - 7.45\text{ eV}) = \boxed{0.33\text{ eV}}$$

CHECK The Part-(*c*) result is greater than zero as expected.

THE COVALENT BOND

A completely different mechanism, the **covalent bond,** is responsible for the bonding of identical or similar atoms to form molecules such as gaseous hydrogen (H_2), nitrogen (N_2), and carbon monoxide (CO). If we calculate the energy needed to form H^+ and H^- ions by the transfer of an electron from one atom to the other and then add this energy to the electrostatic potential energy, we find that there is no separation distance for which the total energy is negative. The bond thus cannot be ionic. Instead, the attraction of two hydrogen atoms can only be explained quantum-mechanically. The decrease in energy when two hydrogen atoms approach each other is due to the sharing of the two electrons by both atoms, which can be explained using the symmetry properties of the wave functions of electrons.

We can gain some insight into covalent bonding by considering a simple, one-dimensional quantum-mechanics problem of two identical finite square wells. We first consider a single electron that is equally likely to be in either well. Because the wells are identical, the probability distribution, which is proportional to $|\psi^2|$, must be symmetric about the midpoint between the wells. Then ψ must be either symmetric or antisymmetric with respect to the two wells. The two possibilities for the ground state are shown in Figure 37-2*a* for the case in which the wells are far apart and in Figure 37-2*b* for the case in which the wells are close together. An important feature of Figure 37-2*b* is that in the region between the wells the symmetric wave function is large and the antisymmetric wave function is small.

Now consider adding a second electron to the two wells. We saw in Section 6 of Chapter 35 that the wave functions for particles that obey the exclusion principle are antisymmetric on exchange of the particles. Thus, the total wave function for the two electrons must be antisymmetric on exchange of the electrons. Note that exchanging the electrons while keeping the wells in place is equivalent to keeping the electrons in place and exchanging the wells. The total wave function for two electrons can be written as a spatial expression and an expression for spin. So, an antisymmetric wave function can be the product of a symmetric spatial expression and an antisymmetric expression for spin or of a symmetric expression for spin and an antisymmetric spatial expression.

To understand the symmetry of the total wave function, we must therefore understand the symmetry of the expression for spin of the wave function. The spin of a single electron can have two possible values for its quantum number m_S: $m_s = +\frac{1}{2}$, which we call spin up, or $m_s = -\frac{1}{2}$, which we call spin down. We will use arrows to designate the spin wave function for a single electron: $\uparrow_1$ or $\uparrow_2$ for electron 1 or electron 2 that both are spin up and $\downarrow_1$ or $\downarrow_2$ for electron 1 or electron 2 that are both spin down. The total spin quantum number for two electrons can be $S = 1$, where $m_S = +1, 0,$ or -1; or $S = 0$, where $m_S = 0$. We use $\phi_{S\,m_S}$ to denote the spin wave function for two electrons. The spin state $\phi_{1\,+1}$, corresponding to $S = 1$ and $m_S = +1$, can be written

$$\phi_{1\,+1} = \uparrow_1\uparrow_2 \qquad S = 1, m_S = +1 \qquad 37\text{-}1$$

Similarly, the spin state for $S = 1, m_S = -1$ is

$$\phi_{1\,-1} = \downarrow_1\downarrow_2 \qquad S = 1, m_S = -1 \qquad 37\text{-}2$$

Note that both of these states are symmetric upon exchange of the electrons. The spin state corresponding to $S = 1$ and $m_S = 0$ is not quite so obvious. It turns out to be proportional to

$$\phi_{10} = \uparrow_1\downarrow_2 + \uparrow_2\downarrow_1 \qquad S = 1, m_S = 0 \qquad 37\text{-}3$$

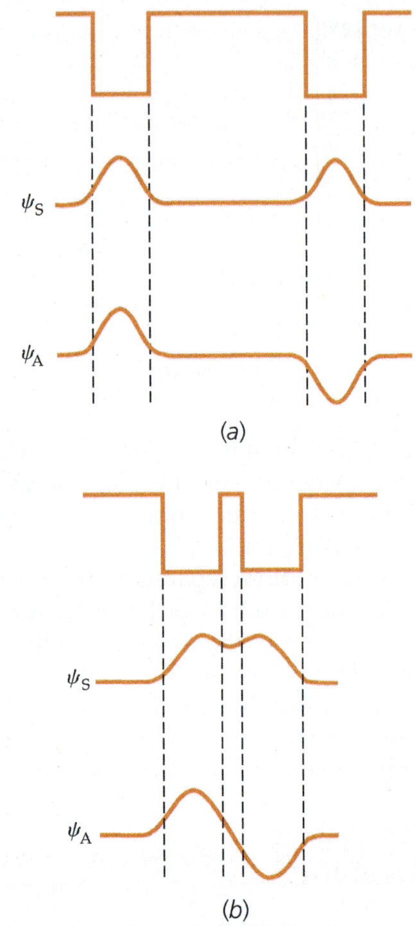

FIGURE 37-2 (*a*) Two square wells far apart. The electron wave function can be either symmetric (ψ_S) or antisymmetric (ψ_A) in space. The probability distributions and energies are the same for the two wave functions when the wells are far apart. (*b*) Two square wells that are close together. Between the wells, the antisymmetric space wave function is approximately zero, whereas the symmetric space wave function is quite large.

This spin state is also symmetric upon exchange of the electrons. The spin state for two electrons with antiparallel spins ($S = 0$) is

$$\phi_{00} = \uparrow_1\downarrow_2 - \uparrow_2\downarrow_1 \qquad S = 0, m_s = 0 \qquad 37\text{-}4$$

This spin state is antisymmetric upon exchange of electrons.

We thus have the important result that the spin part of the wave function is symmetric for parallel spins ($S = 1$) and antisymmetric for antiparallel spins ($S = 0$). Because the total wave function is the product of the spatial expression and the expression for spin, we have the following important result:

> For the total wave function of two electrons to be antisymmetric, the spatial part of the wave function must be antisymmetric for parallel spins ($S = 1$) and symmetric for antiparallel spins ($S = 0$).
>
> SPIN ALIGNMENT AND WAVE-FUNCTION SYMMETRY

We can now consider the problem of two hydrogen atoms. Figure 37-3*a* shows a spatially symmetric wave function ψ_S and a spatially antisymmetric wave function ψ_A for two hydrogen atoms that are far apart, and Figure 37-3*b* shows the same two wave functions for two hydrogen atoms that are close together. The squares of these two wave functions are shown in Figure 37-3*c*. Note that the probability distribution $|\psi|^2$ in the region between the protons is large for the symmetric wave function and small for the antisymmetric wave function. Thus, when the spatial part of the wave function is symmetric ($S = 0$), the electrons are often found in the region between the protons. The electron cloud, as shown in the upper part of Figure 37-3*c*, is concentrated in the space between the protons and the protons are bound together by this negatively charged cloud. Conversely, when the spatial part of the wave function is antisymmetric ($S = 1$), the electrons spend little time between the protons and the atoms do not bond together to form a molecule. In this case, the electron cloud is not concentrated in the space between the protons, as shown in the lower part of Figure 37-3*c*.

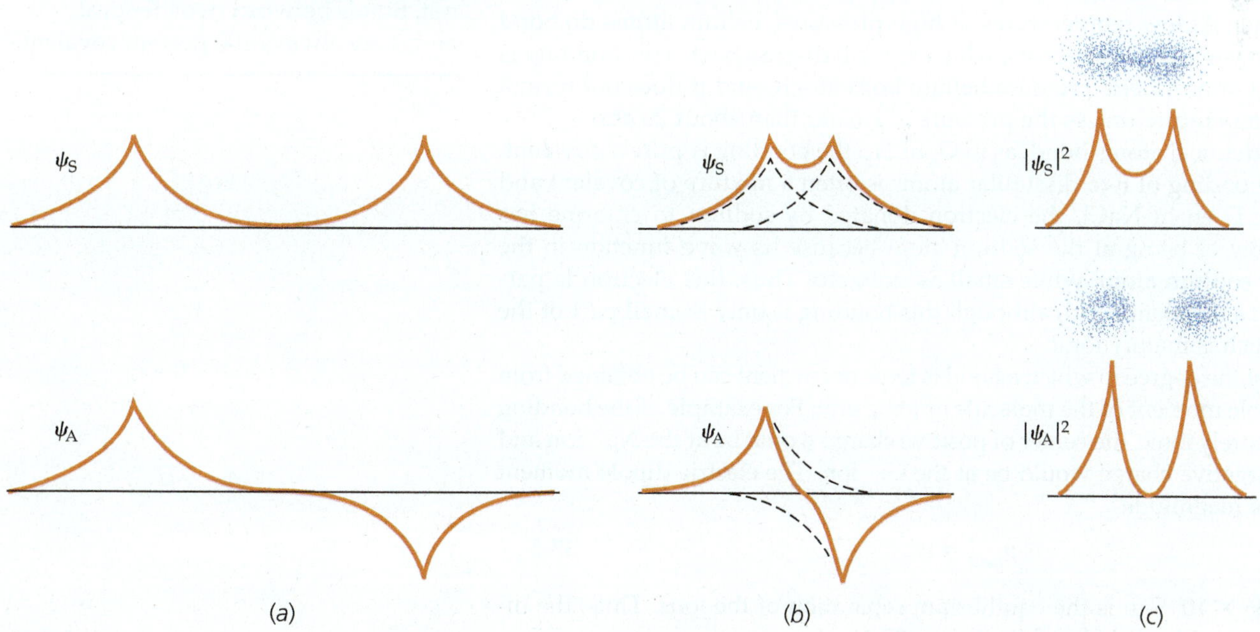

FIGURE 37-3 One-dimensional symmetric and antisymmetric wave functions for two hydrogen atoms (*a*) far apart and (*b*) close together. (*c*) Electron probability distributions ($|\psi|^2$) for the wave functions in Figure 37-3*b*. For the symmetric wave function, the electron charge density is large between the protons. This negative charge density holds the protons together in the hydrogen molecule H_2. For the antisymmetric wave function, the electron charge density is not large between the protons.

The total electrostatic potential energy for the H_2 molecule consists of the positive energy of repulsion of the two electrons and the negative potential energy of attraction of each electron for each proton. Figure 37-4 shows the electrostatic potential energy function U_S for two hydrogen atoms versus separation for the case in which the spatial part of the electron wave function is symmetric, and the electrostatic potential energy function U_A for the case in which the spatial part of the wave function is antisymmetric. We can see that the potential energy for the symmetric state is lower than the potential energy for the antisymmetric state and that the shape of the potential energy curve for the symmetric state is similar to the shape of the potential energy curve for ionic bonding (Figure 37-1). The equilibrium separation for H_2 is $r_0 = 0.074$ nm, and the binding energy is 4.52 eV. For the antisymmetric state, the potential energy is never negative and there is no bonding.

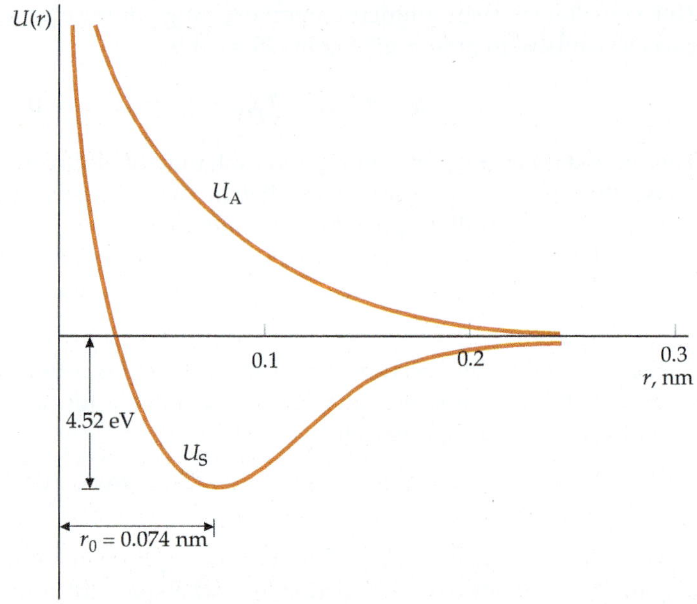

FIGURE 37-4 Potential energy versus separation for two hydrogen atoms. The curve labeled U_S is for a wave function that has a symmetric expression for the spatial part and the curve labeled U_A is for a wave function that has an antisymmetric expression for the spatial part.

We can now see why three hydrogen atoms do not bond to form H_3. If a third hydrogen atom is brought near an H_2 molecule, the third electron cannot be in a 1s state and have its spin antiparallel to the spin of both of the other electrons. If that electron is in an antisymmetric spatial state with respect to exchange with one of the electrons, the repulsion of this atom is greater than the attraction of the other. As the three atoms are pushed together, the third electron is, in effect, forced into a higher quantum-energy state according to the exclusion principle. The bond between two hydrogen atoms is called a **saturated bond** because there is no room for another electron. The two shared electrons essentially fill the 1s states of both atoms.

We can also see why two helium atoms do not normally bond together to form the He_2 molecule. There are no valence electrons that can be shared. The electrons in the filled shells are forced into higher energy states when the two atoms are brought together. At low temperatures or high pressures, helium atoms do bond together due to van der Waals forces, which we will discuss next. This bonding is so weak that at atmospheric pressure helium boils at 4 K, and it does not form a solid at any temperature unless the pressure is greater than about 20 atm.

! Don't think all bonds between atoms are partly ionic. They are not. Bonds between two identical atoms are always 100 percent covalent.

When two identical atoms bond, as in O_2 or N_2, the bonding is purely covalent. However, the bonding of two dissimilar atoms is often a mixture of covalent and ionic bonding. Even in NaCl, the electron donated by sodium to chlorine has some probability of being at the sodium atom because its wave function in the vicinity of the sodium atom, while small, is not zero. Thus, this electron is partially shared in a covalent bond, although this bonding is only a small part of the total bond, which is mainly ionic.

A measure of the degree to which a bond is ionic or covalent can be obtained from the electric dipole moment of the molecule or ionic unit. For example, if the bonding in NaCl were purely ionic, the center of positive charge would be at the Na^+ ion and the center of negative charge would be at the Cl^- ion. The electric dipole moment would have the magnitude

$$p_{\text{ionic}} = er_0 \qquad 37\text{-}5$$

where $r_0 = 2.36 \times 10^{-10}$ m is the equilibrium separation of the ions. Thus, the dipole moment of NaCl would be (from Figure 37-1)

$$\begin{aligned} p_{\text{ionic}} &= er_0 \\ &= (1.60 \times 10^{-19}\ \text{C})(2.36 \times 10^{-10}\ \text{m}) = 3.78 \times 10^{-29}\ \text{C}\cdot\text{m} \end{aligned}$$

The actual measured electric dipole moment of NaCl is

$$p_{\text{measured}} = 3.00 \times 10^{-29}\ \text{C}\cdot\text{m}$$

We can define the ratio of p_{measured} to p_{ionic} as the fractional amount of ionic bonding. For NaCl, this ratio is 3.00/3.78 = 0.79. Thus, the bonding in NaCl is about 79 percent ionic.

PRACTICE PROBLEM 37-1

The equilibrium separation of HCl is 0.128 nm and its measured electric dipole moment is $3.60 \times 10^{-30}\ \text{C}\cdot\text{m}$. What is the percentage of ionic bonding in HCl?

OTHER BONDING TYPES

The van der Waals Bond Any two separated molecules will be attracted to one another by electrostatic forces called *van der Waals forces.* So will any two atoms that do not form ionic or covalent bonds. The **van der Waals bonds** due to these forces are much weaker than the bonds already discussed. At high enough temperatures, these forces are not strong enough to overcome the motion of the atoms or molecules due to thermal energy. At sufficiently low temperatures, these motions become negligible and the van der Waals forces will cause virtually all substances to condense into a liquid and then a solid form.* The van der Waals forces arise from the interaction of the instantaneous electric dipole moments of the molecules.

Figure 37-5 shows how two polar molecules—molecules that have *permanent* electric dipole moments, such as H_2O—can bond. The electric field due to the dipole moment of one molecule orients the other molecule so that the two dipole moments attract. Nonpolar molecules also attract other nonpolar molecules by the van der Waals forces. Although nonpolar molecules have zero electric dipole moments on the average, they have instantaneous dipole moments that are generally not zero because of fluctuations in the positions of the charges. When two nonpolar molecules are near each other, the fluctuations in the instantaneous dipole moments tend to become correlated so as to produce attraction. This is illustrated in Figure 37-6.

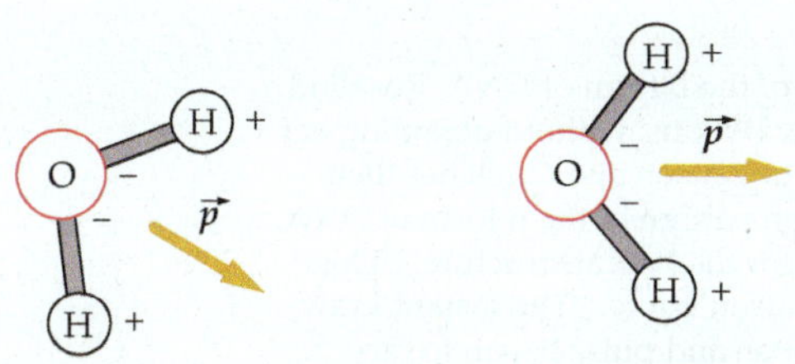

FIGURE 37-5 Bonding of H_2O molecules due to the attraction of the electric dipoles. The dipole moment of each molecule is indicated by $\vec{p}$. The electric field of one dipole orients the other dipole so the two dipole moments tend to be parallel. When the dipole moments are approximately parallel, the center of negative charge of one molecule is closer to the center of positive charge of the other molecule than it is to the center of the negative charge, so the molecules attract.

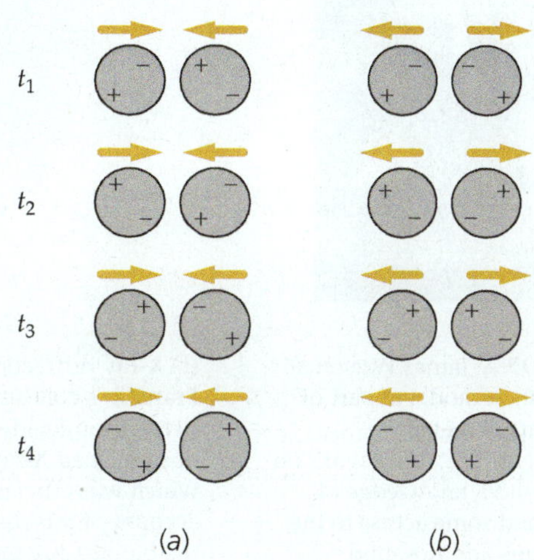

FIGURE 37-6 van der Waals attraction of molecules that have zero permanent dipole moments. (*a*) Possible orientations of instantaneous dipole moments at different times leading to attraction. (*b*) Possible orientations leading to repulsion. The electric field of the instantaneous dipole moment of one molecule tends to polarize the other molecule; thus the orientations leading to attraction (Figure 37-6*a*) are much more likely than those leading to repulsion (Figure 37-6*b*).

* Helium is the only element that does not solidify at any temperature at atmospheric pressure.

The hydrogen bond Another bonding mechanism of great importance is the hydrogen bond, which is formed by the sharing of a proton (the nucleus of the hydrogen atom) between two atoms, frequently two oxygen atoms. This sharing of a proton is similar to the sharing of electrons responsible for the covalent bond already discussed. It is facilitated by the small mass of the proton and by the absence of core electrons in hydrogen. The hydrogen bond often holds groups of molecules together and is responsible for the cross-linking that allows giant biological molecules and polymers to hold their fixed shapes. The well-known helical structure of DNA is due to hydrogen-bond linkages across turns of the helix (Figure 37-7).

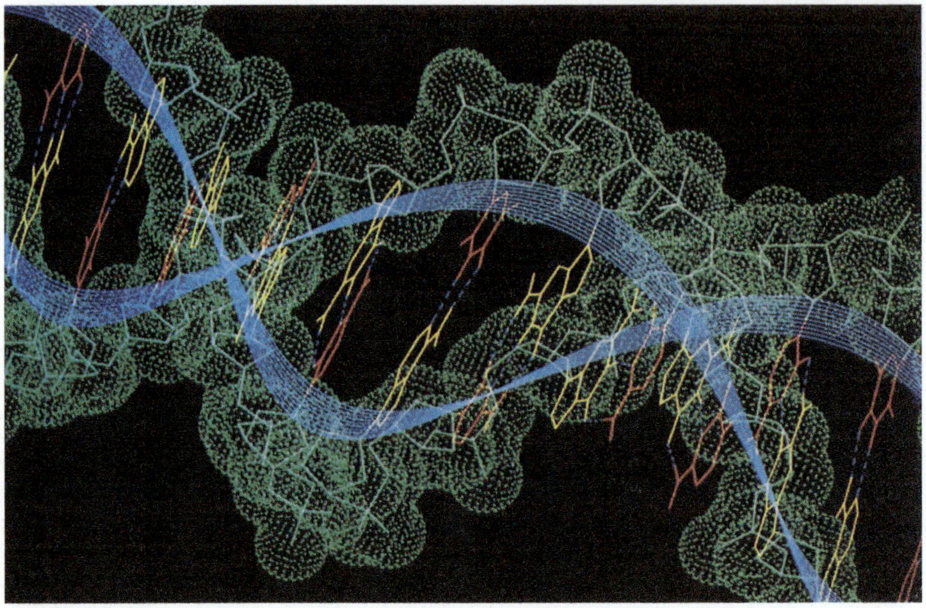

FIGURE 37-7 The DNA molecule. *(© Will and Demi McIntire/Photo Researchers.)*

(*a*) The discoverers of the structure of DNA. James Watson at left and Francis Crick are shown with their model of part of a DNA molecule in 1953. Crick and Watson met at the Cavendish Laboratory, Cambridge, in 1951. Their work on the structure of DNA was performed with a knowledge of Chargaff's ratios of the bases in DNA and some access to the X-ray crystallography of Maurice Wilkins and Rosalind Franklin at King's College London. Combining all of this work led to the deduction that DNA exists as a double helix, thus to its structure. Crick, Watson, and Wilkins shared the 1962 Nobel Prize for Physiology or Medicine; Franklin died from cancer in 1958. *((a) Norman Collection for the History of Molecular Biology.)*

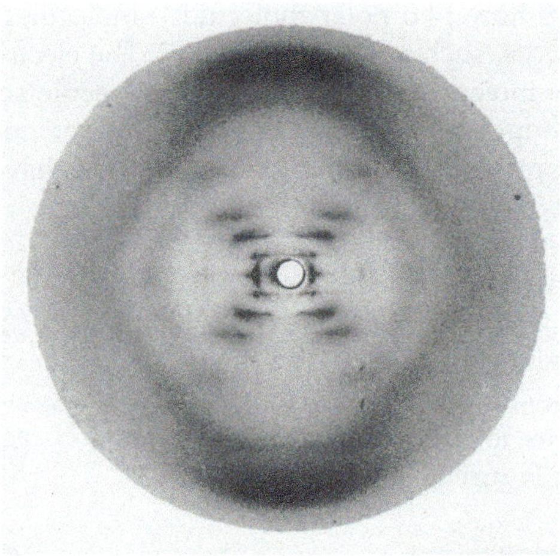

(*b*) X-ray diffraction pattern of the B form of DNA. Rosalind Franklin's colleague Maurice Wilkins, without obtaining her permission, made available to Watson and Crick her then unpublished X-ray diffraction pattern of the B form of DNA, which was crucial evidence for the helical structure. In his account of this discovery, Watson wrote: "The instant I saw the picture, my mouth fell open and pulse began to race. . . . The black cross of reflections which dominated the picture could arise only from a helical structure. . . . Mere inspection of the X-ray picture gave several of the vital helical parameters" (from Stent, Gunther, *The Double Helix*, New York: Norton, 1980). *((b) © A. Barrington Brown/Photo Researchers, NY.)*

The metallic bond In a metal, two atoms do not bond together by exchanging or sharing an electron to form a molecule. Instead, each valence electron is shared by many atoms. The bonding is thus distributed throughout the entire metal. A metal can be thought of as a lattice of positive ions held together by essentially free electrons that roam throughout the solid. In the quantum-mechanical picture, these free electrons form a cloud of negative charge density between the positively charged lattice ions that holds the ions together. In this respect, the metallic bond is somewhat similar to the covalent bond. However, with the metallic bond, there are far more than just two atoms involved, and the negative charge is distributed uniformly throughout the volume of the metal. The number of free electrons per lattice ion varies from metal to metal but is of the order of one free electron per ion.

* 37-2 POLYATOMIC MOLECULES

Molecules that have more than two atoms range from relatively simple molecules such as water, which has a molecular mass number of 18, to such giants as proteins and DNA, which can have molecular mass numbers of hundreds of thousands up to many millions. As with diatomic molecules, the structure of polyatomic molecules can be understood by applying basic quantum mechanics to the bonding of individual atoms. The bonding mechanisms for most polyatomic molecules are the covalent bond and the hydrogen bond. We will discuss only some of the simplest polyatomic molecules—H_2O, NH_3, and CH_4—to illustrate both the simplicity and complexity of the application of quantum mechanics to molecular bonding.

The basic requirement for the sharing of electrons in a covalent bond is that the wave functions of the valence electrons in the individual atoms must overlap as much as possible. As our first example, we will consider the water molecule. The ground-state configuration of the oxygen atom is $1s^2 2s^2 2p^4$. The 1s and 2s electrons are in filled shells and do not contribute to the bonding. The 2p shell has room for six electrons, two in each of the three spatial states (orbitals) corresponding to $\ell = 1$. In an isolated atom, we describe these spatial states by the hydrogen-like wave functions corresponding to $\ell = 1$ and $m_\ell = +1, 0$, and -1. Because the energy is the same for these three spatial states, we could equally well use any linear combination of these wave functions. When an atom participates in molecular bonding, certain combinations of these atomic wave functions are important. These combinations are called **p_x, p_y, and p_z atomic orbitals.** The angular dependence of these orbitals is

$$p_x \propto \sin\theta \cos\phi \qquad 37\text{-}6$$

$$p_y \propto \cos\theta \cos\phi \qquad 37\text{-}7$$

$$p_z \propto \cos\phi \qquad 37\text{-}8$$

The electron charge distribution for these orbitals is maximum along the x, y, or z axis, respectively, as shown in Figure 37-8.

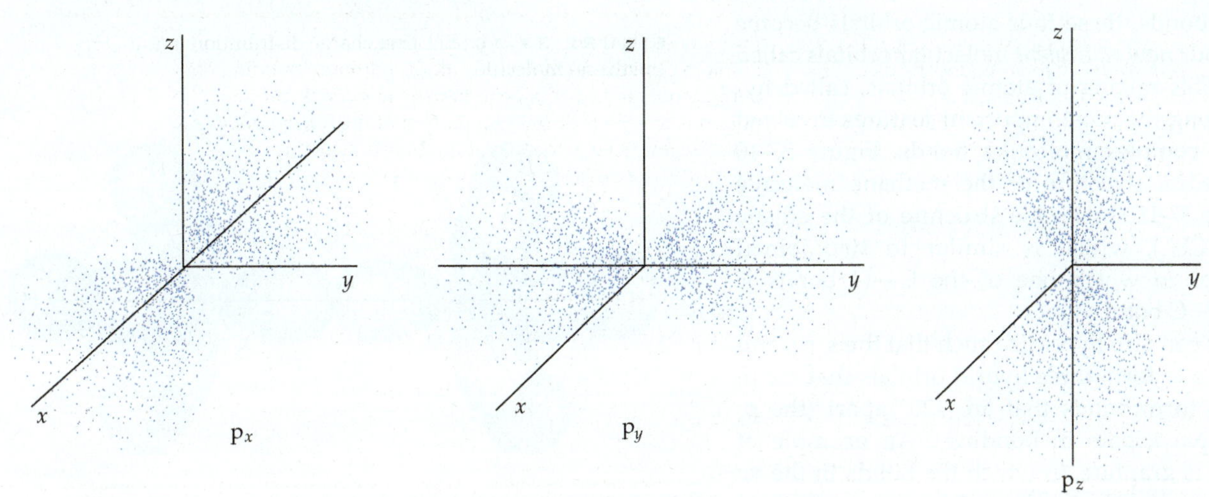

FIGURE 37-8 Computer-generated dot plot illustrating the spatial dependence of the electron charge distribution in the p_x, p_y, and p_z atomic orbitals.

For the oxygen in an H_2O molecule, maximum overlap of the electron wave functions occurs when two of the four 2p electrons are in one of the atomic orbitals (for this example, assume the p_z orbital) with their spins antiparallel, the third 2p electron is in a second orbital (the p_x orbital), and the fourth 2p electron is in the third orbital (the p_y orbital). Each of the unpaired electrons (in the p_x and p_y orbitals for this example) forms a bond with the electron of a hydrogen atom, as shown in Figure 37-9. Because of the repulsion of the two hydrogen atoms, the angle between the O—H bonds is actually greater than 90°. The effect of this repulsion can be calculated, and the result is in agreement with the measured angle of 104.5°.

Similar reasoning leads to an understanding of the bonding in NH_3 (not shown). In the ground state, nitrogen has three electrons in the 2p state. When these three electrons are in the p_x, p_y, and p_z atomic orbitals, they bond to the electrons of hydrogen atoms. Again, because of the repulsion of the hydrogen atoms, the angles between the bonds are somewhat larger than 90°.

The bonding of carbon atoms is somewhat more complicated. Carbon forms single, double and triple bonds, leading to a great diversity in the kinds of organic molecules. The ground-state configuration of carbon is $1s^2 2s^2 2p^2$. From our previous discussion, we might expect carbon to be divalent—that is, bonding only through its two 2p electrons—with the two bonds forming at approximately 90°. However, one of the most important features of the chemistry of carbon is that tetravalent carbon compounds, such as CH_4, are overwhelmingly favored.

The observed valence of 4 for carbon comes about in an interesting way. One of the first excited states of carbon occurs when a 2s electron is excited to a 2p state, giving a configuration of $1s^2 2s^1 2p^3$. In this excited state, we can have four unpaired electrons, one each in the 2s, $2p_x$, $2p_y$, and $2p_z$ atomic orbitals. We might expect there to be three similar bonds corresponding to the three p orbitals and one different bond corresponding to the s orbital. However, when carbon forms tetravalent bonds, these four atomic orbitals become mixed and form four new *equivalent* molecular orbitals called **hybrid orbitals.** This mixing of atomic orbitals, called hybridization, is among the most important features involved in the physics of complex molecular bonds. Figure 37-10 shows the tetrahedral structure of the methane molecule (CH_4), and Figure 37-11 shows the structure of the ethane molecule (CH_3—CH_3), which is similar to two joined methane molecules in which one of the C—H bonds is replaced with a C—C bond.

Carbon orbitals can also hybridize such that the s, p_x, and p_y orbitals combine to form three hybrid orbitals that are in the *xy* plane and form bonds that are 120° apart (the p_z orbital does not participate in bonding). An example of this configuration is graphite, in which the bonds in the *xy* plane provide the strongly layered structure characteristic of the material.

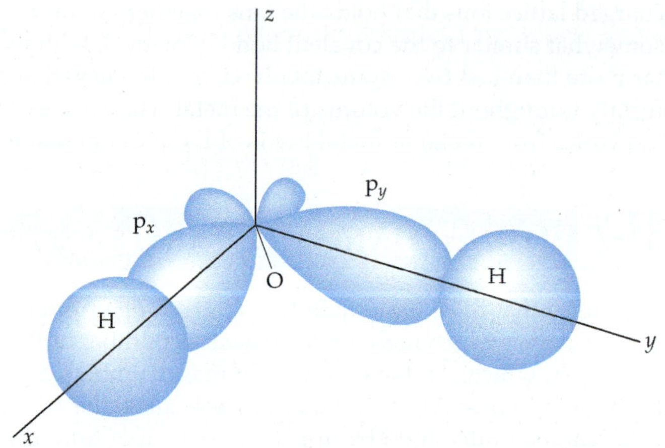

FIGURE 37-9 Electron charge distribution in the H_2O molecule.

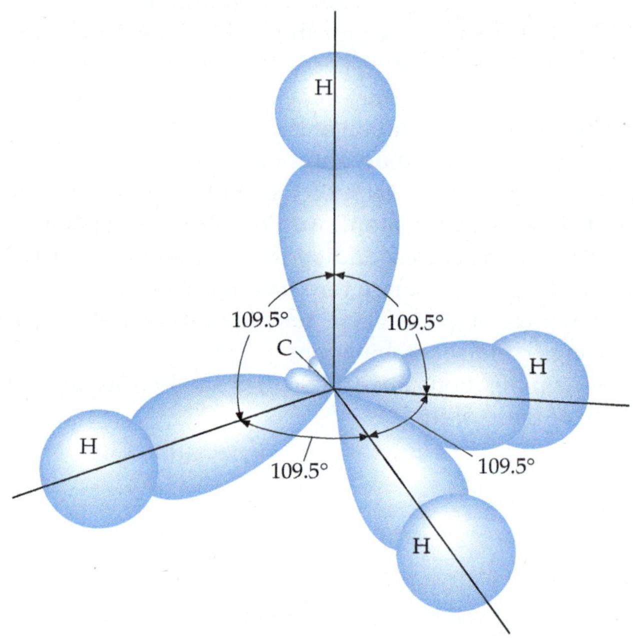

FIGURE 37-10 Electron charge distribution in the CH_4 (methane) molecule.

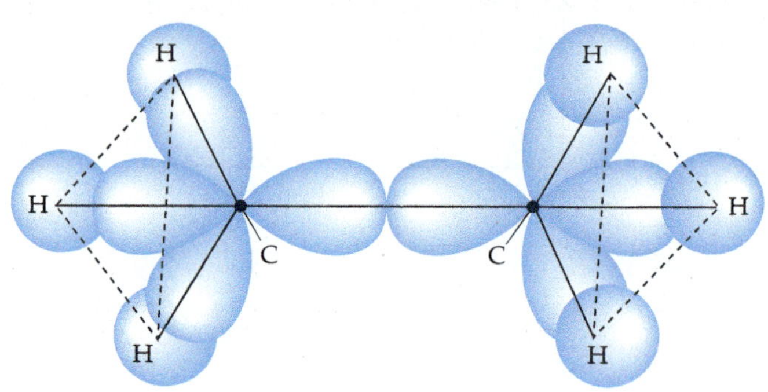

FIGURE 37-11 Electron charge distribution in the CH_3—CH_3 (ethane) molecule.

37-3 ENERGY LEVELS AND SPECTRA OF DIATOMIC MOLECULES

As is the case with an atom, a molecule often emits electromagnetic radiation when it makes a transition from an excited energy state to a state of lower energy. Conversely, a molecule can absorb radiation and make a transition from a lower energy state to a higher energy state. The study of molecular emission and absorption spectra thus provides us with information about the energy states of molecules. For simplicity, we will consider only diatomic molecules here.

The internal energy of a molecule can be conveniently separated into three parts: electronic, due to the excitation of the electrons of the molecule; vibrational, due to the oscillations of the atoms of the molecule; and rotational, due to the rotation of the molecule about its center of mass. The magnitudes of these energies are sufficiently different that they can be treated separately. The energies due to the electronic excitations of a molecule are typically of the order of magnitude of 1 eV, the same as for the electronic excitations of an atom. The energies of the vibrations of the atoms and of the rotation of the molecule are much smaller than the electronic excitation energy.

ROTATIONAL ENERGY LEVELS

Figure 37-12 shows a simple schematic model of a diatomic molecule consisting of particles that have masses of m_1 and m_2, are separated by a distance r and are rotating about the center of mass. Classically, the kinetic energy of rotation (Equation 9-11) is

$$E = \tfrac{1}{2}I\omega^2 \qquad 37\text{-}9$$

where I is the moment of inertia and ω is the angular speed of the rotation motion. If we write this in terms of the angular momentum $L = I\omega$, we have

$$E = \frac{(I\omega)^2}{2I} = \frac{L^2}{2I} \qquad 37\text{-}10$$

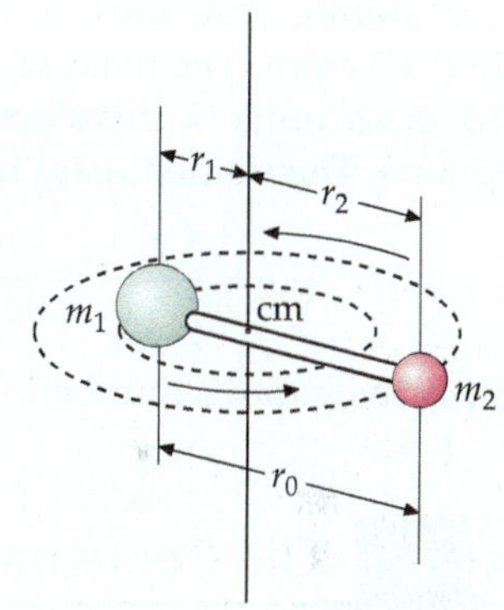

FIGURE 37-12 Diatomic molecule rotating about an axis through its center of mass.

The solution of the Schrödinger equation for rotation leads to quantization of the angular momentum with values given by

$$L^2 = \ell(\ell + 1)\hbar^2 \qquad \ell = 0, 1, 2, \ldots \qquad 37\text{-}11$$

where ℓ is the **rotational quantum number.** This is the same quantum condition on angular momentum that holds for the orbital angular momentum of an electron in an atom. Note, however, that L in Equation 37-10 refers to the angular momentum of the entire molecule rotating about its center of mass. The energy levels of a rotating molecule are therefore given by

$$E_\ell = \frac{\ell(\ell + 1)\hbar^2}{2I} = \ell(\ell + 1)E_{0r} \qquad \ell = 0, 1, 2, \ldots \qquad 37\text{-}12$$

ROTATIONAL ENERGY LEVELS

where E_{0r} is the characteristic rotational energy of a particular molecule, which is inversely proportional to its moment of inertia:

$$E_{0r} = \frac{\hbar^2}{2I} \qquad 37\text{-}13$$

CHARACTERISTIC ROTATIONAL ENERGY

A measurement of the rotational energy of a molecule from its rotational spectrum can be used to determine the moment of inertia of the molecule, which can then be used to find the separation of the atoms in the molecule. The moment of inertia about an axis through the center of mass of a diatomic molecule (see Figure 37-12) is

$$I = m_1 r_1^2 + m_2 r_2^2$$

Using $m_1 r_1 = m_2 r_2$, where r_1 is the distance of atom 1 from the center of mass, r_2 is the distance of atom 2 from the center of mass, and $r_0 = r_1 + r_2$, we can write the moment of inertia (see Problem 26) as

$$I = \mu r_0^2 \qquad \text{37-14}$$

where μ, called the **reduced mass,** is

$$\mu = \frac{m_1 m_2}{m_1 + m_2} \qquad \text{37-15}$$

DEFINITION—REDUCED MASS

If the masses are equal ($m_1 = m_2 = m$), as in H_2 and O_2, the reduced mass is $\mu = \frac{1}{2}m$ and

$$I = \tfrac{1}{2} m r_0^2 \qquad \text{37-16}$$

A unit of mass convenient for discussing atomic and molecular masses is the **unified atomic mass unit,** u, which is defined as one-twelfth the mass of the carbon-12 (^{12}C) atom. The mass of one ^{12}C atom is thus 12 u. The mass of an atom in unified mass units is therefore numerically equal to the molar mass of the atom in grams. The unified mass unit is related to the gram and kilogram by

$$1\text{ u} = \frac{1\text{ g}}{N_A} = \frac{10^{-3}\text{ kg}}{6.0221 \times 10^{23}} = 1.6606 \times 10^{-27}\text{ kg} \qquad \text{37-17}$$

where N_A is Avogadro's number.

Example 37-2 The Reduced Mass of a Diatomic Molecule

Find the reduced mass of the HCl molecule.

PICTURE We find the masses of the hydrogen and chlorine atoms in Appendix C* and use the definition of reduced mass (Equation 37-15).

SOLVE

1. The reduced mass μ is related to the individual masses m_H and m_{Cl}: $\mu = \dfrac{m_H m_{Cl}}{m_H + m_{Cl}}$

2. Find the masses in the periodic table: $m_H = 1.01\text{ u}, \quad m_{Cl} = 35.5\text{ u}$

3. Substitute to calculate the reduced mass: $\mu = \dfrac{m_H m_{Cl}}{m_H + m_{Cl}} = \dfrac{(1.01\text{ u})(35.5\text{ u})}{1.01\text{ u} + 35.5\text{ u}} = \boxed{0.982\text{ u}}$

CHECK The formula for reduced mass is identical to the formula for the equivalent resistance for two resistors in parallel. As expected, the reduced mass is less than either mass.

TAKING IT FURTHER When one atom of a diatomic molecule is much more massive than the other, the center of mass of the molecule is approximately at the center of the more massive atom and the reduced mass is approximately equal to the mass of the lighter atom.

* The masses in the tables are weighted according to the natural isotopic distribution. Thus, the mass of carbon is given as 12.011 rather than 12.000 because natural carbon consists of about 98.9 percent ^{12}C and 1.1 percent ^{13}C. Similarly, natural chlorine consists of about 76 percent ^{35}Cl and 24 percent ^{37}Cl.

Example 37-3 Rotational Kinetic Energy of a Molecule

Estimate the characteristic rotational kinetic energy of an O_2 molecule, assuming that the separation of the atoms is 0.100 nm.

PICTURE The characteristic rotational kinetic energy is given by $E_{0r} = \hbar/(2I)$ (Equation 37-13), where I is the moment of inertia. The moment of inertia is given by $I = \mu r_0^2$ (Equation 37-14), where μ is the reduced mass and r_0 is the average center-to-center separation of the atomic nuclei.

SOLVE

1. The characteristic rotational energy is inversely proportional to the moment of inertia:
$$E_{0r} = \frac{\hbar^2}{2I}$$
2. Calculate the moment of inertia:
$$I = \mu r_0^2 = \tfrac{1}{2} m r_0^2$$
3. Substitute this expression for I into the expression for E_{0r}:
$$E_{0r} = \frac{\hbar^2}{m r_0^2}$$
4. Use $m = 16$ u for the mass of oxygen to calculate E_{0r}:
$$E_{0r} = \frac{\hbar^2}{m r_0^2} = \frac{(1.055 \times 10^{-34}\ \mathrm{J \cdot s})^2}{(16\ \mathrm{u})(10^{-10}\ \mathrm{m})^2} \times \left(\frac{1\ \mathrm{u}}{1.66 \times 10^{-27}\ \mathrm{kg}}\right)$$
$$= 4.19 \times 10^{-23}\ \mathrm{J} = \boxed{2.62 \times 10^{-4}\ \mathrm{eV}}$$

CHECK As expected, the characteristic rotational kinetic energy is small compared with 1 eV (a typical electronic excitation energy).

We can see from Example 37-3 that the rotational energy levels are several orders of magnitude smaller than energy levels due to electron excitation. Transitions within a given set of rotational energy levels yield photons in the microwave region of the electromagnetic spectrum. The rotational energies are also small compared with the typical thermal energy kT at normal temperatures. For $T = 300$ K, for example, kT is about 2.6×10^{-2} eV, which is approximately 100 times the characteristic rotational energy as calculated in Example 37-3 and approximately 1 percent of the typical electronic energy. Thus, at ordinary temperatures, a molecule can be easily excited to the lower rotational energy levels by collisions with other molecules. But such collisions cannot excite the molecule to its electronic energy levels above the ground state.

CONCEPT CHECK 37-1

At room temperature, the molecules of a diatomic gas undergo transitions between rotational states, but the atoms of a monatomic gas do not. Why?

VIBRATIONAL ENERGY LEVELS

The quantization of energy in a simple harmonic oscillator was one of the first problems solved by Schrödinger in his paper proposing his wave equation. Solving the Schrödinger equation for a simple harmonic oscillator gives

$$E_\nu = (\nu + \tfrac{1}{2})hf \qquad \nu = 0, 1, 2, \ldots \qquad 37\text{-}18$$

VIBRATIONAL ENERGY LEVELS

where f is the frequency of the oscillator and ν (lowercase Greek nu) is the **vibrational quantum number.*** An interesting feature of this result is that the energy levels are equally spaced with intervals equal to hf. The frequency of vibration of a diatomic molecule can be related to the force exerted by one atom on the other. Consider two objects of mass m_1 and m_2 connected by a spring of force constant k_F.

* We use ν here rather than n so as not to confuse the vibrational quantum number with the principal quantum number n for electronic energy levels.

The frequency of oscillation of this system (see Problem 32) can be shown to be

$$f = \frac{1}{2\pi}\sqrt{\frac{k_F}{\mu}} \qquad 37\text{-}19$$

where μ is the reduced mass given by Equation 37-15. The effective force constant k_F of a diatomic molecule can thus be determined from a measurement of the frequency of oscillation of the molecule.

A selection rule on transitions between vibrational states (of the same electronic state) requires that the vibrational quantum number ν can change only by ± 1, so the energy of a photon emitted by such a transition is hf and the frequency of the photon is f, the same as the frequency of vibration. There is a similar selection rule that ℓ must change by ± 1 for transitions between rotational states.

A typical measured frequency of a transition between vibrational states is 5×10^{13} Hz, which gives

$$E \approx hf = (4.14 \times 10^{-15}\ \text{eV}\cdot\text{s})(5.0 \times 10^{13}\ \text{s}^{-1}) = 0.2\ \text{eV}$$

and is an estimate for the order of magnitude of vibrational energies. This typical vibrational energy is approximately 1000 times greater than the typical rotational energy E_{0r} of the O_2 molecule we found in Example 37-3 and about 8 times greater than the typical thermal energy $kT = 0.026$ eV at $T = 300$ K. Thus, the vibrational levels are almost never excited by molecular collisions at ordinary temperatures.

Example 37-4 Determining the Force Constant

The frequency of vibration of the CO molecule is 6.42×10^{13} Hz. What is the effective force constant for this molecule?

PICTURE We use $2\pi f = \sqrt{k_F/\mu}$ (Equation 37-19) to relate k_F to the frequency and the reduced mass, and calculate μ from its definition.

SOLVE

1. The effective force constant is related to the frequency and reduced mass by Equation 37-19:

$$f = \frac{1}{2\pi}\sqrt{\frac{k_F}{\mu}}$$

$$k_F = (2\pi f)^2\mu$$

2. Calculate the reduced mass using 12 u for the mass of the carbon atom and 16 u for the mass of the oxygen atom:

$$\mu = \frac{m_1 m_2}{m_1 + m_2} = \frac{(12\ \text{u})(16\ \text{u})}{12\ \text{u} + 16\ \text{u}} = 6.86\ \text{u}$$

3. Substitute this value of μ into the equation for k_F in step 1 and convert to SI units:

$$k_F = (2\pi f)^2\mu$$
$$= 4\pi^2(6.42 \times 10^{13}\ \text{Hz})^2(6.86\ \text{u})$$
$$= 1.12 \times 10^{30}\ \text{u/s}^2 \times \left(\frac{1.66 \times 10^{-27}\ \text{kg}}{1\ \text{u}}\right)$$
$$= \boxed{1.85 \times 10^3\ \text{N/m}}$$

CHECK From Newton's second law we know that 1 kg m/s² = 1 N, so the units of kg/s² that remain after canceling out the u's in step 3 are equal to N/m, which is what is expected for the force constant of a "spring."

EMISSION SPECTRA

Figure 37-13 shows schematically some electronic, vibrational, and rotational energy levels of a diatomic molecule. The vibrational levels are labeled with the quantum number ν and the rotational levels are labeled with ℓ. The lower vibrational levels are evenly spaced, with $\Delta E = hf$. For higher vibrational levels, the approximation that the vibration is simple harmonic is not valid and the levels are not quite evenly spaced.

Note that the potential energy curves representing the force between the two atoms in the molecule do not have exactly the same shape for the electronic ground and excited states. This implies that the fundamental frequency of vibration f is different for different electronic states. For transitions between vibrational states of different electronic states, the selection rule $\Delta\nu = \pm 1$ does not hold. Such transitions result in the emission of photons of wavelengths in or near the visible spectrum, so the emission spectrum of a molecule for electronic transitions is also sometimes called the optical spectrum.

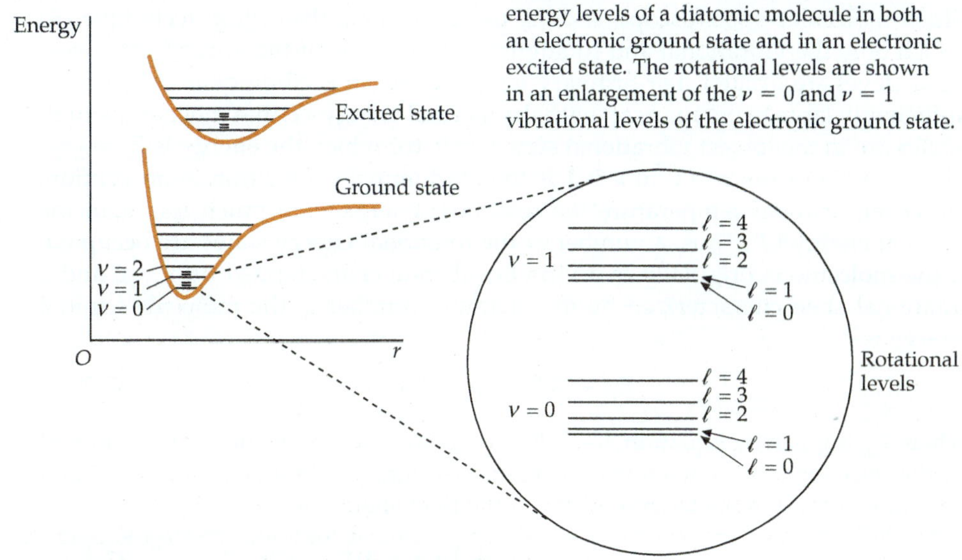

FIGURE 37-13 Vibrational and rotational energy levels of a diatomic molecule in both an electronic ground state and in an electronic excited state. The rotational levels are shown in an enlargement of the $\nu = 0$ and $\nu = 1$ vibrational levels of the electronic ground state.

The spacing of the rotational levels increases with increasing values of ℓ. Because the energies of rotation are so much smaller than those of vibrational excitation or electronic excitation of a molecule, molecular rotation shows up in optical spectra as a fine splitting of the spectral lines. When this fine structure is not resolved, the spectrum appears as bands, as shown in Figure 37-14*a*. Close inspection of these bands reveals that they have a fine structure due to the rotational energy levels, as shown in the enlargement in Figure 37-14*c*.

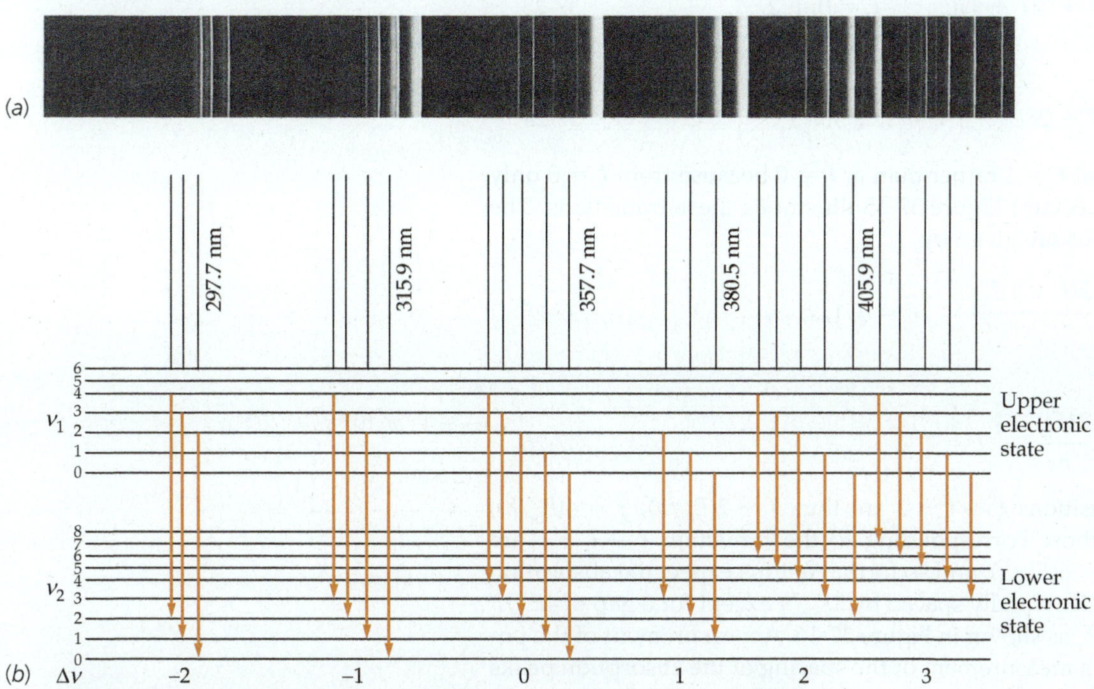

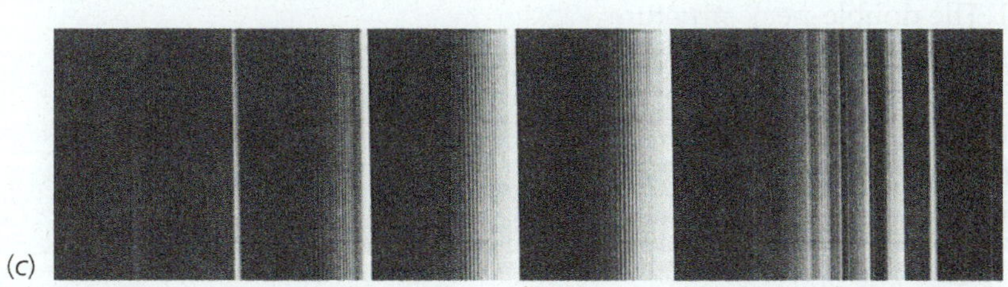

FIGURE 37-14 (*a*) Part of the emission spectrum of N_2. The spectral lines are due to transitions between the vibrational levels of two electronic states, as indicated in the energy level diagram (*b*). (*c*) An enlargement of part of Figure 37-14*a* shows that the apparent lines are in fact bands with structure caused by rotational levels. *(Courtesy of Dr. J. A. Marquissee.)*

ABSORPTION SPECTRA

Much molecular spectroscopy is done using infrared absorption techniques in which only the vibrational and rotational energy levels of the ground-state electronic level are excited. For ordinary temperatures, the vibrational energies are sufficiently large in comparison with the thermal energy kT that most of the molecules are in the lowest vibrational state $\nu = 0$, for which the energy is $E_0 = \frac{1}{2}hf$. The transition from $\nu = 0$ to $\nu = 1$ is the predominant transition in absorption. However, at room temperature the rotational energies are much less than the thermal energy kT. Thus, a number of the rotational energy states are occupied. If the molecule is originally in a vibrational state characterized by $\nu = 0$ and a rotational state characterized by the quantum number ℓ, the molecule's initial energy is

$$E_\ell = \tfrac{1}{2}hf + \ell(\ell + 1)E_{0r} \qquad 37\text{-}20$$

where E_{0r} is given by Equation 37-13. From this state, two transitions are permitted by the selection rules. For a transition to the next higher vibrational state $\nu = 1$ and a rotational state characterized by $\ell + 1$, the final energy is

$$E_{\ell+1} = \tfrac{3}{2}hf + (\ell + 1)(\ell + 2)E_{0r} \qquad 37\text{-}21$$

For a transition to the next higher vibrational state and to a rotational state characterized by $\ell - 1$, the final energy is

$$E_{\ell-1} = \tfrac{3}{2}hf + (\ell - 1)\ell E_{0r} \qquad 37\text{-}22$$

The energy differences therefore are

$$\Delta E_{\ell\to\ell+1} = E_{\ell+1} - E_\ell = hf + 2(\ell + 1)E_{0r} \qquad \ell = 0, 1, 2, \ldots \qquad 37\text{-}23$$

and

$$\Delta E_{\ell\to\ell-1} = E_{\ell-1} - E_\ell = hf - 2\ell E_{0r} \qquad \ell = 1, 2, 3, \ldots \qquad 37\text{-}24$$

(In Equation 37-24, ℓ begins at $\ell = 1$ rather than at $\ell = 0$ because from $\ell = 0$ only the transition $\ell \to \ell + 1$ can occur.) Figure 37-15 illustrates these transitions. The frequencies of these transitions are given by

$$f_{\ell\to\ell+1} = \frac{\Delta E_{\ell\to\ell+1}}{h} = f + \frac{2(\ell + 1)E_{0r}}{h} \qquad \ell = 0, 1, 2, \ldots \qquad 37\text{-}25$$

and

$$f_{\ell\to\ell-1} = \frac{\Delta E_{\ell\to\ell-1}}{h} = f - \frac{2\ell E_{0r}}{h} \qquad \ell = 1, 2, 3, \ldots \qquad 37\text{-}26$$

The frequencies for the transitions $\ell \to \ell + 1$ are thus $f + 2(E_{0r}/h)$, $f + 4(E_{0r}/h)$, $f + 6(E_{0r}/h)$, and so forth; those corresponding to the transition $\ell \to \ell - 1$ are $f - 2(E_{0r}/h)$, $f - 4(E_{0r}/h)$, $f - 6(E_{0r}/h)$, and so forth. We thus expect the absorption spectrum to contain frequencies equally spaced by $2E_{0r}/h$ except for a gap of $4E_{0r}/h$ at the vibrational frequency f, as shown in Figure 37-16. A measurement of the position of the gap gives f and a measurement of the spacing of the absorption peaks gives E_{0r}, which is inversely proportional to the moment of inertia of the molecule.

Figure 37-17 shows the absorption spectrum of HCl. The double-peak structure results from the fact that chlorine occurs naturally in two isotopes, ^{35}Cl and ^{37}Cl, which gives HCl with two different moments of inertia. If all the rotational levels were equally populated initially, we would expect the intensities of each absorption line to be equal. However, the population of a rotational level is proportional to the degeneracy of the level, that is, to the number of states with the same value of ℓ, which is $2\ell + 1$, and to the Boltzmann factor $e^{-E/kT}$, where E is the energy of

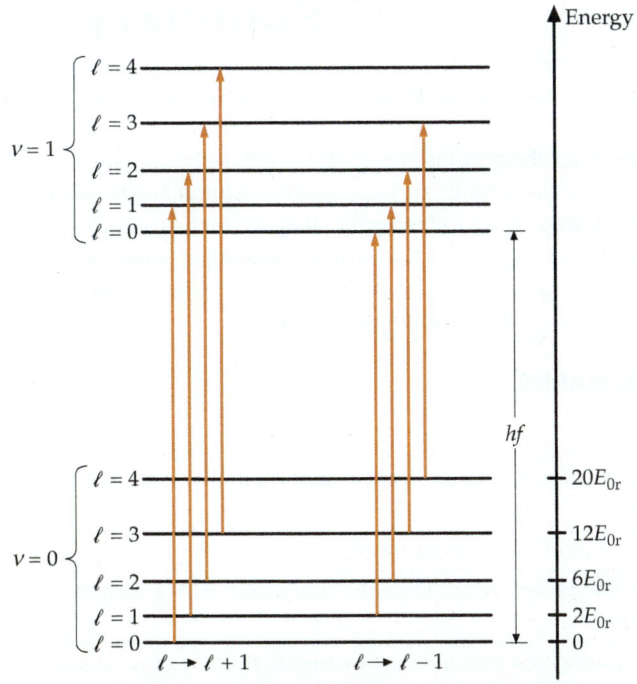

FIGURE 37-15 Absorptive transitions between the lowest vibrational states $\nu = 0$ and $\nu = 1$ in a diatomic molecule. These transitions obey the selection rule $\Delta\ell \pm 1$ and fall into two bands. The energies of the $\ell \rightarrow \ell + 1$ band are $hf + 2E_{0r}$, $hf + 4E_{0r}$, $hf + 6E_{0r}$, and so forth; whereas the energies of the $\ell \rightarrow \ell - 1$ band are $hf - 2E_{0r}$, $hf - 4E_{0r}$, $hf - 6E_{0r}$, and so forth.

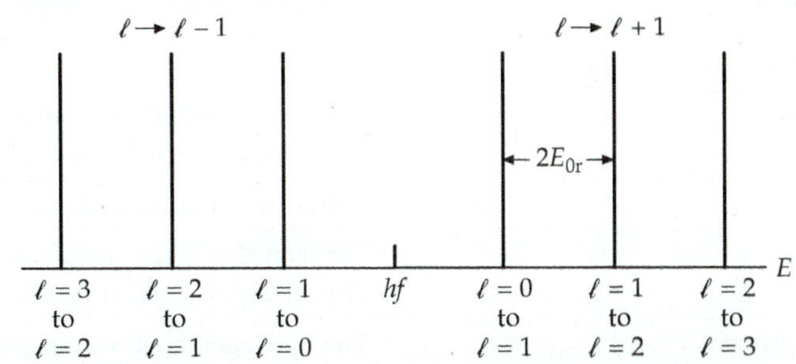

FIGURE 37-16 Expected absorption spectrum of a diatomic molecule. The right branch corresponds to transitions $\ell \rightarrow \ell + 1$ and the left branch corresponds to the transitions $\ell \rightarrow \ell - 1$. The lines are equally spaced by $2E_{0r}$. The energy midway between the branches is hf, where f is the frequency of vibration of the molecule.

the state. (The Boltzmann factor is presented in Chapter 17.) For low values of ℓ, the population increases slightly because of the degeneracy factor, whereas for higher values of ℓ, the population decreases because of the Boltzmann factor. The intensities of the absorption lines therefore increase with ℓ for low values of ℓ and then decrease with ℓ for high values of ℓ, as can be seen from the figure.

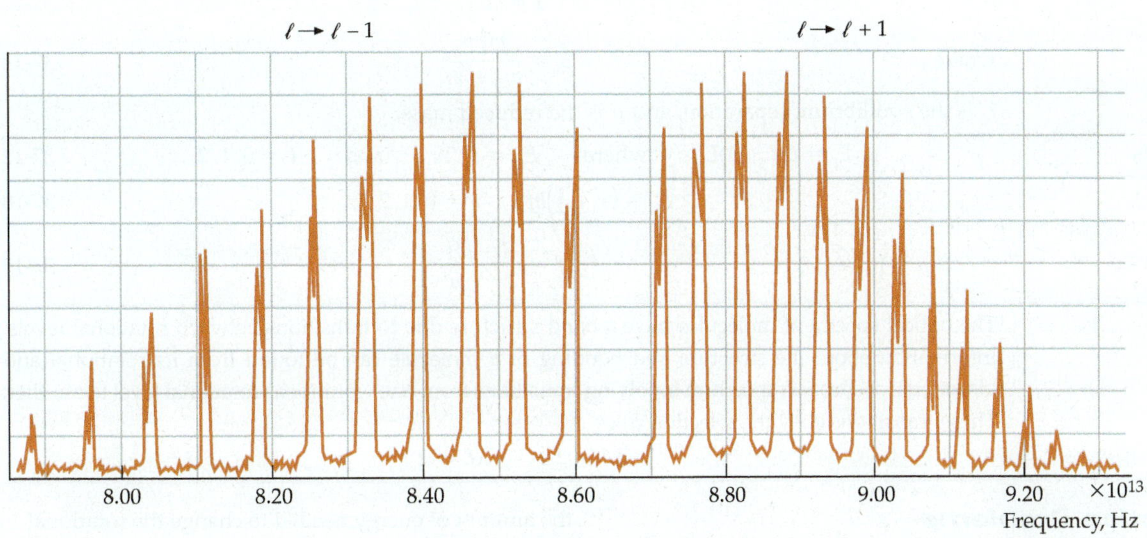

FIGURE 37-17 Absorption spectrum of the diatomic molecule HCl. The double-peak structure results from the two isotopes of chlorine, ^{35}Cl (abundance 75.5 percent) and ^{37}Cl (abundance 24.5 percent). The intensities of the peaks vary because the population of the initial state depends on ℓ.

Summary

1. Atoms are usually found in nature bonded to form molecules or in the lattices of crystalline solids.
2. Ionic bonnds and covalent bonds are the principal mechanisms responsible for forming molecules. Metallic bonds and van der Waals bonds are important in the formation of solids and liquids. Hydrogen bonds enable large biological molecules to maintain their shape.
3. Like atoms, molecules emit electromagnetic radiation when making a transition from a higher energy state to a lower energy state. The internal energy of a molecule can be separated into three parts: electronic, vibrational, and rotational energy.

TOPIC	RELEVANT EQUATIONS AND REMARKS	
1. Molecular Bonding		
Ionic	Ionic bonds result when an electron is transferred from one atom to another, resulting in a positive ion and a negative ion that bond together.	
Covalent	The covalent bond is the sharing of one or more electrons by atoms.	
van der Waals	The van der Waals bonds are weak bonds that result from the interaction of the instantaneous electric dipole moments of molecules.	
Hydrogen	The hydrogen bond results from the sharing of a proton of the hydrogen atom by other atoms.	
Metallic	In the metallic bond, the positive lattice ions of the metal are held together by a cloud of negative charge composed of free electrons.	
Mixed	A diatomic molecule formed from two identical atoms, such as O_2, must bond by covalent bonding. The bonding of two nonidentical atoms is often a mixture of covalent and ionic bonding. The percentage of ionic bonding can be found from the ratio of the magnitude of the measured electric dipole moment to the magnitude of the ionic electric dipole moment defined by $p_{\text{ionic}} = er_0$ where r_0 is the equilibrium separation of the ions.	37-5
2. *Polyatomic Molecules	The shapes of such polyatomic molecules as H_2O and NH_3 can be understood from the spatial distribution of the atomic-orbital or molecular-orbital wave functions. The tetravalent nature of the carbon atom is a result of the hybridization of the 2s and 2p atomic orbitals.	
3. Diatomic Molecules		
Moment of inertia	$I = \mu r_0^2$	37-14
	where $\mu = \dfrac{m_1 m_2}{m_1 + m_2}$	37-15
	r_0 is the equilibrium separation, and μ is the reduced mass.	
Rotational energy levels	$E_\ell = \ell(\ell + 1)E_{0r}$ where $E_{0r} = \hbar/2\pi$ and $\ell = 0, 1, 2, \ldots$	37-12
Vibrational energy levels	$E_\nu = \left(\nu + \frac{1}{2}\right)hf \quad \nu = 0, 1, 2, \ldots$	37-18
Effective force constant k_F	$f = \dfrac{1}{2\pi}\sqrt{\dfrac{k_F}{\mu}}$	37-19
4. Molecular Spectra	The optical spectra of molecules have a band structure due to transitions between rotational levels. Information about the structure and bonding of a molecule can be found from its rotational and vibrational absorption spectrum involving transitions from one vibrational-rotational level to another. These transitions obey the selection rules $\Delta\nu = \pm 1 \quad \Delta\ell = \pm 1$	

Answers to Practice Problems

37-1 17.6 percent

Answers to Concept Check

37-1 The moment of inertia of an atom is much much less than the moment of inertia of a diatomic molecule, so the amount of energy needed to change the rotational state of a single atom is much much larger than the amount needed for a diatomic molecule. At 300 K, the required energy is not available by way of collisions between atoms.

Problems

In a few problems, you are given more data than you actually need; in a few other problems, you are required to supply data from your general knowledge, outside sources, or informed estimate.

Interpret as significant all digits in numerical values that have trailing zeros and no decimal points.

- • Single-concept, single-step, relatively easy
- •• Intermediate-level, may require synthesis of concepts
- ••• Challenging
- SSM Solution is in the *Student Solutions Manual*
- Consecutive problems that are shaded are paired problems.

CONCEPTUAL PROBLEMS

1 • Would you expect NaCl to be polar or nonpolar? SSM

2 • Would you expect N_2 to be polar or nonpolar?

3 • Does neon naturally occur as Ne or Ne_2? Explain your answer.

4 • What type of bonding mechanism would you expect for atoms of (*a*) HF, (*b*) KBr, (*c*) N_2, (*d*) Ag in solid silver?

5 •• The elements in the far right column of the periodic table are sometimes called noble gases, both because they are gases under a wide range of conditions and because atoms of these elements almost never react with other atoms to form molecules or ionic compounds. However, atoms of noble gases can react if the resulting molecule is formed in an electronic excited state. An example is ArF. When it is formed in the excited state, it is written ArF* and is called an excimer (for *ex*cited di*mer*). Refer to Figure 37-13 and discuss how a diagram for the electronic, vibrational, and rotation energy levels of ArF and ArF* would look in which the ArF ground state is unstable and the ArF* excited state is stable. (*Note:* Excimers are used in certain kinds of lasers.) SSM

6 • Find other atoms that have the same subshell electron configurations in their two highest energy orbitals as carbon atoms do. Would you expect the same type of hybridization for these orbitals as for carbon?

7 • How does the value of the effective force constant calculated for a CO molecule in Example 37-4 compare with the value of the force constant of the suspension springs on a typical automobile, which is about 1.5 kN/m?

8 • Explain why the moment of inertia of a diatomic molecule increases slightly with increasing angular momentum.

9 • Why would you expect the separation distance between the two protons to be larger in a H_2^+ ion than in a H_2 molecule?

10 • At room temperature an atom typically absorb radiation only from the ground state, whereas a diatomic molecule typically absorbs radiation from many different rotational states. Why?

11 •• The vibrational energy levels of diatomic molecules are described by a single vibrational frequency f that is the frequency of vibration of the two atoms of the molecule along the line through their centers. Would you expect to see one or more than one vibrational frequency in molecules that have three or more atoms? Consider in particular a water molecule H_2O (Figure 37-9).

ESTIMATION AND APPROXIMATION

12 •• The potential energy for a diatomic molecule has a minimum as shown in Figure 37-13. Near this minimum, the graph for the energy as a function of distance between the atoms may be approximated as a parabola, leading to the harmonic oscillator model for the vibrating molecule. An improved approximation is called the anharmonic oscillator and leads to a modification of the expression for the energy $E_\nu = (\nu + \frac{1}{2})hf$, where $\nu = 0, 1, 2, \ldots$ (Equation 37-18). The modified expression for energy is $E_\nu = (\nu + \frac{1}{2})hf - (\nu + \frac{1}{2})^2 hf\alpha$, where $\nu = 0, 1, 2, \ldots$. For an O_2 molecule, the constants have the values $f = 4.74 \times 10^{13}\,\text{s}^{-1}$ and $\alpha = 7.6 \times 10^{-3}$. Use this formula to estimate the smallest value of the quantum number ν for which the modified expression differs from the original expression by 10 percent.

13 •• To understand why quantum mechanics is not needed to describe many macroscopic systems, estimate the rotational energy quantum number ℓ and spacing between adjacent energy levels for a baseball ($m \sim 300$ g, $r \sim 3$ cm) spinning about its own axis at 20 rev/min. *Hint: Pick ℓ so the quantum energy formula $E_\ell = \ell(\ell + 1)\hbar^2/(2I)$, where $\ell = 0, 1, 2, \ldots$ (Equation 37-12) gives the correct energy for the given system. Then find the energy increase for the next highest energy level.*

14 •• Estimate the quantum number ν and spacing between adjacent energy levels for a 1.0-kg mass attached to spring. The spring has a force constant equal to 1200 N/m and the mass-spring system is vibrating with an amplitude of 3.0 cm. *Hint: Pick ν so that the quantum energy formula $E_\nu = (\nu + \frac{1}{2})hf$, where $\nu = 0, 1, 2, \ldots$ (Equation 37-18) gives the correct energy for the given system. Then find the energy increase for the next highest energy level.*

MOLECULAR BONDING

15 • Calculate the separation of Na^+ and Cl^- ions for which the potential energy of a single ionic unit (one Na^+ ion and one Cl^- ion) is -1.52 eV.

16 • The equilibrium separation of the atoms in a HF molecule is 0.0917 nm and the measured electric dipole moment of the molecule is $6.40 \times 10^{-30}\,\text{C}\cdot\text{m}$. What percentage of the HF bond is ionic?

17 •• The dissociation energy of RbF is 5.12 eV, and the equilibrium separation of RbF is 0.227 nm. The electron affinity of a fluorine atom is -3.40 eV and the ionization energy of rubidium is 4.18 eV. Determine the core-repulsion energy of RbF.

18 •• The equilibrium separation of the K^+ and Cl^- ions in KCl is about 0.267 nm. (*a*) Calculate the potential energy of attraction of the ions. Assume that the ions are point charges at this separation. (*b*) The ionization energy of potassium is 4.34 eV and the electron affinity of chlorine is -3.62 eV. Calculate a value for the dissociation energy using the assumption that the energy of repulsion is negligible. (See Figure 37-1.) (*c*) The measured dissociation energy is 4.49 eV. What is the energy due to repulsion of the ions at the equilibrium separation?

19 •• Indicate an approximate value for the average value of the separation distance r for two vibrational levels on the potential energy curve for a diatomic molecule (one of the curves in Figure 37-13). Your teacher claims that the increase in r_{av} with increases in vibration energy explains why solids expand when heated. Do you agree? If so, give an argument supporting this claim. If not, give an argument opposing this claim.

20 •• Calculate the potential energy of attraction between the Na^+ and Cl^- ions at the equilibrium separation $r_0 = 0.236$ nm. Compare this result with the dissociation energy given in Figure 37-1. What is the energy due to repulsion of the ions at the equilibrium separation?

21 •• The equilibrium separation of the K^+ and F^- ions in KF is about 0.217 nm. (*a*) Calculate the potential energy of attraction of these ions. Assume that the ions are point charges at this separation. (*b*) The ionization energy of potassium is 4.34 eV and the electron affinity of fluorine is −3.40 eV. Find the dissociation energy by neglecting any energy of repulsion. (*c*) The measured dissociation energy is 5.07 eV. Calculate the energy due to repulsion of the ions at the equilibrium separation.

ENERGY LEVELS OF SPECTRA OF DIATOMIC MOLECULES

22 • The characteristic rotational energy E_{0r} for the rotation of a N_2 molecule is 2.48×10^{-4} eV. Using this value, find the separation distance of the 2 nitrogen atoms.

23 • The separation of the two oxygen atoms in a molecule of O_2 is actually slightly greater than the 0.100 nm used in Example 37-3. Furthermore, the characteristic energy of rotation E_{0r} for O_2 is 1.78×10^{-4} eV rather than the result obtained in that example. Use this value to calculate the separation distance of the two oxygen atoms. SSM

24 •• Show that the reduced mass of a diatomic molecule is always smaller than the mass of the molecule. Calculate the reduced mass for (*a*) H_2, (*b*) N_2, (*c*) CO, and (*d*) HCl. Express your answers in unified atomic mass units.

25 •• A CO molecule has a binding energy of approximately 11 eV. Find the vibrational quantum number ν that corresponds to 11 eV. (If a CO molecule actually had this much vibrational energy, it would "shake" apart.)

26 •• Derive the equation $I = \mu r_0^2$ (Equation 37-14) for the moment of inertia in terms of the reduced mass of a diatomic molecule.

27 •• The equilibrium separation between the atoms of a LiH molecule is 0.16 nm. Determine the energy separation between the $\ell = 3$ and $\ell = 2$ rotational levels of the diatomic molecule. SSM

28 •• The equilibrium separation of the K^+ and Cl^- ions in KCl is about 0.267 nm. Use this value together with the reduced mass of KCl to calculate the characteristic rotational energy E_{0r} (Equation 37-13) of KCl.

29 •• The central frequency for the absorption band of HCl shown in Figure 37-17 is at 8.66×10^{13} Hz, and the absorption peaks to either side of the central frequency are separated by about 6×10^{11} Hz. Use this information to find (*a*) the lowest (zero-point) vibrational energy for HCl, (*b*) the moment of inertia of HCl, and (*c*) the equilibrium separation of the two atoms.

30 •• Calculate the effective force constant for HCl from its reduced mass and from the fundamental vibrational frequency obtained from Figure 37-17.

31 •• The equilibrium separation between the atoms of a CO molecule is 0.113 nm. For a molecule, such as CO, that has a permanent electric dipole moment, radiative transitions obeying the selection rule $\Delta\ell = \pm 1$ between two rotational energy levels of the same vibrational level are allowed. (That is, the selection rule $\Delta\nu = \pm 1$ does not hold.) (*a*) Find the moment of inertia of CO and calculate the characteristic rotational energy E_{0r} (in eV). (*b*) Make an energy-level diagram for the rotational levels from $\ell = 0$ to $\ell = 5$ for some vibrational level. Label the energies in electron volts, starting with $E = 0$ for $\ell = 0$. Indicate on your diagram the transitions that obey $\Delta\ell = -1$, and calculate the energies of the photons emitted. (*c*) Find the wavelength of the photons emitted during each transition in (*b*). In what region of the electromagnetic spectrum are those photons? SSM

32 •• Two objects, one of mass m_1 and the other of mass m_2, are connected to opposite ends of a spring of force constant k_F. The objects are released from rest with the spring compressed. (*a*) Show that when the spring is extended and the object of mass m_1 is a distance Δr_1 from its equilibrium position in the center-of-mass reference frame, the force exerted by the spring is given by $F = -k_F(m_1/\mu)\Delta r_1$, where μ is the reduced mass. (*b*) Show that the frequency of oscillation f is related to k_F and μ by $2\pi f = \sqrt{k_F/\mu}$.

33 ••• Calculate the reduced masses μ for $H^{35}Cl$ and $H^{37}Cl$ molecules and the fractional difference $\Delta\mu/\mu$. Show that the mixture of isotopes in HCl leads to a fractional difference in the frequency of a transition from one rotational state to another given by $\Delta f/f = -\Delta\mu/\mu$. Compute $\Delta f/f$ and compare your result with Figure 37-17.

GENERAL PROBLEMS

34 • Show that when one atom of a diatomic molecule is much more massive than the other the reduced mass is approximately equal to the mass of the lighter atom.

35 •• The equilibrium separation between the nuclei of a CO molecule is 0.113 nm. Determine the energy difference between the $\ell = 2$ and $\ell = 1$ rotational energy levels of the molecule.

36 •• The effective force constant for a HF molecule is 970 N/m. Find the frequency of vibration for the molecule.

37 •• The frequency of vibration of a NO molecule is 5.63×10^{13} Hz. Find the effective force constant for NO.

38 •• The effective force constant of the hydrogen bond in a H_2 molecule is 580 N/m. Obtain the energies of the four lowest vibrational levels of the H_2, HD, and D_2 molecules, where H is protium and D is deuterium, and find the wavelengths of photons resulting from transitions between adjacent vibrational levels for each of the molecules.

39 •• The potential energy between two atoms in a molecule separated by a distance r can often be described rather well by the Lenard-Jones (or 6-12) potential function, which can be written as

$$U = U_0\left[\left(\frac{a}{r}\right)^{12} - 2\left(\frac{a}{r}\right)^{6}\right]$$

where U_0 and a are constants. Find the equilibrium separation r_0 in terms of a. *Hint: At the equilibrium separation the potential energy is a minimum.* Find U_{min}, the value of U when $r = r_0$. Use Figure 37-4 to obtain numerical values of r_0 and U_0 for a H_2 molecule, and express your answers in nanometers and electron volts.

40 •• In this problem, you are to determine how the van der Waals force between a polar molecule and a nonpolar molecule depends on the distance between the molecules. Let the polar molecule be at the origin and let its dipole moment be in the $+x$ direction. In addition, let the nonpolar molecule be on the x axis a distance x away. (*a*) How does the electric field strength due to an electric dipole vary with the distance from the dipole in a given direction? (*b*) Use (1) that the potential energy U of an electric dipole of dipole moment $\vec{p}$ in an electric field $\vec{E}$ can be expressed as $U = -\vec{p} \cdot \vec{E}$, and (2) that the induced dipole moment $\vec{p}'$ of the nonpolar molecule is in the direction of $\vec{E}$, and that p' is proportional to E, to determine how the potential energy of interaction of the two molecules depends on the separation distance x. (*c*) Using $F_x = -dU/dx$, determine how the force between the two molecules depends on distance.

41 •• Find the dependence of the force on separation distance between the two polar molecules described in Problem 40.

42 •• Use the infrared absorption spectrum of HCl in Figure 37-17 to obtain (*a*) the characteristic rotational energy E_{0r} (in eV) and (*b*) the vibrational frequency f and the vibrational energy hf (in eV).

43 • The dissociation energy is sometimes expressed in kilocalories per mole (kcal/mol). (*a*) Find the relation between the units eV/molecule and kcal/mol. (*b*) Find the dissociation energy of NaCl in kcal/mol.

CHAPTER 38

IT IS WELL KNOWN THAT ARSENIC IS A POISON. IT IS LESS WELL KNOWN THAT SILICON CRYSTALS THAT HAVE SMALL CONCENTRATIONS OF ARSENIC ATOMS HAVE A MUCH LOWER RESISTIVITY THAN DO CRYSTALS THAT ARE 100 PERCENT SILICON. *(The Natural History Museum/Alamy.)*

? Do you know how many atoms of arsenic it takes to increase the charge-carrier density by a factor of 5 million? (See Example 38-7.)

Solids

The first microscopic model of electric conduction in metals was proposed by Paul K. Drude in 1900 and developed by Hendrik A. Lorentz about 1909. This model successfully predicts that the current is proportional to the potential drop (Ohm's law) and relates the resistivity of conductors to the mean speed and the mean free path* of the free electrons within the conductor. However, when mean speed and mean free path are interpreted classically, there is a disagreement between the calculated values and the measured values of the resistivity, and a similar disagreement between the predicted temperature dependence and the observed temperature dependence that resistivity values have. Thus, the classical theory fails to adequately describe the resistivity of metals. Furthermore, the classical theory says nothing about the most striking

* The mean free path is the average distance traveled between collisions.

property of solids, namely, that some substances are conductors, others are insulators, and still others are semiconductors, which are substances whose resistivity falls between that of conductors and insulators.

When mean speed and mean free path are interpreted using quantum theory, both the magnitude and the temperature dependence of the resistivity are correctly predicted. In addition, quantum theory allows us to determine if a substance will be a conductor, an insulator, or a semiconductor.

In this chapter, we use our understanding of quantum mechanics to discuss the structure of solids and solid-state semiconducting devices. Much of our discussion will be qualitative because, as in atomic physics, the quantum-mechanical calculations are mathematically sophisticated.

38-1 THE STRUCTURE OF SOLIDS

The three phases of matter we observe everyday—gas, liquid, and solid—result from the relative strengths of the attractive forces between atoms and molecules and the thermal energies of the particles. Molecules and atoms in the gas phase have relatively large thermal kinetic energies, and such particles have little influence on one another except during their frequent but brief collisions. (By using the term thermal kinetic energies, we mean the kinetic energies of the molecules and atoms in the center-of-mass reference frame of the gas.) At sufficiently low temperatures, van der Waals forces will cause practically every substance to condense into a liquid and then into a solid. In liquids, the molecules or atoms are close enough—and their thermal kinetic energies are low enough—that they can develop a temporary **short-range order.** As their thermal kinetic energies are further reduced, the molecules or atoms form solids, which are characterized by a lasting order.

If a liquid is cooled slowly so that the kinetic energy of its molecules is reduced slowly, the molecules (or atoms or ions) may arrange themselves in a regular crystalline array, producing the maximum number of bonds and leading to a minimum potential energy. However, if the liquid is cooled rapidly so that its internal energy is removed before the molecules have a chance to arrange themselves, the solid formed is often not crystalline or the arrangement is not regular. Such a solid is called an **amorphous solid.** It displays short-range order but not the long-range order (the order over many molecular, atomic, or ionic diameters) that is characteristic of a crystal. Glass is a typical amorphous solid. A characteristic result of the long-range ordering of a crystal is that it has a well-defined melting point, whereas an amorphous solid merely softens as its temperature is increased. Many substances may solidify into either an amorphous state or a crystalline state depending on how the substances are prepared; others exist only in one such state or the other.

Most common solids are polycrystalline; that is, they consist of many single crystals that meet at *grain boundaries.* The size of a single crystal is typically a fraction of a millimeter. However, large single crystals do occur naturally and can be produced artificially. The most important property of a single crystal is the symmetry and regularity of its structure. It can be thought of as having a single unit structure that is repeated throughout the crystal. This smallest unit of a crystal is called the **unit cell;** its structure depends on the type of bonding—ionic, covalent, metallic, hydrogen, van der Waals—between the atoms, ions, or molecules. If more than one kind of atom is present, the structure will also depend on the relative sizes of the atoms.

Figure 38-1 shows the structure unit cell of crystalline sodium chloride (NaCl). The Na^+ and Cl^- ions are spherically symmetric, and the Cl^- ion is approximately twice as large as the Na^+ ion. The minimum potential energy for this crystal occurs when an ion of either kind has six nearest neighbors of the other kind. This structure is called *face-centered-cubic* (fcc). Note that the Na^+ and Cl^- ions in solid NaCl are *not* paired into NaCl molecules.

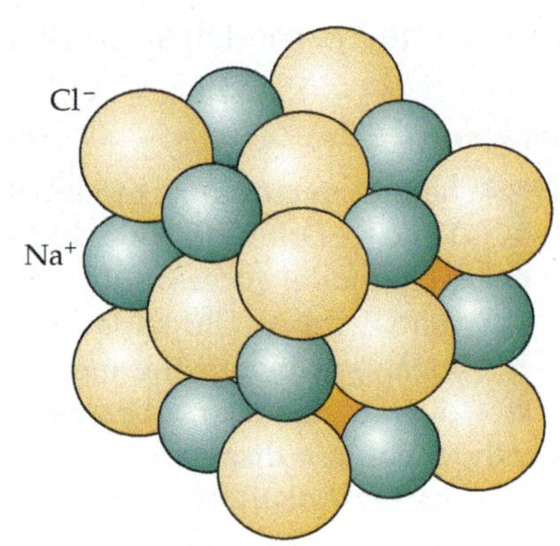

FIGURE 38-1 Face-centered-cubic structure of the NaCl crystal.

The net attractive part of the potential energy of an ion in a crystal can be written

$$U_{\text{att}} = -\alpha \frac{ke^2}{r} \qquad 38\text{-}1$$

where r is the (center-to-center) separation distance between neighboring ions (0.281 nm for the Na^+ and Cl^- ions in crystalline NaCl) and α, called the **Madelung constant,** depends on the geometry of the crystal. If only the six nearest neighbors of each ion in a face-centered-cubic crystalline structure were important, α would be six. However, in addition to the six neighbors of the opposite charge at a distance r, there are twelve ions of the same charge at a distance $\sqrt{2}r$, eight ions of opposite charge at a distance $\sqrt{3}r$, and so on. The Madelung constant is thus an infinite sum:

$$\alpha = 6 - \frac{12}{\sqrt{2}} + \frac{8}{\sqrt{3}} - \cdots \qquad 38\text{-}2$$

The value of the Madelung constant for face-centered-cubic structures is $\alpha = 1.7476$.*

* A large number of terms are needed to calculate the Madelung constant accurately because the sum converges very slowly.

(a)

Crystal structure. (*a*) The hexagonal symmetry of a snowflake arises from a hexagonal symmetry in its lattice of hydrogen atoms and oxygen atoms. (*b*) NaCl (salt) crystals, magnified approximately thirty times. The crystals are built up from a cubic lattice of sodium and chloride ions. In the absence of impurities, an exact cubic crystal is formed. This (false-color) scanning electron micrograph shows that in practice the basic cube is often disrupted by dislocations, giving rise to crystals that have a wide variety of shapes. The underlying cubic symmetry, though, remains evident. (*c*) A crystal of quartz (SiO_2, silicon dioxide), the most abundant and widespread mineral on Earth. If molten quartz solidifies without crystallizing, glass is formed. (*d*) A soldering iron tip, ground down to reveal the copper core within its iron sheath. Visible in the iron is its underlying microcrystalline structure.
((a) Richard Waters 2/89 p. 52 Discover. (b) © Dr. Jeremy Burgess/Science Photo Library/Photo Researchers. (c) © Thomas R. Taylor/Photo Researchers. (d) Courtesy the AT&T Archives.)

(b)

(c)

(d)

When Na^+ and Cl^- ions are very close together, they repel each other because of the overlap of their electron orbitals and the exclusion-principle repulsion discussed in Section 37-1. A simple empirical expression for the potential energy associated with this repulsion that works fairly well is

$$U_{rep} = \frac{A}{r^n}$$

where A and n are constants. The total potential energy of an ion is then

$$U = -\alpha\frac{ke^2}{r} + \frac{A}{r^n} \qquad 38\text{-}3$$

The equilibrium separation $r = r_0$ is that at which the force $F = -dU/dr$ is zero. Differentiating and setting $dU/dr = 0$ at $r = r_0$, we obtain

$$A = \frac{\alpha ke^2 r_0^{n-1}}{n} \qquad 38\text{-}4$$

Substituting for A in Equation 38-3 gives

$$U = -\alpha\frac{ke^2}{r_0}\left[\frac{r_0}{r} - \frac{1}{n}\left(\frac{r_0}{r}\right)^n\right] \qquad 38\text{-}5$$

At $r = r_0$, we have

$$U(r_0) = -\alpha\frac{ke^2}{r_0}\left(1 - \frac{1}{n}\right) \qquad 38\text{-}6$$

If we know the equilibrium separation r_0, the value of n can be found approximately from the *dissociation energy* of the crystal, which is the energy needed to break up the crystal into atoms.

Example 38-1 Separation Distance between Na^+ and Cl^- in NaCl

Calculate the equilibrium separation r_0 for NaCl from the measured density of NaCl, which is $\rho = 2.16\ \text{g/cm}^3$.

PICTURE We consider each ion to occupy a cubic volume of side r_0. The mass of 1 mol of NaCl is 58.4 g, which is the sum of the molar masses of sodium and chlorine. There are $2N_A$ ions in 1 mol of NaCl, where $N_A = 6.02 \times 10^{23}$ is Avogadro's number.

SOLVE

1. We consider each ion to occupy a cubic volume of side r_0. The volume v of one mole of NaCl equals the number of ions multiplied by the volume per ion:

$$v = 2N_A r_0^3$$

2. Relate r_0 to the density ρ and the molar mass M of NaCl:

$$\rho = \frac{M}{v} = \frac{M}{2N_A r_0^3}$$

3. Solve for r_0^3 and substitute the known values:

$$r_0^3 = \frac{M}{2N_A\rho} = \frac{58.4\ \text{g}}{2(6.02 \times 10^{23})(2.16\ \text{g/cm}^3)} = 2.25 \times 10^{-23}\ \text{cm}^3$$

so

$$r_0 = 2.82 \times 10^{-8}\ \text{cm} = \boxed{0.282\ \text{nm}}$$

CHECK In Chapter 36, we found the diameter of the hydrogen atom in the ground state to be about 0.11 nm. Our step 3 result is less than three times larger. Thus, $r_0 = 0.282$ nm is plausible.

The measured dissociation energy of NaCl is 770 kJ/mol. Using $1\ \text{eV} = 1.602 \times 10^{-19}$ J and the fact that 1 mol of NaCl has N_A pairs of ions, we can express the dissociation energy in electron volts per ion pair. The conversion between electron volts per ion pair and kilojoules per mole is

$$1\frac{\text{eV}}{\text{ion pair}} \times \frac{6.022 \times 10^{23}\ \text{ion pairs}}{1\ \text{mol}} \times \frac{1.602 \times 10^{-19}\ \text{J}}{1\ \text{eV}}$$

The result is

$$1\frac{\text{eV}}{\text{ion pair}} = 96.47\frac{\text{kJ}}{\text{mol}} \qquad 38\text{-}7$$

Thus, 770 kJ/mol = 7.98 eV per ion pair. Substituting −7.98 eV for $U(r_0)$, 0.282 nm for r_0, and 1.75 for α in Equation 38-6, we can solve for n. The result is $n = 9.35 \approx 9$.

Most ionic crystals, such as LiF, KF, KCl, KI, and AgCl, have a face-centered-cubic structure. Some elemental solids that have fcc structure are silver, aluminum, gold, calcium, copper, nickel, and lead.

Figure 38-2 shows the structure of CsCl, which is called the *body-centered-cubic* (bcc) structure. In this structure, each ion has eight nearest neighbor ions of the opposite charge. The Madelung constant for these crystals is 1.7627. Elemental solids that have bcc structure include barium, cesium, iron, potassium, lithium, molybdenum, and sodium.

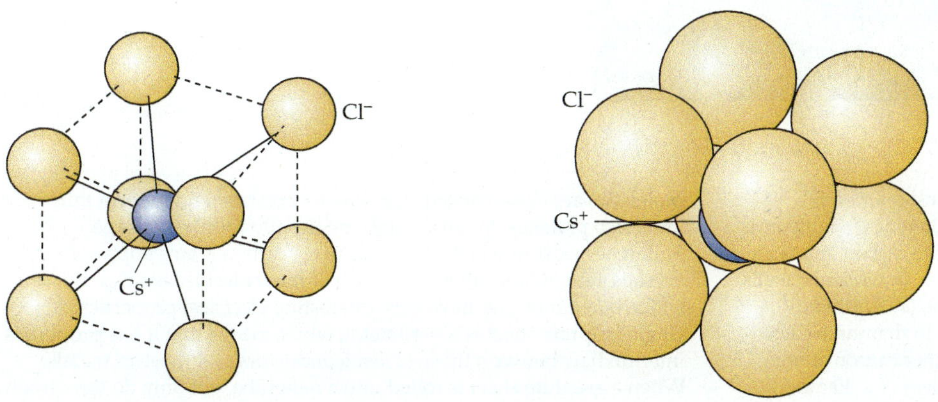

FIGURE 38-2 Body-centered-cubic structure of the CsCl crystal.

Figure 38-3 shows another important crystal structure: the *hexagonal close-packed* (hcp) structure. This structure is obtained by stacking identical spheres, such as bowling balls. In the first layer, each ball touches six others; thus, the name *hexagonal.* In the next layer, each ball fits into a triangular depression of the first layer. In the third layer, each ball fits into a triangular depression of the second layer, so it lies directly over a ball in the first layer. Elemental solids that have hcp structure include beryllium, cadmium, cerium, magnesium, osmium, and zinc.

For solids that have covalent bonding, the crystal structure is determined by the configuration of the bonds. Figure 38-4 illustrates the diamond structure of carbon, in which each atom is bonded to four other atoms as a result of hybridization, which is discussed in Section 37-2. This configuration is also the structure of germanium and silicon.

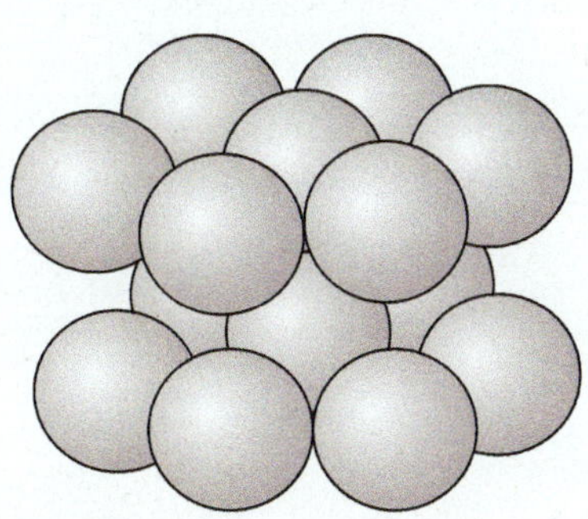

FIGURE 38-3 Hexagonal close-packed crystal structure.

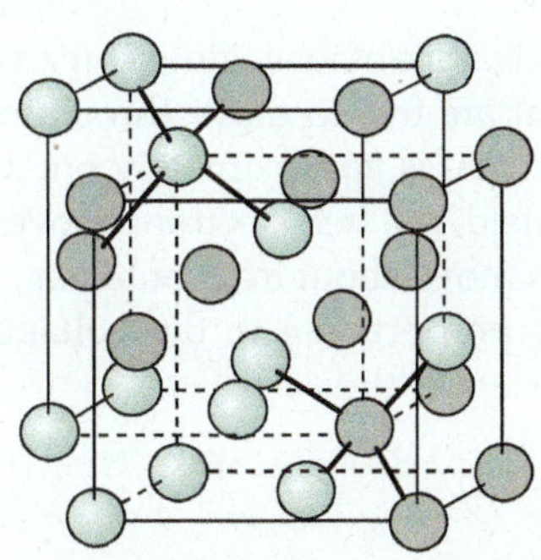

FIGURE 38-4 Diamond crystal structure. This structure can be considered to be a combination of two interpenetrating face-centered-cubic structures.

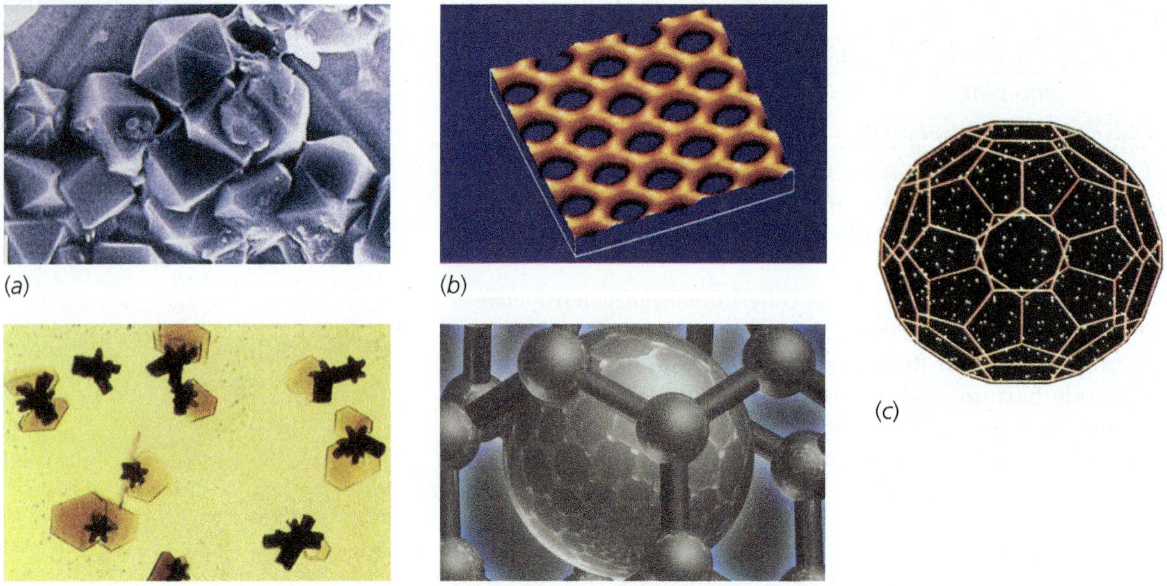

(a) (b) (c) (d) (e)

Carbon exists in three well-defined crystalline forms: diamond, graphite, and fullerenes (short for "buckminsterfullerenes"). Fullerenes were discovered in 1985. The forms differ in how the carbon atoms are packed together in a lattice. A fourth form of carbon, in which no well-defined crystalline form exists, is common charcoal. (*a*) Synthetic diamonds, magnified approximately 75,000 times. In diamond, each carbon atom is centered in a tetrahedron of four other carbon atoms. The strength of these bonds accounts for the hardness of a diamond. (*b*) An atomic-force micrograph of graphite. In graphite, carbon atoms are arranged in sheets, where each sheet is made up of atoms in hexagonal rings. The sheets slide easily across one another, a property that allows graphite to function as a lubricant. (*c*) A single sheet of carbon rings can be closed on itself if certain rings are allowed to be pentagonal, instead of hexagonal. A computer-generated image of the smallest such structure, C_{60}, is shown here. Each of the sixty vertices corresponds to a carbon atom; twenty of the faces are hexagons and twelve of the faces are pentagons. The same geometric pattern is encountered in a soccer ball. (*d*) Fullerene crystals, in which C_{60} molecules are close-packed. The smaller crystals tend to form thin brownish platelets; larger crystals are usually rodlike in shape. Fullerenes exist in which more than sixty carbon atoms appear. In the crystals shown here, about one-sixth of the molecules are C_{70}. (*e*) Carbon nanotubes have very interesting electrical properties. A single graphite sheet is a semimetal, which means that it has properties intermediate between those of semiconductors and those of metals. When a graphite sheet is rolled into a nanotube, not only do the carbon atoms have to line up around the circumference of the tube, but the wave functions of the electrons must also match up. This boundary-matching requirement places restrictions on these wave functions, which affects the motion of the electrons. Depending on exactly how the tube is rolled up, the nanotube can be either a semiconductor or a metal. *((a) Chris Kovach 3/91 p. 69 Discover. (b) Srinivas Manne, University of California, Santa Barbara. (c) Dr. F. A. Quiocho and J. S. Spurlino/Howard Hughes Medical Institute, Baylor College of Medicine. (d) W. Krätschmer/Max-Planck-Institute for Nuclear Physics. (e) © Kenneth Weard/BioGrafx/Science Source/Photo Researchers.)*

38-2 A MICROSCOPIC PICTURE OF CONDUCTION

We consider a metal as a regular three-dimensional lattice of ions filling some volume V and having a large number N of electrons that are free to move throughout the whole metal. The number of free electrons in a metal is approximately one to four electrons per atom. In the absence of an electric field, the free electrons move about the metal randomly, much the way gas molecules move about in a container.

The current in a conducting wire segment is proportional to the voltage drop across the segment:

$$I = \frac{V}{R} \qquad (\text{or } V = IR)$$

The resistance R is proportional to the length L of the wire segment and inversely proportional to the cross-sectional area A:

$$R = \rho\frac{L}{A}$$

where ρ is the resistivity. Substituting $\rho L/A$ for R, and EL for V, we can write the current in terms of the electric field strength E and the resistivity. We have

$$I = \frac{V}{R} = \frac{EL}{\rho L/A} = \frac{1}{\rho}EA$$

Dividing both sides by the area A gives $I/A = (1/\rho)E$, or $J = (1/\rho)E$, where $J = I/A$ is the magnitude of the **current density** vector $\vec{J}$. The current density vector is defined as

$$\vec{J} = qn\vec{v}_{\mathrm{d}} \qquad 38\text{-}8$$

DEFINITION—CURRENT DENSITY

where q, n, and $\vec{v}_{\mathrm{d}}$ are the charge, the number density, and the drift velocity of the charge carrier. (This follows from Equation 25-3.) In vector form, the relation between the current density and the electric field is

$$\vec{J} = \frac{1}{\rho}\vec{E} \qquad 38\text{-}9$$

This relation is the point form of Ohm's law. The reciprocal of the resistivity is called the **conductivity.**

According to Ohm's law, the resistivity is independent of both the current density and the electric field $\vec{E}$. Combining Equations 38-8 and 38-9 gives

$$-en_{\mathrm{e}}\vec{v}_{\mathrm{d}} = \frac{1}{\rho}\vec{E} \qquad 38\text{-}10$$

where $-e$ and n_{e} have been substituted for q and n, respectively. According to Equation 38-10, the drift velocity $\vec{v}_{\mathrm{d}}$ is proportional to $\vec{E}$.

In the presence of an electric field, a free electron experiences a force $-e\vec{E}$. If this were the only force acting, the electron would have a constant acceleration $-e\vec{E}/m_{\mathrm{e}}$. However, Equation 38-10 implies a steady-state situation with a constant drift velocity that is proportional to the field $\vec{E}$. In the microscopic model, it is assumed that a free electron is accelerated for a short time and then makes a collision with a lattice ion. The velocity of the electron immediately after the collision is completely unrelated to the drift velocity. The justification for this assumption is that the magnitude of the drift velocity is extremely small compared with the speeds associated with the thermal kinetic energies of the free electrons.

For a typical free electron, its velocity a time t after its last collision is $\vec{v}_0 - (-e\vec{E}/m_{\mathrm{e}})t$, where $\vec{v}_0$ is its velocity immediately after that collision. Because the direction of $\vec{v}_0$ is random, it does not contribute to the average velocity of the electrons. Thus, the average velocity or drift velocity of the electrons is

$$\vec{v}_{\mathrm{d}} = -\frac{e\vec{E}}{m_{\mathrm{e}}}\tau \qquad 38\text{-}11$$

where τ is the average time since the last collision. Substituting for $\vec{v}_{\mathrm{d}}$ in Equation 38-10, we obtain

$$-n_{\mathrm{e}}e\left(\frac{e\vec{E}}{m_{\mathrm{e}}}\tau\right) = \frac{1}{\rho}\vec{E}$$

so

$$\rho = \frac{m_{\mathrm{e}}}{n_{\mathrm{e}}e^2\tau} \qquad 38\text{-}12$$

The time τ, called the **collision time,** is also the average time between collisions.*

* It is tempting but incorrect to think that if τ is the average time between collisions, the average time since its last collision is $\frac{1}{2}\tau$ rather than τ. If you find this confusing, you may take comfort in the fact that Drude used the incorrect result $\frac{1}{2}\tau$ in his original work.

The average distance an electron travels between collisions is $v_{av}\tau$, which is called the mean free path λ:

$$\lambda = v_{av}\tau \qquad 38\text{-}13$$

where v_{av} is the mean speed of the electrons. (The mean speed is many orders of magnitude greater than the drift speed.) In terms of the mean free path and the mean speed, the resistivity is

$$\rho = \frac{m_e v_{av}}{n_e e^2 \lambda} \qquad 38\text{-}14$$

RESISTIVITY IN TERMS OF v_{AV} AND λ

According to Ohm's law, the resistivity ρ is independent of the electric field $\vec{E}$. Because m_e, n_e, and e are constants, the only quantities that could possibly depend on $\vec{E}$ are the mean speed v_{av} and the mean free path λ. Let us examine these quantities to see if they can possibly depend on the applied field $\vec{E}$.

CLASSICAL INTERPRETATION OF v_{av} AND λ

Classically, at $T = 0$ all the free electrons in a conductor should have zero kinetic energy. As the conductor is heated, the lattice ions acquire an average kinetic energy of $\frac{3}{2}kT$, which is imparted to the free electrons by the collisions between the electrons and the ions. (This is a result of the equipartition theorem studied in Chapters 17 and 18.) The free electrons would then have a Maxwell–Boltzmann distribution just like a gas of molecules. In equilibrium, the electrons would be expected to have a mean kinetic energy of $\frac{3}{2}kT$, which at ordinary temperatures (~300 K) is approximately 0.04 eV. At $T = 300$ K, their root-mean-square (rms) speed,* which is slightly greater than the mean speed, is

$$v_{av} \approx v_{rms} = \sqrt{\frac{3kT}{m_e}} = \sqrt{\frac{3(1.38 \times 10^{-23}\ \text{J/K})(300\ \text{K})}{9.11 \times 10^{-31}\text{kg}}} = 1.17 \times 10^5\ \text{m/s} \qquad 38\text{-}15$$

Note that this is about nine orders of magnitude greater than the typical drift speed of 3.5×10^{-5} m/s, which was calculated in Example 25-1. The very small drift speed caused by the electric field therefore has essentially no effect on the very large mean speed of the electrons, so v_{av} in Equation 38-14 cannot depend on the electric field $\vec{E}$.

The mean free path is related classically to the size of the lattice ions in the conductor and to the number of ions per unit volume. Consider one electron moving with speed v through a region of stationary ions that are assumed to be hard spheres (Figure 38-5). Assume the size of the electron is negligible. The electron will collide with an ion if it comes within a distance r from the center of the ion, where r is the radius of the ion. During some time interval Δt_1, the electron moves a distance vt_1. If there is an ion whose center is in the cylindrical volume $\pi r^2 v\Delta t_1$, the electron will collide with the ion. The electron will then change directions and collide with another ion in time Δt_2 if the center of the ion is in the volume $\pi r^2 vt_2$. Thus, in the total time $\Delta t = \Delta t_1 + \Delta t_2 + \ldots$, the electron will collide with the number of ions whose centers are in the volume $\pi r^2 v\Delta t$. The number of ions in this volume is $n_{ion}\pi r^2 v\Delta t$, where n_{ion} is the number of ions per unit volume.

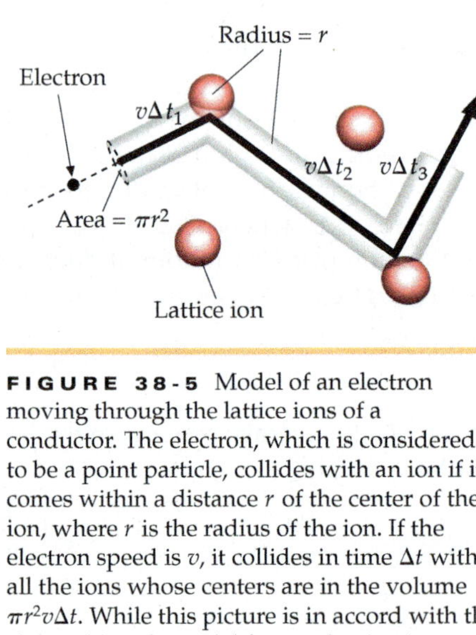

FIGURE 38-5 Model of an electron moving through the lattice ions of a conductor. The electron, which is considered to be a point particle, collides with an ion if it comes within a distance r of the center of the ion, where r is the radius of the ion. If the electron speed is v, it collides in time Δt with all the ions whose centers are in the volume $\pi r^2 v\Delta t$. While this picture is in accord with the classical Drude model for conduction in metals, it is in conflict with the current quantum-mechanical model presented later in this chapter.

* See Equation 17–21.

The total path length divided by the number of collisions is the mean free path:

$$\lambda = \frac{v\Delta t}{n_{\text{ion}}\pi r^2 v\Delta t} = \frac{1}{n_{\text{ion}}\pi r^2} = \frac{1}{n_{\text{ion}}A} \qquad 38\text{-}16$$

where $A = \pi r^2$ is the cross-sectional area of a lattice ion.

SUCCESSES AND FAILURES OF THE CLASSICAL MODEL

Neither n_{ion} nor r depends on the electric field $\vec{E}$, so λ also does not depend on $\vec{E}$. v_{av} and λ do not depend on $\vec{E}$ according to their classical interpretations, so the resistivity ρ does not depend on $\vec{E}$ in accordance with Ohm's law. However, the classical theory gives an incorrect temperature dependence for the resistivity. Because λ depends only on the radius and the number density of the lattice ions, the only quantity in Equation 38-14 that depends on temperature in the classical theory is v_{av}, which is proportional to $\sqrt{T}$. But experiments show that ρ varies linearly with temperature. Furthermore, when ρ is calculated at $T = 300$ K using the Maxwell–Boltzmann distribution for v_{av} and Equation 38-16 for λ, the calculated result is about six times greater than the measured value.

The classical theory of conduction fails because electrons are not classical particles. The wave nature of the electrons must be considered. Because of the wave properties of electrons and the constraints described by the exclusion principle (to be discussed in the following section), the energy distribution of the free electrons in a metal is not even approximately given by the Maxwell–Boltzmann distribution. Furthermore, the collision of an electron with a lattice ion is not similar to the collision of a baseball with a tree. Instead, it involves the scattering of electron waves by the lattice. To understand the quantum theory of conduction, we need a qualitative understanding of the energy distribution of free electrons in a metal. This will also help us understand the origin of contact potentials between two dissimilar metals in contact and the contribution of free electrons to the heat capacity of metals.

38-3 FREE ELECTRONS IN A SOLID

One may want to consider free electrons in a metal to be an *electron gas* in a metal. However, molecules in an ordinary gas, such as air, obey the classical Maxwell–Boltzmann energy distribution, but the free electrons in a metal do not. Instead, they obey a quantum energy distribution called the *Fermi–Dirac distribution*. The main features of a free electron can be understood by considering the electron in a metal to be a particle in a box, a problem whose one-dimensional version we studied extensively in Chapter 34. We discuss the main features of a free electron semiquantitatively in this section and leave the details of the Fermi–Dirac distribution to Section 38-9.

ENERGY QUANTIZATION IN A BOX

In Chapter 34, we found that the wavelength associated with an electron of momentum p is given by the de Broglie relation:

$$\lambda = \frac{h}{p} \qquad 38\text{-}17$$

where h is Planck's constant. When a particle is confined to a finite region of space, such as a box, only certain wavelengths λ_n, where $n = 1, 2, \ldots$, that are specified by standing-wave conditions are allowed. For a one-dimensional box of length L, the standing-wave condition is

$$n\frac{\lambda_n}{2} = L \qquad n = 1, 2, \ldots \qquad 38\text{-}18$$

This results in the quantization of energy:

$$E_n = \frac{p_n^2}{2m} = \frac{(h/\lambda_n)^2}{2m} = \frac{h^2}{2m}\frac{1}{\lambda_n^2} = \frac{h^2}{2m}\frac{1}{(2L/n)^2}$$

or

$$E_n = n^2 E_1 \qquad 38\text{-}19$$

where $E_1 = h^2/(8mL^2)$. The spatial wave function for the nth state is given by

$$\psi_n(x) = \sqrt{\frac{2}{L}}\sin(n\pi x/L) \qquad 38\text{-}20$$

The quantum number n characterizes the wave function for a particular state and the energy of that state. In three-dimensional problems, three quantum numbers arise, one associated with each dimension.

THE EXCLUSION PRINCIPLE

The distribution of electrons among the possible energy states is described by the exclusion principle, which states that no two electrons in an atom can be in the same quantum state; that is, they cannot have the same set of values for their quantum numbers. The exclusion principle applies to all "spin one-half" particles (fermions), which include electrons, protons, and neutrons. These particles have a *spin* quantum number m_s which has two possible values, $+\frac{1}{2}$ and $-\frac{1}{2}$. The quantum state of a particle is characterized by the spin quantum number m_s and the quantum numbers associated with the spatial part of the wave function. Because the spin quantum numbers have just two possible values, the exclusion principle can be stated in terms of the spatial states:

> There can be at most two electrons with the same set of values for their *spatial* quantum numbers.
>
> EXCLUSION PRINCIPLE IN TERMS OF SPATIAL STATES

When there are more than two electrons in a system, such as an atom, only two can be in the lowest energy state. The third and fourth electrons must go into the second-lowest state, and so on.

Example 38-2 Boson-System Energy versus Fermion-System Energy

Compare the total energy of the ground state of five identical bosons of mass m in a one-dimensional box with the total energy of the ground state of five identical fermions of mass m in the same box.

PICTURE The ground state is the lowest possible energy state. The energy levels in a one-dimensional box are given by $E_n = n^2E_1$, where $E_1 = h^2/(8mL^2)$. (This is in accord with Equation 38-19.) The lowest energy for five bosons occurs when all the bosons are in the state $n = 1$, as shown in Figure 38-6*a*. For fermions, the lowest state occurs when two fermions are in the state $n = 1$, two fermions are in the state $n = 2$, and one fermion is in the state $n = 3$, as shown in Figure 38-6*b*.

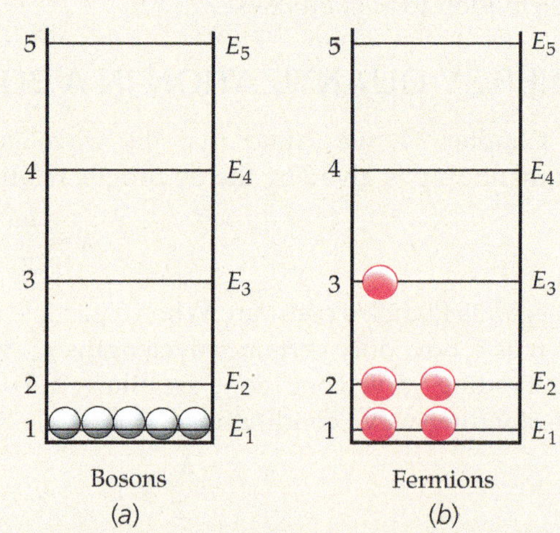

FIGURE 38-6

SOLVE

1. The energy of five bosons in the state $n = 1$ is: $E = 5E_1$

2. The energy of two fermions in the state $n = 1$, two fermions in the state $n = 2$, and one fermion in the state $n = 3$ is:
$$E = 2E_1 + 2E_2 + 1E_3 = 2E_1 + 2(2)^2E_1 + 1(3)^2E_1 = 2E_1 + 8E_1 + 9E_1 = 19E_1$$

3. Compare the total energies: The five identical fermions have 3.8 times the total energy of the five identical bosons.

CHECK The fact that fermions must have different quantum states has a large effect on the total energy of a multiple-particle system, as expected.

THE FERMI ENERGY

When there are many electrons in a box, at $T = 0$ the electrons will occupy the lowest energy states consistent with the exclusion principle. If we have N electrons, we can put two electrons in the lowest energy level, two electrons in the next lowest energy level, and so on. The N electrons thus fill the lowest $N/2$ energy levels (Figure 38-7). The energy of the last filled (or half-filled) level at $T = 0$ is called the Fermi energy E_F. If the electrons moved in a one-dimensional box, the Fermi energy would be given by Equation 38-19, with $n = N/2$:

$$E_F = \left(\frac{N}{2}\right)^2 \frac{h^2}{8m_eL^2} = \frac{h^2}{32m_e}\left(\frac{N}{L}\right)^2 \qquad \text{38-21}$$

FERMI ENERGY AT $T = 0$ IN ONE DIMENSION

FIGURE 38-7 At $T = 0$ the electrons fill up the allowed energy states to the Fermi energy E_F. The levels are so closely spaced that they can be assumed to be continuous.

In a one-dimensional box, the Fermi energy depends on the number of free electrons per unit length of the box.

PRACTICE PROBLEM 38-1

Suppose there is an ion, and therefore a free electron, every 0.100 nm in a one-dimensional box. Calculate the Fermi energy. *Hint: Write Equation 38-21 as*

$$E_F = \frac{(hc)^2}{32m_ec^2}\left(\frac{N}{L}\right)^2 = \frac{(1240\text{ eV}\cdot\text{nm})^2}{32(0.511\text{ MeV})}\left(\frac{N}{L}\right)^2$$

In our model of conduction, the free electrons move in a *three-dimensional* box of volume V. The derivation of the Fermi energy in three dimensions is somewhat difficult, so we will just give the result. In three dimensions, the Fermi energy at $T = 0$ is

$$E_F = \frac{h^2}{8m_e}\left(\frac{3N}{\pi V}\right)^{2/3} \qquad \text{38-22}a$$

FERMI ENERGY AT $T = 0$ IN THREE DIMENSIONS

The Fermi energy depends on the number density of free electrons N/V. Substituting numerical values for the constants gives

$$E_F = (0.3646\text{ eV}\cdot\text{nm}^2)\left(\frac{N}{V}\right)^{2/3} \qquad \text{38-22}b$$

FERMI ENERGY AT $T = 0$ IN THREE DIMENSIONS

Example 38-3 The Fermi Energy for Copper

The number density for electrons in copper was calculated in Example 25-1 and found to be $84.7/\text{nm}^3$. Calculate the Fermi energy at $T = 0$ for copper.

PICTURE The Fermi energy is given by Equations 38-22.

SOLVE

1. The Fermi energy is given by Equation 38-22*b*: $E_F = (0.3646\ \text{eV}\cdot\text{nm}^2)\left(\frac{N}{V}\right)^{2/3}$

2. Substitute the given number density for copper: $E_F = (0.3646\ \text{eV}\cdot\text{nm}^2)(84.7/\text{nm}^3)^{2/3} = \boxed{7.03\ \text{eV}}$

CHECK The Fermi energy (the step-2 result) is much greater than kT at room temperatures as expected. For example, at $T = 300$ K, kT is only about 0.026 eV.

PRACTICE PROBLEM 38-2 Use Equation 38-22*b* to calculate the Fermi energy at $T = 0$ for gold, which has a free-electron number density of $59.0/\text{nm}^3$.

Table 38-1 lists the free-electron number densities and Fermi energies at $T = 0$ for several metals.

The free electrons in a metal are sometimes referred to as a Fermi gas. (They constitute a gas of fermions.) The average energy of a free electron can be calculated from the complete energy distribution of the electrons, which is discussed in Section 38-9. At $T = 0$, the average energy turns out to be

$$E_{av} = \tfrac{3}{5}E_F \qquad 38\text{-}23$$

AVERAGE ENERGY OF ELECTRONS IN A FERMI GAS AT $T = 0$

Table 38-1 Free-Electron Number Densities* and Fermi Energies at $T = 0$ for Selected Elements

	Element	N/V, electrons/nm^3	E_F, eV
Al	Aluminum	181	11.7
Ag	Silver	58.6	5.50
Au	Gold	59.0	5.53
Cu	Copper	84.7	7.03
Fe	Iron	170	11.2
K	Potassium	14.0	2.11
Li	Lithium	47.0	4.75
Mg	Magnesium	86.0	7.11
Mn	Manganese	165	11.0
Na	Sodium	26.5	3.24
Sn	Tin	148	10.2
Zn	Zinc	132	9.46

* Number densities are measured using the Hall effect, discussed in Section 26-4.

For copper, E_{av} is approximately 4 eV. This average energy is huge compared with thermal energies of about $kT \approx 0.026$ eV at a temperature of $T = 300$ K. This result is very different from the classical Maxwell–Boltzmann distribution result that at $T = 0$, $E = 0$, and that at some temperature T, E is of the same order as kT.

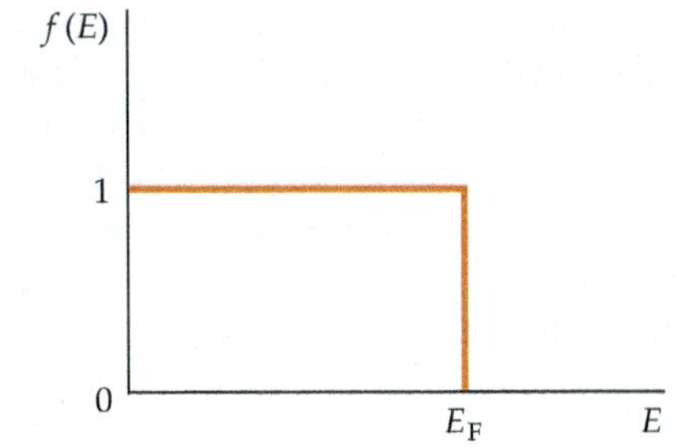

FIGURE 38-8 Fermi factor versus energy at $T = 0$.

THE FERMI FACTOR AT $T = 0$

The probability of an energy state being occupied is called the **Fermi factor,** $f(E)$. At $T = 0$ all the states below E_F are filled, whereas all those above that energy are empty, as shown in Figure 38-8. Thus, at $T = 0$ the Fermi factor is simply

$$f(E) = \begin{cases} 1 & E < E_F \\ 0 & E > E_F \end{cases} \qquad 38\text{-}24$$

THE FERMI FACTOR FOR $T > 0$

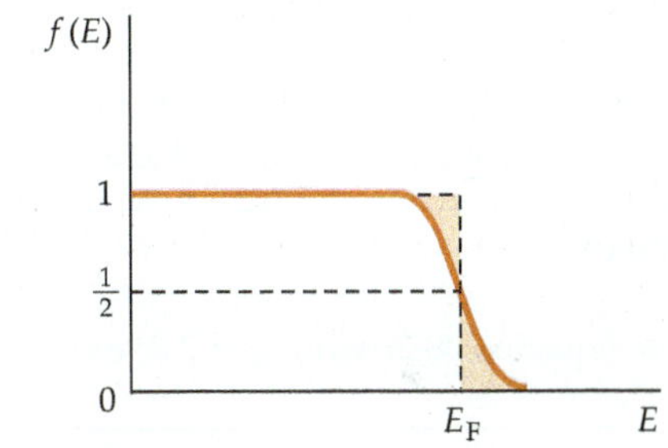

FIGURE 38-9 The Fermi factor for some temperature T. Some electrons that have energies near the Fermi energy are excited, as indicated by the shaded regions. The Fermi energy E_F is that value of E for which $f(E) = \frac{1}{2}$.

At temperatures greater than $T = 0$, some electrons will occupy higher energy states because of thermal energy gained during collisions with the lattice. However, an electron cannot move to a higher or lower state unless it is unoccupied. Because the kinetic energy of the lattice ions is of the order of kT, electrons cannot gain much more energy than kT in collisions with the lattice ions. Therefore, only those electrons that have energies within about kT of the Fermi energy can gain energy as the temperature is increased. At 300 K, kT is only 0.026 eV, so the exclusion principle prevents all but a very few electrons near the top of the energy distribution from gaining energy through random collisions with the lattice ions. Figure 38-9 shows a plot of the Fermi factor for some temperature T. Because for $T > 0$ there is no distinct energy that separates filled levels from unfilled levels, the definition of the Fermi energy must be slightly modified. At temperature T, the Fermi energy is defined to be the energy of the energy state for which the probability of being occupied is $\frac{1}{2}$. For all but extremely high temperatures, the difference between the Fermi energy at temperature T and the Fermi energy at temperature $T = 0$ is very small.

The **Fermi temperature** T_F is defined by

$$kT_F = E_F \qquad 38\text{-}25$$

For temperatures much lower than the Fermi temperature, the average energy of the lattice ions will be much less than the Fermi energy, and the electron energy distribution will not differ greatly from that at $T = 0$.

Example 38-4 The Fermi Temperature for Copper

Find the Fermi temperature for copper.

PICTURE We use Equation 38-25 to find the Fermi temperature. The Fermi energy for copper at $T = 0$, calculated in Example 38-3, is 7.03 eV.

SOLVE

Use $E_F = 7.03$ eV and $k = 8.617 \times 10^{-5}$ eV/K in Equation 38-25:

$$T_F = \frac{E_F}{k} = \frac{7.03 \text{ eV}}{8.617 \times 10^{-5} \text{ eV/K}} = \boxed{81\,600 \text{ K}}$$

CHECK The Fermi temperature is very high, as expected.

TAKING IT FURTHER We can see from this example that the Fermi temperature of copper is much greater than any temperature T for which copper remains a solid.

Because an electric field in a conductor accelerates all of the conduction electrons together, the exclusion principle does not prevent the free electrons in filled states from participating in conduction. Figure 38-10 shows the Fermi factor in one dimension versus *velocity* for an ordinary temperature. The factor is approximately 1 for velocities v_x in the range $-u_F < v_x < u_F$, where the Fermi speed u_F is related to the Fermi energy by $E_F = \frac{1}{2}mu_F^2$. Then

$$u_F = \sqrt{\frac{2E_F}{m_e}} \qquad \text{38-26}$$

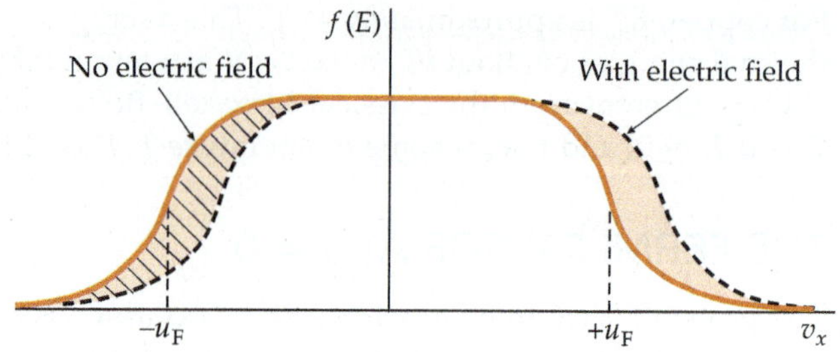

FIGURE 38-10 Fermi factor versus velocity in one dimension with no electric field (solid) and with an electric field in the $-x$ direction (dashed). The difference is greatly exaggerated.

Example 38-5 The Fermi Speed for Copper

Calculate the Fermi speed for copper.

PICTURE We use Equation 38-26 to find the Fermi speed. The Fermi energy for copper at $T = 0$, calculated in Example 38-3, is 7.03 eV.

SOLVE

Use Equation 38-26 with $E_F = 7.03$ eV:

$$u_F = \sqrt{\frac{2(7.03\text{ eV})}{9.11 \times 10^{-31}\text{ kg}}\left(\frac{1.60 \times 10^{-19}\text{ J}}{1\text{ eV}}\right)} = \boxed{1.57 \times 10^6\text{ m/s.}}$$

CHECK As expected, the result (the Fermi speed for copper) is high, but less than the speed of light.

The dashed curve in Figure 38-10 shows the Fermi factor after the electric field has been acting for some time t. Although all of the free electrons have their velocities shifted in the direction opposite to the electric field, the net effect is equivalent to shifting only the electrons near the Fermi energy.

CONTACT POTENTIAL

When two different metals are placed in contact, a potential difference $V_{contact}$ called the **contact potential** develops between them. The contact potential depends on both the work functions of the two metals, ϕ_1 and ϕ_2 (we encountered work functions when the photoelectric effect was introduced in Chapter 34), and the Fermi energies of the two metals. When the metals are in contact, the total energy of the system is lowered if electrons near the boundary move from the metal that has the higher Fermi energy into the metal that has the lower Fermi energy until the Fermi energies of the two metals are the same, as shown in Figure 38-11. When equilibrium is established, the metal that has the lower initial Fermi energy is negatively charged and the other metal is positively charged, so that between them there is a potential difference $V_{contact}$ given by

$$V_{contact} = \frac{\phi_1 - \phi_2}{e} \qquad \text{38-27}$$

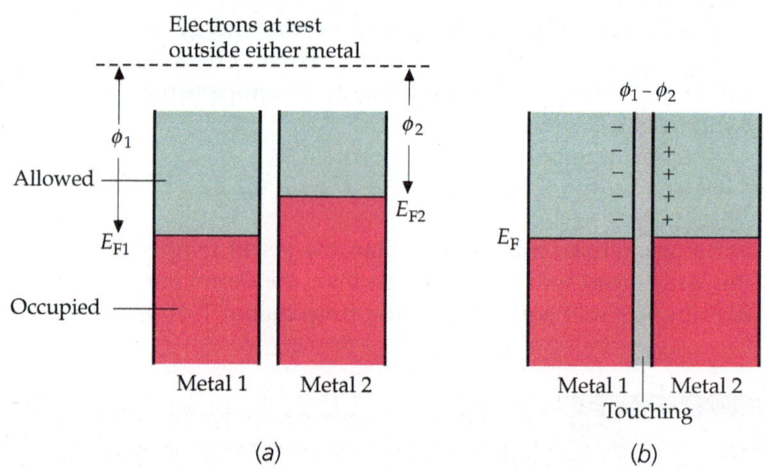

FIGURE 38-11 (*a*) Energy levels for two different metals that have different Fermi energies E_F and work functions ϕ. The work function is the difference between the energy of an electron at rest outside the metal and the Fermi energy within the metal. (*b*) When the metals are in contact, electrons flow from the metal that initially has the higher Fermi energy to the metal that initially has the lower Fermi energy until the Fermi energies are equal.

Table 38-2 lists the work functions for several metals.

Table 38-2 Work Functions for Some Metals

	Metal	ϕ, eV		Metal	ϕ, eV
Ag	Silver	4.7	K	Potassium	2.1
Au	Gold	4.8	Mn	Manganese	3.8
Ca	Calcium	3.2	Na	Sodium	2.3
Cu	Copper	4.1	Ni	Nickel	5.2

Example 38-6 Contact Potential between Silver and Tungsten

The threshold wavelength for the photoelectric effect is 271 nm for tungsten and 262 nm for silver. What is the contact potential developed when silver and tungsten are placed in contact?

PICTURE The contact potential is proportional to the difference in the work functions for the two metals (Equation 38-27). The work function ϕ can be found from the given threshold wavelengths using $\phi = hc/\lambda_t$ (Equation 34-4).

SOLVE

1. The contact potential is given by Equation 38-27:

$$V_{\text{contact}} = \frac{\phi_1 - \phi_2}{e}$$

2. The work function is related to the threshold wavelength (Equation 34-4):

$$\phi = \frac{hc}{\lambda_t}$$

3. Substitute $\lambda_t = 271$ nm for tungsten (the symbol for tungsten is W):

$$\phi_W = \frac{hc}{\lambda_t} = \frac{1240 \text{ eV}\cdot\text{nm}}{271 \text{ nm}} = 4.58 \text{ eV}$$

4. Substitute $\lambda_t = 262$ nm for silver:

$$\phi_{Ag} = \frac{1240 \text{ eV}\cdot\text{nm}}{262 \text{ nm}} = 4.73 \text{ eV}$$

5. The contact potential is thus:

$$V_{\text{contact}} = \frac{\phi_{Ag} - \phi_W}{e} = 4.73 \text{ V} - 4.58 \text{ V} = \boxed{0.15 \text{ V}}$$

CHECK As expected, the contact potential is small (less than one volt). You do not get large potential differences just by putting two metals in contact.

HEAT CAPACITY DUE TO ELECTRONS IN A METAL

The quantum-mechanical description of the electron distribution in metals allows us to understand why the contribution of the free electrons to the heat capacity of a metal is much less that of the ions. According to the classical equipartition theorem, the energy of the lattice ions in n moles of a solid is $3nRT$, and thus the molar specific heat is $c' = 3R$, where R is the universal gas constant (see Section 18-7). In a metal, the number of free electrons is approximately equal to the number of lattice ions. If these electrons obey the classical equipartition theorem, they should have an energy of $\frac{3}{2}nRT$ and contribute an additional $\frac{3}{2}R$ to the molar specific heat. But measured heat capacities of metals are just slightly greater than those of insulators. We can understand this result because at some temperature T, only those electrons that have energies near the Fermi energy can be excited by random collisions with the lattice ions. The number of those electrons is of the order of $(kT/E_F)N$, where N is the total number of free electrons. The energy of those electrons is increased from that at $T = 0$ by an amount that is of the order of kT. So the total increase in thermal energy is of the order of $(kT/E_F)N \times kT$.

We can thus express the energy of N electrons at temperature T as

$$E = NE_{av}(0) + \alpha N\frac{kT}{E_F}kT \tag{38-28}$$

where $E_{av}(0)$ is the average energy at $T = 0$ and α is a constant that we expect to be of the order of 1 if our reasoning is correct. The calculation of α is quite challenging. The result is $\alpha = \pi^2/4$. Using this result and writing E_F in terms of the Fermi temperature, $E_F = kT_F$, we obtain the following for the contribution of the free electrons to the heat capacity at constant volume:

$$C_V = \frac{dE}{dT} = 2\alpha Nk\frac{kT}{E_F} = \frac{1}{2}\pi^2 nR\frac{T}{T_F}$$

where we have written Nk in terms of the gas constant R ($R = Nk/n$). The molar specific heat at constant volume is then

$$c'_V = \frac{1}{2}\pi^2 R\frac{T}{T_F} \tag{38-29}$$

We can see that because of the large value of T_F, the contribution of the free electrons is a small fraction of R at ordinary temperatures. Because $T_F = 81\,600$ K for copper, the molar specific heat of the free electrons at $T = 300$ K is

$$c'_V = \frac{1}{2}\pi^2\frac{300\text{ K}}{81\,600\text{ K}}R \approx 0.02R$$

which is in good agreement with the experiment.

38-4 QUANTUM THEORY OF ELECTRICAL CONDUCTION

We can use Equation 38-14 for the resistivity if we use the Fermi speed u_F (Equation 38-26) in place of v_{av}:

$$\rho = \frac{m_e u_F}{n_e e^2 \lambda} \tag{38-30}$$

We now have two problems. First, because the Fermi speed u_F is approximately independent of temperature, the resistivity given by Equation 38-30 is independent of temperature unless the mean free path should depend on the temperature. The second problem concerns magnitudes. As mentioned earlier, the classical expression for resistivity using v_{av} calculated from the Maxwell–Boltzmann distribution gives values that are about 6 times too large at $T = 300$ K. Because the Fermi speed u_F is about 16 times the Maxwell-Boltzmann value of v_{av}, the magnitude of ρ predicted by Equation 38-30 will be approximately 100 times greater than the experimentally determined value. The resolution of both of these problems lies in the calculation of the mean free path λ.

THE SCATTERING OF ELECTRON WAVES

In Equation 38-16 for the classical mean free path $\lambda = 1/(n_{ion}A)$, the quantity $A = \pi r^2$ is the cross-sectional area of the lattice ion as seen by an electron. In the quantum calculation, the mean free path is related to the scattering of electron waves by the crystal lattice. Detailed calculations show that, for a *perfectly* ordered crystal, $\lambda = \infty$; that is, there is no scattering of the electron waves. The scattering of electron waves arises because of *imperfections* in the crystal lattice, which have nothing to do with the actual cross-sectional area A of the lattice ions.

According to the quantum theory of electron scattering, A depends merely on *deviations* of the lattice ions from a perfectly ordered array and not on the size of the ions. The most common causes of such deviations are thermal vibrations of the lattice ions or impurities.

We can use $\lambda = 1/(n_{ion}A)$ for the mean free path if we reinterpret the area A. Figure 38-12 compares the classical picture and the quantum picture of this area. In the quantum picture, the lattice ions are points that have no size but present an area $A = \pi r_0^2$, where r_0 is the amplitude of thermal vibrations. In Chapter 14, we saw that the energy of vibration in simple harmonic motion is proportional to the square of the amplitude, which is πr_0^2. Thus, the effective area A is proportional to the energy of vibration of the lattice ions. From the equipartition theorem,* we know that the average energy of vibration is proportional to kT. Thus, A is proportional to T, and λ is proportional to $1/T$. Then the resistivity given by Equation 38-14 is proportional to T, in agreement with experiment.

The effective area A due to thermal vibrations can be calculated, and the results give values for the resistivity that are in agreement with experiments. At $T = 300$ K, for example, the effective area turns out to be about 100 times smaller than the actual cross-sectional area of a lattice ion. We see, therefore, that the free-electron model of metals gives a good account of electrical conduction if the classical mean speed v_{av} is replaced by the Fermi speed u_F and if the collisions between electrons and the lattice ions are interpreted in terms of the scattering of electron waves, for which only deviations from a perfectly ordered lattice are important.

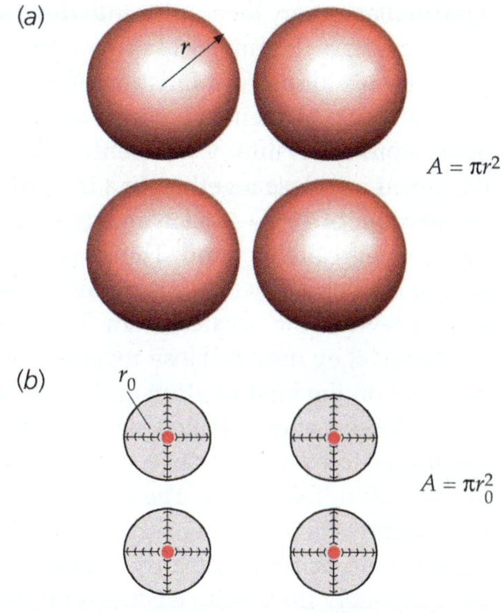

FIGURE 38-12 (*a*) Classical picture of the lattice ions as spherical balls of radius r that each present an area πr^2 to the electrons. (*b*) Quantum-mechanical picture of the lattice ions as points that are vibrating in three dimensions. The area presented to the electrons is πr_0^2, where r_0 is the amplitude of oscillation of the ions.

The presence of impurities in a metal also causes deviations from perfect regularity in the crystal lattice. The effects of impurities on resistivity are approximately independent of temperature. The resistivity of a metal containing impurities can be written $\rho = \rho_t + \rho_i$, where ρ_t is the resistivity due to the thermal motion of the lattice ions and ρ_i is the resistivity due to impurities. Figure 38-13 shows typical resistance versus temperature curves for metals with impurities. As the absolute temperature approaches zero, the resistivity due to thermal motion approaches zero, and the total resistivity approaches the resistivity due to impurities, which is constant.

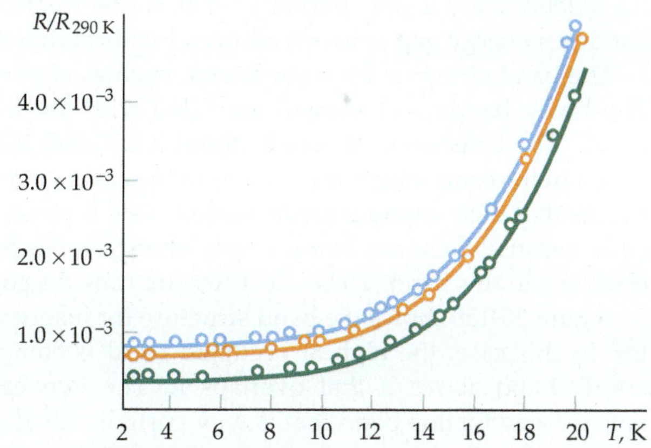

FIGURE 38-13 Relative resistance versus temperature for three samples of sodium. The three curves have the same temperature dependence but different magnitudes because of differing amounts of impurities in the samples.

38-5 BAND THEORY OF SOLIDS

Resistivities vary enormously between insulators and conductors. For a typical insulator, such as quartz, $\rho \sim 10^{16}\ \Omega \cdot \text{m}$, whereas for a typical conductor, $\rho \sim 10^{-8}\ \Omega \cdot \text{m}$. The reason for this enormous variation is the variation in the number density of free electrons n_e. To understand this variation, we consider the effect of the lattice on the electron energy levels.

We begin by considering the energy levels of the individual atoms as they are brought together. The allowed energy levels in an isolated atom are often far apart. For example, in hydrogen, the lowest allowed energy $E_1 = -13.6$ eV is 10.2 eV below the next lowest allowed energy $E_2 = (-13.6 \text{ eV})/4 = -3.4$ eV.† Let us consider two identical atoms and focus our attention on one particular energy level. When the atoms are far apart, the energy of a particular level is the same for each atom. As the atoms are brought closer together, the energy level for each atom changes because of the influence of the other atom. As a result, the level splits into two levels of slightly different energies for the two-atom system. If we bring three atoms close together,

* The equipartition theorem does hold for the lattice ions, which obey the Maxwell–Boltzmann energy distribution.

† The energy levels in hydrogen are discussed in Chapter 36.

a particular energy level splits into three separate levels of slightly different energies. Figure 38-14 shows the energy splitting of two energy levels for six atoms as a function of the separation of the atoms.

If we have N identical atoms, a particular energy level in the isolated atom splits into N different, closely spaced energy levels when the atoms are close together. In a macroscopic solid, N is very large—of the order of 10^{23}—so each energy level splits into a very large number of levels called a **band.** The levels are spaced almost continuously within the band. There is a separate band of levels for each particular energy level of the isolated atom. The bands may be widely separated in energy, they may be close together, or they may even overlap, depending on the kind of atom and the type of bonding in the solid.

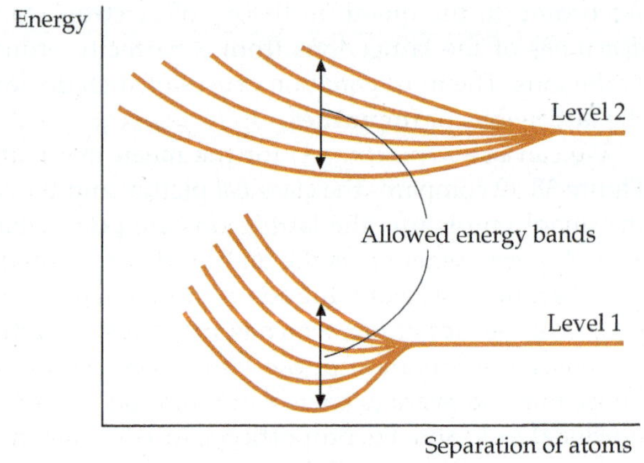

FIGURE 38-14 Energy splitting of two energy levels for six atoms as a function of the separation of the atoms. When there are many atoms, each level splits into a near-continuum of levels called a band.

The lowest energy bands, corresponding to the lowest energy levels of the atoms in the lattice, are filled with electrons that are bound to the individual atoms. The electrons that can take part in conduction occupy the higher energy bands. The highest energy band that contains electrons is called the **valence band.** The valence band may be completely filled with electrons or only partially filled, depending on the kind of atom and the type of bonding in the solid.

We can now understand why some solids are conductors and why others are insulators. If the valence band is only partially filled, there are many available empty energy states in the band, and the electrons in the band can easily be raised to a higher energy state by an electric field. Accordingly, this substance is a good conductor. If the valence band is filled and there is a large energy gap between it and the next available band, an applied electric field may be too weak to excite an electron from the upper energy levels of the filled band across the large gap into the energy levels of the empty band, so the substance is an insulator. The lowest band in which there are unoccupied states is called the **conduction band.** In a conductor, the valence band is only partially filled, so the valence band is also the conduction band. An energy gap between allowed bands is called a **forbidden energy band.**

The band structure for a conductor, such as copper, is shown in Figure 38-15*a*. The lower bands (not shown) are filled with the lower energy electrons of the atoms. The valence band is only about half-filled. When an electric field is established in the conductor, the electrons in the conduction band are accelerated, which means that their energies are increased. This is consistent with the exclusion principle because there are many empty energy states just above those occupied by electrons in this band. These electrons are thus the conduction electrons.

Figure 38-15*b* shows the band structure for magnesium, which is also a conductor. In this case, the highest occupied band is completely filled, but there is an empty band above it that overlaps it. The two bands thus form a combined valence–conduction band that is only partially filled.

Figure 38-15*c* shows the band structure for a typical insulator. At $T = 0$ K, the valence band is completely filled. The next energy band having empty energy states, the conduction band, is separated from the valence band by a large energy gap.

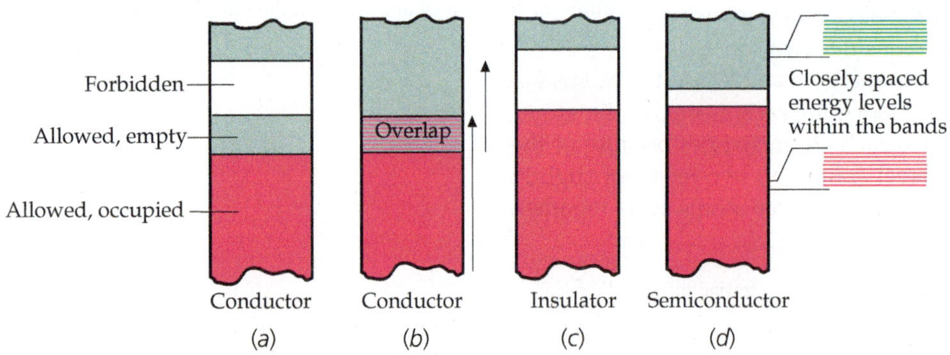

FIGURE 38-15 Four possible band structures for a solid. (*a*) A typical conductor. The valence band is also the conduction band. It is only partially filled, so electrons can be easily excited to nearby energy states. (*b*) A conductor in which the valence band overlaps a conduction band above it. (*c*) A typical insulator. There is a forbidden band that has a large energy gap between the filled valence band and the conduction band. (*d*) A semiconductor. The energy gap between the filled valence band and the conduction band is very small, so some electrons are excited to the conduction band at normal temperatures, leaving holes in the valence band.

At $T = 0$, the conduction band is empty. At ordinary temperatures, a few electrons can be excited to states in that band, but most cannot be excited to states because the energy gap is large compared with the energy an electron might obtain by thermal excitation. Very few electrons can be thermally excited to the nearly empty conduction band, even at fairly high temperatures. When an electric field of ordinary magnitude is established in the solid, electrons cannot be accelerated because there are no empty energy states at nearby energies. We describe this by saying that there are no free electrons. The small conductivity that is observed is due to the very few electrons that are thermally excited into the nearly empty conduction band. When an electric field applied to an insulator is sufficiently strong to cause an electron to be excited across the energy gap to the empty band, dielectric breakdown occurs.

In some substances, the energy gap between the filled valence band and the empty conduction band is very small, as shown in Figure 38-15*d*. At $T = 0$, there are no electrons in the conduction band and the material is an insulator. At ordinary temperatures, however, there are an appreciable number of electrons in the conduction band due to thermal excitation. Such a material is called an **intrinsic semiconductor.** For typical intrinsic semiconductors, such as silicon and germanium, the energy gap is only about 1 eV. In the presence of an electric field, the electrons in the conduction band can be accelerated because there are empty states nearby. Also, for each electron in the conduction band there is a vacancy, or **hole,** in the nearly filled valence band. In the presence of an electric field, electrons in this band can also be excited to a vacant energy level. This contributes to the electric current and is most easily described as the motion of a hole in the direction of the field and opposite to the motion of the electrons. The hole thus acts like a positive charge. To visualize the conduction of holes, think of a two-lane, one-way road that has one lane completely filled with parked cars and the other lane empty. If a car moves out of the completely filled lane into the empty lane, it can move ahead freely. As the other cars move up to occupy the vacated space, the vacated space propagates backward in the direction opposite the motion of the cars. Both the forward motion of the car in the nearly empty lane and the backward propagation of the empty space contribute to a net forward propagation of the cars.

An interesting characteristic of semiconductors is that the resistivity of the substance decreases as the temperature increases, which is contrary to the case for normal conductors. The reason is that as the temperature increases, the number of free electrons increases because there are more electrons in the conduction band. The number of holes in the valence band also increases, of course. In semiconductors, the effect of the increase in the number of charge carriers, both electrons and holes, exceeds the effect of the increase in resistivity due to the increased scattering of the electrons by the lattice ions due to thermal vibrations. Semiconductors therefore have a negative temperature coefficient of resistivity.

38-6 SEMICONDUCTORS

The semiconducting property of intrinsic semiconductors makes them useful as a basis for electronic circuit components whose resistivity can be controlled by application of an external voltage or current. Most such *solid-state devices,* however, such as the semiconductor diode and the transistor, make use of **impurity semiconductors,** which are created through the controlled addition of certain impurities to intrinsic semiconductors. This process is called **doping.** Figure 38-16*a* is a schematic illustration of silicon doped with a small amount of arsenic so that the arsenic atoms replace a few of the silicon atoms in the crystal lattice. The conduction band of pure silicon is virtually empty at ordinary temperatures, so pure silicon is a poor conductor of electricity. However, arsenic has five valence electrons rather than the four valence electrons of silicon. Four of these electrons take part in bonds with the four neighboring silicon atoms, and the fifth electron is very

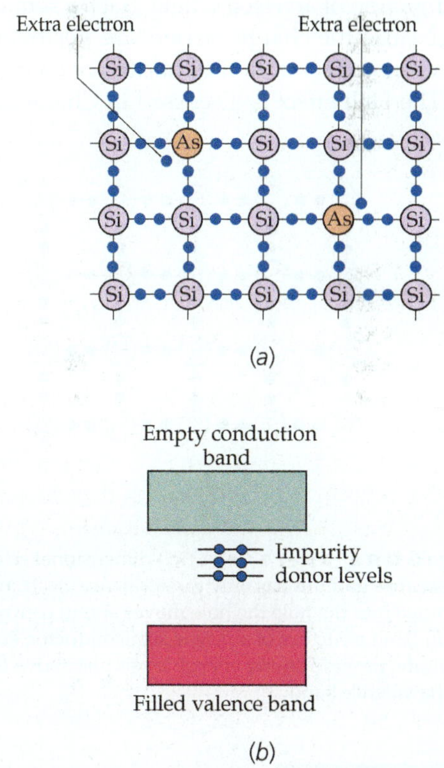

FIGURE 38-16 (*a*) A two-dimensional schematic illustration of silicon doped with arsenic. Because arsenic has five valence electrons, there is an extra, weakly bound electron that is easily excited to the conduction band, where it can contribute to electrical conduction. (*b*) Band structure of an *n*-type semiconductor, such as silicon doped with arsenic. The impurity atoms provide filled energy levels that are just below the conduction band. These levels donate electrons to the conduction band.

loosely bound to the atom. This extra electron occupies an energy level that is just slightly below the conduction band in the solid, and it is easily excited into the conduction band, where it can contribute to electrical conduction.

The effect on the band structure of a silicon crystal achieved by doping it with arsenic is shown in Figure 38-16*b*. The levels shown just below the conduction band are due to the extra electrons of the arsenic atoms. These levels are called **donor levels** because they donate electrons to the conduction band without leaving holes in the valence band. Such a semiconductor is called an ***n*-type semiconductor** because the major charge carriers are negatively charged electrons. The conductivity of a doped semiconductor can be controlled by controlling the amount of impurity added. The addition of just one part per million can increase the conductivity by several orders of magnitude.

Another type of impurity semiconductor can be made by replacing a silicon atom with a gallium atom, which has three valence electrons (Figure 38-17*a*). The gallium atom accepts electrons from the valence band to complete its four covalent bonds, thus creating a hole in the valence band. The effect on the band structure of silicon achieved by doping it with gallium is shown in Figure 38-17*b*. The empty levels shown just above the valence band are due to the holes from the ionized gallium atoms. These levels are called **acceptor levels** because they accept electrons from the filled valence band when those electrons are thermally excited to a higher energy state. This creates holes in the valence band that are free to propagate in the direction of an electric field. Such a semiconductor is called a ***p*-type semiconductor** because the charge carriers are positively charged holes. The fact that conduction is due to the motion of positively charged holes can be verified by the Hall effect. (The Hall effect is discussed in Chapter 26.)

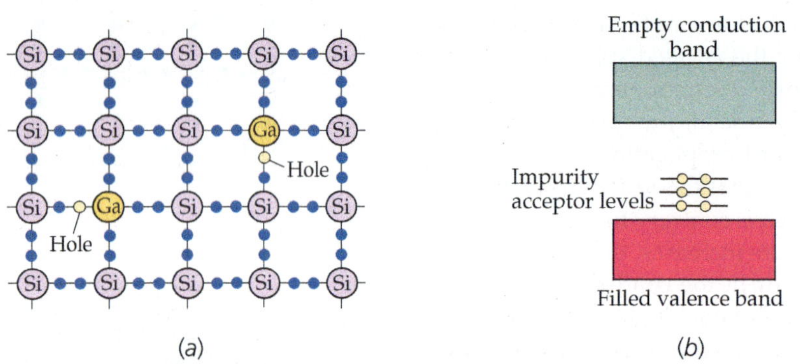

FIGURE 38-17 (*a*) A two-dimensional schematic illustration of silicon doped with gallium. Because gallium has only three valence electrons, there is a hole in one of its bonds. As electrons move into the hole the hole moves about, contributing to the conduction of electrical current. (*b*) Band structure of a *p*-type semiconductor, such as silicon doped with gallium. The impurity atoms provide empty energy levels just above the filled valence band that accept electrons from the valence band.

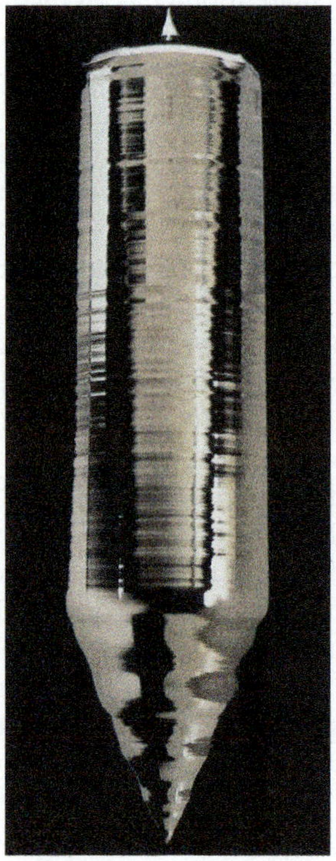

Synthetic crystal silicon is produced beginning with a raw material containing silicon (for instance, common beach sand), separating out the silicon, and melting it. From a seed crystal, the molten silicon grows into a cylindrical crystal, such as the one shown here. The crystals (typically about 1.3 m long) are formed under highly controlled conditions to ensure that they are flawless and the crystals are then sliced into thousands of thin wafers onto which the layers of an integrated circuit are etched. *(Museum of Modern Art.)*

Example 38-7 Number Density of Free Electrons in Arsenic-Doped Silicon

Try It Yourself

The number of free electrons in pure silicon is approximately 10^{10} electrons/cm^3 at ordinary temperatures. If one silicon atom out of every 10^6 atoms is replaced by an arsenic atom, how many free electrons per cubic centimeter are there? (The density of silicon is 2.33 g/cm^3 and its molar mass is 28.1 g/mol.)

PICTURE The number of silicon atoms per cubic centimeter, n_{Si}, can be found from $n_{Si} = \rho N_A/M$. Then, because each arsenic atom contributes one free electron, the number of electrons contributed by the arsenic atoms is $10^{-6}\,n_{Si}$.

SOLVE

Cover the column to the right and try these on your own before looking at the answers.

Steps	Answers
1. Calculate the number of silicon atoms per cubic centimeter.	$n_{Si} = \frac{\rho N_A}{M} = \frac{(2.33\ g/cm^3)(6.02 \times 10^{23}\ atoms/mol)}{28.1\ g/mol} = 4.99 \times 10^{22}\ atoms/cm^3$
2. Multiply by 10^{-6} to obtain the number of arsenic atoms per cubic centimeter, which equals the added number of free electrons per cubic centimeter.	$n_{As} = 10^{-6} n_{Si} = 4.99 \times 10^{16}\ atoms/cm^3$
3. The number of free electrons per cubic centimeter is equal to the number of arsenic atoms per cubic centimeter plus 1×10^{-10} (the number of silicon atoms per cubic centimeter).	$n_e = n_{As} + 1 \times 10^{-10} n_{Si} = 4.99 \times 10^{16}\ cm^{-3} + 1 \times 10^{10}\ cm^{-3} \approx \boxed{5 \times 10^{16}\ electrons/cm^3}$

CHECK As expected, the step-3 result is less than the number density of silicon atoms and more than the number density of conduction electrons in pure silicon.

TAKING IT FURTHER Because silicon has so few free electrons per atom, the number density of conduction electrons is increased by a factor of approximately 5 million per cubic centimeter by doping silicon with just one arsenic atom per million silicon atoms.

PRACTICE PROBLEM 38-3 How many free electrons are there per silicon atom in pure silicon?

* 38-7 SEMICONDUCTOR JUNCTIONS AND DEVICES

Semiconductor devices such as diodes and transistors make use of n-type semiconductors and p-type semiconductors joined together, as shown in Figure 38-18. In practice, the two types of semiconductors are often incorporated into a single silicon crystal doped with donor impurities on one side and acceptor impurities on the other side. The region in which the semiconductor changes from a p-type semiconductor to an n-type semiconductor is called a ***pn* junction.**

When an n-type semiconductor and a p-type semiconductor are placed in contact, the initially unequal concentrations of electrons and holes result in the diffusion of electrons across the junction from the n side to the p side and holes from the p side to the n side until equilibrium is established. The result of this diffusion is a net transport of positive charge from the p side to the n side. Unlike the case when two different metals are in contact, the electrons cannot travel very far from the junction region because the semiconductor is not a particularly good conductor. The diffusion of electrons and holes therefore creates a double layer of charge at the junction similar to that on a parallel-plate capacitor. There is, thus, a potential difference V across the junction, which tends to inhibit further diffusion. In equilibrium, the n side which has a net positive charge will be at a higher potential than the p side which has a net negative charge. In the junction region, between the charge layers, there will be very few charge carriers of either type, so the junction region has a high resistance. Figure 38-19 shows the energy-level diagram for a pn junction. The junction region is also called the **depletion region** because it has been depleted of charge carriers.

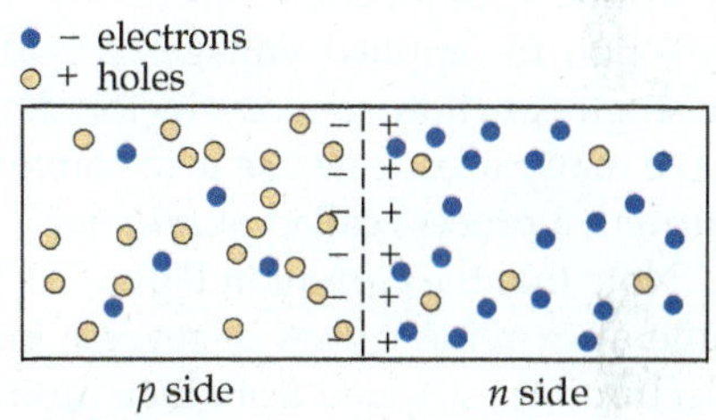

FIGURE 38-18 A pn junction. Because of the difference in their concentrations on either side of the pn junction, holes diffuse from the p side to the n side, and electrons diffuse from the n side to the p side. As a result, there is a double layer of charge at the junction, with the p side being negative and the n side being positive.

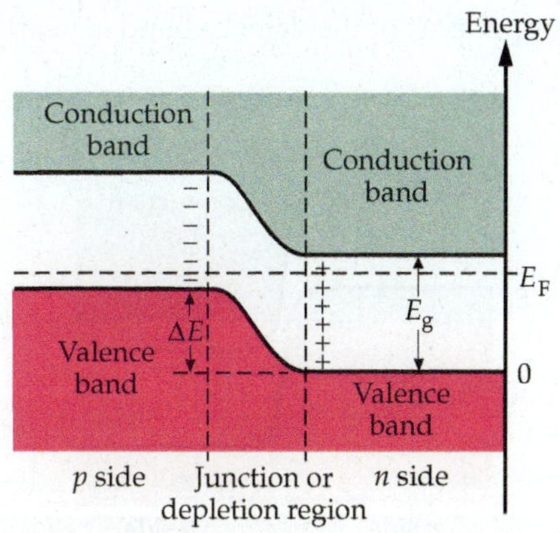

FIGURE 38-19 Electron energy levels for a pn junction.

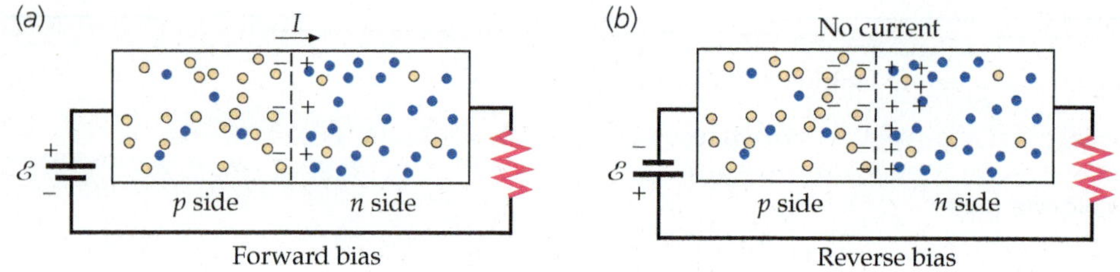

FIGURE 38-20 A *pn*-junction diode. (*a*) Forward-biased *pn* junction. The applied potential difference enhances the diffusion of holes from the *p* side to the *n* side and of electrons from the *n* side to the *p* side, resulting in a current *I*. (*b*) Reverse-biased *pn* junction. The applied potential difference inhibits the further diffusion of holes and electrons across the junction, so there is no current.

*DIODES

In Figure 38-20, an external potential difference has been applied across a *pn* junction by connecting a battery and a resistor to the semiconductor. When the positive terminal of the battery is connected to the *p* side of the junction, as shown in Figure 38-20*a*, the junction is said to be **forward biased.** Forward biasing lowers the potential across the junction. The diffusion of electrons and holes is thereby increased as they attempt to reestablish equilibrium, resulting in a current in the circuit.

If the positive terminal of the battery is connected to the *n* side of the junction, as shown in Figure 38-20*b*, the junction is said to be **reverse biased.** Reverse biasing tends to increase the potential difference across the junction, thereby further inhibiting diffusion. Figure 38-21 shows a plot of current versus voltage for a typical semiconductor junction. Essentially, the junction conducts only in one direction for applied voltages greater than the breakdown voltage. A single-junction semiconductor device is called a **diode.*** Diodes have many uses. One use is to convert alternating current into direct current, a process called *rectification.*

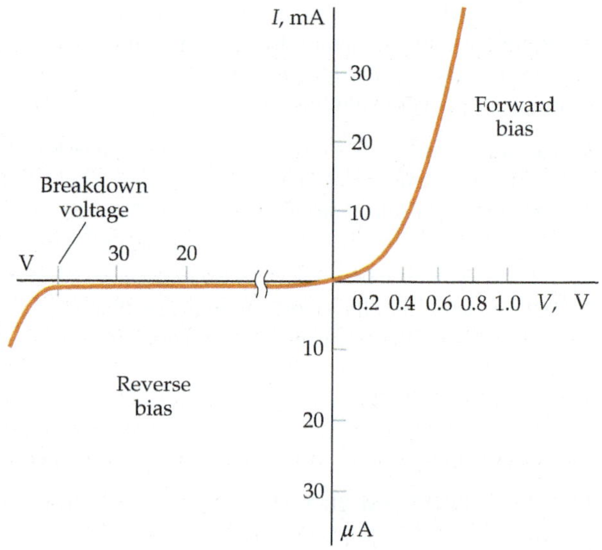

FIGURE 38-21 Plot of current versus applied voltage across a *pn* junction. Note the different scales on both axes for the forward and reverse bias conditions.

Note that the current in Figure 38-21 suddenly increases in magnitude at extreme values of reverse bias. In such large electric fields, electrons are stripped from their atomic bonds and accelerated across the junction. These electrons, in turn, cause others to break loose. This effect is called **avalanche breakdown.** Although such a breakdown can be disastrous in a circuit where it is not intended, the fact that it occurs at a sharply defined voltage makes it of use in a special voltage reference standard known as a **Zener diode.** Zener diodes are also used to protect devices from excessively high voltages.

An interesting effect, one that we discuss only qualitatively, occurs if both the *n* side and the *p* side of a *pn*-junction diode are so heavily doped that the donors on the *n* side provide so many electrons that the lower part of the conduction band is practically filled, and the acceptors on the *p* side accept so many electrons that the upper part of the valence band is nearly empty. Figure 38-22*a* shows the energy-level

FIGURE 38-22 Electron energy levels for a heavily doped *pn*-junction tunnel diode. (*a*) With no bias voltage, some electrons tunnel in each direction. (*b*) With a small bias voltage, the tunneling current is enhanced in one direction, making a sizable contribution to the net current. (*c*) With further increases in the bias voltage, the tunneling current decreases dramatically.

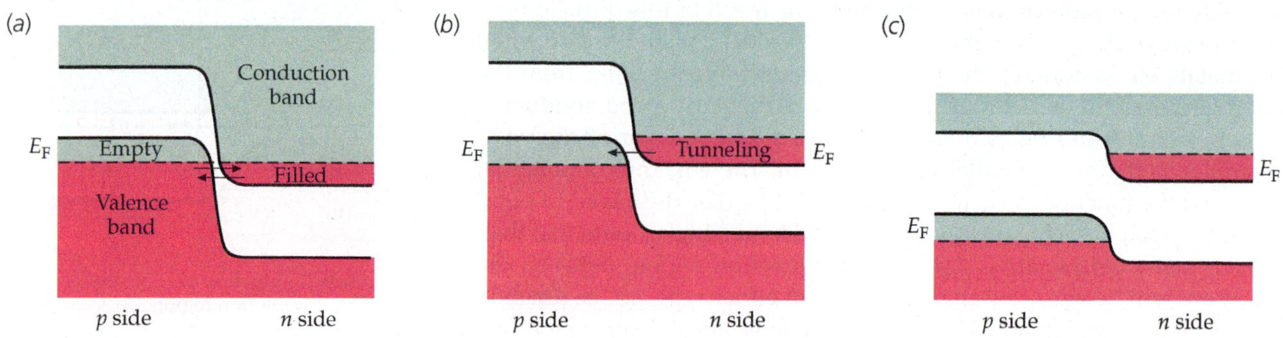

* The name *diode* originates from a vacuum tube device consisting of just two electrodes that also conducts electric current in one direction only.

diagram for this situation. Because the depletion region is now so narrow, electrons can easily penetrate the potential barrier across the junction and tunnel to the other side. The flow of electrons through the barrier is called a **tunneling current,** and such a heavily doped diode is called a **tunnel diode.**

At equilibrium where there is no bias, there is an equal tunneling current in each direction. When a small bias voltage is applied across the junction, the energy-level diagram is as shown in Figure 38-22*b*, and the tunneling of electrons from the *n* side to the *p* side is increased, whereas the tunneling of electrons in the opposite direction is decreased. This tunneling current, in addition to the usual current due to diffusion, results in a considerable net current. When the bias voltage is increased slightly, the energy-level diagram is as shown in Figure 38-22*c*, and the tunneling current is decreased. Although the diffusion current is increased, the net current is decreased. At large bias voltages, the tunneling current is completely negligible, and the total current increases with increasing bias voltage due to diffusion, as in an ordinary *pn*-junction diode. Figure 38-23 shows the current-versus-voltage curve for a tunnel diode. Such diodes are used in electric circuits because of their very fast response time. When operated near the peak in the current-versus-voltage curve, a small change in bias voltage results in a large change in the current.

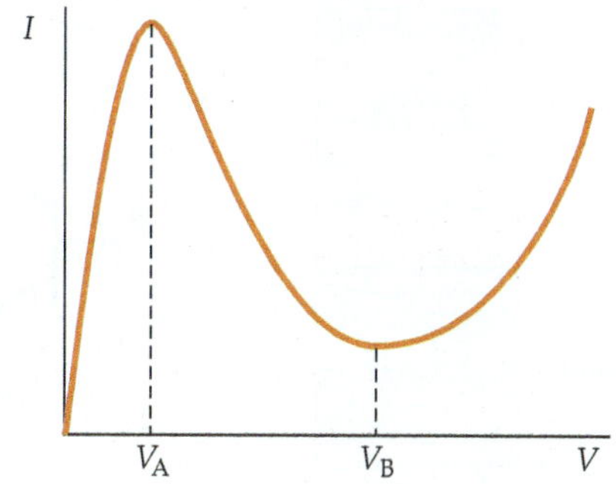

FIGURE 38-23 Current versus applied (bias) voltage V for a tunnel diode. For $V < V_A$, an increase in the bias voltage V enhances tunneling. For $V_A < V < V_B$, an increase in the bias voltage inhibits tunneling. For $V > V_B$, the tunneling is negligible, and the diode behaves like an ordinary *pn*-junction diode.

Another use for the *pn*-junction semiconductor is the **solar cell,** which is illustrated schematically in Figure 38-24. When a photon of energy greater than the gap energy (1.1 eV in silicon) strikes the *p*-type region, it can excite an electron from the valence band into the conduction band, leaving a hole in the valence band. This region is already rich in holes. Some of the electrons created by the photons will recombine with holes, but some will migrate to the junction. From there, they are accelerated into the *n*-type region by the electric field between the double layer of charge. This creates an excess negative charge in the *n*-type region and an excess positive charge in the *p*-type region. The result is a potential difference between the two regions, which in practice is approximately 0.6 V. If a load resistance is connected across the two regions, a charge flows through the resistance. Some of the incident light energy is thus converted into electrical energy. The current in the resistor is proportional to the rate of arrival of incident photons, which is in turn proportional to the intensity of the incident light.

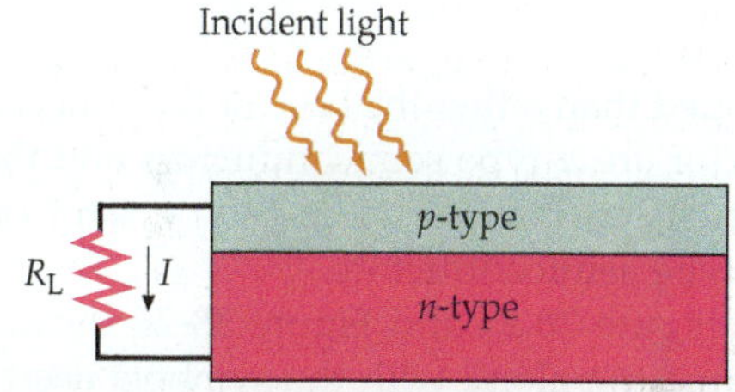

FIGURE 38-24 A *pn*-junction semiconductor as a solar cell. When light strikes the *p*-type region, electron-hole pairs are created, resulting in a current through the load resistance R_L.

There are many other applications of semiconductors with *pn* junctions. Particle detectors, called **surface-barrier detectors,** consist of a *pn*-junction semiconductor that has a large reverse bias so that there is ordinarily no current. When a high-energy particle, such as an electron, passes through the semiconductor, it creates many electron–hole pairs as it loses energy. The resulting current pulse signals the passage of the particle. **Light-emitting diodes** (LEDs) are *pn*-junction semiconductors that have large forward biases that produce large excess concentrations of electrons on the *p* sides and holes on the *n* sides of the junctions. Under these conditions, an LED emits light as the electrons and holes recombine. This is essentially the reverse of the process that occurs in a solar cell, in which electron–hole pairs are created by the absorption of light. LEDs are commonly used as warning indicators and as sources of infrared light beams.

A light-emitting diode (LED). *(© C. Falco/Photo Researchers.)*

*TRANSISTORS

The transistor, a semiconducting device that is used to produce a desired output signal in response to an input signal, was invented in 1948 by William Shockley, John Bardeen, and Walter Brattain and has revolutionized the electronics industry and our everyday world. A *simple bipolar junction transistor** consists of three distinct semiconductor regions called the **emitter,** the **base,** and the **collector.** The base is a very thin region of one type of semiconductor sandwiched between two regions of the opposite type. The emitter semiconductor is much more heavily

* Besides the bipolar junction transistor, there are other categories of transistors, notably, the field–effect transistor.

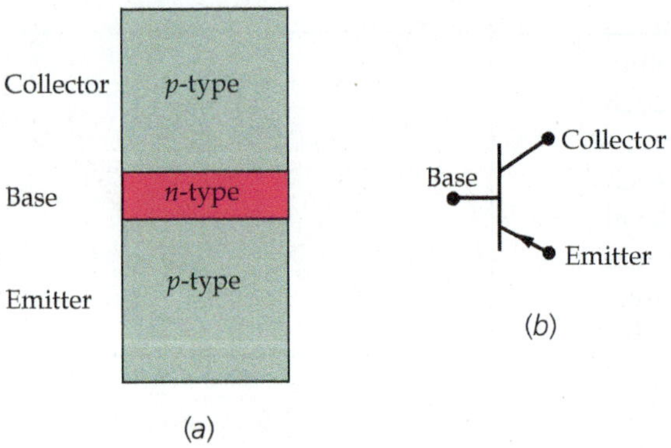

FIGURE 38-25 A *pnp* transistor. (*a*) The heavily doped emitter emits holes that pass through the thin base to the collector. (*b*) Symbol for a *pnp* transistor in a circuit. The arrow points in the direction of the conventional current, which is the same as that of the emitted holes.

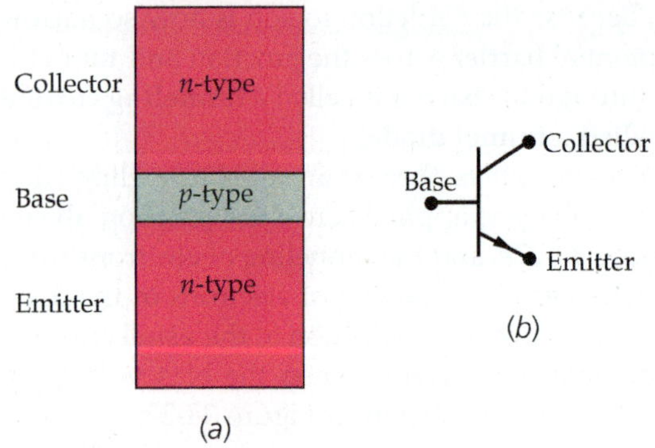

FIGURE 38-26 An *npn* transistor. (*a*) The heavily doped emitter emits electrons that pass through the thin base to the collector. (*b*) Symbol for an *npn* transistor. The arrow points in the direction of the conventional current, which is opposite the direction of the emitted electrons.

doped than either the base or the collector. In an *npn* transistor, the emitter and collector are *n*-type semiconductors and the base is a *p*-type semiconductor; in a *pnp* transistor, the base is an *n*-type semiconductor and the emitter and collector are *p*-type semiconductors.

Figure 38-25 and Figure 38-26 show, respectively, a *npn* transistor and an *npn* transistor, along with the symbols used to represent each transistor in circuit diagrams. We see that either transistor consists of two *pn* junctions. We will discuss the operation of a *npn* transistor. The operation of an *npn* transistor is similar.

In the normal operation of a *pnp* transistor, the emitter-base junction is forward biased, and the base-collector junction is reverse biased, as shown in Figure 38-27. The heavily doped *p*-type emitter emits holes that flow toward the emitter-base junction. This flow constitutes the emitter current I_e. Because the base is very thin, most of the holes flow across the base into the collector. This flow in the collector constitutes a current I_c. However, some of the holes recombine in the base producing a positive charge that inhibits the further flow of charge. To prevent this, some of the holes that do not reach the collector are drawn off the base as a base current I_b in a wire connected to the base. In Figure 38-27, therefore, I_c is almost but not quite equal to I_e, and I_b is much smaller than either I_c or I_e. It is customary to express I_c as

$$I_c = \beta I_b \tag{38-31}$$

where β is called the current gain of the transistor. Transistors can be designed to have values of β as low as ten or as high as several hundred.

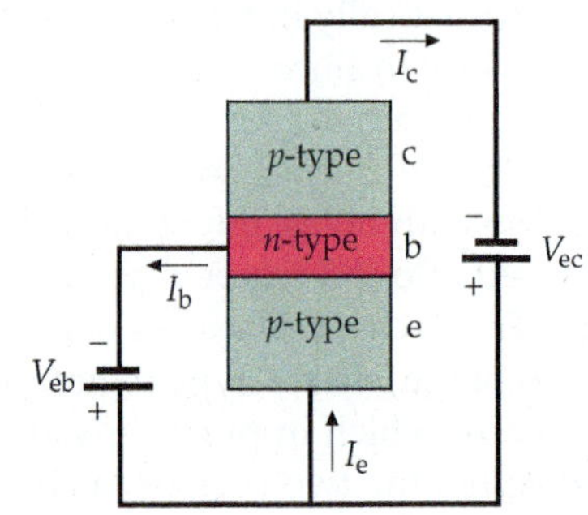

FIGURE 38-27 A *pnp* transistor biased for normal operation. Holes from the emitter can easily diffuse across the base, which is only tens of nanometers thick. Most of the holes flow to the collector, producing the current I_c.

Figure 38-28 shows a simple *pnp* transistor used as an amplifier. A small, time-varying input voltage v_s is connected in series with a constant bias voltage V_{eb}. The base current is then the sum of a steady current I_b produced by the bias voltage V_{eb} and a time-varying current i_b due to the signal voltage v_s. Because v_s may at any instant be either positive or negative, the bias voltage V_{eb} must be large enough to ensure that there is always a forward bias on the emitter-base junction. The collector current will consist of two parts: a constant direct current $I_c = \beta I_b$ and a time-varyng current $i_c = \beta i_b$. We thus have a current amplifier in which the time-varying output current i_c is β multiplied by the input current i_b. In such an amplifier, the steady currents I_c and I_b, although essential to the operation of the transistor, are usually not of interest. The input signal voltage v_s is related to the base current by Ohm's law:

$$i_b = \frac{v_s}{R_b + r_b} \tag{38-32}$$

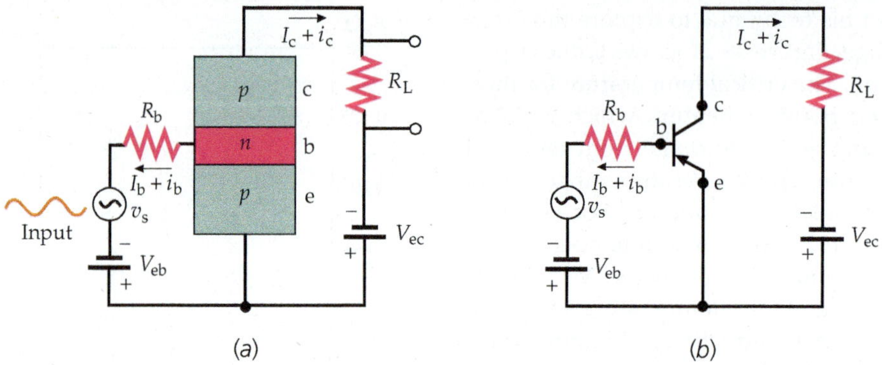

FIGURE 38-28 (*a*) A *pnp* transistor used as an amplifier. A small change i_b in the base current results in a large change i_c in the collector current. Thus, a small signal in the base circuit results in a large signal across the load resistor R_L in the collector circuit. (*b*) The same circuit as in Figure 38-28*a* with the conventional symbol for the transistor.

where r_b is the internal resistance of that part of the transistor between the base and emitter. Similarly, the collector current i_c produces a time-varying voltage v_L across the output or load resistance R_L given by

$$v_L = i_c R_L \qquad 38\text{-}33$$

Using Equation 38-31 and Equation 38-32, we have

$$i_c = \beta i_b = \beta \frac{v_s}{R_b + r_b} \qquad 38\text{-}34$$

The output voltage is thus related to the input voltage by

$$v_L = \beta \frac{v_s}{R_b + r_b} R_L = \beta \frac{R_L}{R_b + r_b} v_s \qquad 38\text{-}35$$

The ratio of the output voltage to the input voltage is the **voltage gain** of the amplifier:

$$\text{Voltage gain} = \frac{v_L}{v_s} = \beta \frac{R_L}{R_b + r_b} \qquad 38\text{-}36$$

A typical amplifier (for example, in a tape player) has several transistors, similar to the one shown in Figure 38-28, connected in series so that the output of one transistor serves as the input for the next. Thus, the very small voltage fluctuations produced by the motion of the magnetic tape past the pickup heads controls the large amounts of power required to drive the loudspeakers. The power delivered to the speakers is supplied by the dc sources connected to each transistor.

The technology of semiconductors extends well beyond individual transistors and diodes. Many of the electronic devices we use every day, such as laptop computers and the processors that govern the operation of vehicles and appliances, rely on large-scale integration of many transistors and other circuit components on a single chip. Large-scale integration combined with advanced concepts in semiconductor theory has created remarkable new instruments for scientific research.

38-8 SUPERCONDUCTIVITY

There are some substances for which the resistivity suddenly drops to zero below a certain temperature T_c, which is called the **critical temperature.** This amazing phenomenon, called **superconductivity,** was discovered in 1911 by the Dutch physicist H. Kamerlingh Onnes, who developed a technique for liquefying

helium (boiling point equal to 4.2 K) and used his technique to explore the properties of substances at temperatures in that range. Figure 38-29 shows Onnes's plot of the resistance of mercury versus temperature. The critical temperature for mercury is approximately the same as the boiling point of helium, which is 4.2 K. Critical temperatures for other superconducting elements range from less than 0.1 K for hafnium and iridium to 9.2 K for niobium. The temperature range for superconductors is much higher for a number of metallic compounds. For example, the superconducting alloy Nb_3Ge, discovered in 1973, has a critical temperature of 25 K, which was the highest known until 1986, when the discoveries of J. Georg Bednorz and K. Alexander Müller launched the era of high-temperature superconductors, now defined as materials that exhibit superconductivity at temperatures above 77 K (the temperature at which nitrogen boils). The highest temperature at which superconductivity has been demonstrated, using thallium-doped $HgBa_2Ca_2Cu_3O_8$ + delta, is 138 K at atmospheric pressure. At extremely high pressures, some materials exhibit superconductivity at temperatures as high as 164 K.

The resistivity of a superconductor is zero. There can be a current in a superconductor even when there is no emf in the superconducting circuit. Indeed, in superconducting rings in which there was no electric field, steady currents have been observed to persist for years without apparent loss. Despite the cost and inconvenience of refrigeration using expensive liquid helium, many superconducting magnets have been built using superconducting materials, because such magnets require no power expenditure to maintain the large current needed to produce a large magnetic field.

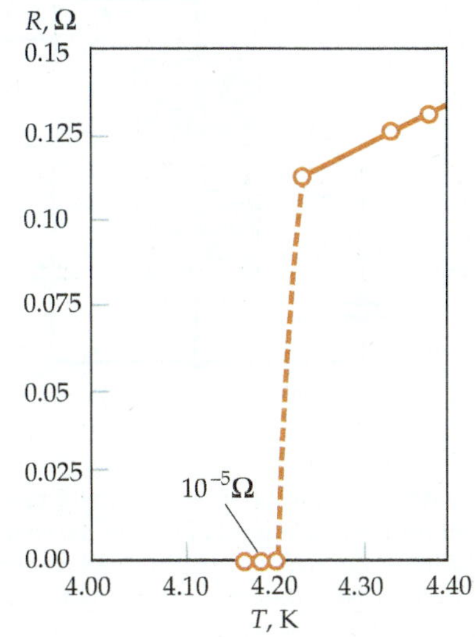

FIGURE 38-29 Plot by H. Kamerlingh Onnes of the resistance of mercury versus temperature, showing the sudden decrease at the critical temperature of $T = 4.2$ K.

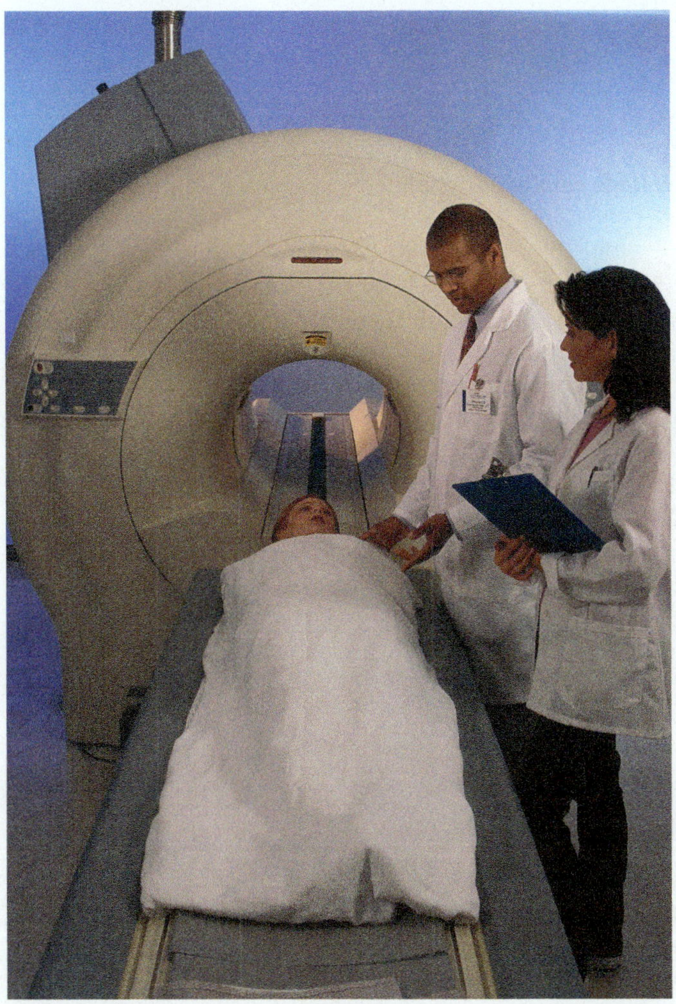

The wires for the magnetic field of a magnetic resonance imaging (MRI) machine carry large currents. To keep the wires from overheating, they are maintained at superconducting temperatures. To accomplish this, they are immersed in liquid helium. *(Corbis.)*

The discovery of high-temperature superconductors has revolutionized the study of superconductivity because relatively inexpensive liquid nitrogen, which boils at 77 K, can be used for a coolant. However, many problems, such as brittleness and the toxicity of the materials, make these new superconductors difficult to use. The search continues for new materials that will be superconductors at even higher temperatures.

THE BCS THEORY

It had been recognized for some time that low temperature superconductivity is due to a collective action of the conducting electrons. In 1957, John Bardeen, Leon Cooper, and Robert Schrieffer published a successful theory of low temperature superconductivity now known by the initials of the inventors as the **BCS theory.** According to this theory, the electrons in a superconductor are coupled in pairs at low temperatures. The coupling comes about because of the interaction between electrons and the crystal lattice. One electron interacts with the lattice and perturbs it. The perturbed lattice interacts with another electron in such a way that there is an attraction between the two electrons that at low temperatures can exceed the Coulomb repulsion between them. The electrons form a bound state called a **Cooper pair.** The electrons in a Cooper pair have equal and opposite spins, so they form a system with zero spin. Each Cooper pair acts as a *single particle* with zero spin, in other words, as a boson. Bosons do not obey the exclusion principle. Any number of Cooper pairs may be in the same quantum state with the same energy. In the ground state of a superconductor (at $T = 0$), all the conduction electrons are in Cooper pairs and all the Cooper pairs are in the same energy state. In the superconducting state, the Cooper pairs are correlated so that they act collectively. An electric current can be produced in a superconductor because all of the electrons in this collective state move together. But energy cannot be dissipated by individual collisions of electron and lattice ions unless the temperature is high enough to break the binding of the Cooper pairs. The required energy is called the *superconducting energy gap* E_g. In the BCS theory, this energy at zero temperature is related to the critical temperature by

$$E_g = \tfrac{7}{2}kT_c \qquad 38\text{-}37$$

The energy gap can be determined by measuring the current across a junction between a normal metal and a superconductor as a function of voltage. Consider two metals separated by a layer of insulating material, such as aluminum oxide, that is only a few nanometers thick. The insulating material between the metals forms a barrier that prevents most electrons from traversing the junction. However, waves can tunnel through a barrier if the barrier is not too thick, even if the energy of the wave is less than that of the barrier.

When the materials on either side of the gap are normal nonsuperconducting metals, the current resulting from the tunneling of electrons through the insulating layer obeys Ohm's law for low applied voltages (Figure 38-30*a*). When one of the metals is a normal metal and the other is a superconductor, there is no current (at absolute zero) unless the applied voltage V is greater than a critical voltage $V_c = E_g/(2e)$, where E_g is the superconductor energy gap. Figure 38-30*b* shows the plot of current versus voltage for this situation. The current escalates rapidly when the energy $2eV$ absorbed by a Cooper pair traversing the barrier approaches $E_g = 2eV_c$, the minimum energy needed to break up the pair. (The small current visible in Figure 38-30*b* before the critical voltage is reached is present because at any temperature above absolute zero some of the electrons in the superconductor are thermally excited above the energy gap and are therefore not paired.) At voltages slightly above V_c, the current-versus-voltage curve becomes that for a normal metal. The superconducting energy gap can thus be measured by measuring the average voltage for the transition region.

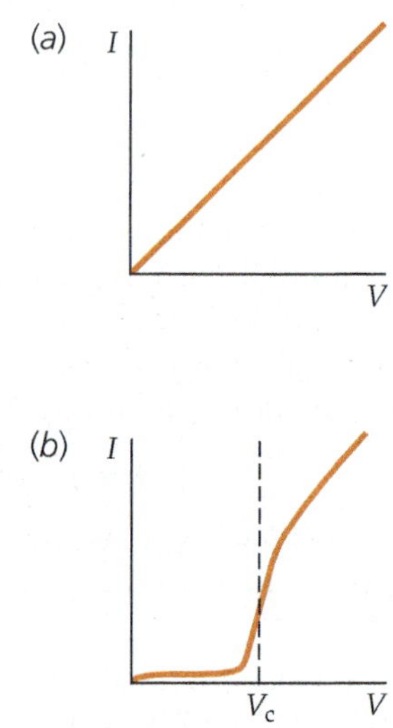

FIGURE 38-30 Tunneling current versus voltage for a junction of two metals separated by a thin oxide layer. (*a*) When both metals are normal metals, the current is proportional to the voltage, as predicted by Ohm's law. (*b*) When one metal is a normal metal and another metal is a superconductor, the current is approximately zero until the applied voltage V approaches the critical voltage $V_c = E_g/(2e)$.

Example 38-8 Superconducting Energy Gap for Mercury

Calculate the superconducting energy gap for mercury ($T_c = 4.2$ K) predicted by the BCS theory.

PICTURE The energy gap is related to the critical temperature by $E_g = 3.5\,kT_c$ (Equation 38-37).

SOLVE

1. The BCS prediction for the energy gap is $E_g = 3.5kT_c$

2. Substitute $T_c = 4.2$ K:

$$E_g = 3.5kT_c$$
$$= 3.5(1.38 \times 10^{-23}\,\text{J/K})(4.2\,\text{K})\left(\frac{1\,\text{ev}}{1.6 \times 10^{-19}\,\text{J}}\right)$$
$$= \boxed{1.3 \times 10^{-3}\,\text{eV}}$$

Note that the energy gap for a typical superconductor is much smaller than the energy gap for a typical semiconductor, which is of the order of 1 eV. As the temperature is increased from $T = 0$, some of the Cooper pairs are broken. Then there are fewer pairs available for each pair to interact with, and the energy gap is reduced until at $T = T_c$ the energy gap is zero (Figure 38-31).

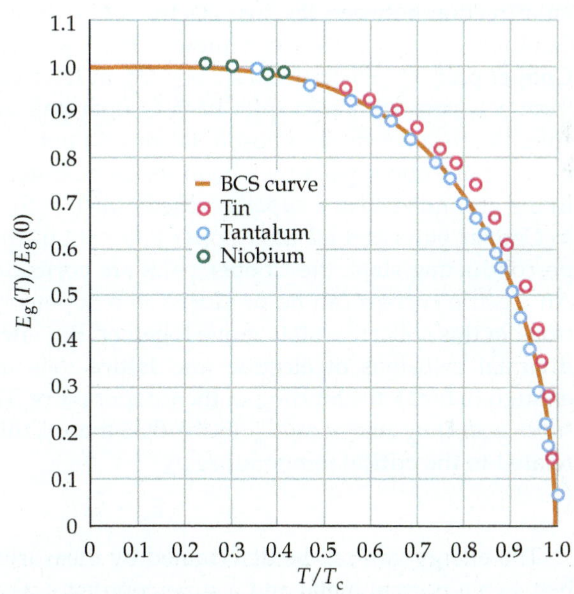

FIGURE 38-31 Ratio of the energy gap at temperature T to that at temperature $T = 0$ as a function of the relative temperature T/T_c. The solid curve is that predicted by the BCS theory.

THE JOSEPHSON EFFECT

When two superconductors are separated by a thin nonsuperconducting barrier (for example, a layer of aluminum oxide a few nanometers thick), the junction is called a **Josephson junction,** based on the prediction in 1962 by Brian Josephson that Cooper pairs could tunnel across such a junction from one superconductor to the other with no resistance. The tunneling of Cooper pairs constitutes a current, which does not require a voltage to be applied across the junction. The current depends on the difference in phase of the wave functions that describe the Cooper pairs. Let ϕ_1 be the phase constant for the wave function of a Cooper pair in one superconductor. All the Cooper pairs in a superconductor act coherently and have the same phase constant. If ϕ_2 is the phase constant for the Cooper pairs in the second superconductor, the current across the junction is given by

$$I = I_{max} \sin(\phi_2 - \phi_1) \qquad 38\text{-}38$$

where I_{max} is the maximum current, which depends on the thickness of the barrier. This result has been observed experimentally and is known as the **dc Josephson effect.**

Josephson also predicted that if a dc voltage V were applied across a Josephson junction, there would be a current that alternates with frequency f given by

$$f = \frac{2e}{h}V \qquad 38\text{-}39$$

This result, known as the **ac Josephson effect,** has been observed experimentally, and careful measurement of the frequency allows a precise determination of the ratio e/h. Because frequency can be measured very accurately, the ac Josephson effect is also used to establish precise voltage standards. The inverse effect, in which the application of an alternating voltage across a Josephson junction results in a dc current, has also been observed.

Example 38-9 Frequency of Josephson Current

Using $e = 1.602 \times 10^{-19}$ C and $h = 6.626 \times 10^{-34}$ J·s, calculate the frequency of the Josephson current if the applied voltage is 1.000 μV.

PICTURE The frequency f is related to the applied voltage V by $hf = 2eV$ (Equation 38-39).

SOLVE

Substitute the given values into Equation 38-39 to calculate f:

$$f = \frac{2e}{h}V = \frac{2(1.602 \times 10^{-19}\text{ C})}{6.626 \times 10^{-34}\text{ J}\cdot\text{s}}(1.000 \times 10^{-6}\text{ V})$$

$$= 4.835 \times 10^{8}\text{ Hz} = \boxed{483.5\text{ MHz}}$$

38-9 THE FERMI–DIRAC DISTRIBUTION

The classical Maxwell–Boltzmann distribution (Equation 17-38) gives the number dN of molecules that have energy E in the range between E and $E + dE$. The number dN is equal to the product of $g(E)\,dE$ where $g(E)$ is the **density of states** (number of energy states in the range dE) and the Boltzmann factor $e^{-E/(kT)}$, which is the probability of a state being occupied. The distribution function for free electrons in a metal is called the **Fermi–Dirac distribution.** The Fermi–Dirac distribution can be written in the same form as the Maxwell–Boltzmann distribution, where the density of states calculated from quantum theory and the Boltzmann factor is replaced by the Fermi factor. Let $n(E)\,dE$ be the number of electrons that have energies between E and $E + dE$. This number is written

$$n(E)dE = f(E)g(E)dE \qquad 38\text{-}40$$

ENERGY DISTRIBUTION FUNCTION

where $g(E)\,dE$ is the number of states that have energies between E and $E + dE$ and $f(E)$ is the probability of a state being occupied, which is the Fermi factor. The density of states in three dimensions is somewhat challenging to calculate, so we just give the result. For electrons in a metal of volume V, the density of states is

$$g(E) = \frac{8\sqrt{2}\pi m_e^{3/2}V}{h^3}E^{1/2} \qquad 38\text{-}41$$

DENSITY OF STATES

As in the classical Maxwell–Boltzmann distribution, the density of states is proportional to $E^{1/2}$.

At $T = 0$, the Fermi factor is given by Equation 38-24:

$$f(E) = \begin{cases} 1 & E < E_F \\ 0 & E > E_F \end{cases}$$

The integral of $n(E)\,dE$ over all energies gives the total number of electrons N. We can derive the equation

$$E_F = \frac{h^2}{8m_e}\left(\frac{3N}{\pi V}\right)^{2/3}$$

(Equation 38-22*a*) for the Fermi energy at $T = 0$ by integrating $n(E)\,dE$ from $E = 0$ to $E = \infty$. We obtain

$$N = \int_0^{\infty} n(E)dE = \int_0^{E_F} n(E)dE + \int_{E_F}^{\infty} n(E)dE$$

$$= \int_0^{E_F} \frac{8\sqrt{2}\pi m_e^{3/2} V}{h^3} E^{1/2}\,dE + 0 = \frac{16\sqrt{2}\pi m_e^{3/2} V}{3h^3} E_F^{3/2}$$

Note that at $T = 0$, $n(E)$ is zero for $E > E_F$. Solving for E_F gives the Fermi energy at $T = 0$:

$$E_F = \frac{h^2}{8m_e}\left(\frac{3N}{\pi V}\right)^{2/3} \qquad 38\text{-}42$$

which is Equation 38-22*a*. In terms of the Fermi energy, the density of states (Equation 38-41) is

$$g(E) = \frac{8\sqrt{2}\pi m_e^{3/2} V}{h^3} E^{1/2} = \frac{3}{2} N E_F^{-3/2} E^{1/2} \qquad 38\text{-}43$$

DENSITY OF STATES IN TERMS OF E_F

which is obtained by solving Equation 38-42 for m_e, and then substituting for m_e in Equation 38-41. The average energy at $T = 0$ is calculated from

$$E_{av} = \frac{\int_0^{E_F} Eg(E)dE}{\int_0^{E_F} g(E)dE} = \frac{1}{N}\int_0^{E_F} Eg(E)dE \qquad 38\text{-}44$$

where $N = \int_0^{E_F} g(E)dE$ is the total number of electrons. Substituting for $g(E)$ from Equation 38-43 and then evaluating the integral in Equation 38-44, we obtain Equation 38-23:

$$E_{av} = \tfrac{3}{5} E_F \qquad 38\text{-}45$$

AVERAGE ENERGY AT $T = 0$

At $T > 0$, the Fermi factor is more complicated. It can be shown to be

$$f(E) = \frac{1}{e^{(E - E_F)/(kT)} + 1} \qquad 38\text{-}46$$

FERMI FACTOR

We can see from this equation that for E greater than E_F, $e^{(E - E_F)/(kT)}$ becomes very large as T approaches zero, so at $T = 0$, the Fermi factor is zero for $E > E_F$. On the other hand, for E less than E_F, $e^{(E - E_F)/(kT)}$ approaches 0 as T approaches zero, so at $T = 0$, $f(E) = 1$ for $E < E_F$. Thus, the Fermi factor given by Equation 38-46 holds for all temperatures. Note also that for any nonzero value of T, $f(E) = \frac{1}{2}$ at $E = E_F$.

The complete Fermi-Dirac distribution function is thus

$$n(E)dE = g(E)f(E)dE = \frac{8\sqrt{2}\pi m_e^{3/2}V}{h^3}E^{1/2}\frac{1}{e^{(E-E_F)/(kT)}+1}dE \qquad \text{38-47}$$

FERMI–DIRAC DISTRIBUTION

We can see that for those few electrons that have energies much greater than the Fermi energy, the Fermi factor approaches $1/e^{(E-E_F)/(kT)} = e^{(E_F-E)/(kT)} = e^{E_F/(kT)}e^{-E/(kT)}$, which is proportional to $e^{-E/(kT)}$. Thus, the high-energy tail of the Fermi–Dirac energy distribution decreases with increasing E as $e^{-E/(kT)}$, just like the classical Maxwell–Boltzmann energy distribution. The reason for this is that in this high-energy region there are many unoccupied energy states and few electrons, so the exclusion principle is not important. Thus, the Fermi–Dirac distribution approaches the classical Maxwell–Boltzmann distribution in the high-energy limit. This result has practical importance because it applies to the conduction electrons in semiconductors.

Example 38-10 Fermi Factor for Copper at 300 K

At what energy is the Fermi factor equal to 0.100 for copper at $T = 300$ K?

PICTURE We set $f(E) = 0.100$ in Equation 38-46, using $T = 300$ K and $E_F = 7.03$ eV from Table 38-1, and solve for E.

SOLVE

1. Solve Equation 38-46 for $e^{(E-E_F)/(kT)}$:

$$f(E) = \frac{1}{e^{(E-E_F)/(kT)}+1}$$

so

$$e^{(E-E_F)/(kT)} = \frac{1}{f(E)} - 1$$

2. Take the logarithm of both sides:

$$\frac{E-E_F}{kT} = \ln\left[\frac{1}{f(E)} - 1\right]$$

3. Solve for E. For E_F, use the value for E_F at $T = 0$ K listed in Table 38-1:

$$E = E_F + \ln\left[\frac{1}{f(E)} - 1\right]kT$$

$$= 7.03\text{ eV} + \ln\left[\frac{1}{0.100} - 1\right](8.62\times10^{-5}\text{ eV/K})(300\text{ K})$$

$$= \boxed{7.09\text{ eV}}$$

CHECK As expected, the energy is slightly above the Fermi energy when the Fermi factor is equal to 0.100.

Example 38-11 Probability of a Higher Energy State Being Occupied

Find the probability that an energy state in copper 0.100 eV above the Fermi energy is occupied at $T = 300$ K.

PICTURE The probability is the Fermi factor given in Equation 38-46, with $E_F = 7.03$ eV and $E = 7.13$ eV.

SOLVE

1. The probability of an energy state being occupied equals the Fermi factor:

$$P = f(E) = \frac{1}{e^{(E-E_F)/(kT)} + 1}$$

2. Calculate the exponent in the Fermi factor (exponents are always dimensionless):

$$\frac{E - E_F}{kT} = \frac{7.13\text{ eV} - 7.03\text{ eV}}{(8.62 \times 10^{-5}\text{ eV/K})(300\text{ K})} = 3.87$$

3. Use this result to calculate the Fermi factor:

$$f(E) = \frac{1}{e^{(E-E_F)/(kT)} + 1} = \frac{1}{e^{3.87} + 1}$$

$$= \frac{1}{48 + 1} = \boxed{0.020}$$

CHECK The probability that an energy state above the Fermi energy is occupied is less than one-half. As expected, the step 4 result is less than one-half.

TAKING IT FURTHER The probability of an electron having an energy 0.100 eV above the Fermi energy at 300 K is only about 2 percent.

Example 38-12 Probability of a Lower Energy State Being Occupied *Try It Yourself*

Find the probability that an energy state in copper 0.10 eV *below* the Fermi energy is occupied at $T = 300$ K.

PICTURE The probability is the Fermi factor given in Equation 38-46, with $E_F = 7.03$ eV and $E = 6.93$ eV.

SOLVE

Cover the column to the right and try these on your own before looking at the answers.

Steps	Answers
1. Write the Fermi factor:	$f(E) = \frac{1}{e^{(E-E_F)/(kT)} + 1}$
2. Calculate the exponent in the Fermi factor:	$\frac{E - E_F}{kT} = \frac{6.93\text{ eV} - 7.03\text{ eV}}{(8.62 \times 10^{-5}\text{ eV/K})(300\text{ K})} = -3.87$
3. Use your result from step 2 to calculate the Fermi factor:	$f(E) = \frac{1}{e^{(E-E_F)/(kT)} + 1} = \frac{1}{e^{-3.87} + 1} = \frac{1}{0.021 + 1} = \boxed{0.98}$

CHECK As expected, the step-3 result is greater than one-half.

TAKING IT FURTHER The probability of an electron having an energy of 0.10 eV *below* the Fermi energy at 300 K is approximately 98 percent.

PRACTICE PROBLEM 38-4 What is the probability of an energy state 0.10 eV below the Fermi energy being unoccupied at 300 K?

Summary

TOPIC	RELEVANT EQUATIONS AND REMARKS	
1. The Structure of Solids	Solids are often found in crystalline form in which a small structure, which is called the unit cell, is repeated over and over. A crystal may have a face-centered-cubic, body-centered-cubic, hexagonal close-packed, or other structure depending on the type of bonding between the atoms, ions, or molecules in the crystal and on the relative sizes of the atoms.	
Potential energy	$U = -\alpha\frac{ke^2}{r} + \frac{A}{r^n}$ where r is the center-to-center separation distance between neighboring ions, α is the Madelung constant, which depends on the geometry of the crystal and is of the order of 1.8, and n is approximately 9.	38-3
2. A Microscopic Picture of Conduction		
Resistivity	$\rho = \frac{m_e v_{av}}{n_e e^2 \lambda}$ where v_{av} is the average speed of the electrons and λ is their mean free path between collisions with the lattice ions.	38-14
Mean free path	$\lambda = \frac{vt}{n_{ion}\pi r^2 vt} = \frac{1}{n_{ion}\pi r^2} = \frac{1}{n_{ion}A}$ where n_{ion} is the number of lattice ions per unit volume, r is their effective radius, and A is their effective cross-sectional area.	38-16
3. Classical Interpretation of v_{av} and λ	v_{av} is determined from the Maxwell–Boltzmann distribution, and r is the actual radius of a lattice ion. (This interpretation is not consistent with measured results.)	
4. Quantum Interpretation of v_{av} and λ	v_{av} is determined from the Fermi–Dirac distribution and is approximately constant independent of temperature. The mean free path is determined from the scattering of electron waves, which occurs only because of deviations from a perfectly ordered array. The radius r is the amplitude of vibration of the lattice ion, which is proportional to $\sqrt{T}$, so A is proportional to T.	
5. Free Electrons		
Fermi energy E_F at $T = 0$	E_F is the energy of the last filled (or half-filled) energy state.	
E_F at $T > 0$	E_F is the energy at which the probability of being occupied is $\frac{1}{2}$.	
Approximate magnitude of E_F	E_F is between 5 eV and 10 eV for most metals.	
Dependence of E_F on the number density (N/V) of free electrons	$E_F = \frac{h^2}{8m_e}\left(\frac{3N}{\pi V}\right)^{2/3}$	38-22*a*
Average energy at $T = 0$	$E_{av} = \frac{3}{5}E_F$	38-23
Fermi factor at $T = 0$	The Fermi factor $f(E)$ is the probability of a state being occupied. $f(E) = \begin{cases} 1 & E < E_F \\ 0 & E > E_F \end{cases}$	38-24
Fermi temperature	$T_F = \frac{E_F}{k}$	38-25
Fermi speed	$u_F = \sqrt{\frac{2E_F}{m_e}}$	38-26

TOPIC	RELEVANT EQUATIONS AND REMARKS	
Contact potential	When two different metals are placed in contact, electrons flow from the metal with the higher Fermi energy to the metal with the lower Fermi energy until the Fermi energies of the two metals are equal. In equilibrium, there is a potential difference between the metals that is equal to the difference in the work function of the two metals divided by the electronic charge e: $$V_{\text{contact}} = \frac{\phi_1 - \phi_2}{e}$$	38-27
Specific heat due to conduction electrons	$$c'_{\text{V}} = \tfrac{1}{2}\pi^2 R \frac{T}{T_{\text{F}}}$$	38-29
6. Band Theory of Solids	When many atoms are brought together to form a solid, the individual energy levels are split into bands of allowed energies. The splitting depends on the type of bonding and the lattice separation. The highest energy band that contains electrons is called the valence band. (The lowest energy band that is not filled with electrons is called the conduction band.) In a conductor, the valence band is only partially filled, so there are many available empty energy states for excited electrons. In an insulator, the valence band is completely filled and there is a large energy gap between it and the next allowed band, the conduction band. In a semiconductor, the energy gap between the filled valence band and the empty conduction band is small; so, at ordinary temperatures, an appreciable number of electrons are thermally excited into the conduction band.	
7. Semiconductors	The conductivity of a semiconductor can be greatly increased by doping. In an n-type semiconductor, the doping adds electrons at energies just below that of the conduction band. In a p-type semiconductor, holes are added at energies just above that of the valence band.	
8. *Semiconductor Junctions and Devices		
**pn* Junctions	Semiconductor devices such as diodes and transistors make use of n-type semiconductors and p-type semiconductors. The two types of semiconductors are typically a single silicon crystal doped with donor impurities on one side and acceptor impurities on the other side. The region in which the semiconductor changes from a p-type semiconductor to an n-type semiconductor is called a junction. Junctions are used in diodes, solar cells, surface barrier detectors, LEDs, and transistors.	
*Diodes	A diode is a single-junction device that carries current in one direction only.	
*Zener diodes	A Zener diode is a diode with a very high reverse bias. It breaks down suddenly at a distinct voltage and can therefore be used as a voltage reference standard.	
*Tunnel diodes	A tunnel diode is a diode that is heavily doped so that electrons tunnel through the depletion barrier. At normal operation, a small change in bias voltage results in a large change in current.	
*Transistors	A transistor consists of a very thin semiconductor of one type sandwiched between two semiconductors of the opposite type. Transistors are used in amplifiers because a small variation in the base current results in a large variation in the collector current.	
9. Superconductivity	In a superconductor, the resistance drops suddenly to zero below a critical temperature T_c. Superconductors with critical temperatures as high as 138 K have been discovered.	
The BCS theory	Superconductivity is described by a theory of quantum mechanics called the BCS theory in which the free electrons form Cooper pairs. The energy needed to break up a Cooper pair is called the superconducting energy gap E_g. When all the electrons are paired, individual electrons cannot be scattered by a lattice ion, so the resistance is zero.	
Tunneling	When a normal conductor is separated from a superconductor by a thin layer of oxide, electrons can tunnel through the energy barrier if the applied voltage across the layer is $E_g/(2e)$, where E_g is the energy needed to break up a Cooper pair. The energy gap E_g can be determined by a measurement of the tunneling current versus the applied voltage.	

TOPIC	RELEVANT EQUATIONS AND REMARKS
Josephson junction	A system of two superconductors separated by a thin layer of nonconducting material is called a Josephson junction.
dc Josephson effect	A dc current is observed to tunnel through a Josephson junction even in the absence of voltage across the junction.
ac Josephson effect	When a dc voltage V is applied across a Josephson junction, an ac current is observed with a frequency $$f = \frac{2e}{h}V \quad (38\text{-}39)$$ Measurement of the frequency of this current allows a precise determination of the ratio e/h.
10. The Fermi–Dirac Distribution	The number of electrons with energies between E and $E + dE$ is given by $$n(E)\,dE = f(E)\,g(E)\,dE \quad (38\text{-}40)$$ where $g(E)$ is the density of states and $f(E)$ is the Fermi factor.
Density of states	$$g(E) = \frac{8\sqrt{2}\pi m_e^{3/2}V}{h^3}E^{1/2} \quad (38\text{-}41)$$
Fermi factor at $T > 0$	$$f(E) = \frac{1}{e^{(E-E_F)/(kT)} + 1} \quad (38\text{-}46)$$

Answers to Practice Problems

38-1 $E_F = 9.40$ eV

38-2 5.53 eV

38-3 2×10^{-13} electrons/atom

38-4 One minus the probability of the energy state being occupied. That is, $1 - 0.98 = 0.02$ or 2 percent.

Problems

In a few problems, you are given more data than you actually need; in a few other problems, you are required to supply data from your general knowledge, outside sources, or informed estimate.

Interpret as significant all digits in numerical values that have trailing zeros and no decimal points.

- • Single-concept, single-step, relatively easy
- •• Intermediate-level, may require synthesis of concepts
- ••• Challenging
- SSM Solution is in the *Student Solutions Manual*
- Consecutive problems that are shaded are paired problems.

CONCEPTUAL PROBLEMS

1 • In the classical model of conduction, the electron loses energy on average during a collision because it loses the drift velocity it had acquired since the last collision. Where does this energy appear?

2 • A metal is a good conductor because the valence energy band for electrons is (*a*) empty, (*b*) partly filled, (*c*) filled, but there is only a small gap to a higher empty band, (*d*) completely filled, (*e*) none of the above.

3 • Thomas refuses to believe that a potential difference can be created simply by bringing two different metals into contact with each other. John talks him into making a small wager and is about to win the bet. (*a*) Which two metals from Table 38-2 would demonstrate his point most effectively? (*b*) What is the value of that contact potential?

4 • (*a*) In Problem 3, which choices of different metals would make the least impressive demonstration? (*b*) What is the value of that contact potential?

5 • When a sample of pure copper is cooled from 300 K to 4 K, its resistivity decreases more than the resistivity of a sample of brass when it is cooled through the same temperature difference. Why? SSM

6 • Insulators are poor conductors of electricity because (*a*) there is a small energy gap between the filled valence band and the next higher band where electrons can exist, (*b*) there is a large energy gap between the completely filled valence band and the next higher band where electrons can exist, (*c*) the valence band has a few vacancies for electrons, (*d*) the valence band is only partly filled, (*e*) none of the above.

7 • How does the sign of the change in the resistivity of a sample of copper compare with the sign of the change in the resistivity of a sample of silicon when the temperatures of both samples increase? SSM

8 • True or false:

(*a*) Solids that are good electrical conductors are usually good heat conductors.

(*b*) At $T = 0$, an intrinsic semiconductor is an insulator.

(*c*) The Fermi energy is the average energy of an electron in a solid.
(*d*) At $T = 0$, the value of the Fermi factor can be either 1 or 0.
(*e*) Semiconductors conduct current in one direction only.
(*f*) The classical free-electron theory adequately explains the heat capacity of metals.
(*g*) The contact potential between two metals is proportional to the difference in the work functions of the two metals.

9 • Which of the following elements are most likely to act as acceptor impurities in germanium? (*a*) bromine, (*b*) gallium, (*c*) silicon, (*d*) phosphorus, (*e*) magnesium

10 • Which of the following elements are most likely to serve as donor impurities in germanium? (*a*) bromine, (*b*) gallium, (*c*) silicon, (*d*) phosphorus, (*e*) magnesium

11 • An electron hole is created when a photon is absorbed by a semiconductor. How does this hole enable the semiconductor to conduct electricity?

12 • Examine the positions of phosphorus, boron, thallium, and antimony in Table 36-1. (*a*) Which of these elements can be used to dope silicon to create an *n*-type semiconductor? (*b*) Which of these elements can be used to dope silicon to create a *p*-type semiconductor?

13 • When photons of visible light strike the *p*-type semiconductor in a *pn* junction solar cell, (*a*) only free electrons are created, (*b*) only positive holes are created, (*c*) both electrons and holes are created, (*d*) protons are created, (*e*) none of the above.

ESTIMATION AND APPROXIMATION

14 • The ratio of the resistivity of the most resistive (least conductive) material to that of the least resistive material (excluding superconductors) is approximately 10^{24}. You can develop a feeling for how remarkable this range is by considering what the ratio is of the largest values to smallest values of other material properties. Choose any three properties of materials, and using tables in this book or some other resource, calculate the ratio of the largest instance of the property to the smallest instance of that property (other than zero) and rank these in decreasing order. Can you find any other property that shows a range as large as that of electrical resistivity?

15 • A device is said to be "ohmic" if a graph of current versus applied voltage is a straight line through the origin. The resistance R of the device is the reciprocal of the slope of this line. A *pn* junction is an example of a nonohmic device, as may be seen from Figure 38-21. For nonohmic devices, it is sometimes convenient to define the *differential resistance* as the reciprocal of the slope of the I-versus-V curve. Using the curve in Figure 38-21, estimate the differential resistance of the *pn* junction at applied voltages of -20 V, $+0.2$ V, $+0.4$ V, $+0.6$ V, and $+0.8$ V.

THE STRUCTURE OF SOLIDS

16 • Calculate the center-to-center separation distance r_0 between the K^+ and the Cl^- ions in KCl. Do this by assuming that each ion occupies a cubic volume of side r_0. The molar mass of KCl is 74.55 g/mol and its density is 1.984 g/cm^3.

17 • The center-to-center separation distance between the Li^+ and Cl^- ions in LiCl is 0.257 nm. Use that value and the molar mass of LiCl (42.4 g/mol) to compute the density of LiCl.

18 • Find the value of n in Equation 38-6 that gives the measured dissociation energy of 741 kJ/mol for LiCl, which has the same structure as NaCl and for which $r_0 = 0.257$ nm.

19 •• (*a*) Use Equation 38-6 and calculate $U(r_0)$ for calcium oxide, CaO, where $r_0 = 0.208$ nm. Assume $n = 8$. (*b*) If n increases from 8 to 10, what is the fractional change in $U(r_0)$?

A MICROSCOPIC PICTURE OF CONDUCTION

20 • A measure of the density of the free electrons in a metal is the distance r_s, which is defined as the radius of the sphere whose volume equals the volume per conduction electron. (*a*) Show that $r_s = [3/(4\pi n)]^{1/3}$, where n is the free-electron number density. (*b*) Calculate r_s for copper in nanometers.

21 • (*a*) Given a mean free path $\lambda = 0.400$ nm and a mean speed $v_{av} = 1.17 \times 10^5$ m/s for the charge flow in copper at a temperature of 300 K, calculate the classical value for the resistivity ρ of copper. (*b*) The classical model suggests that the mean free path is temperature independent and that v_{av} depends on temperature. According to this model, what would ρ be at 100 K?

FREE ELECTRONS IN A SOLID

22 •• Silicon has a molar mass of 28.09 g/mol and a density of 2.41×10^3 kg/m^3. Each atom of silicon has four valence electrons and the Fermi energy of the material is 4.88 eV. (*a*) Given that the electron mean free path at room temperature is $\lambda = 27.0$ nm, estimate the resistivity. (*b*) The accepted value for the resistivity of silicon is 640 $\Omega\cdot$m (at room temperature). How does this accepted value compare to the value calculated in Part (*a*)?

23 • Calculate the number density of free electrons in (*a*) Ag ($\rho = 10.5$ g/cm^3) and (*b*) Au ($\rho = 19.3$ g/cm^3), assuming one free electron per atom, and compare your results with the values listed in Table 38-1.

24 • The density of aluminum is 2.7 g/cm^3. How many free electrons are present per aluminum atom?

25 • The density of tin is 7.3 g/cm^3. How many free electrons are present per tin atom?

26 • Calculate the Fermi temperature for (*a*) Mg, (*b*) Mn, and (*c*) Zn.

27 • What is the speed of a conduction electron whose energy is equal to the Fermi energy E_F for (*a*) Na, (*b*) Au, and (*c*) Sn?

28 • Calculate the Fermi energy for (*a*) Al, (*b*) K, and (*c*) Sn using the number densities given in Table 38-1.

29 • Find the average energy of the conduction electrons at $T = 0$ in (*a*) copper and (*b*) lithium.

30 • Calculate (*a*) the Fermi temperature and (*b*) the Fermi energy at $T = 0$ for iron.

31 •• (*a*) Assuming that each gold atom in a sample of gold metal contributes one free electron, calculate the free-electron density in gold knowing that its atomic mass is 196.97 g/mol and its density is 19.3×10^3 kg/m^3. (*b*) If the Fermi speed for gold is 1.39×10^6 m/s, what is the Fermi energy in electron volts? (*c*) By what factor is the Fermi energy higher than the kT energy at room temperature? (*d*) Explain the difference between the Fermi energy and the kT energy. SSM

32 •• The bulk modulus B of a material can be defined by $B = -V\partial P/\partial V$. (*a*) Use the monatomic ideal-gas relation $PV = \frac{2}{3}NE_{av}$, where E_{av} is the average kinetic energy, Equation 38-22 and Equation 38-23 to show that $P = \frac{2}{5}NE_F/V = CV^{-5/3}$, where C is a constant independent of V. (*b*) Show that the bulk modulus of the free electrons in a solid metal is therefore $B = \frac{5}{3}P = \frac{2}{3}NE_F/V$. (*c*) Compute the bulk modulus in newtons per square meter for the free electrons in a sample of copper and compare your result with the measured value of 140×10^9 N/m^2.

33 •• The pressure of a monatomic ideal gas is related to the average kinetic energy of the gas particles by $PV = \frac{2}{3}NE_{av}$, where n is the number of particles and E_{av} is the average kinetic energy. Use this information to calculate the pressure of the free electrons in a sample of copper in newtons per square meter, and compare your result with atmospheric pressure, which is about 10^5 N/m^2. (*Note:* The units are most easily handled by using the conversion factors 1 N/m^2 = 1 J/m^3 and 1 eV = 1.602×10^{-19} J.) SSM

34 • Calculate the contact potential between (*a*) Ag and Cu, (*b*) Ag and Ni, and (*c*) Ca and Cu.

HEAT CAPACITY DUE TO ELECTRONS IN A METAL

35 •• Gold has a Fermi energy of 5.53 eV. Determine the molar specific heat at constant volume for gold at room temperature. SSM

QUANTUM THEORY OF ELECTRICAL CONDUCTION

36 • The resistivities and Fermi speeds of Na, Au, and Sn at $T = 273$ K are 4.2 $\mu\Omega\cdot$cm, 2.04 $\mu\Omega\cdot$cm, and 10.6 $\mu\Omega\cdot$cm, and 1.07×10^6 m/s, 1.39×10^6 m/s, and 1.89×10^6 m/s, respectively. Use those values to find the mean free paths for the conduction electrons in these elements.

37 •• The resistivity of pure copper increases by approximately 1.0×10^{-8} $\Omega\cdot$m with the addition of 1.0 percent (by number of atoms) of an impurity distributed throughout the metal. The mean free path λ depends on both the impurity and the oscillations of the lattice ions according to the equation $1/\lambda = 1/\lambda_t + 1/\lambda_i$, where λ_t is the mean free path associated with the thermal vibrations of the ions and λ_i is the mean free path associated with the impurities. (*a*) Estimate λ_i using Equation 38-14 and the data given in Table 38-1. (*b*) If r is the effective radius of an impurity lattice ion seen by an electron, the scattering cross section is πr^2. Estimate this area, using the fact that r is related to λ_i by Equation 38-16. SSM

BAND THEORY OF SOLIDS

38 • Electromagnetic radiation is incident on the surface of a semiconductor. The maximum wavelength of this light that is required if electrons are to cross the energy gap between the valence band and the conduction band is 380.0 nm. What is the energy gap, in electron volts, for the semiconductor?

39 • An electron occupies the highest energy level of the valence band in a silicon sample. What is the maximum photon wavelength that will excite the electron across the energy gap if the gap is 1.14 eV? SSM

40 • An electron occupies the highest energy level of the valence band in a germanium sample. What is the maximum photon wavelength that will excite the electron into the conduction band? In germanium, the energy gap between the valence and conduction bands is 0.74 eV.

41 • An electron occupies the highest energy level of the valence band in a diamond sample. What is the maximum photon wavelength that will excite the electron into the conduction band? In diamond, the energy gap between the valence and conduction bands is 7.0 eV.

42 •• A photon of wavelength 3.35 μm has just enough energy to raise an electron from the valence band to the conduction band in a lead sulfide sample. (*a*) Find the energy gap between the bands in lead sulfide. (*b*) Find the temperature T for which kT equals that energy gap.

SEMICONDUCTORS

43 • The donor energy levels in an n-type semiconductor are 0.0100 eV below the conduction band. Find the temperature for which $kT = 0.0100$ eV.

44 •• When a thin slab of semiconducting material is illuminated with monochromatic electromagnetic radiation, most of the radiation is transmitted through the slab if the wavelength is greater than 1.85 mm. For wavelengths less than 1.85 mm, most of the incident radiation is absorbed. Determine the energy gap of the semiconductor.

45 •• The relative binding of the extra electron in the arsenic atom that replaces an atom in silicon or germanium can be understood from a calculation of the first Bohr radius of the electron in these materials. Four of arsenic's valence electrons form covalent bonds, so the fifth electron sees a center of attraction with a charge of $+e$. This model is a modified hydrogen atom. In the Bohr model of the hydrogen atom, the electron moves in free space at a radius a_0 given by $a_0 = 4\pi\epsilon_0\hbar^2/(m_e e^2)$ (Equation 36-12). When an electron moves in a crystal, we can approximate the effect of the other atoms by replacing ϵ_0 with $\kappa\epsilon_0$ and m_e with an effective mass for the electron. For silicon, κ is 12 and the effective mass is approximately $0.2m_e$. For germanium, κ is 16 and the effective mass is approximately $0.1m_e$. Estimate the Bohr radii for the valence electron as it orbits the impurity arsenic atom in silicon and in germanium.

46 •• The ground-state energy of the hydrogen atom is given by $E_1 = -m_e e^4/(8\epsilon_0^2 h^2)$ (Equations 36-15 and 36-16 where $4\pi\epsilon_0$ is substituted for k^{-1}). Modify this equation using information in Problem 45 by replacing ϵ_0 with $\kappa\epsilon_0$ and m_e with an effective mass for the electron to estimate the binding energy of the extra electron of an impurity arsenic atom in (*a*) silicon and (*b*) germanium.

47 •• A doped n-type silicon sample has 1.00×10^{16} electrons per cubic centimeter in the conduction band and has a resistivity of 5.00×10^{23} $\Omega\cdot$m at 300 K. Find the mean free path of the electrons. Use the effective mass of $0.2m_e$ for the mass of the electrons. (See Problem 45.) Compare this mean free path with that of conduction electrons in copper at 300 K.

48 ••• In the Hall effect, the Hall coefficient R_H is the proportionality constant between the transverse electric field and the product of the applied magnetic field and the current density. That is, $E_y = R_H B_z J_x$, where the current density, the transverse electric field, and the applied magnetic field are in the $+x$, $-y$, and $+z$ directions, respectively. (The Hall effect is presented in Chapter 26.) The measured Hall coefficient of a doped silicon sample is 0.0400 V$\cdot$m/(A$\cdot$T) at room temperature. If all the doping impurities have contributed to the total number of charge carriers of the sample, find (*a*) the type of impurity (donor or acceptor) used to dope the sample and (*b*) the concentration of the impurities.

*SEMICONDUCTOR JUNCTIONS AND DEVICES

49 •• Simple theory for the current versus the bias voltage across a pn junction yields the equation $I = I_0(e^{eV_b/kT} - 1)$. Sketch I versus V_b for both positive and negative values of V_b using that equation.

50 • The base current in an npn transistor circuit is 25.0 mA. If 88.0 percent of the electrons entering the base from the emitter reach the collector, what is the base current?

51 •• In Figure 38-28 for the npn-transistor amplifier, suppose $R_b = 2.00$ kΩ and $R_L = 10.0$ kΩ. Suppose further that a 10.0-μA ac base current i_b generates a 0.500-mA ac collector current i_c. What is the voltage gain of the amplifier? SSM

52 •• Germanium can be used to measure the energy of incident photons. Consider a 660-keV gamma ray emitted from ^{137}Cs. (*a*) Given that the band gap in germanium is 0.72 eV, how many electron-hole pairs can be generated as this gamma ray travels through germanium? (*b*) The number N of pairs in Part (*a*) will have statistical fluctuations given by $\pm\sqrt{N}$. What then is the energy resolution of the detector in that photon energy region?

53 •• Make a sketch showing the valence and conduction band edges and Fermi energy of a *pn*-junction diode when biased (*a*) in the forward direction and (*b*) in the reverse direction.

54 •• A good silicon diode has the current-voltage characteristic given by $I = I_0(e^{eV_b/kT} - 1)$. Let $kT = 0.025$ eV (room temperature) and the saturation current $I_0 = 1.0$ nA. (*a*) Show that for small reverse-bias voltages, the resistance is 25 MΩ. *Hint: Do a Taylor-series expansion of the exponential function about $V_b = 0$, or use the expansion for e^x found in Table M-4 of the Math Tutorial.* (*b*) Find the dc resistance for a reverse bias of 0.50 V. (*c*) Find the resistance V/I for a 0.50-V forward bias. What is the current in this case? (*d*) Calculate the differential resistance dV/dI for a 0.50-V forward bias.

55 •• A long slab of silicon of thickness $T = 1.0$ mm and width $w = 1.0$ cm is placed in a magnetic field $B = 0.40$ T. The slab is in the xy plane, where the length of the slab is parallel with the x axis and the magnetic field points in the $+z$ direction. When a current of 0.20 A exists in the sample in the $+x$ direction, a potential difference of 5.0 mV develops across the width of the sample and the electric field in the sample points in the $+y$ direction. Determine the semiconductor type (n or p) and the concentration of charge carriers. (The Hall effect is presented in Chapter 26.)

THE BCS THEORY

56 • (*a*) Use Equation 38-37 to calculate the superconducting energy gap for tin, and compare your result with the measured value of 6.00×10^{-4} eV. (*b*) Use the measured value to calculate the minimum value of the wavelength of a photon that has sufficient energy to break up Cooper pairs in lead ($T_c = 3.72$ K) at $T = 0$.

57 • (*a*) Use Equation 38-37 to calculate the superconducting energy gap for lead, and compare your result with the measured value of 2.73×10^{-3} eV. (*b*) Use the measured value to calculate the minimum value of the wavelength of a photon that has sufficient energy to break up Cooper pairs in tin ($T_c = 7.19$ K) at $T = 0$. SSM

THE FERMI–DIRAC DISTRIBUTION

58 •• The number of electrons in the conduction band of an insulator or intrinsic semiconductor is governed chiefly by the Fermi factor. Because the valence band in these materials is nearly filled and the conduction band is nearly empty, the Fermi energy E_F is generally midway between the top of the valence band and the bottom of the conduction band; that is, it is at $E_g/2$, where E_g is the band gap between the two bands and the energy is measured from the top of the valence band. (*a*) In silicon, $E_g \approx 1.0$ eV. Show that in this case the Fermi factor for electrons at the bottom of the conduction band is given by $\exp(-E_g/2kT)$ and evaluate this factor. Discuss the significance of the result if there are 10^{22} valence electrons per cubic centimeter and the probability of finding an electron in the conduction band is given by the Fermi factor. (*b*) Repeat the calculation in Part (*a*) for an insulator with a band gap of 6.0 eV.

59 •• Approximately how many energy states that have energies between 2.00 eV and 2.20 eV are available to electrons in a cube of silver measuring 1.00 mm on a side?

60 •• Show that at $E = E_F$, the expression for the Fermi factor (Equation 38-24) is equal to 0.5.

61 •• (*a*) Using the equation $E_F = [h^2/(8m_e)][3N/(\pi V)]^{2/3}$ (Equation 38-22*a*), calculate the Fermi energy for silver. (*b*) Determine the average kinetic energy of a free electron and (*c*) find the Fermi speed for silver. SSM

62 •• What is the difference between the energies at which the Fermi factor is 0.9 and 0.1 at 300 K in (*a*) copper, (*b*) potassium, and (*c*) aluminum.

63 •• What is the probability that a conduction electron in silver will have a kinetic energy of 4.90 eV at $T = 300$ K? SSM

64 •• Show that $g(E) = \frac{3}{2}NE_F^{-3/2}E^{1/2}$ (Equation 38-43) follows from Equation 38-41 for $g(E)$, and from Equation 38-22*a* for E_F.

65 •• Carry out the integration $E_{av} = (1/N)\int_0^{E_F} Eg(E)dE$ to show that the average energy at $T = 0$ is $\frac{3}{5}E_F$.

66 •• The density of the electron states in a metal can be written $g(E) = AE^{1/2}$, where A is a constant and E is measured from the bottom of the conduction band. (*a*) Show that the total number of states is $\frac{2}{3}AE_F^{3/2}$. (*b*) Approximately what fraction of the conduction electrons are within kT of the Fermi energy? (*c*) Evaluate that fraction for copper at $T = 300$ K.

67 •• What is the probability that a conduction electron in silver will have a kinetic energy of 5.49 eV at $T = 300$ K?

68 •• Using the expression $g(E) = (8\sqrt{2}\pi m_e^{3/2}V/h^3)E^{1/2}$ (Equation 38-41) for the density of states, estimate the fraction of the conduction electrons in copper that can absorb energy from collisions with the vibrating lattice ions at (*a*) 77 K and (*b*) 300 K.

69 •• In an intrinsic semiconductor, the Fermi energy is about midway between the top of the valence band and the bottom of the conduction band. In germanium, the forbidden energy band has a width of 0.70 eV. Show that at room temperature the distribution function of electrons in the conduction band is given by the Maxwell–Boltzmann distribution function.

70 ••• The root-mean-square (rms) value of a variable is obtained by calculating the average value of the square of that variable and then taking the square root of the result. Use that procedure to determine the rms energy of a Fermi distribution. Express your result in terms of E_F and compare it to the average energy. Why do E_{av} and E_{rms} differ?

GENERAL PROBLEMS

71 • The density of potassium is 0.851 g/cm³. How many free electrons are there per potassium atom in a crystal of potassium?

72 • Calculate the number density of free electrons for (*a*) magnesium, which has a density of 1.74 g/cm³, and (*b*) zinc, which has a density of 7.14 g/cm³. For the calculations assume there are two free electrons per atom, and compare your results with the values listed in Table 38-1.

73 •• Estimate the fraction of free electrons in copper that are in energy states above the Fermi energy at (*a*) 300 K (about room temperature) and (*b*) 1000 K.

74 •• A certain free-electron energy state of manganese has a 10.0 percent chance of being occupied when the temperature of the manganese is $T = 1300$ K. What is the energy of the state?

75 •• The semiconducting compound CdSe is widely used for light-emitting diodes (LEDs). The energy gap in CdSe is 1.80 eV. What is the frequency of the light emitted by a CdSe LED?

76 ••• A 2.00-cm² wafer of pure silicon is irradiated with electromagnetic radiation having a wavelength of 775 nm. The intensity of the radiation is 4.00 W/m² and every photon that strikes the sample is absorbed and creates an electron-hole pair. (*a*) How many electron-hole pairs are produced in one second? (*b*) If the number of electron-hole pairs in the sample is 6.25×10^{11} in the steady state, at what rate do the electron-hole pairs recombine? (*c*) If every recombination event results in the radiation of one photon, at what rate is energy radiated by the sample?

CHAPTER 39

THE ANDROMEDA GALAXY BY MEASURING THE FREQUENCY OF THE LIGHT COMING TO US FROM DISTANT OBJECTS, WE ARE ABLE TO DETERMINE HOW FAST THE OBJECTS ARE APPROACHING TOWARD US OR RECEDING FROM US. *(NASA.)*

? Have you wondered how the frequency of the light enables us to determine the speed of recession of a distant galaxy? (See Example 39-5.)

Relativity

The theory of relativity consists of two rather different theories, the special theory and the general theory. The special theory, developed by Albert Einstein and others in 1905, describes measurements made in different inertial reference frames moving with constant velocity relative to one another. Its consequences, which can be derived with a minimum of mathematics, are applicable in a wide variety of situations encountered in physics and in engineering. On the other hand, the general theory, also developed by Einstein and others around 1916, describes accelerated reference frames and gravity. A thorough understanding of the general theory requires sophisticated mathematics, and the applications of the theory are mainly in the area of gravitation. The general theory is of great importance in cosmology, but it is rarely encountered in other areas of physics or in engineering. The general theory is applied, however, in the engineering of the Global Positioning System (GPS).*

* The satellites used in GPS contain atomic clocks.

In this chapter, we concentrate on the special theory (often referred to as special relativity). General relativity theory (general relativity) will be discussed briefly near the end of the chapter. Special relativity is first presented in Chapter R (which precedes Chapter 11). You should consider reviewing the material in Chapter R before proceeding in this chapter.

39-1 NEWTONIAN RELATIVITY

Newton's first law does not distinguish between a particle at rest and a particle moving with constant velocity. If there is no net external force acting, the particle will remain in its initial state, either at rest or moving with its initial velocity. A particle at rest relative to you is moving with constant velocity relative to an observer who is moving with constant velocity relative to you. How might we distinguish whether you and the particle are at rest and the second observer is moving with constant velocity, or the second observer is at rest and you and the particle are moving?

Let us consider some simple experiments. Suppose we have a railway boxcar moving along a straight, flat track with a constant velocity v. (The velocity v is a signed quantity, and the sign indicates the direction of the motion of the boxcar along the track.) We note that a ball at rest in the boxcar remains at rest relative to the car. If we drop the ball, it falls straight down, relative to the boxcar, with an acceleration g due to gravity. Of course, when viewed from the reference frame of the track, the ball moves along a parabolic path because it has an initial velocity v to the right. No mechanics experiment that we can do—measuring the period of a pendulum, observing the collisions between two objects, or whatever—will tell us whether the boxcar is moving and the track is at rest or the track is moving and the boxcar is at rest. If we have a coordinate system attached to the track and another attached to the boxcar, Newton's laws hold in either system.

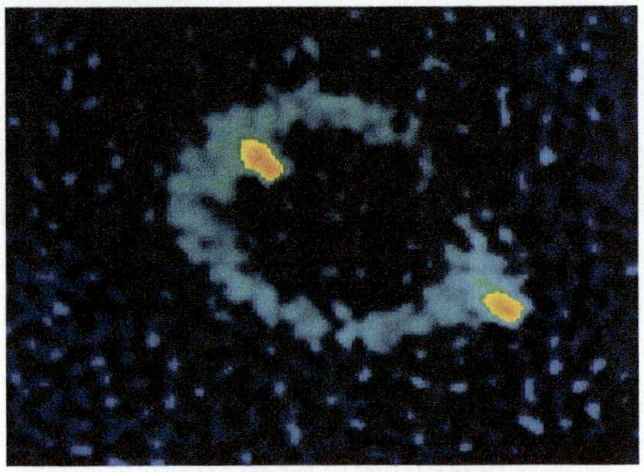

The ringlike structure of the radio source MG1131 + 0456 is thought to be due to *gravitational lensing*, first proposed by Albert Einstein in 1936, in which a source is imaged into a ring by a large, massive object in the foreground. *(NRAO/AUI.)*

A set of coordinate systems at rest relative to each other is called a *reference frame*. A reference frame in which Newton's laws hold is called an *inertial reference frame*.* All reference frames moving at constant velocity relative to an inertial reference frame are also inertial reference frames. If we have two inertial reference frames moving with constant velocity relative to each other, there are no mechanics experiments that can tell us which is at rest and which is moving or if they are both moving. This result is known as the principle of **Newtonian relativity:**

> Absolute motion cannot be detected.
>
> PRINCIPLE OF NEWTONIAN RELATIVITY

This principle was well known by Galileo, Newton, and others in the seventeenth century. By the late nineteenth century, however, this view had changed. It was then generally thought that Newtonian relativity was not valid and that absolute motion could be detected in principle by a measurement of the speed of light.

ETHER AND THE SPEED OF LIGHT

We saw in Chapter 15 that the velocity of a wave depends on the properties of the medium in which the wave travels and not on the velocity of the source of the waves. For example, the velocity of sound relative to still air depends on the temperature of the air. Light and other electromagnetic waves (for example, radio

* Reference frames were first discussed in Section 3-1. Inertial reference frames were also discussed in Section 4-1.

waves and X rays) travel through a vacuum with a speed $c = 3.00 \times 10^8$ m/s that is predicted by James Clerk Maxwell's equations for electricity and magnetism. But what is this speed relative to? What is the equivalent of still air for a vacuum? A proposed medium for the propagation of light was called the *ether;* it was thought to pervade all space. The velocity of light relative to the ether was assumed to be c, as predicted by Maxwell's equations. The velocity of any object relative to the ether was considered to be the absolute velocity of the object.

Albert Michelson, first in 1881 and then again with Edward Morley in 1887, set out to measure the velocity of Earth relative to the ether by an ingenious experiment in which the velocity of light relative to Earth was compared for two light beams, one parallel to the direction of Earth's motion relative to the Sun and the other perpendicular to the direction of Earth's motion. Despite painstakingly careful measurements, they could detect no difference. The experiment has since been repeated under various conditions by a number of people, and no difference has ever been found. The absolute motion of Earth relative to the ether cannot be detected.

39-2 EINSTEIN'S POSTULATES

In 1905, at the age of 26, Albert Einstein published a paper on the electrodynamics of moving bodies.* In this paper, he postulated that absolute motion cannot be detected by any experiment. That is, there is no ether. Earth can be considered to be at rest and the velocity of light will be the same in any direction.† His theory of special relativity can be derived from two postulates. Simply stated, these postulates are as follows:

> Postulate 1: Absolute uniform motion cannot be detected.
>
> Postulate 2: The speed of light is independent of the motion of the source.
>
> EINSTEIN'S POSTULATES

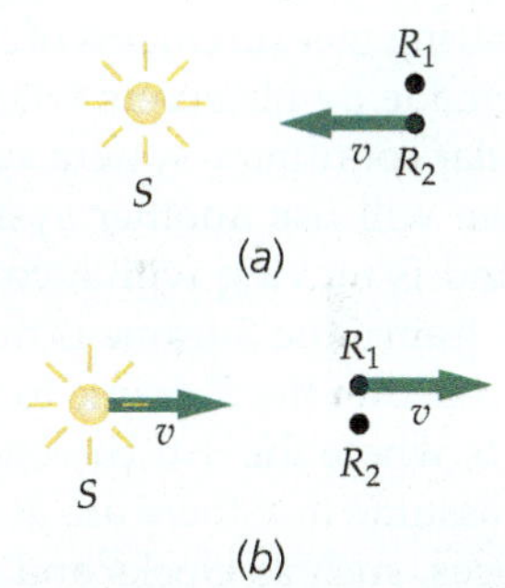

FIGURE 39-1 (*a*) A stationary light source S, a stationary observer R_1, and a second observer R_2 moving toward the source with speed v. (*b*) In the reference frame in which the observer R_2 is at rest, the light source S and observer R_1 move to the right with speed v. If absolute motion cannot be detected, the two views are equivalent. Because the speed of light does not depend on the motion of the source, observer R_2 measures the same value for that speed as observer R_1.

Postulate 1 is merely an extension of the Newtonian principle of relativity to include all types of physical measurements (not just those measurements that are mechanical). Postulate 2 describes a common property of many waves. For example, the speed of sound waves does not depend on the motion of the sound source. The sound waves from a car horn travel through the air with the same speed, relative to the air, independent of whether the car is moving relative to the air or not. The speed of the waves depends only on the properties of the air, such as its temperature.

Although each postulate seems quite reasonable, many of the implications of the two postulates together are quite surprising and contradict what is often called common sense. For example, one important implication of these postulates is that every observer measures the same value for the speed of light independent of the relative motion of the source and the observer. Consider a light source S and two observers, R_1 at rest relative to S and R_2 moving toward S with speed v, as shown in Figure 39-1*a*. The speed of light measured by R_1 is $c = 3.00 \times 10^8$ m/s. What is the speed measured by R_2? The answer is *not* $c + v$. By postulate 1, Figure 39-1*a* is equivalent to Figure 39-1*b*, in which R_2 is at rest and the source S and R_1 are

* *Annalen der Physik*, vol. 17, 1905, p. 841. For a translation from the original German, see W. Perrett and G. B. Jeffery (trans.), *The Principle of Relativity: A Collection of Original Memoirs on the Special and General Theory of Relativity* by H. A. Lorentz, A. Einstein, H. Minkowski, and W. Weyl, Dover, New York, 1923.

† Einstein did not set out to explain the results of the Michelson–Morley experiment. His theory arose from his considerations of the theory of electricity and magnetism and the unusual property of electromagnetic waves that they propagate in a vacuum. In his first paper, which contains the complete theory of special relativity, he made only a passing reference to the Michelson–Morley experiment, and in later years he could not recall whether he was aware of the details of the experiment before he published his theory.

moving with speed v. That is, because absolute motion cannot be detected, it is not possible to say which is really moving and which is at rest. By postulate 2, the speed of light from a moving source is independent of the motion of the source. Thus, looking at Figure 39-1*b*, we see that R_2 measures the speed of light to be c, just as R_1 does. This result is often considered as an alternative to Einstein's second postulate:

> Postulate 2 (alternate): Every observer measures the same value c for the speed of light.

This result contradicts our intuitive ideas about relative velocities. If a car moves at 50 km/h away from an observer and another car moves at 80 km/h in the same direction, the velocity of the second car relative to the first car is 30 km/h. This result is easily measured and conforms to our intuition. However, according to Einstein's postulates, if a light beam is moving in the direction of the cars, observers in both cars will measure the same speed for the light beam. Our intuitive ideas about the combination of velocities are approximations that hold only when the speeds are very small compared with the speed of light. Even in an airplane moving with the speed of sound, to measure the speed of light accurately enough to distinguish the difference between the results c and $c + v$, where v is the speed of the plane, would require a measurement with six-digit accuracy.

39-3 THE LORENTZ TRANSFORMATION

Einstein's postulates have important consequences for measuring time intervals and space intervals, as well as relative velocities. Throughout this chapter, we will be comparing measurements of the positions and times of events (such as lightning flashes) made by observers who are moving relative to each other. We will use a rectangular coordinate system xyz that has an origin O and is called the S reference frame; we will use another system $x'y'z'$ that has an origin O', is called the S' frame, and is moving with a constant velocity $\vec{v}$ relative to the S frame. Relative to the S' frame, the S frame is moving with a constant velocity $-\vec{v}$. For simplicity, we will consider the S' frame to be moving along the x axis in the $+x$ direction relative to S, where the $+x'$ direction is the same as the $+x$ direction. In each frame, we will assume that there are as many observers as are needed who have measuring devices, such as clocks and metersticks, that are identical when compared at rest (see Figure 39-2).

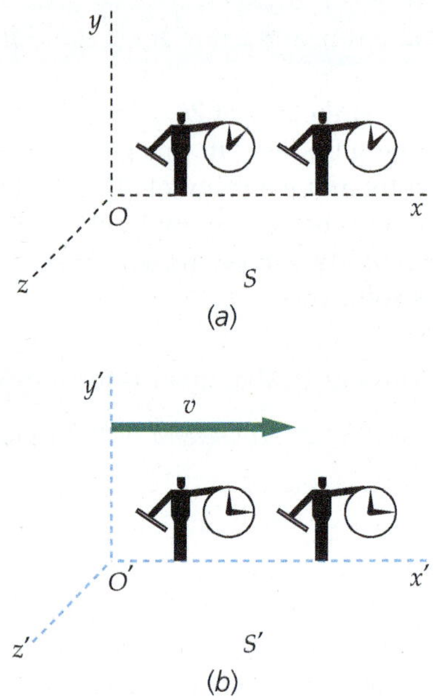

FIGURE 39-2 Coordinate reference frames S and S' moving with relative speed v. In each frame, there are observers who have metersticks and clocks that are identical when compared at rest.

We will use Einstein's postulates to find the general relation between the coordinates x, y, and z and the time t of an event as seen in reference frame S and the coordinates x', y', and z' and the time t' of the same event as seen in reference frame S', which is moving with uniform velocity relative to S. For convenience, we assume that the origins are coincident at time $t = t' = 0$. The classical relation, called the **Galilean transformation,** is

$$x = x' + vt', \quad y = y', \quad z = z', \quad t = t' \qquad \text{39-1}a$$

GALILEAN TRANSFORMATION

The inverse transformation is

$$x' = x - vt, \quad y' = y, \quad z' = z, \quad t' = t \qquad \text{39-1}b$$

These equations are consistent with experimental observations as long as v is much less than c. They lead to the familiar classical rules for velocities. If a particle has velocity $u_x = dx/dt$ in frame S, its velocity in frame S' is

$$u'_x = \frac{dx'}{dt'} = \frac{dx'}{dt} = \frac{d}{dt}(x - vt) = u_x - v \qquad \text{39-2}$$

If we differentiate this equation again, we find that the acceleration of the particle is the same in both frames:

$$a_x = \frac{du_x}{dt} = \frac{du'_x}{dt'} = a'_x$$

It should be clear that the Galilean transformation is not consistent with Einstein's postulates of special relativity. If light moves along the x axis with speed $u'_x = c$ in S', these equations imply that the speed in S' is $u_x = c + v$ rather than $u_x = c$, which is consistent with Einstein's postulates and with experiment. The classical transformation equations must therefore be modified to make them consistent with Einstein's postulates. We will give a brief outline of one method of obtaining the relativistic transformation.

We assume that the relativistic transformation equation for x is the same as the classical equation (Equation 39-1a) except for a constant multiplier on the right side. That is, we assume the equation is of the form

$$x = \gamma(x' + vt') \qquad 39\text{-}3$$

where γ is a constant that can depend on v and c but not on the coordinates. The inverse transformation must look the same except for the plus sign:

$$x' = \gamma(x - vt) \qquad 39\text{-}4$$

Let us consider a light pulse that starts at the origin of S at $t = 0$. Because we have assumed that the two origins are coincident at $t = t' = 0$, the pulse also starts at the origin of S' at $t' = 0$. Einstein's postulates require that the equation for the x component of the wave front of the light pulse is $x = ct$ in frame S and $x' = ct'$ in frame S'. Substituting ct for x and ct' for x' in Equation 39-3 and Equation 39-4, we obtain

$$ct = \gamma(c + v)t' \qquad 39\text{-}5$$

and

$$ct' = \gamma(c - v)t \qquad 39\text{-}6$$

We divide both sides of Equations 39-5 and 39-6 by t, and then eliminate the ratio t'/t from the two equations and determine γ. Thus,

$$\gamma = \frac{1}{\sqrt{1 - (v^2/c^2)}} \qquad 39\text{-}7$$

(Note that γ is always greater than 1, and when v is much less than c, $\gamma \approx 1$.) The relativistic transformation for x and x' is therefore given by Equation 39-3 and Equation 39-4, where γ is given by Equation 39-7. We can obtain equations for t and t' by combining Equation 39-3 with the inverse transformation given by Equation 39-4. Substituting $x = \gamma(x' + vt')$ for x in Equation 39-4, we obtain

$$x' = \gamma[\gamma(x' + vt') - vt] \qquad 39\text{-}8$$

which can be solved for t in terms of x' and t'. The complete relativistic transformation is

$$x = \gamma(x' + vt'), \quad y = y', \quad z = z' \qquad 39\text{-}9$$

$$t = \gamma\left(t' + \frac{vx'}{c^2}\right) \qquad 39\text{-}10$$

LORENTZ TRANSFORMATION

The inverse transformation is

$$x' = \gamma(x - vt), \quad y' = y, \quad z' = z \qquad 39\text{-}11$$

$$t' = \gamma\left(t - \frac{vx}{c^2}\right) \qquad 39\text{-}12$$

The transformation described by Equations 39-9 through 39-12 is called the **Lorentz transformation.** It relates the space and time coordinates x, y, z, and t of an event in frame S to the coordinates x', y', z', and t' of the same event as seen in frame S', which is moving along the x axis with speed v relative to frame S.

We will now look at some applications of the Lorentz transformation.

TIME DILATION

Consider two events, one that occurs on the x' axis at point x'_0 at time t'_1 in frame S' and another that occurs on the x' axis at point x'_0 at time t'_2 in frame S'. (Both events occur at point x'_0 in S'.) We can find the times t_1 and t_2 for the events in S from Equation 39-10. We have

$$t_1 = \gamma\left(t'_1 + \frac{vx'_0}{c^2}\right)$$

and

$$t_2 = \gamma\left(t'_2 + \frac{vx'_0}{c^2}\right)$$

so

$$t_2 - t_1 = \gamma(t'_2 - t'_1)$$

The time between two events that happen at the *same place* in a reference frame is called **proper time Δt_p** between the events. In this case, the time interval $t'_2 - t'_1$ measured in frame S' is proper time. The time interval Δt measured in any other reference frame is always longer than the proper time. This expansion is called **time dilation:**

$$\Delta t = \gamma\,\Delta t_\mathrm{p} \qquad 39\text{-}13$$

TIME DILATION

Example 39-1 Spatial Separation and Temporal Separation of Two Events

Two events occur at the same point x'_0 at times t'_1 and t'_2 in frame S', which is traveling in the $+x$ direction at speed v relative to frame S. (*a*) What is the spatial separation of the events in frame S? (*b*) What is the temporal separation of the events in frame S?

PICTURE The spatial separation in S is $x_2 - x_1$, where x_2 and x_1 are the coordinates of the events in S, which are found using Equation 39-9.

SOLVE

(*a*) 1. The position x_1 in S is given by Equation 39-9 with $x'_1 = x'_0$: $\quad x_1 = \gamma(x'_0 + vt'_1)$

2. Similarly, the position x_2 in S is given by: $\quad x_2 = \gamma(x'_0 + vt'_2)$

3. Subtract to find the spatial separation: $\quad \Delta x = x_2 - x_1 = \gamma v(t'_2 - t'_1) = \boxed{\dfrac{v(t'_2 - t'_1)}{\sqrt{1 - (v^2/c^2)}}}$

(*b*) Using the time dilation formula, relate the two time intervals. The two events occur at the same place in S', so the proper time between the two events is $\Delta t_\mathrm{p} = t'_2 - t'_1$: $\quad \Delta t = t_2 - t_1 = \gamma(t'_2 - t'_1) = \boxed{\dfrac{(t'_2 - t'_1)}{\sqrt{1 - (v^2/c^2)}}}$

CHECK Taking the limits of the Part-(*a*) and Part-(*b*) results as c approaches infinity gives $\Delta x = v(t_2' - t_1')$ and $\Delta t = t_2' - t_1'$, respectively. Combining these expressions gives $\Delta x = v\Delta t$. This is just the classical (nonrelativistic) equation that displacement equals velocity multiplied by time that is developed in Chapter 2 for one-dimensional motion. In addition, the equation $\Delta t = t_2' - t_1'$ is just the classical result that the length of the time between events is the same in both reference frames.

TAKING IT FURTHER Dividing the Part-(*a*) result by the Part-(*b*) result gives $\Delta x/\Delta t = v$. The spatial separation Δx of the two events in S is the distance a fixed point, such as x_0' in S', moves in S during the time interval between the events in S.

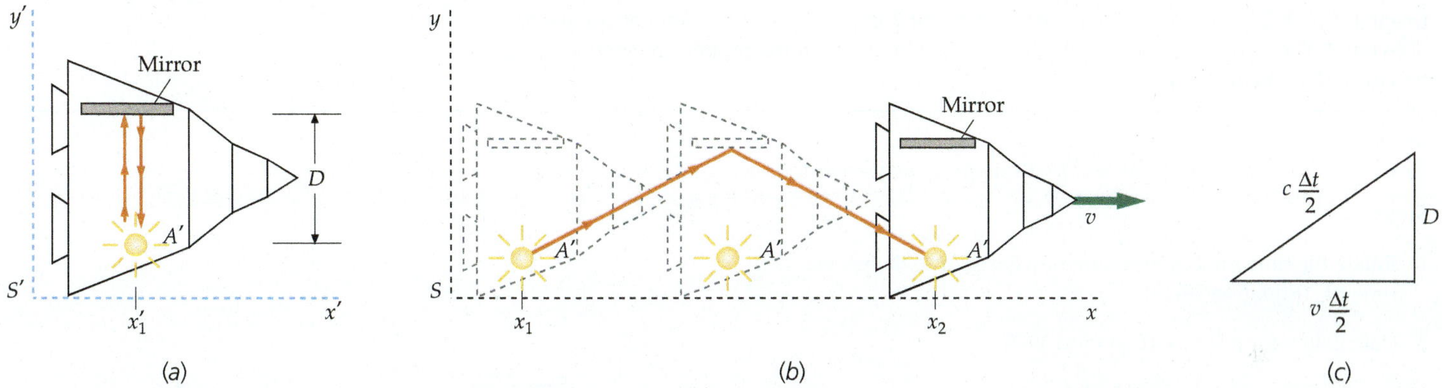

FIGURE 39-3 (*a*) Observer A' and the mirror are in a spaceship at rest in frame S'. The time it takes for the light pulse to reach the mirror and return is measured by A' to be $2D/c$. (*b*) In frame S, the spaceship is moving to the right with speed v. If the speed of light is the same in both frames, the time it takes for the light to reach the mirror and return is longer than $2D/c$ in S because the distance traveled is greater than $2D$. (*c*) A right triangle for computing the time Δt in frame S.

We can understand time dilation directly from Einstein's postulates without using the Lorentz transformation. Figure 39-3*a* shows an observer A' a distance D from a mirror. The observer and the mirror are in a spaceship that is at rest in frame S'. The observer explodes a flash gun and measures the time interval $\Delta t'$ between the original flash (Event 1) and his seeing the return flash from the mirror (Event 2). Because light travels with speed c, this time is

$$\Delta t' = \frac{2D}{c}$$

We now consider the same two events, the original flash of light and the receiving of the return flash, as observed in reference frame S, in which observer A' and the mirror are moving to the right with speed v, as shown in Figure 39-3*b*. Events 1 and 2 happen at positions x_1 and x_2, respectively, in frame S. During the time interval Δt (as measured in S) between the original flash and the return flash, observer A' and his spaceship have moved to the right a distance $v\,\Delta t$. In Figure 39-3, we can see that the path traveled by the light is longer in S than in S'. However, by Einstein's postulates, light travels with the same speed c in frame S as it does in frame S'. Because light travels farther in S at the same speed, it takes longer in S to reach the mirror and return. The time interval in S is thus longer than it is in S'. From the triangle in Figure 39-3*c*, we have

$$\left(\frac{c\,\Delta t}{2}\right)^2 = D^2 + \left(\frac{v\,\Delta t}{2}\right)^2$$

or

$$\Delta t = \frac{2D}{\sqrt{c^2 - v^2}} = \frac{2D}{c}\frac{1}{\sqrt{1 - (v^2/c^2)}}$$

Using $\Delta t' = 2D/c$, we obtain

$$\Delta t = \frac{\Delta t'}{\sqrt{1 - (v^2/c^2)}} = \gamma\,\Delta t'$$

Example 39-2 How Long Is a One-Hour Nap?

Try It Yourself

Astronauts in a spaceship traveling at $v = 0.600c$ relative to Earth sign off from space control, saying that they are going to nap for 1.00 h and then call back. How long does their nap last as measured on Earth?

PICTURE Because the astronauts go to sleep (Event 1) and wake up (Event 2) at the same place in the reference frame of the ship, the time interval for their nap of 1.00 h, as measured by a clock on the ship, is the proper time between the two events. In the reference frame of Earth, they move a considerable distance during the time between the two events. The time interval measured in Earth's frame is measured using two clocks that are stationary relative to Earth. Clock 1 is located at the position of Event 1 and measures the time of occurrence of Event 1. Clock 2 is located at the position of Event 2 and measures the time of occurrence of Event 2. The difference between the two times is longer than the proper time between the two events by the factor γ.

SOLVE

Cover the column to the right and try these on your own before looking at the answers.

Steps	Answers
1. Relate the time interval measured on Earth Δt to the proper time Δt_p (Equation 39-13).	$\Delta t = \gamma\, \Delta t_p$
2. Calculate γ for $v = 0.6c$ (Equation 39-7).	$\gamma = 1.25$
3. Substitute the value for γ to calculate the time of the nap in Earth's frame.	$\Delta t = \gamma\, \Delta t_p = \boxed{1.25\text{ h}}$

CHECK The time interval is longer in the reference frame in which the two events occur at different locations as expected.

PRACTICE PROBLEM 39-1 If the spaceship is moving at $v = 0.800c$, how long would a 1.00 h nap last as measured on Earth?

LENGTH CONTRACTION

A phenomenon closely related to time dilation is **length contraction.** The length of an object measured in the reference frame in which the object is at rest is called its **proper length** L_p. In a reference frame in which the object is moving parallel to its length, the measured length is shorter than its proper length. Consider a rod at rest in frame S' with one end at x_2' and the other end at x_1'. The length of the rod in this frame is its proper length $L_p = x_2' - x_1'$. Some care must be taken to find the length of the rod in frame S. In that frame, the rod is moving to the right with speed v, the speed of frame S'. The length of the rod in frame S is defined as $L = x_2 - x_1$, where x_2 is the position of one end at some time t_2, and x_1 is the position of the other end *at the same time* $t_1 = t_2$ as measured in frame S. To calculate $x_2 - x_1$ at some time t we use Equation 39-11:

$$x_2' = \gamma(x_2 - vt_2)$$

and

$$x_1' = \gamma(x_1 - vt_1)$$

Because $t_2 = t_1$, by subtracting the second equation from the first we obtain

$$x_2' - x_1' = \gamma(x_2 - x_1)$$

Solving for $x_2 - x_1$ gives

$$x_2 - x_1 = \frac{1}{\gamma}(x_2' - x_1') = (x_2' - x_1')\sqrt{1 - \frac{v^2}{c^2}}$$

or

$$L = \frac{1}{\gamma}L_p = L_p\sqrt{1 - \frac{v^2}{c^2}} \qquad 39\text{-}14$$

LENGTH CONTRACTION

Because $1/\gamma$ is less than one, it follows that the length of the rod is smaller when it is measured in a frame in which it is moving parallel to its length. Before Einstein's paper was published, Hendrik A. Lorentz and George F. FitzGerald tried to explain the null result of the Michelson–Morley experiment by assuming that distances in the direction of motion contracted by the amount given in Equation 39-14. This length contraction is now known as the **Lorentz–FitzGerald contraction.**

Example 39-3 The Length of a Moving Meterstick

A stick that has a proper length of 1.00 m moves in a direction parallel to its length with speed v relative to you. The length of the stick as measured by you is 0.914 m. What is the speed v?

PICTURE We can find v directly from Equation 39-14.

SOLVE

1. Equation 39-14 relates the lengths L and L_p and the speed v: $L = L_p\sqrt{1 - \frac{v^2}{c^2}}$

2. Solve for v: $v = c\sqrt{1 - \frac{L^2}{L_p^2}} = c\sqrt{1 - \frac{(0.914\text{ m})^2}{(1.00\text{ m})^2}} = \boxed{0.406c}$

CHECK As expected, the speed is a significant fraction of c.

An interesting example of time dilation or length contraction is the generation of muons as secondary radiation from cosmic rays. Muons decay according to the statistical law of radioactivity:

$$N(t) = N_0 e^{-t/\tau} \qquad 39\text{-}15$$

where N_0 is the number of muons at time $t = 0$, $N(t)$ is the number remaining at time t, and τ is the mean lifetime, which is approximately 2.2 μs for muons at rest. Because muons are generated (from the decay of pions) high in the atmosphere, usually several thousand meters above sea level, few muons should reach sea level. A typical muon moving with speed $0.9978c$ would travel only about 660 m in 2.2 μs. However, the lifetime of the muon measured in Earth's reference frame is increased by the factor $1/\sqrt{1 - (v^2/c^2)}$, which is 15 for this particular speed. The mean lifetime measured in Earth's reference frame is therefore 33 μs, and a muon with speed $0.9978c$ travels approximately 10 000 m during this time. From the muon's point of view, it exists for only 2.2 μs, but the atmosphere is rushing past it with a speed of $0.9978c$. The distance of 10 000 m in Earth's frame is thus contracted to only 660 m in the muon's frame, as indicated in Figure 39-4.

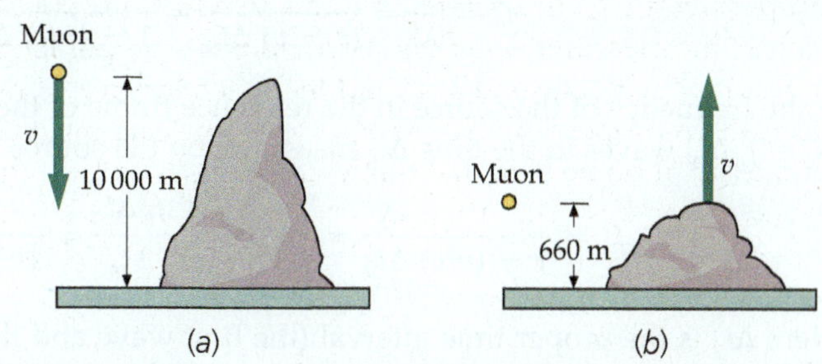

FIGURE 39-4 Although muons are created high above Earth and their mean lifetime is only about 2.2 μs when at rest, many appear at Earth's surface. (*a*) In Earth's reference frame, a typical muon moving at $0.9978c$ has a mean lifetime of 33 μs and travels 10 000 m during that time. (*b*) In the reference frame of the muon, the distance traveled by Earth is only 660 m in the muon's lifetime of 2.2 μs.

It is easy to distinguish experimentally between the classical and relativistic predictions of the observation of muons at sea level. Suppose that we observe 10^8 muons at an altitude of 10 000 m during some time interval with a muon detector. How many would we expect to observe at sea level during the same time interval? According to the nonrelativistic prediction, the time it takes for the muons to travel 10 000 m is $(10\,000 \text{ m})/(0.998c) \approx 33\ \mu\text{s}$, which is 15 lifetimes. Substituting $N_0 = 1.0 \times 10^8$ and $t = 15\tau$ into Equation 39-15, we obtain

$$N = N_0 e^{-t/\tau} = 1.0 \times 10^8 e^{-15} = 31$$

See **Math Tutorial** *for more information on the* **The Exponential Function**

We would thus expect all but about 31 of the original 100 million muons to decay before reaching sea level.

According to the relativistic prediction, Earth must travel only the contracted distance of 660 m in the rest frame of the muon. This trip takes only $2.2\ \mu\text{s} = 1\tau$. Therefore, the number of muons expected at sea level is

$$N = N_0 e^{-t/\tau} = 1.0 \times 10^8 e^{-1} = 37 \times 10^6$$

Thus, relativity predicts that we would observe 37 million muons during the same time interval. Experiments have confirmed the relativistic prediction of 37 million muons.

THE RELATIVISTIC DOPPLER EFFECT

For light or other electromagnetic waves in a vacuum, a distinction between the motion of the source and the motion of the receiver cannot be made. Therefore, the expressions we derived in Chapter 15 for the Doppler effect cannot be correct for light. The reason it is not correct is that in Chapter 15 we assumed the time intervals in the reference frames of the source and receiver to be the same.

Consider a source moving toward a receiver with speed v, relative to the receiver. If the source emits N electromagnetic-wave crests during a time Δt_R (measured in the frame of the receiver), the first crest will travel a distance $c\,\Delta t_R$ and the source will travel a distance $v\,\Delta t_R$ measured in the frame of the receiver. The wavelength in this reference frame will be

$$\lambda' = \frac{c\,\Delta t_R - v\,\Delta t_R}{N}$$

The frequency f' observed by the receiver will therefore be

$$f' = \frac{c}{\lambda'} = \frac{c}{(c - v)}\frac{N}{\Delta t_R} = \frac{1}{1 - (v/c)}\frac{N}{\Delta t_R}$$

If the frequency of the source in the reference frame of the source is f_0, it will emit $N = f_0 \Delta t_S$ waves in the time Δt_S measured by the source. Then

$$f' = \frac{1}{1 - (v/c)}\frac{N}{\Delta t_R} = \frac{1}{1 - (v/c)}\frac{f_0 \Delta t_S}{\Delta t_R} = \frac{f_0}{1 - (v/c)}\frac{\Delta t_S}{\Delta t_R}$$

Here Δt_S is the proper time interval (the first wave and the Nth wave are emitted at the same place in the reference frame of the source). Times Δt_S and Δt_R are related by Equation 39-13 for time dilation:

$$\Delta t_R = \gamma\,\Delta t_S = \frac{\Delta t_S}{\sqrt{1 - (v^2/c^2)}}$$

Thus, when the source and the receiver are moving toward one another we obtain

$$f' = \frac{f_0}{1 - (v/c)}\sqrt{1 - (v^2/c^2)} = \sqrt{\frac{1 + (v/c)}{1 - (v/c)}}\,f_0 \quad \text{approaching} \qquad 39\text{-}16a$$

This differs from our classical equation only in the time-dilation factor. It is left as a problem (Problem 25) for you to show that the same results are obtained if the calculations are done in the reference frame of the receiver.

When the source and the receiver are moving away from one another, the same analysis shows that the observed frequency is given by

$$f' = \frac{f_0}{1 + (v/c)}\sqrt{1 - (v/c^2)} = \sqrt{\frac{1 - (v/c)}{1 + (v/c)}}f_0 \quad \text{receding} \qquad \text{39-16}b$$

An application of the relativistic Doppler effect is the **redshift** observed in the light from distant galaxies. Because the galaxies are moving away from us, the light they emit is shifted toward the longer wavelengths. (Because light that has the longest visible wavelengths appears red, this is referred to as a redshift.) The speed of the galaxies relative to us can be determined by measuring this shift.

Example 39-4 Red Light/Green Light

You are spending the day shadowing two police officers. You have just witnessed the officers pulling over a car that went through a red light. The driver claims that the red light looked green because the car was moving toward the stoplight, which shifted the wavelength of the observed light. You quickly do some calculations to see if the driver has a reasonable case.

PICTURE We can use the Doppler shift formula for approaching objects in Equation 39-16*a*. This will tell us the velocity, but we need to know the frequencies of the light. We can make good guesses for the wavelengths of red light and green light and use $c = f\lambda$ to determine the frequencies.

SOLVE

1. The observer is approaching the light source, so we use the Doppler formula (Equation 39-16*a*) for approaching sources:

$$f' = \sqrt{\frac{1 + (v/c)}{1 - (v/c)}}f_0$$

2. Substitute c/λ for f, then simplify:

$$\frac{c}{\lambda'} = \sqrt{\frac{1 + (v/c)}{1 - (v/c)}}\frac{c}{\lambda_0}$$

$$\left(\frac{\lambda_0}{\lambda'}\right)^2 = \frac{1 + (v/c)}{1 - (v/c)}$$

3. Cross multiply and solve for v/c:

$$(\lambda_0)^2\left(1 - \frac{v}{c}\right) = (\lambda')^2\left(1 + \frac{v}{c}\right)$$

$$(\lambda_0)^2 - (\lambda')^2 = \left[(\lambda_0)^2 + (\lambda')^2\right]\left(\frac{v}{c}\right)$$

$$\frac{v}{c} = \frac{(\lambda_0)^2 - (\lambda')^2}{(\lambda_0)^2 + (\lambda')^2} = \frac{1 - (\lambda'/\lambda_0)^2}{1 + (\lambda'/\lambda_0)^2}$$

4. The values for the wavelengths for the colors of the visible spectrum can be found in Table 30-1. The wavelengths for red are 625 nm or longer, and the wavelengths for green are 530 nm or shorter. Solve for the speed needed to shift the wavelength from 625 nm to 530 nm:

$$\frac{\lambda'}{\lambda_0} = \frac{530\text{ nm}}{625\text{ nm}} = 0.848$$

$$\frac{v}{c} = \frac{1 - 0.848^2}{1 + 0.848^2} = 0.163$$

$$v = 0.163c = 4.90 \times 10^7\text{ m/s} = 1.10 \times 10^8\text{ mi/h}$$

5. This speed is beyond any possible speed for a car:

The driver does not have a plausible case.

CHECK A car cannot travel at relativistic speeds, so the answer to this problem was obvious.

Example 39-5 Finding Speed from the Doppler Shift — *Try It Yourself*

The emission spectrum of hydrogen includes a line that has the wavelength $\lambda_0 = 656$ nm. In light reaching us from a distant galaxy, the wavelength of that spectral line is measured to be $\lambda' = 1458$ nm. Find the speed at which the distant galaxy is receding from Earth.

PICTURE Wavelength is related to frequency by $c = f\lambda$ and the received frequency is related to the unshifted frequency by the Doppler shift equation for a receding source (Equation 39-16*b*).

SOLVE

Cover the column to the right and try these on your own before looking at the answers.

Steps	Answers
1. Use Equation 39-16*b* to relate the speed v to the received frequency f' and the unshifted frequency f_0.	$f' = \sqrt{\dfrac{1-(v/c)}{1+(v/c)}}\,f_0$
2. Substitute $f' = c/\lambda'$ and $f_0 = c/\lambda_0$ and solve for v/c.	$\dfrac{v}{c} = \dfrac{1-(\lambda_0/\lambda')^2}{1+(\lambda_0/\lambda')^2} = 0.664$ $v = \boxed{0.664c}$

CHECK As expected, the result is a significant fraction of c. This result is expected because the wavelength of the received light is large in comparison to the wavelength of the same spectral line in the reference frame of the source.

39-4 CLOCK SYNCHRONIZATION AND SIMULTANEITY

We saw in Section 39-3 that proper time is the time interval between two events that occur at the same point in some reference frame. It can therefore be measured on a single clock. (Remember, in each frame there is, in principle, a stationary clock at each point in space, and the time of an event in a given frame is measured by the clock at that point.) However, in another reference frame moving relative to the first, the same two events occur at different places, so two stationary clocks are needed in this reference frame to record the times. The time of each event is measured on a different clock, and the interval is found by subtraction of the measured times. This procedure requires that the clocks be **synchronized.** We will show in this section that

> Two clocks that are synchronized in one reference frame are typically not synchronized in any other frame moving relative to the first frame.
>
> SYNCHRONIZED CLOCKS

Here is a corollary to this result:

> Two events that are simultaneous in one reference frame typically are not simultaneous in another frame that is moving relative to the first.*
>
> SIMULTANEOUS EVENTS

* This is true unless the x coordinates of the two events are equal, where the x axis is parallel to the relative velocity of the two frames.

Comprehension of these facts usually resolves all relativity paradoxes. Unfortunately, the intuitive (and incorrect) belief that simultaneity is an absolute relation is difficult to overcome.

Suppose we have two clocks at rest, one at point A and the other at point B, where points A and B are a distance L apart in frame S. How can we synchronize the two clocks? If an observer at A looks at the clock at B and sets her clock to read the same time, the clocks will not be synchronized because of the time L/c it takes light to travel from one clock to another. To synchronize the clocks, the observer at A must set her clock ahead by the time L/c. Then she will see that the clock at B reads a time that is L/c behind the time on her clock, but she will calculate that the clocks are synchronized when she allows for the time L/c for the light to reach her. Any other observers in S (except those equidistant from the two clocks) will see the clocks reading different times, but they will also calculate that the clocks are synchronized when they correct for the time it takes the light to reach them. An equivalent method for synchronizing two clocks would be for an observer at point C, a point midway between the clocks, to send a light signal and for the observers at A and B to set their clocks to some prearranged time when they receive the signal.

We now examine the question of **simultaneity.** Suppose observers at A and B agree to explode flashguns at t_0 (having previously synchronized their clocks). The observer at C will see the light from the two flashes at the same time, and because he is equidistant from A and B, he will conclude that the flashes were simultaneous. Other observers in frame S will see the light from A or B first, depending on their location, but after correcting for the time the light takes to reach them, they also will conclude that the flashes were simultaneous. We can thus define simultaneity as follows:

> Two events in a reference frame are simultaneous if light signals from the events reach an observer halfway between the events at the same time.
>
> DEFINITION—SIMULTANEITY

To show that two events that are simultaneous in frame S are not simultaneous in another frame S' moving relative to S, we will use an example introduced by Einstein. A train is moving with speed v past a station platform. We will consider the train to be at rest in S' and the platform to be at rest in S. We have observers A', B', and C' at the front, back, and middle of the train (Figure 39-5). We now suppose that the train and platform are struck by lightning at the front and back of the train and that the lightning bolts are simultaneous in the frame of the platform S. That is, an observer C on the platform halfway between the positions A and B, where the lightning strikes, sees the light from the two strikes at the same time.

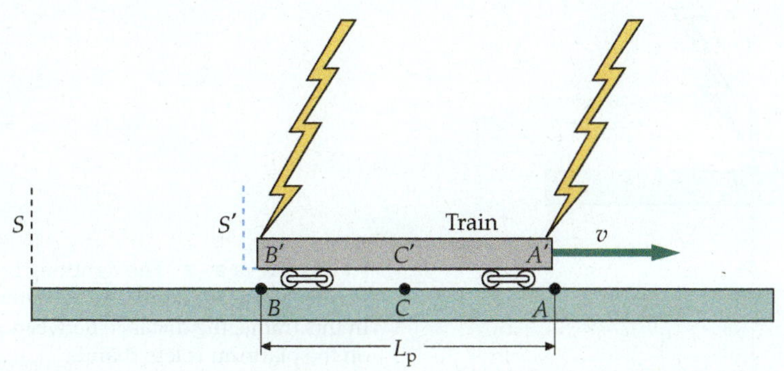

FIGURE 39-5 In frame S attached to the platform, simultaneous lightning bolts strike the ends of a train traveling with speed v. The light from the simultaneous events reaches observer C, standing on the platform midway between the events, at the same time. The distance between the bolts is $L_{\text{p platform}}$.

It is convenient to suppose that the lightning scorches both the train and platform so that the events can be easily located. Because C' is in the middle of the train, halfway between the places on the train that are scorched, the events are simultaneous in S' only if C' sees the flashes at the same time. However, the flash from the front of the train is seen by C' before the flash from the back of the train. We can understand this by considering the motion of C' as seen in frame S (Figure 39-6). By the time the light from the front flash reaches C', C' has moved some distance toward the front flash and some distance away from the back flash. Thus, the light from the back flash has not yet reached C', as indicated in the figure. Observer C' must therefore conclude that the events are not simultaneous and that the front of the train was struck before the back. Furthermore, all observers in S' on the train will agree with C' when they have corrected for the time it takes the light to reach them.

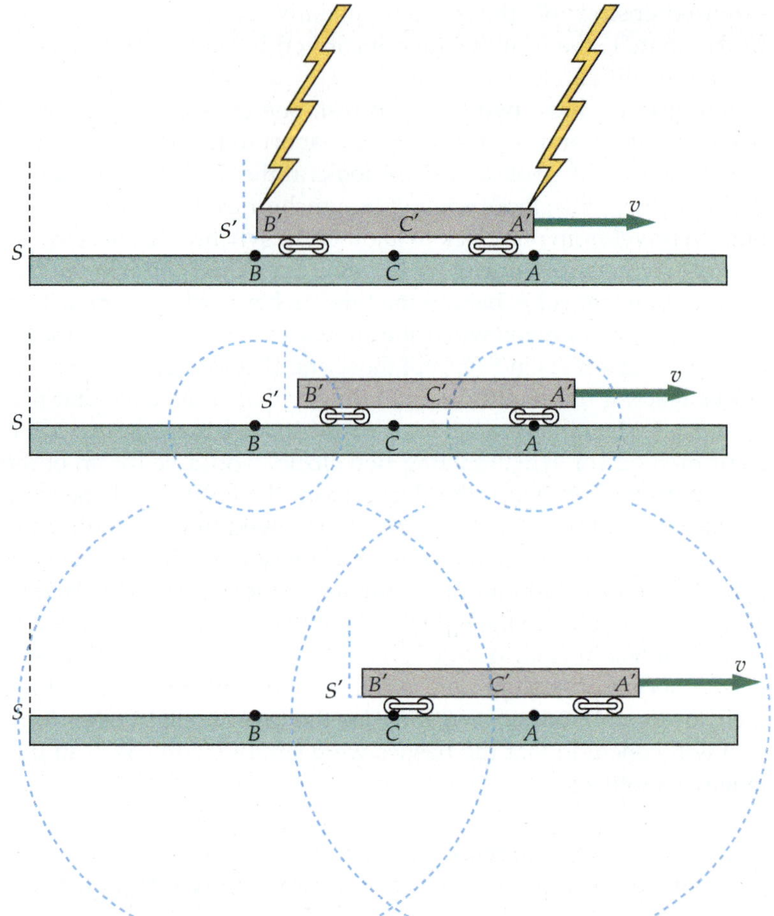

FIGURE 39-6 In frame S attached to the platform, the light from the lightning bolt at the front of the train reaches observer C', standing on the train at its midpoint, before the light from the bolt at the back of the train. Because C' is midway between the events (which occur at the front and rear of the train), the events are not simultaneous for him.

Figure 39-7 shows the events of the lightning bolts as seen in the reference frame of the train (S'). In this frame the platform is moving, so the distance between the scorch marks on the platform is contracted. The platform is shorter than it is in S, and, because the train is at rest, the train is longer than its contracted length in S. When the lightning bolt strikes the front of the train at A', the front of the train is at point A, and the back of the train has not yet reached point B. Later, when the lightning bolt strikes the back of the train at B', the back has reached point B on the platform.

The time discrepancy of two clocks that are synchronized in frame S as seen in frame S' can be found from the Lorentz transformation equations. Suppose we have clocks at points x_1 and x_2 that are synchronized in S. What are the times t_1 and t_2 on the clocks as observed from frame S' at a time t'_0?

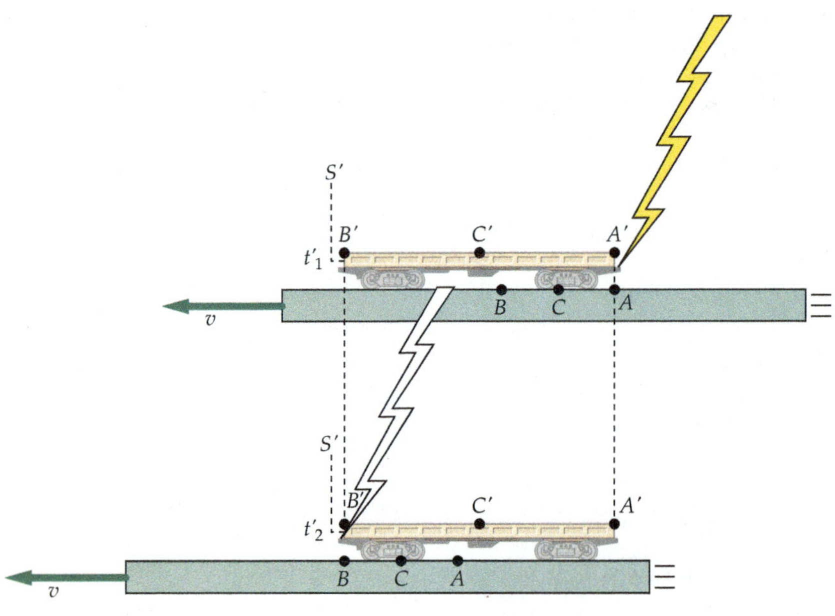

FIGURE 39-7 The lightning bolts of Figure 39-5 as seen in frame S' of the train. In this frame, the distance between A and B on the platform is less than $L_{\text{p platform}}$, and the proper length of the train $L_{\text{p train}}$ is longer than $L_{\text{p platform}}$. The first lightning bolt strikes the front of the train when A' and A are coincident. The second bolt strikes the rear of the train when B' and B are coincident.

From Equation 39-12, when $t'_1 = t'_2 = t'_0$, we have

$$t'_0 = \gamma\left(t_1 - \frac{vx_1}{c^2}\right)$$

and

$$t'_0 = \gamma\left(t_2 - \frac{vx_2}{c^2}\right)$$

Subtracting the first equation from the second, and then rearranging, gives

$$t_2 - t_1 = \frac{v}{c^2}(x_2 - x_1)$$

Note that the chasing clock (at x_2) leads the other (at x_1) by an amount that is proportional to their proper separation $L_p = x_2 - x_1$.

> If two clocks are synchronized in the frame in which they are both at rest, in a frame in which they are moving along the line through both clocks, the chasing clock leads (shows a later time) by an amount
>
> $$\Delta t_S = L_p \frac{v}{c^2} \qquad 39\text{-}17$$
>
> where L_p is the proper distance between the clocks.
>
> CHASING CLOCK SHOWS LATER TIME

A numerical example should help clarify time dilation, clock synchronization, and the internal consistency of these results.

Example 39-6 Synchronizing Clocks

Observer A' in a spaceship that has a flashgun and a mirror is shown in Figure 39-3 (see page 1325). Observer A' is standing next to the flashgun. The distance from the gun to the mirror is 15 light-minutes (written 15 $c \cdot$min) and the spaceship, at rest in frame S', travels with speed $v = 0.80c$ relative to a very long space platform that is at rest in frame S. The platform has two synchronized clocks, one clock at position x_1, the position of the spaceship when the observer explodes the flashgun, and the other clock at position x_2, the position of the spaceship when the light returns to the gun from the mirror. Find the time intervals between the events (exploding the flashgun and receiving the return flash from the mirror) (*a*) in the frame of the spaceship and (*b*) in the frame of the platform. Find (*c*) the distance traveled by the spaceship and (*d*) the amount by which the clocks on the platform are out of synchronization according to observers on the spaceship.

PICTURE The events occur at the same place on the spaceship, so the time between the events in frame S' is the proper time between the events.

SOLVE

(*a*) 1. In the frame of the spaceship, the light travels from the gun to the mirror and back, a total distance $D = 30\ c \cdot \text{min}$. The time required is D/c:

$$\Delta t' = \frac{D}{c} = \frac{30\ c \cdot \text{min}}{c} = \boxed{30\ \text{min}}$$

2. Because the two events happen at the same place in the spaceship, the time interval is proper time:

$$\Delta t_p = \boxed{30\ \text{min}}$$

(*b*) 1. In frame S, the time between the events is longer by the factor γ:

$$\Delta t = \gamma\, \Delta t_p = \gamma(30\ \text{min})$$

2. Calculate γ:

$$\gamma = \frac{1}{\sqrt{1 - (v^2/c^2)}} = \frac{1}{\sqrt{1 - (0.80)^2}} = \frac{1}{\sqrt{0.36}} = \frac{5}{3}$$

3. Use the value of γ to calculate the time between the events as observed in frame S:

$$\Delta t = \gamma\, \Delta t_p = \tfrac{5}{3}(30\ \text{min}) = \boxed{50\ \text{min}}$$

(*c*) In frame *S*, the distance traveled by the spaceship is $v\,\Delta t$: $x_2 - x_1 = v\,\Delta t = (0.80c)(50\text{ min}) = \boxed{40\ c\cdot\text{min}}$

(*d*) 1. The amount that the clocks on the platform are out of synchronization is related to the proper distance between the clocks L_p: $\Delta t_s = L_p\dfrac{v}{c^2}$

2. The Part-(*c*) result is the proper distance between the clocks on the platform: $L_p = x_2 - x_1 = 40\ c\cdot\text{min}$

so

$$\Delta t_s = L_p\frac{v}{c^2} = (40\ c\cdot\text{min})\frac{(0.80c)}{c^2} = \boxed{32\text{ min}}$$

TAKING IT FURTHER Observers on the platform would say that the spaceship's clock is running slow because it records a time of only 30 min between the events, whereas the time measured by observers on the platform is 50 min.

Figure 39-8 shows the situation in Example 39-6 viewed from the spaceship in S'. The platform is traveling past the ship with speed 0.8*c*. There is a clock at point x_1, which coincides with the ship when the flashgun is exploded, and another at point x_2, which coincides with the ship when the return flash is received from the mirror. We assume that the clock at x_1 reads 12:00 noon at the time of the light flash. The clocks at x_1 and x_2 are synchronized in *S* but not in S'. In S', the clock at x_2, which is chasing the one at x_1, leads by 32 min; it would thus read 12:32 to an observer in S'. When the spaceship coincides with x_2, the clock there reads 12:50. The time between the events is therefore 50 min in *S*. Note that according to observers in S', this clock ticks off 50 min − 32 min = 18 min for a trip that takes 30 min in S'. Thus, observers in S' see this clock run slow by the factor 30/18 = 5/3.

Every observer in one frame sees the clocks in the other frame run slow. According to observers in *S*, who measure 50 min for the time interval, the time interval in S' (30 min) is too small, so they see the single clock in S' run too slow by the factor 5/3. According to the observers in S', the observers in *S* measure a time that is too *long* despite the fact that their clocks run too slow because the clocks in *S* are out of synchronization. The clocks tick off only 18 min, but the second clock leads the first clock by 32 min, so the time interval is 50 min.

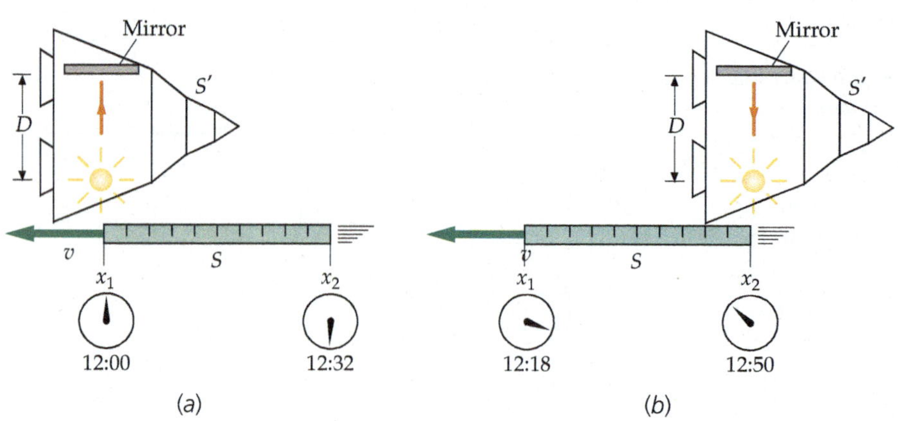

FIGURE 39-8 Clocks on a platform as observed from the spaceship's frame of reference S'. During the time $\Delta t' = 30$ min it takes for the platform to pass the spaceship, the clocks on the platform run slow and tick off (30 min)/γ = 18 min. But the clocks are unsynchronized, with the chasing clock leading by $L_p v/c^2$, which for this case is 32 min. The time it takes for the spaceship to go from x_1 to x_2, as measured on the platform, is therefore 32 min + 18 min = 50 min.

THE TWIN PARADOX

Homer and Ulysses are identical twins. Ulysses travels at high speed to a planet beyond the solar system and returns while Homer remains at home. When they are together again, which twin is older or are they the same age? The correct answer is that Homer, the twin who stays at home, is older. This problem, with variations, has been the subject of spirited debate for decades, though there are very few who disagree with the answer. The problem appears to be a paradox because of the seemingly symmetric roles played by the twins and the asymmetric result in their aging. The paradox is resolved when the asymmetry of the twins' roles is noted. The relativistic result conflicts with common sense based on our strong but incorrect belief in absolute simultaneity. We will consider a particular case with some numerical magnitudes that, though impractical, make the calculations easy.

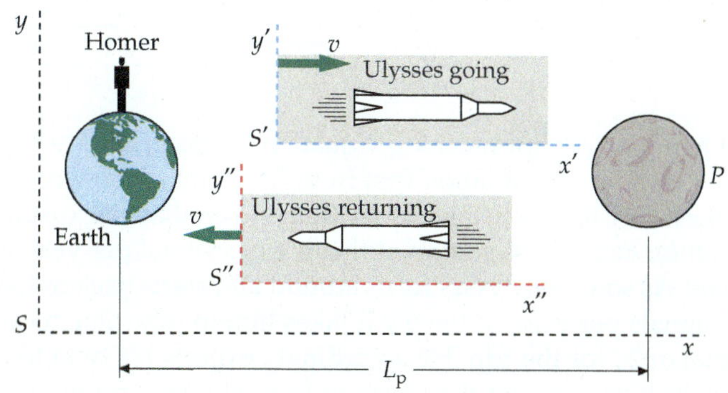

FIGURE 39-9 The twin paradox. Earth and a distant planet are fixed in frame S. Ulysses coasts in frame S' to the planet and then coasts back in frame S''. His twin Homer stays on Earth. When Ulysses returns, he is younger than his twin. The roles played by the twins are not symmetric. Homer remains at rest in one inertial reference frame, but Ulysses must go from being at rest in one inertial reference frame to another if he is to return home.

In reference frame S, Earth, planet P and Homer are at rest and Earth and planet P are a distance L_p apart (Figure 39-9). Homer is on Earth. Reference frames S' and S'' are moving with speed v toward and away from planet P, respectively. Ulysses quickly accelerates to speed v, then coasts, at rest in S', until he reaches the planet, where he quickly decelerates to a stop and is momentarily at rest in S. To return, Ulysses quickly accelerates to speed v toward Earth and then coasts, at rest in S'', until he reaches Earth, where he quickly decelerates to a stop. We can assume that the acceleration (and deceleration) times are negligible compared with the coasting times. We use the following values for illustration: $L_p = 8$ light-years $(8\,c\cdot\text{y})$ and $v = 0.8c$. Then $\sqrt{1-(v^2/c^2)} = 3/5$ and $\gamma = 5/3$.

It is easy to analyze the problem from Homer's point of view on Earth. According to Homer's clock, Ulysses coasts in S' for a time $L_p/v = 10$ y and in S'' for an equal time. Thus, Homer is 20 y older when Ulysses returns. The time interval in S' between Ulysses's leaving Earth and his arriving at the planet is shorter because it is proper time. The time it takes to reach the planet by Ulysses's clock is

$$\Delta t' = \frac{\Delta t}{\gamma} = \frac{10\text{ y}}{5/3} = 6\text{ y}$$

Because the same time is required for the return trip, Ulysses will have recorded 12 y for the round trip and will be 8 years younger than Homer upon his return.

From Ulysses's point of view, the distance between Earth and the planet is contracted and is only

$$L' = \frac{L_p}{\gamma} = \frac{8\,c\cdot\text{y}}{5/3} = 4.8\,c\cdot\text{y}$$

At $v = 0.8c$, it takes only $L'/v = 4.8c\cdot\text{y}/0.8\,c = 6$ y each way.

The real challenge in this problem is for Ulysses to understand why his twin aged 20 y during his absence. If we consider Ulysses as being at rest and Homer as moving away, Homer's clock should run slow and measure only $\frac{3}{5}(6\text{ y}) = 3.6$ y. Then why shouldn't Homer age only 7.2 y during the round trip? This, of course, is the paradox. The difficulty with the analysis from the point of view of Ulysses is that he does not remain in a single inertial reference frame. What happens while Ulysses is stopping and starting? To investigate this problem in detail, we would need to treat accelerated reference frames, a subject dealt with in the study of general relativity and beyond the scope of this book. However, we can get some insight into the problem by having the twins send regular signals to each other so that they can record the other's age continuously. If they arrange to send a signal once a year, each can determine the age of the other merely by counting the signals received. The arrival frequency of the signals will not be 1 per year because of the Doppler shift. The frequency observed will be given by Equations 39-16*a* and 39-16*b*.

Using $v/c = 0.8$ (so $v^2/c^2 = 0.64$), we have for the case in which the twins are receding from each other

$$f' = \frac{f_0}{1 + (v/c)}\sqrt{1 - (v^2/c^2)} = \frac{\sqrt{1 - 0.64}}{1 + 0.8} f_0 = \frac{1}{3} f_0$$

When they are approaching, Equation 39-16*a* gives $f' = 3f_0$.

Consider the situation first from the point of view of Ulysses. During the 6 y it takes him to reach the planet (remember that the distance is contracted in his frame), he receives signals at the rate of $\frac{1}{3}$ signal per year, and so he receives 2 signals. As soon as Ulysses turns around and starts back to Earth, he begins to receive 3 signals per year. In the 6 y it takes him to return he receives 18 signals, giving a total of 20 for the trip. He accordingly expects his twin to have aged 20 years.

We now consider the situation from Homer's point of view. He receives signals at the rate of $\frac{1}{3}$ signal per year not only for the 10 y it takes Ulysses to reach the planet but also for the time it takes for the last signal sent by Ulysses before he turns around to get back to Earth. (He cannot know that Ulysses has turned around until the signals begin reaching him with increased frequency.) Because the planet is 8 light-years away, there is an additional 8 y of receiving signals at the rate of $\frac{1}{3}$ signal per year. During the first 18 y, Homer receives 6 signals. In the final 2 y before Ulysses arrives, Homer receives 6 signals, or 3 per year. (The first signal sent after Ulysses turns around takes 8 y to reach Earth, whereas Ulysses, traveling at $0.8c$, takes 10 y to return and therefore arrives just 2 y after Homer begins to receive signals at the faster rate.) Thus, Homer expects Ulysses to have aged 12 y. In this analysis, the asymmetry of the twins' roles is apparent. When they are together again, both twins agree that the one who has been accelerated will be younger than the one who stayed home.

The predictions of the special theory of relativity concerning the twin paradox have been tested using small particles that can be accelerated to such large speeds that γ is appreciably greater than 1. Unstable particles can be accelerated and trapped in circular orbits in a magnetic field, for example, and their lifetimes can then be compared with those of identical particles at rest. In all such experiments, the accelerated particles live longer on the average than the particles at rest, as predicted. These predictions have also been confirmed by the results of an experiment in which high-precision atomic clocks were flown around the world in commercial airplanes, but the analysis of that experiment is complicated due to the necessity of including gravitational effects treated in the general theory of relativity.

39-5 THE VELOCITY TRANSFORMATION

We can find how velocities transform from one reference frame to another by differentiating the Lorentz transformation equations. Suppose a particle has velocity $u'_x = dx'/dt'$ in frame S', which is moving to the right with speed v relative to frame S. The particle's velocity in frame S is

$$u_x = \frac{dx}{dt}$$

From the Lorentz transformation equations (Equations 39-9 and 39-10), we have

$$dx = \gamma(dx' + v\,dt')$$

and

$$dt = \gamma\left(dt' + \frac{v\,dx'}{c^2}\right)$$

The velocity relative to frame S is thus

$$u_x = \frac{dx}{dt} = \frac{\gamma(dx' + v\,dt')}{\gamma\left(dt' + \frac{v\,dx'}{c^2}\right)} = \frac{\frac{dx'}{dt'} + v}{1 + \frac{v}{c^2}\frac{dx'}{dt'}} = \frac{u'_x + v}{1 + \frac{vu'_x}{c^2}}$$

If a particle has components of velocity along the y or z axes, we can use the same relation between dt and dt', with $dy = dy'$ and $dz = dz'$, to obtain

$$u_y = \frac{dy}{dt} = \frac{dy'}{\gamma\left(dt' + \frac{v\,dx'}{c^2}\right)} = \frac{\frac{dy'}{dt'}}{\gamma\left(1 + \frac{v}{c^2}\frac{dx'}{dt'}\right)} = \frac{u'_y}{\gamma\left(1 + \frac{vu'_x}{c^2}\right)}$$

and

$$u_z = \frac{u'_z}{\gamma\left(1 + \frac{vu'_x}{c^2}\right)}$$

The complete relativistic velocity transformation is

$$u_x = \frac{u'_x + v}{1 + \frac{vu'_x}{c^2}} \qquad \text{39-18}a$$

$$u_y = \frac{u'_y}{\gamma\left(1 + \frac{vu'_x}{c^2}\right)} \qquad \text{39-18}b$$

$$u_z = \frac{u'_z}{\gamma\left(1 + \frac{vu'_x}{c^2}\right)} \qquad \text{39-18}c$$

RELATIVISTIC VELOCITY TRANSFORMATION

The inverse velocity transformation equations are

$$u'_x = \frac{u_x - v}{1 - \frac{vu_x}{c^2}} \qquad \text{39-19}a$$

$$u'_y = \frac{u_y}{\gamma\left(1 - \frac{vu_x}{c^2}\right)} \qquad \text{39-19}b$$

$$u'_z = \frac{u_z}{\gamma\left(1 - \frac{vu_x}{c^2}\right)} \qquad \text{39-19}c$$

These equations differ from the classical and intuitive result $u_x = u'_x + v$, $u_y = u'_y$, and $u_z = u'_z$ because the denominators in the equations are not equal to 1. When v and u'_x are small compared with the speed of light c, $\gamma \approx 1$ and $vu'_x/c^2 \ll 1$. Then the relativistic and classical velocity transformation equations are the same.

Example 39-7 Relative Velocity at Nonrelativistic Speeds

A supersonic plane moves away from you, and in the $+x$ direction, at a speed of 1000 m/s (about 3 times the speed of sound) relative to you. A second plane, traveling in the same direction and ahead of the first plane, moves away from you, and away from the first plane, at a speed of 500 m/s relative to the first plane. How fast is the second plane moving relative to you?

PICTURE The speeds are so small compared with c that we expect the classical equations for combining velocities to be accurate. We show this by calculating the correction term in the denominator of Equation 39-18*a*. Let frame S be your rest frame and frame S' be the rest frame of the first plane. Then v, the velocity of S' relative to S, is 1000 m/s. The second plane has velocity $u'_x = 500$ m/s relative to S'.

SOLVE

1. Let S and S' be the reference frames of you and the first plane, respectively. Also, let u_x and u'_x be the velocities of the second plane relative to S and S', respectively. Equation 39-18*a* can be used to find u_x. The velocity of the second plane relative to you is v:

$$u_x = \frac{u'_x + v}{1 + \dfrac{vu'_x}{c^2}}$$

2. If the correction term in the denominator is negligible (compared to 1), Equation 39-18*a* gives the classical formula for combining velocities. Calculate the value of the correction term:

$$\frac{vu'_x}{c^2} = \frac{(1000)(500)}{(3.00 \times 10^8)^2} = 5.56 \times 10^{-12}$$

3. The correction term is so small that the classical and relativistic results are virtually the same:

$$u_x \approx u'_x + v$$
$$= 500\text{ m/s} + 1000\text{ m/s} = \boxed{1500\text{ m/s}}$$

Example 39-8 Relative Velocity at Relativistic Speeds

Work Example 39-7 if the first plane moves with speed $v = 0.80c$ relative to you and the second plane moves with the same speed $0.80c$ relative to the first plane.

PICTURE These speeds are not small compared with c, so we need to use the relativistic expression (Equation 39-18*a*). We again assume that you are at rest in frame S and the first plane is at rest in frame S' that is moving at $v = 0.80c$ relative to you. The velocity of the second plane relative to S' is $u'_x = 0.80c$.

SOLVE

Use Equation 39-18*a* to calculate the speed of the second plane relative to you:

$$u_x = \frac{u'_x + v}{1 + \dfrac{vu'_x}{c^2}} = \frac{0.80c + 0.80c}{1 + \dfrac{(0.80c)(0.80c)}{c^2}} = \frac{1.60c}{1.64} = \boxed{0.98c}$$

CHECK As expected, the result is less than c.

The result in Example 39-8 is quite different from the classically expected result of $0.80c + 0.80c = 1.60c$. In fact, it can be shown from Equations 39-18*a*–*c* that if the speed of an object is less than c in one frame, it is less than c in all other frames moving relative to that frame with a speed less than c. (See Problem 59.) We will see in Section 39-7 that it takes an infinite amount of energy to accelerate a particle to the speed of light. The speed of light c is thus an upper, unattainable limit for the speed of a particle with mass. (There are massless particles, such as photons, that always move at the speed of light.)

Example 39-9 Relative Speed of a Photon

A photon moves along the x' axis in frame S', with speed $u'_x = c$. What is its speed in frame S?

PICTURE Use Equation 39-18*a* to calculate the speed of the photon in S.

SOLVE

The speed in S is given by Equation 39-18*a*:

$$u_x = \frac{u'_x + v}{1 + \frac{vu'_x}{c^2}} = \frac{c + v}{1 + \frac{vc}{c^2}} = \frac{c + v}{1 + \frac{v}{c}} = \frac{c + v}{\frac{1}{c}(c + v)} = \boxed{c}$$

CHECK The speed in both frames is c, independent of v. This is in accord with Einstein's postulates.

Example 39-10 Rockets Passing in Opposite Directions

Two spaceships, each 100 m long when measured at rest, travel toward each other, each with a speed of $0.85c$ relative to Earth. (*a*) What is the length of each spaceship as measured by someone at rest relative to Earth? (*b*) How fast is each spaceship traveling as measured by an observer at rest relative to the other spaceship? (*c*) What is the length of one spaceship when measured by an observer at rest relative to the other spaceship? (*d*) At time $t = 0$ on Earth, the front ends of the ships are next to each other as they just begin to pass each other. At what time on Earth are their back ends next to each other?

PICTURE (*a*) The length of each spaceship as measured on Earth is the contracted length $\sqrt{1 - (u^2/c^2)}L_p$ (Equation 39-14), where u is the speed of either spaceship relative to Earth. To solve Part (*b*), let Earth be at rest in frame S, and let the spaceship on the left (spaceship 1) be at rest in frame S', which is moving at speed $v = 0.85c$ relative to S. Then the spaceship on the right (spaceship 2) moves with velocity $u_{2x} = -0.85c$ (Figure 39-10). (*c*) The length of spaceship 2 as seen by an observer at rest relative to spaceship 1 is $\sqrt{1 - (u'^2_{2x}/c^2)}L_p$.

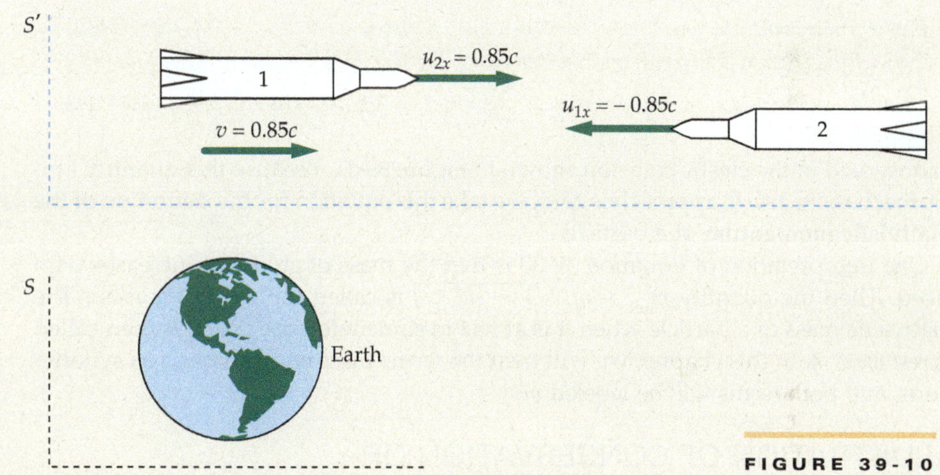

FIGURE 39-10

SOLVE

(*a*) The length of each spaceship in S, the reference frame of Earth, is the proper length divided by γ.

$$L = \frac{1}{\gamma}L_p = \sqrt{1 - \frac{|u_{2x}|^2}{c^2}}L_p = \sqrt{1 - \frac{(0.85c)^2}{c^2}}(100\text{ m}) = \boxed{53\text{ m}}$$

(*b*) Use the velocity transformation formula (Equation 39-19*a*) to find the velocity u'_{2x} of spaceship 2 as seen in frame S':

$$u'_{2x} = \frac{u_{2x} - v}{1 - \frac{vu_{2x}}{c^2}} = \frac{-0.85c - 0.85c}{1 - \frac{(0.85c)(-0.85c)}{c^2}} = \frac{-1.70c}{1.7225} = -0.987c$$

so

$$|u'_{2x}| = \boxed{0.99c}$$

(*c*) In the frame of spaceship 1, spaceship 2 is moving with speed $|u'| = 0.987c$. Use this to calculate the length of spaceship 2 as seen by an observer at rest relative to spaceship 1:

$$L = \frac{1}{\gamma}L_p = \sqrt{1 - \frac{|u'_{2x}|^2}{c^2}}L_p = \sqrt{1 - \frac{(0.987c)^2}{c^2}}(100\text{ m}) = \boxed{16\text{ m}}$$

(*d*) If the front ends of the spaceships are together at $t = 0$ on Earth, their back ends will be together after the time it takes either spaceship to move the length of the spaceship in Earth's frame:

$$t = \frac{L}{u} = \frac{53\text{ m}}{0.85c} = \frac{53\text{ m}}{(0.85)(3.00 \times 10^8\text{ m/s})} = \boxed{2.1 \times 10^{-7}\text{ s}}$$

CHECK As expected, the Part-(*c*) result is less than the Part-(*a*) result, and both results are less than the proper length of 100 m.

39-6 RELATIVISTIC MOMENTUM

We have seen in previous sections that Einstein's postulates require important modifications in our ideas of simultaneity and in our measurements of time and length. Einstein's postulates also require modifications in our concepts of mass, momentum, and energy. In classical mechanics, the momentum of a particle is defined as the product of its mass and its velocity, $m\vec{u}$, where $\vec{u}$ is the velocity. In an isolated system of particles, with no net force acting on the system, the total momentum of the system remains constant.

The reason that the total momentum of a system is important in classical mechanics is that it is conserved when there are no external forces acting on the system, as is the case in collisions. But we have just seen that $\Sigma m_i \vec{u}_i$ is conserved only in the approximation that $u \ll c$. We will define the relativistic momentum $\vec{p}$ of a particle to have the following properties:

1. In collisions, $\vec{p}$ is conserved.
2. As u/c approaches zero, $\vec{p}$ approaches $m\vec{u}$.

We will show that the quantity

$$\vec{p} = \frac{m\vec{u}}{\sqrt{1 - \dfrac{u^2}{c^2}}} \qquad 39\text{-}20$$

RELATIVISTIC MOMENTUM

is conserved in the elastic collision shown in Figure 39-11. Because this quantity also approaches $m\vec{u}$ as u/c approaches zero, we take this equation for the definition of the **relativistic momentum** of a particle.

One interpretation of Equation 39-20 is that the mass of an object increases with speed. Then the quantity $m_{\text{rel}} = m/\sqrt{1 - (u^2/c^2)}$ is called the *relativistic mass*. The relativistic mass of a particle when it is at rest in some reference frame is then called its *rest mass m*. In this chapter, we will treat the terms mass and rest mass as synonymous, and both terms will be labeled m.

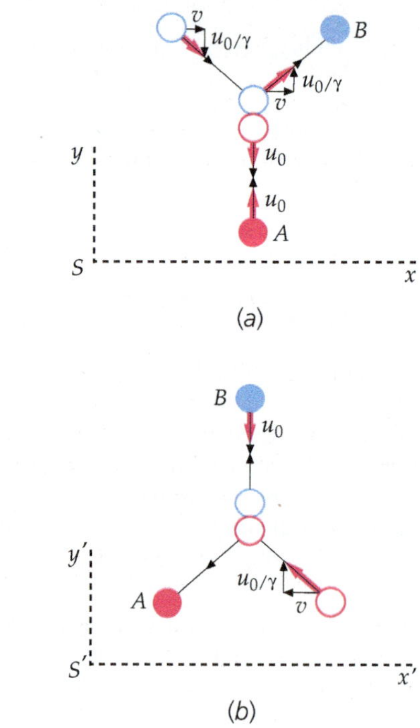

FIGURE 39-11 (*a*) Elastic collision of two identical pucks as seen in frame S. The vertical component of the velocity of puck B is u_0/γ in S if it is u_0 in S'. (*b*) The same collision as seen in S'. In this frame, puck A has a vertical component of velocity equal to u_0/γ.

ILLUSTRATION OF CONSERVATION OF THE RELATIVISTIC MOMENTUM

We consider two observers: one observer at rest in reference frame S, and the other observer at rest in frame S', which is moving to the right in the $+x$ direction with speed v relative to frame S. Each has a puck of mass m that can slide freely across a flat horizontal surface. The two pucks are identical when compared at rest. One observer launches puck A in the $+y$ direction with a speed u_0 relative to himself and the other launches puck B in the $-y$ direction with a speed u_0 relative to himself, so that each puck makes an elastic collision with the other puck, and returns to the person that launched it. Figure 39-11 shows how the collision looks in each reference frame.

We will compute the y component of the relativistic momentum of each puck in the reference frame S for the collision and show that the y component of the total relativistic momentum is zero. The speed of puck A in S is u_0, so the y component of its relativistic momentum is

$$p_{Ay} = \frac{mu_0}{\sqrt{1 - (u_0^2/c^2)}}$$

The speed of puck B in S is more complicated. Its x component is v and its y component is $-u_0/\gamma$ (Equation 39-18*b*). Thus,

$$u_B^2 = u_{Bx}^2 + u_{By}^2 = v^2 + \left[-u_0\sqrt{1 - (v^2/c^2)}\right]^2 = v^2 + u_0^2 - \frac{u_0^2 v^2}{c^2}$$

Using this result to compute $\sqrt{1-(u_B^2/c^2)}$ we obtain

$$1-\frac{u_B^2}{c^2}=1-\frac{v^2}{c^2}-\frac{u_0^2}{c^2}+\frac{u_0^2v^2}{c^4}=\left(1-\frac{v^2}{c^2}\right)\left(1-\frac{u_0^2}{c^2}\right)$$

and

$$\sqrt{1-(u_B^2/c^2)}=\sqrt{1-(v^2/c^2)}\,\sqrt{1-(u_0^2/c^2)}=\left(\frac{1}{\gamma}\right)\sqrt{1-(u_0^2/c^2)}$$

The y component of the relativistic momentum of puck B as seen in S is therefore

$$p_{By}=\frac{mu_{By}}{\sqrt{1-(u_B^2/c^2)}}=\frac{-mu_0/\gamma}{(1/\gamma)\sqrt{1-(u_0^2/c^2)}}=\frac{-mu_0}{\sqrt{1-(u_0^2/c^2)}}$$

Because $p_{By}=-p_{Ay}$, the y component of the total momentum of the two pucks is zero. If the y component of the momentum of each puck is reversed by the collision, the total momentum will remain zero and momentum will be conserved.

39-7 RELATIVISTIC ENERGY

In classical mechanics, the work done by the net force acting on a particle equals the change in the kinetic energy of the particle. In relativistic mechanics, we equate the net force to the rate of change of the relativistic momentum. The work done by the net force can then be calculated and set equal to the change in kinetic energy.

As in classical mechanics, we will define kinetic energy as the work done by the net force in accelerating a particle from rest to some final velocity u_f. Considering one dimension only, we have

$$K=\int_{u=0}^{u=u_f}F_{\text{net}}\,ds=\int_{u=0}^{u=u_f}\frac{dp}{dt}\,ds=\int_{u=0}^{u=u_f}u\,dp=\int_{u=0}^{u=u_f}u\,d\left(\frac{mu}{\sqrt{1-(u^2/c^2)}}\right)\qquad\text{39-21}$$

where we have used $u=ds/dt$. It is left as a problem (Problem 35) for you to show that

$$d\left(\frac{mu}{\sqrt{1-(u^2/c^2)}}\right)=m\left(1-\frac{u^2}{c^2}\right)^{-3/2}du$$

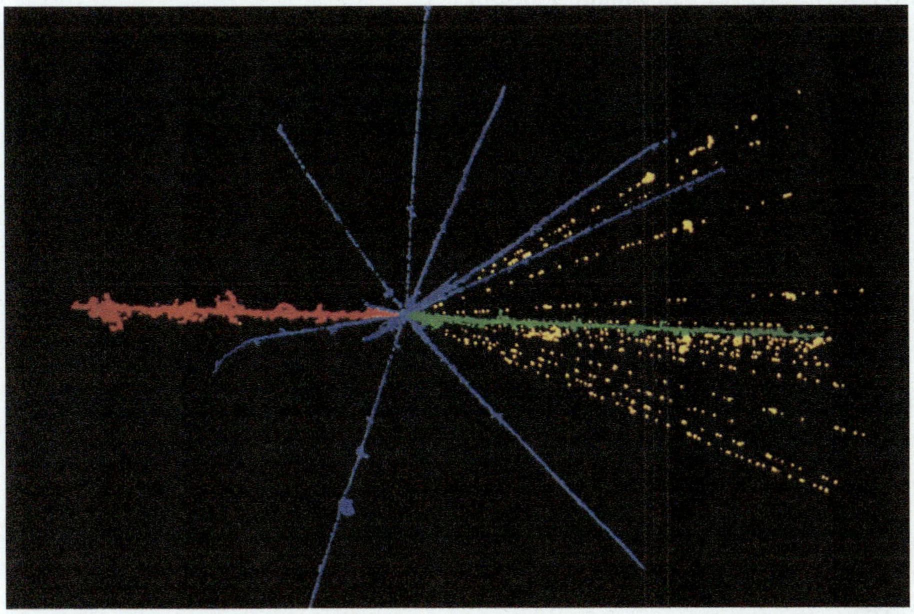

The creation of elementary particles demonstrates the conversion of kinetic energy to rest energy. In this 1950 photograph of a cosmic ray shower, a high-energy sulfur nucleus (red) collides with a nucleus in a photographic emulsion and produces a spray of particles, including a fluorine nucleus (green), other nuclear fragments (blue), and approximately 16 pions (yellow). *(© C. Powell, P. Fowler, and D. Perkins. Science Photo Library/Photo Researchers.)*

If we substitute that expression into the integrand in Equation 39-21, we obtain

$$K = \int_{u=0}^{u=u_f} u\, d\left(\frac{mu}{\sqrt{1-(u^2/c^2)}}\right) = \int_0^{u_f} m\left(1 - \frac{u^2}{c^2}\right)^{-3/2} u\, du$$

$$= mc^2\left(\frac{1}{\sqrt{1-(u_f^2/c^2)}} - 1\right)$$

or

$$K = \frac{mc^2}{\sqrt{1-(u^2/c^2)}} - mc^2 \qquad \text{39-22}$$

RELATIVISTIC KINETIC ENERGY

(In this expression the final speed u_f is arbitrary, so the subscript f is not needed.)

The expression for kinetic energy consists of two terms. The first term depends on the speed of the particle. The second term, mc^2, is independent of the speed. The quantity mc^2 is called the **rest energy** E_0 of the particle. The rest energy is the product of the mass and c^2:

$$E_0 = mc^2 \qquad \text{39-23}$$

REST ENERGY

The total **relativistic energy** E is then defined to be the sum of the kinetic energy and the rest energy:

$$E = K + mc^2 = \frac{mc^2}{\sqrt{1-(u^2/c^2)}} \qquad \text{39-24}$$

RELATIVISTIC ENERGY

Thus, the work done by an unbalanced force increases the energy from the rest energy mc^2 to the final energy $mc^2/\sqrt{1-(u^2/c^2)}$. We can obtain a useful expression for the velocity of a particle by multiplying Equation 39-20 for the relativistic momentum by c^2 and comparing the result with Equation 39-24 for the relativistic energy. We have

$$pc^2 = \frac{mc^2 u}{\sqrt{1-(u^2/c^2)}} = Eu$$

Dividing both sides by cE gives

$$\frac{u}{c} = \frac{pc}{E} \qquad \text{39-25}$$

Energies in atomic and nuclear physics are usually expressed in units of electron volts (eV) or mega-electron volts (MeV):

$$1\text{ eV} = 1.602 \times 10^{-19}\text{ J}$$

A convenient unit for the masses of atomic particles is eV/c^2 or MeV/c^2, which is the rest energy of the particle divided by c^2. The rest energies of some elementary particles and light nuclei are given in Table 39-1.

Table 39-1 Rest Energies of Some Elementary Particles and Light Nuclei

Particle	Symbol	Rest energy, MeV
Photon	γ	0
Electron (positron)	e or $e^-(e^+)$	0.5110
Muon	$\mu^\pm$	105.7
Pion	π^0	135.0
	$\pi^\pm$	139.6
Proton	^{1}H or p	938.272
Neutron	n	939.565
Deuteron	^{2}H or d	1875.613
Triton	^{3}H or t	2808.920
Helion	^{3}He or h	2808.391
Alpha particle	^{4}He or α	3727.379

Example 39-11 Total Energy, Kinetic Energy, and Momentum

An electron (rest energy 0.511 MeV) moves with speed $u = 0.800c$. Find (*a*) its total energy, (*b*) its kinetic energy, and (*c*) the magnitude of its momentum.

PICTURE This problem involves substituting into Equations 39-20 to 39-25.

SOLVE

(*a*) The total energy is given by Equation 39-24:

$$E = \frac{mc^2}{\sqrt{1 - (u^2/c^2)}} = \frac{0.511\text{ MeV}}{\sqrt{1 - 0.64}} = \frac{0.511\text{ MeV}}{0.6} = \boxed{0.852\text{ MeV}}$$

(*b*) The kinetic energy is the total energy minus the rest energy:

$$K = E - mc^2 = 0.852\text{ MeV} - 0.511\text{ MeV} = \boxed{0.341\text{ MeV}}$$

(*c*) The magnitude of the momentum is found from Equation 39-20. We can simplify the momentum expression by multiplying both numerator and denominator by c^2 and using the Part-(*a*) result:

$$p = \frac{mu}{\sqrt{1 - (u^2/c^2)}}$$

$$= \frac{mc^2}{\sqrt{1 - (u^2/c^2)}}\frac{u}{c^2} = (0.852\text{ MeV})\frac{0.8c}{c^2} = \boxed{0.682\text{ MeV}/c}$$

CHECK The kinetic energy is less than the total energy as expected.

TAKING IT FURTHER The technique used to solve Part (*c*) (multiplying numerator and denominator by c^2) is equivalent to using Equation 39-25.

The expression for kinetic energy given by Equation 39-22 does not look much like the classical expression $\frac{1}{2}mu^2$. However, when u is much less than c, we can approximate $1/\sqrt{1 - (u^2/c^2)}$ using the binomial expansion

$$(1 + x)^n = 1 + nx + \frac{n(n-1)}{2}x^2 + \cdots \approx 1 + nx \qquad x \ll 1 \qquad \text{39-26}$$

Then

$$\frac{1}{\sqrt{1 - (u^2/c^2)}} = \left(1 - \frac{u^2}{c^2}\right)^{-1/2} \approx 1 + \frac{1}{2}\frac{u^2}{c^2} \qquad u \ll c$$

See **Math Tutorial** *for more information on the* **Binomial Expansion**

From this result, when u is much less than c, the expression for relativistic kinetic energy becomes

$$K = mc^2\left[\frac{1}{\sqrt{1-(u^2/c^2)}} - 1\right] \approx mc^2\left[1 + \frac{1}{2}\frac{u^2}{c^2} - 1\right] = \frac{1}{2}mu^2 \qquad u \ll c$$

Thus, at low speeds, the relativistic expression is the same as the classical expression.

We note from Equation 39-24 that as the speed u approaches the speed of light c, the energy of the particle becomes very large (because $1/\sqrt{1-(u^2/c^2)}$ becomes very large). At $u = c$, the energy becomes infinite. A simple interpretation of the result is that it takes an infinite amount of energy to accelerate a particle (that has mass) to the speed of light.

In practical applications, the momentum or energy of a particle is often known rather than the speed. Equation 39-20 for the relativistic momentum and Equation 39-24 for the relativistic energy can be combined to eliminate the speed u. The result is

$$E^2 = p^2c^2 + (mc^2)^2 \qquad 39\text{-}27$$

RELATION FOR TOTAL ENERGY, MOMENTUM, AND REST ENERGY

This useful equation can be conveniently remembered from the right triangle shown in Figure 39-12.

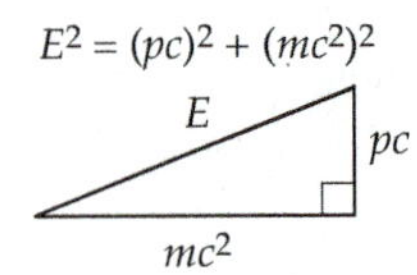

FIGURE 39-12 Right triangle to remember Equation 39-27.

PRACTICE PROBLEM 39-2

A proton (mass equal to 938 MeV/c^2) moving at speed u has a total energy of 1400 MeV. Find (*a*) $1/\sqrt{1-(u^2/c^2)}$, (*b*) the momentum of the proton, and (*c*) the speed u of the proton.

If the energy of a particle is much greater than its rest energy mc^2, the second term on the right side of Equation 39-27 can be neglected, giving the useful approximation

$$E \approx pc \qquad E \gg mc^2 \qquad 39\text{-}28$$

Equation 39-28 is an exact relation between energy and momentum for particles that do not have mass, such as photons.

MASS AND ENERGY

Einstein considered the relation $E_0 = mc^2$ (Equation 39-23) relating the energy of a particle to its mass to be the most significant result of the theory of relativity. Energy and inertia, which were formerly two distinct concepts, are related through this famous equation. As discussed in Chapter 7, the conversion of rest energy to kinetic energy with a corresponding decrease in mass is a common occurrence in radioactive decay and nuclear reactions, including nuclear fission and nuclear fusion. We illustrated this in Section 7-4 with the deuteron, whose mass is 2.22 MeV/c^2 less than the mass of its parts—a proton and a neutron. When a neutron and a proton combine to form a deuteron, 2.22 MeV of energy is released. The breaking up of a deuteron into a neutron and a proton requires 2.22 MeV of energy input. The proton and the neutron are thus bound together in a deuteron by a binding energy of 2.22 MeV. Any stable composite particle, such as a deuteron or an alpha particle (2 neutrons plus 2 protons), that is made up of other particles has a mass and rest energy that are less than the sum of the masses and rest energies of its parts. The difference in these rest energies is the binding energy of the composite particle. The binding energies of atoms and molecules are of the order of a few electron volts, which explains why there is only a negligible difference in mass between the composite particle and its parts. The binding energies of nuclei are of the order of several MeV, which explains why there is a noticeable difference in mass between the composite particle and its parts. Some very heavy nuclei, such as radium, are radioactive and decay into a less massive nucleus plus an alpha particle. In this case, the original nucleus has a rest energy greater than that of the decay particles. The excess energy appears as the kinetic energy of the decay products.

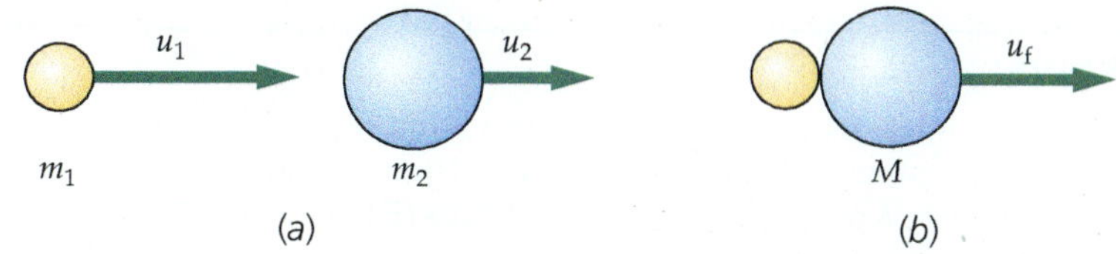

FIGURE 39-13 A perfectly inelastic collision between two particles. One particle of mass m_1 collides with another particle of mass m_2. After the collision, the particles stick together, forming a composite particle of mass M that moves with speed u_f so that relativistic momentum is conserved. Kinetic energy is lost in the process. If we assume that the total energy is conserved, the loss in kinetic energy must equal c^2 multiplied by the increase in the mass of the system.

To further illustrate the connection between mass and energy, we consider a perfectly inelastic collision of two particles. Classically, kinetic energy is lost during such a collision. Relativistically, this loss in kinetic energy shows up as an increase in rest energy of the system; that is, the total energy of the system is conserved. Consider a particle of mass m_1 moving with initial speed u_1 that collides with a particle of mass m_2 moving with initial speed u_2. The particles collide and stick together, forming a particle of mass M that moves with speed u_f, as shown in Figure 39-13. The initial total energy of particle 1 is

$$E_1 = K_1 + m_1c^2$$

where K_1 is its initial kinetic energy. Similarly the initial total energy of particle 2 is

$$E_2 = K_2 + m_2c^2$$

The total initial energy of the system is

$$E_i = E_1 + E_2 = K_1 + m_1c^2 + K_2 + m_2c^2 = K_i + M_ic^2$$

where $K_i = K_1 + K_2$ and $M_i = m_1 + m_2$ are the initial kinetic energy and initial mass of the system. The final total energy of the system is

$$E_f = K_f + M_fc^2$$

If we set the final total energy equal to the initial total energy, we obtain

$$K_f + M_fc^2 = K_i + M_ic^2$$

Rearranging gives $K_f - K_i = -(M_f - M_i)c^2$, which can be expressed

$$\Delta K + (\Delta M)c^2 = 0 \qquad 39\text{-}29$$

where $\Delta M = M_f - M_i$ is the change in mass of the system.

Example 39-12 Totally Inelastic Collision

A particle of mass 2.00 MeV/c^2 and kinetic energy 3.00 MeV collides with a stationary particle of mass 4.00 MeV/c^2. After the collision, the two particles stick together. Find (*a*) the magnitude of the initial momentum of the system, (*b*) the final velocity of the two-particle system, and (*c*) the mass of the two-particle system.

PICTURE (*a*) The initial momentum of the system is the initial momentum of the incoming particle, which can be found from the total energy of the particle. (*b*) The final velocity of the system can be found from its total energy and momentum using $u/c = pc/E$ (Equation 39-25). The energy is found from conservation of energy, and the momentum from conservation of momentum. (*c*) Because the final energy and momentum are known, the final mass can be found using $E^2 = p^2c^2 + (mc^2)^2$.

SOLVE

(*a*) 1. The initial momentum of the system is the initial momentum of the incoming particle. The momentum of a particle is related to its energy and mass (Equation 39-27):

$$E_1^2 = p_1^2c^2 + (m_1c^2)^2$$

$$p_1c = \sqrt{E_1^2 - (m_1c^2)^2}$$

2. The total energy of the moving particle is the sum of its kinetic energy and its rest energy:

$$E_1 = 3.00 \text{ MeV} + 2.00 \text{ MeV} = 5.00 \text{ MeV}$$

3. Use the total energy to calculate the magnitude of the momentum:

$$p_1c = \sqrt{E_1^2 - (m_1c^2)^2} = \sqrt{(5.00 \text{ MeV})^2 - (2.00 \text{ MeV})^2} = \sqrt{21.0} \text{ MeV}$$

$$p_1 = \boxed{4.58 \text{ MeV}/c}$$

(*b*) 1. We can find the final velocity of the system from its total energy E_f and its momentum p_f using Equation 39-25:

$$\frac{u_f}{c} = \frac{p_fc}{E_f}$$

2. By the conservation of total energy, the final energy of the system equals the initial total energy of the two particles:

$$E_f = E_i = E_1 + E_2 = 5.00 \text{ MeV} + 4.00 \text{ MeV} = 9.00 \text{ MeV}$$

3. By the conservation of momentum, the final momentum of the two-particle system equals the initial momentum:

$$p_f = 4.58 \text{ MeV}/c$$

4. Calculate the velocity of the two-particle system from its total energy and momentum using $u/c = pc/E$:

$$\frac{u_f}{c} = \frac{p_fc}{E_f} = \frac{4.58 \text{ MeV}}{9.00 \text{ MeV}} = 0.509$$

$$u_f = \boxed{0.509c}$$

(*c*) We can find the mass M_f of the final two-particle system from Equation 39-27 using $pc = 4.58$ MeV and $E = 9.00$ MeV:

$$E_f^2 = (p_fc)^2 + (M_fc^2)^2$$

$$(9.00 \text{ MeV})^2 = (4.58 \text{ MeV})^2 + (M_fc^2)^2$$

$$M_f = \boxed{7.75 \text{ MeV}/c^2}$$

TAKING IT FURTHER Note that the mass of the system increased from 6.00 MeV/c^2 to 7.75 MeV/c^2. This mass increase, multiplied by c^2, equals the loss in kinetic energy of the system, as you will show in the following exercise.

PRACTICE PROBLEM 39-3 (*a*) Find the final kinetic energy of the two-particle system in Example 39-12. (*b*) Find the loss in kinetic energy, K_{loss}, in the collision. (*c*) Show that $K_{loss} = (\Delta M)c^2$, where ΔM is the change in mass of the system.

Example 39-13 Momentum and Total-Energy Conservation

A 1.00×10^6-kg rocket has 1.00×10^3 kg of fuel on board. The rocket is parked in space when it suddenly becomes necessary to accelerate. The rocket engines ignite, and the 1.00×10^3 kg of fuel are consumed. The exhaust (spent fuel) is ejected during a very short time interval at a speed of 0.500*c* relative to *S*—the inertial reference frame in which the rocket is initially at rest. (*a*) Calculate the change in the mass of the rocket–fuel system. (*b*) Calculate the final speed of the rocket u_R relative to *S*. (*c*) Again, calculate the final speed of the rocket relative to *S*, this time using classical (Newtonian) mechanics.

PICTURE The speed of the rocket and the change in the mass of the system can be calculated using conservation of momentum and conservation of energy. In reference frame *S*, the total momentum of the rocket plus fuel remains zero. After the burn, the magnitude of the momentum of the rocket equals that of the ejected fuel. Let $m_R = 1.00 \times 10^6$ kg be the mass of the rocket, not including the mass of the fuel, let $m_{Fi} = 1.00 \times 10^3$ kg be the mass of the fuel *before* the burn, and let m_{Ff} be the mass of the fuel *after* the burn. The mass of the rocket, m_R, remains fixed, but during the burn the mass of the fuel decreases. (The fuel has less chemical energy after the burn, and so has less mass as well.)

SOLVE

(*a*) 1. The magnitudes of the momentum of the rocket and the momentum of the ejected fuel are equal. For the reasons stated above, the mass of the rocket, not including the 1.00×10^3 kg of fuel, does not change during the burn:

$$p_R = p_F = p$$

$$\frac{m_R u_R}{\sqrt{1-(u_R^2/c^2)}} = \frac{m_{Ff} u_F}{\sqrt{1-(u_F^2/c^2)}} = p$$

$m_R = 1.00 \times 10^6$ kg, $u_F = 0.500c$, and u_R is the final speed of the rocket.

2. The total energy of the system does not change:

$$E_f = E_i$$

3. The initial energy is the rest energy of the rocket and fuel before the burn. The final energy is the energy of the rocket plus energy of the fuel. The energy of each is related to its momentum by Equation 39-27:

$$E_i = m_R c^2 + m_{Fi} c^2 = (m_R + m_{Fi})c^2$$

$$E_{Rf}^2 = p^2c^2 + (m_R c^2)^2$$

$$E_{Ff}^2 = p^2c^2 + (m_{Ff} c^2)^2$$

so

$$E_f = E_{Rf} + E_{Ff}$$

$$E_f = \sqrt{p^2c^2 + (m_R c^2)} + \sqrt{p^2c^2 + (m_{Ff} c^2)^2}$$

4. Equate the initial and final energies:

$$\sqrt{p^2c^2 + (m_R c^2)^2} + \sqrt{p^2c^2 + (m_{Ff} c^2)^2} = (m_R + m_{Fi})c^2$$

5. The step-4 result and $p = \dfrac{m_{Ff} u_F}{1-(u_F^2/c^2)}$ (the step-1 result) constitute two simultaneous equations with unknowns p and m_{Ff}. Solving for m_{Ff} gives:

$$m_{Ff} = 866 \text{ kg}$$

so

$$m_{loss} = m_{Fi} - m_{Ff} = 1000 \text{ kg} - 866 \text{ kg} = \boxed{134 \text{ kg}}$$

(*b*) 1. To solve for u_R, we use Equation 39-25:

$$\frac{u_R}{c} = \frac{pc}{E_{Rf}}$$

2. To solve for p, we substitute the value for m_{Ff} into the Part (*a*), step-1 result:

$$p = \frac{m_{Ff} u_F}{\sqrt{1-(u_F^2/c^2)}} = \frac{(866 \text{ kg})0.500c}{\sqrt{1-0.250}}$$

$$= (5.00 \times 10^2 \text{ kg})c$$

3. We use the value for p to solve for E_{Rf}:

$$E_{Rf}^2 = p^2c^2 + (m_R c^2)^2$$

$$= (5.00 \times 10^2 \text{ kg})^2c^4 + (1.00 \times 10^6 \text{ kg})^2c^4$$

$$= (1.00 \times 10^{12} \text{ kg}^2)c^4$$

so

$$E_{Rf} = (1.00 \times 10^6 \text{ kg})c^2$$

4. Using our Part (*b*), step-1 result, we solve for u_R:

$$u_R = \frac{pc^2}{E_{Rf}} = \frac{(5.00 \times 10^2 \text{ kg})c^3}{(1.00 \times 10^6 \text{ kg})c^2}$$

$$= \boxed{5.00 \times 10^{-4}c = 1.50 \times 10^{-5} \text{ m/s}}$$

(*c*) Equate the magnitude of the classical expressions for the momentum of the rocket and burned fuel and solve for u_R:

$$m_R u_R = m_F u_F$$

$$u_R = \frac{m_F}{m_R} u_F = \frac{1.00 \times 10^3 \text{ kg}}{1.00 \times 10^6 \text{ kg}} 0.500c$$

$$= 5.00 \times 10^{-4}c$$

$$= \boxed{1.50 \times 10^5 \text{ m/s}}$$

CHECK We find the result of the relativistic calculation of the final rocket speed to differ from the classical result. If carried out to five figures, the relativistic calculation gives $u_R = 4.9994 \times 10^{-4}\, c$ for the final speed of the rocket. However, the classical calculation gives $u_R = 5.0000 \times 10^{-4}\, c$. These two values differ by less than one part in 8000.

CONCEPT CHECK 39-1

If the matter being ejected were a 1.00×10^3-kg rigid block launched by a spring with one end attached to the rocket, would the mass of the block change or would the mass of the spring change?

39-8 GENERAL RELATIVITY

The generalization of the theory of relativity to noninertial reference frames by Einstein in 1916 is known as the general theory of relativity. It is much more difficult mathematically than the special theory of relativity, and there are fewer situations in which it can be tested. Nevertheless, its importance calls for a brief qualitative discussion.

The basis of the general theory of relativity is the **principle of equivalence:**

> A homogeneous gravitational field is completely equivalent to a uniformly accelerated reference frame.
>
> PRINCIPLE OF EQUIVALENCE

This principle arises in Newtonian mechanics because of the apparent identity of gravitational mass and inertial mass. In a uniform gravitational field, all objects fall with the same acceleration $\vec{g}$ independent of their masses because the gravitational force is proportional to the (gravitational) mass, whereas the acceleration varies inversely with the (inertial) mass. Consider a compartment in space undergoing a uniform acceleration $\vec{a}$, as shown in Figure 39-14*a*. No mechanics experiment can be performed inside the compartment that will distinguish whether the compartment is actually accelerating in space or is at rest (or is moving with uniform velocity) in the presence of a uniform gravitational field $\vec{g} = -\vec{a}$, as shown in Figure 39-14*b*. If objects are dropped in the compartment, they will fall to the floor with an acceleration $\vec{g} = -\vec{a}$. If people stand on a spring scale, it will read their weight of magnitude $ma = mg$.

Einstein assumed that the principle of equivalence applies to all physics and not just to mechanics. In effect, he assumed that there is no experiment of any kind that can distinguish uniformly accelerated motion from the presence of a gravitational field.

One consequence of the principle of equivalence—the deflection of a light beam in a gravitational field—was one of the first to be tested experimentally. In a region that has no gravitational field, a light beam will travel in a straight line at speed c. The principle of equivalence tells us that a region that has no gravitational field exists only in a compartment that is in free fall. Figure 39-15 shows a beam of light entering a compartment that is accelerating relative to a nearby reference frame in free fall. Successive positions of the compartment at equal time intervals are shown in Figure 39-15*a*. Because the compartment is accelerating, the distance it moves in each time interval increases with time. The path of the beam of light as observed from inside the compartment is therefore a parabola, as shown in Figure 39-15*b*.

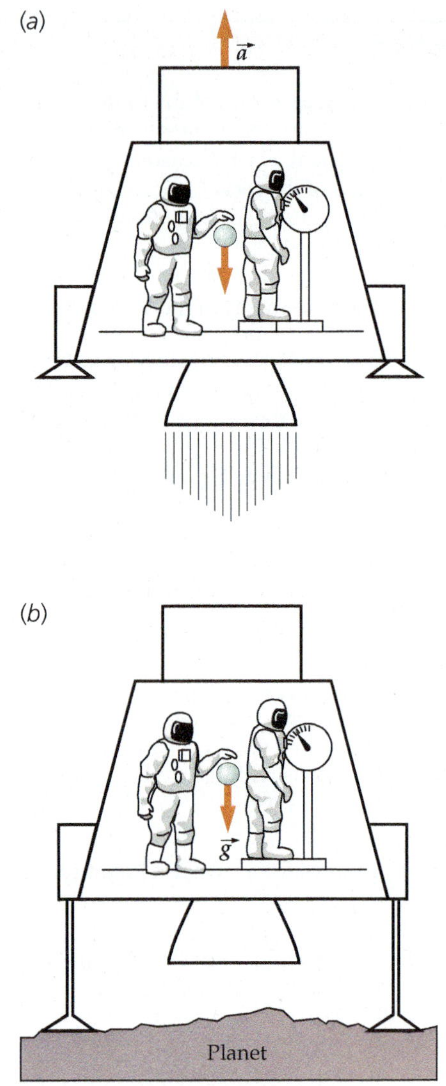

FIGURE 39-14 The results of experiments in a uniformly accelerated reference frame (*a*) cannot be distinguished from those in a uniform gravitational field (*b*) if the acceleration $\vec{a}$ and the gravitational field $\vec{g}$ have the same magnitude.

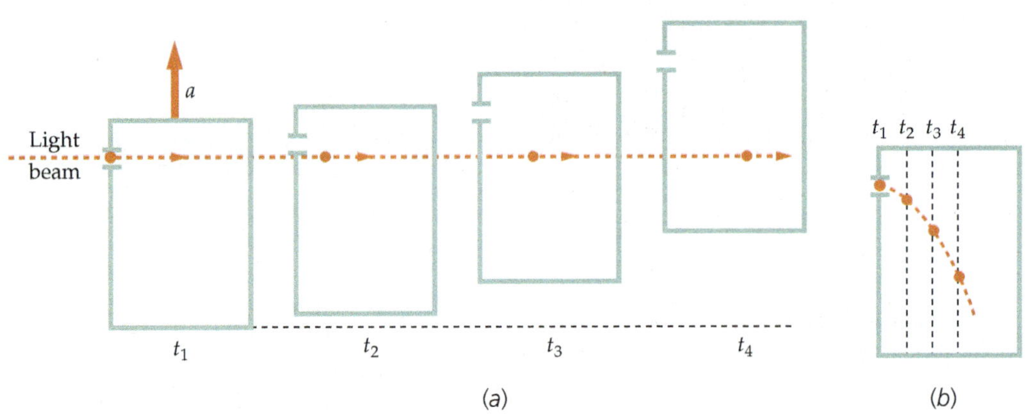

FIGURE 39-15 (*a*) A light beam moving in a straight line through a compartment that is undergoing uniform acceleration relative to a nearby reference frame in free fall. The position of the beam is shown at equally spaced times t_1, t_2, t_3, and t_4. (*b*) In the reference frame of the compartment, the light travels in a parabolic path as a ball would if it were projected horizontally. The vertical displacements are greatly exaggerated for emphasis.

But according to the principle of equivalence, there is no way to distinguish between an accelerating compartment and one moving with uniform velocity in a uniform gravitational field. We conclude, therefore, that a beam of light will accelerate in a gravitational field, just like objects that have mass. For example, near the surface of Earth, light will fall with an acceleration of 9.81 m/s^2. This is difficult to observe because of the enormous speed of light. In a distance of 3000 km, which takes light about 0.01 s to traverse, a beam of light should fall approximately 0.5 mm. Einstein pointed out that the deflection of a light beam in a gravitational field might be observed when light from a distant star passes close to the Sun, as illustrated in Figure 39-16. Because of the brightness of the Sun, this cannot ordinarily be seen. Such a deflection was first observed in 1919 during an eclipse of the Sun. This well-publicized observation brought instant worldwide fame to Einstein.

A second prediction from Einstein's theory of general relativity, which we will not discuss in detail, is the excess precession of the perihelion of the orbit of Mercury of about 0.01° per century. This effect had been known and unexplained for some time, so, in a sense, explaining it constituted an immediate success of the theory.

A third prediction of general relativity concerns the change in time intervals and frequencies of light in a gravitational field. In Chapter 11, we found that the gravitational potential energy between two masses M and m a distance r apart is

$$U = -\frac{GMm}{r}$$

where G is the universal gravitational constant, and the point of zero potential energy has been chosen to be when the separation of the masses is infinite. The potential energy per unit mass near a mass M is called the *gravitational potential* ϕ:

$$\phi = -\frac{GM}{r} \qquad 39\text{-}30$$

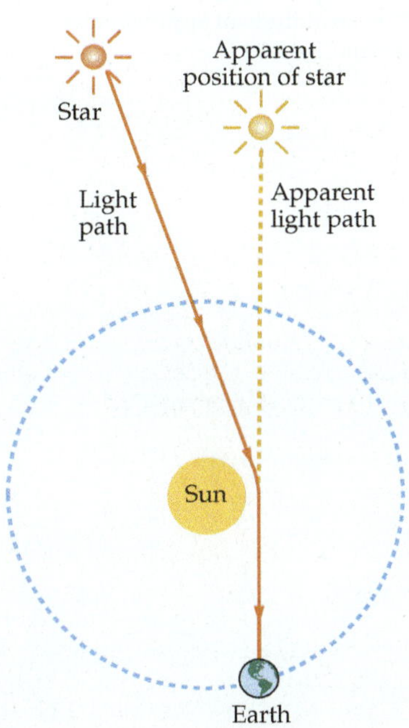

FIGURE 39-16 The deflection (greatly exaggerated) of a beam of light due to the gravitational attraction of the Sun.

The quartz sphere in the top part of the container is probably the world's most perfectly round object. It is designed to spin as a gyroscope in a satellite orbiting Earth. General relativity predicts that the rotation of Earth will cause the axis of rotation of the gyroscope to precess in a circle at a rate of approximately 1 revolution in 100000 years. *(Michael Freeman.)*

According to the general theory of relativity, clocks run more slowly in regions of lower gravitational potential. (Because the gravitational potential is negative, as can be seen from Equation 39-30, the nearer the mass the more negative, and therefore, the lower the gravitational potential.) If Δt_1 is a time interval between two events measured by a clock where the gravitational potential is ϕ_1 and Δt_2 is the interval between the same two events as measured by a clock where the gravitational potential is ϕ_2, general relativity predicts that the fractional difference between the times will be approximately*

$$\frac{\Delta t_2 - \Delta t_1}{\Delta t} = \frac{1}{c^2}(\phi_2 - \phi_1) \qquad \text{39-31}$$

A clock in a region of low gravitational potential will therefore run more slowly than a clock in a region of higher gravitional potential. Because a vibrating atom can be considered to be a clock, the frequency of vibration of an atom in a region of low potential, such as near the Sun, will be lower than the frequency of vibration of the same atom on Earth. This shift toward a lower frequency, and therefore a longer wavelength, is called the **gravitational redshift.**

As our final example of the predictions of general relativity, we mention **black holes,** which were first predicted by J. Robert Oppenheimer and Hartland Snyder in 1939. According to the general theory of relativity, if the density of an object such as a star is great enough, its gravitational attraction will be so great that once inside a critical radius, nothing can escape, not even light or other electromagnetic radiation. (The effect of a black hole on objects outside the critical radius is the same as that of any other mass.) A remarkable property of such an object is that nothing that happens inside it can be communicated to the outside. As sometimes occurs in physics, a simple but incorrect calculation gives the correct results for the relation between the mass and the critical radius of a black hole. In Newtonian mechanics, the speed needed for a particle to escape from the surface of a planet or a star of mass M and radius R is given by Equation 11-21:

$$v_e = \sqrt{\frac{2GM}{R}}$$

If we set the escape speed equal to the speed of light and solve for the radius, we obtain the critical radius R_S, called the **Schwarzschild radius:**

$$R_S = \frac{2GM}{c^2} \qquad \text{39-32}$$

For an object that has a mass equal to five times that of our Sun (theoretically the minimum mass for a black hole) to be a black hole, its radius would have to be approximately 15 km. Because no radiation is emitted from a black hole and its radius is expected to be small, the detection of a black hole is not easy. The best chance of detection occurs in a binary-star system in which a black hole is a close companion to a normal star. Then both stars revolve around their center of mass and the gravitational field of the black hole will pull gas from the normal star into the black hole. However, to conserve angular momentum, the gas does not go straight into the black hole. Instead, the gas orbits around the black hole in a disk, called an accretion disk, while slowly being pulled closer to the black hole. The gas in this disk emits X rays because the temperature of the gas being pulled inward reaches several million kelvins. The mass of a black-hole candidate can often be estimated. An estimated mass of at least five solar masses, along with the emission of X rays, establishes a strong inference that the candidate is, in fact, a black hole. In addition to the black holes just described, there are supermassive black holes that exist at the centers of galaxies. At the center of the Milky Way is a supermassive black hole that has a mass of about two million solar masses.

This extremely accurate hydrogen maser clock was launched in a satellite in 1976, and its time was compared to that of an identical clock on Earth. In accordance with the prediction of general relativity, the clock on Earth, where the gravitational potential was lower, lost approximately 4.3×10^{-10} s each second compared with the clock orbiting Earth at an altitude of approximately 10000 km. *(NASA.)*

---

* Because this shift is usually very small, it does not matter by which interval we divide on the left side of the equation.

Summary

TOPIC	RELEVANT EQUATIONS AND REMARKS	
1. Einstein's Postulates	The special theory of relativity is based on two postulates of Albert Einstein. All of the results of special relativity can be derived from these postulates. Postulate 1: Absolute uniform motion cannot be detected. Postulate 2: The speed of light is independent of the motion of the source. An important implication of these postulates is Postulate 2 (alternate): Every observer measures the same value c for the speed of light.	
2. The Lorentz Transformation	$x = \gamma(x' + vt'), \qquad y = y', \qquad z = z'$	39-9
	$t = \gamma\left(t' + \frac{vx'}{c^2}\right)$	39-10
	$\gamma = \frac{1}{\sqrt{1 - (v^2/c^2)}}$	39-7
Inverse transformation	$x' = \gamma(x - vt), \qquad y' = y, \qquad z' = z$	39-11
	$t' = \gamma\left(t - \frac{vx}{c^2}\right)$	39-12
3. Time Dilation	The time interval measured between two events that occur at the same point in space in some reference frame is called the proper time interval Δt_p between those two events. In another reference frame in which the same two events occur at different places, the time interval Δt between the events is longer by the factor γ.	
	$\Delta t = \gamma\, \Delta t_p$	39-13
4. Length Contraction	The length of an object measured in the reference frame in which the object is at rest is called its proper length L_p. When measured in another reference frame, the length of the object along the direction parallel to the velocity of the object is	
	$L = \frac{L_p}{\gamma}$	39-14
5. The Relativistic Doppler Effect	$f' = \frac{\sqrt{1 - (v^2/c^2)}}{1 - (v/c)} f_0$ approaching	39-16*a*
	$f' = \frac{\sqrt{1 - (v^2/c^2)}}{1 + (v/c)} f_0$ receding	39-16*b*
6. Clock Synchronization and Simultaneity	Two events that are simultaneous in one reference frame typically are not simultaneous in another frame that is moving relative to the first. If two clocks are synchronized in the frame in which they are at rest, they will be out of synchronization in another frame. In the frame in which they are moving, the chasing clock leads by an amount	
	$\Delta t_S = L_p \frac{v}{c^2}$	39-17
	where L_p is the proper distance between the clocks.	
7. The Velocity Transformation	$u_x = \frac{u'_x + v}{1 + (vu'_x/c^2)}$	39-18*a*
	$u_y = \frac{u'_y}{\gamma[1 + (vu'_x/c^2)]}$	39-18*b*
	$u_z = \frac{u'_z}{\gamma[1 + (vu'_x/c^2)]}$	39-18*c*

TOPIC	RELEVANT EQUATIONS AND REMARKS	
Inverse velocity transformation	$u'_x = \dfrac{u_x - v}{1 - (vu_x/c^2)}$	39-19*a*
	$u'_y = \dfrac{u_y}{\gamma[1 - (vu_x/c^2)]}$	39-19*b*
	$u'_z = \dfrac{u_z}{\gamma[1 - (vu_x/c^2)]}$	39-19*c*
8. Relativistic Momentum	$\vec{p} = \dfrac{m\vec{u}}{\sqrt{1 - (u^2/c^2)}}$	39-20
	where m is the mass of the particle.	
9. Relativistic Energy		
Kinetic energy	$K = \dfrac{mc^2}{\sqrt{1 - (u^2/c^2)}} - mc^2$	39-22
Rest energy	$E_0 = mc^2$	39-23
Total Relativistic energy	$E = K + E_0 = \dfrac{mc^2}{\sqrt{1 - (u^2/c^2)}}$	39-24
10. Useful Formulas for Speed, Energy, and Momentum	$\dfrac{u}{p} = \dfrac{pc}{E}$	39-25
	$E^2 = p^2c^2 + (mc^2)^2$	39-27
	$E \approx pc \qquad E \gg mc^2$	39-28

Answer to Concept Check

39-1 Only the rest mass of the spring would change.

Answers for Practice Problems

39-1 1.67 h

39-2 (*a*) 1.49, (*b*) $p = 1.04 \times 10^3$ MeV/*c*, and (*c*) $u = 0.74c$

39-3 (*a*) $K_f = E_f - M_fc^2 = 9.00\text{ MeV} - 7.75\text{ MeV} = 1.25\text{ MeV}$,

(*b*) $K_{loss} = K_i - K_f = 3.00\text{ MeV} - 1.25\text{ MeV} = 1.75\text{ MeV}$, and

(*c*) $(\Delta M)c^2 = (M_f - M_i)c^2 = 7.75\text{ MeV} - (2.00\text{ MeV} + 4.00\text{ MeV}) = 1.75\text{ MeV} = K_{loss}$

Problems

In a few problems, you are given more data than you actually need; in a few other problems, you are required to supply data from your general knowledge, outside sources, or informed estimate.

Interpret as significant all digits in numerical values that have trailing zeros and no decimal points.

- • Single-concept, single-step, relatively easy
- •• Intermediate-level, may require synthesis of concepts
- ••• Challenging
- SSM Solution is in the *Student Solutions Manual*
- Consecutive problems that are shaded are paired problems.

CONCEPTUAL PROBLEMS

1 • The approximate total energy of a particle of mass m moving at speed $u \ll c$ is (*a*) $mc^2 + \frac{1}{2}mu^2$, (*b*) $\frac{1}{2}mu^2$, (*c*) cmu, (*d*) mc^2, (*e*) $\frac{1}{2}cmu$. SSM

2 • A set of twins work in an office building. One twin works on the top floor and the other twin works in the basement. Considering general relativity, which twin will age more quickly? (*a*) They will age at the same rate. (*b*) The twin who works on the top floor will age more quickly. (*c*) The twin who works in the basement will age more quickly. (*d*) It depends on the speed of the office building. (*e*) none of the above

3 • True or false:

(*a*) The speed of light is the same in all reference frames.
(*b*) The time interval between two events is never shorter than the proper time interval between the two events.
(*c*) Absolute motion can be determined by means of length contraction.
(*d*) The light-year is a unit of distance.
(*e*) Simultaneous events must occur at the same place.
(*f*) If two events are not simultaneous in one frame, they cannot be simultaneous in any other frame.
(*g*) The mass of a system that consists of two particles tightly bound together by attractive forces is less than the sum of the masses of the individual particles when separated.

4 • An observer sees a system moving past her that consists of a mass oscillating on the end of a spring and measures the period T of the oscillations. A second observer, who is moving with the mass–spring system, also measures its period. The second observer will find a period that is (*a*) equal to T, (*b*) less than T, (*c*) greater than T, (*d*) either (*a*) or (*b*) depending on whether the system was approaching or receding from the first observer, (*e*) Not enough information is given to answer the question.

5 • The Lorentz transformation for y and z is the same as the classical result: $y = y'$ and $z = z'$. Yet the relativistic velocity transformation does not give the classical result $u_y = u'_y$ and $u_z = u'_z$. Explain why this result occurs.

ESTIMATION AND APPROXIMATION

6 •• The Sun radiates energy at the rate of approximately 4×10^{26} W. Assume that this energy is produced by a reaction whose net result is the fusion of four protons to form a single ^{4}He nucleus and the release of 25 MeV of energy that is radiated into space. Calculate the Sun's loss of mass per day.

7 •• The most distant galaxies that can be seen by the Hubble telescope are moving away from us and have a redshift parameter of about $z = 5$. [The redshift parameter z is defined as $(f - f')/f'$, where f is the frequency measured in the rest frame of the emitter and f' is the frequency measured in the rest frame of the receiver.] (*a*) What is the speed of the galaxies relative to us (expressed as a fraction of the speed of light)? (*b*) *Hubble's law* states that the recession speed is given by the expression $v = Hx$, where v is the speed of recession, x is the distance, and H, the Hubble constant, is equal to 75 km/s/Mpc, where 1 pc = 3.26 $c \cdot$y. (The abbreviation for parsec is pc.) Estimate the distance of such a galaxy from us using the information given. SSM

TIME DILATION AND LENGTH CONTRACTION

8 • The proper mean lifetime of a muon is 2.2 μs. Muons in a beam are traveling through a laboratory at 0.95c. (*a*) What is their mean lifetime as measured in the laboratory? (*b*) How far do they travel, on average, before they decay?

9 •• In the Stanford linear collider, small bundles of electrons and positrons are fired at each other. In the laboratory's frame of reference, each bundle is approximately 1.0 cm long and 10 μm in diameter. In the collision region, each particle has an energy of 50 GeV, and the electrons and the positrons are moving in opposite directions. (*a*) How long and how wide is each bundle in its own reference frame? (*b*) What must be the minimum proper length of the accelerator for a bundle to have both its ends simultaneously in the accelerator in its own reference frame? (The actual proper length of the accelerator is less than 1000 m.) (*c*) What is the length of a positron bundle in a reference frame that moves with the electron bundle?

10 • Use the binomial expansion equation

$$(1 + x)^n = 1 + nx + \frac{n(n-1)}{2}x^2 + \ldots \approx 1 + nx \qquad x \ll 1$$

to derive the following results for the case when v is much less than c.

$$(a)\ \gamma \approx 1 + \frac{1}{2}\frac{v^2}{c^2}$$

$$(b)\ \frac{1}{\gamma} \approx 1 - \frac{1}{2}\frac{v^2}{c^2}$$

$$(c)\ \gamma - 1 \approx 1 - \frac{1}{\gamma} \approx \frac{1}{2}\frac{v^2}{c^2}$$

11 •• Star A and Star B are at rest relative to Earth. Star A is 27 $c \cdot$y from Earth, and as viewed from Earth, Star B is located beyond (behind) Star A. (*a*) A spaceship is making a trip from Earth to Star A at a speed such that the trip from Earth to Star A takes 12 y according to clocks on the spaceship. At what speed, relative to Earth, must the spaceship travel? (Assume that the times for the accelerations are very short compared to the overall trip time.) (*b*) Upon reaching Star A, the spaceship speeds up and departs for Star B at a speed such that the gamma factor, γ, is twice that of Part (*a*). The trip from Star A to Star B takes 5.0 y (spaceship's time). How far, in $c \cdot$y, is Star B from Star A in the rest frame of Earth and the two stars? (*c*) Upon reaching Star B, the spaceship departs for Earth at the same speed as in Part (*b*). It takes it 10 y (spaceship's time) to return to Earth. If you were born on Earth the day the ship left Earth and you remain on Earth, how old are you on the day the ship returns to Earth?

12 • A spaceship travels to a star 35 $c \cdot$y away at a speed of 2.7×10^8 m/s. How long does the spaceship take to get to the star (*a*) as measured on Earth and (*b*) as measured by a passenger on the spaceship?

13 •• Unobtainium (Un) is an unstable particle that decays into normalium (Nr) and standardium (St) particles. (*a*) An accelerator produces a beam of Un that travels to a detector located 100 m away from the accelerator. The particles travel with a velocity of $v = 0.866c$. How long do the particles take (in the laboratory frame) to get to the detector? (*b*) By the time the particles get to the detector, half of the particles have decayed. What is the half-life of Un? (*Note:* half-life as it would be measured in a frame moving with the particles). (*c*) A new detector is going to be used, which is located 1000 m away from the accelerator. How fast should the particles be moving if half of the particles are to make it to the new detector? SSM

14 •• A clock on Spaceship A measures the time interval between two events, both of which occur at the location of the clock. You are on Spaceship B. According to your careful measurements, the time interval between the two events is 1.00 percent longer than that measured by the two clocks on Spaceship A. How fast is Spaceship A moving relative to Spaceship B. *Hint: Use one or more of the results of Problem 10.*

15 •• If a plane flies at a speed of 2000 km/h, how long must the plane fly before its clock loses 1.00 s because of time dilation? *Hint: Use one or more of the results of Problem 10.*

THE LORENTZ TRANSFORMATION, CLOCK SYNCHRONIZATION, AND SIMULTANEITY

16 •• Show that when $v \ll c$ the relativistic transformation equations for x, t, and u_x reduce to the classical transformation equations.

17 •• A spaceship of proper length $L_p = 400$ m moves past a transmitting station at a speed of $0.760c$. (The transmitting station broadcasts signals that travel at the speed of light.) A clock is attached to the nose of the spaceship and a second clock is attached to the transmitting station. The instant that the nose of the spaceship passes the transmitter, the clock attached to the transmitter and the clock attached to the nose of the spaceship are set equal to zero. The instant that the tail of the spaceship passes the transmitter a signal is sent by the transmitter that is subsequently detected by a receiver in the nose of the spaceship. (*a*) When, according to the clock attached to the nose of spaceship, is the signal sent? (*b*) When, according to the clocks attached to the nose of spaceship, is the signal received? (*c*) When, according to the clock attached to the transmitter, is the signal received by the spaceship? (*d*) According to an observer that works at the transmitting station, how far from the transmitter is the nose of the spaceship when the signal is received? SSM

18 •• In frame S, event B occurs 2.0 μs after event A, and event A occurs at the origin whereas event B occurs on the x axis at $x = 1.5$ km. How fast and in what direction must an observer be traveling along the x axis so that events A and B occur simultaneously? Is it possible for event B to precede event A for some observer?

19 •• Observers in reference frame S see an explosion located on the x axis at $x_1 = 480$ m. A second explosion occurs, 5.0 μs later, at $x_2 = 1200$ m. In reference frame S', which is moving along the x axis in the $+x$ direction at speed v, the two explosions occur at the same point in space. What is the separation in time between the two explosions as measured in S'?

20 ••• In reference frame S, events 1 and 2 are separated by a distance $D = x_2 - x_1$ and a time $T = t_2 - t_1$. (*a*) Use the Lorentz transformation to show that in frame S', which is moving along the x axis with speed v relative to S, the time separation is $t_2' - t_1' = \gamma(T - vD/c^2)$. (*b*) Show that the events can be simultaneous in frame S' only if D is greater than cT. (*c*) If one of the events is the *cause* of the other, the separation D must be less than cT, because D/c is the smallest time that a signal can take to travel from x_1 to x_2 in frame S. Show that if D is less than cT, t_2' is greater than t_1' in all reference frames. This shows that if the cause precedes the effect in one frame, it must precede it in all reference frames. (*d*) Suppose that a signal could be sent with speed $c' > c$ so that in frame S the cause precedes the effect by the time $T = D/c'$. Show that there is then a reference frame moving with speed v less than c in which the effect precedes the cause.

21 ••• A rocket that has a proper length of 700 m is moving to the right at a speed of $0.900c$. It has two clocks—one in the nose and one in the tail—that have been synchronized in the frame of the rocket. A clock on the ground and the clock in the nose of the rocket both read zero as they pass by each other. (*a*) At the instant the clock on the ground reads zero, what does the clock in the tail of the rocket read according to observers on the ground? When the clock in the tail of the rocket passes the clock on the ground, (*b*) what does the clock in the tail read according to observers on the ground, and (*c*) what does the clock in the nose read according to observers on the ground, and (*d*) what does the clock in the nose read according to observers on the rocket? (*e*) At the instant the clock in the nose of the rocket reads 1.00 h, a light signal is sent from the nose of the rocket to an observer standing by the clock on the ground. What does the clock on the ground read when the observer on the ground receives the signal? (*f*) When the observer on the ground receives the signal, he immediately sends a return signal to the nose of the rocket. What is the reading of the clock in the nose of the rocket when that signal is received at the nose of the rocket?

THE VELOCITY TRANSFORMATION AND THE RELATIVISTIC DOPPLER EFFECT

22 •• **SPREADSHEET** A spaceship, at rest in a certain reference frame S, is given a speed increase of $0.50c$ (call this increase boost 1). Relative to its new rest frame, the spaceship is given a further $0.50c$ increase 10 seconds later (as measured in its new rest frame; call this increase boost 2). This process is continued indefinitely, at 10-s intervals, as measured in the rest frame of the spaceship. (Assume that the boosts take a very short time compared to 10 s.) (*a*) Using a spreadsheet program, calculate and graph the speed of the spaceship in reference frame S as a function of the boost number for boost 1 to boost 10. (*b*) Graph the gamma factor in the same manner. (*c*) How many boosts does it take until the speed of the ship in S is greater than $0.999c$? (*d*) How far does the spaceship move between boost 1 and boost 6, as measured in reference frame S? What is the average speed of the spaceship between boost 1 and boost 6, as measured in S?

23 • Light is emitted by a sodium sample that is moving toward Earth with speed v. The wavelength of the light is 589 nm in the rest frame of the sample. The wavelength measured in the frame of Earth is 547 nm. Find v.

24 • A distant galaxy is moving away from us at a speed of 1.85×10^7 m/s. Calculate the fractional redshift $(\lambda' - \lambda_0)/\lambda_0$ that we observe the light from the galaxy to have.

25 •• Derive $f' = f_0\sqrt{1 - (v^2/c^2)}/[1 - (v/c)]$ (Equation 39-16*a*) for the frequency received by an observer moving with speed v toward a stationary source of electromagnetic waves.

26 • Show that if v is much less than c, the Doppler shift is given approximately by

$$\Delta f/f \approx \pm v/c$$

27 •• A clock is placed in a satellite that orbits Earth with an orbital period of 90 min. By what time interval will this clock differ from an identical clock on Earth after 1.0 y? (Assume that special relativity applies and neglect general relativity.) SSM

28 •• For light that is Doppler-shifted with respect to an observer, we define the redshift parameter $z = (f - f')/f'$, where f is the frequency of the light measured in the rest frame of the emitter and f' is the frequency measured in the rest frame of the receiver. If the emitter is moving directly away from the receiver, show that the relative velocity between the emitter and the receiver is $v = c(u^2 - 1)/(u^2 + 1)$, where $u = z + 1$.

29 • A light beam moves along the y' axis with speed c in frame S', which is moving in the $+x$ direction with speed v relative to frame S. (*a*) Find the x and y components of the velocity of the light beam in frame S. (*b*) Show that, according to the velocity transformation equations, the magnitude of the velocity of the light beam in S is c.

30 •• A spaceship is moving east at speed $0.90c$ relative to Earth. A second spaceship is moving west at speed $0.90c$ relative to Earth. What is the speed of one spaceship relative to the other spaceship?

31 •• A particle moves with speed $0.800c$ in the $+x''$ direction along the x'' axis of frame S'', which moves with the same speed and in the same direction along the x' axis relative to frame S'. Frame S' moves with the same speed and in the same direction along the x axis relative to frame S. (*a*) Find the speed of the particle relative to frame S'. (*b*) Find the speed of the particle relative to frame S. SSM

RELATIVISTIC MOMENTUM AND RELATIVISTIC ENERGY

32 • A proton that has a rest energy equal to 938 MeV has a total energy of 2200 MeV. (*a*) What is its speed? (*b*) What is its momentum?

33 • If the kinetic energy of a particle equals twice its rest energy, what percentage error is made by using $p = mu$ for the magnitude of its momentum?

34 •• In a certain reference frame, a particle has momentum of 6.00 MeV/c and total energy of 8.00 MeV. (*a*) Determine the mass of the particle. (*b*) What is the total energy of the particle in a reference frame in which its momentum is 4.00 MeV/c? (*c*) What is the relative speed of the two reference frames?

35 •• Show that

$$d\left(\frac{mu}{1-(u^2/c^2)}\right) = m\left(1-\frac{u^2}{c^2}\right)^{-3/2} du$$

(*Note:* This relation was used to derive the relativistically correct expression for kinetic energy (Equation 39-22).)

36 •• The K^0 particle has a mass of 497.7 MeV/c^2. It decays into a π^- and π^+, each having mass 139.6 MeV/c^2. Following the decay of a K^0, one of the pions is at rest in the laboratory. Determine the kinetic energy of the other pion after the decay and of the K^0 prior to the decay.

37 •• In reference frame S', two protons, each moving at $0.500c$, approach each other head-on. (*a*) Calculate the total kinetic energy of the two protons in frame S'. (*b*) Calculate the total kinetic energy of the protons as seen in reference frame S, which is moving with one of the protons. SSM

38 •• An antiproton $\bar{p}$ has the same mass m as a proton p. The antiproton is created during the reaction $p + p \rightarrow p + p + p + \bar{p}$. During an experiment, protons at rest in the laboratory are bombarded with protons of kinetic energy K_L, which must be great enough so that an amount of kinetic energy equal to $2mc^2$ can be converted into the rest energy of the two particles. In the frame of the laboratory, the total kinetic energy cannot be converted into rest energy because of conservation of momentum. However, in the zero-momentum reference frame in which the two initial protons are moving toward each other with equal speed u, the total kinetic energy can be converted into rest energy. (*a*) Find the speed of each proton u so that the total kinetic energy in the zero-momentum frame is $2mc^2$. (*b*) Transform to the laboratory's frame in which one proton is at rest, and find the speed u' of the other proton. (*c*) Show that the kinetic energy of the moving proton in the laboratory's frame is $K_L = 6mc^2$.

39 ••• A particle of mass 1.00 MeV/c^2 and kinetic energy 2.00 MeV collides with a stationary particle of mass 2.00 MeV/c^2. After the collision, the particles stick together. Find (*a*) the speed of the first particle before the collision, (*b*) the total energy of the first particle before the collision, (*c*) the initial total momentum of the system, (*d*) the total kinetic energy after the collision, and (*e*) the mass of the system after the collision.

GENERAL RELATIVITY

40 •• Light traveling in the direction of increasing gravitational potential undergoes a frequency redshift. Calculate the shift in wavelength if a beam of light of wavelength $\lambda = 632.8$ nm is sent up a vertical shaft of height $L = 100$ m.

41 •• Let us revisit a problem from Chapter 3: Two cannons are pointed directly toward each other, as shown in Figure 39-17. When fired, the cannonballs will follow the trajectories shown. Point P is the point where the trajectories cross each other. Ignore any effects due to air resistance. Using the principle of equivalence, show that if the cannons are fired simultaneously (in the rest frame of the cannons), the cannonballs will hit each other at point P.

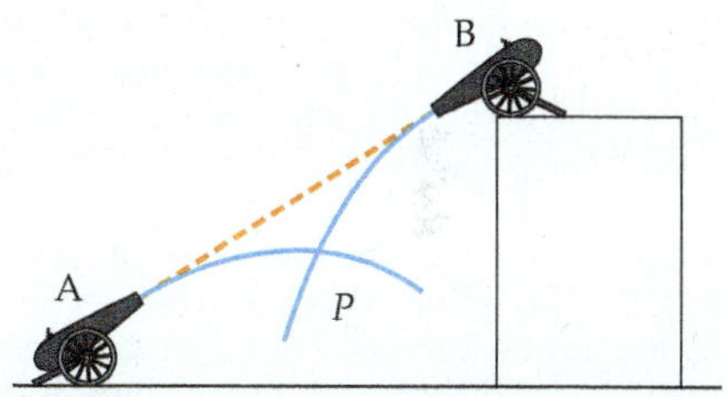

FIGURE 39-17
Problem 41

42 ••• A horizontal turntable rotates with angular speed ω. There is a clock at the center of the turntable and an identical clock mounted on the turntable a distance r from the center. In an inertial reference frame, in which the clock at the center is at rest, the clock at distance r is moving with speed $u = r\omega$. (*a*) Show that from time dilation according to special relativity, the time between ticks, Δt_0 for the clock at rest and Δt_R for the moving clock, are related by

$$\frac{\Delta t_R - \Delta t_0}{\Delta t_0} = -\frac{r^2\omega^2}{2c^2} \qquad r\omega \ll c$$

(*b*) In a reference frame rotating with the table, both clocks are at rest. Show that the clock at distance r experiences a pseudoforce $F_r = mr\omega^2$ in the rotating frame and that this is equivalent to a difference in gravitational potential between r and the origin of $\phi_r - \phi_0 = -\frac{1}{2}r^2\omega^2$. (*c*) Use the difference in gravitational potential given in Part (*b*) to show that in this frame the difference in time intervals is the same as in the inertial frame.

GENERAL PROBLEMS

43 • How fast must a muon travel so that its mean lifetime is 46 μs if its mean lifetime at rest is 2.2 μs?

44 • A distant galaxy is moving away from Earth with a speed that results in each wavelength received on Earth being shifted so that $\lambda' = 2\lambda_0$. Find the speed of the galaxy relative to Earth.

45 •• Frames S and S' are moving relative to each other along the x and x' axes (which superpose). Observers at rest in the two frames set their clocks to $t = 0$ when the two origins coincide. In frame S, event 1 occurs at $x_1 = 1.0\ c\cdot$y and $t_1 = 1.00$ y and event 2 occurs at $x_2 = 2.0\ c\cdot$y and $t_2 = 0.50$ y. The events occur simultaneously in frame S'. (*a*) Find the magnitude and direction of the velocity of S' relative to S. (*b*) At what time do both events occur as measured in S'? SSM

46 •• An interstellar spaceship travels from Earth to a star system 12 light-years away (as measured in Earth's frame). The trip takes 15 y as measured by clocks on the spaceship. (*a*) What is the speed of the spaceship relative to Earth? (*b*) When the spaceship arrives, it sends an electromagnetic signal to Earth. How long after the spaceship leaves Earth will observers on Earth receive the signal?

47 •• The neutral pion π^0 has a mass of 135.0 MeV/c^2. This particle can be created in a proton–proton collision:

$$\text{p} + \text{p} \rightarrow \text{p} + \text{p} + \pi^0$$

Determine the threshold kinetic energy for the creation of a π^0 in a collision of a moving proton and a stationary proton. (See Problem 38.)

48 •• A rocket that has a proper length of 1000 m moves away from a space station and in the $+x$ direction at $0.60c$ relative to an observer on the station. An astronaut stands at the rear of the rocket and fires a dart toward the front of the rocket at $0.80c$ relative to the rocket. How long does it take the dart to reach the front of the rocket (*a*) as measured in the frame of the rocket, (*b*) as measured in the frame of the space station, and (*c*) as measured in the frame of the dart?

49 ••• Using a simple thought experiment, Einstein showed that there is mass associated with electromagnetic radiation. Consider a box of length L and mass M resting on a frictionless surface. Attached to the left wall of the box is a light source that emits a directed pulse of radiation of energy E, which is completely absorbed at the right wall of the box. According to classical electromagnetic theory, the radiation carries momentum of magnitude $p = E/c$ (Equation 30-24). The box recoils when the pulse is emitted by the light source. (*a*) Find the recoil velocity of the box so that momentum is conserved when the light is emitted. (Because p is small and M is large, you may use classical mechanics.) (*b*) When the light is absorbed at the right wall of the box the box stops, so the total momentum of the system remains zero. If we neglect the very small velocity of the box, the time it takes for the radiation to travel across the box is $\Delta t = L/c$. Find the distance moved by the box in that time. (*c*) Show that if the center of mass of the system is to remain at the same place, the radiation must carry mass $m = E/c^2$. SSM

50 ••• Using the relativistic conservation of momentum and energy and the relation between energy and momentum for a photon $E = pc$, prove that a free electron (an electron not bound to an atomic nucleus) cannot absorb or emit a photon.

51 ••• When a moving particle that has a kinetic energy greater than the threshold kinetic energy K_{th} strikes a stationary target particle, one or more particles may be created in the inelastic collision. Show that the threshold kinetic energy of the moving particle is given by

$$K_{\text{th}} = \frac{(\Sigma m_{\text{in}} + \Sigma m_{\text{fin}})(\Sigma m_{\text{fin}} + \Sigma m_{\text{in}})c^2}{2m_{\text{target}}}$$

Here Σm_{in} is the sum of the masses of the particles prior to the collision, Σm_{fin} is the sum of the masses of the particles following the collision, and m_{target} is the mass of the target particle. Use this expression to determine the threshold kinetic energy of protons incident on a stationary proton target for the production of a proton–antiproton pair; compare your result with the result of Problem 38. SSM

52 ••• A particle of mass M decays into two identical particles, each of mass m, where $m = 0.30M$. Prior to the decay, the particle of mass M has a total energy of $4.0mc^2$ in the laboratory reference frame. The velocities of the decay products are along the direction of motion of M. Find the velocities of the decay products in the laboratory reference frame.

53 ••• A rod of proper length L_{p} makes an angle θ with the x axis in frame S. Show that the angle θ' made with the x' axis in frame S', which is moving in the $+x$ direction with speed v, is given by $\tan\theta' = \gamma\tan\theta$ and that the length of the stick in S' is $L' = L_{\text{p}}(\gamma^{-2}\cos^2\theta + \sin^2\theta)^{1/2}$.

54 ••• Show that if a particle moves at an angle θ with the x axis with speed u in frame S, it moves at an angle θ' with the x' axis in S' given by

$$\tan\theta' = \gamma^{-1}\sin\theta/[\cos\theta - (v/u)]$$

55 ••• For the special case of a particle moving with speed u along the y axis in frame S, show that its momentum and energy in frame S', a frame that is moving along the x axis with velocity v, are related to its momentum and energy in S by the transformation equations

$$p'_x = \gamma\left(p_x - \frac{vE}{c^2}\right) \qquad p'_y = p_y \qquad p'_z = p_z \qquad \frac{E'}{c} = \gamma\left(\frac{E}{c} - \frac{vp_x}{c}\right)$$

Compare these equations with the Lorentz transformation equations for x', y', z', and t'. Notice that the quantities p_x, p_y, p_z and E/c transform in the same way as do x, y, z, and ct. SSM

56 ••• The equation for the spherical wavefront of a light pulse that begins at the origin at time $t = 0$ is $x^2 + y^2 + z^2 - (ct)^2 = 0$. Frame S' moves with velocity v along the x axis. Using the Lorentz transformation, show that such a light pulse also has a spherical wavefront in frame S' by showing that $x'^2 + y'^2 + z'^2 - (ct')^2 = 0$.

57 ••• In Problem 56, you showed that the quantity $x^2 + y^2 + z^2 - (ct)^2$ has the same value (zero) in both S and S'. A quantity that has the same value in all inertial frames is called a *Lorentz invariant*. From the results of Problem 55, the quantity $p_x^2 + p_y^2 + p_z^2 - E^2/c^2$ must also be a Lorentz invariant. Show that this quantity has the value $-m^2c^2$ in both the S and S' reference frames.

58 ••• A long rod that is parallel to the x axis is released from rest. Subsequently, it is in free fall with an acceleration of magnitude g in the $-y$ direction. An observer in a rocket ship moving with speed v parallel to the x axis passes by. Using the Lorentz transformations, show that the observer on the rocket ship will measure the rod to be bent into a parabolic shape. Is the parabola concave upward or concave downward?

59 •• Show that if u'_x and v in $u_x = (u'_x + v)/[1 + (vu'_x/c^2)]$ (Equation 39-18*a*) are both positive and less than c, then u_x is positive and less than c. *Hint: Let* $u'_x = (1 - \varepsilon_1)c$ *and* $v = (1 - \varepsilon_2)c$, *where* ε_1 *and* ε_2 *are positive numbers that are less than 1.*

60 ••• In reference frame S, the acceleration of a particle is $\vec{a} = a_x\hat{i} + a_y\hat{j} + a_z\hat{k}$. Derive expressions for the acceleration components a'_x, a'_y, and a'_z of the particle in reference frame S' that is moving relative to S in the x direction with velocity v.

CHAPTER 40

THE DIABLO CANYON NUCLEAR POWER PLANT NEAR SAN LUIS OBISPO, CALIFORNIA. *(Tony Hertz/Alamy.)*

? How much energy is released during the fission of one gram of ^{235}U? (See Example 40-6.)

Nuclear Physics

To many chemists, the atomic nucleus is modeled as a point charge that has most of the mass of the atom. In this chapter, we will look at the nucleus from the physicist's perspective and see how the protons and neutrons that make up the nucleus have played important roles in our everyday life as well as in the history and structure of the universe.

In this chapter, we study the properties of atomic nuclei, examine radioactivity, and explore nuclear reactions. We also discuss fission and fusion. The fission of very heavy nuclei, such as uranium, is a major source of power today, while the fusion of very light nuclei is the energy source that powers the stars, including our Sun, and may hold the key to our energy needs of the future.

40-1 PROPERTIES OF NUCLEI

The nucleus of an atom has just two kinds of particles, protons and neutrons,* which have approximately the same mass (the neutron is approximately 0.2 percent more massive). The proton has a charge of $+e$, and the neutron is uncharged.

* The most prevalent hydrogen nucleus has a single proton.

The number of protons, Z, is the atomic number of the atom, which also equals the number of electrons in the atom. The number of neutrons that a nucleus has, N, is approximately equal to Z for light nuclei. For heavier nuclei, the number of neutrons is increasingly greater than Z. The total number of nucleons* $A = N + Z$ is called the **nucleon number** or **mass number** of the nucleus. A particular nuclear species is called a **nuclide.** Two or more nuclides that have the same atomic number Z but have different values for N and A are called **isotopes.** A particular nuclide is designated by its atomic symbol (for example, H for hydrogen and He for helium) and its mass number A as a superscript. The lightest element, hydrogen, has three isotopes: protium, ^1H, whose nucleus is just a single proton; deuterium, ^2H, whose nucleus is composed of one proton and one neutron; and tritium, ^3H, whose nucleus is composed of one proton and two neutrons. Although the mass of the deuterium atom is about twice the mass of the protium atom and the mass of the tritium atom is about three times the mass of protium, these three atoms have nearly identical chemical properties because they each have one electron. On the average, there are about three stable isotopes for each element, although some atoms have only one stable isotope while others have five or six. The most common isotope of the second lightest element, helium, is ^4He. The ^4He nucleus is also known as an α particle. Another isotope of helium is ^3He, and the ^3He nucleus is also known as helion.

Nucleons exert a strong attractive force on other nucleons. This force, called the **strong nuclear force** or the **hadronic force,** is much stronger than the electrostatic force of repulsion between the protons and is very much stronger than the gravitational forces between the nucleons. (Gravity is so comparably weak that it can always be neglected in nuclear physics.) The strong nuclear force is roughly the same between two neutrons, two protons, or a neutron and a proton. Two protons, of course, also exert a repulsive electrostatic force on each other due to their charges, which tends to weaken the attraction between them somewhat. The strong nuclear force decreases rapidly with distance, and it is negligible when two nucleons are more than a few femtometers apart.

SIZE, SHAPE, AND DENSITY

The size and shape of the nucleus can be determined by bombarding it with high-energy particles and observing the scattering. The results depend somewhat on the kind of experiment. For example, a scattering experiment using electrons measures the charge distribution of the nucleus, whereas a scattering experiment using neutrons determines the region of influence of the strong nuclear force. A wide variety of experiments suggest that most nuclei are approximately spherical, with radii given approximately by

$$R = R_0 A^{1/3} \qquad \text{40-1}$$

NUCLEAR RADIUS

where R_0 is approximately 1.2 fm. The fact that the radius of a spherical nucleus is proportional to $A^{1/3}$ implies that the volume of the nucleus is proportional to A. Because the mass of the nucleus is also approximately proportional to A, the densities of all nuclei are approximately the same. This is analogous to a drop of liquid, which also has constant density independent of its size. The **liquid-drop model** of the nucleus has proved quite successful in explaining nuclear behavior, especially the fission of heavy nuclei.

---

* The word *nucleon* refers to either a neutron or a proton that is part of a nucleus.

N AND *Z* NUMBERS

For light nuclei, the greatest stability is achieved when the numbers of protons and neutrons are approximately equal, $N \approx Z$. For heavier nuclei, instability caused by the electrostatic repulsion between the protons is minimized when there are more neutrons than protons. We can see this by looking at the N and Z numbers for the most abundant isotopes of some representative elements: for $^{16}_{8}O$, $N = 8$ and $Z = 8$; for $^{40}_{20}Ca$, $N = 20$ and $Z = 20$; for $^{56}_{26}Fe$, $N = 30$ and $Z = 26$; for $^{207}_{82}Pb$, $N = 125$ and $Z = 82$; and for $^{238}_{92}U$, $N = 146$ and $Z = 92$. (The atomic number Z has been included here as a subscript of the atomic symbol for emphasis. It is not actually needed because the atomic number is implied by the atomic symbol.)

Figure 40-1 shows a plot of N versus Z for the known stable nuclei. The curve follows the straight line $N = Z$ for small values of N and Z. We can understand this tendency for N and Z to be equal by considering the total energy of A particles in a one-dimensional box. For $A = 8$, Figure 40-2 shows the energy levels for eight neutrons and for four neutrons and four protons. Because of the exclusion principle, only two identical particles (that have opposite spins) can be in the same space state. Because protons and neutrons are not identical, we can put two each in a state, as shown in Figure 40-2*b*. Thus, the total energy for four protons and four neutrons is less than the total energy for eight neutrons (or eight protons), as shown in Figure 40-2*a*. When the Coulomb energy of repulsion, which is proportional to Z^2, is included, this result changes somewhat. For large values of A and Z, the total energy may be increased less by adding two neutrons than by adding one neutron and one proton because of the electrostatic repulsion involved in the latter case. This explains why $N > Z$ for the larger values of A (for the heavier nuclei).

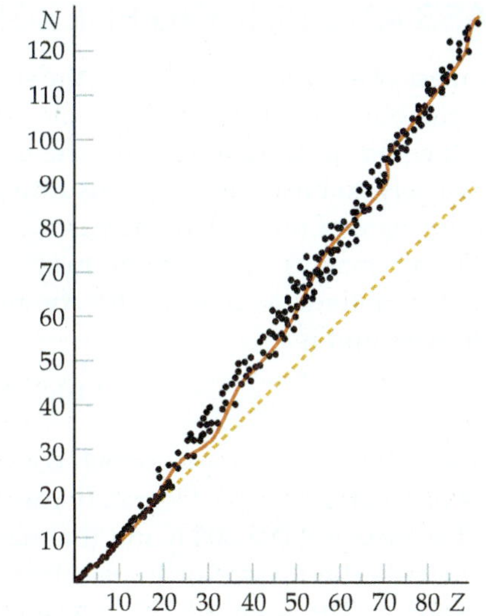

FIGURE 40-1 Plot of number of neutrons N versus number of protons Z for the stable nuclides. The dashed line is $N = Z$.

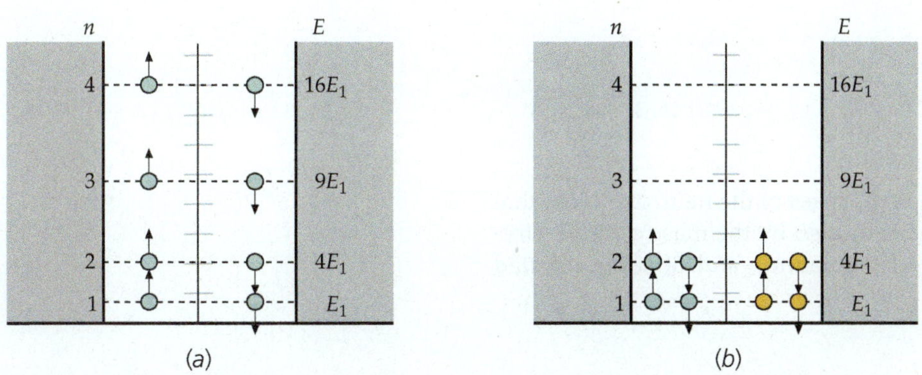

FIGURE 40-2 (*a*) Eight neutrons in a one-dimensional box. In accordance with the exclusion principle, only two neutrons (that have opposite spins) can be in a given energy level. (*b*) Four neutrons and four protons in a one-dimensional box. Because protons and neutrons are not identical particles, two of each can be in the same energy level. The total energy is much less for this case than for the case shown in Figure 40-2*a*.

PRACTICE PROBLEM 40-1

(*a*) Calculate the total energy of the eight neutrons in the one-dimensional box shown in Figure 40-2*a*. (*b*) Calculate the total energy of the four neutrons and four protons in the one-dimensional box shown in Figure 40-2*b*.

MASS AND BINDING ENERGY

The mass of a nucleus is less than the sum of the masses of its parts by E_b/c^2, where E_b is the binding energy and c is the speed of light. When two or more nucleons fuse together to form a nucleus, the total mass decreases and energy is released. Conversely, to break up a nucleus into its parts, energy is absorbed by the system and the mass of the system increases.

Atomic masses and nuclear masses are often given in unified atomic mass units (u), defined as one-twelfth the mass of a ^{12}C atom. The rest energy of one such mass unit is

$$(1\text{ u})c^2 = 931.5\text{ MeV} \qquad \text{40-2}$$

Consider ^{4}He, for example, which consists of two protons and two neutrons. The mass of an atom can be accurately measured in a mass spectrometer. The mass of the ^{4}He atom is 4.002 603 u and the mass of the ^{1}H atom is 1.007 825 u. These values include the masses of the electrons in the atom. The mass of the neutron is 1.008 665 u. The sum of the masses of two ^{1}H atoms and two neutrons is 2(1.007 825 u) + 2(1.008 665 u) = 4.032 980 u, which is greater than the mass of the ^{4}He atom by 0.030 377 u.* We can find the binding energy of the ^{4}He nucleus from this mass difference of 0.030 377 u by using the mass conversion factor $(1\text{ u})c^2 = 931.5$ MeV from Equation 40-2. Then

$$(0.030\,377\text{ u})c^2 = (0.030\,377\text{ u})c^2 \times \frac{931.5\text{ MeV}/c^2}{1\text{ u}} = 28.30\text{ MeV}$$

The total binding energy of ^{4}He is thus 28.30 MeV. In general, the binding energy of a nucleus of an atom of atomic mass M_A having Z protons and N neutrons is found by calculating the difference between the sum of the masses of the nucleons and the mass of the nucleus and then multiplying by c^2:

$$E_b = (ZM_H + Nm_n - M_A)c^2 \qquad \text{40-3}$$

TOTAL NUCLEAR BINDING ENERGY

where M_H is the mass of the ^{1}H atom and m_n is the mass of the neutron. (Note that the mass of the Z electrons in the term ZM_H is canceled by the mass of the Z electrons in the term M_A.†) The atomic masses of the neutron and of some selected isotopes are listed in Table 40-1.

Example 40-1 Binding Energy of the Last Neutron

Find the binding energy of the last neutron in a ^{4}He nucleus.

PICTURE The binding energy is energy equivalent of the mass of a ^{3}He atom plus the mass of a neutron minus the mass of a ^{4}He atom. We find the masses from Table 40-1 and multiply by c^2 to obtain the energy equivalents.

---

* Note that by using the masses of two ^{1}H atoms rather than two protons, the masses of the electrons in the atom are accounted for. We do this because it is atomic masses, not nuclear masses, that are measured directly and listed in mass tables.

† The mass associated with the binding energies of the electrons are not accounted for in this calculation.

SOLVE

1. Add the mass of the neutron to that of ^{3}He:

$$m_{^3\text{He}} + m_\text{n} = 3.016\,030\text{ u} + 1.008\,665\text{ u} = 4.024\,695\text{ u}$$

2. Subtract the mass of ^{4}He from the result:

$$\Delta m = (m_{^3\text{He}} + m_\text{n}) - m_{^4\text{He}} = 4.024\,695\text{ u} - 4.002\,603\text{ u} = 0.022\,092\text{ u}$$

3. Multiply this mass difference by c^2 and convert to MeV:

$$E_\text{b} = (\Delta m)c^2 = (0.022\,092\text{ u})c^2 \times \frac{931.5\text{ MeV}/c^2}{1\text{ u}} = \boxed{20.58\text{ MeV}}$$

CHECK As expected, the step-3 result of 20.58 MeV is less than the total binding energy of a ^{4}He nucleus. (The total binding energy of a ^{4}He nucleus is 28.30 MeV, a value that is calculated preceding Equation 40-3.)

Table 40-1 Atomic Masses of the Neutron and Selected Isotopes*

Element	Symbol	Z	Atomic mass, u
Neutron	n	0	1.008 665
Hydrogen			
Protium	^{1}H	1	1.007 825
Deuterium	^{2}H or D	1	2.014 102
Tritium	^{3}H or T	1	3.016 050
Helium	^{3}He	2	3.016 030
	^{4}He	2	4.002 603
Lithium	^{6}Li	3	6.015 125
	^{7}Li	3	7.016 004
Boron	^{10}B	5	10.012 939
Carbon	^{12}C	6	12.000 000
	^{13}C	6	13.003 354
	^{14}C	6	14.003 242
Nitrogen	^{13}N	7	13.005 738
	^{14}N	7	14.003 074
Oxygen	^{16}O	8	15.994 915
Sodium	^{23}Na	11	22.989 771
Potassium	^{39}K	19	38.963 710
Iron	^{56}Fe	26	55.939 395
Copper	^{63}Cu	29	62.929 592
Silver	^{107}Ag	47	106.905 094
Gold	^{197}Au	79	196.966 541
Lead	^{208}Pb	82	207.976 650
Polonium	^{212}Po	84	211.989 629
Radon	^{222}Rn	86	222.017 531
Radium	^{226}Ra	88	226.025 360
Uranium	^{238}U	92	238.048 608
Plutonium	^{242}Pu	94	242.058 725

* Mass values obtained at <http://physics.nist.gov/PhysRefData/Compositions/index.html>.

Figure 40-3 shows the binding energy per nucleon E_b/A versus A. The mean value is approximately 8.3 MeV. The flatness of the curve for $A > 50$ shows that E_b is approximately proportional to A. This indicates that there is saturation of nuclear forces in the nucleus as would be the case if each nucleon were attracted only to its nearest neighbors. Such a situation also leads to a constant nuclear density consistent with the measurements of the radius. If, for example, there were no saturation and each nucleon bonded to each other nucleon, there would be $A - 1$ bonds for each nucleon and a total of $A(A - 1)$ bonds altogether. The total binding energy, which is a measure of the energy needed to break all these bonds, would then be proportional to $A(A - 1)$, and E_b/A would not be approximately constant. The steep rise in the curve for low A is due to the increase in the number of nearest neighbors and therefore to the increased number of bonds per nucleon. The gradual decrease at high A is due to the Coulomb repulsion of the protons, which increases as Z^2 and decreases the binding energy. For very large A, this Coulomb repulsion is so great that a nucleus that has an A greater than approximately 300 is unstable and undergoes spontaneous fission.

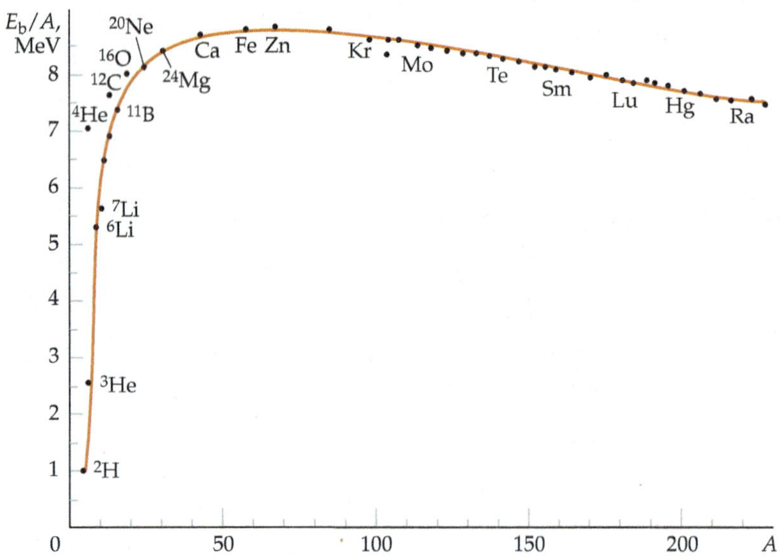

FIGURE 40-3 The binding energy per nucleon versus the nucleon number A. For nuclei that have values of A greater than 50, the curve is approximately constant, indicating that the total binding energy is approximately proportional to A.

40-2 RADIOACTIVITY

Many nuclei are radioactive; that is, they decay into other nuclei by the emission of particles such as photons, electrons, neutrons, or α particles. The terms α decay, β decay, and γ decay were used before it was discovered that α particles are ^{4}He nuclei, β particles are either electrons (β^-) or positrons* (β^+), and γ rays are photons. The rate of decay of a radioactive sample decreases exponentially with increasing time. *This exponential time dependence is characteristic of all radioactivity and indicates that radioactive decay is a statistical process.* Because each nucleus is well shielded from others by the atomic electrons, pressure and temperature changes have little or no effect on the rate of radioactive decay or other nuclear properties.

* The positron has the same mass as an electron and it has a charge of $+e$.

Let N be the number of radioactive nuclei at some time t. If the decay of an individual nucleus is a random event, we expect the number of nuclei that decay in some time interval dt to be proportional both to N and to dt. Because of these decays, the number N will decrease. The change in N between time t and time $t + dt$ is given by

$$dN = -\lambda N\,dt \tag{40-4}$$

where λ is a constant of proportionality called the **decay constant.** The rate of change of N, dN/dt, is proportional to N. This is characteristic of exponential decay. To solve Equation 40-4 for N, we first divide each side by N, thus separating the variables N and t:

$$\frac{dN}{N} = -\lambda\,dt$$

Integrating, we obtain

$$\int_{N_0}^{N'} \frac{dN}{N} = -\lambda \int_0^{t'} dt$$

or

$$\ln\frac{N'}{N_0} = -\lambda t' \tag{40-5}$$

where N' is the number of nuclei that remain at time t'. For convenience, we drop the primes from N' and t'. This introduces no ambiguity because the parameters N and t have been integrated out of the equation. Taking the exponential of each side, we obtain

$$\frac{N}{N_0} = e^{-\lambda t}$$

or

$$N = N_0 e^{-\lambda t} \tag{40-6}$$

See **Math Tutorial** *for more information on*
Exponential Functions

The number of radioactive decays per second is called the **decay rate** R:

$$R = -\frac{dN}{dt} = \lambda N = \lambda N_0 e^{-\lambda t} = R_0 e^{-\lambda t} \tag{40-7}$$

DECAY RATE

where

$$R_0 = \lambda N_0 \tag{40-8}$$

is the decay rate at time $t = 0$. The decay rate R is the quantity that is determined experimentally. The decay rate is also called the **activity** of the sample.

The average or **mean lifetime** τ is equal to the reciprocal of the decay constant (see Problem 40):

$$\tau = \frac{1}{\lambda} \tag{40-9}$$

The mean lifetime is analogous to the time constant in the exponential decrease in the charge on a capacitor in an *RC* circuit that we discussed in Section 25-6. After a time equal to the mean lifetime, the number of radioactive nuclei and the decay rate are each equal to $e^{-1} = 37$ percent of their original values. The **half-life** $t_{1/2}$ is defined as the time it takes for the number of nuclei and the decay rate to decrease by half. Setting $t = t_{1/2}$ and $N = N_0/2$ in Equation 40-6 gives

$$\frac{N_0}{2} = N_0 e^{-\lambda t_{1/2}} \qquad \text{40-10}$$

or

$$e^{+\lambda t_{1/2}} = 2$$

Solving for $t_{1/2}$ gives

$$t_{1/2} = \frac{\ln 2}{\lambda} = (\ln 2)\tau = 0.693\tau \qquad \text{40-11}$$

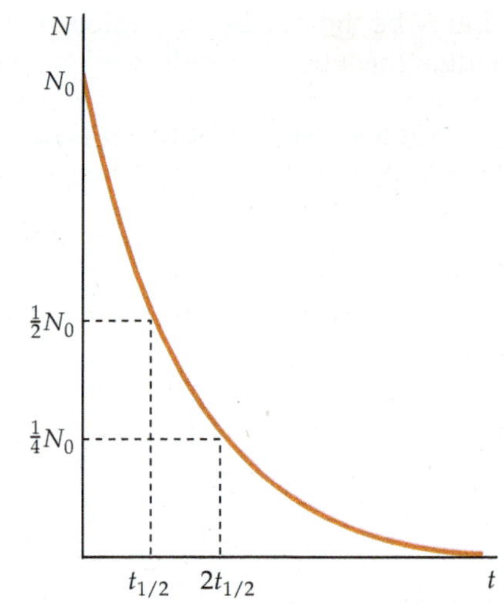

FIGURE 40-4 Exponential radioactive decay. After each half-life $t_{1/2}$, the number of nuclei remaining has decreased by one-half. The decay rate $R = \lambda N$ has the same time dependence as does N.

Figure 40-4 shows a plot of N versus t. If we multiply the numbers on the N axis by λ, this graph becomes a plot of R versus t. After each time interval of one half-life, both the number of nuclei left and the decay rate have decreased to half of their previous values. For example, if the decay rate is R_0 initially, it will be $\frac{1}{2}R_0$ after one half-life, $\left(\frac{1}{2}\right)\left(\frac{1}{2}\right)R_0$ after two half-lives, and so forth. After n half-lives, the decay rate will be

$$R = \left(\tfrac{1}{2}\right)^n R_0 \qquad \text{40-12}$$

The half-lives of radioactive nuclei vary from very small times (less than 1 μs) to very large times (greater than 10^{10} y).

Example 40-2 Counting Rate for Radioactive Decay

A radioactive source has a half-life of 1.0 min. At time $t = 0$, the radioactive source is placed near a detector, and the counting rate (the number of decay particles detected per unit time) is observed to be 2000 counts/s. Find the counting rate at times $t = 1.0$ min, $t = 2.0$ min, $t = 3.0$ min, and $t = 10$ min.

PICTURE The counting rate r is proportional to the decay rate R, and the decay rate is given by $R = \left(\frac{1}{2}\right)^n R_0$ (Equation 40-12), where n is the time divided by 1.0 min.

SOLVE

1. Because the half-life is 1.0 min, the counting rate will be half as great at $t = 1.0$ min as at $t = 0$:

$$r_1 = \tfrac{1}{2}r_0 = \tfrac{1}{2}(2000 \text{ counts/s}) = \boxed{1.0 \times 10^3 \text{ counts/s at } 1.0 \text{ min}}$$

2. At $t = 2.0$ min, the rate is half that at 1 min. It decreases by one-half each minute:

$$r_2 = \left(\tfrac{1}{2}\right)^2 r_0 = \tfrac{1}{4}(2000 \text{ counts/s}) = \boxed{5.0 \times 10^2 \text{ counts/s at } 2.0 \text{ min}}$$

$$r_3 = \left(\tfrac{1}{2}\right)^3 r_0 = \tfrac{1}{8}(2000 \text{ counts/s}) = \boxed{2.5 \times 10^2 \text{ counts/s at } 3.0 \text{ min}}$$

3. At $t = 10$ min, the rate will be $\left(\frac{1}{2}\right)^{10}$ multiplied by the initial rate:

$$r_{10} = \left(\tfrac{1}{2}\right)^{10} r_0 = \tfrac{1}{1024}(2000 \text{ counts/s})$$
$$= 1.95 \text{ counts/s}$$
$$\approx \boxed{2.0 \text{ counts/s at } 10 \text{ min}}$$

CHECK As expected, the counting rate decreases as the number of minutes increases.

CONCEPT CHECK 40-1

A radioactive isotope has a half-life of 10 s. You are observing a sample of this isotope. After approximately one minute of observation, there is only one atom of this isotope left in your sample. How many atoms of this isotope will be left in your sample 15 s later?

Example 40-3 Detection-Efficiency Considerations

If the detection efficiency in Example 40-2 is 20 percent, (*a*) how many radioactive nuclei are there at time $t = 0$ and (*b*) at time $t = 1.0$ min? (c) How many nuclei decay in the first minute?

PICTURE The detection efficiency depends on the probability that a radioactive decay particle will enter the detector and the probability that upon entering the detector it will produce a count. If the efficiency is 20 percent, the decay rate must be five times the counting rate.

SOLVE

(*a*) 1. The number of radioactive nuclei is related to the decay rate R and the decay constant λ:

$$R = \lambda N$$

2. The decay constant is related to the half-life:

$$\lambda = \frac{\ln 2.0}{t_{1/2}} = \frac{0.693}{1.0\text{ min}} = 0.693\text{ min}^{-1}$$

3. Because the detection efficiency is 20 percent, the decay rate is five times the counting rate. Calculate the initial decay rate:

$$R_0 = (5\text{ decays/count}) \times (2000\text{ counts/s}) = 1.0 \times 10^4\text{ decays/s}$$

4. Substitute to calculate the initial number of radioactive nuclei N_0 at $t = 0$:

$$N_0 = \frac{R_0}{\lambda} = \frac{1.0 \times 10^4\text{ s}^{-1}}{0.693\text{ min}^{-1}} \times \frac{60\text{ s}}{1\text{ min}}$$

$$= 8.66 \times 10^5 = \boxed{8.7 \times 10^5}$$

(*b*) At time $t = 1\text{ min} = t_{1/2}$, there are half as many radioactive nuclei as at $t = 0$:

$$N_1 = \tfrac{1}{2}(8.66 \times 10^5) = 4.33 \times 10^5 = \boxed{4.3 \times 10^5}$$

(*c*) The number of nuclei that decay in the first minute is $N_0 - N_1$:

$$\Delta N = N_0 - N_1 = 8.66 \times 10^5 - 4.33 \times 10^5 = \boxed{4.3 \times 10^5}$$

CHECK The results for Parts (*b*) and (*c*) are equal, as expected. At the end of one half-life, half of the nuclei have decayed and the other half remain.

The SI unit of radioactive decay is the **becquerel** (Bq), which is defined as one decay per second:

$$1\text{ Bq} = 1\text{ decay/s} \qquad 40\text{-}13$$

A historical unit that applies to all types of radioactivity is the **curie** (Ci), which is defined as

$$1\text{ Ci} = 3.7 \times 10^{10}\text{ decays/s} = 3.7 \times 10^{10}\text{ Bq} \qquad 40\text{-}14$$

The curie is the rate at which radiation is emitted by 1 g of radium. Because this is a very large unit, the millicurie (mCi) or microcurie (μCi) are often used.

BETA DECAY

Beta decay occurs in nuclei that have too many neutrons or too few neutrons for stability. During β decay, A remains the same while Z either increases by 1 (β^- decay) or decreases by 1 (β^+ decay).

An example of β decay is the decay of a free neutron into a proton and an electron. (The half-life of a free neutron is about 10.8 min.) The energy of β decay is 0.782 MeV, which is the difference between the rest energy of the neutron and the rest energy of the proton and an electron. More generally, during β^- decay, a nucleus of mass number A and atomic number Z decays into a nucleus, referred to as the **daughter nucleus,** of mass number A and atomic number $Z' = Z + 1$ and an electron is emitted. (The original nucleus is called the **parent.**) If the decay energy were shared by only the daughter nucleus and the emitted electron, the energy of the electron would be uniquely determined by the conservation of energy and momentum. Experiments show, however, the energies of the electrons emitted during the β^- decay of a nucleus are observed to vary from zero to the maximum energy available. A typical energy spectrum for the electrons is shown in Figure 40-5.

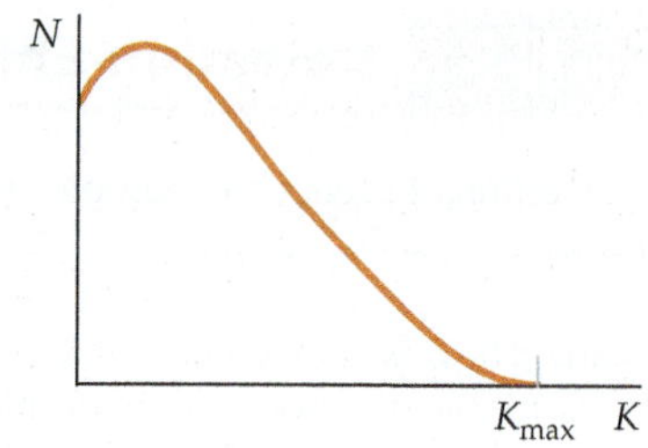

FIGURE 40-5 Number of electrons emitted during β^- decay versus kinetic energy. The fact that all the electrons do not have the same energy K_{max} suggests that another particle, one that shares the energy available for decay, is emitted.

To explain the fact that energy seemed not to be conserved during β decay, Wolfgang Pauli in 1930 suggested that a third particle, which he called the **neutrino,** is also emitted. Because the measured maximum energy of the emitted electrons is equal to the total available for the decay, the rest energy and therefore the mass of the neutrino was assumed to be zero. (It is now known that the mass of the neutrino is very small but not zero.) In 1948, measurements of the momenta of the emitted electron and the recoiling nucleus showed that the neutrino was also needed for the conservation of linear momentum during β decay. The neutrino was first observed experimentally in 1957. It is now known that there are at least three kinds of neutrinos, one (ν_e) associated with electrons, one (ν_μ) associated with muons, and one (ν_τ) associated with the tau particle, τ. Moreover, each neutrino has an antiparticle, written $\bar{\nu}_e$, $\bar{\nu}_\mu$, and $\bar{\nu}_\tau$. It is the electron antineutrino that is emitted during the decay of a neutron, which is written*

$$\mathrm{n} \rightarrow \mathrm{p} + \mathrm{e}^- + \bar{\nu}_\mathrm{e} \qquad 40\text{-}15$$

During β^+ decay, a proton changes into a neutron, and a positron (and a neutrino) is emitted. A free proton cannot decay by positron emission because of conservation of energy (the mass of the neutron and the positron is greater than the mass of the proton); however, because of binding-energy effects, a proton inside a nucleus can decay. A typical β^+ decay is

$$^{13}_{7}\mathrm{N} \rightarrow {}^{13}_{6}\mathrm{C} + \mathrm{e}^+ + \nu_\mathrm{e} \qquad 40\text{-}16$$

! Do not confuse the symbols e^- and e^+ with the symbol e. The symbols e^- and e^+ denote particles (the electron and the positron), whereas the symbol e denotes an amount of charge.

The electrons or the positrons emitted during β decay do not exist inside the nucleus. They are created during the process of decay, just as photons are created when an atom makes a transition from a higher energy state to a lower energy state.

An important example of β decay is that of ^{14}C, which is used in radioactive carbon dating:

$$^{14}\mathrm{C} \rightarrow {}^{14}\mathrm{N} + \mathrm{e}^- + \bar{\nu}_\mathrm{e} \qquad 40\text{-}17$$

The half-life for this decay is 5730 y. The radioactive isotope ^{14}C is produced in the upper atmosphere during nuclear reactions caused by cosmic rays. The chemical reactivity of a carbon atom that has a ^{14}C nucleus is the same as the chemical reactivity of a carbon atom that has a ^{12}C nucleus. For example, atoms that have these nuclei combine with oxygen to form CO_2 molecules. Because living organisms continually exchange CO_2 with the atmosphere, the ratio of ^{14}C to ^{12}C in a living organism is the same as the equilibrium ratio in the atmosphere, which is about 1.3×10^{-12}. After an organism dies, it no longer absorbs ^{14}C from the atmosphere, so the ratio of ^{14}C to ^{12}C continually decreases due to the radioactive decay of ^{14}C. The number of ^{14}C decays per minute per gram of carbon in a living organism can be calculated from the known half-life of ^{14}C and the number of ^{14}C nuclei in a gram of carbon. The result is that there are approximately 15.0 decays per minute per gram of carbon in a living organism. Using this result and the measured number of decays per minute per gram of carbon in a nonliving sample of bone, wood, or other object having carbon, we can determine the age of the sample. For example, if the measured rate were 7.5 decays per minute per gram, the age of the sample would be one half-life = 5730 years.

* This reaction is also written $\mathrm{n} \rightarrow \mathrm{p} + \beta^- + \bar{\nu}_\mathrm{e}$.

Example 40-4 How Old Is the Artifact? *Context-Rich*

You have a summer job working in an archeological research lab. Your supervisor calls to tell you that they found a new bone at their current site and asks you to determine the age of the bone from a sample that she will send you. When the bone sample arrives, you take a section that contains 200 grams of carbon and you find a beta decay rate of 400 decays/min.

PICTURE There are approximately 15.0 decays per minute per gram of carbon in a living organism, and the half-life of carbon-14 is 5730 y. We need to determine the number of half-lives that have occurred since the death of the organism. We do this by using the equality $R_n = (1/2)^n R_0$ (Equation 40-12), where R_n is the current decay rate, R_0 is the initial decay rate, and n is the number of half-lives. We can determine the initial decay rate by multiplying the decay rate per gram by the mass of the carbon of the sample.

SOLVE

1. Write the decay rate after n half-lives in terms of the initial decay rate:
$$R_n = \left(\tfrac{1}{2}\right)^n R_0$$

2. Calculate the initial decay rate (the decay for 200 g of carbon when the organism died):
$$R_0 = [(15\ \text{decays/min})/\text{g}](200\ \text{g}) = 3000\ \text{decays/min}$$

3. Substitute the values for R_0 and R_n into the step-1 equation and solve for n:
$$R_n = \left(\tfrac{1}{2}\right)^n R_0$$
$$400\ \text{decays/min} = \left(\tfrac{1}{2}\right)^n 3000\ \text{decays/min}$$
$$\left(\tfrac{1}{2}\right)^n = \frac{400}{3000}$$
$$2^n = \frac{3000}{400} = 7.5$$

4. We solve for n by taking the logarithm of each side:
$$n \ln 2 = \ln 7.5 \quad \Rightarrow \quad n = \frac{\ln 7.5}{\ln 2} = 2.91$$

5. The age of the bone is $nt_{1/2}$:
$$t = nt_{1/2} = 2.91(5730\ \text{y}) = \boxed{1.67 \times 10^4\ \text{y}}$$

CHECK If the bone were from a recently living organism, we would expect the decay rate to be a steady [(15 decays/min)/g](200 g) = 3000 decays/min. The current decay rate is given as 400 decays/min. Because 400/3000 is roughly 1/8 (actually 1/7.5), the sample must be approximately three half-lives old, which is about 3(5730 y). This is in agreement with the step-5 result of 2.91(5730 y).

PRACTICE PROBLEM 40-2 The Check of Example 40-4 states, "Because 400/3000 is roughly 1/8 (actually 1/7.5), the sample must be approximately three half-lives old. . . ." Explain why this ratio of 1/8 implies an age equal to three half-lives.

GAMMA DECAY

During γ decay, a nucleus in an excited state decays to a lower-energy state by the emission of a photon. This process is the nuclear counterpart of spontaneous emission of photons by atoms and molecules. Unlike β decay or α decay, neither the mass number A nor the atomic number Z change during γ decay. Because the spacing of the nuclear energy levels is of the order of 1 MeV (as compared with spacing of the order of 1 eV in atoms), the wavelengths of the emitted photons are of the order of 1 pm (1 pm = 10^{-12} m):

$$\lambda = \frac{hc}{E} \approx \frac{1240\ \text{eV}\cdot\text{nm}}{1\ \text{MeV}} = 0.00124\ \text{nm} = 1.24\ \text{pm}$$

The mean lifetime for γ decay is often very short. It is usually observed only because it follows either α decay or β decay. For example, if a radioactive parent nucleus decays by β decay to an excited state of the daughter nucleus, the daughter nucleus then decays to its ground state by γ emission. Direct measurements of mean lifetimes as short as approximately 10^{-11} s are possible. Measurements of mean lifetimes shorter than 10^{-11} s are difficult, but they can sometimes be made by indirect methods.

A few γ emitters have very long lifetimes, of the order of hours. The energy states that do have such long lifetimes are called **metastable states.**

ALPHA DECAY

All very heavy nuclei ($Z > 83$) are potentially unstable via α decay because the mass of the original radioactive nucleus is greater than the sum of the masses of the decay products—an α particle and the daughter nucleus. Consider the decay of ^{232}Th ($Z = 90$) into ^{228}Ra ($Z = 88$) and an α particle. This process is written as

$$^{232}\text{Th} \longrightarrow {}^{228}\text{Ra} + \alpha = {}^{228}\text{Ra} + {}^{4}\text{He} \qquad 40\text{-}18$$

The mass of the ^{232}Th atom is 232.038 050 u. The mass of the daughter atom ^{228}Ra is 228.031 064 u. Adding 4.002 603 u (the mass of ^{4}He) to the mass of ^{228}Ra, we get 232.033 667 u for the total mass of the decay products. This value is less than the mass of ^{232}Th by 0.004 383 u, which multiplied by 931.5 MeV/c^2 gives 4.08 MeV/c^2 for the excess mass of ^{232}Th when compared to the total mass of the decay products. The isotope ^{232}Th is therefore potentially unstable to α decay. This decay does in fact occur in nature with the emission of an α particle of kinetic energy 4.08 MeV. (The kinetic energy of the α particle is actually somewhat less than 4.08 MeV because some of the released energy is taken up by the recoiling ^{228}Ra nucleus.)

When a nucleus emits an α particle, both N and Z decrease by 2 and A decreases by 4. The daughter of a radioactive nucleus is often itself radioactive and decays by either α decay or β decay or both. If the original nucleus has a mass number A that is 4 times an integer, the daughter nucleus and all those in the decay chain will also have mass numbers equal to 4 multiplied by an integer. Similarly, if the mass number of the original nucleus is $4n + 1$, where n is an integer, all the nuclei in the decay chain will have mass numbers given by $4n + 1$, where n decreases by one at each α decay. We can see, therefore, that there are four possible α-decay chains, depending on whether A equals $4n$, $4n + 1$, $4n + 2$, or $4n + 3$, where n is an integer. All but one of these decay chains are found on Earth. The $4n + 1$ series is not found because its longest-lived member (other than the stable end product ^{209}Bi) is ^{237}Np, which has a half-life of only 2×10^6 y. Because this period is much less than the age of Earth, this series has disappeared.

Figure 40-6 shows the thorium series, for which $A = 4n$. It begins with an α decay from ^{232}Th to ^{228}Ra. The daughter nuclide of an α decay is on the left or neutron-rich side of the stability curve (the dashed line in the figure), so it often decays by β^- decay. In the thorium series, ^{228}Ra decays by β^- decay to ^{228}Ac, which in turn decays by β^- decay to ^{228}Th. There are then four α decays to ^{212}Pb, which decays by β^- decay to ^{212}Bi. The series branches at ^{212}Bi, which decays either by α decay to ^{208}Tl or by β^- decay to ^{212}Po. The branches meet at the stable lead isotope ^{208}Pb.

The energies of α particles from natural radioactive sources range from approximately 4 MeV to 7 MeV, and the half-lives of the sources range from approximately 10^{-5} s to 10^{10} y. In general, the smaller the energy of the emitted α particle, the longer the half-life. As we discussed in Section 35-4, the enormous variation

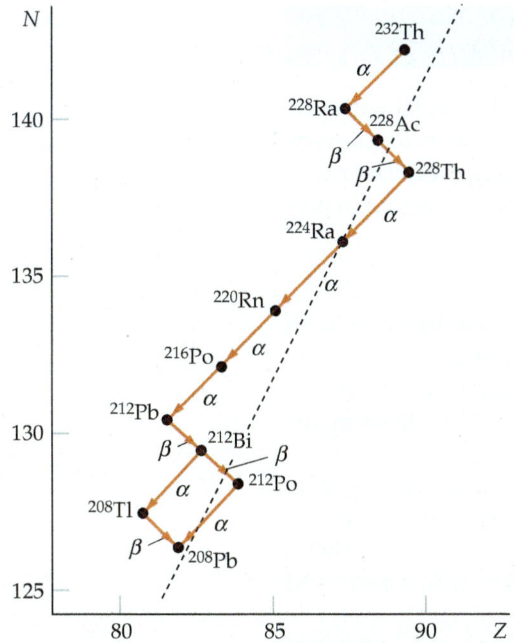

FIGURE 40-6 The thorium (4*n*) α decay series. The dashed line is the curve of stability.

in half-lives was explained by George Gamow in 1928. He considered α decay to be a process in which an α particle is first formed inside a nucleus and then tunnels through the Coulomb barrier (Figure 40-7). A slight increase in the energy of the α particle reduces the relative height $U_{max} - E$ of the barrier and also the thickness $r_1 - R$. Because the probability of penetration is so sensitive to the relative height and thickness of the barrier, a small increase in E leads to a large increase in the probability of barrier penetration and therefore to a significantly shorter lifetime. Gamow was able to derive an expression for the half-life as a function of E that is in excellent agreement with experimental results.

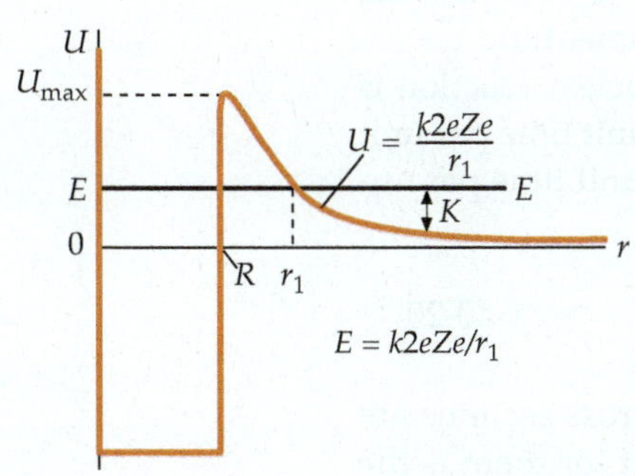

FIGURE 40-7 A model of the potential energy for an α particle and a nucleus. The strong attractive nuclear force that exists for values of r less than the nuclear radius R is indicated by the potential well. Outside the nucleus, the nuclear force is negligible, and the potential energy is given by the Coulomb potential energy function $U = +k2eZe/r$, where Ze is the nuclear charge and $2e$ is the charge of the α particle. The kinetic energy K of the α particle is equal to the energy E when the α particle is far away from the nucleus. A small increase in E reduces the relative height $U_{max} - E$ of the barrier and also reduces its thickness $r_1 - R$, leading to a much greater probability of penetration. An increase in the energy of the emitted α particles by a factor of 2 results in a reduction of the half-life by a factor of more than 10^{20}.

40-3 NUCLEAR REACTIONS

Information about nuclei is typically obtained by bombarding the nuclei with various particles and observing the results. Although the first experiments of this type were limited by the need to use naturally occurring radiation, they produced many important discoveries. In 1932, J. D. Cockcroft and E. T. S. Walton succeeded in producing the reaction

$$\mathrm{p} + {}^{7}\mathrm{Li} \rightarrow {}^{8}\mathrm{Be} \rightarrow {}^{4}\mathrm{He} + {}^{4}\mathrm{He}$$

using artificially accelerated protons. At about the same time, the Van de Graaff electrostatic generator (by R. Van de Graaff in 1931) and the first cyclotron (by E. O. Lawrence and M. S. Livingston in 1932) were built. Since then, enormous advances in the technology for accelerating and detecting particles have been made, and many nuclear reactions have been studied.

When a particle is incident on a nucleus, several different things can happen. The incident particle may be scattered, either elastically or inelastically, or the incident particle may be absorbed by the nucleus, and another particle or particles may be emitted. In inelastic scattering, the nucleus is left in an excited state and subsequently decays by emitting photons (or other particles).

The amount of energy released or absorbed during a reaction (in the center of mass reference frame) is called the **Q value** of the reaction. The Q value equals c^2 multiplied by the mass difference. When energy is released during a reaction, the reaction is said to be an **exothermic reaction.** During an exothermic reaction, the total mass of the incoming particles is greater than the total mass of the outgoing particles, and the Q value is positive. If the total mass of the incoming particles is less than that of the outgoing particles, energy is required for the reaction to take place, and the reaction is said to be an **endothermic reaction.** The Q value of an endothermic reaction is negative. In general, if Δm is the change in mass, the Q value is

$$Q = -(\Delta m)c^2 \qquad 40\text{-}19$$

Q VALUE

An endothermic reaction cannot take place below a specific threshold energy. In the laboratory reference frame in which stationary particles are bombarded by incoming particles, the threshold energy is somewhat greater than $|Q|$ because the outgoing particles must have some kinetic energy to conserve momentum.

A measure of the effective size of a nucleus for a particular nuclear reaction is the **cross section** σ. If I is the number of the incident particles per unit time per unit area (the incident intensity) and R is the number of reactions per unit time per nucleus, the cross section is

$$\sigma = \frac{R}{I} \qquad 40\text{-}20$$

The cross section σ has the dimensions of area. Because nuclear cross sections are of the order of the square of the nuclear radius, a convenient unit for them is the **barn,** which is defined as

$$1\ \text{barn} = 10^{-28}\ \text{m}^2 \qquad 40\text{-}21$$

The cross section for a particular reaction is a function of energy. For an endothermic reaction, it is zero for energies below the threshold energy.

Example 40-5 Exothermic or Endothermic?

Find the Q value of the reaction $\mathrm{p} + {}^{7}\mathrm{Li} \rightarrow {}^{4}\mathrm{He} + {}^{4}\mathrm{He}$ and state whether the reaction is exothermic or endothermic.

PICTURE We find the masses of the atoms from Table 40-1 and calculate the difference in the total mass of the outgoing particles and the incoming particles. The Q value is related to the change in mass Δm by $Q = -(\Delta m)c^2$. If we use the mass of protium rather than the mass of the proton, there will be four electrons on each side of the reaction, so the electron masses will cancel.

SOLVE

1. Find the mass of each atom from Table 40-1:
   ^{1}H 1.007 825 u
   ^{7}Li 7.016 004 u
   ^{4}He 4.002 603 u
2. Calculate the initial mass m_i of the incoming particles:
   $m_i = 1.007\,825\text{ u} + 7.016\,004\text{ u} = 8.023\,829\text{ u}$
3. Calculate the final mass m_f:
   $m_f = 2(4.002\,603\text{ u}) = 8.005\,206\text{ u}$
4. Calculate the change in mass:
   $\Delta m = m_f - m_i = 8.005\,206\text{ u} - 8.023\,829\text{ u}$
   $= -0.018\,623\text{ u}$
5. Calculate the Q value:
   $$Q = -(\Delta m)c^2 = (+0.018\,623\text{ u})c^2 \times \frac{931.5\text{ MeV}}{1\text{ u}}$$
   $$= \boxed{17.35\text{ MeV}}$$
   Q is positive, so the reaction is exothermic.

CHECK Because the initial mass is greater than the final mass, the initial energy is greater than the final energy and the reaction is exothermic, yielding 17.35 MeV.

REACTIONS WITH NEUTRONS

Nuclear reactions that involve neutrons are important for understanding nuclear reactors. The most likely reaction between a nucleus and a neutron that has an energy of more than about 1 MeV is scattering. However, even if the scattering is elastic, the neutron loses some energy to the nucleus because the nucleus recoils. If a neutron is scattered many times in a material, its energy decreases until the neutron is of the order of the energy of thermal motion kT, where k is Boltzmann's constant and T is the absolute temperature. (At ordinary room temperatures, kT is approximately 0.025 eV.) The neutron is then equally likely to gain or lose energy from a nucleus when it is elastically scattered. A neutron that has an energy of the order of kT is called a **thermal neutron.**

At low energies, a neutron is likely to be captured, producing an excited nucleus. A γ ray is then emitted from the excited nucleus. Figure 40-8 shows the neutron-capture cross section for silver as a function of the energy of the neutron. The large

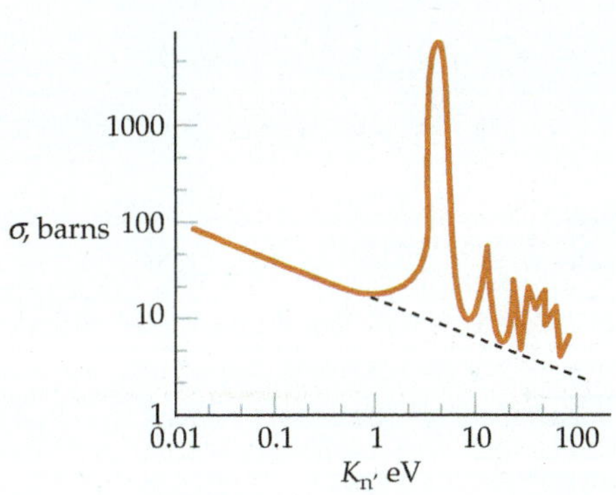

FIGURE 40-8 Neutron-capture cross section for silver as a function of the energy of the incident neutron. The straight line indicates the $1/v$ dependence of the cross section, which is proportional to the time spent by the neutron in the vicinity of the silver nucleus. Superimposed on this dependence are a large resonance and several smaller resonances.

peak in this curve is called a **resonance.** Except for the resonance, the cross section varies fairly smoothly with energy, decreasing with increasing energy roughly as $1/v$, where v is the speed of the neutron. We can understand this energy dependence as follows: Consider a neutron moving with speed v near a nucleus of diameter $2R$. The time it takes the neutron to pass the nucleus is $2R/v$. Thus, the neutron-capture cross section is proportional to the time spent by the neutron in the vicinity of the silver nucleus. The dashed line in Figure 40-8 indicates this $1/v$ dependence. At the maximum of the resonance, the value of the cross section is very large ($\sigma > 5000$ barns) compared with a value of only about 10 barns just past the resonance. Many elements show similar resonances in their neutron-capture cross sections. For example, the maximum cross section for ^{113}Cd is approximately 57 000 barns. This material is thus very useful for shielding against low-energy neutrons.

An important nuclear reaction that involves neutrons is fission, which is discussed in the next section.

40-4 FISSION AND FUSION

Figure 40-9 shows a plot of the nuclear mass difference per nucleon $(M - Zm_{\text{p}} - Nm_{\text{n}})/A$ in units of MeV/c^2 versus A. This curve is just the negative of the binding-energy curve shown in Figure 40-3. From Figure 40-9, we can see that the values for the mass difference per nucleon for both very heavy ($A \approx 200$) and very light ($A \leq 20$) nuclides are greater than the values for nuclides of intermediate mass. Thus, energy is released when a very heavy nucleus, such as ^{235}U, breaks up into two lighter nuclei—during a process called **fission**—or when two very light nuclei, such as ^{2}H and ^{3}H, fuse together to form a nucleus of greater mass—during a process called **fusion.**

The applications of both fission and fusion to the generation of electrical power and the development of nuclear weapons have had a profound effect on our lives since the early twentieth century. The application of these reactions to the development of energy resources may have an even greater effect in the future. We will look at some of the features of fission and fusion that are important for their application in reactors to generate power.

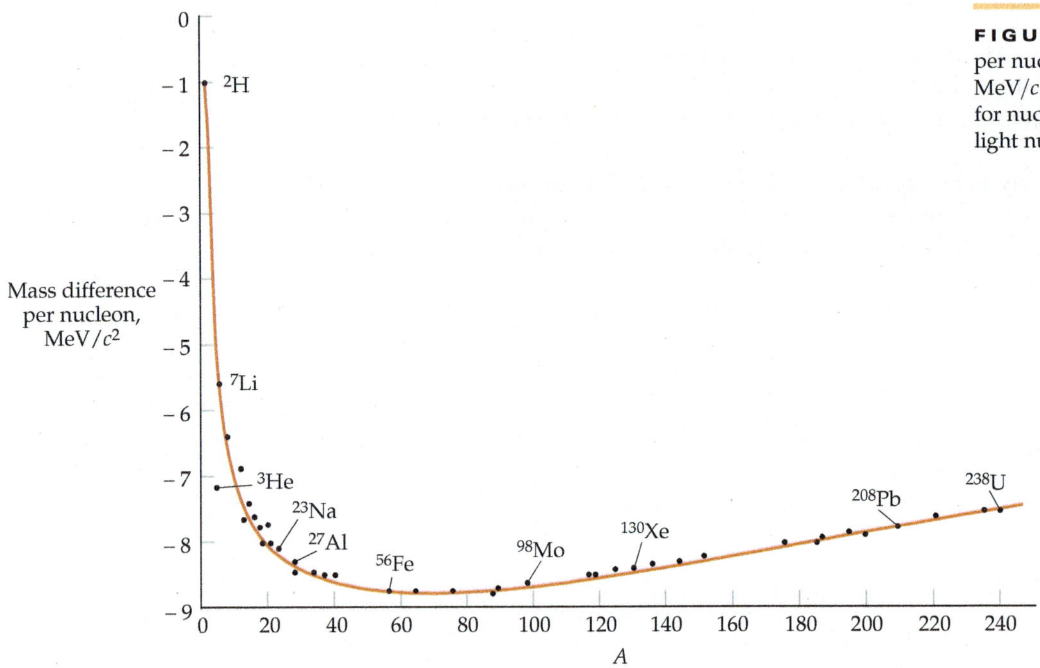

FIGURE 40-9 Plot of mass difference per nucleon $(M - Zm_{\text{p}} - Nm_{\text{n}})/A$ in units of MeV/c^2 versus A. The mass per nucleon is less for nuclei of intermediate mass than for very light nuclei or very heavy nuclei.

Hidden layers in paintings are analyzed by bombarding the painting with neutrons and observing the radiative emissions from nuclei that have captured a neutron. Different elements used in the painting have different half-lives. (*a*) Van Dyck's painting *Saint Rosalie Interceding for the Plague-Stricken of Palermo*. The black-and-white images in (*b*) and (*c*) were formed using a special film sensitive to electrons emitted by the radioactively decaying elements. Image (*b*), taken a few hours after the neutron irradiation, reveals the presence of manganese, found in umber, which is a dark earth pigment used for the painting's base layer. (Blank areas show where modern repairs, free of manganese, have been made.) The image in (*c*) was taken 4 days later, after the umber emissions had died away and when phosphorus, found in charcoal and boneblack, was the main radiating element. Upside down is revealed a sketch of Van Dyck himself. The self-portrait, executed in charcoal, had been overpainted by the artist. *((a) © 1991 by the Metropolitan Museum of Art. (b) and (c) Courtesy of Paintings Conservation Department, Metropolitan Museum of Art.)*

FISSION

Very heavy nuclei ($Z > 92$) are subject to spontaneous fission. They break apart into two nuclei even if the nuclei are not disturbed. We can understand this by considering the analogy of a charged liquid drop. If the drop is not too large, surface tension can overcome the repulsive forces of the charges and hold the drop together. There is, however, a certain maximum size beyond which the drop will be unstable and will spontaneously break apart. Because of spontaneous fission, an upper limit exists on the size of a nucleus and therefore on the number of elements that are possible.

Some heavy nuclei—uranium and plutonium, in particular—can be induced to fission by the capture of neutrons. During the fission of ^{235}U, for example, the uranium nucleus is excited by the capture of a neutron, causing it to split into two nuclei and emit several neutrons. The Coulomb force of repulsion drives the fission fragments apart, with the released energy eventually appearing as thermal energy. Consider, for example, the fission of a nucleus of mass number $A = 200$ into two nuclei of mass number $A = 100$. Because the rest energy for $A = 200$ is about 1 MeV per nucleon greater than that for $A = 100$, approximately 200 MeV per nucleus is released during such a fission. This is a large amount of energy. By contrast, during the chemical reaction of combustion, only about 4 eV of energy is released per molecule of oxygen consumed.

Example 40-6 Energy Released During the Fission of ^{235}U

Calculate the total energy (in kilowatt-hours) released during the fission of 1.00 g of ^{235}U, assuming that 200 MeV is released per fission.

PICTURE We need to find the number of uranium nuclei in one gram of ^{235}U, which we find using the fact that there are Avogadro's number ($N_A = 6.02 \times 10^{23}$) of nuclei in 235 grams.

SOLVE

1. The total energy is the number of nuclei multiplied by the energy per nucleus:

$$E = NE_{\text{nucleus}} = N(200\text{MeV/nucleus})$$

2. Calculate N:

$$N = \frac{6.02 \times 10^{23}\text{ nuclei/mol}}{235\text{ g/mol}} \times 1.00\text{ g}$$
$$= 2.56 \times 10^{21}\text{ nuclei}$$

3. Calculate the energy per gram in eV and convert to kW·h:

$$E = \frac{200 \times 10^{6}\text{ eV}}{1\text{ nucleus}} \times 2.56 \times 10^{21}\text{ nuclei}$$
$$= 5.12 \times 10^{29}\text{ eV} = 8.19 \times 10^{10}\text{ J}$$
$$= 8.19 \times 10^{7}\text{ kW}\cdot\text{s} = \boxed{2.28 \times 10^{4}\text{ kW}\cdot\text{h}}$$

The fission of uranium was discovered in 1938 by Otto Hahn and Fritz Strassmann, who found that medium-mass elements (for example, barium and lanthanum) were produced in the bombardment of uranium with neutrons. The discovery that several neutrons were emitted during the fission process led to speculation concerning the possibility of using those neutrons to cause further

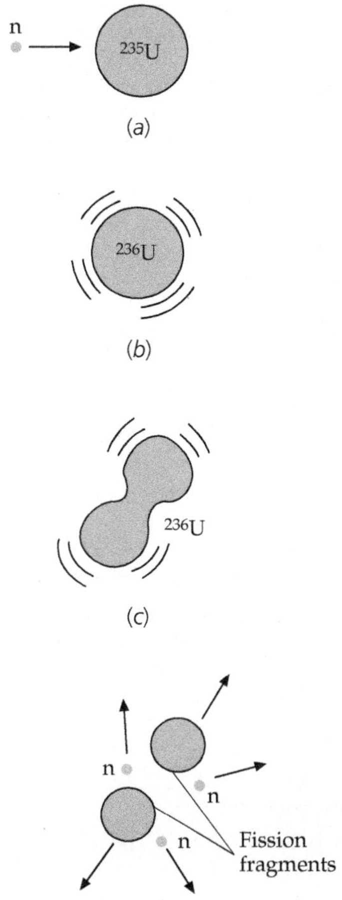

FIGURE 40-10 Schematic illustration of nuclear fission. (*a*) The absorption of a neutron by ^{235}U leads to (*b*) ^{236}U in an excited state. (*c*) The oscillation of ^{236}U has become unstable. (*d*) The nucleus splits apart into two nuclei that are less massive than the original nucleus and emits several neutrons that can produce fission in other nuclei.

fissions, thereby producing a chain reaction. When ^{235}U captures a neutron, the resulting ^{236}U nucleus emits γ rays as it de-excites to the ground state approximately 15 percent of the time and undergoes fission approximately 85 percent of the time. The fission process is somewhat analogous to the oscillation of a liquid drop, as shown in Figure 40-10. If the oscillations are violent enough, the drop splits in two. Using the liquid-drop model, Niels Bohr and John Wheeler calculated the critical energy E_c needed by the ^{236}U nucleus to undergo fission. (^{236}U is the nucleus formed momentarily by the capture of a neutron by ^{235}U.) For this nucleus, the critical energy is 5.3 MeV, which is less than the 6.4 MeV of excitation energy produced when ^{235}U captures a neutron. The capture of a neutron by ^{235}U therefore produces an excited state of the ^{236}U nucleus that has more than enough energy to break apart. On the other hand, the critical energy for fission of the ^{239}U nucleus is 5.9 MeV. The capture of a neutron by a ^{238}U nucleus produces an excitation energy of only 5.2 MeV. Therefore, when a neutron is captured by ^{238}U to form ^{239}U, the excitation energy is not great enough for fission to occur. In this case, the excited ^{239}U nucleus de-excites by γ emission and then decays to ^{239}Np by β decay, and then again to ^{239}Pu by β decay.

A fissioning nucleus can split into a pair of medium-mass nuclei, as shown in Figure 40-11. Depending on the particular reaction, 1, 2, or 3 neutrons may be emitted. The average number of neutrons emitted in the fission of ^{235}U is approximately 2.5. A typical fission reaction is

$$n + {}^{235}U \rightarrow {}^{141}Ba + {}^{92}Kr + 3n$$

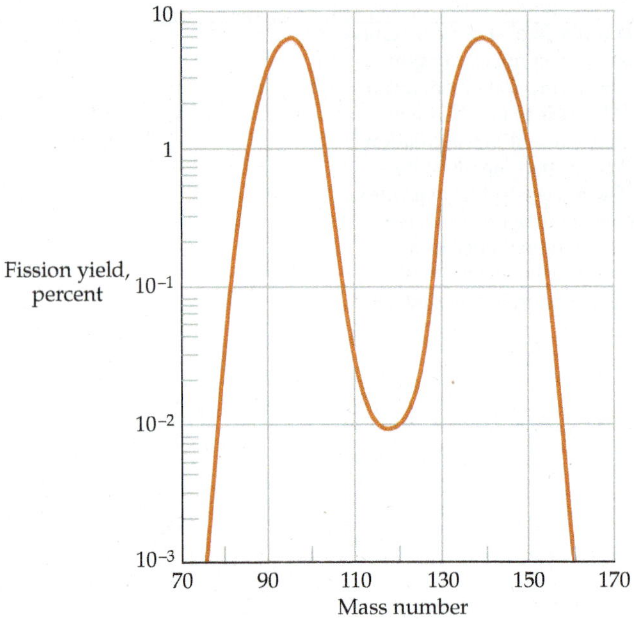

FIGURE 40-11 Distribution of the possible fission fragments of ^{235}U. The splitting of ^{235}U into two fragments of unequal mass is more likely than its splitting into fragments of equal mass.

NUCLEAR FISSION REACTORS

To sustain a chain reaction in a fission reactor, one of the neutrons (on average) that is emitted during and following* the fission of ^{235}U must be captured by another ^{235}U nucleus and cause it to fission. The **reproduction constant** k of a reactor is defined as the average number of neutrons from each fission that cause a subsequent fission. The maximum possible value of k for a uranium reactor is 2.5, but it is normally less than this for two important reasons: (1) Some of the neutrons may escape from the region containing fissionable nuclei and (2) some of the neutrons may be captured by nonfissioning nuclei in the reactor. If k is exactly 1, the reaction will be self-sustaining. If k is less than 1, the reaction will die out. If k is significantly greater than 1, the reaction rate will increase rapidly and become uncontrollable. In the design of nuclear bombs, such a runaway reaction is desired. In power reactors, the value of k must be kept very nearly equal to 1.

Because the neutrons emitted during and following fission have energies of the order of 1 MeV, whereas the chance for neutron capture leading to fission in ^{235}U is largest at small energies, the chain reaction can be sustained only if the neutrons are slowed down before they escape from the reactor. At high energies (1 MeV to 2 MeV), neutrons lose energy rapidly by inelastic scattering from ^{238}U, the principal constituent of natural uranium. (Natural uranium contains 99.3 percent ^{238}U and only 0.7 percent fissionable ^{235}U.) Once the neutron energy is below the excitation energies of the nuclei in the reactor (about 1 MeV), the main process of energy loss is by elastic scattering, in which a fast neutron collides with a nucleus at rest and transfers some of its kinetic energy to that nucleus. Such energy transfers are efficient only if the masses of the two bodies are comparable. A neutron will not transfer much energy in an elastic collision with a heavy uranium nucleus. Such a collision is like one between a marble and a billiard ball. The marble will be deflected by the much more massive billiard ball, and very little of its kinetic energy will be transferred to the billiard ball. A **moderator** consisting of material, such as water or carbon, that has light nuclei is therefore placed around the fissionable material in the core of the reactor to slow down the neutrons.

* Neutrons are sometimes emitted by the fission products. These neutrons are typically emitted a few seconds following the fission.

The inside of a nuclear power plant in Kent, England. A technician is standing on the reactor charge transfer plate, into which uranium fuel rods fit. *(© Jerry Mason/ Photo Researchers.)*

The neutrons are slowed down by elastic collisions with the nuclei of the moderator until they are in thermal equilibrium with the moderator. Because of the relatively large neutron-capture cross section of the hydrogen nucleus in water, reactors that use ordinary water as a moderator cannot easily achieve $k \approx 1$ unless they use enriched uranium, in which the ^{235}U content has been increased from 0.7 percent to between 1 percent and 4 percent. Natural uranium can be used if heavy water (D_2O) is used instead of ordinary (light) water (H_2O) as the moderator. Although heavy water is expensive, most Canadian reactors use heavy water for a moderator to avoid the cost of constructing uranium-enrichment facilities.

Figure 40-12 shows some of the features of a pressurized-water reactor commonly used in the United States to generate electricity. Fission in the core heats the water to a high temperature in the primary loop, which is closed. This water, which also serves as the moderator, is under high pressure to prevent the water from boiling.

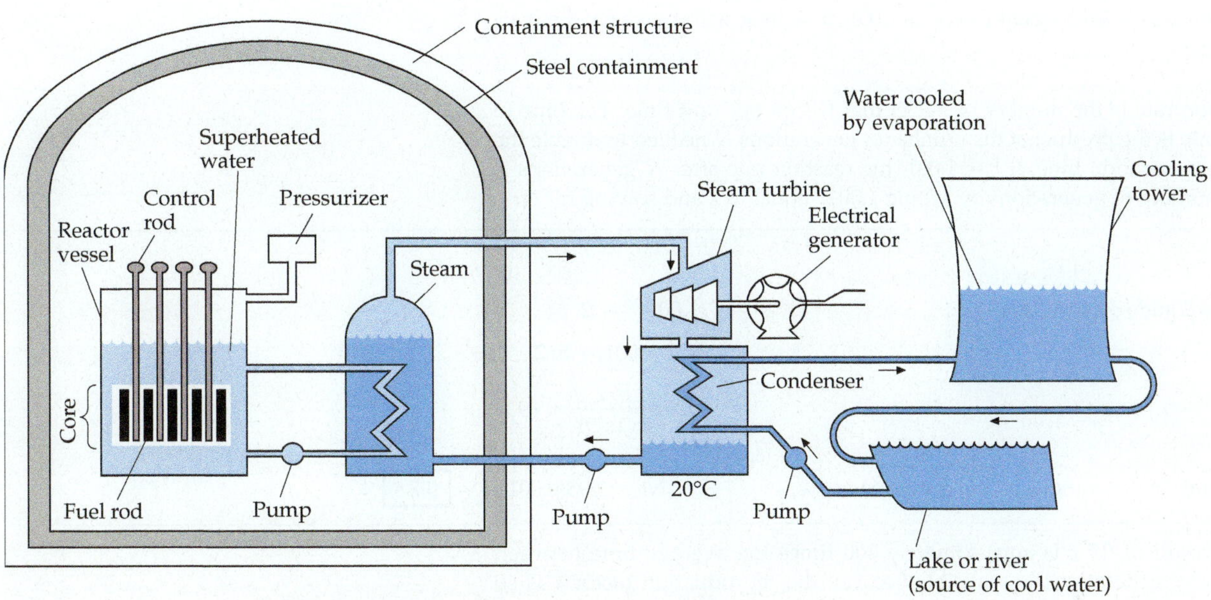

FIGURE 40-12 Simplified drawing of a pressurized-water reactor. The water in contact with the reactor core serves as both the moderator and the heat-transfer material. It is isolated from the water used to produce the steam that drives the turbines. Many features, such as the backup cooling mechanisms, are not shown here.

The hot water is pumped to a heat exchanger, where it heats the water in the secondary loop and converts the water to steam, which is then used to drive the turbines that produce electrical power. Note that the water in the secondary loop is isolated from the water in the primary loop to prevent its contamination by the radioactive nuclei in the reactor core.

The ability to control the reproduction factor k precisely is important if a power reactor is to be operated safely. Both natural negative-feedback mechanisms and mechanical methods of control are used. If k is greater than 1, the reaction rate increases and the temperature of the reactor increases. If water is used as a moderator, its density decreases with increasing temperature and the water becomes a less effective moderator. A second important control method is the use of control rods made of a material, such as cadmium, that has a very large neutron-capture cross section. To decrease the reaction rate, the control rods are inserted so that more neutrons are captured by the rods and k becomes less than 1. To increase the reaction rate, the rods are gradually withdrawn from the reactor; fewer neutrons are captured by the control rods and k becomes greater than 1.

Mechanical control of the reaction rate of a nuclear reactor using control rods is possible only because some of the neutrons emitted during the fission process are **delayed neutrons.** The time needed for a neutron to slow down from 1 MeV or 2 MeV to the thermal-energy level and then be captured is only of the order of a millisecond. If all the neutrons emitted during fission were prompt neutrons, that is, emitted immediately during the fission process, mechanical control would not be possible because the reactor would run away before the rods could be inserted farther. However, approximately 0.65 percent of the neutrons emitted are delayed by an average time of about 14 s. Those neutrons are emitted not during the fission process itself but during the decay of the fission fragments. The effect of the delayed neutrons can be seen in the following examples.

Example 40-7 Doubling Time

If the average time between fission generations (the time it takes for a neutron emitted during one fission to cause another) is $t_1 = 1\text{ ms} = 0.001\text{ s}$ and if the average number of neutrons from each fission that cause a subsequent fission is 1.001, how long will it take for the reaction rate to double?

PICTURE The reaction rate is the number of nuclei that fission per unit time. The time to double the reaction rate is the product of the number of generations N needed to double the reaction rate and the generation time. If $k = 1.001$, the reaction rate after N generations is 1.001^N. We find the number of generations by setting 1.001^N equal to 2 and solving for N.

SOLVE

1. Set 1.001^N equal to 2 and solve for N:

$$(1.001)^N = 2$$

$$N \ln 1.001 = \ln 2$$

$$N = \frac{\ln 2}{\ln 1.001} = 693$$

2. Multiply the number of generations by the generation time:

$$t = Nt_1 = 693(0.001\text{ s}) = \boxed{0.7\text{ s}}$$

CHECK The step-2 result of 0.7 s is approximately 700 times the average time between generations. This many generations is plausible because the reproduction factor k is so close to 1.

TAKING IT FURTHER The doubling time of about 0.7 s is not enough time for insertion of control rods.

Example 40-8 Delayed Neutrons and Control-Rod Insertion

Try It Yourself

Assuming that 0.65 percent of the neutrons emitted are delayed by 14 s, find the average generation time and the doubling time if $k = 1.001$.

PICTURE The doubling time is Nt_{av}, where t_{av} is the average time between generations. Since 99.35 percent of the generation times are 0.001 s and 0.65 percent are 14 s, the average generation time is 0.9935(0.001 s) + 0.0065(14 s).

SOLVE

Cover the column to the right and try these on your own before looking at the answers.

Steps	Answers
1. Compute the average generation time.	$t_{av} = 0.9935(0.001\text{ s}) + 0.0065(14\text{ s}) = 0.092\text{ s}$
2. Use your result to find the time for 693 generations.	$t = 63.8\text{ s} \approx \boxed{60\text{ s}}$

CHECK The number of delayed neutrons is approximately 0.7 percent of the total number of neutrons, but the generation time of 1 ms is approximately 0.007 percent of 14 s. Thus, an increase in the doubling time by a factor of about 100 is plausible.

TAKING IT FURTHER A doubling time of 60 s is plenty of time for mechanical insertion of control rods.

Because of the limited supply of natural uranium, the small fraction of ^{235}U in natural uranium, and the limited capacity of enrichment facilities, reactors based on the fission of ^{235}U cannot meet our energy needs for very long. A promising alternative is the **breeder reactor.** When the relatively plentiful but nonfissionable ^{238}U nucleus captures a neutron, it decays by β decay (with a half-life of 20 min) to ^{239}Np, which in turn decays by β decay (with a half-life of 2.35 days) to the fissionable nuclide ^{239}Pu. Because ^{239}Pu fissions with fast neutrons, no moderator is needed. A reactor initially fueled with a mixture of ^{238}U and ^{239}Pu will breed as much fuel as it uses or more if one or more of the neutrons emitted in the fission of ^{239}Pu is captured by ^{238}U. Practical studies indicate that a typical breeder reactor can be expected to double its fuel supply in 7 to 10 years.

There are two major safety problems inherent with breeder reactors. The fraction of delayed neutrons is only 0.3 percent for the fission of ^{239}Pu, so the time between generations is much less than that for ordinary reactors. Mechanical control is therefore much more difficult. Also, because the operating temperature of a breeder reactor is relatively high and a moderator is not desired, a heat-transfer material, such as liquid sodium metal, is used rather than water (which is the moderator as well as the heat-transfer material in an ordinary reactor). If the temperature of the reactor increases, the resulting decrease in the density of the heat-transfer material leads to positive feedback, because it will absorb fewer neutrons than before. Because of these safety considerations, breeder reactors are not yet in commercial use in the United States. There are, however, several in operation in France, Great Britain, and the former Soviet Union.

FUSION

During fusion, two light nuclei, such as deuterium (^{2}H) and tritium (^{3}H), fuse together to form a heavier nucleus. A typical fusion reaction is

$$^2\text{H} + {}^3\text{H} \rightarrow {}^4\text{He} + \text{n} + 17.6\text{ MeV}$$

The energy released in fusion depends on the particular reaction. For the $^2H + {}^3H$ reaction, the energy released is 17.6 MeV. Although this energy is less than the energy released during a fission reaction, it is a greater amount of energy per unit mass. The energy released during this fusion reaction is (17.6 MeV)/(5 nucleons) = 3.52 MeV per nucleon. This is approximately 3.5 times as great as the 1 MeV per nucleon released in fission.

The production of power from the fusion of light nuclei holds great promise because of the relative abundance of the fuel and the absence of some of the dangers inherent in fission reactors. Unfortunately, the technology necessary to make fusion a practical source of energy has not yet been developed. We will consider the $^2H + {}^3H$ reaction; other reactions present similar problems.

Because of the Coulomb repulsion between the ^2H and ^3H nuclei, very large kinetic energies, of the order of 1 MeV, are needed to get the nuclei close enough together for the attractive nuclear forces to become effective and to cause fusion. Such energies can be obtained in an accelerator, but because the scattering of one nucleus by the other is much more probable than fusion, the bombardment of one nucleus by another in an accelerator requires the input of more energy than is recovered. To obtain energy from fusion, the particles must be heated to a temperature great enough for the fusion reaction to occur as the result of random thermal collisions. Because a significant number of particles have kinetic energies greater than the mean kinetic energy, $\frac{3}{2}kT$, and because some particles can tunnel through the Coulomb barrier, a temperature T corresponding to $kT \approx 10$ keV is adequate to ensure that a reasonable number of fusion reactions will occur if the density of the particles is sufficiently high. The temperature corresponding to $kT = 10$ keV is of the order of 10^8 K. These temperatures occur in the interiors of stars, where such reactions are common. At these temperatures, a gas consists of positive ions and electrons and is called a **plasma.** One of the problems arising in attempts to produce controlled fusion reactions is the problem of confining the plasma long enough for the reactions to take place. In the interior of the Sun, the plasma is confined by the enormous gravitational field of the Sun. In a laboratory on Earth, confinement is a difficult problem.

The energy required to heat a plasma is proportional to the number density of its ions, n, whereas the collision rate is proportional to n^2 (the square of the number density). If τ is the confinement time, the output energy is proportional to $n^2\tau$. If the output energy is to exceed the input energy, we must have

$$C_1 n^2 \tau > C_2 n$$

where C_1 and C_2 are constants. In 1957, the British physicist J. D. Lawson evaluated these constants from estimates of the efficiencies of various hypothetical fusion reactors and derived the following relation between density and confinement time, known as **Lawson's criterion:**

$$n\tau > 10^{20}\ \mathrm{s\cdot particles/m^3} \qquad \text{40-22}$$

LAWSON'S CRITERION

If Lawson's criterion is met and the thermal energy of the ions is great enough ($kT \approx 10$ keV), the energy released by a fusion reactor will just equal the energy input; that is, the reactor will just break even. For the reactor to be practical, much more energy must be released.

Two schemes for achieving Lawson's criterion are currently under investigation. In one scheme, **magnetic confinement,** a magnetic field is used to confine the plasma (see Section 26-2). In the most common arrangement, first developed in the former Soviet Union and called a *tokamak*, the plasma is confined in a large toroid.

The magnetic field is a combination of the doughnut-shaped magnetic field due to the windings of the toroid and the self-field due to the current of the circulating plasma. The break-even point has almost been achieved using magnetic confinement, but we are still a long way from building a practical fusion reactor.

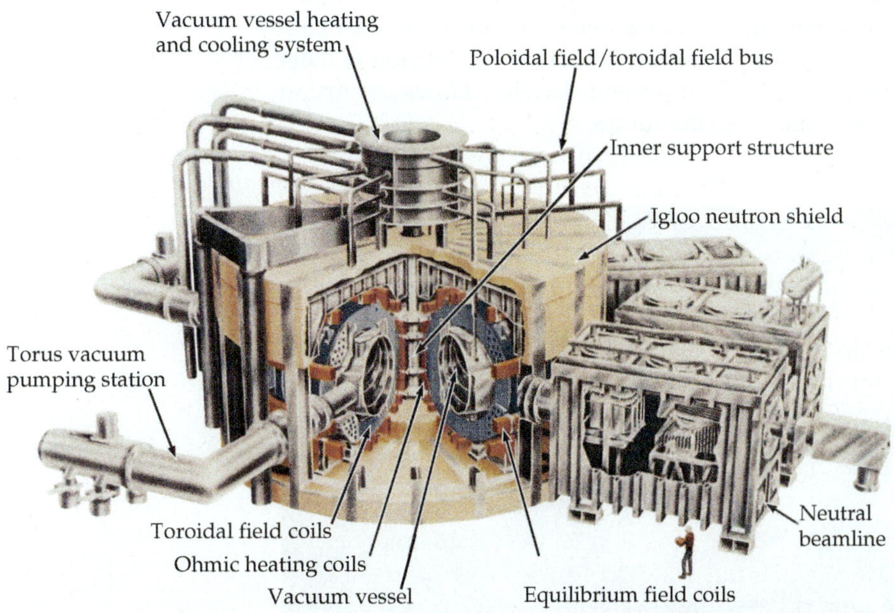

(*a*)

(*b*)

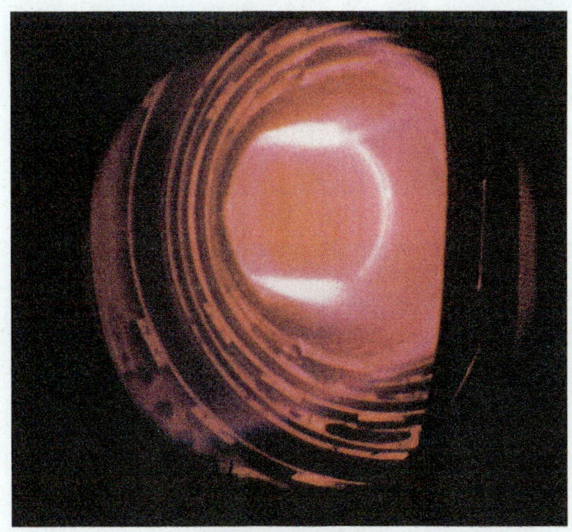

(*c*)

(*a*) Schematic of the Tokamak Fusion Test Reactor (TFTR). The toroidal coils, surrounding the doughnut-shaped vacuum vessel, are designed to conduct current for 3-s pulses, separated by waiting times of 5 min. Pulses peak at 73 000 A, producing a magnetic field of 5.2 T. This magnetic field is the principal means of confining the deuterium–tritium plasma that circulates within the vacuum vessel. Current for the pulses is delivered by converting the rotational energy of two 600-ton flywheels. Sets of poloidal coils, perpendicular to the toroidal coils, carry an oscillating current that generates a current through the confined plasma itself, heating it ohmically. Additional poloidal fields help stabilize the confined plasma. Between four and six neutral-beam injection systems (only one of which is shown in the schematic) are used to inject high-energy deuterium atoms into the deuterium–tritium plasma, heating beyond what could be obtained ohmically, ultimately to the point of fusion. (*b*) The TFTR itself. The diameter of the vacuum vessel is 7.7 m. (*c*) An 800-kA plasma, lasting 1.6 s, as it discharges within the vacuum vessel. *((All) Courtesy of the Princeton Plasma Physics Laboratory.)*

In a second scheme, called **inertial confinement,** a pellet of solid deuterium and tritium is bombarded from all sides by intense pulsed laser beams of energies of the order of 10^4 J lasting about 10^{-8} s. (Intense beams of ions are also used.) Computer simulation studies indicate that the pellet should be compressed to approximately 10^4 times its normal density and heated to a temperature greater than 10^8 K. This should produce approximately 10^6 J of fusion energy in 10^{-11} s, which is so brief that confinement is achieved by inertia alone.

Because the break-even point is just barely being achieved in magnetic-confinement fusion, and because the building of a fusion reactor involves many practical problems that have not yet been solved, the availability of fusion to meet our energy needs is not expected for at least several decades. However, fusion holds great promise as an energy source for the future.

(a)

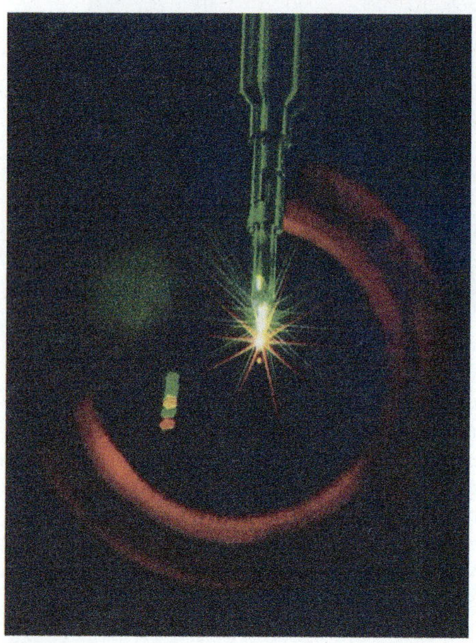

(b)

(*a*) The Nova target chamber, an aluminum sphere approximately 5 m in diameter, inside which 10 beams from the world's most powerful laser converge onto a hydrogen-containing pellet 0.5 mm in diameter. (*b*) The resulting fusion reaction is visible as a tiny star, lasting 10^{-10} s, releasing 10^{13} neutrons. *((All) Courtesy of the Lawrence Livermore National Laboratory/U.S. Department of Energy.)*

Summary

TOPIC	RELEVANT EQUATIONS AND REMARKS	
1. Properties of Nuclei	A nucleus has N neutrons, Z protons, and a mass number $A = N + Z$. For light nuclei, N and Z are approximately equal, whereas for heavy nuclei, N is greater than Z.	
Isotopes	Isotopes are two or more nuclei that have the same atomic number Z but have different values of N and A.	
Size and shape	Most nuclei are approximately spherical in shape and have a volume that is proportional to A. Because the mass is proportional to A, nuclear density is independent of A.	
Radius	$R = R_0 A^{1/3} \approx (1.2\text{ fm})A^{1/3}$	40-1
Mass and binding energy	The mass of a stable nucleus is less than the sum of the masses of its nucleons. The mass difference Δm multiplied by c^2 equals the binding energy E_b of the nucleus. The binding energy is approximately proportional to the mass number A.	
2. Radioactivity	Unstable nuclei are radioactive and decay by emitting α particles (^{4}He nuclei), β particles (electrons or positrons), or γ rays (photons). All radioactivity is statistical in nature and follows an exponential decay law: $N = N_0 e^{-\lambda t}$	40-6
Decay rate	$R = \lambda N = R_0 e^{-\lambda t}$	40-7
Mean lifetime	$\tau = \frac{1}{\lambda}$	40-9
Half-life	$t_{1/2} = \tau \ln 2 = 0.693\tau$ The half-lives of α decay range from a fraction of a second to millions of years. For β decay, the half-lives range up to hours or days. For γ decay, the half-lives are usually less than a microsecond.	40-11
Decay-rate units	The number of decays per second of 1 g of radium is the curie (Ci). $1\text{ Ci} = 3.7 \times 10^{10}\text{ decays/s} = 3.7 \times 10^{10}\text{ Bq}$ $(1\text{ Bq} = 1\text{ decay/s})$	
3. Nuclear Reactions		
Q value	The Q value equals c^2 multiplied by the total mass of the incoming particles less the total mass of the outgoing particles in the center of mass reference frame. If the net mass change is Δm, the Q value is $Q = -(\Delta m)c^2$	40-19
Exothermic reaction	If total mass decreases during a reaction, Q is positive and measures the energy released.	
Endothermic reaction	If total mass increases during a reaction, Q is negative. Then $\lvert Q \rvert$ is the threshold energy for the reaction in the center of mass reference frame.	
4. Fission	Fission occurs when some heavy elements, such as ^{235}U or ^{239}Pu, capture a neutron and split apart into two nuclei. The two nuclei then fly apart because of electrostatic repulsion. A chain reaction is possible because several neutrons are emitted by a nucleus when it undergoes fission. A chain reaction can be sustained in a reactor if, on the average, one of the emitted neutrons is slowed down by scattering in the reactor and is then captured by another fissionable nucleus. Very heavy nuclei ($Z > 92$) are subject to spontaneous fission.	

TOPIC	RELEVANT EQUATIONS AND REMARKS
5. **Fusion**	A large amount of energy is released when two light nuclei, such as ^{2}H and ^{3}H, fuse together. Fusion takes place spontaneously inside the Sun and other stars, where the temperature is great enough (about 10^8 K) for thermal motion to bring the charged hydrogen ions close enough together to fuse. Although controlled fusion holds great promise as a future energy source, practical difficulties have thus far hindered its development.
Lawson criterion	The minimum product of particle density n and confinement time τ to get more energy out of a fusion reactor than is put in is $n\tau > 10^{20}$ s · particles/m^3.

Answer to Concept Check

41-1 The number left can be either one or zero, where zero left is more probable than one left.

Answers to Practice Problems

40-1 (*a*) $60E_1$ (*b*) $20E_1$

40-2 It is because $\frac{1}{8} = \left(\frac{1}{2}\right)^3$, so $n = 3$.

Problems

In a few problems, you are given more data than you actually need; in a few other problems, you are required to supply data from your general knowledge, outside sources, or informed estimate.

Interpret as significant all digits in numerical values that have trailing zeros and no decimal points.

- • Single-concept, single-step, relatively easy
- •• Intermediate-level, may require synthesis of concepts
- ••• Challenging
- SSM Solution is in the *Student Solutions Manual*
- Consecutive problems that are shaded are paired problems.

CONCEPTUAL PROBLEMS

1 • Isotopes of nitrogen, iron and tin have stable isotopes ^{14}N, ^{56}Fe and ^{118}Sn. Give the symbols for two other isotopes of (*a*) nitrogen, (*b*) iron, and (*c*) tin.

2 • Why is the decay chain $A = 4n + 1$ not found in nature?

3 • A decay by α emission is often followed by β decay. When this occurs, it is by β^- and not β^+ decay. Why?

4 • The half-life of ^{14}C is much less than the age of the universe, yet ^{14}C is found in nature. Why?

5 • What effect would a long-term variation in cosmic-ray activity have on the accuracy of ^{14}C dating?

6 • Why does an element that has $Z = 130$ not exist?

7 • Why is a moderator needed in an ordinary nuclear fission reactor?

8 • Explain why water is more effective than lead in slowing down fast neutrons.

9 • The stable isotope of sodium is ^{23}Na. What kind of beta decay would you expect of (*a*) ^{22}Na and (*b*) ^{24}Na?

10 • What is the advantage of a breeder reactor over an ordinary reactor? What are the disadvantages of a breeder reactor?

11 • True or false:

(*a*) In a breeder reactor, fuel can be produced as fast as it is consumed.

(*b*) The atomic nucleus is composed of protons, neutrons, and electrons.

(*c*) The mass of a ^{2}H nucleus is less than the mass of a ^{1}H nucleus plus the mass of a neutron.

(*d*) After two half-lives, all the radioactive nuclei in a given sample have decayed.

12 • Why is it that extreme changes in the temperature or the pressure of a radioactive sample have little or no effect on the radioactivity?

ESTIMATION AND APPROXIMATION

13 • We found in Chapter 25 that the ratio of the resistivity of the most insulating material to the resistivity of the least resistive material (excluding superconductors) is approximately 10^{22}. Few properties of materials show such a wide range of values. Using information in the textbook or other resources, find the ratio of largest to smallest for some nuclear properties of matter. Some examples might be the range of mass densities found in an atom, the half-life of radioactive nuclei, or the range of nuclear masses.

14 •• According to the United States Department of Energy, the U.S. population consumes approximately 10^{20} joules of energy each year. Estimate the mass (in kilograms) of (*a*) uranium that would be needed to produce this much energy using nuclear fission and (*b*) deuterium and tritium that would be needed to produce this much energy using nuclear fusion.

PROPERTIES OF NUCLEI

15 • Calculate the binding energy and the binding energy per nucleon from the masses given in Table 40-1 for (*a*) ^{12}C, (*b*) ^{56}Fe, and (*c*) ^{238}U. SSM

16 • Calculate the binding energy and the binding energy per nucleon from the masses given in Table 40-1 for (*a*) ^{6}Li, (*b*) ^{39}K, and (*c*) ^{208}Pb.

17 • Use the radius formula $R = R_0 A^{1/3}$ (Equation 40-1), where $R_0 = 1.2$ fm, to compute the radii of the following nuclei: (*a*) ^{16}O, (*b*) ^{56}Fe, and (*c*) ^{197}Au.

18 • During a fission process, a ^{239}Pu nucleus splits into two nuclei whose mass number ratio is 3 to 1. Calculate the radii of the nuclei formed during the process.

19 •• The neutron, when isolated from an atomic nucleus, decays into a proton, an electron, and an antineutrino as follows: $n \rightarrow {}^1H + e^- + \bar{\nu}$. The thermal energy of a neutron is of the order of kT, where k is the Boltzmann constant. (*a*) In both joules and electron volts, calculate the energy of a thermal neutron at 25°C. (*b*) What is the speed of that thermal neutron? (*c*) A beam of monoenergetic thermal neutrons is produced at 25°C and has an intensity I. After traveling 1350 km, the beam has an intensity of $\frac{1}{2}I$. Using this information, estimate the half-life of the neutron. Express your answer in minutes. SSM

20 • Use $R = R_0 A^{1/3}$ (Equation 40-1), where $R_0 = 1.2$ fm, for the radius of a spherical nucleus to calculate the density of nuclear matter. Express your answer in grams per cubic centimeter.

21 •• In 1920, 12 years before the discovery of the neutron, Ernest Rutherford argued that proton–electron pairs might exist in the confines of the nucleus in order to explain the mass number, A, being greater than the nuclear charge, Z. He also used this argument to account for the source of beta particles in radioactive decay. Rutherford's scattering experiments in 1910 showed that the nucleus had a diameter of approximately 10 fm. Using this nuclear diameter, the uncertainty principle, and that beta particles have an energy range of 0.02 MeV to 3.40 MeV, show why the hypothetical electrons cannot be confined to a region occupied by the nucleus. SSM

22 •• Consider the following fission process: ${}^{235}_{92}U + n \rightarrow {}^{95}_{37}Rb + {}^{137}_{55}Cs + 4n$. Determine the electrostatic potential energy, in MeV, of the reaction products when the surfaces of the ^{95}Rb nucleus and the ^{137}Cs nucleus are just touching immediately after being formed during the fission process.

RADIOACTIVITY

23 • Homer enters the visitors' chambers, and his Geiger beeper sounds. He shuts off the beeper, removes the device from his shoulder patch, and holds it near the only new object in the room—an orb that is to be presented as a gift from the visiting Cartesians. Pushing a button marked "monitor," Homer reads that the device is reading a counting rate of 4000 counts/s above the background counting rate. After 10 min, the counting rate has dropped to 1000 counts/s above the background rate. (*a*) What is the half-life of the source? (*b*) How high will the counting rate be (above the background counting rate) 20 min after the monitoring device was switched on?

24 • A certain source gives 2000 counts/s at time $t = 0$. Its half-life is 2.0 min. How many counts per second will it give after (*a*) 4.0 min, (*b*) 6.0 min, and (*c*) 8.0 min?

25 • The counting rate from a radioactive source is 8000 counts/s at time $t = 0$, and 10 min later the rate is 1000 counts/s. (*a*) What is the half-life? (*b*) What is the decay constant? (*c*) What is the counting rate after 20 min?

26 • The half-life of radium is 1620 y. Calculate the number of disintegrations per second of 1.00 g of radium and show that the disintegration rate is approximately 1.0 Ci.

27 • A radioactive piece of silver foil ($t_{1/2} = 2.4$ min) is placed near a Geiger counter and 1000 counts/s are observed at time $t = 0$. (*a*) What is the counting rate at $t = 2.4$ min and at $t = 4.8$ min? (*b*) If the counting efficiency is 20 percent, how many radioactive silver nuclei are there at time $t = 0$? At time $t = 2.4$ min? (*c*) At what time will the counting rate be about 30 counts/s?

28 • Use Table 40-1 to calculate the energy release, in MeV, for the α decay of (*a*) ^{226}Ra and (*b*) ^{242}Pu.

29 •• Plutonium is very toxic to the human body. Once it enters the body it collects primarily in the bones, although it also can be found in other organs. Red blood cells are synthesized within the marrow of the bones. The isotope ^{239}Pu is an alpha emitter that has a half-life of 24 360 years. Because alpha particles are an ionizing radiation, the blood-making ability of the marrow is, in time, destroyed by the presence of ^{239}Pu. In addition, many kinds of cancers will also develop in the surrounding tissues because of the ionizing effects of the alpha particles. (*a*) If a person accidentally ingested 2.0 μg of ^{239}Pu and all of it is absorbed by the bones of the person, how many alpha particles are produced per second within the body of the person? (*b*) When, in years, will the activity be 1000 alpha particles per second? SSM

30 •• Consider an alpha-emitting parent nucleus ${}^A_Z X$ initially at rest. The nucleus decays into a daughter nucleus Y and an alpha particle as follows: ${}^A_Z X \rightarrow {}^{A-4}_{Z-2}Y + {}^4_2\alpha + Q$. (*a*) Show that the alpha particle has a kinetic energy of $(A - 4)Q/A$. (*b*) Show that the kinetic energy of the recoiling daughter nucleus is given by $K_Y = 4Q/A$.

31 • The fissile material ^{239}Pu is an alpha emitter. Write the reaction that describes ^{239}Pu undergoing alpha decay. Given that ^{239}Pu, ^{235}U, and an alpha particle have respective masses of 239.052 156 u, 235.043 923 u, and 4.002 603 u, use the relations appearing in Problem 30 to calculate the kinetic energies of the alpha particle and the recoiling daughter nucleus. SSM

32 • Through a friend in the security department at the museum, Angela obtains a sample of a wooden tool handle that contains 175 g of carbon. The decay rate of the ^{14}C in the sample is 8.1 Bq. How long ago was the wood in the handle last alive?

33 • A sample of a radioactive isotope is found to have an activity of 115.0 Bq immediately after it is pulled from the reactor that formed the isotope. Its activity 2 h 15 min later is measured to be 85.2 Bq. (*a*) Calculate the decay constant and the half-life of the sample. (*b*) How many radioactive nuclei were there in the sample initially?

34 •• A 1.00-mg sample of substance that has an atomic mass of 59.934 u and emits β particles has an activity of 1.131 Ci. Find the decay constant for the substance in reciprocal seconds and find the half-life in years.

35 •• Radiation has been used for a long time in medical therapy to control the development and growth of cancer cells. Cobalt-60, a gamma emitter that emits photons that have energies of 1.17 MeV and 1.33 MeV, is used to irradiate and destroy deep-rooted cancers. Small needles made of ^{60}Co of a specified activity are encased in gold and used as body implants in tumors for time periods that are related to tumor size, tumor cell reproductive rate, and the activity of the needle. (*a*) A 1.00-μg sample of ^{60}Co, that has a half-life of

5.27 y and that is used to irradiate a small internal tumor with gamma rays, is prepared in the cyclotron of a medical center. Determine the activity of the sample in curies. (*b*) What will the activity of the sample be 1.75 y from now? SSM

36 •• (*a*) Show that if the decay rate is R_0 at time $t = 0$ and R_1 at some later time $t = t_1$, the decay constant is given by $\lambda = t_1^{-1}\ln(R_0/R_1)$ and the half-life is given by $t_{1/2} = t_1\ln(2)/\ln(R_0/R_1)$. (*b*) Use these results to find the decay constant and the half-life if the decay rate is 1200 Bq at $t = 0$ and 800 Bq at $t_1 = 60.0$ s.

37 •• A wooden casket is thought to be 18 000 years old. How much carbon would have to be recovered from the object to yield a ^{14}C counting rate of no less than 5 counts/min with a detection efficiency of 20 percent?

38 •• A sample of radioactive material is initially found to have an activity of 115.0 decays/min. After 4 d 5 h, its activity is measured to be 73.5 decays/min. (*a*) Calculate the half-life of the material. (*b*) How long (from the initial time) will it take for the sample to reach an activity of 10.0 decays/min?

39 •• The rubidium isotope ^{87}Rb is a β^- emitter that has a half-life of 4.9×10^{10} y. It decays into ^{87}Sr. This nuclear decay is used to determine the age of rocks and fossils. Rocks containing the fossils of early animals have a ratio of ^{87}Sr to ^{87}Rb of 0.0100. Assuming that there was no ^{87}Sr present when the rocks were formed, calculate the age of the fossils.

40 ••• Consider a single nucleus of a radioactive isotope that has a decay rate equal to λ. The nucleus has not decayed at $t = 0$. The probability that the nucleus will decay between time t and time $t + dt$ is equal to $\lambda e^{-\lambda t}dt$. (*a*) Show that this statement is consistent with the fact that the probability is 1 that the nucleus will decay between $t = 0$ and $t = \infty$. (*b*) Show that the expected lifetime of the nucleus is equal to $1/\lambda$. *Hint: The expected lifetime is equal to* $\int_0^\infty t\lambda e^{-\lambda t}dt$ *divided by* $\int_0^\infty \lambda e^{-\lambda t}dt$. (*c*) A sample of material contains a number of these radioactive nuclei at time $t = 0$. What is the mean lifetime of the radioactive nuclei in the sample?

NUCLEAR REACTIONS

41 • Using Table 40-1, find the Q values for the following reactions: (*a*) $^1\text{H} + {}^3\text{H} \rightarrow {}^3\text{He} + \text{n} + Q$ and (*b*) $^2\text{H} + {}^2\text{H} \rightarrow {}^3\text{He} + \text{n} + Q$.

42 • Using Table 40-1, find the Q values for the following reactions: (*a*) $^2\text{H} + {}^2\text{H} \rightarrow {}^3\text{H} + {}^1\text{H} + Q$, (*b*) $^2\text{H} + {}^3\text{He} \rightarrow {}^4\text{He} + {}^1\text{H} + Q$, and (*c*) $^6\text{Li} + \text{n} \rightarrow {}^3\text{H} + {}^4\text{He} + Q$.

43 •• (*a*) Use the values 14.003 242 u and 14.003 074 u for the atomic masses of ^{14}C and ^{14}N, respectively, to calculate the Q value (in MeV) for the β-decay reaction ${}^{14}_{6}\text{C} \rightarrow {}^{14}_{7}\text{N} + e^- + \bar{\nu}_e$. (*b*) Explain why you should not add the mass of the electron to that of atomic ^{14}N for the calculation in Part (*a*). SSM

44 •• (*a*) Use the values 13.005 738 u and 13.003 354 u for the atomic masses of ${}^{13}_{7}$N and ${}^{13}_{6}$C, respectively, to calculate the Q value (in MeV) for the β-decay reaction

$$ {}^{13}_{7}\text{N} \rightarrow {}^{13}_{6}\text{C} + e^+ + \nu_e $$

(*b*) Explain why you need to add twice the mass of an electron to the mass of ${}^{13}_{6}$C during the calculation of the Q value for the reaction in Part (*a*).

FISSION AND FUSION

45 • Assuming an average energy of 200 MeV per fission, calculate the number of fissions per second needed for a 500-MW reactor. SSM

46 • If the reproduction factor in a reactor is 1.1, find the number of generations needed for the power level to (*a*) double, (*b*) increase by a factor of 10, and (*c*) increase by a factor of 100. Find the time needed in each case if (*d*) there are no delayed neutrons, so that the time between generations is 1.0 ms, and (*e*) there are delayed neutrons that make the average time between generations 100 ms.

47 •• Consider the following fission reaction: ${}^{235}_{92}\text{U} + \text{n} \rightarrow {}^{95}_{42}\text{Mo} + {}^{139}_{57}\text{La} + 2\text{n} + Q$. The masses of the neutron, ^{235}U, ^{95}Mo, and ^{139}La are 1.008 665 u, 235.043 923 u, 94.905 842 u, and 138.906 348 u, respectively. Calculate the Q value, in MeV, for the fission reaction. SSM

48 •• In 1989, researchers claimed to have achieved fusion in an electrochemical cell at room temperature. Their now thoroughly discredited claim was that a power output of 4.00 W was produced by deuterium fusion reactions in the palladium electrode of their apparatus. The two most likely reactions are

$$ ^2\text{H} + {}^2\text{H} \rightarrow {}^3\text{He} + \text{n} + 3.27\ \text{MeV} $$

and

$$ ^2\text{H} + {}^2\text{H} \rightarrow {}^3\text{H} + {}^1\text{H} + 4.03\ \text{MeV} $$

Of the deuterium nuclei that participated in these reactions, assume half of the deuterium nuclei participated in the first reaction and the other half participated in the second reaction. How many neutrons per second would we expect to be emitted in the generation of 4.00 W of power?

49 •• A fusion reactor that uses only deuterium for fuel would have the two reactions in Problem 48 taking place in the reactor. The ^{3}H produced in the second reaction reacts immediately with another ^{2}H in the reaction

$$ ^3\text{H} + {}^2\text{H} \rightarrow {}^4\text{He} + \text{n} + 17.6\ \text{MeV} $$

The ratio of ^{2}H to ^{1}H atoms in naturally occurring hydrogen is 1.5×10^{-4}. How much energy would be produced from 4.0 L of water if all of the ^{2}H nuclei undergo fusion?

50 ••• The fusion reaction between ^{2}H and ^{3}H is

$$ ^3\text{H} + {}^2\text{H} \rightarrow {}^4\text{He} + \text{n} + 17.6\ \text{MeV} $$

Using the conservation of momentum and the given Q value, find the final energies of both the ^{4}He nucleus and the neutron, assuming the initial kinetic energy of the system is 1.00 MeV and the initial momentum of the system is zero.

51 ••• Energy is generated in the Sun and other stars by fusion. One of the fusion cycles, the proton–proton cycle, consists of the following reactions:

$$ ^1\text{H} + {}^1\text{H} \rightarrow {}^2\text{H} + e^+ + \nu_e $$

$$ ^1\text{H} + {}^2\text{H} \rightarrow {}^3\text{He} + \gamma $$

followed by

$$ ^1\text{H} + {}^3\text{He} \rightarrow {}^4\text{He} + e^+ + \nu_e $$

(*a*) Show that the net effect of these reactions is

$$ 4\,{}^1\text{H} \rightarrow {}^4\text{He} + 2e^+ + 2\nu_e + \gamma $$

(*b*) Show that 24.7 MeV is released during this cycle (not counting the additional energy of 1.02 MeV that is released when each positron meets an electron and the two annihilate). (*c*) The Sun radiates energy at the rate of approximately 4.0×10^{26} W. Assuming that this is due to the conversion of four protons into helium, γ rays, and neutrinos, which releases 26.7 MeV, what is the rate of proton consumption in the Sun? How long will the Sun last if it continues to radiate at its present level? (Assume that protons constitute about half of the total mass, 2.0×10^{30} kg, of the Sun.)

GENERAL PROBLEMS

52 • (*a*) Show that $ke^2 = 1.44$ MeV·fm, where k is the Coulomb constant and e is the magnitude of the electron charge. (*b*) Show that $hc = 1240$ MeV·fm.

53 • The counting rate from a radioactive source is 6400 counts/s. The half-life of the source is 10 s. Make a plot of the counting rate as a function of time for times up to 1 min. What is the decay constant for the source? SSM

54 • Find the energy needed to remove a neutron from (*a*) ^{4}He and (*b*) ^{7}Li.

55 • The isotope ^{14}C decays according to $^{14}\mathrm{C} \rightarrow {}^{14}\mathrm{N} + \mathrm{e}^- + \bar{\nu}_\mathrm{e}$. The atomic mass of ^{14}N is 14.003 074 u. Determine the maximum kinetic energy of the electron. (Neglect recoil of the nitrogen atom.)

56 • The density of a neutron star is the same as the density of a nucleus. If our Sun were to collapse to a neutron star, what would be the radius of that object?

57 •• Show that the ^{109}Ag nucleus is stable and does not undergo alpha decay, $^{109}_{47}\mathrm{Ag} \rightarrow {}^4_2\mathrm{He} + {}^{105}_{45}\mathrm{Rh} + Q$. The mass of the ^{109}Ag nucleus is 108.904 756 u, and the products of the decay are 4.002 603 u and 104.905 250 u, respectively. SSM

58 •• Gamma rays can be used to induce photofission (fission triggered by the absorption of a photon) in nuclei. Calculate the threshold photon wavelength for the following nuclear reaction: $^2\mathrm{H} + \gamma \rightarrow {}^1\mathrm{H} + \mathrm{n}$. Use Table 40-1 for the masses of the interacting particles.

59 • The relative abundance of ^{40}K (potassium 40) is 1.2×10^{-4}. The isotope ^{40}K has a molar mass of 40.0 g/mol, is radioactive, and has a half-life of 1.3×10^9 y. Potassium is an essential element of every living cell. In the human body the mass of potassium constitutes approximately 0.36 percent of the total mass. Determine the activity of this radioactive source in a student whose mass is 60 kg.

60 •• When a positron makes contact with an electron, the electron–positron pair annihilate by way of the reaction $\beta^+ + \beta^- \rightarrow 2\gamma$. Calculate the minimum total energy, in MeV, of the two photons created when a positron–electron pair annihilate.

61 •• The isotope ^{24}Na is a β emitter and has a half-life of 15 h. A saline solution containing the radioactive isotope has an activity of 600 kBq and is injected into the bloodstream of a patient. Ten hours later, the activity of 1 mL of blood from the individual yields a counting rate of 12 counts/s at a counting efficiency of 20 percent. Determine the volume of blood in the patient.

62 •• (*a*) Determine the distance of closest approach of an 8.0-MeV α particle in a head-on collision with a stationary nucleus of ^{197}Au and with a stationary nucleus of ^{10}B, neglecting the recoil of the struck nuclei. (*b*) Repeat the calculation taking into account the recoil of the struck nuclei.

63 •• Twelve nucleons are in a one-dimensional infinite square well of length $L = 3.0$ fm. (*a*) Using the approximation that the mass of a nucleon is 1.0 u, find the lowest energy of a nucleon in the well. Express your answer in MeV. What is the ground-state energy of the system of 12 nucleons in the well if (*b*) all the nucleons are neutrons so that there can be no more than 2 in each spatial state and (*c*) 6 of the nucleons are neutrons and 6 are protons so that there can be as many as 4 nucleons in each spatial state? (Neglect the energy of Coulomb repulsion of the protons.)

64 •• The helium nucleus or α particle is a very tightly bound system. Nuclei with $N = Z = 2n$, where n is an integer (for example, ^{12}C, ^{16}O, ^{20}Ne, and ^{24}Mg), can be modeled as agglomerates of α particles. (*a*) Use this model to estimate the binding energy of a pair of α particles from the atomic masses of ^{4}He and ^{16}O. Assume that the four α particles in ^{16}O form a regular tetrahedron that has one α particle at each vertex. (*b*) From the result obtained in Part (*a*) determine, on the basis of the model, the binding energy of ^{12}C and compare your result with the result obtained from the atomic mass of ^{12}C.

65 •• Nuclei of a radioactive isotope that has a decay constant of λ are produced in an accelerator at a constant rate R_p. The number of radioactive nuclei N then obeys the equation $dN/dt = R_\mathrm{p} - \lambda N$. (*a*) If N is zero at $t = 0$, sketch N versus t for the situation. (*b*) The isotope ^{62}Cu is produced at a rate of 100 per second by placing ordinary copper (^{63}Cu) in a beam of high-energy photons. The reaction is

$$\gamma + {}^{63}\mathrm{Cu} \rightarrow {}^{62}\mathrm{Cu} + \mathrm{n}$$

The isotope ^{62}Cu decays by β decay and has a half-life of 10 min. After a time long enough so that $dN/dt \approx 0$, how many ^{62}Cu nuclei are there?

66 •• The total energy consumed in the United States in 1 y is approximately 7.0×10^{19} J. How many kilograms of ^{235}U would be needed to provide this amount of energy if we assume that 200 MeV of energy is released by each fissioning uranium nucleus, that all of the uranium atoms undergo fission, and that all of the energy-conversion mechanisms used are 100 percent efficient?

67 •• (*a*) Find the wavelength of a particle in the ground state of a one-dimensional infinite square well of length $L = 2.00$ fm. (*b*) Find the momentum in units of MeV/c for a particle that has this wavelength. (*c*) Show that the total energy of an electron that has this wavelength is approximately $E \approx pc$. (*d*) What is the kinetic energy of an electron in the ground state of the well? This calculation shows that if an electron were confined in a region of space as small as a nucleus, it would have a very large kinetic energy.

68 •• If ^{12}C, ^{11}B, and ^{1}H have respective masses of 12.000 000 u, 11.009 306 u, and 1.007 825 u, determine the minimum energy, Q, in MeV, required to remove a proton from a ^{12}C nucleus.

69 ••• Assume that a neutron decays into a proton and an electron without the emission of a neutrino. The kinetic energy shared by the proton and the electron is then 0.782 MeV. In the rest frame of the neutron, the total momentum is zero, so the momentum of the proton must be equal and opposite the momentum of the electron. This determines the ratio of the kinetic energies of the two particles, but because the electron is relativistic, the exact calculation of these relative kinetic energies is somewhat challenging. (*a*) Assume that the kinetic energy of the electron is 0.782 MeV and calculate the momentum p of the electron in units of MeV/c. *Hint: Use* $E^2 = p^2c^2 + (mc^2)^2$ *(Equation 39-27)*. (*b*) Using your result from Part (*a*), calculate the kinetic energy $p^2/2m_\mathrm{p}$ of the proton. (*c*) Because the total kinetic energy of the electron and the proton is 0.782 MeV, the calculation in Part (*b*) gives a correction to the assumption that the kinetic energy of the electron is 0.782 MeV. What percentage of 0.782 MeV is this correction? SSM

70 ••• In the laboratory reference frame, a neutron of mass m moving with speed v_L makes an elastic head-on collision with a nucleus of mass M that is at rest. (*a*) Show that the speed of the center of mass in the lab frame is $V = mv_\mathrm{L}/(m + M)$. (*b*) What is the speed of the nucleus in the center-of-mass frame before the collision and after the collision? (*c*) What is the speed of the nucleus in the laboratory frame after the collision? (*d*) Show that the energy of the nucleus after the collision in the laboratory frame is

$$\frac{1}{2}M(2V)^2 = \frac{4mM}{(m+M)^2}\left(\frac{1}{2}mv_\mathrm{L}^2\right)$$

(*e*) Show that the fraction of the energy lost by the neutron in the elastic collision is

$$\frac{-\Delta E}{E} = \frac{4mM}{(m+M)^2} = \frac{4(m/M)}{[1 + (m/M)]^2}$$

71 ••• (*a*) Use the result from Part (*e*) of Problem 70 to show that after N head-on collisions of a neutron with carbon nuclei at rest, the energy of the neutron is approximately $(0.714)^N E_0$, where E_0 is its original energy. (*b*) How many head-on collisions are required to reduce the energy of the neutron from 2.0 MeV to 0.020 eV, assuming stationary carbon nuclei?

72 ••• On the average, a neutron loses 63 percent of its energy in a collision with a hydrogen atom and 11 percent of its energy in a collision with a carbon atom. Calculate the number of collisions needed to reduce the energy of a neutron from 2.0 MeV to 0.020 eV if the neutron collides with (*a*) hydrogen atoms and (*b*) carbon atoms.

73 ••• Frequently, the daughter nucleus of a radioactive parent nucleus is itself radioactive. Suppose the parent nucleus, designated by P, has a decay constant λ_P, while the daughter nucleus, designated by D, has a decay constant λ_D. The number of daughter nuclei N_D are then given by the solution to the differential equation

$$dN_D/dt = \lambda_P N_P - \lambda_D N_D$$

where N_P is the number of parent nuclei. (*a*) Justify this differential equation. (*b*) Show that the solution for the equation is

$$N_D(t) = \frac{N_{P0}\lambda_P}{\lambda_D - \lambda_P}(e^{-\lambda_P t} - e^{-\lambda_D t})$$

where N_{P0} is the number of parent nuclei present at $t = 0$ when there are no daughter nuclei. (*c*) Show that the expression for N_D in Part (*b*) gives $N_D(t) > 0$ whether $\lambda_P > \lambda_D$ or $\lambda_D > \lambda_P$. (*d*) Make a plot of $N_P(t)$ and $N_D(t)$ as a function of time when $\tau_D = 3\tau_P$, where τ_D and τ_P are the mean lifetimes of the daughter and parent nuclei, respectively. SSM

74 ••• Suppose isotope A decays to isotope B and has a decay constant λ_A, and isotope B in turn decays and has a decay constant λ_B. Suppose a sample contains, at $t = 0$, only isotope A nuclei. Derive an expression for the time at which the number of isotope B nuclei will be a maximum. (See Problem 73.)

75 ••• An example of the situation discussed in Problem 73 is the radioactive isotope ^{229}Th, an α emitter that has a half-life of 7300 y. Its daughter, ^{225}Ra, is a β emitter that has a half-life of 14.8 d. In this instance, as in many instances, the half-life of the parent is much longer than the half-life of the daughter. Using the expression given in Problem 73, Part (*b*), starting with a sample of pure ^{229}Th containing N_{P0} nuclei, show that the number, N_D, of ^{225}Ra nuclei will, after several years, be given by

$$N_D = \frac{\lambda_P}{\lambda_D} N_P$$

where N_P is the number of ^{229}Th nuclei. The number of daughter nuclei are said to be in *secular equilibrium*.

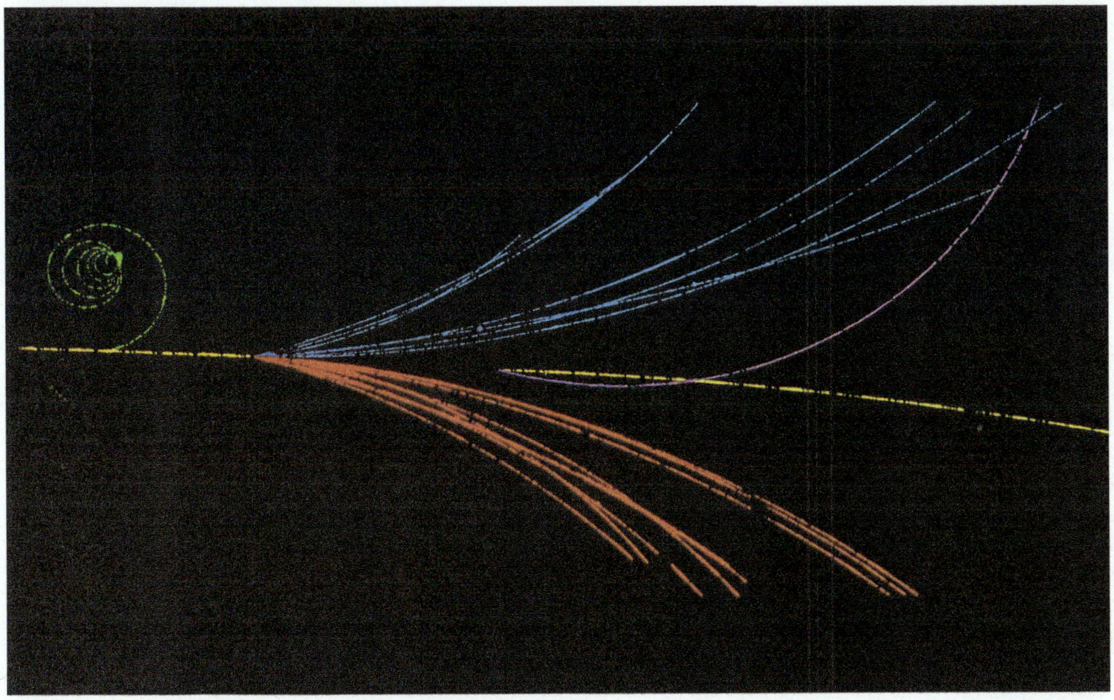

CHAPTER 41

TRACKS IN A BUBBLE CHAMBER PRODUCED BY AN INCOMING HIGH-ENERGY PROTON (YELLOW), INCIDENT FROM THE LEFT, COLLIDING WITH A PROTON AT REST. THE SMALL GREEN SPIRAL IS AN ELECTRON KNOCKED OUT OF AN ATOM. IT CURVES TO THE LEFT BECAUSE OF AN EXTERNAL MAGNETIC FIELD IN THE CHAMBER. THE COLLISION PRODUCES SEVEN NEGATIVE PARTICLES (BLUE), ALL π^-; A NEUTRAL PARTICLE Λ^0 THAT LEAVES NO TRACK; AND NINE POSITIVE PARTICLES (RED) INCLUDING SEVEN π^+, A K^+, AND A PROTON. THE Λ^0 TRAVELS IN THE ORIGINAL DIRECTION OF THE INCOMING PROTON BEFORE DECAYING INTO A PROTON (YELLOW) AND A π^- (PURPLE). *(© Lawrence Livermore Laboratory/Science Photo Library/Photo Researchers.)*

Elementary Particles and the Beginning of the Universe

? How do you determine the energy of particle interactions? (See Example 41-1.)

Items that we encounter in everyday life are made of atoms. In some sense, atoms are the building blocks of nature. However, we know that atoms are not the most fundamental constituents of matter. With the discovery of the electron by J. J. Thomson (1897), the Bohr theory of the nuclear atom (1913), and the discovery of the neutron (1932), it became clear that atoms and even nuclei have considerable structure. Indeed the once simple picture of particle physics in which there were just four "elementary" particles—the proton, neutron, electron, and photon—has become much more complex.

Since the 1950s, enormous sums of money have been spent constructing particle accelerators of greater and greater energies in hopes of finding particles predicted by various theories. At present, we know of several hundred particles that at one time or another have been considered to be elementary, and research teams at the giant accelerator laboratories around the world are searching for and finding new particles. Some of these particles have such short lifetimes (of the order of 10^{-23} s) that they can be detected only indirectly. Many particles are observed only during nuclear reactions using high-energy accelerators. In addition to the usual particle properties of mass, charge, and spin, new properties have been found and given whimsical names such as strangeness, charm, color, topness, and bottomness.

In this chapter, we will first look at the various ways of classifying the multitude of particles that have been found. We will then describe the current theory of elementary particles, called the standard model, *in which all matter in nature—from the exotic particles produced in the giant accelerator laboratories to ordinary grains of sand—is considered to be constructed from just two families of elementary particles, leptons and quarks. In the final section, we will use our knowledge of elementary particles to discuss the big bang theory which describes the origin of the universe.*

41-1 HADRONS AND LEPTONS

All the different forces observed in nature, from ordinary friction to the tremendous forces involved during supernova explosions, can be understood in terms of the four basic interactions: (1) the strong nuclear interaction (also called the hadronic interaction), (2) the electromagnetic interaction, (3) the weak (nuclear) interaction, and (4) the gravitational interaction. The four basic interactions provide a convenient structure for the classification of particles. Some particles participate in all four interactions, whereas other particles participate in only some of the interactions. For example, all particles participate in gravitational interaction, the weakest of the interactions. All particles that have electric charge participate in the electromagnetic interaction.

Particles that interact by the strong interaction are called **hadrons.** There are two kinds of hadrons: **baryons,** which have spin $\frac{1}{2}$ or $\frac{3}{2}$ or $\frac{5}{2}$, etc., and **mesons,** which have spin 0 or 1 or 2, etc. Baryons, which include nucleons, are the most massive of the elementary particles. Mesons have intermediate masses, between the mass of the electron and the mass of the proton. Particles that decay by the strong interaction have very short lifetimes, of the order of 10^{-23} s, which is about the time it takes light to travel a distance equal to the diameter of a nucleus. On the other hand, particles that decay by the weak interaction have much longer lifetimes, of the order of 10^{-10} s. Table 41-1 lists some of the properties of those hadrons that are stable against decay by the strong interaction.

Hadrons are rather complicated entities and have complex structures. If we use the term *elementary particle* to mean a point particle that has no structure and is not constructed from some more elementary entities, then hadrons are not elementary particles. It is now believed that all hadrons are composed of more fundamental entities called *quarks,* which, as far as we know, are truly elementary particles.

Particles that participate in the weak interaction but not in the strong interaction are called **leptons.** These include electrons, muons, and neutrinos, which are all less massive than the lightest hadron. The word *lepton,* meaning "light particle," was chosen to reflect the relatively small mass of the particles. However, the most recently discovered lepton, the *tau,* found by Martin Lewis Perl in 1975, has a mass of 1784 MeV/c^2, nearly twice the mass of the proton (938 MeV/c^2), so we now have a "heavy lepton." In addition, the word *muon,* short for mu-meson, is something of a misnomer. The muon is not now categorized as a meson, so it is best to refer to it

The Super-Kamiokande detector, built in Japan in 1996 as a joint Japanese–American experiment, is essentially a water tank the size of a large cathedral installed in a deep zinc mine 1 mile inside a mountain. When neutrinos pass through the tank, one of the neutrinos occasionally collides with an atom, sending blue light through the water to an array of detectors. This photograph shows the detector wall and top that have approximately 9000 photomultiplier tubes that help detect the neutrinos. Experimental results reported in June 1998 were evidence that the mass of the neutrino cannot be zero. (*ICCR (Institute for Cosmic Ray Research), The University of Tokyo.*)

Table 41-1 Hadrons That Are Stable Against Decay via the Strong Nuclear Interaction

Name	Symbol	Mass MeV/c²	Spin, $\hbar$	Charge, e	Antiparticle	Mean Lifetime, s	Typical Decay Products*
Baryons							
Nucleon	p (proton)	938.3	$\frac{1}{2}$	+1	p^-	Infinite	
	n (neutron)	939.6	$\frac{1}{2}$	0	$\overline{n}$	930	$p + e^- + \overline{\nu}_e$
Lambda	Λ^0	1116	$\frac{1}{2}$	0	$\overline{\Lambda}^0$	2.5×10^{-10}	$p + \pi^-$
Sigma†	Σ^+	1189	$\frac{1}{2}$	+1	$\overline{\Sigma}^-$	0.8×10^{-10}	$n + \pi^+$
	Σ^0	1193	$\frac{1}{2}$	0	$\overline{\Sigma}^0$	10^{-20}	$\Lambda^0 + \gamma$
	Σ^-	1197	$\frac{1}{2}$	−1	$\overline{\Sigma}^+$	1.7×10^{-10}	$n + \pi^-$
Xi	Ξ^0	1315	$\frac{1}{2}$	0	$\overline{\Xi}^0$	3.0×10^{-10}	$\Lambda^0 + \pi^0$
	Ξ^-	1321	$\frac{1}{2}$	−1	$\overline{\Xi}^+$	1.7×10^{-10}	$\Lambda^0 + \pi^-$
Omega	Ω^-	1672	$\frac{3}{2}$	−1	Ω^+	1.3×10^{-10}	$\Xi^0 + \pi^-$
Mesons							
Pion	π^+	139.6	0	+1	π^-	2.6×10^{-8}	$\mu^+ + \nu_\mu$
	π^0	135	0	0	π^0	0.8×10^{-16}	$\gamma + \gamma$
	π^-	139.6	0	−1	π^+	2.6×10^{-8}	$\mu^- + \overline{\nu}_\mu$
Kaon‡	K^+	493.7	0	+1	K^-	1.24×10^{-8}	$\pi^+ + \pi^0$
	K^0	497.7	0	0	$\overline{K}^0$	0.88×10^{-10} and 5.2×10^{-8}	$\pi^+ + \pi^-$ and $\pi^+ + e^- + \overline{\nu}_e$
Eta	η^0	549	0	0		2×10^{-19}	$\gamma + \gamma$

* Other decay modes also occur for most particles.

† The Σ^0 is included here for completeness even though it does decay via the strong interaction.

‡ The K^0 has two distinct lifetimes, sometimes referred to as K^0_{short} and K^0_{long} All other particles have a unique lifetime.

as a muon and not as a mu-meson. As far as we know, leptons are point particles that have no structure and can be considered to be truly elementary in the sense that they are not composed of other particles.

There are six leptons. They are the electron and the electron neutrino, the muon and the muon neutrino, and the tau and the tau neutrino. (Each of the leptons has an antiparticle.) The masses of the electron, the muon, and the tau are quite different. The mass of the electron is 0.511 MeV/c^2, the mass of the muon is 106 MeV/c^2, and the mass of the tau is 1784 MeV/c^2. The standard model predicts that neutrinos, like photons, do not have mass. However, there is now strong evidence that their mass, though very small, is greater than zero. During the late 1990s, experiments using a detector in Japan called the Super-Kamiokande (Super-K) found that neutrinos emitted from the Sun arrived on Earth in much smaller numbers than the numbers that are predicted from the fusion processes in the Sun. This result can be explained if the mass of the neutrino is not zero.* In addition, a neutrino mass as small as a few eV/c^2 would have great cosmological significance. The answer to the question of whether the universe will continue to expand indefinitely or will reach a maximum size and begin to contract depends on the total mass in the universe. Thus, the answer could depend on whether the mass of the neutrino is actually zero or is merely small, because the cosmic density of each species of neutrino is ~100 per cm³.

* The connection between the shortfall of solar-neutrino detections and the mass of the neutrino is elucidated in "On Morphing Neutrinos and Why They Must Have Mass" by Eugene Hecht, *The Physics Teacher* 41 (2003): 164–168.

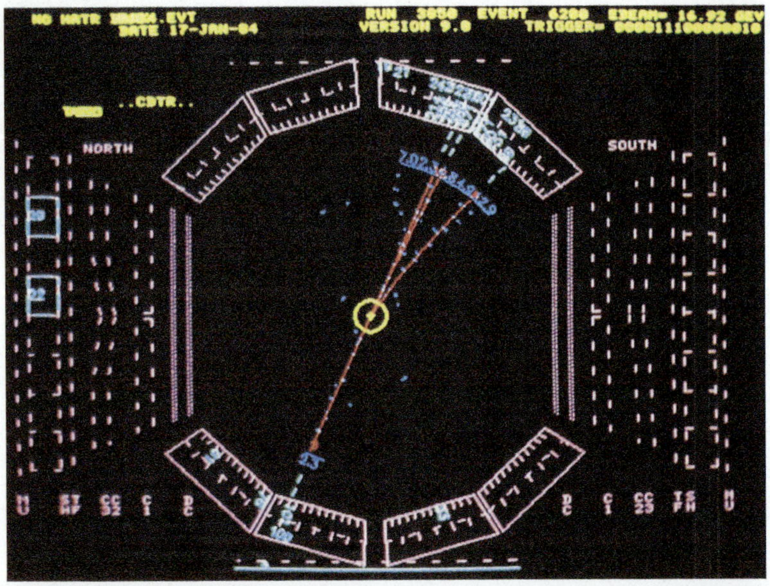

(a)

(b)

(a) A computer display of the production and decay of a τ_1 and τ_2 pair. An electron and a positron annihilate at the center marked by the yellow cross, producing a τ^+ and τ^- pair, which travel in opposite directions, but quickly decay while still inside the beam pipe (yellow circle). The τ^+ decays into two invisible neutrinos and a μ^+, which travels toward the bottom left. Its track in the drift chamber is calculated by a computer and indicated in red. It penetrates the lead–argon counters outlined in purple and is detected at the blue dot near the bottom blue line that marks the end of a muon detector. The τ^- decays into three charged pions (red tracks moving upward) plus invisible neutrinos. (b) The Mark I detector, built by a team from the Stanford Linear Accelerator Center (SLAC) and the Lawrence Berkeley Laboratory, became famous for many discoveries, including the J/ψ meson and the τ lepton. Tracks of particles are recorded by wire spark chambers wrapped in concentric cylinders around the beam pipe extending out to the ring where physicist Carl Friedberg has his right foot. Beyond this are two rings of protruding tubes, housing photomultipliers that view various scintillation counters. The rectangular magnets at the left guide the counterrotating beams that collide in the center of the detector. *((a) Science Photo Library/Photo Researchers. (b) © Lawrence Berkeley Laboratory/ Science Photo Library/Photo Researchers.)*

The observation of electron neutrinos from the supernova 1987A puts an upper limit on the masses of the neutrinos. Because the velocity of a particle that has mass depends on its energy, the arrival time of a burst of neutrinos that have mass from a supernova would be spread out in time. The fact that the electron neutrinos from the 1987 supernova all arrived at Earth within 13 s of one another results in an upper limit of about 16 eV/c^2 for their mass. Note that an upper limit does not imply that the mass is not zero. Measurements of the relative number of muon neutrinos and electron neutrinos entering the huge, underground Super-K detector suggest that at least one type of neutrino can oscillate between types (for example, between a mu neutrino and a tau neutrino). Further measurements of antineutrinos from nuclear reactors strongly show that all three types of neutrinos oscillate between types and thus have mass. Measurements made in Japan, using the *Kam*ioka *L*iquid Scintillator *A*nti-*N*eutrino *D*etector (KamLAND), show that oscillations from one species of neutrino to another species of neutrino can be observed over path lengths as short as 180 km (Figure 41-1).

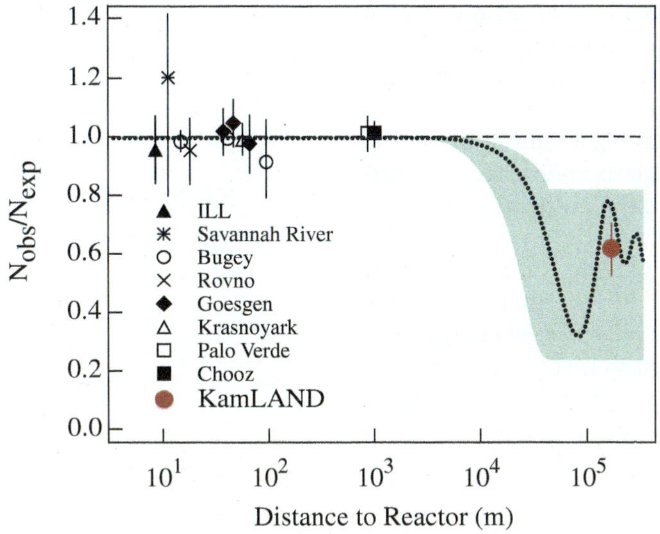

FIGURE 41-1 First evidence for antineutrino disappearance. The ratio of the number of antineutrinos observed N_{obs} to the number that one would expect to observe N_{exp} (assuming no neutrino oscillations) is plotted versus distance to the nearest antineutrino sources. The KamLAND site is 180 km from nearby antineutrino sources (nuclear reactors), while the other eight detector sites are less than 1.0 km from nearby nuclear reactors. For those eight sites, $N_{obs}/N_{exp} = 1.0$, which is what is expected assuming no neutrino oscillations. However, the KamLAND detector found $N_{obs}/N_{exp} = 0.6$. This result is strong evidence that while neutrinos do not oscillate in significant numbers while traveling over path lengths of less than 1.0 km, they do oscillate in significant numbers while traveling over path lengths only a few orders of magnitude longer than 1 km. *(© Lawrence Berkeley Laboratory/ Science Photo Library/Photo Researchers.)*

41-2 SPIN AND ANTIPARTICLES

One important characteristic of a particle is its intrinsic spin angular momentum. We have already discussed the fact that the electron has a quantum number m_s that corresponds to the z component of its intrinsic spin characterized by the quantum number $s = \frac{1}{2}$. Protons, neutrons, neutrinos, and the various other particles that also have an intrinsic spin characterized by the quantum number $s = \frac{1}{2}$ are called **spin-$\frac{1}{2}$ particles.** Particles that have spin $\frac{1}{2}$ (or $\frac{3}{2}$, $\frac{5}{2}$, etc.) are called fermions and obey the exclusion principle. Particles such as pions and other mesons have zero spin or integral spin ($s = 0, 1, 2$, etc.). Those particles are called bosons and do not obey the exclusion principle. That is, any number of those particles can be in the same quantum state.

Spin-$\frac{1}{2}$ particles are described by the Dirac equation, which is an extension of the Schrödinger equation that includes special relativity. One feature of Paul Dirac's theory, proposed in 1927, is the prediction of the existence of antiparticles. In special relativity, the energy of a particle is related to the mass and the momentum of the particle by $E = \pm\sqrt{p^2c^2 + m^2c^4}$ (Equation 39-27). We usually choose the positive solution and dismiss the negative-energy solution with a physical argument. However, the Dirac equation requires the existence of wave functions that correspond to the negative-energy states. Dirac got around this difficulty by postulating that all the negative-energy states were

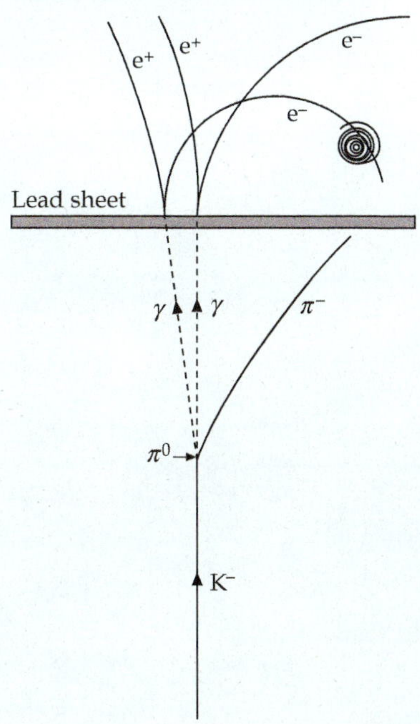

A negative kaon (K^-) enters a bubble chamber from the bottom and decays into a π^-, which moves off to the right, and a π^0, which immediately decays into two photons whose paths are indicated by the dashed lines in the drawing. Each photon interacts in the lead sheet, producing an electron–positron pair. The spiral at the right is another electron that has been knocked out of an atom in the chamber. (Other extraneous tracks have been removed from the photograph.) *(Figure 4 from "First Results from KamLAND: Evidence for Reactor Antineutrino Disappearance" by the KamLAND Collaboration,* Physical Review Letters, *Vol. 90, No. 2, December 17, 2003. Copyright © 2003 The American Physical Society. Reprinted with permission.)*

filled and would therefore not be observable. Only holes in the "infinite sea" of negative-energy states would be observed. For example, a hole in the negative sea of electron energy states would appear as a particle identical to the electron except having positive charge. When such a particle came in the vicinity of an electron the two particles would annihilate, releasing two photons having a minimum total energy of $2m_ec^2$, where m_e is the mass of the electron. This interpretation received little attention until a particle with just those properties, called the positron, was discovered in 1932 by Carl Anderson.

Antiparticles are never created alone but always in particle–antiparticle pairs. In the creation of an electron–positron pair by a photon, the energy of the photon must be at least as great as the rest energy of the electron plus the rest energy of the positron, which is $2m_ec^2 \approx 1.02$ MeV. Although the positron is stable, it has only a short-term existence in our universe because of the large supply of electrons in matter. The fate of a positron is annihilation according to the reaction

$$e^+ + e^- \rightarrow \gamma + \gamma \qquad 41\text{-}1$$

The probability of this reaction is large only if the positron and electron are moving slowly relative to one another. In the center-of-mass reference frame, the momentum of the two particles prior to annihilation is zero, so two photons moving in opposite directions are needed to conserve linear momentum.

The fact that we call electrons *particles* and positrons *antiparticles* does not imply that positrons are less fundamental than electrons. It merely reflects the nature of our universe. If our matter were made up of negative protons and positive electrons, then positive protons and negative electrons would suffer quick annihilation and would be called antiparticles.

An aerial view of the European Laboratory for Particle Physics (CERN) just outside of Geneva, Switzerland. The large circle shows the Large Electron–Positron collider (LEP) tunnel, which is 27 km in circumference. The irregular dashed line is the border between France and Switzerland. *(Richard Ehrlich.)*

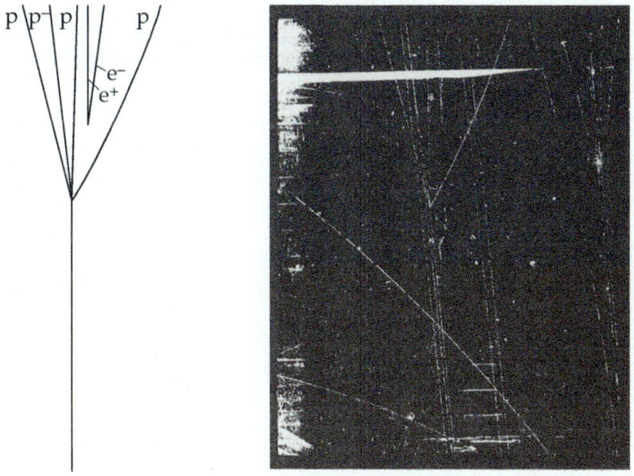

FIGURE 41-2 Bubble-chamber tracks that show the creation of a proton–antiproton pair in the collision of an incident 25-GeV proton with a stationary proton in liquid hydrogen. *(CERN.)*

The antiproton (p^-) was discovered in 1955 by Emilio Segrè and Owen Chamberlain using a beam of protons in the Bevatron at Berkeley to produce the reaction*

$$p^+ + p^+ \rightarrow p^+ + p^+ + p^+ + p^- \qquad 41\text{-}2$$

The creation of a proton–antiproton pair (Figure 41-2) requires kinetic energy of at least $2m_pc^2 = 1877$ MeV $= 1.877$ GeV in the zero-momentum reference frame in which the two protons approach each other with equal and opposite momenta. In the laboratory frame in which one of the protons is initially at rest, the kinetic energy of the incoming proton must be at least $6m_pc^2 = 5.63$ GeV (see Problem 38 of Chapter 39). This energy was not available in laboratories before the development of high-energy accelerators in the 1950s. Antiprotons annihilate with protons to produce two gamma rays in a reaction similar to the reaction in Equation 41-1.

The tunnel of the proton–antiproton collider at CERN. The same bending magnets and focusing magnets can be used for protons or antiprotons moving in opposite directions. The rectangular box in the foreground is a focusing magnet, and the next four boxes are the bending magnets. *(CERN.)*

* The antiproton is sometimes denoted by $\bar{p}$ rather than p^-. For neutral particles, such as the neutron, the bar must be used to denote the antiparticle. Thus, the antineutron is denoted by $\bar{n}$. The electron and proton are often denoted by e and p without the minus sign or plus sign superscripts.

Example 41-1 Proton–Antiproton Annihilation

Context-Rich

You have been reading about nuclear physics and particle interactions. In particular, you have been looking at the reaction $p^+ + p^- \longrightarrow \gamma + \gamma$ (proton–antiproton annihilation). You wonder if the photons produced are visible to the human eye if the two protons are initially at rest. Are the photons visible to the human eye?

PICTURE If the photons are visible, they should have wavelengths in the visible range (400 nm to 800 nm.) Because the proton and the antiproton are at rest, conservation of momentum requires that the two photons created during their annihilation have equal and opposite momenta and therefore equal energies, frequencies, and wavelengths. Conservation of energy implies that the photons have a combined energy equal to the rest energy of the proton plus the rest energy of the antiproton (approximately 938 MeV each).

SOLVE

1. Set the total energy of the two photons, $2E_\gamma$, equal to the rest energy of the proton plus antiproton and solve for E_γ:

$$2E_\gamma = 2m_p c^2$$

so

$$E_\gamma = m_p c^2 = 938 \text{ MeV}$$

2. Set the energy of the photon equal to $hf = hc/\lambda$ and solve for the wavelength λ:

$$E_\gamma = hf = \frac{hc}{\lambda}$$

$$\lambda = \frac{hc}{E_\gamma} = \frac{1240 \text{ eV}\cdot\text{nm}}{938 \text{ MeV}}$$

$$= 1.32 \times 10^{-6} \text{ nm} = 1.32 \text{ fm}$$

3. Compare this wavelength with the wavelengths of visible light:

The photons are *not* in the visible spectrum.

CHECK In Chapter 36, we found that the energies of photons in the visible spectrum are equal to only a few electron volts. It is not a surprise to find that photons that have energies on the order of 10^9 electron volts are not in the visible spectrum.

TAKING IT FURTHER The wavelength of the photons produced by proton–antiproton annihilation is more than eight orders of magnitude less than 400 nm—the shortest wavelength in the visible spectrum.

41-3 THE CONSERVATION LAWS

One adage is "anything that can happen does." If a conceivable decay or reaction does not occur, there must be a reason. The reason is usually expressed in terms of a conservation law. The conservation of energy rules out the decay of any particle for which the total mass of the decay products would be greater than the initial mass of the particle before decay. The conservation of linear momentum requires that when an electron and a positron at rest annihilate, two photons must be emitted. Angular momentum must also be conserved during a reaction or a decay. A fourth conservation law that restricts the possible particle decays and reactions is the conservation of electric charge. The net electric charge before a decay or a reaction must equal the net charge after the decay or the reaction.

There are two additional conservation laws that are important in the reactions and the decays of elementary particles: the conservation of baryon number and the conservation of lepton number. Consider the proposed decay

$$p \rightarrow \pi^0 + e^+$$

where π is the symbol for the pion (pi-meson). This decay would conserve charge, energy, angular momentum, and linear momentum, but it does not occur. It does not conserve either lepton number or baryon number. (The proton p is a baryon, the positron e^+ is a lepton, and the π^0 is a meson.) The conservation of lepton number and baryon number implies that whenever a lepton or a baryon is created, an antiparticle of the same type is also created. We assign the **lepton number** $L = +1$ to all leptons, $L = -1$ to all antileptons, and $L = 0$ to all other particles. Similarly, the **baryon number** $B = +1$ is assigned to all baryons, $B = -1$ to all antibaryons, and $B = 0$ to all other particles. The sum of the baryon numbers and the sum of the lepton numbers cannot change during a reaction or a decay. The conservation of baryon number along with the conservation of energy implies that the least massive baryon, the proton, must be stable.

The conservation of lepton number implies that the neutrino emitted during the β decay of the free neutron is an antineutrino:

$$n \rightarrow p^+ + e^- + \bar{\nu}_e \qquad 41\text{-}3$$

The fact that neutrinos and antineutrinos are different is illustrated by an experiment in which ^{37}Cl is bombarded with an intense antineutrino beam from the decay of reactor neutrons (reactor neutrons are fission products that are produced in nuclear reactors). If neutrinos and antineutrinos were the same, we would expect the following reaction:

$$^{37}_{17}Cl + \bar{\nu}_e \rightarrow {}^{37}_{18}Ar + e^- \qquad 41\text{-}4$$

This reaction is not observed. However, if protons are bombarded with antineutrinos, the reaction

$$p + \bar{\nu}_e \rightarrow n + e^+ \qquad 41\text{-}5$$

is observed. Note that the sum of the lepton numbers is -1 on the left side of the reaction equation in Equation 41-4 and is $+1$ on the right side of the reaction equation. But the sum of the lepton numbers is -1 on both sides of the reaction equation in Equation 41-5.

Not only are neutrinos and antineutrinos distinct particles, but the neutrinos associated with electrons are distinct from the neutrinos associated with muons. Electron-like leptons (e and ν_e), muon-like leptons (μ and ν_μ), and tau-like leptons (τ and ν_τ) are each separately conserved, so we assign separate lepton numbers L_e, L_μ, and L_τ to the particles. The leptons and their lepton numbers are listed in Table 41-2.

CONCEPT CHECK 41-1

Why is it that the conservation of baryon number along with the conservation of energy implies that the proton, which is the least massive baryon, must be stable?

Table 41-2 Lepton Numbers

	L_e	L_μ	L_τ
e^-	+1	0	0
ν_e	+1	0	0
e^+	−1	0	0
$\bar{\nu}_e$	−1	0	0
μ^-	0	+1	0
ν_μ	0	+1	0
μ^+	0	−1	0
$\bar{\nu}_\mu$	0	−1	0
τ^-	0	0	+1
ν_τ	0	0	+1
τ^+	0	0	−1
$\bar{\nu}_\tau$	0	0	−1

Example 41-2 What Laws Are Being Violated? *Conceptual*

What conservation laws (if any) are violated by the following proposed decays: (*a*) $n \rightarrow p + \pi^-$, (*b*) $\Lambda^0 \rightarrow p^- + \pi^+$, and (*c*) $\mu^- \rightarrow e^- + \gamma$? ($\Lambda^0$ is the symbol for the lambda-zero particle.)

PICTURE All reactions must separately conserve energy, electric charge, baryon number, electron lepton number, muon lepton number, and tau lepton number.

SOLVE

(*a*) There are no leptons in this decay, so there is no problem with the conservation of lepton number. The net charge is zero before the decay and after the decay, so charge is conserved. Also, the baryon number is +1 both before and after the decay. However, rest energy of the proton (938.3 MeV) plus the rest energy of the pion (139.6 MeV) is greater than the rest energy of the neutron (939.6 MeV). In the rest frame of the neutron, the energy prior to the reaction (the rest energy of the neutron) is less than the total rest energy following the reaction.

This decay does not conserve energy.

(*b*) Again, there are no leptons involved, and the net charge is zero before the decay and after the decay. Also, the rest energy of the lambda-zero (1116 MeV) is greater than the rest energy of the antiproton (938.3 MeV) plus the rest energy of the pion (139.6 MeV), so in the rest frame of the lambda-zero the energy prior to the reaction (the rest energy of the lambda-zero) is greater than the total rest energy following the reaction. Energy could be conserved, with the loss in rest energy equal to the gain in kinetic energy of the decay products. There are no leptons in the reaction, so all three lepton numbers are conserved. The baryon number is +1 for the lambda particle and −1 for the antiproton and zero for the pi-meson.

This decay does not conserve baryon number.

(*c*) The μ^- has a muon lepton number (L_μ) equal to +1 and an electron lepton number (L_e) equal to 0, the e^- has $L_\mu = 0$ and $L_e = +1$, and the γ has $L_\mu = L_e = 0$.

This reaction does not conserve either muon lepton number or electron lepton number.

TAKING IT FURTHER The muon does decay by $\mu^- \rightarrow e^- + \bar{\nu}_e + \nu_\mu$, which does conserve both muon lepton numbers and electron lepton numbers.

There are some conservation laws that are not universal but apply only to certain kinds of interactions. In particular, there are quantities that are conserved during decays and reactions that occur by the strong interaction but not during decays or reactions that occur by the weak interaction. One of these quantities that is particularly important is **strangeness,** introduced by M. Gell-Mann and K. Nishijima in 1952 to explain the strange behavior of some of the heavy baryons and mesons. Consider the reaction

$$p + \pi^- \rightarrow \Lambda^0 + K^0 \qquad 41\text{-}6$$

where K is the symbol for the kaon (K-meson). The proton and the pion interact by the strong interaction. Both the Λ^0 and K^0 decay into hadrons

$$\Lambda^0 \rightarrow p + \pi^- \qquad 41\text{-}7$$

and

$$K^0 \rightarrow \pi^+ + \pi^- \qquad 41\text{-}8$$

However, the decay times for both the Λ^0 and K^0 are of the order of 10^{-10} s, which is characteristic of the weak interaction, rather than 10^{-23} s, which would be expected for the strong interaction. Other particles showing similar behavior were called **strange particles.** These particles are always produced in pairs, even when

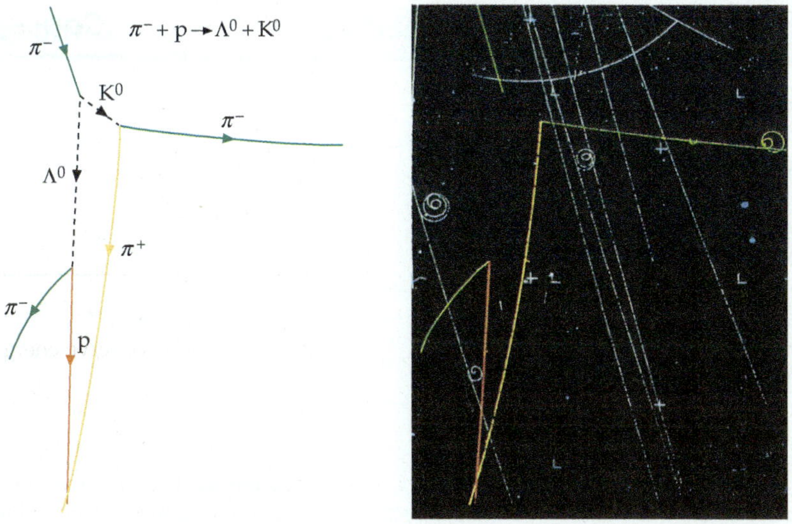

An early photograph of bubble-chamber tracks at the Lawrence Berkeley Laboratory, showing the production and the decay of two particles that have nonzero strangeness, the K^0 and the Λ^0. These neutral particles are identified by the tracks of their electrically charged decay particles. The lambda particle was named because of the similarity of the tracks of its decay particles to the uppercase Greek letter lambda (Λ). (The blue tracks are particles not involved in the reaction of Equation 41-6.) *(© Lawrence Berkeley Laboratory/Science Photo Library/Photo Researchers.)*

all other conservation laws are met. This behavior is described by assigning a new property called strangeness to the particles. During reactions and decays that occur by the strong interaction, strangeness is conserved. During reactions and decays that occur by the weak interaction, the strangeness can only change by ±1. The strangeness of the ordinary hadrons—the nucleons and pions—was arbitrarily taken to be zero. The strangeness of the K^0 was arbitrarily chosen to be +1. The strangeness of the Λ^0 particle must then be −1 so that strangeness is conserved during the reaction described by Equation 41-6. The strangeness of other particles could then be assigned by looking at their various reactions and decays. During reactions and decays that occur by the weak interaction, the strangeness can change by ±1.

Figure 41-3 shows the masses of the baryons and the mesons that are stable against decay by the strong interaction versus strangeness. We can see from this figure that the particles cluster in multiplets of one, two, or three particles of approximately equal mass, and that the strangeness of a multiplet of particles is related to the *center of charge* of the multiplet.

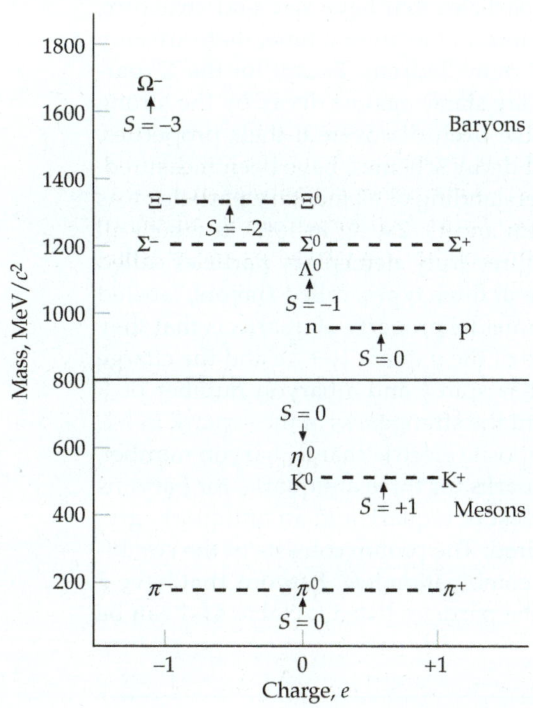

FIGURE 41-3 The strangeness of hadrons shown on a plot of mass versus charge. The strangeness of a baryon–charge multiplet is related to the number of places on the plot that the center of charge of the multiplet is displaced from that of the nucleon doublet. For each "displacement" of *e*, the strangeness changes by ±1. For mesons, the strangeness is related to the number of places the center of charge is displaced from that of the pion triplet. Because of the unfortunate original assignment of +1 for the strangeness of K-mesons (kaons), all of the baryons that are stable against decay by the strong interaction have negative or zero strangeness.

Example 41-3 Strong Interaction, Weak Interaction, or No Interaction *Conceptual*

State whether the following decays can occur by the strong interaction, by the weak interaction, or not at all: (*a*) $\Sigma^+ \rightarrow p + \pi^0$, (*b*) $\Sigma^0 \rightarrow \Lambda^0 + \gamma$, and (*c*) $\Xi^0 \rightarrow n + \pi^0$, where Σ, Λ and Ξ are the symbols for the sigma, lambda and xi particles respectively.

PICTURE We first note that the mass of each decaying particle is greater than the mass of the decay products, so none of the three reactions are in conflict with the principle of conservation of energy. In addition, no leptons are involved in any of the three decays, and charge and baryon number are both conserved during all three reactions. The decay will occur by the strong interaction if strangeness is conserved (if $\Delta S = 0$). If $\Delta S = \pm 1$, the decay will occur via the weak interaction. If $|\Delta S| > 1$, the decay will not occur.

SOLVE

(*a*) From Figure 41-3, we can see that the strangeness of the Σ^+ is -1, whereas the strangeness of both the proton and the pion is zero.

This decay is possible by the weak interaction but not by the strong interaction. It is, in fact, one of the decay modes of the Σ^+ particle which has a lifetime of the order of 10^{-10} s.

(*b*) The strangeness of both the Σ^0 and Λ^0 is -1, whereas the strangeness of the photon is zero.

This decay can proceed by the strong interaction. It is, in fact, the dominant mode of decay of the Σ^0 particle which has a lifetime of approximately 10^{-20} s.

(*c*) The strangeness of the Ξ^0 is -2, whereas the strangeness of both the neutron and the pion is zero.

Because strangeness cannot change by 2 during a decay or a reaction, this decay does not occur.

41-4 QUARKS

Leptons appear to be truly elementary particles in that they do not break down into smaller entities and they seem to have no measurable size or structure. Hadrons, on the other hand, are complex particles that have size and structure, and they decay into other hadrons. Furthermore, at the present time, there are only six known leptons, whereas there are many more hadrons. Except for the Σ^0 particle, Table 41-1 includes only hadrons that are stable against decay by the strong interaction. Hundreds of other hadrons have been discovered; their properties, such as charge, spin, mass, strangeness, and decay schemes, have been measured.

The most important advance in our understanding of elementary particles was the quark model proposed by M. Gell-Mann and G. Zweig in 1963 in which all hadrons consist of combinations of two or three truly elementary particles called **quarks.*** In the original model, quarks came in three types, called **flavors,** labeled *u*, *d*, and *s* (for *up*, *down*, and *strange*). An unusual property of quarks is that they carry fractional electron charges. The charge of the *u* quark is $+\frac{2}{3}e$ and the charge of the *d* and *s* quarks is $-\frac{1}{3}e$. Each quark has spin $\frac{1}{2}$ and a baryon number of $\frac{1}{3}$. The strangeness of the *u* and *d* quark is 0, and the strangeness of the *s* quark is -1. Each quark has an antiquark that has the opposite electric charge, baryon number, and strangeness. Baryons consist of three quarks (or three antiquarks for baryons that are antiparticles), whereas mesons consist of a quark and an antiquark, giving mesons baryon numbers $B = 0$, as required. The proton consists of the combination *uud* and the neutron consists of the combination *udd*. Baryons that have a strangeness $S = -1$ have one *s* quark. All the particles listed in Table 41-1 can be

* The name *quark* was chosen by M. Gell-Mann from a quotation from *Finnegan's Wake* by James Joyce.

constructed from these three quarks and three antiquarks.† The great strength of the quark model is that all the allowed combinations of three quarks or quark–antiquark pairs result in known hadrons. Strong evidence for the existence of quarks inside a nucleon is provided by high-energy scattering experiments called *deep inelastic scattering.* During these experiments, a nucleon is bombarded with electrons, muons, or neutrinos of energies from 15 GeV to 200 GeV. Analyses of particles scattered at large angles indicate that inside the nucleon are three spin-$\frac{1}{2}$ particles of sizes much smaller than that of the nucleon. These experiments are analogous to Rutherford's scattering of α particles by atoms in which the presence of a tiny nucleus in the atom was inferred from the large-angle scattering of the α particles.

In 1967, a fourth quark was proposed to explain some discrepancies between experimental determinations of certain decay rates and calculations based on the quark model. The fourth quark is labeled c for a new property called **charm.** Like strangeness, charm is conserved during strong interactions but changes by ± 1 in weak interactions. In 1975, a new heavy meson called the ψ meson was discovered that has the properties expected of a $c\bar{c}$ combination. Since then, other mesons that have combinations such as $c\bar{d}$ and $\bar{c}d$, as well as baryons having the charmed quark, have been discovered. Two more quarks labeled t and b (for *t*op and *b*ottom) were proposed in the 1970s. In 1977, a massive new meson called the Υ (upsilon) meson or **bottomonium,** which is considered to have the quark combination $b\bar{b}$, was discovered. The top quark was observed in 1995. The properties of the six quarks are listed in Table 41-3.

The six quarks and six leptons (and their antiparticles) are thought to be the fundamental elementary particles of which all matter is composed. Table 41-4 lists the masses of the fundamental particles. In this table, the masses given for neutrinos are upper limits. The masses given for quarks are educated guesses. There is strong experimental evidence for the existence of each of these particles.

Table 41-3 Properties of Quarks and Antiquarks

Flavor	Spin	Charge	Baryon Number	Strangeness	Charm	Topness	Bottomness
Quarks							
u (up)	$\frac{1}{2}\hbar$	$+\frac{2}{3}e$	$+\frac{1}{3}$	0	0	0	0
d (down)	$\frac{1}{2}\hbar$	$-\frac{1}{3}e$	$+\frac{1}{3}$	0	0	0	0
s (strange)	$\frac{1}{2}\hbar$	$-\frac{1}{3}e$	$+\frac{1}{3}$	−1	0	0	0
c (charmed)	$\frac{1}{2}\hbar$	$+\frac{2}{3}e$	$+\frac{1}{3}$	0	+1	0	0
t (top)	$\frac{1}{2}\hbar$	$+\frac{2}{3}e$	$+\frac{1}{3}$	0	0	+1	0
b (bottom)	$\frac{1}{2}\hbar$	$-\frac{1}{3}e$	$+\frac{1}{3}$	0	0	0	+1
Antiquarks							
$\bar{u}$	$\frac{1}{2}\hbar$	$-\frac{2}{3}e$	$-\frac{1}{3}$	0	0	0	0
$\bar{d}$	$\frac{1}{2}\hbar$	$+\frac{1}{3}e$	$-\frac{1}{3}$	0	0	0	0
$\bar{s}$	$\frac{1}{2}\hbar$	$+\frac{1}{3}e$	$-\frac{1}{3}$	+1	0	0	0
$\bar{c}$	$\frac{1}{2}\hbar$	$-\frac{2}{3}e$	$-\frac{1}{3}$	0	−1	0	0
$\bar{t}$	$\frac{1}{2}\hbar$	$-\frac{2}{3}e$	$-\frac{1}{3}$	0	0	−1	0
$\bar{b}$	$\frac{1}{2}\hbar$	$+\frac{1}{3}e$	$-\frac{1}{3}$	0	0	0	−1

† The correct quark combinations of hadrons are not always obvious, because of the symmetry requirements on the total wave function. For example, the π^0 meson is represented by a linear combination of $u\bar{u}$ and $d\bar{d}$.

Table 41-4 Masses of Fundamental Particles

Particle	Mass
Quarks	
u (up)	336 MeV/c^2
d (down)	338 MeV/c^2
s (strange)	540 MeV/c^2
c (charmed)	1500 MeV/c^2
t (top)	174 000 MeV/c^2
b (bottom)	4500 MeV/c^2
Leptons	
e^- (electron)	0.511 MeV/c^2
ν_e (electron neutrino)	$<$ 2.2 eV/c^2
μ^- (muon)	105.659 MeV/c^2
ν_μ (muon neutrino)	$<$ 0.17 MeV/c^2
τ^- (tau)	1784 MeV/c^2
ν_τ (tau neutrino)	$<$ 28 MeV/c^2

Example 41-4 Given the Constituent Quark Species, Identify the Particle — *Conceptual*

What are the properties of the particles made up of the following quarks: (*a*) $u\bar{d}$, (*b*) $\bar{u}d$, (*c*) dds, and (*d*) uss?

PICTURE Baryons are made up of three quarks, whereas mesons consist of a quark and an antiquark. We add the electric charges of the quarks to find the total charge of the hadron. We also find the strangeness of the hadron by adding the strangeness of the quarks.

SOLVE

(*a*) Because $u\bar{d}$ is a quark-antiquark combination, it has baryon number 0 and is therefore a meson. There is no strange quark here (that is, $S = 0$), so the strangeness of the meson is zero. The charge of the up quark is $+\frac{2}{3}e$ and the charge of the antidown quark is $+\frac{1}{3}e$, so the charge of the meson is $+1e$.

The quark combination $u\bar{d}$ is the π^+ meson.

(*b*) The particle $\bar{u}d$ is also a meson that has zero strangeness. Its electric charge is $-\frac{2}{3}e + \left(-\frac{1}{3}e\right) = -1e$.

The quark combination $\bar{u}d$ is the π^- meson.

(*c*) The particle dds is a baryon that has strangeness -1 because it has one strange quark. Its electric charge is $-\frac{1}{3}e - \frac{1}{3}e - \frac{1}{3}e = -1e$.

The quark combination dds is the Σ^- particle.

(*d*) The particle uss is a baryon that has strangeness -2. Its electric charge is $+\frac{2}{3}e - \frac{1}{3}e - \frac{1}{3}e = 0$.

The quark combination uss is the Ξ^0 particle.

QUARK CONFINEMENT

Despite considerable experimental effort, no isolated quark has ever been observed. It is now believed that it is impossible to obtain an isolated quark. Although the force between quarks is not known, it is believed that the potential energy of two quarks increases with increasing separation distance so that an infinite amount of energy would be needed to separate the quarks completely.

This would be true, for example, if the force of attraction between two quarks remains constant or increases with separation distance, rather than decreasing with increasing separation distance as is the case for other fundamental forces, such as the electric force between two charges, the gravitational force between two masses, and the strong nuclear force between two hadrons.

When a large amount of energy is added to a quark system, such as a nucleon, a quark–antiquark pair is created and the original quarks remain confined within the original system. Because quarks cannot be isolated, but are always bound together to form a baryon or a meson, the mass of a quark cannot be accurately known, which is why the masses listed in Table 41-4 are merely educated guesses.

41-5 FIELD PARTICLES

In addition to the six fundamental leptons and six fundamental quarks, there are other particles, called *field particles,* or *field quanta,* that are associated with the forces exerted by one elementary particle on another. In **quantum electrodynamics,** the electromagnetic field of a single charged particle is described by **virtual photons** that are continuously being emitted and reabsorbed by the particle. If we put energy into the system by accelerating the charge, some of these virtual photons are shaken off and become real, observable photons. The photon is said to mediate the electromagnetic interaction. Each of the four basic interactions can be described via mediating field particles.

The field quantum associated with the gravitational interaction, called the **graviton,** has not yet been observed. The gravitational *charge* analogous to electric charge is mass.

The weak interaction is thought to be mediated by three field quanta called **vector bosons:** W^+, W^-, and Z^0. These particles were predicted by Sheldon Glashow, Abdus Salam, and Steven Weinberg in a theory called the *electroweak theory,* which we discuss in the next section. The W and Z particles were first observed in 1983 by a group of over a hundred scientists led by Carlo Rubbia using the high-energy accelerator at CERN in Geneva, Switzerland. The masses of the $W^\pm$ particles (about 80 GeV/c^2) and the Z particle (about 91 GeV/c^2) measured during this experiment were in excellent agreement with those predicted by the electroweak theory. (The W^- particle is the antiparticle of the W^+ particle, so they must have identical masses.)

The field quanta associated with the strong force between quarks are called **gluons.** Isolated gluons have not been observed experimentally. The *charge* responsible for the strong interactions comes in three varieties, labeled *red, green,* and *blue* (analogous with the three primary colors), and the strong charge is called the **color charge.** The field theory for strong interactions, analogous to quantum electrodynamics for electromagnetic interactions, is called **quantum chromodynamics (QCD).**

Table 41-5 lists the bosons responsible for mediating the basic interactions.

Table 41-5 Bosons That Mediate the Basic Interactions

Interaction	Boson	Spin	Mass	Electric Charge
Strong	g (gluon)	1	0	0
Weak	$W^\pm$	1	80.22 GeV/c^2	$\pm 1e$
	Z^0	1	91.19 GeV/c^2	0
Electromagnetic	γ (photon)	1	0	0
Gravitational	Graviton†	2	0	0

† Not yet observed.

41-6 THE ELECTROWEAK THEORY

In the **electroweak theory,** the electromagnetic and weak interactions are considered to be two different manifestations of a more fundamental electroweak interaction. At very high energies ($\gg 100$ GeV), the electroweak interaction would be mediated by four bosons. From symmetry considerations, these would be a triplet consisting of W^+, W^0, and W^-, all of equal mass, and a singlet B^0 of some other mass. Neither the W^0 nor the B^0 would be observed directly, but one linear combination of the W^0 and the B^0 would be the Z^0 and another would be the photon. At ordinary energies, the symmetry is broken. This leads to the separation of the electromagnetic interaction mediated by the massless photon and the weak interaction mediated by the W^+, W^-, and Z^0 particles. The fact that the photon is massless and that the W and Z particles have masses of the order of 100 GeV/c^2 shows that the symmetry assumed in the electroweak theory does not exist at lower energies.

The symmetry-breaking mechanism is called a **Higgs field,** which requires a new boson, the **Higgs boson,** whose rest energy is expected to be of the order of 1 TeV (1 TeV = 10^{12} eV). The Higgs boson has not yet been observed. Calculations show that Higgs bosons (if they exist) should be produced in head-on collisions between protons of energies of the order of a few TeV. Such energies are not presently available, but the Large Hadron Collider, a particle accelerator under construction near Geneva, Switzerland, is scheduled to come on line late in 2007. The LHC is designed to produce head-on proton–proton collisions for which each proton has an energy of 7 TeV.

41-7 THE STANDARD MODEL

The combination of the quark model, electroweak theory, and quantum chromodynamics is called the **standard model.** In this model, the fundamental particles are the leptons and quarks, each of which comes in six flavors, as shown in Table 41-4; the force carriers are the photon, the $W^\pm$ and Z particles, and the gluons (of which there are eight types). The leptons and quarks are all spin-$\frac{1}{2}$ fermions, which obey the exclusion principle, and the force carriers are integral-spin bosons, which do not obey the exclusion principle. Every interaction in nature is due to one of the four basic interactions: strong, electromagnetic, weak, and gravitational. A particle experiences one of the basic interactions if it carries a charge associated with that interaction. Electric charge is the familiar charge that we have studied previously. Weak charge, also called flavor charge, is carried by leptons and quarks. The charge associated with the strong interaction is called color charge and is carried by quarks and gluons but not by leptons. The charge associated with the gravitational force is mass. It is important to note that the photon, which mediates the electromagnetic interaction, does not carry electric charge. Similarly, the $W^\pm$ and Z particles, which mediate the weak interaction, do not carry weak charge. However, the gluons, which mediate the strong interaction, do carry color charge. This fact is related to the confinement of quarks as discussed in Section 41-4.

All matter is made up of leptons or quarks. There are no known composite particles consisting of leptons bound together by the weak force. Leptons exist only as isolated particles. Hadrons (baryons and mesons) are composite particles consisting of quarks bound together by the color charge. A result of QCD theory is that only color-neutral combinations of quarks are allowed. Three quarks of

Table 41-6 Properties of the Basic Interactions

	Gravitational	Weak	Electromagnetic	Strong	
				Fundamental	Residual
Acts on	Mass	Flavor	Electric charge	Color charge	
Particles experiencing	All	Quarks, leptons	Electrically charged	Quarks, gluons	Hadrons
Particles mediating	Graviton	$W^{\pm}, Z$	γ	Gluons	Mesons
Strength for two quarks at 10^{-18} m†	10^{-41}	0.8	1	25	(not applicable)
Strength for two protons in nucleus†	10^{-36}	10^{-7}	1	(not applicable)	20

† Strengths are relative to electromagnetic strength.

different colors can combine to form color-neutral baryons, such as the neutron and the proton. Mesons each have a quark and an antiquark and are also color-neutral. Excited states of hadrons are considered to be different particles. For example, the Δ^+ particle is an excited state of the proton. Both are made up of the *uud* quarks, but the proton is in the ground state and has spin $\frac{1}{2}$ and a rest energy of 938 MeV, whereas the Δ^+ particle is in the first excited state and has spin $\frac{3}{2}$ and a rest energy of 1232 MeV. The two *u* quarks can be in the same spin state in the Δ^+ without violating the exclusion principle, because they have different color. All baryons eventually decay to the lightest baryon, the proton. That the proton does not decay is consistent with the conservation of energy and conservation of baryon number.

The strong interaction has two parts, the fundamental interaction or color interaction and what is called the *residual strong interaction.* The fundamental interaction is responsible for the force exerted by one quark on another quark and is mediated by gluons. The residual strong interaction is responsible for the force between color-neutral nucleons, such as the neutron and the proton. This force is due to the residual strong interactions between the color-charged quarks that make up the nucleons and can be viewed as being mediated by the exchange of mesons. The residual strong interaction between color-neutral nucleons can be thought of as analogous to the residual electromagnetic interaction between neutral atoms that bind them together to form molecules. Table 41-6 lists some of the properties of the basic interactions.

For each particle, there is an antiparticle. A particle and its antiparticle have identical mass and spin but opposite electric charge. For leptons, the lepton numbers L_e, L_μ, and L_τ of the antiparticles are the negatives of the corresponding lepton numbers for the particles. For example, the lepton number for the electron is $L_e = +1$, and the lepton number for the positron is $L_e = -1$. For hadrons, the baryon number, strangeness, charm, topness, and bottomness are the sums of those quantities for the quarks that make up the hadron. The number of each antiparticle is the negative of the number for the corresponding particle. For example, the lambda particle Λ^0, which is made up of the *uds* quarks, has $B = 1$ and $S = -1$, whereas its antiparticle $\overline{\Lambda}^0$, which is made up of the $\overline{u}\overline{d}\overline{s}$ quarks, has $B = -1$ and $S = +1$. A particle such as the photon γ or the Z^0 particle that has zero electric charge; $B = 0$, $L = 0$, $S = 0$; and zero charm, topness, and bottomness is its own antiparticle. Note that the K^0 meson ($d\bar{s}$) has a zero value for all of these quantities except strangeness, which is +1. Its antiparticle, the $\overline{K}^0$ meson ($\bar{d}s$), has strangeness −1, which makes it distinct from the K^0. The π^+ ($u\bar{d}$) and π^- ($\bar{u}d$) are somewhat special in that they have electric charge but zero values for L, B, and S. They are antiparticles of each other, but because there is no conservation law for mesons, it is impossible to say which is the particle and which is the antiparticle. Similarly, the W^+ and W^- are antiparticles of each other.

GRAND UNIFICATION THEORIES

With the success of the electroweak theory, attempts have been made to combine the strong, electromagnetic, and weak interactions in various **grand unification theories (GUTs).** In one of these theories, leptons and quarks are considered to be two aspects of a single class of particles. Under certain conditions, a quark could change into a lepton and vice versa, even though this would appear to violate the conservation of lepton number and baryon number. One of the exciting predictions of this theory is that the proton is not stable but merely has a very long lifetime of the order of 10^{32} y. Such a long lifetime makes proton decay difficult to observe. However, projects are ongoing in which detectors monitor very large numbers of protons in search of an event indicating the decay of a proton.

41-8 THE EVOLUTION OF THE UNIVERSE

In the presently accepted model, the universe began with a singular cataclysmic event called the **big bang** and is expanding. The first evidence that the universe is expanding was the astronomer Edwin Powell Hubble's discovery of the relation between the redshifts in the spectra of galaxies and their distances from us. This relation is illustrated in Figure 41-4 for a group of spiral galaxies used by astronomers for calibrating distances. Provided that the redshift is due to the Doppler effect, the recession velocity v of a galaxy is related to its distance r from us by Hubble's law,

$$v = Hr \tag{41-9}$$

where H is the **Hubble constant.** In principle, the value of H is easy to obtain because it relies on the direct calculation of v from redshift measurements. However, astronomical distances are very challenging to measure, and they have been determined for only a fraction of the 10^{10} or so galaxies in the observable universe.

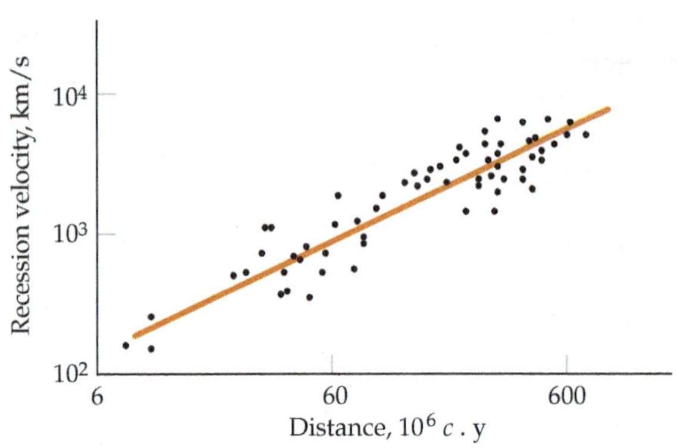

FIGURE 41-4 A plot of the recession velocities of individual galaxies versus distance.

Thus, the value of H changes as distance calibration data are refined. The currently accepted value of the Hubble constant is about

$$H = \frac{23\ \text{km/s}}{10^6\, c \cdot \text{y}} \qquad \text{41-10}$$

Hubble's law tells us that the galaxies are all rushing away from us, and those galaxies that are the farthest away are moving the fastest. However, there is no reason why our location should be special. An observer in any galaxy would make the same observations and compute the same Hubble constant. Thus, Hubble's law suggests that all of the galaxies are receding from each other at an average speed of 23 km/s per $10^6 c \cdot \text{y}$ of separation. In other words, the universe is expanding. Notice that the basic dimension of H is reciprocal time. The quantity $1/H$ is called the **Hubble age** and equals about 1.3×10^{10} y. This would correspond to the age of the universe if the gravitational pull on the receding galaxies were ignored.

Example 41-5 Using Hubble's Law

Redshift measurements of a galaxy in the constellation Virgo yield a recession velocity of 1200 km/s. How far is it to that galaxy?

PICTURE We calculate the distance from Hubble's law.

SOLVE

Use Hubble's law to find r:

$$r = \frac{v}{H} = (1200\ \text{km/s})\frac{10^6\, c \cdot \text{y}}{23\ \text{km/s}} = \boxed{52 \times 10^6\, c \cdot \text{y}}$$

PRACTICE PROBLEM 41-1 Show that $1/H = 1.3 \times 10^{10}$ y.

THE 2.7-*K* BACKGROUND RADIATION

In investigating ways of accounting for the cosmic abundance of atoms that are heavier than hydrogen atoms, cosmologists recognized that nucleosynthesis in stars could explain the abundance of atoms heavier than helium atoms but could not by itself explain the abundance of helium atoms. Helium must therefore have been formed during the big bang. To synthesize an amount of helium sufficient to account for its present abundance, the big bang would have to have occurred at an extremely high initial temperature to provide the necessary reaction rate before fusion was shut down by the decreasing density of the very rapid initial expansion. The high temperature implies a corresponding thermal (blackbody) radiation field that would cool as the expansion progressed. Theoretical analysis predicted that from the estimated time of the big bang to the present, the remnants of the radiation field should have cooled to a temperature of about 3 K, corresponding to a blackbody spectrum with peak wavelength λ_{max} in the microwave region. In 1965, the predicted cosmic background radiation was discovered by Arno Penzias and Robert Wilson at the Bell Labs. Since this landmark discovery, careful analysis has established that the temperature of the background field is 2.7281 K and has shown that it has an isotropic distribution in space.

THE BIG BANG

The singular event that initiated the expansion of the universe is thought to have been a huge explosion. The four interactions of nature (strong, electromagnetic, weak, and gravitational) initially were unified into a single interaction. Physicists have been successful in developing theoretical descriptions that unify the first three interactions, but a theory of quantum gravity, needed for the extreme densities of the single-interaction period, does not yet exist. Consequently, until the cooling universe "froze" or "condensed out" the gravitational interaction at approximately 10^{-43} s after the big bang, when the temperature was still 10^{32} K, we have no means of describing what was occurring. At this point, the average energy of the particles created would have been about 10^{19} GeV. As the universe continued to cool below 10^{32} K, the three interactions other than gravity remained unified and are described by the grand unification theories (GUTs). Quarks and leptons were indistinguishable and particle quantum numbers were not conserved. It was during this period that a slight excess of quarks over antiquarks occurred, roughly 1 part in 10^9, that ultimately resulted in the predominance of matter over antimatter that we now observe in the universe.

At 10^{-35} s, the universe had expanded sufficiently to cool to approximately 10^{27} K, at which point another phase transition occurred as the strong interaction condensed out of the GUTs group, leaving only the electromagnetic and weak interactions still unified as the **electroweak interaction.** During this period, the previously free quarks in the dense mixture of roughly equal numbers of quarks, leptons, their antiparticles, and photons began to combine into hadrons and their antiparticles, including the nucleons. By the time the universe had cooled to approximately 10^{13} K, at about $t = 10^{-6}$ s, the hadrons had mostly disappeared. This is because 10^{13} K corresponds to $kT \sim 1$ GeV, which is the minimum energy needed to create nucleons and antinucleons from the photons present by the reactions

$$\gamma \rightarrow p^+ + p^- \tag{41-11a}$$

and

$$\gamma \rightarrow n + \bar{n} \tag{41-11b}$$

The particle–antiparticle pairs annihilated and there was no new production to replace them. Only the slight earlier excess of quarks over antiquarks led to a slight excess of protons and neutrons over their antiparticles. The annihilations resulted in photons and leptons, and after about $t = 10^{-4}$ s, those particles in roughly equal numbers dominated the universe. This was the **lepton era.** At about $t = 10$ s, the temperature had fallen to 10^{10} K ($kT \sim 1$ MeV). Further expansion and cooling dropped the average photon energy below the energy needed to form an electron–positron pair. Annihilation then removed all of the positrons as it had the antiprotons and antineutrons earlier, leaving only the small excess of electrons arising from charge conservation, and the **radiation era** began. The particles present were primarily photons and neutrinos.

Within a few more minutes, the temperature dropped sufficiently to enable fusing protons and neutrons to form nuclei that were not immediately photodisintegrated. The nuclei of deuterium, helium, and lithium were produced during this **nucleosynthesis period,** but the rapid expansion soon dropped the temperature too low for the fusion to continue and the formation of heavier elements had to await the birth of stars.

A long time later, when the temperature had dropped to about 3000 K as the universe grew to about 1/1000 of its present size, kT dropped below typical atomic ionization energies and atoms were formed. By then, the expansion had redshifted the radiation field so that the total radiation energy was about equal to the energy represented by the remaining mass. As expansion and cooling continued, the energy of the steadily redshifting radiation declined at a steady rate until, at $t = 10^{10}$ y (now), matter came to dominate the universe, with its energy density exceeding that of the 2.7-K radiation remaining from the big bang by a factor of about 1000.

Summary

TOPIC	RELEVANT EQUATIONS AND REMARKS
1. Basic Interactions	There are four basic interactions: strong, electromagnetic, weak, and gravitational.
Strong	The *charge* associated with the strong interaction is called color. Quarks and gluons have color and experience the strong interaction. Hadrons (baryons and mesons) experience a residual strong interaction resulting from the fundamental strong interaction between the quarks that make up the hadrons. Decay times by the strong interaction are typically 10^{-23} s.
Electromagnetic	All particles that have electric charge experience the force due to the electromagnetic interaction.
Weak	The *charge* associated with the weak interaction is called flavor. Quarks and leptons have flavor and experience the weak interaction. Decay times by the weak interaction are typically 10^{-10} s.
Gravitational	The *charge* associated with the gravitational interaction is called mass.
2. Fundamental Particles	There are two families of fundamental particles, leptons and quarks, each having six members. It is thought that these particles have no size and no internal structure.
Leptons	Leptons are spin-$\frac{1}{2}$ fermions: the electron e and its neutrino ν_e, the muon μ and its neutrino ν_μ, and the tau τ and its neutrino ν_τ. The electron, muon, and tau have mass, electric charge, and flavor, but not color; so they participate in the gravitational, electromagnetic, and weak interactions, but not the strong interaction. The neutrinos have flavor but no electric charge and no color. They have a very small mass.
Quarks	There are six quarks, called up u, down d, strange s, charmed c, top t, and bottom b. Each is a spin-$\frac{1}{2}$ fermion. The quarks participate in all of the basic interactions. Because they are always confined in mesons or baryons, their masses can only be estimated.
3. Hadrons	Hadrons are composite particles that are made up of quarks. There are two types of hadrons, baryons and mesons. Baryons, which include the neutron and proton, are fermions of half-integral spin consisting of three quarks. Mesons, which include pions and kaons, have zero or integral spin. Hadrons interact with each other by the residual strong interaction.
4. Field Particles	In addition to the six fundamental leptons and six fundamental quarks, there are field particles that are associated with the basic interactions. *Interaction* — *Field Particle* Gravitational — Graviton (not yet observed) Electromagnetic — Photon Weak — W^+, W^-, Z^0 Strong — Gluons
5. The Conservation Laws	Some quantities, such as energy, linear momentum, electric charge, angular momentum, baryon number, and each of the three lepton numbers, are strictly conserved during all reactions and decays. Others, such as strangeness and charm, are conserved during reactions and decays that proceed by the strong interaction but not in those that proceed by the weak interaction.
6. Particles and Antiparticles	Particles and their antiparticles have identical masses but opposite values for their other properties, such as charge, lepton number, baryon number, and strangeness. Particle–antiparticle pairs can be produced during various nuclear reactions if the energy available is greater than $2mc^2$, where m is the mass of the particle.
7. Hubble's Law	Hubble's law relates the recession velocity of a galaxy, determined from the redshift of its spectrum, to the distance of the galaxy from us: $v = Hr$ (41-9) where the Hubble constant H = 23 km/s per million light-years. From Hubble's law, we conclude that the universe is expanding and that the expansion began approximately $1/H$ years ago.

TOPIC	RELEVANT EQUATIONS AND REMARKS
8. **The Big Bang**	According to the model currently used to describe the evolution of the universe, the universe began with a big bang approximately 10^{10} years ago. The big bang model is supported by substantial experimental observations, including the isotropic, 2.7-K background blackbody radiation spectrum.

Answer to Concept Check

41-1 A proton is a baryon that has a baryon number (B) equal to 1, and all particles that are not baryons have $B = 0$. If a proton decays, conservation of baryon number implies that the decay products must contain a minimum of one baryon. In addition, conservation of energy implies that the rest mass of the decay products cannot be greater than the rest mass of the proton. Because there are no baryons that have a rest mass less than the rest mass of the proton, the proton cannot decay without either violating conservation of baryon number, conservation of energy, or both.

Problems

In a few problems, you are given more data than you actually need; in a few other problems, you are required to supply data from your general knowledge, outside sources, or informed estimate.

Interpret as significant all digits in numerical values that have trailing zeros and no decimal points.

- • Single-concept, single-step, relatively easy
- •• Intermediate-level, may require synthesis of concepts
- ••• Challenging
- SSM Solution is in the *Student Solutions Manual*
- Consecutive problems that are shaded are paired problems.

CONCEPTUAL PROBLEMS

1 • How are baryons and mesons similar? How are they different?

2 • The muon and the pion have nearly the same masses. How do the particles differ?

3 • How can you tell whether a decay proceeds by the strong interaction or the weak interaction? SSM

4 • True or false:

(*a*) All baryons are hadrons.
(*b*) All hadrons are baryons.

5 • True or false: All mesons are spin-$\frac{1}{2}$ particles.

6 • A particle that is made of exactly two quarks is (*a*) a meson, (*b*) a baryon, (*c*) a lepton, (*d*) either a meson or a baryon, but definitely not a lepton.

7 • Have any quark–antiquark combinations whose electric charge is not an integer multiplied by the fundamental charge e been observed?

8 • True or false:

(*a*) A lepton is a combination of three quarks.
(*b*) The typical times for decays by the weak interaction are orders of magnitude longer than the typical times for decays by the strong interaction.
(*c*) The muon and the pion are both mesons.

9 • True or false:

(*a*) Electrons interact with protons by the strong interaction.
(*b*) Strangeness is not conserved in reactions involving the weak interactions.
(*c*) Neutrons have zero charm.

ESTIMATION AND APPROXIMATION

10 •• Grand unification theories predict that the proton has a long but finite lifetime. Current experiments based on detecting the decay of protons in water infer that this lifetime is at least 10^{32} years. Assume 10^{32} years is, in fact, the mean lifetime of the proton. Estimate the expected time between proton decays that occur in the water of a filled Olympic-size swimming pool. An Olympic-size swimming pool is 100 m × 25 m × 2.0 m. Give your answer in days.

11 •• Table 41-6 lists some properties of the four fundamental interactions. To better understand the significance of this table, confirm the ratio of the numerical entries in the second and fourth column of the last row of the table by estimating the ratio of the electromagnetic force to the gravitational force between two protons of a nucleus.

SPIN AND ANTIPARTICLES

12 • Two pions at rest annihilate according to the reaction $\pi^+ + \pi^- \rightarrow \gamma + \gamma$. (*a*) Why must the energies of the two γ rays be equal? (*b*) Find the energy of each γ ray. (*c*) Find the wavelength of each γ ray.

13 • Find the minimum energy of the photon needed for the following pair-production reactions: (*a*) $\gamma \rightarrow \pi^+ + \pi^-$, (*b*) $\gamma \rightarrow p + p^-$, and (*c*) $\gamma \rightarrow \mu^- + \mu^+$.

THE CONSERVATION LAWS

14 • State which of the following decays or reactions violate one or more of the conservation laws, and give the law or laws violated in each case: (*a*) $p^+ \rightarrow n + e^+ + \bar{\nu}_e$, (*b*) $n \rightarrow p^+ + \pi^-$, (*c*) $e^+ + e^- \rightarrow \gamma$, (*d*) $p^+ + p^- \rightarrow \gamma + \gamma$, and (*e*) $\bar{\nu}_e + p^+ \rightarrow n + e^+$.

15 • Determine the change in strangeness in each reaction that follows, and state whether the each decay can proceed by the strong interaction, by the weak interaction, or not at all: (*a*) $\Omega^- \rightarrow \Xi^0 + \pi^-$, (*b*) $\Xi^0 \rightarrow p + \pi^- + \pi^0$, and (*c*) $\Lambda^0 \rightarrow p + \pi^-$.

16 • Determine the change in strangeness for each decay, and state whether each decay can proceed by the strong interaction, by the weak interaction, or not at all: (*a*) $\Omega^- \rightarrow \Lambda^0 + K^-$ and (*b*) $\Xi^0 \rightarrow p + \pi^-$.

17 • Determine the change in strangeness for each decay, and state whether each decay can proceed by the strong interaction, by the weak interaction, or not at all: (*a*) $\Omega^- \rightarrow \Lambda^0 + \bar{\nu}_e + e^-$ and (*b*) $\Sigma^+ \rightarrow p + \pi^0$.

18 • (*a*) Which of the following decays of the τ particle is possible?

$$\tau \rightarrow \mu^- + \bar{\nu}_\mu + \nu_\tau$$
$$\tau \rightarrow \mu^- + \nu_\mu + \bar{\nu}_\tau$$

(*b*) Explain why the other decay is not possible. (*c*) Calculate the kinetic energy of the decay products for the decay that is possible.

19 •• Using Table 41-2 and the laws of conservation of charge number, baryon number, strangeness, and spin, identify the unknown particle, symbolized by (?), in each of the following reactions: (*a*) $p + \pi^- \rightarrow \Sigma^0 + (?)$, (*b*) $p + p \rightarrow \pi^+ + n + K^+ + (?)$, and (*c*) $p + K^- \rightarrow \Xi^- + (?)$ SSM

20 •• Test the following decays for violation of the conservation of energy, electric charge, baryon number, and lepton number: (*a*) $n \rightarrow \pi^+ + \pi^- + \mu^+ + \mu^-$ and (*b*) $\pi^0 \rightarrow e^+ + e^- + \gamma$. Assume that linear momentum and angular momentum are conserved. State which conservation laws (if any) are violated in each decay.

QUARKS

21 • Find the baryon number, charge, and strangeness for the following quark combinations and identify the hadron: (*a*) *uud*, (*b*) *udd*, (*c*) *uus*, (*d*) *dds*, (*e*) *uss*, and (*f*) *dss*.

22 • Find the baryon number, charge, and strangeness for the following quark combinations: (*a*) $u\bar{d}$, (*b*) $\bar{u}d$, (*c*) $u\bar{s}$, and (*d*) $\bar{u}s$.

23 • The Δ^{++} particle is a baryon that decays by the strong interaction. Its strangeness, charm, topness, and bottomness are all zero. What combination of quarks gives a particle that has those properties?

24 • Find a possible combination of quarks that gives the correct values for electric charge, baryon number, and strangeness for (*a*) K^+ and (*b*) K^0.

25 • The D^+ meson has zero strangeness, but it has charm of +1. (*a*) What is a possible quark combination that will give the correct properties for the particle? (*b*) Repeat Part (*a*) for the D^- meson, which is the antiparticle of the D^+ meson.

26 • Find a possible combination of quarks that gives the correct values for electric charge, baryon number, and strangeness for (*a*) K^- (the K^- is the antiparticle of the K^+) and (*b*) $\bar{K}^0$.

27 •• Find a possible quark combination for the following particles: (*a*) Λ^0, (*b*) p^-, and (*c*) Σ^-. SSM

28 •• Find a possible quark combination for the following particles: (*a*) $\bar{n}$, (*b*) Ξ^0, and (*c*) Σ^+.

29 •• Find a possible quark combination for the following particles: (*a*) Ω^- and (*b*) Ξ^-.

30 •• State the properties of the particles made up of the following quarks: (*a*) *ddd*, (*b*) $u\bar{c}$, (*c*) $u\bar{b}$, and (*d*) $\bar{s}\bar{s}\bar{s}$.

THE EVOLUTION OF THE UNIVERSE

31 • A galaxy is receding from Earth at 2.5 percent the speed of light. Estimate the distance from Earth to the galaxy. SSM

32 • Estimate the speed of a galaxy that is $12 \times 10^9\ c \cdot y$ away from us.

33 •• The Doppler frequency shift for a light from a source that is receding from a stationary receiver is given by $f' = f_0\sqrt{(1-\beta)/(1+\beta)}$, where $\beta = v/c$ (Equation 39-16*b*). Show that the Doppler wavelength shift for light is $\lambda' = \lambda_0\sqrt{(1+\beta)/(1-\beta)}$.

34 •• The red line in the spectrum of atomic hydrogen is frequently referred to as the *H*α line, and it has a wavelength of 656.3 nm. Using Hubble's law and the Doppler equation for light from Problem 33, determine the wavelength of the *H*α line in the spectrum emitted from galaxies at distances of (*a*) $5.00 \times 10^6\ c \cdot y$, (*b*) $5.00 \times 10^8\ c \cdot y$, and (*c*) $5.00 \times 10^9\ c \cdot y$ from Earth.

GENERAL PROBLEMS

35 • (*a*) What conditions are necessary for a particle and its antiparticle to be identical? (*b*) Find the quark combination of both the particle and the antiparticle of both the π^0 and the Ξ^0 particles. (*c*) Of the π^0 and the Ξ^0 particles, which, if any, is its own antiparticle?

36 •• The red line in the spectrum of atomic hydrogen is frequently referred to as the *H*α line, and it has a wavelength of 656.3 nm. Light from a distant galaxy shows a redshift of the *H*α line of hydrogen to a wavelength of 1458 nm. (*a*) What is the recessional velocity of the galaxy? (*b*) Estimate the distance to the galaxy.

37 •• (*a*) In terms of the quark model, show that the reaction $\pi^0 \rightarrow \gamma + \gamma$ does not violate any conservation laws. (*b*) Which conservation law is violated by the reaction $\pi^0 \rightarrow \gamma$? SSM

38 •• Test the following decays for violation of the conservation of energy, electric charge, baryon number, and lepton number: (*a*) $\Lambda^0 \rightarrow p + \pi^-$, (*b*) $\Sigma^- \rightarrow n + p^-$, and (*c*) $\mu^- \rightarrow e^- + \bar{\nu}_e + \nu_\mu$. Assume that linear momentum and angular momentum are conserved. State which conservation laws (if any) are violated in each decay.

39 •• Consider the following high-energy particle reaction: $p + p \rightarrow \Lambda^0 + K^0 + p + (?)$, where (?) represents an unknown particle. During this reaction, stationary protons are bombarded with a beam of high-energy protons. (*a*) Use the laws of conservation of charge number, baryon number, strangeness (Table 41-2), and spin to determine the unknown particle. (*b*) Calculate the *Q* value for the reaction. (*c*) The threshold kinetic energy K_{th} for this reaction is given by $K_{th} = -\frac{1}{2}Q(m_p + m_p + M_1 + M_2 + M_3 + M_4)/m_p$, where M_1, M_2, M_3, and M_4 are the masses of the reaction products. Find K_{th}.

40 ••• In this problem, you will calculate the difference in the time of arrival of two neutrinos of different energy from a supernova that is 170 000 light-years away. Let the energies of the neutrinos be $E_1 = 20$ MeV and $E_2 = 5$ MeV, and assume that the mass of a neutrino is $2.0\ \text{eV}/c^2$. Because the total energies of the neutrinos is so much greater than their rest energies, the neutrinos have speeds that are very nearly equal to c and energies that are approximately $E \approx pc$. (*a*) If t_1 and t_2 are the times that the neutrinos with speeds u_1 and u_2 take to travel a distance x, show that $\Delta t = t_2 - t_1 = x(u_1 - u_2)/u_1u_2 \approx (x\,\Delta u)/c^2$. (*b*) The speed of a neutrino of mass m and total energy E can be found from $E = mc^2/[1 - (u^2/c^2)]^{1/2}$ (Equation 39-24). Show that when $E \gg mc^2$, the speed ratio u/c is given approximately by $u/c \approx 1 - \frac{1}{2}(mc^2/E)^2$. (*c*) Use the results from Part (*a*) and Part (*b*) to calculate $u_1 - u_2$ for the energies and mass given, and calculate Δt from the result from Part (*a*) for $x = 170\ 000\ c \cdot \text{y}$. (*d*) Repeat the calculation in Part (*c*) using $20\ \text{eV}/c^2$ for the neutrino mass.

41 ••• A Λ^0 at rest decays by the reaction $\Lambda^0 \rightarrow \text{p} + \pi^-$. (*a*) Calculate the total kinetic energy of the decay products. (*b*) Find the ratio of the kinetic energy of the pion to the kinetic energy of the proton. (*c*) Find the kinetic energies of the proton and the pion for the decay.

42 ••• A Σ^0 particle at rest decays by the reaction $\Sigma^0 \rightarrow \Lambda^0 + \gamma$. (*a*) What is the total energy (total energy includes rest energy) of the decay products? (*b*) Assuming that the kinetic energy of the Λ^0 is negligible compared with the energy of the photon, calculate the approximate momentum of the photon. (*c*) Use your result from Part (*b*) to calculate the kinetic energy of the Λ^0. (*d*) Use your result from Part (*c*) to obtain a better estimate of the momentum and the energy of the photon.

Appendix A

SI Units and Conversion Factors

Base Units*

Length	The *meter* (m) is the distance traveled by light in a vacuum in 1/299,792,458 s.
Time	The *second* (s) is the duration of 9,192,631,770 periods of the radiation corresponding to the transition between the two hyperfine levels of the ground state of the ^{133}Cs atom.
Mass	The *kilogram* (kg) is the mass of the international standard body preserved at Sèvres, France.
Mole	The *mole* (mol) is the amount of substance of a system which contains as many elementary entities as there are atoms in 0.012 kilogram of carbon 12.
Current	The *ampere* (A) is that constant current which, if maintained in two straight parallel conductors of infinite length, of negligible circular cross section, and placed 1 m apart in vacuum would produce between these conductors a force equal to 2×10^{-7} N/m of length.
Temperature	The *kelvin* (K) is 1/273.16 of the thermodynamic temperature of the triple point of water.
Luminous intensity	The *candela* (cd) is the luminous intensity in a given direction, of a source that emits monochromatic radiation of frequency 540×10^{12} hertz and that has a radiant intensity in that direction of 1/683 watt/steradian.

*These definitions are found on the Internet at http://physics.nist.gov/cuu/Units/current.html

Derived Units

Force	newton (N)	$1\ \text{N} = 1\ \text{kg}\cdot\text{m/s}^2$
Work, energy	joule (J)	$1\ \text{J} = 1\ \text{N}\cdot\text{m}$
Power	watt (W)	$1\ \text{W} = 1\ \text{J/s}$
Frequency	hertz (Hz)	$1\ \text{Hz} = \text{cy/s}$
Charge	coulomb (C)	$1\ \text{C} = 1\ \text{A}\cdot\text{s}$
Potential	volt (V)	$1\ \text{V} = 1\ \text{J/C}$
Resistance	ohm (Ω)	$1\ \Omega = 1\ \text{V/A}$
Capacitance	farad (F)	$1\ \text{F} = 1\ \text{C/V}$
Magnetic field	tesla (T)	$1\ \text{T} = 1\ \text{N/(A}\cdot\text{m)}$
Magnetic flux	weber (Wb)	$1\ \text{Wb} = 1\ \text{T}\cdot\text{m}^2$
Inductance	henry (H)	$1\ \text{H} = 1\ \text{J/A}^2$

Conversion Factors

Conversion factors are written as equations for simplicity; relations marked with an asterisk are exact.

Length

1 km = 0.6215 mi

1 mi = 1.609 km

1 m = 1.0936 yd = 3.281 ft = 39.37 in

*1 in = 2.54 cm

*1 ft = 12 in = 30.48 cm

*1 yd = 3 ft = 91.44 cm

1 lightyear = 1 $c \cdot y$ = 9.461×10^{15} m

*1 Å = 0.1 nm

Area

*1 $m^2 = 10^4\ cm^2$

1 $km^2 = 0.3861\ mi^2$ = 247.1 acres

*1 $in^2 = 6.4516\ cm^2$

1 $ft^2 = 9.29 \times 10^{-2}\ m^2$

1 $m^2 = 10.76\ ft^2$

*1 acre = 43 560 ft^2

1 mi^2 = 640 acres = 2.590 km^2

Volume

*1 $m^3 = 10^6\ cm^3$

*1 L = 1000 cm^3 = $10^{-3}\ m^3$

1 gal = 3.785 L

1 gal = 4 qt = 8 pt = 128 oz = 231 in^3

1 $in^3 = 16.39\ cm^3$

1 ft^3 = 1728 $in.^3$ = 28.32 L
= $2.832 \times 10^4\ cm^3$

Time

*1 h = 60 min = 3.6 ks

*1 d = 24 h = 1440 min = 86.4 ks

1 y = 365.24 d = 3.156×10^7 s

Speed

*1 m/s = 3.6 km/h

1 km/h = 0.2778 m/s = 0.6215 mi/h

1 mi/h = 0.4470 m/s = 1.609 km/h

1 mi/h = 1.467 ft/s

Angle and Angular Speed

*π rad = 180°

1 rad = 57.30°

1° = 1.745×10^{-2} rad

1 rev/min = 0.1047 rad/s

1 rad/s = 9.549 rev/min

Mass

*1 kg = 1000 g

*1 tonne = 1000 kg = 1 Mg

1 u = 1.6605×10^{-27} kg
= 931.49 MeV/c^2

1 kg = 6.022×10^{26} u

1 slug = 14.59 kg

1 kg = 6.852×10^{-2} slug

Density

*1 g/cm^3 = 1000 kg/m^3 = 1 kg/L

(1 g/cm^3)g = 62.4 lb/ft^3

Force

1 N = 0.2248 lb = 10^5 dyn

*1 lb = 4.448222 N

(1 kg)g = 2.2046 lb

Pressure

*1 Pa = 1 N/m^2

*1 atm = 101.325 kPa = 1.01325 bars

1 atm = 14.7 lb/in^2 = 760 mmHg
= 29.9 inHg = 33.9 ftH_2O

1 lb/in^2 = 6.895 kPa

1 torr = 1 mmHg = 133.32 Pa

1 bar = 100 kPa

Energy

*1 kW · h = 3.6 MJ

*1 cal = 4.1840 J

1 ft · lb = 1.356 J = 1.286×10^{-3} Btu

*1 L · atm = 101.325 J

1 L · atm = 24.217 cal

1 Btu = 778 ft · lb = 252 cal = 1054.35 J

1 eV = 1.602×10^{-19} J

1 u · c^2 = 931.49 MeV

*1 erg = 10^{-7} J

Power

1 horsepower = 550 ft · lb/s = 745.7 W

1 Btu/h = 2.931×10^{-4} kW

1 W = 1.341×10^{-3} horsepower
= 0.7376 ft · lb/s

Magnetic Field

*1 T = 10^4 G

Thermal Conductivity

1 W/(m·K) = 6.938 Btu·in/(h·ft^2·F°)

1 Btu·in/(h·ft^2·F°) = 0.1441 W/(m·K)

Appendix B
Numerical Data

Terrestrial Data

Free-fall acceleration g	
Standard value (at sea level at 45° latitude)*	9.806 65 m/s^2; 32.1740 ft/s^2
At equator*	9.7804 m/s^2
At poles*	9.8322 m/s^2
Mass of Earth M_E	5.97×10^{24} kg
Radius of Earth R_E, mean	6.37×10^6 m; 3960 mi
Escape speed $\sqrt{2R_E g}$	1.12×10^4 m/s; 6.95 mi/s
Solar constant†	1.37 kW/m^2
Standard temperature and pressure (STP):	
Temperature	273.15 K (0.00°C)
Pressure	101.325 kPa (1.00 atm)
Molar mass of air	28.97 g/mol
Density of air (STP), ρ_{air}	1.217 kg/m^3
Speed of sound (STP)	331 m/s
Heat of fusion of H_2O (0°C, 1 atm)	333.5 kJ/kg
Heat of vaporization of H_2O (100°C, 1 atm)	2.257 MJ/kg

*Measured relative to Earth's surface.

† Average power incident normally on 1 m^2 outside Earth's atmosphere at the mean distance from Earth to the Sun.

Astronomical Data*

Earth	
Distance to moon, mean†	3.844×10^8 m; 2.389×10^5 mi
Distance to the Sun, mean†	1.496×10^{11} m; 9.30×10^7 mi; 1.00 AU
Orbital speed, mean	2.98×10^4 m/s
Moon	
Mass	7.35×10^{22} kg
Radius	1.737×10^6 m
Period	27.32 d
Acceleration of gravity at surface	1.62 m/s^2
Sun	
Mass	1.99×10^{30} kg
Radius	6.96×10^8 m

*Additional solar-system data is available from NASA at <http://nssdc.gsfc.nasa.gov/planetary/planetfact.html>.

† Center to center.

Physical Constants*

Gravitational constant	G	$6.6742(10) \times 10^{-11}\ \mathrm{N \cdot m^2/kg^2}$
Speed of light	c	$2.997\ 924\ 58 \times 10^8\ \mathrm{m/s}$
Fundamental charge	e	$1.602\ 176\ 453(14) \times 10^{-19}\ \mathrm{C}$
Avogadro's number	N_A	$6.022\ 141\ 5(10) \times 10^{23}$ particles/mol
Gas constant	R	$8.314\ 472(15)\ \mathrm{J/(mol \cdot K)}$
		$1.987\ 2065(36)\ \mathrm{cal/(mol \cdot K)}$
		$8.205\ 746(15) \times 10^{-2}\ \mathrm{L \cdot atm/(mol \cdot K)}$
Boltzmann constant	$k = R/N_A$	$1.380\ 650\ 5(24) \times 10^{-23}\ \mathrm{J/K}$
		$8.617\ 343(15) \times 10^{-5}\ \mathrm{eV/K}$
Stefan-Boltzmann constant	$\sigma = (\pi^2/60)k^4/(\hbar^3 c^2)$	$5.670\ 400(40) \times 10^{-8}\ \mathrm{W/(m^2k^4)}$
Atomic mass constant	$m_u = \frac{1}{12}m(^{12}\mathrm{C})$	$1.660\ 538\ 86(28) \times 10^{-27}\ \mathrm{kg} = 1\mathrm{u}$
Magnetic constant (permeability of free space)	μ_0	$4\pi \times 10^{-7}\ \mathrm{N/A^2}$
		$1.256\ 637 \times 10^{-6}\ \mathrm{N/A^2}$
Electric constant (permittivity of free space)	$\epsilon_0 = 1/(\mu_0 C^2)$	$8.854\ 187\ 817 \ldots \times 10^{-12}\ \mathrm{C^2/(N \cdot m^2)}$
Coulomb constant	$k = 1/(4\pi\epsilon_0)$	$8.987\ 551\ 788 \ldots \times 10^9\ \mathrm{N \cdot m^2/C^2}$
Planck's constant	h	$6.626\ 0693(11) \times 10^{-34}\ \mathrm{J \cdot s}$
		$4.135\ 667\ 43(35) \times 10^{-15}\ \mathrm{eV \cdot s}$
	$\hbar = h/2\pi$	$1.054\ 571\ 68(18) \times 10^{-34}\ \mathrm{J \cdot s}$
		$6.582\ 119\ 15(56) \times 10^{-16}\ \mathrm{eV \cdot s}$
Mass of electron	m_e	$9.109\ 382\ 6(16) \times 10^{-31}\ \mathrm{kg}$
		$0.510\ 998\ 918(44)\ \mathrm{MeV}/c^2$
Mass of proton	m_p	$1.672\ 621\ 71(29) \times 10^{-27}\ \mathrm{kg}$
		$938.272\ 029(80) \times \mathrm{MeV}/c^2$
Mass of neutron	m_n	$1.674\ 927\ 28(29) \times 10^{-27}\ \mathrm{kg}$
		$939.565\ 360(81)\ \mathrm{MeV}/c^2$
Bohr magneton	$m_B = eh/2m_e$	$9.274\ 009\ 49(80) \times 10^{-24}\ \mathrm{J/T}$
		$5.788\ 381\ 804(39) \times 10^{-5}\ \mathrm{eV/T}$
Nuclear magneton	$m_n = eh/2m_p$	$5.050\ 783\ 43(43) \times 10^{-27}\ \mathrm{J/T}$
		$3.152\ 451\ 259(21) \times 10^{-8}\ \mathrm{eV/T}$
Magnetic flux quantum	$\phi_0 = h/2e$	$2.067\ 833\ 72(18) \times 10^{-15}\ \mathrm{T \cdot m^2}$
Quantized Hall resistance	$R_K = h/e^2$	$2.581\ 280\ 7449(86) \times 10^4\ \Omega$
Rydberg constant	R_H	$1.097\ 373\ 156\ 8525(73) \times 10^7\ \mathrm{m^{-1}}$
Josephson frequency-voltage quotient	$K_J = 2e/h$	$4.835\ 978\ 79(41) \times 10^{14}\ \mathrm{Hz/V}$
Compton wavelength	$\lambda_C = h/m_e c$	$2.426\ 310\ 238(16) \times 10^{-12}\ \mathrm{m}$

* The values for these and other constants may be found on the Internet at http://physics.nist.gov/cuu/Constants/index.html. The numbers in parentheses represent the uncertainties in the last two digits. (For example, 2.044 43(13) stands for 2.044 43 ± 0.000 13.) Values without uncertainties are exact, including those values with ellipses (such as the value of pi is exactly 3.1415. . .).

For additional data, see the following tables in the text.

Geometry and Trigonometry

$C = \pi d = 2\pi r$	definition of π
$A = \pi r^2$	area of circle
$V = \frac{4}{3}\pi r^3$	spherical volume
$A = dV/dr = 4\pi r^2$	spherical surface area
$V = A_{base}L = \pi r^2 L$	cylindrical volume
$A = \partial V/\partial r = 2\pi r L$	cylindrical surface area

$o = h\sin\theta$
$a = h\cos\theta$

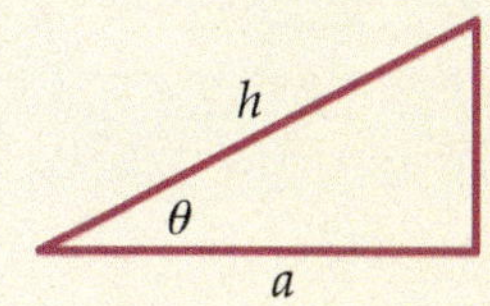

$\sin^2\theta + \cos^2\theta = 1$

$\sin(A \pm B) = \sin A\cos B \pm \cos A\sin B$

$\cos(A \pm B) = \cos A\cos B \mp \sin A\sin B$

$\sin A \pm \sin B = 2\sin[\frac{1}{2}(A \pm B)]\cos[\frac{1}{2}(A \mp B)]$

$\sin\theta \equiv y$
$\cos\theta \equiv x$
$\tan\theta \equiv \frac{y}{x}$

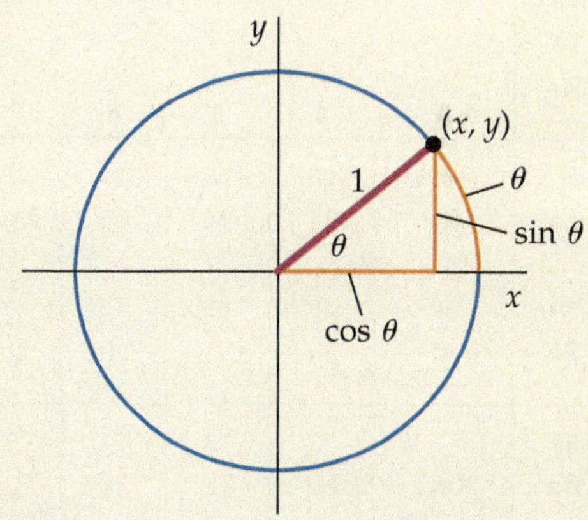

If $|\theta| \ll 1$, then
$\cos\theta \approx 1$ and $\tan\theta \approx \sin\theta \approx \theta$ (θ in radians)

Quadratic Formula

If $ax^2 + bx + c = 0$, then $x = \frac{-b \pm \sqrt{b^2 - 4ac}}{2a}$

Binomial Expansion

If $|x| < 1$, then $(1 + x)^n =$

$$1 + nx + \frac{n(n-1)}{2!}x^2 + \frac{n(n-1)(n-2)}{3!}x^3 + \ldots$$

If $|x| \ll 1$, then $(1 + x)^n \approx 1 + nx$

Differential Approximation

If $\Delta F = F(x + \Delta x) - F(x)$ and if $|\Delta x|$ is small,

then $\Delta F \approx \frac{dF}{dx}\Delta x$.

Appendix C

Periodic Table of Elements*

1	2	3	4	5	6	7	8	9	10	11	12	13	14	15	16	17	18
1 H																	2 He
3 Li	4 Be											5 B	6 C	7 N	8 O	9 F	10 Ne
11 Na	12 Mg											13 Al	14 Si	15 P	16 S	17 Cl	18 Ar
19 K	20 Ca	21 Sc	22 Ti	23 V	24 Cr	25 Mn	26 Fe	27 Co	28 Ni	29 Cu	30 Zn	31 Ga	32 Ge	33 As	34 Se	35 Br	36 Kr
37 Rb	38 Sr	39 Y	40 Zr	41 Nb	42 Mo	43 Tc	44 Ru	45 Rh	46 Pd	47 Ag	48 Cd	49 In	50 Sn	51 Sb	52 Te	53 I	54 Xe
55 Cs	56 Ba	57–71 Rare Earths	72 Hf	73 Ta	74 W	75 Re	76 Os	77 Ir	78 Pt	79 Au	80 Hg	81 Tl	82 Pb	83 Bi	84 Po	85 At	86 Rn
87 Fr	88 Ra	89–103 Actinides	104 Rf	105 Db	106 Sg	107 Bh	108 Hs	109 Mt	110 Ds	111 Rg							

Rare Earths (Lanthanides)	57 La	58 Ce	59 Pr	60 Nd	61 Pm	62 Sm	63 Eu	64 Gd	65 Tb	66 Dy	67 Ho	68 Er	69 Tm	70 Yb	71 Lu
Actinides	89 Ac	90 Th	91 Pa	92 U	93 Np	94 Pu	95 Am	96 Cm	97 Bk	98 Cf	99 Es	100 Fm	101 Md	102 No	103 Lr

*The 1–18 group designation has been recommended by the International Union of Pure and Applied Chemistry (IUPAC). Elements with atomic numbers 112, 114, and 116 have been reported but not fully authenticated as of September 2003. From http://www.iupac.org/reports/periodic_table/IUPAC_Periodic_Table-3Oct05.pdf

Atomic Numbers and Atomic Masses*

Atomic Number	Name	Symbol	Mass	Atomic Number	Name	Symbol	Mass
1	Hydrogen	H	1.00794(7)	57	Lanthanum	La	138.90547(7)
2	Helium	He	4.002602(2)	58	Cerium	Ce	140.116(1)
3	Lithium	Li	6.941(2)	59	Praseodymium	Pr	140.90765(2)
4	Beryllium	Be	9.012182(3)	60	Neodymium	Nd	144.242(3)
5	Boron	B	10.811(7)	61	Promethium	Pm	[145]
6	Carbon	C	12.0107(8)	62	Samarium	Sm	150.36(2)
7	Nitrogen	N	14.0067(2)	63	Europium	Eu	151.964(1)
8	Oxygen	O	15.9994(3)	64	Gadolinium	Gd	157.25(3)
9	Fluorine	F	18.9984032(5)	65	Terbium	Tb	158.92535(2)
10	Neon	Ne	20.1797(6)	66	Dysprosium	Dy	162.500(1)
11	Sodium	Na	22.98976928(2)	67	Holmium	Ho	164.93032(2)
12	Magnesium	Mg	24.3050(6)	68	Erbium	Er	167.259(3)
13	Aluminum	Al	26.9815386(8)	69	Thulium	Tm	168.93421(2)
14	Silicon	Si	28.0855(3)	70	Ytterbium	Yb	173.04(3)
15	Phosphorus	P	30.973762(2)	71	Lutetium	Lu	174.967(1)
16	Sulfur	S	32.065(5)	72	Hafnium	Hf	178.49(2)
17	Chlorine	Cl	35.453(2)	73	Tantalum	Ta	180.94788(2)
18	Argon	Ar	39.948(1)	74	Tungsten	W	183.84(1)
19	Potassium	K	39.0983(1)	75	Rhenium	Re	186.207(1)
20	Calcium	Ca	40.078(4)	76	Osmium	Os	190.23(3)
21	Scandium	Sc	44.955912(6)	77	Iridium	Ir	192.217(3)
22	Titanium	Ti	47.867(1)	78	Platinum	Pt	195.084(9)
23	Vanadium	V	50.9415(1)	79	Gold	Au	196.966569(4)
24	Chromium	Cr	51.9961(6)	80	Mercury	Hg	200.59(2)
25	Manganese	Mn	54.938045(5)	81	Thallium	Tl	204.3833(2)
26	Iron	Fe	55.845(2)	82	Lead	Pb	207.2(1)
27	Cobalt	Co	58.933195(5)	83	Bismuth	Bi	208.98040(1)
28	Nickel	Ni	58.6934(2)	84	Polonium	Po	[209]
29	Copper	Cu	63.546(3)	85	Astatine	At	[210]
30	Zinc	Zn	65.409(4)	86	Radon	Rn	[222]
31	Gallium	Ga	69.723(1)	87	Francium	Fr	[223]
32	Germanium	Ge	72.64(1)	88	Radium	Ra	[226]
33	Arsenic	As	74.92160(2)	89	Actinium	Ac	[227]
34	Selenium	Se	78.96(3)	90	Thorium	Th	232.03806(2)
35	Bromine	Br	79.904(1)	91	Protactinium	Pa	231.03588(2)
36	Krypton	Kr	83.798(2)	92	Uranium	U	238.02891(3)
37	Rubidium	Rb	85.4678(3)	93	Neptunium	Np	[237]
38	Strontium	Sr	87.62(1)	94	Plutonium	Pu	[244]
39	Yttrium	Y	88.90585(2)	95	Americium	Am	[243]
40	Zirconium	Zr	91.224(2)	96	Curium	Cm	[247]
41	Niobium	Nb	92.90638(2)	97	Berkelium	Bk	[247]
42	Molybdenum	Mo	95.94(2)	98	Californium	Cf	[251]
43	Technetium	Tc	[98]	99	Einsteinium	Es	[252]
44	Ruthenium	Ru	101.07(2)	100	Fermium	Fm	[257]
45	Rhodium	Rh	102.90550(2)	101	Mendelevium	Md	[258]
46	Palladium	Pd	106.42(1)	102	Nobelium	No	[259]
47	Silver	Ag	107.8682(2)	103	Lawrencium	Lr	[262]
48	Cadmium	Cd	112.411(8)	104	Rutherfordium	Rf	[261]
49	Indium	In	114.818(3)	105	Dubnium	Db	[262]
50	Tin	Sn	118.710(7)	106	Seaborgium	Sg	[266]
51	Antimony	Sb	121.760(1)	107	Bohrium	Bh	[264]
52	Tellurium	Te	127.60(3)	108	Hassium	Hs	[277]
53	Iodine	I	126.90447(3)	109	Meitnerium	Mt	[268]
54	Xenon	Xe	131.293(6)	110	Darmstadtium	Ds	[271]
55	Cesium	Cs	132.9054519(2)	111	Roentgenium	Rg	[272]
56	Barium	Ba	137.327(7)				

* IUPAC 2005 standard atomic weights (mean relative atomic masses) as approved at the 43rd IUPAC General Assembly in Beijing, China, in August 2005 are listed with uncertainties in the last figure in parentheses. From http://www.iupac.org/reports/periodic_table/IUPAC_Periodic_Table-3Oct05.pdf

Math Tutorial

In this tutorial, we review some of the basic results of algebra, geometry, trigonometry, and calculus. In many cases, we merely state results without proof. Table M-1 lists some mathematical symbols.

M-1 SIGNIFICANT DIGITS

Many numbers we work with in science are the result of measurement and are therefore known only within a degree of uncertainty. This uncertainty should be reflected in the number of digits used. For example, if you have a 1-meter-long rule with scale spacing of 1 cm, you know that you can measure the height of a box to within a fifth of a centimeter or so. Using this rule, you might find that the box height is 27.0 cm. If there is a scale with a spacing of 1 mm on your rule, you might perhaps measure the box height to be 27.03 cm. However, if there is a scale with a spacing of 1 mm on your rule, you might not be able to measure the height more accurately than 27.03 cm because the height might vary by 0.01 cm or so, depending on which part of the box you measure the height at. When you write down that the height of the box is 27.03 cm, you are stating that your best estimate of the height is 27.03 cm, but you are not claiming that it is exactly 27.030000 . . . cm high. The four digits in 27.03 cm are called **significant digits.** Your measured length, 2.703 m, has four significant digits. Significant digits are also called significant figures.

The number of significant digits in an answer to a calculation will depend on the number of significant digits in the given data. When you work with numbers that have uncertainties, you should be careful not to include more digits than the certainty of measurement warrants. *Approximate* calculations (order-of-magnitude estimates) always result in answers that have only one significant digit or none. When you multiply, divide, add, or subtract numbers, you must consider the accuracy of the results. Listed below are some rules that will help you determine the number of significant digits of your results.

Table M-1 Mathematical Symbols

$=$	is equal to
$\neq$	is not equal to
$\approx$	is approximately equal to
$\sim$	is of the order of
$\propto$	is proportional to
$>$	is greater than
$\geq$	is greater than or equal to
$\gg$	is much greater than
$<$	is less than
$\leq$	is less than or equal to
$\ll$	is much less than
Δx	change in x
$\lvert x \rvert$	absolute value of x
n!	$n(n-1)(n-2)\ldots1$
$\sum$	sum
lim	limit
$\Delta t \rightarrow 0$	Δt approaches zero
$\frac{dx}{dt}$	derivative of x with respect to t
$\frac{\partial x}{\partial t}$	partial derivative of x with respect to t
$\int$	integral

1. When multiplying or dividing quantities, the number of significant digits in the final answer is no greater than that in the quantity with the fewest significant digits.
2. When adding or subtracting quantities, the number of decimal places in the answer should match that of the term with the smallest number of decimal places.
3. Exact values have an unlimited number of significant digits. For example, a value determined by counting, such as 2 tables, has no uncertainty and is an exact value. In addition, the conversion factor 0.0254000 . . . m/in is an exact value because 1.000 . . . inches is exactly equal to 0.0254000 . . . meters. (The yard is, by definition, equal to exactly 0.9144 meters, and 0.9144 divided by 36 is exactly equal to 0.0254.)
4. Sometimes zeros are significant and sometimes they are not. If a zero is before a leading nonzero digit, then the zero is not significant. For example, the number 0.00890 has three significant digits. The first three zeroes are not significant digits but are merely markers to locate the decimal point. Note that the zero after the nine is significant.
5. Zeros that are between nonzero digits are significant. For example, 5603 has four significant digits.
6. The number of significant digits in numbers with trailing zeros and no decimal point is ambiguous. For example 31000 could have as many as five significant digits or as few as two significant digits. To prevent ambiguity, you should report numbers by using scientific notation or by using a decimal point.

Example M-1 Finding the Average of Three Numbers

Find the average of 19.90, −7.524, and −11.8179.

PICTURE You will be adding 3 numbers and then dividing the result by 3. The first number has three significant digits, the second number has four, and the third number has five.

SOLVE

1. Sum the three numbers.

$$19.90 + (-7.524) + (-11.8179) = 0.55\textit{81}$$

2. If the problem only asked for the sum of the three numbers, we would round the answer to the least number of decimal places among all the numbers being added. However, we must divide this intermediate result by 3, so we use the intermediate answer with the two extra digits (italicized and red).

$$\frac{0.55\textit{81}}{3} = 0.18\textit{60333}\ldots$$

3. Only two of the digits in the intermediate answer, 0.18*60333*. . . , are significant digits, so we must round this number to get our final answer. The number 3 in the denominator is a whole number and has an unlimited number of significant digits. Thus, the final answer has the same number of significant digits as the numerator, which is 2.

The final answer is $\boxed{0.19}$.

CHECK The sum in step 1 has two significant digits following the decimal point, the same as the number being summed with the least number of significant digits after the decimal point.

PRACTICE PROBLEMS

1. $\dfrac{5.3\ \text{mol}}{22.4\ \text{mol/L}}$
2. 57.8 m/s − 26.24 m/s

M-2 EQUATIONS

An **equation** is a statement written using numbers and symbols to indicate that two quantities, written on either side of an equals sign (=), are equal. The quantity on either side of the equal sign may consist of a single term, or of a sum or difference of two or more **terms.** For example, the equation $x = 1 - (ay + b)/(cx - d)$ contains three terms, x, 1 and $(ay + b)/(cx - d)$.

You can perform the following operations on equations:

1. The same quantity can be added to or subtracted from each side of an equation.
2. Each side of an equation can be multiplied or divided by the same quantity.
3. Each side of an equation can be raised to the same power.

These operations are meant to be applied to each *side* of the equation rather than each term in the equation. (Because multiplication is distributive over addition, operation 2—and only operation 2—of the preceding operations also applies term by term.)

Caution: *Division by zero is forbidden at any stage in solving an equation; results (if any) would be invalid.*

Adding or Subtracting Equal Amounts

To find x when $x - 3 = 7$, add 3 to both sides of the equation: $(x - 3) + 3 = 7 + 3$; thus, $x = 10$.

Multiplying or Dividing by Equal Amounts

If $3x = 17$, solve for x by dividing both sides of the equation by 3; thus, $x = \frac{17}{3}$, or 5.7.

Example M-2 Simplifying Reciprocals in an Equation

Solve the following equation for x:

$$\frac{1}{x} + \frac{1}{4} = \frac{1}{3}$$

Equations containing reciprocals of unknowns occur in geometric optics and in electric circuit analysis—for example, in finding the net resistance of parallel resistors.

PICTURE In this equation, the term containing x is on the same side of the equation as a term not containing x. Furthermore, x is found in the denominator of a fraction.

SOLVE

1. Subtract $\frac{1}{4}$ from each side: $$\frac{1}{x} = \frac{1}{3} - \frac{1}{4}$$
2. Simplify the right side of the equation by using the lowest common denominator: $$\frac{1}{x} = \frac{1}{3} - \frac{1}{4} = \frac{4}{12} - \frac{3}{12} = \frac{4-3}{12} = \frac{1}{12} \quad \text{so} \quad \frac{1}{x} = \frac{1}{12}$$
3. Multiply both sides of the equation by $12x$ to determine the value of x: $$12\cancel{x}\frac{1}{\cancel{x}} = \cancel{12}x\frac{1}{\cancel{12}}$$ $$\boxed{12} = x$$

CHECK Substitute 12 for x in the left side of original equation.

$$\frac{1}{x} + \frac{1}{4} = \frac{1}{12} + \frac{3}{12} = \frac{4}{12} = \frac{1}{3}$$

PRACTICE PROBLEMS Solve each of the following for x:

3. $(7.0\ \text{cm}^3)x = 18\ \text{kg} + (4.0\ \text{cm}^3)x$
4. $\frac{4}{x} + \frac{1}{3} = \frac{3}{x}$

M-3 DIRECT AND INVERSE PROPORTIONS

When we say variable quantities x and y are **directly proportional,** we mean that as x and y change, the ratio x/y is constant. To say that two quantities are proportional is to say that they are directly proportional. When we say variable quantities x and y are **inversely proportional,** we mean that as x and y change, the product xy is constant.

Relationships of direct and inverse proportion are common in physics. Objects moving at the same velocity have momenta directly proportional to their masses. The ideal-gas law ($PV = nRT$) states that pressure P is directly proportional to (absolute) temperature T, when volume V remains constant and is inversely proportional to volume, when temperature remains constant. Ohm's law ($V = IR$) states that the voltage V across a resistor is directly proportional to the electric current in the resistor when the resistance R remains constant.

CONSTANT OF PROPORTIONALITY

When two quantities are directly proportional, the two quantities are related by a *constant of proportionality*. If you are paid for working at a regular rate R in dollars per day, for example, the money m you earn is directly proportional to the time t you work; the rate R is the constant of proportionality that relates the money earned in dollars to the time worked t in days:

$$\frac{m}{t} = R \qquad \text{or} \qquad m = Rt$$

If you earn \$400 in 5 days, the value of R is \$400/(5 days) = \$80/day. To find the amount you earn in 8 days, you could perform the calculation

$$m = (\$80/\text{day})(8 \text{ days}) = \$640$$

Sometimes the constant of proportionality can be ignored in proportion problems. Because the amount you earn in 8 days is $\frac{8}{5}$ times what you earn in 5 days, this amount is

$$m = \frac{8}{5}(\$400) = \$640$$

Example M-3 Painting Cubes

You need 15.4 mL of paint to cover one side of a cube. The area of one side of the cube is 426 cm^2. What is the relation between the volume of paint needed and the area to be covered? How much paint do you need to paint one side of a cube in which the one side has an area of 503 cm^2?

PICTURE To determine the amount of paint for the side whose area is 503 cm^2, you will need to set up a proportion.

SOLVE

1. The volume V of paint needed increases in proportion to the area A to be covered.

   V and A are directly proportional.

   That is, $\frac{V}{A} = k$ or $V = kA$

   where k is the proportionality constant

2. Determine the value of the proportionality constant using the given values $V_1 = 15.4$ mL and $A_1 = 426$ cm^2:

   $$k = \frac{V_1}{A_1} = \frac{15.4 \text{ mL}}{426 \text{ cm}^2} = 0.0361 \text{ mL/cm}^2$$

3. Determine the volume of paint needed to paint a side of a cube whose area is 503 cm^2 using the proportionality constant in step 1:

   $$V_2 = kA_2 = (0.0361 \text{ mL/cm}^2)(503 \text{ cm}^2) = \boxed{18.2 \text{ mL}}$$

CHECK Our value for V_2 is greater than the value for V_1, as expected. The amount of paint needed to cover an area equal to 503 cm^2 should be greater than the amount of paint needed to cover an area of 426 cm^2 because 503 cm^2 is larger than 426 cm^2.

PRACTICE PROBLEMS

5. A cylindrical container holds 0.384 L of water when full. How much water would the container hold if its radius were doubled and its height remained unchanged?
Hint: The volume of a right circular cylinder is given by $V = \pi r^2 h$, *where* r *is its radius and* h *is its height. Thus,* V *is directly proportional to* r^2 *when* h *remains constant.*
6. For the container in Practice Problem 5, how much water would the container hold if both its height and its radius were doubled? How much water would the container hold if its radius were doubled and its height remained unchanged?
Hint: The volume V *of a right circular cylinder is given by* $V = \pi r^2 h$, *where* r *is its radius and* h *is its height.*

M-4 LINEAR EQUATIONS

A **linear equation** is an equation of the form $x + 2y - 4z = 3$. That is, an equation is linear if each term either is constant or is the product of a constant and a variable raised to the first power. Such equations are said to be linear because the plots of these equations form straight lines or planes. The equations of direct proportion between two variables are linear equations.

GRAPH OF A STRAIGHT LINE

A linear equation relating y and x can always be put into the standard form

$$y = mx + b \tag{M-1}$$

where m and b are constants that may be either positive or negative. Figure M-1 shows a graph of the values of x and y that satisfy Equation M-1. The constant b, called the **y intercept,** is the value of y at $x = 0$. The constant m is the **slope** of the line, which equals the ratio of the change in y to the corresponding change in x. In the figure, we have indicated two points on the line, (x_1, y_1) and (x_2, y_2), and the changes $\Delta x = x_2 - x_1$ and $\Delta y = y_2 - y_1$. The slope m is then

$$m = \frac{y_2 - y_1}{x_2 - x_1} = \frac{\Delta y}{\Delta x}$$

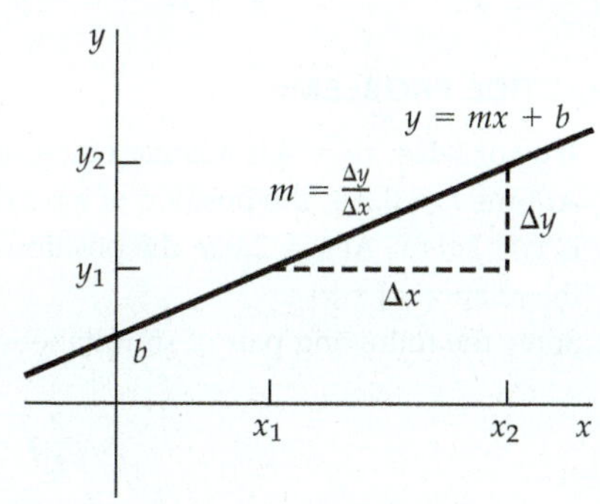

FIGURE M-1 Graph of the linear equation $y = mx + b$, where b is the y intercept and $m = \Delta y/\Delta x$ is the slope.

If x and y are both unknown in the equation $y = mx + b$, there are no unique values of x and y that are solutions to the equation. Any pair of values (x_1, y_1) on the line in Figure M-1 will satisfy the equation. If we have two equations, each with the same two unknowns x and y, the equations can be solved simultaneously for the unknowns. Example M-4 shows how simultaneous linear equations can be solved.

Example M-4 Using Two Equations to Solve for Two Unknowns

Find any and all values of x and y that simultaneously satisfy

$$3x - 2y = 8 \tag{M-2}$$

and

$$y - x = 2 \tag{M-3}$$

PICTURE Figure M-2 shows a graph of the two equations. At the point where the lines intersect, the values of x and y satisfy both equations. We can solve two simultaneous equations by first solving either equation for one variable in terms of the other variable and then substituting the result into the other equation.

SOLVE

1. Solve Equation M-3 for y: $y = x + 2$
2. Substitute this value for y into Equation M-2: $3x - 2(x + 2) = 8$
3. Simplify the equation and solve for x: $3x - 2x - 4 = 8$; $x - 4 = 8$; $x = \boxed{12}$
4. Use your solution for x and one of the given equations to find the value of y: $y - x = 2$, where $x = 12$; $y - 12 = 2$; $y = 2 + 12 = \boxed{14}$

CHECK An alternative method is to multiply one equation by a constant such that one of the unknown terms is eliminated when the equations are added or subtracted. We can multiply through Equation M-3 by 2

$$2(y - x) = 2(2)$$
$$2y - 2x = 4$$

and add the result to Equation M-2 and solve for x:

$$2y - 2x = 4$$
$$3x - 2y = 8$$
$$3x - 2x = 12 \Rightarrow x = 12$$

Substitute into Equation M-3 and solve for y:

$$y - 12 = 2 \Rightarrow y = 14$$

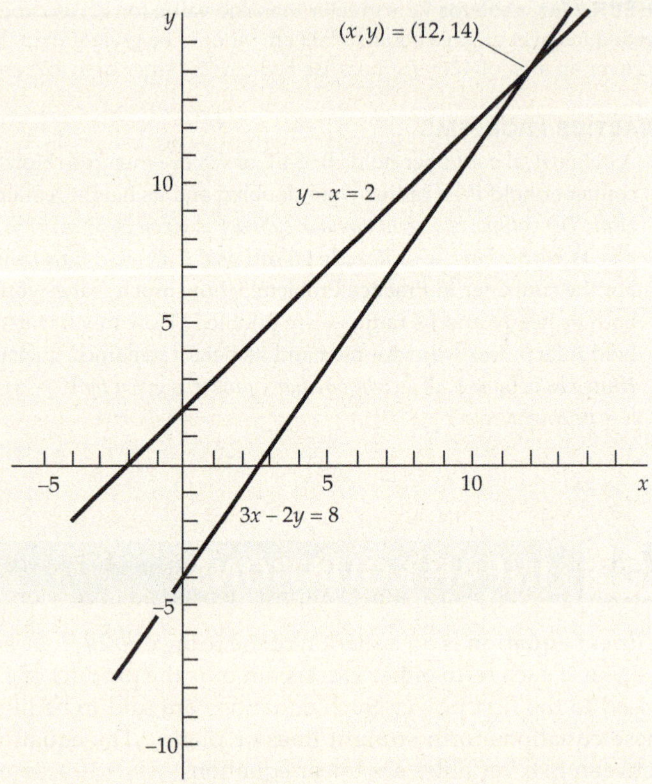

FIGURE M-2 Graph of Equations M-2 and M-3. At the point where the lines intersect, the values of x and y satisfy both equations.

PRACTICE PROBLEMS

7. True or false: $xy = 4$ is a linear equation.
8. At time $t = 0.0$ s, the position of a particle moving along the x axis at a constant velocity is $x = 3.0$ m. At $t = 2.0$ s, the position is $x = 12.0$ m. Write a linear equation showing the relation of x to t.
9. Solve the following pair of simultaneous equations for x and y:

$$\frac{5}{4}x + \frac{1}{3}y = 30$$
$$y - 5x = 20$$

M-5 QUADRATIC EQUATIONS AND FACTORING

A **quadratic equation** is an equation of the form $ax^2 + bxy + cy^2 + ex + fy + g = 0$, where x and y are variables and $a, b, c, e, f,$ and g are constants. In each term of the equation the powers of the variables are integers that sum to 2, 1, or 0. The designation *quadratic equation* usually applies to an equation of one variable that can be written in the standard form

$$ax^2 + bx + c = 0 \qquad \text{M-4}$$

where $a, b,$ and c are constants. The quadratic equation has two solutions or **roots**—values of x for which the equation is true.

FACTORING

We can solve some quadratic equations by **factoring.** Very often terms of an equation can be grouped or organized into other terms. When we factor terms, we look for multipliers and multiplicands—which we now call **factors**—that will yield two or more new terms as a product. For example, we can find the roots of the quadratic equation $x^2 - 3x + 2 = 0$ by factoring the left side, to get $(x - 2)(x - 1) = 0$. The roots are $x = 2$ and $x = 1$.

Factoring is useful for simplifying equations and for understanding the relationships between quantities. You should be familiar with the multiplication of the factors $(ax + by)(cx + dy) = acx^2 + (ad + bc)xy + bdy^2$.

You should readily recognize some typical factorable combinations:

1. Common factor: $2ax + 3ay = a(2x + 3y)$
2. Perfect square: $x^2 - 2xy + y^2 = (x - y)^2$ (If the expression on the left side of a quadratic equation in standard form is a perfect square, the two roots will be equal.)
3. Difference of squares: $x^2 - y^2 = (x + y)(x - y)$

Also, look for factors that are prime numbers (2, 5, 7, etc.) because these factors can help you factor and simplify terms quickly. For example, the equation $98x^2 - 140 = 0$ can be simplified because 98 and 140 share the common factor 2. That is, $98x^2 - 140 = 0$ becomes $2(49x^2 - 70) = 0$, so we have $49x^2 - 70 = 0$.

This result can be further simplified because 49 and 70 share the common factor 7. Thus, $49x^2 - 70 = 0$ becomes $7(7x^2 - 10) = 0$, so we have $7x^2 - 10 = 0$.

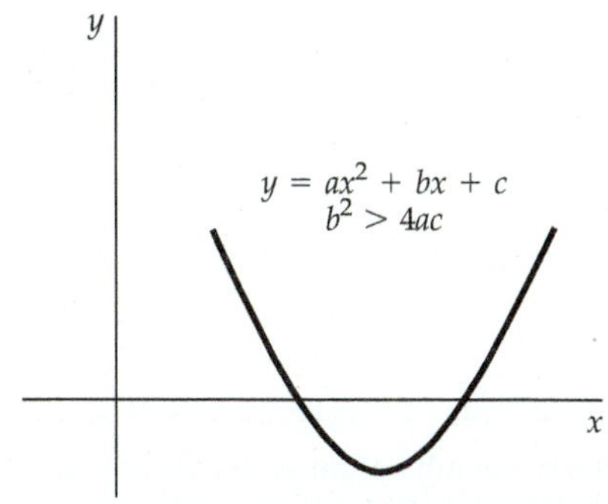

FIGURE M-3 Graph of y versus x when $y = ax^2 + bx + c$ for the case $b^2 > 4ac$. The two values of x for which $y = 0$ satisfy the quadratic equation (Equation M-4).

THE QUADRATIC FORMULA

Not all quadratic equations can be solved by factoring. However, *any* quadratic equation in the standard form $ax^2 + bx + c = 0$ can be solved by the **quadratic formula,**

$$x = \frac{-b \pm \sqrt{b^2 - 4ac}}{2a} = -\frac{b}{2a} \pm \frac{1}{2a}\sqrt{b^2 - 4ac} \qquad \text{M-5}$$

When b^2 is greater than $4ac$, there are two solutions corresponding to the + and − signs. Figure M-3 shows a graph of y versus x where $y = ax^2 + bx + c$. The curve, a **parabola,** crosses the x axis twice. (The simplest representation of a parabola in (x, y) coordinates is an equation of the form $y = ax^2 + bx + c$.) The two roots of this equation are the values for which $y = 0$; that is, they are the *x intercepts.*

When b^2 is less than $4ac$, the graph of y versus x does not intersect the x axis, as is shown in Figure M-4; there are still two roots, but they are not real numbers (see the discussion of complex numbers beginning on page M-19). When $b^2 = 4ac$, the graph of y versus x is tangent to the x axis at the point $x = -b/2a$; the two roots are each equal to $-b/2a$.

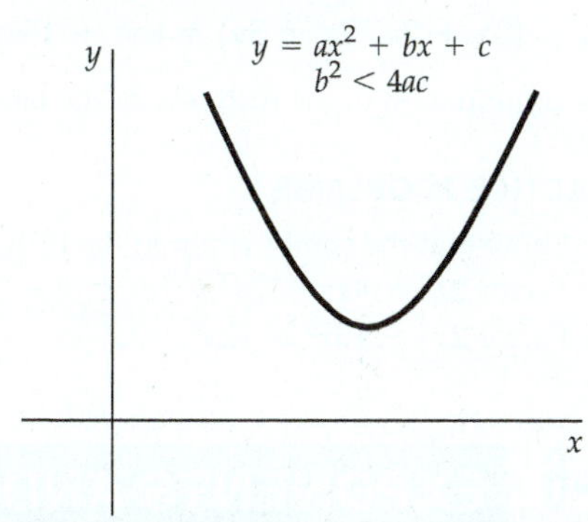

FIGURE M-4 Graph of y versus x when $y = ax^2 + bx + c$ for the case $b^2 < 4ac$. In this case, there are no real values for x for which $y = 0$.

Example M-5 Factoring a Second-Degree Polynomial

Factor the expression $6x^2 + 19xy + 10y^2$.

PICTURE We examine the coefficients of the terms to see whether the expression can be factored without resorting to more advanced methods. Remember that the multiplication $(ax + by)(cx + dy) = acx^2 + (ad + bc)xy + bdy^2$.

SOLVE

1. The coefficient of x^2 is 6 which can be factored two ways: $ac = 6$
   $3 \cdot 2 = 6$ or $6 \cdot 1 = 6$
2. The coefficient of y^2 is 10 which can also be factored two ways: $bd = 10$
   $5 \cdot 2 = 10$ or $10 \cdot 1 = 10$

3. List the possibilities for a, b, c, and d in a table. Include a column for $ad + bc$.

   If $a = 3$, then $c = 2$, and vice versa. In addition, if $a = 6$, then $c = 1$, and vice versa. For each value of a there are four values for b.

a	b	c	d	$ad + bc$
3	5	2	2	16
3	2	2	5	19
3	10	2	1	23
3	1	2	10	32
2	5	3	2	19
2	2	3	5	16
2	10	3	1	32
2	1	3	10	23
6	5	1	2	17
6	2	1	5	32
6	10	1	1	16
6	1	1	10	61
1	5	6	2	32
1	2	6	5	17
1	10	6	1	61
1	1	6	10	16

4. Find a combination such that $ad + bc = 19$. As you can see from the table there are two such combinations, and each gives the same results:

   $ad + bc = 19$

   $3 \cdot 5 + 2 \cdot 2 = 19$

5. Use the combination in the second row of the table to factor the expression in question:

   $6x^2 + 19xy + 10y^2 = (3x + 2y)(2x + 5y)$

CHECK As a check, expand $(3x + 2y)(2x + 5y)$.

$$(3x + 2y)(2x + 5y) = 6x^2 + 15xy + 4xy + 10y^2 = 6x^2 + 19xy + 10y^2$$

The combination in the fifth row of the table also gives the step-4 result.

PRACTICE PROBLEMS

10. Show that the combination in the fifth row of the table also gives the step-4 result.
11. Factor $2x^2 - 4xy + 2y^2$.
12. Factor $2x^4 + 10x^3 + 12x^2$.

M-6 EXPONENTS AND LOGARITHMS

EXPONENTS

The notation x^n stands for the quantity obtained by multiplying x times itself n times. For example, $x^2 = x \cdot x$ and $x^3 = x \cdot x \cdot x$. The quantity n is called the **power,** or the **exponent,** of x (the **base**). Listed below are some rules that will help you simplify terms that have exponents.

1. When two powers of x are multiplied, the exponents are added:

$$(x^m)(x^n) = x^{m+n} \qquad \text{M-6}$$

   Example: $x^2x^3 = x^{2+3} = (x \cdot x)(x \cdot x \cdot x) = x^5$.

2. Any number (except 0) raised to the 0 power is defined to be 1:

$$x^0 = 1 \qquad \text{M-7}$$

3. Based on rule 2,

$$x^n x^{-n} = x^0 = 1$$

$$x^{-n} = \frac{1}{x^n} \qquad \text{M-8}$$

4. When two powers are divided, the exponents are subtracted:

$$\frac{x^n}{x^m} = x^n x^{-m} = x^{n-m} \qquad \text{M-9}$$

5. When a power is raised to another power, the exponents are multiplied:

$$(x^n)^m = x^{nm} \qquad \text{M-10}$$

6. When exponents are written as fractions, they represent the roots of the base. For example,

$$x^{1/2} \cdot x^{1/2} = x$$

so

$$x^{1/2} = \sqrt{x} \qquad (x > 0)$$

Example M-6 Simplifying a Quantity That Has Exponents

Simplify $\frac{x^4 x^7}{x^8}$.

PICTURE According to rule 1, when two powers of x are multiplied, the exponents are added. Rule 4 states that when two powers are divided, the exponents are subtracted.

SOLVE

1. Simplify the numerator $x^4 x^7$ using rule 1. $x^4 x^7 = x^{4+7} = x^{11}$

2. Simplify $\frac{x^{11}}{x^8}$ using rule 4: $\frac{x^{11}}{x^8} = x^{11} x^{-8} = x^{11-8} = x^3$

CHECK Use the value $x = 2$ to determine if your answer is correct.

$$\frac{2^4 2^7}{2^8} = 2^3 = 8$$

$$\frac{2^4 2^7}{2^8} = \frac{(16)(128)}{256} = \frac{2048}{256} = 8$$

PRACTICE PROBLEMS

13. $(x^{1/18})^9 =$
14. $x^6 x^0 =$

LOGARITHMS

Any positive number can be expressed as some power of any other positive number except one. If y is related to x by $y = a^x$, then the number x is said to be the **logarithm** of y to the **base** a, and the relation is written

$$x = \log_a y$$

Thus, logarithms are *exponents*, and the rules for working with logarithms correspond to similar laws for exponents. Listed below are some rules that will help you simplify terms that have logarithms.

1. If $y_1 = a^n$ and $y_2 = a^m$, then

$$y_1 y_2 = a^n a^m = a^{n+m}$$

Correspondingly,

$$\log_a y_1 y_2 = \log_a a^{n+m} = n + m = \log_a a^n + \log_a a^m = \log_a y_1 + \log_a y_2 \qquad \text{M-11}$$

It then follows that

$$\log_a y^n = n \log_a y \qquad \text{M-12}$$

2. Because $a^1 = a$ and $a^0 = 1$,

$$\log_a a = 1 \qquad \text{M-13}$$

and

$$\log_a 1 = 0 \qquad \text{M-14}$$

There are two bases in common use: logarithms to base 10 are called **common logarithms,** and logarithms to base e (where $e = 2.718\ldots$) are called **natural logarithms.**

In this text, the symbol ln is used for natural logarithms and the symbol log, without a subscript, is used for common logarithms. Thus,

$$\log_e x = \ln x \quad \text{and} \quad \log_{10} x = \log x \qquad \text{M-15}$$

and $y = \ln x$ implies

$$x = e^y \qquad \text{M-16}$$

Logarithms can be changed from one base to another. Suppose that

$$z = \log x \qquad \text{M-17}$$

Then

$$10^z = 10^{\log x} = x \qquad \text{M-18}$$

Taking the natural logarithm of both sides of Equation M-18, we obtain

$$z \ln 10 = \ln x$$

Substituting log x for z (see Equation M-17) gives

$$\ln x = (\ln 10)\log x \qquad \text{M-19}$$

Example M-7 Converting Between Common Logarithms and Natural Logarithms

The steps leading to Equation M-19 show that, in general, $\log_b x = (\log_b a)\log_a x$, and thus that conversion of logarithms from one base to another requires only multiplication by a constant. Describe the mathematical relation between the constant for converting common logarithms to natural logarithms and the constant for converting natural logarithms to common logarithms.

PICTURE We have a general mathematical formula for converting logarithms from one base to another. We look for the mathematical relation by exchanging a for b and vice versa in the formula.

SOLVE

1. You have a formula for converting logarithms from base a to base b: $\log_b x = (\log_b a)\log_a x$
2. To convert from base b to base a, exchange all a for b and vice versa: $\log_a x = (\log_a b)\log_b x$
3. Divide both sides of the equation in step 1 by $\log_a x$: $\dfrac{\log_b x}{\log_a x} = \log_b a$
4. Divide both sides of the equation in step 2 by $(\log_a b)\log_a x$: $\dfrac{1}{\log_a b} = \dfrac{\log_b x}{\log_a x}$
5. The results show that the conversion factors $\log_b a$ and $\log_a b$ are reciprocals of one other: $\dfrac{1}{\log_a b} = \log_b a$

CHECK For the value of $\log_{10} e$, your calculator will give 0.43429. For ln 10, your calculator will give 2.3026. Multiply 0.43429 by 2.3026; you will get 1.0000.

PRACTICE PROBLEMS

15. Evaluate $\log_{10}1000$.
16. Evaluate $\log_2 5$.

M-7 GEOMETRY

The properties of the most common **geometric figures**—bounded shapes in two or three dimensions whose lengths, areas, or volumes are governed by specific ratios—are a basic analytical tool in physics. For example, the characteristic ratios within triangles give us the laws of *trigonometry* (see the next section of this tutorial), which in turn give us the theory of vectors, essential in analyzing motion in two or more dimensions. Circles and spheres are essential for understanding, among other concepts, angular momentum and the probability densities of quantum mechanics.

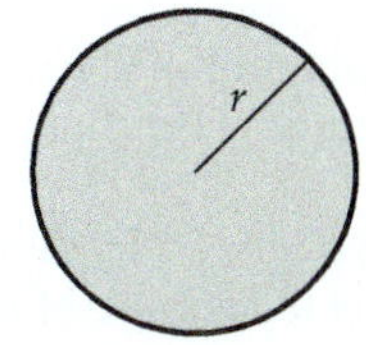

FIGURE M-5 Area of a circle.

BASIC FORMULAS IN GEOMETRY

Circle The ratio of the circumference of a circle to its diameter is a number π, which has the approximate value

$$\pi = 3.141\,592$$

The circumference C of a circle is thus related to its diameter d and its radius r by

$$C = \pi d = 2\pi r \qquad \text{circumference of circle} \qquad \text{M-20}$$

The area of a circle is (Figure M-5)

$$A = \pi r^2 \qquad \text{area of circle} \qquad \text{M-21}$$

Parallelogram The area of a parallelogram is the base b times the height h (Figure M-6):

$$A = bh$$

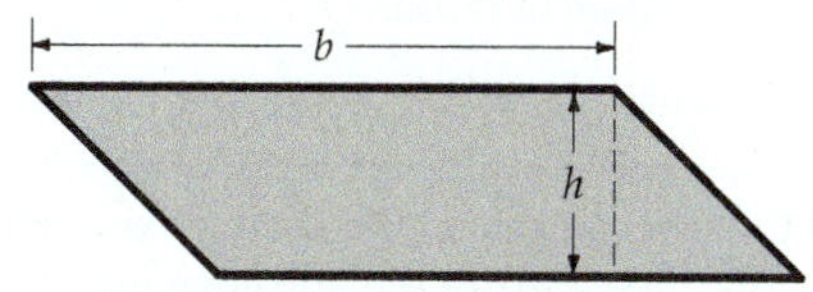

FIGURE M-6 Area of a parallelogram.

Triangle The area of a triangle is one-half the base times the height h (Figure M-7)

$$A = \frac{1}{2}bh$$

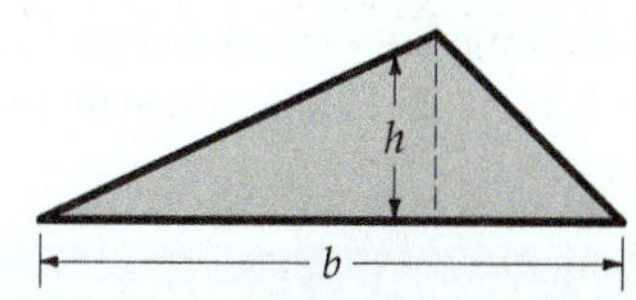

FIGURE M-7 Area of a triangle.

Sphere A sphere of radius r (Figure M-8) has a surface area given by

$$A = 4\pi r^2 \qquad \text{surface area of sphere} \qquad \text{M-22}$$

and a volume given by

$$V = \frac{4}{3}\pi r^3 \qquad \text{volume of sphere} \qquad \text{M-23}$$

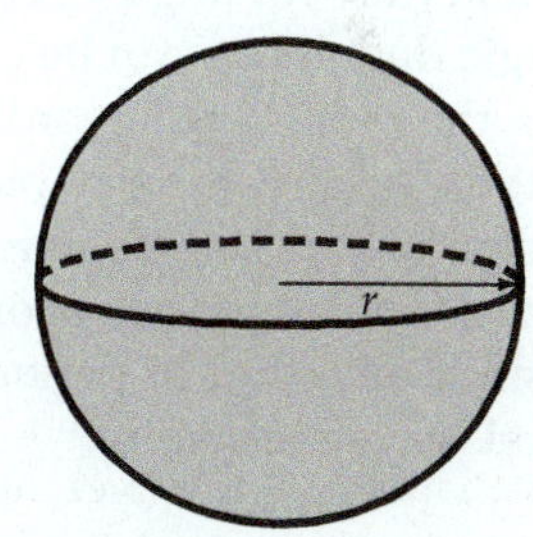

FIGURE M-8 Surface area and volume of a sphere.

Cylinder A cylinder of radius r and length L (Figure M-9) has surface area (not including the end faces) of

$$A = 2\pi rL \qquad \text{surface of cylinder} \qquad \text{M-24}$$

and volume of

$$V = \pi r^2 L \qquad \text{volume of cylinder} \qquad \text{M-25}$$

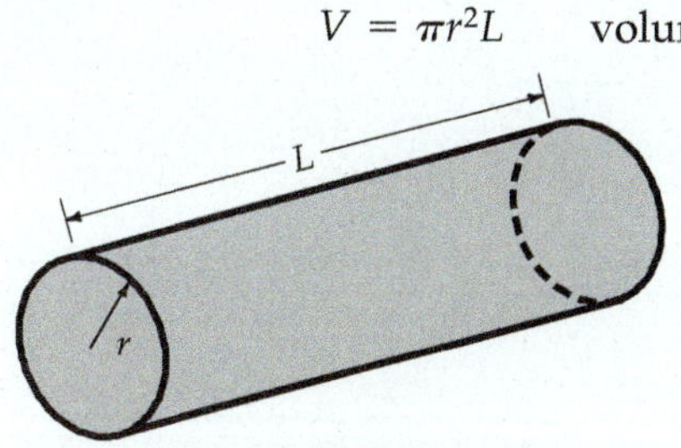

FIGURE M-9 Surface area (not including the end faces) and the volume of a cylinder.

Example M-8 Calculating the volume of a Spherical Shell

An aluminum spherical shell has an outer diameter of 40.0 cm and an inner diameter of 39.0 cm. Find the volume of the aluminum in this shell.

PICTURE The volume of the aluminum in the spherical shell is the volume that remains when we subtract the volume of the inner sphere having $d_i = 2r_i = 39.0$ cm from the volume of the outer sphere having $d_o = 2r_o = 40.0$ cm.

SOLVE

1. Subtract the volume of the sphere of radius r_i from the volume of the sphere of radius r_o: $V = \frac{4}{3}\pi r_o^3 - \frac{4}{3}\pi r_i^3 = \frac{4}{3}\pi(r_o^3 - r_i^3)$
2. Substitute 20.0 cm for r_o and 19.5 cm for r_i: $V = \frac{4}{3}\pi[(20.0\text{ cm})^3 - (19.5\text{ cm})^3] = \boxed{2.45 \times 10^3\text{ cm}^3}$

CHECK The volume of the shell is expected to be the same order of magnitude as the volume of a hollow cube with an outside edge length of 40.0 cm and an inside edge length of 39.0 cm. The volume of such a hollow cube is $(40.0\text{ cm})^3 - (39.0\text{ cm})^3 = 4.68 \times 10^3\text{ cm}^3$. The step-2 result meets the expectation that the volume of the shell is the same order of magnitude as the volume of the hollow cube.

PRACTICE PROBLEMS

17. Find the ratio between the volume V and the surface A of a sphere of radius r.
18. What is the area of a cylinder that has a radius that is 1/3 its length?

M-8 TRIGONOMETRY

Trigonometry, which gets its name from Greek roots meaning "triangle" and "measure," is the study of some important mathematical functions, called **trigonometric functions.** These functions are most simply defined as ratios of the sides of right triangles. However, these right-triangle definitions are of limited use because they are valid only for angles between zero and 90°. However, the validity of the right-triangle definitions can be extended by defining the trigonometric functions in terms of the ratio of the coordinates of points on a circle of unit radius drawn centered at the origin of the xy plane.

In physics, we first encounter trigonometric functions when we use vectors to analyze motion in two dimensions. Trigonometric functions are also essential in the analysis of any kind of periodic behavior, such as circular motion, oscillatory motion, and wave mechanics.

ANGLES AND THEIR MEASURE: DEGREES AND RADIANS

The size of an angle formed by two intersecting straight lines is known as its **measure.** The standard way of finding the measure of an angle is to place the angle so that its **vertex,** or point of intersection of the two lines that form the angle, is at the center of a circle located at the origin of a graph that has Cartesian coordinates and one of the lines extends rightward on the positive x axis. The distance traveled *counterclockwise* on the circumference from the positive x axis to reach the intersection of the circumference with the other line defines the measure of the angle. (Traveling clockwise to the second line would simply give us a negative measure; to illustrate basic concepts, we position the angle so that the smaller rotation will be in the counterclockwise direction.)

The most familiar unit for expressing the measure of an angle is the **degree,** which equals 1/360 of the full distance around the circumference of the circle. For greater precision, or for smaller angles, we either show degrees plus minutes (′)

and seconds ("), with $1' = 1°/60$ and $1'' = 1'/60 = 1°/3600$; or show degrees as an ordinary decimal number.

For scientific work, a more useful measure of an angle is the **radian** (rad). Again, place the angle with its vertex at the center of a circle and measure counterclockwise rotation around the circumference. The measure of the angle in radians is then defined as the length of the circular arc from one line to the other divided by the radius of the circle (Figure M-10). If s is the arc length and r is the radius of the circle, the angle θ measured in radians is

$$\theta = \frac{s}{r} \qquad \text{M-26}$$

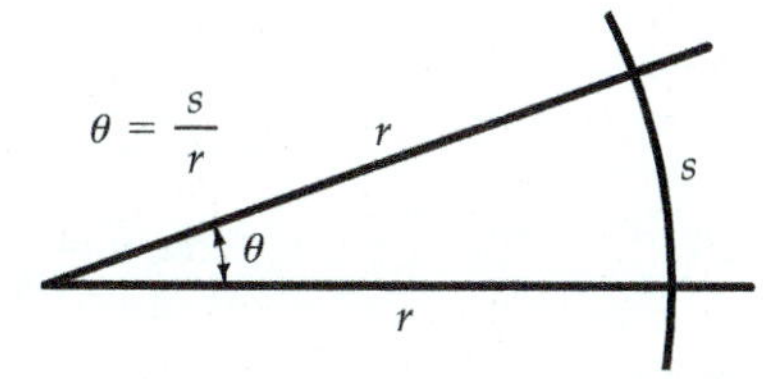

FIGURE M-10 The angle θ in radians is defined to be the ratio s/r, where s is the arc length intercepted on a circle of radius r.

Because the angle measured in radians is the ratio of two lengths, it is dimensionless. The relation between radians and degrees is

$$360° = 2\pi \text{ rad}$$

or

$$1 \text{ rad} = \frac{360°}{2\pi} = 57.3°$$

Figure M-11 shows some useful relations for angles.

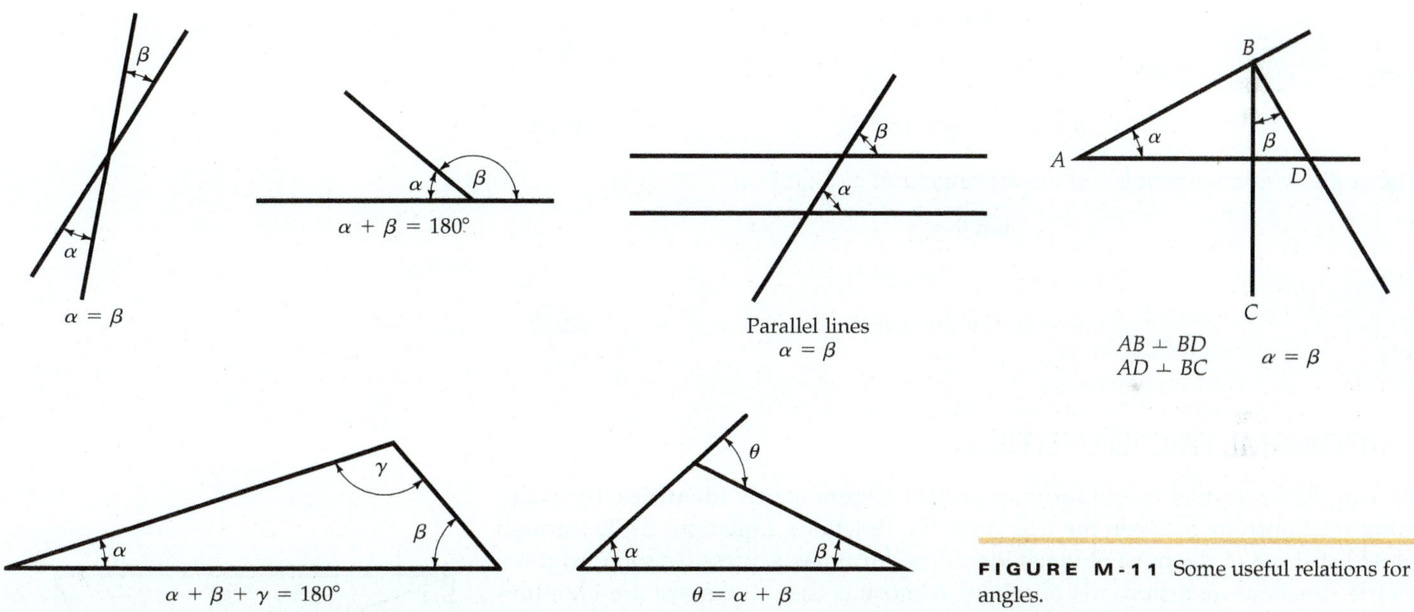

FIGURE M-11 Some useful relations for angles.

THE TRIGONOMETRIC FUNCTIONS

Figure M-12 shows a right triangle formed by drawing the line BC perpendicular to AC. The lengths of the sides are labeled a, b, and c. The right-triangle definitions of the trigonometric functions $\sin\theta$ (the **sine**), $\cos\theta$ (the **cosine**), and $\tan\theta$ (the **tangent**) for an acute angle θ are

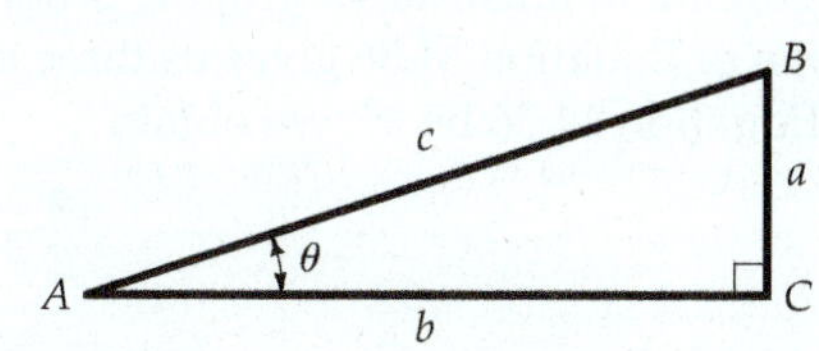

FIGURE M-12 A right triangle with sides of length a and b and a hypotenuse of length c.

$$\sin\theta = \frac{a}{c} = \frac{\text{Opposite side}}{\text{Hypotenuse}} \qquad \text{M-27}$$

$$\cos\theta = \frac{b}{c} = \frac{\text{Adjacent side}}{\text{Hypotenuse}} \qquad \text{M-28}$$

$$\tan\theta = \frac{a}{b} = \frac{\text{Opposite side}}{\text{Adjacent side}} = \frac{\sin\theta}{\cos\theta} \qquad \text{M-29}$$

(**Acute angles** are angles whose positive rotation around the circumference of a circle measures less than 90°, or $\pi/2$.) Three other trigonometric functions—the

secant (sec), the **cosecant** (csc), and the **cotangent** (cot), defined as the reciprocals of these functions—are

$$\sec\theta = \frac{c}{b} = \frac{1}{\cos\theta} \qquad \text{M-30}$$

$$\csc\theta = \frac{c}{a} = \frac{1}{\sin\theta} \qquad \text{M-31}$$

$$\cot\theta = \frac{b}{a} = \frac{1}{\tan\theta} = \frac{\cos\theta}{\sin\theta} \qquad \text{M-32}$$

The angle θ, whose sine is x, is called the arcsine of x, and is written $\sin^{-1}x$. That is, if

$$\sin\theta = x$$

then

$$\theta = \arcsin x = \sin^{-1}x \qquad \text{M-33}$$

The arcsine is the inverse of the sine. The inverse of the cosine and tangent are defined similarly. The angle whose cosine is y is the arccosine of y. That is, if

$$\cos\theta = y$$

then

$$\theta = \arccos y = \cos^{-1}y \qquad \text{M-34}$$

The angle whose tangent is z is the arctangent of z. That is, if

$$\tan\theta = z$$

then

$$\theta = \arctan z = \tan^{-1}z \qquad \text{M-35}$$

TRIGONOMETRIC IDENTITIES

We can derive several useful formulas, called **trigonometric identities,** by examining relationships between the trigonometric functions. Equations M-30 through M-32 list three of the most obvious identities, formulas expressing some trigonometric functions as reciprocals of others. Almost as easy to discern are identities derived from the **Pythagorean theorem,**

$$a^2 + b^2 = c^2 \qquad \text{M-36}$$

(Figure M-13 illustrates a graphic proof of the theorem.) Simple algebraic manipulation of Equation M-36 gives us three more identities. First, if we divide each term in Equation M-36 by c^2, we obtain

$$\frac{a^2}{c^2} + \frac{b^2}{c^2} = 1$$

or, from the definitions of $\sin\theta$ (which is a/c) and $\cos\theta$ (which is b/c)

$$\sin^2\theta + \cos^2\theta = 1 \qquad \text{M-37}$$

Similarly, we can divide each term in Equation M-36 by a^2 or b^2 and obtain

$$1 + \cot^2\theta = \csc^2\theta \qquad \text{M-38}$$

and

$$1 + \tan^2\theta = \sec^2\theta \qquad \text{M-39}$$

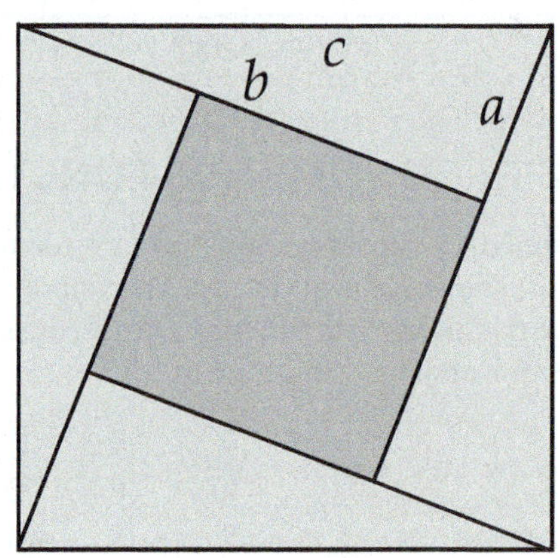

FIGURE M-13 When this figure was first published, the letters were absent and it was accompanied by the single word "Behold!" Using the drawing, establish the Pythagorean theorem ($a^2 + b^2 = c^2$).

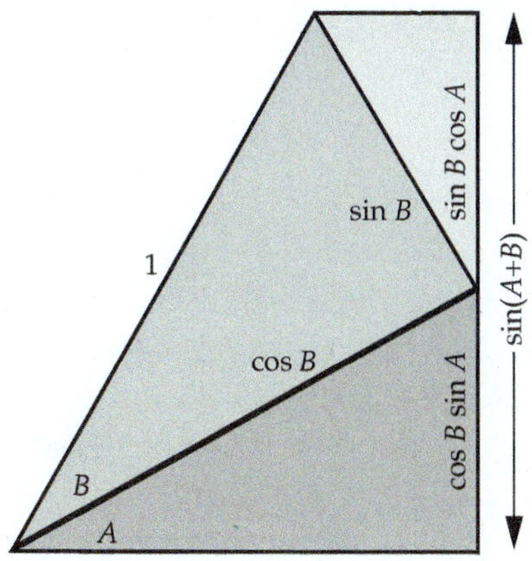

FIGURE M-14 Using this drawing, establish the identity $\sin(A + B) = \sin A \cos B + \cos A \sin B$. You can also use it to establish the identity $\cos(A + B) = \cos A \cos B - \sin A \sin B$. Try it.

Table M-2 Trigonometric Identities

$$\sin(A \pm B) = \sin A \cos B \pm \cos A \sin B$$

$$\cos(A \pm B) = \cos A \cos B \mp \sin A \sin B$$

$$\tan(A \pm B) = \frac{\tan A \pm \tan B}{1 \mp \tan A \tan B}$$

$$\sin A \pm \sin B = 2 \sin\left[\frac{1}{2}(A \pm B)\right]\cos\left[\frac{1}{2}(A \mp B)\right]$$

$$\cos A + \cos B = 2 \cos\left[\frac{1}{2}(A + B)\right]\cos\left[\frac{1}{2}(A - B)\right]$$

$$\cos A - \cos B = 2 \sin\left[\frac{1}{2}(A + B)\right]\sin\left[\frac{1}{2}(B - A)\right]$$

$$\tan A \pm \tan B = \frac{\sin(A \pm B)}{\cos A \cos B}$$

$$\sin^2\theta + \cos^2\theta = 1;\ \sec^2\theta - \tan^2\theta = 1;\ \csc^2\theta - \cot^2\theta = 1$$

$$\sin 2\theta = 2 \sin\theta \cos\theta$$

$$\cos 2\theta = \cos^2\theta - \sin^2\theta = 2\cos^2\theta - 1 = 1 - 2\sin^2\theta$$

$$\tan 2\theta = \frac{2\tan\theta}{1 - \tan^2\theta}$$

$$\sin\frac{1}{2}\theta = \pm\sqrt{\frac{1 - \cos\theta}{2}};\ \cos\frac{1}{2}\theta = \pm\sqrt{\frac{1 + \cos\theta}{2}};\ \tan\frac{1}{2}\theta = \pm\sqrt{\frac{1 - \cos\theta}{1 + \cos\theta}}$$

Table M-2 lists these last three and many more trigonometric identities. Notice that they fall into four categories: functions of sums or differences of angles, sums or differences of squared functions, functions of double angles (2θ), and functions of half angles $\left(\frac{1}{2}\theta\right)$. Notice that some of the formulas contain paired alternatives, expressed with the signs $\pm$ and $\mp$; in such formulas, remember to always apply the formula with either all the "upper" or all the "lower" alternatives. Figure M-14 shows a graphic proof of the first two sum-of-angle identities.

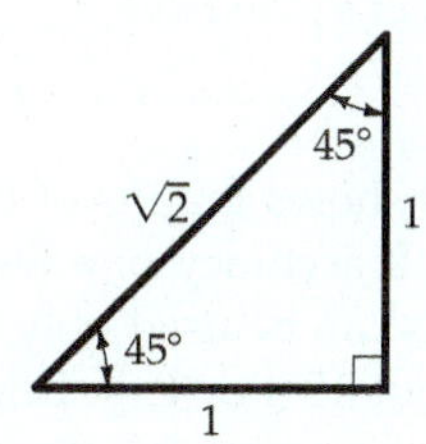

FIGURE M-15 An isosceles right triangle.

SOME IMPORTANT VALUES OF THE FUNCTIONS

Figure M-15 is a diagram of an *isosceles* right triangle (an isosceles triangle is a triangle with two equal sides), from which we can find the sine, cosine, and tangent of 45°. The two acute angles of this triangle are equal. Because the sum of the three angles in a triangle must equal 180° and the right angle is 90°, each acute angle must be 45°. For convenience, let us assume that the equal sides each have a length of 1 unit. The Pythagorean theorem gives us a value for the hypotenuse of

$$c = \sqrt{a^2 + b^2} = \sqrt{1^2 + 1^2} = \sqrt{2} \text{ units}$$

We calculate the values of the functions as follows:

$$\sin 45^\circ = \frac{a}{c} = \frac{1}{\sqrt{2}} = 0.707 \quad \cos 45^\circ = \frac{b}{c} = \frac{1}{\sqrt{2}} = 0.707 \quad \tan 45^\circ = \frac{a}{b} = \frac{1}{1} = 1$$

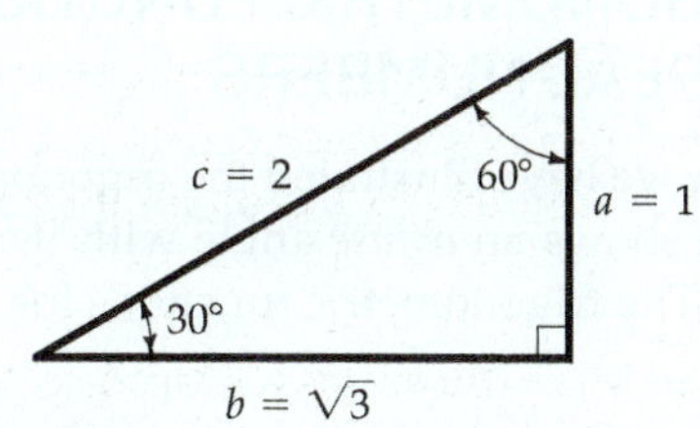

FIGURE M-16 A 30°–60° right triangle.

Another common triangle, a 30°–60° right triangle, is shown in Figure M-16. Because this particular right triangle is in effect half of an *equilateral triangle* (a 60°–60°–60° triangle, or a triangle having three equal sides and three equal angles), we can see that the sine of 30° must be exactly 0.5 (Figure M-17). The equilateral triangle must have all sides equal to c, the hypotenuse of the 30°–60° right triangle. Thus, side a is one-half the length of the hypotenuse, and so

$$\sin 30^\circ = \frac{1}{2}$$

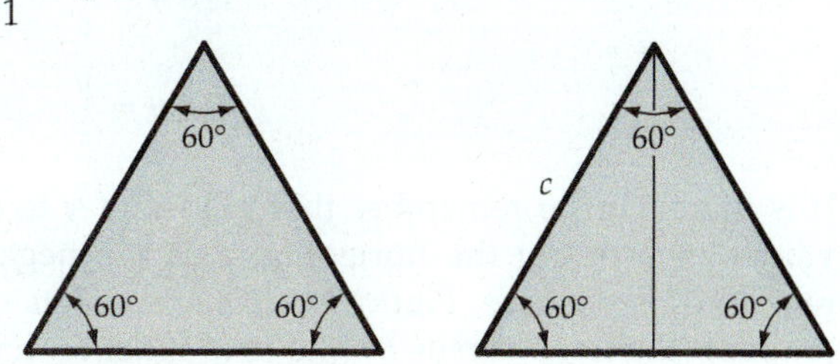

FIGURE M-17 (a) An equilateral triangle. (b) An equilateral triangle that has been bisected to form two 30°–60° right triangles.

To find the other ratios within the 30–60° right triangle, let us assign a value of 1 to the side opposite the 30° angle. Then

$$c = \frac{1}{0.5} = 2 \qquad b = \sqrt{c^2 - a^2} = \sqrt{2^2 - 1^2} = \sqrt{3}$$

$$\cos 30° = \frac{b}{c} = \frac{\sqrt{3}}{2} = 0.866 \qquad \tan 30° = \frac{a}{b} = \frac{1}{\sqrt{3}} = 0.577$$

$$\sin 60° = \frac{b}{c} = \cos 30° = 0.866 \qquad \cos 60° = \frac{a}{c} = \sin 30° = \frac{1}{2}$$

$$\tan 60° = \frac{b}{a} = \frac{\sqrt{3}}{1} = 1.732$$

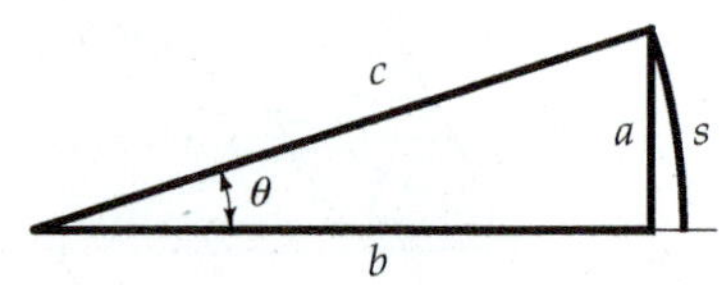

FIGURE M-18 For small angles, $\sin\theta = a/c$, $\tan\theta = a/b$, and the angle $\theta = s/c$ are all approximately equal.

SMALL-ANGLE APPROXIMATION

For small angles, the length a is nearly equal to the arc length s, as can be seen in Figure M-18. The angle $\theta = s/c$ is therefore nearly equal to $\sin\theta = a/c$:

$$\sin\theta \approx \theta \quad \text{for small values of } \theta \qquad \text{M-40}$$

Similarly, the lengths c and b are nearly equal, so $\tan\theta = a/b$ is nearly equal to both θ and $\sin\theta$ for small values of θ:

$$\tan\theta \approx \sin\theta \approx \theta \quad \text{for small values of } \theta \qquad \text{M-41}$$

Equations M-40 and M-41 hold only if θ is measured in radians. Because $\cos\theta = b/c$, and because these lengths are nearly equal for small values of θ, we have

$$\cos\theta \approx 1 \quad \text{for small values of } \theta \qquad \text{M-42}$$

Figure M-19 shows graphs of θ, $\sin\theta$, and $\tan\theta$ versus θ for small values of θ. If accuracy of a few percent is needed, small-angle approximations can be used only for angles of about a quarter of a radian (or about 15°) or less. Below this value, as the angle becomes smaller, the approximation $\theta \approx \sin\theta \approx \tan\theta$ is even more accurate.

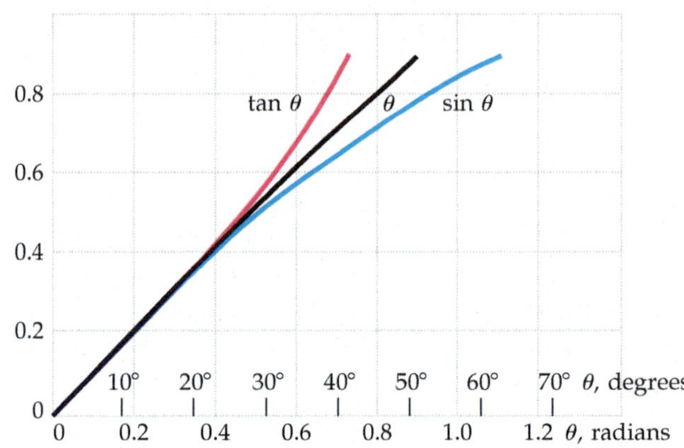

FIGURE M-19 Graphs of $\tan\theta$, θ, and $\sin\theta$ versus θ for small values of θ.

TRIGONOMETRIC FUNCTIONS AS FUNCTIONS OF REAL NUMBERS

So far we have illustrated the trigonometric functions as properties of angles. Figure M-20 shows an *obtuse* angle with its vertex at the origin and one side along the x axis. The trigonometric functions for a "general" angle such as this are defined by

$$\sin\theta = \frac{y}{c} \qquad \text{M-43}$$

$$\cos\theta = \frac{x}{c} \qquad \text{M-44}$$

$$\tan\theta = \frac{y}{x} \qquad \text{M-45}$$

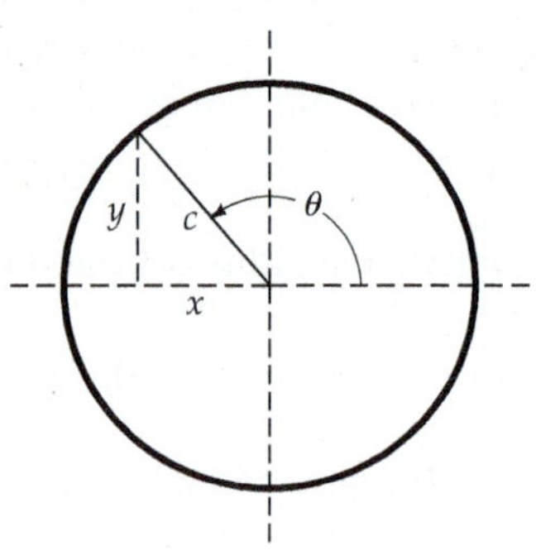

FIGURE M-20 Diagram for defining the trigonometric functions for an obtuse angle.

It is important to remember that values of x to the left of the vertical axis and values of y below the horizontal axis are negative; c in the figure is always regarded as positive. Figure M-21 shows plots of the general sine, cosine, and tangent functions versus θ. The sine function has a period of 2π rad. Thus, for any value of θ, $\sin(\theta + 2\pi) = \sin\theta$, and so forth. That is, when an angle changes by 2π rad, the function returns to its original value. The tangent function has a

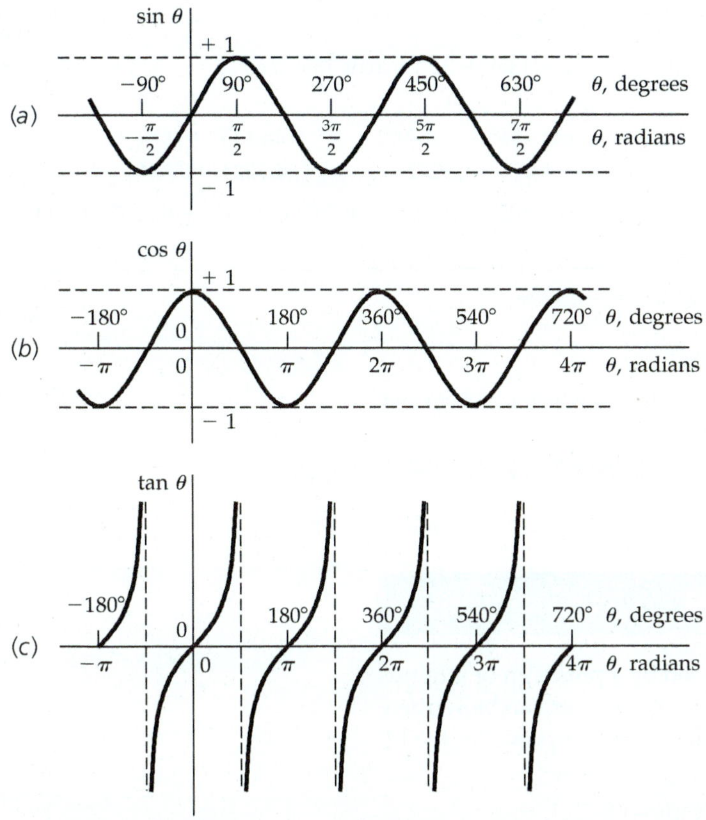

FIGURE M-21 The trigonometric functions sin θ, cos θ, and tan θ versus θ.

period of π rad. Thus, $\tan(\theta + \pi) = \tan\theta$, and so forth. Some other useful relations are

$$\sin(\pi - \theta) = \sin\theta \qquad \text{M-46}$$

$$\cos(\pi - \theta) = -\cos\theta \qquad \text{M-47}$$

$$\sin(\tfrac{1}{2}\pi - \theta) = \cos\theta \qquad \text{M-48}$$

$$\cos(\tfrac{1}{2}\pi - \theta) = \sin\theta \qquad \text{M-49}$$

Because the radian is dimensionless, it is not hard to see from the plots in Figure M-21 that the trigonometric functions are functions of all real numbers. The functions can also be expressed as power series in θ. The series for sin θ and cos θ are

$$\sin\theta = \theta - \frac{\theta^3}{3!} + \frac{\theta^5}{5!} - \frac{\theta^7}{7!} + \cdots \qquad \text{M-50}$$

$$\cos\theta = 1 - \frac{\theta^2}{2!} + \frac{\theta^4}{4!} - \frac{\theta^6}{6!} + \cdots \qquad \text{M-51}$$

When θ is small, good approximations are obtained using only the first few terms in the series.

Example M-9 Cosine of a Sum

Using the suitable trigonometric identity from Table M-2, find cos(135° + 22°). Give your answer in four significant figures.

PICTURE As long as all angles are given in degrees, there is no need to convert to radians, because all operations are numerical values of the functions. Be sure, however, that your calculator is in degree mode. The suitable identity is $\cos(A \pm B) = \cos A \cos B \mp \sin A \sin B$, where the upper signs are appropriate.

SOLVE

1. Write the trigonometric identity for the cosine of a sum, with $A = 135°$ and $B = 22°$:

$\cos(135° + 22°) = (\cos 135°)(\cos 22°) - (\sin 135°)(\sin 22°)$

2. Using a calculator, find cos 135°, sin 135°, cos 22°, and sin 22°:

$\cos 135° = -0.7071$ $\sin 135° = 0.7071$

$\cos 22° = 0.9272$ $\sin 22° = 0.3746$

3. Enter the values in the formula and calculate the answer:

$$\cos(135° + 22°) = (-0.7071)(0.9272) - (0.7071)(0.3746) = -0.9205$$

CHECK The calculator shows that the $\cos(135° + 22°) = \cos(157°) = -0.9205$.

PRACTICE PROBLEMS

19. Find $\sin\theta$ and $\cos\theta$ for the right triangle shown in Figure M-12 in which $a = 4$ cm and $b = 7$ cm. What is the value for θ?
20. Find $\sin\theta$ where $\theta = 8.2°$. Is your answer consistent with the small-angle approximation?

M-9 THE BINOMIAL EXPANSION

A **binomial** is an expression consisting of two terms joined by a plus sign or a minus sign. The **binomial theorem** states that a binomial raised to a power can be written, or *expanded*, as a series of terms. If we raise the binomial $(1 + x)$ to a power n, the binomial theorem takes the form

$$(1 + x)^n = 1 + nx + \frac{n(n-1)}{2!}x^2 + \frac{n(n-1)(n-2)}{3!}x^3 + \cdots \qquad \text{M-52}$$

The series is valid for any value of n if $|x|$ is less than 1. The binomial expansion is very useful for approximating algebraic expressions, because when $|x| < 1$, the higher-order terms in the sum are small. (The order of a term is the power of x in the term. Thus, the terms explicitly shown in Equation M-52 are of order 0, 1, 2, and 3.) The series is particularly useful in situations where $|x|$ is small compared with 1; then each term is *much* smaller than the previous term and we can drop all but the first two or three terms in the expansion. If $|x|$ is much less than 1, we have

$$(1 + x)^n \approx 1 + nx, \qquad |x| \ll 1 \qquad \text{M-53}$$

The binomial expansion is used in deriving many formulas of calculus that are important in physics. A well-known use in physics of the approximation in Equation M-53 is the proof that relativistic kinetic energy reduces to the classic formula when the velocity of a particle is very small compared with the velocity of light c.

Example M-10 Using the Binomial Expansion to Find a Power of a Number

Use Equation M-53 to find an approximate value for the square root of 101.

PICTURE The number 101 readily suggests a binomial, namely, $(100 + 1)$. To approximate the answer using the binomial expansion, we must manipulate the expression to get a binomial consisting of 1 and a term less than 1.

SOLVE

1. Write $(101)^{1/2}$ to give an expression $(1 + x)^n$ in which x is much less than 1:

$$(101)^{1/2} = (100 + 1)^{1/2} = (100)^{1/2}(1 + 0.01)^{1/2} = 10(1 + 0.01)^{1/2}$$

2. Use Equation M-53 with $n = \frac{1}{2}$ and $x = 0.01$ to expand $(1 + 0.01)^{1/2}$:

$$(1 + 0.01)^{1/2} = 1 + \tfrac{1}{2}(0.01) + \frac{\frac{1}{2}\left(-\frac{1}{2}\right)}{2}(0.01)^2 + \cdots$$

3. Because $|x| \ll 1$, we expect the magnitude of terms of order 2 and higher to be significantly smaller than the magnitude of the first-order term. Approximate the binomial (1) by keeping only the zeroth and first-order terms, and (2) by keeping only the first 3 terms:

Keeping only the zeroth and first-order terms gives

$$(1 + 0.01)^{1/2} \approx 1 + \tfrac{1}{2}(0.01) = 1 + 0.005\,000\,0 = 1.005\,000\,0$$

Keeping only the zeroth, first-, and second-order terms gives

$$(1 + 0.01)^{1/2} \approx 1 + \tfrac{1}{2}(0.01) + \frac{\tfrac{1}{2}(-\tfrac{1}{2})}{2}(0.01)^2 \approx 1 + 0.005\,000\,0 - 0.000\,012\,5 = 1.004\,987\,5$$

4. Substitute these results into the equation in step 1:

Keeping only the zeroth and first-order terms gives

$$(101)^{1/2} = 10(1 + 0.01)^{1/2} \approx \boxed{10.050\,000}$$

Keeping only the zeroth, first-, and second-order terms gives

$$(101)^{1/2} = 10(1 + 0.01)^{1/2} \approx \boxed{10.049\,875}$$

CHECK We therefore expect our answer to be correct to within about 0.001%. The value of $(101)^{1/2}$, to eight figures, is 10.049 876. This differs from 10.050 000 by 0.000 124, or about one part in 10^5, and differs from 10.049 875 by about one part in 10^7.

PRACTICE PROBLEMS For the following, calculate the answer keeping the zeroth and first-order terms in the binomial series (Equation M-53), find the answer using your calculator, and show the percentage discrepancy between the two values:

21. $(1 + 0.001)^{-4}$
22. $(1 - 0.001)^{40}$

M-10 COMPLEX NUMBERS

Real numbers are all numbers, from $-\infty$ to $+\infty$, that can be *ordered*. We know that, given two real numbers, one is always equal to, greater than, or less than the other. For example, $3 > 2$, $1.4 < \sqrt{2} < 1.5$, and $3.14 < \pi < 3.15$. A number that *cannot* be ordered is $\sqrt{-1}$; we cannot measure the size of this number, and so it makes no sense to say, for example, that $3 \times \sqrt{-1}$ is greater than or less than $2 \times \sqrt{-1}$. The earliest mathematicians who dealt with numbers containing $\sqrt{-1}$ referred to these numbers as *imaginary* numbers because they could not be used to measure or count something. In mathematics the symbol i is used to represent $\sqrt{-1}$.

Equation M-5, the quadratic formula, applies to equations of the form

$$ax^2 + bx + c = 0$$

The formula shows that there are no real roots when $b^2 < 4ac$. There are, however, still two roots. Each root is a number containing two terms: a real number, and a multiple of $i = \sqrt{-1}$. The multiple of i is called an **imaginary number,** and i is called the **unit imaginary.**

A general **complex number** z can be written

$$z = a + bi \qquad \text{M-54}$$

where a and b are real numbers. The quantity a is called the real part of z, or Re(z), and the quantity b is called the imaginary part of z, or Im(z). We can represent a complex number z as a point in a plane, called the complex plane, as shown in Figure M-22, where the x axis is the **real axis** and the y axis is the **imaginary axis.** We can also use the relations $a = r\cos\theta$ and $b = r\sin\theta$ from Figure M-22 to write the complex number z in **polar coordinates** (a system in which a point is designated by the counterclockwise angle of rotation θ and the distance r in the direction of θ):

$$z = r\cos\theta + ir\sin\theta \qquad \text{M-55}$$

where $r = \sqrt{a^2 + b^2}$ is called the **magnitude** of z.

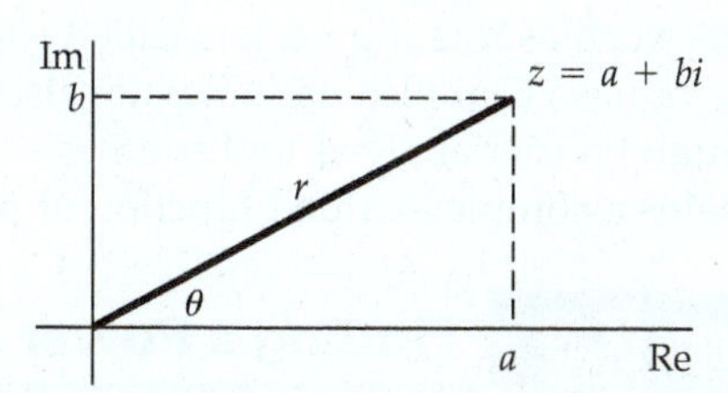

$$z = a + bi = r\cos\theta + (r\sin\theta)i = r(\cos\theta + i\sin\theta)$$

FIGURE M-22 Representation of a complex number in a plane. The real part of the complex number is plotted along the horizontal axis, and the imaginary part is plotted along the vertical axis.

When complex numbers are added or subtracted, the real and imaginary parts are added or subtracted separately:

$$z_1 + z_2 = (a_1 + ib_1) + (a_2 + ib_2) = (a_1 + a_2) + i(b_1 + b_2) \qquad \text{M-56}$$

However, when two complex numbers are multiplied, each part of one number is multiplied by each part of the other number:

$$\begin{aligned} z_1 z_2 &= (a_1 + ib_1)(a_2 + ib_2) = a_1a_2 + i^2b_1b_2 + i(a_1b_2 + a_2b_1) \\ &= a_1a_2 - b_1b_2 + i(a_1b_2 + a_2b_1) \end{aligned} \qquad \text{M-57}$$

where we have used $i^2 = -1$.

The **complex conjugate** z^* of the complex number z is that number obtained by replacing i with $-i$ when writing z. If $z = a + ib$, then

$$z^* = (a + ib)^* = a - ib \qquad \text{M-58}$$

(When a quadratic equation has complex roots, the roots are **conjugate complex numbers,** in the form $a \pm bi$.) The product of a complex number and its complex conjugate equals the square of the magnitude of the number:

$$zz^* = (a + ib)(a - ib) = a^2 + b^2 = r^2 \qquad \text{M-59}$$

A particularly useful function of a complex number is the exponential $e^{i\theta}$. Using an expansion for e^x, we have

$$e^{i\theta} = 1 + i\theta + \frac{(i\theta)^2}{2!} + \frac{(i\theta)^3}{3!} + \frac{(i\theta)^4}{4!} + \cdots$$

Using $i^2 = -1$, $i^3 = -i$, $i^4 = +1$, and so forth, and separating the real parts from the imaginary parts, this expansion can be written

$$e^{i\theta} = \left(1 - \frac{\theta^2}{2!} + \frac{\theta^4}{4!} - \cdots\right) + i\left(\theta - \frac{\theta^3}{3!} + \cdots\right)$$

Comparing this result with Equations M-50 and M-51, we can see that

$$e^{i\theta} = \cos\theta + i\sin\theta \qquad \text{M-60}$$

Using this result, we can express a general complex number as an exponential:

$$z = a + ib = r\cos\theta + ir\sin\theta = re^{i\theta} \qquad \text{M-61}$$

If $z = x + iy$, where x and y are real variables, then z is called a **complex variable.**

COMPLEX VARIABLES IN PHYSICS

Complex variables are often used in formulas describing AC circuits: the impedance of a capacitor or an inductor includes a real part (the resistance) and an imaginary part (the reactance). (There are alternative ways, however, of analyzing AC circuits—such as rotating vectors called *phasors*—that do not require assigning imaginary values.) Complex variables are also important in the study of harmonic waves through Fourier analysis and synthesis. The time-dependent Schrödinger equation contains a complex-valued function of position and time.

Example M-11 Finding a Power of a Complex Number

Calculate $(1 + 3i)^4$ by using the binomial expansion.

PICTURE The expression is of the form $(1 + x)^n$. Because n is a positive integer, the expansion is valid for any value of x, and all terms, other than those of order n or lower must equal zero.

SOLVE

1. Write out the expansion of $(1 + 3i)^4$ to show the terms up through the fouth-order term:

$$1 + 4\cdot 3i + \frac{4(3)}{2!}(3i)^2 + \frac{4(3)(2)}{3!}(3i)^3 + \frac{4(3)(2)(1)}{4!}(3i)^4$$

2. Evaluate each term, remembering that $i^2 = -1$, $i^3 = -i$, and $i^4 = +1$: $1 + 12i - 54 - 108i + 81$

3. Show the result in the form $a + bi$: $(1 + 3i)^4 = \boxed{28 - 96i}$

CHECK We can solve the problem algebraically to show that the answer is correct. We first square $(1 + 3i)$ and then square the result, to get $(1 + 3i)^4$:

$$(1 + 3i)^2 = 1 \cdot 1 + 2 \cdot 1 \cdot 3i + (3i)^2 = 1 + 6i - 9 = -8 + 6i$$

$$(-8 + 6i)^2 = (-8)(-8) + 2(-8)(6i) + (6i)^2 = 64 - 96i - 36 = 28 - 96i$$

PRACTICE PROBLEMS Express in the form $a + bi$:

23. $e^{i\pi}$

24. $e^{i\pi/2}$

M-11 DIFFERENTIAL CALCULUS

Calculus is the branch of mathematics that allows us to deal with instantaneous rates of change of functions and variables. From the equation of a function—say, x as a function of t—we can always find x for a particular t, but with the methods of calculus you can go much further. You can know where x will have certain properties, such as a maximum or a minimum value, without having to try endless values of t. With calculus, if given the proper data, you can find, for example, the location of maximum stress on a beam, or the velocity or position of a falling object at a time t, or the energy a falling object has acquired at the time of impact. The principles of calculus are derived from examining functions at the infinitesimal level—analyzing how, say, x will change when the change in t becomes vanishingly small. We start with **differential calculus,** in which we determine the *limit* of the rate of change of x with respect to t as the change in t becomes closer and closer to zero.

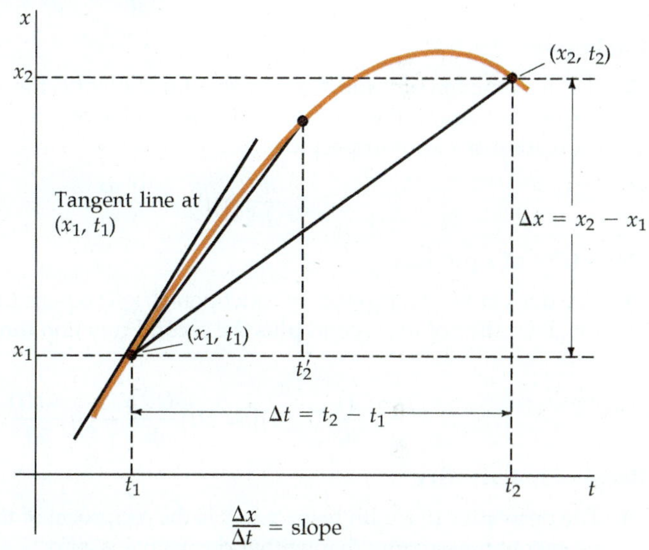

FIGURE M-23 Graph of a typical function $x(t)$. The points (x_1, t_1) and (x_2, t_2) are connected by a straight line. The slope of this line is $\Delta x/\Delta t$. As the time interval beginning at t_1 is decreased, the slope for that interval approaches the slope of the line tangent to the curve at time t_1, which is the derivative of x with respect to t.

Figure M-23 is a graph of x versus t for a typical function $x(t)$. At a particular value $t = t_1$, x has the value of x_1, as indicated. At another value t_2, x has the value x_2. The change in t, $t_2 - t_1$, is written $\Delta t = t_2 - t_1$; and the corresponding change in x is written $\Delta x = x_2 - x_1$. The ratio $\Delta x/\Delta t$ is the slope of the straight line connecting (x_1, t_1) and (x_2, t_2). If we take the limit as t_2 approaches t_1 (as Δt approaches zero) the slope of the line connecting (x_1, t_1) and (x_2, t_2) approaches the slope of the line that is tangent to the curve at the point (x_1, t_1). The slope of this tangent line is equal to the **derivative** of x with respect to t and is written dx/dt:

$$\frac{dx}{dt} = \lim_{\Delta t \to 0} \frac{\Delta x}{\Delta t} \qquad \text{M-62}$$

(When we find the derivative of a function, we say that we are **differentiating** the function; and the very small "dx" and "dt" elements are called **differentials** of x and t, respectively.) The derivative of a function of t is another function of t. If x is a constant and does not change, the graph of x versus t is a horizontal line with zero slope. The derivative of a constant is thus zero. In Figure M-24, x is not constant but is proportional to t:

$$x = Ct$$

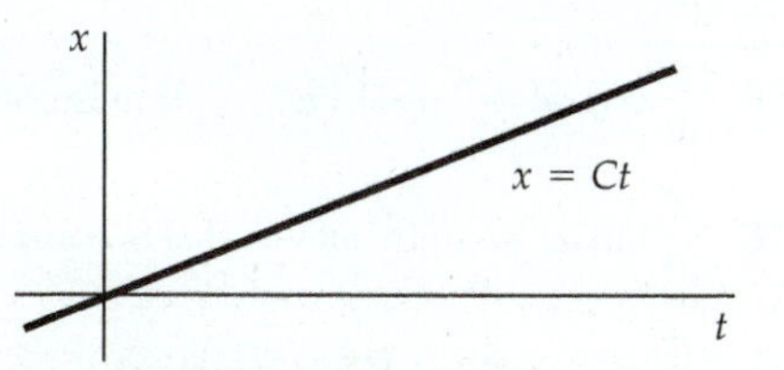

FIGURE M-24 Graph of the linear function $x = Ct$. This function has a constant slope C.

This function has a constant slope equal to C. Thus the derivative of Ct is C. Table M-3 lists some properties of derivatives and the derivatives of some particular functions that occur often in physics. It is followed by comments aimed at making these properties and rules clearer. More detailed discussion can be found in most calculus textbooks.

Table M-3 Properties of Derivatives and Derivatives of Particular Functions

Linearity

1. The derivative of a constant C times a function $f(t)$ equals the constant times the derivative of the function:

$$\frac{d}{dt}[Cf(t)] = C\frac{df(t)}{dt}$$

2. The derivative of a sum of functions equals the sum of the derivatives of the functions:

$$\frac{d}{dt}[f(t) + g(t)] = \frac{df(t)}{dt} + \frac{dg(t)}{dt}$$

Chain rule

3. If f is a function of x and x is in turn a function of t, the derivative of f with respect to t equals the product of the derivative of f with respect to x and the derivative of x with respect to t:

$$\frac{d}{dt}f(x(t)) = \frac{df}{dx}\frac{dx}{dt}$$

Derivative of a product

4. The derivative of a product of functions $f(t)g(t)$ equals the first function times the derivative of the second plus the second function times the derivative of the first:

$$\frac{d}{dt}[f(t)g(t)] = f(t)\frac{dg(t)}{dt} + g(t)\frac{df(t)}{dt}$$

Reciprocal derivative

5. The derivative of t with respect to x is the reciprocal of the derivative of x with respect to t, assuming that neither derivative is zero:

$$\frac{dt}{dx} = \left(\frac{dx}{dt}\right)^{-1} \quad \text{if} \quad \frac{dt}{dx} \neq 0 \quad \text{and} \quad \frac{dx}{dt} \neq 0$$

Derivatives of particular functions

6. If C is a constant, then $dC/dt = 0$.

7. $\frac{d(t^n)}{dt} = nt^{n-1}$ If n is constant.

8. $\frac{d}{dt}\sin\omega t = \omega\cos\omega t$ If ω is constant.

9. $\frac{d}{dt}\cos\omega t = -\omega\sin\omega t$ If ω is constant.

10. $\frac{d}{dt}\tan\omega t = \omega\sin^2\omega t$ If ω is constant.

11. $\frac{d}{dt}e^{bt} = be^{bt}$ If b is constant.

12. $\frac{d}{dt}\ln bt = \frac{1}{t}$ If b is constant.

COMMENTS ON RULES 1 THROUGH 5

Rules 1 and 2 follow from the fact that the limiting process is linear. We can understand rule 3, the chain rule, by multiplying $\Delta f/\Delta t$ by $\Delta x/\Delta x$ and noting that as Δt approaches zero, Δx also approaches zero. That is,

$$\lim_{\Delta t\to 0}\frac{\Delta f}{\Delta t} = \lim_{\Delta t\to 0}\left(\frac{\Delta f}{\Delta t}\frac{\Delta x}{\Delta x}\right) = \lim_{\Delta t\to 0}\left(\frac{\Delta f}{\Delta x}\frac{\Delta x}{\Delta t}\right) = \left(\lim_{\Delta x\to 0}\frac{\Delta f}{\Delta x}\right)\left(\lim_{\Delta t\to 0}\frac{\Delta x}{\Delta t}\right) = \frac{df}{dx}\frac{dx}{dt}$$

where we have used that the limit of the product is equal to product of the limits.

Rule 4 is not immediately apparent. The derivative of a product of functions is the limit of the ratio

$$\frac{f(t+\Delta t)g(t+\Delta t) - f(t)g(t)}{\Delta t}$$

If we add and subtract the quantity $f(t+\Delta t)g(t)$ in the numerator, we can write this ratio as

$$\frac{f(t+\Delta t)g(t+\Delta t) - f(t+\Delta t)g(t) + f(t+\Delta t)g(t) - f(t)g(t)}{\Delta t}$$
$$= f(t+\Delta t)\left[\frac{g(t+\Delta t) - g(t)}{\Delta t}\right] + g(t)\left[\frac{f(t+\Delta t) - f(t)}{\Delta t}\right]$$

As Δt approaches zero, the terms in square brackets become $dg(t)/dt$ and $df(t)/dt$, respectively, and the limit of the expression is

$$f(t)\frac{dg(t)}{dt} + g(t)\frac{df(t)}{dt}$$

Rule 5 follows directly from the definition:

$$\frac{dx}{dt} = \lim_{\Delta t\to 0}\frac{\Delta x}{\Delta t} = \lim_{\Delta x\to 0}\left(\frac{\Delta t}{\Delta x}\right)^{-1} = \left(\frac{dt}{dx}\right)^{-1}$$

COMMENTS ON RULE 7

We can obtain this important result using the binomial expansion. We have

$$f(t) = t^n$$

$$f(t+\Delta t) = (t+\Delta t)^n = t^n\left(1+\frac{\Delta t}{t}\right)^n$$
$$= t^n\left[1 + n\frac{\Delta t}{t} + \frac{n(n-1)}{2!}\left(\frac{\Delta t}{t}\right)^2 + \frac{n(n-1)(n-2)}{3!}\left(\frac{\Delta t}{t}\right)^3 + \cdots\right]$$

Then

$$f(t+\Delta t) - f(t) = t^n\left[n\frac{\Delta t}{t} + \frac{n(n-1)}{2!}\left(\frac{\Delta t}{t}\right)^2 + \cdots\right]$$

and

$$\frac{f(t+\Delta t) - f(t)}{\Delta t} = nt^{n-1} + \frac{n(n-1)}{2!}t^{n-2}\Delta t + \cdots$$

The next term omitted from the last sum is proportional to $(\Delta t)^2$, the following to $(\Delta t)^3$, and so on. Each term except the first approaches zero as Δt approaches zero. Thus

$$\frac{df}{dt} = \lim_{\Delta x\to 0}\frac{f(t+\Delta t) - f(t)}{\Delta t} = nt^{n-1}$$

COMMENTS ON RULES 8 TO 10

We first write $\sin \omega t = \sin\theta$ with $\theta = \omega t$ and use the chain rule,

$$\frac{d\sin\theta}{dt} = \frac{d\sin\theta}{d\theta}\frac{d\theta}{dt} = \omega\frac{d\sin\theta}{d\theta}$$

We then use the trigonometric formula for the sine of the sum of two angles θ and $\Delta\theta$:

$$\sin(\theta + \Delta\theta) = \sin\Delta\theta\cos\theta + \cos\Delta\theta\sin\theta$$

Because $\Delta\theta$ is to approach zero, we can use the small-angle approximations

$$\sin\Delta\theta \approx \Delta\theta \qquad \text{and} \qquad \cos\Delta\theta \approx 1$$

Then

$$\sin(\theta + \Delta\theta) \approx \Delta\theta\cos\theta + \sin\theta$$

and

$$\frac{\sin(\theta + \Delta\theta) - \sin\theta}{\Delta\theta} \approx \cos\theta$$

Similar reasoning can be applied to the cosine function to obtain rule 9.

Rule 10 is obtained by writing $\tan\theta = \sin\theta/\cos\theta$ and applying rule 4 along with rules 8 and 9:

$$\begin{aligned}\frac{d}{dt}(\tan\theta) &= \frac{d}{dt}(\sin\theta)(\cos\theta)^{-1} = \sin\theta\frac{d}{dt}(\cos\theta)^{-1} + \frac{d(\sin\theta)}{dt}(\cos\theta)^{-1}\\ &= \sin\theta(-1)(\cos\theta)^{-2}(-\sin\theta) + (\cos\theta)(\cos\theta)^{-1}\\ &= \frac{\sin^2\theta}{\cos^2\theta} + 1 = \tan^2\theta + 1 = \sec^2\theta\end{aligned}$$

To obtain rule 10, let $\theta = \omega t$ and use the chain rule.

COMMENTS ON RULE 11

Again we use the chain rule

$$\frac{de^\theta}{dt} = \frac{b\,de^\theta}{b\,dt} = b\frac{de^\theta}{d(bt)} = b\frac{de^\theta}{d\theta} \qquad \text{with} \qquad \theta = bt$$

and the series expansion for the exponential function:

$$e^{\theta+\Delta\theta} = e^\theta e^{\Delta\theta} = e^\theta\left[1 + \Delta\theta + \frac{(\Delta\theta)^2}{2!} + \frac{(\Delta\theta)^3}{3!} + \cdots\right]$$

Then

$$\frac{e^{\theta+\Delta\theta} - e^\theta}{\Delta\theta} = e^\theta + e^\theta\frac{\Delta\theta}{2!} + e^\theta\frac{(\Delta\theta)^2}{3!} + \cdots$$

As $\Delta\theta$ approaches zero, the right side of this equation approaches e^θ.

COMMENTS ON RULE 12

Let

$$y = \ln bt$$

Then

$$e^y = bt \Rightarrow t = \frac{1}{b}e^y$$

Then, using rule 11, we obtain

$$\frac{dt}{dy} = \frac{1}{b}e^y \therefore \frac{dt}{dy} = t$$

Then, using rule 5, we obtain

$$\frac{dy}{dt} = \left(\frac{dt}{dy}\right)^{-1} = \frac{1}{t}$$

SECOND- AND HIGHER-ORDER DERIVATIVES; DIMENSIONAL ANALYSIS

Once we have differentiated a function, we can differentiate the resulting derivative as long as terms remain to differentiate. A function such as $x = e^{bt}$ can be differentiated indefinitely: $dx/dt = be^{bt}$ (this function differentiates to give b^2e^{bt}, and so on).

Consider velocity and acceleration. We can define velocity as the rate of change of position of a particle, or dx/dt, and acceleration as the rate of change of velocity, or the *second* derivative of x with respect to t, written dx^2/dt^2. If a particle moves at a constant velocity, then dx/dt will equal a constant. The acceleration, however, will be zero: having constant velocity is the same as having no acceleration, and the derivative of a constant is zero. Now consider a falling object, subject to the constant acceleration of gravity: the velocity itself will be time-dependent, so the *second* derivative, dx^2/dt^2, will be a constant.

The *physical dimensions* of a derivative with respect to a variable are those that would result if the original function of the variable were divided by a value of the variable. For example, the dimension of an equation in which one term is x (for position) is that of length (L); the dimensions of the derivative of x with respect to time t are those of velocity (L/T), and the dimensions of dx^2/dt^2 are those of acceleration (L/T^2).

Example M-12 Position, Velocity, and Acceleration

Find the first and the second derivative of $x = \frac{1}{2}at^2 + bt + c$, where a, b, and c are constants. The function gives the position (in m) of a particle in one dimension, where t is the time (in s), a is acceleration (in m/s^2), b is velocity (in m/s) at a time $t = 0$, and c is the position (in m) of the particle at $t = 0$.

PICTURE Both the first and the second derivatives are sums of terms; for each differentiation we take the derivative of each term separately and add the results.

SOLVE

1. To find the first derivative, first compute the derivative of the first term: $\dfrac{d(\frac{1}{2}at^2)}{dt} = \left(\dfrac{1}{2}a\right)2t^1 = at$
2. Compute the first derivative of the second and third terms: $\dfrac{d(bt)}{dt} = b, \quad \dfrac{d(c)}{dt} = 0$
3. Add these results: $\dfrac{dx}{dt} = at + b$
4. To compute the second derivative, repeat the process for the result in step 3: $\dfrac{d^2x}{dt^2} = a + 0 = a$

CHECK The physical dimensions show that the answer is plausible. The original function is an equation for position; all terms are in meters—the units of t^2 and t cancel the units of s^2 and s in the constants a and b, respectively. In the function for dx/dt, all terms are similarly in m/s: the constant c has differentiated to zero, and the unit for t cancels one of the units for s in the constant a. In the function for dx^2/dt^2, only the acceleration constant remains; as expected, its dimensions are L/T^2.

PRACTICE PROBLEMS

25. Find dy/dx for $y = \frac{5}{8}x^3 - 24x - \frac{5}{8}$.

26. Find dy/dt for $y = ate^{bt}$, where a and b are constants.

SOLVING DIFFERENTIAL EQUATIONS USING COMPLEX NUMBERS

A **differential equation** is an equation in which the derivatives of a function appear as variables. It is an equation in which the variables are related to each other through their derivatives. Consider an equation of the form

$$a\frac{d^2x}{dt^2} + b\frac{dx}{dt} + cx = A\cos\omega t \qquad \text{M-63}$$

that represents a physical process, such as a damped harmonic oscillator driven by a sinusoidal force, or a series *RLC* combination being driven by a sinusoidal potential drop. Although each of the parameters in Equation M-63 is a real number, the time-dependent cosine term suggests that we might find the steady-state solution to this equation by introducing complex numbers. We first construct the "parallel" equation

$$a\frac{d^2y}{dt^2} + b\frac{dy}{dt} + cy = A\sin\omega t \qquad \text{M-64}$$

Equation M-64 has no physical meaning of its own, and we have no interest in solving it. However, it is of use in solving Equation M-63. After multiplying through Equation M-64 by the unit imaginary i, we add Equation M-64 and Equation M-63 to obtain

$$\left(a\frac{d^2x}{dt^2} + ai\frac{d^2y}{dt^2}\right) + \left(b\frac{dx}{dt} + bi\frac{dy}{dt}\right) + (cx + ciy) = A\cos\omega t + Ai\sin\omega t$$

We next combine terms to get

$$a\frac{d^2(x+iy)}{dt^2} + b\frac{d(x+iy)}{dt} + c(x+iy) = A(\cos\omega t + i\sin\omega t) \qquad \text{M-65}$$

which is valid because the derivative of a sum is equal to the sum of the derivatives. We simplify our result by defining $z = x + iy$ and by using the identity $e^{i\omega t} = \cos\omega t + i\sin\omega t$. Substituting these into Equation M-65, we obtain

$$a\frac{d^2z}{dt^2} + b\frac{dz}{dt} + cz = Ae^{i\omega t} \qquad \text{M-66}$$

which we now solve for z. Once z is obtained, we can solve for x using $x = \text{Re}(z)$.

Because we are looking only for the steady-state solution for Equation M-65, we can assume its solution is of the form $x = x_0\cos(\omega t - \phi)$, where ϕ is a constant. This is equivalent to assuming that the solution to Equation M-66 is of the form $z = \eta e^{i\omega t}$, where η, pronounced eta (like beta without the b), is a constant complex number. Then $dz/dt = i\omega z$, $d^2z/dt^2 = -\omega^2 z$, and $e^{i\omega t} = z/\eta$. Substituting these into Equation M-65 gives

$$-a\omega^2 z + i\omega bz + cz = A\frac{z}{\eta}$$

Dividing both sides of this equation by z and solving for η gives

$$\eta = \frac{A}{-a\omega^2 + i\omega b + c}$$

Expressing the denominator in polar form gives

$$(-a\omega^2 + c) + i\omega b = \sqrt{(-a\omega^2 + c)^2 + \omega^2 b^2}\, e^{i\phi}$$

where $\tan\phi = \omega b/(-a\omega^2 + c)$. Thus,

$$\eta = \frac{A}{\sqrt{(-a\omega^2 + c)^2 + \omega^2 b^2}}e^{-i\phi}$$

so

$$z = \eta e^{i\omega t} = \frac{A}{\sqrt{(-a\omega^2 + c)^2 + \omega^2 b^2}} e^{i(\omega t - \phi)}$$

$$= \frac{A}{\sqrt{(-a\omega^2 + c)^2 + \omega^2 b^2}} [\cos(\omega t - \phi) + i \sin(\omega t - \phi)] \qquad \text{M-67}$$

It follows that

$$x = \text{Re}(z) = \frac{A}{\sqrt{(-a\omega^2 + c)^2 + \omega^2 b^2}} \cos(\omega t - \phi) \qquad \text{M-68}$$

THE EXPONENTIAL FUNCTION

An **exponential function** is a function of the form a^{bx}, where $a > 0$ and b are constants. The function is usually written as e^{cx}, where c is constant.

When the rate of change of a quantity is proportional to the quantity itself, the quantity increases or decreases exponentially, depending on the sign of the proportionality constant. An example of an *exponentially* decreasing function is nuclear decay. If N is the number of radioactive nuclei at some time, then the change dN in some very small time interval dt will be proportional to N and to dt:

$$dN = -\lambda N\, dt$$

where λ is the *decay constant* (not to be confused with the decay rate dN/dt, which decreases exponentially). The function N satisfying this equation is

$$N = N_0 e^{-\lambda t} \qquad \text{M-69}$$

where N_0 is the value of N at time $t = 0$. Figure M-25 shows N versus t. A characteristic of exponential decay is that N decreases by a constant factor in a given time interval. The time interval for N to decrease to half its original value is its *half-life* $t_{1/2}$. The half-life is obtained from Equation M-69 by setting $N = \frac{1}{2}N_0$ and solving for the time. This gives

$$t_{1/2} = \frac{\ln 2}{\lambda} = \frac{0.693}{\lambda} \qquad \text{M-70}$$

An example of *exponential increase* is population growth. If the number of organisms is N, the change in N after a very small time interval dt is given by

$$dN = +\lambda N\, dt$$

where λ is now the *growth constant*. The function N satisfying this equation is

$$N = N_0 e^{\lambda t} \qquad \text{M-71}$$

(Note the change of sign in the exponent.) A graph of this function is shown in Figure M-26. An exponential increase can be characterized by a doubling time T_2, which is related to λ by

$$T_2 = \frac{\ln 2}{\lambda} = \frac{0.693}{\lambda} \qquad \text{M-72}$$

Very often, we know population growth as an annual percentage increase and wish to calculate the doubling time. In this case, we find T_2 (in years) from the equation

$$T_2 = \frac{69.3}{r} \qquad \text{M-73}$$

where r is the percent per year. For example, if the population increases by 2 percent per year, the population will double every $69.3/2 \approx 35$ years. Table M-4 lists some useful relations for exponential and logarithmic functions.

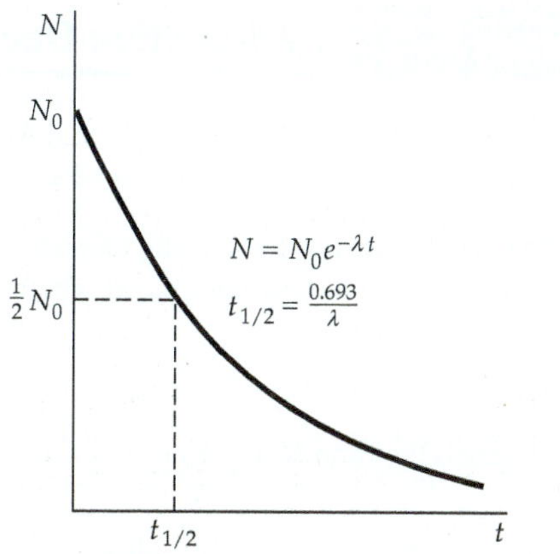

FIGURE M-25 Graph of N versus t when N decreases exponentially. The time $t_{1/2}$ is the time it takes for N to decrease by one-half.

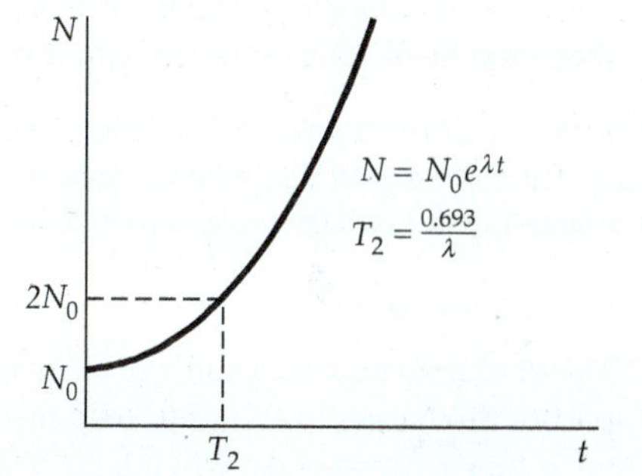

FIGURE M-26 Graph of N versus t when N increases exponentially. The time T_2 is the time it takes for N to double.

Table M-4 Exponential and Logarithmic Functions

$e = 2.718\,28$

$e^0 = 1$

If $y = e^x$, then $x = \ln y$.

$e^{\ln x} = x$

$e^x e^y = e^{(x+y)}$

$(e^x)^y = e^{xy} = (e^y)^x$

$\ln e = 1; \ln 1 = 0$

$\ln xy = \ln x + \ln y$

$\ln \frac{x}{y} = \ln x - \ln y$

$\ln e^x = x; \ln a^x = x \ln a$

$\ln x = (\ln 10) \log x$

$= 2.30\,26 \log x$

$\log x = (\log e) \ln x = 0.434\,29 \ln x$

$e^x = 1 + x + \frac{x^2}{2!} + \frac{x^3}{3!} = \ldots$

$\ln(1 + x) = x - \frac{x^2}{2} + \frac{x^3}{3} - \frac{x^4}{4}$

Example M-13 Radioactive Decay of Cobalt-60

The half-life of cobalt-60 (^{60}Co) is 5.27 y. At $t = 0$ you have a sample of ^{60}Co that has a mass equal to 1.20 mg. At what time t (in years) will 0.400 mg of the sample of ^{60}Co have decayed?

PICTURE When we derived the half-life in exponential decay, we set $N/N_0 = 1/2$. In this example, we are to find the time at which two-thirds of a sample remains, and so the ratio N/N_0 will be 0.667.

SOLVE

1. Express the ratio N/N_0 as an exponential function:
$$\frac{N}{N_0} = 0.667 = e^{-\lambda t}$$
2. Take the reciprocal of both sides:
$$\frac{N_0}{N} = 1.50 = e^{\lambda t}$$
3. Solve for t:
$$t = \frac{\ln 1.50}{\lambda} = \frac{0.405}{\lambda}$$
4. The decay constant is related to the half-life by $\lambda = (\ln 2)/t_{1/2}$ (Equation M-70). Substitute $(\ln 2)/t_{1/2}$ for λ and evaluate the time:
$$t = \frac{\ln 1.5}{\ln 2} t_{1/2} = \frac{\ln 1.5}{\ln 2} \times 5.27\text{ y} = \boxed{3.08\text{ y}}$$

CHECK It takes 5.27 y for the mass of a sample of ^{60}Co to decrease to 50 percent of its initial mass. Thus, we expect it to take less than 5.27 y for the sample to lose 33.3 percent of its mass. Our step-4 result of 3.08 y is less than 5.27 y, as expected.

PRACTICE PROBLEMS

27. The discharge time constant τ of a capacitor in an RC circuit is the time in which the capacitor discharges to e^{-1} (or 0.368) times its charge at $t = 0$. If $\tau = 1$ s for a capacitor, at what time t (in seconds) will it have discharged to 50.0% of its initial charge?
28. If the coyote population in your state is increasing at a rate of 8.0% a decade and continues increasing at the same rate indefinitely, in how many years will it reach 1.5 times its current level?

M-12 INTEGRAL CALCULUS

Integration can be considered the inverse of differentiation. If a function $f(t)$ is *integrated*, a function $F(t)$ is found for which $f(t)$ is the derivative of $F(t)$ with respect to t.

THE INTEGRAL AS AN AREA UNDER A CURVE; DIMENSIONAL ANALYSIS

The process of finding the area under a curve on the graph illustrates integration. Figure M-27 shows a function $f(t)$. The area of the shaded element is approximately $f_i\,\Delta t_i$, where f_i is evaluated anywhere in the interval Δt_i. This approximation is highly accurate if Δt_i is very small. The total area under some stretch of the curve is found by summing all the area elements it covers and taking the limit as each Δt_i approaches zero. This limit is called the **integral** of f over t and is written

$$\int f\,dt = \text{area}_i = \lim_{\Delta t_i \to 0} \sum_i f_i \Delta t_i \qquad \text{M-74}$$

The *physical dimensions* of an integral of a function $f(t)$ are found by multiplying the dimensions of the *integrand* (the function being integrated) and the dimensions of the integration variable t. For example, if the integrand is a velocity function

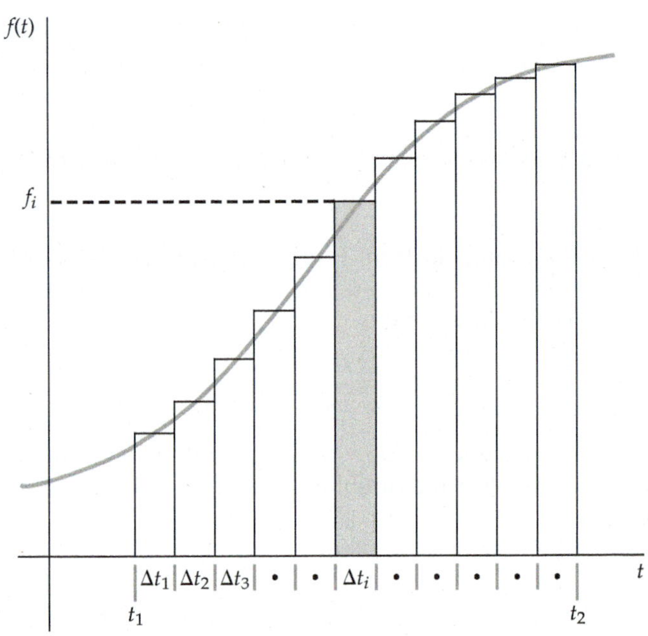

FIGURE M-27 A general function $f(t)$. The area of the shaded element is approximately $f_i\Delta t_i$, where f_i is evaluated anywhere in the interval.

$v(t)$ (dimensions L/T) and the integration variable is time t), the dimension of the integral is L = (L/T) × T. That is, the dimensions of the integral are those of velocity times time.

Let

$$y = \int_{t_1}^{t} f\,dt \qquad \text{M-75}$$

The function y is the area under the f-versus-t curve from t_1 to a general value t. For a small interval Δt, the change in the area Δy is approximately $f\,\Delta t$:

$$\Delta y \approx f\,\Delta t$$

$$f \approx \frac{\Delta y}{\Delta t}$$

If we take the limit as Δt approaches 0, we can see that f is the derivative of y:

$$f = \frac{dy}{dt} \qquad \text{M-76}$$

INDEFINITE INTEGRALS AND DEFINITE INTEGRALS

When we write

$$y = \int f\,dt \qquad \text{M-77}$$

we are showing y as an **indefinite integral** of f over t. To evaluate an indefinite integral, we find the function y whose derivative is f. Because that function could contain a constant term that differentiated to zero, we include as our final term a **constant of integration** C. If we are integrating the function over a known segment—such as t_1 to t_2 in Figure M-27—we can find a **definite integral,** eliminating the unknown constant C:

$$\int_{t_1}^{t_2} f\,dt = y(t_2) - y(t_1) \qquad \text{M-78}$$

Table M-5 lists some important integration formulas. More extensive lists of integration formulas can be found in any calculus textbook or by searching for "table of integrals" on the Internet.

Table M-5 Integration Formulas†

1. $\int A\,dt = At$
2. $\int At\,dt = \frac{1}{2}At^2$
3. $\int At^n\,dt = A\frac{t^{n+1}}{n+1}, \quad n \neq -1$
4. $\int At^{-1}\,dt = A\ln|t|$
5. $\int e^{bt}\,dt = \frac{1}{b}e^{bt}$
6. $\int \cos\omega t\,dt = \frac{1}{\omega}\sin\omega t$
7. $\int \sin\omega t\,dt = -\frac{1}{\omega}\cos\omega t$
8. $\int_0^\infty e^{-ax}\,dx = \frac{1}{a}$
9. $\int_0^\infty e^{-ax^2}\,dx = \frac{1}{2}\sqrt{\frac{\pi}{a}}$
10. $\int_0^\infty xe^{-ax^2}\,dx = \frac{2}{a}$
11. $\int_0^\infty x^2e^{-ax^2}\,dx = \frac{1}{4}\sqrt{\frac{\pi}{a^3}}$
12. $\int_0^\infty x^3e^{-ax^2}\,dx = \frac{4}{a^2}$
13. $\int_0^\infty x^4e^{-ax^2}\,dx = \frac{3}{8}\sqrt{\frac{\pi}{a^5}}$

† In these formulas, A, b, and ω are constants. In formulas 1 through 7, an arbitrary constant C can be added to the right side of each equation. The constant a is greater than zero.

Example M-14 Integrating Equations of Motion

A particle is moving at a constant acceleration a. Write a formula for position x at time t given that the position and velocity are x_0 and v_0 at time $t = 0$.

PICTURE Velocity v is the derivative of x with respect to time t, and acceleration is the derivative of v with respect to t. We should be able to write a function $x(t)$ by performing two integrations.

SOLVE

1. Integrate a with respect to t to find the v as a function of t. The a can be factored from the integrand because a is constant:

$$v = \int a\,dt = a\int dt$$

$$v = at + C_1$$

where C_1 represents a times the constant of integration.

2. The velocity $v = v_0$ when $t = 0$

$$v_0 = 0 + C_1 \Rightarrow C_1 = v_0$$

so $v = v_0 + at$

3. Integrate v with respect to t to find x as a function of t:

$$x = \int v\,dt = \int (v_0 + at)\,dt = \int v_0\,dt + \int at\,dt$$

$$x = v_0 \int dt + a \int t\,dt = v_0 t + \tfrac{1}{2}at^2 + C_2$$

where C_2 represents the combined constants of integration.

4. The position $x = x_0$ when $t = 0$

$$x_0 = 0 + 0 + C_2$$

so $x = x_0 + v_0 t + \tfrac{1}{2}at^2$

CHECK Differentiate the step-4 result twice to get the acceleration

$$v = \frac{dx}{dt} = \frac{d}{dt}(x_0 + v_0 t + \tfrac{1}{2}at^2) = 0 + v_0 + at$$

$$a = \frac{dv}{dt} = \frac{d}{dt}(v_0 + at) = a$$

PRACTICE PROBLEMS

29. $\int_3^6 3\,dx =$

30. $V = \int_5^8 \pi r^2\,dL =$

Answers to Practice Problems

1. 0.24 L
2. 31.6 m/s
3. 6.0 kg/cm^3
4. -3
5. 1.54 L
6. 3.07 L
7. False
8. $x = (4.5\text{ m/s})t + 3.0\text{ m}$
9. $x = 8, y = 60$
11. $2(x - y)^2$
12. $x^2(2x + 4)(x + 3)$
13. $x^{1/2}$
14. x^6
15. 3
16. ~ 2.322
17. $V/A = \frac{1}{3}r$
18. $A = \frac{2}{3}\pi L^2$
19. $\sin\theta = 0.496$, $\cos\theta = 0.868$, $\theta = 29.7°$
20. $\sin 8.2° = 0.1426$, $8.2° = 0.1431$ rad
21. 0.996, 0.996 00, close to 0%
22. 0.96, 0.960 77, $\ll 1\%$
23. $-1 + 0i = -1$
24. $0 + i = i$
25. $dy/dx = \frac{5}{24}x^2 - 24$
26. $dy/dt = ae^{bt}(bt + 1)$
27. 0.693 s
28. 51 y
29. 9
30. $3\pi r^2$

Answers to Odd-Numbered End-of-Chapter Problems

Problem answers are calculated using $g = 9.81\ \text{m/s}^2$ unless otherwise specified. Differences in the last figure can easily result from differences in rounding the input data and are not important.

Chapter 1

1 (*c*)

3 (*c*)

5 1.609×10^5 cm/mi

7 (*e*)

9 False

11

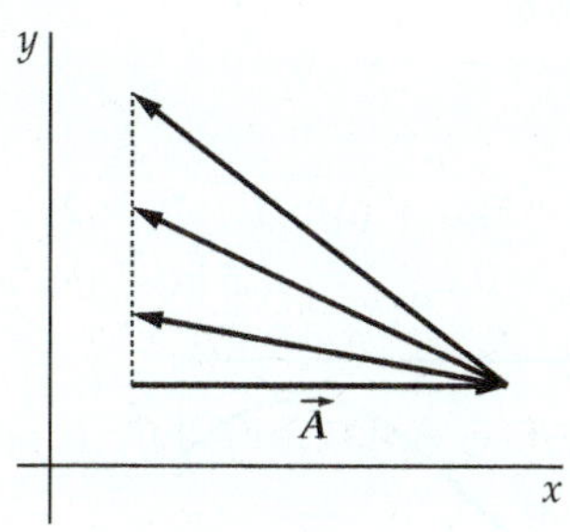

13

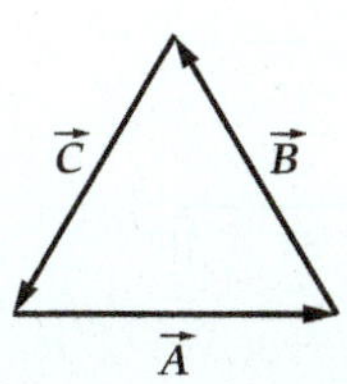

15 2.0×10^{27} molecules

17 (*a*) $\approx 3 \times 10^{10}$ diapers, (*b*) $\approx 2 \times 10^7\ \text{m}^3$, (*c*) $\approx 0.8\ \text{mi}^2$

19 (*a*) 50 MB, (*b*) 7×10^2 novels

21 (*a*) 40×10^{-6} W, (*b*) 4×10^{-9} s, (*c*) 3×10^6 W, (*d*) 25×10^3 m

23 (*a*) C_1 is in m; C_2 is in m/s, (*b*) C_1 is in m/s², (*c*) C_1 is in m/s², (*d*) C_1 is in m; C_2 is in s⁻¹, (*e*) C_1 is in m/s; C_2 is in s⁻¹

25 (*a*) 4.00×10^7 m, (*b*) 6.37×10^6 m, (*c*) 2.49×10^4 mi, 3.96×10^3 mi

27 210 cm

29 1.280 km

31 (*a*) 36.00 km/h·s, (*b*) 10.00 m/s², (*c*) 88 ft/s, (*d*) 27 m/s

33 (*a*) 1.3×10^4 lb (*b*) 4 cases

35 (*a*) m/s², (*b*) s, (*c*) m

37 T^{-1}

39 (*a*) M/T^2, (*b*) $\text{kg}\cdot\text{m}^2/\text{s}^2$

43 M/L^3

45 (*a*) 30 000, (*b*) 0.0062, (*c*) 0.000 004, (*d*) 217 000

47 (*a*) 1.14×10^5, (*b*) 2.25×10^{-8}, (*c*) 8.27×10^3, (*d*) 6.27×10^2

49 3.6×10^6

51 (*a*) 25.8 mm², (*b*) 30.1 mm²

53 (*a*) $A_x = 5.0$ m, $A_y = 8.7$ m, (*b*) $v_x = -19$ m/s, $v_y = -16$ m/s, (*c*) $F_x = 35$ lb, $F_y = 20$ lb

55 You could have gone either 87 m north or 87 m south. The headings of your walk were either 60° north of east or 60° south of east, respectively.

57 (*a*) $40\hat{i} - 50\hat{j}$, (*b*) $-51°$

59 $-0.59\hat{i} - 0.81\hat{j}, 0.92\hat{i} - 0.38\hat{j}, -0.51\hat{i} + 0.86\hat{j}$

61 $\approx 3.3 \times 10^3$ mi/h, $\approx 5.3 \times 10^3$ km/h, $\approx 1.5 \times 10^3$ m/s

63 31.7 y

65 2.0×10^{23}

67 (*a*) $1.4 \times 10^{17}\ \text{kg/m}^3$, (*b*) 2.2×10^2 m

69 (*a*) 4.848×10^{-6} parsec, (*b*) 3.086×10^{16} m, (*c*) 9.461×10^{15} m, (*d*) 6.324×10^4 AU, (*e*) 3.262 light-years

71 (*a*)

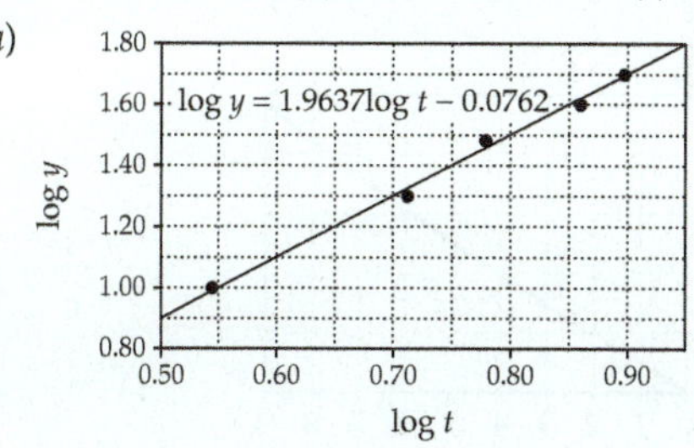

(*c*) $B = 0.84\ \text{m/s}^2$, $C = 2.0$, (*d*) 1.1 s, (*e*) 1.7 m/s²

73 55.4×10^3 tons. The 50,000-ton claim is conservative. The actual weight is closer to 55,000 tons.

75 (*a*)

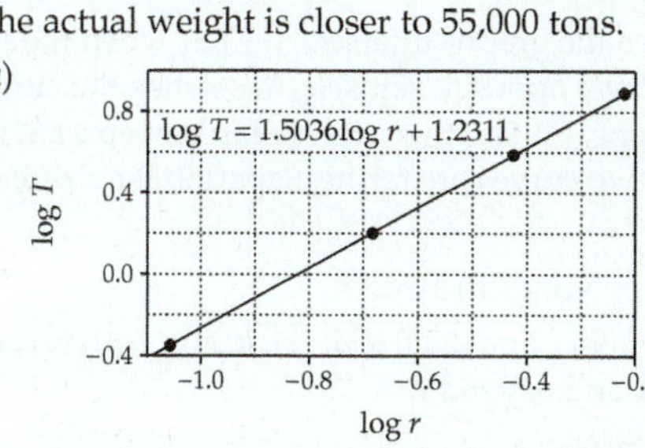

$n = 1.50$, $C = 17.0\ \text{y}/(\text{Gm})^{3/2}$, $T = [17.0\ \text{y}/(\text{Gm})^{3/2}]r^{1.50}$, (*b*) $r = 0.510$ Gm

77 (*a*) $\vec{F}_{\text{Paul}} = (35\text{ lb})\hat{i} + (35\text{ lb})\hat{j}$ $\vec{F}_{\text{Johnny}} = (-53\text{ lb})\hat{i} + (-37.3\text{ lb})\hat{j}$, (*b*) $\vec{F}_{\text{Connie}} = (18\text{ lb})\hat{i} + (1.9\text{ lb})\hat{j}$, $F_{\text{Connie}} = 18\text{ lb}$, $\theta = 6.1°$ N of E

Chapter 2

1 Zero

3 $v_{\text{av 1st half}} = 2H/T$, $v_{\text{av 2nd half}} = -2H/T$

5 (*a*) Your speed increased from zero, stayed constant for a while, and then decreased.

(*b*)

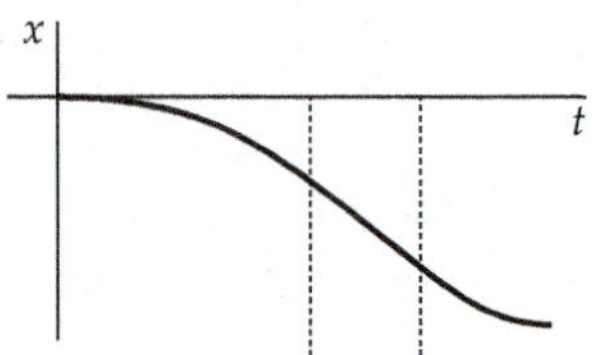

(*c*)

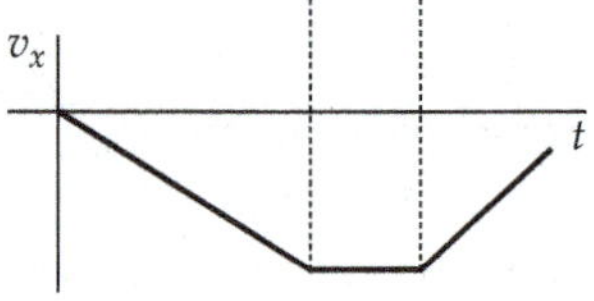

7 True

9 False. If it were true, then any time the initial and final velocities are both zero the average velocity would be zero.

11 (*a*)

13 (*a*) *b*, (*b*) *c*, (*c*) *d*, (*d*) *e*

15 (*a*) B, D, and E, (*b*) A and D, (*c*) C

17 (*a*) True, (*b*) True

19 (*a*) 0, (*b*) $-g$, (*c*) Its acceleration is greater than g in magnitude while the ball is in contact with the ceiling.

21 (*a*) False, (*b*) False, (*c*) True

23 (*a*) *c*

(*b*)

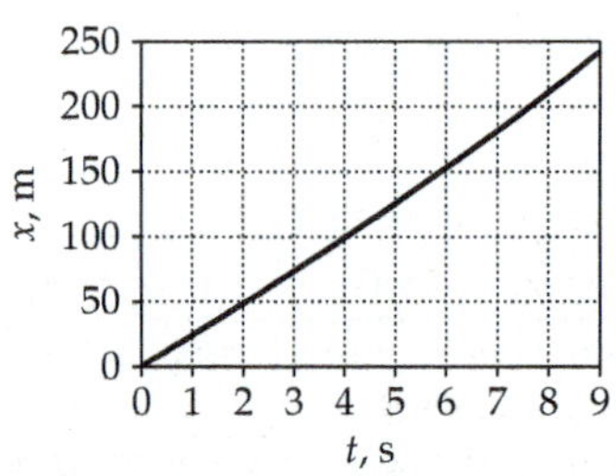

25 B is passing A.

27 (*c*)

29 (*a*) Yes, when the graphs intersect, (*b*) Yes, when the slopes of the curves have opposite signs, (*c*) Yes, when the curves have the same slope, (*d*) The two cars are farthest apart at the instant the two curves are farthest apart in the x direction.

31 $v_J = \frac{1}{2}v_{\max}$

33 (*a*) *d*, (*b*) *b*, (*c*) None, (*d*) *c* and *d*

35 (*a*) *a*, *f*, and *i*, (*b*) *c* and *d*, (*c*) *a*, *d*, *e*, *f*, *h*, and *i*, (*d*) *b*, *c*, and *g*, (*e*) *a* and *i*, *d* and *h*, *f* and *i*

37 $-1.2 \times 10^3\text{ m/s}^2$

39 4.03 m/s^2

41 (*a*) 1.7 km ≈ 1 mi, (*b*) If the uncertainty in your time estimate is less than 1 s (±20%), the uncertainty in the distance estimate will be about 20% of 1.7 km, or approximately 300 m.

43 (*a*) 0.28 km/min, (*b*) −0.083 km/min, (*c*) 0 (*d*) 0.13 km/min

45 (*a*) 2.2 h, (*b*) $(t_{\text{supersonic}}/t_{\text{subsonic}}) \approx 0.45$

47 (*a*) 4.3 y, (*b*) 4.3×10^6 y. Because $4.3 \times 10^6\text{ y} \gg 1000$ y, Gregor does not have to pay.

49 23.5 m/s

51 (*a*) 0, (*b*) 0.3 m/s, (*c*) −2 m/s, (*d*) 1 m/s

53 $v_{\text{av}} = 122$ km/h. $v_{\text{Numerical av}} = 1.04v_{\text{av}}$. The average speed would be equal to one-third the sum of the three speeds if the three speeds were each maintained for the same length of time instead of for the same distance.

55 (*a*)

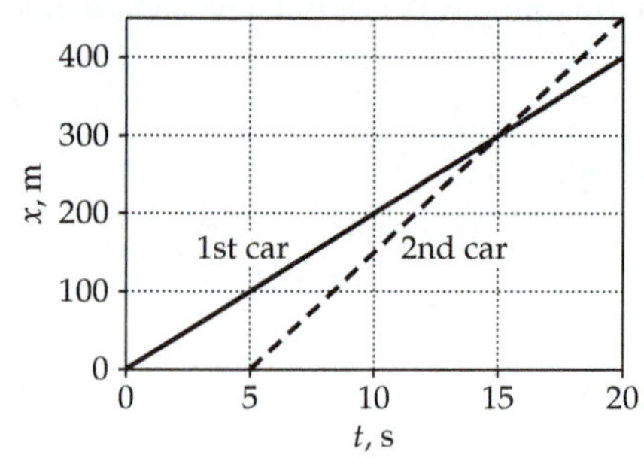

(*b*) 15 s, (*c*) 300 m, (*d*) 100 m

57 15 m/s

59 -2.0 m/s^2

61 (*a*) 2.0 m/s, (*b*) $\Delta x = (2t - 5)\Delta t + (\Delta t)^2$, (*c*) $v(t) = 2t - 5$

63 (*a*) $a_{\text{av AB}} = 3.3\text{ m/s}^2$, $a_{\text{av BC}} = 0$, $a_{\text{av CE}} = -7.5\text{ m/s}^2$, (*b*) 75 m

(*c*)

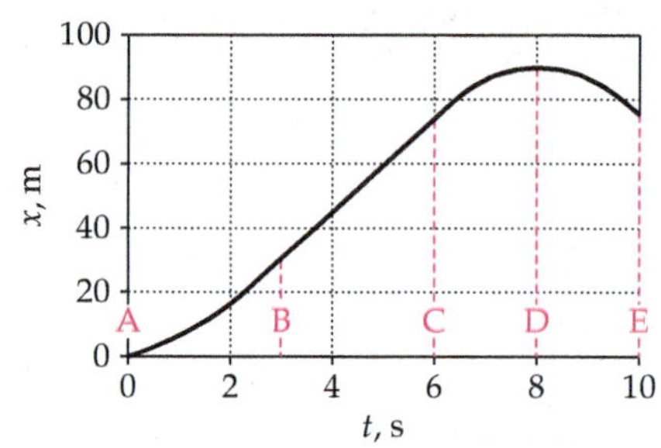

(*d*) At point D, $t = 8$ s, the graph crosses the time axis; therefore $v = 0$.

65 (*a*) 80 m/s, (*b*) 0.40 km, (*c*) 40 m/s

67 16 m/s^2

69 (*a*) 4.1 s, (*b*) 20 m, (*c*) 0.99 s and 3.1 s

71 (*a*)

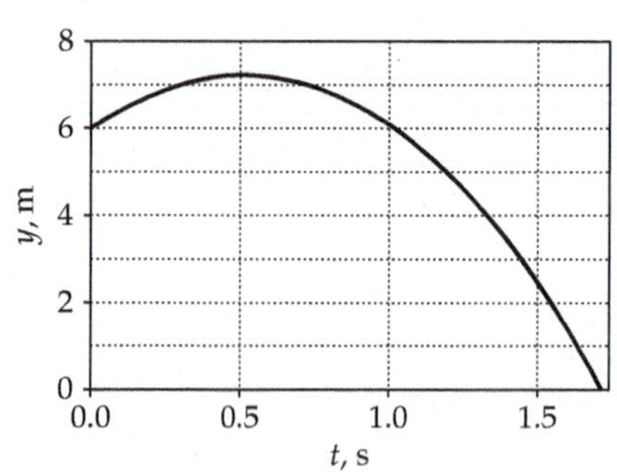

(*b*) 7.3 m, (*c*) 1.7 s, (*d*) 12 m/s

73 44 m

75 68 m/s

77 (*a*) 666 m, (*b*) 14 m/s

79 (*a*) You did not achieve your goal. To go higher, you can increase the acceleration value or the duration of the acceleration. (*b*) 138 s, (*c*) 610 m/s

81 40 cm/s, -6.9 cm/s^2

83 (*a*) 11 mi/h, (*b*) 0.60

85 11 m

87 28 m

89 (*a*) 2.4 m, (*b*) 1.4 s

93 (*a*) 2.1 d, (*b*) 5.8 y

95 4.8 m/s

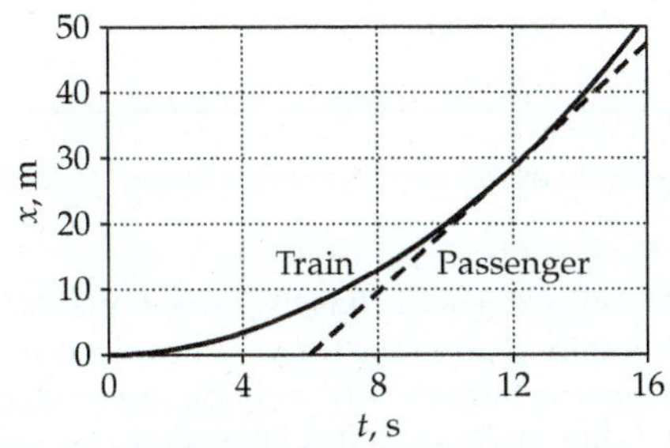

97 $h/3$

99 (*a*) 35 s, (*b*) 1.2 km

(*c*)

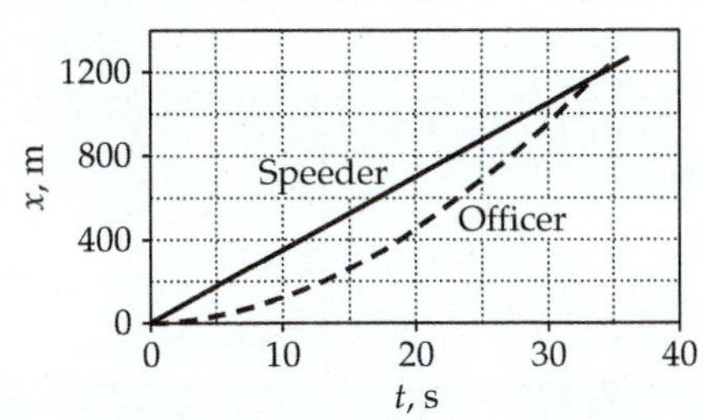

101 (*a*) $2L/3$, (*b*) $\frac{2}{3}t_{\text{fin}}$, (*c*) $\sqrt{4aL/3}$

103 (*a*)

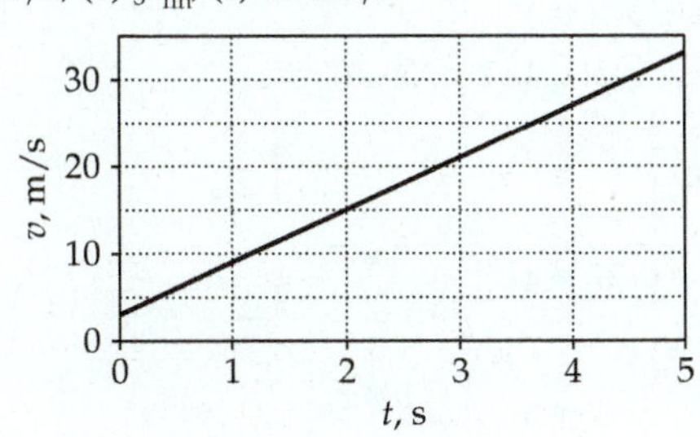

area under the curve = 90 m

(*b*) $x(t) = (3.0\text{ m/s}^2)t^2 + (3.0\text{ m/s})t$, 90 m

105 $x(t) = (2.3\text{ m/s}^3)t^3 - (5.0\text{ m/s})t$

107 (*a*) 0.25 m/s per box, (*b*) 0.93 m/s, 3.0 m/s, 6.0 m/s

(*c*)

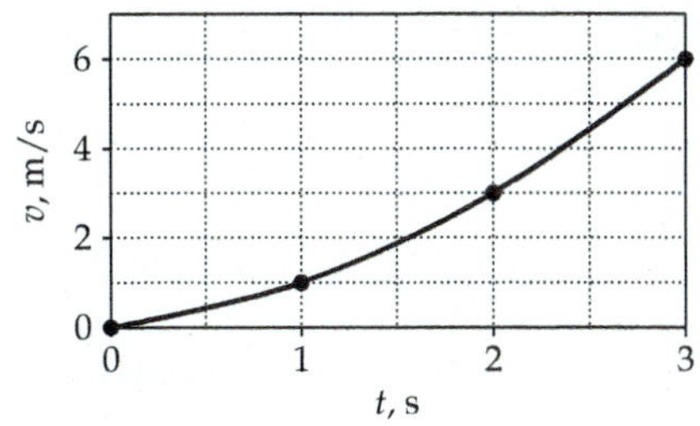

$x(3\text{ s}) = 6.5$ m

109 (*a*)

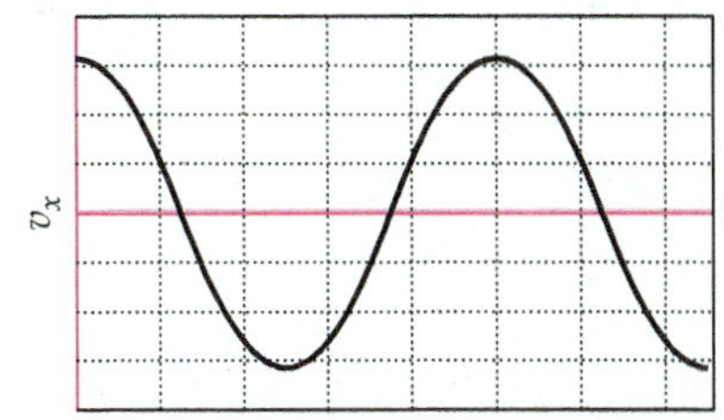

(*b*)

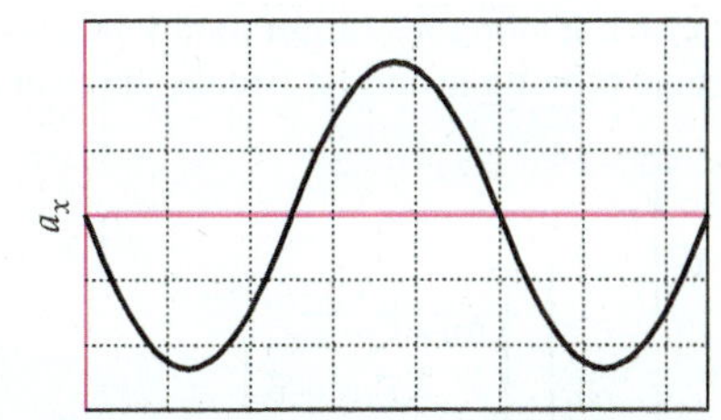

(*c*) The points at the greatest distances from the time axis correspond to turnaround points. The velocity of the body is zero at these points.
(*d*) The velocity is greatest when the slope is greatest, and vice versa. The acceleration is zero when the concavity changes sign, and the acceleration is greatest when the rate of change of the slope with respect to x is greatest.

111 (*a*) $v(t) = (0.10\text{ m/s}^3)t^2 + 9.5\text{ m/s}$,
(*b*) $x(t) = \frac{1}{6}(0.20\text{ m/s}^3)t^3 + (9.5\text{ m/s})t - 5.0\text{ m/s}$,
(*c*) 13 m/s, 15 m/s. v_{av} is not the same as $(v_{\text{i}} + v_{\text{f}})/2$ because the acceleration is not constant.

113 (*b*) 0.452 s, (*c*) 12.0 m/s^2, 22.3%

115 (*a*) The maximum value of the sine function (as in $\sin \omega t$) is 1. Hence, the coefficient $B = v_{\text{max}}$.
(*b*) $a = \omega v_{\text{max}} \cos(\omega t)$. The acceleration is *not* constant.
(*c*) $|a_{\text{max}}| = \omega v_{\text{max}}$, (*d*) $x = x_0 + (v_{\text{max}}/\omega)[1 - \cos(\omega t)]$

117 (*a*) s^{-1}, (*c*) $v_{\text{t}} = g/b$

119 (*b*) 0.762

(*c*)

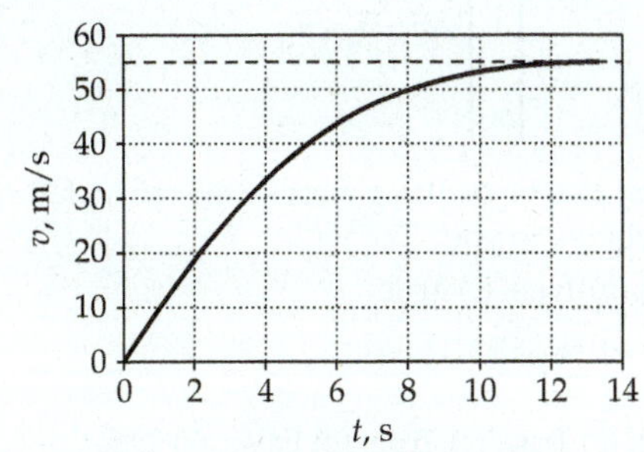

121 You should not contest your ticket.

Chapter 3

1 No. Yes.

3 Zero

5 (*e*)

7 (*c*)

9 (*a*) The velocity vector is tangent to the path.

(*b*)

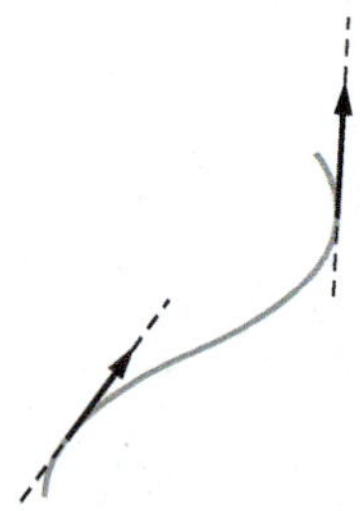

11 (*a*) A car moving along a straight road while slowing down. (*b*) A car moving along a straight road while speeding up. (*c*) A particle moving around a circular track at constant speed.

13 (*a*)

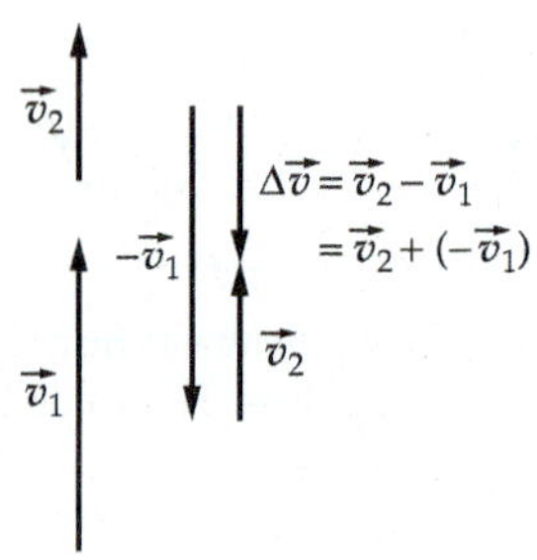

(*b*)

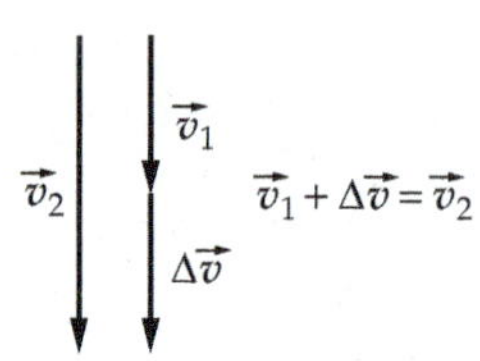

(*c*)

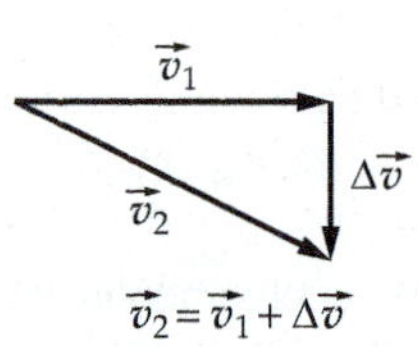

15

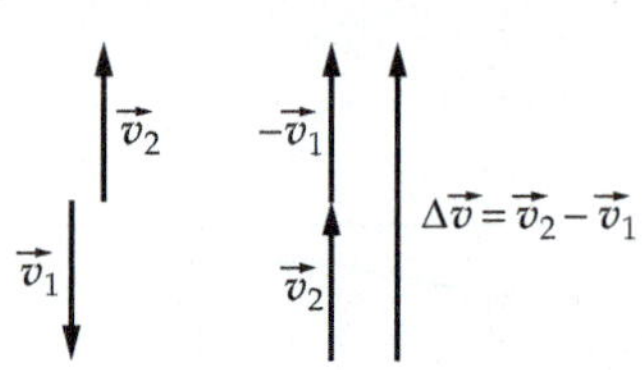

17 You must also be walking west so the rain is falling straight down relative to you.

19 (*a*) True, (*b*) True, (*c*) True

21 (*a*)

23 (*d*)

25 (*a*) False, (*b*) True, (*c*) True, (*d*) False, (*e*) True

27 Assume the $+x$ direction is east and the $+y$ direction is north. Then

(*a*)

Path	Direction of Velocity Vector
AB	North
BC	Northeast
CD	East
DE	Southeast
EF	South

(*b*)

Path	Direction of Acceleration Vector
AB	North
BC	Southeast
CD	0
DE	Southwest
EF	North

(*c*) The magnitude of the acceleration is larger on DE than on BC.

29 The droplet leaving the bottle has the same horizontal velocity as the ship. During the time the droplet is in the air, it is also moving horizontally with the same velocity as the rest of the ship. Because of this, it falls into the vessel, which has the same horizontal velocity. Because you have the same horizontal velocity as the ship does, you see things as if the ship were standing still.

31 (*a*) True, (*b*) False, (*c*) False, (*d*) True

33 (*a*)

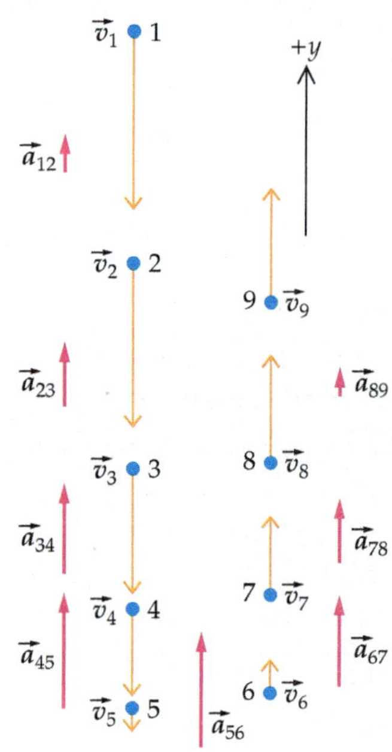

i	v_y (m/s)	Δv_y (m/s)	a_{ave} (m/s^2)
1	−0.78		
2	−0.69	0.09	1.8
3	−0.55	0.14	2.8
4	−0.35	0.20	4.0
5	−0.10	0.25	5.0
6	0.15	0.25	5.0
7	0.35	0.20	4.0
8	0.49	0.14	2.8
9	0.53	0.04	0.8

(*b*) The acceleration vector always points upward, so the sign of its y component does not change. The magnitude of the acceleration vector is greatest when the bungee cord has its maximum extension.

35 $\approx 7 \times 10^4\ m/s^2$

37 15 m/s

39 $\Delta\vec{B} = 0, \Delta\vec{A} = -(0.25\ m)\hat{j} - (0.25\ m)\hat{i}$

41

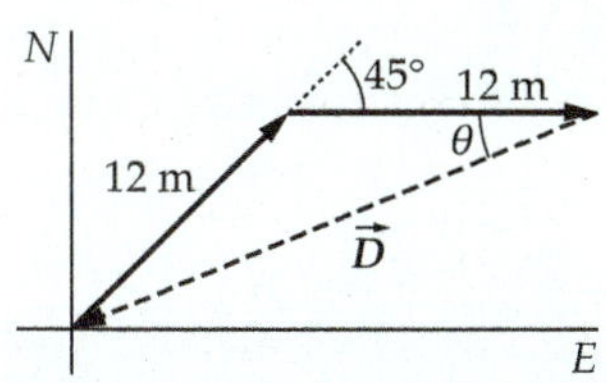

$\vec{D} \approx 22$ m @ 23° south of west

43 (*a*) $\vec{D} = (3.0\ m)\hat{i} + (3.0\ m)\hat{j} + (3.0\ m)\hat{k}$, (*b*) 5.2 m

45 $\vec{v}_{av} = (14\ km/h)\hat{i} + (-4.1\ km/h)\hat{j}$

47 7.2 m/s

49 (*a*) $\vec{v}_{av} = (33\ m/s)\hat{i} + (27\ m/s)\hat{j}$,
(*b*) $\vec{a}_{av} = (-3.0\ m/s^2)\hat{i} + (-1.8\ m/s^2)\hat{j}$

51 $\vec{v} = 30\hat{i} + (40 - 10t)\hat{j}, \vec{a} = (-10\ m/s^2)\hat{j}$

53 (*a*) $\vec{v}_{av} = (20\ m/s)(-\hat{i} + \hat{j})$, (*b*) $\vec{a}_{av} = (-2.0\ m/s^2)\hat{i}$,
(*c*) $\Delta\vec{r} = (600\ m)(-\hat{i} + \hat{j})$

55 (*a*) 16° west of north, (*b*) 280 km/s

57 8.5°, 2.57 h

59 You should fly your plane across the wind.

61 (*a*) $\vec{r}_{AB}(6.0\ s) = (1.2 \times 10^2\ m)\hat{i} + (4.0\ m)\hat{j}$,
(*b*) $\vec{v}_{AB}(6.0\ s) = (-20\ m/s)\hat{i} - (12\ m/s)\hat{j}$, (*c*) $(-2.0\ m/s^2)\hat{j}$

63 (*a*) $\vec{v}_{rel} = (0.80\ m/s)\hat{i} - (1.2\ m/s)\hat{j}$, (*b*) 1.0 m/s

65 $1.5 \times 10^{-6}\ m/s^2$, $1.55 \times 10^{-7}g$

67 (*a*) 463 m/s, 0.343 percent of g, (*b*) From the person toward the center of Earth, (*c*) 380 m/s, $2.76 \times 10^{-2}\ m/s^2$, (*d*) Zero

69 (*a*) 14 s, 1.8 m/s, (*b*) 0.89 m/s, $0.40\ m/s^2$

71 (*a*) 15 cm, (*b*) The range of accelerations is 1300g to 2700g.

73 $h = (v_0^2 \sin^2\theta_0)/2g$

75 34 m/s

77 20.3 m/s, 36.1°

79 69.3°

81 (*a*) 18 m/s, (*b*) 13°

83 (*a*) 8.1 m, (*b*) 23 m/s

85 63.4° below the horizontal

87 (*a*) The kick is short. (*b*) 0.34 m under the bar, (*c*) 5.2 m

89 (*a*) 0.97 s, (*b*) 4.2 m, (*c*) 13 m/s @ 70° below the horizontal

91 (*a*) 485 km, (*b*) 1.70 km/s

93 (*a*) 194 m, (*c*) 219 m, 11 percent

95 (*b*) 80 m, (*c*) 288 m. The approximate solution is 80 m larger. (The approximation ignores higher-order terms, and they are important when the differences are not small.)

99 (*a*) 11 m/s, (*b*) 3.1 s, (*c*) $\vec{v} = (6.5\ m/s)\hat{i} + (-22\ m/s)\hat{j}$

101 (*a*) 21.5 m/s, (*b*) 3.53 s, (*c*) 19.3 m/s

103 (*a*) 7.41 m/s, (*b*) 0.756 s, (*c*) 15.9 m/s, 17.5 m/s, 25.0°

105 (*a*) 0.785 m, (*b*) 105 m

107 (*a*) 1.1 m, (*b*) 3.9 m

109 (*a*) 15 km, (*b*) 54 s

111 806 mi/h, 60.3° north of west

113 4th step

115 (*a*) $v_{min} = \dfrac{x}{\cos\theta}\sqrt{\dfrac{g}{2(x \tan\theta - h)}}$,
(*b*) $v_{min} = 26$ m/s $= 58$ mi/h,

119 (*a*) 26 m, (*b*) 7.8°

121 52.9 km, 52.8° east of north

Chapter 4

1 Yes, there are forces acting on it. They are the same as those that would act on it if it were sitting on your table at home.

3 In the limo you hold one end of the string and suspend the object from the other end. If the string remains vertical, the reference frame of the limo is an inertial reference frame. No, you cannot determine the limo's velocity

5 No. Predicting the direction of the subsequent motion requires additional information.

7 The mass of the probe is constant. However, the solar system will attract the probe with a gravitational force.

9 You and the elevator could be either descending and slowing or ascending and speeding up. In both cases your apparent weight is greater than your actual weight.

11 The most significant force in our everyday world is gravity. It literally keeps us on or near the ground. The other most relevant force is the electromagnetic force. It provides "the glue" to hold solids together and make them rigid. It is of great importance in electric circuits.

13 (*a*) Normal force, contact type, (*b*) Normal, contact, (*c*) Normal, contact, (*d*) Normal, contact, (*e*) Gravitational, action-at-a-distance
The two normal forces that the two blocks exert on each other and the two normal forces that the table and the bottom block exert on each other

15 When the plate is sitting on the floor, the normal force F_n acting upward on it is exerted by the floor and is the same size as the gravitational force F_g on the plate. Hence, the plate does not accelerate. However, to slow the plate down as it hits the floor requires that $F_n > F_g$ (or $F_n \gg F_g$, if the

floor is hard and the plate slows quickly). A large normal force exerted on delicate china can easily break it.

17 (*a*) The normal force of the block on the sprinter, in the forward direction.
(*b*) The frictional force by the ice on the puck, in the opposite direction to the velocity.
(*c*) The gravitation force by Earth on the ball, in the downward direction.
(*d*) The force by the stretched bungee cord on the jumper, in the upward direction.

19 (*a*) (2) 100 N, (*b*) Their accelerations are the same. (*c*) The directions of their acceleration are the same.

21 (*a*)

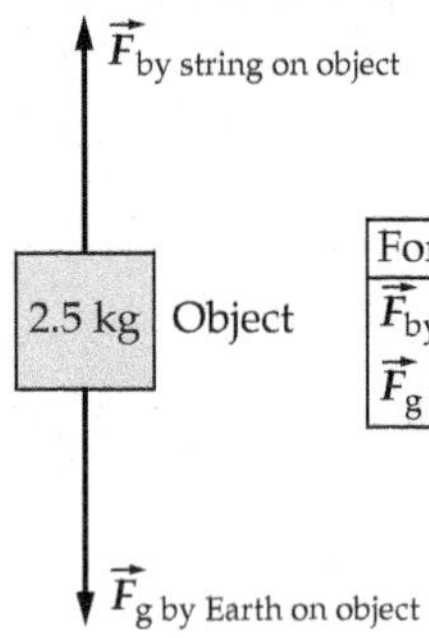

Force	Third-Law Pair
$\vec{F}_{\text{by string on object}}$	$\vec{F}_{\text{by object on string}}$
$\vec{F}_{\text{g by Earth on object}}$	$\vec{F}_{\text{g by object on Earth}}$

(*b*)

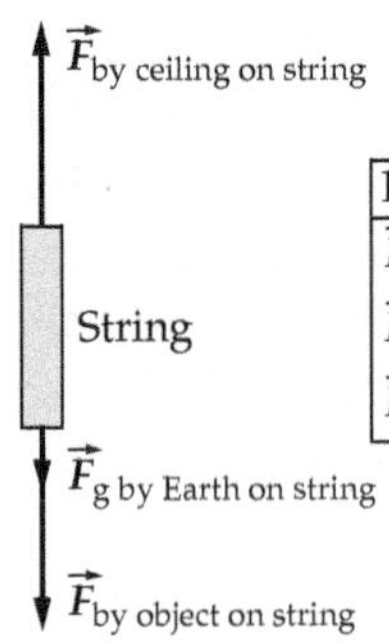

Force	Third-Law Pair
$\vec{F}_{\text{by ceiling on string}}$	$\vec{F}_{\text{by string on ceiling}}$
$\vec{F}_{\text{g by Earth on string}}$	$\vec{F}_{\text{g by string on Earth}}$
$\vec{F}_{\text{by object on string}}$	$\vec{F}_{\text{by string on object}}$

23 (*a*)

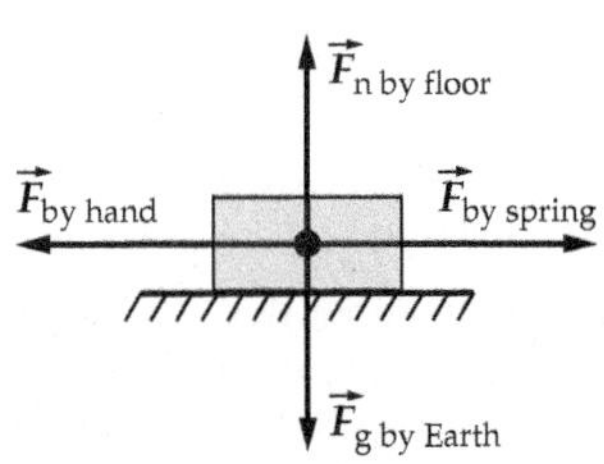

(*b*)

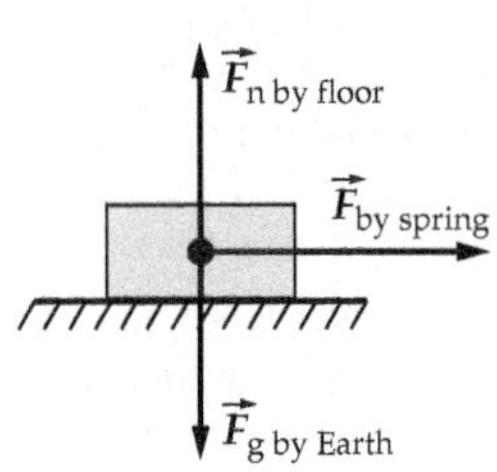

(*c*)

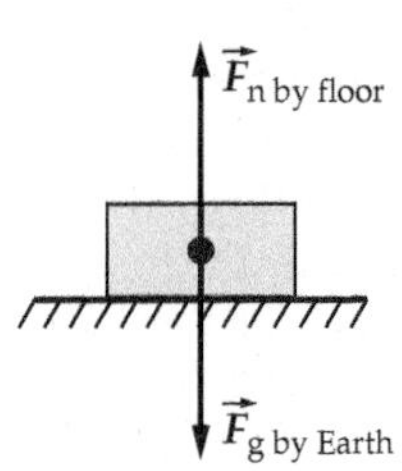

25 (*a*) m_2/m_1, (*b*) m_2/m_1, (*c*) $\Delta x = \frac{1}{2}F(m_2^{-1} - m_1^{-1})(\Delta t)^2$. Because $m_1 > m_2$, the object whose mass is m_2 is ahead.

27 3.6 kN

29 −17 kN

31 (*a*) 6.0 m/s², (*b*) 1/3, (*c*) 2.3 m/s²

33 12 kg

35 (*a*) −3.8 kN, (*b*) 3.00 cm

37 (*a*) 4.2 m/s² @ 45° from each force, (*b*) 8.4 m/s² @ 15° from $2\vec{F}$

39 (*a*) 4.0 m/s², (*b*) 2.4 m/s²

41 (*a*) $\vec{a} = (1.5\text{ m/s}^2)\hat{i} + (-3.5\text{ m/s}^2)\hat{j}$,
(*b*) $\vec{v}(3.0\text{ s}) = (4.5\text{ m/s})\hat{i} + (-11\text{ m/s})\hat{j}$,
(*c*) $\vec{r}(3.0\text{ s}) = (6.8\text{ m})\hat{i} + (-16\text{ m})\hat{j}$

43 (*a*) 5.3×10^2 N, (*b*) 1.2×10^2 lb

45 (*a*) 2.45 kN, (*b*) 409 N, (*c*) 2.45 kN

47 (*a*)

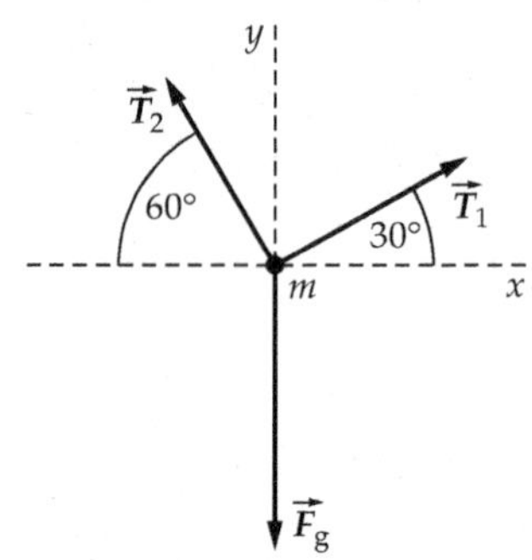

(*b*) T_2 is greater than T_1

49 (*a*) 37°, (*b*) 4.1 N, (*c*) T_1 = 3.4 N, T_2 = 2.4 N, T_3 = 3.4 N

51 $\vec{F}_3 = (-5.0\text{ N})\hat{i} + (-26\text{ N})\hat{j}$

53 (*a*) 3.82 kN, (*b*) 4.30 kN

55 (*a*) If T_H = 10.0 N, then the width of the arch is 9.56 m.
(*b*) If the width of the arch is 8.00 m, then T_H = 3.72 N and the arch is 2.63 m high, tall enough for someone to walk through.

57 56.0 N

59 (*a*) T = 0.42 kN, F_n = 0.25 kN, (*b*) $T = mg\sin\theta$

61 0.55 kN

63 (*a*)

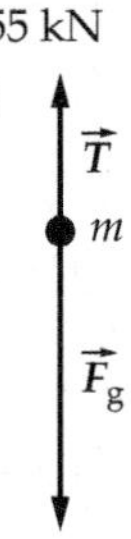

(*b*)

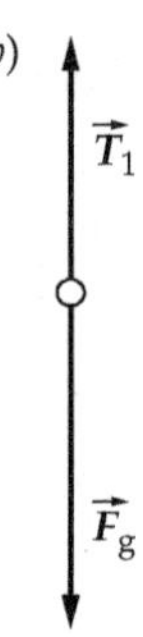

(*c*) No. There is no difference.

65 (*a*) 20 N, (*b*) 20 N, (*c*) 26 N, (*d*) $T_{0\to5.0\text{ s}}$ = 20 N, $T_{5\text{ s}\to9\text{ s}}$ = 15 N

67 (*a*) 1.3 m/s², (*b*) T_1 = 17 N, T_2 = 21 N

69 (*a*) $a = \dfrac{F}{m_1 + m_2}m,\ F_{21} = \dfrac{Fm_1}{m_1 + m_2}$, (*b*) 0.40 m/s², 0.8 N

73 (*a*)

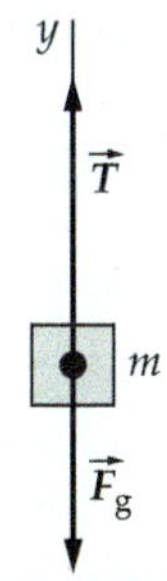

(*b*) 592 N

75 (*a*) $a = \dfrac{g(m_2 - m_1 \sin\theta)}{m_1 + m_2}$, $T = \dfrac{g m_1 m_2 (1 + \sin\theta)}{m_1 + m_2}$,

(*b*) 2.5 m/s^2, 37 N

77 (*a*) 1.4 m/s^2, 61 N, (*b*) $(m_1/m_2) = 1.19$

79 (*a*) 0.40 kN, (*b*) 0.37 kN

81 (*a*) 5.0 cm, (*b*) $a_{20} = 2.5$ m/s^2, $a_5 = 4.9$ m/s^2, $T = 25$ N

83 $m_{\text{2nd mass}} = 1.4$ kg or 1.1 kg

85 $F_{\text{on } m_2} = \left(\dfrac{m^2 + m_1^2 + m_2^2}{m + m_1 + m_2}\right) g$

87 $T_B = 305$ N, $F_{\text{mast on the deck}} = 1.55$ kN

89 (*a*) -0.10 km/s^2, (*b*) 6.1 cm, (*c*) 35 ms

91 (*a*) $a = \dfrac{F}{m_1 + m_2}$, (*b*) $F_{\text{net}} = \dfrac{m_2}{m_1 + m_2} F$, (*c*) $T = \dfrac{m_1}{m_1 + m_2} F$

93 (*a*) 55.0 g, (*b*) 2.45 m/s^2, 2.03 N

95 (*a*) $T = \frac{1}{3}(F_2 + 2F_1)$, (*b*) $t_0 = (3T_0/4C)$

97 (*a*) You should throw your boot in the direction away from the closest shore. (*b*) 420 N, (*c*) 7.52 s

99 (*a*)

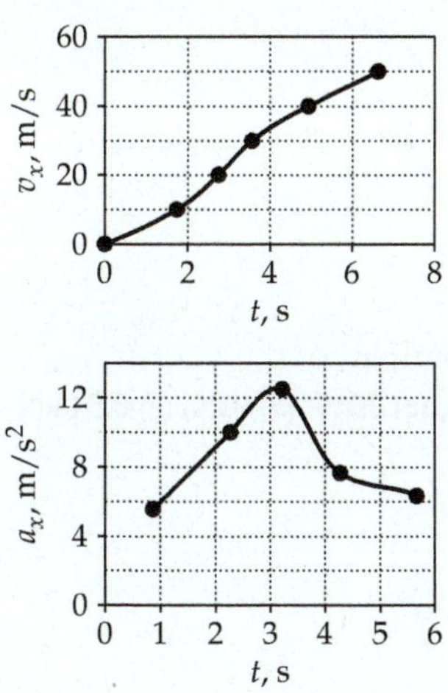

(*b*) 2.8 s to 3.6 s, (*c*) 5500 N, (*d*) 160 m

Chapter 5

1 Static and kinetic frictional forces are responsible for the accelerations. If the coefficient of static friction between the truck bed and the object is sufficiently large, then the object will not slip on the truck bed. The larger the acceleration of the truck, the larger the coefficient of static friction that is needed to prevent slipping.

3 (*d*)

5 (*c*)

7 As the spring is extended, the force exerted by the spring on the block increases. Once that force exceeds the maximum value of the force of static friction, the block will slip. As it does, it will shorten the length of the spring, decreasing the force that the spring exerts. The force of kinetic friction then slows the block to a stop, which starts the cycle over again.

9 (*a*), (*b*), and (*c*)

11 Block 1 will hit the pulley before block 2 hits the wall.

13 Air drag is proportional to the density of air and to the cross-sectional area of the object. On a warm day the air is less dense. The air is also less dense at high altitudes. Pointing his hands results in less area being presented to air drag forces and, hence, reduces them. Rounded and sleek clothing has the same effect as pointing his hands.

15 (*c*)

17 (*a*) The drag force is proportional to the area presented and some power of the speed.The drag force on the feather initially is larger because the feather presents a larger area than does the pebble. As the pebble gains speed the drag force on it increases. The drag force on the pebble eventually exceeds the drag force on the feather because the drag force on the feather cannot exceed the gravitational force on the feather.

(*b*) Terminal speed is much higher for the pebble than for the feather. The acceleration of the pebble will remain high until its speed approaches its terminal speed.

19 $a_{\text{cm}} = (m_1/(m_1 + m_2))a_1$

21 The friction of the road on the tire causes the car to slow down

23 The center of mass moves downward

25 The acceleration of the center of mass is zero.

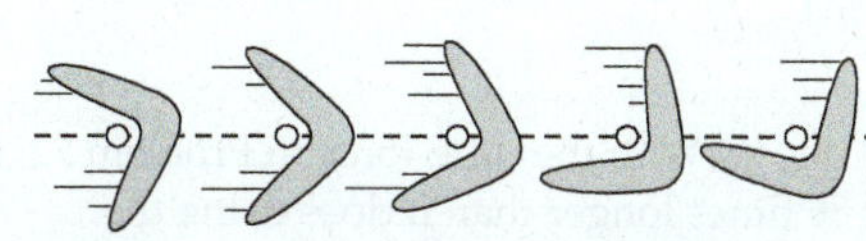

27 (*a*) M/T, kg/s, (*b*) M/L, kg/m, (*c*) ML/T^2, (*d*) 57 m/s, (*e*) 87 m/s

29 $\mu_s \approx 1.4$. This is probably not such a good idea. Tires on asphalt or concrete have a maximum coefficient of static friction of about 1.

31 (*b*)

33 (*a*) 15 N, (*b*) 12 N

35 500 N

37 (*a*) 5.9 m/s^2, (*b*) 76 m

39 (*a*) 49.1 N, (*b*) 123 N

41 (*a*) 4.6°, (*b*) 4.6°

43 0.84 m/s

45 2.4 m/s^2, 37 N

47 (*a*) 4.0 m, (*b*) 0.47

49 (*a*) 2.7 m/s^2, (*b*) 10 s

51 (*a*) 0.96 m/s^2, (*b*) 0.18 N

53 (*a*) The static-frictional force opposes the motion of the object, and the maximum value of the static-frictional force is proportional to the normal force F_N. The normal force is equal to the weight minus the vertical component F_y of the force F. Keeping the magnitude F constant while increasing θ from zero results in a decrease in F_y and thus a corresponding decrease in the maximum static-frictional force, f_{max}. The object will begin to move if the horizontal component F_x of the force F exceeds f_{max}. An increase in θ results in a decrease in F_x. As θ increases from 0, the decrease in F_N is larger than the decrease in F_x, so the object is more and more likely to slip. However, as θ approaches 90°, F_x approaches zero and no movement will be initiated. If F is large enough and if θ increases from 0, then at some value of θ, the block will start to move.

(*b*)

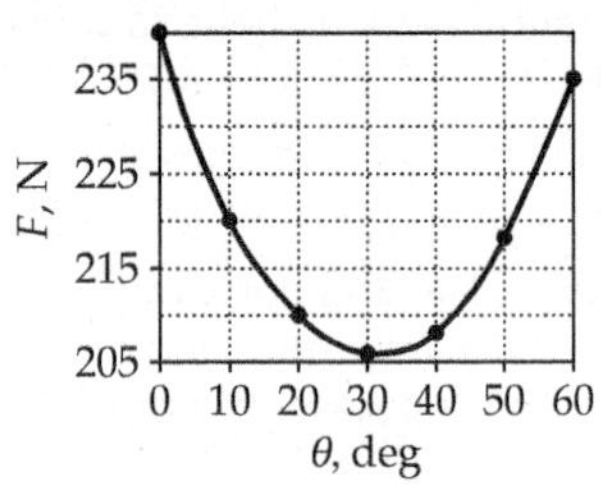

From the graph, we can see that the minimum value for F occurs when $\theta \approx 32°$.

57 (*a*) 0.24, (*b*) 1.4 m/s^2

59 (*a*) 18 N, (*b*) 1.5 m/s^2, 2.9 N, (*c*) $a_1 = 2.0\ m/s^2$, $a_2 = 7.8\ m/s^2$

61 (*a*) 5.7°, (*b*) 1.9 m/s

63 (*a*) $F_{min} = -1.6$ kg, $F_{max} = 84$ N, (*b*) $F_{min} = 5.8$ N and $F_{max} = 37$ N (The $+x$ direction is to the right.)

65 (*b*) 0.30, (*c*) 2.8 m/s

67 2.8×10^{-4} kg/m

69 $d_{n\ filters} = \sqrt{n}\, d_{1\ filter}$

71 25 m/s

73 (*a*) about 39 ms, (*b*) With the drag force in Problem 72, it takes about 86 times longer than it does using the centrifuge.

75 25°

77 (*a*) 1.4 m/s, (*b*) 8.5 N

79 (*a*) 8.33 m/s^2, upward, (*b*) 667 N, upward, (*c*) 1.45 kN, upward

81 $T_1 = [m_2(L_1 + L_2) + m_1L_1](2\pi/T)^2$, $T_1 = m_2(L_1 + L_2)(2\pi/T)^2$

83 (*a*) 53° above horizontal, 0.41 kN, (*b*) 53° below horizontal, 0.41 kN

85 (*a*) 0.40 N, (*b*) 0.644

87 52°

89 12.8 m/s

91 (*a*) 7.3 m/s, (*b*) 0.54

93 22°

95 (*a*) 7.8 kN, (*b*) −0.78 kN

97 $20\ km/h \le v \le 56\ m/h$

99 (*a*) about 60.4 m, (*b*) about 60.6 m, (*c*) about 3.3 s, (*d*) about 3.7 s, (*e*) less than

101 (0.23 cm, 0)

103 (2.0 m, 1.4 m)

105 (1.5 m, 1.4 m)

107 $(\frac{1}{4}L, \frac{1}{4}L)$

113 $\vec{v}_{cm} = (3.0\ m/s)\hat{i} - (1.5\ m/s)\hat{j}$

115 $\vec{a}_{cm} = (2.4\ m/s^2)\hat{i}$

117 (*a*)

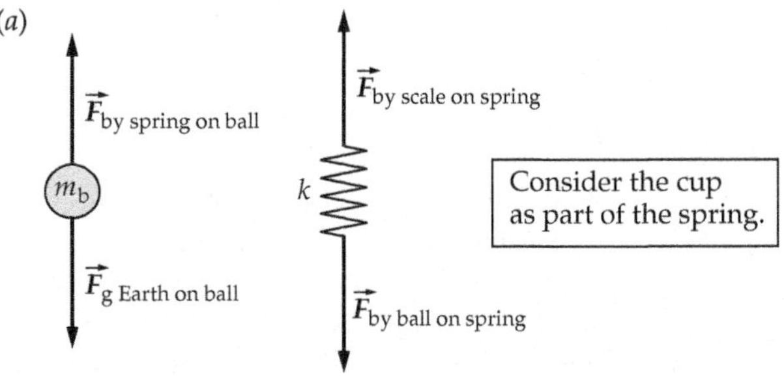

(*c*) $F_{by\ scale} = (m_{ball} + m_{platform})g$

119 (*a*)

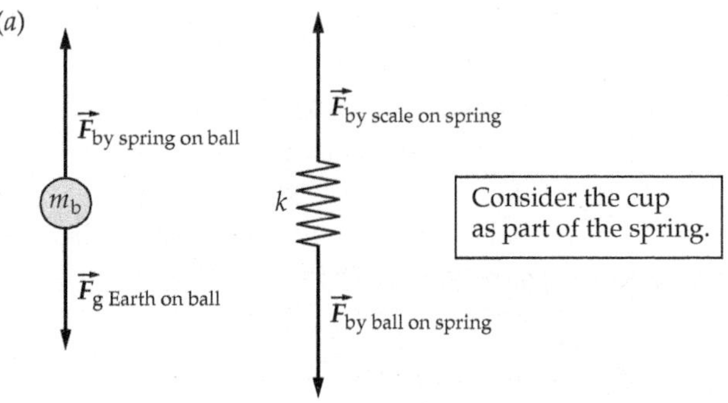

(*b*) $d' = \dfrac{m_b(g + a)}{k} > d$,

(*c*) $F_{by\ scale} = (m_b + m_p)(a + g) > F_{by\ scale\ pblm\ 117}$

121 0.51

123 1.49 kN

125 (*a*) 49 m/s^2, (*b*) 13 rev/min

127 (*a*) 0.19 kN, (*b*) 52 N, (*c*) 35 N, (*d*) 0.24, (*e*) 0.54 kN

129 0.43

131 (*a*)

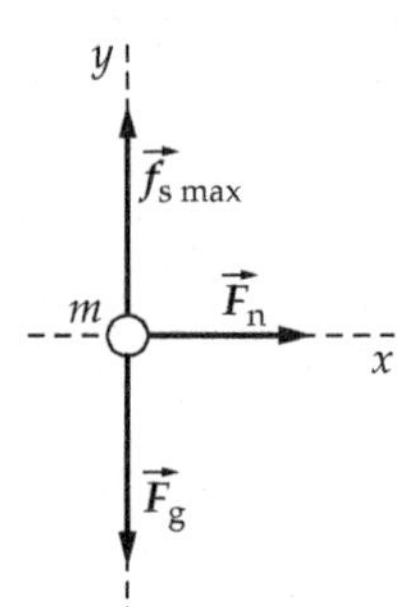

(*b*) 0.74 kN, (*c*) 20 rev/min. This result holds for all patrons, regardless of their mass.

133 Yes. Sally's claim seems to be supported by Liz's calculation.

135 4.67×10^6 m. It is inside the surface of Earth.

137 (*a*) 35 cm, (*b*) 4.7 m/s downward (*c*) 14.2 kN, 9.81 m/s^2, downward

139 (*a*) 4.91 m/s, (*b*) 1.64 m/s downward, (*c*) 1.09 m/s^2 downward

Chapter 6

1 (*a*) True, (*b*) True, (*c*) False, (*e*) True

3 (*a*) False, (*b*) True, (*c*) False, (*d*) True

5 An object moving along a curved path at constant speed has constant kinetic energy, but is accelerating (because its velocity is continually changing direction). No, because if the object is not accelerating, the net force acting on it must be zero and, consequently, its kinetic energy must be constant.

7 The work required to stretch a spring 2.0 cm is greater than the work required to stretch it 1.0 cm by a factor of 4.

9 (*d*)

11 (*a*) False, (*b*) False, (*c*) False, (*d*) False

13 (*a*) False, (*b*) False, (*c*) True, (*d*) True

15 $2\Delta t$

17 The only external force (neglecting air resistance) that does center-of-mass work on the car is the static friction force $\vec{f}_s$ exerted by the road on the tires. The positive center-of-mass work this friction force does is translated into a gain of kinetic energy.

19 (*a*) 4.5×10^{18} J, (*b*) 1%, (*c*) 1.4×10^{11} W

21 21 kJ

23 (*a*) 147 J, (*b*) 266 J

25 11 kJ, 3.5 kW

27 (*a*) 6.0 J, (*b*) 12 J, (*c*) 3.5 m/s

31 (*a*) $m(y) = 20\text{ kg} - (2.5\text{ kg/m})y$, (*b*) 0.59 kJ

33 (*a*)

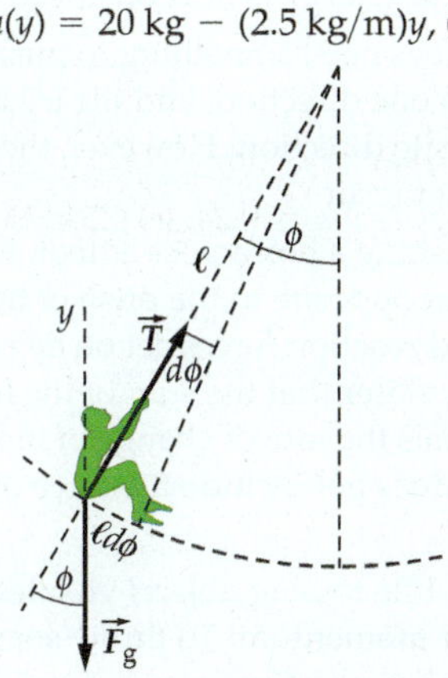

(*b*) $dW_{F_g} = -mg\ell \sin\phi \, d\phi$, (*c*) 2.5 kJ, 7.0 m/s

35 $W_{x_0 \to x} = A\left(\frac{1}{x_0} - \frac{1}{x}\right)$, $K_{x \to \infty} = \frac{A}{x_0}$, $v_{x \to \infty} = \sqrt{\frac{2A}{mx_0}}$

37 180°

39 (*a*) −24, (*b*) −10, (*c*) 0

41 (*a*) 1.0 J, (*b*) 0.21 N

43 (*b*) $6\hat{i} + 8\hat{j}$. Another vector that meets the requirements is $-6\hat{i} - 8\hat{j}$.

45 (*b*) $\vec{p} = (34\text{ kg}\cdot\text{m/s})\hat{i} + (-16\text{ kg}\cdot\text{m/s})\hat{j}$, (*c*) 0.28 kJ, 0.28 kJ

49 (*a*) $P(t) = (3.1\text{ W/s})t$, (*b*) 9.4 W

51 0.15 kW

55 (*a*) 0.381 kg/m, (*b*) 148 mi/h

57 3.2×10^5 m

59 50 kW

61 (*a*) 405 N, (*b*) 19.9 N

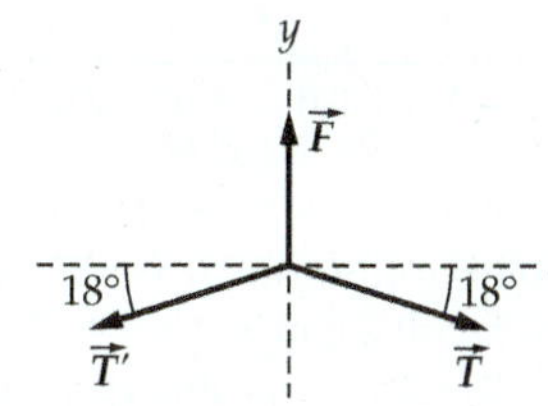

(*c*) 39.7 mJ

63 (*a*) $F(x) = mC^2x$, (*b*) $W = \frac{1}{2}mC^2x_1^2$

65 (*a*) $v = (6t^2 - 8t)$, $a = (12t - 8)$, (*b*) $P = 8mt(9t^2 - 18t + 8)$, (*c*) $W = 2mt_1^2(3t_1 - 4)^2$

67 (*a*) 208 kW, (*b*) 5.74 km

69 (*a*)

x (m)	4.0	−3.0	−2.0	−1.0	0.0	1.0	2.0	3.0	4.0
W (J)	6.0	4.0	2.0	0.5	0.0	0.5	1.5	2.5	3.0

(*b*) 28.0 J

71 (*b*) $W_{1\text{ rev cw}} = (31\text{ m})F_0$, $W_{1\text{ rev ccw}} = (-31\text{ m})F_0$

73 (*a*) $F_x = -kx\left(1 - \frac{y_0}{\sqrt{x^2 + y_0^2}}\right)$, (*c*) $v_f = \frac{L^2}{2y_0}\sqrt{\frac{k}{m}}$

Chapter 7

1 (*d*)

3 (*a*) False, (*b*) False, (*c*) False, (*d*) False

5 (*a*) True, (*b*) True, (*c*) True, (*d*) True, (*e*) False

7 (*a*) False, (*b*) False, (*c*) False, (*d*) False, (*e*) True

9 (*d*)

11 (*a*) Yes, (*b*) No, (*c*) No

13 (*a*) 25 cm, (*b*) −0.12 kJ

15 (*a*) 16 s, (*b*) 6.8 min. Not feasible that you can maintain the pace for 6.8 min.

17 1.5×10^{18} J/y, 3%

19 2.4×10^5 L/s

21 (*a*) 0.39 kJ, (*b*) 2.5 m, 4.9 m/s, (*c*) 24 J, 0.37 kJ, (*d*) 0.39 kJ, 20 m/s

23 (*a*) 10 cm, (*b*) 14 cm

25 (*a*) $F_x = C/x^2$, and $\vec{F} = F_x\hat{i}$
(*b*) If $x > 0$, $\vec{F}$ points away from the origin. If $x < 0$, $\vec{F}$ points toward the origin.
(*c*) Decrease.
(*d*) If $x > 0$, $\vec{F}$ points toward the origin. If $x < 0$, $\vec{F}$ points away from the origin.

27 $U(x) = [(-0.63\text{ kJ}\cdot\text{m})/x] + 0.30\text{ kJ}$

29 (*a*) $F_x = 4x(x + 2)(x - 2)$, (*b*) $x = -2$ m, $x = 0$, and $x = 2$ m, (*c*) Unstable at $x = -2$ m, stable at $x = 0$, unstable at $x = 2$ m

31 (*a*) $x = 0.0$ m and $x = 2.0$ m, (*b*)

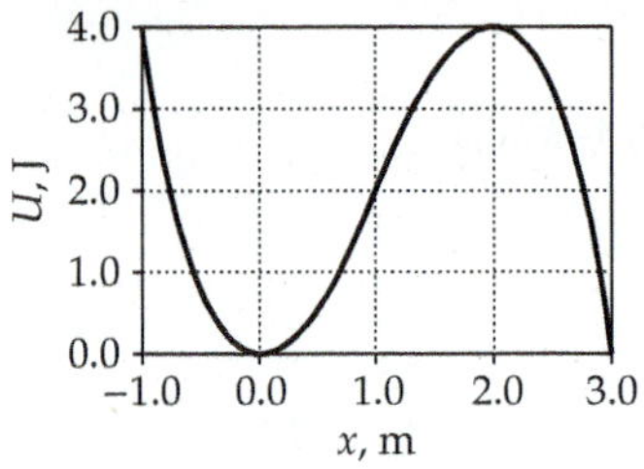

(*c*) Stable equilibrium at $x = 0$, unstable equilibrium at $x - 2.0$ m, (*d*) 2.0 m/s

33 (*a*) $U(\theta) = (m_2\ell_2 - m_1\ell_1)g\sin\theta$, (*b*) U is a minimum at $\theta = -\pi/2$, U is a maximum at $\theta = \pi/2$

35 (*a*) $U(y) = -mgy - 2Mg\left(L - \sqrt{y^2 + d^2}\right)$, (*b*) $y = d\sqrt{m^2/4M^2 - m^2}$, (*d*) This is a point of stable equilibrium.

37 $v = \sqrt{gL/2}$

39 (*a*) 0.858 m, (*b*) The block will retrace its path, rising to a height of 5.00 m.

41 26°

43 $U = \dfrac{[mg(\sin\theta + \mu_s\cos\theta)]^2}{2k}$

45 $6mg$

47 16.7 kN

49 $6mg$

51 (*a*) 31 m, (*b*) 34 m/s

53 (*a*) 0.15 km, (*b*) 45 m/s

55 (*a*) $K_{max} = \frac{5}{2}mgL$, (*b*) $6mg$

57 (*a*) 20°, (*b*) 6.4 m/s

59 $v_2 = L\sqrt{2(g/L)(1 - \cos\theta) + (k/m)(\sqrt{\tfrac{13}{4}} - 3\cos\theta - \tfrac{1}{2})^2}$

61 (*a*) 82 kJ, (*b*) The energy comes from the internal chemical energy in your body. (*c*) 410 kJ, (*d*) 330 kJ,

63 (*a*) 0.10 kJ, (*b*) 70 J, (*c*) 34 J, (*d*) 2.9 m/s

65 (*a*) 7.7 m/s, (*b*) 59 J, (*c*) 0.33

67 (*a*) (14 N)y, (*b*) −(14 N)y, (*c*) 2.0 m/s

69 0.87 m, 2.7 m/s

71 (*a*) 9.0×10^{13} J, (*b*) \$$2.5 \times 10^6$, (*c*) 2.8×10^4 y

73 (*a*) 3.9×10^{31} MeV, (*b*) 4.2×10^8 m/s. As expected, this result (4.2×10^8 m/s) is greater than the speed of light (and thus incorrect). Use of the nonrelativistic expression for kinetic energy is not justified.

75 1.1×10^5 reactions/s

77 0.782 MeV

79 (*a*) 1.1 kg, (*b*) 2.7×10^9 kg

81 (*a*) 6, (*b*) 0.21 eV

83 $\Delta E_{therm} = -mgv\,\Delta t\sin\theta$

85 12 m^2

87 (*a*) 0.208, (*b*) 3.5 MJ

89 (*a*) $x_{1,-} = (2\mu_k mg/k) - x_i$, (*b*) $v_0 = \sqrt{(k/m)x_0^2 - 2\mu_k g x_0}$, (*c*) $\mu_k = kx_i/2mg$

91 (*a*) 11 kW, (*b*) −6.8 kW, (*c*) \$1.81, \$5.43

93 (*a*) 1.61 kJ, (*b*) 0.6 kJ, (*c*) 23 m/s

95 (*b*)

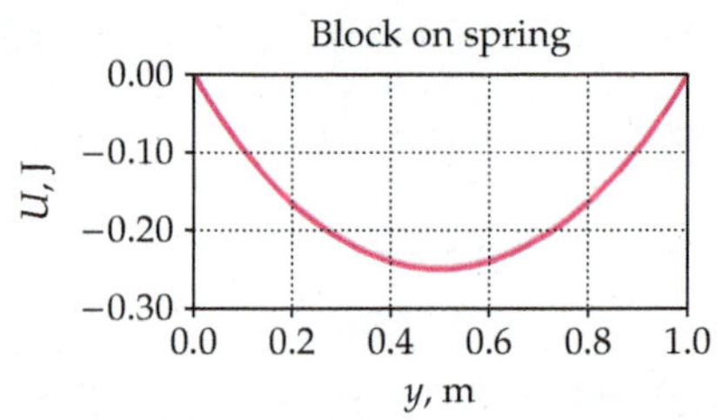

97 (*a*) 17 m, (*b*) 4.91 kN, (*c*) 4.9 m/s^2, (*d*) 13 kN, upward, (*e*) 5.5 kN, 64°, (*f*) 1.4 kN

99 (*a*) $F_{20} = 491$ N, $F_{30} = 981$ N, (*b*) $P_{20} = 9.8$ kW, $P_{30} = 29$ kW, (*c*) 8.8°, (*d*) 6.36 km/L

103 (*a*) $v = \sqrt{2my/(m + M)}$, (*b*) Same as in (*a*)

107 (*a*)

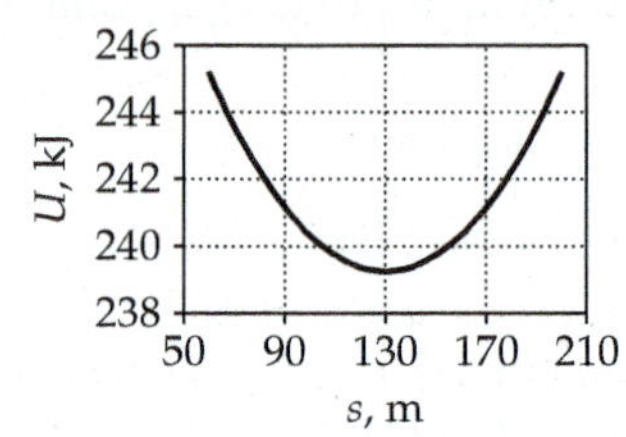

(*b*) 5.4 kJ

Chapter 8

5 The momentum of the bullet–gun system is initially zero. After firing, the bullet's momentum is directed west. Momentum conservation requires that the system's total momentum does not change, so the gun's momentum must be directed east. Kinetic energy is not conserved.

7 In a way, the rocket does need something to push upon. It pushes the exhaust in one direction, and the exhaust pushes it in the opposite direction. However, the rocket does not push against the air.

9 Think of someone pushing a box across a floor. Her push on the box is equal but opposite to the push of the box on her, but the action and reaction forces act on *different objects*. Newton's second law states that the sum of the forces acting on the box equals the rate of change of momentum of the box. This sum does not include the force of the box on her.

11 Hovering in midair while tossing objects violates the conservation of linear momentum! To throw something forward requires being pushed backward. Superheroes are not depicted as experiencing this backward motion that is predicted by conservation of linear momentum. This action does not violate conservation of energy.

13 The road. (The frictional force by the road on the tire causes the car to slow down.)

15 About 10^4

17 (*a*) False, (*b*) True, (*c*) True, (*d*) True

19 (*a*) The loss of kinetic energy is the same in both cases. (*b*) The situation in which the two objects have oppositely directed velocities.

21 (*b*)

23 The water is changing direction when it rounds the corner in the nozzle. Therefore, the nozzle must exert a force on the stream of water to change its momentum, and from Newton's third law, the water exerts an equal but opposite force on the nozzle. This requires a net force in the direction of the momentum change.

25 In the center-of-mass frame the two velocities are equal and opposite, both before and after the collision. In addition, the speed of each puck is the same before and after the collision. The direction of the velocity of each puck changes by some angle during the collision.

27 The downward force of lunar gravity and the upward thrust provided by the rocket combustion products.

29 Think of the sail facing the fan (like the sail on a square rigger might), and think of the stream of air molecules hitting the sail. Imagine that they bounce off the sail elastically—their net change in momentum is then roughly twice the change in momentum that they experienced going through the fan. Thus the change in momentum of the air is backward, so to conserve momentum of the air–fan–boat system the change in momentum of the fan–boat system will be forward.

31 (*a*) 2.34 s, (*b*) 6.7 m/s

33 5.5 m/s

35 4.0 m/s to the right

37 $\vec{v}' = 2v\hat{i} - v\hat{j}$

39 0.084

41 (*a*) 44 J, (*b*) $\vec{v}_{cm} = (1.5\text{ m/s})\hat{i}$, (*c*) $\vec{v}_{1\,rel} = (3.5\text{ m/s})\hat{i}$, $\vec{v}_{2\,rel} = (-3.5\text{ m/s})\hat{i}$, (*d*) 37 J

43 (*a*) 11 N·s, (*b*) 1.3 kN

45 1.81 MN·s, 0.60 MN

47 0.23 kN

49 (*a*) 1.1 N·s directed into the wall, (*b*) 0.36 kN, into the wall, (*c*) 0.48 N·s, away from the wall, (*d*) 3.8 N, away from the wall

51 (*a*) 0.20 s, (*b*) 27 ms, (*c*) Because the collision time is much shorter for the sawdust landing, the average force exerted on the vaulter by the airbag is much less than the average force the sawdust exerts on him.

53 (*a*) 20 m/s, (*b*) Twenty percent of the initial kinetic energy is transformed into thermal energy, acoustic energy, and the deformation of metal.

55 (*a*) 2.0 m/s, (*b*) The collision was not elastic.

57 $v_{p\,f} = -0.25$ km/s, $v_{nuc\,f} = 46$ m/s

59 (*a*) 5.0 m/s, (*b*) 0.25 m, (*c*) $v_{1f} = 0$, $v_{2f} = 7.0$ m/s

61 (*a*) $0.2v_0$, (*b*) $0.4v_0$

63 0.45 km/s

67 $h = (v^2/8g)(m_1/m_2)^2$

69 0.0529

71 (*a*) The meteorite should impact Earth along a line exactly opposite Earth's orbital velocity vector. (*b*) 2.71×10^{-15} percent, (*c*) 1.00×10^{23} kg

73 1.5×10^6 m/s

75 (*a*) $\vec{v}_1 = (312\text{ m/s})\hat{i} + (66.6\text{ m/s})\hat{j}$, (*b*) 5.6 km, (*c*) 35.8 kJ

77 0.91

79 (*a*) 20%, (*b*) 0.89

81 (*a*) 1.7 m/s, (*b*) 0.83

83 (*a*) At room temperature rubber will bounce more when it hits a stick than it will at freezing temperatures. (*b*) 3.8 cm

85 (*a*) 60°, (*b*) $v_{cf} = 2.50$ m/s, $v_8 = 4.33$ m/s

87 (*a*) $v_1 = 1.7$ m/s and $v_2 = 1.0$ m/s, (*b*) The collision was elastic.

89 5.3 m/s, 29°

91 $$K_i = K_f = \frac{p_1^2}{2}\left[\frac{m_1^2 + 6m_1m_2 + m_2^2}{m_1^2m_2 + m_1m_2^2}\right] = \frac{p_1'^2}{2}\left[\frac{m_1^2 + 6m_1m_2 + m_2^2}{m_1^2m_2 + m_1m_2^2}\right]$$
$p_1' = \pm p_1$. If $p_1' = +p_1$, the particles do not collide.

93 (*a*) $\vec{v}_{cm} = 0$, (*b*) $\vec{u}_3 = (-5.0\text{ m/s})\hat{i}$, $\vec{u}_5 = (3.0\text{ m/s})\hat{i}$, (*c*) $\vec{u}_3' = (5.0\text{ m/s})\hat{i}$, $u_5' = 0.75$ m/s, (*d*) $\vec{v}_3' = (5.0\text{ m/s})\hat{i}$, $\vec{v}_5' = (-3.0\text{ m/s})\hat{i}$, (*e*) $K_i = K_f = 60$ J

95 (*a*) 360 kN, (*b*) 120 s, (*c*) 1.72 km/s

97 (*d*) 28

99 0.19 m/s, $K_i = 31$ mJ, $K_f = 12$ mJ

101 (*a*) $\vec{p} = -(1.1 \times 10^5\text{ kg·km/h})\hat{i} + (1.1 \times 10^5\text{ kg·km/h})\hat{j}$, (*b*) 43 km/h. The direction of the wreckage is 46° west of north.

103 (*a*) 6.3 m/s, (*b*) 20 m

105 (*a*) The velocity of the basketball will be equal in magnitude but opposite in direction to the velocity of the baseball. (*b*) 0, (*c*) $2v$

107 (*a*) 30 km/s, (*b*) 8.1. The energy comes from an immeasurably small slowing of Saturn's orbital speed.

109 The driver was not telling the truth.

111 8.9 kg

113 (*b*) 55

115 (*a*) $$v_{2f} = \left(\frac{m_b}{m_2 + m_b}\right)\left(1 + \frac{m_1}{m_1 + m_b}\right)v,$$
$$v_{1f} = -\frac{m_2m_b(2m_1 + m_b)}{(m_1 + m_b)^2(m_2 + m_b)}v,$$
(*b*) $$\Delta K = \frac{1}{2}\frac{m_2m_b^2(2m_1 + m_b)^2}{(m_2 + m_b)^2(m_1 + m_b)^2}\left(1 + \frac{m_1m_2}{(m_1 + m_b)^2}\right)v^2.$$
This additional energy came from chemical energy in the astronaut's bodies.

117 4.53 m/s

Chapter 9

1 (*a*) The point on the rim. (*b*) Both turn through the same angle. (*c*) The point on the rim. (*d*) Both have the same angular speed. (*e*) Both have zero tangential acceleration. (*f*) Both have zero angular acceleration. (*g*) The point on the rim.

3 (*c*)

5 (*a*) Tara, (*b*) Tara, (*c*) Neither

7 By choking up, you are rotating the bat about an axis closer to the center of mass, thus reducing the bat's moment of inertia. The smaller the moment of inertia, the larger the angular acceleration (a quicker bat) for the same torque.

9 (*b*)

11 (*b*)

13 One reason is to maximize the moment arm about the line through the hinge pins for the force exerted by someone pulling or pushing on the knob.

15 (*b*)

17 (*b*)

19 (*a*)

23 12 rev

25 10%

27 Approximately 6

29 (*a*) 16 rad/s, (*b*) 47 rad, (*c*) 7.4 rev, (*d*) 4.7 m/s and 73 m/s^2

31 (*a*) 40 rad/s, (*b*) $a_t = 0.96\ m/s^2$, $a_t = 0.19\ km/s^2$

33 (*a*) 0.59 rad/s^2, (*b*) 4.7 rad/s

35 3.6 rad/s

37 1.0 rad/s, 9.9 rev/min

39 (*a*) 2.94 rad, (*b*) 780 d

41 60 $kg \cdot m^2$

43 (*a*) 28 $kg \cdot m^2$, (*b*) 32 $kg \cdot m^2$

45 2.6 $kg \cdot m^2$

47 (*b*) $I_{cm} = \frac{1}{12}m(a^2 + b^2)$

49 $5.4 \times 10^{-47}\ kg \cdot m^2$

51 $1.4 \times 10^{-2}\ kg \cdot m^2$

55 $I = \frac{3}{10}MR^2$

57 $I_x = 3M((H^2/5) + (R^2/20))$

59 (*a*) 1.9 N · m, (*b*) $1.2 \times 10^2\ rad/s^2$, (*c*) 6.2×10^2 rad/s

61 (*a*) $a_t = g \sin\theta$, (*b*) $mgL \sin\theta$

63 (*a*) $d\tau = (2\mu_k Mg/R^2)r^2\,dr$, (*b*) $\tau = \frac{2}{3}MR\mu_k g$, (*c*) $\Delta t = (3R\omega/4\mu_k g)$

65 (*a*) 85 mJ, (*b*) 72 rev/min

67 (*a*) 19.6 kN, (*b*) 5.9 kN · m, (*c*) 0.27 rad/s, (*d*) 1.6 kW

69 (*a*) 3.6 rad/s, (*b*) 3.6 rad/s

71 3.1 m/s^2, $T_1 = 12$ N, $T_2 = 13$ N

73 30°

75 8.21 m/s

77 (*a*) $a = g/(1 + (2M/5m))$, (*b*) $T = (2mMg)/(5m + 2M)$

79 (*a*) 72 kg, (*b*) 1.4 rad/s^2, 0.29 kN, 0.75 kN

81 (*a*) $a = \dfrac{g \sin\theta}{1 + (m_1/2m_2)}$, (*b*) $T = \dfrac{m_2 g \sin\theta}{1 + (2m_2/m_1)}$,
(*c*) $v = \sqrt{\dfrac{2gh}{1 + (m_1/2m_2)}}$, (*d*) $a = g$, $T = 0$, and $v = \sqrt{2gh}$

83 10 kJ

85 3.1 m/s

87 (*a*) 0.19 m/s^2, (*b*) 0.96 N

89 20°

91 (*a*) $a = \frac{2}{3}g \sin\theta$, (*b*) $f_s = \frac{1}{3}mg \sin\theta$, (*c*) $\theta_{max} = \tan^{-1}(3\mu_s)$

93 $v' = \sqrt{\frac{4}{3}}v$

95 0.22 kJ

97 (*a*) $\alpha = 2F/[R(M + 3m)]$, counterclockwise,
(*b*) $a_C = F/(M + 3m)$, in direction of $\vec{F}$,
(*c*) $a_{CB} = -2F/(M = 3m)$, opposite to direction of $\vec{F}$

99 (*a*) 0.40 rad/s^2, 0.20 rad/s^2, (*b*) 4.0 N, clockwise

101 (*a*) $s_1 = \frac{12}{49}(v_0^2/\mu_k g)$, $t_1 = \frac{2}{7}(v_0/\mu_k g)$, and $v_1 = \frac{5}{7}v_0$, (*b*) $\frac{5}{7}$,
(*c*) $s_1 = 27$ m, $t_1 = 3.9$ s, and $v_1 = 5.7$ m/s

103 $v = (2r\omega_0/7)$

105 (*a*) 0.19 s, (*b*) 0.67 m, (*c*) 2.9 m/s

107 (*a*) $v = \frac{11}{7}v_0$, (*b*) $\Delta t = \frac{4}{7}(v_0/\mu_k g)$, (*c*) $\Delta x = \frac{36}{49}(v_0^2/\mu_k g)$

109 13 cm

111 (*a*) 7.4 m/s^2, (*b*) 15 m/s^2, (*c*) 2.4 m/s

113 (*a*) 7.8×10^2 kJ, (*b*) 90 N · m, 0.15 kN, (*c*) 1.4×10^3 rev

115 (*a*) 15 m, (*b*) 15 rad/s

117 (*a*) $\omega = \sqrt{4g/3r}$, (*b*) $F = \frac{7}{3}Mg$

119 (*a*) 32.2 rad/s, (*b*) 23° (*c*) 24 rad/s^2

121 (*a*) 14.7 m/s^2, (*b*) 66.7 cm

123 42 J

125

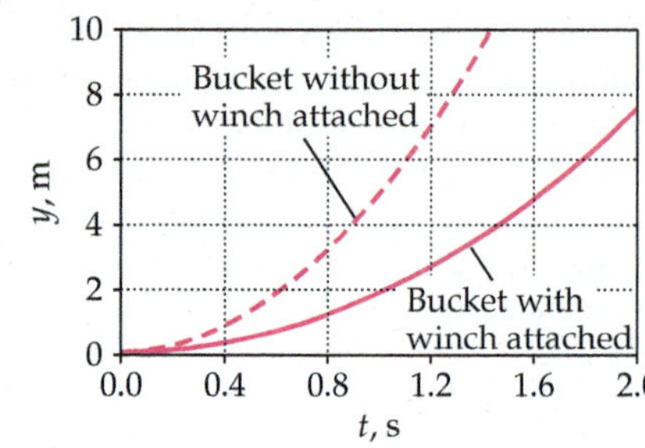

127 (*a*) 26 N, (*b*) 1.1 m/s^2, (*c*) 3.2 kg

Chapter 10

1 (*a*) True, (*b*) False, (*c*) False

3 90°

5 (*a*) *L* is doubled, (*b*) *L* is doubled

7 False. A high diver going from a tucked to a layout position.

9 (*e*)

11 The hardboiled egg is solid inside, so everything rotates with a uniform angular speed. By contrast, when you start an uncooked egg spinning, the yolk will not immediately spin with the shell, and when you stop it from spinning the yolk will continue to spin for a while.

13 (*a*)

15 (*a*) The plane tends to veer to the right. The change in angular momentum $\Delta\vec{L}_{prop}$ for the propeller is up, so the net torque $\vec{\tau}$ on the propeller is up as well. The propeller must exert an equal but opposite torque on the plane. This downward torque exerted by the prop on the plane tends to cause a downward change in the angular momentum of the plane. This means the plane tends to rotate clockwise as viewed from above.
(*b*) The plane tends to veer downward. The change in angular momentum $\Delta\vec{L}_{prop}$ for the propeller is to the right,

so the net torque $\vec{\tau}$ on the propeller is toward the right as well. The propeller must exert an equal but opposite torque on the plane. This leftward directed torque exerted by the prop on the plane tends to cause a leftward-directed change in angular momentum for the plane. This means the plane tends to rotate clockwise as viewed from the right.

17 (*a*) Your kinetic energy decreases. Increasing your moment of inertia I while conserving your angular momentum L decreases your kinetic energy $K = L^2/(2I)$.
(*b*) Extending your arms out to the side increases your moment of inertia, and decreases your angular speed. The angular momentum of the system is unchanged.

19 About 4 rev/s

21 (*a*) 33, (*b*) 33, (*c*) 8, (*d*) 14

23 (*a*) $2.4 \times 10^{-8}\ \text{kg}\cdot\text{m}^2/\text{s}$,
(*b*) $\ell(\ell + 1) = 5.2 \times 10^{52}$, $\ell \approx 2.3 \times 10^{26}$,
(*c*) $\Delta\ell = 2.3 \times 10^{18}$. The quantization of angular momentum is not noticed in macroscopic physics because no experiment can detect a fractional change in ℓ of $10^{-6}\%$.

25 (*a*) 0.331, (*b*) Because experimentally $C < 0.4$, the mass density must be greater near the center of Earth.

27 $\vec{\tau} = FR\hat{k}$

29 (*a*) $24\hat{k}$, (*b*) $-24\hat{j}$, (*c*) $-5\hat{k}$

33 $\vec{B} = 4\hat{j} + 3\hat{k}$

37 (*a*) $54\ \text{kg}\cdot\text{m}^2/\text{s}$, upward, (*b*) $54\ \text{kg}\cdot\text{m}^2/\text{s}$, downward, (*c*) 0

39 (*b*) Downward

41 (*a*) $1.3 \times 10^{-5}\ \text{kg}\cdot\text{m}^2/\text{s}$, away from you,
(*b*) $1.3 \times 10^{-5}\ \text{kg}\cdot\text{m}^2/\text{s}$, away from you,
(*c*) $1.3 \times 10^{-5}\ \text{kg}\cdot\text{m}^2/\text{s}$, away from you,
(*d*) $8.8 \times 10^{-5}\ \text{kg}\cdot\text{m}^2/\text{s}$, pointing toward you

43 (*a*) $-4.9\ \text{N}\cdot\text{m}$. Note that, because L decreases as the particle rotates clockwise, the angular acceleration and the net torque are both upward.
(*b*) $\omega_{\text{orbital}} = 0.48\ \text{rad/s} - (0.19\ \text{rad/s}^2)t$, downward.

45 (*a*) $\tau_{\text{net}} = Rg(m_2 \sin\theta - m_1)$, (*b*) $L = vR((I/R^2) + m_1 + m_2)$,
(*c*) $a = \dfrac{g(m_2 \sin\theta - m_1)}{(I/R^2) + m_1 + m_2}$

49 (*a*) 5.0 rev/s, (*b*) 0.62 kJ, (*c*) Because no external agent does work on the system, the energy comes from your internal energy.

51 10 mm/s

53 (*a*) $r_0 m v_0$, (*b*) $\frac{1}{2}mv_0^2$, (*c*) $m(v_0^2/r_0)$, $-\frac{2}{3}mv_0^2$

55 125°

57 (*a*) $3.46 \times 10^{-47}\ \text{kg}\cdot\text{m}^2$, (*b*) 1.00 meV, 2.01 meV, 6.02 meV

59 (*a*) No, None of the allowed values of E_ℓ are equal to $3E_{0\text{r}}$.
(*b*) 2.5

61 $\vec{v}_{\text{cm}} = \dfrac{m}{M + m}\vec{v}$, $\omega = \left(\dfrac{mMd}{\frac{1}{12}ML^2(M + m) + Mmd^2}\right)v$

63 $v = \sqrt{\dfrac{(0.5M + 0.8m)(\frac{1}{3}ML^2 + 0.64mL^2)g}{0.32Lm^2}}$

65 (*a*) $v_{\text{cm}} = K/M$, (*b*) $V = 4K/M$, (*c*) $V = -2K/M$, (*d*) Yes, one point remains motionless, but only for a very brief time.

67 0.36

69 (*a*) $18\ \text{J}\cdot\text{s}$, (*b*) 0.41 rad/s, (*c*) 15 s, (*d*) $0.079\ \text{J}\cdot\text{s}$

71 (*a*) $\vec{L} = -(48\ \text{kg}\cdot\text{m}^2/\text{s})\hat{k}$, (*b*) $\vec{\tau} = (16\ \text{N}\cdot\text{m})\hat{k}$

73 (*a*) $0.24\ \text{kJ}\cdot\text{s}$, (*b*) 0.31 kJ

75 (*a*) Because $\tau_{\text{net}} \neq 0$, angular momentum is not conserved. (*b*) Because, in this frictionless environment, the net external force acting on the object is the tension force and it acts at right angles to the object's velocity, the energy of the object is conserved. (*c*) v_0

77 Yes. The solution depends only upon conservation of angular momentum of the system, so it depends only upon the initial and final moments of inertia.

81 (*a*) 0.228 rad/s, (*b*) 0.192 rad/s

83 $4.47 \times 10^{22}\ \text{N}\cdot\text{m}$

85 −79.9 cm

Chapter R

1 In the reference frame of the car both events occur at the same location (the location of the car). Thus, your friend's watch measures the proper time between the two events.

3 Yes. Let the initial frame of reference be frame 1. In frame 1 let L be the distance between the events, let T be the time between the events, and let the $+x$ direction be the direction of event B relative to event A. Next, calculate the value of L/T. If L/T is less than c, then consider the two events in a reference frame 2, a frame moving at speed $v = L/T$ in the $+x$ direction. In frame 2 both events occur at the same location.

5 Yes.

7 (*a*)

9 (*b*)

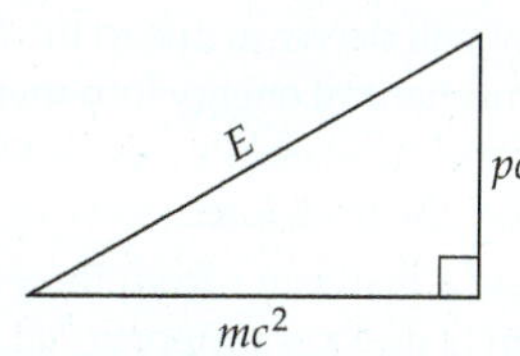

11 (*a*)

13 5.9 ns

15 (*a*) $K/E_0 = 5 \times 10^{-15}$, (*b*) $E/E_0 = 1.0000000000000050$,
(*c*) $K_{\text{non-rel}}/K_{\text{rel}} = 0.9999999999999975$

17 6.6 m

19 (*a*) 599 m

21 (*a*) 1.3×10^2 y, (*b*) 88 y

23 (*a*) 60 cm, (*b*) 2.5 ns

25 $0.80c$

27 (*a*) $4.50 \times 10^{-10}\%$, (*b*) 142 μs less than 1 y, 142 μs

29 36 min

31 12.5 min, 12.5 min

33 60 min

35 $0.400c$. Event B can precede event A provided $v > 0.400c$.

37 (*a*) 11 y, (*b*) 40 y

39 (*a*) 1.01, (*b*) 1.15, (*c*) 1.67, (*d*) 7.09

41 (*a*) $0.155\ mc^2$, (*b*) $1.29\ mc^2$, (*c*) $6.09\ mc^2$

43 2.97 GeV

45 (*b*) $0.866c$, (*c*) $0.999c$

49 (*a*) 0.79%, (*b*) 69%

51 (*a*) 0.943

53 In 100 lifetimes, $d \approx 6600$ km or approximately one Earth radius. This relatively short distance should convince your classmate that the origin of the muons that are observed on Earth is within our atmosphere, and that they certainly are not from the Sun.

55 (*a*) 4.50 km/s, (*b*) 0.334 μs

57 (*a*) 0.75*c*, (*b*) 5.0 ft, (*c*) No. In your rest frame, the back end of the ladder will clear the door before the front end hits the wall of the shed, while in Ernie's rest frame, the front end will hit the wall of the shed while the back end has yet to clear the door.

Chapter 11

1 (*a*) False, (*b*) True, (*c*) True, (*d*) False

3 Earth is closest to the Sun during winter in the northern hemisphere. This is the time of fastest orbital speed. Summer would be the time for minimum orbital speed.

5 To obtain the mass *M* of Venus you need to measure the period *T* and semi-major axis *a* of the orbit of one of the satellites, substitute the measured values into $T^2/a^3 = 4\pi^2/(GM)$ (Kepler's third law) and solve for *M*.

7 (*d*)

9 (*b*)

11 (*b*)

13 You should fire the rocket in a direction to oppose the orbital motion of the satellite. As the satellite gets closer to Earth after the burn, the potential energy will decrease. However, the total mechanical energy will decrease due to the frictional drag forces transforming mechanical energy into thermal energy. The kinetic energy will increase until the satellite enters the atmosphere where the drag forces slow its motion.

15 At a point inside the sphere a distance *r* from its center, the gravitational field strength is directly proportional to the amount of mass within a distance *r* from the center, and inversely proportional to the square of the distance *r* from the center. The mass within a distance *r* from the center is proportional to the cube of *r*. Thus, the gravitational field strength is directly proportional to *r*.

17 $M_{\text{galaxy}} = 1.08 \times 10^{11}\ M_S$

19 About 3.0 km

21 (*a*) 6.28×10^{-4} rad/s, 2.78 h, (*b*) $L_J = 1.93 \times 10^{43}\ \text{kg}\cdot\text{m}^2/\text{s}$, $L_S = 7.85 \times 10^{42}\ \text{kg}\cdot\text{m}^2/\text{s}$, 0.70%, (*c*) $T_{\text{Sun}} = 3.64$ h, $T_{\text{Sun}} = 1.30T_{\text{max}}$

23 84.0 y

25 4.90×10^{11} m = 3.00 AU

27 (*a*) 1.6×10^{11} m = 1.1 AU, (*b*) 2.7×10^{10} m = 0.18 AU, 2.9×10^{11} m = 1.9 AU

29 (*b*) 0.73 AU, (*c*) 0.63 y

31 (*a*) 1.90×10^{27} kg

33 (*a*) 22.7 h, (*b*) 1.22×10^9 m

35 Your weight would be ten times your weight on Earth.

37 2.27×10^4 m/s

39 (*a*) 1.4, (*b*) It is farther from the Sun than Earth. Kepler's third law $[T^2 = Cr_{\text{av}}^3]$ tells us that longer orbital periods together with larger orbital radii means slower orbital speeds, so the speed of objects orbiting the Sun decreases with distance from the Sun. The average orbital speed of Earth, given by $v = 2\pi r_{\text{ES}}/T_{\text{ES}}$, is approximately 30 km/s. Because the given maximum speed of the asteroid is only 20 km/s, the asteroid is further from the Sun.

41 (*a*) 7.37 m, (*b*) 31.9 μm

43 0.605

45 10^9 m

47 2.38 km/s

49 (*a*) 8.7 kW·h, (*b*) $500

51 6.9 km/s

53 19.4 km/s

55 13.8 km/s

57 (*a*) 7.31 h, (*b*) 1.04×10^9 J, (*c*) 8.72×10^{12} J·s

59 1.11×10^{10} J

61 $(4.0\ \text{N/kg})\hat{i}$

63 (*a*) $\vec{g} = \dfrac{Gm}{L^2}\hat{i} + \dfrac{Gm}{L^2}\hat{j}$, (*b*) $|\vec{g}| = \sqrt{2}\dfrac{Gm}{L^2}$

65 (*a*) $(-1.7 \times 10^{-11}\ \text{N/kg})\hat{i}$, (*b*) $(-8.3 \times 10^{-12}\ \text{N/kg})\hat{i}$, (*c*) 2.5 m

67 (*a*) $\frac{1}{2}CL^2$, (*b*) $\vec{g} = -\dfrac{2GM}{L^2}\left[\dfrac{L}{x_0 - L} - \ln\dfrac{x_0}{x_0 - L}\right]\hat{i}$

69 (*a*) 0, (*b*) 0, (*c*) 3.2×10^{-9} N/kg

71 $g_1 = g_2$

73 (*a*) $\dfrac{Gm(M_1 + M_2)}{9a^2}$, (*b*) $\dfrac{GmM_1}{3.61a^2}$, (*c*) 0

77 $g(x) = G\left(\dfrac{4\pi\rho_0 R^3}{3}\right)\left[\dfrac{1}{x^2} - \dfrac{1}{8(x - \frac{1}{2}R)^2}\right]$

79 $\omega = \sqrt{\dfrac{4\pi\rho_0 G}{3}}$

81 1.0 m/s

83 (*a*) $\vec{F} = -\dfrac{GMm}{d^2}\left[1 - \dfrac{d^3/4}{\{d^2 + (R^2/4)\}^{3/2}}\right]\hat{i}$

85 249 y

87 (*a*) $W = GM_E m\left(\dfrac{1}{r_1} - \dfrac{1}{r_2}\right)$

89 (*a*) 3.36×10^9, (*b*) 241

91 1.70 Mm

95 1.60×10^{-4}

97 $v = 1.16\sqrt{\dfrac{GM}{a}}$

99 $$g(r) = \begin{cases} 0 & r < R_1 \\ \dfrac{GM(r^3 - R_1^3)}{r^2(R_2^3 - R_1^3)} & R_1 < r \le R_2 \\ \dfrac{GM}{r^2} & R_2 < r \end{cases}$$

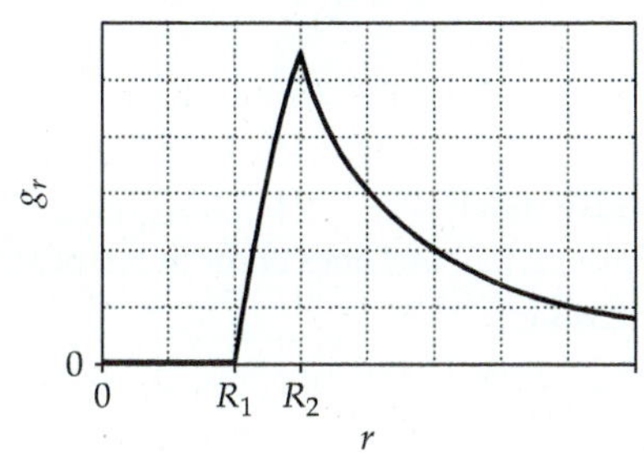

101 $g = (2G\lambda)/r$

103 (*b*) $U = -\frac{GMm_0}{L}\ln\left(\frac{x + L/2}{x - L/2}\right)$, (*c*) $F_x(x) = -\frac{GMm_0}{x^2 - (L/2)^2}$

This answer and the answer given in Example 11-8 are the same.

105 34 pN

107 (*a*) The gravitational force is greater on the lower robot, so if it were not for the cable, its acceleration would be greater than that of the upper robot and they would separate. In opposing this separation, the cable is stressed. (*b*) 2.2×10^5 m

Chapter 12

1 (*a*) False, (*b*) True, (*c*) True, (*d*) False

3 (*b*)

5 The higher the sign, the greater the torque about the horizontal axis through the lowest extremes of the posts for a given wind speed. In addition, the deeper the post hole the greater the maximum opposing torque about the same axis for a given consistency of the dirt. Thus, higher signs require deeper holes for the signposts.

7 The main reason this is done is to lower the center of gravity of the mug. The lower the center of gravity the more stable the mug is.

9 (*b*)

11 (*b*) 2.0 N/cm

13 (*b*) Taking long strides requires a large coefficient of static friction because θ is large for long strides. (*c*) If μ_s is small, that is, there is ice on the surface, θ must be small to avoid slipping.

15 84 cm

17 692 N, 2.54 kN

19 0.728 m

21 $F_1 = \frac{1}{2}Mg$, $F_2 = \frac{\sqrt{3}}{2}Mg$

23 (*a*) $\vec{F} = (30\text{ N})\hat{i} + (30\text{ N})\hat{j}$, (*b*) $\vec{F} = (35\text{ N})\hat{i} + (45\text{ N})\hat{j}$

25 (*a*) $F_n = Mg - F\sqrt{(2R - h)/h}$, (*b*) F,

(*c*) $F\sqrt{(2R - h)/h}$

27 (*a*) 6.87 N, (*b*) 1.7 N·m, (*c*) 8.3 N, 15 N

29 (*a*) 71 N, (*b*) 3.5 m, (*c*) 0.50 kN

31 (*b*) $a/3$, (*c*) $Mg/2$

33 $h = \mu_s L \tan\theta \sin\theta$

35 $\mu_s = \frac{2h}{L \tan\theta \sin\theta}$

37 59°

39 62°

41 (*a*) 42 N, (*b*) 0.14%

43 5.0°

47 (*a*) 1.4×10^6 N/m², (*b*) 7 mJ, (*c*) 28 mJ. There is about 4 times as much energy stored in the rubber when 0.30 kg are hung from it. That is because the stored energy increases quadratically with an increase in the mass.

49 0.69

51 Because $\text{Stress}_{\text{failing}} < \text{Stress}_{\text{cable}}$, the cable will not support the elevator.

55 1.5 kN

57 $m_1 = 0.15$ kg, $m_2 = 0.71$ kg, $m_3 = 0.36$ kg

59 1.8 kg

61 0.15

63 $\mu_s < 0.50$

65 $\mu_s = \frac{1}{2}(\cot\theta + 1)$

67 (*a*) 0.15 kN, (*b*) 3.8 m

69 $\mu_s < 0.50$

71 (*c*)

Cell	Content/Formula	Algebraic Form
B5	B4+1	$i + 1$
C5	C4+$B$1/(2*B5)	$d_i + \frac{L}{2i}$

	A	B	C	D
1	L =	0.20	m	
2		0		
3		*i*	offset	
4		1	0.100	
5		2	0.150	
6		3	0.183	
102		99	0.518	
103		100	0.519	

$d_5 = 15$ cm, $d_{10} = 26$ cm, and $d_{100} = 0.52$ cm

(*d*) No.

73 566 N

75 $F_n = 2mg$, $F = mg\frac{r}{\sqrt{R(2r - R)}}$, $F_W = mg\frac{R - r}{\sqrt{R(2r - R)}}$

Chapter 13

1 (*e*)

3 (*c*)

5 Pressure increases approximately 1 atm every 10 m of depth. To breathe requires creating a pressure of less than 1 atm in your lungs. At the surface you can do this easily, but not at a depth of 10 m.

7 (*b*)

9 False. The buoyant force on a submerged object depends on the weight of the displaced fluid, which, in turn, depends on the volume of the displaced fluid. Because the bricks have the same volume, they will displace the same volume of water.

11 Because the pressure increases with depth, the object will be compressed and its density will increase as its volume decreases. Thus, the object will sink to the bottom.

13

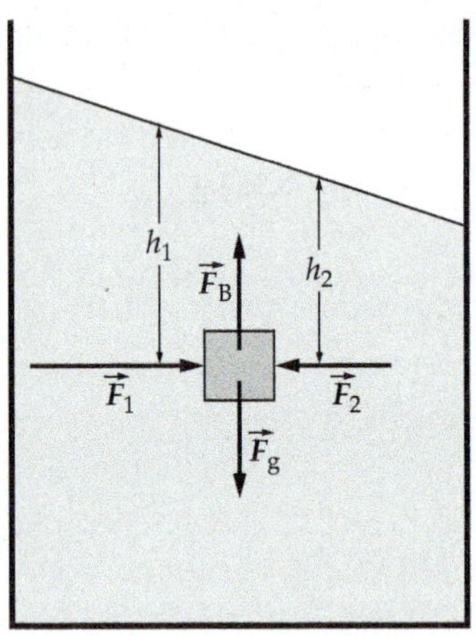

The pictorial representation shows the glass and an element of water in the middle of the glass. As is readily established by a simple demonstration, the surface of the water is not level while the glass is accelerated, showing that there is a pressure gradient (a difference in pressure) due to the differing depths ($h_1 > h_2$, and hence, $F_1 > F_2$) of water on the two sides of the element of water. This pressure gradient results in a net force on the element of water, as shown in the figure. The upward buoyant force is equal in magnitude to the downward gravitational force.

15 (*c*)

17 The mounding around entrance 1 will cause the streamlines to curve concave downward over the entrance. An upward pressure gradient produces the downward centripetal force. This means there is a lowering of the pressure at entrance 1. No such lowering occurs over entrance 2, so the pressure there is higher than the pressure at entrance 1. The air circulates in entrance 2 and out entrance 1. It has been demonstrated that enough air will circulate inside the tunnel even with the slightest breeze outside.

19 1.11 kg/m^3

21 1.0×10^2 kg

23 33.6 kg/m^3

25 0.773

27 29.8 in Hg

29 1.5 N

31 230 N

33 197 atm. Because a depth of only 2 km is required to produce a 1-percent compression, this does occur in our oceans.

35 (*a*) 15 kN, (*b*) 0.34 kg

37 45 cm

39 1.4%

41 4.36 N

43 (*a*) 11×10^3 kg/m^3, (*b*) From Table 13-1, we see that the density of the unknown material is close to that of lead.

45 800 kg/m^3, 1.11

47 250 kg/m^3

49 3.9 kg

51 3.9×10^5 N

53 2.5×10^7 kg

55 (*a*) 12.0 m/s, (*b*) 133 kPa, (*c*) The flow rates are the same.

57 (*a*) 4.6 L/min, (*b*) 7.6×10^{-2} m^2

59 144 kPa

61 0.20 kPa. Because $\Delta P \propto I_V^2$ (a plot of ΔP as a function of I_V is a parabola that opens upward), this pressure difference is the minimum pressure difference.

65 (*b*) $P_{top} = P_{at} - \rho g d$

67 (*a*) 9.28 cm/s, (*b*) 0.331 cm, (*c*) 76 cm

69 (*a*) $x = 2\sqrt{h(H - h)}$, (*c*) $x_{max} = H$

71 1.43 mm

73 90 mi/h. Because most major league pitchers can throw a fastball in the low-to-mid-90s, this abrupt decrease in drag may very well play a role in the game.

75 2.91 L/s

77 2

79 36 kg/m^3

81 0.71 kg

83 12 cm

85 One meter is a plausible diameter for such a pipe.

87 29 s

89 (*a*) 70 m^3, (*b*) 5.2 m/s^2

91 (*b*) 0.13 km^{-1}

93 39 cm^3

Chapter 14

1 (*a*) False, (*b*) True, (*c*) True

3 (*a*)

5 (*c*)

7 (*c*)

9 Neglecting the mass of the spring in your calculations results in your using a value for the mass of the oscillating system that is smaller than its actual value. Hence, your calculated value for the period will be smaller than the actual period of the system and the calculated value for the frequency, which is the reciprocal of the period, will be higher than the actual value.

Because the total energy of the oscillating system depends solely on the amplitude of its motion and on the stiffness of the spring, it is independent of the mass of the system, and so neglecting the mass of the spring would have no effect on your calculation of the system's total energy.

11 (*d*)

13 (*b*)

15 1 matches up with B, 2 matches up with D, and 3 matches up with A.

17 (*c*)

19 (*a*) True, (*b*) False, (*c*) True, (*d*) False, (*e*) True

21 (*b*)

23 (*b*)

25 About 5

27 8π

29 (*a*) 3.00 Hz, (*b*) 0.333 s, (*c*) 7.0 cm, (*d*) 0.0833 s in $-x$ direction.

31 (*a*) $x = (0.25\text{ m})\cos[(4.2\text{ s}^{-1})t]$,
(*b*) $v = -(1.0\text{ m/s})\sin[(4.2\text{ s}^{-1})t]$,
(*c*) $a = -(4.4\text{ m/s}^2)\cos[(4.2\text{ s}^{-1})t]$

33 (*a*) $x = (0.28\text{ m})\cos[(4.2\text{ s}^{-1})t - 0.45]$,
(*b*) $v = -(1.2\text{ m/s})\sin[(4.2\text{ s}^{-1})t - 0.45]$,
(*c*) $a = -(4.9\text{ m/s}^2)\cos[(4.2\text{ s}^{-1})t - 0.45]$

35 (*a*)

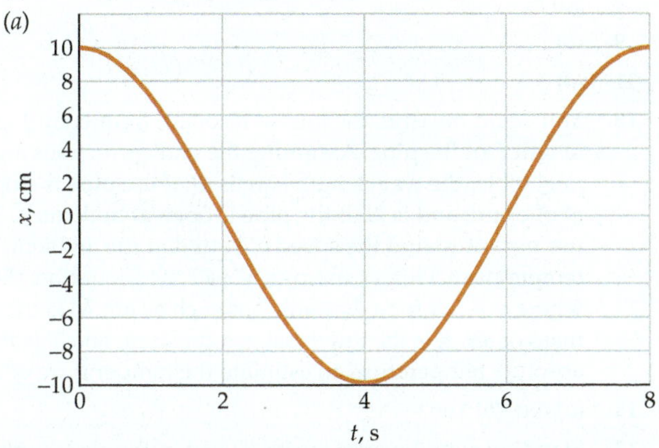

(*b*) 2.9 cm, 7.1 cm, 7.1 cm, 2.9 cm

37 (*a*) 7.9 m/s, 25 m/s², (*b*) 6.3 m/s, −15 m/s²

39 (*a*) 3.1 s, 0.32 Hz, (*b*) $x = (40\text{ cm})\cos[(2.0\text{ s}^{-1})t + (\pi/2)]$

41 23 J

43 (*a*) 0.368 J, (*b*) 3.83 cm

45 1.4 kN/m

47 (*a*) 6.9 Hz, (*b*) 0.15 s, (*c*) 10 cm, (*d*) 4.3 m/s,
(*e*) 1.9×10^2 m/s², (*f*) 36 ms, 0

49 (*a*) 0.68 kN/m, (*b*) 0.42 s, (*c*) 1.5 m/s, (*d*) 23 m/s²

51 (*a*) 3.08 kN/m, (*b*) 4.16 Hz, (*c*) 0.240 s

53 0.262 s

55 (*a*) 1.0 Hz, (*b*) 0.50 s, (*c*) 0.29 N

57 (*a*) 6.7 cm, (*b*) 0.26 s, (*c*) Because $h < 8.0$ cm, the spring is never uncompressed. 77 cm/s

59 44 cm

61 12 s

63 11.7 s

65 $T = 2\pi\sqrt{L/(g\cos\theta)}$

67 1.1 s

69 0.50 kg · m²

71 21.1 cm from the center of the meter stick

75 (*a*) $d = 1.64$ m, (*b*) 2.31 cm

79 (*a*) 0.31, (*b*) -3.1×10^{-2}%

81 (*a*) 1.57%, (*c*) $0.43E_0$

83 (*a*) 5.51 Pa · s, (*b*) 125

85 (*a*) 1.0 Hz, (*b*) 2 Hz, (*c*) 0.35 Hz

87 (*a*) 4.98 cm, (*b*) 14.1 rad/s, (*c*) 35.4 cm, (*d*) 1.00 rad/s

89 (*a*) 0.48 Hz, 2.1 s, (*b*) $v = -(1.2\text{ m/s})\sin\left[(3.0\text{ rad/s})t + \frac{\pi}{4}\right]$,
(*c*) 1.2 m/s

91 The error is greater if the clock is elevated.

93 (*a*) $\mu_s = \dfrac{Ak}{(m_1 + m_2)g}$ (*b*) A is unchanged. E is unchanged. ω is reduced, and T is increased.

95 (*b*) 2.0 cm/s²

101 6.44×10^{13} rad/s

103 $T = 7.78\sqrt{R/g}$

107 (*a*)

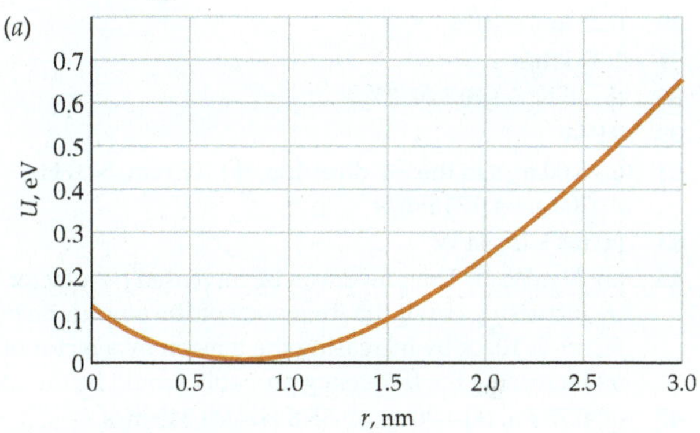

(*b*) $r = r_0$, $k = 2\beta^2 D$, (*c*) $\omega = 2\beta\sqrt{D/m}$

Chapter 15

1 The speed of a transverse wave on a uniform rope increases with increasing tension. The waves on the rope move faster as they move up because the tension increases due to the weight of the rope below.

3 (*b*)

5 The resonant (standing-wave) frequencies on a string are inversely proportional to the square root of the linear density of the string ($f = \sqrt{F_T/\mu}/\lambda$). Thus, extremely high frequencies (which might otherwise require very long strings) can be accommodated on relatively short strings if the strings are linearly denser than the high-frequency strings. High frequencies are not a problem, as they utilize short strings anyway.

7 (*c*)

9 There was only one explosion. Sound travels faster in water than air. Abel heard the sound wave in the water first, then, surfacing, heard the sound wave traveling through the air, which took longer to reach him.

11 (*b*)

13 (*a*)

15 (*b*)

17 (*a*) False, (*b*) True, (*c*) True

19 (*a*)

21 The light from the visible star will be shifted about its mean frequency periodically due to the relative approach toward and recession away from Earth as the star revolves around the common center of mass.

23

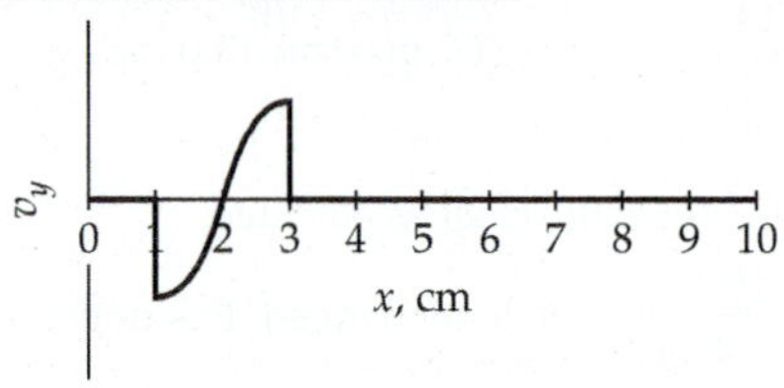

25 Path C. Because the wave speed is highest in the water, and more of path C is underwater than are paths A or B, the sound wave will spend the least time on path C.

27 11 ms

29 0.27 km/s, 21%

31 1.32 km/s

33 (*b*) 40 N, (*c*) 40.8 N, 2%

39 9.9 W

41 (*a*) 5.00 m/s in the $-x$ direction, (*b*) 10.0 cm, 50.0 Hz, 0.0200 s, (*c*) 0.314 m/s

43 (*a*) 6.8 J, (*b*) 44 W

45 (*a*) 79 mW, (*b*) The power can be increased by a factor of 100 by increasing either the frequency or the amplitude by a factor of 10, or by increasing the tension by a factor of 100. (*c*) Increasing the frequency probably would be the easiest.

47 (*a*) 0.75 Pa, (*b*) 4.00 m, (*c*) 85.8 Hz, (*d*) 343 m/s

49 (*a*) 36.4 μm, (*b*) 83.4 mPa

51 (*a*) Zero, (*b*) 3.64 μm

53 (*a*) 0.80 s, (*b*) 30 m, (*c*) 6.8 m

55 (*a*) 50.3 W, (*b*) 2.00 m, (*c*) 4.44 mW/m^2

57 (*a*) 20.0 dB, (*b*) 100 dB

59 90 dB

61 (*a*) 0.10 km, (*b*) 0.13 W

63 (*a*) 80 dB, (*b*) Eliminating the 70 dB and 73 dB sources does not reduce the intensity level significantly.

65 87.8 dB

67 57 dB

73 (*a*) 263 m/s, (*b*) 1.32 m, (*c*) 261 Hz

75 153 Hz

77 2.25×10^8 m/s

79 174 mi/h

81 (*a*) -7.78 kHz, (*b*) -4.44 kHz

83 (*a*) $f_r' = f_s(v - u_r)/(v - u_s)$

85 (*a*) 0.82 kHz, (*b*) 0.85 kHz

87 185 m, 714 Hz

89 $\lambda_{max} = (500\text{ nm})(1 + 4.36 \times 10^{-5})$ and $\lambda_{min} = (500\text{ nm})(1 - 4.36 \times 10^{-5})$

91 529 Hz, 474 Hz

93 7.99 m from the left end of the wire

95 (*a*) 55.6 N/m^2, (*b*) 3.49 W/m^2, (*c*) 0.110 W

97 77 kN

99 206 m

101 0.2 cm

103 (*a*) 10.0 m/s, (*b*) 2.00 m, (*c*) $P_{max} = 1.26 \times 10^{-4}$ kg m/s, (*d*) 3.95 mN

Chapter 16

1

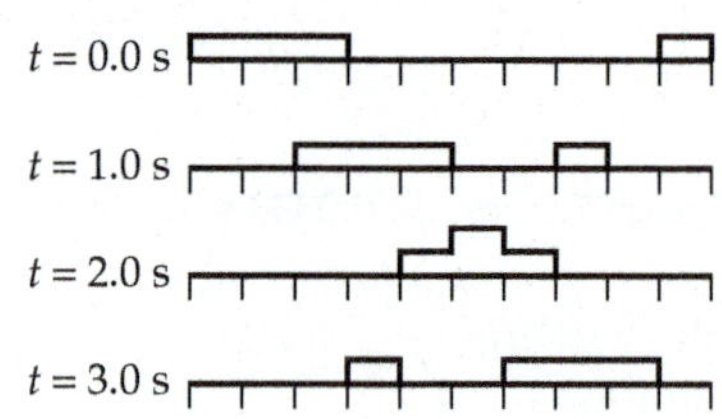

3 (*c*)

5 (*a*)

7 3.0 m

9 (*b*)

11 (*a*)

13 You could measure the lowest resonant frequency f and the length L of the pipe. Assuming the end corrections are negligible, the wavelength equals $4L$ if the pipe is stopped at one end, and is $2L$ if the pipe is open at both ends. Then use $v = f\lambda$ to find the speed of sound at the ambient temperature. Finally, use $v = \sqrt{\gamma RT/M}$ (Equation 15-5), where $\gamma = 1.4$ for a diatomic gas such as air, M is the molar mass of air, R is the universal gas constant, and T is the absolute temperature, to estimate the temperature of the air.

15 (*a*) No, (*b*) Yes

17 Standing sound waves are produced in the air columns above the water. The resonance frequency of the air columns depends on the length of the air column, which depends on how much water is in the glass.

19 The wavelength is determined mostly by the size of the resonant cavity of the mouth; the frequency of sounds he makes is equal to the wave speed divided by the wavelength. Because $v_{He} > v_{air}$ (see Equation 15-5), the resonance frequency is higher if helium is the gas in the cavity.

21 If you do not hear even one beat for the entire time the string and the tuning fork are vibrating, you can be sure that their frequencies, while not exactly the same, are very close. If the sounds of the vibrating string and the tuning fork last for 10 s, it follows that the beat frequency is less than 0.1 Hz. Hence the frequencies of the vibrating string and the tuning fork are within 0.1 Hz of each other.

23 3. The estimated frequencies agree with the observed frequencies to within 14%.

25 7.1 cm

27 (*a*) 89°, (*b*) 1.5*A*

29 (*a*) 0, (*b*) $2I_0$, (*c*) $4I_0$

31 (*a*) 60.0 cm, (*b*) 0.400π rad, (*c*) 24.0 m/s

33 $f_1 = 2.0$ kHz, $f_2 = 5.0$ kHz

35 (*b*)

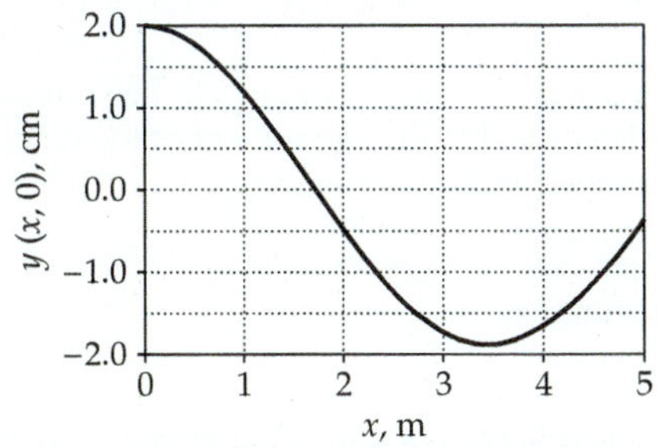

(*c*)

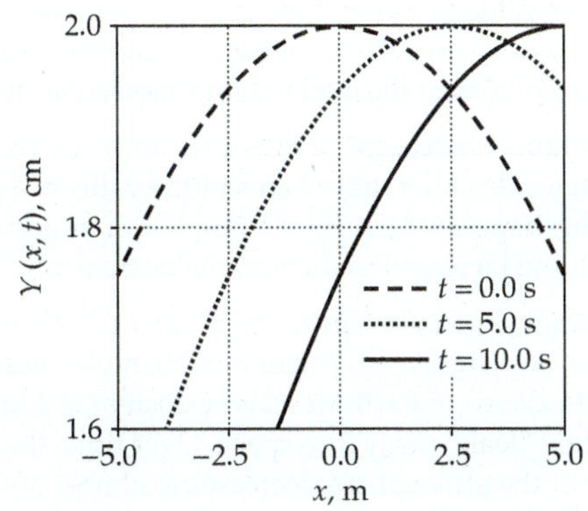

$v_{est} = 46$ cm/s, v_{est} is 92% of $v_{envelope}$

37 1.8 m, 51°

39 (*a*) 0.279 m, (*b*) 1.23 kHz, (*c*) 0.432 rad, 0.592 rad, 0.772 rad, 0.992 rad, and 1.35 rad, (*d*) 0.0698 rad

41 2.0 rad

43 (*b*) 1.2 kHz, (*c*) 15 Hz/(mi/h)

45 (*a*) 2.00 m, 2.50 Hz, (*b*) $y_3(x,t) = (4.00 \text{ mm}) \sin(\pi \text{ m}^{-1})x \cos(50.0\pi \text{ s}^{-1})t$

47 (*a*) 521 m/s, (*b*) 2.80 m, 186 Hz, (*c*) 372 Hz and 558 Hz

49 141 Hz

51 (*a*) 32.4 cm, 47.7 Hz, (*b*) 15.0 m/s, (*c*) 62.8 cm

53 (*a*) 71.5 Hz, (*b*) 5.00 kHz, (*c*) 71

55 452 Hz. Ideally, the pipe should expand so that v/L, where L is the length of the pipe, is independent of temperature.

57 (*a*) 40.0 cm, (*b*) 480 N, (*c*) You should place your finger 9.2 cm from the scroll bridge.

59 (*a*) 75 Hz, (*b*) 5th and 6th, (*c*) 2.0 m

61 (*a*) 0.574 g/m, (*b*) $\mu_E = 1.29$ g/m, $\mu_A = 2.91$ g/m, $\mu_D = 2.91$ g/m, $\mu_G = 6.57$ g/m

63 (*a*) The two sounds produce a beat because the third harmonic of the A string equals the second harmonic of the E string, and the original frequency of the E string is slightly greater than 660 Hz. (*b*) 662 Hz

65 (*a*) Because the frequency is fixed, the wavelength depends only on the tension on the string. This is true because the only parameter that can affect the wave speed on the string is the tension on the string. The tension on the string is provided by the weight hanging from its end. Given that the length of the string is fixed, only certain wavelengths can resonate on the string. Thus, because only certain wavelengths are allowed, only certain wave speeds will work. This, in turn, means that only certain tensions, and therefore weights, will work.
(*b*) Higher frequency modes on the same length of string results in shorter wavelengths. To accomplish this without changing frequency, you need to reduce the wave speed. This is accomplished by reducing the tension in the string. Because the tension is provided by the weight on the end of the string, you must reduce the weight.
(*d*) $w_1 = 19.2$ N, $w_2 = 4.80$ N, $w_3 = 2.13$ N

67 (*a*) $\Delta t = N/f_0$, (*b*) $\lambda \approx \Delta x/N$, (*c*) $k = 2\pi N/\Delta x$, (*d*) N is uncertain because the waveform dies out gradually rather than stopping abruptly at some time; hence, where the pulse starts and stops is not well defined.

69 6.74 m

71 (*a*) 1.9 cm, 3.6 m/s, (*b*) 0, 0, (*c*) −1.2 cm, −2.2 m/s, (*d*) 0, 0

73 98.0 Hz

75 (*a*) The pipe is closed at one end. (*b*) 262 Hz, (*c*) 32.7 cm

77 (*a*) $y_1(x,t) = (0.010 \text{ m})\sin[(\tfrac{1}{2}\pi \text{ m}^{-1})x - (40\pi \text{ s}^{-1})t]$,
$y_2(x,t) = (0.010 \text{ m})\sin[(\tfrac{1}{2}\pi \text{ m}^{-1})x + (40\pi \text{ s}^{-1})t]$,
(*b*) 2.00 m, (*c*) $v_y(1.0 \text{ m},t) = -(2.5 \text{ m/s})\sin[(40\pi \text{ s}^{-1})t]$,
(*d*) $a_y(1.0 \text{ m},t) = -(0.32 \text{ km/s}^2)\cos[(40\pi \text{ s}^{-1})t]$

79 $y_{res}(x,t) = (10.0 \text{ cm})\sin(kx - \omega t)$

81 (*b*) 203.4 Hz, (*c*) 203.4 Hz

83 812 Hz

85 (*a*)

Cell	Content/Formula	Algebraic Form
A6	A5+0.1	$x + \Delta x$
B4	2*B3+1	$2n + 1$
B5	(−1)^B$3*COS(B$4*$A5)/B$4*4/PI()	$\frac{4}{\pi}\sum_{n=0}^{\infty}\frac{(-1)^n \cos((2n+1)x)}{2n+1}$
C5	B5+(−1)^C$3*COS(C$4*$A5)/C$4*4/PI()	$\frac{4}{\pi}\sum_{n=0}^{\infty}\frac{(-1)^n \cos((2n+1)x)}{2n+1}$

	A	B	C	D	K	L
1						
2						
3		0	1	2	9	10
4		1	3	5	19	21
5	0.0	1.2732	0.8488	1.1035	0.9682	1.0289
6	0.1	1.2669	0.8614	1.0849	1.0134	0.9828
134	12.9	1.2030	0.9740	0.9493	0.9691	1.0146
135	13.0	1.1554	1.0422	0.8990	1.0261	0.9685

The solid curve is plotted from the data in columns A and B and is the graph of $f(x)$ for 1 term. The dashed curve is plotted from the data in columns A and F and is the graph of $f(x)$ for 5 terms. The dotted curve is plotted from the data in columns A and K and is the graph of $f(x)$ for 10 terms.

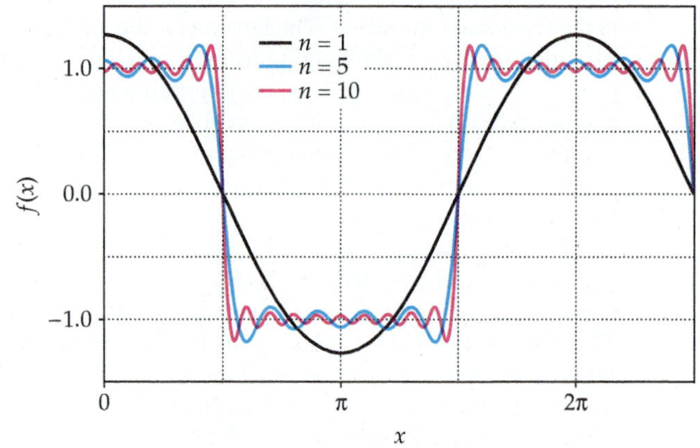

(*b*) It is equivalent to the Leibnitz formula.

87 (*b*)

Cell	Content/Formula	Algebraic Form
B1	90	L
B2	1	r
B3	340	c
B8	B7+1	$n + 1$
C7	2/$B$3*((2*(B7−1)*$B$2)^2+$B$1^2)^0.5	Δt_n

	A	B	C	D
1	L =	90	m	
2	r =	1	m	
3	c =	343	m/s	
6		n	t(n)	delta t(n)
7		1	0.5248	0.0001
8		2	0.5249	0.0004
206		200	2.3739	0.0114

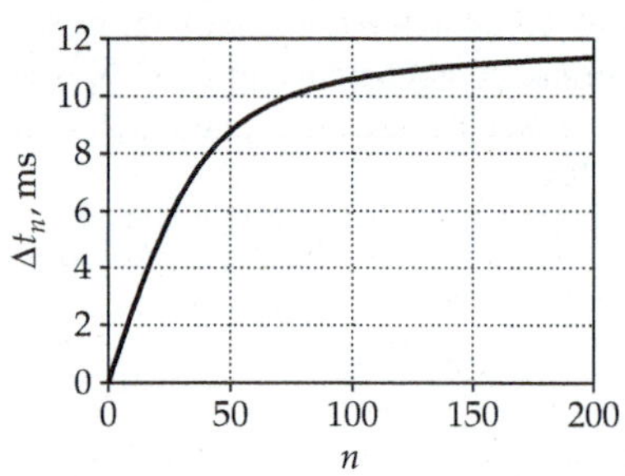

(*c*) The frequency heard at any time is $1/\Delta t_n$, so because Δt_n increases over time, the frequency of the culvert whistler decreases. $f_{highest} = 7.72$ kHz, $f_{lowest} = 85.5$ Hz

Chapter 17

1 (*a*) False, (*b*) False, (*c*) True

3 Mert's room was colder.

5 From the ideal-gas law, we have $P = nRT/V$, and V/T is the slope of the line from the origin to the point (T, V) on the graph. During the process the slope of the line from the origin to (T, V) continuously decreases, so the pressure continuously increases.

7 (*d*)

9 The average kinetic energies are equal. The ratio of their rms speeds is equal to the square root of the reciprocal of the ratio of their molecular masses.

11 False

13 It does not matter.

15 (*b*)

17 The rms speed is always somewhat greater than the speed of sound. However, it is only the component of the molecular velocities in the direction of propagation that is relevant to this issue. In addition, in a gas the mean free path is greater than the average intermolecular distance.

19 If the volume decreases the pressure increases because more molecules hit a unit of area of the walls in a given time. This happens because the number of molecules per unit volume increases as the volume decreases.

21 The average molecular speed of He gas at 300 K is about 1.4 km/s, so a significant fraction of He molecules have speeds in excess of Earth's escape velocity (11.2 km/s). Thus, they "leak" away into space. Over time, the He content of the atmosphere decreases to almost nothing.

23 About 1.2 kg/m^3

25 (*a*) 3600 K, (*b*) 230 K, (*c*) Because hydrogen is lighter than air, it rises to the top of the atmosphere. Because the temperature is high there, a greater fraction of the molecules reach escape speed. (*d*) 160 K, 10 K. Because g is less on the moon, the escape speed is lower. Thus, a larger percentage of the molecules are moving at escape speed.

27 (*a*) 1.23 km/s, (*b*) 310 m/s, (*c*) 264 m/s, (*d*) Because v_e is greater than v_{rms} for O_2, CO_2, and H_2, all three gases should be found on Jupiter.

29 (*a*) 2×10^{11} atm, (*b*) $v_{rms\ protons} = 5 \times 10^5$ m/s, $v_{rms\ electrons} = 2 \times 10^7$ m/s

31 1×10^{-4} g

33 1063°C

35 (*a*)

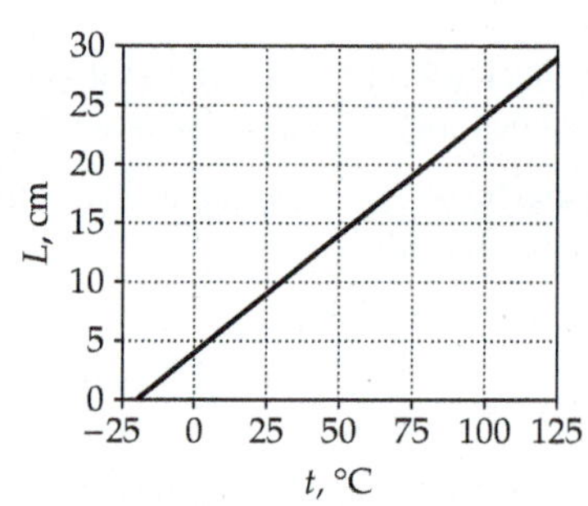

(*b*) 8.40 cm, (*c*) 107°C

37 −320°F

39 (*a*)

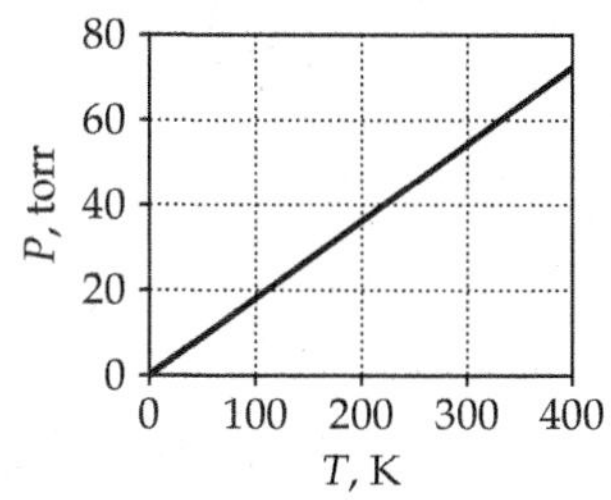

(*b*) 54.9 torr, (*c*) 3.70×10^3 K

41 $-40.0°C = -40.0°F$

43 $-183°C$, $-297°F$

45 (*a*) $R_0 = 3.91 \times 10^{-3}$ K, $B = 3.94 \times 10^3$ K, (*b*) 1.31 kΩ, (*c*) −389 Ω/K, −4.33 Ω/K, (*d*) The thermistor has greater sensitivity at lower temperatures.

47 1.79 mol, 1.08×10^{24} molecules

49 −83 *glips*

51 (*a*) 3.7×10^3 mol, (*b*) 60 mol

53 11.1 atm

55 (*a*) Air will be less dense when its water-vapor content is higher. (*b*) 18 g

57 1.1 kN

59 (*a*) 0.28 km/s, (*b*) 0.87 km/s. The rms speed of argon atoms is slightly less than one-third the rms speed of helium atoms.

61 5.0×10^5 m/s, 2.1×10^{-16} J

65 $K/\Delta U = 7.9 \times 10^4$

69 (*a*) $E_{peak} = \frac{1}{2}kT$, $E_{peak} = \frac{1}{3}E_{av}$

(*b*)

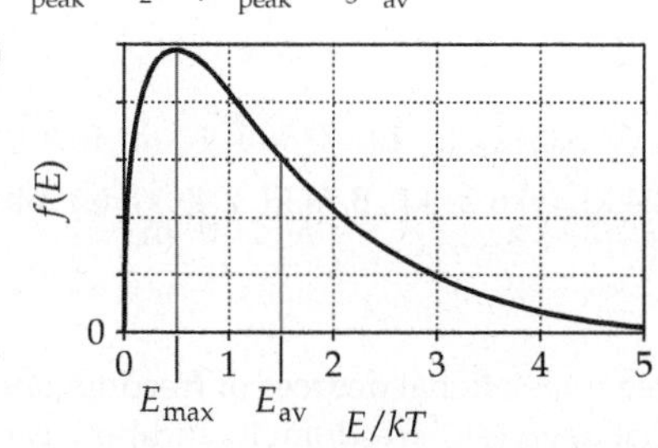

E_{max} is the most probable energy

(*c*) The graph rises from zero to the peak much more rapidly than it falls off to the right of the peak. Because the distribution is so strongly skewed to the right of the peak, the outlying molecules with relatively high energies pull the average ($3kT/2$) far to the right of the most probable value ($kT/2$).

71 (*a*) 1.2×10^2 K, (*b*) 2.4×10^2 K, (*c*) 1.4 atm

73 (*a*) To escape from the surface of a droplet of water, molecules must have enough translational kinetic energy to overcome the attractive forces from their neighbors. Therefore the molecules that escape will be those that are moving faster, leaving the slower molecules behind. The slower molecules have less kinetic energy, so the temperature of the droplet, which is proportional to the average translational kinetic energy per molecule, decreases.
(*b*) As long as the temperature is not too high, the molecules that evaporate from a surface will be only those with the most extreme speeds, at the high-energy "tail" of the Maxwell–Boltzmann distribution. Within this part of the distribution, increasing the temperature only slightly can greatly increase the percentage of molecules with speeds above a certain threshold. For example, suppose that we set an initial threshold at $E = 5kT_1$, then imagine increasing the temperature by 10%, so $T_2 = 1.1T_1$. At the threshold, the ratio of the new energy distribution to the old one is

$$\frac{F(T_2)}{F(T_1)} = \left(\frac{T_1}{T_2}\right)e^{-E/kT_2}e^{+E/kT_1} = (1.1)^{-3/2}e^{-5/1.1}e^5 = 1.365$$

an increase of almost 37%.

75 110 mol of H_2, 55 mol of O_2

77 $4m$

79 (*a*) 142 ms, (*b*) 146 ms

81 (*a*)

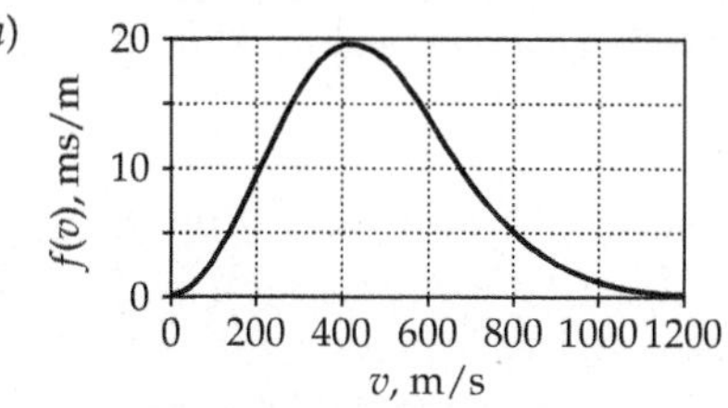

(*b*) As the temperature is increased, the graph spreads out horizontally and gets shorter vertically. More precisely, the horizontal position of the peak moves to the right in proportion to the square root of the temperature, while the height of the peak drops by the same factor, preserving the total area under the graph (which must be 1.0, the total probability of a molecule having any velocity between zero and infinity).
(*c*) A graph of $F(v)$ for nitrogen at 300 K follows. Each number in column C of the spreadsheet [shown in Part (*a*)] is approximately equal to the integral of $f(v)$ from zero up to the corresponding v value. This integral represents the probability of a molecule having a speed greater than or equal to this value of v.

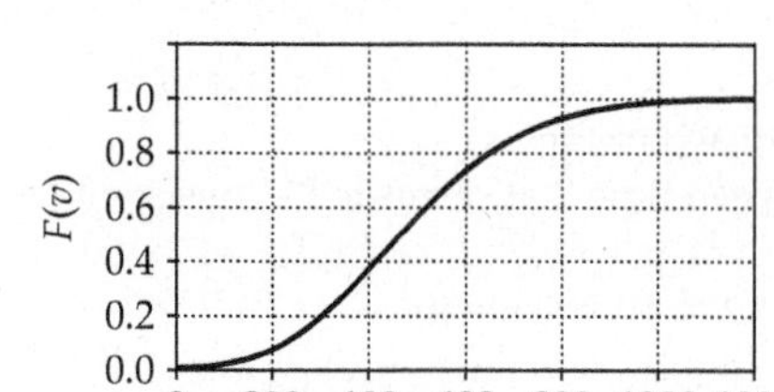

$F(v) = \int_0^v f(v')dv'$

(*d*) About 7%. Note that this value is consistent with the graph of $F(v)$ shown immediately above.
(*e*) A little under 14%.

Chapter 18

1 (*e*)
3 (*c*)
5 (*a*)
7 (*c*)
9 Yes. $\Delta E_{int} = Q_{in} + W_{on}$. If the gas does work at the same rate that it absorbs heat, its internal energy will remain constant.
11 Particles that attract each other have more potential energy the farther apart they are. In a real gas the molecules exert weak attractive forces on each other. These forces increase the internal potential energy during an expansion. An increase in potential energy means a decrease in kinetic energy, and a decrease in kinetic energy means a decrease in translational kinetic energy. Thus, there is a decrease in temperature.
13 Particles that repel each other have more potential energy the closer together they are. The repulsive forces decrease the internal potential energy during an expansion. A decrease in potential energy means an increase in kinetic energy, and an increase in kinetic energy means an increase in translational kinetic energy. Thus, there is an increase in temperature.
15 (*a*)
17 (*a*) False, (*b*) False, (*c*) True, (*d*) True, (*e*) True, (*f*) True
19 (*d*)
21 During a reversible adiabatic process, PV^{γ} is constant, where $\gamma > 1$, and during an isothermal processes, PV is constant. Thus the pressure rise during the compression is greater than the pressure drop during the expansion. The final process could be a constant volume process during which heat is absorbed from the system. A constant-volume cooling will decrease the pressure and return the gas to its original state.

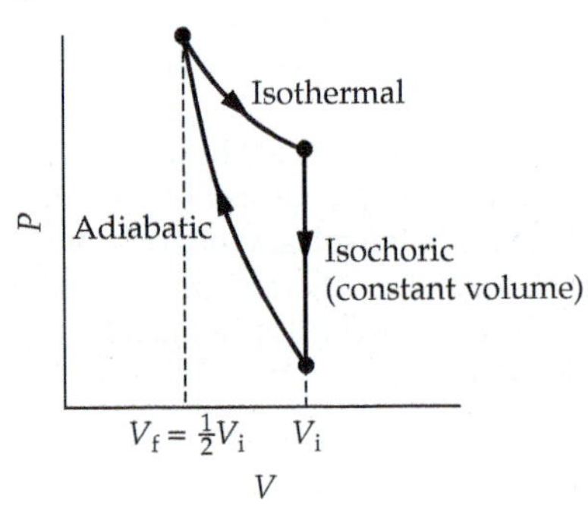

23 The temperature decreases.
25 1.6 min, an elapsed time that seems to be consistent with experience.
27 1.2×10^{-5} (or 1.2×10^{-3} percent)
29 31.3 kJ
31 48.8 g
33 12.1°C
35 4.5×10^2 kg
37 (*a*) 0°C, (*b*) 125 g
39 (*a*) 4.9°C, (*b*) No ice is left.
41 (*a*) 5.26°C, (*b*) 175 g, (*c*) No
43 618°C
45 2.20 kJ
47 54 J
49 (*a*) 6.13 W, (*b*) 38.1 min
51 (*a*) 405 J

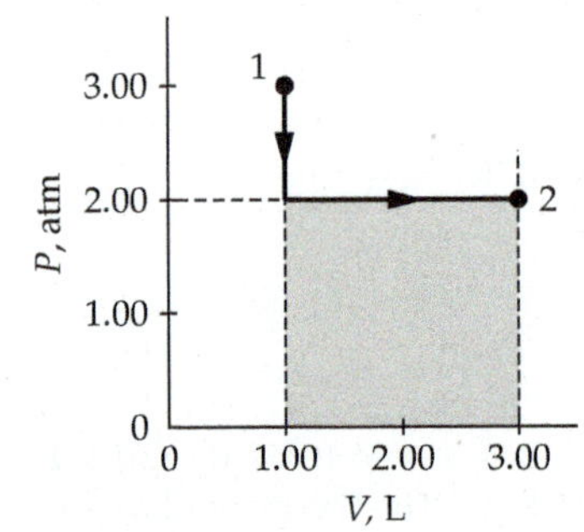

(*b*) 861 J

53 (*a*) 507 J

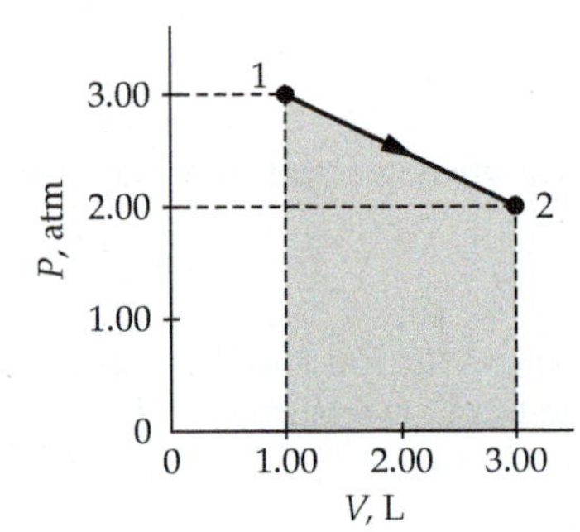

(*b*) 963 J

55 $W_{\text{by gas}} = \frac{3}{2}P_0V_0$
57 (*a*) 555 J, (*b*) 555 J
59 (*a*) 0.495 mol, (*b*) 3.09 kJ, (*c*) 20.8 J/mol·K, (*d*) 10.3 J/K
61 (*a*) 6.24 kJ, 6.24 kJ, (*b*) 6.24 kJ, 8.73 kJ, 2.49 kJ, (*c*) 2.49 kJ
63 59.6 L
65 $\Delta c'_p = -\frac{13}{2}R$
67 There are three translational degrees of freedom and three rotational degrees of freedom. In addition, each of the hydrogen atoms can vibrate against the oxygen atom, resulting in an additional 4 degrees of freedom (2 per atom). $C_{\text{v water}} = 5Nk = 5nR$
69 (*a*) 300 K, 7.80 L, 1.14 kJ, 1.14 kJ, (*b*) 5.40 L, 208 K, 0, 574 J
71 (*a*) 263 K, (*b*) 10.8 L, (*c*) 1.48 kJ, (*d*) −1.48 kJ
73 −0.14 kJ
75

Process	Q_{in} (kJ)	W_{on} (kJ)
D→A	8.98	0
A→B	13.2	−13.2
B→C	−8.98	0
C→D	−6.58	6.58

$W_{\text{by gas total}} = 6.6$ kJ

77 (*a*)

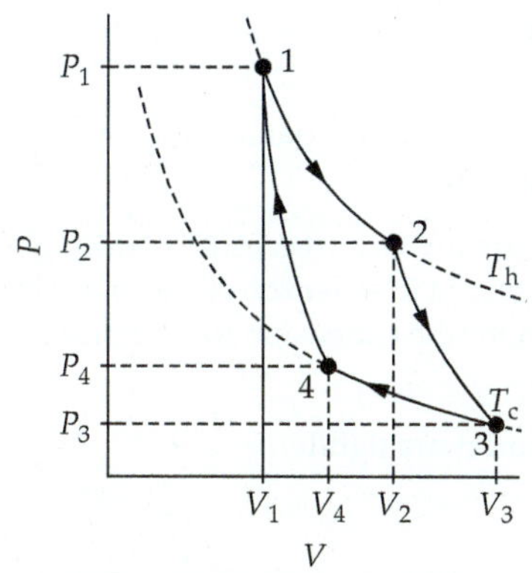

79 (*a*) 65 K, 81 K, (*b*) 1.6 kJ, (*c*) 2.2 kJ

81 (*a*) 81 K, 81 K, (*b*) 2.7 kJ, (*c*) 3.3 kJ

83 256 kcal

85 (*a*) $c(4.00\ \text{K}) = 9.20 \times 10^{-2}\ \text{J/kg}\cdot\text{K}$, (*b*) 0.0584 J/kg

87 (*a*) 2.49 kJ, (*b*) 3.20 kJ

89 171 K

91

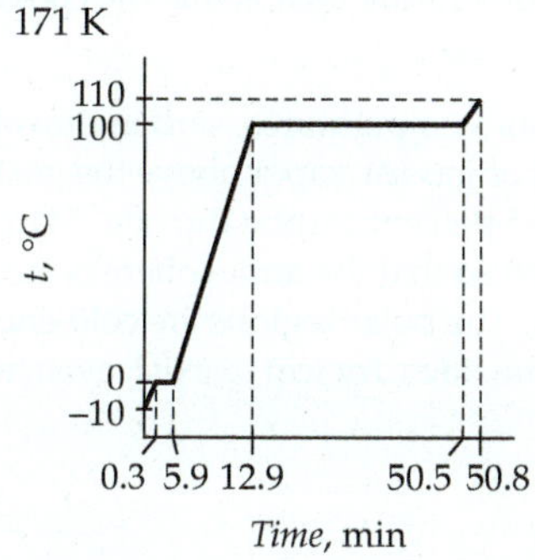

93 396 K

95 (*b*) 4.62 kJ

97 (*a*) $P_2 = \frac{1}{2}P_0$, (*b*) The gas is diatomic. (*c*) During the isothermal process the translational kinetic energy is unchanged. During the adiabatic process the translational kinetic energy increases by a factor of 1.32.

Chapter 19

1 (*c*)

3 (*a*)

5 The COP is defined so as to be a measure of the effectiveness of the device. For a refrigerator or air conditioner, the important quantity is the heat transferred from the already colder interior, Q_c. For a heat pump, the idea is to focus on the heat transferred into the warm interior of the house, Q_h.

7 Increasing the temperature of the steam increases its energy content. In addition, it increases the Carnot efficiency, and generally increases the efficiency of any heat engine.

9 A Carnot-cycle refrigerator is more efficient when the temperatures are close together because it requires less work to extract heat from an already cold interior if the temperature of the exterior is close to the temperature of the interior of the refrigerator. A Carnot-cycle heat engine is more efficient when the temperature difference is large because then more work is done by the engine for each unit of heat absorbed from the hot reservoir.

11 (*c*)

13 (*d*)

15 The cycle is that of the Otto engine (see Figure 19-3).

17

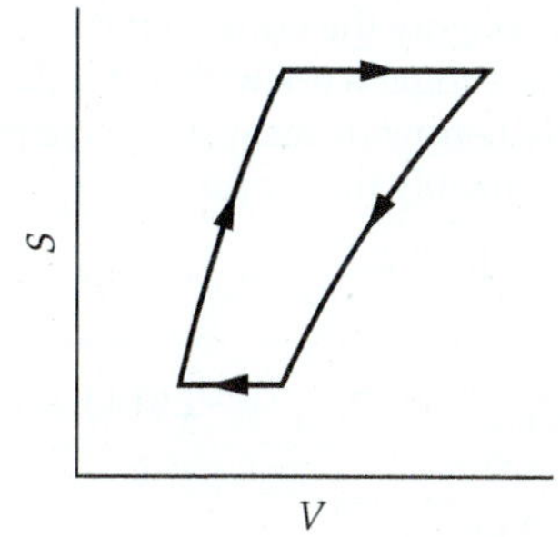

19

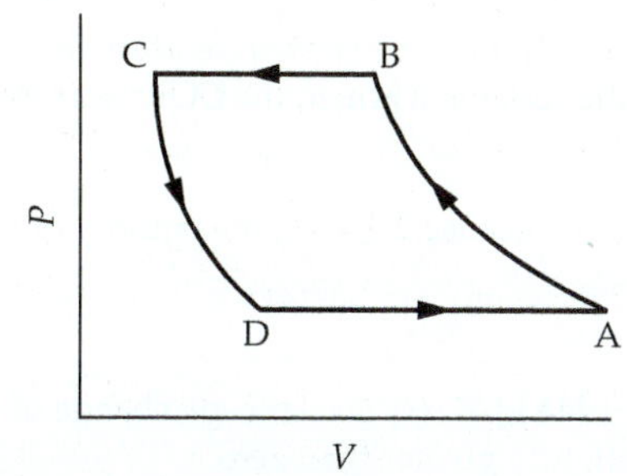

21 An increase of about 47%.

23 56%

25 (*a*) 1.7×10^{17} W, (*b*) 6.02×10^{14} J/(K·s)

27 (*a*) 500 J, (*b*) 400 J

29 (*a*) 40%, (*b*) 80 W

31 (*a*)

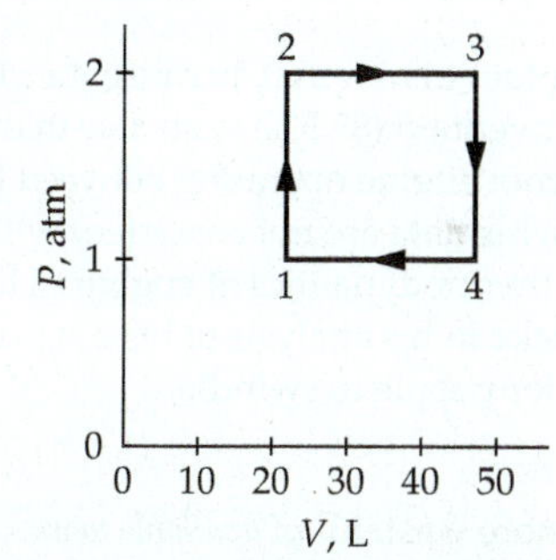

Process	W (kJ)	Q (kJ)	ΔE_{int} (kJ)
1 → 2	0	3.74	3.74
2 → 3	−4.99	12.5	7.5
3 → 4	0	−7.48	−7.48
4 → 1	2.49	−6.24	−3.75

(*b*) 15%

33 13.1%

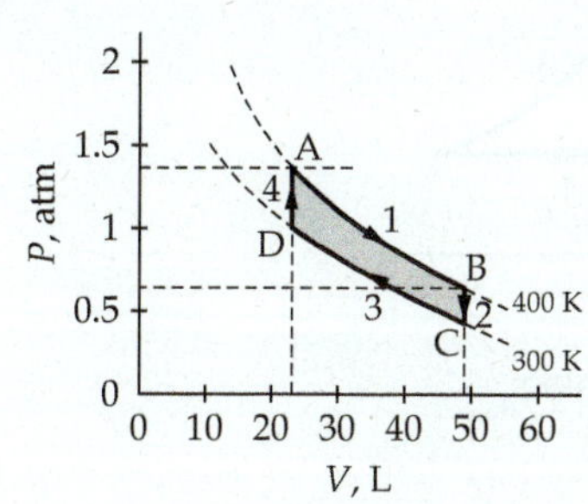

35 (*a*) $T_2 = 600$ K, $T_3 = 1800$ K, $T_4 = 600$ K, (*b*) 15%

37 (*a*) 5.16%. No contradiction. (*b*) Most warm-blooded animals survive under roughly the same conditions as humans. To make a heat engine work with appreciable efficiency, internal body temperatures would have to be maintained at an unreasonably high level.

41 (*a*) 33.3%, (*b*) 33.3 J, (*c*) 67 J, (*d*) 2.0

45 (*a*) 33%

47 (*a*) 100°C, (*b*) 3.12 kJ, $Q_{2\rightarrow3} = 0$, $Q_{3\rightarrow1} = -2.91$ kJ, (*c*) 6.7%, (*d*) 35.5%

49 (*a*) 6.3, (*b*) 3.2 kW, (*c*) 5.3 kW

51 (*a*) 0.17 MJ, (*b*) 0.12 MJ. Because the temperature difference increases when the room is warmer, the COP decreases.

53 6.05 kJ/K

55 $\Delta S_u = 2.40$ J/K and, because $\Delta S_u > 0$, the entropy of the universe increases.

57 (*a*) 0, (*b*) 267 K

59 (*a*) 244 kJ/K, (*b*) −244 kJ/K, (*c*) The entropy change of the universe is just slightly greater than zero.

61 (*a*) −117 J/K, (*b*) 138 J/K, (*c*) 20 J/K

63 (*a*) 0.42 J/K, (*b*) 125 J

65 (*a*) $W_{\text{cycle}} = 20.0$ J, (*b*) $Q_{\text{h cycle}} = 67$ J, $Q_{\text{c cycle}} = 47$ J

67 (*a*) 51%, (*b*) 0.10 MJ, (*c*) 98 kJ

69 113 W/K

71 (*a*) You should explain to him that, because the efficiency he claims for his invention (83.3%) is greater than the efficiency of a Carnot engine operating between the same two temperatures, his data are not consistent with what is known about the thermodynamics of engines. He must have made a mistake in his analysis of his data-or he is a con man looking for people to swindle.

(*b*) 135 W

73 (*a*) Process (2) is more wasteful of *available work*. (*b*) $\Delta S_1 = 1.67$ J/K, $\Delta S_2 = 0.833$ J/K

75 313 K

77 10 W

79 (*a*) 253 kPa, (*b*) 462 K, (*c*) 6.97 kJ

81 (*a*) 253 kPa, (*b*) 416 K, (*c*) 6.59 kJ

83 (*a*)

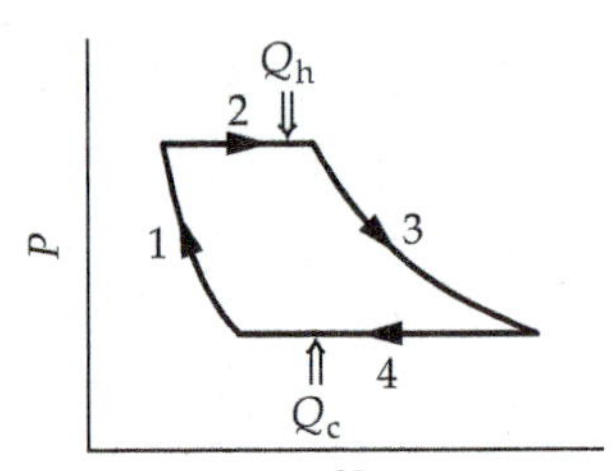

89 $T = 10^{484}$ y, $T \approx 10^{478} T_{\text{Russell}}$

Chapter 20

1 The glass bulb warms and expands first, before the mercury warms and expands.

3 Water expands greatly as it freezes. If a sealed glass bottle full of water is placed in a freezer, as the water freezes there will be no room for the expansion to take place. The bottle will be broken.

5 The strip will curl more tightly.

7 (*c*)

9 (*a*) With increasing altitude, *P* decreases; from curve OC, the temperature *T* of the liquid–gas interface decreases as the pressure decreases, so the boiling temperature decreases. Likewise, from curve OB, the melting temperature increases with increasing altitude.

(*b*) Boiling at a lower temperature means that the cooking time will have to be increased.

11 At very low pressures and temperatures, carbon dioxide can exist only as a solid or gas (or vapor above the gas). The atmosphere of Mars is 95 percent carbon dioxide. Mars, on average, is warm enough so that the atmosphere is mostly gaseous carbon dioxide. The polar regions are cold enough to enable solid carbon dioxide (dry ice) to exist, even at the low pressure.

13 (*a*)

15

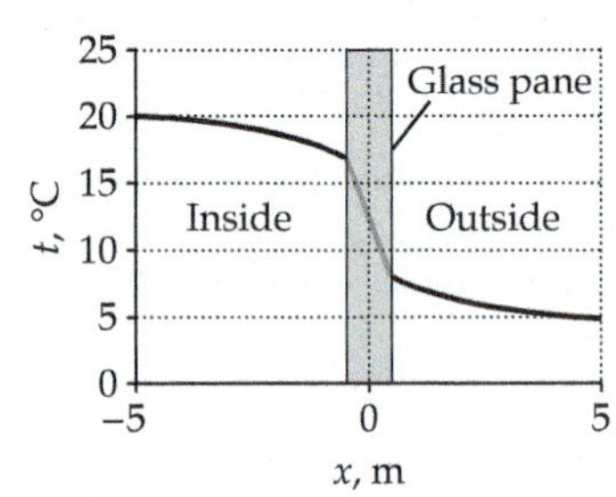

17 Your assumption was not correct and 14 mL of water overflowed.

19 17 mW/(m · K)

21 0.30 kW, 0.1 K/W

23 2.9 nm

25 8(°F · h · ft²)/Btu

27 220°C

29 15×10^{-6} K^{-1}

31 3.7×10^{-12} N/m²

33 (*a*) 90°C, (*b*) 82°C, (*c*) 170 kPa

35 2.1 kBtu/h

37 (*a*) $I_{\text{Cu}} = 0.96$ kW, $I_{\text{Al}} = 0.57$ kW, (*b*) 1.53 kW, (*c*) 0.052 K/W

41 1.3 mm

43 93.5 cm^2

45 1598°C

47 2.1 km

49 5800 K

51 (*b*) The values agree to within 0.3%.

53 1.3×10^{10} kW. About 0.007 percent.

55 142 W

57 $L_2 = L_1$, $\omega_2 \approx (1 - 2\alpha\,\Delta T)\omega_1$, $E_2 = E_1(1 - 2\alpha\,\Delta T)$

59 (*a*) 0.70 cm/h, (*b*) 12 d

Chapter 21

1 The net charge on large objects is always very close to zero. Hence the most obvious force is the gravitational force.

3 (*a*) Coulomb's law is only valid for point particles. The paper bits cannot be modeled as point particles because the paper bits become polarized.

(*b*) No, the attraction does not depend on the sign of the charge on the comb. The induced charge on the paper that is closest to the comb is always opposite in sign to the charge on the comb, and thus the net force on the paper is always attractive.

5 (*a*)

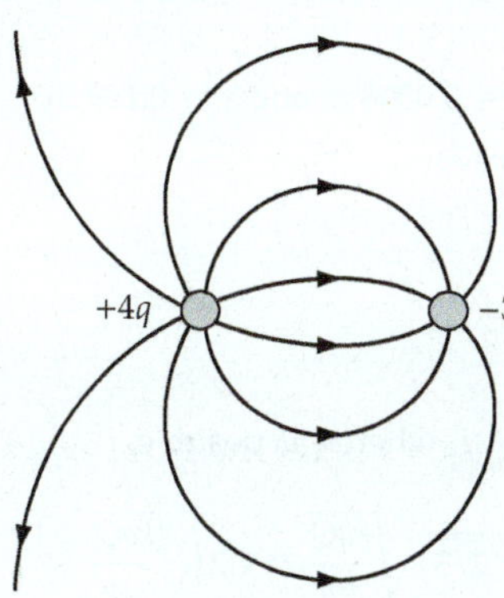

(*b*)

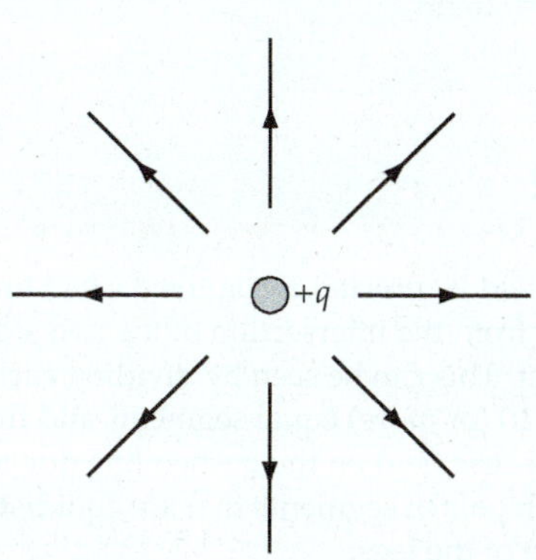

7 Assume that the rod has a negative charge. When the charged rod is brought near the aluminum foil, it induces a redistribution of charges with the side nearer the rod becoming positively charged, and so the ball of foil swings toward the rod. When it touches the rod, some of the negative charge is transferred to the foil, which, as a result, acquires a net negative charge and is now repelled by the rod.

9 (*a*) On the sphere near the positively charged rod, the induced charge is negative and near the rod. On the other sphere, the net charge is positive and on the side far from the rod. This is shown in the diagram.

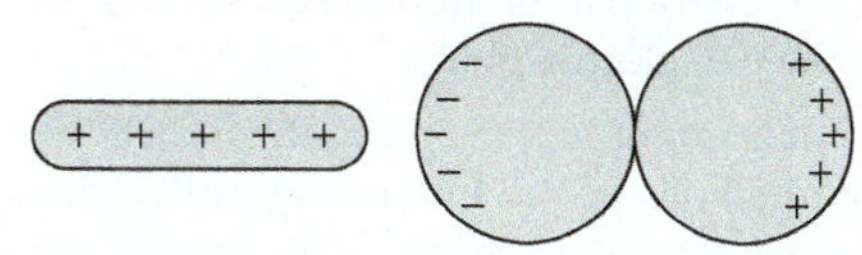

(*b*) When the spheres are separated and far apart and the rod has been removed, the induced charges are distributed uniformly over each sphere. The charge distributions are shown in the diagram.

11 (*a*) False, (*b*) True, (*c*) False, (*d*) Possibly, (*e*) False, (*f*) True

13 (*a*)

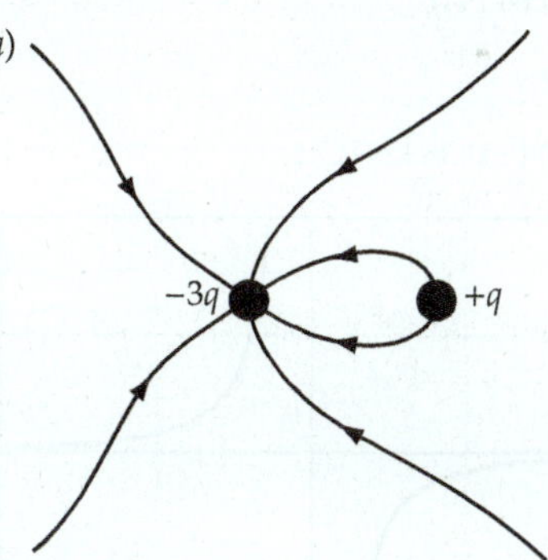

(*b*)

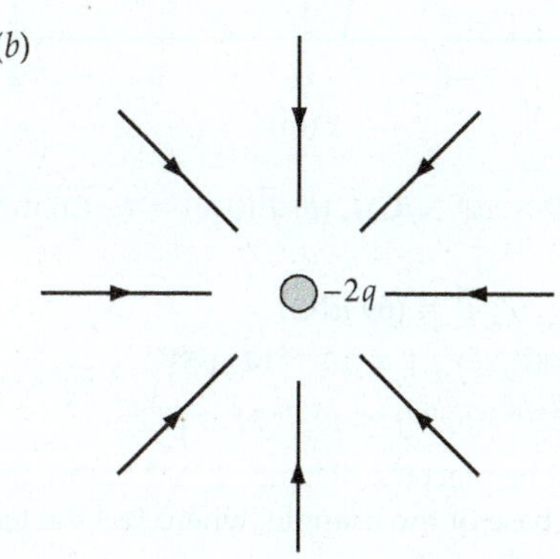

15 The dipole moment rotates back and forth in oscillatory motion. The dipole moment gains angular speed as it rotates toward the direction of the electric field, and loses angular speed as it rotates away from the direction of the electric field.

17

	1	2	3
(*a*)	down	up	up
(*b*)	up	right	left
(*c*)	down	up	up
(*d*)	down	up	up

Figure 21-23 shows the electric field due to a single dipole, where the dipole moment is directed toward the right. The electric field due two a pair of dipoles can be obtained by superposing the two electric fields.

19 Because the can is grounded, the presence of the negatively charged plastic rod induces a positive charge on it. The positive charges induced on the can are attracted, via the Coulomb interaction, to the negative charges on the plastic rod. Unlike charges attract, so the can will roll toward the rod.

21 5.0×10^{12} electrons

23 4.82×10^{7} C

25 (*a*) 2.60 h, (*b*) 2.1×10^{-13} W

27 $\vec{F}_1 = (1.5 \times 10^{-2}\ \text{N})\hat{i}$

29 A distance equal to $0.41L$ from the -2.9-μC charge on the side away from the 4.0-μC charge.

31 $\vec{F}_3 = -(8.65\ \text{N})\hat{j}$

33 $\vec{F}_1 = (0.90\ \text{N})\hat{i} + (1.8\ \text{N})\hat{j}$, $\vec{F}_2 = (-1.3\ \text{N})\hat{i} - (1.2\ \text{N})\hat{j}$, $\vec{F}_3 = (0.4\ \text{N})\hat{i} - (0.64\ \text{N})\hat{j}$

35 $\vec{F}_q = \dfrac{kqQ}{R^2}(1 + \sqrt{2})\hat{i}$

37 (*a*) $(1.0\ \text{kN/C})\hat{i}$, (*b*) $(-0.36\ \text{kN/C})\hat{i}$,

(*c*)

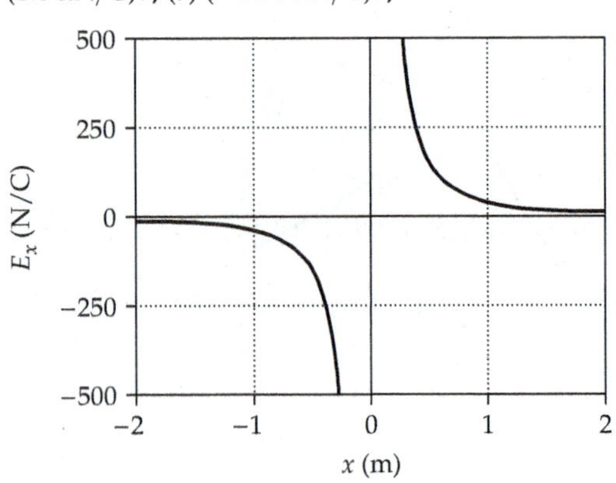

39 (*a*) $\vec{E}(0, 0) = (4.0 \times 10^5\ \text{N/C})\hat{j}$, (*b*) $\vec{F}(0, 0) = (-1.6\ \text{mN})\hat{j}$, (*c*) -40 nC

41 (*a*) 35 kN/C at 0°, (*b*) $\vec{F} = (69\ \mu\text{N})\hat{i}$

43 (*a*) 13 kN/C at 230°, (*b*) 2.1×10^{-15} N at 51°

45 (*a*) 1.9 kN/C at 230°, (*b*) 3.0×10^{-16} N at 230°

47 The charge must be placed a distance $L/\sqrt{3}$ below the midpoint of the base of the triangle, where L is the length of a side of the triangle.

49 (*a*) For a positive test charge, the equilibrium at (0, 0) is unstable for small displacements in either direction along the x axis, and stable for small displacements in either direction along the y axis.
(*b*) For a negative test charge, the equilibrium is stable at (0,0) for displacements along the x axis and unstable for displacements along the y axis.
(*c*) $q_0 = -\frac{1}{4}q$

51 (*a*) 1.76×10^{11} C/kg, (*b*) $1.76 \times 10^{13}\ \text{m/s}^2$ in the direction opposite to the direction of the electric field, (*c*) 0.2 μs, (*d*) 3 mm

53 (*a*) $\vec{a} = (-5.28 \times 10^{13}\ \text{m/s}^2)\hat{j}$, (*b*) 50.0 ns, (*c*) 33.4° in the $-y$ direction

55 800 μC

57 The electron strikes the lower plate 4.1 cm to the right of its initial position.

59 (*a*) $8.0 \times 10^{-18}\ \text{C}\cdot\text{m}$
(*b*)

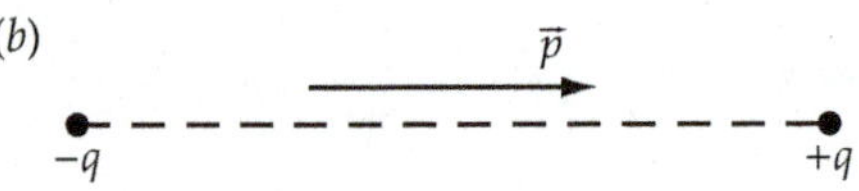

63 (*a*) 1.83×10^6 N/C, (*b*) 1.80×10^6 N/C. The exact and estimated values of E_p agree to within 2%. This difference is this large because the separation of the two charges of the dipole is 20% of the distance from the center of the dipole to point P.

67 (*a*) 1.8×10^{-5} C and 1.8×10^{-4} C
(*b*) -1.4×10^{-5} C and 2.1×10^{-4} C

69 (*a*) 0.225 N, downward, (*b*) 0.112 N·m, counterclockwise, (*c*) 45.8 g, (*d*) 5.00×10^{-7} C

71 (*a*) 28.0 μC and 172 μC
(*b*) 250 N

73 (*a*) $-97.2\ \mu$C, (*b*) $x = 0.0508$ m and $x = 0.169$ m

75 (*a*) 10°, (*b*) 9.9° for each

79 $v = e\sqrt{k/(2mL)}$

83 (*a*) $E_y = \dfrac{2kQy}{[y^2 + \frac{1}{4}a^2]^{3/2}}$,

(*b*) $\vec{F} = \dfrac{2kqQy}{[y^2 + \frac{1}{4}a^2]^{3/2}}\hat{j}$, where q is positive,

(*c*) $v = \sqrt{8\left(1 - \sqrt{2/3}\right)}\sqrt{\dfrac{kqQ}{am}} = 1.21\sqrt{\dfrac{kqQ}{am}}$

85 4.6×10^{-14} m = 46 fm

87 (*b*) 52 μm/s

Chapter 22

1 The resultant field is directed along the dashed line, pointing away from the intersection of the two sides of the L-shaped object. This can be seen by dividing each leg of the object into 10 (or more) equal segments and then drawing the electric field on the dashed line due to the charges on each pair of segments that are equidistant from the intersection of the legs.

3 (*a*) True (assuming there are no charges inside the shell), (*b*) True, (*c*) False

5 (*a*) False, (*b*) True

7 (*a*) False, (*b*) False, (*c*) True, (*d*) False, (*e*) True

9 (*a*) radially inward, (*b*) radially outward, (*c*) radially inward

11 (*a*) radially inward, (*b*) radially inward, (*c*) The field is zero.

13 (*a*) 18 nC, (*b*) 26 N/C, (*c*) 4.4 N/C, (*d*) 2.6 mN/C, (*e*) This result is about 0.01% less than the exact value obtained in (*d*).

15 (*a*) 4.7×10^5 N/C, (*b*) 1.1×10^6 N/C, (*c*) 1.5×10^3 N/C (*d*) 1.5×10^3 N/C. This result agrees exactly, to two significant figures, with the result obtained in Part (*c*).

17 (*a*) $0.189\ kQ/a^2$, (*b*) $0.358\ kQ/a^2$, (*c*) $0.385\ kQ/a^2$, (*d*) $0.354\ kQ/a^2$, (*e*) $0.179\ kQ/a^2$,

(*f*)

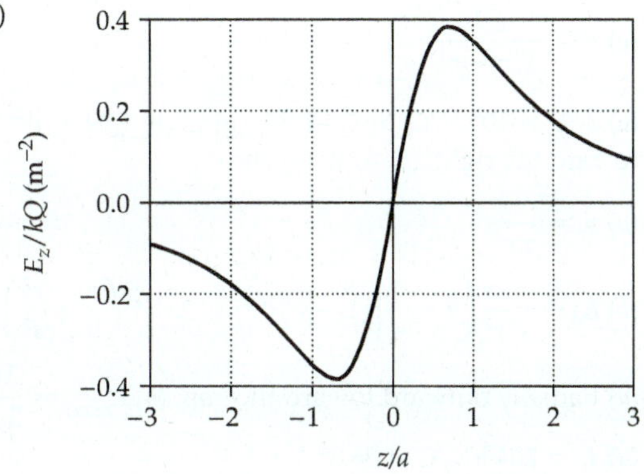

19 (*a*)

Cell	Content/Formula	Algebraic Form
B3	9.00E+09	k
B4	5.00E−10	σ
B5	0.3	r
A8	0	x_0
A9	A8+0.01	$x_0 + 0.01$
B8	2*PI()*$B$3*$B$4*(1−A8/(A8^2+$B$5^2)^2)^0.5)	$2\pi k\sigma\left(1 - \frac{x}{\sqrt{x^2 + a^2}}\right)$
C8	2*PI()*$B$3*$B$4	$2\pi k\sigma$

	A	B	C
1			
2			
3	k =	9.00E+09	N*m2/C2
4	σ =	5.00E−10	C/m2
5	a =	0.300	m
6			
7	x	$E(x)$	E_{sheet}
8	0.00	28.27	28.3
9	0.01	27.33	28.3
77	0.69	2.34	28.3
78	0.70	2.29	28.3

(*b*)

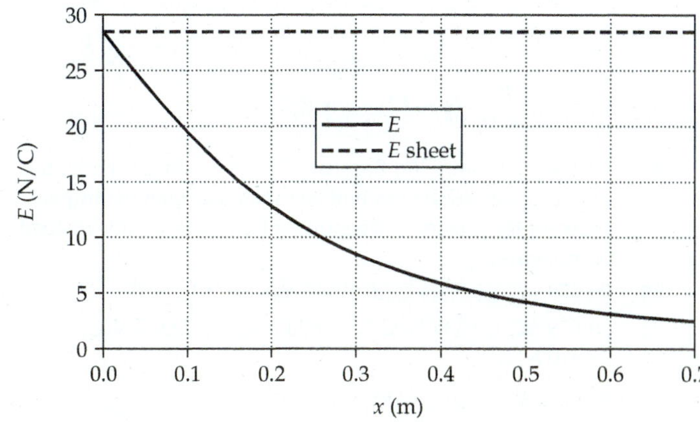

The magnitudes differ by more than 10.0 percent for $x \geq 0.0300$ m.

27 (*a*) 20.0 N·m²/C, (*b*) 17 N·m²/C

29 (*a*) 1.5 N·m²/C, 1.5 N·m²/C, (*b*) 0, (*c*) 3.0 N·m²/C, (*d*) 2.7×10^{-11} C

31 (*a*) 3.14 m², (*b*) 7.19×10^4 N/C, (*c*) 2.26×10^5 N·m²/C, (*d*) No, (*e*) 2.26×10^5 N·m²/C

33 −79.7 nC

35 $\cos\theta$

37 (*a*) $E_{r<R_1} = 0$, $\vec{E}_{R_1<r<R_1} = \frac{kq_1}{r^2}\hat{\boldsymbol{r}}$, $\vec{E}_{r>R_2} = \frac{k(q_1 + q_2)}{r^2}\hat{\boldsymbol{r}}$, (*b*) −1,

(*c*)

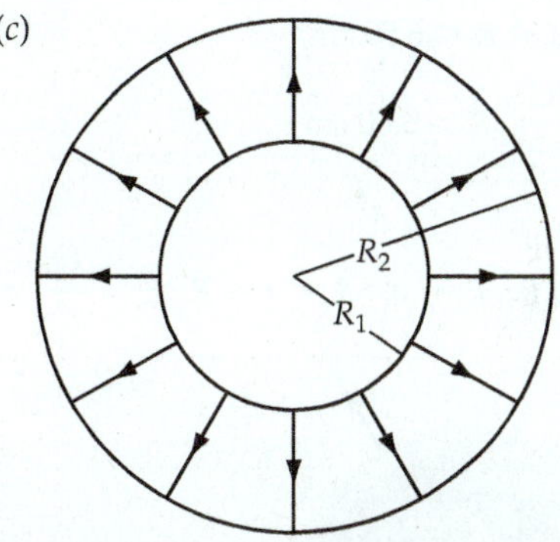

39 (*a*) 0.407 nC, (*b*) 339 N/C, (*c*) 1.00 kN/C, (*d*) 983 N/C, (*e*) 366 N/C

41 (*a*) 2.00 μC/m³, (*b*) 470 N/C

43 (*a*) $Q = 2\pi BR^2$, (*b*) $E_r = \frac{BR^2}{2\epsilon_0 r^2}$ $r > R$, $E_r = \frac{B}{2\epsilon_0}$ $r < R$

(*c*)

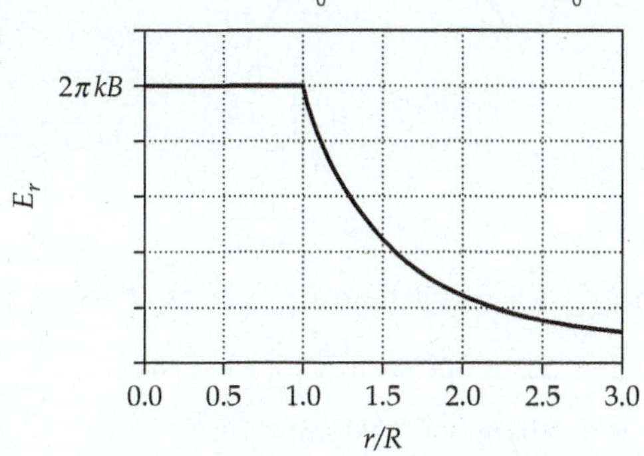

45 (*a*) $Q_{\text{inside}} = \frac{4\pi\rho}{3}(r^3 - R_1^3)$,

(*b*) $E_r = 0 \quad r < R_1, E_r = \frac{\rho}{3\epsilon_0 r^2}(R_2^3 - R_1^3) \quad R_1 < r < R_2,$

$E_r = \frac{\rho}{3\epsilon_0 r^2}(R_2^3 - R_1^3) \quad r > R_2$

47 (*a*) 1.41×10^6 m/s, (*b*) Because of its much larger mass, the impact speed of the ion will be much less than the impact speed of the electron. (The ion will impact the tube instead of the wire.)

49 (*a*) 679 nC, (*b*) 0, (*c*) 0, (*d*) 1.00 kN/C, (*e*) 610 N/C

51 (*a*) 679 nC, (*b*) 339 N/C, (*c*) 1.00 kN/C, (*d*) 1.00 kN/C, (*e*) 610 N/C

53 (*a*) $E_R = 0 \quad r < 1.50$ cm,

$E_R = \frac{(108\ \text{N}\cdot\text{m/C})}{R} \quad 1.50\ \text{cm} < r < 4.50\ \text{cm},$

$E_R = 0 \quad 4.50\ \text{cm} < r < 6.50\ \text{cm},$

$E_R = \frac{156\ \text{N}\cdot\text{m/C}}{R} \quad r > 6.50\ \text{cm},$

(*b*) $\sigma_{\text{inside}} = -21.2\ \text{nC/m}^2$ and $\sigma_{\text{outside}} = 14.7\ \text{nC/m}^2$

55 (*b*) $E_R = \frac{b}{4\epsilon_0}R^3 \quad r < a, E_R = \frac{ba^4}{4R\epsilon_0} \quad r > a$

57 (*a*) 18.8 nC/m, (*b*) $E_R = 22.6$ kN/C $\quad R < 1.50$ cm,

$E_R = \frac{339\ \text{N}\cdot\text{m/C}}{R} \quad 1.50\ \text{cm} < R < 4.50\ \text{cm},$

$E_R = 0 \quad 4.50\ \text{cm} < R < 6.50\ \text{cm},$

$E_R = \frac{339\ \text{N}\cdot\text{m/C}}{R} \quad R > 6.50\ \text{cm}$

59 9.4 kN/C

61 (*a*) $E_r = \frac{kq}{r^2} \quad r < R_1, E_r = 0 \quad R_1 < r < R_2, E_r = \frac{kq}{r^2} \quad r > R_2$

(*b*)

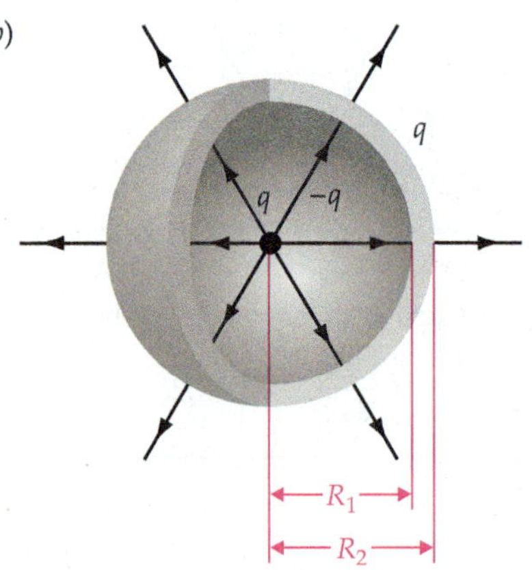

(*c*) $\sigma_{\text{inner}} = -\frac{q}{4\pi R_1^2}, \sigma_{\text{outer}} = \frac{q}{4\pi R_2^2}$

63 (*a*) $\sigma_{\text{inner}} = -0.55\ \mu\text{C/m}^2, \sigma_{\text{outer}} = 0.25\ \mu\text{C/m}^2$

(*b*) $E_r = (2.3 \times 10^4\ \text{N}\cdot\text{m}^2/\text{C})\frac{1}{r^2} \quad r < 60\ \text{cm},$

$E_r = 0 \quad 60\ \text{cm} < r < 90\ \text{cm},$

$E_r = (2.3 \times 10^4\ \text{N}\cdot\text{m}^2/\text{C})\frac{1}{r^2} \quad r > 90\ \text{cm}$

(*c*) $\sigma_{\text{inner}} = -0.55\ \mu\text{C/m}^2, \sigma_{\text{outer}} = 0.59\ \mu\text{C/m}^2,$

$E_r = (2.3 \times 10^4\ \text{N}\cdot\text{m}^2/\text{C})\frac{1}{r^2} \quad r < 60\ \text{cm},$

$E_r = 0 \quad 60\ \text{cm} < r < 90\ \text{cm},$

$E_r = (5.4 \times 10^4\ \text{N}\cdot\text{m}^2/\text{C})\frac{1}{r^2} \quad r > 90\ \text{cm}$

65 (*a*) $Q_{\text{left}} = 15\ \mu\text{C}$ and $Q_{\text{right}} = 65\ \mu\text{C}$,
(*b*) $E_{\text{left}\,x} = -68$ kN/C and $E_{\text{right}\,x} = 294$ kN/C,
$\sigma_{\text{left}} = 0.60\ \mu\text{C/m}^2$ and $\sigma_{\text{right}} = 2.60\ \mu\text{C/m}^2$

67 −115 kN/C

69 (*a*) $E = \frac{Q}{8\pi\epsilon_0 r^2}$, radially outward,

(*b*) $F = \frac{Q^2a^2}{32\pi\epsilon_0 r^4}$, radially outward,

(*c*) $P = \frac{Q^2}{32\pi^2\epsilon_0 r^4}$

71 (*a*) 3.39×10^9 N/C, toward the right, (*b*) 3.39×10^9 N/C, toward the right, (*c*) zero, (*d*) zero

73 (*a*) $\rho_0 = \frac{-e}{\pi a^3}$,

(*b*) $E_r(r) = \frac{ke}{r^2}\left(1 - \frac{1}{4}\left[(1 - e^{-2r/a}) - 2e^{-2r/a}\left(\frac{r}{a} + \frac{r^2}{a^2}\right)\right]\right)$

75 (*a*) Radially outward toward the gap, (*b*) $E_{\text{center}} = \frac{kQ\ell}{2\pi R^3}$

77 (*a*) $\vec{E} = 204$ kN/C at 56.3°,
(*b*) $\vec{E} = 263$ kN/C at 153°

79 (*a*) $v = \sqrt{\frac{2kq\lambda}{m}}$, (*b*) $T = \pi R\sqrt{\frac{2m}{kq\lambda}}$

81 (*a*) 0.997 kg, (*b*) 1.18 Hz

83 (*b*) $\vec{E}_1 = \vec{E}_2 = \frac{\rho b}{3\epsilon_0}\hat{i}$

85 $\vec{E}_1 = \left(\frac{\rho b}{3\epsilon_0} + \frac{Q}{4\pi\epsilon_0 b^2}\right)\hat{i}, \vec{E}_2 = \left(\frac{\rho b}{3\epsilon_0} - \frac{Q}{4\pi\epsilon_0 b^2}\right)\hat{i}$

87 $\frac{1}{2}R$

Chapter 23

1 The proton is moving to a region of higher potential. The proton's electrostatic potential energy is increasing.

3 The electric field is zero throughout the region.

5

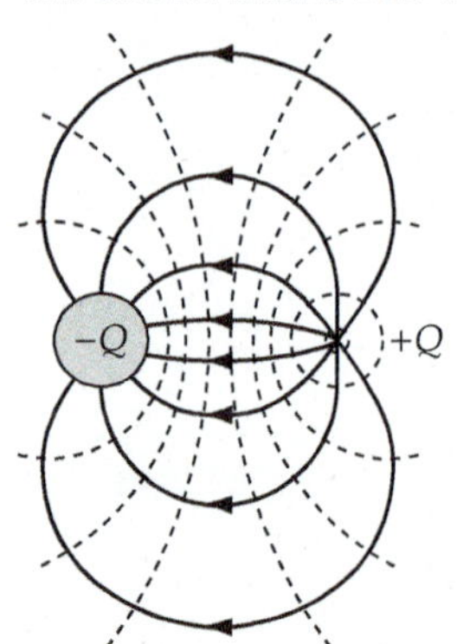

7 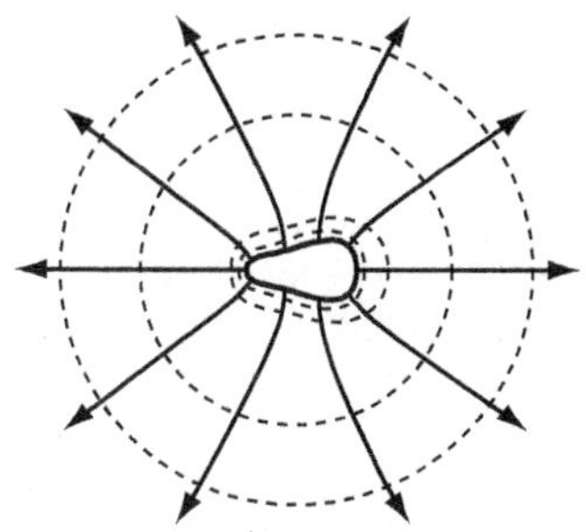

9 (*a*) 2, (*b*) 3

11 No. The local surface charge density is proportional to the normal component of the electric field, not the potential on the surface.

13 3.0×10^9 V

15 0.72 MeV

17 27 μC/m^2

19 (*a*) 4.49 kV, (*b*) 13.5 mJ

21 (*a*) −8.00 kV, (*b*) −24.0 mJ, (*c*) 24.0 mJ, (*d*) $V(x) = -(2.00\ \text{kV/m})x$

23 (*a*) 3.09×10^7 m/s, (*b*) 2.50 MV/m

25 (*a*) $r = kzZe^2/K_i$, (*b*) 46 fm, 25 fm, (*c*) No. The distance of closest approach for a 5-MeV alpha particle found above (45.5 fm) is much larger than the 7 fm radius of a gold nucleus. Hence the scattering was solely the result of the inverse-square Coulomb force.

27 (*a*) 12.9 kV, (*b*) 7.55 kV, (*c*) 4.43 kV

29 (*a*) 135 kV (*b*) 95.3 kV (*c*) Because the two field points are equidistant from all points on the circle, the answers for Parts (*a*) and (*b*) would not change.

31 (*a*) $V(x) = kq\left(\dfrac{1}{|x-a|} + \dfrac{1}{|x+a|}\right)$,

(*b*)

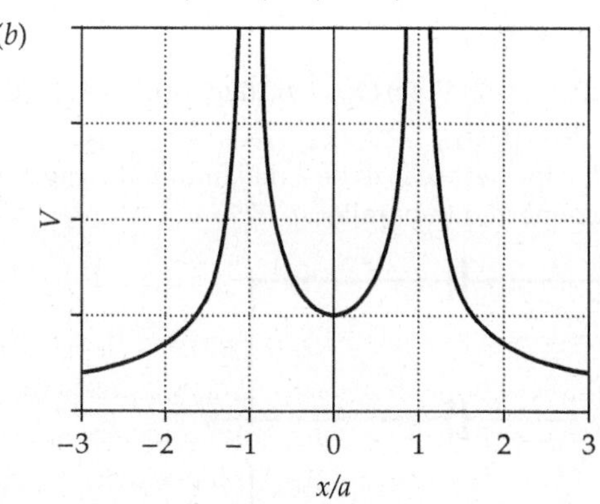

33 (*b*) at points on the *z* axis

35 (*a*) positive, (*b*) 25.0 kV/m

37 (*a*) +668 nC, (*b*) 3.00 kV. The plane at $x = 2.00$ m is at the higher potential.

39 (*a*) $V(x) = kq\left(\dfrac{2}{\sqrt{x^2+a^2}} + \dfrac{1}{|x-a|}\right) \quad x \neq a,$

(*b*) $E_x(x) = \dfrac{2kqx}{(x^2+a^2)^{3/2}} + \dfrac{kq}{(x-a)^2} \quad x > a,$

$E_x(x) = \dfrac{2kqx}{(x^2+a^2)^{3/2}} - \dfrac{kq}{(a-x)^2} \quad x < a$

41 (*a*) 6.02 kV, (*b*) −12.7 kV, (*c*) −42.3 kV

43 ~3×10^{-5} C/m^2

45 $V_a - V_b = \dfrac{2kq}{L}\ln\left(\dfrac{b}{a}\right)$

47 $V_a - V_b = kq\left(\dfrac{1}{a} - \dfrac{1}{b}\right)$

49

Region	$x \le 0$	$0 \le x \le a$	$x \ge a$
Part (*a*)	$\dfrac{\sigma}{\epsilon_0}x$	0	$-\dfrac{\sigma}{\epsilon_0}(x-a)$
Part (*b*)	0	$-\dfrac{\sigma}{\epsilon_0}x$	0

51 (*a*) $V(x,0) = \dfrac{kQ}{L}\ln\left(\dfrac{\sqrt{x^2+\frac{1}{4}L^2}+\frac{1}{2}L}{\sqrt{x^2+\frac{1}{4}L^2}-\frac{1}{4}L^2}\right)$

53 (*a*) $Q = \frac{1}{2}\pi\sigma_0 R^2$, (*b*) $V = \dfrac{2\pi k\sigma_0}{3R^2}\left((R^2 - 2z^2)\sqrt{z^2+R^2} + 2z^3\right)$

55 (*a*) $V(x) = \dfrac{kQ}{L}\ln\left(\dfrac{x+\frac{1}{2}L}{x-\frac{1}{2}L}\right)$

57 (*a*) $dQ = \dfrac{3Q}{R^3}r'^2\,dr'$, (*b*) $dV = \dfrac{3kQ}{R^3}r'\,dr'$, (*c*) $V = \dfrac{3kQ}{2R^3}(R^2 - r^2)$, (*d*) $dV = \left(\dfrac{3kQ}{R^3 r}\right)r'^2\,dr'$, (*e*) $V = \dfrac{kQ}{R^3}r^2$, (*f*) $V = \dfrac{kQ}{2R}\left(3 - \dfrac{r^2}{R^2}\right)$

61 (*a*) The equipotential surfaces are planes parallel to the charged planes. (*b*) The regions to either side of the two charged planes are equipotential regions, so any surface in either of these regions is an equipotential surface.

63 (*a*) 0.224 cm, closer to the wire, (*b*) 0.864 mm, (*c*) The distance between the 700 V and the 725 V equipotentials is 0.0966 mm. This closer spacing of these two equipotential surfaces was to be expected. Close to the central wire, two equipotential surfaces with the same difference in potential should be closer together to reflect the fact that the electric field strength is greater closer to the wire.

65 (*a*) 30.0 mJ, (*b*) −5.99 mJ, (*c*) −18.0 mJ

67 (*a*) 22.3 nC, (*b*) 22.3 μJ

69 $v = q\sqrt{\dfrac{6\sqrt{2}k}{ma}} = 2.91q\sqrt{\dfrac{k}{ma}}$

71 (*a*) 9.61×10^{-20} J, (*b*) 4.59×10^5 m/s, (*c*) Because $2K_{i\,\text{min}} > K_{i\,\text{escape}}$, the electron escapes from the proton with residual kinetic energy.

73 (*a*) $V(x) = \dfrac{2kq}{\sqrt{x^2+a^2}}$, (*b*) $\vec{E}(x) = \dfrac{2kqx}{(x^2+a^2)^{3/2}}\hat{\imath}$

75 (*a*) $V(x,y) = \dfrac{\lambda}{2\pi\epsilon_0}\ln\left(\dfrac{\sqrt{(x-a)^2+y^2}}{\sqrt{(x+a)^2+y^2}}\right)$, $V(0,y) = 0$

(*b*) $y = \pm\sqrt{21.25x - x^2 - 25}$

(*c*)

Cell	Content/Formula	Algebraic Form
A2	1.25	$\frac{1}{4}a$
A3	A2+0.05	$x + \Delta x$
B2	SQRT(21.25*A2−A2^2−25)	$y = \sqrt{21.25x - x^2 - 25}$
B4	−B2	$y = -\sqrt{21.25x - x^2 - 25}$

	A	B	C
1	x	y_{pos}	y_{neg}
2	1.25	0.00	0.00
3	1.30	0.97	−0.97
4	1.35	1.37	−1.37
5	1.40	1.67	−1.67
6	1.45	1.93	−1.93
7	1.50	2.15	−2.15
370	19.65	2.54	−2.54
371	19.70	2.35	−2.35
372	19.75	2.15	−2.15
373	19.80	1.93	−1.93
374	19.85	1.67	−1.67
375	19.90	1.37	−1.37
376	19.95	0.97	−0.97

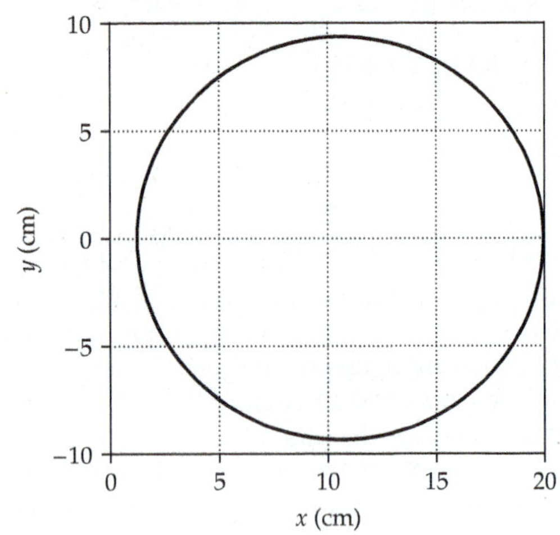

The wires are in the $y = 0$ plane.

77 (*a*) $3.56 \times 10^8\ \text{C/m}^3$,

(*b*) $V(r) = \pi k a^3 \rho_0\left(\frac{1}{a} + \frac{1}{r}\right)e^{-2r/a} = ke\left(\frac{1}{a} + \frac{1}{r}\right)e^{-2r/a}$

79 (*a*) $W_{+Q\to+a} = \frac{kQ^2}{2a}$, (*b*) $W_{-Q\to 0} = \frac{-2kQ^2}{a}$, (*c*) $W_{-Q\to 2a} = \frac{2kQ^2}{3a}$

81 (*a*) 100 eV, (*b*) 1.38×10^5 m/s

83 $R_2 = \frac{2}{3}R_1$

85 7.1 nC

87 (*b*) $\sigma = \frac{qd}{4\pi(d^2 + r^2)^{3/2}}$

89 (*a*) $V(c) = 0$, $V(b) = kQ\left(\frac{1}{b} - \frac{1}{c}\right)$, $V(a) = V(b) = kQ\left(\frac{1}{b} - \frac{1}{c}\right)$,

(*b*) $Q_b = Q$, $V(a) = V(c) = 0$, $Q_a = -Q\frac{a(c-b)}{b(c-a)}$,

$Q_c = -Q\frac{c(b-a)}{b(c-a)}$, $V(b) = kQ\frac{(c-b)(b-a)}{b^2(c-a)}$

93 (*a*) $R' = 0.794R$ (*b*) $\Delta E = -0.370E$

Chapter 24

1 (*c*)

3 False. The electrostatic energy density is not uniformly distributed because the magnitude of the electric field strength is not uniformly distributed.

5 1/3

7 (*a*) True, (*b*) True

9 (*a*) True, (*b*) False. Because $Q = CV$, and C increases, Q must increase. (*c*) True. $E = V/d$, where d is the plate separation. (*d*) False. $U = \frac{1}{2}QV$.

11 (*a*) $U_{\text{parallel}} = 2U_{1\text{ capacitor}}$, (*b*) $U_{\text{series}} = \frac{1}{2}U_{1\text{ capacitor}}$

13 $0.1\ \text{nF/m} \le C/L \le 0.2\ \text{nF/m}$

15 2.3 nF

17 75.0 nF

19 (*a*) 15.0 mJ, (*b*) 45.0 mJ

21 (*a*) 0.625 J, (*b*) 1.88 J

23 (*a*) 100 kV/m, (*b*) 44.3 mJ/m³, (*c*) 88.5 μJ, (*d*) 17.7 nF, (*e*) 88.5 μJ

25 (*a*) 11 nC, (*b*) Because work has to be done to pull the plates farther apart, you would expect the energy stored in the capacitor to increase, (*c*) 0.55 μJ

27 10.00 μF

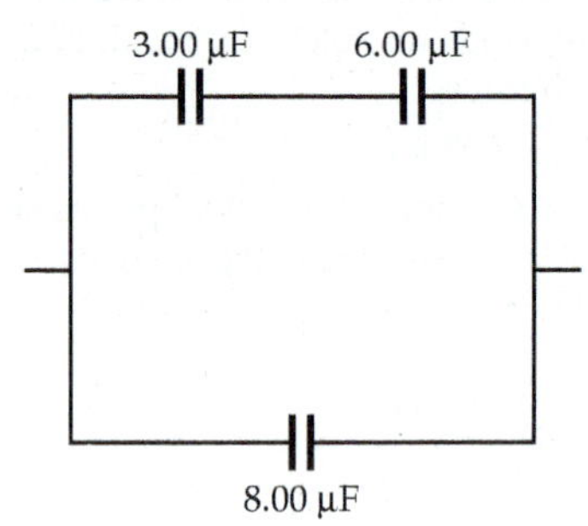

29 (*a*) 30.0 μF, (*b*) 6.00 V, (*c*) $Q_{10} = 60.0\ \mu\text{C}$, $Q_{20} = 120\ \mu\text{C}$, (*d*) $U_{10} = 180\ \mu\text{J}$, $U_{20} = 360\ \mu\text{J}$

31 (*a*) If their capacitance is to be a maximum, the capacitors must be connected in parallel.

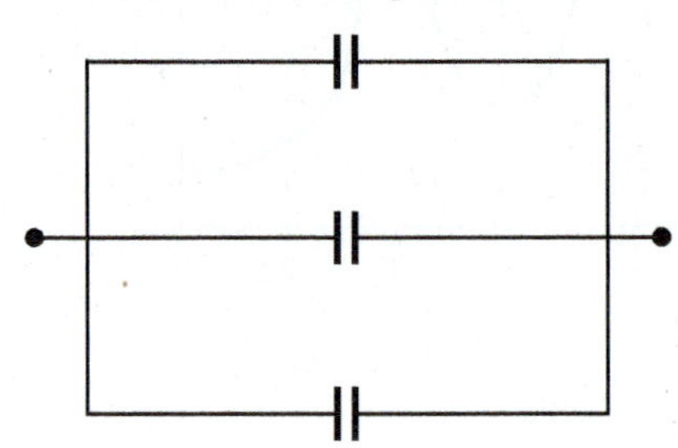

(*b*) (1) 1.67 μF

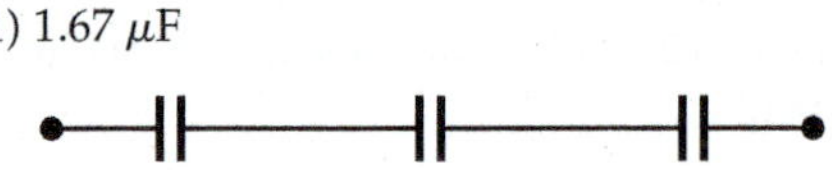

(2) 3.33 μF

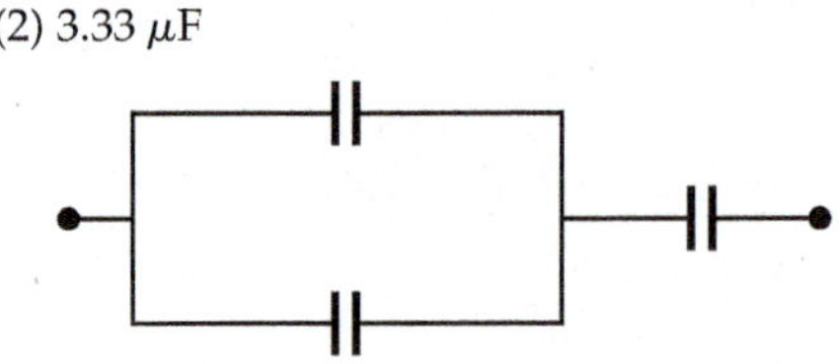

(3) 7.50 μF

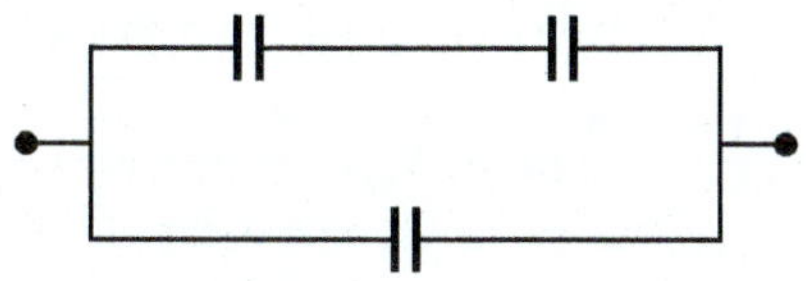

35 (*a*) $2C_0$, (*b*) $11C_0$

37 0.571 μF, 0.667 μF, 0.800 μF, 0.857 μF, 1.33 μF, 1.43 μF, 1.71 μF, 2.33 μF, 2.80 μF, 3.00 μF, 4.67 μF, 5.00 μF, 6.00 μF, 7.00 μF

39 (*a*) 4.80 kV, (*b*) 9.60 mC

41 (*a*) 7.9 m^2, (*b*) 23 V, (*c*) 37 μJ, (*d*) 0.16 J

43 (*a*) 1.55 pF, (*b*) 15.5 nC/m

45 179 pF/m

47 $\Delta C = -2\frac{P}{Y}C$

51 $R' = 2R$

53 (*a*) $V_{100} = V_{400} = 1.20$ kV, (*b*) 640 μJ

55 (*a*) 2.4 μF, (*b*) 0.4 mJ

57 (*a*) $V_{4.00} = V_{12.0} = 6.0$ V, (*b*) $U_i = 1.15$ mJ, $U_f = 0.29$ mJ

59 (*a*) $V_1 = V_2 = V_3 = 200$ V,
(*b*) $Q_1 = -255\ \mu$C, $Q_2 = 145\ \mu$C, $Q_3 = 545\ \mu$C,
(*c*) $V_1 = 127$ V, $V_2 = 36.4$ V, $V_3 = 90.9$ V

61 2.72 nF

63 (*a*) 50 μm, (*b*) 240 cm^2

65 $Q_1 = \frac{2Q}{1+\kappa}$, $Q_2 = \frac{2Q\kappa}{1+\kappa}$

67 (*a*) 16.7 nF, (*b*) 1.17 nC

69 (*a*) 2.1, (*b*) 45 cm^2, (*c*) 5.2 nC

71 A series combination of two of the capacitors connected in parallel with a series combination of the other two capacitors will result in total energy U_0 stored in all four capacitors.

73 2.00 μF

75 (*a*) $\frac{2}{3}C_0$, (*b*) C_0, (*c*) $3C_0$

77 (*a*) $C_{\text{new}} = \frac{\epsilon_0 A}{3d}$, (*b*) $V_{\text{new}} = 3V$, (*c*) $U_{\text{new}} = \frac{3\epsilon_0 AV^2}{2d}$,
(*d*) $W = \frac{\epsilon_0 AV^2}{d}$

79 133 μC, 267 μC

83 (*a*) $U = \frac{Q^2}{2\epsilon_0 A}x$, (*b*) $dU = \frac{Q^2}{2\epsilon_0 A}dx$

85 (*a*) $U = \frac{Q^2 d}{2\epsilon_0 a[(\kappa - 1)x + a]}$, (*b*) $F = \frac{(\kappa - 1)Q^2 d}{2a\epsilon_0[(\kappa - 1)x + a]^2}$,
(*c*) $F = \frac{(\kappa - 1)a\epsilon_0 V^2}{2d}$, (*d*) The force originates from the fringing fields around the edges of the capacitor. The effect of the force is to pull the polarized dielectric into the space between the capacitor plates.

87 (*a*) First show that F is inversely proportional to d for a given V_0. Because F increases as d decreases, a decrease in plate separation will unbalance the system. Hence the balance is unstable. (*b*) $V_0 = d_0\sqrt{\frac{2Mg}{\epsilon_0 A}}$

89 (*a*) $Q_1 = (200\text{ V})C_1$, $Q_2 = (200\text{ V})\kappa C_1$,
(*b*) $U = (2.00 \times 10^4\text{ V}^2)(1 + \kappa)C_1$,
(*c*) $U_f = (1.00 \times 10^4\text{ V}^2)C_1(1 + \kappa)^2$, (*d*) $V_f = 100(1 + \kappa)$ V

91 0.100 μF, 16.0 μC

93 $C = \frac{\epsilon_0 ab}{y_0}\ln(2)$

Chapter 25

1 In earlier chapters the conductors are constrained to be in electrostatic equilibrium. In this chapter, this constraint is no longer in place.

3 (*c*)

5 (*a*)

7 No, it is not necessarily true for a battery. Under normal operating conditions the current in the battery is in the direction away from the negative battery terminal and toward the positive battery terminal. That is, it opposite to the direction of the electric field.

9 (*e*)

11 (*d*)

13 You should decrease the resistance. The heat output is given by $P = V^2/R$. Because the voltage across the resistor is constant, decreasing the resistance will increase P.

15 (*a*)

17 (*a*)

19 (*a*) False, (*b*) True, (*c*) True

21 (*b*)

23 $P_2 = \frac{1}{4}P$ and $P_3 = \frac{1}{2}P$

25 1.9 kA

27 26 m

29 12 gauge

31 0.28 mm/s

33 (*a*) 0.21 mm/s, 0.53 mm/s, (*b*) 0.396

35 (*a*) $1.04 \times 10^8\text{ m}^{-1}$, (*b*) $1.04 \times 10^{14}\text{ m}^{-3}$, (*c*) 5.00 kA/m^2

37 0.86 s

39 (*a*) 33.3 Ω, (*b*) 0.750 A

41 1.9 V

43 63 light-years

45 1.20 Ω

47 31 mΩ

49 $R = \frac{\rho L}{\pi ab}$

51 (*a*) $R = \frac{\rho}{2\pi L}\ln(b/a)$, (*b*) 2.05 A

53 46°C

55 (*a*) 15.0 A, (*b*) 11.1 Ω, (*c*) 1.30 kW

57 (*b*) 3×10^2

59 (*a*) 636 K, (*b*) As the filament heats up, its resistance decreases. This results in more power being dissipated, further heat, higher temperature, etc. If not controlled, this thermal runaway can burn out the filament.

61 0.18 kJ

63 (*a*) 0.24 kW, (*b*) 0.23 kW, (*c*) 1.7 kJ, (*d*) 84 J

65 (*a*) 6.9 MJ, (*b*) 12.8 h

67 (*a*) 26.7 kW, (*b*) 576 kC, (*c*) 69.1 MJ, (*d*) 57.6 km, (*e*) \$0.03 per km

69 $I_4 = 3.00$ A, $I_3 = 4.00$ A, $I_6 = 2.00$ A

71 (*b*) It would not affect it.

73 0.45 kΩ

75 (*a*) 6.00 Ω, (*b*) The current in both the 6.00-Ω and the 12.0-Ω resistor in the upper branch is 667 mA. The current in each 6.00-Ω resistor in the parallel combination in the lower branch is 667 mA. The current in the 6.00-Ω resistor on the right in the lower branch is 1.33 A.

77 8 pieces

79 (*a*) $R_3 = \dfrac{R_1^2}{R_1 + R_2}$, (*b*) 0, (*c*) $R_1 = \dfrac{R_3 + \sqrt{R_3^2 + 4R_2R_3}}{2}$

81 (*a*) 4.00 A, (*b*) 2.00 V, (*c*) 1.00 Ω

83 (*a*)

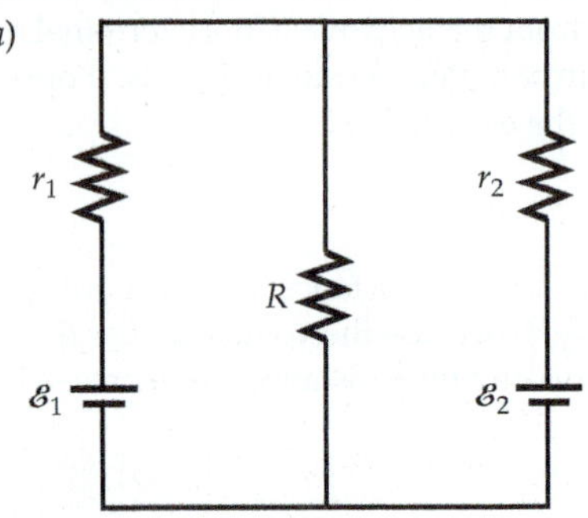

(*b*) $I_1 = -19.0$ A, $I_2 = 25.1$ A, $I_R = 6.17$ A, (*c*) Battery 2 supplies 311 W. Of the 234 W that is delivered to battery 1, 216 W goes into recharging battery 1 and 18.0 W is dissipated by the internal resistance. In addition, 76.2 W is delivered to the 2.00-Ω resistor.

85 (*a*) $I_{4\Omega} = 0.667$ A, $I_{3\Omega} = 0.889$ A, $I_{6\Omega} = 1.56$ A, (*b*) $V_{ab} = 9.33$ V, (*c*) $P_{\text{left}} = 8.00$ W, $P_{\text{right}} = 10.7$ W

87 For the series combination, the power delivered to the load is greater if $R > r$ and is greatest when $R = 2r$. If $r = R$, both arrangements provide the same power to the load. For the parallel combination, the power delivered to the load is greater if $R < r$ and is a maximum when $R = \frac{1}{2}r$.

89 $V_a - V_b = 2.40$ V

91 (*a*) 3.33 V, (*b*) 3.33 V, (*c*) 3.13 V, (*d*) 2.00 V, (*e*) 0.435 V, (*f*) $R_{\max} = 1.67$ MΩ

93 2.5 Ω

95 (*a*) 600 μC, (*b*) 0.200 A, (*c*) 3.00 ms, (*d*) 81.2 μC

97 2.18 MΩ

99 (*a*) 5.69 μC, (*b*) 1.10 μC/s, (*c*) 1.10 μA, (*d*) 6.62 μW, (*e*) 2.44 μW, (*f*) 4.19 μW

103 (*a*) 0.250 A, (*b*) 62.5 mA, (*c*) $I_2(t) = (62.5\text{ mA})(1 - e^{-t/0.750\text{ ms}})$

105 (*a*) 48.0 μA, (*b*) 0.866 s

107 (*a*) (1) The potential drops across R_2 and R_3 are equal, so $I_2 > I_3$. The current in R_1 equals the sum of the currents I_2 and I_3, so I_1 is greater than either I_2 or I_3. (*b*) $I_1 = 1.50$ A, $I_2 = 1.00$ A, $I_3 = 0.50$ A

109 (*a*) 43.9 Ω, (*b*) 300 Ω, (*c*) 3.8 kΩ

111 (*a*) 2.18×10^{13} s^{-1}, (*b*) 210 J/s, (*c*) 27.6 s

113 0.16 L/s

115 (*a*) 10.0 ms, (*c*) 1.00 GΩ, (*d*) 60.9 ps, (*e*) 2.89 kW

119 (*a*) $R_{\text{eq}} = \left(\dfrac{1 + \sqrt{5}}{2}\right)R$, (*b*) $R_{\text{eq}} = \dfrac{R_1 + \sqrt{R_1^2 + 4R_1R_2}}{2}$

Chapter 26

1 (*b*)

3 Because the alternating current running through the filament is changing direction every 1/60 s, the filament experiences a force that changes direction at the frequency of the current. Thus, it oscillates at 60 Hz.

5 (*a*)

9 (*a*) False, (*b*) True, (*c*) True, (*d*) True

11 According to the principle of relativity, this is equivalent to the electron moving from right to left at speed v with the magnet stationary. When the electron is directly over the magnet, the field points directly up, so there is a force directed out of the page on the electron.

13 $I \sim 2 \times 10^3$ A. You should advise him to develop some other act. A current of 2000 A would overheat the wire (which is a gross understatement).

15 (*a*) $-(3.8\ \mu\text{N})\hat{k}$, (*b*) $-(7.5\ \mu\text{N})\hat{k}$, (*c*) 0, (*d*) $(7.5\ \mu\text{N})\hat{j}$

17 0.96 N

19 $-(19\text{ fN})\hat{i} - (13\text{ fN})\hat{j} - (58\text{ fN})\hat{k}$

21 1.5 A

23 $(10\text{ T})\hat{i} + (10\text{ T})\hat{j} - (15\text{ T})\hat{k}$

27 (*a*) 87 ns, (*b*) 4.7×10^7 m/s, (*c*) 11 MeV

29 (*a*) $2v_\alpha = 2v_d = 1v_p$, (*b*) $1K_\alpha = 2K_d = 1K_p$, (*c*) $L_\alpha = 2L_d = 2L_p$

33 (*a*) 24°, 1.3×10^6 m/s, (*b*) 24°, 6.3×10^5 m/s

35 (*a*) 1.6×10^6 m/s, (*b*) 14 keV, (*c*) 7.7 eV

37 7.37 mm

39 (*a*) 63.3 cm, (*b*) 2.58 cm

41 $\Delta t_{58} = 15.7\ \mu$s, $\Delta t_{60} = 16.3\ \mu$s

43 (*a*) 21 MHz, (*b*) 46 MeV, (*c*) $f_{\text{deuterons}} = 11$ MHz, $K_{\text{deuterons}} = 23$ MeV

47 (*a*) 0.30 A·m², (*b*) 0.13 N·m

49 (*a*) 0, (*b*) 2.7×10^{-3} N·m

51 $B_{\min} = \dfrac{mg}{I\pi R}$

53 (*a*) $(0.84\text{ N}\cdot\text{m})\hat{k}$, (*b*) 0, (*c*) 0, (*d*) $(0.59\text{ N}\cdot\text{m})\hat{k}$

55 0.38 A·m², into the page

61 $\mu = \frac{4}{3}\pi\sigma R^4\omega$

63 (*a*) $\tau = \frac{1}{4}\pi\sigma r^4\omega B\sin\theta$, (*b*) $\Omega = \dfrac{\pi\sigma r^2 B}{2m}\sin\theta$

65 (*a*) 3.68×10^{-5} m/s, (*b*) 1.47 μV

67 1.0 mV

69 4

71 (*a*) 1.3 μs, (*b*) 2.4×10^6 m/s, (*c*) 0.12 MeV

75 (*a*) $B = -\dfrac{mg}{IL}\tan\theta$, (*b*) $\vec{a} = g\sin\theta$, up the incline

77 (*a*) $(10\text{ V/m})\hat{j}$, (*b*) The positive end has the lesser y coordinate, (*c*) 20 V

81 1.0×10^{-28} kg

Chapter 27

1 Note that, while the two far fields (the fields far from the dipoles) are the same, the two near fields (the fields near to the dipoles) are not. At the center of the electric dipole, the electric field is antiparallel to the direction of the far field above and below the dipole, and at the center of the magnetic dipole, the magnetic field is parallel to the direction of the far field above and below the dipole. It is especially important to note that while the electric field lines begin and terminate on electric charges, the magnetic field lines are continuous, i.e., they form closed loops.

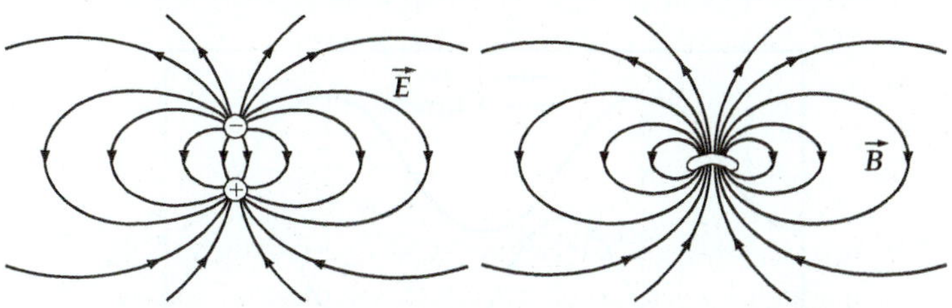

3 (*a*)

5 Both tell you about the respective fluxes through closed surfaces. In the electrical case, the flux is proportional to the net charge enclosed. In the magnetic case, the flux is always zero because there is no such thing as magnetic charge (a magnetic monopole). The source of the magnetic field is NOT the equivalent of electric charge; that is, it is NOT a thing called magnetic charge, but rather it is moving electric charges.

7 Clockwise

9 (*a*) True, (*b*) False, (*c*) True, (*d*) True

11 H_2, CO_2, and N_2 are diamagnetic ($\chi_m < 0$); O_2 is paramagnetic ($\chi_m > 0$).

13 (*a*) $\vec{B}(0, 0) = -(9.0\text{ pT})\hat{k}$, (*b*) $\vec{B}(0, 1.0\text{ m}) = -(36\text{ pT})\hat{k}$, (*c*) $\vec{B}(0, 3.0\text{ m}) = (36\text{ pT})\hat{k}$, (*d*) $\vec{B}(0, 4.0\text{ m}) = (9.0\text{ pT})\hat{k}$

15 (*a*) $\vec{B}(2.0\text{ m}, 2.0\text{ m}) = 0$,
(*b*) $\vec{B}(6.0\text{ m}, 4.0\text{ m}) = -(3.6 \times 10^{-23}\text{ T})\hat{k}$,
(*c*) $\vec{B}(3.0\text{ m}, 6.0\text{ m}) = (4.0 \times 10^{-23}\text{ T})\hat{k}$

17 $\epsilon_0\mu_0 v^2$

19 $\vec{B}(0, 3.0\text{ m}, 4.0\text{ m}) = -(9.6\text{ pT})\hat{i}$

21 (*a*) $B(0) = 54\ \mu\text{T}$, (*b*) $B(0.010\text{ m}) = 46\ \mu\text{T}$, (*c*) $B(0.020\text{ m}) = 31\ \mu\text{T}$, (*d*) $B(0.35\text{ m}) = 34\text{ nT}$

25 (*a*) $\vec{B}(-3.0\text{ cm}) = -(89\ \mu\text{T})\hat{k}$, (*b*) $\vec{B}(0) = 0$
(*c*) $\vec{B}(3.0\text{ cm}) = (89\ \mu\text{T})\hat{k}$, (*d*) $\vec{B}(9.0\text{ cm}) = -(160\ \mu\text{T})\hat{k}$

27 (*a*) $\vec{B}(-3.0\text{ cm}) = -(0.18\text{ mT})\hat{k}$, (*b*) $\vec{B}(0) = -(0.13\text{ mT})\hat{k}$,
(*c*) $\vec{B}(3.0\text{ cm}) = -(0.18\text{ mT})\hat{k}$, (*d*) $\vec{B}(9.0\text{ cm}) = (0.11\text{ mT})\hat{k}$

29 (*a*) $(64\ \mu\text{T})\hat{j}$,
(*b*) $-(48\ \mu\text{T})\hat{k}$

31 (*a*) Because the currents repel, they are antiparallel.
(*b*) 39 mA

33 80 A

35 (*a*) 30 μT, down the page, (*b*) 4.5×10^{-4} N/m, toward the right

37 (*a*) 80 A, (*b*) $\vec{B}(5.0\text{ cm}, 0, 0) = -(0.24\text{ mT})\hat{j}$

39 (*a*) $\dfrac{F}{\ell} = \dfrac{3\sqrt{2}\,\mu_0 I^2}{4\pi a}$, (*b*) $\dfrac{F}{\ell} = \dfrac{\sqrt{2}\,\mu_0 I^2}{4\pi a}$

41 (*a*) 3.3 mT, (*b*) 1.6 mT

45 $B_{\text{inside}} = 0$, $B_{\text{outside}} = \dfrac{\mu_0 I}{2\pi R}$. The direction of the magnetic field is in the direction of the curled fingers of your right hand when you grab the cylinder with your right thumb in the direction of the current.

49 (*a*) $B_{R<a} = 0$, (*b*) $B_{a<R<b} = \dfrac{\mu_0 I}{2\pi R}\dfrac{R^2 - a^2}{b^2 - a^2}$, (*c*) $B_{R>b} = \dfrac{\mu_0 I}{2\pi R}$

51 (*a*) $B(1.10\text{ cm}) = 27.3\text{ mT}$, (*b*) $B(1.50\text{ cm}) = 20.0\text{ mT}$

53 (*a*) $B = B_{\text{app}} = 10.1\text{ mT}$, (*b*) $B_{\text{app}} = 10.1\text{ mT}$, $B = 1.5\text{ T}$

55 -4.0×10^{-5}

57 5.43 A/m

59

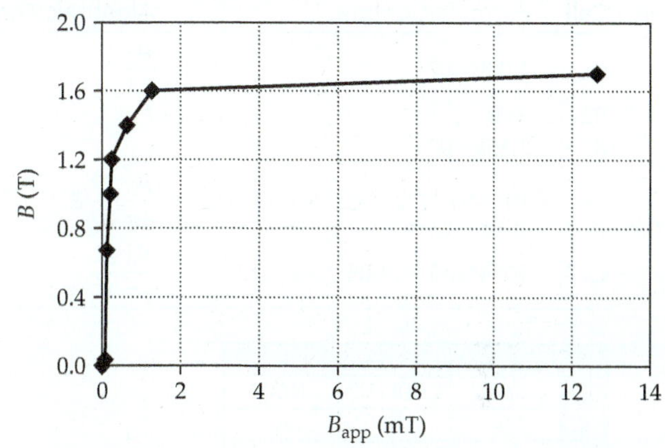

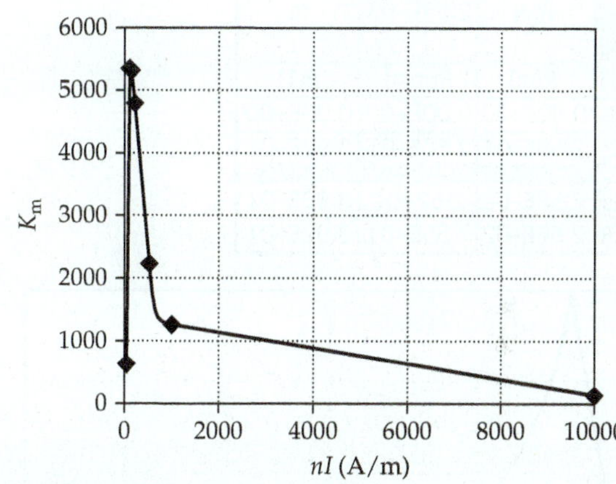

61 $1.69\mu_B$

63 (*b*) 7.46×10^{-4}

65 $B_{\text{app}} = \dfrac{\mu_0 NI}{2\pi a}$, $B = \dfrac{\mu_0 NI}{2\pi a} + \mu_0 M$

67 (*a*) 30.2 mT, (*b*) 6.96 A/m, (*c*) 30.2 mT

69 11.7, $1.48 \times 10^{-5}\text{ N/A}^2$

71 (*a*) 12.6 mT, (*b*) 1.36×10^6 A/m, (*c*) 137

73 (*a*) 1.42×10^6 A/m,
(*b*) $K_m = 90.0$, $\mu = 1.13 \times 10^{-4}\text{ T}\cdot\text{m/A}$, $\chi_m = 89$

75 (*a*) $(8.00\text{ T/m})r$, (*b*) $(3.20 \times 10^{-3}\text{ T}\cdot\text{m})\dfrac{1}{r}$, (*c*) $(8.00 \times 10^{-6}\text{ T}\cdot\text{m})\dfrac{1}{r}$, (*d*) Note that the field in the ferromagnetic region is that which would be produced in a nonmagnetic region by a current of $400I = 1600$ A. The amperian current on the inside of the surface of the ferromagnetic material must therefore be 1600 A − 40 A = 1560 A in the direction of I. On the outside surface there must then be an amperian current of 1560 A in the opposite direction.

77 $\vec{B}_P = \frac{\mu_0 I}{4}\left(\frac{1}{R_1} - \frac{1}{R_2}\right)$, out of the page

79 $\vec{B}_P = \frac{\mu_0}{2\pi}\frac{I}{a}(1 + \sqrt{2})$, out of the page

81 The $+x$ and $+y$ directions are up the page and to the right. (*a*) $\vec{F}_{\text{top}} = -(2.5 \times 10^{-5}\ \text{N})\hat{j}$, $\vec{F}_{\text{left side}} = (1.0 \times 10^{-4}\ \text{N})\hat{i}$, $\vec{F}_{\text{bottom}} = (2.5 \times 10^{-5}\ \text{N})\hat{j}$, $\vec{F}_{\text{right side}} = -(0.29 \times 10^{-4}\ \text{N})\hat{i}$, (*b*) $\vec{F}_{\text{net}} = (0.71 \times 10^{-4}\ \text{N})\hat{i}$

83 7.1 μT, into the page

85

Cell	Formula/Content	Algebraic Form
B1	1.00E−07	$\frac{\mu_0}{4\pi}$
B2	5.00	I
B3	2.55E−03	r_0
C6	10^4*$B$1*2*$B$2*A6/$B$3^2	$\frac{\mu_0}{4\pi}\frac{2I}{R_0^2}R$
C17	10^4*$B$1*2*$B$2*A6/A17	$\frac{\mu_0}{4\pi}\frac{2I}{R}$

	A	B	C
1	μ/4π=	1.00E−07	N/A2
2	I=	5	A
3	R_0=	2.55E−03	m
4			
5	R(m)	R(mm)	B(T)
6	0.00E+00	0.00E+00	0.00E+00
7	2.55E−04	2.55E−01	3.92E−01
105	2.52E−02	2.52+01	3.96E−01
106	2.55E−02	2.55E+01	3.92E−01

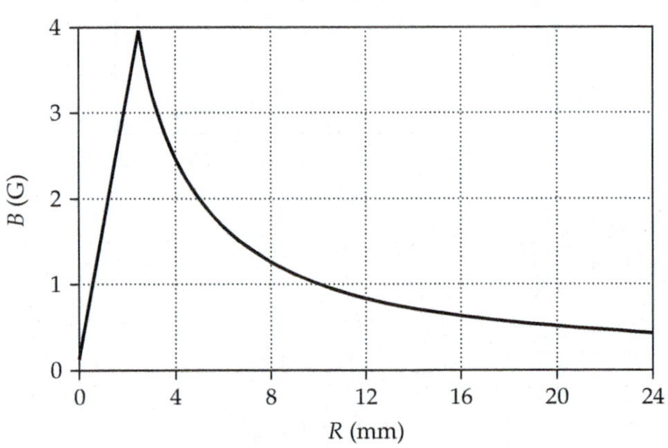

87 (*a*) $5.24 \times 10^{-2}\ \text{A}\cdot\text{m}^2$, (*b*) 7.70×10^5 A/m, (*c*) 23.1 kA

89 (*a*) 15.5 GA, (*b*) Because Earth's magnetic field points down at the north pole, application of the right-hand rule indicates that the current is clockwise when viewed from above the north pole.

91 3.18 cm

93 (*a*) and (*b*) $B(5.0\ \text{cm}) = B(10\ \text{cm}) = 10\ \mu\text{T}$, (*c*) $B(20\ \text{cm}) = 5.0\ \mu\text{T}$

95 2.24 A

97 (*c*) $B_z = \frac{1}{2}\mu_0\omega\sigma\left(\frac{R^2 + 2z^2}{\sqrt{R^2 + z^2}} - 2z\right)$

Chapter 28

1 (*a*) Orient the sheet so the normal to the sheet is both horizontal and perpendicular to the local tangent to the magnetic equator. (*b*) Orient the sheet of paper so the normal to the sheet is perpendicular to the direction of the normal described in the answer to Part (*a*).

5 (*d*)

7 The induced current is clockwise as viewed from the left. The loops repel each other.

9 (*a*) and (*b*)

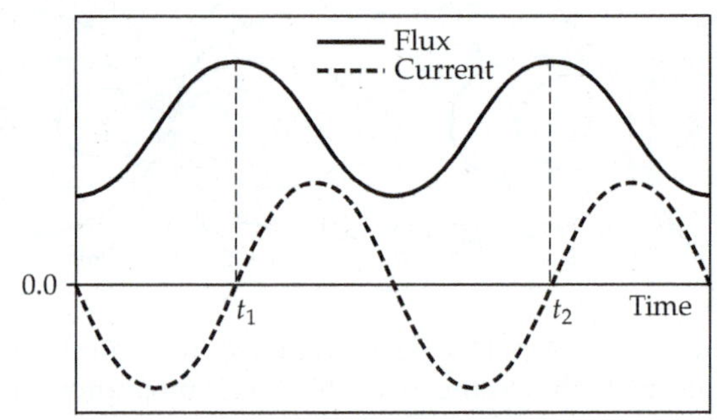

11 The magnetic field of the falling magnet sets up eddy currents in the metal tube. The eddy currents establish a magnetic field that exerts a force on the magnet opposing its motion; thus the magnet is slowed down. If the tube is made of a nonconducting material, there are no eddy currents.

13 (*c*)

15 (*a*) False, (*b*) True, (*c*) False, (*d*) True, (*e*) False

17 $u_m \approx (8 \times 10^3)u_e$

19 (*a*) 0.5 V, (*b*) 7 mV/m

21 (*a*) 0, (*b*) 14 μWb, (*c*) 0, (*d*) 12 μWb

23 $\phi_m = \pm\pi R^2 B$

25 1.68 mWb

27 (*a*) $\phi_m = \mu_0 n I N \pi R_1^2$, (*b*) $\phi_m = \mu_0 n I N \pi R_2^2$

29 $\frac{\phi_m}{L} = \frac{\mu_0 I}{4\pi}$

31 (*a*)

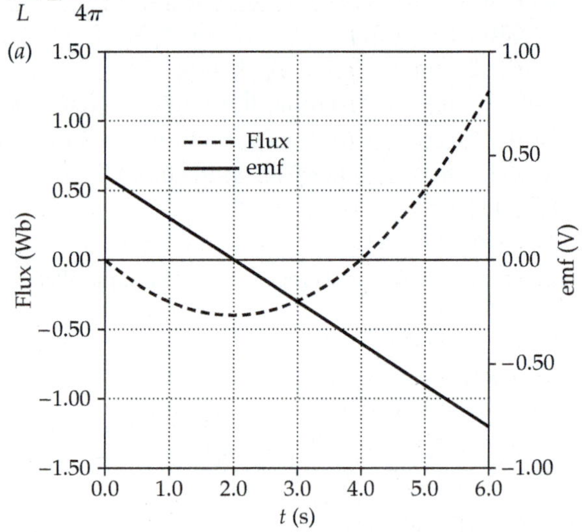

(*b*) The flux is a minimum when $t = 2.0$ s; $V(2.0\text{ s}) = 0$.

(*c*) The flux is zero when $t = 0$ and $t = 4.0$ s; $\mathcal{E}(0) = 0.40$ V and $\mathcal{E}(4.0\text{ s}) = -0.40$ V.

33 (*a*) 1.26 mC, (*b*) 12.6 mA, (*c*) 628 mV

35 79.8 μT

37 400 m/s

39 (*a*)

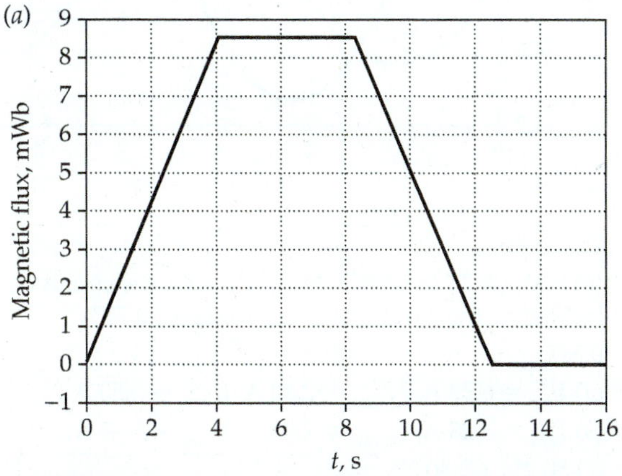

(*b*)

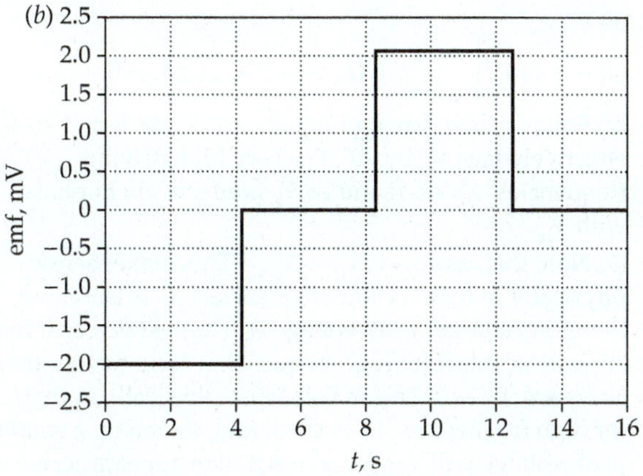

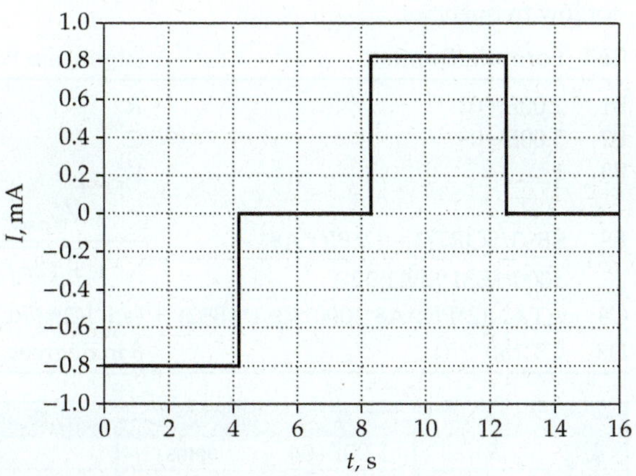

41 (*a*) $F_m = \dfrac{B\ell}{R}(\mathcal{E} - B\ell v)$, (*c*) 0

45 (*a*) 14 V, (*b*) 486 rev/s

47 (*a*) 24.0 Wb, (*b*) −1.60 kV

49 $L = 0$, $R = 162\ \Omega$

51 0.16 μH

57 $\dfrac{dU_m}{dx} = \dfrac{\mu_0 I^2}{16\pi}$

59 (*a*) $I = 0$, $dI/dt = 25.0$ kA/s,

(*b*) $I = 2.27$ A, $dI/dt = 20.5$ kA/s,

(*c*) $I = 7.90$ A, $dI/dt = 9.20$ kA/s,

(*d*) $I = 10.8$ A, $dI/dt = 3.38$ kA/s

61 (*a*) 44.1 W, (*b*) 40.4 W, (*c*) 3.62 W

63 (*a*) 3.00 kA/s, (*b*) 1.50 kA/s, (*c*) 80.0 mA, (*d*) 0.123 ms

65 (*a*) $I_{10\text{-}\Omega} = I_{2\text{-H}} = 1.0$ A, $I_{100\text{-}\Omega} = 0$, (*b*) $V_{2\text{-H}} = 100$ V

(*c*)

Cell	Content/Formula	Algebraic Form
B1	2.0	L
B2	100	R
B3	1	I_0
A6	0	t_0
B6	$B$3*EXP((−$B$2/$B$1)*A6)	$I_0 e^{-\frac{R}{L}t}$

	A	B	C
1	L=	2	H
2	R=	100	ohms
3	I_0=	1	A
4			
5	t	I(t)	V(t)
6	0.000	1.00E+00	100.00
7	0.005	7.79E-01	77.88
8	0.010	6.07E-01	60.65
9	0.015	4.72E-01	47.24
10	0.020	3.68E-01	36.79
11	0.025	2.87E-01	28.65
12	0.030	2.23E-01	22.31
32	0.130	1.50E–03	0.15
33	0.135	1.17E–03	0.12
34	0.140	9.12E–04	0.09
35	0.145	7.10E–04	0.07
36	0.150	5.53E–04	0.06

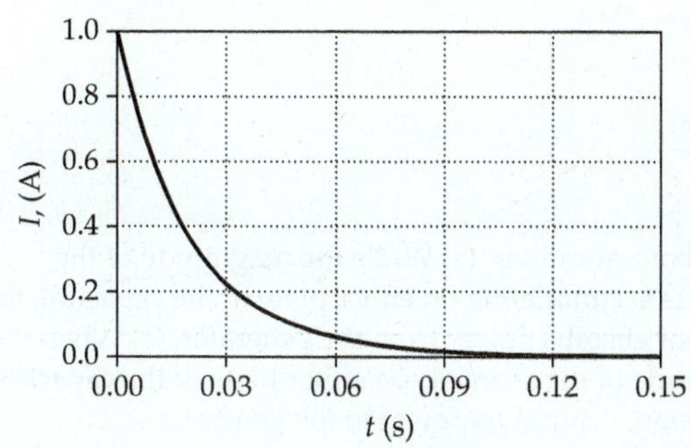

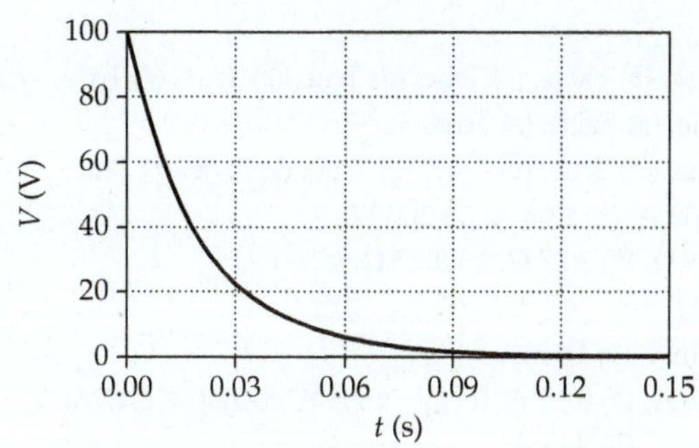

67 (*a*) 88 ms, (*b*) 35 mH

69 (*a*) 3.53 J, (*b*) 1.61 J, (*c*) 1.92 J

71 (*b*) 2.50 krad/s

75 0.28 H

77 (*a*) As the magnet passes through a loop it induces an emf because of the changing flux through the loop. This allows the coil to "sense" when the magnet is passing through it. (*b*) One cannot use a cylinder made of conductive material because eddy currents induced in it by a falling magnet would slow the magnet. (*c*) As the magnet approaches the loop the flux increases, resulting in the negative voltage signal of increasing magnitude. When the magnet is passing a loop, the flux reaches a maximum value and then decreases, so the induced emf becomes zero and then positive. The instant at which the induced emf is zero is the instant at which the magnet is at the center of the loop.

(*d*)

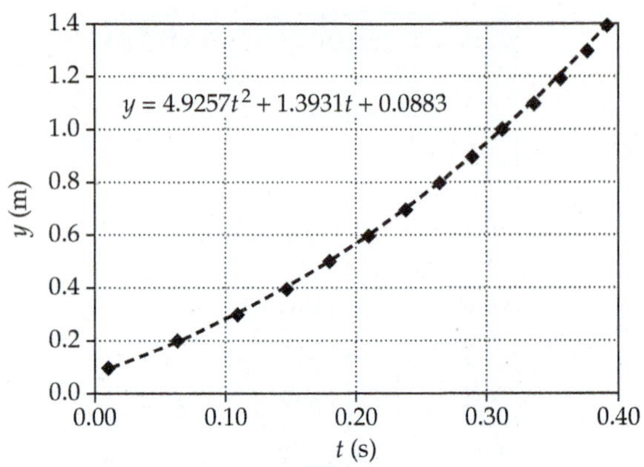

$g = 9.85\ \text{m/s}^2$

79 (*a*) $E_r = -\frac{1}{2} r\mu_0 n I_0 \omega \cos \omega t \quad r < R$,

(*b*) $E_r = -\dfrac{\mu_0 n R^2 I_0 \omega}{2r} \cos \omega t \quad r > R$

Chapter 29

1 8.33 ms

3 (*b*)

5 (*c*)

7 Yes to both questions. (*a*) While the magnitude of the charge is accumulating on either plate of the capacitor, the capacitor absorbs power from the generator. (*b*) When the magnitude of the charge is on either plate of the capacitor is decreasing, it supplies power to the generator.

9 (*a*)

11 (*a*)

13 (*a*) False, (*b*) False, (*c*) True, (*d*) True, (*e*) True, (*f*) True

15 (*a*) True, (*b*) False, (*c*) True

17 (*a*) False, (*b*) True, (*c*) True, (*d*) True, (*e*) True, (*f*) True

19 (*a*) 0.833 A, (*b*) 1.18 A, (*c*) 200 W

21 (*a*) 0.38 Ω, (*b*) 3.77 Ω, (*c*) 37.7 Ω

23 1.6 kHz

25 (*a*) 25 mA, (*b*) 18 mA

27 (*a*) 0.35 A, (*b*) 0.35 A, (*c*) $I = (0.34\ \text{A})\cos(\omega t + 0.17\ \text{rad})$

29 (*a*) 1.3 ms, (*b*) 88 mH

31 (*a*) 2.3 mJ, (*b*) 0.71 kHz, (*c*) 0.67 A

33 (*a*)

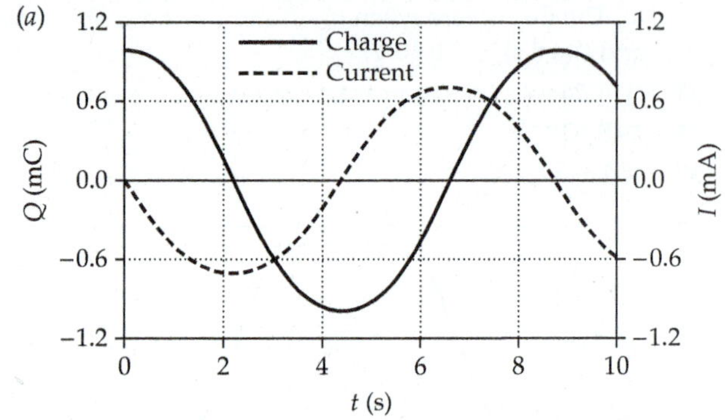

35 29.2 mH

37 (*a*) 0.33, (*b*) 27 Ω, (*c*) 0.20 H, (*d*) Because the circuit is inductive, the current lags the voltage. (*e*) 71°

39 0.397

41 (*a*) $I_{\text{rms}} = 6.23\ \text{A}$, $I_{R_L\,\text{rms}} = 2.80\ \text{A}$, $I_{L\,\text{rms}} = 5.53\ \text{A}$, (*b*) $I_{\text{rms}} = 3.28\ \text{A}$, $I_{R_L\,\text{rms}} = 2.94\ \text{A}$, $I_{L\,\text{rms}} = 1.46\ \text{A}$, (*c*) 50.2%, (*d*) 80.0%

43 60 V

45 (*a*) $\delta = \tan^{-1}\left[-\dfrac{1}{\omega RC}\right]$, (*b*) $\delta \to -90°$, (*c*) $\delta \to 0$,

(*d*) For very low driving frequencies, $X_C \gg R$ and so $\vec{V}_C$ effectively lags $\vec{V}_{\text{in}}$ by 90°. For very high driving frequencies, $X_C \ll R$ and so $\vec{V}_R$ is effectively in phase with $\vec{V}_{\text{in}}$.

51 (*b*) Note that, as $\omega \to 0$, $V_L \to V_{\text{peak}}$. This makes sense physically in that, for low frequencies, X_C is large and, therefore, a larger peak voltage will appear across it than appears across it for high frequencies. Note further that, as $\omega \to \infty$, $V_L \to 0$. This makes sense physically in that, for high frequencies, X_C is small and, therefore, a smaller peak voltage will appear across it than appears across it for low frequencies.

53

Cell	Formula/Content	Algebraic Form
B1	2.00E+03	R
B2	5.00E−09	C
B3	1	$V_{\text{in peak}}$
B8	$B$3/SQRT(1+((2*PI()*A8*1000*$B$1*$B$2)^2))	$\dfrac{V_{\text{in peak}}}{\sqrt{1 + (2\pi f RC)^2}}$
C8	ATAN(2*PI()*A8*1000*$B$1*$B$2)	$\tan^{-1}(2\pi f RC)$
D8	C8*180/PI()	δ in degrees

	A	B	C	D
1	R =	1.00E+04	ohms	
2	C =	5.00E−09	F	
3	$V_{\text{in peak}}$ =	1	V	
4				
5				
6	f(kHz)	V_{out}	δ(rad)	δ(deg)
7	0	1.000	0.000	0.0
8	1	0.954	0.304	17.4
56	49	0.065	1.506	86.3
57	50	0.064	1.507	86.4

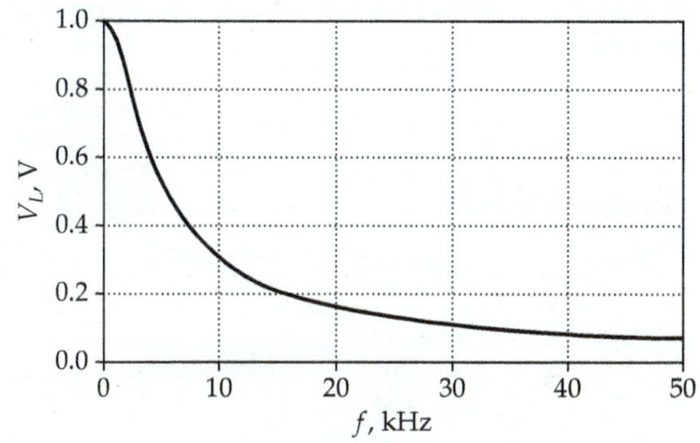

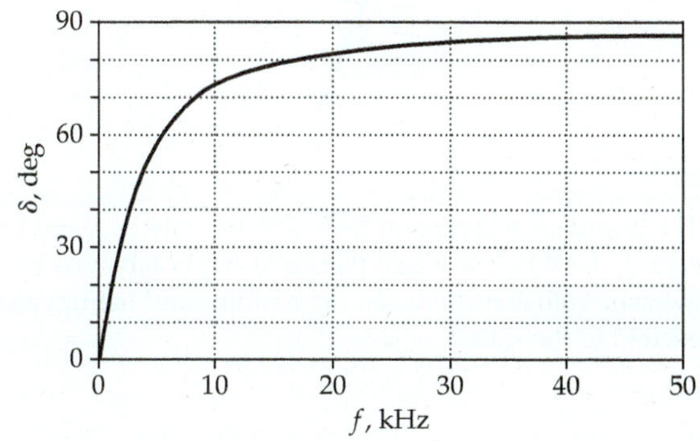

55 (*b*) $\Delta\omega = \frac{R}{L}$

57 33 μF

59 (*a*) $I(t) = -(19\text{ mA})\sin\left(\omega t + \frac{\pi}{4}\right)$, where $\omega = 1250$ rad/s, (*b*) 23 μF,

(*c*) $U_m(t) = (4.9\ \mu\text{J})\sin^2\left(\omega t + \frac{\pi}{4}\right)$,

$U_e(t) = (4.9\ \mu\text{J})\cos^2\left(\omega t + \frac{\pi}{4}\right)$, where $\omega = 1250$ rad/s, $U = 4.9\ \mu$J

61 (*a*) 5.4 fF, (*b*) $f(x) = \frac{70\text{ MHz}}{\sqrt{1 - (4.0\text{ m}^{-1})x}}$

65 (*a*) 14, (*b*) 80 Hz, (*c*) 0.27

67 (*a*) 10 A, (*b*) 53°, (*c*) 0.33 mF, (*d*) 0.13 kV

69 (*a*) 80 V, (*b*) 78 V, (*c*) 0.17 kV, (*d*) 0.11 kV, (*e*) 0.18 kV

71 (*a*)

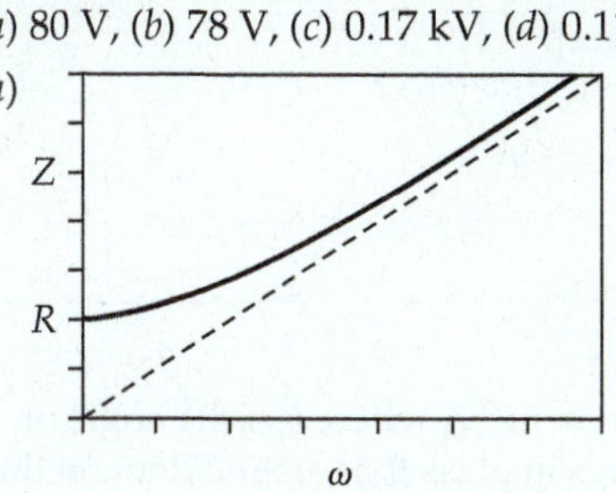

(*b*)

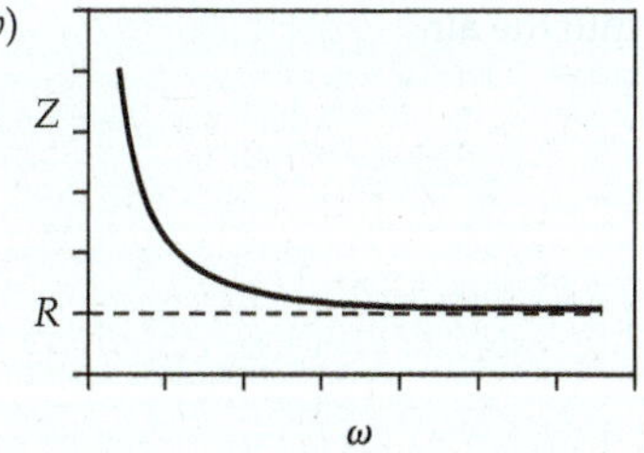

(*c*)

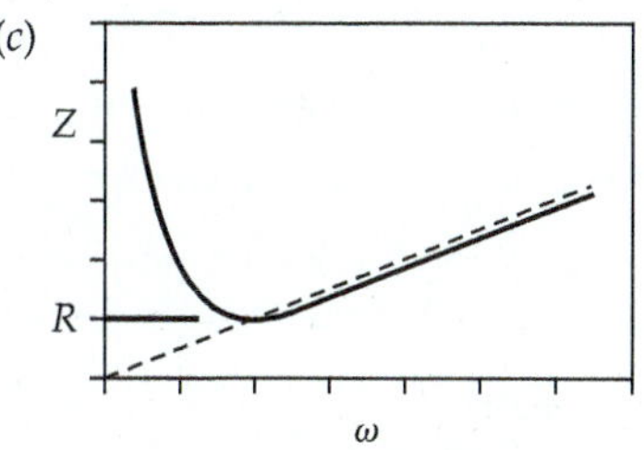

73 (*a*) $Z = 12\ \Omega$, (*b*) $R = 7.2\ \Omega$, $X = 10\ \Omega$, (*c*) If the current leads the emf, the reactance is capacitive.

79 (*a*) 1:5, (*b*) 50 A

81 (*a*) 1.5 A, (*b*) 19

83 3.33×10^3

85 (*a*) 12 V, (*b*) 8.5 V

87 $I_{max} = 1.06$ A, $I_{min} = -0.06$ A, $I_{av} = 0.50$ A, $I_{rms} = 0.64$ A

Chapter 30

1 (*a*) False, (*b*) True, (*c*) True, (*d*) False

3 (*a*) False, (*b*) True, (*c*) True, (*d*) True

5 (*a*) (1) X-rays, (2) green light, (3) red light, (*b*) (1) microwaves, (2) green light, (3) ultraviolet light.

7 (*a*) The electric dipole antenna should be oriented vertically. (*b*) The loop antenna and the electric dipole transmitting antenna should be in the same vertical plane.

9 (*d*)

11 (*d*)

13 2×10^{-7}

15 (*a*) 3.4×10^{14} V/m·s

19 (*a*) 10 A, (*b*) $\frac{dE}{dt} = 2.3 \times 10^{12}$ V/m·s,

(*c*) $\oint_C \vec{B} \cdot d\vec{\ell} = 0.79\ \mu\text{T}\cdot\text{m}$

21 580 nm, 5.17×10^{14} Hz

23 (*a*) 3.00×10^{18} Hz, (*b*) 5.45×10^{14} Hz. Consulting Table 30-1, we see that the color of light that has a wavelength of 550 nm is yellow-green. This result is consistent with those of Problem 21 and is close to the wavelength of the peak output of the Sun. Because we see naturally by reflected sunlight, this result is not surprising.

25 (*a*) 30°, (*b*) 7.1 m

27 4.13 μW/m^2

29 386 nW/m^2

31 (*a*) 283 V/m, (*b*) 943 nT, (*c*) 212 W/m^2, (*d*) 708 nPa

33 (*a*) 40 nN, (*b*) 80 nN

35 (*a*) 45°, (*b*) 5.7°

37 (*a*) $+x$ direction, (*b*) $\lambda = 0.628$ m, $f = 477$ MHz,
(*c*) $\vec{E}(x, t) = (194\text{ V/m})\cos[kx - \omega t]\hat{j}$,
$\vec{B}(x, t) = (647\text{ nT})\cos[kx - \omega t]\hat{k}$,
where $k = 10.0$ rad/m and $\omega = 3.00 \times 10^9$ rad/s.

39 6.10 degrees

41 (*a*) $F_{\text{r Earth}} = 5.83 \times 10^8$ N, $F_{\text{r Earth}} = (1.65 \times 10^{-14})F_{\text{g Earth}}$,
(*b*) $F_{\text{r Mars}} = 7.18 \times 10^7$ N, $F_{\text{r Mars}} = (4.27 \times 10^{-14})F_{\text{g, Mars}}$,
(*c*) Mars

47 2.6 mV

49 (*a*) $I = V_0\left(\frac{1}{R}\sin\omega t + \frac{\epsilon_0 \pi a^2}{d}\cos\omega t\right)$,
(*b*) $B(r) = \frac{\mu_0 V_0}{2\pi r}\left(\frac{1}{R}\sin\omega t + \omega\frac{\epsilon_0 \pi r^2}{d}\cos\omega t\right)$,
(*c*) $\delta = \tan^{-1}\left(\frac{R\omega\epsilon_0\pi a^2}{d}\right)$

51 (*a*) $\vec{S}(x,t) = \frac{1}{\mu_0 c}[E_{10}^2\cos^2(k_1x - \omega_1 t) + 2E_{10}E_{20}\cos(k_1x - \omega_1 t)$
$\cos(k_2x - \omega_2 t + \delta) + E_{20}^2\cos^2(k_2x - \omega_2 t + \delta)]\hat{i}$
(*b*) $\vec{S}_{\text{av}} = \frac{1}{2\mu_0 c}[E_{10}^2 + E_{20}^2]\hat{i}$
(*c*) $\vec{S}(x,t) = \frac{1}{\mu_0 c}[E_{10}^2\cos^2(k_1x - \omega_1 t) - E_{20}^2\cos^2(k_2x + \omega_2 t + \delta)]\hat{i}$,
$\vec{S}_{\text{av}} = \frac{1}{2\mu_0 c}[E_{10}^2 - E_{20}^2]\hat{i}$

53 (*a*) 9.16×10^{-15} T, (*b*) 101 mV, (*c*) 5.49 μV

55 (*a*) $\vec{E} = \frac{I\rho}{\pi a^2}\hat{i}$, where $\hat{i}$ is a unit vector in the direction of the current. (*b*) $\vec{B} = \frac{\mu_0 I}{2\pi a}\hat{\theta}$, where $\hat{\theta}$ is a unit vector perpendicular to $\hat{i}$ and tangent to the surface of the conducting cylinder.
(*c*) $\vec{S} = -\frac{I^2\rho}{2\pi^2 a^3}\hat{r}$, where $\hat{r}$ is a unit vector directed radially outward—away from the axis of the conducting cylinder.
(*d*) $\int S_n\, dA = I^2R$

57 (*a*) 574 nm (*b*) The critical radius is an upper limit and so particles smaller than that radius will be blown out.

59 3.34 mN

Chapter 31

1 (*c*)

3 (*b*)

5 The decrease in the index of refraction *n* of the atmosphere with altitude results in refraction of the light from the Sun, bending it toward the normal to the surfaces on constant *n* (that is, toward the Earth). Consequently the Sun can be seen even after it is just below the horizon.

7 The path of least time is the path through point D.

9 In resonance absorption, the molecules respond to the frequency of the light through the Einstein photon relation $E = hf$. Neither the wavelength nor the frequency of the light within the eyeball depend on the index of refraction of the medium outside the eyeball. Thus, the color appears to be the same in spite of the fact that the wavelength has changed.

11

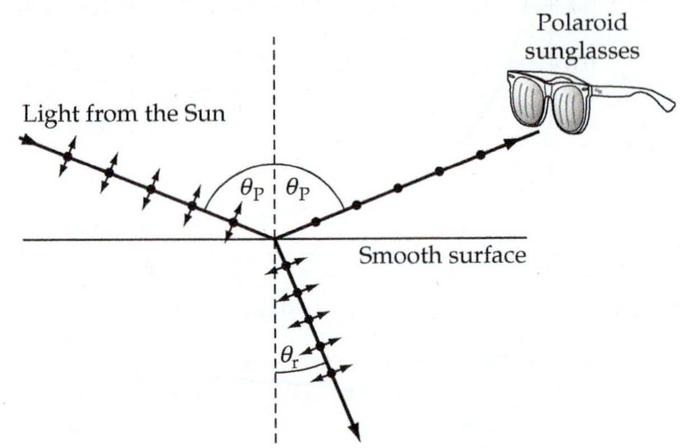

13 (*c*)

15 The population inversion between the state $E_{2\text{ Ne}}$ and the state 1.96 eV below it (see Figure 31-51) is achieved by inelastic collisions between neon atoms and helium atoms excited to the state $E_{2\text{ He}}$.

17 (*d*)

19 3 ps

21 (*a*) 2:00 a.m., September 1, (*b*) 2:08 a.m., September 1

23 (*a*)

25 14 ms

29

	(*a*) speed (m/s)	(*b*) wavelength (nm)	(*c*) frequency (Hz)
air	3.00×10^8	633	4.74×10^{14}
water	2.25×10^8	476	4.74×10^{14}
glass	2.00×10^8	422	4.74×10^{14}

31 (*a*) 50°, (*b*) 39°, (*c*) 26°

33 (*a*) 92%, (*b*) 99%

37 5.1 m wide, 2.2 m deep

39 48.8°

41 1.0×10^2 m^2

43 1.30

45 5°

47 (*a*) 62.5°, (*b*) Yes, if $\theta \geq 41.8°$, where θ is the angle of incidence for the rays in glass that are incident on the glass-water boundary, the rays will leave the glass through the water and pass into the air.

49 1.0°

51 (*a*) 53.1°, (*b*) 56.3°

53 (*a*) $\frac{1}{8}I_0$, (*b*) $\frac{3}{32}I_0$

55 (*a*) 30°, (*b*) 1.7

57 $I_3 = \frac{1}{8}I_0\sin^2 2\omega t$

59
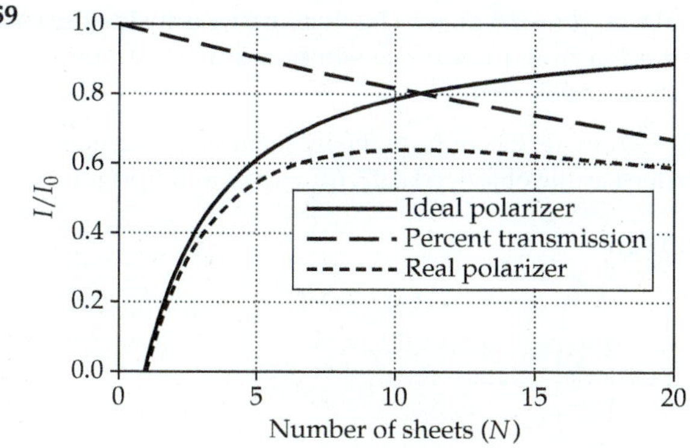

The optimum number of sheets is 11.

61 (*a*) $I_4 = 0.211I_0$, (*b*) For the single sheet between the two end sheets at $\theta = 45°$, $I_3 = 0.125I_0$. The intensity with four sheets at angles of 0°, 30°, 60° and 90° is greater than the intensity of three sheets at angles of 0°, 45° and 90° by a factor of 1.69.

63 (*a*) right circularly polarized, (*b*) $\vec{E} = E_0 \sin(kx + \omega t)\hat{j} - E_0 \cos(kx + \omega t)\hat{k}$

65 (*a*) 435 nm (*b*) 1210 nm

67 (*a*) $\lambda_{max} = 388$ nm, $\lambda_{2\to1} = 1140$ nm, $\lambda_{1\to0} = 588$ nm, (*b*) $\lambda_{1\to0} = 588$ nm, $\lambda_{3\to1} = 554$ nm

69 (*a*) 15 mJ, (*b*) 5.2×10^{16}

71 37°

73 (*a*) 36.8°, (*b*) 38.7°

75 (*a*) $x = -1.00$ m, (*b*) 26.6°, (*c*) 26.6°

77 $\theta_{p\text{ silicate flint}} = 58.3°$, $\theta_{p\text{ borate flint}} = 57.5°$, $\theta_{p\text{ quartz}} = 57.0°$, $\theta_{p\text{ silicate crown}} = 56.5°$

79 (*b*) $\theta_p < \theta_c$

81 (*a*) 1.33, (*b*) 37.2°, (*c*) 48.6°. Because 48.6° is also the angle of incidence at the liquid–air interface and because it is equal to the critical angle for total internal reflection at this interface, no light will emerge into the air.

83 (*c*) 1.67°

Chapter 32

1 Yes. Note that a virtual image is "seen" because the eye focuses the diverging rays to form a real image on the retina. For example, you can photograph the virtual image of yourself in a flat mirror and get a perfectly good picture.

3 (*a*) False, (*b*) False, (*c*) True, (*d*) False

5 (*a*) The mirror will produce an upright image for all object distances.

(*b*) The mirror will produce a virtual image for all object distances.

(*c*) The mirror will produce an image that is smaller than the object for all object distances.

(*d*) The mirror will never produce an enlarged image.

7 (*b*)

9 (*d*)

11 (*a*)

13 The muscles in the eye change the thickness of the lens and thereby change the focal length of the lens to accommodate objects at different distances. A camera lens, on the other hand, has a fixed focal length so that focusing is accomplished by varying the distance between the lens and the light-sensitive surface.

15 The objective lens of a microscope ordinarily produces an image that is larger than the object being viewed (see Figure 32-52), and that image is angularly magnified by the eyepiece. The objective lens of a telescope, on the other hand, ordinarily produces an image that is smaller than the object being viewed (see Figure 32-53), and that image is angularly magnified by the eyepiece. The telescope never produces a real image that is larger than the object.

17 $\frac{1}{2}R_{Earth}$

19
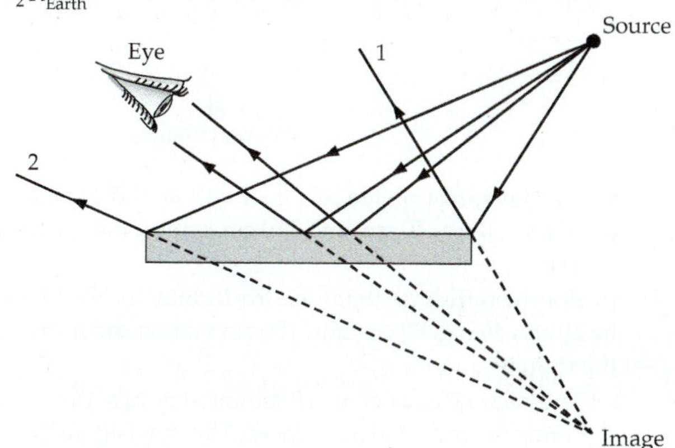

21 (*a*)
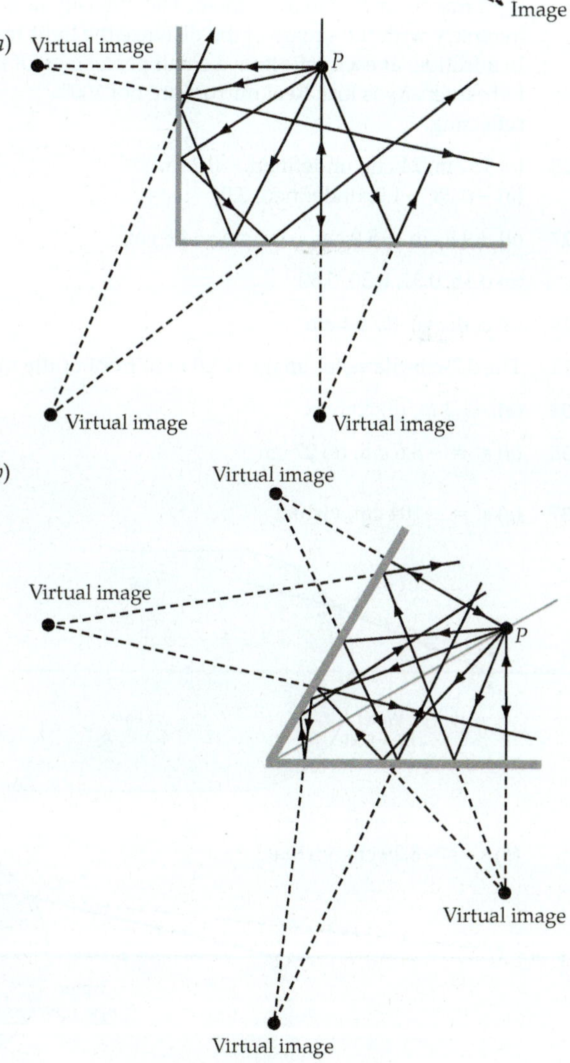

(c)

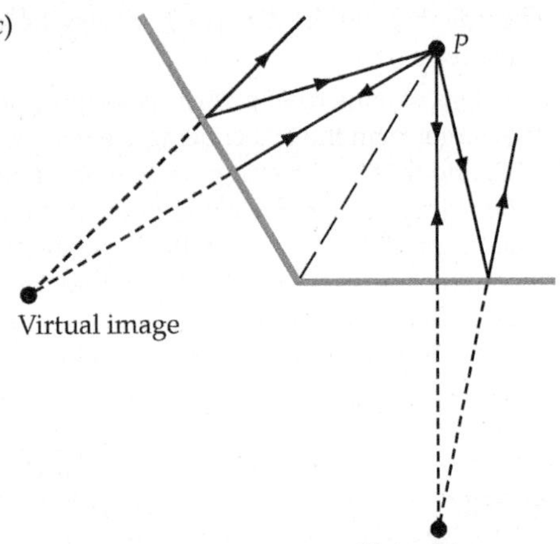

23 (a) For the mirror on the left, the location of the images to be 10 cm, 50 cm, 70 cm, and 110 cm behind the mirror on the left.
(b) For the mirror on the right, the location of the images to be 20 cm, 40 cm, 80 cm, and 100 cm behind the mirror on the right.
(c) The successive images are dimmer because the light travels farther to form them. The intensity falls of inversely with the square of the distance the light travels. In addition, at each reflection a small percentage of the light intensity is lost. Real mirrors are not 100% reflecting.

25 (a) 15 cm, 24 cm, undefined, −0.2 m, (b) −0.28, −1.0, undefined, 3.0

27 (a) −9.9 cm, −8.0 cm, −6.0 cm, −4.8 cm, (b) 0.18, 0.33, 0.50, 0.60

29 (a) concave, (b) 5.1 cm

31 The 3.7-cm-diameter image is 4.0 m in front of the mirror.

33 (a) −1.3 m, (b) convex

35 (a) $s' = -8.6$ cm, (b) 27 cm

39 (a) 0.6 m, (b) −0.8 m, (c) The final image is inside the rod and 0.2 m from the surface whose radius of curvature is 8.00 cm and is virtual.

41 (a) −30 cm, (b) 22 cm from the lens and on the same side of the lens as the object, (c) 0.27, (d) virtual and upright

43 (a) 19 cm

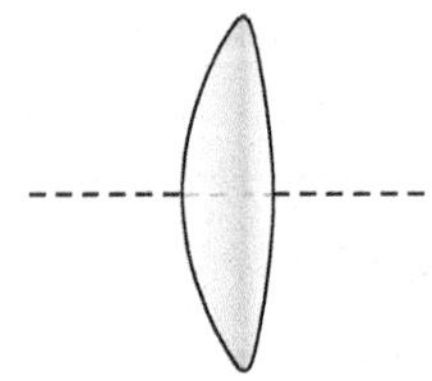

(b) 30 cm

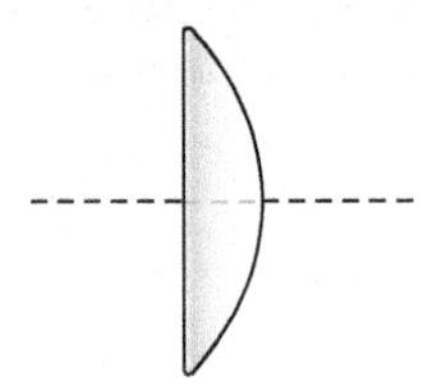

(c) −15 cm

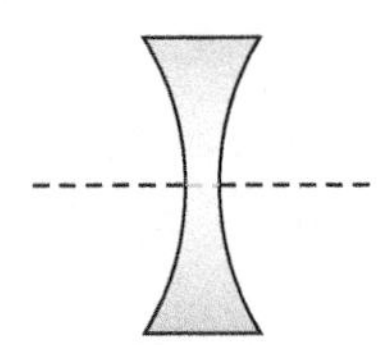

(d) −52 cm

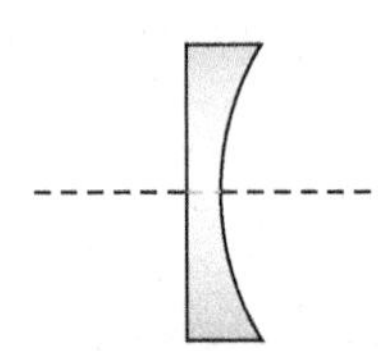

37 (a) $s' = -104$ cm, virtual

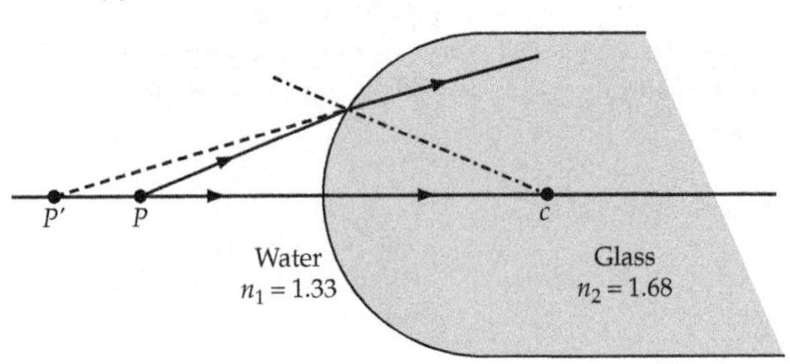

(c) $s' = 64$ cm, real

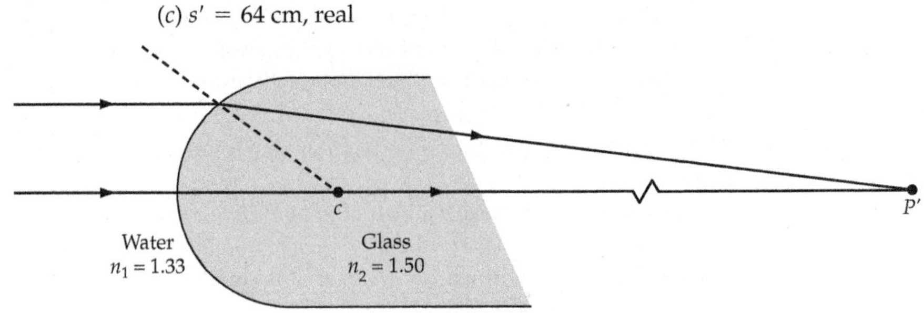

(b) $s' = -8.29$ cm, virtual

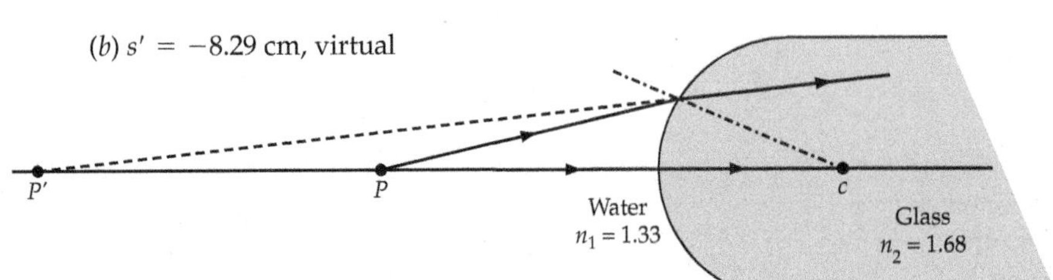

45 (*a*) $s' = 16.7$ cm, $y' = -2.00$ cm. Because $s' > 0$, the image is real and because $y'/y = -0.67$ cm, the image is inverted and diminished. These results confirm those obtained graphically. However, this ray diagram is not to scale.

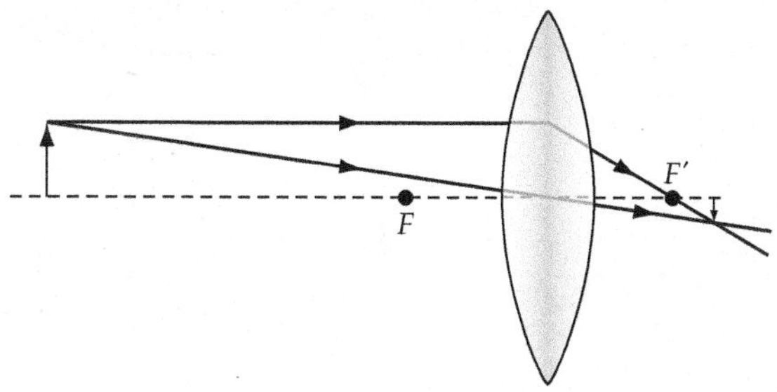

(*b*) $s' = 20.0$ cm, $y' = -3.00$ cm. Because $s' > 0$ the image is real. Because $y' = -3.00$ cm, the image is inverted and the same size as the object. These results confirm those obtained from the ray diagram.

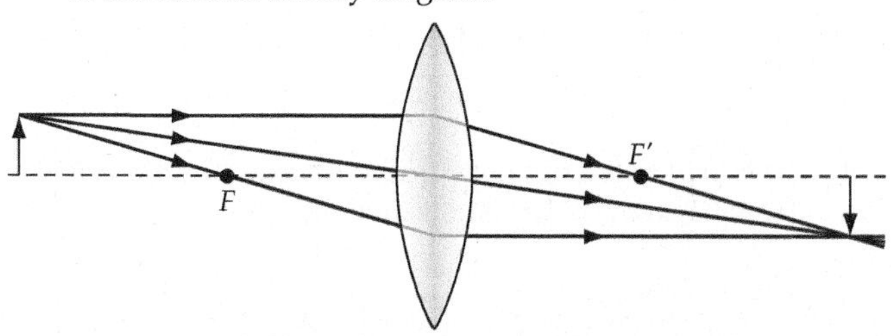

(*c*) $s' = -6.67$ cm, $y' = 1.00$ cm. Because $s' < 0$, the image is virtual. Because $y' = 1.00$ cm, the image is erect and about one-third the size of the object. These results are consistent with those obtained graphically.

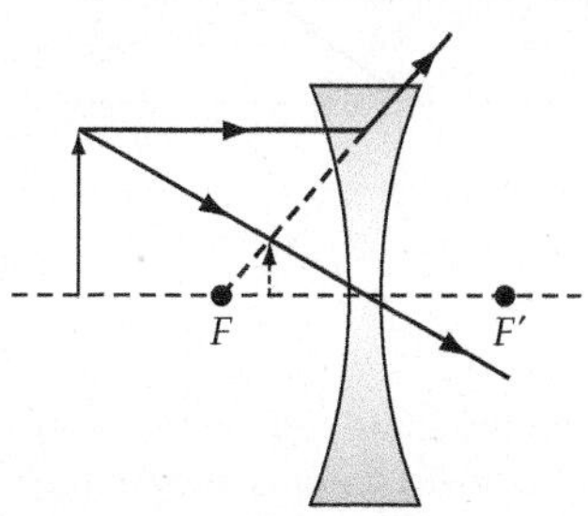

47 $r_1 = -16.2$ cm, $r_2 = \infty$,

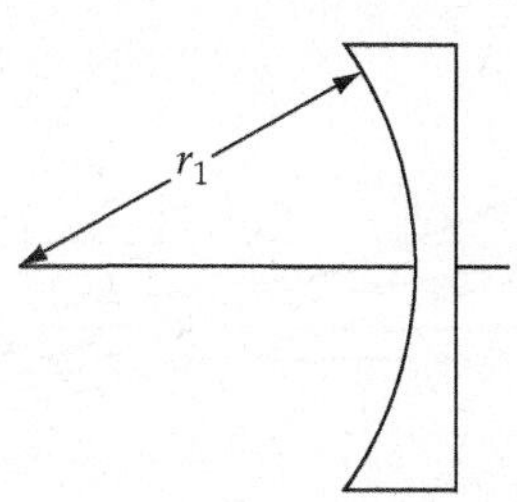

$r_1 = -32.4$ cm, $r_2 = 32.4$ cm,

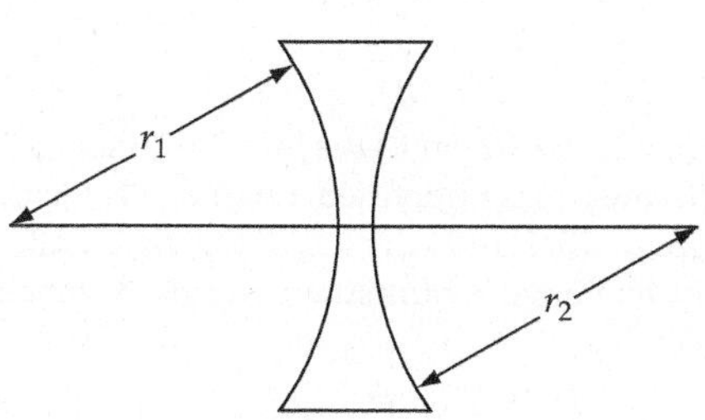

$r_1 = -5.40$ cm, $r_2 = -8.10$ cm

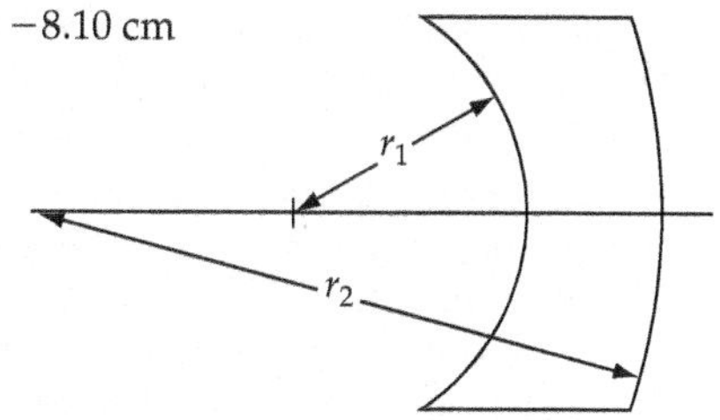

49 (*a*) The final image is 85 cm to the right of the object.

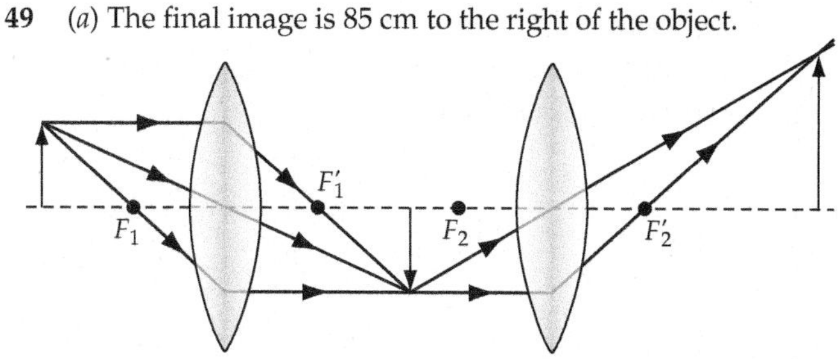

(*b*) Because $s_2' > 0$, the image is real, and because $m = m_1 m_2 = 2.0$, the image is erect and twice the size of the object.

(*c*) 2.0

51 (*b*) 3.70 m

53 (*a*) and (*b*)

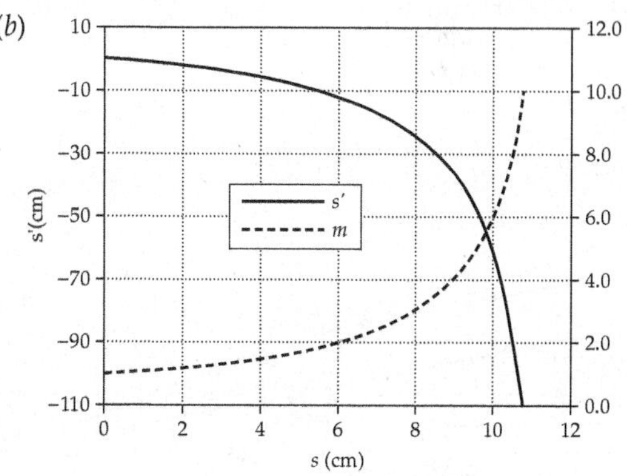

(*c*) The images are virtual and erect for this range of object distances.

(*d*) The equation for the vertical asymptote of the graph of s' versus s is $s = f$. This indicates, that as the object moves toward the second focal point, the magnitude of the image distance becomes large without limit. The equation for the vertical asymptote of the graph of m versus s is $s = f$. This indicates, that as the object moves toward the second focal point, the image becomes the large without limit. In addition, as s approaches zero s' approaches negative infinity and m approaches 1, so as the object moves toward the lens the image becomes the same size as the object and the magnitude of the image distance increases without limit.

55 (*a*) $s_2' = f_2 - 15.0$ cm, $m = m_2 = 1.00$. The final image is 20 cm from the object, virtual, erect, and the same size as the object.

(*b*)

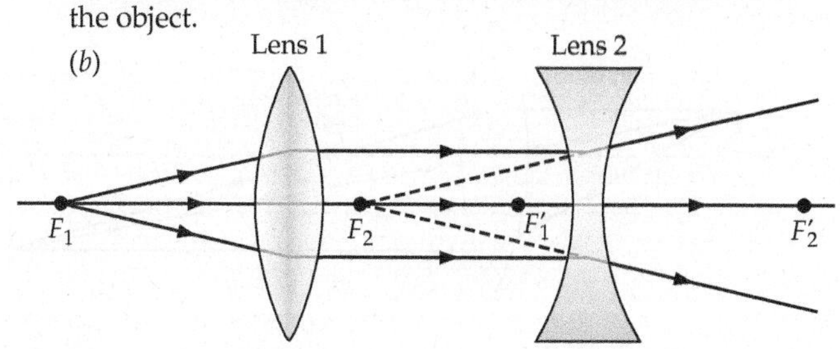

59 (*a*) The final image is 18.7 cm to the right of the second lens. (*b*) −1.53 (*c*) Because $m < 0$, the image is inverted. Because $s'_2 > 0$, the image is real.

61 Reasons for the preference for reflectors include: (1) no chromatic aberrations, (2) less expensive to shape one side of a piece of glass than to shape both sides, (3) reflectors can be more easily supported from rear instead of edges, preventing sagging and focal length changes, and (4) support from rear makes larger sizes easier to handle.

63 −1.77 mm

65 (*a*) 80.0 μrad (*b*) 1.60 mm

67 (*c*) $P_{min} = 40.0$ D, $A = 4.00$ D

69 (*c*) 6.0 D

71 0.444 D

73 3.1 D

75 5.0

77 (*a*) 3.0, (*b*) 4.0

79 (*a*) −19, (*b*) -1.9×10^2

81 −230

83 (*a*) 9.00 mm, (*b*) 0.180 mrad, (*c*) $M = -20.0$

85 (*a*) 25.0, (*b*) −1340

87 (*b*)

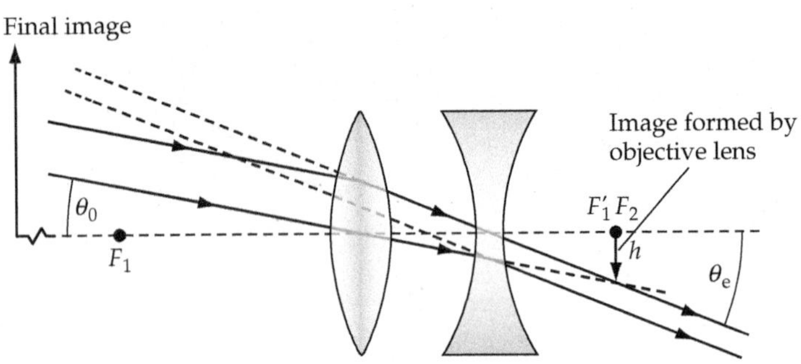

89 -6.67×10^{-3}

91 (*a*) $s = 5.0$ cm, $s' = -10$ cm

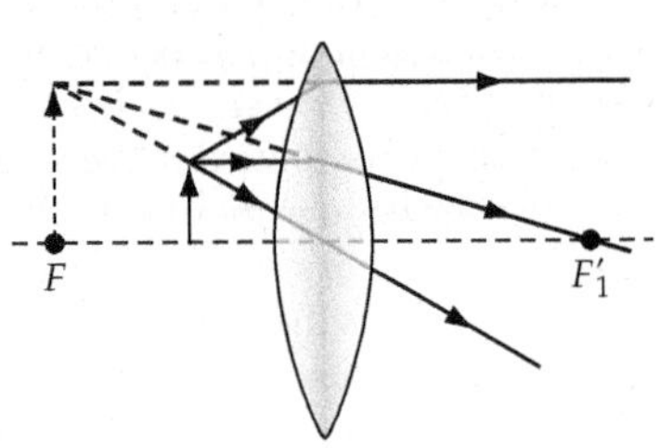

(*b*) $s = 15$ cm, $s' = 30$ cm

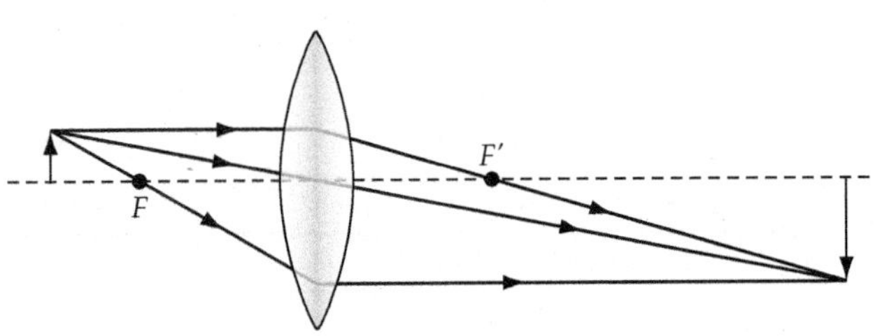

93 (*a*) The lens that has a focal length of 25 mm should be the objective. The two lenses should be separated by 210 mm. The angular magnification is −21.

(*b*)

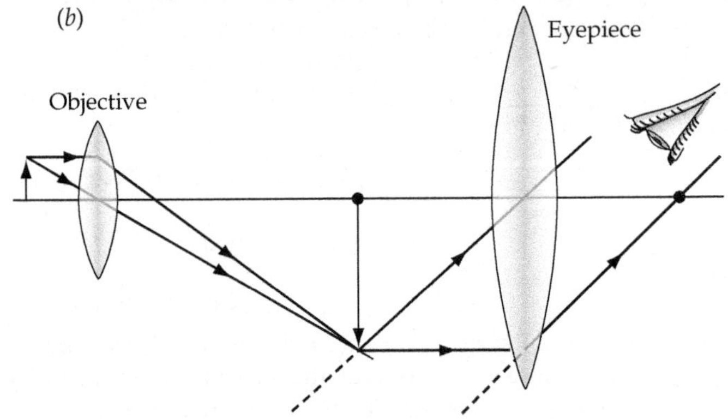

95 3.7 m

97 (*a*) 9.5 cm to the right of the second lens. (*b*) About 20% larger than the object and is inverted.

(*c*)

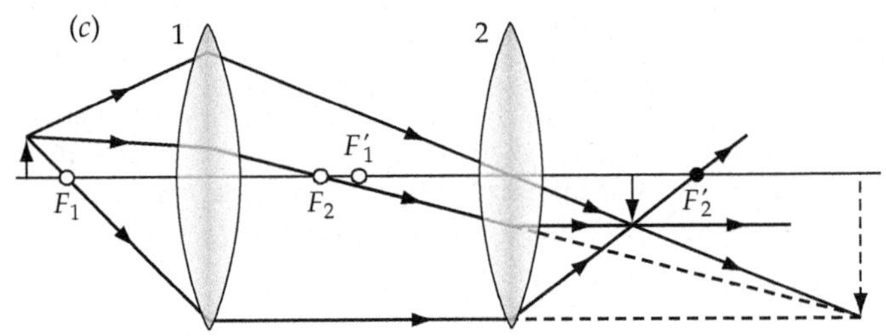

99 9.72 cm/s

101 (*a*) 18 cm from the lens, on the same side as the original object. (*b*) real and upright, (*c*) To see this image the eye must be to the left of image 4.

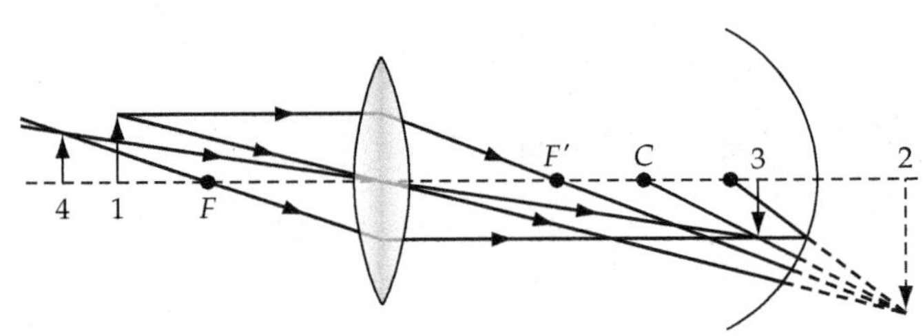

103 37 cm

105 (*a*) The final image is 13.2 cm to the left of the center of the ball. (*b*) Because s'_1 is undefined, no image is formed when the object is 20.0 cm to the left of the glass ball. (Alternatively, an image is formed an infinite distance to the left of the ball.)

Chapter 33

1 (*a*)

3 The thickness of the air space between the flat glass and the lens is approximately proportional to the square of d, the diameter of the ring. Consequently, the separation between adjacent rings is proportional to $1/d$.

5 Colors are observed when the light reflected from the front and back surfaces of the film interfere destructively for some wavelengths and constructively for other wavelengths. For this interference to occur, the phase difference between the light reflected from the front and back surfaces of the film must be constant. This means that twice the thickness of the film must be less than the coherence length of the light. The film is called a thin film if twice its thickness is less than the coherence length of the light.

7 (*d*)

9 (*a*)

11 (*a*)

13 (*a*)

15 (*a*) False, (*b*) True, (*c*) True, (*d*) True, (*e*) True

17 The condition for the resolution of the two sources is given by Rayleigh's criterion: $\alpha_c = 1.22\,\lambda/D$ (Equation 33-25), where α_c is the critical angular separation, D is the diameter of the aperture, and λ is the wavelength of the light illuminating (or emitted by) the objects, in this case headlights, to be resolved. Because the diameter of the pupils of your eyes are larger at night, the critical angle is smaller at night, which means that at night you can resolve the light as coming from two distinct sources when they are at a greater distance.

19 (*a*) 11 km, (*b*) 9.6 km

21 5.9 $c\cdot$y

23 $\approx$2.9 rad

25 5.5 μm $< d <$ 5.8 μm

27 (*a*) 600 nm, (*b*) 720 nm, 514 nm, and 400 nm, (*c*) 720 nm, 514 nm, and 400 nm

29 476 nm

31 (*c*) 68, (*d*) 1.14 cm (*e*) The fringes would become more closely spaced.

33 0.535 mm, 0.926 mm

35 4.95 mm

37 (*a*) 50.0 μm, (*b*) Not with the unaided eye. The separation is too small to be observed with the naked eye. (*c*) 0.500 mm

39 625 nm and 417 nm

41 (*a*) 0.60 mrad, (*b*) 6.0 mrad, (*c*) 60 mrad

43 (*a*) 1.53 km

45 (*a*) 20.0 μm, (*b*) 9

47 8

49 $\vec{E} = 3.6\,A_0 \sin(\omega t - 0.98\text{ rad})\hat{i}$

51 $I/I_0 = 0.0162$

53 (*b*) 6.00 mm. The width for four sources is half the width for two sources.

55 (*a*)

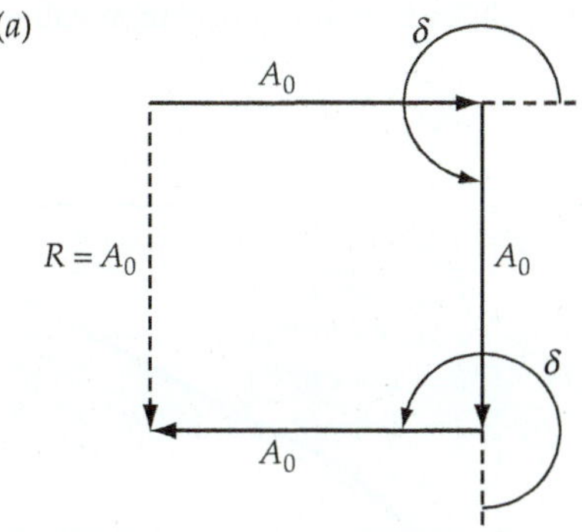

(*b*) 5.56 mW/m^2

57 (*a*) 8.54 mrad, (*b*) 6.83 cm

59 7.00 mm

61 5.00 × 10^9 m

63 (*a*) 86.9 mrad, 82.1 mrad, (*b*) 709 mrad, 662 mrad

65 30.0°

67 One can see the complete spectrum for only the first and second order spectra. That is, only for $m = 1$ and 2. Because 700 nm $< 2 \times 400$ nm, there is no overlap of the second-order spectrum into the first-order spectrum; however, there is overlap of long wavelengths in the second order with short wavelengths in the third-order spectrum.

69 (*a*) 36.4 cm, 80.1 cm, (*b*) 88.4 μm, (*c*) 8000

71 3.09 × 10^5, 5.14 × 10^4 cm^{-1}

73 (*a*) $\phi_m = \frac{1}{2}\sin^{-1}\left(m\frac{\lambda}{d}\right)$, (*b*) 32.1°

75 3.5 μm

77 3.6°, 2.5°

79 (*a*) 15.1 cm, (*b*) 3.33 m^{-1}

81 0.13 mrad

85 (*a*) 97.8 nm, (*b*) No, because 180 nm is not in the visible portion of the spectrum. (*c*) 0.273

87 12 m

Chapter 34

1 (*c*)

3 (*a*)

5 (*a*) True, (*b*) True, (*c*) True

7 (*c*)

9 According to quantum theory, the average value of many measurements of the same quantity will yield the expectation value of that quantity. However, any single measurement may differ from the expectation value.

11 2.48 pm, 2%

13 (*a*)

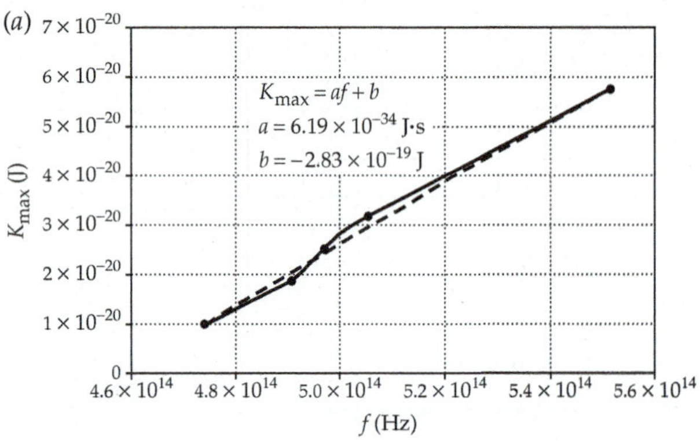

(*b*) 1.77 eV (*c*) cesium

15 (*a*) 4.14×10^{-7} eV, (*b*) 3.72×10^{-9} eV

17 (*a*) 12.4 keV, (*b*) 1.24 GeV

19 $1.95 \times 10^{16}\ s^{-1}$

21 (*a*) 4.13 eV, (*b*) 2.10 eV, (*c*) 0.78 eV, (*d*) 590 nm

23 (*a*) 653 nm, 4.58×10^{14} Hz, (*b*) 3.06 eV, (*c*) 1.64 eV

25 1.2 pm

27 0.18 nm

29 9.32×10^{-24} kg·m/s, 1.80×10^{-23} kg·m/s

31 2.9 nm

33 (*a*) $p_e = 2.09 \times 10^{-22}$ N·s, $p_p = 8.97 \times 10^{-21}$ N·s,

$p_\alpha = 8.97 \times 10^{-21}$ N·s

(*b*) $\lambda_e = 3.17$ pm, $\lambda_p = 73.9$ fm, $\lambda_\alpha = 37.0$ fm

35 20.2 fm

37 0.17 nm

39 4.6 pm

41 (*a*) $E_1 = 205$ MeV, $E_2 = 818$ MeV, $E_3 = 1.84$ GeV

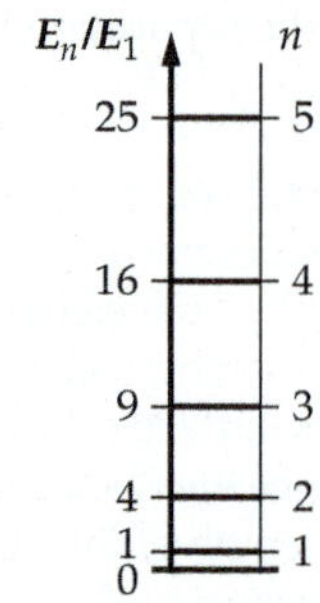

(*b*) $\lambda_{2\to1} = 2.02$ fm,

(*c*) $\lambda_{3\to2} = 1.21$ fm,

(*d*) $\lambda_{3\to1} = 0.758$ fm

43 (*a*) 0, (*b*) 1, (*c*) 0.002

45 (*a*) $L/2$, (*b*) $0.321L^2$

47 (*a*) $1/\sqrt{2}$, (*b*) 0.865

49 (*a*) 0.500, (*b*) 0.402, (*c*) 0.750

51 (*b*) For large values of n, the result agrees with the classical value of $L^2/3$ given in Problem 50.

53 $\langle x \rangle = 0$, $\langle x^2 \rangle = L^2\left[\frac{1}{12} - \frac{1}{2\pi^2}\right]$

55 (*a*) 3.10 eV, (*b*) 6.24×10^{16} eV, (*c*) 2.08×10^{16}

57 (*a*) 1 μm, 10^{-16} kg·m/s, (*b*) 2×10^{11}

59 0.2 keV

61 7×10^3 km

63 (*a*) 92 mW/m², (*b*) 3×10^4

67 1.3 MeV. The energy of the most energetic electron is approximately 2.5 times the rest energy of an electron.

69 1.04 eV, 554 nm

71 (*b*) 0.2% (*c*) Classically, the energy is continuous. For very large values of n, the energy difference between adjacent levels is infinitesimal.

73 (*a*) 6.2×10^{-4} eV/s, (*b*) 53 min

Chapter 35

1 (*a*)

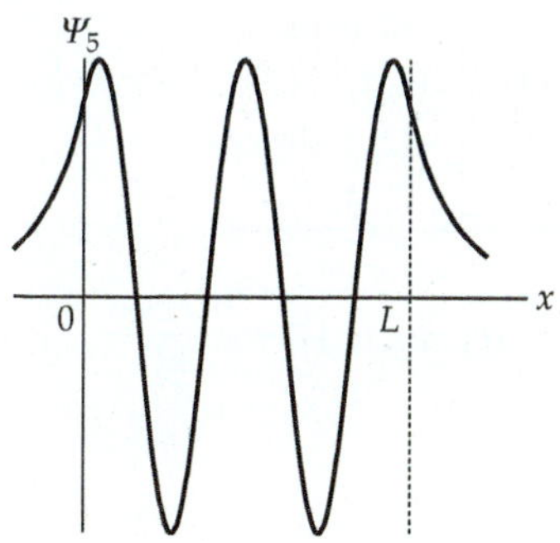

(*b*)

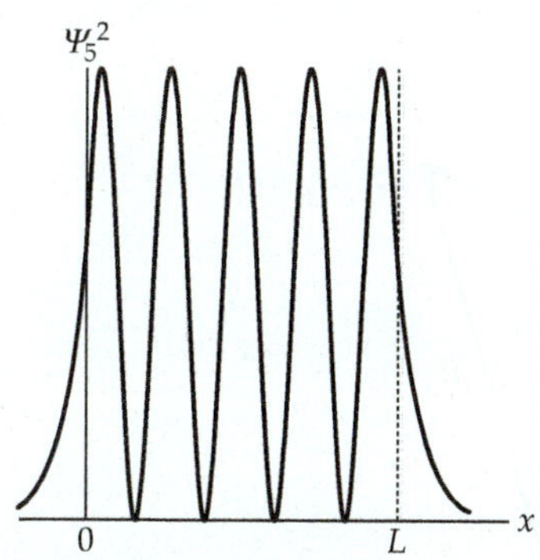

9 (*a*) 9.5 nm, (*b*) 4.1 meV

11 $\Delta x \, \Delta p = \dfrac{\hbar}{2}$

13 (*b*)

Cell	Content/Formula	Algebraic Form
A2	1.0	α
B2	(1−SQRT((A2−1)/A2))/(1+SQRT((A2−1)/A2))²	$\left(\dfrac{1-\sqrt{\dfrac{\alpha-1}{\alpha}}}{1+\sqrt{\dfrac{\alpha-1}{\alpha}}}\right)^2$
C2	1 − B2	$1-\left(\dfrac{1-\sqrt{\dfrac{\alpha-1}{\alpha}}}{1+\sqrt{\dfrac{\alpha-1}{\alpha}}}\right)^2$

	A	B	C
1	α	*R*	*T*
2	1.0	1.000	0.000
3	1.2	0.298	0.702
4	1.4	0.298	0.802
5	1.6	0.149	0.851
18	4.2	0.036	0.964
19	4.4	0.034	0.966
20	4.6	0.032	0.968
21	4.8	0.031	0.969
22	5.0	0.029	0.971

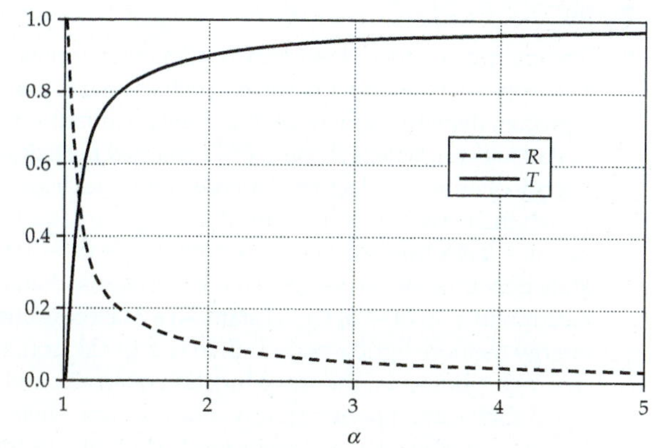

15 (*a*) 10^{-17}, (*b*) 10^{-2}

17 (*a*) $r_{1\,4.0\text{ MeV}} = 66$ fm, $r_{1\,7.0\text{ MeV}} = 38$ fm

(*b*) $T_{4.0\text{ MeV}} \approx 10^{-51}$, $T_{7.0\text{ MeV}} \approx 10^{-38}$

19 (*a*)

n_1	1	1	1	1	1	1	1	1	1	1
n_2	1	1	1	2	1	2	2	1	2	3
n_3	1	2	3	1	4	2	3	5	4	1
E	21	24	29	33	36	36	41	45	48	53

(*b*) (1, 1, 4) and (1, 2, 2)

(*c*) $\psi(1, 1, 4) = A \sin\left(\dfrac{\pi}{L_1}x\right) \sin\left(\dfrac{\pi}{2L_1}y\right) \sin\left(\dfrac{\pi}{L_1}z\right)$

21 (*a*) $\psi(x,y) = A \sin\dfrac{n\pi}{L}x \sin\dfrac{m\pi}{L}y$

(*b*) $E_{nm} = \dfrac{h^2}{8mL^2}\left(n^2 + m^2\right)$

(*c*) (1, 2) and (2, 1)

(*d*) (1, 7), (7, 1), and (5, 5)

23 $E_{110\text{ bosons}} = \dfrac{5h^2}{4mL^2}$

29 $E_0 = \dfrac{5h^2}{mL^2}, E_1 = E_2 = \dfrac{21h^2}{4mL^2}$

31 (*b*) $\langle x^2 \rangle = \dfrac{2}{L}\left(\dfrac{L^3}{24} - \dfrac{L^3}{4n^2\pi^2}\cos n\pi\right)$

35 $A_2 = \sqrt[4]{\dfrac{8m\omega_0}{h}}$

Chapter 36

1 Examination of Figure 36-4 indicates that as n increases, the spacing of adjacent energy levels decreases.

3 (*a*)

5 (*d*)

7 (*a*)

9 The energy of a bound isolated system that consists of two oppositely charged particles, such as an electron and a proton, depends only upon the principle quantum number n. For sodium, which consists of 12 charged particles, the energy of an $n = 3$ electron depends upon the degree to which the wave function of the electron penetrates the $n = 1$ and $n = 2$ electron shells. An electron in a 3s ($n = 3$, $\ell = 0$) state penetrates these shells to a greater degree than does an electron in a 3p ($n = 3$, $\ell = 1$) state, so a 3s electron has less energy (is more tightly bound) than is a 3p electron. In hydrogen, however, the wave function of an electron in the $n = 3$ shell cannot penetrate any other electron shells because no other electron shells exist. Thus, an electron in the 3s state in hydrogen has the same energy as an electron in the 3p state in hydrogen.

11 In conformity with the exclusion principle, the total number of electrons that can be accommodated in states of quantum number n is n^2 (see Problem 48). The fact that closed shells correspond to $2n^2$ electrons indicates that there is another quantum number that can have two possible values.

13 (*a*) phosphorus, (*b*) chromium

15 (*d*)

17 The optical spectrum of any atom is due to the configuration of its outer-shell electrons. Ionizing the next atom in the periodic table gives you an ion with the same number of outer-shell electrons, and almost the same nuclear charge. Hence, the spectra should be very similar.

21 (*a*) 10^5, (*b*) 10^3, (*c*) 5.08×10^4

23 (*a*) 103 nm, (*b*) 97.3 nm

25 (*a*) 1.51 eV, 821 nm
(*b*) 0.661 eV, 1880 nm, 0.967 eV, 1280 nm, 1.13 eV, 1100 nm

6→3	5→3	4→3
1100 nm	1280 nm	1880 nm

27 (*b*) $1.096850 \times 10^7\ \text{m}^{-1}$, $1.097448 \times 10^7\ \text{m}^{-1}$, R_H and $R_{H\,approx}$ agree to three significant figures.
(*c*) 0.0545%

29 (*a*) 1.49×10^{-34} J·s, (*b*) −1, 0, +1

(*c*)

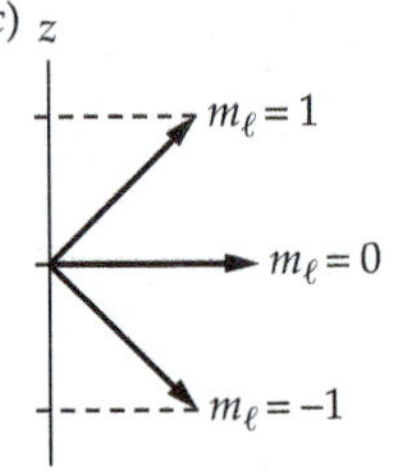

31 (*a*) 0, 1, 2, (*b*) For $\ell = 0$, $m_\ell = 0$. For $\ell = 1$, $m_\ell = -1, 0, +1$. For $\ell = 2$, $m_\ell = -2, -1, 0, +1, +2$. (*c*) 18

33 (*a*) 45.0°, (*b*) 26.6°, (*c*) 8.05°

35 (*a*) $6\hbar^2$, (*b*) $4\hbar^2$, (*c*) $2\hbar^2$

37 (*a*) 4

(*b*)

n	ℓ	m_ℓ	(n, ℓ, m_ℓ)
2	0	0	(2, 0, 0)
2	1	−1	(2, 1, −1)
2	1	0	(2, 1, 0)
2	1	1	(2, 1, 1)

39 (*a*) $\psi_{200}(a_0) = \dfrac{0.0605}{a_0^{3/2}}$, (*b*) $\left[\psi_{200}(a_0)\right]^2 = \dfrac{0.00366}{a_0^3}$

(*c*) $P(a_0) = \dfrac{0.0460}{a_0}$

41 (*a*) 9.20×10^{-4}, (*b*) 0

47 0.323

49 $\ell = 0$ or 1

51

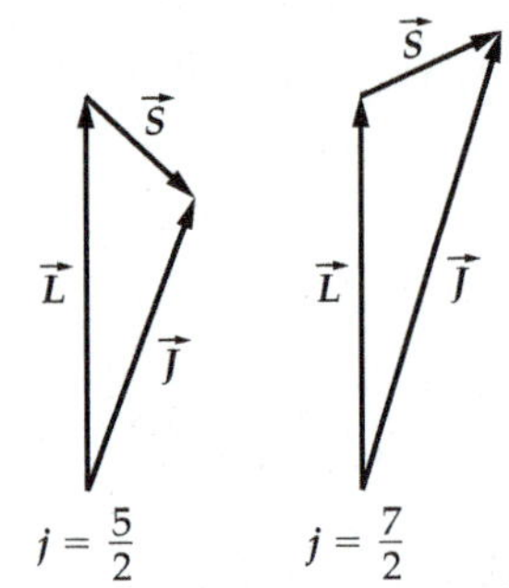

53 (*c*)

55 (*a*) $L_z = -2\hbar, -\hbar, 0, \hbar, 2\hbar$, (*b*) $L_z = -3\hbar, -2\hbar, -\hbar, 0, \hbar, 2\hbar, 3\hbar$

57 (*a*) 2s or 2p, (*b*) $1s^22s^22p^63p$, (*c*) 1s2s

59 (*a*) 0.0610 nm, 0.0578 nm, (*b*) 0.0542 nm

61 (*a*) 1.00 nm, (*b*) 0.155 nm

63 $n_i = 4$ to $n_f = 1$

65

λ, nm	n_i	n_f
164	3	2
230.6	9	3
541	7	4

67 (*a*) 1.6179 eV, 1.6106 eV, (*b*) 0.00730 eV, (*c*) 63.0 T
(*b*) No

71 (*a*) 1.06 GHz, (*b*) 28.4 cm, microwave

73 (*a*) $1.097075 \times 10^7\ \text{m}^{-1}$, (*b*) 0.179 nm

75 (*a*) $1.097074 \times 10^7\ \text{m}^{-1}$, (*b*) 0.0600 nm, (*c*) 0.238 nm

Chapter 37

1 Because the center of charge of the positive Na ion does not coincide with the center of charge for the negative Cl ion, the NaCl molecule has a permanent dipole moment. Hence, it is a polar molecule.

3 Neon occurs naturally as Ne, not Ne_2. Neon is a noble gas. Atoms of noble gases have a closed shell electron configuration.

5 The diagram would consist of a nonbonding ground state with no vibrational or rotational states for ArF (similar to the upper curve in Figure 37-4) but for ArF* there should be a bonding excited state with a definite minimum with respect to internuclear separation and several vibrational states as in the excited state curve of Figure 37-13.

7 The effective force constant from Example 37-4 is 1.85×10^3 N/m. This value is about 25% larger than the given value of the force constant of the suspension springs on a typical automobile.

9 For H_2, the concentration of negative charge between the two protons holds the protons together. In the H_2^+ ion, there is only one electron that is shared by the two positive charges such that most of the electronic charge is again between the two protons. However, in the H_2^+ ion the negative charge between the protons is not as effective as the larger negative charge between them in the H_2 molecule, and the protons should be farther apart. The experimental values support this argument. For H_2, $r_0 = 0.074$ nm, while for H_2^+, $r_0 = 0.106$ nm.

11 For more than two atoms in the molecule, there will be more than just one frequency of vibration because more relative motions are possible. In advanced mechanics, these are known as normal modes of vibration.

13 $\ell \approx 2 \times 10^{30}$, $E_{0r} \approx 5 \times 10^{-65}$ J

15 0.947 nm

17 0.44 eV

19 You should agree. The potential energy curve is shown in the following diagram. The turning points for vibrations of energy E_1 and E_2 are at the values of r where the energies equal $U(r)$. The average value of r for the vibrational levels E_1 and E_2 are labeled $r_{1\,av}$ and $r_{2\,av}$. Note that the estimate of $r_{1\,av}$ is force midway between $r_{1\,min}$ and $r_{1\,max}$. The potential is like a special spring that has a greater force constant for compressions than it has for extensions. The period of a spring-and-mass oscillator is inversely proportional to the square root of the spring constant, so our "special spring" spends more time in extension than in compression. As a result, $r_{1\,av}$ will be greater than the equilibrium radius. This argument can be extended to explain why $r_{2\,av}$ is greater than $r_{1\,av}$. It is because the "force constant" for extension, which can be estimated by taking the average slope of the potential energy curve in the region to the right of the equilibrium position, is greater for $E = E_2$ than for $E = E_1$. It is also because the "force constant" for compression is greater for $E = E_2$ than for $E = E_1$. It follows that $r_{2\,av}$ is greater than $r_{1\,av}$. Because $r_{2\,av}$ is greater than $r_{1\,av}$, it follows that as the vibrational energy of a diatomic molecule increases, the average separation of the atoms of the molecule increases and, hence, the solid expands with heating.

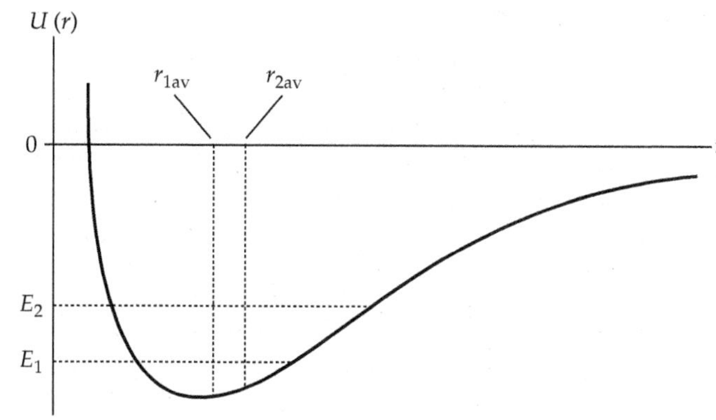

21 (a) $U_e = -6.64$ eV, (b) $E_{d\,calc} = 5.70$ eV, (c) $U_{rep} = 0.63$ eV

23 0.121 nm

25 41

27 5.6 meV

29 (a) 0.179 eV, (b) 3×10^{-47} kg·m², (c) 0.1 nm

31 (a) 1.45×10^{-46} kg·m², 0.239 meV

(b)

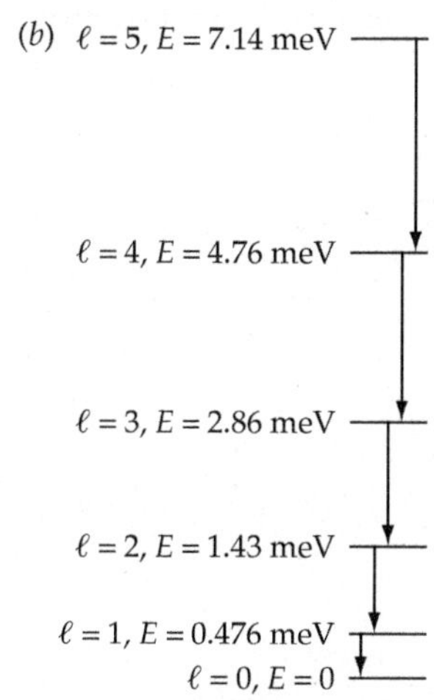

$\Delta E_{54} = 2.38$ meV, $\Delta E_{43} = 1.90$ meV, $\Delta E_{32} = 1.43$ meV, $\Delta E_{21} = 1.25$ meV, $\Delta E_{10} = 0.476$ meV

(c) $\lambda_{10} = 2600\ \mu$m, $\lambda_{21} = 1300\ \mu$m, $\lambda_{32} = 867\ \mu$m, $\lambda_{43} = 650\ \mu$m, $\lambda_{54} = 520\ \mu$m, microwave

33 $\mu_{H^{35}Cl} = 0.972$ u, $\mu_{H^{37}Cl} = 0.974$ u,

$\frac{\Delta\mu}{\mu} = 0.00150$, $\Delta f/f = 0.0012$, in fair agreement (about 20% difference) with the calculated result. Note that Δf is difficult to determine precisely from Figure 37-17.

35 0.955 meV

37 1.55 kN/m

39 $r_0 = a$, $U_{min} = -U_0$, $r_0 = 0.074$ nm, $U_0 = 4.52$ eV

41 $F_x = -\frac{dU}{dx} \propto \frac{1}{x^4}$

43 (a) $\frac{1\text{ eV}}{\text{molecule}} = 23.0$ kcal/mol

(b) 98.2 kcal/mol

Chapter 38

1 The energy lost by the electrons in collision with the ions of the crystal lattice appears as thermal energy throughout the crystal.

3 (*a*) potassium and nickel (*b*) 3.1 V

5 The resistivity of brass at 4 K is almost entirely due to the residual resistance (the resistance due to impurities and other imperfections of the crystal lattice). In brass, the zinc ions act as impurities in copper. In pure copper, the resistivity at 4 K is due to its residual resistance. The residual resistance is very low if the copper is very pure.

7 The resistivity of copper increases with increasing temperature; the resistivity of (pure) silicon decreases with increasing temperature because the number density of charge carriers increases.

9 (*b*)

11 The excited electron is the motion of the electron in the conduction band and contributes to the current. A hole is left in the valence band allowing the positive hole to move through the band, also the motion of the hole contributes to the current.

13 (*c*)

15

V (V)	1/slope (Ω)
−20	∞
+0.2	40
+0.4	20
+0.6	10
+0.8	5

17 2.07 g/cm^3

19 (*a*) −10.6 eV, (*b*) 2.83%

21 (*a*) 0.123 μΩ · m, (*b*) 70.7 nΩ · m

23 (*a*) $n_{Ag} = 5.86 \times 10^{22}$ electrons/cm^3
(*b*) $n_{Ag} = 5.90 \times 10^{22}$ electrons/cm^3. Both these results agree with the values in Table 38-1.

25 4.0

27 (*a*) 1.07×10^6 m/s (*b*) 1.39×10^6 m/s
(*c*) 1.89×10^6 m/s

29 (*a*) 4.22 eV (*b*) 2.85 eV

31 (*a*) 5.90×10^{28} e/m^3, (*b*) 5.50 eV, (*c*) 212, (*d*) The ratio E_F/kT is equal to 212 at $T = 300$ K. The Fermi energy is the energy of the most energetic conduction electron when the crystal is at absolute zero. Because no two conduction electrons can occupy the same state, the Fermi energy is quite high compared with kT. The kT energy is the energy the average conduction electron would have when the crystal is at temperature T if the electrons did not obey the exclusion principle.

33 3.82×10^{10} N/m^2 = 3.77×10^5 atm

35 0.192 J/(mol · K)

37 (*a*) 66 nm, (*b*) 1.8×10^{-4} nm^2

39 1.09 μm

41 180 nm

43 116 K

45 $a_{B\,Si} = 3$ nm, $a_{B\,Ge} = 8$ nm

47 37.1 nm, 38.7 nm. The mean free paths agree to within about 4%.

49

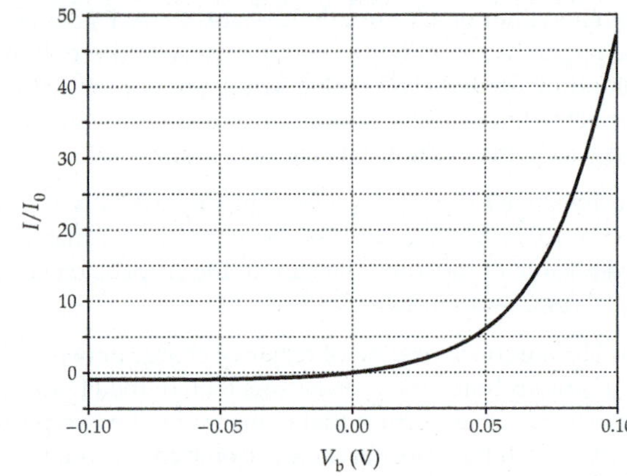

51 250

53 (*a*)

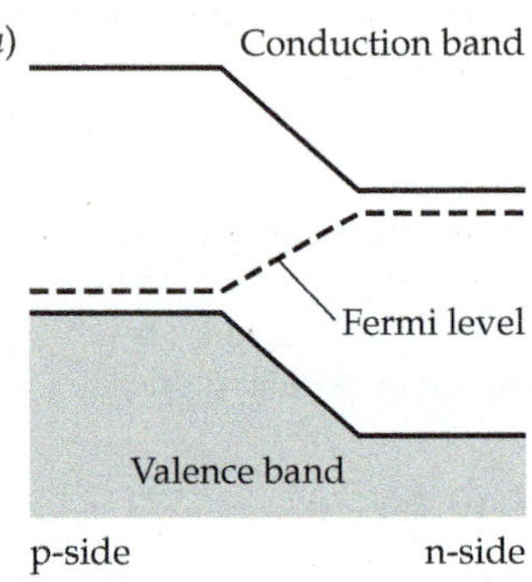

(*b*)

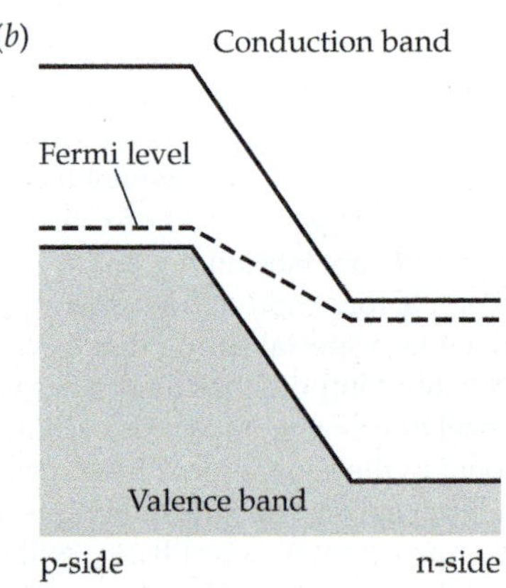

55 The charge carriers are holes and the semiconductor is *p*-type. 1.0×10^{23} m^{-3}

57 (*a*) 2.17 meV, $E_g \approx 0.8E_{g\,measured}$, (*b*) 0.454 mm

59 2.0×10^{18}

61 (*a*) 5.51 eV, (*b*) 3.31 eV, (*c*) 1.08×10^6 m/s

63 1

67 0.60

71 1.07

73 (*a*) 5.51×10^{-3}, (*b*) 1.84×10^{-2}

75 4.35×10^{14} Hz

Chapter 39

1 (*a*)

3 (*a*) True, (*b*) True, (*c*) False, (*d*) True, (*e*) False, (*f*) False, (*g*) True

5 Although $\Delta y = \Delta y'$, $\Delta t \neq \Delta t'$. Consequently, $\Delta y/\Delta t \neq \Delta y'/\Delta t'$.

7 (*a*) 0.946, (*b*) $1.23 \times 10^{10}\ c \cdot y$

9 (*a*) 0.98 km. The width of the beam is unchanged. (*b*) 9.6×10^7 m, (*c*) 0.10 μm

11 (*a*) 0.91 *c*, (*b*) 22 $c \cdot y$, (*c*) 101 y

13 (*a*) 0.385 μs, (*b*) 0.193 μs, (*c*) 0.998 *c*

15 1.85×10^4 y

17 (*a*) 1.76 μs, (*b*) 6.32 μs, (*c*) 3.1 μs, (*d*) 1.70 km

19 4.4 μs

21 (*a*) 2.10 μs, (*b*) 2.59 μs, (*c*) 0.49 μs, (*d*) 2.59 μs, (*e*) 4.36 h, (*f*) 18.8 h

23 2.22×10^7 m/s

27 11 ms

29 (*a*) $u_x = v$ and $u_y = \dfrac{c}{\gamma}$

31 (*a*) 0.976, (*b*) 0.997*c*

33 66.7%

37 (*a*) 290 MeV, (*b*) 629 MeV

39 (*a*) 0.943*c*, (*b*) 3.0 MeV, (*c*) 2.8 MeV/*c*, (*d*) 4.1 MeV/c^2, (*e*) 0.9 MeV

43 0.999*c*

45 (*a*) −0.50*c*, *S*′ moves in the −*x* direction, (*b*) 1.7 y

47 281 MeV

49 (*a*) $v = -\dfrac{E}{Mc}$, (*b*) $d = -\dfrac{LE}{Mc^2}$

51 $K_{th} = 6m_pc^2$ in agreement with Problem 40.

Chapter 40

1 (*a*) ^{15}N, ^{16}N, (*b*) ^{54}Fe, ^{55}Fe, (*c*) ^{117}Sn, ^{119}Sn

3 Generally, decay by α emission leaves the daughter nucleus neutron rich, i.e. above the line of stability. The daughter nucleus therefore tends to decay via β^- emission which converts a nuclear neutron to a proton.

5 It would make the dating unreliable because the current concentration of ^{14}C is not equal to that at some earlier time.

7 The probability for neutron capture by the fissionable nucleus is large only for slow (thermal) neutrons. The neutrons emitted during the fission process are fast (high energy) neutrons and must be slowed to thermal neutrons before they are likely to be captured by another fissionable nucleus.

9 (*a*) β^+, (*b*) β^-

11 (*a*) True (given an unlimited supply of ^{238}U), (b) False, (c) True, (d) False

13

Material Property	Ratio (order of magnitude)
Mass density	10^{15}
Half-life	10^{15}
Nuclear masses	2

15 (*a*) $E_b = 92.2$ MeV, $E_b/A = 7.68$ MeV
(*b*) $E_b = 492$ MeV, $E_b/A = 8.79$ MeV
(*c*) $E_b = 1802$ MeV, $E_b/A = 7.57$ MeV

17 (*a*) 3.0 fm, (*b*) 4.6 fm, (*c*) 7.0 fm

19 (*a*) $E_{thermal} = 4.11 \times 10^{-21}$ J = 25.7 meV, (*b*) 2.22 km/s, (*c*) 10.1 min

23 (*a*) 5 min, (*b*) 250 Bq

25 (*a*) 200 s, (*b*) 3.5×10^{-3} s^{-1}, (*c*) 125 Bq

27 (*a*) 500 Bq, 250 Bq, (*b*) $N_0 = 1.0 \times 10^6$, $N_{2.4\,min} = 5.2 \times 10^5$, (c) 12 min

29 (*a*) $4.5 \times 10^3\ \alpha$/s, (*b*) 5.3×10^4 y

31 $^{239}_{94}\text{Pu} \rightarrow {}^{235}_{92}\text{U} + {}^{4}_{2}\alpha + Q$, $Q = 5.24$ MeV, $K_\alpha = 5.15$ MeV, $K_{^{235}\text{U}} = 89.2$ keV

33 (*a*) $\lambda = 0.133$ h^{-1}, $t_{1/2} = 5.20$ h, (*b*) $N_0 = 3.11 \times 10^6$

35 (*a*) 1.13 mCi, (*b*) 0.898 mCi

37 About 15 g

39 7.0×10^8 y

41 (*a*) −0.764 MeV, (*b*) 3.27 MeV

43 (*a*) 0.156 MeV, (*b*) The masses given are for atoms, not nuclei, so the atomic masses are too large by the atomic number multiplied by the mass of an electron. For the given nuclear reaction, the mass of the carbon atom is too large by $6m_e$ and the mass of the nitrogen atom is too large by $7m_e$. Subtracting $6m_e$ from both sides of the reaction equation leaves an extra electron mass on the right. Not including the mass of the beta particle (electron) is mathematically equivalent to explicitly subtracting $1m_e$ from the right side of the equation.

45 1.56×10^{19} s^{-1}

47 208 MeV

49 3.2×10^{10} J

51 (*c*) 3.7×10^{38} s^{-1}, 5.0×10^{10} y

53 $\lambda = 0.069\ \text{s}^{-1}$

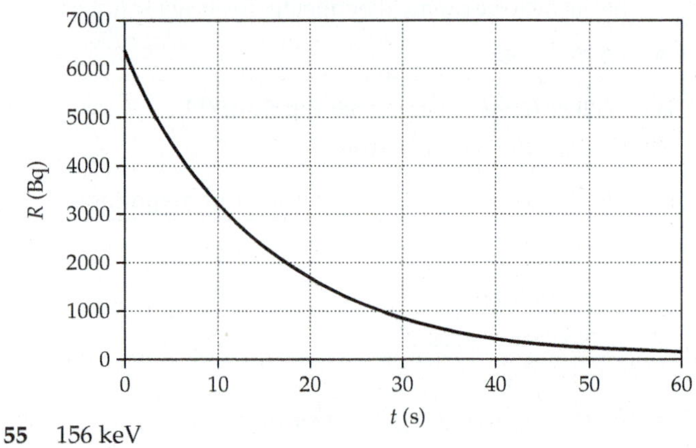

55 156 keV

59 6.7×10^3 Bq

61 6.3 L

63 (*a*) 23 MeV, (*b*) 4.2 GeV, (*c*) 1.3 GeV

65 (*a*)

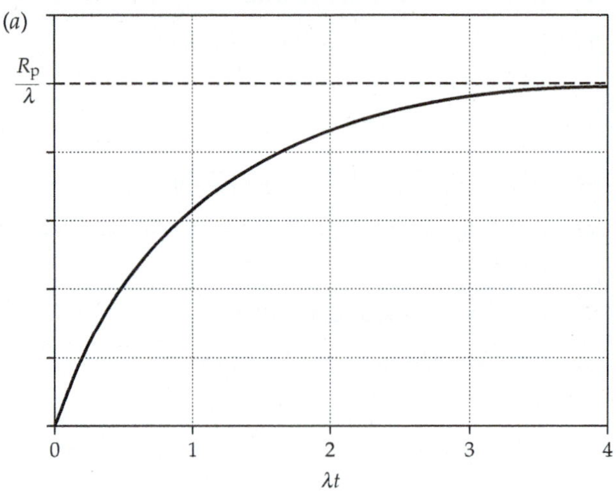

(*b*) 8.7×10^4

67 (*a*) 4.00 fm, (*b*) 310 MeV/*c*, (*d*) 310 MeV

69 (*a*) 1.188 MeV/*c*, (*b*) 752 eV, (*c*) 0.0962%

71 (*b*) 55

73 (*d*)

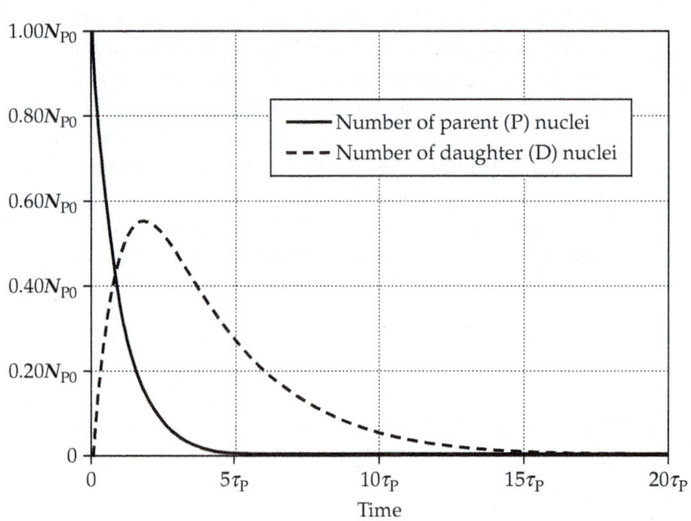

Chapter 41

1

Similarities	Differences
Baryons and mesons are hadrons, i.e., they participate in the strong interaction. Both are composed of quarks.	Baryons consist of three quarks and are fermions. Mesons consist of two quarks and are bosons. Baryons have baryon number +1 or −1. Mesons have baryon number 0.

3 A decay process involving the strong interaction has a very short lifetime ($\sim 10^{-23}$ s), whereas decay processes that proceed via the weak interaction have lifetimes of order 10^{-10} s.

5 False

7 No; from Table 41-3 it is evident that any quark–antiquark combination always results in an integral or zero charge.

9 (*a*) False, (*b*) True, (*c*) True

11 $\frac{F_{\text{em}}}{F_{\text{grav}}} = 1.24 \times 10^{36}$

13 (*a*) 279.2 MeV, (*b*) 1877 MeV, (*c*) 211.3 MeV

15 (*a*) Because $\Delta S = +1$, the reaction can proceed via the weak interaction, (*b*) Because $\Delta S = +2$, the reaction is not allowed, (*c*) Because $\Delta S = +1$, the reaction can proceed via the weak interaction.

17 (*a*) Because $\Delta S = +2$, the reaction is not allowed, (*b*) Because $\Delta S = +1$, the reaction can proceed via the weak interaction.

19 (*a*) K^0, (*b*) Σ^0 or Λ^0, (*c*) K^+

21

	Combination	*B*	*Q*	*S*	Hadron
(*a*)	*uud*	1	+1	0	p^+
(*b*)	*udd*	1	0	0	n
(*c*)	*uus*	1	+1	−1	Σ^+
(*d*)	*dds*	1	−1	−1	Σ^-
(*e*)	*uss*	1	0	−2	Ξ^0
(*f*)	*dss*	1	−1	−2	Ξ^-

23 From Table 41-3 we see that to satisfy the properties of charge number equal to +2 and strangeness, charm, topness, and bottomness all equal to zero, the quark combination must be *uuu*.

25 (*a*) $c\bar{d}$, (*b*) $\bar{c}d$

27 (*a*) *uds*, (*b*) $\bar{u}\bar{u}\bar{d}$, (*c*) *dds*

29 (*a*) *sss*, (*b*) *ssd*

31 $3.3 \times 10^8\, c \cdot \text{y}$

35 (*a*) Baryon number and lepton numbers are conserved quantities. A particle and its antiparticle must have baryon numbers that add to zero and lepton numbers that add to zero. Thus, for a particle and its antiparticle to be identical, its baryon number and all three of its lepton numbers must equal zero. This means it cannot be a lepton or a baryon, so it must be a meson. A particle and its antiparticle have the

complementary quark content. That is, if each quark in a particle is replaced by its antiquark, then the resulting entity is the antiparticle of the particle.
(*b*) The quark combination for the π^0 is a linear combination of $u\overline{u}$ and $d\overline{d}$ and the quark combination for the $\overline{\pi}^0$ is a linear combination of $\overline{u}u$ and $\overline{d}d$. The quark combination for the Ξ^0 is uss and that of the $\overline{\Xi}^0$ is $\overline{u}\,\overline{s}\,\overline{s}$.
(*c*) The π^0 is a meson with quark content of a linear combination of $u\overline{u}$ and $d\overline{d}$, so the π^0 is its own antiparticle.
The Ξ^0 is a baryon. As is explained in the answer to Part (*a*), a baryon cannot be its own antiparticle.

37 (*a*) The u and $\overline{u}$ annihilate, resulting in the photons.
(*b*) Two or more photons are required to conserve linear momentum.

39 (*a*) π^+, (*b*) -815 MeV, (*c*) 1.98 GeV

41 (*a*) 38 MeV, (*b*) 6.72, (*c*) 5 MeV, 33 MeV

Index

If a page number is followed by an *n*, the entry is in a footnote.

Physical Constants*

Atomic mass constant	$m_u = \frac{1}{12}m(^{12}C)$	$1\ u = 1.660\ 538\ 86(28) \times 10^{-27}$ kg
Avogadro's number	N_A	$6.022\ 1415(10) \times 10^{23}$ particles/mol
Boltzmann constant	$k = R/N_A$	$1.380\ 6505(24) \times 10^{-23}$ J/K $8.617\ 343(15) \times 10^{-5}$ eV/K
Bohr magneton	$m_B = e\hbar/(2m_e)$	$9.274\ 009\ 49(80) \times 10^{-24}$ J/T = $5.788\ 381\ 804(39) \times 10^{-5}$ eV/T
Coulomb constant	$k = 1/(4\pi\epsilon_0)$	$8.987\ 551\ 788\ldots \times 10^{9}\ N \cdot m^2/C^2$
Compton wavelength	$\lambda_C = h/(m_e c)$	$2.426\ 310\ 238(16) \times 10^{-12}$ m
Fundamental charge	e	$1.602\ 176\ 53(14) \times 10^{-19}$ C
Gas constant	R	$8.314\ 472(15)$ J/(mol · K) = $1.987\ 2065(36)$ cal/(mol · K) = $8.205\ 746(15) \times 10^{-2}$ L · atm/(mol · K)
Gravitational constant	G	$6.6742(10) \times 10^{-11}\ N \cdot m^2/kg^2$
Mass of electron	m_e	$9.109\ 3826(16) \times 10^{-31}$ kg = $0.510\ 998\ 918(44)\ MeV/c^2$
Mass of proton	m_p	$1.672\ 621\ 71(29) \times 10^{-27}$ kg = $938.272\ 029(80)\ MeV/c^2$
Mass of neutron	m_n	$1.674\ 927\ 28(29) \times 10^{-27}$ kg = $939.565\ 360(81)\ MeV/c^2$
Magnetic constant (permeability of free space)	μ_0	$4\pi \times 10^{-7}\ N/A^2$
Electric constant (permittivity of free space)	ϵ_0	$= 1/(\mu_0 c^2) = 8.854\ 187\ 817\ldots \times 10^{-12}\ C^2/(N \cdot m^2)$
Planck's constant	h	$6.626\ 0693(11) \times 10^{-34}$ J · s = $4.135\ 667\ 43(35) \times 10^{-15}$ eV · s
	$\hbar = h/(2\pi)$	$1.054\ 571\ 68(18) \times 10^{-34}$ J · s = $6.582\ 119\ 15(56) \times 10^{-16}$ eV · s
Speed of light	c	$2.997\ 924\ 58 \times 10^{8}$ m/s
Stefan-Boltzmann constant	σ	$5.670\ 400(40) \times 10^{-8}\ W/(m^2 \cdot K^4)$

*The values for these and other constants can be found in Appendix B as well as on the Internet at http://physics.nist.gov/cuu/Constants/index.html. The numbers in parentheses represent the uncertainties in the last two digits. (For example, 2.044 43(13) stands for 2.044 43 ± 0.000 13.) Values without uncertainties are exact. Values with ellipses are exact (like the number $\pi = 3.1415\ldots$), but are not completely specified.

Derivatives and Definite Integrals

$$\frac{d}{dx}\sin ax = a\cos ax \qquad \int_0^\infty e^{-ax}\,dx = \frac{1}{a} \qquad \int_0^\infty x^2 e^{-ax^2}\,dx = \frac{1}{4}\sqrt{\frac{\pi}{a^3}}$$

$$\frac{d}{dx}\cos ax = -a\sin ax \qquad \int_0^\infty e^{-ax^2}\,dx = \frac{1}{2}\sqrt{\frac{\pi}{a}} \qquad \int_0^\infty x^3 e^{-ax^2}\,dx = \frac{4}{a^2}$$

$$\frac{d}{dx}e^{ax} = ae^{ax} \qquad \int_0^\infty xe^{-ax^2}\,dx = \frac{2}{a} \qquad \int_0^\infty x^4 e^{-ax^2}\,dx = \frac{3}{8}\sqrt{\frac{\pi}{a^5}}$$

The a in the six integrals is a positive constant.

Vector Products

$\vec{A} \cdot \vec{B} = AB\cos\theta$ $\quad$ $\vec{A} \times \vec{B} = AB\sin\theta\,\hat{n}$ ($\hat{n}$ obtained using right-hand rule)

For additional data, see the following tables in the text.

Geometry and Trigonometry

$C = \pi d = 2\pi r$	definition of π
$A = \pi r^2$	area of circle
$V = \frac{4}{3}\pi r^3$	spherical volume
$A = \partial V/\partial r = 4\pi r^2$	spherical surface area
$V = A_{\text{base}}L = \pi r^2 L$	cylindrical volume
$A = \partial V/\partial r = 2\pi r L$	cylindrical surface area

$o = h \sin\theta$

$a = h \cos\theta$

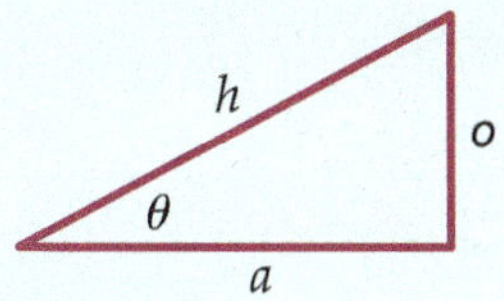

$\sin^2\theta + \cos^2\theta = 1$

$\sin(A \pm B) = \sin A \cos B \pm \cos A \sin B$

$\cos(A \pm B) = \cos A \cos B \mp \sin A \sin B$

$\sin A \pm \sin B = 2 \sin[\frac{1}{2}(A \pm B)] \cos[\frac{1}{2}(A \mp B)]$

$\sin\theta \equiv y$

$\cos\theta \equiv x$

$\tan\theta \equiv \dfrac{y}{x}$

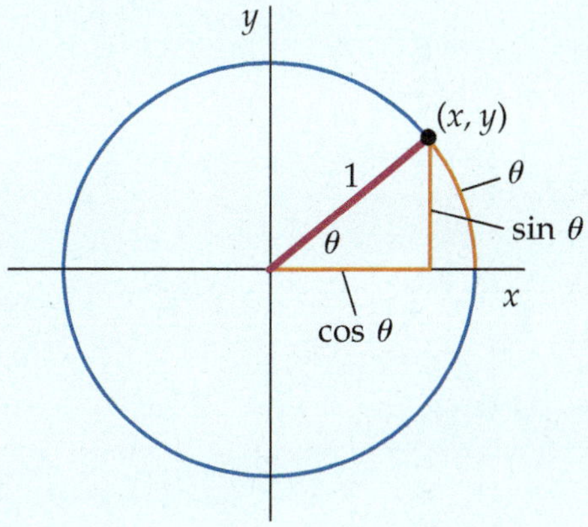

If $|\theta| << 1$, then

$\cos\theta \approx 1$ and $\tan\theta \approx \sin\theta \approx \theta$ (θ in radians)

Quadratic Formula

If $ax^2 + bx + c = 0$, then $x = \dfrac{-b \pm \sqrt{b^2 - 4ac}}{2a}$

Binomial Expansion

If $|x| < 1$, then $(1 + x)^n =$

$$1 + nx + \frac{n(n-1)}{2!}x^2 + \frac{n(n-1)(n-2)}{3!}x^3 + \ldots$$

If $|x| << 1$, then $(1 + x)^n \approx 1 + nx$

Differential Approximation

If $\Delta F = F(x + \Delta x) - F(x)$ and if $|\Delta x|$ is small,

then $\Delta F \approx \dfrac{dF}{dx}\Delta x$.